D1261777

Marks'
**Standard Handbook
for Mechanical Engineers**

Marks'
Standard Handbook
for Mechanical Engineers

Revised by a staff of specialists

EUGENE A. AVALLONE   *Editor*

Consulting Engineer; Professor of Mechanical Engineering, Emeritus
The City College of the City University of New York

THEODORE BAUMEISTER III   *Editor*

Retired Consultant, Information Systems Department
E. I. du Pont de Nemours & Co.

ALI M. SADEGH   *Editor*

Consulting Engineer; Professor of Mechanical Engineering
The City College of the City University of New York

Eleventh Edition

New York   Chicago   San Francisco   Lisbon   London   Madrid
Mexico City   Milan   New Delhi   San Juan   Seoul
Singapore   Sydney   Toronto

The McGraw·Hill Companies

Library of Congress Cataloged The First Issue
of this title as follows:

Standard handbook for mechanical engineers. 1st-ed.;
1916–

New York, McGraw-Hill.

v. Illus. 18–24 cm.

Title varies: 1916–58; Mechanical engineers' handbook.
Editors: 1916–51, L. S. Marks.—1958–   T. Baumeister.
Includes bibliographies.
1. Mechanical engineering—Handbooks, manuals, etc.   I. Marks,
Lionel Simeon, 1871– ed.   II. Baumeister, Theodore, 1897–
ed. III. Title; Mechanical engineers' handbook.
TJ151.S82   502'.4'621   16–12915

Library of Congress Catalog Card Number: 87-641192

MARKS' STANDARD HANDBOOK FOR MECHANICAL ENGINEERS

1 2 3 4 5 6 7 8 9 0 DOW/DOW 0 1 0 9 8 7 6

ISBN-13: 978-0-07-142867-5
ISBN-10: 0-07-142867-4

*The sponsoring editor for this book was Larry S. Hager, the editing supervisor was David E. Fogarty,
and the production supervisor was Richard C. Ruzycka. It was set in Times Roman by International
Typesetting and Composition. The art director for the cover was Anthony Landi.*

Printed and bound by RR Donnelley.

This book is printed on acid-free paper.

*The editors and the publisher will be grateful to readers who notify them of any inaccuracy or impor-
tant omission in this book.*

Contents

For the detailed contents of any section consult the title page of that section.

Index follows Section 20

Contributors

Abraham Abramowitz* *Consulting Engineer; Professor of Electrical Engineering, Emeritus, The City College of The City University of New York* (ILLUMINATION)

Vincent M. Altamuro *President, VMA Inc., Toms River, NJ* (MATERIAL HOLDING, FEEDING, AND METERING. CONVEYOR MOVING AND HANDLING. AUTOMATED GUIDED VEHICLES AND ROBOTS. MATERIAL STORAGE AND WAREHOUSING. METHODS ENGINEERING. AUTOMATIC MANUFACTURING. INDUSTRIAL PLANTS)

Charles A. Amann *Principal Engineer, KAB Engineering* (AUTOMOTIVE ENGINEERING)

Farid M. Amirouche *Professor of Mechanical and Industrial Engineering, University of Illinois at Chicago* (INTRODUCTION TO THE FINITE-ELEMENT METHOD. COMPUTER-AIDED DESIGN, COMPUTER-AIDED ENGINEERING, AND VARIATIONAL DESIGN)

Yiannis Andreopoulos *Professor of Mechanical Engineering, The City College of the City University of New York* (EXPERIMENTAL FLUID MECHANICS)

William Antis* *Technical Director, Maynard Research Council, Inc., Pittsburgh, PA* (METHODS ENGINEERING)

Glenn E. Asauskas *Lubrication Engineer, Chevron Corp.* (LUBRICANTS AND LUBRICATION)

Dennis N. Assanis *Professor of Mechanical Engineering, University of Michigan* (INTERNAL COMBUSTION ENGINES)

Eugene A. Avallone *Consulting Engineer; Professor of Mechanical Engineering, Emeritus, The City College of The City University of New York* (MECHANICAL PROPERTIES OF MATERIALS. GENERAL PROPERTIES OF MATERIALS. PIPE, PIPE FITTINGS, AND VALVES. SOURCES OF ENERGY. STEAM ENGINES. MISCELLANY)

Klemens C. Baczewski *Consulting Engineer* (CARBONIZATION OF COAL AND GAS MAKING)

Glenn W. Baggley* *Former Manager, Regenerative Systems, Bloom Engineering Co., Inc.* (COMBUSTION FURNACES)

Frederick G. Baily *Consulting Engineer; Steam Turbines, General Electric Co.* (STEAM TURBINES)

Robert D. Bartholomew *Associate, Sheppard T. Powell Associates, LLC* (CORROSION)

George F. Baumeister *President, EMC Process Corp., Newport, DE* (MATHEMATICAL TABLES)

John T. Baumeister *Manager, Product Compliance Test Center, Unisys Corp.* (MEASURING UNITS)

E. R. Behnke* *Product Manager, CM Chain Division, Columbus, McKinnon Corp.* (CHAINS)

John T. Benedict* *Retired Standards Engineer and Consultant, Society of Automotive Engineers* (AUTOMOTIVE ENGINEERING)

Bernadette M. Bennett, Esq. *Associate; Carter, DeLuca, Farrell and Schmidt, LLP Melville, NY* (PATENTS, TRADEMARKS, AND COPYRIGHTS)

Louis Bialy *Director, Codes & Product Safety, Otis Elevator Company* (ELEVATORS, DUMBWAITERS, AND ESCALATORS)

Malcolm Blair *Technical and Research Director, Steel Founders Society of America* (IRON AND STEEL CASTINGS)

Omer W. Blodgett *Senior Design Consultant, Lincoln Electric Co.* (WELDING AND CUTTING)

B. Douglas Bode *Engineering Supervisor, Product Customization and Vehicle Enhancement, Construction and Forestry Div., John Deere* (OFF-HIGHWAY VEHICLES AND EARTHMOVING EQUIPMENT)

Donald E. Bolt* *Engineering Manager, Heat Transfer Products Dept., Foster Wheeler Energy Corp.* (POWER PLANT HEAT EXCHANGERS)

G. David Bounds *Senior Engineer, Duke Energy Corp.* (PIPELINE TRANSMISSION)

William J. Bow* *Director, Retired, Heat Transfer Products Department, Foster Wheeler Energy Corp.* (POWER PLANT HEAT EXCHANGERS)

*Contributions by authors whose names are marked with an asterisk were made for the previous edition and have been revised or rewritten by others for this edition. The stated professional position in these cases is that held by the author at the time of his or her contribution.

James L. Bowman* *Senior Engineering Consultant, Rotary-Reciprocating Compressor Division, Ingersoll-Rand Co.* (COMPRESSORS)

Walter H. Boyes, Jr. *Editor-in-Chief/Publisher, Control Magazine* (INSTRUMENTS)

Richard L. Brazill *Technology Specialist, ALCOA Technical Center, ALCOA* (ALUMINUM AND ITS ALLOYS)

Frederic W. Buse* *Chief Engineer, Standard Pump Division, Ingersoll-Rand Co.* (DISPLACEMENT PUMPS)

Charles P. Butterfield *Chief Engineer, National Wind Technology Center, National Renewable Energy Laboratory* (WIND POWER)

C. L. Carlson* *Late Fellow Engineer, Research Labs., Westinghouse Electric Corp.* (NONFERROUS METALS AND ALLOYS. METALS AND ALLOYS FOR NUCLEAR ENERGY APPLICATIONS)

Scott W. Case *Professor of Engineering Science & Mechanics, Virginia Polytechnic Institute and State University* (MECHANICS OF COMPOSITE MATERIALS)

Vittorio (Rino) Castelli *Senior Research Fellow, Retired, Xerox Corp.; Engineering Consultant* (FRICTION. FLUID FILM BEARINGS)

Paul V. Cavallaro *Senior Mechanical Research Engineer, Naval Underwater Warfare Center* (AIR-INFLATED FABRIC STRUCTURES)

Eric L. Christiansen *Johnson Space Center, NASA* (METEOROID/ORBITAL DEBRIS SHIELDING)

Robin O. Cleveland *Associate Professor of Aerospace and Mechanical Engineering, Boston University* (SOUND, NOISE, AND ULTRASONICS)

Gary L. Cloud *Professor, Department of Mechanical Engineering, Michigan State University* (EXPERIMENTAL STRESS AND STRAIN ANALYSIS)

Ashley C. Cockerill *Vice President and Event Coordinator, nanoTech Business, Inc.* (ENGINEERING STATISTICS AND QUALITY CONTROL)

Timothy M. Cockerill *Senior Project Manager, University of Illinois* (ELECTRONICS)

Thomas J. Cockerill *Advisory Engineer, International Business Machines Corp.* (COMPUTERS)

Aaron Cohen *Retired Center Director, Lyndon B. Johnson Space Center, NASA; Zachry Professor, Texas A&M University* (ASTRONAUTICS)

Arthur Cohen *Former Manager, Standards and Safety Engineering, Copper Development Assn.* (COPPER AND COPPER ALLOYS)

D. E. Cole *Director, Office for Study of Automotive Transportation, Transportation Research Institute, University of Michigan* (INTERNAL COMBUSTION ENGINES)

James M. Connolly *Section Head, Projects Department, Jacksonville Electric Authority* (COST OF ELECTRIC POWER)

Alexander Couzis *Professor of Chemical Engineering, The City College of the City University of New York* (INTRODUCTION TO NANOTECHNOLOGY)

Terry L. Creasy *Assistant Professor of Mechanical Engineering, Texas A&M University* (STRUCTURAL COMPOSITES)

M. R. M. Crespo da Silva* *University of Cincinnati* (ATTITUDE DYNAMICS, STABILIZATION, AND CONTROL OF SPACECRAFT)

Richard A. Dahlin *Vice President, Engineering, Walker Magnetics* (LIFTING MAGNETS)

Benjamin C. Davenny *Acoustical Consultant, Acentech Inc., Cambridge, MA* (SOUND, NOISE, AND ULTRASONICS)

William H. Day *President, Longview Energy Associates, LLC; formerly Founder and Board Chairman, The Gas Turbine Association* (GAS TURBINES)

Benjamin B. Dayton *Consulting Physicist, East Flat Rock, NC* (HIGH-VACUUM PUMPS)

Horacio M. de la Fuente *Senior Engineer, NASA Johnson Space Center* (TRANSHAB)

Donald D. Dodge *Supervisor, Retired, Product Quality and Inspection Technology, Manufacturing Development, Ford Motor Co.* (NONDESTRUCTIVE TESTING)

Andrew M. Donaldson *Project Director, Parsons E&C, Reading, PA* (COST OF ELECTRIC POWER)

Joseph S. Dorson *Senior Engineer, Columbus McKinnon Corp.* (CHAIN)

James Drago *Manager, Engineering, Garlock Sealing Technologies* (PACKING, GASKETS, AND SEALS)

Michael B. Duke *Chief, Solar Systems Exploration, Johnson Space Center, NASA* (DYNAMIC ENVIRONMENTS)

F. J. Edeskuty *Retired Associate, Las Alamos National Laboratory* (CRYOGENICS)

O. Elnan* *University of Cincinnati* (SPACE-VEHICLE TRAJECTORIES, FLIGHT MECHANICS, AND PERFORMANCE. ORBITAL MECHANICS)

Robert E. Eppich *Vice President, Technology, American Foundry Society* (IRON AND STEEL CASTINGS)

C. James Erickson* *Retired Principal Consultant, Engineering Department, E. I. du Pont de Nemours & Co.* (ELECTRICAL ENGINEERING)

George H. Ewing* *Retired President and Chief Executive Officer, Texas Eastern Gas Pipeline Co. and Transwestern Pipeline Co.* (PIPELINE TRANSMISSION)

Heimir Fanner *Chief Design Engineer, Ariel Corp.* (COMPRESSORS)

Erich A. Farber *Distinguished Service Professor Emeritus, Director Emeritus of Solar Energy and Energy Conversion Lab., University of Florida* [STIRLING (HOT AIR) ENGINES. SOLAR ENERGY. DIRECT ENERGY CONVERSION]

Raymond E. Farrell, Esq. *Partner; Carter, DeLuca, Farrell and Schmidt, LLP, Melville, NY* (PATENTS, TRADEMARKS, AND COPYRIGHTS)

D. W. Fellenz* *University of Cincinnati* (SPACE-VEHICLE TRAJECTORIES, FLIGHT MECHANICS, AND PERFORMANCE. ATMOSPHERIC ENTRY)

Chuck Fennell *Program Manager, Dalton Foundries* (FOUNDARY PRACTICE AND EQUIPMENT)

Arthur J. Fiehn* *Late Retired Vice President, Project Operations Division, Burns & Roe, Inc.* (COST OF ELECTRIC POWER)

Sanford Fleeter *McAllister Distinguished Professor, School of Mechanical Engineering, Purdue University* (JET PROPULSION AND AIRCRAFT PROPELLERS)

Luc G. Fréchette *Professor of Mechanical Engineering, Université de Sherbrooke, Canada* [AN INTRODUCTION TO MICROELECTROMECHANICAL SYSTEMS (MEMS)]

William L. Gamble *Professor Emeritus of Civil and Environmental Engineering, University of Illinois at Urbana-Champaign* (CEMENT, MORTAR, AND CONCRETE. REINFORCED CONCRETE DESIGN AND CONSTRUCTION)

Robert F. Gambon *Power Plant Design and Development Consultant* (COST OF ELECTRIC POWER)

Burt Garofab *Senior Engineer, Pittston Corp.* (MINES, HOISTS, AND SKIPS. LOCOMOTIVE HAULAGE, COAL MINES)

Siamak Ghofranian *Senior Engineer, Rockwell Aerospace* (DOCKING OF TWO FREE-FLYING SPACECRAFT)

Samuel V. Glorioso *Section Chief, Metallic Materials, Johnson Space Center, NASA* (STRESS CORROSION CRACKING)

Norman Goldberg *Consulting Engineer, Economides and Goldberg* (AIR-CONDITIONING, HEATING, AND VENTILATING)

Andrew Goldenberg *Professor of Mechanical and Industrial Engineering, University of Toronto, Canada; President and CEO of Engineering Service Inc. (ESI) Toronto* (ROBOTICS, MECHATRONICS, AND INTELLIGENT AUTOMATION)

David T. Goldman* *Late Deputy Manager, U.S. Department of Energy, Chicago Operations Office* (MEASURING UNITS)

Frank E. Goodwin *Executive Vice President, ILZRO, Inc.* (BEARING METALS. LOW-MELTING-POINT METALS AND ALLOYS. ZINC AND ZINC ALLOYS)

Don Graham *Manager, Turning Products, Carboloy, Inc.* (CEMENTED CARBIDES)

David W. Green *Supervisory Research General Engineer, Forest Products Lab., USDA* (WOOD)

Leonard M. Grillo *Principal, Grillo Engineering Co.* (MUNICIPAL WASTE COMBUSTION)

Walter W. Guy *Chief, Crew and Thermal Systems Division, Johnson Space Center, NASA* (SPACECRAFT LIFE SUPPORT AND THERMAL MANAGEMENT)

Marsbed Hablanian *Retired Manager of Engineering and R&D, Varian Vacuum Technologies* (HIGH-VACUUM PUMPS)

Christopher P. Hansen *Structures and Mechanism Engineer, NASA Johnson Space Center* (PORTABLE HYPERBARIC CHAMBER)

Harold V. Hawkins* *Late Manager, Product Standards and Services, Columbus McKinnon Corp.* (DRAGGING, PULLING, AND PUSHING. PIPELINE FLEXURE STRESSES)

Keith L. Hawthorne *Vice President—Technology, Transportation Technology Center, Inc.* (RAILWAY ENGINEERING)

V. Terrey Hawthorne *Late Senior Engineer, LTK Engineering Services* (RAILWAY ENGINEERING)

J. Edmund Hay *U.S. Department of the Interior* (EXPLOSIVES)

Terry L. Henshaw *Consulting Engineer, Magnolia, TX* (DISPLACEMENT PUMPS. CENTRIFUGAL PUMPS)

Roland Hernandez *Research Engineer, Forest Products Lab., USDA* (WOOD)

David T. Holmes *Manager of Engineering, Lift-Tech International Div. of Columbus McKinnon Corp.* (MONORAILS. OVERHEAD TRAVELING CRANES)

Hoyt C. Hottel *Late Professor Emeritus, Massachusetts Institute of Technology* (RADIANT HEAT TRANSFER)

Michael W. Hyer *Professor of Engineering Science & Mechanics, Virginia Polytechnic Institute and State University* (MECHANICS OF COMPOSITE MATERIALS)

Timothy J. Jacobs *Research Fellow, Department of Mechanical Engineering, University of Michigan* (INTERNAL COMBUSTION ENGINES)

Michael W. M. Jenkins *Professor, Aerospace Design, Georgia Institute of Technology* (AERONAUTICS)

Peter K. Johnson *Consultant* (POWDERED METALS)

Randolph T. Johnson *Naval Surface Warfare Center* (ROCKET FUELS)

Robert L. Johnston *Branch Chief, Materials, Johnson Space Center, NASA* (METALLIC MATERIALS FOR AEROSPACE APPLICATIONS. MATERIALS FOR USE IN HIGH-PRESSURE OXYGEN SYSTEMS)

Byron M. Jones* *Retired Assistant Professor of Electrical Engineering, School of Engineering, University of Tennessee at Chattanooga* (ELECTRONICS)

Scott K. Jones *Professor, Department of Accounting & MIS, Alfred Lerner College of Business & Economics, University of Delaware* (COST ACCOUNTING)

Robert Jorgensen *Engineering Consultant* (FANS)

Serope Kalpakjian *Professor Emeritus of Mechanical and Materials Engineering, Illinois Institute of Technology* (MACHINING PROCESSES AND MACHINE TOOLS)

Igor J. Karassik* *Late Senior Consulting Engineer, Ingersoll Dresser Pump Co.* (CENTRIFUGAL AND AXIAL FLOW PUMPS)

Jonathan D. Kemp *Vibration Consultant, Acentech, Inc., Cambridge, MA* (SOUND, NOISE, AND ULTRASONICS)

J. Randolph Kissell *President, The TGB Partnership* (ALUMINUM AND ITS ALLOYS)

John B. Kitto, Jr. *Babcock & Wilcox Co.* (STEAM BOILERS)

Andrew C. Klein *Professor of Nuclear Engineering, Oregon State University; Director of Training, Education and Research Partnership, Idaho National Laboratories* (NUCLEAR POWER. ENVIRONMENTAL CONTROL. OCCUPATIONAL SAFETY AND HEALTH. FIRE PROTECTION)

Doyle Knight *Professor of Mechanical and Aerospace Engineering, Rutgers University* (INTRODUCTION TO COMPUTATIONAL FLUID MECHANICS)

Ronald M. Kohner *President, Landmark Engineering Services, Ltd.* (DERRICKS)

Ezra S. Krendel *Emeritus Professor of Operations Research and Statistics, University of Pennsylvania* (HUMAN FACTORS AND ERGONOMICS. MUSCLE GENERATED POWER)

A. G. Kromis* *University of Cincinnati* (SPACE-VEHICLE TRAJECTORIES, FLIGHT MECHANICS, AND PERFORMANCE)

Srirangam Kumaresan *Biomechanics Institute, Santa Barbara, California* (HUMAN INJURY TOLERANCE AND ANTHROPOMETRIC TEST DEVICES)

L. D. Kunsman* *Late Fellow Engineer, Research Labs, Westinghouse, Electric Corp.* (NONFERROUS METALS AND ALLOYS. METALS AND ALLOYS FOR NUCLEAR ENERGY APPLICATIONS)

Colin K. Larsen *Vice President, Blue Giant U.S.A. Corp.* (SURFACE HANDLING)

Stan Lebow *Research Forest Products Technologist, Forest Products Lab., USDA* (WOOD)

John H. Lewis *Engineering Consultant; Formerly Engineering Staff, Pratt & Whitney Division, United Technologies Corp.; Adjunct Associate Professor, Hartford Graduate Center, Rensselaear Polytechnic Institute* (GAS TURBINES)

Jackie Jie Li *Professor of Mechanical Engineering, The City College of the City University of New York* (FERROELECTRICS/PIEZOELECTRICS AND SHAPE MEMORY ALLOYS)

Peter E. Liley *Professor Emeritus of Mechanical Engineering, Purdue University* (THERMODYNAMICS, THERMODYNAMIC PROPERTIES OF SUBSTANCES)

James P. Locke *Flight Surgeon, NASA Johnson Space Center* (PORTABLE HYPERBARIC CHAMBER)

Ernst K. H. Marburg *Manager, Product Standards and Service, Columbus McKinnon Corp.* (LIFTING, HOISTING, AND ELEVATING. DRAGGING, PULLING, AND PUSHING. LOADING, CARRYING, AND EXCAVATING)

Larry D. Means *President, Means Engineering and Consulting* (WIRE ROPE)

Leonard Meirovitch *University Distinguished Professor Emeritus, Department of Engineering Science and Mechanics, Virginia Polytechnic Institute and State University* (VIBRATION)

George W. Michalec *Late Consulting Engineer* (GEARING)

Duane K. Miller *Manager, Engineering Services, Lincoln Electric Co.* (WELDING AND CUTTING)

Patrick C. Mock *Principal Electron Optical Scientist, Science and Engineering Associates, Inc.* (OPTICS)

Thomas L. Moser *Deputy Associate Administrator, Office of Space Flight, NASA Headquarters, NASA* (SPACE-VEHICLE STRUCTURES)

George J. Moshos* *Professor Emeritus of Computer and Information Science, New Jersey Institute of Technology* (COMPUTERS)

Eduard Muljadi *Senior Engineer, National Wind Technology Center, National Renewable Energy Laboratory* (WIND POWER)

Otto Muller-Girard* *Consulting Engineer* (INSTRUMENTS)

James W. Murdock *Late Consulting Engineer* (MECHANICS OF FLUIDS)

Gregory V. Murphy *Professor, Department of Electrical and Computer Engineering, College of Engineering, University of Alabama* (AUTOMATIC CONTROLS)

Joseph F. Murphy *Research General Engineer, Forest Products Lab., USDA* (WOOD)

John Nagy *Retired Supervisory Physical Scientist, U.S. Department of Labor, Mine Safety and Health Administration* (DUST EXPLOSIONS)

B. W. Niebel* *Professor Emeritus of Industrial Engineering, The Pennsylvania State University* (INDUSTRIAL ECONOMICS AND MANAGEMENT)

James J. Noble *Formerly Research Associate Professor of Chemical and Biological Engineering, Tufts University* (RADIANT HEAT TRANSFER)

Charles Osborn *Business Manager, Precision Cleaning Division, PTI Industries, Inc.* (PRECISION CLEANING)

D. J. Patterson *Professor of Mechanical Engineering, Emeritus, University of Michigan* (INTERNAL COMBUSTION ENGINES)

Harold W. Paxton *United States Steel Professor Emeritus, Carnegie Mellon University* (IRON AND STEEL)

Richard W. Perkins *Professor Emeritus of Mechanical, Aerospace, and Manufacturing Engineering, Syracuse University* (WOODCUTTING TOOLS AND MACHINES)

W. R. Perry* *University of Cincinnati* (ORBITAL MECHANICS. SPACE-VEHICLE TRAJECTORIES, FLIGHT MECHANICS, AND PERFORMANCE)

Kenneth A. Phair *Senior Project Engineer, Shaw Stone & Webster* (GEOTHERMAL POWER)

Orvis E. Pigg *Section Head, Structural Analysis, Johnson Space Center, NASA* (SPACE VEHICLE STRUCTURES)

Henry O. Pohl *Chief, Propulsion and Power Division, Johnson Space Center, NASA* (SPACE PROPULSION)

Nicholas R. Rafferty *Retired Technical Associate, E. I. du Pont de Nemours & Co., Inc.* (ELECTRICAL ENGINEERING)

Rama Ramakumar *PSO/Albrecht Naeter Professor and Director, Engineering Energy Laboratory, Oklahoma State University* (WIND POWER)

Pascal M. Rapier* *Scientist III, Retired, Lawrence Berkeley Laboratory* (ENVIRONMENTAL CONTROL. OCCUPATIONAL SAFETY AND HEALTH. FIRE PROTECTION)

James D. Redmond *President, Technical Marketing Services, Inc.* (STAINLESS STEELS)

Darrold E. Roen *Late Manager, Sales & Special Engineering & Government Products, John Deere* (OFF-HIGHWAY VEHICLES)

Ivan L. Ross* *International Manager, Chain Conveyor Division, ACCO* (OVERHEAD CONVEYORS)

Robert J. Ross *Supervisory Research General Engineer, Forest Products Lab., USDA* (WOOD)

J. W. Russell* *University of Cincinnati* (SPACE-VEHICLE TRAJECTORIES, FLIGHT MECHANICS, AND PERFORMANCE. LUNAR- AND INTERPLANETARY FLIGHT MECHANICS)

A. J. Rydzewski *Consultant, DuPont Engineering, E. I. du Pont de Nemours and Company* (MECHANICAL REFRIGERATION)

Ali M. Sadegh *Professor of Mechanical Engineering, The City College of The City University of New York* (MECHANICS OF MATERIALS. NONMETALLIC MATERIALS. MECHANISM. MACHINE ELEMENTS. SURFACE TEXTURE DESIGNATION, PRODUCTION, AND QUALITY CONTROL. INTRODUCTION TO BIOMECHANICS. AIR-INFLATED FABRIC STRUCTURES. RAPID PROTOTYPING.)

Anthony Sances, Jr. *Biomechanics Institute, Santa Barbara, CA* (HUMAN INJURY TOLERANCE AND ANTHROPOMETRIC TEST DEVICES)

C. Edward Sandifer *Professor, Western Connecticut State University, Danbury, CT* (MATHEMATICS)

Erwin M. Saniga *Dana Johnson Professor of Information Technology and Professor of Operations Management, University of Delaware* (OPERATIONS MANAGEMENT)

Adel F. Sarofim *Presidential Professor of Chemical Engineering, University of Utah* (RADIANT HEAT TRANSFER)

Martin D. Schlesinger *Late Consultant, Wallingford Group, Ltd.* (FUELS)

John R. Schley *Manager, Technical Marketing, RMI Titanium Co.* (TITANIUM AND ZIRCONIUM)

Matthew S. Schmidt *Senior Engineer, Rockwell Aerospace* (DOCKING OF TWO FREE-FLYING SPACECRAFT)

Wiliam C. Schneider *Retired Assistant Director Engineering/Senior Engineer, NASA Johnson Space Center; Visiting Professor, Texas A&M University* (ASTRONAUTICS)

James D. Shearouse, III *Late Senior Development Engineer, The Dow Chemical Co.* (MAGNESIUM AND MAGNESIUM ALLOYS)

David A. Shifler *Metallurgist, MERA Metallurgical Services* (CORROSION)

Rajiv Shivpuri *Professor of Industrial, Welding, and Systems Engineering, Ohio State University* (PLASTIC WORKING OF METALS)

James C. Simmons *Senior Vice President, Business Development, Core Furnace Systems Corp.* (ELECTRIC FURNACES AND OVENS)

William T. Simpson *Research Forest Products Technologist, Forest Products Lab., USDA* (WOOD)

Kenneth A. Smith *Edward R. Gilliland Professor of Chemical Engineering, Massachusetts Institute of Technology* (TRANSMISSION OF HEAT BY CONDUCTION AND CONVECTION)

Lawrence H. Sobel* *University of Cincinnati* (VIBRATION OF STRUCTURES)

James G. Speight *Western Research Institute* (FUELS)

Robert D. Steele *Project Manager, Voith Siemens Hydro Power Generation, Inc.* (HYDRAULIC TURBINES)

Stephen R. Swanson *Professor of Mechanical Engineering, University of Utah* (FIBER COMPOSITE MATERIALS)

John Symonds* *Fellow Engineer, Retired, Oceanic Division, Westinghouse Electric Corp.* (MECHANICAL PROPERTIES OF MATERIALS)

Peter L. Tea, Jr. *Professor of Physics Emeritus, The City College of the City University of New York* (MECHANICS OF SOLIDS)

Anton TenWolde *Supervisory Research Physicist, Forest Products Lab., USDA* (WOOD)

W. David Teter *Retired Professor of Civil Engineering, University of Delaware* (SURVEYING)

Michael C. Tracy *Rear Admiral, U.S. Navy* (MARINE ENGINEERING)

John H. Tundermann *Former Vice President, Research and Technology, INCO Intl., Inc.* (METALS AND ALLOYS FOR USE AT ELEVATED TEMPERATURES. NICKEL AND NICKEL ALLOYS)

Charles O. Velzy *Consultant* (MUNICIPAL WASTE COMBUSTION)

Harry C. Verakis *Supervisory Physical Scientist, U.S. Department of Labor, Mine Safety and Health Administration* (DUST EXPLOSIONS)

Arnold S. Vernick *Former Associate, Geraghty & Miller, Inc.* (WATER)

Robert J. Vondrasek* *Assistant Vice President of Engineering, National Fire Protection Assoc.* (COST OF ELECTRIC POWER)

Michael W. Washo *Senior Engineer, Retired, Eastman Kodak, Co.* (BEARINGS WITH ROLLING CONTACT)

Larry F. Wieserman *Senior Technical Supervisor, ALCOA* (ALUMINUM AND ITS ALLOYS)

Robert H. White *Supervisory Wood Scientist, Forest Products Lab., USDA* (WOOD)

John W. Wood, Jr. *Manager, Technical Services, Garlock Klozure* (PACKING, GASKETS, AND SEALS)

The Editors

EUGENE A. AVALLONE, Editor, is Professor of Mechanical Engineering, Emeritus, The City College of the City University of New York. He has been engaged for many years as a consultant to industry and to a number of local and national governmental agencies.

THEODORE BAUMEISTER, III, Editor, is now retired from Du Pont where he was an internal consultant. His specialties are operations research, business decision making, and long-range planning. He has also taught financial modeling in the United States, South America, and the Far East.

ALI M. SADEGH, Editor, is Professor of Mechanical Engineering, The City College of the City University of New York. He is also Director of the Center for Advanced Engineering Design and Development. He is actively engaged in research in the areas of machine design, manufacturing, and biomechanics.

Preface to the Eleventh Edition

The evolutionary trends underlying modern engineering practice are grounded not only on the tried and true principles and techniques of the past, but also on more recent and current advances. Thus, in the preparation of the eleventh edition of "Marks'," the Editors have considered the broad enterprise falling under the rubric of "Mechanical Engineering" and have added to and/or amended the contents to include subject areas that will be of maximum utility to the practicing engineer. That said, the Editors note that the publication of this eleventh edition has been accomplished through the combined and coordinated efforts of contributors, readers, and the Editors.

First, we recognize, with pleasure, the input from our many contributors—past, continuing, and those newly engaged. Their contributions have been prepared with care, and are authoritative, informative, and concise.

Second, our readers, who are practitioners in their own wise, have found that the global treatment of the subjects presented in the "Marks'" permits of great utility and serves as a convenient ready reference. The reading public has had access to "Marks'" since 1916, when Lionel S. Marks edited the first edition. This eleventh edition follows 90 years later. During the intervening years, "Marks'" and "Handbook for Mechanical Engineers" have become synonymous to a wide readership which includes mechanical engineers, engineers in the associated disciplines, and others. Our readership derives from a wide spectrum of interests, and it appears many find the "Marks'" useful as they pursue their professional endeavors.

The Editors consider it a given that every successive edition must balance the requests to broaden the range or depth of subject matter printed, the incorporation of new material which will be useful to the widest possible audience, and the requirement to keep the size of the Handbook reasonable and manageable. We are aware that the current engineering practitioner learns quickly that the revolutionary developments of the recent past soon become standard practice. By the same token, it is prudent to realize that as a consequence of rapid developments, some cutting-edge technologies prove to have a short shelf life and soon are regarded as obsolescent—if not obsolete.

The Editors are fortunate to have had, from time to time, input from readers and reviewers, who have proffered cogent commentary and suggestions; a number are included in this edition. Indeed, the synergy between Editors, contributors, and readers has been instrumental in the continuing usefulness of successive editions of "Marks' Standard Handbook for Mechanical Engineers."

The reader will note that a considerable portion of the tabular data and running text continue to be presented in dual units; i.e., USCS and SI. The date for a projected full transition to SI units is not yet firm, and the "Marks'" reflects that. We look to the future in that regard.

Society is in an era of information technology, as manifest by the practicing engineer's working tools. For example: the ubiquitous personal computer, its derivative use of software programs of a vast variety and number, printers, computer-aided design and drawing, universal access to the Internet, and so on. It is recognized, too, that the great leaps forward which are thereby enhanced still require the engineer to exercise sound and rational judgment as to the reliability of the solutions provided.

Last, the Handbook is ultimately the responsibility of the Editors. The utmost care has been exercised to avoid errors, but if any inadvertently are included, the Publisher and Editors will appreciate being so informed. Corrections will be incorporated into subsequent printings.

Ardsley, NY   EUGENE A. AVALLONE
Newark, DE   THEODORE BAUMEISTER III
Franklin Lakes, NJ   ALI M. SADEGH

Preface to the First Edition*

This Handbook is intended to supply both the practicing engineer and the student with a reference work which is authoritative in character and which covers the field of mechanical engineering in a comprehensive manner. It is no longer possible for a single individual or a small group of individuals to have so intimate an acquaintance with any major division of engineering as is necessary if critical judgment is to be exercised in the statement of current practice and the selection of engineering data. Only by the cooperation of a considerable number of specialists is it possible to obtain the desirable degree of reliability. This Handbook represents the work of fifty specialists.

Each contributor is to be regarded as responsible for the accuracy of his section. The number of contributors required to ensure sufficiently specialized knowledge for all the topics treated is necessarily large. It was found desirable to enlist the services of thirteen specialists for an adequate handling of the "Properties of Engineering Materials." Such topics as "Automobiles," "Aeronautics," "Illumination," "Patent Law," "Cost Accounting," "Industrial Buildings," "Corrosion," "Air Conditioning, "Fire Protection," "Prevention of Accidents," etc., though occupying relatively small spaces in the book, demanded each a separate writer.

A number of the contributions which deal with engineering practice, after examination by the Editor-in-Chief, were submitted by him to one or more specialists for criticism and suggestions. Their cooperation has proved of great value in securing greater accuracy and in ensuring that the subject matter does not embody solely the practice of one individual but is truly representative.

An accuracy of four significant figures has been assumed as the desirable limit; figures in excess of this number have been deleted, except in special cases. In the mathematical tables only four significant figures have been kept.

The Editor-in-Chief desires to express here his appreciation of the spirit of cooperation shown by the Contributors and of their patience in submitting to modifications of their sections. He wishes also to thank the Publishers for giving him complete freedom and hearty assistance in all matters relating to the book from the choice of contributors to the details of typography.

Cambridge, Mass.   LIONEL S. MARKS
April 23, 1916

*Excerpt.

Preface to the First Edition*

Symbols and Abbreviations

For symbols of chemical elements, see Sec. 6; for abbreviations applying to metric weights and measures and SI units, Sec. 1; SI unit prefixes are listed on p. 1–19.

Pairs of parentheses, brackets, etc., are frequently used in this work to indicate corresponding values. For example, the statement that "the cost per kW of a 30,000-kW plant is $86; of a 15,000-kW plant, $98; and of an 8,000-kW plant, $112," is condensed as follows: The cost per kW of a 30,000 (15,000) [8,000]-kW plant is $86 (98) [112].

In the citation of references readers should always attempt to consult the latest edition of referenced publications.

A or Å	Angstrom unit $= 10^{-10}$ m; 3.937×10^{-11} in
A	mass number $= N + Z$; ampere
AA	arithmetical average
AAA	Am. Automobile Assoc.
AAMA	American Automobile Manufacturers' Assoc.
AAR	Assoc. of Am. Railroads
AAS	Am. Astronautical Soc.
ABAI	Am. Boiler & Affiliated Industries
abs	absolute
a.c.	aerodynamic center
a-c, ac	alternating current
ACI	Am. Concrete Inst.
ACM	Assoc. for Computing Machinery
ACRMA	Air Conditioning and Refrigerating Manufacturers Assoc.
ACS	Am. Chemical Soc.
ACSR	aluminum cable steel-reinforced
ACV	air cushion vehicle
A.D.	anno Domini (in the year of our Lord)
AEC	Atomic Energy Commission (U.S.)
a-f, af	audio frequency
AFBMA	Anti-friction Bearings Manufacturers' Assoc.
AFS	Am. Foundrymen's Soc.
AGA	Am. Gas Assoc.
AGMA	Am. Gear Manufacturers' Assoc.
ahp	air horsepower
AIChE	Am. Inst. of Chemical Engineers
AIEE	Am. Inst. of Electrical Engineers (see IEEE)
AIME	Am. Inst. of Mining Engineers
AIP	Am. Inst. of Physics
AISC	American Institute of Steel Construction, Inc.
AISE	Am. Iron & Steel Engineers
AISI	Am. Iron and Steel Inst.
Al. Assn.	Aluminum Association
a.m.	ante meridiem (before noon)
a-m, am	amplitude modulation
Am. Mach.	Am. Machinist (New York)
AMA	Acoustical Materials Assoc.
AMCA	Air Moving & Conditioning Assoc., Inc.
amu	atomic mass unit
AN	ammonium nitrate (explosive); Army-Navy Specification
AN-FO	ammonium nitrate-fuel oil (explosive)
ANC	Army-Navy Civil Aeronautics Committee
ANS	Am. Nuclear Soc.
ANSI	American National Standards Institute
antilog	antilogarithm of
API	Am. Petroleum Inst.
approx	approximately
APWA	Am. Public Works Assoc.
AREA	Am. Railroad Eng. Assoc.
ARI	Air Conditioning and Refrigeration Inst.
ARS	Am. Rocket Soc.
ASCE	Am. Soc. of Civil Engineers
ASHRAE	Am. Soc. of Heating, Refrigerating, and Air Conditioning Engineers
ASLE	Am. Soc. of Lubricating Engineers
ASM	Am. Soc. of Metals
ASME	Am. Soc. of Mechanical Engineers
ASST	Am. Soc. of Steel Treating
ASTM	Am. Soc. for Testing and Materials
ASTME	Am. Soc. of Tool & Manufacturing Engineers
atm	atmosphere
Auto. Ind.	Automotive Industries (New York)
avdp	avoirdupois
avg, ave	average
AWG	Am. Wire Gage
AWPA	Am. Wood Preservation Assoc.
AWS	American Welding Soc.
AWWA	American Water Works Assoc.
b	barns
bar	barometer
B&S	Brown & Sharp (gage); Beams and Stringers
bbl	barrels
B.C.	before Christ
B.C.C.	body centered cubic
Bé	Baumé (degrees)
B.G.	Birmingham gage (hoop and sheet)
bgd	billions of gallons per day
BHN	Brinell Hardness Number
bhp	brake horsepower
BLC	boundary layer control
B.M.	board measure; bench mark
bmep	brake mean effective pressure
B of M, BuMines	Bureau of Mines

BOD	biochemical oxygen demand
bp	boiling point
Bq	bequerel
bsfc	brake specific fuel consumption
BSI	British Standards Inst.
Btu	British thermal units
Btub, Btu/h	Btu per hr
bu	bushels
Bull.	Bulletin
Buweaps	Bureau of Weapons, U.S. Navy
BWG	Birmingham wire gage
c	velocity of light
°C	degrees Celsius (centigrade)
C	coulomb
CAB	Civil Aeronautics Board
CAGI	Compressed Air & Gas Inst.
cal	calories
C-B-R	chemical, biological & radiological (filters)
CBS	Columbia Broadcasting System
cc, cm^3	cubic centimetres
CCR	critical compression ratio
c to c	center to center
cd	candela
c.f.	centrifugal force
cf.	confer (compare)
cfh, ft^3/h	cubic feet per hour
cfm, ft^3/min	cubic feet per minute
C.F.R.	Cooperative Fuel Research
cfs, ft^3/s	cubic feet per second
cg	center of gravity
cgs	centimetre-gram-second
Chm. Eng.	Chemical Eng'g (New York)
chu	centigrade heat unit
C.I.	cast iron
cir	circular
cir mil	circular mils
cm	centimetres
CME	Chartered Mech. Engr. (IMechE)
C.N.	cetane number
coef	coefficient
COESA	U.S. Committee on Extension to the Standard Atmosphere
col	column
colog	cologarithm of
const	constant
cos	cosine of
cos^{-1}	angle whose cosine is, inverse cosine of
cosh	hyperbolic cosine of
$cosh^{-1}$	inverse hyperbolic cosine of
cot	cotangent of
cot^{-1}	angle whose cotangent is (see cos^{-1})
coth	hyperbolic cotangent of
$coth^{-1}$	inverse hyperbolic cotangent of
covers	coversed sine of
c.p.	circular pitch; center of pressure
cp	candle power
cp	coef of performance
CP	chemically pure
CPH	close packed hexagonal
cpm	cycles per minute
cycles/min	
cps, cycles/s	cycles per second
CSA	Canadian Standards Assoc.
csc	cosecant of
csc^{-1}	angle whose cosecant is (see cos^{-1})
csch	hyperbolic cosecant of
$csch^{-1}$	inverse hyperbolic cosecant of
cu	cubic
cyl	cylinder

db, dB	decibel
d-c, dc	direct current
def	definition
deg	degrees
diam. (dia)	diameter
DO	dissolved oxygen
D_2O	deuterium (heavy water)
d.p.	double pole
DP	Diametral pitch
DPH	diamond pyramid hardness
DST	daylight saving time
d^2 tons	breaking strength, d = chain wire diam. in.
DX	direct expansion
e	base of Napierian logarithmic system (= 2.7182+)
EAP	equivalent air pressure
EDR	equivalent direct radiation
EEI	Edison Electric Inst.
eff	efficiency
e.g.	exempli gratia (for example)
ehp	effective horsepower
EHV	extra high voltage
El. Wld.	Electrical World (New York)
elec	electric
elong	elongation
emf	electromotive force
Engg.	Engineering (London)
Engr.	The Engineer (London)
ENT	emergency negative thrust
EP	extreme pressure (lubricant)
ERDA	Energy Research & Development Administration (successor to AEC; see also NRC)
Eq.	equation
est	estimated
etc.	et cetera (and so forth)
et seq.	et sequens (and the following)
eV	electron volts
evap	evaporation
exp	exponential function of
exsec	exterior secant of
ext	external
°F	degrees Fahrenheit
F	farad
FAA	Federal Aviation Agency
F.C.	fixed carbon, %
FCC	Federal Communications Commission; Federal Constructive Council
F.C.C.	face-centered-cubic (alloys)
ff.	following (pages)
fhp	friction horsepower
Fig.	figure
F.I.T.	Federal income tax
f-m, fm	frequency modulation
F.O.B.	free on board (cars)
FP	fore perpendicular
FPC	Federal Power Commission
fpm, ft/min	feet per minute
fps	foot-pound-second system
ft/s	feet per second
F.S.	Federal Specifications
FSB	Federal Specifications Board
fsp	fiber saturation point
ft	feet
fc	foot candles
fL	foot lamberts
ft·lb	foot-pounds
g	acceleration due to gravity
g	grams
gal	gallons

gc	gigacycles per second		J	joule
GCA	ground-controlled approach		J&P	joists and planks
g·cal	gram-calories		*Jour.*	Journal
gd	Gudermannian of		JP	jet propulsion fuel
G.E.	General Electric Co.		*k*	isentropic exponent; conductivity
GEM	ground effect machine		K	degrees Kelvin (Celsius abs)
GFI	gullet feed index		K	Knudsen number
G.M.	General Motors Co.		kB	kilo Btu (1000 Btu)
GMT	Greenwich Mean Time		kc	kilocycles
GNP	gross national product		kcps	kilocycles per second
gpcd	gallons per capita day		kg	kilograms
gpd	gallons per day, grams per denier		kg·cal	kilogram-calories
gpm, gal/min	gallons per minute		kg·m	kilogram-metres
gps, gal/s	gallons per second		kip	1000 lb or 1 kilo-pound
gpt	grams per tex		kips	thousands of pounds
H	henry		km	kilometres
h	Planck's constant $= 6.624 \times 10^{-27}$ org-sec		kmc	kilomegacycles per second
$\hbar$	Planck's constant, $\hbar = h/2\pi$		kmcps	kilomegacycles per second
HEPA	high efficiency particulate matter		kpsi	thousands of pounds per sq in
h-f, hf	high frequency		ksi	one kip per sq in, 1000 psi (lb/in^2)
hhv	high heat value		kts	knots
horiz	horizontal		kVA	kilovolt-amperes
hp	horsepower		kW	kilowatts
h-p	high-pressure		kWh	kilowatt-hours
HPAC	Heating, Piping, & Air Conditioning (Chicago)		L	lamberts
hp·hr	horsepower-hour		l, L	litres
hr, h	hours		£	Laplace operational symbol
HSS	high speed steel		lb	pounds
H.T.	heat-treated		L.B.P.	length between perpendiculars
HTHW	high temperature hot water		lhv	low heat value
Hz	hertz = 1 cycle/s (cps)		lim	limit
IACS	International Annealed Copper Standard		lin	linear
IAeS	Institute of Aerospace Sciences		ln	Napierian logarithm of
ibid.	ibidem (in the same place)		loc. cit.	loco citato (place already cited)
ICAO	International Civil Aviation Organization		log	common logarithm of
ICC	Interstate Commerce Commission		LOX	liquid oxygen explosive
ICE	Inst. of Civil Engineers		l-p, lp	low pressure
ICI	International Commission on Illumination		LPG	liquified petroleum gas
I.C.T.	International Critical Tables		lpw, lm/W	lumens per watt
I.D., ID	inside diameter		lx	lux
i.e.	id est (that is)		L.W.L.	load water line
IEC	International Electrotechnical Commission		lm	lumen
IEEE	Inst. of Electrical & Electronics Engineers (successor to AIEE, *q.v.*)		m	metres
			M	thousand; Mach number; moisture, %
IES	Illuminating Engineering Soc.		mA	milliamperes
i-f, if	intermediate frequency		*Machy.*	Machinery (New York)
IGT	Inst. of Gas Technology		max	maximum
ihp	indicated horsepower		MBh	thousands of Btu per hr
IMechE	Inst. of Mechanical Engineers		mc	megacycles per second
imep	indicated mean effective pressure		m.c.	moisture content
Imp	Imperial		Mcf	thousand cubic feet
in., in	inches		mcps	megacycles per second
in·lb,	inch-pounds		*Mech. Eng.*	Mechanical Eng'g (ASME)
in·lb			mep	mean effective pressure
INA	Inst. of Naval Architects		METO	maximum, except during take-off
Ind. & Eng.	Industrial & Eng'g Chemistry (Easton, PA)		me V	million electron volts
Chem.			MF	maintenance factor
int	internal		mhc	mean horizontal candles
i-p, ip	intermediate pressure		mi	mile
ipm, in/min	inches per minute		MIL-STD	U.S. Military Standard
ipr	inches per revolution		min	minutes; minimum
IPS	iron pipe size		mip	mean indicated pressure
IRE	Inst. of Radio Engineers (see IEEE)		MKS	metre-kilogram-second system
IRS	Internal Revenue Service		MKSA	metre-kilogram-second-ampere system
ISO	International Organization for Standardization		mL	millilamberts
isoth	isothermal		ml, mL	millilitre = 1.000027 cm^3
ISTM	International Soc. for Testing Materials		mlhc	mean lower hemispherical candles
IUPAC	International Union of Pure & Applied Chemistry		mm	millimetres

mm-free	mineral matter free		*Proc.*	Proceedings
mmf	magnetomotive force		PSD	power spectral density, g^2/cps
mol	mole		psi, lb/in^2	lb per sq in
mp	melting point		psia	lb per sq in. abs
MPC	maximum permissible concentration		psig	lb per sq in. gage
mph, mi/h	miles per hour		pt	point; pint
MRT	mean radiant temperature		PVC	polyvinyl chloride
ms	manuscript; milliseconds		Q	10^{18} Btu
msc	mean spherical candles		qt	quarts
MSS	Manufacturers Standardization Soc. of the Valve & Fittings Industry		q.v.	quod vide (which see)
			r	roentgens
mu	micron, micro		*R*	gas constant
MW	megawatts		R	deg Rankine (Fahrenheit abs); Reynolds number
MW day	megawatt day		rad	radius; radiation absorbed dose; radian
MWT	mean water temperature		RBE	see rem
n	polytropic exponent		R-C	resistor-capacitor
N	number (in mathematical tables)		RCA	Radio Corporation of America
N	number of neutrons; newton		R&D	research and development
N$_s$	specific speed		RDX	cyclonite, a military explosive
NA	not available		rem	Roentgen equivalent man (formerly RBE)
NAA	National Assoc. of Accountants		rev	revolutions
NACA	National Advisory Committee on Aeronautics (see NASA)		r-f, rf	radio frequency
NACM	National Assoc. of Chain Manufacturers		RMA	Rubber Manufacturers Assoc.
NASA	National Aeronautics and Space Administration		rms	square root of mean square
nat.	natural		rpm, r/min	revolutions per minute
NBC	National Broadcasting Company		rps, r/s	revolutions per second
NBFU	National Board of Fire Underwriters		RSHF	room sensible heat factor
NBS	National Bureau of Standards (see NIST)		ry.	railway
NCN	nitrocarbonitrate (explosive)		*s*	entropy
NDHA	National District Hearing Assoc.		s	seconds
NEC®	National Electric Code® (National Electrical Code® and NEC® are registered trademarks of the National Fire Protection Association, Inc., Quincy, MA.)		S	sulfur, %; siemens
			SAE	Soc. of Automotive Engineers
			sat	saturated
NEMA	National Electrical Manufacturers Assoc.		SBI	steel Boiler Inst.
NFPA	National Fire Protection Assoc.		scfm	standard cu ft per min
NIST	National Institute of Standards and Technology		SCR	silicon controlled rectifier
NLGI	National Lubricating Grease Institute		sec	secant of
nm	nautical miles		sec^{-1}	angle whose secant is (see cos^{-1})
No. (Nos.)	number(s)		Sec.	Section
NPSH	net positive suction head		sech	hyperbolic secant of
NRC	Nuclear Regulator Commission (successor to AEC; see also ERDA)		sech^{-1}	inverse hyperbolic secant of
			segm	segment
NTP	normal temperature and pressure		SE No.	steam emulsion number
O.D., OD	outside diameter (pipes)		SEI	Structural Engineering Institute
O.H.	open-hearth (steel)		sfc	specific fuel consumption, lb per hphr
O.N.	octane number		sfm, sfpm	surface feet per minute
op. cit.	opere citato (work already cited)		shp	shaft horsepower
OSHA	Occupational Safety & Health Administration		SI	International System of Units (Le Systéme International d'Unites)
OSW	Office of Saline Water			
OTS	Office of Technical Services, U.S. Dept. of Commerce		sin	sine of
oz	ounces		sin^{-1}	angle whose sine is (see cos^{-1})
p. (pp.)	page (pages)		sinh	hyperbolic sine of
Pa	pascal		sinh^{-1}	inverse hyperbolic sine of
P.C.	propulsive coefficient		SME	Society of Manufacturing Engineers (successor to ASTME)
PE	polyethylene			
PEG	polyethylene glycol		SNAME	Soc. of Naval Architects and Marine Engineers
P.E.L.	proportional elastic limit		SP	static pressure
PETN	an explosive		sp	specific
pf	power factor		specif	specification
PFI	Pipe Fabrication Inst.		sp gr	specific gravity
PIV	peak inverse voltage		sp ht	specific heat
p.m.	post meridiem (after noon)		spp	species unspecified (botanical)
PM	preventive maintenance		SPS	standard pipe size
P.N.	performance number		sq	square
ppb	parts per billion		sr	steradian
PPI	plan position indicator		SSF	sec Saybolt Furol
ppm	parts per million		SSU	seconds Saybolt Universal (same as SUS)
press	pressure		std	standard

SUS	Saybolt Universal seconds (same as SSU)	USPHS	U.S. Public Health Service
SWG	Standard (British) wire gage	USS	United States Standard
T	tesla	USSG	U.S. Standard Gage
TAC	Technical Advisory Committee on Weather Design Conditions (ASHRAE)	UTC	Coordinated Universal Time
		V	volt
tan	tangent of	VCF	visual comfort factor
$\tan^{-1}$	angle whose tangent is (see $\cos^{-1}$)	VCI	visual comfort index
tanh	hyperbolic tangent of	VDI	Verein Deutscher Ingenieure
$\tanh^{-1}$	inverse hyperbolic tangent of	vel	velocity
TDH	total dynamic head	vers	versed sine of
TEL	tetraethyl lead	vert	vertical
temp	temperature	VHF	very high frequency
THI	temperature-humidity (discomfort) index	VI	viscosity index
thp	thrust horsepower	viz.	videlicet (namely)
TNT	trinitrotoluol (explosive)	V.M.	volatile matter, %
torr	= 1 mm Hg = 1.332 millibars (1/760) atm = (1.013250/760) dynes per cm^2	vol	volume
		VP	velocity pressure
TP	total pressure	vs.	versus
tph	tons per hour	W	watt
tpi	turns per in	Wb	weber
TR	transmitter-receiver	W&M	Washburn & Moen wire gage
Trans.	Transactions	w.g.	water gage
T.S.	tensile strength; tensile stress	WHO	World Health Organization
tsi	tons per sq in	W.I.	wrought iron
ttd	terminal temperature difference	W.P.A.	Western Pine Assoc.
UHF	ultra high frequency	wt	weight
UKAEA	United Kingdom Atomic Energy Authority	yd	yards
UL	Underwriters' Laboratory	Y.P.	yield point
ult	ultimate	yr	year(s)
UMS	universal maintenance standards	Y.S.	yield strength; yield stress
USAF	U.S. Air Force	z	atomic number, figure of merit
USCG	U.S. Coast Guard	*Zeit.*	Zeitschrift
USCS	U.S. Commercial Standard; U.S. Customary System	Δ	mass defect
USDA	U.S. Dept. of Agriculture	μc	microcurie
USFPL	U.S. Forest Products Laboratory	σ, s	Boltzmann constant
USGS	U.S. Geologic Survey	μ	micro (= 10^{-6}), as in μs
USHEW	U.S. Dept. of Health, Education & Welfare	μm	micrometre (micron) = 10^{-6} m (10^{-3} mm)
USN	U.S. Navy	Ω	ohm
USP	U.S. pharmacopoeia		

MATHEMATICAL SIGNS AND SYMBOLS

+	plus (sign of addition)	$\neq$	not equal to
+	positive	$\rightarrow \doteq$	approaches
−	minus (sign of subtraction)	$\propto$	varies as
−	Negative	∞	infinity
$\pm(\mp)$	plus or minus (minus or plus)	$\sqrt{\ }$	square root of
×	times, by (multiplication sign)	$\sqrt[3]{\ }$	cube root of
·	multiplied by	$\therefore$	therefore
÷	sign of division	$\parallel$	parallel to
/	divided by	( ) [ ] { }	parentheses, brackets and braces; quantities enclosed by them to be taken together in multiplying, dividing, etc.
:	ratio sign, divided by, is to		
::	equals, as (proportion)	$\overline{AB}$	length of line from A to B
<	less than	π	pi (= 3.14159$^+$)
>	greater than	°	degrees
≪	much less than	′	minutes
≫	much greater than	″	seconds
=	equals	$\angle$	angle
≡	identical with	dx	differential of x
∼	similar to	Δ	(delta) difference
≈	approximately equals	Δx	increment of x
≅	approximately equals, congruent	$\partial u/\partial x$	partial derivative of u with respect to x
≤	equal to or less than	$\int$	integral of
≥	equal to or greater than		

$\int_b^a$	integral of, between limits a and b
$\oint$	line integral around a closed path
Σ	(sigma) summation of
$f(x), F(x)$	functions of x
$\exp x = e^x$	[$e = 2.71828$ (base of natural, or Napierian, logarithms)]
∇	del or nabla, vector differential operator
∇^2	Laplacian operator
$\pounds$	Laplace operational symbol

$4!$	factorial $4 = 4 \times 3 \times 2 \times 1$
$|x|$	absolute value of x
$\dot{x}$	first derivative of x with respect to time
$\ddot{x}$	second derivative of x with respect to time
$A \times B$	vector product; magnitude of A times magnitude of B times sine of the angle from A to B; $AB \sin \overline{AB}$
$A \cdot B$	scalar product; magnitude of A times magnitude of B times cosine of the angle from A to B; $AB \cos \overline{AB}$

Marks'
**Standard Handbook
for Mechanical Engineers**

Section **1**

Mathematical Tables and Measuring Units

BY

GEORGE F. BAUMEISTER   *President, EMC Process Co., Newport, DE*
JOHN T. BAUMEISTER   *Manager, Product Compliance Test Center, Unisys Corp.*

1.1   MATHEMATICAL TABLES
by George F. Baumeister

REFERENCES FOR MATHEMATICAL TABLES: Dwight, "Mathematical Tables of Elementary and Some Higher Mathematical Functions," McGraw-Hill. Dwight, "Tables of Integrals and Other Mathematical Data," Macmillan. Jahnke and Emde, "Tables of Functions," B. G. Teubner, Leipzig, or Dover. Pierce-Foster, "A Short Table of Integrals," Ginn. "Mathematical Tables from Handbook of Chemistry and Physics," Chemical Rubber Co. "Handbook of Mathematical Functions," NBS.

Table 1.1.1 Segments of Circles, Given h/c

Given: h = height; c = chord. To find the diameter of the circle, the length of arc, or the area of the segment, form the ratio h/c, and find from the table the value of (diam/c), (arc/c); then, by a simple multiplication,

$$\text{diam} = c \times (\text{diam}/c)$$
$$\text{arc} = c \times (\text{arc}/c)$$
$$\text{area} = h \times c \times (\text{area}/h \times c)$$

The table gives also the angle subtended at the center, and the ratio of h to D.

$\dfrac{h}{c}$	$\dfrac{\text{Diam}}{c}$	Diff	$\dfrac{\text{Arc}}{c}$	Diff	$\dfrac{\text{Area}}{h \times c}$	Diff	Central angle, v	Diff	$\dfrac{h}{\text{Diam}}$	Diff
.00			1.000	0	.6667	0	0.00°	458	.0000	4
1	25.010	12490	1.000	1	.6667	2	4.58	458	.0004	12
2	12.520	*4157	1.001	1	.6669	2	9.16	457	.0016	20
3	8.363	*2073	1.002	2	.6671	4	13.73	457	.0036	28
4	6.290	*1240	1.004	3	.6675	5	18.30	454	.0064	35
.05	5.050	*823	1.007	3	.6680	6	22.84°	453	.0099	43
6	4.227	*586	1.010	3	.6686	7	27.37	451	.0142	50
7	3.641	*436	1.013	4	.6693	8	31.88	448	.0192	58
8	3.205	*337	1.017	4	.6701	9	36.36	446	.0250	64
9	2.868	*268	1.021	5	.6710	10	40.82	442	.0314	71
.10	2.600	*217	1.026	6	.6720	11	45.24°	439	.0385	77
1	2.383	*180	1.032	6	.6731	12	49.63	435	.0462	83
2	2.203	*150	1.038	6	.6743	13	53.98	432	.0545	88
3	2.053	*127	1.044	7	.6756	14	58.30	427	.0633	94
4	1.926	*109	1.051	8	.6770	15	62.57	423	.0727	99
.15	1.817	*94	1.059	8	.6785	16	66.80°	418	.0826	103
6	1.723	*82	1.067	8	.6801	17	70.98	413	.0929	107
7	1.641	*72	1.075	9	.6818	18	75.11	409	.1036	111
8	1.569	*63	1.084	10	.6836	19	79.20	403	.1147	116
9	1.506	56	1.094	9	.6855	20	83.23	399	.1263	116
.20	1.450	50	1.103	11	.6875	21	87.21°	392	.1379	120
1	1.400	44	1.114	10	.6896	22	91.13	387	.1499	123
2	1.356	39	1.124	12	.6918	23	95.00	381	.1622	124
3	1.317	35	1.136	11	.6941	24	98.81	375	.1746	127
4	1.282	32	1.147	12	.6965	24	102.56	370	.1873	127
.25	1.250	28	1.159	12	.6989	25	106.26°	364	.2000	128
6	1.222	26	1.171	13	.7014	27	109.90	358	.2128	130
7	1.196	23	1.184	13	.7041	27	113.48	352	.2258	129
8	1.173	21	1.197	14	.7068	28	117.00	345	.2387	130
9	1.152	19	1.211	14	.7096	29	120.45	341	.2517	130
.30	1.133	17	1.225	14	.7125	29	123.86°	334	.2647	130
1	1.116	15	1.239	15	.7154	31	127.20	328	.2777	129
2	1.101	13	1.254	15	.7185	31	130.48	322	.2906	128
3	1.088	13	1.269	15	.7216	32	133.70	316	.3034	128
4	1.075	11	1.284	16	.7248	32	136.86	311	.3162	127
.35	1.064	10	1.300	16	.7280	34	139.97°	305	.3289	125
6	1.054	8	1.316	16	.7314	34	143.02	299	.3414	124
7	1.046	8	1.332	17	.7348	35	146.01	293	.3538	123
8	1.038	7	1.349	17	.7383	36	148.94	288	.3661	122
9	1.031	6	1.366	17	.7419	36	151.82	282	.3783	119
.40	1.025	5	1.383	18	.7455	37	154.64°	277	.3902	119
1	1.020	5	1.401	18	.7492	38	157.41	271	.4021	116
2	1.015	4	1.419	18	.7530	38	160.12	266	.4137	115
3	1.011	3	1.437	18	.7568	39	162.78	261	.4252	112
4	1.008	2	1.455	19	.7607	40	165.39	256	.4364	111
.45	1.006	3	1.474	19	.7647	40	167.95°	251	.4475	109
6	1.003	1	1.493	19	.7687	41	170.46	245	.4584	107
7	1.002	1	1.512	19	.7728	41	172.91	241	.4691	105
8	1.001	1	1.531	20	.7769	42	175.32	237	.4796	103
9	1.000	0	1.551	20	.7811	43	177.69	231	.4899	101
.50	1.000		1.571		.7854		180.00°		.5000	

* Interpolation may be inaccurate at these points.

Table 1.1.2   Segments of Circles, Given *h/D*

Given: h = height; D = diameter of circle. To find the chord, the length of arc, or the area of the segment, form the ratio h/D, and find from the table the value of (chord/D), (arc/D), or (area/D^2); then by a simple multiplication,

$$\text{chord} = D \times (\text{chord}/D)$$
$$\text{arc} = D \times (\text{arc}/D)$$
$$\text{area} = D^2 \times (\text{area}/D^2)$$

This table gives also the angle subtended at the center, the ratio of the arc of the segment of the whole circumference, and the ratio of the area of the segment to the area of the whole circle.

$\dfrac{h}{D}$	$\dfrac{\text{Arc}}{D}$	Diff	$\dfrac{\text{Area}}{D^2}$	Diff	Central angle, v	Diff	$\dfrac{\text{Chord}}{D}$	Diff	$\dfrac{\text{Arc}}{\text{Circum}}$	Diff	$\dfrac{\text{Area}}{\text{Circle}}$	Diff
.00	0.000	2003	.0000	13	0.00°	2296	.0000	*1990	.0000	*638	.0000	17
1	.2003	*835	.0013	24	22.96	*956	.1990	*810	.0638	*265	.0017	31
2	.2838	*644	.0037	32	32.52	*738	.2800	*612	.0903	*205	.0048	39
3	.3482	*545	.0069	36	39.90	*625	.3412	*507	.1108	*174	.0087	47
4	.4027	*483	.0105	42	46.15	*553	.3919	*440	.1282	*154	.0134	53
.05	.4510	*439	.0147	45	51.68°	*504	.4359	*391	.1436	*139	.0187	58
6	.4949	*406	.0192	50	56.72	*465	.4750	*353	.1575	*130	.0245	63
7	.5355	*380	.0242	52	61.37	*435	.5103	*323	.1705	121	.0308	67
8	.5735	*359	.0294	56	65.72	*411	.5426	*298	.1826	114	.0375	71
9	.6094	*341	.0350	59	69.83	*391	.5724	*276	.1940	108	.0446	74
.10	.6435	*326	.0409	61	73.74°	*374	.6000	*258	.2048	104	.0520	78
1	.6761	*314	.0470	64	77.48	*359	.6258	*241	.2152	100	.0598	82
2	.7075	*302	.0534	66	81.07	*347	.6499	*227	.2252	96	.0680	84
3	.7377	*293	.0600	68	84.54	*335	.6726	*214	.2348	93	.0764	87
4	.7670	*284	.0668	71	87.89	*326	.6940	*201	.2441	91	.0851	90
.15	.7954	276	.0739	72	91.15°	316	.7141	*191	.2532	88	.0941	92
6	.8230	270	.0811	74	94.31	309	.7332	*181	.2620	86	.1033	94
7	.8500	263	.0885	76	97.40	302	.7513	*171	.2706	83	.1127	97
8	.8763	258	.0961	78	100.42	295	.7684	162	.2789	82	.1224	99
9	.9021	252	.1039	79	103.37	289	.7846	154	.2871	81	.1323	101
.20	0.9273	248	.1118	81	106.26°	284	.8000	146	.2952	79	.1424	103
1	0.9521	243	.1199	82	109.10	279	.8146	139	.3031	77	.1527	104
2	0.9764	240	.1281	84	111.89	274	.8285	132	.3108	76	.1631	107
3	1.0004	235	.1365	84	114.63	271	.8417	125	.3184	75	.1738	108
4	1.0239	233	.1449	86	117.34	266	.8542	118	.3259	74	.1846	109
.25	1.0472	229	.1535	88	120.00°	263	.8660	113	.3333	73	.1955	111
6	1.0701	227	.1623	88	122.63	260	.8773	106	.3406	72	.2066	112
7	1.0928	224	.1711	89	125.23	256	.8879	101	.3478	72	.2178	114
8	1.1152	222	.1800	90	127.79	254	.8980	95	.3550	70	.2292	115
9	1.1374	219	.1890	92	130.33	251	.9075	90	.3620	70	.2407	116
.30	1.1593	217	.1982	92	132.84°	249	.9165	85	.3690	69	.2523	117
1	1.1810	215	.2074	93	135.33	247	.9250	80	.3759	69	.2640	119
2	1.2025	214	.2167	93	137.80	245	.9330	74	.3828	68	.2759	119
3	1.2239	212	.2260	95	140.25	242	.9404	70	.3896	67	.2878	120
4	1.2451	210	.2355	95	142.67	241	.9474	65	.3963	67	.2998	121
.35	1.2661	209	.2450	96	145.08°	240	.9539	61	.4030	67	.3119	122
6	1.2870	208	.2546	96	147.48	238	.9600	56	.4097	66	.3241	123
7	1.3078	206	.2642	97	149.86	237	.9656	52	.4163	66	.3364	123
8	1.3284	206	.2739	97	152.23	235	.9708	47	.4229	65	.3487	124
9	1.3490	204	.2836	98	154.58	235	.9755	43	.4294	65	.3611	124
.40	1.3694	204	.2934	98	156.93°	233	.9798	39	.4359	65	.3735	125
1	1.3898	203	.3032	98	159.26	233	.9837	34	.4424	65	.3860	126
2	1.4101	202	.3130	99	161.59	231	.9871	31	.4489	64	.3986	126
3	1.4303	202	.3229	99	163.90	232	.9902	26	.4553	64	.4112	126
4	1.4505	201	.3328	100	166.22	230	.9928	22	.4617	64	.4238	126
.45	1.4706	201	.3428	99	168.52°	230	.9950	18	.4681	64	.4364	127
6	1.4907	201	.3527	100	170.82	230	.9968	14	.4745	64	.4491	127
7	1.5108	200	.3627	100	173.12	229	.9982	10	.4809	64	.4618	127
8	1.5308	200	.3727	100	175.41	230	.9992	6	.4873	63	.4745	128
9	1.5508	200	.3827	100	177.71	229	.9998	2	.4936	64	.4873	127
.50	1.5708		.3927		180.00°		1.0000		.5000		.5000	

* Interpolation may be inaccurate at these points.

Table 1.1.3 Regular Polygons

n = number of sides

$v = 360°/n$ = angle subtended at the center by one side

a = length of one side $= R\left(2\sin\dfrac{v}{2}\right) = r\left(2\tan\dfrac{v}{2}\right)$

R = radius of circumscribed circle $= a\left(\tfrac{1}{2}\csc\dfrac{v}{2}\right) = r\left(\sec\dfrac{v}{2}\right)$

r = radius of inscribed circle $= R\left(\cos\dfrac{v}{2}\right) = a\left(\tfrac{1}{2}\cot\dfrac{v}{2}\right)$

Area $= a^2\left(\tfrac{1}{4}n\cot\dfrac{v}{2}\right) = R^2(\tfrac{1}{2}\,n\sin v) = r^2\left(n\tan\dfrac{v}{2}\right)$

n	v	$\dfrac{\text{Area}}{a^2}$	$\dfrac{\text{Area}}{R^2}$	$\dfrac{\text{Area}}{r^2}$	$\dfrac{R}{a}$	$\dfrac{R}{r}$	$\dfrac{a}{R}$	$\dfrac{a}{r}$	$\dfrac{r}{R}$	$\dfrac{r}{a}$
3	120°	0.4330	1.299	5.196	0.5774	2.000	1.732	3.464	0.5000	0.2887
4	90°	1.000	2.000	4.000	0.7071	1.414	1.414	2.000	0.7071	0.5000
5	72°	1.721	2.378	3.633	0.8507	1.236	1.176	1.453	0.8090	0.6882
6	60°	2.598	2.598	3.464	1.0000	1.155	1.000	1.155	0.8660	0.8660
7	51°.43	3.634	2.736	3.371	1.152	1.110	0.8678	0.9631	0.9010	1.038
8	45°	4.828	2.828	3.314	1.307	1.082	0.7654	0.8284	0.9239	1.207
9	40°	6.182	2.893	3.276	1.462	1.064	0.6840	0.7279	0.9397	1.374
10	36°	7.694	2.939	3.249	1.618	1.052	0.6180	0.6498	0.9511	1.539
12	30°	11.20	3.000	3.215	1.932	1.035	0.5176	0.5359	0.9659	1.866
15	24°	17.64	3.051	3.188	2.405	1.022	0.4158	0.4251	0.9781	2.352
16	22°.50	20.11	3.062	3.183	2.563	1.020	0.3902	0.3978	0.9808	2.514
20	18°	31.57	3.090	3.168	3.196	1.013	0.3129	0.3168	0.9877	3.157
24	15°	45.58	3.106	3.160	3.831	1.009	0.2611	0.2633	0.9914	3.798
32	11°.25	81.23	3.121	3.152	5.101	1.005	0.1960	0.1970	0.9952	5.077
48	7°.50	183.1	3.133	3.146	7.645	1.002	0.1308	0.1311	0.9979	7.629
64	5°.625	325.7	3.137	3.144	10.19	1.001	0.0981	0.0983	0.9968	10.18

Table 1.1.4 Binomial Coefficients

$(n)_0 = 1 \quad (n)_1 = n \quad (n)_2 = \dfrac{n(n-1)}{1\times 2} \quad (n)_3 = \dfrac{n(n-1)(n-2)}{1\times 2\times 3}$ etc. in general $(n)_r = \dfrac{n(n-1)(n-2)\cdots[n-(r-1)]}{1\times 2\times 3\times\cdots\times r}$. Other notations: $nC_r = \dbinom{n}{r} = (n)_r$

n	$(n)_0$	$(n)_1$	$(n)_2$	$(n)_3$	$(n)_4$	$(n)_5$	$(n)_6$	$(n)_7$	$(n)_8$	$(n)_9$	$(n)_{10}$	$(n)_{11}$	$(n)_{12}$	$(n)_{13}$
1	1	1	......	......	......	......	......	......	......	......	......	......	......	......
2	1	2	1	......	......	......	......	......	......	......	......	......	......	......
3	1	3	3	1	......	......	......	......	......	......	......	......	......	......
4	1	4	6	4	1	......	......	......	......	......	......	......	......	......
5	1	5	10	10	5	1	......	......	......	......	......	......	......	......
6	1	6	15	20	15	6	1	......	......	......	......	......	......	......
7	1	7	21	35	35	21	7	1	......	......	......	......	......	......
8	1	8	28	56	70	56	28	8	1	......	......	......	......	......
9	1	9	36	84	126	126	84	36	9	1	......	......	......	......
10	1	10	45	120	210	252	210	120	45	10	1	......	......	......
11	1	11	55	165	330	462	462	330	165	55	11	1	......	......
12	1	12	66	220	495	792	924	792	495	220	66	12	1	......
13	1	13	78	286	715	1287	1716	1716	1287	715	286	78	13	1
14	1	14	91	364	1001	2002	3003	3432	3003	2002	1001	364	91	14
15	1	15	105	455	1365	3003	5005	6435	6435	5005	3003	1365	455	105

NOTE: For $n = 14$, $(n)_{14} = 1$; for $n = 15$, $(n)_{14} = 15$, and $(n)_{15} = 1$.

Table 1.1.5   Compound Interest. Amount of a Given Principal

The amount A at the end of n years of a given principal P placed at compound interest today is $A = P \times x$ or $A = P \times y$, according as the interest (at the rate of r percent per annum) is compounded annually, or continuously; the factor x or y being taken from the following tables.

Values of x (interest compounded annually: $A = P \times x$)

Years	$r = 2$	3	4	5	6	7	8	10	12
1	1.0200	1.0300	1.0400	1.0500	1.0600	1.0700	1.0800	1.1000	1.1200
2	1.0404	1.0609	1.0816	1.1025	1.1236	1.1449	1.1664	1.2100	1.2544
3	1.0612	1.0927	1.1249	1.1576	1.1910	1.2250	1.2597	1.3310	1.4049
4	1.0824	1.1255	1.1699	1.2155	1.2625	1.3108	1.3605	1.4641	1.5735
5	1.1041	1.1593	1.2167	1.2763	1.3382	1.4026	1.4693	1.6105	1.7623
6	1.1262	1.1941	1.2653	1.3401	1.4185	1.5007	1.5869	1.7716	1.9738
7	1.1487	1.2299	1.3159	1.4071	1.5036	1.6058	1.7138	1.9487	2.2107
8	1.1717	1.2668	1.3686	1.4775	1.5938	1.7182	1.8509	2.1436	2.4760
9	1.1951	1.3048	1.4233	1.5513	1.6895	1.8385	1.9990	2.3579	2.7731
10	1.2190	1.3439	1.4802	1.6289	1.7908	1.9672	2.1589	2.5937	3.1058
11	1.2434	1.3842	1.5395	1.7103	1.8983	2.1049	2.3316	2.8531	3.4785
12	1.2682	1.4258	1.6010	1.7959	2.0122	2.2522	2.5182	3.1384	3.8960
13	1.2936	1.4685	1.6651	1.8856	2.1329	2.4098	2.7196	3.4523	4.3635
14	1.3195	1.5126	1.7317	1.9799	2.2609	2.5785	2.9372	3.7975	4.8871
15	1.3459	1.5580	1.8009	2.0789	2.3966	2.7590	3.1722	4.1772	5.4736
16	1.3728	1.6047	1.8730	2.1829	2.5404	2.9522	3.4259	4.5950	6.1304
17	1.4002	1.6528	1.9479	2.2920	2.6928	3.1588	3.7000	5.0545	6.8660
18	1.4282	1.7024	2.0258	2.4066	2.8543	3.3799	3.9960	5.5599	7.6900
19	1.4568	1.7535	2.1068	2.5270	3.0256	3.6165	4.3157	6.1159	8.6128
20	1.4859	1.8061	2.1911	2.6533	3.2071	3.8697	4.6610	6.7275	9.6463
25	1.6406	2.0938	2.6658	3.3864	4.2919	5.4274	6.8485	10.835	17.000
30	1.8114	2.4273	3.2434	4.3219	5.7435	7.6123	10.063	17.449	29.960
40	2.2080	3.2620	4.8010	7.0400	10.286	14.974	21.725	45.259	93.051
50	2.6916	4.3839	7.1067	11.467	18.420	29.457	46.902	117.39	289.00
60	3.2810	5.8916	10.520	18.679	32.988	57.946	101.26	304.48	897.60

NOTE: This table is computed from the formula $x = [1 + (r/100)]^n$.

Values of y (interest compounded continuously: $A = P \times y$)

Years	$r = 2$	3	4	5	6	7	8	10	12
1	1.0202	1.0305	1.0408	1.0513	1.0618	1.0725	1.0833	1.1052	1.1275
2	1.0408	1.0618	1.0833	1.1052	1.1275	1.1503	1.1735	1.2214	1.2712
3	1.0618	1.0942	1.1275	1.1618	1.1972	1.2337	1.2712	1.3499	1.4333
4	1.0833	1.1275	1.1735	1.2214	1.2712	1.3231	1.3771	1.4918	1.6161
5	1.1052	1.1618	1.2214	1.2840	1.3499	1.4191	1.4918	1.6487	1.8221
6	1.1275	1.1972	1.2712	1.3499	1.4333	1.5220	1.6161	1.8221	2.0544
7	1.1503	1.2337	1.3231	1.4191	1.5220	1.6323	1.7507	2.0138	2.3164
8	1.1735	1.2712	1.3771	1.4918	1.6161	1.7507	1.8965	2.2255	2.6117
9	1.1972	1.3100	1.4333	1.5683	1.7160	1.8776	2.0544	2.4596	2.9447
10	1.2214	1.3499	1.4918	1.6487	1.8221	2.0138	2.2255	2.7183	3.3201
11	1.2461	1.3910	1.5527	1.7333	1.9348	2.1598	2.4109	3.0042	3.7434
12	1.2712	1.4333	1.6161	1.8221	2.0544	2.3164	2.6117	3.3201	4.2207
13	1.2969	1.4770	1.6820	1.9155	2.1815	2.4843	2.8292	3.6693	4.7588
14	1.3231	1.5220	1.7507	2.0138	2.3164	2.6645	3.0649	4.0552	5.3656
15	1.3499	1.5683	1.8221	2.1170	2.4596	2.8577	3.3201	4.4817	6.0496
16	1.3771	1.6161	1.8965	2.2255	2.6117	3.0649	3.5966	4.9530	6.8210
17	1.4049	1.6653	1.9739	2.3396	2.7732	3.2871	3.8962	5.4739	7.6906
18	1.4333	1.7160	2.0544	2.4596	2.9447	3.5254	4.2207	6.0496	8.6711
19	1.4623	1.7683	2.1383	2.5857	3.1268	3.7810	4.5722	6.6859	9.7767
20	1.4918	1.8221	2.2255	2.7183	3.3201	4.0552	4.9530	7.3891	11.023
25	1.6487	2.1170	2.7183	3.4903	4.4817	5.7546	7.3891	12.182	20.086
30	1.8221	2.4596	3.3201	4.4817	6.0496	8.1662	11.023	20.086	36.598
40	2.2255	3.3201	4.9530	7.3891	11.023	16.445	24.533	54.598	121.51
50	2.7183	4.4817	7.3891	12.182	20.086	33.115	54.598	148.41	403.43
60	3.3201	6.0496	11.023	20.086	36.598	66.686	121.51	403.43	1339.4

FORMULA: $y = e^{(r/100) \times n}$.

Table 1.1.6   Principal Which Will Amount to a Given Sum

The principal P, which, if placed at compound interest today, will amount to a given sum A at the end of n years $P = A \times x'$ or $P = A \times y'$, according as the interest (at the rate of r percent per annum) is compounded annually, or continuously; the factor x' or y' being taken from the following tables.

Values of x' (interest compounded annually: $P = A \times x'$)

Years	$r = 2$	3	4	5	6	7	8	10	12
1	.98039	.97087	.96154	.95238	.94340	.93458	.92593	.90909	.89286
2	.96117	.94260	.92456	.90703	.89000	.87344	.85734	.82645	.79719
3	.94232	.91514	.88900	.86384	.83962	.81630	.79383	.75131	.71178
4	.92385	.88849	.85480	.82270	.79209	.76290	.73503	.68301	.63552
5	.90573	.86261	.82193	.78353	.74726	.71299	.68058	.62092	.56743
6	.88797	.83748	.79031	.74622	.70496	.66634	.63017	.56447	.50663
7	.87056	.81309	.75992	.71068	.66506	.62275	.58349	.51316	.45235
8	.85349	.78941	.73069	.67684	.62741	.58201	.54027	.46651	.40388
9	.83676	.76642	.70259	.64461	.59190	.54393	.50025	.42410	.36061
10	.82035	.74409	.67556	.61391	.55839	.50835	.46319	.38554	.32197
11	.80426	.72242	.64958	.58468	.52679	.47509	.42888	.35049	.28748
12	.78849	.70138	.62460	.55684	.49697	.44401	.39711	.31863	.25668
13	.77303	.68095	.60057	.53032	.46884	.41496	.36770	.28966	.22917
14	.75788	.66112	.57748	.50507	.44230	.38782	.34046	.26333	.20462
15	.74301	.64186	.55526	.48102	.41727	.36245	.31524	.23939	.18270
16	.72845	.62317	.53391	.45811	.39365	.33873	.29189	.21763	.16312
17	.71416	.60502	.51337	.43630	.37136	.31657	.27027	.19784	.14564
18	.70016	.58739	.49363	.41552	.35034	.29586	.25025	.17986	.13004
19	.68643	.57029	.47464	.39573	.33051	.27651	.23171	.16351	.11611
20	.67297	.55368	.45639	.37689	.31180	.25842	.21455	.14864	.10367
25	.60953	.47761	.37512	.29530	.23300	.18425	.14602	.09230	.05882
30	.55207	.41199	.30832	.23138	.17411	.13137	.09938	.05731	.03338
40	.45289	.30656	.20829	.14205	.09722	.06678	.04603	.02209	.01075
50	.37153	.22811	.14071	.08720	.05429	.03395	.02132	.00852	.00346
60	.30478	.16973	.09506	.05354	.03031	.01726	.00988	.00328	.00111

FORMULA: $x' = [1 + (r/100)]^{-n} = 1/x$.

Values of y' (interest compounded continuously: $P = A \times y'$)

Years	$r = 2$	3	4	5	6	7	8	10	12
1	.98020	.97045	.96079	.95123	.94176	.93239	.92312	.90484	.88692
2	.96079	.94176	.92312	.90484	.88692	.86936	.85214	.81873	.78663
3	.94176	.91393	.88692	.86071	.83527	.81058	.78663	.74082	.69768
4	.92312	.88692	.85214	.81873	.78663	.75578	.72615	.67032	.61878
5	.90484	.86071	.81873	.77880	.74082	.70469	.67032	.60653	.54881
6	.88692	.83527	.78663	.74082	.69768	.65705	.61878	.54881	.48675
7	.86936	.81058	.75578	.70469	.65705	.61263	.57121	.49659	.43171
8	.85214	.78663	.72615	.67032	.61878	.57121	.52729	.44933	.38289
9	.83527	.76338	.69768	.63763	.58275	.53259	.48675	.40657	.33960
10	.81873	.74082	.67032	.60653	.54881	.49659	.44933	.36788	.30119
11	.80252	.71892	.64404	.57695	.51685	.46301	.41478	.33287	.26714
12	.78663	.69768	.61878	.54881	.48675	.43171	.38289	.30119	.23693
13	.77105	.67706	.59452	.52205	.45841	.40252	.35345.	.27253	.21014
14	.75578	.65705	.57121	.49659	.43171	.37531	.32628	.24660	.18637
15	.74082	.63763	.54881	.47237	.40657	.34994	.30119	.22313	.16530
16	.72615	.61878	.52729	.44933	.38289	.32628	.27804	.20190	.14661
17	.71177	.60050	.50662	.42741	.36059	.30422	.25666	.18268	.13003
18	.69768	.58275	.48675	.40657	.33960	.28365	.23693	.16530	.11533
19	.68386	.56553	.46767	.38674	.31982	.26448	.21871	.14957	.10228
20	.67032	.54881	.44933	.36788	.30119	.24660	.20190	.13534	.09072
25	.60653	.47237	.36788	.28650	.22313	.17377	.13534	.08208	.04979
30	.54881	.40657	.30119	.22313	.16530	.12246	.09072	.04979	.02732
40	.44933	.30119	.20190	.13534	.09072	.06081	.04076	.01832	.00823
50	.36788	.22313	.13534	.08208	.04979	.03020	.01832	.00674	.00248
60	.30119	.16530	.09072	.04979	.02732	.01500	.00823	.00248	.00075

FORMULA: $y' = e^{-(r/100) \times n} = 1/y$.

Table 1.1.7   Amount of an Annuity

The amount S accumulated at the end of n years by a given annual payment Y set aside at the end of each year is $S = Y \times v$, where the factor v is to be taken from the following table (interest at r percent per annum, compounded annually).

	Values of v								
Years	$r = 2$	3	4	5	6	7	8	10	12
1	1.0000	1.0000	1.0000	1.0000	1.0000	1.0000	1.0000	1.0000	1.0000
2	2.0200	2.0300	2.0400	2.0500	2.0600	2.0700	2.0800	2.1000	2.1200
3	3.0604	3.0909	3.1216	3.1525	3.1836	3.2149	3.2464	3.3100	3.3744
4	4.1216	4.1836	4.2465	4.3101	4.3746	4.4399	4.5061	4.6410	4.7793
5	5.2040	5.3091	5.4163	5.5256	5.6371	5.7507	5.8666	6.1051	6.3528
6	6.3081	6.4684	6.6330	6.8019	6.9753	7.1533	7.3359	7.7156	8.1152
7	7.4343	7.6625	7.8983	8.1420	8.3938	8.6540	8.9228	9.4872	10.089
8	8.5830	8.8923	9.2142	9.5491	9.8975	10.260	10.637	11.436	12.300
9	9.7546	10.159	10.583	11.027	11.491	11.978	12.488	13.579	14.776
10	10.950	11.464	12.006	12.578	13.181	13.816	14.487	15.937	17.549
11	12.169	12.808	13.486	14.207	14.972	15.784	16.645	18.531	20.655
12	13.412	14.192	15.026	15.917	16.870	17.888	18.977	21.384	24.133
13	14.680	15.618	16.627	17.713	18.882	20.141	21.495	24.523	28.029
14	15.974	17.086	18.292	19.599	21.015	22.550	24.215	27.975	32.393
15	17.293	18.599	20.024	21.579	23.276	25.129	27.152	31.772	37.280
16	18.639	20.157	21.825	23.657	25.673	27.888	30.324	35.950	42.753
17	20.012	21.762	23.698	25.840	28.213	30.840	33.750	40.545	48.884
18	21.412	23.414	25.645	28.132	30.906	33.999	37.450	45.599	55.750
19	22.841	25.117	27.671	30.539	33.760	37.379	41.446	51.159	63.440
20	24.297	26.870	29.778	33.066	36.786	40.995	45.762	57.275	72.052
25	32.030	36.459	41.646	47.727	54.865	63.249	73.106	98.347	133.33
30	40.568	47.575	56.085	66.439	79.058	94.461	113.28	164.49	241.33
40	60.402	75.401	95.026	120.80	154.76	199.64	259.06	442.59	767.09
50	84.579	112.80	152.67	209.35	290.34	406.53	573.77	1163.9	2400.0
60	114.05	163.05	237.99	353.58	533.13	813.52	1253.2	3034.8	7471.6

FORMULA: $v\{[1 + (r/100)]^n - 1\} \div (r/100) = (x - 1) \div (r/100)$.

Table 1.1.8   Annuity Which Will Amount to a Given Sum (Sinking Fund)

The annual payment Y which, if set aside at the end of each year, will amount with accumulated interest to a given sum S at the end of n years is $Y = S \times v'$, where the factor v' is given below (interest at r percent per annum, compounded annually).

	Values of v'								
Years	$r = 2$	3	4	5	6	7	8	10	12
1	1.0000	1.0000	1.0000	1.0000	1.0000	1.0000	1.0000	1.0000	1.0000
2	.49505	.49261	.49020	.48780	.48544	.48309	.48077	.47619	.47170
3	.32675	.32353	.32035	.31721	.31411	.31105	.30803	.30211	.29635
4	.24262	.23903	.23549	.23201	.22859	.22523	.22192	.21547	.20923
5	.19216	.18835	.18463	.18097	.17740	.17389	.17046	.16380	.15741
6	.15853	.15460	.15076	.14702	.14336	.13980	.13632	.12961	.12323
7	.13451	.13051	.12661	.12282	.11914	.11555	.11207	.10541	.09912
8	.11651	.11246	.10853	.10472	.10104	.09747	.09401	.08744	.08130
9	.10252	.09843	.09449	.09069	.08702	.08349	.08008	.07364	.06768
10	.09133	.08723	.08329	.07950	.07587	.07238	.06903	.06275	.05698
11	.08218	.07808	.07415	.07039	.06679	.06336	.06008	.05396	.04842
12	.07456	.07046	.06655	.06283	.05928	.05590	.05270	.04676	.04144
13	.06812	.06403	.06014	.05646	.05296	.04965	.04652	.04078	.03568
14	.06260	.05853	.05467	.05102	.04758	.04434	.04130	.03575	.03087
15	.05783	.05377	.04994	.04634	.04296	.03979	.03683	.03147	.02682
16	.05365	.04961	.04582	.04227	.03895	.03586	.03298	.02782	.02339
17	.04997	.04595	.04220	.03870	.03544	.03243	.02963	.02466	.02046
18	.04670	.04271	.03899	.03555	.03236	.02941	.02670	.02193	.01794
19	.04378	.03981	.03614	.03275	.02962	.02675	.02413	.01955	.01576
20	.04116	.03722	.03358	.03024	.02718	.02439	.02185	.01746	.01388
25	.03122	.02743	.02401	.02095	.01823	.01581	.01368	.01017	.00750
30	.02465	.02102	.01783	.01505	.01265	.01059	.00883	.00608	.00414
40	.01656	.01326	.01052	.00828	.00646	.00501	.00386	.00226	.00130
50	.01182	.00887	.00655	.00478	.00344	.00246	.00174	.00086	.00042
60	.00877	.00613	.00420	.00283	.00188	.00123	.00080	.00033	.00013

FORMULA: $v' = (r/100) \div \{[1 + (r/100)]^n - 1\} = 1/v$.

Table 1.1.9   Present Worth of an Annuity

The capital C which, if placed at interest today, will provide for a given annual payment Y for a term of n years before it is exhausted is $C = Y \times w$, where the factor w is given below (interest at r percent per annum, compounded annually).

Years	\multicolumn{9}{c}{Values of w}
	$r = 2$	3	4	5	6	7	8	10	12
1	.98039	.97087	.96154	.95238	.94340	.93458	.92593	.90909	.89286
2	1.9416	1.9135	1.8861	1.8594	1.8334	1.8080	1.7833	1.7355	1.6901
3	2.8839	2.8286	2.7751	2.7232	2.6730	2.6243	2.5771	2.4869	2.4018
4	3.8077	3.7171	3.6299	3.5460	3.4651	3.3872	3.3121	3.1699	3.0373
5	4.7135	4.5797	4.4518	4.3295	4.2124	4.1002	3.9927	3.7908	3.6048
6	5.6014	5.4172	5.2421	5.0757	4.9173	4.7665	4.6229	4.3553	4.1114
7	6.4720	6.2303	6.0021	5.7864	5.5824	5.3893	5.2064	4.8684	4.5638
8	7.3255	7.0197	6.7327	6.4632	6.2098	5.9713	5.7466	5.3349	4.9676
9	8.1622	7.7861	7.4353	7.1078	6.8017	6.5152	6.2469	5.7590	5.3282
10	8.9826	8.5302	8.1109	7.7217	7.3601	7.0236	6.7101	6.1446	5.6502
11	9.7868	9.2526	8.7605	8.3064	7.8869	7.4987	7.1390	6.4951	5.9377
12	10.575	9.9540	9.3851	8.8633	8.3838	7.9427	7.5361	6.8137	6.1944
13	11.348	10.635	9.9856	9.3936	8.8527	8.3577	7.9038	7.1034	6.4235
14	12.106	11.296	10.563	9.8986	9.2950	8.7455	8.2442	7.3667	6.6282
15	12.849	11.938	11.118	10.380	9.7122	9.1079	8.5595	7.6061	6.8109
16	13.578	12.561	11.652	10.838	10.106	9.4466	8.8514	7.8237	6.9740
17	14.292	13.166	12.166	11.274	10.477	9.7632	9.1216	8.0216	7.1196
18	14.992	13.754	12.659	11.690	10.828	10.059	9.3719	8.2014	7.2497
19	15.678	14.324	13.134	12.085	11.158	10.336	9.6036	8.3649	7.3658
20	16.351	14.877	13.590	12.462	11.470	10.594	9.8181	8.5136	7.4694
25	19.523	17.413	15.622	14.094	12.783	11.654	10.675	9.0770	7.8431
30	22.396	19.600	17.292	15.372	13.765	12.409	11.258	9.4269	8.0552
40	27.355	23.115	19.793	17.159	15.046	13.332	11.925	9.7791	8.2438
50	31.424	25.730	21.482	18.256	15.762	13.801	12.233	9.9148	8.3045
60	34.761	27.676	22.623	18.929	16.161	14.039	12.377	9.9672	8.3240

FORMULA: $w = \{1 - [1 + (r/100)]^{-n}\} \div [r/100] = v/x.$

Table 1.1.10   Annuity Provided for by a Given Capital

The annual payment Y provided for a term of n years by a given capital C placed at interest today is $Y = C \times w'$ (interest at r percent per annum, compounded annually; the fund supposed to be exhausted at the end of the term).

Years	\multicolumn{9}{c}{Values of w'}
	$r = 2$	3	4	5	6	7	8	10	12
1	1.0200	1.0300	1.0400	1.0500	1.0600	1.0700	1.0800	1.1000	1.1200
2	.51505	.52261	.53020	.53780	.54544	.55309	.56077	.57619	.59170
3	.34675	.35353	.36035	.36721	.37411	.38105	.38803	.40211	.41635
4	.26262	.26903	.27549	.28201	.28859	.29523	.30192	.31547	.32923
5	.21216	.21835	.22463	.23097	.23740	.24389	.25046	.26380	.27741
6	.17853	.18460	.19076	.19702	.20336	.20980	.21632	.22961	.24323
7	.15451	.16051	.16661	.17282	.17914	.18555	.19207	.20541	.21912
8	.13651	.14246	.14853	.15472	.16104	.16747	.17401	.18744	.20130
9	.12252	.12843	.13449	.14069	.14702	.15349	.16008	.17364	.18768
10	.11133	.11723	.12329	.12950	.13587	.14238	.14903	.16275	.17698
11	.10218	.10808	.11415	.12039	.12679	.13336	.14008	.15396	.16842
12	.09456	.10046	.10655	.11283	.11928	.12590	.13270	.14676	.16144
13	.08812	.09403	.10014	.10646	.11296	.11965	.12652	.14078	.15568
14	.08260	.08853	.09467	.10102	.10758	.11434	.12130	.13575	.15087
15	.07783	.08377	.08994	.09634	.10296	.10979	.11683	.13147	.14682
16	.07365	.07961	.08582	.09227	.09895	.10586	.11298	.12782	.14339
17	.06997	.07595	.08220	.08870	.09544	.10243	.10963	.12466	.14046
13	.06670	.07271	.07899	.08555	.09236	.09941	.10670	.12193	.13794
19	.06378	.06981	.07614	.08275	.08962	.09675	.10413	.11955	.13576
20	.06116	.06722	.07358	.08024	.08718	.09439	.10185	.11746	.13388
25	.05122	.05743	.06401	.07095	.07823	.08581	.09368	.11017	.12750
30	.04465	.05102	.05783	.06505	.07265	.08059	.08883	.10608	.12414
40	.03656	.04326	.05052	.05828	.06646	.07501	.08386	.10226	.12130
50	.03182	.03887	.04655	.05478	.06344	.07246	.08174	.10086	.12042
60	.02877	.03613	.04420	.05283	.06188	.07123	.08080	.10033	.12013

FORMULA: $w' = [r/100] \div \{1 - [1 + (r/100)]^{-n}\} = 1/w = v' + (r/100).$

Table 1.1.11   Ordinates of the Normal Density Function

$$f(x) = \frac{1}{\sqrt{2\pi}}\, e^{-x^2/2}$$

x	.00	.01	.02	.03	.04	.05	.06	.07	.08	.09
.0	.3989	.3989	.3989	.3988	.3986	.3984	.3982	.3980	.3977	.3973
.1	.3970	.3965	.3961	.3956	.3951	.3945	.3939	.3932	.3925	.3918
.2	.3910	.3902	.3894	.3885	.3876	.3867	.3857	.3847	.3836	.3825
.3	.3814	.3802	.3790	.3778	.3765	.3752	.3739	.3725	.3712	.3697
.4	.3683	.3668	.3653	.3637	.3621	.3605	.3589	.3572	.3555	.3538
.5	.3521	.3503	.3485	.3467	.3448	.3429	.3410	.3391	.3372	.3352
.6	.3332	.3312	.3292	.3271	.3251	.3230	.3209	.3187	.3166	.3144
.7	.3123	.3101	.3079	.3056	.3034	.3011	.2989	.2966	.2943	.2920
.8	.2897	.2874	.2850	.2827	.2803	.2780	.2756	.2732	.2709	.2685
.9	.2661	.2637	.2613	.2589	.2565	.2541	.2516	.2492	.2468	.2444
1.0	.2420	.2396	.2371	.2347	.2323	.2299	.2275	.2251	.2227	.2203
1.1	.2179	.2155	.2131	.2107	.2083	.2059	.2036	.2012	.1989	.1965
1.2	.1942	.1919	.1895	.1872	.1849	.1826	.1804	.1781	.1758	.1736
1.3	.1714	.1691	.1669	.1647	.1626	.1604	.1582	.1561	.1539	.1518
1.4	.1497	.1476	.1456	.1435	.1415	.1394	.1374	.1354	.1334	.1315
1.5	.1295	.1276	.1257	.1238	.1219	.1200	.1182	.1163	.1154	.1127
1.6	.1109	.1092	.1074	.1057	.1040	.1023	.1006	.0989	.0973	.0957
1.7	.0940	.0925	.0909	.0893	.0878	.0863	.0848	.0833	.0818	.0804
1.8	.0790	.0775	.0761	.0748	.0734	.0721	.0707	.0694	.0681	.0669
1.9	.0656	.0644	.0632	.0620	.0608	.0596	.0584	.0573	.0562	.0551
2.0	.0540	.0529	.0519	.0508	.0498	.0488	.0478	.0468	.0459	.0449
2.1	.0440	.0431	.0422	.0413	.0404	.0396	.0387	.0379	.0371	.0363
2.2	.0355	.0347	.0339	.0332	.0325	.0317	.0310	.0303	.0297	.0290
2.3	.0283	.0277	.0270	.0264	.0258	.0252	.0246	.0241	.0235	.0229
2.4	.0224	.0219	.0213	.0208	.0203	.0198	.0194	.0189	.0184	.0180
2.5	.0175	.0171	.0167	.0163	.0158	.0154	.0151	.0147	.0143	.0139
2.6	.0136	.0132	.0129	.0126	.0122	.0119	.0116	.0113	.0110	.0107
2.7	.0104	.0101	.0099	.0096	.0093	.0091	.0088	.0086	.0084	.0081
2.8	.0079	.0077	.0075	.0073	.0071	.0069	.0067	.0065	.0063	.0061
2.9	.0060	.0058	.0056	.0055	.0053	.0051	.0050	.0048	.0047	.0046
3.0	.0044	.0043	.0042	.0040	.0039	.0038	.0037	.0036	.0035	.0034
3.1	.0033	.0032	.0031	.0030	.0029	.0028	.0027	.0026	.0025	.0025
3.2	.0024	.0023	.0022	.0022	.0021	.0020	.0020	.0019	.0018	.0018
3.3	.0017	.0017	.0016	.0016	.0015	.0015	.0014	.0014	.0013	.0013
3.4	.0012	.0012	.0012	.0011	.0011	.0010	.0010	.0010	.0009	.0009
3.5	.0009	.0008	.0008	.0008	.0008	.0007	.0007	.0007	.0007	.0006
3.6	.0006	.0006	.0006	.0005	.0005	.0005	.0005	.0005	.0005	.0004
3.7	.0004	.0004	.0004	.0004	.0004	.0004	.0003	.0003	.0003	.0003
3.8	.0003	.0003	.0003	.0003	.0003	.0002	.0002	.0002	.0002	.0002
3.9	.0002	.0002	.0002	.0002	.0002	.0002	.0002	.0002	.0001	.0001

NOTE: x is the value in left-hand column + the value in top row.
$f(x)$ is the value in the body of the table. *Example:* $x = 2.14$; $f(x) = 0.0404$.

Table 1.1.12   Cumulative Normal Distribution

$$F(x) = \int_{-\infty}^{x} \frac{1}{\sqrt{2\pi}} e^{-t^2/2}\,dt$$

x	.00	.01	.02	.03	.04	.05	.06	.07	.08	.09
.0	.5000	.5040	.5080	.5120	.5160	.5199	.5239	.5279	.5319	.5359
.1	.5398	.5438	.5478	.5517	.5557	.5596	.5636	.5675	.5714	.5735
.2	.5793	.5832	.5871	.5910	.5948	.5987	.6026	.6064	.6103	.6141
.3	.6179	.6217	.6255	.6293	.6331	.6368	.6406	.6443	.6480	.6517
.4	.6554	.6591	.6628	.6664	.6700	.6736	.6772	.6808	.6844	.6879
.5	.6915	.6950	.6985	.7019	.7054	.7088	.7123	.7157	.7190	.7224
.6	.7257	.7291	.7324	.7357	.7389	.7422	.7454	.7486	.7517	.7549
.7	.7580	.7611	.7642	.7673	.7703	.7734	.7764	.7793	.7823	.7852
.8	.7881	.7910	.7939	.7967	.7995	.8023	.8051	.8078	.8106	.8133
.9	.8159	.8186	.8212	.8238	.8264	.8289	.8315	.8340	.8365	.8389
1.0	.8413	.8438	.8461	.8485	.8508	.8531	.8554	.8577	.8599	.8621
1.1	.8643	.8665	.8686	.8708	.8729	.8749	.8770	.8790	.8810	.8830
1.2	.8849	.8869	.8888	.8906	.8925	.8943	.8962	.8980	.8997	.9015
1.3	.9032	.9049	.9066	.9082	.9099	.9115	.9131	.9147	.9162	.9177
1.4	.9192	.9207	.9222	.9236	.9251	.9265	.9279	.9292	.9306	.9319
1.5	.9332	.9345	.9357	.9370	.9382	.9394	.9406	.9418	.9429	.9441
1.6	.9452	.9463	.9474	.9484	.9495	.9505	.9515	.9525	.9535	.9545
1.7	.9554	.9564	.9573	.9582	.9591	.9599	.9608	.9616	.9625	.9633
1.8	.9641	.9649	.9656	.9664	.9671	.9678	.9686	.9693	.9699	.9706
1.9	.9713	.9719	.9726	.9732	.9738	.9744	.9750	.9756	.9761	.9767
2.0	.9772	.9778	.9783	.9788	.9793	.9798	.9803	.9808	.9812	.9817
2.1	.9812	.9826	.9830	.9834	.9838	.9842	.9846	.9850	.9854	.9857
2.2	.9861	.9864	.9868	.9871	.9875	.9878	.9881	.9884	.9887	.9890
2.3	.9893	.9896	.9898	.9901	.9904	.9906	.9909	.9911	.9913	.9916
2.4	.9918	.9920	.9922	.9925	.9927	.9929	.9931	.9932	.9934	.9936
2.5	.9938	.9940	.9941	.9943	.9945	.9946	.9948	.9949	.9951	.9952
2.6	.9953	.9955	.9956	.9957	.9959	.9960	.9961	.9962	.9963	.9964
2.7	.9965	.9966	.9967	.9968	.9969	.9970	.9971	.9972	.9973	.9974
2.8	.9974	.9975	.9976	.9977	.9977	.9978	.9979	.9979	.9980	.9981
2.9	.9981	.9982	.9982	.9983	.9984	.9984	.9985	.9985	.9986	.9986
3.0	.9986	.9987	.9987	.9988	.9988	.9989	.9989	.9989	.9990	.9990
3.1	.9990	.9991	.9991	.9991	.9992	.9992	.9992	.9992	.9993	.9993
3.2	.9993	.9993	.9994	.9994	.9994	.9994	.9994	.9995	.9995	.9995
3.3	.9995	.9995	.9995	.9996	.9996	.9996	.9996	.9996	.9996	.9997
3.4	.9997	.9997	.9997	.9997	.9997	.9997	.9997	.9997	.9997	.9998

NOTE: $x = (a - \mu)/\sigma$ where a is the observed value, μ is the mean, and σ is the standard deviation.
x is the value in the left-hand column + the value in the top row.
$F(x)$ is the probability that a point will be less than or equal to x.
$F(x)$ is the value in the body of the table. *Example:* The probability that an observation will be less than or equal to 1.04 is .8508.
NOTE: $F(-x) = 1 - F(x)$.

Table 1.1.13   Cumulative Chi-Square Distribution

$$F(t) = \int_0^t \frac{x^{(n-2)/2} e^{-x/2} dx}{2^{n/2}[(n-2)/2]!}$$

n \ F	.005	.010	.025	.050	.100	.250	.500	.750	.900	.950	.975	.990	.995
1	.000039	.00016	.00098	.0039	.0158	.101	.455	1.32	2.70	3.84	5.02	6.62	7.86
2	.0100	.0201	.0506	.103	.211	.575	1.39	2.77	4.61	5.99	7.38	9.21	10.6
3	.0717	.155	.216	.352	.584	1.21	2.37	4.11	6.25	7.81	9.35	11.3	12.8
4	.207	.297	.484	.711	1.06	1.92	3.36	5.39	7.78	9.49	11.1	13.3	14.9
5	.412	.554	.831	1.15	1.61	2.67	4.35	· 6.63	9.24	11.1	12.8	15.1	16.7
6	.676	.872	1.24	1.64	2.20	3.45	5.35	7.84	10.6	12.6	14.4	16.8	18.5
7	.989	1.24	1.69	2.17	2.83	4.25	6.35	9.04	12.0	14.1	16.0	18.5	20.3
8	1.34	1.65	2.18	2.73	3.49	5.07	7.34	10.2	13.4	15.5	17.5	20.1	22.0
9	1.73	2.09	2.70	3.33	4.17	5.90	8.34	11.4	14.7	16.9	19.0	21.7	23.6
10	2.16	2.56	3.25	3.94	4.87	6.74	9.34	12.5	16.0	18.3	20.5	23.2	25.2
11	2.60	3.05	3.82	4.57	5.58	7.58	10.3	13.7	17.3	19.7	21.9	24.7	26.8
12	3.07	3.57	4.40	5.23	6.30	8.44	11.3	14.8	18.5	21.0	23.3	26.2	28.3
13	3.57	4.11	5.01	5.89	7.04	9.30	12.3	16.0	19.8	22.4	24.7	27.7	29.8
14	4.07	4.66	5.63	6.57	7.79	10.2	13.3	17.1	21.1	23.7	26.1	29.1	31.3
15	4.60	5.23	6.26	7.26	8.55	11.0	14.3	18.2	22.3	25.0	27.5	30.6	32.8
16	5.14	5.81	6.91	7.96	9.31	11.9	15.3	19.4	23.5	26.3	28.8	32.0	34.3
17	5.70	6.41	7.56	8.67	10.1	12.8	16.3	20.5	24.8	27.6	30.2	33.4	35.7
18	6.26	7.01	8.23	9.39	10.9	13.7	17.3	21.6	26.0	28.9	31.5	34.8	37.2
19	6.84	7.63	8.91	10.1	11.7	14.6	18.3	22.7	27.2	30.1	32.9	36.2	38.6
20	7.43	8.26	9.59	10.9	12.4	15.5	19.3	23.8	28.4	31.4	34.2	37.6	40.0
21	8.03	8.90	10.3	11.6	13.2	16.3	20.3	24.9	29.6	32.7	35.5	38.9	41.4
22	8.64	9.54	11.0	12.3	14.0	17.2	21.3	26.0	30.8	33.9	36.8	40.3	42.8
23	9.26	10.2	11.7	13.1	14.8	18.1	22.3	27.1	32.0	35.2	38.1	41.6	44.2
24	9.89	10.9	12.4	13.8	15.7	19.0	23.3	28.2	33.2	36.4	39.4	43.0	45.6
25	10.5	11.5	13.1	14.6	16.5	19.9	24.3	29.3	34.4	37.7	40.6	44.3	46.9
26	11.2	12.2	13.8	15.4	17.3	20.8	25.3	30.4	35.6	38.9	41.9	45.6	48.3
27	11.8	12.9	14.6	16.2	18.1	21.7	26.3	31.5	36.7	40.1	43.2	47.0	49.6
28	12.5	13.6	15.3	16.9	18.9	22.7	27.3	32.6	37.9	41.3	44.5	48.3	51.0
29	13.1	14.3	16.0	17.7	19.8	23.6	28.3	33.7	39.1	42.6	45.7	49.6	52.3
30	13.8	15.0	16.8	18.5	20.6	24.5	29.3	34.8	40.3	43.8	47.0	50.9	53.7

NOTE: n is the number of degrees of freedom.
Values for t are in the body of the table. *Example:* The probability that, with 16 degrees of freedom, a point will be ≤ 23.5 is .900.

Table 1.1.14   Cumulative "Student's" Distribution

$$F(t) = \int_{-\infty}^{t} \frac{\left(\dfrac{n-1}{2}\right)!}{\left(\dfrac{n-2}{2}\right)! \sqrt{\pi n}\left(1 + \dfrac{x^2}{n}\right)^{(n+1)/2}} dx$$

n	.75	.90	.95	.975	.99	.995	.9995
1	1.000	3.078	6.314	12.70	31.82	63.66	636.3
2	.816	1.886	2.920	4.303	6.965	9.925	31.60
3	.765	1.638	2.353	3.182	4.541	5.841	12.92
4	.741	1.533	2.132	2.776	3.747	4.604	8.610
5	.727	1.476	2.015	2.571	3.365	4.032	6.859
6	.718	1.440	1.943	2.447	3.143	3.707	5.959
7	.711	1.415	1.895	2.365	2.998	3.499	5.408
8	.706	1.397	1.860	2.306	2.896	3.355	5.041
9	.703	1.383	1.833	2.262	2.821	3.250	4.781
10	.700	1.372	1.812	2.228	2.764	3.169	4.587
11	.697	1.363	1.796	2.201	2.718	3.106	4.437
12	.695	1.356	1.782	2.179	2.681	3.055	4.318
13	.694	1.350	1.771	2.160	2.650	3.012	4.221
14	.692	1.345	1.761	2.145	2.624	2.977	4.140
15	.691	1.341	1.753	2.131	2.602	2.947	4.073
16	.690	1.337	1.746	2.120	2.583	2.921	4.015
17	.689	1.333	1.740	2.110	2.567	2.898	3.965
18	.688	1.330	1.734	2.101	2.552	2.878	3.922
19	.688	1.328	1.729	2.093	2.539	2.861	3.883
20	.687	1.325	1.725	2.086	2.528	2.845	3.850
21	.686	1.323	1.721	2.080	2.518	2.831	3.819
22	.686	1.321	1.717	2.074	2.508	2.819	3.792
23	.685	1.319	1.714	2.069	2.500	2.807	3.768
24	.685	1.318	1.711	2.064	2.492	2.797	3.745
25	.684	1.316	1.708	2.060	2.485	2.787	3.725
26	.684	1.315	1.706	2.056	2.479	2.779	3.707
27	.684	1.314	1.703	2.052	2.473	2.771	3.690
28	.683	1.313	1.701	2.048	2.467	2.763	3.674
29	.683	1.311	1.699	2.045	2.462	2.756	3.659
30	.683	1.310	1.697	2.042	2.457	2.750	3.646
40	.681	1.303	1.684	2.021	2.423	2.704	3.551
60	.679	1.296	1.671	2.000	2.390	2.660	3.460
120	.677	1.289	1.658	1.980	2.385	2.617	3.373

NOTE: n is the number of degrees of freedom.
Values for t are in the body of the table. *Example:* The probability that, with 16 degrees of freedom, a point will be ≤ 2.921 is .995.
NOTE: $F(-t) = 1 - F(t)$.

Table 1.1.15   Cumulative *F* Distribution

m degrees of freedom in numerator; *n* in denominator

$$G(F) = \int_0^F \frac{[(m + n - 2)/2]!\, m^{m/2} n^{n/2} x^{(m-2)/2} (n + mx)^{-(m+n)/2}\, dx}{[(m - 2)/2]![(n - 2)/2]!}$$

Upper 5% points ($F_{.95}$)

Degrees of freedom for numerator

	1	2	3	4	5	6	7	8	9	10	12	15	20	24	30	40	60	120	∞
1	161	200	216	225	230	234	237	239	241	242	244	246	248	249	250	251	252	253	254
2	18.5	19.0	19.2	19.2	19.3	19.3	19.4	19.4	19.4	19.4	19.4	19.4	19.4	19.4	19.5	19.5	19.5	19.5	19.5
3	10.1	9.55	9.28	9.12	9.01	8.94	8.89	8.85	8.81	8.79	8.74	8.70	8.66	8.64	8.62	8.59	8.57	8.55	8.53
4	7.71	6.94	6.59	6.39	6.26	6.16	6.09	6.04	6.00	5.96	5.91	5.86	5.80	5.77	5.75	5.72	5.69	5.66	5.63
5	6.61	5.79	5.41	5.19	5.05	4.95	4.88	4.82	4.77	4.74	4.68	4.62	4.56	4.53	4.50	4.46	4.43	4.40	4.37
6	5.99	5.14	4.76	4.53	4.39	4.28	4.21	4.15	4.10	4.06	4.00	3.94	3.87	3.84	3.81	3.77	3.74	3.70	3.67
7	5.59	4.74	4.35	4.12	3.97	3.87	3.79	3.73	3.68	3.64	3.57	3.51	3.44	3.41	3.38	3.34	3.30	3.27	3.23
8	5.32	4.46	4.07	3.84	3.69	3.58	3.50	3.44	3.39	3.35	3.28	3.22	3.15	3.12	3.08	3.04	3.01	2.97	2.93
9	5.12	4.26	3.86	3.63	3.48	3.37	3.29	3.23	3.18	3.14	3.07	3.01	2.94	2.90	2.86	2.83	2.79	2.75	2.71
10	4.96	4.10	3.71	3.48	3.33	3.22	3.14	3.07	3.02	2.98	2.91	2.85	2.77	2.74	2.70	2.66	2.62	2.58	2.54
11	4.84	3.98	3.59	3.36	3.20	3.09	3.01	2.95	2.90	2.85	2.79	2.72	2.65	2.61	2.57	2.53	2.49	2.45	2.40
12	4.75	3.89	3.49	3.26	3.11	3.00	2.91	2.85	2.80	2.75	2.69	2.62	2.54	2.51	2.47	2.43	2.38	2.34	2.30
13	4.67	3.81	3.41	3.18	3.03	2.92	2.83	2.77	2.71	2.67	2.60	2.53	2.46	2.42	2.38	2.34	2.30	2.25	2.21
14	4.60	3.74	3.34	3.11	2.96	2.85	2.76	2.70	2.65	2.60	2.53	2.46	2.39	2.35	2.31	2.27	2.22	2.18	2.13
15	4.54	3.68	3.29	3.06	2.90	2.79	2.71	2.64	2.59	2.54	2.48	2.40	2.33	2.29	2.25	2.20	2.16	2.11	2.07
16	4.49	3.63	3.24	3.01	2.85	2.74	2.66	2.59	2.54	2.49	2.42	2.35	2.28	2.24	2.19	2.15	2.11	2.06	2.01
17	4.45	3.59	3.20	2.96	2.81	2.70	2.61	2.55	2.49	2.45	2.38	2.31	2.23	2.19	2.15	2.10	2.06	2.01	1.96
18	4.41	3.55	3.16	2.93	2.77	2.66	2.58	2.51	2.46	2.41	2.34	2.27	2.19	2.15	2.11	2.06	2.02	1.97	1.92
19	4.38	3.52	3.13	2.90	2.74	2.63	2.54	2.48	2.42	2.38	2.31	2.23	2.16	2.11	2.07	2.03	1.98	1.93	1.88
20	4.35	3.49	3.10	2.87	2.71	2.60	2.51	2.45	2.39	2.35	2.28	2.20	2.12	2.08	2.04	1.99	1.95	1.90	1.84
21	4.32	3.47	3.07	2.84	2.68	2.57	2.49	2.42	2.37	2.32	2.25	2.18	2.10	2.05	2.01	1.96	1.92	1.87	1.81
22	4.30	3.44	3.05	2.82	2.66	2.55	2.46	2.40	2.34	2.30	2.23	2.15	2.07	2.03	1.98	1.94	1.89	1.84	1.78
23	4.28	3.42	3.03	2.80	2.64	2.53	2.44	2.37	2.32	2.27	2.20	2.13	2.05	2.01	1.96	1.91	1.86	1.81	1.76
24	4.26	3.40	3.01	2.78	2.62	2.51	2.42	2.36	2.30	2.25	2.18	2.11	2.03	1.98	1.94	1.89	1.84	1.79	1.73
25	4.24	3.39	2.99	2.76	2.60	2.49	2.40	2.34	2.28	2.24	2.16	2.09	2.01	1.96	1.92	1.87	1.82	1.77	1.71
30	4.17	3.32	2.92	2.69	2.53	2.42	2.33	2.27	2.21	2.16	2.09	2.01	1.93	1.89	1.84	1.79	1.74	1.68	1.62
40	4.08	3.23	2.84	2.61	2.45	2.34	2.25	2.18	2.12	2.08	2.00	1.92	1.84	1.79	1.74	1.69	1.64	1.58	1.51
60	4.00	3.15	2.76	2.53	2.37	2.25	2.17	2.10	2.04	1.99	1.92	1.84	1.75	1.70	1.65	1.59	1.53	1.47	1.39
120	3.92	3.07	2.68	2.45	2.29	2.18	2.09	2.02	1.96	1.91	1.83	1.75	1.66	1.61	1.55	1.50	1.43	1.35	1.25
∞	3.84	3.00	2.60	2.37	2.21	2.10	2.01	1.94	1.88	1.83	1.75	1.67	1.57	1.52	1.46	1.39	1.32	1.22	1.00

Upper 1% points ($F_{.99}$)

Degrees of freedom for numerator

	1	2	3	4	5	6	7	8	9	10	12	15	20	24	30	40	60	120	∞
1	4052	5000	5403	5625	5764	5859	5928	5982	6023	6056	6106	6157	6209	6235	6261	6287	6313	6339	6366
2	98.5	99.0	99.2	99.2	99.3	99.3	99.4	99.4	99.4	99.4	99.4	99.4	99.4	99.5	99.5	99.5	99.5	99.5	99.5
3	34.1	30.8	29.5	28.7	28.2	27.9	27.7	27.5	27.3	27.2	27.1	26.9	26.7	26.6	26.5	26.4	26.3	26.2	26.1
4	21.2	18.0	16.7	16.0	15.5	15.2	15.0	14.8	14.7	14.5	14.4	14.2	14.0	13.9	13.8	13.7	13.7	13.6	13.5
5	16.3	13.3	12.1	11.4	11.0	10.7	10.5	10.3	10.2	10.1	9.89	9.72	9.55	9.47	9.38	9.29	9.20	9.11	9.02
6	13.7	10.9	9.78	9.15	8.75	8.47	8.26	8.10	7.98	7.87	7.72	7.56	7.40	7.31	7.23	7.14	7.06	6.97	6.88
7	12.2	9.55	8.45	7.85	7.46	7.19	6.99	6.84	6.72	6.62	6.47	6.31	6.16	6.07	5.99	5.91	5.82	5.74	5.65
8	11.3	8.65	7.59	7.01	6.63	6.37	6.18	6.03	5.91	5.81	5.67	5.52	5.36	5.28	5.20	5.12	5.03	4.95	4.86
9	10.6	8.02	6.99	6.42	6.06	5.80	5.61	5.47	5.35	5.26	5.11	4.96	4.81	4.73	4.65	4.57	4.48	4.40	4.31
10	10.0	7.56	6.55	5.99	5.64	5.39	5.20	5.06	4.94	4.85	4.71	4.56	4.41	4.33	4.25	4.17	4.08	4.40	3.91
11	9.65	7.21	6.22	5.67	5.32	5.07	4.89	4.74	4.63	4.54	4.40	4.25	4.10	4.02	3.94	3.86	3.78	3.69	3.60
12	9.33	6.93	5.95	5.41	5.06	4.82	4.64	4.50	4.39	4.30	4.16	4.01	3.86	3.78	3.70	3.62	3.54	3.45	3.36
13	9.07	6.70	5.74	5.21	4.86	4.62	4.44	4.30	4.19	4.10	3.96	3.82	3.66	3.59	3.51	3.43	3.34	3.25	3.17
14	8.86	6.51	5.56	5.04	4.70	4.46	4.28	4.14	4.03	3.94	3.80	3.66	3.51	3.43	3.35	3.27	3.18	3.09	3.00
15	8.68	6.36	5.42	4.89	4.56	4.32	4.14	4.00	3.89	3.80	3.67	3.52	3.37	3.29	3.21	3.13	3.05	2.96	2.87
16	8.53	6.23	5.29	4.77	4.44	4.20	4.03	3.89	3.78	3.69	3.55	3.41	3.26	3.18	3.10	3.02	2.93	2.84	2.75
17	8.40	6.11	5.19	4.67	4.34	4.10	3.93	3.79	3.68	3.59	3.46	3.31	3.16	3.08	3.00	2.92	2.83	2.75	2.65
18	8.29	6.01	5.09	4.58	4.25	4.01	3.84	3.71	3.60	3.51	3.37	3.23	3.08	3.00	2.92	2.84	2.75	2.66	2.57
19	8.19	5.93	5.01	4.50	4.17	3.94	3.77	3.63	3.52	3.43	3.30	3.15	3.00	2.92	2.84	2.76	2.67	2.58	2.49
20	8.10	5.85	4.94	4.43	4.10	3.87	3.70	3.56	3.46	3.37	3.23	3.09	2.94	2.86	2.78	2.69	2.61	2.52	2.42
21	8.02	5.78	4.87	4.37	4.04	3.81	3.64	3.51	3.40	3.31	3.17	3.03	2.88	2.80	2.72	2.64	2.55	2.46	2.36
22	7.95	5.72	4.82	4.31	3.99	3.76	3.59	3.45	3.35	3.26	3.12	2.98	2.83	2.75	2.67	2.58	2.50	2.40	2.31
23	7.88	5.66	4.76	4.26	3.94	3.71	3.54	3.41	3.30	3.21	3.07	2.93	2.78	2.70	2.62	2.54	2.45	2.35	2.26
24	7.82	5.61	4.72	4.22	3.90	3.67	3.50	3.36	3.26	3.17	3.03	2.89	2.74	2.66	2.58	2.49	2.40	2.31	2.21
25	7.77	5.57	4.68	4.18	3.86	3.63	3.46	3.32	3.22	3.13	2.99	2.85	2.70	2.62	2.53	2.45	2.36	2.27	2.17
30	7.56	5.39	4.51	4.02	3.70	3.47	3.30	3.17	3.07	2.98	2.84	2.70	2.55	2.47	2.39	2.30	2.21	2.11	2.01
40	7.31	5.18	4.31	3.83	3.51	3.29	3.12	2.99	2.89	2.80	2.66	2.52	2.37	2.29	2.20	2.11	2.02	1.92	1.80
60	7.08	4.98	4.13	3.65	3.34	3.12	2.95	2.82	2.72	2.63	2.50	2.35	2.20	2.12	2.03	1.94	1.84	1.73	1.60
120	6.85	4.79	3.95	3.48	3.17	2.96	2.79	2.66	2.56	2.47	2.34	2.19	2.03	1.95	1.86	1.76	1.66	1.53	1.38
∞	6.63	4.61	3.78	3.32	3.02	2.80	2.64	2.51	2.41	2.32	2.18	2.04	1.88	1.79	1.70	1.59	1.47	1.32	1.00

NOTE: *m* is the number of degrees of freedom in the numerator of *F*; *n* is the number of degrees of freedom in the denominator of *F*.
Values for *F* are in the body of the table.
G is the probability that a point, with *m* and *n* degrees of freedom will be ≤*F*.
Example: With 2 and 5 degrees of freedom, the probability that a point will be ≤13.3 is .99.
SOURCE: "Chemical Engineers' Handbook," 5th edition, by R. H. Perry and C. H. Chilton, McGraw-Hill, 1973. Used with permission.

Table 1.1.16   Standard Distribution of Residuals

a = any positive quantity
y = the number of residuals which are numerically $< a$
r = the probable error of a single observation
n = number of observations

$\dfrac{a}{r}$	$\dfrac{y}{n}$	Diff	$\dfrac{a}{r}$	$\dfrac{y}{n}$	Diff
0.0	.000	54	2.5	.908	13
1	.054	53	6	.921	10
2	.107	53	7	.931	10
3	.160	53	8	.941	9
4	.213	51	9	.950	7
0.5	.264	50	3.0	.957	6
6	.314	49	1	.963	6
7	.363	48	2	.969	5
8	.411	45	3	.974	4
9	.456	44	4	.978	4
1.0	.500	42	3.5	.982	3
1	.542	40	6	.985	2
2	.582	37	7	.987	3
3	.619	36	8	.990	1
4	.655	33	9	.991	2
1.5	.688	31	4.0	.993	6
6	.719	29			
7	.748	27	5.0	.999	
8	.775	25			
9	.800	23			
2.0	.823	20			
1	.843	19			
2	.862	17			
3	.879	16			
4	.895	13			

Table 1.1.17   Factors for Computing Probable Error

n	Bessel $\dfrac{0.6745}{\sqrt{(n-1)}}$	Bessel $\dfrac{0.6745}{\sqrt{n(n-1)}}$	Peters $\dfrac{0.8453}{\sqrt{n(n-1)}}$	Peters $\dfrac{0.8453}{n\sqrt{n-1}}$	n	Bessel $\dfrac{0.6745}{\sqrt{(n-1)}}$	Bessel $\dfrac{0.6745}{\sqrt{n(n-1)}}$	Peters $\dfrac{0.8453}{\sqrt{n(n-1)}}$	Peters $\dfrac{0.8453}{n\sqrt{n-1}}$
2	.6745	.4769	.5978	.4227	30	.1252	.0229	.0287	.0052
3	.4769	.2754	.3451	.1993	31	.1231	.0221	.0277	.0050
4	.3894	.1947	.2440	.1220	32	.1211	.0214	.0268	.0047
5	.3372	.1508	.1890	.0845	33	.1192	.0208	.0260	.0045
6	.3016	.1231	.1543	.0630	34	.1174	.0201	.0252	.0043
7	.2754	.1041	.1304	.0493	35	.1157	.0196	.0245	.0041
8	.2549	.0901	.1130	.0399	36	.1140	.0190	.0238	.0040
9	.2385	.0795	.0996	.0332	37	.1124	.0185	.0232	.0038
10	.2248	.0711	.0891	.0282	38	.1109	.0180	.0225	.0037
11	.2133	.0643	.0806	.0243	39	.1094	.0175	.0220	.0035
12	.2034	.0587	.0736	.0212	40	.1080	.0171	.0214	.0034
13	.1947	.0540	.0677	.0188	45	.1017	.0152	.0190	.0028
14	.1871	.0500	.0627	.0167	50	.0964	.0136	.0171	.0024
15	.1803	.0465	.0583	.0151	55	.0918	.0124	.0155	.0021
16	.1742	.0435	.0546	.0136	60	.0878	.0113	.0142	.0018
17	.1686	.0409	.0513	.0124	65	.0843	.0105	.0131	.0016
18	.1636	.0386	.0483	.0114	70	.0812	.0097	.0122	.0015
19	.1590	.0365	.0457	.0105	75	.0784	.0091	.0113	.0013
20	.1547	.0346	.0434	.0097	80	.0759	.0085	.0106	.0012
21	.1508	.0329	.0412	.0090	85	.0736	.0080	.0100	.0011
22	.1472	.0314	.0393	.0084	90	.0715	.0075	.0094	.0010
23	.1438	.0300	.0376	.0078	95	.0696	.0071	.0089	.0009
24	.1406	.0287	.0360	.0073	100	.0678	.0068	.0085	.0008
25	.1377	.0275	.0345	.0069					
26	.1349	.0265	.0332	.0065					
27	.1323	.0255	.0319	.0061					
28	.1298	.0245	.0307	.0058					
29	.1275	.0237	.0297	.0055					

Table 1.1.18 Decimal Equivalents

From minutes and seconds into decimal parts of a degree				From decimal parts of a degree into minutes and seconds (exact values)				Common fractions — 8ths	16ths	32nds	64ths	Exact decimal values
0′	0°.0000	0″	0°.0000	0°.00	0′	0°.50	30′				1	.01 5625
1	.0167	1	.0003	1	0′ 36″	1	30′ 36″		1		2	.03 125
2	.0333	2	.0006	2	1′ 12″	2	31′ 12″				3	.04 6875
3	.05	3	.0008	3	1′ 48″	3	31′ 48″		1	2	4	.06 25
4	.0667	4	.0011	4	2′ 24″	4	32′ 24″				5	.07 8125
5′	.0833	5″	.0014	0°.05	3′	0°.55	33′			3	6	.09 375
6	.10	6	.0017	6	3′ 36″	6	33′ 36″				7	.10 9375
7	.1167	7	.0019	7	4′ 12″	7	34′ 12″	1	2	4	8	.12 5
8	.1333	8	.0022	8	4′ 48″	8	34′ 48″				9	.14 0625
9	.15	9	.0025	9	5′ 24″	9	35′ 24″			5	10	.15 625
10′	0°.1667	10″	0°.0028	0°.10	6′	0°.60	36′				11	.17 1875
1	.1833	1	.0031	1	6′ 36″	1	36′ 36″		3	6	12	.18 75
2	.20	2	.0033	2	7′ 12″	2	37′ 12″				13	.20 3125
3	.2167	3	.0036	3	7′ 48″	3	37′ 48″			7	14	.21 875
4	.2333	4	.0039	4	8′ 24″	4	38′ 24″				15	.23 4375
15′	.25	15″	.0042	0°.15	9′	0°.65	39′	2	4	8	16	.25
6	.2667	6	.0044	6	9′ 36″	6	39′ 36″				17	.26 5625
7	.2833	7	.0047	7	10′ 12″	7	40′ 12″			9	18	.28 125
8	.30	8	.005	8	10′ 48″	8	40′ 48″				19	.29 6875
9	.3167	9	.0053	9	11′ 24″	9	41′ 24″		5	10	20	.31 25
20′	0°.3333	20″	0°.0056	0°.20	12′	0°.70	42′				21	.32 8125
1	.35	1	.0058	1	12′ 36″	1	42′ 36″			11	22	.34 375
2	.3667	2	.0061	2	13′ 12″	2	43′ 12″				23	.35 9375
3	.3833	3	.0064	3	13′ 48″	3	43′ 48″	3	6	12	24	.37 5
4	.40	4	.0067	4	14′ 24″	4	44′ 24″				25	.39 0625
25′	.4167	25″	.0069	0°.25	15′	0°.75	45′			13	26	.40 625
6	.4333	6	.0072	6	15′ 36″	6	45′ 36″				27	.42 1875
7	.45	7	.0075	7	16′ 12″	7	46′ 12″		7	14	28	.43 75
8	.4667	8	.0078	8	16′ 48″	8	46′ 48″				29	.45 3125
9	.4833	9	.0081	9	17′ 24″	9	47′ 24″			15	30	.46 875
30′	0°.50	30″	0°.0083	0°.30	18′	0°.80	48′				31	.48 4375
1	.5167	1	.0086	1	18′ 36″	1	48′ 36″	4	8	16	32	.50
2	.5333	2	.0089	2	19′ 12″	2	49′ 12″				33	.51 5625
3	.55	3	.0092	3	19′ 48″	3	49′ 48″			17	34	.53 125
4	.5667	4	.0094	4	20′ 24″	4	50′ 24″				35	.54 6875
35′	.5833	35″	.0097	0°.35	21′	0°.85	51′		9	18	36	.56 25
6	.60	6	.01	6	21′ 36″	6	51′ 36″				37	.57 8125
7	.6167	7	.0103	7	22′ 12″	7	52′ 12″			19	38	.59 375
8	.6333	8	.0106	8	22′ 48″	8	52′ 48″				39	.60 9375
9	.65	9	.0108	9	23′ 24″	9	53′ 24″	5	10	20	40	.62 5
40′	0°.6667	40″	0°.0111	0°.40	24′	0°.90	54′				41	.64 0625
1	.6833	1	.0114	1	24′ 36″	1	54′ 36″			21	42	.65 625
2	.70	2	.0117	2	25′ 12″	2	55′ 12″				43	.67 1875
3	.7167	3	.0119	3	25′ 48″	3	55′ 48″		11	22	44	.68 75
4	.7333	4	.0122	4	26′ 24″	4	56′ 24″				45	.70 3125
45′	.75	45″	.0125	0°.45	27′	0°.95	57′			23	46	.71 875
6	.7667	6	.0128	6	27′ 36″	6	57′ 36″				47	.73 4375
7	.7833	7	.0131	7	28′ 12″	7	58′ 12″	6	12	24	48	.75
8	.80	8	.0133	8	28′ 48″	8	58′ 48″				49	.76 5625
9	.8167	9	.0136	9	29′ 24″	9	59′ 24″			25	50	.78 125
50′	0°.8333	50″	0°.0139	0°.50	30′	1°.00	60′				51	.79 6875
1	.85	1	.0142						13	26	52	.81 25
2	.8667	2	.0144	0°.000	0″.0						53	.82 8125
3	.8833	3	.0147	1	3″.6					27	54	.84 375
4	.90	4	.015	2	7″.2						55	.85 9375
55′	.9167	55″	.0153	3	10″.8			7	14	28	56	.87 5
6	.9333	6	.0156	4	14″.4						57	.89 0625
7	.95	7	.0158	0°.005	18″					29	58	.90 625
8	.9667	8	.0161	6	21″.6						59	.92 1875
9	.9833	9	.0164	7	25″.2				15	30	60	.93 75
60′	1.00	60″	0°.0167	8	28″.8						61	.95 3125
				9	32″.4					31	62	.96 875
				0°.010	36″						63	.98 4375

1.2 MEASURING UNITS

by John T. Baumeister

REFERENCES: "International Critical Tables," McGraw-Hill. "Smithsonian Physical Tables," Smithsonian Institution. "Landolt-Börnstein: Zahlenwerte und Funktionen aus Physik, Chemie, Astronomie, Geophysik und Technik," Springer. "Handbook of Chemistry and Physics," Chemical Rubber Co. "Units and Systems of Weights and Measures; Their Origin, Development, and Present Status," NBS LC 1035 (1976). "Weights and Measures Standards of the United States, a Brief History," NBS Spec. Pub. 447 (1976). "Standard Time," Code of Federal Regulations, Title 49. "Fluid Meters, Their Theory and Application," 6th ed., chaps. 1–2, ASME, 1971. H.E. Huntley, "Dimensional Analysis," Richard & Co., New York, 1951. "U.S. Standard Atmosphere, 1962," Government Printing Office. Public Law 89-387, "Uniform Time Act of 1966." Public Law 94-168, "Metric Conversion Act of 1975." ASTM E380-91a, "Use of the International Standards of Units (SI) (the Modernized Metric System)." The International System of Units," NIST Spec. Pub. 330. "Guide for the Use of the International System of Units (SI)," NIST Spec. Pub. 811. "Guidelines for Use of the Modernized Metric System," NBS LC 1120. "NBS Time and Frequency Dissemination Services," NBS Spec. Pub. 432. "Factors for High Precision Conversion," NBS LC 1071. American Society of Mechanical Engineers SI Series, ASME SI 1–9. Jespersen and Fitz-Randolph, "From Sundials to Atomic Clocks: Understanding Time and Frequency," NBS, Monograph 155. ANSI/IEEE Std 268-1992, "American National Standard for Metric Practice."

U.S. CUSTOMARY SYSTEM (USCS)

The USCS, often called the "inch-pound system," is the system of units most commonly used for measures of weight and length (Table 1.2.1). The units are identical for practical purposes with the corresponding English units, but the capacity measures differ from those used in the British Commonwealth, the U.S. gallon being defined as 231 cu in and the bushel as 2,150.42 cu in, whereas the corresponding British Imperial units are, respectively, 277.42 cu in and 2,219.36 cu in (1 Imp gal = 1.2 U.S. gal, approx; 1 Imp bu = 1.03 U.S. bu, approx).

Table 1.2.1 U.S. Customary Units

Units of length	
12 inches	= 1 foot
3 feet	= 1 yard
5½ yards = 16½ feet	= 1 rod, pole, or perch
40 poles = 220 yards	= 1 furlong
8 furlongs = 1,760 yards = 5,280 feet	= 1 mile
3 miles	= 1 league
4 inches	= 1 hand
9 inches	= 1 span
Nautical units	
6,076.11549 feet	= 1 international nautical mile
6 feet	= 1 fathom
120 fathoms	= 1 cable length
1 nautical mile per hr	= 1 knot
Surveyor's or Gunter's units	
7.92 inches	= 1 link
100 links = 66 ft = 4 rods	= 1 chain
80 chains	= 1 mile
33⅓ inches	= 1 vara (Texas)
Units of area	
144 square inches	= 1 square foot
9 square feet	= 1 square yard
30¼ square yards	= 1 square rod, pole, or perch
160 square rods = 10 square chains = 43,560 square feet = 5,645 sq varas (Texas)	= 1 acre
640 acres = 1 square mile =	1 "section" of U.S. government-surveyed land
1 circular inch = area of circle 1 inch in diameter	= 0.7854 sq in
1 square inch	= 1.2732 circular inches
1 circular mil	= area of circle 0.001 in in diam
1,000,000 cir mils	= 1 circular inch
Units of volume	
1,728 cubic inches	= 1 cubic foot
231 cubic inches	= 1 gallon
27 cubic feet	= 1 cubic yard
1 cord of wood	= 128 cubic feet
1 perch of masonry	= 16½ to 25 cu ft

Liquid or fluid measurements

4 gills	= 1 pint
2 pints	= 1 quart
4 quarts	= 1 gallon
7.4805 gallons	= 1 cubic foot

(There is no standard liquid barrel; by trade custom, 1 bbl of petroleum oil, unrefined = 42 gal. The capacity of the common steel barrel used for refined petroleum products and other liquids is 55 gal.)

Apothecaries' liquid measurements

60 minims	= 1 liquid dram or drachm
8 drams	= 1 liquid ounce
16 ounces	= 1 pint

Water measurements

The miner's inch is a unit of water volume flow no longer used by the Bureau of Reclamation. It is used within particular water districts where its value is defined by statute. Specifically, within many of the states of the West the miner's inch is ¹⁄₅₀ cubic foot per second. In others it is equal to ¹⁄₄₀ cubic foot per second, while in the state of Colorado, 38.4 miner's inch is equal to 1 cubic-foot per second. In SI units, these correspond to $.32 \times 10^{-6}$ m³/s, $.409 \times 10^{-6}$ m³/s, and $.427 \times 10^{-6}$ m³/s, respectively.

Dry measures

2 pints	= 1 quart
8 quarts	= 1 peck
4 pecks	= 1 bushel
1 std bbl for fruits and vegetables	= 7,056 cu in or 105 dry qt, struck measure

Shipping measures

1 Register ton	= 100 cu ft
1 U.S. shipping ton	= 40 cu ft = 32.14 U.S. bu or 31.14 Imp bu
1 British shipping ton	= 42 cu ft = 32.70 Imp bu or 33.75 U.S. bu

Board measurements

(Based on nominal not actual dimensions; see Table 12.2.8)

1 board foot = $\begin{cases} 144 \text{ cu in} = \text{volume of board} \\ 1 \text{ ft sq and 1 in thick} \end{cases}$

The international log rule, based upon ¼ in kerf, is expressed by the formula

$$X = 0.904762(0.22\,D^2 - 0.71\,D)$$

where X is the number of board feet in a 4-ft section of a log and D is the top diam in in. In computing the number of board feet in a log, the taper is taken at ½ in per 4 ft linear, and separate computation is made for each 4-ft section.

Weights
(The grain is the same in all systems.)

Avoirdupois weights

16 drams = 437.5 grains	= 1 ounce
16 ounces = 7,000 grains	= 1 pound
100 pounds	= 1 cental
2,000 pounds	= 1 short ton
2,240 pounds	= 1 long ton
1 std lime bbl, small	= 180 lb net
1 std lime bbl, large	= 280 lb net
Also (in Great Britain):	
14 pounds	= 1 stone
2 stone = 28 pounds	= 1 quarter
4 quarters = 112 pounds	= 1 hundredweight (cwt)
20 hundredweight	= 1 long ton

Troy weights

24 grains	= 1 pennyweight (dwt)
20 pennyweights = 480 grains	= 1 ounce
12 ounces = 5,760 grains	= 1 pound

1 assay ton = 29,167 milligrams, or as many milligrams as there are troy ounces in a ton of 2,000 lb avoirdupois. Consequently, the number of milligrams of precious metal yielded by an assay ton of ore gives directly the number of troy ounces that would be obtained from a ton of 2,000 lb avoirdupois.

Apothecaries' weights

20 grains	= 1 scruple ℈
3 scruples = 60 grains	= 1 dram ℨ
8 drams	= 1 ounce ℥
12 ounces = 5,760 grains	= 1 pound

Weight for precious stones
1 carat = 200 milligrams
(Used by almost all important nations)

Circular measures

60 seconds	= 1 minute
60 minutes	= 1 degree
90 degrees	= 1 quadrant
360 degrees	= circumference
57.2957795 degrees	= 1 radian (or angle having
(= 57°17′44.806″)	arc of length equal to radius)

METRIC SYSTEM

In the United States the name **"metric system"** of length and mass units is commonly taken to refer to a system that was developed in France about 1800. The unit of length was equal to 1/10,000,000 of a quarter meridian (north pole to equator) and named the **metre.** A cube 1/10th metre on a side was the **litre,** the unit of volume. The mass of water filling this cube was the **kilogram,** or standard of mass; i.e., 1 litre of water = 1 kilogram of mass. Metal bars and weights were constructed conforming to these prescriptions for the metre and kilogram. One bar and one weight were selected to be the primary representations. The kilogram and the metre are now defined independently, and the litre, although for many years defined as the volume of a kilogram of water at the temperature of its maximum density, 4°C, and under a pressure of 76 cm of mercury, is now equal to 1 cubic decimeter.

In 1866, the U.S. Congress formally recognized metric units as a legal system, thereby making their use permissible in the United States. In 1893, the Office of Weights and Measures (now the National Bureau of Standards), by executive order, fixed the values of the U.S. yard and pound in terms of the meter and kilogram, respectively, as 1 yard = 3,600/3,937 m; and 1 lb = 0.453 592 4277 kg. By agreement in 1959 among the **national standards** laboratories of the **English-speaking nations,** the relations in use now are: 1 yd = 0.9144 m, whence 1 in =

25.4 mm exactly; and 1 lb = 0.453 592 37 kg, or 1 lb = 453.59 g (nearly).

THE INTERNATIONAL SYSTEM OF UNITS (SI)

In October 1960, the Eleventh General (International) Conference on Weights and Measures redefined some of the original metric units and expanded the system to include other physical and engineering units. This expanded system is called, in French, **Le Système International d'Unités** (abbreviated **SI**), and in English, **The International System of Units.**

The **Metric Conversion Act of 1975** codifies the voluntary conversion of the U.S. to the SI system. It is expected that in time all units in the United States will be in SI form. For this reason, additional tables of units, prefixes, equivalents, and conversion factors are included below (Tables 1.2.2 and 1.2.3).

SI consists of **seven base** units, **two supplementary** units, a series of **derived units** consistent with the base and supplementary units, and a series of approved prefixes for the formation of multiples and submultiples of the various units (see Tables 1.2.2 and 1.2.3). Multiple and submultiple prefixes in steps of 1,000 are recommended. (See ASTM E380-91a for further details.)

Base and **supplementary units** are **defined** [NIST Spec. Pub. 330 (2001)] as:

Metre The metre is defined as the length of path traveled by light in a vacuum during a time interval 1/299 792 458 of a second.

Kilogram The kilogram is the unit of mass; it is equal to the mass of the international prototype of the kilogram.

Second The second is the duration of 9,192,631,770 periods of the radiation corresponding to the transition between the two hyperfine levels of the ground state of the cesium 133 atom.

Ampere The ampere is that constant current which, if maintained in two straight parallel conductors of infinite length, of negligible cross section, and placed 1 metre apart in vacuum, would produce between these conductors a force equal to 2×10^{-7} newton per metre of length.

Kelvin The kelvin, unit of thermodynamic temperature, is the fraction 1/273.16 of the thermodynamic temperature of the triple point of water.

Mole The mole is the amount of substance of a system which contains as many elementary entities as there are atoms in 0.012 kilogram of carbon 12. (When the mole is used, the elementary entities must be specified and may be atoms, molecules, ions, electrons, other particles, or specified groups of such particles.)

Candela The candela is the luminous intensity, in a given direction, of a source that emits monochromatic radiation of frequency 540×10^{12} hertz and that has a radiant intensity in that direction of 1/683 watt per steradian.

Radian The unit of measure of a plane angle with its vertex at the center of a circle and subtended by an arc equal in length to the radius.

Steradian The unit of measure of a solid angle with its vertex at the center of a sphere and enclosing an area of the spherical surface equal to that of a square with sides equal in length to the radius.

SI conversion factors are listed in Table 1.2.4 alphabetically (adapted from ASTM E380-91a, "Standard Practice for Use of the International System of Units (SI) (the Modernized Metric System)." Conversion factors are written as a number greater than one and less than ten with six or fewer decimal places. This number is followed by the letter E (for exponent), a plus or minus symbol, and two digits which indicate the power of 10 by which the number must be multiplied to obtain the correct value. For example:

$$3.523\ 907\ E - 02 \text{ is } 3.523\ 907 \times 10^{-2} \text{ or } 0.035\ 239\ 07$$

An asterisk (*) after the sixth decimal place indicates that the conversion factor is exact and that all subsequent digits are zero. All other conversion factors have been rounded off.

Table 1.2.2 SI Units

Quantity	Unit	SI symbol	Formula
Base units*			
Length	metre	m	
Mass	kilogram	kg	
Time	second	s	
Electric current	ampere	A	
Thermodynamic temperature	kelvin	K	
Amount of substance	mole	mol	
Luminous intensity	candela	cd	
Supplementary units*			
Plane angle	radian	rad	
Solid angle	steradian	sr	
Derived units*			
Acceleration	metre per second squared		m/s^2
Activity (of a radioactive source)	disintegration per second		(disintegration)/s
Angular acceleration	radian per second squared		rad/s^2
Angular velocity	radian per second		rad/s
Area	square metre		m^2
Density	kilogram per cubic metre		kg/m^3
Electric capacitance	farad	F	$A \cdot s/V$
Electrical conductance	siemens	S	A/V
Electric field strength	volt per metre		V/m
Electric inductance	henry	H	$V \cdot s/A$
Electric potential difference	volt	V	W/A
Electric resistance	ohm	Ω	V/A
Electromotive force	volt	V	W/A
Energy	joule	J	$N \cdot m$
Entropy	joule per kelvin		J/K
Force	newton	N	$kg \cdot m/s^2$
Frequency	hertz	Hz	1/s
Illuminance	lux	lx	lm/m^2
Luminance	candela per square metre		cd/m^2
Luminous flux	lumen	lm	$cd \cdot sr$
Magnetic field strength	ampere per metre		A/m
Magnetic flux	weber	Wb	$V \cdot s$
Magnetic flux density	tesla	T	Wb/m^2
Magnetic potential difference	ampere	A	
Power	watt	W	J/s
Pressure	pascal	Pa	N/m^2
Quantity of electricity	coulomb	C	$A \cdot s$
Quantity of heat	joule	J	$N \cdot m$
Radiant intensity	watt per steradian		W/sr
Specific heat capacity	joule per kilogram-kelvin		$J/(kg \cdot K)$
Stress	pascal	Pa	N/m^2
Thermal conductivity	watt per metre-kelvin		$W/(m \cdot K)$
Velocity	metre per second		m/s
Viscosity, dynamic	pascal-second		$Pa \cdot s$
Viscosity, kinematic	square metre per second		m^2/s
Voltage	volt	V	W/A
Volume	cubic metre		m^3
Wave number	reciprocal metre		1/m
Work	joule	J	$N \cdot m$
Units in use with the SI†			
Time	minute	min	1 min = 60 s
	hour	h	1 h = 60 min = 3,600 s
	day	d	1 d = 24 h = 86,400 s
Plane angle	degree	°	$1° = \pi/180$ rad
	minute‡	′	$1′ = (1/60)° = (\pi/10,800)$ rad
	second‡	″	$1″ = (1/60)′ = (\pi/648,000)$ rad
Volume	litre	L	$1 L = 1 dm^3 = 10^{-3} m^3$
Mass	metric ton	t	$1 t = 10^3$ kg
	unified atomic mass unit§	u	$1 u = 1.660\ 54 \times 10^{-27}$ kg
Energy	electronvolt§	eV	$1 eV = 1.602\ 18 \times 10^{-19}$ J

* ASTM E380-91a.
  † These units are not part of SI, but their use is both so widespread and important that the International Committee for Weights and Measures in 1969 recognized their continued use with the SI (see NIST Spec. Pub. 330).
  ‡ Use discouraged, except for special fields such as cartography.
  § Values in SI units obtained experimentally. These units are to be used in specialized fields only.

Table 1.2.3 SI Prefixes*

Multiplication factors	Prefix	SI symbol
1 000 000 000 000 000 000 000 000 = 10^{24}	yotta	Y
1 000 000 000 000 000 000 000 = 10^{21}	zetta	Z
1 000 000 000 000 000 000 = 10^{18}	exa	E
1 000 000 000 000 000 = 10^{15}	peta	P
1 000 000 000 000 = 10^{12}	tera	T
1 000 000 000 = 10^{9}	giga	G
1 000 000 = 10^{6}	mega	M
1 000 = 10^{3}	kilo	k
100 = 10^{2}	hecto†	h
10 = 10^{1}	deka†	da
0.1 = 10^{-1}	deci†	d
0.01 = 10^{-2}	centi†	c
0.001 = 10^{-3}	milli	m
0.000 001 = 10^{-6}	micro	μ
0.000 000 001 = 10^{-9}	nano	n
0.000 000 000 001 = 10^{-12}	pico	P
0.000 000 000 000 001 = 10^{-15}	femto	f
0.000 000 000 000 000 001 = 10^{-18}	atto	a
0.000 000 000 000 000 000 001 = 10^{-21}	zepto	z
0.000 000 000 000 000 000 000 001 = 10^{-24}	yocto	y

* ANSI/IEEE Std 268-1992.
† To be avoided where practical.

Table 1.2.4 SI Conversion Factors

To convert from	to	Multiply by
abampere	ampere (A)	1.000 000*E+01
abcoulomb	coulomb (C)	1.000 000*E+01
abfarad	farad (F)	1.000 000*E+09
abhenry	henry (H)	1.000 000*E−09
abmho	siemens (S)	1.000 000*E+09
abohm	ohm (Ω)	1.000 000*E−09
abvolt	volt (V)	1.000 000*E−08
acre-foot (U.S. survey)[a]	metre³ (m³)	1.233 489 E+03
acre (U.S. survey)[a]	metre² (m²)	4.046 873 E+03
ampere, international U.S. (A_{INT-US})[b]	ampere (A)	9.998 43 E−01
ampere, U.S. legal 1948 (A_{US-48})	ampere (A)	1.000 008 E+00
ampere-hour	coulomb (C)	3.600 000*E+03
angstrom	metre (m)	1.000 000*E−10
are	metre² (m²)	1.000 000*E+02
astronomical unit	metre (m)	1.495 98 E+11
atmosphere (normal)	pascal (Pa)	1.013 25 E+05
atmosphere (technical = 1 kg$_f$/cm²)	pascal (Pa)	9.806 650*E+04
bar	pascal (Pa)	1.000 000*E+05
barn	metre² (m²)	1.000 000*E−28
barrel (for crude petroleum, 42 gal)	metre³ (m³)	1.589 873 E−01
board foot	metre³ (m³)	2.359 737 E−03
British thermal unit (International Table)[c]	joule (J)	1.055 056 E+03
British thermal unit (mean)	joule (J)	1.055 87 E+03
British thermal unit (thermochemical)	joule (J)	1.054 350 E+03
British thermal unit (39°F)	joule (J)	1.059 67 E+03
British thermal unit (59°F)	joule (J)	1.054 80 E+03
British thermal unit (60°F)	joule (J)	1.054 68 E+03
Btu (thermochemical)/foot²-second	watt/metre² (W/m²)	1.134 893 E+04
Btu (thermochemical)/foot²-minute	watt/metre² (W/m²)	1.891 489 E+02
Btu (thermochemical)/foot²-hour	watt/metre² (W/m²)	3.152 481 E+00
Btu (thermochemical)/inch²-second	watt/metre² (W/m²)	1.634 246 E+06
Btu (thermochemical) · in/s · ft² · °F (k, thermal conductivity)	watt/metre-kelvin (W/m · K)	5.188 732 E+02
Btu (International Table) · in/s · ft² · °F (k, thermal conductivity)	watt/metre-kelvin (W/m · K)	5.192 204 E+02
Btu (thermochemical) · in/h · ft² · °F (k, thermal conductivity)	watt/metre-kelvin (W/m · K)	1.441 314 E−01
Btu (International Table) · in/h · ft² · °F (k, thermal conductivity)	watt/metre-kelvin (W/m · K)	1.442 279 E−01
Btu (International Table)/ft²	joule/metre² (J/m²)	1.135 653 E+04
Btu (thermochemical)/ft²	joule/metre² (J/m²)	1.134 893 E+04
Btu (International Table)/h · ft² · °F (C, thermal conductance)	watt/metre²-kelvin (W/m² · K)	5.678 263 E+00
Btu (thermochemical)/h · ft² · °F (C, thermal conductance)	watt/metre²-kelvin (W/m² · K)	5.674 466 E+00
Btu (International Table)/pound-mass	joule/kilogram (J/kg)	2.326 000*E+03

Table 1.2.4   SI Conversion Factors   (*Continued*)

To convert from	to	Multiply by
Btu (thermochemical)/pound-mass	joule/kilogram (J/kg)	2.324 444 E+03
Btu (International Table)/lbm · °F (*c*, heat capacity)	joule/kilogram-kelvin (J/kg · K)	4.186 800*E+03
Btu (thermochemical)/lbm · °F (*c*, heat capacity)	joule/kilogram-kelvin (J/kg · K)	4.184 000*E+03
Btu (International Table)/s · ft^2 · °F	watt/metre^2-kelvin (W/m^2 · K)	2.044 175 E+04
Btu (thermochemical)/s · ft^2 · °F	watt/metre^2-kelvin (W/m^2 · K)	2.042 808 E+04
Btu (International Table)/hour	watt (W)	2.930 711 E−01
Btu (thermochemical)/second	watt (W)	1.054 350 E+03
Btu (thermochemical)/minute	watt (W)	1.757 250 E+01
Btu (thermochemical)/hour	watt (W)	2.928 751 E−01
bushel (U.S.)	metre^3 (m^3)	3.523 907 E−02
calorie (International Table)	joule (J)	4.186 800*E+00
calorie (mean)	joule (J)	4.190 02 E+00
calorie (thermochemical)	joule (J)	4.184 000*E+00
calorie (15°C)	joule (J)	4.185 80 E+00
calorie (20°C)	joule (J)	4.181 90 E+00
calorie (kilogram, International Table)	joule (J)	4.186 800*E+03
calorie (kilogram, mean)	joule (J)	4.190 02 E+03
calorie (kilogram, thermochemical)	joule (J)	4.184 000*E+03
calorie (thermochemical)/centimetre^2-minute	watt/metre^2 (W/m^2)	6.973 333 E+02
cal (thermochemical)/cm^2	joule/metre^2 (J/m^2)	4.184 000*E+04
cal (thermochemical)/cm^2 · s	watt/metre^2 (W/m^2)	4.184 000*E+04
cal (thermochemical)/cm · s · °C	watt/metre-kelvin (W/m · K)	4.184 000*E+02
cal (International Table)/g	joule/kilogram (J/kg)	4.186 800*E+03
cal (International Table)/g · °C	joule/kilogram-kelvin (J/kg · K)	4.186 800*E+03
cal (thermochemical)/g	joule/kilogram (J/kg)	4.184 000*E+03
cal (thermochemical)/g · °C	joule/kilogram-kelvin (J/kg · K)	4.184 000*E+03
calorie (thermochemical)/second	watt (W)	4.184 000*E+00
calorie (thermochemical)/minute	watt (W)	6.973 333 E−02
carat (metric)	kilogram (kg)	2.000 000*E−04
centimetre of mercury (0°C)	pascal (Pa)	1.333 22 E+03
centimetre of water (4°C)	pascal (Pa)	9.806 38 E+01
centipoise	pascal-second (Pa · s)	1.000 000*E−03
centistokes	metre^2/second (m^2/s)	1.000 000*E−06
chain (engineer or ramden)	meter (m)	3.048* E+01
chain (surveyor or gunter)	meter (m)	2.011 684 E+01
circular mil	metre^2 (m^2)	5.067 075 E−10
cord	metre^3 (m^3)	3.624 556 E+00
coulomb, international U.S. (C$_{INT-US}$)[b]	coulomb (C)	9.998 43 E−01
coulomb, U.S. legal 1948 (C$_{US-48}$)	coulomb (C)	1.000 008 E+00
cup	metre^3 (m^3)	2.365 882 E−04
curie	becquerel (Bq)	3.700 000*E+10
day (mean solar)	second (s)	8.640 000 E+04
day (sidereal)	second (s)	8.616 409 E+04
degree (angle)	radian (rad)	1.745 329 E−02
degree Celsius	kelvin (K)	$t_K = t_C + 273.15$
degree centigrade	kelvin (K)	$t_K = t_C + 273.15$
degree Fahrenheit	degree Celsius	$t_C = (t_F - 32)/1.8$
degree Fahrenheit	kelvin (K)	$t_K = (t_F + 459.67)/1.8$
deg F · h · ft^2/Btu (thermochemical) (*R*, thermal resistance)	kelvin-metre^2/watt (K · m^2/W)	1.762 280 E−01
deg F · h · ft^2/Btu (International Table) (*R*, thermal resistance)	kelvin-metre^2/watt (K · m^2/ W)	1.761 102 E−01
degree Rankine	kelvin (K)	$t_K = t_R/1.8$
dram (avoirdupois)	kilogram (kg)	1.771 845 E−03
dram (troy or apothecary)	kilogram (kg)	3.887 934 E−03
dram (U.S. fluid)	kilogram (kg)	3.696 691 E−06
dyne	newton (N)	1.000 000*E−05
dyne-centimetre	newton-metre (N · m)	1.000 000*E−07
dyne-centimetre^2	pascal (Pa)	1.000 000*E−01
electron volt	joule (J)	1.602 18 E−19
EMU of capacitance	farad (F)	1.000 000*E+09
EMU of current	ampere (A)	1.000 000*E+01
EMU of electric potential	volt (V)	1.000 000*E−08
EMU of inductance	henry (H)	1.000 000*E−09
EMU of resistance	ohm (Ω)	1.000 000*E−09
ESU of capacitance	farad (F)	1.112 650 E−12
ESU of current	ampere (A)	3.335 6 E−10
ESU of electric potential	volt (V)	2.997 9 E+02
ESU of inductance	henry (H)	8.987 552 E+11

Table 1.2.4   SI Conversion Factors   (*Continued*)

To convert from	to	Multiply by
ESU of resistance	ohm (Ω)	8.987 552 E+11
erg	joule (J)	1.000 000*E−07
erg/centimetre^2-second	watt/metre^2 (W/m^2)	1.000 000*E−03
erg/second	watt (W)	1.000 000*E−07
farad, international U.S. (F_{INT-US})	farad (F)	9.995 05 E−01
faraday (based on carbon 12)	coulomb (C)	9.648 531 E+04
faraday (chemical)	coulomb (C)	9.649 57 E+04
faraday (physical)	coulomb (C)	9.652 19 E+04
fathom (U.S. survey)[a]	metre (m)	1.828 804 E+00
fermi (femtometer)	metre (m)	1.000 000*E−15
fluid ounce (U.S.)	metre^3 (m^3)	2.957 353 E−05
foot	metre (m)	3.048 000*E−01
foot (U.S. survey)[a]	metre (m)	3.048 006 E−01
foot^3/minute	metre^3/second (m^3/s)	4.719 474 E−04
foot^3/second	metre^3/second (m^3/s)	2.831 685 E−02
foot^3 (volume and section modulus)	metre^3 (m^3)	2.831 685 E−02
foot^2	metre^2 (m^2)	9.290 304*E−02
foot^4 (moment of section)[d]	metre^4 (m^4)	8.630 975 E−03
foot/hour	metre/second (m/s)	8.466 667 E−05
foot/minute	metre/second (m/s)	5.080 000*E−03
foot/second	metre/second (m/s)	3.048 000*E−01
foot^2/second	metre^2/second (m^2/s)	9.290 304*E−02
foot of water (39.2°F)	pascal (Pa)	2.988 98 E+03
footcandle	lumen/metre^2 (lm/m^2)	1.076 391 E+01
footcandle	lux (lx)	1.076 391 E+01
footlambert	candela/metre^2 (cd/m^2)	3.426 259 E+00
foot-pound-force	joule (J)	1.355 818 E+00
foot-pound-force/hour	watt (W)	3.766 161 E−04
foot-pound-force/minute	watt (W)	2.259 697 E−02
foot-pound-force/second	watt (W)	1.355 818 E+00
foot-poundal	joule (J)	4.214 011 E−02
ft^2/h (thermal diffusivity)	metre^2/second (m^2/s)	2.580 640*E−05
foot/second^2	metre/second^2 (m/s^2)	3.048 000*E−01
free fall, standard	metre/second^2 (m/s^2)	9.806 650*E+00
furlong	metre (m)	2.011 68 *E+02
gal	metre/second^2 (m/s^2)	1.000 000*E−02
gallon (Canadian liquid)	metre^3 (m^3)	4.546 090 E−03
gallon (U.K. liquid)	metre^3 (m^3)	4.546 092 E−03
gallon (U.S. dry)	metre^3 (m^3)	4.404 884 E−03
gallon (U.S. liquid)	metre^3 (m^3)	3.785 412 E−03
gallon (U.S. liquid)/day	metre^3/second (m^3/s)	4.381 264 E−08
gallon (U.S. liquid)/minute	metre^3/second (m^3/s)	6.309 020 E−05
gamma	tesla (T)	1.000 000*E−09
gauss	tesla (T)	1.000 000*E−04
gilbert	ampere-turn	7.957 747 E−01
gill (U.K.)	metre^3 (m^3)	1.420 653 E−04
gill (U.S.)	metre^3 (m^3)	1.182 941 E−04
grade	degree (angular)	9.000 000*E−01
grade	radian (rad)	1.570 796 E−02
grain (1/7,000 lbm avoirdupois)	kilogram (kg)	6.479 891*E−05
gram	kilogram (kg)	1.000 000*E−03
gram/centimetre^3	kilogram/metre^3 (kg/m^3)	1.000 000*E+03
gram-force/centimetre^2	pascal (Pa)	9.806 650*E+01
hectare	metre^2 (m^2)	1.000 000*E+04
henry, international U.S. (H_{INT-US})	henry (H)	1.000 495 E+00
hogshead (U.S.)	metre^3 (m^3)	2.384 809 E−01
horsepower (550 ft · lbf/s)	watt (W)	7.456 999 E+02
horsepower (boiler)	watt (W)	9.809 50 E+03
horsepower (electric)	watt (W)	7.460 000*E+02
horsepower (metric)	watt (W)	7.354 99 E+02
horsepower (water)	watt (W)	7.460 43 E+02
horsepower (U.K.)	watt (W)	7.457 0 E+02
hour (mean solar)	second (s)	3.600 000*E+03
hour (sidereal)	second (s)	3.590 170 E+03
hundredweight (long)	kilogram (kg)	5.080 235 E+01
hundredweight (short)	kilogram (kg)	4.535 924 E+01
inch	metre (m)	2.540 000*E−02
inch^2	metre^2 (m^2)	6.451 600*E−04
inch^3 (volume and section modulus)	metre^3 (m^3)	1.638 706 E−05
inch^3/minute	metre^3/second (m^3/s)	2.731 177 E−07
inch^4 (moment of section)[d]	metre^4 (m^4)	4.162 314 E−07
inch/second	metre/second (m/s)	2.540 000*E−02
inch of mercury (32°F)	pascal (Pa)	3.386 38 E+03

Table 1.2.4   SI Conversion Factors   (*Continued*)

To convert from	to	Multiply by
inch of mercury (60°F)	pascal (Pa)	3.376 85 E+03
inch of water (39.2°F)	pascal (Pa)	2.490 82 E+02
inch of water (60°F)	pascal (Pa)	2.488 4 E+02
inch/second^2	metre/second^2 (m/s^2)	2.540 000*E−02
joule, international U.S. (J_{INT-US})[b]	joule (J)	1.000 182 E+00
joule, U.S. legal 1948 (J_{US-48})	joule (J)	1.000 017 E+00
kayser	1/metre (1/m)	1.000 000*E+02
kelvin	degree Celsius	$t_C = t_K - 273.15$
kilocalorie (thermochemical)/minute	watt (W)	6.973 333 E+01
kilocalorie (thermochemical)/second	watt (W)	4.184 000*E+03
kilogram-force (kgf)	newton (N)	9.806 650*E+00
kilogram-force-metre	newton-metre (N · m)	9.806 650*E+00
kilogram-force-second^2/metre (mass)	kilogram (kg)	9.806 650*E+00
kilogram-force/centimetre^2	pascal (Pa)	9.806 650*E+04
kilogram-force/metre^3	pascal (Pa)	9.806 650*E+00
kilogram-force/millimetre^2	pascal (Pa)	9.806 650*E+06
kilogram-mass	kilogram (kg)	1.000 000*E+00
kilometre/hour	metre/second (m/s)	2.777 778 E−01
kilopond	newton (N)	9.806 650*E+00
kilowatt hour	joule (J)	3.600 000*E+06
kilowatt hour, international U.S. (kWh$_{INT-US}$)[b]	joule (J)	3.600 655 E+06
kilowatt hour, U.S. legal 1948 (kWh$_{US-48}$)	joule (J)	3.600 061 E+06
kip (1,000 lbf)	newton (N)	4.448 222 E+03
kip/inch^2 (ksi)	pascal (Pa)	6.894 757 E+06
knot (international)	metre/second (m/s)	5.144 444 E−01
lambert	candela/metre^2 (cd/m^2)	3.183 099 E+03
langley	joule/metre^2 (J/m^2)	4.184 000*E+04
league, nautical (international and U.S.)	metre (m)	5.556 000*E+03
league (U.S. survey)[a]	metre (m)	4.828 041 E+03
league, nautical (U.K.)	metre (m)	5.559 552*E+03
light year (365.2425 days)	metre (m)	9.460 54 E+15
link (engineer or ramden)	metre (m)	3.048* E−01
link (surveyor or gunter)	metre (m)	2.011 68* E−01
litre[e]	metre^3 (m^3)	1.000 000*E−03
lux	lumen/metre^2 (lm/m^2)	1.000 000*E+00
maxwell	weber (Wb)	1.000 000*E−08
mho	siemens (S)	1.000 000*E+00
microinch	metre (m)	2.540 000*E−08
micron (micrometre)	metre (m)	1.000 000*E−06
mil	metre (m)	2.540 000*E−05
mile, nautical (international and U.S.)	metre (m)	1.852 000*E+03
mile, nautical (U.K.)	metre (m)	1.853 184*E+03
mile (international)	metre (m)	1.609 344*E+03
mile (U.S. survey)[a]	metre (m)	1.609 347 E+03
mile^2 (international)	metre^2 (m^2)	2.589 988 E+06
mile^2 (U.S. survey)[a]	metre^2 (m^2)	2.589 998 E+06
mile/hour (international)	metre/second (m/s)	4.470 400*E−01
mile/hour (international)	kilometre/hour	1.609 344*E+00
millimetre of mercury (0°C)	pascal (Pa)	1.333 224 E+02
minute (angle)	radian (rad)	2.908 882 E−04
minute (mean solar)	second (s)	6.000 000 E+01
minute (sidereal)	second (s)	5.983 617 E+01
month (mean calendar)	second (s)	2.268 000 E+06
oersted	ampere/metre (A/m)	7.957 747 E+01
ohm, international U.S. (Ω_{INT-US})	ohm (Ω)	1.000 495 E+00
ohm-centimetre	ohm-metre (Ω · m)	1.000 000*E−02
ounce-force (avoirdupois)	newton (N)	2.780 139 E−01
ounce-force-inch	newton-metre (N · m)	7.061 552 E−03
ounce-mass (avoirdupois)	kilogram (kg)	2.834 952 E−02
ounce-mass (troy or apothecary)	kilogram (kg)	3.110 348 E−02
ounce-mass/yard^2	kilogram/metre^2 (kg/m^2)	3.390 575 E−02
ounce (avoirdupois)(mass)/inch^3	kilogram/metre^3 (kg/m^3)	1.729 994 E+03
ounce (U.K. fluid)	metre^3 (m^3)	2.841 306 E−05
ounce (U.S. fluid)	metre^3 (m^3)	2.957 353 E−05
parsec	metre (m)	3.085 678 E+16
peck (U.S.)	metre^3 (m^3)	8.809 768 E−03
pennyweight	kilogram (kg)	1.555 174 E−03
perm (0°C)	kilogram/pascal-second-metre^2 (kg/Pa · s · m^2)	5.721 35 E−11
perm (23 °C)	kilogram/pascal-second-metre^2 (kg/Pa · s · m^2)	5.745 25 E−11

Table 1.2.4   SI Conversion Factors   (*Continued*)

To convert from	to	Multiply by
perm-inch (0°C)	kilogram/pascal-second-metre (kg/Pa · s · m)	1.453 22 E−12
perm-inch (23°C)	kilogram/pascal-second-metre (kg/Pa · s · m)	1.459 29 E−12
phot	lumen/metre² (lm/m²)	1.000 000*E+04
pica (printer's)	metre (m)	4.217 518 E−03
pint (U.S. dry)	metre³ (m³)	5.506 105 E−04
pint (U.S. liquid)	metre³ (m³)	4.731 765 E−04
point (printer's)	metre	3.514 598 E−04
poise (absolute viscosity)	pascal-second (Pa · s)	1.000 000*E−01
poundal	newton (N)	1.382 550 E−01
poundal/foot²	pascal (Pa)	1.488 164 E+00
poundal-second/foot²	pascal-second (Pa · s)	1.488 164 E+00
pound-force (lbf avoirdupois)	newton (N)	4.448 222 E+00
pound-force-inch	newton-metre (N · m)	1.129 848 E−01
pound-force-foot	newton-metre (N · m)	1.355 818 E+00
pound-force-foot/inch	newton-metre/metre (N · m/m)	5.337 866 E+01
pound-force-inch/inch	newton-metre/metre (N · m/m)	4.448 222 E+00
pound-force/inch	newton/metre (N/m)	1.751 268 E+02
pound-force/foot	newton/metre (N/m)	1.459 390 E+01
pound-force/foot²	pascal (Pa)	4.788 026 E+01
pound-force/inch² (psi)	pascal (Pa)	6.894 757 E+03
pound-force-second/foot²	pascal-second (Pa · s)	4.788 026 E+01
pound-mass (lbm avoirdupois)	kilogram (kg)	4.535 924 E−01
pound-mass (troy or apothecary)	kilogram (kg)	3.732 417 E−01
pound-mass-foot² (moment of inertia)	kilogram-metre² (kg · m²)	4.214 011 E−02
pound-mass-inch² (moment of inertia)	kilogram-metre² (kg · m²)	2.926 397 E−04
pound-mass/foot²	kilogram/metre² (kg/m²)	4.882 428 E+00
pound-mass/second	kilogram/second (kg/s)	4.535 924 E−01
pound-mass/minute	kilogram/second (kg/s)	7.559 873 E−03
pound-mass/foot³	kilogram/metre³ (kg/m³)	1.601 846 E+01
pound-mass/inch³	kilogram/metre³ (kg/m³)	2.767 990 E+04
pound-mass/gallon (U.K. liquid)	kilogram/metre³ (kg/m³)	9.977 637 E+01
pound-mass/gallon (U.S. liquid)	kilogram/metre³ (kg/m³)	1.198 264 E+02
pound-mass/foot-second	pascal-second (Pa · s)	1.488 164 E+00
quart (U.S. dry)	metre³ (m³)	1.101 221 E−03
quart (U.S. liquid)	metre³ (m³)	9.463 529 E−04
rad (radiation dose absorbed)	gray (Gy)	1.000 000*E−02
rem (dose equivalent)	sievert (Sv)	1.000 000*E−02
rhe	metre²/newton-second (m²/N · s)	1.000 000*E+01
rod (U.S. survey)[a]	metre (m)	5.029 210 E+00
roentgen	coulomb/kilogram (C/kg)	2.580 000*E−04
second (angle)	radian (rad)	4.848 137 E−06
second (sidereal)	second (s)	9.972 696 E−01
section (U.S. survey)[a]	metre² (m²)	2.589 998 E+06
shake	second (s)	1.000 000*E−08
slug	kilogram (kg)	1.459 390 E+01
slug/foot³	kilogram/metre³ (kg/m³)	5.153 788 E+02
slug/foot-second	pascal-second (Pa · s)	4.788 026 E+01
statampere	ampere (A)	3.335 641 E−10
statcoulomb	coulomb (C)	3.335 641 E−10
statfarad	farad (F)	1.112 650 E−12
stathenry	henry (H)	8.987 552 E+11
statmho	siemens (S)	1.112 650 E−12
statohm	ohm (Ω)	8.987 552 E+11
statvolt	volt (V)	2.997 925 E+02
stere	metre³ (m³)	1.000 000*E+00
stilb	candela/metre² (cd/m²)	1.000 000*E+04
stokes (kinematic viscosity)	metre²/second (m²/s)	1.000 000*E−04
tablespoon	metre³ (m³)	1.478 676 E−05
teaspoon	metre³ (m³)	4.928 922 E−06
ton (assay)	kilogram (kg)	2.916 667 E−02
ton (long, 2,240 lbm)	kilogram (kg)	1.016 047 E+03
ton (metric)	kilogram (kg)	1.000 000*E+03
ton (nuclear equivalent of TNT)	joule (J)	4.184 000*E+09
ton (register)	metre³ (m³)	2.831 685 E+00
ton (short, 2,000 lbm)	kilogram (kg)	9.071 847 E+02
ton (short, mass)/hour	kilogram/second (kg/s)	2.519 958 E−01
ton (long, mass)/yard³	kilogram/metre³ (kg/m³)	1.328 939 E+03
tonne	kilogram (kg)	1.000 000*E+03
torr (mm Hg, 0°C)	pascal (Pa)	1.333 22 E+02
township (U.S. survey)[a]	metre² (m²)	9.323 994 E+07
unit pole	weber (Wb)	1.256 637 E−07

Table 1.2.4   SI Conversion Factors   (*Continued*)

To convert from	to	Multiply by
volt, international U.S. $(V_{INT-US})^b$	volt (V)	1.000 338 E+00
volt, U.S. legal 1948 (V_{US-48})	volt (V)	1.000 008 E+00
watt, international U.S. $(W_{INT-US})^b$	watt (W)	1.000 182 E+00
watt, U.S. legal 1948 (W_{US-48})	watt (W)	1.000 017 E+00
watt/centimetre^2	watt/metre^2 (W/m^2)	1.000 000*E+04
watt-hour	joule (J)	3.600 000*E+03
watt-second	joule (J)	1.000 000*E+00
yard	metre (m)	9.144 000*E−01
yard^2	metre^2 (m^2)	8.361 274 E−01
yard^3	metre^3 (m^3)	7.645 549 E−01
yard^3/minute	metre^3/second (m^3/s)	1.274 258 E−02
year (calendar)	second (s)	3.153 600*E+07
year (sidereal)	second (s)	3.155 815 E+07
year (tropical)	second (s)	3.155 693 E+07

^{a} Based on the U.S. survey foot (1 ft = 1,200/3,937 m).

^{b} In 1948 a new international agreement was reached on absolute electrical units, which changed the value of the volt used in this country by about 300 parts per million. Again in 1969 a new base of reference was internationally adopted making a further change of 8.4 parts per million. These changes (and also changes in ampere, joule, watt, coulomb) require careful terminology and conversion factors for exact use of old information. Terms used in this guide are:

Volt as used prior to January 1948—volt, international U.S. (V_{INT-US})

Volt as used between January 1948 and January 1969—volt, U.S. legal 1948 (V_{INT-48})

Volt as used since January 1969—volt (V)

Identical treatment is given the ampere, coulomb, watt, and joule.

^{c} This value was adopted in 1956. Some of the older International Tables use the value 1.055 04 E+03. The exact conversion factor is 1.055 055 852 62*E+03.

^{d} Moment of inertia of a plane section about a specified axis.

^{e} In 1964, the General Conference on Weights and Measures adopted the name "litre" as a special name for the cubic decimetre. Prior to this decision the litre differed slightly (previous value, 1.000028 dm^3), and in expression of precision, volume measurement, this fact must be kept in mind.

SYSTEMS OF UNITS

The principal units of interest to mechanical engineers can be derived from three base units which are considered to be dimensionally independent of each other. The British "gravitational system," in common use in the United States, uses units of **length, force,** and **time** as base units and is also called the "foot-pound-second system." The metric system, on the other hand, is based on the meter, kilogram, and second, units of **length, mass,** and **time,** and is often designated as the "MKS system." During the nineteenth century a metric "gravitational system," based on a kilogram-force (also called a "kilopond") came into general use. With the development of the International System of Units (SI), based as it is on the original metric system for mechanical units, and the general requirements by members of the European Community that only SI units be used, it is anticipated that the kilogram-force will fall into disuse to be replaced by the newton, the SI unit of force. Table 1.2.5 gives the base units of four systems with the corresponding derived unit given in parentheses.

In the definitions given below, the "standard kilogram body" refers to the international kilogram prototype, a platinum-iridium cylinder kept in the International Bureau of Weights and Measures in Sèvres, just outside Paris. The "standard pound body" is related to the kilogram by a precise numerical factor: 1 lb = 0.453 592 37 kg. This new "unified" pound has replaced the somewhat smaller Imperial pound of the United Kingdom and the slightly larger pound of the United States (see NBS Spec. Pub. 447). The "standard locality" means sea level, 45° latitude, or more strictly any locality in which the acceleration due to gravity has the value 9.80 665 m/s^2 = 32.1740 ft/s^2, which may be called the **standard acceleration** (Table 1.2.6).

The **pound force** is the force required to support the standard pound body against gravity, *in vacuo,* in the standard locality; or, it is the force which, if applied to the standard pound body, supposed free to move, would give that body the "standard acceleration." The word *pound* is used for the unit of both force and mass and consequently is ambiguous. To avoid uncertainty, it is desirable to call the units "pound force" and "pound mass," respectively. The slug has been defined as that mass which will accelerate at 1 ft/s^2 when acted upon by a one pound force. It is therefore equal to 32.1740 pound-mass.

The **kilogram force** is the force required to support the standard kilogram against gravity, *in vacuo,* in the standard locality; or, it is the force which, if applied to the standard kilogram body, supposed free to move, would give that body the "standard acceleration." The word *kilogram* is used for the unit of both force and mass and consequently is ambiguous. It is for this reason that the General Conference on Weights and Measures declared (in 1901) that the kilogram was the unit of mass, a concept incorporated into SI when it was formally approved in 1960.

The **dyne** is the force which, if applied to the standard gram body, would give that body an acceleration of 1 cm/s^2; i.e., 1 dyne = 1/980.665 of a gram force.

The **newton** is that force which will impart to a 1-kilogram mass an acceleration of 1 m/s^2.

Table 1.2.5   Systems of Units

Quantity	Dimensions of units in terms of L/M/F/T	British "gravitational system"	Metric "gravitational system"	CGS system	SI system
Length	L	1 ft	1 m	1 cm	1 m
Mass	M	(1 slug)		1 g	1 kg
Force	F	1 lb	1 kg	(1 dyne)	(1 N)
Time	T	1 s	1 s	1 s	1 s

Table 1.2.6   Acceleration of Gravity

Latitude, deg	g m/s²	g ft/s²	g/g⁰	Latitude, deg	g m/s²	g ft/s²	g/g⁰

Latitude, deg	g m/s²	g ft/s²	g/g^0	Latitude, deg	g m/s²	g ft/s²	g/g^0
0	9.780	32.088	0.9973	50	9.811	32.187	1.0004
10	9.782	32.093	0.9975	60	9.819	32.215	1.0013
20	9.786	32.108	0.9979	70	9.826	32.238	1.0020
30	9.793	32.130	0.9986	80	9.831	32.253	1.0024
40	9.802	32.158	0.9995	90	9.832	32.258	1.0026

NOTE: Correction for altitude above sea level: -3 mm/s² for each 1,000 m; -0.003 ft/s² for each 1,000 ft.
SOURCE: U.S. Coast and Geodetic Survey, 1912.

TEMPERATURE

The SI unit for thermodynamic temperature is the **kelvin, K,** which is the fraction 1/273.16 of the thermodynamic temperature of the triple point of water. Thus 273.16 K is the **fixed (base) point** on the **kelvin scale.**

Another unit used for the measurement of temperature is degrees **Celsius** (formerly **centigrade**), °C. The relation between a thermodynamic temperature T and a Celsius temperature t is

$$t = T - 273.15 \text{ K (the ice point of water)}$$

Thus the unit Celsius degree is equal to the unit kelvin, and a difference of temperature would be the same on either scale.

In the USCS temperature is measured in degrees **Fahrenheit, F.** The relation between the Celsius and the Fahrenheit scales is

$$t_{°C} = (t_{°F} - 32)/1.8$$

(For temperature-conversion tables, see Sec. 4.)

TERRESTRIAL GRAVITY

Standard acceleration of gravity is $g^0 = 9.80665$ m per sec per sec, or 32.1740 ft per sec per sec. This value g^0 is assumed to be the value of g at sea level and latitude 45°.

MOHS SCALE OF HARDNESS

This scale is an arbitrary one which is used to describe the hardness of several mineral substances on a scale of 1 through 10 (Table 1.2.7). The given number indicates a higher relative hardness compared with that of substances below it; and a lower relative hardness than those above it. For example, an unknown substance is scratched by quartz, but it, in turn, scratches feldspar. The unknown has a hardness of between 6 and 7 on the Mohs scale.

Table 1.2.7   Mohs Scale of Hardness

1. Talc	5. Apatite	8. Topaz
2. Gypsum	6. Feldspar	9. Sapphire
3. Calc-spar	7. Quartz	10. Diamond
4. Fluorspar		

TIME

Kinds of Time   Three kinds of time are recognized by astronomers: sidereal, apparent solar, and mean solar time. The **sidereal day** is the interval between two consecutive transits of some fixed celestial object across any given meridian, or it is the interval required by the earth to make one complete revolution on its axis. The interval is constant, but it is inconvenient as a time unit because the noon of the sidereal day occurs at all hours of the day and night. The **apparent solar day** is the interval between two consecutive transits of the sun across any given meridian. On account of the variable distance between the sun and earth, the variable speed of the earth in its orbit, the effect of the moon, etc., this interval is not constant and consequently cannot be kept by any simple mechanisms, such as clocks or watches. To overcome the objection noted above, the **mean solar day** was devised. The mean solar day is

the length of the average apparent solar day. Like the sidereal day it is constant, and like the apparent solar day its noon always occurs at approximately the same time of day. By international agreement, beginning Jan. 1, 1925, the astronomical day, like the civil day, is from midnight to midnight. The hours of the astronomical day run from 0 to 24, and the hours of the civil day usually run from 0 to 12 A.M. and 0 to 12 P.M. In some countries the hours of the civil day also run from 0 to 24.

The Year   Three different kinds of year are used: the sidereal, the tropical, and the anomalistic. The **sidereal year** is the time taken by the earth to complete one revolution around the sun from a given star to the same star again. Its length is 365 days, 6 hours, 9 minutes, and 9 seconds. The **tropical year** is the time included between two successive passages of the vernal equinox by the sun, and since the equinox moves westward 50.2 seconds of arc a year, the tropical year is shorter by 20 minutes 23 seconds in time than the sidereal year. As the seasons depend upon the earth's position with respect to the equinox, the tropical year is the year of civil reckoning. The **anomalistic year** is the interval between two successive passages of the perihelion, viz., the time of the earth's nearest approach to the sun. The anomalistic year is used only in special calculations in astronomy.

The Second   Although the second is ordinarily defined as 1/86,400 of the mean solar day, this is not sufficiently precise for many scientific purposes. Scientists have adopted more precise definitions for specific purposes: in 1956, one in terms of the length of the tropical year 1900 and, more recently, in 1967, one in terms of a specific atomic frequency.

Frequency is the reciprocal of time for 1 cycle; the unit of frequency is the **hertz** (Hz), defined as 1 cycle/s.

The Calendar   The **Gregorian calendar,** now used in most of the civilized world, was adopted in Catholic countries of Europe in 1582 and in Great Britain and her colonies Jan. 1, 1752. The average length of the Gregorian calendar year is $365\frac{1}{4} - \frac{3}{400}$ days, or 365.2425 days. This is equivalent to 365 days, 5 hours, 49 minutes, 12 seconds. The length of the tropical year is 365.2422 days, or 365 days, 5 hours, 48 minutes, 46 seconds. Thus the Gregorian calendar year is longer than the tropical year by 0.0003 day, or 26 seconds. This difference amounts to 1 day in slightly more than 3,300 years and can properly be neglected.

Standard Time   Prior to 1883, each city of the United States had its own time, which was determined by the time of passage of the sun across the local meridian. A system of standard time had been used since its first adoption by the railroads in 1883 but was first legalized on Mar. 19, 1918, when Congress directed the Interstate Commerce Commission to establish limits of the standard time zones. Congress took no further steps until the **Uniform Time Act of 1966** was enacted, followed with an amendment in 1972. This legislation, referred to as "the Act," transferred the regulation and enforcement of the law to the Department of Transportation.

By the legislation of 1918, with some modifications by the Act, the contiguous United States is divided into four **time zones,** each of which, theoretically, was to span 15 degrees of longitude. The first, the **Eastern zone,** extends from the Atlantic westward to include most of Michigan and Indiana, the eastern parts of Kentucky and Tennessee, Georgia, and Florida, except the west half of the panhandle. **Eastern standard time** is based upon the mean solar time of the 75th meridian west of Greenwich, and is 5 hours slower than **Greenwich Mean Time (GMT).** (See also discussion of UTC below.) The second or **Central zone** extends westward to include most of North Dakota, about half of South Dakota and Nebraska, most of Kansas, Oklahoma, and all but the two most westerly counties of Texas. **Central standard time** is based upon the mean solar time of the 90th meridian west of Greenwich, and is 6 hours slower than GMT. The third or **Mountain zone** extends westward to include Montana, most of Idaho, one county of Oregon, Utah, and Arizona. **Mountain standard time** is based upon the mean solar time of the 105th meridian west of Greenwich, and is 7 hours slower than GMT. The fourth or **Pacific zone** includes all of the remaining 48 contiguous states. **Pacific standard time** is based on the mean solar time of the 120th meridian west of Greenwich, and is 8 hours slower than GMT. Exact locations of boundaries may be obtained from the Department of Transportation.

In addition to the above four zones there are four others that apply to the noncontiguous states and islands. The most easterly is the **Atlantic zone,** which includes Puerto Rico and the Virgin Islands, where the time is 4 hours slower than GMT. Eastern standard time is used in the Panama Canal strip. To the west of the Pacific time zone there are the **Yukon,** the **Alaska-Hawaii,** and **Bering zones** where the times are, respectively, 9, 10, and 11 hours slower than GMT. The system of standard time has been adopted in all civilized countries and is used by ships on the high seas.

The Act directs that from the first Sunday in April to the last Sunday in October, the time in each zone is to be advanced one hour for advanced time or daylight saving time (DST). However, any state-by-state enactment may exempt the entire state from using advanced time. By this provision Arizona and Hawaii do not observe advanced time (as of 1973). By the 1972 amendment to the Act, a state split by a time-zone boundary may exempt from using advanced time all that part which is in one zone without affecting the rest of the state. By this amendment, 80 counties of Indiana in the Eastern zone are exempt from using advanced time, while 6 counties in the northwest corner and 6 counties in the southwest, which are in Central zone, do observe advanced time.

Pursuant to its assignment of carrying out the Act, the Department of Transportation has stipulated that municipalities located on the boundary between the Eastern and Central zones are in the Central zone; those on the boundary between the Central and Mountain zones are in the Mountain zone (except that Murdo, SD, is in the Central zone); those on the boundary between Mountain and Pacific time zones are in the Mountain zone. In such places, when the time is given, it should be specified as Central, Mountain, etc.

Standard Time Signals   The National Institute of Standards and Technology broadcasts time signals from **station WWV,** Ft. Collins, CO, and from **station WWVH,** near Kekaha, Kaui, HI. The broadcasts by WWV are on radio carrier frequencies of 2.5, 5, 10, 15, and 20 MHz, while those by WWVH are on radio carrier frequencies of 2.5, 5, 10, and 15 MHz. Effective Jan. 1, 1975, time announcements by both WWV and WWVH are referred to as **Coordinated Universal Time, UTC,** the international coordinated time scale used around the world for most timekeeping purposes. UTC is generated by reference to International Atomic Time (TAI), which is determined by the Bureau International de l'Heure on the basis of atomic clocks operating in various establishments in accordance with the definition of the second. Since the difference between UTC and TAI is defined to be a whole number of seconds, a "leap second" is periodically added to or subtracted from UTC to take into account variations in the rotation of the earth. Time (i.e., clock time) is given in terms of 0 to 24 hours a day, starting with 0000 at midnight at Greenwich zero longitude. The beginning of each 0.8-second-long audio tone marks the end of an announced time interval. For example, at 2:15 P.M., UTC, the voice announcement would be: "At the tone fourteen hours fifteen minutes Coordinated Universal Time," given during the last 7.5 seconds of each minute. The tone markers from both stations are given simultaneously, but owing to propagation interferences may not be received simultaneously.

Beginning 1 minute after the hour, a 600-Hz signal is broadcast for about 45 s. At 2 min after the hour, the standard musical pitch of 440 Hz is broadcast for about 45 s. For the remaining 57 min of the hour, alternating tones of 600 and 500 Hz are broadcast for the first 45 s of each minute (see NIST Spec. Pub. 432). The time signal can also be received via long-distance telephone service from Ft. Collins. In addition to providing the musical pitch, these tone signals may be of use as markers for automated recorders and other such devices.

DENSITY AND RELATIVE DENSITY

Density of a body is its mass per unit volume. With SI units densities are in kilograms per cubic meter. However, giving densities in grams per cubic centimeter has been common. With the USCS, densities are given in pounds per mass cubic foot.

Table 1.2.8   Relative Densities at 60°/60°F Corresponding to Degrees API and Weights per U.S. Gallon at 60°F

$$\left(\text{Calculated from the formula, relative density} = \frac{141.5}{131.5 + \text{deg API}} \right)$$

Degrees API	Relative density	Lb per U.S. gallon	Degrees API	Relative density	Lb per U.S. gallon
10	1.0000	8.328	56	0.7547	6.283
11	0.9930	8.270	57	0.7507	6.249
12	0.9861	8.212	58	0.7467	6.216
13	0.9792	8.155	59	0.7428	6.184
14	0.9725	8.099	60	0.7389	6.151
15	0.9659	8.044	61	0.7351	6.119
16	0.9593	7.989	62	0.7313	6.087
17	0.9529	7.935	63	0.7275	6.056
18	0.9465	7.882	64	0.7238	6.025
19	0.9402	7.830	65	0.7201	5.994
20	0.9340	7.778	66	0.7165	5.964
21	0.9279	7.727	67	0.7128	5.934
22	0.9218	7.676	68	0.7093	5.904
23	0.9159	7.627	69	0.7057	5.874
24	0.9100	7.578	70	0.7022	5.845
25	0.9042	7.529	71	0.6988	5.817
26	0.8984	7.481	72	0.6953	5.788
27	0.8927	7.434	73	0.6919	5.759
28	0.8871	7.387	74	0.6886	5.731
29	0.8816	7.341	75	0.6852	5.703
30	0.8762	7.296	76	0.6819	5.676
31	0.8708	7.251	77	0.6787	5.649
32	0.8654	7.206	78	0.6754	5.622
33	0.8602	7.163	79	0.6722	5.595
34	0.8550	7.119	80	0.6690	5.568
35	0.8498	7.076	81	0.6659	5.542
36	0.8448	7.034	82	0.6628	5.516
37	0.8398	6.993	83	0.6597	5.491
38	0.8348	6.951	84	0.6566	5.465
39	0.8299	6.910	85	0.6536	5.440
40	0.8251	6.870	86	0.6506	5.415
41	0.8203	6.830	87	0.6476	5.390
42	0.8155	6.790	88	0.6446	5.365
43	0.8109	6.752	89	0.6417	5.341
44	0.8063	6.713	90	0.6388	5.316
45	0.8017	6.675	91	0.6360	5.293
46	0.7972	6.637	92	0.6331	5.269
47	0.7927	6.600	93	0.6303	5.246
48	0.7883	6.563	94	0.6275	5.222
49	0.7839	6.526	95	0.6247	5.199
50	0.7796	6.490	96	0.6220	5.176
51	0.7753	6.455	97	0.6193	5.154
52	0.7711	6.420	98	0.6166	5.131
53	0.7669	6.385	99	0.6139	5.109
54	0.7628	6.350	100	0.6112	5.086
55	0.7587	6.316			

NOTE: The weights in this table are weights in air at 60°F with humidity 50 percent and pressure 760 mm.

Table 1.2.9 Relative Densities at 60°/60°F Corresponding to Degrees Baumé for Liquids Lighter than Water and Weights per U.S. Gallon at 60°F

$$\left(\text{Calculated from the formula, relative density } \frac{60°}{60°}F = \frac{140}{130 + \text{deg Baumé}}\right)$$

Degrees Baumé	Relative density	Lb per gallon	Degrees Baumé	Relative density	Lb per gallon
10.0	1.0000	8.328	56.0	0.7527	6.266
11.0	0.9929	8.269	57.0	0.7487	6.233
12.0	0.9859	8.211	58.0	0.7447	6.199
13.0	0.9790	8.153	59.0	0.7407	6.166
14.0	0.9722	8.096	60.0	0.7368	6.134
15.0	0.9655	8.041	61.0	0.7330	6.102
16.0	0.9589	7.986	62.0	0.7292	6.070
17.0	0.9524	7.931	63.0	0.7254	6.038
18.0	0.9459	7.877	64.0	0.7216	6.007
19.0	0.9396	7.825	65.0	0.7179	5.976
20.0	0.9333	7.772	66.0	0.7143	5.946
21.0	0.9272	7.721	67.0	0.7107	5.916
22.0	0.9211	7.670	68.0	0.7071	5.886
23.0	0.9150	7.620	69.0	0.7035	5.856
24.0	0.9091	7.570	70.0	0.7000	5.827
25.0	0.9032	7.522	71.0	0.6965	5.798
26.0	0.8974	7.473	72.0	0.6931	5.769
27.0	0.8917	7.425	73.0	0.6897	5.741
28.0	0.8861	7.378	74.0	0.6863	5.712
29.0	0.8805	7.332	75.0	0.6829	5.685
30.0	0.8750	7.286	76.0	0.6796	5.657
31.0	0.8696	7.241	77.0	0.6763	5.629
32.0	0.8642	7.196	78.0	0.6731	5.602
33.0	0.8589	7.152	79.0	0.6699	5.576
34.0	0.8537	7.108	80.0	0.6667	5.549
35.0	0.8485	7.065	81.0	0.6635	5.522
36.0	0.8434	7.022	82.0	0.6604	5.497
37.0	0.8383	6.980	83.0	0.6573	5.471
38.0	0.8333	6.939	84.0	0.6542	5.445
39.0	0.8284	6.898	85.0	0.6512	5.420
40.0	0.8235	6.857	86.0	0.6482	5.395
41.0	0.8187	6.817	87.0	0.6452	5.370
42.0	0.8140	6.777	88.0	0.6422	5.345
43.0	0.8092	6.738	89.0	0.6393	5.320
44.0	0.8046	6.699	90.0	0.6364	5.296
45.0	0.8000	6.661	91.0	0.6335	5.272
46.0	0.7955	6.623	92.0	0.6306	5.248
47.0	0.7910	6.586	93.0	0.6278	5.225
48.0	0.7865	6.548	94.0	0.6250	5.201
49.0	0.7821	6.511	95.0	0.6222	5.178
50.0	0.7778	6.476	96.0	0.6195	5.155
51.0	0.7735	6.440	97.0	0.6167	5.132
52.0	0.7692	6.404	98.0	0.6140	5.100
53.0	0.7650	6.369	99.0	0.6114	5.088
54.0	0.7609	6.334	100.0	0.6087	5.066
55.0	0.7568	6.300			

Relative density is the ratio of the density of one substance to that of a second (or reference) substance, both at some specified temperature. Use of the earlier term *specific gravity* for this quantity is discouraged. For solids and liquids water is almost universally used as the reference substance. Physicists use a reference temperature of 4°C (= 39.2°F); U.S. engineers commonly use 60°F. With the introduction of SI units, it may be found desirable to use 59°F, since 59°F and 15°C are equivalents.

For gases, relative density is generally the ratio of the density of the gas to that of air, both at the same temperature, pressure, and dryness (as regards water vapor). Because equal numbers of moles of gases occupy equal volumes, the ratio of the molecular weight of the gas to that of air may be used as the relative density of the gas. When this is done, the molecular weight of air may be taken as 28.9644.

The relative density of liquids is usually measured by means of a **hydrometer.** In addition to a scale reading in relative density as defined above, other arbitrary scales for hydrometers are used in various trades and industries. The most common of these are the **API** and **Baumé.** The API (American Petroleum Institute) scale is approved by the American Petroleum Institute, the ASTM, the U.S. Bureau of Mines, and the National Bureau of Standards and is recommended for exclusive use in the U.S. petroleum industry, superseding the Baumé scale for liquids lighter than water. The relation between **API degrees** and relative density (see Table 1.2.8) is expressed by the following equation:

$$\text{Degrees API} = \frac{141.5}{\text{rel dens } 60°/60°F} - 131.5$$

The relative densities corresponding to the indications of the **Baumé hydrometer** are given in Tables 1.2.9 and 1.2.10.

Table 1.2.10 Relative Densities at 60°/60°F Corresponding to Degrees Baumé for Liquids Heavier than Water

$$\left(\text{Calculated from the formula, relative density } \frac{60°}{60°}F = \frac{145}{145 - \text{deg Baumé}}\right)$$

Degrees Baumé	Relative density	Degrees Baumé	Relative density	Degrees Baumé	Relative density
0	1.0000	24	1.1983	48	1.4948
1	1.0069	25	1.2083	49	1.5104
2	1.0140	26	1.2185	50	1.5263
3	1.0211	27	1.2288	51	1.5426
4	1.0284	28	1.2393	52	1.5591
5	1.0357	29	1.2500	53	1.5761
6	1.0432	30	1.2609	54	1.5934
7	1.0507	31	1.2719	55	1.6111
8	1.0584	32	1.2832	56	1.6292
9	1.0662	33	1.2946	57	1.6477
10	1.0741	34	1.3063	58	1.6667
11	1.0821	35	1.3182	59	1.6860
12	1.0902	36	1.3303	60	1.7059
13	1.0985	37	1.3426	61	1.7262
14	1.1069	38	1.3551	62	1.7470
15	1.1154	39	1.3679	63	1.7683
16	1.1240	40	1.3810	64	1.7901
17	1.1328	41	1.3942	65	1.8125
18	1.1417	42	1.4078	66	1.8354
19	1.1508	43	1.4216	67	1.8590
20	1.1600	44	1.4356	68	1.8831
21	1.1694	45	1.4500	69	1.9079
22	1.1789	46	1.4646	70	1.9333
23	1.1885	47	1.4796		

CONVERSION AND EQUIVALENCY TABLES

Note for Use of Conversion Tables (Tables 1.2.11 through 1.2.34)

Subscripts after any figure, 0_s, 9_s, etc., mean that that figure is to be repeated the indicated number of times.

Table 1.2.11   Length Equivalents

Centimetres	Inches	Feet	Yards	Metres	Chains	Kilometres	Miles
1	0.3937	0.03281	0.01094	0.01	0.0_34971	10^{-5}	0.0_66214
2.540	1	0.08333	0.02778	0.0254	0.001263	0.0_4254	0.0_41578
30.48	12	1	0.3333	0.3048	0.01515	0.0_33048	0.0_31894
91.44	36	3	1	0.9144	0.04545	0.0_39144	0.0_35682
100	39.37	3.281	1.0936	1	0.04971	0.001	0.0_36214
2012	792	66	22	20.12	1	0.02012	0.0125
100000	39370	3281	1093.6	1000	49.71	1	0.6214
160934	63360	5280	1760	1609	80	1.609	1

(As used by metrology laboratories for precise measurements, including measurements of surface texture)*

Angstrom units Å	Surface texture (U.S.), microinch μin	Light bands,† monochromatic helium light count ‡	Surface texture foreign, μm	Precision measurements, § 0.0001 in	Close-tolerance measurements, 0.001 in (mils)	Metric unit, mm	USCS unit, in
1	0.003937	0.0003404	0.0001	0.0_43937	0.0_53937	0.0_61	0.0_83937
254	1	0.086	0.0254	0.01	0.001	0.0_4254	0.0_51
2937.5	11.566	1	0.29375	0.11566	0.011566	0.0_329375	0.0_411566
10,000	39.37	3.404	1	0.3937	0.03937	0.001	0.0_43937
25,400	100	8.646	2.54	1	0.1	0.00254	0.0001
254,000	1000	86.46	25.4	10	1	0.0254	0.001
10,000,000	39,370	3404	1000	393.7	39.37	1	0.03937
254,000,000	1,000,000	86,460	25,400	10,000	1000	25.4	1

* Computed by J. A. Broadston.
† One light band equals one-half corresponding wavelength. Visible-light wavelengths range from red at 6,500 Å to violet at 4,100 Å.
‡ One helium light band = 0.000011661 in = 2937.5 Å; one krypton 86 light band = 0.0000119 in = 3,022.5 Å; one mercury 198 light band = 0.00001075 in = 2,730 Å.
§ The designations "precision measurements," etc., are not necessarily used in all metrology laboratories.

Table 1.2.12   Conversion of Lengths*

	Inches to millimetres	Millimetres to inches	Feet to metres	Metres to feet	Yards to metres	Metres to yards	Miles to kilometres	Kilometres to miles
1	25.40	0.03937	0.3048	3.281	0.9144	1.094	1.609	0.6214
2	50.80	0.07874	0.6096	6.562	1.829	2.187	3.219	1.243
3	76.20	0.1181	0.9144	9.843	2.743	3.281	4.828	1.864
4	101.60	0.1575	1.219	13.12	3.658	4.374	6.437	2.485
5	127.00	0.1969	1.524	16.40	4.572	5.468	6.047	3.107
6	152.40	0.2362	1.829	19.69	5.486	6.562	9.656	3.728
7	177.80	0.2756	2.134	22.97	6.401	7.655	11.27	4.350
8	203.20	0.3150	2.438	26.25	7.315	8.749	12.87	4.971
9	228.60	0.3543	2.743	29.53	8.230	9.843	14.48	5.592

* EXAMPLE: 1 in = 25.40 mm.

Common fractions of an inch to millimetres (from 1/64 to 1 in)

64ths	Millimetres	64ths	Millimetres	64th	Millimetres	64ths	Millimetres	64ths	Millimetres	64ths	Millimetres
1	0.397	13	5.159	25	9.922	37	14.684	49	19.447	57	22.622
2	0.794	14	5.556	26	10.319	38	15.081	50	19.844	58	23.019
3	1.191	15	5.953	27	10.716	39	15.478	51	20.241	59	23.416
4	1.588	16	6.350	28	11.112	40	15.875	52	20.638	60	23.812
5	1.984	17	6.747	29	11.509	41	16.272	53	21.034.	61	24.209
6	2.381	18	7.144	30	11.906	42	16.669	54	21.431	62	24.606
7	2.778	19	7.541	31	12.303	43	17.066	55	21.828	63	25.003
8	3.175	20	7.938	32	12.700	44	17.462	56	22.225	64	25.400
9	3.572	21	8.334	33	13.097	45	17.859				
10	3.969	22	8.731	34	13.494	46	18.256				
11	4.366	23	9.128	35	13.891	47	18.653				
12	4.762	24	9.525	36	14.288	48	19.050				

Decimals of an inch to millimetres (0.01 to 0.99 in)

	0	1	2	3	4	5	6	7	8	9
.0		0.254	0.508	0.762	1.016	1.270	1.524	1.778	2.032	2.286
.1	2.540	2.794	3.048	3.302	3.556	3.810	4.064	4.318	4.572	4.826
.2	5.080	5.334	5.588	5.842	6.096	6.350	6.604	6.858	7.112	7.366
.3	7.620	7.874	8.128	8.382	8.636	8.890	9.144	9.398	9.652	9.906
.4	10.160	10.414	10.668	10.922	11.176	11.430	11.684	11.938	12.192	12.446
.5	12.700	12.954	13.208	13.462	13.716	13.970	14.224	14.478	14.732	14.986
.6	15.240	15.494	15.748	16.002	16.256	16.510	16.764	17.018	17.272	17.526
.7	17.780	18.034	18.288	18.542	18.796	19.050	19.304	19.558	19.812	20.066
.8	20.320	20.574	20.828	21.082	21.336	21.590	21.844	22.098	22.352	22.606
.9	22.860	23.114	23.368	23.622	23.876	24.130	24.384	24.638	24.892	25.146

Millimetres to decimals of an inch (from 1 to 99 mm)

	0.	1.	2.	3.	4.	5.	6.	7.	8.	9.
0		0.0394	0.0787	0.1181	0.1575	0.1969	0.2362	0.2756	0.3150	0.3543
1	0.3937	0.4331	0.4724	0.5118	0.5512	0.5906	0.6299	0.6693	0.7087	0.7480
2	0.7874	0.8268	0.8661	0.9055	0.9449	0.9843	1.0236	1.0630	1.1024	1.1417
3	1.1811	1.2205	1.2598	1.2992	1.3386	1.3780	1.4173	1.4567	1.4961	1.5354
4	1.5748	1.6142	1.6535	1.6929	1.7323	1.7717	1.8110	1.8504	1.8898	1.9291
5	1.9685	2.0079	2.0472	2.0866	2.1260	2.1654	2.2047	2.2441	2.2835	2.3228
6	2.3622	2.4016	2.4409	2.4803	2.5197	2.5591	2.5984	2.6378	2.6772	2.7165
7	2.7559	2.7953	2.8346	2.8740	2.9134	2.9528	2.9921	3.0315	3.0709	3.1102
8	3.1496	3.1890	3.2283	3.2677	3.3071	3.3465	3.3858	3.4252	3.4646	3.5039
9	3.5433	3.5827	3.6220	3.6614	3.7008	3.7402	3.7795	3.8189	3.8583	3.8976

Table 1.2.13   Area Equivalents
(1 hectare = 100 ares = 10,000 centiares or square metres)

Square metres	Square inches	Square feet	Square yards	Square rods	Square chains	Roods	Acres	Square miles or sections
1	1550	10.76	1.196	0.0395	0.002471	0.0_39884	0.0_32471	0.0_63861
0.0_36452	1	0.006944	0.0_37716	0.0_42551	0.0_51594	0.0_66377	0.0_61594	0.0_92491
0.09290	144	1	0.1111	0.003673	0.0_32296	0.0_49183	0.0_42296	0.0_53587
0.8361	1296	9	1	0.03306	0.002066	0.0_58264	0.0002066	0.0_63228
25.29	39204	272.25	30.25	1	0.0625	0.02500	0.00625	0.0_59766
404.7	627264	4356	484	16	1	0.4	0.1	0.0001562
1012	1568160	10890	1210	40	2.5	1	0.25	0.0_33906
4047	6272640	43560	4840	160	10	4	1	0.001562
2589988		27878400	3097600	102400	6400	2560	640	1

Table 1.2.14   Conversion of Areas*

	Sq in to sq cm	Sq cm to sq in	Sq ft to sq m	Sq m to sq ft	Sq yd to sq m	Sq m to sq yd	Acres to hectares	Hectares to acres	Sq mi to sq km	Sq km to sq mi
1	6.452	0.1550	0.0929	10.76	0.8361	1.196	0.4047	2.471	2.590	0.3861
2	12.90	0.3100	0.1858	21.53	1.672	2.392	0.8094	4.942	5.180	0.7722
3	19.35	0.4650	0.2787	32.29	2.508	3.588	1.214	7.413	7.770	1.158
4	25.81	0.6200	0.3716	43.06	3.345	4.784	1.619	9.884	10.360	1.544
5	32.26	0.7750	0.4645	53.82	4.181	5.980	2.023	12.355	12.950	1.931
6	38.71	0.9300	0.5574	64.58	5.017	7.176	2.428	14.826	15.540	2.317
7	45.16	1.085	0.6503	75.35	5.853	8.372	2.833	17.297	18.130	2.703
8	51.61	1.240	0.7432	86.11	6.689	9.568	3.237	19.768	20.720	3.089
9	58.06	1.395	0.8361	96.88	7.525	10.764	3.642	22.239	23.310	3.475

* EXAMPLE: 1 in^2 = 6.452 cm^2.

Table 1.2.15   Volume and Capacity Equivalents

Cubic inches	Cubic feet	Cubic yards	U.S. Apothecary fluid ounces	U.S. quarts Liquid	U.S. quarts Dry	U.S. gallons	U.S. bushels	Cubic decimetres or litres
1	0.0_35787	0.0_42143	0.5541	0.01732	0.01488	0.0_44329	0.0_34650	0.01639
1728	1	0.03704	957.5	29.92	25.71	7.481	0.8036	28.32
46656	27	1	25853	807.9	694.3	202.2	21.70	764.6
1.805	0.001044	0.0_43868	1	0.03125	0.02686	0.007812	0.0_38392	0.02957
57.75	0.03342	0.001238	32	1	0.8594	0.25	0.02686	0.9464
67.20	0.03889	0.001440	37.24	1.164	1	0.2909	0.03125	1.101
231	0.1337	0.004951	128	4	3.437	1	0.1074	3.785
2150	1.244	0.04609	1192	37.24	32	9.309	1	35.24
61.02	0.03531	0.001308	33.81	1.057	0.9081	0.2642	0.02838	1

Table 1.2.16   Conversion of Volumes or Cubic Measure*

	Cu in to mL	mL to cu in	Cu ft to cu m	Cu m to cu ft	Cu yd to cu m	Cu m to cu yd	Gallons to cu ft	Cu ft to gallons
1	16.39	0.06102	0.02832	35.31	0.7646	1.308	0.1337	7.481
2	32.77	0.1220	0.05663	70.63	1.529	2.616	0.2674	14.96
3	49.16	0.1831	0.08495	105.9	2.294	3.924	0.4010	22.44
4	65.55	0.2441	0.1133	141.3	3.058	5.232	0.5347	29.92
5	81.94	0.3051	0.1416	176.6	3.823	6.540	0.6684	37.40
6	98.32	0.3661	0.1699	211.9	4.587	7.848	0.8021	44.88
7	114.7	0.4272	0.1982	247.2	5.352	9.156	0.9358	52.36
8	131.1	0.4882	0.2265	282.5	6.116	10.46	1.069	59.84
9	147.5	0.5492	0.2549	317.8	6.881	11.77	1.203	67.32

* EXAMPLE: 1 in^3 = 16.39 mL.

Table 1.2.17   Conversion of Volumes or Capacities*

	Fluid ounces to mL	mL to fluid ounces	Liquid pints to litres	Litres to liquid pints	Liquid quarts to litres	Litres to liquid quarts	Gallons to litres	Litres to gallons	Bushels to hectolitres	Hectolitres to bushels
1	29.57	0.03381	0.4732	2.113	0.9463	1.057	3.785	0.2642	0.3524	2.838
2	59.15	0.06763	0.9463	4.227	1.893	2.113	7.571	0.5284	0.7048	5.676
3	88.72	0.1014	1.420	6.340	2.839	3.170	11.36	0.7925	1.057	8.513
4	118.3	0.1353	1.893	8.454	3.785	4.227	15.14	1.057	1.410	11.35
5	147.9	0.1691	2.366	10.57	4.732	5.284	18.93	1.321	1.762	14.19
6	177.4	0.2092	2.839	12.68	5.678	6.340	22.71	1.585	2.114	17.03
7	207.0	0.2367	3.312	14.79	6.624	7.397	26.50	1.849	2.467	19.86
8	236.6	0.2705	3.785	16.91	7.571	8.454	30.28	2.113	2.819	22.70
9	266.2	0.3043	4.259	19.02	8.517	9.510	34.07	2.378	3.171	25.54

* EXAMPLE: 1 fluid oz = 29.57 mL.

Table 1.2.18   Mass Equivalents

Kilograms	Grains	Ounces — Troy and apoth	Ounces — Avoirdupois	Pounds — Troy and apoth	Pounds — Avoirdupois	Tons — Short	Tons — Long	Tons — Metric
1	15432	32.15	35.27	2.6792	2.205	0.0_21102	0.0_39842	0.001
0.0_46480	1	0.0_22083	0.0_22286	0.0_31736	0.0_31429	0.0_77143	0.0_66378	0.0_66480
0.03110	480	1	1.09714	0.08333	0.06857	0.0_43429	0.0_33061	0.0_43110
0.02835	437.5	0.9115	1	0.07595	0.0625	0.0_43125	0.0_42790	0.0_42835
0.3732	5760	12	13.17	1	0.8229	0.0_44114	0.0_33673	0.0_33732
0.4536	7000	14.58	16	1.215	1	0.0005	0.0_44464	0.0_34536
907.2	140_6	29167	320_3	2431	2000	1	0.8929	0.9072
1016	15680_4	32667	35840	2722	2240	1.12	1	1.016
1000	15432356	32151	35274	2679	2205	1.102	0.9842	1

Table 1.2.19   Conversion of Masses*

	Grains to grams	Grams to grains	Ounces (avdp) to grams	Grams to ounces (avdp)	Pounds (avdp) to kilograms	Kilograms to pounds (avdp)	Short tons (2000 lb) to metric tons	Metric tons (1000 kg) to short tons	Long tons (2240 lb) to metric tons	Metric tons to long tons
1	0.06480	15.43	28.35	0.03527	0.4536	2.205	0.907	1.102	1.016	0.984
2	0.1296	30.86	56.70	0.07055	0.9072	4.409	1.814	2.205	2.032	1.968
3	0.1944	46.30	85.05	0.1058	1.361	6.614	2.722	3.307	3.048	2.953
4	0.2592	61.73	113.40	0.1411	1.814	8.818	3.629	4.409	4.064	3.937
5	0.3240	77.16	141.75	0.1764	2.268	11.02	4.536	5.512	5.080	4.921
6	0.3888	92.59	170.10	0.2116	2.722	13.23	5.443	6.614	6.096	5.905
7	0.4536	108.03	198.45	0.2469	3.175	15.43	6.350	7.716	7.112	6.889
8	0.5184	123.46	226.80	0.2822	3.629	17.64	7.257	8.818	8.128	7.874
9	0.5832	138.89	255.15	0.3175	4.082	19.84	8.165	9.921	9.144	8.858

* EXAMPLE: 1 grain = 0.06480 grams.

Table 1.2.20   Pressure Equivalents

Pascals N/m^2	Bars 10^5 N/m^2	Poundsf per in^2	Atmospheres	Columns of mercury at temperature 0°C and $g = 9.80665$ m/s² — cm	Columns of mercury at temperature 0°C and $g = 9.80665$ m/s² — in	Columns of water at temperature 15°C and $g = 9.80665$ m/s² — cm	Columns of water at temperature 15°C and $g = 9.80665$ m/s² — in
1	10^{-5}	0.000145	0.00001	0.00075	0.000295	0.01021	0.00402
100000	1	14.504	0.9869	75.01	29.53	1020.7	401.8
6894.8	0.068948	1	0.06805	5.171	2.036	70.37	27.703
101326	1.0132	14.696	1	76.000	29.92	1034	407.1
1333	0.0133	0.1934	0.01316	1	0.3937	13.61	5.357
3386	0.03386	0.4912	0.03342	2.540	1	34.56	13.61
97.98	0.0009798	0.01421	0.000967	0.07349	0.02893	1	0.3937
248.9	0.002489	0.03609	0.002456	0.1867	0.07349	2.540	1

Table 1.2.21   Conversion of Pressures*

	Lb/in² to bars	Bars to lb/in²	Lb/in² to atmospheres	Atmospheres to lb/in²	Bars to atmospheres	Atmospheres to bars
1	0.06895	14.504	0.06805	14.696	0.98692	1.01325
2	0.13790	29.008	0.13609	29.392	1.9738	2.0265
3	0.20684	43.511	0.20414	44.098	2.9607	3.0397
4	0.27579	58.015	0.27218	58.784	3.9477	4.0530
5	0.34474	72.519	0.34823	73.480	4.9346	5.0663
6	0.41368	87.023	0.40826	88.176	5.9215	6.0795
7	0.48263	101.53	0.47632	102.87	6.9085	7.0927
8	0.55158	116.03	0.54436	117.57	7.8954	8.1060
9	0.62053	130.53	0.61241	132.26	8.8823	9.1192

* EXAMPLE: 1 lb/in² = 0.06895 bar.

Table 1.2.22   Velocity Equivalents

cm/s	m/s	m/min	km/h	ft/s	ft/min	mi/h	Knots
1	0.01	0.6	0.036	0.03281	1.9685	0.02237	0.01944
100	1	60	3.6	3.281	196.85	2.237	1.944
1.667	0.01667	1	0.06	0.05468	3.281	0.03728	0.03240
27.78	0.2778	16.67	1	0.9113	54.68	0.6214	0.53996
30.48	0.3048	18.29	1.097	1	60	0.6818	0.59248
0.5080	0.005080	0.3048	0.01829	0.01667	1	0.01136	0.00987
44.70	0.4470	26.82	1.609	1.467	88	1	0.86898
51.44	0.5144	30.87	1.852	1.688	101.3	1.151	1

Table 1.2.23   Conversion of Linear and Angular Velocities*

	cm/s to ft/min	ft/min to cm/s	cm/s to mi/h	mi/h to cm/s	ft/s to mi/h	mi/h to ft/s	rad/s to r/min	r/min to rad/s
1	1.97	0.508	0.0224	44.70	0.682	1.47	9.55	0.1047
2	3.94	1.016	0.0447	89.41	1.364	2.93	19.10	0.2094
3	5.91	1.524	0.0671	134.1	2.045	4.40	28.65	0.3142
4	7.87	2.032	0.0895	178.8	2.727	5.87	38.20	0.4189
5	9.84	2.540	0.1118	223.5	3.409	7.33	47.75	0.5236
6	11.81	3.048	0.1342	268.2	4.091	8.80	57.30	0.6283
7	13.78	3.556	0.1566	312.9	4.773	10.27	66.84	0.7330
8	15.75	4.064	0.1790	357.6	5.455	11.73	76.39	0.8378
9	17.72	4.572	0.2013	402.3	6.136	13.20	85.94	0.9425

* EXAMPLE: 1 cm/s = 1.97 ft/min.

Table 1.2.24   Acceleration Equivalents

cm/s²	m/s²	m/(h · s)	km/(h · s)	ft/(h · s)	ft/s²	ft/min²	mi/(h · s)	knots/s
1	0.01	36.00	0.036	118.1	0.03281	118.1	0.02237	0.01944
100	1	3600	3.6	11811	3.281	11811	2.237	1.944
0.02778	0.0002778	1	0.001	3.281	0.0009113	3.281	0.0006214	0.0005400
27.78	0.2778	1000	1	3281	0.9113	3281	0.6214	0.5400
0.008467	0.00008467	0.3048	0.0003048	1	0.0002778	1	0.0001894	0.0001646
30.48	0.3048	1097	1.097	3600	1	3600	0.6818	0.4572
0.008467	0.00008467	0.3048	0.0003048	1	0.0002778	1	0.0001894	0.0001646
44.70	0.4470	1609	1.609	5280	1.467	5280	1	0.8690
51.44	0.5144	1852	1.852	6076	1.688	6076	1.151	1

Table 1.2.25 Conversion of Accelerations*

	$\mathrm{cm/s^2}$ to $\mathrm{ft/min^2}$	$\mathrm{km/(h \cdot s)}$ to $\mathrm{mi/(h \cdot s)}$	$\mathrm{km/(h \cdot s)}$ to knots/s	$\mathrm{ft/s^2}$ to $\mathrm{mi/(h \cdot s)}$	$\mathrm{ft/s^2}$ to knots/s	$\mathrm{ft/min^2}$ to $\mathrm{cm/s^2}$	$\mathrm{mi/(h \cdot s)}$ to $\mathrm{km/(h \cdot s)}$	$\mathrm{mi/(h \cdot s)}$ to knots/s	knots/s to $\mathrm{mi/(h \cdot s)}$	knots/s to $\mathrm{km/(h \cdot s)}$
1	118.1	0.6214	0.5400	0.6818	0.4572	0.008467	1.609	0.8690	1.151	1.852
2	236.2	1.243	1.080	1.364	0.9144	0.01693	3.219	1.738	2.302	3.704
3	354.3	1.864	1.620	2.045	1.372	0.02540	4.828	2.607	3.452	5.556
4	472.4	2.485	2.160	2.727	1.829	0.03387	6.437	3.476	4.603	7.408
5	590.6	3.107	2.700	3.409	2.286	0.04233	8.046	4.345	5.754	9.260
6	708.7	3.728	3.240	4.091	2.743	0.05080	9.656	5.214	6.905	11.11
7	826.8	4.350	3.780	4.772	3.200	0.05927	11.27	6.083	8.056	12.96
8	944.9	4.971	4.320	5.454	3.658	0.06774	12.87	6.952	9.206	14.82
9	1063	5.592	4.860	6.136	4.115	0.07620	14.48	7.821	10.36	16.67

* EXAMPLE: 1 $\mathrm{cm/s^2}$ = 118.1 $\mathrm{ft/min^2}$.

Table 1.2.26 Energy or Work Equivalents

Joules or Newton-metres	Kilogramf-metres	Foot-poundsf	Kilowatt hours	Metric horsepower-hours	Horsepower-hours	Litre-atmospheres	Kilocalories	British thermal units
1	0.10197	0.7376	0.0_62778	0.0_63777	0.0_63725	0.009869	0.0_32388	0.0_39478
9.80665	1	7.233	0.0_52724	0.0_537037	0.0_53653	0.09678	0.002342	0.009295
1.356	0.1383	1	0.0_63766	0.0_651206	0.0_650505	0.01338	0.0_33238	0.001285
3.600×10^6	3.671×10^5	2.655×10^6	1	1.3596	1.341	35528	859.9	3412
2.648×10^6	270000	1.9529×10^6	0.7355	1	0.9863	26131	632.4	2510
2.6845×10^6	2.7375×10^5	1.98×10^6	0.7457	1.0139	1	26493	641.2	2544
101.33	10.333	74.74	0.0_42815	0.0_43827	0.0_43775	1	0.02420	0.09604
4186.8	426.9	3088	0.001163	0.001581	0.001560	41.32	1	3.968
1055	107.6	778.2	0.0_32931	0.0_33985	0.0_33930	10.41	0.25200	1

Table 1.2.27 Conversion of Energy, Work, Heat*

	Ft · lbf to joules	Joules to ft · lbf	Ft · lbf to Btu	Btu to ft · lbf	Kilogramf-metres to kilocalories	Kilocalories to kilogramf-metres	Joules to calories	Calories to joules
1	1.3558	0.7376	0.001285	778.2	0.002342	426.9	0.2388	4.187
2	2.7116	1.4751	0.002570	1,556	0.004685	853.9	0.4777	8.374
3	4.0674	2.2127	0.003855	2,334	0.007027	1,281	0.7165	12.56
4	5.4232	2.9503	0.005140	3,113	0.009369	1,708	0.9554	16.75
5	6.7790	3.6879	0.006425	3,891	0.01172	2,135	1.194	20.93
6	8.1348	4.4254	0.007710	4,669	0.01405	2,562	1.433	25.12
7	9.4906	5.1630	0.008995	5,447	0.01640	2,989	1.672	29.31
8	10.8464	5.9006	0.01028	6,225	0.01874	3,415	1.911	33.49
9	12.2022	6.6381	0.01156	7,003	0.02108	3,842	2.150	37.68

* EXAMPLE: 1 ft · lbf = 1.3558 J.

Table 1.2.28 Power Equivalents

Horsepower	Kilowatts	Metric horsepower	Kgf · m per s	Ft · lbf per s	Kilocalories per s	Btu per s
1	0.7457	1.014	76.04	550	0.1781	0.7068
1.341	1	1.360	102.0	737.6	0.2388	0.9478
0.9863	0.7355	1	75	542.5	0.1757	0.6971
0.01315	0.009807	0.01333	1	7.233	0.002342	0.009295
0.00182	0.001356	0.00184	0.1383	1	0.0_33238	0.001285
5.615	4.187	5.692	426.9	3088	1	3.968
1.415	1.055	1.434	107.6	778.2	0.2520	1

Table 1.2.29   Conversion of Power*

	Horsepower to kilowatts	Kilowatts to horsepower	Metric horsepower to kilowatts	Kilowatts to metric horsepower	Horsepower to metric horsepower	Metric horsepower to horsepower
1	0.7457	1.341	0.7355	1.360	1.014	0.9863
2	1.491	2.682	1.471	2.719	2.028	1.973
3	2.237	4.023	2.206	4.079	3.042	2.959
4	2.983	5.364	2.942	5.438	4.055	3.945
5	3.729	6.705	3.677	6.798	5.069	4.932
6	4.474	8.046	4.412	8.158	6.083	5.918
7	5.220	9.387	5.147	9.520	7.097	6.904
8	5.966	10.73	5.883	10.88	8.111	7.891
9	6.711	12.07	6.618	12.24	9.125	8.877

* EXAMPLE: 1 hp = 0.7457 kW.

Table 1.2.30   Density Equivalents*

Grams per mL	Lb per cu in	Lb per cu ft	Short tons (2,000 lb) per cu yd	Lb per U.S. gal
1	0.03613	62.43	0.8428	8.345
27.68	1	1728	23.33	231
0.01602	0.0$_3$5787	1	0.0135	0.1337
1.187	0.04287	74.7	1	9.902
0.1198	0.004329	7.481	0.1010	1

* EXAMPLE: 1 g per mL = 62.43 lb per cu ft.

Table 1.2.31   Conversion of Density

	Grams per mL to lb per cu ft	Lb per cu ft to grams per mL	Grams per mL to short tons per cu yd	Short tons per cu yd to grams per mL
1	62.43	0.01602	0.8428	1.187
2	124.86	0.03204	1.6856	2.373
3	187.28	0.04805	2.5283	3.560
4	249.71	0.06407	3.3711	4.746
5	312.14	0.08009	4.2139	5.933
6	374.57	0.09611	5.0567	7.119
7	437.00	0.11213	5.8995	8.306
8	499.43	0.12814	6.7423	9.492
9	561.85	0.14416	7.5850	10.679
10	624.28	0.16018	8.4278	11.866

Table 1.2.32   Thermal Conductivity

Calories per cm · s · °C	Watts per cm · °C	Calories per cm · h · °C	Btu · ft per ft^2 · h · °F	Btu · in per ft^2 · day · °F
1	4.1868	3,600	241.9	69,670
0.2388	1	860	57.79	16,641
0.0002778	0.001163	1	0.0672	19.35
0.004134	0.01731	14.88	1	288
0.00001435	0.00006009	0.05167	0.00347	1

Table 1.2.33   Thermal Conductance

Calories per cm^2 · s · °C	Watts per cm^2 · °C	Calories per cm^2 · h · °C	Btu per ft^2 · h · °F	Btu per ft^2 · day · °F
1	4.1868	3,600	7,373	176,962
0.2388	1	860	1,761	42,267
0.0002778	0.001163	1	2.048	49.16
0.0001356	0.0005678	0.4882	1	24
0.000005651	0.00002366	0.02034	0.04167	1

Table 1.2.34   Heat Flow

Calories per cm^2 · s	Watts per cm^2	Calories per cm^2 · h	Btu per ft^2 · h	Btu per ft^2 · day
1	4.1868	3,600	13,272	318,531
0.2388	1	860	3,170	76,081
0.0002778	0.001163	1	3.687	88.48
0.00007535	0.0003154	0.2712	1	24
0.000003139	0.00001314	0.01130	0.04167	1

Section **2**

Mathematics

BY

C. EDWARD SANDIFER *Professor, Western Connecticut State University, Danbury, CT.*
THOMAS J. COCKERILL *Advisory Engineer, International Business Machines, Inc.*

2.1 MATHEMATICS
by C. Edward Sandifer

REFERENCES: Conte and DeBoor, "Elementary Numerical Analysis: An Algorithmic Approach," McGraw-Hill. Boyce and DiPrima, "Elementary Differential Equations and Boundary Value Problems," Wiley. Hamming, "Numerical Methods for Scientists and Engineers," McGraw-Hill. Kreyszig, "Advanced Engineering Mathematics," Wiley.

SETS, NUMBERS, AND ARITHMETIC

Sets and Elements

The concept of a set appears throughout modern mathematics. A **set** is a well-defined list or collection of objects and is generally denoted by capital letters, $A, B, C, \ldots$. The objects composing the set are called **elements** and are denoted by lowercase letters, $a, b, x, y, \ldots$. The notation

$$x \in A$$

is read "x is an element of A" and means that x is one of the objects composing the set A.

There are two basic ways to describe a set. The first way is to list the elements of the set.

$$A = \{2, 4, 6, 8, 10\}$$

This often is not practical for very large sets.

The second way is to describe properties which determine the elements of the set.

$$A = \{\text{even numbers from 2 to 10}\}$$

This method is sometimes awkward since a single set may sometimes be described in several different ways.

In describing sets, the symbol : is read "such that." The expression

$$B = \{x : x \text{ is an even integer}, x > 1, x < 11\}$$

is read "B equals the set of all x such that x is an even integer, x is greater than 1, and x is less than 11."

Two sets, A and B, are equal, written $A = B$, if they contain exactly the same elements. The sets A and B above are equal. If two sets, X and Y, are not equal, it is written $X \neq Y$.

Subsets A set C is a subset of a set A, written $C \subseteq A$, if each element in C is also an element in A. It is also said that C is contained in A. Any set is a subset of itself. That is, $A \subseteq A$ always. A is said to be an "improper subset of itself." Otherwise, if $C \subseteq A$ and $C \neq A$, then C is a proper subset of A.

Two theorems are important about subsets:
(Fundamental theorem of set equality)

$$\text{If } X \subseteq Y \quad \text{and} \quad Y \subseteq X, \quad \text{then } X = Y \qquad (2.1.1)$$

(Transitivity)

$$\text{If } X \subseteq Y \quad \text{and} \quad Y \subseteq Z, \quad \text{then } X \subseteq Z \qquad (2.1.2)$$

Universe and Empty Set In an application of set theory, it often happens that all sets being considered are subsets of some fixed set, say integers or vectors. This fixed set is called the **universe** and is sometimes denoted U.

It is possible that a set contains no elements at all. The set with no elements is called the **empty set** or the **null set** and is denoted $\varnothing$.

Set Operations New sets may be built from given sets in several ways. The **union** of two sets, denoted $A \cup B$, is the set of all elements belonging to A or to B, or to both.

$$A \cup B = \{x : x \in A \quad \text{or} \quad x \in B\}$$

The union has the properties:

$$A \subseteq A \cup B \quad \text{and} \quad B \subseteq A \cup B \qquad (2.1.3)$$

The **intersection** is denoted $A \cap B$ and consists of all elements, each of which belongs to both A and B.

$$A \cap B = \{x : x \in A \quad \text{and} \quad x \in B\}$$

The intersection has the properties

$$A \cap B \subseteq A \quad \text{and} \quad A \cap B \subseteq B \qquad (2.1.4)$$

If $A \cap B = \varnothing$, then A and B are called **disjoint.**

In general, a union makes a larger set and an intersection makes a smaller set.

The **complement** of a set A is the set of all elements in the universe set which are not in A. This is written

$$\tilde{}A = \{x : x \in U, \quad x \notin A\}$$

The difference of two sets, denoted $A - B$, is the set of all elements which belong to A but do not belong to B.

Algebra on Sets The operations of union, intersection, and complement obey certain laws known as **Boolean algebra.** Using these laws, it is possible to convert an expression involving sets into other equivalent expressions. The laws of Boolean algebra are given in Table 2.1.1.

Venn Diagrams To give a pictorial representation of a set, **Venn diagrams** are often used. Regions in the plane are used to correspond to sets, and areas are shaded to indicate unions, intersections, and complements. Examples of Venn diagrams are given in Fig. 2.1.1. The laws of Boolean algebra and their relation to Venn diagrams are particularly important in programming and in the logic of computer searches.

Numbers

Numbers are the basic instruments of computation. It is by operations on numbers that calculations are made. There are several different kinds of numbers.

Natural numbers, or counting numbers, denoted **N,** are the whole numbers greater than zero. Sometimes zero is included as a natural number. Any two natural numbers may be added or multiplied to give another natural number, but subtracting them may produce a negative

Table 2.1.1 Laws of Boolean Algebra

1. Idempotency	
$A \cup A = A$	$A \cap A = A$
2. Associativity	
$(A \cup B) \cup C = A \cup (B \cup C)$	$(A \cap B) \cap C = A \cap (B \cap C)$
3. Commutativity	
$A \cup B = B \cup A$	$A \cap B = B \cap A$
4. Distributivity	
$A \cup (B \cap C) = (A \cup B) \cap (A \cup C)$	$A \cap (B \cup C) = (A \cap B) \cup (A \cap C)$
5. Identity	
$A \cup \varnothing = A$	$A \cap U = A$
$A \cup U = U$	$A \cap \varnothing = \varnothing$
6. Complement	
$A \cup \tilde{}A = U$	$A \cap \tilde{}A = \varnothing$
$\tilde{}(\tilde{}A) = A$	
$\tilde{}U = \varnothing$	
$\tilde{}\varnothing = U$	
7. DeMorgan's laws	
$\tilde{}(A \cup B) = \tilde{}A \cap \tilde{}B$	$\tilde{}(A \cap B) = \tilde{}A \cup \tilde{}B$

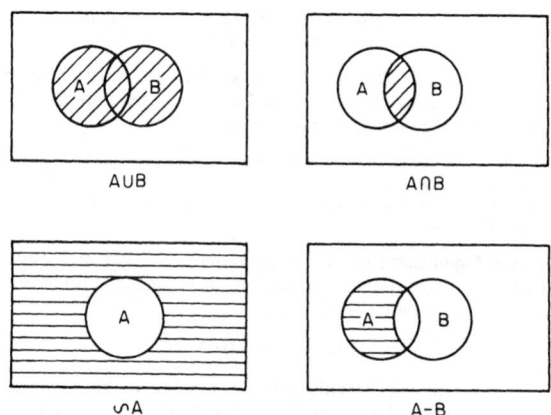

Fig. 2.1.1 Venn diagrams.

number, which is not a natural number, and dividing them may produce a fraction, which is not a natural number. When computers are used, it is important to know what kind of number is being used so the correct data type will be used as well.

Integers, or whole numbers, are denoted by **Z.** They include both positive and negative numbers and zero. Integers may be added, subtracted, and multiplied, but division might not produce an integer.

Real numbers, denoted **R,** are essentially all values which it is possible for a measurement to take, or all possible lengths for line segments. **Rational numbers** are real numbers that are the quotient of two integers, for example, $^{11}/_{78}$. **Irrational numbers** are not the quotient of two integers, for example, π and $\sqrt{2}$. Within the real numbers, it is always possible to add, subtract, multiply, and divide (except division by zero).

Complex numbers, or **imaginary numbers,** denoted **C,** are an extension of the real numbers that include the square root of -1, denoted i. Within the real numbers, only positive numbers have square roots. Within the complex numbers, all numbers have square roots.

Any complex number z can be written uniquely as $z = x + iy$, where x and y are real. Then x is the real part of z, denoted $\mathrm{Re}(z)$, and y is the imaginary part, denoted $\mathrm{Im}(z)$.

The **complex conjugate,** or simply conjugate of a complex number, z is $\bar{z} = x - iy$.

If $z = x + iy$ and $w = u + iv$, then z and w may be manipulated as follows:

$$z + w = (x + u) + i(y + v)$$
$$z - w = (x - u) + i(y - v)$$
$$zw = xu - yv + i(xv + yu)$$
$$\frac{z}{w} = \frac{xu + yv + i(yu - xv)}{u^2 + v^2}$$

As sets, the following relation exists among these different kinds of numbers:

$$\mathbf{N} \subseteq \mathbf{Z} \subseteq \mathbf{R} \subseteq \mathbf{C}$$

Functions

A **function** f is a rule that relates two sets A and B. Given an element x of the set A, the function assigns a unique element y from the set B. This is written

$$y = f(x)$$

The set A is called the **domain** of the function, and the set B is called the **range.** It is possible for A and B to be the same set.

Functions are usually described by giving the rule. For example,

$$f(x) = 3x + 4$$

is a rule for a function with range and domain both equal to **R.** Given a value, say, 2, from the domain, $f(2) = 3(2) + 4 = 10$.

If two functions f and g have the same range and domain and if the ranges are numbers, then f and g may be added, subtracted, multiplied, or divided according to the rules of the range. If $f(x) = 3x + 4$ and $g(x) = \sin(x)$ and both have range and domain equal to **R,** then

$$f + g(x) = 3x + 4 + \sin(x)$$

and

$$\frac{f}{g}(x) = \frac{3x + 4}{\sin x}$$

Dividing functions occasionally leads to complications when one of the functions assumes a value of zero. In the example f/g above, this occurs when $x = 0$. The quotient cannot be evaluated for $x = 0$ although the quotient function is still meaningful. In this case, the function f/g is said to have a **pole** at $x = 0$.

Polynomial functions are functions of the form

$$f(x) = \sum_{i=0}^{n} a_i x^i$$

where $a_n \neq 0$. The domain and range of polynomial functions are always either **R** or **C.** The number n is the degree of the polynomial.

Polynomials of degree 0 or 1 are called **linear;** of degree 2 they are called **parabolic** or **quadratic;** and of degree 3 they are called **cubic.**

The values of f for which $f(x) = 0$ are called the **roots** of f. A polynomial of degree n has at most n roots. There is exactly one exception to this rule: If $f(x) = 0$ is the constant zero function, the degree of f is zero, but f has infinitely many roots.

Roots of polynomials of degree 1 are found as follows: Suppose the polynomial is $f(x) = ax + b$. Set $f(x) = 0$ and solve for x. Then $x = -b/a$.

Roots of polynomials of degree 2 are often found using the quadratic formula. If $f(x) = ax^2 + bx + c$, then the two roots of f are given by the **quadratic formula:**

$$x_1 = \frac{-b + \sqrt{b^2 - 4ac}}{2a} \quad \text{and} \quad x_2 = \frac{-b - \sqrt{b^2 - 4ac}}{2a}$$

Roots of a polynomial of degree 3 fall into two types.

Equations of the Third Degree with Term in x^2 Absent

Solution: After dividing through by the coefficient of x^3, any equation of this type can be written $x^3 = Ax + B$. Let $p = A/3$ and $q = B/2$. The general solution is as follows:

CASE 1. $q^2 - p^3$ positive. One root is real, viz.,

$$x_1 = \sqrt[3]{q + \sqrt{q^2 - p^3}} + \sqrt[3]{q - \sqrt{q^2 - p^3}}$$

The other two roots are imaginary.

CASE 2. $q^2 - p^3$ = zero. Three roots real, but two of them equal.

$$x_1 = 2\sqrt[3]{q} \qquad x_2 = -\sqrt[3]{q} \qquad x_3 = -\sqrt[3]{q}$$

CASE 3. $q^2 - p^3$ negative. All three roots are real and distinct. Determine an angle u between 0 and 180°, such that $\cos u = q/(p\sqrt{p})$. Then

$$x_1 = 2\sqrt{p}\cos(u/3)$$
$$x_2 = 2\sqrt{p}\cos(u/3 + 120°)$$
$$x_3 = 2\sqrt{p}\cos(u/3 + 240°)$$

Graphical Solution: Plot the curve $y_1 = x^3$, and the straight line $y_2 = Ax + B$. The abscissas of the points of intersection will be the roots of the equation.

Equations of the Third Degree (General Case)

Solution: The general cubic equation, after dividing through by the coefficient of the highest power, may be written $x^3 + ax^2 + bx + c = 0$. To get rid of the term in x^2, let $x = x_1 - a/3$. The equation then becomes $x_1^3 = Ax_1 + B$, where $A = 3(a/3)^2 - b$, and $B = -2(a/3)^3 + b(a/3) - c$. Solve this equation for x_1, by the method above, and then find x itself from $x = x_1 - (a/3)$.

Graphical Solution: Without getting rid of the term in x^2, write the equation in the form $x^3 = -a[x + (b/2a)]^2 + [a(b/2a)^2 - c]$, and solve by the graphical method.

Computer Solutions: Equations of degree 3 or higher are frequently solved by using computer algebra systems such as Mathematica or Maple.

Arithmetic

When numbers, functions, or vectors are manipulated, they always obey certain properties, regardless of the types of the objects involved. Elements may be added or subtracted only if they are in the same universe set. Elements in different universes may sometimes be multiplied or divided, but the result may be in a different universe. Regardless of the universe sets involved, the following properties hold true:

1. Associative laws. $a + (b + c) = (a + b) + c$, $a(bc) = (ab)c$
2. Identity laws. $0 + a = a$, $1a = a$
3. Inverse laws. $a - a = 0$, $a/a = 1$
4. Distributive law. $a(b + c) = ab + ac$
5. Commutative laws. $a + b = b + a$, $ab = ba$

Certain universes, for example, matrices, do not obey the commutative law for multiplication.

SIGNIFICANT FIGURES AND PRECISION

Number of Significant Figures   In engineering computations, the data are ordinarily the result of measurement and are correct only to a limited number of significant figures. Each of the numbers 3.840 and 0.003840 is said to be given "correct to four figures"; the true value lies in the first case between 0.0038395 and 0.0038405. The **absolute error** is less than 0.001 in the first case, and less than 0.000001 in the second; but the **relative error** is the same in both cases, namely, an error of less than "one part in 3,840."

If a number is written as 384,000, the reader is left in doubt whether the number of correct significant figures is 3, 4, 5, or 6. This doubt can be removed by writing the number as 3.84×10^5, or 3.840×10^5, or 3.8400×10^5, or 3.84000×10^5.

In any numerical computation, the possible or desirable degree of accuracy should be decided on and the computation should then be so arranged that the required number of significant figures, and no more, is secured. Carrying out the work to a larger number of places than is justified by the data is to be avoided, (1) because the form of the results leads to an erroneous impression of their accuracy and (2) because time and labor are wasted in superfluous computation.

The unit value of the least significant figure in a number is its **precision.** The number 123.456 has six significant figures and has precision 0.001.

Two ways to represent a real number are as **fixed-point** or as **floating-point,** also known as "scientific notation."

In scientific notation, a number is represented as a product of a **mantissa** and a power of 10. The mantissa has its first significant figure either immediately before or immediately after the decimal point, depending on which convention is being used. The power of 10 used is called the **exponent.** The number 123.456 may be represented as either

$$0.123456 \times 10^3 \quad \text{or} \quad 1.23456 \times 10^2$$

Fixed-point representations tend to be more convenient when the quantities involved will be added or subtracted or when all measurements relate to the same precision. Floating-point representations are more convenient for very large or very small numbers or when the quantities involved will be multiplied or divided.

Many different numbers may share the same representation. For example, 0.05 may be used to represent, with precision 0.01, any value between 0.045000 and 0.054999. The largest value a number represents, in this case 0.0549999, is sometimes denoted x^*, and the smallest is denoted x_*.

An awareness of precision and significant figures is necessary so that answers correctly represent their accuracy.

Multiplication and Division   A product or quotient should be written with the smallest number of significant figures of any of the factors involved. The product often has a different precision than the factors, but the significant figures must not increase.

EXAMPLES. $(6.)(8.) = 48$ should be written as 50 since the factors have one significant figure. There is a loss of precision from 1 to 10.

$0.6 \times 0.8 = .048$ should be written as 0.5 since the factors have one significant figure. There is a gain of precision from 0.1 to 0.01.

Addition and Subtraction   A sum or difference should be represented with the same precision as the least precise term involved. The number of significant figures may change.

EXAMPLES. $3.14 + 0.001 = 3.141$ should be represented as 3.14 since the least precise term has precision 0.01.
$3.14 + 0.1 = 3.24$ should be represented as 3.2 since the least precise term has precision 0.1.

Loss of Significant Figures   Addition and subtraction may result in serious loss of significant figures and resultant large relative errors if the sums are near zero. For example,

$$3.15 - 3.14 = 0.01$$

shows a loss from three significant figures to just one. Where it is possible, calculations and measurements should be planned so that loss of significant figures can be avoided.

Mixed Calculations   When an expression involves both products and sums, significant figures and precision should be noted for each term or factor as it is calculated, so that correct significant figures and precision for the result are known. The calculation should be performed to as much precision as is available and should be rounded to the correct precision when the calculation is finished. This process is frequently done incorrectly, particularly when calculators or computers provide many decimal places in their result but provide no clue as to how many of those figures are significant.

Significant Figures in Evaluating Functions   If $y = f(x)$, then the correct number of significant figures in y depends on the number of significant figures in x and on the behavior of the function f in the neighborhood of x. In general, y should be represented so that all of $f(x)$, $f(x^*)$, and $f(x_*)$ are between y^* and y_*.

EXAMPLES.

$$\begin{array}{ll} \text{sqr}\,(2.0) & \text{sqr}\,(1.95) = 1.39642 \\ & \text{sqr}\,(2.00) = 1.41421 \\ & \text{sqr}\,(2.05) = 1.43178 \\ & \text{so } y = 1.4 \end{array}$$

$$\begin{array}{ll} \sin\,(1°) & \sin\,(0.5) = 0.00872 \\ & \sin\,(1.0) = 0.01745 \\ & \sin\,(1.5) = 0.02617 \\ & \text{so } \sin\,(1°) = 0.0 \end{array}$$

$$\begin{array}{ll} \sin\,(90°) & \sin\,(89.5) = 0.99996 \\ & \sin\,(90.0) = 1.00000 \\ & \sin\,(90.5) = 0.99996 \\ & \text{so } \sin\,(90°) = 1.0000 \end{array}$$

Note that in finding sin (90°), there was a gain in significant figures from two to five and also a gain in precision. This tends to happen when $f'(x)$ is close to zero. On the other hand, precision and significant figures are often lost when $f'(x)$ or $f''(x)$ are large.

Rearrangement of Formulas   Often a formula may be rewritten in order to avoid a loss of significant figures.

In using the quadratic formula to find the roots of a polynomial, significant figures may be lost if the $ax^2 + bx + c$ has a root near zero. The quadratic formula may be rearranged as follows:

1. Use the quadratic formula to find the root that is not close to 0. Call this root x_1.
2. Then $x_2 = c/ax_1$.

If $f(x) = \sqrt{x + 1} - \sqrt{x}$, then loss of significant figures occurs if x is large. This can be eliminated by "rationalizing the numerator" as follows:

$$\frac{(\sqrt{x + 1} - \sqrt{x})(\sqrt{x + 1} + \sqrt{x})}{\sqrt{x + 1} + \sqrt{x}} = \frac{1}{\sqrt{x + 1} + \sqrt{x}}$$

and this has no loss of significant figures.

There is an almost unlimited number of "tricks" for rearranging formulas to avoid loss of significant figures, but many of these are very similar to the tricks used in calculus to evaluate limits.

GEOMETRY, AREAS, AND VOLUMES

Geometrical Theorems

Right Triangles $a^2 + b^2 = c^2$. (See Fig. 2.1.2.) $\angle A + \angle B = 90°$. $p^2 = mn$. $a^2 = mc$. $b^2 = nc$.

Oblique Triangles   Sum of angles = 180°. An exterior angle = sum of the two opposite interior angles (Fig. 2.1.2).

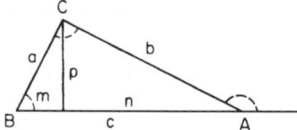

Fig. 2.1.2   Right triangle.

The medians, joining each vertex with the middle point of the opposite side, meet in the center of gravity G (Fig. 2.1.3), which trisects each median.

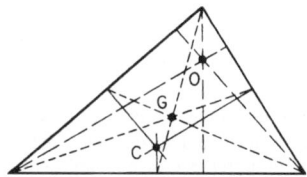

Fig. 2.1.3   Triangle showing medians and center of gravity.

The altitudes meet in a point called the **orthocenter,** O.

The perpendiculars erected at the midpoints of the sides meet in a point C, the center of the circumscribed circle. (In any triangle G, O, and C lie in line, and G is two-thirds of the way from O to C.)

The bisectors of the angles meet in the center of the inscribed circle (Fig. 2.1.4).

Fig. 2.1.4   Triangle showing bisectors of angles.

The largest side of a triangle is opposite the largest angle; it is less than the sum of the other two sides.

Similar Figures   Any two similar figures, in a plane or in space, can be placed in "perspective," i.e., so that straight lines joining corresponding points of the two figures will pass through a common point (Fig. 2.1.5). That is, of two similar figures, one is merely an enlargement of the other. Assume that each length in one figure is k times the corresponding length in the other; then each area in the first figure is k^2 times the corresponding area in the second, and each volume in the first figure is k^3 times the corresponding volume in the second. If two lines are cut by a set of parallel lines (or parallel planes), the corresponding segments are proportional.

Fig. 2.1.5   Similar figures.

The Circle   An angle that is inscribed in a semicircle is a right angle (Fig. 2.1.6).

A tangent is perpendicular to the radius drawn to the point of contact.

Fig. 2.1.6   Angle inscribed in a semicircle.

Fig. 2.1.7   Dihedral angle.

Dihedral Angles   The dihedral angle between two planes is measured by a plane angle formed by two lines, one in each plane, perpendicular to the edge (Fig. 2.1.7). (For solid angles, see Surfaces and Volumes of Solids.)

In a **tetrahedron,** or triangular pyramid, the four medians, joining each vertex with the center of gravity of the opposite face, meet in a point, the center of gravity of the tetrahedron; this point is ¾ of the way from any vertex to the center of gravity of the opposite face.

The Sphere   (See also Surfaces and Volumes of Solids.) If AB is a diameter, any plane perpendicular to AB cuts the sphere in a circle, of which A and B are called the poles. A great circle on the sphere is formed by a plane passing through the center.

Geometrical Constructions

To Bisect a Line AB (Fig. 2.1.8) (1) From A and B as centers, and with equal radii, describe arcs intersecting at P and Q, and draw PQ, which will bisect AB in M. (2) Lay off $AC = BD =$ approximately half of AB, and then bisect CD.

Fig. 2.1.8   Bisectors of a line.

Fig. 2.1.9   Construction of a line parallel to a given line.

To Draw a Parallel to a Given Line l **through a Given Point** A (Fig. 2.1.9) With point A as center draw an arc just touching the line l; with any point O of the line as center, draw an arc BC with the same radius. Then a line through A touching this arc will be the required parallel. Or, use a straightedge and triangle. Or, use a sheet of celluloid with a set of lines parallel to one edge and about ¼ in apart ruled upon it.

To Draw a Perpendicular to a Given Line from a Given Point A **Outside the Line**   (Fig. 2.1.10) (1) With A as center, describe an arc cutting the line at R and S, and bisect RS at M. Then M is the foot of the perpendicular. (2) If A is nearly opposite one end of the line, take any point B of the line and bisect AB in O; then with O as center, and OA or OB as radius, draw an arc cutting the line in M. Or, (3) use a straightedge and triangle.

Fig. 2.1.10   Construction of a line perpendicular to a given line from a point not on the line.

To Erect a Perpendicular to a Given Line at a Given Point P (1) Lay off $PR = PS$ (Fig. 2.1.11), and with R and S as centers draw arcs intersecting at *A*. Then *PA* is the required perpendicular. (2) If *P* is near the end of the line, take any convenient point *O* (Fig. 2.1.12) above the line as center, and with radius *OP* draw an arc cutting the line at *Q*. Produce *QO* to meet the arc at *A*; then *PA* is the required perpendicular. (3) Lay off *PB* = 4 units of any scale (Fig. 2.1.13); from *P* and *B* as centers lay off *PA* = 3 and *BA* = 5; then *APB* is a right angle.

Fig. 2.1.11   Construction of a line perpendicular to a given line from a point on the line.

Fig. 2.1.12   Construction of a line perpendicular to a given line from a point on the line.

To Divide a Line *AB* **into** *n* **Equal Parts**   (Fig. 2.1.14) Through *A* draw a line *AX* at any angle, and lay off *n* equal steps along this line. Connect the last of these divisions with *B*, and draw parallels through the other divisions. These parallels will divide the given line into *n* equal parts. A similar method may be used to divide a line into parts which shall be proportional to any given numbers.

To Bisect an Angle *AOB*   (Fig. 2.1.15) Lay off *OA* = *OB*. From *A* and *B* as centers, with any convenient radius, draw arcs meeting at *M*; then *OM* is the required bisector.

Fig. 2.1.13   Construction of a line perpendicular to a given line from a point on the line.

Fig. 2.1.14   Division of a line into equal parts.

To draw the bisector of an angle when the vertex of the angle is not accessible. Parallel to the given lines *a*, *b*, and equidistant from them, draw two lines *a'*, *b'* which intersect; then bisect the angle between *a'* and *b'*.

To Inscribe a Hexagon in a Circle   (Fig. 2.1.16) Step around the circumference with a chord equal to the radius. Or, use a 60° triangle.

Fig. 2.1.15   Bisection of an angle.

Fig. 2.1.16   Hexagon inscribed in a circle.

To Circumscribe a Hexagon about a Circle   (Fig. 2.1.17) Draw a chord *AB* equal to the radius. Bisect the arc *AB* at *T*. Draw the tangent at *T* (parallel to *AB*), meeting *OA* and *OB* at *P* and *Q*. Then draw a circle with radius *OP* or *OQ* and inscribe in it a hexagon, one side being *PQ*.

To Construct a Polygon of *n* **Sides, One Side** *AB* **Being Given**   (Fig. 2.1.18) With *A* as center and *AB* as radius, draw a semicircle, and divide it into *n* parts, of which *n* − 2 parts (counting from *B*) are to be used. Draw rays from *A* through these points of division, and complete the construction as in the figure (in which *n* = 7). Note that the center of the polygon must lie in the perpendicular bisector of each side.

To Draw a Tangent to a Circle   from an external point *A* (Fig. 2.1.19) Bisect *AC* in *M*; with *M* as center and radius *MC*, draw arc cutting circle in *P*; then *P* is the required point of tangency.

Fig. 2.1.17   Hexagon circumscribed about a circle.

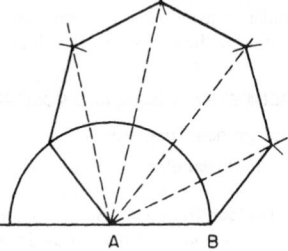

Fig. 2.1.18   Construction of a polygon with a given side.

To Draw a Common Tangent to Two Given Circles   (Fig. 2.1.20) Let *C* and *c* be centers and *R* and *r* the radii (*R* > *r*). From *C* as center, draw two concentric circles with radii *R* + *r* and *R* − *r*; draw tangents to

Fig. 2.1.19   Construction of a tangent to a circle.

Fig. 2.1.20   Construction of a tangent common to two circles.

these circles from *c*; then draw parallels to these lines at distance *r*. These parallels will be the required common tangents.

To Draw a Circle through Three Given Points   *A*, *B*, *C*, or to find the center of a given circular arc (Fig. 2.1.21) Draw the perpendicular bisectors of *AB* and *BC*; these will meet at the center, *O*.

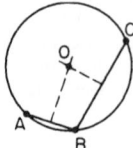

Fig. 2.1.21   Construction of a circle passing through three given points.

To Draw a Circle through Two Given Points *A*, *B*, **and Touching a Given Circle**   (Fig. 2.1.22) Draw any circle through *A* and *B*, cutting the given circle at *C* and *D*. Let *AB* and *CD* meet at *E*, and let *ET* be tangent from *E* to the circle just drawn. With *E* as center, and radius *ET*, draw an arc cutting the given circle at *P* and *Q*. Either *P* or *Q* is the required point of contact. (Two solutions.)

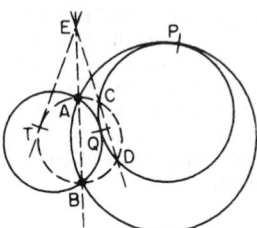

Fig. 2.1.22   Construction of a circle through two given points and touching a given circle.

To Draw a Circle through One Given Point, *A*, **and Touching Two Given Circles**   (Fig. 2.1.23) Let *S* be a center of similitude for the two given circles, i.e., the point of intersection of two external (or internal) common tangents. Through S draw any line cutting one circle at two points, the nearer of which shall be called P, and the other at two points, the more remote of which shall be called Q. Through A, P, Q draw a circle cutting SA at B. Then draw a circle through A and B and touching one of the given circles (see preceding construction). This circle will touch the other given circle also. (Four solutions.)

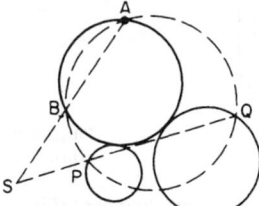

Fig. 2.1.23   Construction of a circle through a given point and touching two given circles.

To Draw an Annulus Which Shall Contain a Given Number of Equal Contiguous Circles   (Fig. 2.1.24) (An annulus is a ring-shaped area enclosed between two concentric circles.) Let $R + r$ and $R - r$ be the inner and outer radii of the annulus, r being the radius of each of the n circles. Then the required relation between these quantities is given by $r = R \sin(180°/n)$, or $r = (R + r)[\sin(180°/n)]/[1 + \sin(180°/n)]$.

Fig. 2.1.24   Construction of an annulus containing a given number of contiguous circles.

Lengths and Areas of Plane Figures

Right Triangle   (Fig. 2.1.25) $a^2 + b^2 = c^2$. Area $= \frac{1}{2}ab = \frac{1}{2}a^2 \cot A = \frac{1}{2}b^2 \tan A = \frac{1}{4}c^2 \sin 2A$.

Equilateral Triangle   (Fig. 2.1.26) Area $= \frac{1}{4}a^2\sqrt{3} = 0.43301a^2$.

Fig. 2.1.25   Right triangle.   **Fig. 2.1.26**   Equilateral triangle.

Any Triangle   (Fig. 2.1.27)

$s = \frac{1}{2}(a + b + c), t = \frac{1}{2}(m1 + m2 + m3)$

$r = \sqrt{(s - a)(s - b)(s - c)/s}$ = radius inscribed circle

$R = \frac{1}{2}a/\sin A = \frac{1}{2}b/\sin B = \frac{1}{2}c/\sin C$ = radius circumscribed circle

Area $= \frac{1}{2}$ base $\times$ altitude $= \frac{1}{2}ah = \frac{1}{2}ab \sin C = rs = abc/4R = \pm\frac{1}{2}\{(x_1 y_2 - x_2 y_1) + (x_2 y_3 - x_3 y_2) + (x_3 y_1 - x_1 y_3)\}$,

where (x_1, y_1), (x_2, y_2), (x_3, y_3) are coordinates of vertices.

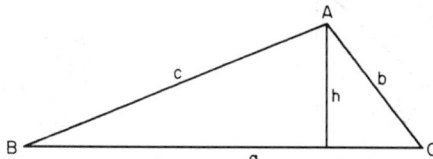

Fig. 2.1.27   Triangle.

Rectangle   (Fig. 2.1.28) Area $= ab = \frac{1}{2}D^2 \sin u$, where u = angle between diagonals D, D.

Rhombus   (Fig. 2.1.29) Area $= a^2 \sin C = \frac{1}{2}D_1D_2$, where C = angle between two adjacent sides; D_1, D_2 = diagonals.

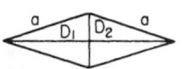

Fig. 2.1.28   Rectangle.   **Fig. 2.1.29**   Rhombus.

Parallelogram   (Fig. 2.1.30) Area $= bh = ab \sin C = \frac{1}{2}D_1D_2 \sin u$, where u = angle between diagonals D_1 and D_2.

Trapezoid   (Fig. 2.1.31) Area $= \frac{1}{2}(a + b) h$ where bases a and b are parallel.

Fig. 2.1.30   Parallelogram.   **Fig. 2.1.31**   Trapezoid.

Any Quadrilateral   (Fig. 2.1.32) Area $= \frac{1}{2}D_1D_2 \sin u$.

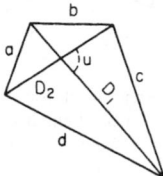

Fig. 2.1.32   Quadrilateral.

Regular Polygons n = number of sides; $v = 360°/n$ = angle subtended at center by one side; a = length of one side $= 2R \sin(v/2) = 2r \tan(v/2)$; R = radius of circumscribed circle $= 0.5 a \csc(v/2) = r \sec(v/2)$; r = radius of inscribed circle $= R \cos(v/2) = 0.5 \cot(v/2)$; area $= 0.25 a^2n \cot(v/2) = 0.5 R^2n \sin(v) = r^2n \tan(v/2)$. Areas of regular polygons are tabulated in Table 1.1.3.

Circle   Area $= \pi r^2 = \frac{1}{2}Cr = \frac{1}{4}Cd = \frac{1}{4}\pi d^2 = 0.785398d^2$, where r = radius, d = diameter, C = circumference $= 2\pi r = \pi d$.

Annulus   (Fig. 2.1.33) Area $= \pi(R^2 - r^2) = \pi(D^2 - d^2)/4 = 2\pi R'b$, where R' = mean radius $= \frac{1}{2}(R + r)$, and $b = R - r$.

Fig. 2.1.33   Annulus.

Sector   (Fig. 2.1.34) Area $= \frac{1}{2}rs = \pi r^2A/360° = \frac{1}{2}r^2$ rad A, where rad A = radian measure of angle A, and s = length of arc $= r$ rad A.

Fig. 2.1.34   Sector.

Segment (Fig. 2.1.35) Area = $\frac{1}{2}r^2(\text{rad } A - \sin A)$ = $\frac{1}{2}[r(s - c) + ch]$, where rad A radian measure of angle A. For small arcs, $s = \frac{1}{3}(8c' - c)$, where c' = chord of half of the arc (Huygens' approximation). Areas of segments are tabulated in Tables 1.1.1 and 1.1.2.

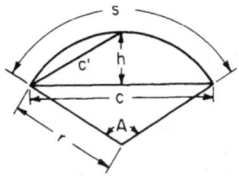

Fig. 2.1.35 Segment.

Ribbon bounded by two parallel curves (Fig. 2.1.36). If a straight line AB moves so that it is always perpendicular to the path traced by its middle point G, then the area of the ribbon or strip thus generated is equal to the length of AB times the length of the path traced by G. (It is assumed that the radius of curvature of G's path is never less than $\frac{1}{2}AB$, so that successive positions of generating line will not intersect.)

Fig. 2.1.36 Ribbon.

Ellipse (Fig. 2.1.37) Area of ellipse = πab. Area of shaded segment = $xy + ab \sin^{-1}(x/a)$. Length of perimeter of ellipse = $\pi(a + b)K$, where $K = (1 + \frac{1}{4}m^2 + \frac{1}{64}m^4 + \frac{1}{256}m^6 + \ldots)$, $m = (a - b)/(a + b)$.

For $m =$	0.1	0.2	0.3	0.4	0.5
$K =$	1.002	1.010	1.023	1.040	1.064
For $m =$	0.6	0.7	0.8	0.9	1.0
$K =$	1.092	1.127	1.168	1.216	1.273

Fig. 2.1.37 Ellipse.

Hyperbola (Fig. 2.1.38) In any hyperbola, shaded area $A = ab \ln[(x/a) + (y/b)]$. In an equilateral hyperbola ($a = b$), area $A = a^2 \sinh^{-1}(y/a) = a^2 \cosh^{-1}(x/a)$. Here x and y are coordinates of point P.

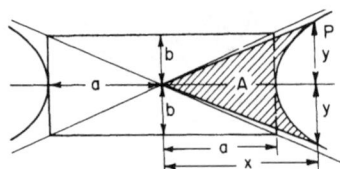

Fig. 2.1.38 Hyperbola.

For lengths and areas of **other curves** see Analytical Geometry.

Surfaces and Volumes of Solids

Regular Prism (Fig. 2.1.39) Volume = $\frac{1}{2}nrah = Bh$. Lateral area = $nah = Ph$. Here n = number of sides; B = area of base; P = perimeter of base.

Right Circular Cylinder (Fig. 2.1.40) Volume = $\pi r^2 h = Bh$. Lateral area = $2\pi rh = Ph$. Here B = area of base; P = perimeter of base.

Fig. 2.1.39 Regular prism.

Fig. 2.1.40 Right circular cylinder.

Truncated Right Circular Cylinder (Fig. 2.1.41) Volume = $\pi r^2 h = Bh$. Lateral area = $2\pi rh = Ph$. Here h = mean height = $\frac{1}{2}(h_1 + h_2)$; B = area of base; P = perimeter of base.

Fig. 2.1.41 Truncated right circular cylinder.

Any Prism or Cylinder (Fig. 2.1.42) Volume = $Bh = Nl$. Lateral area = Ql. Here l = length of an element or lateral edge; B = area of base; N = area of normal section; Q = perimeter of normal section.

Fig. 2.1.42 Any prism or cylinder.

Special Ungula of a Right Cylinder (Fig. 2.1.43) Volume = $\frac{2}{3}r^2 H$. Lateral area = $2rH$. r = radius. (Upper surface is a semiellipse.)

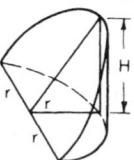

Fig. 2.1.43 Special ungula of a right circular cylinder.

Any Ungula of a right circular cylinder (Figs. 2.1.44 and 2.1.45) Volume = $H(\frac{2}{3}a^3 \pm cB)/(r \pm c) = H[a(r^2 - \frac{1}{3}a^2) \pm r^2c \text{ rad } u]/(r \pm c)$. Lateral area = $H(2ra \pm cs)/(r \pm c) = 2rH(a \pm c \text{ rad } u)/$

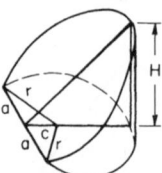

Fig. 2.1.44 Ungula of a right circular cylinder.

Fig. 2.1.45 Ungula of a right circular cylinder.

$(r \pm c)$. If base is greater (less) than a semicircle, use $+ (-)$ sign. $r =$ radius of base; $B =$ area of base; $s =$ arc of base; $u =$ half the angle subtended by arc s at center; rad $u =$ radian measure of angle u.

Regular Pyramid (Fig. 2.1.46) Volume = $\frac{1}{3}$ altitude $\times$ area of base = $\frac{1}{6}hran$. Lateral area = $\frac{1}{2}$ slant height $\times$ perimeter of base = $\frac{1}{2}san$. Here $r =$ radius of inscribed circle; $a =$ side (of regular polygon); $n =$ number of sides; $s = \sqrt{r^2 + h^2}$. Vertex of pyramid directly above center of base.

Fig. 2.1.46   Regular pyramid.

Right Circular Cone Volume = $\frac{1}{3}\pi r^2 h$. Lateral area = πrs. Here $r =$ radius of base; $h =$ altitude; $s =$ slant height = $\sqrt{r^2 + h^2}$.

Frustum of Regular Pyramid (Fig. 2.1.47) Volume = $\frac{1}{6}hran[1 + (a'/a) + (a'/a)^2]$. Lateral area = slant height $\times$ half sum of perimeters of bases = slant height $\times$ perimeter of midsection = $\frac{1}{2}sn(r + r')$. Here $r, r' =$ radii of inscribed circles; $s = \sqrt{(r - r')^2 + h^2}$; $a, a' =$ sides of lower and upper bases; $n =$ number of sides.

Frustum of Right Circular Cone (Fig. 2.1.48) Volume = $\frac{1}{3}\pi r^2 h[1 + (r'/r) + (r'/r)^2] = \frac{1}{3}\pi h(r^2 + rr' + r'^2) = \frac{1}{4}\pi h[r + r')^2 + \frac{1}{3}(r - r')^2]$. Lateral area = $\pi s(r + r')$; $s = \sqrt{(r - r')^2 + h^2}$.

   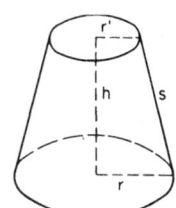

Fig. 2.1.47   Frustum of a regular pyramid.

Fig. 2.1.48   Frustum of a right circular cone.

Any Pyramid or Cone Volume = $\frac{1}{3}Bh$. $B =$ area of base; $h =$ perpendicular distance from vertex to plane in which base lies.

Any Pyramidal or Conical Frustum (Fig. 2.1.49) Volume = $\frac{1}{3}h(B + \sqrt{BB'} + B') = \frac{1}{3}hB[1 + (P'/P) + (P'/P)^2]$. Here B, $B' =$ areas of lower and upper bases; P, $P' =$ perimeters of lower and upper bases.

Fig. 2.1.49   Pyramidal frustum and conical frustum.

Sphere Volume = $V = \frac{4}{3}\pi r^3 = 4.188790r^3 = \frac{1}{6}\pi d^3 = \frac{2}{3}$ volume of circumscribed cylinder. Area = $A = 4\pi r^2 =$ four great circles =

$\pi d^2 =$ lateral area of circumscribed cylinder. Here $r =$ radius; $d = 2r =$ diameter = $\sqrt[3]{6V/\pi} = \sqrt{A/\pi}$.

Hollow Sphere or spherical shell. Volume = $\frac{4}{3}\pi(R^3 - r^3) = \frac{1}{6}\pi(D^3 - d^3) = 4\pi R_1^2 t + \frac{1}{3}\pi t^3$. Here R, $r =$ outer and inner radii; $D, d =$ outer and inner diameters; $t =$ thickness = $R - r$; $R_1 =$ mean radius = $\frac{1}{2}(R + r)$.

Any Spherical Segment. Zone (Fig. 2.1.50) Volume = $\frac{1}{6}\pi h(3a^2 + 3a_1^2 + h^2)$. Lateral area (zone) = $2\pi rh$. Here $r =$ radius of sphere. If the inscribed frustum of a cone is removed from the spherical segment, the volume remaining is $\frac{1}{6}\pi hc^2$, where $c =$ slant height of frustum = $\sqrt{h^2 + (a - a_1)^2}$.

Fig. 2.1.50   Any spherical segment.

Spherical Segment of One Base. Zone (spherical "cap" of Fig. 2.1.51) Volume = $\frac{1}{6}\pi h(3a^2 + h^2) = \frac{1}{3}\pi h^2 (3r - h)$. Lateral area (of zone) = $2\pi rh = \pi(a^2 + h^2)$.

NOTE. $a^2 = h(2r - h)$, where $r =$ radius of sphere.

Spherical Sector (Fig. 2.1.51) Volume = $\frac{1}{3}r \times$ area of cap = $\frac{2}{3}\pi r^2 h$. Total area = area of cap + area of cone = $2\pi rh + \pi ra$.

NOTE. $a^2 = h(2r - h)$.

Spherical Wedge bounded by two plane semicircles and a **lune** (Fig. 2.1.52). Volume of wedge $\div$ volume of sphere = $u/360°$. Area of lune $\div$ area of sphere = $u/360°$. $u =$ dihedral angle of the wedge.

Fig. 2.1.51   Spherical sector.

Fig. 2.1.52   Spherical wedge.

Solid Angles Any portion of a spherical surface subtends what is called a **solid angle** at the center of the sphere. If the area of the given portion of spherical surface is equal to the square of the radius, the subtended solid angle is called a **steradian,** and this is commonly taken as the unit. The entire solid angle about the center is equal to 4π steradians. A so-called "solid right angle" is the solid angle subtended by a quadrantal (or trirectangular) spherical triangle, and a "spherical degree" (now little used) is a solid angle equal to $\frac{1}{90}$ of a solid right angle. Hence 720 spherical degrees = 1 stereogon, or π steradians = 180 spherical degrees. If $u =$ the angle which an element of a cone makes with its axis, then the solid angle of the cone contains $2\pi(1 - \cos u)$ steradians.

Regular Polyhedra $A =$ area of surface; $V =$ volume; $a =$ edge.

Name of solid	Bounded by	A/a^2	V/a^3
Tetrahedron	4 triangles	1.7321	0.1179
Cube	6 squares	6.0000	1.0000
Octahedron	8 triangles	3.4641	0.4714
Dodecahedron	12 pentagons	20.6457	7.6631
Icosahedron	20 triangles	8.6603	2.1917

Ellipsoid (Fig. 2.1.53) Volume = $\frac{4}{3}\pi abc$, where a, b, c = semi-axes.

Torus, or Anchor Ring (Fig. 2.1.54) Volume = $2\pi^2 cr^2$. Area = $4\pi^2 cr$.

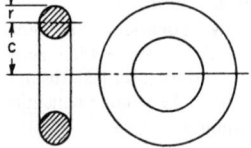

Fig. 2.1.53 Ellipsoid.   **Fig. 2.1.54** Torus.

Volume of a Solid of Revolution (solid generated by rotating an area bounded above by $f(x)$ around the x axis)

$$V = \pi \int_a^b |f(x)|^2 \, dx$$

Area of a Surface of Revolution

$$A = 2\pi \int_a^b y \sqrt{1 + (dy/dx)^2} \, dx$$

Length of Arc of a Plane Curve $y = f(x)$ between values $x = a$ and $x = b$. $s = \int_a^b \sqrt{1 + (dy/dx)^2} \, dx$. If $x = f(t)$ and $y = g(t)$, for $a < t < b$, then

$$s = \int_a^b \sqrt{(dx/dt)^2 + (dy/dt)^2} \, dt$$

PERMUTATIONS AND COMBINATIONS

The product $(1)(2)(3) \ldots (n)$ is written $n!$ and is read "n factorial." By convention, $0! = 1$, and $n!$ is not defined for negative integers.

For large values of n, $n!$ may be approximated by Stirling's formula:

$$n! \approx 2.50663 n^{n+.5} e^{-n}$$

The **binomial coefficient** $C(n, k)$, also written $\binom{n}{k}$, is defined as:

$$C(n, k) \frac{n!}{k!(n-k)!}$$

$C(n, k)$ is read "n choose k" or as "binomial coefficient n-k."

Binomial coefficients have the following properties:

1. $C(n, 0) = C(n, n) = 1$
2. $C(n, 1) = C(n, n - 1) = n$
3. $C(n + 1, k) = C(n, k) + C(n, k - 1)$
4. $C(n, k) = C(n, n - k)$

Binomial coefficients are tabulated in Sec. 1.

Binomial Theorem

If n is a positive integer, then

$$(a + b)^n = \sum_{k=0}^{n} C(n, k) a^k b^{n-k}$$

EXAMPLE. The third term of $(2x + 3)^7$ is $C(7, 4)(2x)^{7-4}3^4 = [7!/(4!3!)](2x)^3 3^4 = (35)(8\,x^3)(81) = 22680x^3$.

Combinations $C(n, k)$ gives the number of ways k distinct objects can be chosen from a set of n elements. This is the number of k-element subsets of a set of n elements.

EXAMPLE. The set of four elements (a, b, c, d) has $C(4, 2) = 6$ two-element subsets, $\{a, b\}$, $\{a, c\}$, $\{a, d\}$, $\{b, c\}$, $\{b, d\}$, and $\{c, d\}$. (Note that $\{a, c\}$ is the same set as $\{c, a\}$.)

Permutations The number of ways k objects may be arranged from a set of n elements is given by

$$P(n, k) = \frac{n!}{(n - k)!}$$

EXAMPLE. Two elements from the set (a, b, c, d) may be arranged in $P(4, 2) = 12$ ways: ab, ac, ad, ba, bc, bd, ca, cb, cd, da, db, and dc. Note that ac is a different arrangement than ca.

Permutations and combinations are examined in detail in most texts on probability and statistics and on discrete mathematics.

If an event can occur in s ways and can fail to occur in f ways, and if all ways are equally likely, then the probability of the event's occurring is $p = s/(s + f)$, and the probability of failure is $q = f/(s + f) = 1 - p$.

The set of all possible outcomes of an experiment is called the **sample space**, denoted S. Let n be the number of outcomes in the sample set. A subset A of the sample space is called an **event**. The number of outcomes in A is s. Therefore $P(A) = s/n$. The probability that A does not occur is $P(\tilde{~}A) = q = 1 - p$.

Always $0 \le p \le 1$ and $P(S) = 1$.

If two events cannot occur simultaneously, then $A \cap B = \varnothing$, and A and B are said to be **mutually exclusive**. Then $P(A \cup B) = P(A) + P(B)$. Otherwise, $P(A \cup B) = P(A) + P(B) - P(A \cap B)$.

Events A and B are **independent** if $P(A \cap B) = P(A)P(B)$.

If E is an event and if $P(E) > 0$, then the probability that A occurs once E has already occurred is called the "conditional probability of A given E," written $P(A|E)$ and defined as

$$P(A|E) = P(A \cap E)/P(E)$$

A and E are independent if $P(A|E) = P(A)$.

If the outcomes in a sample space X are all numbers, then X, together with the probabilities of the outcomes, is called a **random variable**. If x_i is an outcome, then $p_i = P(x_i)$.

The **expected value** of a random variable is

$$E(X) = \Sigma\, e_i p_i$$

The **variance** of X is

$$V(X) = \Sigma[x_i - E(X)]^2 p_i$$

The **standard deviation** is

$$S(X) = \sqrt{[V(X)]}$$

The Binomial, or Bernoulli, Distribution If an experiment is repeated n times and the probability of a success on any trial is p, then the probability of k successes among those n trials is

$$f(n, k, p) = C(n, k) p^k q^{n-k}$$

Geometric Distribution If an experiment is repeated until it finally succeeds, let x be the number of failures observed before the first success. Let p be the probability of success on any trial and let $q = 1 - p$. Then

$$P(x = k) = q^k \cdot p$$

Uniform Distribution If the random variable x assumes the values 1, 2, $\ldots$, n, with equal probabilities, then the distribution is uniform, and

$$P(x = k) = \frac{1}{n}$$

Hypergeometric Distribution—Sampling without Replacement If a finite population of N elements contains x successes and if n items are selected randomly without replacement, then the probability that k

successes will occur among those n samples is

$$h(x; N, n, k) = \frac{C(k, x)C(N - k, n - x)}{C(N, n)}$$

For large values of N, the hypergeometric distribution approaches the binomial distribution, so

$$h(x; N, n, k) \approx f\left(n, k, \frac{x}{N}\right)$$

Poisson Distribution   If the average number of successes which occur in a given fixed time interval is m, then let x be the number of successes observed in that time interval. The probability that $x = k$ is

$$p(k, m) = \frac{e^{-m}m^x}{x!} \quad \text{where } e = 2.71828 \ldots$$

Negative Binomial Distribution   If repeated independent trials have probability of success p, then let x be the trial number upon which success number n occurs. Then the probability that $x = k$ is

$$b*(k; n, p) = C(k - 1, n - 1)p^n q^{k-n}$$

The expected values and variances of these distributions are summarized in the following table:

Distribution	$E(X)$	$V(X)$
Uniform	$(n + 1)/2$	$(n^2 - 1)/12$
Binomial	np	npq
Hypergeometric	nk/N	$[nk(N - n)(1 - k/N)]/[N(N - 1)]$
Poisson	m	m
Geometric	q/p	q/p^2
Negative binomial	nq/p	nq/p^2

LINEAR ALGEBRA

Using linear algebra, it is often possible to express in a single equation a set of relations that would otherwise require several equations. Similarly, it is possible to replace many calculations involving several variables with a few calculations involving vectors and matrices. In general, the equations to which the techniques of linear algebra apply must be linear equations; they can involve no polynomial, exponential, or trigonometric terms.

Vectors

A **row vector** $\mathbf{v}$ is a list of numbers written in a row, usually enclosed by parentheses.

$$\mathbf{v} = (v_1, v_2, \ldots, v_n)$$

A **column vector u** is a list of numbers written in a column:

$$\mathbf{u} = \begin{pmatrix} u_1 \\ u_2 \\ \cdot \\ \cdot \\ \cdot \\ u_n \end{pmatrix}$$

The numbers u_i and v_i may be real or complex, or they may even be variables or functions.

A vector is sometimes called an **ordered n-tuple**. In the case where $n = 2$, it may be called an **ordered pair**.

The numbers v_i are called **components** or **coordinates** of the vector $\mathbf{v}$. The number n is called the **dimension** of $\mathbf{v}$.

Two-dimensional vectors correspond with points in the plane, where v_1 is the x coordinate and v_2 is the y coordinate of the point $\mathbf{v}$. Two-dimensional vectors also correspond with complex numbers, where $z = v_1 + iv_2$.

Three-dimensional vectors correspond to points in space, where v_1, v_2, and v_3 are the x, y, and z coordinates of the point, respectively.

Two- and three-dimensional vectors may be thought of as having a direction and a magnitude. See the section "Analytical Geometry."

Two vectors $\mathbf{u}$ and $\mathbf{v}$ are equal if:
1. $\mathbf{u}$ and $\mathbf{v}$ are the same type (either row or column).
2. $\mathbf{u}$ and $\mathbf{v}$ have the same dimension.
3. Corresponding components are equal; that is, $u_i = v_i$ for $i = 1, 2, \ldots, n$.

Note that the row vectors

$$\mathbf{u} = (1, 2, 3) \quad \text{and} \quad \mathbf{v} = (3, 2, 1)$$

are not equal since the components are not in the same order. Also,

$$\mathbf{u} = (1, 2, 3) \quad \text{and} \quad \mathbf{v} = \begin{pmatrix} 1 \\ 2 \\ 3 \end{pmatrix}$$

are not equal since $\mathbf{u}$ is a row vector and $\mathbf{v}$ is a column vector.

Vector Transpose   If $\mathbf{u}$ is a row vector, then the **transpose** of $\mathbf{u}$, written $\mathbf{u}^T$, is the column vector with the same components in the same order as $\mathbf{u}$. Similarly, the transpose of a column vector is the row vector with the same components in the same order. Note that $(\mathbf{u}^T)^T = \mathbf{u}$.

Vector Addition   If $\mathbf{u}$ and $\mathbf{v}$ are vectors of the same type and the same dimension, then the sum of $\mathbf{u}$ and $\mathbf{v}$, written $\mathbf{u} + \mathbf{v}$, is the vector obtained by adding corresponding components. In the case of row vectors,

$$\mathbf{u} + \mathbf{v} = (u_1 + v_1, u_2 + v_2, \ldots, u_n + v_n)$$

Scalar Multiplication   If a is a number and $\mathbf{u}$ is a vector, then the **scalar product** $a\mathbf{u}$ is the vector obtained by multiplying each component of $\mathbf{u}$ by a.

$$a\mathbf{u} = (au_1, au_2, \ldots, au_n)$$

A number by which a vector is multiplied is called a **scalar.**
The **negative** of vector $\mathbf{u}$ is written $-\mathbf{u}$, and

$$-\mathbf{u} = -1\mathbf{u}$$

The **zero vector** is the vector with all its components equal to zero.

Arithmetic Properties of Vectors   If $\mathbf{u}$, $\mathbf{v}$, and $\mathbf{w}$ are vectors of the same type and dimensions, and if a and b are scalars, then vector addition and scalar multiplication obey the following seven rules, known as the *properties of a vector space:*

1. $(\mathbf{u} + \mathbf{v}) + \mathbf{w} = \mathbf{u} + (\mathbf{v} + \mathbf{w})$	associative law
2. $\mathbf{u} + \mathbf{v} = \mathbf{v} + \mathbf{u}$	commutative law
3. $\mathbf{u} + \mathbf{0} = \mathbf{u}$	additive identity
4. $\mathbf{u} + (-\mathbf{u}) = \mathbf{0}$	additive inverse
5. $a(\mathbf{u} + \mathbf{v}) = a\mathbf{u} + a\mathbf{v}$	distributive law
6. $(ab)\mathbf{u} = a(b\mathbf{u})$	associative law of multiplication
7. $1\mathbf{u} = \mathbf{u}$	multiplicative identity

Inner Product or Dot Product   If $\mathbf{u}$ and $\mathbf{v}$ are vectors of the same type and dimension, then their **inner product** or **dot product**, written $\mathbf{uv}$ or $\mathbf{u} \cdot \mathbf{v}$, is the scalar

$$\mathbf{uv} = u_1 v_1 + u_2 v_2 + \cdots + u_n v_n$$

Vectors $\mathbf{u}$ and $\mathbf{v}$ are **perpendicular** or **orthogonal** if $\mathbf{uv} = 0$.

Magnitude   There are two equivalent ways to define the magnitude of a vector $\mathbf{u}$, written $|\mathbf{u}|$ or $\|\mathbf{u}\|$.

$$|\mathbf{u}| = \sqrt{(\mathbf{u} \cdot \mathbf{u})}$$

or

$$|\mathbf{u}| = \sqrt{(u_1^2 + u_2^2 + \cdots + u_n^2)}$$

Cross Product or Outer Product   If $\mathbf{u}$ and $\mathbf{v}$ are three-dimensional vectors, then they have a **cross product**, also called **outer product** or **vector product.**

$$\mathbf{u} \times \mathbf{v} = (u_2 v_3 - u_3 v_2, v_1 u_3 - v_3 u_1, u_1 v_2 - u_2 v_1)$$

The cross product $\mathbf{u} \times \mathbf{v}$ is a three-dimensional vector that is perpendicular to both $\mathbf{u}$ and $\mathbf{v}$. The cross product is not commutative. In fact,

$$\mathbf{u} \times \mathbf{v} = -\mathbf{v} \times \mathbf{u}$$

Cross product and inner product have two properties involving trigonometric functions. If θ is the angle between vectors $\mathbf{u}$ and $\mathbf{v}$, then

$$\mathbf{uv} = |\mathbf{u}||\mathbf{v}|\cos\theta \quad \text{and} \quad |\mathbf{u} \times \mathbf{v}| = |\mathbf{u}||\mathbf{v}|\sin\theta$$

Matrices

A **matrix** is a rectangular array of numbers. A matrix A with m rows and n columns may be written

$$A = \begin{pmatrix} a_{11} & a_{12} & a_{13} & \cdots & a_{1n} \\ a_{21} & a_{22} & a_{23} & \cdots & a_{2n} \\ a_{31} & a_{32} & a_{33} & \cdots & a_{3n} \\ \cdots & \cdots & \cdots & \cdots & \cdots \\ a_{m1} & a_{m2} & a_{m3} & \cdots & a_{mn} \end{pmatrix}$$

The numbers a_{ij} are called the **entries** of the matrix. The first subscript i identifies the row of the entry, and the second subscript j identifies the column.

Matrices are denoted either by capital letters, A, B, etc., or by writing the general entry in parentheses, (a_{ij}).

The number of rows and the number of columns together define the dimensions of the matrix. The matrix A is an $m \times n$ matrix, read "m by n."

A row vector may be considered to be a $1 \times n$ matrix, and a column vector may be considered as a $n \times 1$ matrix.

The rows of a matrix are sometimes considered as row vectors, and the columns may be considered as column vectors.

If a matrix has the same number of rows as columns, the matrix is called a **square matrix**.

In a square matrix, the entries a_{ii}, where the row index is the same as the column index, are called the diagonal entries.

If a matrix has all its entries equal to zero, it is called a **zero matrix**.

If a square matrix has all its entries equal to zero except its diagonal entries, it is called a **diagonal matrix**.

The diagonal matrix with all its diagonal entries equal to 1 is called the **identity matrix,** and is denoted I, or $I_{n \times n}$ if it is important to emphasize the dimensions of the matrix. The 2×2 and 3×3 identity matrices are:

$$I_{2\times2} = \begin{pmatrix} 1 & 0 \\ 0 & 1 \end{pmatrix} \qquad I_{3\times3} = \begin{pmatrix} 1 & 0 & 0 \\ 0 & 1 & 0 \\ 0 & 0 & 1 \end{pmatrix}$$

The entries of a square matrix a_{ij} where $i > j$ are said to be below the diagonal. Similarly, those where $i < j$ are said to be above the diagonal.

A square matrix with all entries below (resp. above) the diagonal equal to zero is called *upper-triangular* (resp. *lower-triangular*).

Matrix Addition   Matrices A and B may be added only if they have the same dimensions. Then the sum $C = A + B$ is defined by

$$c_{ij} = a_{ij} + b_{ij}$$

That is, corresponding entries of the matrices are added together, just as with vectors. Similarly, matrices may be multiplied by scalars.

Matrix Multiplication   Matrices A and B may be multiplied only if the number of columns in A equals the number of rows of B. If A is an $m \times n$ matrix and B is an $n \times p$ matrix, then the product $C = AB$ is an $m \times p$ matrix, defined as follows:

$$c_{ij} = a_{i1}b_{1j} + a_{i2}b_{2j} + \cdots + a_{in}b_{nj}$$
$$= \sum_{k=1}^{n} a_{ik}b_{kj}$$

The entry c_{ij} may also be defined as the dot product of row i of A with the transpose of column j of B.

EXAMPLE. $\begin{pmatrix} 1 & 2 \\ 5 & 6 \end{pmatrix} \begin{pmatrix} 3 & 4 \\ 7 & 8 \end{pmatrix} =$

$$\begin{pmatrix} 1 \times 3 + 2 \times 7 & 1 \times 4 + 2 \times 8 \\ 5 \times 3 + 6 \times 7 & 5 \times 4 + 6 \times 8 \end{pmatrix} = \begin{pmatrix} 17 & 20 \\ 57 & 68 \end{pmatrix}$$

Matrix multiplication is not commutative. Even if A and B are both square, it is hardly ever true that $AB = BA$. Matrix multiplication does have the following properties:

1. $(AB)C = A(BC)$ associative law
2. $A(B + C) = AB + AC$
3. $(B + C)A = BA + CA$ distributive laws

If A is square, then also

4. $AI = IA = A$ multiplicative identity

If A is square, then powers of A, AA, and AAA are denoted A^2 and A^3, respectively.

The transpose of a matrix A, written A^T, is obtained by writing the rows of A as columns. If A is $m \times n$, then A^T is $n \times m$.

EXAMPLE. $\begin{pmatrix} 1 & 2 & 3 \\ 4 & 5 & 6 \end{pmatrix}^T = \begin{pmatrix} 1 & 4 \\ 2 & 5 \\ 3 & 6 \end{pmatrix}$

The transpose has the following properties:

1. $(A^T)^T = A$
2. $(A + B)^T = A^T + B^T$
3. $(AB)^T = B^T A^T$

Note that in property 3, the order of multiplication is reversed.

If $A^T = A$, then A is called *symmetric*.

Linear Equations

A linear equation in two variables is of the form

$$a_1 x_1 + a_2 x_2 = b \quad \text{or} \quad a_1 x + a_2 y = b$$

depending on whether the variables are named x_1 and x_2 or x and y.

In n variables, such an equation has the form

$$a_1 x_1 + a_2 x_2 + \cdots a_n x_n = b$$

Such equations describe lines and planes. Often it is necessary to solve several such equations simultaneously. A set of m linear equations in n variables is called an $m \times n$ system of simultaneous linear equations.

Systems with Two Variables

1×2 Systems   An equation of the form

$$a_1 x + a_2 y = b$$

has infinitely many solutions which form a straight line in the xy plane. That line has slope $-a_1/a_2$ and y intercept b/a_2.

2×2 Systems   A 2×2 system has the form

$$a_{11}x + a_{12}y = b_1, \qquad a_{21}x + a_{22}y = b_2$$

Solutions to such systems do not always exist.

CASE 1.   The system has exactly one solution (Fig. 2.1.55a). The lines corresponding to the equations intersect at a single point. This occurs whenever the two lines have different slopes, so they are not

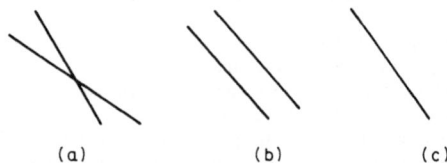

(a)   (b)   (c)

Fig. 2.1.55   Lines corresponding to linear equations. (a) One solution; (b) no solutions; (c) infinitely many solutions.

parallel. In this case,

$$\frac{a_{11}}{a_{21}} \neq \frac{a_{12}}{a_{22}} \quad \text{so} \quad a_{11}a_{22} - a_{21}a_{12} \neq 0$$

CASE 2.   The system has no solutions (Fig. 2.1.55b). This occurs whenever the two lines have the same slope and different y intercepts, so they are parallel. In this case,

$$\frac{a_{11}}{a_{21}} = \frac{a_{12}}{a_{22}}$$

CASE 3.   The system has infinitely many solutions (Fig. 2.1.55c). This occurs whenever the two lines coincide. They have the same slope and y intercept. In this case,

$$\frac{a_{11}}{a_{21}} = \frac{a_{12}}{a_{22}} = \frac{b_1}{b_2}$$

The value $a_{11} a_{22} - a_{21} a_{12}$ is called the **determinant** of the system. A larger $n \times n$ system also has a determinant (see below). A system has exactly one solution when its determinant is not zero.

3 × 2 Systems   Any system with more equations than variables is called **overdetermined.** The only case in which a 3×2 system has exactly one solution is when one of the equations can be derived from the other two.

One basic way to solve such a system is to treat any two equations as a 2×2 system and see if the solution to that subsystem of equations is also a solution to the third equation.

Matrix Form for Systems of Equations   The 2×2 system of linear equations

$$a_{11}x_1 + a_{12}x_2 = b_1 \qquad a_{21}x_1 + a_{22}x_2 = b_2$$

may be written as a matrix equation as follows:

$$\begin{pmatrix} a_{11} & a_{12} \\ a_{21} & a_{22} \end{pmatrix} \begin{pmatrix} x_1 \\ x_2 \end{pmatrix} = \begin{pmatrix} b_1 \\ b_2 \end{pmatrix}$$

or as

$$A\mathbf{x} = \mathbf{b}$$

where A is the 2×2 matrix and $\mathbf{x}$ and $\mathbf{b}$ are two-dimensional column vectors. Then, the determinant of A, written det A or $|A|$, is the same as the determinant of the 2×2 system:

$$\det A = a_{11}a_{22} - a_{21}a_{12}$$

In general, any $m \times n$ system of simultaneous linear equations may be written as

$$A\mathbf{x} = \mathbf{b}$$

where A is an $m \times n$ matrix, $\mathbf{x}$ is an n-dimensional column vector, and $\mathbf{b}$ is an m-dimensional column vector.

An $n \times n$ (square) system of simultaneous linear equations has exactly one solution whenever its determinant is not zero. Then the system and the matrix A are called **nonsingular.** If the determinant is zero, the system is called **singular.**

Elementary Row Operations on a Matrix   There are three operations on a matrix which change the matrix:
1. Multiply each entry in row i by a scalar k (not zero).
2. Interchange row i with row j.
3. Add row i to row j.
Similarly, there are three elementary column operations.
The elementary row operations have the following effects on $|A|$:
1. Multiplying a row (or column) by k multiplies $|A|$ by k.
2. Interchanging two rows (or columns) multiplies $|A|$ by -1.
3. Adding one row (or column) to another does not change $|A|$.

Pivoting, or Reducing, a Column   The process of changing the ij entry of a matrix to 1 and changing the rest of column j to zero, by using elementary row operations, is known as **reducing** column j or as **pivoting**

on the ij entry. Combining pivoting, the properties of the elementary row operations, and the fact:

$$|I_{n \times n}| = 1$$

provides a technique for finding the determinant of $n \times n$ matrices.

EXAMPLE.   Find $|A|$ where

$$A = \begin{pmatrix} 1 & 2 & -4 \\ 5 & -3 & -7 \\ 3 & -2 & 3 \end{pmatrix}$$

First, pivot on the entry in row 1, column 1, in this case, the 1.

Multiplying row 1 by -5, then adding row 1 to row 2, we first multiply the determinant by -5, then do not change it:

$$-5|A| = \begin{vmatrix} -5 & -10 & 20 \\ 5 & -3 & -7 \\ 3 & -2 & 3 \end{vmatrix} = \begin{vmatrix} -5 & -10 & 20 \\ 0 & -13 & 13 \\ 3 & -2 & 3 \end{vmatrix}$$

Next, multiply row 1 by $\frac{3}{5}$ and add row 1 to row 3:

$$-3|A| = \begin{vmatrix} -3 & -6 & 12 \\ 0 & -13 & 13 \\ 3 & -2 & 3 \end{vmatrix} = \begin{vmatrix} -3 & -6 & 12 \\ 0 & -13 & 13 \\ 0 & -8 & 15 \end{vmatrix}$$

Next, divide row 1 by -3:

$$|A| = \begin{vmatrix} 1 & 2 & -4 \\ 0 & -13 & 13 \\ 0 & -8 & 15 \end{vmatrix}$$

Next, pivot on the entry in row 2, column 2. Multiplying row 2 by $-\frac{8}{13}$ and then adding row 2 to row 3, we get:

$$-\frac{8}{13}|A| = \begin{vmatrix} 1 & 2 & -4 \\ 0 & 8 & -8 \\ 0 & -8 & 15 \end{vmatrix} = \begin{vmatrix} 1 & 2 & -4 \\ 0 & 8 & -8 \\ 0 & 0 & 7 \end{vmatrix}$$

Next, divide row 2 by $-\frac{8}{13}$.

$$|A| = \begin{vmatrix} 1 & 2 & -4 \\ 0 & -13 & 13 \\ 0 & 0 & 7 \end{vmatrix}$$

The determinant of a triangular matrix is the product of its diagonal elements, in this case -91.

Inverses   Whenever $|A|$ is not zero, that is, whenever A is nonsingular, then there is another $n \times n$ matrix, denoted A^{-1}, read "A inverse" with the property

$$AA^{-1} = A^{-1}A = I_{n \times n}$$

Then the $n \times n$ system of equations

$$A\mathbf{x} = \mathbf{b}$$

can be solved by multiplying both sides by A^{-1}, so

$$\mathbf{x} = I_{n \times n}\mathbf{x} = A^{-1}A\mathbf{x} = A^{-1}\mathbf{b}$$

so

$$\mathbf{x} = A^{-1}\mathbf{b}$$

The matrix A^{-1} may be found as follows:
1. Make a $n \times 2n$ matrix, with the first n columns the matrix A and the last n columns the identity matrix $I_{n \times n}$.
2. Pivot on each of the diagonal entries of this matrix, one after another, using the elementary row operations.
3. After pivoting n times, the matrix will have in the first n columns the identity matrix, and the last n columns will be the matrix A^{-1}.

EXAMPLE.   Solve the system

$$x_1 + 2x_2 - 4x_3 = -4$$
$$5x_1 - 3x_2 - 7x_3 = 6$$
$$3x_1 - 2x_2 + 3x_3 = 11$$

We must invert the matrix

$$A = \begin{pmatrix} 1 & 2 & -4 \\ 5 & -3 & -7 \\ 3 & -2 & 3 \end{pmatrix}$$

This is the same matrix used in the determinant example above. Adjoin the identity matrix to make a 3 × 6 matrix

$$\begin{pmatrix} 1 & 2 & -4 & 1 & 0 & 0 \\ 5 & -3 & -7 & 0 & 1 & 0 \\ 3 & -2 & 3 & 0 & 0 & 1 \end{pmatrix}$$

Perform the elementary row operations in exactly the same order as in the determinant example.

STEP 1.   Pivot on row 1, column 1.

$$\begin{pmatrix} 1 & 2 & -4 & 1 & 0 & 0 \\ 0 & -13 & 13 & -5 & 1 & 0 \\ 0 & -8 & 15 & -3 & 0 & 1 \end{pmatrix}$$

STEP 2.   Pivot on row 2, column 2.

$$\begin{pmatrix} 1 & 0 & -2 & \frac{3}{13} & \frac{2}{13} & 0 \\ 0 & 1 & -1 & \frac{5}{13} & -\frac{1}{13} & 0 \\ 0 & 0 & 7 & \frac{1}{13} & -\frac{8}{13} & 1 \end{pmatrix}$$

STEP 3.   Pivot on row 3, column 3.

$$\begin{pmatrix} 1 & 0 & 0 & \frac{23}{91} & -\frac{2}{91} & \frac{26}{91} \\ 0 & 1 & 0 & \frac{36}{91} & -\frac{15}{91} & \frac{13}{91} \\ 0 & 0 & 1 & \frac{1}{91} & -\frac{8}{91} & \frac{13}{91} \end{pmatrix}$$

Now, the inverse matrix appears on the right. To solve the equation,

$$\mathbf{x} = A^{-1}\mathbf{b}$$

so,

$$\mathbf{x} = \begin{pmatrix} \frac{23}{91} & -\frac{2}{91} & \frac{26}{91} \\ \frac{36}{91} & -\frac{15}{91} & \frac{13}{91} \\ \frac{1}{91} & -\frac{8}{91} & \frac{13}{91} \end{pmatrix} \begin{pmatrix} -4 \\ 6 \\ 11 \end{pmatrix}$$

$$= \begin{pmatrix} (-4 \times 23 + 6 \times -2 + 11 \times 26)/91 \\ (-4 \times 36 + 6 \times -15 + 11 \times 13)/91 \\ (-4 \times 1 + 6 \times -8 + 11 \times 13)/91 \end{pmatrix} = \begin{pmatrix} 2 \\ -1 \\ 1 \end{pmatrix}$$

The solution to the system is then

$$x_1 = 2 \qquad x_2 = -1 \qquad x_3 = 1$$

Special Matrices   If A is a matrix of complex numbers, then it is possible to take the complex conjugate a_{ij}^* of each entry, a_{ij}. This is called the **conjugate** of A and is denoted A^*.

1. If $a_{ij} = a_{ji}$, then A is **symmetric**.
2. If $a_{ij} = -a_{ji}$, then A is **skew** or **antisymmetric**.
3. If $A^T = A^{-1}$, then A is **orthogonal**.
4. If $A = A^{-1}$, then A is **involutory**.
5. If $A = A^*$, then A is **hermitian**.
6. If $A = -A^*$, then A is **skew hermitian**.
7. If $A^{-1} = A^*$, then A is **unitary**.

Eigenvalues and Eigenvectors   If A is a square matrix and x is a variable, then the matrix $B = A - xI$ is the **characteristic matrix**, or **eigenmatrix**, of A. The determinant $|A - xI|$ is a polynomial of degree n, called the **characteristic polynomial** of A. The roots of this polynomial, x_1, $x_2, \ldots, x_n$, are the **eigenvalues** of A.

Note that some sources define the characteristic matrix as $xI - A$. If n is odd, then this multiplies the characteristic equation by -1, but the eigenvalues are not changed.

EXAMPLE. $A = \begin{vmatrix} -2 & 5 \\ 2 & 1 \end{vmatrix}$ $B = \begin{vmatrix} -2 - x & 5 \\ 2 & 1 - x \end{vmatrix}$

Then the characteristic polynomial is

$$|B| = (-2 - x)(1 - x) - (2)(5)$$
$$= x^2 + x - 2 - 10$$
$$= x^2 + x - 12$$
$$= (x + 4)(x - 3)$$

The eigenvalues are -4 and $+3$.

A nonzero vector $\mathbf{v}$ satisfying

$$(A - x_iI)\mathbf{v} = \mathbf{0}$$

is called an **eigenvector** of A associated with the eigenvalue x_i. Eigenvectors have the special property

$$A\mathbf{v} = x_i\mathbf{v}$$

Any multiple of an eigenvector is also an eigenvector.

A matrix is nonsingular when none of its eigenvalues are zero.

Rank and Nullity   It is possible that the product of a nonzero matrix A and a nonzero vector $\mathbf{v}$ is zero. This cannot happen if A is nonsingular.

The set of all vectors which become zero when multiplied by A is called the **kernel** of A. The **nullity** of A is the dimension of the kernel. It is a measure of how singular a matrix is.

If A is an $m \times n$ matrix, then the **rank** of A is defined as $n -$ nullity. Rank is at most m.

The technique of pivoting is useful in finding the rank of a matrix. The procedure is as follows:

1. Pivot on each diagonal entry in the matrix, starting with a_{11}.
2. If a row becomes all zero, exchange it with other rows to move it to the bottom of the matrix.
3. If a diagonal entry is zero but the row is not all zero, exchange the column containing the entry with a column to the right not containing a zero in that row.

When the procedure can be carried no further, the nullity is the number of rows of zeros in the matrix.

EXAMPLE.   Find the rank and nullity of the 3 × 2 matrix:

$$\begin{pmatrix} 1 & 1 \\ 2 & 1 \\ 4 & 1 \end{pmatrix}$$

Pivoting on row 1, column 1, yields

$$\begin{pmatrix} 1 & 0 \\ 0 & -1 \\ 0 & -3 \end{pmatrix}$$

Pivoting on row 2, column 2, yields

$$\begin{pmatrix} 1 & 0 \\ 0 & 1 \\ 0 & 0 \end{pmatrix}$$

Nullity is therefore 1. Rank is $3 - 1 = 2$.

If the rank of a matrix is n, so that

$$\text{Rank} + \text{nullity} = m$$

the matrix is said to be **full rank**.

TRIGONOMETRY

Formal Trigonometry

Angles or Rotations   An **angle** is generated by the rotation of a ray, as Ox, about a fixed point O in the plane. Every angle has an **initial line** (OA) from which the rotation started (Fig. 2.1.56), and a **terminal line** (OB) where it stopped; and the counterclockwise direction of rotation is taken as positive. Since the rotating ray may revolve as often as desired, angles of any magnitude, positive or negative, may be obtained. Two angles are **congruent** if they may be superimposed so that their initial lines coincide and their terminal lines coincide; i.e., two congruent angles are either equal or differ by some multiple of 360°. Two angles are **complementary** if their sum is 90°; **supplementary** if their sum is 180°.

Fig. 2.1.56   Angle.

(The acute angles of a right-angled triangle are complementary.) If the initial line is placed so that it runs horizontally to the right, as in Fig. 2.1.57, then the angle is said to be an angle in the 1st, 2nd, 3rd, or 4th **quadrant** according as the terminal line lies across the region marked I, II, III, or IV.

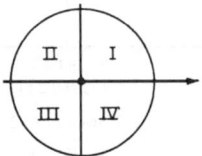

Fig. 2.1.57 Circle showing quadrants.

Units of Angular Measurement

1. *Sexagesimal measure.* (360 degrees = 1 revolution.) Denoted on many calculators by DEG. 1 degree = 1° = 1/90 of a right angle. The degree is usually divided into 60 equal parts called minutes ('), and each minute into 60 equal parts called seconds ("); while the second is subdivided decimally. But for many purposes it is more convenient to divide the degree itself into decimal parts, thus avoiding the use of minutes and seconds.

2. *Centesimal measure.* Used chiefly in France. Denoted on calculators by GRAD. (400 grades = 1 revolution.) 1 grade = 1/100 of a right angle. The grade is always divided decimally, the following terms being sometimes used: 1 "centesimal minute" = 1/100 of a grade; 1 "centesimal second" = 1/100 of a centesimal minute. In reading Continental books it is important to notice carefully which system is employed.

3. *Radian, or circular, measure.* (π radians = 180 degrees.) Denoted by RAD. 1 radian = the angle subtended by an arc whose length is equal to the length of the radius. The radian is constantly used in higher mathematics and in mechanics, and is always divided decimally. Many theorems in calculus assume that angles are being measured in radians, not degrees, and are not true without that assumption.

$$1 \text{ radian} = 57°.30 - = 57°.2957795131$$
$$= 57°17'44".806247 = 180°/\pi$$
$$1° = 0.01745 \ldots \text{ radian} = 0.01745\ 32925 \text{ radian}$$
$$1' = 0.00029\ 08882 \text{ radian}$$
$$1" = 0.00000\ 48481 \text{ radian}$$

Table 2.1.2   Signs of the Trigonometric Functions

If x is in quadrant	I	II	III	IV
sin x and csc x are	+	+	−	−
cos x and sec x are	+	−	−	+
tan x and cot x are	+	−	+	−

Definitions of the Trigonometric Functions   Let x be any angle whose initial line is OA and terminal line OP (see Fig. 2.1.58). Drop a

perpendicular from P on OA or OA produced. In the right triangle OMP, the three sides are MP = "side opposite" O (positive if running upward); OM = "side adjacent" to O (positive if running to the right); OP = "hypotenuse" or "radius" (may always be taken as positive); and the six ratios between these sides are the principal trigonometric

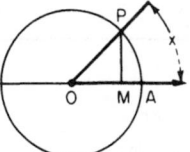

Fig. 2.1.58 Unit circle showing elements used in trigonometric functions.

functions of the angle x; thus:

$$\text{sine of } x = \sin x = \text{opp/hyp} = MP/OP$$
$$\text{cosine of } x = \cos x = \text{adj/hyp} = OM/OP$$
$$\text{tangent of } x = \tan x = \text{opp/adj} = MP/OM$$
$$\text{cotangent of } x = \cot x = \text{adj/opp} = OM/MP$$
$$\text{secant of } x = \sec x = \text{hyp/adj} = OP/OM$$
$$\text{cosecant of } x = \csc x = \text{hyp/opp} = OP/MP$$

The last three are best remembered as the reciprocals of the first three:

$$\cot x = 1/\tan x \qquad \sec x = 1/\cos x \qquad \csc x = 1/\sin x$$

Trigonometric functions, the exponential functions, and complex numbers are all related by the **Euler formula:** $e^{ix} = \cos x + i \sin x$, where $i = \sqrt{-1}$. A special case of this $e^{i\pi} = -1$. Note that here x must be measured in radians.

Variations in the functions as x varies from 0 to 360° are shown in Table 2.1.3. The variations in the sine and cosine are best remembered by noting the changes in the lines MP and OM (Fig. 2.1.59) in the "unit circle" (i.e., a circle with radius = OP = 1), as P moves around the circumference.

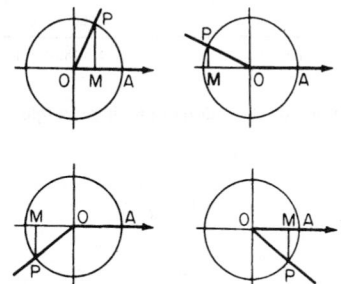

Fig. 2.1.59 Unit circle showing angles in the various quadrants.

Table 2.1.3   Ranges of the Trigonometric Functions

					Values at		
x in DEG	0° to 90°	90° to 180°	180° to 270°	270° to 360°	30°	45°	60°
x in RAD	(0 to $\pi/2$)	($\pi/2$ to π)	(π to $3\pi/2$)	($3\pi/2$ to 2π)	($\pi/6$)	($\pi/4$)	($\pi/3$)
sin x	+0 to +1	+1 to +0	−0 to −1	−1 to −0	1/2	$\frac{1}{2}\sqrt{2}$	$\frac{1}{2}\sqrt{3}$
csc x	+∞ to +1	+1 to +∞	−∞ to −1	−1 to −∞	2	$\sqrt{2}$	$\frac{2}{3}\sqrt{3}$
cos x	+1 to +0	−0 to −1	−1 to −0	+0 to +1	$\frac{1}{2}\sqrt{3}$	$\frac{1}{2}\sqrt{2}$	1/2
sec x	+1 to +∞	−∞ to −1	−1 to −∞	+∞ to +1	$\frac{2}{3}\sqrt{3}$	$\sqrt{2}$	2
tan x	+0 to +∞	−∞ to −0	+0 to +∞	−∞ to −0	$\frac{1}{2}\sqrt{3}$	1	$\sqrt{3}$
cot x	+∞ to +0	−0 to −∞	+∞ to +0	−0 to −∞	$\sqrt{3}$	1	$\frac{1}{3}\sqrt{3}$

To Find Any Function of a Given Angle (Reduction to the first quadrant.) It is often required to find the functions of any angle x from a table that includes only angles between 0 and 90°. If x is not already between 0 and 360°, first "reduce to the first revolution" by simply adding or subtracting the proper multiple of 360° [for any function of (x) = the same function of $(x \pm n \times 360°)$]. Next **reduce to first quadrant** per table below.

$$\tan (x + y) = (\tan x + \tan y)/(1 - \tan x \tan y)$$
$$\cot (x + y) = (\cot x \cot y - 1)/(\cot x + \cot y)$$
$$\sin (x - y) = \sin x \cos y - \cos x \sin y$$
$$\cos (x - y) = \cos x \cos y + \sin x \sin y$$
$$\tan (x - y) = (\tan x - \tan y)/(1 + \tan x \tan y)$$
$$\cot (x - y) = (\cot x \cot y + 1)/(\cot y - \cot x)$$
$$\sin x + \sin y = 2 \sin \tfrac{1}{2}(x + y) \cos \tfrac{1}{2}(x - y)$$

If x is between	90° and 180° ($\pi/2$ and π)	180° and 270° (π and $3\pi/2$)	270° and 360° ($3\pi/2$ and 2π)
Subtract	90° from x ($\pi/2$)	180° from x (π)	270° from x ($3\pi/2$)
Then $\sin x$	$= +\cos (x - 90°)$	$= -\sin (x - 180°)$	$= -\cos (x - 270°)$
$\csc x$	$= +\sec (x - 90°)$	$= -\csc (x - 180°)$	$= -\sec (x - 270°)$
$\cos x$	$= -\sin (x - 90°)$	$= -\cos (x - 180°)$	$= +\sin (x - 270°)$
$\sec x$	$= -\csc (x - 90°)$	$= -\sec (x - 180°)$	$= +\csc (x - 270°)$
$\tan x$	$= -\cot (x - 90°)$	$= +\tan (x - 180°)$	$= -\cot (x - 270°)$
$\cot x$	$= -\tan (x - 90°)$	$= +\cot (x - 180°)$	$= -\tan (x - 270°)$

The "reduced angle" $(x - 90°,$ or $x - 180°,$ or $x - 270°)$ will in each case be an angle between 0 and 90°, whose functions can then be found in the table.

NOTE. The formulas for sine and cosine are best remembered by aid of the unit circle.

To Find the Angle When One of Its Functions Is Given In general, there will be two angles between 0 and 360° corresponding to any given function. The rules showing how to find these angles are tabulated below.

Given	First find an *acute* angle x_0 such that	Then the required angles x_1 and x_2 will be*
$\sin x = + a$	$\sin x_0 = a$	x_0 and $180° - x_0$
$\cos x = + a$	$\cos x_0 = a$	x_0 and $[360° - x_0]$
$\tan x = + a$	$\tan x_0 = a$	x_0 and $[180° + x_0]$
$\cot x = + a$	$\cot x_0 = a$	x_0 and $[180° + x_0]$
$\sin x = - a$	$\sin x_0 = a$	$[180° + x_0]$ and $[360° - x_0]$
$\cos x = - a$	$\cos x_0 = a$	$180° - x_0$ and $[180° + x_0]$
$\tan x = - a$	$\tan x_0 = a$	$180° - x_0$ and $[360° - x_0]$
$\cot x = - a$	$\cot x_0 = a$	$180° - x_0$ and $[360° - x_0]$

* The angles enclosed in brackets lie outside the range 0 to 180 deg and hence cannot occur as angles in a triangle.

Relations Among the Functions of a Single Angle

$$\sin^2 x + \cos^2 x = 1$$

$$\tan x = \frac{\sin x}{\cos x}$$

$$\cot x = \frac{1}{\tan x} = \frac{\cos x}{\sin x}$$

$$1 + \tan^2 x = \sec^2 x = \frac{1}{\cos^2 x}$$

$$1 + \cot^2 x = \csc^2 x = \frac{1}{\sin^2 x}$$

$$\sin x = \sqrt{1 - \cos^2 x} = \frac{\tan x}{\sqrt{1 + \tan^2 x}} = \frac{1}{\sqrt{1 + \cot^2 x}}$$

$$\cos x = \sqrt{1 - \sin^2 x} = \frac{1}{\sqrt{1 + \tan^2 x}} = \frac{\cot x}{\sqrt{1 + \cot^2 x}}$$

Functions of Negative Angles $\sin (-x) = -\sin x; \cos (-x) = \cos x; \tan (-x) = -\tan x.$

Functions of the Sum and Difference of Two Angles

$$\sin (x + y) = \sin x \cos y + \cos x \sin y$$
$$\cos (x + y) = \cos x \cos y - \sin x \sin y$$

$$\sin x - \sin y = 2 \cos \tfrac{1}{2}(x + y) \sin \tfrac{1}{2}(x - y)$$
$$\cos x + \cos y = 2 \cos \tfrac{1}{2}(x + y) \cos \tfrac{1}{2}(x - y)$$
$$\cos x - \cos y = -2 \sin \tfrac{1}{2}(x + y) \sin \tfrac{1}{2}(x - y)$$

$$\tan x + \tan y = \frac{\sin (x + y)}{\cos x \cos y}; \cot x + \cot y = \frac{\sin (x + y)}{\sin x \sin y}$$

$$\tan x - \tan y = \frac{\sin (x - y)}{\cos x \cos y}; \cot x - \cot y = \frac{\sin (x - y)}{\sin x \sin y}$$

$$\sin^2 x - \sin^2 y = \cos^2 y - \cos^2 x = \sin (x + y) \sin (x - y)$$
$$\cos^2 x - \sin^2 y = \cos^2 y - \sin^2 x = \cos (x + y) \cos (x - y)$$
$$\sin (45° + x) = \cos (45° - x)$$
$$\tan (45° + x) = \cot (45° - x)$$
$$\sin (45° - x) = \cos (45° + x)$$
$$\tan (45° - x) = \cot (45° + x)$$

In the following transformations, a and b are supposed to be positive, $c = \sqrt{a^2 + b^2}$, A = the positive acute angle for which $\tan A = a/b$, and B = the positive acute angle for which $\tan B = b/a$:

$$a \cos x + b \sin x = c \sin (A + x) = c \cos (B - x)$$
$$a \cos x - b \sin x = c \sin (A - x) = c \cos (B + x)$$

Functions of Multiple Angles and Half Angles

$$\sin 2x = 2 \sin x \cos x; \sin x = 2 \sin \tfrac{1}{2}x \cos \tfrac{1}{2}x$$
$$\cos 2x = \cos^2 x - \sin^2 x = 1 - 2 \sin^2 x = 2 \cos^2 x - 1$$

$$\tan 2x = \frac{2 \tan x}{1 - \tan^2 x} \qquad \cot 2x = \frac{\cot^2 x - 1}{2 \cot x}$$

$$\sin 3x = 3 \sin x - 4 \sin^3 x; \tan 3x = \frac{3 \tan x - \tan^3 x}{1 - 3 \tan^2 x}$$

$$\cos 3x = 4 \cos^3 x - 3 \cos x$$

$$\sin (nx) = n \sin x \cos^{n-1} x - (n)_3 \sin^3 x \cos^{n-3} x$$
$$+ (n)_5 \sin^5 x \cos^{n-5} x - \cdots$$
$$\cos (nx) = \cos^n x - (n)_2 \sin^2 x \cos^{n-2} x + (n)_4 \sin^4 x \cos^{n-4} x - \cdots$$

where $(n)_2, (n)_3, \ldots$, are the binomial coefficients.

$$\sin \tfrac{1}{2}x = \pm \sqrt{\tfrac{1}{2}(1 - \cos x)}. \quad 1 - \cos x = 2 \sin^2 \tfrac{1}{2}x$$
$$\cos \tfrac{1}{2}x = \pm \sqrt{\tfrac{1}{2}(1 + \cos x)}. \quad 1 + \cos x = 2 \cos^2 \tfrac{1}{2}x$$

$$\tan \tfrac{1}{2}x = \pm \sqrt{\frac{1 - \cos x}{1 + \cos x}} = \frac{\sin x}{1 + \cos x} = \frac{1 - \cos x}{\sin x}$$

$$\tan \left(\frac{x}{2} + 45°\right) = \pm \sqrt{\frac{1 + \sin x}{1 - \sin x}}$$

Here the + or − sign is to be used according to the sign of the left-hand side of the equation.

Approximations for sin x, cos x, and tan x   For small values of x, x measured in radians, the following approximations hold:

$$\sin x \approx x \qquad \tan x \approx x \qquad \cos x \approx 1 - \frac{x^2}{2}$$

The following actually hold:

$$\sin x < x < \tan x \qquad \cos x < \frac{\sin x}{x} < 1$$

As x approaches 0, $\lim [(\sin x)/x] = 1$.

Inverse Trigonometric Functions   The notation $\sin^{-1} x$ (read: arc-sine of x, or inverse sine of x) means the principal angle whose sine is x. Similarly for $\cos^{-1} x$, $\tan^{-1} x$, etc. (The principal angle means an angle between -90 and $+90°$ in case of $\sin^{-1}$ and $\tan^{-1}$, and between 0 and 180° in the case of $\cos^{-1}$.)

Solution of Plane Triangles

The "parts" of a plane triangle are its three sides a, b, c, and its three angles A, B, C (A being opposite a). Two triangles are **congruent** if all their corresponding parts are equal. Two triangles are **similar** if their corresponding angles are equal, that is, $A_1 = A_2$, $B_1 = B_2$, and $C_1 = C_2$. Similar triangles may differ in scale, but they satisfy $a_1/a_2 = b_1/b_2 = c_1/c_2$.

Two different triangles may have two corresponding sides and the angle opposite one of those sides equal (Fig. 2.1.60), and still not be congruent. This is the **angle-side-side theorem.**

Otherwise, a triangle is uniquely determined by any three of its parts, as long as those parts are not all angles. To "solve" a triangle means to find the unknown parts from the known. The fundamental formulas are

$$\text{Law of sines: } \frac{a}{b} = \frac{\sin A}{\sin B}$$

$$\text{Law of cosines: } c^2 = a^2 + b^2 - 2ab \cos C$$

Fig. 2.1.60   Triangles with an angle, an adjacent side, and an opposite side given.

Right Triangles   Use the definitions of the trigonometric functions, selecting for each unknown part a relation which connects that unknown with known quantities; then solve the resulting equations. Thus, in Fig. 2.1.61, if $C = 90°$, then $A + B = 90°$, $c^2 = a^2 + b^2$,

$$\sin A = a/c \qquad \cos A = b/c$$
$$\tan A = a/b \qquad \cot A = b/a$$

If A is very small, use $\tan \frac{1}{2} A = \sqrt{(c - b)/(c + b)}$.

Oblique Triangles   There are four cases. It is highly desirable in all these cases to draw a sketch of the triangle approximately to scale before commencing the computation, so that any large numerical error may be readily detected.

Fig. 2.1.61   Right triangle.

Fig. 2.1.62   Triangle with two angles and the included side given.

CASE 1.   GIVEN TWO ANGLES (provided their sum is $< 180°$) AND ONE SIDE (say a, Fig. 2.1.62). The third angle is known since $A + B + C = 180°$.

To find the remaining sides, use

$$b = \frac{a \sin B}{\sin A} \qquad c = \frac{a \sin C}{\sin A}$$

Or, drop a perpendicular from either B or C on the opposite side, and solve by right triangles.

Check: $c \cos B + b \cos C = a$.

CASE 2.   GIVEN TWO SIDES (say a and b) AND THE INCLUDED ANGLE (C); AND SUPPOSE $a > b$ (Fig. 2.1.63).

Method 1: Find c from $c^2 = a^2 + b^2 - 2ab \cos C$; then find the smaller angle, B, from $\sin B = (b/c) \sin C$; and finally, find A from $A = 180° - (B + C)$.

Check: $a \cos B + b \cos A = c$.

Method 2: Find $\frac{1}{2}(A - B)$ from the law of tangents:

$$\tan \tfrac{1}{2}(A - B) = [(a - b)/(a + b)] \cot \tfrac{1}{2}C$$

and $\frac{1}{2}(A + B)$ from $\frac{1}{2}(A + B) = 90° - C/2$; hence $A = \frac{1}{2}(A + B) + \frac{1}{2}(A - B)$ and $B = \frac{1}{2}(A + B) - \frac{1}{2}(A - B)$. Then find c from $c = a \sin C/\sin A$ or $c = b \sin C/\sin B$.

Check: $a \cos B + b \cos A = c$.

Method 3: Drop a perpendicular from A to the opposite side, and solve by right triangles.

CASE 3.   GIVEN THE THREE SIDES (provided the largest is less than the sum of the other two) (Fig. 2.1.64).

Method 1: Find the largest angle A (which may be acute or obtuse) from $\cos A = (b^2 + c^2 - a^2)/2bc$ and then find B and C (which will always be acute) from $\sin B = b \sin A/a$ and $\sin C = c \sin A/a$.

Check: $A + B + C = 180°$.

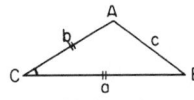

Fig. 2.1.63   Triangle with two sides and the included angle given.

Fig. 2.1.64   Triangle with three sides given.

Method 2: Find A, B, and C from $\tan \frac{1}{2}A = r/(s - a)$, $\tan \frac{1}{2}B = r/(s - b)$, $\tan \frac{1}{2}C = r/(s - c)$, where $s = \frac{1}{2}(a + b + c)$, and $r = \sqrt{(s - a)(s - b)(s - c)/s}$. Check: $A + B + C = 180°$.

Method 3: If only one angle, say A, is required, use

$$\sin \tfrac{1}{2}A = \sqrt{(s - b)(s - c)/bc}$$
or
$$\cos \tfrac{1}{2}A = \sqrt{s(s - a)/bc}$$

according as $\frac{1}{2}A$ is nearer 0° or nearer 90°.

CASE 4.   GIVEN TWO SIDES (say b and c) AND THE ANGLE OPPOSITE ONE OF THEM (B). This is the "ambiguous case" in which there may be two solutions, or one, or none.

First, try to find $C = c \sin B/b$. If $\sin C > 1$, there is no solution. If $\sin C = 1$, $C = 90°$ and the triangle is a right triangle. If $\sin C < 1$, this determines two angles C, namely, an acute angle C_1, and an obtuse angle $C_2 = 180° - C_1$. Then C_1 will yield a solution when and only when $C_1 + B < 180°$ (see Case 1); and similarly C_2 will yield a solution when and only when $C_2 + B < 180°$ (see Case 1).

Other Properties of Triangles   (See also Geometry, Areas, and Volumes.)

$\text{Area} = \frac{1}{2}ab \sin C = \sqrt{s(s - a)(s - b)(s - c)} = rs$ where $s = \frac{1}{2}(a + b + c)$, and $r = $ radius of inscribed circle $= \sqrt{(s - a)(s - b)(s - c)/s}$.

Radius of circumscribed circle $= R$, where

$$2R = a/\sin A = b/\sin B = c/\sin C$$

$$r = 4R \sin\frac{A}{2} \sin\frac{B}{2} \sin\frac{C}{2} = \frac{abc}{4Rs}$$

The length of the bisector of the angle C is

$$z = \frac{2\sqrt{abs(s-c)}}{a+b} = \frac{\sqrt{ab[(a+b)^2 - c^2]}}{a+b}$$

The median from C to the middle point of c is $m = \frac{1}{2}\sqrt{2(a^2 + b^2) - c^2}$.

Hyperbolic Functions

The **hyperbolic sine, hyperbolic cosine,** etc., of any number x, are functions of x which are closely related to the exponential e^x, and which have formal properties very similar to those of the trigonometric functions, sine, cosine, etc. Their definitions and fundamental properties are as follows:

$$\sinh x = \frac{1}{2}(e^x - e^{-x})$$
$$\cosh x = \frac{1}{2}(e^x + e^{-x})$$
$$\tanh x = \sinh x / \cosh x$$

$$\cosh x + \sinh x = e^x$$
$$\cosh x - \sinh x = e^{-x}$$

$$\operatorname{csch} x = 1/\sinh x$$
$$\operatorname{sech} x = 1/\cosh x$$
$$\coth x = 1/\tanh x$$

$$\cosh^2 x - \sinh^2 x = 1$$
$$1 - \tanh^2 x = \operatorname{sech}^2 x$$
$$1 - \coth^2 x = -\operatorname{csch}^2 x$$

$$\sinh(-x) = -\sinh x$$
$$\cosh(-x) = \cosh x$$
$$\tanh(-x) = -\tanh x$$

$$\sinh(x \pm y) = \sinh x \cosh y \pm \cosh x \sinh y$$
$$\cosh(x \pm y) = \cosh x \cosh y \pm \sinh x \sinh y$$
$$\tanh(x \pm y) = (\tanh x \pm \tanh y)/(1 \pm \tanh x \tanh y)$$

$$\sinh 2x = 2 \sinh x \cosh x$$
$$\cosh 2x = \cosh^2 x + \sinh^2 x$$
$$\tanh 2x = (2 \tanh x)/(1 + \tanh^2 x)$$

$$\sinh \tfrac{1}{2}x = \sqrt{\tfrac{1}{2}(\cosh x - 1)}$$
$$\cosh \tfrac{1}{2}x = \sqrt{\tfrac{1}{2}(\cosh x + 1)}$$
$$\tanh \tfrac{1}{2}x = (\cosh x - 1)/(\sinh x) = (\sinh x)/(\cosh x + 1)$$

The hyperbolic functions are related to the rectangular hyperbola, $x^2 - y^2 = a^2$ (Fig. 2.1.66), in much the same way that the trigonometric functions are related to the circle $x^2 + y^2 = a^2$ (Fig. 2.1.65); the analogy, however, concerns not angles but areas. Thus, in either figure, let A

Fig. 2.1.65 Circle.

Fig. 2.1.66 Hyperbola.

represent the shaded area, and let $u = A/a^2$ (a pure number). Then for the coordinates of the point P we have, in Fig. 2.1.65, $x = a \cos u$, $y = a \sin u$; and in Fig. 2.1.66, $x = a \cosh u$, $y = a \sinh u$.

The **inverse hyperbolic sine** of y, denoted by $\sinh^{-1} y$, is the number whose hyperbolic sine is y; that is, the notation $x = \sinh^{-1} y$ means $\sinh x = y$. Similarly for $\cosh^{-1} y$, $\tanh^{-1} y$, etc. These functions are closely related to the logarithmic function, and are especially valuable in the integral calculus.

$$\sinh^{-1}(y/a) = \ln(y + \sqrt{y^2 + a^2}) - \ln a$$
$$\cosh^{-1}(y/a) = \ln(y + \sqrt{y^2 - a^2}) - \ln a$$
$$\tanh^{-1}\frac{y}{a} = \frac{1}{2}\ln\frac{a+y}{a-y}$$
$$\coth^{-1}\frac{y}{a} = \frac{1}{2}\ln\frac{y+a}{y-a}$$

ANALYTICAL GEOMETRY

The Point and the Straight Line

Rectangular Coordinates   (Fig. 2.1.67) Let $P_1 = (x_1, y_1)$, $P_2 = (x_2, y_2)$. Then, distance

$$P_1P_2 = \sqrt{(x_2 - x_1)^2 + (y_2 - y_1)^2}$$

slope of $P_1 P_2 = m = \tan u = (y_2 - y_1)/(x_2 - x_1)$; coordinates of midpoint are $x = \frac{1}{2}(x_1 + x_2)$, $y = \frac{1}{2}(y_1 + y_2)$; coordinates of point 1/nth of the way from P_1 to P_2 are $x = x_1 + (1/n)(x_2 - x_1)$, $y = y_1 + (1/n)(y_2 - y_1)$.

Let m_1, m_2 be the slopes of two lines; then, if the lines are parallel, $m_1 = m_2$; if the lines are perpendicular to each other, $m_1 = -1/m_2$.

Fig. 2.1.67 Graph of straight line.

Fig. 2.1.68 Graph of straight line showing intercepts.

Equations of a Straight Line
1. *Intercept form* (Fig. 2.1.68). $x/a + y/b = 1$. (a, b = intercepts of the line on the axes.)
2. *Slope form* (Fig. 2.1.69). $y = mx + b$. ($m = \tan u$ = slope; b = intercept on the y axis.)
3. *Normal form* (Fig. 2.1.70). $x \cos v + y \sin v = p$. (p = perpendicular from origin to line; v = angle from the x axis to p.)

Fig. 2.1.69 Graph of straight line showing slope and vertical intercept.

Fig. 2.1.70 Graph of straight line showing perpendicular line from origin.

4. *Parallel-intercept form* (Fig. 2.1.71). $\dfrac{y - b}{c - b} = \dfrac{x}{k}$ (b = y intercept, c = intercept on the parallel $x = k$).

Fig. 2.1.71 Graph of straight line showing intercepts on parallel lines.

5. *General form.* $Ax + By + C = 0$. [Here $a = -C/A$, $b = -C/B$, $m = -A/B$, $\cos v = A/R$, $\sin v = B/R$, $p = -C/R$, where $R = \pm\sqrt{A^2 + B^2}$ (sign to be so chosen that p is positive).]
6. *Line through* (x_1, y_1) *with slope* m. $y - y_1 = m(x - x_1)$.

7. *Line through* (x_1, y_1) *and* (x_2, y_2). $y - y_1 = \dfrac{y_2 - y_1}{x_2 - x_1}(x - x_1)$.

8. *Line parallel to x axis.* $y = a$; *to y axis:* $x = b$.

Angles and Distances   If u = angle from the line with slope m_1 to the line with slope m_2, then

$$\tan u = \frac{m_2 - m_1}{1 + m_2 m_1}$$

If parallel, $m_1 = m_2$.
If perpendicular, $m_1 m_2 = -1$.
If u = angle between the lines $Ax + By + C = 0$ and $A'x + B'y + C' = 0$, then

$$\cos u = \frac{AA' + BB'}{\pm \sqrt{(A^2 + B^2)(A'^2 + B'^2)}}$$

If parallel, $A/A' = B/B'$.
If perpendicular, $AA' + BB' = 0$.

The equation of a line through (x_1, y_1) and meeting a given line $y = mx + b$ at an angle u, is

$$y - y_1 = \frac{m + \tan u}{1 - m \tan u}(x - x_1)$$

The distance from (x_0, y_0) to the line $Ax + By + C = 0$ is

$$D = \left| \frac{Ax_0 + By_0 + C}{\sqrt{A^2 + B^2}} \right|$$

where the vertical bars mean "the absolute value of."
The distance from (x_0, y_0) to a line which passes through (x_1, y_1) and makes an angle u with the x axis is

$$D = (x_0 - x_1) \sin u - (y_0 - y_1) \cos u$$

Polar Coordinates   (Fig. 2.1.72) Let (x, y) be the rectangular and (r, θ) the polar coordinates of a given point P. Then $x = r \cos \theta$; $y = r \sin \theta$; $x^2 + y^2 = r^2$.

Fig. 2.1.72   Polar coordinates.

Transformation of Coordinates   If origin is moved to point (x_0, y_0), the new axes being parallel to the old, $x = x_0 + x'$, $y = y_0 + y'$.
If axes are turned through the angle u, without change of origin,

$$x = x' \cos u - y' \sin u \qquad y = x' \sin u + y' \cos u$$

The Circle

The **equation of a circle** with center (a, b) and radius r is

$$(x - a)^2 + (y - b)^2 = r^2$$

If center is at the origin, the equation becomes $x^2 + y^2 = r^2$. If circle goes through the origin and center is on the x axis at point $(r, 0)$, equation becomes $x^2 + y^2 = 2rx$. The **general equation** of a circle is

$$x^2 + y^2 + Dx + Ey + F = 0$$

It has center at $(-D/2, -E/2)$, and radius $= \sqrt{(D/2)^2 + (E/2)^2 - F}$ (which may be real, null, or imaginary).

Equations of Circle in Parametric Form   It is sometimes convenient to express the coordinates x and y of the moving point P (Fig. 2.1.73) in terms of an auxiliary variable, called a **parameter.** Thus, if the parameter be taken as the angle u from the x axis to the radius vector OP, then the equations of the circle in parametric form will be $x = a \cos u$; $y = a$

$\sin u$. For every value of the parameter u, there corresponds a point (x, y) on the circle. The ordinary equation $x^2 + y^2 = a^2$ can be obtained from the parametric equations by eliminating u. The equation of a circle with radius a and center at (h, k) in parametric form will be $x = h + a \cos u$; $y = k + a \sin u$.

Fig. 2.1.73   Parameters of a circle.

The Parabola

The **parabola** is the locus of a point which moves so that its distance from a fixed line (called the **directrix**) is always equal to its distance from a fixed point F (called the **focus**). See Fig. 2.1.74. The point halfway from focus to directrix is the **vertex**, O. The line through the focus, perpendicular to the directrix, is the **principal axis.** The breadth of the curve at the focus is called the **latus rectum**, or **parameter,** $= 2p$, where p is the distance from focus to directrix.

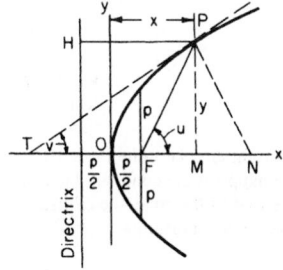

Fig. 2.1.74   Graph of parabola.

NOTE.   Any section of a right circular cone made by a plane parallel to a tangent plane of the cone will be a parabola.

Equation of parabola, principal axis along the x axis, origin at vertex (Fig. 2.1.74): $y^2 = 2px$.
Polar equation of parabola, referred to F as origin and Fx as axis (Fig. 2.1.75): $r = p/(1 - \cos \theta)$.
Equation of parabola with principal axis parallel to y axis: $y = ax^2 + bx + c$. This may be rewritten, using a technique called **completing the square:**

$$y = a \left[x^2 + \frac{b}{a}x + \frac{b^2}{4a^2} \right] + c - \frac{b^2}{4a}$$

$$= a \left[x + \frac{b}{2a} \right]^2 + c - \frac{b^2}{4a}$$

Fig. 2.1.75   Polar plot of parabola.

Fig. 2.1.76   Vertical parabola showing rays passing through the focus.

Then: vertex is the point $[-b/2a, c - b^2/4a]$; latus rectum is $p = 1/2a$; and focus is the point $[-b/2a, c - b^2/4a + 1/4a]$.

A parabola has the special property that lines parallel to its principal axis, when reflected off the inside "surface" of the parabola, will all pass through the focus (Fig. 2.1.76). This property makes parabolas useful in designing mirrors and antennas.

The Ellipse

The **ellipse** (as shown in Fig. 2.1.77), has two **foci**, F and F', and two **directrices**, DH and $D'H'$. If P is any point on the curve, $PF + PF'$ is constant, $= 2a$; and PF/PH (or PF'/PH') is also constant, $= e$, where e is the **eccentricity** ($e < 1$). Either of these properties may be taken as the definition of the curve. The relations between e and the semiaxes a and b are as shown in Fig. 2.1.78. Thus, $b^2 = a^2(1 - e^2)$, $ae = \sqrt{a^2 - b^2}$, $e^2 = 1 - (b/a)^2$. The **semilatus rectum** $= p = a(1 - e^2) = b^2/a$. Note that b is always less than a, except in the special case of the circle, in which $b = a$ and $e = 0$.

Fig. 2.1.77   Ellipse.

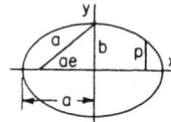

Fig. 2.1.78   Ellipse showing semiaxes.

Any section of a right circular cone made by a plane which cuts all the elements of one nappe of the cone will be an ellipse; if the plane is perpendicular to the axis of the cone, the ellipse becomes a circle.

Equation of ellipse, center at origin:

$$\frac{x^2}{a^2} + \frac{y^2}{b^2} = 1 \qquad \text{or} \qquad y = \pm \frac{b}{a}\sqrt{a^2 - x^2}$$

If $P = (x, y)$ is any point of the curve, $PF = a + ex$, $PF' = a - ex$.

Equations of the ellipse in parametric form: $x = a \cos u$, $y = b \sin u$, where u is the eccentric angle of the point $P = (x, y)$. See Fig. 2.1.81.

Polar equation, focus as origin, axes as in Fig. 2.1.79. $r = p/(1 - e \cos \theta)$.

Equation of the tangent at (x_1, y_1): $b^2 x_1 x + a^2 y_1 y = a^2 b^2$.

The line $y = mx + k$ will be a tangent if $k = \pm \sqrt{a^2 m^2 + b^2}$.

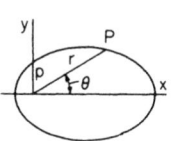

Fig. 2.1.79   Ellipse in polar form.

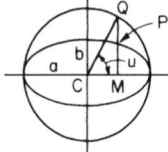

Fig. 2.1.80   Ellipse as a flattened circle.

Ellipse as a Flattened Circle, Eccentric Angle   If the ordinates in a circle are diminished in a constant ratio, the resulting points will lie on an ellipse (Fig. 2.1.80). If Q traces the circle with uniform velocity, the corresponding point P will trace the ellipse, with varying velocity. The angle u in the figure is called the eccentric angle of the point P.

A consequence of this property is that if a circle is drawn with its horizontal scale different from its vertical scale, it will appear to be an ellipse. This phenomenon is common in computer graphics.

The **radius of curvature of an ellipse** at any point $P = (x, y)$ is

$$R = a^2 b^2 (x^2/a^4 + y^2/b^4)^{3/2} = p/\sin^3 v$$

where v is the angle which the tangent at P makes with PF or PF'. At end of major axis, $R = b^2/a = MA$; at end of minor axis, $R = a^2/b = NB$ (see Fig. 2.1.81).

Fig. 2.1.81   Ellipse showing radius of curvature.

The Hyperbola

The **hyperbola** has two **foci**, F and F', at distances $\pm ae$ from the center, and two **directrices**, DH and $D'H'$, at distances $\pm a/e$ from the center (Fig. 2.1.82). If P is any point of the curve, $|PF - PF'|$ is constant, $= 2a$; and PF/PH (or PF'/PH') is also constant, $= e$ (called the **eccentricity**), where $e > 1$. Either of these properties may be taken as the

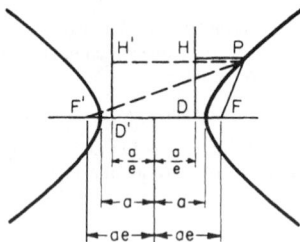

Fig. 2.1.82   Hyperbola.

definition of the curve. The curve has two branches which approach more and more nearly two straight lines called the **asymptotes**. Each asymptote makes with the principal axis an angle whose tangent is b/a. The relations between e, a, and b are shown in Fig. 2.1.83: $b^2 = a^2(e^2 - 1)$, $ae = \sqrt{a^2 + b^2}$, $e^2 = 1 + (b/a)^2$. The semilatus rectum, or ordinate at the focus, is $p = a(e^2 - 1) = b^2/a$.

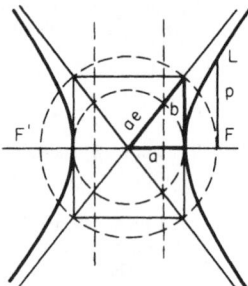

Fig. 2.1.83   Hyperbola showing the asymptotes.

Any section of a right circular cone made by a plane which cuts both nappes of the cone will be a hyperbola.

Equation of the hyperbola, center as origin:

$$\frac{x^2}{a^2} - \frac{y^2}{b^2} = 1 \qquad \text{or} \qquad y = \pm \frac{b}{a}\sqrt{x^2 - a^2}$$

If $P = (x, y)$ is on the right-hand branch, $PF = ex - a$, $PF' = ex + a$. If P is on the left-hand branch, $PF = -ex + a$, $PF' = -ex - a$.

Equations of Hyperbola in Parametric Form   (1) $x = a \cosh u$, $y = b \sinh u$. Here u may be interpreted as A/ab, where A is the area shaded in Fig. 2.1.84. (2) $x = a \sec v$, $y = b \tan v$, where v is an auxiliary angle of no special geometric interest.

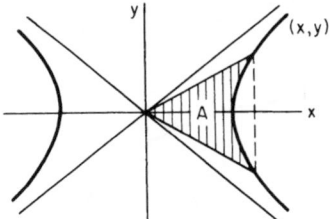

Fig. 2.1.84   Hyperbola showing parametric form.

Polar equation, referred to focus as origin, axes as in Fig. 2.1.85:

$$r = p/(1 - e \cos \theta)$$

Equation of tangent at (x_1, y_1): $b^2 x_1 x - a^2 y_1 y = a^2 b^2$. The line $y = mx + k$ will be a tangent if $k = \pm \sqrt{a^2 m^2 - b^2}$.

Fig. 2.1.85   Hyperbola in polar form.

The triangle bounded by the asymptotes and a variable tangent is of constant area, $= ab$.

Conjugate hyperbolas are two hyperbolas having the same asymptotes with semiaxes interchanged (Fig. 2.1.86). The equations of the hyperbola conjugate to $x^2/a^2 - y^2/b^2 = 1$ is $x^2/a^2 - y^2/b^2 = -1$.

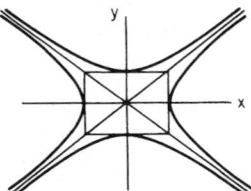

Fig. 2.1.86   Conjugate hyperbolas.

Equilateral Hyperbola   ($a = b$) Equation referred to principal axes (Fig. 2.1.87): $x^2 - y^2 = a^2$.

NOTE. $p = a$ (Fig. 2.1.87). Equation referred to asymptotes as axes (Fig. 2.1.88): $xy = a^2/2$.

Asymptotes are perpendicular. Eccentricity $= \sqrt{2}$. Any diameter is equal in length to its conjugate diameter.

The Catenary

The **catenary** is the curve in which a flexible chain or cord of uniform density will hang when supported by the two ends. Let $w =$ weight of the chain per unit length; $T =$ the tension at any point P; and T_h, $T_v =$ the horizontal and vertical components of T. The horizontal component T_h is the same at all points of the curve.

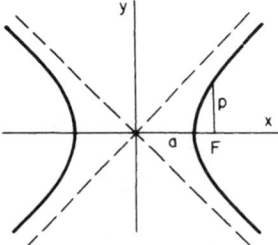

Fig. 2.1.87   Equilateral hyperbola.

The length $a = T_h/w$ is called the **parameter** of the catenary, or the distance from the lowest point O to the **directrix** DQ (Fig. 2.1.89). When a is very large, the curve is very flat.

The rectangular **equation,** referred to the lowest point as origin, is $y = a [\cosh (x/a) - 1]$. In case of very flat arcs (a large), $y = x^2/2a + \cdots$; $s = x + \frac{1}{6} x^3/a^2 + \cdots$, approx, so that in such a case the catenary closely resembles a parabola.

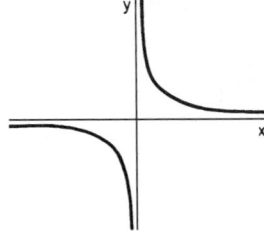

Fig. 2.1.88   Hyperbola with asymptotes as axes.

Calculus properties of the catenary are often discussed in texts on the calculus of variations (Weinstock, "Calculus of Variations," Dover; Ewing, "Calculus of Variations with Applications," Dover).

Problems on the Catenary   (Fig. 2.1.89) When any two of the four quantities, x, y, s, T/w are known, the remaining two, and also the parameter a, can be found, using the following:

$a = x/z$	$s = a \sinh z$
$T = wa \cosh z$	$y/x = (\cosh z - 1)/z$
$s/x = (\sinh z)/z$	$wx/T = z \cosh z$

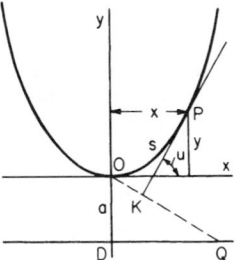

Fig. 2.1.89   Catenary.

NOTE.   If $wx/T < 0.6627$, then there are two values of z, one less than 1.2, and one greater. If $wx/T > 0.6627$, then the problem has no solution.

Given the Length $2L$ of a Chain Supported at Two Points A and B Not in the Same Level, to Find a   (See Fig. 2.1.90; b and c are supposed known.) Let $(\sqrt{L^2 - b^2})/c = s/x$; use $s/x = \sinh z/z$ to find z. Then $a = c/z$.

NOTE.   The coordinates of the midpoint M of AB (see Fig. 2.1.90) are $x_0 = a \tanh^{-1} (b/L)$, $y_0 = (L/\tanh z) - a$, so that the position of the lowest point is determined.

Fig. 2.1.90   Catenary with ends at unequal levels.

Other Useful Curves

The **cycloid** is traced by a point on the circumference of a circle which rolls without slipping along a straight line. Equations of cycloid, in parametric form (axes as in Fig. 2.1.91): $x = a(\text{rad } u - \sin u)$, $y = a(1 - \cos u)$, where a is the radius of the rolling circle, and rad u is the radian measure of the angle u through which it has rolled. The **radius of curvature** at any point P is $PC = 4a \sin (u/2) = 2\sqrt{2ay}$.

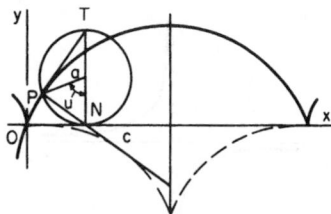

Fig. 2.1.91   Cycloid.

The **trochoid** is a more general curve, traced by any point on a radius of the rolling circle, at distance b from the center (Fig. 2.1.92). It is a prolate trochoid if $b < a$, and a curtate or looped trochoid if $b > a$. The equations in either case are $x = a \text{ rad } u - b \sin u$, $y = a - b \cos u$.

Fig. 2.1.92   Trochoid.

The **epicycloid** (or **hypocycloid**) is a curve generated by a point on the circumference of a circle of radius a which rolls without slipping on the outside (or inside) of a fixed circle of radius c (Fig. 2.1.93 and

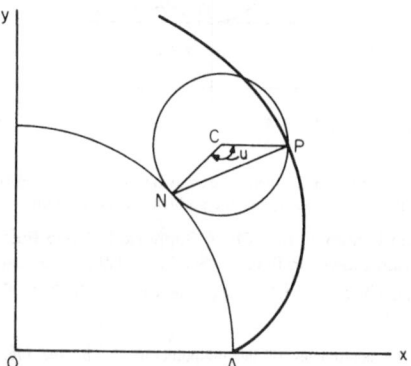

Fig. 2.1.93   Epicycloid.

Fig. 2.1.94). For the **equations,** put $b = a$ in the equations of the epior hypotrochoid, below.

Radius of curvature at any point P is

$$R = \frac{4a(c \pm a)}{c \pm 2a} \times \sin \tfrac{1}{2}u$$

At A, $R = 0$; at D, $R = \dfrac{4a(c \pm a)}{c \pm 2a}$.

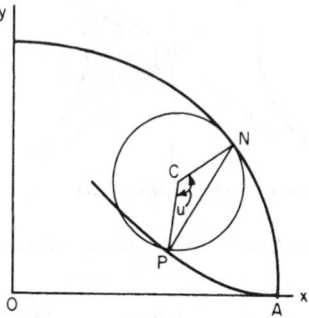

Fig. 2.1.94   Hypocycloid.

Special Cases   If $a = \frac{1}{2}c$, the hypocycloid becomes a straight line, diameter of the fixed circle (Fig. 2.1.95). In this case the hypotrochoid traced by any point rigidly connected with the rolling circle (not necessarily on the circumference) will be an ellipse. If $a = \frac{1}{4}c$, the curve

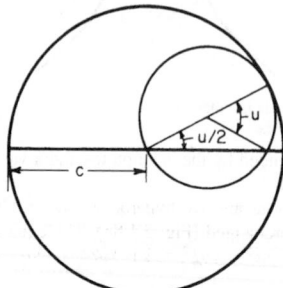

Fig. 2.1.95   Hypocycloid is straight line when the radius of inside circle is half that of the outside circle.

generated will be the four-cusped hypocycloid, or **astroid** (Fig. 2.1.96), whose equation is $x^{2/3} + y^{2/3} = c^{2/3}$. If $a = c$, the epicycloid is the **cardioid,** whose equation in polar coordinates (axes as in Fig. 2.1.97) is $r = 2c(1 + \cos \theta)$. Length of cardioid = $16c$.

The **epitrochoid** (or **hypotrochoid**) is a curve traced by any point rigidly attached to a circle of radius a, at distance b from the center, when this

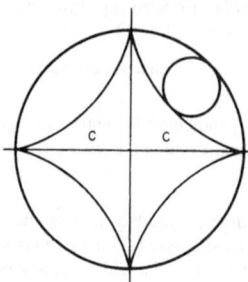

Fig. 2.1.96   Astroid.

circle rolls without slipping on the outside (or inside) of a fixed circle of radius c. The equations are

$$x = (c \pm a) \cos\left(\frac{a}{c}u\right) \pm b \cos\left[\left(1 \pm \frac{a}{c}\right)u\right]$$

$$y = (c \pm a) \sin\left(\frac{a}{c}u\right) - b \sin\left[\left(1 \pm \frac{a}{c}\right)u\right]$$

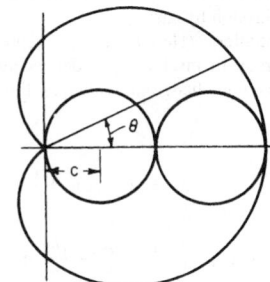

Fig. 2.1.97 Cardioid.

where u = the angle which the moving radius makes with the line of centers; take the upper sign for the epi- and the lower for the hypotrochoid. The curve is called prolate or curtate according as $b < a$ or $b > a$. When $b = a$, the special case of the epi- or hypocycloid arises.

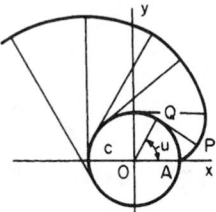

Fig. 2.1.98 Involute of circle.

The **involute of a circle** is the curve traced by the end of a taut string which is unwound from the circumference of a fixed circle, of radius c. If QP is the free portion of the string at any instant (Fig. 2.1.98), QP will be tangent to the circle at Q, and the length of QP = length of arc QA; hence the construction of the curve. The equations of the curve in parametric form (axes as in figure) are $x = c(\cos u + \text{rad } u \sin u)$, $y = c(\sin u - \text{rad } u \cos u)$, where rad u is the radian measure of the angle u which OQ makes with the x axis. Length of arc $AP = \frac{1}{2}c(\text{rad } u)^2$; radius of curvature at P is QP. Polar equations, in terms of parameter

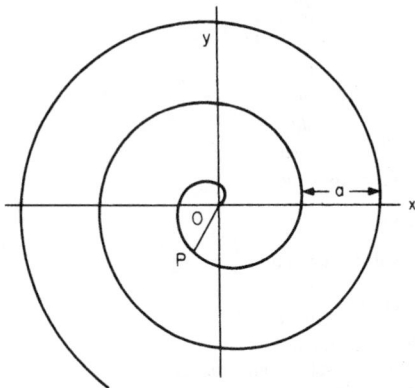

Fig. 2.1.99 Spiral of Archimedes.

$v(= \text{angle } POQ)$, are $r = c \sec v$, rad $\theta = \tan v - \text{rad } v$. Here, $r = OP$, and rad θ = radian measure of angle, AOP (Fig. 2.1.98).

The **spiral of Archimedes** (Fig. 2.1.99) is traced by a point P which, starting from O, moves with uniform velocity along a ray OP, while the ray itself revolves with uniform angular velocity about O. Polar equation: $r = k$ rad θ, or $r = a(\theta°/360°)$. Here $a = 2\pi k$ = the distance measured along a radius, from each coil to the next.

The radius of curvature at P is $R = (k^2 + r^2)^{3/2}/(2k^2 + r^2)$.

The **logarithmic spiral** (Fig. 2.1.100) is a curve which cuts the radii from O at a constant angle v, whose cotangent is m. Polar equation: $r = ae^{m \text{rad} \theta}$. Here a is the value of r when $\theta = 0$. For large negative values of θ, the curve winds around O as an asymptotic point. If PT and PN are the tangent and normal at P, the line TON being perpendicular to OP (not shown in figure), then $ON = rm$, and $PN = r\sqrt{1 + m^2} = r/\sin v$. Radius of curvature at P is PN.

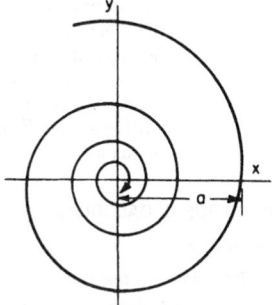

Fig. 2.1.100 Logarithmic spiral.

The **tractrix,** or Schiele's antifriction curve (Fig. 2.1.101), is a curve such that the portion PT of the tangent between the point of contact and the x axis is constant $= a$. Its equation is

$$x = \pm a\left[\cosh^{-1}\frac{a}{y} - \sqrt{1 - \left(\frac{y}{a}\right)^2}\right]$$

or, in parametric form, $x = \pm a(t - \tanh t)$, $y = a/\cosh t$. The x axis is an asymptote of the curve. Length of arc $BP = a \log_e (a/y)$.

Fig. 2.1.101 Tractrix.

The tractrix describes the path taken by an object being pulled by a string moving along the x axis, where the initial position of the object is B and the opposite end of the string begins at O.

Fig. 2.1.102 Lemniscate.

The **lemniscate** (Fig. 2.1.102) is the locus of a point P the product of whose distances from two fixed points F, F' is constant, equal to $\frac{1}{2}a^2$.

The distance $FF' = a\sqrt{2}$. Polar equation is $r = a\sqrt{\cos 2\theta}$. Angle between OP and the normal at P is 2θ. The two branches of the curve cross at right angles at O. Maximum y occurs when $\theta = 30°$ and $r = a/\sqrt{2}$, and is equal to $\frac{1}{4}a\sqrt{2}$. Area of one loop $= a^2/2$.

The helix (Fig. 2.1.103) is the curve of a screw thread on a cylinder of radius r. The curve crosses the elements of the cylinder at a constant angle, v. The pitch, h, is the distance between two coils of the helix, measured along an element of the cylinder; hence $h = 2\pi r \tan v$. Length of one coil $= \sqrt{(2\pi r)^2 + h^2} = 2\pi r/\cos v$. If the cylinder is rolled out on a plane, the development of the helix will be a straight line, with slope equal to $\tan v$.

Fig. 2.1.103 Helix.

DIFFERENTIAL AND INTEGRAL CALCULUS

Derivatives and Differentials

Derivatives and Differentials A **function** of a single variable x may be denoted by $f(x)$, $F(x)$, etc. The value of the function when x has the value x_0 is then denoted by $f(x_0)$, $F(x_0)$, etc. The **derivative** of a function $y = f(x)$ may be denoted by $f'(x)$, or by dy/dx. The value of the derivative at a given point $x = x_0$ is the **rate of change** of the function at that point; or, if the function is represented by a curve in the usual way (Fig. 2.1.104), the value of the derivative at any point shows the **slope of the curve** (i.e., the slope of the tangent to the curve) at that point (positive if the tangent points upward, and negative if it points downward, moving to the right).

Fig. 2.1.104 Curve showing tangent and derivatives.

The **increment** Δy (read: "delta y") in y is the change produced in y by increasing x from x_0 to $x_0 + \Delta x$; i.e., $\Delta y = f(x_0 + \Delta x) - f(x_0)$. The **differential**, dy, of y is the value which Δy would have if the curve coincided with its tangent. (The differential, dx, of x is the same as Δx when x is the independent variable.) Note that the derivative depends only on the value of x_0, while Δy and dy depend not only on x_0 but on the value of Δx as well. The ratio $\Delta y/\Delta x$ represents the secant slope, and dy/dx the slope of tangent (see Fig. 2.1.104). If Δx is made to approach zero, the secant approaches the tangent as a limiting position, so that the derivative is

$$f'(x) = \frac{dy}{dx} = \lim_{\Delta x \to 0}\left[\frac{\Delta y}{\Delta x}\right] = \lim_{\Delta x \to 0}\left[\frac{f(x_0 + \Delta x) - f(x_0)}{\Delta x}\right]$$

Also, $dy = f'(x)\,dx$.

The symbol "lim" in connection with $\Delta x \to 0$ means "the limit, as Δx approaches 0, of ...." (A constant c is said to be the **limit** of a variable u if, whenever any quantity m has been assigned, there is a stage in the variation process beyond which $|c - u|$ is always less than m; or, briefly, c is the limit of u if the difference between c and u can be made to become and remain as small as we please.)

To find the derivative of a given function at a given point: (1) If the function is given only by a curve, measure graphically the slope of the tangent at the point in question; (2) if the function is given by a mathematical expression, use the following rules for differentiation. These rules give, directly, the differential, dy, in terms of dx; to find the derivative, dy/dx, divide through by dx.

Rules for Differentiation (Here u, v, w, ... represent any functions of a variable x, or may themselves be independent variables. a is a constant which does not change in value in the same discussion; $e = 2.71828$.)

1. $d(a + u) = du$
2. $d(au) = a\,du$
3. $d(u + v + w + \cdots) = du + dv + dw + \cdots$
4. $d(uv) = u\,dv + v\,du$

5. $d(uvw\ldots) = (uvw\ldots)\left(\dfrac{du}{u} + \dfrac{dv}{v} + \dfrac{dw}{w} + \cdots\right)$

6. $d\dfrac{u}{v} = \dfrac{v\,du - u\,dv}{v^2}$

7. $d(u^m) = mu^{m-1}\,du$. Thus, $d(u^2) = 2u\,du$; $d(u^3) = 3u^2\,du$; etc.

8. $d\sqrt{u} = \dfrac{du}{2\sqrt{u}}$

9. $d\left(\dfrac{1}{u}\right) = -\dfrac{du}{u^2}$

10. $d(e^u) = e^u\,du$
11. $d(a^u) = (\ln a)a^u\,du$

12. $d\ln u = \dfrac{du}{u}$

13. $d\log_{10} u = \log_{10} e\,\dfrac{du}{u} = (0.4343\ldots)\dfrac{du}{u}$

14. $d\sin u = \cos u\,du$
15. $d\csc u = -\cot u\csc u\,du$
16. $d\cos u = -\sin u\,du$
17. $d\sec u = \tan u\sec u\,du$
18. $d\tan u = \sec^2 u\,du$
19. $d\cot u = -\csc^2 u\,du$

20. $d\sin^{-1} u = \dfrac{du}{\sqrt{1 - u^2}}$

21. $d\csc^{-1} u = -\dfrac{du}{u\sqrt{u^2 - 1}}$

22. $d\cos^{-1} u = \dfrac{du}{\sqrt{1 - u^2}}$

23. $d\sec^{-1} u = \dfrac{du}{u\sqrt{u^2 - 1}}$

24. $d\tan^{-1} u = \dfrac{du}{1 + u^2}$

25. $d\cot^{-1} u = -\dfrac{du}{1 + u^2}$

26. $d\ln\sin u = \cot u\,du$

27. $d\ln\tan u = \dfrac{2\,du}{\sin 2u}$

28. $d\ln\cos u = -\tan u\,du$

29. $d\ln\cot u = -\dfrac{2\,du}{\sin 2u}$

30. $d\sinh u = \cosh u\,du$
31. $d\csch u = -\csch u\coth u\,du$
32. $d\cosh u = \sinh u\,du$
33. $d\sech u = -\sech u\tanh u\,du$

34. $d \tanh u = \mathrm{sech}^2\, u\; du$

35. $d \coth u = -\mathrm{csch}^2\, u\; du$

36. $d \sinh^{-1} u = \dfrac{du}{\sqrt{u^2 + 1}}$

37. $d \,\mathrm{csch}^{-1} u = -\dfrac{du}{u\sqrt{u^2 + 1}}$

38. $d \cosh^{-1} u = \dfrac{du}{\sqrt{u^2 - 1}}$

39. $d \,\mathrm{sech}^{-1} u = -\dfrac{du}{u\sqrt{1 - u^2}}$

40. $d \tanh^{-1} u = \dfrac{du}{1 - u^2}$

41. $d \coth^{-1} u = \dfrac{du}{1 - u^2}$

42. $d(u^v) = (u^{v-1})(u \ln u\; dv + v\; du)$

Derivatives of Higher Orders The derivative of the derivative is called the second derivative; the derivative of this, the third derivative; and so on. If $y = f(x)$,

$$f'(x) = D_x y = \frac{dy}{dx}$$

$$f''(x) = D_x^2 y = \frac{d^2 y}{dx^2}$$

$$f'''(x) = D_x^3 y = \frac{d^3 y}{dx^3} \quad \text{etc.}$$

NOTE. If the notation $d^2 y/dx^2$ is used, this must not be treated as a fraction, like dy/dx, but as an inseparable symbol, made up of a symbol of operation d^2/dx^2, and an operand y.

The geometric meaning of the second derivative is this: if the original function $y = f(x)$ is represented by a curve in the usual way, then at any point where $f''(x)$ is *positive*, the curve is *concave upward*, and at any point where $f''(x)$ is *negative*, the curve is *concave downward* (Fig. 2.1.105). When $f''(x) = 0$, the curve usually has a **point of inflection.**

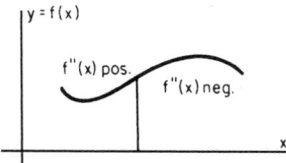

Fig. 2.1.105 Curve showing concavity.

Functions of two or more variables may be denoted by $f(x, y, \ldots)$, $F(x, y, \ldots)$, etc. The derivative of such a function $u = f(x, y, \ldots)$ formed on the assumption that x is the only variable ($y, \ldots$ being regarded for the moment as constants) is called the **partial derivative of** u **with respect to** x, and is denoted by $f_x(x, y)$ or $D_x u$, or $d_x u/dx$, or $\partial u/\partial x$. Similarly, the partial derivative of u with respect to y is $f_y(x, y)$ or $D_y u$, or $d_y u/dy$, or $\partial u/\partial y$.

NOTE. In the third notation, $d_x u$ denotes the differential of u formed on the assumption that x is the only variable. If the fourth notation, $\partial u/\partial x$, is used, this must not be treated as a fraction like du/dx; the $\partial/\partial x$ is a symbol of operation, operating on u, and the "∂x" must not be separated.

Partial derivatives of the second order are denoted by f_{xx}, f_{xy}, f_{yy}, or by $D_x u, D_x(D_y u), D_y^2 u$, or by $\partial^2 u/\partial x^2, \partial^2 u/\partial x\, \partial y, \partial^2 u/\partial y^2$, the last symbols being "inseparable." Similarly for higher derivatives. Note that $f_{xy} = f_{yx}$.

If increments Δx, Δy (or dx, dy) are assigned to the independent variables x, y, the increment, Δu, produced in $u = f(x, y)$ is

$$\Delta u = f(x + \Delta x, y + \Delta y) - f(x, y)$$

while the **differential,** du, i.e., the value which Δu would have if the partial derivatives of u with respect to x and y were constant, is given by

$$du = (f_x) \cdot dx + (f_y) \cdot dy$$

Here the coefficients of dx and dy are the values of the partial derivatives of u at the point in question.

If x and y are functions of a third variable t, then the equation

$$\frac{du}{dt} = (f_x)\frac{dx}{dt} + (f_y)\frac{dy}{dt}$$

expresses the rate of change of u with respect to t, in terms of the separate rate of change of x and y with respect to t.

Implicit Functions If $f(x, y) = 0$, either of the variables x and y is said to be an implicit function of the other. To find dy/dx, either (1) solve for y in terms of x, and then find dy/dx directly; or (2) differentiate the equation through as it stands, remembering that both x and y are variables, and then divide by dx; or (3) use the formula $dy/dx = -(f_x/f_y)$, where f_x and f_y are the partial derivatives of $f(x, y)$ at the point in question.

Maxima and Minima

A **function of one variable**, as $y = f(x)$, is said to have a **maximum** at a point $x = x_0$, if at that point the slope of the curve is zero and the concavity downward (see Fig. 2.1.106); a sufficient condition for a maximum is $f'(x_0) = 0$ and $f''(x_0)$ negative. Similarly, $f(x)$ has a **minimum** if the slope is zero and the concavity upward; a sufficient condition for a minimum is $f'(x_0) = 0$ and $f''(x_0)$ positive. If $f''(x_0) = 0$ and $f'''(x_0) \neq 0$, the point x_0 will be a **point of inflection.** If $f'(x_0) = 0$ and $f''(x_0) = 0$ and $f'''(x_0) = 0$, the point x_0 will be a maximum if $f''''(x_0) < 0$, and a minimum if $f''''(x_0) > 0$. It is usually sufficient, however, in any practical case, to find the values of x which make $f'(x) = 0$, and then decide, from a general knowledge of the curve or the sign of $f'(x)$ to the right and left of x_0, which of these values (if any) give maxima or minima, without investigating the higher derivatives.

Fig. 2.1.106 Curve showing maxima and minima.

A **function of two variables**, as $u = f(x, y)$, will have a **maximum** at a point (x_0, y_0) if at that point $f_x = 0, f_y = 0$, and $f_{xx} < 0, f_{yy} < 0$; and a **minimum** if at that point $f_x = 0, f_y = 0$, and $f_{xx} > 0, f_{yy} > 0$; provided, in each case, $(f_{xx})(f_{yy}) - (f_{xy})^2$ is positive. If $f_x = 0$ and $f_y = 0$, and f_{xx} and f_{yy} have opposite signs, the point (x_0, y_0) will be a "saddle point" of the surface representing the function.

Indeterminate Forms

In the following paragraphs, $f(x)$, $g(x)$ denote functions which approach 0; $F(x)$, $G(x)$ functions which increase indefinitely; and $U(x)$ a function which approaches 1, when x approaches a definite quantity a. The problem in each case is to find the limit approached by certain combinations of these functions when x approaches a. The symbol $\rightarrow$ is to be read "approaches" or "tends to."

CASE 1. "0/0." To find the limit of $f(x)/g(x)$ when $f(x) \rightarrow 0$ and $g(x) \rightarrow 0$, use the theorem that $\lim [f(x)/g(x)] = \lim [f'(x)/g'(x)]$, where $f'(x)$ and $g'(x)$ are the derivatives of $f(x)$ and $g(x)$. This second limit may be easier to find than the first. If $f'(x) \to 0$ and $g'(x) \to 0$, apply the same theorem a second time: $\lim [f'(x)/g'(x)] = \lim [f''(x)/g''(x)]$, and so on.

CASE 2.   "∞/∞." If $F(x) \to \infty$ and $G(x) \to \infty$, then $\lim [F(x)/G(x)] = \lim [F'(x)/G'(x)]$, precisely as in Case 1.

CASE 3.   "$0 \cdot \infty$." To find the limit of $f(x) \cdot F(x)$ when $f(x) \to 0$ and $F(x) \to \infty$, write $\lim [f(x) \cdot F(x)] = \lim\{f(x)/[1/F(x)]\}$ or $= \lim \{F(x)/[1/f(x)]\}$, then proceed as in Case 1 or Case 2.

CASE 4.   The limit of combinations "0^0" or $[f(x)]^{g(x)}$; "1^∞" or $[U(x)]^{F(x)}$; "∞^0" or $[F(x)]^{b(x)}$ may be found since their logarithms are limits of the type evaluated in Case 3.

CASE 5.   "$\infty - \infty$." If $F(x) \to \infty$ and $G(x) \to \infty$, write

$$\lim [F(x) - G(x)] = \lim \frac{1/G(x) - 1/F(x)}{1/[F(x) \cdot G(x)]}$$

then proceed as in Case 1. Sometimes it is shorter to expand the functions in series. It should be carefully noticed that expressions like 0/0, ∞/∞, etc., do not represent mathematical quantities.

Curvature

The **radius of curvature** R of a plane curve at any point P (Fig. 2.1.107) is the distance, measured along the normal, on the concave side of the curve, to the **center of curvature**, C, this point being the limiting position of the point of intersection of the normals at P and a neighboring point Q, as Q is made to approach P along the curve. If the equation of the curve is $y = f(x)$,

$$R = \frac{ds}{du} = \frac{[1 + (y')^2]^{3/2}}{y''}$$

where $ds = \sqrt{dx^2 + dy^2}$ = the differential of arc, $u = \tan^{-1}[f'(x)]$ = the angle which the tangent at P makes with the x axis, and $y' = f'(x)$ and $y'' = f''(x)$ are the first and second derivatives of $f(x)$ at the point P. Note that $dx = ds \cos u$ and $dy = ds \sin u$. The **curvature**, K, at the point P, is $K = 1/R = du/ds$; i.e., the curvature is the rate at which the angle u is changing with respect to the length of arc s. If the slope of the curve is small, $K \approx f''(x)$.

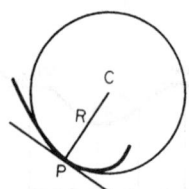

Fig. 2.1.107   Curve showing radius of curvature.

If the equation of the curve in polar coordinates is $r = f(\theta)$, where r = radius vector and θ = polar angle, then

$$R = \frac{[r^2 + (r')^2]^{3/2}}{r^2 - rr'' + 2(r')^2}$$

where $r' = f'(\theta)$ and $r'' = f''(\theta)$.

The **evolute** of a curve is the locus of its centers of curvature. If one curve is the evolute of another, the second is called the **involute** of the first.

Indefinite Integrals

An **integral** of $f(x) \, dx$ is any function whose differential is $f(x) \, dx$, and is denoted by $\int f(x) \, dx$. All the integrals of $f(x) \, dx$ are included in the expression $\int f(x) \, dx + C$, where $\int f(x) \, dx$ is any particular integral, and C is an arbitrary constant. The process of finding (when possible) an integral of a given function consists in recognizing by inspection a function which, when differentiated, will produce the given function; or in transforming the given function into a form in which such recognition

is easy. The most common integrable forms are collected in the following brief table; for a more extended list, see Peirce, "Table of Integrals," Ginn, or Dwight, "Table of Integrals and other Mathematical Data," Macmillan, or "CRC Mathematical Tables."

GENERAL FORMULAS

1. $\int a \, du = a \int du = au + C$

2. $\int (u + v) \, dx = \int u \, dx + \int v \, dx$

3. $\int u \, dv = uv - \int v \, du$ (integration by parts)

4. $\int f(x) \, dx = \int f[F(y)]F'(y) \, dy, x = F(y)$

(change of variables)

5. $\int dy \int f(x, y) \, dx = \int dx \int f(x, y) \, dy$

FUNDAMENTAL INTEGRALS

6. $\int x^n \, dx = \dfrac{x^{n+1}}{n + 1} + C$, when $n \neq -1$

7. $\int \dfrac{dx}{x} = \ln x + C = \ln cx$

8. $\int e^x \, dx = e^x + C$

9. $\int \sin x \, dx = -\cos x + C$

10. $\int \cos x \, dx = \sin x + C$

11. $\int \dfrac{dx}{\sin^2 x} = \int \csc^2 x \, dx = -\cot x + C$

12. $\int \dfrac{dx}{\cos^2 x} = \int \sec^2 x \, dx = \tan x + C$

13. $\int \dfrac{dx}{\sqrt{1 - x^2}} = \sin^{-1} x + C = -\cos^{-1} x + C$

14. $\int \dfrac{dx}{1 + x^2} = \tan^{-1} x + C = -\cot^{-1} x + C$

RATIONAL FUNCTIONS

15. $\int (a + bx)^n \, dx = \dfrac{(a + bx)^{n+1}}{(n + 1)b} + C$

16. $\int \dfrac{dx}{a + bx} = \dfrac{1}{b} \ln (a + bx) + C = \dfrac{1}{b} \ln c(a + bx)$

17. $\int \dfrac{dx}{x^n} = -\dfrac{1}{(n - 1)x^{n-1}} + C$ except when $n = 1$

18. $\int \dfrac{dx}{(a + bx)^2} = -\dfrac{1}{b(a + bx)} + C$

19. $\int \dfrac{dx}{1 - x^2} = \frac{1}{2} \ln \dfrac{1 + x}{1 - x} + C = \tanh^{-1} x + C$, when $x < 1$

20. $\int \dfrac{dx}{x^2 - 1} = \frac{1}{2} \ln \dfrac{x - 1}{x + 1} + C = -\coth^{-1} x + C$, when $x > 1$

21. $\int \dfrac{dx}{a + bx^2} = \dfrac{1}{\sqrt{ab}} \tan^{-1}\left(\sqrt{\dfrac{b}{a}}x\right) + C$

22. $\int \dfrac{dx}{a - bx^2} = \dfrac{1}{2\sqrt{ab}} \ln \dfrac{\sqrt{ab} + bx}{\sqrt{ab} - bx} + C$ $[a > 0, b > 0]$

$\qquad\qquad = \dfrac{1}{\sqrt{ab}} \tanh^{-1}\left(\sqrt{\dfrac{b}{a}}x\right) + C$

23. $\int \dfrac{dx}{a + 2bx + cx^2} =$

$\left. \begin{array}{l} \dfrac{1}{\sqrt{ac - b^2}} \tan^{-1} \dfrac{b + cx}{\sqrt{ac - b^2}} + C \end{array} \right\}$ $[ac - b^2 > 0]$

$\left. \begin{array}{l} = \dfrac{1}{2\sqrt{b^2 - ac}} \ln \dfrac{\sqrt{b^2 - ac} - b - cx}{\sqrt{b^2 - ac} + b + cx} + C \\[2mm] = -\dfrac{1}{\sqrt{b^2 - ac}} \tanh^{-1} \dfrac{b + cx}{\sqrt{b^2 - ac}} + C \end{array} \right\}$ $[b^2 - ac > 0]$

24. $\int \dfrac{dx}{a + 2bx + cx^2} = -\dfrac{1}{b + cx} + C$, when $b^2 = ac$

25. $\int \dfrac{(m + nx)\, dx}{a + 2bx + cx^2} = \dfrac{n}{2c} \ln (a + 2bx + cx^2)$

$+ \dfrac{mc - nb}{c} \int \dfrac{dx}{a + 2bx + cx^2}$

26. In $\int \dfrac{f(x)\, dx}{a + 2bx + cx^2}$, if $f(x)$ is a polynomial of higher than the first degree, divide by the denominator before integrating

27. $\int \dfrac{dx}{(a + 2bx + cx^2)^p} = \dfrac{1}{2(ac - b^2)(p - 1)}$

$\times \dfrac{b + cx}{(a + 2bx + cx^2)^{p-1}}$

$+ \dfrac{(2p - 3)c}{2(ac - b^2)(p - 1)} \int \dfrac{dx}{(a + 2bx + cx^2)^{p-1}}$

28. $\int \dfrac{(m + nx)\, dx}{(a + 2bx + cx^2)^p} = -\dfrac{n}{2c(p - 1)} \times$

$\dfrac{1}{(a + 2bx + cx^2)^{p-1}} + \dfrac{mc - nb}{c} \int \dfrac{dx}{(a + 2bx + cx^2)^p}$

29. $\int x^{m-1}(a + bx)^n\, dx = \dfrac{x^{m-1}(a + bx)^{n+1}}{(m + n)b}$

$- \dfrac{(m - 1)a}{(m + n)b} \int x^{m-2}(a + bx)^n\, dx$

$= \dfrac{x^m(a + bx)^n}{m + n} + \dfrac{na}{m + n} \int x^{m-1}(a + bx)^{n-1}\, dx$

IRRATIONAL FUNCTIONS

30. $\int \sqrt{a + bx}\, dx = \dfrac{2}{3b}(\sqrt{a + bx})^3 + C$

31. $\int \dfrac{dx}{\sqrt{a + bx}} = \dfrac{2}{b}\sqrt{a + bx} + C$

32. $\int \dfrac{(m + nx)\, dx}{\sqrt{a + bx}} = \dfrac{2}{3b^2}(3mb - 2an + nbx)\sqrt{a + bx} + C$

33. $\int \dfrac{dx}{(m + nx)\sqrt{a + bx}}$; substitute $y = \sqrt{a + bx}$, and use 21 and 22

34. $\int \dfrac{f(x, \sqrt[n]{a + bx})}{F(x, \sqrt[n]{a + bx})}\, dx$; substitute $\sqrt[n]{a + bx} = y$

35. $\int \dfrac{dx}{\sqrt{a^2 - x^2}} = \sin^{-1}\dfrac{x}{a} + C = -\cos^{-1}\dfrac{x}{a} + C$

36. $\int \dfrac{dx}{\sqrt{a^2 + x^2}} = \ln(x + \sqrt{a^2 + x^2}) + C = \sinh^{-1}\dfrac{x}{a} + C$

37. $\int \dfrac{dx}{\sqrt{x^2 - a^2}} = \ln(x + \sqrt{x^2 - a^2}) + C = \cosh^{-1}\dfrac{x}{a} + C$

38. $\int \dfrac{dx}{\sqrt{a + 2bx + cx^2}}$

$= \dfrac{1}{\sqrt{c}} \ln(b + cx + \sqrt{c}\sqrt{a + 2bx + cx^2}) + C$, where $c > 0$

$= \dfrac{1}{\sqrt{c}} \sinh^{-1} \dfrac{b + cx}{\sqrt{ac - b^2}} + C$, when $ac - b^2 > 0$

$= \dfrac{1}{\sqrt{c}} \cosh^{-1} \dfrac{b + cx}{\sqrt{b^2 - ac}} + C$, when $b^2 - ac > 0$

$= \dfrac{-1}{\sqrt{-c}} \sin^{-1} \dfrac{b + cx}{\sqrt{b^2 - ac}} + C$, when $c < 0$

39. $\int \dfrac{(m + nx)\, dx}{\sqrt{a + 2bx + cx^2}} = \dfrac{n}{c}\sqrt{a + 2bx + cx^2}$

$+ \dfrac{mc - nb}{c} \int \dfrac{dx}{\sqrt{a + 2bx + cx^2}}$

40. $\int \dfrac{x^m\, dx}{\sqrt{a + 2bx + cx^2}} = \dfrac{x^{m-1}X}{mc} - \dfrac{(m - 1)a}{mc} \int \dfrac{x^{m-2}\, dx}{X}$

$- \dfrac{(2m - 1)b}{mc} \int \dfrac{x^{m-1}}{X}\, dx$ when $X = \sqrt{a + 2bx + cx^2}$

41. $\int \sqrt{a^2 + x^2}\, dx = \dfrac{x}{2}\sqrt{a^2 + x^2} + \dfrac{a^2}{2}\ln(x + \sqrt{a^2 + x^2}) + C$

$= \dfrac{x}{2}\sqrt{a^2 + x^2} + \dfrac{a^2}{2}\sinh^{-1}\dfrac{x}{a} + C$

42. $\int \sqrt{a^2 - x^2}\, dx = \dfrac{x}{2}\sqrt{a^2 - x^2} + \dfrac{a^2}{2}\sin^{-1}\dfrac{x}{a} + C$

43. $\int \sqrt{x^2 - a^2}\, dx = \dfrac{x}{2}\sqrt{x^2 - a^2} - \dfrac{a^2}{2}\ln(x + \sqrt{x^2 - a^2}) + C$

$= \dfrac{x}{2}\sqrt{x^2 - a^2} - \dfrac{a^2}{2}\cosh^{-1}\dfrac{x}{a} + C$

44. $\int \sqrt{a + 2bx + cx^2}\, dx = \dfrac{b + cx}{2c}\sqrt{a + 2bx + cx^2}$

$+ \dfrac{ac - b^2}{2c} \int \dfrac{dx}{\sqrt{a + 2bx + cx^2}} + C$

TRANSCENDENTAL FUNCTIONS

45. $\int a^x\, dx = \dfrac{a^x}{\ln a} + C$

46. $\int x^n e^{ax}\, dx$

$= \dfrac{x^n e^{ax}}{a}\left[1 - \dfrac{n}{ax} + \dfrac{n(n - 1)}{a^2 x^2} - \cdots \pm \dfrac{n!}{a^n x^n}\right] + C$

47. $\int \ln x\, dx = x \ln x - x + C$

48. $\int \dfrac{\ln x}{x^2}\, dx = -\dfrac{\ln x}{x} - \dfrac{1}{x} + C$

49. $\int \dfrac{(\ln x)^n}{x}\, dx = \dfrac{1}{n + 1}(\ln x)^{n+1} + C$

50. $\int \sin^2 x\, dx = -\frac{1}{4}\sin 2x + \frac{1}{2}x + C$

$= -\frac{1}{2}\sin x \cos x + \frac{1}{2}x + C$

51. $\int \cos^2 x\, dx = \frac{1}{4}\sin 2x + \frac{1}{2}x + C$

$= \frac{1}{2}\sin x \cos x + \frac{1}{2}x + C$

52. $\int \sin mx\, dx = -\dfrac{\cos mx}{m} + C$

53. $\int \cos mx\, dx = \dfrac{\sin mx}{m} + C$

54. $\int \sin mx \cos nx\, dx = -\dfrac{\cos(m + n)x}{2(m + n)} - \dfrac{\cos(m - n)x}{2(m - n)} + C$

55. $\int \sin mx \sin nx\, dx = \dfrac{\sin(m - n)x}{2(m - n)} - \dfrac{\sin(m + n)x}{2(m + n)} + C$

56. $\int \cos mx \cos nx\, dx = \dfrac{\sin(m - n)x}{2(m - n)} + \dfrac{\sin(m + n)x}{2(m + n)} + C$

57. $\displaystyle\int \tan x \, dx = -\ln \cos x + C$

58. $\displaystyle\int \cot x \, dx = \ln \sin x + C$

59. $\displaystyle\int \frac{dx}{\sin x} = \ln \tan \frac{x}{2} + C$

60. $\displaystyle\int \frac{dx}{\cos x} = \ln \tan \left(\frac{\pi}{4} + \frac{x}{2} \right) + C$

61. $\displaystyle\int \frac{dx}{1 + \cos x} = \tan \frac{x}{2} + C$

62. $\displaystyle\int \frac{dx}{1 - \cos x} = -\cot \frac{x}{2} + C$

63. $\displaystyle\int \sin x \cos x \, dx = \frac{1}{2} \sin^2 x + C$

64. $\displaystyle\int \frac{dx}{\sin x \cos x} = \ln \tan x + C$

65.* $\displaystyle\int \sin^n x \, dx = -\frac{\cos x \sin^{n-1} x}{n} + \frac{n-1}{n} \int \sin^{n-2} x \, dx$

66.* $\displaystyle\int \cos^n x \, dx = \frac{\sin x \cos^{n-1} x}{n} + \frac{n-1}{n} \int \cos^{n-2} x \, dx$

67. $\displaystyle\int \tan^n x \, dx = \frac{\tan^{n-1} x}{n-1} - \int \tan^{n-2} x \, dx$

68. $\displaystyle\int \cot^n x \, dx = -\frac{\cot^{n-1} x}{n-1} - \int \cot^{n-2} x \, dx$

69. $\displaystyle\int \frac{dx}{\sin^n x} = -\frac{\cos x}{(n-1)\sin^{n-1} x} + \frac{n-2}{n-1} \int \frac{dx}{\sin^{n-2} x}$

70. $\displaystyle\int \frac{dx}{\cos^n x} = \frac{\sin x}{(n-1)\cos^{n-1} x} + \frac{n-2}{n-1} \int \frac{dx}{\cos^{n-2} x}$

71.† $\displaystyle\int \sin^p x \cos^q x \, dx = \frac{\sin^{p+1} x \cos^{q-1} x}{p+q}$

$\displaystyle + \frac{q-1}{p+q} \int \sin^p x \cos^{q-2} x \, dx = -\frac{\sin^{p-1} x \cos^{q+1} x}{p+q}$

$\displaystyle + \frac{p-1}{p+q} \int \sin^{p-2} x \cos^q x \, dx$

72.† $\displaystyle\int \sin^{-p} x \cos^q x \, dx = -\frac{\sin^{-p+1} x \cos^{q+1} x}{p-1}$

$\displaystyle + \frac{p-q-2}{p-1} \int \sin^{-p+2} x \cos^q x \, dx$

73.† $\displaystyle\int \sin^p x \cos^{-q} x \, dx = \frac{\sin^{p+1} x \cos^{-q+1} x}{q-1}$

$\displaystyle + \frac{q-p-2}{q-1} \int \sin^p x \cos^{-q+2} x \, dx$

74. $\displaystyle\int \frac{dx}{a + b \cos x} = \frac{2}{\sqrt{a^2 - b^2}} \tan^{-1} \left(\sqrt{\frac{a-b}{a+b}} \tan \frac{1}{2}x \right) + C,$

when $a^2 > b^2$,

$\displaystyle = \frac{1}{\sqrt{b^2 - a^2}} \ln \frac{b + a \cos x + \sin x \sqrt{b^2 - a^2}}{a + b \cos x} + C,$

when $a^2 < b^2$,

$\displaystyle = \frac{2}{\sqrt{b^2 - a^2}} \tanh^{-1} \left(\sqrt{\frac{b-a}{b+a}} \tan \frac{1}{2}x \right) + C,$ when $a^2 < b^2$

75. $\displaystyle\int \frac{\cos x \, dx}{a + b \cos x} = \frac{x}{b} - \frac{a}{b} \int \frac{dx}{a + b \cos x} + C$

76. $\displaystyle\int \frac{\sin x \, dx}{a + b \cos x} = -\frac{1}{b} \ln (a + b \cos x) + C$

---

* If n is an odd number, substitute $\cos x = z$ or $\sin x = z$.

† If p or q is an odd number, substitute $\cos x = z$ or $\sin x = z$.

---

77. $\displaystyle\int \frac{A + B \cos x + C \sin x}{a + b \cos x + c \sin x} dx = A \int \frac{dy}{a + p \cos y}$

$\displaystyle + (B \cos u + C \sin u) \int \frac{\cos y \, dy}{a + p \cos y}$

$\displaystyle - (B \sin u - C \cos u) \int \frac{\sin y \, dy}{a + p \cos y},$ where $b = p \cos u, c = p$

$\sin u$ and $x - u = y$

78. $\displaystyle\int e^{ax} \sin bx \, dx = \frac{a \sin bx - b \cos bx}{a^2 + b^2} e^{ax} + C$

79. $\displaystyle\int e^{ax} \cos bx \, dx = \frac{a \cos bx + b \sin bx}{a^2 + b^2} e^{ax} + C$

80. $\displaystyle\int \sin^{-1} x \, dx = x \sin^{-1} x + \sqrt{1 - x^2} + C$

81. $\displaystyle\int \cos^{-1} x \, dx = x \cos^{-1} x - \sqrt{1 - x^2} + C$

82. $\displaystyle\int \tan^{-1} x \, dx = x \tan^{-1} x - \frac{1}{2} \ln (1 + x^2) + C$

83. $\displaystyle\int \cot^{-1} x \, dx = x \cot^{-1} x + \frac{1}{2} \ln (1 + x^2) + C$

84. $\displaystyle\int \sinh x \, dx = \cosh x + C$

85. $\displaystyle\int \tanh x \, dx = \ln \cosh x + C$

86. $\displaystyle\int \cosh x \, dx = \sinh x + C$

87. $\displaystyle\int \coth x \, dx = \ln \sinh x + C$

88. $\displaystyle\int \text{sech} \, x \, dx = 2 \tan^{-1}(e^x) + C$

89. $\displaystyle\int \text{csch} \, x \, dx = \ln \tanh (x/2) + C$

90. $\displaystyle\int \sinh^2 x \, dx = \frac{1}{2} \sinh x \cosh x - \frac{1}{2}x + C$

91. $\displaystyle\int \cosh^2 x \, dx = \frac{1}{2} \sinh x \cosh x + \frac{1}{2}x + C$

92. $\displaystyle\int \text{sech}^2 x \, dx = \tanh x + C$

93. $\displaystyle\int \text{csch}^2 x \, dx = -\coth x + C$

Hints on Using Integral Tables   It happens with frustrating frequency that no integral table lists the integral that needs to be evaluated. When this happens, one may (a) seek a more complete integral table, (b) appeal to mathematical software, such as Mathematica, Maple, MathCad or Derive, (c) use numerical or approximate methods, such as Simpson's rule (see section "Numerical Methods"), or (d) attempt to transform the integral into one which may be evaluated. Some hints on such transformation follow. For a more complete list and more complete explanations, consult a calculus text, such as Thomas, "Calculus and Analytic Geometry," Addison-Wesley, or Anton, "Calculus with Analytic Geometry," Wiley. One or more of the following "tricks" may be successful.

TRIGONOMETRIC SUBSTITUTIONS

1. If an integrand contains $\sqrt{(a^2 - x^2)}$, substitute $x = a \sin u$, and $\sqrt{(a^2 - x^2)} = a \cos u$.

2. Substitute $x = a \tan u$ and $\sqrt{(x^2 + a^2)} = a \sec u$.

3. Substitute $x = a \sec u$ and $\sqrt{(x^2 - a^2)} = a \tan u$.

COMPLETING THE SQUARE

4. Rewrite $ax^2 + bx + c = a[x + b/(2a)]^2 + (4ac - b^2)/(4a)$; then substitute $u = x + b/(2a)$ and $B = (4ac - b^2)/(4a)$.

PARTIAL FRACTIONS

5. For a ratio of polynomials, where the denominator has been completely factored into linear factors $p_i(x)$ and quadratic factors $q_j(x)$, and where the degree of the numerator is less than the degree of the denominator, then rewrite $r(x)/[p_1(x) \ldots p_n(x)q_1(x) \ldots q_m(x)] = A_1/p_1(x) + \cdots + A_n/p_n(x) + (B_1x + C_1)/q_1(x) + \cdots + (B_mx + C_m)/q_m(x)$.

INTEGRATION BY PARTS

6. Change the integral using the formula

$$\int u \, dv = uv - \int v \, du$$

where u and dv are chosen so that (a) v is easy to find from dv, and (b) $v \, du$ is easier to find than $u \, dv$.

Kasube suggests ("A Technique for Integration by Parts," *Am. Math. Month.*, vol. 90, no. 3, Mar. 1983): Choose u in the order of preference LIATE, that is, Logarithmic, Inverse trigonometric, Algebraic, Trigonometric, Exponential.

EXAMPLE. Find $\int x \ln x \, dx$. The logarithmic $\ln x$ has higher priority than does the algebraic x, so let $u = \ln(x)$ and $dv = x \, dx$. Then $du = (1/x) \, dx$; $v = x^2/2$, so $\int x \ln x \, dx = uv - \int v \, du = (x^2/2) \ln x - \int (x^2/2)(1/x) \, dx = (x^2/2) \ln x - \int x/2 \, dx = (x^2/2) \ln x - x^2/4 + C$.

Definite Integrals The definite integral of $f(x) \, dx$ from $x = a$ to $x = b$, denoted by $\int_a^b f(x) \, dx$, is the limit (as n increases indefinitely) of a sum of n terms:

$$\int_a^b f(x) \, dx = \lim_{n \to \infty} [f(x_1) \, \Delta x + f(x_2) \, \Delta x + f(x_3) \, \Delta x + \cdots + f(x_n)\Delta x]$$

built up as follows: Divide the interval from a to b into n equal parts, and call each part $\Delta x = (b - a)/n$; in each of these intervals take a value of x (say, $x_1, x_2, \ldots, x_n$), find the value of the function $f(x)$ at each of these points, and multiply it by Δx, the width of the interval; then take the limit of the sum of the terms thus formed, when the number of terms increases indefinitely, while each individual term approaches zero.

Geometrically, $\int_a^b f(x) \, dx$ is the area bounded by the curve $y = f(x)$, the x axis, and the ordinates $x = a$ and $x = b$ (Fig. 2.1.108); i.e., briefly, the "area under the curve, from a to b." The **fundamental theorem** for the evaluation of a definite integral is the following:

$$\int_a^b f(x) \, dx = \left[\int f(x) \, dx \right]_{x=b} - \left[\int f(x) \, dx \right]_{x=a}$$

i.e., the definite integral is equal to the difference between two values of any one of the indefinite integrals of the function in question. In other words, the limit of a sum can be found whenever the function can be integrated.

Fig. 2.1.108 Graph showing areas to be summed during integration.

Properties of Definite Integrals

$$\int_a^b = -\int_b^a; \qquad \int_a^c + \int_c^b = \int_a^b$$

MEAN-VALUE THEOREM FOR INTEGRALS

$$\int_a^b F(x)f(x) \, dx = F(X) \int_a^b f(x) \, dx$$

provided $f(x)$ does not change sign from $x = a$ to $x = b$; here X is some (unknown) value of x intermediate between a and b.

MEAN VALUE. The **mean value** of $f(x)$ with respect to x, between a and b, is

$$\bar{f} = \frac{1}{b - a} \int_a^b f(x) \, dx$$

THEOREM ON CHANGE OF VARIABLE. In evaluating $\int_{x=a}^{x=b} f(x) \, dx$, $f(x) \, dx$ may be replaced by its value in terms of a new variable t and dt, and $x = a$ and $x = b$ by the corresponding values of t, provided that throughout the interval the relation between x and t is a one-to-one correspondence (i.e., to each value of x there corresponds one and only one value of t, and to each value of t there corresponds one and only one value of x). So $\int_{x=a}^{x=b} f(x) \, dx = \int_{t=g(a)}^{t=g(b)} f(g(t)) \, g'(t) \, dt$.

DIFFERENTIATION WITH RESPECT TO THE UPPER LIMIT. If b is variable, then $\int_a^b f(x) \, dx$ is a function b, whose derivative is

$$\frac{d}{db} \int_a^b f(x) \, dx = f(b)$$

DIFFERENTIATION WITH RESPECT TO A PARAMETER

$$\frac{\partial}{\partial c} \int_a^b f(x, c) \, dx = \int_a^b \frac{\partial f(x, c)}{\partial c} \, dx$$

Functions Defined by Definite Integrals The following definite integrals have received special names:

1. Elliptic integral of the first kind $= F(u, k) = \int_0^u \frac{dx}{\sqrt{1 - k^2 \sin^2 x}}$ when $k^2 < 1$.

2. Elliptic integral of the second kind $= E(u, k) = \int_0^u \sqrt{1 - k^2 \sin^2 x} \, dx$, when $k^2 < 1$.

3, 4. Complete elliptic integrals of the first and second kinds; put $u = \pi/2$ in (1) and (2).

5. The probability integral $= \frac{2}{\sqrt{\pi}} \int_0^x e^{-x^2} \, dx$.

6. The gamma function $= \Gamma(n) = \int_0^\infty x^{n-1}e^{-x} \, dx$.

Approximate Methods of Integration. Mechanical Quadrature (See also section "Numerical Methods.")

1. Use Simpson's rule (see also Scarborough, "Numerical Mathematical Analyses," Johns Hopkins Press).

2. Expand the function in a converging power series, and integrate term by term.

3. Plot the area under the curve $y = f(x)$ from $x = a$ to $x = b$ on squared paper, and measure this area roughly by "counting squares."

Double Integrals The notation $\int\int f(x, y) \, dy \, dx$ means $\int[\int f(x, y) \, dy] \, dx$, the limits of integration in the inner, or first, integral being functions of x (or constants).

EXAMPLE. To find the weight of a plane area whose density, w, is variable, say $w = f(x, y)$. The weight of a typical element, $dx \, dy$, is $f(x, y) \, dx \, dy$. Keeping x and dx constant and summing these elements from, say, $y = F_1(x)$ to $y = F_2(x)$, as determined by the shape of the boundary (Fig. 2.1.109), the weight of a typical strip perpendicular to the x axis is

$$dx \int_{y=F_1(x)}^{y=F_2(x)} f(x, y) \, dy$$

Finally, summing these strips from, say, $x = a$ to $x = b$, the weight of the whole area is

$$\int_{x=a}^{x=b} \left[dx \int_{y=F_1(x)}^{y=F_2(x)} f(x, y) \, dy \right] \text{ or, briefly, } \iint f(x, y) \, dy \, dx$$

Fig. 2.1.109 Graph showing areas to be summed during double integration.

Triple Integrals The notation $\iiint f(x, y, z) \, dz \, dy \, dx$ means

$$\iint \left\{ \int \left[\int f(x, y, z) \, dz \right] dy \right\} dx$$

Such integrals are known as **volume integrals.**

EXAMPLE. To find the mass of a volume which has variable density, say, $w = f(x, y, z)$. If the shape of the volume is described by $a < x < b$, $F_1(x) < y < F_2(x)$, and $G_1(x, y) < z < G_2(x, y)$, then the mass is given by

$$\int_a^b \int_{F_1(x)}^{F_2(x)} \int_{G_1(x,y)}^{G_2(x,y)} f(x, y, z) \, dz \, dy \, dx$$

SERIES AND SEQUENCES

Sequences

A **sequence** is an ordered list of numbers, $x_1, x_2, \ldots, x_n, \ldots$.

An infinite sequence is an infinitely long list. A sequence is often defined by a function $f(n)$, $n = 1, 2, \ldots$. The formula defining $f(n)$ is called the **general term** of the sequence.

The variable n is called the **index** of the sequence. Sometimes the index is taken to start with $n = 0$ instead of $n = 1$.

A sequence **converges** to a limit L if the general term $f(n)$ has limit L as n goes to infinity. If a sequence does not have a unique limit, the sequence is said to "diverge." There are two fundamental ways a function can diverge: (1) It may become infinitely large, in which case the sequence is said to be "unbounded," or (2) it may tend to alternate among two or more values, as in the sequence $x_n = (-1)^n$.

A sequence **alternates** if its odd-numbered terms are positive and its even-numbered terms are negative, or vice versa.

Series

A **series** is a sequence of sums. The terms of the sums are another sequence, $x_1, x_2, \ldots$. Then the series is the sequence defined by $s_n = x_1 + x_2 + \cdots + x_n = \sum_{i=1}^{n} x_i$. The sequence s_n is also called the **sequence of partial sums** of the series.

If the sequence of partial sums converges (resp. diverges), then the series is said to converge (resp. diverge). If the limit of a series is S, then the sequence defined by $r_n = S - s_n$ is called the "error sequence" or the "sequence of truncation errors."

Convergence of Series THEOREM. If a series $s_n = x_1 + x_2 + \cdots + x_n$ converges, then it is necessary (but not sufficient) that the sequence

x_n has limit zero. A series of partial sums of an alternating sequence is called an *alternating series.*

THEOREM. An alternating series converges whenever the sequence x_n has limit zero.

A series is a **geometric series** if its terms are of the form ar^n. The value r is called the **ratio** of the series. Usually, for geometric series, the index is taken to start with $n = 0$ instead of $n = 1$.

THEOREM. A geometric series with $x_n = ar^n$, $n = 0, 1, 2, \ldots$, converges if and only if $-1 < r < 1$, and then the limit of the series is $a/(1 - r)$. The partial sums of a geometric series are $s_n = a(1 - r^n)/(1 - r)$.

The series defined by the sequence $x_n = 1/n$, $n = 1, 2, \ldots$, is called the **harmonic series.** The harmonic series diverges.

A series with each term $x_n > 0$ is called a "positive series."

There are a number of tests to determine whether or not a positive series s_n converges.

1. *Comparison test.* If $c_1 + c_2 + \cdots + c_n$ is a positive series that converges, and if $0 < x_n < c_n$, then the series $x_1 + x_2 + \cdots + x_n$ also converges.

If $d_1 + d_2 + \cdots + d_n$ diverges and $x_n > d_n$, then $x_1 + x_2 + \cdots + x_n$ also diverges.

2. *Integral test.* If $f(t)$ is a strictly decreasing function and $f(n) = x_n$, then the series s_n and the integral $\int_1^{\infty} f(t) \, dt$ either both converge or both diverge.

3. *P test.* The series defined by $x_n = 1/n^p$ converges if $p > 1$ and diverges if $p = 1$ or $p < 1$. If $p = 1$, then this is the harmonic series.

4. *Ratio test.* If the limit of the sequence $x_{n+1}/x_n = r$, then the series diverges if $r > 1$, and it converges if $0 < r < 1$. The test is inconclusive if $r = 1$.

5. *Cauchy root test.* If L is the limit of the nth root of the nth term, $\lim x_n^{1/n}$, then the series converges if $L < 1$ and diverges if $L > 1$. If $L = 1$, then the test is inconclusive.

A **power series** is an expression of the form $a_0 + a_1 x + a_2 x^2 + \cdots + a_n x^n + \cdots$ or $\sum_{i=0}^{\infty} a_i x^i$.

The range of values of x for which a power series converges is the **interval of convergence** of the power series.

General Formulas of Maclaurin and Taylor If $f(x)$ and all its derivatives are continuous in the neighborhood of the point $x = 0$ (or $x = a$), then, for any value of x in this neighborhood, the function $f(x)$ may be expressed as a power series arranged according to ascending powers of x (or of $x - a$), as follows:

$$f(x) = f(0) + \frac{f'(0)}{1!} x + \frac{f''(0)}{2!} x^2 + \frac{f'''(0)}{3!} x^3 + \cdots$$
$$+ \frac{f^{(n-1)}(0)}{(n-1)!} x^{n-1} + (P_n) x^n \qquad \text{(Maclaurin)}$$

$$f(x) = f(a) + \frac{f'(a)}{1!} (x - a) + \frac{f''(a)}{2!} (x - a)^2 + \frac{f'''(a)}{3!} (x - a)^3 +$$
$$\cdots + \frac{f^{(n-1)}(a)}{(n-1)!} (x - a)^{n-1} + (Q_n)(x - a)^n \qquad \text{(Taylor)}$$

Here $(P_n)x^n$, or $(Q_n)(x - a)^n$, is called the **remainder term;** the values of the coefficients P_n and Q_n may be expressed as follows:

$$P_n = [f^{(n)}(sx)]/n! = [(1 - t)^{n-1} f^{(n)}(tx)]/(n - 1)!$$
$$Q_n = \{f^{(n)}[a + s(x - a)]\}/n!$$
$$= \{(1 - t)^{n-1} f^{(n)}[a + t(x - a)]\}/(n - 1)!$$

where s and t are certain unknown numbers between 0 and 1; the s form is due to Lagrange, the t form to Cauchy.

The error due to neglecting the remainder term is less than $(\overline{P}_n)x^n$, or $(\overline{Q}_n)(x - a)^n$, where $\overline{P}_n$, or $\overline{Q}_n$, is the largest value taken on by P_n, or

Q_n, when s or t ranges from 0 to 1. If this error, which depends on both n and x, approaches 0 as n increases (for any given value of x), then the general expression with remainder becomes (for that value of x) a convergent infinite series.

The sum of the first few terms of Maclaurin's series gives a good approximation to $f(x)$ for values of x near $x = 0$; Taylor's series gives a similar approximation for values near $x = a$.

The MacLaurin series of some important functions are given below.

Power series may be differentiated term by term, so the derivative of a power series $a_0 + a_1 x + a_2 x^2 + \cdots + a_n x^n$ is $a_1 + 2a_2 x + \cdots + na_n x^{n-1} \ldots$. The power series of the derivative has the same interval of convergence, except that the endpoints may or may not be included in the interval.

Series Expansions of Some Important Functions

The range of values of x for which each of the series is convergent is stated at the right of the series.

Geometrical Series

$$\frac{1}{1 - x} = \sum_{n=0}^{\infty} x^n \qquad -1 < x < 1$$

$$\frac{1}{(1 - x)^m} = 1 + \sum_{n=1}^{\infty} \frac{(m + n - 1)!}{(m - 1)! n!} x^n \qquad -1 < x < 1$$

Exponential and Logarithmic Series

$$e^x = 1 + \frac{x}{1!} + \frac{x^2}{2!} + \frac{x^3}{3!} + \frac{x^4}{4!} + \cdots \qquad [-\infty < x < +\infty]$$

$$a^x = e^{mx} = 1 + \frac{m}{1!} x + \frac{m^2}{2!} x^2 + \frac{m^3}{3!} x^3 + \cdots$$
$$[a > 0, -\infty < x < +\infty]$$

where $m = \ln a = (2.3026)(\log_{10} a)$.

$$\ln (1 + x) = x - \frac{x^2}{2} + \frac{x^3}{3} - \frac{x^4}{4} + \frac{x^5}{5} \cdots \qquad [-1 < x < +1]$$

$$\ln (1 - x) = -x - \frac{x^2}{2} - \frac{x^3}{3} - \frac{x^4}{4} - \frac{x^5}{5} - \cdots \qquad [-1 < x < +1]$$

$$\ln \left(\frac{1 + x}{1 - x} \right) = 2 \left(x + \frac{x^3}{3} + \frac{x^5}{5} + \frac{x^7}{7} + \cdots \right) \qquad [-1 < x < +1]$$

$$\ln \left(\frac{x + 1}{x - 1} \right) = 2 \left(\frac{1}{x} + \frac{1}{3x^3} + \frac{1}{5x^5} + \frac{1}{7x^7} + \cdots \right)$$
$$[x < -1 \text{ or } +1 < x]$$

$$\ln x = 2 \left[\frac{x - 1}{x + 1} + \frac{1}{3} \left(\frac{x - 1}{x + 1} \right)^3 + \frac{1}{5} \left(\frac{x - 1}{x + 1} \right)^5 + \cdots \right]$$
$$[0 < x < \infty]$$

$$\ln (a + x) = \ln a + 2 \left[\frac{x}{2a + x} + \frac{1}{3} \left(\frac{x}{2a + x} \right)^3 \right.$$
$$\left. + \frac{1}{5} \left(\frac{x}{2a + x} \right)^5 + \cdots \right]$$
$$[0 < a < +\infty, -a < x < +\infty]$$

Series for the Trigonometric Functions

In the following formulas, *all angles must be expressed in radians.* If D = the number of degrees in the angle, and x = its radian measure, then $x = 0.017453D$.

$$\sin x = x - \frac{x^3}{3!} + \frac{x^5}{5!} - \frac{x^7}{7!} + \cdots \qquad [-\infty < x < +\infty]$$

$$\cos x = 1 - \frac{x^2}{2!} + \frac{x^4}{4!} - \frac{x^6}{6!} + \frac{x^8}{8!} - \cdots \qquad [-\infty < x < +\infty]$$

$$\tan x = x + \frac{x^3}{3} + \frac{2x^5}{15} + \frac{17x^7}{315} + \frac{62x^9}{2835} + \cdots$$
$$[-\pi/2 < x < +\pi/2]$$

$$\cot x = \frac{1}{x} - \frac{x}{3} - \frac{x^3}{45} - \frac{2x^5}{945} - \frac{x^7}{4725} - \cdots \qquad [-\pi < x < +\pi]$$

$$\sin^{-1} y = y + \frac{y^3}{6} + \frac{3y^5}{40} + \frac{5y^7}{112} + \cdots \qquad [-1 \leq y \leq +1]$$

$$\tan^{-1} y = y - \frac{y^3}{3} + \frac{y^5}{5} - \frac{y^7}{7} + \cdots \qquad [-1 \leq y \leq +1]$$

$$\cos^{-1} y = \tfrac{1}{2}\pi - \sin^{-1} y; \qquad \cot^{-1} y = \tfrac{1}{2}\pi - \tan^{-1} y.$$

Series for the Hyperbolic Functions   (x a pure number)

$$\sinh x = x + \frac{x^3}{3!} + \frac{x^5}{5!} + \frac{x^7}{7!} + \cdots \qquad [-\infty < x < \infty]$$

$$\cosh x = 1 + \frac{x^2}{2!} + \frac{x^4}{4!} + \frac{x^6}{6!} + \cdots \qquad [-\infty < x < \infty]$$

$$\sinh^{-1} y = y - \frac{y^3}{6} + \frac{3y^5}{40} - \frac{5y^7}{112} + \cdots \qquad [-1 < y < +1]$$

$$\tanh^{-1} y = y + \frac{y^3}{3} + \frac{y^5}{5} + \frac{y^7}{7} + \cdots \qquad [-1 < y < +1]$$

ORDINARY DIFFERENTIAL EQUATIONS

An **ordinary differential equation** is one which contains a single independent variable, or argument, and a single dependent variable, or function, with its derivatives of various orders. A **partial differential equation** is one which contains a function of several independent variables, and its partial derivatives of various orders. The order of a differential equation is the order of the highest derivative which occurs in it. A solution of a differential equation is any relation among the variables, involving no derivatives, though possibly involving integrations which, when substituted in the given equation, will satisfy it. The general solution of an ordinary differential equation of the nth order will contain n arbitrary constants.

If specific values of the arbitrary constants are chosen, then a solution is called a **particular solution.** For most problems, all possible particular solutions to a differential equation may be found by choosing values for the constants in a general solution. In some cases, however, other solutions exist. These are called **singular solutions.**

EXAMPLE.   The differential equation $(yy')^2 - a^2 - y^2 = 0$ has general solution $(x - c)^2 + y^2 = a^2$, where c is an arbitrary constant. Additionally, it has the two singular solutions $y = a$ and $y = -a$. The singular solutions form two parallel lines tangent to the family of circles given by the general solution.

The example illustrates a general property of singular solutions; at each point on a singular solution, the singular solution is tangent to some curve given in the general solution.

Methods of Solving Ordinary Differential Equations

DIFFERENTIAL EQUATIONS OF THE FIRST ORDER

1. If possible, separate the variables; i.e., collect all the x's and dx on one side, and all the y's and dy on the other side; then integrate both sides, and add the constant of integration.

2. If the equation is homogeneous in x and y, the value of dy/dx in terms of x and y will be of the form $dy/dx = f(y/x)$. Substituting $y = xt$ will enable the variables to be separated.

Solution: $\log_e x = \displaystyle\int \frac{dt}{f(t) - t} + C.$

3. The expression $f(x, y)\, dx + F(x, y)\, dy$ is an *exact differential* if $\dfrac{\partial f(x, y)}{\partial y} = \dfrac{\partial F(x, y)}{\partial x} (= P,$ say). In this case the solution of $f(x, y)\, dx + F(x, y)\, dy = 0$ is

$$\int f(x, y)\, dx + \int[F(x, y) - \int P\, dx]\, dy = C$$

or

$$\int F(x, y)\, dy + \int[f(x, y) - \int P\, dy]\, dx = C$$

4. Linear differential equation of the first order: $\dfrac{dx}{dx} + f(x) \cdot y = F(x)$.

Solution: $y = e^{-P}[\int e^P F(x)\, dx + C]$, where $P = \int f(x)\, dx$

5. Bernoulli's equation: $\dfrac{dy}{dx} + f(x) \cdot y = F(x) \cdot y^n$. Substituting $y^{1-n} = v$ gives $(dv/dx) + (1 - n)f(x) \cdot v = (1 - n)F(x)$, which is linear in v and x.

6. Clairaut's equation: $y = xp + f(p)$, where $p = dy/dx$. The solution consists of the family of lines given by $y = Cx + f(C)$, where C is any constant, together with the curve obtained by eliminating p between the equations $y = xp + f(p)$ and $x + f'(p) = 0$, where $f'(p)$ is the derivative of $f(p)$.

7. Riccati's equation. $p + ay^2 + Q(x)y + R(x) = 0$, where $p = dy/dx$ can be reduced to a second-order linear differential equation $(d^2u/dx^2) + Q(x)(du/dx) + R(x) = 0$ by the substitution $y = du/dx$.

8. Homogeneous equations. A function $f(x, y)$ is **homogeneous** of degree n if $f(rx, ry) = r^m f(x, y)$, for all values of r, x, and y. In practice, this means that $f(x, y)$ looks like a polynomial in the two variables x and y, and each term of the polynomial has total degree m. A differential equation is homogeneous if it has the form $f(x, y) = 0$, with f homogeneous. $(xy + x^2)\, dx + y^2\, dy = 0$ is homogeneous. $\cos(xy)\, dx + y^2\, dy = 0$ is not.

If an equation is homogeneous, then either of the substitutions $y = vx$ or $x = vy$ will transform the equation into a separable equation.

9. $dy/dx = f[(ax + by + c)/(dx + ey + g)]$ is reduced to a homogeneous equation by substituting $u = ax + by + c$, $v = dx + ey + g$, if $ae - bd = 0$, and $z = ax + by$, $w = dx + ey$ if $ae - bd = 0$.

DIFFERENTIAL EQUATIONS OF THE SECOND ORDER

10. *Dependent variable missing.* If an equation does not involve the variable y, and is of the form $F(x, dy/dx, d^2y/dx^2) = 0$, then it can be reduced to a first-order equation by substituting $p = dy/dx$ and $dp/dx = d^2y/dx^2$.

11. *Independent variable missing.* If the equation is of the form $F(y, dy/dx, d^2y/dx^2) = 0$, and so is missing the variable x, then it can be reduced to a first-order equation by substituting $p = dy/dx$ and $p(dp/dy) = d^2y/dx^2$.

12. $\dfrac{d^2y}{dx^2} = -n^2y$.

Solution: $y = C_1 \sin(nx + C_2)$, or $y = C_3 \sin nx + C_4 \cos nx$.

13. $\dfrac{d^2y}{dx^2} = +n^2y$.

Solution: $y = C_1 \sinh(nx + C_2)$, or $y = C_3 e^{nx} + C_4 e^{-nx}$.

14. $\dfrac{d^2y}{dx^2} = f(y)$.

Solution: $x = \int \dfrac{dy}{\sqrt{C_1 + 2P}} + C_2$, where $P = \int f(y)\, dy$.

15. $\dfrac{d^2y}{dx^2} = f(x)$.

Solution: $y = \int P\, dx + C_1x + C_2$ where $P = \int f(x)\, dx$,

or $y = xP - \int xf(x)\, dx + C_1x + C_2$.

16. $\dfrac{d^2y}{dx^2} = f\left(\dfrac{dy}{dx}\right)$. Putting $\dfrac{dy}{dx} = z, \dfrac{d^2y}{dx^2} = \dfrac{dz}{dx}$,

$$x = \int \dfrac{dz}{f(z)} + C_1, \quad \text{and} \quad y = \int \dfrac{z\, dz}{f(z)} + C_2$$

then eliminate z from these two equations.

17. The equation for damped vibrations: $\dfrac{d^2y}{dx^2} + 2b\dfrac{dy}{dx} + a^2y = 0$.

CASE 1. If $a^2 - b^2 > 0$, let $m = \sqrt{a^2 - b^2}$.
Solution:

$y = C_1e^{-bx}\sin(mx + C_2)$ or $y = e^{-bx}[C_3 \sin(mx) + C_4\cos(mx)]$

CASE 2. If $a^2 - b^2 = 0$, solution is $y = e^{-bx}(C_1 + C_2x)$.
CASE 3. If $a^2 - b^2 < 0$, let $n = \sqrt{b^2 - a^2}$.
Solution:

$y = C_1e^{-bx}\sinh(nx + C_2)$ or $y = C_3e^{-(b+n)x} + C_4e^{-(b-n)x}$

18. $\dfrac{d^2y}{dx^2} + 2b\dfrac{dy}{dx} + a^2y = c$.

Solution: $y = \dfrac{c}{a^2} + y_1$, where $y_1 =$ the solution of the corresponding equation with second member zero [see type 17 above].

19. $\dfrac{d^2y}{dx^2} + 2b\dfrac{dy}{dx} + a^2y = c \sin(kx)$.

Solution: $y = R \sin(kx - S) + y_1$
where $R = c/\sqrt{(a^2 - k^2)^2 + 4b^2k^2}$, $\tan S = 2bk/(a^2 - k^2)$, and $y_1 =$ the solution of the corresponding equation with second member zero [see type 17 above].

20. $\dfrac{d^2y}{dx^2} + 2b\dfrac{dy}{dx} + a^2y = f(x)$.

Solution: $y = R \sin(kx - S) + y_1$
where $R = c/\sqrt{(a^2 - k^2)^2 + 4b^2k^2}$, $\tan S = 2bk/(a^2 - k^2)$, and $y_1 =$ the solution of the corresponding equation with second member zero [see type 17 above].
If $b^2 < a^2$,

$$y_0 = \dfrac{1}{2\sqrt{b^2 - a^2}}\left[e^{m_1x}\int e^{-m_1x}f(x)\, dx - e^{m_2x}\int e^{-m_2x}f(x)\, dx\right]$$

where $m_1 = -b + \sqrt{b^2 - a^2}$ and $m_2 = -b - \sqrt{b^2 - a^2}$.
If $b^2 < a^2$, let $m = \sqrt{a^2 - b^2}$, then

$$y_0 = \dfrac{1}{m}e^{-bx}\left[\sin(mx)\int e^{bx}\cos(mx) \cdot f(x)\, dx\right.$$
$$\left. - \cos(mx)\int e^{bx}\sin(mx) \cdot f(x)\, dx\right]$$

If $b^2 = a^2$, $y_0 = e^{-bx}\left[x\int e^{bx}f(x)\, dx - \int x \cdot e^{bx}f(x)\, dx\right]$.

Types 17 to 20 are examples of linear differential equations with constant coefficients. The solutions of such equations are often found most simply by the use of Laplace transforms. (See Franklin, "Fourier Methods," pp. 198–229, McGraw-Hill or Kreyszig, "Advanced Engineering Mathematics," Wiley.)

Linear Equations

For the linear equation of the nth order

$$A_n(x)\, d^ny/dx^n + A_{n-1}(x)\, d^{n-1}y/dx^{n-1} + \cdots$$
$$+ A_1(x)dy/dx + A_0(x)y = E(x)$$

the general solution is $y = u + c_1u_1 + c_2u_2 + \cdots + c_nu_n$. Here u, the particular integral, is any solution of the given equation, and $u_1, u_2, \ldots, u_n$ form a fundamental system of solutions of the homogeneous equation obtained by replacing $E(x)$ by zero. A set of solutions is fundamental, or independent, if its Wronskian determinant $W(x)$ is not zero, where

$$W(x) = \begin{vmatrix} u_1 & u_2 & \cdots & u_n \\ u_1' & u_2' & \cdots & u_n' \\ \cdot & \cdot & \cdots & \cdot \\ \cdot & \cdot & & \cdot \\ \cdot & \cdot & & \cdot \\ u_1^{(n-1)} & u_2^{(n-1)} & \cdots & u_n^{(n-1)} \end{vmatrix}$$

For any n functions, $W(x) = 0$ if some one u_i is linearly dependent on the others, as $u_n = k_1u_1 + k_2u_2 + \cdots + k_{n-1}u_{n-1}$ with the coefficients k_i constant. And for n solutions of a linear differential equation of the nth order, if $W(x) \neq 0$, the solutions are linearly independent.

Constant Coefficients   To solve the homogeneous equation of the nth order $A_nd^ny/dx^n + A_{n-1}d^{n-1}y/dx^{n-1} + \cdots + A_1dy/dx + A_0y = 0$, $A_n \neq 0$, where $A_n, A_{n-1}, \ldots, A_0$ are constants, find the roots of the auxiliary equation

$$A_np^n + A_{n-1}p^{n-1} + \cdots + A_1p + A_0 = 0$$

For each simple real root r, there is a term ce^{rx} in the solution. The terms of the solution are to be added together. When r occurs twice among the n roots of the auxiliary equation, the corresponding term is $e^{rx}(c_1 + c_2x)$. When r occurs three times, the corresponding term is $e^{rx}(c_1 + c_2x + c_3x^2)$, and so forth. When there is a pair of conjugate complex roots $a + bi$ and $a - bi$, the real form of the terms in the solution is $e^{ax}(c_1 \cos bx + d_1 \sin bx)$. When the same pair occurs twice, the corresponding term is $e^{ax}[(c_1 + c_2 x) \cos bx + (d_1 + d_2 x) \sin bx]$, and so forth.

Consider next the general nonhomogeneous linear differential equation of order n, with constant coefficients, or

$$A_nd^ny/dx^n + A_{n-1}d^{n-1}y/dx^{n-1} + \cdots + A_1dy/dx + A_0y = E(x)$$

We may solve this by adding any particular integral to the complementary function, or general solution, of the homogeneous equation obtained by replacing $E(x)$ by zero. The complementary function may be found from the rules just given. And the particular integral may be found by the methods of the following paragraphs.

Undetermined Coefficients   In the last equation, let the right member $E(x)$ be a sum of terms each of which is of the type k, $k \cos bx$, $k \sin bx$, ke^{ax}, kx, or more generally, kx^me^{ax}, $kx^me^{ax} \cos bx$, or $kx^me^{ax} \sin bx$. Here m is zero or a positive integer, and a and b are any real numbers. Then the form of the particular integral I may be predicted by the following rules.

CASE 1. $E(x)$ **is a single term** T. Let D be written for d/dx, so that the given equation is $P(D)y = E(x)$, where $P(D) = A_nD^n + A_{n-1}D^{n-1} + \cdots + A_1D + A_0y$. With the term T associate the simplest polynomial $Q(D)$ such that $Q(D)T = 0$. For the particular types k, etc., $Q(D)$ will be D, $D^2 + b^2$, $D^2 + b^2$, $D - a$, D^2; and for the general types $kx^m e^{ax}$, etc., $Q(D)$ will be $(D - a)^{m+1}$, $(D^2 - 2aD + a^2 + b^2)^{m+1}$, $(D^2 - 2aD + a^2 + b^2)^{m+1}$. Thus $Q(D)$ will always be some power of a first- or second-degree factor, $Q(D) = F^v$, $F = D - a$, or $F = D^2 - 2aD + a^2 + b^2$.

Use the method described under **Constant Coefficients** to find the terms in the solution of $P(D)y = 0$ and also the terms in the solution of $Q(D)P(D)y = 0$. Then assume the particular integral I is a linear combination with unknown coefficients of those terms in the solution of $Q(D)P(D)y = 0$ which are not in the solution of $P(D)y = 0$. Thus if $Q(D) = F^q$ and F is *not* a factor of $P(D)$, assume $I = (Ax^{q-1} + Bx^{q-2} + \cdots + L)e^{ax}$ when $F = D - a$, and assume $I = (Ax^{q-1} + Bx^{q-2} + \cdots + L)e^{ax} \cos bx + (Mx^{q-1} + Nx^{q-2} + \cdots + R)e^{ax} \sin bx$ when $F = D^2 - 2aD + a^2 + b^2$. When F is a factor of $P(D)$ and the highest power of F which is a divisor of $P(D)$ is F^k, try the I above multiplied by x^k.

CASE 2. $E(x)$ **is a sum of terms**. With each term in $E(x)$, associate a polynomial $Q(D) = F^q$ as before. Arrange in one group all the terms that have the same F. The particular integral of the given equation will be the sum of solutions of equations each of which has one group on the right. For any one such equation, the form of the particular integral is given as for Case 1, with q the highest power of F associated with any term of the group on the right.

After the form has been found in Case 1 or 2, the unknown coefficients follow when we substitute back in the given differential equation, equate coefficients of like terms, and solve the resulting system of simultaneous equations.

Variation of Parameters.   Whenever a fundamental system of solutions $u_1, u_2, \ldots, u_n$ for the homogeneous equation is known, a particular integral of

$$A_n(x)d^ny/dx^n + A_{n-1}(x)d^{n-1}y/dx^{n-1} + \cdots$$
$$+ A_1(x)dy/dx + A_0(x)y = E(x)$$

may be found in the form $y = \Sigma v_ku_k$. In this and the next few summations, k runs from 1 to n. The v_k are functions of x, found by integrating their derivatives v_k', and these derivatives are the solutions of the n simultaneous equations $\Sigma v_k'u_k = 0, \Sigma v_k'u_k' = 0, \Sigma v_k'u''_k = 0, \cdots,$ $\Sigma v_k'u_k^{(n-2)} = 0, A_n(x)\Sigma v_k'u_k^{(n-1)} = E(x)$. To find the v_k from $v_k' = \int v_k' dx + c_k$, any choice of constants will lead to a particular integral.

The special choice $v_k = \int_0^x v_k' dx$ leads to the particular integral having $y, y', y'', \ldots, y^{(n-1)}$ each equal to zero when $x = 0$.

The Cauchy-Euler Equidimensional Equation   This has the form

$$k_nx^nd^ny/dx^n + k_{n-1}x^{n-1}d^{n-1}y/dx^{n-1} + \cdots$$
$$+ k_1x \, dy/dx + k_0y = F(x)$$

The substitution $x = e^t$, which makes

$$x \, dy/dx = dy/dt$$
$$x^kd^ky/dx^k = (d/dt - k + 1) \cdots (d/dt - 2)(d/dt - 1) \, dy/dt$$

transforms this into a linear differential equation with constant coefficients. Its solution $y = g(t)$ leads to $y = g(\ln x)$ as the solution of the given Cauchy-Euler equation.

Bessel's Equation   The general Bessel equation of order n is:

$$x^2y'' + xy' + (x^2 - n^2)y = 0$$

This equation has general solution

$$y = AJ_n(x) + BJ_{-n}(x)$$

when n is not an integer. Here, $J_n(x)$ and $J_{-n}(x)$ are Bessel functions (see section on Special Functions).

In case $n = 0$, Bessel's equation has solution

$$y = AJ_0(x) + B\left[J_0(x) \ln (x) - \sum_{k=1}^{\infty} \frac{(-1)^kH_kx^{2k}}{2^{2k}(k!)^2}\right]$$

where H_k is the kth partial sum of the harmonic series, $1 + \frac{1}{2} + \frac{1}{3} + \cdots + 1/k$.

In case $n = 1$, the solution is

$$y = AJ_1(x) + B\left\{J_1(x) \ln (x) + 1/x \right.$$
$$\left. - \left[\sum_{k=1}^{\infty} \frac{(-1)^k(H_k + H_{k-1})x^{2k-1}}{2^{2k}k!(k - 1)!}\right]\right\}$$

In case $n > 1$, n is an integer, solution is

$$y = AJ_n(x) + B\left\{J_n(x) \ln (x) + \left[\sum_{k=0}^{\infty} \frac{(-1)^{k+1}(n - 1)!x^{2k-n}}{2^{2k+1-n}k!(1 - n)^k}\right]\right.$$
$$\left. + \frac{1}{2}\left[\sum_{k=0}^{\infty} \frac{(-1)^{k+1}(H_k + H_{k+1})x^{2k+n}}{2^{2k+n}k!(k + n)!}\right]\right\}$$

Solutions to Bessel's equation may be given in several other forms, often exploiting the relation between H_k and $\ln (k)$ or the so-called *Euler constant*.

General Method of Power Series   Given a general differential equation $F(x, y, y', \ldots) = 0$, the solution may be expanded as a Maclaurin series, so $y = \Sigma_{n=0}^{\infty}a_nx^n$, where $a_n = f^{(n)}(0)/n!$. The power series for y may be differentiated formally, so that $y' = \sum_{n=1}^{\infty} na_n x^{n-1} = \sum_{n=0}^{\infty}(n+1)a_{n+1}x^n$, and $y'' = \sum_{n=2}^{\infty}n(n-1)a_n x^{n-2} = \sum_{n=0}^{\infty}(n+1)(n+2)a_{n+2}x^n$.

Substituting these series into the equation $F(x, y, y', \ldots) = 0$ often gives useful recursive relationships, giving the value of a_n in terms of previous values. If approximate solutions are useful, then it may be sufficient to take the first few terms of the Maclaurin series as a solution to the equation.

EXAMPLE. Consider $y'' - y' + xy = 0$. The procedure gives $\sum_{n=0}^{\infty}(n+1)(n+2)a_{n+2}x^n - \sum_{n=0}^{\infty}(n+1)a_{n+1}x^n + x\sum_{n=0}^{\infty}a_n x^n = \sum_{n=0}^{\infty}(n+1)(n+2)a_{n+2}x^n - \sum_{n=0}^{\infty}(n+1)a_{n+1}x^n + \sum_{n=1}^{\infty}a_{n-1}x^n = (2a_2 - a_1)x^0 + \sum_{n=1}^{\infty}[(n+1)(n+2)a_{n+2} - (n+1)a_{n+1} + a_{n-1}]x^n = 0$. Thus $2a_2 - a_1 = 0$ and, for $n > 0$, $(n+1)(n+2)a_{n+2} - (n+1)a_{n+1} + a_{n-1} = 0$. Thus, a_0 and a_1 may be determined arbitrarily, but thereafter, the values of a_n are determined recursively.

PARTIAL DIFFERENTIAL EQUATIONS

Partial differential equations (PDEs) arise when there are two or more independent variables. Two notations are common for the partial derivatives involved in PDEs, the "del" or fraction notation, where the first partial derivative of f with respect to x would be written $\partial f/\partial x$, and the subscript notation, where it would be written f_x.

In the same way that ordinary differential equations often involve arbitrary constants, solutions to PDEs often involve arbitrary functions.

EXAMPLE. $f_{xy} = 0$ has as its general solution $g(x) + h(y)$. The function g does not depend on y, so $g_y = 0$. Similarly, $f_x = 0$.

PDEs usually involve boundary or initial conditions dictated by the application. These are analogous to initial conditions in ordinary differential equations.

In solving PDEs, it is seldom feasible to find a general solution and then specialize that general solution to satisfy the boundary conditions, as is done with ordinary differential equations. Instead, the boundary conditions usually play a key role in the solution of a problem. A notable exception to this is the case of linear, homogeneous PDEs since they have the property that if f_1 and f_2 are solutions, then $f_1 + f_2$ is also a solution. The wave equation is one such equation, and this property is the key to the solution described in the section "Fourier Series."

Often it is difficult to find exact solutions to PDEs, so it is necessary to resort to approximations or numerical solutions. (See also "Numerical Methods" below.)

Classification of PDEs

Linear   A PDE is **linear** if it involves only first derivatives, and then only to the first power. The general form of a linear PDE, in two independent variables, x and y, and the dependent variable z, is $P(x, y, z) f_x + Q(x, y, z)f_y = R(x, y, z)$, and it will have a solution of the form $z = f(x, y)$ if its solution is a function, or $F(x, y, z) = 0$ if the solution is not a function.

Elliptic   Laplace's equation $f_{xx} + f_{yy} = 0$ and Poisson's equation $f_{xx} + f_{yy} = g(x, y)$ are the prototypical elliptic equations. They have analogs in more than two variables. They do not explicitly involve the variable time and generally describe steady-state or equilibrium conditions, gravitational potential, where boundary conditions are distributions of mass, electrical potential, where boundary conditions are electrical charges, or equilibrium temperatures, and where boundary conditions are points where the temperature is held constant.

Parabolic $T_t = T_{xx} + T_{yy}$ represents the dynamic condition of diffusion or heat conduction, where $T(x, y, t)$ usually represents the temperature at time t at the point (x, y). Note that when the system reaches steady state, the temperature is no longer changing, so $T_t = 0$, and this becomes Laplace's equation.

Hyperbolic   Wave propagation is described by equations of the type $u_{tt} = c^2(u_{xx} + u_{yy})$, where c is the velocity of waves in the medium.

VECTOR CALCULUS

Vector Fields   A **vector field** is a function that assigns a vector to each point in a region. If the region is two-dimensional, then the vectors assigned are two-dimensional, and the vector field is a two-dimensional vector field, denoted $\mathbf{F}(x, y)$. In the same way, a three-dimensional vector field is denoted $\mathbf{F}(x, y, z)$.

A three-dimensional vector field can always be written:

$$\mathbf{F}(x, y, z) = f_1(x, y, z)\mathbf{i} + f_2(x, y, z)\mathbf{j} + f_3(x, y, z)\mathbf{k}$$

where $\mathbf{i}$, $\mathbf{j}$, and $\mathbf{k}$ are the basis vectors $(1, 0, 0)$, $(0, 1, 0)$, and $(0, 0, 1)$, respectively. The functions f_1, f_2, and f_3 are called **coordinate functions** of $\mathbf{F}$.

Gradient   If $f(x, y)$ is a function of two variables, then the **gradient** is the associated vector field grad$(f) = f_x\mathbf{i} + f_y\mathbf{j}$. Similarly, the gradient of a function of three variables is $f_x\mathbf{i} + f_y\mathbf{j} + f_z\mathbf{k}$. At any point, the direction of the gradient vector points in the direction in which the function is increasing most rapidly, and the magnitude is the rate of increase in that direction.

Parameterized Curves   If C is a curve from a point A to a point B, either in two dimensions or in three dimensions, then a **parameterization** of C is a vector-valued function $\mathbf{r}(t) = r_1(t)\mathbf{i} + r_2(t)\mathbf{j} + r_3(t)\mathbf{k}$, which satisfies $\mathbf{r}(a) = A$, $\mathbf{r}(b) = B$, and $\mathbf{r}(t)$ is on the curve C, for $a \leq t \leq b$. It is also necessary that the function $\mathbf{r}(t)$ be continuous and one-to-one. A given curve C has many different parameterizations.

The derivative of a parameterization $\mathbf{r}(t)$ is a vector-valued function $\mathbf{r}'(t) = r_1'(t)\mathbf{i} + r_2'(t)\mathbf{j} + r_3'(t)\mathbf{k}$. The derivative is the velocity function of the parameterization. It is always tangent to the curve C, and the magnitude is the speed of the parameterization.

Line Integrals   If $\mathbf{F}$ is a vector field, C is a curve, and $\mathbf{r}(t)$ is a parameterization of C, then the **line integral**, or **work integral**, of $\mathbf{F}$ along C is

$$W = \int_C \mathbf{F} \cdot d\mathbf{r} = \int_a^b \mathbf{F}(\mathbf{r}(t)) \cdot \mathbf{r}'(t)\, dt$$

This is sometimes called the work integral because if $\mathbf{F}$ is a force field, then W is the amount of work necessary to move an object along the curve C from A to B.

Divergence and Curl   The **divergence** of a vector field $\mathbf{F}$ is div $\mathbf{F} = f_{1x} + f_{2y} + f_{3z}$. If $\mathbf{F}$ represents the flow of a fluid, then the divergence at a point represents the rate at which the fluid is expanding at that point. Vector fields with div $\mathbf{F} = 0$ are called **incompressible.**

The **curl** of $\mathbf{F}$ is

$$\text{curl } \mathbf{F} = (f_{3y} - f_{2z})\mathbf{i} + (f_{1z} - f_{3x})\mathbf{j} + (f_{2x} - f_{1y})\mathbf{k}$$

If $\mathbf{F}$ is a two-dimensional vector field, then the first two terms of the curl are zero, so the curl is just

$$\text{curl } \mathbf{F} = (f_{2x} - f_{1y})\mathbf{k}$$

If $\mathbf{F}$ represents the flow of a fluid, then the curl represents the rotation of the fluid at a given point. Vector fields with curl $\mathbf{F} = \mathbf{0}$ are called **irrotational.**

Two important facts relate div, grad, and curl:
1. div (curl $\mathbf{F}$) = 0
2. curl (grad f) = $\mathbf{0}$

Conservative Vector Fields   A vector field $\mathbf{F} = f_1\mathbf{i} + f_2\mathbf{j} + f_3\mathbf{k}$ is **conservative** if all of the following are satisfied:

$$f_{1y} = f_{2x} \qquad f_{1z} = f_{3x} \qquad \text{and} \qquad f_{2z} = f_{3y}$$

If $\mathbf{F}$ is a two-dimensional vector field, then the second and third conditions are always satisfied, and so only the first condition must be checked. Conservative vector fields have three important properties:

1. $\int_C \mathbf{F} \cdot d\mathbf{r}$ has the same value regardless of what curve C is chosen that connects the points A and B. This property is called **path independence.**
2. $\mathbf{F}$ is the gradient of some function $f(x, y, z)$.
3. Curl $\mathbf{F} = \mathbf{0}$.

In the special case that $\mathbf{F}$ is a conservative vector field, if $\mathbf{F} = \text{grad}$ (f), then

$$\int_C \mathbf{F} \cdot d\mathbf{r} = f(B) - f(A)$$

THEOREMS ABOUT LINE AND SURFACE INTEGRALS

Two important theorems relate line integrals with double integrals. If R is a region in the plane and if C is the curve tracing the boundary of R in the positive (counterclockwise) direction, and if $\mathbf{F}$ is a continuous vector field with continuous first partial derivatives, line integrals on C are related to double integrals on R by Green's theorem and the divergence theorem.

Green's Theorem

$$\int_C \mathbf{F} \cdot d\mathbf{r} = \int\int_R \text{curl}\,(\mathbf{F}) \cdot d\mathbf{S}$$

The right-hand double integral may also be written as $\int\int_R |\text{curl}\,(\mathbf{F})|\,dA$.

Green's theorem describes the total rotation of a vector field in two different ways, on the left in terms of the boundary of the region and on the right in terms of the rotation at each point within the region.

Divergence Theorem

$$\int_C \mathbf{F} \cdot d\mathbf{N} = \int\int_R \text{div}\,(\mathbf{F})\,dA$$

where $\mathbf{N}$ is the so-called *normal vector field* to the curve C. The divergence theorem describes the expansion of a region in two distinct ways, on the left in terms of the flux across the boundary of the region and on the right in terms of the expansion at each point within the region.

Both Green's theorem and the divergence theorem have corresponding theorems involving surface integrals and volume integrals in three dimensions.

LAPLACE AND FOURIER TRANSFORMS

Laplace Transforms The Laplace transform is used to convert equations involving a time variable t into equations involving a frequency

Table 2.1.4 Laplace Transforms

$f(t)$	$F(s) = \mathcal{L}(f(t))$	Name of function
1. a	a/s	
2. 1	$1/s$	
3. $u_a(t) = \begin{cases} 0 & t < a \\ 1 & t > a \end{cases}$	$a\,e^{-as}/S$	Heaviside or step function
4. $\delta_a(t) = u_a'(t)$	e^{-as}	Dirac or impulse function
5. e^{at}	$1/(s - a)$	
6. $(1/r)e^{-t/r}$	$1/(rs + 1)$	
7. ke^{-at}	$k/(s + a)$	
8. $\sin at$	$a/(s^2 + a^2)$	
9. $\cos at$	$s/(s^2 + a^2)$	
10. $e^{-at}\sin bt$	$b/[(s + a)^2 + b^2]$	
11. $\dfrac{e^{at}}{a - b} - \dfrac{e^{bt}}{a - b}$	$\dfrac{1}{(s - a)(s - b)}$	
12. t	$1/s^2$	
13. t^2	$2/s^3$	
14. t^n	$n!/s^{n+1}$	
15. t^a	$\Gamma(a + 1)/s^{a+1}$	Gamma function (see "Special Functions")
16. $\sinh at$	$a/(s^2 - a^2)$	
17. $\cosh at$	$s/(s^2 - a^2)$	
18. $t^n e^{at}$	$n!/(s - a)^{n+1}$	
19. $t \cos at$	$(s^2 - a^2)/(s^2 + a^2)^2$	
20. $t \sin at$	$2as/(s^2 + a^2)^2$	
21. $\sin at - at \cos at$	$2a^3/(s^2 + a^2)^2$	
22. $\arctan a/s$	$(\sin at)/t$	

Table 2.1.5 Properties of Laplace Transforms

$f(t)$	$F(s) = \mathcal{L}(f(t))$	Name of rule
1. $f(t)$	$\displaystyle\int_0^{\infty} e^{-st}f(t)\,dt$	Definition
2. $f(t) + g(t)$	$F(s) + G(s)$	Addition
3. $kf(t)$	$kF(s)$	Scalar multiples
4. $f'(t)$	$sF(s) - f(0^+)$	Derivative laws
5. $f''(t)$	$s^2F(s) - sf(0^+)$ $- f'(0^+)$	
6. $f'''(t)$	$s^3F(s) - s^2f(0^+)$ $- sf'(0^+) - f''(0^+)$	
7. $\displaystyle\int f(t)\,dt$	$(1/s)F(s)$ $+ (1/s)\displaystyle\int f(t)\,dt\big|_{0^+}$	Integral law
8. $f(bt)$	$(1/b)F(s/b)$	Change of scale
9. $e^{at}f(t)$	$F(s - a)$	First shifting
10. $f * g(t)$	$F(s)G(s)$	Convolution
11. $u_a(t)f(t - a)$	$F(s)e^{-at}$	Second shifting
12. $-tf(t)$	$F'(s)$	Derivative in s

variable s. There are essentially three reasons for doing this: (1) higher-order differential equations may be converted to purely algebraic equations, which are more easily solved; (2) boundary conditions are easily handled; and (3) the method is well-suited to the theory associated with the **Nyquist stability criteria.**

In Laplace-transformation mathematics the following symbols and equations are used (Tables 2.1.4 and 2.1.5):

$f(t) = $ a function of time
$s = $ a complex variable of the form $(\sigma + j\omega)$
$F(s) = $ an equation expressed in the transform variable s, resulting from operating on a function of time with the Laplace integral
$\mathcal{L} = $ an operational symbol indicating that the quantity which it prefixes is to be transformed into the frequency domain
$f(0^+) = $ the limit from the positive direction of $f(t)$ as t approaches zero
$f(0^-) = $ the limit from the negative direction of $f(t)$ as t approaches zero

Therefore, $F(s) = \mathcal{L}[f(t)]$. The Laplace integral is defined as

$$\mathcal{L} = \int_0^{\infty} e^{-st}\,dt. \text{ Therefore, } \mathcal{L}[f(t)] = \int_0^{\infty} e^{-st}f(t)\,dt$$

Direct Transforms

EXAMPLE.

$$f(t) = \sin \beta t$$

$$\mathcal{L}[f(t)] = \mathcal{L}(\sin \beta t) = \int_0^{\infty} \sin \beta t\, e^{-st}\,dt$$

but

$$\sin \beta t = \frac{e^{j\beta t} - e^{-j\beta t}}{2j} \quad \text{where} \quad j^2 = -1$$

$$\mathcal{L}(\sin \beta t) = \frac{1}{2j}\int_0^{\infty}(e^{j\beta t} - e^{-j\beta t})e^{-st}\,dt$$

$$= \frac{1}{2j}\left(\frac{-1}{s - j\beta}\right)e^{(-s+j\beta)t}\Big|_0^{\infty}$$

$$- \frac{1}{2j}\left(\frac{-1}{s + j\beta}\right)e^{(-s-j\beta)t}\Big|_0^{\infty}$$

$$= \frac{\beta}{s^2 + \beta^2}$$

Table 2.1.4 lists the transforms of common time-variable expressions.

Some special functions are frequently encountered when using Laplace methods.

The **Heaviside,** or **step, function** $u_a(t)$ sometimes written $u(t - a)$, is zero for all $t < a$ and 1 for all $t > a$. Its value at $t = a$ is defined differently in different applications, as 0, ½, or 1, or it is simply left undefined. In Laplace applications, the value of a function at a single point does not matter. The Heaviside function describes a force which is "off" until time $t = a$ and then instantly goes "on."

The **Dirac delta function,** or **impulse function,** $\delta_a(t)$, sometimes written $\delta(t - a)$, is the derivative of the Heaviside function. Its value is always zero, except at $t = a$, where its value is "positive infinity." It is sometimes described as a "point mass function." The delta function describes an impulse or an instantaneous transfer of momentum.

The derivative of the Dirac delta function is called the **dipole function.** It is less frequently encountered.

The **convolution** $f * g(t)$ of two functions $f(t)$ and $g(t)$ is defined as

$$f * g(t) = \int_0^t f(u)g(t - u)\, du$$

Laplace transforms are often used to solve differential equations arising from so-called *linear systems.* Many vibrating systems and electrical circuits are linear systems. If an input function $f_i(t)$ describes the forces exerted upon a system and a response or output function $f_o(t)$ describes the motion of the system, then the **transfer function** $T(s) = F_o(s)/F_i(s)$. Linear systems have the special property that the transfer function is independent of the input function, within the elastic limits of the system. Therefore,

$$\frac{F_o(s)}{F_i(s)} = \frac{G_o(s)}{G_i(s)}$$

This gives a technique for describing the response of a system to a complicated input function if its response to a simple input function is known.

EXAMPLE. Solve $y'' + 2y' - 3y = 8e^t$ subject to initial conditions $y(0) = 2$ and $y'(0) = 0$. Let $y = f(t)$ and $Y = F(s)$. Take Laplace transforms of both sides and substitute for $y(0)$ and $y'(0)$, and get

$$s^2Y - 2s + 2(sY - 2) - 3Y = \frac{8}{s - 1}$$

Solve for Y, apply partial fractions, and get

$$Y = \frac{2s^2 + 2s + 4}{(s + 3)(s - 1)^2}$$

$$= \frac{1}{s + 3} + \frac{s + 1}{(s - 1)^2}$$

$$= \frac{1}{(s + 3)} + \frac{1}{(s - 1)} + \frac{2}{(s - 1)^2}$$

Using the tables of transforms to find what function has Y as its transform, we get

$$y = e^{-3t} + e^t + 2te^t$$

EXAMPLE. A vibrating system responds to an input function $f_i(t) = \sin t$ with a response $f_o(t) = \sin 2t$. Find the system response to the input $g_i(t) = \sin 2t$.

Apply the invariance of the transfer function, and get

$$G_o(s) = \frac{F_o G_i}{F_i}$$

$$= \frac{4(s^2 + 1)}{(s^2 + 4)^2}$$

$$= \frac{2(2)}{s^2 + 2^2} - \frac{12}{16}\left[\frac{16}{(s^2 + 2^2)^2}\right]$$

Applying formulas 8 and 21 from Table 2.1.4 of Laplace transforms,

$$g_o(t) = 2 \sin 2t - \tfrac{3}{4} \sin 2t + \tfrac{3}{2} t \cos 2t$$

Inversion When an equation has been transformed, an explicit solution for the unknown may be directly determined through algebraic manipulation. In automatic-control design, the equation is usually the differential equation describing the system, and the unknown is either the

output quantity or the error. The solution gained from the transformed equation is expressed in terms of the complex variable s. For many design or analysis purposes, the solution in s is sufficient, but in some cases it is necessary to retransform the solution in terms of time. The process of passing from the complex-variable (frequency domain) expression to that of time (time domain) is called an **inverse transformation.** It is represented symbolically as

$$\mathcal{L}^{-1}F(s) = f(t)$$

For any $f(t)$ there is only one direct transform, $F(s)$. For any given $F(s)$ there is only one inverse transform $f(t)$. Therefore, tables are generally used for determining inverse transforms. Very complete tables of inverse transforms may be found in Gardner and Barnes, "Transients in Linear Systems." As an example of the inverse procedure consider an equation of the form

$$K = \alpha x(t) + \int \frac{x(t)}{\beta}\, dt$$

It is desired to obtain an expression for $x(t)$ resulting from an instantaneous change in the quantity K. Transforming the last equation yields

$$\frac{K}{s} = X(s)\alpha + \frac{X(s)}{s\beta} + \frac{f^{-1}(0^+)}{s}$$

If $f^{-1}(0)/s = 0$

then

$$X(s) = \frac{K/\alpha}{s + 1/\alpha\beta}$$

$$x(t) = \mathcal{L}^{-1}[X(s)] = \mathcal{L}^{-1} \frac{K/\alpha}{s + 1/\alpha\beta}$$

From Table 2.1.4, $x(t) = \dfrac{K}{\alpha}e^{-t/\alpha\beta}$

Fourier Coefficients Fourier coefficients are used to analyze periodic functions in terms of sines and cosines. If $f(x)$ is a function with period $2L$, then the Fourier coefficients are defined as

$$a_n = \frac{1}{L}\int_{-L}^{L} f(s)\cos\frac{n\pi s}{L}\, ds \qquad n = 0, 1, 2, \ldots$$

$$b_n = \frac{1}{L}\int_{-L}^{L} f(s)\sin\frac{n\pi s}{L}\, ds \qquad n = 1, 2, \ldots$$

Equivalently, the bounds of integration may be taken as 0 to $2L$ instead of $-L$ to L. Many treatments take $L = \pi$ or $L = 1$.

Then the Fourier theorem states that

$$f(x) = \frac{a_0}{2} + \sum_{n=1}^{\infty} a_n \cos\left(\frac{n\pi x}{L}\right) + b_n \sin\left(\frac{n\pi x}{L}\right)$$

The series on the right is called the "Fourier series of the function $f(x)$." The convergence of the Fourier series is usually rapid, so that the function $f(x)$ is usually well-approximated by the sum of the first few sums of the series.

Examples of the Fourier Series If $y = f(x)$ is the curve in Figs. 2.1.110 to 2.1.112, then in Fig. 2.1.110,

$$y = \frac{h}{2} - \frac{4h}{\pi^2}\left(\cos\frac{\pi x}{c} + \frac{1}{9}\cos\frac{3\pi x}{c} + \frac{1}{25}\cos\frac{5\pi x}{c} + \cdots\right)$$

Fig. 2.1.110 Saw-tooth curve.

In Fig. 2.1.111,

$$y = \frac{4h}{\pi}\left(\sin\frac{\pi x}{c} + \frac{1}{3}\sin\frac{3\pi x}{c} + \frac{1}{5}\sin\frac{5\pi x}{c} + \cdots\right)$$

Fig. 2.1.111 Step-function curve or square wave.

In Fig. 2.1.112,

$$y = \frac{2h}{\pi}\left(\sin\frac{\pi x}{c} - \frac{1}{2}\sin\frac{2\pi x}{c} + \frac{1}{3}\sin\frac{3\pi x}{c} - \cdots\right)$$

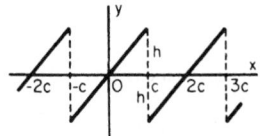

Fig. 2.1.112 Linear-sweep curve.

If the Fourier coefficients of a function $f(x)$ are known, then the coefficients of the derivative $f'(x)$ can be found, when they exist, as follows:

$$a'_n = nb_n \qquad b'_n = -na_n$$

where a'_n and b'_n are the Fourier coefficients of $f'(x)$.

The **complex Fourier coefficients** are defined by:

$$c_n = \tfrac{1}{2}(a_n - ib_n)$$
$$c_0 = \tfrac{1}{2}a_0$$
$$c_n = \tfrac{1}{2}(a_n + ib_n)$$

Then the complex form of the Fourier theorem is

$$f(x) = \sum_{n=-\infty}^{\infty} c_n e^{in\pi x/L}$$

Wave Equation Fourier series are often used in the solution of the wave equation $a^2 u_{xx} = u_{tt}$ where $0 < x < L$, $t > 0$, and initial conditions are $u(x, 0) = f(x)$ and $u_t(x, 0) = g(x)$. This describes the position of a vibrating string of length L, fixed at both ends, with initial position $f(x)$ and initial velocity $g(x)$. The constant a is the velocity at which waves are propagated along the string, and is given by $a^2 = T/p$, where T is the tension in the string and p is the mass per unit length of the string.

If $f(x)$ is extended to the interval $-L < x < L$ by setting $f(-x) = -f(x)$, then f may be considered periodic of period $2L$. Its Fourier coefficients are

$$a_n = 0 \qquad b_n = \int_{-L}^{L} f(x)\sin\frac{n\pi x}{L}\pi\,dx \qquad n = 1, 2, \ldots$$

The solution to the wave equation is

$$u(x, t) = \sum_{n=1}^{\infty} b_n \sin\frac{n\pi x}{L}\cos\frac{n\pi t}{L}$$

Fourier transform A nonperiodic function $f(x)$ requires two functions to describe its Fourier transform:

$$A(w) = \int_{-\infty}^{\infty} f(x)\cos wx\,dx$$

$$B(w) = \int_{-\infty}^{\infty} f(x)\sin wx\,dx$$

Then the **Fourier integral equation** is

$$f(x) = \int_{0}^{\infty} A(w)\cos wx + B(w)\sin wx\,dw$$

The **complex Fourier transform** of $f(x)$ is defined as

$$F(w) = \int_{-\infty}^{\infty} f(x)e^{iwx}\,dx$$

Then the **complex Fourier integral equation** is

$$f(x) = \frac{1}{2\pi}\int_{0}^{\infty} F(w)e^{-iwx}\,dx$$

Heat Equation The Fourier transform may be used to solve the one-dimensional heat equation $u_t(x, t) = u_{xx}(x, t)$, for $t > 0$, given initial condition $u(x, 0) = f(x)$. Let $F(s)$ be the complex Fourier transform of $f(x)$, and let $U(s, t)$ be the complex Fourier transform of $u(x, t)$. Then the transform of $u_t(x, t)$ is $dU(s, t)/dt$.

Transforming $u_t(x, t) = u_{xx}(s, t)$ yields $dU/dt + s^2 U = 0$ and $U(s, 0) = f(s)$. Solving this using the Laplace transform gives $U(s, t) = F(s)e^{s^2 t}$.

Applying the complex Fourier integral equation, which gives $u(x, t)$ in terms of $U(s, t)$, gives

$$u(x, t) = \frac{1}{2\pi}\int_{0}^{\infty} U(s, t)e^{-isx}\,ds$$

$$= \frac{1}{2\pi}\int_{-\infty}^{\infty}\int_{0}^{\infty} f(y)e^{is(y-x)}e^{s^2 t}\,ds\,dy$$

Applying the Euler formula, $e^{ix} = \cos x + i\sin x$,

$$u(x, t) = \frac{1}{2\pi}\int_{-\infty}^{\infty}\int_{0}^{\infty} f(y)\cos\,(s(y - x))e^{s^2 t}\,ds\,dy$$

SPECIAL FUNCTIONS

Gamma Function The gamma function is a generalization of the factorial function. It arises in Laplace transforms of polynomials, in continuous probability, applications involving fractals and fractional derivatives, and in the solution to certain differential equations. It is defined by the improper integral:

$$\Gamma(x) = \int_{0}^{\infty} t^{x-1}e^{-t}\,dt$$

The integral converges for $x > 0$ and diverges otherwise. The function is extended to all negative values, except negative integers, by the relation

$$\Gamma(x + 1) = x\Gamma(x)$$

The gamma function is related to the factorial function by

$$\Gamma(n + 1) = n!$$

for all positive integers n.

An important value of the gamma function is

$$\Gamma(0.5) = \pi^{1/2}$$

Other values of the gamma function are found in CRC Standard Mathematical Tables and similar tables.

Beta Function The beta function is a function of two variables and is a generalization of the binomial coefficients. It is closely related to the gamma function. It is defined by the integral:

$$B(x, y) = \int_0^1 t^{x-1}(1 - t)^{y-1}\, dt \qquad \text{for } x, y > 0$$

The beta function can also be represented as a trigonometric integral, by substituting $t = \sin^2 \theta$, as

$$B(x, y) = 2 \int_0^{\pi/2} (\sin \theta)^{2x-1}(\cos \theta)^{2y-1}\, d\theta$$

The beta function is related to the gamma function by the relation

$$B(x, y) = \frac{\Gamma(x)\Gamma(y)}{\Gamma(x + y)}$$

This relation shows that $B(x, y) = B(y, x)$.

Bernoulli Functions   The Bernoulli functions are a sequence of periodic functions of period 1 used in approximation theory. Note that for any number x, $[x]$ represents the largest integer less than or equal to x. $[3.14] = 3$ and $[-1.2] = -2$. The Bernoulli functions $B_n(x)$ are defined recursively as follows:

1. $B_0(x) = 1$
2. $B_1(x) = x - [x] - \frac{1}{2}$
3. B_{n+1} is defined so that $B'_{n+1}(x) = B_n(x)$ and so that B_{n+1} is periodic of period 1.

B_1 is a special case of the linear-sweep curve (Fig. 2.1.112).

Bessel Functions of the First Kind   Bessel functions of the first kind arise in the solution of Bessel's equation of order v:

$$x^2 y'' + xy' + (x^2 - v^2)y = 0$$

When this is solved using series methods, the recursive relations define the Bessel functions of the first kind of order v:

$$J_v(x) = \sum_{k=0}^{\infty} \frac{(-1)^k}{k!(v + k + 1)}\left(\frac{x}{2}\right)^{v+2k}$$

Chebyshev Polynomials   The Chebyshev polynomials arise in the solution of PDEs of the form

$$(1 - x^2)y'' - xy' + n^2 y = 0$$

and in approximation theory. They are defined as follows:

$$T_0(x) = 1 \qquad T_2(x) = 2x^2 - 1$$
$$T_1(x) = x \qquad T_3(x) = 4x^3 - 3x$$

For $n > 3$, they are defined recursively by the relation

$$T_{n+1}(x) - 2xT_n(x) + T_{n-1}(x) = 0$$

Chebyshev polynomials are said to be orthogonal because they have the property

$$\int_{-1}^1 \frac{T_n(x)T_m(x)}{(1 - x^2)^{1/2}}\, dx = 0 \qquad \text{for } n \neq m$$

NUMERICAL METHODS

Introduction   Classical numerical analysis is based on polynomial approximation of the infinite operations of integration, differentiation, and interpolation. The objective of such analyses is to replace difficult or impossible exact computations with easier approximate computations. The challenge is to make the approximate computations short enough and accurate enough to be useful.

Modern numerical analysis includes Fourier methods, including the *fast Fourier transform* (FFT) and many problems involving the way computers perform calculations. Modern aspects of the theory are changing very rapidly.

Errors   Actual value = calculated value + error. There are several sources of errors in a calculation: mistakes, round-off errors, truncation errors, and accumulation errors.

Round-off errors arise from the use of a number not sufficiently accurate to represent the actual value of the number, for example, using 3.14159 to represent the irrational number π, or using 0.56 to represent $\frac{9}{16}$ or 0.5625.

Truncation errors arise when a finite number of steps are used to approximate an infinite number of steps, for example, the first n terms of a series are used instead of the infinite series.

Accumulation errors occur when an error in one step is carried forward into another step. For example, if $x = 0.994$ has been previously rounded to 0.99, then $1 - x$ will be calculated as 0.01, while its true value is 0.006. An error of less than 1 percent is accumulated into an error of over 50 percent in just one step. Accumulation errors are particularly characteristic of methods involving recursion or iteration, where the same step is performed many times, with the results of one iteration used as the input for the next.

Simultaneous Linear Equations   The matrix equation $Ax = b$ can be solved directly by finding A^{-1}, or it can be solved iteratively, by the method of **iteration in total steps:**

1. If necessary, rearrange the rows of the equation so that there are no zeros on the diagonal of A.
2. Take as initial approximations for the values of x_i:

$$x_1^{(0)} = \frac{b_1}{a_{11}} \qquad x_2^{(0)} = \frac{b_2}{a_{22}} \qquad \cdots \qquad x_n^{(0)} = \frac{b_n}{a_{nn}}$$

3. For successive approximations, take

$$x_i^{(k+1)} = (b_i - a_{i1}x_1^{(k)} - \cdots - a_{in}x_n^{(k)})/a_{ii}$$

Repeat step 3 until successive approximations for the values of x_i reach the specified tolerance.

A property of iteration by total steps is that it is self-correcting: that is, it can recover both from mistakes and from accumulation errors.

Zeros of Functions   An iterative procedure for solving an equation $f(x) = 0$ is the **Newton-Raphson method.** The algorithm is as follows:

1. Choose a first estimate of a root x_0.
2. Let $x_{k+1} = x_k - f(x_k)/f'(x_k)$. Repeat step 2 until the estimate x_k converges to a root r.
3. If there are other roots of $f(x)$, then let $g(x) = f(x)/(x - r)$ and seek roots of $g(x)$.

False Position   If two values x_0 and x_1 are known, such that $f(x_0)$ and $f(x_1)$ are opposite signs, then an iterative procedure for finding a root between x_0 and x_1 is the method of false position.

1. Let $m = [f(x_1) - f(x_0)]/(x_1 - x_0)$.
2. Let $x_2 = x_1 - f(x_1)/m$.
3. Find $f(x_2)$.
4. If $f(x_2)$ and $f(x_1)$ have the same sign, then let $x_1 = x_2$. Otherwise, let $x_0 = x_2$.
5. If x_1 is not a good enough estimate of the root, then return to step 1.

Functional Equalities   To solve an equation of the form $f(x) = g(x)$, use the methods above to find roots of the equation $f(x) - g(x) = 0$.

Maxima   One method for finding the maximum of a function $f(x)$ on an interval $[a, b]$ is to find the roots of the derivative $f'(x)$. The maximum of $f(x)$ occurs at a root or at an endpoint a or b.

Fibonacci Search   An iterative procedure for searching for maxima works if $f(x)$ is unimodular on $[a, b]$. That is, f has only one maximum, and no other local maxima, between a and b. This procedure takes advantage of the so-called *golden ratio*, $r = 0.618034 = (\sqrt{5} - 1)/2$, which arises from the Fibonacci sequence.

1. If a is a sufficiently good estimate of the maximum, then stop. Otherwise, proceed to step 2.
2. Let $x_1 = ra + (1 - r)b$, and let $x_2 = (1 - r)a + rb$. Note $x_1 < x_2$. Find $f(x_1)$ and $f(x_2)$.
   a. If $f(x_1) = f(x_2)$, then let $a = x_1$ and $b = x_2$, and go to step 1.
   b. If $f(x_1) < f(x_2)$, then let $a = x_1$, and go to step 1.
   c. If $f(x_1) > f(x_2)$, then let $b = x_2$, and return to step 1.

In cases b and c, computation is saved since the new value of one of x_1 and x_2 will have been used in the previous step. It has been proved that the Fibonacci search is the fastest possible of the general "cutting" type of searches.

Steepest Ascent   If $z = f(x, y)$ is to be maximized, then the method of steepest ascent takes advantage of the fact that the gradient, grad (f) always points in the direction that f is increasing the fastest.

1. Let (x_0, y_0) be an initial guess of the maximum of f.
2. Let e be an initial step size, usually taken to be small.
3. Let $(x_{k+1}, y_{k+1}) = (x_k, y_k) + e$ grad $f(x_k, y_k)/|$grad $f(x_k, y_k)|$
4. If $f(x_{k+1}, y_{k+1})$ is not greater than $f(x_k, y_k)$, then replace e with $e/2$ (cut the step size in half) and reperform step 3.
5. If (x_k, y_k) is a sufficiently accurate estimate of the maximum, then stop. Otherwise, repeat step 3.

Minimization   The theory of minimization exactly parallels the theory of maximization, since minimizing $z = f(x)$ occurs at the same value of x as maximizing $w = -f(x)$.

Numerical Differentiation   In general, numerical differentiation should be avoided where possible, since differentiation tends to be very sensitive to small errors in the value of the function $f(x)$. There are several approximations to $f'(x)$, involving a "step size" h usually taken to be small:

$$f'(x) = \frac{f(x + h) - f(x)}{h}$$

$$f'(x) = \frac{f(x + h) - f(x - h)}{2h}$$

$$f'(x) = \frac{f(x + 2h) + f(x + h) - f(x - h) - f(x - 2h)}{6h}$$

Other formulas are possible.

If a derivative is to be calculated from an equally spaced sequence of measured data, $y_1, y_2, \ldots, y_n$, then the above formulas may be adapted by taking $y_i = f(x_i)$. Then $h = x_{i+1} - x_i$ is the distance between measurements.

Since there are usually noise or measurement errors in measured data, it is often necessary to smooth the data, expecting that errors will be averaged out. Elementary smoothing is by simple averaging, where a value y_i is replaced by an average before the derivative is calculated. Examples include:

$$y_i \leftarrow \frac{y_{i+1} + y_i + y_{i-1}}{3}$$

$$y_i \leftarrow \frac{y_{i+2} + y_{i+1} + y_i + y_{i-1} + y_{i-2}}{5}$$

More information may be found in the literature under the topics linear filters, digital signal processing, and smoothing techniques.

Numerical Integration

Numerical integration requires a great deal of calculation and is usually done with the aid of a computer. All the methods described here, and many others, are widely available in packaged computer software. There is often a temptation to use whatever software is available without first checking that it really is appropriate. For this reason, it is important that the user be familiar with the methods being used and that he or she ensure that the error terms are tolerably small.

Trapezoid Rule   If an interval $a \leq x \leq b$ is divided into subintervals $x_0, x_1, \ldots, x_n$, then the definite integral

$$\int_a^b f(x)\, dx$$

may be approximated by

$$\sum_{i=1}^n [f(x_i) + f(x_{i-1})]\frac{x_{i+1} - x_i}{2}$$

If the values x_i are equally spaced at distance h and if f_i is written for $f(x_i)$, then the above formula reduces to

$$[f_0 + 2f_1 + 2f_2 + \cdots + 2f_{n-1} + f_n]\frac{h}{2}$$

The error in the trapezoid rule is given by

$$|E_n| \leq \frac{(b - a)^3|f''(t)|}{12n^2}$$

where t is some value $a \leq t \leq b$.

Simpson's Rule   The most widely used rule for numerical integration approximates the curve with parabolas. The interval $a < x < b$ must be divided into $n/2$ subintervals, each of length $2h$, where n is an even number. Using the notation above, the integral is approximated by

$$[f_0 + 4f_1 + 2f_2 + 4f_3 + \cdots + 4f_{n-1} + f_n]\frac{h}{3}$$

The error term for Simpson's rule is given by $|E_n| < nh^5|f^{(4)}(t)|/180$, where $a < t < b$.

Simpson's rule is generally more accurate than the trapezoid rule.

Ordinary Differential Equations

Modified Euler Method   Consider a first-order differential equation $dy/dx = f(x, y)$ and initial condition $y = y_0$ and $x = x_0$. Take x_i equally spaced, with $x_{i+1} - x_i = h$. Then the method is:

1. Set $n = 0$.
2. $y_n' = f(x_n, y_n)$ and $y'' = f_x(x_n, y_n) + y_n'f_y(x_n, y_n)$, where f_x and f_y denote partial derivatives.
3. $y_{n+1}' = f(x_{n+1}, y_{n+1})$.

Predictor steps:

4. For $n > 0$, $y_{n+1}^* = y_{n-1} + 2hy_n'$.
5. $y_{n+1}'^* = f(x_{n+1}, y_{n+1}^*)$.

Corrector steps:

6. $y_n^\# + 1 = y_n + [y_{n+1}'^* + y_n']h/2$.
7. $y_{n+1}'^\# = f(x_{n+1}, y_{n+1}^\#)$.
8. If required accuracy is not yet obtained for y_{n+1} and y_{n+1}', then substitute $y^\#$ for y^*, in all its forms, and repeat the corrector steps. Otherwise, set $n = n + 1$ and return to step 2.

Other predictor-corrector methods are described in the literature.

Runge-Kutta Methods   These make up a family of widely used methods for ordinary differential equations. Given $dy/dx = f(x, y)$ and h = interval size, third-order method (error proportional to h^4):

$$k_0 = hf(x_n)$$

$$k_1 = hf\left(x_n + \frac{h}{2}, y_n + \frac{k_0}{2}\right)$$

$$k_2 = hf(x_n + h, y_n + 2k_1 - k_0)$$

$$y_{n+1} = y_n + \frac{k_0 + 4k_1 + k_2}{6}$$

Higher-order Runge-Kutta methods are described in the literature. In general, higher-order methods yield smaller error terms.

Partial Differential Equations

Approximate solutions to partial differential equations require even more computational effort than ordinary differential equations. A typical problem involves finding the value of a function of two or more variables on the interior of a region, given the values of the function on the boundary of that region and also given a partial differential equation that the function must satisfy.

A typical solution strategy involves representing the continuous problem by its values on a finite set of points, called nodes, within the region, and representing the differential equation by difference equations on the value of the function at the nodes. The boundary value conditions are represented by assigning values to those nodes at or near the boundary of the region.

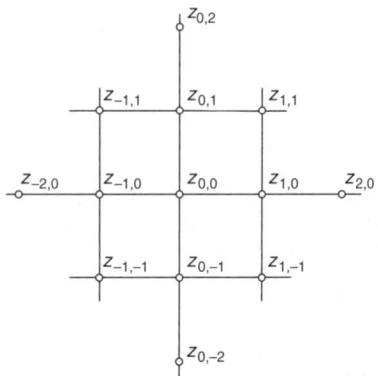

Fig. 2.1.113   Rectangular net.

The pattern of nodes within the region is called a net. A rectangular net is shown in Fig. 2.1.113. There is also an extensive theory of triangular nets. Some techniques use an irregular pattern of nodes, with the nodes more concentrated in more critical parts of the region. Such techniques are called finite element methods.

If the distance between nodes in a net is given by a, then the various partial derivatives at a value $z_{0,0} = f(x_0, y_0)$ may be given by

$$\frac{\partial z}{\partial x} \approx \frac{1}{2a}(z_{1,0} - z_{-1,0})$$

$$\frac{\partial z}{\partial y} \approx \frac{1}{2a}(z_{0,1} - z_{0,-1})$$

$$\frac{\partial^2 z}{\partial x^2} \approx \frac{1}{a^2}(z_{1,0} - 2z_{0,0} + z_{-1,0})$$

$$\frac{\partial^2 z}{\partial x \partial y} \approx \frac{1}{4a^2}(z_{1,1} - z_{-1,1} + z_{-1,-1} - z_{1,-1})$$

$$\frac{\partial^2 z}{\partial y^2} \approx \frac{1}{a^2}(z_{0,1} - 2z_{0,0} + z_{0,-1})$$

Other formulas are possible, and there are also formulas for higher order partial derivatives.

Applying the appropriate difference formulas to each node in a net gives an equation for each node. If there are n nodes, then n linear equations involving n unknowns arise. Though n is likely to be rather large, the system of linear equations is *sparse*, that is, most of the entries in the $n \times n$ matrix that describes the system are zero. Special methods exist for solving sparse systems of simultaneous linear equations relatively efficiently.

EXAMPLE.   Solve the generalized Laplace equation

$$\nabla^2 f = \frac{\partial^2 f}{\partial x^2} + \frac{\partial^2 f}{\partial y^2} = g(x, y)$$

where $g(x, y)$ is a given function of x and y. At each point (x_i, y_j) on the interior of the region, we approximate the partial derivatives as

$$\frac{\partial^2 z}{\partial x^2} \approx \frac{1}{a^2}(z_{i+1,j} - 2z_{i,j} + z_{i-1,j}) \quad \text{and} \quad \frac{\partial^2 z}{\partial y^2} \approx \frac{1}{a^2}(z_{i,j+1} - 2z_{i,j} + z_{i,j-1}).$$

This gives a difference equation

$$\frac{1}{a^2}(z_{i+1,j} + z_{i-1,j} + z_{i,j+1} + z_{i,j-1} - 4z_{i,j}) = g(x_i, y_j)$$

that must be satisfied at each point (x_i, y_j).

Relaxation methods are a family of techniques for finding approximate solutions to such systems of equations. They involve finding the points (x_i, y_j) for which the difference equations are farthest from being true, and adjusting or relaxing the value of $z_{i,j}$ so that the equation is satisfied at that point.

2.2 COMPUTERS
by Thomas J. Cockerill

REFERENCES: Manuals from Computer Manufacturers. Knuth, "The Art of Computer Programming," vols 1, 2, and 3, Addison-Wesley. Yourdon and Constantine, "Structured Design," Prentice-Hall. DeMarco, "Structured Analysis and System Specification," Prentice-Hall. Moshos, "Data Communications," West Publishing. Date, "An Introduction to Database Systems," 4th ed., Addison-Wesley. Wiener and Sincovec, "Software Engineering with Modula-2 and ADA," Wiley. Hamming, "Numerical Methods for Scientists and Engineers," McGraw-Hill. Bowers and Sedore, "SCEPTRE: A Computer Program for Circuit and System Analysis," Prentice-Hall. Tannenbaum, "Operating Systems," Prentice-Hall. Lister, "Fundamentals of Operating Systems, 3d ed., Springer-Verlag. American National Standard Programming Language FORTRAN, ANSI X3.198-1992. Jensen and Wirth, "PASCAL: User Manual and Report," Springer. *Communications, Journal,* and *Computer Surveys,* ACM Computer Society. *Computer, Spectrum,* IEEE.

COMPUTER PROGRAMMING

Machine Types

Computers are machines used for automatically processing information represented by mechanical, electrical, or optical means. They may be classified as analog or digital according to the techniques used to represent and process the information. Analog computers represent information as physically measurable, continuous quantities and process the information by components that have been interconnected to form an analogous model of the problem to be solved. Digital computers, on the other hand, represent information as discrete physical states which have been encoded into symbolic formats, and process the information by sequences of operational steps which have been preplanned to solve the given problem.

When compared to analog computers, digital computers have the advantages of greater versatility in solving scientific, engineering, and commercial problems that involve numerical and nonnumerical information; of an accuracy dictated by significant digits rather than that which can be measured; and of exact reproducibility of results that stay unvitiated by small, random fluctuations in the physical signals. In the past, multiple-purpose analog computers offered advantages of speed and cost in solving a sophisticated class of complex problems dealing with networks of differential equations, but these advantages have disappeared with the advances in solid-state computers. Other than the occasional use of analog techniques for embedding computations as part of a larger system, digital techniques now account almost exclusively for the technology used in computers.

Digital information may be represented as a series of incremental, numerical steps which may be manipulated to position control devices using stepping motors. Digital information may also be encoded into symbolic formats representing digits, alphabetic characters, arithmetic numbers, words, linguistic constructs, points, and pictures which may be processed by a variety of mechanized operators. Machines organized in this manner can handle a more general class of both numerical and nonnumerical problems and so form by far the most common type of digital machines. In fact, the term *computer* has become synonymous with this type of machine.

Digital Machines

Digital machines consist of two kinds of circuits: memory cells, which effectively act to delay signals until needed, and logical units, which perform basic Boolean operations such as AND, OR, NOT, XOR, NAND, and NOR. Memory circuits can be simply defined as units where information can be stored and retrieved on demand. Configurations assembled from the Boolean operators provide the macro operators and functions available to the machine user through encoded instructions.

Both data and the instructions for processing the data can be stored in memory. Each unit of memory has an address at which the contents can be retrieved, or "read." The *read* operation makes the contents at an address available to other parts of the computer without destroying the contents in memory. The contents at an address may be changed by a *write* operation which inserts new information after first nullifying the previous contents. Some types of memory, called read-only memory (**ROM**), can be read from but not written to. They can only be changed at the factory.

Abstractly, the address and the contents at the address serve roles analogous to a variable and the value of the variable. For example, the equation $z = x + y$ specifies that the value of x added to the value of y will produce the value of z. In a similar way, the machine instruction whose format might be:

add, address 1, address 2, address 3

will, when executed, add the contents at address 1 to the contents at address 2 and store the result at address 3. As in the equation where the variables remain unaltered while the values of the variables may be changed, the addresses in the instruction remain unaltered while the contents at the address may change.

Computers differ from other kinds of mechanical and electrical machines in that computers perform work on information rather than on forces and displacements. A common form of information is numbers. Numbers can be encoded into a mechanized form and processed by the four rules of arithmetic ($+$, $-$, $\times$, $\div$). But numbers are only one kind of information that can be manipulated by the computer. Given an encoded alphabet, words and languages can be formed and the computer can be used to perform such processes as information storage and retrieval, translation, and editing. Given an encoded representation of points and lines, the computer can be used to perform such functions as drawing, recognizing, editing, and displaying graphs, patterns, and pictures.

Because computers have become easily accessible, engineers and scientists from every discipline have reformatted their professional activities to mechanize those aspects which can supplant human thought and decision. In this way, mechanical processes can be viewed as augmenting human physical skills and strength, and information processes can be viewed as augmenting human mental skills and intelligence.

COMPUTER DATA STRUCTURES

Binary Notation

Digital computers represent information by strings of digits which assume one of two values: 0 or 1. These units of information are called **bits**, a word contracted from the term *binary digits*. A string of bits may represent either numerical or nonnumerical information.

In order to achieve efficiency in handling the information, the computer groups the bits together into units containing a fixed number of bits which can be referenced as discrete units. By encoding and formatting these units of information, the computer can act to process them. Units of 8 bits, called **bytes**, are common. A byte can be used to encode the basic symbolic characters which provide the computer with input-output information such as the alphabet, decimal digits, punctuation marks, and special characters.

Bit groups may be organized into larger units of 4 bytes (32 bits) called **words**, or even larger units of 8 bytes called *double words;* and sometimes into smaller units of 2 bytes called *half words*. Besides encoding numerical information and other linguistic constructs, these units are used to encode a repertoire of machine instructions. Older machines and special-purpose machines may have other word sizes.

Computers process numerical information represented as binary numbers. The binary numbering system uses a positional notation similar to the decimal system. For example, the decimal number 596.37 represents the value $5 \times 10^2 + 9 \times 10^1 + 6 \times 10^0 + 3 \times 10^{-1} + 7 \times 10^{-2}$. The value assigned to any of the 10 possible digits in the decimal system depends on its position relative to the decimal point (a weight of 10 to zero or positive exponent is assigned to the digits appearing to the left of the decimal point, and a weight of 10 to a negative exponent is applied to digits to the right of the decimal point). In a similar manner, a binary number uses a radix of 2 and two possible digits: 0 and 1. The radix point in the positional notation separates the whole from the fractional part of the number, just as in the decimal system. The binary number 1011.011 represents a value $1 \times 2^3 + 0 \times 2^2 + 1 \times 2^1 + 1 \times 2^0 + 0 \times 2^{-1} + 1 \times 2^{-2} + 1 \times 2^{-3}$.

The operators available in the computer for setting up the solution of a problem are encoded into the instructions of the machine. The instruction repertoire always includes the usual arithmetic operators to handle numerical calculations. These instructions operate on data encoded in the binary system. However, this is not a serious operational problem, since the user specifies the numbers in the decimal system or by mnemonics, and the computer converts these formats into its own internal binary representation.

On occasions when one must express a number directly in the binary system, the number of digits needed to represent a numerical value becomes a handicap. In these situations, a radix of 8 or 16 (called the **octal** or **hexadecimal** system, respectively) constitutes a more convenient system. Starting with the digit to the left or with the digit to the right of the radix point, groups of 3 or 4 binary digits can be easily converted to equivalent octal or hexadecimal digits, respectively. Appending nonsignificant 0s as needed to the rightmost and leftmost part of the number to complete the set of 3 or 4 binary digits may be necessary. Table 2.2.1 lists the conversions of binary digits to their equivalent octal and hexadecimal representations. In the hexadecimal system, the letters A through F augment the set of decimal digits to represent the digits for 10 through 15. The following examples illustrate the conversion

Table 2.2.1 Binary-Hexadecimal and Binary-Octal Conversion

Binary	Hexadecimal	Binary	Octal
0000	0	000	0
0001	1	001	1
0010	2	010	2
0011	3	011	3
0100	4	100	4
0101	5	101	5
0110	6	110	6
0111	7	111	7
1000	8		
1001	9		
1010	A		
1011	B		
1100	C		
1101	D		
1110	E		
1111	F		

Table 2.2.2 Schemes for Encoding Decimal Digits

Decimal digit	BCD	Excess-3	4221 code
0	0000	0011	0000
1	0001	0100	0001
2	0010	0101	0010
3	0011	0110	0011
4	0100	0111	0110
5	0101	1000	1001
6	0110	1001	1100
7	0111	1010	1101
8	1000	1011	1110
9	1001	1100	1111

between binary numbers and octal or hexadecimal numbers using the table.

binary number	011	011	110	101	.	001	111	100
octal number	3	3	6	5	.	1	7	4

binary number	0110	1111	0101	.	0011	1110
hexadecimal number	6	F	5	.	3	E

Formats for Numerical Data

Three different formats are used to represent numerical information internal to the computer: fixed-point, encoded decimal, and floating-point.

A word or half word in fixed-point format is given as a string of 0s and 1s representing a binary number. The program infers the position of the radix point (immediately to the right of the word representing integers, and immediately to the left of the word representing fractions). Algebraic numbers have several alternate forms: 1's complement, 2's complement, and signed-magnitude. Most often 1's and 2's complement forms are adopted because they lead to a simplification in the hardware needed to perform the arithmetic operations. The sign of a 1's complement number can be changed by replacing the 0s with 1s and the 1s with 0s. To change the sign of a 2's complement number, reverse the digits as with a 1's-complement number and then add a 1 to the resulting binary number. Signed-magnitude numbers use the common representation of an explicit + or − sign by encoding the sign in the leftmost bit as a 0 or 1, respectively.

Many computers provide an encoded-decimal representation as a convenience for applications needing a decimal system. Table 2.2.2 gives three out of over 8000 possible schemes used to encode decimal digits in which 4 bits represent each decade value. Many other codes are possible using more bits per decade, but four bits per decimal digit are common because two decimal digits can then be encoded in one byte. The particular scheme selected depends on the properties needed by the devices in the application.

The floating-point format is a mechanized version of the scientific notation ($\pm M \times 10^{\pm E}$, where $\pm M$ and $\pm E$ represent the signed mantissa and signed exponent of the number). This format makes possible the use of a machine word to encode a large range of numbers. The signed mantissa and signed exponent occupy a portion of the word. The exponent is implied as a power of 2 or 16 rather than of 10, and the radix point is implied to the left of the mantissa. After each operation, the machine adjusts the exponent so that a nonzero digit appears in the most significant digit of the mantissa. That is, the mantissa is **normalized** so that its value lies in the range of $1/b \leq M < 1$ where b is the implied base of the number system (e.g.: $1/2 \leq M < 1$ for a radix of 2, and $1/16 \leq M < 1$ for a radix of 16). Since the zero in this notation has many logical representations, the format uses a standard recognizable form for zero, with a zero mantissa and a zero exponent, in order to avoid any ambiguity.

When calculations need greater precision, floating-point numbers use a two-word representation. The first word contains the exponent and mantissa as in the one-word floating point. Precision is increased by appending the extra word to the mantissa. The terms **single precision** and **double precision** make the distinction between the one- and

Fig. 2.2.1 ASCII code set.

two-word representations for floating-point numbers, although *extended precision* would be a more accurate term for the two-word form since the added word more than doubles the number of significant digits.

The equivalent decimal precision of a floating-point number depends on the number n of bits used for the unsigned mantissa and on the implied base b (binary, octal, or hexadecimal). This can be simply expressed in equivalent decimal digits p as: $0.0301 (n - \log_2 b) < p < 0.0301 n$. For example, a 32-bit number using 7 bits for the signed exponent of an implied base of 16, 1 bit for the sign of the mantissa, and 24 bits for the value of the mantissa gives a precision of 6.02 to 7.22 equivalent decimal digits. The fractional parts indicate that some 7-digit and some 8-digit numbers cannot be represented with a mantissa of 24 bits. On the other hand, a double-precision number formed by adding another word of 32 bits to the 24-bit mantissa gives a precision of 15.65 to 16.85 equivalent decimal digits.

The range r of possible values in floating-point notation depends on the number of bits used to represent the exponent and the implied radix. For example, for a signed exponent of 7 bits and an implied base of 16, then $16^{-64} \leq r \leq 16^{63}$.

Formats of Nonnumerical Data

Logical elements, also called **Boolean** elements, have two possible values which simply represent 0 or 1, true or false, yes or no, OFF or ON, etc. These values may be conveniently encoded by a single bit.

A large variety of codes are used to represent the alphabet, digits, punctuation marks, and other special symbols. The most popular ones are the 7-bit ASCII code and the 8-bit EBCDIC code. ASCII and EBCDIC find their genesis in punch-tape and punch-card technologies, respectively, where each character was encoded as a combination of punched holes in a column. Both have now evolved into accepted standards represented by a combination of 0s and 1s in a byte.

Figure 2.2.1 shows the ASCII code. (ASCII stands for American Standard Code for Information Interchange.) The possible 128 bit patterns divide the code into 96 graphic characters (although the codes 0100000 and 1111111 do not represent any printable graphic symbol) and 32 control characters which represent nonprintable characters used in communications, in controlling peripheral machines, or in expanding the code set with other characters or fonts. The graphic codes and the control codes are organized so that subsets of usable codes with fewer bits can be formed and still maintain the pattern.

Data Structure Types

The above types of numerical and nonnumerical data formats are recognized and manipulated by the hardware operations of the computer. Other more complex data structures may be programmed into the computer by building upon these primitive data types. The programmable data structures might include arrays, defined as ordered lists of elements of identical type; **sets,** defined as unordered lists of elements of identical type; **records,** defined as ordered lists of elements that need not be of the same type; **files,** defined as sequential collections of identical records; and **databases,** defined as organized collections of different records or file types.

COMPUTER ORGANIZATION

Principal Components

The principal components of a computer system consist of a central processing unit (referred to as the **CPU** or platform), its working memory, user interface, file storage, and a collection of add-ons and peripheral devices. A computer system can be viewed as a library of collected data and packages of assembled sequences of instructions that can be executed in the prescribed order by the CPU to solve specific problems or perform utility functions for the users. These sequences are variously called programs, subprograms, routines, subroutines, procedures, functions, etc. Collectively they are called **software** and are directly accessible to the CPU through the working memory. The file devices act analogously to a bookshelf—they store information until it is needed. Only after a program and its data have been transferred from the file devices or from peripheral devices to the working memory can the individual instructions and data be addressed and executed to perform their intended functions.

The CPU functions to monitor the flow of data and instructions into and out of memory during program execution, control the order of instruction execution, decode the operation, locate the operand(s) needed, and perform the operation specified. Two characteristics of the memory and storage components dictate the roles they play in the computer system. They are **access time,** defined as the elapsed time between the instant a read or write operation has been initiated and the instant the operation is completed, and size, defined by the number of bytes in a module. The faster the access time, the more costly per bit of memory or storage, and the smaller the module. The principal types of memory and storage components from the fastest to the slowest are registers which operate as an integral part of the CPU, cache and main memory which form the working memory, and mass and archival storage which serve for storing files.

The interrelationships among the components in a computer system and their primary performance parameters will be given in context in the following discussion. However, hundreds of manufacturers of computers and computer products have a stake in advancing the technology and adding new functionality to maintain their competitive edge. In such an environment, no performance figures stay current. With this caveat, performance figures given should not be taken as absolutes but only as an indication of how each component contributes to the performance of the total system.

Throughout the discussion (and in the computer world generally), prefixes indicating large numbers are given by the symbols k for kilo (10^3), M for mega (10^6), G for giga (10^9), and T for tera (10^{12}). For memory units, however, these symbols have a slightly altered meaning. Memories are organized in binary units whereby powers of two form the basis for all addressing schemes. According, k refers to a memory size of 1024 (2^{10}) units. Similarly M refers to 1024^2 (1,048,576), G refers to 1024^3, and T refers to 1024^4. For example, 1-Mbyte memory indicates a size of 1,048,576 bytes.

Memory

The main memory, also known as random access memory (**RAM**), is organized into fixed size bit cells (words, bytes, half words, or double words) which can be located by address and whose contents contain the instructions and data currently being executed. The CPU acts to address the individual memory cells during program execution and transfers their contents to and from its internal registers.

Optionally, the working memory may contain an auxiliary memory, called **cache,** which is faster than the main memory. Cache operates on the premise that data and instructions that will shortly be needed are located near those currently being used. If the information is not found in the cache, then it is transferred from the main memory. The effective average access times offered by the combined configuration of RAM and cache results in a more powerful (faster) computer.

Central Processing Unit

The CPU makes available a repertoire of instructions which the user uses to set up the problem solutions. Although the specific format for instructions varies among machines, the following illustrates the pattern:

name: operator, operand(s)

The name designates an address whose contents contain the operator and one or more operands. The operator encodes an operation permitted by the hardware of the CPU. The operand(s) refer to the entities used in the operation which may be either data or another instruction specified by address. Some instructions have implied operand(s) and use the bits which would have been used for operand(s) to modify the operator.

To begin execution of a program, the CPU first loads the instructions serially by address into the memory either from a peripheral device, or more frequently, from storage. The internal structure of the CPU contains a number of memory registers whose number, while relatively few, depend on the machine's organization. The CPU acts to transfer the instructions one at a time from memory into a designated register where the individual bits can be interpreted and executed by the hardware. The actions of the following steps in the CPU, known as the *fetch-execute cycle,* control the order of instruction execution.

Step 1: Manually or automatically under program control load the address of the starting instruction into a register called the program register (PR).

Step 2: Fetch and copy the contents at the address in PR into a register called the program content register (PCR).

Step 3: Prepare to fetch the next instruction by augmenting PR to the next address in normal sequence.

Step 4: Interpret the instruction in PCR, retrieve the operands, execute the encoded operation, and then return to step 2. Note that the executed instruction may change the address in PR to start a different instruction sequence.

The speed of machines can be compared by calculating the average execution time of an instruction. Table 2.2.3 illustrates a typical instruction mix used in calculating the average. The instruction mix gives the relative frequency each instruction appears in a compiled list of typical programs and so depends on the types of problems one expects the machine to solve (e.g., scientific, commercial, or combination). The equation

$$t = \sum_i w_i t_i$$

expresses the average instruction execution time t as a function of the execution time t_i for instruction i having a relative frequency w_i in the instruction mix. The reciprocal of t measures the processor's performance as the average number of instructions per second (ips) it can execute.

Table 2.2.3   Instruction Mix

i	Instruction type	Weight w_i
1	Add: Floating point	0.07
2	Fixed point	0.16
3	Multiple: Floating point	0.06
4	Load/store register	0.12
5	Shift: One character	0.11
6	Branch: Conditional	0.21
7	Unconditional	0.17
8	Move 3 words in memory	0.10
	Total	1.00

For machines designed to support scientific and engineering calculations, the floating-point arithmetic operations dominate the time needed to execute an average instruction mix. The speed for such machines is given by the average number of floating-point operations which can be executed per second (flops). This measure, however, can be misleading in comparing different machine models. For example, when the machine has been configured with a cluster of processors cooperating through a shared memory, the rate of the configuration (measured in flops) represents the simple sum of the individual processors' flops rates. This does not reflect the amount of parallelism that can be realized within a given problem.

To compare the performance of different machine models, users often assemble and execute a suite of programs which characterize their particular problem load. This idea has been refined so that in 1992 two suites of **benchmark** programs representing typical scientific, mathematical, and engineering applications were standardized: Specint92 for integer operations, and Specfp92 for floating-point operations. Since these benchmark programs were established, many more standardized benchmarks have been created to model critical workloads and applications, and the set is constantly being revised and expanded. Performance ratings for computers are often reported on the basis of results from running a selected subset of the available benchmark programs.

Computer performance depends on a number of interrelated factors used in their design and fabrication, among them compactness, bus size, clock frequency, instruction set size, and number of coprocessors.

The speed that energy can be transmitted through a wire, bounded theoretically at 3×10^{10} cm/s, limits the ultimate speed at which the electronic circuits can operate. The further apart the electronic elements are from each other, the slower the operations. Advances in integrated circuits have produced compact microprocessors operating in the nanosecond range.

The microprocessor's bus size (the width of its data path, or the number of bits that can be sent simultaneously in parallel) affect its performance in two ways: by the number of memory cells that can be directly addressed, and by the number of bits each memory reference can fetch and process at a time. For example, a 16-bit microprocessor can reference 2^{16} 16-bit memory cells and process 16 bits at a time. In order to handle the individual bits, the number of transistors that must be packed into the microprocessor goes up geometrically with the width of the data path. The earliest microprocessors were 8-bit devices, meaning that every memory reference retrieved 8 bits. To retrieve more bits, say 16, 32, or 64 bits, the 8-bit microprocessor had to make multiple references. Microprocessors have become more powerful as the packing technology has improved.

While normally the circuits operate asynchronously, a computer clock times the sequencing of the instructions. Clock speed is given in hertz (Hz, one cycle per second). Each instruction takes an integral number of cycles to complete, with one cycle being the minimum. If an instruction completes its operations in the middle of a cycle, the start of the next instruction must wait for the beginning of the next cycle.

Two schemes are used to implement the computer instruction set in the microprocessors. The more traditional complex instruction set computer (CISC) microprocessors implement by hard-wiring some 300 instruction types. Strange to say, the faster alternate-approach reduced instruction set computer (RISC) implements only about 10 to 30 percent of the instruction types by hard wiring, and implements the remaining more-complex instructions by programming them at the factory into read-only memory. Since the fewer hard-wired instructions are more frequently used and can operate faster in the simpler, more-compact RISC environment, the average instruction time is reduced.

To achieve even greater effectiveness and speed calls for more complex coordination in the execution of instructions and data. In one scheme, microprocessors in a cluster share a common memory to form the machine organization (a multiprocessor or parallel processor). The total work which may come from a single program or independent programs is parceled out to the individual machines which operate independently but are coordinated to work in parallel with the other machines in the cluster. Faster speeds can be achieved when the individual processors can work on different parts of the problem or can be assigned to those parts of the problem solution for which they have been especially designed (e.g., input-output operations or computational

operations). Two other schemes, pipelining and array processing, divide an instruction into the separate tasks that must be performed to complete its execution. A pipelining machine executes the tasks concurrently on consecutive pieces of data. An array processor executes the tasks of the different instructions in a sequence simultaneously and coordinates their completion (which might mean abandoning a partially completed instruction if it had been initiated prematurely). These schemes are usually associated with the larger and faster machines.

User Interface

The user interface is provided to enable the computer user to initiate or terminate tasks, to interrogate the computer to determine the status of tasks during execution, to give and receive instructions, and to otherwise monitor the operation of the system.

The user interface is continuously evolving, but commonly consists of a relatively slow speed keyboard input, a display, and a pointing device. Important display characteristics for usability are the size of the screen and the resolution. The display has its own memory, which refreshes and controls the display. Examples of display devices that have seen common usage are cathode-ray tube (CRT) displays, flat-panel displays, and thin-film displays. For convenience and manual speed, a pointing device such as a mouse, trackball, or touchpad can be used to control the movement of the cursor on the display. The pointing device can also be used to locate and select options available on the display.

File Devices

File devices serve to store libraries of directly accessible programs and data in electronically or optically readable formats. The file devices record the information in large blocks rather than by individual addresses. To be used, the blocks must first be transferred into the working memory. Depending on how selected blocks are located, file devices are categorized as sequential or direct-access. On **sequential** devices the computer locates the information by searching the file from the beginning. Direct-access devices, on the other hand, position the read-write mechanism directly at the location of the needed information. Searching on these devices works concurrently with the CPU and the other devices making up the computer configuration.

Magnetic or **optical disks** that offer a wide choice of options form the commercial direct-access devices. Some file devices are read only and are used to deliver programs or data to be installed or only referenced by the user. Examples of read-only devices are CD-ROM and DVD-ROM. Other file devices can be both read from and written to. Example of read-write devices are magnetic disk drives, CD RJW drives, and DVD-RW drives. The recording surface consists of a platter (or platters) of recording material mounted on a common spindle rotated at high speed. The read-write heads may be permanently positioned along the radius of the platter or may be mounted on a common arm that can be moved radially to locate any specified track of information. Information is recorded on the **tracks** circumferentially using fixed-size blocks called pages or sectors. **Pages** divide the storage and memory space alike into blocks of 4096 bytes so that program transfers can be made without creating unusable space. **Sectors** nominally describe the physical division of the storage space into equal segments for easier positioning of the read-write heads.

The access time for retrieving information from a disk depends on three separately quoted factors, called seek time, latency time, and transfer time. **Seek time** gives the time needed to position the read-write heads from their current track position to the track containing the information. Since the faster fixed-head disks require no radial motion, only latency and transfer time need to be factored into the total access time for these devices. **Latency time** is the time needed to locate the start of the information along the circumferential track. This time depends on the speed of revolution of the disk, and, on average, corresponds to the time required to revolve the platter half a turn. The average latency time can be reduced by repeating the information several times around the track. **Transfer time,** usually quoted as a rate, gives the rate at which information can be transferred to memory after it has been located. There is a large variation in transfer rates depending on the disk system selected.

Computer architects sometimes refer to file storage as **mass storage** or **archival storage,** depending on whether or not the libraries can be kept off-line from the system and mounted when needed. Disk drives with mountable platter(s) and tape drives constitute the archival storage. Sealed disks that often have fixed heads for faster access are the medium of choice for mass storage.

Peripheral Devices and Add-ons

Peripheral devices function as self-contained external units that work on line to the computer to provide or receive information or to control the flow of information. Add-ons are a special class of units whose circuits can be integrated into the circuitry of the computer hardware to augment the basic functionality of the processors. Section 15 covers the electronic technology associated with these devices.

An input device may be defined as any device that provides a machine-readable source of information. For engineering work, the most common forms of input are touch-tone dials, mark sensing, bar codes, and keyboards (usually in conjunction with a printing mechanism or video scope). Many bench instruments have been reconfigured to include digital devices to provide direct input to computers. Because of the datahandling capabilities of the computer, these instruments can be simpler, smaller, and less expensive than the hand instruments they replace. New devices have also been introduced: devices for visual measurement of distance, area, speed, and coordinate position of an object; or for inspecting color or shades of gray for computer-guided vision. Other methods of input that are finding greater acceptance include handwriting recognition, printed character recognition, voice digitizers, and picture digitizers.

Traditionally, output devices play the role of producing displays for the interpretation of results. A large variety of printers, graphical plotters, video displays, and audio sets have been developed for this purpose.

A variety of actuators have been developed for driving control mechanisms. For complex numerical control, programmable controllers (called PLCs) can simultaneously control and update data from multiple tasks. These electronically driven mechanisms and controllers, working with input devices, make possible systems for automatic testing of products, real-time control, and robotics with learning and adaptive capabilities.

Computer Sizes

Computer size refers not only to the physical size but also to the number of electronics elements in the system, and so reflects the performance of the system. Between the two ends of the spectrum from the largest and fastest to the smallest and slowest are machines that vary in speed and complexity. Although no nomenclature has been universally adopted that indicates computer size, the following descriptions illustrate a few generally understood terms used for some common configurations. The choice of which computer is appropriate often requires serious evaluation. In the cases where serious evaluation is required, it is necessary to evaluate common performance benchmarks for the machine. In addition, it may be necessary to evaluate the machine by using a mix of applications designed to simulate current usage of existing machines or the expected usage of the proposed machine.

Personal computers (PCs) have been made possible by the advances in solid-state technology. The name applies to computers that can fit the total complement of hardware on a desktop and operate as stand-alone systems so as to provide immediate dedicated services to an individual user. This popular computer size has been generally credited for spreading computer literacy in today's society. Because of its commercial success, many peripheral devices, add-ons, and software products have been (and are continually being) developed. Laptop PCs are personal computers that have the low weight and size of a briefcase and can easily be transported when peripherals are not immediately needed.

The term **workstation** describes computer systems which have been designed to support complex engineering, scientific, or business applications in a professional environment. Although a top-of-the-line PC or a PC connected as a peripheral to another computer can function like a workstation, one can expect a machine designed as a workstation to offer higher performance than a PC and to support the more specialized peripherals and sophisticated professional software. Nevertheless, the boundary between PCs and workstations changes as the technology advances.

Notebook PCs and the smaller sized **palmtop** PCs are portable, battery-operated machines. These machines find excellent use as portable PCs in some applications and as data acquisition systems. However their undersized keyboards and small displays may limit their usefulness for sustained operations.

Computers larger than a PC or a workstation, called **mainframes** (and sometimes minis or maxis, depending on size), serve to support multiusers and multiapplications. A remotely accessible computing center may house several mainframes which either operate alone or cooperate with each other. Their high speed, large memories, and high reliability allow them to handle complex programs and they have been especially well suited to applications that do not allow for any significant downtime such as banking operations, business transactions, and reservation systems.

At the upper extreme end of the computer spectrum is the **supercomputer,** the class of the fastest machines that can address large, complex scientific/engineering problems which cannot reasonably be transferred to other machines. Obviously this class of computer must have cache and main memory sizes and speeds commensurate with the speed of the platform. While mass memory sizes must also be large, computers which support large databases may often have larger memories than do supercomputers. Large, complex technical problems must be run with high-precision arithmetic. Because of this, performance is measured in double-precision flops. To realize the increasing demand for higher performance, the designers of supercomputers work at the edge of available technology, especially in the use of multiple processors. With multiple processors, however, performance depends as much on the time spent in communication between processors as on the computational speed of the individual processors. In the final analysis, to muster the supercomputer's inherent speed, the development of the software becomes the problem. Some users report that software must often be hand-tailored to the specific problem. The power of the machines, however, can replace years of work in analysis and experimentation.

DISTRIBUTED COMPUTING

Organization of Data Facilities

A distributed computer system can be defined as a collection of computer resources which are remotely located from each other and are interconnected to cooperate in providing their respective services. The resources include both the equipment and the software. Resources distributed to reside near the vicinity where the data is collected or used have an obvious advantage over centralization. But to provide information in a timely and reliable manner, these islands of automation must be integrated.

The size and complexity of an enterprise served by a distributed information system can vary from a single-purpose office to a multipleplant conglomerate. An enterprise is defined as a system which has been created to accomplish a mission in its environment and whose goals involve risk. Internally it consists of organized functions and facilities which have been prepared to provide its services and accomplish its mission. When stimulated by an external entity, the enterprise acts to produce its planned response. An enterprise must handle both the flow of material (goods) and the flow of information. The information system tracks the material in the material system, but itself handles only the enterprise's information.

The technology for distributing and integrating the total information system comes under the industrial strategy known as computer-integrated business (CIB) or computer-integrated manufacturing (CIM). The following reasons have been cited for developing CIB and CIM:

Most data generated locally has only local significance.
Data integrity resides where it is generated.

The quality and consistency of operational decisions demands not only that all parts of the system work with the same data but that they can receive it in a reliable and timely manner.

If a local processor fails, it may disrupt local operations, but the remaining system should continue to function independently.

Small cohesive processors can be best managed and maintained locally.

Through standards, selection of local processes can be made from the best products in a competitive market that can be integrated into the total system.

Obsolete processors can be replaced by processors implemented by more advance technology that conform to standards without the cost of tailoring the products to the existing system.

Figure 2.2.2 depicts the total information system of an enterprise. The database consists of the organized collection of data the processors use in their operations. Because of differences in their communication requirements, the automated procedures are shown separated into those used in the office and those used on the production floor. In a business environment, the front office operations and back office operations make this separation. While all processes have a critical deadline, the production floor handles real-time operations, defined as processes which must complete their response within a critical deadline or else the results of the operations become moot. This places different constraints on the local-area networks (LANs) that serve the communication needs within the office or within the production floor. To communicate with entities outside the enterprise, the enterprise uses a wide-area network (WAN), normally made up from available public network facilities. For efficient and effective operation, the processes must be interconnected by the communications to share the data in the database and so integrate the services provided.

Fig. 2.2.2   Composite view of an enterprise's information system.

Communication Channels

A communication channel provides the connecting path for transmitting signals between a computing system and a remotely located application. Physically the channel may be formed by a wire line using copper, coaxial cable, or optical-fiber cable; or may be formed by a wireless line using radio, microwave, or communication satellites; or may be a combination of these lines.

Capacity, defined as the maximum rate at which information can be transmitted, characterizes a channel independent of the morphic line. Theoretically, an ideal noiseless channel that does not distort the signals has a channel capacity C given by:

$$C = 2W$$

where C is in pulses per second and W is the channel bandwidth. For digital transmission, the Hartley-Shannon theorem sets the capacity of a channel limited by the presence of gaussian noise such as the thermal noise inherent in the components. The formula:

$$C = W \log_2(1 + S/N)$$

gives the capacity C in bits/s in terms of the signal to noise ratio S/N and the bandwidth W. Since the signal to noise ratio is normally given in decibels divisible by 3 (e.g., 12, 18, 21, 24) the following formula provides a workable approximation to the formula above:

$$C = W(S/N)_{db}/3$$

where $(S/N)_{db}$ is the signal-to-noise ratio expressed in decibels. Other forms of noise, signal distortions, and the methods of signal modulation reduce this theoretical capacity appreciably.

Nominal transmission speeds for electronic channels vary from 1000 bits to almost 20 Mbits per second. Fiber optics, however, form an almost noise-free medium. The transmission speed in fiber optics depends on the amount a signal spreads due to the multiple reflected paths it takes from its source to its destination. Advances in fiber technology have reduced this spread to give unbelievable rates. Effectively, the speeds available in today's optical channels make possible the transmission over a common channel, using digital techniques, of all forms of information: text, voice, and pictures.

Besides agreeing on speed, the transmitter and receiver must agree on the mode of transmission and on the timing of the signals. For stations located remotely from each other, transmission occurs by organizing the bits into groups and transferring them, one bit after another, in a serial mode. One scheme, called **asynchronous** or start-stop transmission, uses separate start and stop signals to frame a small group of bits representing a character. Separate but identical clocks at the transmitter and receiver time the signals. For transmission of larger blocks at faster rates, the stations use **synchronous** transmission which embeds the clock information within the transmitted bits.

Communication Layer Model

Figure 2.2.3 depicts two remotely located stations that must cooperate through communication in accomplishing their respective tasks. The communications substructure provides the communication services needed by the application. The application tasks themselves, however, are outside the scope of the communication substructure. The distinction here is similar to that in a telephone system which is not concerned with the application other than to provide the needed communication service. The figure shows the communication facilities packaged into a hierarchical modular layer architecture in which each node contains identical kinds of functions at the same layer level. The layer functions represent abstractions of real facilities, but need not represent specific hardware or software. The entities at a layer provide communication

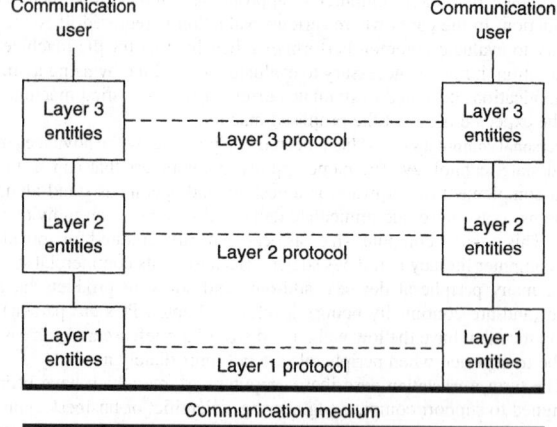

Fig. 2.2.3   Communication layer architecture.

services to the layer above or can request the services available from the layer below. The services provided or requested are available only at service points which are identified by addresses at the boundaries that interface the adjacent layers.

The top and bottom levels of the layered structure are unique. The topmost layer interfaces and provides the communication services to the noncommunication functions performed at a node dealing with the application task (the user's program). This layer also requests communication services from the layer below. The bottom layer does not have a lower layer through which it can request communication services. This layer acts to create and recognize the physical signals transmitted between the bottom entities of the communicating partners (it arranges the actual transmission). The medium that provides the path for the transfer of signals (a wire, usually) connects the service access points at the bottom layers, but itself lies outside the layer structure.

Virtual communication occurs between peer entities, those at the same level. Peer-to-peer communication must conform to layer protocol, defined as the rules and conventions used to exchange information. Actual physical communication proceeds from the upper layers to the bottom, through the communication medium (wire), and then up through the layer structure of the cooperating node.

Since the entities at each layer both transmit and receive data, the protocol between peer layers controls both input and output data, depending on the direction of transmission. The transmitting entities accomplish this by appending control information to each data unit that they pass to the layer below. This control information is later interpreted and removed by the peer entities receiving the data unit.

Communication Standards

Table 2.2.4 lists a few of the hundreds of forums seeking to develop and adopt voluntary standards or to coordinate standards activities. Often users establish standards by agreement that fixes some existing practice. The ISO, however, has described a seven-layer model, called the *Reference Model for Open Systems Interconnection* (OSI), for coordinating and expediting the development of new implementation standards. The term *open systems* refers to systems that allow devices to be interconnected and to communicate with each other by conforming to common implementation standards. The ISO model is not of itself an implementation standard nor does it provide a basis for appraising existing implementations, but it partitions the communication facilities into layers of related function which can be independently standardized by different teams of experts. Its importance lies in the fact that both vendors and users have agreed to provide and accept implementation standards that conform to this model.

Table 2.2.4   Some Groups Involved with Communication Standards

CCITT	Comité Consultatif de Télégraphique et Téléphonique
ISO	International Organization for Standardization
ANSI	American National Standards Institute
EIA	Electronic Industries Association
IEEE	Institute of Electrical and Electronics Engineers
MAP/TOP	Manufacturing Automation Protocols and Technical and Office Protocols Users Group
NIST	National Institute of Standards and Technology

The following lists the names the ISO has given the layers in its ISO model together with a brief description of their roles.

Application layer provides no services to the other layers but serves as the interface for the specialized communication that may be required by the actual application, such as file transfer, message handling, virtual terminal, or job transfer.

Presentation layer relieves the node from having to conform to a particular syntactical representation of the data by converting the data formats to those needed by the layer above.

Session layer coordinates the dialogue between nodes including arranging several sessions to use the same transport layer at one time.

Transport layer establishes and releases the connections between peers to provide for data transfer services such as throughput, transit delays, connection setup delays, error rate control, and assessment of resource availability.

Network layer provides for the establishment, maintenance, and release of the route whereby a node directs information toward its destination.

Data link layer is concerned with the transfer of information that has been organized into larger blocks by creating and recognizing the block boundaries.

Physical layer generates and detects the physical signals representing the bits, and safeguards the integrity of the signals against faulty transmission or lack of synchronization.

The IEEE has formulated several implementation standards for office or production floor LANs that conform to the lower two layers of the ISO model. The functions assigned to the ISO data link layer have been distributed over two sublayers, a logical link control (LLC) upper sublayer that generates and interprets the link control commands, and a medium access control (MAC) lower sublayer that frames the data units and acquires the right to access the medium. From this structure, the IEEE has formulated standards for the MAC sublayer and ISO physical layer combination, and a common standard for the LLC sublayer.

CSMA/CD standardized the access method developed by the Xerox Corporation under its trademark Ethernet. The nodes in the network are attached to a common bus. All nodes hear every message transmitted, but accept only those messages addressed to themselves. When a node has a message to transmit, it listens for the line to be free of other traffic before it initiates transmission. However, more than one node may detect the free line and may start to transmit. In this situation the signals will collide and produce a detectable change in the energy level present in the line. Even after a station detects a collision it must continue to transmit to make sure that all stations hear the collision (all data frames must be of sufficient length to be present simultaneously on the line as they pass each station). On hearing a collision, all stations that are transmitting wait a random length of time and then attempt to retransmit.

The MAP/TOP (Manufacturing Automation Protocols and Technical and Office Protocols) Users Group started under the auspices of General Motors and Boeing Information Systems and now has a membership of many thousands of national and international corporations. The corporations in this group have made a commitment to open systems that will allow them to select the best products through standards, agreed to by the group, that will meet their respective requirements. These standards have also been adopted by NIST for governmentwide use under the title Government Open Systems Interconnections Profile (GOSIP).

The common carriers who offer WAN communication services through their public networks have also developed **packet-switching** networks for public use. Packet switching transmits data in a purely digital format, which, when embellished, can replace the common circuit-switching technology used in analog communications such as voice. A packet is a fixed-sized block of digital data with embedded control information. The network serves to deliver the packets to their destination in an efficient and reliable manner.

Wireless network capabilities for computers based on the IEEE 802.11 set of standards evolved rapidly. The 802.11 standards enabled the implementation of wireless LAN having a performance comparable to that of wired LAN. The wireless LAN network is a set of wireless access points attached to a LAN that allow computers access the LAN. The coverage of the wireless access points can overlap allowing mobility within a region of coverage and allowing transfer from one access point to another if the computer is moved from one coverage region to another.

RELATIONAL DATABASE TECHNOLOGY

Design Concepts

As computer hardware has evolved from small working memories and tape storage to large working memories and large disk storage, so has database technology moved from accessing and processing of a single, sequential file to that of multiple, random-access files. A **relational database** can be defined as an organized collection of interconnected tables or records. The records appear like the flat files of older technology. In each record the information is in columns (fields) which identify attributes, and rows (tuples) which list particular instances of the attributes. One column (or more), known as the **primary key,** identifies each row. Obviously, the primary key must be unique for each row.

If the data is to be handled in an efficient and orderly way, the records cannot be organized in a helter-skelter fashion such as simply transporting existing flat files into relational tables. To avoid problems in maintaining and using the database, redundancy should be eliminated by storing each fact at only one place so that, when making additions or deletions, one need not worry about duplicates throughout the database. This goal can be realized by organizing the records into what is known as the *third normal form.* A record is in the third normal form if and only if all onkey attributes are mutually independent and fully dependent on the primary key.

The advantages of relational databases, assuming proper normalization, are:

Each fact can be stored exactly once.

The integrity of the data resides locally, where it is generated and can best be managed.

The tables can be physically distributed yet interconnected.

Each user can be given his/her own private view of the database without altering its physical structure.

New applications involving only a part of the total database can be developed independently.

The system can be automated to find the best path through the database for the specified data.

Each table can be used in many applications by employing simple operators without having to transfer and manipulate data superfluous to the application.

A large, comprehensive system can evolve from phased design of local systems.

New tables can be added without corrupting everyone's view of the data.

The data in each table can be protected differently for each user (read-only, write-only).

The tables can be made inaccessible to all users who do not have the right to know.

Relational Database Operators

A database system contains the structured collection of data, an on-line catalog and dictionary of data items, and facilities to access and use the data. The system allows users to:

Add new tables

Remove old tables

Insert new data into existing tables

Delete data from existing tables

Retrieve selected data

Manipulate data extracted from several tables

Create specialized reports

As might be expected, these systems include a large collection of operators and built-in functions in addition to those normally used in mathematics. Because of the similarity between database tables and mathematical sets, special set-like operators have been developed to manipulate tables. Table 2.2.5 lists seven typical table operators. The list of functions would normally also include such things as count, sum, average, find the maximum in a column, and find the minimum in a column. A rich collection of report generators offers powerful and flexible capabilities for producing tabular listings, text, graphics (bar charts, pie charts, point plots, and continuous plots), pictorial displays, and voice output.

SOFTWARE ENGINEERING

Programming Goals

Software engineering encompasses the methodologies for analyzing program requirements and for structuring programs to meet the requirements over their life cycle. The objectives are to produce programs that are:

Well documented

Easily read

Proved correct

Bug- (error-) free

Modifiable and maintainable

Implementable in modules

Control-Flow Diagrams

A control-flow diagram, popularly known as a **flowchart,** depicts all possible sequences of a program during execution by representing the control logic as a directed graph with labeled nodes. The theory associated with flowcharts has been refined so that programs can be structured to meet the above objectives. Without loss of generality, the nodes in a flowchart can be limited to the three types shown in Fig. 2.2.4. A **function** may be either a *transformer* which converts input data values into output data values or a *transducer* which converts that data's morphological form. A label placed in the rectangle specifies the function's action. A **predicate** node acts to bifurcate the path through the node. A question labels the diamond representing a predicate node. The answer to the question yields a binary value: 0 or 1, yes or no, ON or OFF. One of the output lines is selected accordingly. A **connector** serves to rejoin separated paths. Normally the circle representing a connector does not contain a label, but when the flowchart is used to document a computer program it may be convenient to label the connector.

Structured programming theory models all programs by their flowcharts by placing minor restrictions on their lines and nodes. Specifically, a flowchart is called a **proper program** if it has precisely one input and one output line, and for every node there exists a path from the input line through the node to the output line. The restriction prohibiting multiple input or output lines can easily be circumvented by funneling the lines through collector nodes. The other restriction simply discards unwanted program structures, since a program with a path that does not reach the output may not terminate.

Table 2.2.5   Relational Database Operators

Operator	Input	Output
Select	A table and a condition	A table of all tuples that satisfy the given condition
Project	A table and an attribute	A table of all values in the specified attribute
Union	Two tables	A table of all unique tuples appearing in one table or the other
Intersection	Two tables	A table of all tuples the given tables have in common
Difference	Two tables	A table of all tuples appearing in the first and not in the second table
Join	Two tables and a condition	A table concatenating the attributes of the tuples that satisfy the given condition
Divide	A table, two attributes, and list of values	A table of values appearing in one specified attribute of the given table when the table has tuples that satisfies every value in the list in the other given attribute

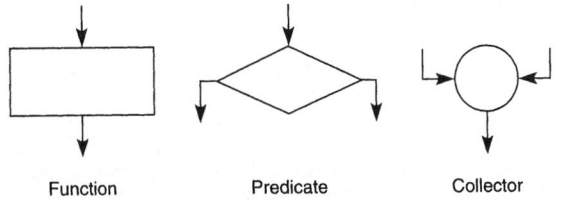

Fig. 2.2.4 Basic flowchart nodes.

Not all proper programs exhibit the desirable properties of meeting the objectives listed above. Figure 2.2.5 lists a group of proper programs whose graphs have been identified as being *well-structured* and useful as basic building blocks for creating other well-structured programs. The name assigned to each of these graph suggests the process each represents. CASE is just a convenient way of showing multiple IFTHENELSES more compactly.

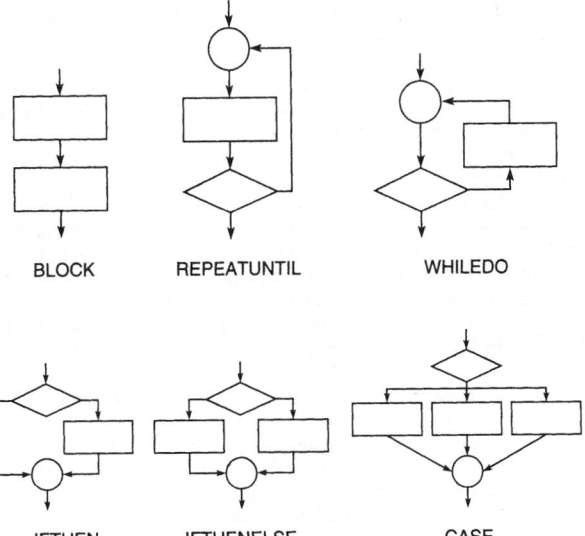

Fig. 2.2.5 Basic flowchart building blocks.

The **structured programming theorem** states: any proper program can be reconfigured to an equivalent program producing the same transformation of the data by a flowchart containing at most the graphs labeled BLOCK, IFTHENELSE, and REPEATUNTIL.

Every proper program has one input line and one output line like a function block. The synthesis of more complex well-structured programs is achieved by substituting any of the three building blocks mentioned in the theorem for a function node. In fact, any of the basic building blocks would do just as well. A program so structured will appear as a block of function nodes with a top-down control flow. Because of the top-down structure, the arrow points are not normally shown.

Figure 2.2.6 illustrates the expansion of a program to find the roots of $ax^2 + bx + c = 0$. The flowchart is shown in three levels of detail.

Data-Flow Diagrams

Data-flow diagrams structure the actions of a program into a network by tracking the data as it passes through the program. They depict the interworkings of a system by the processes performing the work and the communication between the processes. Data-flow diagrams have proved valuable in analyzing existing or new systems to determine the system requirements and in designing systems to meet those requirements.

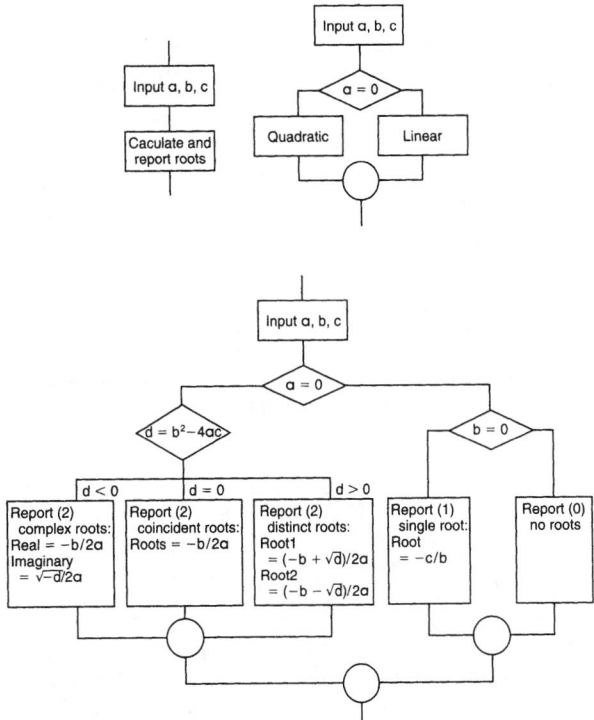

Fig. 2.2.6 Illustration of a control-flow diagram.

Figure 2.2.7 shows the four basic elements used to construct a data-flow diagram. The roles each element plays in the system are:

Rectangular boxes lie outside the system and represent the input data sources or output data sinks that communicate with the system. The sources and sinks are also called *terminators*.

Circles (bubbles) represent processes or actions performed by the system in accomplishing its function.

Twin parallel lines represent a data file used to collect and store data from among the processes or from a process over time which can be later recalled.

Arcs or vectors connect the other elements and represent data flows.

A label placed with each element makes clear its role in the system. The circles contain verbs and the other elements contain nouns. The arcs tie the system together. An arc between a terminator and a process represents input to or output from the system. An arc between two processes represents output from one process which is input to the other. An arc between a process and a file represents data gathered by the process and stored in the file, or retrieval of data from the file.

Analysis starts with a contextual view of the system studied in its environment. The contextual view gives the name of the system, the collection of terminators, and the data flows that provide the system inputs and outputs; all accompanied by a statement of the system objective. Details on the terminators and data they provide may also be described by text, but often the picture suffices. It is understood that the form of the input and output may not be dictated by the designer since they often involve organizations outside the system. Typical inputs in

Fig. 2.2.7 Data-flow diagram elements.

(a) Contextual view

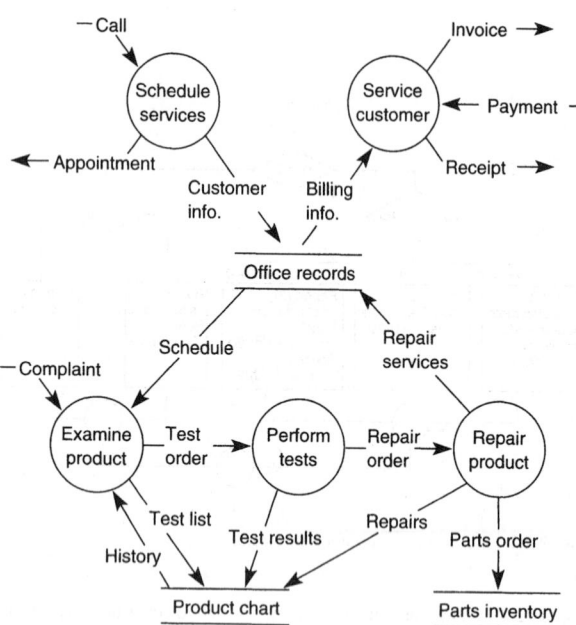

(b) Behavorial view

Fig. 2.2.8 Illustration of a data-flow diagram.

industrial systems include customer orders, payment checks, purchase orders, requests for quotations, etc. Figure 2.2.8a illustrates a context diagram for a repair shop.

Figure 2.2.8b gives many more operational details showing how the parts of the system interact to accomplish the system's objectives. The designer can restructure the internal processors and the formats of the data flows. The bubbles in a diagram can be broken down into further details to be shown in another data-flow diagram. This can be repeated level after level until the processes become manageable and understandable. To complete the system description, each bubble in the dataflow charts is accompanied by a control-flow diagram or its equivalent to describe the algorithm used to accomplish the actions and a data dictionary describing the items in the data flows and in the databases.

The techniques of data-flow diagrams lend themselves beautifully to the analysis of existing systems. In a complex system it would be unusual for an individual to know all the details, but all system participants know their respective roles: what they receive, whence they receive it, what they do, what they send, and where they send it. By carefully structuring interviews, the complete system can be synthesized to any desired level of detail. Moreover, each system component can be verified because what is sent from one process must be received by another and what is received by a process must be used by the process. To automate the total system or parts of the system, control bubbles containing transition diagrams can be implemented to control the timing of the processes.

SOFTWARE SYSTEMS

Software Techniques

Two basic operations form the heart of nonnumerical techniques such as those found in handling large database tables. One basic operation, called **sorting**, collates the information in a table by reordering the items by their key into a specified order. The other basic operation, called **searching**, seeks to find items in a table whose keys have the same or related value as a given argument. The search operation may or may not be successful, but in either case further operations follow the search (e.g., retrieve, insert, replace).

One must recognize that computers cannot do mathematics. They can perform a few basic operations such as the four rules of arithmetic, but even in this case the operations are approximations. In fact, computers represent long integers, long rationals, and all the irrational numbers like π and e only as approximations. While computer arithmetic and the computer representation of numbers exceed the precision one commonly uses, the size of problems solved in a computer and the number of operations that are performed can produce misleading results with large computational errors.

Since the computer can handle only the four rules of arithmetic, complex **functions** must be approximated by polynomials or rational fractions. A rational fraction is a polynomial divided by another polynomial. From these curve-fitting techniques, a variety of weighted-average formulas can be developed to approximate the **definite integral** of a function. These formulas are used in the procedures for solving differential and integral equations. While differentiation can also be expressed by these techniques, it is seldom used, since the errors become unacceptable.

Taking advantage of the machine's speed and accuracy, one can solve **nonlinear equations** by trial and error. For example, one can use the Newton-Raphson method to find successive approximations to the roots of an equation. The computer is programmed to perform the calculations needed in each iteration and to terminate the procedure when it has converged on a root. More sophisticated routines can be found in the libraries for finding real, multiple, and complex roots of an equation.

Matrix techniques have been commercially programmed into libraries of prepared modules which can be integrated into programs written in all popular engineering programming languages. These libraries not only contain excellent routines for solving **simultaneous linear equations** and the **eigenvalues** of characteristic matrices, but also embody procedures guarding against ill-conditioned matrices which lead to large computational errors.

Special matrix techniques called **relaxation** are used to solve partial differential equations on the computer. A typical problem requires setting up a grid of hundreds or thousands of points to describe the region and expressing the equation at each point by finite-difference methods. The resulting matrix is very sparse with a regular pattern of nonzero elements. The form of the matrix circumvents the need for handling large arrays of numbers in the computer and avoids problems in computational accuracy normally found in dealing with extremely large matrices.

The computer is an excellent tool for handling **optimization** problems. Mathematically these problems are formulated as problems in finding the maximum or minimum of a nonlinear equation. The excellent techniques that have been developed can deal effectively with the unique complexities these problems have, such as saddle points which represent both a maximum and a minimum.

Another class of problems, called **linear programming** problems, is characterized by the linear constraint of many variables which plot into regions outlined by multidimensional planes (in the two-dimensional case, the region is a plane enclosed by straight lines). Techniques have been developed to find the optimal solution of the variables satisfying some given value or cost objective function. The solution to the problem proceeds by searching the corners of the region defined by the constraining equations to find points which represent minimum points of a cost function or maximum points of a value function.

The best known and most widely used techniques for solving statistical problems are those of linear **statistics.** These involve the techniques of **least squares** (otherwise known as **regression**). For some problems these techniques do not suffice, and more specialized techniques involving nonlinear statistics must be used, albeit a solution may not exist.

Artificial intelligence (AI) is the study and implementation of programs that model knowledge systems and exhibit aspects of intelligence in problem solving. Typical areas of application are in learning, linguistics, pattern recognition, decision making, and theorem proving. In AI, the computer serves to search a collection of heuristic rules to find a match with a current situation and to make inferences or otherwise reorganize knowledge into more useful forms. AI techniques have been utilized to build sophisticated systems, called **expert systems,** to aid in producing a timely response in problems involving a large number of complex conditions.

Operating Systems

The operating system provides the services that support the needs that computer programs have in common during execution. Any list of services would include those needed to configure the resources that will be made available to the users, to attach hardware units (e.g., memory modules, storage devices, coprocessors, and peripheral devices) to the existing configuration, to detach modules, to assign default parameters to the hardware and software units, to set up and schedule users' tasks so as to resolve conflicts and optimize throughput, to control system input and output devices, to protect the system and users' programs from themselves and from each other, to manage storage space in the file devices, to protect file devices from faults and illegal use, to account for the use of the system, and to handle in an orderly way any exception which might be encountered during program execution. A well-designed operating system provides these services in a user-friendly environment and yet makes itself and the computer operating staff transparent to the user.

The design of a computer operating system depends on the number of users which can be expected. The focus of single-user systems relies on the monitor to provide a user-friendly system through dialog menus with icons, mouse operations, and templets. Table 2.2.6 lists some popular operating systems for PCs by their trademark names. The design of a multiuser system attempts to give each user the impression that he/she is the lone user of the system. In addition to providing the accoutrements of a user-friendly system, the design focuses on the order of processing the jobs in an attempt to treat each user in a fair and equitable fashion. The basic issues for determining the order of processing center on the selection of job queues: the number of queues (a simple queue or a mix of queues), the method used in scheduling the jobs in the queue (first come–first served, shortest job next, or explicit priorities), and the internal handling of the jobs in the queue (batch, multiprogramming, or timesharing).

Table 2.2.6   Some Popular PC Operating Systems

Trademark	Supplier
Windows	Microsoft Corp.
Unix	Unix Systems Laboratory Inc.
Sun/OS	Sun Microsystems Inc.
Macintosh	Apple Computer Inc.
Linux	Multiple suppliers

Batch operating systems process jobs in a sequential order. Jobs are collected in batches and entered into the computer with individual job instructions which the operating system interprets to set up the job, to allocate resources needed, to process the job, and to provide the input/ output. The operating system processes each job to completion in the order it appears in the batch. In the event a malfunction or fault occurs during execution, the operating system terminates the job currently being executed in an orderly fashion before initiating the next job in sequence.

Multiprogramming operating systems process several jobs concurrently. A job may be initiated any time memory and other resources which it needs become available. Many jobs may be simultaneously active in the system and maintained in a partial state of completion. The order of execution depends on the priority assignments. Jobs are executed to completion or put into a wait state until a pending request for service has been satisfied. It should be noted that, while the CPU can execute only a single program at any moment of time, operations with peripheral and storage devices can occur concurrently.

Timesharing operating systems process jobs in a way similar to multiprogramming except for the added feature that each job is given a short slice of the available time to complete its tasks. If the job has not been completed within its time slice or if it requests a service from an external device, it is put into a wait status and control passes to the next job. Effectively, the length of the time slice determines the priority of the job.

Program Preparation Facilities

For the user, the crucial part of a language system is the grammar which specifies the language syntax and semantics that give the symbols and rules used to compose acceptable statements and the meaning associated with the statements. Compared to natural languages, computer languages are more precise, have a simpler structure, and have a clearer syntax and semantics that allows no ambiguities in what one writes or what one means. For a program to be executed, it must eventually be translated into a sequence of basic machine instructions.

The statements written by a user must first be put on some machine-readable medium or typed on a keyboard for entry into the machine. The translator (**compiler**) program accepts these statements as input and translates (compiles) them into a sequence of basic machine instructions which form the executable version of the program. After that, the translated (compiled) program can be run.

During the execution of a program, a run-time program must also be present in the memory. The purpose of the run-time system is to perform services that the user's program may require. For example, in case of a program fault, the run-time system will identify the error and terminate the program in an orderly manner.

Some language systems do not have a separate compiler to produce machine-executable instructions. Instead the run-time system interprets the statements as written, converts them into a pseudo-code, and executes the coded version.

Commonly needed functions are made available as prepared modules, either as an integral part of the language or from stored libraries. The documentation of these functions must be studied carefully to assure correct selection and utilization.

Languages may be classified as procedure-oriented or problem-oriented. With **procedure-oriented** languages, all the detailed steps must be specified by the user. These languages are usually characterized as being more verbose than problem-oriented languages, but are more flexible and can deal with a wider range of problems. **Problem-oriented** languages deal with more specialized classes of problems. The elements of problem-oriented languages are usually familiar to a knowledgeable professional and so are easier to learn and use than procedure-oriented languages.

The most elementary form of a procedure-oriented language is called an **assembler.** This class of language permits a computer program to be written directly in basic computer instructions using mnemonic operators and symbolic operands. The assembler's translator converts these instructions into machine-usable form.

A further refinement of an assembler permits the use of macros. A **macro** identifies, by an assigned name and a list of formal parameters, a sequence of computer instructions written in the assembler's format and stored in its subroutine library. The macroassembler includes these macro instructions in the translated program along with the instructions written by the programmer.

Besides these basic language systems there exists a large variety of other language systems. These are called higher-level language systems since they permit more complex statements than are permitted by a macroassembler. They can also be used on machines produced by different manufacturers or on machines with different instruction repertoires.

One language of historical value is **ALGOL 60**. It is a landmark in the theoretical development of computer languages. It was designed and standardized by an international committee whose goal was to formulate a language suitable for publishing computer algorithms. Its importance lies in the many language features it introduced which are now common in the more recent languages which succeeded it and in the scientific notation which was used to define it.

FORTRAN (FORmula TRANslator) was one of the first languages catering to the engineering and scientific community where algebraic formulas specify the computations used within the program. It has been standardized several times. Each version has expanded the language features and has removed undesirable features which lead to unstructured programs.

The **PASCAL** language couples the ideas of ALGOL 60 to those of structured programming. By allowing only appropriate statement types, it guarantees that any program written in the language will be well structured. In addition, the language introduced new data types and allows programmers to define new complex data structures based on the primitive data types.

The definition of the **Ada** language was sponsored by the Department of Defense as an all-encompassing language for the development and maintenance of very large, software-intensive projects over their life cycle. While it meets software engineering objectives in a manner similar to Pascal, it has many other features not normally found in programming languages. Like other attempts to formulate very large all-inclusive languages, it is difficult to learn and has not found popular favor. Nevertheless, its many unique features make it especially valuable in implementing programs which cannot be easily implemented in other languages (e.g., programs for parallel computations in embedded computers).

By edict, subsets of Ada were forbidden. **Modula-2** was designed to retain the inherent simplicity of PASCAL but include many of the advanced features of Ada. Its advantage lies in implementing large projects involving many programmers. The compilers for this language have rigorous interface cross-checking mechanisms to avoid poor interfaces between components. Another troublesome area is the implicit use of global data. Modula-2 retains the Ada facilities that allow programmers to share data and avoids incorrectly modifying the data in different program units.

The **C** language was developed by AT&T's Bell Laboratories and subsequently standardized by ANSI. It has a reputation for translating programs into compact and fast code, and for allowing program segments to be precompiled. Its strength rests in the flexibility of the language; for example, it permits statements from other languages to be included in-line in a C program and it offers the largest selection of operators that mirror those available in an assembly language. Because of its flexibility, programs written in C can become unreadable. The C^{++} language was also developed by AT&T Bell Laboratories and standardized by ANSI. The C^{++} language was designed as an extension of C so that the concepts required to design and implement a program would be expressed more directly and concisely. Use of classes to embody the design concept is one of the primary concepts of C^{++}.

Problem-oriented languages have been developed for every discipline. A language might deal with a specialized application within an engineering field, or it might deal with a whole gamut of applications covering one or more fields.

A class of problem-oriented languages that deserves special mention are those for solving problems in **discrete simulation**. GPSS, Simscript, and SIMULA are among the most popular. A simulation (another word for *model*) of a system is used whenever it is desirable to watch a succession of many interrelated events or when there is interplay between the system under study and outside forces. Examples are problems in human-machine interaction and in the modeling of business systems. Typical human-machine problems are the servicing of automatic equipment by a crew of operators (to study crew size and assignments, typically), or responses by shared maintenance crews to equipment subject to unpredictable (random) breakdown. Business models often involve

transportation and warehousing studies. A business model could also study the interactions between a business and the rest of the economy such as competitive buying in a raw materials market or competitive marketing of products by manufacturers.

Physical or chemical systems may also be modeled. For example, to study the application of automatic control values in pipelines, the computer model consists of the control system, the valves, the piping system, and the fluid properties. Such a model, when tested, can indicate whether fluid hammer will occur or whether valve action is fast enough. It can also be used to predict pressure and temperature conditions in the fluid when subject to the valve actions.

Another class of problem-oriented languages makes the computer directly accessible to the specialist with little additional training. This is achieved by permitting the user to describe problems to the computer *in terms that are familiar in the discipline of the problem* and for which the language is designed. Two approaches are used. Figures 2.2.9 and 2.2.10 illustrate these.

One approach sets up the computer program directly from the mathematical equations. In fact, problems were formulated in this manner in the past, where analog computers were especially well-suited. Anyone familiar with analog computers finds the transitions to these languages easy. Figure 2.2.9 illustrates this approach using the MIMIC language to write the program for the solution of the initial-value problem:

$$M\ddot{y} + Z\dot{y} + Ky = 1 \quad \text{and} \quad \dot{y}(0) = y(0) = 0$$

MIMIC is a digital simulation language used to solve systems of ordinary differential equations. The key step in setting up the solution is to isolate the highest-order derivative on the left-hand side of the equation and equate it to an expression composed of the remaining terms. For the equation above, this results in:

$$\ddot{y} = (1 - Z\dot{y} - Ky)/M$$

The highest-order derivative is derived by equating it to the expression on the right-hand side of the equation. The lower-order derivatives in the expression are generated successively by integrating the highest-order derivative. The MIMIC language permits the user to write these statements in a format closely resembling mathematical notation.

The alternate approach used in problem-oriented languages permits the setup to be described to the computer directly from the block diagram of the physical system. Figure 2.2.10 illustrates this approach using the SCEPTRE language. SCEPTRE statements are written under headings and subheadings which identify the type of component being described. This language may be applied to network problems of electrical digital-logic elements, mechanical-translation or rotational elements, or transfer-function blocks. The translator for this language develops and sets up the equations directly from this description of the

MIMIC statements	Explanation
$DY2 = (1 - Z * DY1 - K * Y)/M$	Differential equation to be solved. "*" is used for multiplication and $DY2$, $DY1$, and Y are defined mnemonics for $\ddot{y}$, $\dot{y}$, and y.
$DY1 = \text{INT}(DY2,0.)$ $Y = \text{INT}(DY1,0.)$	INT (A, B) is used to perform integration. It forms successive values of $B + \int A\,dt$.
$\text{FIN}(T,10.)$	T is a reserved name representing the independent variable. This statement will terminate execution when $T \geq 10$.
$\text{CON}(M, K, Z)$	Values must be furnished for M, K, and Z. An input with these values must appear after the END card.
$\text{PLO}(T,DY2)$ $\text{PLO}(T,DY1)$ $\text{PLO}(T,Y)$	Three point plots are produced on the line printer; $\ddot{y}$, $\dot{y}$, and y vs. t.
END	Necessary last statement.

Fig. 2.2.9   Illustration of a MIMIC program.

$K1 = 40.$
$D1 = .5$
$M1 = 10.$
$R1 = 7.32$

A node is assigned to:
• ground
• any mass
• point between two elements
The prefix of element name specifies its type, i.e., M for mass, K for spring, D for damper, and R for force.

(a)

SCEPTRE statements	Explanation
MECHANICAL DESCRIPTION ELEMENTS $M1, 1 - 3 = 10.$ $K1, 1 - 2 = 40.$ $D1, 2 - 3 = .5$ $R1, 1 - 3 = 7.32$	Specifies the elements and their position in the diagram using the node numbers.
OUTPUT $SM1, VM1$	Results are listed on the line printer. Prefix on the element specifies the quantity to be listed; S for displacement, V for velocity.
RUN CONTROL STOPTIME $= 10.$	TIME is reserved name for independent variable. Statement will terminate execution of program when TIME is equal to or greater than 10.
END	Necessary statement.

(b)

Fig. 2.2.10 Illustration of SCEPTRE program. (a) Problem to be solved; (b) SCEPTRE program.

network diagram, and so relieves the user from the mathematical aspects of the problem.

Application Packages

An application package differs from a language in that its components have been organized to solve problems in a particular application rather than to create the components themselves. The user interacts with the package by initiating the operations and providing the data. From an operational view, packages are built to minimize or simplify interactions with the users by using a menu to initiate operations and entering the data through templets.

Perhaps the most widely used application package is the **word processor.** The objective of a word processor is to allow users to compose text in an electronically stored format which can be corrected or modified, and from which a hard copy can be produced on demand. Besides the basic typewriter operations, it contains functions to manipulate text in blocks or columns, to create headers and footers, to number pages, to find and correct words, to format the data in a variety of ways, to create labels, and to merge blocks of text together. The better word processors have an integrated dictionary, a spelling checker to find and correct misspelled words, a grammar checker to find grammatical errors, and a thesaurus. They often have facilities to prepare complex mathematical equations and to include and manipulate graphical artwork, including editing color pictures. When enough page- and document formatting capability has been added, the programs are known as **desktop publishing** programs.

One of the programs that contributed to the early acceptance of personal computers was the **spread sheet** program. These programs simulate the common spread sheet with its columns and rows of interrelated data. The computerized approach has the advantage that the equations are stored so that the results of a change in data can be shown quickly after any change is made in the data. Modern spread sheet programs have many capabilities, including the ability to obtain information from other spread sheets, to produce a variety of reports, and to prepare equations which have complicated logical aspects.

Tools for **project management** have been organized into commercially available application packages. The objectives of these programs are in the planning, scheduling, and controlling the time-oriented activities describing the projects. There are two basically similar techniques used in these packages. One, called **CPM** (critical path method), assumes that the project activities can be estimated deterministically. The other, called **PERT** (project evaluation and review technology), assumes that the activities can be estimated probabilistically. Both take into account such items as the requirement that certain tasks cannot start before the completion of other tasks. The concepts of **critical path** and **float** are crucial, especially in scheduling the large projects that these programs are used for. In both cases tools are included for estimating project schedules, estimating resources needed and their schedules, and representing the project activities in graphical as well as tabular form.

A major use of the digital computer is in **data reduction,** data analysis, and visualization of data. In installations where large amounts of data are recorded and kept, it is often advisable to reduce the amount of data by ganging the data together, by averaging the data with numerical filters to reduce the amount of noise, or by converting the data to a more appropriate form for storage, analysis, visualization, or future processing. This application has been expanded to produce systems for evaluation, automatic testing, and fault diagnosis by coupling the data acquisition equipment to special peripherals that automatically measure and record the data in a digital format and report the data as meaningful, nonphysically measurable parameters associated with a mathematical model.

Computer-aided design/computer-aided manufacturing (**CAD/CAM**) is an integrated collection of software tools which have been designed to make way for innovative methods of fabricating customized products to meet customer demands. The goal of modern manufacturing is to process orders placed for different products sooner and faster, and to fabricate them without retooling. CAD has the tools for prototyping a design and setting up the factory for production. Working within a framework of agile manufacturing facilities that features automated vehicles, handling robots, assembly robots, and welding and painting robots, the factory sets itself up for production under computer control. Production starts with the receipt of an order on which customers may pick options such as color, size, shapes, and features. Manufacturing proceeds with greater flexibility, quality, and efficiency in producing an increased number of products with a reduced workforce. Effectively, CAD/CAM provides for the ultimate just-in-time (JIT) manufacturing.

Collaboration software is a set of software tools that have been designed to enhance timely and effective communication within a software preparation group. The use of collaboration software has been an important tool to improve the speed and effectiveness of communication for local teams and especially for geographically dispersed teams. With such software, all team members have access to current information such as financial or experimental data, reports, and presentations. Proposals and updates can be provided rapidly to all team members in response to changing conditions. Working within the framework of the collaboration tools can also provide the team with a history that can be very valuable in addressing new problems and in providing new team members with information required to allow them to quickly become effective members of the team.

Two other types of application package illustrate the versatility of data management techniques. One type ties on-line equipment to a computer for collecting real-time data from the production lines. An animated, pictorial display of the production lines forms the heart of the system, allowing supervision in a central control station to continuously track operations. The other type collects time-series data from the various activities in an enterprise. It assists in what is known as **management by exception.** It is especially useful where the detailed data is so voluminous that it is feasible to examine it only in summaries. The data elements are processed and stored in various levels of detail in a seamless fashion. The system stores the reduced data and connects it to the detailed data from which it was derived. The application package allows management, through simple computer operations, to detect a problem at a higher level and to locate and pinpoint its cause through examination of successively lower levels.

Mechanics of Solids and Fluids

BY

PETER L. TEA, Jr. *Professor of Physics Emeritus, The City College of The City University of New York*
VITTORIO (RINO) CASTELLI *Senior Research Fellow, Retired, Xerox Corp.; Engineering Consultant*
J. W. MURDOCK *Late Consulting Engineer*
LEONARD MEIROVITCH *University Distinguished Professor Emeritus, Department of Engineering Science and Mechanics, Virginia Polytechnic Institute and State University*

3.1 MECHANICS OF SOLIDS
by Peter L. Tea, Jr.

REFERENCES: Beer and Johnston, "Mechanics for Engineers," McGraw-Hill. Ginsberg and Genin, "Statics and Dynamics," Wiley. Higdon and Stiles, "Engineering Mechanics," Prentice-Hall. Holowenko, "Dynamics of Machinery," Wiley. Housnor and Hudson, "Applied Mechanics," Van Nostrand. Meriam, "Statics and Dynamics," Wiley. Mabie and Ocvirk, "Mechanisms and Dynamics of Machinery," Wiley. Synge and Griffith, "Principles of Mechanics," McGraw-Hill. Timoshenko and Young, "Advanced Dynamics," McGraw-Hill. Timoshenko and Young, "Engineering Mechanics," McGraw-Hill.

PHYSICAL MECHANICS

Definitions

Force is the action of one body on another which will cause acceleration of the second body unless acted on by an equal and opposite action counteracting the effect of the first body. It is a vector quantity.

Time is a measure of the sequence of events. In newtonian mechanics it is an absolute quantity. In relativistic mechanics it is relative to the *frames of reference* in which the sequence of events is observed. The common unit of time is the second.

Inertia is that property of matter which causes a resistance to any change in the motion of a body.

Mass is a quantitative measure of *inertia*.

Acceleration of Gravity Every object which falls in a vacuum at a given position on the earth's surface will have the same acceleration g. Accurate values of the acceleration of gravity as measured *relative* to the earth's surface include the effect of the earth's rotation and flattening at the poles. The international gravity formula for the acceleration of gravity at the earth's surface is $g = 32.0881(1 + 0.005288 \sin^2 \phi - 0.0000059 \sin^2 2\phi)$ ft/s^2, where ϕ is latitude in degrees. For extreme accuracy, the local acceleration of gravity must also be corrected for the presence of large water or land masses and for height above sea level. The absolute acceleration of gravity for a nonrotating earth discounts the effect of the earth's rotation and is rarely used, except outside the earth's atmosphere. If g_0 represents the absolute acceleration at sea level, the absolute value at an altitude h is $g = g_0 R^2/(R + h)^2$, where R is the radius of the earth, approximately 3,960 mi (6,373 km).

Weight is the resultant force of attraction on the mass of a body due to a gravitational field. On the earth, units of weight are based upon an acceleration of gravity of 32.1740 ft/s^2 (9.80665 m/s^2).

Linear momentum is the product of mass and the linear velocity of a particle and is a vector. The moment of the linear-momentum vector about a fixed axis is the **angular momentum** of the particle about that fixed axis. For a rigid body rotating about a fixed axis, angular momentum is defined as the product of moment of inertia and angular velocity, each measured about the fixed axis.

An increment of **work** is defined as the product of an incremental displacement and the component of the force vector in the direction of the displacement or the component of the displacement vector in the direction of the force. The increment of work done by a couple acting on a body during a rotation of $d\theta$ in the plane of the couple is $dU = M\,d\theta$.

Energy is defined as the capacity of a body to do work by reason of its motion or configuration (see **Work and Energy**).

A **vector** is a directed line segment that has both magnitude and direction. In script or text, a vector is distinguished from a scalar V by a boldface-type **V**. The magnitude of the scalar is the magnitude of the vector, $V = |\mathbf{V}|$.

A **frame of reference** is a specified set of geometric conditions to which other locations, motion, and time are referred. In newtonian mechanics, the fixed stars are referred to as the **primary (inertial) frame of reference**. Relativistic mechanics denies the existence of a primary reference frame and holds that all reference frames must be described relative to each other.

SYSTEMS AND UNITS OF MEASUREMENTS

In *absolute systems*, the units of **length, mass,** and **time** are considered fundamental quantities, and all other units including that of **force** are derived.

In *gravitational systems*, the units of **length, force,** and **time** are considered fundamental qualities, and all other units including that of **mass** are derived.

In the SI system of units, the unit of mass is the kilogram (kg) and the unit of length is the metre (m). A force of one newton (N) is derived as the force that will give 1 kilogram an acceleration of 1 m/s^2.

In the English engineering system of units, the unit of mass is the pound mass (lbm) and the unit of length is the foot (ft). A force of one pound (1 lbf) is the force that gives a pound mass (1 lbm) an acceleration equal to the standard acceleration of gravity on the earth, 32.1740 ft/s^2 (9.80665 m/s^2). A slug is the mass that will be accelerated 1 ft/s^2 by a force of 1 lbf. Therefore, 1 slug = 32.1740 lbm. When described in the gravitational system, mass is a derived unit, being the constant of proportionality between force and acceleration, as determined by Newton's second law.

General Laws

NEWTON'S LAWS

I. If a balanced force system acts on a particle at rest, it will remain at rest. If a balanced force system acts on a particle in motion, it will remain in motion in a straight line without acceleration.

II. If an unbalanced force system acts on a particle, it will accelerate in proportion to the magnitude and in the direction of the resultant force.

III. When two particles exert forces on each other, these forces are equal in magnitude, opposite in direction, and collinear.

Fundamental Equation The basic relation between mass, acceleration, and force is contained in Newton's second law of motion. As applied to a particle of mass, $\mathbf{F} = m\mathbf{a}$, force = mass × acceleration. This equation is a vector equation, since the direction of $\mathbf{F}$ must be the direction of $\mathbf{a}$, as well as having $\mathbf{F}$ equal in magnitude to $m\mathbf{a}$. An alternative form of Newton's second law states that the resultant force is equal to the time rate of change of momentum, $\mathbf{F} = d(m\mathbf{v})/dt$.

Law of the Conservation of Mass The mass of a body remains unchanged by any ordinary physical or chemical change to which it may be subjected. (True in classical mechanics.)

Law of the Conservation of Energy The principle of conservation of energy requires that the total mechanical energy of a system remain unchanged if it is subjected only to forces which depend on position or configuration.

Law of the Conservation of Momentum The linear momentum of a system of bodies is unchanged if there is no resultant external force on the system. The angular momentum of a system of bodies about a fixed axis is unchanged if there is no resultant external moment about this axis.

Law of Mutual Attraction (Gravitation) Two particles attract each other with a force F proportional to their masses m_1 and m_2 and inversely proportional to the square of the distance r between them, or $F = Gm_1m_2/r^2$, in which G is the gravitational constant. The value of the gravitational constant is $G = 6.673 \times 10^{-11}$ m^3/kg · s^2 in SI or absolute units, or $G = 3.44 \times 10^{-8}$ lbf · ft^2/slug^2 in engineering gravitational units. It should be pointed out that the unit of force F in the SI system is the **newton** and is derived, while the unit force in the gravitational system is the **pound-force** and is a fundamental quantity.

EXAMPLE.   Each of two solid steel spheres 6 in in diam will weigh 32.0 lb on the earth's surface. This is the force of attraction between the earth and the steel sphere. The force of mutual attraction between the spheres if they are just touching is 0.000000136 lbf. (For spheres, center-to-center distance may be used.)

STATICS OF RIGID BODIES

General Considerations

If the forces acting on a rigid body do not produce any acceleration, they must neutralize each other, i.e., form a **system of forces in equilibrium.** Equilibrium is said to be **stable** when the body with the forces acting upon it returns to its original position after being displaced a very small amount from that position; **unstable** when the body tends to move still farther from its original position than the very small displacement; and **neutral** when the forces retain their equilibrium when the body is in its new position.

External and Internal Forces   The forces by which the individual particles of a body act on each other are known as internal forces. All other forces are called external forces. If a body is supported by other bodies while subject to the action of forces, deformations and forces will be produced at the points of support or contact and these internal forces will be distributed throughout the body until equilibrium exists and the body is said to be in a state of tension, compression, or shear. The forces exerted by the body on the supports are known as **reactions.** They are equal in magnitude and opposite in direction to the forces with which the supports act on the body, known as **supporting forces.** The supporting forces are external forces applied to the body.

In considering a body at a definite section, it will be found that all the internal forces act in pairs, the two forces being equal and opposite. The external forces act singly.

General Law   When a body is at rest, the forces acting externally to it must form an equilibrium system. This law will hold for any part of the body, in which case the forces acting at any section of the body become external forces when the part on either side of the section is considered alone. In the case of a **rigid body,** any two forces of the same magnitude, but acting in opposite directions in any straight line, may be added or removed without change in the action of the forces acting on the body, provided the strength of the body is not affected.

Composition, Resolution, and Equilibrium of Forces

The **resultant** of several forces acting at a point is a force which will produce the same effect as all the individual forces acting together.

Forces Acting on a Body at the Same Point   The resultant R of two forces F_1 and F_2 applied to a rigid body at the same point is represented in magnitude and direction by the diagonal of the parallelogram formed by F_1 and F_2 (see Figs. 3.1.1 and 3.1.2).

$$R = \sqrt{F_1^2 + F_2^2 + 2F_1F_2 \cos a}$$

$$\sin a_1 = (F_2 \sin a)/R \qquad \sin a_2 = (F_1 \sin a)/R$$

When $a = 90°$, $R = \sqrt{F_1^2 + F_2^2}$, $\sin a_1 = F_2/R$, and $\sin a_2 = F_1/R$.

When $a = 0°$, $R = F_1 + F_2$ $\Big\}$ Forces act in same
When $a = 180°$, $R = F_1 - F_2$ straight line.

A **force R may be resolved into two component forces** intersecting anywhere on R and acting in the same plane as R, by the reverse of the operation shown by Figs. 3.1.1 and 3.1.2; and by repeating the operation with the components, R may be resolved into any number of component forces intersecting R at the same point and in the same plane.

Resultant of Any Number of Forces Applied to a Rigid Body at the Same Point   Resolve each of the given forces F into components along three rectangular coordinate axes. If A, B, and C are the angles made with XX, YY, and ZZ, respectively, by any force F, the components will be $F \cos A$ along XX, $F \cos B$ along YY, $F \cos C$ along ZZ; add the components of all the forces along each axis algebraically and obtain $\Sigma F \cos A = \Sigma X$ along XX, $\Sigma F \cos B = \Sigma Y$ along YY, and $\Sigma F \cos C = \Sigma Z$ along ZZ.

The resultant $R = \sqrt{(\Sigma X)^2 + (\Sigma Y)^2 + (\Sigma Z)^2}$. The angles made by the resultant with the three axes are A_r with XX, B_r with YY, C_r with ZZ, where

$$\cos A_r = \Sigma X/R \qquad \cos B_r = \Sigma Y/R \qquad \cos C_r = \Sigma Z/R$$

The **direction** of the **resultant** can be determined by plotting the algebraic sums of the components.

If the forces are all in the same plane, the components of each of the forces along one of the three axes (say ZZ) will be 0; i.e., angle $C_r = 90°$ and $R = \sqrt{(\Sigma X)^2 + (\Sigma Y)^2}$, $\cos A_r = \Sigma X/R$, and $\cos B_r = \Sigma Y/R$.

For equilibrium, it is necessary that $R = 0$; i.e., ΣX, ΣY, and ΣZ must each be equal to zero.

General Law   In order that a number of forces acting at the same point shall be in equilibrium, the algebraic sum of their components along any *three* coordinate axes must each be equal to zero. When the forces all act in the same plane, the algebraic sum of their components along any *two* coordinate axes must each equal zero.

When the Forces Form a System in Equilibrium   Three unknown forces can be determined if the lines of action of the forces are *all* known and are in different planes. If the forces are all in the same plane, the lines of action being known, only *two* unknown forces can be determined. If the lines of action of the unknown forces are *not* known, only *one* unknown force can be determined in either case.

Couples and Moments

Couple   Two parallel forces of equal magnitude (Fig. 3.1.3) which act in opposite directions and are not collinear form a couple. **A couple cannot be reduced to a single force.**

Fig. 3.1.3

Displacement and Change of a Couple   The forces forming a couple may be moved about and their magnitude and direction changed, provided they always remain parallel to each other and remain in either the original plane or one parallel to it, and provided the product of one of the forces and the perpendicular distance between the two is constant and the direction of rotation remains the same.

Moment of a Couple   The moment of a couple is the product of the magnitude of one of the forces and the perpendicular distance between the lines of action of the forces. $Fa = $ moment of couple; $a = $ arm of couple. If the forces are measured in pounds and the distance a in feet, **the unit of rotation moment** is the foot-pound. If the force is measured in newtons and the distance in metres, the unit is the newton-metre. In the cgs system the unit of rotation moment is 1 cm-dyne.

Rotation moments of couples acting in the same plane are conventionally considered to be positive for counterclockwise moments and negative for clockwise moments, although it is only necessary to be consistent within a given problem. The magnitude, direction, and sense of rotation of a couple are completely determined by its moment axis, or moment vector, which is a line drawn perpendicular to the plane in which the couple acts, with an arrow indicating the direction from which the couple will appear to have right-handed rotation; the length of the line represents the magnitude of the moment of the couple. Figure 3.1.4a shows a counterclockwise couple, designated $+$. Figure 3.1.4b shows a clockwise couple, designated $-$.

Fig. 3.1.1

Fig. 3.1.2

Composition of Couples   A couple may be represented in space by a single vector (having no specific line of action) perpendicular to the plane of the couple. If you let the relaxed fingers of your right hand represent the "swirl" of the forces in the couple, your stiff right thumb will indicate the direction of the vector. Vector couples may be added vectorially in the same manner by which concurrent forces are added.

Couples lying in the same or parallel planes are added algebraically. Let $+28$ lbf · ft ($+38$ N · m), -42 lbf · ft (-57 N · m), and $+70$ lbf · ft (95 N · m) be the moments of three couples in the same or parallel planes; their resultant is a single couple lying in the same or in a parallel plane, whose moment is $\Sigma M = +28 - 42 + 70 = +56$ lbf · ft ($\Sigma M = +38 - 57 + 95 = 76$ N · m).

Fig. 3.1.4   **Fig. 3.1.5**

If the polygon formed by the moment vectors of several couples closes itself, the couples form an equilibrium system. Two couples will balance each other when they lie in the same or parallel planes and have the same moment in magnitude, but opposite in sign.

Combination of a Couple and a Single Force in the Same Plane (Fig. 3.1.5) Given a force $F = 18$ lbf (80 N) acting as shown at distance x from YY, and a couple whose moment is -180 lbf · ft (244 N · m) in the same or parallel plane, to find the resultant. A couple may be changed to any other couple in the same or a parallel plane having the same moment and same sign. Let the couple consist of two forces of 18 lbf (80 N) each and let the arm be 10 ft (3.05 m). Place the couple in such a manner that one of its forces is opposed to the given force at p. This force of the couple and the given force being of the same magnitude and opposite in direction will neutralize each other, leaving the other force of the couple acting at a distance of 10 ft (3.05 m) from p and parallel and equal to the given force 18 lbf (80 N).

General Rule   The resultant of a couple and a single force lying in the same or parallel planes is a single force, equal in magnitude, in the same direction and parallel to the single force, and acting at a distance from the line of action of the single force equal to the moment of the couple divided by the single force. The moment of the resultant force about any point on the line of action of the given single force must be of the same sense as that of the couple, positive if the moment of the couple is positive, and negative if the moment of the couple is negative. If the moment of the couple in Fig. 3.1.5 had been $+$ instead of $-$, the resultant would have been a force of 18 lbf (80 N) acting in the same direction and parallel to F, but at a distance of 10 ft (3.05 m) to the right of it, making the moment of the resultant about any point on F positive.

To effect a parallel displacement of a single force F over a distance a, a couple whose moment is Fa must be added to the system. The sense of the couple will depend upon which way it is desired to displace force F.

The moment of a force with respect to a point is the product of the force F and the perpendicular distance from the point to the line of action of the force.

The Moment of a Force with Respect to a Straight Line   If the force is resolved into components parallel and perpendicular to the given line, the moment of the force with respect to the line is the product of the magnitude of the perpendicular component and the distance from its line of action to the given line.

Forces with Different Points of Application

Composition of Forces   If each force $\mathbf{F}$ is resolved into components parallel to three rectangular coordinate axes XX, YY, ZZ, the magnitude of the resultant is $R = \sqrt{(\Sigma X)^2 + (\Sigma Y)^2 + (\Sigma Z)^2}$, and its line

of action makes angles A_r, B_r, and C_r with axes XX, YY, and ZZ, where $\cos A_r = \Sigma X/R$, $\cos B_r = \Sigma Y/R$, and $\cos C_r = \Sigma Z/R$; and there are three couples which may be combined by their moment vectors into a single resultant couple having the moment $M_r = \sqrt{(M_x)^2 + (M_y)^2 + (M_z)^2}$, whose moment vector makes angles of A_m, B_m, and C_m with axes XX, YY, and ZZ, such that $\cos A_m = M_x/M_r$, $\cos B_m = M_y/M_r$, $\cos C_m = M_z/M_r$. If this single resulting couple is in the same plane as the single resulting force at the origin or a plane parallel to it, the system may be reduced to a single force $\mathbf{R}$ acting at a distance from $\mathbf{R}$ equal to M_r/R. If the couple and force are not in the same or parallel planes, it is impossible to reduce the system to a single force. If $\mathbf{R} = 0$, i.e., if ΣX, ΣY, and ΣZ all equal zero, the system will reduce to a single couple whose moment is $\mathbf{M}_r$. If $\mathbf{M}_r = 0$, i.e., if M_x, M_y, and M_z all equal zero, the resultant will be a single force $\mathbf{R}$.

When the forces are all in the same plane, the cosine of one of the angles A_r, B_r, or $C_r = 0$, say, $C_r = 90°$. Then $R = \sqrt{(\Sigma X)^2 + (\Sigma Y)^2}$, $M_r = \sqrt{M_x^2 + M_y^2}$, and the final resultant is a force equal and parallel to R, acting at a distance from R equal to M_r/R.

A system of forces in the same plane can always be replaced by either a couple or a single force. If $\mathbf{R} = 0$ and $\mathbf{M}_r · 0$, the resultant is a couple. If $\mathbf{M}_r = 0$ and $\mathbf{R} > 0$, the resultant is a single force.

A rigid body is in equilibrium when acted upon by a system of forces whenever $R = 0$ and $M_r = 0$, i.e., when the following six conditions hold true: $\Sigma X = 0$, $\Sigma Y = 0$, $\Sigma Z = 0$, $M_x = 0$, $M_y = 0$, and $M_z = 0$. When the system of forces is in the same plane, equilibrium prevails when the following three conditions hold true: $\Sigma X = 0$, $\Sigma Y = 0$, $\Sigma M = 0$.

Forces Applied to Support Rigid Bodies

The external forces in equilibrium acting upon a body may be statically determinate or indeterminate according to the number of unknown forces existing. When the forces are all in the same plane and act at a common point, two unknown forces may be determined if their lines of action are known, or one unknown force if its line of action is unknown.

When the forces are all in the same plane and are parallel, two unknown forces may be determined if the lines of action are known, or one unknown force if its line of action is unknown.

When the forces are anywhere in the same plane, three unknown forces may be determined if their lines of action are known, if they are not parallel or do not pass through a common point; if the lines of action are unknown, only one unknown force can be determined.

If the forces all act at a common point but are in different planes, three unknown forces can be determined if the lines of action are known, one if unknown.

If the forces act in different planes but are parallel, three unknown forces can be determined if their lines of action are known, one if unknown.

The first step in the solution of problems in statics is the determination of the supporting forces. The following data are required for the complete knowledge of supporting forces: magnitude, direction, and point of application. According to the nature of the problem, none, one, or two of these quantities are known.

One Fixed Support   The point of application, direction, and magnitude of the load are known. See Fig. 3.1.6. As the body on which the forces act is in equilibrium, the supporting force P must be equal in magnitude and opposite in direction to the resultant of the loads L.

In the case of a **rolling surface,** the point of application of the support is obtained from the center of the connecting bolt A (Fig. 3.1.7), both the direction and magnitude being unknown. The point of application and

Fig. 3.1.6   **Fig. 3.1.7**

line of action of the support at B are known, being determined by the rollers.

When three forces acting in the same plane on the same rigid body are in equilibrium, their lines of action must pass through the same point O. The load L is known in magnitude and direction. The line of action of the support at B is known on account of the rollers. The point of application of the support at A is known. The three forces are in equilibrium and are in the same plane; therefore, the lines of action must meet at the point O.

In the case of the rolling surfaces shown in Fig. 3.1.8, the direction of the support at A is known, the magnitude unknown. The line of action and point of application of the supporting force at B are known, its

Fig. 3.1.8

Fig. 3.1.9

magnitude unknown. The lines of action of the three forces must meet in a point, and the supporting force at A must be perpendicular to the plane XX.

If a member of a truss or frame in equilibrium is pinned at two points and loaded at these two points only, the line of action of the forces exerted on the member or by the member at these two points must be along a line connecting the pins.

If the external forces acting upon a rigid body in equilibrium are all in the same plane, the equations $\Sigma X = 0$, $\Sigma Y = 0$, and $\Sigma M = 0$ must be satisfied. When trusses, frames, and other structures are under discussion, these equations are usually used as $\Sigma V = 0$, $\Sigma H = 0$, $\Sigma M = 0$, where V and H represent vertical and horizontal components, respectively.

In Fig. 3.1.9, the directions and points of application of the supporting forces are known, but all three are of unknown magnitudes. Including load L, there are four forces, and the three unknown magnitudes may be determined by $\Sigma X = 0$, $\Sigma Y = 0$, and $\Sigma M = 0$.

The **supports** are said to be **statically determinate** when the laws of equilibrium are sufficient for their determination. When the conditions are not sufficient for the determination of the supports or other forces, the structure is said to be **statically indeterminate;** the unknown forces can then be determined from considerations involving the deformation of the material.

When several bodies are so connected to one another as to make up a rigid structure, the forces at the points of connection must be considered as internal forces and are not taken into consideration in the determination of the supporting forces for the structure as a whole.

The distortion of any practically rigid structure under its working loads is so small as to be negligible when determining supporting forces. When the forces acting at the different joints in a built-up structure cannot be determined by dividing the structure up into parts, the structure is said to be **statically indeterminate internally.** A structure may be statically indeterminate internally and still be statically determinate externally.

Fundamental Problems in Graphical Statics

A force may be represented by a straight line in a determined position, and its magnitude by the length of the straight line. The direction in which it acts may be indicated by an arrow.

Polygon of Forces The parallelogram of two forces intersecting each other (see Figs. 3.1.4 and 3.1.5) leads directly to the graphic composition by means of the triangle of forces. In Fig. 3.1.10, R is called the **closing side,** and represents the resultant of the forces $\mathbf{F}_1$ and $\mathbf{F}_2$ in magnitude and direction. Its position is given by the point of application O. By means of repeated use of the triangle of forces and by omitting the

closing sides of the individual triangles, the magnitude and direction of the resultant $\mathbf{R}$ of any number of forces in the same plane and intersecting

Fig. 3.1.10

at a single point can be found. In Fig. 3.1.11 the lines representing the forces start from point O, and in the force polygon (Fig. 3.1.12) they are joined in any order, the arrows showing their directions following around the polygon in the same direction. The magnitude of the resultant at the point of application of the forces is represented by the closing side $\mathbf{R}$ of the force polygon; its direction, as shown by the arrow, is counter to that in the other sides of the polygon.

If the forces are in equilibrium, R must equal zero, i.e., **the force polygon must close.**

Fig. 3.1.11

Fig. 3.1.12

If in a closed polygon one of the forces is reversed in direction, this force becomes the resultant of all the others.

Determination of Stresses in Members of a Statically Determinate Plane Structure with Loads at Rest

It will be assumed that the loads are applied at the joints of the structure, i.e., at the points where the different members are connected, and that the connections are pins with no friction. The stresses in the members must then be along lines connecting the pins, unless any member is loaded at more than two points by pin connections.

Equilibrium In order that the whole structure should be in equilibrium, it is necessary that the external forces (loads and supports) shall form a balanced system. Graphical and analytical methods are both of service.

Supporting Forces When the supporting forces are to be determined, it is not necessary to pay any attention to the makeup of the structure under consideration so long as it is practically rigid; the loads may be taken as they occur, or the resultant of the loads may be used instead. When the stresses in the members of the structure are being determined, the loads *must* be distributed at the joints where they belong.

Method of Joints When all the external forces have been determined, any joint at which there are not more than two unknown forces may be taken and these unknown forces determined by the methods of the stress polygon, resolution, or moments. In Fig. 3.1.13, let O be the joint of a structure and $\mathbf{F}$ be the only known force; but let $O1$ and $O2$ be two members of the structure joined at O. Then the lines of action of the unknown forces are known and their magnitude may be determined (1) by a **stress polygon** which, for equilibrium, must close; (2) by resolution into H and V components, using the condition of equilibrium $\Sigma H = 0$, $\Sigma V = 0$; or (3) by moments, using any convenient point on the line of action of $O1$ and $O2$ and the condition of equilibrium $\Sigma M = 0$. No more than *two* unknown forces can be determined. In this manner, proceeding from joint to joint, the stresses in all the members of the truss can usually be determined if the structure is statically determinate internally.

Method of Sections   The structure may be divided into parts by passing a section through it cutting some of its members; one part may then be treated as a rigid body and the external forces acting upon it determined. Some of these forces will be the stresses in the members themselves. For example, let xx (Fig. 3.1.14) be a section taken through a truss loaded at P_1, P_2, and P_3, and supported on rollers at S. As the whole truss is in equilibrium, any part of it must be also, and consequently the part shown to the left of xx must be in equilibrium under the action of the forces acting externally to it. Three of these forces are the

Fig. 3.1.13   **Fig. 3.1.14**

stresses in the members aa, bb, and bc, and are the unknown forces to be determined. They can be determined by applying the condition of equilibrium of forces acting in the same plane but not at the same point. $\Sigma H = 0$, $\Sigma V = 0$, $\Sigma M = 0$. The three unknown forces can be determined only if they are not parallel or do not pass through the same point; if, however, the forces are parallel or meet in a point, two unknown forces only can be determined. Sections may be passed through a structure cutting members in any convenient manner, as a rule, however, cutting not more than three members, unless members are unloaded.

For the determination of stresses in framed structures, see Sec. 12.2.

CENTER OF GRAVITY

Consider a three-dimensional body of any size, shape, and weight, but preferably thin and flat. If it is suspended as in Fig. 3.1.15 by a cord from any point A, it will be in equilibrium under the action of the tension in the cord and the weight W. If the experiment is repeated by suspending the body from point B, it will again be in equilibrium. Use a plumb bob through point A and draw the "first line" on the body. If the lines of action of the weight were marked in each case, they would be concurrent at a point G known as the **center of gravity** or **center of mass**. Whenever the density of the body is uniform, it will be a constant factor and like geometric shapes of different densities will have the same center of gravity. The term **centroid** is used in this case since the location of the center of gravity is of geometric concern only. If densities are nonuniform, like geometric shapes will have the same centroid but different centers of gravity.

Fig. 3.1.15

Centroids of Technically Important Lines, Areas, and Solids

CENTROIDS OF LINES

Straight Lines   The centroid is at its middle point.
Circular Arc AB (Fig. 3.1.16) $x_0 = r \sin c/\text{rad } c$; $y_0 = 2r \sin^2 \frac{1}{2}c/\text{rad } c$. (rad c = angle c measured in radians.)
Circular Arc AC (Fig. 3.1.17) $x_0 = r \sin c/\text{rad } c$; $y_0 = 0$.

Quadrant, AB (Fig. 3.1.18a) $x_0 = y_0 = 2r/\pi = 0.6366r$.
Semicircumference, AC (Fig. 3.1.18b) $y_0 = 2r/\pi = 0.6366r$; $x_0 = 0$.

Fig. 3.1.16   **Fig. 3.1.17**   **Fig. 3.1.18a**

Combination of Arcs and Straight Line (Fig. 3.1.19) AD and BC are two quadrants of radius r. $y_0 = \{(AB)r + 2[0.5\pi r(r - 0.6366r)]\} \div [AB + 2(0.5\pi r)]$ (Symmetrical about YY).

Fig. 3.1.18b   **Fig. 3.1.19**

CENTROIDS OF PLANE AREAS

Triangle   Centroid lies at the intersection of the lines joining the vertices with the midpoints of the sides, and at a distance from any side equal to one-third of the corresponding altitude.
Parallelogram   Centroid lies at the point of intersection of the diagonals.
Trapezoid (Fig. 3.1.20)   Centroid lies on the line joining the middle points m and n of the parallel sides. The distances h_a and h_b are

$$h_a = h(a + 2b)/3(a + b) \qquad h_b = h(2a + b)/3(a + b)$$

Draw $BE = a$ and $CF = b$; EF will then intersect mn at centroid.
Any Quadrilateral   The centroid of any quadrilateral may be determined by the general rule for areas, or graphically by dividing it into two sets of triangles by means of the diagonals. Find the centroid of each of the four triangles and connect the centroids of the triangles belonging to the same set. The intersection of these lines will be centroid of area. Thus, in Fig. 3.1.21, O, O_1, O_2, and O_3 are, respectively, the centroids of the triangles ABD, ABC, BDC, and ACD. The intersection of O_1O_3 with OO_2 gives the centroid of the quadrilateral area.

Fig. 3.1.20   **Fig. 3.1.21**

Segment of a Circle (Fig. 3.1.22) $x_0 = \frac{2}{3}r \sin^3 c/(\text{rad } c - \cos c \sin c)$. A segment may be considered to be a sector from which a triangle is subtracted, and the general rule applied.
Sector of a Circle (Fig. 3.1.23) $x_0 = \frac{2}{3}r \sin c/\text{rad } c$; $y_0 = \frac{4}{3}r \sin^2 \frac{1}{2} c/\text{rad } c$.
Semicircle $x_0 = \frac{4}{3}r/\pi = 0.4244r$; $y_0 = 0$.
Quadrant (90° sector) $x_0 = y_0 = \frac{4}{3}r/\pi = 0.4244r$.

Parabolic Half Segment (Fig. 3.1.24)   Area *ABO*: $x_0 = \frac{3}{5}x_1$; $y_0 = \frac{3}{8}y_1$.
Parabolic Spandrel (Fig. 3.1.24)   Area *AOC*: $x_0' = \frac{3}{10}x_1$; $y_0' = \frac{3}{4}y_1$.

Fig. 3.1.22   **Fig. 3.1.23**

Fig. 3.1.24

Quadrant of an Ellipse (Fig. 3.1.25)   Area *OAB*: $x_0 = \frac{4}{3}(a/\pi)$; $y_0 = \frac{4}{3}(b/\pi)$.

The centroid of a figure such as that shown in Fig. 3.1.26 may be determined as follows: Divide the area *OABC* into a number of parts by lines drawn perpendicular to the axis *XX*, e.g., 11, 22, 33, etc. These parts will be approximately either triangles, rectangles, or trapezoids. The area of each division may be obtained by taking the product of its

Fig. 3.1.25

Fig. 3.1.26

mean height and its base. The centroid of each area may be obtained as previously shown. The sum of the moments of all the areas about *XX* and *YY*, respectively, divided by the sum of the areas will give approximately the distances from the center of gravity of the whole area to the axes *XX* and *YY*. The greater the number of areas taken, the more nearly exact the result.

CENTROIDS OF SOLIDS

Prism or Cylinder with Parallel Bases   The centroid lies in the center of the line connecting the centers of gravity of the bases.
Oblique Frustum of a Right Circular Cylinder (Fig. 3.1.27)   Let 1 2 3 4 be the plane of symmetry. The distance from the base to the plane section is $\frac{1}{2}h + (r^2 \tan^2 c)/8h$, where *c* is the angle of inclination of the oblique section to the base. The distance of the centroid from the axis of the cylinder is $r^2 \tan c/4h$.
Pyramid or Cone   The centroid lies in the line connecting the centroid of the base with the vertex and at a distance of one-fourth of the altitude above the base.
Truncated Pyramid   If *h* is the height of the truncated pyramid and *A* and *B* the areas of its bases, the distance of its centroid from the surface of *A* is

$$h(A + 2\sqrt{AB} + 3B)/4(A + \sqrt{AB} + B)$$

Truncated Circular Cone   If *h* is the height of the frustum and *R* and *r* the radii of the bases, the distance from the surface of the base whose radius is *R* to the centroid is $h(R^2 + 2Rr + 3r^2)/4(R^2 + Rr + r^2)$.

Fig. 3.1.27

Fig. 3.1.28

Segment of a Sphere (Fig. 3.1.28)   Volume *ABC*: $x_0 = 3(2r - h)^2/4(3r - h)$.
Hemisphere $x_0 = 3r/8$.
Hollow Hemisphere $x_0 = 3(R^4 - r^4)/8(R^3 - r^3)$, where *R* and *r* are, respectively, the outer and inner radii.
Sector of a Sphere (Fig. 3.1.28)   Volume *OABCO*: $x_0' = \frac{3}{8}(2r - h)$.
Ellipsoid, with Semiaxes *a*, *b*, and *c*   For each octant, distance from center of gravity to each of the bounding planes = $\frac{3}{8} \times$ length of semiaxis perpendicular to the plane considered.

The formulas given for the determination of the centroid of lines and areas can be used to determine the areas and volumes of surfaces and solids of revolution, respectively, by employing the theorems of Pappus, Sec. 2.1.

Determination of Center of Gravity of a Body by Experiment   The center of gravity may be determined by hanging the body up from different points and plumbing down; the point of intersection of the plumb lines will give the center of gravity. It may also be determined as shown in Fig. 3.1.29. The body is placed on knife-edges which rest on platform scales. The sum of the weights registered on the two scales ($w_1 + w_2$) must equal the weight (*w*) of the body. Taking a moment axis at either end (say, *O*), $w_2A/w = x_0$ = distance from *O* to plane containing the center of gravity. (See also Fig. 3.1.15 and accompanying text.)

Fig. 3.1.29

Graphical Determination of the Centroids of Plane Areas   See Fig. 3.1.40.

MOMENT OF INERTIA

The **moment of inertia of a solid body** with respect to a given axis is the limit of the sum of the products of the masses of each of the elementary particles into which the body may be conceived to be divided and the square of their distance from the given axis.

If $dm = dw/g$ represents the mass of an elementary particle and *y* its distance from an axis, the moment of inertia *I* of the body about this axis will be $I = \int y^2 \, dm = \int y^2 \, dw/g$.

The moment of inertia may be expressed in weight units ($I_w = \int y^2 \, dw$), in which case the moment of inertia in weight units, I_w, is equal to the moment of inertia in mass units, *I*, multiplied by *g*.

If $I = k^2 m$, the quantity k is called the **radius of gyration** or the **radius of inertia.**

If a body is considered to be composed of a number of parts, its moment of inertia about an axis is equal to the sum of the moments of inertia of the several parts about the same axis, or $I = I_1 + I_2 + I_3 + \cdots + I_n$.

The **moment of inertia of an area** with respect to a given axis that lies in the plane of the area is the limit of the sum of the products of the elementary areas into which the area may be conceived to be divided and the square of their distance (y) from the axis in question. $I = \int y^2\, dA = k^2 A$, where $k = $ radius of gyration.

The quantity $\int y^2\, dA$ is more properly referred to as the *second moment of area* since it is not a measure of *inertia* in a true sense.

Formulas for moments of inertia and radii of gyration of various areas follow later in this section.

Relation between the Moments of Inertia of an Area and a Solid
The moment of inertia of a solid of elementary thickness about an axis is equal to the moment of inertia of the area of one face of the solid about the same axis multiplied by the mass per unit volume of the solid times the elementary thickness of the solid.

Moments of Inertia about Parallel Axes   The moment of inertia of an area or solid about any given axis is equal to the moment of inertia about a parallel axis through the center of gravity plus the square of the distance between the two axes times the area or mass.

In Fig. 3.1.30a, the moment of inertia of the area *ABCD* about axis *YY* is equal to I_0 (or the moment of inertia about Y_0Y_0 through the center of gravity of the area and parallel to *YY*) plus $x_0^2 A$, where $A = $ area of *ABCD*. In Fig. 3.1.30b, the moment of inertia of the mass m about $YY = I_0 + x_0^2 m$. Y_0Y_0 passes through the centroid of the mass and is parallel to *YY*.

(a)   (b)

Fig. 3.1.30

Polar Moment of Inertia   The polar moment of inertia (Fig. 3.1.31) is taken about an axis perpendicular to the plane of the area. Referring to Fig. 3.1.31, if I_y and I_x are the moments of inertia of the area A about *YY* and *XX*, respectively, then the polar moment of inertia $I_p = I_x + I_y$, or the polar moment of inertia is equal to the sum of the moments of inertia about any two axes at right angles to each other in the plane of the area and intersecting at the pole.

Product of Inertia   This quantity will be represented by I_{xy}, and is $\iint xy\, dy\, dx$, where x and y are the coordinates of any elementary part into which the area may be conceived to be divided. I_{xy} may be positive or negative, depending upon the position of the area with respect to the coordinate axes *XX* and *XY*.

Relation between Moments of Inertia about Axes Inclined to Each Other   Referring to Fig. 3.1.32, let I_y and I_x be the moments of inertia of the area A about *YY* and *XX*, respectively, I_y' and I_x' the moments about $Y'Y'$ and $X'X'$, and I_{xy} and I_{xy}' the products of inertia for *XX* and *YY*, and $X'X'$ and $Y'Y'$, respectively. Also, let c be the angle between the respective pairs of axes, as shown. Then,

$$I_y' = I_y \cos^2 c + I_x \sin^2 c + I_{xy} \sin 2c$$

$$I_x' = I_x \cos^2 c + I_y \sin^2 c + I_{xy} \sin 2c$$

$$I_{xy}' = \frac{I_x - I_y}{2} \sin 2c + I_{xy} \cos 2c$$

Principal Moments of Inertia   In every plane area, a given point being taken as the origin, there is at least one pair of rectangular axes

          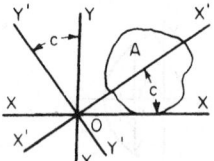

Fig. 3.1.31   **Fig. 3.1.32**

in the plane of the area about one of which the moment of inertia is a maximum, and a minimum about the other. These moments of inertia are called the **principal moments of inertia,** and the axes about which they are taken are the **principal axes of inertia.** One of the conditions for principal moments of inertia is that the product of inertia I_{xy} shall equal zero. **Axes of symmetry** of an area are always principal axes of inertia.

Relation between Products of Inertia and Parallel Axes   In Fig. 3.1.33, X_0X_0 and Y_0Y_0 pass through the center of gravity of the area parallel to the given axes *XX* and *YY*. If I_{xy} is the product of inertia for *XX* and *YY*, and $I_{x_0y_0}$ that for X_0X_0 and Y_0Y_0, then $I_{xy} = I_{x_0y_0} + abA$.

Fig. 3.1.33

Mohr's Circle   The **principal moments of inertia** and the location of the **principal axes of inertia** for any point of a plane area may be established graphically as follows.

Given at any point A of a plane area (Fig. 3.1.34), the moments of inertia I_x and I_y about axes X and Y, and the product of inertia I_{xy} relative to X and Y. The graph shown in Fig. 3.1.34b is plotted on rectangular coordinates with moments of inertia as abscissas and products of inertia

(a)   (b)

Fig. 3.1.34

as ordinates. Lay out $Oa = I_x$ and $ab = I_{xy}$ (upward for positive products of inertia, downward for negative). Lay out $Oc = I_y$ and $cd = $ *negative of* I_{xy}. Draw a circle with bd as diameter. This is **Mohr's circle.** The *maximum* moment of inertia is $I_x' = Of$; the *minimum* moment of inertia is $I_y' = Og$. The principal axes of inertia are located as follows. From axis AX (Fig. 3.1.34a) lay out angular distance $\theta = \frac{1}{2} < bef$. This locates axis AX', one principal axis ($I_x' = Of$). The other principal axis of inertia is AY', perpendicular to AX' ($I_y' = Og$).

The **moment of inertia of any area** may be considered to be made up of the sum or difference of the known moments of inertia of simple figures. For example, the dimensioned figure shown in Fig. 3.1.35 represents the section of a rolled shape with hole *oprs* and may be divided into the semicircle *abc*, rectangle *edkg*, and triangles *mfg* and *hkl*, from which the rectangle *oprs* is to be subtracted. Referring to axis *XX*,

$$I_{xx} = \pi 4^4/8 \text{ for semicircle } abc + (2 \times 11^3)/3 \text{ for rectangle } edkg$$

$$+ 2[(5 \times 3^3)/36 + 10^2(5 \times 3)/2] \text{ for the two triangles } mfg$$

and *hkl*

From the sum of these there is to be subtracted $I_{xx} = [(2 \times 3^2)/12 + 4^2(2 \times 3)]$ for the rectangle *oprs*.

Fig. 3.1.35

If the moment of inertia of the whole area is required about an axis parallel to *XX*, but passing through the center of gravity of the whole area, $I_0 = I_{xx} - x_0^2 A$, where x_0 = distance from *XX* to center of gravity. The **moments of inertia of built-up sections** used in structural work may be found in the same manner, the moments of inertia of the different **rolled sections** being given in Sec. 12.2.

Moments of Inertia of Solids For moments of inertia of solids about parallel axes, $I_x = I_0 + x_0^2 m$.

Moment of Inertia with Reference to Any Axis Let a mass particle *dm* of a body have *x*, *y*, and *z* as coordinates, *XX*, *YY*, and *ZZ* being the coordinate axes and *O* the origin. Let *X'X'* be any axis passing through the origin and making angles of *A*, *B*, and *C* with *XX*, *YY*, and *ZZ*, respectively. The moment of inertia with respect to this axis then becomes equal to

$$I'_x = \cos^2 A \int (y^2 + z^2)\, dm + \cos^2 B \int (z^2 \times x^2)\, dm$$
$$+ \cos^2 C \int (x^2 + y^2)\, dm - 2\cos B \cos C \int yz\, dm$$
$$- 2\cos C \cos A \int zx\, dm - 2\cos A \cos B \int xy\, dm$$

Let the moment of inertia about $XX = I_x = \int (y^2 + z^2)\, dm$, about $YY = I_y = \int (z^2 + x^2)\, dm$, and about $ZZ = I_z = \int (x^2 + y^2)\, dm$. Let the products of inertia about the three coordinate axes be

$$I_{yz} = \int yz\, dm \qquad I_{zx} = \int zx\, dm \qquad I_{xy} = \int xy\, dm$$

Then the moment of inertia I'_x becomes equal to

$$I_x \cos^2 A + I_y \cos^2 B + I_z \cos^2 C - 2I_{yz} \cos B \cos C - 2I_{zx}$$
$$\cos C \cos A - 2I_{xy} \cos A \cos B$$

The moment of inertia of any solid may be considered to be made up of the sum or difference of the moments of inertia of simple solids of which the moments of inertia are known.

Moments of Inertia of Important Solids (Homogeneous)

w = weight per unit of volume of the body
$m = w/g$ = mass per unit of volume of the body
$M = W/g$ = total mass of body
r = radius
I = moment of inertia (mass units)
$I_w = I \times g$ = moment of inertia (weight units)

Solid circular cylinder about its axis: $I = \pi r^4 ma/2 = Mr^2/2$. (*a* = length of axis of cylinder.)

Solid circular cylinder about an axis through the center of gravity and perpendicular to axis of cylinder: $I = M[r^2 + (a^2/3)]/4$.

Hollow circular cylinder about its axis: $I = M(r_1^2 + r_2^2)/2$ (r_1 and r_2 = outer and inner radii).

Thin-walled hollow circular cylinder about its axis: $I = Mr^2$.

Solid sphere about a diameter: $I = 8m\pi r^5/15 = 2Mr^2/5$.

Thin hollow sphere about a diameter: $I = 2Mr^2/3$.

Thick hollow sphere about a diameter: $I = 8m\pi(r_1^5 - r_2^5)/15$. ($r_1$ and r_2 are outer and inner radii.)

Rectangular prism about an axis through center of gravity and perpendicular to a face whose dimensions are *a* and *b*: $I = M(a^2 + b^2)/12$.

Solid right circular cone about an axis through its apex and perpendicular to its axis: $I = 3M[(r^2/4) + h^2]/5$. (*h* = altitude of cone, *r* = radius of base.)

Solid right circular cone about its axis of revolution: $I = 3Mr^2/10$.

Ellipsoid with semiaxes *a*, *b*, and *c*: *I* about diameter 2*c* (*z* axis) = $4m\pi abc\ (a^2 + b^2)/15$. [Equation of ellipsoid: $(x^2/a^2) + (y^2/b^2) + (z^2/c^2) = 1$.]

Ring with Circular Section (Fig. 3.1.36) $I_{yy} = \frac{1}{2}m\pi^2 Ra^2(4R^2 + 3a^2)$; $I_{xx} = m\pi^2 Ra^2[R^2 + (5a^2/4)]$.

Fig. 3.1.36   **Fig. 3.1.37**

Approximate Moments of Inertia of Solids In order to determine the moment of inertia of a solid, it is necessary to know all its dimensions. In the case of a rod of mass *M* (Fig. 3.1.37) and length *l*, with shape and size of the cross section unknown, making the approximation that the weight is all concentrated along the axis of the rod, the moment of inertia about *YY* will be $I_{yy} = \int_0^l (M/l)x^2\, dx = Ml^2/3$

A **thin plate** may be treated in the same way (Fig. 3.1.38): $I_{yy} = \int_0^l (M/l)x^2\, dx = \frac{1}{3}Ml^2$.

Thin Ring, or Cylinder (Fig. 3.1.39) Assume the mass *M* of the ring or cylinder to be concentrated at a distance *r* from *O*. The moment of inertia about an axis through *O* perpendicular to plane of ring or along the axis of the cylinder will be $I = Mr^2$; this will be greater than the exact moment of inertia, and *r* is sometimes taken as the distance from *O* to the center of gravity of the cross section of the rim.

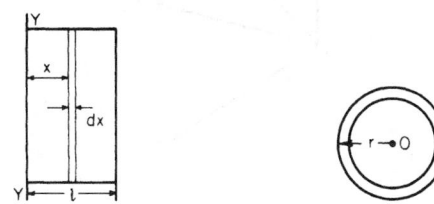

Fig. 3.1.38   **Fig. 3.1.39**

Flywheel Effect The moment of inertia of a solid is often called flywheel effect in the solution of problems dealing with rotating bodies.

Graphical Determination of the Centroids and Moments of Inertia of Plane Areas Required to find the center of gravity of the area *MNP* (Fig. 3.1.40) and its moment of inertia about any axis *XX*.

Draw any line *SS* parallel to *XX* and at a distance *d* from it. Draw a number of lines such as *AB* and *EF* across the figure parallel to *XX*. From *E* and *F* draw *ER* and *FT* perpendicular to *SS*. Select as a pole any

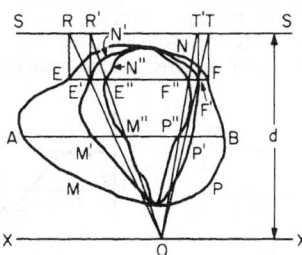

Fig. 3.1.40

point on *XX*, preferably the point nearest the area, and draw *OR* and *OT*, cutting *EF* at *E'* and *F'*. If the same construction is repeated, using other lines parallel to *XX*, a number of points will be obtained, which, if connected by a smooth curve, will give the area *M'N'P'*. Project *E'* and *F'* onto *SS* by lines *E'R'* and *F'T'*. Join *F'* and *T* with *O*, obtaining *E"* and *F"*; connect the points obtained using other lines parallel to *XX* and obtain an area *M"N"P"*. The area *M'N'P'* × *d* = moment of area *MNP* about the line *XX*, and the distance from *XX* to the centroid *MNP* = area *M'N'P'* × *d*/area *MNP*. Also, area *M"N"P"* × d^2 = moment of inertia of *MNP* about *XX*. The areas *M'N'P'* and *M"N"P"* can best be obtained by use of a planimeter.

KINEMATICS

Kinematics is the study of the motion of bodies without reference to the forces causing that motion or the mass of the bodies.

The **displacement** of a point is the directed distance that a point has moved on a geometric path from a convenient origin. It is a **vector**, having both magnitude and direction, and is subject to all the laws and characteristics attributed to vectors. In Fig. 3.1.41, the displacement of the point *A* from the origin *O* is the directed distance *O* to *A*, symbolized by the vector **s**.

The **velocity** of a point is the time rate of change of displacement, or $\mathbf{v} = ds/dt$.

The **acceleration** of a point is the time rate of change of velocity, or $\mathbf{a} = d\mathbf{v}/dt$.

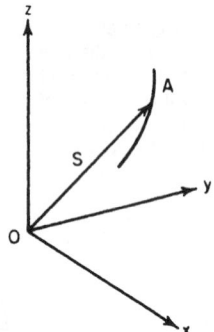

Fig. 3.1.41

The kinematic definitions of velocity and acceleration involve the four variables, *displacement, velocity, acceleration,* and *time*. If we eliminate the variable of time, a third equation of motion is obtained, $ds/v = dt = dv/a$. This differential equation, together with the definitions of velocity and acceleration, make up the *three kinematic equations* of motion, $v = ds/dt$, $a = dv/dt$, and $a\, ds = v\, dv$. These differential equations are usually limited to the scalar form when expressed together, since the last can only be properly expressed in terms of the scalar *dt*. The first two, since they are definitions for velocity and acceleration, are **vector equations.**

A **space-time curve** offers a convenient means for the study of the motion of a point that moves in a straight line. The **slope of the curve** at any point will represent the **velocity** at that time. In Fig. 3.1.42*a* the slope is constant, as the graph is a straight line; the velocity is therefore uniform. In Fig. 3.1.42*b* the slope of the curve varies from point to point, and the velocity must also vary. At *p* and *q* the slope is zero; therefore, the velocity of the point at the corresponding times must also be zero.

Fig. 3.1.42

A **velocity-time curve** offers a convenient means for the study of acceleration. The **slope of the curve** at any point will represent the **acceleration** at that time. In Fig. 3.1.43*a* the slope is constant; so the acceleration must be constant. In the case represented by the full line, the acceleration is positive; so the velocity is increasing. The dotted line shows a negative acceleration and therefore a decreasing velocity. In Fig. 3.1.43*b* the slope of the curve varies from point to point; so the acceleration must also vary. At *p* and *q* the slope is zero; therefore, the acceleration of the point at the corresponding times must also be zero. The area under the velocity-time curve between any two ordinates such as *NL* and *HT* will represent the distance moved in time interval *LT*. In the case of the uniformly accelerated motion shown by the full line in Fig. 3.1.43*a*, the area *LNHT* is ½(*NL* + *HT*) × (*OT* − *OL*) = mean velocity multiplied by the time interval = space passed over during this time interval. In Fig. 3.1.43*b* the mean velocity can be obtained from the equation of the curve by means of the calculus, or graphically by approximation of the area.

Fig. 3.1.43

An **acceleration-time curve** (Fig. 3.1.44) may be constructed by plotting accelerations as ordinates, and times as abscissas. The area under this curve between any two ordinates will represent the total increase in velocity during the time interval. The area *ABCD* represents the total increase in velocity between time t_1 and time t_2.

General Expressions Showing the Relations between Space, Time, Velocity, and Acceleration for Rectilinear Motion

SPECIAL MOTIONS

Uniform Motion If the **velocity** is **constant**, the **acceleration** must be **zero,** and the point has uniform motion. The space-time curve becomes a straight line inclined toward the time axis (Fig. 3.1.42*a*). The velocity-time curve becomes a straight line parallel to the time axis. For this motion $a = 0$, $v =$ constant, and $s = s_0 + vt$.

Uniformly Accelerated or Retarded Motion If the **velocity** is **not uniform** but the **acceleration** is **constant**, the point has uniformly accelerated motion; the acceleration may be either positive or negative. The space-time curve becomes a parabola and the velocity-time curve becomes a straight line inclined toward the time axis (Fig. 3.1.43*a*). The acceleration-time curve becomes a straight line parallel to the time axis. For this motion $a =$ constant, $v = v_0 + at$, $s = s_0 + v_0 t + \frac{1}{2}at^2$.

If the point starts from rest, $v_0 = 0$. Care should be taken concerning the sign + or − for acceleration.

Composition and Resolution of Velocities and Acceleration

Resultant Velocity A velocity is said to be the resultant of two other velocities when it is represented by a vector that is the geometric sum of the vectors representing the other two velocities. This is the **parallelogram of motion.** In Fig. 3.1.45, **v** is the resultant of $\mathbf{v}_1$ and $\mathbf{v}_2$

Fig. 3.1.44

Fig. 3.1.45

and is represented by the diagonal of a parallelogram of which $\mathbf{v}_1$ and $\mathbf{v}_2$ are the sides; or it is the third side of a triangle of which $\mathbf{v}_1$ and $\mathbf{v}_2$ are the other two sides.

Polygon of Motion   The parallelogram of motion may be extended to the polygon of motion. Let $\mathbf{v}_1$, $\mathbf{v}_2$, $\mathbf{v}_3$, $\mathbf{v}_4$ (Fig. 3.1.46a) show the directions of four velocities imparted in the same plane to point O. If the lines $\mathbf{v}_1$, $\mathbf{v}_2$, $\mathbf{v}_3$, $\mathbf{v}_4$ (Fig. 3.1.46b) are drawn parallel to and proportional to the velocities imparted to point O, $\mathbf{v}$ will represent the resultant velocity imparted to O. It will make no difference in what order the velocities are taken in constructing the motion polygon. As long as the arrows showing the direction of the motion follow each other in order about the polygon, the resultant velocity of the point will be represented in magnitude by the closing side of the polygon, but opposite in direction.

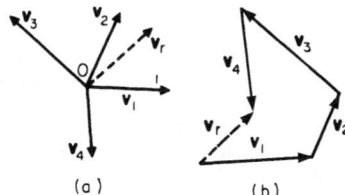

(a)   (b)

Fig. 3.1.46

Resolution of Velocities   Velocities may be resolved into **component velocities** in the same plane, as shown by Fig. 3.1.47. Let the velocity of point O be v_r. In Fig. 3.1.47a this velocity is resolved into two components in the same plane as v_r and at right angles to each other.

$$v_r = \sqrt{(v_1)^2 + (v_2)^2}$$

In Fig. 3.1.47b the components are in the same plane as v_r, but are not at right angles to each other. In this case,

$$v_r = \sqrt{(v_1)^2 + (v_2)^2 + 2v_1v_2\cos B}$$

If the components v_1 and v_2 and angle B are known, the direction of v_r can be determined. $\sin bOc = (v_1/v_r)\sin B$. $\sin cOa = (v_2/v_r)\sin B$. Where v_1 and v_2 are at right angles to each other, $\sin B = 1$.

(a)   (b)

Fig. 3.1.47

Resultant Acceleration   Accelerations may be combined and resolved in the same manner as velocities, but in this case the lines or vectors represent accelerations instead of velocities. If the acceleration had components of magnitude a_1 and a_2, the magnitude of the resultant acceleration would be $a = \sqrt{(a_1)^2 + (a_2)^2 + 2a_1a_2\cos B}$, where B is the angle between the vectors $\mathbf{a}_1$ and $\mathbf{a}_2$.

Curvilinear Motion in a Plane

The linear velocity $v = ds/dt$ of a point in curvilinear motion is the same as for rectilinear motion. Its direction is tangent to the path of the point. In Fig. 3.1.48a, let $P_1P_2P_3$ be the path of a moving point and $\mathbf{V}_1$, $\mathbf{V}_2$, $\mathbf{V}_3$ represent its velocity at points P_1, P_2, P_3, respectively. If O is taken as a pole (Fig. 3.1.48b) and vectors $\mathbf{V}_1$, $\mathbf{V}_2$, $\mathbf{V}_3$ representing the velocities of the point at P_1, P_2, and P_3 are drawn, the curve connecting the terminal points of these vectors is known as the **hodograph** of the motion. This velocity diagram is applicable only to motions all in the same plane.

Acceleration   Tangents to the curve (Fig. 3.1.48b) indicate the directions of the **instantaneous velocities**. The direction of the tangents does not, as a rule, coincide with the direction of the accelerations as represented by tangents to the path. If the acceleration a at some point

in the path is resolved by means of a parallelogram into components tangent and normal to the path, the normal acceleration $a_n = v^2/\rho$, where ρ = radius of curvature of the path at the point in question, and the tangential acceleration $a_t = dv/dt$, where v = velocity tangent to the path at the same point. $a = \sqrt{a_n^2 + a_t^2}$. The normal acceleration is constantly directed toward the concave side of the path.

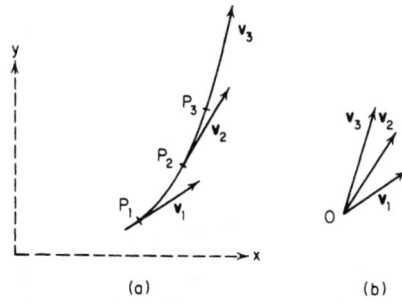

(a)   (b)

Fig. 3.1.48

EXAMPLE.   Figure 3.1.49 shows a point moving in a curvilinear path. At p_1 the velocity is $\mathbf{v}_1$; at p_2 the velocity is $\mathbf{v}_2$. If these velocities are drawn from pole O (Fig. 3.1.49b), $\Delta\mathbf{v}$ will be the difference between $\mathbf{v}_2$ and $\mathbf{v}_1$. The acceleration during travel p_1p_2 will be $\Delta\mathbf{v}/\Delta t$, where Δt is the time interval. The approximation becomes closer to instantaneous acceleration as shorter intervals Δt are employed.

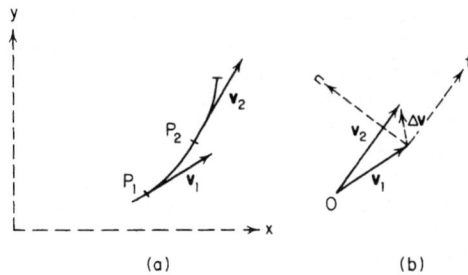

(a)   (b)

Fig. 3.1.49

The acceleration $\Delta\mathbf{v}/\Delta t$ can be resolved into normal and tangential components leading to $\mathbf{a}_n = \Delta v_n/\Delta t$, normal to the path, and $\mathbf{a}_t = \Delta v_t/\Delta t$, tangential to the path.

Velocity and *acceleration* may be expressed in **polar coordinates** such that $v = \sqrt{v_r^2 + v_\theta^2}$ and $a = \sqrt{a_r^2 + a_\theta^2}$. Figure 3.1.50 may be used to explain the r and θ coordinates.

EXAMPLE.   At P_1 the velocity is $\mathbf{v}_1$, with components $\mathbf{v}_{1r}$ in the r direction and $\mathbf{v}_{1\theta}$ in the θ direction. At P_2 the velocity is $\mathbf{v}_2$, with components $\mathbf{v}_{2r}$ in the r direction and $\mathbf{v}_{2\theta}$ in the θ direction. It is evident that the difference in velocities $\mathbf{v}_2 - \mathbf{v}_1 = \Delta\mathbf{v}$ will have components Δv_r and Δv_θ, giving rise to accelerations $\mathbf{a}_r$ and $\mathbf{a}_\theta$ in a time interval Δt.

In polar coordinates, $v_r = dr/dt$, $a_r = d^2r/dt^2 - r(d\theta/dt)^2$, $v_\theta = r(d\theta/dt)$, and $a_\theta = r(d^2\theta/dt^2) + 2(dr/dt)(d\theta/dt)$.

If a point P moves on a circular path of radius r with an angular velocity of ω and an angular acceleration of α, the linear velocity of the point P is $v = \omega r$ and the two components of the linear acceleration are $a_n = v^2/r = \omega^2 r = v\omega$ and $a_t = \alpha r$.

If the angular velocity is constant, the point P travels equal circular paths in equal intervals of time. The projected displacement, velocity, and acceleration of the point P on the x and y axes are sinusoidal functions of time, and the motion is said to be **harmonic motion**. Angular velocity is usually expressed in radians per second, and when the number (N) of revolutions traversed per minute (N/min) by the point P is known, the angular velocity of the radius r is $\omega = 2\pi N/60 = 0.10472N$.

Fig. 3.1.50

Fig. 3.1.51

In Fig. 3.1.51, let the angular velocity of the line OP be a constant ω. Let the point P start at X' and move to P in time t. Then the angle $\theta = \omega t$. If $OP = r$, $OA = s = r\cos\omega t$. The velocity v of the point A on the x axis will equal $ds/dt = -\omega r\sin\omega t$, and the acceleration $a = dv/dt = -\omega^2 r\cos\omega t$. The period τ is the time necessary for the point P to complete one cycle of motion $\tau = 2\pi/\omega$, and it is also equal to the time necessary for A to complete a full cycle on the x axis from X' to X and return.

Curvilinear Motion in Space

If **three dimensions** are used, velocities and accelerations may be resolved into components not in the same plane by what is known as the **parallelepiped of motion**. Three coordinate systems are widely used, cartesian, cylindrical, and spherical. In **cartesian coordinates,** $v = \sqrt{v_x^2 + v_y^2 + v_z^2}$ and $a = \sqrt{a_x^2 + a_y^2 + a_z^2}$. In **cylindrical coordinates,** the radius vector $\mathbf{R}$ of displacement lies in the rz plane, which is at an angle with the xz plane. Referring to (a) of Fig. 3.1.52, the θ coordinate is perpendicular to the rz plane. In this system $v = \sqrt{v_r^2 + v_\theta^2 + v_z^2}$ and $a = \sqrt{a_r^2 + a_\theta^2 + a_z^2}$ where $v_r = dr/dt$, $a_r = d^2r/dt^2 - r(d\theta/dt)^2$, $v_\theta = r(d\theta/dt)$, and $a_\theta = r(d^2\theta/dt^2) + 2(dr/dt)(d\theta/dt)$. In **spherical coordinates**, the three coordinates are the R coordinate, the θ coordinate, and the ϕ coordinate as in (b) of Fig. 3.1.52. The velocity and acceleration are $v = \sqrt{v_R^2 + v_\theta^2 + v_\phi^2}$ and $a = \sqrt{a_R^2 + a_\theta^2 + a_\phi^2}$, where $v_R = dR/dt$, $v_\phi = R(d\phi/dt)$, $v_\theta = R\cos\phi(d\theta/dt)$, $a_R = d^2R/dt^2 - R(d\phi/dt)^2 - R\cos^2\phi(d\theta/dt)^2$, $a_\phi = R(d^2\phi/dt^2) + R\cos\phi\sin\phi\,(d\theta/dt)^2 + 2(dR/dt)(d\phi/dt)$, and $a_\theta = R\cos\phi\,(d^2\theta/dt^2) + 2[(dR/dt)\cos\phi - R\sin\phi\,(d\phi/dt)]\,d\theta/dt$.

(a) Cylindrical coordinates

(b) Spherical coordinates

Fig. 3.1.52

Motion of Rigid Bodies

A body is said to be **rigid** when the distances between all its particles are invariable. Theoretically, rigid bodies do not exist, but materials used in engineering are rigid under most practical working conditions. The motion of a rigid body can be completely described by knowing the **angular motion** of a line on the rigid body and the **linear motion** of a point on this line and relating the motion of all other parts of the rigid body to these motions. If a rigid body moves so that a straight line connecting any two of its particles remains parallel to its original position at all times, it is said to have **translation**. In **rectilinear translation**, all points move in straight lines. In **curvilinear translation**, all points move on congruent curves but without rotation. **Rotation** is defined as angular motion about an axis, which may or may not be fixed. Rigid body motion in which the paths of all particles lie on parallel planes is called **plane motion.**

Angular Motion

Angular displacement is the change in angular position of a given line as measured from a convenient reference line. In Fig. 3.1.53, consider the motion of the lines AB as it moves from its original position $A'B'$. The angle between lines AB and $A'B'$ is the angular displacement of line $\mathbf{AB}$, symbolized as θ. It is a directed quantity and is a vector. The usual notation used to designate angular displacement is a vector normal to

Fig. 3.1.53

the plane in which the angular displacement occurs. The length of the vector is proportional to the magnitude of the angular displacement. For a rigid body moving in three dimensions, the line AB may have angular motion about any three orthogonal axes. For example, the angular displacement can be described in cartesian coordinates as $\theta = \theta_x + \theta_y + \theta_z$, where $\theta = \sqrt{\theta_x^2 + \theta_y^2 + \theta_z^2}$.

Angular velocity is defined as the time rate of change of angular displacement, $\omega = d\theta/dt$. Angular velocity may also have components about any three orthogonal axes.

Angular acceleration is defined as the time rate of change of angular velocity, $\alpha = d\omega/dt = d^2\theta\,dt^2$. Angular acceleration may also have components about any three orthogonal axes.

The *kinematic equations* of angular motion of a line are analogous to those for the motion of a point. In referring to Table 3.1.1, $\omega = d\theta/dt$ $\alpha = d\omega/dt$, and $\alpha\,d\theta = \omega\,d\omega$. Substitute θ for s, ω for v, and α for a.

Motion of a Rigid Body in a Plane

Plane motion is the motion of a rigid body such that the paths of all particles of that rigid body lie on parallel planes.

Table 3.1.1

Variables	$s = f(t)$	$v = f(t)$	$a_2 = f(t)$	$a = f(s, v)$
Displacement		$s = s_0 + \int_{t_0}^{t} v\,dt$	$s = s_0 + \int_{t_0}^{t}\int_{t_0}^{t} a\,dt\,dt$	$s = s_0 + \int_{v_0}^{v} (v/a)\,dv$
Velocity	$v = ds/dt$		$v = v_0 + \int_{t_0}^{t} a\,dt$	$\int_{v_0}^{v} v\,dv = \int_{s}^{s_0} a\,ds$
Acceleration	$a = d^2s/dt^2$	$a = dv/dt$		$a = v\,dv/ds$

Instantaneous Axis   When the axis about which any body may be considered to rotate changes its position, any one position is known as an instantaneous axis, and the line through all positions of the instantaneous axis as the **centrode.**

When the velocity of two points in the same plane of a rigid body having plane motion is known, the instantaneous axis for the body will be at the intersection of the lines drawn from each point and perpendicular to its velocity. See Fig. 3.1.54, in which A and B are two points on the rod AB, $\mathbf{v}_1$ and $\mathbf{v}_2$ representing their velocities. O is the instantaneous axis for AB; therefore point C will have velocity shown in a line perpendicular to OC.

Linear velocities of points in a body rotating about an instantaneous axis are proportional to their distances from this axis. In Fig. 3.1.54, $\mathbf{v}_1 : \mathbf{v}_2 : \mathbf{v}_3 = AO : OB : OC$. If the velocities of A and B were parallel, the lines OA and OB would also be parallel and there would be no instantaneous axis. The motion of the rod would be translation, and all points would be moving with the same velocity in parallel straight lines.

If a body has plane motion, the components of the velocities of any two points in the body along the straight line joining them must be equal. Ax must be equal to By and Cz in Fig. 3.1.54.

EXAMPLE.   In Fig. 3.1.55a, the velocities of points A and B are known—they are $\mathbf{v}_1$ and $\mathbf{v}_2$, respectively. To find the instantaneous axis of the body, perpendiculars AO and BO are drawn. O, at the intersection of the perpendiculars, is the **instantaneous axis** of the body. To find the velocity of any other point, like C, line OC is drawn and v_3 erected perpendicular to OC with magnitude equal to v_1 (CO/AO). The **angular velocity** of the body will be $\omega = v_1/AO$ or v_2/BO or v_3/CO. The instantaneous axis of a wheel rolling on a rack without slipping (Fig. 3.1.55b) lies at the point of contact O, which has zero linear velocity. All points of the wheel will have velocities perpendicular to radii to O and proportional in magnitudes to their respective distances from O.

Another way to describe the plane motion of a rigid body is with the use of **relative motion**. In Fig. 3.1.56 the velocity of point A is $\mathbf{v}_1$. The angular velocity of the line AB is ω_{AB}. The velocity of B relative to A is $\omega_{AB} \times r_{AB}$. Point B is considered to be moving on a circular path around A as a center. The direction of relative velocity of B to A would be tangent to the circular path in the direction that ω_{AB} would make B move. The velocity of B is the vector sum of the velocity A added to the velocity of B relative to A, $\mathbf{v}_B = \mathbf{v}_A + \mathbf{v}_{B/A}$.

The **acceleration** of B is the vector sum of the acceleration of A added to the acceleration of B relative to A, $\mathbf{a}_B = \mathbf{a}_A + \mathbf{a}_{B/A}$. Care must be taken to include the complete relative acceleration of B to A. If B is considered to move on a circular path about A, with a velocity relative to A, it will have an acceleration relative to A that has both normal and tangential components: $\mathbf{a}_{B/A} = (\mathbf{a}_{B/A})_n + (\mathbf{a}_{B/A})_r$

(a) Relative velocity

(b) Relative acceleration

Fig. 3.1.56

If B is a point on a path which lies on the same rigid body as the line AB, a **particle P traveling on the path** will have a velocity $\mathbf{v}_p$ at the instant P passes over point B such that $\mathbf{v}_P = \mathbf{v}_A + \mathbf{v}_{B/A} + \mathbf{v}_{P/B}$, where the velocity $\mathbf{v}_{P/B}$ is the velocity of P relative to point B.

The particle P will have an acceleration $\mathbf{a}_P$ at the instant P passes over the point B such that $\mathbf{a}_P = \mathbf{a}_A + \mathbf{a}_{B/A} + \mathbf{a}_{P/B} + 2\omega_{AB} \times \mathbf{v}_{P/B}$. The term $\mathbf{a}_{P/B}$ is the acceleration of P relative to the path at point B. The last term $2\omega_{AB}\,\mathbf{v}_{P/B}$ is frequently referred to as the **coriolis acceleration.** The direction is always normal to the path in a sense which would rotate the head of the vector $\mathbf{v}_{P/B}$ about its tail in the direction of the angular velocity of the rigid body ω_{AB}.

EXAMPLE.   In Fig. 3.1.57, arm AB is rotating counterclockwise about A with a constant angular velocity of 38 r/min or 4 rad/s, and the slider moves outward with a velocity of 10 ft/s (3.05 m/s). At an instant when the slider P is 30 in (0.76 m) from the center A, the acceleration of the slider will have two components. One component is the normal acceleration directed toward the center A. Its magnitude is $\omega^2 r = 4^2 (30/12) = 40$ ft/s² [$\omega^2 r = 4^2 (0.76) = 12.2$ m/s²]. The second is the coriolis acceleration directed normal to the arm AB, upward and to the left. Its magnitude is $2\omega v = 2(4)(10) = 80$ ft/s² [$2\omega v = 2(4)(3.05) = 24.4$ m/s²].

Fig. 3.1.57

General Motion of a Rigid Body

The general motion of a point moving in a coordinate system which is itself in motion is complicated and can best be summarized by using

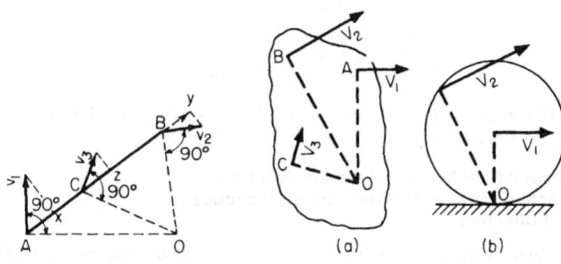

Fig. 3.1.54   **Fig. 3.1.55**

vector notation. Referring to Fig. 3.1.58, let the point P be displaced a vector distance $\mathbf{R}$ from the origin O of a moving reference frame x, y, z which has a velocity $\mathbf{v}_o$ and an acceleration $\mathbf{a}_o$. If point P has a velocity and an acceleration relative to the moving reference plane, let these be $\mathbf{v}_r$ and $\mathbf{a}_r$. The angular velocity of the moving reference fame is ω, and

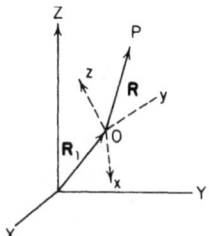

Fig. 3.1.58

the origin of the moving reference frame is displaced a vector distance $\mathbf{R}_1$ from the origin of a primary (fixed) reference frame X, Y, Z. The velocity and acceleration of P are $\mathbf{v}_P = \mathbf{v}_o + \omega \times \mathbf{R} + \mathbf{v}_r$ and $\mathbf{a}_P = \mathbf{a}_o + (d\omega/dt) \times \mathbf{R} + \omega \times (\omega \times \mathbf{R}) + 2\omega \times \mathbf{v}_r + \mathbf{a}_r$.

DYNAMICS OF PARTICLES

Consider a particle of mass m subjected to the action of forces $\mathbf{F}_1$, $\mathbf{F}_2$, $\mathbf{F}_3$, . . . , whose vector resultant is $\mathbf{R} = \Sigma\mathbf{F}$. According to Newton's first law of motion, if $\mathbf{R} = 0$, the body is acted on by a balanced force system, and it will either remain at rest or move uniformly in a straight line. If $\mathbf{R} \neq 0$, Newton's second law of motion states that the body will accelerate in the direction of and proportional to the magnitude of the resultant R. This may be expressed as $\Sigma\mathbf{F} = m\mathbf{a}$. If the resultant of the force system has components in the x, y, and z directions, the resultant acceleration will have proportional components in the x, y, and z direction so that $F_x = ma_x$, $F_y = ma_y$, and $F_z = ma_z$. If the resultant of the force system varies with time, the acceleration will also vary with time.

In **rectilinear motion**, the acceleration and the direction of the unbalanced force must be in the direction of motion. **Forces must be in balance and the acceleration equal to zero in any direction other than the direction of motion.**

EXAMPLE 1. The body in Fig. 3.1.59 has a mass of 90 lbm (40.8 kg) and is subjected to an external horizontal force of 36 lbf (160 N) applied in the direction shown. The coefficient of friction between the body and the inclined plane is 0.1. Required, the velocity of the body at the end of 5 s, if it starts from rest.

Fig. 3.1.59

First determine all the forces acting externally on the body. These are the applied force $F = 36$ lbf (106 N), the weight $W = 90$ lbf (400 N), and the force with which the plane reacts on the body. The latter force can be resolved into component forces, one normal and one parallel to the surface of the plane. Motion will be downward along the plane since a static analysis will show that the body will slide downward unless the static coefficient of friction is greater than 0.269. In the direction normal to the surface of the plane, the forces must be balanced. The normal force is $(3/5)(36) + (4/5)(90) = 93.6$ lbf (416 N). The frictional force is $93.6 \times 0.1 = 9.36$ lbf (41.6 N). The unbalanced force acting on the body along the

plane is $(3/5)(90) - (4/5)(36) - 9.36 = 15.84$ lbf (70.46 N) downward. $F = (W/9)$ $a = (90/g)$ a; therefore, $a = 0.176$ $g = 5.66$ ft/s^2 (1.725 m/s^2). In SI units, $F = ma = 70.46 = 40.8a$; and $a = 1.725$ m/s^2. The body is acted upon by constant forces and starts from rest; therefore, $v = \int_0^5 a\,dt$, and at the end of 5 s, the velocity would be 28.35 ft/s (8.91 m/s).

EXAMPLE 2. The force with which a rope acts on a body is equal and opposite to the force with which the body acts on the rope, and each is equal to the tension in the rope. In Fig. 3.1.60a, neglecting the weight of the pulley and the rope, the tension in the cord must be the force of 27 lbf. For the 18-lb mass, the unbalanced force is $27 - 18 = 9$ lbf in the upward direction, i.e., $27 - 18 = (18/g)a$, and $a = 16.1$ ft/s^2 upward. In Fig. 3.1.60b the 27-lb force is replaced by a 27-lb mass. The unbalanced force is still $27 - 18 = 9$ lbf, but it now acts on two masses so that $27 - 18 = (45a/g)$ and $a = 6.44$ ft/s^2. The 18-lb mass is accelerated upward, and the 27-lb mass is accelerated downward. The tension in the rope is equal to 18 lbf plus the unbalanced force necessary to give it an upward acceleration of $g/5$ or $T = 18 + (18/g)(g/5) = 21.6$ lbf. The tension is also equal to 27 lbf less the unbalanced force necessary to give it a downward acceleration of $g/5$ or $T = 27 - (27/g) \times (g/5) = 21.6$ lbf.

Fig. 3.1.60

In SI units, in Fig. 3.1.60a, the unbalanced force is $120 - 80 = 40$ N, in the upward direction, i.e., $120 - 80 = 8.16a$, and $a = 4.9$ m/s^2 (16.1 ft/s^2). In Fig. 3.1.60b the unbalanced force is still 40 N, but it now acts on the two masses so that $120 - 80 = 20.4a$ and $a = 1.96$ m/s^2 (6.44 ft/s^2). The tension in the rope is the weight of the 8.16-kg mass in newtons plus the unbalanced force necessary to give it an upward acceleration of 1.96 m/s^2, $T = 9.807(8.16) + (8.16)(1.96) = 96$ N (21.6 lbf).

General Formulas for the Motion of a Body under the Action of a Constant Unbalanced Force

Let s = distance, ft; a = acceleration, ft/s^2; v = velocity, ft/s; v_0 = initial velocity, ft/s; h = height, ft; F = force; m = mass; w = weight; g = acceleration due to gravity.

$$\text{Initial velocity} = 0$$
$$F = ma = (w/g)a$$
$$v = at$$
$$s = \tfrac{1}{2}at^2 = \tfrac{1}{2}vt$$
$$v = \sqrt{2as}$$
$$= \sqrt{2gh} \text{ (falling freely from rest)}$$

$$\text{Initial velocity} = v_0$$
$$F = ma = (w/g)a$$
$$v = v_0 + at$$
$$s = v_0t + \tfrac{1}{2}at^2 = \tfrac{1}{2}v_0t + \tfrac{1}{2}vt$$

If a body is to be moved in a straight line by a force, the line of action of this force must pass through its center of gravity.

General Rule for the Solution of Problems When the Forces Are Constant in Magnitude and Direction

Resolve all the forces acting on the body into two components, one in the direction of the body's motion and one at right angles to it. Add the components in the direction of the body's motion algebraically and find the **unbalanced force,** if any exists.

In **curvilinear motion,** a particle moves along a curved path, and the resultant of the unbalanced force system may have components in directions other than the direction of motion. **The acceleration in any given direction is proportional to the component of the resultant in that direction.** It is common to utilize orthogonal coordinate systems such as **cartesian coordinates, polar coordinates,** and **normal and tangential coordinates** in analyzing forces and accelerations.

EXAMPLE.   A conical pendulum consists of a weight suspended from a cord or light rod and made to rotate in a horizontal circle about a vertical axis with a constant angular velocity of N r/min. For any given constant speed of rotation, the angle θ, the radius r, and the height h will have fixed values. Looking at Fig. 3.1.61, we see that the forces in the vertical direction must be balanced, $T \cos \theta = w$. The forces in the direction normal to the circular path of rotation are unbalanced such that $T \sin \theta = (w/g)a_n = (w/g)\omega^2 r$. Substituting $r = l \sin \theta$ in this last equation gives the value of the tension in the cord $T = (w/g)l\omega^2$. Dividing the second equation by the first and substituting $\tan \theta = r/h$ yields the additional relation that $h = g/\omega^2$.

Fig. 3.1.61

An unresisted **projectile** has a motion compounded of the vertical motion of a falling body, and of the horizontal motion due to the horizontal component of the velocity of projection. In Fig. 3.1.62 the only force acting after the projectile starts is gravity, which causes an accelerating downward. The horizontal component of the original velocity v_0 is not changed by gravity. The projectile will rise until the velocity

Fig. 3.1.62

given to it by gravity is equal to the vertical component of the starting velocity v_0, and the equation $v_0 \sin \theta = gt$ gives the time t required to reach the highest point in the curve. The same time will be taken in falling if the surface XX is level, and the projectile will therefore be in flight $2t$ s. The distance $s = v_0 \cos \theta \times 2t$, and the maximum height of ascent $h = (v_0 \sin \theta)^2/2g$. The expressions for the coordinates of any point on the path of the projectile are: $x = (v_0 \cos \theta)t$, and $y = (v_0 \sin \theta)t - \frac{1}{2}gt^2$, giving $y = x \tan \theta - (gx^2/2v_0^2 \cos^2 \theta)$ as the equation for the curve of the path. The radius of curvature of the highest point may be found by using the general expression $v^2 = gr$ and solving for r, v being taken equal to $v_0 \cos \theta$.

Simple Pendulum   The period of oscillation $= \tau = 2\pi \sqrt{l/g}$, where l is the length of the pendulum and the length of the swing is not great compared to l.

Centrifugal and Centripetal Forces   When a body revolves about an axis, some connection must exist capable of applying force enough

to the body to constantly deviate it toward the axis. This deviating force is known as **centripetal force.** The equal and opposite resistance offered by the body to the connection is called the **centrifugal force.** The acceleration toward the axis necessary to keep a particle moving in a circle about that axis is v^2/r; therefore, the force necessary is $ma = mv^2/r = wv^2/gr = w\pi^2N^2r/900g$, where $N = $ r/min. This force is constantly directed toward the axis.

The centrifugal force of a solid body revolving about an axis is the same as if the whole mass of the body were concentrated at its center of gravity. Centrifugal force $= wv^2/gr = mv^2/r = w\omega^2r/g$, where w and m are the weight and mass of the whole body, r is the distance from the axis about which the body is rotating to the center of gravity of the body, ω the angular velocity of the body about the axis in radians, and v the linear velocity of the center of gravity of the body.

Balancing

A rotating body is said to be in **standing balance** when its center of gravity coincides with the axis upon which it revolves. Standing balance may be obtained by resting the axle carrying the body upon two horizontal plane surfaces, as in Fig. 3.1.63. If the center of gravity of the wheel A lies on the axis of the shaft B, there will be no movement, but if the center of gravity does not lie on the axis of the shaft, the shaft will roll until the center of gravity of the wheel comes directly under the

Fig. 3.1.63

axis of the shaft. The center of gravity may be brought to the axis of the shaft by adding or taking away weight at proper points on the diameter passing through the center of gravity and the center of the shaft. Weights may be added to or subtracted from any part of the wheel so long as its center of gravity is brought to the center of the shaft.

A rotating body may be in standing balance and not in **dynamic balance.** In Fig. 3.1.64, AA and BB are two disks whose centers of gravity are at o and p, respectively. The shaft and the disks are in standing balance if the disks are of the same weight and the distances of o and p from the center of the shaft are equal, and o and p lie in the same axial plane but on opposite sides of the shaft. Let the weight of each disk be w and the distances of o and p from the center of the shaft each be equal to r.

Fig. 3.1.64

The force exerted on the shaft by AA is equal to $w\omega^2r/g$, where ω is the angular velocity of the shaft. Also, the force exerted on the shaft by $BB = w\omega^2r/g$. These two equal and opposite parallel forces act at a distance x apart and constitute a couple with a moment tending to rotate the shaft, as shown by the arrows, of $(w\omega^2r/g)x$. A couple cannot be balanced by a single force; so two forces at least must be added to or subtracted from the system to get dynamic balance.

Systems of Particles   The principles of motion for a single particle can be extended to cover a **system of particles.** In this case, **the vector resultant of all external forces acting on the system of particles must equal the total mass of the system times the acceleration of the mass center, and the direction of the resultant must be the direction of the acceleration of the mass center.** This is the **principle of motion of the mass center.**

Rotation of Solid Bodies in a Plane about Fixed Axes

For a rigid body revolving **in a plane** about a fixed axis, **the resultant moment about that axis must be equal to the product of the moment of inertia (about that axis) and the angular acceleration,** $\Sigma M_0 = I_0 \alpha$. This is a general statement which includes the particular case of rotation about an axis that passes through the center of gravity.

Rotation about an Axis Passing through the Center of Gravity The rotation of a body about its center of gravity can only be caused or changed by a **couple.** See Fig. 3.1.65. If a single force F is applied to the wheel, the axis immediately acts on the wheel with an equal force to prevent translation, and the result is a couple (moment Fr) acting on the body and causing rotation about its center of gravity.

Fig. 3.1.65

General formulas for rotation of a body about a fixed axis through the center of gravity, if a constant unbalanced moment is applied (Fig. 3.1.65).

Let θ = angular displacement, rad; ω = angular velocity, rad/s; α = angular acceleration, rad/s^2; M = unbalanced moment, ft · lb; I = moment of inertia (mass); g = acceleration due to gravity; t = time of application of M.

Initial angular velocity = 0	Initial angular velocity = ω_0
$M = I\alpha$	$M = I\alpha$
$\theta = \frac{1}{2}\alpha t^2$	$\theta = \omega_0 t + \frac{1}{2}\alpha t^2$
$\omega = \sqrt{2\alpha\theta}$	$\omega = \sqrt{\omega_0^2 + 2\alpha\theta}$

General Rule for Rotating Bodies Determine all the external forces acting and their moments about the axis of rotation. If these moments are balanced, there will be no change of motion. If the moments are unbalanced, this unbalanced moment, or **torque,** will cause an angular acceleration about the axis.

Rotation about an Axis Not Passing through the Center of Gravity **The resultant force acting on the body must be proportional to the acceleration of the center of gravity and directed along its line of action.** If the axis of rotation does not pass through the center of gravity, the center of gravity will have a resultant acceleration with a component $a_n = \omega^2 r$ directed toward the axis of rotation and a component $a_t = \alpha r$ tangential to its circular path. The resultant force acting on the body must also have two components, one directed normal and one directed tangential to the path of the center of gravity. The line of action of this resultant does not pass through the center of gravity because of the unbalanced moment $M_0 = I_0\alpha$ but at a point Q, as in Fig. 3.1.66. The point of application of this resultant is known as the **center of percussion** and may be **defined** as the point of application of the resultant of all the forces tending to cause a body to rotate about a certain axis. It is the point at which a suspended

Fig. 3.1.66

body may be struck without causing any force on the axis passing through the point of suspension.

Center of Percussion The distance from the axis of suspension to the center of percussion is $q_0 = I/mr_G$, where I = moment of inertia of the body about its axis of suspension and r_G is the distance from the axis of suspension to the center of gravity of the body.

EXAMPLES. 1. Find the center of percussion of the homogeneous rod (Fig. 3.1.67) of length L and mass m, suspended at XX.

$$q_0 = \frac{I}{mr_G}$$

$$I \approx \frac{m}{L}\int_0^L r^2\,dr = mL^2/3 \qquad r_G = L/2 \qquad \therefore q_0 = \frac{mL^2/3}{mL/2} = 2L/3$$

2. Find the center of percussion of a solid cylinder, of mass m, resting on a horizontal plane. In Fig. 3.1.68, the instantaneous center of the cylinder is at O, and serves temporarily as a fixed axis of rotation. The center of percussion will therefore be a height above the plane equal to $q_0 = I/mx_0$. Since $I = (mr^2/2) + mr^2$ and $x_0 = r$, $q_0 = 3r/2$.

Fig. 3.1.67   **Fig. 3.1.68**

Wheel or Cylinder Rolling down a Plane In this case the component of the weight along the plane tends to make it roll down and is treated as a force causing rotation. The forces acting on the body should be resolved into components along the line of motion and perpendicular to it. If the forces are all known, their resultant is at the center of percussion. If one force is to be determined (the exact conditions as regards slipping or not slipping must be known), the center of percussion can be determined and the unknown force found.

Relation between the Center of Percussion and Radius of Gyration $q_0 = I/mr_G = k^2/r_G$ $\therefore$ $k^2 = r_G q_0$ where k = radius of gyration. Therefore, the radius of gyration is a mean proportional between the distance from the axis of oscillation to the center of percussion and the distance from the same axis to the center of gravity.

Interchangeability of Center of Percussion and Axis of Oscillation If a body is suspended from an axis, the center of percussion for that axis can be found. If the body is suspended from this center of percussion as an axis, the original axis of suspension will then become the center of percussion. The center of percussion is sometimes known as the **center of oscillation.**

Period of Oscillation of a Compound Pendulum The length of an equivalent simple pendulum is the distance from the axis of suspension to the center of percussion of the body in question. To find the **period of oscillation** of a body about a given axis, find the distance $q_0 = I/mr_G$ from that axis to the center of percussion of the swinging body. The length of the simple pendulum that will oscillate in the same time is this distance q_0. The period of oscillation for the equivalent single pendulum is $\tau = 2\pi\sqrt{q_0/g}$.

Determination of Moment of Inertia by Experiment To find the moment of inertia of a body, suspend it from some axis not passing through the center of gravity and, by swinging it, determine the period of one complete oscillation in seconds. The known values will then be τ = time of one complete oscillation, r_G = distance from axis to center of gravity, and m = mass of body. The length of the equivalent simple pendulum is $q_0 = I/mr_G$. Substituting this value of q_0 in $\tau = 2\pi\sqrt{q_0/g}$ gives $\tau = 2\pi\sqrt{I/mr_Gg}$, from which $\tau^2 = 4\pi^2 I/mr_Gg$, or $I = mr_Gg\tau^2/4\pi^2$.

Fig. 3.1.69

Plane Motion of a Rigid Body

Plane motion may be considered to be a combination of translation and rotation (see "Kinematics"). For translation, Newton's second law of motion must always be satisfied, and the resultant of the external force system must be equal to the product of the mass times the acceleration of the center of gravity in any system of coordinates. In rotation, the body moving in plane motion will not have a fixed axis. When the methods of relative motion are being used, any point on the body may be used as a reference axis to which the motion of all other points is referred.

The sum of the moments of all external forces about the reference axis must be equal to the vector sum of the centroidal moment of inertia times the angular acceleration and the amount of the resultant force about the reference axis.

EXAMPLE.   Determine the forces acting on the piston pin A and the crankpin B of the connecting rod of a reciprocating engine shown in Fig. 3.1.69 for a position of 30° from TDC. The crankshaft speed is constant at 2,000 r/min. Assume that the pressure of expanding gases on the 4-lbm (1.81-kg) piston at this point is 145 lb/in^2 (10^6 N/m^2). The connecting rod has a mass of 5 lbm (2.27 kg) and has a centroidal radius of gyration of 3 in (0.076 m).

The kinematics of the problem are such that the angular velocity of the crank is $\omega_{OB} = 209.4$ rad/s clockwise, the angular velocity of the connecting rod is $\omega_{AB} = 45.7$ rad/s counterclockwise, and the angular acceleration is $\alpha_{AB} = 5,263$ rad/s^2 clockwise. The linear acceleration of the piston is 7,274 ft/s^2 in the direction of the crank. From the free-body diagram of the piston, the horizontal component of the piston-pin force is $145 \times (\pi/4)(5^2) - P = (4/32.2)(7,274)$, $P = 1,943$ lbf. The acceleration of the center of gravity G is the vector sum of the component accelerations $a_G = a_B + a_{G/B}^n + a_{G/B}^t$ where $a_{G/B}^n = \omega_{GB}^2 \cdot r_{GB} = 3/12(45.7)^2 = 522$ ft/s^2 and $a_{G/B}^t = \alpha_{GB} \cdot r_{GB} = 3/12(5.263) = 1,316$ ft/s^2. The resultant acceleration of the center of gravity is 6,685 ft/s^2 in the x direction and 2,284 ft/s^2 in the negative y direction. The resultant of the external force system will have corresponding components such that $ma_{Gx} = (5/32.2)(6,685) = 1,039$ lbf and $ma_{Gy} = (5/32.2)(2,284) = 355$ lbf. The three remaining unknown forces can be found from the three equations of motion for the connecting rod.

Taking the sum of the forces in the x direction, $\epsilon F = ma_{Gx}$; $P - R_x = ma_{Gx}$, and $R_x = 905.4$ lbf. In the y direction, $\Sigma F = ma_{Gy}$; $R_y - N = ma_{gy}$; this has two unknowns, R_y and N. Taking the sum of the moments of the external forces about the center of mass g, $\Sigma M_G = I_G \alpha AB$; $(N)(5) \cos (7.18°) - (P)(5) \sin (7.18°) + (R_y)(3) \cos (7.18°) - R_x (3) \sin (7.18°) - (5/386.4)(3)^2(5,263)$. Solving for R_y and N simultaneously, $R_y = 494.7$ lbf and $N = 140$ lbf. We could have avoided the solution of two simultaneous algebraic equations by taking the moment summation about end A, which would determine R_y independently, or about end B, which would determine N independently.

In SI units, the kinematics would be identical, the linear acceleration of the piston being 2,217 m/s^2 (7,274 ft/s^2). From the free-body diagram of the piston, the horizontal component of the piston-pin force is $(10^6) \times (\pi/4)(0.127)^2 - P = (1.81)(2,217)$, and $P = 8,640$ N. The components of the acceleration of the center of gravity G are $a_{G/B}^n = 522$ ft/s^2 and $a_{G/B}^t = 1,315$ ft/s^2. The resultant acceleration of the center of gravity is 2,037.5 m/s^2 (6,685 ft/s^2) in the x direction and 696.3 m/s^2 (2,284 ft/s^2) in the negative y direction. The resultant of the external force system will have the corresponding components; $ma_{Gx} = (2.27)(2,037.5) = 4,620$ N; $ma_{Gy} = (2.27)(696.3) = 1,579$ N. $R_x = 4,027$ N, $R_y = 2,201$ N, force $N = 623$ newtons.

WORK AND ENERGY

Work   When a body is displaced against resistance or accelerated, work must be done upon it. An increment of work is defined as the product of an incremental displacement and the component of the force

vector in the direction of the displacement or the product of the component of the incremental displacement and the force in the direction of the force. $dU = F \cdot ds \cos \alpha$, where α is the angle between the vector displacement and the vector force. The increment of work done by a couple M acting in a body during an increment of angular rotation $d\theta$ in the plane of the couple is $dU = M \, d\theta$. In a force-displacement or moment-angle diagram, called a **work diagram** (Fig. 3.1.70), force is plotted as a function of displacement. The area under the curve represents the work done, which is equal to $\int_{s_1}^{s_2} F \, ds \cos \alpha$ or $\int_{\theta_1}^{\theta_2} M \, d\theta$.

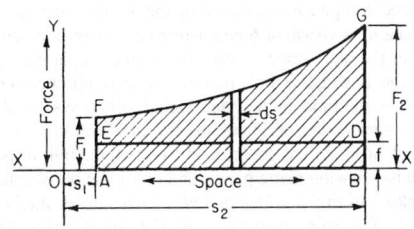

Fig. 3.1.70

Units of Work   When the force of 1 lb acts through the distance of 1 ft, 1 lb · ft of work is done. In SI units, a force of 1 newton acting through 1 metre is 1 joule of work. 1.356 N · m = 1 lb · ft.

Energy   A body is said to possess energy when it can do work. A body may possess this capacity through its **position** or **condition**. When a body is so held that it can do work, if released, it is said to possess energy of position or **potential energy**. When a body is moving with some velocity, it is said to possess energy of motion or **kinetic energy**. An example of potential energy is a body held suspended by a rope; the position of the body is such that if the rope is removed work can be done by the body.

Energy is expressed in the same units as work. The kinetic energy of a particle is expressed by the formula $E = \frac{1}{2}mv^2 = \frac{1}{2}(w/g)v^2$. The kinetic energy of a rigid body in translation is also expressed as $E = \frac{1}{2}mv^2$. Since all particles of the rigid body have the same identical velocity v, the velocity v is the velocity of the center of gravity. The kinetic energy of a rigid body, rotating about a fixed axis is $E = \frac{1}{2}I_0\omega^2$, where I_0 is the mass moment of inertia about the axis of rotation. In plane motion, a rigid body has both translation and rotation. The kinetic energy is the algebraic sum of the translating kinetic energy of the center of gravity and the rotating kinetic energy about the center of gravity, $E = \frac{1}{2}mv^2 + \frac{1}{2}I\omega^2$. Here the velocity v is the velocity of the center of gravity, and the moment of inertia I is the centroidal moment of inertia.

If a force which varies acts through a space on a body of mass m, the work done is $\int_{s_1}^{s} F \, ds$, and if the work is all used in giving kinetic energy to the body it is equal to $\frac{1}{2}m(v_2^2 - v_1^2) =$ **change in kinetic energy**, where v_2 and v_1 are the velocities at distances s_2 and s_1, respectively. This is a specific statement of the law of conservation of energy. **The principle of conservation of energy requires that the mechanical energy of a system remain unchanged if it is subjected only to forces which depend on position or configuration.**

Certain problems in which the velocity of a body at any point in its straight-line path when acted upon by varying forces is required can be easily solved by the use of a **work diagram.**

In Fig. 3.1.70, let a body start from rest at A and be acted upon by a force that varies in accordance with the diagram $AFGBA$. Let the resistance to motion be a constant force $= f$. Find the velocity of the body at point B. The area $AFGBA$ represents the work done upon the body and the area $AEDBA$ ($=$ force $f \times$ distance AB) represents the work that must be done to overcome resistance. The difference of these areas, or $EFGDE$, will represent work done in excess of that required to overcome resistance, and consequently is equal to the increase in kinetic energy. Equating the work represented by the area $EFGDE$ to $\frac{1}{2}wv^2/g$ and solving for v will give the required velocity at B. If the body did not start from rest, this area would represent the change in kinetic energy, and the velocity could be obtained by the formula: Work $= \frac{1}{2}(w/g)$ $(v_1^2 - v_0^2)$, v_1 being the required velocity.

General Rule for Rectilinear Motion   Resolve each force acting on the body into components, one of which acts along the line of motion of the body and the other at right angles to the line of motion. Take the sum of all the components acting in the direction of the motion and multiply this sum by the distance moved through for constant forces. (Take the average force times distance for forces that vary.) This product will be the total work done upon the body. If there is no unbalanced component, there will be no change in kinetic energy and consequently no change in velocity. If there is an unbalanced component, the change in kinetic energy will be this unbalanced component multiplied by the distance moved through.

The **work done by a system of forces acting on a body** is equal to the algebraic sum of the work done by each force taken separately.

Power is the rate at which work is performed, or the number of units of work performed in unit time. In the English engineering system, the units of power are the horsepower, or 33,000 lb $\cdot$ ft/min $= 550$ lb $\cdot$ ft/s, and the kilowatt $= 1.341$ hp $= 737.55$ lb $\cdot$ ft/s. In SI units, the unit of power is the watt, which is 1 newton-metre per second or 1 joule per second.

Friction Brake   In Fig. 3.1.71 a pulley revolves under the band and in the direction of the arrow, exerting a pull of T on the spring. The friction of the band on the rim of the pulley is $(T - w)$, where w is the weight attached to one end of the band. Let the pulley make N r/min; then the work done per minute against friction by the rim of the pulley is $2\pi RN(T - w)$, and the horsepower absorbed by brake $= 2\pi RN(T - w)/33,000$.

Fig. 3.1.71

IMPULSE AND MOMENTUM

The product of force and time is defined as linear impulse. The impulse of a constant force over a time interval $t_2 - t_1$ is $F(t_2 - t_1)$. If the force is not constant in magnitude but is constant in direction, the impulse is $\int_{t_1}^{t_2} F\, dt$. The dimensions of linear impulse are (force) $\times$ (time) in pound-seconds, or newton-seconds.

Impulse is a **vector** quantity which has the direction of the resultant force. Impulses may be added vectorially by means of a vector polygon, or they may be resolved into components by means of a parallelogram. The **moment** of a **linear impulse** may be found in the same manner as

the moment of a force. The linear impulse is represented by a directed line segment, and the moment of the impulse is the product of the magnitude of the impulse and the perpendicular distance from the line segment to the point about which the moment is taken. **Angular impulse** over a time interval $t_2 - t_1$ is a product of the sum of applied moments on a rigid body about a reference axis and time. The dimensions for angular impulse are (force) $\times$ (time) $\times$ (displacement) in foot-pound-seconds or newton-metre-seconds. **Angular impulse and linear impulse cannot be added.**

Momentum is also a vector quantity and can be added and resolved in the same manner as force and impulse. The dimensions of linear momentum are (force) $\times$ (time) in pound-seconds or newton-seconds, and are identical to linear impulse. An alternate statement of Newton's second law of motion is that the resultant of an unbalanced force system must be equal to the time rate of change of linear momentum, $\Sigma F = d(mv)/dt$.

If a variable force acts for a certain time on a body of mass m, the quantity $\int_{t_1}^{t_2} F\, dt = m(v_2 - v_1) =$ the change of momentum of the body.

The **moment of momentum** can be determined by the same methods as those used for the moment of a force or moment of an impulse. The dimensions of the moment of momentum are (force) $\times$ (time) $\times$ (displacement) in foot-pound-seconds, or newton-metre-seconds.

In **plane motion** the angular momentum of a rigid body about a reference axis perpendicular to the plane of motion is the sum of the moments of linear momenta of all particles in the body about the reference axes. Specifically, **the angular momentum of a rigid body in plane motion is the vector sum of the angular momentum about the reference axis and the moment of the linear momentum of the center of gravity about the reference axis,** $H_0 = I_0\omega + d \times mv$.

In three-dimensional rotation about a fixed axis, the angular momentum of a rigid body has components along three coordinate axes, which involve both the moments of inertia about the x, y, and z axes, $I_{0_{xx}}, I_{0_{yy}},$ and $I_{0_{zz}}$, and the products of inertia, $I_{0_{xy}}, I_{0_{zx}},$ and $I_{0_{yz}}$; $H_{0_x} = I_{0_{xx}} \cdot \omega_x - I_{0_{xy}} \cdot \omega_y - I_{0_{zx}} \cdot \omega_z, H_{0_y} = -I_{0_{xy}} \cdot \omega_x + I_{0_{yy}} \cdot \omega_y - I_{0_{yz}} \cdot \omega_z,$ and $H_{0_z} = I_{0_{zx}} \cdot \omega_x - I_{0_{yz}} \cdot \omega_y + I_{0_{zz}} \cdot \omega_z$ where $H_0 = H_{0_x} + H_{0_y} + H_{0_z}$.

Impact

The collision between two bodies, where relatively large forces result over a comparatively short interval of time, is called **impact.** A straight line perpendicular to the plane of contact of two colliding bodies is called the **line of impact.** If the centers of gravity of the two bodies lie on the line of contact, the impact is called **central impact,** in any other case, **eccentric impact.** If the linear momenta of the centers of gravity are also directed along the line of impact, the impact is **collinear or direct central impact.** In any other case impact is said to be **oblique.**

Collinear Impact   When two masses m_1 and m_2, having respective velocities u_1 and u_2, move in the same line, they will collide if $u_2 > u_1$ (Fig. 3.1.72a). During collision (Fig. 3.1.72b), kinetic energy is absorbed

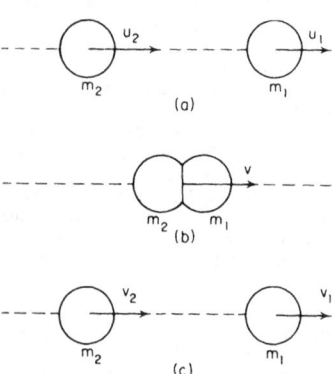

Fig. 3.1.72

in the deformation of the bodies. There follows a period of restoration which may or may not be complete. If complete restoration of the energy of deformation occurs, the impact is **elastic.** If the restoration of energy is incomplete, the impact is referred to as **inelastic.** After collision (Fig. 3.1.72c), the bodies continue to move with changed velocities of v_1 and v_2. Since the contact forces on one body are equal to and opposite the contact forces on the other, the sum of the linear momenta of the two bodies is conserved; $m_1u_1 + m_2u_2 = m_1v_1 + m_2v_2$.

   The law of conservation of momentum states that the linear momentum of a system of bodies is unchanged if there is no resultant external force on the system.

   Coefficient of Restitution   The ratio of the velocity of separation $v_1 - v_2$ to the velocity of approach $u_2 - u_1$ is called the **coefficient of restitution** e, $e = (v_1 - v_2)/(u_2 - u_1)$.

   The value of e will depend on the shape and material properties of the colliding bodies. In elastic impact, the coefficient of restitution is unity and there is no energy loss. A coefficient of restitution of zero indicates perfectly inelastic or plastic impact, where there is no separation of the bodies after collision and the energy loss is a maximum. In **oblique impact,** the coefficient of restitution applies only to those components of velocity along the line of impact or normal to the plane of impact. The coefficient of restitution between two materials can be measured by making one body many times larger than the other so that m_2 is infinitely large in comparison to m_1. The velocity of m_2 is unchanged for all practical purposes during impact and $e = v_1/u_1$. For a small ball dropped from a height H upon an extensive horizontal surface and rebounding to a height h, $e = \sqrt{h/H}$.

   Impact of Jet Water on Flat Plate   When a jet of water strikes a flat plate perpendicularly to its surface, the force exerted by the water on the plate is wv/g, where w is the weight of water striking the plate in a unit of time and v is the velocity. When the jet is inclined to the surface by an angle, A, the force is $(wv/g) \cos A$.

Variable Mass

If the mass of a body is variable such that mass is being either added or ejected, an alternate form of Newton's second law of motion must be used which accounts for changes in mass:

$$\mathbf{F} = m\frac{d\mathbf{v}}{dt} + \frac{dm}{dt}\mathbf{u}$$

The mass m is the instantaneous mass of the body, and dv/dt is the time rate of change of the absolute of velocity of mass m. The velocity u is the velocity of the mass m relative to the added or ejected mass, and dm/dt is the time rate of change of mass. In this case, care must be exercised in the choice of coordinates and expressions of sign. If mass is being added, dm/dt is plus, and if mass is ejected, dm/dt is minus.

Fields of Force—Attraction

The space within which the action of a physical force comes into play on bodies lying within its boundaries is called the **field of the force.**

   The **strength** or **intensity of the field** at any given point is the relation between a force F acting on a mass m at that point and the mass. Intensity of field $= i = F/m$; $F = mi$.

   The **unit of field intensity** is the same as the unit of acceleration, i.e., 1 ft/s^2 or 1 m/s^2. The intensity of a field of force may be represented by a line (or **vector**).

   A field of force is said to be **homogeneous** when the intensity of all points is uniform and in the same direction.

   A field of force is called a **central field of force** with a center O, if the direction of the force acting on the mass particle m in every point of the field passes through O and its magnitude is a function only of the distance r from O to m. A line so drawn through the field of force that its direction coincides at every point with that of the force prevailing at that point is called a **line of force.**

Rotation of Solid Bodies about Any Axis

The general moment equations for three-dimensional motion are usually expressed in terms of the angular momentum. For a reference axis O, which is either a fixed axis of the center of gravity, $M_{0_x} = (dH_{0_x}/dt) - H_{0_z} \cdot \omega_z + H_{0_z} \cdot \omega_y$, $M_{0_y} = (dH_{0_y}/dt) - H_{0_z} \cdot \omega_x + H_{0_x} \cdot \omega_z$, and $M_{0_z} = (dH_{0_z}/dt) - H_{0_y}\omega_y + H_{0_x}\omega_x$. If the coordinate axes are oriented to coincide with the principal axes of inertia, I_{0_x}, I_{0_y}, and I_{0_z}, a similar set of three differential equations results, involving moments, angular velocity, and angular acceleration; $M_{0_x} = I_{0_x}(d\omega_x/dt) + (I_{0_z} - I_{0_y})\omega_y \cdot \omega_z$, $M_{0_y} = I_{0_y}(d\omega_y/dt) + (I_{0_x} - I_{0_z})\omega_z \cdot \omega_x$, and $M_{0_z} = I_{0_z}(d\omega_z/dt) + (I_{0_y} - I_{0_x})\omega_x\omega_y$. These equations are known as **Euler's equations of motion** and may apply to any rigid body.

GYROSCOPIC MOTION AND THE GYROSCOPE

Gyroscopic motion can be explained in terms of Euler's equations. Let I_1, I_2, and I_3 represent the principal moments of inertia of a gyroscope spinning with a constant angular velocity ω, about axis 1, the subscripts 1, 2, and 3 representing a right-hand set of reference axes (Figs. 3.1.73 and 3.1.74). If the gyroscope is precessed about the third axis, a vector moment results along the second axis such that

$$M_2 = I_2(d\omega_2/dt) + (I_1 - I_3)\omega_3\omega_1$$

Where the precession and spin axes are at right angles, the term $(d\omega_2/dt)$ equals the component of $\omega_3 \times \omega_1$ along axis 2. Because of this, in the simple case of a body of symmetry, where $I_2 = I_3$, the gyroscopic

Fig. 3.1.73

moment can be reduced to the common expression $M = I\omega\Omega$, where Ω is the rate of precession, ω the rate of spin, and I the moment of inertia about the spin axis. It is important to realize that these are equations of motion and relate the applied or resulting gyroscopic moment due to forces which act *on* the rotor, as disclosed by a free-body diagram, to the resulting motion of the rotor.

   Physical insight into the behavior of a steady precessing gyro with mutually perpendicular moment, spin, and precession axes is gained by recognizing from Fig. 3.1.74 that the change dH in angular momentum H is equal to the angular impulse $M\,dt$. In time dt, the angular-momentum

Fig. 3.1.74

vector swings from H to H', owing to the velocity of precession ω_3. The vector change dH in angular momentum is in the direction of the applied moment M. This fact is inherent in the basic moment-momentum equation and can always be used to establish the correct spatial relationships between the moment, precessional, and spin vectors. It is seen, therefore, from Fig. 3.1.74 that the spin axis always turns toward the moment axis. Just as the change in direction of the mass-center velocity is in the same direction as the resultant force, so does the change in angular momentum follow the direction of the applied moment.

For example, suppose an airplane is driven by a right-handed propeller (turning like a right-handed screw when moving forward). If a gust of wind or other force turns the machine to the *left,* the gyroscopic action of the propeller will make the forward end of the shaft strive to *rise;* if the wing surface is large, this motion will be practically prevented by the resistance of the air, and the gyroscopic forces become effective merely as internal stresses, whose maximum value can be computed by the formula above. Similarly, if the airplane is dipped *downward,* the gyroscopic action will make the forward end of the shaft strive to turn to the *left.*

Modern **applications** of the gyroscope are based on one of the following properties: (1) a gyroscope mounted in three gimbal rings so as to be entirely free angularly in all directions will retain its direction *in space* in the absence of outside couples; (2) if the axis of rotation of a gyroscope turns or precesses in space, a couple or torque acts on the gyroscope (and conversely on its frame).

Devices operating on the first principle are satisfactory only for short durations, say less than half an hour, because no gyroscope is entirely without outside couple. The friction couples at the various gimbal bearings, although small, will precess the axis of rotation so that after a while the axis of rotation will have changed its direction in space. The chief device based on the first principle is the **airplane compass,** which is a freely mounted gyro, keeping its direction in space during fast maneuvers of a fighting airplane. No magnetic compass will indicate correctly during such maneuvers. After the plane is back on an even keel in steady flight, the magnetic compass once more reads the true magnetic north, and the gyro compass has to be reset to point north again.

An example of a device operating on the second principle is the **automatic pilot** for keeping a vehicle on a given course. This device has been installed on torpedoes, ships, airplanes. When the ship or plane turns from the chosen course, a couple is exerted on the gyro axis, which makes it precess and this operates electric contacts or hydraulic or pneumatic valves. These again operate on the rudders, through relays, and bring the ship back to its course.

Another application is the ship **antirolling** gyroscope. This very large gyroscope spins about a vertical axis and is mounted in a ship so that the axis can be tipped fore and aft by means of an electric motor, the precession motor. The gyro can exert a large torque on the ship about the fore-and-aft axis, which is along the "rolling" axis. The sign of the torque is determined by the direction of rotation of the precession motor, which in turn is controlled by electric contacts operated by a small pilot gyroscope on the ship, which feels which way the ship rolls and gives the signals to apply a countertorque.

The **turn indicator** for airplanes is a gyro, the frame of which is held by springs. When the airplane turns, it makes the gyro axis turn with it, and the resultant couple is delivered by the springs. Thus the elongation of the springs is a measure of the rate of turn, which is suitably indicated by a pointer.

The most complicated and ingenious application of the gyroscope is the **marine compass.** This is a pendulously suspended gyroscope which is affected by gravity and also by the earth's rotation so that the gyro axis is in equilibrium only when it points north, i.e., when it lies in the plane formed by the local vertical and by the earth's north-south axis. If the compass is disturbed so that it points away from north, the action of the earth's rotation will restore it to the correct north position in a few hours.

3.2   FRICTION
by Vittorio (Rino) Castelli

REFERENCES: Bowden and Tabor, "The Friction and Lubrication of Solids," Oxford. Fuller, "Theory and Practice of Lubrication for Engineers," 2nd ed., Wiley. Shigley, "Mechanical Design," McGraw-Hill. Rabinowicz, "Friction and Wear of Materials," Wiley. Ling, Klaus, and Fein, "Boundary Lubrication—An Appraisal of World Literature," ASME, 1969. Dowson, "History of Tribology," Longman, 1979. Petersen and Winer, "Wear Control Handbook," ASME, 1980. M. M. Khonsari and E. R. Booser, "Applied Tribology," Wiley, 2001.

Friction is the resistance that is encountered when two solid surfaces slide or tend to slide over each other. The surfaces may be either dry or lubricated. In the first case, when the surfaces are free from contaminating fluids, or films, the resistance is called **dry friction.** The friction of brake shoes on the rim of a railroad wheel is an example of dry friction.

When the rubbing surfaces are separated from each other by a very thin film of lubricant, the friction is that of **boundary** (or greasy) **lubrication.** The lubrication depends in this case on the strong adhesion of the lubricant to the material of the rubbing surfaces; the layers of lubricant slip over each other instead of the dry surfaces. A journal when starting, reversing, or turning at very low speed under a heavy load is an example of the condition that will cause boundary lubrication. Other examples are gear teeth (especially hypoid gears), cutting tools, wire-drawing dies, power screws, bridge trunnions, and the running-in process of most lubricated surfaces.

When the lubrication is arranged so that the rubbing surfaces are separated by a fluid film, and the load on the surfaces is carried entirely by the hydrostatic or hydrodynamic pressure in the film, the friction is that of **complete** (or viscous) **lubrication.** In this case, the frictional losses are due solely to the internal fluid friction in the film. Oil ring bearings, bearings with forced feed of oil, pivoted shoe-type thrust and journal bearings, bearings operating in an oil bath, hydrostatic oil pads, oil lifts, and step bearings are instances of complete lubrication.

Incomplete lubrication or mixed lubrication takes place when the load on the rubbing surfaces is carried partly by a fluid viscous film and partly by areas of boundary lubrication. The friction is intermediate between that of fluid and boundary lubrication. Incomplete lubrication exists in bearings with drop-feed, waste-packed, or wick-fed lubrication, or on parallel-surface bearings.

STATIC AND KINETIC COEFFICIENTS OF FRICTION

In the absence of friction, the resultant of the forces between the surfaces of two bodies pressing upon each other is normal to the surface of contact. With friction, the resultant deviates from the normal.

If one body is pressed against another by a force P, as in Fig. 3.2.1, the first body will not move, provided the angle a included between the line of action of the force and a normal to the surfaces in contact does not exceed a certain value which depends upon the nature of the surfaces. The reaction force R has the same magnitude and line of action as the force P. In Fig. 3.2.1, R is resolved into two components: a force N

normal to the surfaces in contact and a force F_r parallel to the surfaces in contact. From the above statement it follows that, for motion not to occur,

$$F_r = N \tan a_0 = Nf_0$$

where $f_0 = \tan a_0$ is called the **coefficient of friction of rest** (or of **static friction**) and a_0 is the **angle of friction at rest**.

Fig. 3.2.1

If the normal force N between the surfaces is kept constant, and the tangential force F_r is gradually increased, there will be no motion while $F_r < Nf_0$. A state of *impending motion* is reached when F_r nears the value of Nf_0. If sliding motion occurs, a frictional force F resisting the motion must be overcome. The force F is commonly expressed as $F = fN$, where f is the **coefficient of sliding friction**, or **kinetic friction**. Normally, the coefficients of sliding friction are smaller than the coefficients of static friction. With small velocities of sliding and very clean surfaces, the two coefficients do not differ appreciably.

Table 3.2.4 demonstrates the typical reduction of sliding coefficients of friction below corresponding static values. Figure 3.2.2 indicates results of tests on lubricated machine tool ways showing a reduction of friction coefficient with increasing sliding velocity.

Fig. 3.2.2 Typical relationship between kinetic friction and sliding velocity for lubricated cast iron on cast iron slideways (load, 20 lb/in^2; upper slider, scraped; lower slideway, scraped). *(From Birchall, Kearny, and Moss, Intl. J. Machine Tool Design Research, 1962.)*

This behavior is normal with dry friction, some conditions of boundary friction, and with the break-away friction in ball and roller bearings. This condition is depicted in Fig. 3.2.3, where the friction force decreases with relative velocity. This negative slope leads to locally unstable equilibrium and **self-excited vibrations** in systems such as the one of Fig. 3.2.4. This phenomenon takes place because, for small amplitudes, the oscillatory system displays damping in which the damping

Fig. 3.2.3 Friction force decreases as velocity increases.

factor is equal to the slope of the friction curve and thus is termed **negative damping**. When the slope of the friction force versus sliding velocity is positive (**positive damping**) this type of instability is not possible. This is typical of fluid damping, squeeze films, dash pots, and fluid film bearings in general.

Fig. 3.2.4 Belt friction apparatus with possible self-excited vibrations.

It is interesting to note that these self-excited systems vibrate at close to their natural frequency over a large range of frictional levels and speeds. This symptom is a helpful means of identification. Another characteristic is that the moving body comes periodically to momentary relative rest, that is, zero sliding velocity. For this reason, this phenomenon is also called **stick-slip vibration.** Common examples are violin strings, chalk on blackboard, water-lubricated rubber stern tube ship bearings at low speed, squeaky hinges, and oscillating rolling element bearings, especially if they are supporting large flexible structures such as radar antennas. Control requires the introduction of fluid film bearings, viscous seals, or viscous dampers into the system with sufficient positive damping to override the effects of negative damping.

Under moderate pressures, the frictional force is proportional to the normal load on the rubbing surfaces. It is independent of the pressure per unit area of the surfaces. The direction of the friction force opposing the sliding motion is locally exactly opposite to the local relative velocity. Therefore, it takes very little effort to displace transversally two bodies which have a major direction of relative sliding. This behavior, **compound sliding**, is exploited when easing the extraction of a nail by simultaneously rotating it about its axis, and accounts for the ease with which an automobile may skid on the road or with which a plug gage can be inserted into a hole if it is rotated while being pushed in.

The coefficients of friction for dry surfaces (dry friction) depend on the materials sliding over each other and on the finished condition of the surfaces. With greasy (boundary) lubrication, the coefficients depend both on the materials and conditions of the surfaces and on the lubricants employed.

Coefficients of friction are sensitive to atmospheric dust and humidity, oxide films, surface finish, velocity of sliding, temperature, vibration, and the extent of contamination. In many instances the degree of contamination is perhaps the most important single variable. For example, in Table 3.2.1, values for the static coefficient of friction of steel on steel are listed, and, depending upon the degree of contamination of the specimens, the coefficient of friction varies effectively from ∞ (infinity) to 0.013.

The most effective boundary lubricants are generally those which react chemically with the solid surface and form an adhering film that is attached to the surface with a chemical bond. This action depends upon

Table 3.2.1 Coefficients of Static Friction for Steel on Steel

Test condition	f_0	Ref.
Degassed at elevated temp in high vacuum	∞ (weld on contact)	1
Grease-free in vacuum	0.78	2
Grease-free in air	0.39	3
Clean and coated with oleic acid	0.11	2
Clean and coated with solution of stearic acid	0.013	4

Sources: (1) Bowden and Young, *Proc. Roy. Soc.,* 1951. (2) Campbell, *Trans. ASME,* 1939. (3) Tomlinson, *Phil. Mag.,* 1929. (4) Hardy and Doubleday, *Proc. Roy. Soc.,* 1923.

the nature of the lubricant and upon the reactivity of the solid surface. Table 3.2.2 indicates that a fatty acid, such as found in animal, vegetable, and marine oils, reduces the coefficient of friction markedly only if it can react effectively with the solid surface. Paraffin oil is almost completely nonreactive.

Table 3.2.2   Coefficients of Static Friction at Room Temperature

Surfaces	Clean	Paraffin oil	Paraffin oil plus 1% lauric acid	Degree of reactivity of solid
Nickel	0.7	0.3	0.28	Low
Chromium	0.4	0.3	0.3	Low
Platinum	1.2	0.28	0.25	Low
Silver	1.4	0.8	0.7	Low
Glass	0.9		0.4	Low
Copper	1.4	0.3	0.08	High
Cadmium	0.5	0.45	0.05	High
Zinc	0.6	0.2	0.04	High
Magnesium	0.6	0.5	0.08	High
Iron	1.0	0.3	0.2	Mild
Aluminum	1.4	0.7	0.3	Mild

SOURCE: From Bowden and Tabor, "The Friction and Lubrication of Solids," Oxford.

Values in Table 3.2.4 of sliding and static coefficients have been selected largely from investigations where these variables have been very carefully controlled. They are representative values for smooth surfaces. It has been generally observed that sliding friction between hard materials is smaller than that between softer surfaces.

Effect of Surface Films   Campbell observed a lowering of the coefficient of friction when oxide or sulfide films were present on metal surfaces (*Trans. ASME,* 1939; footnotes to Table 3.2.4). The reductions listed in Table 3.2.3 were obtained with oxide films formed by heating in air at temperatures from 100 to 500°C, and sulfide films produced by immersion in a 0.02 percent sodium sulfide solution.

Table 3.2.3   Static Coefficient of Friction f_0

	Clean and dry	Oxide film	Sulfide film
Steel-steel	0.78	0.27	0.39
Brass-brass	0.88		0.57
Copper-copper	1.21	0.76	0.74

Effect of Sliding Velocity   It has generally been observed that coefficients of friction reduce on dry surfaces as sliding velocity increases. (See results of railway brake-shoe tests below.) Dokos measured this reduction in friction for mild steel on medium steel. Values are for the average of four tests with high contact pressures (*Trans. ASME,* 1946; see footnotes to Table 3.2.4).

Sliding velocity, in/s	0.0001	0.001	0.01	0.1	1	10	100
f	0.53	0.48	0.39	0.31	0.23	0.19	0.18

Effect of Surface Finish   The degree of surface roughness has been found to influence the coefficient of friction. Burwell evaluated this effect for conditions of boundary or greasy friction (*Jour. SAE,* 1942; see footnotes to Table 3.2.4). The values listed in Table 3.2.5 are for sliding coefficients of friction, hard steel on hard steel. The friction coefficient and wear rates of **polymers against metals** are often lowered by decreasing the surface roughness. This is particularly true of composites such as those with polytetrafluoroethylene (PTFE) which function through transfer to the counterface.

Solid Lubricants   In certain applications solid lubricants are used successfully. Boyd and Robertson with pressures ranging from 50,000

to 400,000 lb/in² (344,700 to 2,757,000 kN/m²) found sliding coefficients of friction f for hard steel on hard steel as follows: powdered mica, 0.305; powdered soapstone, 0.306; lead iodide, 0.071; silver sulfate, 0.054; graphite, 0.058; molybdenum disulfide, 0.033; tungsten disulfide, 0.037; stearic acid, 0.029 (*Trans. ASME,* 1945; see footnotes to Table 3.2.4).

Coefficients of Static Friction for Special Cases

Masonry and Earth   Dry masonry on brickwork, 0.6 to 0.7; timber on polished stone, 0.40; iron on stone, 0.3 to 0.7; masonry on dry clay, 0.51; masonry on moist clay, 0.33.

Earth on Earth   Dry sand, clay, mixed earth, 0.4 to 0.7; damp clay, 1.0; wet clay, 0.31; shingle and gravel, 0.8 to 1.1.

Natural Cork   On cork, 0.59; on pine with grain, 0.49; on glass, 0.52; on dry steel, 0.45; on wet steel, 0.69; on hot steel, 0.64; on oiled steel, 0.45; water-soaked cork on steel, 0.56; oil-soaked cork on steel, 0.42.

Coefficients of Sliding Friction for Special Cases

Soapy Wood   Lesley gives for wood on wood, copiously lubricated with tallow, stearine, and soft soap (as used in launching practice), a starting coefficient of friction equal to 0.036, diminishing to an average value of 0.019 for the first 50 ft of motion of the ship. Rennie gives 0.0385 for wood on wood, lubricated with soft soap, under a load of 56 lb/in².

Asbestos-Fabric Brake Material   The coefficient of sliding friction f of asbestos fabric against a cast-iron brake drum, according to Taylor and Holt (*NBS,* 1940) is 0.35 to 0.40 when at normal temperature. It drops somewhat with rise in brake temperature up to 300°F (149°C). With a further increase in brake temperature from 300 to 500°F (149 to 260°C) the value of f may show an increase caused by disruption of the brake surface.

Steel Tires on Steel Rails (Galton)

Speed, mi/h	Start	6.8	13.5	27.3	40.9	54.4	60
Values of f	0.242	0.088	0.072	0.07	0.057	0.038	0.027

Railway Brake Shoes on Steel Tires   Galton and Westinghouse give, for cast-iron brakes, the following values for f, which decrease rapidly with the speed of the rim; the coefficient f decreases also with time, as the temperature of the shoe increases.

Speed, mi/h	10	20	30	40	50	60
f, when brakes were applied	0.32	0.21	0.18	0.13	0.10	0.06
f, after 5 s	0.21	0.17	0.11	0.10	0.07	0.05
f, after 12 s		0.13	0.10	0.08	0.06	0.05

Schmidt and Schrader confirm the marked decrease in the coefficient of friction with the increase of rim speed. They also show an irregular slight decrease in the value of f with higher shoe pressure on the wheel, but they did not find the drop in friction after a prolonged application of the brakes. Their observations are as follows:

Speed, mi/h	20	30	40	50	60
Coefficient of friction	0.25	0.23	0.19	0.17	0.16

Friction of Steel on Polymers   A useful list of friction coefficients between steel and various polymers is given in Table 3.2.6.

Grindstones   The coefficient of friction between coarse-grained sandstone and cast iron is $f = 0.21$ to 0.24; for steel, 0.29; for wrought iron, 0.41 to 0.46, according as the stone is freshly trued or dull; for fine-grained sandstone (wet grinding) $f = 0.72$ for cast iron, 0.94 for steel, and 1.0 for wrought iron.

Honda and Yamada give $f = 0.28$ to 0.50 for carbon steel on emery, depending on the roughness of the wheel.

Table 3.2.4 Coefficients of Static and Sliding Friction
(Reference letters indicate the lubricant used; numbers in parentheses give the sources. See footnote.)

Materials	Static		Sliding	
	Dry	Greasy	Dry	Greasy
Hard steel on hard steel	0.78 (1)	0.11 (1, *a*)	0.42 (2)	0.029 (5, *h*)
		0.23 (1, *b*)		0.081 (5, *c*)
		0.15 (1, *c*)		0.080 (5, *i*)
		0.11 (1, *d*)		0.058 (5, *j*)
		0.0075 (18, *p*)		0.084 (5, *d*)
		0.0052 (18, *h*)		0.105 (5, *k*)
				0.096 (5, *l*)
				0.108 (5, *m*)
				0.12 (5, *a*)
Mild steel on mild steel	0.74 (19)		0.57 (3)	0.09 (3, *a*)
				0.19 (3, *u*)
Hard steel on graphite	0.21 (1)	0.09 (1, *a*)		
Hard steel on babbitt (ASTM No. 1)	0.70 (11)	0.23 (1, *b*)	0.33 (6)	0.16 (1, *b*)
		0.15 (1, *c*)		0.06 (1, *c*)
		0.08 (1, *d*)		0.11 (1, *d*)
		0.085 (1, *e*)		
Hard steel on babbitt (ASTM No. 8)	0.42 (11)	0.17 (1, *b*)	0.35 (11)	0.14 (1, *b*)
		0.11 (1, *c*)		0.065 (1, *c*)
		0.09 (1, *d*)		0.07 (1, *d*)
		0.08 (1, *e*)		0.08 (11, *h*)
Hard steel on babbitt (ASTM No. 10)		0.25 (1, *b*)		0.13 (1, *b*)
		0.12 (1, *c*)		0.06 (1, *c*)
		0.10 (1, *d*)		0.055 (1, *d*)
		0.11 (1, *e*)		
Mild steel on cadmium silver				0.097 (2, *f*)
Mild steel on phosphor bronze			0.34 (3)	0.173 (2, *f*)
Mild steel on copper lead				0.145 (2, *f*)
Mild steel on cast iron		0.183 (15, *c*)	0.23 (6)	0.133 (2, *f*)
Mild steel on lead	0.95 (11)	0.5 (1, *f*)	0.95 (11)	0.3 (11, *f*)
Nickel on mild steel			0.64 (3)	0.178 (3, *x*)
Aluminum on mild steel	0.61 (8)		0.47 93)	
Magnesium on mild steel			0.42 (3)	
Magnesium on magnesium	0.6 (22)	0.08 (22, *y*)		
Teflon on Teflon	0.04 (22)			0.04 (22, *f*)
Teflon on steel	0.04 (22)			0.04 (22, *f*)
Tungsten carbide on tungsten carbide	0.2 (22)	0.12 (22, *a*)		
Tungsten carbide on steel	0.5 (22)	0.08 (22, *a*)		
Tungsten carbide on copper	0.35 (23)			
Tungsten carbide on iron	0.8 (23)			
Bonded carbide on copper	0.35 (23)			
Bonded carbide on iron	0.8 (23)			
Cadmium on mild steel			0.46 (3)	
Copper on mild steel	0.53 (8)		0.36 (3)	0.18 (17, *a*)
Nickel on nickel	1.10 (16)		0.53 (3)	0.12 (3, *w*)
Brass on mild steel	0.51 (8)		0.44 (6)	
Brass on cast iron			0.30 (6)	
Zinc on cast iron	0.85 (16)		0.21 (7)	
Magnesium on cast iron			0.25 (7)	
Copper on cast iron	1.05 (16)		0.29 (7)	
Tin on cast iron			0.32 (7)	
Lead on cast iron			0.43 (7)	
Aluminum on aluminum	1.05 (16)		1.4 (3)	
Glass on glass	0.94 (8)	0.01 (10, *p*)	0.40 (3)	0.09 (3, *a*)
		0.005 (10, *q*)		0.116 (3, *v*)
Carbon on glass			0.18 (3)	
Garnet on mild steel			0.39 (3)	
Glass on nickel	0.78 (8)		0.56 (3)	

(*a*) Oleic acid; (*b*) Atlantic spindle oil (light mineral); (*c*) castor oil; (*d*) lard oil; (*e*) Atlantic spindle oil plus 2 percent oleic acid; (*f*) medium mineral oil; (*g*) medium mineral oil plus ½ percent oleic acid; (*h*) stearic acid; (*i*) grease (zinc oxide base); (*j*) graphite; (*k*) turbine oil plus 1 percent graphite; (*l*) turbine oil plus 1 percent stearic acid; (*m*) turbine oil (medium mineral); (*n*) olive oil; (*p*) palmitic acid; (*q*) ricinoleic acid; (*r*) dry soap; (*s*) lard; (*t*) water; (*u*) rape oil; (*v*) 3-in-1 oil; (*w*) octyl alcohol; (*x*) triolein; (*y*) 1 percent lauric acid in paraffin oil.

SOURCES: (1) Campbell, *Trans. ASME*, 1939; (2) Clarke, Lincoln, and Sterrett, *Proc. API*, 1935; (3) Beare and Bowden, *Phil. Trans. Roy. Soc.*, 1935; (4) Dokos, *Trans. ASME*, 1946; (5) Boyd and Robertson, *Trans. ASME*, 1945; (6) Sachs, *Zeit f. angew. Math. und Mech.*, 1924; (7) Honda and Yamaha, *Jour. I of M*, 1925; (8) Tomlinson, *Phil. Mag.*, 1929; (9) Morin, *Acad. Roy. des Sciences*, 1838; (10) Claypoole, *Trans. ASME*, 1943; (11) Tabor, *Jour. Applied Phys.*, 1945; (12) Eyssen, General Discussion on Lubrication, *ASME*, 1937; (13) Brazier and Holland-Bowyer, General Discussion on Lubrication, *ASME*, 1937; (14) Burwell, *Jour. SAE.*, 1942; (15) Stanton, "Friction," Longmans; (16) Ernst and Merchant, Conference on Friction and Surface Finish, M.I.T., 1940; (17) Gongwer, Conference on Friction and Surface Finish, M.I.T., 1940; (18) Hardy and Bircumshaw, *Proc. Roy. Soc.*, 1925; (19) Hardy and Hardy, *Phil. Mag.*, 1919; (20) Bowden and Young, *Proc. Roy. Soc.*, 1951; (21) Hardy and Doubleday, *Proc. Roy. Soc.*, 1923; (22) Bowden and Tabor, "The Friction and Lubrication of Solids," Oxford; (23) Shooter, *Research*, 4, 1951.

Table 3.2.4 Coefficients of Static and Sliding Friction (*Continued*)
(Reference letters indicate the lubricant used; numbers in parentheses give the sources. See footnote.)

Materials	Static		Sliding	
	Dry	Greasy	Dry	Greasy
Copper on glass	0.68 (8)		0.53 (3)	
Cast iron on cast iron	1.10 (16)		0.15 (9)	0.070 (9, *d*)
				0.064 (9, *n*)
Bronze on cast iron			0.22 (9)	0.077 (9, *n*)
Oak on oak (parallel to grain)	0.62 (9)		0.48 (9)	0.164 (9, *r*)
				0.067 (9, *s*)
Oak on oak (perpendicular)	0.54 (9)		0.32 (9)	0.072 (9, *s*)
Leather on oak (parallel)	0.61 (9)		0.52 (9)	
Cast iron on oak			0.49 (9)	0.075 (9, *n*)
Leather on cast iron			0.56 (9)	0.36 (9, *t*)
				0.13 (9, *n*)
Laminated plastic on steel			0.35 (12)	0.05 (12, *t*)
Fluted rubber bearing on steel				0.05 (13, *t*)

(*a*) Oleic acid; (*b*) Atlantic spindle oil (light mineral); (*c*) castor oil; (*d*) lard oil; (*e*) Atlantic spindle oil plus 2 percent oleic acid; (*f*) medium mineral oil; (*g*) medium mineral oil plus ½ percent oleic acid; (*h*) stearic acid; (*i*) grease (zinc oxide base); (*j*) graphite; (*k*) turbine oil plus 1 percent graphite; (*l*) turbine oil plus 1 percent stearic acid; (*m*) turbine oil (medium mineral); (*n*) olive oil; (*p*) palmitic acid; (*q*) ricinoleic acid; (*r*) dry soap; (*s*) lard; (*t*) water; (*u*) rape oil; (*v*) 3-in-1 oil; (*w*) octyl alcohol; (*x*) triolein; (*y*) 1 percent lauric acid in paraffin oil.
SOURCES: (1) Campbell, *Trans. ASME*, 1939; (2) Clarke, Lincoln, and Sterrett, *Proc. API*, 1935; (3) Beare and Bowden, *Phil. Trans. Roy. Soc.*, 1935; (4) Dokos, *Trans. ASME*, 1946; (5) Boyd and Robertson, *Trans. ASME*, 1945; (6) Sachs, *Zeit f. angew. Math. und Mech.*, 1924; (7) Honda and Yamaha, *Jour. I of M*, 1925; (8) Tomlinson, *Phil. Mag.*, 1929; (9) Morin, *Acad. Roy. des Sciences*, 1838; (10) Claypoole, *Trans. ASME*, 1943; (11) Tabor, *Jour. Applied Phys.*, 1945; (12) Eyssen, General Discussion on Lubrication, *ASME*, 1937; (13) Brazier and Holland-Bowyer, General Discussion on Lubrication, *ASME*, 1937; (14) Burwell, *Jour. SAE.*, 1942; (15) Stanton, "Friction," Longmans; (16) Ernst and Merchant, Conference on Friction and Surface Finish, M.I.T., 1940; (17) Gongwer, Conference on Friction and Surface Finish, M.I.T., 1940; (18) Hardy and Bircumshaw, *Proc. Roy. Soc.*, 1925; (19) Hardy and Hardy, *Phil. Mag.*, 1919; (20) Bowden and Young, *Proc. Roy. Soc.*, 1951; (21) Hardy and Doubleday, *Proc. Roy. Soc.*, 1923; (22) Bowden and Tabor, "The Friction and Lubrication of Solids," Oxford; (23) Shooter, *Research*, 4, 1951.

Table 3.2.5 Coefficient of Friction of Hard Steel on Hard Steel

	Surface					
	Superfinished	Ground	Ground	Ground	Ground	Grit-blasted
Roughness, microinches	2	7	20	50	65	55
Mineral oil	0.128	0.189	0.360	0.372	0.378	0.212
Mineral oil + 2% oleic acid	0.116	0.170	0.249	0.261	0.230	0.164
Oleic acid	0.099	0.163	0.195	0.222	0.238	0.195
Mineral oil + 2% sulfonated sperm oil	0.095	0.137	0.175	0.251	0.197	0.165

Table 3.2.6 Coefficient of Friction of Steel on Polymers
Room temperature, low speeds.

Material	Condition	f
Nylon	Dry	0.4
Nylon	Wet with water	0.15
Plexiglas	Dry	0.5
Polyvinyl chloride (PVC)	Dry	0.5
Polystyrene	Dry	0.5
Low-density (LD) polyethylene, no plasticizer	Dry	0.4
LD polyethylene, no plasticizer	Wet	0.1
High-density (HD) polyethylene, no plasticizer	Dry or wet	0.15
Soft wood	Natural	0.25
Lignum vitae	Natural	0.1
PTFE, low speed	Dry or wet	0.06
PTFE, high speed	Dry or wet	0.3
Filled PTFE (15% glass fiber)	Dry	0.12
Filled PTFE (15% graphite)	Dry	0.09
Filled PTFE (60% bronze)	Dry	0.09
Polyurethane rubber	Dry	1.6
Isoprene rubber	Dry	3–10
Isoprene rubber	Wet (water and alcohol)	2–4

Inflation pressure, lb/in^2	Dry pavement		Wet pavement	
	Static f_0	Sliding f	Static f_0	Sliding f
40	0.90	0.85	0.74	0.69
50	0.88	0.84	0.64	0.58
60	0.80	0.76	0.63	0.56

Tests of the Goodrich Company on wet brick pavement with tires of different treads gave the following values of f:

	Coefficients of friction			
	Static (before slipping)		Sliding (after slipping)	
Speed, mi/h	5	30	5	30
Smooth tire	0.49	0.28	0.43	0.26
Circumferential grooves	0.58	0.42	0.52	0.36
Angular grooves at 60°	0.75	0.55	0.70	0.39
Angular grooves at 45°	0.77	0.55	0.68	0.44

Rubber Tires on Pavement Arnoux gives $f = 0.67$ for dry macadam, 0.71 for dry asphalt, and 0.17 to 0.06 for soft, slippery roads. For a cord tire on a sand-filled brick surface in fair condition. Agg (*Bull.* 88, *Iowa State College Engineering Experiment Station*, 1928) gives the following values of f depending on the inflation of the tire:

Development continues using various manufacturing techniques (bias ply, belted, radial, studs), tread patterns, and rubber compounds, so that it is not possible to present average values applicable to present conditions.

Sleds For **unshod wooden runners** on smooth wood or stone surfaces, $f = 0.07$ (0.15) when tallow (dry soap) is used as a lubricant (= 0.38 when not lubricated); on snow and ice, $f = 0.035$. For **runners**

with metal shoes on snow and ice, $f = 0.02$. Rennie found for steel on ice, $f = 0.014$. However, as the temperature falls, the coefficient of friction will get larger. Bowden cites the following data for brass on ice:

Temperature, °C	f
0	0.025
−20	0.085
−40	0.115
−60	0.14

ROLLING FRICTION

Rolling is substituted frequently for sliding friction, as in the case of wheels under vehicles, balls or rollers in bearings, rollers under skids when moving loads; frictional resistance to the rolling motion is substantially smaller than to sliding motion. The fact that a resistance arises to rolling motion is due to several factors: (1) the contacting surfaces are elastically deflected, so that, on the finite size of the contact, relative sliding occurs, (2) the deflected surfaces dissipate energy due to internal friction (hysteresis), (3) the surfaces are imperfect so that contact takes place on asperities ahead of the line of centers, and (4) surface adhesion phenomena. The coefficient of rolling friction $f_r = P/L$ where L is the load and P is the frictional resistance.

The frictional resistance P to the rolling of a cylinder under a load L applied at the center of the roller (Fig. 3.2.5) is inversely proportional to the radius r of the roller; $P = (k/r)L$. Note that k has the dimensions of length. Quite often k increases with load, particularly for cases involving

Fig. 3.2.5

plastic deformations. Values of k, in inches, are as follows: hardwood on hardwood, 0.02; iron on iron, steel on steel, 0.002; hard polished steel on hard polished steel, 0.0002 to 0.0004.

Data on rolling friction are scarce. Noonan and Strange give, for steel rollers on steel plates and for loads varying from light to those causing a permanent set of the material, the following values of k, in inches:

surfaces well finished and clean, 0.0005 to 0.001; surfaces well oiled, 0.001 to 0.002; surfaces covered with silt, 0.003 to 0.005; surfaces rusty, 0.005 to 0.01.

If a load L is moved on rollers (Fig. 3.2.5) and if k and k' are the respective coefficients of friction for the lower and upper surfaces, the frictional force $P = (k + k')L/d$.

McKibben and Davidson (*Agri. Eng.*, 1939) give the data in Table 3.2.7 on the rolling resistance of various types of wheels for typical road and field conditions. Note that the coefficient f_r is the ratio of resistance force to load.

Moyer found the following average values of f_r for pneumatic rubber tires properly inflated and loaded: hard road, 0.008; dry, firm, and well-packed gravel, 0.012; wet loose gravel, 0.06.

FRICTION OF MACHINE ELEMENTS

Work of Friction—Efficiency In a simple machine or assemblage of two elements, the work done by an applied force P acting through the distance s is measured by the product Ps. The useful work done is less and is measured by the product Ll of the resistance L by the distance l through which it acts. The **efficiency** e of the machine is the ratio of useful work performed to the total work received, or $e = Ll/Ps$. The **work expended in friction** W_f is the difference between the total work received and the useful work, or $W_f = Ps - Ll$. The lost-work ratio $= V = W_f/Ll$, and $e = 1/(1 + V)$.

If a machine consists of a train of mechanisms having the respective efficiencies $e_1, e_2, e_3 \ldots e_n$, the combined efficiency of the machine is equal to the product of these efficiencies.

Efficiencies of Machines and Machine Elements The values for machine elements in Table 3.2.8 are from "Elements of Machine Design," by Kimball and Barr. Those for machines are from Goodman's "Mechanics Applied to Engineering." The quantities given are percentage efficiencies.

Wedges

Sliding in V Guides If a wedge-shaped slide having an angle $2b$ is pressed into a V guide by a force P (Fig. 3.2.6), the total force normal to the wedge faces will be $N = P/\sin b$. A friction force F, opposing motion along the longitudinal axis of the wedge, arises by virtue of the coefficient of friction f between the contacting surface of the wedge and guides: $F = fN = fP/\sin b$. In these formulas, the fact that the elasticity of the materials permits an advance of the wedge into the guide under the load P has been neglected. The common efficiency for V guides is $e = 0.88$ to 0.90.

Table 3.2.7   Coefficients of Rolling Friction f_r for Wheels with Steel and Pneumatic Tires

Wheel	Inflation press, lb/in²	Load, lb	Concrete	Bluegrass sod	Tilled loam	Loose sand	Loose snow 10–14 in deep
2.5 × 36 steel		1,000	0.010	0.087	0.384	0.431	0.106
4 × 24 steel		500	0.034	0.082	0.468	0.504	0.282
4.00–18 4-ply	20	500	0.034	0.058	0.366	0.392	0.210
4 × 36 steel		1,000	0.019	0.074	0.367	0.413	
4.00–30 4-ply	36	1,000	0.018	0.057	0.322	0.319	
4.00–36 4-ply	36	1,000	0.017	0.050	0.294	0.277	
5.00–16 4-ply	32	1,000	0.031	0.062	0.388	0.460	
6 × 28 steel		1,000	0.023	0.094	0.368	0.477	0.156
6.00–16 4-ply	20	1,000	0.027	0.060	0.319	0.338	0.146
6.00–164-ply*	30	1,000	0.031	0.070	0.401	0.387	
7.50–10 4-ply†	20	1,000	0.029	0.061	0.379	0.429	
7.50–16 4-ply	20	1,500	0.023	0.055	0.280	0.322	
7.50–28 4-ply	16	1,500	0.026	0.052	0.197	0.205	
8 × 48 steel		1,500	0.013	0.065	0.236	0.264	0.118
7.50–36 4-ply	16	1,500	0.018	0.046	0.185	0.177	0.0753
9.00–10 4-ply†	20	1,000	0.031	0.060	0.331	0.388	
9.00–16 6-ply	16	1,500	0.042	0.054	0.249	0.272	0.099

* Skid-ring tractor tire.
† Ribbed tread tractor tire.
All other pneumatic tires with implement-type tread.

Table 3.2.8   Efficiencies of Machines and Machine Elements

Common bearing (singly)	96–98
Common bearing, long lines of shafting	95
Roller bearings	98
Ball bearings	99
Spur gear, including bearings	
Cast teeth	93
Cut teeth	96
Bevel gear, including bearings	
Cast teeth	92
Cut teeth	95
Worm gear	
Thread angle, 30°	85–95
Thread angle, 15°	75–90
Belting	96–98
Pin-connected chains (bicycle)	95–97
High-grade transmission chains	97–99
Weston pulley block (½ ton)	30–47
Epicycloidal pulley block	40–45
1-ton steam hoist or windlass	50–70
Hydraulic windlass	60–80
Hydraulic jack	80–90
Cranes (steam)	60–70
Overhead traveling cranes	30–50
Locomotives (drawbar hp/ihp)	65–75
Hydraulic couplings, max	98

Fig. 3.2.6

Taper Keys   In Fig. 3.2.7 if the key is moved in the direction of the force P, the force H must be overcome. The supporting reactions K_1, K_2, and K_3 together with the required force P may be obtained by drawing the force polygon (Fig. 3.2.8). The friction angles of these faces are a_1, a_2, and a_3, respectively. In Fig. 3.2.8, draw AB parallel to H in Fig. 3.2.7, and lay it off to scale to represent H. From the point A, draw AC

Fig. 3.2.7   **Fig. 3.2.8**

parallel to K_1, i.e., making the angle $b + a_1$ with AB; from the other extremity of AB, draw BC parallel to K_2 in Fig. 3.2.7. AC and CB then give the magnitudes of K_1 and K_2, respectively. Now through C draw CD parallel to K_3 to its intersection with AD which has been drawn through A parallel to P. The magnitudes of K_3 and P are then given by the lengths of CD and DA.

By calculation,

$$K_1/H = \cos a_2/\cos (b + a_1 + a_2)$$
$$P/K_1 = \sin (b + a_1 + a_3)/\cos a_3$$
$$P/H = \cos a_2 \sin (b + a_1 + a_3)/\cos a_3 \cos (b + a_1 + a_2)$$

If $a_1 = a_2 = a_3 = a$, then $P = H \tan (b + 2a)$, and efficiency $e = \tan b/\tan (b + 2a)$. Force required to loosen the key $= P_1 = H \tan (2a - b)$.

In order for the key not to slide out when force P is removed, it is necessary that $b < (a_1 + a_3)$, or $b < 2a$.

The forces acting upon the taper key of Fig. 3.2.9 may be found in a similar way (see Fig. 3.2.10).

$$P = 2H \cos a \sin (b + a)/\cos (b + 2a)$$
$$= 2H \tan (b + a)/[1 - \tan a \tan (b + a)]$$
$$= 2H \tan (b + a) \text{ approx}$$

The force to loosen the key is $P_1 = 2H \tan (a - b)$ approx, and the efficiency $e = \tan b/\tan (b + a)$. The key will be self-locking when $b < a$, or, more generally, when $2b < (a_1 + a_3)$.

Fig. 3.2.9   **Fig. 3.2.10**

Screws

Screws with Square Threads (Fig. 3.2.11)   Let r = mean radius of the thread = ½ (radius at root + outside radius), and l = pitch (or lead of a single-threaded screw), both in inches; b = angle of inclination of thread to a plane at right angles to the axis of screw ($\tan b = l/2\pi r$); and f = coefficient of sliding friction = $\tan a$. Then for a screw in uniform

Fig. 3.2.11

motion (friction of the root and outside surfaces being neglected) there is required a force P acting at right angles to the axis at the distance r. $P = L \tan (b \pm a) = L(l \pm 2\pi rf)/(2\pi r \pm fl)$, where the upper signs are for motion in a direction opposed to that of L and the lower for motion in the same direction as that of L. When $b \leq a$, the screw will not "overhaul" (or move under the action of the load L).

The **efficiency** for motion opposed to direction in which L acts = $e = \tan b/\tan(b + a)$; for motion in the same direction in which L acts, $e = \tan (b - a)/\tan b$.

The value of e is a maximum when $b = 45° - \frac{1}{2}a$; e.g., $e_{max} = 0.81$ for $b = 42°$ and $f = 0.1$. Since e increases rapidly for values of b up to 20°, this angle is generally not exceeded; for $b = 20°$, and $f_1 = 0.10$, $e = 0.74$. In presses, where the mechanical advantage is required to be great, b is taken down to 3°, for which value $e = 0.34$ with $f = 0.10$.

Kingsbury found for square-threaded screws running in loose-fitting nuts, the following coefficients of friction: lard oil, 0.09 to 0.25; heavy mineral oil, 0.11 to 0.19; heavy oil with graphite, 0.03 to 0.15.

Ham and Ryan give for screws the following values of coefficients of friction, with medium mineral oil: high-grade materials and workmanship, 0.10; average quality materials and workmanship, 0.12; poor workmanship, 0.15. The use of castor oil as a lubricant lowered f from 0.10 to 0.066. The coefficients of static friction (at starting) were 30 percent higher. Table 3.2.8 gives representative values of efficiency.

Screws with V Threads (Fig.3.2.12)   Let c = half the angle between the faces of a thread. Then, using the same notation as for squarethreaded screws, for a screw in motion (neglecting friction of root and outside surfaces),

$$P = L(l \pm 2\pi rf \sec d)/(2\pi r \pm lf \sec d)$$

d is the angle between a plane normal to the axis of the screw through the point of the resultant thread friction, and a plane which is tangent to the surface of the thread at the same point (see Groat, *Proc. Engs. Soc. West. Penn*, **34**). Sec d = sec $c\sqrt{1 - (\sin b \sin c)^2}$. For small values of b this reduces practically to sec d = sec c, and, for all cases the approximation, $P = L(l \pm 2\pi rf \sec c)/(2\pi r \pm lf \sec c)$ is within the limits of probable error in estimating values to be used for f.

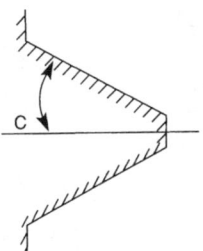

Fig. 3.2.12

The **efficiencies** are: $e = \tan b(1 - f \tan b \sec d)/(\tan b + f \sec d)$ for motion opposed to L, and $e = (\tan b - f \sec d)/\tan b(1 + f \tan b \sec d)$ for motion with L. If we let tan $d' = f \sec d$, these equations reduce, respectively, to $e = \tan b/\tan (b + d')$ and $e = \tan (b - d')/\tan b$. Negative values in the latter case merely mean that the thread will not overhaul. Subtract the values from unity for actual efficiency, considering the external moment and not the load L as being the driver. The efficiency of a V thread is lower than that of a square thread of the same helix angle, since $d' > a$.

For a **V-threaded screw and nut**, let r_1 = outside radius of thread, r_2 = radius at root of thread, $r = (r_1 + r_2)/2$, tan $d' = f \sec d$, r_0 = mean radius of nut seat = $1.5r$ (approx) and f' = coefficient of friction between nut and seat.

To tighten up the nut the turning moment required is $M = Pr + Lr_0 f = Lr[\tan (d' + b) + 1.5f']$. **To loosen** $M = Lr[\tan (d' - b) + 1.5f']$.

The **total tension in a bolt** due to tightening up with a moment M is $T = 2\pi M/(l + fl \sec b \sec d \csc b + f'3\pi r)$. $T \div$ area at root gives unit pure tensile stress induced, S_t. There is also a unit torsional stress: $S_s = 2(M - 1.5rf'T)/\pi r_2^3$. The equivalent combined stress is $S = 0.35S_t + 0.65\sqrt{S_t^2 + 4S_s^2}$.

Kingsbury, from tests on U.S. standard bolts, finds efficiencies for tightening up nuts from 0.06 to 0.12, depending upon the roughness of the contact surfaces and the character of the lubrication.

Toothed and Worm Gearing

The efficiency of **spur and bevel gearing** depends on the material and the workmanship of the gears and on the lubricant employed. For high-speed gears of good quality the efficiency of the gear transmission is 99 percent; with slow-speed gears of average workmanship the efficiency of 96 percent is common. On the average, efficiencies of 97 to 98 percent can be considered normal.

In **helical gears**, where considerable transverse sliding of the meshing teeth on each other takes place, the friction is much greater. If b and c are, respectively, the spiral angles of the teeth of the driving and driven helical gears (i.e., the angle between the teeth and the axis of rotation), $b + c$ is the shaft angle of the two gears, and $f = \tan a$ is the coefficient of sliding friction of the teeth, the efficiency of the gear transmission is $e = [\cos b \cos (c + a)]/[\cos c \cos (b - a)]$.

In the case of **worm gearing** when the shafts are normal to each other $(b + c = 90)$, the efficiency is $e = \tan c/\tan (c + a) = (1 - pf/2\pi r)/(1 + 2\pi rf/p)$, where c is the spiral angle of the worm wheel, or the lead angle of the worm; p the lead, or pitch of the worm thread; and r the mean radius of the worm. Typical values of f are shown in Table 3.2.9.

Journals and Bearings

Friction of Journal Bearings If P = total load on journal, l = journal length, and $2r$ = journal diameter, then $p = P/2rl$ = mean normal pressure on the projected area of the journal. Also, if f_1 is the **coefficient of journal friction**, the **moment of journal friction** for a cylindrical journal is $M = f_1 Pr$. The **work expended in friction** at angular velocity ω is

$$W_f = \omega M = f_1 Pr\omega$$

For the **conical bearing** (Fig. 3.2.13) the mean radius $r_m = (r + R)/2$ is to be used.

Fig. 3.2.13

Values of Coefficient of Friction For very low velocities of rotation (e.g., below 10 r/min), high loads, and with good lubrication, the coefficient of friction approaches the value of greasy friction, 0.07 to 0.15 (see Table 3.2.4). This is also the "pullout" coefficient of friction *on starting* the journal. With higher velocities, a fluid film is established between the journal and bearing, and the values of the coefficient of friction depend on the speed of rotation, the pressure on the bearing, and the viscosity of the oil. For journals running in complete bearing bushings, with a small clearance, i.e., with the diameter of the bushing slightly larger than the diameter of the journal, the experimental data of McKee give approximate values of the coefficient of friction as in Fig. 3.2.14.

Fig. 3.2.14 Coefficient of friction of journal.

If d_1 is the diameter of the bushing in inches, d the diameter of the journal in inches, then $(d_1 - d)$ is the diametral clearance and $m = (d_1 - d)/d$ is the clearance ratio. The diagram of McKee (Fig. 3.2.14) gives the coefficient of friction as a function of the characteristic number

Table 3.2.9   Coefficients of Friction for Worm Gears

	100	200	300	500	800	1200
Rubbing speed of worm, ft/min	100	200	300	500	800	1200
(m/min)	(30.5)	(61)	(91.5)	(152)	(244)	(366)
Phosphor-bronze wheel, polished-steel worm	0.054	0.045	0.039	0.030	0.024	0.020
Single-threaded cast-iron worm and gear	0.060	0.051	0.047	0.034	0.025	

ZN/p, where N is the speed of rotation in revolutions per minute, $p = P/(dl)$ is the average pressure in lb/in² on the projected area of the bearing, P is the load, l is the axial length of the bearing, and Z is the absolute viscosity of the oil in centipoises. Approximate values of Z at 100 (130)°F are as follows: light machine oil, 30 (16); medium machine oil, 60 (25); medium-heavy machine oil, 120 (40); heavy machine oil, 160 (60).

For purposes of design of ordinary machinery with bearing pressures from 50 to 300 lb/in² (344.7 to 2,068 kN/m²) and speeds of 100 to 3,000 rpm, values for the coefficient of journal friction can be taken from 0.008 to 0.020.

Thrust Bearings

Frictional Resistance for Flat Ring Bearing   Step bearings or pivots may be used to resist the end thrust of shafts. Let L = total load in the direction of the shaft axis and f = coefficient of sliding friction.

For a **ring-shaped flat step bearing** such as that shown in Fig. 3.2.15 (or a **collar bearing**), the **moment of thrust friction** $M = \frac{1}{3} fL(D^3 - d^3)/(D^2 - d^2)$. For a **flat circular step bearing**, $d = 0$, and $M = \frac{1}{3} fLD$.

Fig. 3.2.15

The value of the coefficient of sliding friction is 0.08 to 0.15 when the speed of rotation is very slow. At higher velocities when a collar or step bearing is used, $f = 0.04$ to 0.06. If the design provides for the formation of a load carrying oil film, as in the case of the Kingsbury thrust bearing, the coefficient of friction has values $f = 0.001$ to 0.0025.

Where oil is supplied from an external pump with such pressure as to separate the surfaces and provide an oil film of thickness h (Fig. 3.2.15), the frictional moment is

$$M = \frac{Zn(D^4 - d^4)}{67 \times 10^7 h} = \frac{\pi \mu \omega (D^4 - d^4)}{32 h}$$

where D and d are in inches, μ is the absolute viscosity, ω is the angular velocity, h is the film thickness, in, Z is viscosity of lubricant in centipoises, and n is rotation speed, r/min. With this kind of lubrication the frictional moment depends upon the speed of rotation of the shaft and actually approaches zero for zero shaft speeds. The thrust load will be carried on a film of oil regardless of shaft rotation for as long as the pump continues to supply the required volume and pressure (see also Secs. 8 and 14).

EXAMPLE.   A hydrostatic thrust bearing carries 101,000 lb, D is 16 in, d is 10 in, oil-film thickness h is 0.006 in, oil viscosity Z, 30 centipoises at operating temperature, and n is 750 r/min. Substituting these values, the frictional torque M is 310 in · lb (358 cm · kg). The oil supply pressure was 82.5 lb/in² (569 kN/m²); the oil flow, 12.2 gal/min (46.2 l/min).

Frictional Forces in Pin Joints of Mechanisms

In the absence of friction, or when the effect of friction is negligible, the force transmitted by the link b from the driver a to the driven link c (Figs. 3.2.16 and 3.2.17) acts through the centerline OO of the pins connecting the link b with links a and c. With friction, this line of action shifts to the line AA, tangent to small circles of diameter d. The diameter

d of the circle, called the **friction circle**, for each individual joint, is equal to fD, where D is the diameter of the pin and f is the coefficient of friction between the pin and the link. The choice of the proper disposition of the tangent AA with respect to the two friction circles is dictated

Fig. 3.2.16   **Fig. 3.2.17**

by the consideration that friction always opposes the action of the linkage. The force f opposes the motion of a; therefore, with friction it acts on a longer lever than without friction (Figs. 3.2.16 and 3.2.17). On the other hand, the force F drives the link c; friction hinders its action, and the equivalent lever is shorter with friction than without friction; the friction throws the line of action toward the center of rotation of link c.

EXAMPLE.   An engine eccentric (Fig. 3.2.18) is a joint where the friction loss may be large. For the dimensions shown and with a torque of 250 in · lb applied to the rotating shaft, the resultant horizontal force, with no friction, will act through the center of the eccentric and be 250/(2.5 sin 60) or 115.5 lb. With friction coefficient 0.1, the resultant force (which for a long rod remains approximately horizontal) will be tangent to the friction circle of radius 0.1 × 5, or 0.5 in, and have a magnitude of 250/(2.5 sin 60 + 0.5), or 93.8 lb (42.6 kg).

Fig. 3.2.18

Tension Elements

Frictional Resistance   In Fig. 3.2.19, let T_1 and T_2 be the tensions with which a rope, belt, chain, or brake band is strained over a drum, pulley, or sheave, and let the rope or belt be on the point of slipping from T_2 toward T_1 by reason of the difference of tension $T_1 - T_2$. Then $T_1 - T_2$ = circumferential force P transferred by friction must be equal

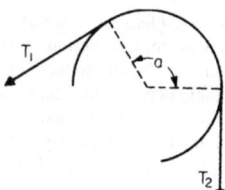

Fig. 3.2.19

to the frictional resistance W of the belt, rope, or band on the drum or pulley. Also, let a = angle subtending the arc of contact between the drum and tension element. Then, disregarding centrifugal forces,

$$T_1 = T_2 e^{fa} \quad \text{and} \quad P = (e^{fa} - 1)T_1/e^{fa} = (e^{fa} - 1)T_2 = W$$

where e = base of the napierian system of logarithms = 2.178 +.

Table 3.2.10 Values of e^{fa}

$\dfrac{a°}{360°}$	f								
	0.1	0.15	0.2	0.25	0.3	0.35	0.4	0.45	0.5
0.1	1.06	1.1	1.13	1.17	1.21	1.25	1.29	1.33	1.37
0.2	1.13	1.21	1.29	1.37	1.46	1.55	1.65	1.76	1.87
0.3	1.21	1.32	1.45	1.60	1.76	1.93	2.13	2.34	2.57
0.4	1.29	1.46	1.65	1.87	2.12	2.41	2.73	3.10	3.51
0.425	1.31	1.49	1.70	1.95	2.23	2.55	2.91	3.33	3.80
0.45	1.33	1.53	1.76	2.03	2.34	2.69	3.10	3.57	4.11
0.475	1.35	1.56	1.82	2.11	2.45	2.84	3.30	3.83	4.45
0.5	1.37	1.60	1.87	2.19	2.57	3.00	3.51	4.11	4.81
0.525	1.39	1.64	1.93	2.28	2.69	3.17	3.74	4.41	5.20
0.55	1.41	1.68	2.00	2.37	2.82	3.35	3.98	4.74	5.63
0.6	1.46	1.76	2.13	2.57	3.10	3.74	4.52	5.45	6.59
0.7	1.55	1.93	2.41	3.00	3.74	4.66	5.81	7.24	9.02
0.8	1.65	2.13	2.73	3.51	4.52	5.81	7.47	9.60	12.35
0.9	1.76	2.34	3.10	4.11	5.45	7.24	9.60	12.74	16.90
1.0	1.87	2.57	3.51	4.81	6.59	9.02	12.35	16.90	23.14
1.5	2.57	4.11	6.59	10.55	16.90	27.08	43.38	69.49	111.32
2.0	3.51	6.59	12.35	23.14	43.38	81.31	152.40	285.68	535.49
2.5	4.81	10.55	23.14	50.75	111.32	244.15	535.49	1,174.5	2,575.9
3.0	6.59	16.90	43.38	111.32	285.68	733.14	1,881.5	4,828.5	12,391
3.5	9.02	27.08	81.31	244.15	733.14	2,199.90	6,610.7	19,851	59,608
4.0	12.35	43.38	152.40	535.49	1,881.5	6,610.7	23,227	81,610	286,744

NOTE: $e^{\pi} = 23.1407$, log $e^{\pi} = 1.3643764$.

f is the static coefficient of friction (f_0) when there is no slip of the belt or band on the drum and the coefficient of kinetic friction (f) when slip takes place. For ease of computation, the values of the quantity e^{fa} are tabulated on Table 3.2.10.

Average values of f_0 for belts, ropes, and brake bands are as follows: for leather belt on cast-iron pulley, very greasy, 0.12; slightly greasy, 0.28; moist, 0.38. For hemp rope on cast-iron drum, 0.25; on wooden drum, 0.40; on rough wood, 0.50; on polished wood, 0.33. For iron brake bands on cast-iron pulleys, 0.18. For wire ropes, Tichvinsky reports coefficients of static friction, f_0, for a ⅝ rope (8 × 19) on a worn-in cast-iron groove: 0.113 (dry); for mylar on aluminum, 0.4 to 0.7.

Belt Transmissions; Effects of Belt Compliance

In the configuration of Fig. 3.2.20, pulley A drives a belt at angular velocity ω_A. Pulley B, here assumed to be of the same radius R as A, is driven at angular velocity ω_B. If the belt is extensible and the resistive torque $M = (T_1 - T_2) R$ is applied at B, ω_B will be smaller than ω_A and power will be dissipated at a rate $W = M(\omega_A - \omega_B)$. Likewise, the surface velocity V_1 of the more stretched belt will be larger than V_2. No slip will take place over the wraps A_T-A_S and B_T-B_S. The slip angles a_A and

a_B can be calculated from

$$a_A = [\ln (T_1/T_2)]/f_A \qquad a_B = [\ln (T_1/T_2)]/f_B$$

where f_A and f_B are the coefficients of friction on pulleys A and B, respectively. To calculate the above values, it is necessary to know the mean tension of the belt, $T = (T_1 + T_2)/2$. Then, $T_1/T_2 = [T + M/(2R)]/[T - M/(2R)]$. In this configuration, when the slip angles become equal to π (180°), complete slip occurs.

It is interesting to note that torque is transmitted only over the slip arcs a_A and a_B since there is no tension variation in the arcs A_T-A_S and B_T-B_S where the belt is in a uniform state of stretch.

Fig. 3.2.20 Pulley transmission with extensible belt.

3.3 MECHANICS OF FLUIDS
by J. W. Murdock

REFERENCES: *Specific.* "Handbook of Chemistry and Physics," Chemical Rubber Company. "Smithsonian Physical Tables," Smithsonian Institution. "Petroleum Measurement Tables," ASTM. "Steam Tables," ASME. "American Institute of Physics Handbook," McGraw-Hill. "International Critical Tables," McGraw-Hill. "Tables of Thermal Properties of Gases," *NBS Circular 564.* Murdock, "Fluid Mechanics and Its Applications," Houghton Mifflin, 1976. "Pipe Friction Manual," Hydraulic Institute. "Flow of Fluids," ASME, 1971. "Fluid Meters," 6th ed., ASME, 1971. "Measurement of Fluid Flow in Pipes Using Orifice, Nozzle, and Venturi," ASME Standard MFC-3M-1984. Murdock, ASME 64-WA/FM-6. Horton, *Engineering News,* 75, 373, 1916. Belvins, ASME 72/WA/FE-39. Staley and Graven, ASME 72PET/30. "Temperature Measurement," PTC 19.3, ASME. Moody, *Trans. ASME,* 1944, pp. 671–684. *General.* Binder, "Fluid Mechanics," Prentice-Hall. Langhaar, "Dimensional Analysis and Theory of Models," Wiley. Murdock, "Fluid Mechanics," Drexel University Press. Rouse, "Elementary Mechanics of Fluids," Wiley. Shames, "Mechanics of Fluids," McGraw-Hill. Streeter, "Fluid Mechanics," McGraw-Hill.

Notation

a = acceleration, area, exponent
A = area
c = velocity of sound
C = coefficient
$\mathbf{C}$ = Cauchy number
$\mathbf{C}_p$ = pressure coefficient
d = diameter, distance
E = bulk modulus of elasticity, modulus of elasticity (Young's modulus), velocity of approach factor, specific energy
$\mathbf{E}$ = Euler number
f = frequency, friction factor
F = dimension of force, force
$\mathbf{F}$ = Froude number
g = acceleration due to gravity
g_c = proportionality constant = 32.1740 lbm/(lbf) (ft/s^2)
G = mass velocity
h = head, vertical distance below a liquid surface
H = geopotential altitude
i = ideal
I = moment of inertia
J = mechanical equivalent of heat, 778.169 ft · lbf
k = isentropic exponent, ratio of specific heats
K = constant, resistant coefficient, weir coefficient
$\mathbf{K}$ = flow coefficient
L = dimension of length, length
m = mass, lbm
$\dot{m}$ = mass rate of flow, lbm/s
M = dimension of mass, mass (slugs)
$\dot{M}$ = mass rate of flow, slugs/s
$\mathbf{M}$ = Mach number
n = exponent for a polytropic process, roughness factor
N = dimensionless number
p = pressure
P = perimeter, power
q = heat added
$\mathbf{q}$ = flow rate per unit width
Q = volumetric flow rate
r = pressure ratio, radius
R = gas constant, reactive force
$\mathbf{R}$ = Reynolds number
R_h = hydraulic radius
s = distance, second
sp. gr. = specific gravity
S = scale reading, slope of a channel
$\mathbf{S}$ = Strouhal number
t = time
T = dimension of time, absolute temperature
u = internal energy
U = stream-tube velocity
v = specific volume
V = one-dimensional velocity, volume
$\mathbf{V}$ = velocity ratio
W = work done by fluid
$\mathbf{W}$ = Weber number
x = abscissa
y = ordinate
Y = expansion factor
z = height above a datum
Z = compressibility factor, crest height
α = angle, kinetic energy correction factor
β = ratio of primary element diameter to pipe diameter
γ = specific weight
δ = boundary-layer thickness
ε = absolute surface roughness
θ = angle
μ = dynamic viscosity
ν = kinematic viscosity
π = 3.14159 . . . , dimensionless ratio
ρ = density
σ = surface tension
τ = unit shear stress
ω = rotational speed

FLUIDS AND OTHER SUBSTANCES

Substances may be classified by their response when at rest to the imposition of a shear force. Consider the two very large plates, one moving, the other stationary, separated by a small distance y as shown in Fig. 3.3.1. The space between these plates is filled with a substance whose surfaces adhere to these plates in such a manner that its upper surface moves at the same velocity as the upper plate and the lower surface is stationary. The upper surface of the substance attains a velocity of U as the result of the application of shear force F_s. As y approaches dy, U approaches dU, and the rate of deformation of the substance becomes dU/dy. The **unit shear stress** is defined by $\tau = F_s/A_s$, where A_s is the shear or surface area. The deformation characteristics of various substances are shown in Fig. 3.3.2.

Fig. 3.3.1 Flow of a substance between parallel plates.

An ideal or **elastic solid** will resist the shear force, and its rate of deformation will be zero regardless of loading and hence is coincident with the ordinate of Fig. 3.3.2. A **plastic** will resist the shear until its yield stress is attained, and the application of additional loading will cause it to deform continuously, or flow. If the deformation rate is directly proportional to the flow, it is called an **ideal plastic**.

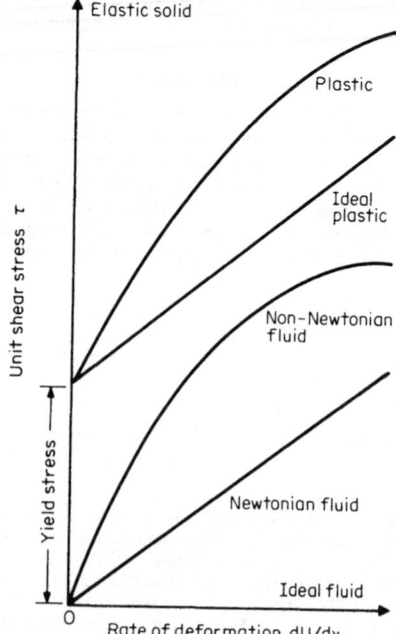

Fig. 3.3.2 Deformation characteristics of substances.

If the **substance** is unable to resist even the slightest amount of shear without flowing, it is a **fluid**. An **ideal fluid** has no internal friction, and hence its deformation rate coincides with the abscissa of Fig. 3.3.2. All real fluids have internal friction so that their rate of deformation is proportional to the applied shear stress. If it is directly proportional, it is called a **Newtonian fluid**; if not, a **non-Newtonian fluid**.

Two kinds of **fluids** are considered in this section, incompressible and compressible. A **liquid** except at very high pressures and/or temperatures may be considered incompressible. **Gases** and **vapors** are compressible fluids, but only **ideal gases** (those that follow the ideal-gas laws) are considered in this section. All others are covered in Secs. 4.1 and 4.2.

FLUID PROPERTIES

The **density** ρ of a fluid is its mass per unit volume. Its dimensions are M/L^3. In **fluid mechanics**, the units are slugs/ft³ and lbf · s²/ft⁴ (515.3788 kg/m³), but in **thermodynamics** (Sec. 4.1), the units are lbm/ft³ (16.01846 kg/m³). Numerical values of densities for selected liquids are shown in Table 3.3.1. The temperature change at 68°F (20°C) required to produce a 1 percent change in density varies from 12°F (6.7°C) for kerosene to 99°F (55°C) for mercury.

The **specific volume** v of a fluid is its volume per unit mass. Its dimensions are L^3/M. The units are ft³/lbm. Specific volume is related to density by $v = 1/\rho g_c$, where g_c is the proportionality constant [32.1740 (lbm/lbf)(ft/s²)]. **Specific volumes** of ideal gases may be computed from the equation of state: $v = RT/p$, where R is the gas constant in ft · lbf/(lbm)(°R) (see Sec. 4.1), T is the temperature in degrees Rankine (°F + 459.67), and p is the pressure in lbf/ft² abs.

The **specific weight** γ of a fluid is its weight per unit volume and has dimensions of F/L^3 or $M/(L^2)(T^2)$. The units are lbf/ft³ or slugs/(ft²)(s²) (157.087 N/m³). Specific weight is related to density by $\gamma = \rho g$, where g is the acceleration of gravity.

The **specific gravity** (sp. gr.) of a substance is a dimensionless ratio of the density of a fluid to that of a reference fluid. Water is used as the reference fluid for solids and liquids, and air is used for gases. Since the density of liquids changes with temperature for a precise definition of **specific gravity**, the temperature of the fluid and the reference fluid should be stated, for example, 60/60°F, where the upper temperature pertains to the liquid and the lower to water. If no temperatures are stated, reference is made to water at its maximum density, which occurs at 3.98°C and atmospheric pressure. The maximum density of water is 1.9403 slugs/ft³ (999.973 kg/m³). See Sec. 1.2 for conversion factors for API and Baumé hydrometers. For gases, it is common practice to use the ratio of the molecular weight of the gas to that of air (28.9644), thus eliminating the necessity of stating the pressure and temperature for ideal gases.

The **bulk modulus of elasticity** E of a fluid is the ratio of the pressure stress to the volumetric strain. Its dimensions are F/L^2. The units are lbf/in² or lbf/ft². E depends upon the thermodynamic process causing the change of state so that $E_x = -v(\partial p/\partial v)_x$, where x is the process. For ideal gases, $E_T = p$ for an isothermal process and $E_s = kp$ for an isentropic process where k is the ratio of specific heats. Values of E_T and E_s for liquids are given in Table 3.3.2. For liquids, a mean value is used by integrating the equation over a finite interval, or $E_{xm} = -v_1(\Delta p/\Delta v)_x = v_1(p_2 - p_1)/(v_1 - v_2)_x$.

EXAMPLE.   What pressure must be applied to ethyl alcohol at 68°F (20°C) to produce a 1 percent decrease in volume at constant temperature?

$$\Delta p = -E_T(\Delta v/v) = -(130,000)(-0.01)$$
$$= 1,300 \text{ lbf/in}^2 \ (9 \times 10^6 \text{ N/m}^2)$$

In a like manner, the pressure required to produce a 1 percent decrease in the volume of mercury is found to be 35,900 lbf/in² (248 × 10⁶ N/m²). For most engineering purposes, liquids may be considered as incompressible fluids.

The **acoustic velocity**, or velocity of sound in a fluid, is given by $c = \sqrt{E_s/\rho}$. For an ideal gas $c = \sqrt{kp/\rho} = \sqrt{kg_c pv} = \sqrt{kg_c RT}$. Values of the speed of sound in liquids are given in Table 3.3.2.

EXAMPLE.   Check the value of the velocity of sound in benzene at 68°F (20°C) given in Table 3.3.2 using the isentropic bulk modulus. $c = \sqrt{E_s/\rho} = \sqrt{144 \times 223,000/1.705} = 4,340$ ft/s (1,320 m/s). Additional information on the velocity of sound is given in Secs. 4, 11, and 12.

Application of shear stress to a fluid results in the continual and permanent distortion known as flow. **Viscosity** is the resistance of a fluid to shear motion—its internal friction. This resistance is due to two phenomena: (1) cohesion of the molecules and (2) molecular transfer from one layer to another, setting up a tangential or shear stress. In liquids, cohesion predominates, and since cohesion decreases with increasing temperature, the viscosity of liquids does likewise. Cohesion is relatively weak in gases; hence increased molecular activity with increasing temperature causes an increase in molecular transfer with corresponding increase in viscosity.

The **dynamic viscosity** μ of a fluid is the ratio of the shearing stress to the rate of deformation. From Fig. 3.3.1, $\mu = \tau/(dU/dy)$. Its dimensions are $(F)(T)/L^2$ or $M/(L)(T)$. The units are lbf · s/ft² or slugs/(ft)(s) [47.88026(N · s)/m²].

In the cgs system, the unit of dynamic viscosity is the **poise**, 2,089 × 10⁻⁶ (lbf · s)/ft² [0.1 (N · s)/m²], but for convenience the **centipoise** (1/100 poise) is widely used. The dynamic viscosity of water at 68°F (20°C) is approximately 1 centipoise.

Table 3.3.3 gives values of dynamic viscosity for selected liquids at atmospheric pressure. Values of viscosity for fuels and lubricants are given in Sec. 6. The effect of pressure on liquid viscosity is generally

Table 3.3.1   Density of Liquids at Atmospheric Pressure

Temp: °C	0	20	40	60	80	100
°F	32	68	104	140	176	212
Liquid	ρ, slugs/ft³ (515.4 kg/m³)					
Alcohol, ethyl[f]	1.564	1.532	1.498	1.463		
Benzene[a,b]	1.746	1.705	1.663	1.621	1.579	
Carbon tetrachloride[a,b]	3.168	3.093	3.017	2.940	2.857	
Gasoline,[c] sp. gr. 0.68	1.345	1.310	1.275	1.239		
Glycerin[a,b]	2.472	2.447	2.423	2.398	2.372	2.346
Kerosene,[c] sp. gr. 0.81	1.630	1.564	1.536	1.508	1.480	
Mercury[b]	26.379	26.283	26.188	26.094	26.000	25.906
Oil, machine,[c] sp. gr. 0.907	1.778	1.752	1.727	1.702	1.677	1.651
Water, fresh[d]	1.940	1.937	1.925	1.908	1.885	1.859
Water, salt[e]	1.995	1.988	1.975			

SOURCES: Computed from data given in:
[a] "Handbook of Chemistry and Physics," 52d ed, Chemical Rubber Company, 1971–1972.
[b] "Smithsonian Physical Tables," 9th rev. ed., 1954.
[c] ASTM-IP, "Petroleum Measurement Tables."
[d] "Steam Tables," ASME, 1967.
[e] "American Institute of Physics Handbook," 3d ed., McGraw-Hill, 1972.
[f] "International Critical Tables," McGraw-Hill.

Table 3.3.2   Bulk Modulus of Elasticity, Ratio of Specific Heats of Liquids and Velocity of Sound at One Atmosphere and 68°F (20°C)

Liquid	E in lbf/in² (6,895 N/m²)		$k = c_p/c_v$	c in ft/s (0.3048 m/s)
	Isothermal E_T	Isentropic E_s		
Alcohol, ethyl[a,e]	130,000	155,000	1.19	3,810
Benzene[a,f]	154,000	223,000	1.45	4,340
Carbon tetrachloride[a,b]	139,000	204,000	1.47	3,080
Glycerin[f]	654,000	719,000	1.10	6,510
Kerosene,[a,e] sp. gr. 0.81	188,000	209,000	1.11	4,390
Mercury[e]	3,590,000	4,150,000	1.16	4,770
Oil, machine,[f] sp. gr. 0.907	189,000	219,000	1.13	4,240
Water, fresh[a]	316,000	319,000	1.01	4,860
Water, salt[a,e]	339,000	344,000	1.01	4,990

SOURCES: Computed from data given in:
[a] "Handbook of Chemistry and Physics," 52d ed., Chemical Rubber Company, 1971–1972.
[b] "Smithsonian Physical Tables," 9th rev. ed., 1954.
[c] ASTM-IP, "Petroleum Measurement Tables."
[d] "Steam Tables," ASME, 1967.
[e] "American Institute of Physics Handbook," 3d ed., McGraw-Hill, 1972.
[f] "International Critical Tables," McGraw-Hill.

Table 3.3.3   Dynamic Viscosity of Liquids at Atmospheric Pressure

Temp: °C	0	20	40	60	80	100
°F	32	68	104	140	176	212
Liquid	μ, (lbf · s)/(ft²) [47.88 (N · s)/(m²)] × 10⁶*					
Alcohol, ethyl[a,e]	37.02	25.06	17.42	12.36	9.028	
Benzene[a]	19.05	13.62	10.51	8.187	6.871	
Carbon tetrachloride[e]	28.12	20.28	15.41	12.17	9.884	
Gasoline,[b] sp. gr. 0.68	7.28	5.98	4.93	4.28		
Glycerin[d]	252,000	29,500	5,931	1,695	666.2	309.1
Kerosene,[b] sp. gr. 0.81	61.8	38.1	26.8	20.3	16.3	
Mercury[a]	35.19	32.46	30.28	28.55	27.11	25.90
Oil, machine,[a] sp. gr. 0.907						
"Light"	7,380	1,810	647	299	164	102
"Heavy"	66,100	9,470	2,320	812	371	200
Water, fresh[c]	36.61	20.92	13.61	9.672	7.331	5.827
Water, salt[d]	39.40	22.61	18.20			

* μ × 10⁶ = tabulated value
μ = tabulated value × 10⁻⁶
SOURCES: Computed from data given in:
[a] "Handbook of Chemistry and Physics," 52d ed., Chemical Rubber Company, 1971–1972.
[b] "Smithsonian Physical Tables," 9th rev. ed., 1954.
[c] "Steam Tables," ASME, 1967.
[d] "American Institute of Physics Handbook," 3d ed., McGraw-Hill, 1972.
[e] "International Critical Tables," McGraw-Hill.

Table 3.3.4   Dynamic Viscosity of Gases at One Atmosphere

Temp: °C	0	20	60	100	200	400	600	800	1000
°F	32	68	140	212	392	752	1112	1472	1832
Gas	μ, (lbf · s)/(ft²) [47.88(N · s)/(m²)] × 10⁸*								
Air[a]	35.67	39.16	41.79	45.95	53.15	70.42	80.72	91.75	100.8
Carbon dioxide[a]	29.03	30.91	35.00	38.99	47.77	62.92	74.96	87.56	97.71
Carbon monoxide[b]	34.60	36.97	41.57	45.96	52.39	66.92	79.68	91.49	102.2
Helium[a]	38.85	40.54	44.23	47.64	55.80	71.27	84.97	97.43	
Hydrogen[a,b]	17.43	18.27	20.95	21.57	25.29	32.02	38.17	43.92	49.20
Methane[a]	21.42	22.70	26.50	27.80	33.49	43.21			
Nitrogen[a,b]	34.67	36.51	40.14	43.55	51.47	65.02	76.47	86.38	95.40
Oxygen[b]	40.08	42.33	46.66	50.74	60.16	76.60	90.87	104.3	116.7
Steam[c]		18.49	21.89	25.29	33.79	50.79	67.79	84.79	

* μ × 10⁸ = tabulated value
μ = tabulated value × 10⁻⁸
SOURCES: Computed from data given in:
[a] "Handbook of Chemistry and Physics," 52d ed., Chemical Rubber Company, 1971–1972.
[b] "Tables of Thermal Properties of Gases," NBS Circular 564, 1955.
[c] "Steam Tables," ASME, 1967.

unimportant in fluid mechanics except in lubricants (Sec. 6). The viscosity of water changes little at pressures up to 15,000 lbf/in², but for animal and vegetable oils it increases about 350 percent and for mineral oils about 1,600 percent at 15,000 lbf/in² pressure.

The dynamic viscosity of gases is primarily a temperature function and essentially independent of pressure. Table 3.3.4 gives values of dynamic viscosity of selected gases.

The **kinematic viscosity** v of a fluid is its dynamic viscosity divided by its density, or $v = \mu/\rho$. Its dimensions are L^2/T. The units are ft²/s (9.290304 × 10⁻² m²/s).

In the cgs system, the unit of kinematic viscosity is the **stoke** (1 × 10⁻⁴ m²/s), but for convenience, the **centistoke** (1/100 stoke) is widely used. The kinematic viscosity of water at 68°F (20°C) is approximately 1 centistoke.

The standard device for **experimental determination of kinematic viscosity** in the United States is the **Saybolt Universal** viscometer. It consists essentially of a metal tube and an orifice built to rigid specifications and calibrated. The time required for a gravity flow of 60 cubic centimeters is called the SSU (Saybolt seconds Universal). Approximate conversions of SSU to stokes may be made as follows:

$$32 < \text{SSU} < 100 \text{ seconds, stokes} = 0.00226 \, (\text{SSU}) - 1.95/(\text{SSU})$$
$$\text{SSU} > 100 \text{ seconds, stokes} = 0.00220 \, (\text{SSU}) - 1.35/(\text{SSU})$$

For viscous oils, the **Saybolt Furol** viscometer is used. Approximate conversions of SSF (Saybolt seconds Furol) may be made as follows:

$$25 < \text{SSF} < 40 \text{ seconds, stokes} = 0.0224 \, (\text{SSF}) - 1.84/(\text{SSF})$$
$$\text{SSF} > 40 \text{ seconds, stokes} = 0.0216 \, (\text{SSF}) - 0.60/(\text{SSF})$$

For exact conversions of Saybolt viscosities, see ASTM D445–71 and Sec. 6.11.

The **surface tension** σ of a fluid is the work done in extending the surface of a liquid one unit of area or work per unit area. Its dimensions are F/L. The units are lbf/ft (14.5930 N/m).

Values of σ for various interfaces are given in Table 3.3.5. Surface tension decreases with increasing temperature. Surface tension is of importance in the formation of bubbles and in problems involving atomization.

Table 3.3.5 Surface Tension of Liquids at One Atmosphere and 68°F (20°C)

| Liquid | δ, lbf/ft (14.59 N/m) × 10³ | | |
	In vapor	In air	In water
Alcohol, ethyl*	1.56	1.53	
Benzene*	2.00	1.98	2.40
Carbon tetrachloride*	1.85	1.83	3.08
Gasoline,* sp. gr. 0.68	1.3–1.6		2.7–3.6
Glycerin*	4.30	4.35	
Kerosene,* sp. gr. 0.81	1.6–2.2		
Mercury*	32.6§	32.8	25.7
Oil, machine,‡ sp. gr. 0.907	2.5	2.6	2.3–3.7
Water, fresh‡		4.99	
Water, salt‡		5.04	

SOURCES: Computed from data given in:
* "International Critical Tables," McGraw-Hill.
† ASTM-IP, "Petroleum Measurement Tables."
‡ "American Institute of Physics Handbook," 3d ed., McGraw-Hill, 1972.
§ In vacuum.

Capillary action is due to surface tension, **cohesion** of the liquid molecules, and the **adhesion** of the molecules on the surface of a solid. This action is of importance in fluid mechanics because of the formation of a meniscus (curved section) in a tube. When the adhesion is greater than the cohesion, a liquid "wets" the solid surface, and the liquid will rise in the tube and conversely will fall if the reverse. Figure 3.3.3 illustrates this effect on manometer tubes. In the reading of a manometer, all data should be taken at the center of the meniscus.

The **vapor pressure** pv of a fluid is the pressure at which its liquid and vapor are in equilibrium at a given temperature. See Secs. 4.1 and 4.2 for further definitions and values.

Fig. 3.3.3 Capillarity in circular glass tubes.

FLUID STATICS

Pressure p is the force per unit area exerted on or by a fluid and has dimensions of F/L^2. In fluid mechanics and in thermodynamic equations, the units are lbf/ft² (47.88026 N/m²), but engineering practice is to use units of lbf/in² (6,894.757 N/m²).

The relationship between **absolute pressure, gage pressure,** and **vacuum** is shown in Fig. 3.3.4. Most fluid-mechanics equations and **all** thermodynamic equations require the use of **absolute pressure,** and unless otherwise designated, a pressure should be understood to be **absolute pressure.** Common practice is to denote absolute pressure as lbf/ft² abs, or psfa, lbf/in² abs or psia; and in a like manner for **gage pressure** lbf/ft² g, lbf/in² g, and psig.

Fig. 3.3.4 Pressure relations.

According to **Pascal's** principle, the pressure in a static fluid is the same in all directions.

The **basic equation of fluid statics** is obtained by consideration of a fluid particle at rest with respect to other fluid particles, all being subjected to body-force accelerations of a_x, a_y, and a_z opposite the directions of x, y, and z, respectively, and the acceleration of gravity in the z direction, resulting in the following:

$$dp = -\rho[a_x \, dx + a_y \, dy + (a_z + g)dz]$$

Pressure-Height Relations For a fluid at rest and subject only to the gravitational force, a_x, a_y, and a_z are zero and the **basic equation for fluid statics** reduces to $dp = -\rho g \, dz = \gamma \, dz$.

Liquids (Incompressible Fluids) The pressure-height equation integrates to $(p_1 - p_2) = \rho g(z_2 - z_1) = \gamma(z_2 - z_1) = \Delta p = \gamma h$, where h is measured from the liquid surface (Fig. 3.3.5).

EXAMPLE. A large closed tank is partly filled with 68°F (20°C) benzene. If the pressure on the surface is 10 lb/in², what is the pressure in the benzene at a depth of 11 ft below the liquid surface?

$$p_1 = \rho g h + p_2 = \frac{1.705 \times 32.17 \times 11}{144} + 10$$

$$= 14.19 \text{ lbf/in}^2 (9.784 \times 10^4 \text{ N/m}^2)$$

Ideal Gases (Compressible Fluids) For problems involving the upper atmosphere, it is necessary to take into account the variation of gravity with altitude. For this purpose, the **geopotential altitude** H is used, defined by $H = Z/(1 + z/r)$, where r is the radius of the earth

Fig. 3.3.5 Pressure equivalence.

($\approx 21 \times 10^6$ ft $\approx 6.4 \times 10^6$ m) and z is the height above sea level. The integration of the pressure-height equation depends upon the thermodynamic process. For an **isothermal process** $p_2/p_1 = e^{-(H_2 - H_1)/RT}$ and for a **polytropic process** ($n \neq 1$)

$$\frac{p_2}{p_1} = \left[1 - \frac{(n-1)(H_2 - H_1)}{nRT_1}\right]^{n/(n-1)}$$

Temperature-height relations for a polytropic process ($n \neq 1$) are given by

$$\frac{n}{1-n} = \frac{H_2 - H_1}{R(T_2 - T_1)}$$

Substituting in the **pressure-altitude** equation,

$$p_2/p_1 = (T_2/T_1)^{(H_2 - H_1)\div R(T_1 - T_2)}$$

EXAMPLE. The U.S. Standard Atmosphere 1962 (Sec. 11) is defined as having a sea-level temperature of 59°F (15°C) and a pressure of 2,116.22 lbf/ft². From sea level to a geopotential altitude of 36,089 ft (11,000 m) the temperature decreases linearly with altitude to −69.70°F (−56.5°C). Check the value of pressure ratio at this altitude given in the standard table.
Noting that $T_1 = 59 + 459.67 = 518.67$, $T_2 = -69.70 + 459.67 = 389.97$, and $R = 53.34$ ft · lbf/(lbm)(°R),

$$p_2/p_1 = (T_2/T_1)^{(H_2 - H_1)/R(T_1 - T_2)}$$
$$= (389.97/518.67)^{(36,089-0)/53.34(518.67-389.97)}$$
$$= 0.2233 \text{ vs. tabulated value of } 0.2234$$

Pressure-Sensing Devices The two principal devices using liquids are the **barometer** and the **manometer.** The barometer senses absolute pressure and the manometer senses pressure differential. For discussion of the barometer and other pressure-sensing devices, refer to Sec. 16.
Manometers are a direct application of the basic equation of fluid statics and serve as a pressure standard in the range of ¹⁄₁₀ in of water to 100 lbf/in². The most familiar type of manometer is the **U tube** shown in Fig. 3.3.6a. Because of the necessity of observing both legs simultaneously, the **well** or **cistern** type (Fig. 3.3.6b) is sometimes used. The **inclined manometer** (Fig. 3.3.6c) is a special form of the well-type manometer designed to enhance the readability of small pressure differentials. Application of the basic equation of fluid statics to each of the types results in the following equations. For the **U tube,** $p_1 - p_2 = (\gamma_m - \gamma_f)h$, where γ_m and γ_f are the specific weights of the manometer and sensed fluids, respectively, and h is the vertical distance between the liquid interfaces. For the **well type,** $p_1 - p_2 = (\gamma_m - \gamma_f)(z_2) \times (1 + A_2/A_1)$, where A_1 and A_2 are as shown in Fig. 3.3.6b and z_2 is the vertical distance from the fill line to the upper interface. Commercial manufacturers of well-type manometers correct for the area ratios so that $p_1 - p_2 = (\gamma_m - \gamma_f)S$, where S is the scale reading and is equal to $z_1(1 + A_2/A_1)$. For this reason, scales should not be interchanged between U type or well type or between well types without consulting the manufacturer. For inclined manometers,

$$p_1 - p_2 = (\gamma_m - \gamma_f)(A_2/A_1 + \sin \theta)R$$

where R is the distance along the inclined tube. Commercial inclined manometers also have special scales so that $p_1 - p_2 = (\gamma_m - \gamma_f)S$, where $S = (A_2/A_1 + \sin \theta)R$.

(a)

(b)

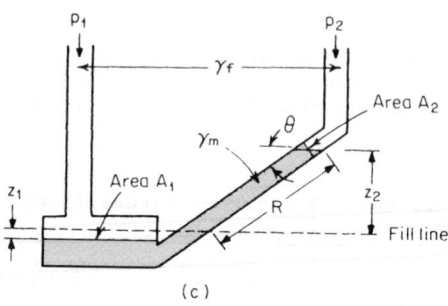

(c)

Fig. 3.3.6 (a) U-tube manometer; (b) well or cistern-type manometer; (c) inclined manometer.

EXAMPLE. A U-tube manometer containing mercury is used to sense the difference in water pressure. If the height between the interfaces is 10 in and the temperature is 68°F (20°C), what is the pressure differential?

$$p_1 - p_2 = (\gamma_m - \gamma_f)h = g(\rho_m - \rho_f)h$$
$$= 32.17(26.283 - 1.937)(10/12)$$
$$= 652.7 \text{ lbf/ft}^2(3.152 \times 10^4 \text{ N/m}^2)$$

Liquid Forces The force exerted by a liquid on a **plane submerged surface** (Fig. 3.3.7) is given by $F = \int p\, dA = \gamma \int h \cdot dA = \gamma h_c A$, where h_c is the distance from the liquid surface to the center of gravity of the surface, and A is the area of the surface. The location of the center of this force is given by

$$s_F = s_c + I_G/s_c A$$

where s_F is the inclined distance from the liquid surface to the center of force, s_c the inclined distance to the center of gravity of the surface, and I_G the moment of inertia around its center of gravity. Values of I_G are given in Sec. 5.2. See also Sec. 3.1. From Fig. 3.3.7, $h = R \sin \theta$, so that the vertical center of force becomes

$$h_F = h_c + I_G(\sin \theta)^2/h_c A$$

EXAMPLE.   Determine the force and its location acting on a rectangular gate 3 ft wide and 5 ft high at the bottom of a tank containing 68°F (20°C) water, 12 ft deep, (1) if the gate is vertical, and (2) if it is inclined 30° from horizontal.

1. *Vertical gate*

$F = \gamma g h_c A = \rho g h_c A$
$\quad = 1.937 \times 32.17(12 - 5/2)(5 \times 3)$
$\quad = 8,800 \ \text{lb}_f(3.914 \times 10^4 \ \text{N})$

$h_F = h_c + I_G(\sin\theta)^2/h_c A$, from Sec. 5.2, I_G for a rectangle = (width)(height)³/12

$h_F = (12 - 5/2) + (3 \times 5^3/12)(\sin 90°)^2/(12 - 5/2)(3 \times 5)$

$h_F = 9.719 \ \text{ft} (2.962 \ \text{m})$

2. *Inclined gate*

$F = \gamma h_c A = \rho g h_c A$
$\quad = 1.937 \times 32.17(12 - 5/2 \sin 30°)(5 \times 3)$
$\quad = 10,048 \ \text{lbf} (4.470 \times 10^4 \ \text{N})$

$h_F = h_c + I_G(\sin\theta)^2/h_c A$
$\quad = (12 - 5/2 \sin 30°) + (3 \times 5^3/12)(\sin 30°)^2/(12 - 5/2 \sin 30°)(3 \times 5)$
$\quad = 10.80 \ \text{ft} (3.291 \ \text{m})$

Fig. 3.3.7   Notation for liquid force on submerged surfaces.

Forces on irregular surfaces may be obtained by considering their horizontal and vertical components. The vertical component F_z equals the weight of liquid above the surface and acts through the centroid of the volume of the liquid above the surface. The horizontal component F_x equals the force on a vertical projection of the irregular surface. This force may be calculated by $F_x = \gamma h_{cx} A_x$, where h_{cx} is the distance from the center of gravity of the horizontal projection, and A_x is the projected area. The forces may be combined by $F = \sqrt{F_z^2 + F_x^2}$.

When fluid masses are **accelerated** without relative motion between fluid particles, the basic equation of fluid statics may be applied. For **translation** of a **liquid** mass due to uniform acceleration, the basic equation integrates to

$$p_2 - p_1 = -\rho[(x_2 - x_1)a_x + (y_2 - y_1)a_y + (z_2 - z_1)(a_z + g)]$$

EXAMPLE.   An open tank partly filled with a liquid is being accelerated up an inclined plane as shown in Fig. 3.3.8. The uniform acceleration is 20 ft/s² and the angle of the incline is 30°. What is the angle of the free surface of the liquid? Noting that on the free surface $p_2 = p_1$ and that the acceleration in the y direction is zero, the basic equation reduces to

$$(x_2 - x_1)a_x + (z_2 - z_1)(a_z + g) = 0$$

Solving for tan θ,

$$\tan\theta = \frac{z_1 - z_2}{x_2 - x_1} = \frac{a_x}{a_z + g} = \frac{a\cos\alpha}{a\sin\alpha + g}$$
$$= (20\cos 30°)/(20\sin 30° + 32.17) = 0.4107$$
$$\theta = 22°20'$$

For **rotation** of liquid masses with uniform rotational acceleration, the basic equation integrates to

$$p_2 - p_1 = \rho\left[\frac{\omega^2}{2}(x_2^2 - x_1^2) - g(z_2 - z_1)\right]$$

where ω is the rotational speed in rad/s and x is the radial distance from the axis of rotation.

Fig. 3.3.8   Notation for translation example.

EXAMPLE.   The closed cylindrical tank shown in Fig. 3.3.9 is 4 ft in diameter and 10 ft high and is filled with 104°F (40°C) benzene. The tank is rotated at 250 r/min about an axis 3 ft from its centerline. Compute the maximum pressure differential in the tank. Analysis of the rotation equation indicates that the maximum

Fig. 3.3.9   Notation for rotation example.

pressure will occur at the maximum rotational radius and the minimum elevation and, conversely, the minimum at the minimum rotational radius and maximum elevation. From Fig. 3.3.9, $x_1 = 3 - 4/2 = 1$ ft, $x_2 = 1 + 4 = 5$ ft, $z_2 - z_1 = -10$ ft, and the rotational speed $\omega = 2\pi N/60 = 2\pi(250)/60 = 26.18$ rad/s. Substituting

into the rotational equation,

$$p_2 - p_1 = \rho \left[\frac{\omega^2}{2}(x_2^2 - x_1^2) - g(z_2 - z_1) \right]$$

$$= \frac{1.663}{144} \left[\frac{(26.18)^2}{2}(5^2 - 1^2) - 32.17(-10) \right]$$

$$= 98.70 \text{ lbf/in}^2 (6.805 \times 10^5 \text{ N/m}^2)$$

Buoyancy Archimedes' principle states that a body immersed in a fluid is buoyed up by a force equal to the weight of the fluid displaced. If an object immersed in a fluid is heavier than the fluid displaced, it will sink to the bottom, and if lighter, it will rise. From the free-body diagram of Fig. 3.3.10, it is seen that for vertical equilibrium,

$$\Sigma F_z = 0 = F_B - F_g - F_D$$

where F_B is the buoyant force, F_g the gravity force (weight of body), and F_D the force required to prevent the body from rising. The buoyant force

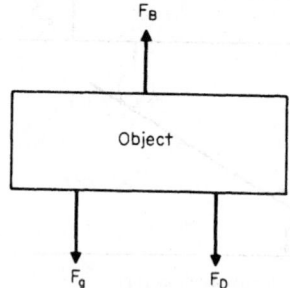

Fig. 3.3.10 Free body diagram of an immersed object.

being the weight of the displaced liquid, the equilibrium equation may be written as

$$F_D = F_B - F_g = \gamma_f V - \gamma_0 V = (\gamma_f - \gamma_0)V$$

where γ_f is the specific weight of the fluid, γ_0 is the specific weight of the object, and V is the volume of the object.

EXAMPLE. An airship has a volume of 3,700,000 ft^3 and is filled with hydrogen. What is its gross lift in air at 59°F (15°C) and 14.696 psia? Noting that $\gamma = p/RT$,

$$F_D = (\gamma_f - \gamma_0)V = \left(\frac{p}{R_a T} - \frac{p}{R_{H_2} T} \right)V$$

$$= \frac{pV}{T}\left(\frac{1}{R_a} - \frac{1}{R_{H_2}} \right)$$

$$= \frac{144 \times 14.696 \times 3,700,000}{59 + 459.7}\left(\frac{1}{53.34} - \frac{1}{766.8} \right)$$

$$= 263,300 \text{ lbf} (1.71 \times 10^6 \text{ N})$$

Flotation is a special case of buoyancy where $F_D = 0$, and hence $F_B = F_g$.

EXAMPLE. A crude **hydrometer** consists of a cylinder of ½ in diameter and 2 in length surmounted by a cylinder of ⅛ in diameter and 10 in long. Lead shot is added to the hydrometer until its total weight is 0.32 oz. To what depth would this hydrometer float in 104°F (40°C) glycerin? For flotation, $F_B = F_g = \gamma_f V = \rho_f g V$ or $V = F_B/\rho_f g = (0.32/16)/(2.423 \times 32.17) = 2.566 \times 10^{-4}$ ft^3. Volume of cylindrical portion of hydrometer $= V_c = \pi D^2 L/4 = \pi (0.5/12)^2 (2/12)/4 = 2.273 \times 10^{-4}$ ft^3. Volume of stem immersed $= V_s = V - V_c = 2.566 \times 10^{-4} - 2.273 \times 10^{-4} = 2.930 = 10^{-5}$ ft^3. Length of immersed stem $= L_s = 4 \ V_s/\pi D^2 = (4 \times 2.930 \times 10^{-5})/\pi (0.125/12)^2 = 0.3438$ ft $= 0.3438 \times 12 = 4.126$ in. Total immersion $= L + L_s = 2 + 4.126 = 6.126$ in (0.156 m).

Static Stability A body is in static equilibrium when the imposition of a small displacement brings into action forces that tend to restore the body to its original position. For **completely submerged bodies,** the center of buoyancy and the center of gravity must lie on the same vertical line and the center of buoyancy must be located above the center of gravity. Figure 3.3.11a shows a balloon and its basket in its normal position with

the center of buoyancy B above and on the same vertical line as the center of gravity G. Figure 3.3.11b shows the balloon displaced from its normal position. In this position, there is a couple $F_B x$ which tends to restore the balloon and its basket to its original position. For **floating bodies,** the center of gravity and the center of buoyancy must lie on the

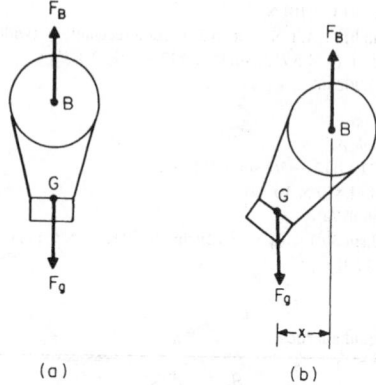

Fig. 3.3.11 Stability of an immersed body.

same vertical line, but the center of buoyancy may be below the center of gravity, as is common practice in surface-ship design. It is required that when displaced, the line of action of the buoyant force intersect the centerline above the center of gravity. Figure 3.3.12a shows a floating body in its normal position with its center of gravity G on the same vertical line and above the center of buoyancy B. Figure 3.3.12b shows the object displaced. The intersection of the line of action of the buoyant force with the centerline of the body at M is called the metacenter. As shown, this is above the center of buoyancy and sets up a restoring couple. When the metacenter is below the center of gravity, the object will capsize (see Sec. 11.3).

Fig. 3.3.12 Stability of a floating body.

FLUID KINEMATICS

Steady and Unsteady Flow If at every point in the fluid stream none of the local fluid properties changes with time, the flow is said to be steady. The mathematical conditions for steady flow are met when $\partial(\text{fluid properties})/\partial t = 0$. While flow is generally unsteady by nature, many real cases of unsteady flow may be treated as steady flow by using average properties or by changing the space reference. The amount of error produced by the averaging technique depends upon the nature of the unsteady flow, but the latter technique is error-free when it can be applied.

Streamlines and Stream Tubes Velocity has both magnitude and direction and hence is a vector. A **streamline** is a line which gives the *direction* of the velocity of a fluid particle at each point in the flow stream. When streamlines are connected by a closed curve in steady flow, they will form a boundary through which the fluid particles cannot pass. The space between the streamlines becomes a **stream tube.** The stream-tube concept broadens the application of fluid-flow principles; for example, it allows treating the flow inside a pipe and the flow around an object with the same laws. A stream tube of decreasing size approaches its own axis, a central streamline; thus equations developed for a stream tube may also be applied to a streamline.

Velocity and Acceleration In the most general case of fluid motion, the resultant **velocity** U along a streamline is a function of both distance s and time t, or $U = f(s, t)$. In differential form,

$$dU = \frac{\partial U}{\partial s} ds + \frac{\partial U}{\partial t} dt$$

An expression for **acceleration** may be obtained by dividing the velocity equation by dt, resulting in

$$\frac{dU}{dt} = \frac{\partial U}{\partial s}\frac{ds}{dt} + \frac{\partial U}{\partial t}$$

for steady flow $\partial U/\partial t = 0$.

Velocity Profile In the flow of real fluids, the individual streamlines will have different velocities past a section. Figure 3.3.13 shows the steady flow of a fluid past a section (A-A) of a circular pipe. The **velocity profile** is obtained by plotting the velocity U of each streamline as it passes A-A. The stream tube that is formed by the space between the streamlines is the annulus whose area is dA, as shown in Fig. 3.3.13 for

Fig. 3.3.13 Velocity profile.

the stream tube whose velocity is U. The volumetric rate of flow Q for the flow past section A-A is $Q = \int U\, dA$. All flows take place between boundaries that are **three-dimensional.** The terms **one-dimensional, two-dimensional,** and **three-dimensional** flow refer to the number of dimensions required to describe the velocity profile. For **three-dimensional** flow, a **volume** (L^3) is required; for example, the flow of a fluid in a circular pipe. For **two-dimensional** flow, an **area** (L^2) is necessary; for example, the flow between two parallel plates. For **one-dimensional** flow, a **line** (L) describes the profile. In cases of two- or three-dimensional flow, $\int U\, dA$ can be integrated either mathematically if the equations are known or graphically if velocity-measurement data are available. In many engineering applications, the **average velocity** V may be used where $V = Q/A = (1/A)\int U\, dA$.

The **continuity equation** is a special case of the general physical law of the conservation of mass. It may be simply stated for a control volume:

Mass rate entering = mass rate of storage + mass rate leaving

This may be expressed mathematically as

$$\rho U\, dA = \left[\frac{\partial}{\partial t}(\rho\, dA\, ds)\right] + \left[\rho U\, dA + \frac{\partial}{\partial s}(\rho U\, dA)\, ds\right]$$

where ds is an incremental distance along the control volume. For steady flow, $\partial/\partial t(\rho\, dA\, ds) = 0$, the general equation reduces to $d(\rho U\, dA) = 0$. Integrating the steady-flow **continuity equation** for the average velocity along a flow passage:

$$\rho VA = \text{a constant} = \rho_1 V_1 A_1 = \rho_2 V_2 A_2 = \cdots = \rho_n V_n A_n = \dot{M}$$

where $\dot{M}$ is the mass flow rate in slugs/s (14.5939 kg/s). In many engineering applications, the flow rate in pounds mass per second is desired,

so that

$$\dot{m} = \frac{V_1 A_1}{v_1} = \frac{V_2 A_2}{v_2} = \cdots = \frac{V_n A_n}{v_n}$$

where $\dot{m}$ is the flow rate in lbm/s (0.4535924 kg/s).

EXAMPLE. Air discharges from a 12-in-diameter duct through a 4-in-diameter nozzle into the atmosphere. The pressure in the duct is 20 lbf/in², and atmospheric pressure is 14.7 lbf/in². The temperature of the air in the duct just upstream of the nozzle is 150°F, and the temperature in the jet is 147°F. If the velocity in the duct is 18 ft/s, compute (1) the mass flow rate in lbm/s and (2) the velocity in the nozzle jet. From the equation of state

$$v = RT/p$$
$$v_D = RT_D/p_D = 53.34\,(150 + 459.7)/(144 \times 20)$$
$$= 11.29 \text{ ft}^3/\text{lbm}$$
$$v_J = RT_J/p_J = 53.34\,(147 + 459.7)/(144 \times 14.7)$$
$$= 15.29 \text{ ft}^3/\text{lbm}$$
$$(1)\ \dot{m} = V_D A_D/v_D = 18[(\pi/4)(12/12)^2]/11.29$$
$$\dot{m}_J = 1.252 \text{ lbm/s } (0.5680 \text{ kg/s})$$
$$(2)\ VJ = mv_J/A_J = (1.252)(15.29)/[(\pi/4)(4/12)^2]$$
$$v_J = 219.2 \text{ ft/s}(66.82 \text{ m/s})$$

FLUID DYNAMICS

Equation of Motion For steady one-dimensional flow, consideration of forces acting on a fluid element of length dL, flow area dA, boundary perimeter in fluid contact dP, and change in elevation dz with a unit shear stress τ moving at a velocity of V results in

$$v\, dp + \frac{V\, dV}{g_c} + \frac{g}{g_c}dz + v\tau\left(\frac{dP}{dA}\right) dL = 0$$

Substituting $v = g/g_c\gamma$ and simplifying,

$$\frac{dp}{\gamma} + \frac{V\, dV}{g} + dz + dh_f = 0$$

where $dh_f = (\tau/\gamma)(dP/dA)\, dL = \tau\, dL/\gamma R_h$.

The expression $1/(dP/dA)$ is the **hydraulic radius** R_h and equals the flow area divided by the perimeter of the solid boundary in contact with the fluid. This perimeter is usually called the "wetted" perimeter. The hydraulic radius of a pipe flowing full is $(\pi D^2/4)/\pi D = D/4$. Values for other configurations are given in Table 3.3.6. Integration of the equation of motion for an incompressible fluid results in

$$\frac{p_1}{\gamma} + \frac{V_1^2}{2g} + z_1 = \frac{p_2}{\gamma} + \frac{V_2^2}{2g} + z_2 + h_{1f2}$$

Each term of the equation is in feet and is equivalent to the height the fluid would rise in a tube if its energy were converted into potential energy. For this reason, in hydraulic practice, each type of energy is referred to as a **head.** The **static pressure head** is p/γ. The **velocity head** is $V^2/2g$, and the **potential head** is z. The energy loss between sections h_{1f2} is called the **lost head** or **friction head.** The **energy grade line** at any point $\Sigma(p/\gamma + V^2/2g + z)$, and the **hydraulic grade line** is $\Sigma(p/\gamma + z)$ as shown in Fig. 3.3.14.

EXAMPLE. A 12-in pipe (11.938 in inside diameter) reduces to a 6-in pipe (6.065 in inside diameter). Benzene at 68°F (20°C) flows steadily through this system. At section 1, the 12-in pipe centerline is 10 ft above the datum, and at section 2, the 6-in pipe centerline is 15 ft above the datum. The pressure at section 1 is 20 lbf/in² and the velocity is 4 ft/s. If the head loss due to friction is 0.05 $V_2^2/2g$, compute the pressure at section 2. Assume $g = g_c$, $\gamma = \rho g = 1.705 \times 32.17 = 54.85$ lbf/ft³. From the continuity equation,

$$M = \rho_1 A_1 V_1 = \rho_2 A_2 V_2 \qquad (p_1 = p_2)$$
$$V_2 = V_1(A_1/A_2) = V_1(\pi D_1^2/4)/(\pi D_2^2/4) = V_1(D_1/D_2)^2$$
$$V_2 = 4(11.938/6.065)^2 = 15.50 \text{ ft/s}$$

From the equation of motion,

$$\frac{p_2}{\gamma} = \frac{p_1}{\gamma} + \frac{V_1^2}{2g} + z_1 - \left(\frac{V_2^2}{2g} + z_2 + h_{1f2}\right)$$
$$\frac{p_2}{\gamma} = \frac{p_1}{\gamma} + \frac{V_1^2 - V_2^2 - 0.05V_2^2}{2g} + z_1 - z_2$$

Table 3.3.6   Values of Flow Area A and Hydraulic Radius R_h for Various Cross Sections

Cross section	Condition			Equations
	Flowing full		$h/D = 1$	$A = \pi D^2/4 \quad R_h = D/4$
	Upper half partly full		$0.5 < h/D < 1$	$\cos(\theta/2) = (2h/D - 1)$ $A = [\pi(360 - \theta) + 180 \sin\theta](D^2/1{,}440)$ $R_h = [1 + (180 \sin\theta)/(\pi\theta)](D/4)$
			$h/D = 0.8128$	$A = 0.6839 D^2 \; R_h \max = 0.3043 D$
	Lower half partly full		$h/D = 0.5$	$A = \pi D^2/8 \quad R_h \max = h/2$
			$0 < h/D < 0.5$	$\cos(\theta/2) = (1 - 2h/D)$ $A = (\pi\theta - 180 \sin\theta)(D^2/1{,}440)$ $R_h = [1 - (180 \sin\theta)/(\pi\theta)](D/4)$
	Flowing full		$h/D = 1$	$A = bD \quad R_h = bD/2(b + D)$
			Square $b = D$	$A = D^2 \quad R_h = D/4$
	Partly full		$h/D < 1$ $h/b = 0.5$ $b \to \infty, h \to 0$	$A = bh \quad R_h = bh/(2h + b)$ $A = b^2/2 \quad R_h \max = h/2$ $R_h \to h$ (wide shallow stream)
	$\alpha \neq \beta$			$R_h \max = h/2$ $A = [b + 1/2h(\cot\alpha + \cot\beta)]h$ $R_h = A/[b + h(\csc\alpha + \csc\beta)]$
	$\alpha = \beta$	$\dfrac{h}{a} = \dfrac{1}{2}$	$\alpha = 26°34'$	$A = (b + 2h)h$ $R_h = (b + 2h)h/(b + 4.472h)$
		$\dfrac{h}{a} = \dfrac{\sqrt{3}}{3}$	$\alpha = 30°$	$A = (b + 1.732h)h$ $R_h = (b + 1.732h)h/(b + 4h)$
		$\dfrac{h}{a} = \dfrac{2}{3}$	$\alpha = 33°41'$	$A = (b + 1.5h)h$ $R_h = (b + 1.5h)h/(b + 3.606h)$
		$\dfrac{h}{a} = 1$	$\alpha = 45°$	$A = (b + h)h$ $R_h = (b + h)h/(b + 2.828h)$
		$\dfrac{h}{a} = \dfrac{3}{2}$	$\alpha = 56°19'$	$A = (b + 0.6667h)h$ $R_h = (b + 0.6667h)h/(b + 2.404h)$
		$\dfrac{h}{a} = \sqrt{3}$	$\alpha = 60°$	$A = (b + 0.5774h)h$ $R_h = (b + 0.5774h)h/(b + 2.309h)$
	$\theta = $ any angle			$A = \tan(\theta/2)h^2 \quad R_h = \sin(\theta/2)h/2$
	$\theta = 30$			$A = 0.2679h^2 \quad R_h = 0.1294h$
	$\theta = 45$			$A = 0.4142h^2 \quad R_h = 0.1913h$
	$\theta = 60$			$A = 0.5774h^2 \quad R_h = 0.2500h$
	$\theta = 90$			$A = h^2 \quad R_h = 0.3536h$

$$\frac{p_2}{\gamma} = \frac{144 \times 20}{54.85} + \frac{4^2 - 1.05(15.50)^2}{2 \times 32.17} + 10 - 15$$

$$\frac{p^2}{\gamma} = 43.83 \text{ ft}$$

$$p_2 = \frac{54.88 \times 43.83}{144} = 16.70 \text{ lbf/in}^2 \; (1.151 \times 10^5 \text{ N/m}^2)$$

Energy Equation   Application of the principles of conservation of energy to a control volume for one-dimensional flow results in the following for steady flow:

$$J \, dq = dW + \frac{V \, dV}{g_c} + \frac{g}{g_c} dz + J \, du + d(pv)$$

where J is the mechanical equivalent of heat, 778.169 ft · lbf/Btu; q is the heat added, Btu/lbm (2,326 J/kg); W is the steady-flow shaft work

done *by* the fluid; and u is the internal energy. Btu/lbm (2,326 J/kg). If the energy equation is integrated for an incompressible fluid,

$$J_1 q_2 = {}_1 W_2 + \frac{V_2^2 - V_1^2}{2g_c} + \frac{g}{g_c}(z_2 - z_1) + J(u_2 - u_1) + v(p_2 - p_1)$$

The equation of motion does not consider thermal energy or steady-flow work; the energy equation has no terms for friction. Subtracting the differential equation of motion from the energy equation and solving for friction results in

$$dh_f = (dW + J \, du + p \, dv - J \, dq)(g_c/g)$$

Integrating for an incompressible fluid ($dv = 0$),

$$h_{1f2} = [{}_1 W_2 + J(u_2 - u_1) - J_1 q_2](g_c/g)$$

In the absence of steady-flow work in the system, the effect of friction is to increase the internal energy and/or to transfer heat from the system.

For steady frictionless, incompressible flow, both the equation of motion and the energy equation reduce to

$$\frac{p_1}{\gamma} + \frac{V_1^2}{2g} + z_1 = \frac{p_2}{\gamma} + \frac{V_2^2}{2g} + z_2$$

which is known as the **Bernoulli equation.**

Fig. 3.3.14   Energy relations.

Area-Velocity Relations   The continuity equation may be written as $\log_e \dot{M} = \log_e V + \log_e A + \log_e \rho$, which when differentiated becomes

$$\frac{dA}{A} = -\frac{dV}{V} - \frac{d\rho}{\rho}$$

For incompressible fluids, $d\rho = 0$, so

$$\frac{dA}{A} = -\frac{dV}{V}$$

Examination of this equation indicates
1. If the area increases, the velocity decreases.
2. If the area is constant, the velocity is constant.
3. There are no critical values.

For the frictionless flow of compressible fluids, it can be demonstrated that

$$\frac{dA}{A} = -\frac{dV}{V}\left[1 - \left(\frac{V}{c}\right)^2\right]$$

Analysis of the above equation indicates:
1. *Subsonic velocity $V < c$.* If the area increases, the velocity decreases. Same as for incompressible flow.
2. *Sonic velocity $V = c$.* Sonic velocity can exist only where the change in area is zero, i.e., at the *end* of a convergent passage or at the *exit* of a constant-area duct.
3. *Supersonic velocity $V > c$.* If area increases, the velocity increases, the reverse of incompressible flow. Also, supersonic velocity can exist only in the **expanding** portion of a passage **after** a constriction where **sonic** velocity existed.

Frictionless adiabatic compressible flow of an ideal gas in a horizontal passage must satisfy the following requirements:

1. *Conservation of mass.* As expressed by the continuity equation $M = \rho_1 A_1 V_1 = \rho_2 A_2 V_2$.
2. *Conservation of energy.* As expressed by the energy equation

$$\frac{V^2}{2g_c} + Ju + pv = \frac{V_1^2}{2g_c} + Ju_1 + p_1v_1 = \frac{V_2^2}{2g_c} + Ju_2 + p_2v_2$$

3. *Process relationship.* For an ideal gas undergoing a frictionless adiabatic (isentropic) process,

$$pv^k = p_1v_1^k = p_2v_2^k$$

4. *Ideal-gas law.* The equation of state for an ideal gas

$$pv = RT$$

In an expanding supersonic flow, a **compression shock wave** will be formed if the requirements for the conservation of mass and energy are not satisfied. This type of wave is associated with large and sudden rises in pressure, density, temperature, and entropy. The shock wave is so thin that for computation purposes it may be considered as a single line. For compressible flow of gases and vapors in passages, refer to Sec. 4.1; for steam-turbine passages, Sec. 9.4; for compressible flow around immersed objects, see Sec. 11.4.

The **impulse-momentum equation** is an application of the principle of conservation of momentum and is derived from Newton's second law. It is used to calculate the forces exerted on a solid boundary by a moving stream. Because velocity and force have both magnitude and direction, they are vectors. The impulse-momentum equation may be written for all three directions:

$$\Sigma F_x = \dot{M}(V_{x2} - V_{x1})$$
$$\Sigma F_y = \dot{M}(V_{y2} - V_{y1})$$
$$\Sigma F_z = \dot{M}(V_{z2} - V_{z1})$$

Figure 3.3.15 shows a free-body diagram of a control volume. The pressure forces shown are those imposed by the boundaries on the fluid and on the atmosphere. The reactive force R is that imposed by the downstream boundary on the fluid for equilibrium. Application of the impulse-momentum equation yields

$$\Sigma F = (F_{p1} + F_{a2}) - (F_{a1} + F_{p2} + R) = \dot{M}(V_2 - V_1)$$

Solving for R,

$$R = (p_1 - p_a)A_1 - (p_2 - p_a)A_2 = \dot{M}(V_2 = V_1)$$

The impulse-momentum equation is often used in conjunction with the continuity and energy equations to solve engineering problems. Because of the wide variety of possible applications, some examples are given to illustrate the methods of attack.

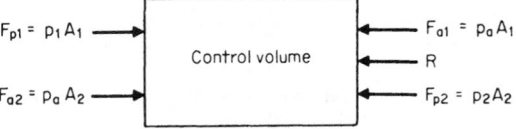

Fig. 3.3.15   Notation for impulse momentum.

EXAMPLE.   **Compressible Fluid in a Duct.** Nitrogen flows steadily through a 6-in (5.761 in inside diameter) straight, horizontal pipe at a mass rate of 25 lbm/s. At section 1, the pressure is 120 lbf/in^2 and the temperature is 100°F. At section 2, the pressure is 80 lbf/in^2 and the temperature is 110°F. Find the friction force opposing the motion. From the equation of state,

$$v = RT/p$$
$$v_1 = 55.16(459.7 + 100)/(144 \times 120) = 1.787 \text{ ft}^3/\text{lbm}$$
$$v_2 = 55.16(459.7 + 110)/(144 \times 80) = 2.728 \text{ ft}^3/\text{lbm}$$

Flow area of pipe $\times \pi D^2/4 = \pi(5.761/12)^2/4 = 0.1810 \text{ ft}^2$

From the continuity equation,

$$v = \dot{m}V/A$$
$$V_1 = (25 \times 1.787)/0.1810 = 246.8 \text{ ft/s}$$
$$V_2 = (25 \times 2.728)/0.1810 = 376.8 \text{ ft/s}$$

Applying the free-body equation for impulse momentum ($A = A_1 = A_2$),

$$R = (p_1 - p_a)A_1 - (p_2 - p_a)A_2 - \dot{M}(V_2 - V_1)$$
$$= (p_1 - p_2)A - M(V_2 - V_1) = 144(120 - 80)\,0.1810$$
$$- (25/32.17)(376.8 - 246.8) = 941.5\ \text{lbf}\ (4.188 \times 10^3\ \text{N})$$

EXAMPLE.   **Water Flow through a Nozzle.** Water at 68°F (20°C) flows through a horizontal 12- by 6-in-diameter nozzle discharging into the atmosphere. The pressure at the nozzle inlet is 65 lbf/in² and barometric pressure is 14.7 lbf/in². Determine the force exerted by the water on the nozzle.

$$A = \pi D^2/4$$
$$A_1 = \pi(12/12)^2/4 = 0.7854\ \text{ft}^2$$
$$A_2 = \pi(6/12)^2/4 = 0.1963\ \text{ft}^2$$
$$\gamma = \rho g = 1.937 \times 32.17 = 62.31\ \text{lb/ft}^3$$

From the continuity equation $\rho_1 A_1 V_1 = \rho_2 A_2 V_2$ for $\rho_1 = \rho_2$, $V_2 = V_1 A_1/A_2 = (0.7854/0.1963)V_1 = 4V_1$. From Bernoulli's equation ($z_1 = z_2$).

$$p_1/\gamma + V_1^2/2g = p_2/\gamma + V_2^2/2g = p_2/\gamma + (4V_1)^2/2g$$

or $V_1 = \sqrt{2g(p_1 - p_2)/15\gamma}$

$$= \sqrt{2 \times 32.17 \times 144(65 - 14.7)/15 \times 62.31} = 22.33\ \text{ft/s}$$
$$V_2 = 4 \times 22.33 = 89.32\ \text{ft/s}$$

Again from the equation of continuity

$$\dot{M} = \rho_1 A_1 V_1 = 1.937 \times 0.7854 \times 22.33 = 33.97\ \text{slugs/s}$$

Applying the free-body equation for impulse momentum,

$$R = (p_1 - p_a)A_1 - (p_2 - p_a)A_2 - \dot{M}(V_2 - V_1)$$
$$= 144(65 - 14.7)0.7854 - 144(14.7 - 14.7)0.1963$$
$$= (33.97)(89.32 - 22.33)$$
$$= 3,413\ \text{lbf}\ (1.518 \times 10^4\ \text{N})$$

EXAMPLE.   **Incompressible Flow through a Reducing Bend.** Carbon tetrachloride flows steadily without friction at 68°F (20°C) through a 90° horizontal reducing bend. The mass flow rate is 4 slugs/s, the inlet diameter is 6 in, and the outlet diameter is 3 in. The inlet pressure is 50 lbf/in² and the barometric pressure is 14.7 lbf/in². Compute the magnitude and direction of the force required to "anchor" this bend.

$$A = \pi D^2/4$$
$$A_1 = (\pi/4)(6/12)^2 = 0.1963\ \text{ft}^2$$
$$A_2 = (\pi/4)(3/12)^2 = 0.04909\ \text{ft}^2$$

From continuity,

$$V = M/\rho A$$
$$V_1 = 4/(3.093)(0.1963) = 6.588\ \text{ft/s}$$
$$V_2 = 4/(3.093)(0.04909) = 26.35\ \text{ft/s}$$

From the Bernoulli equation ($z_1 = z_2$),

$$\frac{p_2}{\gamma} = \frac{p_1}{\gamma} + \frac{V_1^2}{2g} - \frac{V_2^2}{2g} = \frac{144 \times 50}{3.093 \times 32.17} + \frac{(6.588)^2 - (26.35)^2}{2 \times 32.17} = 62.24\ \text{ft}$$
$$p_2 = \frac{(3.093 \times 32.17)(62.24)}{144} = 43.01\ \text{lbf/in}^2$$

From Fig. 3.3.16,

$$\Sigma F_x = (p_1 - p_a)A_1 - (p_2 - p_a)A_2 \cos \alpha - R_x$$
$$= M(V_2 \cos \alpha - V_1)$$

or $R_x = (p_1 - p_a)A_1 - (p_2 - p_a)A_2 \cos \alpha - \dot{M}(V_2 \cos \alpha - V_1)$
$$\Sigma F_y = 0 - (p_2 - p_a)A_2 \sin \alpha + R_y = \dot{M}(V_2 \sin \alpha - 0)$$

or $R_y = (p_2 - p_a)A_2 \sin \alpha + MV_2 \sin \alpha$
$$R_x = 144(50 - 14.7)0.1963 - 144(43.01 - 14.7)(\cos 90°)$$
$$- 4(26.35 \cos 90° - 6.588)$$
$$= 1,024\ \text{lbf}$$
$$R_y = 144(43.01 - 14.7)(0.04909)\sin 90° + 4(26.35)(\sin 90°)$$
$$= 305.5\ \text{lbf}$$
$$R = \sqrt{R_x^2 + R_y^2} = \sqrt{(1,024)^2 + (305.5)^2}$$
$$= 1,068\ \text{lbf}\ (4.753 \times 10^3\ \text{N})$$
$$\theta = \tan^{-1}(F_y/F_x) = \tan^{-1}(305.5/1,024)$$
$$= 16°37'$$

Forces on Blades and Deflectors   The forces imposed on a fluid jet whose velocity is V_J by a blade moving at a speed of V_b away from the jet are shown in Fig. 3.3.17. The following equations were developed from the application of the impulse-momentum equation for an open jet ($p_2 = p_1$) and for **frictionless** flow:

$$F_x = \rho A_J(V_J - V_b)^2(1 - \cos \alpha)$$
$$F_y = \rho A_J(V_J - V_b)^2 \sin \alpha$$
$$F = 2\rho A_J(V_J - V_b)^2 \sin(\alpha/2)$$

EXAMPLE.   In the nozzle-blade system of Fig. 3.3.17, water at 68°F (20°C) enters a 3- by 1½-in-diameter horizontal nozzle with a pressure of 23 lbf/in² and discharges at 14.7 lbf/in² (atmospheric pressure). The blade moves away from the nozzle at a velocity of 10 ft/s and deflects the stream through an angle of 80°. For

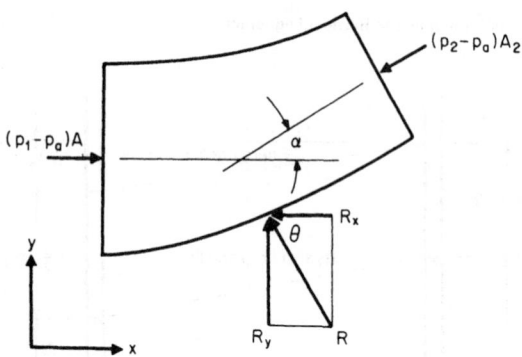

Fig. 3.3.16   Forces on a bend.

frictionless flow, calculate the total force exerted by the jet on the blade. Assume $g = g_c$; then $\gamma = \rho g$. From the continuity equation ($\rho_I = \rho_J$), $\rho_I A_I V_I = \rho_J A_J V_J$, $V_I = (A_J/A_I)V_J$,

$$V_I = \frac{\pi D_J^2/4}{\pi D_I^2/4}V_J = \left(\frac{D_J}{D_I}\right)^2 V_J$$
$$V_I = (1.5/3)^2 V_J = V_J/4$$

From the Bernoulli equation ($z_2 = z_1$),

$$\frac{V_J^2}{2g} = \frac{V_I^2}{2g} + \frac{p_I - p_J}{\rho g}$$
$$\frac{V_J^2 - (V_J/4)^2}{2g} = \frac{(p_I - p_J)}{\rho g}$$
$$V_J = \sqrt{\frac{2(16/15)(p_I - p_J)}{\rho}}$$
$$= \sqrt{\frac{2 \times (16/15)\,144\,(23 - 14.7)}{1.937}} = 36.28\ \text{ft/s}$$

The total force $F = 2\rho A_J(V_J - V_b)^2 \sin(\alpha/2)$

$$F = 2 \times 1.937(\pi/4)(1.5/12)^2(36.28 - 10)^2 \sin(80/2)$$
$$= 21.11\ \text{lbf}\ (93.90\ \text{N})$$

Fig. 3.3.17   Notation for blade study.

Impulse Turbine   In a turbine, the total of the separate forces acting simultaneously on each blade equals that caused by the combined mass flow rate $\dot{M}$ discharged by the nozzle or

$$\Sigma P = \Sigma F_x V_b = \dot{M}(V_J - V_b)(1 - \cos \alpha)V_b$$

The maximum value of power P is found by differentiating P with respect to V_b and setting the result equal to zero. Solving for V_b yields $V_b = V_J/2$, so that maximum power occurs when the velocity of the jet is equal to twice the velocity of the blade. Examination of the power equation also indicates that the angle α for a maximum power results when $\cos \theta = -1$ or $\alpha = 180°$. For theoretical maximum power of a blade, $2V_b = V_J$ and $\alpha = 180°$. It should be noted that in any practical impulse-turbine application, α cannot be $180°$ because the discharge interferes with the next set of blades. Substituting $V_b = V_J/2$, $\alpha = 180°$ in the power equation,

$$\Sigma P_{max} = \dot{M}(V_J - V_J/2)[1 - (-1)]V_J/2 = MV_J^2/2 = \dot{m}V_J^2/2g_c$$

or the maximum power per unit mass is equal to the total power of the jet. Application of the Bernoulli equation between the surface of a reservoir and the discharge of the turbine shows that $\Sigma P_{max} = \dot{M}\sqrt{2g(z_2 - z_1)}$. For design details, see Sec. 9.9.

Flow in a Curved Path   When a fluid flows through a bend, it is also rotated around an axis and the energy required to produce rotation must be supplied from the energy already in the fluid mass. This fluid rotation is called a **free vortex** because it is free of outside energy. Consider the fluid mass $\rho(r_o - r_i)\,dA$ of Fig. 3.3.18 being rotated as it flows through a bend of outer radius r_o, inner radius r_i, with a velocity of V. Application of Newton's second law to this mass results in

$$dF = p_o\,dA - p_i\,dA = [\rho(r_o - r_i)\,dA][V^2/(r_o + r_i)/2]$$

which reduces to

$$p_o - p_i = 2(r_o - r_i)\rho V^2/(r_o + r_i)$$

Because of the difference in fluid pressure between the inner and outer walls of the bend, secondary flows are set up, and this is the primary cause of friction loss of bends. These secondary flows set up turbulence that require 50 or more straight pipe diameters downstream to dissipate.

Fig. 3.3.18   Notation for flow in a curved path.

Thus this loss does not take place in the bend, but in the downstream system. These losses may be reduced by the use of splitter plates which help minimize the secondary flows by reducing $r_o - r_i$ and hence $p_o - p_i$.

EXAMPLE.   104°F (40°C) benzene flows at a rate of 8 ft³/s in a square horizontal duct. This duct makes a 90° turn with an inner radius of 1 ft and an outer radius of 2 ft. Calculate the difference between the walls of this bend. The area of this duct is $(r_o - r_i)^2 = (2 - 1)^2 = 1$ ft². From the continuity equation $V = Q/A = 8/1 = 8$ ft/s. The pressure difference

$$p_o - p_i = 2(r_o - r_i)\rho V^2/(r_o + r_i)$$
$$= 2(2 - 1)1.663\,(8)^2/(2 + 1) = 70.95 \text{ lbf/ft}^2$$
$$= 70.95/144 = 0.4927 \text{ lbf/in}^2\,(3.397 \times 10^3 \text{ N/m}^2)$$

DIMENSIONLESS PARAMETERS

Modern engineering practice is based on a combination of theoretical analysis and experimental data. Often the engineer is faced with the necessity of obtaining practical results in situations where for various reasons, physical phenomena cannot be described mathematically and experimental data must be considered. The generation and use of **dimensionless parameters** provides a powerful and useful tool in (1) reducing the number of variables required for an experimental program,

(2) establishing the principles of model design and testing, (3) developing equations, and (4) converting data from one system of units to another. Dimensionless parameters may be generated from (1) **physical equations**, (2) the principles of **similarity**, and (3) **dimensional analysis.** All **physical equations** must be dimensionally correct so that a dimensionless parameter may be generated by simply dividing one side of the equation by the other. A minimum of two dimensionless parameters will be formed, one being the inverse of the other.

EXAMPLE.   It is desired to generate a series of dimensionless parameters to describe the ratios of static pressure head, velocity head, and potential head to total head for frictionless incompressible flow. From the Bernoulli equation,

$$\frac{p}{\gamma} + \frac{V^2}{2g} + z = \Sigma h = \text{total head}$$

$$N_1 = \frac{p/\gamma + V^2/2g + z}{\Sigma h} = \frac{p/\gamma}{\Sigma h} + \frac{V^2/2g}{\Sigma h} + \frac{z}{\Sigma h} = N_p + N_V + N_z$$

or

$$N_2 = \frac{\Sigma h}{p_2/\gamma + V^2/2g + z} = N_1^{-1}$$

N_1 and N_2 are total energy ratios and N_p, N_V, and N_z are the ratios of the static pressure head, velocity head, and potential head, respectively, to the total head.

Models versus Prototypes   There are times when for economic or other reasons it is desirable to determine the performance of a structure or machine by testing another structure or machine. This type of testing is called model testing. The equipment being tested is called a **model,** and the equipment whose performance is to be predicted is called a **prototype.** A **model** may be smaller than, the same size as, or larger than the **prototype.** Model experiments on aircraft, rockets, missiles, pipes, ships, canals, turbines, pumps, and other structures and machines have resulted in savings that more than justified the expenditure of funds for the design, construction, and testing of the model. In some situations, the model and the prototype may be the same piece of equipment, for example, the laboratory calibration of a flowmeter with water to predict its performance with other fluids. Many manufacturers of fluid machinery have test facilities that are limited to one or two fluids and are forced to test with what they have available in order to predict performance with nonavailable fluids. For towing-tank testing of ship models and for wind-tunnel testing of aircraft and aircraft-component models, see Secs. 11.4 and 11.5.

Similarity Requirements   For complete similarity between a model and its prototype, it is necessary to have **geometric, kinematic,** and **dynamic** similarity. **Geometric similarity** exists between model and prototype when the **ratios** of all corresponding dimensions of the model and prototype are equal. These ratios may be written as follows:

Length:
$$L_{model}/L_{prototype} = L_{ratio}$$
$$= L_m/L_p = L_r$$

Area:
$$L_{model}^2/L_{prototype}^2 = L_{ratio}^2$$
$$= L_m^2/L_p^2 = L_r^2$$

Volume:
$$L_{model}^3/L_{prototype}^3 = L_{ratio}^3$$
$$= L_m^3/L_p^3 = L_r^3$$

Kinematic similarity exists between model and prototype when their streamlines are geometrically similar. The kinematic ratios resulting from this condition are

Acceleration:
$$a_r = a_m/a_p = L_mT_m^{-2}/L_pT_p^{-2}$$
$$= L_r/T_r^{-2}$$

Velocity:
$$V_r = V_m/V_p = L_mT_m^{-1}/L_pT_p^{-1}$$
$$= L_r/T_r^{-1}$$

Volume flow rate:
$$Q_r = Q_m/Q_p = L_m^3T_m^{-1}/L_p^3T_p^{-1}$$
$$= L_r^3/T_r^{-1}$$

Dynamic similarity exists between model and prototype having geometric and kinematic similarity when the ratios of all forces are the same. Consider the model/prototype relations for the flow around the object shown in Fig. 3.3.19. For **geometric similarity** $D_m/D_p = L_m/L_p = L_r$ and for **kinematic similarity** $U_{Am}/U_{Ap} = U_{Bm}/U_{Bp} = V_r = L_rT_r^{-1}$.

Next consider the three forces acting on point C of Fig. 3.3.19 without specifying their nature. From the geometric similarity of their vector polygons and Newton's law, for **dynamic similarity** $F_{1m}/F_{1p} = F_{2m}/F_{2p} = F_{3m}/F_{3p} = M_m a_{Cm}/M_p a_{Cp} = F_r$. For dynamic similarity, these force ratios must be maintained on all corresponding fluid particles

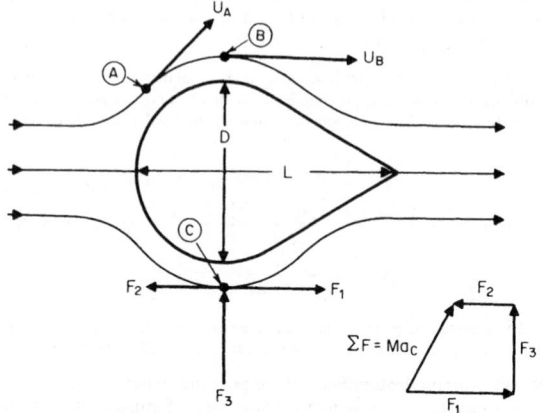

Fig. 3.3.19 Notation for dynamic similarity.

throughout the flow pattern. From the force polygon of Fig. 3.3.19, it is evident that $F_1 + \to F_2 + \to F_3 = Ma_C$. For total model/prototype force ratio, comparisons of force polygons yield

$$F_r = \frac{F_{1m} + \to F_{2m} + \to F_{3m}}{F_{1p} + \to F_{2p} + \to F_{3p}} = \frac{M_m a_{Cm}}{M_p a_{Cp}}$$

Fluid Forces The fluid forces that are considered here are those acting on a fluid element whose mass $= \rho L^3$, area $= L^2$, length $= L$, and velocity $= L/T$.

Inertia force

$$F_i = \text{(mass)(acceleration)}$$
$$= (\rho L^3)(L/T^2) = \rho L(L^2/T^2)$$
$$= \rho L^2 V^2$$

Viscous force

$$F_\mu = \text{(viscous shear stress)(shear area)}$$
$$= \tau L^2 = \mu(dU/dy)L^2 = \mu(V/L)L^2$$
$$= \mu L V$$

Gravity force

$$F_g = \text{(mass)(acceleration due to gravity)}$$
$$= (\rho L^3)(g) = \rho L^3 g$$

Pressure force

$$F_p = \text{(pressure)(area)} = pL^2$$

Centrifugal force

$$F_\omega = \text{(mass)(acceleration)}$$
$$= (\rho L^3)(L/T^2)$$
$$= (\rho L^3)(L\omega^2) = \rho L^4 \omega^2$$

Elastic force

$$F_E = \text{(modulus of elasticity)(area)}$$
$$= EL^2$$

Surface-tension force

$$F_\sigma = \text{(surface tension)(length)} = \sigma L$$

Vibratory force

$$F_f = \text{(mass)(acceleration)}$$
$$= (\rho L^3)(L/T^2)$$
$$= (\rho L^4)(T^{-2}) = \rho L^4 f^2$$

If all fluid forces were acting on a fluid element,

$$F_r = \frac{F_{\mu m} + \to F_{gm} + \to F_{pm} + \to F_{\omega m} + \to F_{Em} + \to F_{\sigma m} + \to F_{fm}}{F_{\mu p} + \to F_{gp} + \to F_{pp} + \to F_{\omega p} + \to F_{Ep} + \to F_{\sigma p} + \to F_{fp}}$$
$$= \frac{F_{im}}{F_{ip}}$$

Examination of the above equation and the force polygon of Fig. 3.3.19 lead to the conclusion that dynamic similarity can be characterized by an equality of force ratios one less than the total number involved. Any force ratio may be eliminated, depending upon the quantities which are desired. Fortunately, in most practical engineering problems, not all of the eight forces are involved because some may not be acting, may be of negligible magnitude, or may be in opposition to each other in such a way as to compensate. In each application of similarity, a good understanding of the fluid phenomena involved is necessary to eliminate the irrelevant, trivial, or compensating forces. When the flow phenomenon is too complex to be readily analyzed, or is not known, only experimental verification with the prototype of results from a model test will determine what forces should be considered in future model testing.

Standard Numbers With eight fluid forces that can act in flow situations, the number of dimensionless parameters that can be formed from

Table 3.3.7 Standard Numbers

Force ratio	Equations	Result	Conventional practice		
			Form	Symbol	Name
$\dfrac{\text{Inertia}}{\text{Viscous}}$	$\dfrac{F_i}{F_\mu} = \dfrac{\rho L^2 V^2}{\mu L V}$	$\dfrac{\rho L V}{\mu}$	$\dfrac{\rho L V}{\mu}$	**R**	Reynolds
$\dfrac{\text{Inertia}}{\text{Gravity}}$	$\dfrac{F_i}{F_g} = \dfrac{\rho L^2 V^2}{\rho L^3 g}$	$\dfrac{V^2}{Lg}$	$\dfrac{V}{\sqrt{Lg}}$	**F**	Froude
$\dfrac{\text{Inertia}}{\text{Pressure}}$	$\dfrac{F_i}{F_p} = \dfrac{\rho L^2 V^2}{\rho L^2}$	$\dfrac{\rho V^2}{p}$	$\dfrac{\rho V^2}{p}$	**E**	Euler
			$\dfrac{2\Delta p}{\rho V^2}$	**Cp**	Pressure coefficient
$\dfrac{\text{Inertia}}{\text{Centrifugal}}$	$\dfrac{F_i}{F_\omega} = \dfrac{\rho L^2 V^2}{\rho L^4 \omega^2}$	$\dfrac{V^2}{L^2 \omega^2}$	$\dfrac{V}{DN}$	**V**	Velocity ratio
$\dfrac{\text{Inertia}}{\text{Elastic}}$	$\dfrac{F_i}{F_E} = \dfrac{\rho L^2 V^2}{EL^2}$	$\dfrac{\rho V^2}{E}$	$\dfrac{\rho V^2}{E}$	**C**	Cauchy
			$\dfrac{V}{\sqrt{E/\rho}}$	**M**	Mach
$\dfrac{\text{Inertia}}{\text{Surface tension}}$	$\dfrac{F_i}{F_\sigma} = \dfrac{\rho L^2 V^2}{\sigma L}$	$\dfrac{\rho L V^2}{\sigma}$	$\dfrac{\rho L V^2}{\sigma}$	**W**	Weber
$\dfrac{\text{Inertia}}{\text{Vibration}}$	$\dfrac{F_i}{F_f} = \dfrac{\rho L^2 V^2}{\rho L^4 f^2}$	$\dfrac{V^2}{L^2 f^2}$	$\dfrac{Lf}{V}$	**S**	Strouhal

SOURCE: Computed from data given in Murdock, "Fluid Mechanics and Its Applications," Houghton Mifflin, 1976.

their ratios is 56. However, conventional practice is to ratio the inertia force to the other fluid forces, usually by division because the inertia force is the vector sum of all the other forces involved in a given flow situation. Results obtained by dividing the inertia force by each of the other forces are shown in Table 3.3.7 compared with the standard numbers that are used in conventional practice.

DYNAMIC SIMILARITY

Vibration   In the flow of fluids around objects and in the motion of bodies immersed in fluids, **vibration** may occur because of the formation of a wake caused by alternate shedding of eddies in a periodic fashion or by the vibration of the object or the body. The **Strouhal number S** is the ratio of the velocity of vibration Lf to the velocity of the fluid V. Since the vibration may be fluid-induced or structure-induced, two frequencies must be considered, the wake frequency $f\omega$ and the natural frequency of the structure f_n. Fluid-induced forces are usually of small magnitude, but as the wake frequency approaches the natural frequency of the structure, the vibratory forces increase very rapidly. When $f\omega = f_n$, the structure will go into resonance and fail. This imposes on the model designer the requirement of matching to scale the natural-frequency characteristics of the prototype. This subject is treated later under Wake Frequency. All further discussions of model/prototype relations are made under the assumption that either vibratory forces are absent or they are taken care of in the design of the model or in the test program.

Incompressible Flow   Considered in this category are the flow of fluids around an object, motion of bodies immersed in incompressible fluids, and the flow of incompressible fluids in conduits. It includes, for example, a submarine traveling under water but not partly submerged, and liquids flowing in pipes and passages when the liquid completely fills them, but not when partly full as in open-channel flow. It also includes aircraft moving in atmospheres that may be considered incompressible. Incompressible flow in rotating machinery is considered separately.

In these situations the gravity force, although acting on all fluid particles, does not affect the flow pattern. Excluding rotating machinery, centrifugal forces are absent. By definition of an incompressible fluid, elastic forces are zero, and since there is no liquid-gas interface, surface-tension forces are absent.

The only forces now remaining for consideration are the inertia, viscous, and pressure. Using standard numbers, the parameters are **Reynolds number** and **pressure coefficient**. The Reynolds number may be converted into a kinematic ratio by noting that by definition $v = \mu/\rho$ and substituting in $\mathbf{R} = \rho LV/\mu = LV/v$. In this form, Reynolds number is the ratio of the fluid velocity V and the "shear velocity" v/L. For this reason, Reynolds number is used to characterize the velocity profile. Forces and pressure losses are then determined by the pressure coefficient.

EXAMPLE.   A submarine is to move submerged through 32°F (0°C) seawater at a speed of 10 knots. (1) At what speed should a 1:20 model be towed in 68°F (20°C) fresh water? (2) If the thrust on the model is found to be 42,500 lbf, what horsepower will be required to propel the submarine?

1. Speed of model for Reynolds-number similarity

$$R_m = R_p = \left(\frac{\rho VL}{\mu}\right)_m = \left(\frac{\rho VL}{\mu}\right)_p$$

$$V_m = V_p(\rho_p/\rho_m)(L_p/L_m)(\mu_m/\mu_p)$$
$$V_m = (10)(1.995/1.937)(20/1)(20.92 \times 10^{-6}/39.40 \times 10^{-6})$$
$$= 109.4 \text{ knots } (56.27 \text{ m/s})$$

2. Prototype horsepower

$$\mathbf{C}_{pp} = \mathbf{C}_{pm} = \left(\frac{2\Delta p}{\rho V^2}\right)_p = \left(\frac{2\Delta p}{\rho V^2}\right)_m$$

$$F = \Delta pL^2, \Delta p = \frac{F}{L^2}, \text{ so that } \left(\frac{2F}{\rho V^2 L^2}\right)_p = \left(\frac{2F}{\rho V^2 L^2}\right)_m$$
$$F_p = F_m(\rho_p/\rho_m)(V_p/V_m)^2(L_p/L_m)^2$$
$$= 42,500(1.995/1.937)(10/109.4)^2(20/1)^2 = 146,300 \text{ lbf}$$

$$P = FV = \left(\frac{146,300}{550}\right)\left(\frac{10 \times 6,076}{3,600}\right)$$
$$= 4,490 \text{ hp } (3.35 \times 10^6 \text{ W})$$

Compressible Flow   Considered in this category are the flow of compressible fluids under the conditions specified for incompressible flow in the preceding paragraphs. In addition to the forces involved in incompressible flow, the elastic force must be added. Conventional practice is to use the square root of the inertia/elastic force ratio or **Mach number.**

Mach number is the ratio of the fluid velocity to its speed of sound and may be written $\mathbf{M} = V/c = V\sqrt{E_s/\rho}$. For an ideal gas, $\mathbf{M} = V/\sqrt{kg_cRT}$. In compressible-flow problems, practice is to use the Mach number to characterize the velocity or kinematic similarity, the Reynolds number for dynamic similarity, and the pressure coefficient for force or pressure-loss determination.

EXAMPLE.   An airplane is to fly at 500 mi/h in an atmosphere whose temperature is 32°F (0°C) and pressure is 12 lbf/in². A 1:20 model is tested in a wind tunnel where a supply of air at 392°F (200°C) and variable pressure is available. At (1) what speed and (2) what pressure should the model be tested for dynamic similarity?

1. Speed for Mach-number similarity

$$\mathbf{M}_m = \mathbf{M}_p = \left(\frac{V}{\sqrt{E/\rho}}\right)_m = \left(\frac{V}{\sqrt{E/\rho}}\right)_p = \left(\frac{V}{\sqrt{kg_cRT}}\right)_m = \left(\frac{V}{\sqrt{kg_cRT}}\right)_p$$
$$V_m = V_p(k_m/k_p)^{1/2}(R_m/R_p)^{1/2}(T_m/T_p)^{1/2}$$

For the same gas $k_m = k_p$, $R_m = R_p$, and

$$V_m = V_p\sqrt{T_m/T_p} = 500\sqrt{(851.7/491.7)} = 658.1 \text{ mi/h}$$

2. Pressure for Reynolds-number similarity

$$\mathbf{R}_m = \mathbf{R}_p = \left(\frac{\rho VL}{\mu}\right)_m = \left(\frac{\rho VL}{\mu}\right)_p$$

$$\rho_m = \rho_p(V_p/V_m)(L_p/L_m)(\mu_m/\mu_p)$$

Since $\rho = p/g_cRT$

$$\left(\frac{p}{g_cRT}\right)_m = \left(\frac{p}{g_cRT}\right)_p\left(\frac{V_p}{V_m}\right)\left(\frac{L_p}{L_m}\right)\left(\frac{\mu_m}{\mu_p}\right)$$

$$p_m = p_p(T_m/T_p)(V_p/V_m)(L_p/L_m)(\mu_m/\mu_p)$$
$$p_m = 12(851.7/491.7)(500/658.1)(20/1)(53.15 \times 10^{-6}/35.67 \times 10^{-6})$$
$$p_m = 470.6 \text{ lbf/in}^2 (3.245 \times 10^6 \text{ N/m}^2)$$

For information about wind-tunnel testing and its limitations, refer to Sec. 11.4.

Centrifugal Machinery   This category includes the flow of fluids in such centrifugal machinery as compressors, fans, and pumps. In addition to the inertia, pressure, viscous, and elastic forces, centrifugal forces must now be considered. Since centrifugal force is really a special case of the inertia force, their ratio as shown in Table 3.3.7 is **velocity ratio** and is the ratio of the fluid velocity to the machine tangential velocity. In model/prototype relations for centrifugal machinery, DN (D = diameter, ft, N = rotational speed) is substituted for the velocity V, and D for L, which results in the following:

$$\mathbf{M} = DN/\sqrt{kg_cRT} \qquad \mathbf{R} = \rho D^2N/\mu \qquad \mathbf{C}_p = 2\Delta p/\rho D^2N^2$$

EXAMPLE.   A centrifugal compressor operating at 100 r/min is to compress methane delivered to it at 50 lbf/in² and 68°F (20°C). It is proposed to test this compressor with air from a source at 140°F (60°C) and 100 lbf/in². Determine compressor speed and inlet-air pressure required for dynamic similarity. Find speed for Mach-number similarity:

$$\mathbf{M}_m = \mathbf{M}_p = (DN/\sqrt{kg_cRT})_m = (DN/\sqrt{kg_cRT})_p$$
$$N_m = N_p(D_p/D_m)\sqrt{(k_m/k_p)(R_m/R_p)(T_m/T_p)}$$
$$= 100(1)\sqrt{(1.40/1.32)(53.34/96.33)(599.7/527.7)}$$
$$= 81.70 \text{ r/min}$$

Find pressure for Reynolds-number similarity:

$$\mathbf{R}_m = \mathbf{R}_p = \left(\frac{\rho D^2N}{\mu}\right)_m = \left(\frac{\rho D^2N}{\mu}\right)_p$$

For an ideal gas $\rho = p/g_c RT$, so that

$$(pD^2N/g_c RT\mu)_m = (pD^2N/g_c RT\mu)_p$$

$$\begin{aligned}p_m &= p_p(D_p/D_m)^2(N_p/N_m)(R_m/R_p)(T_m/T_p)(\mu_m/\mu_p)\\ &= 50(1)^2(100/81.70)(53.34/96.33)(599.7/527.7) \times (41.79\\ &\quad \times 10^{-8}/22.70 \times 10^{-8}) = 70.90 \text{ lbf/in}^2(4.888 \times 10^5 \text{ N/m}^2)\end{aligned}$$

See Sec. 14 for specific information on pump and compressor similarity.

Liquid Surfaces   Considered in this category are ships, seaplanes during takeoff, submarines partly submerged, piers, dams, rivers, open-channel flow, spillways, harbors, etc. Resistance at liquid surfaces is due to surface tension and wave action. Since wave action is due to gravity, the gravity force and surface-tension force are now added to the forces that were considered in the last paragraph. These are expressed as the square root of the inertia/gravity force ratio or **Froude number** $\mathbf{F} = V/\sqrt{Lg}$ and as the inertia/surface tension force ratio or **Weber number** $\mathbf{W} = \rho LV^2/\sigma$. On the other hand, elastic and pressure forces are now absent. Surface tension is a minor property in fluid mechanics and it normally exerts a negligible effect on wave formation *except* when the waves are small, say less than 1 in. Thus the effects of surface tension on the model might be considerable, but negligible on the prototype. This type of "scale effect" must be avoided. For accurate results, the inertia/surface tension force ratio or Weber number should be considered. It is never possible to have complete dynamic similarity of liquid surfaces unless the model and prototype are the same size, as shown in the following example.

EXAMPLE.   An ocean vessel 500 ft long is to travel at a speed of 15 knots. A 1:25 model of this ship is to be tested in a towing tank at design temperature. Determine the model speed required for (1) wave-resistance similarity, (2) viscous or skin-friction similarity, (3) surface-tension similarity, and (4) the model size required for complete dynamic similarity.

1. *Speed for **Froude-number** similarity*

or
$$\mathbf{F}_m = \mathbf{F}_p = (V/\sqrt{Lg})_m = (V/\sqrt{Lg})_p$$
$$V_m = V_p\sqrt{L_m/L_p} = 15\sqrt{1/25} = 3 \text{ knots}$$

2. *Speed for **Reynolds-number** similarity*

$$\mathbf{R}_m = \mathbf{R}_p = (\rho LV/\mu)_m = (\rho LV/\mu)_p$$
$$V_m = V_p(\rho_p/\rho_m)(L_p/L_m)(\mu_m/\mu_p)$$
$$V_m = 15(1)(25/1)(1) = 375 \text{ knots}$$

3. *Speed for **Weber-number** similarity*

$$\mathbf{W}_m = \mathbf{W}_p = (\rho LV^2/\sigma)_m = (\rho LV^2/\sigma)_p$$
$$V_m = V_p\sqrt{(\rho_p/\rho_m)(L_p/L_m)(\sigma_m/\sigma_p)}$$
$$V_m = 15\sqrt{(1)(25)(1)} = 75 \text{ knots}$$

4. *Model size for complete similarity.* First try Reynolds and Froude similarity; let

$$V_m = V_p(\rho_p/\rho_m)(L_p/L_m)(\mu_m/\mu_p) = V_p\sqrt{L_m/L_p}$$

which reduces to

$$L_m/L_p = (\rho_p/\rho_m)^{2/3}(\mu_m/\mu_p)^{2/3}$$

Next try Weber and Froude similarity; let

$$V_m = V_p\sqrt{(\rho_p/\rho_m)(L_p/L_m)(\sigma_m/\sigma_p)} = V_p\sqrt{L_m/L_p}$$

which reduces to

$$L_m/L_p = (\rho_p/\rho_m)^{1/2}(\sigma_m/\sigma_p)^{1/2}$$

For the same fluid at the same temperature, either of the above solves for $L_m = L_p$, or the model must be the same size as the prototype. For use of different fluids and/or the same fluid at different temperatures.

$$L_m/L_p = (\rho_p/\rho_m)^{2/3}(\mu_m/\mu_p)^{2/3} = (\rho_p/\rho_m)^{1/2}(\sigma_m/\sigma_p)^{1/2}$$

which reduces to

$$(\mu^4/\rho\sigma^3)_m = (\mu^4/\rho\sigma^3)_p$$

No practical way has been found to model for complete similarity. Marine engineering practice is to model for wave resistance and correct for skin-friction resistance. See Sec. 11.3.

DIMENSIONAL ANALYSIS

Dimensional analysis is the mathematics of dimensions and quantities and provides procedural techniques whereby the variables that are assumed to be significant in a problem can be formed into dimensionless parameters, the number of parameters being less than the number of variables. This is a great advantage, because fewer experimental runs are then required to establish a relationship between the parameters than between the variables. While the user is not presumed to have any knowledge of the fundamental physical equations, the more knowledgeable the user, the better the results. If any significant variable or variables are omitted, the relationship obtained from dimensional analysis will not apply to the physical problem. On the other hand, inclusion of all possible variables will result in losing the principal advantage of dimensional analysis, i.e., the reduction of the amount of experimental data required to establish relationships. Two formal methods of dimensional analysis are used, the **method of Lord Rayleigh** and **Buckingham's Π theorem.**

Dimensions used in mechanics are mass M, length L, time T, and force F. Corresponding **units** for these dimensions are the slug (kilogram), the foot (metre), the second (second), and the pound force (newton). Any system in mechanics can be defined by three fundamental dimensions. Two systems are used, the force (FLT) and the mass (MLT). In the force system, mass is a derived quantity and in the mass system, force is a derived quantity. Force and mass are related by Newton's law: $F = MLT^{-2}$ and $M = FL^{-1}T^2$. Table 3.3.8 shows common variables and their dimensions and units.

Lord Rayleigh's method uses algebra to determine interrelationships among variables. While this method may be used for any number of variables, it becomes relatively complex and is not generally used for more than four. This method is most easily described by example.

EXAMPLE.   In laminar flow, the unit shear stress τ is some function of the fluid dynamic viscosity μ, the velocity difference dU between adjacent laminae separated by the distance dy. Develop a relationship.

1. Write a functional relationship of the variables:

$$\tau = f(\mu, dU, \mathbf{dy})$$

Assume $\tau = K(\mu^a\, dU^b\, dy^c)$.

2. Write a dimensional equation in either FLT or MLT system:

$$(FL^{-2}) = K(FL^{-2}T)^a(LT^{-1})^b(L)^c$$

3. Solve the dimensional equation for exponents:

		τ	μ	dU	dy
Force	F	$1 =$	$a + 0 + 0$		
Length	L	$-2 =$	$-2a + b + c$		
Time	T	$0 =$	$a - b + 0$		

Solution: $a = 1, b = 1, c = -1$

4. Insert exponents in the functional equation: $\tau = K(\mu^a\, dU^b\, dy^c) = K(\mu^1\, du^1\, dy^{-1})$, or $K = (\mu\, dU/\tau\, dy)$. This was based on the assumption of $\tau = K(\mu^a\, dU^b\, dy^c)$. The general relationship is $K = f(\mu\, dU/\tau\, dy)$. The functional relationship cannot be obtained from dimensional analysis. Only physical analysis and/or experiments can determine this. From both physical analysis and experimental data,

$$\tau = \mu\, dU/dy$$

The Buckingham Π theorem serves the same purpose as the method of Lord Rayleigh for deriving equations expressing one variable in terms of its dependent variables. The Π theorem is preferred when the number of variables exceeds four. Application of the Π theorem results in the formation of dimensionless parameters called π ratios. These π ratios have no relation to 3.14159.... The Π theorem will be illustrated in the following example.

EXAMPLE.   Experiments are to be conducted with gas bubbles rising in a still liquid. Consider a gas bubble of diameter D rising in a liquid whose density is ρ, surface tension σ, viscosity μ, rising with a velocity of V in a gravitational field of g. Find a set of parameters for organizing experimental results.

1. List all the physical variables considered according to type: geometric, kinematic, or dynamic.

Table 3.3.8 Dimensions and Units of Common Variables

Symbol	Variable	Dimensions MLT	FLT	Units USCS*	SI
		Geometric			
L	Length	L		ft	m
A	Area	L^2		ft^2	m^2
V	Volume	L^3		ft^3	m^3
		Kinematic			
t	Time	T		s	s
ω	Angular velocity	T^{-1}		s^{-1}	s^{-1}
f	Frequency				
V	Velocity	LT^{-1}		ft/s	m/s
ν	Kinematic viscosity	L^2T^{-1}		ft^2/s	m^2/s
Q	Volume flow rate	L^3T^{-1}		ft^3/s	m^3/s
α	Angular acceleration	T^{-2}		s^{-2}	s^2
a	Acceleration	LT^{-2}		ft/s^{-2}	m/s^2
		Dynamic			
ρ	Density	ML^{-3}	$FL^{-4}T^2$	slug/ft^3	kg/m^3
M	Mass	M	$FL^{-1}T^2$	slug	kg
I	Moment of inertia	ML^2	FLT^2	slug $\cdot$ ft^3	kg $\cdot$ m^2
μ	Dynamic viscosity	$ML^{-1}T^{-1}$	$FL^{-2}T$	slug/ft $\cdot$ s	kg/m $\cdot$ s
M	Mass flow rate	MT^{-1}	$FL^{-1}T^{-1}$	slug/s	kg/s
MV	Momentum	MLT^{-1}	FT	lbf $\cdot$ s	N $\cdot$ s
Ft	Impulse				
$M\omega$	Angular momentum	ML^2T^{-1}	FLT	slug $\cdot$ ft^2/s	kg $\cdot$ m^2/s
γ	Specific weight	$ML^{-2}T^{-2}$	FL^{-3}	lbf/ft^3	N/m^3
p	Pressure				
τ	Unit shear stress	$ML^{-1}T^{-2}$	FL^{-2}	lbf/ft^2	N/m^2
E	Modulus of elasticity				
σ	Surface tension	MT^{-2}	FL^{-1}	lbf/ft	N/m
F	Force	MLT^{-2}	F	lbf	N
E	Energy				
W	Work	ML^2T^{-2}	FL	lbf $\cdot$ ft	J
FL	Torque				
P	Power	ML^2T^{-3}	FLT^{-1}	lbf $\cdot$ ft/s	W
v	Specific volume	$M^{-1}L^3$	$F^{-1}L^4T^{-2}$	ft^3/lbm	m^3/kg

*United States Customary System.

2. Choose either the *FLT* or *MLT* system of dimensions.
3. Select a "basic group" of variables characteristic of the flow as follows:

  a. B_G, a geometric variable
  b. B_K, a kinematic variable
  c. B_D, a dynamic variable (if three dimensions are used)

4. Assign *A* numbers to the remaining variables starting with A_1.

Type	Symbol	Description	Dimensions	Number
Geometric	D	Bubble diameter	L	B_G
Kinematic	V	Bubble velocity	LT^{-1}	B_K
	g	Acceleration of gravity	LT^{-2}	A_1
Dynamic	ρ	Liquid density	ML^{-3}	B_D
	σ	Surface tension	MT^{-2}	A_2
	μ	Liquid viscosity	$ML^{-1}T^{-1}$	A_3

5. Write the basic equation for each π ratio as follows:

$$\pi_1 = (B_G)^{x1}(B_K)^{y1}(B_D)^{z1}(A_1)$$
$$\pi_2 = (B_G)^{x2}(B_K)^{y2}(B_D)^{z2}(A_2) \ldots \pi_n = (B_G)^{xn}(B_K)^{yn}(B_D)^{zn}(A_n)$$

Note that the number of π ratios is equal to the number of *A* numbers and thus equal to the number of variables less the number of fundamental dimensions in a problem.

6. Write the dimensional equations and use the algebraic method to determine the value of exponents *x*, *y*, and *z* for each π ratio. Note that for all π ratios, the sum of the exponents of a given dimension is zero.

$$\pi_1 = (B_G)^{x1}(B_K)^{y1}(B_D)^{z1}(A_1) = (D)^{x1}(V)^{y1}(\rho)^{z1}(g)$$
$$(M^0L^0T^0) = (L^{x1})(L^{y1}T^{-y1})(M^{z1}L^{-3z1})(LT^{-2})$$

Solution: $x_1 = 1, y_1 = -2, z_1 = 0$

$$\pi_1 = D^1V^{-2}\rho^0g = Dg/V^2$$
$$\pi_2 = (B_G)^{x2}(B_K)^{y2}(B_D)^{z2}(A_2) = (D)^{x2}(V)^{y2}(\rho)^{z2}(\sigma)$$
$$(M^0L^0T^0) = (L^{x2})(L^{y2}T^{-y2})(M^{z2}L^{-3z2})(MT^{-2})$$

Solution: $x_2 = -1, y_2 = -2, z_2 = -1$

$$\pi_2 = D^{-1}V^{-2}\rho^{-1}\sigma = \sigma/DV^2\rho$$
$$\pi_3 = (B_G)^{x3}(B_K)^{y3}(B_D)^{z3}(A_3) = (D)^{x3}(V)^{y3}(\rho)^{z3}(\mu)$$
$$(M^0L^0T^0) = (L^{x3})(L^{y3}T^{-y3})(M^{z3}L^{-3z3})(ML^{-1}T^{-1})$$

Solution: $x_3 = -1, y_3 = -1, z_3 = -1$

$$\pi_3 = D^{-1}V^{-1}\rho^{-1}\mu = \mu/DV\rho$$

7. Convert π ratios to conventional practice. One statement of the Buckingham II theorem is that any π ratio may be taken as a function of all the others, or $f(\pi_1, \pi_2, \pi_3, \ldots, \pi_n) = 0$. This equation is mathematical shorthand for a functional statement. It could be written, for example, as $\pi_2 = f(\pi_1, \pi_3, \ldots, \pi_n)$. This equation states that π_2 is some function of π_1 and π_3 through π_n but is not a statement of *what* function π_2 is of the other π ratios. This can be determined only by physical and/or experimental analysis. Thus we are free to substitute *any* function in the equation; for example, π_1 may be replaced with $2\pi_1^{-1}$ or π_n with $a\pi_n^b$.

The procedures set forth in this example are designed to produce π ratios containing the same terms as those resulting from the application of the principles of similarity so that the physical significance may be understood. However, any other combinations might have been used. The only real requirement for a "basic group" is that it contain the same number of terms as there are dimensions in a problem and that each of these dimensions be represented in it.

The π ratios derived for this example may be converted into conventional practice as follows:

$$\pi_1 = Dg/V^2$$

is recognized as the inverse of the square root of the Froude number $\mathbf{F}$

$$\pi_2 = \sigma/DV^2\rho$$

is the inverse of the Weber number $\mathbf{W}$

$$\pi = \mu/DV\rho$$

is the inverse of the Reynolds number $\mathbf{R}$

Let $\pi_1 = f(\pi_2, \pi_3)$
Then $V = K(Dg)^{1 \div 2}$
where $K = f(\mathbf{W}, \mathbf{R})$

This agrees with the results of the dynamic-similarity analysis of liquid surfaces. This also permits a reduction in the experimental program from variations of six variables to three dimensionless parameters.

FORCES OF IMMERSED OBJECTS

Drag and Lift   When a fluid impinges on an object as shown in Fig. 3.3.20, the undisturbed fluid pressure p and the velocity V change. Writing Bernoulli's equation for two points on the surface of the object, the point S being the most forward point and point A being any other point, we have, for horizontal flow,

$$p + \rho V^2/2 = p_S + \rho V_S^2/2 = p_A + \rho V_A^2/2$$

At point S, $V_S = 0$, so that $p_S = p + \rho V^2/2$. This is called the **stagnation point**, and p_S is the **stagnation pressure**. Since point A is any other point, the result of the fluid impingement is to create a pressure $p_A = p + \rho(V^2 - V_A^2)/2$ acting normal to every point on the surface of the object.

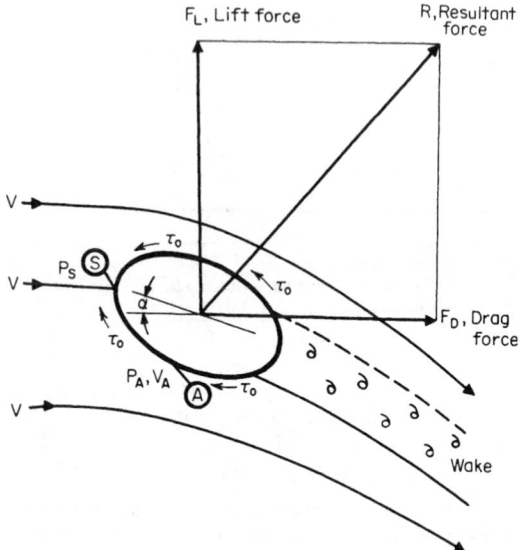

Fig. 3.3.20   Notation for drag and lift.

In addition, a frictional force $F_f = \tau_0 A_s$ tangential to the surface area A_s opposes the motion. The sum of these forces gives the resultant force R acting on the body. The resultant force R is resolved into the **drag** component F_D parallel to the flow and **lift** component F_L perpendicular to the fluid motion. Depending upon the shape of the object, a wake may be formed which sheds eddies with a frequency of f. The angle α is called the angle of attack. (See Secs. 11.4 and 11.5.)

From dimensional analysis or dynamic similarity,

$$f(\mathbf{C}_p, \mathbf{R}, \mathbf{M}, \mathbf{S}) = 0$$

The formation of a wake depends upon the Reynolds number, or $\mathbf{S} = f(\mathbf{R})$. This reduces the functional relation to $f(\mathbf{C}_p, \mathbf{R}, \mathbf{M}) = 0$. Since the drag

and lift forces may be considered independently,

$$F_D = C_D\rho V^2(A)/2$$

where $C_D = f(\mathbf{R}, \mathbf{M})$, and A = characteristic area.

$$F_L = C_L\rho V^2(A)/2$$

where $C_L = f(\mathbf{R}, \mathbf{M})$.

It is evident from Fig. 3.3.20 that C_D and C_L are also functions of the angle of attack. Since the drag force arises from two sources, the pressure or shape drag F_p and the skin-friction drag F_f due to wall shear stress τ_0, the drag coefficient is made up of two parts:

$$F_D = F_p + F_f = C_D\rho AV^2/2 = C_p\rho AV^2/2 + C_f\rho A_s V^2/2$$

or $$C_D = C_p + C_f A_s/A$$

where C_p is the coefficient of pressure, C_f the skin-friction coefficient, and A_s the characteristic area for shear.

Skin-Friction Drag   Figure 3.3.21 shows a fluid approaching a smooth flat plate with a uniform velocity profile of V. As the fluid passes over the plate, the velocity at the plate surface is zero and increases to V at some distance δ from the surface. The region in which the velocity varies from 0 to V is called the **boundary layer**. For some

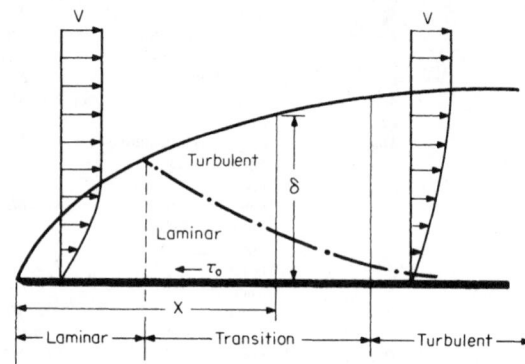

Fig. 3.3.21   Boundary layer along a smooth flat plate.

distance along the plate, the flow within the boundary layer is laminar, with viscous forces predominating, but in the transition zone as the inertia forces become larger, a turbulent layer begins to form and increases as the laminar layer decreases.

Boundary-layer thickness and skin-friction drag for incompressible flow over smooth flat plates may be calculated from the following equations, where $\mathbf{R}_x = \rho VX/\mu$:

Laminar

$$\delta/X = 5.20\,\mathbf{R}_X^{-1/2} \qquad 0 < \mathbf{R}_X < 5 \times 10^5$$
$$C_f = 1.328\,\mathbf{R}_X^{-1/2} \qquad 0 < \mathbf{R}_X < 5 \times 10^5$$

Turbulent

$$\delta/X = 0.377\,\mathbf{R}_X^{-1/5} \qquad\qquad 5 \times 10^4 < \mathbf{R}_X < 10^6$$
$$\delta/X = 0.220\,\mathbf{R}_X^{-1/6} \qquad\qquad 10^6 < \mathbf{R}_X < 5 \times 10^8$$
$$C_f = 0.0735\,\mathbf{R}_X^{-1/5} \qquad\qquad 2 \times 10^5 < \mathbf{R}_X < 10^7$$
$$C_f = 0.455\,(\log_{10}\mathbf{R}_X)^{-2.58} \qquad 10^7 < \mathbf{R}_X < 10^8$$
$$C_f = 0.05863\,(\log_{10}C_f\mathbf{R}_X)^{-2} \qquad 10^8 < \mathbf{R}_x < 10^9$$

Transition   The Reynolds number at which the boundary layer changes depends upon the roughness of the plate and degree of turbulence. The generally accepted number is 500,000, but the transition can take place at Reynolds numbers higher or lower. (Refer to Secs. 11.4 and 11.5.) For transition at any Reynolds number $\mathbf{R}_X$,

$$C_f = 0.455\,(\log_{10}\mathbf{R}_X)^{-2.58} - (0.0735\,\mathbf{R}_t^{4/5} - 1.328\,8\,\mathbf{R}_t^{1/2})\,\mathbf{R}_X^{-1}$$

For $\mathbf{R}_t = 5 \times 10^5$, $C_f = 0.455\,(\log_{10}\mathbf{R}_X)^{-2.58} - 1{,}725\,\mathbf{R}_X^{-1}$.

Pressure Drag Experiments with sharp-edged objects placed perpendicular to the flow stream indicate that their drag coefficients are essentially constant at Reynolds numbers over 1,000. This means that the drag for $R_X > 10^3$ is pressure drag. Values of C_D for various shapes are given in Sec. 11 along with the effects of Mach number.

Wake Frequency An object in a fluid stream may be subject to the downstream periodic shedding of vortices from first one side and then the other. The frequency of the resulting transverse (lift) force is a function of the stream Strouhal number. As the wake frequency approaches the natural frequency of the structure, the periodic lift force increases asymptotically in magnitude, and when resonance occurs, the structure fails. Neglecting to take this phenomenon into account in design has been responsible for failures of electric transmission lines, submarine periscopes, smokestacks, bridges, and thermometer wells. The wake-frequency characteristics of **cylinders** are shown in Fig. 3.3.22. At a Reynolds number of about 20, vortices begin to shed alternately. Behind the cylinder is a staggered stable arrangement of vortices known as the "Kármán vortex trail." At a Reynolds number of about 10^5, the flow changes from laminar to turbulent. At the end of the transition zone ($\mathbf{R} \approx 3.5 \times 10^5$), the flow becomes turbulent, the alternate

Fig. 3.3.22 Flow around a cylinder. (*From Murdock, "Fluid Mechanics and Its Applications," Houghton Mifflin, 1976.*)

shedding stops, and the wake is aperiodic. At the end of the supercritical zone ($\mathbf{R} \approx 3.5 \times 10^6$), the wake continues to be turbulent, but the shedding again becomes alternate and periodic.

The alternating **lift force** is given by

$$F_L(t) = C_L \rho V^2 A \sin (2\pi f t)/2$$

where t is the time. For an analysis of this force in the subcritical zone, see Belvins (Murdock, "Fluid Mechanics and Its Applications," Houghton Mifflin, 1976). For design of steel stacks, Staley and Graven (ASME 72PET/30) recommend $C_L = 0.8$ for $10^4 < \mathbf{R} < 10^5$, $C_L = 2.8 - 0.4 \log_{10} \mathbf{R}$ for $\mathbf{R} = 10^5$ to 10^6, and $C_L = 0.4$ for $10^6 < \mathbf{R} < 10^7$.

The **Strouhal number** is nearly constant to $\mathbf{R} = 10^5$, and a nominal design value of 0.2 is generally used. Above $\mathbf{R} = 10^5$, data from different experimenters vary widely, as indicated by the crosshatched zone of

Fig. 3.3.22. This wide zone is due to experimental and/or measurement difficulties and the dependence on surface roughness to "'trigger" the boundary layer. Examination of Fig. 3.3.22 indicates an inverse relation of Strouhal number to drag coefficient.

Observation of actual structures shows that they vibrate at their natural frequency and with a mode shape associated with their fundamental (first) mode during vortex excitation. Based on observations of actual stacks and wind-tunnel tests, Staley and Graven recommend a constant Strouhal number of 0.2 for all ranges of Reynolds number. The ASME recommends $\mathbf{S} = 0.22$ for thermowell design ("Temperature Measurement," PTC 19.3). Until such time as the value of the Strouhal number above $\mathbf{R} = 10^5$ has been firmly established, designers of structures in this area should proceed with caution.

FLOW IN PIPES

Parameters for Pipe Flow The forces acting on a fluid flowing through and completely filling a horizontal pipe are inertia, viscous, pressure, and elastic. If the surface roughness of the pipe is ε, either similarity or dimensional analysis leads to $\mathbf{C}_p = f(\mathbf{R}, \mathbf{M}, L/D, \varepsilon/D)$, which may be written for incompressible fluids as $\Delta p = \mathbf{C}_p V^2/2 = K\rho V^2/2$, where K is the **resistance coefficient** and ε/D the **relative roughness** of the pipe surface, and the resistance coefficient $K = f(\mathbf{R}, L/D, \varepsilon/D)$. The pressure loss may be converted to the terms of **lost head**: $h_f = \Delta p/\gamma = KV^2/2g$. Conventional practice is to use the **friction factor** f, defined as $f = KD/L$ or $h_f = KV^2/2g = (fL/D)V^2/2g$, where $f = f(\mathbf{R}, \varepsilon/D)$. When a fluid flows into a pipe, the boundary layer starts at the entrance, as shown in Fig. 3.3.23, and grows continuously until it fills the pipe. From the equation of motion $dh_f = \tau \, dL/\gamma R_h$ and for circular ducts $R_h = D/4$. Comparing wall shear stress τ_0 with friction factor results in the following: $\tau_0 = f\rho V^2/8$.

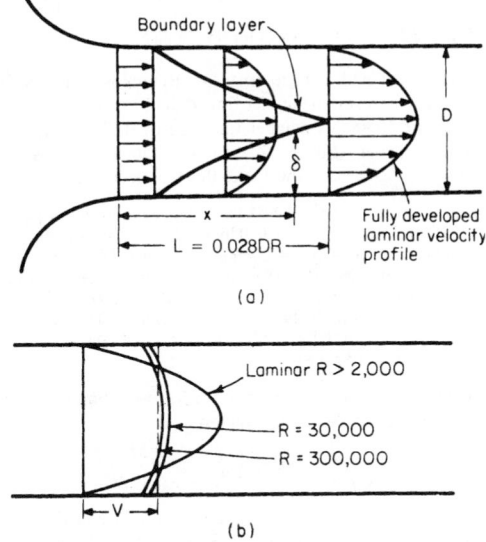

Fig. 3.3.23 Velocity profiles in pipes.

Laminar Flow In this type of flow, the resistance is due to viscous forces only so that it is independent of the pipe surface roughness, or $\tau_0 = \mu \, dU/dy$. Application of this equation to the equation of motion and the friction factor yields $f = 64/\mathbf{R}$. Experiments show that it is possible to maintain laminar flow to very high Reynolds numbers if care is taken to increase the flow gradually, but normally the slightest disturbance will destroy the laminar boundary layer if the value of Reynolds number is greater than 4,000. In a like manner, flow initially turbulent

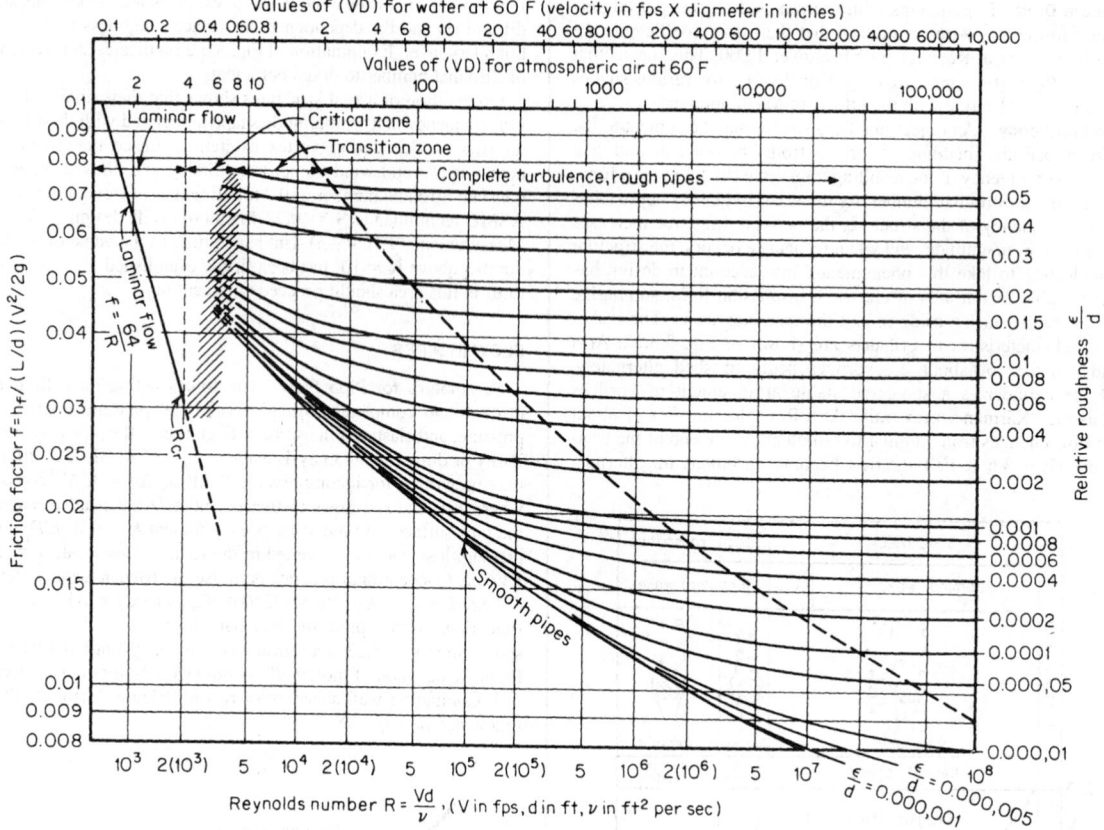

Fig. 3.3.24 Friction factors for flow in pipes.

can be maintained with care to very low Reynolds numbers, but the slightest upset will result in laminar flow if the Reynolds number is less than 2,000. The Reynolds-number range between 2,000 and 4,000 is called the **critical zone** (Fig. 3.3.24). Flow in the zone is **unstable**, and designers of piping systems must take this into account.

EXAMPLE. Glycerin at 68°F (20°C) flows through a horizontal pipe 1 in in diameter and 20 ft long at a rate of 0.090 lbm/s. What is the pressure loss? From the continuity equation $V = Q/A = (m/\rho g)/(\pi D^2/4) = [0.090/(2.447 \times 32.17)]/[(\pi/4)(1/12)^2] = 0.2096$ ft/s. The Reynolds number $\mathbf{R} = \rho VD/\mu = (2.447)(0.2096)(1/12)/(29,500 \times 10^{-6}) = 1.449$. $\mathbf{R} < 2,000$; therefore, flow is laminar and $f = 64/\mathbf{R} = 64/1.449 = 44.17$. $K = fL/D = 44.17 \times 20(1/12) = 10,600$. $\Delta p = K\rho V^2/2 = 10,600 \times 2.447 (0.2096)^2/2 = 569.8$ lbf/ft^2 = 569.8/144 = 3.957 lbf/in^2 (2.728×10^4 N/m^2).

Turbulent Flow   The friction factor for Reynolds number over 4,000 is computed using the Colebrook equation:

$$\frac{1}{\sqrt{f}} = -2 \log_{10}\left(\frac{\varepsilon/D}{3.7} + \frac{2.51}{\mathbf{R}\sqrt{f}}\right)$$

Figure 3.3.24 is a graphical presentation of this equation (Moody, *Trans. ASME*, 1944, pp. 671–684). Examination of the Colebrook equation indicates that if the value of surface roughness ε is small compared with the pipe diameter ($\varepsilon/D \to 0$), the friction factor is a function of Reynolds number only. A **smooth pipe** is one in which the ratio $(\varepsilon/D)/3.7$ is small compared with $2.51/\mathbf{R}\sqrt{f}$. On the other hand, as the Reynolds number increases so that $2.51/\mathbf{R}\sqrt{f} \to 0$, the friction factor becomes a function of relative roughness only and the pipe is called a **rough pipe**. Thus the same pipe may be smooth under one flow condition, and rough under another. The reason for this is that as the Reynolds number increases, the thickness of the laminar sublayer decreases as shown in Fig. 3.3.21, exposing the surface roughness to flow. Values of absolute roughness ε are given in Table 3.3.9. The variation of friction

Table 3.3.9   Values of Absolute Roughness, New Clean Commercial Pipes

Type of pipe or tubing	ε ft (0.3048 m) $\times 10^{-6}$ Range		Design	Probable max variation of f from design, %
Asphalted cast iron	400		400	−5 to + 5
Brass and copper	5		5	−5 to + 5
Concrete	1,000	10,000	4,000	−35 to 50
Cast iron	850		850	−10 to + 15
Galvanized iron	500		500	0 to + 10
Wrought iron	150		150	−5 to 10
Steel	150		150	−5 to 10
Riveted steel	3,000	30,000	6,000	−25 to 75
Wood stave	600	3,000	2,000	−35 to 20

SOURCE: Compiled from data given in "Pipe Friction Manual," Hydraulic Institute, 3d ed., 1961.

factor shown in Fig. 3.3.9 is for new, clean pipes. The change of friction factor with **age** depends upon the chemical properties of the fluid and the piping material. Published data for flow of water through wrought-iron or cast-iron pipes show as much as 20 percent increase after a few months to 500 percent after 20 years. When necessary to allow for service life, a study of specific conditions is recommended. The calculation of friction factor to four significant figures in the examples to follow is only for numerical comparison and should not be construed to mean accuracy.

Engineering Calculations   Engineering pipe computations usually fall into one of the following classes:
1. Determine pressure loss Δp when Q, L, and D are known.
2. Determine flow rate Q when L, D, and Δp are known.
3. Determine pipe diameter D when Q, L, and Δp are known.

Pressure-loss computations may be made to engineering accuracy using an expanded version of Fig. 3.3.24. Greater precision may be obtained by using a combination of Table 3.3.9 and the Colebrook equation, as will be shown in the example to follow. Flow rate may be determined by direct solution of the Colebrook equation. Computation of pipe diameter necessitates the trial-and-error method of solution.

EXAMPLE. *Case 1:* 2,000 gal/min of 68°F (20°C) water flow through 500 ft of cast-iron pipe having an internal diameter of 10 in. At point 1 the pressure is 10 lbf/in^2 and the elevation 150 ft, and at point 2 the elevation 100 ft. Find p_2.

From continuity $V = Q/A = [2,000 \times (231/1,728)/60]/[(\pi/4)(10/12)^2] = 8.170$ ft/s. Reynolds number $\mathbf{R} = \rho V D/\mu = (1.937)(8.170)(10/12)/(20.92 \times 10^{-6}) = 6.304 \times 10^5$. $\mathbf{R} > 4,000$ ∴ flow is turbulent. $\varepsilon/D = (850 \times 10^{-6})/(10/12) = 1.020 \times 10^{-3}$.

Determine f: from Fig. 3.3.24 by interpolation $f = 0.02$. Substituting this value on the right-hand side of the Colebrook equation,

$$\frac{1}{\sqrt{f}} = -2 \log_{10}\left(\frac{\varepsilon/D}{3.7} + \frac{2.51}{\mathbf{R}\sqrt{f}}\right)$$

$$= -2 \log_{10}\left[\frac{1.020 \times 10^{-3}}{3.7} + \frac{2.51}{(6.305 \times 10^5)\sqrt{0.02}}\right]$$

$$\frac{1}{\sqrt{f}} = 7.035 \qquad f = 0.02021$$

Resistance coefficient $K = \dfrac{fL}{D} = \dfrac{0.02021 \times 500}{10/12}$

$$K = 12.13$$
$$h_{1/2} = KV^2/2g = 12.13 \times (8.170)^2/2 \times 32.17$$
$$h_{1/2} = 12.58 \text{ ft}$$

Equation of motion: $p_1/\gamma + V_1^2/2g + z_1 = p_2/\gamma + V_2^2/2g + z_2 + h_{1/2}$. Noting that $V_1 = V_2 = V$ and solving for p_2,

$$p_2 = p_1 + \gamma(z_1 - z_2 - h_{1/2})$$
$$= 144 \times 10 + (1.937 \times 32.17)(150 - 100 - 12.58)$$
$$p_2 = 3,772 \text{ lbf/ft}^2 = 3,772/144 = 26.20 \text{ lbf/in}^2 \ (1.806 \times 10^5 \text{ N/m}^2)$$

EXAMPLE. *Case 2:* Gasoline (sp. gr. 0.68) at 68°F (20°C) flows through a 6-in schedule 40 (ID = 0.5054 ft) welded steel pipe with a head loss of 10 ft in 500 ft. Determine the flow. This problem may be solved directly by deriving equations that do not contain the flow rate Q.

From $h_f = \left(\dfrac{fL}{D}\right)\dfrac{V^2}{2g}$, $\qquad V = (2gh_fD)^{1/2}(fL)^{1/2}$

From $\mathbf{R} = \rho VD/\mu$, $\qquad V = \mathbf{R}\mu/\rho D$

Equating the above and solving,

$$\mathbf{R}\sqrt{f} = (\rho D/\mu)(2gh_fD/L)^{1/2}$$
$$= (1.310 \times 0.5054/5.98 \times 10^{-6})$$
$$\times (2 \times 32.17 \times 10 \times 0.5054/500)^{1/2} = 89,285$$

which is now in a form that may be used directly in the Colebrook equation:

$$\varepsilon/D = 150 \times 10^{-6}/0.5054 = 2.968 \times 10^{-4}$$

From the Colebrook equation,

$$\frac{1}{\sqrt{f}} = -2 \log_{10}\left(\frac{\varepsilon/D}{3.7} + \frac{2.51}{\mathbf{R}\sqrt{f}}\right)$$

$$= -2 \log_{10}\left(\frac{2.968 \times 10^{-4}}{3.7} + \frac{2.51}{89,285}\right)$$

$$\frac{1}{\sqrt{f}} = 7.931 \qquad f = 0.01590$$

$$\mathbf{R} = 89,285/\sqrt{f} = 89,285 \times 7.93 = 7.08 \times 10^5$$
$$\mathbf{R} > 4,000 \ \therefore \text{ flow is turbulent}$$
$$V = \mathbf{R}\mu/\rho D = (7.08 \times 10^5 \times 5.98$$
$$\times 10^{-6})/(1.310 \times 0.5054) = 6.396 \text{ ft/s}$$
$$Q = AV = (\pi/4)(0.5054)^2(6.396)$$
$$Q = 1.283 \text{ ft}^3/\text{s} \ (3.633 \times 10^{-2} \text{ m}^3/\text{s}^1)$$

EXAMPLE. *Case 3;* Water at 68°F (20°C) is to flow at a rate of 500 ft^3/s through a concrete pipe 5,000 ft long with a head loss not to exceed 50 ft. Determine the diameter of the pipe. This problem may be solved by trial and error using methods of the preceding example. First trial: Assume any diameter (say 1 ft).

$$\mathbf{R}\sqrt{f} = (\rho D/\mu)(2gh_fD/L)^{1/2}$$
$$= (1.937D/20.92 \times 10^{-6}) \times (2 \times 32.17 \times 50D/5,000)^{1/2}$$
$$= 74,269D^{3/2} = 74,269(1)^{3/2} = 74,269$$
$$\varepsilon/D_1 = 4,000 \times 10^{-6}/D = 4,000 \times 10^{-6}/(1) = 4,000 \times 10^{-6}$$

$$\frac{1}{\sqrt{f_1}} = -2 \log_{10}\left(\frac{\varepsilon/D_1}{3.7} + \frac{2.51}{\mathbf{R}\sqrt{f_1}}\right)$$

$$= -2 \log_{10}\left(\frac{4,000 \times 10^{-6}}{3.7} + \frac{2.51}{74,269}\right)$$

$$\frac{1}{\sqrt{f_1}} = 5.906 \qquad f_1 = 0.02867$$

$$\mathbf{R}_1 = 74,269/\sqrt{f_1} = 74,269 \times 5.906 = 438,600$$
$$V_1 = \mathbf{R}_1\mu/\rho D_1 = (438,600 \times 20.92 \times 10^{-6})/(1.937 \times 1)$$
$$V_1 = 4.737 \text{ ft/s}$$
$$Q_1 = A_1V_1 = [\pi(1)^2/4]4.737 = 3.720 \text{ ft}^3/\text{s}$$

For the same loss and friction factor,

$$D_2 = D_1(Q/Q_1)^{2/5} = (1)(500/3.720)^{2/5} = 7.102 \text{ ft}$$

For the second trial use $D_2 = 7.102$, which results in $Q = 502.2$ ft^3/s. Since the nearest standard size would be used, additional trials are unnecessary.

Velocity Profile Figure 3.3.23a shows the formation of a laminar velocity profile. As the fluid enters the pipe, the boundary layer starts at the entrance and grows continuously until it fills the pipe. The flow while the boundary is growing is called **generating flow.** When the boundary layer completely fills the pipe, the flow is called **established flow.** The distance required for establishing **laminar flow** is $L/D \approx$ 0.028 **R.** For **turbulent flow,** the distance is much shorter because of the turbulence and not dependent upon Reynolds number, L/D being from 25 to 50.

Examination of Fig. 3.3.23b indicates that as the Reynolds number increases, the velocity distribution becomes "flatter" and the flow approaches **one-dimensional.** The velocity profile for laminar flow is parabolic, $U/V = 2[1 - (r/r_o)^2]$ and for **turbulent flow,** logarithmic (except for the very thin laminar boundary layer), $U/V = 1 + 1.43\sqrt{f} + 2.15\sqrt{f}\log_{10}(1 - r/r_o)$. The use of the average velocity produces an error in the computation of kinetic energy. If α is the **kinetic-energy correction factor,** the true kinetic-energy change per unit mass between two points on a flow system $\Delta KE = \alpha_1 V_1^2/2g_c - \alpha_2 V_2^2/2g_c$, where $\alpha = (1/AV^3)\int U^3 dA$. For **laminar flow,** $\alpha = 2$ and for **turbulent flow,** $\alpha \approx 1 + 2.7f$. Of interest is the **pipe factor** V/U_{max}; for **laminar flow,** $V/U_{max} = 1/2$ and for **turbulent flow,** $V/U_{max} = 1 + 1.43\sqrt{f}$. The location at which the local velocity equals the average velocity for **laminar flow** is $U = V$ at $r/r_o = 0.7071$ and for **turbulent flow** is $U = V$ at $r/r_o = 0.7838$.

Compressible Flow At the present time, there are no true analytical solutions for the computation of actual characteristics of compressible fluids flowing in pipes. In the real flow of a compressible fluid in a pipe, the amount of heat transferred and its direction are dependent upon the amount of insulation, the temperature gradient between the fluid and ambient temperatures, and the heat-transfer coefficient. Each condition requires an individual application of the principles of thermodynamics and heat transfer for its solution.

Conventional engineering practice is to use one of the following methods for flow computation.

1. Assume **adiabatic** flow. This approximates the flow of compressible fluids in short, insulated pipelines.

2. Assume **isothermal** flow. This approximates the flow of gases in long, uninsulated pipelines where the fluid and ambient temperatures are nearly equal.

Adiabatic Flow If the Mach number is less than ¼, results within normal engineering-accuracy requirements may be obtained by considering the fluid to be incompressible. A detailed discussion of and methods for the solution of compressible adiabatic flow are beyond the scope of this section, and any standard gas-dynamics text should be consulted.

Isothermal Flow   The equation of motion for a horizontal piping system may be written as follows:

$$dp + \rho V\, dV + \gamma\, dh_f = 0$$

noting, from the continuity equation, that $\rho V = M/A = G$, where G is the **mass velocity** in slugs/(ft^2)(s), and that $\gamma\, dh_f = [(f/D)\rho V^2/2]\, dL = [(f/D)GV/2]\, dL$. Substituting in the above equation of motion and dividing by $GV/2$ results in

$$\frac{2\rho\, dp}{G^2} + \frac{2dV}{V} + \left(\frac{F}{D}\right) dL = 0$$

Integrating for an isothermal process ($p/\rho = C$) and assuming f is a constant,

$$\frac{\rho_1 p_1}{G^2}\left[\left(\frac{p_2}{p_1}\right)^2 - 1\right] + 2\log_e\left(\frac{V_2}{V_1}\right) + \frac{fL}{D} = 0$$

Noting that $A_1 = A_2$, $V_2/V_1 = \rho_1/\rho_2 = p_1/p_2$, and solving for G,

$$G = \left\{\frac{\rho_1 p_1[1 - (p_2/p_1)^2]}{2\log_e(p_1/p_2) + fL/D}\right\}^{1/2}$$

The Reynolds number may be written as

$$\mathbf{R} = \frac{\rho VD}{\mu} = \frac{GD}{\mu} \quad \text{and} \quad G = \mathbf{R}\frac{\mu}{D}$$

The value of $\mathbf{R}\sqrt{f}$ may be obtained from the simultaneous solution of the two equations for G, assuming that $2\log_e p_1/p_2$ is small compared with fL/D.

$$\mathbf{R}\sqrt{f} \approx \left\{\frac{D^3\rho_1 p_1}{\mu^2 L}\left[1 - \left(\frac{p_2}{p_1}\right)\right]\right\}^{1/2}$$

EXAMPLE.   Air at 68°F (20°C) is flowing isothermally through a horizontal straight standard 1-in steel pipe (inside diameter = 1.049 in). The pipe is 200 ft long, the pressure at the pipe inlet is 74.7 lbf/in^2, and the pressure drop through the pipe is 5 lbf/in^2. Find the flow rate in lbm/s. From the equation of state $\rho_1 = p/g_cRT = (144 \times 74.7)/(32.17 \times 53.34 \times 527.7) = 0.01188$ slugs/ft^3.

$$\mathbf{R}\sqrt{f} = \{[(D^3\rho_1 p_1/\mu^2 L)][1 - (p_2/p_1)^2]\}^{1/2} = \{[(1.049/12)^3(0.01188)$$
$$\times (144 \times 74.7)/(39.16 \times 10^{-8})(200)[1 - (69.7/74.7)^2]\}^{1/2}$$
$$= 18{,}977$$

For steel pipe $\varepsilon = 150 \times 10^{-6}$ ft, $\varepsilon/D = (150 \times 10^{-6})/(1.049/12) = 1.716 \times 10^{-3}$. From the Colebrook equation,

$$\frac{1}{\sqrt{f}} = -2\log_{10}\left(\frac{\varepsilon/D}{3.7} + \frac{2.51}{\mathbf{R}\sqrt{f}}\right)$$
$$= 2\log_{10}[(1.716 \times 10^{-3}/3.7) + (2.51)/(18{,}977)] = 6.449$$
$$f = 0.02404$$
$$\mathbf{R} = (\mathbf{R}\sqrt{f})(1/\sqrt{f}) = (18{,}953)(6.449) = 122{,}200$$
$$\mathbf{R} > 4{,}000 \therefore \text{ flow is turbulent}$$
$$G = \left\{\frac{\rho_1 p_1[1 - (p_2/p_1)^2]}{2\log_e(p_1/p_2) + fL/D}\right\}^{1/2}$$
$$= \left\{\frac{(0.01188)(144 \times 74.7)[1 - (69.7/74.7)^2]}{2\log_e(74.7/69.7) + (0.02404)(200)/(1.049/12)}\right\}^{1/2}$$
$$= 0.5476 \text{ slug/(ft}^2)(\text{s})$$
$$\dot{m} = g_cAG = (32.17)(\pi/4)(1.049/12)^2(0.5476)$$
$$\dot{m} = 0.1057 \text{ lbm/s } (47.94 \times 10^{-3} \text{ kg/s})$$

Noncircular Pipes   For the flow of fluids in noncircular pipes, the **hydraulic diameter** D_h is used. From the definition of hydraulic radius, the diameter of a circular pipe was shown to be four times its **hydraulic radius**; thus $D_h = 4R_h$. The Reynolds number thus may be written as $\mathbf{R} = \rho VD_h/\mu = GD_h/\mu$, the relative roughness as ε/D_h, and the resistance coefficient $K = fL/D_h$. With the above modifications, flows through noncircular pipes may be computed in the same manner as for circular pipes.

EXAMPLE.   Air at 68°F (20°C) and 100 lbf/in^2 enters a rectangular duct 1 by 3 ft at a rate of 720 lbm/s. The duct is horizontal, 100 ft long, and made of galvanized iron. Assuming isothermal flow, estimate the pressure loss due to

friction in this line. From the equation of state, $\rho_1 = p_1/g_cRT_1 = (144 \times 100)/(32.17)(53.34)(527.7) = 0.01590$ slug/ft^3. From Table 3.3.6, $R_h = bD/2(b + D) = 3 \times 1/2(3 + 1) = 0.375$ ft, and $D_h = 4R_h = 4 \times 0.375 = 1.5$ ft. For galvanized iron, $\varepsilon/D_h = 500 \times 10^{-6}/1.5 = 3.333 \times 10^{-4}$

$$G = (\dot{m}/g_c)/A = (720/32.17)/(1 \times 3) = 7.460 \text{ slugs/(ft}^2)(\text{s})$$
$$\mathbf{R} = GD_h/\mu = (7.460)(1.5)/(39.16 \times 10^{-8})$$
$$= 28{,}580{,}000 > 4{,}000 \therefore \text{ flow is turbulent}$$

From Fig. 3.3.24, $f \approx 0.015$

$$\frac{1}{\sqrt{f}} = -2\log_{10}\left(\frac{\varepsilon/D_h}{3.7} + \frac{2.51}{\mathbf{R}\sqrt{f}}\right)$$
$$= -2\log_{10}\left(\frac{3.333 \times 10^{-4}}{3.7} + \frac{2.51}{28{,}580{,}000\sqrt{0.015}}\right)$$
$$f = 0.01530$$

Solving the isothermal equation for p_2/p_1,

$$\frac{p_2}{p_1} = \left\{1 - \left(\frac{G^2}{\rho_1 p_1}\right)\left[2\log_e\left(\frac{p_1}{p_2}\right) + \frac{fL}{D_h}\right]\right\}^{1/2}$$

For first trial, assume $2\log_e(p_1/p_2)$ is small compared with fL/D_h:

$$p_2/p_1 = \{1 - [(7.460)^2/(0.01590)(144 \times 100)][0 + (0.01530)(100)/1.5]\}^{1/2}$$
$$= 0.8672$$

Second trial using first-trial values results in 0.8263. Subsequent trials result in a balance at $p_2/p_1 = 0.8036$, $p_2 = 100 \times 0.8036 = 80.36$ lbf/in^2 (5.541 $\times$ 10^5 N/m^2).

PIPING SYSTEMS

Resistance Parameters   The resistance to flow of a piping system is similar to the resistance of an object immersed in a flow stream and is made up of pressure (inertia) or shape drag and skin-friction (viscous) drag. For long, straight pipes the pressure drag is characterized by the **relative roughness** ε/D and the skin friction by the **Reynolds number R.** For other piping components, two parameters are used to describe the resistance to flow, the **resistance coefficient** $K = fL/D$ and the **equivalent length** $L/D = K/f$. The resistance-coefficient method assumes that the component loss is all due to pressure drag and that the flow through the component is completely turbulent and independent of Reynold's number. The equivalent-length method assumes that resistance of the component varies in the same manner as does a straight pipe. The basic assumption then is that its pressure drag is the same as that for the relative roughness ε/D of the pipe and that the friction drag varies with the Reynolds number $\mathbf{R}$ in the same manner as the straight pipe. Both methods have the inherent advantage of simplicity in application, but neither is correct except in the fully developed turbulent region. Two excellent sources of information on the resistance of piping-system components are the Hydraulic Institute "Pipe Friction Manual," which uses the resistance-coefficient method, and the Crane Company Technical Paper 410 ("Fluid Meters," 6th ed. ASME, 1971), which uses the equivalent-length concept.

For valves, branch flow through tees, and the type of components listed in Table 3.3.10, the pressure drag is predominant, is "rougher" than the pipe to which it is attached, and will extend the completely turbulent region to lower values of Reynolds number. For bends and elbows, the loss is made up of pressure drag due to the change of direction and the consequent secondary flows which are dissipated in 50 diameters or more downstream piping. For this reason, loss through adjacent bends will not be twice that of a single bend.

In long pipelines, the effect of bends, valves, and fittings is usually negligible, but in systems where there is little straight pipe, they are the controlling factor. Under-design will result in the failure of the system to deliver the required capacity. Over-design will result in inefficient operation because it will be necessary to "throttle" one or more of the valves. For estimating purposes, Tables 3.3.10 and 3.3.11 may be used as shown in the examples. When available, the manufacturers' data should be used, particularly for valves, because of the wide variety of designs for the same type. (See also Sec. 12.4.)

Table 3.3.10   Representative Values of Resistance Coefficient K

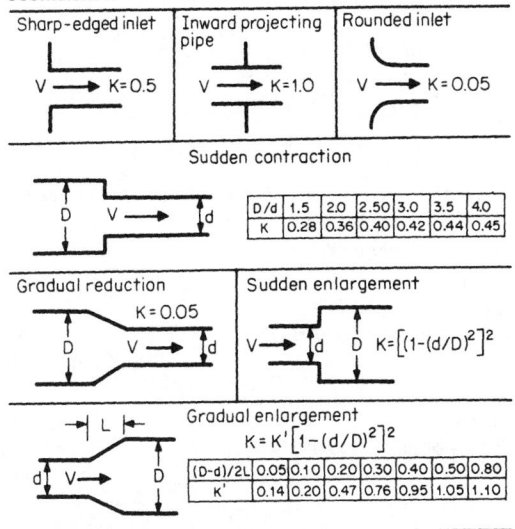

Exit loss = (sharp edged, projecting, Rounded), K=1.0

SOURCE: Compiled from data given in "Pipe Friction Manual," 3d ed., Hydraulic Institute, 1961.

Table 3.3.11   Representative Equivalent Length in Pipe Diameters (L/D) of Various Valves and Fittings

Globe valves, fully open	450
Angle valves, fully open	200
Gate valves, fully open	13
¾ open	35
½ open	160
¼ open	900
Swing check valves, fully open	135
In line, ball check valves, fully open	150
Butterfly valves, 6 in and larger, fully open	20
90° standard elbow	30
45° standard elbow	16
90° long-radius elbow	20
90° street elbow	50
45° street elbow	26
Standard tee:	
Flow through run	20
Flow through branch	60

SOURCE: Compiled from data given in "Flow of Fluids," Crane Company Technical Paper 410, ASME, 1971.

Series Systems   In a single piping system made of various sizes, the practice is to group all of one size together and apply the continuity equation, as shown in the following example.

EXAMPLE.   Water at 68°F (20°C) leaves an open tank whose surface elevation is 180 ft and enters a 2-in schedule 40 steel pipe via a sharp-edged entrance. After 50 ft of straight 2-in pipe that contains a 2-in globe valve, the line enlarges suddenly to an 8-in schedule 40 steel pipe which consists of 100 ft of straight 8-in pipe, two standard 90° elbows and one 8-in angle valve. The 8-in line discharges below the surface of another open tank whose surface elevation is 100 ft. Determine the volumetric flow rate.

$$D_1 = 2.067/12 = 0.1723 \text{ ft} \quad \text{and} \quad D_2 = 7.981/12 = 0.6651 \text{ ft}$$
$$\varepsilon/D_1 = 150 \times 10^{-6}/0.1723 = 8.706 \times 10^{-4}$$
$$\varepsilon/D_2 = 150 \times 10^{-6}/0.6651 = 2.255 \times 10^{-4}$$

For turbulent flow,

$$\frac{1}{\sqrt{f_1}} = -2 \log_{10}\left(\frac{\varepsilon/D}{3.7}\right)$$

$$\frac{1}{\sqrt{f_1}} = -2 \log_{10}\left(\frac{8.706 \times 10^{-4}}{3.7}\right) \qquad f_1 = 0.01899$$

$$\frac{1}{\sqrt{f_1}} = -2 \log_{10}\left(\frac{2.255 \times 10^{-4}}{3.7}\right) \qquad f_2 = 0.01407$$

1. *2-in components*

		K
Entrance loss, sharp-edged	=	0.5
50 ft straight pipe = f_1 (50/0.1723)	=	$290.2f_1$
Globe valve = f_1 (L/D)	=	$450.0f_1$
Sudden enlargement $k = [-(D/D_2)^2]^2$		
= $[1 - (2.067/7.981)^2]^2$	=	0.87
	$\Sigma K_1 =$	$1.37 + 740.2f_1$

2. *8-in components*

		K
100 ft of straight pipe f_2 (100/0.6651)	=	$150.4f_2$
2 standard 90° elbows 2 × $30 f_2$	=	$60 \ f_2$
1 angle valve $200 f_2$	=	$200 \ f_2$
Exit loss	=	1
	$\Sigma K_2 =$	$1 + 410.4f_2$

3. *Apply equation of motion*

$$h_{1f2} = z_1 - z_2 = (\Sigma K_1)\frac{V_1^2}{2g} + (\Sigma K_2)\frac{V_2^2}{2g}$$

From continuity, $\rho_1 A_1 V_1 = \rho_2 A_2 V_2$ for $\rho_1 = \rho_2$

$$V_2 = V_1(A_1/A_2) = V_1(D_1/D_2)^2$$
$$h_{1f2} = z_1 - z_2 = [\Sigma K_1 + \Sigma K_2(D_1/D_2)^4]V_1^2/2g$$
$$V_1 = \{[2g(z_1 - z_2)]/[\Sigma K_1 + \Sigma K_2(D_1/D_2)^4]\}^{1/2}$$
$$V_1 = \left[\frac{2 \times 32.17 \times (180 - 100)}{(1.37 + 740.2f_1) + (1 + 410.4f_2)(2.067/7.981)^4}\right]^{1/2}$$
$$V_1 = \frac{71.74}{(1.374 + 740.2f_1 + 1.846f_2)^{1/2}}$$

4. For first trial assume f_1 and f_2 for complete turbulence

$$V_1 = \frac{71.74}{(1.374 + 740.2 \times 0.01899 + 1.846 \times 0.01407)^{1/2}}$$
$$V_1 = 18.25 \text{ ft/s}$$
$$V_2 = 18.25 (2.067/7.981)^2 = 1.224 \text{ ft/s}$$
$$\mathbf{R}_1 = \rho_1 V_1 D_1/\mu = (1.937)(18.25)(0.1723)/(20.92 \times 10^{-6})$$
$$\mathbf{R}_1 = 291,100 > 4,000 \therefore \text{ flow is turbulent}$$
$$\mathbf{R}_2 = \rho_2 V_2 D_2/\mu_2 = (1.937)(1.224)(0.6651)/(20.92 \times 10^{-6})$$
$$\mathbf{R}_2 = 75,420 > 4,000 \therefore \text{ flow is turbulent}$$

5. For second trial use first trial V_1 and V_2. From Fig. 3.3.24 and the Colebrook equation,

$$\frac{1}{\sqrt{f_1}} = -2 \log_{10}\left(\frac{8.706 \times 10^{-4}}{3.7} + \frac{2.51}{291,100\sqrt{0.020}}\right)$$
$$f_1 = 0.02008$$
$$\frac{1}{\sqrt{f_2}} = -2 \log_{10}\left(\frac{2.255 \times 10^{-4}}{3.7} + \frac{2.51}{75,420\sqrt{0.020}}\right)$$
$$f_2 = 0.02008$$
$$V_1 = \frac{71.74}{(1.374 + 740.2 \times 0.02008 + 1.864 \times 0.02008)^{1/2}}$$
$$V_1 = 17.78$$

A third trial results in $V = 17.77$ ft/s or $Q = A_1 V_1 = (\pi/4)(0.1723)^2(17.77) = 0.4143$ ft³/s (1.173×10^{-2} m³/s).

Parallel Systems   In solution of problems involving two or more parallel pipes, the head loss for all of the pipes is the same as shown in the following example.

EXAMPLE.   Benzene at 68°F (20°C) flows at a rate of 0.5 ft³/s through two parallel, straight, horizontal pipes connecting two pressurized tanks. The pipes are both schedule 40 steel, one being 1 in, the other 2 in. They both are 100 ft long and have connections that project inwardly in the supply tank. If the pressure in the supply tank is maintained at 100 lbf/in², what pressure should be maintained on the receiving tank?

$$D_1 = 1.049/12 = 0.08742 \text{ ft} \quad \text{and} \quad D_2 = 2.067/12 = 0.1723 \text{ ft}$$
$$\varepsilon/D_1 = 150 \times 10^{-6}/0.08742 = 1.716 \times 10^{-3}$$
$$\varepsilon/D_2 = 150 \times 10^{-6}/0.1723 = 8.706 \times 10^{-4}$$

For turbulent flow,

$$\frac{1}{\sqrt{f}} = -2\log_{10}\left(\frac{\varepsilon/D}{3.7}\right)$$

$$\frac{1}{\sqrt{f_1}} = -2\log_{10}\left(\frac{1.716 \times 10^{-3}}{3.7}\right) \qquad f_1 = 0.02249$$

$$\frac{1}{\sqrt{f_2}} = -2\log_{10}\left(\frac{8.706 \times 10^{-4}}{3.7}\right) \qquad f_1 = 0.01899$$

1. *1-in. components*

		K
Entrance loss, inward projection		$= 1.0$
100 ft straight pipe f_1 (100/0.08742)		$= 1,144\,f_1$
Exit loss		$= 1.0$
	$\Sigma K_1 =$	$2.0 + 1,144\,f_1$

2. *2-in components*

Entrance loss, inward projection		$= 1.0$
100 ft straight nine f_2 (100/0.1723)		$= 580.4\,f_2$
Exit loss		$= 1.0$
	$\Sigma K_2 =$	$2.0 + 580.4\,f_2$

$$h_f = \Sigma K_1 V_1^2/2g = \Sigma K_2 V_2^2/2g$$

From the continuity equation, $Q = AV$

$$\Sigma K_1 \frac{Q_1^2}{2gA_1^2} = \Sigma K_2 \frac{Q_2^2}{2gA_2^2}$$

Solving for Q_1/Q_2,

$$\frac{Q_1}{Q_2} = \frac{A_1}{A_2}\sqrt{\frac{\Sigma K_2}{\Sigma K_1}} = \left(\frac{D_1}{D_2}\right)^2\sqrt{\frac{\Sigma K_2}{\Sigma K_1}} = \left(\frac{D_1}{D_2}\right)^2\sqrt{\frac{2.0 + 580.4\,f_2}{2.0 + 1,144\,f_1}}$$

For first trial assume flow is completely turbulent,

$$\frac{Q_1}{Q_2} = \left(\frac{0.08742}{0.1723}\right)^2\sqrt{\frac{2.0 + 580.4 \times 0.01899}{2.0 + 1,144 \times 0.02249}}$$

$$\frac{Q_1}{Q_2} = 0.1764 \qquad Q = Q_1 + Q_2 = 0.1764\,Q_2 + Q_2$$

$$0.5000 = 1.1764\,Q_2 \qquad Q_2 = 0.4250$$
$$Q_1 = 0.5000 - 0.4250 = 0.0750$$

for the second trial use first-trial values,

$$V_1 = Q_1/A_1 = 0.0750/(\pi/4)(0.08742)^2 = 12.50$$
$$V_2 = Q_2/A_2 = 0.4250/(\pi/4)(0.1723)^2 = 18.23$$
$$\mathbf{R}_1 = \rho_1 V_1 D_1/\mu_1 = (1.705)(12.50)(0.08742)/(13.62 \times 10^{-6})$$
$$\mathbf{R}_1 = 136{,}800 > 4{,}000 \therefore \text{flow is turbulent}$$
$$\mathbf{R}_2 = \rho_2 V_2 D_2/\mu_2 = (1.705)(18.23)(0.1723)/(13.62 \times 10^{-6})$$
$$\mathbf{R}_2 = 393{,}200 > 4{,}000 \therefore \text{flow is turbulent}$$

Using the Colebrook equation and Fig. 3.3.24,

$$\frac{1}{\sqrt{f_1}} = -2\log_{10}\left(\frac{1.716 \times 10^{-3}}{3.7} + \frac{2.51}{136{,}800\sqrt{0.024}}\right)$$

$$f_1 = 0.02389$$

$$\frac{1}{\sqrt{f_2}} = -2\log_{10}\left(\frac{8.706 \times 10^{-4}}{3.7} + \frac{2.51}{393{,}200\sqrt{0.020}}\right)$$

$$f_2 = 0.01981$$

$$h_f = \Sigma K_1 \frac{V_1^2}{2g} = \Sigma K_2 \frac{V_2^2}{2g}$$

$$\Sigma K_1 V_1^2/2g = (2.0 + 1,144 \times 0.02389)(12.50)^2/(2 \times 32.17) = 71.23$$
$$\Sigma K_2 V_2^2/2g = (2.0 + 580.4 \times 0.01981)(18.23)^2/(2 \times 32.17) = 69.80$$

$71.23 \approx 69.80$; further trials not justifiable because of accuracy of f, K, L/D. Use average or 70.52, so that $\Delta p = \rho g h_f = (1.705 \times 32.17 \times 70.52)/144 = 26.86$ lbf/in$^2 = p_1 - p_2 = 100 - p_2$, $p_2 = 100 - 26.86 = 73.40$ lbf/in^2 (5.061 $\times$ 10^5 N/m^2).

Branch Flow Problems of a single line feeding several points may be solved as shown in the following example.

EXAMPLE. Ethyl alcohol at 68°F (20°C) flows from tank A, which is maintained at a constant pressure of 100 lb/in^2 through 200 ft of 2-in cast-iron schedule 40 pipe to a Y branch connection ($K = 0.5$) where 100 ft of 2-in pipe goes to tank B, which is maintained at 80 lbf/in^2 and 50 ft of 2-in pipe to tank C, which is also

maintained at 80 lbf/in^2. All tank connections are flush and sharp-edged and are at the same elevation. Estimate the flow rate to each tank.

$$D = 2.067/12 = 0.1723 \text{ ft}$$
$$\varepsilon/D = 850 \times 10^{-6}/0.1723 = 4.933 \times 10^{-3}$$

For turbulent flow, $\dfrac{1}{\sqrt{f}} = -2\log_{10}\left(\dfrac{\varepsilon/D}{3.7}\right)$

$$\frac{1}{\sqrt{f}} = -2\log_{10}\left(\frac{4.933 \times 10^{-3}}{3.7}\right) \qquad f = 0.03025$$

$$h_{AfB} = (p_A - p_B)/\rho g = 144(100 - 80)/(1.532 \times 32.17) = 58.44$$
$$h_{Afc} = (p_A - p_C)/\rho g = h_{AfB} = 58.44$$

Let point X be just before the Y; then

1. *From tank A to Y*

		K
Entrance loss, sharp-edged		$= 0.5$
200 ft straight pipe $= f_{AX}(200/0.1723)$		$= 1,161\,f_{AX}$
	$\Sigma K_{AX} =$	$0.5 + 1,161\,f_{AX}$

2. *From Y to tank B*

Y branch		$= 0.5$
100 ft straight pipe $= f_{XB}(100/0.1723)$		$= 580.4\,f_{XB}$
Exit loss		$= 1.0$
	$\Sigma K_{XB} =$	$1.5 + 580.4\,f_{XB}$

3. *From Y to tank C*

Y branch		$= 0.5$
50 ft straight pipe $= f_{XC}(50/0.1723)$		$= 290.2\,f_{XC}$
Exit loss		$= 1.0$
	$\Sigma K_{XC} =$	$1.5 + 290.2\,f_{XC}$

Balance of flows:

$$Q_{AX} = Q_{XB} + Q_{XC}$$

and from continuity, $(A_{AX} = A_{XB} = A_{XC})$, $V_{AX} = V_{XB} + V_{Xc}$; then

$$h_{AfB} = \Sigma K_{AX}\frac{V_{AX}^2}{2g} + \Sigma K_{XB}\frac{V_{XB}^2}{2g}$$

$$h_{AfC} = \Sigma K_{AX}\frac{V_{AX}^2}{2g} + \Sigma K_{XC}\frac{V_{XC}^2}{2g}$$

For first trial assume completely turbulent flow

$$h_{AfB} = \frac{(0.5 + 1,161\,f_{AX})V_{AX}^2}{2g} + \frac{(1.5 + 580.4\,f_{XB})V_{XB}^2}{2g}$$

$$58.44 = \frac{(0.5 + 1,161 \times 0.03025)V_{AX}^2}{2 \times 32.17} + \frac{(1.5 + 580.4 \times 0.03025)V_{XB}^2}{2 \times 32.17}$$

$$58.44 = 0.5536\,V_{AX}^2 + 0.2962\,V_{XB}^2$$

and in a like manner

$$h_{AfC} = 58.44 = 05536\,V_{XC}^2 + 0.1598\,V_{XC}^2$$

Equating $h_{AfB} = h_{AfC}$,

$$0.5536\,V_{AX}^2 + 0.2962\,V_{XB}^2 = 0.5536\,V_{AX}^2 + 0.1598\,V_{XC}^2$$

or $\qquad V_{XC} = 1.3615\,V_{XB} \quad$ and since $\quad V_{AX} = V_{XB} + V_{XC}$

$$V_{AX} = V_{XB} + 1.3615\,V_{XB} = 2.3615\,V_{XB}$$

so that $\qquad h_{AfB} = 58.44 = 0.5536(2.3615\,V_{XB}^2) + 0.2962\,V_{XB}^2$

$$V_{XB} = 4.156$$
$$V_{XC} = 1.3615(4.156) = 5.658$$
$$V_{AX} = 4.156 + 5.658 = 9.814$$

Second trial,

$$\mathbf{R}_{AX} = \frac{\rho V_{AX} D}{\mu} = \frac{1.532 \times 9.814 \times 0.1723}{25.06 \times 10^{-6}}$$
$$\mathbf{R}_{AX} = 103{,}400 > 4{,}000 \therefore \text{flow is turbulent}$$

In a like manner,

$$\mathbf{R}_{XB} = 43{,}780 \qquad \mathbf{R}_{XC} = 59{,}600$$

Using the Colebrook equation and Fig. 3.3.24,

$$\frac{1}{\sqrt{f_{AX}}} = -2\log_{10}\left(\frac{4.933 \times 10^{-3}}{3.7} + \frac{2.51}{103{,}400\sqrt{0.031}}\right)$$

$$f_{AX} = 0.03116$$

In a like manner,

$$f_{XB} = 0.03231 \qquad f_{XC} = 0.03179$$

$$h_{AfB} = \frac{(0.5 + 1{,}161 \times 0.03116)V_{AX}^2}{2 \times 32.17} + \frac{(1.5 + 580.4 \times 0.03231)V_{XB}^2}{2 \times 32.17}$$

$$h_{AfB} = 0.5700\, V_{AX}^2 + 0.3148\, V_{XB}^2$$

$$h_{AfC} = 0.5700\, V_{AX}^2 + \frac{(1.5 + 290.2 \times 0.03179)V_{XC}^2}{2 \times 32.17}$$

$$h_{AfC} + 0.5700\, V_{AX}^2 + 0.1667\, V_{XC}^2$$

$$0.3148\, V_{XB}^2 = 0.1667\, V_{XC}^2$$

$$V_{XC} = 1.374\, V_{XB}$$

$$V_{AX} = V_{XB} + 1.374\, V_{XB} = 2.374\, V_{XB}$$

so that

$$h_{AfB} = 58.44 = 0.5700\,(2.374\, V_{XB})^2 + 0.3148\, V_{XB}^2$$

$$V_{XB} = 4.070 \qquad V_{XC} = 5.592 \qquad V_{AX} = 9.663$$

Further trials are not justified.

$$A = \pi D^2/4 = (\pi/4)(0.1723)^2 = 0.02332 \text{ ft}^2$$

$$Q_{AB} = V_{AB}A = 4.070 \times 0.02332 = 0.09491 \text{ ft}^3/\text{s } (2.686 \times 10^{-1}\,\text{m}^3/\text{s})$$

$$Q_{XC} = V_{XC}A = 5.592 \times 0.02332 = 0.1304 \text{ ft}^3/\text{s } (3.693 \times 10^{-3}\,\text{m}^3/\text{s})$$

Siphons are arrangements of hose or pipe which cause liquids to flow from one level A in Fig. 3.3.25 to a lower level C over an intermediate summit B. Performance of siphons may be evaluated from the equation of motion between points A and B:

$$\frac{p_A}{\gamma} + \frac{V_A^2}{2g} + z_A = \frac{p_B}{\gamma} + \frac{V_B^2}{2g} + z_B + h_{AfB}$$

Noting that on the surface $V_A = 0$ and the minimum pressure that can exist at point B is the vapor pressure pv, the maximum elevation of point B is

$$z_B - z_A = \frac{p_A}{\gamma} - \left(\frac{p_v}{\gamma} + \frac{V_B^2}{2g} + h_{AfB}\right)$$

The friction loss $h_f = \Sigma K_{AB} V_B^2/2g$, and let $V_B = V$; then

$$z_B - z_A = \frac{p_A - p_v}{\rho g} - (1 - \Sigma K_{AB})\frac{V^2}{2g}$$

Flow under this maximum condition will be uncertain. The air pump or ejector used for priming the pipe (flow will not take place unless the siphon is full of water) might have to be operated occasionally to remove accumulated air and vapor. Values of $z_B - z_A$ less than those calculated by the above equation should be used.

Fig. 3.3.25 Siphon.

EXAMPLE. The siphon shown in Fig. 3.3.25 is composed of 2,000 ft of 6-in schedule 40 cast-iron pipe. Reservoir A is at elevation 800 ft and C at 600 ft. Estimate the maximum height for $z_B - z_A$ if the water temperature may reach 104°F (40°C), and the amount of straight pipe from A to B is 100 ft. For the first bend $L/D = 25$ and the second (at B) $L/D = 50$. Atmospheric pressure is 14.70 lbf/in². For 6-in schedule 40 pipe $D = 6.065/12 = 0.5054$ ft, $\varepsilon/D = 850 \times 10^{-6}/0.5054 = 1.682 \times 10^{-3}$. Turbulent friction factor

$$1/\sqrt{f} = -2 \log_{10}\left(\frac{\varepsilon D}{3.7}\right) = -2 \log_{10}(1.682 \times 10^{-3}/3.7) = 0.02238$$

1. *Components from A to B.* (Note loss in second bend takes place in downstream piping.)

	K
Entrance (inward projection)	= 1.0
100 ft straight pipe $f\,(100/0.5054)$	= 197.9f
First bend	= 25 f
ΣK_{AB}	= 1.0 + 227.9 f

2. *Components from A to C*

ΣK_{AB}	= 1.0 + 2,229 f
1,900 ft of straight pipe $f\,(1{,}900/0.5054)$	= 3,759.4 f
Second bend	= 50 f
Exit loss	= 1
ΣK_{AC}	= 2.0 + 4,032 f

First trial assume complete turbulence. Writing the equation of motion between A and C.

$$\frac{p_A}{\gamma} + \frac{V_A^2}{2g} + z_A = \frac{p_C}{\gamma} + \frac{V_C^2}{2g} + z_C + \Sigma K_{AC}\frac{V^2}{2g}$$

Noting $V_A = V_C = 0$, and $p_A = p_C = 14.7$ lbf/in²,

$$V = \sqrt{\frac{2g(z_A - z_C)}{\Sigma K_{AC}}} = \sqrt{\frac{2g(z_A - z_C)}{2.0 + 4{,}032f}}$$

$$= \sqrt{\frac{2 \times 32.17\,(800 - 600)}{2.0 + 4.032\,f}}$$

$$= \frac{113.44}{\sqrt{2.0 + 4{,}032 \times 0.02238}}$$

$$= 11.81$$

Second trial, use first-trial values,

$$\mathbf{R} = \frac{\rho VD}{\mu} = (1.925)(11.81)(0.5054)/(13.61 \times 10^{-6})$$

$$\mathbf{R} = 846{,}200 > 4{,}000 \therefore \text{ flow is turbulent}$$

From Fig. 3.3.24 and the Colebrook equation,

$$\frac{1}{\sqrt{f}} = -2 \log_{10}\left(\frac{1.682 \times 10^{-3}}{3.7} + \frac{2.51}{844{,}200 \sqrt{0.023}}\right)$$

$$f = 0.02263$$

$$V = \frac{113.44}{\sqrt{2.0 + 4{,}032 \times 0.02263}} = 11.75 \qquad \text{(close check)}$$

From Sec. 4.2 steam tables at 104°F, $p_v = 1.070$ lbf/in², the maximum height

$$z_B - z_A = \frac{p_A - p_v}{\rho g} - (1 + \Sigma K_{AB})\frac{V^2}{2g}$$

$$= \frac{144(14.70 - 1.070)}{1.925 \times 32.17} - (1 + 1 + 227.9$$

$$\times 0.02262)\frac{(11.75)^2}{2 \times 32.17} = 16.58 \text{ ft} (5.053\,\text{m})$$

Note that if a ±10 percent error exists in calculation of pressure loss, maximum height should be limited to ~ 15 ft (5 m).

ASME PIPELINE FLOWMETERS

Parameters Dimensional analysis of the flow of an incompressible fluid flowing in a pipe of diameter D, surface roughness ε, through a primary element (venturi, nozzle or orifice) whose diameter is d with a velocity of V, producing a pressure drop of Δp; sensed by pressure taps located a distance L apart results in $f(\mathbf{C}_p, \mathbf{R}_d, \varepsilon/D, d/D) = 0$, which may be written as $\Delta p = \mathbf{C}_p \rho V^2/2$. Conventional practice is to express the relations as $V = \mathbf{K}\sqrt{2\Delta p/\rho}$, where $\mathbf{K}$ is the **flow coefficient**, $\mathbf{K} = 1\sqrt{\mathbf{C}_p}$, and $\mathbf{K} = f(\mathbf{R}_d, L/D, \varepsilon/d, d/D)$. The ratio of the diameter of the primary element to meter tube (pipe) diameter D is known as the **beta ratio**, where $\beta = d/D$. Application of the continuity equation leads to $Q = \mathbf{K}A_2\sqrt{2\Delta p/\rho}$, where A_2 is the area of the primary element.

Conventional practice is to base flowmeter computations on the assumption of one-dimensional frictionless flow of an incompressible fluid in a horizontal meter tube and to correct for actual conditions by the use of a coefficient for viscous effects and a factor for elastic effects.

Application of the Bernoulli equation for horizontal flow from section 1 (inlet tap) to section 2 (outlet tap) results in $p_1/\rho g + V_1^2/2g = p_2/\rho g + V_2^2/2g$ or $(p_1 - p_2)/\rho = V_2^2 - V_1^2 = \Delta p/\rho$. From the equation of continuity, $Q_i = A_1 V_1 = A_2 V_2$, where Q_i is the **ideal flow rate**. Substituting, $2\Delta p/\rho = Q_i^2/A_1^2 - Q_i^2/A_2^2$, and solving for Q_i, $Q_i = A_2\sqrt{2\Delta p/\rho}/\sqrt{1 - (A_2/A_1)^2}$, noting that $A_2/A_1 = (d/D)^2 = \beta^2$, $Q_i = A_2\sqrt{2\Delta p/\rho}/\sqrt{1 - \beta^4}$. The **discharge coefficient** C is defined as the ratio of the actual flow Q to the ideal flow Q_i, or $C = Q/Q_i$, so that $Q = CQ_i = CA_2\sqrt{2\Delta p/\rho}/\sqrt{1 - \beta^4}$. It is customary to write the volumetric-flow equation as $Q = CEA_2\sqrt{2\Delta p/\rho}$, where $E = 1/\sqrt{1 - \beta^4}$. E is called the **velocity-of-approach factor** because it accounts for the one-dimensional kinetic energy at the upstream tap. Comparing the equation from dimensional analysis with the modified Bernoulli equation, $Q = \mathbf{K}A_2\sqrt{2\Delta p/\rho} = CEA_2\sqrt{2\Delta p/\rho}$, or $\mathbf{K} = CE$ and $C = f(\mathbf{R}_d, L/D, \beta)$.

For compressible fluids, the incompressible equation is modified by the **expansion factor** Y, where Y is defined as the ratio of the flow of a compressible fluid to that of an incompressible fluid at the same value of Reynolds number. Calculations are then based on inlet-tap-fluid properties, and the compressible equation becomes

$$Q_1 = \mathbf{K}YA_2\sqrt{2\Delta p/\rho_1} = CEYA_2\sqrt{2\Delta p/\rho_1}$$

where $Y = f(L/D, \varepsilon/D, \beta, \mathbf{M})$. Reynolds number $\mathbf{R}_d$ is also based on inlet-fluid properties, but on the primary-element diameter or

$$\mathbf{R}_d = \rho_1 V_2 d/\mu_1 = \rho_1(Q_1/A_2)d/\mu_1 = 4\rho_1 Q_1/\pi\, d\mu_1$$

Caution The numerical values of coefficients for flowmeters given in the paragraphs to follow are based on experimental data obtained with long, straight pipes where the velocity profile approaching the primary element was fully developed. The presence of valves, bends, and fittings upstream of the primary element can cause serious errors. For approach and discharge, straight-pipe requirements, "Fluid Meters," (6th ed., ASME, 1971) should be consulted.

Venturi Tubes Figure 3.3.26 shows a typical venturi tube consisting of a cylindrical inlet, convergent cone, throat, and divergent cone. The convergent entrance has an included angle of about 21° and the divergent cone 7 to 8°. The purpose of the divergent cone is to reduce the

Fig. 3.3.26 Venturi tube.

overall pressure loss of the meter; its removal will have no effect on the coefficient of discharge. Pressure is sensed through a series of holes in the inlet and throat. These holes lead to an annular chamber, and the two

chambers are connected to a pressure-differential sensor. Discharge coefficients for venturi tubes as established by the American Society of Mechanical Engineers are given in Table 3.3.12. Coefficients of discharge outside the tabulated limits must be determined by individual calibrations.

EXAMPLE. Benzene at 68°F (20°C) flows through a machined-inlet venturi tube whose inlet diameter is 8 in and whose throat diameter is 3.5 in. The differential pressure is sensed by a U-tube manometer. The manometer contains mercury under the benzene, and the level of the mercury in the throat leg is 4 in. Compute the volumetric flow rate. Noting that $D = 8$ in (0.6667 ft) and $\beta = 3.5/8 = 0.4375$ are within the limits of Table 3.3.12, assume $C = 0.995$, and then check $\mathbf{R}_d$ to verify if it is within limits. For a U-tube manometer (Fig. 3.3.6a), $p_2 - p_1 = (\gamma_m - \gamma_f)h = \Delta p$ and $\Delta p/\rho_1 = (\rho_m g - \rho_f g)h/\rho_f = g(\rho_m/\rho_f - 1)h = 32.17(26.283/1.705 - 1)(4/12) = 154.6$. For a liquid, $Y = 1$ (incompressible fluid), $E = 1/\sqrt{1 - \beta^4} = 1/\sqrt{1 - (0.4375)^4} = 1.019$.

$$\begin{aligned} Q_1 &= CEYA_d\sqrt{2\Delta p/\rho_1} \\ &= (0.995)(1.019)(\pi/4)(3.5/12)^2\sqrt{2 \times 154.6} \\ &= 1.192 \text{ ft}^3/\text{s}\, (3.373 \times 10^{-3} \text{ m}^3/\text{s}) \end{aligned}$$

$\mathbf{R}_d = 4\rho_1 Q_1/\pi\, d\mu_1 = 4(1.705)(1.192)/\pi(3.5/12)(13.62 \times 10^{-6})$
$\mathbf{R}_d = 651,400$, which lies between 200,000 and 1,000,000 of Table 3.3.12 ∴ solution is valid

Flow Nozzles Figure 3.3.27 shows an ASME flow nozzle. This nozzle is built to rigid specifications, and pressure differential may be sensed by either throat taps or pipe-wall taps. Taps are located one pipe diameter upstream and one-half diameter downstream from the nozzle inlet. Discharge coefficients for ASME flow nozzles may be computed from $C = 0.9975 - 0.00653 (10^6/\mathbf{R}_d)^a$, where $a = 1/2$ for $\mathbf{R}_d < 10^6$ and $a = 1/5$ for $\mathbf{R}_d > 10^6$. Most of the data were obtained for D between 2 and 15.75 in, $\mathbf{R}_d$ between 10^4 and 10^6, and beta between 0.15 and 0.75. For values of C within these ranges, a tolerance of 2 percent may be anticipated, and outside these limits, the tolerance may be greater than 2 percent. Because slight variations in form or dimension of either pipe or nozzle may affect the observed pressures, and thus cause the exponent a and the slope term (-0.00653) to vary considerably, nozzles should be individually calibrated.

EXAMPLE. An ASME flow nozzle is to be designed to measure the flow of 400 gal/min of 68°F (20°C) water in a 6-in schedule 40 (inside diameter = 6.065 in) steel pipe. The pressure differential across the nozzle is not to exceed 75 in of water. What should be the throat diameter of the nozzle? $\Delta p = h\rho_1 g$, $\Delta p/\rho_1 = hg = (75/12)(32.17) = 201.1$, $Q = (400/60)(231/1,728) = 0.8912 \text{ ft}^3/\text{s}$. A trial-and-error solution is necessary to establish the values of C and E because they are dependent upon β and $\mathbf{R}_d$, both of which require that d be known. Since $\mathbf{K} = CE \approx 1$, assume for first trial that $CE = 1$. Since a liquid is involved, $Y = 1$, $A_2 = Q_1/(CE)(Y)\sqrt{2\Delta p/\rho_1} = (0.8912)/(1)(1)\sqrt{2 \times 201.1} = 0.04444 \text{ ft}^2$, $d = \sqrt{4A_2/\pi} = \sqrt{4(0.04444)/\pi} = 0.2379 \text{ ft or } d = 0.2379 \times 12 = 2.854 \text{ in}$, $\beta = d/D = 2.854/6.065 = 0.4706$.

For second trial use first-trial value:

$$E = 1/\sqrt{1 - \beta^4} = 1/\sqrt{1 - (0.4706)^4} = 1.025$$

$\mathbf{R}_d = 4\rho_1 Q_1/\pi d\mu_1 = 4(1.937)(0.8912)/\pi(0.2379)(20.92 \times 10^{-6}) = 442,600 < 10^6$ ∴ $a = 1/2$ and $C = 0.9975 - 0.00653(10^6/\mathbf{R}_d)^{1/2}$. $C = 0.9975 - 0.00653(10^6/442,600)^{1/2} = 0.9877$. $A_2 = (0.8912)/(0.9877 \times 1.025)\sqrt{2 \times 201.1} = 0.04389$, $d_2 = \sqrt{4 \times (0.04389/\pi)} = 0.2364$, $d_2 = 0.2364 \times 12 = 2.837$ in $(7.205 \times 10^{-2} \text{ m})$. Further trials are not necessary in view of the ± 2 percent tolerance of C.

Table 3.3.12 ASME Coefficients for Venturi Tubes

Type of inlet cone	Reynolds number $\mathbf{R}_d$		Inlet diam D in $(2.54 \times 10^{-2}$ m)		β		C	Tolerance, %
	Min	Max	Min	Max	Min	Max		
Machined		1×10^6	2	10		0.75	0.995	± 1.0
Rough welded sheet metal	5×10^5	2×10^6	8	48	0.4	0.70	0.985	± 1.5
Rough cast			4	32	0.3	0.75	0.984	± 0.7

Source: Compiled from data given in "Fluid Meters," 6th ed., ASME, 1971.

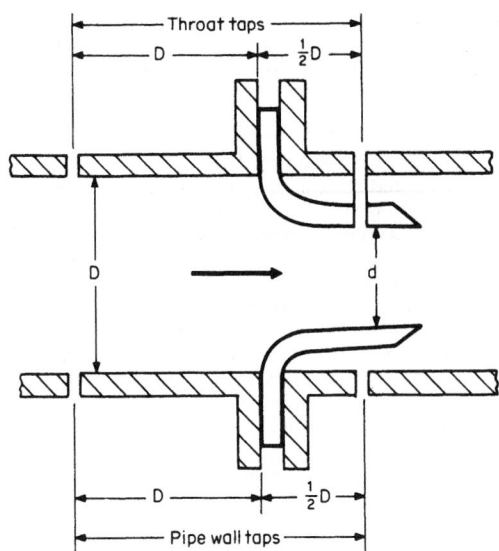

Fig. 3.3.27 ASME flow nozzle.

Compressible Flow—Venturi Tubes and Flow Nozzles   The expansion factor Y is computed based on the assumption of a frictionless adiabatic (isentropic) expansion of an ideal gas from the inlet to the throat of the primary element, resulting in (see Sec. 4.1)

$$Y = \left[\frac{kr^{2/k}(1 - r^{(k-1)/k})(1 - \beta^4)}{(1 - r)(k - 1)(1 - \beta^4 r^{2/k})} \right]^{1/2}$$

where $r = p_2/p_1$.

Maximum flow is obtained when the critical pressure ratio is reached. The critical pressure ratio r_c may be calculated from

$$r^{(1-k)/k} + \frac{k - 1}{2}\beta^4 r^{2/k} = \frac{k + 1}{2}$$

Table 3.3.13 gives selected values of Y_c and r_c.

EXAMPLE. A piping system consists of a compressor, a horizontal straight length of 2-in-inside-diameter pipe, and a 1-in-throat-diameter ASME flow nozzle attached to the end of the pipe, discharging into the atmosphere. The compressor is operated to maintain a flow of air with 115 lbf/in^2 and 140°F (60°C) conditions in the pipe just one pipe diameter before the nozzle inlet. Barometric pressure is 14.7 lbf/in^2. Estimate the flow rate of the air in lbm/s.

From the equation of state, $\rho_1 = p_1/g_c R T_1 = (144 \times 115)/(32.17)(53.34)$ $(140 + 459.7) = 0.01609$ slug/ft^3, $\beta = d/D = 1/2 = 0.5$, $E = 1/\sqrt{1 - \beta^4} =$ $1/\sqrt{1 - (0.5)^4} = 1.033$, $r = p_2/p_1 = 14.7/115 = 0.1278$, but from Table 3.3.13 at $\beta = 0.5$, $k = 1.4$, $r_c = 0.5362$, and $Y_c = 0.6973$, so that because of critical flow the throat pressure $p_c = 115 \times 0.5362 = 61.66$ lbf/in^2. $\Delta p_c/\rho_1 = 144(115 -$ $61.66)/0.01609 = 477{,}375$. A trial-and-error solution is necessary to obtain C. For the first trial assume $10^6/\mathbf{R}_d = 0$ or $C = 0.9975$. Then $Q_1 = CEY_c A_2 \sqrt{2\Delta p_c/\rho_1} =$ $(0.9975)(1.033)(0.6973)(\pi/4)(1.12)^2 \sqrt{2 \times 477{,}375} = 3.829$ ft^3/s, $\mathbf{R}_d = 4\rho_1 Q/$ $\pi\, d\mu_1 = (4)(0.01609)(3.828)/\pi(1/12)(41.79 \times 10^{-8}) = 2{,}252{,}000$.

Second trial, use first-trial values:

$$\mathbf{R} > 10^6, a = 115, C = 0.9975 - (0.00653)(16^6/2{,}252{,}000)^{1/5}$$
$$C = 0.9919, Q_1 = 3.828(0.9919/0.9975) = 3.806 \text{ ft}^3/\text{s}$$

Further trials are not necessary in view of ± 2 percent tolerance on C.

$$\dot{m} = Q_1\rho_1 g = 3.806 \times 0.01609 \times 32.17 = 1.970 \text{ lbm/s (0.8935 kg/s)}$$

Orifice Meters   When a fluid flows through a square-edged thin-plate orifice, the minimum-flow area is found to occur downstream from the orifice plate. This minimum area is called the vena contracta, and its location is a function of beta ratio. Figure 3.3.28 shows the relative pressure difference due to the presence of the orifice plate. Because the location of the pressure taps is vital, it is necessary to specify the exact position of the downstream pressure tap. The jet contraction

Table 3.3.13   Expansion Factors and Critical Pressure Ratios for Venturi Tubes and Flow Nozzles

β	k	Critical values		Expansion factor Y			
		r_c	Y_c	$r = 0.60$	$r = 0.70$	$r = 0.80$	$r = 0.90$
0	1.10	0.5846	0.6894	0.7021	0.7820	0.8579	0.9304
	1.20	0.5644	0.6948	0.7228	0.7981	0.8689	0.9360
	1.30	0.5457	0.7000	0.7409	0.8119	0.8783	0.9408
	1.40	0.5282	0.7049	0.7568	0.8240	0.8864	0.9449
0.20	1.10	0.5848	0.6892	0.7017	0.7817	0.8577	0.9303
	1.20	0.5546	0.6946	0.7225	0.7978	0.8687	0.9359
	1.30	0.5459	0.6998	0.7406	0.8117	0.8781	0.9407
	1.40	0.5284	0.7047	0.7576	0.8237	0.8862	0.9448
0.50	1.10	0.5921	0.6817	0.6883	0.7699	0.8485	0.9250
	1.20	0.5721	0.6872	0.7094	0.7864	0.8600	0.9310
	1.30	0.5535	0.6923	0.7248	0.8007	0.8699	0.9361
	1.40	0.5362	0.6973	0.7440	0.8133	0.8785	0.9405
0.60	1.10	0.6006	0.6729		0.7556	0.8374	0.9186
	1.20	0.5808	0.6784	0.6939	0.7727	0.8495	0.9250
	1.30	0.5625	0.6836	0.7126	0.7875	0.8599	0.9305
	1.40	0.5454	0.6885	0.7292	0.8006	0.8689	0.9352
0.70	1.10	0.6160	0.6570		0.7290	0.8160	0.9058
	1.20	0.5967	0.6624	0.6651	0.7469	0.8292	0.9131
	1.30	0.5788	0.6676	0.6844	0.7626	0.8405	0.9193
	1.40	0.5621	0.6726	0.7015	0.7765	0.8505	0.9247
0.80	1.10	0.6441	0.6277		0.6778	0.7731	0.8788
	1.20	0.6238	0.6331		0.6970	0.7881	0.8877
	1.30	0.6087	0.6383		0.7140	0.8012	0.8954
	1.40	0.5926	0.6433	0.6491	0.7292	0.8182	0.9021

SOURCE: Murdock, "Fluid Mechanics and Its Applications," Houghton Mifflin, 1976.

amounts to about 60 percent of the orifice area; so orifice coefficients are in the order of 0.6 compared with the nearly unity obtained with venturi tubes and flow nozzles.

Three pressure-differential-measuring tap locations are specified by the ASME. These are the flange, vena contracta, and the 1 D and 1/2 D. In the flange tap, the location is always 1 in from either face of the

Fig. 3.3.28   Relative-pressure changes due to flow through an orifice.

orifice plate regardless of the size of the pipe. In the vena contracta tap, the upstream tap is located one pipe diameter from the inlet face of the orifice plate and the downstream tap at the location of the vena contracta. In the 1 D and 1/2 D tap, the upstream tap is located one pipe diameter from the inlet face of the orifice plate and downstream one-half pipe diameter from the inlet face of the orifice plate.

Flange taps are used because they can be prefabricated, and flanges with holes drilled at the correct locations may be purchased as off-the-shelf items, thus saving the cost of field fabrication. The disadvantage of flange taps is that they are not symmetrical with respect to pipe size. Because of this, coefficients of discharge for flange taps vary greatly with pipe size.

Vena contracta taps are used because they give the maximum differential for any given flow. The disadvantage of the vena contracta tap is

that if the orifice size is changed, a new downstream tap must be drilled. The 1 D and 1/2 D taps incorporate the best features of the vena contracta taps and are symmetrical with respect to pipe size.

Discharge coefficients for orifices may be calculated from

$$C = C_o + \Delta C R_d^{-0.75} \qquad (R_d > 10^4)$$

where C_o and ΔC are obtained from Table 3.3.14.

Tolerances for uncalibrated orifice meters are in the order of ± 1 to ± 2 percent depending upon β, D, and R_d.

Compressible Flow through ASME Orifices   As shown in Fig. 3.3.28, the minimum flow area for an orifice is at the vena contracta located downstream of the orifice. The stream of compressible fluid is not restrained as it leaves the orifice throat and is free to expand transversely and longitudinally to the point of minimum-flow area. Thus the contraction of the jet will be less for a compressible fluid than for a liquid. Because of this, the theoretical-expansion-factor equation may not be used with orifices. Neither may the critical-pressure-ratio equation be used, as the phenomenon of critical flow has not been observed during testing of orifice meters.

For orifice meters, the following equation, which is based on experimental data, is used:

$$Y = 1 - (0.41 + 0.35\beta^4)(\Delta p/p_1)/k$$

EXAMPLE. Air at 68°F (20°C) and 150 lbf/in² flows in a 2-in schedule 40 pipe (inside diameter = 2.067 in) at a volumetric rate of 15 ft³/min. A 0.5500-in ASME orifice equipped with flange taps is used to meter this flow. What deflection in inches could be expected on a U-tube manometer filled with 60°F water? From the equation of state, $\rho_1 = p_1/g_cRT_1 = (144 \times 150)/(32.17)(53.34)$ $(68 + 459.7) = 0.02385$ slug/ft³, $\beta = 0.5500/2.067 = 0.2661$. $Q_1 = 15/60 = 0.25$ ft³/s, $A_2 = (\pi/4)(0.5500/12)^2 = 1.650 \times 10^{-3}$ ft². $E = 1/\sqrt{1 - \beta^4} = 1/\sqrt{1 - (0.2661)^4} = 1.003$. $R_d = 4\rho_1Q_1/\pi\, d\mu_1 = 4(0.02385)(0.25)/\pi(0.5500/12)(39.16 \times 10^{-8})$. $R_d = 423,000$.

From Table 3.3.14 at $\beta = 0.2661$, $D = 2.067$-in flange taps, by interpolation, $C_o = 0.5977$, $\Delta C = 9.087$, from orifice-coefficient equation $C = C_o + \Delta C R_d^{-0.75}$. $C = 0.5977 + (9.087)(423,000)^{-0.75} = 0.5982$. A trial-and-error solution is required because the pressure loss is needed in order to compute Y. For the first trial, assume $Y = 1$, $\Delta p = (Q_1/CEYA_2)^2(\rho_1/2) = [(0.25)/(0.5982)(1.003)(Y)(1.650 \times 10^{-3})]^2(0.02385/2) = 760.5/Y^2 = 760.5/(1)^2 = 760.5$ lbf/ft².

Table 3.3.14   Values of C_o and ΔC for Use in Orifice Coefficient Equation

Pipe ID, in	0.20	0.25	0.30	0.35	0.40	0.45	0.50	0.55	0.60	0.65	0.70
					β						
					ΔC, all taps						
All	5.486	8.106	11.153	14.606	18.451	22.675	27.266	32.215	37.513	45.153	49.129
					C_o, vena contracta and 1D and ½D taps						
All	0.5969	0.5975	0.5983	0.5992	0.6003	0.6016	0.6031	0.6045	0.6059	0.6068	0.6069
					C_o, flange taps						
2.0	0.5969	0.5975	0.5982	0.5992	0.6003	0.6016	0.6030	0.6044	0.6056	0.6065	0.6066
2.5	0.5969	0.5975	0.5983	0.5993	0.6004	0.6017	0.6032	0.6046	0.6059	0.6068	0.6068
3.0	0.5969	0.5975	0.5983	0.5993	0.6004	0.6017	0.6031	0.6044	0.6055	0.6061	0.6057
3.5	0.5969	0.5975	0.5983	0.5993	0.6004	0.6016	0.6030	0.6042	0.6052	0.6056	0.6049
4.0	0.5969	0.5976	0.5983	0.5993	0.6004	0.6016	0.6029	0.6041	0.6050	0.6052	0.6043
5.0	0.5969	0.5976	0.5983	0.5993	0.6004	0.6016	0.6028	0.6039	0.6047	0.6047	0.6034
6.0	0.5969	0.5976	0.5983	0.5993	0.6004	0.6016	0.6028	0.6038	0.6045	0.6044	0.6029
8.0	0.5969	0.5976	0.5984	0.5993	0.6004	0.6015	0.6027	0.6037	0.6042	0.6040	0.6022
10.0	0.5969	0.5976	0.5984	0.5993	0.6004	0.6015	0.6026	0.6036	0.6041	0.6037	0.6017
12.0	0.5970	0.5976	0.5984	0.5993	0.6004	0.6015	0.6026	0.6035	0.6040	0.6035	0.6015
16.0	0.5970	0.5976	0.5984	0.5993	0.6003	0.6015	0.6026	0.6035	0.6039	0.6033	0.6011
24.0	0.5970	0.5976	0.5984	0.5993	0.6003	0.6015	0.6025	0.6034	0.6037	0.6031	0.6007
48.0	0.5970	0.5976	0.5984	0.5993	0.6003	0.6014	0.6025	0.6033	0.6036	0.6029	0.6004
∞	0.5970	0.5976	0.5984	0.5993	0.6003	0.6014	0.6025	0.6032	0.6035	0.6027	0.6000

SOURCE: Compiled from data given in ASME Standard MFC-3M-1984 "Measurement of Fluid Flow in Pipes. Using Orifice, Nozzle and Venturi."

For the second trial we use first-trial values.

$$Y = 1 - (0.41 + 0.35\,\beta^4)\frac{\Delta p/p_1}{k}$$

$$= 1 - [0.41 + 0.35(0.2661)^4]\frac{760.5/144 \times 150}{1.4} = 0.9896$$

$$\Delta p = \frac{760.5}{Y^2} = \frac{760.5}{(0.9896)^2} = 776.1 \text{ lbf/ft}^2$$

For the third trial we use second-trial values.

$$Y = 1 - [0.41 + 0.35(0.2661)^4]\frac{776.1/144 \times 150}{1.4} = 0.9894$$

$$\Delta p = \frac{776.1}{(0.9894)^2} = 793.3 \text{ lbf/ft}^2$$

Resubstitution does not produce any further change in Y. From the U-tube-manometer equation:

$$h = \frac{\Delta p}{\gamma_m} = \frac{\Delta p}{g_c(\rho_m - \rho_f)}$$

$$= \frac{793.3}{32.17}(1.937 - 0.02385) = 12.89 \text{ ft}$$

$$= 12.89 \times 12 = 154.7 \text{ in} \ (3.929 \text{ m})$$

PITOT TUBES

Definition A Pitot tube is a device that is shaped in such a manner that it senses stagnation pressure. The name "Pitot tube" has been applied to two general classifications of instruments, the first being a tube that measures the impact or stagnation pressures only, and the second a combined tube that measures both impact and static pressures with a single primary instrument. The combined sensor is called a Pitot-static tube.

Tube Coefficient From Fig. 3.3.29, it is evident that the Pitot tube can sense only the stagnation pressure resulting from the local stream-tube velocity U. The local ideal velocity U_i for an incompressible fluid is obtained by the application of the Bernoulli equation ($z_S = z$), $U_i^2/2g + p/\rho g = U_S^2/2g + p_S/\rho g$. Solving for U_i and noting that by definition

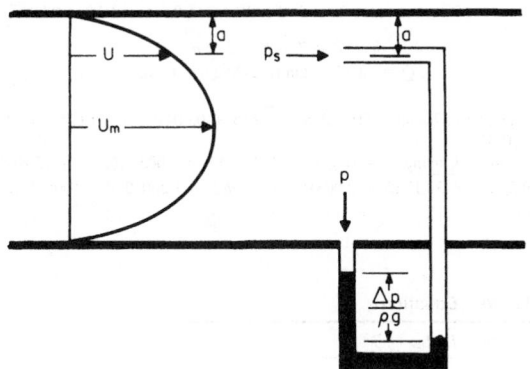

Fig. 3.3.29 Notation for Pitot tube study.

$U_S = 0$, $U_i = \sqrt{2(p_S - p)/\rho}$. Conventional practice is to define the **tube coefficient** C_T as the ratio of the actual stream-tube velocity to the ideal stream-tube velocity, or $C_T = U/U_i$ and $U = C_T U_i = C_T\sqrt{2\Delta p/\rho}$. The numerical value of C_T depends primarily upon its geometry. The value of C_T may be established (1) by calibration with a uniform velocity, (2) from published data for similar geometry, or (3) in the absence of other information, may be assumed to be unity.

Pipe Coefficient For the calculation of volumetric flow rate, it is necessary to integrate the continuity equation, $Q = \int U\,da = AV$. The **pipe coefficient** C_P is defined as the ratio of the average velocity to the stream-tube velocity, or $C_P = V/U$, and $Q = C_P A_1 V = C_P C_T A_1 \sqrt{2\Delta p/\rho}$. The numerical value of C_P is dependent upon the location of the tube and the **velocity profile.** The values of C_P may be established by (1) making a "traverse" by taking data at various points in the flow stream and determining the velocity profile experimentally (see "Fluid Meters," 6th ed., ASME, 1971, for locations of traverse points), (2) using standard velocity profiles, (3) locating the Pitot tube at a point where $U = V$, and (4) assuming one-dimensional flow of $C_P = 1$ **only** in the absence of other data.

Compressible Flow For compressible flow, the **compression factor** Z is based on the assumption of a frictionless adiabatic (isentropic) compression of an ideal gas from the moving stream tube to the stagnation point (see Sec. 4.1), which results in

$$Z = \left[\frac{k}{k-1}\frac{(p_S/p)^{(k-1)/k} - 1}{(p_S/p) - 1}\right]^{1/2}$$

and the volumetric flow rate becomes

$$Q = C_P C_T Z A_1 \sqrt{2\Delta p/\rho}$$

EXAMPLE. Carbon dioxide flows at 68°F (20°C) and 20 lbf/in^2 in an 8-in schedule 40 galvanized-iron pipe. A Pitot tube located on the pipe centerline indicates a pressure differential of 6.986 lbf/in^2. Estimate the mass flow rate. For 8-in schedule 40 pipe $D = 7.981/12 = 0.6651$, $\varepsilon/D = 500 \times 10^{-6}/0.6651 = 7.518 \times 10^{-4}$, $A_1 = \pi D^2/4 = (\pi/4)(0.6651)^2 = 0.3474 \text{ ft}^2$, $p_S = p + \Delta p = 20 + 6.986 = 26.986$ lbf/in^2. From the equation of state, $\rho = p/g_c RT_o = (20 \times 144)/(32.17)(35.11)(68 + 459.7) = 0.004832$,

$$Z = \left[\frac{k}{k-1}\frac{(p_S/p)^{(k-1)/k} - 1}{(p_S/p) - 1}\right]^{1/2}$$

$$= \left\{[1.3/(1.3 - 1)] \times \frac{(26.986/20)^{(1.3-1)/1.3} - 1}{(26.986/20) - 1}\right\}^{1/2} = 0.9423$$

In the absence of other data, C_T may be assumed to be unity. A trial-and-error solution is necessary to determine C_P, since f requires flow rate. For the first trial assume complete turbulence.

$$1/\sqrt{f} = -2\log_{10}(7.518 \times 10^{-4}/3.7) \qquad \sqrt{f} = 0.1354$$

$$C_P = V/U = V/U_{max} = 1/(1 + 1.43\sqrt{f}) = 1/(1 + 1.43 \times 0.1354) = 0.8378$$

$$V = C_P C_T Z\sqrt{2\Delta P/\rho} = (0.8378)(1)(0.9423)\sqrt{2 \times 144(6.987)/(0.004832)}$$

$$V = 509.4 \text{ ft/s}$$

$$R = \rho V D/\mu = (0.004832)(509.4)(0.6651)/(30.91 \times 10^{-18})$$

$$R = 5,296,000 > 4,000 \ \therefore \ \text{flow is turbulent}$$

From the Colebrook equation and Fig. 3.3.24,

$$\frac{1}{\sqrt{f}} = -2\log_{10}\left(\frac{7.518 \times 10^{-4}}{3.7} + \frac{2.51}{5,296,000\sqrt{0.018}}\right)$$

$$\sqrt{f} = 0.1357$$

$$C_P = 1/(1 + 1.43 \times 0.1357) = 0.8375 \qquad \text{(close check)}$$

$$V = 509.4(0.8375/8378) = 509.2 \text{ ft/s}$$

From the continuity equation, $m = \rho A_1 V g_c = (0.004832)(0.3474)(509.2)(32.17) = 27.50$ lbm/s (12.47 kg/s).

ASME WEIRS

Definitions A weir is a dam over which liquids are forced to flow. Weirs are used to measure the flow of liquids in open channels or in conduits which do not flow full; i.e., there is a free liquid surface. Weirs are almost exclusively used for measuring water flow, although small ones have been used for metering other liquids. Weirs are classified according to their notch or opening as follows: (1) **rectangular notch** (original form); (2) V or **triangular notch**; (3) **trapezoidal notch**, which when designed with end slopes one horizontal to four vertical is called the **Cipolletti weir;** (4) the **hyperbolic weir** designed to give a constant coefficient of discharge; and (5) the **parabolic weir** designed to give a linear relationship of head to flow. As shown in Fig. 3.3.30, the top of the weir is the crest and the distance from the liquid surface to the crest h is called the **head.**

The sheet of liquid flowing over the weir crest is called the **nappe.** When the nappe falls downstream of the weir plate, it is said to be free, or **aerated.** When the width of the approach channel L_c is greater than the crest length L_w, the nappe will contract so that it will have a minimum width less than the crest length. For this reason, the weir is known as a **contracted weir.** For the special case where $L_w = L_c$, the contractions do not take place, and such weirs are known as **suppressed weirs.**

Fig. 3.3.30   Notation for weir study.

Parameters   The forces acting on a liquid flowing over a weir are inertia, viscous, surface tension. and gravity. If the weir head produced by the flow is h, the characteristic length of the weir is L_w, and the channel width is L_c, either similarity or dimensional analysis leads to $f(\mathbf{F}, \mathbf{W}, \mathbf{R}, L_w/L_c) = 0$, which may be written as $V = K\sqrt{2gh}$, where K is the **weir coefficient** and $K = f(\mathbf{W}, \mathbf{R}, L_w/L_c)$. Since the weir has been almost exclusively used for metering water flow over limited temperature ranges, the effects of surface tension and viscosity have not been adequately established by experiment.

Caution   The numerical values of coefficients for weirs are based on experimental data obtained from calibration of weirs with long approaches of straight channels. Head measurement should be made at a distance at least three or four times the expected maximum head h. Screens and baffles should be used as necessary to ensure steady uniform flow without waves or local eddy currents. The approach channel should be relatively wide and deep.

Rectangular Weirs   Figure 3.3.31 shows a rectangular weir whose crest width is L_w. The volumetric flow rate may be computed from the continuity equation: $Q = AV = (L_w h)(K\sqrt{2gh}) = KL_w\sqrt{2g}\,h^{3/2}$. The ASME "Fluid Meters" report recommends the following equation for rectangular weirs: $Q = (\frac{2}{3})CL_a\sqrt{2g}\,h_a^{3/2}$, where C is the **coefficient of discharge** $C = f(L_w/L_c, h/Z)$, L_a is the adjusted crest length $L_a = L_w + \Delta L$, and h_a is the adjusted weir head $h_a = h + 0.003$ ft. Values of C and ΔL may be obtained from Table 3.3.15. To avoid the possibility that the liquid drag along the sides of the channel will affect side contractions, $L_c - L_w$ should be at least $4h$. The minimum crest length should be 0.5 ft to prevent mutual interference of the end contractions The minimum head for free flow of the nappe should be 0.1 ft.

EXAMPLE.   Water flows in a channel whose width is 40 ft. At the end of the channel is a rectangular weir whose crest width is 10 ft and whose crest height is 4 ft. The water flows over the weir at a height of 3 ft above the crest of the weir. Estimate the volumetric flow rate. $L_w/L_c = 10/40 = 0.25$, $h/Z = 3/4 = 0.75$, from Table 3.3.15 (interpolated), $C = 0.589$, $\Delta L = 0.008$, $L_a = L_w + \Delta L = 10 + 0.008 = 10.008$ ft, $h_a = h + 0.003 = 3 + 0.003 = 3.003$ ft, $Q = (2/3) CL_a\sqrt{2g}\,h^{3/2}$, $Q = (2/3)(0.589)(10.008)(2 \times 32.17)^{1/2}(3.003)^{3/2}$ $Q = 164.0$ ft³/s (4.644 m³/s).

Fig. 3.3.31   Rectangular weir.

Triangular Weirs   Figure 3.3.32 shows a triangular weir whose notch angle is θ. The volumetric flow rate may be computed from the continuity equation $Q = AV = (h^2 \tan \theta/2)(K\sqrt{2gh}) = K \tan(\theta/2)\sqrt{2g}\,h^{5/2}$. The ASME "Fluid Meters" report recommends the following for triangular weirs: $Q = (8/15)\,C \tan(\theta/2)\sqrt{2g}(h + \Delta h)^{5/2}$, where C is the coefficient of discharge $C = f(\theta)$ and Δh is the correction for head/crest ratio $\Delta h = f(\theta)$. Values of C and Δh may be obtained from Table 3.3.16.

EXAMPLE.   It is desired to maintain a flow of 167 ft³/s in an open channel whose width is 20 ft at a height of 7 ft by locating a triangular weir at the end of the channel. The weir has a crest height of 2 ft. What notch angle is required to maintain these conditions? A trial-and-error solution is required. For the first trial assume $\theta = 60°$ (mean value 20 to 100°); then $C = 0.576$ and $\Delta h = 0.004$.

$$h + Z = 7 = h + 2 \therefore h = 5$$
$$Q = (8/15)\,C \tan(\theta/2)\sqrt{2g}(h + \Delta h)^{5/2}$$

$167 = (8/15)(0.576) \tan(\theta/2)\sqrt{2 \times 32.17}\,(5 + 0.004)^{5/2}$, $\tan^{-1}(\theta/2) = 1.20993$, $\theta = 100°51'$.

Second trial, using $\theta = 100$, $C = 0.581$, $\Delta h = 0.003$, $167 = (8/15)(0.581) \tan(\theta/2)\sqrt{2 \times 32.17}\,(5 + 0.003)^{5/2}$, $\tan^{-1}(\theta/2) = 1.20012$, $\theta = 100°39'$ (close check).

Table 3.3.15   Values of C and ΔL for Use in Rectangular-Weir Equation

| h/Z | \multicolumn{8}{c|}{Crest length/channel width = L_w/L_c} |
	0	0.2	0.4	0.6	0.7	0.8	0.9	1.0
				Coefficient of discharge C				
0	0.587	0.589	0.591	0.593	0.595	0.597	0.599	0.603
0.5	0.586	0.588	0.594	0.602	0.610	0.620	0.631	0.640
1.0	0.586	0.587	0.597	0.611	0.625	0.642	0.663	0.676
1.5	0.584	0.586	0.600	0.620	0.640	0.664	0.695	0.715
2.0	0.583	0.586	0.603	0.629	0.655	0.687	0.726	0.753
2.5	0.582	0.585	0.608	0.637	0.671	0.710	0.760	0.790
3.0	0.580	0.584	0.610	0.647	0.687	0.733	0.793	0.827
				Adjustment for crest length ΔL, ft				
Any	0.007	0.008	0.009	0.012	0.013	0.014	0.013	−0.005

SOURCE: Compiled from data given in "Fluid Meters," ASME, 1971.

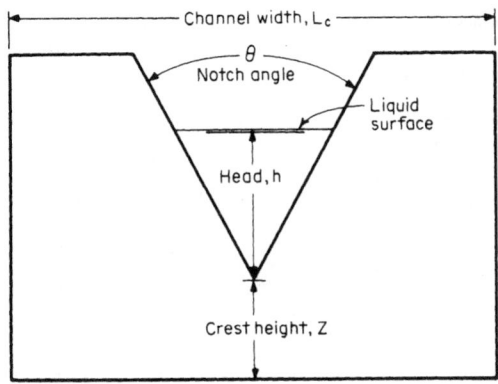

Fig. 3.3.32   Triangular weir.

Table 3.3.16   Values of *C* and Δ*h* for Use in Triangular-Weir Equation

Item	Weir notch angle θ, deg						
	20	30	45	60	75	90	100
C	0.592	0.586	0.580	0.576	0.576	0.579	0.581
Δh, ft	0.010	0.007	0.005	0.004	0.003	0.003	0.003

SOURCE: Compiled from data given in "Fluid Meters." ASME, 1971.

OPEN-CHANNEL FLOW

Definitions   An **open channel** is a conduit in which a liquid flows with a free surface subjected to a constant pressure. Flows of water in natural streams, artificial canals, irrigation ditches, sewers, and flumes are examples where the water surface is subjected to atmospheric pressure. The flow of any liquid in a pipe where there is a free liquid surface is an example of open-channel flow where the liquid surface will be subjected to the pressure existing in the pipe. The **slope** S of a channel is the change in elevation per unit of horizontal distance. For small slopes, this is equivalent to dividing the change in elevation by the distance L measured along the channel bottom between two sections. For steady uniform flow, the velocity distribution is the same at all sections of the channel, so that the energy grade line has the same angle as the bottom of the channel, thus:

$$S = h_f/L$$

The distance between the liquid surface and the bottom of the channel is sometimes called the **stage** and is denoted by the symbol y in Fig. 3.3.33. When the stages between the sections are not uniform, that is, $y_1 \neq y_2$ or the cross section of the channel changes, or both, the flow is said to be **varied**. When a liquid flows in a channel of uniform cross section and the slope of the surface is the same as the slope of the bottom of the channel ($y_1 = y = y_2$), the flow is said to be **uniform**.

Fig. 3.3.33   Notation for open channel flow.

Parameters   The forces acting on a liquid flowing in an open channel are inertia, viscous, surface tension, and gravity. If the channel has a surface roughness of ε, a hydraulic radius of R_h, and a slope of S, either similarity or dimensional analysis leads to $f(\mathbf{F}, \mathbf{W}, \mathbf{R}, \varepsilon/4R_h) = 0$, which may be written as $V = C\sqrt{R_h S}$, where $C = f(\mathbf{W}, \mathbf{R}, \varepsilon/4R_h)$ and is known as the **Chézy coefficient.** The relationship between the Chézy coefficient C and the friction factor may be determined by equating

$$V = \sqrt{8R_h h_f g/fL} = C\sqrt{R_h S} = C\sqrt{(R_h h_f)/L}$$

or $C = (8g/f)^{1/2}$. Although this establishes a relationship between the Chézy coefficient and the friction factor, it should be noted that $f = f(\mathbf{R}, \varepsilon/4R_h)$ and $C = f(\mathbf{W}, \mathbf{R}, \varepsilon/4R_h)$, because in open-channel flow, pressure forces are absent and in pipe flow, surface-tension and gravity forces are absent. For these reasons, data obtained in pipe flow should not be applied to open-channel flow.

Roughness Factors   For open-channel flow, the Chézy coefficient is calculated by the Manning equation, which was developed from examination of experimental results of water tests. The **Manning relation** is stated as

$$C = \frac{1.486}{n} R_h^{1/6}$$

where n is a roughness factor and should be a function of Reynolds number, Weber number, and relative roughness. Since only water-test data obtained at ordinary temperatures support these values, it must be assumed that n is the value for turbulent flow only. Since surface tension is a weak property, the effects of Weber-number variation are negligible, leaving n to be some function of surface roughness. Design values of n are given in Table 3.3.17. Maximum flow for a given slope will take place when R_h is a maximum, and values of R_h max are given in Table 3.3.6.

Table 3.3.17   Values of Roughness Factor *n* for Use in Manning Equation

Surface	n	Surface	n
Brick	0.015	Earth, with stones	0.035
Cast iron	0.015	and weeds	
Concrete, finished	0.012	Gravel	0.029
Concrete, unfinished	0.015	Riveted steel	0.017
Brass pipe	0.010	Rubble	0.025
Earth	0.025	Wood, planed	0.012
		Wood, unplaned	0.013

SOURCE: Compiled from data given in R. Horton, *Engineering News*, 75, 373, 1916.

EXAMPLE.   It is necessary to carry 150 ft³/s of water in a rectangular unplaned timber flume whose width is to be twice the depth of water. What are the required dimensions for various slopes of the flume? From Table 3.3.6, $A = b^2/2$ and $R_h = h/2 = b/4$. From Table 3.3.17, $n = 0.013$ for unplaned wood. From Manning's equation, $C = 1.486/n$, $R_h^{1/6} = (1.486/0.013)(b^{1/6}/(4)^{1/6} = 90.73 b^{1/6}$. From the continuity equation, $V = Q/A = 150/(b^2/2)$, $V = 300/b^2$. From the Chézy equation, $V = C\sqrt{R_h S} = 300/b^2 = 90.73 b^{1/6}\sqrt{(b/4)S}$; solving for b, $b = 2.0308/S^{3/16}$.

Assumed S: 1×10^{-1} 1×10^{-2} 1×10^{-3} 1×10^{-4} 1×10^{-5} 1×10^{-6} ft/ft
Required b: 3.127   4.816   7.416   11.42   17.59   27.08 ft

EXAMPLE.   A rubble-lined trapezoidal canal with 45° sides is to carry 360 ft³/s of water at a depth of 4 ft. If the slope is 9×10^{-4} ft/ft, what should be the dimensions of the canal? From Table 3.3.17, $n = 0.025$ for rubble. From Table 3.3.6 for $\alpha = 45°$, $A = (b + h)h = 4(b + 4)$, and $R_h = (b + h)h/(b + 2.828h) = 4(b + 4)/(b + 11.312)$. From the Manning relation, $C = (1.486/n)(R_h^{1/6}) = (1.486/0.025)R_h^{1/6} = 59.44 R_h^{1/6}$. For the first trial, assume $R_h = R_h$ max $= h/2 = 4/2 = 2$; then $C = 59.44(2)^{1/6} = 66.72$ and $V = C\sqrt{R_h S} = 66.72 \sqrt{2 \times 9 \times 10^{-4}} = 2.831$. From the continuity equation, $A = Q/V = 360/2.831 = 127.2 = 4(b + 4)$; $b = 27.79$ ft. Second trial, use the first trial, $R_h = 4(27.79 + 4)/(27.79 + 11.312)$, $R_h = 3.252$, $V = 59.44(3.252)^{1/6}\sqrt{3.252 \times 9 \times 10^{-4}} = 3.914$. From the equation of continuity, $Q/V = 360/3.914 = 91.97 = 4(b + 4)$, $b = 18.99$. Subsequent trial-and-error solutions result in a balance at $b = 19.93$ ft (6.075 m).

Specific Energy Specific energy is defined as the energy of the fluid referred to the bottom of the channel as the datum. Thus the specific energy E at any section is given by $E = y + V^2/2g$; from the continuity equation $V = Q/A$ or $E = y + (Q/A)^2/2g$. For a rectangular channel whose width is b, $A = by$; and if $\mathbf{q}$ is defined as the flow rate per unit width, $\mathbf{q} = Q/b$ and $E = y + (\mathbf{q}b/by)^2/2g = y + (\mathbf{q}/y)^2/2g$.

Critical Values For rectangular channels, if the specific-energy equation is differentiated and set equal to zero, critical values are obtained; thus $dE/dy = d/dy[y + (\mathbf{q}/y)^2/2g] = 0 = 1 - \mathbf{q}^2/y^3g$ or $\mathbf{q}_c^2 = y_c^3 g$. Substituting in the specific-energy equation, $E = y_c + y_c^3 g/2gy_c^2 = 3/2y_c$. Figure 3.3.34 shows the relation between depth and specific energy for a constant flow rate. If the depth is greater than critical, the flow is *subcritical;* at critical depth it is *critical* and at depths below critical the flow is *supercritical.* For a given specific energy, there is a maximum unit flow rate that can exist.

Fig. 3.3.34 Specific energy diagram, constant flow rate.

The **Froude number** $\mathbf{F} = V/\sqrt{gy}$, when substituted in the specific-energy equation, yields $E = y + (\mathbf{F}^2 gy)/2g = y(1 + \mathbf{F}^2/2)$ or $E/y = 1 + \mathbf{F}^2/2$. For critical flow, $E_c/y_c = 3/2$. Substituting $E_c/y_c = 3/2 = 1 + \mathbf{F}_c^2/2$, or $\mathbf{F} = 1$,

$$\mathbf{F} < 1 \qquad \text{Flow is subcritical}$$
$$\mathbf{F} = 1 \qquad \text{Flow is critical}$$
$$\mathbf{F} > 1 \qquad \text{Flow is supercritical}$$

It is seen that for open-channel flow the Froude number determines the type of flow in the same manner as Mach number for compressible flow.

EXAMPLE.   Water flows at a ate of 600 ft³/s in a rectangular channel 10 ft wide at a depth of 4 ft. Determine (1) specific energy and (2) type of flow.
1. from the continuity equation,

$$V = Q/A = 600/(10 \times 4) = 15 \text{ ft/s}$$
$$E = y + V^2/2g = 4 + (15)^2/2(2 \times 32.17) = 7.497 \text{ ft}$$

2. $\mathbf{F} = V/\sqrt{gy} = 15/\sqrt{32.17 \times 4} = 1.322$; $\mathbf{F} > 1$ ∴ flow is supercritical.

FLOW OF LIQUIDS FROM TANK OPENINGS

Steady State Consider the jet whose velocity is V discharging from an open tank through an opening whose area is a, as shown in Fig. 3.3.35. The liquid height above the centerline is h, and the cross-sectional area of the tank at h is A. The ideal velocity of the jet is $V_i = \sqrt{2gh}$. The ratio of the actual velocity V to the ideal velocity V_i is the **coefficient of velocity** Cv, or $V = C_v V_i = C_v \sqrt{2gh}$. The ratio of the actual opening a to the minimum area of the jet a_c is the **coefficient of contraction** C_c, or $a = C_c a_c$. The ratio of the actual discharge Q to the ideal discharge Q_i is

the **coefficient of discharge** C, or $Q = CQ_i = C_a V_i = C_c C_v a \sqrt{2gh}$, and $C = C_c C_v$. Nominal values of coefficients for various openings are given in Fig. 3.3.36.

Fig. 3.3.35 Notation for tank flow.

Unsteady State If the rate of liquid entering the tank Q_{in} is different from that leaving, the level h in the tank will change because of the change in storage. For liquids, the conservation-of-mass equation may be written as $Q_{in} - Q_{out} = Q_{stored}$; for a time interval dt, $(Q_{in} - Q_{out}) dt = A\, dh$,

Type	Coefficient		
	c	c_c	c_v
Sharp-edged orifice	0.61	0.62	0.98
Rounded-edged orifice	0.98	1.00	0.98
Short tube $\frac{L}{D} \sim 1$	0.80	1.00	0.80
Borda	0.51	0.52	0.98

Fig. 3.3.36 Nominal coefficients of orifices.

neglecting fluid acceleration,

$$Q_{out}\, dt = Ca\sqrt{2gh}\, dt, \text{ or } (Q_{in} - Ca\sqrt{2gh})\, dt$$

$$= A\, dh, \text{ or } \int_{t_1}^{t_2} dt = \int_{h_1}^{h_2} \frac{A\, dh}{Q_{in} - Q_{out}}$$

$$= \int_{h_1}^{h_2} \frac{A\, dh}{Q_{in} - Ca\sqrt{2gh}}$$

EXAMPLE.   An open cylindrical tank is 6 ft in diameter and is filled with water to a depth of 10 ft. A 4-in-diameter sharp-edged orifice is installed on the bottom of the tank. A pipe on the top of the tank supplies water at the rate of 1 ft³/s. Estimate (1) the steady-state level of this tank, (2) the time required to reduce the tank level by 2 ft.

1. *Steady-state level.* From Fig. 3.3.36, $C = 0.61$ for a sharp-edged orifice, $a = (\pi/4)d^2 = (\pi/4)(4/12)^2 = 0.08727 \text{ ft}^2$. For steady state, $Q_{in} = Q_{out} = Ca\sqrt{2gh} = 1 = (0.61)(0.08727)(2 \times 32.17h)^{1/2}$; $h = 5.484$ ft.

2. Time required to lower level 2 ft, $A = (\pi/4)D^2 = (\pi/4)(6)^2 = 28.27 \text{ ft}^2$

$$t_2 - t_1 = \int_{h_1}^{h_2} \frac{A\, dh}{Q_{in} - Ca\sqrt{2gh}}$$

This equation may be integrated by letting $Q = Ca\sqrt{2g}h^{1/2}$; then $dh = 2Q\, dQ/(Ca\sqrt{2g})^2$; then

$$t_2 - t_1 = \frac{2A}{(Ca\sqrt{2g})^2}\left[Q_{in}\log_e\left(\frac{Q_{in} - Q_1}{Q_{in} - Q_2}\right) + Q_1 - Q_2\right]$$

At t_1: $Q_1 = 0.61 \times 0.08727\sqrt{2 \times 32.17 \times 10} = 1.350 \text{ ft}^3/\text{s}$

At t_2: $Q_2 = 0.61 \times 0.08727\sqrt{2 \times 32.17 \times 8} = 1.208 \text{ ft}^3/\text{s}$

$$t_2 - t_1 = \frac{2 \times 28.27}{(0.61 \times 0.08727\sqrt{2 \times 32.17})^2}$$
$$\times\left[(1)\log_e\left(\frac{1 - 1.350}{1 - 1.208}\right) + 1.350 - 1.208\right]$$

$$t_2 - t_1 = 205.4 \text{ s}$$

WATER HAMMER

Equations   Water hammer is the series of shocks, sounding like hammer blows, produced **by suddenly reducing the flow** of a fluid in a pipe. Consider a fluid flowing frictionlessly in a rigid pipe of uniform area A with a velocity V. The pipe has a length L, and inlet pressure p_1 and a pressure p_2 at L. At length L, there is a valve which can suddenly reduce the velocity at L to $V - \Delta V$. The equivalent mass rate of flow of a pressure wave traveling at sonic velocity c, $M = \rho Ac$. From the impulse-momentum equation, $M(V_2 - V_1) = p_2A_2 - p_1A_1$; for this application, $(\rho Ac)(V - \Delta V - V) = p_2A - p_1A$, or the increase in pressure $\Delta p = -\rho c\Delta V$. When the liquid is flowing in an elastic pipe, the

equation for pressure rise must be modified to account for the expansion of the pipe; thus

$$c = \sqrt{\frac{E_s}{\rho[1 + (E_s/E_p)(D_o + D_i)/(D_o - D_i)]}}$$

where ρ = mass density of the fluid, E_s = bulk modulus of elasticity of the fluid, E_p = modulus of elasticity of the pipe material, D_o = outside diameter of pipe, and D_i = inside diameter of pipe.

Time of Closure   The time for a pressure wave to travel the length of pipe L and return is $t = 2L/c$. If the time of closure $t_c \leq t$, the approximate pressure rise $\Delta p \approx -2\,\rho V(L/t_c)$. When it is not feasible to close the valve slowly, **air chambers** or **surge tanks** may be used to absorb all or most of the pressure rise. Water hammer can be very dangerous. See Sec. 9.9.

EXAMPLE.   Water flows at 68°F (20°C) in a 3-in steel schedule 40 pipe at a velocity of 10 ft/s. A valve located 200 ft downstream is suddenly closed. Determine (1) the increase in pressure considering pipe to be rigid, (2) the increase considering pipe to be elastic, and (3) the maximum time of valve closure to be considered "sudden."

For water, $\rho = -1.937 \text{ slugs/ft}^3 = 1.937 \text{ lb} \cdot \text{sec}^2/\text{ft}^4$; $E_s = 319,000 \text{ lb/in}^2$; $E_p = 28.5 \times 10^6 \text{ lb/in}^2$ (Secs. 5.1 and 6); $c = 4,860$ ft/s; from Sec. 8.7, $D_o = 3.5$ in, $D_i = 3.068$ in.

1. *Inelastic pipe*

$$\Delta p = -\rho c\Delta V = -(1.937)(4,860)(-10) = 94,138 \text{ lbf/ft}^2$$
$$= 94,138/144 = 653.8 \text{ lbf/in}^2\,(4.507 \times 10^6 \text{ N/m}^2)$$

2. *Elastic pipe*

$$c = \sqrt{\frac{E_s}{\rho[1 + (E_s/E_p)(D_o + D_i)/(D_o - D_i)]}}$$
$$= \sqrt{\frac{319,000 \times 144}{1.937\left[1 + \frac{(319,000/28.5 \times 10^6)(3.500 + 3.067)}{(3.500 - 3.067)}\right]}}$$
$$= 4,504$$

$$\Delta p = -(1.937)(4,504)(-10)$$
$$= 87,242 \text{ lbf/ft}^2 = 605.9 \text{ lbf/in}^2\,(4.177 \times 10^6 \text{ N/m}^2)$$

3. *Maximum time for closure*

$$t = 2L/c = 2 \times 200/4,860 = 0.08230 \text{ s or less than } 1/10 \text{ s}$$

COMPUTATIONAL FLUID DYNAMICS (CFD)

The partial differential equations of fluid motion are very difficult to solve. CFD methods are utilized to provide discrete approximations for the solution of those equations. A brief introduction to CFD is included in Section 20.6. The references therein will guide the reader further.

3.4 VIBRATION
by Leonard Meirovitch

REFERENCES: Harris, "Shock and Vibration Handbook," 3d ed., McGraw-Hill. Thomson, "Theory of Vibration with Applications," Prentice Hall. Meirovitch, "Fundamentals of Vibration," McGraw-Hill. Meirovitch, "Principles and Techniques of Vibrations," Prentice-Hall.

SINGLE-DEGREE-OF-FREEDOM SYSTEMS

Discrete System Components   A system is defined as an aggregation of components acting together as one entity. The components of a vibratory mechanical system are of three different types, and they relate forces to displacements, velocities, and accelerations. The component relating forces to displacements is known as a spring (Fig. 3.4.1a). For a **linear spring** the force F_s is proportional to the elongation $\delta = x_2 - x_1$, or

$$F_s = k\delta = k(x_2 - x_1) \tag{3.4.1}$$

where k represents the **spring constant,** or the **spring stiffness,** and x_1 and x_2 are the displacements of the end points. The component relating

forces to velocities is called a **viscous damper** or a **dashpot** (Fig. 3.4.1b). It consists of a piston fitting loosely in a cylinder filled with liquid so that the liquid can flow around the piston when it moves relative to the cylinder. The relation between the damper force and the velocity of the piston relative to the cylinder is

$$F_d = c(\dot{x}_2 - \dot{x}_1) \tag{3.4.2}$$

in which c is the **coefficient of viscous damping**; note that dots denote derivatives with respect to time. Finally, the relation between forces and accelerations is given by Newton's second law of motion:

$$F_m = m\ddot{x} \tag{3.4.3}$$

where m is the **mass** (Fig. 3.4.1c).

The spring constant k, coefficient of viscous damping c, and mass m represent physical properties of the components and are the **system parameters.** By implication, these properties are concentrated at points, thus they are **lumped,** or **discrete, parameters.** Note that springs and dampers are assumed to be massless and masses are assumed to be rigid.

Springs can be arranged in parallel and in series. Then, the proportionality constant between the forces and the end points is known as an

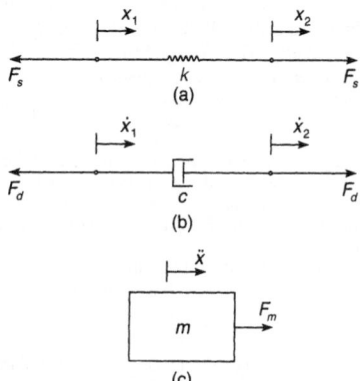

Fig. 3.4.1

Table 3.4.1　Equivalent Spring Constants

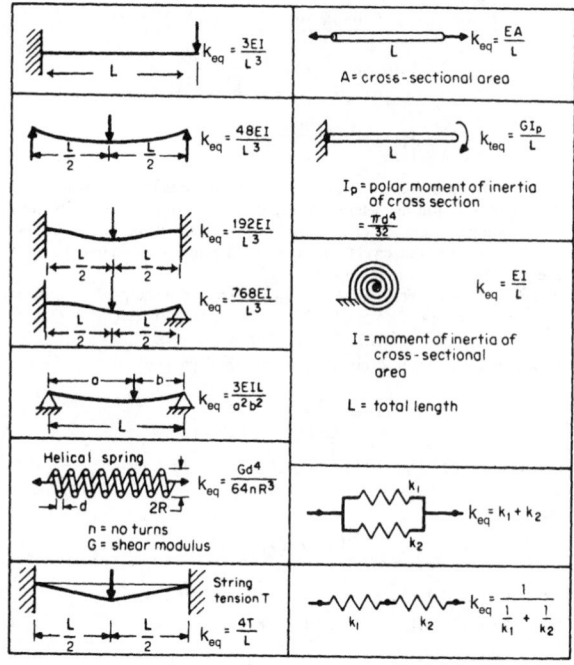

equivalent spring constant and is denoted by k_{eq}, as shown in Table 3.4.1. Certain elastic components, although distributed over a given line segment, can be regarded as lumped with an equivalent spring constant given by $k_{eq} = F/\delta$, where δ is the deflection at the point of application of the force F. A similar relation can be given for springs in torsion. Table 3.4.1 lists the equivalent spring constants for a variety of components.

Equation of Motion　The dynamic behavior of many engineering systems can be approximated with good accuracy by the **mass-damper-spring model** shown in Fig. 3.4.2. Using Newton's second law in conjunction with Eqs. (3.4.1)–(3.4.3) and measuring the displacement $x(t)$ from the static equilibrium position, we obtain the differential equation of motion

$$m\ddot{x}(t) + c\dot{x}(t) + kx(t) = F(t) \qquad (3.4.4)$$

which is subject to the **initial conditions** $x(0) = x_0$, $\dot{x}(0) = v_0$, where x_0 and v_0 are the **initial displacement** and **initial velocity,** respectively. Equation (3.4.4) is in terms of a single coordinate. namely $x(t)$; the system of Fig. 3.4.2 is therefore said to be a **single-degree-of-freedom system.**

Free Vibration of Undamped Systems　Assuming zero damping and external forces and dividing Eq. (3.4.4) through by m, we obtain

$$\ddot{x} + \omega_n^2 x = 0 \qquad \omega_n = \sqrt{k/m} \qquad (3.4.5)$$

In this case, the vibration is caused by the initial excitations alone. The solution of Eq. (3.4.5) is

$$x(t) = A \cos(\omega_n t - \phi) \qquad (3.4.6)$$

which represents **simple sinusoidal,** or **simple harmonic oscillation** with **amplitude** A, **phase angle** ϕ, and **frequency**

$$\omega_n = \sqrt{k/m} \qquad \text{rad/s} \qquad (3.4.7)$$

Systems described by equations of the type (3.4.5) are called **harmonic oscillators.** Because the frequency of oscillation represents an inherent property of the system, independent of the initial excitation, ω_n is called the **natural frequency.** On the other hand, the amplitude and

phase angle do depend on the initial displacement and velocity, as follows:

$$A = \sqrt{x_0^2 + (v_0/\omega_n)^2} \qquad \phi = \tan^{-1} v_0/x_0\omega_n \qquad (3.4.8)$$

The time necessary to complete one cycle of motion defines the **period**

$$T = 2\pi/\omega_n \qquad \text{seconds} \qquad (3.4.9)$$

The reciprocal of the period provides another definition of the **natural frequency,** namely,

$$f_n = \frac{1}{T} = \frac{\omega_n}{2\pi} \qquad \text{Hz} \qquad (3.4.10)$$

where Hz denotes *hertz* [1 Hz = 1 cycle per second (cps)].

A large variety of vibratory systems behave like harmonic oscillators, many of them when restricted to small amplitudes. Table 3.4.2 shows a variety of harmonic oscillators together with their respective natural frequency.

Free Vibration of Damped Systems　Let $F(t) = 0$ and divide through by m. Then, Eq. (3.4.4) reduces to

$$\ddot{x}(t) + 2\zeta\omega_n\dot{x}(t) + \omega_n^2 x(t) = 0 \qquad (3.4.11)$$

where

$$\zeta = c/2m\omega_n \qquad (3.4.12)$$

is the **damping factor,** a nondimensional quantity. The nature of the motion depends on ζ. The most important case is that in which $0 < \zeta < 1$.

Fig. 3.4.2

Table 3.4.2   Harmonic Oscillators and Natural Frequencies

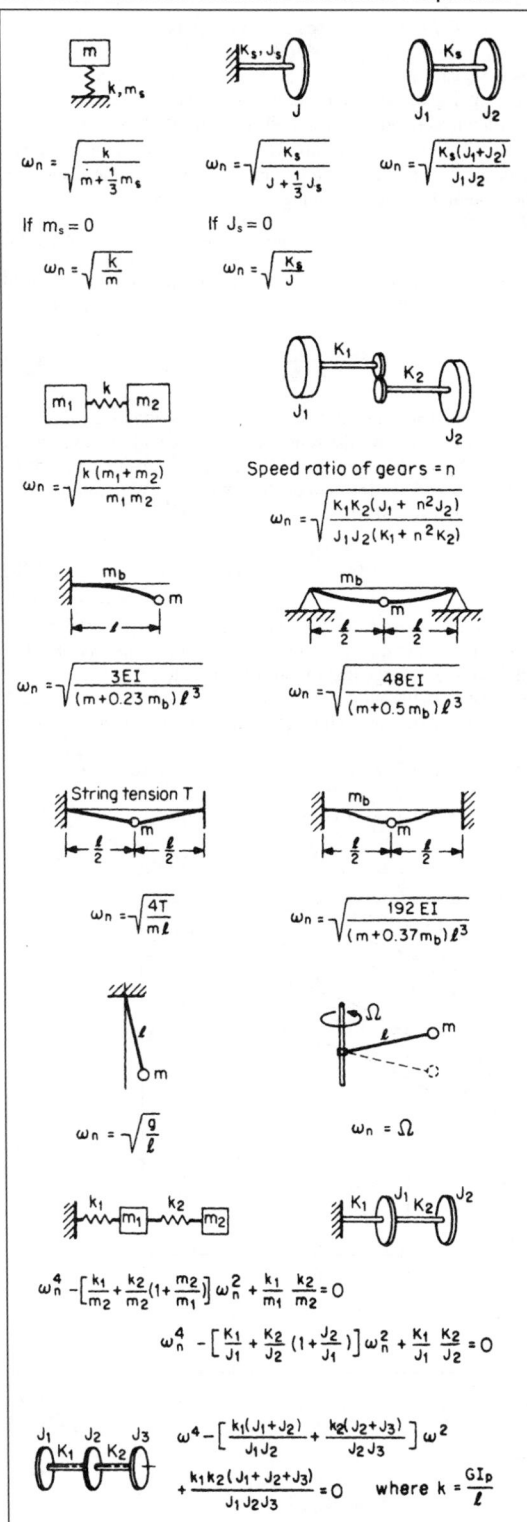

In this case, the system is said to be **underdamped** and the solution of Eq. (3.4.11) is

$$x(t) = Ae^{-\zeta\omega_n t}\cos(\omega_d t - \phi) \qquad (3.4.13)$$

where

$$\omega_d = (1 - \zeta^2)^{1/2}\omega_n \qquad (3.4.14)$$

is the **frequency of damped free vibration** and

$$T = 2\pi/\omega_d \qquad (3.4.15)$$

is the **period of damped oscillation.** The amplitude and phase angle depend on the initial displacement and velocity, as follows:

$$A = \sqrt{x_0^2 + (\zeta\omega_n x_0 + v_0)^2/\omega_d^2} \quad \phi$$
$$= \tan^{-1}(\zeta\omega_n x_0 + v_0)/x_0\omega_d \qquad (3.4.16)$$

The motion described by Eq. (3.4.13) represents **decaying oscillation,** where the term $Ae^{-\zeta\omega_n t}$ can be regarded as a time-dependent amplitude, providing an envelope bounding the harmonic oscillation.

When $\zeta \geq 1$, the solution represents **aperiodic decay.** The case $\zeta = 1$ represents **critical damping,** and

$$c_c = 2m\omega_n \qquad (3.4.17)$$

is the **critical damping coefficient,** although there is nothing critical about it. It merely represents the borderline between oscillatory decay and aperiodic decay. In fact, c_c is the smallest damping coefficient for which the motion is aperiodic. When $\zeta > 1$, the system is said to be **overdamped.**

Logarithmic Decrement   Quite often the damping factor is not known and must be determined experimentally. In the case in which the system is underdamped, this can be done conveniently by plotting $x(t)$ versus t (Fig. 3.4.3) and measuring the response at two different times

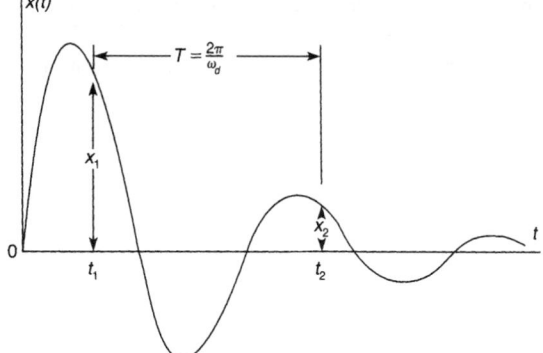

Fig. 3.4.3

separated by a complete period. Let the times be t_1 and $t_1 + T$, introduce the notation $x(t_1) = x_1$, $x(t_1 + T) = x_2$, and use Eq. (3.4.13) to obtain

$$\frac{x_1}{x_2} = \frac{Ae^{-\zeta\omega_n t_1}\cos(\omega_d t_1 - \phi)}{Ae^{-\zeta\omega_n(t_1+T)}\cos[\omega_d(t_1 + T) - \phi]} = e^{\zeta\omega_n T} \qquad (3.4.18)$$

where $\cos[\omega_d(t_1 + T) - \phi] = \cos(\omega_d t_1 - \phi + 2\pi) = \cos(\omega_d t_1 - \phi)$. Equation (3.4.18) yields the **logarithmic decrement**

$$\delta = \ln\frac{x_1}{x_2} = \zeta\omega_n T = \frac{2\pi\zeta}{\sqrt{1 - \zeta^2}} \qquad (3.4.19)$$

which can be used to obtain the damping factor

$$\zeta = \frac{\delta}{\sqrt{(2\pi)^2 + \delta^2}} \qquad (3.4.20)$$

For small damping, the logarithmic decrement is also small, and the damping factor can be approximated by

$$\zeta \approx \frac{\delta}{2\pi} \qquad (3.4.21)$$

Response to Harmonic Excitations Consider the case in which the excitation force $F(t)$ in Eq. (3.4.4) is harmonic. For convenience, express $F(t)$ in the form $kA \cos \omega t$, where k is the spring constant, A is an amplitude with units of displacement and ω is the **excitation frequency.** When divided through by m, Eq. (3.4.4) has the form

$$\ddot{x} + 2\zeta\omega_n\dot{x} + \omega_n^2 x = \omega_n^2 A \cos \omega t \qquad (3.4.22)$$

The solution of Eq. (3.4.22) can be expressed as

$$x(t) = A|G(\omega)| \cos(\omega t - \phi) \qquad (3.4.23)$$

where

$$|G(\omega)| = \frac{1}{\sqrt{[1 - (\omega/\omega_n)^2]^2 + (2\zeta\omega/\omega_n)^2}} \qquad (3.4.24)$$

is a **nondimensional magnitude factor*** and

$$\phi(\omega) = \tan^{-1} \frac{2\zeta\omega/\omega_n}{1 - (\omega/\omega_n)^2} \qquad (3.4.25)$$

is the phase angle; note that both the magnitude factor and phase angle depend on the excitation frequency ω.

Equation (3.4.23) shows that the response to harmonic excitation is also harmonic and has the same frequency as the excitation, but different amplitude $A|G(\omega)|$ and phase angle $\phi(\omega)$. Not much can be learned by plotting the response as a function of time, but a great deal of information can be gained by plotting $|G|$ versus ω/ω_n and ϕ versus ω/ω_n. They are shown in Fig. 3.4.4 for various values of the damping factor ζ.

In Fig. 3.4.4, for low values of ω/ω_n, the nondimensional magnitude factor $|G(\omega)|$ approaches unity and the phase angle $\phi(\omega)$ approaches zero. For large values of ω/ω_n, the magnitude approaches zero (see accompanying footnote about magnification factor) and the phase angle approaches $180°$. The magnitude experiences peaks for $\omega/\omega_n =$

* The term $|G(\omega)|$ is often referred to as *magnification factor,* but this is a misnomer, as we shall see shortly.

$\sqrt{1 - 2\zeta^2}$, provided $\zeta < 1/\sqrt{2}$. The peak values are $|G(\omega)|_{\max} = 1/2\zeta\sqrt{1 - \zeta^2}$. For small ζ, the peaks occur approximately at $\omega/\omega_n = 1$ and have the approximate values $|G(\omega)|_{\max} = Q \approx 1/2\zeta$, where Q is known as the **quality factor.** In such cases, the phase angle tends to $90°$. Clearly, for small ζ the system experiences large-amplitude vibration, a condition known as **resonance.** The points P_1 and P_2, where $|G|$ falls to $Q/\sqrt{2}$, are called **half-power points.** The increment of frequency associated with the half-power points P_1 and P_2 represents the **bandwidth** $\Delta\omega$ of the system. For small damping, it has the value

$$\Delta\omega = \omega_2 - \omega_1 \approx 2\zeta\omega_n \qquad (3.4.26)$$

The case $\zeta = 0$ deserves special attention. In this case, referring to Eq. (3.4.22), the response is simply

$$x(t) = \frac{A}{1 - (\omega/\omega_n)^2} \cos \omega t \qquad (3.4.27)$$

For $\omega/\omega_n < 1$, the displacement is in the same direction as the force, so that the phase angle is zero; the **response is in phase** with the excitation. For $\omega/\omega_n > 1$, the displacement is in the direction opposite to the force, so that the phase angle is $180°$ **out of phase** with the excitation. Finally, when $\omega = \omega_n$ the response is

$$x(t) = \frac{A}{2} \omega_n t \sin \omega_n t \qquad (3.4.28)$$

This is typical of the **resonance** condition, when the response increases without bounds as time increases. Of course, at a certain time the displacement becomes so large that the spring ceases to be linear, thus violating the original assumption and invalidating the solution. In practical terms, unless the excitation frequency varies, passing quickly through $\omega = \omega_n$, the system can break down.

When the excitation is $F(t) = kA \sin \omega t$, the response is

$$x(t) = A|G(\omega)| \sin(\omega t - \phi) \qquad (3.4.29)$$

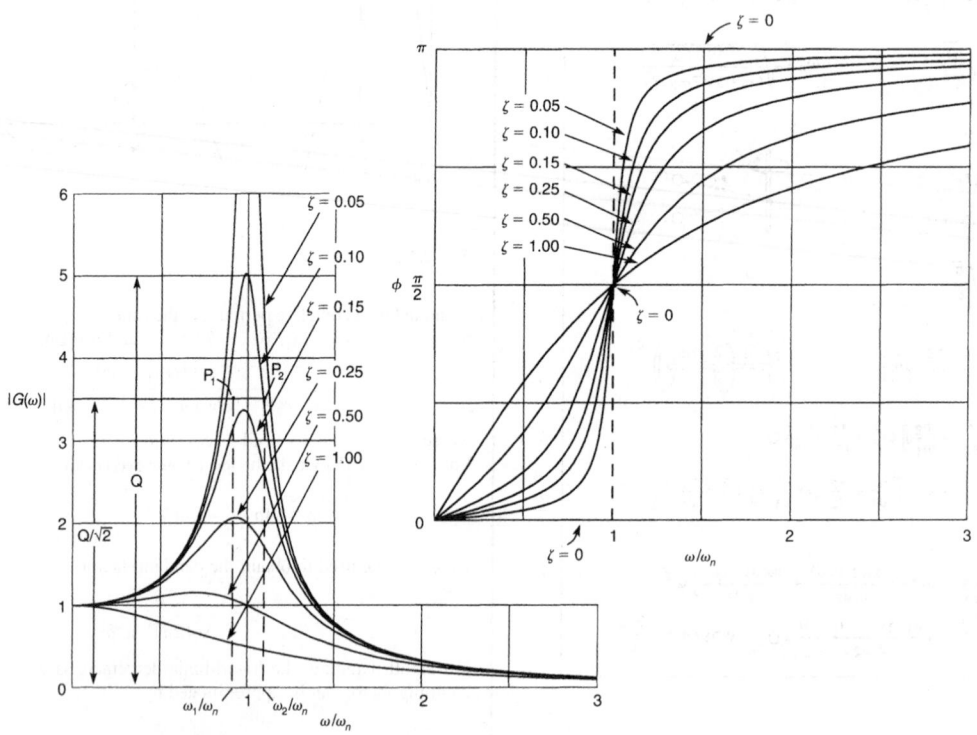

Fig. 3.4.4 Frequency response plots.

One concludes that in **harmonic response,** time plays a secondary role to the frequency. In fact, the only significant information is extracted from the magnitude and phase angle plots of Fig. 3.4.4. They are referred to as **frequency-response plots.**

Since time plays no particular role, the harmonic response is called **steady-state response.** In general, for linear systems with constant parameters, such as the mass-damper-spring system under consideration, the response to the initial excitations is added to the response to the excitation forces. The response to initial excitations, however, represents **transient response.** This is due to the fact that every system possesses some amount of damping, so that the response to initial excitations disappears with time. In contrast, steady-state response persists with time. Hence, in the case of harmonic excitations, it is meaningless to add the response to initial excitations to the harmonic response.

Vibration Isolation   A problem of great interest is the magnitude of the force transmitted to the base by a system of the type shown in Fig. 3.4.2 subjected to harmonic excitation. This force is a combination of the spring force kx and the dashpot force $c\dot{x}$. Recalling Eq. (3.4.23), write

$$kx = kA|G|\cos(\omega t - \phi)$$
$$c\dot{x} = -c\omega A|G|\sin(\omega t - \phi)$$
$$= c\omega A|G|\cos\left(\omega t - \phi + \frac{\pi}{2}\right) \qquad (3.4.30)$$

so that the dashpot force is $90°$ out of phase with the spring force. Hence, the magnitude of the force is

$$F_{tr} = \sqrt{(kA|G|)^2 + (c\omega A|G|)^2} = kA|G|\sqrt{1 + (c\omega/k)^2}$$
$$= kA|G|\sqrt{1 + (2\zeta\omega/\omega_n)^2} \qquad (3.4.31)$$

Let the magnitude of the harmonic excitation be $F_0 = kA$; the measure of the force transmitted to the base is then

$$T = \frac{F_{tr}}{F_0} = |G|\sqrt{1 + (2\zeta\omega/\omega_n)^2}$$
$$= \sqrt{\frac{1 + (2\zeta\omega/\omega n)^2}{[1 - (\omega/\omega_n)^2]^2 + (2\zeta\omega/\omega_n)^2}} \qquad (3.4.32)$$

which represents a nondimensional ratio called **transmissibility.** Figure 3.4.5 plots F_{tr}/F_0 versus ω/ω_n for various values of ζ.

Fig. 3.4.5

The transmissibility is less than 1 for $\omega/\omega_n > \sqrt{2}$, and decreases as ω/ω_n increases. Hence, for **an isolator to perform well,** its natural frequency must be much smaller than the excitation frequency. However, for very low natural frequencies, difficulties can be encountered in isolator design. Indeed, the natural frequency is related to the static deflection δ_{st} by $\omega_n = \sqrt{k/m} = \sqrt{g/\delta_{st}}$, where g is the gravitational constant. For the natural frequency to be sufficiently small, the static deflection may have to be impractically large. The relation between the excitation frequency f measured in rotations per minute and the static deflection δ_{st} measured in inches is

$$f = 187.7\sqrt{\frac{2 - R}{\delta_{st}(1 - R)}} \qquad \text{rpm} \qquad (3.4.33)$$

where $R = 1 - T$ represents **the percent reduction** in vibration. Figure 3.4.6 shows a logarithmic plot of f versus δ_{st} with R as a parameter.

Fig. 3.4.6

Figure 3.4.7 depicts two types of isolators. In Fig. 3.4.7a, isolation is accomplished by means of springs and in Fig. 3.4.7b by rubber rings supporting the bearings. Isolators of all shapes and sizes are available commercially.

Fig. 3.4.7

Rotating Unbalanced Masses   Many appliances, machines, etc., involve components spinning relative to a main body. A typical example is the clothes dryer. Under certain circumstances, the mass of the spinning component is not symmetric relative to the center of rotation, as when the clothes are not spread uniformly in the spinning drum, giving rise to harmonic excitation. The behavior of such systems can be simulated adequately by the single-degree-of-freedom model shown in Fig. 3.4.8, which consists of a main mass $M - m$, supported by two springs of combined stiffness k and a dashpot with coefficient of viscous damping c, and two eccentric masses $m/2$ rotating in opposite sense with the constant angular velocity ω. Although there are three masses, the motion of the eccentric masses relative to the main mass is prescribed, so that there is only one degree of freedom. The equation of motion for the system is

$$M\ddot{x} + c\dot{x} + kx = m/\omega^2 \sin \omega t \qquad (3.4.34)$$

Fig. 3.4.8

Using the analogy with Eq. (3.4.29), the solution of Eq. (3.4.34) is

$$x(t) = \frac{m}{M} l \left(\frac{\omega}{\omega_n}\right)^2 |G(\omega)| \sin(\omega t - \phi) \qquad \omega_n^2 = \frac{k}{M} \quad (3.4.35)$$

The magnitude factor in this case is $(\omega/\omega_n)^2 |G(\omega)|$, where $|G(\omega)|$ is given by Eq. (3.4.24); it is plotted in Fig. 3.4.9. On the other hand, the phase angle remains as in Fig. 3.4.4.

Fig. 3.4.9

Whirling of Rotating Shafts   Many mechanical systems involve rotating shafts carrying disks. If the disk has some eccentricity, then the centrifugal forces cause the shaft to bend, as shown in Fig. 3.4.10a. The rotation of the plane containing the bent shaft about the bearing axis is called **whirling.** Figure 3.4.10b shows a disk with the body axes x, y rotating about the origin O with the angular velocity ω. The geometrical

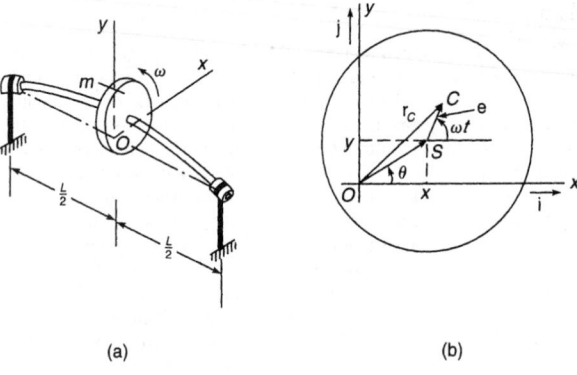

(a)   (b)

Fig. 3.4.10

center of the disk is denoted by S and the mass center by C. The distance between the two points is the eccentricity e. The shaft is massless and of stiffness k_{eq} and the disk is rigid and of mass m. The x and y components of the displacement of S relative to O are independent from one another and, for no damping, satisfy the equations of motion

$$\ddot{x} + \omega_n^2 x = e\omega^2 \cos \omega t \qquad \ddot{y} + \omega_n^2 y = e\omega^2 \sin \omega t \qquad \omega_n^2 = k_{eq}/m$$
$$(3.4.36)$$

where, assuming that the shaft is simply supported (see Table 3.4.1), $k_{eq} = 48EI/L^3$, in which E is the modulus of elasticity, I the cross-sectional area moment of inertia, and L the length of the shaft. By analogy with Eq. (3.4.27), Eqs. (3.4.36) have the solution

$$x(t) = \frac{e(\omega/\omega_n)^2}{1 - (\omega/\omega_n)^2} \cos \omega t \qquad y(t) = \frac{e(\omega/\omega_n)^2}{1 - (\omega/\omega_n)^2} \sin \omega t \quad (3.4.37)$$

Clearly, resonance occurs when the whirling angular velocity coincides with the natural frequency. In terms of rotations per minute, it has the value

$$f_c = \frac{60}{2\pi} \omega_n = \frac{60}{2\pi} \sqrt{\frac{48EI}{mL^3}} \qquad \text{rpm} \qquad (3.4.38)$$

where f_c is called the **critical speed.**

Structural Damping   Experience shows that energy is dissipated in all real systems, including those assumed to be undamped. For example, because of internal friction, energy is dissipated in real springs undergoing cyclic stress. This type of damping is called **structural damping** or **hysteretic damping** because the energy dissipated in one cycle of stress is equal to the area inside the hysteresis loop. Systems possessing structural damping and subjected to harmonic excitation with the frequency ω can be treated as if they possess viscous damping with the equivalent coefficient

$$c_{eq} = \alpha/\pi\omega \qquad (3.4.39)$$

where α is a material constant. In this case, the equation of motion is

$$m\ddot{x} + \frac{\alpha}{\pi\omega}\dot{x} + kx = kA \cos \omega t \qquad (3.4.40)$$

The solution of Eq. (3.4.40) is

$$x(t) = A|G| \cos(\omega t - \phi) \qquad (3.4.41)$$

where this time the magnitude factor and phase angle have the values

$$G = \frac{1}{\sqrt{[1 - (\omega/\omega_n)^2]^2 + \gamma^2}} \qquad \phi = \tan^{-1} \frac{\gamma\omega_n^2}{\omega[1 - (\omega/\omega_n)^2]} \quad (3.4.42)$$

in which

$$\gamma = \frac{\alpha}{\pi k} \qquad (3.4.43)$$

is known as the **structural damping factor.** One word of caution is in order: **the analogy between structural and viscous damping is valid only for harmonic excitation.**

Balancing of Rotating Machines   Machines such as electric motors and generators, turbines, compressors, etc. contain rotors with journals supported by bearings. In many cases, the rotors rotate relative to the bearings at very high rates, reaching into tens of thousands of revolutions per minute. Ideally the rotor is rigid and the axis of rotation coincides with one of its principal axes; by implication, the rotor center of mass lies on the axis of rotation. Such a rotor does not wobble and the only forces exerted on the bearings are due to the weight of the rotor. Such a rotor is said to be **perfectly balanced.** These ideal conditions are seldom realized, and in practice the mass center lies at a distance e (eccentricity) from the axis of rotation, so that there is a net centrifugal force $F = me\omega^2$ acting on the rotor, where m is the mass of the rotor and ω is the rotational speed. This centrifugal force is balanced by reaction forces in the bearings, which tend to wear out the bearings with time.

The rotor unbalance can be divided into two types, static and dynamic. Static unbalance can be detected by placing the rotor on a pair of parallel rails. Then, the mass center will settle in the lowest position in a vertical plane through the rotation axis and below this axis. To balance the rotor statically, it is necessary to add a mass m' in the same plane at a distance r from the rotation axis and above this axis, where m' and r must be such that $m'r = me$. In this manner, the net centrifugal force on the rotor is zero. The net result of **static balancing** is to cause the mass center to coincide with the rotation axis, so that the rotor will remain in any position placed on the rails. However, unless the mass m' is placed on a line containing m and at right angles with the bearings axis, the centrifugal forces on m and m' will form a couple (Fig. 3.4,11), Static balancing is suitable when the rotor is in the form of a thin disk, in which case the couple tends to be small. Automobile tires are at times balanced statically, although strictly speaking they are neither thin nor rigid.

Fig. 3.4.11

In general, for practical reasons, the mass m' cannot be placed on an axis containing m and perpendicular to the bearing axis. Hence, although in static balancing the mass center lies on the rotation axis, the rotor principal axis does not coincide with the bearing axis, as shown in Fig. 3.4.12, causing the rotor to wobble during rotation. In this case, the rotor is said to be **dynamically unbalanced.** Clearly, it is highly desirable to place the mass m' so that the rotor is both statically and dynamically balanced. In this regard, note that the end planes of the rotor are

Fig. 3.4.12

convenient locations to place correcting masses. In Fig. 3.4.13, if the mass center is at a distance a from the right end, then **dynamic balance** can be achieved by placing masses $m'a/L$ and $m'(L - a)/L$ on the intersection of the plane of unbalance and the rotor left end plane and right end plane, respectively. In this manner, the resultant centrifugal force is zero

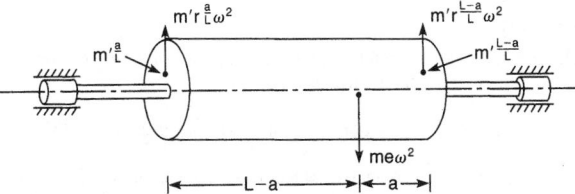

Fig. 3.4.13

and the two couples thus created are equal in value to $m'a(L - a)\omega^2/L$ and opposite in sense, so that they cancel each other. This results in a rotor completely balanced, i.e., balanced statically and dynamically.

The task of determining the magnitude and position of the unbalance is carried out by means of a balancing machine provided with elastically supported bearings permitting the rotor to spin (Fig. 3.4.14). The unbalance causes the bearings to oscillate laterally so that electrical pickups and stroboflash light can measure the amplitude and phase of the rotor with respect to an arbitrary rotor.

In cases in which the rotor is very long and flexible. the position of the unbalance depends on the elastic configuration of the rotor, which in turn depends on the speed of rotation. temperature, etc. In such cases, it is necessary to balance the rotor under normal operating conditions by means of a portable balancing instrument.

Inertial Unbalance of Reciprocating Engines   The crank-piston mechanism of a reciprocating engine produces dynamic forces capable of causing undesirable vibrations. Rotating parts, such as the crankshaft,

Fig. 3.4.14

can be balanced. However, translating parts, such as the piston, cannot be easily balanced, and the same can be said about the connecting rod, which executes a more complex motion of combined rotation and translation.

In the calculation of the unbalanced forces in a single-cylinder engine, the mass of the moving parts is divided into a reciprocating mass and a rotating mass. This is done by apportioning some of the mass of the connecting rod to the piston and some to the crank end. In general, this division of the connecting rod into two lumped masses tends to cause errors in the moment of inertia, and hence in the torque equation. On the other hand, the force equation can be regarded as being accurate. (See also Sec. 8.2.)

Assuming that the rotating mass is counterbalanced, only the reciprocating mass is of concern, and the inertia force for a single-cylinder engine is

$$F = m_{\text{rec}}r\omega^2 \cos \omega t + m_{\text{rec}}\frac{r^2}{L}\omega^2 \cos 2\omega t \qquad (3.4.44)$$

where m_{rec} is the reciprocating mass, r the radius of the crank, ω the angular velocity of the crank, and L the length of the connecting rod. The first component on the right side, which alternates once per revolution, is denoted by X_1 and referred to as the **primary force,** and the second component, which is smaller and alternates twice per revolution, is denoted by X_2 and is called the **secondary force.**

In addition to the inertia force, there is an unbalanced torque about the crankshaft axis due to the reciprocating mass. However, this torque is considered together with the torque created by the power stroke, and the torsional oscillations resulting from these excitations can be mitigated by means of a pendulum-type absorber (see "Centrifugal Pendulum Vibration Absorbers" below) or a torsional damper.

The analysis for the single-cylinder engine can be extended to multicylinder in-line and V-block engines by superposition. For the in-line engine or one block of the V engine, the inertia force becomes

$$F = m_{\text{rec}}r\omega^2\sum_{j=1}^{n} \cos (\omega t + \phi_j)$$
$$+ m_{\text{rec}}\frac{r^2}{L}\omega^2\sum_{j=1}^{n} \cos 2(\omega t + \phi_j) \qquad (3.4.45)$$

where ϕ_j is a phase angle corresponding to the crank position associated with cylinder j and n is the number of cylinders. The vibration force can be eliminated by proper spacing of the angular positions $\phi_j(j = 1, 2, \ldots, n)$.

Even if $F = 0$, there can be pitching and yawing moments due to the spacing of the cylinders. Table 3.4.3 gives the inertial unbalance and pitching of the primary and secondary forces for various crank-angle arrangements of n-cylinder engines.

Centrifugal Pendulum Vibration Absorbers   For a rotating system, such as the crank mechanism just discussed, the exciting torques are proportional to the rotational speed ω, which varies over a wide range. Hence, for a vibration absorber to be effective, its natural frequency must be proportional to ω. The centrifugal pendulum shown in Fig. 3.4.15 is ideally suited to this task. Strictly speaking, the system of Fig. 3.4.15 represents a two-degree-of-freedom nonlinear system. However, assuming that the motion of the wheel consists of a steady rotation ω and a small harmonic oscillation at the frequency Ω, or

$$\theta(t) = \omega t + \theta_0 \sin \Omega t \qquad (3.4.46)$$

Table 3.4.3   Inertial Unbalance of Four-Stroke-per-Cycle Engines

No. n of cylinders	Crank phase angle ϕ_j	Unbalanced forces		Unbalanced pitching moments about 1st cylinder	
		Primary	Secondary	Primary	Secondary
1		X_1	X_2	—	—
2	$0-180°$	0	$2X_2$	ℓX_1	$2\ell X_2$
4	$0-180°-180°-0$	0	$4X_2$	0	$6\ell X_2$
4	$0-90°-270°-180°$	0	0	$\ell X_1\sqrt{1+3^2}$	0
6	$0-120°-240°-$ $240°-120°-0$	0	0	0	0
8	$0-180°-90°-270°-$ $270°-90°-180°-0$	0	0	0	0
90° V-8	$0-90°-270°-180°$	0	0	Rotating primary couple of constant magnitude $\sqrt{10}\ell X_1$ which may be completely counterbalanced	

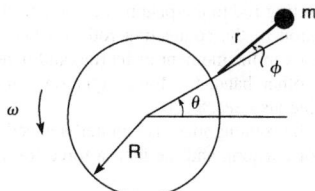

Fig. 3.4.15

and that the pendulum angle ϕ is relatively small, then the equation of motion of the pendulum reduces to the linear single-degree-of-freedom system

$$\ddot{\phi} + \omega_n^2\phi = \frac{R+r}{r}\Omega^2\theta_0\sin\Omega t \qquad (3.4.47)$$

where

$$\omega_n = \omega\sqrt{R/r} \qquad (3.4.48)$$

is the natural frequency of the pendulum. The torque exerted by the pendulum on the wheel is

$$T = -\frac{m(R+r)^2}{1-r\Omega^2/R\omega^2}\ddot{\theta} \qquad (3.4.49)$$

so that the system behaves like a wheel with the effective mass moment of inertia

$$J_{\text{eff}} = -\frac{m(R+r)^2}{1-r\Omega^2/R\omega^2} \qquad (3.4.50)$$

which becomes infinite when Ω is equal to the natural frequency ω_n. To suppress disturbing torques of frequency Ω several times larger than the rotational speed ω, the ratio r/R must be very small, which requires a short pendulum. The **bifilar pendulum** depicted in Fig. 3.4.16, which consists of a U-shaped counterweight that fits loosely and rolls on two pins of radius r_2 within two larger holes of equal radius r_1, represents a suitable design whereby the effective pendulum length is $r = r_1 - r_2$.

Fig. 3.4.16

Response to Periodic Excitations   A problem of interest in mechanical vibrations concerns the response $x(t)$ of the cam and follower system shown in Fig. 3.4.17. As the cam rotates at a constant angular rate, the follower undergoes the periodic displacement $y(t)$, where $y(t)$ has the period T. The equation of motion is

$$m\ddot{x} + (k_1 + k_2)x = k_2 y \qquad (3.4.51)$$

Fig. 3.4.17

Any periodic function can be expanded in a series of harmonic components in the form of the Fourier series

$$y(t) = \frac{1}{2}a_0 + \sum_{p=1}^{\infty}(a_p\cos\omega_0 t + b_p\sin p\omega_0 t) \quad \omega_0 = 2\pi/T \qquad (3.4.52)$$

where ω_0 is called the **fundamental harmonic** and $p\omega_0$ $(p = 1, 2, \ldots)$ are called **higher harmonics,** in which p is an integer. The coefficients have the expressions

$$a_p = \frac{2}{T}\int_0^T y(t)\cos p\omega_0 t \qquad p = 0, 1, 2, \ldots$$
$$ \qquad (3.4.53)$$
$$b_p = \frac{2}{T}\int_0^T y(t)\sin p\omega_0 t \qquad p = 1, 2, \ldots$$

Note that the limits of integration can be changed, as long as the integration covers one complete period. From Eq. (3.4.27), and a companion equation for the sine counterpart, the response is

$$x(t) = \frac{k_2}{k_1+k_2}\left[\frac{1}{2}a_0 + \sum_{p=1}^{\infty}\frac{1}{1-(p\omega_0/\omega_n)^2}\right.$$
$$\left. \times (a_p\cos p\omega_0 t + b_p\sin p\omega_0 t)\right] \qquad (3.4.54)$$

where

$$\omega_n = \sqrt{(k_1+k_2)/m} \qquad (3.4.55)$$

is the natural frequency of the system. Equation (3.5.54) describes a steady-state response, so that a description in terms of time is not very informative. More significant information can be extracted by plotting the amplitudes of the harmonic components versus the harmonic number. Such plots are called **frequency spectra,** and there is one for the excitation and one for the response. Equation (3.4.54) leads to the conclusion that **resonance occurs for** $p\omega_0 = \omega_n$.

As an example, consider the periodic excitation shown in Fig. 3.4.18 and use Eqs. (3.4.53) to obtain the coefficients

$$a_0 = 2A, a_p = 0, b_p = \begin{cases} 4B/p\pi & p \text{ odd} \\ 0 & p \text{ even} \end{cases} \quad (3.4.56)$$

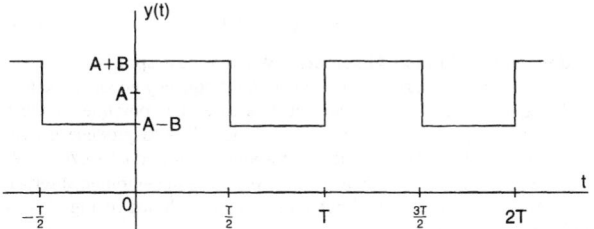

Fig. 3.4.18 Example of periodic excitation.

The excitation and response frequency spectra are displayed in Figs. 3.4.19a and b, the latter for the case in which $\omega_n = 4\omega_0$.

(a)

(b)

Fig. 3.4.19 (a) Excitation frequency spectrum; (b) response frequency spectrum for the periodic excitation of Fig. 3.4.18.

Unit Impulse and Impulse Response Harmonic and periodic forces represent **steady-state excitations;** they **persist indefinitely.** The response to such forces is also steady state. An entirely different class of forces consists of **arbitrary, or transient, forces.** The term *transient* is not entirely appropriate, as some of these forces can also persist indefinitely. Concepts pivotal to the response to arbitrary forces are the unit impulse and the impulse response. The **unit impulse,** denoted by $\delta(t - a)$, represents a function of very high amplitude and defined over a very small time interval at $t = a$ such that the area enclosed is equal to 1 (Fig. 3.4.20). The **impulse response,** denoted by $g(t)$, is defined as the response of a system to a unit impulse applied at $t = 0$, with the initial

conditions being equal to zero. For the mass-damper-spring system of Fig. 3.4.2, the impulse response is

$$g(t) = \frac{1}{m\omega_d} e^{-\zeta\omega_n t} \sin \omega_d t \quad t > 0 \quad (3.4.57)$$

Fig. 3.4.20

Convolution Integral An arbitrary force $F(t)$ as shown in Fig. 3.4.21 can be regarded as a superposition of impulses of magnitude $F(\tau)d\tau$ and applied at $t = \tau$. Hence, the response to an arbitrary force can be regarded as a superposition of impulse responses $g(t - \tau)$ of magnitude $F(\tau)d\tau$, or

$$x(t) = \int_0^t F(\tau)g(t - \tau)\,d\tau$$

$$= \frac{1}{m\omega_d} \int_0^t F(\tau)e^{-\zeta\omega_n(t-\tau)} \sin \omega_d(t - \tau)\,d\tau \quad (3.4.58a)$$

which is called the **convolution integral** or the **superposition integral;** it can also be written in the form

$$x(t) = \int_0^t F(t - \tau)g(\tau)d\tau$$

$$= \frac{1}{m\omega_d} \int_0^t F(t - \tau)e^{-\zeta\omega_n\tau} \sin \omega_d\tau\,d\tau \quad (3.4.58b)$$

Fig. 3.4.21

Shock Spectrum Many systems are subjected on occasions to large forces applied suddenly and over periods of time that are short compared to the natural period. Such forces are capable of inflicting serious damage on a system and are referred to as **shocks.** The severity of a shock is commonly measured in terms of the maximum value of the response of a mass-spring system. The plot of the peak response versus the natural frequency is called the **shock spectrum** or **response spectrum.**

A shock $F(t)$ is characterized by its maximum value F_0, its duration T, and its shape. It is common to approximate the force by the half-sine pulse

$$F(t) = \begin{cases} F_0 \sin \omega t & \text{for } 0 < t < T = \pi/\omega \\ 0 & \text{for } t < 0 \text{ and } t > T \end{cases} \quad (3.4.59)$$

Using the convolution integral, Eq. (3.4.58b) with $\zeta = 0$, the response of a mass-spring system *during the duration of the pulse* is

$$x(t) = \frac{F_0}{k[1 - (\omega/\omega_n)^2]} \left(\sin \omega t - \frac{\omega}{\omega_n} \sin \omega_n t \right)$$

$$0 < t < \pi/\omega \quad (3.4.60)$$

The maximum response is obtained when $\dot{x} = 0$ and has the value

$$x_{max} = \frac{F_0}{k(1 - \omega/\omega_n)} \sin \frac{2i\pi}{1 + \omega_n/\omega}$$

$$i = 1, 2, \ldots ; i < \frac{1}{2}\left(1 + \frac{\omega_n}{\omega}\right) \quad (3.4.61)$$

On the other hand, the response *after the termination of the pulse* is

$$x(t) = \frac{F_0\omega_n^2/\omega}{k[1 - (\omega_n/\omega)^2]} [\cos \omega_n t + \cos \omega_n(t - T)] \quad (3.4.62)$$

which has the maximum value

$$x_{max} = \frac{2F_0\omega_n/\omega}{k[1 - (\omega_n/\omega)^2]} \cos \frac{\pi\omega_n}{2\omega} \quad (3.4.63)$$

The shock spectrum is the plot x_{max} versus ω_n/ω. For $\omega_n < \omega$, the maximum response is given by Eq. (3.4.63) and for $\omega_n > \omega$ by Eq. (3.4.61). The shock spectrum is shown in Fig. 3.4.22 in the form of the nondimensional plot $x_{max}k/F_0$ versus ω_n/ω.

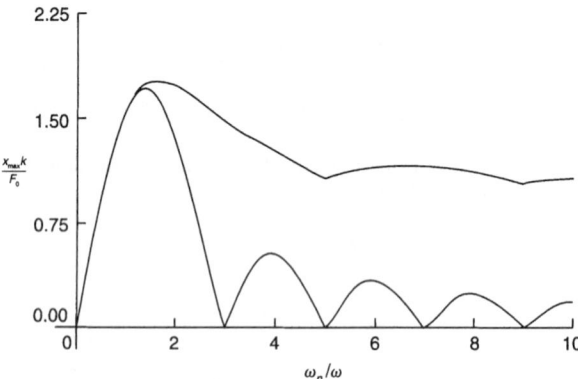

Fig. 3.4.22

MULTI-DEGREE-OF-FREEDOM SYSTEMS

Equations of Motion   Many vibrating systems require more elaborate models than a single-degree-of-freedom system, such as the **multi-degree-of-freedom system** shown in Fig. 3.4.23. By using Newton's second law for each of the n masses m_i $(i = 1, 2, \ldots, n)$, the equations of motion can be written in the form

$$m_i\ddot{x}_i(t) + \sum_{j=1}^{n} c_{ij}\dot{x}_j(t) + \sum_{j=1}^{n} k_{ij}x_j(t) = F_i(t)$$
$$i = 1, 2, \ldots, n \quad (3.4.64)$$

where $x_i(t)$ is the displacement of mass m_i, $F_i(t)$ is the force acting on m_i, and c_{ij} and k_{ij} are **damping** and **stiffness coefficients**, respectively. The matrix form of Eqs. (3.4.64) is

$$M\ddot{\mathbf{x}}(t) + C\dot{\mathbf{x}}(t) + K\mathbf{x}(t) = \mathbf{F}(t) \quad (3.4.65)$$

in which $\mathbf{x}(t)$ is the n-dimensional displacement vector, $\mathbf{F}(t)$ the corresponding force vector, M the **mass matrix**, C the **damping matrix**, and K

the **stiffness matrix,** all three symmetric matrices. (In the present case the mass matrix is diagonal, but in general it is not, although it is symmetric.)

Response of Undamped Systems to Harmonic Excitations   Let the harmonic excitation have the form

$$\mathbf{F}(t) = \mathbf{F}_0 \sin \omega t \quad (3.4.66)$$

where $\mathbf{F}_0$ is a constant vector and ω is the excitation, or driving frequency. The response to the harmonic excitation is a **steady-state** response and can be expressed as

$$\mathbf{x}(t) = Z^{-1}(\omega)\mathbf{F}_0 \sin \omega t \quad (3.4.67)$$

where $Z^{-1}(\omega)$ is the inverse of the **impedance matrix** $Z(\omega)$. In the absence of damping, the impedance matrix is

$$Z(\omega) = K - \omega^2 M \quad (3.4.68)$$

Undamped Vibration Absorbers   When a mass-spring system m_1, k_1 is subjected to a harmonic force with the frequency equal to the natural frequency, resonance occurs. In this case, it is possible to add a second mass-spring system m_2, k_2 so designed as to produce a two-degree-of-freedom system with the response of m_1 equal to zero. We refer to m_1, k_1 as the **main system** and to m_2, k_2 as the **vibration absorber.** The resulting two-degree-of-freedom system is shown in Fig. 3.4.24 and has the impedance matrix

$$Z(\omega)\begin{bmatrix} k_1 + k_2 - \omega^2 m_1 & -k_2 \\ -k_2 & k_2 - \omega^2 m_2 \end{bmatrix} \quad (3.4.69)$$

Fig. 3.4.24

Inserting Eq. (3.4.69) into Eq. (3.4.67), together with $F_1(t) = F_1 \sin \omega t$, $F_2(t) = 0$, write the steady-state response in the form

$$x_1(t) = X_1(\omega) \sin \omega t \quad (3.4.70a)$$
$$x_2(t) = X_2(\omega) \sin \omega t \quad (3.4.70b)$$

Fig. 3.4.23

where the amplitudes are given by

$$X_1(\omega) = \frac{[1 - (\omega/\omega_a)^2]x_{st}}{[1 + \mu(\omega_a/\omega_n)^2 - (\omega/\omega_n)^2][1 - (\omega/\omega_a)^2] - \mu(\omega_a/\omega_n)^2}$$

$$(3.4.71a)$$

$$X_2(\omega) = \frac{x_{st}}{[1 + \mu(\omega_a/\omega_n)^2 - (\omega/\omega_n)^2][1 - (\omega/\omega_a)^2] - \mu(\omega_a/\omega_n)^2}$$

$$(3.4.71b)$$

in which

$\omega_n = \sqrt{k_1/m_1}$ = the natural frequency of the main system alone

$\omega_a = \sqrt{k_2/m_2}$ = the natural frequency of the absorber alone

$x_{st} = F_1/k_1$ = the static deflection of the main system

$\mu = m_2/m_1$ = the ratio of the absorber mass to the main mass

From Eqs. (3.4.70a) and (3.4.71a), we conclude that if we choose m_2 and k_2 such that $\omega_a = \omega$, the response $x_1(t)$ of the main mass is zero. Moreover, from Eqs. (3.4.70b) and (3.4.71b),

$$x_2(t) = -\left(\frac{\omega_n}{\omega_a}\right)^2 \frac{x_{st}}{\mu} \sin \omega t = -\frac{F_1}{k_2} \sin \omega t \qquad (3.4.72)$$

so that the force in the absorber spring is

$$k_2 x_2(t) = -F_1 \sin \omega t \qquad (3.4.73)$$

Hence, *the absorber exerts a force on the main mass balancing exactly the applied force $F_1 \sin \omega t$.*

A vibration absorber designed for a given operating frequency ω can perform satisfactorily for operating frequencies that vary slightly from ω. In this case, the motion of m_1 is not zero, but its amplitude tends to be very small, as can be verified from a frequency response plot $X_1(\omega)/x_{st}$ versus ω/ω_n; Fig. 3.4.25 shows such a plot for $\mu = 0.2$ and $\omega_n = \omega_a$. The shaded area indicates the range in which the performance can be regarded as satisfactory. Note that the thin line in Fig. 3.4.25 represents the frequency response of the main system alone. Also note that the system resulting from the combination of the main system and the absorber has two resonance frequencies, but they are removed from the operating frequency $\omega = \omega_n = \omega_a$.

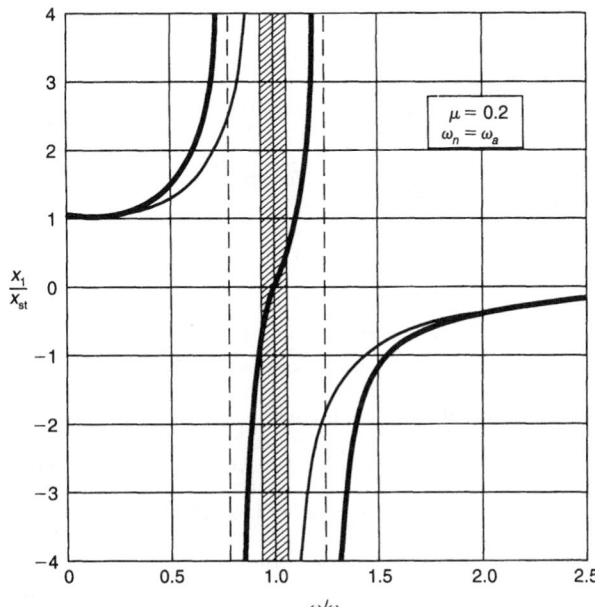

Fig. 3.4.25

Natural Modes of Vibration   In the absence of damping and external forces, Eq. (3.4.65) reduces to the free-vibration equation

$$M\ddot{\mathbf{x}}(t) + K\mathbf{x}(t) = \mathbf{0} \qquad (3.4.74)$$

which has the harmonic solution

$$\mathbf{x}(t) = \mathbf{u} \cos(\omega t - \phi) \qquad (3.4.75)$$

where $\mathbf{u}$ is a constant vector, ω a frequency of oscillation, and ϕ a phase angle. Introduction of Eq. (3.4.75) into Eq. (3.4.74) and division through by $\cos(\omega t - \phi)$ results in

$$K\mathbf{u} = \omega^2 M\mathbf{u} \qquad (3.4.76)$$

which represents a set of n simultaneous algebraic equations known as the **eigenvalue problem.** It has n solutions consisting of the **eigenvalues** ω_r^2; the square roots represent the **natural frequencies** ω_r ($r = 1, 2, \ldots, n$). Moreover, to each natural frequency ω_r there corresponds a vector $\mathbf{u}_r$ ($r = 1, 2, \ldots, n$) called **eigenvector,** or **modal vector,** or **natural mode.** The modal vectors possess the **orthogonality property,** or

$$\mathbf{u}_s^T M \mathbf{u}_r = 0 \qquad (3.4.77a)$$
$$\mathbf{u}_s^T K \mathbf{u}_r = 0 \qquad (3.4.77b)$$

(for $r, s = 1, 2, \ldots, n; r \neq s$), in which $\mathbf{u}_s^T$ is the **transpose** of $\mathbf{u}_s$, a row vector. It is convenient to adjust the magnitude of the modal vectors so as to satisfy

$$\mathbf{u}_r^T M \mathbf{u}_r = 1 \qquad (3.4.78a)$$
$$\mathbf{u}_r^T K \mathbf{u}_r = \omega_r^2 \qquad (3.4.78b)$$

(for $r = 1, 2, \ldots, n$), a process known as normalization, in which case $\mathbf{u}_r$ are called **normal modes.** Note that the normalization process involves Eq. (3.4.78a) alone, as Eq. (3.4.78b) follows automatically. The solution of the eigenvalue problem can be obtained by a large variety of computational algorithms (Meirovitch, "Principles and Techniques of Vibrations," Prentice-Hall). Commercially, they are available in software packages for numerical computations, such as MATLAB.

The actual solution of Eq. (3.4.74) is obtained below in the context of the transient response.

Transient Response of Undamped Systems   From Eq. (3.4.65), the vibration of undamped systems satisfies the equation

$$M\ddot{\mathbf{x}}(t) + K\mathbf{x}(t) = \mathbf{F}(t) \qquad (3.4.79)$$

where $\mathbf{F}(t)$ is an arbitrary force vector. In addition, the displacement and velocity vectors must satisfy the initial conditions $\mathbf{x}(0) = \mathbf{x}_0$, $\dot{\mathbf{x}}(0) = \mathbf{v}_0$. The solution of Eq. (3.4.79) has the form

$$\mathbf{x}(t) = \sum_{r=1}^{n} \mathbf{u}_r q_r(t) \qquad (3.4.80)$$

in which $\mathbf{u}_r$ are the modal vectors and $q_r(t)$ are associated **modal coordinates.** Inserting Eq. (3.4.80) into Eq. (3.4.79), premultiplying the result by $\mathbf{u}_s^T$, and using Eqs. (3.4.77) and (3.4.78) we obtain the modal equations

$$\ddot{q}_r(t) + \omega_r^2 q_r(t) = Q_r(t) \qquad r = 1, 2, \ldots, n \qquad (3.4.81)$$

where

$$Q_r(t) = \mathbf{u}_r^T \mathbf{F}(t) \qquad r = 1, 2, \ldots, n \qquad (3.4.82)$$

are **modal forces.** Equations (3.4.81) resemble the equation of a single-degree-of-freedom system and have the solution

$$q_r(t) = \frac{1}{\omega_r} \int_0^t Q_r(t - \tau) \sin \omega_r \tau \, d\tau + q_r(0) \cos \omega_r t + \frac{\dot{q}_r(0)}{\omega_r} \sin \omega_r t$$

$$r = 1, 2, \ldots, n \quad (3.4.83)$$

where

$$q_r(0) = \mathbf{u}_r^T M \mathbf{x}_0 \qquad (3.4.84a)$$
$$\dot{q}_r(0) = \mathbf{u}_r^T M \mathbf{v}_0 \qquad (3.4.84b)$$

(for $r = 1, 2, \ldots, n$) are **initial modal displacements and velocities,** respectively. The solution to both external forces and initial excitations is obtained by inserting Eqs. (3.4.83) into Eq. (3.4.80).

Systems with Proportional Damping When the system is damped, the response does not in general have the form of Eq. (3.4.80), and a more involved approach is necessary (Meirovitch, "Fundamentals of Vibration," McGraw-Hill). In the special case in which the damping matrix C is proportional to the mass matrix M or the stiffness matrix K, or is a linear combination of M and K, the preceding approach yields the modal equations

$$\ddot{q}_r(t) + 2\zeta_r\omega_r\dot{q}_r(t) + \omega_r^2 q_r(t) = Q_r(t) \qquad r = 1, 2, \ldots, n \quad (3.4.85)$$

where ζ_r are **modal damping factors**. Equations (3.4.85) have the solution

$$q_r(t) = \frac{1}{\omega_{dr}}\int_0^t Q_r(t-\tau)e^{-\zeta_r\omega_r\tau}\sin\omega_{dr}\tau\,d\tau$$
$$+ \frac{q_r(0)}{\sqrt{1-\zeta_r^2}}e^{-\zeta_r\omega_r t}\cos(\omega_{dr}t - \psi_r) + \frac{\dot{q}_r(0)}{\omega_{dr}}\sin\omega_{dr}t$$
$$r = 1, 2, \ldots, n \quad (3.4.86)$$

in which

$$\omega_{dr} = \omega_r\sqrt{1-\zeta_r^2} \qquad r = 1, 2, \ldots, n \quad (3.4.87)$$

is the damped frequency in the rth mode and

$$\psi_r = \tan^{-1}\frac{\zeta_r}{\sqrt{1-\zeta_r^2}} \qquad r = 1, 2, \ldots, n \quad (3.4.88)$$

is a phase angle associated with the rth mode. The quantities $Q_r(t)$, $q_r(0)$, and $\dot{q}_r(0)$ remain as defined by Eqs. (3.4.82), (3.4.84a), and (3.4.84b), respectively.

DISTRIBUTED-PARAMETER SYSTEMS

Vibration of Rods, Shafts, and Strings The axial vibration of rods is described by the equation

$$-\frac{\partial}{\partial x}\left[EA(x)\frac{\partial u(x, t)}{\partial x}\right] + m(x)\frac{\partial^2 u(x, t)}{\partial t^2} = f(x, t)$$
$$0 < x < L \quad (3.4.89)$$

where $u(x, t)$ is the axial displacement, $f(x, t)$ the axial force per unit length, E the modulus of elasticity, $A(x)$ the cross-sectional area, and $m(x)$ the mass per unit length. The solution $u(x, t)$ is subject to one boundary condition at each end.

Before attempting to solve Eq. (3.4.89), consider the free vibration problem, $f(x, t) = 0$. The solution of the latter problem is harmonic and can be expressed as

$$u(x, t) = U(x)\cos(\omega t - \phi) \quad (3.4.90)$$

where $U(x)$ is the amplitude, ω the frequency, and ϕ an inconsequential phase angle. Inserting Eq. (3.4.90) into Eq. (3.4.89) with $f(x, t) = 0$ and dividing through by $\cos(\omega t - \phi)$, we conclude that $U(x)$ and ω must satisfy the *eigenvalue problem*

$$-\frac{d}{dx}\left[EA(x)\frac{dU(x)}{dx}\right] = \omega^2 m(x)U(x) \qquad 0 < x < L \quad (3.4.91)$$

where $U(x)$ must satisfy one boundary condition at each end. At a fixed end the displacement U must be zero and at a free end the axial force $EA\,dU/dx$ is zero.

Exact solutions of the eigenvalue problem are possible in only a few cases, mostly for uniform rods, in which case Eq. (3.4.91) reduces to

$$\frac{d^2U(x)}{dx^2} + \beta^2 U(x) = 0 \qquad \beta^2 = \frac{\omega^2 m}{EA} \qquad 0 < x < L \quad (3.4.92)$$

whose solution is

$$U(x) = A\sin\beta x + B\cos\beta x \quad (3.4.93)$$

where A and B are constants of integration, determined from specified boundary conditions. In the case of a **fixed-free** rod, the boundary conditions are

$$U(0) = 0 \quad (3.4.94a)$$

$$EA\frac{dU}{dx}\bigg|_{x=L} = 0 \quad (3.4.94b)$$

Condition (3.4.94a) gives $B = 0$ and condition (3.4.94b) yields the **characteristic equation**

$$\cos\beta L = 0 \quad (3.4.95)$$

which has the infinity of solutions

$$\beta_r L = \frac{(2r-1)\pi}{2} \qquad r = 1, 2, \ldots \quad (3.4.96)$$

where β_r represent the *eigenvalues;* they are related to the *natural frequencies* ω_r by

$$\omega_r = \beta_r\sqrt{\frac{EA}{m}} = \frac{(2r-1)\pi}{2}\sqrt{\frac{EA}{mL^2}} \qquad r = 1, 2, \ldots \quad (3.4.97)$$

From Eq. (3.4.93), the **normal modes** are

$$U_r(x) = \sqrt{\frac{2}{mL}}\sin\frac{(2r-1)\pi x}{2L} \qquad r = 1, 2, \ldots \quad (3.4.98)$$

For a **fixed-fixed** rod, the natural frequencies and normal modes are

$$\omega_r = r\pi\sqrt{\frac{EA}{mL^2}} \qquad U_r(x) = \sqrt{\frac{2}{mL}}\sin\frac{r\pi x}{L}$$
$$r = 1, 2, \ldots \quad (3.4.99)$$

and for a **free-free** rod they are

$$\omega_0 = 0 \qquad U_0 = \sqrt{\frac{1}{mL}} \quad (3.4.100a)$$

$$\omega_r = r\pi\sqrt{\frac{EA}{mL^2}} \qquad U_r(x) = \sqrt{\frac{2}{mL}}\cos\frac{r\pi x}{L}$$
$$r = 1, 2, \ldots \quad (3.4.100b)$$

Note that U_0 represents a **rigid-body mode**, with zero natural frequency.

In every case the modes are orthogonal, satisfying the conditions

$$\int_0^L mU_s(x)U_r(x)\,dx = 0 \quad (3.4.101a)$$

$$-\int_0^L U_s(x)\frac{d}{dx}\left[EA\frac{dU_r(x)}{dx}\right]dx = 0 \quad (3.4.101b)$$

(for $r, s = 0, 1, 2, \ldots, r \neq s$) and have been normalized to satisfy the relations

$$\int_0^L mU_r^2(x)\,dx = 1 \quad (3.4.102a)$$

$$-\int_0^L U_r(x)\frac{d}{dx}\left[EA\frac{dU_r(x)}{dx}\right]dx = \omega_r^2 \quad (3.4.102b)$$

(for $r = 0, 1, 2, \ldots$). Note that the orthogonality of the normal modes extends to the rigid-body mode.

The response of the rod has the form

$$u(x, t) = \sum_{r=1}^{\infty} U_r(x)q_r(t) \quad (3.4.103)$$

Introducing Eq. (3.4.103) into Eq. (3.4.89), multiplying through by $U_s(x)$, integrating over the length of the rod, and using Eqs. (3.4.101) and (3.4.102) we obtain the **modal equations**

$$\ddot{q}_r(t) + \omega_r^2 q_r(t) = Q_r(t) \qquad r = 1, 2, \ldots \quad (3.4.104)$$

where

$$Q_r(t) = \int_0^L U_r(x)f(x, t)\,dx \qquad r = 1, 2, \ldots \quad (3.4.105)$$

Table 3.4.4   Analogous Quantities for Rods, Shafts, and Strings

	Rods	Shafts	Strings
Displacement	Axial—$u(x, t)$	Torsional—$\theta(x, t)$	Transverse—$w(x, t)$
Inertia (per unit length)	Mass—$m(x)$	Mass polar moment of inertia—$I(x)$	Mass—$\rho(x)$
Stiffness	Axial—$EA(x)$ E = Young's modulus $A(x)$ = cross-sectional area	Torsional—$GJ(x)$ G = shear modulus $J(x)$ = area polar moment of inertia	Tension—$T(x)$
Load (per unit length)	Force—$f(x, t)$	Moment—$m(x, t)$	Force—$f(x, t)$

are the **modal forces**. Equations (3.4.104) resemble Eqs. (3.4.81) entirely; their solution is given by Eqs. (3.4.83). The displacement of the rod is obtained by inserting Eqs. (3.4.83) into Eq. (3.4.103).

As an example, consider the response of a uniform fixed-free rod to the uniformly distributed impulsive force

$$f(x, t) = \hat{f}_0 \delta(t) \qquad (3.4.106)$$

Inserting Eqs. (3.4.98) and (3.4.106) into Eq. (3.4.105), we obtain the modal forces

$$Q_r(t) = \sqrt{\frac{2}{mL}} \int_0^L \sin \frac{(2r-1)\pi x}{2L} \hat{f}_0 \delta(t)\, dx$$

$$= \frac{2}{(2r-1)\pi} \sqrt{\frac{2L}{m}} \hat{f}_0 \delta(t) \qquad r = 1, 2, \ldots \quad (3.4.107)$$

so that, from Eqs. (3.4.83), the modal displacements are

$$q_r(t) = \frac{1}{\omega_r} \frac{2}{(2r-1)\pi} \sqrt{\frac{2L}{m}} \hat{f}_0 \int_0^t \delta(t-\tau) \sin \omega_r \tau \, d\tau$$

$$= \left[\frac{2}{(2r-1)\pi}\right]^2 \sqrt{\frac{2L^3}{EA}} \hat{f}_0 \sin \frac{(2r-1)\pi}{2} \sqrt{\frac{EA}{mL^2}} t$$

$$r = 1, 2, \ldots \quad (3.4.108)$$

Finally, from Eq. (3.4.103), the response is

$$u(x, t) = \frac{8\hat{f}_0 L}{\pi^2} \sqrt{\frac{1}{mEA}} \sum_{r=1}^{\infty} \frac{1}{(2r-1)^2} \sin \frac{(2r-1)\pi x}{2L}$$

$$\times \sin \frac{(2r-1)\pi}{2} \sqrt{\frac{EA}{mL^2}} t \quad (3.4.109)$$

The torsional vibration of shafts and the transverse vibration of strings are described by the same differential equation and boundary conditions as the axial vibration of rods, except that the nature of the displacement, inertia and stiffness parameters, and external excitations differs, as indicated in Table 3.4.4.

Bending Vibration of Beams   The procedure for evaluating the response of beams in transverse vibration is similar to that for rods, the main difference arising in the stiffness term. The differential equation for beams in bending is

$$\frac{\partial^2}{\partial x^2}\left[EI(x)\frac{\partial^2 w(x, t)}{\partial x^2}\right] + m(x)\frac{\partial^2 w(x, t)}{\partial t^2}$$

$$= f(x, t) \qquad 0 < x < L \quad (3.4.110)$$

in which $w(x, t)$ is the transverse displacement, $f(x, t)$ the force per unit length, $I(x)$ the cross-sectional area moment of inertia, and $m(x)$ the mass per unit length. The solution $w(x, t)$ must satisfy two boundary conditions at each end.

The eigenvalue problem is described by the differential equation

$$\frac{d^2}{dx^2}\left[EI(x)\frac{d^2 W(x)}{dx^2}\right] = \omega^2 m(x) W(x) \qquad 0 < x < L \quad (3.4.111)$$

and two boundary conditions at each end, depending on the type of support. Some possible boundary conditions are given in Table 3.4.5. The solution of the eigenvalue problem consists of the natural frequencies ω_r and natural modes $W_r(x)$ $(r = 1, 2, \ldots)$. The first five normalized natural frequencies of uniform beams with six different boundary conditions are listed in Table 3.4.6. The normal modes for the hinged-hinged beam are

$$W_r(x) = \sqrt{\frac{2}{mL}} \sin \frac{r\pi x}{L} \qquad r = 1, 2, \ldots \quad (3.4.112)$$

The normal modes for the remaining beam types are more involved and they involve both trigonometric and hyperbolic functions (Meirovitch, "Fundamentals of Vibrations"). The modes for every beam type are orthogonal and can be used to obtain the response $w(x, t)$ in the form of a series similar to Eq. (3.4.103).

Table 3.4.5   Quantities Equal to Zero at Boundary

Boundary type	Displacement W	Slope dW/dx	Bending moment EId^2W/dx^2	Shearing force $d(EId^2W/dx^2)/dx$
Hinged	✓		✓	
Clamped	✓	✓		
Free			✓	✓

Vibration of Membranes   A **membrane** is a very thin sheet of material stretched over a two-dimensional domain enclosed by one or two nonintersecting boundaries. It can be regarded as the two-dimensional counterpart of the string. Like a string, it derives the ability to resist transverse displacements from tension, which acts in all directions in the plane of the membrane and at all its points. It is commonly assumed that the tension is uniform and does not change as the membrane deflects.

Table 3.4.6   Normalized Natural Frequencies for Various Beams

Beam type	$\omega_1 \sqrt{mL^4/EI}$	$\omega_2 \sqrt{mL^4/EI}$	$\omega_3 \sqrt{mL^4/EI}$	$\omega_4 \sqrt{mL^4/EI}$	$\omega_5 \sqrt{mL^2/EI}$
Hinged-hinged	π^2	$4\pi^2$	$9\pi^2$	$16\pi^2$	$25\pi^2$
Clamped-free	1.875^2	4.694^2	7.855^2	10.996^2	14.137^2
Free-free	0	0	$(1.506\pi)^2$	$(2.500\pi)^2$	$(3.500\pi)^2$
Clamped—clamped	$(1.506\pi)^2$	$(2.500\pi)^2$	$(3.500\pi)^2$	$(4.500\pi)^2$	$(5.500\pi)^2$
Clamped-hinged	3.927^2	7.069^2	10.210^2	13.352^2	16.493^2
Hinged-free	0	3.927^2	7.069^2	10.210^2	13.352^2

The general procedure for calculating the response of membranes remains the same as for rods and beams, but there is one significant new factor, namely, the shape of the boundary, which dictates the type of coordinates to be used. For rectangular membranes cartesian coordinates must be used, and for circular membranes polar coordinates are indicated.

The differential equation for the transverse vibration of membranes is

$$-T\nabla^2 w + \rho\frac{\partial^2 w}{\partial t^2} = f \qquad (3.4.113)$$

which must be satisfied at every interior point of the membrane, where w is the transverse displacement, f the transverse force per unit area, T the tension, and ρ the mass per unit area. Moreover, ∇^2 is the Laplacian operator, whose expression depends on the coordinates used. The solution w must satisfy one boundary condition at every boundary point. Using the established procedure, the eigenvalue problem is described by the differential equation

$$-T\nabla^2 W = \omega^2 \rho W \qquad (3.4.114)$$

where W is the displacement amplitude; it must satisfy one boundary condition at every point of the boundary.

Consider a **rectangular membrane fixed** at $x = 0, a$ and $y = 0, b$, in which case the Laplacian operator in terms of the cartesian coordinates x and y has the form

$$\nabla^2 = \frac{\partial^2}{\partial x^2} + \frac{\partial^2}{\partial y^2} \qquad (3.4.115)$$

The boundary conditions are $W(0, y) = W(a, y) = W(x, 0) = W(x, b) = 0$. The natural frequencies are

$$\omega_{mn} = \pi\sqrt{\left[\left(\frac{m}{a}\right)^2 + \left(\frac{n}{b}\right)^2\right]\frac{T}{\rho}}$$
$$m, n = 1, 2, \ldots \quad (3.4.116)$$

and the normal modes are

$$W_{mn}(x, y) = \frac{2}{\sqrt{\rho ab}}\sin\frac{m\pi x}{a}\sin\frac{n\pi y}{b} \qquad m, n = 1, 2, \ldots \quad (3.4.117)$$

The modes satisfy the orthogonality conditions

$$\int_0^a\int_0^b \rho W_{mn}(x, y)W_{rs}(x, y)\, dx\, dy = 0,$$
$$m \neq r \text{ and/or } n \neq s \quad (3.4.118a)$$

$$-\int_0^a\int_0^b W_{mn}(x, y)T\nabla^2 W_{rs}(x, y)\, dx\, dy = 0,$$
$$m \neq r \text{ and/or } n \neq s \quad (3.4.118b)$$

and have been normalized so that $\int_0^a\int_0^b \rho W_{mn}^2(x, y)\, dx\, dy = 1 (m, n = 1, 2, \ldots)$. Note that, because the problem is two-dimensional, it is necessary to identify the natural frequencies and modes by two subscripts. With this exception, the procedure for obtaining the response is the same as for rods and beams.

Next, consider a **uniform circular membrane fixed at** $r = a$. In this case, the Laplacian operator in terms of the polar coordinates r and θ is

$$\nabla^2 = \frac{\partial^2}{\partial r^2} + \frac{1}{r}\frac{\partial}{\partial r} + \frac{1}{r^2}\frac{\partial^2}{\partial\theta^2} \qquad (3.4.119)$$

The natural modes for circular membranes are appreciably more involved than for rectangular membranes. They are products of Bessel functions of $\omega_{mn}r$ and trigonometric functions of $m\theta$, where $m = 0, 1, 2, \ldots$ and $n = 1, 2, \ldots$. The modes are given in Meirovitch, "Principles and Techniques of Vibrations," Prentice-Hall. Table 3.4.7 gives the normalized natural frequencies $\omega_{mn}^* = (\omega_{mn}/2\pi)\sqrt{\rho a^2/T}$ corresponding to $m = 0, 1, 2$ and $n = 1, 2, 3$. The modes satisfy the orthogonality relations

$$\int_0^a\int_0^{2\pi} \rho W_{mn}(r, \theta)W_{rs}(r, \theta)r\, dr\, d\theta = 0$$
$$m \neq r \text{ and/or } n \neq s \quad (3.4.120a)$$

Table 3.4.7   Circular Membrane Normalized Natural Frequencies
$\omega_{mn}^* = (\omega_{mn}/2\pi)\sqrt{\rho a^2/T}$

m	n		
	1	2	3
0	0.3827	0.8786	1.3773
1	0.6099	1.1165	1.6192
2	0.8174	1.3397	1.8494

$$-\int_0^a\int_0^{2\pi} W_{mn}(r, \theta)T\nabla^2 W_{rs}(r, \theta)r\, dr\, d\theta = 0$$
$$m \neq r \text{ and/or } n \neq s \quad (3.4.120b)$$

The response of circular membranes is obtained in the usual manner.

Bending Vibration of Plates Consider **plates** whose behavior is governed by the elementary plate theory, which is based on the following assumptions: (1) deflections are small compared to the plate thickness; (2) the normal stresses in the direction transverse to the plate are negligible; (3) there is no force resultant on the cross-sectional area of a plate differential element: the middle plane of the plate does not undergo deformations and represents a neutral plane, and (4) any straight line normal to the middle plane remains so during bending. Under these assumptions, the differential equation for the bending vibration of plates is

$$D\nabla^4 w + m\frac{\partial^2 w}{\partial t^2} = f \qquad (3.4.121)$$

and is to be satisfied at every interior point of the plate, where w is the transverse displacement, f the transverse force per unit area, m the mass per unit area, $D = Eh^3/12(1 - v^2)$ the plate flexural rigidity, E Young's modulus, h the plate thickness, and v Poisson's ratio. Moreover, ∇^4 is the biharmonic operator. The solution w must satisfy two boundary conditions at every point of the boundary. The eigenvalue problem is defined by the differential equation

$$D\nabla^4 W = \omega^2 mW \qquad (3.4.122)$$

and corresponding boundary conditions.

Consider a **rectangular plate simply supported** at $x = 0, a$ and $y = 0, b$. Because of the shape of the plate, we must use cartesian coordinates, in which case the biharmonic operator has the expression

$$\nabla^4 = \nabla^2\nabla^2 = \left(\frac{\partial^2}{\partial x^2} + \frac{\partial^2}{\partial y^2}\right)\left(\frac{\partial^2}{\partial x^2} + \frac{\partial^2}{\partial y^2}\right)$$
$$= \frac{\partial^4}{\partial x^4} + 2\frac{\partial^4}{\partial x^2\partial y^2} + \frac{\partial^4}{\partial y^4} \qquad (3.4.123)$$

Moreover, the boundary conditions are $W = 0$ and $\partial^2 W/\partial x^2 = 0$ for $x = 0, a$ and $W = 0$ and $\partial^2 W/\partial y^2 = 0$ for $y = 0, b$. The natural frequencies are

$$\omega_{mn} = \pi^2\left[\left(\frac{m}{a}\right)^2 + \left(\frac{n}{b}\right)^2\right]\sqrt{\frac{D}{m}}$$
$$m, n = 1, 2, \ldots \quad (3.4.124)$$

and no confusion should arise because the same symbol is used for one of the subscripts and for the mass per unit area. The corresponding normal modes are

$$W_{mn}(x, y) = \frac{2}{\sqrt{mab}}\sin\frac{m\pi x}{n}\sin\frac{n\pi y}{b} \qquad m, n = 1, 2, \ldots \quad (3.4.125)$$

and they are recognized as being the same as for rectangular membranes fixed at all boundaries.

A **circular plate** requires use of polar coordinates, so that the biharmonic operator has the form

$$\nabla^4 = \nabla^2\nabla^2 = \left(\frac{\partial^2}{\partial r^2} + \frac{1}{r}\frac{\partial}{\partial r} + \frac{1}{r^2}\frac{\partial^2}{\partial\theta^2}\right)\left(\frac{\partial^2}{\partial r^2} + \frac{1}{r}\frac{\partial}{\partial r} + \frac{1}{r^2}\frac{\partial^2}{\partial\theta^2}\right)$$
$$(3.4.126)$$

Table 3.4.8 Circular Plate Normalized Natural Frequencies
$\omega_{mn}^* = \omega_{mn}(a/\pi)^2 \sqrt{m/D}$

	n		
m	1	2	3
0	1.015^2	2.007^2	3.000^2
1	1.468^2	2.483^2	3.490^2
2	1.879^2	2.992^2	4.000^2

Consider a **plate clamped** at $r = a$, in which case the boundary conditions are $W(r, \theta) = 0$ and $\partial W(r, \theta)/\partial r = 0$ at $r = a$. In addition, the solution must be finite at every interior point in the plate, and in particular at $r = 0$. The natural modes have involved expressions; they are given in Meirovitch, "Principles and Techniques of Vibrations," Prentice-Hall. Table 3.4.8 lists the normalized natural frequencies $\omega_{mn}^* = \omega_{mn}(a/\pi)^2 \sqrt{m/D}$ corresponding to $m = 0, 1, 2$ and $n = 1, 2, 3$.

The natural modes of the plates are orthogonal and can be used to obtain the response to both initial and external excitations.

APPROXIMATE METHODS FOR DISTRIBUTED SYSTEMS

Rayleigh's Energy Method   The eigenvalue problem contains vital information concerning vibrating systems, namely, the natural frequencies and modes. In the majority of practical cases, exact solutions to the eigenvalue problem for distributed systems are not possible, so that the interest lies in **approximate solutions.** This is often the case when the mass and stiffness are distributed nonuniformly and/or the boundary conditions cannot be satisfied, the latter in particular for two-dimensional systems with irregularly shaped boundaries.

When the objective is to estimate the lowest natural frequency, Rayleigh's energy method has few equals. As discussed earlier, free vibration of undamped systems is harmonic and can be expressed as

$$w(x, t) = W(x) \cos(\omega t - \phi) \qquad (3.4.127)$$

where $W(x)$ is the displacement amplitude, ω the free vibration frequency, and ϕ an inconsequential phase angle. The kinetic energy represents an integral involving the velocity squared. Hence, using Eq. (3.4.127), the kinetic energy can be written in the form

$$T(t) = \frac{1}{2}\int_0^L m(x)\left[\frac{\partial w(x, t)}{\partial t}\right]^2 dx = \omega^2 T_{\text{ref}} \sin^2(\omega t - \phi) \qquad (3.4.128)$$

where

$$T_{\text{ref}} = \frac{1}{2}\int_0^L m(x)W^2(x)\, dx \qquad (3.4.129)$$

is called the **reference kinetic energy.** The form of the potential energy is system-dependent, but in general is an integral involving the square of the displacement and of its derivatives with respect to the spatial coordinates (see Table 3.4.9). It can be expressed as

$$V(t) = V_{\max} \cos^2(\omega t - \phi) \qquad (3.4.130)$$

where $V_{\max}$ is the maximum potential energy, which can be obtained by simply replacing $w(x, t)$ by $W(x)$ in $V(t)$. Using the principle of conservation of energy in conjunction with Eqs. (3.4.128) and (3.4.130), we can write

$$E = T + V = T_{\max} + 0 = 0 + V_{\max} \qquad (3.4.131)$$

in which

$$T_{\max} = \omega^2 T_{\text{ref}} \qquad (3.4.132)$$

It follows that

$$\omega^2 = \frac{V_{\max}}{T_{\text{ref}}} \qquad (3.4.133)$$

Equation (3.4.133) represents **Rayleigh's quotient,** which has the remarkable property that it has a minimum value for $W(x) = W_1(x)$, the minimum value being ω_1^2. Rayleigh's energy method amounts to selecting a trial function $W(x)$ reasonably close to the lowest natural mode $W_1(x)$, inserting this function into Rayleigh's quotient, and carrying out the indicated integrations. Then, ω^2 will be one order of magnitude closer

to the lowest eigenvalue ω_1^2 than $W(x)$ is to $W_1(x)$, thus providing a good estimate ω of the lowest natural frequency ω_1. Quite often, the static deformation of the system acted on by loads proportional to the mass distribution is a good choice. In some cases, the lowest mode of a related simpler system can yield good results.

As an example, estimate the lowest natural frequency of a uniform bar in axial vibration with a mass M attached at $x = L$ (Fig. 3.4.26) for the three trial functions (1) $U(x) = x/L$; (2) $U(x) = (1 + M/mL)(x/L) - (x/L)^2/2$, representing the static deformation; and (3) $U(x) = \sin \pi x/2L$, representing the lowest mode of the bar without the mass M. The Rayleigh quotient for this bar is

$$\omega^2 = \frac{\displaystyle\int_0^L EA(x)[dU(x)/dx]^2\, dx}{\displaystyle\int_0^L m(x)U^2(x)\, dx + MU^2(L)} \qquad (3.4.134)$$

Fig. 3.4.26

The results are:

1. $\displaystyle \omega^2 = \frac{\displaystyle\int_0^L EA(1/L)^2\, dx}{\displaystyle\int_0^L m(x/L)^2\, dx + M} = \frac{EA}{(M + mL/3)L}$

2. $\dfrac{\displaystyle\int_0^L EA(1 + M/mL - x/L)^2(1/L)^2\, dx}{\displaystyle\int_0^L m[(1 + M/mL)(x/L) - (x/L)^2/2]^2\, dx + M(1 + 2M/mL)^2/4}$

   $= \dfrac{\left(\dfrac{M}{mL}\right)^2 + \dfrac{M}{mL} + \dfrac{1}{3}}{\dfrac{1}{3}\left(\dfrac{M}{mL}\right)^2 + \dfrac{5}{12}\dfrac{M}{mL} + \dfrac{2}{15} + \dfrac{M}{mL}\left(\dfrac{1}{2} + \dfrac{M}{mL}\right)^2}\dfrac{EA}{mL^2}$ (3.4.135)

3. $\displaystyle \omega^2 = \frac{\displaystyle\int_0^L EA\left(\frac{\pi}{2L}\right)^2 \cos^2\frac{\pi x}{2L}\, dx}{\displaystyle\int_0^L m \sin^2\frac{\pi x}{2L}\, dx + M} = \frac{\pi^2}{8\left(\dfrac{1}{2} + \dfrac{M}{mL}\right)}\dfrac{EA}{mL^2}$

Table 3.4.9 Potential Energy for Various Systems

System	Potential energy* $V(t)$
Rods (also shafts and strings)	$\dfrac{1}{2}\displaystyle\int_0^L EA(x)[\partial u(x, t)/\partial x]^2 dx$
Beams	$\dfrac{1}{2}\displaystyle\int_0^L EI(x)[\partial^2 w(x, t)/\partial x^2]^2 dx$
Beams with axial force	$\dfrac{1}{2}\displaystyle\int_0^L \{EI(x)[\partial^2 w(x, t)/\partial x^2]^2 + P(x)[\partial \omega(x, t)/\partial x]^2\}\, dx$
Membranes	$\dfrac{1}{2}\displaystyle\int_{\text{Area}} T\{[\partial w(x, y, t)/\partial x]^2 + [\partial w(x, y, t)/\partial y]^2\}\, dx\, dy$
Plates	$\dfrac{1}{2}\displaystyle\int_{\text{Area}} D\{\nabla^2 w(x, y, t)^2 + 2(1 - v)[\partial^2 w(x, y, t)/\partial x\, \partial y]^2 - [\partial^2 w(x, y, t)/\partial x^2][\partial^2 w(x, y, t)/\partial y^2]\}dx\, dy$

*If the distributed system has a spring at the boundary point a, then add a term $kw^2(a, t)/2$.

For comparison purposes, let $M = mL$, which yields the following estimates for the lowest natural frequency:

$$1. \quad \omega = 0.8660\sqrt{\frac{EA}{mL^2}}$$

$$2. \quad \omega = 0.8629\sqrt{\frac{EA}{mL^2}} \qquad (3.4.136)$$

$$3. \quad \omega = 0.9069\sqrt{\frac{EA}{mL^2}}$$

The best estimate is the lowest one, which corresponds to case 2, with the trial function in the form of the static displacement. Note that the estimate obtained in case 1 is also quite good. It corresponds to the first case in Table 3.4.2, representing a mass-spring system in which the mass of the spring is included.

Rayleigh-Ritz Method Rayleigh's quotient, Eq. (3.4.133), corresponding to any trial function $W(x)$ is always larger than the lowest eigenvalue ω_1^2, and it takes the minimum value of ω_1^2 when $W(x)$ coincides with the lowest natural mode $W_1(x)$. However, this possibility must be ruled out by virtue of the assumption that W_1 is not available. The **Rayleigh-Ritz method** is a procedure for minimizing Rayleigh's quotient by means of a sequence of approximate solutions converging to the actual solution of the eigenvalue problem. The minimizing sequence has the form

$$W(x) = a_1\phi_1(x)$$

$$W(x) = a_1\phi_1(x) + a_2\phi_2(x) = \sum_{j=1}^{2} a_j\phi_j(x)$$

$$\cdots \qquad (3.4.137)$$

$$W(x) = a_1\phi_1(x) + a_2\phi_2(x) + \cdots + a_n\phi_n(x) = \sum_{j=1}^{n} a_j\phi_j(x)$$

where a_j are undetermined coefficients and $\phi_j(x)$ are suitable trial functions satisfying all, or at least the geometric boundary conditions. The coefficients $a_j (j = 1, 2, \ldots, n)$ are determined so that Rayleigh's quotient has a minimum. With Eqs. (3.4.137) inserted into Eq. (3.4.133), Rayleigh's quotient becomes

$$\Omega^2 = \frac{\displaystyle\sum_{i=1}^{n}\sum_{j=1}^{n} k_{ij}a_ia_j}{\displaystyle\sum_{i=1}^{n}\sum_{j=1}^{n} m_{ij}a_ia_j} \qquad n = 1, 2, \ldots \qquad (3.4.138)$$

where $k_{ij} = k_{ji}$ and $m_{ij} = m_{ji} (i, j = 1, 2, \ldots, n)$ are symmetric stiffness and mass coefficients whose nature depends on the potential energy and kinetic energy, respectively. The special case in which $n = 1$ represents Rayleigh's energy method. For $n \geq 2$, minimization of Rayleigh's quotient leads to the solution of the eigenvalue problem

$$\sum_{j=1}^{n} k_{ij}a_j = \Omega^2 \sum_{j=1}^{n} m_{ij}a_j$$

$$i = 1, 2, \ldots, n; n = 2, 3, \ldots \qquad (3.4.139)$$

Equations (3.4.139) can be written in the matrix form

$$K\mathbf{a} = \Omega^2 M\mathbf{a} \qquad (3.4.140)$$

in which $K = [k_{ij}]$ is the symmetric stiffness matrix and $M = [m_{ij}]$ is the symmetric mass matrix. Equation (3.4.140) resembles the eigenvalue problem for multi-degree-of-freedom systems, Eq. (3.4.76), and its solutions possess the same properties. The eigenvalues Ω_r^2 provide approximations to the actual eigenvalues ω_r^2, and approach them from above as n increases. Moreover, the eigenvectors $\mathbf{a}_r = [a_{r1}\ a_{r2} \ldots a_{rn}]^T$ can be used to obtain the approximate natural modes by writing

$$W_r(x) = a_{r1}\phi_1(x) + a_{r2}\phi_2(x) + \cdots + a_{rn}\phi_n(x) = \sum_{j=1}^{n} a_{rj}\phi_j(x)$$

$$r = 1, 2, \ldots, n; n = 2, 3, \ldots \qquad (3.4.141)$$

As an illustration, consider the same rod in axial vibration used to demonstrate Rayleigh's energy method. Insert Eqs. (3.4.137) with $W(x)$ replaced by $U(x)$ into the numerator and denominator of Eq. (3.4.134) to obtain

$$\int_0^L EA(x)\left[\frac{dU(x)}{dx}\right]^2 dx$$

$$= \sum_{i=1}^{n}\sum_{j=1}^{n}\left[\int_0^L EA(x)\frac{d\phi_i(x)}{dx}\frac{d\phi_j(x)}{dx}\,dx\right]a_ia_j \qquad (3.4.142a)$$

$$\int_0^L m(x)U^2(x)\,dx + MU^2(L)$$

$$= \sum_{i=1}^{n}\sum_{j=1}^{n}\left[\int_0^L m(x)\phi_i(x)\phi_j(x)\,dx + M\phi_i(L)\phi_j(L)\right]a_ia_j \qquad (3.4.142b)$$

so that the stiffness and mass coefficients are

$$k_{ij} = \int_0^L EA(x)\frac{d\phi_i(x)}{dx}\frac{d\phi_j(x)}{dx}\,dx \qquad i, j = 1, 2, \ldots, n \qquad (3.4.143a)$$

$$m_{ij} = \int_0^L m(x)\phi_i(x)\phi_j(x)\,dx + M\phi_i(L)\phi_j(L)$$

$$i, j = 1, 2, \ldots, n \qquad (3.4.143b)$$

respectively. As trial functions, use

$$\phi_j(x) = (x/L)^j \qquad j = 1, 2, \ldots, n \qquad (3.4.144)$$

which are zero at $x = 0$, thus satisfying the geometric boundary condition. Hence, the stiffness and mass coefficients are

$$k_{ij} = \frac{EAij}{L^{i+j}}\int_0^L x^{i-1}x^{j-1}\,dx = \frac{ij}{i+j-1}\frac{EA}{L}$$

$$i, j = 1, 2, \ldots, n \qquad (3.4.145a)$$

$$m_{ij} = \frac{m}{L^{i+j}}\int_0^L x^ix^j\,dx + M = \frac{mL}{i+j+1} + M$$

$$i, j = 1, 2, \ldots, n \qquad (3.4.145b)$$

so that the stiffness and mass matrices are

$$K = \frac{EA}{L}\begin{bmatrix} 1 & 1 & 1 & \cdots & 1 \\ 1 & 4/3 & 3/2 & \cdots & 2n/(n+1) \\ 1 & 3/2 & 9/5 & \cdots & 3n/(n+2) \\ \cdots\cdots\cdots\cdots\cdots\cdots\cdots\cdots \\ 1 & 2n/(n+1) & 3n/(n+2) & \cdots & n^2/(2n-1) \end{bmatrix} \qquad (3.4.146a)$$

$$M = mL\begin{bmatrix} 1/3 & 1/4 & 1/5 & \cdots & 1/(n+2) \\ 1/4 & 1/5 & 1/6 & \cdots & 1/(n+3) \\ 1/5 & 1/6 & 1/7 & \cdots & 1/(n+4) \\ \cdots\cdots\cdots\cdots\cdots\cdots\cdots\cdots \\ 1/(n+2) & 1/(n+3) & 1/(n+4) & \cdots & 1/(2n+1) \end{bmatrix}$$

$$+ M\begin{bmatrix} 1 & 1 & 1 & \cdots & 1 \\ 1 & 1 & 1 & \cdots & 1 \\ 1 & 1 & 1 & \cdots & 1 \\ \cdots\cdots\cdots\cdots\cdots \\ 1 & 1 & 1 & \cdots & 1 \end{bmatrix} \qquad (3.4.146b)$$

For comparison purposes, consider the case in which $M = mL$. Then, for $n = 2$, the eigenvalue problem is

$$\begin{bmatrix} 1 & 1 \\ 1 & 4/3 \end{bmatrix}\begin{bmatrix} a_1 \\ a_2 \end{bmatrix} = \lambda\begin{bmatrix} 4/3 & 5/4 \\ 5/4 & 6/5 \end{bmatrix}\begin{bmatrix} a_1 \\ a_2 \end{bmatrix}$$

$$\lambda = \Omega^2\frac{mL^2}{EA} \qquad (3.4.147)$$

which has the solutions

$$\lambda_1 = 0.7407 \qquad \mathbf{a}_1 = [1 \quad -0.1667]^T$$
$$\lambda_2 = 12.0000 \qquad \mathbf{a}_2 = [1 \quad -1.0976]^T \qquad (3.4.148)$$

Hence, the computed natural frequencies and modes are

$$\Omega_1 = 0.8607\sqrt{\frac{EA}{mL^2}} \qquad U_1(x) = \frac{x}{L} - 0.1667\left(\frac{x}{L}\right)^2$$

$$\Omega_2 = 3.4641\sqrt{\frac{EA}{mL^2}} \qquad U_2(x) = \frac{x}{L} - 1.0976\left(\frac{x}{L}\right)^2$$

(3.4.149)

Comparing Eqs. (3.4.149) with the estimates obtained by Rayleigh's energy method, Eqs. (3.4.136), note that the Rayleigh-Ritz method has produced a more accurate approximation for the lowest natural frequency. In addition, it has produced a first approximation for the second lowest natural frequency, as well as approximations for the two lowest modes, which Rayleigh's energy method is unable to produce. The approximate solutions can be improved by letting $n = 3, 4, \ldots$.

Finite Element Method In the Rayleigh-Ritz method, the trial functions extend over the entire domain of the system and tend to be complicated and difficult to work with. More importantly, they often cannot be produced, particularly for two-dimensional problems. Another version of the Rayleigh-Ritz method, the **finite element method,** does not suffer from these drawbacks. Indeed, the trial functions extending only over small subdomains, referred to as **finite elements,** are known low-degree polynomials and permit easy computer coding. As in the Rayleigh-Ritz method, a solution is assumed in the form of a linear combination of trial functions, known as **interpolation functions,** multiplied by undetermined coefficients. In the finite element method the coefficients have physical meaning, as they represent "nodal" displacements, where "nodes" are boundary points between finite elements. The computation of the stiffness and mass matrices is carried out for each of the elements separately and then the element stiffness and mass matrices are assembled into global stiffness and mass matrices. One disadvantage of the finite element method is that it requires a large number of degrees of freedom.

To illustrate the method, and for easy visualization, consider the transverse vibration of a string fixed at $x = 0$ and with a spring of stiffness K attached at $x = L$ (Fig. 3.4.27) and divide the length L into n elements of width h, so that $nh = L$. Denote the displacements of the nodal points x_e by a_e and assume that the string displacement is linear between any two nodal points. Figure 3.4.28 shows a typical element e.

Introducing Eq. (3.4.151) into Eqs. (3.4.152) and considering the boundary conditions, we obtain the element stiffness and mass matrices

$$K_1 = \frac{EA}{h} \qquad K_e = \frac{EA}{h}\begin{bmatrix} 1 & -1 \\ -1 & 1 \end{bmatrix} \qquad e = 2, 3, \ldots, n-1$$

$$K_n = \frac{EA}{h}\begin{bmatrix} 1 & -1 \\ -1 & Kh/EA \end{bmatrix} \qquad M_1 = \frac{hm}{3}$$

$$M_e = \frac{hm}{6}\begin{bmatrix} 2 & 1 \\ 1 & 2 \end{bmatrix} \qquad e = 2, 3, \ldots, n$$

(3.4.153)

where K_1 and M_1 are really scalars, because the left end of the first element is fixed, so that the displacement is zero. Then, since the nodal

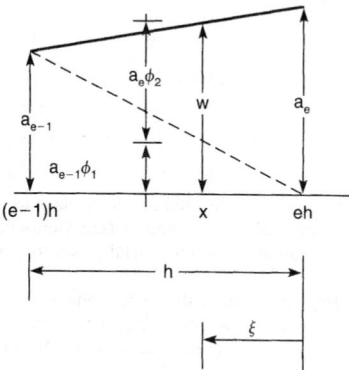

Fig. 3.4.28

displacement a_e is shared by elements e and $e + 1$ ($e = 1, 2, \ldots, n-2$), the element stiffness and mass matrices can be assembled into the global stiffness and mass matrices

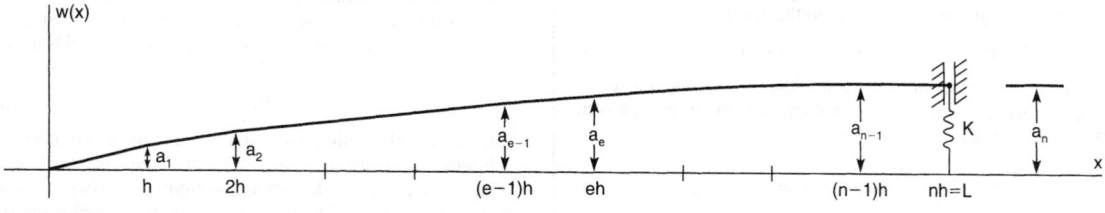

Fig. 3.4.27

The process can be simplified greatly by introducing the nondimensional local coordinate $\xi = j - x/h$. Then, considering the two linear interpolation functions

$$\phi_1(\xi) = \xi \qquad \phi_2(\xi) = 1 - \xi$$

(3.4.150)

the displacement at point ξ can be expressed as

$$\omega(\xi) = a_{e-1}\phi_1(\xi) + a_e\phi_2(\xi)$$

(3.4.151)

where a_{e-1} and a_e are the nodal displacements for element e. Using Eqs. (3.4.143) and changing variables from x to ξ, we can write the element stiffness and mass coefficients

$$k_{eij} = \frac{1}{h}\int_0^1 EA\frac{d\phi_i}{d\xi}\frac{d\phi_j}{d\xi}d\xi \qquad m_{eij} = h\int_0^1 m\phi_i\phi_j\,d\xi,$$
$$i, j = 1, 2 \quad (3.4.152)$$

$$K = \frac{EA}{h}\begin{bmatrix} 2 & -1 & 0 & \cdots & 0 & 0 \\ -1 & 2 & -1 & \cdots & 0 & 0 \\ 0 & -1 & 2 & \cdots & 0 & 0 \\ \multicolumn{6}{c}{\dotfill} \\ 0 & 0 & 0 & \cdots & 2 & -1 \\ 0 & 0 & 0 & \cdots & -1 & Kh/EA \end{bmatrix}$$

(3.4.154)

$$M = \frac{hm}{6}\begin{bmatrix} 4 & 1 & 0 & \cdots & 0 & 0 \\ 1 & 4 & 1 & \cdots & 0 & 0 \\ 0 & 1 & 4 & \cdots & 0 & 0 \\ \multicolumn{6}{c}{\dotfill} \\ 0 & 0 & 0 & \cdots & 4 & 1 \\ 0 & 0 & 0 & \cdots & 1 & 2 \end{bmatrix}$$

For beams in bending, the displacements consist of one translation and one rotation per node; the interpolation functions are the Hermite cubics

$$\phi_1(\xi) = 3\xi^2 - 2\xi^3, \phi_2(\xi) = \xi^2 - \xi^3$$
$$\phi_3(\xi) = 1 - 3\xi^2 + 2\xi^3, \phi_4(\xi) = -\xi + 2\xi^2 - \xi^3 \quad (3.4.155)$$

and the element stiffness and mass coefficients are

$$k_{eij} = \frac{1}{h^3} \int_0^1 EI \frac{d^2\phi_i}{d\xi^2} \frac{d^2\phi_j}{d\xi^2} d\xi \qquad m_{eij} = h \int_0^1 m\phi_i\phi_j \, d\xi$$
$$i, j = 1, 2, 3, 4 \quad (3.4.156)$$

yielding typical element stiffness and mass matrices

$$K_e = \frac{EI}{h^3} \begin{bmatrix} 12 & 6 & -12 & 6 \\ 6 & 4 & -6 & 2 \\ -12 & -6 & 12 & -6 \\ 6 & 2 & -6 & 4 \end{bmatrix}$$

$$M_e = \frac{hm}{420} \begin{bmatrix} 156 & 22 & 54 & -13 \\ 22 & 4 & 13 & -3 \\ 54 & 13 & 156 & -22 \\ -13 & -3 & -22 & 4 \end{bmatrix} \quad (3.4.157)$$

The treatment of two-dimensional problems, such as for membranes and plates, is considerably more complex (see Meirovitch, "Principles and Techniques of Vibration," Prentice-Hall) than for one-dimensional problems.

The various steps involved in the finite element method lend themselves to ready computer programming. There are many computer codes available commercially; one widely used is NASTRAN.

VIBRATION-MEASURING INSTRUMENTS

Typical quantities to be measured include acceleration, velocity, displacement, frequency, damping, and stress. Vibration implies motion, so that there is a great deal of interest in transducers capable of measuring motion relative to the inertial space. The basic transducer of many vibration-measuring instruments is a mass-damper-spring enclosed in a case together with a device, generally electrical, for measuring the displacement of the mass relative to the case, as shown in Fig. 3.4.29. The equation for the displacement $z(t)$ of the mass relative to the case is

$$m\ddot{z}(t) + c\dot{z}(t) + kz(t) = -m\ddot{y}(t) \quad (3.4.158)$$

where $y(t)$ is the displacement of the case relative to the inertial space. If this displacement is harmonic, $y(t) = Y \sin \omega t$, then by analogy with Eq. (3.4.35) the response is

$$z(t) = Y\left(\frac{\omega}{\omega_n}\right)^2 |G(\omega)| \sin(\omega t - \phi)$$
$$= Z(\omega) \sin(\omega t - \phi) \quad (3.4.159)$$

so that the magnitude factor $Z(\omega)/Y = (\omega/\omega_n)^2 |G(\omega)|$ is as plotted in Fig. 3.4.9 and the phase angle ϕ is as in Fig. 3.4.4. The plot $Z(\omega)/Y$

Fig. 3.4.29

versus ω/ω_n is shown again in Fig. 3.4.30 on a scale more suited to our purposes.

Accelerometers are high-natural-frequency instruments. Their usefulness is limited to a frequency range well below resonance. Indeed, for small values of ω/ω_n, Eq. (3.4.159) yields the approximation

$$Z(\omega) \approx \frac{1}{\omega_n^2} \omega^2 Y \quad (3.4.160)$$

so that the signal amplitude is proportional to the amplitude of the acceleration of the case relative to the inertial space. For $\xi = 0.7$, the accelerometer can be used in the range $0 \leq \omega/\omega_n \leq 0.4$ with less than 1 percent error, and the range can be extended to $\omega/\omega_n \leq 0.7$ if proper corrections, based on instrument calibration, are made.

Commonly used accelerometers are the compression-type piezoelectric accelerometers. They consist of a mass resting on a piezoelectric ceramic crystal, such as quartz, tourmaline, or ferroelectric ceramic, with the crystal acting both as spring and sensor. Piezoelectric actuators have negligible damping, so that their range must be smaller, such as $0 < \omega/\omega_n < 0.2$. In view of the fact, however, that the natural frequency is very high, about 30,000 Hz, this is a respectable range.

Displacement-Measuring Instruments These are low-natural-frequency devices and their usefulness is limited to a frequency range well above resonance. For $\omega/\omega_n \gg 1$, Eq. (3.4.159) yields the approximation

$$Z(\omega) \approx Y \quad (3.4.161)$$

so that the signal amplitude is proportional to the amplitude of the case displacement. Instruments with low natural frequency compared to the excitation frequency are known as **seismometers.** They are commonly used to measure ground motions, such as those caused by earthquakes or underground nuclear explosions. The requirement of low natural frequency dictates that the mass, referred to as **seismic mass,** be very large and the spring very soft, so that essentially the mass remains

Fig. 3.4.30

stationary in an inertial space while the case attached to the ground moves relative to the mass.

Seismometers tend to be considerably larger in size than accelerometers. If a large-size instrument is undesirable, or even if size is not an issue, displacements in harmonic motion, as well as velocities, can be obtained from accelerometer signals by means of electronic integrators.

Some other transducers, not mass-damper-spring transducers, are as follows (Harris, "Shock and Vibration Handbook," 3d ed., McGraw-Hill):

Differential-transformer pickups: They consist of a core of magnetic material attached to the vibrating structure, a primary coil, and two secondary coils. As the core moves, both the inductance and induced voltage of one secondary coil increase while those of the other decrease. The output voltage is proportional to the displacement over a wide range. Such pickups are used for very low frequencies, up to 60 Hz.

Strain gages: They consist of a grid of fine wires which exhibit a change in electrical resistance proportional to the strain experienced. Their flimsiness requires that strain gages be either mounted on or bonded to some carrier material.

Section 4

Heat

BY

PETER E. LILEY *Professor Emeritus of Mechanical Engineering, Purdue University.*
HOYT C. HOTTEL *Late Professor Emeritus, Massachusetts Institute of Technology.*
ADEL F. SAROFIM *Presidential Professor of Chemical Engineering, University of Utah.*
JAMES J. NOBLE *Formerly Research Associate Professor of Chemical and Biological Engineering, Tufts University.*
KENNETH A. SMITH *Edward R. Gilliland Professor of Chemical Engineering, Massachusetts Institute of Technology.*

4.1 THERMODYNAMICS
by Peter E. Liley

NOTE: References are placed throughout the text for the reader's convenience. (No material is presented relating to the calibration of thermometers at fixed points, etc. Specific details of the measurement of temperature, pressure, etc. are found in Benedict, "Fundamentals of Temperature, Pressure and Flow Measurements," 3d ed. Measurement of other properties is reviewed in Maglic et al., "Compendium of Thermophysical Property Measurement Methods," vol. 1, Plenum Press. The periodical *Metrologia* presents latest developments, particularly for work of a definitive caliber.)

Thermodynamic properties of a variety of other specific materials are listed also in Secs. 4.2, 6.1, and 9.8.

THERMOMETER SCALES

Let F and C denote the readings on the Fahrenheit and Celsius (or centigrade) scales, respectively, for the same temperature. Then

$$C = \frac{5}{9}(F - 32) \qquad F = \frac{9}{5}C + 32$$

If the pressure readings of a constant-volume hydrogen thermometer are extrapolated to zero pressure, it is found that the corresponding temperature is $-273.15°C$, or $-459.67°F$. An **absolute temperature scale** was formerly used on which zero corresponding with zero pressure on the hydrogen thermometer. The basis now used is to define and give a numerical value to the temperature at a single point, the triple point of water, defined as $0.01°C$. The scales are:

Kelvins (K) = degrees Celsius + 273.15
Degrees Rankine (°R) = degrees Fahrenheit + 459.67

EXPANSION OF BODIES BY HEAT

Coefficients of Expansion The **coefficient of linear expansion** a' of a solid is defined as the increment of length in a unit of length for a rise in temperature of 1 deg. Likewise, the **coefficient of cubical expansion** a''' of a solid, liquid, or gas is the increment of volume of a unit volume for a rise of temperature of 1 deg. Denoting these coefficients by a' and a''', respectively, we have

$$a' = \frac{1}{l}\frac{dl}{dt} \qquad a''' = \frac{1}{V}\frac{dV}{dt}$$

in which l denotes length, V volume, and t temperature. For homogeneous solids $a''' = 3a'$, and the **coefficient of area expansion** $a'' = 2a'$.

The coefficients of expansion are, in general, dependent upon the temperature, but for ordinary ranges of temperature, constant mean values may be taken. If lengths, areas, and volumes at $32°F$ ($0°C$) are taken as standard, then these magnitudes at other temperatures t_1 and t_2 are related as follows:

$$\frac{l_1}{l_2} = \frac{1 + a't_1}{1 + a't_2} \qquad \frac{A_1}{A_2} = \frac{1 + a''t_1}{1 + a''t_2} \qquad \frac{V_1}{V_2} = \frac{1 + a'''t_1}{1 + a'''t_2}$$

Since for solids and liquids the expansion is small, the preceding formulas for these bodies become approximately

$$l_2 - l_1 = a'l_1(t_2 - t_1)$$
$$A_2 - A_1 = a''A_1(t_2 - t_1)$$
$$V_2 - V_1 = a'''V_1(t_2 - t_1)$$

The coefficients of cubical expansion for different gases at ordinary temperatures are about the same. From 0 to $212°F$ and at atmospheric pressure, the values multiplied by 1,000 are as follows: for NH_3, 2.11; CO, 2.04; CO_2, 2.07; H_2, 2.03; NO, 2.07. For an ideal gas, the coeffi-

cient at any temperature is the reciprocal of the (absolute) temperature. (See also Table 6.1.10.)

UNITS OF FORCE AND MASS

Force mass, length, and time are related by Newton's second law of motion, which may be expressed as

$$F \sim ma$$

In order to write this as an equality, a constant must be introduced which has magnitude and dimensions. For convenience, in the fps system, the constant may be designated as $1/g_c$. Thus,

$$F = \frac{ma}{g_c}$$

Since this equation must be homogeneous insofar as the dimensions are concerned, the units for g_c are $mL/(t^2F)$. Consider a 1-lb mass, lbm, in the earth's gravitational field, where the acceleration is 32.1740 ft/s². The force exerted on the pound mass will be defined as the pound force, lbf. This system of units gives for g_c the following magnitude and dimensions:

$$1 \text{ lbf} = \frac{(1 \text{ lbm})(32.174 \text{ ft/s}^2)}{g_c}$$

hence

$$g_c = 32.174 \text{ lbm} \cdot \text{ft/(lbf} \cdot \text{s}^2)$$

Note that g_c may be used with other units, in which case the numerical value changes. The numerical value of g_c for four systems of units is

$$g_c = 32.174 \frac{\text{lbm} \cdot \text{ft}}{\text{lbf} \cdot \text{s}^2} = 1 \frac{\text{slug} \cdot \text{ft}}{\text{lbf} \cdot \text{s}^2} = 1 \frac{\text{lbm} \cdot \text{ft}}{\text{pdl} \cdot \text{s}^2} = 1 \frac{\text{g} \cdot \text{cm}}{\text{dyn} \cdot \text{s}^2}$$

In SI, the constant is chosen to be unity and $F(N) = m(\text{kg})a(\text{m/s}^2)$. There are four possible constants, and all have been used. (See Blackman, "SI Units in Engineering," Macmillan.)

Consider now the relationship which involves weight, a gravitational force, and mass by applying the basic equation for a body of fixed mass acted upon by a gravitational force g and no other forces. The acceleration of the mass caused by the gravitational force is the acceleration due to gravity g.

Substituting gives the relationship between weight and mass

$$w = \frac{mg}{g_c}$$

If the gravitational acceleration is constant, the weight and mass are in a fixed proportion to each other; hence for accounting purposes in mass balances they can be used interchangeably. This is not possible if g is a variable.

We may now write the relation between mass m and weight w as

$$w = m\frac{g}{g_c}$$

The constant g_c is used throughout the following paragraphs. (An extensive table of conversion factors from customary units to SI units is found in Sec. 1.)

The SI unit of pressure is the newton per square metre. It is a very small pressure, as normal atmospheric pressure is 1.01325×10^5 N/m². While some use has been made of the pressure expressed in kN/m² or kPa (1 Pa = 1 N/m²) and in MN/m² or MPa, the general technical usage now seems to favor the bar = 10^5 N/m^2 = 10^5 Pa so that 1 atm = 1.01325 bar. For many approximate calculations the atmosphere and the bar can be equated.

Many representative accounts of the measurement of low and high pressure have appeared. (See, for example, Lawrance, *Chem. Eng. Progr.*, **50**, 1954, p. 155; Leck, "Pressure Measurement in Vacuum Systems," Inst. Phys., London, 1957; Peggs, "High Pressure Measurement Techniques," Appl. Sci. Publishers, Barking, Essex.)

MEASUREMENT OF HEAT

Units of Heat   Many units of heat have been dependent on the experimentally determined properties of some substance. To eliminate experimental variations, the unit of heat may be defined in terms of fundamental units. The International Steam Table Conference (London, 1929) defines the Steam Table (IT) calorie as 1/860 of a watthour. One British thermal unit (Btu) is defined as 251.996 IT cal, 778.26 ft · lb.

Previously, the Btu was defined as the heat necessary to raise one pound of water one degree Fahrenheit at some arbitrarily chosen temperature level. Similarly, the calorie was defined as the heat required to heat one gram of water one degree Celsius at 15°C (or at 17.5°C). These units are roughly the same in value as those mentioned above. In SI, the **joule** is the heat unit, and the newton-metre the work unit of energy. The two are equal, so that 1 J = 1 N · m; that is, in SI, the mechanical equivalent of heat is unity.

Heat Capacity and Specific Heat   The **heat capacity** of a material is the amount of heat transferred to raise a unit mass of a material 1 deg in temperature. The ratio of the amount of heat transferred to raise unit mass of a material 1 deg to that required to raise unit mass of water 1 deg at some specified temperature is the **specific heat** of the material. For most engineering purposes, heat capacities may be assumed numerically equal to specific heats. Two heat capacities are generally used, that at constant pressure c_p and that at constant volume c_v. For unit mass, the instantaneous heat capacities are defined as

$$\left(\frac{\partial h}{\partial t}\right)_p = c_p \qquad \left(\frac{\partial u}{\partial t}\right)_v = c_v$$

Over a range in temperature, the mean heat capacities are given by

$$c_{pm} = \frac{1}{t_2 - t_1}\int_{t1}^{t2} c_p\, dt \qquad c_{vm} = \frac{1}{t_2 - t_1}\int_{t1}^{t2} c_v\, dt$$

Denoting by c the heat capacity, the heat required to raise the temperature of m lb of a substance from t_1 to t_2 is $Q = mc(t_2 - t_1)$, provided c is a constant.

In general, c varies with the temperature, though for moderate temperature ranges a constant mean value may be taken. If, however, c is taken as varable, $Q = m\int_{t1}^{t2} c\, dt$. The *mean* heat capacity from 0 to t deg is given by $c_m = \frac{1}{t}\int_0^t c\, dt$. If $c = a_1 + a_2 t + a_3 t^2 + \cdots$

$$c_m = a_1 + \tfrac{1}{2}a_2 t + \tfrac{1}{3}a_3 t^2 + \cdots$$

Data for the specific heat of some solids, liquids, and gases are found in Tables 4.2.22 and 4.2.27.

Specific Heat of Solids   For eleraents near room temperature, the specific heat may be approximated by the rule of Dulong and Petit, that the specific heat at constant volume approaches $3R$. At lower temperatures, Debye's theory leads to the equation

$$\frac{C_v}{R} = 3\left(\frac{T}{\Theta}\right)\int_0^{\Theta_{max}/T}\left(\frac{\Theta}{T}\right)^4 \frac{\exp(\Theta/T)}{[\exp(\Theta/T) - 1]^2}\, d\left(\frac{\Theta}{T}\right)$$

commonly known as Debye's function. Figure 4.1.1 shows the variation of c_v/R with T/Θ, as predicted from the equation above. The principal difficulties in using the Debye equation arise from (1) the difficulty in finding unique values of Θ for any given material and (2) the need to consider other contributions to the specific heat. For further references on the Debye equation, see Harrison and Neighbours. *U.S. Naval Post-*

graduate Lab. Rept. 49, Dec. 1964; Overton and Hancock, *Naval Research Lab. Rept. 5502*, 1960; Hilsenrath and Zeigler, *NBS Monograph 49*, 1962. For a thorough discussion of electronic, lattice, and magnetic contributions to specific heat, see Gopal, "Specific Heats at Low Temperatures," Plenum Press, New York. See also Table 6.1.11.

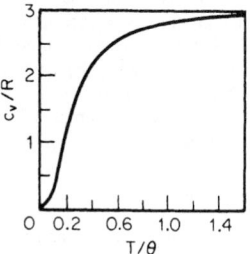

Fig. 4.1.1

For solid compounds at about room temperature, Kopp's approximation is often useful. This states that the specific heat of a solid compound at room temperature is equal to the sum of the specific heats of the atoms forming the compound.

SPECIFIC HEAT OF LIQUIDS

No general theory of any simple practical utility seems to exist for the specific heat of liquids. In "Thermophysical Properties of Refrigerants," ASHRAE, Atlanta, 1976, the interpolation device was a polynomial in temperature, usually up to T^3.

SPECIFIC HEAT OF GASES

The following table summarizes results of kinetic theory for specific heats of gases:

Gas Type	c_p/R	c_v/R	c_p/c_v
Monatomic	5/2	3/2	5/3
Diatomic	7/2	5/2	7/5
n degrees of freedom	$(n + 2)/2$	$n/2$	$1 + 2/n$

Determination of the effective number of degrees of freedom limits extending this method to more comlex gases.

Properties of gases are, usually, most readily correlated on the mol basis. A **pound mol** is the mass in pounds equal to the molecular weight. Thus 1 pound mol of oxygen is 32 lb. At the same pressure and temperature, the volume of 1 mol is the same for all perfect gases, i.e., following the gas laws. Experimental findings led Avogadro (1776–1856) to formulate the microscopic hypothesis now known as **Avogadro's principle,** which states that 1 mol of any perfect gas contains the same number of molecules. The number is known as the **Avogadro number** and is equal to

$$N = 6.02214 \times 10^{26} \text{ molecules/(kg·mol)}$$
$$= 2.73160 \times 10^{26} \text{ molecules/(lb·mol)}$$

For perfect gases, $Mc_p - Mc_v = MR = 1.987$.

$$c_i = R/(k - 1) \qquad c_p = Rk(k - 1)$$

Passut and Danner [*Ind. Eng. Chem. Process Design & Dev.* (**11,** 1972, p. 543)] developed a set of thermodynamically consistent polynomials for estimating ideal gas enthalpy, entropy, and heat capacity, fittings being given for 89 compounds. The same journal reported 2 years later an extension of the work to another 57 compounds, by Huang and Daubert. Fittings of a cubic-in-temperature polynomial for 408 hydrocarbons and related compounds in the ideal gas state were reported by Thinh et al. in *Hydrocarhon Processing,* Jan. 1971. pp. 98–104. On p. 153 of a later issue (Aug. 1976), they claimed that the function $C_p = A + B\exp(-c/T^n)$ fitted for 221 hydrocarbons: graphite and hydrogen gave a more accurate fit. A cubic polynomial in temperature was also fitted for more than 700 compounds from 273 to 1,000 K by Seres et al. in *Acta Phys. Chem.*, Univ. Szegediensis (Hungary), **23**, 1977, pp. 433–468. A 1975 formula of Wilhoit was fitted for 62 substances by A. Harmens in *Proc. Conf. Chemical Thermodynamic Data on Fluids,* IPC Sci. Tech. Press, Guildford, U.K., pp. 112–120. A cubic polynomial fitting for 435 substances appeared in *J. Chem. Eng., Peking*(**2**, 1979, pp. 109–132). The reader is reminded that specific heat at constant pressure values can readily be calculated from tabulated enthalpy-temperature tables, for any physical state. Table 4.2.22 gives values for liquids and gases, while Tables 4.2.27 to 4.2.29 provide similar information for selected solids.

SPECIFIC HEAT OF MIXTURES

If w_1 lb of a substance at temperature t_1 and with specific heat c_1 is mixed with w_2 lb of a second substance at temperature t_2 and with specific heat c_2, provided chemical reaction, heat evolution, or heat absorption does not occur, the specific heat of the mixture is

$$c_m = (w_1c_1 + w_2c_2)/(w_1 + w_2)$$

and the temperature of the mixture is

$$t_m = (w_1c_1t_1 + w_2c_2t_2)/(w_1c_1 + w_2c_2)$$

In general, $t_m = \Sigma wct / \Sigma wc$.

To raise the temperature of w_1 lb of a substance having a specific heat c_1 from t_1 to t_m, the weight w_2 of a second substance required is

$$w_2 = w_1c_1(t_m - t_1)/c_2(t_2 - t_m)$$

SPECIFIC HEAT OF SOLUTIONS

For aqueous solutions of salts, the specific heat may be estimated by assuming the specific heat of the solution equal to that of the water alone. Thus, for a 20 percent by weight solution of sodium chloride in water, the specific heat would be approximately 0.8.

Although approximate calculations of mixture properties often consist simply of multiplying the mole fraction of each constituent by the property of each constituent, more accurate calculations are possible. (See "Technical Data Book—Petroleum Refining" API, Washington, DC, 1984; Daubert, "Chemical Engineering Thermodynamics," McGraw-Hill; "The Properties of Gases and Liquids," 3d ed., McGraw-Hill; Perry, "Chemical Engineers Handbook," McGraw-Hill; Walas, "Phase Equilibria in Chemical Engineering," Butterworth.)

LATENT HEAT

For pure substances, the heat effects accompanying changes in state at constant pressure are known as latent effects, because no temperature change is evident. Heat of fusion, vaporization, sublimation, and change in crystal form are exarnples.

The values for the heat of fusion and latent heat of vaporization are presented in Tables 4.2.21 and 4.2.28.

GENERAL PRINCIPLES OF THERMODYNAMICS

Notation

B = availability (by definition, $B = H - T_0S$)
c_p = specific heat at constant pressure
c_v = specific heat at constant volume
E, e = total energy associated with system
g = local acceleration of gravity, ft/s^2
g_c = a dimensional constant
H, h = enthalpy, Btu (by definition $h = u + pv$)
J = mechanical equivalent of heat = 778.26 ft · lb/Btu = 4.1861 J/cal
$k = c_p/c_v$
m = mass of substance under consideration, lbm

M = molecular weight
p = absolute pressure, lb/ft^2
Q, q = quantity of heat absorbed by system from surroundings, Btu
R = ideal gas constant
R_u = universal gas constant
S, s = entropy
t = temperature, °F
$T = t + 459.69$ = absolute temperature = °R
T_0 = sink or discard temperature
U, u = internal energy
$\bar{v}$ = linear velocity
v = volume
V = total volume
w = weight of substance under consideration, lb
W = external work performed on surroundings during change of state, ft · lb
$$Y = \left(\frac{p_1}{p_2}\right)^{(k-1)k} - 1$$
z = distance above or below chosen datum
g = free energy (by definition, $g = h - Ts$)
f = Helmholtz free energy (by definition, $f = u - Ts$)
ϵ = exergy

In thermodynamics, unless otherwise noted, the convention followed is that the change in any property $\psi = \Delta\psi =$ final value $-$ initial value $= \psi_2 - \psi_1$.

In this notation, small letters usually denote magnitudes referred to a unit mass of the substance, capital letters corresponding magnitudes referred to m units of mass. Thus, v denotes the volume of 1 lb, and $V = mv$, the volume of m lb. Similarly, $U = mu$, $S = ms$, etc. Subscripts are used to indicate different states; thus, p_1, v_1, T_1, u_1, s_1 refer to state 1; p_2, v_2, T_2, u_2, s_2 refer to state 2; Q_{12} is used to denote the heat transferred during the change from state 1 to state 2, and W_{12} denotes the external work done during the same change.

Thermodynamics is the study which deals with energy, the various concepts and laws describing the conversion of one form of energy to another, and the various systems employed to effect the conversions. Thermodynamics deals in general with systems in equilibrium. By means of its fundamental concepts and basic laws, the behavior of an engineering system may be described when the various variables are altered. Thermodynamics covers a very broad field and includes many systems, for example, those dealing with chemical, thermal, mechanical, and electrical force fields and potentials. The quantity of matter within a prescribed boundary under consideration is called the **system,** and everything external to the system is spoken of as the **surroundings.** With a **closed system** there is no interchange of matter between system and surroundings; with an **open system** there is such an interchange. Any change that the system may undergo is known as a **process.** Any process or series of processes in which the system returns to its original condition or state is called a **cycle.**

Heat is energy **in transit** from one mass to another because of a temperature difference between the two. Whenever a force of any kind acts through a distance, **work** is done. Like heat, work is also energy in transit. Work is to be differentiated from the capacity of a quantity of energy to do work.

The two fundamental and general laws of thermodynamics are: (1) energy may be neither created nor destroyed, (2) it is impossible to bring about any change or series of changes the sole net result of which is transfer of energy as heat from a low to a high temperature; in other words, heat will not of itself flow from low to high temperatures.

The **first law of thermodynamics,** one of the very important laws of nature, is the law of conservation of energy. Although the law has been stated in a variety of ways, all have essentially the same meaning. The following are examples of typical statements: Whenever energy is transformed from one form to another, energy is always conserved; energy can neither be created nor destroyed; the sum total of all energy remains constant. The energy conservation hypothesis was stated by a number of investigators; however, experimental evidence was not available until the famous work of J. P. Joule. Transformation of matter into energy ($E = mc^2$), as in nuclear reactions, is ignored; within the realm of thermodynamics discussed here, mass is conserved.

It has long been the custom to designate the law of conservation of energy, the first law of thermodynamics, when it is used in the analysis of engineering systems involving heat transfer and work. Statements of the first law may be written as follows: Heat and work are mutually convertible; or, since energy can neither be created nor destroyed, the total energy associated with an energy conversion remains constant.

Before the first law may be applied to the analysis of engineering systems, it is necessary to express it in some form of expression. Thus it may be stated for an **open system** as

$$\begin{bmatrix} \text{Net amount of} \\ \text{energy added to} \\ \text{system as heat} \\ \text{and all forms} \\ \text{of work} \end{bmatrix} + \begin{bmatrix} \text{stored} \\ \text{energy} \\ \text{of mass} \\ \text{entering} \\ \text{system} \end{bmatrix} - \begin{bmatrix} \text{stored} \\ \text{energy} \\ \text{of mass} \\ \text{leaving} \\ \text{system} \end{bmatrix} = \begin{bmatrix} \text{net in-} \\ \text{crease} \\ \text{in stored} \\ \text{energy of} \\ \text{system} \end{bmatrix}$$

For an open system with fluid entering only at section 1 and leaving only at section 2 and with no electrical, magnetic, or surface-tension effects, this equation may be written as

$$Q + W + \int \left(h_1 + \frac{v_1^2}{2g_c} + \frac{gz_1}{g_c} \right) \delta m_1$$
$$- \int \left(h_2 + \frac{v_2^2}{2g_c} + \frac{gz_2}{g_c} \right) \delta m_2 = U_f - U_i$$
$$+ \frac{m_f v_f^2 - m_i v_i^2}{2g_c} + \frac{g}{g_c}(m_f z_f - m_i z_i)$$

Note that the same sign is given to both heat and work transfers. Heat and work added to the system are given a positive sign; heat lost and work output are given a negative sign. The subscripts i and f refer to entire systems before and after the process occurs, and δm refers to a differential quantity of matter.

It must be remembered that all terms in the first-law equation must be expressed in the same units.

For a **closed stationary system,** the first-law expression reduces to

$$Q + W = U_2 - U_1$$

For an **open system** fixed in position but undergoing **steady flow,** e.g., a turbine or reciprocating steam engine, for a mass flow rate of m is

$$\dot{Q} + \dot{W} = \dot{m} \left[(h_2 - h_1) + \frac{v_2^2 - v_1^2}{2g_c} + \frac{8}{g_c}(z_2 - z_1) \right]$$

In a **steady-flow** process, the mass rate of flow into the apparatus is equal to the mass rate of flow out; in addition, at any point in the apparatus, the conditions are unchanging with time.

This condition is usually called the **continuity equation** and is written as

$$\dot{m} = \frac{dm}{dt} = \frac{A\bar{v}}{v} = \frac{A_1\bar{v}}{v_1} = \frac{A_2\bar{v}_2}{v_2}$$

where mass flow rate $\dot{m}$ is related to volume flow rate $\dot{V}$ by $\dot{V} = \dot{m}v$ and A is the cross-sectional area.

Since for many processes the last two terms are often negligible, they will be omitted for simplicity except when such omission would introduce appreciable error.

Work done in overcoming a fluid pressure is measured by $W = -\int p\, dv$, where p is the pressure *effectively* applied to the surroundings for doing work and dv represents the change in volume of the system.

Reversible and Irreversible Processes   A **reversible** process is one in which both the system and the surroundings may be returned to their original states. After an **irreversible** process, this is not possible. No process involving friction or an unbalanced potential can be reversible. No loss in ability to do work is suffered because of a reversible process,

but there is always a loss in ability to do work because of an irreversible process. All actual processes are irreversible. Any series of reversible processes that starts and finishes with the system in the same state is called a **reversible cycle.**

Steady-Flow Processes   With **steady flow,** the conditions at any point in an apparatus through which a fluid is flowing do not change progressively with time. Steady-flow processes involving only mechanical effects are equivalent to similar nonflow processes occurring between two weightless frictionless diaphragms or pistons moving at constant pressure with the system as a whole in motion. Under these circumstances, the total work done by or on a unit amount of fluid is made up of that done on the two diaphragms $p_2 v_2 - p_1 v_1$ and that done on the rest of the surroundings $-\int p\, dv - p_2 v_2 + p_1 v_1$. Differentiating, $-p\, dv - d(p)v = v\, dp$. The net, useful flow work done on the surroundings is $\int v\, dp$. This is often called the shaft work. The net, useful or shaft work differs from the total work by $p_2 v_2 - p_1 v_1$. The first-law equation may be written to indicate this result for a unit mass flow rate as

$$q + W_{\text{net}} = u_2 - u_1 + p_2 v_2 - p_1 v_1 + \frac{1}{2g_c}(\bar{v}_2^2 - \bar{v}_1^2) + \frac{g}{g_c}(z_2 - z_1)$$

or, since by definition

$$u + pv = h$$
$$q + W_{\text{net}} = h_2 - h_1 + \frac{1}{2g_c}(\bar{v}_2^2 - \bar{v}_1^2) + \frac{g}{g_c}(z_2 - z_1)$$

If all net work effects are mechanical,

$$q + \int v\, dp = h_2 - h_1 + \frac{1}{2g_c}(\bar{v}_2^2 - \bar{v}_1^2) + \frac{g}{g_c}(z_2 - z_1)$$

Since in evaluating $\int v\, dp$ the pressure is that *effectively* applied to the surroundings, the integration cannot usually be performed except for reversible processes.

If a fluid is passed adiabatically through a conduit (i.e., without heat exchange with the conduit), without doing any net or useful work, and if velocity and potential effects are negligible, $h_2 = h_1$. A process of the kind indicated is the Joule-Thomson flow, and the ratio $(\partial T/\partial p)_h$ for such a flow is the Joule-Thomson coefficient.

If a fluid is passed through a nonadiabatic conduit without doing any net or useful work and if velocity and potential effects are negligible, $q = h_2 - h_1$. This equation is important in the calculation of heat balances on flow apparatus, e.g., condensers, heat exchangers, and coolers.

In many engineering processes the movement of materials is not independent of time; hence the steady-flow equations do not apply. For example, the process of oxygen discharging from a storage bottle represents a transient condition. The pressure within the bottle changes as the amount of oxygen in the tank decreases. The analysis of some transient processes is very complex; however, in order to show the general approach, a simple case will be considered.

The quantity of material flowing into and out of the engineering system in Fig. 4.1.2 varies with time. The amount of work and the heat transfer crossing the system boundary are likewise dependent upon time. According to the law of conservation of mass, the rate of change of mass within the system is equal to the rate of mass flow into and out

Fig. 4.1.2   Variable-flow system.

of the system. Hence, in terms of mass flow rates,

$$\frac{dm_s}{d\tau} = \frac{dm_1}{d\tau} - \frac{dm_2}{d\tau}$$

For a finite period of time, this relation may be expressed as

$$\Delta m_s = \Delta m_1 - \Delta m_2$$

The first law may be written as follows:

$$\frac{dU_s}{d\tau} = \frac{dQ}{d\tau} + \frac{dW}{d\tau} + \left(h_1 + \frac{\bar{v}_1^2}{2g_c} + \frac{g}{g_c} z_1 \right) \frac{dm_1}{d\tau}$$

$$- \left(h_2 + \frac{\bar{v}_2^2}{2g_c} + \frac{g}{g_c} z_2 \right) \frac{dm_2}{d\tau}$$

Under non-steady-flow conditions the variables h, $\bar{v}$, z may change with time as well as flow rate, in which case the solution is very involved.

If steady-flow conditions prevail, then ΔU_s is equal to 0 and the integrands are independent of time, in which case the above equation reduces to the familiar steady-flow relation.

The **second law of thermodynamics** is a statement that conversion of heat to work is limited by the temperature at which conversion occurs. It may be shown that:

1. No cycle can be more efficient than a reversible cycle operating between given temperature limits.

2. The efficiency of all reversible cycles absorbing heat only at a single constant higher temperature T_1 and rejecting heat only at a single constant lower temperature T_2 must be the same.

3. For all such cycles, the efficiency is

$$\eta = \frac{W}{Q_1} = \frac{T_1 - T_2}{T_1}$$

This is usually called the **Carnot cycle efficiency**. By the first law $W = Q_1 - Q_2$,

$$(Q_1 - Q_2)/Q_1 = (T_1 - T_2)/T_1$$

By algebraic rearrangement,

$$Q_1/T_1 = Q_2/T_2$$

Clapeyron Equation

$$\frac{dp}{dT} = \frac{Q}{TV_{12}}$$

This important relation is useful in calculations relating to constant-pressure evaporation of pure substances. In that case the equation may be written

$$v_{fg} = \frac{h_{fg}}{T\,(dp/dT)}$$

where v_{fg} and h_{fg} are defined on p. 4-14.

ENTROPY

For reversible cyclical processes in which the temperature varies during heat absorption and rejection, i.e., for any *reversible cycle*, $\int (dQ/T) = 0$. Consequently, for any *reversible process*, $\int (dQ/T)$ is not a function of the particular *reversible path* followed. This integral is called the entropy change, or $\int_1^2 (dQ_{rev}/T) = S_2 - S_1 = S_{12}$. The entropy of a substance is dependent only on its state or condition. Mathematically, dS is a complete or perfect differential and S is a point function in contrast with Q and W which are path functions. For any reversible process, the change in entropy of the system and surroundings is zero, whereas for any irreversible process, the net entropy change is positive.

All actual processes are irreversible and therefore occur with a decrease in the amount of energy available for doing work, i.e., with an increase in unavailable energy. The increase in unavailable energy is the product of two factors, T_0 the lowest available temperature for heat discard (practically always the temperature of the atmosphere) and the net change in entropy. The increase in unavailable energy is $T_0 \Delta S_{net}$. Any process that occurs of itself (any **spontaneous process**) will proceed in such a direction as to result in a net increase in entropy. This is an important concept in the application of thermodynamics to chemical processes.

Three important **potentials** used in the Maxwell relations are:

1. The familiar potential, known as **enthalpy**,

$$h = u + pv$$

2. The **free energy** or the **Helmholtz function** is defined by the following relation:

$$f = u - Ts$$

3. The **free enthalpy** or the **Gibbs function** is defined by

$$g = h - Ts = f + pv = u + pv - Ts$$

The names used for these potentials have not gained universal acceptance. In particular, the name **free energy** is used for g in many textbooks on chemical thermodynamics. One should be very cautious when referring to different books or technical papers and should verify by definition, rather than rely on the name of the potential.

Availability of a system or quantity of energy is defined as $g = h - T_0 s$. In this equation, all quantities except T_0 refer to the system irrespective of the state of the surroundings. T_0 is the lowest temperature available for heat discard. The preceding definition assumes the absence of velocity, potential, and similar effects. When these are not negligible, proper allowance must be made, for example, $g = h - T_0 s + \bar{v}^2/(2g_c) + (g/g_c)z$. By substitution of $Q = T_0(S_2 - S_1)$ in the appropriate first-law expressions, it may be shown that for any steady-flow process, or for any constant-pressure nonflow process, decrease in availability is equal to the maximum possible (reversible) net work effect with sink for heat discard at T_0.

The availability function g is of particular value in the thermodynamic analysis of changes occurring in the stages of a turbine and is of general utility in determining thermodynamic efficiencies, i.e., the ratio of actual work performed during a process to that which theoretically should have been performed.

EXERGY

In the following, the general concept of exergy is explained, equations for a few simple processes are given, and three sets of references are added. The first is for information on the basics, the second lists thermodynamic texts, and the final lists references to the exergy of specific topics.

Exergy is a term first suggested by Rant.[13] An alternative term is *availability*, but exergy is now the preferred usage. Exergy is the maximum useful work obtainable from a system and its surroundings.

For a system exchanging heat Q with a reservoir whose temperature T may vary, the exergy is

$$\epsilon = Q - T_0 \int \left(\frac{dQ}{T} \right)$$

where the subscript 0 refers to the surroundings. For a system with enthalpy h and entropy s the exergy is

$$\epsilon = (h + \text{kinetic energy} + \text{potential energy}) - [h_0 + T_0(s - s_0)].$$

The loss of exergy in a system is called the *irreversibility I*, where $I = \epsilon_{in} - \epsilon_{out} -$ useful work.[1,24] The useful work is the total work minus the work done against the atmosphere. As $H = U + PV$, we define an exergy function $\phi = U + P_0V - T_0S$, $\Delta \epsilon = \phi_2 - \phi_1$.

The stream exergy is the exergy transported by unit incoming mass. Neglecting kinetic and potential energies and using the Darrieus function[1,24] $b = h - T_0 s$, we see that the stream exergy is $b - b_0$. The irreversibility of the process is $I = T_0 \Sigma \Delta s$, where Δs is for all systems including the atmosphere. This is the Guoy-Stodola equation.[3,15] (See also Refs. 24 and 25.) This is also called the loss of availability. If $T = T_0$ and $V = V_0$, then $\epsilon = A - A_0$, where $A = U - TS$. If $T = T_0$ and $P = P_0$, then $\epsilon = G - G_0$, where $G = H - TS$.

One example of exergy change is the heat transfer between two identical solid blocks, each of mass m and constant specific heat c. Block A is initially at temperature T_a and block B at T_b. The surrounding atmosphere is at P_0, T_0. Finally, the two temperatures become equal. The change in exergy is

$$\Delta\epsilon = -mc\,T_0 \ln\left[\frac{(T_a + T_b)^2}{4T_aT_b}\right]$$

for $T_a \neq T_b$, $\Delta\epsilon < 0$. The irreversibility I is $-\Delta\epsilon$. The exchange of heat between systems at different temperatures is irreversible, and hence there is a destruction of exergy. In a reversible process the loss in exergy is exactly equal to the useful work done. If the process is irreversible, the loss in exergy exceeds the useful work done.

The exergy of a vacuum of volume V in an atmosphere at P_0, T_0 is P_0V while that of an ideal gas at P, V, T_0 in an atmosphere at P_0, T_0 is

$$P_0V\left\{1 - \left(\frac{P}{P_0}\right)\left[1 - \ln\left(\frac{P}{P_0}\right)\right]\right\}$$

If air from an atmosphere at P_0, T_0 slowly flows into an initially exhausted insulted tank of volume V until the pressure becomes P_0, the irreversibility of the process is

$$I = \left(\frac{P_0V}{k-1}\right)\ln k$$

where $k = c_p/c_v$.

The Gibbs function is of particular importance in processes where chemical changes occur. For reversible isothermal steady-flow processes, or for reversible constant-pressure isothermal nonflow processes, change in free energy is equal to net work.

Helmholtz free energy, $f = u - Ts$, is equal to the work during a constant-volume isothermal reversible nonflow process.

All these functions g and f are point functions, and like E, h, and s their differentials are complete or perfect.

REFERENCES FOR ENERGY
Fundamentals
1. Darrieus, *Sci. Ind.,* **157,** no. 206, 1931, pp. 122–126; *Rev. Gen. Electr.,* **27,** no. 25, 1930, pp. 963–968.
2. Gallo and Milanex, *Energy,* **15,** no. 2, 1990, pp. 113–121 (choice of reference states).
3. Guoy, *J. Physique,* pt. II, **8,** 1889, pp. 501–518.
4. Haywood, "Equilibrium Thermodynamics," Wiley, New York, 1975, pp. 186–223.
5. Haywood, "Analysis of Engineering Cycles," Pergamon, New York, 1991, Appendix A, pp. 263–279.
6. Kotas, *Chem. Eng. Res. Des.,* **64,** 1986, pp. 212–229 (basics, plants, process synthesis and thermoeconomics).
7. Kotas et al., *Proc. Inst. Mech. Eng.,* **209A,** 1995, pp. 275–280 (suggested nomenclature).
8. McGovern, *Proc. Inst. Mech. Eng.,* **183C,** 1989, pp. 59–67 (compressor rational efficiency).
9. McGovern, *Proc. Inst. Mech. Eng.,* **204A,** 1990, pp. 253–262, 263–268.
10. Moran, "Availability Analysis," Prentice-Hall, Englewood Cliffs, N.J., 1982.
11. Mumsch et al., *Int. Chem. Eng.,* **33,** no. 2, 1993, pp. 197–201 (choice of reference states).
12. O'Toole, *Proc. Inst. Mech. Eng.,* **204A,** 1990, pp. 329–340.
13. Rant, *Forsch. Ing. Wes.,* **22,** no. 1, 1956, pp. 36–37.
14. Sato, "Chemical Energy and Exergy," Elsevier, Amsterdam, 2004.
15. Stodola, *Zeits. V.D.I.,* **32,** no. 38, 1898, pp. 1086–1091.
16. Szargut et al., "Exergy Analysis of Thermal, Chemical and Metallurgical Processes," Hemisphere, New York, 1988. (262 ref. This is the definitive survey for work up to 1984.)
Textbooks
17. Bejan, "Advanced Engineering Thermodynamics," Wiley, New York, 1988, pp. 111–146, 7 problems.
18. Black and Hartley, "Thermodynamics," HarperCollins, New York, 2002, pp. 322–374, 92 problems.
19. Bosnjakovic, "Technical Thermodynamics," Holt, Rinehart and Winston, New York, 1965.
20. Callen, "Thermodynamics," Wiley, New York, 1960.
21. Cengel and Boles, "Thermodynamics," McGraw-Hill, New York, 2002, pp. 391–450, 143 problems.
22. Gyftopoulos and Beretta, "Thermodynamics: Fundamentals and Applications," MacMillan, New York, 1991, pp. 413–427, 21 problems.
23. Hsieh, "Engineering Thermodynamics," Prentice-Hall, Englewood Cliffs, N.J., 1993, pp. 265–316, 61 problems.
24. Jones and Dugan, "Engineering Thermodynamics," Prentice-Hall, Englewood Cliffs, N.J., 1996, pp. 433–483, 87 problems.
25. Kirillin et al., "Engineering Thermodynamics," Mir, Moscow, 1982 (ISBN 0-8285-2056-9).
26. Moran and Shapiro, "Fundamentals of Engineering Thermodynamics," Wiley, New York, 2000, pp. 313–371, 134 problems.
27. Wark, "Advanced Thermodynamics for Engineers," McGraw-Hill, New York, 1995.
Specific References
28. Demirel, *Int. J. Exergy,* **1,** no. 1, 2004, pp. 128–146 (biological).
29. Horlock and Haywood, *Proc. Inst. Mech. Eng.,* **199C,** no. 1, 1985, pp. 11–17 (cogeneration).
30. Rosen et al., *Int. J. Exergy,* **1,** no. 1, 2004, pp. 94–110 (cogeneration).
31. Dincer, *Energy Policy,* **30,** 2002, pp. 137–149 (energy policy).
32. Wang et al., *Energy,* **30,** 2005, pp. 85–95 (exergy transfer).
33. Dincer, *Int. J. Hydrogen Energy,* **27,** 2002, pp. 265–285 (fuel cells).
34. Borgert and Velasquez, *Int. J. Exergy,* **1,** no. 1, 2004, pp. 18–28 (Kalina cycle).
35. Cengel et al., *ECOS Symp.,* Istanbul, Turkey, 2001, pp. 91–98 (mixing).
36. Dunbar and Lior, *Int. J. M.E. Educ.,* **25,** no. 1, 1996, pp. 13–31 (power plants).
37. Liley, *Exergy,* **2,** 2002, pp. 55–57 (psychrometrics).
38. Soufi et al., *Int. J. Exergy,* **1,** no. 1, 2004, pp. 29–46 (Rankine cycle).
39. Thirumaleshwar, *Cryogenics,* **19,** 1979, pp. 355–361. (cryogenic refrigerators).
40. Nikolaidis and Probert, *Appl. Energy,* **43,** 1992, pp. 201–220. (refrigerators).
41. Somasumdaram et al., *Int. J. Exergy,* **1,** no. 1, 2001, pp. 80–81 (cascade refrigeration).
42. Reistad et al., "Second Law Analysis," (AES, vol. 25/HTD, vol. 191), American Society of Mechanical Engineers, New York, 1991 (15 technical papers).

PERFECT DIFFERENTIALS. MAXWELL RELATIONS

If z is some function of x and y, in general

$$dz = \left(\frac{\partial z}{\partial x}\right)_y dx + \left(\frac{\partial z}{\partial y}\right)_x dy$$

Substituting M for $(\partial z/\partial x)_y$ and N for $(\partial z/\partial y)_x$,

$$dz = M\,dx + N\,dy$$

But $\dfrac{\partial}{\partial y}\left(\dfrac{\partial z}{\partial x}\right) = \dfrac{\partial}{\partial x}\left(\dfrac{\partial z}{\partial y}\right)$ or $\dfrac{\partial M}{\partial y} = \dfrac{\partial N}{\partial x}$. This is Euler's criterion for integrability. A perfect differential has the characteristics of dz stated above. Many important thermodynamic relations may be derived from the appropriate point function by the use of this relation; see Table 4.1.1.

From the third column of the bottom half of the table, by equating various of the terms which are equal, one may obtain

$$\left(\frac{\partial u}{\partial s}\right)_v = \left(\frac{\partial h}{\partial s}\right)_p \qquad \left(\frac{\partial u}{\partial v}\right)_s = \left(\frac{\partial f}{\partial v}\right)_T$$

$$\left(\frac{\partial h}{\partial p}\right)_s = \left(\frac{\partial g}{\partial p}\right)_T \qquad \left(\frac{\partial g}{\partial T}\right)_p = \left(\frac{\partial f}{\partial T}\right)_v$$

By mathematical manipulation of equations previously given, the following important relations may be formulated:

$$c_v = \left(\frac{\partial q}{\partial T}\right)_v = T\left(\frac{\partial s}{\partial T}\right)_v = \left(\frac{\partial u}{\partial T}\right)_v$$

$$c_p = \left(\frac{\partial q}{\partial T}\right)_p = T\left(\frac{\partial s}{\partial T}\right)_p = \left(\frac{\partial h}{\partial T}\right)_p$$

$$c_p - c_v = T\left(\frac{\partial v}{\partial T}\right)_p\left(\frac{\partial p}{\partial T}\right)_v$$

$$\left(\frac{\partial c_v}{\partial v}\right)_T = T\left(\frac{\partial^2 p}{\partial T^2}\right)_v \qquad \left(\frac{\partial c_p}{\partial T^2}\right)_p = -T\left(\frac{\partial^2 v}{\partial T^2}\right)_p$$

Table 4.1.1   Maxwell Relations

Function	Differential	Maxwell relation
$\Delta u = q + W$	$du = T ds - p dv$	$\left(\dfrac{\partial T}{\partial v}\right)_s = -\left(\dfrac{\partial p}{\partial s}\right)_v$
$h = u + pv$	$dh = T ds + v dp$	$\left(\dfrac{\partial T}{\partial p}\right)_s = \left(\dfrac{\partial v}{\partial s}\right)_p$
$f = u - Ts$	$df = -s dT - p dv$	$\left(\dfrac{\partial s}{\partial v}\right)_T = \left(\dfrac{\partial p}{\partial T}\right)_v$
$g = h - Ts$	$dg = -s dT + v dp$	$\left(\dfrac{\partial s}{\partial p}\right)_T = -\left(\dfrac{\partial v}{\partial T}\right)_p$

By holding certain variables constant, a second set of relations is obtained:

Differential	Independent variable held constant	Relation
$du = T ds - p dv$	s	$\left(\dfrac{\partial u}{\partial v}\right)_s = -p$
	v	$\left(\dfrac{\partial u}{\partial s}\right)_v = T$
$dh = T ds + v dp$	s	$\left(\dfrac{\partial h}{\partial p}\right)_s = v$
	p	$\left(\dfrac{\partial h}{\partial s}\right)_p = T$
$df = -s dT - p dv$	T	$\left(\dfrac{\partial f}{\partial v}\right)_T = -p$
	v	$\left(\dfrac{\partial f}{\partial T}\right)_v = -s$
$dg = -s dT + v dp$	T	$\left(\dfrac{\partial g}{\partial p}\right)_T = v$
	p	$\left(\dfrac{\partial g}{\partial T}\right)_p = -s$

Relations involving q, u, h, and s:

$$dq = c_v dT + T\left(\frac{\partial p}{\partial T}\right)_v dv = c_p dT - T\left(\frac{\partial v}{\partial T}\right)_p dp$$

$$du = c_v dT + \left[T\left(\frac{\partial p}{\partial T}\right)_v - p\right]dv$$

$$dh = c_p dT - \left[T\left(\frac{\partial v}{\partial T}\right)_p - v\right]dp$$

$$ds = c_v \frac{dT}{T} + \left(\frac{\partial p}{\partial T}\right)_v dv = c_p \frac{dT}{T} - \left(\frac{v}{\partial T}\right)_p dp$$

Since $q + W = du$ and $h = u + pv$, for reversible processes,

$$du = T\, ds - p\, dv \qquad \text{and} \qquad dh = du + p\, dv + v\, dp$$

it follows that

$$v = T\left(\frac{\partial s}{\partial p}\right)_T + \left(\frac{\partial h}{\partial p}\right)_T$$

But from the Maxwell relations,

$$\left(\frac{\partial v}{\partial T}\right)_p = -\left(\frac{\partial s}{\partial p}\right)_T$$

Therefore,

$$\left(\frac{\partial h}{\partial p}\right)_T = v - T\left(\frac{\partial v}{\partial T}\right)_p$$

Similarly,

$$\left(\frac{\partial u}{\partial v}\right)_T = -\left[p - T\left(\frac{\partial p}{\partial T}\right)_v\right]$$

These last two equations give in terms of p, v, and T the necessary relations that must hold for any system, however complex. An equation in p, v, and T for the properties of a substance is called an equation of state. These two equations applicable to any substance or system are known as **thermodynamic equations of state.**

Presentation of Thermal Properties   Before the laws of thermodynamics can be applied and quantitative results obtained in the analysis of an engineering system, it is necessary to have available the properties of the system, some of which are temperature, pressure, internal energy entropy, and enthalpy. In general, the property of a pure substance under equilibrium conditions may be expressed as a function of two other properties. This is based on the assumption that certain effects, such as gravitational and magnetic, are not important for the condition under investigation. The various properties of a pure substance under equilibrium conditions may be expressed by an equation of state, which in general form follows:

$$p = f(T, v)$$

In this relation the pressure is shown to be a function of both the temperature and the specific volume. Many special forms of equations of state are used in the analysis of engineering systems. Plots of the properties of various pure substances are very useful in studies dealing with thermodynamics. Two-dimensional plots, such as $p - v$, $p - h$, $p - T$, $T - s$, etc., show phase relations and are important in the analysis of cycles.

The constants in the equations of state are usually based on experimental data. The properties may be presented in many different ways, some of which are:

1. As equations of state, e.g., the perfect gas laws and the van der Waals equation.
2. As charts or graphs.
3. As tables.
4. As approximations which may be useful when more reliable data are not available.

IDEAL GAS LAWS

At low pressures and high enough temperatures, in the absence of chemical reaction, all gases approach a condition such that their P-V-T properties may be expressed by the simple relation

$$pv = RT$$

If v is expressed as volume per unit weight, the value of the constant R will be different for different gases. If v is expressed as the volume of one molecular weight of gas, then R_u is the same for all gases in any chosen system of units. Hence $R = R_u/M$.

In general, for any amount of gas, the ideal gas equation becomes

$$pV = nMRT = nR_u T = \frac{m}{M}R_u T$$

where V is now the total gas volume, n is the number of moles of gas in the volume V, M is the molecular weight, and $R_u = MR$ the universal gas constant. An alternative ideal gas equation of state is $pv = R_u T/M$. It is different from the preceding in the use of specific volume v rather than total volume V.

For all ideal gases, $R_u = MR$ in lb · ft is 1,546. One pound mol of any perfect gas occupies a volume of 359 ft³ at 32°F and 1 atm.

For many engineering purposes, use of the gas laws is permissible up to pressures of 100 to 200 lb/in² if the absolute temperatures are at least twice the critical temperatures. Below the critical temperature, errors introduced by use of the gas laws may usually be neglected up to 15 lb/in² pressure although errors of 5 percent are often met when dealing with saturated vapors.

The **van der Waals equation of state,** $p = BT/(v - b) - a/v^2$, is a modification of the ideal gas law which is sometimes useful at high pressures. The quantities B, a, and b are constants.

Many empirical or semiempirical equations of state have been proposed to represent the real variation of pressure with volume and temperature. The Benedict-Webb-Rubin equation is among them; see Perry, "Chemical Engineers Handbook," 8th ed., McGraw-Hill. Computer programs have been devised for the purpose; see Deutsch, "Microcomputer Programs for Chemical Engineers," McGraw-Hill. For computer programs and output for steam and other fluids, including air tables, see Irvine and Liley, "Steam and Gas Tables with Computer Programs," Academic Press.

Approximate P-V-T Relations   For many gases, P-V-T data are not available. An approximation useful under such circumstances is based on the observation of van der Waals that in terms of reduced properties most gases approximate a common **reduced equation of state.** The reduced quantities are the actual ones divided by the corresponding critical quantities, e.g., the reduced temperature $T_R = T_{actual}/T_{critical}$, the reduced volume $v_R = v_{actual}/v_{critical}$, the reduced pressure $p_R = p_{actual}/p_{critical}$. The gas laws may be made to apply to any nonperfect gas by the introduction of a correction factor

$$pV = ZNR_uT$$

When the gas laws apply, $Z = 1$ and on a molal basis $Z = pV/(R_uT)$. If on a plot of Z versus p_R lines of constant T_R are drawn, for different substances these are found to fall in narrow bands. Single T_R lines may be drawn to represent approximately the various bands. This has been done in Fig. 4.1.3. To use the chart, only the critical pressure and temperature of the gas need be known.

EXAMPLE.   Find the volume of 1 lb of steam at 5,500 psia and 1200°F (by steam tables, $v = 0.1516$ ft^3/lb).

For water, critical temperature = 705.4°F; critical pressure = 3,206.4 psia; reduced temp = 1660/1165 = 1.43; reduced pressure = 5,500/3,206.4 = 1.72; μ (see Fig. 4.1.3) = 0.83, v = 0.83 (1,546)(1,660)/(18)(5,500)(144) = 0.149 ft^3. Error = 100(0.152–0.149)/0.152 = 1.7 percent. If the gas laws had been used, the error would have been 17 percent.

No entirely satisfactory method for calculation for gaseous mixtures has been developed, but the use of average critical constants as proposed by Kay (*Ind. Eng. Chem.,* **28**, 1936, p. 1014) is easy and gives satisfactory results under conditions considerably removed from the critical. He assumes the gaseous mixture can be treated as if it were a single pure gas with a pseudocritical pressure and temperature estimated by a method of molar averaging.

$$(T_c)_{mixture} = (T_c)_a y_a + (T_c)_b y_b + (T_c)_c y_c + \cdots$$
$$(p_c)_{mixture} = (p_c)_a y_a + (p_c)_b y_b + (p_c)_c y_c + \cdots$$

where $(T_c)_a$ is the critical temperature of pure a, etc.; $(p_c)_a$ is the critical pressure of pure a, etc.; and y_a is the mole fraction of a, etc. For a gaseous mixture made up of gases, a, b, c, etc., the pseudocritical constants having been determined, the gaseous mixture is handled on the μ charts as if it were a single pure gas.

IDEAL GAS MIXTURES

Many of the fluids involved in engineering systems are physical mixtures of the permanent gases or one or more of these with superheated or saturated vapors. For example, normal atmospheric air is a mixture of oxygen and nitrogen with traces of other gases, plus superheated or saturated water vapor, or at times saturated vapor and liquid. If the properties of each constituent of a mixture would have to be considered individually during an analysis of a system, the procedures would be very complex. Experience has demonstrated that a mixture of gases may be regarded as an equivalent gas, the properties of which depend upon the kind and proportion of each of the constituents. The general relations applicable to a mixture of perfect gases will be presented. Let V denote the total volume of the mixture, $m_1, m_2, m_3, \ldots$ the masses of the constituent gases, $R_1, R_2, R_3, \ldots$ the corresponding gas constants, and R_m the constant for the mixture. The **partial pressures** of the constituents, i.e., the pressures that the constituents would have if occupying the total volume V, are $p_1 = m_1R_1T/V$, $p_2 = m_2R_2T/V$, etc.

According to Dalton's law, the **total pressure p of the mixture** is the sum of the partial pressures; i.e., $p = p_1 + p_2 + p_3 + \cdots$. Let $m = m_1 + m_2 + m_3 + \cdots$ denote the total mass of the mixture; then $pV = mR_mT$ and $R_m = \Sigma(m_iR_i)/m$. Also $p_1/p = m_1R_1/(mR_m)$, $p_2/p = m_2R_2/(mR_m)$, etc.

Let $V_1, V_2, V_3 + \ldots$ denote the volumes that would be occupied by the constituents at pressure p and temperature T (these are given by the volume composition of the gas). Then $V = V_1 + V_2 + V_3 + \cdots$ and the apparent molecular weight m_m of the mixture is $m_m = \Sigma(m_iV_i)/V$. Then $R_m = 1,546/m_m$. The subscript i denotes an individual constituent.

Volume of 1 lb at 32°F and atm pressure = $359/m_m$.
Mass of 1 ft^3 at 32°F and atm pressure = $0.002788m_m$.

The specific **heats of the mixture** are, respectively,

$$c_p = \Sigma(m_ic_{pi})/m \qquad c_v = \Sigma(m_ic_{vi})/m$$

Internal Energy, Enthalpy, and Entropy of an Ideal Gas   If an ideal gas with constant specific heats changes from an initial state p_1, V_1, T_1 to a final state p_2, V_2, T_2, the following equations hold:

$$u_2 - u_1 = mc_v(T_2 - T_1) = (p_2v_2 - p_1v_1)(k - 1)$$

$$h_2 - h_1 = mc_p(T_2 - T_1) = k\frac{(p_2v_2 - p_1v_1)}{k - 1}$$

$$s_2 - s_1 = m\left(c_v\ln\frac{T_2}{T_1} + R\ln\frac{v_2}{v_1}\right)$$

$$= m\left(c_p\ln\frac{T_2}{T_1} - R\ln\frac{p_2}{p_1}\right) = m\left(c_p\ln\frac{v_2}{v_1} + c_v\ln\frac{p_2}{p_1}\right)$$

Fig. 4.1.3   Compressibility factors for gases and vapors. (*From Hougen and Watson, "Chemical Process Principles,"* Wiley.)

In general, the energy per unit mass is $u = c_v T + u_0$, the enthalpy is $h = c_p T + h_0$, and the entropy is $s = c_v \ln T + R \ln v + s_0 = c_p \ln T - R \ln p + s_0' = c_p \ln v + c_v \ln p = s_0''$.

The two fundamental equations for ideal gases are

$$dq = c_v\, dT + p\, dv \qquad dq = c_p\, dT - v\, dp$$

SPECIAL CHANGES OF STATE FOR IDEAL GASES

(Specific heats assumed constant)

In the following formulas, the subscripts 1 and 2 refer to the initial and final states, respectively.

1. *Constant volume:* $p_2/p_1 = T_2/T_1$.

$$Q_{12} = U_2 - U_1 = mc_v(t_2 - t_1) = V(p_2 - p_1)/(k - 1)$$
$$W_{12} = 0 \qquad s_2 - s_1 = mc_v \ln(T_2/T_1)$$

2. *Constant pressure:* $V_2/V_1 = T_2/T_1$.

$$W_{12} = -p(V_2 - V_1) = -mR(t_2 - t_1)$$
$$Q_{12} = mc_p(t_2 - t_1) = kW_{12}/(k - 1)$$
$$s_2 - s_1 = mc_p \ln(T_2/T_1)$$

3. *Isothermal (constant temperature):* $p_2/p_1 = V_1/V_2$.

$$U_2 - U_1 = 0 \qquad W_{12} = -mRT \ln(V_2/V_1) = -p_1 V_1 \ln(V_2/V_1)$$
$$Q_{12} = -W_{12} \qquad s_2 - s_1 = Q_{12}/T = mR \ln(V_2/V_1)$$

4. *Reversible adiabatic, isentropic:* $p_1 V_1^k = p_2 V_2^k$.

$$T_2/T_1 = (V_1/V_2)^{k-1} = (p_2/p_1)^{(k-1)/k}$$
$$W_{12} = U_1 - U_2 = mc_v(t_1 - t_2)$$
$$Q_{12} = 0 \qquad s_2 - s_1 = 0$$
$$W_{12} = (p_2 V_2 - p_1 V_1)/(k - 1)$$
$$= -p_1 V_1 [(p_2/p_1)^{(k-1)/k} - 1]/(k - 1)$$

5. *Polytropic:* This name is given to the change of state which is represented by the equation $pV^n =$ const. A polytropic curve usually represents actual expansion and compression curves in motors and air compressors for pressures up to a few hundred pounds. By giving n different values and assuming specific heats constant, the preceding changes may be made special cases of the polytropic change, thus,

For $n = 1$,	$pv = $ const	isothermal
$n = k$,	$pv^k = $ const	isentropic
$n = 0$,	$p = $ const	constant pressure
$n = \infty$,	$v = $ const	constant volume

For a polytropic change of an ideal gas (for which c_v is constant), the specific heat is given by the relation $c_n = c_v(n - k)(n - 1)$; hence for $1 < n < k$, c_n is negative. This is approximately the case in air compression up to a few hundred pounds pressure. The following are the principal formulas:

$$p_1 V_1^n = p_2 V_2^n$$
$$T_2/T_1 = (V_1/V_2)^{n-1} = (p_2/p_1)^{(n-1)/n}$$
$$W_{12} = (p_2 V_2 - p_1 V_1)/(n - 1)$$
$$= -p_1 V_1 [(p_2/p_1)^{(n-1)/n} - 1]/(n - 1)$$
$$Q_{12} = mc_n(t_2 - t_1)$$
$$W_{12} : U_2 - U_1 : Q_{12} = k - 1 : 1 - n : k - n$$

The quantity $(p_2/p_1)^{(k-1)/k} - 1$ occurs frequently in calculations for perfect gases.

Determination of Exponent n   If two representative points $(p_1, V_1$ and $p_2, V_2)$ be chosen, then

$$n = (\log p_1 - \log p_2)/(\log V_2 - \log V_1)$$

Several pairs of points should be used to test the constancy of n.

Changes of State with Variable Specific Heat   In case of a considerable range of temperature, the assumption of constant specific heat is not permissible, and the equations referring to changes of state must be suitably modified. (This statement does not apply to inert or monatomic gases.) Experiments on the specific heat of various gases show that the specific heat may sometimes be taken as a linear function of the temperature: thus, $c_v = a + bT$; $c_p = a' + b'T$. In that case, the following expressions apply for the change of internal energy and entropy, respectively:

$$U_2 - U_1 = m[a(T_2 - T_1 + 0.5b(T_2^2 - T_1^2)]$$
$$S_2 - S_1 = m[a \ln(T_2/T_1) + b(T_2 - T_1) + R \ln(V_2/V_1)]$$

and for an isentropic change,

$$W_{12} = U_2 - U_1$$
$$R \ln(V_1/V_2) = a \ln(T_2/T_1) + b(T_2 - T_1)$$

GRAPHICAL REPRESENTATION

The change of state of a substance may be shown graphically by taking any two of the six variables p, V, T, S, U, H as independent coordinates and drawing a curve to represent the successive values of these two variables as the change proceeds. While any pair may be chosen, there are three systems of graphical representation that are specially useful.

1. p and V. The curve (Fig. 4.1.4) represents the simultaneous values of p and V during the change (reversible) from state 1 to state 2. The

Fig. 4.1.4

area between the curve and the axis OV is given by the integral $\int_{V_1}^{V_2} p\, dV$ and therefore represents the external work W_{12} done by the gas during the change. The area included by a closed cycle represents the work of the cycle (as in the indicator diagram of the steam engine).

2. T and S (Fig. 4.1.5). The absolute temperature T is taken as the ordinate, the entropy S as the abscissa. The area between the curve of change of state and the S axis is given by the integral $\int_{S_1}^{S_2} T\, dS$, and it therefore represents the heat Q_{12} absorbed by the substance from external sources provided there are no irreversible effects. On the T-S diagram, an isothermal is a straight line, as AB, parallel to the S axis; a *reversible* adiabatic is a straight line, as CD, parallel to the T axis.

Fig. 4.1.5

In the case of internal generation of heat through friction, as in steam turbines, the increase of entropy is given by $\int_{T_1}^{T_2} (dQ'/T)$ and the area under the curve represents the heat Q' thus generated. In this case, an adiabatic is *not* a straight line parallel to the T axis.

3. *H* and *S*. In the system of representation devised by Dr. Mollier, the enthalpy *H* is taken as the ordinate and the entropy *S* as the abscissa. If on this diagram (Fig. 4.1.6) a line of constant pressure, as 12, be drawn, the heat absorbed during the change at consiant pressure is given by $Q_{12} = H_2 - H_1$, and this is represented by the line segment 23. The **Mollier diagram** is specially useful in problems that involve the flow of fluids, throttling, and the action of steam in turbines.

Fig. 4.1.6

IDEAL CYCLES WITH PERFECT GASES

Gases are used as heat mediums in several important types of machines. In air compressors, air engines, and air refrigerating machines, atmospheric air is the medium. In the internal-combustion engine, the medium is a mixture of products of combustion. Engines using gases are operated in certain well-defined cycles, which are described below. In the analyses given, ideal conditions that cannot be attained by actual motors are assumed. However, conclusions derived from such analyses are usually approximately valid for the modified actual cycle.

In the following, the subscripts 1, 2, 3, etc., refer to corresponding points shown in the figures. The work of the cycle is denoted by *W* and the net heat absorbed by *Q*.

Carnot Cycle   The Carnot cycle (Fig. 4.1.7) is of historic interest. It consists of two isothermals and two isentropics. The heat absorbed

Fig. 4.1.7   Carnot cycle.

along the upper isothermal 12 is $Q_{12} = mRT \ln(V_2/V_1)$, and the heat transformed into work, represented by the cycle area, is $W = Q_{12}(1 - T_0/T)$.

$$W = -mR(T - T_0)\ln\left(\frac{V_2}{V_1}\right)$$

If the cycle is traversed in the reverse sense, $Q_{43} = mRT_0 \ln (V_3/V_4)$ is the heat absorbed from the cold body (brine), and the ratio $Q_{43} : (W) = T_0 : (T - T_0)$ is the **coefficient of performance** of the refrigerating machine.

Leff (*Amer. J. Phys.,* **55,** no. 7, 1987, pp. 602–610) showed that the **thermal efficiency** of a heat engine producing the maximum possible work per cycle consistent with its operating temperature range resulted in efficiencies equal to or well approximated by $\eta = 1 - \sqrt{T_c/T_h}$, where *c* = cold and *h* = hot, as found by Curzon and Ahlborn (*Amer. J. Phys.,* **43,** no. 1, 1975, pp. 22–24) for maximum power output. If the work output per cycle is kept fixed, the thermal efficiency can be increased by operating the heat engine at less than maximum work output per cycle, the limit being an engine of infinite size having a Carnot efficiency. Leff's paper considers Otto, Brayton-Joule, Diesel, and Atkinson cycles. Figure 4.1.8 illustrates the difference between maximum power and maximum efficiency. Detailed discussion of **finite time thermodynamics**

appears in Sieniutcyz and Salamon, "Finite Time Thermodynamics and Thermoeconomies," *Advan. Thermo.,* **4,** 306 pp., 1990, Taylor & Francis, London.

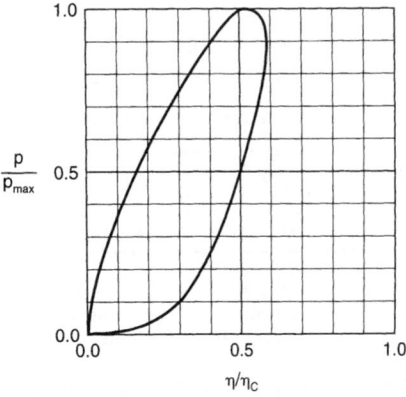

Fig. 4.1.8   Ratio of actual power to maximum power as a function of ratio of actual thermal efficiency to Carnot efficiency.

Finite time thermodynamics is a term applied to the consideration that, for any finite energy transfer, a finite time must occur. A common statement in the literature is that the analysis started from the work of Curzon and Ahlborn (*Amer. J. Phys.,* **43,** 1975, pp. 22–24). According to Bejan (*Amer. J. Phys.,* **62,** no. 1, Jan. 1994, pp. 11–12), this statement is not true, and the original analysis was by Novikov (*At. Energy,* **3,** 1957, p. 409, and *Nucl. Energy,* pt. II, **7,** 1958, pp. 125–148). Wu (*Energy Convsn. Mgmt.,* **34,** no. 12, 1993, pp. 1239–1247) discusses the *endoreversible Carnot heat engine* being one in which all the losses are associated with the transfer of heat to and from the engine, there being no internal losses within the engine itself and refers to Wu and Kiang (*Trans. J. Eng. Gas Turbines & Power,* **113,** 1991, p. 501) for a detailed literature survey.

Otto and Diesel Cycles   The ideal cycles usually employed for internal-combustion engines may be classified in two groups: (1) explosive—Otto (the fluid is introduced in gaseous form), (2) nonexplosive—Diesel, Joule (the fluid is introduced in liquid form).

Otto Cycle   (Fig. 4.1.9 for pressure-volume plane, Fig. 4.1.10 for temperature-entropy plane) Isentropic compression 12 is followed by ignition and rapid heating at constant volume 23. This is followed by isentropic expansion, 34. Assuming constant specific heats the following relations hold:

$$\frac{T_2}{T_1} = \frac{T_3}{T_4} = \left(\frac{p_2}{p_1}\right)^{k-1/k} = \left(\frac{p_3}{p_4}\right)^{k-1/k} = \left(\frac{V_1}{V_2}\right)^{k-1}$$

$$Q_{23} = mc_v(T_1 - T_2)$$

$$W = Q_{23}[1 - (T_1/T_2)] = mc_v(T_3 - T_4 - T_2 + T_1)$$

$$\text{Efficiency} = 1 - \frac{T_1}{T_2} = 1 - \left(\frac{V_2}{V_1}\right)^{k-1} = 1 - \left(\frac{p_1}{p_2}\right)^{(k-1)/k}$$

Fig. 4.1.9   Otto cycle.

Fig. 4.1.10   Otto cycle.

If the compression and expansion curves are polytropics with the same value of n, replace k by n in the first relation above. In this case,

$$W = [(p_3V_3 - p_4V_4) - (p_2V_2 - p_1V_1)]/(n - 1)$$
$$= mR(T_3 - T_4 - T_2 + T_1)/(n - 1)$$

The **mean effective pressure** of the diagram is given by

$$p_m = ap_1(p_3/p_2 - 1)$$

where a has the values given in the following table.

	$p_2/p_1 = 3$	4	5	6	8	10	12	14	16
($n = 1.4$)	$a = 1.70$	1.94	2.13	2.31	2.62	2.88	3.10	3.31	3.50
($n = 1.3$)	$a = 1.69$	1.92	2.11	2.28	2.57	2.81	3.03	3.22	3.39
($n = 1.2$)	$a = 1.68$	1.90	2.08	2.25	2.51	2.74	2.94	3.12	3.27

A later paper by Wu and Blank (*Energy Convsn. Mgmt.,* **34,** no. 12, 1993, pp. 1255–1269) considered optimization of the endoreversible Otto cycle with respect to both power and mean effective pressure.

Diesel Cycle   In the diesel oil engine, air is compressed to a high pressure. Fuel is then injected into the air, which is at a temperature above the ignition point, and it burns at nearly constant pressure (23, in Fig. 4.1.11). Isentropic expansion of the products of combustion is followed by exhaust and suction of fresh air, as in the Otto cycle.

Fig. 4.1.11   Diesel cycle.

The work obtained is

$$W = m[c_p(T_3 - T_2) - c_v(T_4 - T_1)]$$

and the efficiency of the ideal cycle is

$$1 - [(T_4 - T_1)/k(T_3 - T_2)]$$

The **Joule cycle,** also called the **Brayton cycle** (Fig. 4.1.12), consists of two isentropics and two constant-pressure lines. The following relations hold:

$$V_3/V_2 = V_4/V_1 = T_3/T_2 = T_4/T_1$$
$$\frac{T_2}{T_1} = \frac{T_3}{T_4} = \left(\frac{V_1}{V_2}\right)^{k-1} = \left(\frac{V_4}{V_3}\right)^{k-1} = \left(\frac{p_2}{p_1}\right)^{k-1/k}$$
$$W = mc_p(T_3 - T_2 - T_4 + T_1)$$
$$\text{Efficiency} = W/Q_{23} = 1 - T_1/T_2$$

Fig. 4.1.12   Joule or Brayton cycle.

The Joule cycle has assumed renewed importance as a basis for analysis of gas turbine operation.

For additional information on internal combustion engines, see Campbell, "Thermodynamic Analysis of Internal Combustion Engines," Wiley; Taylor, "The Internal Combustion Engine in Theory and Practice," MIT Press. New designs for internal-combustion engines were reviewed by Wallace (*Sci. Progr., Oxford,* **75,** 1991, pp. 15–32).

Stirling Cycle   The Stirling engine may be visualized as a cylinder with a piston at each end. Between the pistons is a regenerator. The cylinder is assumed to be insulated except for a contact with a hot reservoir at one end and a contact with a cold reservoir at the other end.

Starting with state 1, Fig. 4.1.13, heat from the hot reservoir is added to the gas at T_H (or $T_H - dT$). During the reversible isothermal process, the left piston moves outward, doing work as the system volume increases and the pressure falls. Both pistons are then moved to the right at the same rate to keep the system volume constant (process 2–3). No

Fig. 4.1.13   Stirling cycle.

heat transfer occurs with either reservoir. As the gas passes through the regenerator, heat is transferred from the gas to the regenerator, causing the gas temperature to fall to T_L by the time the gas leaves the right end of the regenerator. For this heat-transfer process to be reversible, the temperature of the regenerator at each point must equal the gas temperature at that point. Hence there is a temperature gradient through the regenerator from T_H at the left end to T_L at the right end. No work is accomplished during this process. During the path 3–4, heat is removed from the gas at T_L (or $T_L + dT$) to the reservoir at T_L. To hold the gas temperature constant, the right piston is moved inward—doing work on the gas with a resulting increase in pressure. During process 4–1, both pistons are moved to the left at the same rate to keep the system volume constant. The pistons are closer together during this process than they were during process 2–3, since $V_4 = V_1 < V_2 = V_3$. No heat is transferred to either reservoir. As the gas passes back through the re-generator, the energy stored in the regenerator during 2–3 is returned to the gas. The gas emerges from the left end of the regenerator at the tempe rature T_H No work is performed during this process since the volume remains constant. Thus the cycle is completed and is externally reversible. The system exchanges a net amount of heat with only the two energy reservoirs T_H and T_L. Two types of Stirling engines are shown in Fig. 4.1.14. Extensive research-and-development effort has been devoted to the Stirling engines for future use as prime movers in space power systems operating on solar energy. (See also Sec. 9.6.)

More information can be found in Meijer, *De Ingenieur,* **81,** nos. 18 and 19, 1969; Reader and Hooper, "Stirling Engines," Spon, London; Sternlicht, *Chem. Tech.,* **13,** 1983, pp. 28–36; Walker, "Stirling Engines," Oxford Univ. Press.

Fig. 4.1.14 Two main types of Stirling engine: (1) left, double-cylinder two piston; (2) right, single-cylinder, piston plus displacer. Each has two variable-volume working spaces filled with the working fluid—one for expansion and one for compression of the gas. Spaces are at different temperatures—the extreme temperatures of the working cycle—and are connected by a duct, which holds the regenerators and heat exchangers. (*Intl. Science and Technology,* May 1962.)

AIR COMPRESSION

It is assumed that the compressor works under ideal reversible conditions without clearance and without friction losses and that the changes are over ranges where the gas laws are applicable. Where the gas laws cannot be used, analysis in terms of Z charts is convenient. If the compression from p_1 to p_2 (Fig. 4.1.15) follows the law $pV^n = $ const, the work represented by the indicator diagram is

$$W = n(p_2V_2 - p_1V_1)/(n - 1)$$
$$= np_1V_1[(p_2/p_1)^{(n-1)/n} - 1]/(n - 1)$$

The temperature at the end of compression is given by $T_2/T_1 = (p_2/p_1)^{(n - 1)/n}$. The work W is smaller the smaller the value on n, and the purpose of the water jackets is to reduce n from the isentropic value 1.4. Under usual working conditions, n is about 1.3.

Fig. 4.1.15 Air compressor cycle.

When the pressure p_2 is high, it is advantageous to divide the process into two or more stages and cool the air between the cylinders. The saving effected is best shown on the T-S plane (Fig. 4.1.16). With single-stage compression, 12 represents the compression from p_1 to p_2, and if the constant-pressure line 23 is drawn cutting the isothermal through point 1 in point 3, the area 1'1233' represents the work W. When two stages are used, 14 represents the compression from p_1 to an intermediate pressure p', 45 cooling at constant pressure in the inter-

Fig. 4.1.16 Air compressor cycle

cooler between the cylinders, and 56 the compression in the second stage. The area under 14563 represents the work of the two stages and

the area 2456 the saving effected by compounding. This saving is a maximum when $T_4 = T_6$, and this is the case when the intermediate pressure p' is given by $p' = \sqrt{p_1p_2}$ (see Sec. 14.3).

The total work in two-stage compression is

$$np_1V_1[(p'/p_1)^{(n-1)/n} + (p_2/p')^{(n-1)/n} - 2]/(n - 1)$$

Gas Turbine The Brayton cycle, also called the Joule or constant pressure cycle, employs an air engine, a compressor, and a combustion chamber. Air enters the compressor wherein the pressure is increased. Fuel burning in the combustion chamber raises the temperature of the compressed air under constant-pressure conditions. The resulting high-temperature gases are then introduced to the engine where they expand and perform work. The excess work of the engine over that required to compress the air is available for operating other devices, such as a generator.

Basically, the simple gas-turbine cycle is the same as the Brayton cycle, except that the air compressor and engine are replaced by an axial flow compressor and gas turbine. Air is compressed in the compressor, after which it enters a combustion chamber where the temperature is increased while the pressure remain constant. The resulting high-temperature air then enters the turbine, thereby performing work.

Boyce ("Gas Turbine Engineering Handbook," Gulf) gives numerous examples of ideal and actual gas-turbine cycles. The graphs in this source as well as in a review by Dharmadhikari (*Chemical Engineer* (*London*), Feb. 1989, pp. 16–20) show the same relation between work output and thermal efficiency as the general graph of Gordon (*Amer. J. Phys.,* **59,** no. 6, 1991, pp. 551–555). MacDonald (ASME Paper 89-GT-103, Toronto Gas Turbine Exposition, 1989) describes the increasing use of heat exchangers in gas-turbine plants and reviews the use of recuperators (i.e., regenerators) (in *Heat Recovery Systs. & CHP.* **10,** no. 1, 1990, pp. 1–30). Gas turbines as the topping cycle with steam in the bottoming cycle were described by Huang (ASME Paper 91-GT-186 and *J. Eng. Gas Turbines & Power,* **112,** Jan. 1990, pp. 117–121) and by Cerri (*Trans. ASME,* **109,** Jan. 1987, pp. 46–54). The use of steam injection in gas-turbine cycles has received renewed attention; see, e.g., Consonni (45th Congr. Nat. Assoc. Termotecnica Ital., **IIID,** 1990, pp. 49–60), Ediss (City Univ. London Conf. Paper, Nov. 1991), Lundberg (*ASME Cogen-Turbo IGTI,* **6,** 1991, pp. 9–18). Fraize and Kinney (*J. Eng. Power,* **101,** 1979, pp. 217–227), and Larson and Williams (*J. Eng. Gas Turbines & Power,* **109,** Jan. 1987, pp. 55–63).

Analysis of closed-cycle gas-turbine plant for maximum and zero power output and for maximum efficiency was made by Woods et al. (*Proc. Inst. Mech. Eng.,* **205A,** 1991, pp. 59–66). See also pp. 287–291 and ibid., **206A,** 1992, pp. 283–288. A series of papers by Najjar appeared in *Int. J. M. E. Educ.* (**15,** no. 4, 1987, pp. 267–286); *High Temp. Technol.* (**8,** no. 4, 1990, pp. 283–289); *Heat Recovery Systems & CHP* (**6,** no. 4, 1986, pp. 323–334). For more detailed information regarding the actual gas-turbine cycles see Sec. 9.

VAPORS

General Characteristics of Vapors Let a gas be compressed at constant temperature; then, provided this temperature does not exceed a certain critical value, the gas begins to liquefy at a definite pressure, which depends upon the temperature. At the beginning of liquefaction, a unit mass of gas will also have a definite volume v_g, depending on the temperature. In Fig. 4.1.17, AB represents the compression and the

Fig. 4.1.17

point B gives the **saturation** pressure and volume. If the compression is continued, the pressure remains constant with the temperature, as indicated by BC, until at C the substance is in the liquid state with the volume v_f.

The curves v_f and v_g giving the volumes for various temperatures at the end and beginning of liquefaction, respectively, may be called the **limit curves**. A point B on curve v_g represents the state of **saturated vapor**; a point C on the curve v_f represents the saturated liquid state; and a point M between B and C represents a mixture of vapor and liquid of which the part $x = MC/BC$ is vapor and the part $1 - x = BM/BC$ is liquid. The ratio x is called the **quality of the mixture**. The region between the curves v_f and v_g is thus the region of liquid and vapor mixtures. The region to the right of curve v_g is the region of **superheated vapor**. The curve v_g dividing these regions represents the so-called **saturated vapor**.

For saturated vapor, saturated liquid, or a mixture of vapor and liquid, the pressure is a function of the temperature only, and the volume of the mixture depends upon the temperature and quality x. That is, $p = f(t)$, $v = F(t, x)$.

For the vapor in the superheated state, the volume depends on pressure and temperature [$v = F(p, t)$], and these may be varied independently.

Critical State   If the temperature of the gas lies above a definite temperature t_c called the **critical temperature**, the gas cannot be liquefied by compression alone. The saturation pressure corresponding to t_c is the **critical pressure** and is denoted by p_c. At the critical states, the limit curves v_f and v_g merge; hence for temperatures above t_c, it is impossible to have a mixture of vapor and liquid. Table 4.2.21 gives the critical data for various gases; also the boiling temperature t_b corresponding to atmospheric pressure. Study of the critical region is becoming a specialized topic. NBS Misc. Publ. 273, 1966, contains 33 papers on phenomena near critical points. The ASME symposia on thermophysical properties proceedings contain numerous papers on the subject.

Vapor Pressures   At a specified temperature, a pure liquid can exist in equilibrium contact with its vapor at but one pressure, its vapor pressure. A plot of these pressures against the corresponding temperatures is known as a vapor pressure curve. As noted by Martin, "Thermodynamic and Transport Properties of Gases, Liquids and Solids," ASME, New York, p. 112, the true shape of a log vapor pressure versus reciprocal absolute temperature curve is an S shape. But if the curvature (often slight) is neglected, the equation of the curve becomes $\ln P = A + B/T$. In terms of any two pairs of values (P_1, T_1), (P_2, T_2), $A = (T_2 \ln P_2 - T_1 \ln P_1)/(T_2 - T_1)$ and $B = T_1 T_2/(T_2 - T_1) \ln (P_1/P_2)$. (Note that B is always negative.) Once the values of A and B have been determined, the equation can be used to determine P_3 at $T = T_3$ or T_3 at $P = P_3$. Algebraically, A and B can be eliminated to yield $\ln (P_3/P_1) = [T_3(T_3 - T_1)/T_3(T_2 - T_1)] \ln (P_2/P_1)$, and at any temperature T the slope of the vapor pressure curve is $dP/dT = (1/T^2)[T_1 T_2/(T_1 - T_2)] \ln (P_2/P_1)$.

A classic survey of equations for estimating vapor pressures was given by Miller in *Ind. Eng. Chem.*, **56**, 1964, pp. 46–57. The comprehensive tables of Stull in *Ind. Eng. Chem.*, **39**, 1947, pp. 517–550, are useful though slightly dated. Table 4.2.24 gives $T(K)$ for various P(bar) for 50 substances; P(bar) tables for various $T(K)$ for 16 elements are given in Table 4.2.29. Boublik et al., "The Vapor Pressure of Pure Substances," Elsevier, presents an extensive collection of data.

THERMAL PROPERTIES OF SATURATED VAPORS AND OF VAPOR AND LIQUID MIXTURES

Notation

v_f, v_g = specific volume of 1 lb of saturated liquid and vapor, respectively

c_f, c_g = specific heat of saturated liquid and vapor, respectively

h_f, h_g = specific enthalpy of saturated liquid and vapor, respectively

u_f, u_g = specific internal energy of saturated liquid and vapor, respectively

s_f, s_g = specific entropy of saturated liquid and vapor, respectively

$v_{fg} = vg - v_f$ = increase of volume during vaporization

$h_{fg} = h_g - h_f$ = heat of vaporization, or heat required to vaporize a unit mass of liquid at constant pressure and temperature

Note that r may be used for h_{fg} when several heats of vaporization (as r_1, r_2, r_3, etc.) are under consideration.

$u_{fh} = u - u_f$ = increase of internal energy during vaporization

$s_{fg} = s_g - s_f = h_{fg}/T$ = increase of entropy during vaporization

pv_{fg} = work performed during vaporization

The energy equation applied to the vaporization process is

$$h_{fg} = u_{fg} + pv_{fg}$$

The properties of a unit mass of a mixture of liquid and vapor of quality x are given by the following expressions:

$$v = v_f + xv_{fg}$$
$$h = h_f + xh_{fg}$$
$$u = u_f + xu_{fg}$$
$$s = s_f + xs_{fg}$$

Any property ψ can be expressed as a function of the property of the saturated liquid, ψ_f, that of the saturated vapor, ψ_g, and the quality x by three entirely equivalent equations:

$$\psi = (1 - x)\psi_f + x\psi_g$$
$$= \psi_f + x\psi_{fg}$$
$$= \psi_g - (1 - x)\psi_{fg}$$

where $\psi_{fg} - \psi_g - \psi_f$. Tables of superheated vapor usually give values of v, h, and s per unit mass. If not tabulated, the internal energy u per unit mass can be found from the equation

$$u = h - pv$$

CHARTS FOR SATURATED AND SUPERHEATED VAPORS

Certain properties of vapor mixtures and superheated vapors may be shown graphically by means of charts. Such charts show the behavior of vapors and have a practical application in the solution of certain problems.

Temperature-Entropy Chart   Figures 4.2.10 and 4.2.11 show the temperature-entropy chart for water vapor. The liquid curve is obtained by plotting corresponding values of T and s_f, and the saturation curve by plotting values of T and s_g. The values are taken from Tables 4.2.17 to 4.2.20. The two curves merge into each other at the critical temperature $T = 1,165.4°R$ (647 K). Between these two curves, constant pressure lines are also lines of constant temperature; but at the saturation curve the constant pressure lines show a sharp break with rising temperature. The constant quality lines $x = 0.2, 0.4$, etc., are equally spaced between the liquid and saturation curves.

Figure 4.2.1 is a temperature-entropy chart for air.

Enthalpy-Entropy Chart (Mollier Chart)   In this chart, the enthalpy h is taken as the ordinate and the entropy s as the abscissa.

Enthalpy–Log Pressure Chart   Previously, a chart with coordinates of enthalpy and pressure was termed a pressure-enthalpy chart. In this edition these charts are called enthalpy–log pressure charts, to more correctly identify the scale plotted for pressure. This follows modern usage. Charts with pressure per se as a coordinate have a greatly different scale and appearance.

For examples of the enthalpy–log-pressure chart, see Sec. 4.2. For enthalpy–log-pressure charts for various fluids, see "Engineering Data Book," 9th ed., Gas Processors Suppliers Assoc., Tulsa, OK; Reynolds, "Thermodynamic Properties in SI," Mech. Eng. Stanford Univ. Publication.

The energy-temperature diagram reported by Bucher (*Amer. J. Phys.*, **54**, 1986, pp. 850–851) for reversible cycles and by Wallingford (*Amer. J. Phys.*, **57**, 1989, pp. 379–381) for irreversible cycles was claimed by

Bejan (*Amer. J. Phys.*, **62**, no. 1, Jan. 1994, pp. 11–12) to have been first reported at an earlier date (Bejan, *Mech. Eng. News,* May 1977, pp. 26–28).

CHANGES OF STATE. SUPERHEATED VAPORS AND MIXTURES OF LIQUID AND VAPOR

Isothermal   In the only important cases, the fluid is a mixture of liquid and vapor in both initial and final states.

$$t = \text{const} \qquad p = \text{const}$$
$$x_1, x_2 = \text{initial and final qualities}$$
$$Q_{12} = mh_{fg}(x_2 - x_1)$$
$$W_{12} = mpv_{fg}(x_2 - x_1)$$
$$U_2 - U_1 = mu_{fg}(x_2 - x_1)$$
$$S_2 - S_1 = Q_{12}/T$$

Constant Pressure   If the fluid is a mixture at the beginning and end of the change, the constant pressure change is also isothermal. If the initial state is in the mixture region and the final state is that of a superheated vapor, the following are the equations for Q_{12}, etc. Let h_2, u_2, v_2, and s_2 be the properties of 1 lb of superheated vapor in the final state 2; then

$$Q_{12} = m(h_2 - h_1)$$
$$U_2 - U_1 = m(u_2 - u_1)$$
$$S_2 - S_1 = m(s_2 - s_1)$$
$$W_{12} = -mp(v_2 - v_1)$$
$$h_1 = h_{f1} + x_1 h_{fg1}$$
$$u_1 = u_{f1} + x_1 u_{fg1}$$
$$s = s_{f1} + x_1 s_{fg1}$$
$$v_1 = v_{f1} + x_1 v_{fg1}$$

Constant Volume   Since v_f the liquid volume is nearly constant,

$$x_1 v_{fg1} = x_2 v_{fg2}$$
$$x_2 = x_1 v_{fg1}/v_{fg2} \quad \text{or} \quad x_2 = x_1 v_{g1}/v_{g2} \quad \text{approx}$$
$$Q_{12} = U_2 - U_1 = m(u_2 - u_1) \qquad W_{12} = 0$$

Isentropic $s = \text{const.}$ If the fluid is a mixture in the initial and final states,

$$s_{f1} + x_1 s_{fg1} = s_{f2} + x_2 s_{fg2}$$

If the initial state is that of superheated vapor,

$$s_1 + s_{f2} + x_2 s_{fg2}$$

in which s_1 is read from the table of superheated vapor. The final value x_2 is determined from one of these equations, and the final internal energy u_2 is then

$$u_{f2} + x_2 u_{fg2} \qquad Q_{12} = 0 \qquad W_{12} = U_2 - U_1 = m(u_2 - u_1)$$

For water vapor, the relation between p and v during an isentropic change may be represented approximately by the equation $pv^n = \text{const.}$ The exponent n is not constant, but varies with the initial quality and initial pressure, as shown in Table 4.1.2.

The isentropic expansion of superheated steam is fairly represented by $pv^n = \text{const}$, with $n = 1.315$.

The volume at the end of expansion (or compression) is $V_2 = V_1(p_1/p_2)^{1/n}$, and the external work is

$$W_{12} = (p_2 V_2 - p_1 V_1)/(n - 1)$$
$$= -p_1 V_1[1 - (p_2/p_1)^{(n-1)/n}]/(n - 1)$$

If the initial state is in the region of superheat and final state in the mixture region, two values of n must be used: $n = 1.315$ for the expansion to the state of saturation, and the appropriate value from the first row of Table 4.1.2 for the expansion of the mixture.

MIXTURES OF AIR AND WATER VAPOR

Atmospheric Humidity   The atmosphere is a mixture of air and water vapor. Dalton's law of partial pressures (for the mixture) and the ideal gas law (for each constituent) may safely be assumed to apply. The **total pressure** p_t (barometric pressure) is the sum of the **vapor pressure** p_v and the **air pressure** p_a.

The temperature of the atmosphere, as indicated by an ordinary thermometer, is the **dry-bulb temperature** t_d. If the atmosphere is cooled under constant total pressure, the partial pressures remain constant until a temperature is reached at which condensation of vapor begins. This temperature is the **dew point** t_c (condensation temperature) and is the saturation temperature, or boiling point, corresponding to the actual vapor pressure p_v. If a thermometer bulb is covered with absorbent material, e.g., linen, wet with distilled water and exposed to the atmosphere, evaporation will cool the water and the thermometer bulb to the **wet-bulb temperature** t_w. This is the temperature given by a psychrometer. The wet-bulb temperature lies between the dry-bulb temperature and the dew point. These three temperatures are distinct except for a saturated atmosphere, for which they are identical. For each of these temperatures, there is a corresponding vapor pressure. The actual vapor pressure p_v corresponds with the dew point t_c. The vapor pressures p_d and p_w, corresponding with t_d and t_w, do not represent pressures actually appearing in the atmosphere but are used in computations.

Relative humidity r is the ratio of the actual vapor pressure to the pressure of saturated vapor at the prevailing dry-bulb temperature $r = p_v/p_d$. Within the limits of usual accuracy, this equals the ratio of actual vapor density to the density of saturated vapor at dry-bulb temperature, $r = p_v/p_d$. It is to be noted that relative humidity is a property of the vapor alone; it has nothing to do with the fact that the vapor is mixed with air. It is a method of expressing the departure of the vapor from saturation. (See "ASHRAE Handbook" for information on industrial applications of relative humidity.)

Molal humidity f is the mass of water vapor in mols per 1 mol of air. The laws of Dalton and Avogadro state that the molal composition of a mixture is proportional to the distribution of partial pressures. or $f = p_v/p_a = p_v/(p_t - P_v)$.

Specific humidity (humidity ratio) W is the mass of water vapor (pounds or grains) per pound of dry air. Mass in pounds equals mass in moles multiplied by the molecular weight. The molecular weight of water is 18, and the equivalent molecular weight of air is 28.97. The ratio 28.97/18 = 1.608, or 1.61 with ample accuracy. Thus $W = f/1.61$.

Air density ρ_a is the pounds of air in one cubic foot. **Vapor density** ρ_v is the pounds of vapor in one cubic foot. **Mixture density** ρ_m is the sum of these, i.e., the pounds of air plus vapor in one cubic foot.

Table 4.1.2   Values of *n* (Water Vapor)

Initial quality	\multicolumn{12}{c}{Initial pressure, psia}											
	20	40	60	80	100	120	140	160	180	200	220	240
1.00	1.131	1.132	1.133	1.134	1.136	1.137	1.138	1.139	1.141	1.142	1.143	1.145
0.95	1.127	1.128	1.129	1.130	1.131	1.131	1.132	1.133	1.134	1.135	1.136	1.137
0.90	1.123	1.123	1.124	1.124	1.125	1.125	1.126	1.126	1.127	1.127	1.128	1.129
0.85	1.119	1.119	1.119	1.119	1.120	1.120	1.120	1.120	1.120	1.120	1.120	1.121
0.80	1.115	1.115	1.114	1.114	1.114	1.114	1.113	1.113	1.113	1.113	1.112	1.112
0.75	1.111	1.110	1.110	1.109	1.109	1.108	1.107	1.106	1.106	1.105	1.104	1.104

Notation   The subscripts a, v, m, and f apply to air, vapor, mixture, and liquid water, respectively. The subscripts d and w apply to conditions pertaining to the dry- and wet-bulb temperature, respectively.

HUMIDITY MEASUREMENTS

Many methods are in use: (1) the **dew point** method measures the temperature at which condensation begins; water-vapor pressure can then be found from steam tables. Dew point apparatus can either cool a surface or compress and expand moist air. (2) **Hygrometers** measure relative humidity, often by using the change in dimensions of a hygroscopic material such as human hair, wood, or paper; these instruments are simple and inexpensive but require frequent calibration. The electrical resistance of an electrolytic film can also be used as an indication of relative humidity. (3) The wet- and dry-bulb **psychrometer** is widely used. Humidity measurements of air flowing in ducts can be made with psychrometers that use mercury-in-glass thermometers, thermocouples, or resistance thermometers. Humidity measurements of still air can be made with sling psychrometers as aspiration psychrometers. Psychrometric wet-bulb temperatures must be corrected to obtain thermodynamic wet-bulb temperatures, or there must be adequate air motion past the wet-bulb thermometer, 800 to 900 ft/min (with duct walls at air temperature), to ensure a proper balance between radiation and convection. (4) **Chemical analysis** by the use of desiccants such as sulfuric acid, phosphorus pentoxide, lithium chloride, or silica gel can be used as primary standards of humidity measurement.

The following equations give various properties in terms of pressure in inches Hg and temperature in degrees Fahrenheit.

Relative humidity: $r = p_v/p_d$
Specific humidity: $W = p_v/1.61(p_t - p_v)$ lb/lb dry air

Volume of mixture per pound of dry air:

$$v_a = \frac{1}{\rho_a} = 0.754(t_d + 460)/(p_t - rp_d) \qquad \text{ft}^3$$

Volume of mixture per pound of mixture:

$$v_m = \frac{1}{\rho m} = v_d/(1 + W) \qquad \text{ft}^3$$

The **enthalpy** of a mixture of dry air and steam, when each constituent is assumed to be an ideal gas, in Btu per pound of dry air, is the sum of the enthalpy of 1 lb of dry air and the enthalpy of the W lb of steam mixed with that air. The specific enthalpy of dry air (above 0°F) is $h_a = 0.240t_d$ (up to 130°F, the specific heat of dry air is 0.240; at higher temperatures, it is larger). The specific enthalpy of low-pressure steam (saturated or superheated) is nearly independent of the vapor pvessure and depends only on t_d. An empirical equation for the specific enthalpy of low-pressure steam for the range of temperatures from −40 to 250°F is

$$h_v = 1,062 + 0.44t_d \qquad \text{Btu/lb}$$

The enthalpy of a mixture of air and steam is

$$h_m = 0.240t_d + W(1,062 + 0.44t_d)$$

The specific heat of a mixture of dry air and steam per pound of dry air may be called **humid specific heat** and is $0.240 + 0.44W$ Btu/lb dry air. For a steady-flow process without change of specific humidity, heat transfers per pound of dry air may be computed as the product of humid specific heat and change in dry-bulb temperature.

Thermodynamic Wet-Bulb Temperature (Temperature of Adiabatic Saturation)   The thermodynamic wet-bulb temperature t^* is an important property of state of mixtures of dry air and superheated steam; it is the temperature at which water (or ice), by evaporating into a mixture of air and steam, will bring the mixture to saturation at the same temperature in a steady-flow process in the absence of external heat transfer. For a mixture of dry air and saturated steam only, $t^* = t_d$; where $r < 1$, $t^* < t_d$. By writing energy and mass balances for the process of

adiabatic saturation with water supplied at t^*, the following equation may be derived:

$$W = W^* - \frac{(0.240 + 0.44W^*)(t_d - t^*)}{1,094 + 0.44t_d - t^*}$$

where W^* = specific humidity for saturation at the total pressure of p_t.

The enthalpy of a mixture of dry air and **saturated** steam at the total pressure p_t and thermodynamic wet-bulb temperature t^* exceeds the enthalpy of a mixture of dry air and **superheated** steam at the same p_t and t^* for

$$h_m^* = h_m + (W^* - W)h_f^*$$

A property of the mixture that remains constant for constant p_t and t^* has been called the Σ **function,** for

$$\Sigma^* = h_m^* - W^*h_f^* = \Sigma = h_m - Wh_f^*$$

EXAMPLE.   A mixture of dry air and **saturated** steam; $p_t = 24$ inHg; $t_d = 76$°F. Partial pressure of water vapor from tables:

$$p_v = p_d = 0.905 \text{ inHg}$$

Partial pressure of dry air: $p_a = p_t - p_v = 23.095$ inHg. Specific humidity:

$$W = 0.905/1.61(23.095) = 0.0243 \text{ lb/lb dry air}$$

Volume of mixture per pound of dry air:

$$v_a = 0.754(536)/23.095 = 17.5 \text{ ft}^3$$

Volume of mixture per pound of mixture:

$$v_m = 17.5/1.0243 = 17.1 \text{ ft}^3$$

Enthalpy of mixture:

$$h_m = 0.240(76) + 0.0243(1,095) = 44.85 \text{ Btu/lb dry air}$$

EXAMPLE.   A mixture of dry air and **superheated** steam; $p_t = 24$ inHg; $t_d = 76$°F; $t_w = t^* = 62$°F. Pressure of saturated steam at $t^* = 0.560$ inHg (from tables):

$$W^* = 0.560/1.61(23.44) = 0.01484 \text{ lb/lb dry air}$$

Specific humidity:

$$W = 0.01484 - \frac{0.2465(14)}{1,065.4} = 0.0116 \text{ lb/lb dry air}$$

Partial pressure of water vapor:

$$0.0116 = p_v/[1.61(24 - p_v)] \qquad \text{and} \qquad p_v = 0.44 \text{ inHg}$$

Relative humidity: $r = 0.44/0.905 = 0.486$.
Volume of mixture per pound of dry air:

$$v_a = 0.754(536)/23.56 = 17.2 \text{ ft}^3$$

Volume of mixture per pound of mixture:

$$v_m = 17.2/1.0116 = 17.0 \text{ ft}^3$$

Enthalpy of mixture:

$$h_m = 0.240(76) + 0.0116(1,095) = 30.95 \text{ Btu/lb dry air}$$

PSYCHROMETRIC CHARTS

For occasional use, algebraic equations are less confusing and more reliable; for frequent use, a **psychrometric chart** may be preferable. A disadvantage of charts is that each applies for only one value of barometric pressure, usually 760 mm or 30 inHg. Correction to other barometric readings is not simple. The equations have the advantage that the actual barometric pressure is taken into account. The equations are often more convenient for equal accuracy or more accurate for equal convenience.

Psychrometric charts are usually plotted, as indicated by Fig. 4.1.18, with dry-bulb temperature as abscissa and specific humidity as ordinate. Since the specific humidity is determined by the vapor pressure and the barometric pressure (which is constant for a given chart), and is nearly proportional to the vapor pressure, a second ordinate scale, departing slightly from uniform graduations, will give the vapor pressure. The

Fig. 4.1.18   Skeleton humidity chart.

saturation curve ($r = 1.0$) gives the specific humidity and vapor pressure for a mixture of air and saturated vapor. Similar curves below it give results for various values of relative humidity. Inclined lines of one set carry fixed values of the wet-bulb temperature, and those of another set carry fixed values of v_a, cubic feet per pound of air. Many charts carry additional scales of enthalpy or Σ function.

Any two values will locate the point representing the state of the atmosphere, and the desired values can be read directly.

Psychrometric charts at different temperatures and barometric pressures are useful in solving problems that fall outside the normal range indicated in Fig. 4.1.18. A collection ("trial set") of 17 different psychro-metric charts in both USCS and SI units, for low, normal, and high temperatures, at sea level and at four elevations above sea level, is available from the Carrier Corp., Syracuse, NY.

AIR CONDITIONING

Air-conditioning processes alter the temperature and specific humidity of the atmosphere. The weight of dry air remains constant and consequently computations are best based upon 1 lb of dry air.

Liquid water may enter or leave the apparatus. Its weight m_f lb of air is often merely the difference between the specific humidities of the entering and leaving atmospheres. Its specific enthalpy at the observed or assumed temperature of supply or removal t_f is

$$h_f = t_f - 32 \text{ Btu/lb of liquid}$$

Because most air conditioning involves steady-flow processes, thermal results are computed by the steady flow equation, written for 1 lb of air. Using subscript 1 for entering atmosphere and liquid water, and for heat supplied; and 2 for departing atmosphere and water, and for heat abstracted; the equation becomes (in the absence of work)

$$h_{m1} + m_{f1}h_{f1} + q_1 = h_{m2} + m_{f2}h_{f2} + q_2 \qquad \text{Btu/lb air}$$

Either or both values of m_f or q may be zero.

In terms of the sigma function, the steady-flow equation becomes

$$\Sigma_1 + W_1(t_{w1} + 32) + m_{f1}h_{f1} + q_1 = \Sigma_2 + W_2(t_{w2} - 32)$$
$$+ m_{f2}h_{f2} + q_2 \qquad \text{Btu/lb air}$$

Unit processes involved in air conditioning include heating and cooling an atmosphere above its dew point, cooling below the dew point, adiabatic saturation, and mixing of two atmospheres. These, in various sequences, make it possible to start with any given atmosphere and produce an atmosphere of any required characteristics.

Heating and cooling above the dew point entail no condensation of vapor. Barometric pressure and composition being unaltered, partial pressures remain constant. The process is represented in Fig. 4.1.19.

EXAMPLE.   Initial conditions: $p_t = 28$ inHg; $t_a = 60°F$; $t_w = 50°F$; $p_v = 0.26$ inHg; $V = 1,200$ ft³.

Final conditions: $t_d = 82°F$.

Initial computed values: $r = 0.50$; $W = 0.0058$ lb vapor/lb air; $\rho_a = 0.0707$ lb air/ft³; $m_a = V \times \rho_a = 1,200 \times 0.0707 = 84.9$ lb air; $h_m = 20.7$ Btu/lb air.

Final computed values: p_v, W, and m_a unaltered; $r = 0.24$; $\rho_a = 0.0679$ lb air/ft³; $V = m_a/\rho_a = 84.9/0.0679 = 1,250$ ft³; $h_m = 26.1$ Btu/lb air.

Heat added: $q = h_{m2} - h_{m1} = 26.1 - 20.7 = 5.4$ Btu/lb air; $Q = q \times m_a = 5.4 \times 84.9 = 458$ Btu.

Fig. 4.1.19   **Fig. 4.1.20**

Cooling below the dew point, and **dehumidification,** entails condensation of vapor; the final atmosphere will be saturated, liquid will appear (see Fig. 4.1.20).

EXAMPLE.   Initial conditions: $p_t = 29$ inHg; $t_d = 75°F$; $t_w = 65°F$; $V = 1,500$ ft³.

Final condition: $t_d = 45°F$.

Initial computed values: $W = 0.0113$ lb vapor/lb air; $\rho_a = 0.0706$ lb air/ft³; $m_a = 1,500 \times 0.0706 = 106.0$ lb air; $h_m = 30.4$ Btu/lb air; $t_c = 60°F$.

Final computed values: $t_d = 45°F$; $p_v = 0.30$ inHg; $r = 1.0$; $W = 0.0065$ lb vapor/lb air; $\rho_a = 0.0754$ lb air/ft³; $V = 106.0/0.0754 = 1,406$ ft³; $h_m = 17.8$ Btu/lb air.

Liquid formed: $m_f = W_1 - W_2 = 0.0113 - 0.0065 = 0.0048$ lb liqud/lb air; $h_f = 50 - 32 = 18$ Btu/lb liquid (assuming that the liquid is drained out at an average temperature $t_f = 50°F$).

Heat abstracted: $q = h_{m1} - h_{m2} - m_1 h_f = 30.4 - 17.8 - 0.0048 \times 18 = 12.5$ Btu/lb air; $Q = q \times m_a = 12.5 \times 106.0 = 1,325$ Btu.

Dehumidification may be accomplished in a **surface cooler,** in which the air passes over tubes cooled by brine or refrigerant flowing through them. The solution of this type of problem is most easily handled on the chart (see Fig. 4.1.21). Locate the point representing the state of the entering atmosphere, and draw a straight line to a point on the saturation curve ($r = 1.0$) at the temperature of the cooling surface. The final state of the issuing atmosphere is approximated by a point on this line whose position on the line is determined by the heat abstracted by the cooling medium. This depends upon the extent of surface and the coefficient of heat transfer.

Fig. 4.1.21

Adiabatic saturation (humidification) may be conducted in a spray chamber through which atmosphere flows. A large excess of water is recirculated through spray nozzles, and evaporation is made up by a suitable water supply. After the process has been operating for some time, the water in the spray chamber will have been cooled to the temperature of adiabatic saturation, which differs from the wet-bulb temperature only because of radiation and velocity errors that affect the wet-bulb thermometer. No heat is added or abstracted; the process is adiabatic. The heat of vaporization for the water that is evaporated is supplied by the cooling of the air passing through the chamber. The wet-bulb temperature of the atmosphere is constant throughout the chamber (Fig. 4.1.22). If the chamber is sufficiently large, the issuing atmosphere will be saturated at the wet-bulb temperature of the entering atmosphere; i.e., as the atmosphere passes through the chamber, t_w remains constant, t_d is reduced from its initial value to t_w. In a chamber of commercial size, the action may terminate somewhat short of this,

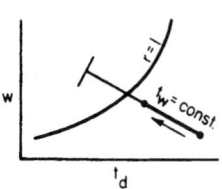

Fig. 4.1.22

the precise endpoint being determined by the duration and effectiveness of contact between air and spray water. In any case, the weight of water evaporated equals the increase in the specific humidity of the atmosphere.

EXAMPLE.   Initial conditions $p_t = 30$ inHg; $t_d = 78°F$; $t_w = 55°F$; $r = 0.20$; $W = 28$ grains vapor/lb air.
Final conditions: $t_d = t_w = 55°F$; $r = 1.0$; $W = 64$ grains vapor/lb air.
Water evaporated: $W_2 - W_1 = 64 - 28 = 36$ grains water/lb air

The design of the spray chamber to produce this result is necessarily based upon experience with like apparatus previously built.

In practice, the spray chamber is preceded and followed by heating coils, the first to warm the entering atmosphere to the desired value of t_w, determined by the prescribed final specific humidity, the second to warm the issuing atmosphere to the desired temperature, and simultaneously to reduce its relative humidity to the desired value.

The **spray chamber** that is used for adiabatic saturation (humidification) in winter may be used for dehumidification in summer by supplying the spray nozzles with refrigerated water instead of recirculated water. In this case, the issuing atmosphere will be saturated at the temperature of the spray water, which will be held at the desired dew point. Subsequent heating of the atmosphere to an acceptable temperature will simultaneously reduce the relative humidity to the desired value.

Mixing Two Atmospheres   In recirculating ventilation systems, two atmospheres (1 and 2) are mixed to form a third (3). The state of the final atmosphere is readily found graphically on the psychrometric chart (see Fig. 4.1.23). Locate the points 1 and 2 representing the states of the initial atmospheres. Connect these points by a straight line. Locate a point that divides this line into segments inversely proportional to the weights of air in the respective atmospheres. The division point represents the state of the final mixture, so long as it falls below the saturation curve ($r = 1$). If the final point falls above the saturation curve, as in Fig. 4.1.24, condensation will ensue, and the true final point 4 is found

Fig. 4.1.23

Fig. 4.1.24

by drawing a line from the apparent point 3, parallel to the lines of constant wet-bulb temperature, to its intersection with the saturation curve. From all the points involved, readings of specific humidity may be taken, including point 3 when it falls above the saturation curve, and in this case the difference between W_3 and W_4 will be the weight of condensate, pounds per pounds air.

If the chart is sectional and the two points do not fall in the same section, or in any case in which it is preferred, the same method may be carried out arithmetically.

For adiabatic mixing in a steady-flow process of two masses of "moist" air, each at the total pressure of p_t,

$$m_{a3} = m_{a1} + m_{a2}$$

In the absence of condensation,

and
$$m_{a3}W_3 = m_{a1}W_1 + m_{a2}W_2$$
$$m_{a3}h_{m3} = m_{a1}h_{m1} + m_{a2}h_{m2}$$

When condensation occurs, assume that the condensate is removed at the final temperature t_4 and that the final mixture consists of dry air and saturated water vapor at this same temperature. The weight of condensate is

$$m_c = m_{a1}W_1 + m_{a2}W_2 - m_{a3}W_4$$

where W_4 is the specific humidity for saturation at temperature t_4 and total pressure p_t. Also

$$m_{a1}h_{m1} + m_{a2}h_{m2} = m_{a2}h_{m4} + m_c h_{f4}$$

In the case of condensation, a trial solution is necessary to find the temperature t_4 that will satisfy these relations.

EXAMPLE.   Two thousand ft³ of air per min at $t_{d1} = 80°F$ and $t_{w1} = 65°F$ are mixed in an adiabatic, steady-flow process with 1,000 ft³ of air per min at $t_{d2} = 95°F$ and $t_{w2} = 75°F$; the total pressure of each mixture is 29 inHg.
By computation, $m_{a1} = 140$ lb dry air/min; $W_1 = 0.010$ lb/lb dry air; $m_{a2} = 67.6$ lb dry air/min; $W_2 = 0.0146$ lb/lb dry air.
$m_{a3} = 207.6$ lb dry air/min.
$W_3 = 0.0116$ lb/lb dry air.
$h_{m1} = 30.3$ Btu/lb dry air and $h_{m2} = 38.9$ Btu/lb dry air.
$h_{m3} = 33.1$ Btu/lb dry air.
$t_{d3} = 84.9°F$.

EXAMPLE.   Fifteen hundred ft³ of air per min at $t_{d1} = 0°F$ and $r_1 = 0.8$ are mixed in an adiabatic, steady-flow process with 1,000 ft³ of air per min at $t_{d2} = 100°F$ and $r_2 = 0.9$; the total pressure of each mixture is 30 inHg.
By computation, $m_{a1} = 129.6$ lb of dry air/min; $W_1 = 0.000626$ lb/lb dry air; $m_{a2} = 66.9$ lb of dry air/min; $W_2 = 0.03824$ lb/lb dry air; $h_{m1} = 0.09$ Btu/lb dry air; $h_{m2} = 66.29$ Btu/lb dry air.
The three equations that must be satisfied by a choice of the terminal temperature, $t_4 = t_{d4} = t_{w4}$, are

$$m_c = 2.64 - 196.5W_4$$
$$4551 = 196.5h_{m4} + m_c h_{f4}$$
$$W_4 = p_{v4}/1.61(30 - p_{v4}) \quad \text{for } r_4 = 1$$

The value of t_4 that satisfies these equations is 55°F; condensation amounts to 0.84 lb/min.

The **cooling tower** is a chamber in which outdoor atmosphere flows through a spray of entering hot water, which is to be cooled. The temperature of the water is reduced in part by the warming of the air, and in greater part by the evaporation of a portion of the water. The atmosphere enters at given conditions and emerges at a higher temperature and usually saturated ($r = 1$). It is commonly possible to cool the water below the temperature of the entering air, often to about halfway between t_d and t_w. The volume of atmosphere per pound of entering water and the weight of water evaporated are to be computed.

EXAMPLE.   A cooling tower is to receive water at 120°F and atmosphere at $t_d = 90$, $t_w = 80$, whence $p_v = 0.92$, $W = 0.0196$ lb vapor/lb air, $\rho_a = 0.0702$ lb air/ft³, and $h_m = 43.2$. The water is to be cooled to 85°F. What volume of atmosphere must be passed through the tower, and what weight of water will be lost by evaporation?
The issuing atmosphere will be assumed to be saturated at 115°F. Then $t_d = 115°F$, $p_v = 3.0$ inHg, $W = 0.0690$ lb vapor/lb air, $\rho_a = 0.0623$ lb air/ft³, and/ $h_m = 104.4$ Btu/lb air.
The two unknowns are the weight of air to be passed through the tower and the weight of water to be evaporated. The two equations are the water-weight balance and the enthalpy balance (the steady-flow equation for zero heat transfer to or from outside). Assume that 1 lb water enters, of which x lb are evaporated. The water-weight balance $1 + m_a W_1 = 1 - x + m_a W_2$ becomes $x = m_a(W_2 - W_1) = m_a(0.0690 - 0.0196) = 0.0494m_a$. The enthalpy balance $1 \times (120 - 32) + m_a h_{m1} = (1 - x)(85 - 32) + m_a h_{m2}$ becomes $88 + 43.2m_a = 53(1 - x) + 104.4m_a$; whence $53x = 53 - 88 + m_a(104.4 - 43.2) = -35 + 61.2m_a$.
Solving these simultaneous equations, $x = 0.0295$ lb water evaporated per pound of water entering and $m_a = 0.597$ lb air per pound water entering.

In an **evaporative condenser,** vapor is condensed within tubes that are cooled by the evaporation of water flowing over the outside of the tubes; the water evaporates into the atmosphere. The computation of results is similar to that for the cooling tower.

REFRIGERATION

Vapor Compression Machines   The essential parts of a vapor-compression system are the same as in the system using air, except that the expansion cylinder is replaced by an expansion value through which the liquefied medium flows from the high-pressure condensing coils to the low-pressure brine coils. The cycle of operation is best shown on the *T-S* plane (Fig. 4.1.25). The point *B* represents the state of the refrigerating medium leaving the brine coils and entering the compressor. Usually in this state the fluid is nearly dry saturated vapor; i.e., point *B* is near the saturation curve S_g. *BC* represents the assumed reversible adiabatic compression, during which the fluid is usually superheated. In the state *C*, the superheated vapor passes into the cooling coils and is cooled at constant pressure, as indicated by *CD*, and then condensed at temper-

Fig. 4.1.25   Vapor compression refrigeration cycle.

ature T_2, as shown by *DE*. The liquid now flows through the expansion valve into the brine coils. This is a throttling process, and the final-state point *A* is located on the T_1-line in such a position as to make the enthalpy for state *A* (= area $OHGAA_1$) equal to the enthalpy at *E* (= area $OHEE_1$). The mixture of liquid and vapor now absorbs heat from the brine and vaporizes, as indicated by *AB*.

The **heat absorbed from the brine**, represented by area A_1ABC_1, is

$$Q_1 = h_b - h_a = h_b - h_e$$

The **heat rejected to the cooling water**, represented by area C_1CDEE_1, is

$$Q_2 = h_c - h_e = c_p(T_c - T_d) + r_2 \qquad \text{approx}$$

where r_2 denotes the enthalpy of vaporization at the upper temperature T_2, and c_p the specific heat of the superheated vapor. The **work** that must be **supplied per pound of fluid circulated** is $W = Q_2 - Q_1 = h_c - h_b$. The ratio $Q_1/W = (h_b - h_e)/(h_c - h_b)$ is sometimes called the **coefficient of performance**.

If *Q* denotes the heat to be absorbed from the brine per hour, then the **quantity of fluid circulated per hour** is $m = Q/(h_b - h_a)$; or, if *B* is taken on the saturation curve, $m = Q/(h_{g1} - h_{f2})$.

The work per hour is $W = m(h_c - h_{g1}) = Q(h_c - h_{g1})/(h_{g1} - h_{f2})$ ft · lb, and the **horsepower required** is $H = Q(h_c - h_{g1})/2544(h_{g1} - h_{f2})$. The (U.S.) ton of refrigeration represents a cooling rate of 200 Btu/min, which is closely equivalent to that of 50 kcal/min, 210 kJ/min, or 3.5 kW. (Extensive tables of refrigerant properties are found in the "ASHRAE Handbook.")

If v_{g1} is the volume of the saturated vapor at the temperature T_1 in the brine coils, and *n* the number of working strokes per minute, the **displacement volume** of the **compressor cylinder** is $V = mv_{g1}/(60n)$.

The work necessary for operating a refrigerator, although usually supplied through the compressor, may be supplied in other ways. Thus in **absorption refrigerators** (see Secs. 12 and 19) an absorbent, usually water, absorbs the refrigerant, usually ammonia. The water, by its affinity for the ammonia, has, in a thermodynamic sense, ability to do work. Having absorbed the ammonia and thereby lost its ability to do work, the water may have its work capacity restored by passing the ammonia-water solution through a rectifying column from which water and ammonia emerge. With operation under a suitable pressure, the ammonia is

condensed to a liquid. This, in turn, may be evaporated, yielding refrigeration, the ammonia vapors being once again absorbed in the water. Thermodynamically the analysis for these absorption cycles is similar to that for compression cycles. See Wood, "Applications of Thermodynamics," 2d ed., Addison-Wesley, for an in-depth discussion of the absorption refrigeration cycle.

Other papers on absorption refrigeration or **heat pump systems** include Chen, *Heat Recovery Systs. & CHP*, **30**, 1988, pp. 37–51; Narodoslawsky et al., *Heat Recovery Systs. & CHP*, **8**, no. 5, 1988, pp. 459–468, and **8**, no. 3, 1988, pp. 221–233; Egrican, **8**, no. 6, 1988, pp. 549–558; and for the aqua ammonia system, Ataer and Goguy, *Int. J. Refrig.*, **14**, Mar. 1991, pp. 86–92. Kalina (*J. Eng. Gas Turbines & Power*, **106**, 1984, pp. 737–742) proposed a new cycle using an ammonia-water solution as a bottoming cycle system. The proper selection of the composition and parameters of the working fluid was stated to be critical in the cycle design. Absorption replaces condensation of the working fluid after expansion in the turbine. Special care is also needed to regulate pressure drops between turbine stages. Chuang et al. [*AES (A.S.M.E.)*, **10**, no. 3, 1989, pp. 73–77] evaluated exergy changes in the cycle while Kouremenos and Rogdaikis [*AES (A.S.M.E.)*, **19**, 1990, pp. 13–19] developed a computer code for use with *h-s* and *T-s* diagrams. For heat pumps using **binary mixtures** see Hihara and Sato, *ASME/JSME Thermal Eng. Proc.*, **3**, 1991, p. 297. For **supercritical heat pump** cycles see, e.g., Angelino and Invernizzi, *Int. J. Refrig.*, **17**, no. 8, 1991; pp. 543–554.

An analysis of industrial gas separation to yield minimum overall cost, i.e., taking into account both energy and capital cost, for the processes of distillation, absorption, adsorption, and membranes was given by Haselden, *Gas Separation & Purification,* **3**, Dec. 1989, pp. 209–216. Analysis of the thermodynamic regenerator cycle with compressed-gas throttling, called the **Linde cycle,** was made by Lavrenchenko, *Cryogenics, 33*, no. 11, 1993, pp. 1040–1045.

Orifice pulse tube refrigerators are receiving more attention. Kittel (*Cryogenics, 32*, no. 9, 1992, pp. 843–844) examines their thermodynamic efficiency and refers to Radebaugh (*Adv. Cryog. Eng.*, **35B**, 1990, pp. 1192–1205) for a review of these devices.

de Rossi et al. (*Proc Mtg. IIR-IIF Comm.* B1, Tel Aviv, 1990) gave an interactive computer code for refrigerant thermodynamic properties which was evaluated for 20 different refrigerants in vapor compression and a reversed Rankine cycle (*Appl. Energy, 38*, 1991, pp. 163–180). The evaluation included an exergy analysis. A similar publication was *AES (A.S.M.E.), 3*, no. 2, 1987, pp. 23–31. Other papers include Kumar et al. (*Heat Recovery Systems & CHP, 9*, no. 2, 1989, pp. 151–157), Alefeld (*Int. J. Refrig., 10*, Nov. 1987, pp. 331–341), and Nikolaidis and Probert (*Appl. Energy, 43*, 1992, pp. 201–220).

The problem of deciding which refrigerants will be used in the future is complex. Although recommendations exist as to the phasing out of existing substances, one cannot predict the extent to which the recommendations will be followed. The blends proposed by seveial manufacturers have yet to receive extensive testing. One must bear in mind that in addition to the thermodynamic suitability considerations of material compatibility, ozone depletion potential, etc. have to be taken into account. According to Dr. McLinden of the National Institute of Standards and Technology, Boulder, CO (private communication, March 1995), R 11, R 12, R 22, R 123, R 134a as well as ammonia (R 717) and propane/isobutane (R 290/R 600a) blends are likely to be important for many years. The presentation of the tables in Sec. 4.2 has been revised to include some of these plus a few other compounds used in ternary blends. The best single source for further information is the "ASHRAE Handbook—Fundamentals," 1993 or latest edition.

STEAM CYCLES

Rankine Cycle   The ideal Rankine cycle is generally employed by engineers as a standard of reference for comparing the performance of actual steam engines and steam turbines. Figure 4.1.26 shows this cycle on the *T-S* and *p-V* planes. *AB* represents the heating of the water in the boiler, BC represents evaporation (and superheating if there is any), *CD* the assumed isentropic expansion in the engine cylinder, and *DA* condensation in the condenser.

Fig. 4.1.26   Rankine cycle.

Let h_a, h_b, h_c, h_d represent the enthalpy per unit mass of steam in the four states *A*, *B*, *C*, and *D*, respectively. Then the energy transformed into work, represented by the area *ABCD*, is $h_c - h_d$ (enthalpies in Btu/lbm).

The energy expended on the fluid is $h_c - h_a$, hence the **Rankine cycle efficiency** is $e_t = (h_c - h_d)/(h_c - h_a)$.

The **steam consumption** of the ideal Rankine engine in pounds per horsepower-hour is $N_r = 2{,}544/(h_c - h_d)$. Expressed in pounds per kilowatthour, the steam consumption of the ideal Rankine cycle is $3{,}412.7/(h_c - h_d)$.

The performance of an engine is frequently stated in terms of the heat used per horsepower-hour. For the ideal Rankine engine, this is

$$Q_r = 2{,}544/e_t = 2{,}544(h_c - h_a)/(h_c - h_d)$$

Efficiency of the Actual Engine   Let *Q* denote the heat transformed into work per pound of steam by the actual engine; then if Q_1 is the heat furnished by the boiler per pound of steam, the **thermal efficiency of the engine** is $e_t = Q/Q_1$.

The efficiency thus defined is misleading, as it takes no account of the conditions of boiler and condenser pressure, superheat, or quality of steam. It is customary therefore to define the efficiency as the ratio Q/Q_a, where Q_a is the **available heat,** or the heat that could be transformed under ideal conditions. For steam engines and turbines, the Rankine cycle is usually taken as the ideal, and the quantity $Q/Q_a = Q(h_c - h_d)$ is called the **engine efficiency.** For engines and turbines, this efficiency ranges from 0.50 to 0.85. The engine efficiency *e* may also be expressed in terms of steam consumed; thus, if N_a is the steam consumption of the actual engine and N_r is the steam consumption of the ideal Rankine engine under similar conditions, then $e = N_r/N_a$.

EXAMPLE.   Suppose the boiler pressure to be 180 psia, superheat 150°F, and the condenser pressure 3 in of mercury. From the steam tables or diagiarn, the following values are found: $h_c = 1{,}283.3$, $h_d = 942$, $h_a = 82.99$. The available heat is $Q_a = 1{,}283.3 - 942 = 341.3$ Btu, and the thermodynamic efficiency of the cycle is $341.3/(1{,}283.3 - 82.99) = 0.284$. The steam consumption per horsepower-hour is $2{,}544/341.3 = 7.46$ lb, and the heat used per horsepower-hour is $2{,}544/0.284 = 8{,}960$ Btu. If an actual engine working under the same conditions has a steam consumption of 11.4 lb/(hp · h), its efficiency is $7.46/11.4 = 0.655$, and its heat consumption per horsepower-hour is $8{,}960/0.655 = 13{,}680$ Btu.

Reheating Cycle   Let the steam after expansion from p_1 to an intermediate pressure p_2 (*cd*, Fig. 4.1.27) be reheated at constant pressure p_2, as indicated by *de*. Then follows the isentropic expansion to pressure p_3, represented by e_f.

The energy absorbed by 1 lb of steain is $h_c - h_a$ from the boiler, and $h_e - h_d$ from the reheating. The work done, neglecting the energy required to operate the boiler feed pump, etc., is $h_c - h_d + h_e - h_f$. Hence the efficiency of the cycle is

$$e_t = \frac{h_c - h_d + h_e - h_f}{h_c - h_a + h_e - h_d}$$

Bleeding Cycle   In the **regenerative** or bleeding cycle, steam is drawn from the turbine at one or more stages and used to heat the feed water. Figure 4.1.28 shows a diagrammatic arrangement for bleeding at one stage. Entering the turbine is $1 + w$ lb of steam at p_1, t_1, and

enthalpy h_1. At the bleeding point *w* lb at p_2, t_2, h_2 enters the feedwater heater. The remaining 1 lb passes through the turbine and condenser and enters the feedwater heater as water at temperature t_3. Let t' denote the temperature of the water leaving the heater, and h' the correspond-

Fig. 4.1.27   Reheating cycle.

ing enthalpy of the liquid. Then the equation for the interchange of heat in the heater is

$$w(h_2 - h') = h' - h_{f3}$$

The work done by the bled steam is $w(h_1 - h_2)$ and that by the 1 lb of steam going completely through the turbine is $h_1 - h_3$. Total work $= w(h_1 - h_2) + (h_1 - h_3)$ if work to the pumps is neglected. The heat supplied between feedwater heater and turbine is $(1 + w)(h_1 - h')$. Hence the ideal efficiency of the cycle is

$$e_t = \frac{w(h_1 - h_2) + h_1 - h_3}{(1 + w)(h - h')}$$

A **computer program** which is claimed to model the thermodynamic performance of any steam power system has been described by Thelen and Somerton [*AES, (A.S.M.E.), 33*, 1994, pp. 167–175], extending an earlier analysis.

Fig. 4.1.28   Regenerative feedwater heating.

The use of selected fluid mixtures in Rankine cycles was proposed by Radermacher (*Int. J. Ht. Fluid Flow, 10*, no. 2, June 1989, p. 90). Lee and Kim (*Energy Convsn. Mgmt., 33*, no. 1, 1992, pp. 59–67) describe the finite time optimization of a modified Rankine heat engine. For a steam Wankel engine see Badr et al. (*Appl. Energy, 40*, 1991, pp. 157–190).

Heat from **nuclear reactors** can be used for heating services or, through thermodynamic cycles, for power purposes. The reactor coolant transfers the heat generated by fission so as to be used directly, or through an intermediate heat-exchange system, avoiding radioactive contamination. Steam is the preferred thermodynamic fluid in practice so that the Rankine-cycle performance standards with regenerative and reheat variations prevail. Adaptation of gas-turbine cycles, using various gases, can be expected as allowable reactor temperatures are raised.

Many engineers and scientists are actively engaged in research dealing with the location, production, utilization, transmission, storage, and distribution of new forms of energy. Examples are the study of the energy released in the fusion of hydrogen nuclei and research in solar energy. Considerable effort is being expended in studying the feasibility of combining the gas turbine with a steam-generating plant. The possibility of efficiency improvement over the steam cycle is due to higher inlet temperatures associated with the gas turbine. Significant advances are predicted in the near future in expanding our energy sources and reserves.

THERMODYNAMICS OF FLOW OF COMPRESSIBLE FLUIDS

Important examples of the flow of compressible fluids are the following: (1) the flow of air and steam through orifices and short tubes or nozzles, as in the steam turbine, (2) the flow of compressed air, steam, and illuminating gas in long mains, (3) the flow of low-pressure gases, as furnace gases in ducts and chimneys or air in ventilating ducts, and (4) the flow of gases in moving channels, as in the centrifugal fan.

Notation

Let A = area of section, ft^2
$\quad C$ = empirically determined coefficient of discharge
$\quad D$ = inside diameter of pipe, ft
$\quad d = 12D$ = inside diameter of pipe, in
$\quad F_{12}$ = energy expended in overcoming internal and external friction between sections A_1 and A_2
$\quad F'$ = energy used in overcoming friction, ft · lb/lb of fluid flowing
$\quad f$ = friction factor = $4f'$
$\quad g$ = 32.2 = local acceleration of gravity, ft/s^2
$\quad g_c$ = a dimensional constant
$\quad h$ = enthalpy, Btu/lb
$\quad J$ = 778.3 ft · lb/Btu
$\quad k = c_p/c_v$
$\quad L$ = equivalent length of pipe, ft
$\quad m$ = mass of fluid flowing past a given section per s, lb
$\quad \mu$ = viscosity, cP
$\quad P$ = pressure, lb/in^2 abs
$\quad \Delta P$ = differential pressure across nozzle, lb/in^2
$\quad p$ = pressure of fluid at given section, lb/ft^2 abs
$\quad p_m$ = critical flow pressure
$\quad Q_{12}$ = heat entering the flowing fluid between sections A_1 and A_2
$\quad q$ = volume of fluid flowing past section, ft^3/min
$\quad R$ = ideal gas constant
$\quad \rho$ = density, lb/ft^3
$\quad T$ = temperature, °R
$\quad \bar{v}$ = mean velocity at the given section, ft/s
$\quad v$ = specific volume
$\quad w$ = weight of fluid power flowing past a given section per s, lb
$\quad z$ = height from center of gravity of flow to fixed base level, ft

The cross sections of the tube or channel are denoted by A_1, A_2, etc. (Fig. 4.1.29), and the various magnitudes pertaining to these sections are denoted by corresponding subscripts. Thus, at section A_1, the velocity, specific volume, and pressure are, respectively, $\bar{v}_1$, v_1, p_1; at section A_2, they are $\bar{v}_2$, v_2, p_2.

Fig. 4.1.29

Fundamental Equations   In the interpretation of fluid-flow phenomena, three fundamental equations are of importance.

1. The continuity equation, or material balance,

$$\frac{A_1\bar{v}_1}{v_1} = \frac{A_2\bar{v}_2}{v_2} \quad \text{or} \quad \frac{dv}{v} = \frac{dA}{A} + \frac{d\bar{v}}{\bar{v}}$$

2. The first law of energy balance for steady flow,

$$q = (h_2 - h_1) + \frac{\bar{v}_2^2 - \bar{v}_1^2}{2g_c} + \frac{g}{g_c}(z_2 - z_1)$$

3. The available energy balance for a steady-flow process, based on unit weight, is

$$v\,dp + \frac{\bar{v}\,d\bar{v}}{g} + dF + dz = 0$$

In the process here discussed, no net external or shaft work is performed.

For most actual processes, the third equation cannot be integrated because the actual path is not known. Usually, adiabatic flow is assumed, but occasionally the assumption of isothermal conditions may be more nearly correct.

For **adiabatic flow of imcompressible fluids,** the last equation above can be written in the more familiar form known as Bernoulli's equation:

$$(p_2 - p_1)/\rho + (\bar{v}_2^2 - \bar{v}_1^2)/2 + g/g_c(z_2 - z_1) = 0$$

Flow through Orifices and Nozzles   As a compressible fluid passes through a nozzle, drop in pressure and simultaneous increase in velocity result. By assuming the type of flow, e.g., adiabatic, it is possible to calculate from the properties of the fluid the required area for the cross section of the nozzle at any point in order that the flowing fluid may just fill the provided space. From this calculation, it is found that for all compressible fluids the nozzle form must first be converging but eventually, if the pressure drops sufficiently, a place is reached where to accommodate the increased volume due to the expansion the nozzle must become diverging in form. The smallest cross section of the nozzle is called the throat, and the pressure at the throat is the **critical flow pressure** (not to be confused with the critical pressure). If the nozzle is cut off at the throat with no diverging section and the pressure at the discharge end is progressively decreased, with fixed inlet pressure, the amount of fluid passing increases until the discharge pressure equals the critical, but further decrease in discharge pressure does not result in increased flow. This is not true for thin plate orifices. For any particular gas, the ratio of critical to inlet pressure is approximately constant. For gases, $p_m/p_1 = 0.53$ approx; for saturated steam the ratio is about 0.575; and for moderately superheated steam it is about 0.55.

Formulas for Orifice Computations   The general fundamental relation is given by the energy balance $(\bar{v}_2^2 - \bar{v}_1^2)/(2g_c) = -h_{12}$. Referring to Fig. 4.1.30, let section 2 be taken at the orifice, section 3 is somewhat

Fig. 4.1.30

beyond the orifice on the downstream side, and section 1 is before the orifice on the upstream side. Then

$$\bar{v}_2 = C\sqrt{2g_c(h_1 - h_2)} \Big/ \sqrt{1 - \left(\frac{A_2}{A_1}\right)^2\left(\frac{v_1}{v_2}\right)^2}$$

The coefficient of discharge C is discussed in Secs. 3 and 16. The volume of gas passing is $\bar{v}_2 A_2$ ft^3/s, and the quantity is $\bar{v}_2 A_2 p$. For ideal gases, assuming reversible adiabatic expansion through the orifice,

$$\bar{v}_2 = C \frac{\sqrt{2g_c p_1 \bar{v}_1 \dfrac{k}{k-1}\left[1 - \left(\dfrac{p_2}{p_1}\right)^{(k-1)/k}\right]}}{\sqrt{1 - \left(\dfrac{A_2}{A_1}\right)^2\left(\dfrac{p_2}{p_1}\right)^{2/k}}}$$

$$m = CA_2 p_2 \frac{\sqrt{\dfrac{2g_c}{RT_1}\dfrac{k}{k-1}\left(\dfrac{p_1}{p_2}\right)^{(k-1)/k}\left[\left(\dfrac{p_1}{p_2}\right)^{(k-1)/k} - 1\right]}}{\sqrt{1 - \left(\dfrac{A_2}{A_1}\right)^2\left(\dfrac{p_2}{p_1}\right)^{2/k}}}$$

Often $\bar{v}_1$ is small compared with $\bar{v}_2$, and under these conditions the denominators in the preceding equations become approximately equal

Table 4.1.3   Values for Y'

	k = 1.40					k = 1.35					k = 1.30				
	d_n/d_1					d_n/d_1					d_n/d_1				
P_2/P_1	0	0.2	0.3	0.4	0.5	0	0.2	0.3	0.4	0.5	0	0.2	0.3	0.4	0.5
1.00	1.000	1.001	1.004	1.013	1.033	1.00	1.001	1.004	1.013	1.033	1.000	1.001	1.004	1.013	1.033
0.99	0.995	0.995	0.999	1.007	1.027	0.994	0.995	0.999	1.007	1.027	0.994	0.995	0.998	1.007	1.026
0.98	0.989	0.990	0.993	1.002	1.021	0.989	0.990	0.993	1.001	1.020	0.988	0.989	0.992	1.001	1.020
0.97	0.984	0.985	0.988	0.996	1.015	0.983	0.984	0.987	0.995	1.014	0.983	0.983	0.986	0.995	1.013
0.96	0.978	0.979	0.982	0.990	1.009	0.978	0.978	0.981	0.990	1.008	0.977	0.977	0.980	0.989	1.007
0.95	0.973	0.974	0.977	0.985	1.002	0.972	0.973	0.976	0.984	1.001	0.971	0.972	0.974	0.982	1.000
0.94	0.967	0.968	0.971	0.979	0.996	0.966	0.967	0.970	0.978	0.995	0.965	0.966	0.968	0.976	0.993
0.93	0.962	0.963	0.965	0.973	0.990	0.961	0.961	0.964	0.972	0.989	0.959	0.960	0.962	0.970	0.987
0.92	0.956	0.957	0.960	0.967	0.984	0.955	0.955	0.958	0.966	0.982	0.953	0.954	0.956	0.964	0.980
0.91	0.951	0.951	0.954	0.961	0.978	0.949	0.950	0.952	0.960	0.976	0.947	0.948	0.950	0.957	0.973
0.90	0.945	0.946	0.948	0.956	0.971	0.943	0.944	0.946	0.953	0.969	0.941	0.942	0.944	0.951	0.966
0.89	0.939	0.940	0.943	0.950	0.965	0.937	0.938	0.940	0.947	0.963	0.935	0.935	0.938	0.945	0.959
0.88	0.934	0.934	0.937	0.944	0.959	0.931	0.932	0.934	0.941	0.956	0.929	0.929	0.932	0.938	0.953
0.87	0.928	0.928	0.931	0.938	0.953	0.925	0.926	0.926	0.935	0.950	0.922	0.923	0.926.	0.932	0.946
0.86	0.922	0.923	0.925	0.932	0.946	0.919	0.920	0.922	0.929	0.943	0.916	0.917	0.919	0.926	0.939
0.85	0.916	0.917	0.919	0.926	0.940	0.913	0.914	0.916	0.923	0.936	0.910	0.911	0.913	0.923	0.932
0.84	0.910	0.911	0.913	0.920	0.933	0.907	0.908	0.910	0.916	0.930	0.904	0.904	0.907	0.919	0.925
0.83	0.904	0.905	0.907	0.913	0.927	0.901	0.902	0.904	0.910	0.923	0.897	0.898	0.900	0.916	0.918
0.82	0.898	0.899	0.901	0.907	0.920	0.895	0.895	0.898	0.904	0.917	0.891	0.891	0.894	0.900	0.911
0.81	0.892	0.893	0.895	0.901	0.914	0.889	0.889	0.891	0.897	0.910	0.885	0.885	0.887	0.893	0.904
0.80	0.886	0.887	0.889	0.895	0.907	0.883	0.883	0.885	0.891	0.903	0.878	0.879	0.880	0.886	0.897
0.79	0.880	0.881	0.883	0.889	0.901	0.876	0.877	0.879	0.834	0.896	0.872	0.872	0.874	0.880	0.890
0.78	0.874	0.875	0.877	0.882	0.894	0.870	0.870	0.872	0.878	0.889	0.865	0.865	0.868	0.873	0.883
0.77	0.868	0.869	0.871	0.876	0.887	0.864	0.864	0.866	0.871	0.882	0.859	0.859	0.861	0.866	0.876
0.76	0.862	0.862	0.864	0.869	0.881	0.857	0.858	0.859	0.865	0.876	0.852	0.852	0.854	0.859	0.869
0.75	0.856	0.856	0.858	0.863	0.874	0.851	0.851	0.853	0.858	0.869	0.845	0.846	0.848	0.852	0.862
0.74	0.849	0.850	0.852	0.857	0.867	0.844	0.845	0.846	0.851	0.862	0.839	0.839	0.841	0.845	0.855
0.73	0.843	0.844	0.845	0.850	0.860	0.838	0.838	0.840	0.845	0.855	0.832	0.832	0.834	0.838	0.848
0.72	0.837	0.837	0.839	0.844	0.854	0.831	0.831	0.833	0.838	0.848	0.825	0.825	0.827	0.831	0.841
0.71	0.820	0.831	0.832	0.837	0.847	0.825	0.825	0.827	0.831	0.840	0.818	0.819	0.820	0.824	0.834
0.70	0.824	0.824	0.826	0.830	0.840	0.818	0.818	0.820	0.824	0.833	0.811	0.812	0.813	0.817	0.826

If the velocity of approach is zero (as with a nozzle taking in air from the outside), d_1 is infinite and d_n/d_1 is zero.

to unity. For air, assuming $R = 53.3$, $k = 1.3937$, and $\bar{v}_1$ negligible,

$$m = 2.05 C A_2 p_2 \sqrt{(1/T_1)(p_1/p_2)^{0.283}[(p_1/p_2)^{0.283} - 1]}$$

Although the preceding formulas are generally applicable under the assumed conditions, it must be remembered that irrespective of the value of p_3, p_2 cannot become less than p_m. When p_3 is less than p_m, the flow rate becomes independent of the downstream pressure; for ideal gases,

$$m = C A_2 p_1 \sqrt{\frac{g_c}{RT_1} k \left(\frac{2}{k+1}\right)^{(k+1)(k-1)}}$$

or for air

$$m = 0.53 C p_1 \frac{A_2}{\sqrt{T_1}}$$

The following formula is useful for calculating the flow rate, in cubic feet per minute, of any gas (provided no condensation occurs) through a nozzle for pressure drops less than the critical range:

$$q_1 = \frac{31.5 C d_n^2 Y'}{\rho_1} \sqrt{\rho_1 \Delta P}$$

In this equation,

$$Y' = \left(\frac{k}{k-1}\right)^{1/2} \left(\frac{P_2}{P_1}\right)^{1/k}$$

$$\times \sqrt{\left[1 - \left(\frac{P_2}{P_1}\right)^{(k-1)/k}\right] \Big/ \left(1 - \frac{P_2}{P_1}\right)\left[1 - \left(\frac{d_n}{d_1}\right)^4 \left(\frac{P_2}{P_1}\right)^{2/k}\right]}$$

where P_1 = static pressure on upstream side of nozzle, psia; P_2 = static pressure on downstream side of nozzle, psia; d_1 = diameter of pipe upstream of nozzle, in; d_n = nozzle throat diameter, in; ρ_1 = specific

weight of gas at upstream side of nozzle, lb/ft³. Values of Y' are given in Table 4.1.3.

Where the pressure drop through the orifice is small, the hydraulic formulas applicable to incompressible fluids may be employed for gases and other compressible fluids.

In general, the formulas of the preceding section are applicable to nozzles. When so used, however, the proper value of the discharge coefficient must be employed. For steam nozzles, this may be as high as 0.94 to 0.96, although for many orifice installations it is as low as 0.50 to 0.60. Steam nozzles constitute a most important type, and calculations for these are best carried out with the aid of a Mollier or similar chart.

Formulas for Discharge of Steam   When the back pressure p_3 is less than the critical pressure p_m, the discharge depends upon the area of orifice A_2 and reservoir pressure p_1. There are three formulas widely used to express, approximately, the discharge m of **saturated** steam in terms of A_2 and p_1 as follows:

1. Napier's equation, $m = A_2 p_1/70$.
2. Grashof's formula, $m = 0.0165 A_2 p_1^{0.97}$.
3. Rateau's formula, $m = A_2 p_1 (16.367 - 0.96 \log p_1)/1,000$.

In these formulas, A_2 is to be taken in square inches, p_1 in pounds per square inch. Napier's formula is merely convenient as a rough check. Formulas 2 and 3 are applicable to well-rounded convergent orifices, in which case the coefficient of discharge may be taken as 1; i.e., no correction is required.

When the back pressure p_2 is greater than the critical flow pressure p_m, the velocity and discharge are found most conveniently from the general formulas of flow. From the steam tables or from the **Mollier chart,** find the initial enthalpy h_1 and the enthalpy h_2 after isentropic expansion; also the specific volume v_2 (see Fig. 4.1.31). Then

$$\bar{v}_2 = 223.7 \sqrt{h_1 - h_2} \quad \text{and} \quad m = A_2 \bar{v}_2/v_2$$

The same method is used in the case of steam initially superheated.

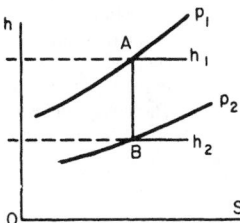

Fig. 4.1.31

EXAMPLE.   Required the discharge through an orifice ½ in diam of steam at 140 psi superheated 110°F, back pressure, 90 psia.

From the Mollier chart and the steam tables, $h_1 = 1,255.7$, $h_2 = 1,214$, $v_2 = 5.30$ ft^3.

$$\bar{v}_2 = 233.7\sqrt{1,255.7 - 1,214} = 1,455$$
$$A_2 = 0.1964 \text{ in}^2 = (0.1964/144) \text{ ft}^2$$
$$m = A_2\bar{v}_2/v_2 = (0.1964/144) \times (1,455/5.30) = 0.372 \text{ lb/s}$$

This calculation assumes ideal conditions, and the results must be multiplied by the correct coefficient of discharge to get actual results.

Flow through Converging-Diverging Nozzles   At the throat, or smallest cross section of the nozzle (Fig. 4.1.32), the pressure of saturated steam takes the value $p_m = 0.57p_1$. The quantity discharged is fixed by the area A_2 of the throat and the initial pressure p_1. For saturated steam, Grashof's or Rateau's formula (see above) may be used. The diverging part of the nozzle permits further expansion to the break pressure p_3, the velocity of the jet meanwhile increasing from $\bar{v}_m (=\bar{v}_2)$, the critical velocity at the throat, to $\bar{v}_3$ given by the fundamental equation $\bar{v}_3 = 223.7\sqrt{h_1 - h_3}$.

Fig. 4.1.32

The frictional resistances in the nozzle have the effect of decreasing the jet energy $\bar{v}_3^2/(2g_c)$ and correspondingly increasing the enthalpy of the flowing fluid. Thus, if h_3 is the enthalpy in the final state with frictionless expansion, $h_3' (>h_3)$ is the enthalpy when friction is taken into account; hence $(\bar{v}_3')^2/(2g_c) = (h_1 - h_3')$ is less than $\bar{v}_3^2/g_c = h_1 - h_3$. The loss of kinetic energy, in Btu, is $h_3' - h_3$, and the ratio of this loss to the available kinetic energy, i.e., $(h_3' - h_3)/(h_1 - h_3)$, is denoted by y.

The **design of a nozzle for a given discharge** m with pressures p_1 and p_3 is most conveniently effected with the aid of the Mollier chart. Determine p_m, the critical pressure, and h_1, h_m, h_3, assuming frictionless flow. Then

$$\bar{v}_m = 223.7\sqrt{h_1 - h_m}$$

and

$$\bar{v}_3' = 223.7\sqrt{(1 - y)(h_1 - h_3)}$$

Next find v_m and v_3'. Then, from the equation of continuity,

$$A_m = mv_m/\bar{v}_m$$

and

$$A_3' = mv_3'/\bar{v}_3'$$

The following example illustrates the method.

EXAMPLE.   Required the throat and end sections of a nozzle to deliver 0.7 lb of steam per second. The initial pressure is 160 psia, the back pressure 15 psia, and the steam is initially superheated 100°F; $y = 0.15$.

The critical pressure is $160 \times 0.55 = 88$ lb. On the Mollier chart (Fig. 4.1.33), the point A representing the initial state is located, and line of constant entropy (frictionless adiabatic) is drawn from A. This cuts the curves $p = 88$ and $p = 15$ in the points B and C, respectively. The three values of h are found to be $h_1 = 1,253$, $h_m = 1,199$, $h_3 = 1,067$. Of the available drop in enthalpy, $h_1 - h_3 = 185.5$ Btu,

Fig. 4.1.33

15 percent or 27.9 Btu is lost through friction. Hence, $CD = 27.9$ is laid off and D is projected horizontally to point C' on the curve $p = 15$. Then C' represents the final state of the steam, and the quality is found to be $x = 0.943$. The specific volume in the state C' is $26.29 \times 0.943 = 24.8$ ft^3. Likewise, the specific volume for the state B is found to be 5.29 ft^3/lb.

For the velocities at throat and end sections,

$$\bar{v}_m = 223.7\sqrt{1,253 - 1,199} = 1,643 \text{ ft/s}$$
$$\bar{v}_3 = 223.7\sqrt{185.5 - 27.9} = 2,813 \text{ ft/s}$$
$$A_m = (0.7 \times 5.29)/1,643 = 0.00225 \text{ ft}^2 = 0.324 \text{ in}^2$$
$$A_3 = (0.7 \times 24.8)/2,813 = 0.00617 \text{ ft}^2 = 0.89 \text{ in}^2$$

The diameters are $d_m = 0.643$ in and $d_3 = 1.064$ in.

Divergence of Nozzles   Figure 4.1.34 gives, for various ratios of expansion, the required "divergence" of nozzle, i.e., the ratio of the area of any section to the throat area. Thus in the case of saturated steam, if the final pressure is 1/15 of the initial pressure the ratio of the areas is 3.25. The curves apply to frictionless flow; the effect of friction is to increase the divergence.

Fig. 4.1.34

Theory of Supersaturation   Certain discrepancies between the discharge of saturated steam through an orifice as calculated from the preceding theory and the discharge actually observed are explained by a hypothesis first advanced by Martin, viz., that steam when expanded rapidly, as in turbine nozzles, becomes supersaturated; in other words, the condensation required by the ordinary theory of adiabatic expansion does not occur on account of the rapidity of the expansion.

The effect of supersaturation in turbines is a loss of energy, the amount of which may be 1.5 to 3 percent of the available energy of the steam.

Flow of Wet Steam   When the steam entering a nozzle is wet, the speed of the water particles at exit is not the same as the speed of the steam. Denoting by $\bar{v}$ the speed of the steam, the speed of the water drop is $f\bar{v}$, and f may vary perhaps 0.20 to 0.05 or less, depending on the pressure. The actual velocity $\bar{v}$ of the steam is greater than the velocity $\bar{v}_0$ calculated on the usual assumption that steam and water have the same velocity. If x is the quality of the steam, the ratio of these velocities is

$$\bar{v}/\bar{v}_0 = 1/\sqrt{x + f^2(1 - x)}$$

Thus with $x = 0.92$, $f = 0.15$, $\bar{v}/\bar{v}_0 = 1.036$. Since the discharge is practically proportional to the steam velocity, the actual discharge in this case is 3.6 percent greater than the discharge computed on the usual assumptions.

Velocity Coefficients, Loss of Energy y On account of friction losses, the actual velocity $\bar{v}$ attained by the jet is less than the velocity $\bar{v}_0$ calculated under ideal conditions. That is, $\bar{v} = x\bar{v}_0$, where x (<1) is a velocity coefficient. The coefficient x is connected with the coefficient y, giving the loss of energy, by the relation, $y = 1 - x^2$.

The elaborate and accurate experiments of the General Electric Co. on turbine nozzles give convergent nozzles values of x in excess of 0.98, with a corresponding loss of energy $y = 0.025$ to 0.04. For similar nozzles, the experiments of the Steam Nozzles Research Committee (of England) by a different method give values of x around 0.96, or $y = 0.08$. In the case of divergent nozzles, the velocity coefficient may be somewhat lower.

FLOW OF FLUIDS IN CIRCULAR PIPES

The fundamental equation as previously given on a unit weight bases, assuming the pipe horizontal, is

$$(\bar{v}\,d\bar{v}/g) + v\,dp + dF = 0$$

The friction term dF includes not only losses due to frictional flow along the pipe but also those due to fittings, valves, etc., as well as losses occasioned by any enlargement or contraction of the pipe as, for instance, the loss occurring when a fluid passes from a pipe into a tank. For long straight pipes of uniform diameter, dF is approximately equal to $2f'[\bar{v}^2 dL/(gD)]$. It is usual to express friction due to fittings, etc., in terms of additional length of pipe, adding this to the actual pipe length to get the equivalent pipe length.

Integration of the fundamental equation leads to two sets of formulas.

1. For pressure drops, small relative to the initial pressure, the specific volume v and the velocity $\bar{v}$ may be assumed constant. Then approximately

$$p_1 - p_2 = 2f'\,\bar{v}^2 L/(vgD)$$

Expressing pressure in pounds per square inch, p', the diameter in inches, and $\bar{v}$ as a function of wv/d^2, this equation becomes

$$p_1' - p_2' = 174.2 f' w^2 v L/d^5$$

2. For considerable pressure drops, when dealing with approximately isothermal flow of gases and vapors to which the gas laws are applicable, the fundamental equation on a weight basis may be integrated to give

$$p_1^2 - p_2^2 = \frac{2w^2 RT}{gA^2}\ln\frac{v_2}{v_1} + \frac{4f'RTw^2 L}{gA^2 D}$$

Coefficients of Friction   The coefficient of friction f is not a constant but is a function of the dimensionless expression $\mu/(\rho\bar{v}a)$ or $\mu v/(\bar{v}d)$, which is the reciprocal of the Reynolds number. McAdams and Sherwood formulate the expression

$$f' = 0.0054 + 0.375[\mu v/(\bar{v}d)]$$

This formula is applicable to water and other fluids. For high-pressure steam, the second term in the expression is small and f' is approximately equal to 0.0054. Babcock has suggested the approximation $f' = 0.0027(1 + 3.6/d)$ for steam.

Values of $f = 4f'$ as a function of pipe surface are given in Sec. 3.3.

For predicting the capacity of a given pipe operating on a chosen fluid with fixed pressure drop, the use of Fig. 4.1.35 eliminates the trial-and-error methods usually involved.

Resistances due to fittings, expressed in terms of L/D, are as follows: 90° elbows, 1–2½ (3–6) [7–10] in, 30 (40) [50]; 90° curves, radius of centerline of curve 2–8 pipe diameters, 10; globe valves, 1–2½ (3–6) [7–10] in, 45 (60) [75]; tees, 1–4 in, 60. The resistance in energy units, due to sudden enlargement in a pipe, is approximately $(\bar{v}_1 - \bar{v}_2)^2/(2g)$, For sudden contraction it is $1.5(1 - r)\bar{v}_2^2/[2g(3 - r)]$, where $r = A_2/A_1$. (See "Camerons Hydraulic Data," latest edition, Ingersoll Rand Co., Woodcliff Lake, NJ; Warring, "Hydraulic Handbook," 8th ed., Gulf, Houston and Trade & Tech. Press, Morden, Surrey; Houghton and

Brock, "Tables for the Compressible Flow of Dry Air," 3d ed., E. Arnold, London, with a review of basic equations and tabular data for the isentropic flow of dry air with Prandtl-Meyer expansion angles, Rayleigh flow, Fanno flow, and plane normal and oblique shock wave tables; Shapiro, "The Dynamics and Thermodynamics of Compressible Fluid Flow," Ronald Press, New York; Blevins, "Applied Fluid Dynamics Handbook," Van Nostrand Reinhold.)

Fig. 4.1.35   Chart for estimating rate of flow from the pressure gradient.

THROTTLING

Throttling or Wire Drawing   When a fluid flows from a region of higher pressure into a region of lower pressure through a valve or constricted passage, it is said to be throttled or wire-drawn. Examples are seen in the passage of steam through pressure-reducing valves, in the flow through ports and passages in the steam engine, and in the expansion valve of the refrigerating machine.

The **general equation applicable to throttling processes** is

$$(\bar{v}_2^2 - \bar{v}_1^2)/(2g_c) = h_1 - h_2$$

The velocities $\bar{v}_2$ and $\bar{v}_1$ are practically equal, and it follows that $h_1 = h_2$; i.e., in a throttling process there is no change in enthalpy.

For a mixture of liquid and vapor, $h = h_f + xh_{fg}$; hence the equation of throttling is $h_{f1} + x_1 h_{fg1} = h_{f2} + x_2 h_{fg2}$. In the case of a perfect gas, $h = c_p T + h_0$; hence the equation of throttling is $c_p T_1 + h_0 = c_p T_2 + h_0$, or $T_1 = T_2$.

Joule-Thomson Effect   The investigations of Joule and Lord Kelvin showed that a gas drops in temperature when throttled. This is not universally true. For some gases, notably hydrogen, the temperature rises for throttling processes over ordinary ranges of temperature and pressure. Whether there is a rise or fall in temperature depends on the particular range of pressure and temperature over which the change occurs. For each gas, there are different values of pressure and temperature at which no temperature change occurs during a Joule-Thomson expansion. That temperature is called the inversion temperature. Below this temperature, a gas cools on throttling; above it, its temperature rises. The ratio of the observed drop in temperature to the drop in pressure, i.e., dT/dP, is the Joule-Thomson coefficient. In actual design, the effect of heat leaks must be carefully evaluated before the theoretical Joule-Thomson coefficients are applied.

Figure 4.1.36 shows the variation of inversion temperature with pressure and temperature; the exact values are shown in Table 4.1.4. The inversion locus for air is shown in Table 4.1.5.

The cooling **effect produced by throttling** has been applied to the liquefaction of gases.

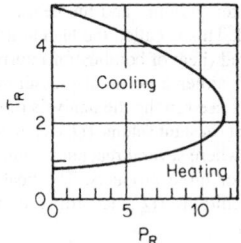

Fig. 4.1.36 Inversion curve.

Loss due to Throttling A throttling process in a cycle of operations always introduces a loss of efficiency. If T_0 is the temperature corresponding to the back pressure, the loss of available energy is the product of T_0 and the increase of entropy during the throttling process. The following example illustrates the calculation in the case of ammonia passing through the expansion valve of a refrigerating machine.

EXAMPLE. The liquid ammonia at a temperature of 70°F passes through the valve into the brine coil in which the temperature is 20 deg and the pressure is 48.21 psia. The initial enthalpy of the liquid ammonia is $h_{f1} = 120.5$, and therefore the final enthalpy is $h_{f2} + x_2 h_{fg2} = 64.7 + 553.1x_2 = 120.5$, whence $x_2 = 0.101$. The initial entropy is $s_{f1} = 0.254$. The final entropy is $s_{f2} + (x_2 h_{fg2}/T_2) = 0.144 + 0.101 \times 1.153 = 0.260$. $T_0 = 20 + 460 = 480$; hence the loss of refrigerating effect is $480 \times (0.260 - 0.254) = 2.9$ Btu.

Table 4.1.4 Approximate Inversion-Curve Locus in Reduced Coordinates ($T_R = T/T_C$; $P_R = P/P_C$)*

P_r	0	0.5	1	1.5	2	2.5	3	4
T_{RL}	0.782	0.800	0.818	0.838	0.859	0.880	0.903	0.953
T_{RU}	4.984	4.916	4.847	4.777	4.706	4.633	4.550	4.401

P_r	5	6	7	8	9	10	11	11.79
T_{RL}	1.01	1.08	1.16	1.25	1.35	1.50	1.73	2.24
T_{RU}	4.23	4.06	3.88	3.68	3.45	3.18	2.86	2.24

* Calculated from the best three-constant equation recommended by Miller, *Ind. Eng. Chem. Fundam.* **9**, 1970, p. 585. T_{RL} refers to the lower curve and T_{RU} to the upper curve.

Table 4.1.5 Approximate Inversion-Curve Locus for Air

P_r, bar	0	25	50	75	100	125	150	175	200	225
T_L, K	(112)*	114	117	120	124	128	132	137	143	149
T_U, K	653	641	629	617	606	594	582	568	555	541

P, bar	250	275	300	325	350	375	400	425	432
T_L, K	156	164	173	184	197	212	230	265	300
T_U, K	526	509	491	470	445	417	386	345	300

* Hypothetical low-pressure limit.

COMBUSTION

REFERENCES: Chigier, "Energy, Combustion and Environment," McGraw-Hill. Campbell, "Thermodynamic Analysis of Combustion Engines," Wiley. Glassman, "Combustion," Academic Press. Lefebvre, "Gas Turbine Combustion," McGraw-Hill. Strehlow, "Combustion Fundamentals," McGraw-Hill. Williams et al., "Fundamental Aspects of Solid Propellant Rockets," *Agardograph*, **116**, Oct. 1969. Basic thermodynamic table type information needed in this area is found in Glushko et al., "Thermodynamic and Thermophysical Properties of Combustion Products," Moscow, and IPST translation; Gordon, NASA Technical Paper 1906, 1982; "JANAF Thermochemical Tables," NSRDS-NBS-37, 1971.

Fuels For special properties of various fuels, see Sec. 7. In general, fuels may be classed under three headings: (1) gaseous fuels, (2) liquid fuels, and (3) solid fuels.

The combustible elements that characterize fuels are carbon, hydrogen, and, in some cases, sulfur. The complete combustion of carbon gives, as a product, carbon dioxide, CO_2; the combustion of hydrogen gives water, H_2O.

Combustion of Gaseous and Liquid Fuels

Combustion Equations The approximate molecular weights of the important elements and compounds entering into combustion calculations are:

Material	C	H_2	O_2	N_2	CO	CO_2	H_2O
Molecular weight	12	2	32	28	28	44	18

Material	CH_4	C_2H_4	C_2H_6O	S	NO	NO_2	SO_2
Molecular weight	16	28	46	32	30	46	64

For the elements C and H, the equations of complete combustion are

$$C + O_2 = CO_2 \qquad H_2 + \tfrac{1}{2}O_2 = H_2O$$
$$12\ lb + 32\ lb = 44\ lb \qquad 2\ lb + 16\ lb = 18\ lb$$

For a combustible compound, as CH_4, the equation may be written

$$CH_4 + x \cdot O_2 = y \cdot CO_2 + z \cdot H_2O$$

Taking, as a basis, 1 molecule of CH_4 and making a balance of the atoms on the two sides of the equation, it is seen that

$$y = 1 \qquad z = 2 \qquad 2x = 2y + z \qquad \text{or} \qquad x = 2$$

Hence,

$$CH_4 + 2O_2 = CO_2 + 2H_2O$$
$$16\ lb + 64\ lb = 44\ lb + 36\ lb$$

The coefficients in the combustion equation give the combining volumes of the gaseous components. Thus, in the last equation 1 ft³ of CH_4 requires for combustion 2 ft³ of oxygen and the resulting gaseous products of combustion are 1 ft³ of CO_2 and 2 ft³ of H_2O. The coefficients multiplied by the corresponding molecular weights give the combining weights. These are conveniently referred to 1 lb of the fuel. In the combustion of CH_4, for example, 1 lb of CH_4 requires 64/16 = 4 lb of oxygen for complete combustion and the products are 44/16 = 2.75 lb of CO_2 and 36/16 = 2.25 lb of H_2O.

Air Required for Combustion The composition of air is approximately 0.232 O_2 and 0.768 N_2 on a pound basis, or 0.21 O_2 and 0.79 N_2 by volume. For exact analyses, it may be necessary sometimes to take account of the water vapor mixed with the air, but ordinarily this may be neglected.

The minimum amount of air required for the combustion of 1 lb of a fuel is the quantity of oxygen required, as found from the combustion equation, divided by 0.232. Likewise, the minimum volume of air required for the combustion of 1 ft³ of a fuel gas is the volume of oxygen divided by 0.21. For example, in the combustion of CH_4 the air required per pound of CH_4 is 4/0.232 = 17.24 lb and the volume of air per cubic foot of CH_4 is 2/0.21 = 9.52 ft³. Ordinarily, more air is provided than is required for complete combustion. Let a denote the minimum amount required and xa the quantity of air admitted; then $x - 1$ is the **excess coefficient.**

Products of Combustion The products arising from the complete combustion of a fuel are CO_2, H_2O, and if sulfur is present, SO_2. Accompanying these are the nitrogen brought in with the air and the oxygen in the excess of air. Hence the products of complete combustion are principally CO_2, H_2O, N_2, and O_2. The **presence of CO indicates incomplete combustion.** In simple calculations the reaction of nitrogen with oxygen to form noxious oxides, often termed NO_x, such as nitric oxide (NO), nitrogen peroxide (NO_2), etc., is neglected. In practice, an automobile engine is run at a lower compression ratio to reduce NO, formation. The reduced pollution is bought at the expense of reduced operating efficiency. The composition of the products of combustion is readily calculated from the combustion equations, as shown by the following illustrative example. (See also Table 4.1.7.)

EXAMPLE.   A producer gas having the volume composition given is burned with 20 percent excess of air; required the volume composition of the exhaust gases.

	V
H_2	0.08
CO	0.22
CH_4	0.024
CO_2	0.066
N_2	0.61
	1.0

Coefficients in reaction equations			Coefficients multiplied by V		
O_2	CO_2	H_2O	O_2	CO_2	H_2O
0.5	0	1	0.04	0	0.08
0.5	1	0	0.11	0.22	0
2	1	2	0.048	0.024	0.048
0	1	0	0	0.066	0
0	0	0	0	0	0
			0.198	0.31	0.128

For 1 ft³ of the producer gas, 0.198 ft³ of O_2 is required for complete combustion. The minimum volume of air required is 0.198/0.21 = 0.943 ft³ and with 20 percent excess the air supplied is 0.943 × 1.2 = 1.132 ft³. Of this, 0.238 ft³ is oxygen and 0.894 ft³ is N_2. Consequently, for 1 ft³ of the fuel gas, the exhaust gas contains

CO_2		0.31 ft³
H_2O		0.128 ft³
N_2	0.61 + 0.894 =	1.504 ft³
O_2 (excess)	0.238 − 0.198 =	0.040 ft³
		1.982 ft²

or

CO_2	15.7 percent
H_2O	6.5 percent
N_2	75.8 percent
O_2	2.0 percent
	100.0 percent

Volume Contraction   As a result of chemical action, there is often a change of volume; for example, in the reaction $2H_2 + O_2 = 2H_2O$, three volumes (two of H_2 and one of O_2) contract to two volumes of water vapor. In the example just given, the volume of producer gas and air supplied is 1 ft³ gas + 1.132 ft³ air = 2.132 ft³, and the corresponding volume of the exhaust gas is 1.982 ft³, showing a contraction of about 7 percent. For a hydrocarbon having the composition C_mH_n, the relative volume contraction is $1 − n/4$; thus for CH_4 and C_2H_4 there is no change of volume, for C_2H_2 the contraction is half the volume, and for C_2H_6 there is an increase of one-half in volume.

The change of volume accompanying a chemical reaction, such as a combustion, causes a corresponding change in the gas constant R. Let R' denote the constant for the mixture of gas and air (1 lb of gas and xa lb of air) before combustion, and R'' the constant of the mixture of resulting products of combustion. Then, if y is the resulting contraction of volume, $R''/R' = (1 + xa − y)/(1 + xa)$.

Heat of Combustion   Usually, a chemical change is accompanied by the generation or absorption of heat. The union of a combustible with oxygen produces heat, and the heat thus generated when 1 lb of combustible is completely burned is called the **heat of combustion** or the **heat value** of the combustible. Heat values are determined experimentally by calorimeters in which the products of combustion are cooled to the initial temperature and the heat absorbed by the cooling medium is measured. This is called the **high heat value.**

The heat transferred (heat of combustion) during a combustion reaction is computed on either a constant-pressure or a constant-volume basis. The first law is used in the the analysis of either process.

1. The **heat value at constant volume** (Q_v). Consider a constant-volume combustion process where several reactants combine under proper conditions to form one or more products. The heat of combustion under constant-volume conditions (Q_v) according to the first law may be expressed as follows:

$$Q_v = \Sigma(Nu)_P - \Sigma(Nu)_R$$

The term N refers to the amount of material, and the symbol u signifies the internal energy per unit quantity of material. The subscripts P and R refer to the products and reactants, respectively. Hence, it may be concluded that Q_v is equal to the change in internal energy. The heat of combustion under constant-volume conditions may also be described as the quantity of heat transferred from a calorimeter to the external surroundings when the termperature and volume of the combustion products are brought to the temperature and volume, respectively, of the gaseous mixture before burning.

2. The **heat value at constant pressure** (Q_p). For a constant-pressure process the first law may be expressed as

$$Q_p = \Sigma(Nu)_P - \Sigma(Nu)_R + pV_P - pV_R$$

Here the symbols p and V refer to the pressure and total volume, respectively. Usually in combustion reactions that part of the change in internal energy resulting from a volume change is small in comparison to the total change; hence it may usually be neglected. Assuming therefore that the internal energy change for a constant-volume reaction is approximately equal to that for a constant-pressure change. the following equation results:

$$Q_p = Q_v + p(V_P - V_R)$$

Since Q_v is equal to the change in internal energy, this relation may be changed to enthalpy values, from which it may be concluded that Q_v is equal to the change in enthalpy. The heat of combustion under constant pressure conditions may also be described as the heat transferred from a calorimeter when the pressure and temperature of the products are brought back to the pressure and temperature, respectively, of the gaseous mixture before burning.

If the reactants and products are assumed to be ideal gases, then the relation for (Q_p) may be expressed as follows, where ΔN represents the change in number of moles and R the universal gas constant:

$$Q_p = Q_v + \Delta N RT$$

From this relation the heat transferred (heat of combustion) at constant pressure may be found from the heat of reaction at constant volume, or vice versa, if the temperature and molar-volume change are known.

If there is no change of volume due to the combustion, the heat values Q_p and Q_v are the same. When there is a contraction of volume, Q_p exceeds Q_v by the heat equivalent of the work done on the gas during the contraction. For example, in the burning of CO according to the equation $CO + \frac{1}{2}O_2 = CO_2$, there is a contraction of ½ volume. Taking 62°F as the temperature, the volume of 1 lb CO at atmospheric pressure is 13.6 ft³; hence the equivalent of the work done at atmospheric pressure is ½ × 13.6 × 2,116/778 = 18.5 Btu, which is about 0.4 percent of the heat value of CO. Since the difference between Q_p and Q_v is small in most fuels, it is usually neglected.

It is also to be noted that heat values vary with the initial temperature (which is also the final temperature), but the variation is usually negligible.

Heat Value per Unit Volume   Since the consumption of a fuel gas is more easily measured by volume than by mass, it is convenient to express heat values in terms of volumes. For this purpose, a standard temperature and pressure must be assumed. It is customary to take atmospheric pressure (14.70 psi) as standard, but there is diversity of

Table 4.1.6 Heats of Combustion

Fuel	Chemical symbol	High heat value		Low heat value	
		Btu/lb	*Btu/ft^3	Btu/lb	*Btu/ft^3
Carbon to CO_2	C	14,096			
Carbon to CO	C	3,960			
CO to CO_2	CO	4,346	316.0		
Sulfur to SO_2	S	3,984			
Hydrogen	H_2	61,031	319.4	51,593	270.0
Methane	CH_4	23,890	994.7	21,518	896.0
Ethane	C_2H_6	22,329	1,742.6	20,431	1,594.5
Propane	C_3H_8	21,670	2,480.1	19,944	2,282.6
Butane	C_4H_{10}	21,316	3,215.6	19,679	2,968.7
Pentane	C_5H_{12}	21,095	3,950.2	19,513	3,654.0
Hexane (liquid)	C_6H_{14}	20,675	....	19,130	
Octane (liquid)	C_8H_{18}	20,529	....	19,029	
n-Decane (liquid)	$C_{10}H_{22}$	20,371	....	19,175	
Ethylene	C_2H_4	21,646	1,576.1	20,276	1,477.4
Propene (propylene)	C_3H_6	21,053	2,299.4	19,683	2,151.3
Acetylene (ethyne)	C_2H_2	21,477	1,451.4	20,734	1,402.0
Benzene	C_6H_6	18,188	3,687.5	17,446	3,539.3
Toluene (methyl benzene)	C_7H_8	18,441	4,410.1	17,601	4,212.6
Methanol (methyl alcohol. liquid)	CH_4O	9,758	....	8,570	
Ethanol (ethyl alcohol, liquid)	C_2H_6O	12,770	....	11,531	
Naphthalene (solid)	$C_{10}H_8$	17,310	....	13,110	

* Measured as a gas at 68°F and 14.70 psia. Multiply by 1.0154 for 60°F and 14.70 psia.

practice in the matter of a **standard temperature.** The temperature of 68°F (20°C) is generally accepted in metric countries and has been recommended by the American delegates to the meeting of the International Committee of Weights and Measures and also by the ASME Power Test Codes Committee. The American Gas Assoc. uses 60°F as the standard temperature of reference. Conversion of density and heat values from 68 to 60°F of dry (saturated) gas is obtained by multiplying by the factor 1.0154 (1.0212). Conversion of specific volumes of dry (saturated) gas is obtained by multiplying by the factor 0.9848 (0.9792).

If the gas is at some other pressure and temperature, say p_1 psia and T_1 °R, the heat value per cubic foot is found by multiplying the heat value per cubic foot under standard conditions by $35.9p_1/T_1$.

The heat values of a few of the more common fuels per pound and per cubic foot are given in Table 4.1.6.

Heat Value per Unit Volume of Mixture Let a denote the volume of air required for the combustion of 1 ft^3 of fuel gas and xa the value of air actually admitted, $(x - 1)a$ being therefore the excess. Then the volume of the mixture of fuel gas and air is $1 + xa$, and the quotient $Q/(1 + xa)$ may be called the heat value per cubic foot of mixture. This magnitude is useful in comparing the relative volumes of mixture required with different fuel gases. Thus a lean gas, as blast-furnace gas or producer gas, has a low heat value Q, but the value of a is correspondingly low. On the other hand, a rich gas, like natural gas, has a high heat value but requires a large volume of air for combustion (see Table 4.1.7).

Low and High Heat Values Any fuel containing hydrogen yields water as one product of combustion. At atmospheric pressure, the partial pressure of the water vapor in the resulting combustion gas mixture will usually be sufficiently high to cause water to condense out if the temperature is allowed to fall below 120 to 140°F. This causes liberation of the heat of vaporization of any water condensed. The low heat value is evaluated assuming no water vapor condensed, whereas the high heat value is calculated assuming all water vapor condensed.

To facilitate calculations of the temperature attained by combustion, it is desirable to make use of the **low heat value.** The necessity of taking into account the heat of vaporization of the water vapor and the difference between the specific heats of liquid water and of water vapor is

thus avoided. The high heat of combustion exceeds the low heat of combustion by the difference between the heat actually given up on cooling the products to the initial temperature and that which would have been given up if the products had remained in the gaseous state. A bomb calorimeter (constant volume) gives practically correct values of the high heat value; a gas calorimeter (constant pressure) gives values which, for the usual fuels, may be incorrect by a fraction of 1 percent. The quantity to be subtracted from the high heat value to obtain the low heat value will vary with the composition of the fuel; an approximate value is $1,050m$, where m is the number of pounds of H_2O formed per pound of fuel burned.

In Germany, the low heat value of the fuel is used in calculating efficiencies of internal-combustion engines. In the United States, the high value is specified by the ASME Power Test Codes.

Heat of Formation The change in enthalpy resulting when a compound is formed from its elements isothermally and at constant pressure is numerically equal to, but of opposite sign to, the heat of formation, $\Delta H_f = -Q_f$. It is equal to the difference between the heats of combustion of the constituents forming the compound and the heat of combustion of the compound itself. The following values for heats of formation are in Btu per pound of the compound. The elements before the change and the compounds formed are assumed in their ordinary stable states at 65°F and 1 atm. A plus sign indicates heat evolved on forming the compound, a minus sign heat absorbed from the surroundings.

Fuels Methane, CH_4 (gas), 2,001.4; ethane, C_2H_6 (gas), 1,206.1; propane, C_3H_8 (vapor), 1,008.5; acetylene, C_2H_2 (gas), -3747; ethylene, C_2H_4 (gas), -805.3; benzene, C_6H_6 (vapor), -459; toluene, C_7H_8 (vapor), -234.9; methyl alcohol, CH_3OH (liquid), 3,227.3; ethyl alcohol, C_2H_5OH (liquid), 2,623.3.

Inorganic Compounds Al_2O_3, 6,710; CaO, 4,869; $CaCO_3$, 5,206; FeO, 1,611; Fe_2O_3, 2,238; Fe_3O_4, 2,075; FeS_2, 532.7; HCl (gas), 1,089; HNO_3 (liquid), 1,190; H_2O (liquid), 6,827; H_2S (gas), 279.9; H_2SO_4 (liquid), 3,555.8; K_2O, 164.7; MgO, 6,522; MnO, 2,449; NO, $-1,296$; N_2O, -803.5; Na_2O, 2,888; NH_3, 1,163; NH_4CL, 1,480; NiO, 1,407; P_2O_5, 5,394; PbO (red), 423.0; PbO_2, 489.1; SO_2, 1,933; SO_3, 2,112; SnO, 904.7; ZnO, 1,847.

Table 4.1.7 Products of Combustion

Fuel	Chemical formula	Molecular weight $O_2 = 32$	Specific weight, lb/ft³ at 68°F and 14.70 lb/in²	Volume of air necessary for combustion of unit volume of fuel at same temperature and pressure	Products of combustion of 1 ft³ of fuel in theoretical amount of air, ft³			Weight of air necessary for combustion of unit weight of fuel	Products of combustion of 1 lb of fuel in theoretical amount of air, lb		
					CO_2	H_2O	N_2		CO_2	H_2O	N_2
Oxygen	O_2	32	0.0831								
Nitrogen	N_2	28.08	0.0727								
Air			0.0753								
Hydrogen	H_2	2.016	0.0052	2.39	0	1	1.89	34.2	0.0	8.94	26.28
Steam	H_2O	18.016									
Carbon monoxide	CO	28.00	0.0727	2.39							
Carbon dioxide	CO_2	44.00	0.1142								
Methane	CH_4	16.03	0.0416	9.55	1	2	7.55	17.21	2.75	2.248	13.22
Ethane	C_2H_6	30.05	0.0779	16.71	2	3	13.21	16.07	2.93	1.799	12.34
Propane	C_3H_8	44.06	0.1142	23.87	3	4	18.87	15.65	3.00	1.635	12.02
Butane	C_4H_{10}	58.1	0.1506	30.94	4	5	24.53	15.44	3.03	1.551	11.86
Pentane	C_5H_{12}	72.1	0.1869	38.08	5	6	30.2	15.31	3.05	1.499	11.76
Hexane	C_6H_{14}	86.1	0.2232	45.3	6	7	35.8	15.22	3.07	1.465	11.69
Heptane	C_7H_{16}	100.1	0.2596	52.5	7	8	41.5	15.15	3.08	1.439	11.64
Octane	C_8H_{18}	114.1	0.2959	59.7	8	9	47.2	15.11	3.08	1.421	11.60
Nonane	C_9H_{20}	128.2	0.3323	66.8	9	10	52.8	15.07	3.09	1.406	11.57
Benzene	C_6H_6	78.0	0.2025	35.8	6	3	28.3	13.26	3.38	0.693	10.18
Toluene	C_7H_8	92.1	0.2388	42.9	7	4	34.0	13.50	3.35	0.783	10.36
Xylene	C_8H_{10}	106.2	0.2752	50.1	8	5	39.6	13.57	3.31	0.845	10.42
Cyclohexane	C_6H_{12}	84.0	0.2180	43.0	6	6	34.0	14.76	3.14	1.285	11.34
Ethylene	C_2H_4	28.03	0.0728	14.32	2	2	11.32	14.76	3.14	1.285	11.34
Propylene	C_3H_6	42.0	0.1090	21.48	3	3	16.98	14.76	3.14	1.285	11.34
Butylene	C_4H_8	64.1	0.1454	28.64	4	4	22.64	14.76	3.14	1.285	11.34
Acetylene	C_2H_2	26.02	0.0675	11.93	2	1	9.43	13.26	3.38	0.693	10.18
Allylene	C_3H_4	40.0	0.1038	19.09	3	2	15.09	13.78	3.30	0.900	10.59
Naphthalene	$C_{10}H_8$	128.1	0.3322	57.3	10	4	45.28	12.93	3.44	0.563	9.93
Methyl alcohol	CH_4O	32.0	0.0830	7.16	1	2	5.66	6.46	1.37	1.125	4.96
Ethyl alcohol	C_2H_6O	46.0	0.1194	14.32	2	3	11.32	8.99	1.91	1.174	6.90

SOURCE: Marks, "The Airplane Engine."

INTERNAL ENERGY AND ENTHALPY OF GASES

Table 4.1.8 gives the internal energy of various common gases in Btu/(lb · mol) measured above 520°R (60°F). The corresponding values of the enthalpy are obtained by adding the value of Apv from the last column.

TEMPERATURE ATTAINED BY COMBUSTION

Excluding the effect of dissociation, the temperature attained at the end of combustion may be calculated by a simple energy balance. The heat of combustion less the heat lost by conduction and radiation during the process is equal to the increase in internal energy of the products mixture if the combustion is at constant volume; or, if the combustion is at constant pressure, the difference is equal to the increase in enthalpy of the products mixture.

EXAMPLE. To calculate the temperature of combustion of a fuel gas having the composition $H_2 = 0.50$, CO = 0.46, $CO_2 = 0.04$. The gas is burned with 15 percent excess air at constant volume, and the initial temperature is 62°F; i.e., $T = 522°R$.

The volume compositions of the initial mixture of fuel gas and air and of the mixture of products are, respectively,

Initial: H_2, 0.50; CO, 0.46; CO_2, 0.04; O_2, 0.552; N_2, 2.098
Products: H_2O, 0.50; CO_2, 0.50; O_2, 0.072; N_2, 2.098

Since a volume composition is also a mol composition, the products mixture may be regarded as made up of 0.5 mol each of H_2O and CO_2, 0.072 mol of O_2, and 2.098 mols of N_2. If values are taken from Tables 4.1.6 and 4.1.7, the heat generated by combustion of the fuel mixture is $0.50 \times 2 \times 51,593 + 0.46 \times 28 \times 4,346 = 107,569$ Btu. The internal energy u of the products mixture at $T = 522$ is now calculated (Table 4.1.8). For 0.5 mol H_2O + 0.5 mol CO_2 + 0.072 mol O_2 + 2.098 mol N_2 this is $6.1 + 6.95 + 0.72 + 20.34 = 34.11$ Btu.

The energy u of the mixture is next calculated for various assumed temperatures, the proper values being taken from Table 4.1.8.

If the heat of combustion, 107,569 Btu, is entirely used in the increase of energy, the temperature attained lies somewhere between 5,000 and 5,100; by interpolation, the value 5,073° is obtained.

Loss of heat during combustion may readily be taken into account; thus if 10 percent of the heat of combustion is lost, the amount available for increasing the energy of the products is $107,569 \times 0.90 = 96,812$ Btu, and this increase gives $T_2 = 4,671°$. If the fuel is burned at constant pressure, Q_p is used instead of Q_v and values of h are determined from Table 4.1.8 instead of values of u.

T_2 assumed	4,700	4,800	4,900	5,000	5,100
Energy 0.5 mol H_2O	18,308	18,851	19,396	19,943	20,429
Energy 0.5 mol CO_2	23,742	24,388	25,035	25,683	26,331
Energy 0.072 mol O_2	1,973	2,026	2,079	2,132	2,185
Energy 2.098 mol N_2	54,596	55,018	56,447	57,882	59,321
$u_2 =$	97,619	100,283	102,957	105,640	108,329
$u_1 =$	34	34	34	34	34
	97,585	100,249	102,923	105,606	108,295

Table 4.1.8 Internal Energy of Gases
Btu/(lb · mol) above 520°R

Temp. °R	O_2	N_2	Air	CO_2	H_2O	H_2	CO	Apv
520	0	0	0	0	0	0	0	1,033
540	100	97	97	139	122	96	97	1,072
560	200	196	196	280	244	193	196	1,112
580	301	295	295	424	357	291	295	1,152
600	402	395	395	570	490	390	396	1,192
700	920	896	897	1,320	1,110	887	896	1,390
800	1,449	1,399	1,403	2,120	1,734	1,386	1,402	1,589
900	1,989	1,905	1,915	2,965	2,366	1,886	1,913	1,787
1,000	2,539	2,416	2,431	3,852	3,009	2,387	2,430	1,986
1,100	3,101	2,934	2,957	4,778	3,666	2,889	2,954	2,185
1,200	3,675	3,461	3,492	5,736	4,399	3,393	3,485	2,383
1,300	4,262	3,996	4,036	6,721	5,030	3,899	4,026	2,582
1,400	4,861	4,539	4,587	7,731	5,740	4,406	4,580	2,780
1,500	5,472	5,091	5,149	8,764	6,468	4,916	5,145	2,979
1,600	6,092	5,652	5,720	9,819	7,212	5,429	5,720	3,178
1,700	6,718	6,224	6,301	10,896	7,970	5,945	6,305	3,376
1,800	7,349	6,805	6,889	11,993	8,741	6,464	6,899	3,575
1,900	7,985	7,393	7,485	13,105	9,526	6,988	7,501	3,773
2,000	8,629	7,989	8,087	14,230	10,327	7,517	8,109	3,972
2,100	9,279	8,592	8,698	15,368	11,146	8,053	8,722	4,171
2,200	9,934	9,203	9,314	16,518	11,983	8,597	9,339	4,369
2,300	10,592	9,817	9,934	17,680	12,835	9,147	9,961	4,568
2,400	11,252	10,435	10,558	18,852	13,700	9,703	10,588	4,766
2,500	11,916	11,056	11,185	20,033	14,578	10,263	11,220	4,965
2,600	12,584	11,682	11,817	21,222	15,469	10,827	11,857	5,164
2,700	13,257	12,313	12,453	22,419	16,372	11,396	12,499	5,362
2,800	13,937	12,949	13,095	23,624	17,288	11,970	13,144	5,561
2,900	14,622	13,590	13,742	24,836	18,217	12,549	13,792	5,759
3,000	15,309	14,236	14,394	26,055	19,160	13,133	14,443	5,958
3,100	16,001	14,888	15,051	27,281	20,117	13,723	15,097	6,157
3,200	16,693	15,543	15,710	28,513	21,086	14,319	15,754	6,355
3,300	17,386	16,199	16,369	29,750	22,066	14,921	16,414	6,554
3,400	18,080	16,855	17,030	30,991	23,057	15,529	17,078	6,752
3,500	18,776	17,512	17,692	32,237	24,057	16,143	17,744	6,951
3,600	19,475	18,171	18,356	33,487	25,067	16,762	18,412	7,150
3,700	20,179	18,833	19,022	34,741	26,085	17,385	19,082	7,348
3,800	20,887	19,496	19,691	35,998	27,110	18,011	19,755	7,547
3,900	21,598	20,162	20,363	37,258	28,141	18,641	20,430	7,745
4,000	22,314	20,830	21,037	38,522	29,178	19,274	21,107	7,944
4,100	23,034	21,500	21,714	39,791	30,221	19,911	21,784	8,143
4,200	23,757	22,172	22,393	41,064	31,270	20,552	22,462	8,341
4,300	24,482	22,845	23,073	42,341	32,326	21,197	23,140	8,540
4,400	25,209	23,519	23,755	43,622	33,389	21,845	23,819	8,738
4,500	25,938	24,194	24,437	44,906	34,459	22,497	24,499	8,937
4,600	26,668	24,869	25,120	46,193	35,535	23,154	25,179	9,136
4,700	27,401	25,546	25,805	47,483	36,616	23,816	25,860	9,334
4,800	28,136	26,224	26,491	48,775	37,701	24,480	26,542	9,533
4,900	28,874	26,905	27,180	50,069	38,791	25,418	27,226	9,731
5,000	29,616	27,589	27,872	51,365	39,885	25,819	27,912	9,930
5,100	30,361	28,275	28,566	52,663	40,983	26,492	28,600	10,129
5,200	31,108	28,961	29,262	53,963	42,084	27,166	29,289	10,327
5,300	31,857	29,648	29,958	55,265	43,187	27,842	29,980	10,526
5,400	32,607	30,337	30,655	56,569	44,293	28,519	30,674	10,724

SOURCE: L. C. Lichty, "Internal Combustion Engines," p. 582, derived from data given by Hershey, Eberhardt, and Hottel, *Trans. SAE*, **31**, 1936, p. 409.

Table 4.1.9   Flame Temperatures, Deg R, at 14.7 psia, Allowing for Dissociation

Fuel	Percent of theoretical air				
	80	90	100	120	140
Hydrogen	4,210	4,330	4,390	4,000	3,670
Carbon monoxide	4,280	4,370	4,320	4,140	3,850
Methane	4,050	. . . . .	4,010	3,660	3,330
Carbureted water gas	3,940	. . . . .	4,150	3,820	3,510
Coal gas	3,920	. . . . .	4,050	3,780	3,440
Natural gas	4,010	. . . . .	4,180	3,840	3,520
Producer gas	3,040	. . . . .	3,330	3,130	2,970
Blast furnace gas	2,810	. . . . .	3,060	2,920	2,750

SOURCE: Satterfield, "Generalized Thermodynamics of High-Temperature Combustion," Sc.D. thesis, M.I.T., 1946.

The volumetric compositions of the fuels of Table 4.1.9 are given below:

Fuels	CO	H_2	CH_4	C_2H_4	Illuminants (assumed C_2H_4)	CO_2	O_2	N_2
H_2	. . . .	100.0						
CO	100.0							
CH_4	. . . .	. . . .	100.0					
Carbureted water gas	24.1	32.5	9.0	2.2	10.3	4.6	0.6	16.7
Coal gas	5.9	53.2	29.6	. . . .	2.7	1.4	0.7	6.5
Natural gas	. . . .	. . . .	78.8	14.0	. . . . . .	0.4	. . . . .	6.8
Producer gas	26.0	3.0	0.5	. . . .	. . . . . .	2.50	. . . . .	56.0
Blast furnace gas	26.5	3.5	0.2	. . . .	. . . . . .	12.8	0.1	56.9

EFFECT OF DISSOCIATION

The maximum temperature that can be obtained by the combustion of any fuel is limited by the dissociation of the products formed. The dissociation and equilibriums involved in high-temperature combustion are exceedingly complex, involving such chemical species as CO_2, CO, H_2O, H_2, H, OH, N_2, NO, N, O_2, and O. The equilibrium reached is a direct consequence of the second law of thermodynamics. However, the calculation of the equilibrium constant even for simple reactions is tedious. For all possible reactions $aA + bB \rightarrow cC + dD$ $(a, b, c, d \leq 2)$, the excellent tables of the equilibrium constant k_p contained in the "American Institute of Physics Handbook," 3d edition, McGraw-Hill, pp. 4-31 and 4-32, are recommended to save time. Papers describing the calculation of k_p for multicomponent reacting gases are contained in the first *ASME Symposium on Thermophysical Properties Proceedings*.

Calculated flame temperatures, allowing for dissociation, for gaseous fuels with stated amounts of air present are given in Table 4.1.9. The combustion is assumed to be adiabatic and at 14.7 lb/in^2 absolute.

In the case of **explosion** in the **internal-combustion engine**, the figures in Table 4.1.9 will be somewhat changed. The effect of compression is to increase both the initial temperature and the initial pressure. The resulting increase in the explosion temperature will tend to increase the dissociation; the increase of pressure will tend to reduce it. The net effect will be a small reduction.

COMBUSTION OF LIQUID FUELS

For properties of fuel oils, heat values, etc., see Sec. 7. Calculations for the burning of liquid fuels are fundamentally the same as for gaseous fuels. Liquid fuels are almost always gasified before or during actual combustion.

COMBUSTION OF SOLID FUELS

For **properties of solid fuels**, heat values, etc., see Sec. 7.

Air Required for Combustion   Let c, h, and o, denote, respectively, the parts of carbon, hydrogen, and oxygen in 1 lb of the fuel. Then the **minimum amount of oxygen** required for complete combustion is $2.67c + 8h - o$ lb, and the **minimum quantity of air** required is $a = (2.67c + 8h - o)/0.23 = 11.6[c + 3(h - o/8)]$ lb.

With air at 62°F and at atmospheric pressure, the minimum volume of air required is $v_m = 147[c + 3(h - o/8]$ ft^3. In practice, an excess of air over that required for combustion is admitted to the furnace. The actual quantity admitted per pound of fuel may be denoted by xa. Then x = amount admitted ÷ minimum amount.

Combustion Products   If v_m is the minimum volume of air required for complete combustion and xv_m the actual volume supplied, then the products will contain per pound of fuel, $O_2 = 0.21v_m(x - 1)$ ft^3, $N_2 = 0.79xv_m$ ft^3.

From the reaction equation $C + O_2 = CO_2$, the volume of CO_2 formed is equal to the volume of oxygen required for the carbon constituent alone; hence volume of $CO_2 = 0.21v_mc/[c + 3(h - 0.125o)]$.

Of the *dry* gaseous products (i.e., without water), the CO_2 content by volume is therefore given by the expression

$$CO_2 = 0.21c/[xc + (x - 0.21)3(h - 0.125o)]$$

The combined CO_2 and O_2 content is

$$CO_2 + O_2 = 0.21\left\{1 - 0.79 \middle/ \left[\frac{x + cx}{3(h - 0.125o) - 0.21}\right]\right\}$$

If the fuel is all carbon, the combined CO_2 and O_2 is by volume 21 percent of the gaseous products. The more hydrogen contained in the fuel, the smaller is the $CO_2 + O_2$ content. The CO_2 content depends in the first instance on the excess of air. Thus, for pure carbon, it is $CO_2 = 0.21/x$.

The excess of air may be calculated from the composition of the gases and that of the fuel. Thus

$$x = 0.21\left[\frac{c}{[CO_2]} + 3(h - 0.125o)\right]\middle/[c + 3(h - 0.125o)]$$

in which $[CO_2]$ denotes the percent by volume of the CO_2 in the dry gas.

The temperature of combustion is calculated by the same method as for gaseous fuels.

Loss due to Incomplete Combustion   The loss due to incomplete combustion of the carbon in the fuel, in Btu/lb of fuel, is

$$L = 10,136C \times CO/(CO + CO_2)$$

where 10,136 = difference in heat evolved in burning 1 lb of carbon to CO_2 and to CO; CO and CO_2 = percentages by volume of carbon monoxide and carbon dioxide as found by analysis; and C = fraction of quantity of carbon in the fuel which is actually burned and passes up the stack, either as CO or CO_2. The presence of 1 percent of CO in the flue gases will represent a decrease in the boiler efficiency of 4.5 percent. An additional loss is caused by passage through the grate to the ashpit of any unburned or partly burned fuel.

It is generally assumed that high CO_2 readings are indicative of good combustion and, hence, of high efficiencies. Such readings are not satisfactory when considered apart from the CO determination. The best percentage of CO_2 to maintain varies with different fuels and is lower for those with a high hydrogen content than for fuel mainly composed of carbon.

Hydrogen in a fuel increases the nitrogen content of the flue gases. This is due to the fact that the water vapor formed by the combustion of hydrogen will condense at the temperature at which the analysis is made, while the nitrogen which accompanied the oxygen maintains its gaseous form and passes in that form into the sampling apparatus. For this reason, where higbly volatile coals containing considerable hydrogen are burned, the flue gas contains an apparently increased amount of nitrogen. The effect is even more pronounced when burning gaseous or liquid hydrocarbon fuels.

The amount of flue gases per pound of fuel, including moisture formed by the hydrogen component, is approximately $3.02[N/(CO_2 + CO)]C + (1 - A)$, where A = percent of ash found in test. The quantity of dry flue gases per pound of fuel may be approximated from the formula $W_2 = C[11CO_2 + 8O + 7(CO + N)]/3(CO_2 + CO)$. In these formulas. the amount of gas is per pound of dry or moist fuel as the percentage of C is referred to a dry or moist basis.

The ratio of air suppplied per pound of fuel to the air theoretically required is

$$\frac{W_1}{W} = \frac{3.02C[N/(CO_2 + CO)]}{34.56(C/3 + H - O/8)}$$

The ratio of air supplied per pound of combustible to that theoretically required is $N/[N - 3.782(O - \frac{1}{2}CO),]$ on the assumption that all the nitrogen in the flue gas comes from the air supplied. Figure 4.1.37 gives the value of this ratio for varying flue-gas analyses where there is no CO present.

For petroleum fuels with hydrogen content from 9 to 16 percent, the excess air can be determined from the CO_2 content of the flue gases (with no CO present) by the use of Fig. 4.1.38. The curves are based on the assumption of 0.4 percent sulfur in the oil.

Fig. 4.1.37   Ratio of air supplied per pound of combustible to that theoretically required.

Fig. 4.1.38   Relation of CO_2 to excess air for fuels.

4.2 THERMODYNAMIC PROPERTIES OF SUBSTANCES
by Peter E. Liley

NOTE: Thermodynamic properties of a variety of other specific materials are listed also in Secs. 4.1, 6.1, and 9.8.

Table 4.2.1   Enthalpy and Psi Functions for Ideal-Gas Air*

T, K	h, kJ/kg	Ψ	T, K	h, kJ/kg	Ψ	T, K	h, kJ/kg	Ψ
200	200.0	−0.473	800	821.9	1.679	1,400	1,515	2.653
220	220.0	−0.329	820	844.0	1.720	1,420	1,539	2.679
240	240.1	−0.197	840	866.1	1.760	1,440	1,563	2.705
260	260.1	−0.076	860	888.3	1.800	1,460	1,587	2.730
280	280.1	0.037	880	910.6	1.838	1,480	1,612	2.755
300	300.2	0.142	900	933.0	1.876	1,500	1,636	2.779
320	320.3	0.240	920	955.4	1.914	1,520	1,660	2.803
340	340.4	0.332	940	978.0	1.950	1,540	1,684	2.827
360	360.6	0.419	960	1,000.6	1.987	1,560	1,709	2.851
380	380.8	0.502	980	1,023.3	2.022	1,580	1,738	2.875
400	401.0	0.580	1,000	1,046.1	2.057	1,600	1,758	2.898
420	421.3	0.655	1,020	1,068.9	2.091	1,620	1,782	2.921
440	441.7	0.727	1,040	1,091.9	2.125	1,640	1,806	2.944
460	462.1	0.795	1,060	1,114.9	2.158	1,660	1,831	2.966
480	482.5	0.861	1,080	1,138.0	2.190	1,680	1,855	2.988
500	503.1	0.925	1,100	1,161.1	2.223	1,700	1,880	3.010
520	523.7	0.986	1,120	1,184.3	2.254	1,720	1,905	3.032
540	544.4	1.045	1,140	1,207.6	2.285	1,740	1,929	3.054
560	565.2	1.102	1,160	1,230.9	2.316	1,760	1,954	3.075
580	586.1	1.158	1,180	1,254.3	2.346	1,780	1,979	3.096
600	607.0	1.211	1,200	1,278	2.376	1,800	2,003	3.117
620	628.1	1.264	1,220	1,301	2.406	1,840	2,053	3.158
640	649.2	1.314	1,240	1,325	2.435	1,880	2,102	3.198
660	670.5	1.364	1,260	1,349	2.463	1,920	2,152	3.238
680	691.8	1.412	1,280	1,372	2.491	1,960	2,202	3.277
700	713.3	1.459	1,300	1,396	2.519	2,000	2,252	3.215
720	734.8	1.505	1,320	1,420	2.547	2,050	2,315	3.262
740	756.4	1.550	1,340	1,444	2.574	2,100	2,377	3.408
760	778.2	1.594	1,360	1,467	2.601	2,150	2,440	3.453
780	800.0	1.637	1,380	1,491	2.627	2,200	2,504	3.496

* Values rounded off from Chappell and Cockshutt, Nat. Res. Counc. Can. Rep. NRC LR 759 (NRC No. 14300), 1974. This source tabulates values of seven thermodynamic functions at 1-K increments from 200 to 2,200 K in SI units and at other increments for two other unit systems. An earlier report (NRC LR 381, 1963) gives a more detailed description of an earlier fitting from 200 to 1,400 K. In the above table h = specific enthalpy, kJ/kg, and $\Psi_2 - \Psi_1 = \log (P_2/P_1)$, for an isentrope. In terms of the Keenan and Kaye function ϕ, $\Psi = [\log (e/R)]\phi$.

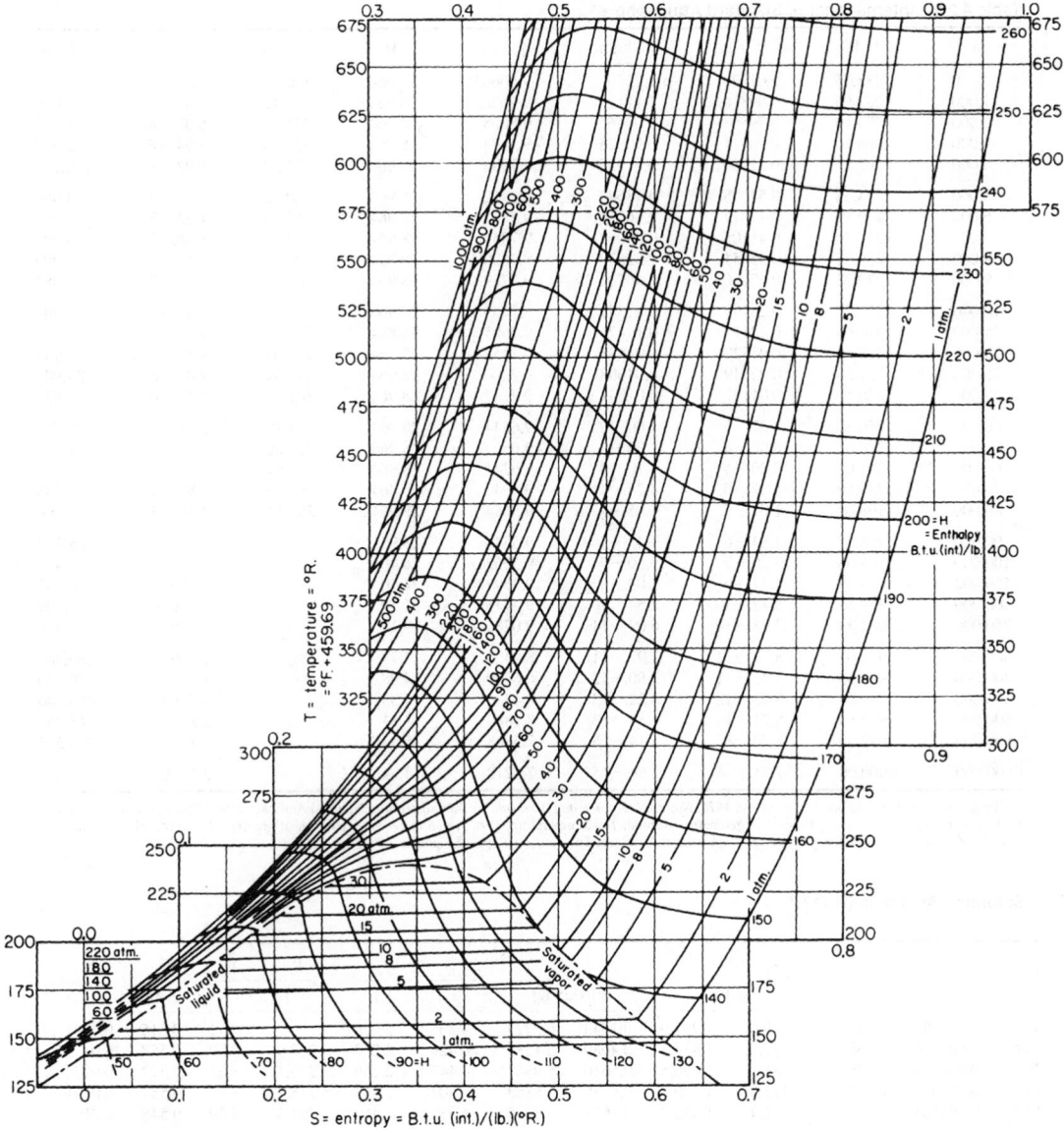

Fig. 4.2.1   Temperature-entropy diagram for air. (*Landsbaum et al, AIChE J.,* **1,** no. 3, 1955, p. 303.) *(Reproduced by permission of the authors and editor, AIChE.)*

Table 4.2.2   International ρ, Standard Atmosphere*

Z, m	T, K	P, bar	ρ, kg/m³	g, m/s²	M	a, m/s	λ, m	H, m
0	288.15	1.01325	1.2250	9.80665	28.964	340.29	6.63.−8	0
1,000	281.65	0.89876	1.1117	9.8036	28.964	336.43	7.31.−8	1,000
2,000	275.15	0.79501	1.0066	9.8005	28.964	332.53	8.07.−8	1,999
3,000	268.66	0.70121	0.90925	9.7974	28.964	328.58	8.94.−8	2,999
4,000	262.17	0.61660	0.81935	9.7943	28.964	324.59	9.92.−8	3,997
5,000	255.68	0.54048	0.73643	9.7912	28.964	320.55	1.10.−7	4,996
6,000	249.19	0.47217	0.66011	9.7882	28.964	316.45	1.23.−7	5,994
7,000	242.70	0.41105	0.59002	9.7851	28.964	312.31	1.38.−7	6,992
8,000	236.22	0.35651	0.52579	9.7820	28.964	308.11	1.55.−7	7,990
9,000	229.73	0.30800	0.46706	9.7789	28.964	303.85	1.74.−7	8,987
10,000	223.25	0.26499	0.41351	9.7759	28.964	299.53	1.97.−7	9,984
15,000	216.65	0.12111	0.19476	9.7605	28.864	295.07	4.17.−7	14,965
20,000	216.65	0.05529	0.08891	9.7452	28.964	295.07	9.14.−7	19,937
25,000	221.55	0.02549	0.04008	9.7300	28.964	298.39	2.03.−6	24,902
30,000	226.51	0.01197	0.01841	9.7147	28.964	301.71	4.42.−6	29,859
40,000	250.35	2.87.−3	4.00.−3	9.6844	28.964	317.19	2.03.−5	39,750
50,000	270.65	8.00.−4	1.03.−3	9.6542	28.964	329.80	7.91.−5	49,610
60,000	247.02	2.20.−4	3.10.−4	9.6241	28.964	315.07	2.62.−4	59,439
70,000	219.59	5.22.−5	8.28.−5	9.5942	28.964	297.06	9.81.−4	69,238
80,000	198.64	1.05.−5	1.85.−5	9.5644	28.964	282.54	4.40.−3	79,006
90,000	186.87	1.84.−6	3.43.−6	9.5348	28.95		2.37.−2	88,744
100,000	195.08	3.20.−7	5.60.−7	9.5052	28.40		0.142	98,451
150,000	634.39	4.54.−9	2.08.−9	9.3597	24.10		33	146,542
200,000	854.56	8.47.−10	2.54.−10	9.2175	21.30		240	193,899
250,000	941.33	2.48.−10	6.07.−11	9.0785	19.19		890	240,540
300,000	976.01	8.77.−11	1.92.−11	8.9427	17.73		2,600	286,480
400,000	995.83	1.45.−11	2.80.−12	8.6799	15.98		1.6.+4	376,320
500,000	999.24	3.02.−12	5.22.−13	8.4286	14.33		7.7.+4	463,540
600,000	999.85	8.21.−13	1.14.−13	8.1880	11.51		2.8.+5	548,252
800,000	999.99	1.70.−13	1.14.−14	7.7368	5.54		1.4.+6	710,574
1,000,000	1,000.00	7.51.−14	3.56.−15	7.3218	3.94		3.1.+6	864,071

* Extracted from *U.S. Standard Atmosphere, 1976, National Oceanic and Atmospheric Administration,* National Aeronautics and Space Administration and the U.S. Air Force, Washington, 1976. Z = geometric altitude, T = temperature, P = pressure, g = acceleration of gravity, M = molecular weight, a = velocity of sound, λ = mean free path, and H = geopotential altitude. The notation 1.79. −5 signifies 1.79 × 10⁻⁵.

Table 4.2.3   Saturated Ammonia (R 717)*
See Note.

P, bar	T, °C	v_f	v_g	h_f	h_g	s_f	s_g	c_{pf}	c_{pg}	μ_f	μ_g	k_f	k_g	Pr_f	Pr_g
		m³/kg		kJ/kg		kJ/(kg·K)		kJ/(kg · K)		μPa · s		W/(m · K)			
0.5	−46.5	0.001438	2.175	−9.0	1,397.9	0.1643	6.3723	4.366	2.126		7.71	0.615	0.0161		
1	−33.6	0.001466	1.138	47.9	1,418.3	0.4080	6.1286	4.429	2.233	262.9	8.09	0.588	0.0175	1.98	1.032
1.5	−25.2	0.001488	0.779	86.1	1,430.5	0.5610	5.9867	4.447	2.266	236.1	8.33	0.572	0.0184	1.84	1.046
2	−18.9	0.001507	0.595	113.8	1,439.2	0.6745	5.8863	4.507	2.393	218.4	8.52	0.554	0.0191	1.78	1.060
2.5	−13.7	0.001523	0.482	137.4	1,445.9	0.7658	5.8085	4.535	2.460	205.4	8.69	0.548	0.0199	1.70	1.074
3	−9.2	0.001536	0.406	157.5	1,451.3	0.8426	5.7449	4.561	2.521	195.0	8.81	0.539	0.0204	1.65	1.089
4	−1.9	0.001560	0.309	191.3	1,459.8	0.9680	5.6443	4.605	2.630	179.5	9.03	0.524	0.0215	1.59	1.104
5	4.1	0.001580	0.250	219.2	1,466.1	1.0692	5.5660	4.663	2.728	168.0	9.21	0.512	0.0225	1.54	1.118
6	9.3	0.001598	0.210	243.2	1,471.0	1.1546	5.5017	4.678	2.818	158.8	9.38	0.501	0.0234	1.48	1.129
8	17.9	0.001630	0.160	283.7	1,478.2	1.2946	5.3994	4.741	2.983	147.4	9.59	0.487	0.0247	1.43	1.160
10	24.9	0.001658	0.1285	317.4	1,483.0	1.4080	5.3189	4.798	3.133	134.6	9.86	0.469	0.0263	1.38	1.174
15	38.7	0.001719	0.0862	384.7	1,489.5	1.6258	5.1683	4.929	3.479	115.6	10.34	0.438	0.0292	1.34	1.233
20	49.4	0.001773	0.0644	437.9	1,491.1	1.7909	5.0564	5.057	3.809	104.9	10.68	0.417	0.0315	1.30	1.292
25	58.2	0.001823	0.0512	483.0	1,489.9	1.9259	4.9651	5.192	4.142	96.2	11.01	0.398	0.0335	1.26	1.360
30	65.8	0.001871	0.0421	522.6	1,486.7	2.0415	4.8864	5.340	4.488	89.3	11.31	0.381	0.0356	1.25	1.426
35	72.4	0.001918	0.03564	558.4	1,482.0	2.1434	4.8161	5.505	4.856	83.7	11.61	0.366	0.0375	1.26	1.50
40	78.4	0.001965	0.03069	591.5	1,476.1	2.2354	4.7516	5.692	5.255	78.8	11.90	0.352	0.0397	1.28	1.57
45	83.9	0.002012	0.02680	622.4	1,469.0	2.3198	4.6912	5.904	5.692	74.6	12.20	0.338	0.0419	1.30	1.66
50	88.9	0.002060	0.02364	651.7	1,461.0	2.3985	4.6338	6.148	6.181	70.8	12.49	0.326	0.0441	1.33	1.75
60	97.9	0.002161	0.01883	706.8	1,442.0	2.5431	4.5244	6.764	7.375	64.3	13.14	0.302	0.0489	1.36	1.98
80	112.9	0.002406	0.01253	810.6	1,390.7	2.8052	4.3076	9.005	11.548	53.4	14.78				
100	125.2	0.002793	0.00826	920.3	1,309.8	3.0715	4.4131	17.08	26.04	42.9	17.76				
113.4†	132.3	0.004260	0.00426	1,105.5	1,105.5	3.5006	3.5006								

* The T, P, v, h, and s values interpolated, rounded, and converted from "ASHRAE Handbook—Fundamentals," 1993. The c_p, μ, and k values from Liley and Desai, CINDAS Rep. 106, 1992. Similar values can be found in "ASHRAE Handbook—Fundamentals," 1993.
  † Critical point.
  NOTE: The reference point used for the data in this table is different from that used in the chart. Use property data from the table or from the chart, but *not* both.

Fig. 4.2.3 Enthalpy–log pressure diagram for ammonia (R717). The reference point used for the data in this chart is different from that used in the table. Use property data from the chart or from the table, but *not* both.

Fig. 4.2.2 Enthalpy–log pressure diagram for air.

Table 4.2.4   Saturated Carbon Dioxide*

T, K	P, bar	v_f, m³/kg	v_g, m³/kg	h_f, kJ/kg	h_g, kJ/kg	s_f, kJ/(kg·K)	s_g, kJ/(kg·K)	c_{pf}, kJ/(kg·K)
216.6	5.180	8.484.−4	0.0712	386.3	731.5	2.656	4.250	1.707
220	5.996	8.574.−4	0.0624	392.6	733.1	2.684	4.232	1.761
225	7.357	8.710.−4	0.0515	401.8	735.1	2.723	4.204	
230	8.935	8.856.−4	0.0428	411.1	736.7	2.763	4.178	1.879
235	10.75	9.011.−4	0.0357	402.5	737.9	2.802	4.152	
240	12.83	9.178.−4	0.0300	430.2	738.9	2.842	4.128	1.933
245	15.19	9.358.−4	0.0253	440.1	739.4	2.882	4.103	
250	17.86	9.554.−4	0.0214	450.3	739.6	2.923	4.079	1.992
255	20.85	9.768.−4	0.0182	460.8	739.4	2.964	4.056	
260	24.19	1.000.−3	0.0155	471.6	738.7	3.005	4.032	2.125
270	32.03	1.056.−3	0.0113	494.4	735.6	3.089	3.981	2.410
275	36.59	1.091.−3	0.0097	506.5	732.8	3.132	3.954	
280	41.60	1.130.−3	0.0082	519.2	729.1	3.176	3.925	2.887
290	53.15	1.241.−3	0.0058	547.6	716.9	3.271	3.854	3.724
300	67.10	1.470.−3	0.0037	585.4	690.2	3.393	3.742	
304.2†	73.83	2.145.−3	0.0021	636.6	636.6	3.558	3.558	∞

* The notation 8.484. − 4 signifies 8.484×10^{-4}.
† Critical point.

Table 4.2.5   Superheated Carbon Dioxide*

P, bar		Temperature, K								
	300	350	400	450	500	600	700	800	900	1,000
v	0.5639	0.6595	0.7543	0.8494	0.9439	1.1333	1.3324	1.5115	1.7005	1.8894
1 h	809.3	853.1	899.1	947.1	997.0	1,102	1,212	1,327	1,445	1,567
s	4.860	4.996	5.118	5.231	5.337	5.527	5.697	5.850	5.990	6.120
v	0.1106	0.1304	0.1498	0.1691	0.1882	0.2264	0.2645	0.3024	0.3403	0.3782
5 h	805.5	850.3	897.0	945.5	995.8	1,101	1,211	1,326	1,445	1,567
s	4.548	4.686	4.810	4.925	5.031	5.222	5.392	5.546	5.685	5.814
v	0.0539	0.0642	0.0742	0.0841	0.0938	0.1131	0.1322	0.1513	0.1703	0.1893
10 h	800.7	846.9	894.4	943.5	994.1	1,100	1,211	1,326	1,445	1,567
s	4.405	4.548	4.674	4.790	4.897	5.089	5.260	5.414	5.555	5.683
v	0.0255	0.0311	0.0364	0.0416	0.0466	0.0564	0.0661	0.0757	0.0853	0.0948
20 h	790.2	839.8	889.3	939.4	990.8	1,098	1,209	1,325	1,444	1,567
s	4.249	4.402	4.534	4.653	4.762	4.955	5.127	5.282	5.423	5.551
v	0.0159	0.0201	0.0238	0.0274	0.0309	0.0375	0.0441	0.0505	0.0570	0.0633
30 h	778.5	832.4	883.8	935.2	987.3	1,096	1,208	1,324	1,444	1,566
s	4.144	4.341	4.447	4.569	4.679	4.876	5.049	5.204	5.346	5.474
v	0.0110	0.0146	0.0175	0.0203	0.0230	0.0281	0.0331	0.0379	0.0428	0.0476
40 h	764.9	824.6	878.3	931.1	984.3	1,094	1,205	1,323	1,443	1,566
s	4.055	4.239	4.380	4.507	4.619	4.818	4.993	5.148	5.291	5.419
v	0.0080	0.0112	0.0138	0.0161	0.0183	0.0224	0.0265	0.0304	0.0343	0.0382
50 h	748.2	816.3	872.6	926.9	981.1	1,091	1,205	1,322	1,443	1,566
s	3.968	4.179	4.330	4.457	4.572	4.773	4.948	5.104	5.247	5.377
v	0.0058	0.0090	0.0113	0.0133	0.0151	0.0187	0.0221	0.0254	0.0286	0.0318
60 h	726.9	807.7	866.9	922.7	977.8	1,089	1,204	1,321	1,442	1,565
s	3.878	4.126	4.314	4.416	4.532	4.736	4.912	5.069	5.212	5.341
v		0.0062	0.0081	0.0097	0.0112	0.0140	0.0166	0.0191	0.0216	0.0240
80 h		788.4	855.1	914.2	971.3	1,085	1,201	1,320	1,441	1,565
s		4.029	4.208	4.347	4.468	4.675	4.854	5.011	5.155	5.286

Table 4.2.5   Superheated Carbon Dioxide*   (*Continued*)

P, bar		Temperature, K								
		300	350	400	450	500	600	700	800	900	1,000
100	v	0.0045	0.0062	0.0076	0.0089	0.0111	0.0133	0.0153	0.0173	0.0193
	h	766.2	843.0	905.7	964.9	1,081	1,198	1,318	1,440	1,564
	s	3.936	4.144	4.290	4.417	4.627	4.808	4.967	5.111	5.241
150	v	0.0023	0.0038	0.0049	0.0058	0.0074	0.0089	0.0103	0.0117	0.0130
	h	704.5	811.9	884.8	949.4	1,072	1,192	1,314	1,437	1,562
	s	3.716	4.005	4.177	4.313	4.536	4.722	4.884	5.030	5.162
200	v	0.0017	0.0027	0.0035	0.0043	0.0056	0.0067	0.0078	0.0088	0.0099
	h	670.0	783.2	865.2	934.9	1,063	1,186	1,310	1,435	1,561
	s	3.591	3.894	4.088	4.234	4.468	4.668	4.824	4.970	5.104
300	v		0.0018	0.0023	0.0029	0.0038	0.0046	0.0053	0.0060	0.0067
	h		745.3	834.0	910.6	1,047	1,176	1,303	1,431	1,559
	s		3.747	3.956	4.118	4.367	4.573	4.743	4.886	5.021
400	v		0.0015	0.0018	0.0022	0.0029	0.0035	0.0041	0.0047	0.0052
	h		728.1	814.6	893.3	1,035	1,168	1,298	1,428	1,558
	s		3.663	3.867	4.033	4.292	4.497	4.671	4.824	4.960
500	v			0.0016	0.0018	0.0024	0.0029	0.0034	0.0038	0.0043
	h			803.5	881.9	1,027	1,162	1,294	1,426	1,557
	s			3.805	3.970	4.234	4.443	4.620	4.774	4.913

* Interpolated and rounded from Vukalovich and Altunin, "Thermophysical Properties of Carbon Dioxide," Atomizdat, Moscow, 1965; and Collett, England, 1968. Note: v, h, and s units are the same as in Table 4.2.4.

Table 4.2.6   Saturated Iso-Butane (R 600a)*

P, bar	T, °C	v_f	v_g	h_f	h_g	s_f	s_g	c_{pf}	c_{pg}	μ_f	μ_g	k_f	k_g	Pr_f	Pr_g
		m³/kg		kJ/kg		kJ/(kg · K)		kJ/(kg · K)		μPa · s		W/(m · K)			
1	−12.13	0.001683	0.3601	288.2	655.5	3.4552	4.8626	2.24	1.56	229	6.63	0.112	0.0125	4.58	0.827
1.5	−1.42	0.001720	0.2468	312.8	668.2	3.5470	4.8615	2.30	1.63	203	6.93	0.108	0.0136	4.31	0.831
2	6.82	0.001746	0.1886	332.3	681.2	3.6166	4.8631	2.35	1.68	184	7.18	0.104	0.0145	4.16	0.832
2.5	13.60	0.001771	0.1528	348.5	690.5	3.6734	4.8658	2.38	1.73	171	7.38	0.101	0.0153	4.03	0.834
3	19.38	0.001793	0.1290	362.6	698.3	3.7214	4.8689	2.42	1.78	160	7.57	0.098	0.0159	3.95	0.847
4	29.17	0.001834	0.0978	382.3	711.5	3.8020	4.8757	2.49	1.86	144	7.89	0.094	0.0170	3.81	0.863
5	37.32	0.001870	0.0785	407.6	722.3	3.8688	4.8824	2.54	1.93	132	8.18	0.090	0.0181	3.73	0.872
6	44.28	0.001904	0.0657	425.7	731.5	3.9254	4.8889	2.60	1.99	122	8.44	0.087	0.0190	3.65	0.884
8	56.08	0.001966	0.0490	456.6	746.7	4.0213	4.9008	2.70	2.11	108	8.91	0.083	0.0206	3.51	0.913
10	65.88	0.002026	0.0389	484.2	758.8	4.1010	4.9112	2.76	2.23	97	9.34	0.079	0.0221	3.39	0.942
15	85.29	0.002220	0.0222	556.3	786.1	4.3020	4.9345	3.04	2.56	78	10.4	0.072	0.0252	3.29	1.057
20	100.38	0.002332	0.0175	588.7	795.1	4.3878	4.9405	3.38	3.01	64	11.4	0.067	0.0284	2.99	1.208
25	112.83	0.002522	0.0135	631.9	802.6	4.4980	4.9403	3.92	3.79	54	12.7	0.063	0.0326	3.36	1.476
30	123.33	0.002786	0.0095	673.7	802.6	4.6008	4.9251	6.3	7.4	44	14.3	0.061	0.0414	4.54	2.556
35	132.33	0.003312	0.0064	720.8	782.0	4.7155	4.8663			33	17.6	0.075	0.0723		
35.5†	134.85	0.004464	0.0045	752.5	752.4	4.791	4.791								

* P, T, v, h, and s are interpolated and rounded from "ASHRAE Handbook—Fundamentals," 1993. c_p, μ, and k from Liley and Desai, CINDAS Rep. 106, 1992. Substantially similar values appear in the "ASHRAE Thermophysical Properties of Refrigerants," 1993.
 † Critical point.

Table 4.2.7   Saturated Normal Hydrogen*

T	P	v_f	v_g	h_f	h_g	s_f	s_g	c_{pf}	c_{pg}
13.95	0.072	0.0130	7.974	218.3	565.4	14.08	46.64	6.36	10.52
14	0.074	0.0130	7.205	219.6	669.3	14.17	46.30	6.47	10.54
15	0.127	0.0132	4.488	226.4	678.2	14.64	44.76	6.91	10.67
16	0.204	0.0133	2.954	233.8	686.7	15.10	43.42	7.36	10.85
17	0.314	0.0135	2.032	241.6	694.7	15.57	42.23	7.88	11.07
18	0.461	0.0137	1.449	249.9	702.1	16.03	41.16	8.42	11.34
19	0.654	0.0139	1.064	258.8	708.8	16.50	40.19	8.93	11.66
20	0.901	0.0141	0.802	268.3	714.8	16.97	39.30	9.45	12.04
21	1.208	0.0143	0.618	278.4	720.2	17.44	38.49	10.13	12.49
22	1.585	0.0146	0.483	289.2	724.4	17.92	37.71	10.82	13.03
23	2.039	0.0148	0.383	300.8	727.6	18.41	36.97	11.69	13.69
24	2.579	0.0151	0.307	313.3	729.8	18.90	36.27	12.52	14.49
25	3.213	0.0155	0.243	326.7	730.7	19.41	35.58	13.44	15.52
26	3.950	0.0159	0.203	341.2	730.2	19.93	34.90	14.80	16.85
27	4.800	0.0164	0.167	357.0	728.0	20.47	34.22	16.17	18.66
28	5.770	0.0170	0.137	374.3	723.7	21.04	33.52	18.48	21.24
29	6.872	0.0177	0.113	393.6	716.6	21.65	32.80	22.05	25.19
30	8.116	0.0185	0.092	415.4	705.9	22.31	32.00	26.59	31.99
31	9.510	0.0198	0.074	441.3	689.7	23.08	31.09	36.55	46.56
32	11.068	0.0217	0.057	474.7	663.2	24.03	29.93	65.37	87.02
33.18^c	13.130	0.0318	0.032	565.4	565.4	26.68	26.68		

* T = temperature, K; P = pressure, bar; c = critical point; v = specific volume, m³/kg; h = specific enthalpy, kJ/kg; s = specific entropy, kJ/(kg · K); c_p = specific heat at constant pressure, kJ/(kg · K); subscript f represents saturated liquid and subscript g represents saturated vapor.

Table 4.2.8   Saturated Propane (R290)*

P, bar	T, °C	v_f	v_g	h_f	h_g	s_f	s_g	c_{pf}	c_{pg}	μ_f	μ_g	k_f	k_g	Pr_f	Pr_g
		m³/kg		kJ/kg		kJ/(kg · K)		kJ/(kg · K)		μPa · s		W/(m · K)			
0.5	−56.95	0.001674	0.8045	388.5	831.5	3.7263	5.7747	2.181	1.374	233	6.05	0.139	0.0101	3.66	0.823
1	−42.38	0.001721	0.4186	420.9	849.0	3.8705	5.7258	2.246	1.457	198	6.46	0.130	0.0113	3.42	0.833
1.5	−32.83	0.001755	0.2871	442.8	860.5	3.9636	5.7015	2.294	1.517	178	6.74	0.124	0.0122	3.29	0.838
2	−25.48	0.001783	0.2194	460.1	869.2	4.0341	5.6861	2.336	1.568	164	6.97	0.119	0.0130	3.22	0.841
2.5	−19.43	0.601807	0.1778	474.5	876.3	4.0916	5.6753	2.371	1.610	154	7.16	0.116	0.0136	3.15	0.848
3	−14.23	0.001828	0.1498	487.1.	882.4	4.1404	5.6672	2.406	1.652	146	7.33	0.113	0.0141	3.11	0.859
4	−5.53	0.001867	0.1138	508.5	892.3	4.2213	5.6556	2.467	1.723	134	7.61	0.108	0.0151	3.06	0.868
5	1.66	0.001901	0.0918	526.6	900.4	4.2875	5.6477	2.522	1.787	124	7.86	0.104	0.0160	3.01	0.878
6	7.82	0.001932	0.0771	542.4	907.1	4.3437	5.6418	2.574	1.847	116	8.08	0.101	0.0168	2.96	0.888
8	18.20	0.001990	0.0580	569.7	917.9	4.4379	5.6333	2.674	1.961	104	8.48	0.096	0.0182	2.90	0.914
10	26.86	0.002044	0.04609	593.1	926.4	4.5161	5.6270	2.769	2.072	95	9.04	0.092	0.0195	2.86	0.961
15	43.84	0.002173	0.03009	641.6	941.1	4.6697	5.6148	3.013	2.363	79	9.63	0.084	0.0225	2.83	1.011
20	57.14	0.002304	0.02165	682.3	949.9	4.7923	5.6026	3.290	2.717	67	10.4	6.078	0.0256	2.83	1.104
25	68.15	0.002450	0.01642	719.0	954.1	4.8979	5.5872	3.665	3.216	58	11.6	0.073	0.0295	2.91	1.265
30	77.67	0.002627	0.01269	753.8	953.8	4.9950	5.5654	4.270	4.041	50	12.4	0.071	0.0355	3.01	1.412
35	85.99	0.002866	0.00978	788.7	947.5	5.0895	5.5318	5.594	5.848	44	13.6	0.073	0.0412	3.37	1.93
40	93.38	0.00336	0.00685	830.0	928.9	5.200	5.470	12.12	14.25	33	17.8	0.084	0.0728	4.76	3.48
42.4†	96.65	0.00457	0.00457	879.2	879.2	5.330	5.330								

* The T, P, v, h, and s values are interpolated, rounded, and converted from "ASHRAE Handbook—Fundamentals," 1993. The c_p, μ, and k values are from Liley and Desai, CINDAS Rep. 106, 1992. Similar values can be found in "ASHRAE Handbook—Fundamentals," 1993.
† Critical point.

Table 4.2.9   Saturated Refrigerant 11*
See Note.

P, bar	T, °C	v_f	v_g	h_f	h_g	s_f	s_g	c_{pf}	c_{pg}	μ_f	μ_g	k_f	k_g	Pr_f	Pr_g
		m³/kg		kJ/kg		kJ/(kg · K)		kJ/(kg · K)		μPa · s		W/(m · K)			
0.5	5.18	0.000657	0.3298	204.5	392.4	1.0162	1.6916	0.873	0.560	537	10.3	0.094	0.0083	4.35	0.695
1	23.55	0.000676	0.1731	220.3	401.8	1.0711	1.6833	0.887	0.580	440	11.0	0.090	0.0088	4.01	0.725
1.5	35.26	0.000689	0.1186	230.9	407.9	1.1060	1.6800	0.896	0.594	385	11.4	0.089	0.0091	3.80	0.736
2	44.42	0.000700	0.0908	239.1	412.5	1.1322	1.6783	0.905	0.604	341	11.8	0.085	0.0096	3.63	0.740
2.5	51.90	0.000709	0.0737	245.9	416.3	1.1532	1.6774	0.913	0.614	315	12.0	0.083	0.0099	3.47	0.745
3	58.37	0.000718	0.0620	251.8	419.4	1.1711	1.6768	0.920	0.623	288	12.3	0.081	0.0102	3.27	0.751
4	69.18	0.000733	0.0470	261.9	424.7	1.2008	1.6764	0.934	0.637	252	12.8	0.079	0.0107	2.98	0.762
5	78.07	0.000746	0.0380	270.2	428.9	1.2247	1.6763	0.947	0.652	228	13.1	0.077	0.0111	2.80	0.770
6	85.76	0.000759	0.0317	277.6	432.4	1.2482	1.6764	0.959	0.669	211	13.5	0.076	0.0115	2.66	0.779
8	98.59	0.000781	0.0239	290.1	438.0	1.2790	1.6768	0.983	0.689	187	14.1	0.073	0.0122	2.52	0.792
10	109.3	0.000802	0.0190	300.8	442.3	1.3069	1.6771	1.008	0.713	170	14.6	0.070	0.0129	2.45	0.807
15	130.3	0.000853	0.0124	322.6	449.9	1.3614	1.6770	1.076	0.783	138	15.7	0.065	0.0143	2.28	0.860
20	146.6	0.000903	0.0090	340.5	454.5	1.4038	1.6754	1.153	0.876	118	16.9	0.062	0.0158	2.19	0.937
25	160.2	0.000959	0.0068	356.4	457.0	1.4399	1.6721	1.256	1.021	101	18.1	0.059	0.0174	2.15	1.062
30	171.9	0.001024	0.0053	371.1	457.6	1.4722	1.6670	1.384	1.317	86	19.3	0.058	0.0193	2.13	1.317
35	182.2	0.001105	0.0042	385.5	456.1	1.5032	1.6583	1.82	1.84						
40	191.3	0.001246	0.0031	401.1	451.3	1.5352	1.6432	2.95	2.31						
44.1†	198.0	0.00181	0.0018	428.6	428.6	1.5933	1.5933								

* The T, P, v, h, and s values are interpolated, converted, and rounded from "ASHRAE Handbook—Fundamentals," 1993. The c_p, μ, and k are from Liley and Desai, CINDAS Rep. 106, 1992. Similar values appear in "ASHRAE Handbook—Fundamentals," 1993.
† Critical point.
NOTE: The reference point used for the data in this table is different from that used in the chart. Use property data from the table or from the chart, but *not* both.

Fig. 4.2.4   Enthalpy–log pressure diagram for refrigerant 11.1 MPa = 10 bar. *(Copyright 1981 by ASHRAE and reproduced by permission.)* The reference point used for the data in this chart is different from that used in the table. Use property data from the chart or from the table, but *not* both.

Table 4.2.10   Saturated Refrigerant 12*
See Note.

$P,$ bar	$T,$ °C	v_f	v_g	h_f	h_g	s_f	s_g	c_{pf}	c_{pg}	μ_f	μ_g	k_f	k_g	Pr_f	Pr_g
		m³/kg		kJ/kg		kJ/(kg · K)		kJ/(kg · K)		μPa · s		W/(m · K)			
0.5	−45.24	0.000653	0.3072	159.5	331.5	0.8386	1.5936	0.883	0.545	420	9.8	0.0940	0.0062	4.08	0.861
1	−30.11	0.000672	0.1611	172.7	338.8	0.8947	1.5780	0.895	0.575	358	10.4	0.0883	0.0070	3.63	0.854
1.5	−20.15	0.000686	0.1103	181.6	345.3	0.9304	1.5702	0.906	0.595	322	11.0	0.0846	0.0075	3.45	0.873
2	−12.52	0.000697	0.0843	188.5	347.0	0.9571	1.5654	0.913	0.612	297	11.4	0.0817.	0.0079	3.32	0.883
2.5	−6.22	0.000707	0.0682	194.2	349.9	0.9788	1.5619	0.921	0.627	277	11.7	0.0793	0.0083	3.22	0.834
3	−0.84	0.000715	0.0574	199.2	352.3	0.9972	1.5594	0.929	0.640	262	12.0	0.0774	0.0086	3.13	0.893
4	8.19	0.000731	0.0436	207.7	356.3	1.0275	1.5556	0.945	0.663	238	12.4	0.0739	0.0091	3.04	0.903
5	15.64	0.000744	0.0351	214.8	359.4	1.0522	1.5530	0.959	0.683	221	12.8	0.0714	0.0095	2.97	0.920
6	22.01	0.000757	0.0294	220.9	362.0	1.0731	1.5510	0.969	0.702	207	13.2	0.0692	0.0098	2.90	0.946
8	32.79	0.000780	0.0221	231.7	366.2	1.1082	1.5479	0.995	0.738	186	13.9	0.0653	0.0105	2.83	0.977
10	41.70	0.000802	0.0176	240.8	369.4	1.1370	1.5455	1.021	0.773	170	14.5	0.0621	0.0111	2.80	1.01
15	59.30	0.000854	0.0141	259.6	374.7	1.1938	1.5400	1.107	0.868	143	15.9	0.0558	0.0125	2.84	1.10
20	72.99	0.000907	0.0082	275.2	377.5	1.2386	1.5341	1.225	0.993	124	17.3	0.0512	0.0137	2.97	1.25
25	84.33	0.000967	0.0062	289.2	378.4	1.2770	1.5265	1.36	1.029	108	18.9	0.0469	0.0151	3.13	1.40
30	94.05	0.001040	0.0048	302.4	377.3	1.3120	1.5162	1.51	1.55	92	20.7	0.0429	0.0167	3.24	1.92
35	102.6	0.001141	0.0036	315.7	373.5	1.3437	1.4975		2.50	75	23.2	0.0389	0.0191	3.04	3.04
40	110.1	0.001360	0.0025	332.3	362.5	1.3871	1.4659		10.9	55	28.2	0.0346	0.0222		
41.2†	111.8	0.001771	0.0018	347.4	347.4	1.4272	1.4272								

* The T, P, v, h, and s values are interpolated, converted, and rounded from "ASHRAE Handbook—Fundamentals," 1993. The c_p, μ, and k values are from Liley and Desai, CINDAS Rep. 106, 1992. Similar values appear in "ASHRAE Handbook—Fundamentals," 1993.
† Critical point.
NOTE: The reference point used for the data in this table is different from that used in the chart. Use property data from the table or from the chart, but *not* both.

Fig. 4.2.5 Enthalpy–log pressure diagram for refrigerant 12. Prepared at the Center for Applied Thermodynamic Studies, University of Idaho, Moscow. *(Copyright by ASHRAE and reproduced by permission.)* The reference point used for the data in this chart is different from that used in the table. Use property data from the chart or from the table, but *not* both.

Table 4.2.11   Saturated Refrigerant 22*
See Note.

P, bar	T, °C	v_f	v_g	h_f	h_g	s_f	s_g	c_{pf}	c_{pg}	μ_f	μ_g	k_f	k_g	Pr_f	Pr_g
		m³/kg		kJ/kg		kJ/(kg · K)		kJ/(kg · K)		μPa · s		W/(m · K)			
0.5	−54.80	0.000690	0.4264	138.7	381.1	0.7510	1.8619	1.080	0.574			0.121	0.0060		
1	−41.39	0.000709	0.2153	153.6	387.6	0.8173	1.8256	1.092	0.605			0.114	0.0069		
1.5	−32.07	0.000723	0.1472	163.6	393.5	0.8591	1.8056	1.104	0.630			0.110	0.0075		
2	−25.19	0.000734	0.1125	171.2	394.7	0.8902	1.7919	1.115	0.650			0.107	0.0079		
2.5	−19.52	0.000743	0.0910	177.5	397.2	0.9155	1.7814	1.126	0.669	258.6	11.02	0.105	0.0083	2.78	0.888
3	−14.66	0.000752	0.0766	183.1	399.2	0.9368	1.7730	1.136	0.686	245.9	11.21	0.103	0.0086	2.71	0.894
4	−6.57	0.000767	0.0582	192.3	402.4	0.9718	1.7599	1.155	0.716	225.6	11.54	0.099	0.0091	2.63	0.907
5	0.11	0.000780	0.0469	200.1	404.9	1.0005	1.7498	1.171	0.745	209.9	11.80	0.096	0.0095	2.56	0.924
6	5.85	0.000789	0.0392	206.9	405.7	1.0327	1.7417	1.189	0.771	197.2	11.97	0.094	0.0099	2.50	0.936
8	15.44	0.000815	0.0295	218.5	410.0	1.0650	1.7287	1.221	0.819	177.7	12.42	0.090	0.0104	2.42	0.974
10	23.39	0.000835	0.0236	228.3	412.3	1.0981	1.7185	1.252	0.871	163.0	12.82	0.086	0.0109	2.36	1.026
15	39.07	0.000883	0.0155	248.5	415.7	1.1628	1.6985	1.332	1.000	137.7	13.9	0.080	0.0118	2.29	1.12
20	51.23	0.000929	0.0113	265.0	417.1	1.2132	1.6822	1.426	1.149						
25	61.33	0.000978	0.0087	279.6	417.0	1.2560	1.6670	1.550	1.341						
30	70.05	0.001030	0.0068	301.3	415.7	1.2942	1.6517	1.613	2.070						
35	77.70	0.001087	0.0056	305.0	413.7	1.3275	1.6371	2.03	2.05						
40	84.53	0.001174	0.0044	318.9	408.3	1.3648	1.6150	2.67	2.96						
45	90.67	0.001326	0.0033	335.8	397.9	1.4100	1.5810	4.47	5.19						
49.9†	96.14	0.001909	0.0019	366.6	366.6	1.4918	1.4918								

* Values are interpolated and rounded from "ASHRAE Handbook—Fundamentals," 1993.
† Critical point.
NOTE: The reference point used for the data in this table is different from that used in the chart. Use property data from the table or from the chart, but *not* both.

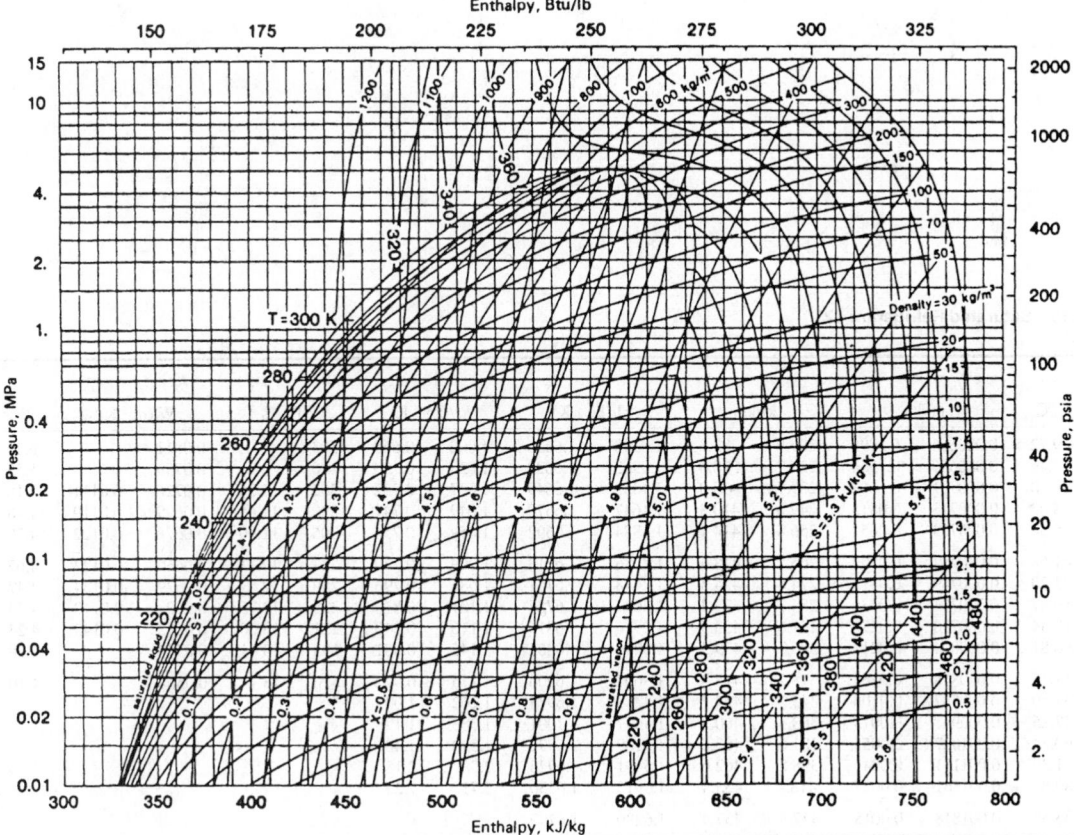

Fig. 4.2.6   Enthalpy–log pressure diagram for refrigerant 22. 1 MPa = 10 bar. *(Copyright 1981 by ASHRAE and reproduced by permission.)* The reference point used for the data in this chart is different from that used in the table. Use property data from the chart or from the table, but *not* both.

Table 4.2.12 Saturated Refrigerant 32*

P, bar	T, °C	v_f	v_g	h_f	h_g	s_f	s_g	c_{pf}	c_{pg}	μ_f	μ_g	k_f	k_g	Pr_f	Pr_g
		m³/kg		kJ/kg		kJ/(kg·K)		kJ/(kg·K)		μPa·s		W/(m·K)			
1	−51.68	0.000832	0.3361	114.6	497.5	0.6565	2.3855	1.559	0.873	278.5	10.50	0.189	0.0082	2.30	1.12
1.5	−43.66	0.000847	0.2394	127.2	501.5	0.7123	2.3435	1.576	0.911	251.6	10.61	0.181	0.0085	2.19	1.14
2	−37.35	0.000860	0.1773	137.3	504.5	0.7555	2.3127	1.601	0.955	226.6	10.70	0.175	0.0088	2.10	1.16
2.5	−32.16	0.000870	0.1433	145.7	506.8	0.7906	2.2888	1.615	1.003	218.8	10.77	0.171	0.0091	2.06	1.18
3	−27.74	0.000880	0.1205	152.7	508.6	0.8202	2.2693	1.627	1.020	207.7	10.83	0.167	0.0094	2.02	1.18
4	−20.39	0.000897	0.0914	165.2	511.3	0.8689	2.2383	1.653	1.074	192.7	10.93	0.161	0.0100	1.98	1.18
5	−14.34	0.000912	0.0736	175.3	513.3	0.9084	2.2140	1.678	1.123	180.8	11.02	0.156	0.0104	1.94	1.19
6	−9.16	0.000925	0.0616	184.2	514.7	0.9418	2.1940	1.701	1.169	170.1	11.10	0.152	0.0109	1.90	1.19
8	−0.51	0.000950	0.0463	199.1	516.7	0.9968	2.1616	1.743	1.255	154.8	11.32	0.145	0.0117	1.86	1.21
10	6.63	0.000972	0.0369	211.7	517.8	1.0415	2.1356	1.784	1.337	142.4	11.53	0.139	0.0124	1.83	1.24
15	20.64	0.001023	0.0242	237.0	518.3	1.1282	2.0855	1.895	1.541	120.7	12.08	0.128	0.0141	1.79	1.32
20	31.45	0.001072	0.0176	257.5	516.7	1.1949	2.0460	2.009	1.761	105.7	12.67	0.119	0.0156	1.78	1.43
25	40.36	0.001122	0.0136	275.3	513.7	1.2506	2.0112	2.151	2.026	94.0	13.29	0.112	0.0171	1.81	1.57
30	48.00	0.001175	0.0102	291.4	509.4	1.2997	1.9786	2.314	2.352	84.9	13.98	0.107	0.0186	1.84	1.77
35	54.69	0.001232	0.0088	306.6	503.9	1.3447	1.9463	2.524	2.791	77.7	14.74	0.101	0.0200	1.94	2.06
40	60.66	0.001299	0.0072	322.1	496.7	1.3876	1.9128	2.744	3.367	71.4	15.60	0.095	0.0215	2.06	2.44
45	66.05	0.001380	0.0060	336.5	487.8	1.4304	1.8763		4.49	66.0	16.61	0.089	0.0191		3.90
50	70.95	0.001490	0.0048	352.8	475.6	1.4759	1.8328			61.8	17.85	0.082	0.0167		
58.6†	78.41	0.002383	0.0024	413.8	413.8	1.6465	1.6465								

* The P, T, v, h, and s values are interpolated and rounded from "ASHRAE Handbook—Fundamentals," 1993; c_p values are interpolated and converted from Defbaugh et al., *J. Chem. Eng. Data*, **39**, 1994, pp. 333–340; μ_f and μ_g are interpolated from Oliveira and Wakeham, *Int. J. Thermophys.*, **14**, no. 6, 1993, pp. 1131–1143.
† Critical point.

Table 4.2.13 Saturated Refrigerant 123*
See Note.

P, bar	T, °C	v_f	v_g	h_f	h_g	s_f	s_g	c_{pf}	c_{pg}	μ_f	μ_g	k_f	k_g	Pr_f	Pr_g
		m³/kg		kJ/kg		kJ/(kg·K)		kJ/(kg·K)		μPa·s		W/(m·K)			
0.5	9.72	0.000666	0.2995	209.9	387.3	1.0348	1.6626	1.002	0.668	503		0.0811		6.21	
1	27.46	0.000686	0.1564	227.7	398.0	1.0963	1.6629	1.023	0.700	410		0.0759		5.53	
1.5	39.10	0.000701	0.1068	239.7	405.0	1.1353	1.6649	1.037	0.722	362	11.36	0.0726	0.0111	5.16	0.738
2	48.05	0.000713	0.0813	249.0	410.3	1.1647	1.6670	1.049	0.741	330	11.65	0.0699	0.0119	4.78	0.728
2.5	55.38	0.000723	0.0657	256.8	414.7	1.1884	1.6692	1.059	0.756	306	11.90	0.0678	0.0122	4.78	0.736
3	61.68	0.000732	0.0552	263.5	418.4	1.2086	1.6713	1.069	0.770	287	12.11	0.0660	0.0127	4.65	0.739
4	73.20	0.000749	0.0417	274.8	424.5	1.2418	1.6751	1.086	0.796	259	12.47	0.0629	0.0134	4.47	0.742
5	80.87	0.000764	0.0335	284.3	429.4	1.2687	1.6784	1.101	0.818	238	12.75	0.0604	0.0140	4.33	0.746
6	88.34	0.000778	0.0280	292.6	433.5	1.2916	1.6815	1.116	0.840	221	13.00	0.0582	0.0145	4.23	0.754
8	100.81	0.000804	0.0217	306.7	439.3	1.3246	1.6865	1.145	0.881	195	13.41	0.0547		4.08	
10	111.15	0.000828	0.0165	318.7	445.5	1.3607	1.6906	1.175	0.922	176	13.74	0.0504		4.10	
15	131.50	0.000887	0.0106	343.2	454.7	1.4218	1.6974	1.262	1.037						
20	147.25	0.000951	0.0075	363.4	460.3	1.4697	1.7001	1.383	1.198						
25	160.24	0.001027	0.0055	381.6	463.0	1.5109	1.6990	1.590	1.481						
30	171.30	0.001131	0.0041	398.8	462.6	1.5491	1.6926	2.08	2.18						
35	180.88	0.001361	0.0027	418.4	455.4	1.5915	1.6730	5.71	7.22						
36.6†	183.68	0.001818	0.0018	437.4	437.4	1.6290	1.6290								

* The P, T, v, h, s, and c_p values are interpolated and rounded from Younglove and McLinden, *J. Phys. Chem. Ref. Data*, **23**, no. 5, 1994, pp. 731–779; μ and k interpolated from "ASHRAE Handbook—Fundamentals," 1993.
† Critical point.
NOTE: The reference point used for the data in this table is different from that used in the chart. Use property data from the table or from the chart, but *not* both.

Fig. 4.2.7 Enthalpy–log pressure diagram for refrigerant 123. 1 MPa = 10 bar. *(Reprinted by permission of the ASHRAE from the 1993 "ASHRAE Handbook—Fundamentals.")* The reference point used for the data in this chart is different from that used in the table. Use property data from the chart or from the table, but *not* both.

Table 4.2.14   Saturated Refrigerant 134a*
See Note.

P, bar	T, K	Spec. vol., m³/kg		Enthalpy, kJ/kg		Entropy, kJ/(kg · K)		Spec. ht. const., P, kJ/(kg · K)	
		Liquid	Vapor	Liquid	Vapor	Liquid	Vapor	Liquid	Vapor
0.5	232.7	0.000706	0.3572	148.4	374.1	0.7966	1.7640	1.254	0.748
1	246.8	0.000726	0.1926	165.4	382.6	0.8675	1.7474	1.279	0.793
1.5	256.0	0.000741	0.1313	177.4	388.3	0.9148	1.7388	1.299	0.826
2	263.1	0.000753	0.0999	186.6	392.6	0.9502	1.7334	1.315	0.854
2.5	268.9	0.000764	0.0807	194.3	396.1	0.9789	1.7295	1.330	0.878
3	273.8	0.000774	0.0677	200.9	399.0	1.0032	1.7267	1.343	0.900
4	282.1	0.000791	0.0512	212.1	403.7	1.0433	1.7225	1.367	0.940
5	288.9	0.000806	0.0411	221.5	407.5	1.0759	1.7196	1.389	0.976
6	294.7	0.000820	0.0343	229.7	410.6	1.1037	1.7184	1.411	1.010
8	304.5	0.000846	0.0256	243.6	415.5	1.1497	1.7140	1.453	1.075
10	312.5	0.000870	0.0203	255.5	419.2	1.1876	1.7112	1.495	1.139
15	328.4	0.000928	0.0131	279.8	425.2	1.2621	1.7048	1.611	1.313
20	340.6	0.000989	0.00929	300.0	428.3	1.3208	1.6975	1.761	1.539
25	350.7	0.001057	0.00694	317.8	429.0	1.3711	1.6880	1.983	1.647
30	359.4	0.001141	0.00528	334.7	427.3	1.4170	1.6748	2.388	2.527
35	366.9	0.001263	0.00237	351.9	422.2	1.4626	1.6549	3.484	4.292
40	373.5	0.001580	0.00256	375.6	405.4	1.5247	1.6045	26.33	37.63
40.6†	374.3	0.001953	0.00195	389.6	389.6	1.5620	1.5620		

* Values are rounded, converted, and interpolated from Tillner-Roth and Baehr, *J. Phys. Chem. Ref. Data,* **23,** no. 5, 1994, pp. 657–730. Liquid enthalpy and entropy at 0°C = 273.15 K are 200 kJ/kg and 1.0000 kJ/kg · K, respectively.
† Critical point.
NOTE: The reference point used for the data in this table is different from that used in the chart. Use property data from the table or from the chart, but *not* both.

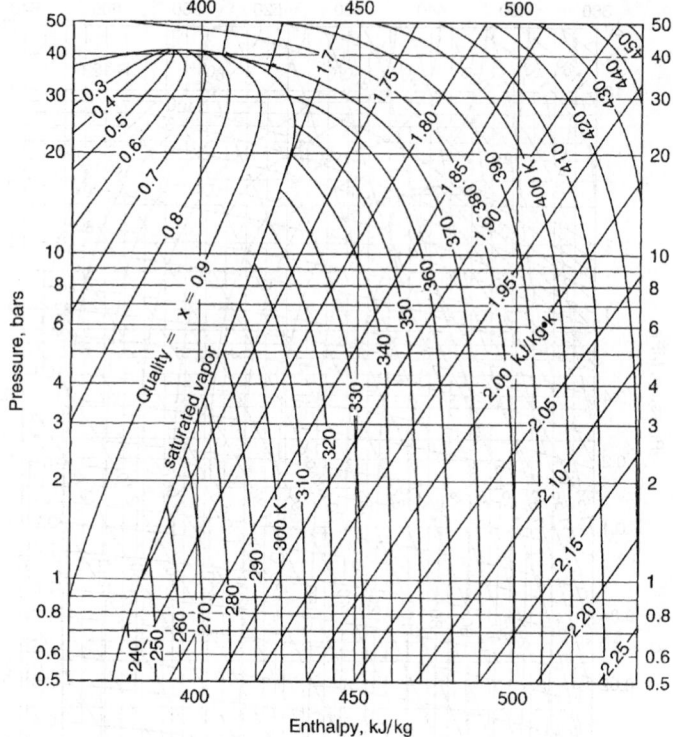

Fig. 4.2.8 Enthalpy–log pressure diagram for refrigerant 134a. The reference point used for the data in this chart is different from that used in the table. Use property data from the chart or from the table, but *not* both.

Table 4.2.15   Saturated Refrigerant 143a*

P, bar	T, °C	v_f	v_g	h_f	h_g	s_f	s_g	c_{pf},	μ_f,	k_f,	Pr_f
		m³/kg		kJ/kg		kJ/(kg · K)		kJ/(kg · K)	μPa · s	W/(m · K)	
0.5	−61.06	0.000840	0.4160	115.8	352.1	0.6541	1.769	1.291	314.6	0.1214	3.34
1	−47.49	0.000861	0.1977	136.4	361.9	0.7474	1.738	1.327	262.4	0.1121	3.11
1.5	−38.59	0.000875	0.1486	145.5	366.1	0.7865	1.728	1.346	243.4	0.1080	3.03
2	−31.77	0.000889	0.1113	154.7	370.2	0.8253	1.718	1.366	225.6	0.1039	2.97
2.5	−26.16	0.000903	0.0908	162.4		0.8567	1.710	1.384	212.5	0.1005	2.93
3	−21.35	0.000915	0.0755	169.2	376.3		1.706	1.401	201.8	0.0976	2.90
4	−13.34	0.000936	0.0566	180.5	380.8	0.9277	1.698	1.430	185.6	0.0928	2.86
5	−6.72	0.000955	0.0461	190.1	384.3	0.9636	1.693	1.459	173.7	0.0889	2.85
6	−1.04	0.000973	0.0382	198.5	387.2	0.9944	1.688	1.486	163.6	0.0856	2.84
8	8.47	0.001005	0.0284	212.8	391.8	1.0457	1.681	1.538	147.0	0.0800	2.83
10	16.34	0.001036	0.0224	225.1	395.2	1.088	1.675	1.590	134.2	0.0755	2.83
15	31.80	0.001112	0.0144	250.6	400.3	1.171	1.664	1.741	113.0	0.0667	2.95
20	43.75	0.001191	0.0102	271.8	402.4	1.238	1.651	1.943	99.1	0.0598	3.22
25	53.83	0.001288	0.0074	291.4	401.5	1.297	1.634	2.369	87.0	0.0538	3.83
30	61.96	0.001405	0.0056	308.9	398.0	1.350	1.616	2.93			
35	69.26	0.001616	0.0040	329.8	338.9	1.403	1.576				
38.3†	73.60	0.002311	0.0023	360.6	360.6	1.471	1.471				

* The P, T, v, h, and s values are iterpolated from a tabulation as a function of temperature supplied by Dr. Friend, NIST, Boulder, CO, based on REFROP 5.
† Critical point.

Table 4.2.16 Saturated Refrigerant 152a*

See Note.

P, bar	T, °C	v_f	v_g	h_f	h_g	s_f	s_g	c_{pf}	c_{pg}	μ_f	μ_g	k_f,	Pr_f
		m³/kg		kJ/kg		kJ/(kg · K)		kJ/(kg · K)		μPa · s		W/(m · K)	
0.5	−38.88	0.000960	0.5737	135.9	478.5	0.7477	2.2103	1.593	0.951	346.9			
1	−24.29	0.000989	0.2997	159.4	489.3	0.8448	2.1710	1.632	1.030	286.1	8.65	0.1280	3.65
1.5	−13.05	0.001012	0.1923	178.0	497.5	0.9177	2.1464	1.665	1.098	249.2	8.98	0.1231	3.37
2	−7.68	0.001025	0.1569	187.1	501.4	0.9523	2.1360	1.682	1.134	233.8	9.14	0.1199	3.28
2.5	−1.60	0.001039	0.1267	197.3	505.6	0.9900	2.1254	1.702	1.175	218.3	9.32	0.1164	3.18
3	3.59	0.001052	0.1057	206.2	509.1	1.0222	2.1169	1.720	1.211	205.9	9.47	0.1134	3.10
4	12.12	0.001074	0.0800	221.8	514.7	1.0747	2.1042	1.751	1.275	187.5	9.67	0.1084	3.03
5	19.14	0.001094	0.0644	233.4	519.1	1.1173	2.0947	1.781	1.331	174.2	9.91	0.1046	2.97
6	25.17	0.001113	0.0546	244.3	522.7	1.1527	2.0871	1.809	1.383	163.6	10.12	0.1010	2.93
8	35.25	0.001146	0.0402	262.8.	528.4	1.2140	2.0752	1.862	1.314	147.4	10.48	0.0953	2.88
10	43.57	0.001177	0.0320	278.5	532.6	1.2635	2.0658	1.915	1.567	135.5	10.81	0.0907	2.86
15	59.95	0.001251	0.0207	310.7	539.2	1.3608	2.0470	2.055	1.800	114.6	11.56	0.0820	2.87
20	72.61	0.001325	0.0148	337.3	542.2	1.4371	2.0299	2.224	2.083		12.32	0.0758	
25	83.07	0.001405	0.0112	360.7	542.5	1.5021	2.0125	2.456	2.476		13.18		
30	92.03	0.001498	0.0087	382.6	540.3	1.5605	1.9927	2.814	3.101				
35	99.87	0.001615	0.0068	403.8	535.0	1.6161	1.9699	3.42	4.32				
40	106.76	0.001840											
45.2†	113.26	0.002717	0.0027	476.7	476.7	1.8019	1.8019						

† The P, T, v, h, s, c_p, μ, and k values are interpolated from "ASHRAE Handbook—Fundamentals," 1993.
† Critical point.
NOTE: The reference point used for the data in this table is different from that used in the chart. Use property data from the table or from the chart, but *not* both.

Table 4.2.17 Saturated Water Substance

T, K	P, bar	v_c, m³/kg	v_g, m³/kg	h_c, kJ/kg	h_g, kJ/(kg · K)	s_c, kJ/(kg · K)	s_g, kJ/(kg · K)
250	0.00076	1.087.−3	1520	−381.5	2,459	−1.400	9.954
260	0.00196	1.088.−3	612.2	−360.5	2,477	−1.323	9.590
270	0.00469	1.090.−3	265.4	−339.6	2,496	−1.296	9.255
273.15	0.00611	1.091.−3	206.3	−333.5	2,502	−1.221	9.158
273.15	0.00611	1.000.−3	206.3	0.0	2,502	0.000	9.158
280	0.00990	1.000.−3	130.4	28.8	2,514	0.104	8.980
290	0.01917	1.001.−3	69.7	70.7	2,532	0.251	8.740
300	0.03531	1.003.−3	39.13	112.5	2,550	0.393	8.520
310	0.06221	1.007.−3	22.93	154.3	2,568	0.530	8.318
320	0.1053	1.011.−3	13.98	196.1	2,586	0.649	8.151
330	0.1719	1.016.−3	8.82	237.9	2,604	0.791	7.962
340	0.2713	1.021.−3	5.74	279.8	2,622	0.916	7.804
350	0.4163	1.027.−3	3.846	321.7	2,639	1.038	7.657
360	0.6209	1.034.−3	2.645	363.7	2,655	1.156	7.521
370	0.9040	1.041.−3	1.861	405.8	2,671	1.271	7.394
373.15	1.0133	1.044.−3	1.679	419.1	2,676	1.307	7.356
380	1.2869	1.049.−3	1.337	448.0	2,687	1.384	7.275
390	1.794	1.058.−3	0.980	490.4	2,702	1.494	7.163
400	2.455	1.067.−3	0.731	532.9	2,716	1.605	7.058
420	4.370	1.088.−3	0.425	618.6	2,742	1.810	6.865
440	7.333	1.110.−3	0.261	705.3	2,764	2.011	6.689
460	11.71	1.137.−3	0.167	793.5	2,782	2.205	6.528
480	17.90	1.167.−3	0.111	883.4	2,795	2.395	6.377
500	26.40	1.203.−3	0.0766	975.6	2,801	2.581	6.233
520	37.70	1.244.−3	0.0525	1,071	2,801	2.765	6.093
540	52.38	1.294.−3	0.0375	1,170	2,792	2.948	5.953
560	71.08	1.355.−3	0.0269	1,273	2,772	3.132	5.808
580	94.51	1.433.−3	0.0193	1,384	2,737	3.321	5.654
600	123.5	1.541.−3	0.0137	1,506	2,682	3.520	5.480
620	159.1	1.705.−3	0.0094	1,647	2,588	3.741	5.259
647.3*	221.2	3.170.−3	0.0032	2,107	2,107	4.443	4.443

Above the solid line the condensed phase is solid; below it is liquid. The notation 1.087. −3 signifies 1.087 × 10⁻³. * Critical temperature

Fig. 4.2.9 Enthalpy–log pressure diagram for refrigerant 152a. *(Reprinted by permission of the ASHRAE from the 1993 "ASHRAE Handbook—Fundamentals.")* The reference point used for the data in this chart is different from that used in the table. Use property data from the chart or from the table, but *not* both.

Fig. 4.2.10   Temperature-entropy diagram for water substance, SI units.

Table 4.2.18 Compressed Steam*

Temperature, K		Pressure, bar												Temperature, K
		1	10	20	40	60	80	100	200	400	600	800	1,000	
350	v	1.027.−3	1.027.−3	1.026.−3	1.025.−3	1.024.−3	1.023.−3	1.023.−3	1.018.−3	1.009.−3	1.002.−3	9.937.−4	9.865.−4	350
	h	231.8	322.5	323.3	324.9	326.4	328.1	329.7	337.7	353.8	369.7	385.7	401.7	
	s	1.037	1.037	1.036	1.035	1.034	1.032	1.031	1.025	1.013	1.001	0.991	0.979	
400	v	1.827	1.067.−3	1.066.−3	1.065.−3	1.064.−3	1.063.−3	1.061.−3	1.056.−3	1.045.−3	1.035.−3	1.027.−3	1.018.−3	400
	h	2,730	533.4	534.1	535.4	536.8	538.2	539.6	546.5	560.6	574.9	589.3	603.8	
	s	7.502	1.600	1.599	1.597	1.595	1.593	1.592	1.583	1.565	1.549	1.533	1.518	
450	v	2.063	1.124.−3	1.123.−3	1.121.−3	1.119.−3	1.118.−3	1.116.−3	1.108.−3	1.094.−3	1.082.−3	1.070.−3	1.059.−3	450
	h	2,830	749.0	749.8	750.8	751.9	753.0	754.1	759.5	771.0	783.0	795.3	807.9	
	s	7.736	2.110	2.107	2.105	2.102	2.099	2.097	2.085	2.061	2.039	2.019	2.002	
500	v	2.298	0.221	0.104	1.201.−3	1.198.−3	1.196.−3	1.193.−3	1.181.−3	1.160.−3	1.142.−3	1.126.−3	1.112.−3	500
	h	2,929	2,891.2	2,839.4	975.9	976.3	976.8	977.3	980.3	987.4	995.9	1005.3	1015.4	
	s	7.944	6.823	6.422	2.578	2.575	2.571	2.567	2.549	2.517	2.488	2.461	2.437	
600	v	2.76	0.271	0.133	0.0630	0.0396	0.0276	0.0201	1.483.−3	1.392.−3	1.337.−3	1.296.−3	1.265.−3	600
	h	3,129	3,109	3,087	3,036	2,976	2,906	2,820	1,489	1,462	1,452	1,447	1,447	
	s	8.309	7.223	6.875	6.590	6.224	5.997	5.775	3.469	3.379	3.316	3.266	3.223	
700	v	3.23	0.319	0.158	0.0769	0.0500	0.0346	0.0283	1.157.−2	2.630.−3	1.831.−3	1.639.−3	1.536.−3	700
	h	2,334	3,322	3,307	3,278	3,247	3,214	3,179	2,965	2,233	2,021	1,962	1,931	
	s	8.625	7.550	7.215	6.864	6.644	6.431	6.334	5.770	4.554	4.192	4.058	3.972	
800	v	3.69	0.367	0.182	0.0689	0.0589	0.0436	0.0343	1.575.−2	6.391.−3	3.496.−3	2.484.−3	2.072.−3	800
	h	3,546	3,537	3,526	3,506	3,485	3,464	3,442	3,325	3,047	2,734	2,567	2,465	
	s	8.908	7.837	7.507	7.151	6.965	6.809	6.685	6.252	5.654	5.175	4.864	4.701	
900	v	4.15	0.414	0.206	0.102	0.0674	0.0501	0.0398	1.899.−2	8.619.−3	5.257.−3	3.704.−3	2.907.−3	900
	h	3,764	3,757	3,750	3,737	3,719	3,704	3,688	3,609	3,440	3,269	3,113	2,995	
	s	9.165	8.097	7.770	7.462	7.237	7.092	6.975	6.587	6.119	5.780	5.510	5.305	
1,000	v	4.15	0.414	0.206	0.102	0.0674	0.0501	0.0398	2.186.−2	1.038.−2	6.605.−3	4.792.−3	3.763.−3	1,000
	h	3,990	3,984	3,978	3,967	3,955	3,944	3,935	3,874	3,756	3,640	3,532	3,435	
	s	9.402	8.336	8.011	7.682	7.486	7.345	7.233	6.867	6.453	6.172	5.951	5.727	
1,500	v	6.92	0.692	0.341	0.1730	0.1153	0.0865	0.0692	0.0346	0.0173	0.0116	0.00871	0.00700	1,500
	h	5,227	5,224	5,221	5,217	5,212	5,207	5,203	5,198	5,171	5,144	5,120	5,095	
	s	10.40	9.34	9.015	8.693	8.503	8.368	8.262	7.936	7.597	7.391	7.239	7.118	
2,000	v	9.26	0.925	0.462	0.231	0.1543	0.1157	0.0926	0.0465	0.0234	0.0157	0.0119	0.0096	2,000
	h	6,706	6,649	6,639	6,629	6,623	6,619	6,616	6,610	6,599	6,590	6,581	6,574	
	s	11.25	10.15	9.828	9.503	9.313	9.178	9.073	8.748	8.418	8.222	8.082	7.971	
2,500	v	11.90	1.171	0.583	0.291	0.1942	0.1457	0.1166	0.0584	0.0294	0.0197	0.0149	0.0120	2,500
	h	9,046	8,504	8,413	8,342	8,307	8,285	8,269	8,269	8,267	8,261	8,250	8,240	
	s	12.28	10.80	10.62	10.26	10.06	9.920	9.810	9.468	9.129	8.930	8.788	8.677	

* v = specific volume, m³/kg; h = specific enthalpy, kJ/kg; s = specific entropy, kJ/(kg · K). The notation 1.027.−3 signifies 1.027×10^{-3}.

Table 4.2.19   Saturated Water Substance, fps Units

Abs press, lb/in²	Temp., °F	Specific volume, ft³/lb		Enthalpy, Btu/lb			Entropy, Btu/(lb · R)			Internal energy, evap., Btu/lb
		Liquid	Vapor	Liquid	Evap.	Vapor	Liquid	Evap.	Vapor	
0.08866	32.02	0.016022	3,302	0.01	1,075.4	1,075.4	0.00000	2.1869	2.1869	1,021.2
1.0	101.70	0.016136	333.6	69.74	1,036.0	1,105.8	0.13266	1.8453	1.9779	974.3
1.5	115.65	0.016187	227.7	83.65	1,028.0	1,111.7	0.15714	1.7867	1.9438	964.8
2	126.04	0.016230	173.75	94.02	1,022.1	1,116.1	0.17499	1.7448	1.9198	957.8
3	141.43	0.016300	118.72	109.39	1,013.1	1,122.5	0.20089	1.6852	1.8861	947.2
4	152.93	0.016358	90.64	120.89	1,006.4	1,127.3	0.21983	1.6426	1.8624	939.3
5	162.21	0.016407	73.53	130.17	1,000.9	1,131.0	0.23486	1.6093	1.8441	932.9
10	193.19	0.016590	38.42	161.23	982.1	1,143.3	0.28358	1.5041	1.7877	911.0
14.696	211.99	0.016715	26.80	180.15	970.4	1,150.5	0.31212	1.4446	1.7567	897.5
15	213.03	0.016723	26.29	181.19	969.7	1,150.9	0.31367	1.4414	1.7551	896.8
20	227.96	0.016830	20.09	196.26	960.1	1,156.4	0.33580	1.3962	1.7320	885.8
25	240.08	0.016922	16.306	208.52	952.2	1,160.7	0.35345	1.3607	1.7142	876.9
30	250.34	0.017004	13.748	218.93	945.4	1,164.3	0.36821	1.3314	1.6996	869.2
35	259.30	0.017078	11.900	228.04	939.3	1,167.4	0.38093	1.3064	1.6873	862.4
40	267.26	0.017146	10.501	236.16	933.8	1,170.0	0.39214	1.2845	1.6767	856.2
45	274.46	0.017209	9.403	243.51	928.8	1,172.3	0.40218	1.2651	1.6673	850.7
50	281.03	0.017269	8.518	250.24	924.2	1,174.4	0.41129	1.2476	1.6589	845.5
55	287.10	0.017325	7.789	256.46	919.9	1,176.3	0.41963	1.2317	1.6513	840.8
60	292.73	0.017378	7.177	262.25	915.8	1,178.0	0.42733	1.2170	1.6444	836.3
65	298.00	0.017429	6.657	267.67	911.9	1,179.6	0.43450	1.2035	1.6380	832.1
70	302.96	0.017478	6.209	272.79	908.3	1,181.0	0.44120	1.1909	1.6321	828.1
75	307.63	0.017524	5.818	277.61	904.8	1,182.4	0.44749	1.1790	1.6265	824.3
80	312.07	0.017570	5.474	282.21	901.4	1,183.6	0.45344	1.1679	1.6214	820.6
85	316.29	0.017613	5.170	286.58	898.2	1,184.8	0.45907	1.1574	1.6165	817.1
90	320.31	0.017655	4.898	290.76	895.1	1,185.9	0.46442	1.1475	1.6119	813.8
95	324.16	0.017696	4.654	294.76	892.1	1,186.9	0.46952	1.1380	1.6076	810.6
100	327.86	0.017736	4.434	298.61	889.2	1,187.8	0.47439	1.1290	1.6034	807.5
150	358.48	0.018089	3.016	330.75	864.2	1,194.9	0.51422	1.0562	1.5704	781.0
200	381.86	0.018387	2.289	355.6	843.7	1,199.3	0.5440	1.0025	1.5464	759.6
250	401.04	0.018653	1.8448	376.2	825.8	1,202.1	0.5680	0.9594	1.5274	741.4
300	417.13	0.018896	1.5442	394.1	809.8	1,203.9	0.5883	0.9232	1.5115	725.1
350	431.82	0.019124	1.3267	409.9	795.0	1,204.9	0.6060	0.8917	1.4978	710.3
400	444.70	0.019340	1.1620	424.2	781.2	1,205.5	0.6218	0.8638	1.4856	696.7
450	456.39	0.019547	1.0326	437.4	768.2	1,205.6	0.6360	0.8385	1.4746	683.9
500	467.13	0.019748	0.9283	449.5	755.8	1,205.3	0.6490	0.8154	1.4645	671.7
550	477.07	0.019943	0.8423	460.9	743.9	1,204.8	0.6611	0.7941	1.4551	660.2
600	486.33	0.02013	0.7702	471.7	732.4	1,204.1	0.6723	0.7742	1.4464	649.1
700	503.23	0.02051	0.6558	491.5	710.5	1,202.0	0.6927	0.7378	1.4305	628.2
800	518.36	0.02087	0.5691	509.7	689.6	1,199.3	0.7110	0.7050	1.4160	608.4
900	532.12	0.02123	0.5009	526.6	669.5	1,196.0	0.7277	0.6750	1.4027	589.6
1,000	544.75	0.02159	0.4459	542.4	650.0	1,192.4	0.7432	0.6471	1.3903	571.5
1,500	596.39	0.02346	0.2769	611.5	557.2	1,168.7	0.8082	0.5276	1.3359	486.9
2,000	636.00	0.02565	0.18813	671.9	464.4	1,136.3	0.8623	0.4238	1.2861	404.2
2,500	668.31	0.02860	0.13059	730.9	360.5	1,091.4	0.9131	0.3196	1.2327	313.4
3,000	695.52	0.03431	0.08404	802.5	213.0	1,015.5	0.9732	0.1843	1.1575	185.4
3,203.6	705.44	0.05053	0.05053	902.5	0	902.5	1.0580	0	1.0580	0

SOURCE: Abstracted from Keenan, Keyes, Hill, and Moore, "Steam Tables," 1969.

Table 4.2.20   Compressed Water Substance, fps Units

Pressure, psia (saturation temp., °F)		Temperature of steam, °F							
		200	300	400	500	600	800	1,000	1,200
10	v	38.85	44.99	51.03	57.04	63.03	74.98	86.91	98.84
(193.19)	h	1,146.6	1,193.7	1,240.5	1,287.7	1,335.5	1,433.3	1,534.6	1,639.4
	s	1.7927	1.8592	1.9171	1.9690	2.0164	2.1009	2.1755	2.2428
50	v	0.01663		8.772	10.061	11.305	14.949	17.352	19.747
(281.03)	h	168.1	1,332.8	1,184.4	1,235.0	1,284.0	1,431.7	1,533.5	1,638.7
	s	0.2940	1.8371	1.6722	1.7348	1.7887	1.9225	1.9975	2.0650
100	v	0.01663	0.01745	4.934	5.587	6.216	7.445	8.657	9.861
(327.86)	h	168.2	269.7	1,227.5	1,279.1	1,329.3	1,429.6	1,532.1	1,637.7
	s	0.2939	0.4372	1.6517	1.7085	1.7582	1.8449	1.9204	1.9882
150	v	0.01663	0.01744	3.221	3.679	4.111	4.944	5.759	6.566
(358.48)	h	168.4	269.8	1,219.5	1,274.1	1,325.7	1,427.5	1,530.7	1,454.5
	s	0.2938	0.4371	1.5997	1.6598	1.7110	1.7989	1.8750	1636.7
200	v	0.01662	0.01744	2.361	2.724	3.058	3.693	4.310	4.918
(381.86)	h	168.5	269.9	1,210.8	1,268.8	1,322.1	1,425.3	1,529.3	1,635.7
	s	0.2938	0.4370	1.5600	1.6239	1.6767	1.7660	1.8425	1.9109
300	v	0.01662	0.01743	0.01863	1.7662	2.004	2.442	2.860	3.270
(417.43)	h	168.8	270.1	375.2	1,257.5	1,314.5	1,421.0	1,526.5	1,633.8
	s	0.2936	0.4368	0.5665	1.5701	1.6266	1.7187	1.7964	1.8653
400	v	0.01661	0.01742	0.01862	1.2843	1.4760	1.8163	2.136	2.446
(444.70)	h	169.0	270.3	375.3	1,245.2	1,306.6	1,416.6	1,523.6	1,631.8
	s	0.2935	0.4366	0.5662	1.5282	1.5892	1.6844	1.7632	1.8327
500	v	0.01661	0.01741	0.01861	.9924	1.1583	1.4407	1.7008	1.9518
(467.13)	h	169.2	270.5	375.4	1,231.5	1,298.3	1,412.1	1,520.7	1,629.8
	s	0.2934	0.4364	0.5660	1.4923	1.5585	1.6571	1.7371	1.8072
600	v	0.01660	0.01740	0.01860	.7947	.9456	1.1900	1.4108	1.6222
(486.33)	h	169.4	270.7	375.5	1,216.2	1,289.5	1,407.6	1,517.8	1,627.8
	s	0.2933	0.4362	0.5657	1.4592	1.5320	1.6343	1.7155	1.7861
700	v	0.01660	0.01740	0.01859	0.02042	.7929	1.0109	1.2036	1.3868
(503.23)	h	169.6	270.9	375.6	487.6	1,280.2	1,402.9	1,514.9	1,625.8
	s	0.2932	0.4360	0.5655	0.6887	1.5081	1.6145	1.6970	1.7682
800	v	0.01659	0.01739	0.01857	0.02040	.6776	.8764	1.0482	1.2102
(518.36)	h	169.9	271.1	375.8	487.6	1,270.4	1,398.2	1,511.9	1,623.8
	s	0.2930	0.4359	0.5652	0.6883	1.4861	1.5969	1.6807	1.7526
900	v	0.01658	0.01739	0.01856	0.02038	.5871	.7717	.9273	1.0729
(532.14)	h	170.1	271.3	375.9	487.5	1,260.0	1,393.4	1,508.9	1,621.7
	s	0.2929	0.4355	0.5650	0.6878	1.4652	1.5810	1.6662	1.7386
1,000	v	0.01658	0.01738	0.01855	0.02036	0.5140	0.6878	0.8305	0.9630
(544.75)	h	170.3	271.5	376.0	487.5	1,248.8	1,388.5	1,505.9	1,619.7
	s	0.2928	0.4355	0.5647	0.6874	1.4450	1.5664	1.6530	1.7261
1,500	v	0.01655	0.01734	0.01849	0.02024	0.2816	0.4350	0.5400	0.6334
(596.39)	h	171.5	272.4	376.6	487.4	1,174.8	1,362.5	1,490.3	1,609.3
	s	0.2922	0.4346	0.5634	0.6853	1.3416	1.5058	1.6001	1.6765
2,000	v	0.01653	0.01731	0.01844	0.02014	0.02330	0.3071	0.3945	0.4685
(636.00)	h	172.6	273.3	377.2	487.3	614.0	1,333.8	1,474.1	1,598.6
	s	0.2916	0.4338	0.5622	0.6832	0.8046	1.4562	1.5598	1.6398
2,500	v	0.01650	0.01727	0.01839	0.02004	0.02300	0.2291	0.3069	0.3696
(668.31)	h	173.8	274.3	377.8	487.3	611.6	1,301.7	1,457.2	1,587.7
	s	0.2910	0.4329	0.5609	0.6813	0.8043	1.4112	1.5262	1.6101
3,000	v	0.01648	0.01724	0.01833	0.01994	0.02274	0.17572	0.2485	0.3036
(695.52)	h	174.9	275.2	378.5	487.3	609.6	1,265.2	1,439.6	1,576.6
	s	0.2905	0.4321	0.5597	0.6794	0.8004	1.3675	1.4967	1.5848
4,000	v	0.01643	0.01717	0.01824	0.01977	0.02229	0.1052	0.1752	0.2213
	h	177.2	277.2	379.9	487.5	606.5	1,172.9	1,402.6	1,553.9
	s	0.2931	0.4304	0.5573	0.6758	0.7936	1.2740	1.4449	1.5423
5,000	v	0.01638	0.01711	0.01814	0.01960	0.02191	0.05932	0.1312	0.1720
	h	179.5	279.1	381.3	487.9	604.2	1,042.1	1,363.4	1,530.8
	s	0.2882	0.4288	0.5551	0.6724	0.7876	1.1583	1.3988	1.5066

v = specific volume, ft³/lb; h = enthalpy, Btu/lb; s = entropy, Btu/(lb · R).

Fig. 4.2.11   Temperature-entropy diagram for water substance, fps units. *(Data from Keenan and Keyes, "Thermodynamic Properties of Steam," Wiley.)*

Table 4.2.21   Phase Transition and Other Data for 100 Fluids*

Name	Formula	M	T_m, K	Δh_{fus}, kJ/kg	T_b, K	Δh_{vap}, kJ/kg	P_c, bar	v_c, m³/kg	T_c, K	Z_c
Acetaldehyde	C_2H_4O	44.053	149.7	73.2	293.7	584.0	55.4	0.00382	461	0.243
Acetic acid	$C_2H_4O_2$	60.053		195.2	391.2	404.7	57.9	0.00285	594.5	0.200
Acetone	C_3H_6O	58.080	178.5	98.5	329.3	500.9	47.2	0.00360	508.2	0.232
Acetylene	C_2H_2	26.038	179.0	96.5	189.2	687.0	62.4	0.00435	308.3	0.276
Air	Mixed	28.966	60.0		79, 82	206.5	37.7	0.00320	132.6	0.263
Ammonia	NH_3	17.031	195.4	331.9	239.7	1,368	112.9	0.00427	405.7	0.244
Aniline	C_6H_7N	93.129	266.8	113.2	457.5	484.9	53.0	0.00340	698.8	0.289
Argon	A	39.948	83.8	29.4	87.5	163.2	48.7	0.00187	150.8	0.290
Benzene	C_6H_6	78.114	278.7	125.9	353.3	394.0	49.2	0.00332	562.1	0.273
Bromine	Br_2	159.81	264.9	66.2	331.6	187.7	103.4	0.00079	584.2	0.270
Butane, n	C_4H_{10}	58.124	113.7	78.2	261.5	366.4	36.5	0.00452	408.1	0.283
Butane, iso	C_4H_{10}	58.124	137.0	80.2	272.7	385.5	38.0	0.00439	425.2	0.274
Butanol	$C_4H_{10}O$	74.123	183.9	121.2	390.8	593.2	44.1	0.00370	562.9	0.259
Batylene	C_4H_8	56.108	87.8	68.6	266.9	391.0	40.2	0.00428	419.6	0.277
Carbon dioxide	CO_2	44.010	216.6	18.4	194.7	573.2	73.8	0.00214	304.1	0.274
Carbon disulfide	CS_2	76.131	161.1	57.7	319.4	351.6	79.0	0.00223	552.0	0.293
Carbon monoxide	CO	28.010	68.1	29.8	81.6	215.1	35.0	0.00332	132.9	0.295
Carbon tetrachloride	CCl_4	153.82	250.3	16.3	349.8	195.0	45.6	0.00170	556.4	0.258
Carbon tetrafluoride	CF_4	88.005	89.5	8.0	145.2	138	37.4	0.00156	228.0	0.272
Cesium	Cs	132.91	301.6	16.4	942.4	494.3	153.7	0.00230	2,043	0.240
Chlorine	Cl_2	70.906	172.2	90.4	238.6	287.5	77.0	0.00175	417.2	0.278
Chloroform	$CHCl_3$	119.38	209.7	77.1	334.5	248.5	54.5	0.00202	536.0	0.294
o-Cresol	C_7H_8O	108.14	303.8		464.1		50.0	0.00291	697.6	0.271
Cyclohexane	C_6H_{12}	84.162	279.8	31.7	353.9	357.4	40.7	0.00366	553.4	0.273
Cyclopropane	C_3H_6	42.081	145.5	129.4	240.3	477	55.0	0.00387	397.8	0.271
Decane	$C_{10}H_{22}$	142.29	243.4	201.8	447.3	276.2	21.0	0.00424	617.5	0.247
Deuterium	D_2	4.028	18.71	48.9	23.7	304.4	16.7	0.00143	38.34	0.301
Diphenyl	$C_{12}H_{10}$	154.21	342.4	120.6	527.6	317.3	38.5	0.00326	789.0	0.295
Ethane	C_2H_6	30.070	89.9	45.0	184.6	488.4	48.8	0.00486	305.4	0.281
Ethanol	C_2H_6O	46.069	159.0	109.0	351.5	840.9	63.8	0.00362	516.2	0.248
Ethyl acetate	$C_4H_8O_2$	88.107	189.4	119.0	350.3	366.2	38.3	0.00325	523.2	0.252
Ethyl bromide	C_2H_5Br	108.97	153.5	54.0	311.5	218	62.3	0.00197	503.9	0.320
Ethyl chloride	C_2H_5Cl	64.515	134.9	68.9	285.4	382.5	52.0	0.00309	460.4	0.270
Ethyl ether	$C_4H_{10}O$	74.123	150.0	93.1	307.8	358.9	36.8	0.00377	466.8	0.262
Ethyl formate	$C_3H_6O_2$	74.080	193.8	124.3	327.4	405.8	47.4	0.00310	508.5	0.257
Ethylene	C_2H_4	28.054	104.0	119.5	169.5	480.0	50.5	0.00455	283.1	0.274
Ethylene oxide	C_2H_4O	44.054	160.6	117.5	283.6	580.0	71.9	0.00318	468.9	0.258
Fluorine	F_2	37.997	53.5	13.4	85.1	172.1	52.2	0.00174	144.3	0.288
Helium 4	He	4.003			4.3	20.6	2.3	0.0144	5.189	0.303
Heptane	C_7H_{16}	100.20	182.5	139.9	371.6	316.3	27.4	0.0043	540.2	0.263
Hexane	C_6H_{14}	86.178	177.8	151.2	341.9	334.8	30.3	0.0043	507.6	0.265
Hydrazine	N_2H_4	32.045	274.7	395.0	386.7	1,207	147.0		653.2	0.284
Hydrogen	H_2	2.016	14.0	58.0	20.4	454.0	13.0	0.0323	33.3	0.305
Hydrogen bromide	HBr	80.912	186.3	37.4	206.4	217.5	85.5	0.00124	363.2	0.284
Hydrogen chloride	HCl	36.461	160.0	54.7	188.1	443.0	83.1	0.00022	324.7	0.249
Hydrogen fluoride	HF	20.006	181.8	196.3	272.7	374.3	64.9	0.00345	461.2	0.120
Hydrogen iodide	HI	127.91	222.4	22.4	237.8	154.0	83.1	0.00106	423.9	0.318
Hydrogen sulfide	H_2S	34.076	187.5	69.8	213.0	248.0	89.4	0.00289	373.1	0.283

Table 4.2.21   Phase Transition and Other Data for 100 Fluids*   (Continued)

Name	Formula	M	T_m, K	Δh_{fus}, kJ/kg	T_b, K	Δh_{vap}, kJ/kg	P_c, bar	v_c, m³/kg	T_c, K	Z_c
Iodine	I_2	253.81	387.0	62.1	457.5	164.3	117.5	0.00054	785.0	0.248
Krypton	Kr	83.80	116.0	19.5	121.4	107.9	55.0	0.00109	209.4	0.288
Lithium	Li	6.940	453.8		1,615	1,945			3,750	
Mercury	Hg	200.59	234.3	11.4	630.1	295.6	1,510	0.00018	1,763	
Methane	CH_4	16.043	90.7	58.4	111.5	511.8	46.0	0.00617	190.5	0.287
Methanol	CH_4O	32.042	175.5	98.9	337.7	1,104	79.5	0.00368	512.6	0.220
Methyl acetate	$C_3H_6O_2$	74.080	175		330.3	410.0	46.9	0.00308	506.9	0.254
Methyl bromide	CH_3Br	94.939	179.5	62.8	276.7	252.0		0.00173	467.2	
Methyl chloride	CH_3Cl	50.49	175.4	127.4	249.4	428.5	66.8	0.00270	416.3	0.277
Methyl formate	$C_2H_4O_2$	60.053	173.4	125.5	304.7	481.2	60.0	0.00287	487.2	0.255
Methylene chloride	CH_2Cl_2	84.922	176.5	54.4	312.9	328	61.3	0.00197	510.	0.255
Naphthalene	$C_{10}H_8$	128.17	353.4	148.1	491.1	341	40.5	0.00321	748.4	0.270
Neon	Ne	20.179	24.5	16.6	27.3	91.3	27.6	0.00207	44.4	0.311
Nitric oxide	NO	30.006	111	76.6	121.4	460	64.9	0.00192	180	0.249
Nitrogen	N_2	28.013	63.1	25.7	77.3	197.6	34.0	0.00318	126.2	0.287
Nitrogen peroxide	NO_2	46.006	263	159.5	294.5	414.4	101.3	0.00180	431.4	0.233
Nitrous oxide	N_2O	44.013	176	148.6	184.7	376.0	72.4	0.00221	309.6	0.274
Octane	C_8H_{18}	114.23	216.4	161.6	398.9	302.7	25.0	0.00426	508.9	0.258
Oxygen	O_2	31.999	54.4	13.9	90.0	212.5	50.4	0.00229	154.6	0.288
Pentane, iso	C_5H_{12}	72.151	113.7	71.1	301.0	341.0	33.5	0.00427	460.4	0.270
Pentane, n	C_5H_{12}	72.151	143.7	116.6	309.2	357.2	33.7	0.00431	469.6	0.268
Potassium	K	39.098	336.4	59.8	1,030		167		2,265	
Propane	C_3H_8	44.097	86	80.0	231.1	425.7	42.6	0.00453	369.8	0.277
Propanol	C_3H_8O	60.096	147.0	86.5	370.4	695.8	51.7	0.00364	536.7	0.253
Propylene	C_3H_6	42.081	87.9	71.4	225.5	437.5	46.0	0.00429	365.1	0.275
Radon	Rn	224	201	12.3	211	82.8	65.5	0.00063	377.0	0.293
Refrigerant 11	$CFCl_3$	137.37	162.2	50.2	296.9	180.2	44.1	0.00181	471.2	0.280
Refrigerant 12	CF_2Cl_2	120.91	115.4	34.3	243.4	165.1	41.2	0.00179	385.2	0.278
Refrigerant 13	CF_3Cl	104.46	92.1		191.7	148.4	38.7	0.00173	302.0	0.279
Refrigerant 13B1	CF_3Br	148.91	105.4		215.4	118.7	39.6	0.00134	340.2	0.280
Refrigerant 21	$CHFCl_2$	102.91	138.2		282.1	242.1	51.7	0.00192	451.4	0.271
Refrigerant 22	CHF_2Cl	86.469	113.2	47.6	232.4	233.6	49.8	0.00191	369.2	0.267
Refrigerant 23	CHF_3	70.014	118.0	58.0	191.2	239	48.4	0.00190	299.1	0.259
Refrigerant 113	$C_2F_3Cl_3$	187.38	238.2		320.8	146.8	34.1	0.00174	487.3	0.274
Refrigerant 114	$C_2F_4Cl_2$	170.92	179.2		276.7	136.0	32.6	0.00172	418.9	0.275
Refrigerant 115	C_2F_5Cl	154.47	171	12.2	234.0	124.1	31.6	0.00163	353.1	0.271
Refrigerant 142b	$C_2F_2H_3Cl$	100.50	142.4	26.7	263.9	223	41.5	0.00232	410.0	0.279
Refrigerant 152a	$C_2F_2H_4$	66.051	156		248	326	45.0	0.00274	386.7	0.253
Refrigerant 216	$C_3F_6Cl_2$	220.93			308	117.3	27.5	0.00174	453.2	0.281
Refrigerant C318	C_4F_8	200.03	233.0		267	116	27.8	0.00161	388.5	0.272
Refrigerant 500	Mix	99.303	114.3		239.7	201.1	44.3	0.00201	378.7	0.281
Refrigerant 502	Mix	111.63			237	172.2	40.7	0.00178	355.3	0.275
Refrigerant 503	Mix	87.267			184	172.9			293	
Refrigerant 504	Mix	79.240			216	242.9			339	
Refrigerant 505	Mix	103.43			243.6		47.3		391	
Refrigerant 506	Mix	93.69			260.7		51.6		416	
Rubidium	Rb	85.468	312.6		959.4	811.3		0.00288	2,083	
Sodium	Na	22.990	371.0		1,155	3,880			2,730	
Sulfur dioxide	SO_2	64.059	197.8		268.4	368.3	78.8	0.00190	430.7	0.269
Sulfur hexafluoride	SF_6	146.051					37.8	0.00137	318.7	0.285
Toluene	C_7H_8	92.141	178.2		383.8	339.0	40.5	0.00345	594.0	0.260
Water	H_2O	18.015	273.2		373.2	2,256	221.2	0.00315	647.3	0.234
Xenon	Xe	131.36	161.5		165	99.2	58.7	0.00091	290	0.290

* M = molecular weight, T_m = normal melting temperature, Δh_{fus} = enthalpy (or latent heat) of fusion, T_b = normal boiling point temperature, Δh_{vap} = enthalpy (or latent heat) of vaporization, T_c = critical temperature, P_c = critical pressure, v_c = critical volume, Z_c = critical compressibility factor.
Source: Prepared by the author and abstracted from Rohsenow et al., "Handbook of Heat Transfer Fundamentals," McGraw-Hill.

Table 4.2.22   Specific Heat at Constant Pressure [kJ/(kg · K)] of Liquids and Gases

Substance	200	225	250	275	300	325	350	375	400	425	450	475	500
Acetylene	1.457	1.517	1.575	1.635	1.695	1.747							
Air	1.048	1.036	1.029	1.025	1.021	1.021	1.020	1.021	1.022	1.024	1.027	1.030	1.035
Ammonia	4.605	4.360	2.210	2.176	2.169	2.183	2.210	2.247	2.289	2.331	2.381	2.429	2.477
Argon	0.524	0.523	0.522	0.522	0.522	0.521	0.521	0.521	0.521	0.521	0.521	0.521	0.521
Butane, i	1.997	2.099	2.207	1.590	1.694	1.810	1.921	2.035	2.149	2.258	2.367	2.436	2.571
Butane, n	2.066	2.134	2.214	1.642	1.731	1.835	1.941	2.047	2.154	2.253	2.361	2.461	2.558
Carbon dioxide		0.763	0.791	0.818	0.845	0.870	0.894	0.917	0.938	0.958	0.978	0.995	1.014
Ethane	1.419	1.492	1.574	1.658	1.762	1.871	1.968	2.081	2.184	2.287	2.393	2.492	2.590
Ethylene	1.296	1.340	1.397	1.475	1.560	1.656	1.748	1.839	1.928	2.014	2.097	2.179	2.258
Fluorine	0.785	0.793	0.803	0.815	0.827								
Helium	5.193	5.193	5.193	5.193	5.193	5.193	5.193	5.193	5.193	5.193	5.193	5.193	5.193
Hydrogen, n	13.53	13.83	14.05	14.20	14.31	14.38	14.43	14.46	14.48	14.49	14.50	14.50	14.51
Krypton	0.252	0.251	0.250	0.249	0.249	0.249	0.249	0.249	0.249	0.249	0.249	0.249	0.248
Methane	2.105	2.122	2.145	2.184	2.235	2.297	2.375	2.454	2.534	2.617	2.709	2.797	2.892
Neon	1.030	1.030	1.030	1.030	1.030	1.030	1.030	1.030	1.030	1.030	1.030	1.030	1.030
Nitrogen	1.043	1.042	1.042	1.041	1.041	1.041	1.042	1.043	1.045	1.047	1.050	1.053	1.056
Oxygen	0.915	0.914	0.915	0.917	0.920	0.924	0.929	0.935	0.942	0.949	0.956	0.964	0.972
Propane	2.124	2.220	1.500	1.596	1.695	1.805	1.910	2.030	2.140	2.249	2.355	2.458	2.558
Propylene	2.094	2.132	1.436	1.481	1.536	1.625	1.709	1.793	1.890	1.981	2.070	2.153	2.245
Refrigerant 12			0.549	0.578	0.602	0.625	0.647	0.666	0.685	0.701	0.716	0.728	0.741
Refrigerant 21	0.973	0.982	0.991	1.015	0.594	0.617	0.641	0.661	0.683	0.701	0.720	0.735	0.752
Refrigerant 22	1.065	1.085	0.588	0.617	0.647	0.676	0.704	0.729	0.757	0.782	0.806	0.826	0.848
Water substance	1.545	1.738	1.935	4.211	4.179	4.182	4.195	2.035	1.996	1.985	1.981	1.980	1.983
Xenon	0.165	0.163	0.162	0.161	0.160	0.160	0.160	0.159	0.159	0.159	0.159	0.159	0.159

Values for the saturated liquid are tabulated up to the normal boiling point. Higher temperature values are for the dilute gas. Values for water substance below 275 K are for ice.

Table 4.2.23   Specific Heat Ratio c_p/c_v for Liquids and Gases at Atmospheric Pressure

Substance	200	225	250	275	300	325	350	375	400	425	450	475	500
Acetylene	1.313	1.289	1.269	1.250	1.234	1.219	1.205						
Air	1.399	1.399	1.399	1.399	1.399	1.399	1.398	1.397	1.395	1.393	1.391	1.388	1.386
Ammonia					1.327	1.302	1.295	1.285	1.278	1.269	1.262	1.256	1.249
Argon	1.663	1.665	1.666	1.666	1.666	1.666	1.666	1.666	1.666	1.666	1.666	1.666	1.666
Butane, i	1.357	1.356	1.359	1.116	1.103	1.094	1.086	1.080	1.075	1.070	1.066	1.063	1.060
Butane, n	1.418	1.412	1.407	1.114	1.103	1.094	1.086	1.080	1.075	1.071	1.067	1.063	1.061
Carbon dioxide		1.344	1.323	1.302	1.290	1.279	1.269	1.260	1.252	1.245	1.239	1.233	1.229
Carbon monoxide	1.405	1.404	1.403	1.402	1.401	1.401	1.400	1.398	1.396	1.394	1.392	1.390	1.387
Ethane		1.246	1.226	1.210	1.193	1.180	1.167	1.157	1.148	1.139	1.132	1.126	1.120
Ethylene			1.275	1.254	1.236	1.220	1.206	1.194	1.183	1.173	1.165	1.158	1.151
Fluorine	1.393	1.386	1.377	1.370	1.362								
Helium	1.667	1.667	1.667	1.667	1.667	1.667	1.667	1.667	1.667	1.667	1.667	1.667	1.667
Hydrogen, n	1.439	1.426	1.415	1.410	1.406	1.403	1.401	1.400	1.399	1.398	1.398	1.397	1.397
Krypton	1.649	1.655	1.658	1.660	1.662	1.662	1.662	1.662	1.662	1.663	1.664	1.665	1.667
Methane	1.337	1.332	1.325	1.316	1.306	1.295	1.282	1.271	1.258	1.247	1.237	1.228	1.219
Neon	1.667	1.667	1.667	1.667	1.667	1.667	1.667	1.667	1.667	1.667	1.667	1.667	1.667
Nitrogen	1.399	1.399	1.399	1.399	1.399	1.399	1.399	1.398	1.397	1.396	1.394	1.393	1.391
Oxygen	1.398	1.397	1.396	1.395	1.394	1.391	1.388	1.385	1.381	1.377	1.373	1.369	1.365
Propane	1.513	1.504	1.164	1.148	1.135	1.124	1.114	1.107	1.100	1.094	1.089	1.085	1.081
Propylene			1.171	1.160	1.150	1.140	1.131	1.123	1.116	1.110	1.105	1.100	1.096
Refrigerant 12			1.165	1.114	1.101	1.088	1.077	1.065	1.055	1.044	1.034	1.025	1.071
Refrigerant 21					1.179	1.164	1.152	1.144	1.137	1.132	1.127	1.124	1.120
Refrigerant 22			1.207	1.190	1.172	1.164	1.155	1.148	1.143	1.138	1.133	1.129
Steam								1.322	1.319	1.317	1.314	1.312	1.309
Xenon		1.623	1.634	1.642	1.650	1.655	1.659	1.662	1.662	1.662	1.662	1.662	1.662

Table 4.2.24   Saturation Temperature, in Kelvins, of Selected Substances

Pressure, bar	Acetone C_3H_6O	Acetylene, C_2H_2	Air, sat. liquid	Air, sat. vapor	Ammonia, NH_3	Aniline C_6H_7N	Argon, Ar	Benzene, C_6H_6	Butanol, $C_4H_{10}O$	n-Butane, C_4H_{10}	Carbon dioxide, CO_2	Carbon tetrachloride, CCl_4	Carbon tetrafluoride, CF_4
0.010	237.8					337.2			299.1			255.1	
0.015	243.8					344.2			305.4	196.5		259.7	
0.020	247.1					350.0			310.0	200.3		263.2	106.4
0.025	251.6					353.6			313.8	203.1		266.0	107.9
0.030	254.3					357.1			316.9	205.7			109.3
0.04	259.0					363.1		276.5	321.8	209.8		270.9	111.5
0.05	262.7					368.1		280.0	325.6	212.9		275.0	113.2
0.06	266.8					372.3		283.2	328.8	215.8		278.5	114.7
0.08	270.9				198.9	379.0		288.6	333.9	220.3		284.6	117.2
0.10	275.1		62.3	66.3	201.9	385.2		293.0	337.8	224.1		289.1	119.1
0.15	283.2		64.6	68.5	207.6	397.1		301.7	345.8	231.0		298.1	122.9
0.20	289.6		66.4	70.2	211.8	404.5		308.3	351.8	236.3		304.8	125.8
0.25	294.1		67.9	71.5	215.2	410.6		313.7	356.8	240.7		310.2	128.1
0.30	298.0		69.1	72.7	218.1	415.8		318.2	360.8	244.4		314.9	130.1
0.40	304.7		71.2	74.6	222.8	424.8		325.8	367.4	250.5		322.2	133.3
0.5	310.1		72.8	76.2	226.6	432.1		331.8	372.8	255.4		327.8	136.0
0.6	315.0		74.3	77.5	229.9	438.2		337.1	377.2	259.6		333.1	138.2
0.8	322.8		76.6	79.7	235.2	448.8	85.1	345.7	384.3	266.6		342.0	142.0
1.0	329.0		78.6	81.6	239.6	456.5	87.2	352.7	390.0	272.2		349.2	145.0
1.5	341.0	195.1	82.3	85.3	247.9	472.1	91.2	366.5	402.6	283.4		363.8	151.0
2.0	350.1	200.1	85.2	88.1	254.3	484.2	94.3	377.0	412.0	291.9		374.7	155.6
2.5	358.9	204.8	87.6	90.4	259.5	494.0	96.8	385.7	419.8	299.0		383.0	159.3
3.0	365.9	208.8	89.7	92.4	263.9	502.0	99.0	393.1	426.1	305.0		390.2	162.1
4.0	377.1	215.3	93.1	95.7	271.3	516.0	102.6	405.6	436.8	315.1		402.0	168.0
5.0	386.1	220.9	96.0	98.4	277.3	527.0	105.8	415.8	445.1	323.4		412.8	172.5
6	393.9	225.8	98.4	100.8	282.4	536.2	108.4	424.7	452.9	330.6	220.0	422.0	175.5
8	406.4	233.9	102.7	104.8	291.0	551.5	110.8	439.9	465.4	342.6	227.1	433.0	182.9
10	417.3	240.3	106.1	108.1	298.1	564.7	116.6	451.6	475.8	352.6	233.0	449.5	187.7
15	435.9	252.5	113.1	114.7	311.9	593.0	124.0	475.8	496.0	372.2	244.6	474.3	199.0
20	452.0	262.3	118.5	119.9	322.5	613.6	129.8	494.6	510.1	387.6	253.6	495.0	207.4
25	465.8	270.3	123.0	124.1	331.3	631.2	134.6	510.1		400.2	260.6	511.2	214.3
30	476.9	277.6	127.0	227.8	338.9	645.8	138.7	523.5		410.9	269.6	524.0	220.2
40	496.0	289.5			351.6	672.0	145.7	545.8			278.5	547.6	
50		299.0			362.1	693.0					287.4		
60		306.9			371.1						295.1		
80					386.1								
100					398.4								

Table 4.2.24   Saturation Temperature, in Kelvins, of Selected Substances   (*Continued*)

Pressure, bar	Chlorine, Cl_2	Cyclohexane, C_6H_{12}	Cyclopropane, C_3H_6	Diphenyl, $C_{12}H_{10}$	Ethane, C_2H_6	Ethanol C_2H_6O	Ethyl chloride C_2H_5Cl	Ethylene, C_2H_4	Fluorine, F_2	Helium, He	Heptane, C_7H_{16}	Hexane, C_6H_{14}	Hydrazine, N_2H_4
0.010			172.9	384.3	127.6	266.7		117.4	58.2		266.9	244.2	285.9
0.015	173.4		176.9	393.5	131.1	272.7	207.9	120.4	59.8		273.2	250.0	293.2
0.020	176.6		179.9	400.1	133.6	277.1	212.0	122.8	61.0		277.9	254.2	298.3
0.025	179.1		182.3	405.9	135.7	280.6	215.1	124.7	62.0		281.8	257.7	302.4
0.030	181.2		184.4	410.2	137.3	283.1	217.6	126.1	62.7		285.0	260.7	306.0
0.04	184.7		187.7	418.0	140.4	287.7	221.9	128.8	64.1		290.1	265.6	311.7
0.05	186.4		190.1	423.8	142.5	291.2	225.0	130.8	65.1		294.4	269.7	315.9
0.06	189.8	282.2	192.5	428.9	144.6	294.3	227.8	132.6	66.1	2.25	298.1	273.0	319.7
0.08	193.8	288.0	196.2	436.8	147.6	299.2	232.4	136.9	67.6	2.38	304.2	278.4	325.5
0.10	197.0	292.6	199.2	443.6	150.4	303.1	236.1	138.0	68.8	2.49	309.0	282.6	330.1
0.15	203.0	301.4	205.0	456.3	155.3	310.4	243.4	143.0	71.1	2.71	318.3	291.5	339.1
0.20	207.7	308.0	209.5	466.1	158.9	315.7	248.8	146.6	72.9	2.82	325.1	298.3	345.9
0.25	211.4	313.6	213.2	473.9	161.9	320.1	253.1	149.3	74.3	3.03	330.5	304.0	350.8
0.30	214.8	318.1	216.1	480.4	164.6	323.9	256.9	151.7	75.5	3.15	335.0	308.6	355.2
0.40	219.9	325.7	221.2	491.5	168.8	330.1	263.1	155.3	77.6	3.37	343.0	316.6	362.3
0.5	224.0	331.8	225.1	499.6	172.4	335.2	267.9	158.3	79.2	3.55	349.2	321.4	368.1
0.6	227.7	337.3	228.7	506.6	175.3	339.2	272.1	161.0	80.6	3.71	354.8	326.4	372.8
0.8	233.8	346.9	234.6	518.2	180.2	345.9	279.5	165.6	82.9	3.98	363.9	334.6	380.2
1.0	238.8	353.7	239.3	528.7	184.3	351.4	285.1	169.2	84.8	4.21	371.0	341.3	386.9
1.5	248.3	368.3	249.0	548.0	192.2	362.3	297.6	176.5	88.6	4.67	385.3	355.2	399.1
2.0	255.9	378.9	260	562.2	198.2	370.6	306.9	181.9	91.4	5.03	396.7	361.0	408.1
2.5	262.1	387.7	269	573.8	203.1	377.5	314.6	186.3	93.7		406.0	373.7	415.6
3.0	267.2	395.0	275	583.8	207.4	382.9	319.6	190.6	95.8		413.5	381.7	422.0
4.0	275.9	407.9	282	600.3	214.5	392.0	329.4	197.7	99.2		426.0	393.3	432.4
5.0	283.0	418.6	287	614.3	220.4	399.1	337.0	202.9	102.0		431.9	404.0	441.2
6	289.2	427.8	292	626.6	225.5	405.0	343.5	207.2	104.4		441.5	412.2	449.0
8	290.6	443.0	301	646.6	234.0	415.6	354.7	215.0	108.4		462.1	427.0	462.5
10	307.8	455.2	309	664.2	241.1	424.1	364.1	221.2	111.8		474.9	438.9	473.7
15	325.0	481.1	327	697.8	255.1	442.4	385.9	234.1	118.5		498.9	462.0	494.3
20	338.0	499.9	342	722.5	266.0	456.0	400.4	243.9	123.7		518.9	479.1	510.2
25	349.4	516.3	355	744.2	275.1	467.0	412.8	252.3	128.1		535.0	495.0	522.2
30	359.2	528.8	365	763.4	282.9	476.0	422.2	259.9	131.8			506.3	531.7
40	376.0	551.1	379	794.0	296.0	490.4	442.4	272.7	138.1				547.8
50	389.0			818.2		502.8	456.5	282.7	143.3				561.8
60	400.7					513.4							574.0
80													598.0
100													616.2

Table 4.2.24 Saturation Temperature, in Kelvins, of Selected Substances (Continued)

Pressure, bar	n-Hydrogen, H_2	Hydrochloric acid, HCl	Hydrogen sulfide, H_2S	Mercury, Hg	Methane, CH_4	Methanol	Methyl chloride, CH_3Cl	Naphthalene	Nitrogen, N_2	Octane, C_8H_{18}	Oxygen, O_2	Pentane, C_5H_{12}	Potassium, K
0.010				448.6		252.6	173.2	353.0		286.4	62.2	219.1	699
0.015				460.5		258.2	178.4	361.2		293.8	63.0	225.0	719
0.020				468.9		262.3	182.6	367.6		299.4	64.3	228.8	732
0.025				475.3		265.6	185.9	372.8		303.7	65.2	232.0	743
0.030				481.6		268.4	188.6	376.9		306.8	66.2	234.7	752
0.04				491.2		272.9	192.7	384.3		312.4	67.6	239.2	764
0.05				498.2		276.5	196.0	389.9		316.9	68.8	242.8	776
0.06				504.7		278.8	198.6	394.8		320.8	69.8	245.9	788
0.08	14.0			515.2		284.4	202.8	402.7		327.1	71.4	250.8	811
0.10	14.4			523.6		288.4	206.0	408.9		332.0	72.7	254.9	829
0.15	15.2	161.0		538.9	92.6	295.9	211.9	420.9	64.1	341.6	75.2	263.0	861
0.20	15.8	164.5		551.0	95.0	301.4	216.5	429.5	65.8	348.7	77.1	269.1	883
0.25	16.3	167.2	189.8	561.0	97.0	305.9	220.5	437.1	67.2	354.6	78.7	274.0	899
0.30	16.7	169.3	192.5	569.0	98.6	309.7	223.8	443.3	68.3	359.7	80.0	277.9	916
0.40	17.5	172.7	197.0	582.2	101.4	315.9	229.1	454.0	70.2	368.3	82.2	284.9	940
0.5	18.1	175.2	200.4	592.7	103.7	320.8	233.4	462.1	71.8	374.9	83.9	290.0	960
0.6	18.6	177.5	203.1	601.8	105.6	325.0	237.0	469.4	73.2	380.6	85.5	294.7	977
0.8	19.5	181.0	208.0	616.8	108.8	331.9	243.1	481.1	75.4	390.2	88.0	302.1	1,006
1.0	20.2	187.8	212.5	628.9	111.5	337.5	248.0	490.2	77.2	398.0	90.1	308.6	1,029
1.5	21.7	195.6	222.0	652.1	116.6	348.1	257.9	509.0	80.8	413.3	94.1	321.6	1,074
2.0	22.8	201.5	227.9	670.0	120.6	356.1	265.4	523.3	83.6	424.6	97.2	331.1	1,108
2.5	23.8	206.1	234.3	684.5	123.9	362.6	271.5	535.0	85.9	434.4	99.8	339.2	1,137
3.0	24.6	210.0	237.9	696.7	126.7	368.1	276.6	544.8	87.9	442.1	102.0	345.7	1,176
4.0	26.0	216.7	245.1	717.1	131.4	377.2	285.5	560.3	91.2	456.3	105.7	356.7	1,203
5.0	27.1	222.0	250.9	733.5	135.3	384.6	292.6	573.2	94.0	467.4	108.8	365.4	1,238
6	28.1	226.2	256.1	747.8	138.7	390.9	298.9	584.7	96.4	477.0	111.4	373.6	1,268
8	29.8	233.6	264.9	770.8	144.4	401.4	309.4	605.0	100.4	494.0	115.9	387.6	1,318
10	31.2	241.6	272.1	790.2	149.1	410.0	318.1	620.7	103.8	507.0	119.6	398.2	1,359
15		254.2	283.3	827.5	158.8	426.5	332.7	652.6	110.4	532.4	127.0	420.7	1,440
20		263.7	299.8	857.2	165.8	439.1	348.6	677.9	115.6	553.6	132.8	436.8	1,505
25		272.3	309.4	879.2	172.0	449.4	359.4	699.0	119.9		137.6	450.4	1,558
30		279.9	316.7	901.0	177.2	458.2	368.6	718.4	123.6		141.7	462.1	1,606
40		291.9	330.0	934.4	186.1	473.0	384.3				148.6		1,689
50		301.1	340.4	962.8		484.9	397.2				154.4		1,759
60		309.5	350.3	987.6		495.3	408.5						1,819
80		322.9	366.9	1,028		511.9							1,920
100				1,065									2,000

Table 4.2.24   Saturation Temperature, in Kelvins, of Selected Substances   (Continued)

Pressure, bar	Propane, C_3H_8	Propanol, C_3H_8O	Propylene, C_3H_6	Refrigerant 11, $CFCl_3$	Refrigerant 12, CF_2Cl_2	Refrigerant 13, CF_3Cl	Refrigerant 21, $CHFCl_2$	Refrigerant 22, CHF_2Cl	Sodium, Na	Sulfur dioxide, SO_2	Toluene, C_7H_8	Water, H_2O
0.010	162.0	284.0	157.8	209.9	171.5	134.1	201.8	165.7	804	190.6	275.1	280.1
0.015	166.2	289.8	161.9	215.0	175.8	137.6	206.6	170.0	825	195.9	282.0	286.1
0.020	169.4	294.0	165.1	219.1	179.1	140.3	210.5	173.1	841	199.6	286.3	290.6
0.025	171.9	297.3	167.4	222.2	181.8	142.4	212.9	175.6	854	202.6	290.0	294.2
0.030	174.1	299.9	169.6	224.7	184.1	144.2	215.5	177.8	865	204.8	293.4	297.2
0.04	177.4	304.2	172.8	229.0	187.8	147.1	219.6	180.9	883	208.7	298.9	302.1
0.05	180.1	307.8	175.7	232.6	190.3	149.4	222.9	183.5	898	211.8	303.3	306.0
0.06	182.4	310.7	177.9	235.5	193.0	151.3	225.7	185.8	910	214.3	307.1	309.3
0.08	186.2	315.7	181.7	240.7	197.0	154.6	230.2	189.6	930	218.6	313.7	314.7
0.10	189.4	319.7	184.9	244.7	200.1	157.1	233.9	192.6	946	222.0	318.6	319.0
0.15	195.6	327.4	190.7	252.1	206.3	162.1	241.0	198.5	977	228.2	328.5	327.3
0.20	200.1	333.2	195.2	257.8	211.1	165.9	246.3	202.9	1,000	232.9	335.1	333.2
0.25	203.9	337.8	198.7	262.4	214.9	169.0	250.6	206.6	1,019	236.2	341.6	338.1
0.30	207.0	341.7	201.7	266.4	218.2	171.7	254.2	209.6	1,034	239.8	346.3	342.3
0.40	212.1	348.1	206.8	273.1	223.5	176.0	260.2	214.7	1,061	244.8	354.1	349.0
0.5	216.3	352.9	211.0	278.2	227.9	179.5	265.1	218.6	1,082	248.7	360.3	354.5
0.6	219.9	357.3	214.5	282.8	231.7	182.5	269.2	220.2	1,099	252.2	366.1	359.1
0.8	225.9	364.6	221.4	290.3	237.9	187.1	276.1	227.7	1,129	258.0	375.6	366.7
1.0	230.7	370.1	225.2	296.6	243.0	191.4	281.7	232.2	1,153	262.8	383.2	372.8
1.5	240.3	381.4	234.5	308.6	253.0	199.3	292.6	241.2	1,199	272.6	398.2	384.5
2.0	247.7	389.9	241.6	317.9	260.6	205.4	299.4	248.1	1,235	279.6	409.3	393.4
2.5	253.2	396.9	247.3	325.5	266.9	210.4	307.8	253.8	1,264	285.4	419.0	400.6
3.0	258.9	402.3	252.6	331.8	272.3	214.6	313.6	258.6	1,289	290.0	427.0	406.7
4.0	267.6	411.9	262.6	342.7	281.3	221.8	323.5	266.5	1,330	298.6	441.7	416.8
5.0	274.8	419.8	267.2	351.1	288.8	227.1	331.6	273.0	1,364	305.3	451.4	425.0
6	281.0	426.8	273.9	358.8	295.2	232.8	338.5	279.0	1,393	311.1	460.9	432.0
8	291.4	438.5	284.0	372.2	306.0	241.4	350.2	289.1	1,430	320.7	476.8	445.6
10	300.0	447.8	292.3	382.7	314.9	248.5	361.0	297.1	1,480	328.6	490.0	453.0
15	317.0	466.6	308.9	403.7	332.6	262.7	379.1	312.3	1,556	345.2	515.8	472.0
20	330.3	481.4	321.8	419.2	346.3	273.7	394.0	324.9	1,623	357.8	535.5	485.5
25		493.6	332.5	432.1	357.5	282.8	406.4	335.1	1,676	368.2	552.3	497.1
30		503.8	341.7	444.8	367.2	290.7	417.1	343.4	1,720	377.0	566.4	507.0
40		511.3	357.0	463.9	383.3		434.7	358.3	1,795	391.6	589.6	523.5
50		536.1					449.2		1,859	403.8		537.1
60									1,913	414.5		548.7
80									2,010			568.1
100									2,085			584.1

Table 4.2.25   Color Scale of Temperature for Iron or Steel

Color	Temperature °F	K
Dark blood red, black red	1,000	810
Dark red, blood red, low red	1,050	840
Dark cherry red	1,175	910
Medium cherry red	1,250	950
Cherry, full red	1,375	1,020
Light cherry, light red	1,550	1,120
Orange, free scaling heat	1,650	1,170
Light orange	1,725	1,210
Yellow	1,825	1,270
Light yellow	1,975	1,350
White	2,200	1,475

Table 4.2.26   Melting Points of Refractory Materials

Material	Temperature °F	K
Aluminum nitride, AlN	4,060	2,500
Aluminum oxide, Al_2O_3	3,720	2,320
Aluminum oxide–beryllium oxide, Al_2O_3-BeO	3,400	2,140
Beryllium carbide, Be_2C	3,810	2,370
Beryllium nitride, Be_3N_4	4,000	2,480
Beryllium oxide, BeO	4,570	2,790
Beryllium silicide, $2BeO \cdot SiO_2$	3,630	2,270
Borazon, BN	5,430	3,270
Calcia (lime), CaO	4,660	2,840
Graphite, C	6,700	3,980
Hafnia, HfO_2	5,090	3,090
Magnesia, MgO	5,070	3,070
Niobium carbide, NbC	6,330	3,770
Silica, SiO_2	3,110	1,980
Silicon carbide, SiC	3,990	2,470
Thoria, ThO_2	5,830	3,490
Titanium carbide, TiC	5,680	3,410
Zirconium aluminide, $ZrAl_2$	3,000	1,920
Zirconium beryllide, $ZrBe_{13}$	3,180	2,020
Zirconium carbide, ZrC	6,400	3,810
Zirconium disilicide, $ZrSi_2$	3,090	1,970
Zirconium nitride, ZrN	5,400	3,260
Zirconium oxide, ZrO_2	4,900	2,980
Zirconium silicides, $Zr_3Si_2, Zr_4Si_3, Zr_6Si_5$	$\approx$4,050	$\approx$2,500

Table 4.2.27   Mean Specific Heats of Various Solids (32–212°F, 273–373 K)

Solid	c, Btu/(lb · °F)	c, kJ/(kg · K)
Alumina	0.183	0.77
Asbestos	0.20	0.84
Ashes	0.20	0.84
Bakelite	$\approx$0.35	1.50
Basalt (lava)	0.20	0.84
Bell metal	0.086	0.36
Bismuth-tin	0.043	0.18
Borax	0.229	0.96
Brass, yellow	0.088	0.37
Brass, red	0.090	0.38
Brick	0.22	0.92
Bronze	0.104	0.44
Carbon-coke	0.203	0.85
Chalk	0.215	0.90
Charcoal	0.20	0.84
Cinders	0.18	0.75
Coal	$\approx$0.30	$\approx$1.25
Concrete	0.156	0.65
Constantan	0.098	0.41
Cork	0.485	2.03
Corundum	0.198	0.83
D'Arcet's metal	0.050	0.21
Dolomite	0.222	0.93
Ebonite	0.33	1.38
German silver	0.095	0.40
Glass, crown	0.16	0.70
Glass, flint	0.12	0.50
Glass, normal	0.20	0.84
Gneiss	0.18	0.75
Granite	0.20	0.84
Graphite	0.20	0.84
Gypsum	0.26	1.10
Hornblende	0.20	0.84
Humus (soil)	0.44	1.80
India rubber (para)	$\approx$0.37	1.50
Kaolin	0.224	0.94
Lead oxide (PbO)	0.055	0.23
Limestone	0.217	0.91
Lipowitz's metal	0.040	0.17
Magnesia	0.222	0.93
Magnesite (Fe_3O_4)	0.168	0.70
Marble	0.210	0.88
Nickel steel	0.109	0.46
Paraffin wax	0.69	2.90
Porcelain	0.22	0.92
Quartz	$\approx$0.23	0.96
Quicklime	0.217	0.91
Rose's metal	0.050	0.21
Salt, rock	0.21	0.88
Sand	0.195	0.82
Sandstone	0.22	0.92
Serpentine	0.25	1.05
Silica	0.191	0.80
Soda	0.231	0.97
Solders (Pb + Sn)	0.043	0.18
Sulfur	0.180	0.75
Talc	0.209	0.87
Tufa	0.33	1.40
Type metal	0.039	0.16
Vulcanite	0.331	1.38
Wood, fir	0.65	2.70
Wood, oak	0.57	2.40
Wood, pine	0.67	2.80
Wood's metal	0.040	0.17

Table 4.2.28 Phase Transition and Other Data for the Elements

Name	Symbol	Formula weight	T_m, K	Δh_{fus}, kJ/kg	T_b, K	Δh_{vap}, kJ/kg	T_c, K	P_c, bar	T_{tr}, K
Actinium	Ac	227.028	1,323	63	3,475	1,750			
Aluminum	Al	26.9815	933.5	398	2,750	10,875	7,850	4,800	
Antimony	Sb	121.75	903.9	163	1,905		5,700	3,200	368, 686
Argon	Ar	39.948	83	30	87.2	163	151	50	
Arsenic	As	74.9216				1,703	2,100		
Barium	Ba	137.33	1,002	55.8		1,099	4,450	720	643
Beryllium	Be	9.01218	1,560	1,355	2,750	32,450	6,200	4,600	1,530
Bismuth	Bi	208.980	544.6	54.0	1,838	725	4,450	1,400	
Boron	B	10.81	2,320	1,933	4,000		3,300		1,473
Bromine	Br	159.808	266	66.0	332	188	584		
Cadmium	Cd	112.41	594	55.1	1,040	886	2,690	1,680	
Calcium	Ca	40.08	1,112	213.1	1,763	3.833	4,300	1,000	720
Carbon	C	12.011	3,810		4,275		7,200	11,500	
Cerium	Ce	140.12	1,072	390		2,955	9,750	3,350	103, 263, 1003
Cesium	Cs	132.905	301.8	16.4	951	496	2,015	125	
Chlorine	Cl$_2$	70.906	172	180.7	239	576	417		
Chromium	Cr	51.996	2,133	325.6	2,950	6.622	5,500		2,113
Cobalt	Co	58.9332	1,766	274.7	3,185	6,390	6,300		700, 1,400
Copper	Cu	63.546	1,357	206.8	2,845	4,726	8,280	7,400	
Dysprosium	Dy	162.50	1,670	68.1	2,855	1,416	6,925	2,500	1,659
Erbium	Er	167.26	1,795	119.1	3,135	1,563	7,250		1,643
Europium	Eu	151.96	1,092	60.6	1,850	944	4,350	690	
Fluorine	F$_2$	37.997	53.5	13.4	85.0	172	144		46
Gadolinium	Gd	157.25	1,585	63.8	3,540	2,285	8,670		1,537
Gallium	Ga	69.72	303	80.1	2,500	3,688	7,125	4,150	276
Germanium	Ge	72.59	1,211	508.9	3,110	4,558	8,900	5,300	
Gold	Au	196.967	1,337	62.8	3,130	1,701	7,250	5,450	
Hafnium	Hf	178.49	2,485	134.8	4,885	3,211	10,400		2,000
Helium	He	4.00260	3.5	2.1	4.22	21	5.2	2.3	2.2
Holmium	Ho	164.930	1,744	73.8	2,968	1,461	7,575		1,703
Hydrogen	H$_2$	2.0159	14.0		20.4				
Indium	In	114.82	430	28.5	2,346	2,019	6,150	2,550	
Iodine	I$_2$	253.809	387	125.0	457		785		
Iridium	Ir	192.22	2,718	13.7	4,740	3,185	7,800		
Iron	Fe	55.847	1,811	247.3	3,136	6,259	8,500	10,000	1,183, 1,671
Krypton	Kr	83.80	115.8	19.6	119.8	108	209.4	55	
Lanthanum	La	138.906	1,194	44.6	3,715	2,978	10,500		550, 1,134
Lead	Pb	207.2	601	23.2	2,025	858	5,500	1,650	
Lithium	Li	6.941	454		1,607	21,340	3,700	1,000	80
Lutetium	Lu	174.967	1,937	106.6	3.668	2,034			
Magnesium	Mg	24.305	922	368.4	1,364	5,252	3,850	1,750	
Manganese	Mn	54.9380	1,518	219.3	2,334	4,112	4,325	560	1,374, 1,447
Mercury	Hg	200.59	234.6	11.4	630	293	1,720	1,500	194
Molybdenum	Mo	95.94	2,892	290.0	4,900		1,450	12,000	
Neodymium	Nd	144.24	1,290	49.6	3,341	1,891	7,900		1,132, 1,297
Neon	Ne	20.179	24.5	16.4	27.1	89	44.5	26.6	
Neptunium	Np	237.048	910		4,160		12,000		551, 847
Nickel	Ni	58.70	1,728	297.6	3,190	6,308	8,000	11,100	631
Niobium	Nb	92.9064	2,740	283.7	5,020	7,341	12,500		
Nitrogen	N$_2$	28.013	63.2	25.7	77.3	198	126.2	34.0	35.6
Osmium	Os	190.2	3,310	150.0	5,300	3,310	12,700		
Oxygen	O$_2$	31.9988	54.4	13.8	90.2	213	154.8		23.8, 43.8
Palladium	Pd	106.4	1,826	165.0	3,240	3,358	7,700	7,100	
Phosphorus	P	30.9738	317		553		995	81	196, 298
Platinum	Pt	195.09	2,045	101	4,100	2,612	10,700	11,000	
Plutonium	Pu	244	913	11.7	3,505	1,409	10,500	3,250	395, 480, 588, 730
Potassium	K	39.0983	336.4	60.1	1,032	2,052	2,210	170	
Praseodymium	Pr	140.908	1,205	49	3,785	2,05	8,900		1,066
Promethium	Pm	145	1,353		2,730				
Protactinium	Pa	231	1,500	64.8	4,300	2,036			
Radium	Ra	226.025	973		1,900				
Radon	Rn	222	202	12.3	211	83	377	66	
Rhenium	Re	186.207	3,453	177.8	5,920	3,842	18,900	14,900	
Rhodium	Rh	102.906	2,236	209.4	3,980	4,798	7,000		
Rubidium	Rb	85.4678	312.6	26.4	964	810	2,070	168	

Table 4.2.28   Phase Transition and Other Data for the Elements   (Continued)

Name	Symbol	Formula weight	T_m, K	Δh_{fus}, kJ/kg	T_b, K	Δh_{vap}, kJ/kg	T_c, K	P_c, bar	T_{tr}, K
Ruthenium	Ru	101.07	2,525	256.3	4,430	5,837	9,600		1,300, 1,475, 1,775
Samarium	Sm	150.4	1,345	57.3	2,064	1,107	5,050	1,780	1,190
Scandium	Sc	44.9559	1,813	313.6	3,550	6,989	6,410	3,750	1,608
Selenium	Se	78.96	494	66.2	958	1,210	1,810	320	398, 425
Silicon	Si	28.0855	1,684	1,802	3,540	14,050	5,160	540	
Silver	Ag	107.868	1,234	104.8	2,435	2,323	6,400	4,450	
Sodium	Na	22.9898	371	113.1	1,155	4,263	2,500	370	
Strontium	Sr	87.62	1,043	1,042	1,650	1,585	4,275	375	505, 893
Sulfur	S	32.06	388	53.4	718		1,210	130	369, 374
Tantalum	Ta	180.948	3,252	173.5	5,640	4,211	16,500	12,000	
Technetium	Tc	98	2,447	232	4,550	5,830	11,500		
Tellurium	Te	127.60	723	137.1	1,261	895	2,330		
Terbium	Tb	158.925	1,631	67.9	3,500	2,083	8,470		228, 1,575
Thallium	Tl	204.37	577	20.1	1,745	806	4,550	1,700	507
Thorium	Th	232.038	2,028	69.4	5,067	2,218	14,400	6,165	1,670
Thulium	Tm	168.934	1,819	99.6	2,220	1,129	6,450		
Tin	Sn	118.69	505	58.9	2,890	2,496	7,700	2,250	286, 476
Titanium	Ti	47.90	1,943	323.6	3,565	8,787	5,850		1,162, 1,353
Tungsten	W	183.85	3,660	192.5	5,890	4,483	15,500	15,000	
Uranium	U	238.029	1,406	35.8	4,422	1,949	12,500	5,000	938, 1,046
Vanadium	V	50.9415	2,191	410.7	3,680	8,870	11,300	10,300	
Xenon	Xe	131.30	161.3	17.5	164.9	96	290	58	
Ytterbium	Yb	173.04	1,098	44.2	1,467	745	4,080	1,150	1,050
Yttrium	Y	88.9059	1,775	128.2	3,610	4,485	8,950		1,758
Zinc	Zn	65.38	692.7	113.0	1,182	1,768			
Zirconium	Zr	91.22	2,125	185.3	4,681	6,376	10,500		1,135

Formula weight = molecular weight, T_m = normal melting temperature, Δh_{fus} = enthalpy (or latent heat) of fusion, T_b = normal boiling-point temperature, Δh_{vap} = enthalpy (or latent heat) of vaporization, T_c = critical temperature, P_c = critical pressure, T_{tr} = transition temperature.

SOURCE: Prepared by the author and abstracted from Rohsenow et al., "Handbook of Heat Transfer Fundamentals," McGraw-Hill.

Table 4.2.29   Thermophysical Properties of Selected Solid Elements

Element	Property	Temperature, K							
		100	200	300	400	500	600	800	1,000
Al	P, bar			2.1. − 43	4.9. − 31	1.1. − 23	1.0. − 18	1.4. − 12	6.6. − 9
	ρ, kg/m^3	2,732	2,719	2,701	2,681	2,661	2,639	2,591	2,365
	h, kJ/kg			4,623	4,716	4,812	4,913	5,131	5,768
	s, kJ/(kg · K)			1.056	1.323	1.539	1.723	2.035	2.728
	c_p, kJ/(kg · K)	0.481	0.797	0.902	0.949	0.997	1.042	1.134	0.921
	λ, W/(m · K)	300	237	237	240	236	231	218	
	α, m^2/s	2.3. − 4	1.1. − 4	9.7. − 5	9.4. − 5	8.9. − 5	8.4. − 5	7.4. − 5	6.6. − 5
Cr	P, bar			4.6. − 62	8.9. − 45	2.1. − 34	1.6. − 27	6.3. − 19	9.1. − 14
	ρ, kg/m^3	7,155	7,145	7,135	7,120	7,110	7,080	7,040	7,000
	h, kJ/kg			4,113	4,160	4,210	4,263	4,375	4,495
	s, kJ/(kg · K)			0.457	0.591	0.703	0.800	0.962	1.094
	c_p, kJ/(kg · K)	0.190	0.382	0.450	0.501	0.537	0.565	0.611	0.653
	λ, W/(m · K)	160	110	94	91	86	81	71	65
	α, m^2/s	1.2. − 4	4.1. − 5	2.9. − 5	2.6. − 5	2.3. − 5	2.0. − 5	1.7. − 5	1.4. − 5
Cu	P, bar			1.1. − 52	9.0. − 38	5.5. − 29	3.8. − 23	7.6. − 16	1.7. − 11
	ρ, kg/m^3	9,009	8,973	8,930	8,884	8,837	8,787	8,686	8,568
	h, kJ/kg			5,067	5,106	5,146	5,188	5,273	5,361
	s, kJ/(kg · K)			0.524	0.637	0.726	0.802	0.924	1.022
	c_p, kJ/(kg · K)	0.254	0.357	0.386	0.396	0.406	0.431	0.448	0.466
	λ, W/(m · K)	480	413	401	393	386	379	366	352
	α, m^2/s	2.2. − 4	1.3. − 4	1.2. − 4	1.1. − 4	1.1. − 4	1.0. − 4	9.0. − 5	8.0. − 5
Au	P, bar			6.7. − 58	6.3. − 42	2.8. − 32	3.6. − 25	6.5. − 18	3.7. − 13
	ρ, kg/m^3	19,460	19,380	19,300	19,210	19,130	19,040	18,860	18,660
	h, kJ/kg			6,046	6,059	6,072	6,086	6,113	6,142
	s, kJ/(kg · K)			0.241	0.279	0.309	0.333	0.373	0.404
	c_p, kJ/(kg · K)	0.109	0.124	0.129	0.131	0.133	0.136	0.141	0.147
	λ, W/(m · K)	327	323	317	311	304	298	284	270
	α, m^2/s	1.5. − 4	1.34. − 4	1.27. − 4	1.23. − 4	1.19. − 4	1.15. − 4	1.07. − 4	9.8. − 5
Fe	P, bar			3.1. − 65	6.3. − 54	3.9. − 47	1.5. − 36	6.6. − 20	1.5. − 14
	ρ, kg/m^3	7,900	7,880	7,860	7,830	7,800	7,760	7,690	7,650
	h, kJ/kg			4,523	4,570	4,621	4,676	4,801	4,958
	s, kJ/(kg · K)			0.491	0.626	0.740	0.840	1,018	1.193
	c_p, kJ/(kg · K)	0.216	0.384	0.450	0.491	0.524	0.555	0.692	1.034
	λ, W/(m · K)	134	94	80	70	61	55	43	32
	α, m^2/s	8.2. − 5	3.1. − 5	2.2. − 5	1.8. − 5	1.5. − 5	1.3. − 5	1.1. − 5	1.0. − 5
Pb	P, bar			6.3. − 29	1.8. − 20	2.2. − 15	5.4. − 12	6.2. − 8	1.6. − 5
	ρ, kg/m^3	11,520	11,430	11,330	11,230	11,130	11,010	10,430	10,190
	h, kJ/kg			6,929	6,942	6,955	6,969	7,022	7,050
	s, kJ/(kg · K)			0.314	0.351	0.381	0.406	0.487	0.519
	c_p, kJ/(kg · K)	0.118	0.125	0.129	0.132	0.137	0.142	0.145	0.142
	λ, W/(m · K)	39.7	36.7	35.3	34.0	32.8	31.4		
	α, m^2/s	2.9. − 5	2.6. − 5	2.4. − 5	2.3. − 5	2.2. − 5	2.0. − 5	1.3. − 5	1.5. − 5
Li	P, bar								
	ρ, kg/m^3	546	541	533	526	492	482	462	442
	h, kJ/kg			4,615	4,919	5,844	6,274	7,113	7,945
	s, kJ/(kg · K)			4.214	5.289	7.182	7.967	9.173	10.102
	c_p, kJ/(kg · K)	1.923	3.105	3.54	3.76	4.34	4.26	4.17	4.15
	λ, W/(m · K)	105	90	85	80				
	α, m^2/s	1.0. − 4	5.4. − 5	4.5. − 5	3.2. − 5	2.1. − 5	2.3. − 5	2.8. − 5	3.3. − 5
Ni	P, bar			1.1. − 67	5.8. − 49	9.8. − 38	2.7. − 30	5.5. − 21	2.1. − 15
	ρ, kg/m^3	8,960	8,930	8,900	8,860	8,820	8,780	8,690	8,610
	h, kJ/kg			4,837	4,883	4,934	4,990	5,099	5,207
	s, kJ/(kg · K)			0.512	0.645	0.758	0.859	1.017	1.137
	c_p, kJ/(kg · K)	0.232	0.383	0.444	0.490	0.540	0.590	0.530	0.556
	λ, W/(m · K)	165	105	91	80	72	66	68	72
	α, m^2/s	8.0. − 5	3.1. − 5	2.3. − 5	1.9. − 5	1.5. − 5	1.3. − 5	1.4. − 5	1.5. − 5
Pt	P, bar			3.2. − 91	1.3. − 66	7.3. − 52	5.0. − 42	9.7. − 30	2.3. − 22
	ρ, kg/m^3	21,550	21,500	21,450	21,380	21,330	21,270	21,140	21,010
	h, kJ/kg			5,837	5,850	5,864	5,878	5,907	5,937
	s, kJ/(kg · K)			0.214	0.253	0.283	0.309	0.350	0.383
	c_p, kJ/(kg · K)	0.101	0.127	0.134	0.136	0.138	0.140	0.146	0.152
	λ, W/(m · K)	78	73	72	72	72	73	76	79
	α, m^2/s	3.6. − 5	2.7. − 5	2.5. − 5	2.5. − 5	2.5. − 5	2.5. − 5	2.5. − 5	2.5. − 5

Table 4.2.29   Thermophysical Properties of Selected Solid Elements   (Continued)

Element	Property	Temperature, K							
		100	200	300	400	500	600	800	1,000
Rh	P, bar			5.5. − 89	6.5. − 65	1.8. − 50	7.1. − 41	7.0. − 29	1.1. − 21
	ρ, kg/m³	12,480	12,460	12,430	12,400	12,360	12,330	12,250	12,170
	h, kJ/kg			4,921	4,946	4,972	4,999	5,055	5,115
	s, kJ/(kg · K)			0.308	0.380	0.437	0.487	0.568	0.635
	c_p, kJ/(kg · K)	0.147	0.220	0.246	0.257	0.265	0.274	0.290	0.307
	λ, W/(m · K)	190	154	150	146	141	136	127	121
	α, m²/s	1.0. − 4	5.6. − 5	4.9. − 5	4.6. − 5	4.3. − 5	4.0. − 5	3.6. − 5	3.2. − 5
Ag	P, bar			2.1. − 43	4.9. − 31	1.1. − 23	1.0. − 18	1.4. − 12	6.6. − 9
	ρ, kg/m³	10,600	10,550	10,490	10,430	10,360	10,300	10,160	10,010
	h, kJ/kg			5,791	5,815	5,839	5,864	5,915	5,968
	s, kJ/(kg · K)			0.3959	0.4641	0.5180	0.5630	0.6365	0.6964
	c_p, kJ/(kg · K)	0.187	0.225	0.236	0.240	0.245	0.251	0.264	0.276
	λ, W/(m · K)	450	430	429	425	419	412	396	379
	α, m²/s	2.3. − 4	1.8. − 4	1.7. − 4	1.7. − 4	1.7. − 4	1.6. − 4	1.5 − 4	1.3. − 4
Ti	P, bar			1.0. − 74	4.6. − 54	5.2. − 42	7.2. − 35	1.2. − 23	1.4. − 17
	ρ, kg/m³	4,530	4,520	4,510	4,490	4,480	4,470	4,440	4,410
	h, kJ/kg			4,857	4,911	4,967	5,025	5,147	5,278
	s, kJ/(kg · K)			0.643	0.797	0.922	1.028	1.205	1.350
	c_p, kJ/(kg · K)	0.295	0.464	0.525	0.555	0.578	0.597	0.627	0.670
	λ, W/(m · K)	31	25	21	20	20	19	19	21
	α, m²/s								
W	P, bar			3.2. − 141	2.9. − 104	4.3. − 82	2.7. − 67	8.7. − 49	1.1. − 37
	ρ, kg/m³	19,310	19,290	19,270	19,240	19,220	19,190	19,130	19,080
	h, kJ/kg			6,255	6,268	6,282	6,296	6,325	6,354
	s, kJ/(kg · K)			0.178	0.217	0.248	0.273	0.315	0.347
	c_p, kJ/(kg · K)	0.089	0.125	0.135	0.137	0.139	0.140	0.144	0.148
	λ, W/(m · K)	208	185	174	159	146	137	125	118
	α, m²/s								
V	P, bar			3.0. − 82	7.1. − 60	1.9. − 46	1.6. − 37	2.4. − 26	1.2. − 19
	ρ, kg/m³	6,074	6,062	6,050	6,030	6,010	6,000	5,960	5,920
	h, kJ/kg			4,740	4,790	4,843	4,896	5,006	5,121
	s, kJ/(kg · K)			0.571	0.716	0.832	0.930	1.088	1.216
	c_p, kJ/(kg · K)	0.257	0.434	0.483	0.512	0.528	0.540	0.563	0.598
	λ, W/(m · K)	36	31	31	31	32	33	36	38
	α, m²/s	2.3. − 5	1.2. − 5	1.1. − 5	1.0. − 5	1.0. − 5	1.0. − 5	1.1. − 5	1.1. − 5
Zn	P, bar			3.7. − 17	1.6. − 11	3.7. − 8	6.7. − 6	3.4. − 3	1.2. − 1
	ρ, kg/m³	7,260	7,200	7,135	7,070	7,000	6,935	6,430	6,260
	h, kJ/kg			5,690	5,730	5,771	5,813	5,970	6,114
	s, kJ/(kg · K)	0.295		0.639	0.753	0.844	0.922	1.216	1.323
	c_p, kJ/(kg · K)		0.366	0.389	0.404	0.419	0.435	0.479	0.479
	λ, W/(m · K)	117	118	116	111	107	103		
	α, m²/s	5.5. − 5	4.7. − 5	4.1. − 5	3.9. − 5	3.7. − 5	3.4. − 5	1.8. − 5	2.2. − 5
Zr	P, bar			2.8. − 99	8.6. − 73	6.6. − 57	2.7. − 46	4.6. − 33	4.1. − 25
	ρ, kg/m³	6,535	6,525	6,515	6,510	6,490	6,480	6,450	6,420
	h, kJ/kg			5,540	5,569	5,600	5,632	5,698	5,768
	s, kJ/(kg · K)			0.429	0.513	0.590	0.640	0.735	0.813
	c_p, kJ/(kg · K)	0.120	0.126	0.130	0.136	0.143	0.153	0.153	0.153
	λ, W/(m · K)	33	25	23	22	21	21	21	23
	α, m²/s								

P = saturation vapor pressure; ρ = density; h = enthalpy; s = entropy; c_p = specific heat at constant pressure; λ = thermal conductivity; α = thermal diffusivity.

4.3 RADIANT HEAT TRANSFER

by Hoyt C. Hottel, Adel F. Sarofim, and James J. Noble

REFERENCE: Baukal, "The John Zink Combustion Handbook," CRC Press. Blokh, "Heat Transfer in Steam Boiler Furnaces," Taylor & Francis. "Radiative Heat Transfer," 2d ed., Academic Press, 2003. Noble, *Int. J. Heat Mass Transfer*, **18**, 1975, pp. 261–269. Siegel and Howell, "Thermal Radiation Heat Transfer," 4th ed., Taylor & Francis. Stultz and Kitto, "Steam," 41st ed., Babcock and Wilcox. Hottel and Sarofim, "Radiative Transfer," McGraw-Hill.

A heated body loses energy continuously by radiation, at a rate dependent on the shape, the size, and, particularly, the temperature of the body. In contrast to conductive energy transport, such emitted radiation is capable of passage to a distant body, where it may be absorbed, reflected, scattered, or transmitted.

Consider a pencil of radiation, defined as all the rays passing through each of two small, widely separated areas dA_1 and dA_2. The rays at dA_1 will have a solid angle of divergence $d\Omega_1$; equal to the apparent area of dA_2 viewed from dA_1, divided by the square of the separating distance. Let the normal to dA_1 make the angle θ_1 with the pencil. The flux density q [energy/(time)(area normal to beam)] per unit solid angle of divergence is called the **intensity I**, and the flux dQ_1 (energy/time) through area dA_1 (of apparent area $dA_1 \cos \theta_1$ normal to the beam) is therefore given by

$$dQ_1 = dA_1 \cos \theta_1 q_1 = I \, dA_1 \cos \theta_1 \, d\Omega_1 \qquad (4.3.1)$$

The intensity I along a pencil, in the absence of absorption or scatter is constant (unless the beam passes into a medium of different refractive index n; $I_1/n_1^2 = I_2/n_2^2$). The **emissive power*** of a surface is the flux density [energy/(time)(surface area)] due to emission from it throughout a hemisphere. If the intensity I of emission from a surface is independent of the angle of emission, Eq. (4.3.1) may be used to show that the surface emissive power is πI, though the emission is throughout 2π steradians.

BLACKBODY RADIATION

Engineering calculations of thermal radiation from surfaces are best keyed to the radiation characteristics of the **blackbody, or ideal radiator.** The characteristic properties of a blackbody are that it absorbs all the radiation incident on its surface and that the quality and intensity of the radiation it emits are completely determined by its temperature. The total radiative flux throughout a hemisphere from a black surface of area A and absolute temperature T is given by the **Stefan-Boltzmann law:** $Q = A\sigma T^4$ or $q = \sigma T^4$. The Stefan-Boltzmann constant σ has the value 5.67×10^{-8} W/m^2(K)^4, 0.1713×10^{-8} Btu/(ft)^2(h)($^\circ$R)^4 or 1.356×10^{-12} cal/(cm)^2(s)(K)^4. From the above definition of emissive power, σT^4 is the total emissive power of a blackbody, called E; and the intensity I_B of emission from a blackbody is E/π, or $\sigma T^4/\pi$.

The spectral distribution of energy flux from a blackbody is expressed by Planck's law

$$E_\lambda \, d\lambda = \frac{2\pi hc^2 n^2 \lambda^{-5}}{e^{hc/(k\lambda T)} - 1} \, d\lambda \equiv \frac{n^2 c_1 \lambda^{-5}}{e^{c_2/(\lambda T)} - 1} \qquad (4.3.2)$$

wherein $E_\lambda \, d\lambda$ is the hemispherical flux density in W/m^2 lying in the wavelength range λ to $\lambda + d\lambda$; h is Planck's constant, 6.6262×10^{-34} J $\cdot$ s; c is the velocity of light in vacuo, 2.9979×10^8 m/s; k is the Boltzmann constant, 1.3807×10^{-23} J/K; λ is the wavelength measured in vacuo, m; n is the refractive index of the emitter; c_1 and c_2, the first and second Planck's law constants, are 3.7418×10^{-16} W $\cdot$ m^2 and

* Variously called, in the literature, **emittance, total hemispherical intensity, radiant flux density** or **exitance.**

1.4388×10^{-2} m $\cdot$ K. To show how E_λ varies with wavelength or temperature, Planck's law may be cast in the form

$$\frac{E_\lambda}{n^2 T^5} = \frac{c_1 (\lambda T)^{-5}}{e^{c_2/(\lambda T)} - 1} \qquad (4.3.3)$$

i.e., when $n \cong 1$ (e.g., in a gas), E_λ/T^5 is a unique function of λT. And E_λ is a maximum at $\lambda T = 2,898$ μm $\cdot$ K (Wien's displacement law). A more useful displacement law: Half of blackbody radiation lies on either side of $\lambda T = 4,107$ μm $\cdot$ K. Another: The maximum intensity per unit fractional change in wavelength *or* frequency is at $\lambda T = 3,670$ μm $\cdot$ K. Integration of E_λ over λ shows that the fraction f of blackbody radiation lying at wavelengths below λ depends only on λT. Values of f versus λT appear in Table 4.3.1. A twofold range of λT geometrically centered on $\lambda T = 3,670$ μm $\cdot$ K spans about half the energy.

A limiting form of the Planck equation as $\lambda T \to 0$ is $E_\lambda = n^2 c_1 \lambda^{-5} e^{-c_2/(\lambda T)}$, the Wien equation, less than 1 percent in error when λT is less than 3,000 μm $\cdot$ K. This is useful for optical pyrometry (red screen $\lambda = 0.65$ μm) when $T < 4,800$ K.

Table 4.3.1 Fraction f of Blackbody Radiation below λT
$\lambda T = \mu$m $\cdot$ K

λT	1,200	1,600	1,800	2,000	2,200	2,400	2,600	2,800
f	0.002	0.020	0.039	0.067	0.101	0.140	0.183	0.228
λT	3,000	3,200	3,400	3,600	3,800	4,000	4,200	4,500
f	0.273	0.318	0.362	0.404	0.443	0.480	0.516	0.564
λT	4,800	5,100	5,500	6,000	6,500	7,000	7,600	8,400
f	0.608	0.646	0.691	0.738	0.776	0.808	0.839	0.871
λT	10,000	12,000	14,000	20,000	50,000			
f	0.914	0.945	0.963	0.986	0.999			

RADIATIVE EXCHANGE BETWEEN SURFACES OF SOLIDS

The ratio of the total radiating power of a real surface to that of a black surface at the same temperature is called the **emittance** of the surface (for a perfectly plane surface, the **emissivity**), designated by ε. Subscripts λ, θ, and n may be assigned to differentiate monochromatic, directional, and surface-normal values, respectively, from the total hemispherical value. If radiation is incident on a surface, the fraction absorbed is called the **absorptance (absorptivity),** a term in which two subscripts may be appended, the first to identify the temperature of the surface and the second to identify the quality of the incident radiation. According to **Kirchhoff's law,** the emissivity and absorptivity of a surface *in surroundings at its own temperature* are the same, for both monochromatic and total radiation. When the temperatures of the surface and its surroundings differ, the total emissivity and absorptivity of the surface are found often to be different, but because absorptivity is substantially independent of irradiation density, the monochromatic emissivity and absorptivity of surfaces are for all practical purposes the same. The difference between total emissivity and absorptivity depends on the variation, with wavelength, of ε_λ and on the difference between the emitter temperature and the effective source temperature.

Consider radiative exchange between a body of area A_1 and temperature T_1 and its black surroundings at T_2. The net interchange is given by

$$\dot{Q}_{1 \rightleftharpoons 2} = A_1 \int_0^\infty [\varepsilon_\lambda E_\lambda(T_1) - \alpha_\lambda E_\lambda(T_2)] \, d\lambda$$

$$= A_1(\varepsilon_1 \sigma T_1^4 - \alpha_{12} \sigma T_2^4) \qquad (4.3.4)$$

where

$$\varepsilon_1 = \int_0^1 \varepsilon_\lambda \, df_{\lambda T_1} \quad \text{and} \quad \alpha_{12} = \int_0^1 \varepsilon_\lambda \, df_{\lambda T_2} \quad (4.3.5)$$

i.e., ε_1 (or α_{12}) is the area under a curve of ε_λ versus f, read as a function of λT at T_1 (or T_2) from Table 4.3.1. If ε_λ does not change with wavelength, the surface is called **gray**, and $\varepsilon_1 = \alpha_{12} = \varepsilon_\lambda$. A selective surface is one whose ε_λ changes dramatically with wavelength. If this change is monotonic, ε_1, and α_{12} are, according to Eqs. (4.3.4) and (4.3.5), markedly different when the absolute temperature ratio is far from 1; e.g., when $T_1 = 293$ K (ambient temperature) and $T_2 = 5,800$ K (effective solar temperature), $\varepsilon_1 = 0.9$ and $\alpha_{12} = 0.1$–0.2 for a white paint, but ε_1 can be as low as 0.12 and α_{12} above 0.9 for a thin layer of copper oxide on bright aluminum, or of chromic oxide on bright nickel.

Although values of emittances and absorptances depend in very complex ways on the real and imaginary components of the refractive index and on the geometric structure of the surface layer, some generalizations are possible.

Polished Metals   (1) ε_λ is quite low in the infrared and, for $\lambda > 8$ μm, can be adequately approximated by $0.00365 \sqrt{r/\lambda}$, where r is the resistivity in ohm · cm and λ is in micrometres; at shorter wavelengths, ε_λ increases and, for many metals, has values of 0.4 to 0.8 in the visible

Fig. 4.3.1   Variation of absorptivity with temperature of radiation source. (1) Slate composition roofing; (2) linoleum, red-brown; (3) asbestos slate (asbestos use is obsolete, but may be encountered in existing construction); (4) soft rubber, gray; (5) concrete; (6) porcelain; (7) vitreous enamel, white; (8) red brick; (9) cork; (10) white Dutch tile; (11) white chamotte; (12) MgO, evaporated; (13) anodized aluminum; (14) aluminum paint; (15) polished aluminum; (16) graphite. The two dashed lines bound the limits of data for gray paving brick, asbestos paper (asbestos use is obsolete, but may be encountered in existing construction), wood, various cloths, plaster of paris, lithopone, and paper.

(0.4–0.7 μm). ε_λ is approximately proportional to the square root of the absolute temperature ($\varepsilon_\lambda \propto \sqrt{r}$ and $r \propto T$) in the far infrared ($\lambda > 8$ μm), is temperature insensitive in the near infrared (0.7–1.5 μm) and, in the visible, decreases slightly as temperature increases. (2) Total emittance is substantially proportional to absolute temperature; at moderate temperature, $\varepsilon_n = 0.58T\sqrt{r_0/T_0}$, where T is in kelvins. (3) Total absorptance of a metal at T_1 for radiation from a black or gray source at T_2 is equal to the emissivity evaluated at the geometric mean of T_1 and T_2. (4) The ratio of hemispherical to normal emittance (absorptance) varies from 1.33 at very lower ε's (α's) to about 1.03 at an ε (α) of 0.4.

Unless extraordinary pains are taken to prevent oxidation, however, a metallic surface may exhibit several times the emittance or absorptance of a polished specimen. The emittance of iron and steel, for example, varies widely with degree of oxidation and roughness—clean metallic surfaces have an emittance of from 0.05–0.45 at ambient temperatures to 0.4–0.7 at high temperatures; oxidized and/or rough surfaces range from 0.6–0.95 at low temperatures to 0.9–0.95 at high temperatures.

Refractory Materials   Grain size and concentration of trace impurities are important. (1) Most refractory materials have an ε_λ of 0.8 to 1.0 at wavelengths beyond 2 to 4 μm; ε_λ decreases rapidly toward shorter wavelengths for materials that are white in the visible but retains its high value for black materials such as FeO and Cr_2O_3. Small concentrations of FeO and Cr_2O_3 or other colored oxides can cause marked increases in the emittance of materials that are normally white. ε_λ for refractory materials varies little with temperature. (2) Refractory materials generally have a total emittance which is high (0.7 to 1.0) at ambient temperatures and decreases with increase in temperature; a change from 1,000 to 1,600°C may cause a decrease in ε of one-fourth to one-third. (3) The emittance and absorptance increase with increase in grain size over a grain-size range of 1–200 μm. (4) The ratio $\varepsilon/\varepsilon_n$ of hemispherical to normal emissivity of polished surfaces varies with refractive index from 1 at $n = 1$ to 0.95 at $n = 1.5$ (common glass) and back to 0.98 at $n = 3.5$. (5) The ratio $\varepsilon/\varepsilon_n$ for a surface composed of particulate matter which scatters isotropically varies with ε from 1 when $\varepsilon = 1$ to 0.8 when $\varepsilon = 0.07$. (6) The total absorptance shows a decrease with increase in temperature of the radiation source similar to the decrease in emittance with increase in the specimen temperature. Figure 4.3.1 shows the effect of the temperature of the radiation source on the absorptance of surfaces of various materials at room temperature. It will be noted that polished aluminum (line 15) and anodized aluminum (line 13), representative of metals and nonmetals, respectively, respond oppositely to a change in the temperature of the radiation source. The absorptance of surfaces for sunlight may be read from the right of Fig. 4.3.1, assuming sunlight to consist of blackbody radiation from a source at 10,440°R (5,800 K).

When T_2 is not too different from T_1, α_{12} may be expressed as $\varepsilon_1(T_2/T_1)^n$, with n determined from Fig. 4.3.1. For this case, Eq. (4.3.4) becomes

$$\dot{Q}_{1,\text{net}} = \sigma A_1 \varepsilon_{AV}(1 + n/4)(T_1^4 - T_2^4) \quad (4.3.6)$$

where ε_{AV} is evaluated at the arithmetic mean of T_1 and T_2.

Table 4.3.2 gives the emittance of various surfaces and emphasizes the variation possible in a single material. The values in the table apply, with a few exceptions, to normal radiation from the surface.

For opaque materials, the **reflectance** ρ is the complement of the absorptance. The directional distribution of the reflected radiation depends on the material, its degree of roughness or grain size, and if a metal, its state of oxidation. Polished surfaces of homogeneous materials reflect specularly. In contrast, the intensity of the radiation reflected from a **perfectly diffuse**, or **Lambert**, surface is independent of direction. The directional distribution of reflectance of many oxidized metals, refractory materials, and natural products approximates that of a perfectly diffuse reflector. A better model, adequate for many calculational purposes, is achieved by assuming that the total reflectance ρ is the sum of diffuse and specular components ρ_D and ρ_S (Hottel and Sarofim, p. 180).

Black Surface Enclosures   When several surfaces are present, the need arises for evaluating a geometric factor F, called the **direct-view factor**. Restriction is temporarily to black surfaces, the intensity from which is independent of angle of emission. Define F_{12} as the fraction of

Table 4.3.2   Emissivity of Surfaces

Surface	Temp.,* °C	Emissivity*
Metals and their oxides		
Aluminum:		
Highly polished	230–580	0.039–0.057
Polished	23	0.040
Rough plate	26	0.055–0.07
Oxidized at 600°C	200–600	0.11–0.19
Oxide	280–830	0.63–0.26
Alloy 75ST	24	0.10
75ST, repeated heating	230–480	0.22–0.16
Brass:		
Highly polished	260–380	0.03–0.04
Rolled plate, natural	22	0.06
Rolled, coarse-emeried	22	0.20
Oxidized at 600°C	200–600	0.61–0.59
Chromium	40–540	0.08–0.26
Copper:		
Electrolytic, polished	80	0.02
Comm'l plate, polished	20	0.030
Heated at 600°C	200–600	0.57–0.57
Thick oxide coating	25	0.78
Cuprous oxide	800–1,100	0.66–0.54
Molten copper	1,080–1,280	0.16–0.13
Dow metal, cleaned, heated	230–400	0.24–0.20
Gold, highly polished	230–630	0.02–0.04
Iron and steel:		
Pure Fe, polished	180–980	0.05–0.37
Wrought iron, polished	40–250	0.28
Smooth sheet iron	700–1,040	0.55–0.60
Rusted plate	20	0.69
Smooth oxidized iron	130–530	0.78–0.82
Strongly oxidized	40–250	0.95
Molten iron and steel	1,500–1,770	0.40–0.45
Lead:		
99.96%, unoxidized	130–230	0.06–0.08
Gray, oxidized	24	0.28
Oxidized at 190°C	190	0.63
Mercury, pure clean	0–100	0.09–0.12
Molybdenum filament	730–2590	0.10–0.29
Monel metal, K5700		
Washed, abrasive soap	24	0.17
Repeated heating	230–875	0.46–0.65
Nickel and alloys:		
Electrolytic, polished	23	0.05
Electroplated, not polished	20	0.11
Wire	190–1,010	0.10–0.19
Plate, oxid. at 600°C	200–600	0.37–0.48
Nickel oxide	650–1,250	0.59–0.86
Copper-nickel, polished	100	0.06
Nickel-silver, polished	100	0.14
Nickelin, gray oxide	21	0.26
Nichrome wire, bright	50–1,000	0.65–0.79
Nichrome wire, oxide	50–500	0.95–0.98
ACI-HW (60Ni, 12Cr); firm black ox, coat	270–560	0.89–0.82
Platinum, polished plate	230–1,630	0.05–0.17
Silver, pure polished	230–630	0.02–0.03
Stainless steels:		
Type 316, cleaned	24	0.28
316, repeated heating	230–870	0.57–0.66
304, 42 h at 520°C	220–530	0.62–0.73
310, furnace service	220–530	0.90–0.97
Allegheny #4, polished	100	0.13
Tantalum filament	1,330–3,000	0.194–0.33
Thorium oxide	280–830	0.58–0.21
Tin, bright	24	0.04–0.06
Tungsten, aged filament	25–3,320	0.03–0.35
Zinc, 99.1%, comm'l, polished	230–330	0.05
Galv. iron, bright	28	0.23
Galv. iron, gray oxide	24	0.28

Table 4.3.2   Emissivity of Surfaces   (Continued)

Surface	Temp.,* °C	Emissivity*
Refractories, building materials, paints, misc.		
Alumina	260–680	0.6–0.33
Alumina, 50-μm grain size	1,010–1,570	0.39–0.28
Alumina-silica, cont'g	1,010–1,570	
0.4% Fe_2O_3		0.61–0.43
1.7% Fe_2O_3		0.73–0.62
2.9% Fe_2O_3		0.78–0.68
Al paints (vary with amount of lacquer body, age)	100	0.27–0.67
Asbestos	40–370	0.93–0.95
Calcium oxide	750–1,100	0.29–0.28
Candle soot; lampblack-waterglass	20–370	0.95 ± 0.01
Carbon plate, heated	130–630	0.81–0.79
Ferric oxide (Fe_2O_3)	500–900	0.8–0.43
Magnesium oxide, 1 μm	260–760	0.67–0.41
Oil layers		
Lube oil, 0.01 in on pol. Ni	20	0.82
Linseed, 1–2 coats on A1	20	0.56–0.57
Rubber, soft gray reclaimed	24	0.86
Silica, 3 μm	260–740	0.7–0.5
Misc. I: shiny black lacquer, planed oak, white enamel, serpentine, gypsum, white enamel paint, roofing paper, lime plaster, black matte shellac	21	0.87–0.91
Misc. II: glazed porcelain, white paper, fused quartz, polished marble, rough red brick, smooth glass, hard glossy rubber, flat black lacquer, water, electrographite	21	0.92–0.96

* When two temperatures and two emissivities are given they correspond, first to first and second to second, and linear interpolation is suggested.

the radiation leaving surface A_1 in all directions which is intercepted by surface A_2. Since the net interchange between A_1 and A_2 must be zero when their temperatures are alike, it follows that $A_1F_{12} = A_2F_{21}$. From the definition of F and Eq. (4.3.1),

$$A_1F_{12} = \int_{A1} \int_{\Omega} \frac{dA_1 \cos \theta_1 \, d\Omega_1}{\pi}$$

$$= \int_{A1} \int_{A2} \frac{dA_1 \cos \theta_1 \, dA_2 \cos \theta_2}{\pi r^2} \qquad (4.3.7)$$

where $dA \cos \theta$ is the projection of dA normal to r, the line connecting dA_1 and dA_2. The product A_1F_{12}, having the dimensions of area, will be called the **direct-interchange area** and be designated by $\overline{s_1 s_2}$, sometimes for brevity by $\overline{12} \ (\equiv \overline{21})$. Clearly, $\overline{11} + \overline{12} + \overline{13} + \cdots = A_1$; and when A_1 cannot "see" itself, $\overline{11} = 0$. Values of F or $\overline{ss}$ have been calculated for various surface arrangements.

Direct-View Factors and Direct Interchange Areas

CASE 1.   Directly opposed parallel rectangles of equal dimensions, and with lengths of sides X and Y divided by separating distance z:

$$\overline{s_1 s_2} \ (\equiv A_1F_{12} \equiv A_2F_{21}) = \frac{z^2}{\pi}\left[\ln\left\{ \frac{(1 + X^2)(1 + Y^2)}{1 + X^2 + Y^2} \right\}^{1/2} \right.$$
$$+ X\sqrt{1 + Y^2} \tan^{-1}\frac{X}{\sqrt{1 + Y^2}} + Y\sqrt{1 + X^2} \tan^{-1}\frac{Y}{\sqrt{1 + X^2}}$$
$$\left. - X \tan^{-1}X - Y \tan^{-1}Y \right]$$

See also Fig. 4.3.2.

CASE 2.   Parallel circular disks with centers on a common normal and with radii R_1 and R_2 divided by separating distance z:

$$\overline{s_1 s_2} \ (\equiv A_1F_{12} \equiv A_2F_{21}) = \frac{\pi z^2}{2}\left[1 + R_1^2 + R_2^2 \right.$$
$$\left. - \sqrt{(1 + R_1^2 + R_2^2)^2 - 4R_1^2 R_2^2} \right]$$

Fig. 4.3.2   Variation of the factor F or $\overline{F}$ for parallel planes directly opposed.

CASE 3.   Rectangles in perpendicular planes of area A_1 and A_2, with a common edge l and with other dimension divided by $l = W_1$ and W_2:

$$\overline{s_1 s_2}\ (\equiv A_1 F_{12} \equiv A_2 F_{21})$$

$$= \frac{l^2}{\pi}\left(\frac{1}{4}\ln\left\{ \frac{(1 + W_1^2)(1 + W_2^2)}{1 + W_1^2 + W_2^2}\left[\frac{W_1^2(1 + W_1^2 + W_2^2)}{(1 + W_1^2)(W_1^2 + W_2^2)} \right]^{W_1^2} \right.\right.$$

$$\left. \times \left[\frac{W_2^2(1 + W_1^2 + W_2^2)}{(1 + W_2^2)(W_1^2 + W_2^2)} \right]^{W_2^2} \right\}$$

$$\left. + W_1\tan^{-1}\frac{1}{W_1} + W_2\tan^{-1}\frac{1}{W_2} - \sqrt{W_1^2 + W_2^2}\ \tan^{-1}\frac{1}{\sqrt{W_1^2 + W_2^2}} \right)$$

CASE 4.   Circular cylinder of radius r_1 surrounded by cylinder of radius r_2, both of equal length l and on a common axis:

$$A_1 = 2\pi r_1 l \qquad A_2 = 2\pi r_2 l$$

Let $r_1/l = R_1$; $r_2/l = R_2$; $[1/R_1^2 - (R_2/R_1)^2 + 1] = B$; $[1/R_1^2 + (R_2/R_1)^2 + 1] = D$; and $r_2/r_1 = R$.

$$\overline{s_1 s_2}\ (\equiv A_1 F_{12} \equiv A_2 F_{21})$$

$$= l^2\left\{ R_1^2\left[\sqrt{(D + 2)^2 - 4R^2}\cos^{-1}\frac{B}{DR} + B\sin^{-1}\frac{1}{R} - \frac{\pi}{2}D \right] \right.$$

$$\left. + 2R_1\left(\pi - \cos^{-1}\frac{B}{D} \right) \right\}$$

$$\overline{s_2 s_2}\ (\equiv A_2 F_{22}) = l^2 R_1\left\{ 2\pi(R - 1) + 4\tan^{-1}(2R_1\sqrt{R^2 - 1}) \right.$$

$$- \sqrt{4R^2 + \frac{1}{R_1^2}}\left(\frac{\pi}{2} + \sin^{-1}\frac{4(R^2 - 1) + 1/[R_1^2(1 - 2/R^2)]}{4(R^2 - 1) + 1/R_1^2} \right)$$

$$\left. + \frac{1}{R_1}\left[\sin^{-1}\left(1 - \frac{2}{R^2} \right) + \frac{\pi}{2} \right] \right\}$$

CASE 5.   Two closed surfaces, one enclosing the other and neither having any negative curvature; A_1 is inside. Since $F_{12} = 1$,

$$\overline{s_1 s_2}\ (\equiv A_1 F_{12} \equiv A_2 F_{21}) \equiv A_1$$

$$F_{21} = \frac{A_1}{A_2} \qquad F_{22} = 1 - \frac{A_1}{A_2}$$

CASE 6.   Sphere of total inside area A_T; radiative exchange between sphere segments of areas A_1 and A_2. Application of Eq. (4.3.7) shows that, independent of relative position,

$$\overline{s_1 s_2}\ (\equiv A_1 F_{12} \equiv A_2 F_{21}) = \frac{A_1 A_2}{A_T} \qquad F_{12} = \frac{A_2}{A_T}$$

CASE 7.   Two dimensional surfaces A_1 and A_2 per unit length normal to cross section, with each area defined by the length of string, stretched on inside face, between ends (i.e., elimination of negative curvature). Graphical exact solution: $\overline{s_1 s_2}$ (per unit normal length) = sum of lengths of crossed stretched strings between ends of A_1 and A_2 minus sum of uncrossed strings, all divided by 2. If an obstruction lies between A_1 and A_2, there may be two sets of strings to represent views on both sides of the obstruction, with results added. The relations for cases 8, 9, and 10 are the results of three among many applications of this principle.

CASE 8.   Exchange among inside surfaces of hollow triangular shape of infinite length and areas A_1, A_2, and A_3:

$$\overline{s_1 s_2}\ (\equiv A_1 F_{12} \equiv A_2 F_{21}) = \frac{A_1 + A_2 - A_3}{2} \qquad F_{12} = \frac{A_1 + A_2 - A_3}{2A_1}$$

CASE 9.   Exchange between two long parallel circular tubes of diameter D and center-to-center distance C, having areas A_{1a} and A_{1b} per unit length:

$$\overline{s_{1a} s_{1b}}\ (\text{per unit length}) = D\left[\sin^{-1}\frac{D}{C} + \sqrt{\left(\frac{C}{D}\right)^2 - 1} - \frac{C}{D} \right]$$

$$F_{1a\to 1b} = \frac{1}{\pi}\left[\sin^{-1}\frac{D}{C} + \sqrt{\left(\frac{C}{D}\right)^2 - 1} - \frac{C}{D} \right]$$

CASE 10.   Exchange between a row of tubes and a plane parallel to it. Consider a unit length along tube axes, with single tube area $A_{1a} = \pi D$ and associated plane area $A_p = C$. A tube sees two tubes and two plane areas:

$$A_{1a} = 2\overline{s_{1a} s_{1b}} + 2\overline{s_{1a} s_p}$$

$$\overline{s_{1a} s_p}\ (\equiv \overline{s_p s_{1a}} \equiv A_p F_{p1}) = \frac{A_{1a}}{2} - \overline{s_{1a} s_{1b}}$$

Substituting from previous example (case 9) yields

$$F_{p1} = 1 - \frac{D}{C}\left[\sin^{-1}\frac{D}{C} + \sqrt{\left(\frac{C}{D}\right)^2 - 1} - \frac{\pi}{2} \right]$$

The value from case 10 appears as line 1 of Fig. 4.3.3. The same figure gives the fraction going to the second row. Additional curves in Fig. 4.3.3 can be obtained by considering the refractory backing as radiatively adiabatic, i.e., by assuming that the radiation that is not absorbed directly is reflected or reradiated, undergoing the same fractional absorption as the incoming beam. In a furnace chamber one zone of which is one or two rows of tubes backed by a refractory, one may visualize the zone as a continuous plane of area A_p at a temperature T_T, the tube surface-temperature, and having an effective absorptivity or

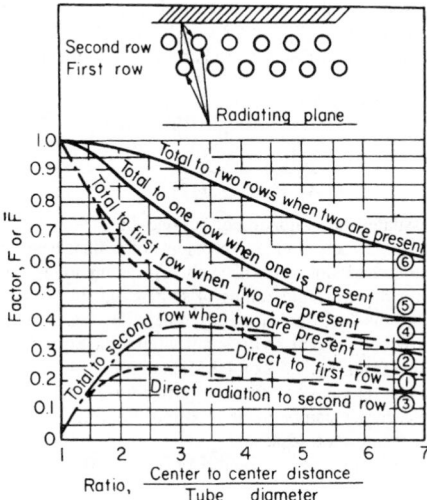

Fig. 4.3.3 Values of F or $\overline{F}$ for a plane parallel to rows of tubes.

emissivity $\varepsilon\,(= \mathscr{F}_{pT})$ that is equal to the value read from Fig. 4.3.3, line 5 or 6; in total exchange area nomenclature (see below), it is $\overline{(S_pS_T)}_R/A_p$. Its complement is headed back toward the emitter, which is whatever faces the replaced tube zone—radiating gas or surfaces or a mixture of them.

When the tubes are gray,

$$\frac{A_p}{\overline{(S_pS_T)}_R} = \frac{A_p}{\overline{(S_pS_T)}_R}\Bigg)_{\substack{\text{black}\\\text{tubes}}} + \frac{\rho_T}{\varepsilon_T} \qquad (4.3.8)$$

When $C/D = 2$, the treatment of a single tube row system with the tubes divided into two zones, front and rear half, reduces $\overline{(S_pS_T)}_R$ or $\mathscr{F}_{pT}$ below the value given by Eq. (4.3.8) by only 1.7 percent (3 percent) when ε_T is 0.8 (0.6).

For other cases, see References.

The view factor F may often be evaluated from that for simpler configurations by the application of three principles: that of reciprocity, $A_iF_{ij} = A_jF_{ji}$; that of conservation, $\Sigma F_{ij} = 1$; and that due to Yamauti, showing that the exchange areas AF between two pairs of surfaces are equal when there is a one-to-one correspondence for all sets of symmetrically placed pairs of elements in the two surface combinations (Hottel and Sarofim, p. 60).

EXAMPLE. The exchange area between the two squares 1 and 4 of Fig. 4.3.4 is to be evaluated. The following exchange areas may be obtained from the values of F for common-side rectangles (case 3, direct-view factors): $\overline{13} = 0.24$, $\overline{24} = 2 \times 0.29 = 0.58$, $\overline{(1+2)(3+4)} = 3 \times 0.32 = 0.96$. Expression of $\overline{(1+2)(3+4)}$ in terms of its components yields $\overline{(1+2)(3+4)} = \overline{13} + \overline{14} + \overline{23} + \overline{24}$. And by the Yamauti principle $\overline{14} = \overline{23}$, since for every pair of elements in 1 and 4, there is a corresponding pair in 2 and 3. Therefore,

$$\overline{14} = [\overline{(1+2)(3+4)} - \overline{13} - \overline{24}]/2 = 0.07$$

Case 1 may be modified in the same way. Another example is the evaluation of AF for exchange between the outside of the smaller of two coaxial cylinders and the inside of the larger when they are *not* coextensive, given the view factor for coextensive cylinders (case 4).

Fig. 4.3.4 Illustration of the Yamauti principle.

Enclosures Containing Gray Source and Sink Surfaces, Refractory Surfaces, and No Absorbing Gas (*Note:* An alternative formulation of the material to follow using matrix algebra is provided by Noble (see References) and in Sec. 5, "Perry's Handbook," 8th ed., McGraw-Hill. The matrix method readily addresses large systems using digital computation, and it can also be used to treat isotropic scatter in the gas volumes.) The calculation of interchange between a source and a sink under conditions involving successive multiple reflections from other source-sink surfaces in the enclosure, as well as reradiation from refractory surfaces, can become complicated. Let a **zone** of a furnace enclosure be an area small enough to make all elements of itself have substantially equivalent "views" of the rest of the enclosure. (In a furnace containing a symmetry plane, parts of a single zone would lie on either side of the plane.) Zones are of two classes, source-sink surfaces, designated by numerical subscripts and having areas $A_1, A_2, \ldots$ and emissivities $\varepsilon_1, \varepsilon_2, \ldots$; and surfaces at which the net radiant-heat flux is zero (fulfilled by the average refractory wall where difference between internal convection and external loss is minute compared with incident radiation), designated by letter subscripts starting with r, and having areas $A_r, A_s, \ldots$. It may be shown that the net radiation interchange between source-sink zones i and j is given by

$$\dot{Q}_{i=j} = \overline{S_iS_j}\sigma T_i^4 - \overline{S_jS_i}\sigma T_j^4 \qquad (4.3.9)$$

The term $\overline{S_iS_j}$ is called the **total interchange area** shared by areas A_i and A_j and depends on the shape of the enclosure and the emissivity and absorptivity of the source and sink zones. It is sometimes called $A_i\mathscr{F}_{ij}$. Restriction here is to gray source-sink zones, for which $\overline{S_iS_j} = \overline{S_jS_i}$; the more general case is treated elsewhere (Hottel and Sarofim, Chaps. 3 and 5).

Evaluation of the $\overline{SS}$'s that characterize an enclosure involves solution of a system of radiation balances on the surfaces. If at a surface the total **leaving flux density**, emitted plus reflected, is denoted by W, radiation balances take the form for source-sink surface j:

$$A_j\varepsilon_jE_j + \rho_j\sum_i\overline{(ij)}W_i = A_jW_j \qquad (4.3.10)$$

and for adiabatic surface r:

$$\sum_i\overline{(ir)}W_i = A_rW_r \qquad (4.3.11)$$

where ρ is reflectance and the summation is over all surfaces in the enclosure. These equations apply to surfaces which emit and reflect diffusely (i.e., their leaving intensity W_i/π is independent of its direction). Most nonmetallic, tarnished, or rough metal surfaces correspond reasonably well to this restriction (but see p. 4-64). In matrix notation, Eqs. (4.3.10) and (4.3.11) become

$$\begin{bmatrix} \overline{11} - \dfrac{A_1}{\rho_1} & \overline{12} & \cdots & \overline{1r} & \overline{1s} & \cdots \\ \overline{12} & \overline{22} - \dfrac{A_2}{\rho_2} & \cdots & \overline{2r} & \overline{2s} & \cdots \\ \cdots & \cdots & \cdots & \cdots & \cdots \\ \overline{1r} & \overline{2r} & \cdots & \overline{rr} - A_r & \overline{rs} & \cdots \\ \overline{1s} & \overline{2s} & \cdots & \overline{rs} & \overline{ss} - A_s & \cdots \\ \cdots & & & & & \end{bmatrix} \begin{bmatrix} W_1 \\ W_2 \\ \cdots \\ W_r \\ W_s \\ \cdots \end{bmatrix} = \begin{bmatrix} -\dfrac{A_1\varepsilon_1}{\rho_1}E_1 \\ -\dfrac{A_2\varepsilon_2}{\rho_2}E_2 \\ \cdots \\ 0 \\ 0 \\ \cdots \end{bmatrix}$$

$$(4.3.12)$$

This represents a system of simultaneous equations equal in number to the number of rows of the square matrix. Each equation consists, on the left, of the sum of the products of the members of a row of the square matrix and the corresponding members of the W-column matrix, and, on the right, of the member of that row in the third matrix. With the above set of equations solved for W_i, the net flux at any surface A_i is given by

$$\dot{Q}_{i,\text{net}} = \frac{A_1\varepsilon_i}{\rho_i}(E_i - W_i) \qquad (4.3.13)$$

Refractory temperature is obtained from $W_r = E_r = \sigma T_r^4$.

The more general use of Eq. (4.3.12) is to obtain the set of total-interchange areas $\overline{SS}$ which constitute a complete description of the effect of shape, size, and emissivity on radiative flux, independent of the presence or absence of other transfer mechanisms. It may be shown that

$$\overline{S_i S_j} \equiv \overline{S_j S_i} \equiv A_i \mathcal{F}_{ij} = \frac{A_i \varepsilon_i}{\rho_i} \frac{A_j \varepsilon_j}{\rho_j} \left(-\frac{D_{ij}{}'}{D} \right) \tag{4.3.14}$$

where D is the determinant of the square coefficient matrix in Eq. (4.3.12) and $D_{ij}{}'$ is the cofactor of its ith row and jth column, or -1^{i+j} times the minor of D formed by crossing out the ith row and jth column.

As an example, consider radiation between two surfaces A_1 and A_2 which together form a complete enclosure. Equation (4.3.12) takes the form

$$A_1 \mathcal{F}_{12} = \frac{A_1 \varepsilon_1}{\rho_1} \frac{A_2 \varepsilon_2}{\rho_2} \frac{\overline{12}}{\begin{vmatrix} \overline{11} - \dfrac{A_1}{\rho_1} & \overline{12} \\ \\ \overline{12} & \overline{22} - \dfrac{A_2}{\rho_2} \end{vmatrix}} \tag{4.3.15}$$

Only one direct-view factor F_{12} or direct exchange area $\overline{12}$ is needed because F_{11} equals $1 - F_{12}$ and F_{22} equals $1 - F_{21}$ or $1 - F_{12}A_1/A_2$. Then $\overline{11}$ equals $A_1 - \overline{12}$, and $\overline{22}$ equals $A_2 - \overline{21}$. With the above substitutions, Eq. (4.3.15) becomes

$$A_1 \mathcal{F}_{12} = \frac{A_1}{1/F_{12} + 1/\varepsilon_1 - 1 + (A_1/A_2)(1/\varepsilon_2 - 1)} \tag{4.3.16}$$

Special cases include:

1. Parallel plates, large compared to clearance. Substitution of $F_{12} = 1$ and $A_1 = A_2$ gives

$$A_1 \mathcal{F}_{12} = \frac{A_1}{1/\varepsilon_1 + 1/\varepsilon_2 - 1} \tag{4.3.17}$$

2. Sphere of area A_1 concentric with surrounding sphere of area A_2. $F_{12} = 1$. Then

$$A_1 \mathcal{F}_{12} = \frac{A_1}{1/\varepsilon_1 + (A_1/A_2)(1/\varepsilon_2 - 1)} \tag{4.3.18}$$

3. Body of surface A_1 having no negative curvature, surrounded by very much larger surface A_2. $F_{12} = 1$ and $A_1/A_2 \to 0$. Then

$$\mathcal{F}_{12} = \varepsilon_1 \tag{4.3.19}$$

Many furnace problems are adequately handled by dividing the enclosure into but two source-sink zones A_1 and A_s, and any number of no-flux zones, $A_r, A_s, \ldots$. For this case Eq. (4.3.14) yields

$$\frac{1}{\overline{S_1 S_2}} \left(\equiv \frac{1}{\overline{S_2 S_1}} \right) = \frac{1}{A_1} \left(\frac{1}{\varepsilon_1} - 1 \right)$$
$$+ \frac{1}{A_2} \left(\frac{1}{\varepsilon_2} - 1 \right) + \frac{1}{(\overline{S_1 S_2})_B} \tag{4.3.20}$$

Here the expression $(\overline{S_1 S_2})_B [\equiv (\overline{S_2 S_1})_B]$ represents the total interchange area for the limiting case of a black source and black sink (the refractory emissivity is of no moment). The factor $(\overline{S_1 S_2})_B/A_1$, called $\overline{F}_{12}$, is known exactly for a few geometrically simple cases and may be approximated for others. If A_1 and A_2 are equal parallel disks, squares, or rectangles, connected by nonconducting but reradiating refractory walls, then $\overline{F}$ is given by Fig. 4.3.2, lines 5 to 8. If A_1 represents an infinite plane and A_2 is one or two rows of infinite parallel tubes in a parallel plane, and if the only other surface is a refractory surface behind the tubes, $\overline{F}_{12}$ is given by line 5 or 6 of Fig. 4.3.3. If an enclosure may be divided into several radiant-heat sources or sinks A_1, A_2, etc., and the rest of the enclosure (reradiating refractory surface) may be lumped together as A_r at a uniform temperature T_r, then the total interchange area for zone pairs in the black system is given by

$$(\overline{S_1 S_2})_B (\equiv A_1 \overline{F}_{12}) = \overline{12} + \frac{(\overline{1r})(\overline{r2})}{A_r - \overline{rr}} \tag{4.3.21}$$

For the two-source-sink-zone system to which Eq. (4.3.20) applies, Eq. (4.3.21) simplifies to $(\overline{S_1 S_2})_B = \overline{12} + 1/[1/\overline{1r} + 1/(\overline{2r})]$; and if A_1 and A_2 each can see none of itself, there is further simplification to

$$(\overline{S_1 S_2})_B = \overline{12} + \frac{1}{1/(A_1 - \overline{12}) + 1/(A_2 - \overline{12})}$$
$$= \frac{A_1 A_2 - (\overline{12})^2}{A_1 + A_2 - 2(\overline{12})} \tag{4.3.22}$$

which necessitates the evaluation of but one direct-view factor F.

Equation (4.3.20) covers many problems of radiant heat interchange between source and sink in furnace enclosures involving no radiating gas. The error due to single zoning of source and sink is small even if the "views" of the enclosure from different parts of each zone are quite different, provided the emissivity is fairly high; the error $\overline{F}$ in is zero if it is obtainable from Fig. 4.3.2 or 4.3.3, small if Eq. (4.3.21) is used and the variation in temperature over the refractory is small. Approach to any desired accuracy can be made by use of Eq. (4.3.14) with division of the surfaces into more zones.

From the definitions of F, $\overline{F}$, and $\mathcal{F}$ or of $\overline{ss}$, $(\overline{SS})_B$, and $\overline{SS}$ it is to be noted that

$$\left. \begin{array}{l} F_{11} + F_{12} + \cdots + F_{1r} + F_{1s} + \cdots = 1 \\ F_{11} + F_{12} + \cdots \qquad\qquad\qquad = 1 \\ \mathcal{F}_{11} + \mathcal{F}_{12} + \cdots \qquad\qquad\qquad = \varepsilon_1 \end{array} \right\}$$

or

$$\begin{array}{l} \overline{s_1 s_2} + \overline{s_1 s_2} + \cdots + \overline{s_1 s_r} + \overline{s_1 s_s} + \cdots = A_1 \\ (\overline{S_1 S_1})_B + (\overline{S_1 S_2})_B + \cdots = A_1 \\ \overline{S_1 S_1} + \overline{S_1 S_2} + \cdots = A_1 \varepsilon_1 \end{array}$$

EXAMPLE.   A furnace chamber of rectangular parallelepipedal form is heated by the combustion of gas inside vertical radiant tubes lining the side walls. The tubes are on centers 2.4 diameters apart. The stock forms a continuous plane on the hearth. Roof and end walls are refractory. Dimensions are shown in Fig. 4.3.5. The radiant tubes and stock are gray bodies having emissivities 0.8 and 0.9, respectively. What is the net rate of heat transmission to the stock by radiation when the mean temperature of the tube surface is $1,500°F$ ($1,089$ K) and that of the stock is $1,200°F$ (922 K)?

Fig. 4.3.5   Dimensions of a furnace chamber.

This problem must be broken up into two parts, first considering the walls with their refractory-backed tubes. To imaginary planes A_2 of area 6×10 ft located parallel to and inside the rows of radiant tubes, the tubes emit radiation $\sigma T_1^4 A_1 \mathcal{F}_{12}$, which equals $\sigma T_1^4 A_2 \mathcal{F}_{21}$. To find $\mathcal{F}_{21}$ use Fig. 4.3.3, line 5, from which $F_{21} = 0.81$. Then from Eq. (4.3.20).

$$\mathcal{F}_{21} = 1/[(1/0.81) + (1/1 - 1) + (2.4/\pi)(1/0.8 - 1)] = 0.702$$

This amounts to saying that the system of refractory-backed tubes is equal in radiating power to a continuous plane A_2 replacing the tubes and refractory back of them, having a temperature equal to that of the tubes and an equivalent or effective emissivity of 0.702.

The new simplified furnace now consists of an enclosure formed by two 6×10 ft radiating side walls (area A_2, of emissivity 0.702), a 5×10 ft receiving plane on the floor (A_3), and refractory surfaces (A_R) to complete the enclosure (ends, roof, and floor side strips); the desired heat transfer is

$$q_{2\to3} = \sigma(T_1^4 - T_3^4) A_2 \mathcal{F}_{23}$$

To evaluate $\mathcal{F}_{23}$, start with the direct interchange factor F_{23}. $F_{23} = F$ from A_2 to $(A_3 + \text{a strip of } A_R \text{ alongside } A_3 \text{ which has a common edge with } A_2)$ minus F from

A_2 to the strip only. These two F's may be evaluated from case 3 for direct-view factors. For the first F, $Y/X = 6/10$, $Z/X = 6.5/10$, $F = 0.239$; for the second F, $Y/X = 6/10$, $Z/X = 1.5/10$, $F = 0.100$. Then $F_{23} = 0.239 - 0.10 = 0.139$. Now $\overline{F}$ may be evaluated. From Eq. (4.3.21) et seq.,

$$A_2\overline{F}_{23} = \overline{23} + \frac{1}{1/\overline{2r} + 1/\overline{3r}}$$

$$\overline{F}_{23} = F_{23} + \frac{1}{1/F_{2r} + (A_2/A_3)(1/F_{3r})}$$

Since A_2 "sees" A_r, A_3, and some of itself (the plane opposite), $F_{2r} = 1 - F_{22} - F_{23}$. F_{22}, the direct interchange factor between parallel 6×10 ft rectangles separated by 8 ft, may be taken as the geometric mean of the factors for 6-ft squares separated by 8 ft, and 10-ft squares separated by 8 ft. These come from Fig. 4.3.2, line 2, according to which $F_{22} = \sqrt{0.13 \times 0.255} = 0.182$. Alternatively, the first of the 10 cases listed above under "Direct-View Factors" may be used. Then $F_{2r} = 1 - 0.182 - 0.139 = 0.679$. The other required direct factor is $F_{3r} = 1 - F_{32} = 1 - F_{25}A_2/A_3 = 1 - 0.139 \times {}^{120}\!/_{50} = 0.666$. Then $\overline{F}_{23} = 0.139 + \{1/[(1/0.679) + (120/50)(1/0.666)]\} = 0.336$. Having $\overline{F}_{23}$, we may now evaluate the factor $\mathscr{F}_{23}$ using Eq. (4.3.20) with $A_1 \to A_2$, $A_2 \to A_3$, and $[\overline{S_1S_2}]_{1B} \to A_2\overline{F}_{23}$.

$$\mathscr{F}_{23} = \frac{1}{1/0.336 + 1/0.702 - 1 + (120/50)(1/0.9 - 1)}$$
$$= 0.273$$
$$\dot{Q}_{net} = \sigma(T_1^4 - T_3^4)A_2\mathscr{F}_{23} = 0.171(19.6^4 - 16.6^4)(120)(0.273)$$
$$= 402{,}000 \text{ Btu/h}$$

In SI units

$$\dot{Q}_{net} = 5.67(10.89^4 - 9.22^4)(120 \times .3048^2)(0.273) = 118{,}000 \text{ W}$$

A result of interest is obtained by dividing the term $A_2\mathscr{F}_{23}$ (120×0.273, or 32.7 ft^2) by the actual area A_1 of the radiating tubes [$(\pi/2.4) \times 60 \times 2 = 157$ ft^2]. This is $32.7/157 = 0.208$; i.e., the net radiation from a tube to the stock is 20.8 percent as much as if the tube were black and completely surrounded by black stock.

Enclosures of Surfaces That Are Not Diffuse Reflectors   The total interchange-area concept has been generalized to include surfaces the reflectance ρ of which can be divided into a diffuse, or Lambert-reflecting, component ρ_D and a specular component ρ_S independent of angle of incidence, with $\varepsilon + \rho_S + \rho_D = 1$. In application to concentric spheres or infinite cylinders, with A_1 the inner surface, the method yields (Hottel and Sarofim, p. 181)

$$A_1\mathscr{F}_{12} \equiv \overline{S_1S_2}$$

$$= \frac{1}{\dfrac{1}{A_1\varepsilon_1} + \dfrac{1}{A_2}\left(\dfrac{1}{\varepsilon_2} - 1\right) + \dfrac{\rho_{s2}}{1 - \rho_{s2}}\left(\dfrac{1}{A_1} - \dfrac{1}{A_2}\right)} \quad (4.3.23)$$

When there is no specular reflectance, the third term in the denominator drops out, in agreement with Eq. (4.3.18). When the reflectance is exclusively specular, the denominator becomes $1/(A_1\varepsilon_1) + \rho_{S2}/[A_1(1 - \rho_{S2})]$, easily derivable from first principles.

RADIATION FROM FLAMES, COMBUSTION PRODUCTS, AND PARTICLE CLOUDS

The radiation from a flame consists of (1) radiation throughout the spectrum from burning soot particles of microscopic and submicroscopic dimensions, from suspended larger particles of coal, coke, or ash, all contributing to what is spoken of as flame luminosity, (2) infrared radiation, mostly from the water vapor and carbon dioxide in the hot gaseous combustion products, and (3) nonequilibrium radiation associated with the combustion process itself, called chemiluminescence and not a significant contributor to the total radiation. A major problem is the effect of the shape of the emitting volume on the radiative flux; this will be considered first.

Mean Beam Lengths   Evaluation of radiation from a nonisothermal volume is beyond the scope of this section (see Hottel and Sarofim, Chap. 11). If a volume emitter is isothermal and at a temperature T, the ratio of the emission from an element of its volume subtending the solid angle $d\Omega$ at a receiver element dA, and making the angle θ with the

normal thereto, to blackbody radiation arriving from within the same solid angle is called the **gas emissivity**. Clearly, ε depends on the path length L through the volume to dA. A hemispherical volume radiating to a spot on the center of its base represents the case in which L is independent of direction. Flux at that spot relative to hemispherical blackbody flux is thus an alternative way to visualize emissivity. The flux density to an area of interest on the envelope of an emitter volume of any shape can be matched by that at the base of a hemispherical volume of some radius L, which will be called the **mean beam length**. It is found that, although the ratio of L to a characteristic dimension D of the shape varies with opacity, the variation is small enough for most engineering purposes to permit use of a constant ratio, L_M/D, where L_M is the **average mean beam length**. L_M can be defined to apply either to a spot on the envelope or to any finite portion of its area. An important limiting case is that of opacity approaching zero ($pD \to 0$, where $p =$ partial pressure of the emitter constituent). For this case, L (called L_0) equals $4 \times$ ratio of gas volume to bounding area when interest is in radiation to the entire envelope. For the range of pD encountered in practice, L (now L_M) is always less. For various shapes, 0.8 to 0.95 times L_0 has been found optimum (see Table 4.3.3); for shapes not reported in Table 4.3.3, a factor of 0.88 (or $L_M = 0.88L_0 = 3.5V/A$) is recommended.

Luminous Flames   Luminosity conventionally refers to soot radiation. Soot is formed in locally fuel-rich portions of flames in amounts that usually correspond, at atmospheric pressure, to amounts less than 1 percent of the carbon in the fuel. Because soot particles are small relative to the wavelength of the radiation of interest in flames (soot primary particle diameters on the order of 20 nm compared to wavelengths of interest of 500 to 8000 nm), the incident radiation permeates the particles and the absorption is proportional to the volume of the particles. In the limit of $r/\lambda \ll 1$, the so-called *Rayleigh limit*, the monochromatic emissivity ε_λ is given by

$$\varepsilon_\lambda = 1 - e^{-Kf_vL/\lambda} \quad (4.3.24)$$

where f_v is the volume of soot per unit volume of space, L is the path length in the same units as λ, and K is dimensionless. The value K will vary with fuel type, experimental conditions, and the temperature history of the soot. The values from a wide range of systems are within a factor of about 2 of one another. The single most important factor governing the value of K is the hydrogen/carbon ratio of the soot with the value of K increasing as the H/C ratio decreases. A value of 9.9 is recommended on the basis of seven studies involving 29 fuels (Mulholland and Croarkin, *Fire and Materials* **24**, 2000, pp. 227–230).

The total emissivity of soot ε_s can be shown to be

$$\varepsilon_s = \int_0^1 \varepsilon_\lambda \, df(\lambda T) = 1 - \frac{15}{4}\left[\Psi^{(3)}\left(1 + \frac{Kf_vLT}{c_2}\right)\right] \quad (4.3.25)$$

where $\Psi^{(3)}(x)$ is the pentagamma function of x. An excellent approximation to Eq. (4.3.25) is

$$\varepsilon_s \cong 1 - \left(1 + \frac{Kf_vLT}{c_2}\right)^{-4} \quad (4.3.26)$$

Expression of ε_s in an e power for combination with the gas emissivity correlation is feasible. In that form

$$\varepsilon_s = 1 - e^{-3.83\,Kf_vLT/c_2} \quad (4.3.27)$$

The approximate relation is good to better than 5 percent for arguments giving values of ε_s less than 0.5.

The biggest uncertainty at present in estimating soot emissivities lies in the estimation of the soot volume fraction f_v. Soot forms in the fuel-rich zones of flames. Soot formation rates are a function of fuel type, mixing rate, local equivalence ratio, temperature, and pressure. Soot formation increases with the aromaticity or C/H ratio of fuels with benzene, α-methyl naphthalene, and acetylene having a high propensity to form soot and methane having a low soot formation propensity; oxygenated fuels, such as alcohols, emit little soot. In practical turbulent diffusion flames, soot forms on the fuel side of the flame front; in premixed flames, at a given temperature, the rate of soot formation increases rapidly as the equivalence ratio exceeds 2.0. Soot burns out rapidly,

Table 4.3.3   Mean Beam Lengths for Volume Radiation

Shape	Characteristic dimension D	L_0/D	L_M/D
Sphere	Diameter	0.67	0.63
Infinite cylinder	Diameter	1	0.94
Semi-infinite cylinder, radiating to:			
Center of base	Diameter	1	0.90
Entire base	Diameter	0.81	0.65
Right-circle cylinder, ht = diam, radiating to:			
Center of base	Diameter	0.76	0.71
Whole surface	Diameter	0.67	0.60
Right-circle cylinder, ht = 0.5 diam, radiating to:			
End	Diameter	0.47	0.43
Side	Diameter	0.52	0.46
Total surface	Diameter	0.50	0.45
Right-circle cylinder, ht = 2 × diam, radiating to:			
End	Diameter	0.73	0.60
Side	Diameter	0.82	0.76
Total surface	Diameter	0.80	0.73
Infinite cylinder, half-circle cross section, radiating to spot on middle of flat side	Radius		1.26
Rectangular parallelepipeds			
1:1:1 (cube)	Edge	0.67	0.60
1:1:4, radiating to:			
1 × 4 face	Shortest edge	0.90	0.82
1 × 1 face	Shortest edge	0.86	0.71
Whole surface	Shortest edge	0.89	0.81
1:2:6, radiating to:			
2 × 6 face	Shortest edge	1.18	
1 × 6 face	Shortest edge	1.24	
1 × 2 face	Shortest edge	1.18	
Whole surface	Shortest edge	1.2	
Infinite parallel planes	Clearance	2.00	1.76
Space outside infinite bank of tubes, centers on equilateral triangles; tube diam = clearance	Clearance	3.4	2.8
Same, except tube diam = 0.5 clearance	Clearance	4.45	3.8
Same, except tube centers on squares, diam = clearance	Clearance	4.1	3.5

in less than 0.1s, under fuel-lean conditions (equivalence ratios less than 1) at temperature above 1,500 K. Because of this rapid soot burnout, soot is usually localized in a relatively small fraction of a furnace or combustor volume. Long, poorly mixed diffusion flames promote soot formation while highly backmixed combustors can burn soot-free. The peak volumetric soot concentration f_v in a typical atmospheric pressure flame is of the range of 1×10^{-7} to 1×10^{-6}, corresponding to a conversion to soot of about 1.5 to 15 percent of the carbon in the fuel. (In calculating f_v at high pressures, allowance needs to be made for the significant increase with pressure of soot formation and for the inverse proportionality of f_v and pressure.) Great progress has been made in the ability to calculate soot in premixed flames [see, for examples, the comparisons of predictions of soot concentration in a well-stirred reactor operated over a wide range of equivalent ratios and temperatures in Brown, Revzan, and Frenklach, *Twenty-seventh Symposium (International) on Combustion*, 1998, pp. 1573–1580] and in a turbulent diffusion flame (Zucca, Marchisio, Barresi, and Fox, "Chemical Engineering Science," 2005).

The importance of soot radiation varies widely between combustors: In large boilers the soot is confined to a small volume and is of only local importance; in gas turbines minimizing the heat flux to the combustor lining is of paramount importance so that only small incremental soot radiation is of concern; in high-temperature glass tanks, the added 0.1 to 0.2 due to soot emissivity is valued in oil-fired flames, and efforts, mostly unsuccessful, had been made to augment the emissivity of natural gas–fired flames; soot contributions to the radiation of pool fires are often dominant and impact the safe separation distances from dikes around oil tanks and the distance from a rig where a flare is located.

Clouds of Large Black Particles   The emissivity of a cloud of particles depends on their area projected along the line of sight. The projected area per unit volume of space is the projected area A of a particle

times the particle number concentration c, or the volume fraction f_v of space occupied by particles times b/d, the projected-surface/volume ratio, where d is the characteristic dimension. [For any randomly oriented particles without dimples, A/(total area) is $\frac{1}{4}$; for spheres, $b = \frac{3}{2}$.] The emissivity of a particle cloud is then given by the alternative formulations

$$\varepsilon = 1 - e^{-bf_v L/d} = 1 - e^{-cAL} \qquad (4.3.28)$$

As an example, consider heavy fuel oil ($CH_{1.5}$, s.g. 0.95) atomized to a surface mean particle diameter of d μm, burned with 20 percent excess air to produce coke residue particles having the original drop diameter, and suspended in combustion products at 1,500 K. From stoichiometry, $f_v = 1.27 \times 10^{-5}$. For spherical particles $b = \frac{3}{2}$, and the flame emissivity due to the particles along a path L will be $1 - e^{-1.9 \times 10^{-2} L/d}$. With 200-$\mu$m particles and an L of 3 m, the particle contribution to emissivity will be 0.25. Soot luminosity will increase this; particle burnout will decrease it. The combined emissivity due to several kinds of emitters will be treated later. The correction for nonblackness of the particles is complicated by multiple scatter of the radiation reflected by each particle. The emissivity ε_M of a cloud of gray particles of individual surface emissivity ε_1 can be estimated by the use of Eq. (4.3.28) with its exponent multiplied by ε_1 if the optical thickness cAL does not exceed about 2.

Gaseous Combustion Products   Radiation from water vapor and carbon dioxide occurs in spectral bands in the infrared. Its magnitude is 3 to 10 times that of convection at furnace temperatures. It depends on gas temperature T_G, on the partial pressure-beam length products $p_w L$ and $p_c L$ (subscripts w and c refer to water vapor and carbon dioxide), and to a much lesser extent on total and partial pressure. The gas emissivity ε_G is the sum of the separate contributions due to H_2O and CO_2, corrected for pressure broadening of the spectral bands and for band overlap (Hottel and Sarofim, Chap. 6). The elaborate calculations can be combined for a restricted set of conditions, here taken to be the practically important cases of 1-atm total pressure and partial pressures representative of fossil fuel combustion in air. In the range of furnace operating conditions the product $\overline{\varepsilon_G T_G}$ varies much less than ε_G with T_G, and $\overline{\varepsilon_G T_G}$ depends primarily on $(p_w + p_c)L$, much less on $p_w/(p_w + p_c)$, and so little on T_G as to permit linear interpolation between widely separated T_G's. An equation of the form

$$\log \overline{\varepsilon_G T_G} = a_0 + a_1 \log pL + a_2 \log^2 pL + a_3 \log^3 pL \quad (4.3.29)$$

where p is the sum of partial pressures $p_w + p_c$ atm and L is the mean beam length, has been found capable of fitting emissivity data over a 1000-fold range of pL, from 0.01 to 10 m · atm (0.03 to 30 ft · atm).

Table 4.3.4, section 2, gives values of the constants representing the results of an averaging of all the available total and integrated spectral data on CO_2 and H_2O, together with corrections for spectral band broadening and overlap. Equation (4.3.29) represents the original data with a precision greater than their accuracy. The constants are given for

computation in either metres and kelvins or feet and degrees Rankine for mixtures, in nonradiating gases, of water vapor alone, CO_2 alone, and four p_w/p_c mixtures. Four suffice, since a change halfway from one mixture ratio to the adjacent one changes the emissivity by a maximum of only 5 percent; linear interpolation may be used if necessary. The constants are given for only three temperatures, which is adequate for linear interpolation since $\overline{\varepsilon_G T}$ changes a maximum of only one-sixth due to a change from one temperature base halfway to the adjacent one. Based on metre atmospheres and kelvins, the interpolation relation, with T_H and T_L representing the higher and lower base temperatures bracketing T, and with the brackets in the term $[A(x)]$ indicating that the parentheses refer not to a multiplier but to an argument, is

$$\overline{\varepsilon_G T_G} = \frac{[\overline{\varepsilon_G T_H}(pL)](T_G - T_L) + [\overline{\varepsilon_G T_L}(pL)](T_H - T_G)}{500} \quad (4.3.30)$$

Extrapolation to a temperature which is above the highest or below the lowest of the three base temperatures in Table 4.3.4 uses the same

Table 4.3.4   Emissivity of ε_G of H_2O-CO_2 Mixtures
Section 1: Limited range for furnaces, valid over 25-fold range of $p_{w+c}L$, 0.046–1.15 m · atm (0.15–3.75 ft · atm)

p_w/p_c	0	0.5	1	2	3	∞
$\dfrac{p_w}{p_w + p_c}$	0(0–0.2)	⅓(0.2–0.42)	½ (0.42–0.6)	⅔(0.6–0.7)	¾(0.7–0.8)	1(0.8–1.0)
	CO_2 only	Corresponding to $(CH)_x$, covering coal, heavy oils, ptich	Corresponding to $(CH_2)_x$, covering distillate oils, paraffins, olefines	Corresponding to CH_4, covering natural gas and refinery gas	Corresponding to $(CH_6)_x$, covering future high-H_2 fuels	H_2O only

Constant b and n of equation $\overline{\varepsilon_G T} = b(pL - 0.015)^n$, pL in m·atm, T in K

T, K	b	n	b	n	b	n	b	n	b	n	b	n
1,000	188	0.209	384	0.33	416	0.34	444	0.34	455	0.35	416	0.400
1,500	252	0.256	448	0.38	495	0.40	540	0.42	548	0.42	548	0.523
2,000	267	0.316	451	0.45	509	0.48	572	0.51	594	0.52	632	0.640

Constants b and n of equation $\overline{\varepsilon_G T} = b(pL - 0.05)^n$, pL in ft·atm, T in °R

T, °R	b	n	b	n	b	n	b	n	b	n	b	n
1,800	264	0.209	467	0.33	501	0.34	534	0.34	541	0.35	466	0.400
2,700	335	0.256	514	0.38	555	0.40	591	0.42	600	0.42	530	0.523
3,600	330	0.316	476	0.45	519	0.48	563	0.51	577	0.52	532	0.640

Section 2: Full range, vaild over 2000-fold range of $p_{w+c}L$, 0.005–10.0 m · atm (0.016–32.0 ft · atm)
Constants of equation, $\log \overline{\varepsilon_G T} = a_0 + a_1 \log pL + a_2 \log^2 pL + a_3 \log^3 pL$

$\dfrac{p_w}{p_c}$	$\dfrac{p_w}{p_w + p_c}$		pL in m · atm, T in K					pL in ft · atm, T in °R			
		T, K	a_0	a_1	a_2	a_3	T, °R	a_0	a_1	a_2	a_3
0	0	1,000	2.2661	0.1742	−0.0390	0.0040	1,800	2.4206	0.2176	−0.0452	0.0040
		1,500	2.3954	0.2203	−0.0433	0.00562	2,700	2.5248	0.2695	−0.0521	0.00562
		2,000	2.4104	0.2602	−0.0651	−0.00155	3,600	2.5143	0.3261	−0.0627	−0.00155
$\dfrac{1}{2}$	$\dfrac{1}{3}$	1,000	2.5754	0.2792	−0.0648	0.0017	1,800	2.6691	0.3474	−0.0674	0.0017
		1,500	2.6461	0.3418	−0.0685	−0.0043	2,700	2.7064	0.4091	−0.0618	−0.0043
		2,000	2.6504	0.4279	−0.0674	−0.0120	3,600	2.6686	0.4879	−0.0489	−0.0120
1	$\dfrac{1}{2}$	1,000	2.6090	0.2799	−0.0745	−0.0006	1,800	2.7001	0.3563	−0.0736	−0.0006
		1,500	2.6862	0.3450	−0.0816	−0.0039	2,700	2.7423	0.4261	−0.0756	−0.0039
		2,000	2.7029	0.4440	−0.0859	−0.0135	3,600	2.7081	0.5219	−0.0650	−0.0135
2	$\dfrac{2}{3}$	1,000	2.6367	0.2723	−0.0804	0.0030	1,800	2.7296	0.3577	−0.0850	0.0030
		1,500	2.7178	0.3386	−0.0990	−0.0030	2,700	2.7724	0.4384	−0.0944	−0.0030
		2,000	2.7482	0.4464	−0.1086	−0.0139	3,600	2.7461	0.5474	−0.0871	−0.0139
3	$\dfrac{3}{4}$	1,000	2.6432	0.2715	−0.0816	0.0052	1,800	2.7359	0.3599	−0.0896	0.0052
		1,500	2.7257	0.3355	−0.0981	0.0045	2,700	2.7811	0.4403	−0.1051	0.0045
		2,000	2.7592	0.4372	−0.1122	−0.0065	3,600	2.7599	0.5478	−0.1021	−0.0065
∞	1	1,000	2.5995	0.3015	−0.0961	0.0119	1,800	2.6720	0.4102	−0.1145	0.0119
		1,500	2.7083	0.3969	−0.1309	0.00123	2,700	2.7238	0.5330	−0.1328	0.00123
		2,000	2.7709	0.5099	−0.1646	−0.0165	3,600	2.7215	0.6666	−0.1391	−0.0165

NOTE: Values of $p_w/(p_w + p_c)$ of ⅓, ½, ⅔, ¾ may be used to cover the ranges 0.2–0.42, 0.42–0.6, 0.6–0.7, and 0.7–0.8, respectively, with a maximum error in ε_G of 5 percent at $pL = 6.5$ m · atm, less at lower pL's. Linear interpolation reduces the error generally to less than 1 percent. Linear interpolation or extrapolation on T introduces an error generally below 2 percent, less than the accuracy of the original data.

formulation, but one of the terms becomes negative. Linearization on the constants a_0 to a_3 rather than on $\overline{\varepsilon_G T}$ may be preferable if fuel quality is unchanging.

When pL lies in the 25-fold range of 0.046 to 1.15 m · atm (0.15 to 3.75 ft · atm), adequate for furnaces, a much simpler two-constant relation is adequate.

$$\overline{\varepsilon_G T} = \begin{cases} b(pL - 0.015)^n & \text{with } T = \text{K}, pL = \text{m} \cdot \text{atm} \\ b(pL - 0.05)^n & \text{with } T = °\text{R}, pL = \text{ft} \cdot \text{atm} \end{cases}$$

The constants are given in Table 4.3.4, section 1.

Combined Radiation from Gases and Suspended Solids

The total emissivity of gases and suspended solids is less than the sum of the separate contributions because of interference between overlapping spectral emissions. The spectral overlap of H_2O and CO_2 radiation has been taken into account by the constants of Table 4.3.4 used for obtaining ε_G. Additional overlap occurs when soot emissivity ε_s is added. If the emission bands of water vapor and CO_2 were randomly placed in the spectrum and soot radiation were gray, the combined emissivity would be $\varepsilon_G + \varepsilon_s$ minus an overlap correction $\varepsilon_G \varepsilon_s$. Monochromatic soot emissivity rises as the wavelength gets shorter, and in a highly sooted flame at 1,500 K half the soot emission lies below 2.5 μm where H_2O and CO_2 emission is negligible. Then the correction $\varepsilon_G \varepsilon_s$ must be reduced, and the following is recommended:

$$\varepsilon_{G+s} = \varepsilon_G + \varepsilon_s - M\varepsilon_G \varepsilon_s \qquad (4.3.31)$$

where M depends mostly on T_G and to a much less extent on the optical density $K f_v L$. Values that have been calculated from this simple model can be represented with acceptable error by

$$M = 1.07 + 2.7 \times 10^5 f_v L - 0.27(T/1,000) \qquad (0 < M < 1)$$

If, in addition to gas and soot, massive particles such as fly ash, coal char, or carbonaceous cenospheres from heavy fuel oil of emissivity ε_M are present, it is recommended that the total emissivity be approximated by

$$\varepsilon_{\text{total}} = \varepsilon_{G+s} + \varepsilon_M - \varepsilon_{G+s}\varepsilon_M \qquad (4.3.32)$$

Radiant interchange between a gas and a *completely bounding black surface* at T_1 produces a surface flux density q given by

$$q = \sigma(\varepsilon_G T_G^4 - \alpha_{G1} T_1^4) \qquad (4.3.33)$$

where α_{G1} is the **absorptivity** of the gas at T_G for radiation from a surface at T_1. The absorptivity of water vapor–CO_2 mixtures may also be obtained from the constants for emissivities. The product $\overline{\alpha_{G1} T_1}$—the absorptivity of gas at T_G for black radiation at T_1 times the surface temperature—is the product $\overline{\varepsilon_G T_1}$ with ε_G evaluated at surface temperature T_1 instead of T_G and at pLT_1/T_G instead of pL, then multiplied by $(T_G/T_1)^{0.5}$, or

$$\overline{\alpha_{G1} T_1} = [\overline{\varepsilon_G T_1}(pLT_1/T_G)](T_G/T_1)^{0.5} \qquad (4.3.34)$$

The exponent 0.5 is an adequate average of the exponents for the pure components. The interpolation relation for absorptivity is

$$\overline{\alpha_{G1} T_1} = \left[\overline{\varepsilon_G T_H}\left(\frac{pLT_H}{T_G}\right)\right]\left(\frac{T_G}{T_H}\right)^{0.5}\frac{T_1 - T_L}{500}$$
$$+ \left[\overline{\varepsilon_G T_L}\left(\frac{pLT_L}{T_G}\right)\right]\left(\frac{T_G}{T_L}\right)^{0.5}\frac{T_H - T_1}{500} \qquad (4.3.35)$$

The base temperature pair T_H and T_L can be different for the evaluation of ε_G and α_{G1} if T_G and T_1 are far enough apart. Extrapolation from the lowest T_G in Eq. (4.3.35) to a much lower T_1 to obtain α_{G1} may yield too high a value for it. That occurs, however, only when $T_1 \ll T_G$, and the fourth-power temperature relation makes the error in q negligible.

If the surface is not black, the right-hand side of Eq. (4.3.33) must be modified. If the surface is gray, multiplication by $\alpha_1 (\equiv \varepsilon_1)$ allows for reduction in the primary beam from gas to surface and surface to gas, but some of the gas radiation initially reflected from the surface has further opportunity for absorption at the surface because the gas is but

incompletely opaque to the reflected beam. Consequently, the factor to allow for surface grayness lies between absorptance α_1 and unity, nearer the latter the more transparent the gas (low pL) and the more convoluted the surface. In the absorptance range of most industrial surfaces, 0.7 to 1.0, an adequate approximation consists in use of an effective absorptance α_1' halfway between the actual value and unity. If the surface is not gray, q depends much more on surface absorptance, which modifies $\varepsilon_G T_G^4$, than on emittance, which modifies $\alpha_{G1} T_1^4$. Absorption is treated more rigorously later in the section.

EXAMPLE. Flue gas containing 9.5 percent CO_2 and 7.1 percent H_2O, wet basis, flows through a bank of tubes of 1.5-in OD on equilateral triangular centers 4.5 in apart. In a section in which the gas and tube surface temperatures are 1,700 and 1,000°F, what is the heat transfer rate per square foot of tube area, due to gas radiation only? Tube surface absorptance = 0.8.
$T_G = 2,160°\text{R}$ (1,200 K); $T_s = 1,460°\text{R}$ (811 K)
$p_w/(p_w + p_c) = 7.1/16.6 = 0.428$; use 0.5
$pL = 0.166 [3.8(4.5 - 1.5)/12] = 0.158$ ft · atm (0.0480 m · atm)
$pL(Ts/T_G) = 0.1580(1,460/2,160) = 0.1066$ ft · atm (0.0325 m · atm)
From Table 4.3.4, for $T_G = 1,500$ K and $pL = 0.0480$ and 0.0325, $\varepsilon T = 125$ and 101 K, and for $T_G = 1,000$ K and $pL = 0.0480$ and 0.0325, $\varepsilon T = 129$ and 107 K. Then

$$\varepsilon_G = \frac{1}{1,200}\frac{125(1,200 - 1,000) + 129(1,500 - 1,200)}{1,500 - 1,000} = 0.106$$

$$\alpha_{G1} = \frac{1}{811}\frac{101(811 - 1,000) + 107(1,500 - 811)}{1,500 - 1,000} = 0.135$$

The effective surface absorptance factor $\alpha_1 = (0.8 + 1)/2 = 0.9$. From Eq. (4.3.33), modified,

$$q = 0.9 \times 0.1713(0.106 \times 21.6^4 - 0.135 \times 14.6^4)$$
$$= 2,612 \text{ Btu/(ft}^2 \cdot \text{h)}$$

or

$$q = 0.9 \times 5.67(0.106 \times 12^4 - 0.135 \times 8.111^4) = 8,235 \text{ W/m}^2$$

This is equivalent to a convection coefficient of 2,612/700 or 3.73 Btu/(ft^2 · F · h) or 21.2 W/(m^2 · K). The emissivity of an equivalent gray flame is $(0.106 \times 21.6^4 - 0.135 \times 14.6^4)/(21.6^4 - 14.6^4) = 0.098$.

RADIATIVE EXCHANGE IN ENCLOSURES OF RADIATING GAS

The so-called *radiant section* of a furnace presents a heat-transfer problem in which there enters the combined action of direct radiation from the flame to the stock or heat sink and radiation from the flame to refractory surfaces and thence back through the flame (with partial absorption) to the sink, convection, and external losses. Solutions of the problem based on varying degrees of simplification are available, including allowance for temperature variation in both gas and refractory walls (Hottel and Sarofim, Chap. 14). A less rigorous treatment suffices, however, for handling many problems. There are two limiting cases: the long chamber with gas temperature varying only in the direction of gas flow and the compact chamber containing a gas or flame at a uniform temperature. The latter, with variations, will be considered first.

Total Exchange Areas $\overline{SS}$ and $\overline{GS}$ The arguments leading to the development of the interchange factor $A_i\mathscr{F}_{ij}(=\overline{S_iS_j})$ between surface zones [Eq. (4.3.14) et seq.] apply to the case of absorption within the gas volume if, in the evaluation of the direct exchange area, allowance is made for attenuation of the radiant beam through the gas. This necessitates nothing more than the redefinition, in Eqs. (4.3.7) to (4.3.22), of every term $\overline{ij}(\equiv\overline{s_is_j}\equiv A_iF_{ij})$ to represent, per unit of black emissive power, flux from A_i through an absorbing gas to A_j; that is, the prior F_{ij} must be multiplied by a mean transmittance τ_{ij} of the gas (= $1 - \alpha_{ij} = 1 - \varepsilon_G$ for a gray gas). In a system containing an isothermal gas and source-sink boundaries of areas $A_1, A_2, \ldots, A_n$, the total emission from A_1 per unit of its black emissive power is $A_1\varepsilon_1$, of which $\overline{S_1S_1} + \overline{S_1S_2} + \cdots + \overline{S_1S_n}$ is absorbed in the surfaces by all mechanisms, direct and indirect. The difference has been absorbed in the gas; it is called the **gas surface total exchange area** $\overline{GS_1}$:

$$\overline{GS_1} = A_1\varepsilon_1 - \sum_i \overline{S_1S_i} \qquad (4.3.36)$$

The letters identifying total exchange areas are, of course, commutative; $\overline{GS_1} \cong \overline{S_1G}$. Note that although $\overline{S_1S_1}$ is never used in calculating radiative interchange, its value is needed for use of Eq. (4.3.36) in calculating $\overline{GS_1}$. $\overline{GS_1}$ embraces the full effect of radiation complexities on radiative exchange between gas and A_1, including multiple reflection at all surfaces, and it is capable of including the effects of gas nongrayness and of assistance given by refractory surfaces to gas-A_1 interchange. It is but mildly temperature-sensitive and is independent of any changes in conduction, convection, mass flow, and energy balances except for their effect on the temperature used in evaluating it.

If the gas volume is not isothermal, the principles used here can be extended to setting up balances on a zoned gas volume (see, e.g., Hottel and Sarofim, "Radiative Transfer," McGraw-Hill, Chap. 11).

Systems with a Single Gas Zone and Two Surface Zones

An enclosure consisting of but one isothermal gas zone and two gray surface zones, when properly specified, can model so many industrially important radiation problems as to merit detailed presentation. One can evaluate the total radiation flux between any two of the three zones, including multiple reflection at all surfaces.

$$\dot{Q}_{G \leftrightarrow 1} = \overline{GS_1}\sigma(T_G^4 - T_1^4)$$
$$\dot{Q}_{1 \leftrightarrow 2} = \overline{S_1S_2}\sigma(T_1^4 - T_2^4) \qquad (4.3.37)$$

The total exchange area takes a relatively simple closed form, even when important allowance is made for gas radiation not being gray and when a reduction of the number of system parameters is introduced by assuming that one of the surface zones, if refractory, is radiatively adiabatic (see later). Before allowance is made for these factors, the case of a **gray gas** enclosed by two source-sink surface zones will be presented. Modification of Eqs. (4.3.7) to (4.3.22), discussed previously, combined with the assumption that a single mean beam length applies to all transfers, i.e., that there is but one gas transmittance τ $(= 1 - \varepsilon_G)$, gives

$$\overline{S_1S_2} = \frac{A_1\varepsilon_1\varepsilon_2 F_{12}}{1/\tau + \tau\rho_1\rho_2(1 - F_{12}/C_2) - \rho_1(1 - F_{12}) - \rho_2(1 - F_{21})}$$

$$(4.3.38)$$

$$\overline{S_1S_1} = \frac{A_1\varepsilon_1^2[F_{11} + \rho_2\tau(F_{12}/C_2 - 1)]}{\text{same denominator}} \qquad (4.3.39)$$

$$\overline{GS_1} = \frac{A_1\varepsilon_1\varepsilon_G[1/\tau + \rho_2(F_{12}/C_2 - 1)]}{\text{same denominator}} \qquad (4.3.40)$$

(Here C is the area expressed as a ratio to the total enclosure area A_T; $C_1 = A_1/A_T$, $C_2 = A_2/A_T$; $C_1 + C_2 = 1$.) The three equations above suffice to formulate total exchange areas for gas-enclosing arrangements which include, e.g., the four geometric cases illustrated in Table 4.3.5, to be discussed later.

An additional surface arrangement of importance is a single zone surface fully enclosing gas. With the gas assumed gray, the simplest derivation of $\overline{GS_1}$ is to note that the emission from surface A_1 per unit of its blackbody emissive power is $A_1\varepsilon_1$, of which the fractions ε_G and $(1 - \varepsilon_G)\varepsilon_1$ are absorbed by the gas and the surface, respectively, and the surface reflected residue always repeats this distribution. Therefore,

$$\overline{GS}_{\substack{\text{single surface} \\ \text{zone surrounding} \\ \text{gray gas}}} \equiv \overline{GS_1} = A_1\varepsilon_1 \frac{\varepsilon_G}{\varepsilon_G + (1 - \varepsilon_G)\varepsilon_1} = \frac{A_1}{1/\varepsilon_G + 1/\varepsilon_1 - 1}$$

$$(4.3.41)$$

Alternatively, $\overline{GS_1}$ could be obtained from case 1 of Table 4.3.5 by letting plane area A_1 approach 0, leaving A_2 as the sole surface zone.

Although departure of gas from grayness has a marked effect on radiative transfer, the subject is complex and will be presented in stages, as the cases shown in Table 4.3.5 are discussed.

Partial Allowance for the Effect of Gas Nongrayness on Total Exchange Areas

A radiating gas departs from grayness in two ways: (1) Gas emissivity ε_G and absorptivity α_{G1} are not the same unless T_1 equals T_G. (2) The fractional transmittance τ of radiation through successive path lengths L_m due to surface reflection, instead of being constant, keeps increasing because at the wavelengths of high absorption the incremental absorption decreases with increasing path length. The first of these effects is sufficiently straightforward to be introduced at this point, coupled with allowance for refractory surfaces being substantially radiatively adiabatic. The second, much more complicated effect will be introduced later; it sometimes changes the computed flux significantly.

In the simplest case of gas-surface radiative exchange—a gas at T_G completely enclosed by a black surface at T_1—the net flux $\dot{Q}_{G \leftrightarrow 1}$ is given by

$$\dot{Q}_{G \leftrightarrow 1} = \sigma(\varepsilon_G T_G^4 - \alpha_{G1}T_1^4) \equiv \sigma\varepsilon_{G,e}(T_G^4 - T_1^4)$$

The evaluation of the absorptivity α_G was covered in Eqs. (4.3.34) and (4.3.35). The second form of the above equation defines $\varepsilon_{G,e}$, the equivalent gray-gas emissivity

$$\varepsilon_{G,e} = \frac{\varepsilon_G - \alpha_{G1}(T_1^4/T_G^4)}{1 - (T_1/T_G)^4} \qquad (4.3.42)$$

Although this introduction of $\varepsilon_{G,e}$ has added no information, the evaluation of $\dot{Q}_{G \leftrightarrow 1}$ in terms of $\varepsilon_{G,e}$ rather than ε_G and α_{G1} gives a better structure for trial-and-error solutions of problems in which either T_G or T_1 is not known and a second energy relation is available.

With partial allowance for gas nongrayness having been made, the evaluation of radiative flux $\dot{Q}_{G \leftrightarrow 1}$ or $\dot{Q}_{1 \leftrightarrow 2}$ [Eq. (4.3.37)] for cases falling in one of the categories of Table 4.3.5 is straightforward if both A_1 and A_2 are source-sink surfaces. Wherever ε_G or τ appears in the table, or in Eqs. (4.3.38) to (4.3.41), use $\varepsilon_{G,e}$ or $1 - \varepsilon_{G,e}$ instead.

EXAMPLE (FIRST APPROXIMATION TO NONGRAYNESS).   Methane is burned to completion with 20 percent excess air (air half saturated with water vapor at 289 K (60°F), 0.0088 mol H_2O/mol dry air) in a furnace chamber with floor dimensions of 3 × 10 m and 5 m high. The whole surface is a gray energy sink of emissivity 0.8 at 1,000 K, surrounding gas at 1,500 K, well stirred. Find the effective gas emissivity $\varepsilon_{G,e}$ and the surface radiative flux density, assuming that the only correction necessary for gas nongrayness is use of $\varepsilon_{G,e}$ rather than ε_G.

SOLUTION.   Combustion is 1 CH_4 + 2 × 1.2 O_2 + 1.2 × (79/21)N_2 + 2 × 1.2 × 100/21 × 0.0088 H_2O going to 1 CO_2 + [2 + 2 × 1.2 × (100/21) × 0.0088] H_2O + 0.4 O_2 + 9.03 N_2 = 12.53 mol/mol of CH_4. And $P_C + P_W = (1 + 2.1)/12.53 = 0.2474$ atm. The mean beam length $L_m = 0.88 \times 4V/A_T = 0.88 \times 4(10 \times 3 \times 5)/\{2[2 \times (10 \times 3 + 10 \times 5 + 3 \times 5)]\} = 2.779$ m. And $pL_m = 0.2474 \times 2.779 = 0.6875$ m · atm. From emissivity Table 4.3.4, $b(1,500) = 540$; $n(1,500) = 0.42$; $b(1,000) = 444$; $n(1,000) = 0.34$. Also $\varepsilon_G(pL) = 540(0.6875 - 0.015)^{0.42}/1,500 = 0.3047$, and $\alpha_{G1}(pL) = 444(0.6875 \times 1,000/1,500 - 0.015)^{0.34}(1,500/1,000)^{0.5}/1,000 = 0.4124$. Then $\varepsilon_{G,e}(pL) = [0.3047 - 0.4124(1,000/1,500)^4]/[1 - (1,000/1,500)^4] = 0.2782$. From Eq. (4.3.41), with ε_G replaced by $\varepsilon_{G,e}$, $(\overline{GS_1}/A_1) = 1/(1/0.2782 + 1/0.8-1) = 0.2601$. Then $\dot{Q}_{G \leftrightarrow 1}/A_1 = 56.7 \times 0.260[(1,500/1,000)^4 - (1,000/1,000)^4] = 59.91$ kW/m²[18,990 Btu/(ft² · h)].

Refractory Surfaces   If one of the surfaces A_r of an enclosure of gas is refractory, an extra temperature T_{refr} and an extra heat transfer equation are needed to determine the fluxes *unless A_r* can be assumed to be radiatively adiabatic. Consider the facts that irradiation of A_r plus convection from gas to it must equal back radiation plus conduction through it if steady state exists, and irradiation is enormous compared to convection. It then follows that the difference between convection and conduction is so minute compared to irradiation or back radiation as to make A_r substantially radiatively adiabatic; assume that A_1 is a sourcesink zone and A_2 a radiatively adiabatic zone, and call it A_r. The condition for adiabaticity of A_r is

$$\overline{GS_r}(T_G^4 - T_r^4) = \overline{S_rS_1}(T_r^4 - T_1^4)$$

or, to eliminate T_r,

$$\frac{T_G^4 - T_r^4}{1/\overline{GS_r}} = \frac{T_r^4 - T_1^4}{1/\overline{S_rS_1}} = \frac{T_G^4 - T_1^4}{1/\overline{GS_r} + 1/\overline{S_rS_1}} \qquad (4.3.43)$$

The net flux from gas G is $\overline{GS_1}\sigma(T_G^4 - T_1^4) + \overline{GS_r}\sigma(T_G^4 - T_r^4)$ which, with replacement of the last term, using Eq. (4.3.43), gives the single

Table 4.3.5 Total Exchange Areas for Four Arrangements of Two-Zone-Surface Enclosures of a Gray Gas*

Case 1	Case 2	Case 3	Case 4

Case 1

A plane surface A_1, and a surface A_2 complete the enclosure

$F_{12} = 1$

$$\frac{\overline{S_1S_2}}{A_1} = \frac{\varepsilon_1\varepsilon_2}{D_1}$$

$$\frac{\overline{GS_1}}{A_1} = \frac{\varepsilon_1\varepsilon_G(1/\tau + \rho_2C_1/C_2)}{D_1}$$

$$\frac{\overline{GS_2}}{A_2} = \frac{\varepsilon_2\varepsilon_G(1/\tau + \rho_1C_1/C_2)}{D_1}$$

$$\frac{\overline{S_1S_1}}{A_1} = \frac{\varepsilon_1^2\tau\rho_2C_1/C_2}{D_1}$$

$$D_1 \equiv \frac{1}{\tau} - \rho_2\left[1 - \frac{C_1}{C_2}(1-\tau\rho_1)\right]$$

$$\tau = 1 - \varepsilon_G$$

Case 2

Infinite parallel planes

$F_{12} = F_{21} = 1$

$$\frac{\overline{S_1S_2}}{A_1} = \frac{\varepsilon_1\varepsilon_2}{D_2}$$

$$\frac{\overline{GS_1}}{A_1} = \frac{\varepsilon_1\varepsilon_G(1/\tau + \rho_2)}{D_2}$$

$$\frac{\overline{S_1S_1}}{A_1} = \frac{\varepsilon_1^2\rho_2\tau}{D_2}$$

$$D_2 \equiv \frac{1}{\tau} - \tau\rho_1\rho_2$$

Case 3

Concentric spherical or infinite cylindrical surface zones, A_1 inside

$$F_{12} = 1 \qquad F_{21} = \frac{A_1}{A_2} = \frac{C_1}{C_2}$$

$$\frac{\overline{S_1S_2}}{A_1} = \frac{\varepsilon_1\varepsilon_2}{D_3}$$

$$\frac{\overline{GS_1}}{A_1} = \frac{\varepsilon_1\varepsilon_G(1/\tau + \rho_2C_1/C_2)}{D_3}$$

$$\frac{\overline{GS_2}}{A_2} = \frac{\varepsilon_2\varepsilon_G(1/\tau + \rho_1C_1/C_2)}{D_3}$$

$$\frac{\overline{S_1S_1}}{A_1} = \frac{\varepsilon_1^2\rho_2\tau C_1/C_2}{D_3}$$

$$D_3 \equiv \frac{1}{\tau} - \rho_2\left[1 - \frac{C_1}{C_2}(1-\tau\rho_1)\right]$$

Case 4

Two-surface-zone enclosure, each zone in one or more parts, any shape

Assume that enclosing surface is a speckled enclosure, or spherical

$$F_{12} = F_{22} = C_2$$
$$F_{21} = F_{11} = C_1$$

$$\frac{\overline{S_1S_2}}{A_1} = \frac{\varepsilon_1\varepsilon_2C_2}{D_4}$$

$$\frac{\overline{GS_1}}{A_1} = \frac{\varepsilon_1\varepsilon_G/\tau}{D_4}$$

$$\frac{\overline{S_1S_1}}{A_1} = \frac{\varepsilon_1^2C_1}{D_4}$$

$$D_4 \equiv \frac{1}{\tau} - \rho_1C_1 - \rho_2C_2$$

* All equations above come from Eqs. (4.3.38) to (4.3.40), with substitutions for view factor F given before equations.

term multiplying a fourth-power temperature difference:

$$\dot{Q}_{G \leftrightarrow 1} = \sigma(T_G^4 - T_1^4)\left[\overline{GS_1} + \frac{1}{1/\overline{GS_r} + 1/\overline{S_rS_1}}\right]$$

$$\equiv (\overline{GS_1})_R \sigma(T_G^4 - T_1^4) \quad (4.3.44)$$

The bracketed term is called $(\overline{GS_1})_R$, the total exchange area from G to A_1 with assistance from a refractory surface. Table 4.3.5 supplies the forms for the three total exchange area terms needed to formulate $(\overline{GS_1})_R$, with A_r substituted for A_2 and with ε_G or τ replaced by $\varepsilon_{G,e}$ or $1 - \varepsilon_{G,e}$.

A general expression for a gray gas enclosure of two surfaces, one of which is radiatively adiabatic, comes from Eq. (4.3.44), in which $(\overline{GS_1})_R$ becomes $\overline{GS_1}$ because $\overline{S_rS_1}$ and $\overline{GS_r}$ are zero, and then from Eq. (4.3.36), which becomes $\overline{GS_1}[=(\overline{GS_1})_R] = A_1\varepsilon_1 - \overline{S_1S_1}$. With $\overline{S_1S_1}$ coming from Eq. (4.3.39), one finally obtains

$$\frac{(\overline{GS_1})_R}{A_1} = \frac{1}{\rho_1/\varepsilon_1 + 1/\{\varepsilon_G[1 + 1/(C_1/C_2 + \varepsilon_G/\tau F_{1r})]\}} \quad (4.3.45)$$

Note that since the first denominator term is zero when A_1 is black, the denominator of the second term is $(\overline{GS_1})_R/A_1$ for a black surface.

The above equation is perfectly general when the **gas is gray** and the two enclosing surfaces are a sink and a radiatively adiabatic surface. When A_1 is a plane (simulation of slab or billet heating furnaces and glass tanks), $(\overline{GS_1})_R/A_1$ is Eq. (4.3.45) with $F_{1r} = 1$. A different but completely equivalent form is

$$\left[\frac{(\overline{GS_1})_R}{A_1}\right]_{A_1 \text{ is plane}} = \frac{1}{1/\varepsilon_1 + (1/\varepsilon_G - 1)^2/[1/(C_1\varepsilon_G) - 1]} \quad (4.3.46)$$

For most refinery **processing furnaces,** with sink and refractory assumed to form a **speckled** enclosure, $(\overline{GS_1})_R/A_1$ is Eq. (4.3.45) with $F_{1r} = C_r$. A better but completely equivalent form is

$$\left[\frac{(\overline{GS_1})_R}{A_1}\right]_{\substack{\text{surface is} \\ \text{speckled}}} = \frac{1}{1/\varepsilon_1 + C_1(1/\varepsilon_G - 1)} \quad (4.3.47)$$

As previously stated, partial allowance for the gas not being gray is made by evaluating $\overline{GS}$ or $(\overline{GS})_R$ with use of $\varepsilon_{G,e}$ rather than ε_G, and $1 - \varepsilon_{G,e}$ rather than τ, in Eqs. (4.3.45) to (4.3.47). A slightly better but more tedious allowance for partial **nongrayness** in evaluating $\dot{Q}_{rad}$ is to replace $\overline{GS_1}\sigma(T_G^4 - T_1^4)$, with $\overrightarrow{GS_1}$ evaluated by using $\varepsilon_{G,e}$, by $\overrightarrow{GS_1}T_G^4 - \overleftarrow{GS_1}T_1^4$, where $\overrightarrow{GS}$, and $\overleftarrow{GS}$, are evaluated by using ε_G and $\alpha_{G,1}$, respectively.

This completes the presentation of procedures for evaluating the total exchange area between gas and sink surface when the gas is gray or by making approximate allowance for the nongrayness by using ε_G and $\alpha_{G,1}$. These exchange areas can be used in the formulations below on furnace chamber performance. Methods will be presented first for a more rigorous treatment of gas nongrayness.

Full Allowance for Gas Nongrayness The above paragraphs failed to allow for the previously discussed change in gas transmittance on successive passages of reflected radiation through the gas. In many radiative transfer problems, interchange between many different pairs of radiators creates a system of simultaneous equations to be solved for the energy fluxes, and a shift from use of the Stefan-Boltzmann to the Planck equation would enormously increase the difficulty of solution. Use will be made of the fact that the total emissivity of a real gas, the spectral emissivity and absorptivity ε_λ of which vary in any way with λ, can be expressed *rigorously* as the a-weighted mean of a suitable number, n, of gray gas emissivity or absorptivity terms $\varepsilon_{G,i}$ or $\alpha_{G,i}$ representing the gray gas emissivity or absorptivity in the energy fractions a_i of the blackbody spectrum. Then

$$\varepsilon_G = \sum_0^n a_i \varepsilon_{G,i} = \sum_0^n a_i(1 - e^{-kpL}) \quad (4.3.48)$$

The linearity between flux $\dot{Q}$ and blackbody emissive power E_B allows the above relations to be used for converting $\overline{GS}$ or $\overline{SS}$ for a gray gas

to a form allowing for nongrayness. For a real gas $\overline{GS}$ or $\overline{SS}$ is the a_i-weighted sum of its values based on each of the $\varepsilon_{G,i}$ values of Eq. (4.3.48). Into an expression for $\overline{GS}_{gray}$ replace ε_G by $\varepsilon_{G,i}$, or τ by $1 - \varepsilon_{G,i}$, multiply the result by a_i and sum the resultant $\overline{GS}$ values to obtain $\overline{GS}_{real\ gas}$. Obviously, the number of terms n should be as small as possible while consistent with small error. Consider an n of 2, with the gas modeled as the sum of one gray gas plus a clear gas, with the gray gas of absorption coefficient k occupying the energy fraction a of the blackbody spectrum and the clear gas ($k = 0$) the fraction $1 - a$. Then, at path lengths L and $2L$,

$$[\varepsilon_G(pL)] = a(1 - e^{-kpL}) + (1 - a)(0)$$
$$[\varepsilon_G(2pL)] = a(1 - e^{-2kpL}) + (1 - a)(0) \quad (4.3.49)$$

Solution of these gives

$$a = \frac{\varepsilon_G(pL)}{2 - \varepsilon_G(2pL)/\varepsilon_G(pL)}$$

$$kpL = -\ln\left[1 - \frac{\varepsilon_G(pL)}{a}\right] \quad (4.3.50)$$

The equivalent gray gas emissivity in the spectral range a is $1 - e^{-kpL}$, from Eq. (4.3.48), and from Eq. (4.3.49) that is $\varepsilon_G(pL)/a$; in the spectral range $1 - a$, the equivalent gray gas emissivity is zero. This simple model will be correct for the contribution of the direct gas emission from path length L and for that of the once-reflected emission (path length $2L$); and the added contributions due to increasing numbers of reflections will be attenuated sufficiently by surface reflections to make errors in them unimportant. Note that a and k are not general constants; they are specific to the **subject mean beam length** L_m and come from basic data, such as Table 4.3.4. Note also that when full allowance for non-grayness is to be made by replacing ε_G by $\varepsilon_{G,e}$, then a is also changed to a_e, which comes from Eq. (4.3.50), with $\varepsilon_{G,e}$ replacing ε_G.

Conversion of gray gas total exchange areas $\overline{GS}$ and $\overline{SS}$ to their non-gray forms is carried out as follows when the nongray model is gray-plus-clear gas: From Eq. (4.3.49) the equivalent gray gas emissivity in the spectral energy fraction a_e is $\varepsilon_{G,e}(pL)/a_e$, which replaces ε_G wherever it or its complement τ occurs in $\overline{GS}$; the result is then multiplied by a_e. There is no contribution from the clear gas energy fraction. Conversion of $\overline{SS}$ from gray to gray plus clear involves making the same substitution, but for $\overline{SS}$ another term must be added. For the clear gas contribution, 0 and 1 are substituted for ε_G and τ, and the result is multiplied by the weighting factor $1 - a_e$; this $\overline{SS}$ is added to the preceding one to give $\overline{SS}_{g+c}$.

The simplest application of this gray-plus-clear model of gas radiation is the case of a single gas zone surrounded by a single surface zone, the case covered for a gray gas by Eq. (4.3.51) and illustrated in the last numerical example, where radiation from methane combustion products is surrounded by a single-zone sink surface. That example will be repeated using the gray-plus-clear gas model, for which the total exchange area is

$$\frac{\overline{GS_1}}{A_1} = \frac{a_e}{a_e/\varepsilon_{G,e} + 1/\varepsilon_1 - 1} \quad (4.3.51)$$

EXAMPLE (GRAY-PLUS-CLEAR GAS RADIATION FROM METHANE COMBUS-TION PRODUCTS). The computations of the previous example, radiative flux from methane combustion products, will be repeated with the more rigorous treatment of nongrayness, and the results will be compared with the more approximate cal-culations for the case of a wall emissivity ε_1 of 0.4 and 0.8.

SOLUTION. Repeat the calculations given, in the earlier example, of $\varepsilon_G(pL)$, $\alpha_{G1}(pL)$ and $\varepsilon_{G,e}(pL)$ for $pL = 2 \times 0.6875$, to give $\varepsilon_G(2pL) = 0.4096$, $\alpha_{G1}(2pL) = 0.5250$, and $\varepsilon_{G,e}(2pL) = 0.3812$. Then $a_e = 0.2782/(2 - 0.3812/0.2782) = 0.4418$, and the emissivity substitute for the gray gas portion of the gray-plus-clear gas model is $0.2782/0.4418 = 0.6297$. For a single enveloping surface zone, the total exchange area comes from Eq. (4.3.51): $\overline{GS_1}/A_1 = a_e/(a_e/\varepsilon_{G,e} + 1/\varepsilon_1 - 1) = 0.4418/(0.4418/0.2782 + 1/0.8 - 1) = 0.2404$. The flux density is $\dot{Q}/A = q = (\overline{GS_1}/A)\sigma(T_G^4 - T_1^4) = 0.2404 \times 56.7 \times [(1,500/1,000)^4 - (1,000/1,000)^4] = 55.37\ kW/m^2$ [17,550 Btu/(ft$^2 \cdot$ h)]. This is 7.6 percent lower than it is when only the difference between ε_G and $\alpha_{G,1}$ is allowed for in finding the effect of gas nongrayness. In some problems the difference is as high as 20 percent. (Note that allowing for average humidity in air adds 5 percent to H_2O and about 2 percent to the gas emissivity.) Changing ε_1 from 0.8 to 0.4 changes the approximate solution for $\overline{GS_1}/A_1$ from 0.2501 to 0.1963 and the gray-plus-clear treatment from 0.2404 to 0.1431.

The procedures for introducing the nongray gas model can be used to convert the total exchange areas for the basic one-gas two-surface model, Eqs. (4.3.38) to (4.3.40), as used to evaluate the cases in Table 4.3.5, to the following gray-plus-clear-gas model forms:

$$\frac{\overline{S_1 S_2}}{A_1} = F_{12} \varepsilon_1 \varepsilon_2 \left(\frac{a_e}{D_a} + \frac{1 - a_e}{D_b} \right) \tag{4.3.52}$$

$$\frac{\overline{S_1 S_2}}{A_1} = F_{12} \varepsilon_1^2 \left\{ \frac{a_e[1 - F_{12} + \rho_2(1 - \varepsilon_{G,e}/a_e)(F_{12}/C_2 - 1)]}{D_a} \right.$$
$$\left. + \frac{(1 - a_e)[1 - F_{12} + \rho_2(F_{12}/C_2 - 1)]}{D_b} \right\} \tag{4.3.53}$$

$$\frac{\overline{GS_1}}{A_1} = \frac{\varepsilon_1 \varepsilon_{G,e}[1/(1 - \varepsilon_{G,e}/a_e) + \rho_2(F_{12}/C_2 - 1)]}{D_a} \tag{4.3.54}$$

$$D_a = \frac{1}{1 - \varepsilon_{G,e}/a_e} + \left(1 - \frac{\varepsilon_{G,e}}{a_e}\right) \rho_1 \rho_2 (1 - F_{12} C_2)$$
$$- \rho_1(1 - F_{12}) - \rho_2(1 - F_{21})$$

$$D_b = 1 + \rho_1 \rho_2 \left(1 - \frac{F_{12}}{C_2}\right) - \rho_1(1 - F_{12}) - \rho_2(1 - F_{21})$$

$$\text{or} \qquad = \varepsilon_1 \varepsilon_2 + F_{12} \left[\varepsilon_2 + \frac{\varepsilon_1(C_1 - \varepsilon_2)}{C_2} \right]$$

The above relations, with the view factor F_{12} specified, may be used to convert the geometric cases of Table 4.3.5 to their more nearly correct forms with gray gas replaced by the gray-plus-clear gas model. That has been done in Table 4.3.6, which covers a moderate idealization of many practical industrial systems.

Effect of Gas Nongrayness on Refractory Zones Full allowance for the effect of gas nongrayness on enclosures in which part of the enclosing surface is radiatively adiabatic is straightforward but sometimes tedious. The term of Eq. (4.3.44) must be evaluated. It is tempting to use Eq. (4.3.45), but that is invalid because, although total radiative interchange at zone A_r is 0, the gas nongrayness makes A_r a net absorber in the spectral energy fraction a (or a_e) and a net emitter in the clear gas fraction $1 - a$. It is necessary, then, to use the basic equation

$$(\overline{GS_1})_R = \left(\overline{GS_1} + \frac{1}{1/\overline{GS_r} + 1/\overline{S_r S_1}} \right) \tag{4.3.55}$$

evaluating each of the right-hand members of a geometric system of interest, such as found in Table 4.3.5 (where A_r is A_2). As previously discussed, $\varepsilon_{G,e}/a_e$ is substituted for ε_G (or its complement for τ), and the result is weighted by the factor a_e; and for $\overline{S_r S_1}$ an additional term based on ε_G being replaced by 0 or τ by 1, with weighting $1 - a_e$, is added.

Of the cases covered in Table 4.3.5, only two will be evaluated to make A_2 represent the radiatively adiabatic zone A_r. The first is for the case of heat sink A_1 in a plane—the simulation of a slab-heating furnace. Insertion into Eq. (4.3.44) of the gray-plus-clear terms $\overline{GS_1}$, $\overline{GS_r}$, and $\overline{S_r S_1}$ from Table 4.3.5 (with subscript r replacing subscript 2) and rearrangement gives:

$$\left[\frac{(\overline{GS_1})_R}{A_1} \right]_{\substack{A_1 \text{ in a} \\ \text{plane}}} = \frac{\varepsilon_G}{D_1} \left[\varepsilon_1 \left(\rho_r \frac{C_1}{C_r} + \frac{1}{1 - \varepsilon_{G/a}} \right) \right.$$

$$\left. + \frac{\varepsilon_r}{\dfrac{1}{\rho_1 + \dfrac{C_r}{C_1}\left(1 - \dfrac{\varepsilon_G}{a}\right)} + \dfrac{\varepsilon_G/\varepsilon_1}{a + \dfrac{(1 - a)D_1}{\varepsilon_r + \rho_r \varepsilon_1 C_1/C_r}}} \right] \tag{4.3.56}$$

where $D_1 = 1/(1 - \varepsilon_G/a) - \rho_r[1 - (C_1/C_r)(\varepsilon_1 + \rho_1 \varepsilon_G/a)]$. Although ε_G and a are used here, $\varepsilon_{G,e}$ and a_e should be used if allowance is to be made for the difference between gas emissivity and absorptivity. Com-

parison of Eq. (4.3.56) with it gray gas equivalent, Eq. (4.3.46), shows the complexity introduced by allowance for gas nongrayness. $[(\overline{GS_1})_R$ for the gray-plus-clear gas model is about 15 percent higher than for gray gas when $\varepsilon_1 = 0.8$, $\varepsilon_G = 0.3$, $C_1 = \frac{1}{3}$, $a = 0.4$, and $\varepsilon_r = 0.6$, but only 1 percent higher when $\varepsilon_r = 1$.]

The second conversion of $\overline{GS}$ to $(\overline{GS_1})_R$ will be case 4 of Table 4.3.5, the two-surface-zone enclosure with the computation simplified by assuming that the direct-view factor from any spot to a surface equals the fraction of the whole enclosure which the surface occupies (the speckled furnace model). This case can be considered an idealization of many processing furnaces such as distilling and cracking coil furnaces, with parts of the enclosure tube-covered and part left refractory. (The refractory under the tubes is not to be classified as part of the refractory zone.) Again, one starts with substitution, into Eq. (4.3.44), of the terms $\overline{GS_1}$, $\overline{GS_r}$, and $\overline{S_r S_1}$, from Table 4.3.5, case 4, with all terms first converted to their gray-plus-clear form. To indicate the procedure, one of the components, $\overline{S_r S_1}$, will be formulated.

$$\frac{\overline{S_r S_1}}{A_1} = a \frac{C_r \varepsilon_1 \varepsilon_r}{D_4'} + (1 - a) \frac{C_r \varepsilon_1 \varepsilon_r}{1 - \rho_1 C_1 - \rho_r C_r}$$

$$= \frac{C_r \varepsilon_1 \varepsilon_r}{D_4'} \left[1 + \frac{\varepsilon_G (1 - a)(a - \varepsilon_G)}{1 - \rho_1 C_1 - \rho_r C_r} \right]$$

With $D_4' = 1/(1 - \varepsilon_G/a) - \rho_1 C_1 - \rho_r C_r$, the result of the full substitution simplifies to

$$\frac{(\overline{GS_1})_R}{A_1} = \frac{1}{C_1 \left(\dfrac{1}{\varepsilon_G} - \dfrac{1}{a} \right) + \dfrac{1}{\varepsilon_1} + \dfrac{1/a - 1}{\varepsilon_1 + \varepsilon_r(C_r/C_1)}} \tag{4.3.57}$$

For a gray gas ($a = 1$) the above becomes

$$\frac{(\overline{GS_1})_R}{A_1} = \frac{1}{C_1(1/\varepsilon_G - 1) + 1/\varepsilon_1} \tag{4.3.58}$$

Equation (4.3.57) has wide applicability.

The beginning of this subsection mentions compact chambers (just treated) and long chambers as limiting cases. The latter will now be treated.

The Long Combustion Chamber If a chamber is long enough in the x direction compared to its mean hydraulic radius, the local flux from gas to wall sink comes substantially from gas at its local temperature, with $\overline{GS_1}$ [or $(\overline{GS_1})_R$] calculated by methods just described but based on a two-dimensional structure; i.e., the opposed upstream and downstream fluxes through the flow cross section will substantially cancel. That limiting case will be considered, with $(\overline{GS_1})_R/A_1$ evaluated by using local mean values of T_G and T_1. The local $(\overline{GS_1})_R$ applicable to a surface element of length dx and perimeter P is then $[(\overline{GS_1})_R/A_1]P \, dx$. Let $T_{G,\text{in}}$, $T_{G,\text{out}}$, $T_{1,\text{in}}$, and $T_{1,\text{out}}$ be specified; furnace length L is to be determined. Assume a constant sink temperature T_1. The equation of heat transfer in the furnace length element $P \, dx$ is then

$$-\dot{m} C_p \, dT_G = P \, dx \left[\frac{(\overline{GS_1})_R}{A_1} \right] [\sigma(T_G^4 - T_1^4) + h(T_G - T_1)] \tag{4.3.59}$$

The second of the heat-transfer terms is an order of magnitude smaller than the first, and to permit ready integration, $h(T_G - T_1)$ will be set equal to $b\sigma(T_G^4 - T_1^4)$, from which

$$b\sigma = \frac{h}{T_G^3 + T_G^2 T_1 + T_G T_1^2 + T_1^3} \cong \frac{h}{4 T_{G1}^3}$$

T_{G1} is the mean value of T_G and T_1, and a 10 percent error in T_1 will make but a 1 percent error in the calculated heat transfer. Then Eq. (4.3.59) becomes

$$-\dot{m} C_p \, dT_G = P \, dx \left[\frac{(\overline{GS_1})_R}{A_1} + \frac{h}{4\sigma T_{G1}^3} \right] \sigma(T_G^4 - T_1^4) \tag{4.3.60}$$

Table 4.3.6 Conversion of Total Exchange Areas for Cases of Table 4.3.5 to Their Gray-plus-Clear Values

Case 1: Plane slab A_1 and surface A_2 completing an enclosure of gas; $F_{12} = 1$

$$\frac{\overline{S_1 S_2}}{A_1} = \frac{a\varepsilon_1\varepsilon_2}{D_1} + \frac{(1-a)\varepsilon_1\varepsilon_2}{1 - \rho_2(1 - \varepsilon_1 C_1/C_2)}$$

where

$$D_1 = \frac{1}{(1 - \varepsilon_G/a)} - \rho_2\left[1 - \frac{C_1}{C_2}\left(\varepsilon_1 + \frac{\rho_1\varepsilon_G}{a}\right)\right]$$

$$\frac{\overline{GS_1}}{A_1} = \varepsilon_1\varepsilon_G\left[\frac{1/(1 - \varepsilon_G/a) + \rho_2 C_1/C_2}{D_1}\right]$$

$$\frac{\overline{GS_2}}{A_2} = \varepsilon_2\varepsilon_G\left[\frac{1/(1 - \varepsilon_G/a) + \rho_1 C_1/C_2}{D_1}\right]$$

Case 2: Infinite parallel planes, gas between; $F_{12} = F_{21} = 1$

$$\frac{\overline{S_1 S_2}}{A_1} = \frac{a\varepsilon_1\varepsilon_2}{D_2} + \frac{(1-a)\varepsilon_1\varepsilon_2}{1 - \rho_1\rho_2}$$

where

$$D_2 = \frac{1}{1 - \varepsilon_G/a} - \left(1 - \frac{\varepsilon_G}{a}\right)\rho_1\rho_2$$

$$\frac{\overline{GS_1}}{A_1} = \varepsilon_1\varepsilon_G\left[\frac{1/(1 - \varepsilon_G/a) + \rho_2}{D_2}\right]$$

Case 3: Concentric spherical or infinite cylindrical surface zones, A_1 inside; $F_{12} = 1$; $F_{21} = A_1/A_2 \equiv C_1/C_2$

$$\frac{\overline{S_1 S_2}}{A_1} = \frac{a\varepsilon_1\varepsilon_2}{D_3} + \frac{(1-a)\varepsilon_1\varepsilon_2}{1 - \rho_2(1 - \varepsilon_1 C_1/C_2)}$$

where

$$D_3 = \frac{1}{1 - \varepsilon_G/a} - \rho_2\left[1 - \left(\frac{C_1}{C_2}\right)\left(\varepsilon_1 + \frac{\rho_1\varepsilon_G}{a}\right)\right]$$

$$\frac{\overline{GS_1}}{A_1} = \varepsilon_1\varepsilon_G\left[\frac{1/(1 - \varepsilon_G/a) + \rho_2 C_1/C_2}{D_3}\right]$$

$$\frac{\overline{GS_2}}{A_2} = \varepsilon_2\varepsilon_G\left[\frac{1/(1 - \varepsilon_G/a) + \rho_2 C_1/C_2}{D_3}\right]$$

Case 4: Spherical enclosure of two surface zones *or* speckled $A_1:A_2$ enclosure; $F_{12} = F_{22} = C_2$; $F_{21} = F_{11} = C_1$

$$\frac{\overline{S_1 S_2}}{A_1} = \frac{a\varepsilon_1\varepsilon_2 C_2}{D_4} + \frac{(1-a)\varepsilon_1\varepsilon_2 C_2}{1 - \rho_1 C_1 - \rho_2 C_2}$$

where

$$D_4 = \frac{1}{1 - \varepsilon_G/a} - \rho_1 C_1 - \rho_2 C_2$$

$$C_1 = \frac{A_1}{A_1 + A_2}$$

$$\frac{\overline{GS_1}}{A_1} = \frac{\varepsilon_1\varepsilon_G/(1 - \varepsilon_G/a)}{D_4}$$

Integration of T_G from $T_{G,\text{in}}$ to $T_{G,\text{out}}$ and of x from 0 to L, and solution for L give

$$L = \frac{\dot{m}\overline{C}_p\left(\tan^{-1}\dfrac{T_{G,\text{out}}}{T_1} - \tan^{-1}\dfrac{T_{G,\text{in}}}{T_1} - \dfrac{1}{2}\ln\dfrac{T_{G,\text{out}} - T_1 T_{G,\text{in}} + T_1}{T_{G,\text{out}} + T_1 T_{G,\text{in}} - T_1}\right)}{2PT_1^3\sigma\left[\dfrac{(\overline{GS_1})_R}{A_1} + \dfrac{h}{4\sigma T_{G1}^3}\right]}$$

(4.3.61)

Trial and error are necessary if L is specified and $T_{G,\text{out}}$ is to be found. If L is not long, axial radiative flux becomes important and a much more complex treatment is necessary. Use of a multigas zone system is one possibility.

Partially Stirred Model of Furnace Chamber Performance An equation representing an energy balance on a combustion chamber of two surface zones—a heat sink A_1 at temperature T_1 and a refractory surface A_r assumed radiatively adiabatic at T_r—is most simply solved if the total enthalpy input H is expressed as $\dot{m}\overline{C}_p(T_F - T_0)$; $\dot{m}$ is the mass rate of fuel plus air, and T_F is a pseudo-adiabatic flame temperature based on a mean specific heat from base temperature T_0 up to the gas

exit temperature T_E rather than up to T_F. Assume that enough stirring occurs in the chamber to produce two temperatures—the heat-transfer temperature T_G and the leaving gas enthalpy temperature T_E—the two differing by an empirical amount, zero if the stirring were perfect.

Of the many ways tried to introduce this empiricism, the best is to assume that $T_G - T_E$, expressed as a ratio to T_F, is a constant Δ. Although Δ will vary with burner type, the effects of excess air and firing rate are small. At the limits of very high and low firing densities, Δ tends toward zero. These conditions excepted and in the absence of performance data on the subject furnace type, assume $\Delta = 0.08$, or

$$\frac{T_G - T_E}{T_F} = \Delta = 0.08$$

This assumption bypasses complex allowance for temperature variations in the chamber gas and for the effects of fluid mechanics and combustion kinetics, but at the cost of not permitting evaluation of **flux distribution** over the surface. The heat-transfer rate $\dot{Q}$ out of the gas is then $\dot{H} = \dot{m}\overline{C}_p(T_E - T_0)$ or $\dot{m}\overline{C}_p(T_F - T_E)$. A combination of energy balance and heat transfer, with the ambient temperature taken as the

enthalpy base temperature T_0, gives

$$(\dot{Q} =) \dot{H} - \dot{m}\overline{C}_p(T_E - T_0) = \dot{m}\,\overline{C}_p(T_F - T_E)$$
$$= (\overline{GS_1})_R\sigma(T_G^4 - T_1^4) + h_1A_1(T_G - T_1) \quad (4.3.62)$$

To make the relations dimensionless, divide through by $(\overline{GS_1})_R\sigma T_F^4$, and let all temperatures, expressed as ratios to T_F, be called T^*. For simplification, the following definitions of D and N_c are introduced:

$$\dot{m}\overline{C}_p/(\overline{GS_1})_R\sigma T_F^3[\equiv \dot{H}/(\overline{GS_1})_R\sigma_F^4(1 - T_0^*)]$$

$$= \text{dimensionless firing density } D$$

After the division, the left-hand side term of Eq. (4.3.62) = $D(1 - T_E^*)$ and the first right-hand side term = $T_G^{*4} - T_1^{*4}$.

$$\frac{h_1A_1}{(\overline{GS_1})_R\sigma T_F^3} = N_c, \text{ convection number (dimensionless)}$$

The equation then becomes

$$D(1 - T_E^*) = T_G^{*4} - T_1^{*4} + N_c(T_G^* - T_1^*) \quad (4.3.63)$$

The two unknowns T_G^* and T_E^* are reduced to one by expressing T_E^* in terms T_G^* of and Δ. Equation (4.3.63), with coefficients of T_G^{*4} and T_G^* collected, then becomes

$$T_G^{*4} + (D + N_c)T_G^* - [T_1^{*4} + N_cT_1^* + D(1 + \Delta)] = 0 \quad (4.3.64)$$

Although Eq. (4.3.64) is a quartic equation, it is capable of explicit solution. Trial and error enter, however, because $(\overline{GS_1})_R$ and $\overline{C}_p$ are mild functions of T_G and related T_E, respectively, and a preliminary guess of T_G is necessary. Ambiguity can exist in the interpretation of terms. If part of the enclosure surface consists of screen tubes over the chamber gas exit to a convection section, radiative transfer to those tubes is included in the chamber energy balance but convection is not, because it has no effect on the chamber gas temperature. Although the results must be considered approximations, depending as they do on the empirical Δ, the equation may be used to find the effect of firing rate, excess air, and air preheat on efficiency. With some performance data available, the small effect of various factors on Δ may be found.

For the commonly encountered case of the sink consisting of a row of tubes mounted on a refractory wall, A_1 is the area of the whole plane in which the tubes lie, T_1 is tube surface temperature, and ε_1 is the effective emissivity of the tube-row-refractory-wall combination, as in the earlier numerical example associated with Fig. 4.3.6, where $\varepsilon_1 = 0.702$. A further simplification is to replace A_1/A_T by C, the "cold" fraction of the wall ($A_T \equiv A_1 + A_r$).

With T_G^* known, the chamber efficiency η_G based on heat transfer from the gas is given by

$$\eta_G = \frac{(1 - T_G^* + \Delta)}{1 - T_O^*} \quad (4.3.65)$$

In the absence of heat losses, this is equal to the efficiency based on energy to the sink. It is expressed as

$$\eta_G = \frac{(GS_1)_R\sigma(T_G^4 - T_1^4) + h_1A_1(T_G - T_1)}{\dot{H}} \quad (4.3.66)$$

Solution of Eq. (4.3.63) gives the relation between T_G^* and D, and substituting the value of T_G^* into either Eq. (4.3.65) or Eq. (4.3.66) provides the dependence of η_G on D.

Furnace Chamber Performance—General Although the chamber efficiency η depends on D, N_c, and T_1^*, the reduced firing density D is the dominant factor; it makes allowance for such operating variables as fuel type, excess air or air preheat—which affect flame temperature or gas emissivity, for fractional occupancy of the walls by sink surfaces, and for sink emissivity. Variation in the normalized sink temperature T_1^* has little effect until it exceeds 0.3; N_c, though significant, is secondary.

Solution of Eq. (4.3.63) gives the relation between D and T_G^*, and Eqs. (4.3.65) and (4.3.66) give the relation between η_G and D. As an example, Fig. 4.3.6 gives D versus η_G. Approximate operating regimes of various classes of furnaces are shown in Table 4.3.7. Note the significant properties of the functions presented: (1) As firing rate D goes down, η_G rises; T_G^* approaches T_1^* in the limit as D decreases to zero. (2) Changing T_1^* has a large effect only when it exceeds about 0.4. (3) As the furnace walls approach complete coverage by a black sink ($C = \varepsilon_1 = 1$), $\overline{GS_1}$ becomes ε_GA_T and $D \propto 1/\varepsilon_G$. Thus, at very high firing rates where η_G approaches inverse proportionality to D, the efficiency of heat transfer varies directly as ε_G (gas-turbine chambers), but at low firing rates ε_G has relatively little effect. (4) When $C\varepsilon_1 \ll 1$ because of a nonblack sink or much refractory surface, the effect of changing flame emissivity is to produce a much less than proportional effect on heat flux. The regimes of operation of different practical combustion systems are given in Table 4.3.7.

Equations (4.3.63) to (4.3.66) predict the effects of excess air, air preheat, and fuel quality on performance, through the effect on T_F; through $\overline{GS_1}$ they show the effect of gas and sink emissivity and the fraction of the chamber walls occupied by heat sink. They serve as a framework for correlating the performance of furnaces with flow patterns—plug flow, parabolic profile, and recirculatory flow—differing from

Fig. 4.3.6 Reduced gas-side furnace efficiency versus effective firing density for well-stirred combustion chamber model. $\Delta = 0$; $T^* = 0.0, 0.4, 0.5, 0.6, 0.7, 0.8, 0.9$.

Table 4.3.7   Operational Domains for Representative Process Furnaces and Combustors

Domain	Furnace or combustor type	Dimensionless sink temperature	Dimensionless firing density
A	Oil processing furnaces; radiant section of oil tube stills and cracking coils	$T^* \approx 0.4$	$0.1 < D_{\text{eff}} < 1.0$
B	Domestic boiler combustion chambers	$T^* \approx 0.2$	$0.5 < D_{\text{eff}} < 1.1$
C	Glass furnaces	$0.7 < T^* < 0.8$	$0.035 < D_{\text{eff}} < 0.8$
D	Soaking pits	$T^* \approx 0.6$	$0.7 < D_{\text{eff}} < 1.1$
E	Gas-turbine combustors	$0.4 < T^* < 0.7$	$4.0 < D_{\text{eff}} < 25.0$

the well-stirred model (Hottel and Sarofim, Chap. 14). As expected, plug-flow furnaces show somewhat higher efficiency, mild recirculation types somewhat lower efficiency, and strong recirculation furnaces an efficiency closely similar to that of the well-stirred model.

Refractory Temperature T_r Though the average value of T_r in a combustion chamber is not involved in the evaluation of T_G or η, it is sometimes of interest. It assumes a mean value between T_G and T_1, given by

$$\left(\frac{T_r}{T_G}\right)^4 = \frac{1 + E(T_1/T_G)^4}{1 + E}$$

For the speckled furnace gray gas model

$$E = C\varepsilon_1\left(\frac{1}{\varepsilon_G} - 1\right)$$

Allowance is made for the difference between emissivity and absorptivity by changing ε_G to $\varepsilon_{G,e}$ in the above equation. For the gray-plus-clear-gas model,

$$E = C\varepsilon_1\left[\frac{1}{\varepsilon_G} - \frac{1}{a} + \frac{1/a - 1}{C\varepsilon_1 + (1 + C)\varepsilon_R}\right]$$

4.4   TRANSMISSION OF HEAT BY CONDUCTION AND CONVECTION
by Kenneth A. Smith

REFERENCES: McAdams, "Heat Transmission," McGraw-Hill. Eckert and Drake, "Analysis of Heat and Mass Transfer," McGraw-Hill. Carslaw and Jaeger, "Conduction of Heat in Solids," Oxford. Jakob, "Heat Transfer," vols. I and II, Wiley. Kays, Crawford, and Weigand, "Convective Heat and Mass Transfer," 4th ed., McGraw-Hill. Wilkes, "Heat Insulation," Wiley. Kays and London, "Compact Heat Exchangers," McGraw-Hill. "Thermophysical Properties Data Book," Purdue University. Gebhart, Jaluria, Mahajan, and Sammakia, "Buoyancy–Induced Flows and Transport," Hemisphere.

Notation and Units   The units are based on feet, pounds, hours, degrees Fahrenheit, and Btu. Any other consistent set may be used in the dimensionless relations given, but for the dimensional equations the units of this table must be used.

A = area of heat-transfer surface, ft^2

A_i = inside area

A_o = outside area

A_m = average value of A, ft^2

a = empirical constant

C_p = specific heat at constant pressure, Btu/lb · °F

D = diameter, ft

D_o = outside diameter, ft

D_i = inside diameter, ft

D' = diameter, in

D'_o = outside diameter, in

D'_i = inside diameter, in

G = mass velocity, equals w/S, lb/h · ft^2 of cross section occupied by fluid

G_{max} = mass velocity through minimum free area in a row of pipes normal to fluid stream, lb/h · ft^2

g_c = conversion factor, equal to 4.18×10^8 (mass lb)(ft)/ (force lb)(h)^2

g_L = local acceleration due to gravity, 4.18×10^8 ft/h^2 at sea level

h = local individual coefficient of heat transfer, equals $dq/dA\ \Delta t$, Btu/h(ft^2)(°F)diff

$h_c + h_r$ = combined coefficient by conduction, convection, and radiation between surface and surroundings

h_m = mean value of h for entire surface, based on $(\Delta t)_m$

$h_{\text{a.m.}}$ = average h, arbitrarily based on arithmetic-mean temperature difference

h_s = heat-transfer coefficient through scale deposits

J = mechanical equivalent of heat, 778 ft · lb/Btu

K = empirical constant

k = thermal conductivity, Btu/h · ft · °F

$$k_m = -\frac{1}{t_1 - t_2}\int_1^2 k\,dt$$

k_f = k at the "film" temperature, $t_f = (t + t_w)/2$

l = thickness of material normal to heat flow, ft

L = length of heat-transfer surface, heated length, ft

N = number of rows of tubes

N_{Gr} = Grashof number, $L_c^3\rho_f^2 g_L\beta_f(\Delta t)_s/\mu_f^2$

q = total rate of heat flow, Btu/h

$\dot{\mathbf{q}}$ = heat-flux vector, Btu/h · ft^2

$\dot{\mathbf{q}}_x$ = x component of heat flux vector

R = thermal resistances, $1/(UA)$, $1/(hA)$, $1/(h_c + h_r)A_0$

$\mathcal{R}$ = recovery factor

r = radius, ft

S = cross section, filled by fluid, in plane normal to direction of fluid flow, ft^2

T = temperature, °R = $t + 460$

T_1, T_2 = inlet and outlet bulk temperatures, respectively, of warmer fluid, °F

t = bulk temperature (based on heat balance), °F

$t_{\text{a.w.}}$ = temperature of adiabatic wall

t_∞ = temperature at infinity

t_w = wall temperature, °F
t_1, t_2 = inlet and outlet bulk temperatures of colder fluid, °F
t_i, t_o = temperatures of fluid inside and outside, °F
$t_f = (t + t_w)/2$
t_{sat} = saturation temperature, °F
U = overall coefficient of heat transfer, Btu/h · ft² · °F; U_i, U_o based on inside and outside surface, respectively
V = mean velocity, ft/h
V_s = average velocity, volumetric rate divided by cross section filled by fluid, ft/s
V_{sm} = maximum velocity, through minimum cross section, ft/s
x = one of the axes of a Cartesian reference frame, ft
$X = (t_2 - t_1)/(T_1 - t_1)$
w = mass rate of flow per tube, lbm/h/tube
$Z = (T_1 - T_2)/(t_2 - t_1)$
β = volumetric coefficient of thermal expansion, °F⁻¹
Γ = mass rate of flow, lbm/(h) (ft of wetted periphery measured on a plane normal to direction of fluid flow); = $w/\pi D$ for a vertical and $w/2L$ for a horizontal tube
γ = ratio of specific heats, c_p/c_v; 1.4 for air
$\mathcal{R}$ = gradient operator
Δt = temperature difference, °F
$(\Delta t)_{ave}$, $(\Delta t)_{l.m.}$ = arithmetic and logarithmic means of terminal temperature differences, respectively, °F
$(\Delta t)_m$ = true mean value of the terminal temperature differences, °F
$(\Delta t)_o$ = overall temperature difference, °F
$(\Delta t)_s$ = temperature difference between surface and surroundings, °F
λ = latent heat (enthalpy) of vaporization, Btu/lb
μ = viscosity at bulk temperature, lbm/h · ft; equals 2.42 times centipoises; equals 116,000 times viscosity in (lb force)(s)/ft²
μ_f = viscosity, lbm/h · ft, at arithmetic mean of wall and fluid temperatures

μ_w = viscosity at wall temperature, lbm/h · ft
ρ = density, lbm/ft³
σ = surface tension, lb force/ft

Subscripts:

l = liquid
v = vapor

Preliminary Statements The transfer of heat is usually considered to occur by three processes:

1. **Conduction** is the transfer of heat from one part of a body to another part or to another body by short-range interaction of molecules and/or electrons.

2. **Convection** is the transfer of heat by the combined mechanisms of fluid mixing and conduction.

3. **Radiation** is the emission of energy in the form of electromagnetic waves. All bodies above absolute zero temperature radiate. Radiation incident on a body may be absorbed, reflected, and transmitted. (See Sec. 4.3.)

CONDUCTION

See Tables 4.4.1 to 4.4.7 and 4.4.10

The **basic Fourier conduction law** for an isotropic material is

$$\dot{\mathbf{q}} = -k\nabla t \qquad (4.4.1)$$

In cartesian coordinates, the x component of this equation is $\dot{\mathbf{q}}_x = -k(\partial t/\partial x)$, and if the heat flow is essentially unidimensional, $\mathbf{q} = \dot{q}A(x) = -kA(x)(dt/dx)$. This states that the steady-state rate of heat conduction q is proportional to the cross-sectional area $A(x)$ normal to the direction of flow and to the temperature gradient —partial—t/—partial—x along the conduction path. The proportionality constant k is called the **"true" thermal conductivity** of the material.

The thermal conductivity of a given material varies with temperature, and the mean thermal conductivity is defined by

$$k_m = \frac{1}{t_0' - t_i'} \int_{t_0}^{t_i'} k \, dt$$

Table 4.4.1   Thermal Conductivities of Metals*

k = Btu/h · ft · °F
$k_t = k_{t_0} - a(t - t_0)$

Substance	Temp range, °F	k_{t_0}	a	Substance	Temp range, °F	k_{t_0}	a
Metals				Tin	60–212	36	0.0135
Aluminum	70–700	130	0.03	Titanium	70–570	9	0.001
Antimony	70–212	10.6	0.006	Tungsten	70–570	92	0.02
Beryllium	70–700	80	0.027	Uranium	70–770	14	−0.007
Cadmium	60–212	53.7	0.01	Vanadium	70	20	—
Cobalt	70	28	—	Zinc	60–212	65	0.007
Copper	70–700	232	0.032	Zirconium	32	11	—
Germanium	70	34	—	Alloys:			
Gold	60–212	196	—	Admiralty metal	68–460	58.1	−0.054
Iron, pure	70–700	41.5	0.025	Brass	−265–360	61.0	−0.066
Iron, wrought	60–212	34.9	0.002	(70% Cu, 30% Zn)	360–810	84.6	0
Steel (1% C)	60–212	26.2	0.002	Bronze, 7.5% Sn	130–460	34.4	−0.042
Lead	32–500	20.3	0.006	7.7% Al	68–392	39.1	−0.038
Magnesium	32–370	99	0.015	Constantan	−350–212	12.7	−0.0076
Mercury	32	4.8	—	(60% Cu, 40% Ni)	212–950	10.1	−0.019
Molybdenum	32–800	79	0.016	Dural 24S (93.6% Al, 4.4% Cu,	−321–550	63.8	−0.083
Nickel	70–560	36	0.0175	1.5% Mg, 0.5% Mn)	550–800	130.	+0.038
Palladium	70	39	—	Inconel X (73% Ni, 15% Cr, 7%	27–1,070	7.62	−0.0068
Platinum	70–800	41	0.0014	Fe, 2.5% Ti)			
Plutonium	70	5	—	Manganin (84% Cu, 12% Mn,	1,070–1,650	3.35	−0.0111
Rhodium	70	88	—	4% Ni)	−256–212	11.5	−0.015
Silver	70–600	242	0.058	Monel (67.1% Ni, 29.2% Cu,	−415–1,470	12.0	−0.008
Tantalum	212	32	—	1.7% Fe, 1.0% Mn)			
Thallium	32	29	—	Nickel silver (64% Cu, 17% Zn,	68–390	18.1	−0.0156
Thorium	70–570	17	−0.0045	18% Ni)			

* For refractories see Sec. 6; for pipe coverings, Sec. 8; for building materials, Sec. 4. Conversion factors for various units are given in Sec. 1. Tables 4.4.3–4.4.7 were revised by G. B. Wilkes.

Over a moderate range, k varies linearly with t, and hence k_m is the value of k at the arithmetic mean of t_i' and t_0'.

Thermal Conductivity of Nickel-Chromium Alloys with Iron

$k_t = k_{t0} - a(t - t_0)$

ANSI number	Temp, range, °F	k_{t0}	a
301, 302, 303, 304 (303 Se, 304 L)	95–1,650	8.08	−0.0052
310 (3105)	32–1,650	6.85	−0.0072
314	80–572	10.01	−0.00124
	572–1,650	8.20	−0.0045
316 (316 L)	−60–1,750	7.50	−0.0042
321, 347 (348)	−100–1,650	8.22	−0.0050
403, 410 (416, 416 Se, 420)	−100–1,850	15.0	0
430 [430 F, 430 F (Se)]	122–1,650	12.60	−0.0012
440 C	212–932	12.77	−0.0043
446	32–1,850	12.96	−0.0050
501, 502	80–1,520	21.4	+0.0037

For unidimensional heat flow through a material of thickness l

$$q \int_0^l \frac{dx}{A(x)} = \int_{t_i'}^{t_0'} k \, dt = k_m(t_i' - t_0') \qquad (4.4.2)$$

With an obvious definition for the mean area:

$$\frac{q}{A_m} = k_m(t_i' - t_0')/l$$

For flat plates, $A_m = A_i = A_0$; for hollow cylinders, $A_m = (A_0 - A_i)/\ln(A_0/A_i)$; for hollow spheres, $A_m = \sqrt{A'A_0}$. For more complex shapes, Eq. (4.4.1) must be employed. For other configurations, mean areas may often be found elsewhere, e.g., Kutateladze and Borishanskei, "A Concise Encyclopedia of Heat Transfer," Pergamon Press, pp. 36–44.

Extended Surfaces   Fin efficiency is defined as the ratio of the mean temperature difference from surface to fluid divided by the temperature difference from fin to fluid at the base or root of the fin. Graphs of fin efficiency for extended surfaces of various types are given by Gardner (*Trans. ASME*, **67**, pp. 621–628, 1945) and in numerous texts, e.g., by Eckert and Drake, pp. 92–93. Heat-transfer coefficients for various extended surfaces are given by Kays and London.

CONDUCTION AND CONVECTION

Phenomena of Heat Transmission   In many practical cases of heat transmission—e.g., boilers, condensers, the cooling of engine cylinders—heat is transmitted from one fluid to another through a wall separating the two. The processes occurring in the fluids may be extremely complex. However, to facilitate discussion, it is convenient to imagine that most of the fluid offers no resistance to heat transmission but that a thin film of fluid adjacent to the wall offers considerable resistance. This situation is depicted in Fig. 4.4.1. Then, by definition,

$$q = h_i A_i(t_i - t_i') = \frac{k}{l} A_m(t_i' - t_0) = h_0 A_0(t_0' - t_0)$$

The terms h_i and h_0 are the **film coefficients,** or **unit conductances,** of the films f_1 and f_2, respectively, and k is the thermal conductivity of the wall. Since q, A, $t_i - t_i'$, and $t_0' - t_0$ are susceptible to direct measurement, h_i and h_0 are simply defined quantities and the propriety of the above equation does not rest upon the heuristic film concept. Indeed, for laminar flow, the film concept is a gross misrepresentation, and yet the definition of a film coefficient (or heat-transfer coefficient) remains convenient and valid.

Fig. 4.4.1   Temperature gradients in heat flow through a wall.

Properties of Molten Metals*

Metal (melting point)	Temperature, °F	k, Btu / (h)(ft)(°F)	ρ, lb / cu ft	c_p, Btu / (lb)(°F)	μ, lb / (ft)(h)
Bismuth (520°F)	600	9.5	625	0.0345	3.92
	1,000	9.0	608	0.0369	2.66
	1,400	9.0	591	0.0393	1.91
Lead (621°F)	700	10.5	658	0.038	5.80
	900	11.4	650	0.037	4.65
	1,300	—	633	—	3.31
Mercury (−38°F)	50	4.7	847	0.033	3.85
	300	6.7	826	0.033	2.66
	600	8.1	802	0.032	2.09
Potassium (147°F)	300	26.0	50.4	0.19	0.90
	800	22.8	46.3	0.18	0.43
	1,300	19.1	42.1	0.18	0.31
Sodium (208°F)	200	49.8	58.0	0.33	1.69
	700	41.8	53.7	0.31	0.68
	1,300	34.5	48.6	0.30	0.43
Na, 56 wt % K, 44 wt % (66.2°F)	200	14.8	55.4	0.270	1.40
	700	15.9	51.3	0.252	0.570
	1,300	16.7	46.2	0.249	0.389
Na, 22 wt % K, 78 wt % (12°F)	200	14.1	53.0	0.226	1.19
	750	15.4	48.4	0.210	0.500
	1,400	—	43.1	0.211	0.353
Pb, 44.5 wt % Bi, 55.5 wt % (257°F)	300	5.23	657	0.035	
	700	6.85	639	0.035	3.71
	1,200	—	614	—	2.78

* Based largely on "Liquid-Metals Handbook," 2d ed., Government Printing Office, Washington.

Table 4.4.2 Thermal Conductivities of Liquids and Gases

Substance	Temp, °F	k	Substance	Temp, °F	k
Liquids:			Gases:		
Acetone	68	0.103	Air (see below)	32	0.0140
Ammonia	45	0.29	Ammonia, vapor	32	0.0126
Aniline	32	0.104	Ammonia	212	0.0192
Benzol	86	0.089	Argon	32	0.00915
Carbon bisulfide	68	0.0931	Carbon dioxide	32	0.0084
Ethyl alcohol	68	0.105		212	0.0128
Ether	68	0.0798	Carbon monoxide	32	0.0135
Glycerin, USP, 95%	68	0.165	Chlorine	32	0.0043
Kerosene	68	0.086	Ethane	32	0.0106
Methyl alcohol	68	0.124	Ethylene	32	0.0101
n-Pentane	68	0.0787	Helium	32	0.0818
Petroleum ether	68	0.0758	n-Hexane	32	0.0072
Toluene	86	0.086	Hydrogen	32	0.0966
Water	32	0.343		212	0.124
	140	0.377	Methane	32	0.0175
Oil, castor	39	0.104	Neon	32	0.0267
Oil, olive	39	0.101	Nitrogen	32	0.0140
Oil, turpentine	54	0.0734	Nitrous oxide	32	0.0088
Vaseline	59	0.106		212	0.0090
			Nitric oxide	32	0.0138
			Oxygen	32	0.0142
			n-Pentane	32	0.0074
			Sulphur dioxide	32	0.005

Thermal Conductivities of Air and Steam

Temperature, °F	32	200	400	600	800	1,000
Air, 1 atm	0.0140	0.0181	0.0225	0.0266	0.0303	0.0337
Steam, 1 lb/in² absolute	—	0.0132	0.0184	0.0238	0.0292	0.0345

SOURCE: F. G. Keyes, *Tech. Rept. 37,* Project Squid (Apr. 1, 1952).

If t_i' and t_0' are eliminated from the above equation, a relation is obtained for steady flow through several resistances in series:

$$q = \frac{t_i - t_0}{1/(h_i A_i) + 1/(k A_m) + 1/(h_0 A_0)} \qquad (4.4.3)$$

Each of the terms in the denominator represents a resistance to heat transfer. There may also be a resistance, $1/(h_s A_s)$, due to the presence of a scale deposit on the surface. Thus, if the overall heat transfer is given by $q = UA(t_i - t_0)$, then the total thermal resistance is given by

$$1/(UA) = 1/(h_i A_i) + 1/(k A_m) + 1/(h_0 A_0) + 1/(h_s A_s) \qquad (4.4.4)$$

Coefficients for scale deposits are given in Table 4.4.9.

Mean Temperature Difference The basic equation for any steadily operated heat exchanger is $dq = U(\Delta t)_o\, dA$, in which U is the overall coefficient [Eq. (4.4.4)], $(\Delta t)_o$ is the overall temperature difference between hot and cold fluids, and dq/dA is the local rate of flow per unit surface. In order to apply this relation to a finite exchanger, it is necessary to integrate it. The assumptions usually made are constant U, constant mass rates of flow, no changes in phase, constant specific heats, and negligible heat losses. The resulting equation for parallel or countercurrent flow of fluids is

$$q = UA(\Delta t)_m = UA[(\Delta t)_{01} - (\Delta t)_{02}]/\ln\,[(\Delta t)_{01}/(\Delta t)_{02}] \qquad (4.4.5a)$$

in which $(\Delta t)_m$ is the logarithmic mean of the terminal temperature differences, $(\Delta t)_{01}$ and $(\Delta t)_{02}$, between hot and cold fluid. The value of UA is evaluated from the resistance concept of Eq. (4.4.4) and the values of h are obtained from the following pages. For more complicated flow geometries, the logarithmic mean is not appropriate, and the true mean temperature difference may be obtained from Fig. 4.4.2, where

$$Y = \text{ordinate}$$
$$= \frac{\text{true mean temp difference}}{\text{logarithmic mean temp difference for counterflow}}$$

For the other symbols see p. 4-79. (From *Trans. ASME,* **62,** 1940, pp. 283–294.)

The above discussion focuses on the concepts of an overall coefficient and a mean-temperature difference. An alternative approach focuses on the concepts of effectiveness and the number of transfer units. The alternatives are basically equivalent, but one or the other may enjoy a computational advantage. The latter method is presented in detail by Kays and London and by Mickley and Korchak (*Chem. Eng.,* **69,** 1962, pp. 181–188 and 239–242).

EXAMPLE. Assume an exchanger in which the hot fluid enters at 400°F and leaves at 327°F; the cold fluid enters at 100°F and leaves at 283°F. Assuming U independent of temperature, what will be the true mean temperature difference from hot to cold fluid, (1) for counterflow and (2) for a reversed current apparatus with one well-baffled pass in the shell and two equal passes in the tubes?

1. With counterflow, the terminal differences are $400 - 283 = 117°F$ and $327 - 100 = 227°F$; the logarithmic mean difference is $110/0.662 = 166°F$.

2. $Z = (400 - 327)/(283 - 100) = 0.4$; $X = (283 - 100)/(400 - 100) = 0.61$; from section A of Fig. 4.4.2, $Y = 0.9 = (\Delta t)_m/166$; $(\Delta t)_m = 149°F$.

If one of the temperatures remains constant, as in a condenser or in an evaporative cooler, Eq. (4.4.5a) applies for parallel flow, counterflow, reversed current, and cross flow.

If U varies considerably with temperature, the apparatus should be considered to be divided into stages, in each of which the variation of U with temperature or temperature difference is linear. Then for parallel or counterflow operation, the following relation may be applied to each stage:

$$q = \frac{A[U_2(\Delta t)_{01} - U_1(\Delta t)_{02}]}{\ln[U_2(\Delta t)_{01}/U_1(\Delta t)_{02}]} \qquad (4.4.5b)$$

Colburn (*Ind. & Eng. Chem.,* **25,** 1933, pp. 873–877) has shown that this equation is exact if U varies linearly with temperature.

Fig. 4.4.2 (A) One shell pass and two tube passes; (B) two shell passes and four tube passes; (C) three shell passes and six tube passes; (D) four shell passes and eight tube passes; (E) cross flow, one shell pass and one tube pass, both fluids mixed; (F) single-pass cross-flow exchanger, both fluids unmixed; (G) single-pass cross-flow exchanger, one fluid mixed, the other unmixed; (H) two-pass cross-flow exchanger, shell fluid mixed, tube fluid unmixed, shell fluid first crossing the second tube pass; (I) same as (H), but shell fluid first crosses the first tube pass.

FILM COEFFICIENTS

The important physical properties which affect film coefficients (see Sec. 4.1) are thermal conductivity, viscosity, density, and specific heat. Factors within the control of the designer include fluid velocity and shape and arrangement of the heating surface. With forced flow of gases or water, under the conditions usually met in practice, the flow is turbulent (see Sec. 3) and under these conditions the film coefficient can be greatly increased by increasing the velocity of the fluid at the expense of a greater power requirement. For a given velocity and fluid, the film coefficient depends upon the direction of flow of fluid relative to the heating surface. With free or natural convection for a given arrangement of surface, the film coefficient depends on an additional fluid property, the coefficient of thermal expansion, on the temperature difference between surface and fluid, and on the local gravitational acceleration. With forced convection at low rates of flow, particularly with viscous fluids such as oils, laminar motion may prevail and the film coefficient depends on thermal conductivity, specific heat, mass rate of flow per tube, and length and diameter of the tube. In any event, the film coefficients h are correlated in terms of dimensionless groups of the controlling factors.

Turbulent Flow inside Clean Tubes (No Change in Phase), $DG/\mu_f > 7,000$

$$\frac{h_m}{C_p G}\left(\frac{C_p \mu_f}{k_f}\right)^{2/3} = \frac{0.023}{(DG/\mu_f)^{0.2}} \tag{4.4.6a}$$

For L/D less than 60, multiply the right-hand side of Eqs. (4.4.6a), (4.4.6b), and (4.4.6c) by $1 + (D/L)^{0.7}$.

Turbulent Flow of Gases inside Clean Tubes, $DG/\mu_f > 7,000$

$$h_m = 0.024 C_p G^{0.8}/(D_i')^{0.2} \tag{4.4.6b}$$

Turbulent Flow of Water inside Clean Tubes, $DG/\mu_f > 7,000$

$$h_m = 160(1 + 0.012 t_f)V_s^{0.8}/(D_i')^{0.2} \tag{4.4.6c}$$

Turbulent Flow of Liquid Metals inside Clean Tubes, $C_p\mu/k < 0.05$ The equation of Sleicher and Tribus ("Recent Advances in Heat Transfer," p. 281, McGraw-Hill, 1961) is recommended for isothermal tube walls:

$$\frac{h_m D}{k} = 6.3 + 0.016\left(\frac{DGC_p}{k}\right)^{0.91}\left(\frac{C_p\mu}{k}\right)^{0.3} \tag{4.4.6d}$$

Turbulent Flow of Gases or Water in Annuli Use Eq. (4.4.6b) or (4.4.6c), with D' taken as the clearance, inches. If the clearance is comparable to the diameter of the inner tube, see Kays, Crawford, and Weigand.

Water in Coiled Pipes Multiply h_m for the staight pipe by the term $(1 + 3.5 D_i/D_c)$, where D_i is the inside diameter of the pipe and D_c is that of the coil.

Turbulent Boundary Layer on a Flat Plate, $V_\infty \rho_f x/\mu_f > 4 \times 10^5$, no pressure gradient

$$\frac{h}{\rho_f C_p V_\infty}\left(\frac{C_p\mu}{k}\right)^{2/3} = \frac{0.0148}{(\rho_f V_\infty x/\mu_f)^{0.2}} \tag{4.4.6e}$$

$$\frac{h_m}{\rho_f C_p V_\infty}\left(\frac{C_p\mu}{k}\right)_f^{2/3} = \frac{0.0185}{(\rho_f V_\infty L/\mu_f)^{0.2}} \tag{4.4.6f}$$

Fluid Flow Normal to a Single Tube, $D_o G/\mu_f$ from 1,000 to 50,000

$$\frac{h_m D_o}{k_f} = 0.26\left(\frac{D_o G}{\mu_f}\right)^{0.6}\left(\frac{C_p\mu}{k}\right)_f^{0.3} \tag{4.4.7}$$

Gas Flow Normal to a Single Tube, $D_o G/u_f$ from 1,000 to 50,000

$$h_m = 0.30 C_p G^{0.6}/(D_o')^{0.4} \tag{4.4.7a}$$

Fluid Flow Normal to a Bank of Staggered Tubes, $D_o G_{max}/\mu_f$ from 2,000 to 40,000

$$\frac{h_m D_o}{k_f} = K\left(\frac{C_p\mu}{k}\right)_f^{1/3}\left(\frac{D_o G_{max}}{\mu_f}\right)^{0.6} \tag{4.4.8}$$

Values of K are given in Table 4.4.8.

Water Flow Normal to a Bank of Staggered Tubes, $D_o G_{max}/\mu_f$ from 2,000 to 40,000

$$h_m = 370(1 + 0.0067 t_f)V_{sm}^{0.6}/(D_o')^{0.4} \tag{4.4.8a}$$

For baffled exchangers, to allow for leakage of fluids around the baffles, use 60 percent of the values of h_m from Eq. (4.4.8); for tubes in line, deduct 25 percent from the values of h_m given by Eq. (4.4.8).

Water Flow in Layer Form over Horizontal Tubes, $4\Gamma/\mu < 2,100$

$$h_{a.m.} = 150(\Gamma/D_o')^{1/3} \tag{4.4.9}$$

for Γ ranging from 100 to 1,000 lb of water per h per ft (each side).

Water Flow in Layer down Vertical Tubes, $w/\pi D > 500$

$$h_m = 120\Gamma^{1/3} \tag{4.4.9a}$$

Table 4.4.3   Thermal Conductivities of Miscellaneous Solid Substances*
Values of k are to be regarded as rough average values for the temperature range indicated

Material	Bulk density, lb/ft³	Temp, °F	k	Material	Bulk density, lb/ft³	Temp, °F	k
Asbestos board, compressed asbestos and cement	123.	86.	0.225	Quartz, crystal, parallel to C axis	...	−300.	25.0
						0.	8.3
						300.	4.2
Asbestos millboard	60.5	86.	0.070	Rubber, hard	74.3	100.	0.092
Asbestos wool	25.	212.	0.058	Rubber, soft, vulcanized	68.6	86.	0.08
Ashes, soft wood	12.5	68.	0.018				
Ashes, volcanic	51.	300.	0.123	Sand, dry	94.8	68.	0.188
Carbon black	12.	133.	0.012	Sawdust, dry	13.4	68.	0.042
Cardboard, corrugated	...	...	0.037	Silica, fused	...	200.	0.83
Celluloid	87.3	86.	0.12	Silica gel, powder	32.5	131.	0.049
Cellulose sponge, du Pont	3.4	82.	0.033	Soil, dry	...	68.	0.075
Concrete, sand, and gravel	142.	75.	1.05	Soil, dry, including stones	127.	68.	0.30
Concrete, cinder	97.	75.	0.41	Snow	7–31	32.	0.34–1.3
Charcoal, powder	11.5	63.	0.029	Titanium oxide, finely ground	52.	1000.	0.041
Cork, granulated	5.4	23.	0.028	Wool, pure	5.6	86.	0.021
Cotton wool	5.0	100.	0.035	Zirconia grain	113.	600.	0.11
Diamond	151.	70.	320.	Woods, oven dry, across grain†:			
Earth plus 42% water	108.	0.	0.62	Aspen	26.	85.	0.069
Fiber, red	80.5	68.	0.27	Bald cypress	24.	85.	0.063
Flotofoam (U.S. Rubber Co.)	1.6	92.	0.017	Balsa	10.	85.	0.034
				Basswood	24.	85.	0.058
Glass, pyrex	139	200.	0.59	Douglas Fir	29.	85.	0.063
Glass, soda lime	...	200.	0.59	Elm, rock	48.	85.	0.097
Graphite, solid	93.5	122.	87.	Fir, white	26.	85.	0.069
Gravel	116.	68.	0.22	Hemlock	29.	85.	0.066
Gypsum board	51.	99.	0.062	Larch, western	36.	85.	0.078
Ice	57.5	...	1.26	Maple, sugar	43.	85.	0.094
Kaolin wool	10.6	800.	0.059	Oak, red	42.	85.	0.099
Leather, sole	62.4	...	0.092	Pine, southern yellow	35.	85.	0.078
Mica	122.	...	0.25	Pine, white	25.	85.	0.060
Pearlite, Arizona, spherical shell of siliceous material	9.1	112.	0.035	Red cedar, western	21.	85.	0.053
Polystyrene, expanded "Styrofoam"	1.7	...	0.021	Redwood	25.	85.	0.062
				Spruce	21.	85.	0.052
Pumice, powdered	49.	300.	0.11				
Quartz, crystal, perpendicular to C axis	...	−300.	12.5				
		0.	4.3				
		300.	2.3				

* The thermal conductivity of different materials varies greatly. For metals and alloys k is high, while for certain insulating materials, such as glass wool, cork, and kapok, it is very low. In general, k varies with the temperature, but in the case of metals, the variation is relatively small. With most other substances, k increases with rising temperatures, but in the case of many crystalline materials, the reverse is true.
† With heat flow parallel to the grain, k may be 2 to 3 times that with heat flow perpendicular to the grain, the values for wool are taken chiefly from J. D. MacLean, *Trans. ASHRAE,* **47,** 1941, p. 323.

Heat Transfer to Gases Flowing at Very High Velocities If a nonreactive gas stream is brought to rest adiabatically, as at the true stagnation point of a blunt body, the temperature rise will be

$$t_s - t_\infty = V^2/(2g_c JC_p) \qquad (4.4.9b)$$

where t_s is the stagnation temperature and t_∞ is the temperature of the free stream moving at velocity V. At every other point on the body, the gas temperature is influenced by convection, viscous dissipation, and thermal conduction. At an insulated surface, the gas temperature will therefore be neither the free-stream temperature nor the stagnation temperature. In general, the rise in gas temperature will be given by

$$t_{aw} - t_\infty = \Re(t_s - t_\infty) = \Re V^2/(2g_c JC_p) \qquad (4.4.9c)$$

where t_{aw} is the gas temperature at the adiabatic wall and $\Re$ is the recovery factor.

If a given point on the surface of a body is not at the temperature t_{aw} given by Eq. (4.4.9c) with the proper local value of $\Re$ inserted, there will be a transfer of heat to or from the body. This suggests defining the coefficient of heat transfer in the usual way, except that the difference $t_w - t_{aw}$ should be used:

$$q/A = h(t_w - t_{aw}) = h\{t_w - [t + \Re V^2/(2g_c JC_p)]\} \qquad (4.4.9d)$$

where t_w is the surface temperature of the heated wall. With this modification, it is found that the correlations for h are nearly independent of Mach number; e.g., Eq. (4.4.6a) may be used for turbulent, compressible flow in a pipe. Obviously, $\Re = 1.0$ at a forward stagnation point. For flows parallel to surfaces which have little or no curvature in the direction of flow, the following are recommended:

Laminar flow $\qquad \Re = \left(\dfrac{C_p\mu}{k}\right)^{1/2}$

Turbulent flow $\qquad \Re = \left(\dfrac{C_p\mu}{k}\right)^{1/3}$

Very little is presently known about point values of the recovery factor for flow over more complex shapes. Thus, special thermocouples

Table 4.4.4 Thermal Conductivities for Building Insulation

Material	Bulk density, lb/ft³	Temp, °F	k
Balsam wool, blanket	3.6	70.	0.021
Cabot's Quilt, eelgrass	15.6	86.	0.027
Glass wool, blanket	3.25	100.	0.022
Hairfelt, blanket	11.0	86.	0.022
Insulating boards, Insulite, Celotex, etc.	12–19	100.	0.027–0.031
Kapok, DryZero, blanket	1.6	75.	0.019
Redwood bark, loose, shredded, Palco Bark	4.0	100.	0.025
Rock wool, loose	7.	117.	0.024
Sil-O-Cel powder	10.6	86.	0.026
Vermiculite, loose, Zonolite	8.2	60.	0.038

Table 4.4.5 Thermal Conductivities of Material for Refrigeration and Extreme Low Temperatures

Material	Bulk density, lb/ft³	Temp, °F	k
Corkboard	6.9	100	0.022
		−100	0.018
		−300	0.010
Fiberglas with asphalt coating (board)	11.0	100	0.023
		−100	0.014
		−300	0.007
Glass blocks, expanded cellular glass	8.5	100	0.033
		−100	0.024
		−300	0.016
Mineral wool board, Rockcork	14.3	100	0.024
		−100	0.017
		−300	0.008
Silica aerogel, powder, Santocel	5.3	100	0.013
		0	0.012
		−100	0.010
Vegetable fiberboard, asphalt coating	14.4	100	0.028
		−100	0.021
		−300	0.013
Foams:			
Polystyrene*	2.9	−100	0.015
Polyurethane†	5.0	−100	0.019

* Test space pressure, 1.0 atm; $k = 0.0047$ at 10^{-5} mmHg.
† Test space pressure, 1.0 atm; $k = 0.007$ at 10^{-3} mmHg.

should be used to measure the temperature of high-velocity gas streams (Hottel and Kalitinsky, *Jour. Applied Mechanics*, 1945, pp. A25–A32; and Franz, *Jahrb* 1938 *deut. Luftfahrt-Forsch* II, pp. 215–218). Eckert (*Trans. ASME,* **78,** 1956, pp. 1273–1283) recommends that all property values be evaluated at a film temperature defined by

$$t_f = (t_\infty + t_w)/2 + 0.22(t_{aw} - t_\infty) \qquad (4.4.9e)$$

Nielsen (*NACA Wartime Rep.* L-179) gives graphs for predicting the heat transfer and pressure drop for airflow at Mach numbers up to 1.0, in tubes having a uniform wall temperature.

Heat transfer from a reacting gas to a surface is treated by Lees ("Recent Advances in Heat and Mass Transfer," p. 161, McGraw-Hill).

LAMINAR FLOW

PIPE FLOW, $DG/\mu < 2{,}100$. Use the Sieder-tate modification of the Graetz equation for isothermal tube walls and $wC_p/kL > 10$:

$$h_{a.m.}D/k = 2.0(wC_p/kL)^{1/3}(\mu/\mu_w)^{0.14} \qquad (4.4.10)$$

or

$$(h_{a.m.}/C_pG)(C_p\mu/k)^{2/3}(\mu_w/\mu)^{0.14} = 1.85(D/L)^{1/3}(DG/\mu)^{-2/3} \qquad (4.4.10a)$$

As shown in Fig. 4.4.3, as DG/μ increases from 2,100 to 7,000, the effect of L/D diminishes and finally becomes negligible for $L/D > 60$.

Fig. 4.4.3 Heating and cooling of viscous oils flowing inside tubes. [The curves for D_G/μ below 2,100 are based on Eq. (4.4.10).]

LAMINAR BOUNDARY LAYER ON A FLAT PLATE. $\rho V_\infty x/\mu < 4 \times 10^5$, isothermal plate, no pressure gradient

$$\frac{h}{\rho_f C_p V_\infty}\left(\frac{C_p\mu}{k}\right)_f^{2/3} = \frac{0.332}{(\rho_f V_\infty x/\mu_f)^{1/2}} \qquad (4.4.10b)$$

$$\frac{h_m}{\rho_f C_p V_\infty}\left(\frac{C_p\mu}{k}\right)_f^{2/3} = \frac{0.664}{(\rho_f V_\infty L/\mu_f)^{1/2}}$$

Table 4.4.6 Thermal Conductivities of Insulating Materials for High Temperatures

Material	Bulk density, lb/ft³	Max temp, °F	100°F	300°F	500°F	1,000°F	1,500°F	2,000°F
Asbestos paper, laminated	22.	400	0.038	0.042				
Asbestos paper, corrugated	16.	300	0.031	0.042				
Diatomaceous earth, silica, powder	18.7	1,500	0.037	0.045	0.053	0.074		
Diatomaceous earth, asbestos and bonding material	18.	1,600	0.045	0.049	0.053	0.065		
Fiberglas block, PF612	2.5	500	0.023	0.039				
Fiberglas block, PF614	4.25	500	0.021	0.033				
Fiberglas block, PF617	9.	500	0.020	0.033				
Fiberglas, metal mesh blanket, #900	......	1,000	0.020	0.030	0.040			
Cellular glass blocks, ave. value	8.5	900	0.033	0.045	0.062			
Hydrous calcium silicate, "Kaylo"	11.	1,200	0.032	0.038	0.045			
85% magnesia	12.	600	0.029	0.035				
Micro-quartz fiber, blanket	3.	3,000	0.021	0.028	0.042	0.075	0.108	0.142
Potassium titanate, fibers	71.5	......	......	0.022	0.024	0.030		
Rock wool, loose	8–12	......	0.027	0.038	0.049	0.078		
Zirconia grain	113.	3,000	......	......	0.108	0.129	0.163	0.217

Table 4.4.7   Thermal Conductance across Airspaces
Btu/(h)(ft^2)—Reflective insulation

Airspace, in	Direction of heat flow	Temp diff, °F	Mean temp, °F	Aluminum surfaces, $\varepsilon = 0.05$	Ordinary surfaces, nonmetallic, $\varepsilon = 0.90$
Horizontal, ¾–4 across	Upward	20.	80.	0.60	1.35
Vertical, ¾–4 across	Across	20.	80.	0.49	1.19
Horizontal, ¾ across	Downward	20.	75.	0.30	1.08
Horizontal, 4 across	Downward	20.	80.	0.19	0.93

Table 4.4.8   Values of K for N Rows Deep

N	1	2	3	4	5	6	7	10
K	0.24	0.25	0.27	0.29	0.30	0.31	0.32	0.33

Table 4.4.9   Heat-Transfer Coefficients h_s for Scale Deposits from Water*
For use in Eq. (4.4.4)

	Temp of heating medium			
	Up to 240°F		240 to 400°F	
	Temp of water			
	125°F or less		Above 125°F	
	Water velocity, ft/s			
	3 and less	Over 3	3 and less	Over 3
Distilled	2,000	2,000	2,000	2,000
Sea water	2,000	2,000	1,000	1,000
Treated boiler feedwater	1,000	2,000	500	1,000
Treated makeup for cooling tower	1,000	1,000	500	500
City, well, Great Lakes	1,000	1,000	500	500
Brackish, clean river water	500	1,000	330	500
River water, muddy, silty†	330	500	250	330
Hard (over 15 grains per gal)	330	330	200	200
Chicago Sanitary Canal	130	170	100	130

Miscellaneous cases: Refrigerating liquids, brine clean petroleum distillates, organic vapors, 1,000; refrigerant vapor, 500; vegetable oils, 330; fuel oil (topped crude), 200.
* From standards of Tubular Exchanger Manufacturers Assoc., 1952.
† Delaware, East River (NY), Mississippi, Schuylkill, and New York Bay.

Natural Convection   Heat transfer by natural convection is governed by relations of the form

$$\frac{h_m L_c}{k_f} = f[L_c^3 \rho_f^2 g_L \beta_f (\Delta t)_s / \mu_f^2, \ (C_p \mu/k)_f] \qquad (4.4.11)$$

where β_f is defined by the equation $\rho_f = \rho_\infty[1 - \beta_f(\Delta t)_s]$. For perfect gases, $\beta_f = 1/T_\infty$. The dimensionless group $L_c^3 \rho_f^2 g_L \beta_f (\Delta t)_s / \mu_f^2 \equiv N_{Gr}$ represents the ratio of the product (inertial force times buoyant force) to (viscous force squared).

If the flow is of the laminar boundary layer type and if $(C_p \mu/k)_f > 1$, an effective correlation is

$$\frac{h_m L_c}{k_f} = B_1 [N_{Gr} (C_p \mu/k)_f]^{0.25} \qquad (4.4.11a)$$

where B_1 is a weak function of $(C_p \mu/k)_f$. Similarly, for $(C_p \mu/k)_f < 1$,

$$\frac{h_m L_c}{k_f} = B_2 [N_{Gr} (C_p \mu/k)_f^2]^{0.25} \qquad (4.4.11b)$$

VERTICAL FLAT PLATES.   For this case, $L_c = L$ and the flow of the laminar boundary layer type will be laminar if

$$(C_p \mu/k)_f > 1 \qquad 10^9 > N_{Gr}(C_p \mu/k)_f > 10^4$$
$$(C_p \mu/k)_f < 1 \qquad ? > N_{Gr}(C_p \mu/k)_f^2 > 10^4$$

Lefevre (*Rept. Heat* 113, National Engineering Laboratory, Great Britain, Aug. 1956) gives an interpolation formula which contains the proper limiting forms and is in complete agreement with existing numerical results:

$$\frac{h_m L_c}{k_f} = \left[\frac{N_{Gr}(C_p \mu/k)^2}{2.435 + 4.884(C_p \mu/k)_f^{1/2} + 4.953(C_p \mu/k)_f} \right]^{0.25} \qquad (4.4.11c)$$

If $(C_p \mu/k)_f$ is in the vicinity of unity and if $N_{Gr} (C_p \mu/k)_f > 10^9$, the boundary layer will be turbulent and

$$\frac{hL}{k_f} = 0.13[N_{Gr}(C_p \mu/k)_f]^{1/3} \qquad (4.4.11d)$$

HORIZONTAL CYLINDERS.   Replace L in the vertical flat plate formulas by $\pi D_o/2$.

HEATED HORIZONTAL PLATES FACING UPWARD OR COOLED HORIZONTAL PLATES FACING DOWNWARD.

$$2 \times 10^7 > N_{Gr}(C_p \mu/k)_f > 10^5$$
$$\frac{h_m L}{k_f} = 0.54[N_{Gr}(C_p \mu/k)_f]^{0.25} \qquad (4.4.11e)$$

$$N_{Gr}(C_p \mu/k)_f > 2 \times 10^7$$
$$\frac{h_m L}{k_f} = 0.14[N_{Gr}(C_p \mu/k)_f]^{1/3} \qquad (4.4.11f)$$

Table 4.4.10   Typical Values of h_m for Heating and Cooling, Forced Convection
$D_0' = 1.31$ in, $D_i' = 1.05$ in

Fluid and arrangement	t_f, °F	Velocity* ft/s	lb/(h · ft^2)	Btu/(h · ft^2 · °F)	Eq. no.
Air inside tubes	. . . . .	$V_s = 31.8, G = 8,600$		8.0	4.4.6b
Air normal to staggered tubes	170	$V_s = 8.92, G_m = 2,000$		7.5	4.4.8
Water inside tubes	100	$V_s = 5.0, G = 1.12 \times 10^6$		1260	4.4.6c
Water normal to staggered tubes	100	$V_s = 2.0, G_m = 0.448 \times 10^6$		800	4.4.8a
Trickle cooler, water	. . . . .	$\Gamma = 100$ lb/(h · ft)		640	4.4.9
Falling water film vertical tube	. . . . .	$\Gamma = 1,000$ lb/(h · ft)		1200	4.4.9a

* Velocity in ft/s at 70°F and 1 atm = $G/3,600\rho$.

Table 4.4.11 Maximum Flux and Corresponding Overall Temperature Difference for Liquids Boiled at 1 atm with a Submerged Horizontal Steam-Heated Tube

Liquid	Aluminum		Copper		Chromium-plated copper		Steel	
	q/A 1,000	$(\Delta t)_o$	q/A 1,000	$(\Delta t)_o$	q/A 1,000	$(\Delta t)_o$	q/A 1,000	$(\Delta t)_o$
Ethyl acetate	41	70	61	55	77	55		
Benzene	51	80	58	70	73	100	82	100
Ethyl alcohol	55	80	85	65	124	65		
Methyl alcohol	.....	.....	100	95	110	110	155	110
Distilled water	.....	.....	230	85	350	75	410	150

HEATED HORIZONTAL PLATES FACING DOWNWARD OR COOLED HORIZONTAL PLATES FACING UPWARD.

$$3 \times 10^{10} > N_{Gr}(C_p\mu/k)_f > 3 \times 10^5$$

$$\frac{hL}{k_f} = 0.27[N_{Gr}(C_p\mu/k)_f]^{0.25} \qquad (4.4.11g)$$

Equations (4.4.11e) to (4.4.11g) should not be considered reliable if $(C_p\mu/k)_f$ differs greatly from unity.

For more complex systems, it is best to consult experimental data (Gebhart et al.).

For any particular fluid, the above equations may be greatly simplified. For air which is at room temperature and atmospheric pressure and is subjected to the gravitational attraction at sea level:

VERTICAL PLATES.

$$10^3 > L^3(\Delta t)_s > 10^{-2}$$
$$h_m = 0.28[(\Delta t)_s/L]^{0.25} \qquad (4.4.12a)$$
$$L^3(\Delta t)_s > 10^3$$
$$h_m = 0.19(\Delta t)_s^{1/3} \qquad (4.4.12b)$$

HORIZONTAL CYLINDERS.

$$10^2 > D^3(\Delta t)_s > 10^{-3}$$
$$h_m = 0.25[(\Delta t)_s/D]^{0.25} \qquad (4.4.12c)$$
$$D^3(\Delta t)_s > 10^2$$
$$h_m = 0.19(\Delta t)_s^{1/3} \qquad (4.4.12d)$$

HEATED HORIZONTAL PLATES FACING UPWARD OR COOLED HORIZONTAL PLATES FACING DOWNWARD.

$$10 > L^3(\Delta t)_s > 0.1$$
$$h_m = 0.27[(\Delta t)_s/L]^{0.25} \qquad (4.4.12e)$$
$$10^4 > L^3(\Delta t)_s > 10$$
$$h_m = 0.22(\Delta t)_s^{1/3} \qquad (4.4.12f)$$

HEATED HORIZONTAL PLATES FACING DOWNWARD OR COOLED HORIZONTAL PLATES FACING UPWARD.

$$10^4 > L^3(\Delta t)_s > 0.1 \qquad (4.4.12g)$$
$$h_m = 0.12[(\Delta t)_s/L]^{0.25}$$

Condensing Vapors If the condensate of a single pure vapor, saturated or supersaturated, wets the surface, film-type condensation is obtained. The rate of heat transfer equals $h_m(\Delta t)_m$, where $(\Delta t)_m$ is the mean difference between the saturation temperature and the temperature of the surface. As long as the condensate flow is laminar ($4\Gamma/\mu_f < 2,100$), the following dimensionless equations may be used:

For **horizontal tubes**,

$$h_m D/k = 0.73[D^3\rho^2\lambda g_L/k\mu_f N(\Delta t)_m]^{0.25}$$
$$= 0.76(D^3\rho^2 g_L/\mu_f\Gamma)^{1/3} \qquad (4.4.13)$$

For **vertical tubes**,

$$h_m L/k = 0.94[L^3\rho^2\lambda g_L/k\mu_f(\Delta t)_m]^{0.25}$$
$$= 0.93(L^3\rho^2 g_L/\mu_f\Gamma)^{1/3} \qquad (4.4.13a)$$

The equations show that a tube of given dimensions, for the usual case where $L/(ND)$ is greater than 2.76, is more effective in a horizontal than in a vertical position. Thus for $L/(ND) = 100$, a horizontal tube gives an average h which is 2.5 times that for a vertical tube. Since there is but little variation in the thermal conductivity or viscosity of the condensate at the condensing temperature at 1 atm, there is little variation in h_m. With horizontal tubes, use h_m from 200 to 400 Btu/(h · ft² · °F) for the following vapors condensing at atmospheric pressure: benzene, carbon tetrachloride, dichlormethane, dichlordifluoromethane, diphenyl ethyl alcohol, heptane, hexane, methyl alcohol, octane, toluene, and xylene. Ammonia gives h_m of 1,000, and mixtures of steam and organic vapors, forming immiscible condensates, give h_m ranging from 250 to 750, increasing with increasing proportion of steam. With film-type condensation of clean steam on horizontal tubes, h_m ranges from 1,000 to 3,000; see Eq. (4.4.13). With vertical tubes 10 to 20 ft long, ripples form in the film; values of h_m from Eq. (4.4.13a) should be increased 20 percent.

For long vertical tubes, $4\Gamma/\mu_f$ may exceed 2,100; in that case:

$$h_m(\mu_f^2/k_f^3\rho_f^2 g_L)^{1/3} = 0.0077(4\Gamma/\mu_f)^{0.4} \qquad (4.4.13b)$$

The presence of **noncondensible gas**, such as air, seriously reduces h, and consequently all vapor-heated apparatus should be well vented.

With steam, small traces of certain promoters (Nagle, U.S. Patent 1,995,361) such as oleic acid and benzyl mercaptan become adsorbed in a very thin layer on the surface of the tubes, preventing the condensate from wetting the metal and inducing dropwise condensation, which gives much higher values of h_m (7,000 to 70,000) than film-type condensation. However, with dirty or corroded surfaces, it is difficult to maintain dropwise condensation. Figure 4.4.4 shows overall coefficients U_o for condensing steam at 1 atm on a vertical 10-ft length of copper tube, ⅝ in OD, 0.049-in wall, at various water velocities.

Fig. 4.4.4 Overall coefficients between condensing steam and water. Curve 1, chromium-plated copper, oleic acid; curve 2, copper, benzyl mercaptan; curve 3, copper, oleic acid; curve 4, admiralty metal, no promoter.

Boiling Liquids The nature of the heat transfer from a submerged heater to a pool of boiling water is shown in Fig. 4.4.5. Other liquids exhibit the same qualitative features. In the range AB, heat transfer to the liquid occurs solely by natural convection, and evaporation occurs at the free surface of the pool. In the range BC, **nucleate boiling** occurs. Bubbles form at active nuclei on the heating surface, detach, and rise to the pool surface. At point C, the heat flux passes through a maximum at

Fig. 4.4.5 Boiling of water at 212°F on a platinum surface.

a temperature difference called the **critical** Δt. In the range CD, transitional boiling occurs. At point D, the transition is complete and the heating surface is completely blanketed by a vapor film. This is the point of minimum heat flux, or the **Leidenfrost point**. In the range DE, the heating surface continues to be blanketed by a vapor film.

The range AB is adequately correlated by the usual natural-convection equations. No truly adequate correlation is available for the range BC because the complex processes of nucleation and interfacial interaction are only partially understood. However, the relation due to Rohsenow (*Trans. ASME,* **74,** 1952, pp. 969–976) is one of the best and can be reliably used for modest extrapolations of existing data.

$$\frac{C_{p,l}(t_w - t_{\text{sat}})}{\lambda} = C_{fs} \left[\frac{q/A}{\mu_l \lambda} \sqrt{\frac{g_c \sigma}{g_L(\rho_l - \rho_v)}} \right]^{1/3} \left(\frac{C_{p,l}\mu_l}{k_l} \right)^{1.7} \quad (4.4.14a)$$

The value of the constant C_{fs} is intimately dependent on the nature of the particular fluid-solid pair and must be determined by experiment. It usually assumes values in the range $0.003 < C_{fs} < 0.05$ and is not affected by moderate subcooling or the shape of the heating surface.

Zuber (*USAEC Rep.* AECU-4439, June, 1959) has presented a theoretical equation for the maximum heat flux from a flat, horizontal surface. The analysis is based on considerations of hydrodynamic stability. For saturated liquids,

$$\left(\frac{q}{A} \right)_{\text{max}} = K_1 \rho_v \lambda \left[\frac{\sigma g_L g_c(\rho_l - \rho_v)}{\rho_v^2} \right]^{1/4} \left(\frac{\rho_l}{\rho_l + \rho_v} \right)^{1/2} \quad (4.4.14b)$$

$$0.12 < K_1 < 0.157 \quad \text{(theoretical)}$$

Berenson (Sc.D. thesis, Mechanical Engineering Department, MIT, 1960) used a similar analysis and obtained a relation which is identical for $\rho_l \gg \rho_v$, but he found that $K_1 = 0.18$ gives better agreement with the data. The theoretical basis of this equation has been subject to attack, but the correlation appears to be the best available. Zuber also performed an analysis for subcooled liquids and proposed a modification which is also in excellent agreement with experiment:

$$\left(\frac{q}{A} \right)_{\text{max}} = K_1 \rho_v [\lambda + C_{p,l}(t_{\text{sat}} - t_l)] \left[\frac{\sigma g_L g_c(\rho_l - \rho_v)}{\rho_v^2} \right]^{0.25}$$

$$\times \left(\frac{\rho_l}{\rho_l + \rho_v} \right)^{1/2}$$

$$\times \left\{ 1 + \frac{5.33(\rho_l C_{p,l} k_l)^{1/2}(t_{\text{sat}} - t_l)}{\rho_v [\lambda + C_{p,l}(t_{\text{sat}} - t_l)]} \left[\frac{g_L(\rho_l - \rho_v)\rho_v^2}{\sigma^3 g_c^3} \right]^{1/8} \right\} \quad (4.4.14c)$$

Zuber's hydrodynamic analysis of the Leidenfrost point yields

$$(q/A)_{\text{min}} = K_2 \lambda \rho_v \left[\frac{\sigma g_L g_c(\rho_l - \rho_v)}{\rho_l^2} \right]^{1/4}$$

$$0.144 < K_2 < 0.177 \quad (4.4.14d)$$

Berenson finds better agreement with the data if $K_2 = 0.09$. For very small wires, the heat flux will exceed that predicted by this flat-plate formula. A reliable prediction of the critical temperature is not available.

For nucleate boiling accompanied by forced convection, the heat flux may be approximated by the sum of the heat flux for pool boiling alone and the heat flux for forced convection alone. This procedure will not be satisfactory at high qualities, and no satisfactory correlation exists for the maximum heat flux.

For a given liquid and system pressure, the nature of the surface may substantially influence the flux at a given (Δt), Table 4.4.11. These data may be used as rough approximations for a bank of submerged tubes. Film coefficients for scale deposits are given in Table 4.4.9.

For **forced-circulation evaporators,** vapor binding is also encountered. Thus with liquid benzene entering a 4-pass steam-jacketed pipe at 0.9 ft/s, up to the point where 60 percent by weight was vaporized, the maximum flux of 60,000 Btu/h · ft² was obtained at an overall temperature difference of 60°F; beyond this point, the coefficient and flux decreased rapidly, approaching the values obtained in superheating vapor, see Eq. (4.4.6b). For comparison, in a natural convection evaporator, a maximum flux of 73,000 Btu/h · ft² was obtained at (Δt)$_v$ of 100°F.

Combined Convection and Radiation Coefficients   In some cases of heat loss, such as that from bare and insulated pipes, where loss is by convection to the air and radiation to the walls of the enclosing space, it is convenient to use a combined convection and radiation coefficient $h_c + h_r$. The rate of heat loss thus becomes

$$q = (h_c + h_r)A(\Delta t)_s \quad (4.4.15)$$

where (Δt)$_s$ is the temperature difference, °F, between the surface of the hot body and the walls of the space. In evaluating ($h_c + h_r$), h_c should be calculated by the appropriate convection formula [see Eqs. (4.4.11c) to (4.4.11g)] and h_r from the equation

$$h_r = 0.00685\varepsilon(T_{\text{av}}/100)^3$$

where ε is the emissivity of the radiating surface (see Sec. 4.3). T_{av} is the average temperature of the surface and the enclosing walls, °R. For oxidized bare steel pipe, the sum $h_c + h_r$ may be taken directly from Table 4.4.12.

Table 4.4.12   Values of $h_c + h_r$
For horizontal bare or insulated standard steel pipe of various sizes and for flat plates in a room at 80°F

Nominal pipe diam, in	(Δt)$_s$, temperature difference, °F, from surface to room														
	50	100	150	200	250	300	400	500	600	700	800	900	1,000	1,100	1,200
½	2.12	2.48	2.76	3.10	3.41	3.75	4.47	5.30	6.21	7.25	8.40	9.73	11.20	12.81	14.65
1	2.03	2.38	2.65	2.98	3.29	3.62	4.33	5.16	6.07	7.11	8.25	9.57	11.04	12.65	14.48
2	1.93	2.27	2.52	2.85	3.14	3.47	4.18	4.99	5.89	6.92	8.07	9.38	10.85	12.46	14.28
4	1.84	2.16	2.41	2.72	3.01	3.33	4.02	4.83	5.72	6.75	7.89	9.21	10.66	12.27	14.09
8	1.76	2.06	2.29	2.60	2.89	3.20	3.88	4.68	5.57	6.60	7.73	9.05	10.50	12.10	13.93
12	1.71	2.01	2.24	2.54	2.82	3.13	3.83	4.61	5.50	6.52	7.65	8.96	10.42	12.03	13.84
24	1.64	1.93	2.15	2.45	2.72	3.03	3.70	4.48	5.37	6.39	7.52	8.83	10.28	11.90	13.70
Flat plates															
Vertical	1.82	2.13	2.40	2.70	2.99	3.30	4.00	4.79	5.70	6.72	7.86	9.18	10.64	12.25	14.06
HFU*	2.00	2.35	2.65	2.97	3.26	3.59	4.31	5.12	6.04	7.07	8.21	9.54	11.01	12.63	14.45
HFD*	1.58	1.85	2.09	2.36	2.63	2.93	3.61	4.38	5.27	6.27	7.40	8.71	10.16	11.76	13.57

* HFU = horizontal facing upward; HFD = horizontal facing downward.

Heat Transmission through Pipe Insulation (McMillan, *Trans. ASME,* 1915) For any number of layers of insulation on any size of pipe, Eqs. (4.4.2), (4.4.4), and (4.4.15) combine to give

$$\frac{q_o}{A_o} = \frac{(\Delta t)_o}{\dfrac{r_o}{k_1} \ln \dfrac{r_2}{r_1} + \dfrac{r_o}{k_2} \ln \dfrac{r_3}{r_2} + \cdots + \dfrac{1}{h_c + h_r}} \qquad (4.4.16)$$

where q_o/A_o is the Btu/(h · ft²) of outer surface of the last layer; $(\Delta t)_o$ is the overall temperature difference (°F) between pipe and air; r_o is the radius, feet, of the outer surface; r_1 is the outside radius (ft) of the pipe, $r_2 = r_1$ + thickness of first layer of insulation, foot; $r_3 = r_2$ plus the thickness of second layer, etc.; and k_1, k_2, k_3, etc., are the conductivities of the respective layers. For average indoor conditions, $h_c + h_r$ is often taken as 2 as an approximation, since a substantial error in $h_c + h_r$ will have but little effect on the overall loss of heat. Figure 4.4.6 shows the variation in U_o with pipe size and thickness of insulation (for $k = 0.042$) for pipe and air temperatures of 375 and 75°F, respectively.

Fig. 4.4.6 Variation with pipe size of overall coefficient U_o for a given thickness of insulation, for $k = 0.042$.

Section **5**

Strength of Materials

BY

JOHN SYMONDS *Fellow Engineer (Retired), Oceanic Division, Westinghouse Electric Corporation*
ALI M. SADEGH *Professor of Mechanical Engineering, The City College of The City University of New York*
HAROLD V. HAWKINS *Late Manager, Product Standards and Services, Columbus McKinnon Corporation*
DONALD D. DODGE *Supervisor (Retired), Product Quality and Inspection Technology, Manufacturing Development, Ford Motor Company*
GARY L. CLOUD *Professor of Mechanical Engineering, Michigan State University*
MICHAEL W. HYER *Professor of Engineering Science and Mechanics, Virginia Polytechnic Institute and State University*
SCOTT W. CASE *Professor of Engineering Science and Mechanics, Virginia Polytechnic Institute and State University*

5.1 MECHANICAL PROPERTIES OF MATERIALS
by John Symonds, Revised by E. A. Avallone

REFERENCES: Davis et al., "Testing and Inspection of Engineering Materials," McGraw-Hill. Timoshenko, "Strength of Materials," pt. II, Van Nostrand. Richards, "Engineering Materials Science," Wadsworth. Nadai, "Plasticity," McGraw-Hill. Tetelman and McEvily, "Fracture of Structural Materials," Wiley. "Fracture Mechanics," ASTM STP-833. McClintock and Argon (eds.), "Mechanical Behavior of Materials," Addison-Wesley. Dieter, "Mechanical Metallurgy," McGraw-Hill. "Creep Data," ASME. ASTM Standards, ASTM. Blaznynski (ed.), "Plasticity and Modern Metal Forming Technology," Elsevier Science.

STRESS-STRAIN DIAGRAMS

The Stress-Strain Curve The engineering tensile stress-strain curve is obtained by **static loading** of a standard specimen, that is, by applying the load slowly enough that all parts of the specimen are in equilibrium at any instant. The curve is usually obtained by controlling the loading rate in the tensile machine. ASTM Standards require a loading rate not exceeding 100,000 lb/in² (70 kgf/mm²)/min. An alternate method of obtaining the curve is to specify the strain rate as the independent variable, in which case the loading rate is continuously adjusted to maintain the required strain rate. A strain rate of 0.05 in/in/(min) is commonly used. It is measured usually by an extensometer attached to the gage length of the specimen. Figure 5.1.1 shows several stress-strain curves.

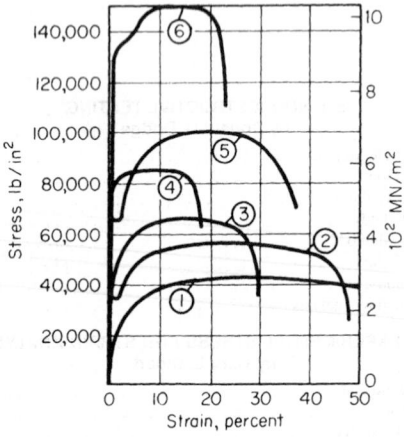

Fig. 5.1.1 Comparative stress-strain diagrams. (1) Soft brass; (2) low carbon steel; (3) hard bronze; (4) cold rolled steel; (5) medium carbon steel, annealed; (6) medium carbon steel, heat treated.

For most engineering materials, the curve will have an initial linear elastic region (Fig. 5.1.2) in which deformation is reversible and time-independent. The slope in this region is **Young's modulus** E. The **proportional elastic limit** (PEL) is the point where the curve starts to deviate from a straight line. The **elastic limit** (frequently indistinguishable from PEL) is the point on the curve beyond which plastic deformation is present after release of the load. If the stress is increased further, the stress-strain curve departs more and more from the straight line. Unloading the specimen at point X (Fig. 5.1.2), the portion XX' is linear and is essentially parallel to the original line OX''. The horizontal distance OX' is called the **permanent set** corresponding to the stress at X. This is the basis for the construction of the arbitrary **yield strength.** To determine the yield strength, a straight line XX' is drawn parallel to the initial elastic line OX'' but displaced from it by an arbitrary value of permanent strain. The permanent strain commonly used is 0.20 percent of the original gage length. The intersection of this line with the curve determines the stress value called the yield strength. In reporting the yield strength, the amount of permanent set should be specified. The arbitrary yield strength is used especially for those materials not exhibiting a natural yield point such as nonferrous metals; but it is not limited to these. Plastic behavior is somewhat time-dependent, particularly at high temperatures. Also at high temperatures, a small amount of time-dependent reversible strain may be detectable, indicative of **anelastic** behavior.

Fig. 5.1.2 General stress-strain diagram.

The **ultimate tensile strength** (UTS) is the maximum load sustained by the specimen divided by the original specimen cross-sectional area. The percent **elongation at failure** is the plastic extension of the specimen at failure expressed as (the change in original gage length × 100) divided by the original gage length. This extension is the sum of the **uniform** and **nonuniform** elongations. The uniform elongation is that which occurs prior to the UTS. It has an unequivocal significance, being associated with uniaxial stress, whereas the nonuniform elongation which occurs during localized extension (necking) is associated with triaxial stress. The nonuniform elongation will depend on geometry, particularly the ratio of specimen gage length L_0 to diameter D or square root of cross-sectional area A. ASTM Standards specify test-specimen geometry for a number of specimen sizes. The ratio $L_0/\sqrt{A}$ is maintained at 4.5 for flat- and round-cross-section specimens. The original gage length should always be stated in reporting elongation values.

The specimen percent **reduction in area** (RA) is the contraction in cross-sectional area at the fracture expressed as a percentage of the original area. It is obtained by measurement of the cross section of the broken specimen at the fracture location. The RA along with the load at fracture can be used to obtain the **fracture stress,** that is, fracture load divided by cross-sectional area at the fracture. See Table 5.1.1.

The type of fracture in tension gives some indications of the quality of the material, but this is considerably affected by the testing temperature, speed of testing, the shape and size of the test piece, and other conditions. Contraction is greatest in tough and ductile materials and least in brittle materials. In general, fractures are either of the **shear** or of the **separation** (loss of cohesion) type. Flat tensile specimens of ductile metals often show shear failures if the ratio of width to thickness is greater than 6:1. A completely shear-type failure may terminate in a chisel edge, for a flat specimen, or a point rupture, for a round specimen. Separation failures occur in brittle materials, such as certain cast irons. Combinations of both shear and separation failures are common on round specimens of ductile metal. Failure often starts at the axis in a necked region and produces a relatively flat area which grows until the material shears along a cone-shaped surface at the outside of the specimen, resulting in what is known as the cup-and-cone fracture.

Table 5.1.1   Typical Mechanical Properties at Room Temperature
(Based on ordinary stress-strain values)

Metal	Tensile strength, 1,000 lb/in²	Yield strength, 1,000 lb/in²	Ultimate elongation, %	Reduction of area, %	Brinell no.
Cast iron	18–60	8–40	0	0	100–300
Wrought iron	45–55	25–35	35–25	55–30	100
Commercially pure iron, annealed	42	19	48	85	70
Hot-rolled	48	30	30	75	90
Cold-rolled	100	95			200
Structural steel, ordinary	50–65	30–40	40–30		120
Low-alloy, high-strength	65–90	40–80	30–15	70–40	150
Steel, SAE 1300, annealed	70	40	26	70	150
Quenched, drawn 1,300°F	100	80	24	65	200
Drawn 1,000°F	130	110	20	60	260
Drawn 700°F	200	180	14	45	400
Drawn 400°F	240	210	10	30	480
Steel, SAE 4340, annealed	80	45	25	70	170
Quenched, drawn 1,300°F	130	110	20	60	270
Drawn 1,000°F	190	170	14	50	395
Drawn 700°F	240	215	12	48	480
Drawn 400°F	290	260	10	44	580
Cold-rolled steel, SAE 1112	84	76	18	45	160
Stainless steel, 18-S	85–95	30–35	60–55	75–65	145–160
Steel castings, heat-treated	60–125	30–90	33–14	65–20	120–250
Aluminum, pure, rolled	13–24	5–21	35–5		23–44
Aluminum-copper alloys, cast	19–23	12–16	4–0		50–80
Wrought, heat-treated	30–60	10–50	33–15		50–120
Aluminum die castings	30		2		
Aluminum alloy 17ST	56	34	26	39	100
Aluminum alloy 51ST	48	40	20	35	105
Copper, annealed	32	5	58	73	45
Copper, hard-drawn	68	60	4	55	100
Brasses, various	40–120	8–80	60–3		50–170
Phosphor bronze	40–130		55–5		50–200
Tobin bronze, rolled	63	41	40	52	120
Magnesium alloys, various	21–45	11–30	17–0.5		47–78
Monel 400, Ni-Cu alloy	79	30	48	75	125
Molybdenum, rolled	100	75	30		250
Silver, cast, annealed	18	8	54		27
Titanium 6–4 alloy, annealed	130	120	10	25	352
Ductile iron, grade 80-55-06	80	55	6		225–255

NOTE: Compressive strength of cast iron, 80,000 to 150,000 lb/in².
Compressive yield strength of all metals, except those cold-worked = tensile yield strength.
Stress 1,000 lb/in² × 6.894 = stress, MN/m².

Double cup-and-cone and rosette fractures sometimes occur. Several types of tensile fractures are shown in Fig. 5.1.3.

Annealed or hot-rolled mild steels generally exhibit a **yield point** (see Fig. 5.1.4). Here, in a constant strain-rate test, a large increment of extension occurs under constant load at the elastic limit or at a stress just below the elastic limit. In the latter event the stress drops suddenly from the **upper yield point** to the **lower yield point.** Subsequent to the drop, the yield-point extension occurs at constant stress, followed by a rise to the UTS. Plastic flow during the yield-point extension is discontinuous;

Fig. 5.1.3 Typical metal fractures in tension.

successive zones of plastic deformation, known as **Luder's bands** or **stretcher strains,** appear until the entire specimen gage length has been uniformly deformed at the end of the yield-point extension. This behavior causes a banded or stepped appearance on the metal surface. The exact form of the stress-strain curve for this class of material is

sensitive to test temperature, test strain rate, and the characteristics of the tensile machine employed.

The plastic behavior in a uniaxial tensile test can be represented as the **true stress-strain curve.** The **true stress** σ is based on the instantaneous

Fig. 5.1.4 Yielding of annealed steel.

cross section A, so that $\sigma = $ load$/A$. The instantaneous **true strain** increment is $-dA/A$, or dL/L prior to necking. Total true strain ε is

$$\int_{A_0}^{A} -\frac{dA}{A} = \ln\left(\frac{A_0}{A}\right)$$

or $\ln (L/L_0)$ prior to necking. The true stress-strain curve or flow curve obtained has the typical form shown in Fig. 5.1.5. In the part of the test subsequent to the maximum load point (UTS), when necking occurs, the true strain of interest is that which occurs in an infinitesimal length at the region of minimum cross section. True strain for this element can still be expressed as $\ln (A_0/A)$, where A refers to the minimum cross

Fig. 5.1.5   True stress-strain curve for 20°C annealed mild steel.

section. Methods of constructing the true stress-strain curve are described in the technical literature. In the range between initial yielding and the neighborhood of the maximum load point the relationship between plastic strain ε_p and true stress often approximates

$$\sigma = k\varepsilon_p^n$$

where k is the **strength coefficient** and n is the **work-hardening exponent.** For a material which shows a yield point the relationship applies only to the rising part of the curve beyond the lower yield. It can be shown that at the maximum load point the slope of the true stress-strain curve equals the true stress, from which it can be deduced that for a material obeying the above exponential relationship between ε_p and n, $\varepsilon_p = n$ at the maximum load point. The exponent strongly influences the spread between YS and UTS on the engineering stress-strain curve. Values of n and k for some materials are shown in Table 5.1.2. A point on the flow curve indentifies the **flow stress** corresponding to a certain strain, that is, the stress required to bring about this amount of plastic deformation. The concept of true strain is useful for accurately describing large amounts of plastic deformation. The linear strain definition $(L - L_0)/L_0$ fails to correct for the continuously changing gage length, which leads to an increasing error as deformation proceeds.

During extension of a specimen under tension, the change in the specimen cross-sectional area is related to the elongation by **Poisson's ratio** μ, which is the ratio of strain in a transverse direction to that in the longitudinal direction. Values of μ for the elastic region are shown in Table 5.1.3. For plastic strain it is approximately 0.5.

Table 5.1.2   Room-Temperature Plastic-Flow Constants for a Number of Metals

Material	Condition	k, 1,000 in^2 (MN/m^2)	n
0.40% C steel	Quenched and tempered at 400°F (478K)	416 (2,860)	0.088
0.05% C steel	Annealed and temper-rolled	72 (49.6)	0.235
2024 aluminum	Precipitation-hardened	100 (689)	0.16
2024 aluminum	Annealed	49 (338)	0.21
Copper	Annealed	46.4 (319)	0.54
70–30 brass	Annealed	130 (895)	0.49

SOURCE: Reproduced by permission from "Properties of Metals in Materials Engineering," ASM, 1949.

Table 5.1.3   Elastic Constants of Metals
(Mostly from tests of R. W. Vose)

Metal	E Modulus of elasticity (Young's modulus). 1,000,000 lb/in^2	G Modulus of rigidity (shearing modulus). 1,000,000 lb/in^2	K Bulk modulus. 1,000,000 lb/in^2	μ Poisson's ratio
Cast steel	28.5	11.3	20.2	0.265
Cold-rolled steel	29.5	11.5	23.1	0.287
Stainless steel 18–8	27.6	10.6	23.6	0.305
All other steels, including high-carbon, heat-treated	28.6–30.0	11.0–11.9	22.6–24.0	0.283–0.292
Cast iron	13.5–21.0	5.2–8.2	8.4–15.5	0.211–0.299
Malleable iron	23.6	9.3	17.2	0.271
Copper	15.6	5.8	17.9	0.355
Brass, 70–30	15.9	6.0	15.7	0.331
Cast brass	14.5	5.3	16.8	0.357
Tobin bronze	13.8	5.1	16.3	0.359
Phosphor bronze	15.9	5.9	17.8	0.350
Aluminum alloys, various	9.9–10.3	3.7–3.9	9.9–10.2	0.330–0.334
Monel metal	25.0	9.5	22.5	0.315
Inconel	31	11		0.27–0.38
Z-nickel	30	11		±0.36
Beryllium copper	17	7		±0.21
Elektron (magnesium alloy)	6.3	2.5	4.8	0.281
Titanium (99.0 Ti), annealed bar	15–16	6.5		0.34
Zirconium, crystal bar	11–14			
Molybdenum, arc-cast	48–52			

The general effect of increased strain rate is to increase the resistance to plastic deformation and thus to raise the flow curve. Decreasing test temperature also raises the flow curve. The effect of strain rate is expressed as **strain-rate sensitivity** m. Its value can be measured in the tension test if the strain rate is suddenly increased by a small increment during the plastic extension. The flow stress will then jump to a higher value. The strain-rate sensitivity is the ratio of incremental changes of $\log \sigma$ and $\log \dot{\varepsilon}$

$$m = \left(\frac{\delta \log \sigma}{\delta \log \dot{\varepsilon}} \right)_{\varepsilon}$$

For most engineering materials at room temperature the strain rate sensitivity is of the order of 0.01. The effect becomes more significant at elevated temperatures, with values ranging to 0.2 and sometimes higher.

Compression Testing   The compressive stress-strain curve is similar to the tensile stress-strain curve up to the yield strength. Thereafter, the progressively increasing specimen cross section causes the compressive stress-strain curve to diverge from the tensile curve. Some ductile metals will not fail in the compression test. Complex behavior occurs when the direction of stressing is changed, because of the **Bauschinger effect,** which can be described as follows: If a specimen is first plastically strained in tension, its yield stress in compression is reduced and vice versa.

Combined Stresses   This refers to the situation in which stresses are present on each of the faces of a cubic element of the material. For a given cube orientation the applied stresses may include shear stresses over the cube faces as well as stresses normal to them. By a suitable rotation of axes the problem can be simplified: applied stresses on the new cubic element are equivalent to three mutually orthogonal **principal stresses** σ_1, σ_2, σ_3 alone, each acting normal to a cube face. Combined stress behavior in the elastic range is described in Sec. 5.2, Mechanics of Materials.

Prediction of the conditions under which plastic yielding will occur under combined stresses can be made with the help of several empirical theories. In the **maximum-shear-stress theory** the criterion for yielding is that yielding will occur when

$$\sigma_1 - \sigma_3 = \sigma_{ys}$$

in which σ_1 and σ_3 are the largest and smallest principal stresses, respectively, and σ_{ys} is the uniaxial tensile yield strength. This is the simplest theory for predicting yielding under combined stresses. A more accurate prediction can be made by the **distortion-energy theory,** according to which the criterion is

$$(\sigma_1 - \sigma_2)^2 + (\sigma_1 - \sigma_3)^2 + (\sigma_2 - \sigma_3)^2 = 2(\sigma_{ys})^2$$

Stress-strain curves in the plastic region for combined stress loading can be constructed. However, a particular stress state does not determine a unique strain value. The latter will depend on the stress-state path which is followed.

Plane strain is a condition where strain is confined to two dimensions. There is generally stress in the third direction, but because of mechanical constraints, strain in this dimension is prevented. Plane strain occurs in certain metalworking operations. It can also occur in the neighborhood of a crack tip in a tensile loaded member if the member is sufficiently thick. The material at the crack tip is then in triaxial tension, which condition promotes brittle fracture. On the other hand, ductility is enhanced and fracture is suppressed by triaxial compression.

Stress Concentration   In a structure or machine part having a notch or any abrupt change in cross section, the maximum stress will occur at this location and will be greater than the stress calculated by elementary formulas based upon simplified assumptions as to the stress distribution. The ratio of this maximum stress to the nominal stress (calculated by the elementary formulas) is the **stress-concentration factor** K_t. This is a constant for the particular geometry and is independent of the material, provided it is isotropic. The stress-concentration factor may be determined experimentally or, in some cases, theoretically from the mathematical theory of elasticity. The factors shown in Figs. 5.1.6 to 5.1.13 were determined from both photoelastic tests and the theory of

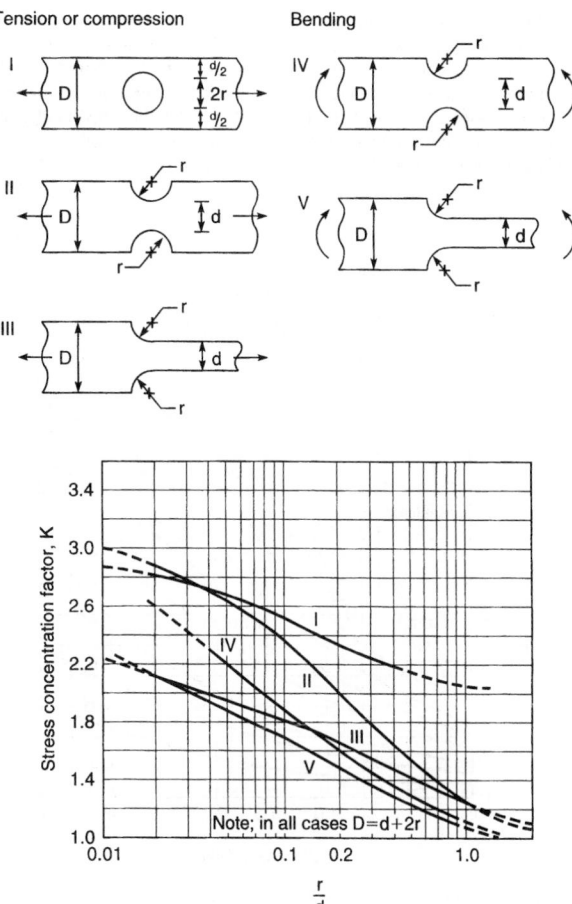

Fig. 5.1.6   Flat plate with semicircular fillets and grooves or with holes. I, II, and III are in tension or compression; IV and V are in bending.

Fig. 5.1.7   Flat plate with grooves, in tension.

elasticity. Stress concentration will cause failure of brittle materials if the concentrated stress is larger than the ultimate strength of the material. In ductile materials, concentrated stresses higher than the yield strength will generally cause local plastic deformation and redistribution of stresses (rendering them more uniform). On the other hand, even with ductile materials areas of stress concentration are possible sites for fatigue if the component is cyclically loaded.

Fig. 5.1.8   Flat plate with fillets, in tension.

Fig. 5.1.9   Flat plate with grooves, in bending.

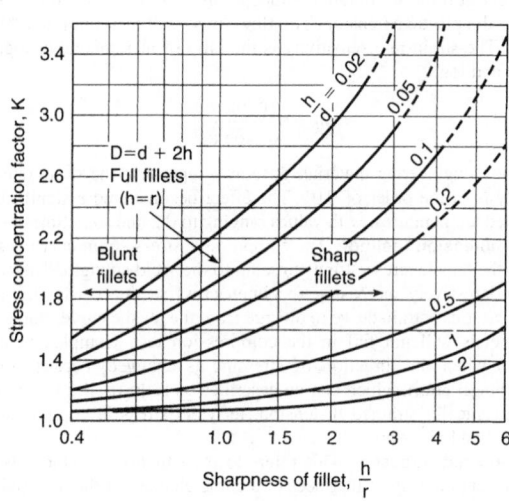

Fig. 5.1.10   Flat plate with fillets, in bending.

Fig. 5.1.11   Flat plate with angular notch, in tension or bending.

Fig. 5.1.12   Grooved shaft in torsion.

Fig. 5.1.13 Filleted shaft in torsion.

FRACTURE AT LOW STRESSES

Materials under tension sometimes fail by rapid fracture at stresses much below their strength level as determined in tests on carefully prepared specimens. These **brittle, unstable,** or **catastrophic failures** originate at preexisting stress-concentrating flaws which may be inherent in a material.

The transition-temperature approach is often used to ensure fracture-safe design in structural-grade steels. These materials exhibit a characteristic temperature, known as the **ductile brittle transition** (DBT) temperature, below which they are susceptible to brittle fracture. The transition-temperature approach to fracture-safe design ensures that the transition temperature of a material selected for a particular application is suitably matched to its intended use temperature. The DBT can be detected by plotting certain measurements from tensile or impact tests against temperature. Usually the transition to brittle behavior is complex, being neither fully ductile nor fully brittle. The range may extend over 200°F (110 K) interval. The **nil-ductility temperature** (NDT), determined by the **drop weight test** (see ASTM Standards), is an important reference point in the transition range. When NDT for a particular steel is known, temperature-stress combinations can be specified which define the limiting conditions under which catastrophic fracture can occur.

In the **Charpy V-notch** (CVN) impact test, a notched-bar specimen (Fig. 5.1.26) is used which is loaded in bending (see ASTM Standards). The energy absorbed from a swinging pendulum in fracturing the specimen is measured. The pendulum strikes the specimen at 16 to 19 ft (4.88 to 5.80 m)/s so that the specimen deformation associated with fracture occurs at a rapid strain rate. This ensures a conservative measure of toughness, since in some materials, toughness is reduced by high strain rates. A CVN impact energy vs. temperature curve is shown in Fig. 5.1.14, which also shows the transitions as given by percent brittle fracture and by percent lateral expansion. The CVN energy has no analytical significance. The test is useful mainly as a guide to the fracture behavior of a material for which an empirical correlation has been established between impact energy and some rigorous fracture criterion. For a particular grade of steel the CVN curve can be correlated with NDT. (See ASME Boiler and Pressure Vessel Code.)

Fracture Mechanics This analytical method is used for ultra-high-strength alloys, transition-temperature materials below the DBT temperature, and some low-strength materials in heavy section thickness.

Fracture mechanics theory deals with crack extension where plastic effects are negligible or confined to a small region around the crack tip. The present discussion is concerned with a through-thickness crack in a tension-loaded plate (Fig. 5.1.15) which is large enough so that the crack-tip stress field is not affected by the plate edges. Fracture mechanics theory states that unstable crack extension occurs when the work required for an increment of crack extension, namely, surface energy and energy consumed in local plastic deformation, is exceeded by the elastic-strain energy released at the crack tip. The elastic-stress field

Fig. 5.1.14 CVN transition curves. (*Data from Westinghouse R & D Lab.*)

surrounding one of the crack tips in Fig. 5.1.15 is characterized by the **stress intensity** K_I, which has units of $(lb\sqrt{in})/in^2$ or $(N/\sqrt{m})/m^2$. It is a function of applied nominal stress σ, crack half-length a, and a geometry factor Q:

$$K_I^2 = Q\sigma^2\pi a \qquad (5.1.1)$$

for the situation of Fig. 5.1.15. For a particular material it is found that as K_I is increased, a value K_c is reached at which unstable crack

Fig. 5.1.15   Through-thickness crack geometry.

propagation occurs. K_c depends on plate thickness B, as shown in Fig. 5.1.16. It attains a constant value when B is great enough to provide plane-strain conditions at the crack tip. The low plateau value of K_c is an important material property known as the **plane-strain critical stress intensity** or **fracture toughness** K_{Ic}. Values for a number of materials are shown in Table 5.1.4. They are influenced strongly by processing and small changes in composition, so that the values shown are not necessarily typical. K_{Ic} can be used in the critical form of Eq. (5.1.1):

$$(K_{Ic})^2 = Q\sigma^2\pi a_{cr} \qquad (5.1.2)$$

to predict failure stress when a maximum flaw size in the material is known or to determine maximum allowable flaw size when the stress is set. The predictions will be accurate so long as plate thickness B satisfies the **plane-strain criterion:** $B \geq 2.5(K_{Ic}/\sigma_{ys})^2$. They will be conservative if a plane-strain condition does not exist. A big advantage of the fracture mechanics approach is that stress intensity can be calculated by

Fig. 5.1.16   Dependence of K_c and fracture appearance (in terms of percentage of square fracture) on thickness of plate specimens. Based on data for aluminum 7075-T6. (*From Scrawly and Brown, STP-381, ASTM.*)

Table 5.1.4   Room-Temperature K_{Ic} Values on High-Strength Materials*

Material	0.2% YS, 1,000 in² (MN/m²)	K_{Ic}, 1,000 in² $\sqrt{in}$ (MN m^{1/2}/m²)
18% Ni maraging steel	300 (2,060)	46 (50.7)
18% Ni maraging steel	270 (1,850)	71 (78)
18% Ni maraging steel	198 (1,360)	87 (96)
Titanium 6-4 alloy	152 (1,022)	39 (43)
Titanium 6-4 alloy	140 (960)	75 (82.5)
Aluminum alloy 7075-T6	75 (516)	26 (28.6)
Aluminum alloy 7075-T6	64 (440)	30 (33)

* Determined at Westinghouse Research Laboratories.

equations analogous to (5.1.1) for a wide variety of geometries, types of crack, and loadings (Paris and Sih, "Stress Analysis of Cracks," STP-381, ASTM, 1965). Failure occurs in all cases when K_I reaches K_{Ic}. Fracture mechanics also provides a framework for predicting the occurrence of **stress-corrosion cracking** by using Eq. (5.1.2) with K_{Ic} replaced by K_{Iscc}, which is the material parameter denoting resistance to stress-corrosion-crack propagation in a particular medium.

Two standard test specimens for K_{Ic} determination are specified in ASTM standards, which also detail specimen preparation and test procedure. Recent developments in fracture mechanics permit treatment of crack propagation in the ductile regime. (See "Elastic-Plastic Fracture," ASTM.)

FATIGUE

Fatigue is generally understood as the gradual deterioration of a material which is subjected to repeated loads. In fatigue testing, a specimen is subjected to periodically varying constant-amplitude stresses by means of mechanical or magnetic devices. The applied stresses may alternate between equal positive and negative values, from zero to maximum positive or negative values, or between unequal positive and negative values. The most common loading is alternate tension and compression of equal numerical values obtained by rotating a smooth cylindrical specimen while under a bending load. A series of fatigue tests are made on a number of specimens of the material at different stress levels. The stress endured is then plotted against the number of cycles sustained. By choosing lower and lower stresses, a value may be found which will not produce failure, regardless of the number of applied cycles. This stress value is called the **fatigue limit.** The diagram is called the stress-cycle diagram or *S-N* **diagram.** Instead of recording the data on cartesian coordinates, either stress is plotted vs. the logarithm of the number of cycles (Fig. 5.1.17) or both stress and cycles are plotted to logarithmic scales. Both diagrams show a relatively sharp bend in the curve near the fatigue limit for ferrous metals. The fatigue limit may be established for most steels between 2 and 10 million cycles. Nonferrous metals usually show no clearly defined fatigue limit. The *S-N* curves in these cases indicate a continuous decrease in stress values to several hundred million cycles, and both the stress value and the number of cycles sustained should be reported. See Table 5.1.5.

The mean stress (the average of the maximum and minimum stress values for a cycle) has a pronounced influence on the stress range (the algebraic difference between the maximum and minimum stress values). Several empirical formulas and graphical methods such as the "modified Goodman diagram" have been developed to show the influence of the mean stress on the stress range for failure. A simple but conservative approach (see Soderberg, Working Stresses, *Jour. Appl. Mech.,* **2,** Sept. 1935) is to plot the variable stress S_v (one-half the stress range) as ordinate vs. the mean stress S_m as abscissa (Fig. 5.1.18). At zero mean stress, the ordinate is the fatigue limit under completely reversed stress. Yielding will occur if the mean stress exceeds the yield stress S_o, and this establishes the extreme right-hand point of the diagram. A straight line is drawn between these two points. The coordinates of any other point along this line are values of S_m and S_v which may produce failure.

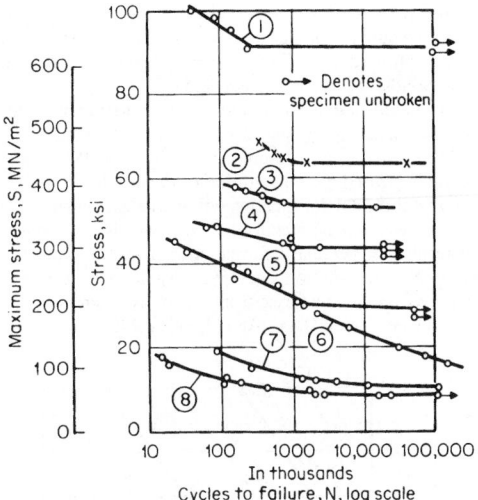

Fig. 5.1.17 The *S-N* diagrams from fatigue tests. (1) 1.20% C steel, quenched and drawn at 860°F (460°C); (2) alloy structural steel; (3) SAE 1050, quenched and drawn at 1,200°F (649°C); (4) SAE 4130, normalized and annealed; (5) ordinary structural steel; (6) Duralumin; (7) copper, annealed; (8) cast iron (reversed bending).

Surface defects, such as roughness or scratches, and notches or shoulders all reduce the fatigue strength of a part. With a notch of prescribed geometric form and known concentration factor, the reduction in strength is appreciably less than would be called for by the concentration factor itself, but the various metals differ widely in their susceptibility to the effect of roughness and concentrations, or **notch sensitivity.**

For a given material subjected to a prescribed state of stress and type of loading, notch sensitivity can be viewed as the ability of that material to resist the concentration of stress incidental to the presence of a notch. Alternately, notch sensitivity can be taken as a measure of the degree to which the geometric stress concentration factor is reduced. An attempt is made to rationalize notch sensitivity through the equation $q = (K_f - 1)/(K - 1)$, where q is the notch sensitivity, K is the geometric stress concentration factor (from data similar to those in Figs. 5.1.5 to 5.1.13 and the like), and K_f is defined as the ratio of the strength of unnotched material to the strength of notched material. Ratio K_f is obtained from laboratory tests, and K is deduced either theoretically or from laboratory tests, but both must reflect the same state of stress and type of loading. The value of q lies between 0 and 1, so that (1) if $q = 0$, $K_f = 1$ and the material is not notch-sensitive (soft metals such as copper, aluminum, and annealed low-strength steel); (2) if $q = 1$, $K_f = K$, the material is fully notch-sensitive and the full value of the geometric stress concentration factor is not diminished (hard, high-strength steel). In practice, q will lie somewhere between 0 and 1,

but it may be hard to quantify. Accordingly, the pragmatic approach to arrive at a solution to a design problem often takes a conservative route and sets $q = 1$. The exact material properties at play which are responsible for notch sensitivity are not clear.

Further, notch sensitivity seems to be higher, and ordinary fatigue strength lower in large specimens, necessitating full-scale tests in many cases (see Peterson, Stress Concentration Phenomena in Fatigue of

Fig. 5.1.18   Effect of mean stress on the variable stress for failure.

Metals, *Trans. ASME,* **55,** 1933, p. 157, and Buckwalter and Horger, Investigation of Fatigue Strength of Axles, Press Fits, Surface Rolling and Effect of Size, *Trans. ASM,* **25,** Mar. 1937, p. 229). **Corrosion** and **galling** (due to rubbing of mating surfaces) cause great reduction of fatigue strengths, sometimes amounting to as much as 90 percent of the original endurance limit. Although any corroding agent will promote severe corrosion fatigue, there is so much difference between the effects of "sea water" or "tap water" from different localities that numerical values are not quoted here.

Overstressing specimens above the fatigue limit for periods shorter than necessary to produce failure at that stress reduces the fatigue limit in a subsequent test. Similarly, **understressing** below the fatigue limit may increase it. Shot peening, nitriding, and cold work usually improve fatigue properties.

No very good overall correlation exists between fatigue properties and any other mechanical property of a material. The best correlation is between the fatigue limit under completely reversed bending stress and the ordinary tensile strength. For many ferrous metals, the fatigue limit is approximately 0.40 to 0.60 times the tensile strength if the latter is below 200,000 lb/in². Low-alloy high-yield-strength steels often show higher values than this. The fatigue limit for nonferrous metals is approximately to 0.20 to 0.50 times the tensile strength. The fatigue limit in reversed shear is approximately 0.57 times that in reversed bending.

In some very important engineering situations components are cyclically stressed into the plastic range. Examples are thermal strains resulting from temperature oscillations and notched regions subjected to secondary stresses. Fatigue life in the plastic or **"low-cycle" fatigue** range has been found to be a function of plastic strain, and low-cycle fatigue testing is done with strain as the controlled variable rather than stress. Fatigue life N and cyclic plastic strain ε_p tend to follow the relationship

$$N\varepsilon_p^2 = C$$

where C is a constant for a material when $N < 10^5$. (See Coffin, A Study of Cyclic-Thermal Stresses in a Ductile Material, *Trans. ASME,* **76,** 1954, p. 947.)

Table 5.1.5   Typical Approximate Fatigue Limits for Reversed Bending

Metal	Tensile strength, 1,000 lb/in²	Fatigue limit, 1,000 lb/in²	Metal	Tensile strength, 1,000 lb/in²	Fatigue limit, 1,000 lb/in²
Cast iron	20–50	6–18	Copper	32–50	12–17
Malleable iron	50	24	Monel	70–120	20–50
Cast steel	60–80	24–32	Phosphor bronze	55	12
Armco iron	44	24	Tobin bronze, hard	65	21
Plain carbon steels	60–150	25–75	Cast aluminum alloys	18–40	6–11
SAE 6150, heat-treated	200	80	Wrought aluminum alloys	25–70	8–18
Nitralloy	125	80	Magnesium alloys	20–45	7–17
Brasses, various	25–75	7–20	Molybdenum, as cast	98	45
Zirconium crystal bar	52	16–18	Titanium (Ti-75A)	91	45

Note: Stress, 1,000 lb/in² × 6.894 = stress, MN/m².

The type of physical change occurring inside a material as it is repeatedly loaded to failure varies as the life is consumed, and a number of stages in fatigue can be distinguished on this basis. The early stages comprise the events causing **nucleation of a crack** or flaw. This is most likely to appear on the surface of the material; **fatigue failures** generally originate at a surface. Following nucleation of the crack, it grows during the **crack-propagation** stage. Eventually the crack becomes large enough for some rapid terminal mode of failure to take over such as ductile rupture or brittle fracture. The rate of crack growth in the crack-propagation stage can be accurately quantified by fracture mechanics methods. Assuming an initial flaw and a loading situation as shown in Fig. 5.1.15, the rate of crack growth per cycle can generally be expressed as

$$da/dN = C_0(\Delta K_I)^n \tag{5.1.3}$$

where C_0 and n are constants for a particular material and ΔK_I is the range of stress intensity per cycle. K_I is given by (5.1.1). Using (5.1.3), it is possible to predict the number of cycles for the crack to grow to a size at which some other mode of failure can take over. Values of the constants C_0 and n are determined from specimens of the same type as those used for determination of K_{Ic} but are instrumented for accurate measurement of slow crack growth.

Constant-amplitude fatigue-test data are relevant to many rotary-machinery situations where constant cyclic loads are encountered. There are important situations where the component undergoes variable loads and where it may be advisable to use **random-load testing.** In this method, the load spectrum which the component will experience in service is determined and is applied to the test specimen artificially.

CREEP

Experience has shown that, for the design of equipment subjected to sustained loading at elevated temperatures, little reliance can be placed on the usual short-time tensile properties of metals at those temperatures. Under the application of a constant load it has been found that materials, both metallic and nonmetallic, show a gradual flow or **creep** even for stresses below the proportional limit at elevated temperatures. Similar effects are present in low-melting metals such as lead at room temperature. The deformation which can be permitted in the satisfactory operation of most high-temperature equipment is limited.

In metals, creep is a plastic deformation caused by slip occurring along crystallographic directions in the individual crystals, together with some flow of the grain-boundary material. After complete release of load, a small fraction of this plastic deformation is recovered with time. Most of the flow is nonrecoverable for metals.

Since the early creep experiments, many different types of tests have come into use. The most common are the **long-time creep test** under constant tensile load and the **stress-rupture test.** Other special forms are the **stress-relaxation test** and the **constant-strain-rate test.**

The **long-time creep test** is conducted by applying a dead weight to one end of a lever system, the other end being attached to the specimen surrounded by a furnace and held at constant temperature. The axial deformation is read periodically throughout the test and a curve is plotted of the strain ε_0 as a function of time t (Fig. 5.1.19). This is repeated for various loads at the same testing temperature. The portion of the curve OA in Fig. 5.1.19 is the region of **primary creep,** AB the region of **secondary creep,** and BC that of **tertiary creep.** The strain rates, or the slopes of the curve, are decreasing, constant, and increasing, respectively,

in these three regions. Since the period of the creep test is usually much shorter than the duration of the part in service, various extrapolation procedures are followed (see Gittus, "Creep, Viscoelasticity and Creep Fracture in Solids," Wiley, 1975). See Table 5.1.6.

In practical applications the region of constant-strain rate (secondary creep) is often used to estimate the probable deformation throughout the life of the part. It is thus assumed that this rate will remain constant during periods beyond the range of the test data. The working stress is chosen so that this total deformation will not be excessive. An **arbitrary creep strength,** which is defined as the stress which at a given temperature will result in 1 percent deformation in 100,000 h, has received a certain amount of recognition, but it is advisable to determine the proper stress for each individual case from diagrams of stress vs. creep rate (Fig. 5.1.20) (see "Creep Data," ASTM and ASME).

Fig. 5.1.20   Creep rates for 0.35% C steel.

Additional temperatures (°F) and stresses (in 1,000 lb/in²) for stated creep rates (percent per 1,000 h) for wrought nonferrous metals are as follows:

60-40 Brass: Rate 0.1, temp. 350 (400), stress 8 (2); rate 0.01, temp 300 (350) [400], stress 10 (3) [1].
Phosphor bronze: Rate 0.1, temp 400 (550) [700] [800], stress 15 (6) [4] [4]; rate 0.01, temp 400 (550) [700], stress 8 (4) [2].
Nickel: Rate 0.1, temp 800 (1000), stress 20 (10).
70 Cu, 30 NI. Rate 0.1, temp 600 (750), stress 28 (13–18); rate 0.01, temp 600 (750), stress 14 (8–9).
Aluminum alloy 17 S (Duralumin): Rate 0.1, temp 300 (500) [600], stress 22 (5) [1.5].
Lead pure (commercial) (0.03 percent Ca): At 110°F, for rate 0.1 percent the stress range, lb/in², is 150–180 (60–140) [200–220]; for rate of 0.01 percent, 50–90 (10–50) [110–150].
Stress, 1,000 lb/in² × 6.894 = stress, MN/m², $t_k = \frac{5}{9}(t_F + 459.67)$.

Structural changes may occur during a creep test, thus altering the metallurgical condition of the metal. In some cases, premature rupture appears at a low fracture strain in a normally ductile metal, indicating that the material has become embrittled. This is a very insidious condition and difficult to predict. The **stress-rupture test** is well adapted to study this effect. It is conducted by applying a constant load to the specimen in the same manner as for the long-time creep test. The nominal stress is then plotted vs. the time for fracture at constant temperature on a log-log scale (Fig. 5.1.21).

The stress reaction is measured in the **constant-strain-rate test** while the specimen is deformed at a constant strain rate. In the **relaxation test,** the decrease of stress with time is measured while the total strain (elastic + plastic) is maintained constant. The latter test has direct application to the loosening of turbine bolts and to similar problems. Although some correlation has been indicated between the results

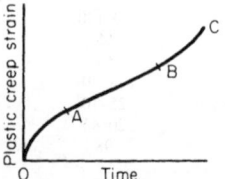

Fig. 5.1.19   Typical creep curve.

Table 5.1.6 Stresses for Given Creep Rates and Temperatures*

Material temp, °F	Creep rate 0.1% per 1,000 h					Creep rate 0.01% per 1,000 h				
	800	900	1,000	1,100	1,200	800	900	1,000	1,100	1,200
Wrought steels:										
SAE 1015	17–27	11–18	3–12	2–7	1	10–18	6–14	3–8	1	
0.20 C , 0.50 Mo	26–33	18–25	9–16	2–6	1–2	16–24	11–22	4–12	2	1
0.10–0.25 C , 4–6 Cr + Mo	22	15–18	9–11	3–6	2–3	14–17	11–15	4–7	2–3	1–2
SAE 4140	27–33	20–25	7–15	4–7	1–2	19–28	12–19	3–8	2–4	1
SAE 1030–1045	8–25	5–15	5	2	1	5–15	3–7	2–4	1	
Commercially pure iron	7		4		3	5		2		
0.15 C, 1–2.5 Cr, 0.50 Mo	25–35	18–28	8–20	6–8	3–4	20–30	12–18	3–12	2–5	1–2
SAE 4340	20–40	15–30	2–12	1–3		8–20		1–6		
SAE X3140	7–10		5–4			3–8		1–2		
0.20 C, 4–6 Cr	30	10–20	7–10	1				3–5		
0.25 C, 4–6 Cr + W	30	10–15	4–10	2–8			6–11	2–7		
0.16 C, 1.2 Cu		18	10–15	3	1		10–18	7–12		
0.20 C, 1 Mo	35	27	12			25	12	6		
0.10–0.40 C, 0.2–0.5 Mo, 1–2 Mn	30–40	12–20	4–14			25–28	8–15	2–8		0.5
SAE 2340	7–12	5	2							
SAE 6140	30	12	4			7	6	1		
SAE 7240	30	21	6–15	2		30	11	3–9	1	
Cr + Va + W, various	20–70	14–30	5–15			18–50	8–18	2–13		
Temp, °F	1,100	1,200	1,300	1,400	1,500	1,000	1,100	1,200	1,300	1,400
Wrought chrome-nickel steels:										
18–8†	10–18	5–11	3–10	2–5	2.5	11–16	5–12	2–10		1–2
10–25 Cr, 10–30 Ni‡	10–20	5–15	3–10	2–5			6–15	3–10	2–8	1–3
Temp, °F	800	900	1,000	1,100	1,200	800	900	1,000	1,100	1,200
Cast steels:										
0.20–0.40 C	10–20	5–10	3			8–15		1		
0.10–0.30 C, 0.5–1 Mo	28	20–30	6–12	2		20	10–15	2–5		
0.15–0.30 C, 4–6 Cr + Mo	25–30	15–25	8–15	8		20–25	9–15	2–7	2	
18–8§			20–25	15	10			20	15	8
Cast iron	20	8	4			10		2		
Cr Ni cast iron			9					3		

* Based on 1,000-h tests. Stresses in 1,000 lb/in².
† Additional data. At creep rate 0.1 percent and 1,000 (1,600)°F the stress is 18–25 (1); at creep rate 0.01 percent at 1,500°F, the stress is 0.5.
‡ Additional data. At creep rate 0.1 percent and 1,000 (1,600)°F the stress is 10–30 (1).
§ Additional data. At creep rate 0.1 percent and 1,600°F the stress is 3; at creep rate 0.01 and 1,500°F, the stress is 2–3.

of these various types of tests, no general correlation is yet available, and it has been found necessary to make tests under each of these special conditions to obtain satisfactory results.

The interrelationship between strain rate and temperature in the form of a velocity-modified temperature (see MacGregor and Fisher, A Velocity-modified Temperature for the Plastic Flow of Metals, *Jour. Appl. Mech.*, Mar. 1945) simplifies the creep problem in reducing the number of variables.

Superplasticity Superplasticity is the property of some metals and alloys which permits extremely large, uniform deformation at elevated temperature, in contrast to conventional metals which neck down and subsequently fracture after relatively small amounts of plastic deformation. Superplastic behavior requires a metal with small equiaxed grains, a slow and steady rate of deformation (strain rate), and a temperature elevated to somewhat more than half the melting point. With such metals, large plastic deformation can be brought about with lower external

Fig. 5.1.21 Relation between time to failure and stress for a 3% chromium steel. (1) Heat treated 2 h at 1,740°F (950°C) and furnace cooled; (2) hot rolled and annealed 1,580°F (860°C).

loads; ultimately, that allows the use of lighter fabricating equipment and facilitates production of finished parts to near-net shape.

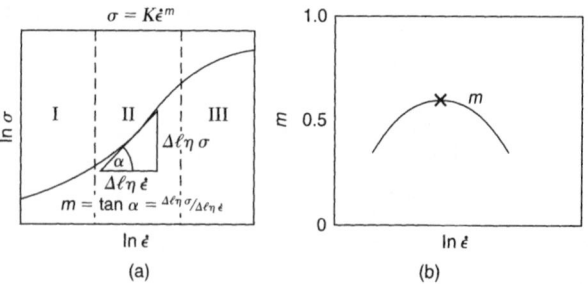

Fig. 5.1.22 Stress and strain rate relations for superplastic alloys. (*a*) Log-log plot of $\sigma = K\dot{\epsilon}^m$; (*b*) *m* as a function of strain rate.

Stress and strain rates are related for a metal exhibiting superplasticity. A factor in this behavior stems from the relationship between the applied stress and strain rates. This factor *m*—the strain rate sensitivity index—is evaluated from the equation $\sigma = K\dot{\epsilon}^m$, where σ is the applied stress, K is a constant, and $\dot{\epsilon}$ is the strain rate. Figure 5.1.22*a* plots a stress/strain rate curve for a superplastic alloy on log-log coordinates. The slope of the curve defines *m*, which is maximum at the point of inflection. Figure 5.1.22*b* shows the variation of *m* versus ln $\dot{\epsilon}$. Ordinary metals exhibit low values of *m*—0.2 or less; for those behaving superplastically, $m = 0.6$ to $0.8 +$. As *m* approaches 1, the behavior of the metal will be quite similar to that of a newtonian viscous solid, which elongates plastically without necking down.

In Fig. 5.1.22*a*, in region I, the stress and strain rates are low and creep is predominantly a result of diffusion. In region III, the stress and strain rates are highest and creep is mainly the result of dislocation and slip mechanisms. In region II, where superplasticity is observed, creep is governed predominantly by grain boundary sliding.

Shape Memory

Some particular alloys have a remarkable property which may be quite useful. The **shape memory effect** inherent in an alloy is a thermoelastic property that results from a reversible martensitic transformation when heated or cooled within a range of temperatures below the melting point. In essence, a given geometry at elevated temperature is cooled and further deformed. Upon being heated to the previous elevated temperature, the deformation imparted at the lower temperature is negated, and the former elevated-temperature geometry is restored. A number of useful applications have been found for those alloys which have this property.

HARDNESS

Hardness has been variously defined as resistance to local penetration, to scratching, to machining, to wear or abrasion, and to yielding. The multiplicity of definitions, and corresponding multiplicity of hardness-measuring instruments, together with the lack of a fundamental definition, indicates that hardness may not be a fundamental property of a material but rather a composite one including yield strength, work hardening, true tensile strength, modulus of elasticity, and others.

Scratch hardness is measured by **Mohs scale** of minerals (Sec. 1.2) which is so arranged that each mineral will scratch the mineral of the next lower number. In recent mineralogical work and in certain microscopic metallurgical work, jeweled scratching points either with a set load or else loaded to give a set width of scratch have been used. Hardness in its relation to machinability and to wear and abrasion is generally dealt with in direct machining or wear tests, and little attempt is made to separate hardness itself, as a numerically expressed quantity, from the results of such tests.

The resistance to localized penetration, or **indentation hardness**, is widely used industrially as a measure of hardness, and indirectly as an indicator of other desired properties in a manufactured product. The indentation tests described below are essentially nondestructive, and in most applications may be considered nonmarring, so that they may be applied to each piece produced; and through the empirical relationships of hardness to such properties as tensile strength, fatigue strength, and impact strength, pieces likely to be deficient in the latter properties may be detected and rejected.

Brinell hardness is determined by forcing a hardened sphere under a known load into the surface of a material and measuring the diameter of the indentation left after the test. The **Brinell hardness number,** or simply the **Brinell number,** is obtained by dividing the load used, in kilograms, by the actual surface area of the indentation, in square millimetres. The result is a pressure, but the units are rarely stated.

$$\text{BHN} = P \left/ \left[\frac{\pi D}{2}(D - \sqrt{D^2 - d^2}) \right] \right.$$

where BHN is the Brinell hardness number; P the imposed load, kg; D the diameter of the spherical indenter, mm; and d the diameter of the resulting impression, mm.

Hardened-steel bearing balls may be used for hardness up to 450, but beyond this hardness specially treated steel balls or jewels should be used to avoid flattening the indenter. The standard-size ball is 10 mm and the standard loads 3,000, 1,500, and 500 kg, with 100, 125, and 250 kg sometimes used for softer materials. If for special reasons any other size of ball is used, the load should be adjusted approximately as follows: for iron and steel, $P = 30D^2$; for brass, bronze, and other soft metals, $P = 5D^2$; for extremely soft metals, $P = D^2$ (see "Methods of Brinell Hardness Testing," ASTM). Readings obtained with other than the standard ball and loadings should have the load and ball size appended, as such readings are only approximately equal to those obtained under standard conditions.

The size of the specimen should be sufficient to ensure that no part of the plastic flow around the impression reaches a free surface, and in no case should the thickness be less than 10 times the depth of the impression. The load should be applied steadily and should remain on for at least 15 s in the case of ferrous materials and 30 s in the case of most nonferrous materials. Longer periods may be necessary on certain soft materials that exhibit creep at room temperature. In testing thin materials, it is not permissible to pile up several thicknesses of material under the indenter, as the readings so obtained will invariably be lower than the true readings. With such materials, smaller indenters and loads, or different methods of hardness testing, are necessary.

In the standard Brinell test, the diameter of the impression is measured with a low-power hand microscope, but for production work several testing machines are available which automatically measure the depth of the impression and from this give readings of hardness. Such machines should be calibrated frequently on test blocks of known hardness.

In the **Rockwell method** of hardness testing, the depth of penetration of an indenter under certain arbitrary conditions of test is determined. The indenter may be either a steel ball of some specified diameter or a spherical-tipped conical diamond of 120° angle and 0.2-mm tip radius, called a "Brale." A *minor load* of 10 kg is first applied which causes an initial penetration and holds the indenter in place. Under this condition, the dial is set to zero and the major load applied. The values of the latter are 60, 100, or 150 kg. Upon removal of the major load, the reading is taken while the minor load is still on. The hardness number may then be read directly from the scale which measures penetration, and this scale is so arranged that soft materials with deep penetration give low hardness numbers.

A variety of combinations of indenter and major load are possible; the most commonly used are R_B using as indenter a $\frac{1}{16}$-in ball and a major load of 100 kg and R_C using a Brale as indenter and a major load of 150 kg (see "Rockwell Hardness and Rockwell Superficial Hardness of Metallic Materials," ASTM).

Compared with the Brinell test, the Rockwell method makes a smaller indentation, may be used on thinner material, and is more rapid, since hardness numbers are read directly and need not be calculated. However, the Brinell test may be made without special apparatus and is somewhat more widely recognized for laboratory use. There is also a **Rockwell superficial hardness test** similar to the regular Rockwell, except that the indentation is much shallower.

The **Vickers** method of hardness testing is similar in principle to the Brinell in that it expresses the result in terms of the pressure under the indenter and uses the same units, kilograms per square millimetre. The indenter is a diamond in the form of a square pyramid with an apical angle of 136°, the loads are much lighter, varying between 1 and 120 kg, and the impression is measured by means of a medium-power compound microscope.

$$V = P/(0.5393d^2)$$

where V is the Vickers hardness number, sometimes called the **diamond-pyramid hardness** (DPH); P the imposed load, kg; and d the diagonal of indentation, mm. The Vickers method is more flexible and is considered to be more accurate than either the Brinell or the Rockwell, but the equipment is more expensive than either of the others and the Rockwell is somewhat faster in production work.

Among the other hardness methods may be mentioned the **Scleroscope,** in which a diamond-tipped "hammer" is dropped on the surface and the rebound taken as an index of hardness. This type of apparatus is seriously affected by the resilience as well as the hardness of the material and has largely been superseded by other methods. In the **Monotron** method, a penetrator is forced into the material to a predetermined depth and the load required is taken as the indirect measure of the hardness. This is the reverse of the Rockwell method in principle, but the loads and indentations are smaller than those of the latter. In the **Herbert pendulum,** a 1-mm steel or jewel ball resting on the surface to be tested acts as the fulcrum for a 4-kg compound pendulum of 10-s period. The swinging of the pendulum causes a rolling indentation in the material, and from the behavior of the pendulum several factors in hardness, such as **work hardenability,** may be determined which are not revealed by other methods. Although the Herbert results are of considerable significance, the instrument is suitable for laboratory use only (see Herbert, The Pendulum Hardness Tester, and Some Recent Developments in Hardness Testing, *Engineer,* **135,** 1923, pp. 390, 686). In the **Herbert cloudburst** test, a shower of steel balls, dropped from a predetermined height, dulls the surface of a hardened part in proportion to its softness and thus reveals defective areas. A variety of **mutual indentation methods,** in which crossed cylinders or prisms of the material to be tested are forced together, give results comparable with the Brinell test. These are particularly useful on wires and on materials at high temperatures.

The relation among the scales of the various hardness methods is not exact, since no two measure exactly the same sort of hardness, and a relationship determined on steels of different hardnesses will be found only approximately true with other materials. The **Vickers-Brinell relation** is nearly linear up to at least 400, with the Vickers approximately 5 percent higher than the Brinell (actual values run from + 2 to + 11 percent) and nearly independent of the material. Beyond 500, the values become more widely divergent owing to the flattening of the Brinell ball. The **Brinell-Rockwell relation** is fairly satisfactory and is shown in Fig. 5.1.23. Approximate relations for the **Shore Scleroscope** are also given on the same plot.

Fig. 5.1.23 Hardness scales.

The **hardness of wood** is defined by the ASTM as the load in pounds required to force a ball 0.444 in in diameter into the wood to a depth of 0.222 in, the speed of penetration being ¼ in/min. For a summary of the work in hardness see Williams, "Hardness and Hardness Measurements," ASM.

TESTING OF MATERIALS

Testing Machines Machines for the mechanical testing of materials usually contain elements (1) for gripping the specimen, (2) for deforming it, and (3) for measuring the load required in performing the deformation. Some machines (ductility testers) omit the measurement of load and substitute a measurement of deformation, whereas other machines include the measurement of both load and deformation through apparatus either integral with the testing machine (stress-strain recorders) or auxiliary to it (strain gages). In most general-purpose testing machines, the deformation is controlled as the independent variable and the resulting load measured, and in many special-purpose machines, particularly those for light loads, the load is controlled and the resulting deformation is measured. Special features may include those for constant rate of loading (pacing disks), for constant rate of straining, for constant load maintenance, and for cyclical load variation (fatigue).

In modern testing systems, the load and deformation measurements are made with load-and-deformation-sensitive transducers which generate electrical outputs. These outputs are converted to load and deformation readings by means of appropriate electronic circuitry. The readings are commonly displayed automatically on a recorder chart or digital meter, or they are read into a computer. The transducer outputs are typically used also as feedback signals to control the test mode (constant loading, constant extension, or constant strain rate). The load transducer is usually a load cell attached to the test machine frame, with electrical output to a bridge circuit and amplifier. The load cell operation depends on change of electrical resistivity with deformation (and load) in the transducer element. The deformation transducer is generally an extensometer clipped on to the test specimen gage length, and operates on the same principle as the load cell transducer: the change in electrical resistance in the specimen gage length is sensed as the specimen deforms. Optical extensometers are also available which do not make physical contact with the specimen. Verification and classification of extensometers is controlled by ASTM Standards. The application of load and deformation to the specimen is usually by means of a screw-driven mechanism, but it may also be applied by means of hydraulic and servohydraulic systems. In each case the load application system responds to control inputs from the load and deformation transducers. Important features in test machine design are the methods used for reducing friction, wear, and backlash. In older testing machines, test loads were determined from the machine itself (e.g., a pressure reading from the machine hydraulic pressure) so that machine friction made an important contribution to inaccuracy. The use of machine-independent transducers in modern testing has eliminated much of this source of error.

Grips should not only hold the test specimen against slippage but should also apply the load in the desired manner. Centering of the load is of great importance in compression testing, and should not be neglected in tension testing if the material is brittle. Figure 5.1.24 shows the theoretical errors due to off-center loading; the results are directly applicable to compression tests using swivel loading blocks. Swivel (ball-and-socket) holders or compression blocks should be used with all except the most ductile materials, and in compression testing of brittle materials (concrete, stone, brick), any rough faces should be smoothly capped with plaster of paris and one-third portland cement. Serrated grips may be used to hold ductile materials or the shanks of other holders in tension; a taper of 1 in 6 on the wedge faces gives a self-tightening action without excessive jamming. Ropes are ordinarily held by wet eye splices, but braided ropes or small cords may be given several turns over a fixed pin and then clamped. Wire ropes should be zinced into forged sockets (solder and lead have insufficient strength). Grip selection for tensile testing is described in ASTM standards.

Accuracy and Calibration ASTM standards require that commercial machines have errors of less than 1 percent within the "loading range" when checked against acceptable standards of comparison at at least five suitably spaced loads. The "loading range" may be any range through which the preceding requirements for accuracy are satisfied, except that it shall not extend below 100 times the least load to which the machine will respond or which can be read on the indicator. The use of calibration plots or tables to correct the results of an otherwise inaccurate machine is not permitted under any circumstances. Machines with errors less than 0.1 percent are commercially available (Tate-Emery and others), and somewhat greater accuracy is possible in the most refined research apparatus.

Fig. 5.1.24 Effect of centering errors on brittle test specimens.

Dead loads may be used to check machines of low capacity; accurately calibrated proving levers may be used to extend the range of available weights. Various elastic devices (such as the Morehouse proving ring) made of specially treated steel, with sensitive disortion-measuring devices, and calibrated by dead weights at the NIST (formerly Bureau of Standards) are among the most satisfactory means of checking the higher loads.

Two **standard forms of test specimens** (ASTM) are shown in Figs. 5.1.25 and 5.1.26. In wrought materials, and particularly in those which

Fig. 5.1.25 Test specimen, 2-in (50-mm) gage length, ½-in (12.5-mm) diameter. Others available for 0.35-in (8.75-mm) and 0.25-in (6.25-mm) diameters. (ASTM).

Fig. 5.1.26 Charpy V-notch impact specimens. (ASTM.)

have been cold-worked, different properties may be expected in different directions with respect to the direction of the applied work, and the test specimens, often called **coupons,** should be cut out from the parent material in such a way as to give the strength in the desired direction. With the exception of fatigue specimens and specimens of extremely brittle materials, surface finish is of little practical importance, although extreme roughness tends to decrease the ultimate elongation.

5.2 MECHANICS OF MATERIALS
by Ali M. Sadegh

REFERENCES: Timoshenko and MacCullough, "Elements of Strength of Materials," Van Nostrand. Seeley, "Advanced Mechanics of Materials," Wiley. Timoshenko and Goodier, "Theory of Elasticity," McGraw-Hill. Phillips, "Introduction to Plasticity," Ronald. Van Den Broek, "Theory of Limit Design," Wiley. Hetényi, "Handbook of Experimental Stress Analysis," Wiley. Dean and Douglas, "Semi-Conductor and Conventional Strain Gages," Academic. Robertson and Harvey, "The Engineering Uses of Holography," University Printing House, London. Sellers, "Basic Training Guide to the New Metrics and SI Units," National Tool, Die and Precision Machining Association. Roark and Young, "Formulas for Stress and Strain," McGraw-Hill. Perry and Lissner, "The Strain Gage Primer," McGraw-Hill. Donnell, "Beams, Plates, and Sheets," Engineering Societies Monographs, McGraw-Hill. Griffel, "Beam Formulas" and "Plate Formulas," Ungar. Durelli et al., "Introduction to the Theoretical and Experimental Analysis of Stress and Strain," McGraw-Hill. "Stress Analysis Manual," Department of Commerce, Pub. no. AD 759 199, 1969. Blodgett, "Welded Structures," Lincoln Arc Welding Foundation. "Characteristics and Applications of Resistance Strain Gages," Department of Commerce, NBS Circ. 528, 1954.

NOTE: The almost universal availability and utilization of computers in engineering practice has led to the development of many forms of software individually tailored to the solution of specific design problems in the area of mechanics of materials. Their use will permit the reader to amplify and supplement a good portion of the formulary and tabular collection in this section, as well as utilize those powerful computational tools in newer and more powerful techniques to facilitate solutions to problems. Many of the approximate methods, involving laborious iterative mathematical schemes, have been supplanted by the computer.

Developments along those lines continue apace and bid fair to expand the types of problems handled, all with greater confidence in the results obtained thereby.

Main Symbols

UNIT STRESS

S = apparent stress
S_v or S_s = pure shearing
T = true (ideal) stress
S_p = proportional elastic limit
S_y = yield point
S_M = ultimate strength, tension
S_c = ultimate compression
S_v = vertical shear in beams
S_R = modulus of rupture

MOMENT

M = bending
M_t = torsion

EXTERNAL ACTION

P = force
G = weight of body
W = weight of load
V = external shear

MODULUS OF ELASTICITY

E = longitudinal
G = shearing
K = bulk
U_p = modulus of resilience
U_R = ultimate resilience

GEOMETRIC

l = length
A = area
V = volume
ν = velocity
r = radius of gyration
I = rectangular moment of inertia
I_P or J = polar moment of inertia

DEFORMATION

e, e' = gross deformation
$\varepsilon, \varepsilon'$ = unit deformation; strain
d or α = unit, angular
s' = unit, lateral
μ = Poisson's ratio
n = reciprocal of Poisson's ratio
r = radius
f = deflection

SIMPLE STRESSES AND STRAINS

Deformations are changes in form produced by external forces or loads that act on nonrigid bodies. Deformations are **longitudinal,** e, a lengthening (+) or shortening (−) of the body; and **angular,** α, a change of angle between the faces.

Unit deformation (strain) (dimensionless number) is the deformation in unit distance. Unit longitudinal deformation (longitudinal strain), $\varepsilon = e/l$ (Fig. 5.2.1). Unit angular-deformation tan α equals α approx (Fig. 5.2.2).

The accompanying lateral deformation results in unit lateral deformation (lateral strain) $\varepsilon' = e'/l'$ (Fig. 5.2.1). For homogeneous, isotropic material operating in the elastic region, the ratio $-\varepsilon'/\varepsilon$ is a constant and is a definite property of the material; this ratio is called **Poisson's ratio** μ.

A fundamental relation among the **three interdependent constants** E, G, and μ for a given material is $E = 2G(1 + \mu)$. Note that μ cannot be larger than 0.5; thus the shearing modulus G is always smaller than the elastic modulus E. At the extremes, for example, $\mu \approx 0.5$ for rubber and paraffin; $\mu \approx 0$ for cork. For concrete, μ varies from 0.10 to 0.20 at working stresses and can reach 0.25 at higher stresses; μ for ordinary glass is about 0.25. In the absence of definitive data, μ for most structural metals can be taken to lie between 0.25 and 0.35. Extensive listings of Poisson's ratio are found in other sections; see Tables 5.1.3 and 6.1.9.

Fig. 5.2.1

Stress is an internal distributed force, or, force per unit area; it is the internal mechanical reaction of the material accompanying deformation. Stresses always occur in pairs. Stresses are **normal** [tensile stress (+) and compressive stress (−)]; and **tangential,** or **shearing.**

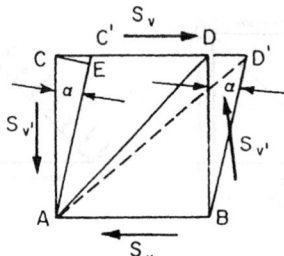

Fig. 5.2.2

Intensity of stress, or **unit stress,** S, lb/in^2 (kgf/cm^2), is the amount of force per unit of area (Fig. 5.2.3). P is the load acting through the center of gravity of the area. The uniformly distributed normal stress is

$$S = P/A$$

When the stress is not uniformly distributed, $S = dP/dA$.

A **long rod** will stretch under its own weight G and a terminal load P (see Fig. 5.2.4). The total elongation e is that due to the terminal load plus that due to one-half the weight of the rod considered as acting at the end.

$$e = (Pl + Gl/2)/(AE)$$

The maximum stress is at the upper end.

Fig. 5.2.3   **Fig. 5.2.4**   **Fig. 5.2.5**

When a **load** is **carried by several paths** to a support, the different paths take portions of the load in proportion to their stiffness, which is controlled by material *(E)* and by design.

EXAMPLE.   Two pairs of bars rigidly connected (with the same elongation) carry a load P_0 (Fig. 5.2.5). A_1, A_2 and E_1, E_2 and P_1, P_2 and S_1, S_2 are cross sections, moduli of elasticity, loads, and stresses of the bars, respectively; e = elongation.

$$e = P_1 l/(E_1 A_1) = P_2 l/(E_2 A_2)$$

$$P_0 = 2P_1 + 2P_2$$

$$S_2 = P_2/A_2 = \tfrac{1}{2}[P_0 E_2/(E_1 A_1 + E_2 A_2)]$$

$$S_1 = \tfrac{1}{2}[P_0 E_1/(E_1 A_1 + E_2 A_2)]$$

Shearing stresses (Fig. 5.2.2) act tangentially to surface of contact and do not change length of sides of elementary volume; they **change the angle between faces** and the **length of diagonal.** Two pairs of shearing stresses must act together. **Shearing stress intensities are of equal magnitude on all four faces of an element.** $S_v = S'_v$ (Fig. 5.2.6).

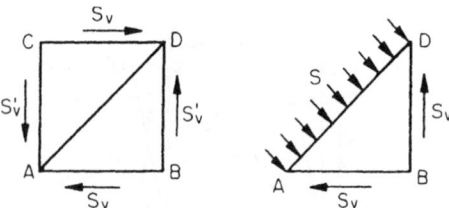

Fig. 5.2.6

In the presence of **pure shear** on external faces (Fig. 5.2.6), the **resultant stress** S on one diagonal plane at 45° is pure tension and on the other diagonal plane pure compression; $S = S_v = S'_v$. Failure under pure shear is difficult to produce experimentally, except under torsion and in certain special cases. Figure 5.2.7 shows an ideal case, and Fig. 5.2.8 a common form of test piece that introduces bending stresses.

Fig. 5.2.7   **Fig. 5.2.8**

Let Fig. 5.2.9 represent the symmetric section of area A with a shearing force V acting through its centroid. If pure shear exists, $S_v = V/A$, and this shear would be uniformly distributed over the area A. When this **shear is accompanied by bending (transverse shear in beams),** the unit

Fig. 5.2.9

shear S_v increases from the extreme fiber to its maximum, which may or may not be at the neutral axis OZ. The unit shear parallel to OZ at a point d distant from the neutral axis (Fig. 5.2.9) is

$$S_v = \frac{V}{Ib} \int_d^e yz \, dy$$

where $z =$ the section width at distance y; and I is the moment of inertia of the **entire** section about the neutral axis OZ. Note that $\int_d^e yz \, dy$ is the first moment of the area above d with respect to axis OZ. For a **rectangular cross section** (Fig. 5.2.10a),

$$S_v = \frac{3}{2} \frac{V}{bh} \left[1 - \left(\frac{2y}{h} \right)^2 \right]$$

$$S_v(\text{max}) = \frac{3}{2} \frac{V}{bh} = \frac{3}{2} \frac{V}{A} \qquad \text{for } y = 0$$

For a **circular cross section** (Fig. 5.2.10b),

$$S_v = \frac{4}{3} \frac{V}{\pi r^2} \left[1 - \left(\frac{y}{r} \right)^2 \right]$$

$$S_v(\text{max}) = \frac{4}{3} \frac{V}{\pi r^2} = \frac{4}{3} \frac{V}{A} \qquad \text{for } y = 0$$

For a **circular ring** (thickness small in comparison with the major diameter), $S_v(\text{max}) = 2V/A$, for $y = 0$.

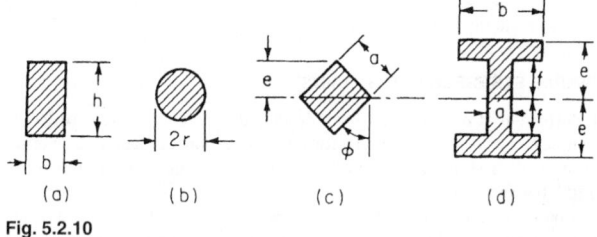

Fig. 5.2.10

For a **square cross section** (diagonal vertical, Fig. 5.2.10c),

$$S_v = \frac{V\sqrt{2}}{a^2} \left[1 + \frac{y\sqrt{2}}{a} - 4 \left(\frac{y}{2} \right)^2 \right]$$

$$S_v(\text{max}) = 1.591 \frac{V}{A} \qquad \text{for } y = \frac{e}{4}$$

For an **I-shaped cross section** (Fig. 5.2.10d),

$$S_v(\text{max}) = \frac{3}{4} \frac{V}{a} \left[\frac{be^2 - (b - a)f^2}{be^3 - (b - a)f^3} \right] \qquad \text{for } y = 0$$

Consider a beam of uniform cross section that is subjected to a bending moment M. **Bending stress,** or the flexure formula, of the beam is $S = -My/I$, where y is the distance from the neutral axis and I is the moment of inertia of the cross section. For a circular cross section beam that is subjected to **torsion** M_t, pure shear stress in the beam is $S_s = M_t r/J$, where r is the radius of the cross section.

Elasticity is the ability of a material to return to its original dimensions after the removal of stresses. The **elastic limit** S_p is the limit of stress within which the deformation completely disappears after the removal of stress; i.e., no set remains.

Hooke's law states that, within the elastic limit, deformation produced is proportional to the stress. Unless modified, the deduced formulas of mechanics apply only within the elastic limit. Beyond this, they are modified by experimental coefficients, as, for instance, the modulus of rupture.

The **modulus of elasticity,** lb/in² (kgf/cm²), is the ratio of the increment of unit stress to increment of unit deformation within the elastic limit.

The modulus of elasticity in tension, or **Young's modulus,**

$$E = \text{unit stress/unit deformation} = Pl/(Ae)$$

The modulus of elasticity in compression is similarly measured.

The **modulus of elasticity in shear** or **coefficient of rigidity,** $G = S_v/\alpha$ where α is expressed in radians (see Fig. 5.2.2).

The **bulk modulus of elasticity** K is the ratio of normal stress, applied to all six faces of a cube, to the change of volume.

Change of volume under normal stress is so small that it is rarely of significance. For example, given a body with length l, width b, thickness d, Poisson's ratio μ, and longitudinal strain ε, $V = lbd = $ original volume. The deformed volume $= (1 + \varepsilon)l(1 - \mu\varepsilon)b(1 - \mu\varepsilon)d$. Neglecting powers of ε, the deformed volume $= (1 + \varepsilon - 2\mu\varepsilon)V$. The change in volume is $\varepsilon(1 - 2\mu)V$; the **unit volumetric strain** is $\varepsilon(1 - 2\mu)$. Thus, a steel rod ($\mu = 0.3$, $E = 30 \times 10^6$ lb/in^2) compressed to a stress of 30,000 lb/in^2 will experience $\varepsilon = 0.001$ and a unit volumetric strain of 0.0004, or 1 part in 2,500.

The following relationships exist between the modulus of elasticity in tension or compression E, modulus of elasticity in shear G, bulk modulus of elasticity K, and Poisson's ratio μ:

$$E = 2G(1 + \mu)$$

$$G = E/[2(1 + \mu)]$$

$$\mu = (E - 2G)/(2G)$$

$$K = E/[3(1 - 2\mu)]$$

$$\mu = (3K - E)/(6K)$$

Resilience U (in · lb)[(cm · kgf)] is the potential energy stored up in a deformed body. The amount of resilience is equal to the work required to deform the body from zero stress to stress S, when S does not exceed the elastic limit. For normal stress, resilience = work of deformation = average force times deformation = $\tfrac{1}{2}Pe = \tfrac{1}{2}AS \times Sl/E = \tfrac{1}{2} S^2V/E$.

Modulus of resilience U_p (in · lb/in^3) [(cm · kgf/cm^3)], or **unit resilience,** is the **elastic energy** stored up in a cubic inch of material at the elastic limit. For normal stress,

$$U_p = \tfrac{1}{2}S_p^2/E$$

The unit resilience for any other kind of stress, as shearing, bending, torsion, is a constant times one-half of the square of the stress divided by the appropriate modulus of elasticity. For values, see Table 5.2.1.

Unit rupture work U_R, sometimes called **ultimate resilience,** is measured by the area of the stress-deformation diagram to rupture.

$$U_R = \tfrac{1}{3} e_u(S_y + 2S_M) \qquad \text{approx}$$

where e_u is the total deformation at rupture.

For structural steel, $U_R = \tfrac{1}{3} \times {}^{27}\!/_{100} \times [35,000 + (2 \times 60,000)] = 13,950$ in · lb/in^3 (982 cm · kgf/cm^3).

EXAMPLE 1.   A load $P = 40,000$ lb compresses a wooden block of cross-sectional area $A = 10$ in^2 and length = 10 in, in an amount $e = {}^4\!/_{100}$ in. Stress $S = {}^1\!/_{10} \times 40,000 = 4,000$ lb/in^2. Unit elongation $s = {}^4\!/_{100} \div 10 = {}^1\!/_{250}$. Modulus of elasticity $E = 4,000 \div {}^1\!/_{250} = 1,000,000$ lb/in^2. Unit resilience $U_p = \tfrac{1}{2} \times 4,000 \times 4,000/1,000,000 = 8$ in · lb/in^3 (0.563 cm · kgf/cm^3).

EXAMPLE 2.   A weight $G = 5,000$ lb falls through a height $h = 2$ ft; $V =$ number of cubic inches required to absorb the shock without exceeding a stress of 4,000 lb/in^2. Work done by falling weight = $Gh = 5,000 \times 2 \times 12$ in · lb (2,271 $\times$ 61 cm · kgf) Resilience of block = $V \times 8$ in · lb = $5,000 \times 2 \times 12$. Therefore, $V = 15,000$ in^3 (245,850 cm^3).

Thermal Stresses   A bar will change its length when its temperature is raised (or lowered) by the amount $\Delta l_0 = \alpha l_0(t_2 - 32)$. The linear **coefficient of thermal expansion** α is assumed constant at normal temperatures and l_0 is the length at 32°F (273.2 K, 0°C). If this expansion (or contraction) is prevented, a **thermal-time stress** is developed, equal to $S = E\alpha(t_2 - t_1)$, as the temperature goes from t_1 to t_2. In thin flat plates the stress becomes $S = E\alpha(t_2 - t_1)/(1 - \mu)$; μ is Poisson's ratio. Such stresses can occur in castings containing large and small sections. Similar stresses also occur when heat flows through members because of the difference in temperature between one point and another. The heat flowing across a length b as a result of a linear drop in temperature Δt equals $Q = kA \, \Delta t/b$ Btu/h (cal/h). The thermal conductivity k is in Btu/(h)(ft^2)(°F)/(in of thickness) [cal/(h)(m^2)(k)/(m)]. The **thermal-flow stress** is then $S = E\alpha Qb/(kA)$. Note, when Q is substituted the stress becomes $S = E\alpha \, \Delta t$ as above, only t is now a function of distance rather than time. For coefficient of thermal expansion α, see Sec. 4.

EXAMPLE.   A cast-iron plate 3 ft square and 2 in thick is used as a fire wall. The temperature is 330°F on the hot side and 160°F on the other. What is the thermal-flow stress developed across the plate?

$$S = E\alpha \, \Delta t = 13 \times 10^6 \times 6.5 \times 10^{-6} \times 170$$

$$= 14,360 \text{ lb/in}^2 \ (1,010 \text{ kgf/cm}^2)$$

or

$$Q = 2.3 \times 9 \times 170/2 = 1,760 \text{ Btu/h}$$

and

$$S = 13 \times 10^6 \times 6.5 \times 10^{-6} \times 1,760 \times 2/2.3 \times 9$$

$$= 14,360 \text{ lb/in}^2 \ (1,010 \text{ kgf/cm}^2)$$

COMBINED STRESSES

In the discussion that follows, the element is subjected to stresses lying in one plane; this is the case of **plane stress,** or **two-dimensional stress.**

Simple stresses, defined as such by the flexure and torsion theories, lie in planes normal or parallel to the line of action of the forces. Normal, as well as shearing, stresses may, however, exist in other directions. A particle out of a loaded member will contain normal and shearing stresses as shown in Fig. 5.2.11. Note that the four shearing stresses must be of the same magnitude, if equilibrium is to be satisfied.

If the particle is "cut" along the plane AA, equilibrium will reveal that, in general, normal as well as shearing stresses act upon the plane

Table 5.2.1   Resilience per Unit of Volume U_p

(S = longitudinal stress; S_v = shearing stress; E = tension modulus of elasticity; G = shearing modulus of elasticity)

Tension or compression	$\tfrac{1}{2}S^2/E$	Torsion	
Shear	$\tfrac{1}{2}S_v^2/G$	Solid circular	$\tfrac{1}{4}S_v^2/G$
Beams (free ends)			
Rectangular section, bent in arc of circle; no shear	$\tfrac{1}{6}S^2/E$	Hollow, radii R_1 and R_2	$\dfrac{R_1^2 + R_2^2}{R_1^2} \dfrac{1}{4} \dfrac{S_v^2}{G}$
Ditto, circular section	$\tfrac{1}{8}S^2/E$	Springs	
Concentrated center load; rectangular cross section	$\tfrac{1}{18}S^2/E$	Carriage	$\tfrac{1}{6}S^2/E$
		Flat spiral, rectangular section	$\tfrac{1}{24}S^2/E$
Ditto, circular cross section	$\tfrac{1}{24}S^2/E$	Helical: axial load, circular wire	$\tfrac{1}{4}S_v^2/G$
Uniform load, rectangular cross section	$\tfrac{5}{36}S^2/E$	Helical: axial twist	$\tfrac{1}{8}S^2/E$
1-beam section, concentrated center load	$\tfrac{3}{32}S^2/E$	Helical: axial twist, rectangular section	$\tfrac{1}{6}S^2/E$

AC (Fig. 5.2.12). The normal stress on plane AC is labeled S_n, and shearing S_s. The application of equilibrium yields

$$S_n = \frac{S_x + S_y}{2} + \frac{S_x - S_y}{2}\cos 2\theta + S_{xy}\sin 2\theta$$

and

$$S_s = \frac{S_x - S_y}{2}\sin 2\theta - S_{xy}\cos 2\theta$$

Fig. 5.2.11   Fig. 5.2.12

A *sign convention* must be used. A tensile stress is positive while compression is negative. A shearing stress is positive when directed as on plane AB of Fig. 5.2.12; i.e., when the shearing stresses on the vertical planes form a clockwise couple, the stress is positive.

The planes defined by $\tan 2\theta = 2S_{xy}/S_x - S_y$, the *principal planes*, contain the **principal stresses**—the maximum and minimum normal stresses. These stresses are

$$S_M, S_m = \frac{S_x + S_y}{2} \pm \sqrt{\left(\frac{S_x - S_y}{2}\right)^2 + S_{xy}^2}$$

The maximum and minimum shearing stresses are represented by the quantity

$$S_{sM,m} = \pm\sqrt{\left(\frac{S_x - S_y}{2}\right)^2 + S_{xy}^2}$$

and they act on the planes defined by

$$\tan 2\theta = -\frac{S_x - S_y}{2S_{xy}}$$

EXAMPLE.   The steam in a boiler subjects a particular particle on the outer surface of the boiler shell to a circumferential stress of 8,000 lb/in² and a longitudinal stress of 4,000 lb/in² as shown in Fig. 5.2.13. Find the stresses acting on the plane XX, making an angle of 60° with the direction of the 8,000 lb/in² stress. Find the principal stresses and locate the principal planes. Also find the maximum and minimum shearing stresses.

Fig. 5.2.13

$$60°\ S_n = \frac{4,000 + 8,000}{2} + \frac{4,000 - 8,000}{2}(-0.5000) + 0$$

$$= 7,000\ \text{lb/in}^2$$

$$60°\ S_s = \frac{4,000 - 8,000}{2}(0.8660) - 0 = -1,732\ \text{lb/in}^2$$

$$S_{M,m} = \frac{4,000 + 8,000}{2} \pm \sqrt{\left(\frac{4,000 - 8,000}{2}\right)^2 + 0}$$

$$= 6,000 \pm 2,000$$

$$= 8,000 \text{ and } 4,000\ \text{lb/in}^2\ (564 \text{ and } 282\ \text{kgf/cm}^2)$$

at $\tan 2\theta = \dfrac{2 \times 0}{4,000 - 8,000} = 0$ or $\theta = 90°$ and $0°$

$$S_{sM,m} = \pm\sqrt{\left(\frac{4,000 - 8,000}{2}\right)^2 + 0}$$

$$= \pm 2,000\ \text{lb/in}^2\ (\pm 141\ \text{kgf/cm}^2)$$

Mohr's Stress Circle   The biaxial stress field with its combined stresses can be represented graphically by the Mohr stress circle. For instance, for the particle given in Fig. 5.2.11, Mohr's circle is as shown in Fig. 5.2.14. The stress *sign convention* previously defined must be adhered to. Furthermore, in order to locate the point (on Mohr's circle) that yields the stresses on a plane $\theta°$ from the vertical side of the particle (such as plane AA in Fig. 5.2.11), $2\theta°$ must be laid off in the same

Fig. 5.2.14

direction from the radius to (S_x, S_{xy}). For the previous example, Mohr's circle becomes Fig. 5.2.15.

Eight special stress fields are shown in Figs. 5.2.16 to 5.2.23, along with Mohr's circle for each.

Fig. 5.2.15

Combined Loading   Combined flexure and torsion arise, for instance, when a shaft twisted by a torque M_t is bent by forces produced by belts or gears. An element on the surface, such as $ABCD$ on the shaft of Fig. 5.2.24, is subjected to a **flexure stress** $S_x = Mc/I = 8Fl/(\pi d^3)$ and

Fig. 5.2.16

Fig. 5.2.17

Fig. 5.2.18

Fig. 5.2.19

Fig. 5.2.20

Fig. 5.2.21

Fig. 5.2.22

Fig. 5.2.23

a **torsional shearing stress** $S_{xy} = M_t c/J = 16 M_t (\pi d^3)$. These stresses will induce combined stresses. The maximum combined stresses will be

$$S_n = \tfrac{1}{2}(S_x \pm \sqrt{S_x^2 + 4S_{xy}^2})$$

and

$$S_s = \pm \tfrac{1}{2}\sqrt{S_x^2 + 4S_{xy}^2}$$

The above situation appllies to any case of normal stress with shear, as when a bolt is under both tension and shear. A beam subjected to both flexure and transverse shear is another case.

Fig. 5.2.24

Combined torsion and longitudinal loads exist on a propeller shaft. An element on this shaft will contain a tensile stress computed using $S = F/A$ and a torsion shearing stress equal to $S_s = M_t c/J$. The free body of an element on the surface of a vertical turbine shaft is subjected to direct compression and torsion.

Fig. 5.2.25

When combined loading results in stresses of the same type and direction, the addition is algebraic. Such a situation exists on an offset link like that of Fig. 5.2.25.

Mohr's Strain Circle Strain equations can also be derived for plane-strain fields. Strains e_x and e_y are the extensional strains (tension or

compression) occurring at a point in two right-angle directions, and the change of the angle between them is γ_{xy}. The strain e at the point in any direction a at an angle θ with the x direction derives as

$$e_a = \frac{e_x + e_y}{2} + \frac{e_x - e_y}{2}\cos 2\theta + \frac{\gamma_{xy}}{2}\sin 2\theta$$

Similarly, the shearing strain γ_{ab} (change in the original right angle between directions a and b) is defined by

$$\gamma_{ab} = (e_x - e_y)\sin 2\theta + \gamma_{xy}\cos 2\theta$$

Inspection easily reveals that the above equations for e_a and γ_{ab} are mathematically identical to those for S_n and S_s. Thus, once a sign convention is established, a Mohr circle for strain can be constructed and used as the stress circle is used. The strain e is positive when an extension and negative when a contraction. If the direction associated with the first subscript a rotates counterclockwise during straining with respect to the direction indicated by the second subscript b, the shearing strain is positive; if clockwise, it is negative. In constructing the circle, positive extensional strains will be plotted to the right as abscissas and positive **half-shearing** strains will be plotted upward as ordinates.

For the strains shown in Fig. 5.2.26a, Mohr's strain circle becomes that shown in Fig. 5.2.26b. The extensional strain in the direction a, making an angle of θ_a with the x direction, is e_a, and the shearing strain is γ_{ab} counterclockwise. The strain 90° away is e_b. The maximum principal strain is e_M at an angle θ_M clockwise from the x direction. The other principal or minimum strain is e_m 90° away.

Fig. 5.2.26

PLASTIC DESIGN

Early efforts in stress analysis were based on limit loads, that is, loads which stress a member "wholly" to the yield strength. Euler's famous paper on column action ("Sur la Force des Colonnes," Academie des Sciences de Berlin, 1757) deals with the column problem this way. More recently, the concept of limit loads, referred to as **limit**, or **plastic**, **design**, has found strong application in the design of certain structures. The theory presupposes a ductile material, absence of stress raisers, and fabrication free of embrittlement. Local load overstress is allowed, provided the structure does not deform appreciably.

To visualize the limit-load approach, consider a simple beam of uniform section subjected to a concentrated load of midspan, as depicted in Fig. 5.2.27a. According to elastic theory, the outermost fiber on each side and at midspan—the section of maximum bending moment—will first reach the yield-strength value, Across the depth of the beam, the stress distribution will, of course, follow the triangular pattern, becoming zero at the neutral axis. If the material is ductile, the stress in the outermost fibers will remain at the yield value until every other fiber reaches the same value as the load increases. Thus the stress distribution assumes the rectangular pattern before the *plastic hinge* forms and failure ensues.

The problem is that of finding the final limit load. Elastic-flexure theory gives the maximum load—triangular distribution—as

$$F_y = \frac{2S_y bh^2}{3l}$$

For the rectangular stress distribution, the limit load becomes

$$F_L = \frac{S_y bh^2}{l}$$

The ratio $F_L/F_y = 1.50$—an increase of 50 percent in load capability. The ratio F_L/F_y has been named **shape factor** (Jenssen, Plastic Design in Welded Structures Promises New Economy and Safety, *Welding Jour.*, Mar. 1959). See Fig. 5.2.27b for shape factors for some other sections. The shape factor may also be determined by dividing the first moment of area about the neutral axis by the section modulus.

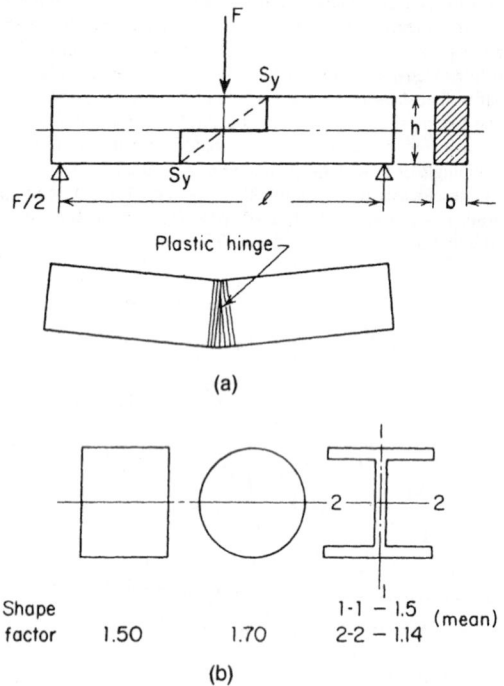

Fig. 5.2.27

A constant-section beam with both ends fixed, supporting a uniformly distributed load, illustrates another application of the plastic-load approach. The bending-moment diagram based on the elastic theory drawn in Fig. 5.2.28 (broken line) shows a moment at the center equal to one-half the moment at either end. A preferable situation, it might be argued, is one in which the moments are the same at the three stations—solid line. Thus, applying equilibrium to, say, the left half of the beam yields a bending moment at each of the three plastic hinges of

$$M_L = \frac{wl^2}{16}$$

DESIGN STRESSES

If a machine part is to safely transmit loads acting upon it, a permissible maximum stress must be established and used in the design. This is the allowable stress, the working stress, or preferably, the **design stress.** The design stress should not waste material, yet should be large enough to prevent failure in case loads exceed expected values, or other uncertainties react unfavorably.

The design stress is determined by dividing the applicable material property—yield strength, ultimate strength, fatigue strength—by a **factor of safety.** The factor should be selected only after all **uncertainties** have

Fig. 5.2.28

been thoroughly considered. Among these are the uncertainty with respect to the magnitude and kind of operating load, the reliability of the material from which the component is made, the assumptions involved in the theories used, the environment in which the equipment might operate, the extent to which localized and fabrication stresses might develop, the uncertainty concerning causes of possible failure, and the endangering of human life in case of failure. Factors of safety vary from industry to industry, being the result of accumulated experience with a class of machines or a kind of environment. Many codes, such as the ASME code for power shafting, recommend design stresses found safe in practice.

In general, the **ductility** of the material determines the property upon which the factor should be based. Materials having an elongation of over 5 percent are considered ductile. In such cases, the factor of safety is based upon the yield strength or the endurance limit. For materials with an elongation under 5 percent, the ultimate strength must be used because these materials are **brittle** and so fracture without yielding.

Factors of safety based on yield are often taken **between 1.5 and 4.0.** For more reliable materials or well-defined design and operating conditions, the lower factors are appropriate. In the case of untried materials or otherwise uncertain conditions, the larger factors are safer. The same values can be used when loads vary, but in such cases they are applied to the fatigue or endurance strength. When the ultimate strength determines the design stress (in the case of brittle materials), the factors of safety can be doubled.

Thus, under static loading, the design stress for, say, SAE 1020, which has a yield strength of 45,000 lb/in² (3,170 kgf/cm²) may be taken at 45,000/2, or 22,500 lb/in² (1,585 kgf/cm²), if a reasonably certain design condition exists. A Class 30 cast-iron part might be designed at 30,000/5 or 6,000 lb/in² (423 kgf/cm²). A 2017S-0 aluminum-alloy component (13,000 lb/in² endurance strength) could be computed at a design stress of 13,000/2.5 or 5,200 lb/in² (366 kgf/cm²) in the usual fatigue-load application.

BEAMS

For properties of structural steel and wooden beams, see Sec. 12.2.

Notation

I = rectangular moment of inertia
I_p = polar moment of inertia
I/c = section modulus
M = bending moment
P, W' = concentrated load
Q or V = total vertical shear
R = reaction
S = unit normal stress

S_s or S_v = transverse shearing stress
W = total distributed load
f = deflection
i = slope
l = distance between supports
r = radius of gyration
r_c = radius of curvature
w = distributed load per longitudinal unit

A **simple beam** rests on supports at its ends which permit rotation. A **cantilever beam** is fixed (no rotation) at one end. When computing reactions and moments, distributed loads may be replaced by their resultants acting at the center of gravity of the distributed-load area.

Reactions are the forces and/or couples acting at the supports and holding the beam in place. In general, the weight of the beam should be accounted for.

The **bending moment** (pound-feet or pound-inches) (kgf · m) at any section is the algebraic sum of the external forces and moments acting on the beam on one side of the section. It is also equal to the moment of the internal-stress forces at the section, $M = \int s\, dA/y$. A bending moment that bends a beam convex downward (tensile stress on bottom fiber) is considered **positive,** while convex upward (compression on bottom) is **negative.**

The **vertical shear** V (lb) (kgf) effective on a section is the algebraic sum of all the forces acting parallel to and on one side of the section, $V = \Sigma F$. It is also equal to the sum of the transverse shear stresses acting on the section, $V = \int S_s\, dA$.

Moment and **shear diagram** may be constructed by plotting to scale the particular entity as the ordinate for each section of the beam. Such diagrams show in continuous form the variation along the length of the beam.

Moment-Shear Relation The shear V is the first derivative of moment with respect to distance along the beam, $V = dM/dx$. This relationship does not, however, account for any sudden changes in moment.

Fig. 5.2.29

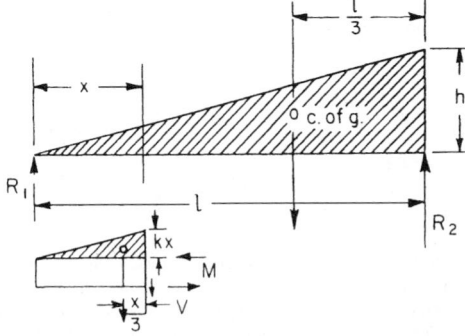

Fig. 5.2.30

EXAMPLES. Figure 5.2.29 illustrates a simple beam subjected to a uniform load. $M = R_1 x - wx \times \dfrac{x}{2} = \dfrac{w/x}{2} - \dfrac{wx^2}{2}$ and $V = R_1 - wx = \dfrac{wl}{2} - wx$. Note

also that $V = \dfrac{d}{dx}\left(\dfrac{wlx}{2} - \dfrac{wx^2}{2} \right) = \dfrac{wl}{2} - wx$.

Figure 5.2.30 is a simple beam carrying a uniformly varying load; $M =$

$R_1 x - h\dfrac{x}{1} \times \dfrac{x}{2} \times \dfrac{x}{3} = \dfrac{hlx}{6} - \dfrac{hx^3}{6l}$, if h is in pounds per foot and weight

of beam is neglected. The vertical shear $V = R_1 - \dfrac{hx}{1} \times \dfrac{x}{2} = \dfrac{hl}{6} - \dfrac{hx^2}{2l}$.

Note again that $V = \dfrac{d}{dx}\left(\dfrac{hlx}{6} - \dfrac{hx^3}{6l} \right) = \dfrac{hl}{6} - \dfrac{hx^2}{2l}$.

Table 5.2.2 gives the reactions, bending-moment equations, vertical shear equations, and the deflection of some of the more common types of beams.

Maximum Safe Load on Steel Beams See Table 5.2.3   To obtain maximum safe load (or maximum deflection under maximum safe load) for any of the conditions of loading given in Table 5.2.5, multiply the corresponding coefficient in that table by the greatest safe load (or deflection) for distributed load for the particular section under consideration as given in Table 5.2.4.

The following approximate factors for reducing the load should be used when beams are long in comparison with their breadth:

Ratio of unsupported (lateral) length to flange width or breadth	20	30	40	50	60	70
Ratio of greatest safe load to calculated load	1	0.9	0.8	0.7	0.6	0.5

Theory of Flexure A bent beam is shown in Fig. 5.2.31. The concave side is in compression and the convex side is in tension. These are divided by the **neutral plane** of zero stress $A'B'BA$. The intersection of the neutral plane with the face of the beam is in the **neutral line or elastic curve** AB. The intersection of the neutral plane with the cross section is the **neutral axis** NN'.

Fig. 5.2.31

It is assumed that a beam is prismatic, of a length at least 10 times its depth, and that the external forces are all at right angles to the axis of the beam and in a plane of symmetry, and that flexure is slight. Other assumptions are: (1) That the material is homogeneous, and obeys Hooke's law. (2) That **stresses are within the elastic limit.** (3) That every layer of material is free to expand and contract longitudinally and laterally under stress as if separate from other layers. (4) That the tensile and compressive moduli of elasticity are equal. (5) That the cross section remains a plane surface. (The assumption of plane cross sections is strictly true only when the shear is constant or zero over the cross section, and when the shear is constant throughout the length of the beam.) It follows then that: (1) The internal forces are in horizontal balance. (2) The **neutral axis contains the center of gravity** of the cross section, where there is no resultant axial stress. (3) The stress intensity varies directly with the distance from the neutral axis.

The moment of the elastic forces about the neutral axis, i.e., the **stress moment** or **moment of resistance,** is $M = SI/c$, where S is an elastic unit stress at outer fiber whose distance from the neutral axis is c; and I is the rectangular moment of inertia about the neutral axis. I/c is the **section modulus.**

This formula is for the **strength of beams.** For rectangular beams, $M = \frac{1}{6}\, Sbh^2$, where b = breadth and h = depth; i.e., the elastic **strength of beam sections** varies as follows: (1) for equal width, as the square of the depth; (2) for equal depth, directly as the width; (3) for equal depth

Table 5.2.2   Beams of Uniform Cross Section, Loaded Transversely

and width, directly as the strength of the material; (4) if span varies, then for equal depth, width, and material, inversely as the span.

If a beam is cut in halves vertically, the two halves laid side by side each will carry only one-half as much as the original beam.

Tables 5.2.6 to 5.2.8 give the properties of various beam cross sections. For properties of structural-steel shapes, see Sec. 12.2.

Oblique Loading   It should be noted that Table 5.2.6 includes certain cases for which the horizontal axis is not a neutral axis, assuming the common case of vertical loading. The rectangular section with the diagonal as a horizontal axis (Table 5.2.6) is such a case. These cases must be handled by the principles of oblique loading.

Every section of a beam has two principal axes passing through the center of gravity, and these two axes are always at right angles to each other. The principal axes are axes with respect to which the moment of inertia is, respectively, a maximum and a minimum, and for which the product of inertia is zero. For symmetrical sections, axes of symmetry are always principal axes. For unsymmetrical sections, like a **rolled angle** section (Fig. 5.2.32), the inclination of the principal axis with the X axis may be found from the formula $\tan 2\theta = 2I_{xy}/(I_y - I_x)$, in which θ = angle of inclination of the principal axis to the X axis, I_{xy} = the product of inertia of the section with respect to the X and Y axes, I_y = moment of inertia of the section with respect to the Y axis, I_x = moment of inertia

Table 5.2.2   Beams of Uniform Cross Section, Loaded Transversely   (*Continued*)

$$R_2 = W = wl$$

$$M_x = -\frac{wx^2}{2}$$

$$M_{max} = -\frac{wl^2}{2}, \ (x = l)$$

$$Q_x = -wx$$

$$Q_{max} = -wl, \ (x = l)$$

$$f = \frac{W}{EI}\frac{l^3}{8} \ (\text{max})$$

$$R_1 = \frac{W}{2} = \frac{wl}{2}$$

$$R_2 = \frac{W}{2} = \frac{wl}{2}$$

$$M_x = \frac{wx}{2}(l - x)$$

$$M_{max} = \frac{wl^2}{8}, (x = \tfrac{1}{2}l)$$

$$Q_x = \frac{wl}{2} - wx$$

$$Q_{max} = \frac{wl}{2}, (x = 0)$$

$$f = \frac{W}{EI}\frac{5l^3}{384} \ (\text{max})$$

$$R_1 = \frac{3}{8}W = \frac{3}{8}wl$$

$$R_2 = \frac{5}{8}W = \frac{5}{8}wl$$

$$M_x = \frac{wx}{2}\left(\frac{3}{4}l - x\right)$$

$$M_{max} = \frac{9}{128}wl^2, \left(x = \frac{3}{8}l\right)$$

$$M_{max} = -\frac{wl^2}{8}, (x = l)$$

$$Q_x = \frac{3}{8}wl - wx$$

$$Q_{max} = -\frac{5}{8}wl$$

$$f = \frac{W}{EI}\frac{l^3}{185} \ (\text{max})$$

$$R_1 = \frac{W}{2} = \frac{wl}{2}, \ R_2 = \frac{W}{2} = \frac{wl}{2}$$

$$M_x = -\frac{wl^2}{2}\left(\frac{1}{6} - \frac{x}{l} + \frac{x^2}{l^2}\right)$$

$$M_{max} = -\frac{1}{12}wl^2, (x = 0, \text{ or } x = l)$$

$$Q_x = \frac{wl}{2} - wx$$

$$Q_{max} = \pm\frac{wl}{2}$$

$$f = \frac{W}{EI}\frac{l^3}{384} \ (\text{max})$$

$$R_2 = W = \text{total load}$$

$$M_x = -\frac{W}{3}\frac{x^3}{l^2}$$

$$M_{max} = -\frac{Wl}{3}$$

$$Q_x = -\frac{Wx^2}{l^2}$$

$$Q_{max} = -W$$

$$f = \frac{W}{EI}\frac{l^3}{15} \ (\text{max})$$

$$R_1 = \frac{1}{3}W, \ R_2 = \frac{2}{3}W$$

$$M_x = \frac{Wx}{3}\left(1 - \frac{x^2}{l^2}\right)$$

$$M_{max} = \frac{2}{9\sqrt{3}}Wl, \left(x = \frac{1}{\sqrt{3}}\right)$$

$$Q_x = W\left(\frac{1}{3} - \frac{x^2}{l^2}\right)$$

$$Q_{max} = -\frac{2}{3}W, (x = l)$$

$$f = 0.01304\frac{Wl^3}{EI} \ (\text{max})$$

Table 5.2.2 Beams of Uniform Cross Section, Loaded Transversely (*Continued*)

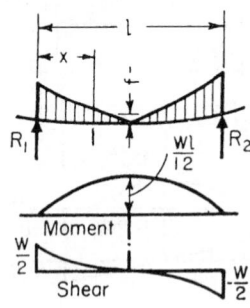

$$R_1 = \frac{W}{2}, R_2 = \frac{W}{2}$$

$$M_x = Wx\left(\frac{1}{2} - \frac{x}{l} + \frac{2x^2}{3l^2}\right)$$

$$M_{max} = \frac{Wl}{12}, \left(x = \frac{1}{2}l\right)$$

$$Q_x = W\left(\frac{1}{2} - \frac{2x}{l} + \frac{2x^2}{l^2}\right)$$

$$Q_{max} = \pm\frac{W}{2}, (x = 0)$$

$$f = \frac{W}{EI}\frac{3l^3}{320} \text{ (max)}$$

$$R_1 = \frac{W}{2}, R_2 = \frac{W}{2}$$

$$M_x = Wx\left(\frac{1}{2} - \frac{2x^2}{3l^2}\right)$$

$$M_{max} = \frac{Wl}{6}, \left(x = \frac{1}{2}l\right)$$

$$Q_x = W\left(\frac{1}{2} - \frac{2x^2}{l^2}\right)$$

$$Q_{max} = \pm\frac{W}{2}, (x = 0)$$

$$f = \frac{W}{EI}\frac{l^3}{60} \text{ (max)}$$

$$R_1 = \frac{W}{5}, R_2 = \frac{4W}{5}$$

$$M_x = Wx\left(\frac{1}{5} - \frac{x^2}{3l^2}\right)$$

$$M_{max} = -\frac{2}{15}Wl \text{ at support 2}$$

$$Q_x = W\left(\frac{1}{5} - \frac{x^2}{l^2}\right)$$

$$Q_{max} = -\frac{4W}{5}$$

$$f = \frac{16Wl^3}{1,500\sqrt{5}EI}$$

$$= \frac{0.00477\,Wl^3}{EI} \text{ (max)}$$

$$R_1 = \frac{W}{2} = \frac{wl}{2}, R_2 = \frac{W}{2} = \frac{wl}{2}$$

$$M_x = \frac{Wx}{2}\left(1 - \frac{c}{x} - \frac{x}{l}\right), (x > c)$$

$$M_x = -\frac{Wx^2}{2l}, (x \leq c)$$

$$M_{max} = \frac{Wl}{4}\left(\frac{1}{2} - \frac{2c}{l}\right), c \leq \left(\frac{\sqrt{2}-1}{2}\right)l$$

$$Q_x = \frac{W}{2} - wx\,(x > c)$$

$$Q_x = -wx\,(x \leq c)$$

Concentrated load W'
Uniformly dist. load $W = wl$

$$R_1 = W'\frac{c_1^2(3c + 2c_1)}{2l^3} + \frac{3}{8}W$$

$$R_2 = W'\frac{(2c^2 + 6cc_1 + 3c_1^2)c}{2l^3} + \frac{5}{8}W$$

$$M_2 = W'\frac{cc_1(2c + c_1)}{2l^2} + W\left(\frac{1}{8}\right)$$

$$M_{W'} = W'\frac{cc_1^2(3c + 2c_1)}{2l^3} + W\frac{(3c_1 - c)c}{8l}$$

(a) $\dfrac{W'}{W} < \dfrac{l^2}{4c_1^2}\dfrac{5c - 3c_1}{3c + 2c_1}$

$$M_{c\,max} = \frac{R_1^2}{2W}l, \left(x = \frac{R_1 l}{W}\right)$$

(b) $\dfrac{W'}{W} < \dfrac{l^2(3c_1 - 5c)}{4c(2c^2 + 6cc_1 + 3c_1^2)}$

$$M_{c_1\,max} = W'c + \frac{(R_1 - W')^2}{2W}l, \left(x = \frac{R_1 - W'}{W}l\right)$$

Deflection under W'

$$f = \frac{W'}{EI}\frac{c^2c_1^3(4c + 3c_1)}{12l^3} + \frac{W}{EI}\frac{cc_1^2(3c + c_1)}{48l}$$

Table 5.2.2   Beams of Uniform Cross Section, Loaded Transversely   (*Continued*)

Concentrated load W'
Uniformly dist. load $W = wl$; $c < c_1$

$$R_1 = W'\frac{c_1}{l} + \frac{W_2}{2}$$

$$R_2 = W'\frac{c}{l} + \frac{W}{2}$$

(a) $\dfrac{W'}{W} < \dfrac{c_1 - c}{2c}$

$$M_{max} = R_2\frac{x_1}{2} = \frac{R_2^2 l}{2W'}\left(x_1 = \frac{R_2 l}{W}\right)$$

(b) $\dfrac{W'}{W} > \dfrac{c_1 - c}{2c}$

$$M_{max} = \left(W' + \frac{W}{2}\right)\frac{cc_1}{l}, (x_1 = c_1)$$

Deflection of beam under W':

$$f = \left(W' + \frac{l^2 + cc_1}{8cc_1}W\right)\frac{c^2c_1^2}{3EIl}$$

$c < c_1$

$$R_1 = W'\frac{(3c + c_1)c_1^2}{l^3} + \frac{W}{2}$$

$$R_2 = W'\frac{(c + 3c_1)c^2}{l^3} + \frac{W}{2}$$

$$M_{max} = M_1 = W'\frac{cc_1^2}{l^2} + \frac{Wl}{12}$$

Deflection under W'

$$f = \frac{1}{EI}\left(W'\frac{c^3c_1^3}{3l^3} + W\frac{c^2c_1^2}{24l}\right)$$

Table 5.2.3   Uniformally Distributed Loads on Simply Supported Rectangular Beams 1-in Wide*
(Laterally Supported Sufficiently to Prevent Buckling)
[Calculated for unit fiber stress at 1,000 lb/in² (70 kgf/cm²): nominal size]
Total load in pounds (kgf)† including the weight of beam

Span, ft (m)‡	Depth of beam, in (cm)§										
	6	7	8	9	10	11	12	13	14	15	16
5	800	1,090	1,420	1,800	2,220	2,690	3,200	3,750	4,350	5,000	5,690
6	670	910	1,180	1,500	1,850	2,240	2,670	3,130	3,630	4,170	4,740
7	570	780	1,010	1,290	1,590	1,920	2,280	2,680	3,110	3,570	4,060
8	500	680	890	1,120	1,390	1,680	2,000	2,350	2,720	3,130	3,560
9	440	600	790	1,000	1,230	1,490	1,780	2,090	2,420	2,780	3,160
10	400	540	710	900	1,110	1,340	1,600	1,880	2,180	2,500	2,840
11	360	490	650	820	1,010	1,220	1,450	1,710	1,980	2,270	2,590
12	330	450	590	750	930	1,120	1,330	1,560	1,810	2,080	2,370
13	310	420	550	690	850	1,030	1,230	1,440	1,680	1,920	2,190
14	290	390	510	640	790	960	1,140	1,340	1,560	1,790	2,030
15	270	360	470	600	740	900	1,070	1,250	1,450	1,670	1,900
16	250	340	440	560	690	840	1,000	1,170	1,360	1,560	1,780
17	230	320	420	530	650	790	940	1,100	1,280	1,470	1,670
18	220	300	400	500	620	750	890	1,040	1,210	1,390	1,580
19	210	290	380	470	590	710	840	990	1,150	1,320	1,500
20	200	270	360	450	560	670	800	940	1,090	1,250	1,420
22	180	250	320	410	500	610	730	850	990	1,140	1,290
24	160	230	290	370	460	560	670	780	910	1,040	1,180
26	150	210	270	340	420	520	610	720	840	960	1,090
28	140	190	250	320	390	480	570	670	780	890	1,010
30	130	180	240	300	370	450	530	630	730	830	950

* This table is convenient for wooden beams. For any other fiber stress S', multiply the values in table by $S'/1,000$. See Sec. 12.2 for properties of wooden beams of commercial sizes.
† To change to kgf, multiply by 0.454.
‡ To change to m, multiply by 0.305.
§ To change to cm, multiply by 2.54.

of the section with respect to the X axis. When this principal axis has been found, the other principal axis is at right angles to it.

Calling the moments of inertia with respect to the principal axes I'_x and I'_y, the unit stress existing anywhere in the section at a point whose coordinates are x and y (Fig. 5.2.33) is $S = My\cos\alpha/I'_x + Mx\sin\alpha/I'_y$, in which M = bending moment with respect to the section in question, α = the angle which the plane of bending moment or the plane of the loads makes with the y axis, $M\cos\alpha$ = the component of bending moment causing bending about the principal axis which has been designated as the X axis, $M\sin\alpha$ = the component of bending moment causing bending about the principal axis which has been designated as the Y axis. The sign of the two terms for unit stress may be determined by inspection in the usual way, and the result will be tension or compression as determined by the algebraic sum of the two terms.

Table 5.2.4 Approximate Safe Loads in Pounds (kgf) on Steel Beams,* Simply Supported, Single Span

Allowable fiber stress for steel, 16,000 lb/in^2 (1,127 kgf/cm^2) (basis of table)

Beams simply supported at both ends.

L = distance between supports, ft (m) a = interior area, in^2 (cm^2)
A = sectional area of beam, in^2 (cm^2) d = interior depth, in (cm)
D = depth of beam, in (cm) w = total working load, net tons (kgf)

| | Greatest safe load, lb | | Deflection, in | |
Shape of section	Load in middle	Load distributed	Load in middle	Load distributed
Solid rectangle	$\dfrac{890AD}{L}$	$\dfrac{1,780AD}{L}$	$\dfrac{wL^3}{32AD^2}$	$\dfrac{wL^3}{52AD^2}$
Hollow rectangle	$\dfrac{890(AD - ad)}{L}$	$\dfrac{1,780(AD - ad)}{L}$	$\dfrac{wL^3}{32(AD^2 - ad^2)}$	$\dfrac{wL^3}{52(AD^2 - ad^2)}$
Solid cylinder	$\dfrac{667AD}{L}$	$\dfrac{1,333AD}{L}$	$\dfrac{wL^3}{24AD^2}$	$\dfrac{wL^3}{38AD^2}$
Hollow cylinder	$\dfrac{667(AD - ad)}{L}$	$\dfrac{1,333(AD - ad)}{L}$	$\dfrac{wL^3}{24(AD^2 - ad^2)}$	$\dfrac{wL^3}{38(AD^2 - ad^2)}$
I beam	$\dfrac{1,795AD}{L}$	$\dfrac{3,390AD}{L}$	$\dfrac{wL^3}{58AD^2}$	$\dfrac{wL^3}{93AD^2}$

Fig. 5.2.32

Fig. 5.2.33

In general, it may be stated that when the plane of the bending moment coincides with one of the principal axes, the other principal axis is the neutral axis. This is the ordinary case, in which the ordinary formula for unit stress may be applied. When the plane of the bending moment does not coincide with one of the principal axes, the above formula for oblique loading may be applied.

Internal Moment Beyond the Elastic Limit

Ordinarily, the expression $M = SI/c$ is used for stresses above the elastic limit, in which case S becomes an experimental coefficient S_R, the **modulus of rupture,** and the formula is empirical. The true relation is obtained by applying to the cross section a stress-strain diagram from a tension and compression test, as in Fig. 5.2.34. Figure 5.2.34 shows the side of a beam of depth d under flexure beyond its elastic limit; line 1–1

Table 5.2.5 Coefficients for Correcting Values in Table 5.2.4 for Various Methods of Support and of Loading, Single Span

Conditions of loading	Max relative safe load	Max relative deflection under max relative safe load
Beam supported at ends:		
Load uniformly distributed over span	1.0	1.0
Load concentrated at center of span	½	0.80
Two equal loads symmetrically concentrated	$l/4c$	
Load increasing uniformly to one end	0.974	0.976
Load increasing uniformly to center	¾	0.96
Load decreasing uniformly to center	½	1.08
Beam fixed at one end, cantilever:		
Load uniformly distributed over span	¼	2.40
Load concentrated at end	⅛	3.20
Load increasing uniformly to fixed end	⅜	1.92
Beam continuous over two supports equidistant from ends:		
Load uniformly distributed over span		
1. If distance $a > 0.2071l$	$l^2/(4a^2)$	
2. If distance $a < 0.2071l$	$\dfrac{l}{l - 4a}$	
3. If distance $a = 0.2071l$	5.83	
Two equal loads concentrated at ends	$l/(4a)$	

NOTE: l = length of beam; c = distance from support to nearest concentrated load; a = distance from support to end of beam.

Table 5.2.6 Properties of Various Cross Sections*

(I = moment of inertia; I/c = section modulus; $r = \sqrt{I/A}$ = radius of gyration)

Rectangle (vertical):
$$I = \frac{bh^3}{12}$$
$$\frac{I}{c} = \frac{bh^2}{6}$$
$$r = \frac{h}{\sqrt{12}} = 0.289h$$

Rectangle (N.A. at base):
$$\frac{bh^3}{3}$$
$$\frac{bh^2}{3}$$
$$\frac{h}{\sqrt{3}} = 0.577h$$

Rectangle (diagonal axis):
$$\frac{b^3 h^3}{6(b^2 + h^2)}$$
$$\frac{b^2 h^2}{6\sqrt{b^2 + h^2}}$$
$$\frac{bh}{\sqrt{6(b^2 + h^2)}}$$

Rectangle (inclined at angle a):
$$\frac{bh}{12}(h^2\cos^2 a + b^2\sin^2 a)$$
$$\frac{bh}{6}\left(\frac{h^2\cos^2 a + b^2\sin^2 a}{h\cos a + b\sin a}\right)$$
$$\sqrt{\frac{h^2\cos^2 a + b^2\sin^2 a}{12}}$$

Hollow rectangle:
$$I = \frac{b}{12}(H^3 - h^3)$$
$$\frac{I}{c} = \frac{b}{6}\frac{H^3 - h^3}{H}$$
$$r = \sqrt{\frac{H^3 - h^3}{12(H - h)}}$$

Hollow square:
$$\frac{H^4 - h^4}{12}$$
$$\frac{1}{6}\frac{H^4 - h^4}{H}$$
$$\sqrt{\frac{H^2 + h^2}{12}}$$

Hollow square (diagonal):
$$\frac{H^4 - h^4}{12}$$
$$\frac{\sqrt{2}}{12}\frac{H^4 - h^4}{H}$$
$$\sqrt{\frac{H^2 + h^2}{12}}$$

Triangle (apex up):
$$\frac{bh^3}{36}; \quad c = \frac{2}{3}h$$
$$\frac{bh^2}{24}$$
$$\frac{h}{\sqrt{18}}$$

Triangle (N.A.):
$$I = \frac{bh^3}{12}$$
$$\frac{I}{c} = \frac{bh^2}{12}$$
$$r = \frac{h}{\sqrt{6}}$$

Hexagon (flat sides, width R):
$$\frac{5}{8}R^3$$

Hexagon:
$$\frac{5\sqrt{3}}{16}R^4$$
$$\frac{5\sqrt{3}}{16}R^3$$
$$\sqrt{\frac{5}{24}}R$$

Octagon:
$$\frac{1 + 2\sqrt{2}}{6}R^4$$
$$0.6906R^3$$
$$0.475R$$

NOTE: Square, axis same as first rectangle, side = h; $I = h^4/12$; $I/c = h^3/6$; $r = 0.289h$.
Square, diagonal taken as axis: $I = h^4/12$; $I/c = 0.1179h^3$; $r = 0.289h$.

Fig. 5.2.34

shows the distorted cross section; line 3–3, the usual rectilinear relation of stress to strain; and line 2–2, an actual stress-strain diagram, applied to the cross section of the beam, compression above and tension below. The neutral axis is then below the gravity axis. The **outer material** may be expected to develop **greater ultimate strength** than in simple stress, because of the reinforcing action of material nearer the neutral axis that is not yet overstrained. This leads to an **equalization of stress** over the cross section. S_R exceeds the ultimate strength S_M in tension as follows: for cast iron, $S_R = 2S_M$; for sandstone, $S_R = 3S_M$; for concrete, $S_R = 2.2S_M$; for wood (green), $S_R = 2.3S_M$.

In the case of steel I beams, failure begins practically when the elastic limit in the compression flange is reached.

Because of the support of adjoining material, the **elastic limit in flexure** S_p is also greater than in tension, depending upon the relation of breadth to depth of section. For the same breadth, the difference decreases with

Table 5.2.6 Properties of Various Cross Sections* *(Continued)*

Equilateral Polygon A = area R = rad circumscribed circle r = rad inscribed circle n = no sides a = length of side Axis as in preceding section of octagon	$I = \dfrac{A}{24}(6R^2 - a^2)$ $= \dfrac{A}{48}(12r^2 + a^2)$ $= \dfrac{AR^2}{4}$ (approx)	$\dfrac{I}{c} = \dfrac{I}{r}$ $= \dfrac{I}{R\cos\dfrac{180°}{n}}$ $= \dfrac{AR}{4}$ (approx)	$\sqrt{\dfrac{1}{A}} = \sqrt{\dfrac{6R^2 - a^2}{24}} \approx \dfrac{R}{2}$ $= \sqrt{\dfrac{12r^2 + a^2}{48}}$
	$I = \dfrac{6b^2 + 6bb_1 + b_1^2}{36(2b + b_1)}h^3$ $c = \dfrac{1}{3}\dfrac{3b + 2b_1}{2b + b_1}h$	$\dfrac{I}{c} = \dfrac{6b^2 + 6bb_1 + b_1^2}{12(3b + 2b_1)}h^2$	$r = \dfrac{h\sqrt{12b^2 + 12bb_1 + 2b_1^2}}{6(2b + b_1)}$
		$I = \dfrac{BH^3 + bh^3}{12}$ $\dfrac{I}{c} = \dfrac{BH^3 + bh^3}{6H}$	$r = \sqrt{\dfrac{BH^3 + bh^3}{12(BH + bh)}}$
		$I = \dfrac{BH^3 - bh^3}{12}$ $\dfrac{I}{c} = \dfrac{BH^3 - bh^3}{6h}$	$r = \sqrt{\dfrac{BH^3 - bh^3}{12(BH - bh)}}$
	$I = \frac{1}{3}(BC_1^3 - B_1h^3 + bc_3^3 - b_1h_1^3)$ $c_1 = \dfrac{1}{2}\dfrac{aH^2 + B_1d^2 + b_1d_1(2H - d_1)}{aH + B_1d + b_1d_1}$		$r = \sqrt{\dfrac{I}{Bd + bd_1 + a(h + h_1)}}$
	$I = \frac{1}{3}(Bc_1^3 - bh^3 + ac_2^3)$ $c_1 = \dfrac{1}{2}\dfrac{aH^2 + bd^2}{aH + bd}$ $c_2 = H - c_1$		$r = \sqrt{\dfrac{I}{Bd + a(H - d)}}$
	$I = \dfrac{\pi d^4}{64} = \dfrac{\pi r^4}{4} = \dfrac{A}{4}r^2$ $= 0.05d^4$ (approx)	$\dfrac{I}{c} = \dfrac{\pi d^3}{32} = \dfrac{\pi r^3}{4} = \dfrac{A}{4}r$ $= 0.1d^3$ (approx)	$\sqrt{\dfrac{I}{A}} = \dfrac{r}{2} = \dfrac{d}{4}$

increase of height. No difference will occur in the case of an I beam, or with hard materials.

Wide plates will not expand and contract freely, and the value of E will be increased on account of side constraint. As a consequence of lateral contraction of the fibers of the tension side of a beam and lateral swelling of fibers at the compression side, the cross section becomes distorted to a trapezoidal shape, and the neutral axis is at the center of gravity of the trapezoid. Strictly, this shape is one with a curved perimeter, the radius being r_c/μ, where r_c is the radius of curvature of the neutral line of the beam, and μ is Poisson's ratio.

Deflection of Beams

When a beam is subjected to bending, the fibers on one side elongate, while the fibers on the other side shorten (Fig. 5.2.35). These changes in length cause the beam to deflect. All points on the beam except those directly over the support fall below their original position, as shown in Figs. 5.2.31 and 5.2.35.

The **elastic curve** is the curve taken by the neutral axis. The radius of curvature at any point is

$$r_c = EI/M$$

Table 5.2.6   Properties of Various Cross Sections*   (*Continued*)

$$I = \frac{64}{105}(b_1 h_1^2 - b_2 h_2^3), \text{ where}$$

$h_1 = \frac{1}{2}(H + t)$	$b_1 = \frac{1}{4}(B + 2.6t)$
$h_2 = \frac{1}{2}(H - t)$	$b_2 = \frac{1}{4}(B - 2.6t)$

$$\frac{I}{c} = \frac{2I}{H + t}$$

$$r = \sqrt{\frac{3I}{t(2B + 5.2H)}}$$

Corrugated sheet iron, parabolically curved

Approximate values of *least* radius of gyration r

$r =$	$0.36336D$	$0.295D$	$D/4.58$ (I-beam)	$D/3.54$ (Channel) $D/6$ (Deck beam)

$r =$	$D/4.74$ (T-beam)	$D/5$ (Angle Equal legs)	$BD/[2.6(B + D)]$ (Angle Unequal legs) $D/4.74$ (Cross)

* Some of the cross sections depicted in this table will be encountered most often in machinery as castings, forgings, or individual sections assembled and joined mechanically (or welded). A number of the sections shown are obsolete and will be encountered mainly in older equipment and/or building structures.

expressed in radians. A second integration gives the vertical deflection of any point of the elastic curve from its original position.

EXAMPLE.   In the cantilever beam shown in Fig. 5.2.35, the bending moment at any section $= -P(l - x) = EI\, d^2y/(dx)^2$. Integrate and determine constant by the condition that when $x = 0$, $dy/dx = 0$. Then $EI\, dy/dx = -Plx + \frac{1}{2}Px^2$. Integrate again; and determine constant by the condition that when $x = 0$, $y = 0$. Then $EIy = -\frac{1}{2}Plx^2 + Px^3/6$. This is the equation of the elastic curve. When $x = l$, $y = f = -Pl^3/(3EI)$. In general, the two constants of integration must be determined simultaneously.

Deflection in general, f, may be expressed by the equation $f = Pl^3/(mEI)$, where m is a coefficient. See Tables 5.2.2 and 5.2.4 for values of f for beams of various sections and loadings. For coefficients of deflection of wooden beams and structural steel shapes, see Sec. 12.2.

Since I varies as the cube of the depth, the **stiffness,** or inverse deflection, of various **beams** varies, other factors remaining constant, inversely as the load, inversely as the cube of the span, and directly as the cube of the depth. This deflection is due to bending moment only. In general, however, the bending of beams involves transverse shearing stresses which cause **shearing strains** and thus **add to the total deflection.** The contribution of shearing strain to overall deflection becomes significant only when the beam span is very short. These strains may affect substantially the strength as well as the deflection of beams. When deflection due to transverse shear is to be accounted for, the differential equation of the elastic curve takes the form

$$EI\frac{d^2y}{dx^2} = EI\left(\frac{d^2y_b}{dx^2} + \frac{d^2y_s}{dx^2}\right) = M - \frac{kEI}{AG} \times \frac{d^2M}{dx^2}$$

where k is a factor dependent upon the beam cross section. Sergius Sergev, in "The Effect of Shearing Forces on the Deflection and Strength of Beams" (*Univ. Wash. Eng. Exp. Stn. Bull.* 114) gives $k = 1.2$ for rectangular sections, 10/9 for circular sections, and 2.4 for I beams. He also points out that in the case of a deep, rectangular-section cantilever, carrying a concentrated load at the free end, the deflection due to shear may be up to 3.1 percent of that due to bending moment; if this beam supports a uniformly distributed load, it may be up to 4.1 percent. A deep, simple beam deflection may increase up to 15.6 percent when carrying a uniformly distributed load and up to 12.5 percent when the load is concentrated at midspan.

Design of beams may be based on **strength** (stress) or on **stiffness** if deflection must be limited. Deflection rather than stress becomes the criterion

for design, e.g., of machine tools, for which the relative positions of tool and workpiece must be maintained while the cutting loads are applied during operation. Similarly, large steam-turbine shafts supported on two end bearings must maintain alignment and tight critical clearances between the rotating blade assemblies and the stationary stator blades during operation. When more than one beam shares a load, each beam will assume a portion of the load that is proportional to its stiffness. **Superposition** may be used in connection with both stresses and deflections.

EXAMPLE.   (Fig. 5.2.36). Two wooden stringers—one *(A)* 8 × 16 in in cross section and 20 ft in span, the other *(B)* 8 in × 8 in × 16 ft—carrying the center load $P_0 = 22,000$ lb are required, the load carried by each stringer. The deflections f of the two stringers must be equal. Load on $A = P_1$, and on $B = P_2$. $f = P_1 l_1^3/(48EI_1) = P_2 l_2^3/(48EI_2)$. Then $P_1/P_2 = l_2^3 I_1/(l_1^3 I_2) = 4$. $P_0 = P_1 + P_2 = 4P_2 + P_2$, whence $P_2 = 22,000/5 = 4,400$ lb (1,998 kgf) and $P_1 = 4 \times 4,400 = 17,600$ lb (7,990 kgf).

Fig. 5.2.36

Relation between Deflection and Stress

Combine the formula $M = SI/c = Pl/n$, where n is a constant, $P =$ load, and $l =$ span, with formula $f = Pl^3/(mEI)$, where m is a constant. Then

$$f = C''SI^2/(Ec)$$

where C'' is a new constant $= n/m$. Other factors remaining the same, the **deflection varies directly as the stress and inversely as** E. If the span is

Table 5.2.7

Beam	Load	n	m	C''
Cantilever	Concentrated at end	1	3	$\frac{1}{3}$
Cantilever	Uniform	2	8	$\frac{1}{4}$
Simple	Concentrated at center	4	48	$\frac{1}{12}$
Simple	Uniform	8	384/5	$\frac{5}{48}$
Fixed ends	Concentrated at center	8	192	$\frac{1}{24}$
Fixed ends	Uniform	12	384	$\frac{1}{32}$
One end fixed, one end supported	Concentrated at center	16/3	768/7	$\frac{7}{144}$
One end fixed, one end supported	Uniform	128/9	185	$\frac{1}{13}$
Simple	Uniformly varying, maximum at center	6	60	$\frac{1}{10}$

constant, a shallow beam will submit to greater deformations than a deeper beam without exceeding a safe stress. If depth is constant, a beam of double span will attain a given deflection with only one-quarter the stress. Values of n, m, and C'' are given in Table 5.2.7 (for other values, see Table 5.2.2).

Graphical Relations

Referring to Fig. 5.2.37, the shear V acting at any section is equal to the total load on the right of the section, or

$$V = \int w\,dx$$

Since $w\,dx$ is the product of w, a loading intensity (which is expressed as a vertical height in the load diagram), and dx, an elementary length along the horizontal, evidently $w\,dx$ is the area of a small vertical strip of the **load diagram**. Then $\int w\,dx$ is the summation of all such vertical strips between two indefinite points. Thus, to obtain the shear in any

Fig. 5.2.37

section mn, find the area of the load diagram up to that section, and draw a second diagram called the **shear diagram**, any ordinate of which is proportional to the shear, or to the area in the load diagram to the right of mn. Since $V = dM/dx$,

$$\int V\,dx = M$$

By similar reasoning, a **moment diagram** may be drawn, such that the ordinate at any point is proportional to the area of the shear diagram to the right of that point. Since $M = EI\, d^2f/dx^2$,

$$\int M\,dx = EI\,(df/dx + C) = EI(i + C)$$

if I is constant. Here C is a constant of integration. Thus i, the slope or grade of the elastic curve at any point, is proportional to the area of the

moment diagram $\int M\,dx$ up to that point; and a **slope diagram** may be derived from the moment diagram in the same manner as the moment diagram was derived from the shear diagram.

If I is not constant, draw a new curve whose ordinates are M/I and use these M/I ordinates just as the M ordinates were used in the case where I was constant; that is, $\int (M/I)\,dx = E(i + C)$. The ordinate at any point of the slope curve is thus proportional to the area of the M/I curve to the right of that point. Again, since $iE = E\,df/dx$.

$$\int iE\,dx = \int E\,df = E(f + C)$$

and thus the ordinate f to the elastic curve at any point is proportional to the area of the slope diagram $\int i\,dx$ up to that point. The equilibrium polygon may be used in drawing the **deflection curve** directly from the M/I diagram.

Thus, the five curves of load, shear, moment, slope, and deflection are so related that each curve is derived from the previous one by a process of graphical integration, and with proper regard to scales the deflection is thereby obtained.

The vertical distance from any point A (Fig. 5.2.38) on the elastic curve of a beam to the tangent at any other point B equals the moment of the area of the $M/(EI)$ diagram from A to B about A. This distance, the **tangential deviation** t_{AB}, may be used with the slope-area relation and the geometry of the elastic curve to obtain deflections. These theorems, together with the equilibrium equations, can be used to compute reactions in the case of statically **indeterminate beams**.

Fig. 5.2.38

EXAMPLE. The deflections of points B and D (Fig. 5.2.38) are

$$y_B = -t_{AB} = \text{moment area}\ \frac{M}{EI}\Big|_A^B$$

$$= -\frac{1}{EI} \times \frac{Pl}{4} \times \frac{l}{4} \times \frac{l}{3} = -\frac{Pl^3}{48EI}$$

$$\theta_c = \nabla\theta\Big|_B^C = \text{area}\ \frac{M}{EI}\Big|_B^C = \frac{1}{EI} \times \frac{Pl}{4} \times \frac{l}{4} = \frac{Pl^2}{16EI}$$

$$y_D = -\left(\theta_c \times \frac{l}{4} - t_{DC}\right)$$

$$= -\frac{Pl^2}{16EI} \times \frac{l}{4} \times \frac{l}{EI} \times \frac{Pl}{8} \times \frac{l}{8} \times \frac{l}{12} = -\frac{11Pl^3}{768EI}$$

Resilience of Beams

The external work of a load gradually applied to a beam, and which increases from zero to P, is $\frac{1}{2}Pf$ and equals the **resilience** U. But, from the formulas $P = nSl/(cl)$ and $f = nSl^2/(mcE)$, where n and m are constants that depend upon loading and supports, S = fiber stress, c = distance from neutral axis to outer fiber, and l = length of span. Substitute for P and f, and

$$U = \frac{n^2}{m}\left(\frac{k}{c}\right)^2 \frac{S^2 V}{2E}$$

where k is the radius of gyration and V the volume of the beam. For values of U, see Table 5.2.1.

The resilience of beams of similar cross section at a given stress is proportional to their volumes. The **internal resilience,** or the elastic deformation energy in the material of a beam in a length x is dU, and

$$U = \frac{1}{2}\int M^2\, dx/(EI) = \frac{1}{2}\int M\, di$$

where M is the moment at any point x, and di is the angle between the tangents to the elastic curve at the ends of dx. The values of resilience and deflection in special cases are easily developed from this equation.

Rolling Loads

Rolling or **moving loads** are those loads which may change their position on a beam. Figure 5.2.39 represents a beam with two equal concentrated moving loads, such as two wheels on a crane girder, or the wheels of a

Fig. 5.2.39

truck on a bridge. Since the maximum moment occurs where the shear is zero, it is evident from the shear diagram that the maximum moment will occur under a wheel. $x < a/2$:

$$R_1 = P\left(1 - \frac{2x}{l} + \frac{a}{l}\right)$$

$$M_2 = \frac{Pl}{2}\left(1 - \frac{a}{l} + \frac{2x}{l}\frac{a}{l} - \frac{4x^2}{l^2}\right)$$

$$R_2 = P\left(1 + \frac{2x}{l} - \frac{a}{l}\right)$$

$$M_1 = \frac{Pl}{2}\left(1 - \frac{a}{l} - \frac{2a^2}{l^2} + \frac{2x}{l}\frac{3a}{l} - \frac{4x^2}{l^2}\right)$$

$$M_2 \text{ max when } x = \tfrac{1}{4}a$$
$$M_1 \text{ max when } x = \tfrac{3}{4}a$$

$$M_{\max} = \frac{Pl}{2}\left(1 - \frac{a}{2l}\right)^2 = \frac{P}{2l}\left(l - \frac{a}{2}\right)^2$$

EXAMPLE.   Two wheel loads of 3,000 lb each, spaced on 5-ft centers, move on a span of $l = 15$ ft, $x = 1.25$ ft, and $R_2 = 2,500$ lb. $\therefore M_{\max} = M_2 = 2,500 \times 6.25$ $(1,135 \times 1.90) = 15,600$ lb $\cdot$ ft (2,159 kgf $\cdot$ m).

Figure 5.2.40 shows the condition when two equal loads are equally distant on opposite sides of the center. The moment is equal under the two loads.

If the **two moving loads** are **of unequal weight,** the condition for **maximum moment** is that the maximum moment will occur under the heavy wheel when the center of the beam bisects the distance between the resultant of the loads and the heavy wheel. Figure 5.2.41 shows the position and the shear and moment diagrams.

When **several wheel loads** constituting a system occur, the several suspected wheels must be examined in turn to determine which will cause the greatest moment. The **position for the greatest moment** that can occur under a given wheel is, as stated, when the center of the span bisects the distance between the wheel in question and the resultant of all the loads then on the span. The **position for maximum shear** at the support will be when one wheel is passing off the span.

Fig. 5.2.40

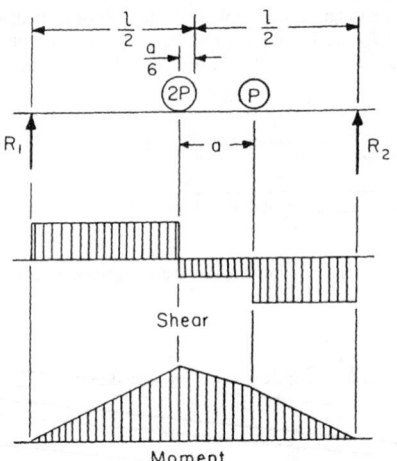

Fig. 5.2.41

Constrained Beams

Constrained beams are those so held or "built in" at one or both ends that the tangent to the elastic curve remains fixed in direction. These beams are held at the ends in such a manner as to allow free horizontal motion, as illustrated by Fig. 5.2.42. A constrained beam is stiffer than a simple beam of the same material, because of the modification of the moment by an end resisting moment. Figure 5.2.43 shows the two most common cases of constrained beams. See also Table 5.2.2.

Fig. 5.2.42   **Fig. 5.2.43**

Continuous Beams

A continuous beam is one resting upon several supports which may or may not be in the same horizontal plane. The general discussion for beams holds for continuous beams. $S_yA = V$, $SI/c = M$, and $d^2f/dx^2 = M/(EI)$. The **shear** at any section is equal to the algebraic sum of the components parallel to the section of all external forces on either side of the section. The bending moment at any section is equal to the moment of all external forces on either side of the section. The relations stated above between shear and moment diagrams hold true for continuous beams. The bending moment at any section is equal to the bending moment at any other section, plus the shear at that section times its arm, plus the product of all the intervening external forces times their respective arms. To illustrate (Fig. 5.2.44):

$$V_x = R_1 + R_2 + R_3 - P_1 - P_2 - P_3$$
$$M_x = R_1(l_1 + l_2 + x) + R_2(l_2 + x) + R_3x$$
$$\quad - P_1(l_2 + c + x) - P_2(b + x) - P_3a$$
$$M_x = M_3 + V_3x - P_3a$$

Table 5.2.8 gives the value of the moment at the various supports of a uniformly loaded continuous beam over equal spans, and it also gives the values of the shears on each side of the supports. Note that the shear is of opposite sign on either side of the supports and that the sum of the two shears is equal to the reaction.

Fig. 5.2.44

Figure 5.2.45 shows the relation between the moment and shear diagrams for a uniformly loaded continuous beam of four equal spans (see Table 5.2.8). Table 5.2.8 also gives the **maximum bending moment** which will occur **between supports**, and in addition the position of this moment and the points of inflection (see Fig. 5.2.46).

Figure 5.2.46 shows the values of the functions for a uniformly loaded continuous beam resting on three equal spans with four supports.

Continuous beams are **stronger and much stiffer than simple beams.** However, a small, unequal subsidence of piers will cause serious

Fig. 5.2.45

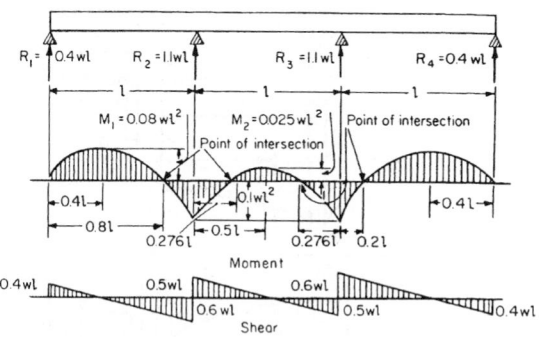

Fig. 5.2.46

Table 5.2.8 Uniformly Loaded Continuous Beams over Equal Spans
(Uniform load per unit length $= w$; length of equal span $= l$)

Number of supports	Notation of support of span	Shear on each side of support. L = left, R = right. Reaction at any support is $L + R$		Moment over each support	Max moment in each span	Distance to point of max moment, measured to right from support	Distance to point of inflection, measured to right from support
		L	R				
2	1 or 2	0	$\frac{1}{2}$	0	0.125	0.500	None
3	1	0	$\frac{3}{8}$	0	0.0703	0.375	0.750
	2	$\frac{5}{8}$	$\frac{5}{8}$	$\frac{1}{8}$	0.0703	0.625	0.250
4	1	0	$\frac{4}{10}$	0	0.080	0.400	0.800
	2	$\frac{6}{10}$	$\frac{5}{10}$	$\frac{1}{10}$	0.025	0.500	0.276, 0.724
5	1	0	$\frac{11}{28}$	0	0.0772	0.393	0.786
	2	$\frac{17}{28}$	$\frac{15}{28}$	$\frac{3}{28}$	0.0364	0.536	0.266, 0.806
	3	$\frac{13}{28}$	$\frac{13}{28}$	$\frac{2}{28}$	0.0364	0.464	0.194, 0.734
6	1	0	$\frac{15}{38}$	0	0.0779	0.395	0.789
	2	$\frac{23}{38}$	$\frac{20}{38}$	$\frac{4}{38}$	0.0332	0.526	0.268, 0.783
	3	$\frac{18}{38}$	$\frac{19}{38}$	$\frac{3}{38}$	0.0461	0.500	0.196, 0.804
7	1	0	$\frac{41}{104}$	0	0.0777	0.394	0.788
	2	$\frac{63}{104}$	$\frac{55}{104}$	$\frac{11}{104}$	0.0340	0.533	0.268, 0.790
	3	$\frac{49}{104}$	$\frac{51}{104}$	$\frac{8}{104}$	0.0433	0.490	0.196, 0.785
	4	$\frac{53}{104}$	$\frac{53}{104}$	$\frac{9}{104}$	0.0433	0.510	0.215, 0.804
8	1	0	$\frac{56}{142}$	0	0.0778	0.394	0.789
	2	$\frac{86}{142}$	$\frac{75}{142}$	$\frac{15}{142}$	0.0338	0.528	0.268, 0.788
	3	$\frac{67}{142}$	$\frac{70}{142}$	$\frac{11}{142}$	0.0440	0.493	0.196, 0.790
	4	$\frac{72}{142}$	$\frac{71}{142}$	$\frac{12}{142}$	0.0405	0.500	0.215, 0.785
Values apply to:		wl	wl	wl^2	wl^2	l	l

NOTE: The numerical values given are coefficients of the expressions at the foot of each column.

Table 5.2.9   Beams of Uniform Strength (in Bending)

Beam	Cross section	Elevation and plan	Formulas
1. Fixed at one end, load P concentrated at other end			
	Rectangle: width (b) constant, depth (g) variable	Elevation: 1, top, straight line; bottom, parabola. 2, complete parabola. Plan: rectangle	$y^2 = \dfrac{6P}{bS_s}x$ $h = \sqrt{\dfrac{6Pl}{bS_s}}$ Deflection at A: $f = \dfrac{8P}{bE}\left(\dfrac{l}{h}\right)^3$
	Rectangle: width (y) variable, depth (h) constant	Elevation: rectangle. Plan: triangle	$y = \dfrac{6P}{h^2 S_s}x$ $b = \dfrac{6Pl}{h^2 S_s}$ Deflection at A: $f = \dfrac{6P}{bE}\left(\dfrac{l}{h}\right)^3$
	Rectangle: width (z) variable, depth (y) variable $\dfrac{z}{y} = k$(const.)	Elevation: cubic parabola. Plan: cubic parabola	$y^3 = \dfrac{6P}{kS_s}x$ $z = ky$ $h = \sqrt[3]{\dfrac{6Pl}{kS_s}}$ $b = kh$
	Circle: diam (y) variable	Elevation: cubic parabola. Plan: cubic parabola	$y^3 = \dfrac{32P}{\pi S_s}x$ $d = \sqrt[3]{\dfrac{32Pl}{\pi S_s}}$
2. Fixed at one end, load of total magnitude P uniformly distributed			
	Rectangle width (b) constant, depth (y) variable	Elevation: triangle. Plan: rectangle	$y = x\sqrt{\dfrac{3P}{b/S}}$ $h = \sqrt{\dfrac{3Pl}{bS_s}}$ $f = 6\dfrac{P}{bE}\left(\dfrac{l}{h}\right)^3$
	Rectangle: width (y) variable, depth (h) constant	Elevation: rectangle. Plan: two parabolic curves with vertices at free end	$y = \dfrac{3Px^2}{lS_s h^2}$ $b = \dfrac{3Pl}{S_s h^2}$ Deflection at A: $f = \dfrac{3P}{bE}\left(\dfrac{l}{h}\right)^3$

Table 5.2.9   Beams of Uniform Strength (in Bending)   (*Continued*)

Beam	Cross section	Elevation and plan	Formulas

2. Fixed at one end, load of total magnitude P uniformly distributed

| | Rectangle: width (z) variable, depth (y) variable, $\dfrac{z}{y} = k$ | Elevation: semi-cubic parabola

Plan: semicubic parabola | $y^3 = \dfrac{3Px^2}{kS_sl}$
$z = ky$
$h = \sqrt[3]{\dfrac{3Pl}{kS_s}}$
$b = kh$ |

3. Supported at both ends, load P concentrated at point C

| | Rectangle: width (b) constant, depth (y) variable | Elevation: two parabolas, vertices at points of support

Plan: rectangle | $y = \sqrt{\dfrac{3P}{S_sb}}\,x$
$h = \sqrt{\dfrac{3Pl}{2bS_s}}$
$f = \dfrac{P}{2Eb}\left(\dfrac{l}{h}\right)^3$ |
| | Rectangle: width (y) variable, depth (h) constant | Elevation: rectangle
Plan: two triangles, vertices at points of support | $y = \dfrac{3P}{S_sh^2}\,x$
$b = \dfrac{3Pl}{2S_sh^2}$
$f = \dfrac{3Pl^3}{8Ebh^3}$ |

Load P moving across span

| | Rectangle: width (b) constant, depth (y) variable | Elevation: ellipse
Major axis $= l$
Minor axis $= 2h$
Plan: rectangle | $\dfrac{x^2}{\left(\dfrac{l}{2}\right)^2} + \dfrac{y^2}{\dfrac{3Pl}{2bS_s}} = 1$
$h = \sqrt{\dfrac{2Pl}{2bS_s}}$ |

4. Supported at both ends, load of total magnitude P uniformly distributed

| | Rectangle: width (b) constant, depth (y) variable | Elevation: ellipse

Plan: rectangle | $\dfrac{x^2}{\left(\dfrac{l}{2}\right)^2} + \dfrac{y^2}{\dfrac{3Pl}{4bS_s}} = 1$
$h = \sqrt{\dfrac{3Pl}{4bS_s}}$ |

Deflection at O:

$$f = \dfrac{1}{64}\dfrac{Pl^3}{EI}$$
$$= \dfrac{3}{16}\dfrac{P}{bE}\left(\dfrac{l}{h}\right)^3$$

changes in sign and magnitude of the bending stresses, reactions, and shears.

Maxwell's Theorem   When a number of loads rest upon a beam, the deflection at any point is equal to the sum of the deflections at this point due to each of the loads taken separately. Maxwell's theorem states that if unit loads rest upon a beam at two points A and B, the deflection at A due to the unit load at B equals the deflection at B due to the unit load at A.

Castigliano's theorem states that the deflection of the point of application of an external force acting on a beam is equal to the partial derivative of the work of deformation with respect to this force. Thus, if P is the force, f the deflection, and U the work of deformation, which equals the resilience,

$$dU/dP = f$$

According to the **principle of least work,** the deformation of any structure takes place in such a manner that the work of deformation is a minimum.

Similarly, angular rotation of a point in a structure θ can be determined by $dU/dM_t = \theta$, where M_t is the bending or torsional moment.

In applying Castigliano's theorem, the strain energy must be expressed as a function of the load. If the structure is subjected to combination of loads, for example, axial force N, bending moment M, shearing force V, and torsional moment M_t, the strain energy is

$$U = \int \frac{N^2\,dx}{2EA} + \int \frac{M^2\,dx}{2EI} + \int \frac{f_s V^2\,dx}{2GA} + \int \frac{M_t^2\,dx}{2GJ}$$

Note that the last integral is valid only for a circular cross section. Using Castigliano's theorem, the displacement f is given by

$$f = \frac{1}{EA}\int N\frac{\partial N}{\partial P}dx + \frac{1}{EI}\int M\frac{\partial M}{\partial P}dx + \frac{1}{GA}\int f_s V\frac{\partial V}{\partial P}dx + \frac{1}{GJ}\int M_t\frac{\partial M_t}{\partial P}dx$$

Beams of Uniform Strength

Beams of uniform strength vary in section so that the unit stress S remains constant, the I/c varies as M. For **rectangular beams,** of breadth b and depth d, $I/c = bd^2/6$; and $M = Sbd^2/6$. Thus, for a cantilever beam of rectangular cross section, under a load P, $Px = Sbd^2/6$. If b is constant, d^2 varies with x, and the profile of the shape of the beam will be a parabola, as Fig. 5.2.47. If d is constant, b will vary as x and the beam will be triangular in plan, as shown in Fig. 5.2.48.

Shear at the end of a beam necessitates a modification of the forms determined above. The area required to resist shear will be P/S_v in a cantilever and R/S_v in a simple beam. The dotted extensions in Figs. 5.2.47 and 5.2.48 show the changes necessary to enable these cantilevers to resist shear. The extra material and cost of fabrication, however, make many of the forms impractical.

Table 5.2.9 shows some of the simple **sections of uniform strength.** In none of these, however, is shear taken into account.

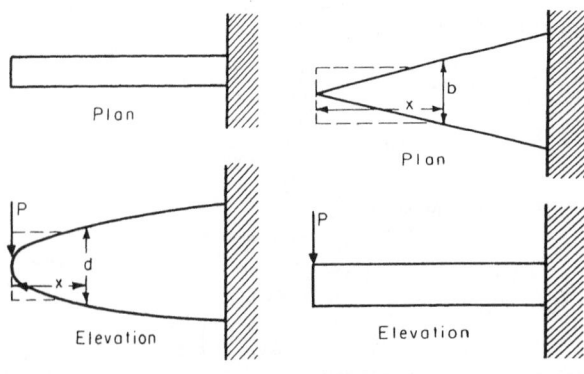

Fig. 5.2.47   **Fig. 5.2.48**

TORSION

Under torsion, a bar (Fig. 5.2.49) is twisted by a couple of magnitude Pp. Elements of the surface becomes helices of angle d, and a radius rotates through an angle θ in a length l, both d and θ being expressed in radians. S_v = shearing unit stress at distance r from center; I_p = *polar moment of inertia;* G = shearing modulus of elasticity. It is assumed that the cross sections remain plane surfaces. The strain on the cross section is wholly tangential, and is zero at the center of the section. Note that $ld = r\theta$.

In the case of a **circular cross section,** the stress S_v increases directly as the distance of the strained element from the center.

The polar moment of inertia I_p for any section may be obtained from $I_p = I_1 + I_2$, where I_1 and I_2 are the rectangular moments of inertia of the section about any two lines at right angles to each other, through the center of gravity.

The **external twisting moment** M_t is balanced by the internal resisting moment.

For strength, $M_t = S_v I_p/r$.
For stiffness, $M_t = aGI_p/l$.
The torsional resilience $U = \tfrac{1}{2}Ppa = S_v^2 I_p l/(2r^2 G) = a^2 GI_p/(2l)$.

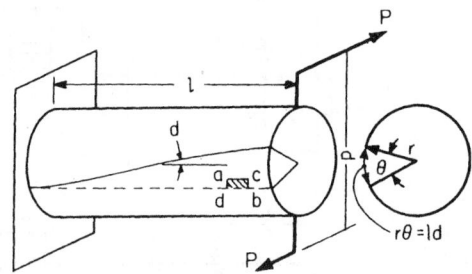

Fig. 5.2.49

The state of stress on an element taken from the surface of the shaft, as in Fig. 5.2.50, is pure shear. Pure tension exists at right angles to one 45° helix and pure compression at right angles to the opposite helix.

Reduced **formulas for shafts of various sections** are given in Table 5.2.11. When the ratio of shaft length to the largest lateral dimension in the cross section is less than approximately 2, the **end effects** may drastically affect the torsional stresses calculated.

Failure under torsion in brittle materials is a tensile failure at right angles to a helical element on the surface. Plastic materials twist off squarely. Fibrous materials separate in long strips.

Torsion of Noncircular Sections   When a section is not circular, the unit stress no longer varies directly as the distance from the center. Cross sections become warped, and the greatest unit stress usually occurs at a point on the perimeter of the cross section *nearest* the axis of twist; thus, there is no stress at the corners of square and rectangular sections. The analyses become complex for noncircular sections, and the methods for solution of design problems using them most often admit only of approximations.

Fig. 5.2.50

Torsion problems have been solved for many different noncircular cross sections by utilizing the **membrane analogy,** due to Prandtl, which makes use of the fact that the mathematical treatment of a twisted bar is governed by the same equations as for a membrane stretched over a hole which has the same shape as the bar and then inflated. The resulting three-dimensional surface provides the following: (1) The torque transmitted is proportional to twice the volume under the inflated membrane, and (2) the shear stress at any point is proportional to the slope of the curve measured perpendicular to that slope. In recent years, several other mathematical techniques have become widely used, especially with the aid of faster computational methods available from electronic computers.

By using **finite-difference methods,** the differential operators of the governing equations are replaced with difference operators which are related to the desired unknown values at a gridwork of points in the outline of the cross section being investigated.

The **finite-element method,** commonly referred to as **FEM,** deals with a spatial approximation of a complex shape which is then analyzed to determine deformations, stresses, etc. By using FEM, the exact structure is replaced with a set of simple structural elements interconnected at a finite number of nodes. The governing equations for the approximate structure can be solved exactly. Note that inasmuch as there is an exact solution for an approximate structure, the end result must be viewed, and the results thereof used, as approximate solutions to the real structure.

Using a finite-difference approach to Poisson's partial differential equation, which defines the stress functions for solid and hollow shafts with generalized contours, along with Prandl's membrane analogy, Isakower has developed a series of practical design charts (ARRAD-COMMISD Manual UN 80-5, January 1981, Department of the Army). Dimensionless charts and tables for transmitted torque and maximum shearing stress have been generated. Information for circular shafts with rectangular and circular keyways, external splines and milled flats, as well as rectangular and X-shaped torsion bars, is presented.

Assuming the stress distribution from the point of maximum stress to the corner to be parabolic, Bach derived the approximate expression, $S_s M = 9M_t/(2b^2 h)$ for a rectangular section, b by h, where $h > b$. For closer results, the shearing stresses for a **rectangular section** (Fig. 5.2.51) may be expressed $S_A = M_t/(\alpha_A b^2 h)$ and $S_B = M_t/(\alpha_B b^2 h)$. The angle of twist for these shafts is $\theta = M_t l/(\beta G b^3 h)$. The factors α_A, α_B, and β are functions of the ratio h/b and are given in Table 5.2.10.

In the case of **composite sections,** such as a tee or angle, the torque that can be resisted is $M_t = G\,\theta \Sigma \beta h b^3$; the summation applies to each of the rectangles into which the section can be divided. The maximum stress occurs on the component rectangle having the largest b value. It is computed from

$$S_A = M_t \beta_A b_A/(\alpha_A \Sigma \beta h b^3)$$

Torque, deflection, and work relations for some additional sections are given in Table 5.2.11.

Fig. 5.2.51

COLUMNS

Members subjected to direct compression can be grouped into three classes. **Compression blocks** are so short (slenderness ratios below 30) that bending of member is unlikely. At the other limit, columns so slender that bending is primary, are the **long columns** defined by Euler's theory. The intermediate columns, quite common in practice, are called **short columns.**

Long columns and the more slender short columns usually fail by **buckling** when the **critical load** is reached. This is a matter of **instability;** that is, the column may continue to yield and deflect even though the load is not increased above critical. The **slenderness ratio** is the unsupported length divided by the least radius of gyration, parallel to which it can bend.

Long columns are handled by **Euler's column formula,**

$$P_{cr} = n\pi^2 EI/l^2 = n\pi^2 EA/(l/r)^2$$

The **coefficient** n accounts for **end conditions.** When the column is pivoted at both ends, $n = 1$; when one end is fixed and other rounded, $n = 2$; when both are fixed, $n = 4$; and when one end is fixed with the other free, $n = \frac{1}{4}$. The slenderness ratio that separates long columns from short ones depends upon the modulus of elasticity and the yield strength of the column material. When Euler's formula results in $(P_{cr}/A) > S_y$, strength rather than buckling causes failure, and the column ceases to be long. In round numbers, this **critical slenderness ratio** falls between 120 and 150. Table 5.2.12 gives additional facts concerning long columns.

Short Columns The stress in a short column may be considered partly due to compression and partly due to bending. A theoretical equation has not been derived. Empirical, though rational, expressions are, in general, based on the assumption that the permissible stress must be reduced below that which could be permitted were it due to compression only. The manner in which this reduction is made determines the type of equation as well as the slenderness ratio beyond which the **equation does not apply.** Figure 5.2.52 illustrates the situation. Some typical formulas are given in Table 5.2.13.

EXAMPLE. A machine member unsupported for a length of 15 in has a square cross section 0.5 in on a side. It is to be subjected to compression. What maximum safe load can be applied centrally, according to the AISC formula? At the computed load, what size section (also square) would be needed, if it were to be designed according to the AREA formula?

$$l/r = 15/0.5/\sqrt{12} = 104 \qquad \therefore \text{short column}$$

$$P/A = 17,000 - 0.485(104)^2 = 11,730$$

or

$$P = 0.25 \times 11,730 = 2,940 \text{ lb } (1,335 \text{ kgf})$$

and

$$\frac{2,940}{a^2} \le 15,000 - 50\left(\frac{15la}{\sqrt{12}}\right) = 15,000 - \frac{2,600}{a}$$

thus

$$a^2 - 0.173a - 0.196 = 0 \quad \text{or} \quad a = 0.536 \text{ in } (1.36 \text{ cm})$$

Combined Flexure and Longitudinal Force Figure 5.2.53 shows a bar under flexure due to transverse and longitudinal loads. The maximum fiber stress S is made up of S_0, due to the direct action of load P, and S_b, due to the entire bending moment M. M is the algebraic sum of two bending moments, M_1 due to longitudinal load (+ for compression and − for tension), and M_2 due to transverse load. $M = M_2 \pm M_1$. Here $M_1 = Pf$ and $f = CS_b l^2/(Ec)$.

Table 5.2.10 Factors for Torsion of Rectangular Shafts (Fig. 5.2.51)

$\dfrac{h}{b}$	1.00	1.50	1.75	2.00	2.50	3.00	4.00	5.00	6.00	8.00	10.0	∞
α_A	0.208	0.231	0.239	0.246	0.258	0.267	0.282	0.291	0.299	0.307	0.312	0.333
α_B	0.208	0.269	0.291	0.309	0.336	0.355	0.378	0.392	0.402	0.414	0.421	
β	0.141	0.196	0.214	0.229	0.249	0.263	0.281	0.291	0.299	0.307	0.312	0.333

Table 5.2.11 Torsion of Shafts of Various Cross Sections
(For strength and stiffness of shafts, see Sec. 8.2)

Cross section	Torsional resisting moment M_t	Angular twist, θ_1 (length = 1 in, radius = 1 in) In terms of torsional moment	Angular twist, θ_1 In terms of max shear	Work of torsion (V = volume)
	$\dfrac{\pi}{16} d^3 S_v$	$\dfrac{M_t}{GI_P} = \dfrac{32}{\pi d^4}\dfrac{M_t}{G}$	$2\dfrac{S_{v_{max}}}{G}\dfrac{1}{d}$	$\dfrac{1}{4}\dfrac{S^2_{v_{max}}}{G} V$ (Note 1)
	$\dfrac{\pi}{16}\dfrac{D^4 - d^4}{D} S_v$	$\dfrac{32}{\pi(D^4 - d^4)}\dfrac{M_t}{G}$	$2\dfrac{S_{v_{max}}}{G}\dfrac{1}{D}$	$\dfrac{1}{4}\dfrac{S^2_{v_{max}}}{G}\dfrac{D^2 + d^2}{D^2} V$ (Note 2)
	$\dfrac{\pi}{16} b^2 h S_v$ ($h > b$)	$\dfrac{16}{\pi}\dfrac{b^2 + h^2}{b^3 h^3}\dfrac{M_t}{G}$	$\dfrac{S_{v_{max}}}{G}\dfrac{b^2 + h^2}{bh^2}$	$\dfrac{1}{8}\dfrac{S^2_{v_{max}}}{G}\dfrac{b^2 + h^2}{h^2} V$ (Note 3)
	$\tfrac{2}{9} b^2 h S_v$ ($h > b$)	$3.6* \dfrac{b^2 + h^2}{b^3 h^3}\dfrac{M_t}{G}$	$0.8* \dfrac{S_{v_{max}}}{G}\dfrac{b^2 + h^2}{bh^2}$	$\dfrac{4}{45}\dfrac{S^2_{v_{max}}}{G}\dfrac{b^2 + h^2}{h^2} V$ (Note 4)
	$\tfrac{2}{9} h^3 S_v$	$7.2\dfrac{1}{h^4}\dfrac{M_t}{G}$	$1.6\dfrac{S_{v_{max}}}{G}\dfrac{1}{h}$	$\dfrac{8}{45}\dfrac{S^2_{v_{max}}}{G} V$ (Note 5)
	$\dfrac{b^2}{20} S_v$	$46.2\dfrac{1}{b^4}\dfrac{M_t}{G}$	$2.31\dfrac{S_{v_{max}}}{G}\dfrac{1}{b}$	
	$\dfrac{b^3}{1.09} S_v$	$0.967\dfrac{1}{b^4}\dfrac{M_t}{G}$	$0.9\dfrac{S_{v_{max}}}{G}\dfrac{1}{b}$	

* When h/b =   1   2   4   8
Coefficient 3.6 becomes = 3.56   3.50   3.35   3.21
Coefficient 0.8 becomes = 0.79   0.78   0.74   0.71
NOTES: (1) $S_{v_{max}}$ at circumference. (2) $S_{v_{max}}$ at outer circumference. (3) $S_{v_{max}}$ at A; $S_{vB} = 16M_t/\pi bh^2$. (4) $S_{v_{max}}$ at middle of side h; in middle of b, $S_v = 9M_t/2bh^2$.
(5) $S_{v_{max}}$ at middle of side.

Table 5.2.12 Strength of Round-ended Columns according to Euler's Formula*

Material	Cast iron	Wrought iron[†]	Low-carbon steel	Medium-carbon steel
Ultimate compressive strength, lb/in^2	107,000	53,400	62,600	89,000
Allowable compressive stress, lb/in^2 (maximum)	7,100	15,400	17,000	20,000
Modulus of elasticity	14,200,000	28,400,000	30,600,000	31,300,000
Factor of safety	8	5	5	5
Smallest I allowable at worst section, in^4	$\dfrac{Pl^2}{17,500,000}$	$\dfrac{Pl^2}{56,000,000}$	$\dfrac{Pl^2}{60,300,000}$	$\dfrac{Pl^2}{61,700,000}$
Limit of ratio, l/r	50.0	60.6	59.4	55.6
Rectangle ($r = b\sqrt{1/12}$), $l/b >$	14.4	17.5	17.2	16.0
Circle ($r = \frac{1}{4}d$), $l/d >$	12.5	15.2	14.9	13.9
Circular ring of small thickness ($r = b\sqrt{1/18}$), $l/d >$	17.6	21.4	21.1	19.7

* P = allowable load, lb; l = length of column, in; b = smallest dimension of a rectangular section, in; d = diameter of a circular section, in; r = least radius of gyration of section.
† This material is no longer manufactured but may be encountered in existing structures and machinery.

Table 5.2.13 Typical Short-Column Formulas

Formula	Material	Code	Slenderness ratio
$S_w = 17{,}000 - 0.48\left(\dfrac{l}{r}\right)^2$	Carbon steels	AISC	$l/r < 120$
$S_w = 16{,}000 - 70\,(l/r)$	Carbon steels	Chicago	$l/r < 120$
$S_w = 15{,}000 - 5\left(\dfrac{l}{r}\right)$	Carbon steels	AREA	$l/r < 150$
$S_w = 19{,}000 - 100\,(l/r)$	Carbon steels	Am. Br. Co.	$60 < \dfrac{l}{r} < 120$
$S_{cr}^{*} = 135{,}000 - \dfrac{15.9}{c}\left(\dfrac{l}{r}\right)^2$	Alloy-steel tubing	ANC	$\dfrac{1}{\sqrt{cr}} < 65$
$S_w = 9{,}000 - 40\left(\dfrac{l}{r}\right)$	Cast iron	NYC	$\dfrac{l}{r} < 70$
$S_{cr}^{*} = 34{,}500 - \dfrac{245}{\sqrt{c}}\left(\dfrac{l}{r}\right)$	2017ST Aluminum	ANC	$\dfrac{1}{\sqrt{cr}} < 94$
$S_{cr}^{*} = 5{,}000 - \dfrac{0.5}{c}\left(\dfrac{l}{r}\right)^2$	Spruce	ANC	$\dfrac{1}{\sqrt{cr}} < 72$
$S_{cr}^{*} = S_y\left[1 - \dfrac{S_y}{4n\pi^2 E}\left(\dfrac{l}{r}\right)^2\right]$	Steels	Johnson	$\dfrac{l}{r} < \sqrt{\dfrac{2n\pi^2 E}{S_y}}$
$S_{cr}^{*}\dagger = \dfrac{S_y}{1 + \dfrac{ec}{r^2}\sec\left(\dfrac{l}{r}\sqrt{\dfrac{P}{4AE}}\right)}$	Steels	Secant	$\dfrac{l}{r} <$ critical

* S_{cr} = theoretical maximum, c = end fixity coefficient,
   $c = 2$, both ends pivoted; $c = 2.86$, one pivoted, other fixed;
   $c = 4$, both ends fixed; $c = 1$, one end fixed, one end free.
† e is initial eccentricity at which load is applied to center of column cross section.

Fig. 5.2.52

FOR THE CASE OF LONGITUDINAL COMPRESSION. $S_b l/c = M_2 + CP S_b l^2/(Eo)$, or $S_b = M_2 c(I - CP l^2/E)$. The maximum stress is $S = S_b + S_0$ compression. The constant C for the case of Fig. 5.2.53 is derived from the equations $P'l/4 = S_b l/c$ and $f = P'l^3/(48EI)$. Solving for f; $f = \frac{1}{12} S_b l^2/(Ec)$, or $C = \frac{1}{12}$. For a beam supported at the ends and uniformly loaded, $C = \frac{5}{48}$. Other cases can be similarly calculated.

FOR THE CASE OF LONGITUDINAL TENSION. $M = M_2 - Pf$, and $S_b = M_2 c/(I + CP l^2/E)$. The maximum stress is $S = S_b + S_0$, tension.

Fig. 5.2.53

ECCENTRIC LOADS

When **short blocks** are **loaded eccentrically** in compression or in tension, i.e., not through the center of gravity (cg), a combination of axial and bending stress results. The maximum unit stress S_M is the algebraic sum of these two unit stresses.

In Fig. 5.2.54 a load P acts in a line of symmetry at the distance e from cg; r = radius of gyration. The unit stresses are (1) S_c, due to P, as

Fig. 5.2.54

if it acted through cg, and (2) S_b, due to the bending moment of P acting with a leverage of e about cg. Thus unit stress S at any point y is

$$S = S_c \pm S_b$$
$$= P/A \pm Pey/I$$
$$= S_c(1 \pm et/r^2)$$

y is positive for points on the same side of cg as P, and negative on the opposite side. For a **rectangular cross section** of width b, the maximum

stress $S_M = S_c (1 + 6e/b)$. When P is outside the middle third of width b and is a compressive load, tensile stresses occur.

For a **circular cross section** of diameter d, $S_M = S_c(1 + 8e/d)$. The stress due to the weight of the solid will modify these relations.

NOTE. In these formulas e is measured from the gravity axis, and gives tension when e is greater than one-sixth the width (measured in the same direction as e), for rectangular sections; and when greater than one-eighth the diameter for solid circular sections.

If, as in certain classes of masonry construction, the **material cannot withstand tensile stress** and thus no tension can occur, the center of moments (Fig. 5.2.55) is taken at the center of stress. For a **rectangular section,** P acts at distance k from the nearest edge. Length under compression $= 3k$, and $S_M = \frac{2}{3}P/(hk)$. For a **circular section,** $S_M = [0.372 + 0.056(k/r)]P/k \sqrt{rk}$, where r = radius and k = distance of P from circumference. For a **circular ring,** S = average compressive stress on cross section produced by P; e = eccentricity of P; z = length of diameter under compression (Fig. 5.2.56). Values of z/r and of the ratio of S_{max} to average S are given in Tables 5.2.14 and 5.2.15.

CHIMNEY PROBLEM. Weight of chimney = 563,000 lb; e = 1.56 ft; OD of chimney = 10 ft 8 in; ID = 6 ft 6 ½ in. Overturning moment = Pe = 878,000 ft· lb, r_1/r = 0.6. e/r = 0.29. This gives $z/r > 2$. Therefore, the entire area of the base is under compression. Area under compression = 55.8 ft²; I = 546; S = 563,000/55.8 ± (878,000 × 5.33)/546 = 18,700 (max) and 1,500 (min) lb compression per ft². From Table 5.2.15, by interpolation, S_{max}/S_{av} = 1.85. ∴ S_{max} = (563,000/55.8) × 1.85 = 18,685 lb/ft² (91,313 kgf/m²).

The **kern** is the area around the center of gravity of a cross section within which any load applied will produce stress of only one sign throughout the entire cross section. Outside the kern, a load produces stresses of different sign. Figure 5.2.57 shows kerns (shaded) for various sections.

Fig. 5.2.55   **Fig. 5.2.56**

Fig. 5.2.57

Table 5.2.14   Values of the Ratio z/r (Fig. 5.2.56)

$\frac{e}{r}$	$\frac{r_1}{r}$							$\frac{e}{r}$
	0.0	0.5	0.6	0.7	0.8	0.9	1.0	
0.25	2.00							0.25
0.30	1.82							0.30
0.35	1.66	1.89	1.98					0.35
0.40	1.51	1.75	1.84	1.93				0.40
0.45	1.37	1.61	1.71	1.81	1.90			0.45
0.50	1.23	1.46	1.56	1.66	1.78	1.89	2.00	0.50
0.55	1.10	1.29	1.39	1.50	1.62	1.74	1.87	0.55
0.60	0.97	1.12	1.21	1.32	1.45	1.58	1.71	0.60
0.65	0.84	0.94	1.02	1.13	1.25	1.40	1.54	0.65
0.70	0.72	0.75	0.82	0.93	1.05	1.20	1.35	0.70
0.75	0.59	0.60	0.64	0.72	0.85	0.99	1.15	0.75
0.80	0.47	0.47	0.48	0.52	0.61	0.77	0.94	0.80
0.85	0.35	0.35	0.35	0.36	0.42	0.55	0.72	0.85
0.90	0.24	0.24	0.24	0.24	0.24	0.32	0.49	0.90
0.95	0.12	0.12	0.12	0.12	0.12	0.12	0.25	0.95

Table 5.2.15   Values of the Ratio S_{max}/S_{avg} (Fig. 5.2.56)
(In determining S average, use load P divided by total area of cross section)

$\dfrac{e}{r}$	\multicolumn{7}{c}{$\dfrac{r_1}{r}$}	$\dfrac{e}{r}$
	0.0	0.5	0.6	0.7	0.8	0.9	1.0	
0.00	1.00	1.00	1.00	1.00	1.00	1.00	1.00	0.00
0.05	1.20	1.16	1.15	1.13	1.12	1.11	1.10	0.05
0.10	1.40	1.32	1.29	1.27	1.24	1.22	1.20	0.10
0.15	1.60	1.48	1.44	1.40	1.37	1.33	1.30	0.15
0.20	1.80	1.64	1.59	1.54	1.49	1.44	1.40	0.20
0.25	2.00	1.80	1.73	1.67	1.61	1.55	1.50	0.25
0.30	2.23	1.96	1.88	1.81	1.73	1.66	1.60	0.30
0.35	2.48	2.12	2.04	1.94	1.85	1.77	1.70	0.35
0.40	2.76	2.29	2.20	2.07	1.98	1.88	1.80	0.40
0.45	3.11	2.51	2.39	2.23	2.10	1.99	1.90	0.45
0.50	3.55	2.80	2.61	2.42	2.26	2.10	2.00	0.50
0.55	4.15	3.14	2.89	2.67	2.42	2.26	2.17	0.55
0.60	4.96	3.58	3.24	2.92	2.64	2.42	2.26	0.60
0.65	6.00	4.34	3.80	3.30	2.92	2.64	2.42	0.65
0.70	7.48	5.40	4.65	3.86	3.33	2.95	2.64	0.70
0.75	9.93	7.26	5.97	4.81	3.93	3.33	2.89	0.75
0.80	13.87	10.05	8.80	6.53	4.93	3.96	3.27	0.80
0.85	21.08	15.55	13.32	10.43	7.16	4.50	3.77	0.85
0.90	38.25	30.80	25.80	19.85	14.60	7.13	4.71	0.90
0.95	96.10	72.20	62.20	50.20	34.60	19.80	6.72	0.95
1.00	∞	∞	∞	∞	∞	∞	∞	1.00

For **a circular ring,** the radius of the kern $r = D[1 + (d/D)^2]/8$.

For a **hollow square** (H and h = lengths of outer and inner sides), the kern is a square similar to Fig. 5.2.57a, where

$$r_{min} = \frac{H}{6}\,\frac{1}{\sqrt{2}}\left[1 + \left(\frac{h}{H}\right)^2\right] = 0.1179H\left[1 + \left(\frac{h}{H}\right)^2\right]$$

For a **hollow octagon** R_a and R_i = radii of circles circumscribing the outer and inner sides; thickness of wall = $0.9239(R_a - R_i)$, the kern is an octagon similar to Fig. 5.2.57c, where $0.2256R$ becomes $0.2256R_a[1 + (R_i/R_a)^2]$.

CURVED BEAMS

The application of the flexure formula for a straight beam to the case of a curved beam results in error. When all "fibers" of a member have the same center of curvature, the **concentric** or common type of curved beam exists (see Fig. 5.2.58). Such a beam is defined by the Winkler-Bach theory. The stress at a point y units from the centroidal axis is

$$S = \frac{M}{AR}\left[1 + \frac{y}{Z(R + y)}\right]$$

M is the bending moment, positive when it increases curvature; Y is positive when measured toward the convex side; A is the cross-sectional area; R is the radius of the centroidal axis; Z **is a cross-section property** defined by

$$Z = -\frac{1}{A}\int \frac{y}{R + y}\,dA$$

Analytical expressions for Z of certain sections are given in Table 5.2.16. Also Z can be found by **graphical** integration methods (see any advanced

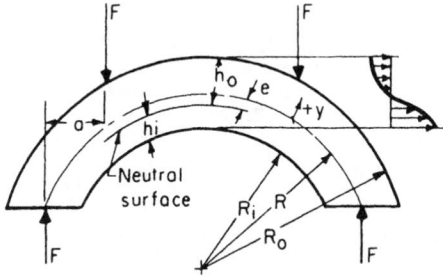

Fig. 5.2.58

strength book). The **neutral surface** shifts toward the center of curvature, or inside fiber, an amount equal to $e = ZR/(Z + 1)$. The Winkler-Bach theory, though practically satisfactory, disregards radial stresses as well as lateral deformations and assumes pure bending. The **maximum stress** occurring on the inside fiber is $S = Mh_i/(AeR_i)$, while that on the outside fiber is $S = Mh_0/(AeR_0)$.

EXAMPLE.   A split steel ring of rectangular cross section is subjected to a diametral force of 1,000 lb as shown in Fig. 5.2.59a. Compute the stress at the point 0.5 in from the outside fiber on plane mm. Also compute the maximum stress.

$$Z = -1 + \frac{R}{h}\ln\frac{R + C}{R - C}$$

$$= -1 + \frac{10}{4}\ln\frac{10 + 2}{10 - 2} = 0.0133$$

$$S_{1.5} = \frac{M}{AR}\left[1 + \frac{y}{Z(R + y)}\right] + \frac{F}{A}$$

$$= \frac{-1,000 \times 10}{8 \times 10}\left[1 + \frac{1.5}{0.0133(10 + 1.5)}\right] + \frac{1,000}{8}$$

$$= -1,250 + 125 = -1,125 \text{ lb/in}^2 \text{ (compr.)} (79 \text{ kgf/cm}^2)$$

$$S_M = \frac{-1,000}{8}\left[1 + \frac{-2}{0.0133(10-2)}\right] + \frac{1,000}{8}$$

$$= 2,230 + 125 = 2,335 \text{ lb/in}^2 \, (166 \text{ kgf/cm}^2)$$

or

$$e = \frac{ZR}{Z+1} = \frac{0.0133 \times 10}{0.0133 + 1} = 0.131$$

and

$$S_M = \frac{Mh_i}{AeR_i} + \frac{F}{A} = \frac{1,000 \times 10 \times 1.87}{8 \times 0.131 \times 8} + \frac{1,000}{8}$$

$$= 2,355 \text{ lb/in}^2 \, (166 \text{ kgf/cm}^2)$$

The **deflection** in curved beams can be computed by means of the moment-area theory. If the origin of axes is taken at the point whose deflection is wanted, it can be shown that the component displacements in the x and y directions are

$$\Delta_x = \int_0^s \frac{My \, ds}{EI} \qquad \text{and} \qquad \Delta_y = \int_0^s \frac{Mx \, ds}{EI}$$

The resultant deflection is then equal to $\Delta_0 = \sqrt{\Delta_x^2 + \Delta_y^2}$ in the direction defined by $\tan \theta = \Delta_y/\Delta_x$. Deflections can also be found conveniently by use of **Castigliano's theorem.** It states that in an elastic system the displacement in the direction of a force (or couple) and due to that force (or couple) is the partial derivative of the strain energy with respect to the force (or couple). Stated mathematically, $\Delta_z = \partial U/\partial F_z$. If a force does not exist at the point and/or in the direction desired, a dummy force may be applied. This force must then be eliminated by equating it to zero at the end.

EXAMPLE. A quadrant of radius R is fixed at one end as shown in Fig. 5.2.59b. The force F is applied in the radial direction at the free end B. Find the deflection of B.

By moment area:

$$y = R \sin \theta \qquad x = R(1 - \cos \theta)$$
$$ds = R \, d\theta \qquad m = FR \sin \theta$$

$$_B\Delta_x = \frac{FR^3}{EI}\int_0^{\pi/2} \sin^2 \theta \, d\theta = \frac{\pi FR^3}{4EI}$$

$$_B\Delta_y = \frac{FR^3}{EI}\int_0^{\pi/2} \sin \theta (1 - \cos \theta) \, d\theta = -\frac{FR^3}{2EI}$$

and

$$\Delta_B = \frac{FR}{2EI}\sqrt{1 + \frac{\pi^2}{4}}$$

at

$$\theta_x = \tan^{-1}\left(-\frac{FR^3}{2EI} \times \frac{4EI}{\pi FR^3}\right) = \tan^{-1}\frac{2}{\pi} = 32.5°$$

By Castigliano:

$$_B\Delta_x = \frac{\partial U}{\partial F} = \frac{\partial}{\partial F_x}\int_0^{\pi/2}\frac{F^2R^3}{2EI}\sin^2 \theta \, d\theta = \frac{\pi FR^3}{4EI}$$

$$_B\Delta_y = \frac{\partial U}{\partial F_y} = \frac{\partial}{\partial F_y}\int_0^{\pi/2}\frac{[FR \sin \theta - F_yR(1 - \cos \theta)]^2 R \, d\theta}{2EI}$$

$$= -\frac{FR^3}{2EI}$$

The F_y, assumed downward, is equated to zero, after the integration and differentiation are performed to find $_B\Delta_y$. The remainder of the computation is exactly as in the moment-area method.

Fig. 5.2.59

Eccentrically Curved Beams These beams (Fig. 5.2.60) are bounded by arcs having different centers of curvature. In addition, it is possible for either radius to be the larger one. The one in which the section depth shortens as the central section is approached may be called the **arch beam.** When the central section is the largest, the beam is of the crescent type. **Crescent I** denotes the beam of larger outside radius and **crescent II** of larger inside radius. The stress at the **central section** of such beams may be found from $S = KMC/I$. In the case of rectangular cross section, the equation becomes $S = 6KM/(bh^2)$ where M is the bending moment, b is the width of the beam section, and h its height. The **stress factors** K for the **inner boundary**, established from photoelastic data, are given in Table 5.2.17. The outside radius is denoted by R_o and the inside by R_i. The geometry of crescent beams is such that the stress can be larger in **off-center sections.** The stress at the central section determined above must then be multiplied by the **position factor** k, given in Table 5.2.18. As in the concentric beam, the **neutral surface** shifts slightly toward the inner boundary (see Vidosic, Curved Beams with Eccentric Boundaries, *Trans. ASME*, 79, pp., 1317–1321).

Table 5.2.16   Analytical Expressions for Z

Section	Expression	Section	Expression
	$Z = -1 + \frac{R}{h}\ln\frac{R+C}{R-C}$		$Z = -1 + \frac{R}{A}\left[t \ln(R+C_1) + (b-t)\ln(R-C_3) - b \ln(R-C_3)\right]$ $A = tC_1 - (b-t)C_3 + bC_2$
	$Z = -1 + 2\left(\frac{R}{r}\right)\left[\frac{R}{r} - \sqrt{\left(\frac{R}{r}\right)^2 - 1}\right]$		$Z = -1 + \frac{R}{A}\left[b \ln\frac{R+C_2}{R-C_2} + (t-b)\ln\frac{R+C_1}{R-C_1}\right]$ $A = 2[(t-b)C_1 + bC_2]$

Fig. 5.2.60

Table 5.2.17   Stress Factors for Inner Boundary at Central Section (See Fig. 5.2.60)

1. For the arch-type beams

(a) $K = 0.834 + 1.504 \dfrac{h}{R_o + R_i}$ if $\dfrac{R_o + R_i}{h} < 5$.

(b) $K = 0.899 + 1.181 \dfrac{h}{R_o + R_i}$ if $5 < \dfrac{R_o + R_i}{h} < 10$.

(c) In the case of larger section ratios use the equivalent beam solution.

2. For the crescent I-type beams

(a) $K = 0.570 + 1.536 \dfrac{h}{R_o + R_i}$ if $\dfrac{R_o + R_i}{h} < 2$.

(b) $K = 0.959 + 0.769 \dfrac{h}{R_o + R_i}$ if $2 < \dfrac{R_o + R_i}{h} < 20$.

(c) $K = 1.092 \left(\dfrac{h}{R_o + R_i}\right)^{0.0298}$ if $\dfrac{R_o + R_i}{h} > 20$.

3. For the crescent II-type beams

(a) $K = 0.897 + 1.098 \dfrac{h}{R_o + R_i}$ if $\dfrac{R_o + R_i}{h} < 8$.

(b) $K = 1.119 \left(\dfrac{h}{R_o + R_i}\right)^{0.0378}$ if $8 < \dfrac{R_o + R_i}{h} > 20$.

(c) $K = 1.081 \left(\dfrac{h}{R_o + R_i}\right)^{0.0270}$ if $\dfrac{R_o + R_i}{h} > 20$.

Table   5.2.18 Crescent-Beam Position Stress Factors (See Fig. 5.2.60)

Angle θ, deg	k	
	Inner	Outer
10	$1 + 0.055H/h$	$1 + 0.03H/h$
20	$1 + 0.164H/h$	$1 + 0.10H/h$
30	$1 + 0.365H/h$	$1 + 0.25H/h$
40	$1 + 0.567H/h$	$1 + 0.467H/h$
50	$1.521 - \dfrac{(0.5171 - 1.382H/h)^{\frac{1}{2}}}{1.382}$	$1 + 0.733H/h$
60	$1.756 - \dfrac{(0.2416 - 0.6506H/h)^{\frac{1}{2}}}{0.6506}$	$1 + 1.123H/h$
70	$2.070 - \dfrac{(0.4817 - 1.298H/h)^{\frac{1}{2}}}{0.6492}$	$1 + 1.70H/h$
80	$2.531 - \dfrac{(0.2939 - 0.7084H/h)^{\frac{1}{2}}}{0.3542}$	$1 + 2.383H/h$
90		$1 + 3.933H/h$

NOTE: All formulas are valid for $0 < H/h \le 0.325$. Formulas for the inner boundary, except for 40 deg, may be used to $H/h \le 0.36$. H = distance between centers.

IMPACT

A force or stress is considered **suddenly applied** when the duration of load application is less than one-half the **fundamental natural period** of vibration of the member upon which the force acts. Under impact, a compression wave propagates through the member at a velocity $c = \sqrt{E/\rho}$, where ρ is the mass density. As this compression wave travels back and forth by reflection from one end of the bar to the other, a maximum stress is produced which is many times larger than what it would be statically. An exact determination of this stress is most difficult. However, if conservation of kinetic and strain energies is applied, the **impact stress** is found to be

$$S' = S\sqrt{\frac{W}{W_b}\left(\frac{3W}{3W + W_b}\right)}$$

The weight of the striking mass is here denoted by W, that of the struck bar by W_b, while S is the static stress, W/A (A is the cross-sectional area of the bar). Above is the case of **sudden impact.** When the ratio W/W_b is small, the stress computed by the above equation may be erroneous. A better solution of this problem may result from

$$S' = S + S\sqrt{\frac{W}{W_b} + \frac{2}{3}}$$

If a weight W falls a distance h before striking a bar of mass W_b, energy conservation will yield the relation

$$S' = S\left(1 + \sqrt{1 + \frac{2h}{e} \times \frac{3W}{3W + W_b}}\right)$$

The elongation $e = \varepsilon l = Sl/E$. When the striking mass W is assumed rigid, the elasticity factor is taken equal to 1. Thus the equation becomes

$$S' = S(1 + \sqrt{1 + 2h/e})$$

If, in addition, h is taken equal to zero (sudden impact), the radical equals 1, and so the stress becomes $S' = 2S$. Since Hooke's law is applicable, the relations

$$e' = e(1 + \sqrt{1 + 2h/e}) \qquad \text{and} \qquad e' = 2e$$

are also true for the same conditions.

The expression may be converted, by using $v^2 = 2gh$, to

$$S' = S[1 + \sqrt{1 + v^2/(eg)}]$$

This might be called the **energy impact** form. If the **natural frequency** f_n of the bar is used, the stress equation is

$$S' = S(1 + \sqrt{1 + 0.204hf_n^2})$$

In general, the **maximum impact stress** in a **beam** and a **shaft** can be approximated from the simplified falling-weight equation. It is necessary, though, to substitute the maximum **deflection** y for e, in the case of beams, and for the **angle of twist** θ in the case of shafts. Of course $S = Mc/I$ and M_tc/J, respectively. Thus

$$S' = S\left[1 + \sqrt{1 + \frac{2h}{y(\text{or } \theta)}}\right]$$

For a more exact solution, elastic yield in each member must be considered. The theory then yields

$$S' = S\left[1 + \sqrt{1 + \frac{2h}{y}\left(\frac{35W}{35W + 17W_b}\right)}\right]$$

for a **simply supported beam** struck in the middle by a weight W.

THEORY OF ELASTICITY

For loaded members in which the stress distribution cannot be estimated by elementary strength-of-material methods, the more advanced mathematical principles of the **theory of elasticity** must be applied. When this is not possible, experimental stress analysis has to be used. Because of the complexity of solution, only some of the more practical problems have been solved by the theory of elasticity. The more general concepts and methods are presented.

Two kinds of forces may act on a body. **Surface forces** are distributed over the surface as the result of, for instance, the pressure of one body on another. Forces due to gravity, inertia, magnetism, etc., which act over the entire volume of a body, are called **body forces**. Both surface and body forces can be best handled if resolved into three orthogonal components. Surface forces are thus designated $\overline{X}$, $\overline{Y}$, and $\overline{Z}$, while body forces are labeled X, Y, and Z.

In general, there exists a normal stress σ and a shearing stress τ at each point of a loaded member. It is convenient to deal with components of each of these stresses on each of six orthogonal planes that bound the point element. Thus there are at each point **six stress components**, σ_x, σ_y, σ_z, $\tau_{yx} = \tau_{xy}$, $\tau_{xz} = \tau_{zx}$, and $\tau_{yz} = \tau_{zy}$. Similarly, if the normal unit strain is designated by the letter ε and shearing unit strain by γ, the **six components of strain** are defined by

$$\varepsilon_x = \partial u/\partial x \qquad \varepsilon_y = \partial v/\partial y \qquad \varepsilon_z = \partial w/\partial z$$

$$\gamma_{xy} = \partial u/\partial y + \partial v/\partial x \qquad \gamma_{yz} = \partial v/\partial z + \partial w/\partial y$$

and $\qquad \gamma_{xz} = \partial u/\partial z + \partial w/\partial x$

The elastic **displacements** of particles on the body in the x, y, and z directions are identified as the u, v, and w **components**, respectively.

Since metals have the usually assumed elastic as well as isotropic properties, Hooke's law holds. Therefore, the interrelationships between stress and strain can easily be obtained.

$$\varepsilon_x = \frac{1}{E}[\sigma_x - \mu(\sigma_y + \sigma_z)]$$

$$\varepsilon_y = \frac{1}{E}[\sigma_y - \mu(\sigma_x + \sigma_z)]$$

$$\varepsilon_z = \frac{1}{E}[\sigma_z - \mu(\sigma_x + \sigma_y)]$$

$$\gamma_{xy} = \tau_{xy}/G \qquad \gamma_{xz} = \tau_{xz}/G$$

and $\qquad \gamma_{yz} = \tau_{yz}/G$

The general case of strain can be obtained by superposing the elongation strains upon the shearing strains.

Problems depending upon theories of elasticity are considerably simplified if the stresses are all parallel to one plane or if all deformations occur in planes perpendicular to the length of the member. The first case is one of **plane stress**, as when a thin plate of uniform thickness is subjected to central, boundary forces parallel to the plane of the plate. The second is a case of **plane strain**, such as a gate subjected to hydrostatic pressure, the intensity of which does not vary along the gate's length. All particles therefore displace at right angles to the length, and so cross sections remain plane. In plane-stress problems, three of the six stress components vanish, thus leaving only σ_x, σ_y, and τ_{xy}. Similarly, in plane strain, only ε_x, ε_y, and γ_{xy} will not equal zero; thus the same three stresses σ_x, σ_y, and τ_{xy} remain to be considered. Plane problems can thus be represented by the element shown in Fig. 5.2.61. Equilibrium considerations applied to this particle result in the **differential equations of equilibrium** which reduce to

$$\frac{\partial \sigma_x}{\partial x} + \frac{\partial \tau_{xy}}{\partial y} + X = 0$$

and $\qquad \dfrac{\partial \sigma_y}{\partial y} + \dfrac{\partial \tau_{xy}}{\partial x} + Y = 0$

Since the two differential equations of equilibrium are insufficient to find the three stresses, a third equation must be used. This is the **compatibility equation** relating the three strain components. It is

$$\frac{\partial^2 \varepsilon_x}{\partial y^2} + \frac{\partial^2 \varepsilon_y}{\partial x^2} = \frac{\partial^2 \gamma_{xy}}{\partial x \, \partial y}$$

If strains are expressed in terms of the stresses, the compatibility equation becomes

$$\left(\frac{\partial^2}{\partial x^2} + \frac{\partial^2}{\partial y^2} \right) (\sigma_x + \sigma_y) = 0$$

Now, in any **two-dimensional** problem, the compatibility equation along with the differential equilibrium equations must be simultaneously solved for the three unknown stresses. This is accomplished using **stress functions**, which permit the integration and satisfy boundary conditions in each particular situation.

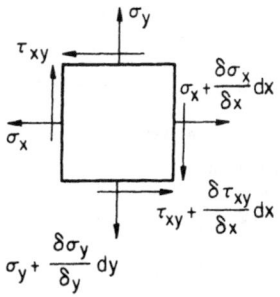

Fig. 5.2.61

In **three-dimensional** problems, the third dimension must be considered. This results in three differential equations of equilibrium, as well as three compatibility equations. The six stress components can thus be found. The complexity involved in the solution of these equations is such, however, that only a few special cases have been solved.

In certain problems, such as rotating circular disks, **polar coordinates** become more convenient. In such cases, the stress components in a two-dimensional field are the **radial stress** σ_r, the **tangential stress** σ_θ, and the **shearing stress** $\tau_{r\theta}$. In terms of these stresses the **polar differential equations** become

$$\frac{\partial \sigma_r}{\partial r} + \frac{1}{r} \frac{\partial \tau_{r\theta}}{\partial \theta} + \frac{\sigma_r - \sigma_\theta}{r} + R = 0$$

and $\qquad \dfrac{1}{r} \dfrac{\partial \sigma_\theta}{\partial \theta} + \dfrac{\partial \tau_{r\theta}}{\partial r} + \dfrac{2\tau_{r\theta}}{r} = 0$

The body force per unit volume is represented by R. The **compatibility equation** in polar coordinates is

$$\left(\frac{\partial^2}{\partial r^2} + \frac{1}{r} \frac{\partial}{\partial r} + \frac{1}{r^2} \frac{\partial^2}{\partial \theta^2} \right) \left(\frac{\partial^2 \phi}{\partial r^2} + \frac{1}{r} \frac{\partial \phi}{\partial r} + \frac{1}{r^2} \frac{\partial^2 \phi}{\partial \theta^2} \right) = 0$$

ϕ is again a stress function of r and θ that will provide a solution of the differential equations and satisfy boundary conditions. As an example, the exact solution of a **simply supported beam** carrying a uniformly distributed load w yields

$$\sigma_x = \frac{w}{2I}(l^2 - x^2)y + \frac{w}{2I}\left(\frac{2y^3}{3} - \frac{2c^2 y}{5} \right)$$

The origin of coordinates is at the center of the beam, $2c$ is the beam depth, and $2l$ is the span length. Thus the maximum stress at $x = 0$ and $y = c$ is $\sigma_x = \dfrac{wl^2 c}{2I} + \dfrac{2}{15}\dfrac{wc^3}{I}$. The first term represents the stress as obtained by the elementary flexure theory; the second is a correction. The second term becomes negligible when c is small compared to l.

The important case of a **flat plate** of unit width with a **circular hole** of diameter $2a$ at its center, subjected to a uniform tensile load, has been solved using polar coordinates. If S is the uniform stress at some distance from the hole, r is measured from the center of the hole, and θ is the angle of r with respect to the longitudinal axis of the member, the stresses are

$$\sigma_r = \frac{S}{2}\left(1 - \frac{a^2}{r^2}\right) + \frac{S}{2}\left(1 + \frac{3a^4}{r^4} - \frac{4a^2}{r^2}\right)\cos 2\theta$$

$$\sigma_\theta = \frac{S}{2}\left(1 + \frac{a^2}{r^2}\right) - \frac{S}{2}\left(1 + \frac{3a^4}{r^4}\right)\cos 2\theta$$

$$\tau_r\theta = -\frac{S}{2}\left(1 - \frac{3a^4}{r^4} + \frac{2a^2}{r^2}\right)\sin 2\theta$$

CYLINDERS AND SPHERES

A **thin-wall cylinder** has a wall thickness such that the assumption of constant stress across the wall results in negligible error. Cylinders having internal-diameter-to-thickness (D/t) ratios greater than 10 are usually considered thin-walled. Boilers, drums, tanks, and pipes are often treated as such. Equilibrium equations reveal the circumferential, or hoop, stress to be $S = pr/t$ under an internal pressure p (see Fig. 5.2.62). If the cylinder is closed at the ends, a longitudinal stress of $pr/(2t)$ is developed. The tensile stress developed in a thin hollow sphere subjected to internal pressure is also $pr/(2t)$.

Fig. 5.2.62

When thin-walled cylinders, such as vacuum tanks and submarines, are subjected to **external pressure, collapse** becomes the mode of failure. The shell is assumed perfectly round and of uniform thickness, the material obeys Hooke's law, the radial stress is negligible, and the normal stress distribution is linear. Other, lesser assumptions are also made. Using the theory of elasticity, R. G. Sturm (*Univ. Ill. Eng. Exp. Stn. Bull.*, no 12, Nov. 11, 1941) derived the **collapsing pressure** as

$$W_c = KE\left(\frac{t}{D}\right)^3 \quad \text{lb/in}^2$$

The factor K, a numerical coefficient, depends upon the L/R and D/t ratios (D is outside-shell diameter), the kind of end support, and whether pressure is applied radially only, or at the ends as well. Figures 5.2.63 to 5.2.66, reproduced from the bulletin, supply the K values. N on these charts indicates the number of lobes into which the shell collapses. These values are for materials having Poisson's ratio $\mu = 0.3$. It may also be pointed out that in the case of long cylinders (infinitely long, theoretically) the value of K approaches $2/(1 - \mu^2)$.

When the **cylinder** is **stiffened with rings,** the shell may be assumed to be divided into a series of shorter shells, equal in length to the ring spacing. The previous equation can then be applied to a ring-to-ring length of cylinder. However, the flexural rigidity of the combined stiffener and shell EI_c necessary to withstand the pressure is $EI_c = W_s D^3 L_s/24$. W_s is the pressure, L_s the length between rings, and I_c the combined moment of inertia of the ring and that portion of the shell assumed acting with the ring.

Fig. 5.2.63   Radial external pressure with simply supported edges.

In some instances, cylinders collapse only after a stress in excess of the elastic limit has been reached; that is, **plastic range** stresses are present. In such cases the same equation applies, but the modulus of elasticity must be modified.

Fig. 5.2.64   Radial external pressure with fixed edges.

When the average stress S_a is less than the proportional limit S_p, and the maximum stress (direct, plus bending) is S, the modified modulus

$$E' = E\left[1 - \frac{1}{4}\left(\frac{S - S_l}{S_u - S_a}\right)^2\right]$$

S_u is the modulus of rupture. When the average stress is larger than the proportional limit, the modified modulus is taken as the tangent at the average stress.

In **thick-walled cylinders** (Fig. 5.2.67a) the circumferential, hoop, or tangential stress S_t is not uniform. In addition a radial stress S_r is present. When equilibrium is applied to the annulus taken out of Fig. 5.2.67a and

Fig. 5.2.65 Radial and end external pressure with simply supported edges.

shown in Fig. 5.2.67*b* and the equation is integrated, the general tangential and radial stress relations, called the **Lamé** equations, are derived.

$$S_t = \frac{r_1^2 p_1 - r_2^2 p_2 + (p_1 - p_2) r_1^2 r_2^2 / r^2}{r_2^2 - r_1^2}$$

and

$$S_r = \frac{r_1^2 p_1 - r_2^2 p_2 - (p_1 - p_2) r_1^2 r_2^2 / r^2}{r_2^2 - r_1^2}$$

When the external pressure $p_2 = 0$, the equations reduce to

$$S_t = \frac{r_1^2 p_1}{r_2^2 - r_1^2}\left(1 + \frac{r_2^2}{r^2}\right)$$

and

$$S_r = \frac{r_1^2 p_1}{r_2^2 - r_1^2}\left(1 + \frac{r_2^2}{r^2}\right)$$

Fig. 5.2.66 Radial and end external pressure with fixed edges.

At the inner boundary the tangential elongation ε_t is equal to

$$\varepsilon_t = (S_t - \mu S_r)/E$$

The increase in the bore radius Δr_1 resulting therefrom is

$$\Delta r_1 = \frac{r_1 p_1}{E_h}\left(\frac{1 + r_1^2/r_2^2}{1 - r_1^2/r_2^2} + \mu\right)$$

Similarly a solid shaft of r radius under external pressure p_2 will have its radius decreased by the amount

$$\Delta r = -\frac{r p_2}{E_s}(1 - \mu)$$

In the case of a **press or shrink fit**, $p_1 = p_2 = p$. The sum of Δr_1 and Δr_1 absolute is the radial interference; twice this sum is the **diametral interference** Δ or

$$\Delta = 2 r_1 p\left[\frac{1}{E_h}\left(\frac{1 + r_1^2/r_2^2}{1 - r_1^2/r_2^2} + \mu\right) + \frac{1 - \mu}{E_s}\right]$$

If the hub and shaft materials are the same, $E_h = E_s = E$, and

$$\Delta = \frac{4 r_1 r_2^2 p}{E(r_2^2 - r_1^2)}$$

If the equation is solved for p and this value is substituted in Lamé's equation, the maximum tangential stress on the inner surface of the hub is found to be

$$S_t = \frac{E\Delta}{4 r_1 r_2^2}(r_2^2 + r_1^2)$$

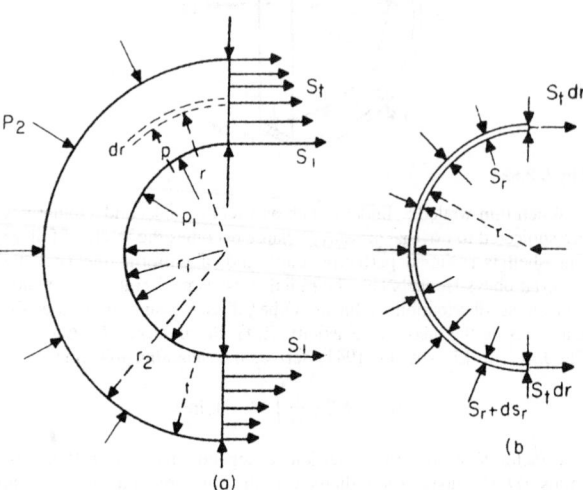

Fig. 5.2.67

EXAMPLE.   The barrel of a field gun has an outside diameter of 9 in and a bore of 4.7 in. An internal pressure of 16,000 lb/in² is developed during firing. What maximum stress occurs in the barrel? An investigation of Lamé's equations for internal pressure reveals the maximum stress to be the tangential one on the inner surface. Thus,

$$S_t = p_1\frac{r_1^2 + r_2^2}{r_2^2 - r_1^2} = \frac{16,000(2.35^2 + 4.5^2)}{4.5^2 - 2.35^2}$$

$$= 28,000 \text{ lb/in}^2 \ (1,972 \text{ kgf/cm}^2)$$

Oval Hollow Cylinders   In Fig. 5.2.68, let a and b be the semiminor and semimajor axes. The bending moments at A and C will then be

$$M_0 = pa^2/2 - pI_x/(2S) - pI_y/(2S)$$
$$M_1 = M_0 - p(a^2 - b^2)/2$$

where I_x and I_y are the moments of inertia of the arc AC about the x and y axes, respectively. The **bending moment at any point** will be

$$M = M_0 - pa^2/2 + px^2/2 + py^2/2$$

Thick Hollow Spheres   With an **internal pressure** p, where $p < T/0.65$.

$$r_2 = r_1[(T + 0.4p)/(T - 0.65p)]^{1/3}$$

The maximum tensile stress is on the inner surface, in the direction of the circumference. With an **external pressure** p, where $p < T/1.05$,

$$r_2 = r_1[T/(T - 1.05p)]^{1/3}$$

In both cases T is the true stress.

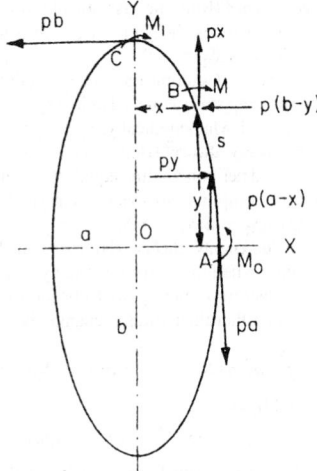

Fig. 5.2.68

PRESSURE BETWEEN BODIES WITH CURVED SURFACES (CONTACT STRESSES)

(See Hertz, "Gesammelte Werke," vol. 1, pp. 159 *et seq.*, Barth.)

Two Spheres   The radius A of the compressed area is obtained from the formula $A^3 = 0.68P(c_1 + c_2)/(1/r_1 + 1/r_2)$, in which P is the compressing force, c_1 and c_2 ($= 1/E_1$ and $1/E_2$) are reciprocals of the respective moduli of elasticity, and r_1 and r_2 are the radii. (Reciprocal of Poisson's ratio is assumed to be $n = 10/3$.) The **greatest contact pressure** in the middle of the compressed surface will be $S_{max} = 1.5(P/\pi A^2)$, and

$$S_{max}^3 = 0.235P(1/r_1 + 1/r_2)^2/(c_1 + c_2)^2$$

The **total deformation** of the two spheres will be Y, which is obtained from

$$Y^3 = 0.46P^2(c_1 + c_2)^2(1/r_1 + 1/r_2)$$

For $c_1 = c_2 = 1/E$, i.e., **two spheres with the same modulus of elasticity**, it follows that $A^3 = 1.36P/E(1/r_1 + 1/r_2)$, $S_{max}^3 = 0.059PE^2(1/r_1 + 1/r_2)^2$, and $Y^3 = 1.84P^2(1/r_1 + 1/r_2)/E^2$. If the radii of these spheres are also equal, $A^3 = 0.68Pr/E = 0.34Pd/E$; $S_{max}^3 = 0.235PE^2/r^2 = 0.94PE^2/d^2$; and $Y^3 = 3.68P^2/(E^2r) = 7.36P^2/(E^2d)$.

Sphere and Flat Plate   In this case $r_1 = r$ and $r_2 = \infty$, and the above formulas become $A^3 = 0.68Pr(c_1 + c_2) = 1.36Pr/E$, and

$$S_{max}^3 = 0.235P/[r^2(c_1 + c_2)^2] = 0.059PE^2/r^2$$
$$Y^3 = 0.46P^2(c_1 + c_2)^2/r = 1.84P^2/(E^2r)$$

Two Cylinders   The width b of the rectangular pressure surface is obtained from $(b/4)^2 = 0.29P(c_1 + c_2)/l[(1/r_1 + (1/r_2)]$, where r_1 and r_2 are the radii, and l the length

$$S_{max}^2 = [4P/(\pi bl)]^2 = 0.35P(1/r_1 + 1/r_2)/l(c_1 + c_2)]$$

For cylinders with the same moduli of elasticity, $c_1 = c_2 = 1/E$, and $(b/4)^2 = 0.58P/El[(1/r_1) + (1/r_2)]$; and $S_{max}^3 = 0.175PE(1/r_1 + 1/r_2)/l$. When $r_1 = r_2 = r$, $(b/4)^2 = 0.29Pr/(El)$, and $S_{max}^2 = 0.35PE/(lr)$.

Cylinder and Flat Plate   Here $r_1 = r$, $r_2 = \infty$, and the above formulas reduce to $(b/4)^2 = 0.29Pr(c_1 + c_2)/l = 0.58Pr/(El)$, and

$$S_{max}^2 = 0.35P/[lr(c_1 + c_2)] = 0.175PE/(lr)$$

For application to ball and roller bearings and to gear teeth, see Sec. 8.

FLAT PLATES

The analysis of flat plates subjected to **lateral loads** is very involved because plates bend in all vertical planes. Strict mathematical derivations have therefore been accomplished only in some special cases. Most of the available formulas contain some amount of rational empiricism. Plates may be classified as (1) **thick plates,** in which transverse shear is important; (2) **average-thickness plates,** in which flexure stress predominates; (3) **thin plates,** which depend in part upon direct tension; and (4) **membranes,** which are subject to direct tension only. However, exact lines of demarcation do not exist.

The flat-plate formulas given apply primarily to symmetrically loaded **average-thickness plates** of constant thickness. They are valid only if the maximum deflection is small relative to the plate thickness; usually, $y \leq 0.4t$. In the mathematical analyses, allowance for stress redistribution, because of slight local yielding, is usually not made. Since this yielding, especially in ductile materials, is beneficial, the formulas generally err on the side of safety. Certain cases of symmetrically loaded **circular and rectangular plates** are presented in Figs. 5.2.69 and 5.2.70. The maximum stresses are calculated from

$$S_M = k\frac{wR^2}{t^2} \qquad S_M = k\frac{P}{t^2} \qquad \text{or} \qquad S_M = k\frac{C}{t^2}$$

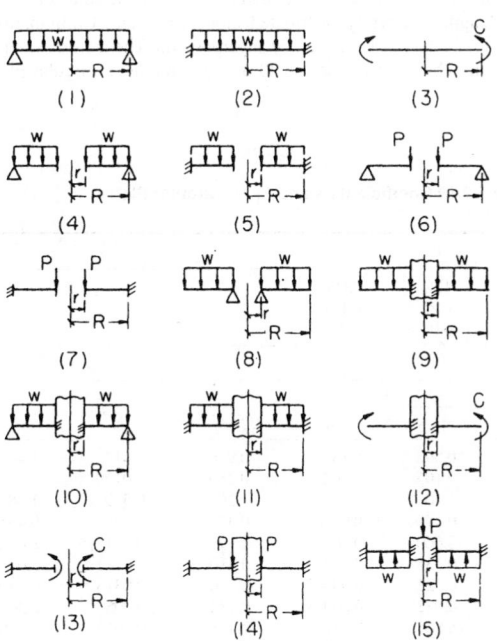

Fig. 5.2.69   Circular plates. Cases (4), (5), (6), (7), (8), and (13) have central hole of radius r; cases (9), (10), (11), (12), (14), and (15) have a central piston of radius r to which the plate is fixed.

(16) Rect., simply supported on all edges.

(17) Rect., fixed on all edges.

(18) Rect., simply supported on three edges; one edge R free

(19) Rect., fixed on one edge r, simply supported on other three edges.

(20) Rect., fixed on two opposite edges r, and simply supported on other two edges R.

(21) Same as (20) but r > R

(22) Rect., simply supported on all edges.

(23) Same as (22) but r > R

(24) Elliptical, 2R major axis, 2r minor axis, uniformly loaded, simply supported along edge.

(25) Same as (24) but fixed along edge.

Fig. 5.2.70 Rectangular and elliptical plates. [R is the longer dimension except in cases (21) and (23).]

The first equation is for a uniformly distributed load w, lb/in²; the second supports a concentrated load P, lb; and the third a couple C, per unit length, uniformly distributed along the edge. Combinations of these loadings may be treated by superposition. The factors k and k_1 are given in Tables 5.2.19 and 5.2.20; R is the radius of circular plates or

one side of rectangular plates, and t is the plate thickness. [In Figs. 5.2.69 and 5.2.70, r = radius of hole or piston for circular plates and r = smaller side of rectangular plates.]

The **maximum deflection** for the same cases is given by

$$y_M = k_1 \frac{wR^4}{Et^3} \qquad y_M = k_1 \frac{PR^2}{Et^3} \quad \text{and} \quad y_M = k_1 \frac{CR^2}{Et^3}$$

The factors k_1 are also given in the tables. For additional information, including shells, refer to ASME Handbook, "Metals Engineering: Design," McGraw-Hill.

THEORIES OF FAILURE

Material properties are usually determined from tests in which specimens are subjected to **simple stresses** under static or fluctuating loads. The attempt to apply these data to **bi- or triaxial stress fields** has resulted in the proposal of various theories of failure. Figure 5.2.71 shows the principal stresses on a triaxially stressed element. It is assumed, for simplicity. that $S_1 > S_2 > S_3$. Compressive stresses are negative.

1. **Maximum-stress theory** (Rankine) assumes failure occurs when the largest principal stress reaches the yield stress in a tension (or compression) specimen. That is, $S_1 = \pm S_y$.

2. **Maximum-shear theory** (Coulomb) assumes yielding (failure) occurs when the maximum shearing stress equals that in a simple tension (or compression) specimen at yield. Mathematically, $S_1 - S_3 = \pm S_y$.

3. **Maximum-strain-energy theory** (Beltrami) assumes failure occurs when the energy absorbed per unit volume equals the strain energy per unit volume in a tension (or compression) specimen at yield. Mathematically, $S_1^2 + S_2^2 + S_3^2 - 2\mu(S_1 S_2 + S_2 S_3 + S_3 S_1) = S_y^2$.

4. **Maximum-distortion-energy theory** (Huber, von Mises, Hencky) assumes yielding occurs when the distortion energy equals that in simple tension at yield. The distortion energy, that portion of the total energy which causes distortion rather than volume change, is

$$U_d = \frac{1 + \mu}{3E}(S_1^2 + S_2^2 + S_3^2 - S_1 S_2 - S_2 S_3 - S_3 S_1)$$

Thus failure is defined by

$$S_1^2 + S_2^2 + S_3^2 - (S_1 S_2 + S_2 S_3 + S_3 S_1) = S_y^2$$

5. **Maximum-strain theory** (Saint-Venant) claims failure occurs when the maximum strain equals the strain in simple tension at yield or $S_1 - \mu(S_2 + S_3) = S_y$.

Table 5.2.19 Coefficients k and k_1 for Circular Plates
($\mu = 0.3$)

Case	k	k_1
1	1.24	0.696
2	0.75	0.171
3	6.0	4.2

| | R/r | | | | | | | | | | | |
| | 1.25 | | 1.5 | | 2 | | 3 | | 4 | | 5 | |
Case	k	k_1	k	k_1	k	k_1	k	k_1	k	k_1	k	k_1
4	0.592	0.184	0.976	0.414	1.440	0.664	1.880	0.824	2.08	0.830	2.19	0.813
5	0.105	0.0025	0.259	0.0129	0.481	0.057	0.654	0.130	0.708	0.163	0.730	0.176
6	1.10	0.341	1.26	0.519	1.48	0.672	1.88	0.734	2.17	0.724	2.34	0.704
7	0.195	0.0036	0.320	0.024	0.455	0.081	0.670	0.171	1.00	0.218	1.30	0.238
8	0.660	0.202	1.19	0.491	2.04	0.902	3.34	1.220	4.30	1.300	5.10	1.310
9	0.135	0.0023	0.410	0.0183	1.04	0.0938	2.15	0.293	2.99	0.448	3.69	0.564
10	0.122	0.00343	0.336	0.0313	0.740	0.1250	1.21	0.291	1.45	0.417	1.59	0.492
11	0.072	0.00068	0.1825	0.005	0.361	0.023	0.546	0.064	0.627	0.092	0.668	0.112
12	6.865	0.2323	7.448	0.6613	8.136	1.493	8.71	2.555	8.930	3.105	9.036	3.418
13	6.0	0.196	6.0	0.485	6.0	0.847	6.0	0.940	6.0	0.801	6.0	0.658
14	0.115	0.00129	0.220	0.0064	0.405	0.0237	0.703	0.062	0.933	0.092	1.13	0.114
15	0.090	0.00077	0.273	0.0062	0.710	0.0329	1.54	0.110	2.23	0.179	2.80	0.234

Table 5.2.20   Coefficients k and k_1 for Rectangular and Elliptical Plates
($\mu = 0.3$)

	R/r									
	1.0		1.5		2.0		3.0		4.0	
Case	k	k_1	k	k_1	k	k_1	k	k_1	k	k_1
16	0.287	0.0443	0.487	0.0843	0.610	0.1106	0.713	0.1336	0.741	0.1400
17	0.308	0.0138	0.454	0.0240	0.497	0.0277	0.500	0.028	0.500	0.028
18	0.672	0.140	0.768	0.160	0.792	0.165	0.798	0.166	0.800	0.166
19	0.500	0.030	0.670	0.070	0.730	0.101	0.750	0.132	0.750	0.139
20	0.418	0.0209	0.626	0.0582	0.715	0.0987	0.750	0.1276	0.750	
21*	0.418	0.0216	0.490	0.0270	0.497	0.0284	0.500	0.0284	0.500	0.0284
22	0.160	0.0221	0.260	0.0421	0.320	0.0553	0.370	0.0668	0.380	0.0700
23*	0.160	0.0220	0.260	0.0436	0.340	0.0592	0.430	0.0772	0.490	0.0908
24	1.24	0.70	1.92	1.26	2.26	1.58	2.60	1.88	2.78	2.02
25	0.75	0.171	1.34	0.304	1.63	0.379	1.84	0.419	1.90	0.431

* Length ratio is r/R in cases 21 and 23.

6. **Internal-friction theory** (Mohr). When the ultimate strengths in tension and compression are the same, this theory reduces to that of maximum shear. For principal stresses of opposite sign, failure is defined by $S_1 - (S_{uc}/S_u) S_2 = -S_{uc}$; if the signs are the same $S_1 = S_u$ or $-S_{uc}$, where S_{uc} is the ultimate strength in compression. If the principal stresses are both either tension or compression, then the larger one, say S_1, must equal S_u when S_1 is tension or S_{uc} when S_1 is compression.

A **graphical representation** of the first four theories applied to a biaxial stress field is presented in Fig. 5.2.72. Stresses outside the bounding lines in the case of each theory mean failure (yield or fracture). A comparison with experimental data proves the distortion-energy theory (4) best for ductile materials of equal tension-compression properties. When these properties are unequal, the internal friction theory (6) appears best. In practice, judging by some accepted codes, the maximum-shear theory (2) is generally used for ductile materials, and the maximum-stress theory (1) for brittle materials.

Fatigue failures cannot be related, theoretically, to elastic strength and thus to the theories described. However, experimental results justify this, at least to a limited extent. Therefore, the theory evaluation given above holds for **fluctuating stresses,** provided that principal stresses at the maximum load are used and the **endurance strength** in simple bending is substituted for the yield strength.

EXAMPLE.   A steel shaft, 4 in diameter, is subjected to a bending moment of 120,000 in · lb, as well as a torque. If the yield strength in tension is 40,000 lb/in², what maximum torque can be applied under the (1) maximum-shear theory and (2) the distortion-energy theory?

$$S_x = \frac{M_c}{I} = \frac{120,000 \times 2}{12.55}$$

$$= 19,100 \text{ lb/in}^2 \qquad S_{xy} = \frac{TC}{J} = \frac{T \times 2}{25.1} = 0.0798T$$

and

$$S_{M,m} = \frac{S_x}{2} \pm \sqrt{\left(\frac{S_x}{2}\right)^2 + S_{xy}^2}$$

$$S_M - S_m = S_y \qquad \text{or} \qquad 2\sqrt{\left(\frac{19,100}{2}\right)^2 + (0.0798T)^2} \qquad (1)$$

$$= 40,000$$

or

$$T = 221,000 \text{ in} \cdot \text{lb}(254,150 \text{ cm} \cdot \text{kgf})$$

$$S_M^2 + S_m^2 - S_M S_m = S_y^2 \qquad (2)$$

substituting and simplifying,

$$(9,550)^2 + 3\sqrt{\left(\frac{19,100}{2}\right)^2 + (0.0798T)^2} = (40,000)^2$$

or

$$T = 255,000 \text{ in} \cdot \text{lb}(293,250 \text{ cm} \cdot \text{kgf})$$

PLASTICITY

The reaction of materials to stress and strain in the plastic range is not fully defined. However, some concepts and theories have been proposed.

Ideally, a **purely elastic material** is one complying explicitly with Hooke's law. In a **viscous material,** the shearing stress is proportional to the shearing strain. The **purely plastic material** yields indefinitely, but only after reaching a certain stress. Combinations of these are the **elastoviscous** and the **elastoplastic** materials.

Engineering materials are not ideal, but usually contain some of the elastoplastic characteristics. The **total strain** ε_t is the sum of the **elastic strain** ε_o plus the **plastic strain** ε_p, as shown in Fig. 5.2.73, where the stress-strain curve is approximated by two straight lines. The **natural strain,** which is at the same time the total strain, is $\varepsilon = \int_{l_o}^{l} dl/l =$ ln (l/l_o). In this equation, l is the instantaneous length, while l_o is the original length. In terms of the normal strain, the natural strain becomes $\bar{\varepsilon} = \ln(1 + \varepsilon_o)$. Since it is assumed that the volume remains constant, $l/l_o = A_o/A$, and so the natural stress becomes $\bar{S} = P/A = (P/A_o)(1 + \varepsilon_o)$. A_o is the original cross-sectional area. If the natural stress is plotted against strain on log-log paper, the graph is very nearly a straight line. The plastic-range relation is thus approximated by $\bar{S} = K\bar{\varepsilon}^n$, where the proportionality factor K and the **strain-hardening coefficient** n are determined from best fits to experimental data. Values of K and n

Fig. 5.2.71

Maximum stress
Maximum shear

Maximum strain energy
Maximum distortion energy

Fig. 5.2.72

Table 5.2.21   Constants *K* and *n* for Sheet Materials

Material	Treatment	K, lb/in^2	n
0.05%C rimmed steel	Annealed	77,100	0.261
0.05%C killed steel	Annealed and tempered	73,100	0.234
Decarburized 0.05%C steel	Annealed in wet H$_2$	75,500	0.284
0.05/0.07% phos. low C	Annealed	93,330	0.156
SAE 4130	Annealed	169,400	0.118
SAE 4130	Normalized and tempered	154,500	0.156
Type 430 stainless	Annealed	143,000	0.229
Alcoa 24-S	Annealed	55,900	0.211
Reynolds R-301	Annealed	48,450	0.211

determined by Low and Garofalo (*Proc. Soc. Exp. Stress Anal.*, vol. IV, no. 2,1947) are given in Table 5.2.21.

Fig. 5.2.73

The geometry of Fig. 5.2.73 can be used to arrive at a second approximate relation

$$\bar{S} = S_o + (\varepsilon_p - \varepsilon_o)\tan\theta = S_o\left(1 - \frac{H}{E}\right) + \varepsilon_p H$$

where $H = \tan\theta$ is a kind of **plastic modulus**.

The **deformation theory of plastic flow** for the general case of combined stress is developed using the above concepts. Certain additional assumptions involved include: principal plastic-strain directions are the same as principal stress directions; the elastic strain is negligible compared to plastic strain; and the ratios of the three principal shearing strains—$(\bar{\varepsilon}_1 - \bar{\varepsilon}_2)$, $(\bar{\varepsilon}_2 - \bar{\varepsilon}_3)$, $(\bar{\varepsilon}_3 - \bar{\varepsilon}_1)$—to the principal shearing stresses—$(\bar{S}_1 - \bar{S}_2)/2$, $(\bar{S}_2 - \bar{S}_3)/2$, $(\bar{S}_3 - \bar{S}_1)/2$—are equal. The relations between the principal strains and stresses in terms of the simple tension quantities become

$$\bar{\varepsilon}_1 = \bar{\varepsilon}/\bar{S}[\bar{S}_1 - (\bar{S}_2 + \bar{S}_3)/2]$$
$$\bar{\varepsilon}_2 = \bar{\varepsilon}/\bar{S}[\bar{S}_2 - (\bar{S}_3 + \bar{S}_1)/2]$$
$$\bar{\varepsilon}_3 = \bar{\varepsilon}/\bar{S}[\bar{S}_3 - (\bar{S}_1 + \bar{S}_2)/2]$$

If these equations are added, the plastic-flow theory is expressed:

$$\frac{\bar{S}}{\bar{\varepsilon}} = \sqrt{\frac{[(\bar{S}_1 - \bar{S}_2)^2 + (\bar{S}_2 - \bar{S}_3)^2 + (\bar{S}_3 - \bar{S}_1)^2]/2}{2(\bar{\varepsilon}_1^2 + \bar{\varepsilon}_2^2 + \bar{\varepsilon}_3^2)/3}}$$

In the above equation

$$\sqrt{[(\bar{S}_1 - \bar{S}_2)^2 + (\bar{S}_2 - \bar{S}_3)^2 + (\bar{S}_3 - \bar{S}_1)^2]/2} = \bar{S}_e$$

and

$$\sqrt{2(\bar{\varepsilon}_1^2 + \bar{\varepsilon}_2^2 + \bar{\varepsilon}_3^2)/3} = \varepsilon_e$$

are the effective, or significant, stress and strain, respectively,

EXAMPLE.   An annealed, stainless-steel type 430 tank has a 41-in inside diameter and has a wall 0.375 in thick. The ultimate strength of the stainless steel is 85,000 lb/in^2. Compute the maximum strain as well as the pressure at fracture.

The tank constitutes a biaxial stress field where $S_1 = pd/(2t)$, $S_2 = pd/(4t)$, and $S_3 = 0$. Taking the power stress-strain relation

$$\bar{S}_e = K\bar{\varepsilon}_e^n \quad \text{or} \quad \bar{\varepsilon}/\bar{S} = \bar{S}_e^{(1-n)/n}/K^{1/n}$$

thus

$$\bar{\varepsilon}_1 = \frac{\bar{S}_e^{(1-n)/n}}{K^{1/n}}\left(\frac{3}{4}\bar{S}_1\right)\bar{\varepsilon} = 0$$

and

$$\bar{\varepsilon}_3 = \frac{\bar{S}_e^{(1-n)/n}}{K^{1/n}}\left(-\frac{3}{4}\bar{S}_t\right) = -\bar{\varepsilon}_1$$

The maximum-shear theory, which is applicable to a ductile material under combined stress, is acceptable here. Thus rupture will occur at

$$\bar{S}_1 - \bar{S}_3 = \bar{S}_u, \text{ and}$$

$$\bar{S}_e = \sqrt{\frac{1}{2}\left[\left(\frac{\bar{S}_1}{2}\right)^2 + \left(\frac{\bar{S}_1}{2}\right)^2 + \bar{S}_1^2\right]} = \sqrt{\frac{3}{4}\bar{S}_1^2} = \left(\frac{3}{4}\right)^{1/2}\bar{S}_u$$

$$\bar{\varepsilon}_1 = \frac{[(3/4)]^{1/2}\bar{S}_u]^{(1-n)/n}}{K^{1/n}}\left(\frac{3}{4}\bar{S}_u\right) = \left(\frac{3}{4}\right)^{1+0.229/0.458}\left(\frac{85,000}{143,000}\right)^{1/0.225}$$

$$= 0.0475 \text{ in/in } (0.0475 \text{ cm/cm})$$

since

$$\bar{S}_u = S_1 = \frac{pd}{2t}, \text{ then } p = \frac{2t\bar{S}_u}{d}$$

or

$$p = \frac{2 \times 0.375 \times 85,000}{41} = 1,550 \text{ lb/in}^2 \text{ (109 kgf/cm}^2)$$

ROTATING DISKS

Rotating circular disks may be of various profiles, of constant or variable thickness, with or without centrally and noncentrally located holes, and with radial, tangential, and shearing stresses.

Solution starts with the differential equations of equilibrium and compatibility and the subsequent application of appropriate boundary conditions for the derivation of working-stress equations.

If the disk thickness is small compared with the diameter, the variation of stress with thickness can be assumed to be negligible, and symmetry eliminates the shearing stress. In the rotating case, the disk weight is neglected, but its inertia force becomes the body-force term in the equilibrium equations.

Thus solved, the stress components in a solid disk become

$$\sigma_r = \frac{3 + \mu}{8}\rho\omega^2(R^2 - r^2)$$

$$\sigma_\theta = \frac{3 + \mu}{8}\rho\omega^2 R^2 - \frac{1 + 3\mu}{8}\rho\omega^2 r^2$$

where μ = Poisson's ratio; ρ = mass density, lb $\cdot$ s²/in⁴; ω = angular speed, rad/s; R = outside disk radius; and r = radius to point in question.

The largest stresses occur at the center of the solid disk and are

$$\sigma_r = \sigma_\theta = \frac{3 + \mu}{8}\rho\omega^2 R^2$$

A disk with a central hole of radius r_h (no external forces) is subjected to the following stresses:

$$\sigma_r = \frac{3 + \mu}{8}\rho\omega^2\left(R^2 + r_h^2 - \frac{R^2 r_h^2}{r^2} - r^2\right)$$

$$\sigma_\theta = \frac{3 + \mu}{8}\rho\omega^2\left(R^2 + r_h^2 + \frac{R^2 r_h^2}{r^2} - \frac{1 + 3\mu}{3 + \mu}r^2\right)$$

The maximum radial stress $\sigma_r|_M$ occurs at $r = \sqrt{Rr_h}$, and

$$\sigma_r|_M = \frac{3 + \mu}{8}\rho\omega^2(R - r_h)^2$$

The largest tangential stress $\sigma_\theta|_M$ exists at the inner boundary, and

$$\sigma_\theta|_M = \frac{3 + \mu}{4}\rho\omega^2\left(R^2 + \frac{1 - \mu}{3 + \mu}r_h^2\right)$$

As the hole radius r_h approaches zero, the tangential stress assumes a value twice that at the center of a rotating solid disk, given above.

5.3 PIPELINE FLEXURE STRESSES
by Harold V. Hawkins

EDITOR'S NOTE: The almost universal availability and utilization of personal computers in engineering practice has led to the development of many competing and complementary forms of piping stress analysis software. Their use is widespread, and individual packaged software allows analysis and design to take into account static and dynamic conditions, restraint conditions, aboveground and buried configurations, etc. The reader is referred to the technical literature for the most suitable and current software available for use in solving the immediate problems at hand.

The brief discussion in this section addresses the fundamental concepts entailed and sets forth the solution of simple systems as an exercise in application of the principles.

REFERENCES: Shipman, Design of Steam Piping to Care for Expansion, *Trans. ASME*, 1929. Wahl, Stresses and Reactions in Expansion Pipe Bends, *Trans. ASME*, 1927. Hovgaard, The Elastic Deformation of Pipe Bends, *Jour. Math. Phys.*, Nov. 1926, Oct. 1928, and Dec. 1929. M. W. Kellog Co., "The Design of Piping Systems," Wiley.

For details of pipe and pipe fittings see Sec. 8.7.

Nomenclature (see Figs. 5.3.1 and 5.3.2)

M_0 = end moment at origin, in $\cdot$ lb (N $\cdot$ m)
M = max moment, in $\cdot$ lb (N $\cdot$ m)
F_x = end reaction at origin in x direction, lb (N)
F_y = end reaction at origin in y direction, lb (N)
S_l = $(Mr/I)\alpha$ = max unit longitudinal flexure stress, lb/in² (N/m²)
S_t = $(Mr/I)\beta$ = max unit transverse flexure stress, lb/in² (N/m²)
S_s = $(Mr/I)\gamma$ = max unit shearing stress, lb/in² (N/m²)
Δx = relative deflection of ends of pipe parallel to x direction caused by either temperature change or support movement, or both, in (m)

Δy = same as Δx but parallel to y direction, in. Note that Δx and Δy are positive if under the change in temperature the end opposite the origin tends to move in a positive x or y direction, respectively.
t = wall thickness of pipe, in (m)
r = mean radius of pipe cross section, in (m)
λ = constant = tR/r^2
I = moment of inertia of pipe cross section about pipe centerline, in⁴ (m⁴)
E = modulus of elasticity of pipe at *actual working* temperature, lb/in² (N/m²)
K = flexibility index of pipe. K = 1 for all straight pipe sections, K = $(10 + 12\lambda^2)/(1 + 12\lambda^2)$ for all curved pipe sections where $\lambda > 0.335$ (see Fig. 5.3.3)
a, β, γ = ratios of actual max longitudinal flexure, transverse flexure, and shearing stresses to Mr/I for curved sections of pipe (see Fig. 5.3.3)

Fig. 5.3.1   **Fig. 5.3.2**

Fig. 5.3.3 Flexure constants of initially curved pipes.

A, B, C, F, G, H = constants given by Table 5.3.2

θ = angle of intersection between tangents to direction of pipe at reactions

$\Delta\theta$ = change in θ caused by movements of supports, or by temperature change, or both, rad

ds = an infinitesimal element of length of pipe

s = length of a particular curved section of pipe, in (m)

R = radius of curvature of pipe centerline, in (m)

General Discussion

Under the effect of changes in temperature of the pipeline, or of movement of support reactions (either translation or rotation), or both, the determination of stress distribution in a pipe becomes a statically indeterminate problem. In general the problem may be solved by a slight modification of the standard arch theory: $\Delta x = -K\int My\,ds(EI)$, $\Delta y = K\int Mx\,ds/(EI)$, and $\Delta\theta = K\int M\,ds/(EI)$ where the constant K is introduced to correct for the increased flexibility of a curved pipe, and where the integration is over the entire length of pipe between supports. In Table 5.3.1 are given equations derived by this method for moment and thrust at one reaction point for pipes in one plane that are fully fixed, hinged at both ends, hinged at one end and fixed at the

other, or partly fixed. If the reactions at one end of the pipe are known, the moment distribution in the entire pipe then can be obtained by simple statics.

Since an initially curved pipe is more flexible than indicated by its moment of inertia, the constant K is introduced. Its value may be taken from Fig. 5.3.3, or computed from the equation given below. $K = 1$ for all straight pipe sections, since they act according to the simple flexure theory.

In Fig. 5.3.3 are given the flexure constants K, α, β, and γ for initially curved pipes as functions of the quantity $\lambda = tR/r^2$. The flexure constants are derived from the equations.

$$K = (10 + 12\lambda^2)/(1 + 12\lambda^2) \qquad \text{when } \lambda > 0.335$$

$$\alpha = \tfrac{2}{3}K\sqrt{(5 + 6\lambda^2)/18} \qquad \lambda \leq 1.472$$

$$\alpha = K(6\lambda^2 - 1)/(6\lambda^2 + 5) \qquad \lambda > 1.472$$

$$\beta = 18\lambda/(1 + 12\lambda^2)$$

$$\gamma = [8\lambda - 36\lambda^3 + (32\lambda^2 + 20/3)$$
$$\times \sqrt{(4/3)\lambda^2 + 5/18}] \div (1 + 12\lambda^2) \qquad \text{when } \lambda < 0.58$$
$$= (12\lambda^2 + 18\lambda - 2)/(1 + 12\lambda^2) \qquad \text{when } \lambda > 0.58$$

Table 5.3.1   General Equations for Pipelines in One Plane (See Figs. 5.3.1 and 5.3.2)

Type of supports	Unsymmetric	Symmetric about y-axis
Both ends fully fixed	$M_0 = \dfrac{EI\,\Delta x(CF - AB) + EI\,\Delta y(BF - AG)}{2ABF + CGH - B^2H - A^2G - CF^2}$	$M_0 = \dfrac{EI\,\Delta x\,F}{GH - F^2}$
	$F_x = \dfrac{EI\,\Delta x(CH - A^2) + EI\,\Delta y(BH - AF)}{2ABF + CGH - B^2H - A^2G - CF^2}$	$F_x = \dfrac{EI\,\Delta x\,H}{GH - F^2}$
	$F_y = \dfrac{EI\,\Delta x(BH - AF) + EI\,\Delta y(GH - F^2)}{2ABF + CGH - B^2H - A^2G - CF^2}$	$F_y = 0$
	$\Delta\theta = 0$	$\Delta\theta = 0$
Both ends hinged	$M_0 = 0$	$M_0 = 0$
	$F_x = \dfrac{EI\,\Delta x\,C + EI\,\Delta y\,B}{CG - B^2}$	$F_x = \dfrac{EI\,\Delta x}{G}$
	$F_y = \dfrac{EI\,\Delta x\,B + EI\,\Delta y\,G}{CG - B^2}$	$F_y = 0$
	$\Delta\theta = \dfrac{\Delta x(AB - CF) + \Delta y(AG - BF)}{CG - B^2}$	$\Delta\theta = \dfrac{-\Delta x\,F}{G}$
Origin end only hinged, other end fully fixed	$M_0 = 0$	
	$F_x = \dfrac{EI\,\Delta x\,C + EI\,\Delta y\,B}{CG - B^2}$	
	$F_y = \dfrac{EI\,\Delta x\,B + EI\,\Delta y\,G}{CG + B^2}$	
	$\Delta\theta = \dfrac{\theta x(AB - CF) + \Delta y(AG - BF)}{CG - B^2}$	
In general for any specific rotation $\Delta\theta$ and movement Δx and Δy ...	$M_0 = \dfrac{EI\,\Delta x(CF - AB) + EI\,\Delta y(BF - AG) + EI\,\Delta\theta(CG - B^2)}{2ABF + CGH - A^3G - CF^2 - B^2H}$	
	$F_x = \dfrac{EI\,\Delta x(CH - A^2) + EI\,\Delta y(BH - AF) + EI\,\Delta\theta(CF - AB)}{2ABF + CGH - A^2G - CF^2 - B^2H}$	
	$F_y = \dfrac{EI\,\Delta x(BH - AF) + EI\,\Delta y(GH - F^2) + EI\,\Delta\theta(BF - AG)}{2ABF + CGH - A^2G - CF^2 - B^2H}$	

The increased flexibility of the curved pipe is brought about by the tendency of its cross section to flatten. This flattening causes a transverse flexure stress whose maximum is S_r. Because the maximum longitudinal and maximum transverse stresses do not occur at the same point in the pipe's cross section, the resulting maximum shear is not one-half the difference of S_l and S_t; it is S_s. In the straight sections of the pipe, $\alpha = 1$, the transverse stress disappears, and $\lambda = \frac{1}{2}$. This discussion of S_s does not include the uniform transverse or longitudinal tension stresses induced by the internal pressure in the pipe; their effects should be added if appreciable.

Table 5.3.2 gives values of the constants A, B, C, F, G, and H for use in equations listed in Table 5.3.1. The values may be used (1) for the solution of any pipeline or (2) for the derivation of equations for standard shapes composed of straight sections and arcs of circles as of Fig. 5.3.5. Equations for shapes not given may be obtained by algebraic addition of those given. All measurements are from the left-hand end of the pipeline. Reactions and stresses are greatly influenced by end conditions. Formulas are given to cover the extreme conditions. The following suggestions and comments should be considered when laying out a pipeline:

Avoid expansion bends, and design the entire pipeline to take care of its own expansion.

The movement of the equipment to which the ends of the pipeline are attached must be included in the Δx and Δy of the equations.

Maximum flexibility is obtained by placing **supports** and **anchors** so that they will not interfere with the natural movement of the pipe.

That shape is most efficient in which the **maximum length of pipe is working** at the maximum safe stress.

Excessive **bending moment at joints** is more likely to cause trouble than excessive stresses in pipe walls. Hence, keep pipe joints away from points of high moment.

Reactions and stresses are greatly influenced by **flattening of the cross section** of the curved portions of the pipeline.

It is recommended that **cold springing** allowances be discounted in stress calculations.

Application to Two- and Three-Plane Pipelines   Pipelines in more than one plane may be solved by the successive application of the preceding data, dividing the pipeline into two or more one-plane lines.

EXAMPLE 1.   The unsymmetric pipeline of Fig. 5.3.4 has fully fixed ends. From Table 5.3.2 use $K = 1$ for all sections, since only straight segments are involved.

Upon introduction of $a = 120$ in (3.05 m), $b = 60$ in (1.52 m), and $c = 180$ in (4.57 m), into the preceding relations (Table 5.3.3) for A, B, C, F, G, H, the equations for the reactions at 0 from Table 5.3.1 become

$$M_0 = EI\,\Delta x(-7.1608 \times 10^{-5}) + EI\,\Delta y(-8.3681 \times 10^{-5})$$

$$F_x = EI\,\Delta x(+1.0993 \times 10^{-5}) + EI\,\Delta y(+3.1488 \times 10^{-6})$$

$$F_y = EI\,\Delta x(+3.1488 \times 10^{-6}) + EI\,\Delta y(+1.33717 \times 10^{-6})$$

Also it follows that

$$M_1 = M_0 + F_y a = EI\,\Delta x(+3.0625 \times 10^{-4}) + EI\,\Delta y(+7.6799 \times 10^{-5})$$

Fig. 5.3.4

$$M_2 = M_1 - F_x b = EI\,\Delta x(-3.5333 \times 10^{-4}) + EI\,\Delta y(-1.1215 \times 10^{-4})$$

$$M_3 = M_2 + F_y c = EI\,\Delta x(+2.1345 \times 10^{-4}) + EI\,\Delta y(+1.2854 \times 10^{-4})$$

Thus the maximum moment M occurs at 3.

The total maximum longitudinal fiber stress ($a = 1$ for straight pipe)

$$S_l = \frac{F_x}{2\pi r t} \pm \frac{M_3 r}{I}$$

There is no transverse flexure stress since all sections are straight. The maximum shearing stress is either (1) one-half of the maximum longitudinal fiber stress as given above, (2) one-half of the hoop-tension stress caused by an internal radial pressure that might exist in the pipe, or (3) one-half of the difference of the maximum longitudinal fiber stress and hoop-tension stress, whichever of these three possibilities is numerically greatest.

EXAMPLE 2.   The equations of Table 5.3.1 may be employed to develop the solution of generalized types of pipe configurations for which Fig. 5.3.5 is a typical example. If only temperature changes are considered, the reactions for the right-angle pipeline (Fig. 5.3.5) may be determined from the following equations:

$$M_0 = C_1 EI\,\Delta x/R^2$$

$$F_x = C_2 EI\,\Delta x/R^3$$

$$F_y = C_3 EI\,\Delta x/R^3$$

In these equations, Δx is the x component of the deflection between reaction points caused by temperature change only. The values of C_1, C_2, and C_3 are given in Fig. 5.3.6 for $K = 1$ and $K = 2$. For other values of K, interpolation may be employed.

Fig. 5.3.5   Right angle pipeline.

EXAMPLE 3.   With $a/R = 20$ and $b/R = 3$, the value of C_1 is 0.185 for $K = 1$ and 0.165 for $K = 2$. If $K = 1.75$, the interpolated value of C_1 is 0.175.

Elimination of Flexure Stresses   Pipeline flexure stresses that normally would result from movement of supports or from the tendency of the pipes to expand under temperature change often may be avoided entirely through the use of expansion joints (Sec. 8.7). Their use may simplify both the design of the pipeline and the support structure. When using expansion joints, the following suggestions should be considered: (1) select the expansion joint carefully for maximum temperature range (and deflection) expected so as to prevent damage to expansion fitting; (2) provide guides to limit movement at the expansion joint to direction permitted by joint; (3) provide adequate anchors at one end of each straight section or along its midlength, forcing movement to occur at the expansion joint yet providing adequate support for the pipeline; (4) mount expansion joints adjacent to an anchor point to prevent sagging of the pipeline under its own weight and do not depend upon the expansion joint for stiffness—it is intended to be flexible; (5) give consideration to effects of corrosion, since the corrugated character of expansion joints makes cleaning difficult.

Table 5.3.2 Values of A, B C, F, G, and H for Various Piping Elements

Element	$A = K\int x\,ds$	$B = K\int xy\,ds$	$C = K\int x^2\,ds$	$F = K\int y\,ds$	$G = K\int y^2\,ds$	$H = K\int ds$
(straight horizontal) $(x_1,y)\ (x_2,y)$, length s	$\dfrac{s}{2}(x_1 + x_2)$	Ay	$s\left(\dfrac{s^2}{3} + x_1 x_2\right)$	sy	Fy	s
(straight vertical) $(x,y_2)\ (x,y_1)$, length s	sx	$\dfrac{A}{2}(y_1 + y_2)$	Ax	$\dfrac{s}{2}(y_1 + y_2)$	$s\left(\dfrac{s^2}{3} + y_1 y_2\right)$	s
(diagonal) $(x_2,y_2)\ (x_1,y_1)$, length s	$\dfrac{s}{2}(x_1 + x_2)$	$\dfrac{A}{3}(y_1 + y_2)$ $+ \dfrac{s}{6}(x_1 y_1 + x_2 y_2)$	$\dfrac{s}{3}(x_1^2 + x_1 x_2 + x_2^2)$	$\dfrac{s}{2}(y_1 + y_2)$	$\dfrac{s}{3}(y_1^2 + y_1 y_2 + y_2^2)$	s
(semicircle) (x,y)	$\pi K R x$	$A\left(y + \dfrac{2R}{\pi}\right)$	$A\left(x + \dfrac{R^2}{2x}\right)$	$(\pi y + 2R)KR$	$Fy + \left(2y + \dfrac{\pi}{2}R\right)KR^2$	πKR
(semicircle, second) (x,y)		$A\left(y - \dfrac{2R}{\pi}\right)$		$(\pi y - 2R)KR$	$Fy - \left(2y - \dfrac{\pi}{2}R\right)KR^2$	
(quarter circle) (x,y)	$\left(\dfrac{\pi x}{2} - R\right)KR$	$Ay + \left(x - \dfrac{R}{2}\right)KR^2$	$Ax + \left(\dfrac{\pi R}{4} - x\right)KR^2$	$\left(\dfrac{\pi y}{2} + R\right)KR$	$Fy + \left(\dfrac{\pi R}{4} + y\right)KR^2$	$\dfrac{\pi KR}{2}$

Shape						
Quarter arc, (x,y)	$\left(\dfrac{\pi x}{2} + R\right) KR$	$Ay - \left(x + \dfrac{R}{2}\right) KR^2$	$Ax + \left(\dfrac{\pi R}{4} + x\right) KR^2$	$\left(\dfrac{\pi y}{2} + R\right) KR$	$Fy + \left(\dfrac{\pi R}{4} + y\right) KR^2$	$\dfrac{\pi KR}{2}$
Quarter arc, (x,y)	$\left(\dfrac{\pi x}{2} - R\right) KR$	$Ay - \left(x + \dfrac{R}{2}\right) KR^2$	$Ax + \left(\dfrac{\pi R}{4} - x\right) KR^2$	$\left(\dfrac{\pi y}{2} - R\right) KR$	$Fy + \left(\dfrac{\pi R}{4} - y\right) KR^2$	
Quarter arc, (x,y)		$Ay - \left(x - \dfrac{R}{2}\right) KR^2$		$\left(\dfrac{\pi y}{2} - R\right) KR$	$Fy + \left(\dfrac{\pi R}{4} - y\right) KR^2$	
$45°$ arc, (x,y)	$\left(\dfrac{\pi x}{4} - \dfrac{R}{\sqrt{2}}\right) KR$	$Ay + \left[\left(1 - \dfrac{\sqrt{2}}{2}\right) x - \dfrac{R}{4}\right] KR^2$	$Ax + \left[\left(\dfrac{\pi}{8} + \dfrac{1}{4}\right) R - \dfrac{x\sqrt{2}}{2}\right] KR^2$	$\left[\dfrac{\pi y}{4} + \left(1 - \dfrac{\sqrt{2}}{2}\right) R\right] KR$	$Fy + \left[\left(1 - \dfrac{\sqrt{2}}{2}\right) R + \left(\dfrac{\pi}{8} - \dfrac{1}{4}\right) R\right] KR^2$	$\dfrac{\pi KR}{4}$
$45°$ arc, (x,y)		$Ay - \left[\left(1 - \dfrac{\sqrt{2}}{2}\right) x - \dfrac{R}{4}\right] KR^2$		$\left[\dfrac{\pi y}{4} - \left(1 - \dfrac{\sqrt{2}}{2}\right) R\right] KR$	$Fy - \left[\left(1 - \dfrac{\sqrt{2}}{2}\right) R - \left(\dfrac{\pi}{8} - \dfrac{1}{4}\right) R\right] KR^2$	
$45°$ arc, (x,y)	$\left(\dfrac{\pi x}{4} - \dfrac{R}{\sqrt{2}}\right) KR$	$Ay + \left[\left(1 - \dfrac{\sqrt{2}}{2}\right) x + \dfrac{R}{4}\right] KR^2$	$Ax + \left[\left(\dfrac{\pi}{8} + \dfrac{1}{4}\right) R - \dfrac{x\sqrt{2}}{2}\right] KR^2$	$\left[\dfrac{\pi y}{4} + \left(1 - \dfrac{\sqrt{2}}{2}\right) R\right] KR$	$Fy + \left[\left(1 - \dfrac{\sqrt{2}}{2}\right) R + \left(\dfrac{\pi}{8} - \dfrac{1}{4}\right) R\right] KR^2$	
$45°$ arc, (x,y)		$Ay - \left[\left(1 - \dfrac{\sqrt{2}}{2}\right) x + \dfrac{R}{4}\right] KR^2$		$\left[\dfrac{\pi y}{4} - \left(1 - \dfrac{\sqrt{2}}{2}\right) R\right] KR$	$Fy - \left[\left(1 - \dfrac{\sqrt{2}}{2}\right) R - \left(\dfrac{\pi}{8} - \dfrac{1}{4}\right) R\right] KR^2$	
General arc θ_1,θ_2, (x,y)	$[r(\theta_2 - \theta_1) - R(\sin\theta_2 - \sin\theta_1)] KR$	$Ay - \left[x(\cos\theta_2 - \cos\theta_1) + \dfrac{R}{2}(\sin^2\theta_2 - \sin^2\theta_1)\right] KR^2$	$Ax - \left[x(\sin\theta_2 - \sin\theta_1) - \dfrac{R}{4}(\sin 2\theta_2 - \sin 2\theta_1) - \dfrac{R}{2}(\theta_2 - \theta_1)\right] KR^2$	$[y(\theta_2 - \theta_1) - R(\cos\theta_2 - \cos\theta_1)] KR$	$Fy - \left[y(\cos\theta_2 - \cos\theta_1) + \dfrac{R}{4}(\sin 2\theta_2 - \sin 2\theta_1) - \dfrac{R}{2}(\theta_2 - \theta_1)\right] KR^2$	$(\theta_2 - \theta_1)KR$

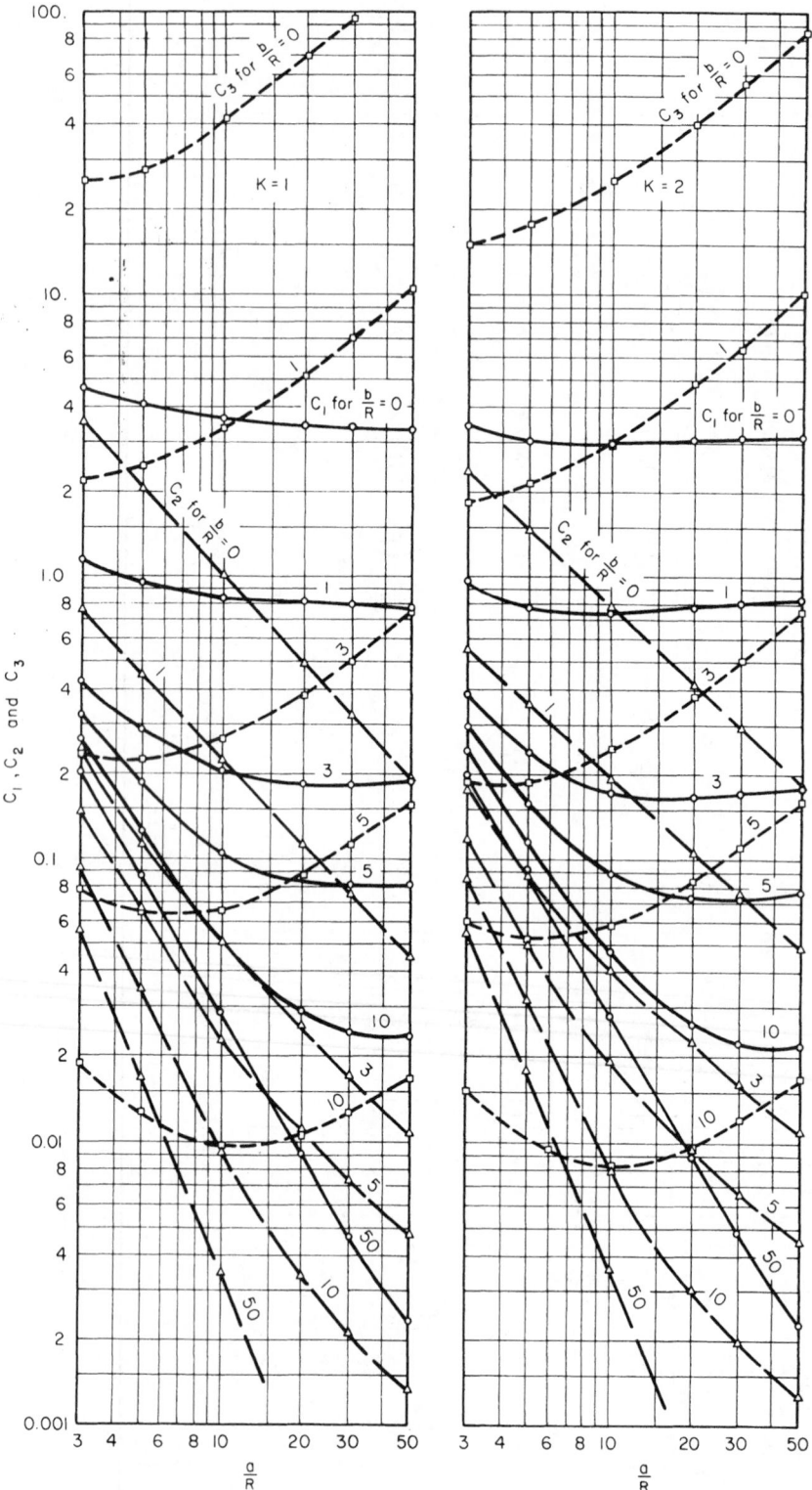

Fig. 5.3.6 Reactions for right-angle pipelines.

Table 5.3.3 Example 1 Showing Determination of Integrals

Part of pipe	Values of integrals					
	A	B	C	F	G	H
0–1	$\dfrac{a^2}{2}$	0	$\dfrac{a^3}{3}$	0	0	a
1–2	ab	$\dfrac{ab^2}{2}$	a^2b	$\dfrac{b^2}{2}$	$\dfrac{b^3}{3}$	b
2–3	$\dfrac{c}{2}(2a+c)$	$\dfrac{bc}{2}(2a+c)$	$\dfrac{c^3}{3}+ac(a+c)$	bc	b^2c	c
Total 0–3	$\dfrac{a^2}{2}+ab$ $+\dfrac{c}{2}(2a+c)$	$\dfrac{ab^2}{2}+\dfrac{bc}{2}(2a+c)$	$\dfrac{a^3}{3}+a^2b+\dfrac{c^3}{3}$ $+ac(a+c)$	$\dfrac{b^2}{2}+bc$	$\dfrac{b^3}{3}+b^2c$	$a+b+c$

5.4 NONDESTRUCTIVE TESTING
by Donald D. Dodge

REFERENCES: Various authors, "Nondestructive Testing Handbook," 8 vols., American Society for Nondestructive Testing. Boyer, "Metals Handbook," vol. 11, American Society for Metals. Heuter and Bolt, "Sonics," Wiley. Krautkramer, "Ultrasonic Testing of Materials," Springer-Verlag. Spanner, "Acoustic Emission: Techniques and Applications," Intex, American Society for Nondestructive Testing. Crowther, "Handbook of Industrial Radiography," Arnold. Wiltshire, "A Further Handbook of Industrial Radiography," Arnold. "Standards," vol. 03.03, ASTM, Boiler and Pressure Vessel Code, Secs. III, V, XI, ASME. ASME Handbook, "Metals Engineering—Design," McGraw-Hill. SAE Handbook, Secs. J358, J359, J420, J425-J428, J1242, J1267, SAE. Materials Evaluation, *Jour. Am. Soc. Nondestructive Testing.*

Nondestructive tests are those tests that determine the usefulness, serviceability, or quality of a part or material without limiting its usefulness. Nondestructive tests are used in machinery maintenance to avoid costly unscheduled loss of service due to fatigue or wear; they are used in manufacturing to ensure product quality and minimize costs. Consideration of test requirements early in the design of a product may facilitate testing and minimize testing cost. Nearly every form of energy is used in nondestructive tests, including all wavelengths of the electromagnetic spectrum as well as vibrational mechanical energy. Physical properties, composition, and structure are determined; flaws are detected; and thickness is measured. These tests are here divided into the following basic methods: **magnetic particle, penetrant, radiographic, ultrasonic, eddy current, acoustic emission, microwave, and infrared.** Numerous techniques are utilized in the application of each test method. Table 5.4.1 gives a summary of many nondestructive test methods.

MAGNETIC PARTICLE METHODS

Magnetic particle testing is a nondestructive method for detecting discontinuities at or near the surface in **ferromagnetic materials.** After the test object is properly magnetized, finely divided magnetic particles are applied to its surface. When the object is properly oriented to the induced magnetic field, a discontinuity creates a leakage flux which attracts and holds the particles, forming a visible indication. Magnetic-field direction and character are dependent upon how the magnetizing force is applied and upon the type of current used. For best sensitivity, the **magnetizing current** must flow in a direction parallel to the principal direction of the expected defect. Circular fields, produced by passing current through the object, are almost completely contained within the test object. Longitudinal fields, produced by coils or yokes, create external poles and a general-leakage field. Alternating, direct, or half-wave direct current may be used for the location of surface defects. Half-wave direct current is most effective for locating subsurface defects. Magnetic particles may be applied dry or as a wet suspension in a liquid such as kerosene or water. Colored **dry powders** are advantageous when testing for subsurface defects and when testing objects that have rough surfaces, such as castings, forgings, and weldments. **Wet particles** are preferred for detection of very fine cracks, such as fatigue, stress corrosion, or grinding cracks. Fluorescent wet particles are used to inspect objects with the aid of ultraviolet light. Fluorescent inspection is widely used because of its greater sensitivity. Application of particles while magnetizing current is on (continuous method) produces stronger indications than those obtained if the particles are applied after the current is shut off (residual method). Interpretation of subsurface-defect indications requires experience. Demagnetization of the test object after inspection is advisable.

Magnetic flux leakage is a variation whereby leakage flux due to flaws is detected electronically via a Hall-effect sensor. Computerized signal interpretation and data imaging techniques are employed.

Electrified particle testing indicates minute cracks in nonconducting materials. Particles of calcium carbonate are positively charged as they are blown through a spray gun at the test object. If the object is metal-backed, such as porcelain enamel, no preparation other than cleaning is necessary. When it is not metal-backed, the object must be dipped in an aqueous penetrant solution and dried. The penetrant remaining in cracks provides a mobile electron supply for the test. A readily visible powder indication forms at a crack owing to the attraction of the positively charged particles.

PENETRANT METHODS

Liquid penetrant testing is used to locate flaws open to the surface of **nonporous materials.** The test object must be thoroughly cleaned before testing. Penetrating liquid is applied to the surface of a test object by a brush, spray, flow, or dip method. A time allowance (1 to 30 min) is required for liquid penetration of surface flaws. Excess penetrant is then carefully removed from the surface, and an absorptive coating, known as developer, is applied to the object to draw penetrant out of flaws, thus showing their location, shape, and approximate size. The developer is typically a fine powder, such as talc usually in suspension in a liquid. Penetrating-liquid types are (1) for test in visible light, and (2) for test under ultraviolet light (3,650 Å). Sensitivity of penetrant testing is greatest when a fluorescent penetrant is used and the object is

Table 5.4.1   Nondestructive Test Methods*

Method	Measures or detects	Applications	Advantages	Limitations
Acoustic emission	Crack initiation and growth rate Internal cracking in welds during cooling Boiling or cavitation Friction or wear Plastic deformation Phase transformations	Pressure vessels Stressed structures Turbine or gearboxes Fracture mechanics research Weldments Sonic-signature analysis	Remote and continuous surveillance Permanent record Dynamic (rather than static) detection of cracks Portable Triangulation techniques to locate flaws	Transducers must be placed on part surface Highly ductile materials yield low-amplitude emissions Part must be stressed or operating Interfering noise needs to be filtered out
Acoustic-impact (tapping)	Debonded areas or delaminations in metal or nonmetal composites or laminates Cracks under bolt or fastener heads Cracks in turbine wheels or turbine blades Loose rivets or fastener heads Crushed core	Brazed or adhesive-bonded structures Bolted or riveted assemblies Turbine blades Turbine wheels Composite structures Honeycomb assemblies	Portable Easy to operate May be automated Permanent record or positive meter readout No couplant required	Part geometry and mass influences test results Impactor and probe must be repositioned to fit geometry of part Reference standards required Pulser impact rate is critical for repeatability
D-Sight (Diffracto)	Enhances visual inspection for surface abnormalities such as dents protrusions, or waviness Crushed core Lap joint corrosion Cold-worked holes Cracks	Detect impact damage to composites or honeycomb corrosion in aircraft lap joints Automotive bodies for waviness	Portable Fast, flexible Noncontact Easy to use Documentable	Part surface must reflect light or be wetted with a fluid
Eddy current	Surface and subsurface cracks and seams Alloy content Heat-treatment variations Wall thickness, coating thickness Crack depth Conductivity Permeability	Tubing Wire Ball bearings "Spot checks" on all types of surfaces Proximity gage Metal detector Metal sorting Measure conductivity in % IACS	No special operator skills required High speed, low cost Automation possible for symmetric parts Permanent-record capability for symmetric parts No couplant or probe contact required	Conductive materials Shallow depth of penetration (thin walls only) Masked or false indications caused by sensitivity to variations such as part geometry Reference standards required Permeability variations
Magneto-optic eddy-current imager	Cracks Corrosion thinning in aluminum	Aluminum aircraft Structure	Real-time imaging Approximately 4-in area coverage	Frequency range of 1.6 to 100 kHz Surface contour Temperature range of 32 to 90°F Directional sensitivity to cracks
Eddy-sonic	Debonded areas in metal-core or metal-faced honeycomb structures Delaminations in metal laminates or composites Crushed core	Metal-core honeycomb Metal-faced honeycomb Conductive laminates such as boron or graphite-fiber composites Bonded-metal panels	Portable Simple to operate No couplant required Locates far-side debonded areas Access to only one surface required May be automated	Specimen or part must contain conductive materials to establish eddy-current field Reference standards required Part geometry
Electric current	Cracks Crack depth Resistivity Wall thickness Corrosion-induced wall thinning	Metallic materials Electrically conductive materials Train rails Nuclear fuel elements Bars, plates other shapes	Access to only one surface required Battery or dc source Portable	Edge effect Surface contamination Good surface contact required Difficult to automate Electrode spacing Reference standards required
Electrified particle	Surface flaws in nonconducting material Through-to-metal pinholes on metal-backed material Tension, compression, cyclic cracks Brittle-coating stress cracks	Glass Porcelain enamel Nonhomogeneous materials such as plastic or asphalt coatings Glass-to-metal seals	Portable Useful on materials not practical for penetrant inspection	Poor resolution on thin coatings False indications from moisture streaks or lint Atmospheric conditions High-voltage discharge

Table 5.4.1   Nondestructive Test Methods*   (Continued)

Method	Measures or detects	Applications	Advantages	Limitations
Filtered particle	Cracks Porosity Differential absorption	Porous materials such as clay, carbon, powdered metals, concrete Grinding wheels High-tension insulators Sanitary ware	Colored or fluorescent particles Leaves no residue after baking part over 400°F Quickly and easily applied Portable	Size and shape of particles must be selected before use Penetrating power of suspension medium is critical Particle concentration must be controlled Skin irritation
Infrared (radiometry) (thermography)	Hot spots Lack of bond Heat transfer Isotherms Temperature ranges	Brazed joints Adhesive-bonded joints Metallic platings or coatings; debonded areas or thickness Electrical assemblies Temperature monitoring	Sensitive to 0.1°F temperature variation Permanent record or thermal picture Quantitative Remote sensing; need not contact part Portable	Emissivity Liquid-nitrogen-cooled detector Critical time-temperature relationship Poor resolution for thick specimens Reference standards required
Leak testing	Leaks: Helium Ammonia Smoke Water Air bubbles Radioactive gas Halogens	Joints: Welded Brazed Adhesive-bonded Sealed assemblies Pressure or vacuum chambers Fuel or gas tanks	High sensitivity to extremely small, light separations not detectable by other NDT methods Sensitivity related to method selected	Accessibility to both surfaces of part required Smeared metal or contaminants may prevent detection Cost related to sensitivity
Magnetic particle	Surface and slightly subsurface flaws; cracks, seams, porosity, inclusions Permeability variations Extremely sensitive for locating small tight cracks	Ferromagnetic materials; bar, plate, forgings, weldments, extrusions, etc.	Advantage over penetrant is that it indicates subsurface flaws, particularly inclusions Relatively fast and low-cost May be portable	Alignment of magnetic field is critical Demagnetization of parts required after tests Parts must be cleaned before and after inspection Masking by surface coatings
Magnetic field (also magnetic flux leakage)	Cracks Wall thickness Hardness Coercive force Magnetic anisotropy Magnetic field Nonmagnetic coating thickness on steel	Ferromagnetic materials Ship degaussing Liquid-level control Treasure hunting Wall thickness of nonmetallic materials Material sorting	Measurement of magnetic material properties May be automated Easily detects magnetic objects in nonmagnetic material Portable	Permeability Reference standards required Edge effect Probe lift-off
Microwave (300 MHz–300 GHz)	Cracks, holes, debonded areas, etc., in nonmetallic parts Changes in composition, degree of cure, moisture content Thickness measurement Dielectric constant Loss tangent	Reinforced plastics Chemical products Ceramics Resins Rubber Wood Liquids Polyurethane foam Radomes	Between radio waves and infrared in electromagnetic spectrum Portable Contact with part surface not normally required Can be automated	Will not penetrate metals Reference standards required Horn-to-part spacing critical Part geometry Wave interference Vibration
Liquid penetrants (dye or fluorescent)	Flaws open to surface of parts; cracks, porosity, seams, laps, etc. Through-wall leaks	All parts with nonabsorbing surfaces (forgings, weldments, castings, etc.). Note: Bleed-out from porous surfaces can mask indications of flaws	Low cost Portable Indications may be further examined visually Results easily interpreted	Surface films such as coatings, scale, and smeared metal may prevent detection of flaws Parts must be cleaned before and after inspection Flaws must be open to surface
Fluoroscopy (cinefluorography) (kinefluorography)	Level of fill in containers Foreign objects Internal components Density variations Voids, thickness Spacing or position	Flow of liquids Presence of cavitation Operation of valves and switches Burning in small solid-propellant rocket motors	High-brightness images Real-time viewing Image magnification Permanent record Moving subject can be observed	Costly equipment Geometric unsharpness Thick specimens Speed of event to be studied Viewing area Radiation hazard

Table 5.4.1   Nondestructive Test Methods*   (*Continued*)

Method	Measures or detects	Applications	Advantages	Limitations
Neutron radiology (thermal neutrons from reactor, accelerator, or Californium 252)	Hydrogen contamination of titanium or zirconium alloys Defective or improperly loaded pyrotechnic devices Improper assembly of metal, nonmetal parts Corrosion products	Pyrotechnic devices Metallic, nonmetallic assemblies Biological specimens Nuclear reactor fuel elements and control rods Adhesive-bonded structures	High neutron absorption by hydrogen, boron, lithium, cadmium, uranium, plutonium Low neutron absorption by most metals Complement to X-ray or gamma-ray radiography	Very costly equipment Nuclear reactor or accelerator required Trained physicists required Radiation hazard Nonportable Indium or gadolinium screens required
Gamma radiology (cobalt 60, iridium 192)	Internal flaws and variations, porosity, inclusions, cracks, lack of fusion, geometry variations, corrosion thinning Density variations Thickness, gap, and position	Usually where X-ray machines are not suitable because source cannot be placed in part with small openings and/or power source not available Panoramic imaging	Low initial cost Permanent records; film Small sources can be placed in parts with small openings Portable Low contrast	One energy level per source Source decay Radiation hazard Trained operators needed Lower image resolution Cost related to source size
X-ray radiology	Internal flaws and variations; porosity, inclusions, cracks, lack of fusion, geometry variations, corrosion Density variations Thickness, gap, and position Misassembly Misalignment	Castings Electrical assemblies Weldments Small, thin, complex wrought products Nonmetallics Solid-propellant rocket motors Composites Container contents	Permanent records; film Adjustable energy levels (5 kV–25 meV) High sensitivity to density changes No couplant required Geometry variations do not affect direction of X-ray beam	High initial costs Orientation of linear flaws in part may not be favorable Radiation hazard Depth of flaw not indicated Sensitivity decreases with increase in scattered radiation
Radiometry X-ray, gamma ray, beta ray (transmission or backscatter)	Wall thickness Plating thickness Variations in density or composition Fill level in cans or containers Inclusions or voids	Sheet, plate, strip, tubing Nuclear reactor fuel rods Cans or containers Plated parts Composites	Fully automatic Fast Extremely accurate In-line process control Portable	Radiation hazard Beta ray useful for ultrathin coatings only Source decay Reference standards required
Reverse-geometry digital X-ray	Cracks Corrosion Water in honeycomb Carbon epoxy honeycomb Foreign objects	Aircraft structure	High-resolution 10^6 pixel image with high contrast	Access to both sides of object Radiation hazard
X-ray computed tomography (CT)	Small density changes Cracks Voids Foreign objects	Solid-propellant rocket motors Rocket nozzles Jet-engine parts Turbine blades	Measures X-ray opacity of object along many paths	Very expensive Trained operator Radiation hazard
Shearography electronic	Lack of bond Delaminations Plastic deformation Strain Crushed core Impact damage Corrosion in Al honeycomb	Composite-metal honeycomb Bonded structures Composite structures	Large area coverage Rapid setup and operation Noncontacting Video image easy to store	Requires vacuum thermal, ultrasonic, or microwave stressing of structure to cause surface strain
Thermal (thermochromic paint, liquid crystals)	Lack of bond Hot spots Heat transfer Isotherms Temperature ranges Blockage in coolant passages	Brazed joints Adhesive-bonded joints Metallic platings or coatings Electrical assemblies Temperature monitoring	Very low initial cost Can be readily applied to surfaces which may be difficult to inspect by other methods No special operator skills	Thin-walled surfaces only Critical time-temperature relationship Image retentivity affected by humidity Reference standards required
Sonic (less than 0.1 MHz)	Debonded areas or delaminations in metal or nonmetal composites or laminates Cohesive bond strength under controlled conditions Crushed or fractured core Bond integrity of metal insert fasteners	Metal or nonmetal composite or laminates brazed or adhesive-bonded Plywood Rocket-motor nozzles Honeycomb	Portable Easy to operate Locates far-side debonded areas May be automated Access to only one surface required	Surface geometry influences test results Reference standards required Adhesive or core-thickness variations influence results

Table 5.4.1 Nondestructive Test Methods* (*Continued*)

Method	Measures or detects	Applications	Advantages	Limitations
Ultrasonic (0.1–25 MHz)	Internal flaws and variations; cracks, lack of fusion, porosity, inclusions, delaminations, lack of bond, texturing Thickness or velocity Poisson's ratio, elastic modulus	Metals Welds Brazed joints Adhesive-bonded joints Nonmetallics In-service parts	Most sensitive to cracks Test results known immediately Automating and permanent-record capability Portable High penetration capability	Couplant required Small, thin, or complex parts may be difficult to inspect Reference standards required Trained operators for manual inspection Special probes
Thermoelectric probe	Thermoelectric potential Coating thickness Physical properties Thompson effect *P-N* junctions in semiconductors	Metal sorting Ceramic coating thickness on metals Semiconductors	Portable Simple to operate Access to only one surface required	Hot probe Difficult to automate Reference standards required Surface contaminants Conductive coatings

* From Donald J. Hagemaier, "Metal Progress Databook," Douglas Aircraft Co., McDonnell-Douglas Corp., Long Beach, CA.

observed in a semidarkened location. After testing, the penetrant and developer are removed by washing with water, sometimes aided by an emulsifier, or with a solvent.

In **filtered particle** testing, cracks in **porous objects** (100 mesh or smaller) are indicated by the difference in absorption between a cracked and a flaw-free surface. A liquid containing suspended particles is sprayed on a test object. If a crack exists, particles are filtered out and concentrate at the surface as liquid flows into the additional absorbent area created by the crack. Fluorescent or colored particles are used to locate flaws in unfired dried clay, certain fired ceramics, concrete, some powdered metals, carbon, and partially sintered tungsten and titanium carbides.

RADIOGRAPHIC METHODS

Radiographic test methods employ X-rays, gamma rays, or similar penetrating radiation to reveal flaws, voids, inclusions, thickness, or structure of objects. Electromagnetic energy wavelengths in the range of 0.01 to 10 Å (1 Å = 10^{-8} cm) are used to examine the interior of opaque materials. Penetrating radiation proceeds from its source in straight lines to the test object. Rays are differentially absorbed by the object, depending upon the energy of the radiation and the nature and thickness of the material.

X-rays of a variety of wavelengths result when high-speed electrons in a vacuum tube are suddenly stopped. An X-ray tube contains a heated filament (cathode) and a target (anode); radiation intensity is almost directly proportional to filament current (mA). tube voltage (kV) determines the penetration capability of the rays. As tube voltage increases, shorter wavelengths and more intense X-rays are produced. When the energy of penetrating radiation increases, shorter wavelengths and more intense X-rays are produced. Also, when the energy of penetrating radiation increases, the difference in attenuation between materials decreases. Consequently, more film-image contrast is obtained at lower voltage, and a greater range of thickness can be radiographed at one time at higher voltage.

Gamma rays of a specific wavelength are emitted from the disintegrating nuclei of natural radioactive elements, such as radium, and from a variety of artificial radioactive isotopes produced in nuclear reactors. Cobalt 60 and iridium 192 are commonly used for industrial radiography. The half-life of an isotope is the time required for half of the radioactive material to decay. This time ranges from a few hours to many years.

Radiographs are photographic records produced by the passage of penetrating radiation onto a film. A void or reduced mass appears as a darker image on the film because of the lesser absorption of energy and the resulting additional exposure of the film. The quantity of X-rays absorbed by a material generally increases as the atomic number increases.

A radiograph is a shadow picture, since X-rays and gamma rays follow the laws of light in shadow formation. Four factors determine the best geometric sharpness of a picture: (1) The effective focal-spot size of the radiation source should be as small as possible. (2) The source-to-object distance should be adequate for proper definition of the area of the object farthest from the film. (3) The film should be as close as possible to the object. (4) The area of interest should be in the center of and perpendicular to the X-ray beams and parallel to the X-ray film.

Radiographic films vary in speed, contrast, and grain size. Slow films generally have smaller grain size and produce more contrast. Slow films are used where optimum sharpness and maximum contrast are desired. Fast films are used where objects with large differences in thickness are to be radiographed or where sharpness and contrast can be sacrificed to shorten exposure time. Exposure of a radiographic film comes from direct radiation and scattered radiation. **Direct radiation** is desirable, image-forming radiation; **scattered radiation,** which occurs in the object being X-rayed or in neighboring objects, produces undesirable images on the film and loss of contrast. **Intensifying screens** made of 0.005- or 0.010-in- (0.13-mm or 0.25-mm) thick lead are often used for radiography at voltages above 100 kV. The lead filters out much of the low-energy scatter radiation. Under action of X-rays or gamma rays above 88 kV, a lead screen also emits electrons which, when in intimate contact with the film, produce additional coherent darkening of the film. Exposure time can be materially reduced by use of intensifying screens above and below the film.

Penetrameters are used to indicate the contrast and definition which exist in a radiograph. The type generally used in the United States is a small rectangular plate of the same material as the object being X-rayed. It is uniform in thickness (usually 2 percent of the object thickness) and has holes drilled through it. ASTM specifies hole diameters 1, 2, and 4 times the thickness of the penetrameter. Step, wire, and bead penetrameters are also used. (See ASTM Materials Specification E94.)

Because of the variety of factors that affect the production and measurements of an X-ray image, operating factors are generally selected from reference tables or graphs which have been prepared from test data obtained for a range of operating conditions.

All materials may be inspected by radiographic means, but there are limitations to the configurations of materials. With optimum techniques, wires 0.0001 in (0.003 mm) in diameter can be resolved in small electrical components. At the other extreme, welded steel pressure vessels with 20-in (500-mm) wall thickness can be routinely inspected by use of high-energy accelerators as a source of radiation. **Neutron radiation** penetrates extremely dense materials such as lead more readily than X-rays or gamma rays but is attenuated by lighter-atomic-weight materials such as plastics, usually because of their hydrogen content.

Radiographic standards are published by ASTM, ASME, AWS, and API, primarily for detecting lack of penetration or lack of fusion in welded objects. Cast-metal objects are radiographed to detect conditions such as shrink, porosity, hot tears, cold shuts, inclusions, coarse structure, and cracks.

The usual method of utilizing penetrating radiation employs film. However, Geiger counters, semiconductors, phosphors (fluoroscopy), photoconductors (xeroradiography), scintillation crystals, and vidicon tubes (image intensifiers) are also used. Computerized digital radiography is an expanding technology.

The **dangers** connected with exposure of the human body to X-rays and gamma rays should be fully understood by any person responsible for the use of radiation equipment. NIST is a prime source of information concerning radiation safety. NRC specifies maximum permissible exposure to be a 1.25 R/¼ year.

ULTRASONIC METHODS

Ultrasonic nondestructive test methods employ high-frequency mechanical vibrational energy to detect and locate structural discontinuities or differences and to measure thickness of a variety of materials. An electric pulse is generated in a test instrument and transmitted to a transducer, which converts the electric pulse into mechanical vibrations. These low-energy-level vibrations are transmitted through a coupling liquid into the test object, where the ultrasonic energy is attenuated, scattered, reflected, or resonated to indicate conditions within material. Reflected, transmitted, or resonant sound energy is reconverted to electrical energy by a transducer and returned to the test instrument, where it is amplified. The received energy is then usually displayed on a cathode-ray tube. The presence, position, and amplitude of echoes indicate conditions of the test-object material.

Materials capable of being tested by ultrasonic energy are those which transmit vibrational energy. Metals are tested in dimensions of up to 30 ft (9.14 m). Noncellular plastics, ceramics, glass, new concrete, organic materials, and rubber can be tested. Each material has a characteristic sound velocity, which is a function of its density and modulus (elastic or shear).

Material characteristics determinable through ultrasonics include structural discontinuities, such as flaws and unbonds, physical constants and metallurgical differences, and thickness (measured from one side). A common application of ultrasonics is the inspection of welds for inclusions, porosity, lack of penetration, and lack of fusion. Other applications include location of unbond in laminated materials, location of fatigue cracks in machinery, and medical applications. Automatic testing is frequently performed in manufacturing applications.

Ultrasonic systems are classified as either **pulse-echo,** in which a single transducer is used, or **through-transmission,** in which separate sending and receiving transducers are used. Pulse-echo systems are more common. In either system, ultrasonic energy must be transmitted into, and received from, the test object through a **coupling medium,** since air will not efficiently transmit ultrasound of these frequencies. Water, oil, grease, and glycerin are commonly used couplants. Two types of testing are used: contact and immersion. In contact testing, the transducer is placed directly on the test object. In immersion testing, the transducer and test object are separated from one another in a tank filled with water or by a column of water or by a liquid-filled wheel. Immersion testing eliminates transducer wear and facilitates scanning of the test object. Scanning systems have paper-printing or computerized video equipment for readout of test information.

Ultrasonic transducers are piezoelectric units which convert electric energy into acoustic energy and convert acoustic energy into electric energy of the same frequency. Quartz, barium titanate, lithium sulfate, lead metaniobate, and lead zirconate titanate are commonly used transducer crystals, which are generally mounted with a damping backing in a housing. Transducers range in size from ¹⁄₁₆ to 5 in (0.15 to 12.7 cm) and are circular or rectangular. Ultrasonic beams can be focused to improve resolution and definition. Transducer characteristics and beam patterns are dependent upon frequency, size, crystal material, and construction.

Test frequencies used range from 40 kHz to 200 MHz. Flaw-detection and thickness-measurement applications use frequencies between 500 kHz and 25 MHz, with 2.25 and 5 MHz being most commonly employed for flaw detection. Low frequencies (40 kHz to 1.0 MHz) are used on materials of low elastic modulus or large grain size. High frequencies (2.25 to 25 MHz) provide better resolution of smaller defects and are used on fine-grain materials and thin sections. Frequencies above 25 MHz are employed for investigation and measurement of physical properties related to acoustic attenuation.

Wave-vibrational modes other than longitudinal are effective in detecting flaws that do not present a reflecting surface to the ultrasonic beam, or other characteristics not detectable by the longitudinal mode. They are useful also when large areas of plates must be examined. Wedges of plastic, water, or other material are inserted between the transducer face and the test object to convert, by refraction, to shear, transverse, surface, or Lamb vibrational modes. As in optics, Snell's law expresses the relationship between incident and refracted beam angles; i.e., the ratio of the sines of the angle from the normal, of the incident and refracted beams in two mediums, is equal to the ratio of the mode acoustic velocities in the two mediums.

Limiting conditions for ultrasonic testing may be the test-object shape, surface roughness, grain size, material structure, flaw orientation, selectivity of discontinuities, and the skill of the operator. Test sensitivity is less for cast metals than for wrought metals because of grain size and surface differences.

Standards for acceptance are published in many government, national society, and company specifications (see references above). Evaluation is made by comparing (visually or by automated electronic means) received signals with signals obtained from reference blocks containing flat bottom holes between ¹⁄₆₄ and ⁸⁄₆₄ in (0.40 and 0.325 cm) in diameter, or from parts containing known flaws, drilled holes, or machined notches.

EDDY CURRENT METHODS

Eddy current nondestructive tests are based upon correlation between electromagnetic properties and physical or structural properties of a test object. Eddy currents are induced in metals whenever they are brought into an ac magnetic field. These eddy currents create a secondary magnetic field, which opposes the inducing magnetic field. The presence of discontinuities or material variations alters eddy currents, thus changing the apparent impedance of the inducing coil or of a detection coil. Coil impedance indicates the magnitude and phase relationship of the eddy currents to their inducing magnetic-field current. This relationship is dependent upon the mass, conductivity, permeability, and structure of the metal and upon the frequency, intensity, and distribution of the alternating magnetic field. Conditions such as heat treatment, composition, hardness, phase transformation, case depth, cold working, strength, size, thickness, cracks, seams, and inhomogeneities are indicated by eddy current tests. Correlation data must usually be obtained to determine whether test conditions for desired characteristics of a particular test object can be established. Because of the many factors which cause variation in electromagnetic properties of metals, care must be taken that the instrument response to the condition of interest is not nullified or duplicated by variations due to other conditions.

Alternating-current **frequencies** between 1 and 5,000,000 Hz are used for eddy current testing. Test frequency determines the depth of current penetration into the test object, owing to the ac phenomenon of "skin effect." One "standard depth of penetration" is the depth at which the eddy currents are equal to 37 percent of their value at the surface. In a plane conductor, depth of penetration varies inversely as the square root of the product of conductivity, permeability, and frequency. High-frequency eddy currents are more sensitive to surface flaws or conditions while low-frequency eddy currents are sensitive also to deeper internal flaws or conditions.

Test coils are of three general types: the **circular coil,** which surrounds an object; the **bobbin coil,** which is inserted within an object; and the **probe coil,** which is placed on the surface of an object. Coils are further classified as **absolute,** when testing is conducted without direct comparison with a reference object in another coil; or **differential,** when comparison is made through use of two coils connected in series opposition. Many variations of these coil types are utilized. Axial length of a circular test coil should not be more than 4 in (10.2 cm), and its shape should correspond closely to the shape of the test object for best results. Coil diameter should be only slightly larger than the test-object diameter for consistent and useful results. Coils may be of the air-core or magnetic-core type.

Instrumentation for the analysis and presentation of electric signals resulting from eddy current testing includes a variety of means, ranging from meters to oscilloscopes to computers. Instrument meter or alarm circuits are adjusted to be sensitive only to signals of a certain electrical phase or amplitude, so that selected conditions are indicated while others are ignored. Automatic and automated testing is one of the principal advantages of the method.

Thickness measurement of metallic and nonmetallic coatings on metals is performed using eddy current principles. Coating thicknesses measured typically range from 0.0001 to 0.100 in (0.00025 to 0.25 cm). For measurement to be possible, coating conductivity must differ from that of the base metal.

MICROWAVE METHODS

Microwave test methods utilize electromagnetic energy to determine characteristics of nonmetallic substances, either solid or liquid. Frequencies used range from 1 to 3,000 GHz. Microwaves generated in a test instrument are transmitted by a waveguide through air to the test object. Analysis of reflected or transmitted energy indicates certain material characteristics, such as moisture content, composition, structure, density, degree of cure, aging, and presence of flaws. Other applications include thickness and displacement measurement in the range of 0.001 in (0.0025 cm) to more than 12 in (30.4 cm). Materials that can be tested include most solid and liquid nonmetals, such as chemicals, minerals, plastics, wood, ceramics, glass, and rubber.

INFRARED METHODS

Infrared nondestructive tests involve the detection of infrared electromagnetic energy emitted by a test object. Infrared radiation is produced naturally by all matter at all temperatures above absolute zero.

Materials emit radiation at varying intensities, depending upon their temperature and surface characteristics. A **passive** infrared system detects the natural radiation of an unheated test object, while an **active** system employs a source to heat the test object, which then radiates infrared energy to a detector. Sensitive indication of temperature or temperature distribution through infrared detection is useful in locating irregularities in materials, in processing, or in the functioning of parts. Emission in the infrared range of 0.8 to 15 μm is collected optically, filtered, detected, and amplified by a test instrument which is designed around the characteristics of the detector material. Temperature variations on the order of 0.1°F can be indicated by meter or graphic means. Infrared theory and instrumentation are based upon radiation from a blackbody; therefore, **emissivity correction** must be made electrically in the test instrument or arithmetically from instrument readings.

ACOUSTIC SIGNATURE ANALYSIS

Acoustic signature analysis involves the analysis of sound energy emitted from an object to determine characteristics of the object. The object may be a simple casting or a complex manufacturing system. A passive test is one in which **sonic** energy is transmitted into the object. In this case, a mode of resonance is usually detected to correlate with cracks or structure variations, which cause a change in effective modulus of the object, such as a nodular iron casting. An active test is one in which the object emits sound as a result of being struck or as a result of being in operation. In this case, characteristics of the object may be correlated to damping time of the sound energy or to the presence or absence of a certain frequency of sound energy. Bearing wear in rotating machinery can often be detected prior to actual failure, for example. More complex analytical systems can monitor and control manufacturing processes, based upon analysis of emitted sound energy.

Acoustic emission is a technology distinctly separate from acoustic signature analysis and is one in which strain produces bursts of energy in an object. These are detected by **ultrasonic** transducers coupled to the object. Growth of microcracks, and other flaws, as well as incipient failure, is monitored by counting the pulses of energy from the object or recording the time rate of the pulses of energy in the ultrasonic range (usually a discrete frequency between 1 kHz and 1 MHz).

5.5   EXPERIMENTAL STRESS AND STRAIN ANALYSIS
by Gary L. Cloud

REFERENCES: Cloud, *Optical Methods of Engineering Analysis,* revised second printing, Cambridge University Press, New York, 1998. Dally and Riley *Experimental Stress Analysis,* 4th ed., College House Enterprises, Knoxville, Tenn. 2005. Kobayashi (ed.), *Handbook on Experimental Mechanics,* Prentice-Hall, Englewood Cliffs, N.J., 1987.

Experimental stress and strain analysis, otherwise called **experimental mechanics,** is the quantitative measurement of the response of bodies and structures to stimuli, often in the form of applied loads. Applications of experimental mechanics are numerous and include, e.g., biomechanics, geomechanics, fracture mechanics, infrastructure, materials characterization, manufacturing processes, transducer design, motion analysis, mechanical design, and nondestructive testing.

Many problems faced by analysts and designers stretch the limitations of analytical and numerical techniques, so experimental solutions are needed. Hybrid approaches wherein experimental data are used as input to a numerical solution that is later subjected to experimental validation are becoming more commonplace. The major problem that must be faced in measurement of strain in structures is that,

for most engineering materials, the strain induced by loads is very small, the change in length per unit gage length being on the order of parts per million (microstrain), so a sensitivity of 1 to 10 microstrain is required. A direct way to measure stress does not exist, except, possibly, by photoelasticity. To determine stress, the strain must be measured and converted to stress by use of the stress-strain relations for the material.

Many techniques are available to attack problems in experimental mechanics, and engineers should be familiar with a spectrum of methods so that the most appropriate one for a given situation might be chosen. Some techniques have evolved, through long usage, into well-known standard approaches, while others are evolving or have not yet been thoroughly accepted by industrial practitioners. The former class includes resistance strain gages, photoelasticity, geometric moiré, and brittle coatings. Those less known or still in evolution include holographic interferometry, speckle interferometry, speckle shearography, acoustic emission, thermography, digital photoelasticity, digital image correlation, and moiré interferometry. This section describes in some detail three of the standard methods and the basic functions of several of the more advanced methods. Systems, supplies, information, and software for implementing these techniques are available from vendors.

RESISTANCE STRAIN GAGES

A resistance strain gage consists of a length of a fine conducting wire or foil, often configured as a grid, that is intimately attached to the specimen. As the specimen is deformed, the gage is deformed by the same amount, so that its resistance changes. If the relationship between resistance change and strain is known for the gage material, then the observed resistance change can be converted to specimen strain at the gage location. The relationship between strain and resistance change is

$$\varepsilon = \frac{1}{F}\left(\frac{\Delta R}{R}\right)$$

where ε is the strain along the gage axis, R is the gage resistance, ΔR is the change of gage resistance with strain, and F is the **gage factor** for the given gage. Gage factor F is determined by the gage manufacturer through batch testing in a known uniaxial stress field and is provided to the gage user, as is the gage resistance.

Strain gages are employed in two distinct ways. First, they are used for direct measurement of strain. Second, they are commonly used in transducing devices in which strain measurements are used as indicators of some other physical parameter such as load, pressure, or acceleration.

Strain gages of many types, sizes, and configurations are available. The most common type is the **bonded metallic gage** in which the sensing element is a thin foil that is mounted on a plastic substrate. Strain gages must be carefully selected for an application in accordance with manufacturers' recommendations. Perfect bonding between gage and specimen is necessary, and rigorous but simple cleaning, bonding, wiring, and protective coating practices must be followed.

Gage factors for common metallic gages are about 2; so since the strain is small, the resistance change per unit gage resistance is also small. An important part of a strain gage installation is the **interface circuit** that converts the small resistance change to a voltage that can be amplified and measured. Two basic interface circuits are commonly used, usually with additions that add capability and flexibility.

Wheatstone Bridge The Wheatstone bridge interface circuit is shown in Fig. 5.5.1. In the usual unbalanced operational mode, the variable resistor R_2 is first adjusted to give 0-V signal output E_O. Then as the resistance of the gage R_1 is changed by ΔR_1, the signal voltage changes proportionately. Often, more than one of and perhaps all the resistors in the circuit are strain gages. For four identical active gages and where E is the supply voltage and F is the gage factor, the relationship between output voltage and gage strains is

$$E_O = \frac{EF}{4}(-\varepsilon_1 + \varepsilon_2 - \varepsilon_3 + \varepsilon_4)$$

This voltage is usually amplified and recorded by a voltmeter, an oscilloscope, or a digital DAQ system. Dedicated strain gage readout devices

Fig. 5.5.1 Wheatstone bridge circuit for resistance strain gages.

and/or the use of circuit calibration allows direct indication in strain units. The output signal is proportional to the excitation voltage applied to the bridge, so a stable power supply must be used to limit noise and to maintain constant circuit sensitivity. Either dc or ac excitation can be used; the latter facilitates the use of filtering and carrier amplification to reduce noise.

An advantage of the bridge circuit is that more than one active strain gage can be used to multiply output or to cancel unwanted inputs such as temperature effects. For example, consider the case where surface strain in a beam is to be measured. Two gages can be used. The one on the top of the beam might be subjected to tensile strain, and the one on the bottom will experience compressive strain. The gages are wired into the bridge circuit in adjacent arms, e.g., R_1 and R_4. The bridge output will then be twice what will be seen if only one gage is used. Any effect of axial loads on the beam will be canceled as will temperature effects, because they have the same sign. The unbalanced bridge circuit can be used to measure either static or dynamic strain. It is the most widely used basic circuit, being implemented with many extra features in most commercial strain indicators and DAQ systems.

Potentiometer Circuit The potentiometer circuit, also called the ballast circuit or the voltage divider circuit, is shown in its basic form in Fig. 5.5.2. This circuit is particularly simple to fabricate. The function of the ballast resistor in the circuit is to limit the current through the strain gage while still giving usable output. Usually, this circuit is used with only one strain gage, so care must be taken that unwanted inputs, such as temperature effects, are eliminated in other ways.

A disadvantage of the potentiometer circuit is that there is no way to initially balance it so that the starting output voltage is zero. The small

Fig. 5.5.2 Potentiometer circuit for resistance strain gages.

change in output voltage that is caused by the strain gage will be superimposed on a relatively large bias voltage, meaning that one is faced with the problem of measuring accurately the small change in a large quantity. This fact limits the use of the circuit to problems involving rapidly varying or **dynamic strains.** A capacitor placed in the circuit acts as a high-pass filter so as to block the steady bias voltage and pass the time-varying component of interest to an amplifier. The circuit is dc-powered, and the power supply must be stable and noise-free. In spite of these limitations, this circuit is exceptionally useful and easy to use, e.g., in measuring strains in machinery.

Circuit Calibration The output equations for the interface circuits are useful for basic design of the measuring system, but not for accurate calibration of the entire system for measuring strain. This calibration is best accomplished by creating a known resistance change at the input. Implementation involves placing a large shunt calibration resistor R_{cal} in parallel with the strain gage R_g. The equivalent calibration strain ε_{cal} is given by

$$\varepsilon_{cal} = \frac{1}{F}\left(\frac{R_g}{R_g + R_{cal}}\right)$$

The relationship between circuit observable output (e.g., oscilloscope trace deflection) and input strain is thus established.

Temperature Effects Strain gages exhibit a resistance change with temperature that cannot be distinguished from the effects of strain. Large errors can result if the gage temperature changes during an experiment and steps are not taken to eliminate these errors. Two approaches are used singly or together.

Manufacturers of strain gages provide gages that are self-compensated for broad classes of materials (e.g., brass, steel). These gages can be used without any other temperature compensation with good, but not perfect, results. They are especially useful for single-gage applications over short periods of time where mechanical strains are relatively large.

The output voltage equation for the Wheatstone bridge can be exploited to eliminate temperature effects in two related ways. In the first, an inactive dummy gage that is the same as the active gage is mounted on an unstressed coupon of the same material. It is wired into an arm of the circuit adjacent to the measuring gage. False strains from temperature are canceled. Another approach is to design the experiment so that two active measuring gages are placed in adjacent bridge arms. The desired strain signal is multiplied, and the false thermal strain is eliminated. Similar approaches are used to eliminate temperature effects on gage lead wires, leading, e.g., to the familiar "three-wire hookup" for single-gage installations.

Transverse Sensitivity   Resistance strain gages exhibit sensitivity to the strain component that is perpendicular to the gage axis. If the stress field is uniaxial, this transverse sensitivity does not cause errors because the gage factor is determined in such a field. In any other biaxial stress field, ignoring the effects of transverse sensitivity can lead to errors that range from negligible to very large, depending on the nature of the field and the strain component being measured. The **transverse sensitivity** K_T is determined by the gage manufacturer by batch testing and is provided to the gage user. For typical foil gages, K_T is less than 2 percent.

The errors caused by transverse sensitivity are eliminated by using two gages in a biaxial stress field, even if only one strain component is sought. That is, for accurate determination of one strain in an unknown biaxial stress field, two gages are required. The two gages are mounted perpendicular to each other. Usually the gages are identical. The common gage factor F as given by the manufacturer is set on the strain indicator. The strain indicator gives an apparent strain for each gage, but the readings are not true strains. The readings are corrected as follows. Let the apparent indicated strain for gage A be Q_A and the reading for the second gage B be Q_B. The true or actual strain ε_A along the axis of gage A is

$$\varepsilon_A = \frac{1 - \mu_0 K_T}{1 - K_T^2}(Q_A - K_T Q_B)$$

Note that if $Q_B \gg Q_A$, the error caused by ignoring K_T will be large. The μ_0 in the equation is the Poisson ratio for the material that was used by the manufacturer in determining the gage factor and the transverse sensitivity factor. It is not always available, but may be taken to be 0.285 with reasonable safety. The true strain for gage B is found by permuting the subscripts in the above equation.

Strain Rosettes   In a general biaxial strain field, there are three unknowns—the two principal strains and the orientations of the principal strain axes. If complete knowledge of the field is sought, then three strain measurements are required. Strain gages do not respond to shear strain, so three determinations of normal strain are used. Three ordinary gages can be employed, but gage manufacturers provide standard combinations of gages, called strain rosettes, that contain two or more gage elements in a single package which can be applied just as a single gage would. The two most common configurations of gage elements in rosettes are (1) the **rectangular rosette,** having gage elements at 0°, 45°, and 90°, and (2) the **delta (equiangular) rosette,** having elements at 0°, 60°, and 120°.

The analysis of rosette data involves the strain transformation equations and the transverse sensitivity correction that are described above. Consider the case in which a strain-indicating system is used, so the output is in strain units. Set the usual manufacturer's gage factor F on the instrument. Take Q_n to be the apparent strain reading from gage n. The true strain for gage n is ε_n. Recall that the transverse sensitivity of each gage is K_T. The true strains for each gage element in the rectangular rosette are then found to be $\varepsilon_1 = Q_1 - K_T Q_3$; $\varepsilon_2 = (1 + K_T)Q_2 - K_T(Q_1 + Q_3)$; $\varepsilon_3 = Q_3 - K_T Q_1$. If gage elements 1 and 3 establish an xy coordinate system, then the strain transformation equations yield $\varepsilon_x = \varepsilon_1$, $\varepsilon_y = \varepsilon_3$, and shear strain $\varepsilon_{xy} = \varepsilon_1 + \varepsilon_3 - 2\varepsilon_2$. With strains in the xy coordinate system known,

the principal strains and the principal axes can be found from the transformation equations or Mohr's circle. Similar data reduction equations for other types of rosettes are easily found in basic strain gage literature.

Strain Gage Transducers   Resistance strain gages are used extensively in constructing transducers to measure physical quantities such as force, torque, pressure, and acceleration. Simple devices of this type can be easily fabricated in the lab for specific applications, or they may be purchased. As an example, consider the situation mentioned above in which two gages are mounted on the top and bottom of a beam to measure the strain in the beam. If the output of the circuitry is calibrated in terms of the load on the beam, then the device becomes a load-measuring transducer. Likewise, a common wrench is instrumented with strain gages and calibrated so that the output is indicative of the torque transmitted by the wrench.

Critical factors in designing such transducers involve the design of the elastic member such that the physical quantity to be measured is converted to a detectable strain, the proper mounting of the gages to detect the strain, the design of the circuit so that it responds to the quantity of interest and rejects unwanted inputs, and the calibration of the device.

Because the strain in the elastic member is not of quantitative interest in a transducer, circuit calibration is not useful or appropriate. Instead, direct calibration is accomplished by applying known values of the desired physical quantity to the transducer and relating these known values to the output. For example, if a beam is used as a load transducer, then known dead weights are applied to the beam. The circuit output is then adjusted to indicate directly the values of the applied known load.

GEOMETRIC MOIRÉ

The geometric moiré or **mechanical moiré** effect is the mechanical occlusion of light by superimposed patterns to produce another pattern that is called a **moiré fringe pattern.** It is not an optical interference effect. Geometric moiré provides a basis for a family of methods for measuring deformation, strain, shape, and contour. Advanced moiré methods, including moiré interferometry, are easily understood if the principles of geometric moiré are known.

This section describes the basic moiré phenomenon and its implementation for measurement of displacement and strain in a deformable body. Then two techniques for using moiré to perform measurements of out-of-plane shape and deformation are outlined.

In-Plane Moiré   Figure 5.5.3 illustrates the basic moiré phenomenon that appears when one system of closely spaced straight lines, a grill, is laid over a similar set. The two sets have different line spacings, and one set is also rotated slightly with respect to the other. At certain locations, the lines from one set block the gaps in the other set so that no light is allowed through. If the line spacings are sufficiently small, then the areas where the light is blocked appear to join up to form another system of broad lines, called a moiré fringe pattern. Simple experimentation with

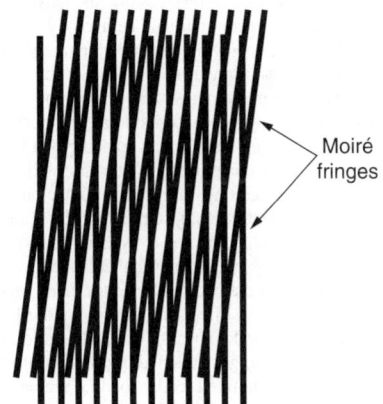

Fig. 5.5.3   Moiré effect from superimposing two line arrays.

such a system shows that the moiré fringes undergo large movement when one of the grills is shifted only a small amount relative to the other grill. Clearly, the moiré pattern is a motion magnifier that can be exploited to give sensitive measurements of relative motion.

Figure 5.5.4 shows a cross-sectional view of how the moiré effect is used to measure displacement and strain. Light is projected through two grills that were originally identical but that now have slightly different

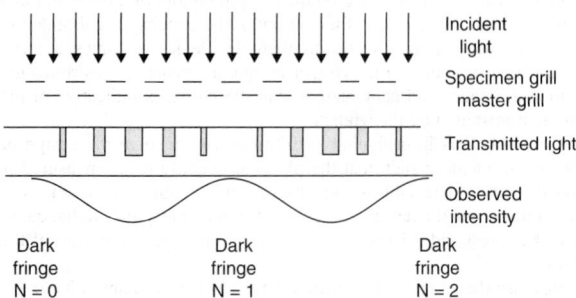

Fig. 5.5.4   Moiré effect as used for displacement measurement.

spacings because one grill has been stretched relative to the other. Consider that this grill, called the specimen grill, has been stretched so that 10 lines fill the space occupied by 12 lines on the second grill, called the master grill. The widths of the cracks between the lines of the gratings vary slowly from zero to a maximum. Recall that the grill lines are actually close together, so the eye or camera does not see the light passing through the individual cracks; rather, local averages are seen. The local average seems to oscillate twice from dark to light in the space shown, meaning 2 orders of moiré fringes are present. The extension or stretch in the specimen grill is $u = Np$, where N is the moiré fringe order and p is the line spacing or pitch of the master grill. Applying the strain-displacement relationship results in the following:

$$\varepsilon = \frac{\partial u}{\partial x} = \frac{\partial (Np)}{\partial x} = p\,\frac{\partial N}{\partial x}$$

which means that the strain along an axis perpendicular to the grill lines is the grill pitch times the gradient of the moiré fringe order (or fringe spacings) along the same axis. Similar relationships can be derived for strain in other directions and for the shear strain, so the complete strain map over the extent of the specimen can be determined. Fringes caused by small relative rotations of the grills are oriented perpendicular to the grating lines, so they do not affect the measurement of normal strain.

Specimen and Master Gratings   Grills or gratings must be attached to the surface of the specimen and superimposed with a master grating using one of several observing techniques. For measurement of strain in metals, the gratings must be very fine, having about 10 to 100 lines per millimetre. A considerable body of art and technology has developed for creating and attaching specimen gratings. Superb photographically reproduced gratings for moiré work are available from vendors. Another good approach is to obtain fine photomechanical screens that are used by the printing industry. These screens are available as photographic negatives or as fine metallic mesh. They may be cut up and glued to the specimen using techniques similar to those used for attaching strain gages. More commonly, the purchased gratings are kept as intact masters and are copied onto the specimen and as submaster gratings by photographic contact printing, by stencil techniques, or by vacuum deposition. Contact printing into a thin coating of the photoresist that is used for making printed circuit boards has been a method of choice for many practitioners.

Observing Techniques   A direct and simple method of measuring displacements and strains in flat specimens involves the use of a transparent model. A grill or grating is attached or printed onto the specimen surface. The specimen is mounted in the loading frame, and a master grill is put into direct contact with the specimen grill and initially oriented to bring the moiré fringe orders to a minimum. The master

grating is held in position with, e.g., a spring clothespin. A camera set up at some distance from the specimen-master combination is used to photograph the moiré fringes as the specimen is loaded.

If a nontransparent specimen is used, the specimen grating must be made to have high visibility in reflected light. The direct superposition method can still be used, but fringe visibility will suffer. An alternative viewing method is to image the specimen grating in the back of a view camera. The master grating replaces the ground glass in the camera back, and the moiré pattern will be visible there. Direct photography of the specimen grating provides two other approaches. The first approach is to use double exposure, one exposure before loading and the other after loading. The moiré fringe pattern is then contained in the photograph. The second approach is to photograph the specimen grating for each loading state. These photographic replicas can be superimposed directly with a submaster or with each other at any later time, which is often an advantage in dealing with nonreversible processes, e.g., those involving plastic deformation. A disadvantage of any procedure that involves imaging of the specimen grating is that the optical system must be capable of resolving finely spaced lines and very accurate focus must be established. Even so, the moiré fringes might not be of good contrast. In these cases, optical Fourier processing can be implemented to improve fringe visibility and reduce noise, while offering the possibility of enhancing strain sensitivity. Figure 5.5.5 shows a moiré fringe pattern obtained with these techniques.

Fig. 5.5.5   Portion of moiré pattern for measuring strain near a plastically deformed hole in aluminum.

Data Reduction   The objective is to create a map of strain distribution from a moiré fringe pattern, so the derivative of the moiré fringe order with respect to position coordinates must be found. Conceptually, this is easily done by drawing an array of axes across the fringe pattern with the axes oriented perpendicular to the grill direction. The location of each moiré fringe pattern relative to some fiducial origin along a chosen axis is measured and plotted as fringe order versus distance in specimen space. This graph is differentiated to obtain the fringe gradient, which is then multiplied by the grill pitch to obtain normal strain as a function of position along the axis. The process is repeated for each of the axes to complete the strain map. Fringe digitization and computer processing are useful, but manual processing serves well for small jobs, since only local fringe spacings are required.

Shadow Moiré The shadow method of geometric moiré utilizes the superimposition of a master grating with its own shadow. The fringes are loci of constant out-of-plane elevation, so they form an absolute contour map of the object being examined. This method and related laser scanning techniques are used, e.g., in manufacturing and medical diagnosis.

To create a shadow moiré pattern, a master grill or grating of pitch p is placed in front of the specimen object. For large objects, the master grating can be created by stretching strings across a frame. The combination is illuminated with a collimated beam at incidence angle α. Observation is at normal incidence by means of a field lens that focuses the light to a point, at which is placed a camera that is focused on the object. The incident illumination creates a shadow of the grating on the surface of the specimen. The grating shadows on the specimen are elongated or foreshortened by a factor that depends on the inclination of the surface, and they are shifted laterally by an amount that depends on the incidence angle and the distance w from the master grating to the specimen. The camera superimposes the image of the master grating with its own distorted shadow, thereby forming a moiré fringe pattern. The fringes are numbered from some origin. If N is the moiré fringe order, then the elevation is $w = Np/\tan\alpha$.

Projection Moiré Another useful method for using moiré for out-of-plane displacement measurement or determining changes of contour, as opposed to absolute contour, involves projecting the reference grating onto the specimen by means of, e.g., a slide projector. This method is especially useful when the shapes of two objects must be compared, a situation that arises in, e.g., examining sheet metal stampings. An advantage is that only a small transparency of the specimen grating is required. Alternatively, a projected grating can be created by oblique interference of collimated beams of coherent light. Another advantage is that changes of contour are determined directly, and subtraction of absolute contours is not required. A disadvantage is that the imaging system must be able to resolve the individual grating lines, as with double-exposure in-plane moiré.

To create the fringe pattern, the grating is projected at incidence angle α onto the specimen surface in its initial state, or onto a master reference surface. The projected grating image is photographed by a camera along the normal to the surface. The specimen is then distorted, or else the second surface is inserted if contour difference is sought. This grating is photographed over the first image. The developed double-exposure image will contain moiré fringes that are numbered as usual. At any point in the image where the moiré fringe order is N, the elevation difference w between the two states of the specimen is given by $w = Np/\sin\alpha$.

OPTICAL INTERFEROMETRY

Many important methods of experimental mechanics are based on the phenomenon of optical interference. Approaches in this class include, e.g., photoelasticity, hologram interferometry, speckle interferometry, and moiré interferometry. Brief study of interference, one of the cornerstones of optics along with diffraction, facilitates quick understanding and use of all such methods.

Collinear Interference Light is considered to be electromagnetic waves that are characterized by an electric vector that oscillates in time and space. The magnitude of the electric vector may be taken to be a simple harmonic plane wave described as $E = A\cos{(2\pi/\lambda)(z - vt)}$, where A is amplitude, λ is the wavelength of the light, z is the distance along the optical path, v is the wave velocity, and t is time. Consider two waves that are of equal amplitude, identically polarized and *coherent,* meaning that they originate from the same source, such as a laser. These two waves travel along the same path, but one of the waves leads the other by some distance or phase difference r. The two waves will interfere with each other to produce a resultant wave whose intensity I is a function of the phase difference; $I = 4A^2\cos^2{(\pi r/\lambda)}$.

The phase lag between two light waves, which is not visible to the eye or detectable by any other means, is converted, through the interference process, to an intensity difference, which can be measured. The transformation of invisible phase difference to visible intensity difference is the basis of all interferometry methods of measurement. It allows determination of physical quantities such as motion to within a fraction of the wavelength of light, giving sensitivities on the order of a few nanometres.

Generic Interferometer Figure 5.5.6 shows a conceptual model that contains all the important elements of an interferometric system for measurement. Light comes from a coherent source such as a laser and is split into two parts by a **beam splitter.** The two waves have identical phases as they leave this component, but they travel along different optical paths, one or both of which might contain optical elements or specimens. The two waves are out of phase as they are directed into a **combiner,** where they interfere to produce an intensity that depends on the **phase difference.** A **detector** such as a photocell, an eye, or a camera records the intensity, from which the phase difference can be inferred. The phase difference is indicative of the difference in *optical path lengths* between the two paths.

Fig. 5.5.6 The generic interferometer model.

The conceptual model seems to imply that only pointwise measurements can be made with one wave, but the process can be executed for every point in a broad field simultaneously if a beam of light is used and appropriate optical elements such as lenses, mirrors, and imaging devices are employed. The interference process produces a map of light and dark patches, called **interference fringes,** that are superimposed on the image of the object being studied. This whole-field capability is a great strength of optical methods in general.

Interferometry can be employed in various modes. The generic model suggests that the difference between the two paths can be obtained directly with one observation. More commonly, one path is held constant, and two observations are taken for two different states of the specimen path. When they are subtracted, the change in only the specimen path is determined.

PHOTOELASTIC STRESS ANALYSIS

Photoelastic stress analysis is a method whereby **polarized light** is used to obtain, through interferometry, the stress state in a loaded transparent model or in a coating that is glued to the surface of a prototype.

Stress Birefringence The speed of light in a vacuum is divided by the speed of light in a material to give the **refractive index** of the material. Optical path length is the physical path length (distance) traveled times the index of refraction. Changes or differences in optical path length cause phase lags or retardations in optical waves as they traverse a material.

In certain materials, including many plastics, the index of refraction at any point is closely related to the state of stress at that point, and such materials are said to exhibit stress birefringence. Experiments on these materials show that

1. The index of refraction depends on the plane of polarization of light that passes through the material; i.e., the index is direction-dependent.

2. The state of the index of refraction can be described in terms of principal values that are associated with principal directions.

3. The principal values of the refractive index are proportional to the principal stresses.

4. The principal axes of refractive index coincide with the principal stress axes.

A conclusion is that if the directions and magnitudes of the principal refractive indexes can be ascertained, then the principal stresses and principal directions can be determined, provided that the relation between stress and refractive index is known for the material. Determination of stress by interferometric observation of refractive index changes is accomplished by the method of photoelasticity.

Photoelasticity Experiment and theory show that **birefringent materials** can transmit light only if the light is polarized in the principal directions. If a polarized wave falls on a birefringent material, it is immediately divided into two polarized components, one in each principal direction, so the beam splitter function that is necessary for interferometry is served by the surface of the birefringent material. The two waves created traverse the material at differing speeds, depending on the two principal values of the refractive index. As these two waves exit the material, one will lag behind the other by an amount which is called the **relative retardation** R that depends on the local principal stresses and a material stress-optic coefficient C_σ according to the equation $R = C_\sigma(\sigma_1 - \sigma_2)d$, where σ_1 and σ_2 are the principal stresses and d is the thickness of the specimen. Figure 5.5.7 illustrates this process. To determine R, the two wave components must be recombined interferometrically. This task is accomplished by a second polarizer in the system, known as the analyzer, which extracts the portion of each wave that lies parallel to the transmission axis of the analyzer to create the conditions for collinear interference. The intensity produced through interference is directly related to the relative retardation as described above.

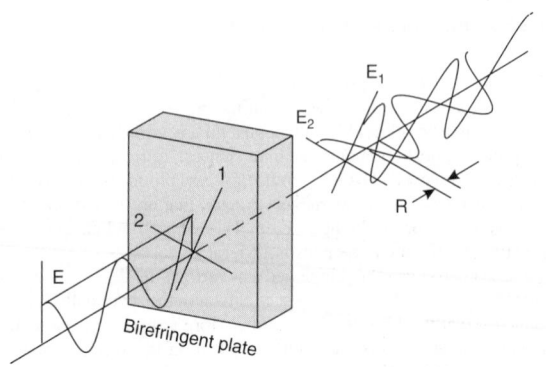

Fig. 5.5.7 Effect of birefringent plate on incident polarized light.

If this interferometric process occurs simultaneously over the entire field and a photograph is made of the result, **interference fringes** will be observed. Attention for the moment is limited to the case in which the polarization axis for the light entering the specimen is perpendicular to the analyzer axis. For linear polarization, two separate families of fringes are produced, as shown in Fig. 5.5.8. These families are

1. Isochromatic fringes that are loci of points having constant relative retardation R, meaning they are lines of constant $\sigma_1 - \sigma_2$. From the isochromatic family, the magnitudes of stress in the specimen can be found.

2. Isoclinic fringes that are loci of points where the principal stress axes are parallel to the axes of the polarizer and analyzer. This family is used to establish the principal stress directions for the entire specimen.

Circular Polarization A problem with the photoelastic measurements described above is that an isoclinic fringe always masks some of the isochromatics. The isoclinics can be eliminated by inserting two quarter-wave plates into the system, one on either side of the specimen.

A quarter-wave plate is a birefringent slab that produces a quarter-wavelength of relative retardation over its extent. The plates are inserted into the photoelastic measuring system with their principal axes

Fig. 5.5.8 Isoclinic and isochromatic fringes in compressed disk.

oriented at $\pm45°$ to the crossed axes of the polarizer and analyzer. The entering plane-polarized wave is divided into two equal orthogonal components, and upon exit from the quarter-wave plate, one component will lag behind the other by $\lambda/4$. The resultant of the two out-of-phase waves will describe a circular helix in space, hence the term **circularly polarized.** The physical effect of using circular polarization is that the directional dependence of the interference is removed. The second $\lambda/4$ plate essentially cancels the effect of the first one, but does not restore the directional dependence.

Photoelastic Polariscope A combination of optical devices that is used for photoelasticity is called a polariscope. Many polariscope configurations are used. One setup, called the diffused-light polariscope, is shown in Fig. 5.5.9. In this system, a diffuser scatters light in all directions, only some of which passes through the system parallel to the opti-

Fig. 5.5.9 Diffused-light photoelastic polariscope.

cal axis. The polarizer, quarter-wave plates, specimen, and analyzer are arranged as described above. These elements are usually mounted in calibrated rings so that they can be rotated relative to one another and to the specimen. A field lens is used to direct the light to a point, where a camera lens is placed to create an image of the specimen on film or a CCD array. The distance from the field lens to the camera aperture must be made equal to the focal length of the field lens; otherwise, the image is made with light that has not traveled parallel to the optical axis, and errors will be introduced. For quantitative measurements, a narrowband color filter is used to create nearly monochromatic light (single wavelength), and/or a spectral lamp is used as the light source.

Material Calibration Accurate determination of the **stress-optic coefficient C_σ** for the photoelastic model material is prerequisite to good quantitative results. Calibration is accomplished by testing a model in which the stress field is known, such as a tensile specimen, a beam in pure bending, or a disc in diametral compression. The simple tension test is the most generally valid, since the stress state in tension does not depend on any assumptions of linearity. The coefficient is dependent not only on the material, but also on time and wavelength, and these factors must be taken into account. Several different calibration procedures are valid; only a basic one is described here.

A "dog bone" specimen of the model material is placed in a loading frame within the polariscope. A load is applied, and the isochromatic order in the central section is carefully estimated by direct observation at a certain time t_1, say 10 min, after load. The load is increased by an increment, and the isochromatic order again is estimated at t_1 after the load increase. The process is repeated for several loads, with the fringe order being determined always at the same time after changing the load. A plot of fringe order versus stress in the tensile specimen is constructed. The slope of this plot gives the stress-optic coefficient for time t_1 after load.

When the unknown specimen of the same material is tested to determine its stress state, the isochromatic orders are recorded for the same time t_1 after loading that was used in the calibration. Sometimes this is not possible. To avoid repeating the calibration testing, the isochromatic orders are often observed at a sequence of times after load so that a family of calibration results is obtained. Note that the definition of stress-optic coefficient is not standardized. Care is required in using calibration data provided by others.

Recording and Interpreting Isochromatics The specimen and polariscope are arranged as described above, first with the polarizer and analyzer axes crossed. The quarter-wave plates are also crossed, and their axes are set to produce circularly polarized light. No light transits the system, so this setup is known as the **dark-field** arrangement. The specimen is loaded, and after waiting the prescribed time, the isochromatics are photographed. Immediately, the analyzer is rotated by 90° so as to produce a **light-field** arrangement, meaning that the background around the specimen will be bright. The new light-field isochromatic pattern is photographed. The reason that both light-field and dark-field patterns are recorded is that twice as many data are obtained, which aids in interpretation. A typical dark-field pattern is shown in Fig. 5.5.10.

Fig. 5.5.10 Isochromatic pattern indicating stresses in compressed disk and arch.

The dark fringes in the dark-field picture are whole-order isochromatics along which the relative retardation is an integral multiple of the wavelength. The isochromatics are numbered in order. Assigning correct orders is not a trivial task. One good technique is to keep track of them as the load is applied. Along these whole-order fringes, the relationship between stress and fringe order is found by combining equations already presented, to give

$$\sigma_1 - \sigma_2 = 2\tau_{\max} = \frac{m\lambda}{C_\sigma d}$$

where m is the fringe order. The dark fringes in the light-field photograph are half-order isochromatics that lie between the whole orders.

The **isochromatic fringes** give for every point inside the specimen only the difference between the principal stresses, which is the diameter of Mohr's circle and equals twice the maximum shear stress. Many methods to obtain the individual stresses have been devised. The problem is simplified along a boundary where there is no load and $\sigma_2 = 0$. For practical purposes, a **boundary stress plot** can be constructed to

show stress concentrations. Normals to the boundary are erected with their lengths proportional to the local σ_1, and the tips of the normals are connected with a smooth curve.

Often, interpolation between whole and half orders of isochromatics does not give sufficient precision for critical areas. One of several precise methods of determining fractional fringe orders must then be used.

Recording and Interpreting Isoclinics For observation of isoclinics, a second model of a material that is optically less sensitive than that used for isochromatic study is often employed so that the isochromatic orders will be small and not mixed with the isoclinics. The model is placed in the polariscope, the quarter-wave plates are removed, and the polarizer and analyzer are set so their axes are horizontal and vertical, as are the axes of the specimen. The 0° isoclinic fringe appears, and it is recorded and labeled. Recording of isoclinics can be accomplished by direct tracing from a view camera back or a video monitor. They can also be photographed if heavy exposures are used to narrow the fringes.

The polarizer and analyzer are then rotated by an increment of, say, 10° in the same direction, and the isoclinic for that setting is recorded and labeled. This process is repeated incrementally until 90° is reached, where the isoclinic will duplicate the first one. The direction of rotation of the polarizers is arbitrary, but the direction used must be noted, usually on the isoclinic tracing. The result will be a complete pattern of isoclinic fringes that show the principal directions everywhere in the specimen.

Visualizing the stress direction map from the isoclinics is rather difficult, so an overlay orthogonal network of lines that are everywhere tangent to the principal directions, called the **stress trajectory map,** is often constructed from the isoclinic pattern.

Scaling Stresses from Model to Prototype To use stresses obtained from a photoelasticity model, they must be transferred to the prototype component or structure, which likely is of a different material, of a different size, and subjected to different loads.

Two aspects of the problem must be given attention: the difference of material properties and the differences in size and load. If the component being studied is elastic, simply connected (no holes), or multiply connected (with holes) with no unequilibrated forces on a single boundary, and if the body forces are negligible, and if no displacement boundary conditions apply, then the stress distribution is completely independent of material properties. If displacement conditions are specified, then the modulus of elasticity is important. For the other cases, only the Poisson ratio is a factor, and the error in ignoring this effect will not exceed about 7 percent. In other words, for most problems, material properties are not a factor in transferring measured stresses from model to prototype.

The laws of dimensional similitude, developed from the Buckingham theorem, are used to attend to size and load differences. The resulting laws of dimensional similarity are almost intuitively obvious. Geometric similarity requires that the model be only scaled up or down relative to the prototype, meaning its shape and proportions cannot change. A similar rule applies for load similarity; the ratios of like loads between model and prototype must be constant. If displacement boundary conditions are specified, then the absolute load magnitudes must be scaled also, but this case rarely arises. If the similarity conditions are satisfied, then the stresses in model and prototype satisfy the following relationship, which shows how the stresses are scaled,

$$\sigma_p = \sigma_m \left(\frac{P_p}{P_m} \right) \left(\frac{a_m}{a_p} \right) \left(\frac{d_m}{d_p} \right)$$

where subscript m means model and subscript p means prototype. Here P_p/P_m is the **load scale factor,** a_m/a_p is the **dimension scale factor,** and d_m/d_p is the **thickness scale factor.** This relationship applies only to two-dimensional problems, and it allows the thickness to be conveniently scaled differently from the lateral dimensions.

Reflection Photoelasticity The strain distribution in nontransparent prototype components and structures can be determined by an extension of the photoelasticity methods described above. The method requires that a thin layer of birefringent material be attached to the surface of the component, using a cement that contains a light-scattering medium such as aluminum powder. The polariscope is then folded in the middle. The polarized light (circular or linear) falls on the photoelastic coating and undergoes a relative retardation as it traverses the coating twice. The scattered (reflected) light travels through the second quarter-wave plate and the analyzer to the eye or a camera. As the prototype is loaded, isoclinic and isochromatic fringes will be obtained.

Interpretation of isochromatic fringes differs from before in that the surface strains, not stresses, are transferred to the birefringent coating by the cement. The material is calibrated in terms of principal strains, and the isochromatics are so interpreted. If stress data are required, then the prototype material properties and the stress-strain relations are used.

Reflection photoelasticity is highly developed because of its practicality, especially in the industrial setting.

Three-Dimensional Photoelasticity   Photoelasticity can be employed to determine stress states in three-dimensional objects by several techniques. All such methods basically slice the three-dimensional model into a stack of two-dimensional cases. Two such approaches are described briefly here.

The **embedded polariscope technique** uses a thin slice cut from the birefringent model. The slice is then glued back into the model, with polarizer and quarter-wave sheets replacing the saw kerfs. The result is a polariscope encased within the object with what is essentially a two-dimensional model between the polarizers. The model is loaded, and the fringe patterns for the slice are obtained by passing unpolarized light through the model. Studies have shown that cutting the model and gluing it back together with the polarizers in place does not significantly affect the stress state if care is taken to use a matching cement. This method works very well for obtaining the stresses in the plane of a single slice of the object.

The **stress freezing and slicing method** requires that the model be brought to its second-order transition point by careful heating. A small deformation is then applied by loading, and this deformation is maintained while the model is cooled. The model is permanently deformed, and the birefringence, which is related to the strains in this case, is "locked in." The model can be sliced into two-dimensional cases for study in an ordinary polariscope without affecting the strain state. Again, the principal stress components in the plane of each slice are obtained. Additional observations, as by oblique incidence, are needed to completely characterize the three-dimensional state of stress.

HOLOGRAPHIC INTERFEROMETRY

Holography is a method of storing and regenerating all the amplitude and phase information contained in the light that is scattered from a body that has been illuminated by a laser. Because all the information is reproduced, the reconstructed object beam is, ideally, indistinguishable from the original, so a full three-dimensional image is created. Since it is possible to record the exact shape and position of the body in two different states, it is possible to compare the two records interferometrically to obtain precise measurement of movement or deformation. This measurement technique is called holographic interferometry.

The making of a hologram is by oblique interference of two beams. The setup for creating this interference follows the model of the generic interferometer described above. A laser beam is split into two parts. One part, called the reference beam, is expanded and made to fall on a high-resolution photographic plate or film. The other part, called the object beam, is expanded and made to illuminate the object being studied. The object scatters this illumination, so some of it falls on the same photographic film. The reference and object beams interfere at the film to create a very fine complex grating, which is stored in the film emulsion. It can be shown that this grating contains all information about the object shape, size, and position, although there will be no recognizable image stored.

Reconstruction of the holographic image is through diffraction by the exposed and developed film. The film is placed back into its original position and illuminated by only the reference beam. Grating diffraction will divide the beam into three parts, one of which, in the usual setup, recreates the object beam. If the eye is placed to receive this beam, a perfect virtual image of the object will appear in its original state.

Holographic interferometry is implemented in three different ways. In the double-exposure method, also called the **frozen-fringe technique,** the initial state of the specimen is recorded as outlined above, but the film is not developed at this point. The object is then loaded or moved, and a second exposure of the object is recorded on the same film. The film is developed and replaced in the setup so as to reconstruct the images. The two superimposed images will be seen as one, and it will contain interferometric fringes that indicate the deformation or motion of the specimen. Alternative approaches are the **real-time (live fringe) technique** and **time-average technique** for vibrating objects, neither of which is treated here.

Holographic interferometry is primarily used for nondestructive testing of, e.g., tires and turbine components. In this application, subsurface voids, disbonds, and damage cause local perturbations in the otherwise smooth holographic fringes.

SPECKLE INTERFEROMETRY

The technique offers a way to achieve interferometric sensitivities for very rapid, whole-field, noncontacting measurements of displacement and strain. No special specimen preparation is required, and video imaging can be utilized.

The basic phenomenon employed in this class of techniques is that an object viewed or photographed when illuminated with laser light will exhibit a dense pattern of bright and dark granular spots. This pattern is called a speckle pattern, and it is caused by local interference of the many waves that are scattered from every point on the illuminated object and that subsequently fall on the receiving sensor array. In order that the speckles have correct properties for measurement of deformation, two beams must be mixed at the sensor. In the first case, a reference beam is used as in holography. In the second instance, two object beams are used.

To utilize these speckles in electronic or digital speckle interferometry, a first digital image is captured to computer memory. The specimen is then moved or deformed, and a second image is recorded. The second digital image is subtracted from the first, pixel by pixel. This subtraction of two intensity patterns can be done in real time on a desktop computer. The resulting video image will exhibit fringes that indicate the local displacements of the surface of the specimen.

If a reference beam is used, if the illuminated object is photographed at near-normal incidence, and if the object beam is nearly parallel with the reference beam at the sensor, then the fringes indicate only the out-of-plane motion of the object. If two object beams are used, then they are set up to illuminate the object at equal and opposite angles of incidence, and viewing is along the normal. In this case, the fringes correspond to in-plane motion or deformation. The measured in-plane displacements are then locally differentiated to obtain strain distribution.

The speckle methods just described produce speckle correlation fringes that are most commonly used in qualitative applications such as nondestructive testing. Phase-shifting techniques extend the method for quantitative measurement. A common implementation is to incorporate into one of the optical paths a mirror that can be moved by small amounts so as to change that path length. At least three images are captured for a matching number of different path length differences. Processing yields a whole-field phase change map that indicates both the sign and the magnitude of displacement. The displacement map is locally differentiated in the computer. Thus, whole-field strain measurements with high sensitivity and small gage lengths can be accomplished in only a few seconds.

SPECKLE SHEAROGRAPHY

This approach uses the speckle effect described above in a different way but with much of the same equipment. An extra optical element, such as a Fresnel biprism or a split lens, creates two images of the object, one very slightly displaced from the other, on the camera sensor. A first image pair is captured, the specimen is loaded, and the second image pair is captured and subtracted from the first, typically in real time. The resulting image pair exhibits fringes that indicate the in-plane derivative of the out-of-plane displacement (i.e., the change in slope), at each point of the object surface.

Electronic speckle shearography may be made highly quantitative by use of phase shifting as outlined above, but it is primarily used without phase shifting as a very important nondestructive evaluation tool that is simple, quick, and easy to implement. As with holographic interferometry, the presence of flaws is indicated by local perturbations of the otherwise smooth fringe pattern.

MOIRÉ INTERFEROMETRY

To implement this extension of the moiré method, a grating having spatial frequencies on the order of several hundreds of lines per millimetre is created by oblique interference of two collimated laser beams. This grating is transferred to the surface of the specimen by one of several techniques. The specimen is then replaced in the same setup that was used to make the grating, and the specimen is loaded. Reflective diffraction of the incident beams from the specimen grating creates two slightly diverging beams, where the local divergence angle is related to the local change of spatial frequency of the specimen grating that is caused by strain. The two diffracted beams are collected by a camera lens and brought together to form two coincident images that appear as one. This image will contain interference fringes resulting from local oblique interference. Interpretation of the fringe pattern is the same as for geometric moiré. This important method overcomes the sensitivity limitation of geometric moiré. Its main disadvantage is that significant specimen preparation is required.

DIGITAL IMAGE CORRELATION

Digital image correlation (DIC) is a noncontacting measurement method that uses digital signal processing algorithms to track the motion in the image plane of subsets of either a coherent or an incoherent random speckle pattern. Two-dimensional DIC (2D-DIC) uses a single camera to convert images of a nominally planar object that deforms predominantly in-plane into a full field of in-plane displacements (u, v). Three-dimensional DIC (3D-DIC) converts images obtained from cameras located at multiple viewpoints into a full field of the actual, three-dimensional displacements (u, v, w) of all points in a planar or nonplanar object. Both methods require that a calibration process be performed so that the image motions are converted to displacements with optimal accuracy. The process can be performed either before or after completion of the experiment. When photographic-quality lenses are used to acquire images, calibration in 2D-DIC is relatively simple, requiring only an image of a ruler or similar metric object. For 3D-DIC using photographic-quality lenses, pairs of images of a calibration object undergoing a set of 3D rotations are required to obtain the orientations and positions of the cameras.

THERMOELASTICITY

Thermoelasticity is a noncontacting analysis technique that utilizes the minute local changes of temperature that are caused by stress. When a body is subjected to tensile stress, the temperature of the body reduces slightly; conversely, when the stress is compressive, a slight increase in temperature occurs. In the elastic regime, this temperature change is directly proportional to the change in the sum of principal stresses on the component surface. The only surface preparation required is to ensure uniform emissivity of the component surface at the infrared wavelengths used by the detector, and this requirement can usually be met by the application of a thin layer of matte black paint. Since the detectors are measuring photon emission from a surface, the technique can also be used obliquely, with satisfactory results obtainable at up to 60° from the normal to the surface.

5.6 MECHANICS OF COMPOSITE MATERIALS
by Michael W. Hyer and Scott W. Case

REFERENCES: *General references:* "The Composite Materials Handbook MIL-17," American Society for Testing and Materials, West Conshohocken, Penn. Hyer, "Stress Analysis of Fiber-Reinforced Composite Materials," WCB/McGraw-Hill, New York, 1998. Jones, "Mechanics of Composite Materials," 2d ed., Taylor and Francis, Philadelphia, 1999. Herakovich, "Mechanics of Fibrous Composites," Wiley, New York, 1998. Daniel and Ishai, "Engineering Mechanics of Composite Materials," Oxford University Press, New York, 1994. Gibson, "Principles of Composite Material Mechanics," McGraw-Hill, New York, 1994. Swanson, "Introduction to Design and Analysis with Advanced Composite Materials," Prentice-Hall, Upper Saddle River, N.J., 1997. Vinson and Sierakowski, "The Behavior of Structures Composed of Composite Materials," 2d ed., Martinus Nijhoff Publishers, Boston, 2002. Wolff, "Introduction to the Dimensional Stability of Composite Materials," DEStech Publications, Lancaster, Penn., 2004. *References dealing primarily with damage tolerance and failure:* Hinton, Soden, and Kaddour, "Failure Criteria in Fibre-Reinforced-Polymer Composites," Elsevier Publishers, 2004. Reifsnider and Case, "Damage Tolerance and Durability of Material Systems," Wiley, New York, 2002.

This section deals with some of the basic issues that must be addressed in applying the classic methods of mechanics of materials to predict the mechanical behavior of **fiber-reinforced composite** materials. The section deals exclusively with the mechanical behavior of materials that consist of stiff and strong **fibers,** such as glass, carbon, or aramid, embedded in a softer material, such as a thermosetting or thermoplastic polymer, known as the matrix material, or simply, the **matrix.** The role of the fiber is to carry the load, while the role of the matrix is to keep the fibers aligned, transfer the load from fiber to fiber, and protect the fibers from direct mechanical contact and exposure to environmental factor. Other fibers, such as boron, have been employed, and composite materials that use metal, such as aluminum or titanium, for the matrix have been developed. Composite materials that utilize boron fibers or a metal matrix are much less common and are quite expensive. The mechanical behavior of metal matrix composite materials can differ significantly from that of polymer matrix composite materials in that metals yield before failing, whereas the polymers used in composite materials are brittle by comparison and generally fail by cracking. This difference in the two matrix materials can lead to different failure criteria and therefore different analysis techniques to predict whether a particular structural component will fail or not. Here emphasis will be on polymer matrix composite materials. However, some of the assumptions and calculations are applicable to metal matrix composite materials if the stresses in the matrix are below the yield level.

The application of the classical methods of the mechanics of materials to the analysis and design of structural components such as beams, plates, curved panels, and cylinders fabricated from fiber-reinforced composite materials requires considerably more steps than the analysis and design of a similar components made of conventional materials such as aluminum or steel. The primary reasons for this are (1) the layered nature of components fabricated of fiber-reinforced composite materials and (2) the fact that fibers in the layers result in stiffness and strength properties that are directionally dependent, not only for that layer, but also for the overall structural component. When material behavior is not directionally dependent, the material behavior is said to be *isotropic.* Except for slight rolling or other effects due to fabrication or consolidation, which are generally ignored, conventional materials such as aluminum or steel exhibit isotropic behavior. When material properties are directionally dependent, material behavior is said to be *anisotropic,* or nonisotropic. There are special classes of anisotropic material behavior, and one which will be defined shortly is *orthotropic* material behavior. To reach a point where the only two differences between a component fabricated from a fiber-reinforced composite material and one fabricated from a conventional material are the layered nature and anisotropic behavior, several steps and assumptions have to be considered. Specifically, the fiber and matrix materials each have their own stiffness and strength properties. When these two constituents are combined, the resulting mixture, or composite, has its own stiffness and strength properties, called **composite properties.** Composite properties depend not only on the properties of each constituent, but also on the quantity of each constituent in the composite material. A key measure of the quantity of a constituent is the **volume fraction** of that constituent in a given volume of composite material. There are models for predicting the properties of a composite material based on the properties and volume fraction of each constituent, in particular, the volume fraction of fibers and the volume fraction of matrix material. These models are referred to as **micromechanical models.** Several micromechanical models will be introduced here for computing such properties as overall stiffness or the overall thermal expansion properties of a layer of composite material. These models are based on assumptions regarding the mechanical behavior of the individual fibers, the matrix surrounding them, and importantly, assumptions regarding the interaction between the fibers and matrix. The result is a collection of models each with its own assumptions and each representing the combined stiffness and thermal expansion effects of all the fibers and the accompanying matrix material. A layer of composite material is assumed to exhibit these combined effects, and it is assumed the combined effects are uniformly distributed throughout the volume of the layer. The assumption of having a uniform distribution of composite properties, despite the fibers and matrix having individual properties, is the concept of **homogenization,** or **smearing,** of constituent properties. Once the smeared composite properties are predicted, the individual fibers and the matrix material lose their identities and a layer is treated as a single material with a single set of properties. Each layer may have different properties, because one layer may use glass fibers and another layer may use carbon fibers, or there may be a mixture of glass and carbon fibers within one layer; but within a layer, the properties are assumed to be uniform. Homogenization is an important concept. If this concept were not accepted, it would be impossible to study the mechanical behavior of fiber-reinforced composite materials, as it would be an insurmountable task to keep track of the behavior of each fiber and the matrix material surrounding each fiber.

After the effects of the fibers and matrix have been homogenized, before design and analysis can proceed, the concept of homogenization must be utilized a second time to represent the combined effects of all the layers on the overall behavior of a structural component. This second step is necessary because a component could be constructed of many, many layers of material. The assemblage of all the layers is referred to as a **laminate.** It becomes a large chore to keep track of the behavior of each layer in a laminate when interest may center on the overall behavior of the laminated component, e.g., such as its lowest natural frequency or its buckling load. The steps for homogenization at this level also depend on assumptions regarding the behavior of the individual layers and assumptions regarding how each layer interacts with the surrounding layers. The study of the individual layers in a laminate and how they interact is often referred to as **macromechanics.** The behavior of a laminate when loading, support, and other conditions are considered is often referred to as **composite structural mechanics.** Many of the tools of mechanics apply to composite materials and structures because the tools are based on principles that do not depend on material behavior. Ultimately, however, material properties must be defined, and it is at this step that the design and analysis of composite structures and of isotropic structures diverge.

MICROMECHANICS OF A LAYER

The practice of using fiber and matrix properties to estimate the homogenized response of a composite material has received lots of attention, and some treatment is given in almost all introductory composite materials texts. Practical reasons for such an investigation include the ability to perform tradeoff studies, such as varying the amount and type of fiber used, and the ability to estimate, during the initial design process, quantities that are difficult to determine experimentally. Three principal approaches have been employed in developing these models for homogenization: (1) mechanics of materials techniques, (2) elasticity solutions, and (3) empirical or semiempirical techniques. In the mechanics of materials approaches, simplified geometries and simplified material behavior models are usually employed. In general, these models do not capture the details of the stress and strain distributions within the fiber and matrix, and as a result, details of the actual fiber arrangement within the composite are unimportant. In the elasticity solutions, an attempt is made to determine the actual stresses and strains in the fiber and matrix materials, and as a result, the fiber packing arrangement within the composite becomes important. In such a case, the actual fiber packing is often represented by idealized packings, as shown in Fig. 5.6.1.

Fig. 5.6.1 Photomicrograph of actual composite cross section (left), and idealized representations of fiber packing: hexagonal (center) and square (right).

It is important to keep in mind, however, that these models are estimates of the actual material behavior, and that there is no substitute for direct experimental measurement of the desired properties. As a result, the models presented in this section are selected for the balance between ease of use and accuracy.

Before we discuss the micromechanics models in detail, it is necessary to begin with a description of the thermal-mechanical response of a single layer of unidirectional composite material. In discussing this response, it is convenient to use an orthogonal coordinate system that has one axis aligned with the direction of the fiber reinforcement, a second axis that lies in the plane of the layer, and a third axis through the thickness of the layer, as illustrated in Fig. 5.6.2. Such a coordinate system is referred to as the **principal material coordinate system** (or often the 1-2-3 coordinate system). In this case, the 1 direction is the fiber direction, while the 2 and 3 directions are transverse directions, often

Fig. 5.6.2 Layer of composite material illustrating principal material coordinate system.

referred to as matrix directions. At this point, it is clear that the material properties will not be the same in all directions. For example, it is reasonable to expect the composite to be stronger and stiffer in the 1 direction than in either the 2 or 3 directions. It might also be expected that the 2 and 3 directions have similar—but not necessarily identical—properties because they are both directions perpendicular to the fiber direction. Accordingly, the response of an element of material at a point within the

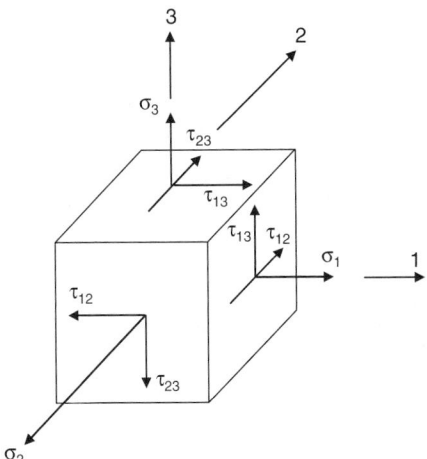

Fig. 5.6.3 Illustration of stresses acting on the small element of composite material from Fig. 5.6.2.

layer that has different properties in the different directions is to be described. Such an element can be acted on by six stress components, as shown in Fig. 5.6.3.

In Fig. 5.6.3, σ_1, σ_2, and σ_3 are the normal stresses, in the principal material directions; τ_{12} is the in-plane shear stress; and τ_{13} and τ_{23} are through-thickness shear stresses. The corresponding normal strains are given as ε_1, ε_2, and ε_3; the engineering shear strains are given as γ_{12}, γ_{13}, and γ_{23}. The strains can then be related to the stresses as

$$
\begin{Bmatrix} \varepsilon_1 \\ \varepsilon_2 \\ \varepsilon_3 \\ \gamma_{23} \\ \gamma_{13} \\ \gamma_{12} \end{Bmatrix} =
\begin{bmatrix}
\dfrac{1}{E_1} & -\dfrac{\nu_{21}}{E_2} & -\dfrac{\nu_{31}}{E_3} & 0 & 0 & 0 \\
-\dfrac{\nu_{12}}{E_1} & \dfrac{1}{E_2} & -\dfrac{\nu_{32}}{E_3} & 0 & 0 & 0 \\
-\dfrac{\nu_{13}}{E_1} & -\dfrac{\nu_{23}}{E_2} & \dfrac{1}{E_3} & 0 & 0 & 0 \\
0 & 0 & 0 & \dfrac{1}{G_{23}} & 0 & 0 \\
0 & 0 & 0 & 0 & \dfrac{1}{G_{13}} & 0 \\
0 & 0 & 0 & 0 & 0 & \dfrac{1}{G_{12}}
\end{bmatrix}
\begin{Bmatrix} \sigma_1 \\ \sigma_2 \\ \sigma_3 \\ \tau_{23} \\ \tau_{13} \\ \tau_{12} \end{Bmatrix}
\tag{5.6.1}
$$

In Eq. (5.6.1), E_1 is the extensional modulus in the fiber direction, E_2 is the extensional modulus in the transverse 2 direction, and E_3 is the extensional modulus in the transverse 3 direction of the composite material. Poisson's ratios ν_{ij} are defined so that $-(\nu_{ij}/E_i)\sigma_i$ is the strain in the j direction due to a stress applied in the i direction. The three shear moduli are G_{23}, G_{13}, and G_{12}. The extensional moduli, the shear moduli, and Poisson's ratios are collectively referred to as *engineering properties*. The square 6-by-6 matrix in Eq. (5.6.1) is referred to as the **compliance matrix** and is normally denoted by S. The stress-strain relations, in terms of S, are then given as

$$
\begin{Bmatrix} \varepsilon_1 \\ \varepsilon_2 \\ \varepsilon_3 \\ \gamma_{23} \\ \gamma_{13} \\ \gamma_{12} \end{Bmatrix} =
\begin{bmatrix}
S_{11} & S_{12} & S_{13} & 0 & 0 & 0 \\
S_{21} & S_{22} & S_{23} & 0 & 0 & 0 \\
S_{31} & S_{32} & S_{33} & 0 & 0 & 0 \\
0 & 0 & 0 & S_{44} & 0 & 0 \\
0 & 0 & 0 & 0 & S_{55} & 0 \\
0 & 0 & 0 & 0 & 0 & S_{66}
\end{bmatrix}
\begin{Bmatrix} \sigma_1 \\ \sigma_2 \\ \sigma_3 \\ \tau_{23} \\ \tau_{13} \\ \tau_{12} \end{Bmatrix}
\tag{5.6.2}
$$

A direct comparison of Eqs. (5.6.1) and (5.6.2) shows that the compliances may be calculated in terms of the engineering properties as

$$
\begin{aligned}
S_{11} = \frac{1}{E_1} \qquad & S_{12} = -\frac{\nu_{21}}{E_2} \qquad S_{13} = -\frac{\nu_{31}}{E_3} \\[4pt]
S_{21} = -\frac{\nu_{12}}{E_1} \qquad & S_{22} = \frac{1}{E_2} \qquad S_{23} = -\frac{\nu_{32}}{E_3} \\[4pt]
S_{31} = -\frac{\nu_{13}}{E_1} \qquad & S_{32} = -\frac{\nu_{23}}{E_2} \qquad S_{33} = \frac{1}{E_3} \\[4pt]
S_{44} = \frac{1}{G_{23}} \qquad & S_{55} = \frac{1}{G_{13}} \qquad S_{66} = \frac{1}{G_{12}}
\end{aligned}
\tag{5.6.3}
$$

The inverse of the compliance matrix is called the **stiffness matrix** and is normally denoted by C. Once this inverse has been defined, the stress-strain relations can be written as

$$
\begin{Bmatrix} \sigma_1 \\ \sigma_2 \\ \sigma_3 \\ \tau_{23} \\ \tau_{13} \\ \tau_{12} \end{Bmatrix} =
\begin{bmatrix}
C_{11} & C_{12} & C_{13} & 0 & 0 & 0 \\
C_{21} & C_{22} & C_{23} & 0 & 0 & 0 \\
C_{31} & C_{32} & C_{33} & 0 & 0 & 0 \\
0 & 0 & 0 & C_{44} & 0 & 0 \\
0 & 0 & 0 & 0 & C_{55} & 0 \\
0 & 0 & 0 & 0 & 0 & C_{66}
\end{bmatrix}
\begin{Bmatrix} \varepsilon_1 \\ \varepsilon_2 \\ \varepsilon_3 \\ \gamma_{23} \\ \gamma_{13} \\ \gamma_{12} \end{Bmatrix}
\tag{5.6.4}
$$

The forms of Eqs. (5.6.2) and (5.6.4) suggest that 12 compliance or 12 stiffness components are required to describe the stress-strain response of an element of composite material. However, it can be shown that the compliance matrix is symmetric, i.e., $S_{ij} = S_{ji}$. For that to be true, from Eq. (5.6.3) the following relations must hold:

$$
\frac{\nu_{12}}{E_1} = \frac{\nu_{21}}{E_2} \qquad \frac{\nu_{13}}{E_1} = \frac{\nu_{31}}{E_3} \qquad \frac{\nu_{23}}{E_2} = \frac{\nu_{32}}{E_3}
\tag{5.6.5}
$$

The relations in Eq. (5.6.5) are referred to as the **reciprocity relations.** Thus, only nine independent engineering properties are needed to describe the elastic response of a layer of composite material.

In addition to the strains due to the stresses, expansional strains due to temperature changes and moisture absorption are often present. For such a case, the strains are related to the stresses by a modification of Eq. (5.6.1) as

$$
\begin{Bmatrix} \varepsilon_1 \\ \varepsilon_2 \\ \varepsilon_3 \\ \gamma_{23} \\ \gamma_{13} \\ \gamma_{12} \end{Bmatrix} =
\begin{bmatrix}
\dfrac{1}{E_1} & -\dfrac{\nu_{21}}{E_2} & -\dfrac{\nu_{31}}{E_3} & 0 & 0 & 0 \\
-\dfrac{\nu_{12}}{E_1} & \dfrac{1}{E_2} & -\dfrac{\nu_{32}}{E_3} & 0 & 0 & 0 \\
-\dfrac{\nu_{13}}{E_1} & -\dfrac{\nu_{23}}{E_2} & \dfrac{1}{E_3} & 0 & 0 & 0 \\
0 & 0 & 0 & \dfrac{1}{G_{23}} & 0 & 0 \\
0 & 0 & 0 & 0 & \dfrac{1}{G_{13}} & 0 \\
0 & 0 & 0 & 0 & 0 & \dfrac{1}{G_{12}}
\end{bmatrix}
\begin{Bmatrix} \sigma_1 \\ \sigma_2 \\ \sigma_3 \\ \tau_{23} \\ \tau_{13} \\ \tau_{12} \end{Bmatrix}
+ \begin{Bmatrix} \alpha_1 \\ \alpha_2 \\ \alpha_3 \\ 0 \\ 0 \\ 0 \end{Bmatrix} \Delta T
+ \begin{Bmatrix} \beta_1 \\ \beta_2 \\ \beta_3 \\ 0 \\ 0 \\ 0 \end{Bmatrix} \Delta M
\tag{5.6.6}
$$

In the above equation, α_1, α_2, and α_3 are coefficients of thermal expansion; β_1, β_2, and β_3 are coefficients of moisture expansion; ΔT is the change in temperature from a reference temperature; and ΔM is the change in moisture content within the material from a reference moisture content. (This reference moisture content is normally taken as zero.) Note that there are no thermally induced or moisture-induced shear strains in the principal material coordinate system. The form of the stress-strain relations in Eq. (5.6.6), and Eqs. (5.6.1), (5.6.2), and (5.6.4), defines **orthotropic material** behavior. Such behavior is characterized by (1) a decoupling of shear and normal responses, resulting in many of the off-diagonal terms being zero, and (2) different material properties in the three mutually perpendicular directions. Specifically, the 0s represent the fact that with orthotropic material behavior, normal strains and normal stresses are related; each particular shear strain is related to the corresponding shear stress, but there are no relations between normal strains and shear stresses, or between shear strains and normal stresses. Also, as noted, there are no thermal- or moisture-induced shear strains. The form of the stress-strain relations for isotropic materials is the same as the form of Eq. (5.6.6), the difference being that the material properties are the same in all directions, and only two engineering properties are needed (e.g., the extensional modulus E and Poisson's ratio ν) to fully describe the response. That is, for the engineering properties $E_1 = E_2 = E_3 = E$, $\nu_{23} = \nu_{13} = \nu_{12} = \nu$, $G_{23} = G_{13} = G_{12} = G = E/2(1 + \nu)$, while for the expansional properties $\alpha_1 = \alpha_2 = \alpha_3 = \alpha$ and $\beta_1 = \beta_2 = \beta_3 = \beta$.

Equation (5.6.6) represents the general response of a layer of composite material subjected to a three-dimensional stress state. In many practical situations involving the use of composite materials, it may be assumed that the composite is in a state of *plane stress,* a state in which three of the stress components are so much smaller than the other three that they may be assumed to be equal to zero for subsequent stress analysis. Some situations in which the plane-stress assumption may be employed arise when you are considering beams, plates, and cylinders in which one dimension is at least an order of magnitude greater than the other dimensions. As an example, consider a plate in which the thickness direction, which is aligned with the 3 direction, is much less than the in-plane dimensions. The out-of-plane stress components σ_3, τ_{23}, and τ_{13} will then be much smaller than the in-plane stress components σ_1, σ_2, and τ_{12}. As a result, Eq. (5.6.6) can be simplified considerably.

Before we examine the simplified stress-strain relations for the case of a plane stress state, a number of cautions are in order. First, the plane stress assumption can lead to inaccuracies. For example, delamination of laminates often initiates at free edges due to the presence of out-of-plane stresses—exactly the stress components ignored in a plane stress analysis. Second, locations of discontinuities in structures, such as at bonded joints, stiffeners, or ply drops, result in three-dimensional stress states. Finally, in many cases the through-thickness stresses are quite small in comparison to the in-plane stresses (particularly σ_1), but may still be large enough in comparison to the corresponding strength values that failure may occur.

To see why the plane stress assumption is important, the effects of this assumption on the stress-strain relations given in Eq. (5.6.6) can be examined. For the case of plane stress in which σ_3, τ_{23}, and τ_{13} are taken to be identically zero, the stress-strain relationship reduces to

$$\begin{Bmatrix} \varepsilon_1 \\ \varepsilon_2 \\ \gamma_{12} \end{Bmatrix} = \begin{bmatrix} \dfrac{1}{E_1} & -\dfrac{\nu_{12}}{E_1} & 0 \\[2mm] \dfrac{\nu_{12}}{E_1} & \dfrac{1}{E_2} & 0 \\[2mm] 0 & 0 & \dfrac{1}{G_{12}} \end{bmatrix} \begin{Bmatrix} \sigma_1 \\ \sigma_2 \\ \tau_{12} \end{Bmatrix}$$

$$+ \begin{Bmatrix} \alpha_1 \\ \alpha_2 \\ 0 \end{Bmatrix} \Delta T + \begin{Bmatrix} \beta_1 \\ \beta_2 \\ 0 \end{Bmatrix} \Delta M \qquad (5.6.7)$$

The 3-by-3 matrix is referred to as the **reduced compliance matrix,** although the expressions for the elements of the reduced compliance matrix in terms of the engineering properties are identical to those in the full 6-by-6 compliance matrix. In addition, it is important to remember that while the normal stress σ_3 is equal to zero for the plane stress condition, the corresponding normal strain ε_3 is nonzero and is given by

$$\varepsilon_3 = -\frac{\nu_{13}}{E_1} \sigma_1 - \frac{\nu_{23}}{E_2} \sigma_2 + \alpha_3 \Delta T + \beta_3 \Delta M \qquad (5.6.8)$$

Thus, to analyze the plane stress response of a layer of composite material to mechanical, thermal, and moisture loads, it is necessary to determine four engineering properties as well as the in-plane coefficients of thermal and moisture expansion. To determine these, the approach taken is to make use of models based primarily on the mechanics of materials, and use other models when the mechanics of materials models do not yield results of sufficient accuracy (or when the more accurate models result in expressions nearly as simple as the mechanics of materials results).

The basic starting point for the mechanics of materials model is the representative volume element shown in Fig. 5.6.4. This volume element consists of enough fiber and matrix that the volume fraction of fiber within the volume is the same as the average value in the composite. A model for the fiber-direction extensional modulus E_1 can be developed by assuming that when a load is applied in the 1 direction, the strains in the fiber and matrix are the same. This **isostrain** assumption leads to the result

$$E_1 = V_f E_1^f + (1 - V_f)E_1^m \qquad (5.6.9)$$

where V_f is the fiber volume fraction, E_1^f is the extensional modulus of the fiber along its length (the 1 direction), and E_1^m is the matrix extensional modulus in the 1 direction. Equation (5.6.9) is often referred to as the **rule-of-mixtures (ROM) model** for E_1 because the result is simply the fractional amount of fiber multiplied by the fiber extensional modulus added to the fractional amount of matrix multiplied by its extensional modulus. Likewise, if an isostrain model is used to estimate Poisson's ratio of the composite material, the resulting expression is

$$\nu_{12} = V_f \nu_{12}^f + (1 - V_f)\nu_{12}^m \qquad (5.6.10)$$

where ν_{12}^f is the fiber Poisson ratio and ν_{12}^m is Poisson's ratio of the matrix. Despite representing a round fiber with a square cross section, numerical results to be presented later in this section will justify recommending the use of rule-of-mixtures estimates for E_1 and ν_{12}.

An inspection of the representative volume shown in Fig. 5.6.4 reveals that the isostrain assumption is not appropriate for estimating the extensional modulus E_2. Rather, a first estimate would be that the

$$V_f = \frac{\text{Volume fiber}}{\text{Total volume}} = \frac{\text{Area fiber}}{\text{Total area}} = \frac{w_f}{w}$$

Fig. 5.6.4   Representative volume element for micromechanics analysis.

stress σ_2 is the same in the fiber and the matrix, and is uniform, when a transverse load is applied. This **isostress assumption** results in an estimate for E_2 of the composite given by

$$\frac{1}{E_2} = \frac{V_f}{E_2^f} + \frac{1 - V_f}{E_2^m} \qquad (5.6.11)$$

where E_2^f and E_2^m are the fiber and matrix transverse extensional modulus values, respectively. While this result based on the isostress assumption is often presented in introductory composite materials textbooks, it produces relatively inaccurate results, as will be shown later. An improved estimate may be developed by considering the representative volume element shown in Fig. 5.6.5. The analysis of this unit element is conducted by assuming that the central fiber/matrix element has an effective extensional modulus that may be calculated by using the rule of mixtures. From there, an analysis of the deformation of the total element leads to the result

$$\frac{1}{E_2} = \frac{1 - \sqrt{V_f}}{E_2^m} + \frac{\sqrt{V_f}}{E_2^f \sqrt{V_f} + E_2^m(1 - \sqrt{V_f})} \qquad (5.6.12)$$

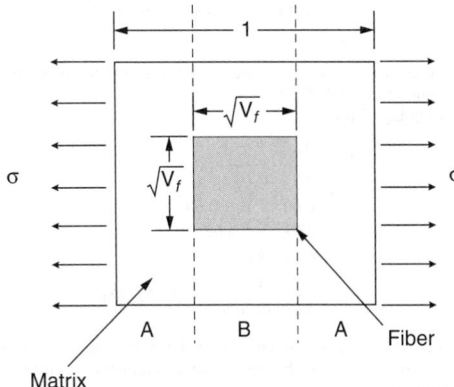

Fig. 5.6.5 Representative volume element for improved mechanics of materials model.

This **improved mechanics of materials model** yields results that are in much better agreement with exact elasticity analyses than those in Eq. (5.6.11), and it is the recommended model for calculating E_2.

For the case of shear loading, the simple mechanics of materials analysis with an isostress assumption gives

$$\frac{1}{G_{12}} = \frac{V_f}{G_{12}^f} + \frac{1 - V_f}{G_{12}^m} \qquad (5.6.13)$$

As with the composite extensional modulus in the 2 direction, often referred to as the transverse extensional modulus, this isostress model gives poor estimates of the composite shear modulus G_{12}. It is possible to develop an analogous expression to Eq. (5.6.12) for the shear modulus. However, there is an **approximate elasticity solution** for a simplified geometry (concentric cylinders of fiber and matrix) available in closed form. The resulting expression for the composite shear modulus is

$$G_{12} = G_{12}^m \frac{(G_{12}^m + G_{12}^f) - V_f(G_{12}^m - G_{12}^f)}{(G_{12}^m + G_{12}^f) + V_f(G_{12}^m - G_{12}^f)} \qquad (5.6.14)$$

This model gives results that are extremely accurate in comparison to exact results for hexagonally and square-packed arrays of fibers.

In addition to these models that describe the response of a layer of composite materials to mechanical loading conditions, it is often useful to have micromechanics models for the expansional coefficients [such as the thermal expansion coefficients and moisture expansion coeffi-

cients that appear in Eq. (5.6.7)]. By using the mechanics of materials representative volume element of Fig. 5.6.4, it is possible to develop expressions for the coefficient of thermal expansion in the 1 direction α_1 as

$$\alpha_1 = \frac{\alpha_1^f E_1^f V_f + \alpha_1^m E_1^m (1 - V_f)}{E_1^f V_f + E_1^m (1 - V_f)} \qquad (5.6.15)$$

The expression for the coefficient of moisture expansion β_1 is developed by analogy. The coefficient of thermal expansion in the 2 direction α_2 is given by

$$\alpha_2 = \left[\alpha_2^f - \left(\frac{E_1^m}{E_1} \right) \nu_{12}^f (\alpha_1^m - \alpha_1^f)(1 - V_f) \right] V_f$$
$$+ \left[\alpha_2^m + \left(\frac{E_1^f}{E_1} \right) \nu_{12}^m (\alpha_1^m - \alpha_1^f) V_f \right] (1 - V_f) \qquad (5.6.16)$$

where E_1 is the rule-of-mixtures expression for the fiber direction extensional modulus given by Eq. (5.6.9). Once again the coefficient of moisture expansion in the 2 direction may be developed by analogy.

In addition to the mechanics of materials, improved mechanics of materials, and elasticity models described above, semiempirical models have been developed for calculating engineering properties of a layer based upon fiber and matrix properties. The most widely used are the **Halpin-Tsai equations**

$$\frac{M}{M_m} = \frac{1 + \eta \xi V_f}{1 - \eta V_f} \qquad (5.6.17)$$

in which

$$\eta = \frac{M_f/M_m - 1}{M_f/M_m + \xi} \qquad (5.6.18)$$

where M = composite modulus E_2 or G_{12}
M_f = corresponding fiber modulus E_2^f or G_{12}^f
M_m = corresponding matrix modulus E_m or ν_m

and ξ is a parameter having admissible values $0 \le \xi < \infty$, which may be different for E_2 and G_{12}. A simplified form of Eqs. (5.6.17) and (5.6.18), obtained by substitution of Eq. (5.6.18) into Eq. (5.6.17), is given by

$$M = M_m \frac{(M_f + \xi M_m) - \xi V_f(M_m - M_f)}{(M_f + \xi M_m) + V_f(M_m - M_f)} \qquad (5.6.19)$$

A value of $\xi = 1$ gives excellent results for G_{12}—an identical result to that given by Eq. (5.6.14). A value of $\xi = 2$ has been found to give good results for E_2.

An assessment of the accuracy of each of the models presented in this section may be obtained by comparing the results from these simple models to results from exact elasticity solutions. Such a comparison requires properties for the fiber and matrix materials. Representative properties for glass fibers, graphite fibers, and polymer matrix materials are taken as the values given in Table 5.6.1. Tables 5.6.2 and 5.6.3

Table 5.6.1   Representative Properties for Graphite Fiber, Glass Fiber, and Polymeric Matrix Materials

	Graphite fiber	Glass fiber*	Polymer matrix*
E_1 (GPa)	233	73.1	4.62
E_2 (GPa)	23.1	73.1	4.62
G_{12} (GPa)	8.96	30.0	1.699
ν_{12}	0.200	0.22	0.36
α_1 (/°C)	-0.540×10^{-6}	5.04×10^{-6}	41.4×10^{-6}
α_2 (/°C)	10.10×10^{-6}	5.04×10^{-6}	41.4×10^{-6}

* Assumed to be isotropic.

Table 5.6.2   Micromechanics Model Results for Graphite-Polymer Composite Material

	Fiber volume fraction							
	0.0	0.1	0.2	0.3	0.4	0.5	0.6	0.7
E_1 (GPa), Eq. (5.6.9)	4.62	27.5	50.3	73.1	96.0	118.8	141.6	164.5
E_1 (GPa), elasticity	4.62	27.5	50.3	73.2	96.0	118.9	141.7	164.5
ν_{12}, Eq. (5.6.10)	0.360	0.344	0.328	0.312	0.296	0.280	0.264	0.248
ν_{12}, elasticity	0.360	0.343	0.327	0.311	0.297	0.282	0.269	0.256
E_2 (GPa), Eq. (5.6.11)	4.62	5.02	5.50	6.08	6.79	7.70	8.88	10.50
E_2 (GPa), Eq. (5.6.12)	4.62	5.61	6.48	7.40	8.45	9.67	11.15	12.98
E_2 (GPa), Eq. (5.6.19)	4.62	5.46	6.41	7.49	8.73	10.16	11.85	13.86
E_2 (GPa), elasticity	4.62	5.78	6.54	7.40	8.43	9.70	11.29	13.27
G_{12} (GPa), Eq. (5.6.13)	1.699	1.848	2.03	2.24	2.51	2.86	3.31	3.93
G_{12} (GPa), Eq. (5.6.14)	1.699	1.947	2.23	2.57	2.97	3.45	4.05	4.80
G_{12} (GPa), elasticity	1.699	1.947	2.23	2.57	2.97	3.46	4.05	4.82
α_1 (/°C) $\times$ 10^6, Eq. (5.6.15)	41.4	5.81	2.54	1.315	0.671	0.275	7.17×10^{-3}	−0.1866
α_1 (/°C) $\times$ 10^6, elasticity	41.4	5.96	2.68	1.440	0.779	0.365	7.87×10^{-2}	−0.1323
α_2 (/°C) $\times$ 10^6, Eq. (5.6.16)	41.4	49.7	46.2	42.0	37.6	33.1	28.5	23.9
α_2 (/°C) $\times$ 10^6, elasticity	41.4	48.9	44.9	40.3	35.7	31.2	26.7	22.4

provide a comparison of composite properties as a function of fiber volume fraction using the various relations discussed.

EXAMPLE 1.   Evaluate the engineering and thermal expansion properties of a layer of glass-fiber-reinforced polymer having a fiber volume fraction of 60 percent.

APPROACH.   Make use of the properties given in Table 5.6.1 along with the micromechanics models described above.

SOLUTION.   The fiber properties are obtained from Table 5.6.1 as

$$E_1^f = E_2^f = 73.1\,\text{GPa}$$
$$G_{12}^f = 30.0\,\text{GPa}$$
$$\nu_{12}^f = 0.22$$
$$\alpha_1^f = \alpha_2^f = 5.04 \times 10^{-6}/°\text{C}$$

The matrix properties are obtained from Table 5.6.1 as

$$E_1^m = E_2^m = 4.62\,\text{GPa}$$
$$G_{12}^m = 1.699\,\text{GPa}$$
$$\nu_{12}^m = 0.36$$
$$\alpha_1^m = \alpha_2^m = 41.4 \times 10^{-6}/°\text{C}$$

The fiber volume fraction is given as $V_f = 0.60$.
For E_1, using Eq. (5.6.9),

$$E_1 = V_f E_1^f + (1 - V_f)E_1^m$$
$$= (0.60)(73.1) + (1 - 0.60)(4.62)$$
$$= 45.7\,\text{GPa}$$

This value compares favorably with the exact value of 45.7 GPa found by using an elasticity solution (see Table 5.6.3).

For ν_{12}, using Eq. (5.6.10),

$$\nu_{12} = V_f \nu_{12}^f + (1 - V_f)\nu_{12}^m$$
$$= (0.60)(0.22) + (1 - 0.60)(0.36)$$
$$= 0.276$$

This value compares favorably with the exact value of 0.269 found by using an elasticity solution (see Table 5.6.3).
For E_2, using Eq. (5.6.11),

$$\frac{1}{E_2} = \frac{V_f}{E_2^f} + \frac{1 - V_f}{E_2^m}$$
$$= \frac{0.60}{73.1} + \frac{1 - 0.60}{4.62}$$
$$= 0.09478\,\text{GPa}^{-1}$$
$$E_2 = \frac{1}{0.09478\,\text{GPa}^{-1}} = 10.55\,\text{GPa}$$

This value is in relatively poor agreement with the exact value of 15.56 GPa found by using the elasticity solution. For the improved mechanics of materials model, using Eq. (5.6.12),

$$\frac{1}{E_2} = \frac{1 - \sqrt{V_f}}{E_2^m} + \frac{\sqrt{V_f}}{E_2^f \sqrt{V_f} + E_2^m(1 - \sqrt{V_f})}$$
$$= \frac{1 - \sqrt{0.60}}{4.62} + \frac{\sqrt{0.60}}{73.1\sqrt{0.60} + 4.62(1 - \sqrt{0.60})}$$
$$= 0.0622\,\text{GPa}^{-1}$$
$$E_2 = 16.07\,\text{GPa}$$

Table 5.6.3   Micromechanics Model Results for Glass-Polymer Composite Material

	Fiber volume fraction							
	0.0	0.1	0.2	0.3	0.4	0.5	0.6	0.7
E_1 (GPa), Eq. (5.6.9)	4.62	11.47	18.32	25.2	32.0	38.9	45.7	52.6
E_1 (GPa), elasticity	4.62	11.48	18.33	25.2	32.0	38.9	45.7	52.6
ν_{12}, Eq. (5.6.10)	0.360	0.346	0.332	0.318	0.304	0.290	0.276	0.262
ν_{12}, elasticity	0.360	0.343	0.327	0.311	0.297	0.282	0.269	0.256
E_2 (GPa), Eq. (5.6.11)	4.62	5.10	5.69	6.43	7.39	8.69	10.55	13.42
E_2 (GPa), Eq. (5.6.12)	4.62	6.25	7.56	9.02	10.78	13.03	16.07	20.5
E_2 (GPa), Eq. (5.6.19)	4.62	5.88	7.39	9.23	11.53	14.49	18.42	23.9
E_2 (GPa), elasticity	4.62	5.77	6.84	8.14	9.82	12.16	15.56	20.7
G_{12} (GPa), Eq. (5.6.13)	1.699	1.875	2.09	2.37	2.73	3.21	3.91	5.00
G_{12} (GPa), Eq. (5.6.14)	1.699	2.03	2.44	2.94	3.59	4.44	5.62	7.36
G_{12} (GPa), elasticity	1.699	2.03	2.44	2.94	3.58	4.44	5.64	7.47
α_1 (/°C) $\times$ 10^6, Eq. (5.6.15)	41.4	18.22	12.38	9.71	8.19	7.20	6.51	6.00
α_1 (/°C) $\times$ 10^6, elasticity	41.4	18.56	12.75	10.06	8.50	7.46	6.72	6.16
α_2 (/°C) $\times$ 10^6, Eq. (5.6.16)	41.4	45.0	42.2	38.2	33.8	29.1	24.4	19.62
α_2 (/°C) $\times$ 10^6, elasticity	41.4	43.9	40.3	35.8	31.1	26.4	21.8	17.38

This model gives better agreement with the exact value for this case (and others as well) and is the recommended model for E_2. The semiempirical Halpin-Tsai expression, Eq. (5.6.19), with $\xi = 2$ results in

$$E_2 = E_2^m \frac{(E_2^f + \xi E_2^m) - \xi V_f(E_2^m - E_2^f)}{(E_2^f + \xi E_2^m) + V_f(E_2^m - E_2^f)}$$

$$= (4.62) \frac{[73.1 + (2)(4.62)] - (2)(0.60)(4.62 - 73.1)}{[73.1 + (2)(4.62)] + (0.60)(4.62 - 73.1)}$$

$$= 18.42 \text{ GPa}$$

For G_{12}, using Eq. (5.6.13),

$$\frac{1}{G_{12}} = \frac{V_f}{G_{12}^f} + \frac{1 - V_f}{G_{12}^m}$$

$$= \frac{0.60}{30.0} + \frac{1 - 0.60}{1.699}$$

$$= 0.255 \text{ GPa}^{-1}$$

$$G_{12} = 3.91 \text{ GPa}$$

This value compares poorly with the exact value of 5.64 GPa obtained from the elasticity solution. The concentric cylinders approximate elasticity model, Eq. (5.6.14), gives

$$G_{12} = G_{12}^m \frac{(G_{12}^m + G_{12}^f) - V_f(G_{12}^m - G_{12}^f)}{(G_{12}^m + G_{12}^f) + V_f(G_{12}^m - G_{12}^f)}$$

$$= 1.699 \frac{(1.699 + 30.0) - 0.60(1.699 - 30.0)}{(1.699 + 30.0) + 0.60(1.699 - 30.0)}$$

$$= 5.62 \text{ GPa}$$

This model gives good agreement with that of the exact solution, and it is the recommended model for G_{12}.

For α_1, using Eq. (5.6.15),

$$\alpha_1 = \frac{\alpha_1^f E_1^f V_f + \alpha_1^m E_1^m (1 - V_f)}{E_1^f V_f + E_1^m (1 - V_f)}$$

$$= \frac{(5.04 \times 10^{-6})(73.1)(0.60) + (41.4 \times 10^{-6})(4.62)(1 - 0.60)}{73.1(0.60) + 4.62(1 - 0.60)}$$

$$= 6.51 \times 10^{-6}/°C$$

This value compares favorably with the exact result of $6.72 \times 10^{-6}/°C$.

For α_2, using Eq. (5.6.16),

$$\alpha_2 = \left[\alpha_2^f - \left(\frac{E_1^m}{E_1} \right) \nu_{12}^f (\alpha_1^m - \alpha_1^f)(1 - V_f) \right] V_f$$

$$+ \left[\alpha_2^m + \left(\frac{E_1^f}{E_1} \right) \nu_{12}^m (\alpha_1^m - \alpha_1^f) V_f \right] (1 - V_f)$$

$$= \left[5.04 \times 10^{-6} - \left(\frac{4.62}{45.7} \right)(0.22)(41.4 \times 10^{-6} \right.$$

$$\left. - 5.04 \times 10^{-6})(1 - 0.60) \right](0.60)$$

$$+ \left[41.4 \times 10^{-6} + \left(\frac{73.1}{45.7} \right)(0.36)(41.4 \times 10^{-6} \right.$$

$$\left. - 5.04 \times 10^{-6})(0.60) \right](1 - 0.60)$$

$$= 24.4 \times 10^{-6}/°C$$

This value compares favorably with the exact result of $21.8 \times 10^{-6}/°C$.

NOTE: It is important to note that in the above examples and all to follow we assume that the fiber and matrix properties are quantified exactly by using three significant figures, as in Table 5.6.1. All subsequent numerical calculations are carried out by starting with these properties, and these subsequent numerical calculations will be reported rounded to three significant figures. Also note that in all numerical examples to follow, a glass-reinforced composite with 60 percent fiber volume fraction is used. Because of their simplicity and accuracy, the engineering properties are computed by employing Eqs. (5.6.9), (5.6.10), (5.6.12), (5.6.14), (5.6.15), and (5.6.16).

Additional example results may be constructed by examining different fiber volume fraction or changing the fiber type, and comparing the results with the values given in Table 5.6.2 or Table 5.6.3, as appropriate for the fiber type chosen. For the graphite fibers, care must be taken to use the appropriate extensional moduli and expansion coefficients, as these fibers are orthotropic.

Although it has not been necessary to this point to describe the thickness of a layer, in the subsequent numerical examples a layer thickness h of 0.000125 m, or 0.125 mm, will be used.

DESCRIBING A LAMINATE

With details of the properties of a layer considered, we turn our attention to describing the behavior of a number of layers, bonded together, one next to the other, to form a laminate. An important part of the description of a laminate is the specification of its **stacking sequence.** The stacking sequence is a list that defines the through-thickness locations and fiber orientations of all the layers in a laminate. Referring to Fig. 5.6.6, which shows the exploded view of a laminate and the **laminate coordinate system,** the xyz coordinate system, the stacking sequence lists the layer fiber orientations relative to the $+x$ axis of the laminate coordinate system, starting with the layer at the negativemost z position. The stacking sequence $[\pm 30/0]_S$ describes the six-layer laminate shown in Fig. 5.6.6 which has $+30°$ layers on the outside, $-30°$ layers inside of those, and two $0°$ layers in the center. The stacking sequence could have been written as $[+30/-30/0/0/-30/+30]_T$, where T signifies that the total laminate has been described, but the shorthand notation with the subscript S is more convenient and implies that the fiber orientation arrangement is symmetric with respect to the xy plane, i.e., the $z = 0$ location. The formal definition of a symmetric laminate will be given shortly. Other shorthand notation can be used, such as $[(\pm 30/0)_3/(90/0)_2]_S$ to describe a 26-layer symmetric laminate which has two subsequences that repeat, one 3 times, the other twice. The notation $[\pm 30]_{2S}$ is interpreted to mean $[(\pm 30)_2]_S$. Of course it is necessary to specify the material properties and thickness of each layer to completely describe a laminate.

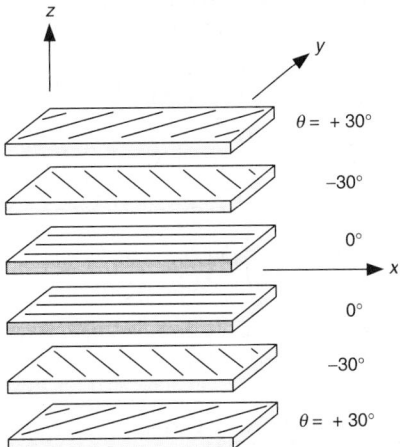

Fig. 5.6.6 Exploded view of a $[\pm 30/0]_S$ laminate and laminate coordinate system.

STRESS-STRAIN BEHAVIOR OF A LAYER: PRINCIPAL MATERIAL COORDINATE SYSTEM

Within the context of the macromechanical theory presented here to study laminate deformations and stresses, referred to as **classical lamination theory,** or **CLT,** a layer is assumed to be in a state of plane stress. Such a condition was previously described. Accordingly, from Eq. (5.6.7), the stress-strain behavior of a layer written in terms of the stresses and strains in the principal material, or 1-2-3, coordinate system, including the effects of thermal expansion, can be written as

$$\begin{Bmatrix} \sigma_1 \\ \sigma_2 \\ \tau_{12} \end{Bmatrix} = \begin{pmatrix} Q_{11} & Q_{12} & 0 \\ Q_{12} & Q_{22} & 0 \\ 0 & 0 & Q_{66} \end{pmatrix} \begin{Bmatrix} \varepsilon_1 - \alpha_1 \Delta T \\ \varepsilon_2 - \alpha_2 \Delta T \\ \gamma_{12} \end{Bmatrix} \quad (5.6.20)$$

where

$$Q_{11} = \frac{E_1}{1 - \nu_{12}\nu_{21}} \qquad Q_{12} = \frac{\nu_{12}E_2}{1 - \nu_{12}\nu_{21}}$$

$$Q_{22} = \frac{E_2}{1 - \nu_{12}\nu_{21}} \qquad Q_{66} = G_{12} \tag{5.6.21}$$

The Q matrix is the **reduced stiffness matrix.** The term *reduced* is used to distinguish the 3-by-3 stiffness matrix of Eq. (5.6.20) from the full 6-by-6 stiffness matrix of Eq. (5.6.4) that must be used when all six components of stress and strain are considered. It is important to recognize that while the numerical values in the reduced compliance matrix are identical to those in the full 6-by-6 compliance matrix, the numerical values in the reduced stiffness matrix are different from those in the full 6-by-6 stiffness matrix. Again, the form of the reduced stiffness matrix in Eq. (5.6.20) is said to represent orthotropic material behavior. For simplicity, moisture content effects have not been included in Eq. (5.6.20). They can be included in the same manner as thermal expansion effects.

EXAMPLE 2. Compute the numerical values of the reduced stiffness matrix for a glass-fiber-reinforced composite.

APPROACH. Substitute the engineering properties directly into Eq. (5.6.21).
[As noted, all examples are based on using a 60 percent volume fraction of glass fibers and, because of their simplicity and accuracy, employing Eqs. (5.6.9), (5.6.10), (5.6.12), (5.6.14), (5.6.15), and (5.6.16) to calculate the material properties, as presented in Table 5.6.3.]

SOLUTION

$$Q_{11} = \frac{E_1}{1 - \nu_{12}\nu_{21}} = \frac{45.7 \times 10^9}{1 - (0.276)(0.0970)} = 47.0 \times 10^9 = 47.0\,\text{GPa}$$

$$Q_{12} = \frac{\nu_{12}E_2}{1 - \nu_{12}\nu_{21}} = \frac{0.276(16.07 \times 10^9)}{1 - (0.276)(0.0970)} = 4.56\,\text{GPa}$$

$$Q_{22} = \frac{E_2}{1 - \nu_{12}\nu_{21}} = \frac{16.07 \times 10^9}{1 - (0.276)(0.0970)} = 16.51\,\text{GPa}$$

$$[Q] = \begin{pmatrix} Q_{11} & Q_{12} & 0 \\ Q_{12} & Q_{22} & 0 \\ 0 & 0 & Q_{66} \end{pmatrix} = \begin{pmatrix} 47.0 & 4.56 & 0 \\ 4.56 & 16.51 & 0 \\ 0 & 0 & 5.62 \end{pmatrix} \times 10^9$$

$$= \begin{pmatrix} 47.0 & 4.56 & 0 \\ 4.56 & 16.51 & 0 \\ 0 & 0 & 5.62 \end{pmatrix} \text{GPa}$$

The reciprocity relation of Eq. (5.6.5) was used to compute ν_{21}. We emphasize that in nearly all cases $\nu_{12} \neq \nu_{21}$.

STRESS-STRAIN BEHAVIOR OF A LAYER: LAMINATE COORDINATE SYSTEM

Since there can be many fiber orientations within a laminate, with each orientation having its own principal material coordinate system, it is more convenient to describe laminate behavior in terms of one coordinate system, namely, the laminate, or *xyz*, coordinate system shown in Fig. 5.6.6. The stresses and strains in the laminate coordinate system, including thermal strains, can be expressed in terms of those in the principal material coordinate system by **transformation relations,** namely,

$$\begin{Bmatrix} \sigma_x \\ \sigma_y \\ \tau_{xy} \end{Bmatrix} = [T]^{-1} \begin{Bmatrix} \sigma_1 \\ \sigma_2 \\ \tau_{12} \end{Bmatrix} \qquad \begin{Bmatrix} \varepsilon_x \\ \varepsilon_y \\ \frac{1}{2}\gamma_{xy} \end{Bmatrix} = [T]^{-1} \begin{Bmatrix} \varepsilon_1 \\ \varepsilon_2 \\ \frac{1}{2}\gamma_{12} \end{Bmatrix}$$

$$\begin{Bmatrix} \alpha_x \\ \alpha_y \\ \frac{1}{2}\alpha_{xy} \end{Bmatrix} \Delta T = [T]^{-1} \begin{Bmatrix} \alpha_1 \\ \alpha_2 \\ 0 \end{Bmatrix} \Delta T \tag{5.6.22}$$

where

$$[T]^{-1} = \begin{bmatrix} T_{11} & T_{12} & T_{13} \\ T_{21} & T_{22} & T_{23} \\ T_{31} & T_{32} & T_{33} \end{bmatrix}^{-1} = \begin{bmatrix} m^2 & n^2 & -2mn \\ n^2 & m^2 & 2mn \\ mn & -mn & m^2 - n^2 \end{bmatrix} \tag{5.6.23}$$

and where $m = \cos\theta$ and $n = \sin\theta$, with θ being the angle measured from the $+x$ axis to the $+1$ axis of the layer. In Eq. (5.6.23), the 3-by-3 matrix, denoted as T^1, is referred to as the **inverse transformation matrix.** The word *inverse* is used because the stresses and strains of importance within a laminate are those parallel and perpendicular to the fibers, namely, those in the principal material coordinate system. The principal material coordinate system stresses and strains are used to determine at what load level a layer will fail. The relationship in Eq. (5.6.22) transforms those important stresses and strains into stresses and strains in the *xyz* coordinate system. The latter quantities are neither parallel nor perpendicular to the fibers, so it is more difficult to evaluate whether the stresses and strains are high enough to cause, e.g., failure of a layer due to excess stress in the fiber direction. Note well in the above relations to factor of ½ associated with the sheer strain.

The above relations can be inverted and rewritten as

$$\begin{Bmatrix} \sigma_1 \\ \sigma_2 \\ \tau_{12} \end{Bmatrix} = [T] \begin{Bmatrix} \sigma_x \\ \sigma_y \\ \tau_{xy} \end{Bmatrix} \qquad \begin{Bmatrix} \varepsilon_1 \\ \varepsilon_2 \\ \frac{1}{2}\gamma_{12} \end{Bmatrix} = [T] \begin{Bmatrix} \varepsilon_x \\ \varepsilon_y \\ \frac{1}{2}\gamma_{xy} \end{Bmatrix}$$

$$\begin{Bmatrix} \alpha_1 \\ \alpha_2 \\ 0 \end{Bmatrix} \Delta T = [T] \begin{Bmatrix} \alpha_x \\ \alpha_y \\ \frac{1}{2}\alpha_{xy} \end{Bmatrix} \Delta T \tag{5.6.24}$$

where

$$[T] = \begin{bmatrix} T_{11} & T_{12} & T_{13} \\ T_{21} & T_{22} & T_{23} \\ T_{31} & T_{32} & T_{33} \end{bmatrix} = \begin{bmatrix} m^2 & n^2 & 2mn \\ n^2 & m^2 & -2mn \\ -mn & mn & m^2 - n^2 \end{bmatrix} \tag{5.6.25}$$

The above 3-by-3 matrix, denoted as T, is referred to as the **transformation matrix.**

As a result of the fibers being at an angle relative to the laminate coordinate system, stresses and strains measured in the laminate coordinate system are quite different from those measured in the principal material coordinate system. Example 3 demonstrates the interesting strain response of a single layer caused by a temperature change, but measured in the laminate coordinate system.

EXAMPLE 3. The temperature of an element cut from a single layer of glass epoxy with its fibers at $+30°$ is decreased by $-100°$C. What shear strain is measured in the laminate coordinate system due to this temperature decrease? Note that this temperature decrease is similar in magnitude to the decrease from the curing temperature of the material to room temperature.

APPROACH. The thermally induced strains in the laminate coordinate system can be computed from the known values of the thermally induced strains in the principal material coordinate system from Table 5.6.3 and the transformation relations of Eqs. (5.6.22) and (5.6.23).

SOLUTION. Since $\theta = 30°$, $m = \sqrt{3/2}$, and $n = \frac{1}{2}$,

$$\begin{Bmatrix} \alpha_x \\ \alpha_y \\ \frac{1}{2}\alpha_{xy} \end{Bmatrix} \Delta T = [T(30°)]^{-1} \begin{Bmatrix} \alpha_1 \\ \alpha_2 \\ 0 \end{Bmatrix} \Delta T$$

$$= \begin{bmatrix} \frac{3}{4} & \frac{1}{4} & \frac{-\sqrt{3}}{2} \\ \frac{1}{4} & \frac{3}{4} & \frac{\sqrt{3}}{2} \\ \frac{\sqrt{3}}{4} & \frac{-\sqrt{3}}{4} & \frac{1}{2} \end{bmatrix} \begin{Bmatrix} 6.51 \times 10^{-6} \\ 24.4 \times 10^{-6} \\ 0 \end{Bmatrix} (-100)$$

or

$$\alpha_x \Delta T = -1,099 \times 10^{-6} \qquad \alpha_y \Delta T = -1,099 \times 10^{-6}$$

$$\alpha_{xy} \Delta T = 1,551 \times 10^{-6}$$

DISCUSSION. What is important to note is that although there is no shear strain in the principal material coordinate system, i.e., $\alpha_{12} \Delta T = 0$, a shear strain as well as contraction strains in the x and y directions are measured in the laminate coordinate system. Since polymer matrix composites are cured at an elevated temperature and then cooled, when a layer is part of a laminate, the thermally induced strains in a layer caused by cooling the laminate from its curing temperature can produce residual thermal stresses. If the element is square and $0.1\,\text{m} \times 0.1\,\text{m}$ in dimensions before the temperature change, after cooling the length in the x and y directions would be less. In addition, the original right angles would be changed by 1,551 microradians, denoted as 1,551 μrad, or 0.0888 deg, as shown in exaggerated fashion in Fig. 5.6.7, with the right angles of two opposite corners increasing and the right angles of the other two opposite corners decreasing. This may not seem like much of an angle change, but it can cause significant stresses if resisted. The strains in the x and y directions are −1,099 and −1,994 microstrain, denoted, respectively, as −1,099 and −1,994 $\mu\varepsilon$. The deformed element would have dimensions 0.0999 and 0.0998 m in the x and y directions, respectively.

Fig. 5.6.7 Cooling deformations of an element with fibers as oriented at $+30°$.

As it is convenient to transform the stresses and strains from each of the principal material coordinate systems of the various layers to the single laminate coordinate system, it is convenient to transform the stress-strain relation of each layer to the laminate coordinate system. This can be accomplished by starting with the stress-strain relation of Eq. (5.6.20) and applying the transformation relations of Eq. (5.6.22). As a result, the stress-strain relation expressed in the laminate coordinate system is written as

$$\begin{Bmatrix} \sigma_x \\ \sigma_y \\ \tau_{xy} \end{Bmatrix} = \begin{pmatrix} \overline{Q}_{11} & \overline{Q}_{12} & \overline{Q}_{16} \\ \overline{Q}_{12} & \overline{Q}_{22} & \overline{Q}_{26} \\ \overline{Q}_{16} & \overline{Q}_{26} & \overline{Q}_{66} \end{pmatrix} \begin{Bmatrix} \varepsilon_x - \alpha_x \Delta T \\ \varepsilon_y - \alpha_x \Delta T \\ \gamma_{xy} - \alpha_{xy} \Delta T \end{Bmatrix} \quad (5.6.26)$$

where

$$\overline{Q}_{11} = Q_{11}m^4 + 2(Q_{12} + 2Q_{66})m^2n^2 + Q_{22}n^4$$
$$\overline{Q}_{12} = (Q_{11} + Q_{22} - 4Q_{66})n^2m^2 + Q_{12}(n^4 + m^4)$$
$$\overline{Q}_{16} = (Q_{11} - Q_{12} - 2Q_{66})nm^3 + (Q_{12} - Q_{22} + 2Q_{66})n^3m \quad (5.6.27)$$
$$\overline{Q}_{22} = Q_{11}n^4 + 2(Q_{12} + 2Q_{66})m^2n^2 + Q_{22}m^4$$
$$\overline{Q}_{26} = (Q_{11} - Q_{12} - 2Q_{66})n^3m + (Q_{12} - Q_{22} + 2Q_{66})nm^3$$
$$\overline{Q}_{66} = (Q_{11} + Q_{22} - 2Q_{12} - 2Q_{66})n^2m^2 + Q_{66}(n^4 + m^4)$$

As mentioned, $m = \cos\theta$ and $n = \sin\theta$, so the $\overline{Q}$ terms are functions of θ. The 3-by-3 $\overline{Q}$ matrix is referred to as the **transformed reduced stiffness matrix**. As can be observed, because of fiber orientation, compared to conventional isotropic materials or the orthotropic behavior of Eq. (5.6.20), unusual relations between stresses and strains occur. Specifically, the shear stress is related to the two normal strains, and the two normal stresses are related to the shear strain. There are no zeros in the $\overline{Q}$ matrix as there are with the Q matrix in Eq. (5.6.20). The behavior represented by Eq. (5.6.27) is referred to as **anisotropic behavior.** A word of caution: Note carefully the factors of $\frac{1}{2}$ and 2 that occur in some relations but not in others. This is due to the use of engineering shear strains γ_{12} and γ_{xy}, as opposed to tensor shear strains, which are given by $\frac{1}{2}\gamma_{12}$ and $\frac{1}{2}\gamma_{xy}$.

EXAMPLE 4. A normal stress of 50 MPa is applied in the x direction to a square element of composite material with its fibers at $30°$ relative to the $+x$ axis while the temperature is kept constant, i.e., $\Delta T = 0$. No other stresses are present. What strains result from this applied stress?

APPROACH AND SOLUTION. Substituting directly into the stress-strain relation expressed in the laminate coordinate system, Eq. (5.6.26), gives

$$\begin{Bmatrix} \sigma_x \\ \sigma_y \\ \tau_{xy} \end{Bmatrix} = \begin{Bmatrix} 50 \times 10^6 \\ 0 \\ 0 \end{Bmatrix} = \begin{pmatrix} \overline{Q}_{11}(30°) & \overline{Q}_{12}(30°) & \overline{Q}_{16}(30°) \\ \overline{Q}_{12}(30°) & \overline{Q}_{22}(30°) & \overline{Q}_{26}(30°) \\ \overline{Q}_{16}(30°) & \overline{Q}_{26}(30°) & \overline{Q}_{66}(30°) \end{pmatrix} \begin{Bmatrix} \varepsilon_x \\ \varepsilon_y \\ \gamma_{xy} \end{Bmatrix}$$

where the $\overline{Q}$'s are computed by using Eq. (5.6.27) with $\theta = 30°$, namely,

$$\begin{pmatrix} \overline{Q}_{11}(30°) & \overline{Q}_{12}(30°) & \overline{Q}_{16}(30°) \\ \overline{Q}_{12}(30°) & \overline{Q}_{22}(30°) & \overline{Q}_{26}(30°) \\ \overline{Q}_{16}(30°) & \overline{Q}_{26}(30°) & \overline{Q}_{66}(30°) \end{pmatrix} = \begin{pmatrix} 33.4 & 10.54 & 10.05 \\ 10.54 & 18.15 & 3.14 \\ 10.05 & 3.14 & 11.60 \end{pmatrix} \times 10^9\,\text{N/m}^2$$

resulting in

$$\begin{pmatrix} 33.4 & 10.54 & 10.05 \\ 10.54 & 18.15 & 3.14 \\ 10.05 & 3.14 & 11.60 \end{pmatrix} \times 10^9 \begin{Bmatrix} \varepsilon_x \\ \varepsilon_y \\ \gamma_{xy} \end{Bmatrix} = \begin{Bmatrix} 50 \times 10^6 \\ 0 \\ 0 \end{Bmatrix}$$

which leads to

$$\begin{Bmatrix} \varepsilon_x \\ \varepsilon_y \\ \gamma_{xy} \end{Bmatrix} = \begin{Bmatrix} 2,370 \\ -1,069 \\ -1,760 \end{Bmatrix} \times 10^{-6}$$

As shown in Fig. 5.6.8, the element of material stretches in the x direction and contracts in the y direction, as an element of conventional material such as aluminum or steel would, the contraction being due to Poisson effects. Unlike an element of conventional material, however, the element experiences a shear strain. This results in corner angles that are not right angles when the square element deforms.

Fig. 5.6.8 Deformations due to application of 50-MPa normal stress in x direction.

To be noted is the fact that there is distinct difference between stating that the element of material is *stressed* in the x direction and stating that the element of material is *strained* in the x direction. If the element is strained in the x direction by, e.g., 1,000 $\mu\varepsilon$, with the element experiencing no other strains, then the stresses required are given by again using Eq. (5.6.26), namely,

$$\begin{Bmatrix} \sigma_x \\ \sigma_y \\ \tau_{xy} \end{Bmatrix} = \begin{pmatrix} \overline{Q}_{11}(30°) & \overline{Q}_{12}(30°) & \overline{Q}_{16}(30°) \\ \overline{Q}_{12}(30°) & \overline{Q}_{22}(30°) & \overline{Q}_{26}(30°) \\ \overline{Q}_{16}(30°) & \overline{Q}_{26}(30°) & \overline{Q}_{66}(30°) \end{pmatrix} \begin{Bmatrix} 1,000 \\ 0 \\ 0 \end{Bmatrix} \times 10^{-6}$$

or $$\begin{Bmatrix} \sigma_x \\ \sigma_y \\ \tau_{xy} \end{Bmatrix} = \begin{Bmatrix} 33.4 \\ 10.54 \\ 10.05 \end{Bmatrix} \times 10^6\,\text{Pa} = \begin{Bmatrix} 33.4 \\ 10.54 \\ 10.05 \end{Bmatrix}\,\text{MPa}$$

It can be concluded that to produce this rather simple state of deformation, a tensile stress in the y direction and a positive shear stress, in addition to a tensile stress in the x direction, are required. These stresses are illustrated in Fig. 5.6.9. For a conventional material, although a tensile stress in the y direction would be required, the shear stress would not be required because there is no tendency for the corner right angle to change. If the fiber angle were $-30°$, the same tensile stresses in the x and y directions would be required, but the shear stress required would be -10.05 MPa, a negative shear stress.

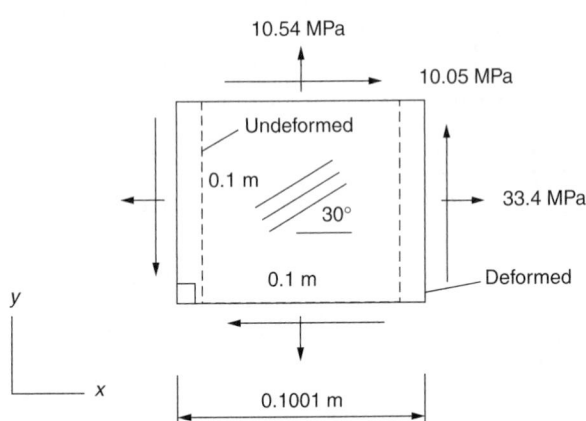

Fig. 5.6.9 Stresses required to produce strain of $1{,}000$ $\mu\varepsilon$ in the x direction.

MACROMECHANICS OF A LAMINATE

When layers are combined to form a laminate, the unusual behavior in the various layers can interact to produce results that are even more unusual than the behavior of a single layer. To study the behavior of a laminate, it is assumed that within a laminate the layers are perfectly bonded together and do not slip relative to one another. Only small deformations are considered, much the same as in studying the mechanics of conventional materials. Like a layer, a laminate occupies a volume in three-dimensional space, so the analysis and the design of laminates are truly a three-dimensional problem. However, the plane stress assumption eliminates some of the three-dimensionality of the problem, as does one other key assumption, namely, the **Kirchhoff hypothesis**. As a direct result of the Kirchhoff hypothesis, the strains $\varepsilon_x, \varepsilon_y,$ and γ_{xy} vary linearly through the thickness of the laminate. If the no-slip condition were not true, the Kirchhoff hypothesis certainly would not be valid. The Kirchhoff hypothesis assumes that when the laminate deflects out of plane in the z direction, all points through the thickness of the laminate at a given x and y location deflect out of plane the same amount. That is, the out-of-plane displacement of a point within a laminate is independent of the z location of the point. The assumed deformation characteristics of a laminate are further illustrated in Fig. 5.6.10, where the deformation is viewed in the xz plane. The figure shows that straight line AA' through the thickness of the laminate before deformation remains straight after deformation, a result that leads to a linear variation of strains through the thickness of the laminate. Specifically, the displacements in the x, y, and z directions of a generic point P at location x, y, z within the laminate are assumed to have the following forms:

$$u(x,y,z) = u^o(x, y) - z\frac{\partial w^o(x,y)}{\partial x}$$

$$v(x,y,z) = v^o(x, y) - z\frac{\partial w^o(x,y)}{\partial y} \qquad (5.6.28)$$

$$w(x,y,z) = w^o(x,y)$$

where the superscript o denotes the displacements of corresponding point P^o on the **geometric midplane** of the laminate, which is also referred to as the **reference surface**. Only small displacements and rotations are considered in the development of Eq. (5.6.28). As will be seen, if the deformation of the reference surface is known, i.e., if the

strains and curvatures of the reference surface are known, then the strains, and subsequently the stresses, in every layer can be determined. The reference surface is somewhat like the neutral axis in simple beam analysis, except with composite materials the stresses are not necessarily zero at the reference surface location.

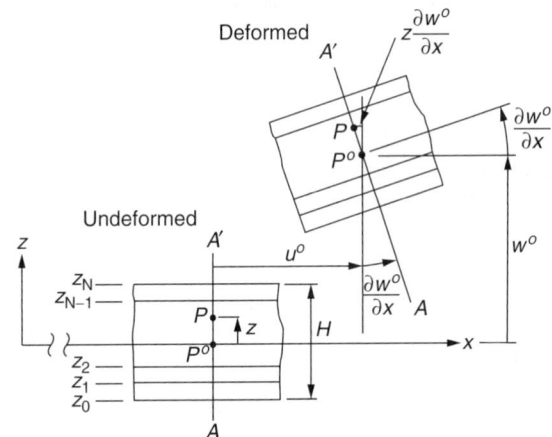

Fig. 5.6.10 Implications of Kirchhoff hypothesis for laminate deformations.

To be noted in Fig. 5.6.10 is the notation $z_0, z_1, z_2, \ldots$ which describes the z locations of the interfaces between layers, or alternatively, the z locations of the lower and upper surfaces of the layers. Accordingly, $z_0 = -H/2$, with H being the total laminate thickness, and the thickness of the kth layer h_k is given by $z_k - z_{z-1}$.

The strains at any location within the laminate can be written in terms of the displacements as

$$\varepsilon_x = \frac{\partial u}{\partial x} = \frac{\partial u^o}{\partial x} - z\frac{\partial^2 w^o}{\partial x^2} = \varepsilon_x^o + z\kappa_x^o$$

$$\varepsilon_y = \frac{\partial v}{\partial y} = \frac{\partial v^o}{\partial y} - z\frac{\partial^2 w^o}{\partial y^2} = \varepsilon_y^o + z\kappa_y^o \qquad (5.6.29)$$

$$\gamma_{xy} = \frac{\partial u}{\partial y} + \frac{\partial v}{\partial x} = \frac{\partial u^o}{\partial y} + \frac{\partial v^o}{\partial x} - 2z\frac{\partial^2 w^o}{\partial x \partial y} = \gamma_{xy}^o + z\kappa_{xy}^o$$

where

$$\left\{ \begin{array}{c} \varepsilon_x^o \\ \varepsilon_y^0 \\ \gamma_{xy}^o \end{array} \right\} = \left\{ \begin{array}{c} \dfrac{\partial u^o}{\partial x} \\ \dfrac{\partial v^o}{\partial y} \\ \dfrac{\partial u^o}{\partial y} + \dfrac{\partial v^o}{\partial x} \end{array} \right\} \qquad \left\{ \begin{array}{c} \kappa_x^o \\ \kappa_y^o \\ \kappa_{xy}^o \end{array} \right\} = \left\{ \begin{array}{c} -\dfrac{\partial^2 w^o}{\partial x^2} \\ -\dfrac{\partial^2 w^o}{\partial y^2} \\ -2\dfrac{\partial^2 w^o}{\partial x \partial y} \end{array} \right\} \qquad (5.6.30)$$

The quantities $\varepsilon_x^o, \varepsilon_y^o,$ and γ_{xy}^o are referred to as the **midplane** or **reference surface strains**. The quantities $\kappa_x^o, \kappa_y^o,$ and κ_{xy}^o are the **midplane** or **reference surface curvatures** and are the bending curvatures of the x and y directions, and the twist curvature, respectively, of the reference surface. By knowing the midplane strains and curvatures, the state of strain at any z location within any layer can be computed though the use of Eq. (5.6.29). These strains, in turn, can be transformed to determine the strains in the principal material coordinate system of the layer at this z location by using the second equation of Eq. (5.6.24). The stresses in the laminate coordinate system at this z location can then be computed using Eq. (5.6.20). As can be seen, then, knowing the midplane strains and curvatures is key to determining the stresses anywhere through the thickness of the laminate. The plane stress assumption and the Kirchhoff hypothesis are keys to the analysis simplifying to this degree. Without these assumptions, the design and analysis of composite laminates would be intractable.

Since the midplane strains and curvatures are the keys to determining the stresses within a laminate, it is important to be able to compute these midplane quantities in terms of the loads applied to a laminate. In the application of composite materials, most often loads, in the form of pressures, forces, or bending moments, are specified rather than midplane strains and curvatures, or rather than the stresses. A relationship between the specified loads and the midplane strains and curvatures is therefore necessary. So far all that has been discussed are strains and stresses. Therefore, relationships between *loads* and the stresses or between *loads* and the midplane strains and curvatures need to be developed to be able to proceed with studying the behavior of laminates in practical applications. To develop these relations, the loads are defined in terms of the stresses as follows:

$$N_x = \int_{-H/2}^{H/2} \sigma_x\, dz \qquad N_y = \int_{-H/2}^{H/2} \sigma_y\, dz \qquad N_{xy} = \int_{-H/2}^{H/2} \tau_{xy}\, dz$$

$$M_x = \int_{-H/2}^{H/2} \sigma_x z\, dz \qquad M_y = \int_{-H/2}^{H/2} \sigma_y z\, dz \qquad M_{xy} = \int_{-H/2}^{H/2} \tau_{xy} z\, dz \tag{5.6.31}$$

The Ns are the *force resultants* and the Ms are the *moment resultants*. The units of the force and moment resultants are force *per unit length* and moment *per unit length*, respectively. Collectively the six quantities are the *stress resultants*. The introduction of the stress resultants further reduces the three-dimensional nature of the problem. The *per unit length* part of the definition of the stress resultants is very important. The resultant N_x is the force in the x direction per unit length of the laminate in the y direction; N_y is the force in the y direction per unit length of laminate in the x direction. Similar interpretations are associated with the moment resultants M_x and M_y. Since it is based on stress component τ_{xy}, the force resultant N_{xy} is the force in the x direction per unit length of laminate in the y direction. However, it is also the force in the y direction per unit length of laminate in the x direction, depending on which edge of the laminate is being discussed. A similar dual definition is associated with M_{xy}. It should be emphasized that it is impossible from equilibrium considerations to have N_{xy} and M_{xy} acting on one edge of an element of material and not have the same values of N_{xy} and M_{xy} acting on the adjacent edge. It is important to be aware of the positive sense of the stress resultants, which are illustrated in Fig. 5.6.11.

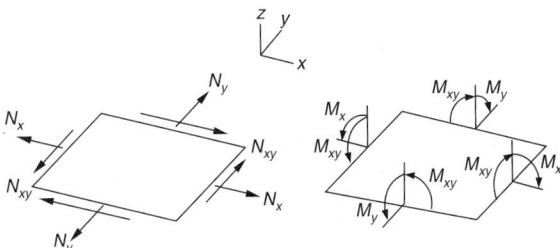

Fig. 5.6.11 Illustration of positive sense of force and moment resultants.

If the three components of stress from Eq. (5.6.26) are substituted into the integrands of the definitions of the six stress resultants, the integration with respect to z can be carried out. Considering N_x as an example, substitution of the first equation of the stress-strain relation from Eqs. (5.6.26) into the first equation of Eqs. (5.6.31) results in

$$N_x = \int_{-H/2}^{H/2} [\overline{Q}_{11}(\varepsilon_x^o + z\kappa_x^o - \alpha_x \Delta T) + \overline{Q}_{12}(\varepsilon_y^o + z\kappa_y^o - \alpha_y \Delta T)$$
$$+ \overline{Q}_{16}(\gamma_{xy}^o + z\kappa_{xy}^o - \alpha_{xy}\Delta T)]\, dz \tag{5.6.32}$$

Regrouping gives

$$N_x = \int_{-H/2}^{H/2} [\overline{Q}_{11}(\varepsilon_x^o + z\kappa_x^o) + \overline{Q}_{12}(\varepsilon_y^o + z\kappa_y^o) + \overline{Q}_{16}(\gamma_{xy}^o + z\kappa_{xy}^o)]\, dz$$
$$- \int_{-H/2}^{H/2} (\overline{Q}_{11}\alpha_x + \overline{Q}_{12}\alpha_y + \overline{Q}_{16}\alpha_{xy})\Delta T\, dz \tag{5.6.33}$$

The second integral is special. It is defined as the **effective thermal force resultant** in the x direction, i.e.,

$$N_x^T = \int_{-H/2}^{H/2} (\overline{Q}_{11}\alpha_x + \overline{Q}_{12}\alpha_y + \overline{Q}_{16}\alpha_{xy})\Delta T\, dz \tag{5.6.34}$$

where the superscript T identifies that the resultant is due to thermal effects. If the temperature change is not a function of z, i.e., if the temperature change is uniform through the thickness of the laminate, then ΔT can be removed from the integral to yield

$$N_x^T = \left[\int_{-H/2}^{H/2} (\overline{Q}_{11}\alpha_x + \overline{Q}_{12}\alpha_y + \overline{Q}_{16}\alpha_{xy})\, dz \right] \Delta T \tag{5.6.35}$$

The integral within the brackets is called the *unit effective thermal force resultant*. It is only a function of material properties and the laminate geometry and is defined as

$$\hat{N}_x^T = \int_{-H/2}^{H/2} (\overline{Q}_{11}\alpha_x + \overline{Q}_{12}\alpha_y + \overline{Q}_{16}\alpha_{xy})\, dz \tag{5.6.36}$$

Note that in the definition of the unit effective thermal force resultant, the coefficients of thermal expansion combine with the reduced stiffnesses. This indicates that the magnitude of the thermal effects depends not only on the thermal expansion or contraction of the individual layers due to temperature changes, but also on how resistant the layers are to deformation, i.e., how stiff they are. By using the nomenclature of Eq. (5.6.36), Eq. (5.6.33) can be written as

$$N_x = \left(\int_{-H/2}^{H/2} \overline{Q}_{11}\, dz \right)\varepsilon_x^o + \left(\int_{-H/2}^{H/2} \overline{Q}_{12}\, dz \right)\varepsilon_y^o + \left(\int_{-H/2}^{H/2} \overline{Q}_{16}\, dz \right)y_{xy}^o$$
$$+ \left(\int_{-H/2}^{H/2} \overline{Q}_{11}z\, dz \right)\kappa_x^o + \left(\int_{-H/2}^{H/2} \overline{Q}_{12}z\, dz \right)\kappa_y^o$$
$$+ \left(\int_{-H/2}^{H/2} \overline{Q}_{16}z\, dz \right)\kappa_{xy}^o - \hat{N}_x^T\Delta T \tag{5.6.37}$$

where the reference surface strains and curvatures have been removed from the integral because they are not functions of z and are therefore not involved in the integration with respect to z. This is a very important point, and it is a direct result of the Kirchhoff hypothesis. As a final step, Eq. (5.6.37) is written as

$$N_x = A_{11}\varepsilon_x^o + A_{12}\varepsilon_y^o + A_{16}y_{xy}^o + B_{11}\kappa_x^o$$
$$+ B_{12}\kappa_y^o + B_{16}\kappa_{xy}^o - \hat{N}_x^T\Delta T \tag{5.6.38}$$

The As and Bs are defined as the integrals of the reduced stiffnesses through the thickness of the laminate, and hence they represent **smeared, integrated,** or **overall stiffnesses** of the laminate. The reduced stiffnesses of every layer contribute to the smeared stiffnesses of the laminate. In the same spirit, the effective thermal force resultant represents smeared, integrated, or overall effects. Note that Eq. (5.6.38) relates the three reference surface strains and three reference surface curvatures to the force resultant and to thermal effects. This relationship and the other five to follow are quite important and absolutely necessary if the stresses within the various layers are to be computed from the loads. Before proceeding, it should be noted, as was stated at the onset, that smearing of properties is used for a second time with the definition of the As, Bs, and the effective thermal force resultant. The smearing of properties, be it at the micromechanics level or at the macromechanics, or laminate, level, is essential if progress is to be made in modeling and predicting behavior.

If the other five integrals in Eq. (5.6.31) are treated in a similar fashion and the results combined with Eq. (5.6.38), the following relationship results:

$$
\begin{Bmatrix} N_x \\ N_y \\ N_{xy} \\ M_x \\ M_y \\ M_{xy} \end{Bmatrix} = \begin{bmatrix} A_{11} & A_{12} & A_{16} & B_{11} & B_{12} & B_{16} \\ A_{12} & A_{22} & A_{26} & B_{12} & B_{22} & B_{26} \\ A_{16} & A_{26} & A_{66} & B_{16} & B_{26} & B_{66} \\ B_{11} & B_{12} & B_{16} & D_{11} & D_{12} & D_{16} \\ B_{12} & B_{22} & B_{26} & D_{12} & D_{22} & D_{26} \\ B_{16} & B_{26} & B_{66} & D_{16} & D_{26} & D_{66} \end{bmatrix} \begin{Bmatrix} \varepsilon_x^o \\ \varepsilon_y^o \\ \gamma_{xy}^o \\ \kappa_x^o \\ \kappa_y^o \\ \kappa_{xy}^o \end{Bmatrix}
$$

$$
- \begin{Bmatrix} \hat{N}_x^T \\ \hat{N}_y^T \\ \hat{N}_{xy}^T \\ \hat{M}_x^T \\ \hat{M}_y^T \\ \hat{M}_{xy}^T \end{Bmatrix} \Delta T \qquad (5.6.39)
$$

The 6-by-6 matrix is referred to as the **laminate stiffness matrix** or, more simply, the *ABD matrix*. This matrix relates the stress resultants, or loads, to the deformations, or strains and curvatures, of the reference surface. Since the reduced stiffnesses do not vary through the thickness of any particular layer, the integrals that define the elements of the *ABD* matrix reduce to summations, namely,

$$
A_{ij} = \sum_{k=1}^{N} \overline{Q}_{ij_k}(z_k - z_{k-1}) \qquad B_{ij} = \frac{1}{2}\sum_{k=1}^{N} \overline{Q}_{ij_k}(z_k^2 - z_{k-1}^2)
$$
$$
D_{ij} = \frac{1}{3}\sum_{k=1}^{N} \overline{Q}_{ij_k}(z_k^3 - z_{k-1}^3) \qquad (5.6.40)
$$

where the z_k are defined in Fig. 5.6.10 and N is the number of layers in the laminate. Since the coefficients of thermal expansion also do not vary through the thickness of any particular layer, the unit effective thermal stress resultants become summations given by

$$
\hat{N}_x^T = \sum_{k=1}^{N} (\overline{Q}_{11k}\alpha_{xk} + \overline{Q}_{12k}\alpha_{yk} + \overline{Q}_{16k}\alpha_{xyk})(z_k - z_{k-1})
$$
$$
\hat{N}_y^T = \sum_{k=1}^{N} (\overline{Q}_{12k}\alpha_{xk} + \overline{Q}_{22k}\alpha_{yk} + \overline{Q}_{26k}\alpha_{xyk})(z_k - z_{k-1})
$$
$$
\hat{N}_{xy}^T = \sum_{k=1}^{N} (\overline{Q}_{16k}\alpha_{xk} + \overline{Q}_{26k}\alpha_{yk} + \overline{Q}_{66k}\alpha_{xyk})(z_k - z_{k-1})
$$
$$
\hat{M}_x^T = \frac{1}{2}\sum_{k=1}^{N} (\overline{Q}_{11k}\alpha_{xk} + \overline{Q}_{12k}\alpha_{yk} + \overline{Q}_{16k}\alpha_{xyk})(z_k^2 - z_{k-1}^2)
$$
$$
\hat{M}_y^T = \frac{1}{2}\sum_{k=1}^{N} (\overline{Q}_{12k}\alpha_{xk} + \overline{Q}_{22k}\alpha_{yk} + \overline{Q}_{26k}\alpha_{xyk})(z_k^2 - z_{k-1}^2)
$$
$$
\hat{M}_{xy}^T = \frac{1}{2}\sum_{k=1}^{N} (\overline{Q}_{16k}\alpha_{xk} + \overline{Q}_{26k}\alpha_{yk} + \overline{Q}_{66k}\alpha_{xyk})(z_k^2 - z_{k-1}^2)
$$

$$(5.6.41)$$

An inspection of the various elements of the *ABD* matrix reveals that not only do their definitions depend on the reduced stiffnesses of every layer, but also their definitions include the location within the laminate of every layer by virtue of z_k and z_{k-1}. The elements of the *ABD* matrix are thus weighted and smeared stiffness quantities, the weighting factor being the thickness and location of the various layers within the laminate. Strictly speaking, the values of the As involve only the layer thicknesses, i.e., $z_k - z_{k-1} = h_k$, the thickness of the kth layer, while the values of the Bs and Ds involve layer thicknesses and locations. Physically, the As are the in-plane stiffnesses of the laminate, the Ds are the bending stiffnesses, and the Bs are unique to composite materials

and are referred to as the *bending-stretching stiffnesses*. This nomenclature refers to the fact that through these terms, bending effects, namely, the curvatures, κ_x^o, κ_y^o, and κ_{xy}^o, are coupled to in-plane force resultants N_x, N_y, and N_{xy}, and conversely, the in-plane strains ε_x^o, ε_y^o, and γ_{xy}^o are coupled to the bending moment resultants M_x, M_y, and M_{xy}. The unit effective thermal stress resultants are also functions of stiffnesses and coefficients of thermal expansion weighted by geometry. With the elements of the *ABD* matrix and the unit effective thermal stress resultants depending on layer thickness and location, a layer can have greater or less influence on the overall behavior of a laminate, depending on its location within the stacking sequence and its thickness. It is important to realize that once the location, material properties, and fiber orientation of each layer are known, the sums defining the *ABD* matrix and the unit effective thermal stress resultants can be computed and numerical values obtained.

Special Cases of Laminate Stacking Sequences

For so-called **symmetric laminates**, all elements of the B matrix are 0, as are the three unit effective thermal moment resultants if the temperature change ΔT is uniform through the thickness of the laminate. A symmetric laminate is defined as follows: If, for every layer at a specific location and with a specific thickness, material properties, and fiber orientation located on one side of the geometric midplane, there is a layer with identical thickness, material properties, and fiber orientation at the mirror-image location on the other side of the geometric midplane, then the laminate is said to be symmetric. For **balanced laminates**, A_{16}, A_{26}, and $\hat{N}_{xy}^T$ are zero. A balanced laminate is defined as follows: If, for every layer with a specific thickness, material properties, and fiber orientation, there is a layer *somewhere* within the laminate with the same specific thickness and material properties, but with the *opposite* fiber orientation, then the laminate is said to be balanced. Symmetric and balanced laminates are very common, and the relation of Eq. (5.6.39) simplifies considerably to

$$
\begin{Bmatrix} N_x \\ N_y \\ N_{xy} \\ M_x \\ M_y \\ M_{xy} \end{Bmatrix} = \begin{bmatrix} A_{11} & A_{12} & 0 & 0 & 0 & 0 \\ A_{12} & A_{22} & 0 & 0 & 0 & 0 \\ 0 & 0 & A_{66} & 0 & 0 & 0 \\ 0 & 0 & 0 & D_{11} & D_{12} & D_{16} \\ 0 & 0 & 0 & D_{12} & D_{22} & D_{26} \\ 0 & 0 & 0 & D_{16} & D_{26} & D_{66} \end{bmatrix} \begin{Bmatrix} \varepsilon_x^o \\ \varepsilon_y^o \\ \gamma_{xy}^o \\ \kappa_x^o \\ \kappa_y^o \\ \kappa_{xy}^o \end{Bmatrix}
$$

$$
- \begin{Bmatrix} \hat{N}_x^T \\ \hat{N}_y^T \\ 0 \\ 0 \\ 0 \\ 0 \end{Bmatrix} \Delta T \qquad (5.6.42)
$$

It is important to note that D_{16} and D_{26} are not zero. It is also important to remember that if the temperature is not uniform through the thickness of the laminate, even though the laminate is symmetric, Eq. (5.6.42) does not apply. Because of the simplifications, Eq. (5.6.42) can be written as two separate equations, namely,

$$
\begin{Bmatrix} N_x \\ N_y \\ N_{xy} \end{Bmatrix} = \begin{Bmatrix} A_{11} & A_{12} & 0 \\ A_{12} & A_{22} & 0 \\ 0 & 0 & A_{66} \end{Bmatrix} \begin{Bmatrix} \varepsilon_x^o \\ \varepsilon_y^o \\ \gamma_{xy}^o \end{Bmatrix} - \begin{Bmatrix} \hat{N}_x^T \\ \hat{N}_y^T \\ 0 \end{Bmatrix} \Delta T
$$

$$
= [A]\begin{Bmatrix} \varepsilon_x^o \\ \varepsilon_y^o \\ \gamma_{xy}^o \end{Bmatrix} - \begin{Bmatrix} \hat{N}_x^T \\ \hat{N}_y^T \\ 0 \end{Bmatrix} \Delta T \qquad (5.6.43)
$$

$$
\begin{Bmatrix} M_x \\ M_y \\ M_{xy} \end{Bmatrix} = \begin{Bmatrix} D_{11} & D_{12} & D_{16} \\ D_{12} & D_{22} & D_{26} \\ D_{16} & D_{26} & D_{66} \end{Bmatrix} \begin{Bmatrix} \kappa_x^o \\ \kappa_y^o \\ \kappa_{xy}^o \end{Bmatrix} = [D]\begin{Bmatrix} \kappa_x^o \\ \kappa_y^o \\ \kappa_{xy}^o \end{Bmatrix} \qquad (5.6.44)
$$

where it is seen that for symmetric laminates, bending effects (M_x, M_y, M_{xy} and κ_x^o, κ_y^o, κ_{xy}^o) and in-plane effects (N_x, N_y, N_{xy} and ε_x^o, ε_y^o, γ_{xy}^o) decouple. Actually, Eq. (5.6.43) can be further simplified to two other equations as

$$\begin{Bmatrix} N_x \\ N_y \end{Bmatrix} = \begin{bmatrix} A_{11} & A_{12} \\ A_{12} & A_{22} \end{bmatrix} \begin{Bmatrix} \varepsilon_x^o \\ \varepsilon_y^o \end{Bmatrix} - \begin{Bmatrix} \hat{N}_x^T \\ \hat{N}_y^T \end{Bmatrix} \Delta T \qquad (5.6.45)$$

$$N_{xy} = A_{66}\gamma_{xy}^o \qquad (5.6.46)$$

Cross-ply laminates are laminates constructed of layers with fiber orientations of only 0 or 90°. For such laminates, because by Eq. (5.6.27) $\overline{Q}_{16}$ and $\overline{Q}_{26}$ are zero for every layer, all 16 and 26 terms and $\hat{N}_{xy}^T$ and $\hat{M}_{xy}^T$ in Eq. (5.6.39) are zero. For a symmetric cross-ply laminate, the relations of Eq. (5.6.39) simplify even more to give what can be written as four separate equations:

$$\begin{Bmatrix} N_x \\ N_y \end{Bmatrix} = \begin{bmatrix} A_{11} & A_{12} \\ A_{12} & A_{22} \end{bmatrix} \begin{Bmatrix} \varepsilon_x^o \\ \varepsilon_y^o \end{Bmatrix} - \begin{Bmatrix} \hat{N}_x^T \\ \hat{N}_y^T \end{Bmatrix} \Delta T \qquad (5.6.47)$$

$$N_{xy} = A_{66}\gamma_{xy}^o \qquad (5.6.48)$$

$$\begin{Bmatrix} M_x \\ M_y \end{Bmatrix} = \begin{bmatrix} D_{11} & D_{12} \\ D_{12} & D_{22} \end{bmatrix} \begin{Bmatrix} \kappa_x^o \\ \kappa_y^o \end{Bmatrix} \qquad (5.6.49)$$

$$M_{xy} = D_{66}\kappa_y^o \qquad (5.6.50)$$

The existence of the four separate relations signifies a high degree of uncoupling of various stress resultants and reference surface deformations. Again as a reminder, if the temperature is not uniform through the thickness of the laminate, then Eqs. (5.6.47) to (5.6.50) do not apply. If the layers are all isotropic, or for a single isotropic layer, the form of Eq. (5.6.39) reduces to the form of Eqs. (5.6.47) to (5.6.50).

To follow are numerical examples of the A, B, and D matrices for a number of common laminates. The numerical examples are based upon the properties for a glass-fiber-reinforced polymer having a fiber volume fraction of 60 percent, as given in Example 2.

EXAMPLE 5 [±45/0/90]$_S$. The subscript S signifies that the laminate is symmetric, and because there is a $-45°$ layer for each $+45°$ layer, the laminate is balanced. Neither the 0 nor 90° layers contribute to any of the 16 terms or the unit effective thermal stress resultants.

$$\begin{bmatrix} A_{11} & A_{12} & A_{16} \\ A_{12} & A_{22} & A_{26} \\ A_{16} & A_{26} & A_{66} \end{bmatrix} = \begin{bmatrix} 27.8 & 8.55 & 0 \\ 8.55 & 27.8 & 0 \\ 0 & 0 & 9.60 \end{bmatrix} \frac{\text{MN}}{\text{m}}$$

$$\begin{bmatrix} D_{11} & D_{12} & D_{16} \\ D_{12} & D_{22} & D_{26} \\ D_{16} & D_{26} & D_{66} \end{bmatrix} = \begin{bmatrix} 2.18 & 0.961 & 0.1784 \\ 0.961 & 1.945 & 0.1784 \\ 0.1784 & 0.1784 & 1.050 \end{bmatrix} \text{N} \cdot \text{m}$$

$$\begin{Bmatrix} \hat{N}_x^T \\ \hat{N}_y^T \end{Bmatrix} = \begin{Bmatrix} 425 \\ 425 \end{Bmatrix} \frac{\text{N/m}}{°\text{C}}$$

This eight-layer laminate is called a *quasi-isotropic* laminate because the in-plane extensional stiffnesses A_{11} and A_{22} are equal, but the in-plane shear stiffness A_{66} is not related to A_{11} ($=A_{22}$), as it would be for a truly isotropic material. In addition, the bending stiffness D_{11} and D_{22} are not equal; nor are D_{16} and D_{26} zero, as they would be for a truly isotropic material, but they are equal. Note also that $\hat{N}_x^T$ and $\hat{N}_y^T$ are equal. Herein, as stated before, the thickness of a layer is taken to be 0.000125 m (0.125 mm).

EXAMPLE 6 [±60/0]$_S$

$$\begin{bmatrix} A_{11} & A_{12} & A_{16} \\ A_{12} & A_{22} & A_{26} \\ A_{16} & A_{26} & A_{66} \end{bmatrix} = \begin{bmatrix} 20.8 & 6.41 & 0 \\ 6.41 & 20.8 & 0 \\ 0 & 0 & 7.20 \end{bmatrix} \frac{\text{MN}}{\text{m}}$$

$$\begin{bmatrix} D_{11} & D_{12} & D_{16} \\ D_{12} & D_{22} & D_{26} \\ D_{16} & D_{26} & D_{66} \end{bmatrix} = \begin{bmatrix} 0.675 & 0.363 & 0.0491 \\ 0.363 & 1.151 & 0.1570 \\ 0.0491 & 0.1570 & 0.400 \end{bmatrix} \text{N} \cdot \text{m}$$

$$\begin{Bmatrix} \hat{N}_x^T \\ \hat{N}_y^T \end{Bmatrix} = \begin{Bmatrix} 319 \\ 319 \end{Bmatrix} \frac{\text{N/m}}{°\text{C}}$$

This six-layer laminate is also quasi-isotropic. The bending stiffnesses D_{16} and D_{26} are not zero.

EXAMPLE 7 [±30/0]$_S$

$$\begin{bmatrix} A_{11} & A_{12} & A_{16} \\ A_{12} & A_{22} & A_{26} \\ A_{16} & A_{26} & A_{66} \end{bmatrix} = \begin{bmatrix} 28.4 & 6.41 & 0 \\ 6.41 & 13.20 & 0 \\ 0 & 0 & 7.20 \end{bmatrix} \frac{\text{MN}}{\text{m}}$$

$$\begin{bmatrix} D_{11} & D_{12} & D_{16} \\ D_{12} & D_{22} & D_{26} \\ D_{16} & D_{26} & D_{66} \end{bmatrix} = \begin{bmatrix} 1.191 & 0.363 & 0.1570 \\ 0.363 & 0.636 & 0.0491 \\ 0.1570 & 0.0491 & 0.400 \end{bmatrix} \text{N} \cdot \text{m}$$

$$\begin{Bmatrix} \hat{N}_x^T \\ \hat{N}_y^T \end{Bmatrix} = \begin{Bmatrix} 315 \\ 323 \end{Bmatrix} \frac{\text{N/m}}{°\text{C}}$$

It is interesting to note the equality of some of the terms in this laminate and the [±60/0]$_S$ laminate above.

EXAMPLE 8 [±30]$_{2S}$. This eight-layer laminate is a symmetric and balanced *angle-ply* laminate.

$$\begin{bmatrix} A_{11} & A_{12} & A_{16} \\ A_{12} & A_{22} & A_{26} \\ A_{16} & A_{26} & A_{66} \end{bmatrix} = \begin{bmatrix} 33.4 & 10.54 & 0 \\ 10.54 & 18.15 & 0 \\ 0 & 0 & 11.60 \end{bmatrix} \frac{\text{MN}}{\text{m}}$$

$$\begin{bmatrix} D_{11} & D_{12} & D_{16} \\ D_{12} & D_{22} & D_{26} \\ D_{16} & D_{26} & D_{66} \end{bmatrix} = \begin{bmatrix} 2.78 & 0.878 & 0.314 \\ 0.878 & 1.512 & 0.0981 \\ 0.314 & 0.0981 & 0.966 \end{bmatrix} \text{N} \cdot \text{m}$$

$$\begin{Bmatrix} \hat{N}_x^T \\ \hat{N}_y^T \end{Bmatrix} = \begin{Bmatrix} 421 \\ 429 \end{Bmatrix} \frac{\text{N/m}}{°\text{C}}$$

It is interesting to compare the properties of this laminate with the [±30/0]$_S$ above, which differs in thickness and the presence of 0° layers.

EXAMPLE 9 [±60/0/90]$_S$

$$\begin{bmatrix} A_{11} & A_{12} & A_{16} \\ A_{12} & A_{22} & A_{26} \\ A_{16} & A_{26} & A_{66} \end{bmatrix} = \begin{bmatrix} 24.9 & 7.55 & 0 \\ 7.55 & 32.6 & 0 \\ 0 & 0 & 8.61 \end{bmatrix} \frac{\text{MN}}{\text{m}}$$

$$\begin{bmatrix} D_{11} & D_{12} & D_{16} \\ D_{12} & D_{22} & D_{26} \\ D_{16} & D_{26} & D_{66} \end{bmatrix} = \begin{bmatrix} 1.773 & 0.816 & 0.0736 \\ 0.816 & 2.65 & 0.235 \\ 0.0736 & 0.235 & 0.904 \end{bmatrix} \text{N} \cdot \text{m}$$

$$\begin{Bmatrix} \hat{N}_x^T \\ \hat{N}_y^T \end{Bmatrix} = \begin{Bmatrix} 427 \\ 423 \end{Bmatrix} \frac{\text{N/m}}{°\text{C}}$$

EXAMPLE 10 [0/90]$_{2S}$. This is an eight-layer symmetric cross-ply laminate.

$$\begin{bmatrix} A_{11} & A_{12} & A_{16} \\ A_{12} & A_{22} & A_{26} \\ A_{16} & A_{26} & A_{66} \end{bmatrix} = \begin{bmatrix} 31.7 & 4.56 & 0 \\ 4.56 & 31.7 & 0 \\ 0 & 0 & 5.62 \end{bmatrix} \frac{\text{MN}}{\text{m}}$$

$$\begin{bmatrix} D_{11} & D_{12} & D_{16} \\ D_{12} & D_{22} & D_{26} \\ D_{16} & D_{26} & D_{66} \end{bmatrix} = \begin{bmatrix} 3.12 & 0.380 & 0 \\ 0.380 & 2.17 & 0 \\ 0 & 0 & 0.468 \end{bmatrix} \text{N} \cdot \text{m}$$

$$\begin{Bmatrix} \hat{N}_x^T \\ \hat{N}_y^T \end{Bmatrix} = \begin{Bmatrix} 425 \\ 425 \end{Bmatrix} \frac{\text{N/m}}{°\text{C}}$$

This laminate could be considered quasi-isotropic also. It is interesting to note the difference in D_{11} and D_{22}. The outer 0° layers provide most of the contribution to D_{11}, while the 90° layers provide most of the contribution to D_{22}. Since the 90° layers are closer to the reference surface than the 0° layers, the bending stiffness in the y direction, D_{22}, is less than the bending stiffness in the x direction, D_{11}. On the other hand, the four $0°$ layers and the four $90°$ layers provide the same resistance to extension in the x and y directions, respectively, so $A_{22} = A_{11}$.

EXAMPLE 11 $[0_4/90_4]_T$.   This an eight-layer *antisymmetric* cross-ply laminate.

$$
\begin{bmatrix}
A_{11} & A_{12} & A_{16} & B_{11} & B_{12} & B_{16} \\
A_{12} & A_{22} & A_{26} & B_{12} & B_{22} & B_{26} \\
A_{16} & A_{26} & A_{66} & B_{16} & B_{26} & B_{66} \\
B_{11} & B_{12} & B_{16} & D_{11} & D_{12} & D_{16} \\
B_{12} & B_{22} & B_{26} & D_{12} & D_{22} & D_{26} \\
B_{16} & B_{26} & B_{66} & D_{16} & D_{26} & D_{66}
\end{bmatrix}
$$

$$
=
\begin{bmatrix}
31.7\ \tfrac{MN}{m} & 4.56\ \tfrac{MN}{m} & 0 & -3{,}810\,N & 0 & 0 \\
4.56\ \tfrac{MN}{m} & 31.7\ \tfrac{MN}{m} & 0 & 0 & 3{,}810\,N & 0 \\
0 & 0 & 5.62\ \tfrac{MN}{m} & 0 & 0 & 0 \\
-3{,}810\,N & 0 & 0 & 2.65\ \tfrac{N}{m} & 0.380\ \tfrac{N}{m} & 0 \\
0 & 3{,}810\,N & 0 & 0.380\ \tfrac{N}{m} & 2.65\ \tfrac{N}{m} & 0 \\
0 & 0 & 0 & 0 & 0 & 0.468\ \tfrac{N}{m}
\end{bmatrix}
$$

$$
\begin{Bmatrix}
\hat{N}_x^T \\
\hat{N}_y^T \\
\hat{N}_{xy}^T \\
\hat{M}_x^T \\
\hat{M}_y^T \\
\hat{M}_{xy}^T
\end{Bmatrix}
=
\begin{Bmatrix}
425\ \tfrac{N/m}{°C} \\
425\ \tfrac{N/m}{°C} \\
0 \\
19.77 \times 10^{-4}\,\tfrac{N}{°C} \\
-19.77 \times 10^{-4}\,\tfrac{N}{°C} \\
0
\end{Bmatrix}
$$

The only terms in the B matrix are B_{11}, which equals $-B_{22}$, but this single coupling term can cause interesting effects. There are also thermally induced moments, which cause this laminate to curl out of plane when cooled from its elevated curing temperature. The two unit equivalent thermal moment resultants are equal in magnitude and opposite in sign. Note that the A matrix is the same as for the previous cross-ply laminate, Example 10.

INVERSE RELATIONS

Generally the force and moment resultants and temperature change are known for an element of laminate, and it is of interest to compute the strains and stresses in individual layers, or throughout the entire thickness of the laminate if a complete stress analysis is desired. It is convenient to invert the relations previously discussed so the reference surface strains and curvatures can be computed more easily. For a general laminate, from Eq. (5.6.39), the inverse relation is

$$
\begin{bmatrix}
\varepsilon_x^o \\
\varepsilon_y^o \\
\gamma_{xy}^o \\
\kappa_x^o \\
\kappa_y^o \\
\kappa_{xy}^o
\end{bmatrix}
=
\begin{bmatrix}
a_{11} & a_{12} & a_{16} & b_{11} & b_{12} & b_{16} \\
a_{12} & a_{22} & a_{26} & b_{21} & b_{22} & b_{26} \\
a_{16} & a_{26} & a_{66} & b_{61} & b_{62} & b_{66} \\
b_{11} & b_{21} & b_{61} & d_{11} & d_{12} & d_{16} \\
b_{12} & b_{22} & b_{62} & d_{12} & d_{22} & d_{26} \\
b_{16} & b_{26} & b_{66} & d_{16} & d_{26} & d_{66}
\end{bmatrix}
\left(
\begin{Bmatrix}
N_x \\
N_y \\
N_{xy} \\
M_x \\
M_y \\
M_{xy}
\end{Bmatrix}
+
\begin{Bmatrix}
\hat{N}_x^T \\
\hat{N}_y^T \\
\hat{N}_{xy}^T \\
\hat{M}_x^T \\
\hat{M}_y^T \\
\hat{M}_{xy}^T
\end{Bmatrix}
\Delta T
\right)
\tag{5.6.51}
$$

where

$$
\begin{bmatrix}
a_{11} & a_{12} & a_{16} & b_{11} & b_{12} & b_{16} \\
a_{12} & a_{22} & a_{26} & b_{21} & b_{22} & b_{26} \\
a_{16} & a_{26} & a_{66} & b_{61} & b_{62} & b_{66} \\
b_{11} & b_{21} & b_{61} & d_{11} & d_{12} & d_{16} \\
b_{12} & b_{22} & b_{62} & d_{12} & d_{22} & d_{26} \\
b_{16} & b_{26} & b_{66} & d_{16} & d_{26} & d_{66}
\end{bmatrix}
=
\begin{bmatrix}
A_{11} & A_{12} & A_{16} & B_{11} & B_{12} & B_{16} \\
A_{12} & A_{22} & A_{26} & B_{12} & B_{22} & B_{26} \\
A_{16} & A_{26} & A_{66} & B_{16} & B_{26} & B_{66} \\
B_{11} & B_{12} & B_{16} & D_{11} & D_{12} & D_{16} \\
B_{12} & B_{22} & B_{26} & D_{12} & D_{22} & D_{26} \\
B_{16} & B_{26} & B_{66} & D_{16} & D_{26} & D_{66}
\end{bmatrix}^{-1}
\tag{5.6.52}
$$

is the **laminate compliance matrix** or **ABD inverse matrix.**

For a symmetric balanced laminate, the inverses of Eqs. (5.6.44) to (5.6.46) are

$$
\begin{Bmatrix}
\varepsilon_x^o \\
\varepsilon_y^o
\end{Bmatrix}
=
\begin{bmatrix}
a_{11} & a_{12} \\
a_{12} & a_{22}
\end{bmatrix}
\begin{Bmatrix}
N_x + \hat{N}_x^T \Delta T \\
N_y + \hat{N}_y^T \Delta T
\end{Bmatrix}
\tag{5.6.53}
$$

$$
\gamma_{xy}^o = a_{66} N_{xy}
\tag{5.6.54}
$$

$$
\begin{Bmatrix}
\kappa_x^o \\
\kappa_y^o \\
\kappa_{xy}^o
\end{Bmatrix}
=
\begin{bmatrix}
d_{11} & d_{12} & d_{16} \\
d_{12} & d_{22} & d_{26} \\
d_{16} & d_{26} & d_{66}
\end{bmatrix}
\begin{Bmatrix}
M_x \\
M_y \\
M_{xy}
\end{Bmatrix}
= [d]
\begin{Bmatrix}
M_x \\
M_y \\
M_{xy}
\end{Bmatrix}
\tag{5.6.55}
$$

where

$$
\begin{bmatrix}
a_{11} & a_{12} & 0 \\
a_{12} & a_{22} & 0 \\
0 & 0 & a_{66}
\end{bmatrix}
=
\begin{bmatrix}
A_{11} & A_{12} & 0 \\
A_{12} & A_{22} & 0 \\
0 & 0 & A_{66}
\end{bmatrix}^{-1}
$$

$$
\begin{bmatrix}
d_{11} & d_{12} & d_{16} \\
d_{12} & d_{22} & d_{26} \\
d_{16} & d_{26} & d_{66}
\end{bmatrix}
=
\begin{bmatrix}
D_{11} & D_{12} & D_{16} \\
D_{12} & D_{22} & D_{26} \\
D_{16} & D_{26} & D_{66}
\end{bmatrix}^{-1}
\tag{5.6.56}
$$

The inverse matrices, namely the compliance matrices, are actually more useful than the stiffness matrices for computing the reference surface strains and curvatures from the force and moment resultants. For a symmetric cross-ply laminate, the inverse relations are determined from Eqs. (5.6.47) to (5.6.50) as

$$
\begin{Bmatrix}
\varepsilon_x^o \\
\varepsilon_y^o
\end{Bmatrix}
=
\begin{bmatrix}
a_{11} & a_{12} \\
a_{12} & a_{22}
\end{bmatrix}
\begin{Bmatrix}
N_x + \hat{N}_x^T \Delta T \\
N_y + \hat{N}_y^T \Delta T
\end{Bmatrix}
\tag{5.6.57}
$$

$$
\gamma_{xy}^o = a_{66} N_{xy}
\tag{5.6.58}
$$

$$
\begin{Bmatrix}
\kappa_x^o \\
\kappa_y^o
\end{Bmatrix}
=
\begin{bmatrix}
d_{11} & d_{12} \\
d_{12} & d_{22}
\end{bmatrix}
\begin{Bmatrix}
M_x \\
M_y
\end{Bmatrix}
\tag{5.6.59}
$$

$$
\kappa_{xy}^o = d_{66} M_{xy}
\tag{5.6.60}
$$

where the compliance matrices are given by

$$
\begin{bmatrix}
a_{11} & a_{12} & 0 \\
a_{12} & a_{22} & 0 \\
0 & 0 & a_{66}
\end{bmatrix}
=
\begin{bmatrix}
A_{11} & A_{12} & 0 \\
A_{12} & A_{22} & 0 \\
0 & 0 & A_{66}
\end{bmatrix}^{-1}
$$

$$
\begin{bmatrix}
d_{11} & d_{12} & 0 \\
d_{12} & d_{22} & 0 \\
0 & 0 & d_{66}
\end{bmatrix}
=
\begin{bmatrix}
D_{11} & D_{12} & 0 \\
D_{12} & D_{22} & 0 \\
0 & 0 & D_{66}
\end{bmatrix}^{-1}
\tag{5.6.61}
$$

The numerical values of the compliance matrices for the laminates discussed in Examples 5 to 11 above are as follows:

EXAMPLE 12 $[\pm45/0/90]_S$

$$
\begin{bmatrix}
a_{11} & a_{12} & a_{16} \\
a_{12} & a_{22} & a_{26} \\
a_{16} & a_{26} & a_{66}
\end{bmatrix}
=
\begin{bmatrix}
39.8 & -12.26 & 0 \\
-12.26 & 39.8 & 0 \\
0 & 0 & 104.1
\end{bmatrix}
\tfrac{m}{GN}
$$

$$
\begin{bmatrix}
d_{11} & d_{12} & d_{16} \\
d_{12} & d_{22} & d_{26} \\
d_{16} & d_{26} & d_{66}
\end{bmatrix}
=
\begin{bmatrix}
0.588 & -0.286 & -0.0514 \\
-0.286 & 0.662 & -0.0638 \\
-0.0514 & -0.0638 & 0.972
\end{bmatrix}
\tfrac{1}{N \cdot m}
$$

where, for example,

$$
39.8\ \frac{m}{GN} = 39.8\ \frac{m}{10^9\,N} = 39.8 \times 10^{-9}\ \frac{m}{N}
$$

EXAMPLE 13 $[\pm60/0]_S$

$$
\begin{bmatrix}
a_{11} & a_{12} & a_{16} \\
a_{12} & a_{22} & a_{26} \\
a_{16} & a_{26} & a_{66}
\end{bmatrix}
=
\begin{bmatrix}
53.1 & -16.34 & 0 \\
-16.34 & 53.1 & 0 \\
0 & 0 & 138.8
\end{bmatrix}
\tfrac{m}{GN}
$$

$$\begin{bmatrix} d_{11} & d_{12} & d_{16} \\ d_{12} & d_{22} & d_{26} \\ d_{16} & d_{26} & d_{66} \end{bmatrix} = \begin{bmatrix} 1.782 & -0.562 & 0.001835 \\ -0.562 & 1.095 & -0.361 \\ 0.001835 & -0.361 & 2.64 \end{bmatrix} \frac{1}{N \cdot m}$$

EXAMPLE 14 $[\pm 30/0]_S$

$$\begin{bmatrix} a_{11} & a_{12} & a_{16} \\ a_{12} & a_{22} & a_{26} \\ a_{16} & a_{26} & a_{66} \end{bmatrix} = \begin{bmatrix} 39.5 & -19.18 & 0 \\ -19.18 & 85.1 & 0 \\ 0 & 0 & 138.8 \end{bmatrix} \frac{m}{GN}$$

$$\begin{bmatrix} d_{11} & d_{12} & d_{16} \\ d_{12} & d_{22} & d_{26} \\ d_{16} & d_{26} & d_{66} \end{bmatrix} = \begin{bmatrix} 1.062 & -0.579 & -0.346 \\ -0.579 & 1.903 & -0.00631 \\ -0.346 & -0.00631 & 2.64 \end{bmatrix} \frac{1}{N \cdot m}$$

EXAMPLE 15 $[\pm 30]_{2S}$

$$\begin{bmatrix} a_{11} & a_{12} & a_{16} \\ a_{12} & a_{22} & a_{26} \\ a_{16} & a_{26} & a_{66} \end{bmatrix} = \begin{bmatrix} 36.7 & -21.3 & 0 \\ -21.3 & 67.5 & 0 \\ 0 & 0 & 86.2 \end{bmatrix} \frac{m}{GN}$$

$$\begin{bmatrix} d_{11} & d_{12} & d_{16} \\ d_{12} & d_{22} & d_{26} \\ d_{16} & d_{26} & d_{66} \end{bmatrix} = \begin{bmatrix} 0.454 & -0.256 & -0.1215 \\ -0.256 & 0.810 & 0.000881 \\ -0.1215 & 0.000881 & 1.074 \end{bmatrix} \frac{1}{N \cdot m}$$

EXAMPLE 16 $[\pm 60/0/90]_S$

$$\begin{bmatrix} a_{11} & a_{12} & a_{16} \\ a_{12} & a_{22} & a_{26} \\ a_{16} & a_{26} & a_{66} \end{bmatrix} = \begin{bmatrix} 43.1 & -10.00 & 0 \\ -10.00 & 33.0 & 0 \\ 0 & 0 & 116.2 \end{bmatrix} \frac{m}{GN}$$

$$\begin{bmatrix} d_{11} & d_{12} & d_{16} \\ d_{12} & d_{22} & d_{26} \\ d_{16} & d_{26} & d_{66} \end{bmatrix} = \begin{bmatrix} 0.657 & -0.203 & -0.000718 \\ -0.203 & 0.450 & -0.1006 \\ -0.000718 & -0.1006 & 1.132 \end{bmatrix} \frac{1}{N \cdot m}$$

EXAMPLE 17 $[0/90]_{2S}$

$$\begin{bmatrix} a_{11} & a_{12} & a_{16} \\ a_{12} & a_{22} & a_{26} \\ a_{16} & a_{26} & a_{66} \end{bmatrix} = \begin{bmatrix} 32.2 & -4.62 & 0 \\ -4.62 & 32.2 & 0 \\ 0 & 0 & 178.0 \end{bmatrix} \frac{m}{GN}$$

$$\begin{bmatrix} d_{11} & d_{12} & d_{16} \\ d_{12} & d_{22} & d_{26} \\ d_{16} & d_{26} & d_{66} \end{bmatrix} = \begin{bmatrix} 0.327 & -0.0573 & 0 \\ -0.0573 & 0.471 & 0 \\ 0 & 0 & 2.14 \end{bmatrix} \frac{1}{N \cdot m}$$

EXAMPLE 18 $[0_4/90_4]_T$

$$\begin{bmatrix} a_{11} & a_{12} & a_{16} & b_{11} & b_{12} & b_{16} \\ a_{12} & a_{22} & a_{26} & b_{21} & b_{22} & b_{26} \\ a_{16} & a_{26} & a_{66} & b_{61} & b_{62} & b_{66} \\ b_{11} & b_{21} & b_{61} & d_{11} & d_{12} & d_{16} \\ b_{12} & b_{22} & b_{62} & d_{12} & d_{22} & d_{26} \\ b_{16} & b_{26} & b_{66} & d_{16} & d_{26} & d_{66} \end{bmatrix}$$

$$= \begin{bmatrix} 39.1 \frac{m}{GN} & -5.61 \frac{m}{GN} & 0 & 5.62 \times 10^{-5} \frac{1}{N} & 0 & 0 \\ -5.61 \frac{m}{GN} & 39.1 \frac{m}{GN} & 0 & 0 & -5.62 \times 10^{-5} \frac{1}{N} & 0 \\ 0 & 0 & 178.0 \frac{m}{GN} & 0 & 0 & 0 \\ 5.62 \times 10^{-5} \frac{1}{N} & 0 & 0 & 0.469 \frac{m}{N} & -0.0673 \frac{m}{N} & 0 \\ 0 & -5.62 \times 10^{-5} \frac{1}{N} & 0 & -0.0673 \frac{m}{N} & 0.469 \frac{m}{N} & 0 \\ 0 & 0 & 0 & 0 & 0 & 2.14 \frac{m}{N} \end{bmatrix}$$

It is interesting to note that while a_{16} and a_{26} are always negative, the sign and existence of d_{16} and d_{26} vary with the laminate. This has implications regarding the directions of the curvatures produced by applied moments, particularly the twist quantities.

LAMINATE STRESS ANALYSIS

With what has been presented, it is now possible to compute the stresses in an element of a laminate that is loaded by known applied force and moment resultants and that is subjected to a temperature change. The material properties, fiber orientation, and z coordinates of each layer lead to numerical values for the elements of the A, B, and D matrices and unit effective thermal stress resultants. Given the magnitude and direction of the applied loads and the dimensions of the element, the force and moment resultants can be determined. The strains and curvatures of the reference surface can be computed by using the various

inverse relations of the previous section, depending on the particular laminate. The strains at every z location through the thickness of the laminate can be determined from Eq. (5.6.29). These strains, in turn, can be transformed by using Eq. (5.6.24) to determine the strains in the principal material coordinate system at every z location. These strains can then be used in Eq. (5.6.20) to compute the stresses in the principal material coordinate system at every z location. The approach is very systematic, the procedures to employ at each step are straightforward and well defined, and numerical values can be determined. It goes without saying that these steps are best executed by programming a calculator or computer to deal with the many algebraic relations. Executing the many algebraic steps by hand is very error-prone. However, any program should be thoroughly checked to be sure it is computing what is intended before the results of the program are used for analysis and design. Once checked for accuracy, the program can be very useful.

Below are several examples that demonstrate the use of the steps described above to compute the stresses within a laminate.

EXAMPLE 19. Consider a flat six-layer $[\pm 30/0]_S$ laminate with in-plane dimensions 0.35 m by 0.25 m subjected to a force of 15,000 N in the x direction uniformly distributed along opposite edges, as shown in Fig. 5.6.12. The temperature change from the cure condition is $\Delta T = -100°C$. Compute the stresses in each layer.

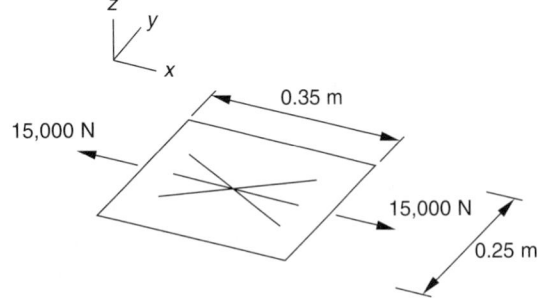

Fig. 5.6.12 Laminate loaded in x direction.

APPROACH.   The relations between stress resultants and deformations and the deformations and stresses, and the transformation relations, all outlined above are applied to the problem to determine the desired answers. Note that the laminate is symmetric and balanced, so Eqs. (5.6.53) to (5.6.55) apply, and only a single force is applied. This should considerably simplify the calculations.

SOLUTION.   Except for the effective thermal stress resultants, N_x is the only stress resultant, and since the force is uniformly distributed, it is given by

$$N_x = \frac{15,000 \text{ N}}{0.25 \text{ m}} = 60,000 \text{ N/m}$$

The unit effective thermal stress resultants consist of just two components given by Example 7, namely,

$$\begin{Bmatrix} \hat{N}_x^T \\ \hat{N}_y^T \\ \hat{N}_{xy}^T \end{Bmatrix} = \begin{Bmatrix} 315 \\ 323 \\ 0 \end{Bmatrix} \frac{\text{N/m}}{°C}$$

Since there are no moment resultants, the reference surface curvatures are zero. The two reference surface strains are given by direct application of Eqs. (5.6.53) and (5.6.54), with numerical values from Example 14, to yield

$$\begin{Bmatrix} \varepsilon_x^o \\ \varepsilon_y^o \\ \gamma_{xy}^o \end{Bmatrix} = \begin{bmatrix} a_{11} & a_{12} & 0 \\ a_{12} & a_{22} & 0 \\ 0 & 0 & a_{66} \end{bmatrix} \begin{Bmatrix} N_x + \hat{N}_x^T \Delta T \\ N_y + \hat{N}_y^T \Delta T \\ N_{xy} + \hat{N}_{xy}^T \Delta T \end{Bmatrix}$$

$$= \begin{bmatrix} 39.5 & -19.18 & 0 \\ -19.18 & 85.1 & 0 \\ 0 & 0 & 138.8 \end{bmatrix} \times 10^{-9} \begin{Bmatrix} 60,000 + 315(-100) \\ 323(-100) \\ 0 \end{Bmatrix}$$

or $\begin{Bmatrix} \varepsilon_x^o \\ \varepsilon_y^o \\ \gamma_{xy}^o \end{Bmatrix} = \begin{Bmatrix} 2,370 \\ -1,151 \\ 0 \end{Bmatrix} \times 10^{-6} + \begin{Bmatrix} -625 \\ -2,140 \\ 0 \end{Bmatrix} \times 10^{-6} = \begin{Bmatrix} 1,745 \\ -3,290 \\ 0 \end{Bmatrix} \times 10^{-6}$

The first term after the first equals sign is the strains due to the applied force, and the second term is the strains due to the decrease in temperature. Due to the applied force, the element of laminate stretches in the direction of the applied force, contracts perpendicular to that direction, and experiences no shear strain. Due to thermal effects, the element of laminate contracts in both directions, but more in the y direction. The net effect is stretching in the direction of the applied force and considerable contraction perpendicular to that direction, mostly due to thermal effects.

Transforming the strains to the principal material coordinate system for each fiber orientation from Eq. (5.6.24) results in

$$\begin{Bmatrix} \varepsilon_1^o(30°) \\ \varepsilon_2^o(30°) \\ \frac{1}{2}\gamma_{12}^o(30°) \end{Bmatrix} = [T(30°)] \begin{Bmatrix} \varepsilon_x^o \\ \varepsilon_y^o \\ \frac{1}{2}\gamma_{xy}^o \end{Bmatrix}$$

$$= \begin{bmatrix} \frac{3}{4} & \frac{1}{4} & \frac{\sqrt{3}}{2} \\ \frac{1}{4} & \frac{3}{4} & -\frac{\sqrt{3}}{2} \\ -\frac{\sqrt{3}}{4} & \frac{\sqrt{3}}{4} & \frac{1}{2} \end{bmatrix} \begin{Bmatrix} 1,745 \\ -3,290 \\ 0 \end{Bmatrix} \times 10^{-6}$$

$$= \begin{Bmatrix} 486 \\ -2,030 \\ -2,180 \end{Bmatrix} \times 10^{-6}$$

$$\begin{Bmatrix} \varepsilon_1^o(-30°) \\ \varepsilon_2^o(-30°) \\ \frac{1}{2}\gamma_{12}^o(-30°) \end{Bmatrix} = [T(-30°)] \begin{Bmatrix} \varepsilon_x^o \\ \varepsilon_y^o \\ \frac{1}{2}\gamma_{xy}^o \end{Bmatrix}$$

$$= \begin{bmatrix} \frac{3}{4} & \frac{1}{4} & -\frac{\sqrt{3}}{2} \\ \frac{1}{4} & \frac{3}{4} & \frac{\sqrt{3}}{2} \\ \frac{\sqrt{3}}{4} & -\frac{\sqrt{3}}{4} & \frac{1}{2} \end{bmatrix} \begin{Bmatrix} 1,745 \\ -3,290 \\ 0 \end{Bmatrix} \times 10^{-6}$$

$$= \begin{Bmatrix} 486 \\ -2,030 \\ 2,180 \end{Bmatrix} \times 10^{-6}$$

$$\begin{Bmatrix} \varepsilon_1^o(0°) \\ \varepsilon_2^o(0°) \\ \frac{1}{2}\gamma_{12}^o(0°) \end{Bmatrix} = [T(0°)] \begin{Bmatrix} \varepsilon_x^o \\ \varepsilon_y^o \\ \frac{1}{2}\gamma_{xy}^o \end{Bmatrix}$$

$$= \begin{bmatrix} 1 & 0 & 0 \\ 0 & 1 & 0 \\ 0 & 0 & 1 \end{bmatrix} \begin{Bmatrix} 1,745 \\ -3,290 \\ 0 \end{Bmatrix} \times 10^{-6} = \begin{Bmatrix} 1,745 \\ -3,290 \\ 0 \end{Bmatrix} \times 10^{-6}$$

In the principal material coordinate system there are substantial shear strains in the $\pm 30°$ layers, and they are of opposite sign. The extensional strains are identical. Of course, no transformation is necessary for the $0°$ layers. Substituting the strains for each layer into the stress-strain relation of Eq. (5.6.20) provides the desired stresses:

$$\begin{Bmatrix} \sigma_1(+30°) \\ \sigma_2(+30°) \\ \tau_{12}(+30°) \end{Bmatrix} = \begin{pmatrix} Q_{11} & Q_{12} & 0 \\ Q_{12} & Q_{22} & 0 \\ 0 & 0 & Q_{66} \end{pmatrix} \begin{Bmatrix} \varepsilon_1(+30°) - \alpha_1\,\Delta T \\ \varepsilon_2(+30°) - \alpha_2\,\Delta T \\ \gamma_{12}(+30°) \end{Bmatrix}$$

$$\begin{Bmatrix} \sigma_1(+30°) \\ \sigma_2(+30°) \\ \tau_{12}(+30°) \end{Bmatrix} = \begin{pmatrix} 47.0 & 4.56 & 0 \\ 4.56 & 16.51 & 0 \\ 0 & 0 & 5.62 \end{pmatrix}$$

$$\times 10^9 \begin{Bmatrix} 486 - 6.51(-100) \\ -2,030 - 24.4(-100) \\ -4,360 \end{Bmatrix} \times 10^{-6} = \begin{Bmatrix} 55.3 \\ 11.94 \\ -24.5 \end{Bmatrix} \text{ MPa}$$

$$\begin{Bmatrix} \sigma_1(-30°) \\ \sigma_2(-30°) \\ \tau_{12}(-30°) \end{Bmatrix} = \begin{pmatrix} Q_{11} & Q_{12} & 0 \\ Q_{12} & Q_{22} & 0 \\ 0 & 0 & Q_{66} \end{pmatrix} \begin{Bmatrix} \varepsilon_1(-30°) - \alpha_1\,\Delta T \\ \varepsilon_2(-30°) - \alpha_2\,\Delta T \\ \gamma_{12}(-30°) \end{Bmatrix}$$

$$\begin{Bmatrix} \sigma_1(-30°) \\ \sigma_2(-30°) \\ \tau_{12}(-30°) \end{Bmatrix} = \begin{pmatrix} 47.0 & 4.56 & 0 \\ 4.56 & 16.51 & 0 \\ 0 & 0 & 5.62 \end{pmatrix}$$

$$\times 10^9 \begin{Bmatrix} 486 - 6.51(-100) \\ -2,030 - 24.4(-100) \\ 4,360 \end{Bmatrix} \times 10^{-6} = \begin{Bmatrix} 55.3 \\ 11.94 \\ 24.5 \end{Bmatrix} \text{ MPa}$$

$$\begin{Bmatrix} \sigma_1(0°) \\ \sigma_2(0°) \\ \tau_{12}(0°) \end{Bmatrix} = \begin{pmatrix} Q_{11} & Q_{12} & 0 \\ Q_{12} & Q_{22} & 0 \\ 0 & 0 & Q_{66} \end{pmatrix} \begin{Bmatrix} \varepsilon_1(0°) - \alpha_1\,\Delta T \\ \varepsilon_2(0°) - \alpha_2\,\Delta T \\ \gamma_{12}(0°) \end{Bmatrix}$$

$$\begin{Bmatrix} \sigma_1(0°) \\ \sigma_2(0°) \\ \tau_{12}(0°) \end{Bmatrix} = \begin{pmatrix} 47.0 & 4.56 & 0 \\ 4.56 & 16.51 & 0 \\ 0 & 0 & 5.62 \end{pmatrix}$$

$$\times 10^9 \begin{Bmatrix} 1,745 - 6.51(-100) \\ -3,290 - 24.4(-100) \\ 0 \end{Bmatrix} \times 10^{-6} = \begin{Bmatrix} 108.7 \\ -3.11 \\ 0 \end{Bmatrix} \text{ MPa}$$

As might be expected, the stress in the fiber direction in the $0°$ layers is the highest. The $\pm 30°$ layers experience a shear stress. If thermal effects were ignored, ΔT would be taken to be zero, resulting in

$$\begin{Bmatrix} \sigma_1(+30°) \\ \sigma_2(+30°) \\ \sigma_{12}(+30°) \end{Bmatrix} = \begin{Bmatrix} 68.7 \\ 2.33 \\ -17.12 \end{Bmatrix} \text{ MPa} \qquad \begin{Bmatrix} \sigma_1(-30°) \\ \sigma_2(-30°) \\ \sigma_{12}(-30°) \end{Bmatrix} = \begin{Bmatrix} 68.7 \\ 2.33 \\ 17.12 \end{Bmatrix} \text{ MPa}$$

$$\begin{Bmatrix} \sigma_1(0°) \\ \sigma_2(0°) \\ \tau_{12}(0°) \end{Bmatrix} = \begin{Bmatrix} 106.1 \\ -8.20 \\ 0 \end{Bmatrix} \text{ MPa}$$

It can be seen that residual thermal effects increase some stress components and decrease others. Even though some stress components are small compared to others, e.g., σ_2 compared to σ_1, the material is much weaker perpendicular to the fibers than in the fiber direction. The small stress may be approaching levels to cause failure perpendicular to the fibers.

Figure 5.6.13 shows the variation through the thickness of the three components of stress in the principal material coordinate system. The stresses are constant within each layer, but the overall distributions through the thickness are discontinuous, a characteristic which makes composite materials different from

Fig. 5.6.13 Through-thickness distribution of stresses: σ_1, solid; σ_2, dotted; τ_{12}, dashed.

conventional materials such as aluminum or steel. For aluminum or steel, for example, the stress distributions through the thickness would be continuous and there would be no shear stress. Note that there is no net shear strain for the laminate, although there are shear strains, but of opposite sign, for the $\pm 30°$ layers. This characteristic is due to the balanced nature of the laminate.

EXAMPLE 20. A flat eight-layer $[\pm 45/0/90]_S$ quasi-isotropic laminate 0.1 by 0.4 m is loaded by a 0.7 N · m moment, as shown in Fig. 5.6.14. The through-thickness distribution of the stresses in the principal material coordinate system is of interest. The temperature change from the cure temperature condition is $-100°$C.

APPROACH. The symmetric and balanced nature of the laminate, along with the fact that only a single moment is applied, simplifies the computations considerably. Equations (5.6.53) to (5.6.55) and the material properties from Examples 5 and 12 apply.

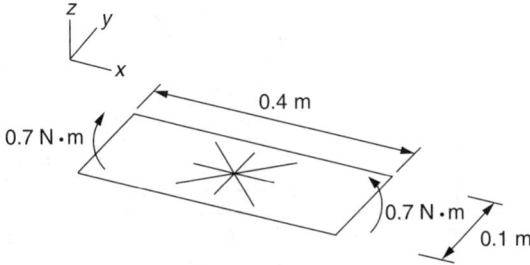

Fig. 5.6.14 Laminate loaded by moment.

SOLUTION.   If it is assumed that the applied moment is uniformly distributed along the opposite edges, the moment resultant is given by

$$M_x = \frac{-0.7}{0.1} = -7 \text{ N} \cdot \text{m/m}$$

The values of M_y and M_{xy} are zero. There are no unit equivalent thermal moment resultants, and the unit equivalent thermal force resultants are given in Example 5 as

$$\begin{Bmatrix} \hat{N}_x^T \\ \hat{N}_y^T \\ \hat{N}_{xy}^T \end{Bmatrix} = \begin{Bmatrix} 425 \\ 425 \\ 0 \end{Bmatrix} \frac{\text{N/m}}{^\circ\text{C}}$$

The reference surface strains are due only to the equivalent thermal force resultants and are computed as follows by way of Eqs. (5.6.53) and (5.6.54), with numerical values from Example 12, as

$$\begin{Bmatrix} \varepsilon_x^o \\ \varepsilon_y^o \\ \gamma_{xy}^o \end{Bmatrix} = \begin{bmatrix} a_{11} & a_{12} & 0 \\ a_{12} & a_{22} & 0 \\ 0 & 0 & a_{66} \end{bmatrix} \begin{Bmatrix} \hat{N}_x^T \Delta T \\ \hat{N}_y^T \Delta T \\ \hat{N}_{xy}^T \Delta T \end{Bmatrix}$$

$$= \begin{bmatrix} 39.8 & -12.26 & 0 \\ -12.26 & 39.8 & 0 \\ 0 & 0 & 104.1 \end{bmatrix} \times 10^{-9} \begin{Bmatrix} 425(-100) \\ 425(-100) \\ 0 \end{Bmatrix}$$

$$= \begin{Bmatrix} -1{,}171 \\ -1{,}171 \\ 0 \end{Bmatrix} \times 10^{-6}$$

The reference surface curvatures are due to the applied moment and are computed by way of Eq. (5.6.55) and Example 12 as

$$\begin{Bmatrix} \kappa_x^o \\ \kappa_y^o \\ \kappa_{xy}^o \end{Bmatrix} = \begin{bmatrix} d_{11} & d_{12} & d_{16} \\ d_{12} & d_{22} & d_{26} \\ d_{16} & d_{26} & d_{66} \end{bmatrix} \begin{Bmatrix} M_x \\ M_y \\ M_{xy} \end{Bmatrix}$$

$$= \begin{bmatrix} 0.588 & -0.286 & -0.0514 \\ -0.286 & 0.662 & -0.0638 \\ -0.0514 & -0.0638 & 0.972 \end{bmatrix} \begin{Bmatrix} -7 \\ 0 \\ 0 \end{Bmatrix} \frac{1}{m} = \begin{Bmatrix} -4.12 \\ 2.00 \\ 0.360 \end{Bmatrix} \frac{1}{m}$$

Note that in addition to producing curvature component κ_x^o, as expected, the single moment M_x produces curvature components κ_y^o and κ_{xy}^o. The former curvature would exist if this were a conventional material such as aluminum or steel, and it is called the *anticlastic* curvature. It is due to Poisson's ratio, which is reflected in the value of d_{12} and is related to $\overline{Q}_{12}$. The twist curvature κ_{xy}^o is due to the existence of d_{16}, which can be related to the existence of nonzero values of $\overline{Q}_{16}$ in the $\pm 45°$ layers. Twist curvature would not exist if this were a conventional material.

The strain as a function of z in the laminate coordinate system is given by Eq. (5.6.29) as

$$\begin{Bmatrix} \varepsilon_x \\ \varepsilon_y \\ \gamma_{xy} \end{Bmatrix} = \begin{Bmatrix} \varepsilon_x^o \\ \varepsilon_y^o \\ \gamma_{xy}^o \end{Bmatrix} + z \begin{Bmatrix} \kappa_x^o \\ \kappa_y^o \\ \kappa_{xy}^o \end{Bmatrix} = \begin{Bmatrix} -1{,}171 \\ -1{,}171 \\ 0 \end{Bmatrix} \times 10^{-6} + z \begin{Bmatrix} -4.12 \\ 2.00 \\ 0.360 \end{Bmatrix}$$

These relations are valid for the full range of z, $-0.000500 \text{ m} \le z \le +0.000500 \text{ m}$. Transforming these strains to the principal material coordinate system by way of Eq. (5.6.24) for the various fiber orientations gives the following:

For $-0.000500 \text{ m} \le z \le -0.000375 \text{ m}$ and $0.000375 \text{ m} \le z \le 0.000500 \text{ m}$,

$$\begin{Bmatrix} \varepsilon_1(45°) \\ \varepsilon_2(45°) \\ \frac{1}{2}\gamma_{12}(45°) \end{Bmatrix} = [T(45°)] \begin{Bmatrix} \varepsilon_x \\ \varepsilon_y \\ \frac{1}{2}\gamma_{xy} \end{Bmatrix}$$

$$= \begin{bmatrix} \frac{1}{2} & \frac{1}{2} & 1 \\ \frac{1}{2} & \frac{1}{2} & -1 \\ -\frac{1}{2} & \frac{1}{2} & 0 \end{bmatrix} \begin{Bmatrix} -1{,}171 \times 10^{-6} - 4.12z \\ -1{,}171 \times 10^{-6} + 2.00z \\ 0.180z \end{Bmatrix}$$

$$= \begin{Bmatrix} -1{,}171 \times 10^{-6} - 0.878z \\ -1{,}171 \times 10^{-6} - 1.238z \\ 3.06z \end{Bmatrix}$$

For $-0.000375 \text{ m} \le z \le -0.000250 \text{ m}$ and $0.000250 \text{ m} \le z \le 0.000375 \text{ m}$,

$$\begin{Bmatrix} \varepsilon_1(-45°) \\ \varepsilon_2(-45°) \\ \frac{1}{2}\gamma_{12}(-45°) \end{Bmatrix} = \begin{Bmatrix} -1{,}171 \times 10^{-6} - 1.238z \\ -1{,}171 \times 10^{-6} - 0.878z \\ -3.06z \end{Bmatrix}$$

For $-0.000250 \text{ m} \le z \le -0.000125 \text{ m}$ and $0.000125 \text{ m} \le z \le 0.000250 \text{ m}$,

$$\begin{Bmatrix} \varepsilon_1(0°) \\ \varepsilon_2(0°) \\ \frac{1}{2}\gamma_{12}(0°) \end{Bmatrix} = \begin{Bmatrix} -1{,}171 \times 10^{-6} - 4.12z \\ -1{,}171 \times 10^{-6} + 2.00z \\ 0.1799z \end{Bmatrix}$$

For $-0.000125 \text{ m} \le z \le 0.000125 \text{ m}$,

$$\begin{Bmatrix} \varepsilon_1(90°) \\ \varepsilon_2(90°) \\ \frac{1}{2}\gamma_{12}(90°) \end{Bmatrix} = \begin{Bmatrix} -1{,}171 \times 10^{-6} + 2.00z \\ -1{,}171 \times 10^{-6} - 4.12z \\ -0.1799z \end{Bmatrix}$$

Note that each equation is valid for two ranges of z, the ranges depending on the z location of the layers with the different fiber orientations. The stresses in the various layers in the principal material coordinate system can be computed directly from Eq. (5.6.20) by using the above strains. Again there is a range of z for each relation, depending on the layer. The stresses in the principal material coordinate system are as follows:

For $-0.000500 \text{ m} \le z \le -0.000375 \text{ m}$ and $0.000375 \text{ m} \le z \le 0.000500 \text{ m}$,

$$\begin{Bmatrix} \sigma_1(45°) \\ \sigma_2(45°) \\ \tau_{12}(45°) \end{Bmatrix} = \begin{bmatrix} Q_{11} & Q_{12} & 0 \\ Q_{12} & Q_{22} & 0 \\ 0 & 0 & Q_{66} \end{bmatrix} \begin{Bmatrix} \varepsilon_1(45°) - \alpha_1 \Delta T \\ \varepsilon_2(45°) - \alpha_2 \Delta T \\ \gamma_{12}(45°) \end{Bmatrix}$$

$$\begin{Bmatrix} \sigma_1(45°) \\ \sigma_2(45°) \\ \tau_{12}(45°) \end{Bmatrix} = \begin{bmatrix} 47.0 & 4.56 & 0 \\ 4.56 & 16.51 & 0 \\ 0 & 0 & 5.62 \end{bmatrix}$$

$$\times 10^9 \begin{Bmatrix} -1{,}171 \times 10^{-6} - 0.878z - 6.51 \times 10^{-6}(-100) \\ -1{,}171 \times 10^{-6} - 1.238z - 24.4 \times 10^{-6}(-100) \\ 6.12z \end{Bmatrix}$$

$$\begin{Bmatrix} \sigma_1(45°) \\ \sigma_2(45°) \\ \tau_{12}(45°) \end{Bmatrix} = \begin{Bmatrix} -18.62 - 46{,}900z \\ 18.62 - 24{,}400z \\ 34{,}400z \end{Bmatrix} \text{ MPa}$$

For $-0.000375 \text{ m} \le z \le -0.000250 \text{ m}$ and $0.000250 \text{ m} \le z \le 0.000375 \text{ m}$,

$$\begin{Bmatrix} \sigma_1(-45°) \\ \sigma_2(-45°) \\ \tau_{12}(-45°) \end{Bmatrix} = \begin{Bmatrix} -18.62 - 62{,}100z \\ 18.62 - 20{,}100z \\ -34{,}400z \end{Bmatrix} \text{ MPa}$$

For $-0.000250 \text{ m} \le z \le -0.000125 \text{ m}$ and $0.000125 \text{ m} \le z \le 0.000250 \text{ m}$,

$$\begin{Bmatrix} \sigma_1(0°) \\ \sigma_2(0°) \\ \tau_{12}(0°) \end{Bmatrix} = \begin{Bmatrix} -18.62 - 184{,}300z \\ 18.62 + 14{,}310z \\ 2{,}020z \end{Bmatrix} \text{ MPa}$$

For -0.000125 m $\leq z \leq 0.000125$ m,

$$\begin{Bmatrix} \sigma_1(90°) \\ \sigma_2(90°) \\ \tau_{12}(90°) \end{Bmatrix} = \begin{Bmatrix} -18.62 + 75{,}300z \\ 18.62 - 58{,}900z \\ -2{,}020z \end{Bmatrix} \text{MPa}$$

The first term in the expressions for the stresses are the stresses due to thermal effects. Because of the quasi-isotropic nature of the laminate, every layer experiences identical thermally induced stresses, and the stress perpendicular to the fiber direction σ_2 is equal to but opposite in value to the stress in the fiber direction σ_1. There are no thermally induced shear stresses. The second terms, the terms linear in z, are the stresses due to the applied moment. These stresses vary linearly with z within each layer, but are discontinuous from layer to layer. The through-thickness distributions of the thermally induced and moment-induced stresses are plotted separately in Figs. 5.6.15 and 5.6.16, respectively.

Fig. 5.6.15 Through-thickness distribution of stresses due to thermal effects: σ_1, solid; σ_2, dotted; $\tau_{12} = 0$.

It is seen that the largest tensile and compressive stresses in the principal material coordinate system due to the applied moment occur in layers 3 and 6, respectively, not in the outer layers. This is very much unlike the stress distributions in a conventional material such as aluminum or steel, where the stresses with the largest magnitude occur at $z = \pm H/2$. This characteristic of layered materials is often overlooked.

ENGINEERING PROPERTIES OF A LAMINATE

Often there is interest in what can be considered smeared engineering properties for a laminate. For example, it may be desirable to treat a laminate as a bar or beam made of a homogeneous material with an equivalent modulus. Since a bar is usually associated with tension, or compression, and a beam is usually associated with bending, the equivalent modulus for a laminated composite bar could be different from the equivalent modulus for a laminated composite beam. This makes sense, since the As depend differently on the thicknesses and locations of the layers than the Ds. Given a bar and a one-dimensional state of stress, the stress-strain behavior can be described by

$$\sigma_x = E_x \varepsilon_x \tag{5.6.62}$$

where E_x is the extensional modulus of the material. If a material is homogeneous, then the stress will be uniform through the thickness. However, it has been shown in the examples that, in general, for a laminated material the stress is not uniform through the thickness. Therefore, consider the stress in Eq. (5.6.62) to be the *average* stress $\overline{\sigma}_x$. Then, considering that under tension or compression the strain in a bar constructed of a symmetric balanced laminate would be that given by the geometric midsurface, or reference surface, strain, Eq. (5.6.62) can be rewritten as

$$\overline{\sigma}_x = E_x \varepsilon_x^o \tag{5.6.63}$$

The integral definition of the stress resultant N_x in Eq. (5.6.31) would reduce to

$$N_x = \overline{\sigma}_x H \tag{5.6.64}$$

Fig. 5.6.16 Through-thickness distribution of stresses due to applied moment: σ_1, solid; σ_2, dotted; τ_{12}, dashed.

where, recall, H is the thickness of the laminate. For a bar N_y and N_{xy} would be zero, resulting in, from Eq. (5.6.53),

$$\varepsilon_x^o = a_{11} N_x \tag{5.6.65}$$

Combining Eqs. (5.6.63) to (5.6.65) leads to

$$\varepsilon_x^o = \frac{\overline{\sigma}_x}{E_x} = a_{11} N_x = a_{11} \overline{\sigma}_x H \tag{5.6.66}$$

By equating the second and fourth parts of the relationship, the equivalent extensional modulus can thus be defined as

$$(E_x)_{\text{extension}} = \frac{1}{a_{11} H} \tag{5.6.67}$$

In a similar fashion, if, instead of tension, bending is considered, then for a homogeneous material the well-known moment-curvature relation for bending of a beam in the x direction can be written as

$$M_x b = -E_x I \frac{d^2 w^o}{dx^2} = E_x I \kappa_x^o \tag{5.6.68}$$

where I is the second moment of area of the rectangular cross section beam and is given by

$$I = \frac{1}{12} b H^3 \tag{5.6.69}$$

The quantity b is the width of the beam (dimension in the y direction in Fig. 5.6.11) that must be included with the moment since the definition of moment M_x as used herein is moment per unit in-plane dimension in the y direction. For bending of a beam, M_y and M_{xy} would be zero, so the curvature in the x direction for a symmetric balanced laminate is given by, from Eq. (5.6.55),

$$\kappa_x^o = d_{11} M_x \tag{5.6.70}$$

Combining Eqs. (5.6.68) to (5.6.70) results in

$$\kappa_x^o = \frac{M_x b}{E_x I} = \frac{M_x b}{E_x \left(\frac{1}{12} b H^3\right)} = d_{11} M_x \tag{5.6.71}$$

Using the third and fourth parts of the relationship provides the definition of the bending modulus as

$$(E_x)_{\text{bending}} = \frac{12}{d_{11} H^3} \tag{5.6.72}$$

EXAMPLE 21.   Compute the extensional and bending modulus in the x direction for an eight-layer $[\pm 45/0/90]_s$ laminate.

APPROACH.   Use the results of Example 12 directly in Eqs. (5.6.67) and (5.6.72).

SOLUTION

$$(E_x)_{\text{extension}} = \frac{1}{a_{11} H} = \frac{1}{39.8 \times 10^{-9}(8 \times 0.000125)} = 25.1 \text{ GPa}$$

$$(E_x)_{\text{bending}} = \frac{12}{d_{11} H^3} = \frac{12}{0.588(8 \times 0.000125)^3} = 20.4 \text{ GPa}$$

Note the bending modulus is lower than the extensional modulus because the $0°$ layers are near the center of the laminate and have less influence on resistance to bending than on resistance to extension.

FAILURE OF COMPOSITE MATERIALS

Generally, the calculation of the stresses within a laminate is carried out so that the values of the stresses can be compared with values known to cause failure. Knowing the levels of the stresses without having a metric to compare the stresses to is generally not useful. Most often it is of interest to know what value of applied load for a particular laminate in a particular loading situation causes failure. As might be expected, failure of fiber-reinforced composite materials is a complex process, and historically no aspect of the behavior of composite materials has been studied and debated more than the issue of failure. The topic is complex because of the multiple modes of failure and because the strength of a composite material in tension is considerably different from the strength in compression. A composite can fail due to excess stress in the fiber direction, and the strength in the fiber direction in tension is greater than the strength in compression. Alternatively, a composite material can fail because of excess stress perpendicular to the fibers. The strength perpendicular to the fibers in tension is considerably less than the strength in compression. And finally, a composite can fail due to excess shear stress. The strength in shear is considerable less than the strength in tension in the fiber direction, although the strength in shear does not depend on the sign of the shear stress. These issues are further complicated by the fact that when the fibers are oriented at an angle relative to the loading direction, through the transformation relations, there is a stress component in the fiber direction, a component perpendicular to the fiber direction, and a shear stress. Which component of stress leads to failure? Or is it a combination of the components? If it is a combination, how should the stresses be combined to predict failure?

There are many issues related to failure of composite materials and there are a number of failure theories. Some theories are quite straightforward, while others are complicated and require a considerable number of experiments to determine all the parameters necessary to implement the theory. No one theory, or criterion, works perfectly for all materials in all situations. It is better to think of failure theories as *indicators* of failure rather than *predictors* of failure. Also, failure in one layer due to excess stress perpendicular to the fibers, for example, does not necessarily mean catastrophic failure of the entire laminate. The other layers may be able to assume a greater portion of the load to compensate for the failed layer. Thus the issue of failure really becomes one of damage accumulation. With this viewpoint, failure of a laminate will occur when there is sufficient damage accumulation that not enough undamaged material remains to support the applied load. The level of **damage accumulation** that can be accepted depends on the application. For example, excess tensile stress perpendicular to the fibers generally causes cracking of a composite material, the cracks being parallel with the fibers. If some cracking is acceptable, then the prediction of cracking does not equate to failure of the laminate. However, the existence of any cracking at all is equivalent to the accumulation of some degree of damage. How much cracking can be tolerated depends very much on the application.

One failure theory that is easy to implement is the **maximum stress failure criterion.** The maximum stress failure criterion independently treats tension failure in the fiber direction, compression failure in the fiber direction, tension failure perpendicular to the fiber direction, compression failure perpendicular to the fiber direction, and failure in shear. There are thus five possible modes of failure for a plane stress state. A known level of failure stress for each mode is required. Failure in tension in the fiber direction is a direct result of fibers fracturing or otherwise breaking. Failure in compression in the fiber direction is due to the kinking of fibers, generally due to the lack of support of the matrix material surrounding the fibers and the development of shear, or kink, bands within the fiber. Failure in tension perpendicular to the fibers is due to failure of the matrix material between the fibers, failure of the

fibers, failure of the bond between the matrix material and the fibers, or a combination of all three failures. Failure in compression perpendicular to the fibers is generally due to crushing of the matrix material. Finally, failure in shear is also due to failure of the matrix material between the fibers, failure of the fibers, failure of the bond between the matrix material and the fibers, or a combination of all three. It is important to realize there is a variability to failure strengths, so the approach to take is to employ allowable stress levels for each mode that adequately account for the variability. If these definitions for failure stresses in each of the modes are used

σ_1^T = tensile failure stress in the 1 direction

σ_1^C = compression failure stress in the 1 direction (a negative number)

σ_2^T = tensile failure stress in the 2 direction

σ_2^C = compression failure stress in the 2 direction (a negative number)

τ_{12}^F = shear failure stress in 1-2 plane (a positive number)

then the maximum stress failure criterion states that failure occurs at a point in a laminate if any of the following six equalities are satisfied:

$$\begin{array}{ll} \sigma_1 = \sigma_1^T & \sigma_1 = \sigma_1^C \\ \sigma_2 = \sigma_2^T & \sigma_2 = \sigma_2^C \\ \tau_{12} = \tau_{12}^F & \tau_{12} = -\tau_{12}^F \end{array} \qquad (5.6.73)$$

The last two equations can be replaced by one, namely,

$$|\tau_{12}| = \tau_{12}^F \qquad (5.6.74)$$

EXAMPLE 22.   A thin-walled cylindrical pressure vessel with radius $R = 0.25$ m is constructed using eight layers of glass-fiber-reinforced material and a stacking sequence of $[\pm 60/0/90]_S$. The fiber angles are specified relative to the axial direction of the cylinder. The end caps of the vessel are suitably designed and reinforced, so the concern is with failure away from the end caps. The change in temperature due to curing is $-100°C$. The level of internal pressure is increased from zero. In what layer, or layers, does the first failure occur, what is the mode of failure, and what is the pressure level?

APPROACH.   Since interest centers on the stresses away from the end caps, the stress resultants acting within the cylinder wall due to the internal pressure can be determined by considering the equilibrium of a section of cylinder away from the end caps, just as is done with conventional materials. These stress resultants will be a function of the cylinder radius as well as the pressure. By using the stress resultants, a stress analysis of the laminate can be conducted.

SOLUTION.   As shown in Fig. 5.6.17, by considering equilibrium in the Y direction of the section of cylinder of length ΔL, and equilibrium in the X direction of the flat end cap, the axial and circumferential stress resultants acting on the cylinder wall can be computed as a function of pressure and cylinder radius.

Fig. 5.6.17   Forces acting on section of cylinder away from end caps and on end caps.

Summing forces in the X direction on the end cap and in the Y direction of the section cylinder results in

$$\sum F_X = 0 : p\pi R^2 - N_x 2\pi R = 0 \rightarrow N_x = \tfrac{1}{2} pR$$

$$\sum F_Y = 0 : 2pR\,\Delta L - 2N_\theta \,\Delta L = 0 \rightarrow N_\theta = pR$$

or

$$\begin{Bmatrix} N_x \\ N_\theta \\ N_{x\theta} \end{Bmatrix} = \begin{Bmatrix} \tfrac{1}{2} \\ 1 \\ 0 \end{Bmatrix} pR$$

where, for generality, the results are left in term of p and R. These are the only two nonzero stress resultants acting within the cylinder wall away from the end caps due to the internal pressure. There are no moment resultants. There are, of course, equivalent thermal force resultants from Example 9 given by

$$\begin{Bmatrix} \hat{N}_x^T \\ \hat{N}_\theta^T \\ 0 \end{Bmatrix} = \begin{Bmatrix} 427 \\ 423 \\ 0 \end{Bmatrix} \dfrac{\text{N/m}}{{}^\circ\text{C}}$$

As can be seen, the subscript y in the nomenclature has been replaced by the subscript θ for this particular problem. This problem is like Example 19 as there are no moment resultants, and therefore no reference surface curvatures, only reference surface strains. However, the problem is unlike Example 19 in that the applied load level, in this case the internal pressure level, is not known. This is not a problem. The unknown pressure can be carried along symbolically in the calculations. Accordingly, from Eqs. (5.6.53) and (5.6.54), with numerical values from Examples 9 and 16, the reference surface strains are given by

$$[a]\begin{Bmatrix} N_x + \hat{N}_x^T \,\Delta T \\ N_y + \hat{N}_y^T \,\Delta T \\ N_{xy} + \hat{N}_{xy}^T \,\Delta T \end{Bmatrix}$$

$$= \begin{bmatrix} 43.1 & -10.00 & 0 \\ -10.00 & 33.0 & 0 \\ 0 & 0 & 116.2 \end{bmatrix} \times 10^{-9} \begin{Bmatrix} \tfrac{1}{2}pR + 427(-100) \\ pR + 423(-100) \\ 0\,pR \end{Bmatrix}$$

For convenience, the pressure p in pascals is converted to atmospheres by using the relation

$$p = 101,400\, p_a$$

where p_a is the internal pressure in atmospheres. The reference surface strains become

$$\begin{Bmatrix} \varepsilon_x^o \\ \varepsilon_\theta^o \\ \gamma_{x\theta}^o \end{Bmatrix} = \begin{Bmatrix} -1,418 + 1,172\, p_a R \\ -970 + 2,840\, p_a R \\ 0\, p_a R \end{Bmatrix} \times 10^{-6}$$

The strains in the principal material coordinate system for the layers with $60°$ fiber orientation are computed from the transformation relation of Eq. (5.6.24) as

$$\begin{Bmatrix} \varepsilon_1(60°) \\ \varepsilon_2(60°) \\ \tfrac{1}{2}\gamma_{12}(60°) \end{Bmatrix} = [T(60°)]\begin{Bmatrix} \varepsilon_x^o \\ \varepsilon_\theta^o \\ \gamma_{x\theta}^o \end{Bmatrix}$$

$$= \begin{bmatrix} \tfrac{1}{4} & \tfrac{3}{4} & \tfrac{\sqrt{3}}{2} \\ \tfrac{3}{4} & \tfrac{1}{4} & -\tfrac{\sqrt{3}}{2} \\ -\tfrac{\sqrt{3}}{4} & \tfrac{\sqrt{3}}{4} & -\tfrac{1}{2} \end{bmatrix} \begin{bmatrix} -1,418 + 1,172\, p_a R \\ -970 + 2,840\, p_a R \\ 0\, p_a R \end{bmatrix} \times 10^{-6}$$

Carrying out the algebra and doing the same for the other layer orientations give

$$\begin{Bmatrix} \varepsilon_1(60°) \\ \varepsilon_2(60°) \\ \gamma_{12}(60°) \end{Bmatrix} = \begin{Bmatrix} -1,082 + 2,420\, p_a R \\ -1,306 + 1,589\, p_a R \\ 388 + 1,446\, p_a R \end{Bmatrix} \times 10^{-6}$$

$$\begin{Bmatrix} \varepsilon_1(-60°) \\ \varepsilon_2(-60°) \\ \gamma_{12}(-60°) \end{Bmatrix} = \begin{Bmatrix} -1,082 + 2,420\, p_a R \\ -1,306 + 1,589\, p_a R \\ -388 - 1,446\, p_a R \end{Bmatrix} \times 10^{-6}$$

$$\begin{Bmatrix} \varepsilon_1(0°) \\ \varepsilon_2(0°) \\ \gamma_{12}(0°) \end{Bmatrix} = \begin{Bmatrix} -1,418 + 1,172\, p_a R \\ -970 + 2,840\, p_a R \\ 0 \end{Bmatrix} \times 10^{-6}$$

$$\begin{Bmatrix} \varepsilon_1(90°) \\ \varepsilon_2(90°) \\ \gamma_{12}(90°) \end{Bmatrix} = \begin{Bmatrix} -970 + 2,840\, p_a R \\ -1,418 + 1,172\, p_a R \\ 0\, p_a R \end{Bmatrix} \times 10^{-6}$$

The stresses in the various layers can be computed using Eq. (5.6.20) and the above strains as follows:

$$\begin{Bmatrix} \sigma_1(60°) \\ \sigma_2(60°) \\ \tau_{12}(60°) \end{Bmatrix} = [Q]\begin{Bmatrix} \varepsilon_1(60°) - \alpha_1\,\Delta T \\ \varepsilon_2(60°) - \alpha_2\,\Delta T \\ \gamma_{12}(60°) \end{Bmatrix}$$

$$= \begin{bmatrix} 47.0 & 4.56 & 0 \\ 4.56 & 16.51 & 0 \\ 0 & 0 & 5.62 \end{bmatrix}$$

$$\times 10^9 \begin{Bmatrix} -1,082 + 2,420 p_a R - 6.51(-100) \\ -1,306 + 1,589 p_a R - 24.4(-100) \\ 388 + 1,446\, p_a R \end{Bmatrix} \times 10^{-6}$$

Carrying out the algebra, substituting for the radius with $R = 0.25$ m, and doing similar calculations for the other three layer angles result in

$$\begin{Bmatrix} \sigma_1(60°) \\ \sigma_2(60°) \\ \tau_{12}(60°) \end{Bmatrix} = \begin{Bmatrix} -15.08 + 30.3\, p_a \\ 16.78 + 9.32\, p_a \\ 2.18 + 2.03\, p_a \end{Bmatrix} \text{MPa}$$

$$\begin{Bmatrix} \sigma_1(-60°) \\ \sigma_2(-60°) \\ \tau_{12}(-60°) \end{Bmatrix} = \begin{Bmatrix} -15.08 + 30.3\, p_a \\ 16.78 + 9.32\, p_a \\ -2.18 - 2.03\, p_a \end{Bmatrix} \text{MPa}$$

$$\begin{Bmatrix} \sigma_1(0°) \\ \sigma_2(0°) \\ \tau_{12}(0°) \end{Bmatrix} = \begin{Bmatrix} -29.3 + 17.00\, p_a \\ 20.8 + 13.07\, p_a \\ 0\, p_a \end{Bmatrix} \text{MPa}$$

$$\begin{Bmatrix} \sigma_1(90°) \\ \sigma_2(90°) \\ \tau_{12}(90°) \end{Bmatrix} = \begin{Bmatrix} -10.34 + 34.7\, p_a \\ 15.45 + 8.08\, p_a \\ 0\, p_a \end{Bmatrix} \text{MPa}$$

By using Eq. (5.6.73), the stresses in each layer are equated to the failure levels for each failure mode, and the value of p_a is computed for each mode. Interest is in positive values of p_a, corresponding to the internal pressure as the problem is formulated here. However, for the sake of completeness, all six failure equations will be considered for the four fiber angles, resulting in 24 values of p_a. The smallest positive value is the value of internal pressure that first causes failure as the pressure is increased from zero. The corresponding layer and the corresponding mode of failure can be determined from the particular equation that leads to this lowest positive value of p_a. For convenience, the six equations of Eq. (5.6.73) are expressed as

$$\begin{Bmatrix} \sigma_1 \\ \sigma_2 \\ \tau_{12} \end{Bmatrix} = \begin{Bmatrix} \sigma_1^T \\ \sigma_2^T \\ \tau_{12}^F \end{Bmatrix} = \begin{Bmatrix} 1,000 \\ 30 \\ 70 \end{Bmatrix} \text{MPa} \qquad \begin{Bmatrix} \sigma_1 \\ \sigma_2 \\ \tau_{12} \end{Bmatrix} = \begin{Bmatrix} \sigma_1^C \\ \sigma_2^C \\ -\tau_{12}^F \end{Bmatrix} = \begin{Bmatrix} -600 \\ -120 \\ -70 \end{Bmatrix} \text{MPa}$$

where typical failure stress levels have been used. For the $60°$ layers, using megapascals, the above six failure equations are

$$\begin{Bmatrix} -15.08 + 30.3\, p_a \\ 16.78 + 9.32\, p_a \\ 2.18 + 2.03\, p_a \end{Bmatrix} = \begin{Bmatrix} 1,000 \\ 30 \\ 70 \end{Bmatrix} \qquad \begin{Bmatrix} -15.08 + 30.3\, p_a \\ 16.78 + 9.32\, p_a \\ 2.18 + 2.03\, p_a \end{Bmatrix} = \begin{Bmatrix} -600 \\ -120 \\ -70 \end{Bmatrix}$$

For the $-60°$ layers, the six failure equations are

$$\begin{Bmatrix} -15.08 + 30.3\, p_a \\ 16.78 + 9.32\, p_a \\ -2.18 - 2.03\, p_a \end{Bmatrix} = \begin{Bmatrix} 1,000 \\ 30 \\ 70 \end{Bmatrix} \qquad \begin{Bmatrix} -15.08 + 30.3\, p_a \\ 16.78 + 9.32\, p_a \\ 2.18 - 2.03\, p_a \end{Bmatrix} = \begin{Bmatrix} -600 \\ -120 \\ -70 \end{Bmatrix}$$

For the $0°$ layers, the six failure equations are

$$\begin{Bmatrix} -29.3 + 17.00\, p_a \\ 20.8 + 13.07\, p_a \\ 0\, p_a \end{Bmatrix} = \begin{Bmatrix} 1,000 \\ 30 \\ 70 \end{Bmatrix} \qquad \begin{Bmatrix} -29.3 + 17.00\, p_a \\ 20.8 + 13.07\, p_a \\ 0\, p_a \end{Bmatrix} = \begin{Bmatrix} -600 \\ -120 \\ -70 \end{Bmatrix}$$

For the 90° layers, the six failure equations are

$$\begin{Bmatrix} -10.34 + 34.7p_a \\ 15.45 + 8.08p_1 \\ 0\,p_a \end{Bmatrix} = \begin{Bmatrix} 1{,}000 \\ 30 \\ 70 \end{Bmatrix} \qquad \begin{Bmatrix} -10.34 + 34.7p_a \\ 15.45 + 8.08p_a \\ 0\,p_a \end{Bmatrix} = \begin{Bmatrix} -600 \\ -120 \\ -70 \end{Bmatrix}$$

From these 24 equations there are 10 positive values of p_a, 10 negative values of p_a, and 4 values of infinity. The infinite values come from the two shear equations for the 0° layers and the two shear equations for the 90°layers. From these multiple values of p_a, the lowest level of the internal pressure is $p_a = 0.704$ atm. This value is obtained from the solution of the second equation for the positive failure stress levels for the 0° layer, i.e.,

$$20.8 + 13.07p_a = 30$$

This is interpreted to mean that at $p_a = 0.704$ atm cracks parallel to the fibers occur in the 0° layers. If matrix cracking is to be avoided, only pressures below this level are allowed. The answer to the original question is that failures occur when $p_a = 0.704$ atm, the mode of failure is excess tensile stress perpendicular to the fibers, and it occurs in the 0° layers. To go one step beyond what was asked for, however, a negative pressure of magnitude 10.78 atm is obtained from the solution of the second equation for negative failure stress levels for the 0° layers, i.e.,

$$20.8 + 13.07p_a = -120$$

This value is the negative value of p_a with the lowest magnitude. It is interpreted to mean that an *external* pressure of 10.78 atm will cause failure of the 0° layers due to excessive compressive stress perpendicular to the fibers, assuming that the cylinder has not buckled due to the external pressure.

Section **6**

Materials of Engineering

BY

EUGENE A. AVALLONE   *Consulting Engineer; Professor of Mechanical Engineering, Emeritus, The City College of The City University of New York*
HAROLD W. PAXTON   *United States Steel Professor Emeritus, Carnegie Mellon University*
JAMES D. REDMOND   *Principal, Technical Marketing Resources, Inc.*
MALCOLM BLAIR   *Technical and Research Director, Steel Founders Society of America*
ROBERT E. EPPICH   *Vice President, Technology, American Foundry Society*
L. D. KUNSMAN   *Late Fellow Engineer, Research Labs, Westinghouse Electric Corp.*
C. L. CARLSON   *Late Fellow Engineer, Research Labs, Westinghouse Electric Corp.*
J. RANDOLPH KISSEL   *President, The TGB Partnership*
LARRY F. WIESERMAN   *Senior Technical Supervisor, ALCOA*
RICHARD L. BRAZILL   *Technology Specialist, ALCOA*
FRANK E. GOODWIN   *Executive Vice President, ILZRO, Inc.*
DON GRAHAM   *Manager, Turning Products, Carboly, Inc.*
ARTHUR COHEN   *Formerly Manager, Standards and Safety Engineering, Copper Development Assn.*
JOHN H. TUNDERMANN   *Formerly Vice President, Research and Technology, INCO International, Inc.*
JAMES D. SHEAROUSE, III   *Late Senior Development Engineer, The Dow Chemical Co.*
PETER K. JOHNSON   *Director, Marketing and Public Relations, Metal Powder Industries Federation*
JOHN R. SCHLEY   *Manager, Technical Marketing, RMI Titanium Co.*
ROBERT D. BARTHOLOMEW   *Associate, Sheppard T. Powell Associates, LLC*
DAVID A. SHIFLER   *MERA Metallurgical Services*
DAVID W. GREEN   *Supervisory Research General Engineer, Forest Products Lab, USDA*
ROLAND HERNANDEZ   *Research Engineer, Forest Products Lab, USDA*
JOSEPH F. MURPHY   *Research General Engineer, Forest Products Lab, USDA*
ROBERT J. ROSS   *Supervisory Research General Engineer, Forest Products Lab, USDA*
WILLIAM T. SIMPSON   *Research Forest Products Technologist, Forest Products Lab, USDA*
ANTON TENWOLDE   *Supervisory Research Physicist, Forest Products Lab, USDA*
ROBERT H. WHITE   *Supervisory Wood Scientist, Forest Products Lab, USDA*
STAN LEBOW   *Research Forest Products Technologist, Forest Products Lab, USDA*
ALI M. SADEGH   *Professor of Mechanical Engineering, The City College of The City University of New York*
WILLIAM L. GAMBLE   *Professor Emeritus of Civil and Environmental Engineering, University of Illinois at Urbana-Champaign*
ARNOLD S. VERNICK   *Formerly Associate, Geraghty & Miller, Inc.*
GLENN E. ASAUSKAS   *Lubrication Engineer, Chevron Corp.*
STEPHEN R. SWANSON   *Professor of Mechanical Engineering, University of Utah*

6.1 GENERAL PROPERTIES OF MATERIALS
Revised by E. A. Avallone

REFERENCES: "International Critical Tables," McGraw-Hill. "Smithsonian Physical Tables," Smithsonian Institution. Landolt, "Landolt-Börnstein, Zahlenwerte und Funktionen aus Physik, Chemie, Astronomie, Geophysik und Technik," Springer. "Handbook of Chemistry and Physics," Chemical Rubber Co. "Book of ASTM Standards," ASTM. "ASHRAE Refrigeration Data Book," ASHRAE. Brady, "Materials Handbook," McGraw-Hill. Mantell, "Engineering Materials Handbook," McGraw-Hill. International Union of Pure and Applied Chemistry, Butterworth Scientific Publications. "U.S. Standard Atmosphere," Government Printing Office. Tables of Thermodynamic Properties of Gases, *NIST Circ.* 564, ASME Steam Tables.

Thermodynamic properties of a variety of other specific materials are listed also in Secs. 4.1, 4.2, and 9.8. Sonic properties of several materials are listed in Sec. 12.6.

CHEMISTRY

Every **elementary substance** is made up of exceedingly small particles called **atoms** which are all alike and which cannot be further subdivided or broken up by **chemical processes.** It will be noted that this statement is virtually a definition of the term elementary substance and a limitation of the term chemical process. There are as many different classes or families of atoms as there are **chemical elements.** See Table 6.1.1.

Two or more atoms, either of the same kind or of different kinds, are, in the case of most elements, capable of uniting with one another to form a higher order of distinct particles called **molecules.** If the molecules or atoms of which any given material is composed are all exactly alike, the material is a **pure substance.** If they are not all alike, the material is a **mixture.**

If the atoms which compose the molecules of any pure substances are all of the same kind, the substance is, as already stated, an **elementary substance.** If the atoms which compose the molecules of a pure chemical substance are not all of the same kind, the substance is a **compound substance.** The atoms are to be considered as the smallest particles which occur separately in the structure of molecules of either compound or elementary substances, so far as can be determined by ordinary chemical analysis. The molecule of an element consists of a definite (usually small) number of its atoms. The molecule of a compound consists of one or more atoms of each of its several elements, the numbers of the

Table 6.1.1 Chemical Elements[a]

Element	Symbol	Atomic no.	Atomic weight[b]	Valence
Actinium	Ac	89		
Aluminum	Al	13	26.9815	3
Americium	Am	95		
Antimony	Sb	51	121.75	3, 5
Argon[c]	Ar	18	39.948	0
Arsenic[d]	As	33	74.9216	3, 5
Astatine	At	85		
Barium	Ba	56	137.34	2
Berkelium	Bk	97		
Beryllium	Be	4	9.0122	2
Bismuth	Bi	83	208.980	3, 5
Boron[d]	B	5	10.811[l]	3
Bromine[e]	Br	35	79.904[m]	1, 3, 5
Cadmium	Cd	48	112.40	2
Calcium	Ca	20	40.08	2
Californium	Cf	98		
Carbon[d]	C	6	12.01115[l]	2, 4
Cerium	Ce	58	140.12	3, 4
Cesium[k]	Cs	55	132.905	1
Chlorine[f]	Cl	17	35.453[m]	1, 3, 5, 7
Chromium	Cr	24	51.996[m]	2, 3, 6
Cobalt	Co	27	58.9332	2, 3
Columbium (see Niobium)				
Copper	Cu	29	63.546[m]	1, 2
Curium	Cm	96		
Dysprosium	Dy	66	162.50	3
Einsteinium	Es	99		
Erbium	Er	68	167.26	3
Europium	Eu	63	151.96	2, 3
Fermium	Fm	100		
Fluorine[g]	F	9	18.9984	1
Francium	Fr	87		
Gadolinium	Gd	64	157.25	3
Gallium[k]	Ga	31	69.72	2, 3
Germanium	Ge	32	72.59	2, 4
Gold	Au	79	196.967	1, 3
Hafnium	Hf	72	178.49	4
Helium[c]	He	2	4.0026	0
Holmium	Ho	67	164.930	3
Hydrogen[h]	H	1	1.00797[i]	1
Indium	In	49	114.82	1, 2, 3
Iodine[d]	I	53	126.9044	1, 3, 5, 7

Table 6.1.1   Chemical Elements^a (Continued)

Element	Symbol	Atomic no.	Atomic weight^b	Valence
Iridium	Ir	77	192.2	2, 3, 4, 6
Iron	Fe	26	55.847^m	2, 3
Krypton^c	Kr	36	83.80	0
Lanthanum	La	57	138.91	3
Lead	Pb	82	207.19	2, 4
Lithium^i	Li	3	6.939	1
Lutetium	Lu	71	174.97	3
Magnesium	Mg	12	24.312	2
Manganese	Mn	25	54.9380	2, 3, 4, 6, 7
Mendelevium	Md	101		
Mercury^e	Hg	80	200.59	1, 2
Molybdenum	Mo	42	95.94	3, 4, 5, 6
Neodymium	Nd	60	144.24	3
Neon^c	Ne	10	20.183	0
Neptunium	Np	93		
Nickel	Ni	28	58.71	2, 3, 4
Niobium	Nb	41	92.906	2, 3, 4, 5
Nitrogen^f	N	7	14.0067	3, 5
Nobelium	No	102		
Osmium	Os	76	190.2	2, 3, 4, 6, 8
Oxygen^f	O	8	15.9994^l	2
Palladium	Pd	46	106.4	2, 4
Phosphorus^d	P	15	30.9738	3, 5
Platinum	Pt	78	195.09	2, 4
Plutonium	Pu	94		
Polonium	Po	84		2, 4
Potassium	K	19	39.102	1
Praseodymium	Pr	59	140.907	3
Promethium	Pm	61		5
Protactinium	Pa	91		
Radium	Ra	88		2
Radon^j	Rn	86		0
Rhenium	Re	75	186.2	1, 4, 7
Rhodium	Rh	45	102.905	3, 4
Rubidium	Rb	37	85.47	1
Ruthenium	Ru	44	101.07	3, 4, 6, 8
Samarium	Sm	62	150.35	3
Scandium	Sc	21	44.956	3
Selenium^d	Se	34	78.96	2, 4, 6
Silicon^d	Si	14	28.086^l	4
Silver	Ag	47	107.868^m	1
Sodium	Na	11	22.9898	1
Strontium	Sr	38	87.62	2
Sulfur^d	S	16	32.064^l	2, 4, 6
Tantalum	Ta	73	180.948	4, 5
Technetium	Tc	43		
Tellurium^d	Te	52	127.60	2, 4, 6
Terbium	Tb	65	158.924	3
Thallium	T1	81	204.37	1, 3
Thorium	Th	90	232.038	3
Thulium	Tm	69	168.934	3
Tin	Sn	50	118.69	2, 4
Titanium	Ti	22	47.90	3, 4
Tungsten	W	74	183.85	3, 4, 5, 6
Uranium	U	92	238.03	4, 6
Vanadium	V	23	50.942	1, 2, 3, 4, 5
Xenon^c	Xe	54	131.30	0
Ytterbium	Yb	70	173.04	2, 3
Yttrium	Y	39	88.905	3
Zinc	Zn	30	65.37	2
Zirconium	Zr	40	91.22	4

^{a} All the elements for which atomic weights listed are metals, except as otherwise indicated. No atomic weights are listed for most radioactive elements, as these elements have no fixed value.

^{b} The atomic weights are based upon nuclidic mass of $C^{12} = 12$.

^{c} Inert gas. ^{d} Metalloid. ^{e} Liquid. ^{f} Gas. ^{g} Most active gas. ^{h} Lightest gas. ^{i} Lightest metal. ^{j} Not placed. ^{k} Liquid at 25 °C.

^{l} The atomic weight varies because of natural variations in the isotopic composition of the element. The observed ranges are boron, $\pm$ 0.003; carbon, $\pm$ 0.00005; hydrogen, $\pm$ 0.00001; oxygen, $\pm$ 0.0001; silicon, $\pm$ 0.001; sulfur, $\pm$ 0.003.

^{m} The atomic weight is believed to have an experimental uncertainty of the following magnitude: bromine, $\pm$ 0.001; chlorine, $\pm$ 0.001; chromium, $\pm$ 0.001; copper, $\pm$ 0.001; iron, $\pm$ 0.003; silver, $\pm$ 0.001. For other elements, the last digit given is believed to be reliable to $\pm$ 0.5.

Source: Table courtesy IUPAC and Butterworth Scientific Publications.

various kinds of atoms and their arrangement being definite and fixed and determining the character of the compound. This notion of molecules and their constituent atoms is useful for interpreting the observed fact that chemical reactions—e.g., the analysis of a compound into its elements, the synthesis of a compound from the elements, or the changing of one or more compounds into one or more different compounds— take place so that the masses of the various substances concerned in a given reaction stand in definite and fixed ratios.

It appears from recent researches that some substances which cannot by any available means be decomposed into simpler substances and which must, therefore, be defined as elements, are continually undergoing spontaneous changes or **radioactive transformation** into other substances which can be recognized as physically and chemically different from the original substance. Radium is an element by the definition given and may be considered as made up of atoms. But it is assumed that these atoms, so called because they resist all efforts to break them up and are, therefore, apparently indivisible, nevertheless split up spontaneously, at a rate which scientists have not been able to influence in any way, into other atoms, thus forming other elementary substances of totally different properties. See Table 6.1.3.

The view generally accepted at present is that the atoms of all the chemical elements, including those not yet known to be radioactive, consist of several kinds of still smaller particles, three of which are known as **protons, neutrons,** and **electrons.** The protons are bound together in the atomic nucleus with other particles, including neutrons, and are positively charged. The neutrons are particles having approximately the mass of a proton but are uncharged. The electrons are negatively charged particles, all alike, external to the nucleus, and sufficient in number to neutralize the nuclear charge in an atom. The differences between the atoms of different chemical elements are due to the different numbers of these smaller particles composing them. According to the original Bohr theory, an ordinary atom is conceived as a stable system of such electrons revolving in closed orbits about the nucleus like the planets of the solar system around the sun. In a hydrogen atom, there is 1 proton and 1 electron; in a radium atom, there are 88 electrons surrounding a nucleus 226 times as massive as the hydrogen nucleus. Only a few, in general the outermost or valence electrons of such an atom, are subject to rearrangement within, or ejection from, the atom, thereby enabling it, because of its increased energy, to combine with other atoms to form molecules of either elementary substances or compounds. The **atomic number** of an

Table 6.1.2 Solubility of Inorganic Substances in Water
(Number of grams of the anhydrous substance soluble in 1,000 g of water. The common name of the substance is given in parentheses)

Substance	Composition	Temperature, °F (°C)		
		32 (0)	122 (50)	212 (100)
Aluminum sulfate	$Al_2(SO_4)_3$	313	521	891
Aluminum potassium sulfate (potassium alum)	$Al_2K_2(SO_4)_4 \cdot 24H_2O$	30	170	1,540
Ammonium bicarbonate	NH_4HCO_3	119		
Ammonium chloride (sal ammoniac)	NH_4Cl	297	504	760
Ammonium nitrate	NH_4NO_3	1,183	3,440	8,710
Ammonium sulfate	$(NH_4)_2SO_4$	706	847	1,033
Barium chloride	$BaCl_2 \cdot 2H_2O$	317	436	587
Barium nitrate	$Ba(NO_3)_2$	50	172	345
Calcium carbonate (calcite)	$CaCO_3$	0.018*		0.88
Calcium chloride	$CaCl_2$	594		1,576
Calcium hydroxide (hydrated lime)	$Ca(OH)_2$	1.77		0.67
Calcium nitrate	$Ca(NO_3)_2 \cdot 4H_2O$	931	3,561	3,626
Calcium sulfate (gypsum)	$CaSO_4 \cdot 2H_2O$	1.76	2.06	1.69
Copper sulfate (blue vitriol)	$CuSO_4 \cdot 5H_2O$	140	334	753
Ferrous chloride	$FeCl_2 \cdot 4H_2O$	644§	820	1,060
Ferrous hydroxide	$Fe(OH)_2$	0.0067‡		
Ferrous sulfate (green vitriol or copperas)	$FeSO_4 \cdot 7H_2O$	156	482	
Ferric chloride	$FeCl_3$	730	3,160	5,369
Lead chloride	$PbCl_2$	6.73	16.7	33.3
Lead nitrate	$Pb(NO_3)_2$	403		1,255
Lead sulfate	$PbSO_4$	0.042†		
Magnesium carbonate	$MgCO_3$	0.13‡		
Magnesium chloride	$MgCl_2 \cdot 6H_2O$	524		723
Magnesium hydroxide (milk of magnesia)	$Mg(OH)_2$	0.009‡		
Magnesium nitrate	$Mg(NO_3)_2 \cdot 6H_2O$	665	903	
Magnesium sulfate (Epsom salts)	$MgSO_4 \cdot 7H_2O$	269	500	710
Potassium carbonate (potash)	K_2CO_3	893	1,216	1,562
Potassium chloride	KCl	284	435	566
Potassium hydroxide (caustic potash)	KOH	971	1,414	1,773
Potassium nitrate (saltpeter or niter)	KNO_3	131	851	2,477
Potassium sulfate	K_2SO_4	74	165	241
Sodium bicarbonate (baking soda)	$NaHCO_3$	69	145	
Sodium carbonate (sal soda or soda ash)	$NaCO_3 \cdot 10H_2O$	204	475	452
Sodium chloride (common salt)	$NaCl$	357	366	392
Sodium hydroxide (caustic soda)	$NaOH$	420	1,448	3,388
Sodium nitrate (Chile saltpeter)	$NaNO_3$	733	1,148	1,755
Sodium sulfate (Glauber salts)	$Na_2SO_4 \cdot 10H_2O$	49	466	422
Zinc chloride	$ZnCl_2$	2,044	4,702	6,147
Zinc nitrate	$Zn(NO_3)_2 \cdot 6H_2O$	947		
Zinc sulfate	$ZnSO_4 \cdot 7H_2O$	419	768	807

* 59°F.
† 68°F.
‡ In cold water.
§ 50°F.

Table 6.1.3 Periodic Table of the Elements

Legend:
Atomic Number → Element
Symbol
Valence
Atomic weight based on $C^{12} = 12.00$
( ) denotes mass number of most stable known isotope

Group categories: Light metals · Brittle metals · Ductile metals · Low melting · Nonmetallic elements · Inert gases

Z	Element	Symbol	Atomic weight	Valence
1	Hydrogen	H	1.00797	1
2	Helium	He	4.0026	0
3	Lithium	Li	6.939	1
4	Beryllium	Be	9.0122	2
5	Boron	B	10.811	3
6	Carbon	C	12.01115	2, 4
7	Nitrogen	N	14.0067	3, 5
8	Oxygen	O	15.9994	2
9	Fluorine	F	18.9984	1
10	Neon	Ne	20.183	0
11	Sodium	Na	22.9898	1
12	Magnesium	Mg	24.312	2
13	Aluminum	Al	26.9815	3
14	Silicon	Si	28.086	4
15	Phosphorus	P	30.9738	3, 5
16	Sulphur	S	32.064	2, 4, 6
17	Chlorine	Cl	35.453	1, 3, 5, 7
18	Argon	Ar	39.948	0
19	Potassium	K	39.102	1
20	Calcium	Ca	40.08	2
21	Scandium	Sc	44.956	3
22	Titanium	Ti	47.90	3, 4
23	Vanadium	V	50.942	1, 2, 3, 4, 5
24	Chromium	Cr	51.996	2, 3, 6
25	Manganese	Mn	54.938	2, 3, 4, 6, 7
26	Iron	Fe	55.847	2, 3
27	Cobalt	Co	58.9332	2, 3
28	Nickel	Ni	58.71	2, 3, 4
29	Copper	Cu	63.546	1, 2
30	Zinc	Zn	65.37	2
31	Gallium	Ga	69.72	2, 3
32	Germanium	Ge	72.59	2, 4
33	Arsenic	As	74.9216	3, 5
34	Selenium	Se	78.96	2, 4, 6
35	Bromine	Br	79.904	1, 3, 5
36	Krypton	Kr	83.80	0
37	Rubidium	Rb	85.47	1
38	Strontium	Sr	87.62	2
39	Yttrium	Y	88.905	3
40	Zirconium	Zr	91.22	4
41	Niobium	Nb	92.906	2, 3, 4, 5
42	Molybdenum	Mo	95.94	3, 4, 5, 6
43	Technetium	Tc	(99)	
44	Ruthenium	Ru	101.07	3, 4, 6, 8
45	Rhodium	Rh	103.905	3, 4
46	Palladium	Pd	106.4	2, 4
47	Silver	Ag	107.868	1
48	Cadmium	Cd	112.40	2
49	Indium	In	114.82	1, 2, 3
50	Tin	Sn	118.69	2, 4
51	Antimony	Sb	121.75	3, 5
52	Tellurium	Te	127.60	2, 4, 6
53	Iodine	I	126.9044	1, 3, 5, 7
54	Xenon	Xe	131.30	0
55	Cesium	Cs	132.905	1
56	Barium	Ba	137.34	2
57	Lanthanum	La	137.91	3
72	Hafnium	Hf	178.49	4
73	Tantalum	Ta	180.948	4, 5
74	Tungsten	W	183.85	3, 4, 5, 6
75	Rhenium	Re	186.2	1, 4, 7
76	Osmium	Os	190.2	2, 3, 4, 6, 8
77	Iridium	Ir	192.2	2, 3, 4, 6
78	Platinum	Pt	195.09	2, 4
79	Gold	Au	196.967	1, 3
80	Mercury	Hg	200.59	1, 2
81	Thallium	Tl	204.37	1, 3
82	Lead	Pb	207.19	2, 4
83	Bismuth	Bi	208.98	3, 5
84	Polonium	Po	(210)	2, 4
85	Astatine	At	(210)	
86	Radon	Rn	(212)	0
87	Francium	Fr	(223)	
88	Radium	Ra	(227)	2
89	Actinium	Ac	(227)	3
71	Lutetium	Lu	174.97	3
103	Lawrencium	Lw	(257)	

Lanthanide Series (Rare earth elements)

Z	Element	Symbol	Atomic weight	Valence
58	Cerium	Ce	140.12	3, 4
59	Praseodymium	Pr	140.907	3
60	Neodymium	Nd	144.24	3
61	Promethium	Pm	(147)	5
62	Samarium	Sm	150.35	3
63	Europium	Eu	151.96	2, 3
64	Godalinium	Gd	157.25	3
65	Terbium	Tb	158.924	3
66	Dysprosium	Dy	162.50	3
67	Holmium	Ho	164.93	3
68	Erbium	Er	167.26	3
69	Thulium	Tm	168.934	3
70	Ytterbium	Yb	173.04	2, 3

Actinide Series (Transuranium elements)

Z	Element	Symbol	Atomic weight	Valence
90	Thorium	Th	232.038	4
91	Protactinium	Pa	(231)	
92	Uranium	U	238.03	3, 4, 5, 6
93	Neptunium	Np	(237)	
94	Plutonium	Pu	(242)	
95	Americium	Am	(243)	
96	Curium	Cm	(245)	
97	Berkelium	Bk	(249)	
98	Californium	Cf	(249)	
99	Einsteinium	Es	(254)	
100	Fermium	Fm	(252)	
101	Mendelevium	Md	(256)	
102	Nobelium	No	(254)	

Table 6.1.4 Solubility of Gases in Water
(By volume at atmospheric pressure)

	t, °F (°C)				t, °F (°C)		
	32 (0)	68 (20)	212 (100)		32 (0)	68 (20)	212 (100)
Air	0.032	0.020	0.012	Hydrogen	0.023	0.020	0.018
Acetylene	1.89	1.12		Hydrogen sulfide	5.0	2.8	0.87
Ammonia	1,250	700		Hydrochloric acid	560	480	
Carbon dioxide	1.87	0.96	0.26	Nitrogen	0.026	0.017	0.0105
Carbon monoxide	0.039	0.025		Oxygen	0.053	0.034	0.185
Chlorine	5.0	2.5	0.00	Sulfuric acid	87	43	

element is the number of excess positive charges on the nucleus of the atom. The essential feature that distinguishes one element from another is this charge of the nucleus. It also determines the position of the element in the periodic table. Modern researches have shown the existence of **isotopes,** that is, two or more species of atoms having the same atomic number and thus occupying the same place in the periodic system, but differing somewhat in atomic weight. These isotopes are chemically identical and are merely different species of the same chemical element. Most of the ordinary inactive elements have been shown to consist of a mixture of isotopes. This convenient atomic model should be regarded as only a working hypothesis for coordinating a number of phenomena about which much yet remains to be known.

Calculation of the Percentage Composition of Substances Add the atomic weights of the elements in the compound to obtain its molecular weight. Multiply the atomic weight of the element to be calculated by the number of atoms present (indicated in the formula by a subscript number) and by 100, and divide by the molecular weight of the compound. For example, hematite iron ore (Fe_2O_3) contains 69.94 percent of iron by weight, determined as follows: Molecular weight of Fe_2O_3 = $(55.84 \times 2) + (16 \times 3) = 159.68$. Percentage of iron in compound = $(55.84 \times 2) \times 100/159.68 = 69.94$.

SPECIFIC GRAVITIES AND DENSITIES AND OTHER PHYSICAL DATA

Table 6.1.5 Approximate Specific Gravities and Densities
(Water at 39°F and normal atmospheric pressure taken as unity) For more detailed data on any material, see the section dealing with the properties of that material. Data given are for usual room temperatures.

Substance	Specific gravity	Avg density lb/ft³	Avg density kg/m³
Metals, alloys, ores*			
Aluminum, cast-hammered	2.55–2.80	165	2,643
Brass, cast-rolled	8.4–8.7	534	8,553
Bronze, aluminum	7.7	481	7,702
Bronze, 7.9–14% Sn	7.4–8.9	509	8,153
Bronze, phosphor	8.88	554	8,874
Copper, cast-rolled	8.8–8.95	556	8,906
Copper ore, pyrites	4.1–4.3	262	4,197
German silver	8.58	536	8,586
Gold, cast-hammered	19.25–19.35	1,205	19,300
Gold coin (U.S.)	17.18–17.2	1,073	17,190
Iridium	21.78–22.42	1,383	22,160
Iron, gray cast	7.03–7.13	442	7,079
Iron, cast, pig	7.2	450	7,207
Iron, wrought	7.6–7.9	485	7,658
Iron, spiegeleisen	7.5	468	7,496
Iron, ferrosilicon	6.7–7.3	437	6,984
Iron ore, hematite	5.2	325	5,206
Iron ore, limonite	3.6–4.0	237	3,796
Iron ore, magnetite	4.9–5.2	315	5,046
Iron slag	2.5–3.0	172	2,755
Lead	11.34	710	11,370
Lead ore, galena	7.3–7.6	465	7,449
Manganese	7.42	475	7,608
Manganese ore, pyrolusite	3.7–4.6	259	4,149
Mercury	13.546	847	13,570

Table 6.1.5 Approximate Specific Gravities and Densities   (Continued)

Substance	Specific gravity	Avg density lb/ft³	Avg density kg/m³
Monel metal, rolled	8.97	555	8,688
Nickel	8.9	537	8,602
Platinum, cast-hammered	21.5	1,330	21,300
Silver, cast-hammered	10.4–10.6	656	10,510
Steel, cold-drawn	7.83	489	7,832
Steel, machine	7.80	487	7,800
Steel, tool	7.70–7.73	481	7,703
Tin, cast-hammered	7.2–7.5	459	7,352
Tin ore, cassiterite	6.4–7.0	418	6,695
Tungsten	19.22	1,200	18,820
Uranium	18.7	1,170	18,740
Zinc, cast-rolled	6.9–7.2	440	7,049
Zinc, ore, blende	3.9–4.2	253	4,052
Various solids			
Cereals, oats, bulk	0.41	26	417
Cereals, barley, bulk	0.62	39	625
Cereals, corn, rye, bulk	0.73	45	721
Cereals, wheat, bulk	0.77	48	769
Cordage (natural fiber)	1.2–1.5	85	1,360
Cordage (plastic)	0.9–1.3	69	1,104
Cork	0.22–0.26	15	240
Cotton, flax, hemp	1.47–1.50	93	1,491
Fats	0.90–0.97	58	925
Flour, loose	0.40–0.50	28	448
Flour, pressed	0.70–0.80	47	753
Glass, common	2.40–2.80	162	2,595
Glass, plate or crown	2.45–2.72	161	2,580
Glass, crystal	2.90–3.00	184	1,950
Glass, flint	3.2–4.7	247	3,960
Hay and straw, bales	0.32	20	320
Leather	0.86–1.02	59	945
Paper	0.70–1.15	58	929
Plastics (see Sec. 6.12)			
Potatoes, piled	0.67	44	705
Rubber, caoutchouc	0.92–0.96	59	946
Rubber goods	1.0–2.0	94	1,506
Salt, granulated, piled	0.77	48	769
Saltpeter	2.11	132	2,115
Starch	1.53	96	1,539
Sulfur	1.93–2.07	125	2,001
Wool	1.32	82	1,315
Timber, air-dry			
Apple	0.66–0.74	44	705
Ash, black	0.55	34	545
Ash, white	0.64–0.71	42	973
Birch, sweet, yellow	0.71–0.72	44	705
Cedar, white, red	0.35	22	352
Cherry, wild red	0.43	27	433
Chestnut	0.48	30	481
Cypress	0.45–0.48	29	465
Fir, Douglas	0.48–0.55	32	513
Fir, balsam	0.40	25	401
Elm, white	0.56	35	561

Table 6.1.5   Approximate Specific Gravities and Densities   (*Continued*)

Substance	Specific gravity	Avg density lb/ft³	kg/m³
Hemlock	0.45–0.50	29	465
Hickory	0.74–0.80	48	769
Locust	0.67–0.77	45	722
Mahogany	0.56–0.85	44	705
Maple, sugar	0.68	43	689
Maple, white	0.53	33	529
Oak, chestnut	0.74	46	737
Oak, live	0.87	54	866
Oak, red, black	0.64–0.71	42	673
Oak, white	0.77	48	770
Pine, Oregon	0.51	32	513
Pine, red	0.48	30	481
Pine, white	0.43	27	433
Pine, Southern	0.61–0.67	38–42	610–673
Pine, Norway	0.55	34	541
Poplar	0.43	27	433
Redwood, California	0.42	26	417
Spruce, white, red	0.45	28	449
Teak, African	0.99	62	994
Teak, Indian	0.66–0.88	48	769
Walnut, black	0.59	37	593
Willow	0.42–0.50	28	449
Various liquids			
Alcohol, ethyl (100%)	0.789	49	802
Alcohol, methyl (100%)	0.796	50	809
Acid, muriatic, 40%	1.20	75	1,201
Acid, nitric, 91%	1.50	94	1,506
Acid, sulfuric, 87%	1.80	112	1,795
Chloroform	1.500	95	1,532
Ether	0.736	46	738
Lye, soda, 66%	1.70	106	1,699
Oils, vegetable	0.91–0.94	58	930
Oils, mineral, lubricants	0.88–0.94	57	914
Turpentine	0.861–0.867	54	866
Water			
Water, 4°C, max density	1.0	62.426	999.97
Water, 100°C	0.9584	59.812	958.10
Water, ice	0.88–0.92	56	897
Water, snow, fresh fallen	0.125	8	128
Water, seawater	1.02–1.03	64	1,025
Ashlar masonry			
Granite, syenite, gneiss	2.4–2.7	159	2,549
Limestone	2.1–2.8	153	2,450
Marble	2.4–2.8	162	2,597
Sandstone	2.0–2.6	143	2,290
Bluestone	2.3–2.6	153	2,451
Rubble masonry			
Granite, syenite, gneiss	2.3–2.6	153	2,451
Limestone	2.0–2.7	147	2,355
Sandstone	1.9–2.5	137	2,194
Bluestone	2.2–2.5	147	2,355
Marble	2.3–2.7	156	2,500
Dry rubble masonry			
Granite, syenite, gneiss	1.9–2.3	130	2,082
Limestone, marble	1.9–2.1	125	2,001
Sandstone, bluestone	1.8–1.9	110	1,762
Brick masonry			
Hard brick	1.8–2.3	128	2,051
Medium brick	1.6–2.0	112	1,794
Soft brick	1.4–1.9	103	1,650
Sand-lime brick	1.4–2.2	112	1,794
Concrete masonry			
Cement, stone, sand	2.2–2.4	144	2,309
Cement, slag, etc.	1.9–2.3	130	2,082
Cement, cinder, etc.	1.5–1.7	100	1,602

Table 6.1.5   Approximate Specific Gravities and Densities   (*Continued*)

Substance	Specific gravity	Avg density lb/ft³	kg/m³
Various building materials			
Ashes, cinders	0.64–0.72	40–45	640–721
Cement, portland, loose	1.5	94	1,505
Portland cement	3.1–3.2	196	3,140
Lime, gypsum, loose	0.85–1.00	53–64	849–1,025
Mortar, lime, set	1.4–1.9	103	1,650
		94	1,505
Mortar, portland cement	2.08–2.25	135	2,163
Slags, bank slag	1.1–1.2	67–72	1,074–1,153
Slags, bank screenings	1.5–1.9	98–117	1,570–1,874
Slags, machine slag	1.5	96	1,538
Slags, slag sand	0.8–0.9	49–55	785–849
Earth, etc., excavated			
Clay, dry	1.0	63	1,009
Clay, damp, plastic	1.76	110	1,761
Clay and gravel, dry	1.6	100	1,602
Earth, dry, loose	1.2	76	1,217
Earth, dry, packed	1.5	95	1,521
Earth, moist, loose	1.3	78	1,250
Earth, moist, packed	1.6	96	1,538
Earth, mud, flowing	1.7	108	1,730
Earth, mud, packed	1.8	115	1,841
Riprap, limestone	1.3–1.4	80–85	1,282–1,361
Riprap, sandstone	1.4	90	1,441
Riprap, shale	1.7	105	1,681
Sand, gravel, dry, loose	1.4–1.7	90–105	1,441–1,681
Sand, gravel, dry, packed	1.6–1.9	100–120	1,602–1,922
Sand, gravel, wet	1.89–2.16	126	2,019
Excavations in water			
Sand or gravel	0.96	60	951
Sand or gravel and clay	1.00	65	1,041
Clay	1.28	80	1,281
River mud	1.44	90	1,432
Soil	1.12	70	1,122
Stone riprap	1.00	65	1,041
Minerals			
Asbestos	2.1–2.8	153	2,451
Barytes	4.50	281	4,504
Basalt	2.7–3.2	184	2,950
Bauxite	2.55	159	2,549
Bluestone	2.5–2.6	159	2,549
Borax	1.7–1.8	109	1,746
Chalk	1.8–2.8	143	2,291
Clay, marl	1.8–2.6	137	2,196
Dolomite	2.9	181	2,901
Feldspar, orthoclase	2.5–2.7	162	2,596
Gneiss	2.7–2.9	175	2,805
Granite	2.6–2.7	165	2,644
Greenstone, trap	2.8–3.2	187	2,998
Gypsum, alabaster	2.3–2.8	159	2,549
Hornblende	3.0	187	2,998
Limestone	2.1–2.86	155	2,484
Marble	2.6–2.86	170	2,725
Magnesite	3.0	187	2,998
Phosphate rock, apatite	3.2	200	3,204
Porphyry	2.6–2.9	172	2,758
Pumice, natural	0.37–0.90	40	641
Quartz, flint	2.5–2.8	165	2,645
Sandstone	2.0–2.6	143	2,291
Serpentine	2.7–2.8	171	2,740
Shale, slate	2.6–2.9	172	2,758
Soapstone, talc	2.6–2.8	169	2,709
Syenite	2.6–2.7	165	2,645
Stone, quarried, piled			
Basalt, granite, gneiss	1.5	96	1,579
Limestone, marble, quartz	1.5	95	1,572
Sandstone	1.3	82	1,314

Table 6.1.5   Approximate Specific Gravities and Densities   (Continued)

Substance	Specific gravity	Avg density lb/ft³	Avg density kg/m³
Shale	1.5	92	1,474
Greenstone, hornblend	1.7	107	1,715
Bituminous substances			
Asphaltum	1.1–1.5	81	1,298
Coal, anthracite	1.4–1.8	97	1,554
Coal, bituminous	1.2–1.5	84	1,346
Coal, lignite	1.1–1.4	78	1,250
Coal, peat, turf, dry	0.65–0.85	47	753
Coal, charcoal, pine	0.28–0.44	23	369
Coal, charcoal, oak	0.47–0.57	33	481
Coal, coke	1.0–1.4	75	1,201
Graphite	1.64–2.7	135	2,163
Paraffin	0.87–0.91	56	898
Petroleum	0.87	54	856
Petroleum, refined (kerosene)	0.78–0.82	50	801
Petroleum, benzine	0.73–0.75	46	737
Petroleum, gasoline	0.70–0.75	45	721
Pitch	1.07–1.15	69	1,105
Tar, bituminous	1.20	75	1,201
Coal and coke, piled			
Coal, anthracite	0.75–0.93	47–58	753–930
Coal, bituminous, lignite	0.64–0.87	40–54	641–866
Coal, peat, turf	0.32–0.42	20–26	320–417
Coal, charcoal	0.16–0.23	10–14	160–224
Coal, coke	0.37–0.51	23–32	369–513
Gases (see Sec. 4)			

* See also Sec. 6.4.

Compressibility of Liquids

If v_1 and v_2 are the volumes of the liquids at pressures of p_1 and p_2 atm, respectively, at any temperature, the coefficient of compressibility b is given by the equation

$$b = \frac{1}{v_1}\frac{v_1 - v_2}{p_2 - p_1}$$

The value of $b \times 10^6$ for oils at low pressures at about 70°F varies from about 55 to 80; for mercury at 32°F, it is 3.9; for chloroform at 32°F, it is 100 and increases with the temperature to 200 at 140°F; for ethyl alcohol, it increases from about 100 at 32°F and low pressures to 125 at 104°F; for glycerin, it is about 24 at room temperature and low pressure.

Table 6.1.6   Average Composition of Dry Air between Sea Level and 90-km (295,000-ft) Altitude

Element	Formula	% by Vol.	% by Mass	Molecular weight
Nitrogen	N_2	78.084	75.55	28.0134
Oxygen	O_2	20.948	23.15	31.9988
Argon	Ar	0.934	1.325	39.948
Carbon Dioxide	CO_2	0.0314	0.0477	44.00995
Neon	Ne	0.00182	0.00127	20.183
Helium	He	0.00052	0.000072	4.0026
Krypton	Kr	0.000114	0.000409	83.80
Methane	CH_4	0.0002	0.000111	16.043

From 0.0 to 0.00005 percent by volume of nine other gases.
Average composite molecular weight of air 28.9644.
SOURCE: "U.S. Standard Atmosphere," Government Printing Office.

Table 6.1.7   Specific Gravity and Density of Water at Atmospheric Pressure*
(Weights are in vacuo)

Temp, °C	Specific gravity	Density lb/ft³	Density kg/m³	Temp, °C	Specific gravity	Density lb/ft³	Density kg/m³
0	0.99987	62.4183	999.845	40	0.99224	61.9428	992.228
2	0.99997	62.4246	999.946	42	0.99147	61.894	991.447
4	1.00000	62.4266	999.955	44	0.99066	61.844	990.647
6	0.99997	62.4246	999.946	46	0.98982	61.791	989.797
8	0.99988	62.4189	999.854	48	0.98896	61.737	988.931
10	0.99973	62.4096	999.706	50	0.98807	61.682	988.050
12	0.99952	62.3969	999.502	52	0.98715	61.624	987.121
14	0.99927	62.3811	999.272	54	0.98621	61.566	986.192
16	0.99897	62.3623	998.948	56	0.98524	61.505	985.215
18	0.99862	62.3407	998.602	58	0.98425	61.443	984.222
20	0.99823	62.3164	998.213	60	0.98324	61.380	983.213
22	0.99780	62.2894	997.780	62	0.98220	61.315	982.172
24	0.99732	62.2598	997.304	64	0.98113	61.249	981.113
26	0.99681	62.2278	996.793	66	0.98005	61.181	980.025
28	0.99626	62.1934	996.242	68	0.97894	61.112	978.920
30	0.99567	62.1568	995.656	70	0.97781	61.041	977.783
32	0.99505	62.1179	995.033	72	0.97666	60.970	976.645
34	0.99440	62.0770	994.378	74	0.97548	60.896	975.460
36	0.99371	62.0341	993.691	76	0.97428	60.821	974.259
38	0.99299	61.9893	992.973	78	0.97307	60.745	973.041

* See also Secs. 4.2 and 6.10.

Table 6.1.8   Volume of Water as a Function of Pressure and Temperature

Temp, °F (°C)	Pressure, atm 0	500	1,000	2,000	3,000	4,000	5,000	6,500	8,000
32 (0)	1.0000	0.9769	0.9566	0.9223	0.8954	0.8739	0.8565	0.8361	
68 (20)	1.0016	0.9804	0.9619	0.9312	0.9065	0.8855	0.8675	0.8444	0.8244
122 (50)	1.0128	0.9915	0.9732	0.9428	0.9183	0.8974	0.8792	0.8562	0.8369
176 (80)	1.0287	1.0071	0.9884	0.9568	0.9315	0.9097	0.8913	0.8679	0.8481

SOURCE: "International Critical Tables."

Table 6.1.9 Basic Properties of Several Metals
(Staff contribution)*

Material	Density,† g/cm³	Coefficient of linear thermal expansion,‡ in/(in·°F) × 10⁻⁶	Thermal conductivity, Btu/(h·ft·°F)	Specific heat,‡ Btu/(lb·°F)	Approx melting temp, °F	Modulus of elasticity, lb/in² × 10⁶	Poisson's ratio	Ultimate Yield stress, lb/in² × 10³	stress, lb/in² × 10³	Elongation, %
Aluminum 2024-T3	2.77	12.6	110	0.23	940	10.6	0.33	50	70	18
Aluminum 6061-T6	2.70	13.5	90	0.23	1,080	10.6	0.33	40	45	17
Aluminum 7079-T6	2.74	13.7	70	0.23	900	10.4	0.33	68	78	14
Beryllium, QMV	1.85	6.4–10.2	85	0.45	2,340	40–44	0.024–0.030	27–38	33–51	1–3.5
Copper, pure	8.90	9.2	227	0.092	1,980	17.0	0.32		18	30
Gold, pure	19.32		172	0.031	1,950	10.8	0.42		See "Metals Handbook"	
Lead, pure	11.34	29.3	21.4	0.031	620	2.0	0.40–0.45	1.3	2.6	20–50
Magnesium AZ31B-H24 (sheet)	1.77	14.5	55	0.25	1,100	6.5	0.35	22	37	15
Magnesium HK31A-H24	1.79	14.0	66	0.13	1,100	6.4	0.35	29	37	8
Molybdenum, wrought	10.3	3.0	83	0.07	4,730	40.0	0.32	80	120–200	Small
Nickle, pure	8.9	7.2	53	0.11	2,650	32.0	0.31§		See "Metals Handbook"	35–40
Platinum	21.45	5.0	40	0.031	3,217	21.3	0.39		20–24	Small
Plutonium, alpha phase	19.0–19.7	30.0	4.8	0.034	1,184	14.0	0.15–0.21	40	60	48
Silver, pure	10.5	11.0	241	0.056	1,760	10–11	0.37		18	36
Steel, AISI C1020 (hot-worked)	7.85	6.3	27	0.10	2,750	29–30	0.29	48	65	65
Steel, AISI 304 (sheet)	8.03	9.9	9.4	0.12	2,600	28	0.29	39	87	1–40
Tantalum	16.6	3.6	31	0.03	5,425	27.0	0.35		50–145	34
Thorium, induction melt	11.6	6.95	21.7	0.03	3,200	7–10	0.27	21	32	9
Titanium, B 120VCA (aged)	4.85	5.2	4.3	0.13	3,100	14.8	0.3	190	200	1–3
Tungsten	19.3	2.5	95	0.033	6,200	50	0.28		18–600	4
Uranium D-38	18.97	4.0–8.0	17	0.028	2,100	24	0.21	28	56	

Room-temperature properties are given. For further information, consult the "Metals Handbook" or a manufacturer's publication.
* Compiled by Anders Lundberg, University of California, and reproduced by permission.
† To obtain the preferred density units, kg/m³, multiply these values by 1,000.
‡ See also Tables 6.1.10 and 6.1.11.
§ At 25°C.

Table 6.1.10   Coefficient of Linear Thermal Expansion for Various Materials
[Mean values between 32 and 212°F except as noted; in/(in · °F) $\times 10^{-4}$]

Metals		Other Materials		Paraffin:	
Aluminum bronze	0.094	Bakelite, bleached	0.122	32–61°F	0.592
Brass, cast	0.104	Brick	0.053	61–100°F	0.724
Brass, wire	0.107	Carbon—coke	0.030	100–120°F	2.612
Bronze	0.100	Cement, neat	0.060	Porcelain	0.02
Constantan (60 Cu, 40 Ni)	0.095	Concrete	0.060	Quarts:	
German silver	0.102	Ebonite	0.468	Parallel to axis	0.044
Iron:		Glass:		Perpend. to axis	0.074
Cast	0.059	Thermometer	0.045	Quarts, fused	0.0028
Soft forged	0.063	Hard	0.033	Rubber	0.428
Wire	0.080	Plate and crown	0.050	Vulcanite	0.400
Magnalium (85 Al, 15 Mg)	0.133	Flint	0.044	Wood (ǁ to fiber):	
Phosphor bronze	0.094	Pyrex	0.018	Ash	0.053
Solder	0.134	Granite	0.04–0.05	Chestnut and maple	0.036
Speculum metal	0.107	Graphite	0.044	Oak	0.027
Steel:		Gutta percha	0.875	Pine	0.030
Bessemer, rolled hard	0.056	Ice	0.283	Across the fiber:	
Bessemer, rolled soft	0.063	Limestone	0.023–0.05	Chestnut and pine	0.019
Nickel (10% Ni)	0.073	Marble	0.02–0.09	Maple	0.027
Type metal	0.108	Masonry	0.025–050	Oak	0.030

Table 6.1.11   Specific Heat of Various Materials
[Mean values between 32 and 212°F; Btu/(lb · °F)]

Solids					
Alloys:		Granite	0.195	Wood:	
Bismuth-tin	0.040–0.045	Graphite	0.201	Fir	0.65
Bell metal	0.086	Gypsum	0.259	Oak	0.57
Brass, yellow	0.0883	Hornblende	0.195	Pine	0.67
Brass, red	0.090	Humus (soil)	0.44	**Liquids**	
Bronze	0.104	Ice:		Acetic acid	0.51
Constantan	0.098	−4°F	0.465	Acetone	0.544
German silver	0.095	32°F	0.487	Alcohol (absolute)	0.58
Lipowits's metal	0.040	India rubber (Para)	0.27–0.48	Aniline	0.49
Nickel steel	0.109	Kaolin	0.224	Bensol	0.40
Rose's metal	0.050	Limestone	0.217	Chloroform	0.23
Solders (Pb and Sn)	0.040–0.045	Marble	0.210	Ether	0.54
Type metal	0.0388	Oxides:		Ethyl acetate	0.478
Wood's metal	0.040	Alumina (Al_3O_2)	0.183	Ethylene glycol	0.602
40 Pb + 60 Bi	0.0317	Cu_2O	0.111	Fusel oil	0.56
25 Pb + 75 Bi	0.030	Lead oxide (PbO)	0.055	Gasoline	0.50
Asbestos	0.20	Lodestone	0.156	Glycerin	0.58
Ashes	0.20	Magnesia	0.222	Hydrochloric acid	0.60
Bakelite	0.3–0.4	Magnetite (Fe_3O_4)	0.168	Kerosene	0.50
Basalt (lava)	0.20	Silica	0.191	Naphthalene	0.31
Borax	0.229	Soda	0.231	Machine oil	0.40
Brick	0.22	Zinc oxide (ZnO)	0.125	Mercury	0.033
Carbon-coke	0.203	Paraffin wax	0.69	Olive oil	0.40
Chalk	0.215	Porcelain	0.22	Paraffin oil	0.52
Charcoal	0.20	Quarts	0.17–0.28	Petroleum	0.50
Cinders	0.18	Quicklime	0.21	Sulfuric acid	0.336
Coal	0.3	Salt, rock	0.21	Sea water	0.94
Concrete	0.156	Sand	0.195	Toluene	0.40
Cork	0.485	Sandstone	0.22	Turpentine	0.42
Corundum	0.198	Serpentine	0.25	Molten metals:	
Dolomite	0.222	Sulfur	0.180	Bismuth (535–725°F)	0.036
Ebonite	0.33	Talc	0.209	Lead (590–680°F)	0.041
Glass:		Tufa	0.33	Sulfur (246–297°F)	0.235
Normal	0.199	Vulcanite	0.331	Tin (460–660°F)	0.058
Crown	0.16				
Flint	0.12				

6.2 IRON AND STEEL
by Harold W. Paxton

REFERENCES: "Metals Handbook," ASM International, latest ed., ASTM Standards, pt. 1. SAE Handbook. "Steel Products Manual," AISI. "Making, Shaping and Treating of Steel," AISE, latest ed.

CLASSIFICATION OF IRON AND STEEL

Iron (Fe) is not a high-purity metal commercially but contains other chemical elements which have a large effect on its physical and mechanical properties. The amount and distribution of these elements are dependent upon the method of manufacture. The most important commercial forms of iron are listed below.

Pig iron is the product of the blast furnace and is made by the reduction of iron ore.

Cast iron is an alloy of iron containing enough carbon to have a low melting temperature and which can be cast to close to final shape. It is not generally capable of being deformed before entering service.

Gray cast iron is an iron which, as cast, has combined carbon (in the form of cementite, Fe_3C) not in excess of a eutectoid percentage—the balance of the carbon occurring as graphite flakes. The term "gray iron" is derived from the characteristic gray fracture of this metal.

White cast iron contains carbon in the combined form. The presence of cementite or iron carbide (Fe_3C) makes this metal hard and brittle, and the absence of graphite gives the fracture a white color.

Malleable cast iron is an alloy in which all the combined carbon in a special white cast iron has been changed to free or temper carbon by suitable heat treatment.

Nodular (ductile) cast iron is produced by adding alloys of magnesium or cerium to molten iron. These additions cause the graphite to form into small nodules, resulting in a higher-strength, ductile iron.

Ingot iron, electrolytic iron (an iron-hydrogen alloy), and **wrought iron** are terms for low-carbon materials which are no longer serious items of commerce but do have considerable historical interest.

Steel is an alloy predominantly of iron and carbon, usually containing measurable amounts of manganese, and often readily formable.

Carbon steel is steel that owes its distinctive properties chiefly to the carbon it contains.

Alloy steel is steel that owes its distinctive properties chiefly to some element or elements other than carbon, or jointly to such other elements and carbon. Some alloy steels necessarily contain an important percentage of carbon, even as much as 1.25 percent. There is no complete agreement about where to draw the line between the alloy steels and the carbon steels.

Basic oxygen steel and **electric-furnace steel** are steels made by the basic oxygen furnace and electric furnace processes, irrespective of carbon content; the effective individual alloy content in engineering steels can range from 0.05 percent up to 3 percent, with a total usually less than 5 percent. Open-hearth and Bessemer steelmaking are no longer practiced in the United States.

Iron ore is reduced in a blast furnace to form **pig iron,** which is the raw material for practically all iron and steel products. Formerly, nearly 90 percent of the iron ore used in the United States came from the Lake Superior district; the ore had the advantages of high quality and the cheapness with which it could be mined and transported by way of the Great Lakes. With the rise of global steelmaking and the availability of high-grade ores and pellets (made on a large scale from low-grade ores) from many sources, the choice of feedstock becomes an economic decision.

The modern **blast furnace** consists of a vertical shaft up to 10 m or 40 ft in diameter and over 30 m (100 ft) high containing a descending column of iron ore, coke, and limestone and a large volume of ascending hot gas. The gas is produced by the burning of coke in the hearth of the furnace and contains about 34 percent carbon monoxide. This gas reduces the iron ore to metallic iron, which melts and picks up considerable quantities of carbon, manganese, phosphorus, sulfur, and silicon. The **gangue** (mostly silica) of the iron ore and the ash in the coke combine with the limestone to form the blast-furnace slag. The pig iron and slag are drawn off at intervals from the hearth through the iron notch and cinder notch, respectively. Some of the larger blast furnaces produce around 10,000 tons of pig iron per day. The blast furnace produces a liquid product for one of three applications: (1) the huge majority passes to the steelmaking process for refining; (2) pig iron is used in foundries for making castings; and (3) ferroalloys, which contain a considerable percentage of another metallic element, are used as addition agents in steelmaking. Compositions of commercial pig irons and two ferroalloys (ferromanganese and ferrosilicon) are listed in Table 6.2.1.

Physical Constants of Unalloyed Iron Some physical properties of iron and even its dilute alloys are sensitive to small changes in composition, grain size, or degree of cold work. The following are reasonably accurate for "pure" iron at or near room temperature; those with an asterisk are sensitive to these variables perhaps by 10 percent or more. Those with a dagger (†) depend measurably on temperature; more extended tables should be consulted.

Specific gravity, 7.866; melting point, 1,536°C (2,797°F); heat of fusion 277 kJ/kg (119 Btu/lbm); thermal conductivity 80.2 W/(m · C) [557 Btu/(h · ft^2 · in · °F)*†; thermal coefficient of expansion 12×10^{-6}/°C (6.7×10^{-6}/°F)†; electrical resistivity 9.7 $\mu\Omega$ · cm*†; and temperature coefficient of electrical resistance 0.0065/°C (0.0036/°F).†

Mechanical Properties Representative mechanical properties of annealed low-carbon steel (often similar to the former ingot iron) are as follows: yield strength 130 to 150 MPa (20 to 25 ksi); tensile strength 260 to 300 MPa (40 to 50 ksi); elongation 20 to 45 percent in 2 in; reduction in area of 60 to 75 percent; Brinell hardness 65 to 100. These figures are at best approximate and depend on composition (especially trace additives) and processing variables. For more precise data, suppliers or broader databases should be consulted.

Young's modulus for ingot iron is 202,000 MPa (29,300,000 lb/in^2) in both tension and compression, and the shear modulus is 81,400 MPa (11,800,000 lb/in^2). Poisson's ratio is 0.28. The **effect of cold rolling** on the tensile strength, yield strength, elongation, and shape of the stressstrain curve is shown in Fig. 6.2.1, which is for Armco ingot iron but would not be substantially different for other low-carbon steels.

Uses Low-carbon materials weld evenly and easily in all processes, can be tailored to be readily paintable and to be enameled, and with other treatments make an excellent low-cost soft magnetic material with high permeability and low coercive force for mass-produced motors and transformers. Other uses, usually after galvanizing, include culverts, flumes, roofing, siding, and housing frames; thin plates can be used in oil and water tanks, boilers, gas holders, and various nondemanding pipes; enameled sheet retains a strong market in ranges, refrigerators, and other household goods, in spite of challenges from plastics.

World Production From about 1970 to 1995, the annual world production of steel was remarkably steady at some 800,000,000 tons. Reductions in some major producers (United States, Japan, some European countries, and the collapsing U.S.S.R.) were balanced by the many smaller countries which began production. The struggle for markets led to pricing that did not cover production costs and led to international political strains. A major impact began in the 1990s as China pushed its own production from a few tens of million tons toward 300,000,000 tons in 2004, pushing world production to over 1 billion tons.

Table 6.2.1 Types of Pig Iron for Steelmaking and Foundry Use

Designation	Chemical composition, %*				Principal use
	Si	P	Mn	C†	
Basic pig, northern	1.50 max	0.400 max	1.01–2.00	3.5–4.4	Basic oxygen steel
In steps of	0.25		0.50		
Foundry, northern	3.50 max	0.301–0.700	0.50–1.25	3.0–4.5	A wide variety of castings
In steps of	0.25		0.25		
Foundry, southern	3.50 max	0.700–0.900	0.40–0.75	3.0–4.5	Cast-iron pipe
In steps of	0.25		0.25		
Ferromanganese (3 grades)	1.2 max	0.35 max	74–82	7.4 max	Addition of manganese to steel or cast iron
Ferrosilicon (silvery pig)	5.00–17.00	0.300 max	1.00–2.00	1.5 max	Addition of silicon to steel or cast iron

* Excerpted from "The Making, Shaping and Treating of Steel," AISE, 1984; further information in "Steel Products Manual," AISI, and ASTM Standards, Pt. 1.
† Carbon content not specified—for information only. Usually S is 0.05 max (0.06 for ferrosilicon) but S and P for basic oxygen steel are typically much lower today.

Fig. 6.2.1 Effect of cold rolling on the stress-strain relationship of Armco ingot iron. *(Kenyon and Burns.)*

The world raw material infrastructure (iron ore, coke, and scrap) was not able to react quickly to support this increase, and prices soared; steel prices followed. Recently (2006) long-term ore contracts reflected large price increases, so it appears that for some time **steel prices** will remain high by historical comparisons. The effects of material substitution are not yet clear, but as end users examine their options more closely, the marketplace will most likely settle the matter.

STEEL

Steel Manufacturing

Steel is produced by the removal of impurities from pig iron in a basic oxygen furnace or an electric furnace.

Basic Oxygen Steel This steel is produced by blowing pure (99 percent) oxygen either vertically under high pressure (1.2 MPa or 175 lb/in²) onto the surface of molten pig iron (BOP) or through tuyeres in the base of the vessel (the Q-BOP process). Some facilities use a combination depending on local circumstances and product mix. This is an autogenous process that requires no external heat to be supplied. The furnaces are similar in shape to the former Bessemer converters but range in capacity to 275 metric tons (t) (300 net tons) or more. The barrel-shaped furnace or vessel may or may not be closed on the bottom, is open at the top, and can rotate in a vertical plane about a horizontal axis for charging and for pouring the finished steel. Selected scrap is charged into the vessel first, up to 30 percent by weight of the total charge. Molten pig iron (often purified from the raw blast-furnace hot metal to give lower sulfur, phosphorus, and sometimes silicon) is poured into the vessel. In the Q-BOP, oxygen must be flowing through the bottom tuyeres at this time to prevent clogging; further flow serves to refine the charge and carries in fluxes as powders. In the BOP process, oxygen is introduced through a water-cooled lance introduced through the top of the vessel. Within seconds after the oxygen is turned on, some iron in the charge is converted to ferrous oxide, which reacts rapidly with the impurities of the charge to remove them from the metal. As soon as reaction starts, limestone is added as a flux. Blowing is continued until the desired degree of purification is attained. The reactions take place very rapidly, and blowing of a heat is completed in

about 20 min in a 200-net-ton furnace. Because of the speed of the process, a computer is used to calculate the charge required for making a given heat of steel, the rate and duration of oxygen blowing, and to regulate the quantity and timing of additions during the blow and for finishing the steel. Production rates of well over 270 t per furnace hour (300 net tons) can be attained. The comparatively low investment cost and low cost of operation have already made the basic oxygen process the largest producer of steel in the world, and along with electric furnaces, it almost completely replaces the basic open hearth as the major steelmaking process. No open hearths operate in the United States today.

Electric Steel The biggest change in steelmaking over the last 20 years is the fraction of steel made by remelting scrap in an electric furnace (EF), originally to serve a relatively nondemanding local market, but increasingly moving up in quality and products to compete with mills using the blast-furnace/oxygen steelmaking route. The economic competition is fierce and has served to improve choices for customers. In the last decade the fraction of steel made in the EF in the United States has gone above 50 percent by a combination of contraction in blast-furnace (BF) production (no new ones built, retirement of some, and a shortage of suitable coke) and an increase of total production, which has been met with EFs. At one time, there was concern that some **undesirable elements in scrap** that are not eliminated in steelmaking, notably copper and zinc, would increase with remelting cycles. However, the development of alternative **iron units** (direct reduced iron, iron carbide, and even pig iron) to dilute scrap additions has at least postponed this as an issue.

Early processes used three-phase alternating current, but increasingly the movement is to a single dc electrode with a conducting hearth. The high-power densities necessitate water cooling and improved basic refractory linings. Scrap is charged into the furnace, which usually contains some of the last heat to improve efficiency. Older practices often had a second slag made after the first meltdown and refining by oxygen blowing, but today, final refining takes place outside the melting unit in a ladle furnace, which allows refining, temperature control, and alloying additions to be made without interfering with the next heat. The materials are **continuously cast** including slabs only 50 mm (2 in) thick and casting directly to sheet in the 1- to 2-mm (0.04- to 0.08-in) thickness range is passing through the pilot stage.

The degree to which electric melting can replace more conventional methods is of great interest and depends in large part on the availability of sufficiently pure scrap at an attractive price and some improvements in surface quality to be able to make the highest-value products. Advances in EF technology are countered aggressively by new developments and cost control in traditional steelmaking; it may well be a decade or more before the pattern clarifies.

The **induction furnace** is simply a fairly small melting furnace to which the various metals are added to make the desired alloy, usually quite specialized. When steel scrap is used as a charge, it will be a high-grade scrap the composition of which is well known (see also Sec. 7).

Ladle Metallurgy One of the biggest contributors to quality in steel products is the concept of refining liquid steel outside the first melting unit—BOP, Q-BOP, or EF, none of which is well designed to perform the final refining function. In this separate unit, gases in solution (oxygen, hydrogen, and, to a lesser extent, nitrogen) can be reduced by vacuum treatment, carbon can be adjusted to desirable very low levels by reaction with oxygen in solution, alloy elements can be added, the temperature can be adjusted, and the liquid steel can be stirred by inert gases to float out inclusions and provide a homogeneous charge to the continuous casters which are now virtually ubiquitous. Reducing oxygen in solution means a "cleaner" steel (fewer nonmetallic inclusions) and a more efficient recovery of active alloying elements added with a purpose and which otherwise might end up as oxides.

Steel Ingots   With the advent of continuous casters, ingot casting is now generally reserved for the production of relatively small volumes of material such as heavy plates and forgings which are too big for current casters. Ingot casting, apart from being inefficient in that the large volume change from liquid to solid must be handled by discarding the large void space usually at the top of the ingot (the **pipe**), also has several other undesirable features caused by the solidification pattern in a large volume, most notably significant differences in composition throughout the piece (**segregation**) leading to different properties, **inclusions** formed during solidification, and surface flaws from poor mold surfaces, splashing and other practices, which if not properly removed lead to defects in finished products (**seams, scabs, scale,** etc.). Some defects can be removed or attenuated, but others cannot; in general, with the exception of some very specialized tool and bearing steels, products from ingots are no longer state-of-the-art unless they are needed for size.

Continuous Casting   This concept, which began with Bessemer in the 1850s, began to be a reliable production tool around 1970 and since then has replaced basically all ingot casting. Industrialized countries all continuously cast close to 100 percent of their production. Sizes cast range from 2-m (80-in)—or more—by 0.3-m (12-in) **slabs** down to 0.1-m (4-in) square or round **billets.** Multiple strands are common where production volume is important. Many heats of steel can be cast in a continuous string with changes of width possible during operation. Changes of composition are possible in succeeding ladles with a discard of the short length of mixed composition.

By intensive process control, it is often possible to avoid cooling the cast slabs to room temperature for inspection, enabling energy savings since the slabs require less reheating before hot rolling. If for some reason the slabs are cooled to room temperature, any surface defects which might lead to quality problems can be removed—usually by **scarfing** with an oxyacetylene torch or by grinding. Since this represents a yield loss, there is a real economic incentive to avoid the formation of such defects by paying attention to casting practices.

Mechanical Treatment of Steel

Cast steel, in the form of slabs, billets, or bars (these latter two differ somewhat arbitrarily in size) is treated further by various combinations of hot and cold deformation to produce a finished product for sale from the mill. Further treatments by fabricators usually occur before delivery to the final customer. These treatments have three purposes: (1) to change the shape by deformation or metal removal to desired tolerances; (2) to break up—at least partially—the segregation and large grain sizes inevitably formed during the solidification process and to redistribute the nonmetallic inclusions which are present; and (3) to change the properties. For example, these may be functional—strength or toughness—or largely aesthetic, such as reflectivity.

These purposes may be separable or in many cases may be acting simultaneously. An example is hot-rolled sheet or plate in which often the rolling schedule (reductions and temperature of each pass, and the cooling rate after the last reduction) is a critical path to obtain the properties and sizes desired and is often known as "heat treatment on the mill." The development of controls to do this has allowed much higher tonnages at attractive prices, and increasing robustness of rolls has allowed steel to be hot-rolled down to the 1 mm (0.04-in) range, where it can compete with the more expensive "cold-rolled" material.

Most steels are reduced after appropriate heating (to above 1,000°C) in various multistand hot rolling mills to produce **sheet, strip, plate, tubes,**

shaped sections, or bars. More specialized deformation, e.g., by **hammer forging,** can result in working in more than one direction, with a distribution of inclusions which is not extended in one direction. Rolling, e.g., more readily imparts anisotropic properties. **Press forging** at slow strain rates changes the worked structure to greater depths and is preferred for high-quality products. The degree of reduction required to eliminate the cast structure varies from 4:1 to 10:1; clearly smaller reductions would be desirable but are currently not usual.

The slabs, blooms, and billets from the caster must be reheated in an atmosphere-controlled furnace to the working temperature, often from room temperature, but if practices permit, they may be charged hot to save energy. Coupling the hot deformation process directly to slabs at the continuous caster exit is potentially more efficient, but practical difficulties currently limit this to a small fraction of total production.

The steel is oxidized during heating to some degree, and this oxidation is removed by a combination of light deformation and high-pressure water sprays before the principal deformation is applied. There are differences in detail between processes, but as a representative example, the conventional production of wide "hot-rolled sheet" [>1.5 m (60 in)] will be discussed.

The slab, about 0.3 m (12 in) thick at about 1200°C is passed through a *scale breaker* and high-pressure water sprays to remove the oxide film. It then passes through a set of *roughing passes* (possibly with some modest width reduction) to reduce the thickness to just over 25 mm (1 in), the ends are sheared perpendicular to the length to remove irregularities, and finally they are fed into a series of up to seven roll stands each of which creates a reduction of 50 to 10 percent passing along the train. Process controls allow each mill stand to run sufficiently faster than the previous one to maintain tension and avoid pileups between stands. The temperature of the sheet is a balance between heat added by deformation and that lost by heat transfer, sometimes with interstand water sprays. Ideally the temperature should not vary between head and tail of the sheet, but this is hard to accomplish.

The deformation encourages recrystallization and even some grain growth between stands; even though the time is short, temperatures are high. Emerging from the last stand between 815 and 950°C, the **austenite*** may or may not recrystallize, depending on the temperature. At higher temperatures, when austenite does recrystallize, the grain size is usually small (often in the 10- to 20-μm range). At lower exit temperatures austenite grains are rolled into "pancakes" with the short dimension often less than 10 μm. Since several **ferrite*** grains nucleate from each austenite grain during subsequent cooling, the ferrite grain size can be as low as 3 to 6 μm (ASTM 14 to 12). We shall see later that small ferrite grain sizes are a major contributor to the superior properties of today's carbon steels, which provide good strength and superior toughness simultaneously and economically.

Some of these steels also incorporate strong **carbide*** and **nitride*** formers in small amounts to provide extra strength from precipitation hardening; the degree to which these are undissolved in austenite during hot rolling affects recrystallization significantly. The subject is too complex to treat briefly here; the interested reader is referred to the ASM "Metals Handbook."

After the last pass, the strip may be cooled by programmed water sprays to between 510 and 730°C so that during coiling, any desired precipitation processes may take place in the coiler. The finished coil, usually 2 to 3 mm (0.080 to 0.120 in) thick and sometimes 1.3 to 1.5 mm (0.052 to 0.060 in) thick, which by now has a light oxide coating, is taken off line and either shippped directly or retained for further processing to make higher value-added products. Depending on composition, typical values of yield strength are from 210 up to 380 MPa (30 to 55 ksi), UTS in the range of 400 to 550 MPa (58 to 80 ksi), with an elongation in 200 mm (8 in) of about 20 percent. The higher strengths correspond to low-alloy steels.

About half the sheet produced is sold directly as **hot-rolled sheet.** The remainder is further cold-worked after scale removal by **pickling** and either is sold as cold-worked to various **tempers** or is recrystallized to form a very formable product known as **cold-rolled and annealed,**

---

* See Fig. 6.2.4 and the discussion thereof.

or more usually as **cold-rolled,** sheet. Strengthening by cold work is common in sheet, strip, wire, or bars. It provides an inexpensive addition to strength but at the cost of a serious loss of ductility, often a better surface finish, and a finished product held to tighter tolerances. It improves springiness by increasing the yield strength, but does not change the elastic moduli. Examples of the effect of cold working on carbon-steel drawn wires are shown in Figs. 6.2.2 and 6.2.3.

Fig. 6.2.2 Increase of tensile strength of plain carbon steel with increasing amounts of cold working by drawing through a wire-drawing die.

Fig. 6.2.3 Reduction in ductility of plain carbon steel with increasing amounts of cold working by drawing through a wire-drawing die.

To make the highest class of formable sheet is a very sophisticated operation. After pickling, the sheet is again reduced in a multistand (three, four, or five) mill with great attention paid to tolerances and surface finish. Reductions per pass range from 25 to 45 percent in early passes to 10 to 30 percent in the last pass. The considerable heat generated necessitates an oil-water mixture to cool and to provide the necessary lubrication. The finished coil is degreased prior to annealing.

The purpose of **annealing** is to provide, for the most demanding applications, pancake-shaped grains after recrystallization of the cold-worked ferrite, in a matrix with a very sharp crystal texture containing little or no carbon or nitrogen in solution. The exact metallurgy is complex but well understood. Two types of annealing are possible: slow heating, holding, and cooling of coils in a hydrogen atmosphere (box annealing) lasting several days, or continuous feeding through a furnace with a computer-controlled time-temperature cycle. The latter is much quicker but very capital-intensive and requires careful and complex process control.

As requirements for formability are reduced, production controls can be relaxed. In order of increasing cost, the series is commercial quality (CQ), drawing quality (DQ), deep drawing quality (DDQ), and extra deep drawing quality (EDDQ). Even more formable steels are possible, but often are not commercially necessary.

Some other deformation processes are occasionally of interest, such as wire drawing, usually done cold, and extrusion, either hot or cold. **Hot extrusion** for materials that are difficult to work became practical through the employment of a glass lubricant. This method allows the hot extrusion of highly alloyed steels and other exotic alloys subjected to service at high loads and/or high temperatures.

Constitution and Structure of Steel

As a result of the methods of production, the following elements are always present in steel: carbon, manganese, phosphorus, sulfur, silicon, and traces of oxygen, nitrogen, and aluminum. Various alloying elements are frequently deliberately added, such as nickel, chromium, copper, molybdenum, niobium (columbium), and vanadium. The most important of the above elements in steel is carbon, and it is necessary to understand the effect of carbon on the internal structure of steel to understand the heat treatment of carbon and low-alloy steels.

The iron–iron carbide equilibrium diagram in Fig. 6.2.4 shows the phases that are present in steels of various carbon contents over a range of temperatures under equilibrium conditions. Pure iron when heated to 910°C (1,670°F) changes its internal crystalline structure from a body-centered cubic arrangement of atoms, **alpha iron,** to a face-centered

Fig. 6.2.4 Iron–iron carbide equilibrium diagram, for carbon content up to 5 percent. (Dashed lines represent equilibrium with cementite, or iron carbide; adjacent solid lines indicate equilibrium with graphite.)

cubic structure, **gamma iron.** At 1,390°C (2,535°F), it changes back to the body-centered cubic structure, **delta iron,** and at 1,539°C (2,802°F) the iron melts. When carbon is added to iron, it is found that it has only slight solid solubility in alpha iron (much less than 0.001 percent at room temperature at equilibrium). These small amounts of carbon, however, are critically important in many high-tonnage applications where formability is required. On the other hand, gamma iron will hold up to 2.0 percent carbon in solution at 1,130°C (2,066°F). The alpha iron containing carbon or any other element in solid solution is called **ferrite,** and the gamma iron containing elements in solid solution is called **austenite.** Usually when not in solution in the iron, the carbon forms a compound Fe_3C (iron carbide) which is extremely hard and brittle and is known as **cementite.** Other carbides of iron exist but are only of interest in rather specialized instances.

The temperatures at which the phase changes occur are called **critical points** (or temperatures) and, in the diagram, represent equilibrium conditions. In practice there is a lag in the attainment of equilibrium, and the critical points are found at lower temperatures on cooling and at higher temperatures on heating than those given, the difference increasing with the rate of cooling or heating.

The various critical points have been designated by the letter A; when obtained on cooling, they are referred to as Ar, on the heating as Ac. The subscripts r and c refer to *refroidissement* and *chauffage,* respectively, and reflect the early French contributions to heat treatment. The various critical points are distinguished from each other by numbers after the letters, being numbered in the order in which they occur as the temperature increases. Ac_1 represents the beginning of transformation of ferrite to austenite on heating; Ac_3 the end of transformation of ferrite to austenite on heating, and Ac_4 the change from austenite to delta iron on heating. On cooling, the critical points would be referred to as Ar_4, Ar_3, and Ar_1, respectively. The subscript 2, not mentioned here, refers to a magnetic transformation. The error came about through a misunderstanding of what rules of thermodynamics apply in phase diagrams. It must be remembered that the diagram represents the pure iron-iron carbide system at equilibrium. The varying amounts of impurities in commercial steels affect to a considerable extent the position of the curves and especially the lateral position of the eutectoid point.

Carbon steel in equilibrium at room temperature will have present both ferrite and cementite. The physical properties of ferrite are approximately those of pure iron and are characteristic of the metal. Cementite is itself hard and brittle; its shape, amount, and distribution control many of the mechanical properties of steel, as discussed later. The fact that the carbides can be dissolved in austenite is the basis of the heat treatment of steel, since the steel can be heated above the A_1 critical temperature to dissolve all the carbides, and then suitable cooling through the appropriate range will produce a wide and predictable range of the desired size and distribution of carbides in the ferrite.

If austenite with the **eutectoid** composition at 0.76 percent carbon (Fig. 6.2.4) is cooled slowly through the critical temperature, ferrite and cementite are rejected simultaneously, forming alternate plates or lamellae. This microstructure is called **pearlite,** since when polished and etched it has a pearly luster. When examined under a high-power optical microscope, however, the individual plates of cementite often can be distinguished easily. If the austenite contains less carbon than the eutectoid composition (i.e., **hypoeutectoid** compositions), free ferrite will first be rejected on slow cooling through the critical temperature until the remaining austenite reaches eutectoid composition, when the simultaneous rejection of both ferrite and carbide will again occur, producing pearlite. A hypoeutectoid steel at room temperature will be composed of areas of free ferrite and areas of pearlite; the higher the carbon percentage, the greater the amount of pearlite present in the steel. If the austenite contains more carbon than the eutectoid composition (i.e., **hypereutectoid** composition) and is cooled slowly through the critical temperature, then cementite is rejected and appears at the austenitic grain boundaries, forming a continuous cementite network until the remaining austenite reaches eutectoid composition, at which time pearlite is formed. A hypereutectoid steel, when slowly cooled, will exhibit areas of pearlite surrounded by a thin network of cementite, or iron carbide.

As the cooling rate is increased, the spacing between the pearlite lamellae becomes smaller; with the resulting greater dispersion of carbide preventing slip in the iron crystals, the steel becomes harder. Also, with an increase in the rate of cooling, there is less time for the separation of excess ferrite or cementite, and the equilibrium amount of these constituents will not be precipitated before the austenite transforms to pearlite. Thus with a fast rate of cooling, pearlite may contain less or more carbon than given by the eutectoid composition. When the cooling rate becomes very rapid (as obtained by quenching), the carbon does not have sufficient time to separate out in the form of carbide, and the austenite transforms to a highly elastically stressed structure supersaturated with carbon called **martensite.** This structure is exceedingly hard but brittle and requires **tempering** to increase the ductility. Tempering consists of heating martensite to some temperature below the critical temperature, causing the carbide to precipitate in the form of small spheroids, or especially in alloy steels, as needles or platelets. The higher the tempering temperature, the larger the carbide particle size, the greater the ductility of the steel, and the lower the hardness.

In a carbon steel, it is possible to have a structure consisting either of parallel plates of carbide in a ferrite matrix, the distance between the plates depending upon the rate of cooling, or of carbide spheroids in a ferrite matrix, the size of the spheroids depending upon the temperature to which the hardened steel was heated. (Some spheroidization occurs when pearlite is heated, but only at high temperatures close to the critical temperature range.)

Heat-Treating Operations

The following definitions of terms have been adopted by the ASTM, SAE, and ASM in substantially identical form.

Heat Treatment   An operation, or combination of operations, involving the heating and cooling of a metal or an alloy in the solid state, for the purpose of obtaining certain desirable conditions or properties.

Quenching   Rapid cooling by immersion in liquids or gases or by contact with metal.

Hardening   Heating and quenching certain iron-base alloys from a temperature either within or above the critical range for the purpose of producing a hardness superior to that obtained when the alloy is not quenched. Usually restricted to the formation of martensite.

Annealing   A heating and cooling operation implying usually a relatively slow cooling. The purpose of such a heat treatment may be (1) to remove stresses; (2) to induce softness; (3) to alter ductility, toughness, electrical, magnetic, or other physical properties; (4) to refine the crystalline structure; (5) to remove gases; or (6) to produce a definite microstructure. The temperature of the operation and the rate of cooling depend upon the material being heat-treated and the purpose of the treatment. Certain specific heat treatments coming under the comprehensive term annealing are as follows:

Full Annealing   Heating iron base alloys above the critical temperature range, holding above that range for a proper period of time, followed by slow cooling to below that range. The annealing temperature is usually about 50°C ($\approx$100°F) above the upper limit of the critical temperature range, and the time of holding is usually not less than 1 h for each 1-in section of the heaviest objects being treated. The objects being treated are ordinarily allowed to cool slowly in the furnace. They may, however, be removed from the furnace and cooled in some medium that will prolong the time of cooling as compared with unrestricted cooling in the air.

Process Annealing   Heating iron-base alloys to a temperature below or close to the lower limit of the critical temperature range followed by cooling as desired. This heat treatment is commonly applied in the sheet and wire industries, and the temperatures generally used are from 540 to 705°C (about 1,000 to 1,300°F).

Normalizing   Heating iron base alloys to approximately 40°C (about 100°F) above the critical temperature range followed by cooling to below that range in still air at ordinary temperature.

Patenting   Heating iron base alloys above the critical temperature range followed by cooling below that range in air or a molten mixture of nitrates or nitrites maintained at a temperature usually between 425 and

565°C (about 800 to 1,050°F), depending on the carbon content of the steel and the properties required of the finished product. This treatment is applied in the wire industry to medium- or high-carbon steel as a treatment to precede further wire drawing.

Spheroidizing Any process of heating and cooling steel that produces a rounded or globular form of carbide. The following spheroidizing methods are used: (1) Prolonged heating at a temperature just below the lower critical temperature, usually followed by relatively slow cooling. (2) In the case of small objects of high-carbon steels, the spheroidizing result is achieved more rapidly by prolonged heating to temperatures alternately within and slightly below the critical temperature range. (3) Tool steel is generally spheroidized by heating to a temperature of 750 to 805°C (about 1,380 to 1,480°F) for carbon steels and higher for many alloy tool steels, holding at heat from 1 to 4 h, and cooling slowly in the furnace.

Tempering (also termed Drawing) Reheating hardened steel to some temperature below the lower critical temperature, followed by any desired rate of cooling. Although the terms *tempering* and *drawing* are practically synonymous as used in commercial practice, the term *tempering* is preferred.

Transformation Reactions in Carbon Steels Much of, but not all, the heat treatment of steel involves heating into the region above Ac_3 to form austenite, followed by cooling at a preselected rate. If the parts are large, heat flow may limit the available cooling rates. As an example selected for simplicity rather than volume of products, we may follow the possible transformations in a eutectoid steel over a range of temperature. (The reactions to produce ferrite in hypoeutectoid steels, which are by far the most common, do not differ in principle; the products are, of course, softer.) The curve is a derivative of the TTT (time-temperature-transformation) curves produced by a systematic study of austenite transformation rates isothermally on specimens thin enough to avoid heat flow complications. The data collected for many steels are found in the literature.

Figure 6.2.5 summarizes the rates of decomposition of a eutectoid carbon steel over a range of temperatures. Various cooling rates are shown diagrammatically, and it will be seen that the faster the rate of cooling, the lower the temperature of transformation, and the harder the product formed. At around 540°C (1,000°F), the austenite transforms rapidly to fine pearlite; to form martensite it is necessary to cool very rapidly through this temperature range to avoid the formation of pearlite before the specimen reaches the temperature at which the formation of martensite begins (M_s). The minimum rate of cooling that is required to form a fully martensite structure is called the **critical cooling rate.** No matter at what rate the steel is cooled, the only products of transformation of this steel will be pearlite or martensite. However, if the steel can be given an interrupted quench in a molten bath at some temperature between 205 and 540°C (about 400 and 1,000°F), an acicular structure, called **bainite,** of considerable toughness, combining high strength with

Fig. 6.2.5 Influence of cooling rate on the product of transformation in a eutectoid carbon steel.

high ductility, is obtained, and this heat treatment is known as austempering. A somewhat similar heat treatment called martempering can be utilized to produce a fully martensitic structure of high hardness, but free of the cracking, distortion, and residual stresses often associated with such a structure. Instead of quenching to room temperature, the steel is quenched to just above the martensitic transformation temperature and held for a short time to permit equalization of the temperature gradient throughout the piece. Then the steel may be cooled relatively slowly through the martensitic transformation range without superimposing thermal stresses on those introduced during transformation. The limitation of austempering and martempering for carbon steels is that these two heat treatments can be applied only to articles of small cross section, since the rate of cooling in salt baths is not sufficient to prevent the formation of pearlite in samples with diameter of more than ½ in.

The maximum hardness obtainable in a high-carbon steel with a fine pearlite structure is approximately 400 Brinell, although a martensitic structure would have a hardness of approximately 700 Brinell. Besides being able to obtain structures of greater hardness by forming martensite, a spheroidal structure will have considerably higher proof stress (i.e., stress to cause a permanent deformation of 0.01 percent) and ductility than a lamellar structure of the same tensile strength and hardness. It is essential, therefore, to form martensite when optimum properties are desired in the steel. This can be done with a piece of steel having a small cross section by heating the steel above the critical and quenching in water; but when the cross section is large, the cooling rate at the center of the section will not be sufficiently rapid to prevent the formation of pearlite. The characteristic of steel that determines its capacity to harden throughout the section when quenched is called **hardenability.** In the discussion that follows, it is pointed out that hardenability is significantly affected by most alloying elements. This term should not be confused with the ability of a steel to attain a certain hardness. The intensity of hardening, i.e., the maximum hardness of the martensite formed, is very largely dependent upon the carbon content of the steel.

Determination of Hardenability A long-established test for hardenability is the **Jominy test** which performs controlled water cooling on one end of a standard bar. Since the thermal conductivity of steel does not vary significantly, each distance from the quenched end (DQE) corresponds to a substantially unique cooling rate, and the structure obtained is a surrogate for the TTT curve on cooling. For a detailed account of the procedure, see the SAE Handbook. The figures extracted from the Jominy test can be extended to many shapes and quenching media. In today's practices, the factors discussed below can often be used to calculate hardenability from chemical composition with considerable confidence, leaving the actual test as a referee or for circumstances where a practice is being developed.

Three main factors affect the hardenability of steel: (1) austenite composition; (2) austenite grain size; and (3) amount, nature, and distribution of undissolved or insoluble particles in the austenite. All three determine the rate of decomposition, in the range of 540°C (about 1,000°F). The slower the rate of decomposition, the larger the section that can be hardened throughout, and therefore the greater the hardenability of the steel. Everything else being equal, the higher the carbon content, the greater the hardenability; this approach, however, is often counterproductive in that other strength properties may be affected in undesirable ways. The question of **austenitic grain size** is of considerable importance in any steel that is to be heat-treated, since it affects the properties of the steel to a considerable extent. When a steel is heated to just above the critical temperature, small polyhedral grains of austenite are formed. With increase in temperature, there is an increase in the size of grains, until at temperatures close to the melting point the grains are very large. Since the transformation of austenite to ferrite and pearlite usually starts at grain boundaries, a fine-grained steel will transform more rapidly than a coarse-grained steel because the latter has much less surface area bounding the grains than a steel with a fine grain size.

The grain size of austenite at a particular temperature depends primarily on the "pinning" of the boundaries by undissolved particles. These particles, which can be aluminum nitride from the deoxidation or various carbides and/or nitrides (added for their effect on final properties), dissolve as temperature increases, allowing grain growth. While hardenability is increased by large austenite grain size, this is not usually favored since some properties of the finished product can be seriously downgraded. Small particles in the austenite will act as nuclei for the beginning of transformation in a manner similar to grain boundaries, and therefore the presence of a large number of small particles (sometimes submicroscopic in size) will also result in low hardenability.

Determination of Austenitic Grain Size The subject of austenite grain size is of considerable interest because of the fact noted above that the grain size developed during heat treatment has a large effect on the physical properties of the steel. In steels of similar chemical analysis, the steel developing the finer austenitic grain size will have a lower hardenability but will, in general, have greater toughness, show less tendency to crack or warp on quenching, be less susceptible to grinding cracks, have lower internal stresses, and retain less austenite than coarse-grained steel. There are several methods of determining the grain-size characteristics of a steel.

The **McQuaid-Ehn test** (ASTM E112), which involves the outlining of austenite grains by cementite after a specific carburizing treatment, is still valid. It has been largely replaced, however, by quenching from the austenitizing treatment under investigation and observing the grain size of the resulting martensite after light tempering and etching with Vilella's reagent. There are several ways to report the grain size observed under the microscope, the one used most extensively being the ASTM index numbers. In fps or English units, the numbers are based on the formula: number of grains per square inch at $100x = 2^{N-1}$, in which N is the grain-size index. The usual range in steels will be from 1 to 128 grains/in^2 at $100x$, and the corresponding ASTM numbers will be 1 to 8, although today grain sizes up to 12 to 14 are common. Whereas at one time "coarse" grain sizes were 1 to 4, and "fine" grain sizes 5 to 8, these would not serve modern requirements. Most materials in service would be no coarser than 7 or 8, and the ferritic low-alloy high-strength steels routinely approach 12 or higher. Grain-size relationships in SI units are covered in detail in Designation E112 of ASTM Standards. For further information on grain size, refer to the ASM "Metals Handbook."

EFFECT OF ALLOYING ELEMENTS ON THE PROPERTIES OF STEEL

When relatively large amounts of alloying elements are added to steel, the characteristic behavior of carbon steels is not lost. Most **alloy steel** is medium- or high-carbon steel to which various elements have been added to modify its properties to an appreciable extent; the alloys as a minimum allow the properties characteristic of the carbon content to be fully realized even in larger sections, and in some cases may provide additional benefits. The percentage of alloy element required for a given purpose ranges from a few hundredths of 1 percent to possibly as high as 5 percent.

When ready for service, these steels will usually contain only two constituents, ferrite and carbide. The only way that an alloying element can affect the properties of the steel is to change the dispersion of carbide in the ferrite, change the properties of the ferrite, or change the characteristics of the carbide. The effect on the distribution of carbide is the most important factor, since in sections amenable to close control of structure, carbon steel is only moderately inferior to alloy steel. However, in large sections where carbon steels will fail to harden throughout the section even on a water quench, the hardenability of the steel can be increased at a cost by the addition of any alloying element (except possibly cobalt). The increase in hardenability permits the hardening of a larger section of alloy steel than of plain carbon steel. The quenching operation does not have to be so drastic. Consequently, there is a smaller difference in temperature between the surface and center during quenching, and cracking and warping resulting from sharp temperature gradients in a steel during hardening can be avoided. The elements most effective in increasing the hardenability of steel are manganese, silicon, and chromium, or combinations of small amounts of several elements such as chromium, nickel, and molybdenum in SAE 4340, where the joint effects are greater than alloys acting singly.

Elements such as molybdenum, tungsten, and vanadium are effective in increasing the hardenability when dissolved in the austenite, but not when present in the austenite in the form of carbides. When dissolved in austenite, and thus contained in solution in the resulting martensite, they can modify considerably the rate of coarsening of carbides in tempered martensite. Tempering relieves the internal stresses in the hardened steel in part by precipitating various carbides of iron at fairly low temperature, which coarsen as the tempering temperature is increased. The increasing particle separation results in a loss of hardness and strength accompanied by increased ductility. See Fig. 6.2.6. Alloying elements can cause slower coarsening rates or, in some cases at temperatures from 500 to 600°C, can cause dissolution of cementite and the precipitation of a new set of small, and thus closely spaced, alloy carbides which in some cases can cause the hardness to actually rise again with no loss in toughness or ductility. This is especially important in tool steels. The presence of these stable carbide-forming elements enables higher tempering temperatures to be used without sacrificing strength. This permits these alloy steels to have a greater ductility for a given strength, or, conversely, greater strength for a given ductility, than plain carbon steels.

The third factor which contributes to the strength of alloy steel is the presence of the alloying element in the ferrite. Any element in solid solution in a metal will increase the strength of the metal, so that these elements will materially contribute to the strength of hardened and tempered steels. The elements most effective in strengthening the ferrite are phosphorus, silicon, manganese, and, to a lesser extent, nickel, molybdenum, and chromium. Carbon and nitrogen are very strong ferrite strengtheners but generally are not present in interstitial solution in significant amounts, and there are other processing reasons to actively keep the amount in solution small by adding strong carbide and/or nitride formers to give interstitial-free (IF) steels.

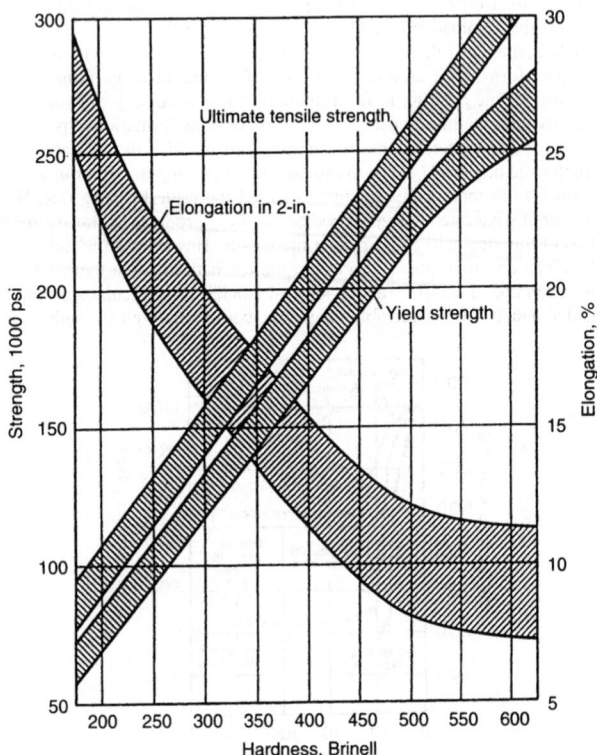

Fig. 6.2.6 Range of tensile properties in several quenched and tempered steels at the same hardness values. (*Janitzky and Baeyertz. Source: ASM; reproduced by permission.*)

Table 6.2.2   Trends of Influence of Some Alloying Elements

Element	Strengthening as dissolved in ferrite	Hardenability effects if dissolved in austenite	Effect on grain coarsening in austenite if undissolved as compound	Effects on tempered hardness, strength, and toughness
Al	*	*	†	None
Cr	*	†	†	*
Co	†	Negative	None	None
Cu	†	*	None	None
Mn	†	*	*	*
Mo	*	†	†	None
Nb (Cb)	None	†‡	†	†
Ni	*	*	None	None
P	†	*	None	None
Si	*	*	None	None
Ta	‡	†‡	†	†
Ti	†	†	†	‡
W	*	†	†	†
V	*	†	†	†

* Effects are moderate at best.
† Effects are strong to very strong.
‡ Effects not clear, or not used significantly.

A final important effect of alloying elements discussed above is their influence on the austenitic grain size. Martensite formed from a fine-grained austenite has considerably greater resistance to shock than when formed from a coarse-grained austenite.

In Table 6.2.2, a summary of the effects of various alloying elements is given. Remember that this table indicates only the trends of the various elements, and the fact that one element has an important influence on one factor does not prevent it from having a completely different influence or another one.

PRINCIPLES OF HEAT TREATMENT OF IRON AND STEEL

When heat-treating a steel for a given part, certain precautions have to be taken to develop optimum mechanical properties in the steel. Some of the major factors that have to be taken into consideration are outlined below.

Heating   The first step in the heat treatment of steel is the heating of the material to above the critical temperature to make it fully austenitic. The **heating rate** should be sufficiently slow to avoid injury to the material through excessive thermal and transformational stresses. In general, hardened steel should be heated more slowly and uniformly than is necessary for soft stress-free materials. Large sections should not be placed in a hot furnace, the allowable size depending upon the carbon and alloy content. For high-carbon steels, care should be taken in heating sections as small as 50-mm (2-in) diameter, and in medium-carbon steels precautions are required for sizes over 150-mm (6-in) diameter. The **maximum temperature** selected will be determined by the chemical composition of the steel and its grain-size characteristics. In hypoeutectoid steel, a temperature about 25 to 50°C above the upper critical range is used, and in hypereutectoid steels, a temperature between the lower and the upper critical temperature is generally used to retain enough carbides to keep the austenite grain size small and preserve what is often limited toughness. Quenching temperatures are usually a little closer to the critical temperature for hypoeutectoid steels than to those for normalizing; annealing for softening is carried out just below Ac_1 for steels up to 0.3 percent C and just above for higher-carbon steels. Tables of suggested temperatures can be found in the ASM "Metals Handbook," or a professional heat treater may be consulted.

The **time** at maximum temperature should be such that a uniform temperature is obtained throughout the cross section of the steel. Care should be taken to avoid undue length of time at temperature, since this will result in undesirable grain growth, scaling, or decarburization of the surface. A practical figure often given for the total time in the hot furnace is 12 min/cm (about ½ h/in) of cross-sectional thickness. When

the steel has attained a uniform temperature, the **cooling rate** must be such as to develop the desired structure; slow cooling rates (furnace or air cooling) to develop the softer pearlitic structures and high cooling rates (quenching) to form the hard martensitic structures. In selecting a **quenching medium** (see ASM "Metals Handbook"), it is important to select the quenching medium for a particular job on the basis of size, shape, and allowable distortion before choosing the steel composition. It is convenient to classify steels in two groups on the basis of depth of hardening: shallow hardening and deep hardening. **Shallow-hardening steels** may be defined as those which, in the form of 25-mm- (1-in-) diameter rounds, have, after brine quenching, a completely martensitic shell not deeper than 6.4 mm (¼ in). The shallow-hardening steels are those of low or no alloy content, whereas the deep-hardening steels have a substantial content of those alloying elements that increase penetration of hardening, notably chromium, manganese, and nickel. The high cooling rates required to harden shallow-hardening steel produce severe distortion and sometimes quench cracking in all but simple, symmetric shapes having a low ratio of length to diameter or thickness. Plain carbon steels cannot be used for complicated shapes where distortion must be avoided. In this case, water quenching must be abandoned and a less active quench used which materially reduces the temperature gradient during quenching. Certain oils are satisfactory but are incapable of hardening shallow-hardening steels of substantial size. A change in steel composition is required with a change from water to an oil quench. Quenching in oil does not entirely prevent distortion. When the degree of distortion produced by oil quenching is objectionable, recourse is taken to air hardening. The cooling rate in air is very much slower than in oil or water; so an exceptionally high alloy content is required. This means that a high price is paid for the advantage gained, in terms of both metal cost and loss in machinability, though it may be well justified when applied to expensive tools or dies. In this case, danger of cracking is negligible.

Liquids for Quenching Shallow-Hardening Steels   Shallow-hardening steels require extremely rapid surface cooling in the quench, particularly in the temperature range around 550°C (1,020°F). A submerged water spray will give the fastest and most reproducible quench practicable. Such a quench is limited in application to simple short objects which are not likely to warp. Because of difficulty in obtaining symmetric flow of the water relative to the work, the spray quench is conducive to warping. The ideal practical quench is one that will give the required surface cooling without agitation of the bath. The addition of ordinary salt, sodium chloride, greatly improves the performance of water in this respect, the best concentration being around 10 percent. Most inorganic salts are effective in suppressing the formation of vapor at the surface of the steel and thus aid in cooling steel uniformly and eliminating the formation of soft spots. To minimize the formation of vapor, water-base quenching liquids must be kept cool, preferably under 20°C (about 70°F). The addition of some other soluble materials to water such as soap is extremely detrimental because of increased formation of vapor.

Liquids for Quenching Deeper-Hardening Steels   When oil quenching is necessary, use a steel of sufficient alloy content to produce a completely martensitic structure at the surface over the heaviest section of the work. To minimize the possibility of cracking, especially when hardening tool steels, keep the quenching oil warm, preferably between 40 and 65°C (about 100 and 150°F). If this expedient is insufficient to prevent cracking, the work may be removed just before the start of the hardening transformation and cooled in air. Whether or not transformation has started can be determined with a permanent magnet, the work being completely nonmagnetic before transformation.

The cooling characteristics of quenching oils are difficult to evaluate and have not been satisfactorily correlated with the physical properties of the oils as determined by the usual tests. The standard tests are important with regard to secondary requirements of quenching oils. Low viscosity assures free draining of oil from the work and therefore low oil loss. A high flash and fire point assures a high boiling point and reduces the fire hazard which is increased by keeping the oil warm. A low carbon residue indicates stability of properties with continued use and little sludging. The steam-emulsion number should be low to ensure low water content, water being objectionable because of its vapor-film-forming tendency and high cooling power. A low saponification number assures that the oil is of mineral base and not subject to organic deterioration of fatty oils which give rise to offensive odors. Viscosity index is a valuable property for maintenance of composition.

In recent years, polymer-water mixtures have found application because of their combination of range of heat abstraction rates and relative freedom from fire hazards and environmental pollution. In all cases, the balance among productivity, danger of distortion and cracking, and minimum cost to give adequate hardenability is not simple; even though some general guides are available (e.g., "Metals Handbook," vol. 1, 10th ed.), consultation with experienced professionals is recommended.

Effect of the Condition of Surface   The factors that affect the depth of hardening are the hardenability of the steel, the size of specimen, the quenching medium, and finally the condition of the surface of the steel before quenching. Steel that carries a heavy coating of scale will not cool so rapidly as a steel that is comparatively scale-free, and soft spots may be produced; or, in extreme cases, complete lack of hardening may result. It is therefore essential to minimize scaling as much as possible. Decarburization can also produce undesirable results such as nonuniform hardening and thus lowers the resistance of the material to alternating stresses (i.e., fatigue).

Tempering, as noted above, relieves quenching stresses and offers the ability to obtain useful combinations of properties through selection of tempering temperature. The ability of alloying elements to slow tempering compared to carbon steel allows higher temperatures to be used to reach a particular strength. This is accompanied by some usually modest increases in ductility and toughness. Certain high-hardenability steels are subject to delayed cracking after quenching and should be tempered without delay. Data on tempering behavior are available from many sources, such as ASM "Metals Handbook" or Bain and Paxton, "Functions of the Alloying Elements in Steel," 3d ed., ASM International.

Relation of Design to Heat Treatment

Care must be taken in the design of a machine part to prevent cracking or distortion during heat treatment. With proper design the entire piece may be heated and cooled at approximately the same rate during the heat-treating operation. A light section should never be joined to a heavy section. Sharp reentrant angles should be avoided. Sharp corners and inadequate fillets produce serious stress concentration, causing the actual service stresses to build up to a point where they amount to two to five times the normal working stress calculated by the engineer in the

original layout. The use of generous fillets is especially desirable with all high-strength alloy steels.

The modulus of elasticity of all commercial steels, either carbon or alloy, is the same so far as practical designing is concerned. The deflection under load of a given part is, therefore, entirely a function of the section of the part and is not affected by the composition or heat treatment of the steel. Consequently if a part deflects excessively, a change in design is necessary; either a heavier section must be used or the points of support must be increased.

COMPOSITE MATERIALS

For some applications, it is not necessary or even desirable that the part have the same composition throughout. The oldest method of utilizing this concept is to produce a high-carbon surface on low-carbon steel by **carburizing,** a high-temperature diffusion treatment, which after quenching gives a wear-resistant case 1 or 2 mm (0.04 or 0.08 in) thick on a fairly shock-resistant core. Clearly, this is an attractive process for gears and other complex machined parts. Other processes which produce a similar product are **nitriding** (which can be carried out at lower temperatures) and **carbonitriding** (a hybrid), or where corrosion resistance is important, by **chromizing.** Hard surfaces can also be obtained on a softer core by using selective heating to produce surface austenite (induction, flames, etc.) before quenching, or by depositing various hard materials on the surface by welding.

Many other types of surface treatment which provide corrosion protection and sometimes aesthetic values are common, beginning with paint or other polymeric films and ranging through enamels (a glass film); films such as zinc and its alloys which can be applied by dipping in molten baths or can be deposited electrolytically on one or both sides (galvanizing); tinplate for cans and other containers; chromium plating; or a light coherent scale of iron oxide. These processes are continually being improved, and they may be used in combinations, e.g., paint on a galvanized surface for exposed areas of automobiles. (See Sec. 6.6.)

Carburizing   Various methods are available depending on the production volume. Pack carburizing can handle diverse feedstock by enclosing the parts in a sealed heat-resistant alloy box with carbonates and carbonaceous material, and by heating for several hours at about 925°C. The carbon dioxide evolved reacts with carbon to form carbon monoxide as the carrier gas, which does the actual carburizing. While the process can be continuous, much of it is done as a batch process, with consequent high labor costs and uncertain quality to be balanced against the flexibility of custom carburizing.

For higher production rates, furnaces with controlled atmospheres involving hydrocarbons and carbon monoxide provide better controls, low labor costs, shorter times to produce a given case depth [4 h versus 9 h for a 1-mm (0.04-in) case], and automatic quenching. Liquid carburizing using cyanide mixtures is even quicker for thin cases, but often it is not as economical for thicker ones; its great advantage is flexibility and control in small lots.

To provide the inherent value in this material of variable composition, it must be treated to optimize the properties of case and core by a double heat treatment. In many cases, to avoid distortion of precision parts, the material is first annealed at a temperature above the carburizing temperature and is cooled at a rate to provide good machinability. The part is machined and then carburized; through careful steel selection, the austenite grain size is not large at this point, and the material can be quenched directly without danger of cracking or distortion. Next the part is tempered; any retained austenite from the low M_s of the high-carbon case has a chance to precipitate carbides, raise its M_s, and transform during cooling after tempering. Care is necessary to avoid internal stresses if the part is to be ground afterward. An alternative approach is to cool the part reasonably slowly after carburizing and then to heat-treat the case primarily by suitable quenching from a temperature typical for hypereutectoid steels between A_1 and A_{cm}. This avoids some retained austenite and is helpful in high-production operations. The core is often carbon steel, but if alloys are needed for hardenability, this must be recognized in the heat treatment.

Nitriding A very hard, thin case can be produced by exposing an already quenched and tempered steel to an ammonia atmosphere at about 510 to 540°C, but unfortunately for periods of 50 to 90 h. The nitrogen diffuses into the steel and combines with strong nitride formers such as aluminum and chromium, which are characteristically present in steels where this process is to be used. The nitrides are small and finely dispersed; since quenching is not necessary after nitriding, dimensional control is excellent and cracking is not an issue. The core properties do not change since tempering at 550°C or higher temperature has already taken place. A typical steel composition is as follows: C, 0.2 to 0.3 percent; Mn, 0.04 to 0.6 percent; Al, 0.9 to 1.4 percent; Cr, 0.9 to 1.4 percent; and Mo, 0.15 to 0.25 percent.

Carbonitriding (Cyaniding) An interesting intermediate which rapidly adds both carbon and nitrogen to steels can be obtained by immersing parts in a cyanide bath just above the critical temperature of the core followed by direct quenching. A layer of about 0.25 mm (0.010 in) can be obtained in 1 h. The nitrides add to the wear resistance.

Local Surface Hardening For some parts which do not readily fit in a furnace, the surface can be hardened preferentially by local heating using flames, induction coils, electron beams, or lasers. The operation requires skill and experience, but in proper hands it can result in very good local control of structure, including the development of favorable surface compressive stresses to improve fatigue resistance.

Clad steels can be produced by one of several methods, including simple cladding by rolling a sandwich out of contact with air at a temperature high enough to bond (1,200°C); by explosive cladding where the geometry is such that the energy of the explosive causes a narrow molten zone to traverse along the interface and provide a good fusion bond; and by various casting and welding processes which can deposit a wide variety of materials (ranging from economical, tough, corrosion-resistant, or high-thermal-conductivity materials to hard and stable carbides in a suitable matrix). Many different product shapes lend themselves to these practices.

Chromizing Chromizing of low-carbon steel is effective in improving corrosion resistance by developing a surface containing up to 40 percent chromium. Some forming operations can be carried out on chromized material. Most chromizing is accomplished by packing the steel to be treated in a powdered mixture of chromium and alumina and then heating to above 1,260°C (2,300°F) for 3 or 4 h in a reducing atmosphere. Another method is to expose the parts to be treated to gaseous chromium compounds at temperatures above 845°C (about 1,550°F). Flat rolled sheets for corrosive applications such as auto mufflers can thus be chromized in open-coil annealing facilities.

THERMOMECHANICAL TREATMENT

The effects of mechanical treatment and heat treatment on the mechanical properties of steel have been discussed earlier in this section. Thermomechanical treatment consists of combining controlled (sometimes large) amounts of plastic deformation with the heat-treatment cycle to achieve improvements in yield strength beyond those attainable by the usual rolling practices alone or rolling following by a separate heat treatment. The tensile strength, of course, is increased at the same time as the yield strength (not necessarily to the same degree), and other properties such as ductility, toughness, creep resistance, and fatigue life can be improved. However, the high strength and hardness of thermomechanically treated steels limit their usefulness to the fabrication of components that require very little cold forming or machining, or very simple shapes such as strip and wire that can be used as part of a composite structure. Although the same yield strength may be achieved in a given steel by different thermomechanical treatments, the other mechanical properties (particularly the toughness) are not necessarily the same.

There are many possible combinations of deformation schedules and time-temperature relationships in heat treatment that can be used for thermomechanical treatment, and individual treatments will not be discussed here. Table 6.2.3 classifies broadly thermomechanical treatments into three principal groups related to the time-temperature

dependence of the transformation of austenite discussed earlier under heat treatment. The names in parentheses following the subclasses in the table are those of some types of thermomechanical treatments that have been used commercially or have been discussed in the literature. At one time it was thought that these procedures would grow in importance, but in fact they are still used in a very minor way with the important exception of class 1*a* and class 1*c*, which are critically important in large tonnages.

Table 6.2.3 Classification of Thermomechanical Treatments

Class I. Deformation before austenite transformation
    a. Normal hot-working processes
        (hot/cold working)
    b. Deformation before transformation to martensite
        (ausforming, austforming, austenrolling, hot-cold working, marworking, warm working)
    c. Deformation before transformation to ferrite-carbide aggregates (austen-tempering)
Class II. Deformation during austenite transformation
    a. Deformation during transformation to martensite (Zerolling and Ardeform processes)
    b. Deformation during transformation to ferrite-carbide aggregates (flow tempering of bainite and isoforming)
Class III. Deformation after austenite transformation
    a. Deformation of martensite followed by tempering
    b. Deformation of tempered martensite followed by aging (flow tempering, marstraining, strain tempering, tempforming, warm working)
    c. Deformation of isothermal transformation products (patenting, flow tempering, warm working)

Source: Radcliffe Kula, Syracuse University Press, 1964.

COMMERCIAL STEELS

The wide variety of applications of steel for engineering purposes is due to the range of mechanical properties obtainable by changes in carbon content and heat treatment. Some typical applications of carbon steels are given in Table 6.2.4. Carbon steels can be subdivided roughly into three groups: (1) low-carbon steel, 0.01 to 0.25 percent carbon, for use where only moderate strength is required together with considerable plasticity; (2) machinery steels, 0.30 to 0.55 percent carbon, which can be heat-treated to develop high strength; and (3) tool steels, containing from 0.60 to 1.30 percent carbon (this range also includes rail and spring steels).

Table 6.2.4 Some Typical Applications of Carbon Steels

Percent C	Uses
0.01–0.10	Sheet, strip, tubing, wire nails
0.10–0.20	Rivets, screws, parts to be case-hardened
0.20–0.35	Structural steel, plate, forgings such as camshafts
0.35–0.45	Machinery steel—shafts, axles, connecting rods, etc.
0.45–0.55	Large forgings—crankshafts, heavy-duty gears, etc.
0.60–0.70	Bolt-heading and drop-forging dies, rails, setscrews
0.70–0.80	Shear blades, cold chisels, hammers, pickaxes, band saws
0.80–0.90	Cutting and blanking punches and dies, rock drills, hand chisels
0.90–1.00	Springs, reamers, broaches, small punches, dies
1.00–1.10	Small springs and lathe, planer, shaper, and slotter tools
1.10–1.20	Twist drills, small taps, threading dies, cutlery, small lathe tools
1.20–1.30	Files, ball races, mandrels, drawing dies, razors

The **product mix** has changed noticeably over the last 20 years as consumption rose from about 100 million to 120 to 130 million tpy, some of which were imports, both semifinished (slabs, billets, etc.) and finished. Rails, in Carnegie's time, " the" product, dropped enormously as track maintenance slipped; tool steels fell, reflecting competition from carbides and oxide tools. Sheet was the main gainer with a large contribution from coated sheet (largely galvanized, reflecting cosmetic influences in the auto industry). Hot-rolled sheet displaced some cold-rolled applications as noted above. As an indication of current (2006) volumes, both of these are approaching 20 million tpy in the United States, from 10 to 12 million tpy in the early 1990s; cold-rolled has remained virtually constant at 12 million tpy. However, some of the specialized forms of steel products remain very important, and their capabilities are sometimes unique. The combination of property, performance, and price must be evaluated for each case.

The chemical compositions and mechanical and physical properties of many of the steels whose uses are listed in Table 6.2.4 are covered by specifications adopted by the American Society for Testing and Materials (ASTM), the Society of Automotive Engineers (SAE), and the American Society of Mechanical Engineers (ASME); other sources include government specifications for military procurement, national specifications in other industrial countries, and smaller specialized groups with their own interests in mind. The situation is more complex than necessary because of mixtures of chemical composition ranges and property and geometric ranges. Simplification will take a great deal of time to develop, and understanding among users, when and if it is reached, will require a major educational endeavor. For the time being, most of the important specifications and equivalents are shown in the ASM "Metals Handbook," 10th ed., vol. 1.

Understanding Some Mechanical Properties of Steels

Before we give brief outlines of some of the more important classes of steels, a discussion of the important factors influencing selection of steels and an appreciation of those for relevent competitive materials may be helpful. The easy things to measure for a material, steel or otherwise, are chemical composition and a stress-strain curve, from which one can extract such familiar quantities as yield strength, tensile strength, and ductility expressed as elongation and reduction of area. Where the application is familiar and the requirements are not particularly crucial, this is still appropriate and, in fact, may be overkill; if the necessary properties are provided, the chemical composition can vary outside the specification with few or no ill effects. However, today's designs emphasize total life-cycle cost and performance and often will require consideration of several other factors such as manufacturing and assembly methods; costs involving weldability (or more broadly joinability), formability, and machinability; corrosion resistance in various environments; fatigue, creep, and fracture; dimensional tolerances; and acceptable disposability at the end of the life cycle. These factors interact with each other in such complex ways that merely using tabulated data can invite disaster in extreme cases. It is not possible to cover all these issues in detail, but a survey of important factors for some of the most common ones may be helpful.

Yield Strength   In technical terms the yield stress, which measures the onset of plastic deformation (for example, 0.1 or 0.2 percent permanent extension), occurs when significant numbers of "dislocations" move in the crystal lattice of the major phase present, usually some form of ferrite. Dislocations are line discontinuities in the lattice which

allow crystal planes to slide over each other; they are always present at levels of 10^4 to 10^6 per square centimeter in undeformed metals, and this dislocation density can increase steadily to 10^{10} to 10^{12} per square centimeter after large deformations. Any feature which interferes with dislocation movement increases the yield stress. Such features include the following:

1. The lattice itself provides a lower limit (the **Peierls' stress**) by exhibiting an equivalent viscosity for movement. As a practical matter, this may be up to about 20 ksi (135 MPa) and is the same for all iron-based materials.

2. Dislocations need energy to cut through each other; thus, the stress necessary to continue deformation rises continuously as the dislocation density increases (work hardening). This is an important, inexpensive source of strength as various tempers are produced by rolling or drawing (see Table 6.2.5); it is accompanied by a significant loss of ductility.

3. Precipitates interfere with dislocation movement. The magnitude of interference depends sensitively on precipitate spacing, rising to a maximum at a spacing from 10 to 30 nm for practical additions. If a given heat treatment cannot develop these fine spacings (e.g., because the piece is so large that transformations take place at high temperatures), the strengthening is more limited. Precipitates are an important source of strength; martensite is used as an intermediate structure for carbon and alloy steels precisely because the precipitate spacing can be accurately controlled by tempering.

4. Small grain sizes, by interfering with the passage of dislocations across the boundaries, result in important increases in yield strength. To a very good approximation, the increase in yield is proportional to (grain diameter)$^{-1/2}$. A change from ASTM 8 to 12, for example, increases the yield strength by about 100 MPa.

5. Elements in solid solution also cause local lattice strains and make dislocation motion more difficult. The effect depends on the element; 1 percent P, for example, can double the hardness of iron or low-carbon steel. It is not used to this degree because of deleterious effects on toughness.

To a first approximation, these effects are additive; Fig. 6.2.7 depicts the effects of the iron lattice itself, solid solution strengthening, and grain size. Working the material would move all curves up the y axis by adding a work-hardening term. Figure 6.2.8 shows typical effects in an LAHS steel (see later) on the separated effects of manganese on yield strength. Manganese provides a little solid solution hardening and by its effect on ferrite grain size provides nonlinear strengthening. Figure 6.2.9 is a schematic of some combinations to obtain desired yield strengths and toughness simultaneously.

Tensile Properties   Materials fail in tension when the increment of nominal stress from work hardening can no longer support the applied load on a decreasing diameter. The load passes through a maximum [the **ultimate tensile stress** (UTS)], and an instability (**necking**) sets in at that point. The triaxial stresses thus induced encourage the formation of internal voids nucleated at inclusions or, less commonly, at other particles

Table 6.2.5   Approximate Mechanical Properties for Various Tempers of Cold-Rolled Carbon Strip Steel

Temper	Tensile strength MPa	Tensile strength 1,000 lb/in²	Elongation in 50 mm or 2 in for 1.27-mm (0.050-in) thickness of strip, %	Remarks
No. 1 (hard)	621 ± 69	90 ± 10		A very stiff cold-rolled strip intended for flat blanking only, and not requiring ability to withstand cold forming
No. 2 (half-hard)	448 ± 69	65 ± 10	10 ± 6	A moderately stiff cold-rolled strip intended for limited bending
No. 3 (quarter-hard)	379 ± 69	55 ± 10	20 ± 7	A medium-soft cold-rolled strip intended for limited bending, shallow drawing, and stamping
No. 4 (skin-rolled)	331 ± 41	48 ± 6	32 ± 8	A soft ductile cold-rolled strip intended for deep drawing where no stretcher strains or fluting are permissible
No. 5 (dead-soft)	303 ± 41	44 ± 6	39 ± 6	A soft ductile cold-rolled strip intended for deep drawing where stretcher strains or fluting are permissible. Also for extrusions

SOURCE: ASTM A109. Complete specification should be consulted in ASTM Standards (latest edition).

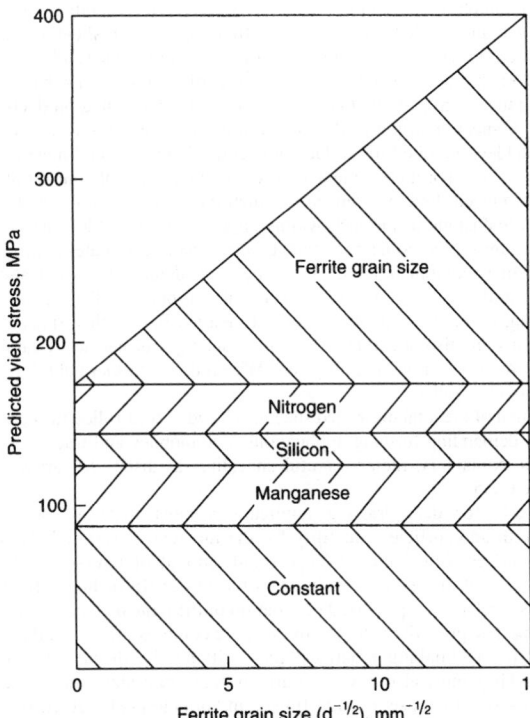

Fig. 6.2.7 The components of yield stress predicted for an air-cooled carbon-manganese steel containing 1.0 percent manganese, 0.25 percent silicon, and 0.01 percent nitrogen. *(Source: Union Carbide Corp.; reproduced by permission.)*

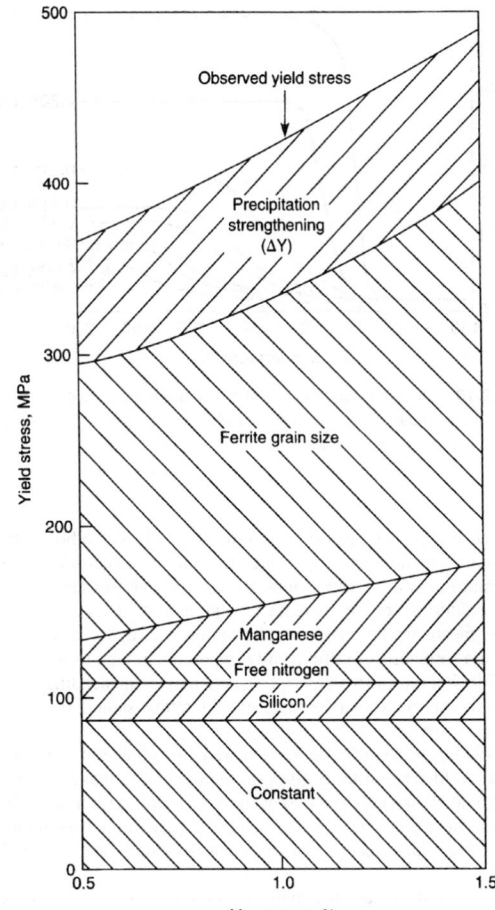

Fig. 6.2.8 The effect of increasing manganese content on the components of the yield stress of steels containing 0.2 percent carbon, 0.2 percent silicon, 0.15 percent vanadium, and 0.015 percent nitrogen, normalized from 900°C (1650°F). *(Source: Union Carbide Corp., reproduced by permission.)*

Fig. 6.2.9 Combinations of yield strength and impact transition temperature available in normalized high strength, low alloy (HSLA) steels. *(Source: Union Carbide Corp.; reproduced by permission.)*

such as precipitates. With increasing strain, these voids grow until they join and ultimately lead to a ductile failure.

Some plastic behavior of steels is sensitive to the number and type of inclusions. In a tensile test the UTS and **elongation to failure** are not affected much by increasing the number of inclusions (although the **uniform ductility** prior to necking is), but the **reduction in area** is; of greater importance, the energy absorbed to propagate a crack (related to the **fracture toughness**) is very sensitive to inclusion content. This relates directly to steelmaking practices, and to ladle metallurgy in particular, which has proved extremely effective in control of deleterious inclusions (not all inclusions are equally bad).

Toughness A full treatment of this topic is not possible here, but we note the following:

1. As strength increases, toughness falls in all cases except where strengthening arises from grain-size reduction. Fine grain size thus is a double blessing; it will increase strength and toughness simultaneously.

2. The energy involved in crack growth depends on carbon content (Fig. 6.2.10). High-carbon materials have not only much lower propagation energies (e.g., the "shelf" in a Charpy test, or the value of K_c), but also higher **impact transition temperatures** (ITTs). Below the ITT the energy for crack propagation becomes very small, and we may loosely describe the steel as "brittle." There is, then, a real incentive to keep carbon in steel as low as possible, especially because pearlite does not increase the yield strength, but does increase the ITT, often to well above room temperature.

3. **Welding** is an important assembly technique. Since the hardenability of the steel can lead to generally undesirable martensite in the **heat-affected zone** (HAZ) and possibly to cracks in this region, the toughness of welds necessitates using steels with the lowest possible carbon equivalent. **Carbon equivalent** is based on a formula which includes C and strong hardenability agents such as Mn, Ni, and Cr (see Sec. 6.3). Furthermore, the natural tendency of austenite to grow to a

Fig. 6.2.10 The effect of carbon, and hence the pearlite content on impact transition temperature curves of ferrite-pearlite steels. *(Source: Union Carbide Corp.; reproduced by permission.)*

very coarse grain size near the weld metal may be desirably restrained by adding elements such as Ti, which as undissolved carbides, interfere with grain growth. In critical applications, finer grain sizes after transformation during cooling increase toughness in the HAZ.

Other mechanical properties such as fatigue strength, corrosion resistance, and formability are frequently important enough to warrant consideration, but too short a description here may be misleading; standard texts and/or professional journals and literature should be consulted when necessary.

Steel is not necessarily the only material of choice in today's competitive world. The issue revolves on the method to make a legitimate and rational choice between two or more materials with many different properties and characteristics. Consider a very simple example (see Ashby, "Materials Selection in Mechanical Design," Pergamon Press, Oxford).

A furniture designer conceives of a lightweight table—a flat sheet of toughened glass supported on slender unbraced cylindrical legs. The legs must be solid, as light as possible, and must support the tabletop and whatever is placed on it without buckling. This involves consideration of the minimum weight and maximum aspect ratio of the legs. Examination of the mechanics shows that for minimum weight, we need a maximum value of the quantity $M_1 = E^{1/2}/\rho$, where E is Young's modulus and ρ is the density. For resistance to buckling, we need the maximum value of $M_2 = E$. If we plot E versus ρ of several materials, we can evaluate regions where both M_1 and M_2 are large. In doing so, we find attractive candidates are carbon-fiber-reinforced polymers (CFRPs) and certain engineering ceramics. They win out over wood, steel, and aluminum by their combination of high modulus and light weight. If we had other constraints such as cost, the CFRPs would be eliminated; as for toughness, the engineering ceramics would be unsuitable. Whenever many constraints must be satisfied simultaneously, their priority must be established. The net result is a compromise—but one which is better than an off-the-cuff guess!

This simple example can be generalized and used to develop a short list of very different alternatives in a reasonably quantitative and nonjudgmental manner for a wide variety of necessary properties. Ashby (op. cit.) provides a large collection of data.

Low-Carbon Steels   Of the many low-carbon-steel products, sheet and strip steels are becoming increasingly important. The consumption of steel in the sheet and tinplate industry has accounted for approximately 60 percent of the total steel production in the United States. This large production has been made possible by refinement of continuous sheet and strip rolling mills. Applications in which large quantities of sheet are employed are tinplate for food containers; black, galvanized, and terne-coated sheets for building purposes; and high-quality sheets

for automobiles, furniture, refrigerators, and countless other stamped, formed, and welded products. The difference between **sheet** and **strip** is based on width and is arbitrary. Cold working produces a better surface finish, improves the mechanical properties, and permits the rolling of thinner-gage material than hot rolling. Some hot-rolled products are sold as sheet or strip directly after coiling from the hot mill. Coils may be pickled and oiled for surface protection. Thicker coiled material up to 2 m (80 in) or more wide may be cut into plates, while thicker plates are produced directly from mills which may involve rolling a slab in two dimensions to produce widths approaching 5 m (200 in). Some plates are subject to heat treatment, occasionally elaborate, to produce particular combinations of properties for demanding applications. Structurals (beams, angles, etc.) are also produced directly by hot rolling, normally without further heat treating. Bars with various cross sections are also hot-rolled; some may undergo further heat treatment depending on service requirements. Wire rods for further cold drawing are also hot-rolled.

Much of the total hot-rolled sheet produced is cold-rolled further; the most demanding forming applications (automobiles and trucks, appliances, containers) require careful annealing, as discussed earlier. See Table 6.2.5.

Steels for deep-drawing applications must have a low yield strength and sufficient ductility for the intended purpose. This ductility arises from having a very small amount of interstitial atoms (carbon and nitrogen) in solution in the ferrite (**IF** or **interstitial-free** steels) and a sharp preferred orientation of the pancake-shaped ferrite grains. The procedures to obtain these are complex and go all the way back to steelmaking practices; they will not be discussed further here. They must also have a relatively fine grain size, since a large grain size will cause a rough finish, an "orange-peel" effect, on the deep drawn article. The sharp yield point characteristic of conventional low-carbon steel must be eliminated to prevent sudden local elongations in the sheet during forming, which result in strain marks called **stretcher strains** or **Lüders lines.** This can be done by cold working (Fig. 6.2.1), a reduction of only 1 percent in thickness usually being sufficient. This cold reduction is usually done by cold rolling, known as **temper rolling,** followed by alternate bending and reverse bending in a roller leveler. Temper rolling must always precede roller leveling because soft annealed sheets will "break" (yield locally) in the roller leveler. An important phenomenon in these temper-rolled low-carbon sheets is the return, partial or complete, of the sharp yield point after a period of time, by interstitial diffusion back to dislocations. This is known as **aging** in steel. The return of the yield point is accompanied by an increase in hardness and a loss in ductility. While a little more expensive, IF steels do not display a yield point and would not really require temper rolling, although they normally receive about 0.25 percent to improve surface finish. (The 1 percent typically used for non-IF steels would cause harmful changes in the stress-strain curve.) With these options available, aging is no longer the problem it used to be.

High-strength hot-rolled, cold-rolled, and galvanized sheets are now available with specified yield strengths. By the use of small alloy additions of niobium, vanadium, and sometimes copper, it is possible to meet the requirements of ASTM A607-70 for hot-rolled and cold-rolled sheets—345 MPa (50,000 lb/in²) yield point, 483 MPa (70,000 lb/in²) tensile strength, and 22 percent elongation in 2 in. By further alloy additions, sheets are produced to a minimum of 448 MPa (65,000 lb/in²) yield point, 552 MPa (80,000 lb/in²) tensile strength, and 16 percent elongation in 2 in. The sheets are available in coil form and are used extensively for metallic buildings and for welding into tubes for construction of furniture, etc.

Structural Carbon Steels   Bridges and buildings historically have been constructed with structural carbon steel meeting the requirements of ASTM A36 (Table 6.2.6). This steel has a minimum yield point of 248 MPa (36,000 lb/in²) and was developed to fill the need for a higher-strength structural carbon steel than the steels formerly covered by ASTM A7 and A373. The controlled composition of A36 steel provides good weldability and furnishes a significant improvement in the

Table 6.2.6 Mechanical Properties of Some Constructional Steels*

ASTM designation	Thickness range, mm (in)	Yield point, min		Tensile strength		Elongation in 200 mm (8 in) min, %	Suitable for welding?
		MPa	1,000 lb/in²	MPa	1,000 lb/in²		
Structural carbon-steel plates							
ASTM A36	To 100 mm (4 in), incl.	248	36	400–552	58–80	20	Yes
Low- and intermediate-tensile-strength carbon-steel plates							
ASTM A283	(structural quality)						
Grade A	All thicknesses	165	24	310	45	28	Yes
Grade B	All thicknesses	186	27	345	50	25	Yes
Grade C	All thicknesses	207	30	379	55	22	Yes
Grade D	All thicknesses	228	33	414	60	20	Yes
Carbon-silicon steel plates for machine parts and general construction							
ASTM A284							
Grade A	To 305 mm (12 in)	172	25	345	50	25	Yes
Grade B	To 305 mm (12 in)	159	23	379	55	23	Yes
Grade C	To 305 mm (12 in)	145	21	414	60	21	Yes
Grade D	To 200 mm (8 in)	145	21	414	60	21	Yes
Carbon-steel pressure-vessel plates							
ASTM A285							
Grade A	To 50 mm (2 in)	165	24	303–379	44–55	27	Yes
Grade B	To 50 mm (2 in)	186	27	345–414	50–60	25	Yes
Grade C	To 50 mm (2 in)	207	30	379–448	55–65	23	Yes
Structural steel for locomotives and railcars							
ASTM A113							
Grade A	All thicknesses	228	33	414–496	60–72	21	No
Grade B	All thicknesses	186	27	345–427	50–62	24	No
Grade C	All thicknesses	179	26	331–400	48–58	26	No
Structural steel for ships							
ASTM A131 (all grades)		221	32	400–490	58–71	21	No
Heat-treated constructional alloy-steel plates							
ASTM A514	To 64 mm (2½ in), incl.	700	100	800–950	115–135	18†	Yes
	Over 64 to 102 mm (2½ to 4 in), incl.	650	90	750–950	105–135	17†	Yes

* See appropriate ASTM documents for properties of other plate steels, shapes, bars, wire, tubing, etc.
† Elongation in 50 mm (2 in), min.

economics of steel construction. The structural carbon steels are available in the form of plates, shapes, sheet piling, and bars, all in the hot-rolled condition. A uniform strength over a range of section thickness is provided by adjusting the amount of carbon, manganese, and silicon in the A36 steel.

High-Strength Low-Alloy Steels While A36 still finds wide use, requirements for weight reduction, energy use in manufacture, and environmental savings have led to a steady, if slow, use of **higher-strength steels.** Slow adoption comes from several sources—the time necessary for specification changes, the education and general confidence level of designers, and inertia due to unfamiliarity with new materials.

The original development of inexpensive steels with higher yield stress began around 1930 when lighter railcars allowing higher payload were an issue. U.S. Steel with COR-TEN® and Bethlehem with Mayari-R™ used solid solution hardening by Cu, Ni, and P to achieve 350-MPa (50-Ksi) yield, and a modest market developed. Some 20 years later, architects noted that under proper conditions, an attractive patina developed and new uses of "weathering steels" developed. The general corrosion resistance was improved by Cu, although care was (and is!) necessary where localized corrosion may be a concern.

In the 1960s a new concept arose of using alloy carbonitrides to produce precipitation hardening. In the days before ladle metallurgy, reproducibility was not always adequate and usage was slow to develop.

However, with ladle metallurgy and the contemporary development of controlled rolling and austenite recrystallization on the hot mill giving fine ferrite grain size, the growth of this class of materials has been encouraging. Much remains, to be done in this regard. These steels have in the past been referred to as "high-tensile steels" and "low-alloy steels," but the name **high-strength low-alloy steels,** abbreviated **HSLA steels,** is now the generally accepted designation.

HSLA steels are a group of steels, intended for general structural and miscellaneous applications, that have minimum yield strengths from 280 to 550 MPa (40 to 80 ksi) with greatest usage around 350 to 420 MPa (50 to 60 ksi). Many opportunities exist for increased usage of the readily weldable, reasonably priced, and tough material in the range of 500 MPa (70 ksi) and up; however, there are not many current prototypes which would encourage a movement to this strength level. These steels typically contain small amounts of alloying elements to achieve their strength in the hot-rolled or normalized condition. Among the elements used in small amounts, singly or in combination, are niobium, titanium, vanadium, manganese, copper, and phosphorus. A complete listing of HSLA steels available from producers in the United States and Canada shows many brands or variations, often not covered by ASTM or other specifications. These steels generally are available as sheet, strip, plates, bars, and shapes and often are sold as proprietary grades.

Some of the steels have improved resistance to corrosion, so that the necessary equal service life in a thinner section or longer life in the same section is obtained in comparison with that of a structural carbon steel member. Good resistance to repeated loading and good abrasion resistance in service may be other characteristics of some of the steels. While high strength is a common characteristic of all HSLA steels, the other properties mentioned above may or may not be, singly or in combination, exhibited by any particular steel.

HSLA steels have found increasing, but still perhaps too limited, acceptance in many fields, among which are the construction of railroad cars, trucks, automobiles, trailers, and buses; welded steel bridges; television and power-transmission towers and lighting standards; columns in highrise buildings; portable liquefied petroleum gas containers; ship construction; oil storage tanks; air conditioning equipment; agricultural and earthmoving equipment.

Dual-Phase Steels   The good properties of HSLA steels do not provide sufficient cold formability for automotive components which involve stretch forming. **Dual-phase steels** generate a microstructure of ferrite plus islands of austenite-martensite by quenching from a temperature between A_1 and A_3. This leads to continuous yielding (rather than a sharp yield point) and a high work-hardening rate with a larger elongation to fracture. Modest forming strains can give a yield strength in the deformed product of 350 MPa (about 50 ksi), comparable to HSLA steels.

Quenched and Tempered Low-Carbon Constructional Alloy Steels   These steels, having yield strengths at the 689-MPa (100,000 lb/in^2) level, are covered by ASTM A514, by military specifications, and for pressure-vessel applications, by ASME Code Case 1204. They are available in plates, shapes, and bars and are readily welded. Since they are heat-treated to a tempered martensitic structure, they retain excellent toughness at temperatures as low as $-45°C$ ($-50°F$). Major cost savings have been effected by using these steels in the construction of pressure vessels, in mining and earthmoving equipment, and for major members of large steel structures.

Ultraservice Low-Carbon Alloy Steels (Quenched and Tempered)   The need for high-performance materials with higher strength-to-weight ratios for critical military needs, for hydrospace explorations, and for aerospace applications has led to the development of quenched and tempered ultraservice alloy steels. Although these steels are similar in many respects to the quenched and tempered low-carbon constructional alloy steels described above, their significantly higher notch toughness at yield strengths up to 965 MPa (140,000 lb/in^2) distinguishes the ultraservice alloy steels from the constructional alloy steels. These steels are not included in the AISI-SAE classification of alloy steels. There are numerous proprietary grades of ultraservice steels in addition to those covered by ASTM designations A543 and A579. Ultraservice steels may be used in large welded structures subjected to unusually high loads, and must exhibit excellent weldability and toughness. In some applications, such as hydrospace operations, the steels must have high resistance to fatigue and corrosion (especially stress corrosion) as well.

Maraging Steels   For performance requiring high strength and toughness and where cost is secondary, age-hardening low-carbon martensites have been developed based on the essentially carbon-free iron-nickel system. The as-quenched martensite is soft and can be shaped before an aging treatment. Two classes exist: (1) a nominal 18 percent nickel steel containing cobalt, molybdenum, and titanium with yield strengths of 1,380 to 2,070 MPa (200 to 300 ksi) and an outstanding resistance to stress corrosion cracking (normally a major problem at these strengths) and (2) a nominal 12 percent nickel steel containing chromium, molybdenum, titanium, and aluminum adjusted to yield strengths of 1,034 to 1,379 MPa (150 to 200 ksi). The toughness of this series is the best available of any steel at these yield strengths.

Cryogenic-Service Steels   For the economical construction of cryogenic vessels operating from room temperature down to the temperature of liquid nitrogen ($-195°C$ or $-320°F$) a 9 percent nickel alloy steel has been developed. The mechanical properties as specified by ASTM A353 are 517 MPa (75,000 lb/in^2) minimum yield strength and 689 to 827 MPa (100,000 to 120,000 lb/in^2) minimum tensile strength. The minimum Charpy impact requirement is 20.3 J (15 ft · lbf) at $-195°C$ ($-320°F$). For lower temperatures, it is necessary to use austenitic stainless steel.

Machinery Steels   A large variety of carbon and alloy steels is used in the automotive and allied industries. Specifications are published by SAE on all types of steel, and these specifications should be referred to for detailed information. A note on **specifications** may be in order. Many bodies set specifications on materials in their sphere of interest although efforts continue to rationalize the complexities. In a perfect world, a customer would specify expectations of properties. These would typically involve the range of mechanical properties, but also other matters of interest such as size, shape, surface finish, and weldability. However, composition often serves as the surrogate, and current specifications have considerable overlap. Relations between compositions and properties, preferably seen through the microstructure created by processing, are reasonably well known but not simple—many customers and suppliers arrive at suitable choices by mutual consultation, and by trial and error.

AISI, which for a long time coordinated its composition specifications with SAE, has discontinued its practice of designating steels, and the new arbiter is vol. 1 ("Materials") of "SAE Handbook." ASTM has a mixture of compositions, sizes, and properties required. The Unified Numbering System (UNS), which has been developed by a number of groups, is a designation of chemical composition and not a specification. It is described in SAE J1086 and ASTM E 527.

A numerical index is used to identify the compositions of AISI (and SAE) steels. Most SAE alloy steels are made by the basic oxygen or basic electric furnace processes; a few steels that at one time were made in the electric furnace carry the prefix E before their number, i.e., E52100. However, with the almost complete use of ladle furnaces, this distinction is rarely necessary today. Steels are "melted" in the BOP or EF and "made" in the ladle. A series of four numerals designates the composition of the AISI steels; the first two indicate the steel type, and the last two indicate, as far as feasible, the average carbon content in "points" or hundredths of 1 percent. Thus 1020 is a carbon steel with a carbon range of 0.18 to 0.23 percent, probably made in the basic oxygen furnace, and E4340 is a nickel-chromium molybdenum steel with 0.38 to 0.43 percent carbon made in the electric-arc furnace. A group of steels known as **H steels,** which are similar to the standard SAE steels, are being produced with a specified Jominy hardenability; these steels are identified by a suffix H added to the conventional series number. In general, these steels have a somewhat greater allowable variation in chemical composition but a smaller variation in hardenability than would be normal for a given grade of steel. This smaller variation in hardenability results in greater reproducibility of the mechanical properties of the steels on heat treatment; therefore, H steels have become increasingly important in machinery steels.

Boron steels are designated by the letter B inserted between the second and third digits, e.g., 50B44. The effectiveness of boron in increasing hardenability was a discovery of the late thirties, when it was noticed that heats treated with complex deoxidizers (containing boron) showed exceptionally good hardenability, high strength, and ductility after heat treatment. It was found that as little as 0.0005 percent of boron increased the hardenability of steels with 0.15 to 0.60 carbon, whereas boron contents of over 0.005 percent had an adverse effect on hot workability. Boron steels achieve special importance in times of alloy shortages, for they can replace such critical alloying elements as nickel, molybdenum, chromium, and manganese and, when properly heat-treated, possess physical properties comparable to the alloy grades they replace. Additional advantages for the use of boron in steels are a decrease in susceptibility to flaking, formation of less adherent scale, greater softness in the unhardened condition, and better machinability. It is also useful in low-carbon bainite and acicular ferrite steels.

Specific applications of these steels cannot be given, since the selection of a steel for a given part must depend upon an intimate knowledge of factors such as the availability and cost of the material, the detailed design of the part, and the severity of the service to be imposed. However, the mechanical properties desired in the part to be heat-treated will determine to a large extent the carbon and alloy content of the steel. Table 6.2.7 gives a résumé of mechanical properties that can be expected on heat-treating SAE steels, and Table 6.2.8 gives an indication of the effect of size on the mechanical properties of heat-treated steels.

Table 6.2.7   Mechanical Properties of Certain SAE Steels with Various Heat Treatments
Sections up to 40 mm (or 1½ diam or thickness)

Tempering temp		Tensile strength		Yield point		Reduction of area, %	Elongation in 50 mm (2 in), %	Brinell hardness
°C	°F	MPa	1,000 lb/in²	MPa	1,000 lb/in²			

Tempering temp		Tensile strength		Yield point		Reduction of area, %	Elongation in 50 mm (2 in), %	Brinell hardness
°C	°F	MPa	1,000 lb/in²	MPa	1,000 lb/in²			
SAE 1040 quenched in water from 815°C (1,500°F)								
315	600	862	125	717	104	46	11	260
425	800	821	119	627	91	53	13	250
540	1,000	758	110	538	78	58	15	220
595	1,100	745	108	490	71	60	17	216
650	1,200	717	104	455	66	62	20	210
705	1,300	676	98	414	60	64	22	205
SAE 1340 normalized at 865°C (1,585°F), quenched in oil from 845°C (1,550°F)								
315	600	1,565	227	1,420	206	43	11	448
425	800	1,248	181	1,145	166	51	13	372
540	1,000	966	140	834	121	58	17.5	297
595	1,100	862	125	710	103	62	20	270
650	1,200	793	115	607	88	65	23	250
705	1,300	758	110	538	78	68	25.5	234
SAE 4042 normalized at 870°C (1,600°F), quenched in oil from 815°C (1,500°F)								
315	600	1,593	231	1,448	210	41	12	448
425	800	1,207	175	1,089	158	50	14	372
540	1,000	966	140	862	125	58	19	297
595	1,100	862	125	758	110	62	23	260
650	1,200	779	113	683	99	65	26	234
705	1,800	724	105	634	92	68	30	210

Table 6.2.8   Effect of Size of Specimen on the Mechanical Properties of Some SAE Steels

Diam of section		Tensile strength		Yield point		Reduction of area, %	Elongation in 50 mm (2 in), %	Brinell hardness
in	mm	MPa	1,000 lb/in²	MPa	1,000 lb/in²			
SAE 1040, water-quenched, tempered at 540°C (1,000°F)								
1	25	758	110	538	78	58	15	230
2	50	676	98	448	65	49	20	194
3	75	641	93	407	59	48	23	185
4	100	621	90	393	57	47	24.5	180
5	125	614	89	372	54	46	25	180
SAE 4140, oil-quenched, tempered at 540°C (1,000°F)								
1	25	1,000	145	883	128	56	18	297
2	50	986	143	802	125	58	19	297
3	75	945	137	814	118	59	20	283
4	100	869	126	758	110	60	18	270
5	125	841	122	724	105	59	17	260
SAE 8640, oil-quenched, tempered at 540°C (1,000°F)								
1	25	1,158	168	1,000	145	44	16	332
2	50	1,055	153	910	132	45	20	313
3	75	951	138	807	117	46	22	283
4	100	889	129	745	108	46	23	270
5	125	869	126	724	105	45	23	260

The low-carbon SAE steels are used for carburized parts, cold-headed bolts and rivets, and for similar applications where high quality is required. The SAE 1100 series are low-carbon **free-cutting** steels for high-speed screw-machine stock and other machining purposes. These steels have high sulfur present in the steel in the form of manganese sulfide inclusions causing the chips to break short on machining. Manganese and phosphorus harden and embrittle the steel, which also contributes toward free machining. The high manganese contents are intended to ensure that all the sulfur is present as manganese sulfide. Lead was a common additive, but there are environmental problems; experiments have been conducted with selenium, tellurium, and bismuth as replacements.

Cold-finished carbon-steel bars are used for bolts, nuts, typewriter and cash register parts, motor and transmission power shafting, piston pins, bushings, oil-pump shafts and gears, etc. Representative mechanical properties of cold-drawn steel are given in Table 6.2.9. Besides

Table 6.2.9   Representative Average Mechanical Properties of Cold-Drawn Steel

SAE no.	Tensile strength		Yield strength		Elongation in 50 mm (2 in), %	Reduction of area, %	Brinell hardness
	MPa	1,000 lb/in²	MPa	1,000 lb/in²			
1010	462	67	379	55.0	25.0	57	137
1015	490	71	416	60.3	22.0	55	149
1020	517	75	439	63.7	20.0	52	156
1025	552	80	469	68.0	18.5	50	163
1030	600	87	509	73.9	17.5	48	179
1035	634	92	539	78.2	17.0	45	187
1040	669	97	568	82.4	16.0	40	197
1045	703	102	598	86.7	15.0	35	207
1117	552	80	469	68.0	19.0	51	163
1118	569	82.5	483	70.1	18.5	50	167
1137	724	105	615	89.2	16.0	35	217
1141	772	112	656	95.2	14.0	30	223

Sizes 16 to 50 mm (⅝ to 2 in) diam, test specimens 50 × 13 mm (2 × 0.505 in).
SOURCE: ASM "Metals Handbook."

improved mechanical properties, cold-finished steel has better machining properties than hot-rolled products. The surface finish and dimensional accuracy are also greatly improved by cold finishing.

Forging steels, at one time between 0.30 and 0.40 percent carbon and used for axles, bolts, pins, connecting rods, and similar applications, can now contain up to 0.7 percent C with microalloying additions to refine the structure (for better toughness) and to deliver precipitation strengthening. In some cases, air cooling can be used, thereby saving the cost of alloy steels. These steels are readily forged and, after heat treatment, develop considerably better mechanical properties than low-carbon steels. For heavy sections where high strength is required, such as in crankshafts and heavy-duty gears, the carbon may be increased and sufficient alloy content may be necessary to obtain the desired hardenability.

TOOL STEELS

The application of tool steels can generally be fitted into one of the following categories or types of operations: cutting, shearing, forming, drawing, extruding, rolling, and battering. Each of these operations requires in the tool steel a particular physical property or a combination of such metallurgical characteristics as hardness, strength, toughness, wear resistance, and resistance to heat softening, before optimum performance can be realized. These considerations are of prime importance in tool selection; but hardenability, permissible distortion, surface decarburization during heat treatment, and machinability of the tool steel are a few of the additional factors to be weighed in reaching a final decision. In actual practice, the final selection of a tool steel represents a compromise of the most desirable physical properties with the best overall economic performance. Tool steels have been identified and classified by the SAE into six major groups, based upon quenching methods, applications, special characteristics, and use in specific industries. These six classes are water-hardening, shock-resisting, cold-work, hot-work, high-speed, and special-purpose tool steels. A simplified classification of these six basic types and their subdivisions is given in Table 6.2.10.

Water-hardening tool steels, containing 0.60 to 1.40 percent carbon, are widely used because of their low cost, good toughness, and excellent machinability. They are shallow-hardening steels, unsuitable for nondeforming applications because of high warpage, and possess poor resistance to softening at elevated temperatures. Water-hardening tool steels have the widest applications of all major groups and are used for files, twist drills, shear knives, chisels, hammers, and forging dies.

Shock-resisting tool steels, with chromium-tungsten, silicon-molybdenum, or silicon-manganese as the dominant alloys, combine good hardenability with outstanding toughness. A tendency to distort easily is their greatest disadvantage. However, oil quenching can minimize this characteristic.

Table 6.2.10   Simplified Tool-Steel Classification*

Major grouping	Symbol	Types
Water-hardening tool steels	W	
Shock-resisting tool steels	S	
Cold-work tool steels	O	Oil hardening
	A	Medium-alloy air hardening
	D	High-carbon, high-chromium
Hot-work tool steels	H	H10–H19 chromium base
		H20–H39 tungsten base
		H40–H59 molybdenum base
High-speed tool steels	T	Tungsten base
	M	Molybdenum base
Special-purpose tool steels	F	Carbon-tungsten
	L	Low-alloy
	P	Mold steels
		P1–P19 low carbon
		P20–P39 other types

* Each subdivision is further identified as to type by a suffix number which follows the letter symbol.

Cold-work tool steels are divided into three groups: oil-hardening, medium-alloy air-hardening, and high-carbon, high-chromium. In general, this class possesses high wear resistance and hardenability, develops little distortion, but is only average in toughness and in resistance to heat softening. Machinability ranges from good in the oil-hardening grade to poor in the high-carbon, high-chromium steels.

Hot-work tool steels are either chromium- or tungsten-based alloys possessing good nondeforming, hardenability, toughness, and resistance to heat-softening characteristics, with fair machinability and wear resistance. Either air or oil hardening can be employed. Applications are blanking, forming, extrusion, and casting dies where temperatures may rise to 540°C (1,000°F).

High-speed tool steels, the best-known tool steels, possess the best combination of all properties except toughness, which is not critical for high-speed cutting operations, and are either tungsten or molybdenumbase types. Cobalt is added in some cases to improve the cutting qualities in roughing operations. They retain considerable hardness at a red heat. Very high heating temperatures are required for the heat treatment of high-speed steel and, in general, the tungsten-cobalt high-speed steels require higher quenching temperatures than the molybdenum steels. High-speed steel should be tempered at about 595°C (1,100°F) to increase the toughness; owing to a secondary hardening effect, the hardness of the tempered steels may be higher than as quenched.

Special-purpose tool steels are composed of the low-carbon, low-alloy, carbon-tungsten, mold, and other miscellaneous types.

SPRING STEEL

For small springs, steel is often supplied to spring manufacturers in a form that requires no heat treatment except perhaps a low-temperature anneal to relieve forming strains. Types of previously treated steel wire for small helical springs are **music wire** which has been given a special heat treatment called patenting and then cold-drawn to develop a high yield strength, **hard-drawn wire** which is of lower quality than music wire since it is usually made of lower-grade material and is seldom patented, and **oil-tempered wire** which has been quenched and tempered. The wire usually has a Brinell hardness between 352 and 415, although this will depend on the application of the spring and the severity of the forming operation. Steel for small flat springs has either been cold-rolled or quenched and tempered to a similar hardness.

Steel for both helical and flat springs which is hardened and tempered after forming is usually supplied in an annealed condition. Plain carbon steel is satisfactory for small springs; for large springs it is necessary to use alloy steels such as chrome-vanadium or silicon-manganese steel in order to obtain a uniform structure throughout the cross section. It is especially important for springs that the surface of the steel be free from all defects and decarburization, which lowers fatigue strength.

SPECIAL ALLOY STEELS

Many steel alloys with compositions tailored to specific requirements are reported periodically. Usually, the compositions and/or treatments are patented, and they are most likely to have registered trademarks and trade names. They are too numerous to be dealt with here in any great detail, but a few of the useful properties exhibited by some of those special alloys are mentioned.

Iron-silicon alloys with minimum amounts of both carbon and other alloying elements have been used by the electrical equipment industry for a long time; often these alloys are known as **electrical sheet steel.** Iron-nickel alloys with high proportions of nickel, and often with other alloying elements, elicit properties such as nonmagnetic behavior, high permeability, low hysteresis loss, low coefficient of expansion, etc.; iron-cobalt alloys combined with other alloying elements can result in materials with low resistivity and high hysteresis loss. The reader interested in the use of materials with some of these or other desired properties is directed to the extensive literature and data available.

STAINLESS STEELS
by James D. Redmond

REFERENCES: "Design Guidelines for the Selection and Use of Stainless Steel," Specialty Steel Institute of North America (SSINA), Washington, DC. Publications of the Nickel Development Institute, Toronto, Ontario, Canada. "Metals Handbook," 10th ed., ASM International. ASTM Standards.

When the chromium content is increased to about 11 percent in an iron-chromium alloy, the resulting material is generally classified as a stainless steel. With that minimum quantity of chromium, a thin, protective, passive film forms spontaneously on the steel. This passive film acts as a barrier to prevent corrosion. Further increases in chromium content strengthen the passive film and enable it to repair itself if it is damaged in a corrosive environment. Stainless steels are also heat-resistant because exposure to high temperatures (red heat and above) causes the formation of a tough oxide layer which retards further oxidation.

Other alloying elements are added to stainless steels to improve corrosion resistance in specific environments or to modify or optimize mechanical properties or characteristics. Nickel changes the crystal structure to improve ductility, toughness, and weldability. Nickel improves corrosion resistance in reducing environments such as sulfuric acid. Molybdenum increases pitting and crevice corrosion resistance in chloride environments. Carbon and nitrogen increase strength. Aluminum and silicon improve oxidation resistance. Sulfur and selenium are added to improve machinability. Titanium and niobium (columbium) are added to prevent **sensitization** by preferentially combining with carbon and nitrogen, thereby preventing intergranular corrosion.

Stainless Steel Grades   There are more than 200 different **grades** of stainless steel. Each is alloyed to provide a combination of corrosion resistance, heat resistance, mechanical properties, or other specific characteristics. There are five **families** of stainless steels: austenitic, ferritic, duplex, martensitic, and precipitation hardening. Table 6.2.11 is a representative list of stainless steel grades by family and chemical composition as typically specified in ASTM Standards. The mechanical properties of many of these grades are summarized in Table 6.2.12. Note that the number designation of a stainless steel does not describe its performance when utilized to resist corrosion.

Austenitic stainless steels are the most widely used, and while most are designated in the 300 series, some of the highly alloyed grades, though austenitic, have other identifying grade designations, e.g., alloy 20, 904L, and the 6 percent molybdenum grades. These latter are often known by their proprietary designation. These stainless steel grades are available in virtually all wrought product forms, and many are employed as castings. Some austenitic grades in which manganese and nitrogen are substituted partially for nickel are also designated in the 200 series. Types 304 (18 Cr, 8 Ni) and 316 (17 Cr, 10 Ni, 2 Mo) are the workhorse grades, and they are utilized for a broad range of equipment and structures. Austenitic grades can be hardened by cold work but not by heat treatment; cold work increases strength with an accompanying decrease in ductility. They provide excellent corrosion resistance, respond very well to forming operations, and are readily welded. When fully annealed, they are not magnetic, but may become slightly magnetic when cold-worked. Compared to carbon steel, they have higher coefficients of thermal expansion and lower thermal conductivities. When austenitic grades are employed to resist corrosion, their performance may vary over a wide range, depending on the particular corrosive media encountered. As a general rule, increased levels of chromium, molybdenum, and nitrogen result in increased resistance to pitting and crevice corrosion in chloride environments. Type 304 is routinely used for atmospheric corrosion resistance and to handle lowchloride potable water. At the other end of the spectrum are the 6% Mo austenitic stainless steels, which have accumulated many years of service experience handling seawater in utility steam condensers and in piping systems on offshore oil and gas platforms. Highly alloyed austenitic stainless steels are subject to precipitation reactions in the range of 1,500 to 1,800°F (815 to 980°C), resulting in a reduction of their corrosion resistance and ambient-temperature impact toughness.

Low-carbon grades, or **L grades,** are restricted to very low levels of carbon (usually 0.030 wt % maximum) to reduce the possibility of sensitization due to chromium carbide formation either during welding or when exposed to a high-temperature thermal cycle. When a sensitized stainless steel is exposed subsequently to a corrosive environment, intergranular corrosion may occur. Other than improved resistance to intergranular corrosion, the low-carbon grades have the same resistance to pitting, crevice corrosion, and chloride stress corrosion cracking as the corresponding grade with the higher level of carbon (usually 0.080 wt % maximum).

There is a trend to dual-certify some pairs of stainless-steel grades. For example, an 18 Cr–8 Ni stainless steel low enough in carbon to meet the requirements of an L grade and high enough in strength to qualify as the standard carbon version may be **dual-certified.** ASTM specifications allow such a material to be certified and marked, for example, 304/304L and S30400/S30403.

Ferritic stainless steels are iron-chromium alloys with 11 to 30 percent chromium. They can be strengthened slightly by cold working. They are magnetic. They are difficult to produce in plate thicknesses, but are readily available in sheet and bar form. Some contain molybdenum for improved corrosion resistance in chloride environments. They exhibit excellent resistance to chloride stress corrosion cracking. Resistance to pitting and crevice corrosion is a function of the total chromium and molybdenum content. The coefficient of thermal expansion is similar to that of carbon steel. The largest quantity produced is Type 409 (11 Cr), used extensively for automobile catalytic converters, mufflers, and exhaust system components. The most highly alloyed ferritic stainless steels have a long service history in seawater-cooled utility condensers.

Table 6.2.11 ASTM/AWS/AMS Chemical Composition (Wt. Pct)[a]

UNS number	Grade	C	Cr	Ni	Mo	Cu	N	Other
				Austenitic stainless steels				
S20100	201	0.15	16.00–18.00	3.50–5.50	—[b]	—	0.25	Mn: 5.50–7.50
S20200	202	0.15	17.00–19.00	4.00–6.00	—	—	0.25	Mn: 7.50–10.00
S30100	301	0.15	16.00–18.00	6.00–8.00	—	—	0.50	—
S30200	302	0.15	17.00–19.00	8.00–10.00	—	—	0.10	—
S30300	303	0.15	17.00–19.00	8.00–10.00	—	—	—	S: 0.15 min
S30323	303Se	0.15	17.00–19.00	8.00–10.00	—	—	—	Se: 0.15 min
S30400	304	0.08	18.00–20.00	8.00–10.50	—	—	—	—
S30403	304L	0.030	18.00–20.00	8.00–12.00	—	—	0.10	—
S30409	304H	0.04–0.10	18.00–20.00	8.00–10.50	—	—	0.10	—
S30451	304N	0.08	18.00–20.00	8.00–10.50	—	—	0.10–0.16	—
S30500	305	0.12	17.00–19.00	10.50–13.00	—	—	—	—
S30800	308	0.08	19.00–21.00	10.00–12.00	—	—	—	—
S30883	308L	0.03	19.50–22.00	9.00–11.00	0.05	0.75	—	Si: 0.39–0.65
S30908	309S	0.08	22.00–24.00	12.00–15.00	—	—	—	—
S31008	310S	0.08	24.00–26.00	19.00–22.00	—	—	—	—
S31600	316	0.08	16.00–18.00	10.00–14.00	2.00–3.00	—	0.10	—
S31603	316L	0.030	16.00–18.00	10.00–14.00	2.00–3.00	—	0.10	—
S31609	316H	0.04–0.10	16.00–18.00	10.00–14.00	2.00–3.00	—	0.10	—
S31703	317L	0.030	18.00–20.00	11.00–15.00	3.00–4.00	—	0.10	—
S31726	317LMN	0.030	17.00–20.00	13.50–17.50	4.0–5.0	—	0.10–0.20	—
S32100	321	0.080	17.00–19.00	9.00–12.00	—	—	0.10	Ti: 5(C + N) min; 0.70 max
S34700	347	0.080	17.00–19.00	9.00–13.00	—	—	—	Nb: 10C min; 1.00 max; Nb + Ta: 8C min; 1.00 max
N08020	Alloy 20	0.07	19.00–21.00	32.00–38.00	2.00–3.00	3.00–4.00	—	—
N08904	904L	0.020	19.00–23.00	23.00–28.00	4.0–5.0	1.0–2.0	0.10	—
S31254	254 SMQ[c]	0.020	19.50–20.50	17.50–18.50	6.00–6.50	0.50–1.00	0.18–0.22	—
N08367	AL-6XN[d]	0.030	20.00–22.00	23.50–25.50	6.00–7.00	0.75	0.18–0.25	—
N08926	25-6MO/1925hMo[f]	0.020	19.00–21.00	24.00–26.00	6.0–7.0	0.5–1.0	0.15–0.25	—
S32654	654 SMO[c]	0.020	24.00–25.00	21.00–23.00	7.00–8.00	0.30–0.60	0.45–0.55	Mn: 2.00–4.00
				Duplex stainless steels				
S31260	DP-3[g]	0.030	24.0–26.0	5.50–7.50	2.50–3.50	0.20–0.80	0.10–0.30	W 0.10–0.50
S31803		0.030	21.0–23.0	4.50–6.50	2.50–3.50	—	0.08–0.20	—
S32003	AL-2003[d]	0.030	19.5–22.5	3.0–4.0	1.50–2.00	—	0.14–0.20	—
S32101	LDX 2101[c]	0.040	21.0–22.0	1.35–1.70	0.10–0.80	—	0.20–0.25	—
S32205	2205	0.030	21.0–23.0	4.50–6.50	2.50–3.50	—	0.14–0.20	—
S32304	2304	0.030	21.5–24.5	3.00–5.00	0.05–0.60	0.05–0.50	0.05–0.20	—
S32550	Ferralium 255[h]	0.040	24.0–27.0	4.50–6.50	2.90–3.90	1.5–2.5	0.10–0.25	—
S32750	2507	0.030	24.0–26.0	6.00–8.00	3.00–5.00	0.50	0.24–0.32	—
S32900	329	0.08	23.00–28.00	2.50–5.00	1.00–2.00	—	—	—

Ferritic stainless steels

UNS	Alloy	C	Cr	Ni	Mo	Cu	N	Other
S40500	405	0.08	11.50–14.50	0.60	—	—	—	Al: 0.10–0.30
S40900	409	0.08	10.50–11.75	0.50	—	—	—	Ti: 6xC Min; 0.75 max
S43000	430	0.12	16.00–18.00	—	—	—	—	—
S43020	430F	0.12	16.00–18.00	—	—	—	—	S: 0.15 min
S43035	439	0.07	17.00–19.00	0.5	—	—	0.04	Ti: 0.20 + 4(C + N) min; 1.10 max Al: 0.15 max
S43400	434	0.12	16.00–18.00	—	0.75–1.25	—	—	—
S44400	444 (18 Cr-2 Mo)	0.025	17.5–19.5	1.00	1.75–2.50	—	0.035	Ti + Nb: 0.20 + 4(C + N) min; 0.80 max
S44600	446	0.20	23.00–27.50	—	—	—	0.25	Nb: 0.05–0.20
S44627	E-BRITE 26-1[d]	0.01	25.00–27.00	—	0.75–1.50	0.20	0.015	Ti + Nb: 0.20–1.00 and 6(C + N) min
S44660	SEA-CURE[i]	0.030	25.00–28.00	1.0–3.50	3.00–4.00	—	0.040	Ti + Nb: 0.20–1.00 and 6(C + N) min
S44735	AL 29-4C[c]	0.030	28.00–30.00	1.00	3.60–4.20	—	0.045	Ti + Nb: 0.20–1.00 and 6(C + N) min

Martensitic stainless steels

UNS	Alloy	C	Cr	Ni	Mo	Cu	N	Other
S40300	403	0.15	11.50–13.50	—	—	—	—	—
S41000	410	0.15	11.50–13.50	0.75	—	—	—	—
S41008	410S	0.08	11.50–13.50	0.60	—	—	—	—
S41600	416	0.15	12.00–14.00	—	—	—	—	—
S41623	416Se	0.15	12.00–14.00	—	—	—	—	Se: 0.15 min
S42000	420	0.15 min	12.00–14.00	—	0.60	—	—	—
S42020	420F	0.08	12.00–14.00	0.50	—	—	—	S: 0.15 min
S43100	431	0.20	15.00–17.00	1.25–2.50	—	—	—	—
S44002	440A	0.60–0.75	16.00–18.00	—	0.75	—	—	—
S44003	440B	0.75–0.95	16.00–18.00	—	0.75	—	—	—
S44004	440C	0.95–1.20	16.00–18.00	—	0.75	—	—	—
S44020	440F	0.95–1.20	16.00–18.00	0.50	0.6	—	—	S: 0.15 min

Precipitation-hardening stainless steels

UNS	Alloy	C	Cr	Ni	Mo	Cu	N	Other
S13800	XM-13/13-8Mo PH	0.05	12.25–13.25	7.50–8.50	2.00–2.50	—	0.01	Al: 0.90–1.35
S15700	632/15-7PH	0.09	14.00–16.00	6.50–7.75	2.00–3.00	—	—	Al: 0.75–1.00
S17400	630/17-4PH	0.07	15.00–17.50	3.00–5.00	—	3.00–5.00	—	Nb + Ta: 0.15–0.45
S17700	631/17-7PH	0.09	16.00–18.00	6.50–7.75	—	—	—	Al: 0.75–1.00
S35000	AMS 350	0.07–0.11	16.00–17.00	4.00–5.00	2.50–3.25	—	0.07–0.13	—
S35500	634/AMS 355	0.10–0.15	15.00–16.00	4.00–5.00	2.50–3.25	—	0.07–0.13	Mn: 0.50–1.25
S45000	XM-25/Custom 450[j]	0.05	14.00–16.00	5.00–7.00	0.50–1.00	1.25–1.75	—	Nb: 8C min
S45500	XM-16/Custom 455[j]	0.03	11.00–12.50	7.50–9.50	0.5	1.50–2.50	—	Ti: 0.90–1.40

[a] Maximum unless range or minimum is indicated.
[b] None required in the specification.
[c] Trademark of Outokumu.
[d] Trademark of Allegheny Ludlum Corp.
[e] Trademark of Special Metals.
[f] Trademark of Thyssen Krupp VDM.
[g] Trademark of Sumitomo Metals.
[h] Trademark of Langley Alloys Ltd.
[i] Trademark of Crucible Materials Corp.
[j] Trademark of Carpenter Technology Corp.

Duplex stainless steels combine some of the best features of the austenitic and the ferritic grades. Duplex stainless steels have excellent chloride stress corrosion cracking resistance and can be produced in the full range of product forms typical of the austenitic grades. Duplex stainless steels are comprised of approximately 50 percent ferrite and 50 percent austenite. They are magnetic. Their yield strength is about double that of the 300-series austenitic grades, so economies may be achieved from reduced piping and vessel wall thicknesses. The coefficient of thermal expansion of duplex stainless steels is similar to that of carbon steel and about 30 to 40 percent less than that of the austenitic grades. The general-purpose duplex grade is 2205 (22 Cr, 5 Ni, 3 Mo, 0.15 N). Because they suffer embrittlement with prolonged high-temperature exposure, ASME constructions using the duplex grades are limited to a maximum 600°F (315°C) service temperature. Duplex stainless steel names often reflect their chemical composition or are proprietary designations.

Martensitic stainless steels are straight chromium grades with relatively high levels of carbon. The martensitic grades can be strengthened by heat treatment. To achieve the best combination of strength, corrosion resistance, ductility, and impact toughness, they are tempered in the range of 300 to 700°F (150 to 370°C). They have a 400-series designation; 410 is the general-purpose grade. They are magnetic. The martensitic grades containing up to about 0.15 wt % carbon—e.g., grades 403, 410, and 416 (a free-machining version) can be hardened to about 45 RC. The high-carbon martensitics, such as Types 440A, B, and C, can be hardened to about 60 RC. The martensitics typically exhibit excellent wear or abrasion resistance but limited corrosion resistance. In most environments, the martensitic grades have less corrosion resistance than Type 304.

Precipitation-hardening stainless steels can be strengthened by a relatively low-temperature heat treatment. The low-temperature heat treatment minimizes distortion and oxidation associated with higher-temperature heat treatments. They can be heat-treated to strengths greater than can the martensitic grades. Most exhibit corrosion resistance superior to the martensitics and approach that of Type 304. While some precipitation-hardening stainless steels have a 600-series designation, they are most frequently known by names which suggest their chemical composition, for example, 17–4PH, or by proprietary names.

The commonly used stainless steels have been approved for use in ASME boiler and pressure vessel construction. Increasing the carbon content increases the maximum **allowable stress values.** Nitrogen additions are an even more powerful strengthening agent. This may be seen by comparing the allowable stresses for 304L (0.030 C maximum), 304H (0.040 C minimum), and 304N (0.10 N minimum) in Table 6.2.13. In general, increasing the total alloy content—especially chromium, molybdenum, and nitrogen—increases the allowable design stresses. However, precipitation reactions which can occur in the most highly alloyed grades limit their use to a maximum of about 750°F (400°C). The duplex grade 2205 has higher allowable stress values than even the most highly alloyed of the austenitic grades, up to a maximum service temperature of 600°F (315°C). The precipitation-hardening grades exhibit some of the highest allowable stress values but also are limited to about 650°F (345°C).

Copious amounts of information and other technical data regarding physical and mechanical properties, examples of applications and histories of service lives in specific corrosive environments, current costs, etc. are available in the references and elsewhere in the professional and trade literature.

Table 6.2.12   ASTM Mechanical Properties (A 240, A 276, A 479, A 564, A 582)

UNS no.	Grade	Condition*	Tensile strength, min ksi	Tensile strength, min MPa	Yield strength, min ksi	Yield strength, min MPa	Elongation in 2 in or 50 mm min, %	Hardness, max Brinell	Hardness, max Rockwell B†
Austenitic stainless steels									
S20100	201	Annealed	95	655	38	260	40.0	—‡	95
S20200	202	Annealed	90	620	38	260	40.0	241	—
S30100	301	Annealed	75	515	30	205	40.0	217	95
S30200	302	Annealed	75	515	30	205	40.0	201	92
S30300	303	Annealed	—	—	—	—	—	262	—
S30323	303Se	Annealed	—	—	—	—	—	262	—
S30400	304	Annealed	75	515	30	205	40.0	201	92
S30403	304L	Annealed	70	485	25	170	40.0	201	92
S30409	304H	Annealed	75	515	30	205	40.0	201	92
S30451	304N	Annealed	80	550	35	240	30.0	201	92
S30500	305	Annealed	75	515	30	205	40.0	183	88
S30908	309S	Annealed	75	515	30	205	40.0	217	95
S31008	310S	Annealed	75	515	30	205	40.0	217	95
S31600	316	Annealed	75	515	30	205	40.0	217	95
S31603	316L	Annealed	70	485	25	170	40.0	217	95
S31609	316H	Annealed	75	515	30	205	40.0	217	95
S31703	317L	Annealed	75	515	30	205	40.0	217	95
S31726	317LMN	Annealed	80	550	35	240	40.0	223	96
S32100	321	Annealed	75	515	30	205	40.0	217	95
S34700	347	Annealed	75	515	30	205	40.0	201	92
N08020	Alloy 20	Annealed	80	551	35	241	30.0	217	95
N08904	904L	Annealed	71	490	31	215	35.0	—	—
S31254	254 SMO	Annealed	94	650	44	300		223	96
N08367	AL-6XN	Annealed	95	655	45	310	30.0	233	—
N08926	25-6MO/1925hMo	Annealed	94	650	43	295	35.0	—	—
S32654	654 SMO	Annealed	109	750	62	430	40.0	250	—
Duplex stainless steels									
S31260	DP-3	Annealed	100	690	70	485	20.0	290	—
S31803	—	Annealed	90	620	65	450	25.0	293	31HRC
S32003	AL 2003	Annealed	90	620	65	450	25.0	293	31HRC
S32101	LDX 2101	Annealed	94	650	65	450	30.0	290	—
S32205	2205	Annealed	95	655	65	450	25.0	293	31HRC

Table 6.2.12   ASTM Mechanical Properties (A 240, A 276, A 479, A 564, A 582)   (*Continued*)

UNS no.	Grade	Condition*	Tensile strength, min		Yield strength, min		Elongation in 2 in or 50 mm min, %	Hardness, max	
			ksi	MPa	ksi	MPa		Brinell	Rockwell B†
S32304	2304	Annealed	87	600	58	400	25.0	290	32HRC
S32550	Ferralium 255	Annealed	110	760	80	550	15.0	302	32HRC
S32750	2507	Annealed	116	795	80	550	15.0	310	32HRC
S32900	329	Annealed	90	620	70	485	15.0	269	28HRC
			Ferritic stainless steels						
S40500	405	Annealed	60	415	25	170	20.0	179	88
S40900	409	Annealed	55	380	25	205	20.0	179	88
S43000	430	Annealed	65	450	30	205	22.0	183	89
S43020	430F	Annealed	—	—	—	—	—	262	—
S43035	439	Annealed	60	415	30	205	22.0	183	89
S43400	434	Annealed	65	450	35	240	22.0	—	—
S44400	444(18 Cr–2 Mo)	Annealed	60	415	40	275	20.0	217	96
S44600	446	Annealed	65	515	40	275	20.0	217	96
S44627	E-BRITE 26-1	Annealed	65	450	40	275	22.0	187	90
S44660	SEA-CURE	Annealed	85	585	65	450	18.0	241	100
S44735	AL 29-4C	Annealed	80	550	60	415	18.0	255	25HRC
			Martensitie stainless steels						
S40300	403	Annealed	70	485	40	275	20.0	223	—
S41000	410	Annealed	65	450	30	205	20.0	217	96
S41008	410S	Annealed	60	415	30	205	22.0	183	89
S41600	416	Annealed	—	—	—	—	—	262	—
S41623	416Se	Annealed	—	—	—	—	—	262	—
S42000	420	Annealed	—	—	—	—	—	241	—
S42020	420F	Annealed	—	—	—	—	—	262	—
S43100	431	Annealed	—	—	—	—	—	285	—
S44002	440A	Annealed	—	—	—	—	—	269	—
S44003	440B	Annealed	—	—	—	—	—	269	—
S44004	440C	Annealed	—	—	—	—	—	269	—
S44020	440F	Annealed	—	—	—	—	—	285	—
			Precipitation-hardening stainless steels						
S13800	XM-13/13-8Mo PH	Annealed	—	—	—	—	—	363	38HRC
		H950	220	1,520	205	1,410	10.0	430	45HRC
		H1000	205	1,410	190	1,310	10.0	400	43HRC
		H1050	175	1,210	165	1,140	12.0	372	40HRC
		H1150	135	930	90	620	14.0	283	30HRC
S15700	632/15-7PH	Annealed	—	—	—	—	—	269	100
		RH950	200	1,380	175	1,210	7.0	415	—
S17400	630/17-4PH	Annealed	—	—	—	—	—	363	38HRC
		H900	190	1,310	170	1,170	10.0	388	40HRC
		H925	170	1,170	155	1,070	10.0	375	38HRC
		H1025	155	1,070	145	1,000	12.0	331	35HRC
		H1100	140	965	115	795	14.0	302	31HRC
S17700	631/17-7PH	Annealed	—	—	—	—	—	229	98
		RH950	185	1,275	150	1,035	6.0	388	41HRC
S35500	634/AMS 355	Annealed	—	—	—	—	—	363	—
		H1000	170	1,170	155	1,070	12.0	341	37HRC
S45000	XM-25/Custom 450	Annealed	130	895	95	655	10.0	321	32HRC
		H900	180	1,240	170	1,170	10.0	363	39HRC
		H950	170	1,170	160	1,100	10.0	341	37HRC
		H1000	160	1,100	150	1,030	12.0	331	36HRC
		H1100	130	895	105	725	16.0	285	30HRC
S45500	XM-16/Custom 455	Annealed	—	—	—	—	—	331	36HRC
		H900	235	1,620	220	1,520	8.0	444	47HRC
		H950	220	1,520	205	1,410	10.0	415	44HRC
		H1000	205	1,410	185	1,280	10.0	363	40HRC

* Condition defined in applicable ASTM specification.
† Rockwell B unless hardness Rockwell C (HRC) is indicated.
‡ None required in ASTM specifications.

Table 6.2.13   Maximum Allowable Stress Values (ksi), Plate: ASME Section I; Section III, Classes 2 and 3; Section VIII, Division 1

UNS no.	Grade	−20 to 100°F	300°F	400°F	500°F	600°F	650°F	700°F	750°F	800°F	850°F	900°F
					Austenitic stainless steels							
S30403	304L	16.7	12.8	11.7	10.9	10.3	10.5	10.0	9.8	9.7	—*	—
S30409	304H	18.8	14.1	12.9	12.1	11.4	11.2	11.1	10.8	10.6	10.4	10.2
S30451	304N	20.0	16.7	15.0	13.9	13.2	13.0	12.7	12.5	12.3	12.1	11.8
S31603	316L	16.7	12.7	11.7	10.9	10.4	10.2	10.0	9.8	9.6	9.4	—
N08904	904L	17.8	15.1	13.8	12.7	12.0	11.7	11.4	—	—	—	—
S31254	254 SMO	23.5	21.4	19.9	18.5	17.9	17.7	17.5	17.3	—	—	—
					Duplex stainless steel							
S31803	—	22.5	21.7	20.9	20.4	20.2	—	—	—	—	—	—
S32101	LDX 2101	26.9	25.6	24.7	24.7	24.7	—	—	—	—	—	—
S32205	2205	25.7	24.8	23.9	23.3	23.1	—	—	—	—	—	—
					Ferritic stainless steels							
S40900	409	13.8	12.7	12.2	11.8	11.4	11.3	11.1	10.7	10.2	—	—
S43000	430	16.3	15.0	14.4	13.9	13.5	13.3	13.1	12.7	12.0	11.3	10.5
					Precipitation-hardening stainless steel							
S17400	630/17-4	35.0	35.0	34.1	33.3	32.8	32.6	—	—	—	—	—

* No allowable stress values at this temperature; material not recommended for service at this temperature.

6.3   IRON AND STEEL CASTINGS
by Malcolm Blair and Robert E. Eppich

REFERENCES: "Metals Handbook," ASM. "Iron Casting Handbook," Iron Casting Society. "Malleable Iron Castings," Malleable Founders' Society. ASTM Specifications. "Ductile Iron Handbook," American Foundry Society. "Modern Casting," American Foundry Society. "Ductile Iron Data for Design Engineers," Quebec Iron and Titanium (QIT). "History Cast in Metal," American Foundry Society. "Steel Castings Handbook," Steel Founders' Society of America.

CLASSIFICATION OF CASTINGS

Cast-Iron Castings   The term **cast iron** covers a wide range of iron-carbon-silicon alloys containing from 2.0 to 4.0 percent carbon and from 0.5 to 3.0 percent silicon. The alloy typically also contains varying percentages of manganese, sulfur, and phosphorus along with other alloying elements such as chromium, molybdenum, copper, and titanium. For a type and grade of cast iron, these elements are specifically and closely controlled. Cast iron is classified into five basic types; each type is generally based on graphite morphology as follows:

Gray iron
Ductile iron
Malleable iron
Compacted graphite iron
White iron

Even though chemistry is important to achieve the type of cast iron, the molten metal processing and cooling rates play major roles in developing each type. Different mechanical properties are generally associated with each of the five basic types of cast iron.

Figure 6.3.1 illustrates the range of carbon and silicon for each type. The chemistry for a specific grade within each type is usually up to the foundry producing the casting. There are situations where the chemistry for particular elements is specified by ASTM, SAE, and others; this should always be fully understood prior to procuring the casting.

Steel Castings   There are two main classes of steel castings: carbon and low-alloy steel castings, and high-alloy steel castings. These classes may be broken down into the following groups: (1) **low-carbon steels** (C < 0.20 percent), (2) **medium-carbon steels** (C between 0.20 and

Fig. 6.3.1   Approximate ranges of carbon and silicon for steels and various cast irons. (*QIT-Fer Titane, Inc.*)

0.50 percent), (3) **high-carbon steels** (C > 0.50 percent), (4) **low-alloy steels** (total alloy ≤ 8 percent), and (5) **high-alloy steels** (total alloy > 8 percent). The tensile strength of cast steel varies from 60,000 to 250,000 lb/in² (400 to 1,700 MPa) depending on the composition and heat treatment. Steel castings are produced weighing from ounces to over 200 tons. They find universal application where strength, toughness, and reliability are essential.

CAST IRON

The engineering and physical properties of cast iron vary with the type of iron; the designer must match the engineering requirements with all the properties of the specific type of cast iron being considered. Machinability, e.g., is significantly affected by the type of cast iron specified. Gray cast iron is the most machinable; the white cast irons are the least machinable.

Composition The properties of cast iron are controlled primarily by the graphite morphology and the quantity of graphite. Also to be considered is the matrix microstructure, which may be established either during cooling from the molten state (as-cast condition) or as a result of heat treatment. Except where previous engineering evaluations have established the need for specific chemistry or the cast iron is to be produced to a specific ASTM specification, such that chemistry is a part of that specification, the foundry normally chooses the composition that will meet the specified mechanical properties and/or graphite morphology.

Types of Cast Iron

Gray Iron Gray iron castings have been produced since the sixth century B.C. During solidification, when the composition and cooling rate are appropriate, carbon will precipitate in the form of graphite flakes that are interconnected within each eutectic cell. Figure 6.3.2 is a photomicrograph of a typical gray iron structure. The flakes appear black in the photograph. The sample is unetched; therefore the matrix microstructure does not show. Gray iron fractures primarily along the graphite. The fracture appears gray; hence the name *gray iron*. This graphite morphology establishes the mechanical properties. The cooling rate of the casting while it solidifies plays a major role in the size and shape of the graphite flakes. Thus the mechanical properties in a single casting can vary by as much as 10,000 lb/in² with thinner sections having higher strength (Fig. 6.3.3).

The flake graphite imparts unique characteristics to gray iron, such as excellent machinability, ability to resist galling, excellent wear resistance (provided the matrix microstructure is primarily pearlite), and excellent damping capacity. This same flake graphite exhibits essentially zero ductility and very low impact strength.

Mechanical Properties Gray iron **tensile strength** normally ranges from 20,000 to 60,000 lb/in² and is a function of chemistry, graphite morphology, and matrix microstructure. ASTM A48 recognizes nine grades based on the tensile strength of a separately cast test bar whose dimensions are selected to be compatible with the critical controlling section

Fig. 6.3.3 The influence of casting section thickness on the tensile strength and hardness for a series of gray irons classified by their strength as cast in 1.2-in- (30-mm-) diameter, B bars. *(Steel Founders Society of America.)*

size of the casting. These dimensions are summarized in Table 6.3.1. Test bars are machined to the specified size prior to testing. Table 6.3.2 summarizes the various specifications under ASTM A48. Most engineering applications utilize gray iron with tensile strengths of 30,000 to 35,000 lb/in² (typically a B bar). An alternate method of specifying gray iron is to utilize SAE J431. This specification, shown in Table 6.3.3, is based on hardness (BHN) rather than tensile test bars.

Yield strength is not normally shown and is not specified because it is not a well-defined property of the material. There is no abrupt change in the stress-strain relation as is normally observed in other material. A proof stress or load resulting in 0.1 percent permanent strain will range form 65 to 80 percent of the tensile strength.

Table 6.3.4 summarizes the mechanical properties of gray iron as they are affected by a term called the **carbon equivalent.** Carbon equivalent is often represented by the equation

$$CE = \text{carbon (C)} + \tfrac{1}{3}\,\text{silicon (Si)} + \tfrac{1}{3}\,\text{phosphorus (P)}$$

Typical applications for gray iron are engine blocks, compressor housings, transmission cases, and valve bodies.

Ductile Iron Whereas gray iron has been used as an engineering material for hundreds of years, ductile iron was developed recently. It was invented in the late 1940s and patented by the International Nickel Company. By adding magnesium to the molten metal under controlled conditions, the graphite structure becomes spherical or nodular. The material resulting has been variously named *spheroidal graphite iron, sg iron, nodular iron,* and *ductile iron.* **Ductile iron** is preferred since it also describes one of its characteristics. The structure is shown in Fig. 6.3.4. As would be expected, this nodular graphite structure, when compared to the graphite flakes in gray iron, shows significant improvement in mechanical properties. However, the machinability and thermal conductivity of ductile iron are lower than those for gray iron.

The generally accepted grades of ductile iron are shown in Table 6.3.5, which summarizes ASTM A-536. In all grades the nodularity of the

Fig. 6.3.2 Gray iron with flake graphite. Unetched, 100 ×.

Table 6.3.1   ASTM A48: Separately Cast Test Bars for Use when a Specific Correlation Has Not Been Established between Test Bar and Casting

Thickness of wall of controlling section of casting, in (mm)	Test bar diameter (See Table 6.3.2)
Under 0.25 (6)	S
0.25 to 0.50 (6 to 12)	A
0.51 to 1.00 (13 to 25)	B
1.01 to 2 (26 to 50)	C
Over 2 (50)	S

SOURCE: Abstracted from ASTM data, by permission.

Table 6.3.2   ASTM A48: Requirements for Tensile Strength of Gray Cast Irons in Separately Cast Test Bars

Class no.	Tensile strength, min, ksi (MPa)	Nominal test bar diameter, in (mm)
20 A	20 (138)	0.88 (22.4)
20 B		1.2 (30.5)
20 C		2.0 (50.8)
20 S		Bars S*
25 A	25 (172)	0.88 (22.4)
25 B		1.2 (30.5)
25 C		2.0 (50.8)
25 S		Bars S*
30 A	30 (207)	0.88 (22.4)
30 B		1.2 (30.5)
30 C		2.0 (50.8)
30 S		Bars S*
35 A	35 (241)	0.88 (22.4)
35 B		1.2 (30.5)
35 C		2.0 (50.8)
35 S		Bars S*
40 A	40 (276)	0.88 (22.4)
40 B		1.2 (30.5)
40 C		2.0 (50.8)
40 S		Bars S*
45 A	45 (310)	0.88 (22.4)
45 B		1.2 (30.5)
45 C		2.0 (50.8)
45 S		Bars S*
50 A	50 (345)	0.88 (22.4)
50 B		1.2 (30.5)
50 C		2.0 (50.8)
50 S		Bars S*
55 A	55 (379)	0.88 (22.4)
55 B		1.2 (30.5)
55 C		2.0 (50.8)
55 S		Bars S*
60 A	60 (414)	0.88 (22.4)
60 B		1.2 (30.5)
60 C		2.0 (50.8)
60 S		Bars S*

* All dimensions of test bar S shall be as agreed upon between the manufacturer and the purchaser.
SOURCE: Abstracted from ASTM data, by permission.

Fig. 6.3.4 Ductile iron with spherulitic graphite. Unetched, 100 ×.

as cast, but annealing is sometimes used to enhance these properties. Grades 60-40-18 and 60-45-12 have a ferrite microstructure. Both major elements (C and S) and minor elements (Mn, P, Cr, and others) are controlled at the foundry in order to achieve the specified properties. Figure 6.3.5 illustrates the mechanical properties and impact strength as they are affected by temperature and test conditions.

An important new material within the nodular iron family has been developed within the last 20 years. This material, **austempered ductile iron (ADI),** exhibits tensile strengths up to 230,000 lb/in^2, but of greater significance is higher elongation, approximately 10 percent, compared to 2 percent for conventional ductile iron at equivalent tensile strength. The material retains the spherical graphite nodules characteristic of ductile iron, but the matrix microstructure of the heat-treated material now consists of acicular ferrite in a high-carbon austenite. Since 1990, ADI has been specified per ASTM 897. Properties associated with this specification are shown in Table 6.3.6 and are summarized in Fig. 6.3.6. Significant research has resulted in excellent documentation of other important properties such as fatigue resistance (Fig. 6.3.7), abrasion resistance (Fig. 6.3.8), and heat-treating response to specific compositions.

Malleable Iron   Whiteheart malleable iron was invented in Europe in 1804. It was cast as white iron and heat-treated for many days, resulting in a material with many properties of steel. Blackheart malleable iron was invented in the United States by Seth Boyden. This material contains no free graphite in the as-cast condition, but when subjected to an extended annealing cycle (6 to 10 days), the graphite is present as

graphite remains the same. Thus the properties are controlled by the casting microstructure, which is influenced by both the chemistry and the cooling rate after solidification. The grades exhibiting high elongation with associated lower tensile and yield properties can be produced

Table 6.3.3   SAE J431: Grades of Automotive Gray Iron Castings Designed by Brinell Hardness

SAE grade	Specified hardness BHN*	Minimum tensile strength (for design purposes)		Other requirements
		lb/in^2	MPa	
G1800	187 max	18,000	124	
G2500	170–229	25,000	173	
G2500a†	170–229	25,000	173	3.4% min C and microstructure specified
G3000	187–241	30,000	207	
G3500	207–255	35,000	241	
G3500b†	207–255	35,000	241	3.4% min C and microstructure specified
G3500c†	207–255	35,000	241	3.5% min C and microstructure specified
G4000	217–269	40,000	276	

* Hardness at a designated location on the castings.
† For applications such as brake drums, disks, and clutch plates to resist thermal shock.
SOURCE: Abstracted from SAE data, by permission.

Table 6.3.4   Effect of Carbon Equivalent on Mechanical Properties of Gray Cast Irons

Carbon equivalent	Tensile strength, lb/in²*	Modulus of elasticity in tension at ½ load, lb/in² × 10⁶†	Modulus of rapture, lb/in²*	Deflection in 18-in span, in‡	Shear strength, lb/in²*	Endurance limit, lb/in²*	Compressive strength, lb/in²*	Brinell hardness, 3,000-kg load	Izod impact, unnotched, ft · lb
4.8	20,400	8.0	48,300	0.370	29,600	10,000	72,800	146	3.6
4.6	22,400	8.7	49,200	0.251	33,000	11,400	91,000	163	3.6
4.5	25,000	9.7	58,700	0.341	35,500	11,800	95,000	163	4.9
4.3	29,300	10.5	63,300	0.141	37,000	12,300	90,900	179	2.2
4.1	32,500	13.6	73,200	0.301	44,600	16,500	128,800	192	4.2
4.0	35,100	13.3	77,000	0.326	47,600	17,400	120,800	196	4.4
3.7	40,900	14.8	84,200	0.308	47,300	19,600	119,100	215	3.9
3.3	47,700	20.0	92,000	0.230	60,800	25,200	159,000	266	4.4

* × 6.89 = kPa. † × 0.00689 = MPa. ‡ × 25.4 = mm.

Fig. 6.3.5 Effect of temperature on low-temperature properties of ferritic ductile iron. *(QIT-Fer Titane, Inc.)*

Fig. 6.3.6 Austempered ductile iron mechanical properties. *(QIT-Fer Titane, Inc.)*

irregularly shaped nodules called **temper carbon** (Fig. 6.3.9). All malleable iron castings produced currently are of blackheart malleable iron; the terms blackheart and whiteheart are obsolete.

The properties of several grades of malleable iron are shown in Table 6.3.7. The properties are similar to those of ductile iron, but the annealing cycles still range from 16 to 30 h. Production costs are higher than for ductile iron, so that virtually all new engineered products are designed to be of ductile iron. Malleable iron accounts for only 3 percent of total cast-iron production and is used mainly for pipe and electrical fittings and replacements for existing malleable-iron castings.

Compacted Graphite Iron   Compacted graphite iron was developed in the mid-1970s. It is a controlled class of iron in which the graphite takes a form between that in gray iron and that in ductile iron. Figure 6.3.10

shows the typical graphite structure that is developed during casting. It was desired to produce a material that, without extensive alloying, would be stronger than the highest-strength grades of gray iron but would be more machinable than ductile iron. These objectives were met. The properties of the various grades of compacted iron are summarized in Table 6.3.8, which summarizes ASTM A842. Several other properties of this material should be considered. For example, its galling resistance is superior to that of ductile iron, and it is attractive for hydraulic applications; its thermal conductivity is superior to that of ductile iron but inferior to that of gray iron. Compacted graphite iron properties have resulted in its applications to satisfy a number of unique requirements.

White Iron   When white iron solidifies, virtually all the carbon appears in the form of carbides. White irons are hard and brittle, and they break with a white fracture. These irons are usually alloyed with chromium and nickel; Table 6.3.9 summarizes ASTM A532. Their hardness is in the range of 500 to 600 BHN. The specific alloying that is required is a function of section size and application; there must be

Table 6.3.5   ASTM A536: Ductile Irons. Tensile Requirements

	Grade 60-40-18	Grade 65-45-12	Grade 80-55-06	Grade 100-70-03	Grade 120-90-02
Tensile strength, min, lb/in²	60,000	65,000	80,000	100,000	120,000
Tensile strength, min, MPa	414	448	552	689	827
Yield strength, min, lb/in²	40,000	45,000	55,000	70,000	90,000
Yield strength, min, MPa	276	310	379	483	621
Elongation in 2 in or 50 mm, min, %	18	12	6.0	3.0	2.0

SOURCE: Abstracted from ASTM data, by permission.

Table 6.3.6   ASTM A897: Austempered Ductile Irons. Mechanical Property Requirements

	Grade 125/80/10	Grade 150/100/7	Grade 175/125/4	Grade 200/155/1	Grade 230/185/—
Tensile strength, min, ksi	125	150	175	200	230
Yield strength, min, ksi	80	100	125	155	185
Elongation in 2 in, min %	10	7	4	1	*
Impact energy, ft · lb†	75	60	45	25	*
Typical hardness, BHN, kg/mm²‡	269–321	302–363	341–444	388–477	444–555

* Elongation and impact requirements are not specified. Although grades 200/155/1 and 230/185/— are both primarily used for gear and wear resistance applications, Grade 200/155/1 has applications where some sacrifice in wear resistance is acceptable in order to provide a limited amount of ductility and toughness.

† Unnotched charpy bars tested at $72 \pm 7°F$. The values in the table are a minimum for the average of the highest three test values of the four tested samples.

‡ Hardness is not mandatory and is shown for information only.
Source: Abstracted from ASTM data, by permission.

Fig. 6.3.7   Fatigue properties of notched and unnotched ADI. *(QIT-Fer Titane, Inc.)*

and 6.3.13. Castings made from these alloys exhibit performance comparable to many stainless steels.

Fig. 6.3.8   Comparison of pin abrasion test results of ADI, ductile iron, and two abrasion-resistant steels. *(QIT-Fer Titane, Inc.)*

Fig. 6.3.9   Temper carbon form of graphite in malleable iron. Unetched, 100 ×.

coordination between designer and foundry. These irons exhibit outstanding wear resistance and are used extensively in the mining industry for ballmill shell liners, balls, impellers, and slurry pumps.

Specialty Irons   There are a number of highly alloyed gray and ductile iron grades of cast iron. These grades typically contain between 20 and 30 percent nickel along with other elements such as chromium and molybdenum. They have the graphite morphology characteristic of conventional gray or ductile iron, but alloying imparts significantly enhanced performance when wear resistance or corrosion resistance is required. Tables 6.3.10 and 6.3.11 summarize ASTM A436 for austenitic gray iron and ASTM A439 for austenitic ductile iron. This family of alloys is best known by its trade name *Ni-Resist*. Corrosion resistance of several of *Ni-Resist* alloys is summarized in Tables 6.3.12

Table 6.3.7   SAE J158a: Malleable Irons. Mechanical Property Requirements

Grade	Casting hardness	Heat treatment
M 3210	156 BHN max.	Annealed
M 4504	163–217 BHN	Air-quenched and tempered
M 5003	187–241 BHN	Air-quenched and tempered
M 5503	187–241 BHN	Liquid-quenched and tempered
M 7002	229–269 BHN	Liquid-quenched and tempered
M 8501	269–302 BHN	Liquid-quenched and tempered

Source: Abstracted from SAE data, by permission.

Several other specialty irons utilize silicon or aluminum as the major alloying elements. These also find specific application in corrosion or elevated-temperature environments.

Fig. 6.3.10 Compacted graphite iron. Unetched, 100 ×.

Table 6.3.8   ASTM A842: Compacted Graphite Irons. Mechanical Property Requirements

	Grade* 250	Grade 300	Grade 350	Grade 400	Grade† 450
Tensile strength, min, MPa	250	300	350	400	450
Yield strength, min, MPa	175	210	245	280	315
Elongation in 50 mm, min, %	3.0	1.5	1.0	1.0	1.0

* The 250 grade is a ferritic grade. Heat treatment to attain required mechanical properties and microstructure shall be the option of the manufacturer.
† The 450 grade is a pearlitic grade usually produced without heat treatment with addition of certain alloys to promote pearlite as a major part of the matrix.
SOURCE: Abstracted from ASTM data, by permission.

Allowances for Iron Castings

Dimension allowances that must be applied in the production of castings arise from the fact that different metals contract at different rates during solidification and cooling. The shape of the part has a major influence on the as-cast dimensions of the casting. Some of the principal allowances are discussed briefly here.

Shrinkage allowances for iron castings are of the order of ⅛ in/ft (0.1 mm/cm). The indiscriminate adoption of this shrinkage allowance can lead to problems in iron casting dimensions. It is wise to discuss shrinkage allowances with the foundry that makes the castings before patterns are made.

Casting finish allowances (unmachined) and **machine finish allowances** are based on the longest dimension of the casting, although the weight of the casting may also influence these allowances. Other factors which affect allowances include pattern materials and the method of pattern construction. A useful guide is the international standard, ISO 8062

Table 6.3.9   ASTM A532: White Irons. Chemical Analysis Requirements

Class	Type	Designation	Carbon	Manganese	Silicon	Nickel	Chromium	Molybdenum	Copper	Phosphorus	Sulfur
I	A	Ni-Cr-Hc	2.8–3.6	2.0 max	0.8 max	3.3–5.0	1.4–4.0	1.0 max	—	0.3 max	0.15 max
I	B	Ni-Cr-Lc	2.4–3.0	2.0 max	0.8 max	3.3–5.0	1.4–4.0	1.0 max	—	0.3 max	0.15 max
I	C	Ni-Cr-GB	2.5–3.7	2.0 max	0.8 max	4.0 max	1.0–2.5	1.0 max	—	0.3 max	0.15 max
I	D	Ni-HiCr	2.5–3.6	2.0 max	2.0 max	4.5–7.0	7.0–11.0	1.5 max	—	0.10 max	0.15 max
II	A	12% Cr	2.0–3.3	2.0 max	1.5 max	2.5 max	11.0–14.0	3.0 max	1.2 max	0.10 max	0.06 max
II	B	15% Cr-Mo	2.0–3.3	2.0 max	1.5 max	2.5 max	14.0–18.0	3.0 max	1.2 max	0.10 max	0.06 max
II	D	20% Cr-Mo	2.0–3.3	2.0 max	1.0–2.2	2.5 max	18.0–23.0	3.0 max	1.2 max	0.10 max	0.06 max
III	A	25% Cr	2.0–3.3	2.0 max	1.5 max	2.5 max	23.0–30.0	3.0 max	1.2 max	0.10 max	0.06 max

SOURCE: Abstracted from ASTM data, by permission.

Table 6.3.10   ASTM A436: Austenitic Gray Irons. Chemical Analysis and Mechanical Property Requirements

Element	Composition, %							
	Type 1	Type 1b	Type 2	Type 2b	Type 3	Type 4	Type 5	Type 6
Carbon, total, max	3.00	3.00	3.00	3.00	2.60	2.60	2.40	3.00
Silicon	1.00–2.80	1.00–2.80	1.00–2.80	1.00–2.80	1.00–2.00	5.00–6.00	1.00–2.00	1.50–2.50
Manganese	0.5–1.5	0.5–1.5	0.5–1.5	0.5–1.5	0.5–1.5	0.5–1.5	0.5–1.5	0.5–1.5
Nickel	13.50–17.50	13.50–17.50	18.00–22.00	18.00–22.00	28.00–32.00	29.00–32.00	34.00–36.00	18.00–22.00
Copper	5.50–7.50	5.50–7.50	0.50 max	0.50 max	0.50 max	0.50 max	0.50 max	3.50–5.50
Chromium	1.5–2.5	2.50–3.50	1.5–2.5	3.00–6.00*	2.50–3.50	4.50–5.50	0.10 max	1.00–2.00
Sulfur, max	0.12	0.12	0.12	0.12	0.12	0.12	0.12	0.12
Molybdenum, max	—	—	—	—	—	—	—	1.00

	Type 1	Type 1b	Type 2	Type 2b	Type 3	Type 4	Type 5	Type 6
Tensile strength, min, ksi (MPa)	25 (172)	30 (207)	25 (172)	30 (207)	25 (172)	25 (172)	20 (138)	25 (172)
Brinell hardness (3,000 kg)	131–183	149–212	118–174	171–248	118–159	149–212	99–124	124–174

* Where some matching is required, the 3.00 to 4.00 percent chromium range is recommended.
SOURCE: Abstracted from ASTM data, by permission.

Table 6.3.11   ASTM A439: Austenitic Ductile Irons. Chemical Analysis and Mechanical Property Requirements

					Type				
	D-2*	D-2B	D-2C	D-3*	D-3A	D-4	D-5	D-5B	D-5S
Element					Composition, %				
Total carbon, max	3.00	3.00	2.90	2.60	2.60	2.60	2.40	2.40	2.30
Silicon	1.50–3.00	1.50–3.00	1.00–3.00	1.00–2.80	1.00–2.80	5.00–6.00	1.00–2.80	1.00–2.80	4.90–5.50
Manganese	0.70–1.25	0.70–1.25	1.80–2.40	1.00 max†	1.00 max†	1.00 max†	1.00 max†	1.00 max†	1.00 max
Phosphorus, max	0.08	0.08	0.08	0.08	0.08	0.08	0.08	0.08	0.08
Nickel	18.00–22.00	18.00–22.00	21.00–24.00	28.00–32.00	28.00–32.00	28.00–32.00	34.00–36.00	34.00–36.00	34.00–37.00
Chromium	1.75–2.75	2.75–4.00	0.50 max†	2.50–3.50	1.00–1.50	4.50–5.50	0.10 max	2.00–3.00	1.75–2.25

					Type				
	D-2	D-2B	D-2C	D-3	D-3A	D-4	D-5	D-5B	D-5S
Element					Properties				
Tensile strength, min, ksi (MPa)	58 (400)	58 (400)	58 (400)	55 (379)	55 (379)	60 (414)	55 (379)	55 (379)	65 (449)
Yield strength (0.2% offset), min, ksi (MPa)	30 (207)	30 (207)	28 (193)	30 (207)	30 (207)	—	30 (207)	30 (207)	30 (207)
Elongation in 2 in or 50 mm, min, %	8.0	7.0	20.0	6.0	10.0	—	20.0	6.0	10
Brinell hardness (3,000 kg)	139–202	148–211	121–171	139–202	131–193	202–273	131–185	139–193	131–193

* Additions of 0.7 to 1.0% of molybdenum will increase the mechanical properties above 800°F (425°C).
† Not intentionally added.
SOURCE: Abstracted from ASTM data, by permission.

Table 6.3.12   Corrosion Resistance of Ductile Ni-Resist Irons D-2 and D-2C

Corrosive media	Penetration, in/yr	
	Type D-2C	Type D-2
Ammonium chloride solution: 10% NH₄Cl, pH 5.15, 13 days at 30°C (86°F), 6.25 ft/min	0.0280	0.0168
Ammonium sulfate solution: 10% (NH₄)2SO₄, pH 5.7, 15 days at 30°C (86°F), 6.25 ft/min	0.0128	0.0111
Ethylene vapors & splash: 38% ethylene glycol, 50% diethylene glycol, 4.5% H₂O, 4% Na₂SO₄, 2.7% NaCl, 0.8% Na₂CO₃ + trace NaOH, pH 8 to 9, 85 days at 275–300°F	0.0023	0.0019*
Fertilizer: commercial "5-10-5", damp, 290 days at atmospheric temp	—	0.0012
Nickel chloride solution: 15% NiCl₂, pH 5.3, 7 days at 30°C (86°F), 6.25 ft/min	0.0062	0.0040
Phosphoric acid, 86%, aerated at 30°C (86°F), velocity 16 ft/min, 12 days	0.213	0.235
Raw sodium chloride brine, 300 gpl of chlorides, 2.7 gpl CaO, 0.06 gpl NaOH, traces of NH3 & H₂S, pH 6–6.5, 61 days at 50°F, 0.1 to 0.2 fps	0.0023	0.0020*
Seawater at 26.6°C (80°F), velocity 27 ft/s, 60-day test	0.039	0.018
Soda & brine: 15.5% NaCl, 9.0% NaOH, 1.0% Na₂SO₄, 32 days at 180°F	0.0028	0.0015
Sodium bisulfate solution: 10% NaHSO₄, pH 1.3, 13 days at 30°C (86°F), 6.25 ft/min	0.0431	0.0444
Sodium chloride solution: 5% NaCl, pH 5.6, 7 days at 30°C (86°F), 6.25 ft/min	0.0028	0.0019
Sodium hydroxide:		
50% NaOH + heavy conc. of suspended NaCl, 173 days at 55°C (131°F), 40 gal/min.	0.0002	0.0002
50% NaOH saturated with salt, 67 days at 95°C (203°F), 40 gal/min	0.0009	0.0006
50% NaOH, 10 days at 260°F, 4 days at 70°F	0.0048	0.0049
30% NaOH + heavy conc. of suspended NaCl, 82 days at 85°C (185°F)	0.0004	0.0005
74% NaOH, 19³/4 days, at 260°F	0.005	0.0056
Sodium sulfate solution: 10% Na₂SO₄, pH 4.0, 7 days at 30°C (86°F), 6.25 ft/min	0.0136	0.0130
Sulfuric acid: 5%, at 30°C (86°F) aerated, velocity 14 ft/min, 4 days	0.120	0.104
Synthesis of sodium bicarbonate by Solvay process: 44% solid NaHCO₃ slurry plus 200 gpl NH₄Cl, 100 gpl NH₄HCO₃, 80 gpl NaCl, 8 gpl NaHCO₃, 40 gpl CO₂, 64 days at 30°C (86°F)	0.0009	0.0003
Tap water aerated at 30°C, velocity 16 ft/min, 28 days	0.0015	0.0023
Vapor above ammonia liquor: 40% NH₃, 9% CO₂, 51% H₂O, 109 days at 85°C (185°F), low velocity	0.011	0.025
Zinc chloride solution: 20% ZnCl₂, pH 5.25, 13 days at 30°C (86°F), 6.25 ft/min	0.0125	0.0064

* Contains 1 % chromium.
SOURCE: QIT-Fer Titane, Inc.

(Castings—System of Dimensional Tolerances and Machining Allowances). It is strongly recommended that designers discuss dimensional and machine finish allowance requirements with the foundry to avoid unnecessary costs.

STEEL CASTINGS

Composition   There are five classes of commercial steel castings: low-carbon steels (C < 0.20 percent), medium-carbon steels (C between 0.20 and 0.50 percent), high-carbon steels (C > 0.50 percent), low-alloy steels (total alloy ≤ 8 percent), and high-alloy steels (total alloy > 8 percent).

Carbon Steel Castings   Carbon steel castings contain less than 1.00 percent C, along with other elements which may include Mn (0.50 to 1.00 percent), Si (0.20 to 0.70 percent), max P (0.05 percent), and max S (0.06 percent). In addition, carbon steels, regardless of whether they are cast or wrought, may contain small percentages of other elements as residuals from raw materials or additives incorporated in the steelmaking process.

Low-Alloy Steel Castings   In a low-alloy steel casting, the alloying elements are present in percentages greater than the following: Mn, 1.00; Si, 0.70; Cu, 0.50; Cr, 0.25; Mo, 0.10; V, 0.05; W, 0.05; and Ti, 0.05. Limitations on phosphorus and sulfur contents apply to low-alloy steels as well as carbon steels unless they are specified to be different

Table 6.3.13   Oxidation Resistance of Ductile Ni-Resist, Ni-Resist, Conventional and High-Silicon Ductile Irons, and Type 309 Stainless Steel

	Oxidation resistance	
	Penetration (in/yr)	
	Test 1	Test 2
Ductile iron (2.5 Si)	0.042	0.50
Ductile iron (5.5 Si)	0.004	0.051
Ductile Ni-Resist type D-2	0.042	0.175
Ductile Ni-Resist type D-2C	0.07	—
Ductile Ni-Resist type D-4	0.004	0.0
Conventional Ni-Resist type 2	0.098	0.30
Type 309 stainless steel	0.0	0.0

Test 1—Furnace atmosphere—air, 4,000 h at 1,300°F.
Test 2—Furnace atmosphere—air, 600 h at 1,600–1,700°F, 600 h between 1,600–1,700°F, and 800–900°F, 600 h at 800–900°F.
SOURCE: QIT-Fer Titane, Inc.

for the purpose of producing some desired effect, e.g., free machining. Carbon and low-alloy steels account for approximately 85 percent of the steel castings produced in the United States.

High-Alloy Steel Castings   Steel castings with total alloy content greater than 8 percent are generally considered to be high-alloy, and usually the steel castings industry requires that the composition contain greater than 11 percent Cr. This Cr content requirement eliminates the potential confusion arising from including austenitic manganese steels (Hadfield) in this group. High-alloy steels include corrosion-resistant, heat-resistant, and duplex stainless steels. Many of the cast corrosion-resistant and duplex stainless steels are similar to the wrought stainless steels, but their performance may not be the same. Table 6.3.14 lists cast high-alloy steels and similar wrought grades.

Weight Range   Steel castings weigh from a few ounces to over 200 tons. Steel castings may be made in any thickness down to 0.25 in (6 mm), and in special processes a thickness of 0.080 in (2 mm) has been achieved. In investment castings, wall thicknesses of 0.060 in (1.5 mm) with sections tapering down to 0.030 in (0.75 mm) are common.

Table 6.3.14   ACI Designations for Heat- and Corrosion-Resistant High-Alloy Castings
(See also Secs. 6.2 and 6.4.)

Cast desig.	UNS	Wrought alloy type*	Cast desig.	UNS	Wrought alloy type*
CA15	J91150	410	CW2M	N26455	C4
CA15M	J91151	—	CW6M	N30107	—
CA28MWV	J91422	422	CW6MC	N26625	625
CA40	J91153	420	CW12MW	N30002	C
CA6NM	J91540	F6NM	CX2MW	N26022	C22
CA6N	J91650	—	CY5SnBiM	N26055	—
CB30	J91803	431	HA		—
CB7Cu-1	J92180	17-4	HC		—
CB7Cu-2	J92110	15-5	HD		—
CC50	J92615	446	HE		—
CD3MWCuN	J93380	Zeron 100	HF		—
CD3MN	J92205	2205	HH		—
CD4MCu	J93370	255	HI		—
CD6MN	J93371	—	HK		—
CE3MN	J93404	—	HL		—
CE30	J93423	—	HN		—
CE8MN	J93345	—	HP		—
CF3	J92500	304NL	HP		—
CF8	J92600	304	HT		—
CF10SMnN	J92972	Nitronic 60	HU		—
CF20	J92602	302	HW		—
CF3M	J92800	316l	HX		—
CF3MN	J92804	316LN	M25S	N24025	—
CF8M	J92900	316	M30C	N24130	—
CF8C	J92710	347	M30H	N24030	—
CF16F	J92701	303	M35-1	N24135	400
CG6MMN	J93790	Nitronic 50	M35-2	N04020	400
CG12	J93001	—	N7M	N30007	B2
CG8M	J93000	317	N12MV	N30012	B
CH20	J93402	309			
CK20	J94202	310			
CK3MCuN	J93254	254SMO			
CN3MN	J94651	AL6XN			
CN7M	.N08007	—			
CN7MS	N02100	—			
CU5MCuC	N08826	825			

* Wrought alloy type references are listed only for the convenience of those who want corresponding wrought and cast grades. This table does not imply that the performance of the corresponding cast and wrought grades will be the same. Because the cast alloy chemical composition ranges are not the same as the wrought composition ranges, buyers should use cast alloy designations for proper identification of castings.
† Common description, formerly used by AISI.
‡ Proprietary trademark: ‡¹ Armco, Inc, ‡² Avesta Sheffield, ‡³ Allegheny Ludlum Corp.
§ Common name used by two or more producers; not a trademark.
¶ ASTM designation.
NOTE: Most of the standard grades listed are covered for general applications by ASTM specifications.
SOURCE: *Steel Founders Society of America Handbook,* 5th ed.

Mechanical Properties   The outstanding mechanical properties of cast steel are high strength, ductility, resistance to impact, stiffness, endurance strength, and resistance to both high and low temperatures. It is also weldable. The mechanical properties of carbon steel castings are shown in Figs. 6.3.11 and 6.3.12; those of low-alloy cast steels are shown in Figs. 6.3.13 and 6.3.14.

Tensile and Yield Strength   Ferritic steels of a given hardness or hardenability have the same tensile strength whether cast, wrought, or forged regardless of alloy content. For design purposes involving tensile and yield properties, cast, wrought, and forged properties are considered the same.

Ductility   If the ductility values of steels are compared with the hardness values, the cast, wrought, and forged values are almost identical. The longitudinal properties of wrought and forged steels are slightly higher than those for castings. The transverse properties of the wrought

Fig. 6.3.11   Yield strength and elongation versus carbon content for cast carbon steels. (1) Water quenched and tempered at 1,200°F; (2) normalized; (3) normalized and tempered at 1,200°F; (4) annealed.

Fig. 6.3.12   Tensile properties of carbon steel as a function of hardness.

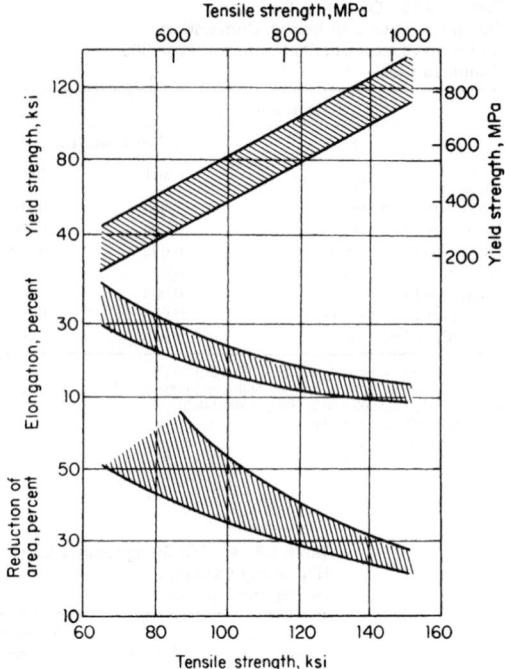

Fig. 6.3.13   Properties of normalized and tempered low-alloy cast steels.

Fig. 6.3.14   Properties of quenched and tempered low-alloy cast steels.

and forged steels are lower, by an amount that depends on the degree of working. Since most service conditions involve several directions of loading, the isotropic behavior of cast steels is particularly advantageous (Fig. 6.3.15).

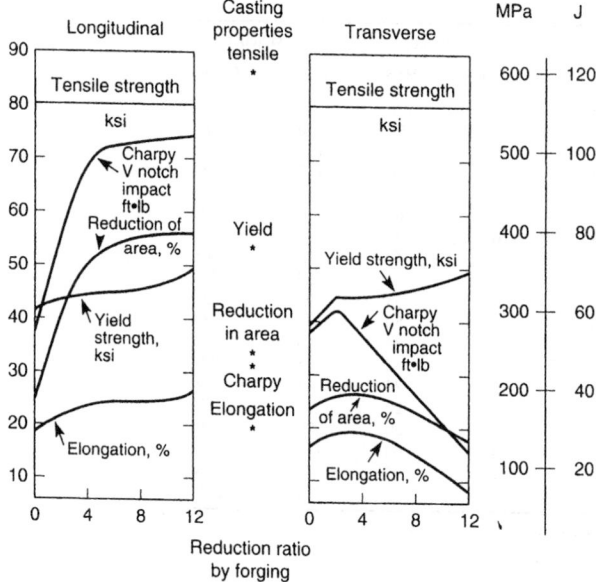

Fig. 6.3.15 Influence of forging reduction on anisotropy for a 0.35 percent carbon wrought steel. Properties for a cast of 0.35 percent carbon steel are shown with an asterisk for comparison purposes. *(Steel Founders Society of America Handbook, 5th ed.)*

Impact The notched-bar impact test is used often as a measure of the toughness of materials. Cast steels have excellent impact properties at normal and low temperatures. Generally wrought steels are tested in the direction of rolling, show higher impact values than cast steels of similar composition. Transverse impact values will be 50 to 70 percent lower than these values. Cast steels do not show directional properties. If the directional properties are averaged for wrought steels, the values obtained are comparable to the values obtained for cast steels of similar composition. The **hardenability** of cast steels is influenced by composition and other variables in the same manner as the hardenability of wrought steels. The ratio of the **endurance limit** to the tensile strength for cast steel varies from 0.42 to 0.50, depending somewhat on the composition and heat treatment of the steel. The notch-fatigue ratio varies from 0.28 to 0.32 for cast steels and is the same for wrought steels (Fig. 6.3.16).

Fig. 6.3.16 Variations of endurance limit with tensile strength for comparable carbon and low-alloy cast and wrought steels.

In **wear testing,** cast steels react similarly to wrought steels and give corresponding values depending on composition, structure, and hardness. Carbon cast steels of approximately 0.50 percent C and low-alloy cast steels of the chromium, chromium-molybdenum, nickel-chromium, chromium-vanadium, and medium-manganese types, all of which contain more than 0.40 percent C, exhibit excellent resistance to wear in service.

Corrosion Resistance Cast and wrought steels of similar composition and heat treatment appear to be equally resistant to corrosion in the same environments. Small amounts of copper in cast steels increase the resistance of the steel to atmospheric corrosion. High-alloy steels of chromium, chromium-nickel, and chromium-nickel-molybdenum types are normally used for corrosion service.

Heat Resistance Although not comparable to the high-alloy steels of the nickel-chromium types developed especially for heat resistance, the 4.0 to 6.5 percent Cr cast steels, particularly with additions of 0.75 to 1.25 percent W, or 0.40 to 0.70 percent Mo, and 0.75 to 1.00 percent Ti, show good strength and considerable resistance to scaling up to 1,000°F (550°C).

Many cast high-alloy nickel-chromium steels have **creep** and **rupture** properties superior to those of wrought materials. In many cases wrought versions of the cast grades are not available due to hot-working difficulties. These cast materials may be used in service conditions up to 2,000°F (1,100°C), with some grades capable of withstanding temperatures of 2,200°F (1,200°C).

Machinability of carbon and alloy cast steels is comparable to that of wrought steels having equivalent strength, ductility, hardness, and similar microstructure. Factors influencing the machinability of cast steels are as follows: (1) Microstructure has a definite effect on machinability of cast steels. In some cases it is possible to improve machinability by as much as 100 to 200 percent through heat treatments which alter the microstructure. (2) Generally speaking, hardness alone cannot be taken as the criterion for predicting tool life in cutting cast steels. (3) In general, for a given structure, plain carbon steels machine better than alloy steels. (4) When machining cast carbon steel (1040) with carbides, tool life varies with the ratio of ferrite to pearlite in the microstructure of the casting, the 60 : 40 ratio resulting in optimum tool life. For equal tool life, the skin of a steel casting should be machined at approximately one-half the cutting speed recommended for the base metal. The machinability of various carbon and low-alloy cast steels is given in Table 6.3.15.

Welding steel castings presents the same problems as welding wrought steels. It is interesting to note that cast low-alloy steels show greater resistance to underbead cracking than their wrought counterparts.

Purchase Specifications for Steel Castings Many steel castings are purchased according to mechanical property specifications, although more frequent use of specifications which indicate the use of steel castings may be more helpful. Note that it is now possible to order steel castings as cast equivalents of the SAE/AISI chemical composition ranges using ASTM A915. It is preferable to order steel castings to conform with national standards such as ASTM. ASTM standards are prepared by purchasers and suppliers and are well understood by steel foundries. The property levels specified in ASTM specifications A27 and A148 are shown in Table 6.3.16. Table 6.3.17 lists current applicable ASTM steel castings standards.

Steel Melting Practice Steel for steel castings is produced commercially by processes similar to those used for wrought steel production, but electric arc and electric induction melting furnaces predominate. Some steel foundries also use secondary refining processes such as **argon oxygen decarburization** (AOD) and/or **vacuum oxygen degassing** (VOD). Wire injection is also used selectively depending on service and process requirements for the steel casting. In addition to the materials listed in the ASTM standards, the steel casting industry will produce compositions tailored especially for particular service applications, and even in small quantity.

Allowances for Steel Castings

Dimensional allowances that must be applied in the production of castings are due to the fact that different metals contract at different rates

Table 6.3.15   Machinability Index for Cast Steels

Steel	BHN	Conventional		Metcut*	
		Carbide	HSS	HSS	Carbide
B1112 Free machining steel (wrought)	179	. . .	100		
1020 Annealed	122	10	90	160	400
1020 Normalized	134	6	75	135	230
1040 Double normalized	185	11	70	130	400
1040 Normalized and annealed	175	10	75	135	380
1040 Normalized	190	6	65	120	325
1040 Normalized and oil-quenched	225	6	45	80	310
1330 Normalized	187	2	40	75	140
1330 Normalized and tempered	160	3	65	120	230
4130 Annealed	175	4	55	95	260
4130 Normalized and spheroidized	175	3	50	90	200
4340 Normalized and annealed	200	3	35	60	210
4340 Normalized and spheroidized	210	6	55	95	290
4340 Quenched and tempered	300	2	25	45	200
4340 Quenched and tempered	400	½	20	35	180
8430 Normalized and tempered at 1,200°F (660°C)	200	3	50	90	200
8430 Normalized and tempered at 1,275°F (702°C)	180	4	60	110	240
8630 Normalized	240	2	40	75	180
8630 Annealed	175	5	65	120	290

* The Metcut speed index number is the actual cutting speed (surface ft/min) which will give 1 h tool life in turning.

Table 6.3.16   ASTM Requirements for Steel Castings in A27 and A148

Grade	Tensile strength, min, ksi (MPa)	Yield point min, ksi (MPa)	Elongation, min, %	Reduction of area, min, %
		ASTM A27		
U-60-30	60 (415)	30 (205)	22	30
60-30	60 (415)	30 (205)	24	35
65-35	65 (450)	35 (240)	24	35
70-36	70 (485)	36 (250)	22	30
70-40	70 (485)	40 (275)	22	30
		ASTM A148		
80–40	80 (550)	40 (275)	18	30
80–50	80 (550)	50 (345)	22	35
90–60	90 (620)	60 (415)	20	40
105–85	105 (725)	85 (585)	17	35
115–95	115 (795)	95 (655)	14	30
130–115	130 (895)	115 (795)	11	25
135–125	135 (930)	125 (860)	9	22
150–135	150 (1,035)	135 (930)	7	18
160–145	160 (1,105)	145 (1,000)	6	12
165–150	165 (1,140)	150 (1,035)	5	20
165–150L	165 (1,140)	150 (1,035)	5	20
210–180	210 (1,450)	180 (1,240)	4	15
210–180L	210 (1,450)	180 (1,240)	4	15
260–210	260 (1,795)	210 (1,450)	3	6
260–210L	260 (1,795)	210 (1,450)	3	6

SOURCE: Abstracted from ASTM data, by permission.

during solidification and cooling. The shape of the part has a major influence on the as-cast dimensions of the casting. Some of the principal allowances are discussed briefly here.

Shrinkage allowances for steel castings vary from ⁹⁄₃₂ to ¹⁄₁₆ in/ft (0.03 to 0.005 cm/cm). The amount most often used is ¼ in/ft (0.021 cm/cm), but if it is applied indiscriminately, it can lead to problems in steel casting dimensions. The best policy is to discuss shrinkage allowances with the foundry that makes the castings before patterns are made.

Minimum Section Thickness   The fluidity of steel when compared to other metals is low. In order that all sections may be completely filled, a minimum value of section thickness must be adopted that is a function of the largest dimension of the casting. These values are suggested for design use:

Min section thickness		Max length of section	
in	mm	in	cm
¼	6	12	30
½	13	50	125

Casting finish allowances (unmachined) are based on the longest dimension of the casting, but the casting weight may also influence these allowances. Other factors which affect allowances include pattern materials and the method of pattern construction. Average values of allowance as a function of pattern type are shown in Table 6.3.18; other sources of similar data are "Steel Castings Handbook," Supplement 3, "Tolerances," SFSA, and ISO 8062 (Castings—System of Dimensional Tolerances and Machining Allowances). It is strongly recommended that designers discuss allowance requirements with the foundry to avoid unnecessary costs.

Machining finish allowances added to the casting dimensions for machining purposes will depend entirely on the casting design. Definite values are not established for all designs, but Table 6.3.19 presents a brief guide to machining allowances on gears, wheels, and circular, flat castings.

Table 6.3.17 A Summary of ASTM Specifications for Steel and Alloy Castings

A 27 Steel Castings, Carbon, for General Application	A 451 Centrifugally Cast Austenitic Steel Pipe for High-Temperature Service
A 128 Steel Castings, Austenitic Manganese	A 487 Steel Castings, Suitable for Pressure Service
A 148 Steel Castings, High Strength, for Structural Purposes	A 494 Castings, Nickel and Nickel Alloy
A 216 Steel Castings, Carbon, Suitable for Fusion Welding, for High-Temperature Service	A 560 Castings, Chromium-Nickel Alloy
	A 660 Centrifugally Cast Carbon Steel Pipe for High-Temperature Service
A 217 Steel Castings, Martensitic Stainless and Alloy, for Pressure-Containing Parts, Suitable for High-Temperature Service	A 703 Steel Castings, General Requirements, for Pressure-Containing Parts
A 297 Steel Castings, Iron-Chromium and Iron-Chromium-Nickel, Heat-Resistant, for General Application	A 732 Castings, Investment, Carbon and Low-Alloy Steel for General Application, and Cobalt Alloy for High Strength at Elevated Temperatures
A 351 Steel Castings, Austenitic, Austenitic-Ferritic (Duplex), for Pressure-Containing Parts	A 743 Castings, Iron-Chromium, Iron-Chromium-Nickel, Corrosion-Resistant, for General Application
A 352 Steel Castings, Ferritic and Martensitic, for Pressure-Containing Parts, Suitable for Low-Temperature Service	A 744 Castings, Iron-Chromium, Iron-Chromium-Nickel, Corrosion-Resistant, for Severe Service
A 356 Steel Castings, Carbon, Low-Alloy, and Stainless Steel, Heavy-Walled for Steam Turbines	A 747 Steel Castings, Stainless, Precipitation Hardening
	A 757 Steel Castings, Ferritic and Martensitic, for Pressure-Containing and Other Applications, for Low-Temperature Service
A 389 Steel Castings, Alloy, Specially Heat-Treated, for Pressure-Containing Parts, Suitable for High-Temperature Service	A 781 Castings, Steel and Alloy, Common Requirements, for General Industrial Use
A 426 Centrifugally Cast Ferritic Alloy Steel Pipe for High-Temperature Service	A 890 Castings, Iron-Chromium-Nickel-Molybdenum, Corrosion-Resistant, Duplex (Austenitic/Ferritic) for General Application
A 447 Steel Castings, Chromium-Nickel-Iron Alloy (25-12 Class), for High-Temperature Service	A 915 Steel Castings, Carbon and Alloy, Chemical Requirements Similar to Standard Wrought Grades

SOURCE: Abstracted from ASTM data, by permission.

Table 6.3.18 Dimensional Tolerances for Steel Castings*

(Deviation from the design dimension)

	Drawing dimension, in†			
Pattern type	0–3.0	3.1–7.0	7.1–20.0	20.1–100.0
Metal match plate	$+\frac{1}{32}, -\frac{1}{16}$	$+\frac{3}{32}, -\frac{1}{16}$	$+\frac{1}{8}, -\frac{1}{16}$	$+\frac{1}{8}, -\frac{1}{8}$
Metal pattern mounted on cope and drag boards	$+\frac{1}{16}, -\frac{1}{16}$	$+\frac{3}{32}, -\frac{3}{32}$	$+\frac{1}{8}, -\frac{3}{32}$	$+\frac{7}{32}, -\frac{1}{8}$
Hardwood pattern mounted on cope and drag boards	$+\frac{3}{32}, -\frac{1}{16}$	$+\frac{1}{8} -\frac{3}{32}$	$+\frac{1}{8}, -\frac{3}{32}$	$+\frac{1}{4}, -\frac{3}{32}$

* Surfaces that are not to be machined. † × 2.54 = cm.

Table 6.3.19 Machining Allowances for Steel Castings

Specific dimension or reference line (plane) distance, in		Greatest dimension of casting							
		10 in		10–20 in		20–100 in		Over 100 in*	
Greater than	But not exceeding	0.xx in*	X/32 in*	0.xx in*	X/32 in*	0.xx in*	X/32 in*	0.xx in*	X/32 in*
		+	+	+	+	+	+	+	+
0	2	0.187	6	0.218	7	0.25	8	0.312	10
2	5	0.25	8	0.281	9	0.312	10	0.437	14
5	10	0.312	10	0.344	11	0.375	12	0.531	17
10	20			0.406	13	0.468	15	0.625	20
20	50					0.562	18	0.75	24
50	75					0.656	21	0.875	28
75	100					0.75	24	1.00	32
100	500							1.25	40

		Machine tolerances on section thickness	
		0.XXX in*	X/32 in*
		+	+
0	½	0.062	2
½	½	0.092	3
1½	4	0.187	6
4	7	0.25	8
7	10	0.344	11
10		0.50	16

* × 2.54 = cm.

Use of Steel Castings

All industries use steel castings. They are used as wheels, sideframes, bolsters, and couplings in the railroad industry; rock crushers in the mining and cement industries; components in construction vehicles; suspension parts and fifth wheels for trucks; valves; pumps; armor. High-alloy steel castings are used in highly corrosive environments in the chemical, petrochemical, and paper industries. Generally, steel castings are applied widely when strength, fatigue, and impact properties are required, as well as weldability, corrosion resistance, and high-temperature strength. Often, overall economy of fabrication will dictate a steel casting as opposed to a competing product.

Casting Design Maximum service and properties can be obtained from castings only when they are properly designed. Design rules for castings have been prepared in detail to aid design engineers in preparing efficient designs. Engineers are referred to the following organizations for the latest publications available:

Steel Founders' Society of America, Crystal lake, IL, 60014.
American Foundry Society, Des Plaines, IL 60016.
Investment Casting Institute, Dallas, TX 75206.

6.4 NONFERROUS METALS AND ALLOYS; METALLIC SPECIALTIES

REFERENCES: Current edition of "Metals Handbook," American Society for Materials. Current listing of applicable ASTM, AISI, and SAE Standards. Publications of the various metal-producing companies. Publications of the professional and trade associations which contain educational material and physical property data for the materials promoted.

INTRODUCTION

by L. D. Kunsman and C. L. Carlson; Amended by Staff

Seven **nonferrous metals** are of primary commercial importance: copper, zinc, lead, tin, aluminum, nickel, and magnesium. Some 40 other elements are frequently alloyed with these to make the commercially important alloys. There are also about 15 minor metals that have important specific uses. The properties of these elements are given in Table 6.4.1. (See also Secs. 4, 5, and 6.1.)

Metallic Properties Metals are substances that characteristically are opaque crystalline solids of high reflectivity having good electrical and thermal conductivities, a positive chemical valence, and, usually, the important combination of considerable strength and the ability to flow before fracture. These characteristics are exhibited by the metallic elements (e.g., iron, copper, aluminum, etc.), or **pure metals**, and by combinations of elements (e.g., steel, brass, dural), or **alloys**. Metals are composed of many small **crystals**, which grow individually until they fill the intervening spaces by abutting neighboring crystals. Although the external shape of these crystals and their orientation with respect to each other are usually random, within each such **grain**, or crystal, the atoms are arranged on a regular three-dimensional lattice. Most metals are arranged according to one of the three common types of **lattice**, or **crystal structure**: face-centered cubic, body-centered cubic, or hexagonal close-packed. (See Table 6.4.1.)

The several properties of metals are influenced to varying degrees by the testing or service **environment** (temperature, surrounding medium) and the **internal structure** of the metal, which is a result of its chemical composition and previous history such as casting, hot rolling, cold extrusion, annealing, and heat treatment. These relationships, and the discussions below, are best understood in the framework of the several phenomena that may occur in metals processing and service and their general effect on metallic structure and properties.

Ores, Extractions, Refining All metals begin with the mining of ores and are successively brought through suitable physical and chemical processes to arrive at commercially useful degrees of purity. At the higher-purity end of this process sequence, scrap metal is frequently combined with that derived from ore. Availability of suitable ores and the extracting and refining processes used are specific for each metal and largely determine the price of the metal and the impurities that are usually present in commercial metals.

Melting Once a metal arrives at a useful degree of purity, it is brought to the desired combination of shape and properties by a series of physical processes, each of which influences the internal structure of

the metal. The first of these is usually melting, during which several elements can be combined to produce an alloy of the desired composition. Depending upon the metal, its container, the surrounding atmosphere, the addition of alloy-forming materials, or exposure to vacuum, various chemical reactions may be utilized in the melting stage to achieve optimum results.

Casting The molten metal of desired composition is poured into some type of mold in which the heat of fusion is dissipated and the melt becomes a solid of suitable shape for the next stage of manufacture. Such castings either are used directly (e.g., sand casting, die casting, permanent-mold casting) or are transferred to subsequent operations (e.g., ingots to rolling, billets to forging or extrusion). The principal advantage of castings is their design flexibility and low cost. The typical casting has, to a greater or lesser degree, a large crystal size (**grain size**), extraneous **inclusions** from slag or mold, and **porosity** caused by gas evolution and/or shrinking during solidification. The foundryman's art lies in minimizing the possible defects in castings while maximizing the economies of the process.

Metalworking (See also Sec. 13.) The major portion of all metals is subjected to additional shape and size changes in the solid state. These metalworking operations substantially alter the internal structure and eliminate many of the defects typical of castings. The usual sequence involves first **hot working** and then, quite often, **cold working**. Some metals are only hot-worked, some only cold-worked; most are both hotand cold-worked. These terms have a special meaning in metallurgical usage.

A cast metal consists of an aggregate of variously oriented grains, each one a single crystal. Upon deformation, the grains flow by a process involving the slip of blocks of atoms over each other, along definite crystallographic planes. The metal is hardened, strengthened, and rendered less ductile, and further deformation becomes more difficult. This is an important method of increasing the strength of nonferrous metals. The effect of progressive **cold rolling** on brass (Fig. 6.4.1) is typical. Terms such as "soft," "quarter hard," "half hard," and "full hard" are frequently used to indicate the degree of hardening produced by such working. For most metals, the hardness resulting from cold working is stable at room temperature, but with lead, zinc, or tin, it will decrease with time.

Upon **annealing, or heating the cold-worked metal,** the first effect is to relieve macrostresses in the object without loss of strength; indeed, the strength is often increased slightly. Above a certain temperature, softening commences and proceeds rapidly with increase in temperature (Fig. 6.4.2). The cold-worked distorted metal undergoes a change called **recrystallization.** New grains form and grow until they have consumed the old, distorted ones. The temperature at which this occurs in a given time, called the **recrystallization temperature,** is lower and the resulting grain size finer, the more severe the working of the original piece. If the original working is carried out above this temperature,

Table 6.4.1 Physical Constants of the Principal Alloy-Forming Elements

	Atomic no.	Atomic weight	Density, lb/in³	Melting point, °F	Boiling point, °F	Specific heat*	Latent heat of fusion, Btu/lb	Linear coef of thermal exp, per °F × 10⁻⁶	Thermal conductivity (near 68°F), Btu/(ft²·h·°F/in)	Electrical resistivity, μΩ·cm	Modulus of elasticity (tension), lb/in² × 10⁶	Crystal structure†	Transition temp, °F	Symbol
Aluminum	13	26.97	0.09751	1,220.4	4,520	0.215	170	13.3	1,540	2.655	10	FCC		Al
Antimony	51	121.76	0.239	1,166.9	2,620	0.049	68.9	4.7 – 7.0	131	39.0	11.3	Rhom		Sb
Arsenic	33	74.91	0.207	1,497	1,130	0.082	159	2.6		35	11	Rhom		As
Barium	56	137.36	0.13	1,300	2,980	0.068		(10)		50	1.8	BCC		Ba
Beryllium	4	9.02	0.0658	2,340	5,020	0.52	470	6.9	1,100	5.9	37	HCP		Be
Bismuth	83	209.00	0.354	520.3	2,590	0.029	22.5	7.4	58	106.8	4.6	Rhom		Bi
Boron	5	10.82	0.083	3,812	4,620	0.309		4.6		1.8×10^{22}		O		B
Cadmium	48	112.41	0.313	609.6	1,409	0.055	23.8	16.6	639	6.83	8	HCP		Cd
Calcium	20	40.08	0.056	1,560	2,625	0.149	100	12	871	3.43	3	FCC/BCC	867	Ca
Carbon	6	12.010	0.0802	6,700	6,730	0.165		0.3 – 2.4	165	1,375	0.7	Hex/D		C
Cerium	58	140.13	0.25	1,460	4,380	0.042	27.2	3.4		78		HCP/FCC	572/1328	Ce
Chromium	24	52.01	0.260	3,350	4,500	0.11	146	6.8	464	13	36	BCC/FCC	3344	Cr
Cobalt	27	58.94	0.32	2,723	6,420	0.099	112	4.0	479	6.24	30	HCP/FCC/HCP	783/2048	Co
Columbium (niobium)	41	92.91	0.310	4,380	5,970	0.065				13.1	15	BCC		Cb
Copper	29	63.54	0.324	1,981.4	4,700	0.092	91.1	9.2	2,730	1.673	16	FCC		Cu
Gadolinium	64	156.9	0.287									HCP		Gd
Gallium	31	69.72	0.216	85.5	3,600	0.0977	34.5	10.1	232	56.8	1	O		Ga
Germanium	32	72.60	0.192	1,756	4,890	0.086	205.7	(3.3)		60×10^{6}	11.4	D		Ge
Gold	79	197.2	0.698	1,945.4	5,380	0.031	29.0	7.9	2,060	2.19	12	FCC		Au
Hafnium	72	178.6	0.473	3,865	9,700	0.0351		(3.3)		32.4	20	HCP/BCC	3540	Hf
Hydrogen	1	1.0030	3.026×10^{-6}	-434.6	-422.9	3.45	27.0		1.18			Hex		H
Indium	49	114.76	0.264	313.5	3,630	0.057	12.2	18	175	8.37	1.57	FCT		In
Iridium	77	193.1	0.813	4,449	9,600	0.031	47	3.8	406	5.3	75	FCC		Ir
Iron	26	55.85	0.284	2,802	4,960	0.108	117	6.50	523	9.71	28.5	BCC/FCC/BCC	1663/2554	Fe
Lanthanum	57	138.92	0.223	1,535	8,000	0.0448				59	5	HCP/FCC/?	662/1427	La
Lead	82	207.21	0.4097	621.3	3,160	0.031	11.3	16.3	241	20.65	2.6	FCC		Pb
Lithium	3	6.940	0.019	367	2,500	0.79	286	31	494	11.7	1.7	BCC		Li
Magnesium	12	24.32	0.0628	1,202	2,030	0.25	160	14	1,100	4.46	6.5	HCP		Mg
Manganese	25	54.93	0.268	2,273	3,900	0.115	115	12		185	23	CCX/CCX/FCT	1340/2010/2080	Mn

Table 6.4.1 Physical Constants of the Principal Alloy-Forming Elements (Continued)

Element	Atomic no.	Atomic weight	Density, lb/in³	Melting point, °F	Boiling point, °F	Specific heat*	Latent heat of fusion, Btu/lb	Linear coef of thermal exp., per °F × 10^{-6}	Thermal conductivity (near 68°F), Btu/(ft²·h·°F/in)	Electrical resistivity, μΩ·cm	Modulus of elasticity (tension), lb/in² × 10^6	Crystal structure†	Transition temp., °F	Symbol
Mercury	80	200.61	0.4896	-37.97	675	0.033	4.9	33.8	58	94.1		Rhom		Hg
Molybdenum	42	95.95	0.369	4,750	8,670	0.061	126	3.0	1,020	5.17	50	BCC		Mo
Nickel	28	58.69	0.322	2,651	4,950	0.105	133	7.4	639	6.84	30	FCC		Ni
Niobium (see Columbium)														Nb
Nitrogen	7	14.008	0.042×10^{-3}	-346	-320.4	0.247	11.2		0.147			Hex		N
Osmium	76	190.2	0.813	4,900	9,900	0.031		2.6		9.5	80	HCP		Os
Oxygen	8	16.000	0.048×10^{-3}	-361.8	-297.4	0.218	5.9		0.171			C		O
Palladium	46	106.7	0.434	2,829	7,200	0.058	69.5	6.6	494	10.8	17	FCC		Pd
Phosphorus	15	30.98	0.0658	111.4	536	0.177	9.0	70		10^{17}		C		P
Platinum	78	195.23	0.7750	3,224.3	7,970	0.032	49	4.9	494	9.83	21	FCC		Pt
Plutonium	94	239	0.686	1,229				50-65	1.83	150		MC	6 forms	Pu
Potassium	19	39.096	0.031	145	1,420	0.177	26.1	46	697	6.15	0.5	BCC		K
Radium	88	226.05	0.18	1,300								?		Ra
Rhenium	75	186.31	0.765	5,733	10,700	0.0326	76		610	21	75	HCP		Re
Rhodium	45	102.91	0.4495	3,571	8,100	0.059		4.6		4.5	54	FCC		Rh
Selenium	34	78.96	0.174	428	1,260	0.084	29.6	21	3	12	8.4	MC/Hex	248	Se
Silicon	14	28.06	0.084	2,605	4,200	0.162	607	1.6-4.1	581	10^5	16	D		Si
Silver	47	107.88	0.379	1,760.9	4,010	0.056	45.0	10.9	2,900	1.59	11	FCC		Ag
Sodium	11	22.997	0.035	207.9	1,638	0.295	49.5	39	929	4.2		BCC		Na
Sulfur	16	32.066	0.0748	246.2	832.3	0.175	16.7	36		2×10^{-23}	1.3	Rhom/FCC	204	S
Tantalum	73	180.88	0.600	5,420	9,570	0.036		3.6	377	12.4	27	BCC		Ta
Tellurium	52	127.61	0.225	840	2,530	0.047	13.1	9.3	41	2×10^{-5}	6	Hex		Te
Thallium	81	204.39	0.428	577	2,655	0.031	9.1	16.6	2,700	18	1.2	HCP/BCC	446	Tl
Thorium	90	232.12	0.422	3,348	8,100	0.0355	35.6	6.2		18.6	11.4	FCC		Th
Tin	50	118.70	0.264	449.4	4,120	0.054	26.1	13	464	11.5	6	D/BCT	55	Sn
Titanium	22	47.90	0.164	3,074	6,395	0.139	187	4.7	119	47.8	16.8	HCP/BCC	1650	Ti
Tungsten	74	183.92	0.697	6,150	10,700	0.032	79	2.4	900	5.5	50	BCC		W
Uranium	92	238.07	0.687	2,065	7,100	0.028	19.8	11.4	186	29	29.7	O/Tet/BCC	1229/1427	U
Vanadium	23	50.95	0.217	3,452	5,430	0.120		4.3	215	26	18.4	BCC	2822	V
Zinc	30	65.38	0.258	787	1,663	0.092	43.3	9.4-22	784	5.92	12	HCP		Zn
Zirconium	40	91.22	0.23	3,326	9,030	0.066		3.1	116	41.0	11	HCP/BCC	1585	Zr

NOTE: See Sec. 1 for conversion factors to SI units.
*Cal/(g · °C) at room temperature equals Btu/(lB · °F) at room temperature.
†FCC = face-centered cubic; BCC = body centered cubic; C = cubic; HCP = hexagonal closet packing; Rhomb = rhombohedral; Hex = hexagonal; FCT = face-centered tetragonal; O = orthorhombic; FCO = face-centered orthorhombic; CCX = cubic complex; D = diamond cubic; BCT = body-centered tetragonal; MC = monoclinic.
SOURCE: ASM "Metals Handbook," revised and supplemented where necessary from Hampel's "Rare Metals Handbook" and elsewhere.

it does not harden the piece, and the operation is termed **hot working**. If the annealing temperature is increased beyond the recrystallization temperature or if the piece is held at that temperature for a long time, the average grain size increases. This generally softens and decreases the strength of the piece still further.

Fig. 6.4.1   Effect of cold rolling on annealed brass (Cu 72, Zn 28).

Hot working occurs when deformation and annealing proceed simultaneously, so that the resulting piece of new size and shape emerges in a soft condition roughly similar to the annealed condition.

Alloys   The above remarks concerning structure and property control through casting, hot working, cold working, and annealing gener-

Fig. 6.4.2   Effect of annealing on cold-rolled brass (Cu 72, Zn 28).

ally apply both to pure metals and to alloys. The addition of alloying elements makes possible other means for controlling properties. In some cases, the addition to a metal A of a second element B simply results in the appearance of some new crystals of B as a **mixture** with crystals of A; the resulting properties tend to be an average of A and B. In other cases, an entirely new substance will form—the **intermetallic compound AB**, having its own set of distinctive properties (usually hard and brittle). In still other cases, element B will simply dissolve in element A to form the **solid solution A(B)**. Such solid solutions have the characteristics of the solvent A modified by the presence of the **solute B,** usually causing increased hardness, strength, electrical resistance, and recrystallization temperature. The most interesting case involves the combination of solid solution A(B) and the precipitation of a second constituent, either B or AB, brought about by the precipitation-hardening heat treatment, which is particularly important in the major nonferrous alloys.

Precipitation Hardening   Many alloys, especially those of aluminum but also some alloys of copper, nickel, magnesium, and other metals, can be hardened and strengthened by heat treatment. The heat treatment is usually a two-step process which involves (1) a solution heat treatment followed by rapid quenching and (2) a precipitation or aging treatment to cause separation of a second phase from solid solution and thereby cause hardening. After a solution treatment, these alloys are comparatively soft and consist of homogeneous grains of solid solution generally indistinguishable microscopically from a pure metal. If very slowly cooled from the solution treatment temperature, the alloy will deposit crystals of a second constituent, the amount of which increases as the temperature decreases. Rapid cooling after a solution treatment will retain the supersaturated solution at room temperature, but if the alloy is subsequently reheated to a suitable temperature, fine particles

of a new phase will form and in time will grow to a microscopically resolvable size. At some stage in this precipitation process the hardness, the tensile strength, and particularly the yield strength of the alloy, will be increased considerably. If the reheating treatment is carried out too long, the alloy will overage and soften. The temperature and time for both solution and precipitation heat treatments must be closely controlled to obtain the best results. To some extent, precipitation hardening may be superimposed upon hardening resulting from cold working. Precipitation-hardened alloys have an unusually high ratio of proportional limit to tensile strength, but the endurance limit in fatigue is not increased to nearly the same extent.

Effect of Environment   The properties of metals under "normal" conditions of 70°F and 50 percent relative humidity in clean air can be markedly changed under other conditions, with the various metals differing greatly in their degree of response to such changing conditions. The topic of **corrosion** is both important and highly specific (see Sec. 6.5). **Oxidation** is a chemical reaction specific to each metal and, wherever important, is so treated. The **effect of temperature,** however, can be profitably considered in general terms. The primary consequence of increase in temperature is increased atom movement, or **diffusion.** In the discussion above, the effects of recrystallization, hot working, solution treatment, and precipitation were all made possible by increased diffusion at elevated temperatures. The temperature at which such atom movements become appreciable is roughly proportional to the melting point of the metal. If their melting points are expressed on an absolutetemperature scale, then various metals can be expected to exhibit comparable effects at about the same fraction of their melting points. Thus the recrystallization temperatures of lead, zinc, aluminum, copper, nickel, molybdenum, and tungsten are successively higher. The consequent rule of thumb is that alloys do not have useful structural strength at temperatures above about 0.5 of their absolute melting temperatures.

ALUMINUM AND ITS ALLOYS
by J. Randolph Kissell, Larry F. Wieserman, and Richard L. Brazill

REFERENCES: Kissell and Ferry, "Aluminum Structure," Wiley, 2002. Aluminum Association publications: "Aluminum Standards and Data 2003." "Aluminum Design Manual 2005." "Aluminum with Food and Chemicals." "Designation System for Aluminum Finishes," 9th ed. "Aluminum for Automotive Body Sheet Panels." "Care of Aluminum," 5th ed. "Guidelines for In-Plant Handling of Aluminum Sheet and Plate." "Guidelines for Minimizing Water Staining of Aluminum." "Aluminum: Technology, Applications, and Environment—A Profile of a Modern Metal," 6th ed. "Standards for Aluminum Sand and Permanent Mold Castings," 2000. "Welding Aluminum: Theory and Practice." ASM International publications: "Aluminum and Aluminum Alloys." "Corrosion of Aluminum and Aluminum Alloys." "The Surface Treatment and Finishing of Aluminum and Its Alloys." "Properties of Aluminum Alloys: Tensile, Creep, and Fatigue Data at High and Low Temperatures." "Fracture Resistance of Aluminum Alloys: Notch Toughness, Tear Toughness, and Fracture Toughness." "Aluminum Handbook," vol. one, "Fundamentals and Materials." "Aluminum Handbook," vol. 2, "Forming Casting, Surface Treatment, Recycling, and Ecology." Minford, "Handbook of Aluminum Bonding Technology and Data," Marcel Dekker, 1993.

Introduction

Aluminum owes most of applications to its low fabrication costs, low density [about 0.1 lb/in³ (2,700 kg/m³)], and relatively high strength of its alloys. Other uses rely on its corrosion resistance, electrical and thermal conductivity, reflectivity, and toughness at low temperatures. Designs utilizing aluminum should take into account its relatively low modulus of elasticity [10,000 ksi (70,000 MPa)] and high coefficient of thermal expansion [$13 \times 10^{-6}/°F$ ($23 \times 10^{-6}/°C$)]. Commercially **pure aluminum** contains at least 99 percent aluminum and is a ductile metal used where strength is not required. **Aluminum alloys,** on the other hand, possess better casting and machining characteristics and better mechanical properties than the pure metal and, therefore, are used for applications such as transportation (aircraft, trains, boats, cars), building products (roofing and siding, window and door frames, canopy and space frames), food and pharmaceutical packaging (cans and foil), sports equipment, heat exchangers, cryogenic containers, and optical components. A great variety of aluminum alloys and products are available from producers; a useful guide to those that are commonly available is the Aluminum Association publication "Aluminum Standards and Data."

Aluminum alloys are divided into two classes: wrought alloys and cast alloys, each with its own designation system specified in ANSI H35.1, "Alloy and Temper Designation Systems for Aluminum," maintained by the Aluminum Association.

Aluminum Alloy Designation Systems

Wrought aluminum and **aluminum alloys** are designated by a four-digit number. The first digit indicates the alloy group according to the main alloying element as follows:

	Designation no.
Aluminum, 99.00 percent and greater	1xxx
Aluminum alloys grouped by major alloying elements	
Copper	2xxx
Manganese	3xxx
Silicon	4xxx
Magnesium	5xxx
Magnesium and silicon	6xxx
Zinc	7xxx
Other element	8xxx
Unused series	9xxx

The main alloying element is a good indicator of the properties of the alloy. For example, aluminum-copper alloys have greater strength but less corrosion resistance than pure aluminum. The second digit indicates modifications of the original alloy or impurity limits. For example, 2319 is identical to 2219 except for additional titanium to improve welding characteristics. The last two digits of the 1xxx series alloys indicate the aluminum purity. "Aluminum Standards and Data" provides the chemical compositions of commonly used alloys.

Some common wrought alloys (by major alloying elements) and applications are as follows:

1xxx Series: Aluminum of 99 percent or greater is used in the electrical and chemical industries for its high conductivity, corrosion resistance, and workability. It is available extruded or rolled and is strengthened by cold working but not by heat treatment (see tempers below).

2xxx Series: These alloys are heat-treatable and some attain strengths comparable to those of steel alloys. They are less corrosion resistant than other aluminum alloys and thus are often clad with pure aluminum or an alloy of the 6xxx series (see alclad aluminum below). Alloys 2014 and 2024 are popular, 2024 being perhaps the most widely used aircraft alloy. Many of the 2xxx alloys, including 2014, have poor welding characteristics.

3xxx Series: Alloys in this series are strengthened only by cold working. Alloys 3003, 3004, and 3105 are popular general-purpose alloys with moderate strengths and good workability, and they are often used for sheet metal applications such as gutters, downspouts, roofing, and siding.

4xxx Series: Silicon added to alloys of this group lowers the melting point, making these alloys suitable for use as weld filler wire (such as 4043) and brazing alloys.

5xxx Series: These alloys attain moderate to high strength by cold working and the addition to magnesium. They usually have the highest welded strengths of aluminum alloys and good corrosion resistance. Alloys 5083, 5086, 5154, 5454, and 5456 are used in welded structures, including pressure vessels and ships. Alloy 5052 is a popular, low-cost sheet metal alloy. For marine applications, 5xxx alloys of the -H116 and -H321 tempers are used for their corrosion resistance in freshwater and saltwater.

6xxx Series: Although these alloys usually are not as strong as those in the other heat-treatable series (2xxx and 7xxx), they offer a good combination of strength and corrosion resistance. Alloys 6061 and 6063 are used widely in structural applications, with 6061 providing greater strength at a slightly greater cost. Both alloys have good weldability.

7xxx Series: Heat-treating alloys in this series produces some of the highest-strength aluminum alloys, frequently used in aircraft, such as 7050, 7075, 7178, and 7475. The 7178-T651 plate has a minimum tensile strength of 84 ksi (580 MPa), the greatest of any commonly used aluminum alloy. Most alloys of this group contain copper and so are not readily welded, with 7005 being an exception.

Aluminum-lithium alloys have higher strength and modulus of elasticity than those of any of the other alloys, but cost about 4 times as much.

Cast aluminum and **aluminum alloys** are also designated by a four-digit number, but with a decimal point between the last two digits. The first digit indicates the alloy group according to the main alloying element as follows:

	Designation no.
Aluminum, 99.00 percent and greater	1xx.x
Aluminum alloys grouped by major alloying elements	
Copper	2xx.x
Silicon, with copper and/or magnesium	3xx.x
Silicon	4xx.x
Magnesium	5xx.x
Zinc	7xx.x
Tin	8xx.x
Other element	9xx.x
Unused series	6xx.x

For the 1xx.x series the second two digits indicate the aluminum purity. The last digit indicates whether the product form is a casting (designated by 0) or ingot (designated by 1 or 2) and is often dropped in common usage for castings. A modification of the original alloy or impurity limits is indicated by a serial letter prefix before the numerical designation. An example is A356, a modification of 356 with slightly tighter controls on the minor alloying elements iron, copper, manganese, and zinc. The serial letters are assigned in alphabetical order but omitting I, O, Q, and X.

Tempers

The **temper designation system** applies to all aluminum and aluminum alloys, wrought or cast, except ingot. The temper designation follows the alloy designation, separated by a hyphen. The basic tempers are designated by letters as follows:

F   as fabricated (no control over thermal conditions or strain hardening)
O   annealed (the lowest strength temper of an alloy)
H   strain-hardened (applies to wrought products only)
T   thermally treated

The H and T tempers are followed by one or more numbers, e.g., T6 or H14, indicating specific sequences of treatments. The properties of alloys of H and T tempers are discussed further below.

The alloys listed in Table 6.4.2 are divided into two classes: those which may be strengthened by cold working (also called strain hardening) only (non-heat-treatable, designated by the H temper) and those which may be strengthened by thermal treatment (heat-treatable, designated by the T temper).

Heat treatment (see Table 6.4.3) begins with a solution heat treatment to put the soluble alloying elements into solid solution. For example, alloy 6061 is heated to 990°F (530°C). This is followed by rapidly dropping the temperature, usually by quenching in water. At this point, the alloy is very workable; but if it is held at room temperature or above, strength gradually increases and ductility decreases as precipitation of constituents of the supersaturated solution begins. This gradual change in properties is called natural aging. Alloys that have received solution heat treatment

Table 6.4.2 Composition and Typical Room-Temperature Properties of Wrought Aluminum Alloys

Aluminum Assoc. alloy designation*	Nominal composition, % (balance aluminum)					Temper†	Density, lb/in³	Electrical conductivity, % IACS	Brinell hardness, 500-kg load, 10-mm ball	Mechanical properties		
	Cu	Si	Mn	Mg	Other elements					Tensile yield strength,‡ ksi	Tensile strength, ksi	Elong, %, in 2 in§
							Work-hardenable alloys					
1100	0.12					O	0.098	59	23	5	13	35
						H14			32	17	18	9
						H18		57	44	22	24	5
3003	0.12		1.2			O	0.099	50	28	6	16	30
						H14		41	40	21	22	8
						H18		40	55	27	29	4
5052				2.5	0.25 Cr	O	0.097	35	47	13	28	25
						H34			68	31	38	10
						H38		35	77	37	42	7
5056			0.12	5.0	0.12 Cr	O	0.095	29	65	22	42	35
						H18		27	105	59	63	10
							Heat-treatable alloys					
2011	5.5				0.4 Pb	T3	0.102	39	95	43	55	15
					0.4 Bi	T8		45	100	45	59	12
2014	4.4	0.8	0.8	0.5		O	0.101	50	45	14	27	18
						T4, 451		34	105	42	62	20
						T6, 651		40	135	60	70	13
Alclad 2014						O	0.101	40		10	25	21
						T4, 451				37	61	22
						T6, 651				60	68	10
2017	4.0	0.5	0.7	0.6		O	0.101	50	45	10	26	22
						T4, 451		34	105	40	62	22
2024	4.4		0.6	1.5		O	0.100	50	47	11	27	20
						T3		30	120	50	70	18
						T4, 351		30	120	47	68	20
Alclad 2024						O	0.100			11	26	20
						T3				45	65	18
						T4, 351				42	64	19
2025	4.4	0.8	0.8			T6	0.101	40	110	37	58	19
2219	6.3		0.3		0.18 Zr	O		44		11	25	18
					0.10 V	T351				36	52	17
					0.06 Ti	T851				51	66	10
6053		0.7	1.2		0.25 Cr	O	0.097	45	26	8	16	35
						T6		42	80	32	37	13
6061	0.28	0.6		1.0	0.20 Cr	O	0.098	47	30	8	18	25
						T6, 651		43	95	40	45	12
6063		0.4		0.7		O	0.097	58	25	7	13	25
						T6		53	73	31	35	12
7075	1.6			2.5	5.6 Zn	O	0.101		60	15	33	17
					0.23 Cr	T6, 651		33	150	73	83	11
Alclad 7075						O	0.101			14	32	17
						T6, 651				67	76	11
7178	2.0			2.8	0.23 Cr, 6.8 Zn	T6, 651	0.102	31		78	88	10

NOTE: See Sec. 1 for conversion factors for SI units.
* Aluminum Association Standardized System of Alloy Designation adopted October 1954.
† Standard temper designations: O = fully annealed. H14 and H34 correspond to half-hard and H18 to H38 to hard strain-hardened tempers. T3 = solution treated and then cold-worked; T4 = solution heat-treated; T6 = solution treated and then artificially aged; T-51 = stretcher stress-relieved; T8 = solution treated, cold-worked, and artificially aged.
‡ At 0.2% offset.
§ Percent in 2 in, $\frac{1}{16}$-in thick specimen.

Table 6.4.3 Conditions for Heat Treatment of Aluminum Alloys

Alloy	Product	Solution heat treatment			Precipitation heat treatment		
		Temp, °F	Temper designation		Temp, °F	Time of aging	Temper designation
2014	Extrusions	925–945	T4		315–325	18 h	T6
2017	Rolled or cold finished wire, rod and bar	925–945	T4		Room	4 days	
2024	Coiled Sheet	910–930	T4		Room	4 days	
6063	Extrusions	940–960	T4		340–360	8 h	T6
6061	Sheet	980–1,000	T4		310–330	18 h	T6
7075	Forgings	860–880	W		215–235 }*	6–8 h	T73
					340–360 ∫	8–10 h	

°C = (°F − 32)/1.8.
* This is a two-stage treatment.

only are designated by a W or T temper and can be more readily cold-worked in this condition. Alloys that naturally age to a substantially stable condition are designated with a T1 through T4 temper, depending on the shaping process and degree of cold work imparted. Some materials that are to receive severe forming operations, such as cold-driven rivets, are held in this workable condition by storing below 32°F (0°C).

By applying a second heat treatment, precipitation heat treatment or artificial aging, at slightly elevated temperatures for a controlled time, further strengthening is achieved and properties are stabilized. For example, 6061 sheet is artificially aged when held at 320°F (160°C) for 18 h. Material so treated is designated T6 temper. Artificial aging changes the shape of the stress-strain curve above yield, thus affecting the tangent modulus of elasticity. For this reason, inelastic buckling strengths are determined differently for artificially aged alloys and alloys that have not received such treatment.

Strain hardening, also called cold working, is used to strengthen non-heat-treatable alloys, which cannot be strengthened by heat treatment. Non-heat-treatable alloys may be given a heat treatment to stabilize mechanical properties.

Annealing reduces an alloy's strength to its lowest level but increases ductility. All wrought alloys may be annealed by heating to a prescribed temperature. For example, 6061 is annealed when held at 775°F (415°C) for approximately 2 to 3 h. Strengths are degraded whenever strain-hardened or artificially aged material is heated above about 250°F (120°C). The loss in strength is proportional to the time at elevated temperature. For example, 6061-T6 heated to 400°F (200°C) for no longer than 30 min or 300°F (150°C) for 1,000 h exhibits less than a 5 percent decrease in strength. Table 6.4.4 shows the effect of elevated temperatures on the strength of some alloy tempers.

Minimum **mechanical properties** (usually tensile ultimate and yield strengths and elongation) are specified for most wrought alloy tempers by the Aluminum Association in "Aluminum Standards and Data." These minimum properties are also listed in ASTM specifications, but are grouped by ASTM by product (such as sheet and plate) rather than by alloy. The minimum properties are established at levels at which 99 percent of the material is expected to exceed at a confidence level of 95 percent. The strengths of aluminum members subjected to axial force, bending, shear, and torsion under static and fatigue loads given in the Aluminum Association's "Aluminum Design Manual" are calculated using these minimum properties.

Product Forms

Extrusions, produced by pushing heated metal through a die opening, are among the most useful aluminum products. A great variety of custom shapes, as well as standard shapes such as I beams, angles, channels, pipe, rectangular tube, tees, and zees, are extruded. Extrusion cross sections may be as large as those that fit with a 31-in (790-mm) circle, but commonly available shapes are limited to about an 18-in (450-mm) circle. Alloys extruded include 1100, 1350, 2014, 2024, 2219, 3003, 5083, 5086, 5154, 5454, 6005, 6061, 6063, 6082, 6101, 6105, 6351, 6463, 7005, 7050, 7075, 7178, and 7475. The most common are the 6xxx series, especially 6061 and 6063. Rod, bar, and wire are also produced rolled or cold-finished as well as extruded. Tubes may be drawn or extruded.

Flat-rolled products include foil, sheet, and plate. **Foil** is defined as rolled product less than 0.006 in (through 0.15 mm) thick. **Sheet** is less than 0.25 in (through 6.3 mm) thick but not less than 0.006 in thick; aluminum sheet gauge thicknesses are different from those used for steel, and decimal thicknesses are preferred when ordering. Sheet is available in flat panels (solid, perforated, or expanded metal) and coiled. **Plates** are 0.25 in thick and greater, up to about 8 in (200 mm) thick. Minimum bend radii for sheet and plate depend on alloy and temper, but are generally greater than the bend radii for mild carbon steel. Commercial roofing and siding sheet are available in a number of profiles, including corrugated, ribbed, and V-beam.

Aluminum **forgings** are produced by open-die and closed-die methods. Minimum mechanical properties are not published for open-die forgings, so structural applications usually require closed-die forgings.

Table 6.4.4 Typical Tensile Properties of Aluminum Alloys at Elevated Temperatures

Alloy and temper	Property *	Temp, °F*				
		75	300	400	500	700
Wrought alloys						
1100-H18	T.S.	24	18	6	4	2.1
	Y.S.	22	14	3.5	2.6	1.6
	El.	15	20	65	75	85
2017-T4	T.S.	62	40	16	9	4.3
	Y.S.	40	30	13	7.5	3.5
	El.	22	15	35	45	70
2024-T4	T.S.	68	45	26	11	5
	Y.S.	47	36	19	9	4
	El.	19	17	27	55	100
3003-H18	T.S.	29	23	14	7.5	2.8
	Y.S.	27	16	9	4	1.8
	El.	10	11	18	60	70
3004-H38	T.S.	41	31	22	12	5
	Y.S.	36	27	15	7.5	3
	El.	6	15	30	50	90
5052-H34	T.S.	38	30	24	12	5
	Y.S.	31	27	15	7.5	3.1
	El.	16	27	45	80	130
6061-T6	T.S.	45	34	19	7.5	3
	Y.S.	40	31	15	5	1.8
	El.	17	20	28	60	95
6063-T6	T.S.	35	21	9	4.5	2.3
	Y.S.	31	20	6.5	3.5	2
	El.	18	20	40	75	105
7075-T6	T.S.	83	31	16	11	6
	Y.S.	73	27	13	9	4.6
	El.	11	30	55	65	70

		Temp, °F*				
		75	300	400	500	600
Sand castings						
355-T51	T.S.	28	24	14	10	6
	Y.S.	23	19	10	5	3
	El.	2	3	8	16	36
356-T6	T.S.	33	23	12	8	4
	Y.S.	24	20	9	5	3
	El.	4	6	18	35	60
Permanent-mold castings						
355-T51	T.S.	30	23	15	10	6
	Y.S.	24	20	10	5	3
	El.	2	4	9	33	98
356-T6	T.S.	38	21	12	8	4
	Y.S.	27	17	9	5	3
	El.	5	10	30	55	70

*T.S., tensile strength ksi. Y.S., yield strength, ksi, 0.2 percent offset. El , elongation in 2 in, percent.

ksi × 6.895 = MPa °C = (°F − 32)/1.8.

Tensile tests made on pieces maintained at elevated temperatures for 10,000 h under no load. Stress applied at 0.05 in/(in/min) strain rate.

Like castings, forgings may be produced in complex shapes, but have more uniform properties and better ductility than castings and are used for products such as wheels and aircraft components requiring high strength-to-weight ratios. For some forging alloys, minimum mechanical properties are slightly lower in directions other than parallel to the grain flow. Common forging alloys are 2014, 2219, 5083, 6061, 7050, and 7175.

Castings are used for parts of complex shapes and are produced by sand casting, permanent mold casting, and die casting. The compositions and minimum mechanical properties of some cast aluminum alloys are given in Tables 6.4.5a to 6.4.5d. Castings generally exhibit greater variation in

Table 6.4.5a Composition Limits of Aluminum Casting Alloys (Percent)

Alloy	Product*	Silicon	Iron	Copper	Manganese	Magnesium	Chromium	Nickel	Zinc	Titanium	Tin	Others Each	Total
201.0	S	0.10	0.15	4.0–5.2	0.20–0.50	0.15–0.55	—	—	—	0.15–0.35	—	0.05	0.10
204.0	S&P	0.20	0.35	4.2–5.0	0.10	0.15–0.35	—	0.05	0.10	0.15–0.30	—	0.05	0.15
208.0	S&P	2.5–3.5	1.2	3.5–4.5	0.50	0.10	—	0.35	1.0	0.25	—	—	0.50
222.0	S&P	2.0	1.5	9.2–10.7	0.50	0.15–0.35	—	0.50	0.8	0.25	—	—	0.35
242.0	S&P	0.7	1.0	3.5–4.5	0.35	1.2–1.8	0.25	1.7–2.3	0.35	0.25	—	0.05	0.15
295.0	S	0.7–1.5	1.0	4.0–5.0	0.35	0.03	—	—	0.35	0.25	—	0.05	0.15
296.0	P	2.0–3.0	1.2	4.0–5.0	0.35	0.05	—	0.35	0.50	0.25	—	—	0.35
308.0	P	5.0–6.0	1.0	4.0–5.0	0.50	0.10	—	—	1.0	0.25	—	—	0.50
319.0	S&P	5.5–6.5	1.0	3.0–4.0	0.50	0.10	—	0.35	1.0	0.25	—	—	0.50
328.0	S	7.5–8.5	1.0	1.0–2.0	0.20–0.6	0.20–0.6	0.35	0.25	1.5	0.25	—	—	0.50
332.0	P	8.5–10.5	1.2	2.0–4.0	0.50	0.50–1.5	—	0.50	1.0	0.25	—	—	0.50
333.0	P	8.0–10.0	1.0	3.0–4.0	0.50	0.05–0.50	—	0.50	1.0	0.25	—	—	0.50
336.0	P	11.0–13.0	1.2	0.50–1.5	0.35	0.7–1.3	—	2.0–3.0	0.35	0.25	—	0.05	—
354.0	S&P	8.6–9.4	0.20	1.6–2.0	0.10	0.40–0.6	—	—	0.10	0.20	—	0.05	0.15
355.0	S&P	4.5–5.5	0.6	1.0–1.5	0.50	0.40–0.6	0.25	—	0.35	0.25	—	0.05	0.15
C355.0	S&P	4.5–5.5	0.20	1.0–1.5	0.10	0.40–0.6	—	—	0.10	0.20	—	0.05	0.15
356.0	S&P	6.5–7.5	0.6	0.25	0.35	0.25–0.45	—	—	0.35	0.25	—	0.05	0.15
A356.0	S&P	6.5–7.5	0.20	0.20	0.10	0.25–0.45	—	—	0.10	0.20	—	0.05	0.15
357.0	S&P	6.5–7.5	0.15	0.05	0.03	0.45–0.6	—	—	0.05	0.20	—	0.05	0.15
A357.0	S&P	6.5–7.5	0.20	0.20	0.10	0.40–0.7	—	—	0.10	0.04–0.20	—	0.05	0.15
359.0	S&P	8.5–9.5	0.20	0.20	0.10	0.50–0.7	—	—	0.10	0.20	—	0.05	0.15
360.0	D	9.0–10.0	2.0	0.6	0.35	0.40–0.6	—	0.50	0.50	—	0.15	—	0.25
A360.0	D	9.0–10.0	1.3	0.6	0.35	0.40–0.6	—	0.50	0.50	—	0.15	—	0.25
380.0	D	7.5–9.5	2.0	3.0–4.0	0.50	0.10	—	0.50	3.0	—	0.35	—	0.50
A380.0	D	7.5–9.5	1.3	3.0–4.0	0.50	0.10	—	0.50	3.0	—	0.35	—	0.50
383.0	D	9.5–11.5	1.3	2.0–3.0	0.50	0.10	—	0.30	3.0	—	0.15	—	0.50
384.0	D	10.5–12.0	1.3	3.0–4.5	0.50	0.10	—	0.50	3.0	—	0.35	—	0.50
390.0	D	16.0–18.0	1.3	4.0–5.0	0.10	0.45–0.65	—	—	0.10	0.20	—	—	0.20
B390.0	D	16.0–18.0	1.3	4.0–5.0	0.50	0.45–0.65	—	0.10	1.5	0.10	—	—	0.20
392.0	D	18.0–20.0	1.5	0.40–0.80	0.20–0.60	0.80–1.20	—	0.50	0.50	0.20	0.30	—	0.50
413.0	D	11.0–13.0	2.0	1.0	0.35	0.10	—	0.50	0.50	—	0.15	—	0.25
A413.0	D	11.0–13.0	1.3	1.0	0.35	0.10	—	0.50	0.50	—	0.15	—	0.25
C433.0	D	4.5–6.0	2.0	0.6	0.35	0.10	—	0.50	0.50	—	0.15	—	0.25
443.0	S&P	4.5–6.0	0.8	0.6	0.50	0.05	0.25	—	0.50	0.25	—	0.05	0.35
B443.0	S&P	4.5–6.0	0.8	0.15	0.35	0.05	—	—	0.35	0.25	—	0.05	0.15
A444.0	P	6.5–7.5	0.20	0.10	0.10	0.05	—	—	0.10	0.20	—	0.05	0.15
512.0	S	1.4–2.2	0.6	0.35	0.8	3.5–4.5	0.25	—	0.35	0.25	—	0.05	0.15
513.0	P	0.30	0.40	0.10	0.30	3.5–4.5	—	—	1.4–2.2	0.20	—	0.05	0.15
514.0	S	0.35	0.50	0.15	0.35	3.5–4.5	—	—	0.15	0.25	—	0.05	0.15
518.0	D	0.35	1.8	0.25	0.35	7.5–8.5	—	0.15	0.15	—	0.15	—	0.25
520.0	S	0.25	0.30	0.25	0.15	9.5–10.6	—	—	0.15	0.25	—	—	0.15
535.0	S&P	0.15	0.15	0.05	0.10–0.25	6.2–7.5	—	—	—	0.10–0.25	—	0.05	0.15
705.0	S&P	0.20	0.8	0.20	0.40–0.6	1.4–1.8	0.20–0.40	—	2.7–3.3	0.25	—	0.05	0.15
707.0	S&P	0.20	0.8	0.20	0.40–0.6	1.8–2.4	0.20–0.40	—	4.0–4.5	0.25	—	0.05	0.15
710.0	S	0.15	0.50	0.35–0.65	0.05	0.6–0.8	—	—	6.0–7.0	0.25	—	0.05	0.15
711.0	P	0.30	0.7–1.4	0.35–0.65	0.05	0.25–0.45	—	—	6.0–7.0	0.20	—	0.05	0.20
712.0	S	0.30	0.50	0.25	0.10	0.50–0.65	0.40–0.6	—	5.0–6.5	0.15–0.25	—	0.05	0.20
713.0	S&P	0.25	1.1	0.40–1.0	0.6	0.20–0.50	0.35	0.15	7.0–8.0	0.25	—	0.10	0.25
771.0	S	0.15	0.15	0.10	0.10	0.8–1.0	0.06–0.20	—	6.5–7.5	0.10–0.20	—	0.05	0.15
850.0	S&P	0.7	0.7	0.7–1.3	0.10	0.10	—	0.7–1.3	—	0.20	—	—	0.30
851.0	S&P	2.0–3.0	0.7	0.7–1.3	0.10	0.10	—	0.30–0.7	—	0.20	—	—	0.30
852.0	S&P	0.40	0.7	1.7–2.3	0.10	0.6–0.9	—	0.9–1.5	—	0.20	—	—	0.30

* S = sand casting, P = permanent mold casting, D = die casting.

Table 6.4.5b   Mechanical Properties of Aluminum Sand Castings

Alloy	Temper	Minimum tensile ultimate strength, ksi	Minimum tensile yield strength, ksi	Minimum percent elongation in 2 in or 4D	Typical Brinell hardness (500-kgf load, 10-mm ball)
201.0	T7	60.0	50.0	3.0	—
204.0	T4	45.0	28.0	6.0	—
242.0	O	23.0	A	A	70
242.0	T61	32.0	20.0	A	105
A242.0	T75	29.0	A	1.0	75
295.0	T4	29.0	13.0	6.0	60
295.0	T6	32.0	20.0	3.0	75
295.0	T62	36.0	28.0	A	95
295.0	T7	29.0	16.0	3.0	70
319.0	F	23.0	13.0	1.5	70
319.0	T5	25.0	A	A	80
319.0	T6	31.0	20.0	1.5	80
355.0	T51	25.0	18.0	A	65
355.0	T6	32.0	20.0	2.0	80
355.0	T71	30.0	22.0	A	75
C355.0	T6	36.0	25.0	2.5	—
356.0	F	19.0	9.5	2.0	55
356.0	T51	23.0	16.0	A	60
356.0	T6	30.0	20.0	3.0	70
356.0	T7	31.0	A	A	75
356.0	T71	25.0	18.0	3.0	60
A356.0	T6	34.0	24.0	3.5	80
A356.0	T61	35.0	26.0	1.0	—
443.0	F	17.0	7.0	3.0	40
B443.0	F	17.0	6.0	3.0	40
512.0	F	17.0	10.0	—	50
514.0	F	22.0	9.0	6.0	50
520.0	T4	42.0	22.0	12.0	75
535.0	F	35.0	18.0	9.0	70
705.0	T5	30.0	17.0 B	5.0	65
707.0	T7	37.0	30.0 B	1.0	80
710.0	T5	32.0	20.0	2.0	75
712.0	T5	34.0	25.0 B	4.0	75
713.0	T5	32.0	22.0	3.0	75
771.0	T5	42.0	38.0	1.5	100
771.0	T51	32.0	27.0	3.0	85
771.0	T52	36.0	30.0	1.5	85
771.0	T6	42.0	35.0	5.0	90
771.0	T71	48.0	45.0	2.0	120
850.0	T5	16.0	A	5.0	45
851.0	T5	17.0	A	3.0	45
852.0	T5	24.0	18.0	A	60

A = not required; B = to be determined only when specified by the purchaser. These properties are for separately cast test specimens, not the castings themselves.

strength and less ductility than wrought products. Tolerances, draft requirements, heat treatments, and quality standard are given in the Aluminum Association's "Standards for Aluminum Sand and Permanent Mold castings." Tolerance and the quality and frequency of inspection must be specified by the user if desired.

Sand castings (ASTM B26) are for larger parts—up to 7,000 lb (3,200 kg)—in relatively small quantities at slow solidification rates. The sand mold is used only once. Tolerance and minimum thicknesses for sand and castings are greater than those for other castings types.

Permanent mold castings (ASTM B108) are produced by pouring the molten metal into a reusable mold, sometimes with a vacuum to assist the flow. While more expensive than sand castings, permanent mold castings can be used for parts with wall thicknesses as thin as about 0.09 in (2.3 mm).

In die casting (ASTM B85), aluminum is injected into a reusable steel mold, or die, at high velocity; fast solidification rates are achieved.

The lowest-cost, general-purpose casting alloy is 356-T6; A356-T6 is common in more structurally demanding applications. Alloy A444-T4 provides excellent ductility, exceeding that of many wrought alloys. These three alloys rank among the most weldable of the casting alloys.

The cast alloys with copper (2xx.x) generally offer the higher strengths at elevated temperatures. In addition to end use, alloys selection should account for fluidity, resistance to hot cracking, and pressure tightness.

Fabrication

Machining (see Sec. 13)   Many aluminum alloys are easily machined without special technique at faster cutting speeds than for other metals. Pure aluminum and alloys of aluminum-magnesium (5xxx) are harder to machine than alloys of aluminum-copper (2xxx) and aluminum-zinc (7xxx). The most machinable wrought alloy is 2011 in the T3, T4, T6, and T8 tempers, producing small broken chips and excellent finish; it is used where physical properties are subordinate to high machinability, such as for screw machine products. Castings are also machined; aluminum-copper (2xx.x), aluminum-magnesium (5xx.x), aluminum-zinc (7xx.x), and aluminum-tin (8xx.x) are among the best choices for good machinability.

Joining   Mechanical fasteners (including rivets, bolts, and screws) are the most common method of joining aluminum parts, because the application of heat during welding decreases the strength of all but annealed material. Aluminum **bolts** (2024-T4, 6061-T6, or 7075-T73)

Table 6.4.5*c*   Mechanical Properties of Aluminum Permanent Mold Castings

Alloy	Temper	Minimum tensile ultimate strength, ksi	Minimum tensile yield strength, ksi	Minimum percent elongation in 2 in or 4*D*	Typical Brinnell hardness (500-kgf load, 10-mm ball)
204.0	T4 separately cast specimens	48.0	29.0	8.0	—
242.0	T571	34.0	—	—	105
242.0	T61	40.0	—	—	110
319.0	F	27.0	14.0	2.5	95
332.0	T5	31.0	—	—	105
333.0	F	28.0	—	—	90
333.0	T5	30.0	—	—	100
333.0	T6	35.0	—	—	105
333.0	T7	31.0	—	—	90
336.0	T551	31.0	—	—	105
336.0	T65	40.0	—	—	125
354.0	T61 separately cast specimens	48.0	37.0	3.0	
354.0	T61 castings, designated area	47.0	36.0	3.0	
354.0	T61 castings, no location designated	43.0	33.0	2.0	
354.0	T62 separately cast specimens	52.0	42.0	2.0	
354.0	T62 castings, designated area	50.0	42.0	2.0	
354.0	T62 castings, no location designated	43.0	33.0	2.0	
355.0	T51	27.0	—	—	75
355.0	T62	42.0	—	—	105
355.0	T7	36.0	—	—	90
355.0	T71	34.0	27.0	—	80
C355.0	T61 separately cast specimens	40.0	30.0	3.0	85–90
C355.0	T61 castings, designated area	40.0	30.0	3.0	
C355.0	T61 castings, no location designated	37.0	30.0	1.0	85
356.0	F	21.0	10.0	3.0	
356.0	T6	33.0	22.0	3.0	85
356.0	T71	25.0	—	3.0	70
A356.0	T61 separately cast specimens	38.0	26.0	5.0	80–90
A356.0	T61 castings, designated area	33.0	26.0	5.0	
A356.0	T61 castings, no location designated	28.0	26.0	3.0	
357.0	T6	45.0	—	3.0	—
A357.0	T61 separately cast specimens	45.0	36.0	3.0	100
A357.0	T61 castings, designated area	46.0	36.0	3.0	—
A357.0	T61 castings, no location designated	41.0	31.0	3.0	—
359.0	T61 separately cast specimens	45.0	34.0	4.0	90
359.0	T61 castings, designated area	45.0	34.0	4.0	
359.0	T61 castings, no location designated	40.0	30.0	3.0	
359.0	T62 separately cast specimens	47.0	38.0	3.0	100
359.0	T62 castings, designated area	47.0	38.0	3.0	
359.0	T62 castings, no location designated	40.0	30.0	3.0	
443.0	F	21.0	7.0	2.0	45
B443.0	F	21.0	6.0	2.5	45
A444.0	T4 separately cast specimens	20.0	—	20	—
A444.0	T4 castings, designated area	20.0	—	20	—
513.0	F	22.0	12.0	2.5	60
535.0	F	35.0	18.0	8.0	—
705.0	T1 or T5	37.0	17.0	10.0	
707.0	T1	42.0	25.0	4.0	
707.0	T7	45.0	35.0	3.0	
711.0	T1	28.0	18.0	7.0	70
713.0	T1 or T5	32.0	22.0	4.0	
850.0	T5	18.0	—	8.0	
851.0	T5	17.0	—	3.0	
851.0	T6	18.0	—	8.0	
852.0	T5	27.0	—	3.0	

A = not required;
B = to be determined only when specified by the purchaser.
When not stated otherwise, properties are for separately cast test specimens, not the castings themselves.

are available in diameters from ¼ in (6.4 mm) to 1 in (25 mm), with properties meeting ASTM F468. Mechanical properties are given in Table 6.4.6. Aluminum nuts (2024-T4, 6061-T6, or 6262-T9) are also available (ASTM F467). Galvanized steel and 300 series stainless steel bolts are also used to join aluminum. Hole diameter usually exceeds bolt diameter by 1/16 in (1.6 mm) or less. **Rivets** are used to resist shear loads only; they do not develop sufficient clamping force between the parts joined and so cannot reliably resist tensile loads. In general, rivets of composition similar to the base metal are used; specialized aerospace applications sometimes use titanium and nickel alloy fasteners. Table 6.4.7 lists common aluminum rivet alloys and their minimum ultimate shear strengths. Hole diameter for cold-driven rivets should be no more than 4 percent larger than the nominal rivet diameter. **Screws** of 2024-T4, 7075-T73, or stainless steel are often used to fasten aluminum sheet.

Table 6.4.5*d*   Mechanical Properties of Aluminum Die Castings

Alloy	Typical tensile ultimate strength,[*] ksi	Typical tensile yield strength,[*] ksi	Typical elongation in 2 in (%)
360.0	44	25	2.5
A360.0	46	24	3.5
380.0	46	23	2.5
A380.0	47	23	3.5
383.0	45	22	3.5
384.0	48	24	2.5
390.0	40.5	35.0	< 1
B390.0	46.0	36.0	< 1
392.0	42.0	39.0	< 1
413.0	43	21	2.5
A413.0	42	19	3.5
C443.0	33	14	9.0
518.0	45	28	5

[*] These are properties of separately cast test bars, not the castings themselves.

Table 6.4.6   Mechanical Properties of Aluminum Bolts

Alloy temper	Minimum tensile strength, ksi	Minimum shear strength, ksi
2024-T4	62	37
6061-T6	42	25
7075-T73	68	41

ksi × 6.895 = MPa

Table 6.4.7   Minimum Shear Strength of Aluminum Rivets

Alloy temper before driving	Minimum shear strength, ksi
1100-H14	10
2017-T4	33
2024-T4	37
2117-T4	26
5056-H32	25
6053-T61	20
6061-T6	25
7050-T7	39

ksi × 6.895 = MPa

Holes for fasteners may be punched, drilled, or reamed, but not punched if the metal thickness is greater than the diameter of the hole. **Adhesives** are also used to join aluminum and are a good replacement for seal welding.

Welding (see Sec. 13)   Most wrought aluminum alloys are weldable by experienced operators using either the fusion or the resistance method. Fusion welding is typically by **gas tungsten arc welding** (GTAW), commonly called TIG (for tungsten inert gas) welding, or **gas metal arc welding** (GMAW), referred to as MIG (metal inert gas) welding. TIG welding is usually used to join parts from about $\frac{1}{32}$ to $\frac{1}{8}$ in (0.8 to 3.2 mm) thick; MIG welding is usually used to weld thicker parts. The American Welding Society Publication D1.2 "Structural Welding Code—Aluminum" provides specifications for structural applications of aluminum fusion welding methods. Filler rod alloys are chosen for strength, corrosion resistance, and compatibility with the parent alloys to be welded. Cast alloys may also be welded, but are more susceptible to cracking than wrought alloys. All aluminum tempers other than annealed tempers suffer a reduction in strength in the heat-affected weld zone (known as the HAZ); this reduction is less in

some of the aluminum-magnesium (5xxx) alloys. **Postweld heat treatment** may be used to counter the reduction in strength caused by welding, but care must be taken to avoid embrittling or warping the weldment. **Resistance welding** includes spot welding, often used for lap joints, and seam welding. Friction stir welding, electron beam welding, and laser beam welding are more recent welding methods that may reduce the extent of heat-affected zones. Methods of **nondestructive testing** of aluminum welds include dye-penetrant methods to detect flaws accessible to the surface and ultrasonic and radiographic (X-ray) inspection.

Brazing is also used to join aluminum alloys with relatively high melting points, such as 1050, 1100, 3003, and 6063, using aluminum-silicon alloys such as 4047 and 4145. Special brazing alloys dominate the heat exchanger market, especially for air conditioners and radiators for trucks and cars. Aluminum may also be **soldered**, but corrosion resistance of soldered joints is inferior to that of welded, brazed, or mechanically fastened joints.

Corrosion Resistance

Although aluminum is chemically active, the presence of a rapidly forming and firmly adherent self-healing oxide surface coating inhibits corrosive action except under conditions that tend to remove this surface film. Concentrated nitric and acetic acids are handled in aluminum not only because of its resistance to attack but also because any resulting corrosion products are colorless. For the same reason, aluminum is employed in the preparation and storage of foods and beverages. Hydrochloric acid and most alkalies dissolve the protective surface film and result in fairly rapid attack. Moderately alkaline soaps and the like can be used with aluminum if a small amount of sodium silicate is added. Aluminum is very resistant to sulfur and most of its gaseous products. Except in special applications, aluminum is not recommended in solutions of pH of less than 4 or greater than 8. Aluminum reacts with many organic compounds and should be used with appropriate corrosion inhibitors in such exposure.

Galvanic corrosion may occur when aluminum is electrically connected by an electrolyte to another metal. Aluminum is more anodic than most metals and will be sacrificed while protecting the other metal, a principal called cathodic protection. Consequently, aluminum is usually isolated from other metals such as steel (but not stainless steel) where moisture is present. Large castings of special aluminum alloys are used to protect steel in offshore oil platforms against corrosion.

Another form of corrosion is **exfoliation**, a delamination or peeling of layers of metal in planes approximately parallel to the metal surface, caused by the formation of corrosion product. Alloys with more than 3 percent magnesium (such as 5083, 5086, 5154, and 5456) and held at temperatures above 150°F (65°C) for extended periods are susceptible to this form of attack.

Care should be taken to store aluminum in a manner to avoid trapping water between adjacent flat surfaces, which causes **water stains**. These stains, which vary in color from dark gray to white, do not compromise strength but are difficult to remove and may be cosmetically unacceptable.

Ordinary atmospheric corrosion is resisted by aluminum and most of its alloys, and they may be used outdoors without any protective coating. (An exception is 2014, which is usually painted when exposed to the elements.) The pure metal is most resistant to attack, and additions of alloying elements decrease corrosion resistance, particularly after heat treatment. Under severe conditions of exposure such as may prevail in marine environments or where the metal is continually in contact with wood or other absorbent material in the presence of moisture, a protective coat of paint on the aluminum is advisable.

Alclad Aluminum   The corrosion resistance of aluminum alloys may by augmented by coating the material with a surface layer of high-purity aluminum or a corrosion-resistant aluminum alloy. Such products are referred to as **alclad.** This cladding becomes an integral part of the material, is metallurgically bonded, and provides cathodic protection in a manner similar to zinc galvanizing on steel. Because the cladding usually has lower strength than the base metal, alclad products have slightly less strength than uncoated material. Cladding thickness varies from 1.5 to 10 percent. Products available clad are 3003 tube, 5056 wire, and 2014, 2024, 2219, 3003, 3004, 6061, 7075, 7178, and 7475 sheet and plate.

Finishes

Anodizing is used to improve the corrosion resistance and appearance of aluminum and is achieved by making the parts to be treated the anode in an electrolytic bath such as sulfuric acid. This process produces a tough, adherent coating of aluminum oxide, usually 0.4 mil (0.01 mm) thick or greater. "The Surface Treatment and Finishing of Aluminum Alloys," by P. G. Sheasby and R. Pinner, provides useful information on anodizing. Any welding should be performed before anodizing, and filler alloys should be chosen judiciously for good anodized color match with the base alloy. The film is colorless on pure aluminum and tends to be gray or colored on alloys containing silicon, copper, or other constituents. To provide a consistent color appearance after anodizing, AQ (anodizing quality) grade may be specified in certain alloys. Where appearance is the overriding concern, 5005 sheet and 6063 extrusions are preferred for anodizing. If a colored finish is desired, the electrolytically oxidized article may be treated with a dye solution. Care must be taken in selecting the dye when the part will be exposed to weather, for not all have proved colorfast. Two-step electrolytic coloring, produced by first clear anodizing and then electrolytically depositing another metal oxide, can produce shades of bronze, burgundy, and blue. Anodized aluminum is used for lighting fixtures, automotive trim, cosmetic cases, and aluminum wheels.

Painting   It is important to properly prepare the surface prior to the application of paint or lacquers. A thin anodic film makes an excellent paint base; alternatively, the aluminum surface may be chemically treated with a dilute phosphoric acid solution. Abrasion blasting may be used on parts thick enough to resist perforation or warping. Zinc chromate is frequently used as a primer, especially for corrosive environments, or another conversion coating or special primer may be used. Most paints are baked on; this can affect the strength of the metal if the bake temperature is high enough or applied long enough to cause annealing. Sometimes the paint baking process is used as the artificial aging heat treatment. Aluminum sheet is available with factory-baked paint finish; minimum mechanical properties must be obtained from the supplier. High-, medium-, and low-gloss paint finishes are available and are determined in accordance with ASTM D523.

Porcelain enameled finishes are ceramic coatings fired to coat the surface of aluminum. They are particularly useful for high-temperature applications such as appliances or cookware or for increased corrosion resistance to acids and alkalies.

The Aluminum Association's publication "Designation System for Aluminum Finishes" describes mechanical, chemical, and anodic finishes; "Care of Aluminum" describes aluminum finishes, cleaning, handling, storage, cleaners, and coatings.

Aluminum Conductors   On a weight basis, aluminum has twice the electrical conductivity of copper; on a volume basis, the conductivity of aluminum is about 62 percent that of copper. For electrical applications, a special, high-purity grade of aluminum is often used (1350, also referred to as EC), or aluminum is alloyed to improve strength with minimum sacrifice in conductivity. Table 6.4.8 lists common conductor alloys and gives their strengths and conductivities in various forms and treatments. In power transmission lines, the necessary strength for long spans is obtained by stranding aluminum wires about a core wire of steel (ACSR) or a higher-strength aluminum alloy. Since many soils are corrosive and electrically conductive, aluminum electrical cables are shrouded in nonpermeable coatings when used underground.

Table 6.4.8   Aluminum Electrical Conductors

Product, alloy, temper (size)	Tensile ultimate strength, ksi*	Tensile yield strength, ksi*	Min. electrical conductivity, % IACS[†]
Drawing Stock (Rod)			
1350-O (0.375- to 1.000-in diameter)	8.5–14.0		61.8
1350-H12, H22 (0.375- to 1.000-in diameter)	12.0–17.0		61.5
1350-H14, H24 (0.375- to 1.000-in diameter)	15.0–20.0		61.4
1350-H16, H26 (0.375- to 1.000-in diameter)	17.0–22.0		61.3
5005-O (0.375-in diameter)	14.0–20.0		54.3
5005-H12, H22 (0.375-in diameter)	17.0–23.0		54.0
5005-H14, H24 (0.375-in diameter)	20.0–26.0		53.9
5005-H16, H26 (0.375-in diameter)	24.0–30.0		53.8
8017-H12, H22 (0.375-in diameter)	16.0–22.0		58.0
8030-H12 (0.375-in diameter)	16.0–20.5		60.0
8176-H14 (0.375-in diameter)	16.0–20.0		59.0
Wire			
1350-H19 (0.0801 to 0.0900 in diameter)	26.0–27.5		61.0
5005-H19 (0.0801 to 0.0900 in diameter)	37.0–39.0		53.5
6201-T81 (0.0612 to 0.1327 in diameter)	46.0–48.0		52.5
8176-H24 (0.0500 to 0.2040 in diameter)	15.0–17.0		61.0
Extrusions			
1350-H111 (all)	8.5 min	3.5 min	61.0
6101-H111 (0.250 to 2.000 in)	12.0 min	8.0 min	59.0
6101-T6 (0.125 to 0.500 in)	29.0 min	25.0 min	55.0
6101-T61 (0.125 to 0.749 in)	20.0 min	15.0 min	57.0
6101-T61 (0.750 to 1.499 in)	18.0 min	11.0 min	57.0
6101-T61 (1.500 to 2.000 in)	15.0 min	8.0 min	57.0
Rolled Bar			
1350-H12 (0.125 to 1.000 in)	12.0 min	8.0 min	61.0
Sawed Plate			
1350-H112 (0.125 to 0.499 in)	11.0 min	6.0 min	61.0
1350-H112 (0.500 to 1.000 in)	10.0 min	4.0 min	61.0
1350-H112 (1.001 to 1.500 in)	9.0 min	3.5 min	61.0
Sheet			
1350-O (0.006 to 0.125 in)	8.0–14.0		61.8

*ksi × 6.895 = MPa.
† % IACS × 0.581 = MS/m.

BEARING METALS
by Frank E. Goodwin

REFERENCES: Current edition of ASM "Metals Handbook." Publications of the various metal producers. Trade association literature containing material properties. Applicable current ASTM and SAE Standards.

Babbitt metal is a general term used for soft tin and lead-base alloys which are cast as bearing surfaces on steel, bronze, or cast-iron shells. Babbitts have excellent **embeddability** (ability to embed foreign particles in itself) and **conformability** (ability to deform plastically to compensate for irregularities in bearing assembly) characteristics. These alloys may be run satisfactorily against a soft-steel shaft. The limitations of Babbitt alloys are the tendency to spread under high, steady loads and to fatigue under high, fluctuating loads. These limitations apply more particularly at higher temperatures; e.g., increase in temperature between 68 and 212°F (20 and 100°C) reduces the metal's strength by 50 percent. These limitations can be overcome by properly designing the thickness and rigidity of the backing material, properly choosing the Babbitt alloy for good mechanical characteristics, and ensuring a good bond between backing and bearing materials.

The important tin- and lead-base (Babbitt) bearing alloys are listed in Table 6.4.9. Alloys 1 and 15 have been used in internal-combustion engines, but the desire of automobile manufacturers to eliminate lead has resulted in the substitution of aluminum-based alloys. Alloy 1 performs satisfactorily at low temperatures, but alloy 15, an arsenic alloy, provides superior performance at elevated temperatures by virtue of its better high-temperature hardness, ability to support higher loads, and longer fatigue life. Alloys 2 and 3 contain more antimony, are harder, and are less likely to pound out. Alloys 7 and 8 are lead-base Babbitts which will function satisfactorily under moderate conditions of load and speed. Alloy B has been used for diesel engine bearings. In general, increasing the lead content in tin-based Babbitt provides higher hardness, greater ease of casting, but lower strength values.

Silver lined bearings have an excellent record in heavy-duty applications in aircraft engines and diesels. For reciprocating engines, silver bearings normally consist of electrodeposited silver on a steel backing with an overlay of 0.001 to 0.005 in of lead. An indium flash on top of the lead overlay is used to increase corrosion resistance of the material.

Aluminum and **zinc** alloys are used for high-load, low-speed applications but have not replaced Babbits for equipment operating under a steady high-speed load. The Al 20 to 30, Sn 3 copper alloy is bonded to a steel bearing shell. The Al 6.25, Sn 1, Ni 1, copper alloy can be either used as-cast or bonded to a bearing shell. The Al 3 cadmium alloy with varying amounts of Si, Cu, and Ni can also be used in either of these two ways. The Zn 11, Al 1, Cu 0.02, magnesium (ZA-12) and Zn 27, Al 2, Cu 0.01 magnesium (ZA-27) alloys are used in cast form, especially in continuous-cast hollow-bar form.

Mention should be made of **cast iron** as a bearing material. The flake graphite in cast iron develops a glazed surface which is useful at surface speeds up to 130 ft/min and at loads up to 150 lb/in² approx. Because of the poor conformability of cast iron, good alignment and freedom from dirt are essential.

Copper-base bearing alloys have a wide range of bearing properties that fit them for many applications. Used alone or in combination with steel, Babbitt (white metal), and graphite, the bronzes and copper-leads meet the conditions of load and speed given in Table 6.4.10. **Copper-lead alloys** are cast onto steel backing strips in very thin layers (0.02 in) to provide bearing surfaces. Lead-based bearing alloys are still used in diesel engines; no lead-free alloy can withstand diesel firing loads.

Three families of copper alloys are used for bearing and wear-resistant alloys in cast form: **phosphor bronzes** (Cu-Sn), Cu-Sn-Pb alloys, and **manganese bronze, aluminum bronze,** and **silicon bronze.** Typical compositions and applications are listed in Table 6.4.10. Phosphor bronzes have residual phosphorus ranging from 0.1 to 1 percent. Hardness increases with phosphorus content. Cu-Sn-Pb alloys have high resistance to wear, high hardness, and moderate strength. The high lead compositions are well suited for applications where lubricant may be deficient. The Mn, Al, and Si bronzes have high tensile strength, high hardness, and good resistance to shock. They are suitable for a wide range of bearing applications.

Porous bearing materials are used in light- and medium-duty applications as small-sized bearings and bushings. Since they can operate for long periods without an additional supply of lubricant, such bearings are useful in inaccessible or inconvenient places where lubrication would be difficult. Porous bearings are made by pressing mixtures of copper and tin (bronze), and often graphite, Teflon, or iron and graphite, and sintering these in a reducing atmosphere without melting. Iron-based, oil-impregnated sintered bearings are often used. By controlling the conditions under which the bearings are made, porosity may be adjusted so that interconnecting voids of up to 35 percent of the total volume may be available for impregnation by lubricants. Applicable specifications for these bearings are given in Tables 6.4.11 and 6.4.12.

Miscellaneous A great variety of materials, e.g., rubber, wood, phenolic, carbon-graphite, ceramets, ceramics, and plastics, are in use for special applications. Carbon-graphite is used where contamination by oil or grease lubricants is undesirable (e.g., textile machinery, pharmaceutical equipment, milk and food processing) and for elevated-temperature applications. Composite bushings of carbon graphite with metals, including copper and iron, are capable of service temperatures up to 1,000°F (537°C). Notable among plastic materials are Teflon and nylon, the polycarbonate Lexan, and the acetal Delrin. Since Lexan and Delrin can be injection-molded easily, bearings can be formed quite economically from these materials.

CEMENTED CARBIDES
by Don Graham

REFERENCES: Schwartzkopf and Kieffer, "Refractory Hard Metals," Macmillan. German, "Powder Metallurgy Science," 2d ed., Metal Powder Industries Federation. "Powder Metallurgy Design Manual," 2d ed., Metal Powder Industries Federation, Princeton, NJ. Goetzal, "Treatise on Powder Metallurgy,"

Table 6.4.9   Compositions and Properties of Some Babbitt Alloys

ASTM B23-00 grade	Composition, %					Compressive ultimate strength		Brinell hardness	
	Sn	Sb	Pb	Cu	As	68°F, lb/in²	20°C, MPa	68°F (20°C)	212°F (100°C)
1	91.0	4.5	0.35	4.5	—	12.9	88.6	17	8
2	89.0	7.5	0.35	3.5	—	14.9	103	24.5	12
3	84.0	8.0	0.35	8.0	—	17.6	121	27.0	14.5
7	10.0	15.0	74.5	0.5	0.45	15.7	108	22.5	10.5
8	5.0	15.0	79.5	0.5	0.45	15.6	108	20.0	9.5
15	1.0	16.0	82.5	0.5	1.10	—	—	21.0	13.0
Other alloys									
B	0.8	12.5	83.3	0.1	3.05	—	—		
ASTM B102, alloy Py 1815A	65	15	18	2	0.15	15	103	23	10

SOURCE: ASTM, reprinted with permission.

Table 6.4.10   Compositions, Properties, and Applications of Some Copper-Base Bearing Metals

| Specifications | Composition, % | | | | | | Minimum tensile strength | | Applications |
	Cu	Sn	Pb	Zn	P	Other	ksi	MPa	
C86100, SAE J462	64	—	—	24	—	3 Fe, 5 Al, 4 Mn	119	820	Extra-heavy-duty bearings
C87610, ASTM B30	90	—	0.2	5	—	4 Si	66	455	High-strength hearings
C90700, ASTM B505	89	11	0.3	—	0.2	—	44	305	Wormgears and wheels; high-speed low-load bearings
C91100	84	16	—	—	—	—	35	240	Heavy load, low-speed bearings
C91300	81	19	—	—	—	—	35	240	Heavy load, low-speed bearings
C93200, SAE J462	83	7	7	3	—	—	35	240	General utility bearings, automobile fittings
C93700, ASTM B22	80	10	10	—	—	—	35	240	High-speed, high-pressure bearings, good corrosion resistant
C93800, ASTM B534	78	7	15	—	—	—	30	205	General utility bearings, backing for Babbit bearings
C94300, ASTM B584	70	5	25	—	—	—	27	185	Low-load, high-speed bearings

SOURCE: ASTM, reprinted with permisstion.

Interscience. Schwartzkopf, "Powder Metallurgy," Macmillan. Davis (ed.), Tool Materials, "ASM Specialty Handbook," ASM, Metals Park, OH, 1995. Graham and Hale, Coated Cutting Tools, *Carbide and Tool J.*, May–June 1982, p. 34. Graham, Trends in Coatings to the Year 2000, "Cutting Tool Technology Conference Proceedings," SME, 1997.

Cemented carbides are a commercially and technically important class of composite material. Comprised primarily of hard tungsten carbide (WC) grains cemented together with a cobalt (Co) binder, they can contain major alloying additions of titanium carbide (TiC), titanium carbonitride (TiCN), tantalum carbide (TaC), niobium carbide (NbC), chromium carbide (CrC), etc., and minor additions of other elements. Because of the extremely high hardness, stiffness, and wear resistance of these materials, their primary use is in cutting tools for the material (usually metal, wood, composite, or rock) removal industry. Secondary applications are as varied as pen balls, food processing equipment, fuel pumps, wear surfaces, fishing line guide rings, large high-pressure components, and even wedding rings. Since their introduction in the mid-1920s, a very large variety of grades of carbides has been developed and put into use in a diverse number of applications. They are ubiquitous throughout industry, particularly so in the high-production metalworking sector.

Most cemented carbide parts are produced by **powder metallurgical processes.** Powders of WC, Co, and sometimes TiC and TaC combined with NbC are blended by either ball or attritor milling. The mixed powder is then **compacted** under very high pressure in appropriately shaped molds. Pressed parts are **sintered** at temperatures between 1,250 and 1,500°C, depending on the composition of the powder. During the

Table 6.4.11   Oil-Impregnated Iron-Base Sintered Bearings (ASTM Standard B439-83)

| Element | Composition, % | | | |
	Grade 1	Grade 2	Grade 3	Grade 4
Copper			7.0–11.0	18.0–22.0
Iron	96.25 min	95.9 min	Remainder*	Remainder*
Total other elements by difference, max	3.0	3.0	3.0	3.0
Combined carbon† (on basis of iron only)	0.25 max	0.25–0.60	—	—
Silicon, max	0.3	0.3	—	—
Aluminum, max	0.2	0.2	—	—

* Total of iron plus copper shall be 97% min.
† The combined carbon may be a metallographic estimate of the carbon in the iron.
SOURCE: ASTM, reprinted with permission.

Table 6.4.12   Oil-Impregnated Sintered Bronze Bearings (ASTM Standard B438-83)

| Element | Composition, % | | | |
| | Grade 1 | | Grade 2 | |
	Class A	Class B	Class A	Class B
Copper	87.5–90.5	87.5–90.5	82.6–88.5	82.6–88.5
Tin	9.5–10.5	9.5–10.5	9.5–10.5	9.5–10.5
Graphite, max	0.1	1.75	0.1	1.75
Lead	*	*	2.0–4.0	2.0–4.0
Iron, max	1.0	1.0	1.0	1.0
Total other elements by difference, max	0.5	0.5	1.0	1.0

* Included in other elements.
SOURCE: ASTM, reprinted with permission.

sintering process, cobalt melts and wets the carbide particles. On cooling, cobalt "cements" the carbide grains together. Essentially complete densification takes place during sintering. A unique feature of cemented carbides is the fact that, although the part shrinks by as much as 40 percent by volume when the porosity is eliminated by the molten cobalt, the pressed part retains its geometric shape very accurately. This is an important feature since further shaping of this extremely hard material can be done accurately only by grinding with diamond wheels or electrodischarge machining. Lasers can be used to cut cemented carbide but not to tight tolerances. While binder metals such as iron or nickel are sometimes used, cobalt is preferred because of its ability to wet the tungsten carbide.

The metal removal industry consumes most of the cemented carbide that is produced. Tips having unique geometries, called **inserts,** are pressed and sintered. Many of these pressed and sintered parts are ready for use directly after sintering. Others are finish-ground either to improve flatness on the seating surface or to produce very tight dimensional tolerances. Although the cutting edges on some inserts are left in the sharp condition, most are honed slightly to improve chippage resistance and reliability. Sharp edges typically are used for machining such materials as nonferrous metals and alloys, high-temperature alloys, alloys used for medical components (e.g., hip and knee replacement parts) and nonmetals. Honed inserts are used for all other applications, including those in the commercially important iron and steels industries.

The first cemented carbides developed consisted only of WC and Co. As cutting speeds were increased to take advantage of the high hardness of this material, temperatures at the tool/workpiece interface increased. In the presence of iron or steel at high temperatures, carbide tools interact chemically with the work materials and result in **cratering,** which, somewhat simplistically, is the dissolution of WC at the cutting edge. This results in a dimple on the surface of the insert that resembles a crater. Additions of secondary carbides such as TiC and TaC minimize cratering—and thereby improve tool life—by virtue of their greater chemical stability. Grades containing these secondary carbides were introduced during the 1940s and were able to machine steel at speeds up to 600 surface feet per minute (sfpm).

With the ever-increasing desire for even greater metal removal rates, higher cutting speeds (and accompanying higher temperatures) were used. **Chemical vapor deposition** (CVD) overlay coatings were developed which were far more effective at preventing cratering than TiC and TaC additions to the carbide. These hard coatings have the additional benefit of increased abrasion resistance.

Titanium carbide coatings became available in 1969; titanium nitride (TiN) coatings followed shortly, Aluminum oxide (Al_2O_3) coatings for even higher-speed applications were introduced in 1973. Today the large majority of coated carbide grades incorporate all three materials in distinct layers. Over the years dramatic improvements in deposition techniques and structural control have led to equally dramatic improvements in performance. Coatings typically result in tool life increases of at least 50 percent although 10-fold improvements are common. Extreme cases of 10,000 percent tool life improvements over uncoated carbide have been reported.

Each coating composition has its own specific advantages. For example, TiC and TiCN are very hard at lower temperatures and are ideal for very abrasive applications. Hence carbide grades designed for machining ductile irons, austempered ductile iron, compacted graphite iron, malleable iron, alloy steels, etc. incorporate relatively thick layers of these coatings.

TiN, along with other commercially available nitride coatings such as hafnium nitride (HfN) and zirconium nitride (ZrN), exhibits excellent metal buildup resistance. This feature is particularly important when machining softer, gummy alloys such as stainless steel, low-carbon steel, and many high-temperature alloys. Metal buildup occurs when fragments of the work material become welded to the cutting edge. This change in the geometry of the cutting edge results in a poor surface finish and, eventually, chippage of the cutting edge. Since metal buildup is particularly problematic at low cutting speeds, grades designed for lower speeds almost always feature this coating. Even grades designed

for higher speeds often use TiN for a top coat, not only to prevent metal buildup but also, because it is gold in color, to allow for easy determination of which insert edges have been used. Such detection is more difficult if the insert is black.

Aluminum oxide is very stable chemically and retains its hardness at elevated temperatures. Consequently, this coating is especially effective at elevated speeds. Grades designed for high-productivity ferrous applications almost always feature a thick layer of Al_2O_3. Although clear, Al_2O_3 appears black due to the adhesive that bonds this coating to the substrate or underlying coating layers.

Another commercially important coating technique is **physical vapor deposition** (PVD). For certain applications, this coating technique provides certain advantages over the CVD technique. First, PVD coatings are thinner and therefore replicate the cutting edge condition more accurately. This means that very sharp edges can be preserved with these coatings. Second, there is no loss in strength associated with the deposition of a PVD coating. The presence of a CVD coating, by comparison, usually reduces the edge toughness of a carbide insert by about 10 percent. As a consequence, PVD coated carbides are used most often for the turning of high-temperature alloys, stainless steels at low to medium cutting speeds, and for a wide variety of drilling, threading, and even milling of a broad range of materials. PVD coated carbide is also used in a number of mold, die, punch, and wear resistance applications.

The primary disadvantage of PVD coatings is their lack of thickness and consequent low wear and crater resistance. Typical CVD coated inserts have coating thicknesses ranging from 10 to 20 μm; PVD coatings are usually 2.5 to 4 μm thick.

The most common PVD coatings are TiN, titanium aluminum nitride (TiAlN), and TiCN. Of these, TiAlN is the hardest and most stable and useful at relatively high speeds (although not as high as CVD coatings allow). Other coatings that can be obtained include ZrN, CrN, and various carbon coatings. The PVD field is very dynamic; improved versions of the listed coatings as well as brand new coatings are always being introduced.

Design Considerations

Cemented carbide, in common with all **brittle materials,** may fragment in service, particularly under conditions of impact or upon release from high compressive loading. Additionally, if other failure modes—chipping, cratering, deformation, wear, thermal cracking—encountered during use are allowed to progress too far, fracture can occur. Precautionary measures must be taken to ensure that personnel and equipment are protected from flying fragments and sharp edges when working with brittle tools.

Failure of brittle materials frequently occurs as a result of tensile stress at or near the surface; thus the strength of brittle materials is very dependent on surface conditions. Chips, scratches, thermal cracks, and grinding marks, which decrease the strength and breakage resistance of cemented carbide parts, should be avoided. Even electric-discharge machined (EDM) surfaces should be ground or lapped to a depth of 0.05 mm to remove damaged material.

Special care must be taken when one is grinding cemented carbides. Adequate ventilation of grinding spaces must comply with existing government regulations applicable to the health and safety of the workplace environment. During the fabrication of cemented carbide parts, particularly during grinding of tools and components, adequate provisions must be made for the removal and collection of generated dusts and cutting fluid mists that contain microscopic metal particles, even though those concentrations may be very low. Although the elements contained in these alloys are not radioactive, note that they may become radioactive when exposed to a sufficiently strong radiation source. The cobalt in the cemented carbide alloys, in particular, could be made radioactive.

Physical and Mechanical Properties

Typical property data are shown in Table 6.4.13. In addition to being an effective "cement," cobalt is the primary component that determines mechanical properties. For example, as cobalt content increases,

Table 6.4.13 Properties of Cemented Carbides

Composition, wt. %	Grain size, μm	Hardness, Ra	Abrasion resistance, l/vol. loss cm³	Density, g/cm³	Transverse rupture strength, 1,000 lb/in²	Ultimate compressive strength, 1,000 lb/in²	Ultimate tensile strength, 1,000 lb/in²	Modulus of elasticity, 10⁶ lb/in²	Proportional limit, 1,000 lb/in²	Ductility, % elong	Fracture toughness, (lb/in²)($\sqrt{\text{in}}$)	Thermal expansion 75 to 400°F, in/(in·°F) × 10⁻⁶	Thermal conductivity, cal/(s·°C·cm)	Electrical resistivity, μΩ·cm
WC-3% Co	1.7	93	60	15.3	290	850		98	350			2.2	0.3	17
WC-6% Co	0.8	93.5	62	15										
WC-6% Co	1.1	92.8	60	15	335	860	160	92	370	0.2	9,200	2.9	0.25	17
WC-6% Co	2.1	92	35	15	380	790		92	280		11,500	2.5	0.25	17
WC-6% Co	3	91	15	15	400	750	220	92	210		13,000	2.4	0.25	
WC-9% Co	4	89.5	10	14.7	425	660		87	140		11,500	2.7		
WC-10% Co	1.8	91	13	14.6	440	750		85	230					
WC-10% Co	5.2	89	7	14.5	460	630		85	130			2.5	0.25	17
WC-13% Co	4	88.2	4	14.2	500	600		81	140	0.3	14,500	3	0.02	17
WC-16% Co	4	86.8	3	13.9	500	560		77	100	0.4	15,800	3.2		
WC-25% Co	3.7	84	2	13	440	450	270	67	60		21,000	3.5	0.2	18
Other materials*														
Tool steel (T-g)		85 (66 R_c)	2	8.4	575	600	300	34				6.5	0.12	20
Carbon steel (1095)		79 (55 R_c)		7.8	105			30						
Cast iron				7.3				15–30						
Copper				8.9								9.2	0.94	1.67

*Data for other materials are included to aid in comparing carbide properties with those of referenced materials.

6-61

toughness increases but hardness and wear resistance decrease. This tradeoff is illustrated in Table 6.4.13. A secondary factor that affects mechanical properties is WC grain size. Finer grain size leads to increased hardness and wear resistance but lower toughness.

Micrograin Carbides   An important segment of the cemented carbides is micrograin carbide. As the name implies, these grades feature grains that are less than 1 μm in diameter. As the **carbide grain size decreases,** hardness increases but toughness decreases. However, once the WC grain size diameter drops below approximately 1 μm, this trade-off becomes much more favorable. Further decreases in grain size result in continued increases in hardness but are accompanied by a much smaller drop in toughness. As a result, **micrograin cemented carbides** have outstanding hardness, wear resistance, and compressive strength combined with surprising toughness relative to conventional carbides. This combination of properties makes these grades particularly useful in machining nickel-, cobalt-, and iron-based superalloys and other difficult to machine alloys, such as refractories and titanium alloys. They also find favor machining nonferrous materials.

Mechanical Properties   The outstanding features of cemented carbides are their hardness, high compressive strength, stiffness, and wear resistance. Unfortunately, but as might be expected of hard materials, the toughness and ductility of carbides are low. Table 6.4.13 lists some **mechanical** and **physical property data** for typical and popular cemented carbide compositions.

Hardness and Wear Resistance   Most applications of cemented carbide alloys involve wear and abrasive conditions. In general, the wear resistance of these materials can be estimated based on hardness. Note that wear is a complex process and can occur by means other than normal abrasion. Should wear occur by microchipping, fracture, chemical attack, or metallurgical interactions, it will be obvious that performance does not correlate directly with hardness.

The data presented in Table 6.4.13 under the heading Abrasion Resistance were determined by using an abrasive wheel. Results from this simplified test should be used only as a rough guide in the selection of those alloys, since abrasive wear that occurs under actual service conditions is usually complex and varies greatly with particular circumstances. The performance characteristics of a given grade are, of course, best determined under actual service conditions.

In many applications, cemented carbides are subjected to relatively high operating temperatures. For example, the temperatures of the interface between the chip and the cutting edge of a carbide cutting tool may reach between 500 and 1,100°C. Wear parts may have point contact temperatures in this range. The extent to which the cemented carbide maintains its hardness at the elevated temperatures encountered in use is an important consideration. The cemented carbides not only have high room-temperature hardness, but also maintain hardness at elevated temperatures better than do steels and cast alloys.

Corrosion Resistance   The corrosion resistance of the various cemented carbide alloys is fairly good when compared with other materials, and they may be employed very advantageously in some corrosive environments where outstanding wear resistance is required. Generally they are not employed where corrosion resistance alone is the requirement because other materials usually can be found which are either cheaper or more corrosion-resistant.

Corrosion may cause strength deterioration due to the preferential attack on the binder matrix phase. This results in the creation of microscopic surface defects to which cemented carbides, in common with other brittle materials, are sensitive.

Special corrosion-resistant grades have been developed in which the cobalt binder has been modified or replaced by other metals or alloys, resulting in improved resistance to acid attack. Usually these grades feature a nickel (Ni) binder, perhaps alloyed with small amounts of chromium (Cr). Other factors such as temperature, pressure, and surface condition may significantly influence corrosive behavior, so that specific tests under actual operating conditions should be performed when possible to select a cemented carbide grade for service under corrosive conditions.

Cermets

Another class of composite material that fits in this general category is **cermet**. These alloys are more complex in terms of both their chemistry and their structure, but usually consist of TiCN or TiC grains and a nickel (Ni) or nickel-molybdenum (Ni-Mo) alloy binder. Other alloying elements such as Co and W are usually present. Cermets usually find application in the high-speed finish machining of steels, stainless steels, and occasionally cast irons and high-temperature alloys. They are also very useful for machining powder metallurgical materials. As a rule, cermets provide better surface finishes and are more wear-resistant but less tough than traditional cemented tungsten carbides. Recent processing improvements, such as sinter HIP (hot isostatic pressing) and a better understanding of the alloying characteristics of cermets, have helped better their toughness immensely, so that cermets tough enough for milling are now available.

COPPER AND COPPER ALLOYS
by Arthur Cohen

REFERENCES: ASM "Metals Handbook," Applicable current ASTM and SAE Standards. Publications of the Copper Development Association Standards promulgated by industrial associations related to specific types of products.

Copper and **copper alloys** constitute one of the major groups of commercial metals. They are widely used because of their excellent electrical and thermal conductivities, outstanding corrosion resistance, and ease of fabrication. They are generally nonmagnetic and can be readily joined by conventional soldering, brazing, and welding processes (gas and resistance).

Primary Copper

Copper used in the manufacture of fabricated brass mill or wire and cable products may originate in the ore body of open-pit or underground mines or from rigidly controlled recycled high-grade copper scrap.

Traditional sulfide ores require conventional crushing, milling, concentration by flotation, smelting to form a matte, conversion to blister copper, and final refining by either the **electrolytic** or the **electrowinning** processes to produce copper cathode, which is fully described in ASTM B115 Electrolytic Cathode Copper. This **electrochemical refining** step results in the deposition of virtually pure copper, with the major impurity being oxygen.

Probably the single most important innovation in the copper industry in the past 30 years has been the introduction of **continuous cast wire rod** technology. Within this period, the traditional 250-lb wire bar has been almost completely replaced domestically (and largely internationally) with continuous cast wire rod product produced to meet the requirements of ASTM B49 Copper Redraw Rod for Electrical Purposes.

Generic Classification of Copper Alloys

The most common way to categorize wrought and cast copper and copper alloys is to divide them into six families: coppers, high-copper alloys, brasses, bronzes, copper nickels, and nickel silvers. They are further subdivided and identified via the **Unified Numbering System** (UNS) for metals and alloys.

The designation system is an orderly method of defining and identifying the alloys. The copper section is administered by the Copper Development Association, but it is not a specification. It eliminates the limitations and conflicts of alloy designations previously used and at the same time provides a workable method for the identification marking of mill and foundry products.

In the designation system, numbers from C10000 through C79999 denote **wrought alloys.** Cast alloys are numbered from C80000 through C999999. Within these two categories, compositions are further grouped into families of coppers and copper alloys shown in Table 6.4.14.

Coppers   These metals have a designated minimum copper content of 99.3 percent or higher.

High-Copper Alloys   In wrought form, these are alloys with designated copper contents less than 99.3 percent but more than 96 percent

Table 6.4.14   Generic Classification of Copper Alloys

Generic name	UNS Nos.	Composition
Wrought alloys		
Coppers	C10100–C15815	> 99% Cu
High-copper alloys	C16200–C19900	> 96% Cu
Brasses	C21000–C28000	Cu-Zn
Leaded brasses	C31200–C38500	Cu-Zn-Pb
Tin brasses	C40400–C48600	Cu-Zn-Sn-Pb
Phosphor bronzes	C50100-C52400	Cu-Sn-P
Leaded phosphor bronzes	C53200–C54400	Cu-Sn-Pb-P
Copper-phosphorus and copper-silver-phosphorus alloys	C55180–C55284	Cu-P-Ag
Aluminum bronzes	C60800–C64210	Cu-Al-Ni-Fe-Si-Sn
Silicon bronzes	C64700–C66100	Cu-Si-Sn
Other copper-zinc alloys	C66400–C69710	—
Copper nickels	C70100–C72950	Cu-Ni-Fe
Nickel silvers	C73500–C79800	Cu-Ni-Zn
Cast alloys		
Coppers	C80100–C81200	> 99% Cu
High-copper alloys	C81400–C82800	> 94% Cu
Red and leaded red brasses	C83300–C84800	Cu-Zn-Sn-Pb (75–89% Cu)
Yellow and leaded yellow brasses	C85200–C85800	Cu-Zn-Sn-Pb (57–74% Cu)
Manganese bronzes and leaded manganese bronzes	C86100–C86800	Cu-Zn-Mn-Fe-Pb
Silicon bronzes, silicon brasses	C87300–C87800	Cu-Zn-Si
Tin bronzes and leaded tin bronzes	C90200–C94500	Cu-Sn-Zn-Pb
Nickel-tin bronzes	C94700–C94900	Cu-Ni-Sn-Zn-Pb
Aluminum bronzes	C95200–C95900	Cu-Al-Fe-Ni
Copper nickels	C96200–C96900	Cu-Ni-Fe
Nickel silvers	C97300–C97800	Cu-Ni-Zn-Pb-Sn
Leaded coppers	C98200–C98840	Cu-Pb
Miscellaneous alloys	C99300–C99750	—

which do not fall into any other copper alloy group. The cast high-copper alloys have designated copper contents in excess of 94 percent, to which silver may be added for special properties.

Brasses   These alloys contain zinc as the principal alloying element with or without other designated alloying elements such as iron, aluminum, nickel, and silicon. The wrought alloys comprise three main families of brasses: copper-zinc alloys; copper-zinc-lead alloys (leaded brasses); and copper-zinc-tin alloys (tin brasses). The cast alloys comprise four main families of brasses: copper-tin-zinc (red, semired, and yellow brasses); **manganese bronze** alloys (high-strength yellow brasses); leaded manganese bronze alloys (leaded high-strength yellow brasses); and copper-zinc-silicon alloys (silicon brasses and bronzes). Ingot for remelting for the manufacture of castings may vary slightly from the ranges shown.

Bronzes   Broadly speaking, bronzes are copper alloys in which the major element is not zinc or nickel. Originally *bronze* described alloys with tin as the only or principal alloying element. Today, the term generally is used not by itself, but with a modifying adjective. For wrought alloys, there are four main families of bronzes: copper-tinphosphorus alloys (phosphor bronzes); copper-tin-lead-phosphorus alloys (leaded phosphor bronzes); copper-aluminum alloys (aluminum bronzes); and copper-silicon alloys (silicon bronzes).

The cast alloys comprise four main families of bronzes: copper-tin alloys (tin bronzes); copper-tin-lead alloys (leaded and high leaded tin bronzes); copper-tin-nickel alloys (nickel-tin bronzes); and copper-aluminum alloys (aluminum bronzes).

The family of alloys known as **manganese bronzes,** in which zinc is the major alloying element, is included in the brasses.

Copper-Nickels   These are alloys with nickel as the principal alloying element, with or without other designated alloying elements.

Copper-Nickel-Zinc Alloys   Known commonly as **nickel silvers,** these are alloys which contain zinc and nickel as the principal and secondary alloying elements, with or without other designated elements.

Leaded Coppers   These comprise a series of cast alloys of copper with 20 percent or more lead, sometimes with a small amount of silver, but without tin or zinc.

Miscellaneous Alloys   Alloys whose chemical compositions do not fall into any of previously described categories are combined under "miscellaneous alloys."

Temper Designations

Temper designations for wrought copper and copper alloys were originally specified on the basis of cold reduction imparted by the rolling of sheet or drawing of wire. Designations for rod, seamless tube, welded tube, extrusions, castings and heat-treated products were not covered. In 1974, ASTM B601 Standard Practice for Temper Designations for Copper and Copper Alloys—Wrought and Cast, based on an alphanumeric code, was created to accommodate this deficiency. The general temper designation codes listed in ASTM B601 by process and product are shown in Table 6.4.15.

Coppers

Coppers include the **oxygen-free coppers** (C10100 and C10200), made by melting prime-quality cathode copper under nonoxidizing conditions. These coppers are particularly suitable for applications requiring high electrical and thermal conductivities coupled with exceptional ductility and freedom from hydrogen embrittlement.

The most common copper is C11000—**electrolytic tough pitch.** Its electrical conductivity exceeds 100 percent IACS and is invariably selected for most wire and cable applications. Selective properties of wire produced to ASTM B1 Hard Drawn Copper Wire, B2 Medium Hard Drawn Copper Wire, and B3 Soft or Annealed Copper Wire are shown in Table 6.4.16.

Where resistance to softening along with improved fatigue strength is required, small amounts of silver are added. This permits the silver-containing coppers to retain the effects of cold working to a higher temperature than pure copper (about 600°F versus about 400°F), a property particularly useful where comparatively high temperatures are to be withstood, as in soldering operations or for stressed conductors designed to operate at moderately elevated temperatures.

If superior **machinability** is required, C14500 (tellurium copper), C14700 (sulfur copper), or C18700 (leaded copper) can be selected.

Table 6.4.15 ASTM B601 Temper Designation Codes for Copper and Copper Alloys

Temper designation	Former temper name or material conditioned	Temper designation	Former temper name or material condition	Temper designation	Former temper name or material condition
Cold-worked tempers[a]		Annealed tempers[d]		Mill-hardened tempers	
H00	1/8 hard	O10	Cast and annealed (homogenized)	TM00	AM
H01	1/4 hard	O11	As cast and precipitation heat-treated	TM01	1/4 HM
H02	1/2 hard	O20	Hot-forged and annealed	TM02	1/2 HM
H03	3/4 hard	O25	Hot-rolled and annealed	TM04	HM
H04	Hard	O30	Hot-extruded and annealed	TM06	XHM
H06	Extra hard	O31	Extruded and precipitation heat-treated	TM08	XHMS
H08	Spring	O40	Hot-pierced and annealed	Quench-hardened tempers	
H10	Extra spring	O50	Light anneal	TQ00	Quench-hardened
H12	Special spring	O60	Soft anneal	TQ30	Quench-hardened and tempered
H13	Ultra spring	O61	Annealed	TQ50	Quench-hardened and temper-annealed
H14	Super spring	O65	Drawing anneal	TQ55	Quench-hardened and temper-annealed, cold-drawn and stress-relieved
Cold-worked tempers[b]		O68	Deep-drawing anneal	TQ75	Interrupted quench-hardened
H50	Extruded and drawn	O70	Dead soft anneal	Precipitation-hardened, cold-worked; and thermal-stress-relieved tempers	
H52	Pierced and drawn	O80	Annealed to temper-1/8 hard	TR01	TL01 and stress-relieved
H55	Light drawn	O81	Annealed to temper-1/4 hard	TR02	TL02 and stress-relieved
H58	Drawn general-purpose	O82	Annealed to temper-1/2 hard	TR04	TL04 and stress-relieved
H60	Cold heading; forming	Annealed tempers[e]		Tempers of welded tubing[f]	
H63	Rivet	OS005	Nominal Avg. grain size, 0.005 mm	WH00	Welded and drawn: 1/8 hard
H64	Screw	OS010	Nominal Avg. grain size, 0.010 mm	WH01	Welded and drawn: 1/4 hard
H66	Bolt	OS015	Nominal Avg. grain size, 0.015 mm	WH02	Welded and drawn: 1/2 hard
H70	Bending	OS025	Nominal Avg. grain size, 0.025 mm	WH03	Welded and drawn: 3/4 hard
H80	Hard-drawn	OS035	Nominal Avg. grain size, 0.035 mm	WH04	Welded and drawn: full hard
H85	Medium-hard-drawn electrical wire	OS050	Nominal Avg. grain size, 0.050 mm	WH06	Welded arid drawn: extra hard
H86	Hard-drawn electrical wire	OS060	Nominal Avg. grain size, 0.060 mm	WM00	As welded from H00 (1/8-hard) strip
H90	As-finned	OS070	Nominal Avg. grain size, 0.070 mm	WM01	As welded from H01 (1/4-hard) strip
Cold-worked and stress-relieved tempers		OS100	Nominal Avg. grain size, 0.100 mm	WM02	As welded from H02 (1/2-hard) strip
HR01	H01 and stress-relieved	OS120	Nominal Avg. grain size, 0.120 mm	WM03	As welded from H03 (3/4-hard) strip
HR02	H02 and stress-relieved	OS150	Nominal Avg. grain size, 0.150 mm	WM04	As welded from H04 (hard) strip
HR04	H04 and stress-relieved	OS200	Nominal Avg. grain size, 0.200 mm	WM06	As welded from H06 (extra hard) strip
HR08	H08 and stress-relieved	Solution-treated temper		WM08	As welded from H08 (spring) strip
HR10	H10 and stress-relieved	TB00	Solution heat-treated (A)	WM10	As welded from H10 (extra spring) strip
HR20	As-finned	Solution-treated and cold-worked tempers		WM15	WM50 and stress-relieved
HR50	Drawn and stress-relieved	TD00	TB00 cold-worked: 1/8 hard	WM20	WM00 and stress-relieved
Cold-rolled and order-strengthened tempers[c]		TD01	TB00 cold-worked: 1/4 hard	WM21	WM01 and stress-relieved
HT04	H04 and treated	TD02	TB00 cold-worked: 1/2 hard	WM22	WM02 and stress-relieved
HT08	H08 and treated	TD03	TB00 cold-worked: 3/4 hard	WM50	As welded from annealed strip
As-manufactured tempers		TD04	TB00 cold-worked: hard	WO50	Welded and light-annealed
M01	As sand cast	Solution treated and precipitation-hardened temper		WR00	WM00: drawn and stress-relieved
M02	As centrifugal cast	TF00	TB00 and precipitation-hardened	WR01	WM01: drawn and stress-relieved
M03	As plaster cast	Cold-worked and precipitation-hardened tempers		WR02	WM02: drawn and stress-relieved
M04	As pressure die cast	TH01	TD01 and precipitation-hardened	WR03	WM03: drawn and stress-relieved
M05	As permanent mold cast	TH02	TD02 and precipitation-hardened	WR04	WM04: drawn and stress-relieved
M06	As investment cast	TH03	TD03 and precipitation-hardened	WR06	WM06: drawn and stress-relieved
M07	As continuous cast	TH04	TD04 and precipitation-hardened		
M10	As hot-forged and air-cooled	Precipitation-hardened or spinodal heat treated and cold-worked tempers			
M11	As forged and quenched	TL00	TF00 cold-worked: 1/8 hard		
M20	As hot-rolled	TL01	TF00 cold-worked: 1/4 hard		
M30	As hot-extruded	TL02	TF00 cold-worked: 1/2 hard		
M40	As-hot-pierced	TL04	TF00 cold-worked: hard		
M45	As hot-pierced and rerolled	TL08	TF00 cold-worked: spring		
		TL10	TF00 cold-worked: extra spring		

[a] Cold-worked tempers to meet standerd requirements based on cold rolling or cold drawing.
[b] Cold-worked tempers to meet standerd requirements based on temper names applicable to specific products.
[c] Tempers produced by controlled amounts of cold work followed by a thermal treatment to produce order strengthening.
[d] Annealed to meet mechanical properties.
[e] Annealed to meet nominal average grain size.
[f] Tempers of fully finished tubing that has been drawn or annealed to produce specified mechanical properties or that has been annealed to produce a prescribed nominal average grain size are commonly identified by the property H, O, or OS temper designation.
SOURCE: ASTM, reprinted with permission.

Table 6.4.16   Mechanical Properties of Copper Wire

Diameter		Hard-drawn			Medium hard			Soft or annealed
		Tensile strength (nominal)		Elongation (nominal) in 10 in (250 mm), % min	Tensile strength		Elongation in 10 in (250 mm), % min	Elongation (nominal) in 10 in (250 mm), % min
in	mm	ksi	MPa		ksi	MPa		
0.460	11.7	49.0	340	3.8	42.0–49.0	290–340	3.8	35
0.325	8.3	54.5	375	2.4	45.0–52.0	310–360	3.0	35
0.229	5.8	59.0	405	1.7	48.0–55.0	330–380	2.2	30
0.162	4.1	62.1	430	1.4	49.0–56.0	340–385	1.5	30
0.114	2.9	64.3	445	1.2	50.0–57.0	345–395	1.3	30
0.081	2.05	65.7	455	1.1	51.0–58.0	350–400	1.1	25
0.057	1.45	66.4	460	1.0	52.0–59.0	360–405	1.0	25
0.040	1.02	67.0	460	1.0	53.0–60.0	365–415	1.0	25

SOURCE: ASTM, abstracted with permission.

With these coppers, superior machinability is gained at a modest sacrifice in electrical conductivity.

Similarly, chromium and zirconium are added to increase elevated-temperature strength with little decrease in conductivity.

Copper-beryllium alloys are **precipitation-hardening alloys** that combine moderate conductivity with very high strengths. To achieve these properties, a typical heat treatment would involve a solution heat treatment for 1 h at 1,450°F (788°C) followed by water quenching, then a precipitation heat treatment at 600°F (316°C) for 3 h.

Wrought Copper Alloys (Brasses and Bronzes)

There are approximately 230 wrought brass and bronze compositions. The most widely used is alloy C26000, which corresponds to a 70:30 copper-zinc composition and is most frequently specified unless high corrosion resistance or special properties of other alloys are required. For example, alloy C36000—free-cutting brass—is selected when extensive machining must be done, particularly on automatic screw machines. Other alloys containing aluminum, silicon, or nickel are specified for their outstanding corrosion resistance. The other properties of greatest importance include mechanical strength, fatigue resistance, ability to take a good finish, corrosion resistance, electrical and thermal conductivities, color, ease of fabrication, and machinability.

The bronzes are divided into five alloy families: phosphor bronzes, aluminum bronzes, silicon bronzes, copper nickels, and nickel silvers.

Phosphor Bronzes   Three tin bronzes, commonly referred to as **phosphor bronzes,** are the dominant alloys in this family. They contain 5, 8, and 10 percent tin and are identified, respectively, as alloys C51000, C52100, and C52400. Containing up to 0.4 percent phosphorus, which improves the casting qualities and hardens them somewhat, these alloys have excellent elastic properties.

Aluminum Bronzes   These alloys with 5 and 8 percent aluminum find application because of their high strength and corrosion resistance, and sometimes because of their golden color. Those with 10 percent aluminum content or higher are very plastic when hot and have exceptionally high strength, particularly after heat treatment.

Silicon Bronzes   There are three dominant alloys in this family in which silicon is the primary alloying agent but which also contain appreciable amounts of zinc, iron, tin, or manganese. These alloys are as corrosion-resistant as copper (slightly more so in some solutions) and possess excellent hot workability with high strengths. Their outstanding characteristic is that of ready weldability by all methods. The alloys are extensively fabricated by arc or acetylene welding into tanks and vessels for hot-water storage and chemical processing.

Copper Nickels   These alloys are extremely malleable and may be worked extensively without annealing. Because of their excellent corrosion resistance, they are used for condenser tubes for the most severe service. Alloys containing nickel have the best high-temperature properties of any copper alloy.

Nickel Silvers   Nickel silvers are white and are often applied because of this property. They are tarnish resistant under atmospheric conditions. Nickel silver is the base for most silver-plated ware.

Overall, the primary selection criteria can be met satisfactorily by one or more of the alloys listed in Table 6.4.17.

Cast Copper-Base Alloys

Casting makes it possible to produce parts with shapes that cannot be achieved easily by fabrication methods such as forming or machining. Often it is more economical to produce a part as a casting than to fabricate it by other means.

Copper alloy castings serve in applications that require superior corrosion resistance, high thermal or electrical conductivity, good bearing surface qualities, or other special properties.

All copper alloys can be successfully **sand-cast,** for this process allows the greatest flexibility in casting size and shape and is the most economical and widely used casting method, especially for limited production quantities. Permanent mold casting is best suited for tin, silicon, aluminum, and manganese bronzes, as well as yellow brasses. Most copper alloys can be cast by the centrifugal process.

Brass die castings are made when great dimensional accuracy and/or a better surface finish is desired. While inferior in properties to hot-pressed parts, die castings are adaptable to a wider range of designs, for they can be made with intricate coring and with considerable variation in section thickness.

The estimated market distribution of the dominant copper alloys by end-use application is shown in Table 6.4.18. Chemical compositions and selective mechanical and physical properties of the dominant alloys used to produce sand castings are listed in Table 6.4.19.

Test results obtained on **standard test bars** (either attached to the casting or separately poured) indicate the quality of the metal used but not the specific properties of the casting itself because of variations of thickness, soundness, and other factors. The ideal casting is one with a fairly uniform metal section with ample fillets and a gradual transition from thin to thick parts.

Properties and Processing of Copper Alloys

Mechanical Properties   Cold working copper and copper alloys increases both tensile and yield strengths, with the more pronounced increase imparted to yield strength. For most alloys, tensile strength of the hardest cold-worked temper is approximately twice the tensile strength of the annealed temper, whereas the yield strength of the hardest coldworked temper can be up to 5 to 6 times that of the annealed temper. While hardness is a measure of temper, it is not an accurate one, because the determination of hardness is dependent upon the alloy, its strength level, and the method used to test for hardness.

All the brasses may be **hot-worked** irrespective of their lead content. Even for alloys having less than 60 percent copper and containing the beta phase in the microstructure, the process permits more extensive changes in shape than cold working because of the plastic (ductile) nature of the beta phase at elevated temperatures, even in the presence of lead. Hence a single hot-working operation can often replace a sequence of forming and annealing operations. Alloys for **extrusion, forging,** or **hot pressing** contain the beta phase in varying amounts.

Table 6.4.17 Composition and Properties of Selected Wrought Copper and Copper Alloys

Alloy no. (and name)	Nominal composition, %	Commercial forms[a]	Tensile strength ksi	Tensile strength MPa	Yield strength ksi	Yield strength MPa	Elongation in 2 in (50 mm), %[b]
C10200 (oxygen-free copper)	99.95 Cu	F, R, W, T, P, S	32–66	221–455	10–53	69–365	55–4
C11000 (electrolytic tough pitch copper)	99.90 Cu, 0.04 O	F, R, W, T, P, S	32–66	221–455	10–53	69–365	55–4
C12200 (phosphorus-deoxidized copper, high residual phosphorus)	99.90 Cu, 0.02 P	F, R, T, P	32–55	221–379	10–50	69–345	45–8
C14500 (phosphorus-deoxidized tellurium-bearing copper)	99.5 Cu, 0.50 Te, 0.008 P	F, R, W, T	32–56	221–386	10–51	69–352	50–3
C14700 (sulfur-bearing copper)	99.6 Cu, 0.40 S	R, W	32–57	221–393	10–55	69–379	52–8
C15000 (zirconium copper)	99.8 Cu, 0.15 Zr	R, W	29–76	200–524	6–72	41–496	54–1.5
C17000 (beryllium copper)	99.5 Cu, 1.7 Be, 0.20 Co	F, R	70–190	483–1,310	32–170	221–1,172	45–3
C17200 (beryllium copper)	99.5 Cu, 1.9 Be, 0.20 Co	F, R, W, T, P, S	68–212	469–1,462	25–195	172–1,344	48–1
C18200 (chromium copper)	99.0 Cu[c], 1.0 Cr	F, W, R, S, T	34–86	234–593	14–77	97–531	40–5
C18700 (leaded copper)	99.0 Cu, 1.0 Pb	R	32–55	221–379	10–50	69–345	45–8
C19400	97.5 Cu, 2.4 Fe, 0.13 Zn, 0.03 P	F	45–76	310–524	24–73	165–503	32–2
C21000 (gilding, 95%)	95.0 Cu, 5.0 Zn	F, W	34–64	234–441	10–58	69–400	45–4
C22000 (commercial bronze, 90%)	90.0 Cu, 10.0 Zn	F, R, W, T	37–72	255–496	10–62	69–427	50–3
C23000 (red brass, 85%)	85.0 Cu, 15.0 Zn	F, W, T, P	39–105	269–724	10–63	69–434	55–3
C24000 (low brass, 80%)	80.0 Cu, 20.0 Zn	F, W	42–125	290–862	12–65	83–448	55–3
C26000 (cartridge brass, 70%)	70.0 Cu, 30.0 Zn	F, R, W, T	44–130	303–896	11–65	76–448	66–3
C26800 C27000 (yellow brass)	65.0 Cu, 35.0 Zn	F, R, W	46–128	317–883	14–62	97–427	65–3
C28000 (Muntz metal)	60.0 Cu, 40.0 Zn	F, R, T	54–74	372–510	21–55	145–379	52–10
C31400 (leaded commercial bronze)	89.0 Cu, 1.8 Pb, 9.2 Zn	F, R	37–60	255–414	12–55	83–379	45–10
C33500 (low-leaded brass)	65.0 Cu, 0.5 Pb, 34.5 Zn	F	46–74	317–510	14–60	97–414	65–8
C34000 (medium-leaded brass)	65.0 Cu, 1.0 Pb, 34.0 Zn	F, R, W, S	47–88	324–607	15–60	103–414	60–7
C34200 (high-leaded brass)	64.5 Cu, 2.0 Pb, 33.5 Zn	F, R	49–85	338–586	17–62	117–427	52–5
C35000 (medium-leaded brass)	62.5 Cu, 1.1 Pb, 36.4, Zn	F, R	45–95	310–655	13–70	90–483	66–1
C35300 (high-leaded brass)	62.0 Cu, 1.8 Pb, 36.2 Zn	F, R	49–85	338–586	17–62	117–427	52–5
C35600 (extra-high-leaded brass)	63.0 Cu, 2.5 Pb, 34.5 Zn	F	49–74	338–510	17–60	117–414	50–7
C36000 (free-cutting brass)	61.5 Cu, 3.0 Pb, 35.5 Zn	F, R, S	49–68	338–469	18–45	124–310	53–18
C36500 to C36800 (leaded Muntz metal)[c]	60.0 Cu[e], 0.6 Pb, 39.4 Zn	F	54	372	20	138	45
C37000 (free-cutting Muntz metal)	60.0 Cu, 1.0 Pb, 39.0 Zn	T	54–80	372–552	20–60	138–414	40–6
C37700 (forging brass)[d]	59.0 Cu, 2.0 Pb, 39.0 Zn	R, S	52	359	20	138	45
C38500 (architectural bronze)[d]	57.0 Cu, 3.0 Pb, 40.0 Zn	R, S	60	414	20	138	30
C40500	95.0 Cu, 1.0 Sn, 4.0 Zn	F	39–78	269–538	12–70	83–483	49–3
C41300	90.0 Cu, 1.0 Sn, 9.0 Zn	F, R, W	41–105	283–724	12–82	83–565	45–2
C43500	81.0 Cu, 0.9 Sn, 18.1 Zn	F, T	46–80	317–552	16–68	110–469	46–7
C44300, C44400, C44500 (inhibited admiralty)	71.0 Cu, 28.0 Zn, 1.0 Sn	F, W, T	48–55	331–379	18–22	124–152	65–60
C46400 to C46700 (naval brass)	60.0 Cu, 39.3 Zn, 0.7 Sn	F, R, T, S	55–88	379–607	25–66	172–455	50–17
C48200 (naval brass, medium-leaded)	60.5 Cu, 0.7 Pb, 0.8 Sn, 38.0 Zn	F, R, S	56–75	386–517	25–53	172–365	43–15
C48500 (leaded naval brass)	60.0 Cu, 1.8 Pb, 37.5 Zn, 0.7 Sn	F, R, S	55–77	379–531	25–53	172–365	40–15
C51000 (phosphor bronze, 5% A)	95.0 Cu, 5.0 Sn, trace P	F, R, W, T	47–140	324–965	19–80	131–552	64–2
C51100	95.6 Cu, 42 Sn, 0.2 P	F	46–103	317–710	50–80	345–552	48–2
C52100 (phosphor bronze, 8% C)	92.0 Cu, 8.0 Sn, trace P	F, R, W	55–140	379–965	24–80	165–552	70–2
C52400 (phosphor bronze, 10% D)	99.0 Cu, 10.0 Sn, trace P	F, R, W	66–147	455–1,014	28	193	70–3
C54400 (free-cutting phosphor bronze)	88.0 Cu, 4.0 Pb, 4.0 Zn, 4.0 Sn	F, R	44–75	303–517	19–63	131–434	50–16
C60800 (aluminum bronze, 5%)	95.0 Cu, 5.0 Al	T	60	414	27	186	55
C61000	92.0 Cu, 8.0 Al	R, W	70–80	483–552	30–55	207–379	65–25
C61300	92.7 Cu, 0.3 Sn, 7.0 Al	F, R, T, P, S	70–85	483–586	30–58	207–400	42–35
C61400 (aluminum bronze, D)	91.0 Cu, 7.0 Al, 2.0 Fe	F, R, W, T, P, S	76–89	524–614	33–60	228–414	45–32
C63000	82.0 Cu, 3.0 Fe, 10.0 Al, 5.0 Ni	F, R	90–118	621–814	50–75	345–517	20–15
C63200	82.0 Cu, 4.0 Fe, 9.0 Al, 5.0 Ni	F, R	90–105	621–724	45–53	310–365	25–20
C64200	91.2 Cu, 7.0 Al	F, R	75–102	517–703	35–68	241–469	32–22
C65100 (low-silicon bronze, B)	98.5 Cu, 1.5 Si	R, W, T	40–95	276–655	15–69	103–476	55–11
C65500 (high-silicon bronze, A)	97.0 Cu, 3.0 Si	F, R, W, T	56–145	386–1,000	21–70	145–483	63–3
C67500 (manganese bronze, A)	58.5 Cu, 1.4 Fe, 39.0 Zn, 1.0 Sn, 0.1 Mn	R, S	65–84	448–579	30–60	207–414	33–19
C68700 (aluminum brass, arsenical)	77.5 Cu, 20.5 Zn, 2.0 Al, 0.1 As	T	60	414	27	186	55
C70600 (copper nickel, 10%)	88.7 Cu, 1.3 Fe, 10.0 Ni	F, T	44–60	303–414	16–57	110–393	42–10
C71500 (copper nickel, 30%)	70.0 Cu, 30.0 Ni	F, R, T	54–75	372–517	20–70	138–483	45–15
C72500	88.2 Cu, 9.5 Ni, 2.3 Sn	F, R, W, T	55–120	379–827	22–108	152–745	35–1
C74500 (nickel silver, 65–10)	65.0 Cu, 25.0 Zn, 10.0 Ni	F, W	49–130	338–896	18–76	124–524	50–1
C75200 (nickel silver, 65–18)	65.0 Cu, 17.0 Zn, 18.0 Ni	F, R, W	56–103	386–710	25–90	172–621	45–3
C77000 (nickel silver, 55–18)	55.0 Cu, 27.0 Zn, 18.0 Ni	F, R, W	60–145	414–1,000	27–90	186–621	40–2
C78200 (leaded nickel silver, 65–8–2)	65.0 Cu, 2.0 Pb, 25.0 Zn, 8.0 Ni	F	53–91	365–627	23–76	159–524	40–3

[a] F, flat products; R, rod; W, wire; T, tube; P, pipe; S, shapes.
[b] Ranges are from softest to hardest commercial forms. The strength of the standard copper alloys depends on the temper (annealed grain size or degree of cold work) and the section thickness of the mill product. Ranges cover standard tempers for each alloy.
[c] Values are for as-hot-rolled material.
[d] Values are for as-extruded material.
[e] Rod, 61.0 Cu min.
SOURCE: Copper Development Association Inc.

| Machinability rating, %[c] | Melting point | | Density, lb/in³ | Specific gravity | Coefficient of thermal expansion $\times 10^{-6}$ at 77–572°F (25–300°C) | Electrical conductivity (annealed) % IACS | Thermal conductivity Btu · ft/(h · ft² · °F) [cal · cm/(s · cm² · °C)] |
	Solidus, °F (°C)	Liquidus, °F (°C)					
20	1,981 (1,083)	1,981 (1,083)	0.323	8.94	9.8 (17.7)	101	226 (0.934)
20	1,949 (1,065)	1,981 (1,083)	0.321–0.323	8.89–8.94	9.8 (17.7)	101	226 (0.934)
20	1,981 (1,083)	1,981 (1,083)	0.323	8.94	9.8 (17.7)	85	196 (0.81)
85	1,924 (1,051)	1,967 (1,075)	0.323	8.94	9.9 (17.8)	93	205 (0.85)
85	1,953 (1,067)	1,969 (1,076)	0.323	8.94	9.8 (17.7)	95	216 (0.89)
20	1,796 (980)	1,976 (1,080)	0.321	8.89	9.8 (17.7)	93	212 (0.876)
20	1,590 (865)	1,800 (980)	0.304	8.41	9.9 (17.8)	22	62–75 (0.26–0.31)
20	1,590 (865)	1,800 (980)	0.298	8.26	9.9 (17.8)	22	62–75 (0.26–0.31)
20	1,958 (1,070)	1,967 (1,075)	0.321	8.89	9.8 (17.6)	80	187 (0.77)
85	1,747 (953)	1,976 (1,080)	0.323	8.94	9.8 (17.6)	96	218 (0.93)
20	1,980 (1,080)	1,990 (1,090)	0.322	8.91	9.8 (17.9)	65	150 (0.625)
20	1,920 (1,050)	1,950 (1,065)	0.320	8.86	10.0 (18.1)	56	135 (0.56)
20	1,870 (1,020)	1,910 (1,045)	0.318	8.80	10.2 (18.4)	44	109 (0.45)
30	1,810 (990)	1,880 (1,025)	0.316	8.75	10.4 (18.7)	37	92 (0.38)
30	1,770 (965)	1,830 (1,000)	0.313	8.67	10.6 (19.1)	32	81 (0.33)
30	1,680 (915)	1,750 (955)	0.308	8.53	11.1 (19.9)	28	70 (0.29)
30	1,660 (905)	1,710 (930)	0.306	8.47	11.3 (20.3)	27	67 (0.28)
40	1,650 (900)	1,660 (905)	0.303	8.39	11.6 (20.8)	28	71 (0.29)
80	1,850 (1,010)	1,900 (1,040)	0.319	8.83	10.2 (18.4)	42	104 (0.43)
60	1,650 (900)	1,700 (925)	0.306	8.47	11.3 (20.3)	26	67 (0.28)
70	1,630 (885)	1,700 (925)	0.306	8.47	11.3 (20.3)	26	67 (0.28)
90	1,630 (885)	1,670 (910)	0.306	8.47	11.3 (20.3)	26	67 (0.28)
70	1,640 (895)	1,680 (915)	0.305	8.44	11.3 (20.3)	26	67 (0.28)
90	1,630 (885)	1,670 (910)	0.306	8.47	11.3 (20.3)	26	67 (0.28)
100	1,630 (885)	1,660 (905)	0.307	8.50	11.4 (20.5)	26	67 (0.28)
100	1,630 (885)	1,650 (900)	0.307	8.50	11.4 (20.5)	26	67 (0.28)
60	1,630 (885)	1,650 (900)	0.304	8.41	11.6 (20.8)	28	71 (0.29)
70	1,630 (885)	1,650 (900)	0.304	8.41	11.6 (20.8)	27	69 (0.28)
80	1,620 (880)	1,640 (895)	0.305	8.44	11.5 (20.7)	27	69 (0.28)
90	1,610 (875)	1,630 (890)	0.306	8.47	11.6 (20.8)	28	71 (0.29)
20	1,875 (1,025)	1,940 (1,060)	0.319	8.83	— (—)	41	95 (0.43)
20	1,850 (1,010)	1,900 (1,038)	0.318	8.80	10.3 (18.6)	30	77 (0.32)
30	1,770 (965)	1,840 (1,005)	0.313	8.66	10.8 (19.4)	28	— (—)
30	1,650 (900)	1,720 (935)	0.308	8.53	11.2 (20.2)	25	64 (0.26)
30	1,630 (885)	1,650 (900)	0.304	8.41	11.8 (21.2)	26	67 (0.28)
50	1,630 (885)	1,650 (900)	0.305	8.44	11.8 (21.2)	26	67 (0.28)
70	1,630 (885)	1,650 (900)	0.305	8.44	11.8 (21.2)	26	67 (0.28)
20	1,750 (950)	1,920 (1,050)	0.320	8.86	9.9 (17.8)	15	40 (0.17)
20	1,785 (975)	1,945 (1,060)	0.320	8.86	9.9 (17.8)	20	48.4 (0.20)
20	1,620 (880)	1,880 (1,020)	0.318	8.80	10.1 (18.2)	13	36 (0.15)
20	1,550 (845)	1,830 (1,000)	0.317	8.78	10.2 (18.4)	11	29 (0.12)
80	1,700 (930)	1,830 (1,000)	0.321	8.89	9.6 (17.3)	19	50 (0.21)
20	1,920 (1,050)	1,945 (1,063)	0.295	8.17	10.0 (18.1)	17	46 (0.19)
20	— —	1,905 (1,040)	0.281	7.78	9.9 (17.9)	15	40 (0.17)
30	1,905 (1,040)	1,915 (1,045)	0.287	7.95	9.0 (16.2)	12	32 (0.13)
20	1,905 (1,040)	1,915 (1,045)	0.285	7.89	9.0 (16.2)	14	39 (0.16)
30	1,895 (1,035)	1,930 (1,054)	0.274	7.58	9.0 (16.2)	7	22.6 (0.09)
30	1,905 (1,040)	1,940 (1,060)	0.276	7.64	9.0 (16.2)	7	20 (0.086)
60	1,800 (985)	1,840 (1,005)	0.278	7.69	10.0 (18.1)	8	26 (0.108)
30	1,890 (1,030)	1,940 (1,060)	0.316	8.75	9.9 (17.9)	12	33 (0.14)
30	1,780 (970)	1,880 (1,025)	0.308	8.53	10.0 (18.0)	7	21 (0.09)
30	1,590 (865)	1,630 (890)	0.302	8.36	11.8 (21.2)	24	61 (0.26)
30	1,740 (950)	1,765 (965)	0.296	8.33	10.3 (18.5)	23	58 (0.24)
20	2,010 (1,100)	2,100 (1,150)	0.323	8.94	9.5 (17.1)	9.0	26 (0.11)
20	2,140 (1,170)	2,260 (1,240)	0.323	8.94	9.0 (16.2)	4.6	17 (0.07)
20	1,940 (1,060)	2,065 (1,130)	0.321	8.89	9.2 (16.5)	11	31 (0.13)
20	— (—)	1,870 (1,020)	0.314	8.69	9.1 (16.4)	9.0	26 (0.11)
20	1,960 (1,070)	2,030 (1,110)	0.316	8.73	9.0 (16.2)	6.0	19 (0.08)
30	— (—)	1,930 (1,055)	0.314	8.70	9.3 (16.7)	5.5	17 (0.07)
60	1,780 (970)	1,830 (1,000)	0.314	8.69	10.3 (18.5)	10.9	28 (0.11)

Table 6.4.18 Estimated Casting Shipment Distribution by Alloy for End-Use Application

End use	Dominant copper alloys—Estimated
Plumbing and heating	C84400 (75%), C83800/C84800 (25%)
Industrial valves and fittings	C83600 (45%), C83800/C84400 (15%)
	C90300 (5%), C92200 (15%), Misc. (20%)
Bearings and bushings	C93200 (80%), C93700 (20%)
Water meters, hydrants, and water system components	C83600 (25%), C84400 (75%)
Electrical connectors and components	C83300 (50%), C83600 (20%), C95400 (15%), C94600 (15%)
Pumps, impellers, and related components	C92200 (20%), C95400/C95500 (50%), Misc. (30%)
Outdoor sprinkler systems	C84400 (100%)
Heavy equipment components	C86300 (80%), C95500 (20%)
Hardware, plaques, and giftware	C83600 (15%), C85200/C85400/C85700 (50%), C92200 (35%)
Marine	C86500 (30%), C95800 (50%), C96200 (5%), C96400 (15%)
Locksets	C85200 (90%), Misc. (10%)
Railroad journal bearings and related parts	C93200 (90%), Misc. (10%)
Other	C87300 (35%), C87610 (65%)

Extruded sections of many copper alloys are made in a wide variety of shapes. In addition to architectural applications of extrusions, extrasion is an important production process since many objects such as hinges, pinions, brackets, and lock barrels can be extruded directly from bars.

While the copper-zinc alloys (brasses) may contain up to 40 percent zinc, those with 30 to 35 percent zinc find the greatest application for they exhibit high ductility and can be readily cold-worked. With decreasing zinc content, the mechanical properties and corrosion resistance of the alloys approach those of copper. The properties of these alloys are listed in Table 6.4.20.

Heat Treating Figures 6.4.3 and 6.4.4 show the progressive effects of cold rolling and annealing of alloy C26000 flat products. **Cold rolling** clearly increases the hardness and the tensile and yield strengths while concurrently decreasing the ductility.

Fig. 6.4.3 Effect of cold working on annealed brass (alloy C26000).

Fig. 6.4.4 Effect of annealing on cold-rolled brass (alloy C26000).

Annealing below a certain minimum temperature has practically no effect, but when the temperature is in the recrystallization range, a rapid decrease in strength and an increase in ductility occur. With a proper anneal, the effects of cold working are almost entirely removed. Heating beyond this point results in grain growth and comparatively little further increase in ductility. Figures 6.4.5 and 6.4.6 show the variation of properties of various brasses after annealing at the temperatures indicated.

Fig. 6.4.5 Tensile strengths of copper-zinc alloys.

Fig. 6.4.6 Percent of elongation in 2 in of copper-zinc alloys.

Work-hardened metals are restored to a soft state by annealing. Single-phase alloys are transferred into unstressed crystals by **recovery, recrystallization,** and **grain growth.** In severely deformed metal, recrystallization occurs at lower temperatures than in lightly deformed metal. Also, the grains are smaller and more uniform in size when severely deformed metal is recrystallized.

Grain size can be controlled by proper selection of cold-working and annealing practices. Large amounts of prior cold work, rapid heating to annealing temperature, and short annealing times foster formation of fine grains. Larger grains are normally produced by a combination of limited deformation and long annealing times.

Practically all wrought copper alloys are used in the cold-worked condition to gain additional strength. Articles are often made from annealed stock and depend on the cold work of the forming operation to shape and harden them. When the cold work involved is too small to do this, brass rolled to a degree of temper consistent with final requirements should be used.

Brass for springs should be rolled as hard as is consistent with the subsequent forming operations. For articles requiring sharp bends, or

Table 6.4.19 Composition and Properties of Selected Cast Copper Alloys

Alloy Number (and name)	Nominal composition, %	Mechanical properties					Melting point		Density, lb/in³	Specific gravity	Coefficient of thermal expansion × 10⁻⁶ at 77-572°F (25-300°C)	Electrical conductivity (annealed), % IACS	Thermal conductivity, Btu · ft/(h · ft2 · °F) [cal · cm/(s · cm² · °C)]
		Tensile strength		Yield strength		Elongation in 50 mm (2 in), %[b]	Solidus °F (°C)	Liquidus °F (°C)					
		ksi	MPa	ksi	MPa								
C83600	85 Cu, 5 Sn, 5 Pfa, 5 Zn	37	255	17	117	30	1,570 (855)	1,850 (1,010)	0.318	8.83	10.0 (18.0)	15	41.6 (0.172)
C83800	83 Cu, 4 Sn, 6 Pb, 7 Zn	35	240	16	110	25	1,550 (845)	1,840 (1,005)	0.312	8.64	10.0 (18.0)	15	41.9 (0.173)
C84400	81 Cu, 3 Sn, 7 Pb, 9 Zn	34	235	15	105	26	1,540 (840)	1,840 (1,005)	0.314	8.70	10.4 (18.0)	16.4	41.9 (0.173)
C84800	76 Gu, 3 Sn, 6 Pb, 15 Zn	37	255	14	97	35	1,530 (832)	1,750 (954)	0.310	8.58	10.4 (18.7)	16.4	41.6 (0.172)
C85200	72 Cu, 1 Sn, 3Pb; 24 Zn	38	260	13	90	35	1,700 (925)	1,725 (940)	0.307	8.50	11.5 (21)	18.6	48.5 (0.20)
C85400	67 Cu, 1 Sn, 3 Pb, 29 Zn	34	235	12	83	35	1,700 (925)	1,725 (940)	0.305	8.45	11.2 (20.2)	19.6	51 (0.21)
C86300	62 Cu, 26 Zn, 3 Fe, 6 Al, 3 Mn	119	820	67	460	18	1,625 (885)	1,693 (923)	0.278	7.69	12 (22)	9	21 (0.085)
C87300	95 Cu, 1 Mn, 4 Si	55	380	25	170	30	1,680 (916)	1,510 (821)	0.302	8.36	10.9 (19.6)	6.7	28 (0.113)
C87610	92 Cu, 4 Zn, 4 Si	55	380	25	170	30	— (—)	— (—)	0.302	8.36	— (—)	6.0	— (—)
C90300	88 Cu, 8 Sn, 4 Zn	45	310	21	145	30	1,570 (854)	1,830 (1,000)	0.318	8.80	10.0 (18.0)	12	43 (0.18)
C92200	88 Cp, 6 Sn, 1.5 Pb, 3.5 Zn	40	275	20	140	30	1,520 (825)	1,810 (990)	0.312	8.64	10.0 (18.0)	14.3	40 (0.17)
C93200	83 Cu, 7 Sn, 7 Pb, 3 Zn	35	240	18	125	20	1,570 (855)	1,790 (975)	0.322	8.93	10.0 (18.0)	12	34 (0.14)
C93700	80 Cu, 10 Sn, 10 Sn	35	240	18	125	20	1,403 (762)	1,705 (930)	0.323	3.95	10.3 (18.5)	10	27 (0.11)
C95400	88.5 Cu, 4 Fe, 10.5 Al	75	515	30	205	12	1,880 (1,025)	1,900 (1,040)	0.269	7.45	9.0 (16.2)	13	34 (0.14)
C95400 (TQ50)	88.5 Cu, 4 Fe, 10.5 Al	90	620	45	310	6	1,880 (1025)	1,900 (1,040)	0.269	7.45	9.0 (16.2)	13	34 (0.14)
C95500	81 Cu, 4 Fe, 11 Al, 4 Ni	90	620	40	275	6	1,930 (1,055)	1,900 (1,040)	0.272	7.53	9.0 (16.2)	8.5	24 (0.10)
C95500 (TQ50)	81 Cu, 4 Fe, 11 Al, 4 Ni	110	760	60	415	5	1,930 (1,055)	1,900 (1,040)	0.272	7.53	9.0 (16.2)	8.5	24 (0.10)
C95800	81.5 Cu, 4 Fe, 9 Al, 4 Ni, 1.5 Mn	85	585	35	240	15	1,910 (1,045)	1,940 (1,060)	0.276	7.64	9.0 (16.2)	7.1	21 (0.085)
C96200	88.5 Cu, 1.5 Fe, 10 Ni	45	310	25	172	20	2,010 (1,100)	2,100 (1,150)	0.323	8.94	9.5 (17.3)	11	26 (0.11)
C96400	69 Cu, 1 Fe, 30 Ni	68	470	27	255	28	2,140 (1,170)	2,260 (1,240)	0.323	8.94	9.0 (16.2)	5	17 (0.068)
C97600	64 Ca, 4 Pb, 20 Ni, 4 Sn, 8 Zn	45	310	24	165	20	2,027 (1,108)	2,089 (1,543)	0.321	8.90	9.3 (17)	5	13 (0.075)

Table 6.4.20 Mechanical Properties of Rolled Yellow Brass (Alloy C26800)

Standard temper designation (ASTM B 601)*	Nominal grain size, mm	Tensile strength ksi	Tensile strength MPa	Approximate Rockwell hardness† F	Approximate Rockwell hardness† 30T
OS120	0.120	—	—	50–62	21 max
OS070	0.070	—	—	52–67	3–27
OS050	0.050	—	—	61–73	20–35
OS035	0.035	—	—	65–76	25–38
OS025	0.025	—	—	67–79	27–42
OS015	0.015	—	—	72–85	33–50

Standard temper designation (ASTM B 601)*	Former temper designation	Tensile strength ksi	Tensile strength MPa	B scale 0.020 (0.058) to 0.036 in (0.914 mm) incl	B scale Over 0.036 in (0.914 mm)	Superficial 30-T 0.012 (0.305) to 0.028 in (0.711 mm) incl	Superficial 30-T Over 0.028 in (0.711 mm)
M20	As hot-rolled	40–50	275–345	—	—	—	—
H01	Quarter-hard	49–59	340–405	40–61	44–65	43–57	46–60
H02	Half-hard	55–65	380–450	57–71	60–74	54–64	56–66
H03	Three-quarters hard	62–72	425–495	70–77	73–80	65–69	67–71
H04	Hard	68–78	470–540	76–82	78–84	68–72	69–73
H06	Extra hard	79–89	545–615	83–87	85–89	73–75	74–76
H08	Spring	86–95	595–655	87–90	89–92	75–77	76–78
H10	Extra spring	90–99	620–685	88–91	90–93	76–78	77–79

* Refer to Table 6.4.15 for definition of temper designations.
† Rockwell hardness values apply as follows: The F scale applies to metal 0.020 in (0.508 mm) in thickness and over; the 30-T scale applies to metal 0.015 in (0.381 mm) in thickness and over.
‡ Rockwell hardness values apply as follows: The B scale values apply to metal 0.020 in (0.508 mm) and over in thickness, and the 30-T scale values apply to metal 0.012 in (0.305 mm) and over in thickness.
SOURCE: ASTM, abstracted with permission.

for deep-drawing operations, annealed brass must be used. In general, smaller grain sizes are preferred.

Table 6.4.20 summarizes the ASTM B36 specification requirements for alloy C26800 (65:35 copper-zinc) for various rolled and annealed tempers.

Effect of Temperature Copper and all its alloys increase in strength slightly and uniformly as temperature decreases from room temperature. No low-temperature brittleness is encountered. Copper is useless for prolonged-stress service much above 400°F (204°C), but some of its alloys may be used up to 550°F (287°C). For restricted service above this temperature, only copper-nickel and copper-aluminum alloys have satisfactory properties.

For specific data, see Elevated-Temperature Properties of Copper Base Alloys, ASTM STP 181; Low-Temperature Properties of Copper and Selected Copper Alloys, *NBS Mono.* 101, 1967; and Wilkens and Bunn, "Copper and Copper Base Alloys," McGraw-Hill, 1943.

Electrical and Thermal Conductivities The brasses (essentially copper-zinc alloys) have relatively good **electrical and thermal conductivities**, although the levels are somewhat lower than those of the coppers because of the effect of solute atoms on the copper lattice. For this reason, copper and high-copper alloys are preferred over the brasses containing more than a few percent total alloy content when high electrical or thermal conductivity is required for the application.

Machinability Machinability of copper alloys is governed by their metallurgical structure and related properties. In that regard, they are divided into three subgroups. All leaded brasses are yellow, have moderate electrical conductivity and fair hot-working qualities, but are relatively poor with respect to fabrication when compared with C26000.

While they all have excellent to satisfactory machinability ratings, the most important alloy for machined products is C36000, **free-cutting brass,** which is the standard material for automatic screw machine work where the very highest machinability is necessary.

ASTM B16 Free-Cutting Brass Rod, Bar, and Shapes for Use in Screw Machines is the dominant specification. Use of this material will often result in considerable savings over the use of steel. Most copper alloys are readily machined by usual methods using standard tools

designed for steel, but machining is done at higher cutting speeds. Consideration of the wide range of characteristics presented by various types of copper alloys and the adaptation of machining practice to the particular material concerned will improve the end results to a marked degree.

Group A is composed of alloys of homogeneous structure: copper, wrought bronzes up to 10 percent tin, brasses and nickel silvers up to 37 percent zinc, aluminum bronzes up to 8 percent aluminum, silicon bronzes, and copper nickel. These alloys are all tough and ductile and form long, continuous chips. When they are severely cold-worked, they approach the group B classification in their characteristics.

Group B includes lead-free alloys of duplex structure, some cast bronzes, and most of the high-strength copper alloys. They form continuous but brittle chips by a process of intermittent shearing against the tool edge. Chatter will result unless work and tool are rigid.

Many of the basic **Group C** brasses and bronzes are rendered particularly adaptable for machining operations by the addition of 0.5 to 3.0 percent lead, which resides in the structure as minute, uniformly distributed droplets. They serve to break the chip and lubricate the tool cutting edge. Chips are fine, almost needlelike, and are readily removed. Very little heat is evolved, but the tendency to chatter is greater than for Group A alloys. Lead additions may be made to most copper alloys, but its low melting point makes hot working impossible. Tellurium and sulfur additions are used in place of lead when the combination of hot workability and good machinability is desired.

For **turning operations,** the tough alloys of Group A need a sharp top rake angle (20° to 30° for copper and copper nickel; 12° to 16° for the brasses, bronzes, and silicon bronzes with high-speed tools; 8° to 12° with carbide tools, except for copper, for which 16° is recommended). Type C (leaded) materials require a much smaller rake angle to minimize chatter, a maximum of 8° with high-speed steels and 3° to 6° with carbide. Type B materials are intermediate, working best with 6° to 12° rake angle with high-speed tools and 3° to 8° with carbide, the higher angle being used for the tougher materials. Side clearance angle should be 5° to 7° except for tough, "sticky" materials like copper and copper nickel, where a side rake of 10° to 15° is better.

Many copper alloys will **drill** satisfactorily with standard helix angle drills. Straight fluted tools (helix angle 0°) are preferable for the Group C leaded alloys. A helix angle of 10° is preferred for Group B and 40° for copper and copper nickel. Feed rates generally are 2 to 3 times faster than those used for steel. With Type B alloy, a fairly course feed helps to break the chip, and with Type A, a fine feed and high speed give best results, provided that sufficient feed is used to prevent rubbing and work hardening.

Corrosion Resistance For over half a century, copper has been the preferred material for tubing used, to convey potable water in domestic plumbing systems because of its outstanding corrosion resistance. In outdoor exposure, however, copper alloys containing less than 20 percent zinc are preferred because of their resistance to stress corrosion cracking, a form of corrosion originally called **season cracking.** For this type of corrosion to occur, a combination of tensile stress (either residual or applied) and the presence of a specific chemical reagent such as ammonia, mercury, mercury compounds, and cyanides is necessary. Stress relief annealing after forming alleviates the tendency for stress corrosion cracking. Alloys that are either zinc-free or contain less than 15 percent zinc generally are not susceptible to stress corrosion cracking.

Arsenical copper (C14200), arsenical admiralty metal (C44300), and arsenical aluminum brass (C68700) were once popular **condenser tube alloys** for freshwater power plant applications. They have been successfully replaced largely with alloy 706 (90:10 copper-nickel). In seawater or brackish water, copper, admiralty metal, and aluminum brass are even less suitable because of their inability to form protective films.

Caution must be exercised, for copper and its alloys are not suitable for use in oxidizing acidic solutions or in the presence of moist ammonia and its compounds. Specific alloys are also limited to maximum flow velocities.

Fabrication Practically any of the copper alloys listed in Table 6.4.17 can be obtained in sheet, rod, and wire form, and many can be obtained as tubes. Most, in the annealed condition, will withstand extensive amounts of cold work and may be shaped to the desired form by deep drawing, flanging, forming, bending, and similar operations. If extensive cold work is planned, the material should be purchased in the annealed condition, and it may need intermediate annealing either to avoid metal failure or to minimize power consumption. Annealing is done at 900 to 1,300°F (482 to 704°C), depending on the alloy, and is usually followed by air cooling. Because of the ready workability of brass, it is often less costly to use than steel. Brass may be drawn at higher speeds than ferrous metals and with less wear on the tools. In cupping operations, a take-in of 45 percent is usual and on some jobs, it may be larger. Brass hardened by cold working is softened by annealing at about 1,100 F (593°C).

Joining by Welding, Soldering, and Brazing

Welding Deoxidized copper will **weld** satisfactorily by the oxy-acetylene method. Sufficient heat input to overcome its high thermal conductivity must be maintained by the use of torches considerably more powerful than those customary for steel, and preferably by additional preheating. The filler rod must be deoxidized. Gas-shielded arc welding is preferred. Tough-pitch copper will not result in high-strength welds because of embrittlement due to the oxygen content. Copper may be arc-welded, using shielded metal arc, gas metal arc (MIG), or gas tungsten arc (TIG) welding procedures using experienced operators. Filler rods of phosphor bronze or silicon bronze will give strong welds more consistently and are used where the presence of a weld of different composition and corrosion-resistance characteristics is not harmful. Brass may be welded by the oxyacetylene process but not by arc welding. A filler rod of about the same composition is used, although silicon is frequently added to prevent zinc fumes.

Copper-silicon alloys respond remarkably well to welding by all methods. The conductivity is not too high, and the alloy is, to a large extent, self-fluxing.

Applicable specifications for joining copper and copper alloys by welding include

ANSI/AWS A5.6 Covered Copper and Copper Alloy Arc Welding Electrodes
ANSI/AWS A5.7 Copper and Copper Alloy Bare Welding Rods and Electrodes
ANSI/AWS A5.27 Copper and Copper Alloy Rods for Oxyfuel Gas Welding

The **fluxes** used for brazing copper joints are water-based and function by dissolving and removing residual oxides from the metal surface; they protect the metal from reoxidation during heating and promote wetting of the joined surfaces by the brazing filler metal.

Soldering Soldered joints with **capillary fittings** are used in plumbing for water and sanitary drainage lines. Such joints depend on capillary action drawing free-flowing molten solder into the gap between the fitting and the tube.

Flux acts as a cleaning and wetting agent which permits uniform spreading of the molten solder over the surfaces to be joined.

Selection of a solder depends primarily on the operating pressure and temperature of the system. Lead-free solders required for joining copper tube and fittings in potable water systems are covered by ASTM B32 Solder Metal.

As in brazing, the functions of soldering flux are to remove residual oxides, promote wetting, and protect the surfaces being soldered from oxidation during heating.

Fluxes best suited for soldering copper and copper alloy tube should meet the requirements of ASTM B813 Liquid and Paste Fluxes for Soldering Applications of Copper and Copper Alloy Tube with the joining accomplished per ASTM B828 Making Capillary Joints by Soldering of Copper and Copper Alloy Tube and Fittings.

JEWELRY METALS
Staff Contribution

REFERENCES: ASM "Metals Handbook." Publications of the metal-producing companies. Standards applicable to classes of metals available from manufacturers' associations. Trade literature pertinent to each specific metal.

Gold is used primarily as a monetary standard; the small amount put to metallurgical use is for jewelry or decorative purposes, in dental work, for fountain-pen nibs, and as an electrodeposited protecting coating. Electroplated gold is widely used for electronic junction points (transistors and the like) and at the mating ends of telephone junction wires. An alloy with palladium has been used as a platinum substitute for laboratory vessels, but its present price is so high that it does not compete with platinum.

Silver has the highest electrical conductivity of any metal, and it found some use in bus bars during World War II, when copper was in short supply. Since its density is higher than that of copper and since government silver frequently has deleterious impurities, it offers no advantage over copper as an electrical conductor. Heavy-duty electrical contacts are usually made of silver. It is used in aircraft bearings and solders. Its largest commercial use is in tableware as sterling silver, which contains 92.5 percent silver (the remainder is usually copper). United States coinage used to contain 90 percent silver, 10 percent copper, but coinage is now manufactured from a debased alloy consisting of copper sandwiched between thin sheets of nickel.

Platinum has many uses because of its high melting point, chemical inertness, and catalytic activity. It is the standard catalyst for the oxidation of sulfur dioxide in the manufacture of sulfuric acid. Because it is inert toward most chemicals, even at elevated temperatures, it can be used for laboratory apparatus. It is the only metal that can be used for an electric heating element about 2,300°F without a protective atmosphere. Thermocouples of platinum with platinum-rhodium alloy are standard for high temperatures. Platinum and platinum alloys are used in large amounts in feeding mechanisms of glass-working equipment to ensure constancy of the orifice dimensions that fix the size of glass products. They are also used for electrical contacts, in dental work, in aircraft spark-plug electrodes, and as jewelry. It has been used for a long time as a catalyst in refining of gasoline and is in widespread use in automotive catalytic converters as a pollution control device.

Palladium follows platinum in importance and abundance among the platinum metals and resembles platinum in most of its properties. Its density and melting point are the lowest of the platinum metals, and it forms an oxide coating at a dull-red heat so that it cannot be heated in air above 800°F (426°C) approx. In the finely divided form, it is an excellent hydrogenation catalyst. It is as ductile as gold and is beaten into leaf as thin as gold leaf. Its hardened alloys find some use in dentistry, jewelry, and electrical contacts.

Iridium is one of the platinum metals. Its chief use are as a hardener for platinum jewelry alloys and as platinum contacts. Its alloys with osmium are used for tipping fountain-pen nibs. Isotope Ir 192 is one of the basic materials used in radiation therapy and is widely employed as a radioactive implant in oncological surgical procedures. It is the most corrosion-resistant element known.

Rhodium is used mainly as an alloying addition to platinum. It is a component of many of the pen-tipping alloys. Because of its high reflectivity and freedom from oxidation films, it is frequently used as an electroplate for jewelry and for reflectors for motion-picture projectors, aircraft searchlights, and the like.

LOW-MELTING-POINT METALS AND ALLOYS
by Frank E. Goodwin

REFERENCES: Current edition of ASM "Metals Handbook." Current listing of applicable ASTM and AWS standards. Publications of the various metal-producing companies.

Metals with low-melting temperatures offer a diversity of industrial applications. In this field, much use is made of the **eutectic**-type alloy, in which two or more elements are combined in proper proportion so as to have a minimum melting temperature. Such alloys melt at a single, fixed temperature, as does a pure metal, rather than over a range of temperatures, as with most alloys.

Liquid Metals   A few metals are used in their liquid state. **Mercury** [mp, -39.37°F (-40°C)] is the only metal that is liquid below room temperature. In addition to its use in thermometers, scientific instruments, and electrical contacts, it is a constituent of some very low-melting alloys. Its application in dental amalgams is unique. Regulatory restrictions on mercury have resulted in substitution of alternatives because of health concerns. It has been used as a heat-exchange fluid, as have **sodium** and the sodium-potassium alloy **NaK** (see Sec. 9.8).

Moving up the temperature scale, **tin** melts at 449.4°F (232°C) and finds its largest single use in coating steel to make tinplate. Tin may be applied to steel, copper, or cast iron by hot-dipping, although for steel this method has been replaced largely by electrodeposition onto continuous strips of rolled steel. Typical tin coating thicknesses are 40 μin (1 μm). **Solders** constitute an important use of tin. Sn-Pb solders are

widely used, although Sn-Sb and Sn-Ag solders can be used in applications requiring lead-free compositions. Tin is alloyed in bronzes and battery grid material. Alloys of 12 to 25 percent Sn, with the balance lead, are applied to steel by hot-dipping and are known as **terneplate.**

Modern **pewter** is a tarnish-resistant alloy used only for ornamental ware and is composed of 91 to 93 percent Sn, 1 to 8 percent Sb, and 0.25 to 3 percent Cu. Alloys used for casting are lower in copper than those used for spinning hollowware. Pewter does not require intermediate annealing during fabrication. Lead was a traditional constituent of pewter but has been eliminated in modern compositions because of the propensity for lead to leach out into fluids contained in pewter vessels.

Pure **lead** melts at 620°F (327°C), and four grades of purities are recognized as standard; see Table 6.4.21. Corroding lead and common lead are 99.94 percent pure, while chemical lead and copper-bearing lead are 99.9 percent pure. Corroding lead is used primarily in the chemical industry, with most used to manufacture white lead or litharge (lead monoxide, PbO). Chemical lead contains residual copper that improves both its corrosion resistance and stiffness, allowing its extensive employment in chemical plants to withstand corrosion, particularly from sulfuric acid. Copper-bearing lead also has high corrosion resistance because of its copper content. Common lead is used for alloying and for battery oxides. Lead is strengthened by alloying with antimony at levels between 0 and 15 percent. Lead-tin alloys with tin levels over the entire composition range are widely used.

The largest use of lead is in **lead-acid batteries.** Deep cycling batteries typically use Pb 6–7 antimony, while Pb 0.1 Ca 0.3 tin alloys are used for maintenance-free batteries found in automobiles. Battery alloys usually have a minimum tensile strength of 6,000 lb/in² (41 MPa) and 20 percent elongation. Pb 0.85 percent antimony alloy is also used for sheathing of high-voltage power cables. Other cable sheathing alloys are based on Sn-Cd, Sn-Sb, or Te-Cu combined with lead. The lead content in cable sheathing alloys always exceeds 99 percent.

Lead sheet for construction applications, primarily in the chemical industries, is based on either pure lead or Pb 6 antimony. When cold-drolled, this alloy has a tensile strength of 4,100 lb/in² (28.3 MPa) and an elongation of 47 percent. Lead sheet is usually $\frac{3}{64}$ in (1.2 mm) thick and weighs 3 lb/ft² (15 kg/m²). **Architectural uses for lead** are numerous, primarily directed to applications where water is to be contained (shower and tiled bathing areas overlaid by ceramic tile). Often, copper is lead-coated to serve as roof flashing or valleys; in those applications, the otherwise offending verdigris coating which would develop on pure copper sheet is inhibited.

Ammunition for both sports and military purposes uses alloys containing percentages up to 8 Sb and 2 As. **Bullets** have somewhat higher alloy content than shot.

Type metals are the hardest and strongest lead alloys. Type metals are obsolescent by virtue of being superceded by newer printing techniques. Their use is restricted to existing printing machinery. Typical compositions are shown in Table 6.4.22. A small amount of copper may be added to increase hardness. Electrotype metal contains the lowest alloy content because it serves only as a backing to the shell and need

Table 6.4.21   Composition Specifications for Lead, According to ASTM B-29–03

Element	Composition, wt %			
	Corroding lead	Common lead	Chemical lead	Copper-bearing lead
Silver, max	0.015	0.005	0.020	0.020
Silver, min	—	—	0.002	—
Copper, max	0.0015	0.0015	0.080	0.080
Copper, min	—	—	0.040	0.040
Silver and copper together, max	0.025	—	—	—
Arsenic, antimony, and tin together, max	0.002	0.002	0.002	0.002
Zinc, max	0.001	0.001	0.001	0.001
Iron, max	0.002	0.002	0.002	0.002
Bismuth, max	0.050	0.050	0.005	0.025
Lead (by difference), min	99.94	99.94	99.90	99.90

SOURCE: ASTM, reprinted with permission.

Table 6.4.22 Compositions and Properties of Type Metal*

Service	Composition, %			Liquidus		Solidus		Brinell hardness†
	Sn	Sb	Pb	°F	°C	°F	°C	
Electrotype	4	3	93	561	294	473	245	12.5
Linotype	5	11	84	475	246	462	239	22
Stereotype	6.5	13	80.5	485	252	462	239	22
Monotype	8	17	75	520	271	462	239	27

* The use of type metals is obsolescent in the printing industry.
† 0.39-in (10-mm) ball, 550-lb (250-kg) load
SOURCE: "Metals Handbook," ASM International, 10th ed.

not be hard. Linotype, or slug casting metal, must be fluid and capable of rapid solidification for its use in composition of newspaper type; thus metal of nearly eutectic composition is used. It is rarely used as the actual printing surface and therefore need not be as hard as stereotype and monotype materials.

Fusible Alloys (See Table 6.4.23.)   These alloys are used typically as fusible links in sprinkler heads, as electric cutouts, as fire-door links, for making castings, for patterns in making match plates, for making electroforming molds, for setting punches in multiple dies, and for dyeing cloth. Some fusible alloys can be cast or sprayed on wood, paper, and other materials without damaging the base materials, and many of these alloys can be used for making hermetic seals. Since some of these alloys melt below the boiling point of water, they can be used in bending tubing. The properly prepared tubing is filled with the molten alloy and allowed to solidify, and after bending, the alloy is melted out by immersing the tube in boiling water. The volume changes during the solidification of a fusible alloy are, to a large extent, governed by the bismuth content of the alloy. As a general rule, alloys containing more than about 55 percent bismuth expand and those containing less than about 48 percent bismuth contract during solidification; those containing 48 to 55 percent bismuth exhibit little change in volume. The change in volume due to cooling of the solid metal is a simple linear shrinkage, but some of the fusible alloys owe much of their industrial importance to other volume changes, caused by change in structure of the solid alloy, which permit the production of castings having dimensions equal to, or greater than, those of the mold in which the metal was cast.

For **fire-sprinkler heads,** with a rating of 160°F (71°C) **Wood's metal** is used for the fusible-solder-alloy link. Wood's metal gives the most suitable degree of sensitivity at this temperature, but in tropical countries and in situations where industrial processes create a hot atmosphere (e.g., baking ovens, foundries), solders having a higher melting point must be used. Alloys of eutectic compositions are used since they melt sharply at a specific temperature.

Fusible alloys are also used as molds for thermoplastics, for the production of artificial jewelry in pastes and plastic materials, in foundry patterns, chucking glass lenses, as hold-down bolts, and inserts in plastics and wood.

Solders are filler metals that produce coalescence of metal parts by melting the solder while heating the base metals below their melting point. To be termed *soldering* (versus brazing), the operating temperature must not exceed 840°F (450°C). The filler metal is distributed between the closely fitting faying surfaces of the joint by capillary action. Compositions of solder alloys are shown in Table 6.4.24. The Pb 63 tin eutectic composition has the lowest melting point—361°F (183°C)—of the binary tin-lead solders. For joints in copper pipe and cables, a wide melting range is needed. Solders containing less than 5 percent Sn are used to seal precoated containers and in applications where the surface temperatures exceed 250°F (120°C). Solders with 10 to 20 percent Sn are used for repairing automobile radiators and bodies, while compositions containing 40 to 50 percent Sn are used for manufacture of automobile radiators, electrical and electronic connections, roofing seams, and heating units. Silver is added to tin-lead solders for electronics applications to reduce the dissolution of silver from the substrate. Sb-containing solders should not be used to join base metals containing zinc, including galvanized iron and brass. A 95.5 percent Sn 4 Cu 0.5 Ag solder has supplanted lead-containing solders in potable water plumbing.

Brazing is similar to soldering but is defined as using filler metals which melt at or above 840°F (450°C). As in soldering, the base metal is not melted. Typical applications of brazing filler metals classified by AWS and ASTM are listed in Table 6.4.25; their compositions and melting points are given in Table 6.4.26.

METALS AND ALLOYS FOR USE AT ELEVATED TEMPERATURES
by John H. Tundermann

REFERENCES: "High Temperature High Strength Alloys," AISI. Simmons and Krivobok, Compilation of Chemical Compositions and Rupture Strength of Super-strength Alloys, *ASTM Tech. Pub.* 170-A. "Metals Handbook," ASM. Smith, "Properties of Metals at Elevated Temperatures," McGraw-Hill. Clark,

Table 6.4.23 Compositions, Properties, and Applications of Fusible Alloys

Composition, %					Solidus		Liquidus		Typical application
Sn	Bi	Pb	Cd	In	°F	°C	°F	°C	
8.3	44.7	22.6	5.3	19.10	117	47	117	47	Dental models, part anchoring, lens chucking
13.3	50	26.7	10.0	—	158	70	158	70	Bushings and locators in jigs and fixtures, lens chucking, reentrant tooling, founding cores and patterns, light sheet-metal embossing dies, tube bending
—	55.5	44.5	—	—	255	124	255	124	Inserts in wood, plastics, bolt anchors, founding cores and patterns, embossing dies, press-form blocks, duplicating plaster patterns, tube bending, hobbyist parts
12.4	50.5	27.8	9.3	—	163	73	158	70	Wood's metal-sprinkler heads
14.5	48.0	28.5	(9% Sb)	—	440	227	217	103	Punch and die assemblies, small bearings, anchoring for machinery, tooling, forming blocks, stripper plates in stamping dies
60	40	—	—	—	338	170	281	138	Locator members in tools and fixtures, electroforming cores, dies for lost-wax patterns, plastic casting molds, prosthetic development work, encapsulating avionic components, spray metallizing, pantograph tracer molds

SOURCE: "Metals Handbook," ASM International, 10th ed.

Table 6.4.24   Compositions and Properties of Selected Solder Alloys

Composition, %		Solidus temperature		Liquidus temperature		
Tin	Lead	°C	°F	°C	°F	Uses
2	98	316	601	322	611	Side seams for can manufacturing
5	95	305	581	312	594	Coating and joining metals
10	90	268	514	302	576	Sealing cellular automobile radiators, filling seams or dents
15	85	227	440	288	550	Sealing cellular automobile radiators, filling seams or dents
20	80	183	361	277	531	Coating and joining metals, or filling dents or seams in automobile bodies
25	75	183	361	266	511	Machine and torch soldering
30	70	183	361	255	491	
35	65	183	361	247	477	General-purpose and wiping solder
40	60	183	361	238	460	Wiping solder for joining lead pipes and cable sheaths; also for automobile radiator cores and heating units
45	55	183	361	227	441	Automobile radiator cores and roofing seams
50	50	183	361	216	421	Most popular general-purpose solder
60	40	183	361	190	374	Primarily for electronic soldering applications where low soldering temperatures are required
63	37	183	361	183	361	Lowest-melting (eutectic) solder for electronic applications
40	58 (2% Sb)	185	365	231	448	General-purpose, not recommended on zinc-containing materials
95	0 (5% Sb)	232	450	240	464	Joints on copper, electrical plumbing, heating, not recommended on zinc-containing metals
0	97.5 (2.5% Ag)	304	580	308	580	Copper, brass, not recommended in humid environments
95.5	0 (4% Cu, 0.5% Ag)	226	440	260	500	Potable water plumbing

SOURCE: "Metals Handbook," vol. 2, 10th ed., p. 553.

Table 6.4.25   Brazing Filler Metals

AWS-ASTM filler-metal classification	Base metals joined
BAlSi (aluminum-silicon)	Aluminum and aluminum alloys
BCuP (copper-phosphorus)	Copper and copper alloys; limited use on tungsten and molybdenum; should not be used on ferrous or nickel-base metals
BAg (silver)	Ferrous and non-ferrous metals except aluminum and magnesium; iron, nickel, cobalt-base alloys; thin-base metals
BCu (copper)	Ferrous and non-ferrous metals except aluminum and magnesium
RBCuZn (copper-zinc)	Ferrous and non-ferrous metals except aluminum and magnesium; corrosion resistance generally inadequate for joining copper, silicon, bronze, copper, nickel, or stainless steel
BMg (magnesium)	Magnesium-base metals
BNi (nickel)	AISI 300 and 400 stainless steels; nickel- and cobalt-base alloys; also carbon steel, low-alloy steels, and copper where specific properties are desired

Table 6.4.26   Compositions and Melting Ranges of Brazing Alloys

AWS designation	Nominal composition, %	Melting temp, °F
BAg-1	45 Ag, 15 Cu, 16 Zn, 24 Cd	1,125–1,145
BNI-7	13 Cr, 10 P, bal. Ni	1,630
BAu-4	81.5 Au, bal. Ni	1,740
BAlSi-2	7.5 Si, bal. Al	1,070–1,135
BAlSi-5	10 Si, 4 Cu, 10 Zn, bal. Al	960–1,040
BCuP-5	80 Cu, 15 Ag, 5 P	1,190–1,475
BCuP-2	93 Cu, 7 P	1,310–1,460
BAg-1a	50 Ag, 15.5 Cu, 16.5 Zn, 18 Cd	1,160–1,175
BAg-7	56 Ag, 22 Cu, 17 Zn, 5 Sn	1,145–1,205
RBCuZn-A	59 Cu, 40 Zn, 0.6 Sn	1,630–1,650
RBCuZn-D	48 Cu, 41 Zn, 10 Ni, 0.15 Si, 0.25 P	1,690–1,715
BCu-1	99.90 Cu, min	1,980
BCu-1a	99.9 Cu, min	1,980
BCu-2	86.5 Cu, min	1,980

°C = (°F − 32)/1.8.
SOURCE: "Metals Handbook," ASM.

"High-Temperature Alloys," Pittman. Cross, Materials for Gas Turbine Engines, *Metal Progress,* March 1965. "Heat Resistant Materials," ASM Handbook, vol. 3, 9th ed., pp. 187–350. Lambert, "Refractory Metals and Alloys," ASM Handbook, vol. 2, 10th ed., pp. 556–585. Watson et al., "Electrical Resistance Alloys," ASM Handbook, vol. 2, 10th ed., pp. 822–839.

Some of data presented here refer to materials with names which are proprietary and registered trademarks. They include Inconel, Incoloy, and Nimonic (Inco Alloys International); Hastelloy (Haynes International); MAR M (Martin Marietta Corp.); and René (General Electric Co.).

Metals are used for an increasing variety of applications at elevated temperatures, "elevated" being a relative term that depends upon the specific metal and the specific service environment. Elevated-temperature properties of the common metals and alloys are cited in the several subsections contained within this section. This subsection deals with metals and alloys whose prime use is in **high-temperature applications.** [Typical operating temperatures are compressors, 750°F (399°C); steam turbines, 1,100°F (593°C); gas turbines, 2,000°F (1,093°C); resistance-heating elements, 2,400°F (1,316°C); electronic vacuum tubes, 3,500°F (1,926°C), and lamps, 4,500°F (2,482°C).] In general, alloys for high-temperature service must have melting points above the operating temperature, low vapor pressures at that temperature, resistance to attack (oxidation, sulfidation, corrosion) in the operating environment, and sufficient strength to withstand the applied load for the service life without deforming beyond permissible limits. At high temperatures, atomic diffusion becomes appreciable, so that time is an important factor with respect to surface chemical reactions, to **creep** (slow deformation under constant load), and to internal changes within the alloy during service. The effects of time and temperature are conveniently combined by the empirical **Larson-Miller parameter** $P = T(C + \log t) \times 10^{-3}$, where T = test temp in °R (°F + 460) and t = test time, h. The constant C depends upon the material but is frequently taken to be 20.

Many alloys have been developed specifically for such applications. The selection of an alloy for a specific high-temperature application is strongly influenced by service conditions (stress, stress fluctuations, temperature, heat shock, atmosphere, service life), and there are hundreds of alloys from which to choose. The following illustrations, data, and discussion should be regarded as examples. Vendor literature and more extensive references should be consulted.

Figure 6.4.7 indicates the general stress-temperature range in which various alloy types find application in elevated-temperature service.

Fig. 6.4.7 Stress-temperature application ranges for several alloy types; stress to produce rupture in 1,000 h.

Fig. 6.4.8 Effect of time on rupture strength of type 304 stainless steel and alloy M252.

Figure 6.4.8 indicates the important effect of time on the strength of alloys at high temperatures, comparing a familiar stainless steel (type 304) with **superalloy** (M252).

Common Heat-Resisting Alloys A number of alloys containing large amounts of **chromium and nickel** are available. These have excellent oxidation resistance at elevated temperatures. Several of them have been developed as electrical-resistance heating elements; others are

modifications of stainless steels, developed for general corrosion resistance. Selected data on these alloys are summarized in Table 6.4.27. The maximum temperature value given is for resistance to oxidation with a reasonable life. At higher temperatures, failure will be rapid because of scaling. At lower temperatures, much longer life will be obtained. At the maximum useful temperature, the metal may be very weak and frequently must be supported to prevent sagging. Under load, these alloys are generally useful only at considerably lower temperatures, say up to 1,200°F (649°C) max, depending upon permissible creep rate and the load.

Superalloys were developed largely to meet the needs of aircraft gas turbines, but they have also been used in other applications demanding high strength at high temperatures. These alloys are based on nickel and/or cobalt, to which are added (typically) chromium for oxidation resistance and a complex of other elements which contribute to hot strength, both by solid solution hardening and by forming relatively stable dispersions of fine particles. Hardening by cold work, hardening by precipitation-hardening heat treatments, and hardening by deliberately arranging for slow precipitation during service are all methods used to enhance the properties of these alloys. **Fabrication** of these alloys is difficult since they are designed to resist distortion even at elevated temperatures. Forging temperatures of about 2,300°F (1,260°C) are used with small reductions and slow rates of working. Many of these alloys are fabricated by precision casting. Cast alloys that are given a strengthening heat treatment often have better properties than wrought alloys, but the shapes that can be made are limited.

Vacuum melting is an important factor in the production of superalloys. The advantages which result from its application include the ability to melt higher percentages of reactive metals, improved mechanical properties (particularly fatigue strength), decreased scatter in mechanical properties, and improved billet-to-bar stock-conversion ratios in wrought alloys.

Table 6.4.28 to 6.4.31 list compositions and properties of some superalloys, and Fig. 6.4.9 indicates the temperature dependence of the rupture strength of a number of such alloys.

Cast tool alloys are another important group of materials having high-temperature strength and wear resistance. They are principally alloys of cobalt, chromium, and tungsten. They are hard and brittle, and they must be cast and ground to shape. Their most important application is for hard facing, but they compete with high-speed steels and cemented carbides for many applications, in certain instances being superior to both. (See Cemented Carbides, Sec. 6.4.) Typical compositions are given in Table 6.4.32; typical properties are: elastic modulus $= 35 \times 10^6$ lb/in^2 (242 GPa); specific gravity = 8.8; hardness, BHN at room temp = 660; 1,500°F (815°C), 435, and 2,000°F (1,093°C) 340.

For still higher-temperature applications, the possible choices are limited to a few metals with high melting points, all characterized by

Fig. 6.4.9 Temperature dependence of strengths of some high-temperature alloys; stress to produce rupture in 1,000 h.

Table 6.4.27   Properties of Common Heat-Resisting Alloys

Alloy UNS no.	Name	Max. temp for oxidation resistance °F	°C	Nominal chemical composition, % C	Cr	Ni	Fe	Other	Specific gravity	Coef of thermal expansion per °F × 10⁻⁶ (0–1,200°F)	Stress to rupture in 1,000 h, ksi 1,200°F (649°C)	1,500°F (816°C)	1,800°F (982°C)	0.2 percent offset yield strength, ksi Room temp	1,000°F (538°C)	1,200°F (649°C)
N 06004	62 Ni, 15 Cr	1,700	926		15	62	Bal		8.19	9.35						
N 06003	80 Ni, 20 Cr	2,100	1,148		20	80			8.4	9.8						
	Kanthal	2,450	1,371		25		Bal	3 Co, 5 Al	7.15					63		
N 06600	Inconel alloy 600	2,050	1,121		13	79	Bal		8.4		14.5	3.7		36	22	22
G 10150	1,015	1,000	537	0.15			Bal		7.8	8.36	2.7			42	20	10
S 50200	502	1,150	621	0.12	5		Bal	0.5 Mo	7.8	7.31	6	1.5		27	18	12
S 44600	446	2,000	1,093	0.12	26	0.3	Bal		7.6	6.67	4	1.2				
S 30400	304	1,650	871	0.06	18	9	Bal		7.9	10.4	11	3.5		32	14	11
S 34700	347	1,650	871	0.08	18.5	11.5	Bal	0.8 Nb	8.0	10.7	20			41	31	26
S 31600	316	1,650	873	0.07	18	13	Bal	2.5 Mo	8.0	10.3	25	7		41	22	21
S 31000	310	2,000	1,093	0.12	25	20	Bal		7.9	9.8	13.2	3.0	2.7	34	28	25
S 32100	321	1,650	899	0.06	18	10	Bal	0.5 Ti	8.0	10.7	17.5	3.7		39	27	25
	NA 22H	2,200	1,204	0.5	28	48	Bal	5W		8.6	30	18	3.6			
N 06617	Inconel alloy 617	2,050	1,121	0.1	22	Bal	3	9 Mo, 12.5 Co, 1.2 Al, 0.6 Ti	8.36	6.4	52	14	4	43	29	25
N 08800	Inconel alloy 800	1,800	982	0.1	21	32	Bal		7.94	7.9	24	7	2	36	26	26
N 08330	330	1,850	1,010	0.05	19	36	Bal		8.08	8.3		5	1.3	40	26	21

limited availability and by difficulty of extraction and fabrication. Advanced alloys of metals such as chromium, niobium, molybdenum, tantalum, and tungsten are still being developed. Table 6.4.33 cites properties of some of these alloys. Wrought tool steel materials are described in "Tool Steels," ASM Handbook, vol. 3, 9th ed., pp. 421–447.

Chromium with room-temperature ductility has been prepared experimentally as a pure metal and as strong high-temperature alloys. It is not generally available on a commercial basis. Large quantities of chromium are used as melt additions to steels, stainless steels, and nickel-base alloys.

Molybdenum is similar to tungsten in most of its properties. It can be prepared in the massive form by powder metallurgy techniques, by inert-atmosphere or vacuum-arc melting. Its most serious limitation is its ready formation of a volatile oxide at temperatures of 1,400°F

approx. In the worked form, it is inferior to tungsten in melting point, tensile strength, vapor pressure, and hardness, but in the recrystallized condition, the ultimate strength and elongation are higher. Tensile strengths up to 350,000 lb/in² (2,413 MPa) have been reported for hard-drawn wire, and to 170,000 lb/in² (1,172 MPa) for soft wire. In the hard-drawn condition, molybdenum has an elongation of 2 to 5 percent, but after recrystallizing, this increases to 10 to 25 percent. Young's modulus is 50×10^6 lb/in² (345 GPa). It costs about the same as tungsten, per pound, but its density is much less. It has considerably better forming properties than tungsten and is extensively used for anodes, grids, and supports in vacuum tubes, lamps, and X-ray tubes. Molybdenum is generally used for winding electric furnaces for temperatures up to 3,000°F (1,649°C). As it must be protected against oxidation, such furnaces are usually operated in hydrogen. It is the common material for cathodes for radar devices, heat radiation shields,

Table 6.4.28   Wrought Superalloys, Compositions*

Common designation	Chemical composition, % C	Mn	Si	Cr	Ni	Co	Mo	W	Nb	Ti	Al	Fe	Other
19-9D L	0.2	0.5	0.6	19	9	....	1.2	1.25	0.3	0.3	....	Bal	
Timken	0.1	0.5	0.5	16	25	....	6	...	...	...	....	Bal	0.15 N
L.C. N-155	0.1	0.5	0.5	20	20	20	3	2	1	...	....	Bal	0.15 N
S-590	0.4	0.5	0.5	20	20	20	4	4	4	...	....	Bal	
S-816	0.4	0.5	0.5	20	20	Bal	4	4	4	...	....	4	
Nimonic alloy 80A	0.05	0.7	0.5	20	76	....	....	....	...	2.3	1.0	0.5	
K-42-B	0.05	0.7	0.7	18	42	22	....	....	...	2.0	0.2	14	
Hastelloy alloy B	0.1	0.5	0.5	..	65	....	28	....	...	....	....	6	0.4 V
M252†	0.10	1.0	0.7	19	53.5	10	10	....	...	2.5	0.75	2	
J1570	0.20	0.1‡	0.2‡	20	29	37.5	....	7	...	4.1	....	2	
HS-R235	0.10	....	....	16	Bal	1.5	6		...	3.0	1.8	8	
Hastelloy alloy X	0.10	....	....	21	48	2.0	9	1.0	...	...	....	18	
HS 25 (L605)	0.05	1.5	1.0‡	20	10	53	....	15	...	...	....	1.0	
A-286	0.05	....		15	26	....	1.3	....	...	2.0	0.35	55	0.3 V
Inconel alloy 718	0.05	0.2	0.2	19	52.5	....	3	....	5	0.9	0.5	18.5	
Inconel alloy X-750	0.1	1.0	0.5	15	72	1.0	....	....	...	2.5	0.7	7	

* For a complete list of superalloys, both wrought and cast, experimental and under development, see *ASTM Spec. Pub.* 170, "Compilation of Chemical Compositions and Rupture Strengths of Superstrength Alloys."

† Waspaloy has a similar composition (except for lower Mo content), and similar properties.

‡ Max.

Table 6.4.29 Wrought Superalloys, Properties

		Max temp under load		Coef of thermal expansion, $in/(in \cdot °F) \times 10^{-6} (70–1,200°F)$	Specific gravity	Stress to rupture, ksi			Short-time tensile properties, ksi							
						At 1,200°F	At 1,500°F		Yield strength 0.2% offset				Tensile strength			
Alloy						1,000 h	100 h	1,000 h	Room	1,200°F	1,500°F	1,800°F	Room	1,200°F	1,500°F	1,800°F
UNS	Name	°F	°C													
K63198	19-9 DL	1,200	649	9.7	7.75	38	17	10	115	39	30		140	75	33	13
	Timken	1,350	732	9.25	8.06	36	13.5	9	96*	70*	16*		134	90	40	18
R30155	L.C. N-155	1,400	760	9.4	8.2	48	20.0	15	53	40	33	12	115	53	35	19
	S-590	1,450	788	8.0	8.34	40	19.0	15	78*	70	47	20	140	82	60	22
R30816	S-816	1,500	816	8.3	8.66	53	29.0	18	63*	45	41		140	112	73	25
N07080	Nimonic alloy 80A	1,450	788	7.56		56	24	15	80*	73	47		150	103	69	
—	K-42-B	1,400	760	8.5	8.23	38	22	15	105	84	52		158	117	54	
N10001	Hastelloy alloy B	1,450	788	6.9	9.24	36	17	10	58	42			135	94	66	24
N07252	M 252	1,500	816	7.55	8.25	70	26	18	90	75	65		160	140	80	20
—	J 1570	1,600	871	8.42	8.66	84	34	24	81	70	71	11	152	135	82	20
—	HS-R235	1,600	871	8.34	7.88	70	35	23	100	90	80	22	170	145	83	25
N06002	Hastelloy alloy X	1,350	732			30	14	10	56	41	37		113	83	52	15
R30605	HS 25 (L605)	1,500	816			54	22	18	70	35				75	50	23
K66286	A-286	1,350	732			45	13.8	7.7								
N07718	Inconel alloy 718	1,400	760	7.2	8.19	54	19	13	152	126			186	149		
N07750	Inconel alloy X-750	1,300	704	7.0	8.25				92	82	45		162	120	52	

* 0.02 percent offset.

rocket nozzles and other missile components, and hot-working tools and die-casting cores in the metalworking industry. Its principal use is still as an alloying addition to steels, especially tool steels and high-temperature steels.

Molybdenum alloys have found increased use in aerospace and commercial structural applications, where their high stiffness, microstructural stability, and creep strength at high temperatures are required. They are generally stronger than niobium alloys but are not ductile in the welded condition. Molybdenum alloys are resistant to alkali metal corrosion.

Niobium (also known as **columbium**) is available as a pure metal and as the base of several niobium alloys. The alloys are used in nuclear applications at temperatures of 1,800 to 2,200°F (980 to 1,205°C) because of their low thermal neutron cross section, high strength, and good corrosion resistance in liquid or gaseous alkali metal atmospheres. Niobium alloys are used in leading edges, rocket nozzles, and guidance structures for hypersonic flight vehicles. A niobium-46.5 percent titanium alloy is

used as a low-temperature superconductor material for magnets in magnetic resonance imaging machines.

Tantalum has a melting point that is surpassed only by tungsten. Its early use was as an electric-lamp filament material. It is more ductile than molybdenum or tungsten; the elongation for annealed material may be as high as 40 percent. The tensile strength of annealed sheet is about 50,000 lb/in² (345 MPa). In this form, it is used primarily in electronic capacitors. It is also used in chemical-processing equipment, where its high rate of heat transfer (compared with glass or ceramics) is particularly important, although it is equivalent to glass in corrosion resistance. Its corrosion resistance also makes it attractive for surgical implants. Like tungsten and molybdenum, it is prepared by powder metallurgy techniques; thus, the size of the piece that can be fabricated is limited by the size of the original pressed compact. Tantalum carbide is used in cemented-carbide tools, where it decreases the tendency to seize and crater. The stability of the anodic oxide film on tantalum leads to rectifier and capacitor applications. Tantalum and its **alloys** become

Table 6.4.30 Cast Superalloys, Nominal Composition

	Chemical composition, %										
Common designation	C	Mn	Si	Cr	Ni	Co	Mo	W	Nb	Fe	Other*
Vitallium (Haynes 21)	0.25	0.6	0.6	27	2	Bal	6	—	—	1	—
61 (Haynes 23)	0.4	0.6	0.6	26	1.5	Bal	—	5	—	1	—
422-19 (Haynes 30)	0.4	0.6	0.6	26	16	Bal	6	—	—	1	—
X-40 (Haynes 31)	0.4	0.6	0.6	25	10	Bal	—	7	—	1	—
S-816	0.4	0.6	0.6	20	20	Bal	4	4	4	5	—
HE 1049	0.45	0.7	0.7	25	10	45	—	15	—	1.5	0.4B
Hastelloy alloy C	0.10	0.8	0.7	16	56	1	17	4	—	5.0	0.3V
B 1900 + Hf	0.1	—	—	8	Bal	10	6	—	—	—	6 Al 4 Ta 1 Ti 1 5 Hf
MAR-M 247	0.15	—	—	8.5	Bal	10	0.5	10	—	—	5.5 Al 3Ta 1 Ti 1.5 Hf
René 80	0.15	—	—	14	Bal	9.5	4	4	—	—	3 Al 5 Ti
MO-RE 2	0.2	0.3	0.3	32	48	—	—	15	—	Bal	1 Al

* Trace elements not shown.

Table 6.4.31 Cast Superalloys, Properties

Common designation	Max temp under load °F	Max temp under load °C	Coef of Thermal expansion in/(in · °F) 3 10⁻⁶ (70–1,200°F)	Specific gravity	Stress to rupture in 1,000 h, ksi 1,200°F	1,500°F	1,800°F	Yield strength 0.2% offset Room 68°F (20°C)	1,200°F (650°C)	1,500°F (816°C)	1,800°F (982°C)	Tensile strength Room 68°F (20°C)	1,200°F (650°C)	1,500°F (816°C)	1,800°F (982°C)
Vitallium (Haynes 21)	1,500	816	8.35	8.3	44	14	7	82	71	49	33	101	89	59	33
61(Haynes 23)	1,500	816	8.5	8.53	47	22	5.5	58	74	40		105	97	58	45
422-19 (Haynes 30)	1,500	816	8.07	8.31		21	7	55	37*	48		98	59*	64	37
N-40 (Haynes 31)	1,500	816	8.18	8.60	46	23	9.8	74	37*	44		101	77*	59	29
S-816	1,500	816	8.27	8.66	46	21	9.8					112			
HE 1049	1,650	899		8.9	75	35	7	80	72	62	36	90	82	81	51
Hastelloy alloy C	1,800	982		8.91	42.5	14.5	1.4	54	50			130	87	51	19
B 1900 + Hf	1,700	927	7.73	8.22		55	15	120	134	109	60	140	146	126	80
MAR-M 247	1,700	927	7.7	8.53			18	117	117	108	48	140	152	130	
René 80				8.16			15	124	105	90		149	149	123	76

* Specimen not aged.

Table 6.4.32 Typical Compositions of Cast Tool Alloys

Name	C	Cr	W	Co	Fe	V	Ta	B	Other
Rexalloy	3	32	20	45					
Stellite 98 M2	3	28	18	35	9	4	0.1	0.1	3 Ni
Tantung G-2	3	15	21	40	...	...	19	0.2	
Borcoloy no. 6	0.7	5	18	20	Bal	1.3	....	0.7	6 Mo
Colmonoy WCR 100	...	10	15	...	Bal	...	....	3	

Table 6.4.33 Properties of Selected Refractory Metals

Alloy	Melting temp, °F	Density, lb/in³ at 75°F	Elastic modulus at 75°F	Ductile-to-brittle transition temp, °F	Tensile strength, ksi (recrystallized condition)				
					75°F	1,800°F	2,200°F	3,000°F	3,500°F
Chromium	3,450	0.760	42	625		12	8		
Columbium	4,474	0.310	16	−185	45	12	10		
F48 (15 W, 5 Mo, 1 Zr, 0.05 C)			25		121	75	50		
Cb74 or Cb752 (10 W, 2 Zr)		0.326			84	60	36		
Molybdenum	4,730	0.369	47	85	75	34	22	6	3
Mo (1/2 Ti)					110	70	48	8	4
TZM (0.5 Ti, 0.08 Zr)					130	85	70	14	5
Tantalum	5,425	0.600	27	< −320	30	22	15	8	4
Ta (10 W)					65	35	12	8	
Rhenium	5,460	0.759	68	< 75	170	85	60	25	11
Tungsten	6,170	0.697	58	645	85	36	32	19	10
W (10 Mo)								28	10
W (2 ThO₂)							40	30	25

competitive with niobium at temperatures above 2,700°F (1,482°C). As in the case of niobium, these materials are resistant to liquid or vapor metal corrosion and have excellent ductility even in the welded condition.

Tungsten has a high melting point, which makes it useful for high-temperature structural applications. The massive metal is usually prepared by powder metallurgy from hydrogen-reduced powder. As a metal, its chief use is as filaments in incandescent lamps and electronic tubes, since its vapor pressure is low at high temperatures. Tensile strengths over 600,000 lb/in² (4,136 MPa) have been reported for fine tungsten wires; in larger sizes (0.040 in), tensile strength is only 200,000 lb/in² (1,379 MPa). The hard-drawn wire has an elongation of 2 to 4 percent, but the recrystallized wire is brittle. Young's modulus is 60 million lb/in² (415 GPa). It can be sealed directly to hard glass and so is used for lead-in wires. A considerable proportion of tungsten rod and sheet is used for electrical contacts in the form of disks cut from rod or sheet and brazed to supporting elements. The major part of the tungsten used is made into ferroalloys for addition to steels or into tungsten carbide for cutting tools. Other applications include elements of electronic tubes, X-ray tube anodes, and arc-welding electrodes. Alloys of tungsten that are commercially available are W-3 Re, W-25 Re, and thoriated tungsten. The rhenium-bearing alloys are more ductile than unalloyed tungsten at room temperature. Thoriated tungsten is stronger than unalloyed tungsten at temperatures up to the recrystallization temperature.

METALS AND ALLOYS FOR NUCLEAR ENERGY APPLICATIONS
by L.D. Kunsman and C.L. Carlson; Amended by Staff

REFERENCES: Glasstone, "Principles of Nuclear Reactor Engineering," Van Nostrand. Hausner and Roboff, "Materials for Nuclear Power Reactors," Reinhold. Publications of the builders of reactors (General Electric Co., Westinghouse Corp., General Atomic Co.) and suppliers of metal used in reactor components. See also Sec. 9.8 commentary and other portions of Sec. 6.4.

The advent of atomic energy not only created a demand for new metals and alloys but also focused attention on certain properties and combinations of properties which theretofore had been of little consequence.

Reactor technology requires special materials for fuels, fuel cladding, moderators, reflectors, controls, heat-transfer mediums, operating mechanisms, and auxiliary structures. Some of the properties pertinent to such applications are given in a general way in Table 6.4.34. In general, outside the reflector, only normal engineering requirements need be considered.

Nuclear Properties A most important consideration in the design of a nuclear reactor is the control of the number and speed of the neutrons resulting from fission of the fuel. The designer must have knowledge of the effectiveness of various materials in slowing down neutrons or in capturing them. The slowing-down power depends not only on the relative energy loss per atomic collision but also on the number of collisions per second per unit volume. The former will be larger, the lower the atomic weight, and the latter larger, the greater the atomic density and the higher the probability of a scattering collision. The effectiveness of a moderator is frequently expressed in terms of the moderating ratio, the ratio of the slowing-down power to the capture cross section. The capturing and scattering tendencies are measured in terms of nuclear cross section in barns (10^{-24} cm²). Data for some of the materials of interest are given in Tables 6.4.35 to 6.4.37. The absorption data apply to slow, or thermal, neutrons; entirely different cross sections obtain for fast neutrons, for which few materials have significantly high capture affinity. Fast neutrons have energies of about 10^6 eV, while slow, or thermal, neutrons have energies of about 2.5×10^{-2} eV. Special consideration must frequently be given to the presence of small amounts of high cross-section elements such as Co or W which are either normal incidental impurities in nickel alloys and steels or important components of high-temperature alloys.

Effects of Radiation Irradiation affects the properties of solids in a number of ways: dimensional changes; decrease in density; increase in hardness, yield, and tensile strengths; decrease in ductility; decrease in electrical conductivity; change in magnetic susceptibility. Another consideration is the activation of certain alloying elements by irradiation. Tantalum[181] and Co[60] have moderate radioactivity but long half-lives; isotopes having short lives but high-activity levels include Cr[51], Mn[56], and Fe[59].

Metallic Coolants The need for the efficient transfer of large quantities of heat in a reactor has led to use of several metallic coolants. These have raised new problems of pumping, valving, and corrosion. In

Table 6.4.34   Requirements of Materials for Nuclear Reactor Components

Component	Neutron absorption cross section	Effect in slowing neutrons	Strength	Resistance to radiation damage	Thermal conductivity	Corrosion resistance	Cost	Other	Typical material
Moderator and reflector	Low	High	Adequate	—	—	High	Low	Low atomic wt	H_2O graphite, Be
Fuel*	Low	—	Adequate	High	High	High	Low	—	U, Th, Pu
Control rod	High	—	Adequate	Adequate	High	—	—	—	Cd, B_4C
Shield	High	High	High	—	—	—	—	High γ radiation absorption	Concrete
Cladding	Low	—	Adequate	—	High	High	Low	—	Al, Zr, stainless steel
Structural	Low	—	High	Adequate	—	High	—	—	Zr, stainless steel
Coolant	Low	—	—	—	High	—	—	Low corrosion rate, high heat capacity	H_2O, Na, NaK, CO_2 He

U, Th, and Pu are used as fuels in the forms of metals, oxides, and carbides.

Table 6.4.35   Moderating Properties of Materials

Moderator	Slowing-down power, cm^{-1}	Moderating ratio
H_2O	1.53	70
D_2O	0.177	21,000
He	1.6×10^{-5}	83
Be	0.16	150
BeO	0.11	180
C (graphite)	0.063	170

SOURCE: Adapted from Glass, "Principles of Nuclear Reactor Engineering, Van Nostrand.

Table 6.4.36   Slow-Neutron Absorption by Structural Materials

Material	Relative neutron absorption per cm^3 $\times 10^3$	Relative neutron absorpton for pipes of equal strength, 68°F (20°C)	Melting point °F	°C
Magnesium	3.5	10	1,200	649
Aluminum	13	102	1,230	666
Stainless steel	226	234	2,730	1,499
Zirconium	12.6	16	3,330	1,816

SOURCE: Leeser, *Materials & Methods*, **41**, 1955, p. 98.

Table 6.4.37   Slow-Neutron Absorption Cross Sections

Low		Intermediate		High	
Element	Cross section, barns	Element	Cross section, barns	Element	Cross section, barns
Oxygen	0.0016	Zinc	1.0	Manganese	12
Carbon	0.0045	Columbium	1.2	Tungsten	18
Beryllium	0.009	Barium	1.2	Tantalum	21
Fluorine	0.01	Strontium	1.3	Chlorine	32
Bismuth	0.015	Nitrogen	1.7	Cobalt	35
Magnesium	0.07	Potassium	2.0	Silver	60
Silicon	0.1	Germanium	2.3	Lithium	67
Phosphorus	0.15	Iron	2.4	Gold	95
Zirconium	0.18	Molybdenum	2.4	Hafnium	100
Lead	0.18	Gallium	2.8	Mercury	340
Aluminum	0.22	Chromium	2.9	Iridium	470
Hydrogen	0.32	Thallium	3.3	Boron	715
Calcium	0.42	Copper	3.6	Cadmium	3,000
Sodium	0.48	Nickel	4.5	Samarium	8,000
Sulphur	0.49	Tellurium	4.5	Gadolinium	36,000
Tin	0.6	Vanadium	4.8		
		Antimony	5.3		
		Titanium	5.8		

SOURCE: Leeser, *Materials & Methods*, **41**, 1955, p. 98.

Table 6.4.38 Resistance of Materials to Liquid Sodium and NaK*

	Good	Limited	Poor
< 1,000°F	Carbon steels, low-alloy steels, alloy steels, stainless steels, nickel alloys, cobalt alloys, refractory metals, beryllium, aluminum oxide, magnesium oxide, aluminum bronze	Gray cast iron, copper, aluminum alloys, magnesium alloys, glasses	Sb, Bi, Cd, Ca, Au, Pb, Se, Ag, S, Sn, Teflon
1,000–1,600°F	Armco iron, stainless steels, nickel alloys,† cobalt alloys, refractory metals‡	Carbon steels, alloy steels, Monel, titanium, zirconium, beryllium, aluminum oxide, magnesium oxide	Gray cast iron, copper alloys, Teflon, Sb, Bi, Cd, Ca, Au, Pb, Se, Ag, S, Sn, Pt, Si, Magnesium alloys

* For more complete details, see "Liquid Metals Handbook."
† Except Monel.
‡ Except titanium and zirconium.

addition to their thermal and flow properties, consideration must also be given the nuclear properties of prospective coolants. Extensive thermal, flow, and corrosion data on metallic coolants are given in the "Liquid Metals Handbook," published by USNRC. The resistances of common materials to liquid sodium and NaK are given qualitatively in Table 6.4.38. Water, however, remains the most-used coolant; it is used under pressure as a single-phase liquid, as a boiling two-phase coolant, or as steam in a superheat reactor.

Fuels   There are, at present, only three fissionable materials, U^{233}, U^{235}, and Pu^{239}. Of these, only U^{235} occurs naturally as an isotopic "impurity" with natural uranium U^{238}. Uranium^{233} may be prepared from natural thorium, and Pu^{239} from U^{238} by neutron bombardment. Both uranium and thorium are prepared by conversion of the oxide to the tetrafluoride and subsequent reduction to the metal.

The properties of **uranium** are considerably affected by the three allotropic changes which it undergoes, the low-temperature forms being highly anisotropic. The strength of the metal is low (see Table 6.4.39) and decreases rapidly with increasing temperature. The corrosion resistance is also poor. Aluminum-base alloys containing uranium-aluminum intermetallic compounds have been used to achieve improved properties. **Thorium** is even softer than uranium but is very ductile. Like uranium it corrodes readily, particularly at elevated temperatures. Owing to its crystal structure, its properties are isotropic. **Plutonium** is unusual in possessing six allotropic forms, many of which have

anomalous physical properties. The electrical resistivity of the alpha phase is greater than that for any other metallic element. Because of the poor corrosion resistance of nuclear-fuel metals, most fuels in power reactors today are in the form of oxide UO_2, ThO_2, or PuO_2, or mixtures of these. The carbides UC and UC_2 are also of interest in reactors using liquid-metal coolants.

Control-Rod Materials   Considerations of neutron absorption cross sections and melting points limit the possible control-rod materials to a very small group of elements, of which only four—boron, cadmium, hafnium, and gadolinium—have thus far been prominent. **Boron** is a very light metal of high hardness (~3,000 Knoop) prepared by thermal or hydrogen reduction of BCl_3. Because of its high melting point, solid shapes of boron are prepared by powder-metallurgy techniques. Boron has a very high electrical resistivity at room temperature but becomes conductive at high temperatures. The metal in bulk form is oxidation-resistant below 1,800°F (982°C) but reacts readily with most halogens at only moderate temperatures. Rather than the elemental form, boron is generally used as boron steel or as the carbide, oxide, or nitride. **Cadmium** is a highly ductile metal of moderate hardness which is recovered as a by-product in zinc smelting. Its properties greatly resemble those of zinc. The relatively low melting point renders it least attractive as a control rod of the four metals cited. **Hafnium** metal is reduced from hafnium tetrachloride by sodium and subsequently purified by the iodide hot-wire process. It is harder and less readily worked than

Table 6.4.39 Mechanical Properties of Metals for Nuclear Reactors

Material	Room temp					Charpy impact ft · lb	Elevated temp*		
	Longitudinal			Transverse					
	Yield strength, ksi	Ult strength, ksi	Elong,† %	Ult str, ksi	Elong,† %		°F	Ult str, ksi	Elong. %
Beryllium:									
Cast, extracted and annealed		40	1.82	16.6	0.18		392	62	23.5
Flake, extruded and annealed		63.7	5.0	25.5	0.30		752	43	29
Powder, hot-extruded	39.5	81.8	15.8	45.2	2.3	4.1	1,112	23	8.5
Powder, vacuum hot-pressed	32.1	45.2	2.3	45.2	2.3	0.8	1,472	5.2	10.5
Zirconium:									
Kroll-50% CW		82.6				14.8	250	32	
Kroil-annealed		49.0					500	23	
Iodide	15.9	35.9	31			2.5–6.0	700	17	
							900	12	
							1,500	3	
Uranium	25	53	<10			15	302	27	
							1,112	12	
Thorium	27	37.5	40				570	22	
							900	17.5	

* All elevated-temperature data on beryllium for hot-extruded powder; on zirconium for iodide material.
† Beryllium is extremely notch-sensitive. The tabulated data have been obtained under very carefully controlled conditions, but ductility values in practice will be found in general to be much lower and essentially zero in the transverse direction.

zirconium, to which it is otherwise very similar in both chemical and physical properties. Hafnium reacts easily with oxygen, and its properties are sensitive to traces of most gases. The very high absorption cross section of **gadolinium** renders it advantageous for fast-acting control rods. This metal is one of the rare earths and as yet is of very limited availability. It is most frequently employed as the oxide.

Beryllium   Great interest attaches to beryllium because it is unique among the metals with respect to its very low neutron-absorption cross section and high neutron-slowing power. It may also serve as a source of neutrons when subjected to alpha-particle bombardment. Beryllium is currently prepared almost entirely by magnesium reduction of the fluoride, although fused-salt electrolysis is also practicable and has been used. The high affinity of the metal for oxygen and nitrogen renders its processing and fabrication especially difficult. At one time, beryllium was regarded as almost hopelessly brittle. Special techniques, such as vacuum hot pressing, or vacuum casting followed by hot extrusion of clad slugs, led to material having marginally acceptable, though highly directional, mechanical properties. The extreme toxicity of beryllium powder necessitates special precautions in all operations.

Zirconium   Zirconium's importance in nuclear technology derives from its low neutron-absorption cross section, excellent corrosion resistance, and high strength at moderate temperatures. The metal is produced by magnesium reduction of the tetrachloride (Kroll process). The Kroll product is usually converted to ingot form by consumable-electrode-arc melting. An important step in the processing for many applications is the difficult chemical separation of the 1½ to 3 percent hafnium with which zirconium is contaminated. Unless removed, this small hafnium content results in a prohibitive increase in absorption cross section from 0.18 to 3.5 barns. The mechanical properties of zirconium are particularly sensitive to impurity content and fabrication technique. In spite of the high melting point, the mechanical properties are poor at high temperatures, principally because of the allotropic transformation at 1,585°F (863°C). A satisfactory annealing temperature is 1,100°F (593°C). The low-temperature corrosion behavior is excellent but is seriously affected by impurities. The oxidation resistance at high temperatures is poor. Special alloys (Zircoloys) have been developed having greatly improved oxidation resistance in the intermediate-temperature range. These alloys have a nominal content of 1.5 percent Sn and minor additions of iron-group elements.

Stainless Steel   Although stainless steel has a higher thermal neutron-absorption cross section than do the zirconium alloys, its good corrosion resistance, high strength, low cost, and ease of fabrication make it a strong competitor with zirconium alloys as a fuel-cladding material for water-cooled power-reactor applications. Types 348 and 304 stainless are used in major reactors.

MAGNESIUM AND MAGNESIUM ALLOYS
by James D. Shearouse, III

REFERENCES: "Metals Handbook," ASM. Beck, "Technology of Magnesium and Its Alloys." Annual Book of Standards, ASTM. Publications of the Dow Chemical Co.

Magnesium is the lightest metal of structural importance [108 lb/ft^3 (1.740 g/cm^3)]. The principal uses for **pure magnesium** are in aluminum alloys, steel desulfurization, and production of nodular iron. Principal uses for **magnesium alloys** are die castings (for automotive and computer applications), wrought products, and, to a lesser extent, gravity castings (usually for aircraft and aerospace components). Because of its chemical reactivity, magnesium can be used in pyrotechnic material and for sacrificial galvanic protection of other metals. Since magnesium in its molten state reacts with the oxygen in air, a protective atmosphere containing sulfur hexafluoride is employed as a controlled atmosphere.

Commercially pure magnesium contains 99.8 percent magnesium and is produced by extraction from seawater or by reduction from magnesite and dolomite ores. Chief impurities are iron, silicon, manganese, and aluminum. The major use of magnesium, consuming about 50 percent of total magnesium production, is as a component of aluminum alloys for beverage can stock. Another use for primary magnesium is in steel desulfurization. Magnesium, usually mixed with lime or calcium carbide, is injected beneath the surface of molten steel to reduce the sulfur content.

A growing application for **magnesium** alloys is die cast automotive components. Their light weight can be used to help achieve reduced fuel consumption, while high-ductility alloys are used for interior components which must absorb impact energy during collisions. These alloys can be cast by hot or cold chamber die-casting methods, producing a diverse group of components ranging from gear cases to steering wheel frames. Some vehicles use magnesium die castings as structural members to serve as instrument panel support beams and steering columns. Designs employing magnesium must account for the relatively low value of the modulus of elasticity (6.5 × 10^6 lb/in^2) and high thermal coefficient of expansion [14 × 10^{-6} per °F at 32°F (0°C) and 16 × 10^{-6} for 68 to 752°F (20 to 400°C)]. (See Tables 6.4.40 and 6.4.41.)

The development of jet engines and high-velocity aircraft, missiles, and spacecraft led to the development of magnesium alloys with improved **elevated-temperature properties.** These alloys employ the addition of some combination of rare earth elements, yttrium, manganese, zirconium, or, in the past, thorium. Such alloys have extended the useful temperatures at which magnesium can serve in structural applications to as high as 700 to 800°F.

Magnesium alloy forgings are used for applications requiring properties superior to those obtainable in castings. They are generally **pressforged.** Alloy AZ61 is a general-purpose alloy, while alloys AZ80A-T5 and ZK60A-T5 are used for the highest-strength press forgings of simple design. These alloys are aged to increase strength.

A wide range of extruded shapes are available in a number of compositions. Alloys AZ31B, AZ61A, AZ80A-T5, and ZK60A-T5 increase in cost and strength in the order named. **Impact extrusion** produces smaller, symmetric tubular parts.

Sheet is available in several alloys (see Table 6.4.41), in both the soft (annealed) and the hard (cold-rolled) tempers. Magnesium alloy sheet usually is hot-formed at temperatures between 400 and 650°F, although simple bends of large radius are made cold.

Joining   Magnesium alloys are joined by **riveting** or **welding.** Riveting is most common. Aluminum alloy rivets are used; alloy 5052 is preferred, to minimize contact corrosion. Other aluminum alloy rivets can be used, but they are not as effective in inhibiting contact corrosion. Rivets should be anodized to prevent contact corrosion. Adhesive bonding is another accepted method for joining magnesium. It saves weight and improves fatigue strength and corrosion resistance.

Arc welding with inert-gas (helium or argon) shielding of the molten metal produces satisfactory joints. Butt joints are preferred, but any type of welded joint permissible for mild steel can be used. After welding, a stress relief anneal is necessary. Typical stress relief anneal times are 15 min at 500°F (260°C) for annealed alloys and 1 h at 400°F (204°C) for cold-rolled alloys.

Machining   Magnesium in all forms is a free-machining metal. Standard tools such as those used for brass and steel can be used with slight modification. Relief angles should be from 7° to 12°; rake angles from 0° to 15°. High-speed steel is satisfactory and is used for most drills, taps, and reamers. Suitable grades of cemented carbides are better for production work and should be used where the tool design permits it. Finely divided magnesium constitutes a fire hazard, and good housekeeping in production areas is essential. (see also Sec. 13.4.)

Corrosion Resistance and Surface Protection   All magnesium alloys display good resistance to ordinary inland atmospheric exposure, to most alkalies, and to many organic chemicals. High-purity alloys introduced in the mid-1980s demonstrate excellent corrosion resistance and are used in under-vehicle applications without coatings. However, galvanic couples formed by contact with most other metals, or by contamination of the surface with other metals during fabrication, can cause rapid attack of the magnesium when exposed to salt water. Protective treatments or coated fasteners can be used to isolate galvanic couples and prevent this type of corrosion.

Table 6.4.40 Typical Mechanical Properties of Magnesium Casting Alloys

| Alloy | Condition or temper* | Nominal composition, % (balance Mg plus trace elements) | | | | | Tensile strength, ksi | Tensile yield strength, ksi | Elongation, %, in 2 in | Shear strength, ksi | Strength, ksi | | Hardness BHN | Electrical conductivity, % IACS‡ |
		Al	Zn	Mn, min	Zr	Other					Tensile	Yield		
Permanent mold and sand casting alloys:														
AM100A	−T6	10.0		0.10			40	22	1	22			70	14
AZ63A	−F	6.0	3.0	0.15			29	14	6	18	60	40	50	14
	−T4						40	13	12	17	60	44	55	12
	−T5						30	14	4	17	60	40	55	
	−T6						40	19	5	20	75	52	73	15
AZ81A	−T4	7.5	0.7	0.13			40	12	15	17	60	44	55	12
AZ91E	− F	8.7	0.7	0.17			24	14	2	18	60	40	52	13
	−T4						40	12	14	17	60	44	53	11
	−T5						26	17	3					
	−T6						40	19	5	20	75	52	66	13
AZ92A	−F	9.0	2.0	0.10			24	14	2	18	50	46	65	12
	−T4						40	14	9	20	68	46	63	10
	−T5						26	16	2	19	50	46	70	
	−T6						40	21	2	22	80	65	84	14
EZ33A	−T5		2.5		0.8	3.5 RE†	23	15	3	22	57	40	50	25
K1A	−F				0.7		25	7	19	8				31
QE22A	−T6				0.7	2.0 RE,† 2.5 Ag	40	30	4	23				25
ZE41A	−T5		4.0		0.7	1.3 RE†	30	20	3.5	22	70	51	62	31
ZK51A	−T5		4.5		0.8		40	24	8	22	72	47	65	27
ZK61A	−T6		6.0		0.8		45	28	10	26			70	27
Die casting alloys:														
A291D	−F	9.0	0.7	0.15			34	23	3	20			75	
AM60B	−F	6.0	0.22 max	0.24			32	19	6–8				62	
AM50A	−F	5.0	0.22 max	0.26			32	18	8–10				57	
AS41B	−F	4.2	0.12 max	0.35		1.0 Si	31	18	6				75	
AE42X1	−F	4.0	0.22 max	0.25		2.4 RE†	33	20	8-10				57	

ksi × 6.895 = MPa
* −F = as cast; −T4 = artificially aged; −T5 = solution heat-treated; −T6 = solution heat-treated.
† RE = rare-earth mixture.
‡ Percent electrical conductivity/100 approximately equals thermal conductivity in cgs units.

POWDERED METALS

by Peter K. Johnson

REFERENCES: German, "Powder Metallurgy Science," 2d ed., Metal Powder Industries Federation, Princeton, NJ. "Powder Metallurgy Design Manual," 2d ed., Metal Powder Industries Federation.

Powder metallurgy (PM) is an automated manufacturing process to make precision metal parts from metal powders or particulate materials. Basically a net or near-net shape metalworking process, a PM operation usually results in a finished part containing more than 97 percent of the starting raw material. Most PM parts weigh less than 5 lb (2.26 kg), although parts weighing as much as 35 lb (15.89 kg) can be fabricated in conventional PM equipment.

In contrast to other metal-forming techniques, PM parts are shaped directly from powders, whereas castings originate from molten metal, and wrought parts are shaped by deformation of hot or cold metal, or by machining (Fig. 6.4.10). The PM process is cost-effective in manufacturing simple or complex shapes at, or very close to, final dimensions in production rates which can range from a few hundred to several thousand parts per hour. Normally, only a minimum amount of machining is required.

Powder metallurgy predates melting and casting of iron and other metals. Egyptians made iron tools, using PM techniques from at least 3000 B.C. Ancient Inca Indians made jewelry and artifacts from precious metal powders. The first modern PM product was the tungsten filament for electric light bulbs, developed in the early 1900s. Oil-impregnated PM bearings were introduced for automotive use in

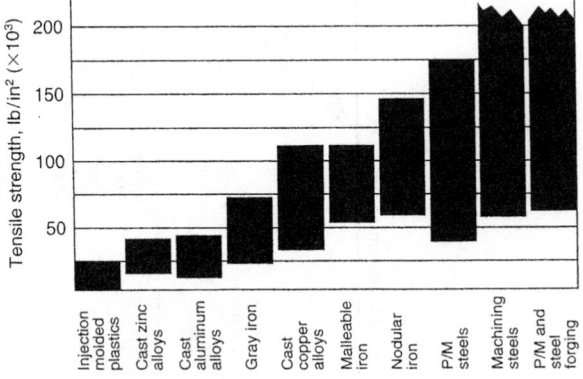

Fig. 6.4.10 Comparison of material strengths.

the 1920s. This was followed by tungsten carbide cutting tool materials in the 1930s, automobile parts in the 1960s and 1970s, aircraft gas-turbine engine parts in the 1980s, and **powder forged** (PF) and **metal injection molding** (MIM) parts in the 1990s.

PM parts are used in a variety of end products such as lock hardware, garden tractors, snowmobiles, automobile automatic transmissions and engines, auto antilock brake systems (ABS), washing machines, power tools, hardware, firearms, copiers, off-road equipment, hunting knives,

Table 6.4.41 Properties of Wrought Magnesium Alloys

Alloy and temper*	Al	Mn	Zn	Zr	Density, lb/in³	Thermal conductivity, cgs units, 68°F (20°C)	Electrical resistivity, μΩ·cm, 68°F (20°C)	Tensile strength, ksi	Tensile yield strength, ksi	Elongation in 2 in, %	Compressive yield strength, ksi	Shear strength, ksi	Bearing strength, ksi Tensile	Bearing strength, ksi Yield	Hardness BHN
Extruded bars, rods, shapes															
AZ31B-F	3.0		1.0		0.0639	0.18	9.2	38	29	15	14	19	56	33	49
AZ61A-F	6.5		1.0		0.0647	0.14	12.5	45	33	16	19	22	68	40	60
AZ80A-T5	8.5		0.5		0.0649	0.12	14.5	55	40	7	35	24	60	58	82
ZK60A-F			5.7	0.5	0.0659	0.28	6.0	49	38	14	33	24	76	56	75
-T5			5.7	0.5	0.0659	0.29	5.7	53	44	11	36	26	79	59	82
Extruded tube															
AZ31B-F	3.0		1.0		0.0639	0.18	9.2	36	24	16	12				46
AZ61A-F	6.5		1.0		0.0647	0.14	12.5	41	24	14	16				50
ZK60A-F			5.7	0.5	0.0659	0.28	6.0	47	35	13	25				75
-T5			5.7	0.5	0.0659	0.29	5.7	50	40	11	30				82
M1A		1.2			0.0635	0.31	5.4	37	26	11	12				
Sheet and plate															
AZ31B-H24	3.0		1.0		0.0639	0.18	9.2	42	32	15	26	29	77	47	73
								40	29	17	23	28	72	45	
								39	27	19	19	27	70	40	
AZ31B-0	3.0		1.0		0.0639	0.18	9.2	37	22	21	16				
M1A		1.2			0.0635	0.31	5.4	37	26	11	12	26	66	37	56
Tooling plate															
AZ31B	3.0		1.0		0.0639	0.18	9.2	35	19	12	10				
Tread plate															
AZ31B	3.0		1.0		0.0639	0.18	9.2	35	19	14	11				52

Nominal composition, % (balance Mg plus trace elements) — columns Al, Mn, Zn, Zr.
Physical properties — Density, Thermal conductivity, Electrical resistivity.
Room-temperature mechanical properties (typical) — Tensile strength, Tensile yield strength, Elongation, Compressive yield strength, Shear strength, Bearing strength, Hardness.

See Sec. 1 for conversion factors to SI units.

NOTE: For all above alloys: coefficient of thermal expansion = 0.0000145; modulus of elasticity = 6,500,000 lb/in²; Position's ratio = 0.35.

* Temper: −F = as fabricated; −H24 = strain-hardened, then partially annealed; −O = fully annealed; −T5 = artificially aged.

Table 6.4.42   Properties of Powdered Ferrous Metals

Material	Density, g/cm³	Tensile strength* MPa	Tensile strength* lb/in²	Comments
Carbon steel				
MPIF F-0008	6.9	370†	54,000†	Often cost-effective
0.8% combined carbon (c.c.)	6.9	585‡	85,000‡	
Copper steel				
MPIF FC-0208	6.7	415†	60,000†	Good sintered strength
2% Cu, 0.8% c.c.	6.7	585‡	85,000‡	
Nickel steel				
MPIF FN-0205	6.9	345†	50,000†	Good heat-treated strength,
2% Ni, 0.5% c.c.	6.9	825‡	120,000‡	impact energy
MPIF FN-0405	6.9	385†	56,000†	
4% Ni, 0.5% c.c.	6.9	845‡	123,000‡	
Infiltrated steel				
MPIF FX-1005	7.3	530†	77,000†	Good strength, closed-off
10% Cu, 0.5% c.c.	7.3	825‡	120,000‡	internal porosity
MPIF FX-2008	7.3	550†	80,000†	
20% Cu, 0.8% c.c.	7.3	690‡	100,000‡	
Low-alloy steel, prealloyed Ni, Mo, Mn				
MPIF FL-4605 HT	6.95	895‡	130,000‡	Good hardenability, con-
MPIF FL-4405 HT	7.10	1100‡	160,000‡	sistency in heat treatment
Stainless steels				
MPIF SS-316 N2 (316 stainless)	6.5	415†	60,000†	Good corrosion resistance, appearance
MPIF SS-410 HT (410 stainless)	6.5	725‡	105,000‡	

* Reference: MPIF Standard 35, Materials Standards for P/M Structural Parts. Strength and density given as typical values.
† As sintered.
‡ Heat-treated.

hydraulic assemblies, X-ray shielding, oil and gas drilling well-head components, and medical equipment. The typical five- or six-passenger car contains about 44 1b of PM parts, a figure that could increase within the next several years. Iron powder is used as a carrier for toner in electrostatic copying machines. People in the United States consume about 2 million 1b of iron powder annually in ironenriched cereals and breads. Copper powder is used in electromagnetic imaging (EMI) shielding, environmentally friendly green bullets replacing lead, and metallic pigmented inks for packaging and printing. Aluminum powder is used in solid-fuel booster rockets for the space shuttle program. The spectrum of applications of powdered metals is very wide indeed.

There are five major processes that consolidate metal powders into precision shapes: conventional powder metallurgy (PM) (the dominant sector of the industry), metal injection molding (MIM), powder forging (PF), warm compaction, hot isostatic pressing (HIP), and cold isostatic pressing (CIP).

PM processes use a variety of alloys, giving designers a wide range of material properties (Tables 6.4.42 to 6.4.44). Major alloy groups include

Table 6.4.43   Properties of Powdered Nonferrous Metals

Material	Use and characteristics
Copper	Electrical components
Bronze	Self-lubricating bearings: @ 6.6 g/cm³ oiled
MPIF CTG-1001	density, 160 MPa (23,000 lb/in²) K strength,*
10% tin, 1% graphite	17% oil content by volume
Brass	Electrical components, applications requiring good corrosion resistance, appearance, and ductility
MPIF CZ-1000	@ 7.9 g/cm³, yield strength 75 MPa (11,000
10% zinc	lb/in²)
MPIF CZ-3000	@7.9 g/cm³, yield strength 125 MPa (18,300
30% zinc	lb/in²)
Nickel silver	Improved corrosion resistance, toughness @ 7.9
MPIF CNZ-1818	g/cm³, yield strength 140 MPa (20,000 lb/in²)
18% nickel, 18% zinc	
Aluminum alloy	Good corrosion resistance, lightweight, good electrical and thermal conductivity
Titanium	Good strength/mass ratio, corrosion resistance

* Radial crashing constant. See MPIF Standard 35, Materials Standards for P/M Self-Lubricating Bearings.

Table 6.4.44   Properties of Powdered Soft Magnetic Metals

Material	Density, g/cm³	Maximum magnetic induction, kG*	Coercive field, Oe
Iron, low-density	6.6	9.5	1.8
Iron, high-density	7.2	12.0	1.8
Phosphorous iron, 0.45% P	7.0	11.5	1.5
Silicon iron, 3% Si	7.0	12.0	1.2
Nickel iron, 50% Ni	7.0	9.0	0.3
410 Stainless steel	7.1	10.0	2.0
430 Stainless steel	7.1	10.0	2.0

* Magnetic field—15 oersteds (Oe).

powdered iron and alloy steels, stainless steel, bronze, and brass. Powders of aluminum, copper, tungsten carbide, tungsten, and heavymetal alloys, tantalum, molybdenum, superalloys, titanium, high-speed tool steels, and precious metals are successfully fabricated by PM techniques.

A PM microstructure can be designed to have controlled microporosity. This inherent advantage of the PM process can often provide special useful product properties. Sound and vibration damping can be enhanced. Components are impregnated with oil to function as self-lubricating bearings, resin impregnated to seal interconnecting microporosity, infiltrated with a lower-melting-point metal to increase strength and impact resistance, and steam-treated to increase corrosion resistance and seal microporosity. The amount and characteristics of the microporosity can be controlled within limits through powder characteristics, powder composition, and the compaction and sintering processes.

Powder metallurgy offers the following advantages to the designer:

1. PM eliminates or minimizes machining.
2. It eliminates or minimizes material losses.
3. It maintains close dimensional tolerances.
4. It offers the possibility of utilizing a wide variety of alloyed materials.
5. PM produces good surface finishes.
6. It provides components that may be heat-treated for increased strength and wear resistance.
7. It provides part-to-part reproducibility.
8. It provides controlled microporosity for self-lubrication or filtration.

9. PM facilitates the manufacture of complex or unique shapes that would be impractical or impossible with other metalworking processes.

10. It is suitable for moderate- to high-volume production.

11. It offers long-term performance reliability for parts in critical applications.

Metal powders are materials precisely engineered to meet a wide range of performance requirements. Major metal powder production processes include atomization (water, gas, centrifugal); chemical (reduction of oxides, precipitation from a liquid, precipitation from a gas) and thermal; and electrolytic. **Particle shape,** size, and size distribution strongly influence the characteristics of powders, particularly their behavior during die filling, compaction, and sintering. The range of shapes covers highly spherical, highly irregular, flake, dendritic (needlelike), and sponge (porous). Metal powders are classified as elemental, partially alloyed, or prealloyed.

The three basic steps for producing conventional-density PM parts are mixing, compacting, and sintering.

Elemental or prealloyed metal powders are mixed with lubricants and/or other alloy additions to produce a homogeneous mixture.

In compacting, a controlled amount of mixed powder is automatically gravity-fed into a precision die and is compacted, usually at room temperatures at pressures as low as 10 or as high as 60 or more tons/in^2 (138 to 827 MPa), depending on the density requirements of the part. Normally compacting pressures in the range of 30 to 50 tons/in^2 (414 to 690 MPa) are used. Special mechanical or hydraulic presses are equipped with very rigid dies designed to withstand the extremely high loads experienced during compaction. Compacting loose powder produces a *green compact* which, with conventional pressing techniques, has the size and shape of the finished part when ejected from the die and is sufficiently rigid to permit in-process handling and transport to a sintering furnace. Other specialized compacting and alternate forming methods can be used, such as powder forging, isostatic pressing, extrusion, injection molding, and spray forming.

During **sintering**, the green part is placed on a wide mesh belt and moves slowly through a controlled-atmosphere furnace. The parts are heated to below the melting point of the base metal, held at the sintering temperature, and then cooled. Basically a solid-state diffusion process, sintering transforms compacted mechanical bonds between the powder particles to metallurgical bonds. Sintering gives the PM part its primary functional properties. PM parts generally are ready for use after sintering. If required, parts can also be repressed, impregnated, machined, tumbled, plated, or heat-treated. Depending on the material and processing technique, PM parts demonstrate tensile strengths exceeding 200,000 lb/in^2 (1,379 MPa).

Warm compaction and **surface densification** are growing in use to improve the density and mechanical properties of PM parts. In warm compaction, powders and tooling members are preheated above room temperature. Surface densification by transverse rolling improves the performance of PM parts and highly loaded gears.

Powder forging (P/F) has become a standard technique for forming automotive connecting rods. More than 500 million P/F rods have been made since 1986. The use of MIM parts is growing rapidly. Applications include firearms, surgical tools, orthodontic appliances, electronics, and telecommunications. The **MIM process,** similar to plastic injection molding, injects a mixture of fine powder (less than 20 μm) and a binder into highly complex parts weighing less than 400 g. Rapid prototyping or desktop manufacturing is another PM technique that combines a three-dimensional computer model of the product and a laser sintering system to produce prototype and production parts. Powder is deposited in layers that form three-dimensional shapes. Two major processes are laser free-form manufacturing and selective laser sintering.

Development continues on PM and particulate materials covering conventional steels, stainless steels, high-speed steels, titanium, aluminum, nickel- and cobalt-based superalloys, refractory metals, composites, aluminides, and nanoscale materials.

NICKEL AND NICKEL ALLOYS
by John H. Tundermann

REFERENCES: Mankins and Lamb, Nickel and Nickel Alloys, "ASM Handbook," vol. 2, 10th ed., pp. 428–445. Tundermann et al., Nickel and Nickel Alloys, "Kirk Othmer Encyclopedia of Chemical Technology," vol. 17. Frantz, Low Expansion Alloys, "ASM Handbook," vol. 2, 10th ed., pp. 889–896.

Nickel is refined from sulfide and lateritic (oxide) ores using pyrometallurgical, hydrometallurgical, and other specialized processes. Sulfide ores are used to produce just over 50 percent of the nickel presently used in the world. High-purity nickel (99.7$^+$ percent) products are produced via electrolytic, carbonyl, and powder processes.

Electroplating accounts for about 10 percent of the total annual consumption of nickel. Normal nickel electroplate has properties approximating those of wrought nickel, but special baths and techniques can give much harder plates. Bonds between nickel and the base metal are usually strong. The largest use of nickel in plating is for corrosion protection of iron and steel parts and zinc-base die castings for automotive use. A 0.04- to 0.08-mm (1.5- to 3-mil) nickel plate is covered with a chromium plate only about one hundredth as thick to give a bright, tarnish-resistant, hard surface. Nickel electrodeposits are also used to facilitate brazing of chromium-containing alloys, to reclaim worn parts, and for electroformed parts.

Wrought nickel and nickel alloys are made by several melting techniques followed by hot and cold working to produce a wide range of product forms such as plate, tube, sheet, strip, rod, and wire.

The nominal composition and typical properties of selected commercial nickel and various nickel alloys are summarized in Tables 6.4.45 to 6.4.48.

Commercially pure nickels, known as Nickel 200 and 201, are available as sheet, rod, wire, tubing, and other fabricated forms. They are used where the thermal or electrical properties of nickel are required and where corrosion resistance is needed in parts that have to be worked extensively.

Commercial nickel may be forged or rolled at 871 to 1,260°C (1,600 to 2300°F). It becomes increasingly harder below 871°C (1,600°F) but

Table 6.4.45 Nominal Compositions, %, of Selected Nickel Alloys

Alloy	Ni	Cu	Fe	Cr	Mo	Al	Si	Mn	W	C	Co	Nb	Ti	Y$_2$O$_3$
Nickel 200	99.5	0.05	0.1				0.05	0.25		0.05				
Duranickel alloy 301	93.5	0.2	0.2			4.4	0.5	0.3		0.2			0.5	
Monel alloy 400	65.5	31.5	1.5				0.25	1.0		0.15				
Monel alloy K-500	65.5	29.5	1.0			3.0	0.15	0.5		0.15				
Hastelloy alloy C-276	56		5	15.5	16		0.05	1.0	4	0.02	2.5			
Inconel alloy 600	75.5	0.5	8	15.5			0.2	0.5		0.08				
Inconel alloy 625	62		2.5	22	9	0.2	0.2	0.2		0.05		3.5	0.2	
Inconel alloy 718	52.5		18.5	19	3	0.5	0.2	0.2		0.04		5	0.9	
Inconel alloy X-750	71	0.5	7	15.5		0.7	0.5	1.0		0.08	1.0		2.5	
Inconel alloy MA 754	78.5			20		0.3				0.05			0.5	0.6
Incoloy alloy 825	42	2.2	30	21	3	0.1	0.25	0.5		0.03			0.9	
Incoloy alloy 909	38		42				0.4			0.1	13.0	4.7	1.5	
Hastelloy alloy X	45		19.5	22	9		1.0	1.0	0.6	0.1	1.5			

Table 6.4.46   Typical Physical Properties of Selected Alloys

Alloy	UNS no.	Density, g/cm^3	Elastic modulus, GPa	Specific heat (20°C), J/(kg · K)	Thermal expansion (20°–93°C), μm/(m · K)	Thermal conductivity (20°C), W/(m · K)	Electrical resistivity (annealed), μΩ · m
Nickel 200	N02200	8.89	204	456	13.3	70	0.096
Duranickel alloy 301	N03301	8.25	207	435	13.0	23.8	0.424
Monel alloy 400	N04400	8.80	180	427	13.9	21.8	0.547
Monel alloy K-500	N05500	8.44	180	419	13.9	17.5	0.615
Hastelloy alloy C-276	N10276	8.89	205	427	11.2	9.8	1.29
Inconel alloy 600	N06600	8.47	207	444	13.3	14.9	1.19
Inconel alloy 625	N06625	8.44	207	410	12.8	9.8	1.29
Inconel alloy 718	N07718	8.19	211	450	13.0	11.4	1.25
Inconel alloy X-750	N07750	8.25	207	431	12.6	12.0	1.22
Inconel alloy MA 754	N07754	8.30	160	440	12.2	14.3	1.075
Incoloy alloy 825	N08825	8.14	206	440	14.0	11.1	1.13
Incoloy alloy 909	N19909	8.30	159	427	7.7	14.8	0.728
Hastelloy alloy X	N06002	8.23	205	461	13.3	11.6	1.16

Table 6.4.47   Typical Mechanical Properties of Selected Nickel Alloys

Alloy	Yield strength, MPa	Tensile strength, MPa	Elongation, % in 2 in	Hardness
Nickel 201	150	462	47	110 HB
Duranickel alloy 301	862	1,170	25	35 HRC
Monel alloy 400	240	550	40	130 HB
Monel alloy K-500	790	1,100	20	300 HB
Hastelloy alloy C-276	355	790	60	90 HRB
Inconel alloy 600	310	655	40	36 HRC
Inconel alloy 625	520	930	42	190 HB
Inconel alloy 718	1,036	1,240	12	45 HRC
Inconel alloy X-750	690	1,137	20	330 HB
Inconel alloy MA 754	585	965	22	25 HRC
Incoloy alloy 825	310	690	45	75 HRB
Incoloy alloy 909	1,035	1,275	15	38 HRC
Hastelloy alloy X	355	793	45	90 HRB

The addition of certain elements to nickel renders it responsive to precipitation or aging treatments to increase its strength and hardness. In the unhardened or quenched state, Duranickel alloy 301 fabricates almost as easily as pure nickel, and when finished, it may be hardened by heating for about 8 h at 538 to 593°C (1,000 to 1,100°F). Intermediate anneals during fabrication are at 900 to 954°C (1,650 to 1,750°F). The increase in strength due to aging may be superimposed on that due to cold work.

Nickel castings are usually made in sand molds by using special techniques because of the high temperatures involved. The addition of 1.5 percent silicon and lesser amounts of carbon and manganese is necessary to obtain good casting properties. Data on nickel and nickel alloy castings can be found in Spear, Corrosion-Resistant Nickel Alloy Castings, "ASM Handbook," vol. 3, 9th ed, pp. 175–178.

Nickel has good to excellent resistance to corrosion in caustics and nonoxidizing acids such as hydrochloric, sulfuric, and organic acids, but poor resistance to strongly oxidizing acids such as nitric acid. Nickel is resistant to corrosion by chlorine, fluorine, and molten salts.

Nickel is used as a catalyst for the hydrogenation of organic fats and for many industrial processes. Porous nickel electrodes are used for battery and fuel cell applications.

Nickel-Copper Alloys   Alloys containing less than 50 percent nickel are discussed under "Copper and Copper Alloys." **Monel** alloy 400

has no brittle range. The recrystallization temperature of cold-worked pure nickel is about 349°C (660°F), but commercial nickel recrystallizes at about 538°C (1,100° F) and is usually annealed at temperatures between 538 and 954°C (1,100 and 1,750°F).

Table 6.4.48   Short-Time High-Temperature Properties of Hot-Rolled Nickel and Its Alloys*

	Temperature, °C							
	21	316	427	540	650	816	982	1,093
Nickel 200†								
Tensile strength, MPa	505	575	525	315	235	170	55	
Yield strength 0.2% offset, MPa	165	150	145	115	105			
Elongation, % in 2 in	40	50	52	55	57	65	91	
Monel alloy 400								
Tensile strength, MPa	560	540	490	350	205	110	55	
Yield strength 0.2% offset, MPa	220	195	200	160	125	60		
Elongation, % in 2 in	46	51	52	29	34	58	45	
Inconel alloy 600								
Tensile strength, MPa	585	545	570	545	490	220	105	75
Yield strength 0.2% offset, MPa	385	185	195	150	150			
Elongation, % in 2 in	50	51	50	21	5	23	51	67
Inconel alloy 718								
Tensile strength, MPa	1,280			1,140	1,030			
Yield strength 0.2% offset, MPa	1,050			945	870			
Elongation, % in 2 in	22			26	15			

* See also data on metals and alloys for use at elevated temperature elsewhere in Sec. 6.4.
† Nickel 200 is not recommended for use above 316°C; low-carbon nickel 201 is the preferred substitute.

(see Tables 6.4.45 to 6.4.47) is a nickel-rich alloy that combines high strength, high ductility, and excellent resistance to corrosion. It is a homogeneous solid-solution alloy; hence its strength can be increased by cold working alone. In the annealed state, its tensile strength is about 480 MPa (70 ksi), and this may be increased to 1,170 MPa (170 ksi) in hard-drawn wire. It is available in practically all fabricated forms. Alloy 400 is hot-worked in the range 871 to 1,177°C (1,600 to 2,150°F) after rapid heating in a reducing, sulfur-free atmosphere. It can be cold-worked in the same manner as mild steel, but requires more power. Very heavily cold-worked alloy 400 may commence to recrystallize at 427°C (800°F), but in normal practice no softening will occur below 649°C (1,200°F). Annealing can be done for 2 to 5 h at about 760°C (1,400°F) or for 2 to 5 min at about 940°C (1,725°F). Nonscaling, sulfur-free atmospheres are required.

Because of its toughness, alloy 400 must be machined with high-speed tools with slower cutting speeds and lighter cuts than mild steel. A special grade containing sulfur, Monel alloy 405, should be used where high cutting speeds must be maintained. This alloy is essentially the same as the sulfur-free alloy in mechanical properties and corrosion resistance, and it can be hot-forged.

The short-time tensile strengths of alloy 400 at elevated temperatures are summarized in Table 6.4.48.

The fatigue endurance limit of alloy 400 is about 240 MPa (34 ksi) when annealed and 325 MPa (47 ksi) when hard-drawn. The action of corrosion during fatigue is much less drastic on alloy 400 than on steels of equal or higher endurance limit.

Alloy 400 is highly resistant to atmospheric action, seawater, steam, foodstuffs, and many industrial chemicals. It deteriorates rapidly in the presence of moist chlorine and ferric, stannic, or mercuric salts in acid solutions. It must not be exposed when hot to molten metals, sulfur, or gaseous products of combustion containing sulfur.

Small additions of aluminum and titanium to the alloy 400 base nickel-copper alloy produces precipitation-hardenable Monel alloy K-500. This alloy is sufficiently ductile in the annealed state to permit drawing, forming, bending, or other cold-working operations but workhardens rapidly and requires more power than mild steel. It is hotworked at 927 to 1,177°C (1,700 to 2,150°F) and should be quenched from 871°C (1,600°F) if the metal is to be further worked or hardened. Heat treatment consists of quenching from 871°C (1,600°F), cold working if desired, and reheating for 10 to 16 h at 538 to 593°C (1,000 to 1,100°F). If no cold working is intended, the quench may be omitted on sections less than 50 mm (2 in) thick and the alloy hardened at 593°C (1,100°F). The properties of the heat-treated alloy remain quite stable, at least up to 538°C (1,000°F). It is nonmagnetic down to −101°C (−150°F).

Heat-Resistant Nickel-Chromium and Nickel-Chromium-Iron Alloys See "Metals and Alloys for Use at Elevated Temperatures" for further information on high-temperature properties of nickel alloys.

The addition of chromium to nickel improves strength and corrosion resistance at elevated temperature. Nickel chromium alloys such as 80 Ni 20 Cr are extensively used in electrical resistance heating applications. Typically, additions of up to 4 percent aluminum and 1 percent yttrium are made to these alloys to increase hot oxidation and corrosion resistance. Incorporating fine dispersions of inert oxides to this system significantly enhances high-temperature properties and microstructural stability. **Inconel** alloy MA 754, which contains 0.6 percent Y_2O_3, exhibits good fatigue strength and corrosion resistance at 1,100°C (2,010°F) and is used in gas-turbine engines and other extreme service applications.

Inconel alloy 600 is a nickel-chromium-iron alloy. Alloy 600 is a high-strength nonmagnetic [at −40°C (−40°F)] alloy which is used widely for corrosion- and heat-resisting applications at temperatures up to 1,204°C (2,200°F) in sulfidizing atmospheres. In sulfidizing atmospheres, the maximum recommended temperature is 816°C (1,500°F). Inconel alloy X-750 is an age-hardenable modification suitable for stressed applications at temperatures up to 649 to 816°C (1,200 to 1,500°F). The addition of molybdenum to this system provides alloys, e.g., Hastelloy alloy X, with additional solid-solution strengthening and good oxidation resistance up to 1,200°C (2,200°F).

The combination of useful strength and oxidation resistance makes nickel alloys frequent choices for high-temperature service. Table 6.4.48 indicates the changes in mechanical properties with temperature.

Corrosion-Resistant Nickel-Molybdenum and Nickel-Iron-Chromium Alloys A series of solid-solution nickel alloys containing molybdenum exhibit superior resistance to corrosion by hot concentrated acids such as hydrochloric, sulfuric, and nitric acids. Hastelloy alloy C-276, which also contains chromium, tungsten, and cobalt, is resistant to a wide range of chemical process environments including strong oxidizing acids, chloride solutions, and other acids and salts. Incoloy alloy 825, a nickel-iron-chromium alloy with additions of molybdenum, is especially resistant to sulfuric and phosphoric acids. These alloys are used in pollution control, chemical processing, acid production, pulp and paper production, waste treatment, and oil and gas applications. Although these types of alloys are used extensively in aqueous environments, they also have good elevated-temperature properties.

Nickel-Iron Alloys Nickel is slightly ferromagnetic but loses its magnetism at a temperature of 368°C (695°F) when pure. For commercial nickel, this temperature is about 343°C (650°F). Monel alloy 400 is lightly magnetic and loses all ferromagnetism above 93°C (200°F). The degree of ferromagnetism and the temperature at which ferromagnetism is lost are very sensitive to variations in composition and mechanical and thermal treatments.

An important group of soft magnetic alloys is the nickel irons and their modifications, which exhibit high initial permeability, high maximum magnetization, and low residual magnetization. Nickel-iron alloys also exhibit low coefficients of thermal expansion that closely match those of many glasses and are therefore often used for glass-sealing applications. Nickel-iron alloys such as Incoloy alloy 909, which contains cobalt, niobium, and titanium, have found wide applications in gas-turbine and rocket engines, springs and instruments, and as controlled-expansion alloys designed to provide high strength and low coefficients of thermal expansion up to 650°C (1,200°F).

Low-Temperature Properties of Nickel Alloys Several nickel-base alloys have very good properties at low temperatures, in contrast to ferrous alloys whose impact strength (the index of brittleness) falls off very rapidly with decreasing temperature. The impact strength remains nearly constant with most nickel-rich alloys, while the tensile and yield strengths increase as they do in the other alloys. Specific data on lowtemperature properties may be found in White and Siebert, "Literature Survey of Low-Temperature Properties of Metals," Edwards, and "Mechanical Properties of Metals at Low Temperatures," *NBS Circ.* 520, 1952.

Welding Alloys Nickel alloys can be welded with similar-composition welding materials and basic welding processes such as gas tungsten arc welding (GTAW), gas metal arc welding (GMAW), shielded metal arc welding (SMAW), brazing, and soldering. The procedures used are similar to those for stainless steels. Nickel-base welding products are also used to weld dissimilar materials. Nickel-base filler metals, especially with high molybdenum contents, are typically used to weld other alloys to ensure adequate pitting and crevice corrosion resistance in final weld deposits. Nickel and nickel iron welding electrodes are used to weld cast irons.

(*Note:* Inconel, Incoloy, and Monel are trademarks of the Inco family of companies. Hastelloy is a trademark of Haynes International.)

TITANIUM AND ZIRCONIUM
by John R. Schley

REFERENCES: Donachie, "Titanium, A Technical Guide," ASM International, Metals Park, OH. "Titnanium, The Choice," rev. 1990, Titanium Development Association, Boulder, CO. ASM Metals Handbook, "Corrosion," vol. 13, 1987 edition, ASM International, Metals Park, OH. Schemel, "ASTM Manual on Zirconium and Hafnium," ASTM STP-639, Philadelphia, PA. "Corrosion of Zirconium Alloys," ASTM STP 368, Philadelphia. "Zircadyne Properties and Applications," Teledyne Wah Chang Albany, Albany, OR. "Basic Design Facts about Titanium," RMI Titanium Company, Niles, OH.

Titanium

The metal titanium is the ninth most abundant element in the earth's surface and the fourth most abundant structural metal after aluminum, iron, and magnesium. It is a soft, ductile, silvery-gray metal with a density about 60 percent of that of steel and a melting point of 1,675°C (3,047°F). A combination of low density and high strength makes titanium attractive for structural applications, particularly in the aerospace industry. These characteristics, coupled with excellent corrosion resistance, have led to the widespread use of titanium in the chemical processing industries and in oil production and refining.

Like iron, titanium can exist in two crystalline forms: hexagonal close-packed (hcp) below 883°C (1,621°F) and body-centered cubic (bcc) at higher temperatures up to the melting point. Titanium combines readily with many common metals such as aluminum, vanadium, zirconium, tin, and iron and is available in a variety of alloy compositions offering a broad range of mechanical and corrosion properties. Certain of these compositions are heat-treatable in the same manner as steels.

Production Processes

The commercial production of titanium begins with an ore, either ilmenite ($FeTiO_3$) or rutile (TiO_2), more often the latter, which passes through a series of chemical reduction steps leading to elemental titanium particles, referred to as **titanium sponge.** The sponge raw material then follows a sequence of processing operations resembling those employed in steelmaking operations. Beginning with the melting of titanium or titanium alloy ingots, the process follows a hot-working sequence of ingot breakdown by forging, then rolling to sheet, plate, or bar product. Titanium is commercially available in all common mill product forms such as billet and bar, sheet and plate, as well as tubing and pipe. It also is rendered into castings.

Mechanical Properties

Commercial titanium products fall into three structural categories corresponding to the allotropic forms of the metal. The alpha structure is present in **commercially pure** (CP) products and in certain alloys containing alloying elements that promote the alpha structure. The alpha-beta form occurs in a series of alloys that have a mixed structure at room temperature, and the beta class of alloys contains elements that stabilize the beta structure down to room temperature. Strength characteristics of the alpha-beta and beta alloys can be varied by heat treatment.

Typical alloys in each class are shown in Table 6.4.49, together with their nominal compositions and tensile properties in the annealed condition. Certain of these alloys can be solution-heat-treated and aged to higher strengths.

Corrosion Properties

Titanium possesses outstanding **corrosion resistance** to a wide variety of environments. This characteristic is due to the presence of a protective, strongly adherent oxide film on the metal surface. This film is normally transparent, and titanium is capable of healing the film almost instantly in any environment where a trace of moisture or oxygen is present. It is very stable and is only attacked by a few substances, most notably hydrofluoric acid. Titanium is unique in its ability to handle certain chemicals such as chlorine, chlorine chemicals, and chlorides. It is essentially inert in seawater, for example. The unalloyed CP grades typically are used for corrosion applications in industrial service, although the alloyed varieties are used increasingly for service where higher strength is required.

Fabrication

Titanium is cast and forged by conventional practices. Both means of fabrication are extensively used, particularly for the aerospace industry. Wrought titanium is readily fabricated in the same manner as and on the same equipment used to fabricate stainless steels and nickel alloys. For example, cold forming and hot forming of sheet and plate are done on press brakes and hydraulic presses with practices modified to accommodate the forming characteristics of titanium. Titanium is machinable by all customary methods, again using practices recommended for titanium. Welding is the most common method used for joining titanium, for the metal is readily weldable. The standard welding process is **gas tungsten arc welding** (GTAW), or **TIG.** Two other welding processes, **plasma arc welding** (PAW) and **gas metal arc welding** (GMAW), or MIG, are used to a lesser extent. Since titanium reacts readily with the atmospheric gases oxygen and nitrogen, precautions must be taken to use an inert-gas shield to keep air away from the hot weld metal during welding. Usual standard practices apply to accomplish this. Titanium can be torch-cut by **oxyacetylene flame** or by **plasma torch,** but care must be taken to remove contaminated surface metal. For all titanium fabricating practices cited above, instructional handbooks and other reference materials are available. Inexperienced individuals are well advised to consult a titanium supplier or fabricator.

Table 6.4.49 Typical Commercial Titanium Alloys

Designation	Tensile strength MPa	ksi	0.2% Yield strength MPa	ksi	Elonga-tion, %	Al	V	Sn	Zr	Mo	Other	N (max)	C (max)	H (max)	Fe (max)	O (max)
Unalloyed (CP grades:																
ASTM grade, 1	240	35	170	25	24	—	—	—	—	—	—	0.03	0.10	0.015	0.20	0.18
ASTM grade 2	340	50	280	40	20	—	—	—	—	—	—	0.03	0.10	0.015	0.30	0.25
ASTM grade 7	340	50	280	40	20	—	—	—	—	—	0.2 Pd	0.03	0.10	0.015	0.30	0.25
Alpha and near-alpha alloys:																
Ti-6 Al-2 Sn-4 Zr-2 Mo	900	130	830	120	10	6.0	—	2.0	4.0	2.0	—	0.05	0.05	0.0125	0.25	0.15
Ti-8 Al-1 Mo-1 V	930	135	830	120	10	8.0	1.0	—	—	1.0	—	0.05	0.08	0.012	0.30	0.12
Ti-5 Al-2 5 Sn	830	120	790	115	10	5.0	—	2.5	—	—	—	0.05	0.08	0.02	0.50	0.20
Alpha-beta alloys:																
Ti-3 Al-2 5V.	620	90	520	75	15	3.0	2.5	—	—	—	—	0.015	0.05	0.015	0.30	0.12
Ti-6 Al-2 V*	900	130	830	120	10	6.0	4.0	—	—	—	—	0.05	0.10	0.0125	0:30	0.20
Ti-6 Al-6 V-2 Sn*	1,030	150	.970	140	8	6.0	6.0	2.0	—	—	0.75 Cu	0.04	0.05	0.015	1.0	0.20
Ti-6 Al-2 Sn-4 Zr-6 Mo†	1,170	170	1,100	160	8	6.0	—	2.0	4.0	6.0	—	0.04	0,04	0.0125	0.15	0.15
Beta alloys:																
Ti-10 V-2 Fe-3 Al†	1,170	180	1,100	170	8	3.0	10.0	—	—	—	—	0.05	0.05	0.015	2.2	0.13
Ti-15V-3 Al-3 Sn-3 Cr†	1,240	180	1,170	170	10	3.0	15.0	3.0	—	—	3.0 Cr	0.03	0.03	0,015	0.30	0.13
Ti-3 Al-8 V-6 Cr-4 Mo-4. Zr*	900	130	860	125	16	3.0	8.9	—	4.0	4.0	6.0 Cr	0.03	0.05	0.020	0.25	0.12

* Mechanical properties given for annealed condition; may be solution-treated and aged to increase strength.
† Mechanical properties given for solution-treated and aged condition.

Applications

Titanium as a material of construction offers the unique combination of wide-ranging corrosion resistance, low density, and high strength. For this reason, titanium is applied broadly for applications that generally are categorized as aerospace and nonaerospace, the latter termed *industrial*. The **aerospace applications** consume about 70 percent of the annual production of titanium, and the titanium alloys rather than the CP grades predominate since these applications depend primarily on titanium's high strength/weight ratio. This category is best represented by **aircraft gas-turbine engines,** the largest single consumer of titanium, followed by **airframe structural members** ranging from fuselage frames, wing beams, and landing gear to springs and fasteners. **Spacecraft** also take a growing share of titanium. Nonaerospace or **industrial applications** more often use titanium for its corrosion resistance, sometimes coupled with a requirement for high strength. The unalloyed, commercially pure grades predominate in most corrosion applications and are typically represented by equipment such as heat exchangers, vessels, and piping.

A recent and growing trend is the application of titanium in the offshore oil industry for service on and around offshore platforms. As applied there, titanium uses are similar to those onshore, but with a growing trend toward large-diameter pipe for subsea service. Strength is a major consideration in these applications, and titanium alloys are utilized increasingly.

In the automotive field, the irreversible mandate for lightweight, fuel-efficient vehicles appears to portend a large use of titanium and its alloys.

Zirconium

Zirconium was isolated as a metal in 1824 but was only developed commercially in the late 1940s, concurrently with titanium. Its first use was in nuclear reactor cores, and this is still a major application. Zirconium bears a close relationship to titanium in physical and mechanical properties, but has a higher density, closer to that of steel. Its melting point is 1,852°C (3,365°F) and, like titanium, it exhibits two allotropic crystalline forms. The pure metal is body-centered cubic (bcc) above 865°C (1,590°F) and hexagonal close-packed (hcp) below this temperature. Zirconium is available commercially in unalloyed form and in several alloyed versions, combined most often with small percentages of niobium or tin. It is corrosion-resistant in a wide range of environments.

Zirconium is derived from naturally occurring zircon sand processed in a manner similar to titanium ores and yielding a **zirconium sponge**. The sponge is converted to mill products in the same manner as titanium sponge. The principal commercial compositions intended for corrosion applications together with corresponding mechanical properties are listed in Tables 6.4.50 and 6.4.51.

Zirconium is a reactive metal that owes its **corrosion resistance** to the formation of a dense, stable, and highly adherent oxide on the metal surface. It is resistant to most organic and mineral acids, strong alkalies, and some molten salts. A notable corrosion characteristic is its resistance to both strongly reducing hydrochloric acid and highly oxidizing nitric acid, relatively unique among metallic corrosion-resistant materials.

Zirconium and its alloys can be machined, formed, and welded by conventional means and practices much like those for stainless steels and identical to those for titanium. Cast shapes are available for equipment such as pumps and valves, and the wrought metal is produced in all standard mill product forms. Persons undertaking the fabrication of zirconium for the first time are well advised to consult a zirconium supplier.

The traditional application of zirconium in the nuclear power industry takes the form of fuel element cladding for reactor fuel bundles. The zirconium material used here is an alloy designated **Zircaloy,** containing about 1.5 percent tin and small additions of nickel, chromium, and iron. It is selected because of its low neutron cross section coupled with the necessary strength and resistance to water corrosion at reactor operating temperatures. Most zirconium is employed here. An interesting noncorrosion zirconium application in photography had been its use as a foil in flash bulbs, but the advent of the electronic flash has curtailed this. An active and growing corrosion application centers on fabricated chemical processing equipment, notably heat exchangers, but also a variety of other standard equipment.

ZINC AND ZINC ALLOYS
by Frank E. Goodwin

REFERENCES: Porter, "Zinc Handbook," Marcel Dekker, New York. ASM "Metals Handbook," 10th ed., vol. 2. ASTM, "Annual Book of Standards." "NADCA Product Specification Standards for Die Casting," Die Casting Development Council.

Zinc, one of the least expensive nonferrous metals, is produced from sulfide, silicate, or carbonate ores by a process involving concentration and roasting followed either by thermal reduction in a zinc-lead blast furnace or by leaching out the oxide with sulfuric acid and electrolyzing the solution after purification. Zinc distilled from blast-furnace production typically contains Pb, Cd, and Fe impurities that may be eliminated by

Table 6.4.50   Chemical Compositions of Zirconium Alloys (Percent)

Grade (ASTM designation)	R60702	R60704	R60705	R60706
Zr + Hf, min	99.2	97.5	95.5	95.5
Hafnium, max	4.5	4.5	4.5	4.5
Fe + Cr	max 0.20	0.2–0.4	max 0.2	max 0.2
Tin	—	1.0–2.0	—	—
Hydrogen, max	0.005	0.005	0.005	0.005
Nitrogen, max	0.025	0.025	0.025	0.025
Carbon, max	0.05	0.05	0.05	0.05
Niobium	—	—	2.0–3.0	2.0–3.0
Oxygen, max	0.16	0.18	0.18	0.16

Table 6.4.51   Minimum ASTM Requirements for the Mechanical Properties of Zirconium at Room Temperature (Cold-Worked and Annealed)

Grade (ASTM designation)	R60702	R60704	R60705	R60706
Tensile strength, min, ksi (MPa)	55 (379)	60 (413)	80 (552)	74 (510)
Yield strength, min, ksi (MPa)	30 (207)	35 (241)	55 (379)	50 (345)
Elongation (0.2% offset), min, percent	16	14	16	20
Bend test radius*	5*T*	5*T*	3*T*	2.5*T*

* Bend tests are not applicable to material over 0.187 in (4.75 mm) thick, and *T* equals the thickness of the bend test sample.

fractional redistillation to produce zinc of 99.99 + percent purity. Metal of equal purity can be directly produced by the electrolytic process. Zinc ingot range from cast balls weighing a fraction of a pound to a 1.1-ton [1 metric ton (t)] "jumbo" blocks. Over 20 percent of zinc metal produced each year is from recycled scrap.

The three standard grades of zinc available in the United States are described in ASTM specification B-6 (see Table 6.4.52). **Special high grade** is overwhelmingly the most commercially important and is used in all applications except as a coating for steel articles galvanized after fabrication. In other applications, notably die casting, the higher levels of impurities present in grades other than special high grade can have harmful effects on corrosion resistance, dimensional stability, and formability. Special high grade is also used as the starting point for brasses containing zinc.

Galvanized Coatings

Protective coatings for steel constitute the largest use of zinc and rely upon the galvanic or sacrificial property of zinc relative to steel. Lead, antimony, or zinc can be added to produce the solidified surface pattern called **spangle** preferred on unpainted articles. The addition of aluminum in amounts of 5 and 55 percent results in coatings with improved corrosion resistance termed **Galfan** and **Galvalume**, respectively.

Zinc Die Castings

Zinc alloys are particularly well suited for making die castings since the melting point is reasonably low, resulting in long die life even with ordinary steels. High accuracies and good surface finish are possible. Alloys currently used for die castings in the United States are covered by ASTM Specifications B-86 and B-240. Nominal compositions and typical properties of these compositions are given in Table 6.4.53. The low limits of impurities are necessary to avoid disintegration of the casting by intergranular corrosion under moist atmospheric conditions. The presence of magnesium or nickel prevents this if the impurities are not higher than the specification values. The mechanical properties shown in Table 6.4.53 are from die-cast test pieces of 0.25-in (6.4-mm) section

Table 6.4.52 ASTM Specification B6-03 for Slab Zinc

	Composition, %			
Grade	Lead, max	Iron, max	Cadmium, max	Zinc, min, by difference
Special high grade*	0.003	0.003	0.003	99.990
High grade	0.03	0.02	0.02	99.90
Prime western†	1.4	0.05	0.20	98.0

* Tin in special high grade shall not exceed 0.001%.
† Aluminum in prime western zinc shall not exceed 0.05%.
SOURCE: ASTM, reprinted with permission.

thickness. Zinc die castings can be produced with section thicknesses as low as 0.03 in (0.75 mm) so that considerable variations in properties from those listed must be expected. These die-casting alloys generally increase in strength with increasing aluminum content. The 8, 12, and 27 percent alloys can also be gravity-cast by other means.

Increasing copper content results in growth of dimensions at elevated temperatures. A measurement of the expansion of the die casting after exposure to water vapor at 203°F (95°C) for 10 days is a suitable index of not only stability but also freedom from susceptibility to intergranular corrosion. When held at room temperature, the copper-containing alloys begin to shrink immediately after removal from the dies; total **shrinkage** will be approximately two-thirds complete in 5 weeks. The maximum extent of this **shrinkage** is about 0.001 in/in (10 μm/cm). The copper-free alloys do not exhibit this effect. Stabilization may be effected by heating the alloys for 3 to 6 h at 203°F (95°C), followed by air cooling to room temperature.

Wrought Zinc

Zinc rolled in the form of sheet, strip, or plate of various thicknesses is used extensively for automobile trim, dry-cell battery cans, fuses, and plumbing applications. Compositions of standard alloys are shown in Table 6.4.54. In addition, a comparatively new series of alloys containing titanium, exhibiting increased strength and creep resistance together

Table 6.4.53 Composition and Typical Properties of Zinc-Base Die Casting Alloys

	Composition, %					
Element	UNS Z33521 (AG40A) Alloy 3	UNS Z33522 (AG40B) Alloy 7	UNS Z35530 (AC41A) Alloy 5	UNS Z25630 ZA-8	UNS Z35630 ZA-12	UNS Z35840 ZA-27
Copper	0.25 max	0.25 max	0.75–1.25	0.8–1.3	0.5–1.25	2.0–2.5
Aluminum	3.5–4.3	3.5–4.3	3.5–4.3	8.0–8.8	10.5–11.5	25.0–28.0
Magnesium	0.020–0.05	0.005–0.020	0.03–0.08	0.015–0.030	0.015–0.030	0.010–0.020
Iron, max	0.100	0.075	0.100	0.10	0.075	0.10
Lead, max	0.005	0.0030	0.005	0.004	0.004	0.004
Cadmium, max	0.004	0.0020	0.004	0.003	0.003	0.003
Tin, max	0.003	0.0010	0.003	0.002	0.002	0.002
Nickel	—	0.005–0.020	—	—	—	—
Zinc	Rest	Rest	Rest	Rest	Rest	Rest
Yield strength						
lb/in^2	32	32	39	41–43	45–48	52–55
MPa	221	221	269	283–296	310–331	359–379
Tensile strength						
lb/in^2	41.0	41.0	47.7	54	59	62
MPa	283	283	329	372	400	426
Elongation in 2 in (5 cm), %	10	14	7	6–10	4–7	2.0–3.5
Charpy impact on square specimens						
ft·lb	43	43	48	24–35	15–27	7–12
J	58	58	65	32–48	20–37	9–16
Brinell Hardness number on square specimens, 500-kg load, 10-mm ball, 30 s	82	80	91	100–106	95–105	116–122

SOURCE: ASTM, reprinted with permission.
Compositions from ASTM Specifications B-86–04 and B-240–04. Properties from NADCA Product Specification Standards for Die Castings.

Table 6.4.54   Typical Composition of Rolled Zinc (ASTM Specification B69–01), %

Lead	Iron, max	Cadmium	Copper	Magnesium	Zinc
0.05 max	0.010	0.005 max	0.001 max	—	Remainder
0.05–0.12	0.012	0.005 max	0.001 max	—	Remainder
0.30–0.65	0.020	0.20–0.35	0.005 max	—	Remainder
0.05–0.12	0.012	0.005 max	0.65–1.25	—	Remainder
0.05–0.12	0.015	0.005 max	0.75–1.25	0.007–0.02	Remainder

SOURCE: ASTM, reprinted with permission.

with low thermal expansion, has become popular in Europe. A typical analysis is 0.5 to 0.8 percent Cu, 0.08 to 0.16 percent Ti, and maximum values of 0.2 percent Pb, 0.015 percent Fe, 0.01 percent Cd, 0.01 percent Mn, and 0.02 percent Cr. All alloys are produced by hot rolling followed by cold rolling when some stiffness and temper are required. Deep-drawing or forming operations are carried out on the softer grades, while limited forming to produce architectural items, plumbing, and automotive trim can be carried out using the harder grades.

Alloys containing 0.65 to 1.025 percent Cu are significantly stronger than unalloyed zinc and can be work-hardened. Since 1982, the United States one-cent coin has been made from a zinc-copper alloy that is copper-plated before coining. The addition of 0.01 percent Mg allows design stresses up to 10,000 lb/in² (69 MPa). The Zn-Cu-Ti alloy is much stronger, with a typical tensile strength of 25,000 lb/in² (172 MPa).

Rolled zinc may be easily formed by all standard techniques. The deformation behavior of rolled zinc and its alloys varies with direction; its crystal structure renders it nonisotropic. In spite of this, it can be formed into parts similar to those made of copper, aluminum, and brass, and usually with the same tools, provided the temperature is not below

70°F (21°C). More severe operations can best be performed at temperatures up to 125°F (52°C). When a cupping operation is performed, a take-in of 40 percent on the first draw is usual. Warm, soapy water is widely used as a lubricant. The soft grades are self-annealing at room temperature, but harder grades respond to deformation better if they are annealed between operations. The copper-free zincs are annealed at 212°F (100°C) and the zinc-copper alloys at 440°C (220°C). Welding is possible with a wire of composition similar to the base metals. Soldering with typical tin-lead alloys is exceptionally easy. The impact extrusion process is widely used for producing battery cups and similar articles.

Effect of Temperature

Properties of zinc and zinc alloys are very sensitive to temperature. **Creep resistance** decreases with increasing temperature, and this must be considered in designing articles to withstand continuous load. When steel screws are used to fasten zinc die castings, maximum long-term clamping load will be reached if an engagement length 4 times the diameter of the screw is used along with cut (rather than rolled) threads on the fasteners.

6.5   CORROSION
by Robert D. Bartholomew and David A. Shifler

REFERENCES: Uhlig, "The Corrosion Handbook," Wiley, New York, 1948. R. Winston Revie (ed.) "Uhlig's Corrosion Handbook," 2d ed., Wiley Interscience, New York, 2000. Evans, "An Introduction to Metallic Corrosion," 3d ed., Edward Arnold & ASM International, Metals Park, OH, 1981. Van Delinder (ed.), "Corrosion Basics—An Introduction," 2d ed., NACE International, Houston, TX, 1984. Wranglen, "An Introduction to Corrosion and Protection of Metals," Chapman & Hall, London, 1986. Fontana, "Corrosion Engineering," 3d ed., McGraw-Hill, New York, 1986. Jones, "Principles and Prevention of Corrosion," Macmillan, New York, 1992. Jones, "Principles and Prevention of Corrosion," 2d ed., Prentice-Hall, Upper Saddle River, NJ, 1996. Scully, "The Fundamentals of Corrosion" 3d ed., Pergamon, New York, 1990. Kaesche, "Metallic Corrosion," 2d ed., Rapp (trans.), NACE International, Houston, TX, 1985. Sheir (ed.), "Corrosion," 2d ed., Newnes-Butterworths, London, 1976. Sheir and Jarman. (eds.), "Corrosion," 3d ed., Butterworth-Heinemann, Oxford, 1994. "Corrosion," vol. 13, "Metals Handbook," 9th ed., ASM International, Metals Park, OH, 1987. Cramer and Covino (eds.), "Corrosion: Fundamentals, Testing, and Protection," ASM International, Materials Park, OH, 2003. Gellings, "Introduction to Corrosion Prevention and Control," Delft University Press, Delft, The Netherlands, 1985. Sedriks, "Corrosion of Stainless Steels," 2d ed., Wiley Interscience, New York, 1996. Pourbaix, "Atlas of Electrochemical Equilibria in Aqueous Solutions," NACE International, Houston, TX, 1974. Prentice, "Electrochemical Engineering Principles," Prentice-Hall, Englewoods Cliffs, NJ, 1991. Newman, and Thomas-Alyea, "Electrochemical Systems," 3d ed., Wiley Interscience, New York, 2004. Bockris and Reddy, "Modern Electrochemistry," Plenum, New York, 1970. Bockris and Khan, "Surface Electrochemistry—A Molecular Level Approach," Plenum, New York, 1993.

INTRODUCTION

Corrosion is the deterioration of a material or its properties because of its reaction with the environment. Materials may include metals and alloys, nonmetals, woods, ceramics, plastics, composites, etc. Although corrosion as a science is barely 150 years old, its effects have impacted humankind for thousands of years. The importance of understanding the causes, initiation, and propagation of corrosion and the methods for controlling its degradation is threefold. First, corrosion generates an **economic impact** through materials losses and failures of various structures and components. A recent study estimated that the combined annualized economic losses from 1999 to 2001 in the United States from corrosion was $276 billion (Payer, et al., "Corrosion Costs and Preventative Strategies in the United States," Publication no. FHWA-RD-01-156, U.S. Federal Highway Administration, 2001; Cost of Corrosion Study Unveiled, *Mater. Perf.*, supplement, NACE International, July 2002). Studies of corrosion costs in other countries have determined that 3 to 4 percent of GNP is related to economic losses from corrosion. These estimates consider only the direct economic costs. Indirect costs are difficult to assess but can include plant downtime, loss of products or services, loss of efficiency, contamination, and overdesign of structures and components. Application of current corrosion control technology (discussed later) could recover or avoid 15 to 20 percent of the costs due to corrosion.

Second, an improved consideration of corrosion prevention or control is improved **safety**. Catastrophic degradation and failures of pressure vessels, petrochemical plants, boilers, aircraft sections, tanks containing toxic materials, and automotive parts have led to thousands of personal injuries and death, which often result in subsequent litigation. If the safety of individual workers or the public is endangered in some manner by selection of a material or use of a corrosion control method, that approach should be abandoned. Once assurance is given that this criterion is met through proper materials selection and improved design and corrosion control methods, other factors can be evaluated to provide the optimum solution.

Third, understanding factors leading to corrosion can **conserve resources.** The world's supply of easily extractable raw material is limited, and corrosion control can minimize the use of precious resources. Corrosion also causes a wastage of energy, water resources, and raw or processed materials.

Corrosion can occur through chemical or electrochemical reactions and is usually an interfacial process between a material and its environment. Deterioration solely by physical means such as erosion, galling, or fretting is not generally considered corrosion. **Chemical corrosion** may be a heterogeneous reaction which occurs at the metal/environment interface and involves the material (metal) itself as one of the reactants. Such occurrences may include metals in strong acidic or alkaline solution, high-temperature oxidation metal/gas reactions in which the oxide or compound is volatile, the breakdown of chemical bonds in polymers, dissolution of solid metals or alloys in a liquid metal (e.g., aluminum in mercury), or the dissolution of metals in fused halides or nonaqueous solutions.

However, even in acidic or alkaline solutions and in most high-temperature environments, corrosion by electrochemical processes consisting of two or more partial reactions involves the transfer of electrons or charges. An electrochemical reaction requires: (1) an anode; (2) a cathode; (3) an electrolyte; and (4) a complete electric circuit. In the reaction of a metal (M) in hydrochloric acid, the metal is oxidized and the electrons are generated at the anode [sec Eq.(6.5.1)]. At the same time, hydrogen cations are reduced and electrons consumed to generate hydrogen gas at the cathode, the **hydrogen evolution reaction** (HER) as described in Eq. (6.5.2).

Anodic (oxidation) reaction: $M \rightarrow M^{+n} + ne^-$ (6.5.1)

Cathodic (reduction) reaction: $2H^+ + 2e^- \rightarrow H_2$ (6.5.2)

Generally, oxidation occurs at the anode, while reduction occurs at the cathode. Corrosion is usually involved at the anode where the metal or alloy is oxidized; this causes dissolution, wastage, and penetration. Cathodic reactions significant to corrosion are few. These may include, in addition the HER, the following:

Oxygen reduction (acidic solutions): $O_2 + 4H^+ + 4e^- \rightarrow 2H_2O$

(6.5.3)

Oxygen reduction (neutral or basic): $O_2 + 2H_2O + 4e^- \rightarrow 4OH^-$

(6.5.4)

Metal ion reduction: $Fe^{3+} + e^- \rightarrow Fe^{2+}$ (6.5.5)

Metal deposition: $Cu^{2+} + 2e^- \rightarrow Cu$ (6.5.6)

THERMODYNAMICS OF CORROSION

For an anodic oxidation reaction of metal to occur, a simultaneous reduction must take place. In corroding metal systems, the anodic and cathodic half-reactions are mutually dependent on each other and form a galvanic or spontaneous cell. A cell in which electrons are driven by an external energy source in the direction counter to the spontaneous half-reactions is termed an **electrolytic cell.** A metal establishes a potential or emf with respect to its environment; it is dependent on the ionic strength and composition of the electrolyte, the temperature, the metal or alloy itself, and other factors. The potential of a metal at the anode in solution arise from the release of positively charged metal cations together with the creation of a negatively charged metal. The standard potential of a metal is defined by fixing the equilibrium concentration of its ions at unit activity and under reversible (zero net current) and standard conditions (one atm, 101 kPa; 25°C). At equilibrium, the net current density ($\mu A/cm^2$) of an electrochemical reaction is zero. The nonzero anodic and cathodic currents are equal and opposite, and the absolute magnitude of either current at equilibrium is equal to the exchange current density (i.e., $i_{an} = i_{cath} = i_o$).

The potential, a measure of the driving influence of an electrochemical reaction, cannot be evaluated in absolute terms, but is determined by the difference between it and another reference electrode. Common **standard reference electrodes** are saturated calomel electrode (SCE, 0.2416 V), Ag/AgCl in 0.1N KCl (0.2881 V), and saturated Cu/CuSO$_4$ (0.316 V) whose potentials are measured relative to **standard hydrogen electrode** (SHE), which by definition is 0.000 V under standard conditions. A luggin capillary is often used to minimize the ohmic drop between the working electrode (the material being measured) and the reference electrode. Positive electrochemical potentials of half-cell reactions versus SHE (written as a reduction reaction) are more easily reduced and noble, while negative values signify half-cell reactions that are more difficult to reduce and, conversely, more easily oxidized or active than the SHE. The signs of the potentials are reversed if the SHE and half-cell reactions of interest are written as oxidation reactions. The potential of a galvanic cell is the sum of the potentials of the anodic and cathodic half-cells in the environment surrounding them.

From thermodynamics, the potential of an electrochemical reaction is a measure of the **Gibbs free energy** $\Delta G = -nFE$, where n is the number of electrons participating in the reaction, F is Faraday's constant (96,480 C/mol), and E is the electrode potential. The potential of the galvanic cell will depend on the concentrations of the reactants and products of the respective partial reactions and on the pH in aqueous solutions. The potential is related to the Gibbs free energy by the Nernst equation

$$\Delta E = \Delta E^\circ + \frac{2.3RT}{nF} \log \frac{(ox)^x}{(red)^y}$$

where ΔE° is the standard electrode potential, (ox) is the activity of an oxidized species, (red) is the activity of the reduced species; and x and y are stoichiometric coefficients involved in the respective half-cell reactions. Corrosion will not occur unless the spontaneous direction of the reaction (i.e., $\Delta G < 0$) indicates metal oxidation. The application of thermodynamics to corrosion phenomena has been generalized by the use of **potential-pH plots** or **Pourbaix diagrams.** Such diagrams are constructed from calculations based on the Nernst equation and solubility data for various metal compounds. As shown in Fig. 6.5.1, it is possible to differentiate regions of potential pH in which iron is either immune or will potentially passivate from regions in which corrosion will thermodynamically occur. The main uses of these diagrams are for (1) predicting the spontaneous directions of reactions; (2) estimating the composition of corrosion products; and (3) predicting environmental changes that will prevent or reduce corrosion. The major limitations of

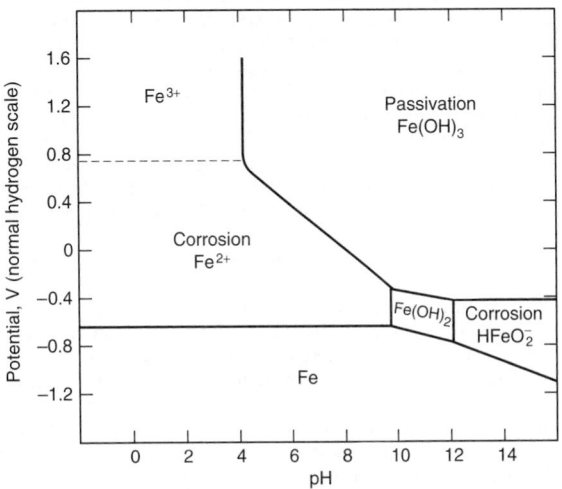

Fig. 6.5.1 Simplified potential-pH diagram for the Fe-H$_2$O system. (*M. Pourbaix, "Atlas of Electrochemical Equilibria in Aqueous Solutions."*)

Pourbaix diagrams are that (1) only pure metals, and not alloys, are usually considered; (2) pH is assumed to remain constant, whereas HER may alter the pH; (3) they do not provide information on metastable films; (4) possible aggressive solutions containing such as Cl⁻, Br⁻, I⁻, or NH_4^+ are not usually considered; and (5) they do not predict the kinetics or corrosion rates of the different electrochemical reactions. The predictions possible are based on calculations of both the metal and solution compositions and concentrations and the temperature designated in the diagrams. Higher-temperature potential-pH diagrams have been developed ("Computer-Calculated Potential-pH Diagrams to 300°C," NP-3137, vol. 2, Electric Power Research Institute, Palo Alto, CA, 1983). **Ellingham diagrams** provide thermodynamically derived data for pure metals in gaseous environments to predict stable phases, although they also do not predict the kinetics of these reactions.

CORROSION KINETICS

Corrosion is thermodynamically possible for most environmental conditions. It is of primary importance to assess the kinetics of corrosion. While free energy and electrode potential are thermodynamic parameters of an electrochemical reaction, the equilibrium exchange current density i_o is a fundamental kinetic property of the reaction. It is dependent on the material, surface properties, environment, and temperature.

Anodic or cathodic electrochemical reactions such as established in Eqs. (6.5.1) to (6.5.6) often proceed at finite rates. When a cell is short-circuited, net oxidation and reduction occur at the anode and cathode, respectively. The potentials of these electrodes will no longer be at equilibrium. This deviation form **equilibrium potential** is termed **polarization** and results from the flow of a net current. Overvoltage or overpotential is a measure of polarization. Deviations from the equilibrium potential may be caused by (1) concentration polarization, (2) activation polarization, (3) resistance polarization, or (4) mixed polarization. **Concentration polarization** η_c is caused by the limiting diffusion rate of the electrolyte of the electrode surface and is generally important only for cathodic reactions. For example, at high reduction rates of the HER, the cathode surface will become depleted of hydrogen ions. If the reduction rate is increased further, the diffusion rate of hydrogen ions to the cathode will approach a limiting diffusion current density i_L. This limiting current density can be increased by agitation, higher temperatures, and higher reactant concentrations.

Activation polarization η_a refers to the electrochemical process that is controlled by a slow rate-determining step at the metal/electrolyte interface. Activation polarization is characteristic of cathodic reactions such as HER and metal/ion deposition or dissolution. The relationship between activation polarization and the rate of reaction i_A at the anode is $\eta_{a(A)} = \beta_A \log (i_A/i_o)$ where β_A represents the Tafel slope (~0.60 to 0.120 V per decade of current) of the anodic half-reaction. A similar expression can be written for the cathodic half-reaction.

Resistance polarization η_R includes the ohmic potential drop through a portion of the electrolyte surrounding the electrode, or the ohmic resistance in a metal-reaction product film on the electrode surface, or both. High-resistivity solutions and insulating films deposited at either the cathode or anode restrict or completely block contact between the metal and the solution and will promote a high resistance polarization. **Mixed polarization** occurs in most systems and is the sum of η_c, η_a, and η_R. Given β, i_L, and i_o, the kinetics of almost any corrosion reaction can be described.

When polarization occurs mostly at the anodes of a cell, the corrosion reaction is anodically controlled. When polarization develops mostly at the cathode, the corrosion rate is cathodically controlled. Resistance control occurs when the electrolyte resistance is so high that the resultant current is insufficient to polarize either the anode or the cathode. Mixed control is common in most systems in which both the anodes and cathodes are polarized to some degree.

Hydrogen overpotential is a dominant factor in controlling the corrosion rate of many metals either in deaerated water or in nonoxidizing acids at cathodic areas; the overpotential is dependent on the metal,

temperature, and surface roughness. The result of hydrogen overpotential often will be a surface film of hydrogen atoms. Decreasing pH usually will promote the HER reaction and corrosion rate; acids and acidic salts create a corrosive environment for many metals. Conversely, high-pH conditions may increase hydrogen overpotential and decrease corrosion, and may provide excellent protection even without film formation, such as when caustic agents, ammonia, or amines are added to condensate or to demineralized, or completely softened, water.

Oxygen dissolved in water reacts with the atomic hydrogen film on cathodic regions of the metal surface by depolarization and enables corrosion to continue. The rate of corrosion is approximately limited by the rate at which oxygen diffuses toward the cathode (**oxygen polarization**); thus, extensive corrosion occurs near or at the water line of a partially immersed or filled steel specimen. In systems where both hydrogen ions and oxygen are present, the initial predominant cathodic reaction will be the one that is most thermodynamically and kinetically favorable.

Corrosion rates have been expressed in a variety of ways. Mils per year (mpy) and micrometers per year (μm/yr) (SI) are desirable forms in which to measure corrosion or penetration rates and to predict the life of a component from the weight loss data of a corrosion coupon test, assuming that the measured weight corresponds to a uniform corrosion rate, as shown by Eqs. (6.5.7a) and (6.5.7b).

$$\text{mpy} = \frac{534W}{DAT} \qquad (6.5.7a)$$

$$\mu\text{m/yr} = \frac{87,600W}{DAT} \qquad (6.5.7b)$$

where W = weight loss, mg; D = density of the specimen, g/cm³; A = area of specimen, mpy, in² (μm/yr, cm²); T = exposure, h. The corrosion rate considered detrimental for a metal or an alloy will depend on its initial cost, design life, cost of materials replacement, environment, and temperature.

During metallic corrosion, the sum of the anodic currents must equal the sum of the cathodic currents. The corrosion potential of the metal surface is defined by the intersection of the anodic and cathodic polarization curves, where anodic and cathodic currents are equal. Figure 6.5.2 illustrates an idealized polarization curve common for most metals, while Fig. 6.5.3 displays typical anodic dissolution behavior of **active/passive metals** such as iron, chromium, cobalt, molybdenum, nickel, titanium, and alloys containing major amounts of these elements. Anodic currents increase with higher anodic potentials until a

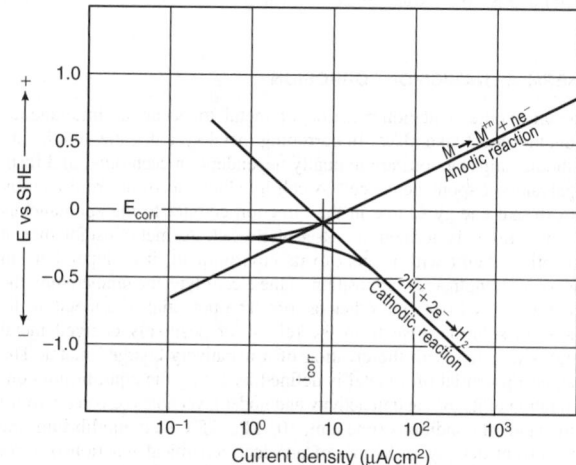

Fig. 6.5.2 Idealized polarization common for most metals, at E_{corr}, where anodic ($i_{corr} = i_a$) and cathodic currents are equal.

Fig. 6.5.3 Idealized anodic polarization curve for active/passive metals that exhibit passivity. Three different potential regions are identified.

critical anodic current density i_{pp} is reached at the primary passive potential E_{pp}. Higher anodic potentials will result in formation of a **passive film** until the transpassion potential is reached.

Generally, metals and alloys (with the possible exception of gold) are not in their most stable thermodynamic state. Corrosion tends to convert these pure materials back to their stable thermodynamic form. The major factor controlling corrosion of metals and alloys is the nature of the **protective, passive film** is its environment. **Passivity** may be defined as the loss of chemical reactivity under certain environmental conditions. A second definition is illustrated by Fig. 6.5.3; a metal becomes passive if, on increasing its potential to more anodic or positive values, the rate of anodic dissolution decreases (even though the rate should increase as the potential increases). A metal can passivate by (1) chemisorption of the solvent, (2) salt film formations, (3) oxide/oxyhydroxide formation (Hoar et al., *Corr. Sci.*, **5**, 1965, p. 279; Agladze, et al., *Prot. of Metals*, **22**, 1987, p. 404), and (4) electropolymerization (Shifler et al., *Electrochim. Acta*, **38**, 1993, p. 881). The stability of the passive film will depend on its chemical, ionic, and electronic properties; its degree of crystallinity; the film flexibility; and the film mechanical properties in a given environment Frankenthal and Kruger (eds.), "Passivity of Metals—Proc. of the Fourth International Symposium on Passivity," The Electrochemical Society, Pennington, NJ, 1978). The **oxide/oxyhydroxide** film is the most common passive film at ambient temperatures in most environments. The passive layer thickness, whether considered as an adsorbed oxygen structure or an oxide film, is generally 50 nm or less.

FACTORS INFLUENCING CORROSION

A number of factors influence the stability and breakdown of the **passive film** and include (1) surface finish, (2) metallurgy, (3) stress, (4) heat treatment, and (5) environment. Polished surfaces generally resist corrosion initiation better than rough surfaces. Surface roughness tends to increase the kinetics of the corrosion reaction by increasing the exchange current density of the HER or the oxygen reduction at the cathodes.

Metallurgical structures and properties often have major effects on corrosion. Regions of varying composition exist along the surface of most metal or alloys. These local compositional changes have different potentials that may initiate local-action cells. Nonmetallic inclusions, particularly sulfide inclusions, are known to initiate corrosion on carbon and stainless steels. The size and distribution of sulfide inclusions in 304 stainless steel have a large impact on dissolution kinetics and pitting susceptibility and growth. Elements such as chromium and nickel improve the corrosion resistance of carbon steel by improving the stability of the passive oxide film. Copper (0.2 percent) added to carbon steel improves the atmospheric corrosion resistance of weathering steels "H. H. Uhlig

Memorial Symposium—Corrosion Science: From Theory to Practice," (Mansfeld et al. (eds.), PV 94-26, The Electrochemical Society, Pennington, NJ, 1995).

Stresses, particularly tensile stresses, affect corrosion behavior. These stresses may be either applied or residual. **Residual stresses** may arise from either dislocation pileups or stacking faults due to deforming or cold-working the metal that may arise from forming, heat-treating, machining, welding, and fabrication operations. Cold-working increases the applied stresses to the individual grains by distorting the crystals. Corrosion at cold-worked sites is not increased in natural waters, but increases several-fold in acidic solutions. Possible segregation of carbon and nitrogen occurs during cold-working. Welding can induce residual stresses and provide sites that are subject to preferential corrosion and cracking. The welding of two dissimilar metals that possess different thermal expansion coefficients may restrict expansion of one member and induce applied stresses during service. Thermal fluctuations of fabrication bends or attachments welded to components may cause applied stresses to develop that may lead to cracking.

Improper **heat treatment** or welding can influence the microstructure of different alloys by causing either precipitation of deleterious phases at alloy grain boundaries or (as is the case of austenitic stainless steels) depletion of chromium in grain boundary zones which decreases the local corrosion resistance. **Welding** also can cause phase transformations and formation of secondary precipitates, and it can induce stresses in and around the weld. Rapid quenching of steels from austenizing temperatures may form martensite, a distorted tetragonal structure, that often suffers from preferential corrosion.

The nature of the **environment** can affect the rate and form of corrosion. Environments include (1) natural and treated waters, (2) the atmosphere, (3) soil, (4) microbiological organisms, and (5) high temperature. Corrosivity in freshwater varies with oxygen content, hardness, chloride and sulfur content, temperature, and velocity. Water may contain colloidal or suspended matter, dissolved solids, and dissolved gases. All these constituents may stimulate or suppress corrosion by affecting either the cathodic or the anodic reaction, or by forming a protective barrier. Oxygen is probably the most significant constituent affecting corrosion in neutral and alkaline solution; hydrogen ions are more significant in acidic solutions. **Freshwater** can be hard or soft. In **hard waters,** calcium carbonate often deposits on the metal surface and protects it; pitting may occur if the calcareous coating is not complete. **Soft waters** are usually more corrosive because protective deposits do not form. Several saturation indices are used to provide scaling tendencies. Nitrates and chlorides increase aqueous conductivity and reduce the effectiveness of natural protective films. The ratio of chloride/bicarbonate has been observed to predict the probability of dezincification. Sulfides in polluted waters tend to cause pitting. Deposits on metal surfaces may lead to local stagnant conditions which may lead to pitting. High velocities usually increase corrosion rates by removing corrosion products that otherwise might suppress the anodic reaction and by stimulating the cathodic reaction by providing more oxygen. **High-purity water,** used in nuclear and high-pressure power units, decreases corrosion by increasing the electrical resistance of the fluid. Temperature effects in aqueous systems are complex, depending on the nature of the cathodic and anodic reactions.

Seawater is roughly equivalent to 3½% sodium chloride, but also contains a number of major constituents and traces of almost all naturally occurring elements. Seawater has a higher conductivity than freshwater, which alone can increase the corrosion of many metals. The high conductivity permits larger areas to participate in corrosion reactions. The high chloride cuntent in seawater can increase localized breakdown of oxide films. The pH of seawater is usually 8.1 to 8.3. Plant photosynthesis and decomposition of marine organisms can raise or lower the pH, respectively. Because of the relatively high pH, the most important cathodic reaction in seawater corrosion processes is oxygen reduction. Highly aerated water, such as tidal splash zones, is usually a region of very high corrosion rates. Barnacles attached to metal surfaces can cause localized attack if present as discontinuous barriers. Copper and copper alloys have the natural ability to suppress barnacles and other microfouling organisms. Sand, salt, and abrasive particles suspended in seawater can aggravate erosion corrosion. Increased seawater temperatures generally increase corrosivity. [Laque, "Marine Corrosion-Causes and Prevention," Wiley Interscience, New York, 1975; Shifler and Aylor, Considerations for the Testing of Materials and Components in Seawater, *Corrosion/2002*, Paper 217, Houston, TX, 2002; Shifler et al. (eds.), "Corrosion and Corrosion Control in Saltwater Environments," PV 99–26, The Electrochemical Society, Pennington, NJ, 2000; Shifler et al. (eds.), "Corrosion and Corrosion Control in Saltwater Environments II," PV 2004–14, The Electrochemical Society, Pennington, NJ, 2005].

In general, **atmospheric corrosion** is the result of the conjoint action of oxygen and water, although contaminants such as sodium chloride and sulfur dioxide accelerate corrosion rates. In the absence of moisture (below 60 to 70 percent relative humidity), most metals corrode very slowly at ambient temperatures. Water is required to provide an electrolyte for charge transfer. Damp corrosion requires moisture from the atmosphere; wet corrosion occurs when water pockets or visible water layers are formed on the metal surface due to salt spray, rain, or dew formation. The solubility of the corrosion products can affect the corrosion rate during wet corrosion; soluble corrosion products usually increase corrosion rates.

Time of wetness is a critical variable that determines the duration of the electrochemical corrosion processes in atmospheric corrosion. Temperature, climatic conditions, relative humidity, and surface shape and conditions that affect time of wetness also influence the corrosion rate. Metal surfaces that retain moisture generally corrode faster than surfaces exposed to rain. Atmospheres can be classified as rural, marine, or industrial. Rural atmospheres tend to have the lowest corrosion rates. The presence of NaCl near coastal shores increases the aggressiveness of the atmosphere. Industrial atmospheres are more corrosive than rural atmospheres because of sulfur compounds. Under humid conditions SO_2 can promote the formation of sulfurous or sulfuric acid. Other contaminants include nitrogen compounds, H_2S, and dust particles. Dust particles adhere to the metal surface and absorb water, prolonging the time of wetness; these particles also may include chlorides that tend to break down passive films [Ailor (ed.), "Atmospheric Corrosion," Wiley-Interscience, New York, 1982; Leygraf and Graedel, "Atmospheric Corrosion," Wiley Interscience, New York, 2000].

Soil is a complex, dynamic environment that changes continuously, both chemically and physically, with the seasons for the year. Characterizing the **corrosivity of soil** is difficult at best. **Soil resistivity** is a measure of the concentration and mobility of ions required to migrate through the soil electrolyte to a metal surface for a corrosion reaction to continue. Soil resistivity is an important parameter in underground corrosion; high resistivity values often suggest low corrosion rates. The mineralogical composition and earth type affect the gain size, effective surface area, and pore size which, in turn, effect the soil corrosivity. Further, soil corrosivity can be strongly affected by certain chemical species, microorganisms, and soil acidity or alkalinity. A certain water content in soil is an essential requirement for corrosion to occur. Oxygen also is generally required for corrosion processes, although steel corrosion can occur under oxygen-free, anaerobic conditions in the presence of **sulfate-reducing bacteria** (SRB). Soil water can regulate the oxygen supply and its transport [Romanoff, Underground Corrosion, *NBS Circ.* **579,** 1957; NACE International, Houston, TX, 1989; Escalante (ed.), "Underground Corrosion," STP 741, ASTM, Philadelphia, 1980].

Almost all commercial alloys are affected by **microbiological influenced corrosion** (MIC). Most MIC involves localized corrosion. Biological organisms are present in virtually all natural aqueous environments (freshwater, brackish water, seawater, or industrial water) and in some soils. In these environments, the tendency is for the microorganisms to attach to and grow as a biofilm on the surface of the structural materials. Environmental variables (pH, velocity, oxidizing power, temperature, electrode polarization, and concentration) under a biofilm can be vastly different from the bulk environment. MIC can occur under aerobic or anaerobic conditions. Biofilms may cause corrosion under conditions that otherwise would not cause dissolution in the environment, change the mode of corrosion, increase or decrease the corrosion rate, or may not influence corrosion at all. MIC may actively or passively cause corrosion. Active MIC directly accelerates or establishes new electrochemical corrosion reactions. In passive MIC, the biomass acts as any dirt or surface deposit under which concentration cells can initiate and propagate. Several forms of bacteria are linked to accelerated corrosion by MIC: (1) SRB; (2) sulfur/sulfide oxidizing bacteria; (3) acid-producing bacteria and fungi; (4) iron-oxidizing bacteria; (5) manganese-fixing bacteria; (6) acetate-oxidizing bacteria; (7) acetate-producing bacteria; and (8) slime formers [Dexter (ed.), "Biological Induced Corrosion," NACE-8, NACE International, Houston, TX, 1986; Kobrin (ed.), "A Practical Manual on Microbiological Influenced Corrosion," NACE International, Houston, TX, 1993; Borenstein, "Microbiological Influenced Corrosion Handbook," Industrial Press, New York, 1994; Videla, "Manual of Biocorrosion," CRC–Lewis Publishers, New York, 1996; Dexter, Microbiologically Influenced Corrosion, in Cramer and Covino (eds.), "Corrosion: Fundamentals, Testing, and Protection," p. 398, ASM International, Materials Park, OH, 2003].

Passive alloys such as stainless steels, nickel-base alloys, and titanium can experience ennoblement of several hundred millivolts with exposure time in natural seawater, thus magnifying the potential differences that may exist between dissimilar alloys. Ennoblement can accelerate galvanic or crevice corrosion (discussed later) [LaFontaine and Dexter, Effect of Marine Biofilm on Galvanic Corrosion, *Proc. of the 1997 Tri-Service Conference on Corrosion*, Session 6, Tri-Service Committee on Corrosion, Wrightsville Beach, NC, November 1997]. Ennoblement of the **open circuit potential** (OCP) occurs in fresh, brackish, and sea waters as long as corrosion initiation does not interfere. The rate of ennoblement is slower in freshwater than in seawater, but the amount of ennoblement is inversely related to the salinity of the electrolyte (Dexter and Zhang, Effect of Biofilms on Corrosion Potential of Stainless Alloys in Estuarine Water, *Proc. 11th* International Corrosion Congress, Florence, Italy, p. 4.333, April 1990). The most noble potentials are consistently found in freshwater. Ennoblement is likely caused by the formation of microbiological films, which increases the kinetics of the cathodic reaction. Many mechanisms have been proposed for OCP ennoblement. Most of the early mechanisms involved enhancement of the cathodic oxygen reduction reaction. This has been proposed to take place by (1) organometallic catalysis, (2) catalysis by bacterially produced enzymes, (3) a decrease in pH, and (4) combinations of pH, peroxide, and oxygen. Some dissolved oxygen at the metal surface within the biofilm matrix is a prerequisite for OCP ennoblements, as Chandrasekaran and Dexter showed that ennoblement disappeared upon depletion of oxygen in the system. They also showed that ennoblement was not sustained under oxygen-saturated conditions. Other mechanisms that have been proposed for OCP ennoblement involve siderophores, biomineralization of dissolved Mn compounds to MnO_2, and oxidation of organic matter (Scotto et al., The Influence of Marine Aerobic Microbial Film on Stainless-Steel Corrosion Behavior, *Corrosion Sci.* **25**, no. 3, 1985, pp. 185–194; Eashwar et al., *Corrosion Sci.* **37**, no. 8, 1995, pp. 1169–1176; Dexter and Maruthamuthu, Response of Passive Alloys with N- and p-Type Passive Films to Manganese in Biofilms, Paper 01256, *Corrosion/2001*, NACE International, Houston, TX, 2001; Chandrasekaran and Dexter, Thermodynamic and Kinetic Factors Influenced by Biofilm Chemistry Prior to Passivity Breakdown, Paper 482, *Corrosion/94*, NACE International, Houston, TX, 1994; Salvago and Magagnin, Biofilms' Effect on the Cathodic and Anodic Processes on Stainless Steel in Seawater near the Corrosion Potential—Part 1: Corrosion Potential, *Corrosion,* **57**, no. 8, 2001, pp. 680–692; Salvago and Magagnin, Biofilms' Effect on the Cathodic and Anodic Proceses on Stainless Steel in Seawater near the corrosion Potential—Part 2: Oxygen Reduction on Passive Metal, *Corrosion* **57**, no. 9, 2001, pp. 759–767).

High-temperature service ($\geq 100°C$) is especially damaging to many metals and alloys because of the exponential increase in reaction rate with temperature. Hot gases, steam, molten or fused salts, molten metals, or refractories, ceramics, and glasses can affect metals and alloys at high temperatures. The most common reactant is oxygen; therefore all gas/metal reactions are usually referred to as oxidations, whether the reaction involves oxygen, steam, hydrogen sulfide, or combustion gases. Diffusion is a fundamental property involved in the oxidation reactions. The corrosion rate can be influenced considerably by the presence of contaminants, particularly if they are adsorbed on the metal surface. Pressure generally has little effect on the corrosion rate unless the dissociation pressure of the oxide or scale constituent lies within the pressure range involved. Stress may be important when attack is intergranular. Differential mechanical properties between a metal and its scale may cause periodic scale cracking which leads to accelerated oxidation. **Thermal cycling** can lead to cracking and flaking of scale layers. A variety of molten salt baths are used in industrial processes. Salt mixtures of nitrates, carbonates, or halides of alkaline or alkaline-earth metals may adsorb on a metal surface and cause beneficial or deleterious effects.

There are three major types of **cells** that transpire in corrosion reactions. The first is the **dissimilar electrode cell.** The is derived from potential differences that exist between a metal containing separate electrically conductive impurity phases on the surface and the metal itself, different metals or alloys connected to one another, cold-worked metal in contact with the same metal annealed, and a new iron pipe in contact with an old iron pipe.

The second cell type, the **concentration cell**, involves having identical electrodes each in contact with an environment of differing composition. One kind of concentration cell involves solutions of differing salt levels. Anodic dissolution or corrosion will tend to occur in the more-dilute salt or lower-pH solution, while the cathodic reaction will tend to occur in the more-concentrated salt or higher-pH solution. Identical pipes may corrode in soils of one composition (clay) while little corrosion occurs in sandy soil. This is due in part to differences in soil composition and in part to differential aeration effects. Corrosion will be experienced in regions low in oxygen; cathodes will exist in high-oxygen areas.

The third type of cell, the **thermogalvanic cell,** involves electrodes of the same metal that are exposed, initially, in electrolytes of the same composition but at different temperatures. Electrode potentials change with temperature, but temperature also may affect the kinetics of dissolution. Depending on other aspects of the environment, thermogalvanic cells may accelerate or slow the rate of corrosion. Denickelification of copper-nickel alloys may occur in hot areas of a heat exchange in brackish water, but are unaffected in colder regions.

FORMS OF CORROSION

It is convenient to classify corrosion by the various forms in which it exists, preferably by visual examination, although in many cases analysis may require more sophisticated methods. An arbitrary list of **corrosion types** is (1) uniform corrosion, (2) galvanic corrosion, (3) crevice corrosion, (4) pitting, (5) intergranular corrosion, (6) dealloying or selective leaching, (7) erosion corrosion, (8) environmentally induced cracking, and (9) high-temperature corrosion.

Uniform corrosion is the most common form of corrosion. It is typified by an electrochemical or chemical attack that effects the entire exposed surface. The metal becomes thinner and eventually fails. Unlike many of the other forms of corrosion, uniform corrosion can be easily measured by weight loss tests and by using equations such as Eqs. (6.5.7a) and (6.5.7b). The life of structures and components suffering from uniform corrosion then can be accurately estimated.

Stray-current corrosion or **stray-current electrolysis** is caused by externally induced electric currents and it usually independent of environmental factors such as oxygen concentration or pH. Stray currents are escaped currents that follow paths other than their intended circuit via a low-resistance path through soil, water, or any suitable electrolyte. Direct currents from electric mass transit systems, welding machines, and implied cathodic protection systems (discussed later) are major sources of stray-current corrosion. At the point where stray currents enter the unintended structure, the sites will become cathodic because of changes in potential. Areas where the stray currents leave the metal structure become anodic and become sites where serious corrosion can occur. Alternating current (ac) generally cause less severe damage than direct current (dc). Stray-current corrosion may vary over short time periods, and its appearance is sometimes similar to that of galvanic corrosion (Szeliga (ed.), "Stray Current Corrosion: The Past, Present, and Future of Rail Transit Systems," NACE International, Houston, TX, 1994).

Galvanic corrosion occurs when a metal or alloy is electrically coupled to another metal, alloy, or conductive nonmetal either directly or through an external path in a common, conductive medium. A potential difference usually exists between dissimilar metals, which causes a flow of electrons between them. The corrosion rate of the less corrosion-resistant (active) anodic metal is increased, while the corrosion rate of the more corrosion-resistant (noble) cathodic metal or alloy is decreased. Galvanic corrosion may be affected by many factors including (1) potential difference between two or more alloys; (2) reaction kinetics; (3) alloy composition; (4) protective film characteristics in the given environment; (5) mass transport (migration, diffusion, and/or convection); (6) bulk solution properties (oxygen content, pH, conductivity, pollutant/contamination level); (7) cathode/anode ratio; (8) polarization; (9) volume of solution; (10) temperature; (11) type of joint; (12) flow rate; and (13) geometry of coupling members [Oldfield, Electrochemical Theory of Galvanic Corrosion, "Galvanic Corrosion," STP 978, Hack (ed.), ASTM Philadelphia, 1988, p. **5,** Francis, *British Corrosion Jour.* **29,** no. 53, 1994; Francis, "Galvanic Corrosion: A Practical Guide for Engineers," NACE International, Houston, TX, 2001].

A driving force for corrosion or current flow is the potential difference formed between dissimilar metals. The extent of accelerated corrosion resulting from galvanic corrosion is related to potential differences between dissimilar metals or alloys, the nature of the environment, the polarization behavior of metals involved in the galvanic couple, and the geometric relationship of the component metals. A **galvanic series of metals and alloys** is useful for predicting the possibility for galvanic corrosion (see. Fig. 6.5.4) (Baboian, "Galvanic and Pitting Corrosion—Field and Laboratory Studies," STP 576, ASTM, Philadelphia, 1976). A galvanic series must be arranged according to the potentials of the metals or alloys in a specific electrolyte at a given temperature and conditions. Separation of two metals in the galvanic series may indicate the possible magnitude of the galvanic corrosion. Since corrosion product films and other changes in the surface composition may occur in different environments, no singular value may be assigned for a particular metal or alloy. This is important since polarity reversal may occur. For example, at ambient temperatures, zinc is more active than steel and is used to cathodically protect steel; however, above 60°C (such as found in hot water heaters), corrosion products formed on zinc may cause the steel to become anodic and corrode preferentially to zinc.

The magnitude of the potential difference between the dissimilar materials does not indicate the degree of galvanic corrosion because potentials are a function of the thermodynamics and not of the reaction kinetics that occur. The relative surface kinetics of the galvanic member ultimately will determine the severity of galvanic corrosion. A difference of 50 mV between select dissimilar materials can lead to severe corrosion while a potential difference of 800 mV has been successfully coupled (Francis, "Galvanic Corrosion: A Practical Guide for Engineers," NACE International, Houston, TX, 2001).

Corrosion from galvanic effects is usually worse near the couple junction and decreases with distance from the junction. The distance influenced by galvanic corrosion will be affected by solution conductivity. In low-conductivity solutions, galvanic attack may be observed as a sharp localized groove at the junction, while in high-conductivity solutions, galvanic corrosion may be shallow and spread over a relatively long distance. Another important factor in galvanic corrosion is the cathodic-to-anodic area ratio. Since anodic and cathodic currents (in microamperes) must be equal in a corrosion reaction, an unfavorable area ratio is composed of a large cathode and a small anode. The current

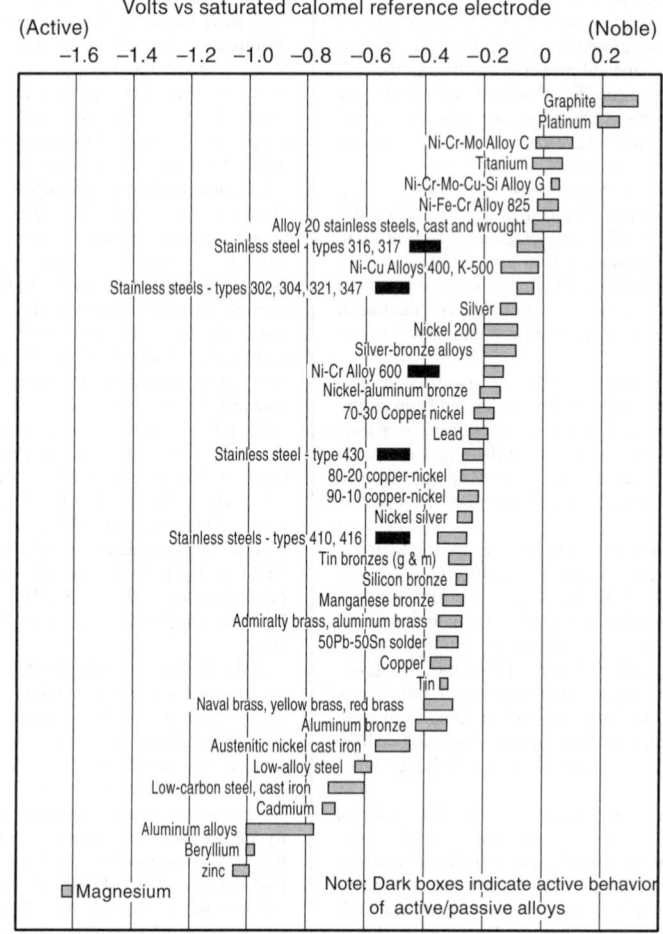

Volts vs saturated calomel reference electrode

Fig. 6.5.4 Galvanic series for seawater, Flowing seawater at 2.4 to 4.0 m/s for 5 to 15 days at 5 to 30°C.

density [in microamperes per square centimeter ($\mu A/cm^2$)] and corrosion rate are greater for the small anode than for the large cathode. Galvanic corrosion rates for an unfavorable cathode-to-anode ratio may be 100 to 1,000 times greater than if the cathode and anode areas were equal. The method by which dissimilar metals or alloys are joined may affect the degree of galvanic corrosion. Welding, in which a gradual transition from one material to another often exists, could react differently in a system where two materials are insulated by a gasket but electrically connected elsewhere through the solution, or differently still in a system connected by fasteners.

Galvanic corrosion may be minimized by (1) selecting combinations of metals or alloys near each other in the galvanic series; (2) avoiding unfavorably large cathode/anode area ratios; (3) completely insulating dissimilar metals whenever possible to disrupt the electric circuit; (4) adding inhibitors to decrease the aggressiveness of the environment; (5) avoiding threaded joints connecting materials far apart in the galvanic series; (6) designing or readily replaceable anodic parts; and (7) adding a third metal that is anodic to both metals or alloys in the galvanic couple [Hack (ed.), "Galvanic Corrosion," STP 978, ASTM, Philadelphia, 1988].

Crevice corrosion is the intensive localized corrosion that may occur because of the presence of narrow openings or gaps between metal-to-metal (under bolts, lap joints, or rivet heads) or nonmetal-to-metal (under gaskets, surface deposits, or dirt) components and small volumes of stagnant solution. Crevice corrosion sometimes is referred to as **underdeposit corrosion** or **gasket corrosion**. Resistance to crevice corrosion

can vary from one alloy/environment system to another. Crevice corrosion may range from near uniform attack to severe localized dissolution at the metal surface within the crevice and may occur in a variety of metals and alloys ranging from relatively noble metals such as silver and copper to active metals such as aluminum and titanium. Stainless steels are susceptible to crevice corrosion. Fibrous gasket materials that draw solution into a crevice by capillary action are particularly effective in promoting crevice corrosion. The environment can be any neutral or acidic aggressive solution, but solutions containing chlorides are particularly conducive to crevice corrosion. In cases where the bulk environment is particularly aggressive, general corrosion may deter localized corrosion at a crevice site.

A condition for crevice corrosion is a crevice or gap that is wide enough to permit solution entry, but is sufficiently narrow to maintain a stagnant zone of solution within the crevice to restrict entry of cathodic reactants and removal of corrosion products through diffusion and migration. Crevice gaps are usually 0.025 to 0.1 mm (0.001 to 0.004 in) wide. The critical depth for crevice corrosion is dependent on the gap width, and it may be only 15 to 40 μm deep. Most cases of crevice corrosion occur in near-neutral solutions in which dissolved oxygen is the cathodic reactant. In seawater, localized corrosion of copper and its alloys (due to differences in Cu^{2+} concentration) is different from that of the stainless steel group because the attack occurs outside of the crevice rather than within. In acidic solutions where hydrogen ion is the cathodic reactant, crevice corrosion actually occurs at the exposed surface near the crevice. Decreasing crevice width or increasing crevice depth, bulk chloride concentration, and/or acidity will increase the potential for crevice corrosion. In some cases, deep crevices may restrict propagation because of the voltage drop through the crevice solution (Lee et al., *Corrosion/83*, Paper 69, NACE International, Houston, TX, 1983).

A condition common to crevice corrosion is the development of an occluded localized environment that differs considerably from the bulk solution. The basic overall mechanism of crevice corrosion is the dissolution of a metal [Eq. (6.5.1)] and the reduction of oxygen to hydroxide ions [Eq. (6.5.4)]. Initially, these reactions take place uniformly, both inside and outside the crevice. However, after a short time, oxygen within the crevice is depleted because of restricted convection, and oxygen reduction ceases in this area. Because the area within the crevice is much smaller than the external area, oxygen reduction remains virtually unchanged. Once oxygen reduction ceases within the crevice, the continuation of metal dissolution tends to produce an excess of positive charge within the crevice. To maintain charge neutrality, chloride or possibly hydroxide ions migrate into the crevice. This results in an increased concentration of metal chlorides within the crevice, which hydrolyzes in water according to Eq. (6.5.8) to form an insoluble hydroxide and a free acid.

$$M^+Cl^- + H_2O \rightarrow MOH\downarrow + H^+Cl^- \qquad (6.5.8)$$

The increased acidity boosts the dissolution rates of most metals and alloys, which increases migration and results in a rapidly accelerating and autocatalytic process. The chloride concentration within crevices exposed to neutral chloride solutions is typically 3 to 10 times higher than in the bulk solution. The pH within the crevice is 2 to 3. The pH drop with time and critical crevice solution that will cause breakdown in stainless steels can be calculated [Oldfield and Sutton, *Brit. Corrosion J.,* **13**, no. 13, 1978; IJsseling (ed.), "Survey of Literature on Crevice Corrosion (1979–1998)," European Federation of Corrosion Publications, no. 30, IOM Communications, London, 2000].

Optimum crevice corrosion resistance can be achieved with an active/passive metal that possesses (1) a narrow active/passive transition, (2) a small critical current density, and (3) an extended passive region. Crevice corrosion may be minimized by avoiding riveted, lap, or bolted joints in favor of properly welded joints, designing vessels for complete drainage and removal of sharp corners and stagnant areas, removing deposits frequently, and using solid, **nonabsorbent gaskets** (such as Teflon) wherever possible.

Pitting is a form of extremely localized attack that results in a cavity or hole in the metal or alloy. Deterioration by pitting is one of the most dangerous types of localized corrosion, but its unanticipated occurrences and propagation rates are difficult to consider in practical engineering designs. Pits are often covered by corrosion products. Depth of pitting is sometimes expressed by the **pitting factor,** which is the ratio of deepest metal penetration to average metal penetration as determined by weight loss measurements. A pitting factor of 1 denotes uniform corrosion.

Pitting is usually associated with the breakdown of a passive metal. Breakdown involves the existence of a critical potential E_b, induction time at a potential $> E_b$, presence of aggressive species (Cl^-, Br^-, ClO_4^-, etc.), and discrete sites of attack. Pitting is associated with local imperfections in the passive layer. Nonmetallic inclusions, second-phase precipitates, grain boundaries, scratch lines, dislocations, and other surface inhomogeneities can become initiation sites for pitting (Szklarska-Smialowska, "Pitting of Metals," NACE International, Houston, TX, 1986; Szklarska-Smialowska, "Pitting and Crevice Corrosion," NACE International, Houston TX, 2005). The induction time before pits initiate may range from days to years. Induction time depends on the metal, aggressiveness of the environment, and the potential, and tends to decrease if either the potential or the concentration of aggressive species increases.

Once pits are initiated, they continue to grow through an autocatalytic process (i.e., the corrosion conditions within the pit are both stimulating and necessary for continuing pit growth). Pit growth is controlled by the rate of depolarization at the cathodic areas. Oxygen in aqueous environments and ferric chloride in various industrial environments are effective depolarizers. The propagation of pits is virtually identical to crevice corrosion. Factors contributing to localized corrosion such as crevice

corrosion and pitting of various materials are found in other publications "Localized Corrosion—Cause of Metal Failure," STP 516, ASTM, Philadelphia, 1972; Brown et al. (eds.), "Localized Corrosion," NACE-3, NACE International, Houston, TX, 1974; Isaacs et al., "Advances in Localized Corrosion," NACE-9, NACE International, Houston, TX, 1990; Frankel and Newman (eds.), "Critical Factors in Localized Corrosion," PV 92-9, The Electrochemical Society, Pennington, NJ, 1992; Natishan et al. (eds.), "Critical Factors in Localized Corrosion II," PV 95-15, The Electrochemical Society, Pennington, NJ, 1996; Natishan et al. (eds.), "Critical Factors in Localized Corrosion III, A symposium in Honor of the 70th Birthday of Jerome Kruger," PV 98-17, The Electrochemical Society, Pennington, NJ, 1999; Virtanen et al. (eds.), "Critical Factors in Localized Corrosion IV: A Symposium in Honor of the 65th Birthday of Hans Bohni," PV 2002-24, The Electrochemical Society, Pennington, NJ, 2002].

Copper and its alloys suffer localized corrosion by nodular pitting in which the attacked areas are covered with small mounds or nodules composed of corrosion product and $CaCO_3$ precipitated from the water. The pit interior is covered with CuCl which prevents formation of a protective Cu_2O layer. Pitting occurs most often in cold, moderately hard to hard waters, but also happens in soft waters above 60°C. Pitting is associated with the presence of very thin carbon (cathodic) films from lubricants used during tube manufacture. Copper pitting is favored by a high sulfate/chloride ratio. Copper-nickel tubing suffers from accelerated corrosion in seawater due to the presence of both sulfides and oxygen. Sulfides prevent the formation of a protective oxide layer, which allows the anodic reaction to proceed unabated supported by the oxygen reduction reaction.

Caustic corrosion, frequently referred to as **caustic gouging** or **ductile gouging,** is a form of localized corrosion that occurs as a result of fouled heat-transfer surfaces of water-cooled tubes and the presence of an active corrodent in the water of high-pressure boilers. Porous deposits that form and grow in these high-heat-input areas produce conditions that allow sodium hydroxide to permeate the deposits by a process called **wick boiling,** and concentrate to extremely high levels (e.g., 100 ppm bulk water concentrated to 220,000 ppm NaOH in water film under these deposits) which dissolves the protective magnetite and then reacts directly with the steel to cause rapid corrosion. A smooth, irregular thinning occurs, usually downstream of flow disruptions or in horizontal or inclined tubing (Port and Herro, "The Nalco Guide to Boiler Tube Analysis," McGraw-Hill, New York, 1991, pp. 57–70; Shifler and Wilkes, Historical Perspectives of Intergranular Corrosion of Boiler Metal by Water. *Corrosion/93,* Paper 44, NACE International, Houston, TX, 1993).

Intergranular corrosion is localized attack that follows a narrow path along the grain boundaries of a metal or an alloy. Intergranular corrosion is caused by the presence of impurities, precipitation of one of the alloying elements, or depletion of one of these elements at or along the grain boundaries. The driving force of intergranular corrosion is the difference in corrosion potential that exists between a thin grain boundary zone and the bulk of immediately adjacent grains. Intergranular corrosion most commonly occurs in aluminum or copper alloys and austenitic stainless steels.

When austenitic stainless steels such as 304 are heated or placed in service at temperatures of 950 to 1450°C, they become sensitized and susceptible to intergranular corrosion, because chromium carbide ($Cr_{26}C_6$) is formed. This effectively removes chromium from the vicinity of the grain boundaries, thereby approaching the corrosion resistance of carbon steel in these chromium-depleted areas. Methods to avoid **sensitization** and control intergranular corrosion of austenitic stainless steels (1) employ high-temperature heat treatment (solution quenching) at 1950 to 2050°C, (2) add elements that are stronger carbide formers [stabilizers such as columbium (niobium), niobium plus tantalum, or titanium] than chromium, and (3) lowering the carbon content below 0.03 percent. Stabilized austenitic stainless steels can experience intergranular corrosion if heated above 2250°C, where columbium carbides dissolve, as may occur during welding. Intergranular corrosion of stabilized stainless steels is called **knife-line attack.** This attack transpires in a narrow band in the parent metal adjacent to a weld. Reheating the steel to around 1950 to 2250°C reforms the stabilized carbides.

Surface carburization also can cause intergranular corrosion in cast austenitic stainless steel. Overaging of aluminum alloys can form precipitates such as $CuAl_2$, $FeAl_3$, $MgAl_3$, $MgZn_2$, and $MnAl_6$ along grain boundaries and promote intergranular corrosion. Some nickel alloys can form intergranular precipitates of carbides and intermetallic phases during heat treatment or welding. Alloys with these intergranular precipitates are subject to intergranular corrosion.

Dealloying involves the selective removal of the most electrochemically active constituent metal in the alloy. Selective removal of one metal can result in either localized attack leading to possible perforation (plug dealloying) or a more uniform attack (layer dealloying) resulting in loss of component strength. Although selective removal of metal such as Al, Fe, Si, Co, Ni, and Cr from their alloys has occurred, the most common dealloying is the selective removal of zinc from brasses, called **dezincification.** Dezincification of brasses (containing 15 percent more zinc) takes place in soft waters, especially if the carbon dioxide content is high. Other factors that increase the susceptibility of brasses to dealloying are high temperatures, waters with high chloride content, low water velocity, crevices, and deposits on the metal surface. Dezincification is experienced over a wide pH range. Layer dezincification is favored when the environment is acidic and the brass has a high zinc content. Plug dezincification tends to develop when the environment is slightly acidic, neutral, or alkaline and the zinc content is relatively low. It has been proposed either that zinc is selectively dissolved from the alloy, leaving a porous structure of metallic copper in situ within the brass lattice, or that both copper and zinc are dissolved initially, but copper immediately redeposits at sites close to where the brass dissolved.

Dezincification can be minimized by reducing oxygen in the environment or by cathodic protection. Dezincification of α-brass (70 percent Cu, 30 percent Zn) can be minimized by alloy additions of 1 percent tin. Small amounts (~0.05 percent) of arsenic, antimony, or phosphorus promote an effective inhibitor of dezincification for 70/30 α-brass (inhibited Admiralty brass). Brasses containing less than 15 percent zinc have not been reported to suffer dezincification.

Gray cast iron selectively leaches iron in mild environments, particularly in underground piping systems. Graphite is cathodic to iron, which sets up a galvanic cell. The dealloying of iron from gray cast iron leaves a porous network of rust, voids, and graphite termed **graphitization** or **graphitic corrosion.** The cast iron loses both its strength and its metallic properties, but experiences no apparent dimensional changes. Graphitization does not occur in nodular or malleable cast irons because a continuous graphite network is not present. White cast iron also is immune to graphitization since it has essentially no available free carbon.

Dealloying also can occur at high temperatures. Exposure of stainless steels to low-oxygen atmospheres at 1800°F (980°C) results in the selective oxidation of chromium and the creation of a more protective scale, and the depletion of chromium under the scale in the substrate metal. Decarburization can occur in carbon steel tubing if heated above 1,600°F (870°C). Denickelification of austenitic stainless steels occurs in liquid sodium.

Corrosion rates are often dependent on fluid flow and availability of appropriate species required to drive electrochemical reactions. An increase in fluid flow may increase corrosion rates by removing protective films or increasing the diffusion or migration of deleterious species. An increase in fluid flow may also decrease corrosion rates by eliminating aggressive ion concentration or enhancing passivation or inhibition by transporting the protective species to the fluid/metal interface. There are a number of mechanisms caused by the conjoint action of flow and corrosion that result in several types of flow-influenced corrosion: (1) mass transport-controlled corrosion, (2) phase transport-controlled corrosion, (3) erosion corrosion, and (4) cavitation (Heitz, *Corrosion*, **47**, no. 135, 1991; also *Corrosion/90*, Paper no. 1, NACE International, Houston, TX, 1990). Mass transport-controlled corrosion implies that the rate of corrosion is dependent on the convective mass transfer processes at the metal/fluid interface. Corrosion by mass transport will often be streamlined and smooth. Phase transport-controlled corrosion suggests that the wetting of the metal surface by a corrosive phase is flow-dependent. This may occur from the separation of one liquid phase from another or by the

formation of a second phase from a liquid. An example of the second mechanism is the formation of discrete bubbles or a vapor phase from boiler water in horizontal or inclined tubes in high-heat-flux areas under low-flow conditions (departure from nucleate boiling). The distribution and morphology of the corrosive attack will be related to the distribution of the corrosive phase—the corroded sites will frequently display rough, irregular surfaces and be coated with thick, porous corrosion deposits (Shifler, Flow-Influenced Corrosion in a Waste Energy Sparge Line, *Mater. Perf.* **37**, 1998, pp. 38–44).

Erosion corrosion is the acceleration or increase of attack because of the relative movement between a corrosive fluid and the metal surface. Mechanically, the conjoint action of erosion and corrosion damages not only the protective film but also the metal or alloy surface itself by the abrasive action of a fluid (gas or liquid) at high velocity. The inability to maintain a protective passive film on the metal or alloy surface raises the corrosion rate. Erosion corrosion is characterized by grooves, gullies, waves, rounded holes, and valleys, usually exhibiting in a directional manner. Components exposed to moving fluids and subject to erosion corrosion include bends, elbows, tees, valves, pumps, blowers, propellers, impellers, agitators, condenser tubing, orifices, turbine blades, nozzles, wear plates, baffles, boiler tubes, and ducts.

The hardness of the alloy or metal and/or the mechanical or protective nature of the passive film will affect the resistance to erosion corrosion in a particular environment. Copper and brasses are particularly susceptible to erosion corrosion. Models and experimental results have shown that flow stream velocity and increases of particle size, sharpness, density, and concentration increase the erosion corrosion rate. Increases in fluid viscosity, density, target material hardness, and/or pipe diameter decrease the corrosion rate. Often, an escalation in the fluid velocity increases the attack by erosion corrosion. The effect may be nil until a critical velocity is reached, above which attack may increase very rapidly. The critical velocity is dependent on the alloy, passive layer, temperature, and nature of the fluid environment. The increased attack may be attributed, in part, to increasing cathodic reactants to the metal surface. Increased velocity may decrease attack by increasing the effectiveness of inhibitors or by preventing silt or dirt from depositing and causing crevice corrosion.

Erosion corrosion includes impingement, cavitation, and fretting corrosion. **Impingement attack** occurs when a fluid strikes the metal surface at high velocity at directional change sections such as bends, tees, turbine blades, and inlets of pipes or tubes. In the majority of cases involving impingement attack, a geometric feature of the system results in turbulence. Air bubbles or solids present in turbulent flow will accentuate attack.

Cavitation damage is caused by the formation and collapse of vapor bubbles in a liquid near a metal surface. Cavitation occurs where high-velocity water and hydrodynamic pressure changes are encountered, such as in ship propellers or pump impellers. If pressure on a liquid is lowered drastically by flow divergence, the water vaporizes and forms bubbles. These bubbles generally collapse rapidly, producing shock waves with pressures as high as 60,000 lb/in² (410 MPa), which destroys the passive film. The newly exposed metal surface corrodes, reforms a passive film that is destroyed again by another cavitation bubble, and so forth. As many as 2,000,000 bubbles may collapse over a small area in 1 s. Cavitation damage causes both corrosion and mechanical effects.

Fretting corrosion is a form of damage caused at the interface of two closely fitting surfaces under load when subjected to vibration and slip. Fretting is explained as:(1) the surface oxide is ruptured at localized sites prompting reoxidation; or (2) material wear and friction cause local oxidation. The degree of plastic deformation is greater for softer materials than for hard; seizing and galling may often occur. Lubrication, increased hardness, increased friction, and the use of gaskets can reduce fretting corrosion.

Environmentally induced cracking is a brittle fracture of an otherwise ductile material due to the conjoint action of tensile stresses in a specific environment over a period of time. Environmentally induced cracking includes (1) stress corrosion cracking, (2) hydrogen damage, (3) corrosion fatigue, (4) liquid-metal embrittlement, and (5) solid-metal induced embrittlement.

Stress corrosion cracking (SCC) refers to service failures due to the joint interaction of tensile stress with a specific corrodent and a susceptible normally ductile material. The observed crack propagation is the result of mechanical stress and corrosion reactions. Stress can be externally applied, but most often is the result of residual tensile stresses. SCC can initiate and propagate with little outside evidence of corrosion or macroscopic deformation and usually provides no warning of impending failure. Cracks often initiate at surface flaws that were preexisting or formed during service. Branching is frequently associated with SCC; the cracks tend to be very fine and tight and visible only with special techniques or by microscopic examination. SCC can be intergranular or transgranular. Intergranular SCC is often associated with grain boundary precipitation or grain boundary segregation of alloy constituents such as found in sensitized austenitic stainless steels. Transgranular SCC is related to crystal structure, anisotropy, grain size and shape, dislocation density and geometry, phase composition, yield strength, ordering, and stacking fault energies. The specific corrodents of SCC for different alloys include the following: (1) carbon steels—sodium hydroxide; (2) stainless steels—NaOH or chlorides; (3) α-brass—ammoniacal solutions; (4) aluminum alloys—aqueous Cl^-, Br^-, I^- solutions; (5) titanium alloys—aqueous Cl^-, Br^-, I^- solutions, organic liquids; (6) high nickel alloys—high-purity steam [Gangloff and Ives (eds.), "Environment-Induced Cracking of Metals," NACE-10, NACE International, Houston, TX, 1990; Bruemmer et al. (eds.), "Parkins Symposium on Fundamental Aspects of Stress Corrosion Cracking," The Minerals, Metals, & Materials Society, Warrendale, PA, 1992; Jones, "Stress Corrosion Cracking: Materials Performance and Evaluation," ASM International, Materials Park, OH, 1992; Jones (ed.), "Chemistry and Electrochemistry of Corrosion and Stress Corrosion Cracking: A Symposium Honoring the Contributions of R. W. Staehle." The Minerals, Metals, & Materials Society, Warrendale, PA, 2001].

Hydrogen interactions can affect most engineering metals and alloys. Terms to describe these interactions are not universally agreed upon. Hydrogen damage or hydrogen attack is caused by the diffusion of atomic hydrogen through carbon and low-alloy steels. Hydrogen reacts with carbon, either in elemental form or as carbides, to form methane gas. Methane accumulates at grain boundaries and can cause local pressures to exceed the yield strength of the material. Hydrogen damage causes intergranular, discontinuous cracks and decarburization of the steel. Once cracking occurs, the damage is irreversible. In utility boilers, hydrogen damage has been associated with fouled heat-transfer surfaces. Hydrogen damage is temperature-dependent with a threshold temperature of about 400°F (204°C). Hydrogen damage has been observed in boilers operating at pressures of 450 to 2700 psi (3.1 to 18.6 MPa) and tube metal temperatures of 600 to 950°F (316 to 510°C). Hydrogen damage may occur in either acidic or alkaline conditions, although this contention is not universally accepted.

Hydrogen embrittlement results from penetration and adsorption of atomic hydrogen into an alloy matrix. This process results in a decrease of toughness and ductility of alloys due to hydrogen ingress. Hydrogen may be removed by baking at 200 to 300°F (93 to 149°C), which returns the alloy mechanical properties very near to those existing before hydrogen entry. Hydrogen embrittlement can affect most metals and alloys. Hydrogen blistering results from atomic hydrogen penetration at internal defects near the surface such as laminations or nonmetallic inclusions where molecular hydrogen forms. Pressure from H can cause local plastic deformation of surface exfoliation [Moody and Thompson (eds.), "Hydrogen Effects on Materials Behavior," The Minerals, Metals, & Materials Society, Warrendale, PA, 1990; Shifler and Wilkes, Historical Perspectives of Intergranluar Corrosion of Boiler Metal by Water, *Corrosion/93,* Paper 44, NACE International, Houston, TX, 1993].

For higher-alloyed steels, the higher the strength of the alloy, the more susceptible the alloy will be to internal hydrogen embrittlement and to hydrogen environmental embrittlement. Hydrogen solubility in an ultrahigh-strength steel (UHHS)/environment system is controlled by trapping and the binding energies of the particular trap sites. Understanding hydrogen interactions and transport in UHSS microstructures is necessary to control hydrogen embrittlement (Li et al.,

Hydrogen Trap States in Ultrahigh-Strength AERMET 100 Steel, *Metallurg. Mater. Trans. A,* **35A,** no. 849, 2004; Thomas et al., Internal Hydrogen Embrittlement of Ultrahigh-Strength AERMET 100 Steel, *Metallurg. Mater. Trans. A,* **34A,** no. 327, 2003).

When a metal or an alloy is subjected to cyclic, changing stresses, cracking and fracture can occur through fatigue at stresses lower than the yield stress. A clear relationship exists between stress amplitude and number of cyclic loads; an endurance limit is the stress level below which fatigue will not occur indefinitely. Fracture solely by fatigue is not viewed as an example of environmental cracking. However, the conjoint action of a corrosive medium and cyclic stresses on a material is termed **corrosion fatigue.** Corrosion fatigue failures are usually "thick-walled" brittle ruptures showing little local deformation. An endurance limit often does not exist in corrosion fatigue failures. Cracking often begins at pits, notches, surface irregularities, welding defects, or sites of intergranular corrosion. Surface features of corrosion fatigue cracking vary with alloys and specific environmental conditions. Crack paths can be transgranular (carbon steels, aluminum alloys, etc.) or intergranular (copper, copper alloys, etc.); exceptions to the norm are found in each alloy system. "Oyster shell" markings may denote corrosion fatigue failures, but corrosion products and corrosion often cover and obscure this feature. Cracks usually propagate in the direction perpendicular to the principal tensile stress; multiple, parallel cracks are usually present near the principal crack which led to failure. Applied cyclic stresses are caused by mechanical restraints or thermal fluctuations; susceptible areas may also include structures that possess residual stresses from fabrication or heat treatment. Low pH, and high stresses, stress cycles, and oxygen content or corrosive species concentrations will increase the probability of corrosion fatigue cracking [McEvily and Staehle (eds.), "Corrosion Fatigue-Mechanics, Metallurgy, Electrochemistry, and Engineering," STP 801, ASTM, Philadelphia, 1983; [McEvily and Staehle (eds.), "Corrosion Fatigue-Chemistry, Mechanics, and Microstructure," NACE International, Houston, TX, 1986].

Liquid-metal embrittlement (LME) or liquid metal–induced embrittlement is the catastrophic brittle fracture of a normally ductile metal when coated by a thin film of liquid metal and subsequently stressed in tension. Cracking may either be intergranular or transgranular. Usually, a solid that has little or no solubility in the liquid and forms no intermetallic compounds with the liquid constitutes a couple with the liquid. Fracture can occur well below the yield stress of the metal or alloy (Jones, Eng. Failure Anal. **1,** no. 51, 1994). A partial list of examples of LME includes (1) aluminum (by mercury, gallium, indium, tin-zinc, lead-tin, sodium, lithium, and inclusions of lead, cadmium, or bismuth in aluminum alloys); (2) copper and select copper alloys (mercury, antimony, cadmium, lead, sodium, lithium, and bismuth); (3) zinc (mercury, gallium, indium, Pb-Sn solder); (4) titanium or titanium alloys (mercury, zinc, and cadmium); (5) iron and carbon steels (aluminum, antimony, cadmium, copper, gallium, indium, lead, lithium, zinc, and mercury); and (6) austenitic and nickel-chromium steels (tin and zinc).

Solid metal–induced embrittlement (SMIE) occurs below the melting temperature T_m of the solid in various LME couples. Many instances of loss of ductility or strength, and brittle fracture have taken place with electroplated metals or coatings and inclusions of low-melting alloys below T_m. Cadmium-plated steel, leaded steels, and titanium embrittled by cadmium, silver, or gold are materials affected by SMIE. The prerequisites for SMIE and LME are similar, although multiple cracks are formed in SMIE and SMIE crack propagation is 100 to 1,000 times slower than LME cracking. More detailed descriptions of LME and SMIE are given elsewhere [Kambar (ed.), "Embrittlement of Metals," American Institute of Mining, Metallurgical and Petroleum Engineers, St. Louis, MO, 1984].

High-temperature corrosion may occur in many environments and is affected by numerous factors such as temperature, alloy or protective coating composition, time, and gas composition. The rate by which a material may degrade in a high-temperature environment will depend on the reaction or range of operative reactions, the reaction thermodynamics, kinetics, and chemical physical properties of the material exposed to this environment. A material can degrade at high temperatures upon exposure to gaseous, liquid (molten ash, salts, or metal/alloys), or solid-state reaction processes such as diffusion and interdiffusion, precipitation, and creep.

High-temperature corrosion/degradation of materials may occur through a number of potential processes: (1) oxidation, (2) carburization and metal dusting, (3) nitridation, (4) sulfidation, (5) hot corrosion, (6) chloridation and other halogenization reactions, (7) hydrogen interactions, (8) molten metals, (7) molten salts, (8) aging reactions such as sensitization, (9) creep, and (10) erosion corrosion and wear. Various materials (alloys, high-temperature coatings, refractories, glasses) are exposed to high-temperature environments and experience varying degrees of degradation (Pond and Shifler, High Temperature Corrosion Failures, in vol. 11: "Failure-Analysis, Metals Handbook," ASM International, Materials Park, OH, 2002; Shifler, High Temperature Gaseous Corrosion Testing, vol. 13A, "Corrosion Principles, ASM Handbook," ASM International, Materials Park, OH, 2003).

Oxide growth advances with time by parabolic, linear, cubic, or logarithmic rates. The corrosion rate can be influenced considerably by the presence of contaminants, particularly if they are adsorbed on the metal surface. Each alloy has an upper temperature limit at which the oxide scale, based on its chemical composition and mechanical properties, cannot protect the underlying alloy. *Catastrophic oxidation* refers to metal-oxygen reactions that occur at continuously increasing rates or break away from protective behavior very rapidly. Molybdenum, tungsten, vanadium, osmium, rhenium, and alloys containing Mo or V may oxidize catastrophically. Internal oxidation may take place in copper- or silver-based alloys containing small levels of Al, Zn, Cd, or Be when more stable oxides are possible below the surface of the metal/scale interface. Sulfidation forms sulfide phases with less stable-base metals, rather than oxide phases, when H_2S, S_2, SO_2, and other gaseous sulfur species have a sufficiently high concentration. In corrosion-resistant alloys, sulfidation usually occurs when Al or Cr ties up with sulfides which interferes with the process of developing an oxide. Carburization of high-temperature alloys is possible in reducing carbon-containing environments. Hot corrosion involves the high-temperature, liquid-phase attack of alloys by eutectic alkali-metal sulfates that solubilize a protective alloy oxide, and exposes the bare metal, usually iron, to oxygen which forms more oxide and promotes subsequent metal loss at temperatures of 1,000 to 1,300°F (538 to 704°C). Vanadium pentoxide and sodium oxide can also form low-melting-point [~1,000°F (538°C)] liquids capable of causing alloy corrosion. These are usually from contaminants in fuels, either coal or oil.

Alloy strength decreases rapidly when metal temperatures approach 900 to 1,000°F and above. Creep entails a time-dependent deformation involving grain boundary sliding and atom movements at high temperatures. When sufficient strain has developed at the grain boundaries, voids and microcracks develop, and grow and coalesce to form larger cracks until failure occurs. Creep rate will increase and the projected time to failure will decrease when stress and/or tube metal temperature is increased [Rapp (ed.), "High Temperature Corrosion," NACE-10, NACE International Houston, TX, 1983; Lai, "High Temperature Corrosion of Engineering Alloys," ASM International, Materials Park, OH, 1990; Gleeson, High-Temperature Corrosion of Metallic Alloys and Coatings, in Schütze (ed.), "Corrosion and Environmental Degradation of Materials," vol. 19 of the series "Materials Science and Technology," Wiley-VCH, Germany, 2000, pp. 173–228; Khanna, "Introduction to High Temperature Oxidation and Corrosion," ASM International, Materials Park, OH, 2002; Tortovelli et al. (eds.), "John Stringer Symposium on High Temperature Corrosion," ASM International, Materials Park, OH, 2003; Opila et al., "High Temperature Corrosion and Materials Chemistry IV," PV 2003-12, The Electrochemical Society, Pennington, NJ, 2003; Edwards et al., "Proceedings of the Workshop on Materials and Practice to Improve Resistance to Fuel Derived Environmental Damage in Land and Sea Based Turbines." October 22–23, 2002, EPRI Publication, Electric Power Research Institute, Palo Alto, CA, 2003].

CORROSION TESTING

Corrosion testing can reduce cost, improve safety, and conserve resources in industrial, commercial, and personal components, processes, and applications. Corrosion evaluation includes laboratory, pilot-plant, and field monitoring and testing. The selection of materials for corrosion resistance in specific industrial services is best effected by field testing and actual service performance. However, uncontrollable environmental factors may skew precise comparison of potential alloys or metals, and test times may be unrealistically long. Accelerated corrosion tests in the laboratory can provide practical information to assess potential material candidates in a given environment, predict the service life of a product or component, evaluate new alloys and processes, assess the effects of environmental variations or conditions on corrosion and corrosion control methods, provide quality control, and study corrosion mechanisms.

Careful planning is essential to obtain meaningful, representative, and reliable test results. Metallurgical factors, environmental variables, statistical treatment, and proper interpretation and correlation of accelerated test results to actual field conditions are considerations in planning and conducting corrosion tests. The chemical composition, fabrication and metallurgical history, identification, and consistency of materials should be known. Although, ideally, the surface condition of the test specimen should mirror the plant conditions, a standard test condition should be selected and maintained. This often requires prescribed procedures for cutting, surface finishing, polishing, cleaning, chemical treatment or passivation, measuring and weighing, and handling. The exposure technique should provide easy assessment of the environment to the sample and should not directly cause conditions for test corrosion (e.g., crevice corrosion) unrelated to field conditions.

The type, interval, and duration of the tests should be seriously considered. Corrosion rates of the material may decrease with time because of the formation of protective films or the removal of a less resistant surface metal. Corrosion rates may increase with time if corrosion-inducting salts or scales are produced or a more resistant metal is removed. The environment may change during testing because of the decrease or increase in concentration of a corrosive species or inhibitor, or the formation of autocatalytic products or other metal-catalyzed variations in the solution. These test solution variations should be recognized and related to field condition changes during time, if possible. Testing conditions should prescribe the presence or absence of oxygen in the system. Aeration, or the presence of oxygen, can influence the corrosion rate of an alloy in solution. Aeration may increase or decrease the corrosion rate based on the influence of oxygen in the corrosion reaction or the nature of the passive films which develop on different metals and alloys. Temperature is an extremely important parameter in corrosion reactions. The temperature at the specimen surface should be known; a rough rule of thumb is that the corrosion rate doubles for each 10°C rise in temperature (Shifler and Aylor, Considerations for the Testing of Materials and Components in Seawater, *Corrosion/2002*, Paper 217, Houston, TX, 2002).

Accelerated laboratory corrosion tests are used to predict corrosion behavior when service history is unavailable and cost and time prohibit simulated field testing. Laboratory tests also can provide screening evaluations before field testing. Such tests include nonelectrochemical and sophisticated electrochemical tests. Nonelectrochemical laboratory techniques include immersion and various salt spray tests to evaluate the corrosion of ferrous and nonferrous metals and alloys, as well as the degree of protection afforded by both organic and inorganic coatings. Since these are accelerated tests by design, the results must be interpreted cautiously. **Standardized test methods** are very effective for both specifications and routine tests used to evaluate experimental or candidate alloys, inhibitors, coatings, and other materials. Many such tests are described in the "Annual Book of ASTM Standards" (vol. 3.02, "Metal Corrosion, Erosion, and Wear") and in standard test methods from the NACE International Book of Standards.

Electrochemical techniques are attractive because they allow a direct method of accelerating corrosion processes without changing the environment, can be used as a nondestructive tool to evaluate corrosion rates, and offer in situ (field) or ex situ (laboratory) investigations. Most typical forms of corrosion can be investigated by electrochemical techniques. Electrochemical tests may include linear polarization resistance, ac electrochemical impedance, electrochemical noise, cyclic voltammetry, cyclic potentiodynamic polarization, potentiodynamic and potentiostatic polarization, and scratch-repassivation. AC impedance has been used to monitor coating integrity and evaluate the effectiveness of cathodic protection and coating systems in underground piping. There are complications involved in various electrochemical test methods which must be resolved or eliminated before interpretation is possible. The methods used above employ a potentiostat, an instrument that regulates the electrode potential and measures current as a function of potential. A galvanostat varies potential as a function of current.

Corrosion coupons are generally easy to install, reflect actual environmental conditions, and can be used for long exposure tests. Coupons can evaluate inhibitor programs and are designed to measure specific forms of corrosion. A simple embrittlement detector covered by ASTM D807 permits concentration of boiler water on a stressed steel specimen to detect caustic cracking. Field coupon testing has several limitations: (1) It cannot detect brief process upsets; (2) it cannot guarantee initiation of localized corrosion rates before the coupons are removed; (3) it cannot directly correlate the calculated coupon corrosion rate to equipment or component corrosion; and (4) it cannot detect certain forms of corrosion. Corrosion rate sensors created to monitor the corrosion of large infrastructural components in real time have been developed for aqueous, soil, and concrete environments using linear polarization resistance (Ansuini et al., *Corrosion/95*, Paper 14, NACE International, Houston, TX, 1995). Ultrasonic thickness, radiography, or eddy current measurements can evaluate corrosion rates in situ. The component surfaces must be clean (free of dirt, paint, corrosion products) to make measurements. Ultrasonic attenuation techniques have been applied to evaluate the extent of hydrogen damage. A noncontact nondesctructive inspection technique for pitting, SCC, and crevice corrosion has been developed using digital speckle correlation (Jin and Chiang, *Corrosion/95*, Paper 535, NACE International, Houston, TX, 1995). ASTM [Baboian (ed.), "Manual 20—Corrosion Tests and Standards: Application and Interpretation," ASTM, Philadelphia, 1995; Baboian (ed.), "Manual 20—Corrosion Tests and Standards: Application and Interpretation," ASTM, International, West Conshohocken, PA, 2005] provides a detailed comprehensive reference for field and laboratory testing in different environments of metals and alloys, coatings, and composites. It includes testing for different corrosion forms and for different industrial applications.

CORROSION PROTECTION METHODS

The basis of **corrosion protection** methods involves restricting or controlling anodic and/or cathodic portions of corrosion reactions, changing the environment variables, or breaking the electrical contact between anodes and cathodes. Corrosion protection methods comprise (1) proper materials selection, (2) design, (3) coatings, (4) use of inhibitors, (5) anodic protection, and (6) cathodic protection [Baboian (ed.), "NACE Corrosion Engineer's Reference Book," 3d ed., NACE International, Houston, TX, 2002].

The most common way to reduce corrosion is to select the right material for the environmental conditions and applications. Ferrous and nonferrous metals and alloys, thermoplastics, nonmetallic linings, resin coatings, composites, glass, concrete, and nonmetal elements are a few of the potential materials available for selection (Mangonon, "The Principles of Materials Selection for Engineering Design," Prentice-Hall, New York, 1999). Experience and data, either in-house or from outside vendors and fabricators, may assist in the materials selection process. Many materials can be eliminated by service conditions (temperature, pressure, strength, chemical compatibility). Corrosion data for a particular chemical or environment may be obtained through a literature survey ("Corrosion Abstracts," NACE International; "Corrosion Data Survey—Metals Section," 6th ed., 1985; "Handbook of Corrosion Data," 2d ed., ASM International, Materials Park, OH, 1994; "Corrosion Data Survey—Nonmetals Section," 5th ed., NACE International, Houston, TX, 1975) or from expert systems or databases. Different organizations (ASTM, ASM International, ASME, NACE International) may offer references to help solve problems and make predictions about the corrosion behavior of candidate materials. The material should be cost-effective and resistant to the various forms of corrosion (discussed earlier) that may be encountered in its application. General rules may be applied to determine resistance of metals and alloys. For reducing or nonoxidizing environments (such as air-free acids and aqueous solutions) nickel, copper, and their alloys are usually satisfactory. For oxidizing conditions, chromium alloys are used, while in extremely powerful oxidizing environments, titanium and its alloys have shown superior resistance. The corrosion resistance of chromium and titanium can be enhanced in hot, concentrated, oxidizer-free acid by small additions of platinum, palladium, or rhodium.

Many costs related to corrosion could be eliminated by **proper design.** The designer should have a credible knowledge of corrosion or should work in cooperation with a materials and corrosion engineer. The design should avoid gaps or structures where dirt or deposit could easily form crevices; horizontal faces should slope for easy drainage. Tanks should be properly supported and employ the use of drip skirts and dished, fatigue-resistant bottoms ("Guidelines for the Welded Fabrication of Nickel Alloys for Corrosion-Resistant Service," Pt. 3, Nickel Development Institute, Toronto, Canada, 1994). Connections should avoid dissimilar metals or alloys widely separated in the galvanic series for lap joints or fasteners, if possible. Welded joints should avoid crevices, microstructural segregation, and high residual or applied stresses. Positioning of parts should account for prevailing environmental conditions. The design should avoid local differences in concentration, temperature, and velocity or turbulence. All parts requiring maintenance or replacement must be easily accessible. All parts to be coated must be accessible for the coating application.

Metal and alloys degrade by several known corrosion mechanisms. Relatively thin coatings of metallic, organic, and inorganic materials can provide an effective barrier between metals/alloys and their environment. **Protective coatings** include metallic, chemical conversion, inorganic nonmetallic, and organic coatings. Before the application of an effective coating on metals and alloys, it is necessary to clean the surface carefully to remove, dirt, grease, salts, and oxides such as mill scale and rust. Metallic coatings may be applied by cladding, electrodeposition (copper, cadmium, nickel, tin, chromium, silver, zinc, gold), hot dipping (tin, zinc/galvanizing, lead, and aluminum), diffusion (chromium/chromizing, zinc/sherardizing, boron, silicon, aluminum/calorizing or aluminizing, tin, titanium, molybdenum), metallizing or flame spraying (zinc, aluminum, lead, tin, stainless steels), vacuum evaporation (aluminum), ion implantation, and other methods. All commercially prepared coatings are porous to a degree. Corrosion or galvanic action may influence the performance of the metal coating. Noble (determined by the galvanic series) coatings must be thicker and have a minimum of pores and small pore size to delay entry of a deleterious fluid. Some pores of noble metal coatings are filled with an organic lacquer, or a second, lower-melting metal is diffused into the initial coating. Sacrificial or cathodic coatings such as cadmium, zinc, and in some environments, tin and aluminum, cathodically protect the base metal. Porosity of sacrificial coatings is not critical as long as cathodic protection of the base metal continues. Higher solution conductivity allows larger defects in the sacrificial coating; thicker coatings provide longer times of effective cathodic protection.

Conversion coatings are produced by electrochemical reaction of the metal surface to form adherent protective corrosion products. Anodizing aluminum forms a protective film of Al_2O_3. Phosphate and chromate treatments provide temporary corrosion resistance and usually provide a base for painting. Chromates are extremely effective and widely used as corrosion inhibitors for high-strength Al alloys in aerospace applications. The combined properties of storage, release, migration, and irreversible reduction provided by chromate coatings underlie their outstanding corrosion protection. Of special note is the action of Cr^{VI} as a "suicidal inhibitor" in which reduction to Cr^{III} locks in the inhibiting film by converting mobile Cr^{VI} to insoluble form and substituting inert Cr^{III}, as discovered in the university research program for the U.S. Air Force (Frankel, Mechanism of Al Alloy Corrosion and the Role of Chromate Inhibitors, AFOSR contact no. F49620-96-1-0479, Final Report, Dec. 1, 2001). Inorganic nonmetallic coatings include vitreous enamels, glass linings, cement, or porcelain enamels bonded on metals.

Susceptibility to mechanical damage and cracking by thermal shock are major disadvantage of these coatings. An inorganic coating must be perfect and defect-free unless cathodic protection is applied.

Paints, lacquers, coal-tar or asphalt enamels, waxes, and varnishes are typical organic coatings. Converted epoxy, epoxy polyester, moisture-cured polyurethane, vinyl, chlorinated rubber, epoxy ester, oil-modified phenolics, and zinc-rich organic coatings are other barriers used to control corrosion. Paints, which are a mixture of insoluble particles of pigments (e.g., TiO_2, Pb_3O_4, Fe_2O_3, $ZnCrO_4$) in a continuous organic or aqueous vehicle such as linseed or tung oil, are the most common organic coatings. All paints are permeable to water and oxygen to a degree and are subject to mechanical damage and eventual breakdown. Inhibitor and antifouling agents are added to improve corrosion protection. Underground pipelines and tanks should be covered with thicker coatings of asphalt or bituminous paints, in conjunction with cloth or plastic wrapping. Polyester and polyethylene have been used for tank linings and as coatings for tank bottoms [Munger, "Corrosion Prevention by Protective Coatings," NACE International, Houston, TX, 1984; Munger and Vincent, "Corrosion Prevention by Protective Coatings," 2d ed., NACE International, Houston, TX, 1999; Fedrizzi and Bonora (eds.), Organic and Inorganic Coatings for Corrosion Prevention—Research and Experiences, European Federation of Corrosion Report No. 20, Institute of Materials, London, 1997].

Inhibitors, when added above a threshold concentration, decrease the corrosion rate of a material in an environment. If the level of an inhibitor is below the threshold concentration, corrosion could be accelerated more than if the inhibitor were completely absent. Inhibitors may be organic or inorganic and fall into several different classes: (1) passivators or oxidizers; (2) precipitators; (3) cathodic or anodic; (4) organic adsorbents; (5) vapor phase; and (6) slushing compounds. Any one inhibitor may be grouped into one or more classifications. Factors such as temperature, fluid velocity, pH, salinity, cost, solubility, interfering species, and metal or alloy may determine both the effectiveness and the choice of inhibitor. Federal EPA regulations may limit inhibitor choices (such as chromate use) based on its toxicity and its disposal requirements.

Passivators in contact with the metal surface act as depolarizers, initiating high anodic current densities and shifting the potential into the passivation range of active/passive metal and alloys. However, if present in concentrations below 10^{-3} to 10^{-4} M, these inhibitors can cause pitting. Passivating inhibitors are anodic inhibitors and may include oxidizing anions such as chromate, nitrite, and nitrate. Phosphate, molybdate, and tungstate require oxygen to passivate steel and are considered nonoxidizing. Nonoxidizing sodium benzoate, cinnamate, and polyphosphate compounds effectively passivate iron in the near-neutral range by facilitating oxygen adsorption. Alkaline compounds (NaOH, Na_3PO_4, $Na_2B_4O_7$, and $Na_2O-nSiO_4$) indirectly assist iron passivation by enhancing oxygen adsorption.

Precipitators are film-forming compounds which create a general action over the metal surface and subsequently indirectly interfere with both anodes and cathodes. Silicates and phosphates in conjunction with oxygen provide effective inhibition by forming deposits. Calcium and magnesium interfere with inhibition by silicates; 2 to 3 ppm of polyphosphates are added to overcome this. The levels of calcium and phosphate must be balanced for effective calcium phosphate inhibition. Addition of a zinc salt often improves polyphosphate inhibition.

Cathodic inhibitors (cathodic poisons, cathodic precipitates, oxygen scavengers) are generally cations which migrate toward cathode surfaces where they are selectively precipitated either chemically or electrochemically to increase circuit resistance and restrict diffusion of reducible species to the cathodes. Cathodic poisons interfere with and slow the HER reaction, thereby slowing the corrosion process. Depending on pH, arsenic, bismuth, antimony, sulfides, and selenides are useful cathodic poisons. Sulfides and arsenic can cause hydrogen blistering and hydrogen embrittlement. Cathodic precipitation-type inhibitors such as calcium and magnesium carbonates and zinc sulfate in natural waters require adjustment of the pH to be effective. Oxygen scavengers help inhibit corrosion, either alone or with another inhibitor, to prevent cathodic depolarization. Sodium sulfite, hydrazine, carbohydrazide, hydroquinone, methylethylketoxime, and diethylhydroxylamine are oxygen scavengers.

Organic inhibitors, in general, affect the entire surface of a corroding metal and affect both the cathode and anode to differing degrees, depending on the potential and the structure or size of the molecule. Cationic, positively charged inhibitors such as amines and anionic, negatively charged inhibitors such as sulfonates will be adsorbed preferentially depending on whether the metal surface is either negatively or positively charged. Soluble organic inhibitors form a protective layer only a few molecules thick; if an insoluble organic inhibitor is added, the film may become about 0.003 in (76 μm) thick. Thick films show good persistence by continuing to inhibit even when the inhibitor is no longer being injected into the system.

Vapor-phase inhibitors consist of volatile aliphatic and cyclic amines and nitrites that possess high vapor pressures. Such inhibitors are placed in the vicinity of the metal to be protected (e.g., inhibitor-impregnated paper) which transfers the inhibiting species to the metal by sublimation or condensation where they adsorb on the surface and retard the cathodic, anodic, or both corrosion processes. This inhibitor protects against water and/or oxygen. Vapor-phase inhibitors are usually effective only if used in closed spaces and are used primarily to retard atmospheric corrosion. Slushing compounds are polar inorganic or organic additives in oil, grease, or wax that adsorb on the metal surface to form a continuous protective film. Suitable additives include organic amines, alkali and alkaline-earth metal salts of sulfonated oils, sodium nitrite, and organic chromates [Nathan (ed.), "Corrosion Inhibitors," NACE International, Houston, TX, 1973; Raman and Labine (eds.), "Reviews of Corrosion Inhibitor Science and Technology," vol. 1, NACE International, Houston, TX, 1993; Raman and Labine (eds.), "Reviews of Corrosion Inhibitor Science and Technology," vol. 2, NACE International Houston, TX, 1996].

Environmental concerns and regulations over using some traditional corrosion control materials have led to research and implementation of alternative corrosion control technologies for (1) volatile organic compounds (VOCs), (2) chromates, and (3) cadmium. By far the most important environmental impact from paints and coatings is the release of VOCs from drying. Virtually everything but the solids in a typical coating formulation is released to the air around the surface being coated. In an enclosed system, such as a paint booth, some of this emission may be captured before release to the atmosphere. New coating formulations with higher proportions of solids meet current environmental requirements, eventually, 100 percent solids coating formulations will be common.

Hexavalent chromium is a cancer hazard that can damage lungs, skin, eyes, and nasal passages. Substitutes for hard chromium coatings are trivalent chromium, nickel-based binary and ternary plating chemicals. Alternatives to electroplating are pressure-controlled atomization processes (Thermal Spray) high-velocity oxygen fuels (HVOF) processes, amorphous nanocrystalline composite depositions, flexible preceramic coatings deposition, organic sealants application, inductive-coupled radio-frequency plasma torch, and cold spray. Cadmium plating has been used to protect steel components in marine environments. Cadmium has been used because of its hardness, close dimensional tolerance, lubricity, and ability to restrict hydrogen permeation into or out of high-strength steels. Cadmium plating is frequently used for fasteners and other very tight tolerance parts because of the dual qualities of lubricity at minimal thickness and superior sacrificial corrosion protection. However, cadmium is highly toxic to personnel and the environment, thereby necessitating the search for alternative corrosion control coating systems. Research is vigorous to develop cadmium replacement coatings with mechanical and sacrificial properties and corrosion performance equivalent to or exceeding those of cadmium. Coatings that show some promise are Zn-Ni, Zn-Sn, and Ni-Zn-P; Al-Co-Ce alloys; but further research and testing will be required.

The degradation of materials in marine environments depends on the expected materials demands and service requirements for the component or systems. Corrosion is dependent on a number of factors, and understanding these factors can optimize the ability of modeling to predict galvanic interactions in various environments and under various operating conditions.

The computational, storage, and analytical instrumentations are readily available to organize, compile, and verify the many **corrosion models** developed from previous materials-based research and empirical investigations related to materials performance in the various environments. This is an arduous task for there are many parameters and factors, which affect passivity and the breakdown of passivity of materials. These are limited corrosion models that can describe, formulate, and improve our understanding of materials interactions leading to protection or corrosion in marine environments. Yet, small changes, additions, or subtractions of various environmental parameters may alter (1) the corrosion resistance of materials, (2) the precise mechanism(s) of breakdown, and (3) the accuracy of corrosion models to explain the mode of breakdown. It is important that computational or mechanistic models not be used beyond their capabilities. In the future, fundamental research on the atomic level will help clarify many aspects of current corrosion models, modify earlier empirically based models, or develop and verify new corrosion models. Fundamentally based research will also improve our understanding of critical materials properties and the effects that various marine environmental interactions have on materials. Better physical and computational models will improve our understanding of corrosion and corrosion control that can lead to predictive and prognostic capabilities to better structural design [Shifler, Galvanic Corrosion Modeling in Flowing Systems, "Proc. ELECTROCOR 2005: Simulation of Electrochemical Processes," Brebbia et al. (eds.), WITPress, Southampton, UK, 2005, p. 59; Shifler, Understanding Materials Interactions in Marine Environments to Promote Extended Structural Life, *Corrosion Sci.* 2005].

Anodic protection is based on formation of a protective film on metals by externally applied anodic currents. Figure 6.5.5 illustrates the effect. The applied anodic current density is equal to the difference between the total oxidation and reduction rates of the system, or $i_{app} = i_{oxid} - i_{red}$. The potential range in which anodic protection is achieved is the protection range or passive region. At the optimum potential E_A, the applied current is approximately 1 μA/cm^2. Anodic protection is limited to active/passive metals and alloys and can be utilized in environments ranging from weak to very aggressive. The applied current is usually equal to the corrosion of the protected system, which can be used to monitor the instantaneous corrosion rate. Operating conditions in the field usually can be accurately and quickly determined by electrochemical laboratory tests (Riggs and Locke, "Anodic Protection—Theory and Practice in the Prevention of Corrosion," Plenum Press, New York, 1981).

Cathodic protection (CP) controls corrosion by supplying electrons to a metal structure, thereby suppressing metal dissolution and increasing hydrogen evolution. Figure 6.5.5 shows that when an applied cathodic current density ($i_{app} = i_{red} - i_{oxid}$) of 10,000 μA/cm^2 on a bare metal surface shifts the potential in the negative or active direction to E_c, the corrosion rate has been reduced to one μA/cm^2. CP is applicable to all metals and alloys and is common for use in aqueous and soil environments.

CP can be accomplished by (1) impressed current from an external power source through an inert anode or (2) use of galvanic couplings by sacrificial anodes. Sacrificial anodes include zinc, aluminum, magnesium, and alloys containing these metals. The total protective current is directly proportional to the surface area of the metal being protected; hence, CP is combined with surface coatings where only the coating pores or holidays and damaged spots need be cathodically protected [e.g., for 10 mi of a bare pipe, a current of 500 A would be required, while for 10 mi of a superior coated pipe (5×10^6 Ω/ft^2) 0.03 A would be required]. The polarization potential can be measured versus a reference, commonly saturated Cu/CuSO$_4$. The criterion for proper CP is -0.85 V versus this reference electrode. Overprotection by CP (denoted by potentials <-0.85 V) can lead to blistering and debonding of the pipe coatings and hydrogen embrittlement of steel from hydrogen gas evolution. Another problem with CP involves stray currents to unintended structures. The proper CP for a system is determined empirically and must resolve a number of factors to be effective [Bianchetti (ed.), "Peabody's Control of Pipeline Corrosion," 2d ed., NACE International, Houston, TX, 2001; Morgan, "Cathodic Protection," 2d ed., NACE International, Houston, TX, 1987; von Baeckmann et al. (eds.), "Handbook of Cathodic Corrosion Protection: Theory and Practice of Electrochemical Protection Processes," 3d ed., Gulf Publishing, 1997].

CORROSION IN INDUSTRIAL AND POWER PLANT STEAM-GENERATING SYSTEMS

Types of Steam Generators

The type of steam generator will dictate the types of corrosion-related damage experienced. Most steam generators can be grouped into the following categories: heat recovery steam generators (HRSG), direct-fired boilers (i.e., a fuel is burned in the furnace), nuclear steam generators, and electric boilers. This section focuses on the corrosion of two categories of steam generators: direct-fired boilers and HRSG units. Generally the term *steam generator* is used here when referring to both. Details about the design of steam generators are discussed in Sec. 9 and the references [Stultz and Kitto (eds.), "Steam: Its Generation and Use," 40th ed., Babcock and Wilcox, 1992].

Steam/Water Cycle Components

Figure 6.5.6 presents an example of a steam/water cycle for a power plant with a direct-fired boiler. The flow is basically in a clockwise direction in the figure. Major components in the cycle include the condenser, **deaerator,** boiler (including the economizer, steam generator, superheater, and reheater tubes), and turbine.

Heat recovery steam generator units have simpler condensate/feedwater systems (often with no external feedwater heaters or deaerators) and more complicated drum and steam systems (often two or three drums operating at different pressures with superheated steam from the intermediate-pressure drum feeding the high-pressure cold reheat lines and superheated steam from the low-pressure (LP) drum feeding the LP turbine). For industrial and cogeneration facilities, a portion of the steam produced would be used in the process, and process condensate normally would be treated in a condensate polisher and returned to the condensate/feedwater system.

Steam/Water Cycle Corrosion

Table 6.5.1 provides a summary of the corrosion mechanisms experienced in steam/water cycle components. Corrosion can occur on the internal or external surfaces of the components. In most components, *internal pitting* refers to the inner diameter (ID) of the tube, and *external pitting* refers to the outer diameter (OD) of heat-transfer tubes. Underdeposit corrosion includes caustic gouging and acid phosphate corrosion as well as concentration of other salts under deposits (e.g., in steam generator tubes during condenser leaks or other contamination events). A detailed (370 pages) discussion of corrosion of steam/water cycle materials was presented in "ASME Handbook on Water Technology for Thermal Power Systems," Chap. 9, ASME, New York, 1989. Chapter 17, Management of Cycle Chemistry, of this ASME

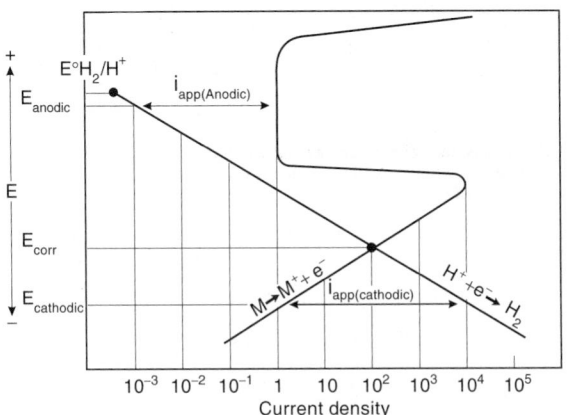

Fig. 6.5.5 Effect of applied anodic and cathodic currents on behavior of an active passive alloy necessary for anodic and cathodic protection, respectively.

Fig. 6.5.6   Example of a steam/water cycle for a direct-fired boiler.

Handbook is scheduled to be rewritten to reflect the changes in understanding regarding corrosion, carryover, monitoring, and chemistry control in steam/water cycle components. A number of other types of failure mechanisms are experienced by steam/water cycle components, including thermal and/or vibration fatigue, erosion, overheating, manufacturing defects, and maintenance problems. Some of the actual failure mechanisms listed are combinations of mechanical and chemical mechanisms, such as corrosion fatigue and stress corrosion cracking (SCC). The identification, cause(s), and corrective action(s) for most types of

failure mechanism were discussed in "The Boiler Tube Failure Metallurgical Guide," vol. 1: "Technical Report," TR-102433-VI, EPRI, Palo Alto, CA, 1993, and in a later version of this document (Dooley and McNaughton, "Boiler Tube Failures: Theory and Practice," vols. 1–3, TR105261, EPRI, Palo Alto, CA, 1996). Corrosion mechanisms which have been more widely recognized since these documents were prepared include stress-assisted corrosion (Labuda and Bartholomew, Stress Assisted Corrosion: Case Histories, *Power Plant Chem.*, March 2005). For information on corrosion in nuclear steam

Table 6.5.1   Summary of Corrosion Mechanisms for Steam/Water Cycle Components

Corrosion mechanism	Condenser	Feedwater heaters	Economizer	Steam generator	Superheater	Reheater	Steam turbine
Internal pitting	X	X	X	X	X	X	X
External pitting	X	X	X	X			
Stress corrosion cracking	X	X			X	X	X
Stress-assisted corrosion			X	X	X		
Corrosion fatigue			X	X			X
Dealloying	X	X					
Ammonia grooving	X						
Ennoblement	X						
Microbiologically influenced corrosion (MIC)	X	While rare, MIC can occur due to improper handling and storage before components are fully assembled or installed.					
Flow-accelerated corrosion (FAC)		X	X	X (HRSG)			
Erosion corrosion	X	X	X				
Dew point corrosion	*		X				
Underdeposit corrosion				X			
Hydrogen damage	X (Rare)			X			
Galvanic corrosion	X						
Intergranular corrosion	X				X	X	
Sulfide corrosion	X			X			
Liquid ash corrosion				X	X	X	
Halogenation					X	X	
Crevice corrosion	X						

* Air-cooled condenser.

generators, a detailed government document on corrosion mechanisms and corrosion predictions is available via the Internet ("Incorporating Aging Effects into Probabilistic Risk Assessment—A Feasibility Study Utilizing Reliability Physics Models," NUREG/CR-5632, Office of Nuclear Regulatory Research, Washington, DC, August 2001).

Deaerators can experience oxygen pitting and corrosion fatigue. NACE Standard RP0590-96, "Recommended Practice for Prevention, Detection, and Correction of Deaerator Cracking" (1996), was largely written in response to deaerator failures from corrosion fatigue cracking. The interconnecting piping in the feedwater system can experience flow-accelerated corrosion (FAC)—sometimes with catastrophic consequences. Corrosion also can be experienced in other steam generator facility components (e.g., flue gas desulfurization system, gas turbines, wet ash handling systems, and makeup and wastewater treatment systems).

Control of Waterside/Steamside Corrosion in the Steam/Water Cycle

To minimize corrosion associated with the water and steam sides of steam generators and other steam/water cycle components, a comprehensive chemistry program should be instituted. A wide variety of chemical treatment programs are utilized for these systems depending on the materials of construction, operating temperatures, pressures, heat fluxes, contaminant levels, and purity criteria for components using the steam.

Very general chemistry guidelines for feedwater, boiler water, and steam for industrial boilers were presented in "Consensus on Operating Practices for the Control of Feedwater and Boiler Water Chemistry in Modern Industrial Boilers," ASME, New York, 1994. This does not indicate the details of a complete chemistry program. Updated consensus documents for chemistry monitoring guidance ("Consensus on Operating Practices for the Sampling and Monitoring of Feedwater and Boiler Water Chemistry in Modern Industrial Boilers," ASME, New York) and for controlling water chemistry ("Consensus on Operating Practices for Control of Water and Steam Chemistry in Combined Cycle and Cogeneration Power Plants") are prepared by ASME to provide guidance for heat recovery steam generators.

For the electric power industry, a series of detailed chemistry guidelines are available from the Electric Power Research Institute (EPRI). EPRI is a research organization funded by its utility members. Its first chemistry guides were published in 1982 and 1986 for pressurized water reactor (PWR) nuclear steam generators and fossil-fuel-fired steam generators (respectively) in the electric utility industry. A number of facilities experienced failures in boilers and other cycle chemistry components despite adherence to these guidelines. EPRI continues to update both documents to reflect the changes in understanding regarding corrosion and corrosion product transport. For example, as of this document, the latest chemistry guideline document for the secondary water system of PWR nuclear steam generators was in its sixth version (2004). EPRI currently has a number of chemistry guidelines tailored for different types of steam generators (fossil-fuel-fired boilers, HRSG, fluidized-bed boilers, etc.) For EPRI member organizations, it is suggested that you obtain the latest guideline for your type of steam generator.

A good chemistry control program requires a high-purity makeup water for the steam/water cycle. For high-pressure electric utility boilers, two-stage demineralization (e.g., cation/anion/mixed bed ion exchange; reverse osmosis/membrane decarbonation/continuous electrodeionization, and combinations of these) is generally utilized. Consult the cited reference for information on design of makeup treatment systems (Sheppard T. Powell Associated, LLC, "Revised Guidelines for Makeup Water Treatment," TR113692, EPRI, Palo Alto, CA, 1999). For industrial boilers, one- or two-stage demineralization or softening (sometimes with dealkalization) is utilized for **makeup water treatment**—depending on the boiler pressure. Unsoftened makeup water has been utilized in small, low-pressure (e.g., ≤ 15 psig, ≤ 100 kPa) boiler systems which receive large amounts of condensate return, but this practice increases the potential for corrosion and deposition. Some makeup systems also include vacuum degasifiers for the removal of dissolved oxygen from the makeup water prior to its entrance into the steam/water cycle.

Condensate returns often require purification or polishing. **Condensate polishing** can be achieved with specially designed filters, softeners, or mixed-bed ion exchangers. Industrial boilers typically use softeners and/or filters (e.g., electromagnetic or cartridge filtration). Once-through boilers typically use full-flow, mixed-bed polishers (preferred) or precoat (with crushed ion exchange resin) filters. Mixed-bed condensate polishers also are strongly advised for drum-type steam generators with condenser cooling water having a high content of chloride and/or a high ratio of hardness to alkalinity. These systems are more subject to underdeposit corrosion of steam generator tubes in the event of a condenser tube leak.

Minimizing the transport of corrosion products to the steam generator is essential for corrosion control in the steam generator. Corrosion products from the feedwater deposit on steam generator tubes, inhibit cooling of tube surfaces, and provide an evaporative-type concentration mechanism of dissolved salts under boiler tube deposits. Therefore, the chemistry of water in contact with the tube surfaces is often much worse than that in the bulk water.

There are two basic philosophies of **feedwater treatment.** The first type of feedwater treatment, which is used for industrial boilers and many high-pressure power boilers, relies primarily on raising the feedwater pH to 8.5 to 10 (each plant will have a smaller control range) and maintaining a reducing environment by eliminating dissolved oxygen (to <5 to 10 ppb) and applying a reducing agent. In this environment, the predominant oxide formed on steel is a mixed metal oxide composed primarily of magnetite (Fe_3O_4). The mixed oxide layer on steel formed in deoxygenated solutions resists dissolution or erosion and thus controls iron transport to the boiler and corrosion of the underlying metal. Copper oxides are maintained in a reduced state (Cu_2O) which is less subject to transport to the steam generator (or carryover into the steam). However, excessive application of reducing agents can result in low dissolved oxygen levels (<2 ppb) and low oxidizing reduction potentials (ORPs) which render steel susceptible to flow-accelerated corrosion.

FAC is the gradual dissolution of magnetite layers from steel surfaces in flowing systems. It only occurs in areas of high flow or turbulence, and the metal surface often has the scalloped surface morphology, which is sometimes referred to as erosion corrosion. FAC occurs between 100°C (212°F) and 250°C (482°F) with a maximum attack around 150°C (302°F) for single-phase FAC and 140 to 180°C (284 to 356°F) for two-phase (i.e., steam/water mixtures) FAC. FAC is decreased by raising the pH, raising the dissolved oxygen level (as little as 1 to 2 ppb sometimes can have a significant effect), reducing or eliminating the feed of oxygen scavengers (reducing agents), changing to an alloy with greater chromium content (e.g., T22 or P22 provide substantial resistance), and reducing the flow-induced erosive forces (e.g., decreasing fluid velocity and avoiding pressure reductions or other sources of flashing and two-phase flow).

FAC has caused gradual thinning and catastrophic failures of feedwater lines (several resulting in fatalities), economizer tubes, evaporator tubes (particularly in low-pressure HRSG evaporator sections), heater drain lines, and attemperation water lines in power plants. Facilities feeding oxygen scavengers should evaluate their system for areas likely to have FAC and should check the thickness of metal in those locations. Modeling programs are available for selecting areas for routine evaluation of FAC, as discussed in "Flow-Accelerated Corrosion in Power Plants," TR-106611, EPRI, Palo Alto, CA, 1996.

An adapted version of the deoxygenated (<5 to 10 ppb) **feedwater treatment** program for systems without any copper present relies solely on mechanical deaeration (no reducing agent is applied). This results in a slightly more oxidizing solution (typically 1 to 7 ppb of oxygen) and a partially oxidized oxide layer on steel (in the feedwater system) which has been found to be more resistant to FAC than steel with purely a magnetite layer.

The second basic type of **feedwater treatment** program is **oxygenated treatment** (OT). While used for many years in Germany and Russia, OT was not generally utilized in the United States until the 1990s. It involves a completely different approach from the other treatment programs since oxygen is fed and it relies on the formation of a gamma ferric oxide (Fe_2O_3) or gamma ferric oxide hydrate (FeOOH) layer on the steel surfaces in the preboiler cycle.

Originally, oxygenated treatment was based on neutral pH values (7 to 8) and oxygen levels of 50 to 250 ppb in very pure feedwater (cation conductivities <0.15 μS/cm). Later, a program based on slightly alkaline pH values (8.0 to 8.5) and 30 to 150 ppb of dissolved oxygen was developed. This treatment program was originally referred to as **combined water treatment,** but it is now commonly referred to as oxygenated treatment in the United States. Its application was primarily limited to once-through boilers with all-steel steam/water cycles (if full-flow **condensate polishing** is provided, as recommended, copper alloys are permitted in the condensers). The primary benefit of OT has been to minimize deposit-induced pressure drop through the boiler during service. The reduction in iron oxide transport into and deposition in the boiler can also significantly reduce the frequency of chemical cleanings. However, OT requires extremely pure feedwater, and pitting can result from contaminated feedwater. The details of oxygenated treatment programs for once-through boilers were presented in "Cycle Chemistry Guidelines for Fossil Plants: Oxygenated Treatment," TR-102285, EPRI, Palo Alto, CA, 1994. This document also presented guidance for utilizing OT for drum-type boilers and HRSG units, although OT in drum-type units incurs much greater risk because feedwater contaminants are concentrated in the drums.

As indicated in Fig. 6.5.6, a **deaerator** is included in many steam/water cycles for the removal of dissolved oxygen and other noncondensable gases. Deaerators are direct contact steam heaters which spray feedwater into fine droplets and strip the dissolved oxygen from the water with steam. This is often referred to as *mechanical deaeration*. Deaerators normally operate at pressures of 5 to 180 psig (35 to 1,241 kPa), although some old deaerators operated at 1 to 2 psig (6.9 to 13.8 kPa). Most modern deaerators reliably remove dissolved oxygen to ≤5 to 7 ppb, provided they are properly maintained and operated. While they are sometimes rated for 7 ppb, the performance of deaerating condensers currently is more variable in industry and involves maintaining proper vacuum, controlling air in-leakage, minimizing condensate subcooling, and deaerating or installing spray headers for the makeup.

Following mechanical deaeration, chemical reducing agents called **oxygen scavengers** often are applied to minimize residual dissolved oxygen and/or to inhibit pitting (enhance passivation). This is recommended for all systems containing copper in feedwater heaters or steam-fed heat exchangers (copper corrosion is minimized at ≤2 ppb) and for steel systems with elevated levels of contamination.

A wide variety of oxygen scavengers are utilized. For boilers below 700 to 900 psig (4.8 to 6.2 MPa), sodium sulfite (often catalyzed with cobalt) is commonly used although other nonvolatile or volatile oxygen scavengers also may be effective. In industrial boiler systems, severe oxygen tuberculation and pitting and failures of economizer tubes have occurred due to excessive oxygen and inadequate feed of sodium sulfite. A slight excess of sodium sulfite is applied to the feedwater, resulting in a residual of sulfite in the boiler water. The exact level of sulfite maintained depends primarily on the boiler pressure and oxygen concentrations in the feedwater, although other factors such as boiler water pH, steam/condensate purity, and feedwater temperatures are involved. Sodium sulfite should not be used in boilers operating above 900 psig (6.2 MPa), as a significant amount of the sulfite decomposes to volatile sulfur species at the elevated temperatures. The resultant acidic species decrease steam and condensate pH values, contributing to increased corrosion in steam/water cycle components. Sulfur species in steam may contribute to SCC. Also, sulfite cannot be used in systems that directly inject feedwater into steam for attemperation (desuperheating).

Commonly used volatile oxygen scavengers (reducing agents) include hydrazine, carbohydrazide, hydroquinone, diethylhydroxylamine (DEHA), and methylethylketoxime. These are normally applied to maintain a free residual of the scavenger in the feedwater at the economizer inlet to the boiler. Due to the slower reaction rates, routine dissolved oxygen monitoring usually is more crucial for boilers using volatile oxygen scavengers than for boilers using sulfite. It is possible for dissolved oxygen and these volatile reducing agents to coexist due

to slow reaction rates. A benefit and a disadvantage of some of the volatile oxygen scavengers (e.g., hydrazine, carbohydrazide, DEHA) are their ability to enhance metal passivation. This has been useful in systems with poorly operating **deaerators** because they maintain a magnetite layer on the steel and appear to inhibit oxygen pitting, despite the presence of elevated dissolved oxygen levels. However, excessive feed of these volatile reducing agents in systems with demineralized makeup must be avoided as they can lead to FAC.

Sulfite is an effective oxygen scavenger, but has little effect on passivation beyond the effect of oxygen removal. Therefore, it will not convert oxide films on steel and does not cause FAC. However, coexistence of sulfite and oxygen does not protect the steel against operational oxygen pitting. Additional information on oxygen scavengers can be found in the NACE technical committee report 3A194 ("Oxygen Scavengers in Steam Generating Systems and in Oil Production," NACE, Houston, TX, 1994).

Aqueous solutions of ammonia or volatile neutralizing amines are usually applied to the feedwater to control feedwater, steam, and condensate pH. For all steel cycles these pH levels can range from 9.0 to 10.0 with the higher pH values resulting in lower iron oxide transport. The higher pH and amines levels result in higher baseline conductivities, which obscures the ability to detect system contamination (which increases the potential for contamination-causing deposits or scale in the steam generator). In systems containing both steel and copper alloys, pH levels in the feedwater, steam, and condensate are typically controlled in the 8.5 to 9.4 range. The exact pH control limits usually are a much tighter range depending on the water purity, amine used, temperature, and type of copper alloys present.

Neutralizing amines commonly utilized in steam/water cycles include cyclohexylamine, morpholine, ethanolamine, diethanolamine, diethylaminoethanol, dimethyl isopropanolamine, and methoxypropylamine. For large steam distribution systems, a blend of two to four amines with different distribution coefficients (ratio of amine in steam to condensate) is often applied to provide pH control throughout the steam/condensate system. Amines tend to decompose in high-pressure, high-temperature systems and produce organic acids which may contribute to corrosion.

Filming amines are high-molecular-weight amines applied to develop thin films on surfaces to serve as a barrier against corrosion. Applications date at least back to the middle of the twentieth century. Various types of filming amines used up through the 1990s in the United States resulted in fouling of equipment (e.g., steam traps, condensate polishers) when the waxy films and associated iron oxides sloughed off surfaces. New high-molecular-weight amines (currently called polyamines) continue to be introduced and applied. Caution is advised in the application of this technology.

In addition to the condensate/**feedwater treatment** program, chemistry must be controlled in the steam generator (boiler water) to minimize corrosion. Currently, the most common types of steam generator treatment programs in use in the United States are coordinated **phosphate treatment,** congruent phosphate treatment, equilibrium phosphate treatment, phosphate continuum treatment, phosphate treatment with caustic, caustic treatment, all-volatile treatment (AVT), and **oxygenated treatment** (for once-through steam generators). With the exception of oxygenated treatment (OT, discussed earlier) and some AVT programs, all these treatment programs are based on no oxygen (<5 ppb) in the steam generator and the formation of a mixed metal oxide of predominantly magnetite (Fe_3O_4).

The variety of phosphate and caustic treatments is designed to provide a reserve of alkalinity to avoid acid attack and minimize pH fluctuations. Caustic or acidic materials concentrating underneath deposits in boiler tubing contribute to most types of waterside corrosion. It is generally desired to control the pH in a range that controls the level of free caustic alkalinity and acidic constituents within acceptable values. Phosphate also precipitates a portion of the hardness contamination entering with the feedwater as a hardness phosphate sludge in the steam drum (or generation bank downcomer tubes) before reaching the higher heat evaporator sections.

The type of **phosphate treatment** mainly depends on the operating pressure and water purity; low-pressure boilers on softened makeup generally operate with substantial amounts of free caustic alkalinity in the boiler water. Industrial boilers operating up to about 900 psig (6.2 MPa) receiving demineralized makeup generally limit free sodium hydroxide (caustic) in the steam generator to ≤1 to 10 ppm, and intermediate- and high-pressure boilers and steam generators usually limit caustic to ≤0 to 1 ppm. These limits are primarily set to avoid caustic corrosion in the boiler. Facilities with free caustic and carryover also are susceptible to stress corrosion cracking of stainless steel steam system components. These facilities favor the use of coordinated phosphate treatment with a sodium/phosphate mole ratio (actually moles of sodium alkalinity divided by moles of phosphate) in the boiler water up to a maximum of 3.0, which corresponds to trisodium phosphate. The minimum ratio varied, but typically sodium/phosphate ratios were controlled at 2.4 to 3.0 for boilers operating at less than 1,000 psig (at 6.9 MPa) and 2.6 to 3.0 for higher-pressure boilers. The minimum ratio for higher-pressure boilers has been raised as indicated in the next paragraph.

Another series of captive alkalinity programs called *congruent treatment* used lower sodium/phosphate ratios (e.g., range of 2.2 to 2.6), but these were found to lead to acid phosphate corrosion in high-pressure boilers (mainly >1,800 psig, >12.4 MPa). Acid phosphate corrosion results from reaction of sodium phosphates with magnetite to form acidic compounds on tube surfaces with the simultaneous generation of caustic in solution. During load reductions, these deposits can redissolve and contribute to corrosion. Research studies determined the minimum sodium/phosphate ratio to avoid acid phosphate corrosion as a function of temperature or drum pressure (Tremaine et al., Phosphate Interactions with Metal Oxides under High-Performance Boiler Hideout Conditions, Paper no. 35, International Water Conference, Pittsburgh, PA, 1993; Tremaine et al., Solubility and Thermodynamics of Sodium Phosphate Reaction Products under Hideout Conditions in High-Pressure Boilers, IWC-96-21, International Water Conference, Pittsburgh, PA, (1996). For high-pressure boilers (especially ≥2,000 psig or ≥13.8 MPa), equilibrium phosphate (which restricts the maximum level of phosphate applied and maintains sodium alkalinity to phosphate ratios above 3.0 with up to 1 ppm of caustic) currently is considered to be one of the best forms of **phosphate treatment** for boilers and HRSG units [Stodola, Fifteen Years of Equilibrium Phosphate Treatment (Correct Use of Phosphates in Drum Boilers), *Power Plant Chem.*, **5**, no. 2, 2003]. While they may not be optimal for the boiler, slightly higher phosphate levels and lower minimum ratios (to 2.7 at 1,500 to 2,000 psig or 10.3 to 13.8 MPa, and down to 2.4 at <1,000 psig or <6.9 MPa) have been safely used in some lower-pressure boilers with lower heat flux, minimal phosphate hideout, and more variable chemistry control. With phosphate or caustic treatment the steam generator pH should usually be greater than the feedwater pH to ensure nonvolatile alkali is present.

AVT is an extension of the **feedwater treatment** programs. Steam generator water pH control is maintained by the portion of amines or ammonia from the feedwater which remains in the steam generator water. Since these materials are volatile, the steam generator pH normally is 0.2 to 0.3 pH units less than the feedwater pH. Due to the low concentrations and poor buffering capacity of ammonia or amines in steam generator water, strict purity limits are maintained on the feedwater and steam generator water. AVT is generally only utilized in high-pressure steam generators (>1,800 psig, 12.4 kPa) with condensate polishers, although some lower-pressure boilers (850 psig, 5.9 kPa) also utilize this treatment. Many all-steel HRSG units also rely solely on AVT due to problems with phosphate hideout in the steam generator and carryover of the water from the drums into the steam.

Control of Fireside Corrosion of Steam Generators

Corrosive constituents in fuel at appropriate metal temperatures may promote fireside corrosion in boiler tube steel. For coal, lignite, oil, or refuse, the corrosive ingredients can form liquids (liquid ash corrosion) that solubilize the oxide film on tubing and react with the underlying metal to reduce the tube wall thickness. This can occur in the combustion zone (e.g., waterwall tubes) or convection pass (e.g., superheaters). Technologies to reduce NO_x emissions from the boiler also result in more reducing conditions in the furnace, which greatly increases the potential

for liquid ash corrosion. This has been alleviated by installing air ports along the walls to create an air blanket and installing stainless and high-alloy weld overlays. For black liquor recovery boilers, sulfur or carbon (char) induced corrosion can occur through the direct reaction with the metal oxide layer. To combat the high corrosion rates of carbon steel in the lower furnace of black liquor recovery boilers, stainless steel and nickel-alloy overlays and composite tubes have been used. Cracking of these composite tubes around air ports and in floor tubes currently is the focus of much investigation. High-nickel alloys also have been used in flue gas desulfurization systems to combat the severe corrosion resulting from the hot, wet environment with low pH values and high dissolved solids levels.

Fluidized-bed boilers can experience severe fireside wastage of boiler tubes due to erosion. Fireside erosion-related mechanisms also can cause severe damage to tubes in the convection pass. Dew point corrosion may occur on the fireside surfaces of the economizer and other low-temperature surfaces (e.g., near drum corrosion in generation bank tubes for black liquor recovery boilers) when condensation can form acidic products (Reid, "External Corrosion and Deposits—Boiler and Gas Turbine," American Elsevier, New York, 1971; Labossiere and Henry, Chromizing for Near-Drum Corrosion Protection, *TAPPI Jour.*, Sept. 1999).

In the preceding discussion, we indicated a few examples of corrective measures for controlling fireside corrosion. To summarize, fireside corrosion often can be mitigated by (1) material selection (e.g., stainless or nickel alloys); (2) purchase of fuels with low impurity levels (limit levels of sulfur, chloride, sodium, vanadium, etc.); (3) purification of fuels (e.g., coal washing); (4) application of fuel additives (e.g., magnesium oxide); (5) adjustment of operating conditions (e.g., increasing percent of excess air, percent of solids in recovery boilers, etc.); and (6) modification of layup practices (e.g., dehumidification systems).

Control of Steam Turbine Corrosion

Steam turbine corrosion is primarily controlled by maintaining steam purity. As the pressure and temperature decrease through the turbine, contaminants in the steam deposit on the turbine. Fortunately, current researchers indicate that no more than one percent of materials in excess of their steam solubility actually deposits in turbines due to the high velocities. Nevertheless, the deposits which happen to form can lead to pitting (particularly in the case of sodium chloride) during a subsequent shutdown or out-of-service period in which oxygen levels will be elevated in the condensate in contact with turbine surfaces. Once initiated, corrosion sites can lead to cracking and stress corrosion cracking during subsequent service. EPRI has various publications on turbine damage ("Turbine Steam Path Damage: Theory and Practice," vols. 1 and 2, EPRI, Palo Alto, CA, 1999; "Turbine Steam Chemistry and Corrosion," 1006283, EPRI, Palo Alto, CA, Sept. 2001) and is seeking to model and predict turbine damage based on *damage function analysis* (DFA) and the *point defect model* (PDM) of corrosion (Engelhardt et al., Deterministic Prediction of Corrosion Damage in Low Pressure Steam Turbines, *Power Plant Chem.*, **6**, no. 11, Nov. 2004).

The source of steam contamination for once-through steam generators is the feedwater and for drum-type and other subcritical steam generators is a combination of carryover into the saturated steam and contamination in the attemperation water (which is added to reduce the temperature of the steam). There are several methods of desuperheating to control steam temperatures. The most common in the power industry is the spray of feedwater directly into the steam. A common, and the most preferred, means of desuperheating for industrial boilers is to use a "sweet water condenser" which condenses saturated steam (with feedwater) for injection into the steam. Some older boilers have drum-type attemperators where the steam is condensed in tube bundles immersed in boiler water in the mud drum. Unfortunately, these tubes can leak and significantly contaminate the steam.

CORROSION IN HEATING AND COOLING WATER SYSTEMS AND COOLING TOWERS

Control of corrosion in the water sides of cooling and heating water systems begins with proper preconditioning of the system. This usually

involves alkaline detergent cleaning, extensive flushing, and initial application of elevated corrosion inhibitor levels (2 to 3 times normal). Galavanized towers need to be treated differently, as discussed in "White Rust: An Industry Update and Guide Paper," Association of Water Technologies, 2002. Objectives of the subsequent cooling water treatment program are as follows:
- Prevent corrosion of metals in the system.
- Prevent the formation of scale (e.g., calcium carbonate, calcium sulfate, calcium phosphate).
- Minimize the amount and deposition of suspended solids.
- Minimize biological growth.

Substantial emphasis in a cooling tower treatment program is placed on preventing deposits of scale, suspended solids, or biological matter from forming on system components. The primary reason for this emphasis is to maintain the performance of the cooling system. However, minimizing deposition is integral to corrosion control to avoid problems with underdeposit corrosion or MIC attack. Stainless steels are particularly susceptible to MIC: Stainless steel condensers have failed in less than a year of operation due to MIC related to stagnant flow conditions. Automatic ball cleaning or brush cleaning systems are usually recommended for stainless steel condenser tubes. Suspended particles can lead to erosion of tubing and piping, particularly copper tubing. In HVAC systems, installation of sidestream filtration or full-flow strainers has been used to minimize suspended matter in the cooling water. Sidestream softening also has been used to reduce scale-forming constituents in cooling tower water system at zero liquid discharge facilities.

Depending on the inherent corrosivity of the cooling water, corrosion control in open, recirculating cooling water systems may consist merely of controlling the pH, conductivity, and alkalinity levels. However, in systems with extensive piping networks, corrosion inhibitors and deposit inhibitors (dispersants) are often utilized to protect materials in the system. The corrosion inhibitors can include the following compounds: zinc salts, molybdate salts, triazoles, orthophosphates, polyphosphates, phosphonates, and polysilicates. Often, two or more of these corrosion inhibitors are used together. Deposit inhibitors usually are called *dispersants* and consist of water-soluble organic compounds such as polyacrylates, phosphonates, and various copolymers and terpolymers.

Oxidizing or nonoxidizing biocides also are required in cooling tower systems to inhibit biological activity. Oxidizing biocides result in increased general corrosion rates for copper and copper alloys and to a lesser extent for ferrous metals. However, oxidizing biocides are sometimes required to provide adequate control of slime-forming bacteria and/or *Legionella* bacteria. Control of *Legionella*, applications of oxidizing biocides, and monitoring requirements for open cooling water systems are covered in the references ("Legionella 2003: An Update and Statement by the Association of Water Technologies (AWT)," McLean, VA, 2003; "Application of Oxidizing Biocides," WTP-141, Cooling Technology Institute, 1994; Bartholomew, "Bromine Based Biocides for Cooling Water Systems: A Literature Review," International Water Conference, Pittsburgh, PA, 1998; Standard Recommended Practice, "Online Monitoring of Cooling Waters," RP0189–2002, NACE, Houston, TX, 2002).

For closed cooling and heating water systems, corrosion inhibitors are usually applied and pH levels are controlled at 8 to 11. Corrosion inhibitors utilized in closed cooling and heating water systems include the inhibitors listed for open, recirculating cooling water systems except for zinc salts. Also nitrite salts are often used in closed water systems.

A **closed system** is defined by the loss of 5 percent of the system volume per year. Corrosion control in closed cooling systems is greatly facilitated by minimizing water loss from the system. Makeup water introduced to replace leakage introduces oxygen, organic matter to support biological activity, and scale constituents. Utilization of pumps with mechanical seals (with filtered seal water) rather than packing seals can greatly reduce water losses from closed systems. Control of oxygen levels by minimizing water leakage and aeration is probably the single most important factor for reducing corrosion in these systems. Glycol, intermittently used for freeze protection of out-of-service components, often is not able to be adequately purged from the system. The glycol is converted to organic acids, which contributes to low pH values and elevated corrosion. Whenever possible, other means of freeze protection are suggested.

In heating and cooling water systems, concerns regarding corrosion control generally focus on the water side or interior of piping and components. However, substantial corrosion can occur on the exterior of these piping systems underneath insulation. External corrosion of piping generally requires a source of external moisture. In chill water and secondary water systems, moisture can be introduced through exposure to the air and subsequent condensation of moisture on the pipe surface.

While chill water piping systems generally are installed with vapor barriers, breaks in the vapor barrier can be present due to inadequate installation or subsequent maintenance activities. Secondary water system piping is sometimes not installed with vapor barriers because the cooling water temperature is designed to be above the expected dew point. However, humidity levels in the building can sometimes be elevated and lead to condensation at higher temperatures than initially expected.

Leaks from piping at fittings or joints can saturate the insulation with moisture and lead to corrosion. This type of moisture source can be particularly troublesome in hot water piping, as the moisture can be evaporated to dryness and wet/dry corrosion can result. Corrosion rates during this transition from wetness to dryness can be many times higher than those of a fully wetted surface [Pollock and Steely (eds.), "Corrosion under Wet Thermal Insulation," NACE International, Houston, TX 1990]. If the moisture can be eliminated, corrosion underneath insulation should be negligible. However, if the moisture cannot be eliminated, coatings may need to be applied to the metal surface. Inhibitors are also available to mitigate corrosion underneath insulation due to condensation, but they may not be effective at controlling corrosion due to leakage because the inhibitors can be washed away. Preferably, insulated systems would have all welded joints in the piping system to prevent joint leakage and corrosion under insulation.

CORROSION IN THE CHEMICAL PROCESS INDUSTRY

Corrosion control in the chemical process industry is important to provide continuous operation in order to meet production schedules and to maintain purity of manufactured products. Also, protection of personnel and the environment is of primary importance to the chemical process industry. Many of the chemicals utilized have a high toxicity, and their release due to unexpected corrosion failures of piping and vessels is not permitted. Corrosion prevention and control in the chemical process industry are largely focused on material selection. Nondestructive evaluation methods are also routinely used to evaluate material integrity. The American Petroleum Institute (API) has at least four guidelines relating to evaluating equipment integrity and fitness for service of components (e.g., API Recommended Practice 579, "Fitness for Service," American Petroleum Institute, Englewood, CO, 2000).

Due to the wide range of chemicals and operating conditions utilized in the chemical process industry, the types of corrosion cannot be adequately addressed here. The following references address the corrosivity of the various environments and appropriate material selection and operational and maintenance practices (Moniz and Pollock, "Process Industries Corrosion," NACE Houston, TX, 1986; Degnan, "Corrosion in the Chemical Processing Industry, Metals Handbook," 9th ed., vol. 13, "Corrosion," ASM International, Metals Park, OH, 1987; Aller et al., "First International Symposium on Process Industry Piping," Materials Technology Institute, Houston, TX, 1993; Dillon, "Materials Selection for the Chemical Process Industries," McGraw-Hill, 1992; Dillon, "Corrosion Control in the Chemical Process Industries," McGraw-Hill, New York, 1986).

6.6 PAINTS AND PROTECTIVE COATINGS
Revised by Staff

REFERENCES: Keane (ed.). "Good Painting Practice," Steel Structures Painting Council. Levinson and Spindel, "Recent Developments in Architectural and Maintenance Painting." Levinson, "Electrocoat, Powder Coat, Radiate." Reprints, *Journal of Coatings Technology,* 1315 Walnut St., Philadelphia. Roberts, "Organic Coatings—Their Properties, Selection and Use," Government Printing Office. "Paint/Coatings Dictionary," Federation of Societies for Coatings Technology. Weismantel (ed.), "Paint Handbook," McGraw-Hill. Publications of the National Paint and Coatings Association.

Paint is a mixture of filmogen (film-forming material, binder) and pigment. The pigment imparts color, and the filmogen, continuity; together, they create opacity. Most paints require volatile thinner to reduce their consistencies to a level suitable for application. An important exception is the powder paints made with fusible resin and pigment.

In **conventional oil-base paint,** the filmogen is a vegetable oil. Driers are added to shorten drying time. The thinner is usually petroleum spirits.

In **water-thinned paints,** the filmogen may be a material dispersible in water, such as solubilized linseed oil or casein, an emulsified polymer, such as butadiene-styrene or acrylate, a cementitious material, such as portland cement, or a soluble silicate.

Varnish is a blend of resin and drying oil, or other combination of filmogens, in volatile thinner. A solution of resin alone is a **spirit varnish,** e.g., shellac varnish. A blend of resin and drying oil is an **oleoresinous varnish,** e.g., spar varnish. A blend of nonresinous, nonoleaginous filmogens requiring a catalyst to promote the chemical reaction necessary to produce a solid film is a **catalytic coating.**

Enamel is paint that dries relatively harder, smoother, and glossier than the ordinary type. These changes come from the use of varnish or synthetic resins instead of oil as the liquid portion. The varnish may be oleoresinous, spirit, or catalytic.

Lacquer is a term that has been used to designate several types of painting materials; it now generally means a spirit varnish or enamel, based usually on cellulose nitrate or cellulose acetate butyrate, acrylate or vinyl.

PAINT INGREDIENTS

The ingredients of paints are drying and semidrying vegetable oils, resins, plasticizers, thinners (solvents), driers or other catalysts, and pigments.

Drying oils dry (become solid) when exposed in thin films to air. The drying starts with a chemical reaction of the oil with oxygen. Subsequent or simultaneous polymerization completes the change. The most important drying oil is linseed oil. Addition of small percentages of driers shortens the drying time.

Soybean oil, with poorer drying properties than linseed oil, is classed as a **semidrying oil. Tung oil** dries much faster than linseed oil, but the film wrinkles so much that it is reminiscent of frost; heat treatment eliminates this tendency. Tung oil gels after a few minutes cooking at high temperatures. It is more resistant to alkali than linseed oil. Its chief use is in oleoresinous varnishes. **Oiticica oil** resembles tung oil in many of its properties. **Castor oil** is nondrying, but heating chemically "dehydrates" it and converts it to a drying oil. The dehydrated oil resembles tung oil but is slower drying and less alkali-resistant. **Nondrying oils,** like coconut oil, are used in some baking enamels. An oxidizing hydrocarbon oil, made from petroleum, contains no esters or fatty acids.

The drying properties of oils are linked to the amount and nature of unsaturated compounds. By molecular distillation or solvent extraction the better drying portions of an oil can be separated. Drying is also improved by changing the structure of the unsaturated

compounds (isomerization) and by modifying the oils with small amounts of such compounds as phthalic acid and maleic acid.

Driers are oil-soluble compounds of certain metals, mainly lead, manganese, and cobalt. They accelerate the drying of coatings made with oil. The metals are introduced into the coatings by addition of separately prepared compounds. Certain types of synthetic nonoil coatings (catalytic coatings) dry by baking or by the action of a **catalyst** other than conventional drier. Drier action begins only after the coating has been applied. Catalyst action usually begins immediately upon its addition to the coating; hence addition is delayed until the coating is about to be used.

Resins Both natural and synthetic resins are used. **Natural resins** include fossil types from trees now extinct, recent types (rosin, Manila, and dammar), lac (secretion of an insect), and asphalts (gilsonite). **Synthetic resins** include ester gum, phenolic, alkyd, urea, melamine, amide, epoxy, urethane, vinyl, styrene, rubber, petroleum, terpene, cellulose nitrate, cellulose acetate, and ethyl cellulose.

Latex filmogens contain a stable aqueous dispersion of synthetic resin produced by emulsion polymerization as the principal constituent.

Plasticizers In an oleoresinous varnish the oil acts as a softening agent or plasticizer for the resin. There are other natural compounds and substances, and many synthetic ones, for plasticizing film formers such as the cellulosics. Plasticizers include dibutyl phthalate, tricresyl phosphate, dibutyl sebacate, dibutyltartrate, tributyl citrate, methyl abietate, and chlorinated biphenyl.

Thinners As the consistency of the film-forming portion of most paints and varnishes is too high for easy application, thinners (**solvents**) are needed to reduce it. If the nonvolatile portion is already liquid, the term thinner is used; if solid, the term solvent is used. **Petroleum** or **mineral spirits** has replaced turpentine for thinning paints in the factory. The coal-tar products—**toluene, xylene,** and **solvent naphtha**—are used where solvency better than that given by petroleum spirits is needed. Esters, alcohols, and ketones are standard for cellulosic or vinyl lacquers. Finally, water is used to thin latex paints, to be thinned at point of use.

Pigments may be natural or synthetic, organic or inorganic, opaque or nonopaque, white or colored, chemically active or inert. Factors entering into the selection of a pigment are color, opacity, particle size, compatibility with other ingredients, resistance to light, heat, alkali, and acid, and cost. The most important pigments are: **white**—zinc oxide, leaded zinc oxide, titanium dioxide; **red**—iron oxides, red lead* (for rust prevention rather than color), chrome orange, molybdate orange, toluidine red, para red; **yellow**—iron oxide, chrome yellow, zinc yellow (for rust prevention rather than color), Hansa; **green**—chrome green, chrome oxide; **blue**—iron, ultramarine, phthalocyanine; **extenders** (nonopaque), i.e., magnesium silicate, calcium silicate, calcium carbonate, barium sulfate, aluminum silicate.

Miscellany Paint contains many other ingredients, minor in amount but important in function, such as emulsifiers, antifoamers, leveling agents, and thickeners.

Ecological Considerations Many jurisdictions restrict the amounts of volatile organic compounds that may be discharged into the atmosphere during coating operations. Users should examine container labels for applicable conditions.

PAINTS

Aluminum paint is a mixture of aluminum pigment and varnish, from 1 to 2 lb of pigment per gal of varnish. The aluminum is in the form of thin flakes. In the paint film, the flakes "leaf"; i.e., they overlap like

---

* The use of lead pigment (red lead) is prohibited by law. (See discussion below under Painting Steel.)

leaves fallen from trees. Leafing gives aluminum paint its metallic appearance and its impermeability to moisture. Aluminum paint ranks high as a reflector of the sun's radiation and as a retainer of heat in hot-air or hot-water pipes or tanks.

Bituminous Paint   Hard asphalts, like gilsonite, cooked with drying oils, and soft asphalts and coal tar, cut back with thinner, may be used to protect metal and masonry wherever their black color is not objectionable.

Chemical-Resistant Paint   Chemical resistance is obtained by use of resins such as vinyls, epoxies and urethanes. Vinyls are air-drying types. Epoxies are catalytically cured with acids or amino resins. The latter are two-component paints to be mixed just prior to use. Urethanes can be two-component also, or single-component paints that react with moisture in the air to cure.

Emulsion paints are emulsions in water of oil or resin base mixed with pigment. The **latex paints** (acrylic, butadiene-styrene, polyvinylacetate, etc.) belong to this class. Emulsion or latex paints are designed for both exterior and interior use.

Strippable coatings are intended to give temporary protection to articles during storage or shipment.

Powder coatings are 100 percent solids coatings applied as a dry powder mix of resin and pigment and subsequently formed into a film with heat. The solid resin binder melts upon heating, binds the pigment, and results in a pigment coating upon cooling. The powder is applied either by a electrostatic spray or by passing the heated object over a fluidized bed of powder with subsequent oven heating to provide a smooth continuous film.

Marine Paint   Ocean-going steel ships require antifouling paint over the anticorrosive primer. Red lead or zinc yellow is used in the primer. **Antifouling** paints contain ingredients, such as cuprous oxide and mercuric oxide, that are toxic to barnacles and other marine organisms. During World War II, the U.S. Navy developed **hot plastic** ship-bottom paint. It is applied by spray gun in a relatively thick film and dries or sets as it cools. **Copper powder paint** is suitable for small wooden craft.

Fire-Retardant Paint   Most paint films contain less combustible matter than does wood. To this extent, they are fire-retardant. In addition they cover splinters and fill in cracks. However, flames and intense heat will eventually ignite the wood, painted or not. The most that ordinary paint can do is to delay ignition. Special compositions fuse and give off flame-smothering fumes, or convert to spongy heat-insulating masses, when heated. The relatively thick film and low resistance to abrasion and cleaning make these coatings unsuited for general use as paint, unless conventional decorative paints are applied over them. However, steady improvement in paint properties is being made.

Traffic paint is quick-drying, so that traffic is inconvenienced as little as possible. Rough texture and absence of gloss increase visibility. Tiny glass beads may be added for greater night visibility. Catalytically cured resinous coating are also a recent development for fast-drying traffic paints.

Heat-Reflecting Paint   Light colors reflect more of the sun's radiation than do dark colors. White is somewhat better than aluminum, but under some conditions, aluminum may retain its reflecting power longer.

Heat-Resistant Paint   Silicone paint is the most heat-resistant paint yet developed. Its first use was as an insulation varnish for electric motors, but types for other uses, such as on stoves and heaters, have been developed. Since it must be cured by baking, it is primarily a product finish.

Fungicides should be added to paint that is to be used in bakeries, breweries, sugar refineries, dairies, and other places where fungi flourish. Until ecological considerations forbade their use, mercury compounds were important fungicides. Their place has been filled by chlorinated phenols and other effective chemical compounds. Fungi-infected surfaces should be sterilized before they are painted. Scrub them with mild alkali; if infection is severe, add a disinfectant.

Wood preservatives of the paintable type are used extensively to treat wood windows, doors, and cabinets. They reduce rotting that may be promoted by water that enters at joints. They comprise solutions of compounds, such as chlorinated phenols, and copper and zinc naphthenates in petroleum thinner. Water repellants, such as paraffin wax, are sometimes added but may interfere with satisfactory painting.

Clear **water repellents** are treatments for masonry to prevent wetting by water. Older types are solutions of waxes, drying oils, or metallic soaps. Silicone solutions, recently developed, are superior, both in repelling water and in effect on appearance.

Zinc-Rich Primer   Anticorrosive primers for iron and steel, incorporating zinc dust in a concentration sufficient to give electrical conductivity in the dried film, enable the zinc metal to corrode preferentially to the substrate, i.e., to give cathodic protection. The binders can be organic resins or inorganic silicates.

Painting

Exterior Architectural Painting   Surfaces must be clean and dry, except that wood to be painted with emulsion paint will tolerate a small amount of moisture and masonry to be painted with portland cement-base paint should be damp. Scrape or melt the resin from the knots in wood, or scrub it off with paint thinner or alcohol. As an extra precaution, seal the knots with shellac varnish, aluminum paint, or proprietary knot sealer. When painting wood for the first time, fill nail holes and cracks with putty, after the priming coat has dried. If the wood has been painted before, remove loose paint with a scraper, wire brush, or sandpaper. Prime all bare wood. In bad cases of cracking and scaling, remove all the paint by dry scraping or with paint and varnish remover or with a torch.

Paint on new wood should be from 4 to 5 mils (0.10 to 0.13 mm) thick. This usually requires three coats. A system consisting of nonpenetrating primer and special finish coat may permit the minimum to be reached with two coats. Repainting should be frequent enough to preserve a satisfactory thickness of the finish coat. Nonpenetrating primers stick on some types of wood better than regular house paint does.

Interior Architectural Painting   Interior paints are used primarily for decoration, illumination, and sanitation. Enamels usually are used for kitchen and bathroom walls, where water resistance and easy cleaning are needed; semiglossy and flat types are used on walls and ceilings where it is desired to avoid glare. Wet plaster will eventually destroy oil-base paints. Fresh plaster must be allowed to dry out. Water from leaks and condensation must be kept away from aged plaster. Several coats of paint with low permeability to water vapor make an effective vapor barrier.

Painting Concrete and Masonry   Moisture in concrete and masonry brings alkali to the surface where it can destroy oil paint. Allow 2 to 6 months for these materials to dry out before painting; a year or more, if they are massive. Emulsion or latex paints may be used on damp concrete or masonry if there is reason to expect the drying to continue. Portland-cement paint, sometimes preferred for masonry, is especially suitable for first painting of porous surfaces such as cinder block. Before applying this paint, wet the surface so that capillarity will not extract the water from the paint. Scrubbing the paint into the surface with a stiff brush of vegetable fiber gives the best job, but good jobs are also obtained with usual brushes or by spraying. Keep the paint damp for 2 or 3 days so that it will set properly. Latex and vinyl paints are much used on concrete, because of their high resistance to alkali.

Painting Steel   Preparation of steel for painting, type of paint, and condition of exposure are closely related. Methods of preparation, arranged in order of increasing thoroughness, include (1) removal of oil with solvent; (2) removal of dirt, loose rust, and loose mill scale with scraper or wire brush; (3) flame cleaning; (4) sandblasting; (5) pickling; (6) phosphating. Exposure environments, arranged in order of increasing severity, include (*a*) dry interiors, or arid regions; (*b*) rural or light industrial areas, normally dry; (*c*) frequently wet; (*d*) continuously wet; (*e*) corrosive chemical. Paint systems for condition (*a*) often consist of a single coat of low-cost paint. For other conditions, the systems comprise one or two coats of rust-inhibitive primer, such as a zinc-rich primer, and one or more finish coats, selected according to severity of conditions.

The primer contains one or more rust-inhibitive pigments, selected mainly from red lead (found mostly in existing painted surfaces), zinc yellow, and zinc dust. It may also contain zinc oxide, iron oxide, and extender pigments. Of equal importance is the binder, especially for the top coats. For above-water surfaces, linseed oil and alkyd varnish give good service; for underwater surfaces, other binders, like phenolic and vinyl resin, are better. Chemical-resistant binders include epoxy, synthetic rubber, chlorinated rubber, vinyl, urethanes, and neoprene.

Paint for structural steel is normally air-drying. Large percentages for factory-finished steel products are catalytic-cured, or are baked.

By virtue of the health problems associated with lead-based paint, this product has effectively been proscribed by law from further use. A problem arises in the matter of removal and repainting areas which have previously been coated with lead-based paints; this is particularly true in the case of steel structures such as bridges, towers, and the like.

Removal of leaded paints from such structures requires that extreme care be exercised in the removal of the paint, due diligence being paid to the protection of not only the personnel involved therein but also the general public and that, further, the collection and disposal of the hazardous waste be effectuated in accordance with prescribed practice. Reference to current applicable OSHA documents will provide guidance in this regard.

The removal of leaded paints can be accomplished by chipping, sand blasting, use of solvents (strippers), and so forth. In all cases, stringent precautions must be emplaced to collect all the residue of old paint as well as the removal medium. Alternatively, recourse may be had to topcoating the existing leaded paint, effectively sealing the surface of the old paint and relying on the topcoating medium to keep the leaded undercoat from flaking away. In this type of application, urethane coatings, among others, have proved suitable. While the topcoating procedure will alleviate current problems in containment of existing leadpainted surfaces, it is not likely that this procedure will prove to be the final solution. Ultimately, as with any painted surface, there will come a time when the leaded-paint subcoat can no longer be contained in situ and must be removed. It is expected that better and more economical procedures will evolve in the near future, so that the extant problems associated with removal of leaded paint will become more tractable. The AISI and AASHTO have collaborated in this regard and have issued a guide addressing these problems.

Galvanized Iron   Allow new galvanized iron to weather for 6 months before painting. If there is not enough time, treat it with a proprietary etching solution. For the priming coat, without pretreatment, a paint containing a substantial amount of zinc dust or portland cement should be used. If the galvanizing has weathered, the usual primers for steel are also good.

Painting Copper   The only preparation needed is washing off any grease and roughening the surface, if it is a polished one. Special primers are not needed. Paint or varnish all copper to prevent corrosion products from staining the adjoining paint.

Painting Aluminum   The surface must be clean and free from grease. Highly polished sheet should be etched with phosphoric acid or chromic acid. Zinc-yellow primers give the best protection against corrosion. The only preparation needed for interior aluminum is to have the surface clean; anticorrosive primers are not needed.

Magnesium and its alloys corrode readily, especially in marine atmospheres. Red-lead primers must not be used. For factory finishing, it is customary to chromatize the metal and then apply a zinc-yellow primer.

Water-Tank Interiors   Among the best paints for these are asphaltic compositions. For drinking-water tanks, select one that imparts no taste. Zinc-dust paints made with phenolic varnish are also good.

Wood Products   Finishes may be lacquer or varnish, or their corresponding enamels. High-quality clear finishes may require many operations, such as sanding, staining, filling, sealing, and finishing. Furniture finishing is done mostly by spray. Small articles are often finished by tumbling; shapes like broom handles, by squeegeeing.

Plastics   Carefully balanced formulations are necessary for satisfactory adhesion, to avoid crazing and to prevent migration of plasticizer.

Paint Deterioration   Heavy dew, hot sun, and marine atmospheres shorten the life of paint. Industrial zones where the atmosphere is contaminated with hygroscopic and acidic substances make special attention to painting programs necessary. Dampness within masonry and plaster walls brings alkali to the surface where it can destroy oil-base paints. Interior paints on dry surfaces endure indefinitely; they need renewal to give new color schemes or when it becomes impractical to wash them. Dry temperate climates are favorable to long life of exterior paints.

Application   Most industrial finishing is done with spray guns. In electrostatic spraying, the spray is charged and attracted to the grounded target. Overspray is largely eliminated. Other methods of application include dipping, electrocoating, flowing, tumbling, doctor blading, rolling, fluid bed, and screen stenciling.

An increasing proportion of maintenance painting is being done with spray and hand roller. The spray requires up to 25 percent more paint than the brush, but the advantage of speed is offsetting. The roller requires about the same amount of paint as the brush.

Dry finishing is done by flame-spraying powdered pigment-filmogen compositions or by immersing heated articles in a **fluid bed** of the powdered composition.

Spreading Rates   When applied by brush, approximate spreading rates for paints on various surfaces are as shown in Table 6.6.1.

OTHER PROTECTIVE AND DECORATIVE COATINGS

Porcelain enamel as applied to metal or clay products is a vitreous, inorganic coating which is set by firing. Enameling is done in several different ways, but the resulting surface qualities and properties are largely similar. The coating finds utility in imparting to the product a surface which is hard and scratch-resistant, possesses a remarkable resistance to general corrosion, and perhaps even has the constituents of the coating material tailored for the enamel to be resistant to specific working environments.

Table 6.6.1   Spreading Rates for Brushed Paints

Surface	Type	First coat ft²/gal	First coat m²/L	Second and third coats ft²/gal	Second and third coats m²/L
Wood	Oil emulsion	400–500	10–12	500–600	12–15
Wood, primed	Emulsion	500–600	12–15	500–600	12–15
Structural steel	Oil	450–600	11–15	650–900	16–22
Sheet metal	Oil	500–600	12–15	550–650	13–16
Brick, concrete	Oil	200–300	5–7	400–500	10–12
Brick, concrete	Cement	100–125	2.5–3	125–175	3–4.5
Brick, concrete	Emulsion	200–300	5–7	400–500	10–12
Smooth plaster	Oil	550–650	13–16	550–650	13–16
Smooth plaster	Emulsion	400–500	10–12	500–600	12–15
Concrete floor	Enamel	400–500	10–12	550–650	13–16
Wood floor	Varnish	550–650	13–16	550–650	13–16

Frit is the basic unfired raw coating material, composed of silica, suitable fluxes, and admixtures to impart either opacity or color.

The surface to be enameled must be rendered extremely clean, after which a tightly adhering undercoat is applied. The surface is covered with powdered frit, and the product finally is fired and subsequently cooled. Despite its demonstrated advantageous properties, porcelain enamel has effectively zero tolerance for deformation, and it will craze or spall easily when mechanically distressed.

Fused dry resin coatings employing polymers as the active agent can be bonded to most metals, with the resin applied in a fluidized bed or by means of electrostatic deposition. The surface to be treated must be extremely clean. Resinous powders based on vinyl, epoxy, nylon, polyethylene, and the like are used. Intermediate bonding agents may be required to foster a secure bond between the metal surface and the resin. After the resin is applied, the metal piece is heated to fuse the resin into a continuous coating. The tightly bonded coating imparts corrosion resistance and may serve decorative functions.

Bonded phosphates provide a corrosion-resistant surface coat to metals, principally steel. Usually, the phosphated surface layer exhibits a reduced coefficient of friction and for this reason can be applied to sheet steel to be deep-drawn, resulting in deeper draws per die pass. Passing the raw sheet metal through an acid phosphate bath results in the deposition of an insoluble crystalline phosphate on the surface.

Hot dipping permits the deposition of metal on a compatible substrate and serves a number of purposes, the primary ones being to impart corrosion resistance and to provide a base for further surface treatment, e.g., paint. In **hot dip galvanizing,** a ferrous base metal, usually in the form of sheet, casting, or fabricated assembly, is carefully cleaned and then passed through a bath of molten zinc. The surface coating is metallographically very complex, but the presence of zinc in the uppermost surface layer provides excellent corrosion resistance. In the event that the zinc coating is damaged, leaving the base metal exposed to corrosive media, galvanic action ensues and zinc will corrode preferentially to the base metal. This sacrificial behavior thereby protects the base metal. (See Secs. 6.4 and 6.5.)

Automotive requirements in recent years have fostered the use of galvanized sheet steel in the fabrication of many body components, often with the metal being galvanized on one side only.

Hardware and fittings exposed to salt water historically have been hot dip galvanized to attain maximum service life.

Terne, a lead coating applied to steel (hence, **terne sheet),** is applied by hot dipping. The sheet is cleaned and then passed through a bath of molten lead alloyed with a small percentage of tin. The product is widely applied for a variety of fabricated pieces and is employed architecturally when its dull, though lasting, finish is desired. It exhibits excellent corrosion resistance, is readily formed, and lends itself easily to being soldered.

Lead-coated copper is used largely as a substitute for pure copper when applied as an architectural specialty for roof flashing, valleys, and the like, especially when the design requires the suppression of green copper corrosion product (verdigris). Its corrosion resistance is effectively equivalent to that of pure lead. It is produced by dipping copper sheet into a bath of molten lead.

Electroplating, or **electrodeposition,** employs an electric current to coat one metal with another. The method can be extended to plastics, but requires that the plastic surface be rendered electrically conductive by some means. In the basic electroplating circuit, the workpiece is made the cathode (negative), the metal to be plated out is made the anode (positive), and both are immersed in an electrolytic bath containing salts of the metal to be plated out. The circuit is completed with the passage of a direct current, whence ions migrate from anode to cathode, where they lose their charge and deposit as metal on the cathode. (In essence, the anode becomes sacrificial to the benefit of the cathode, which accretes the metal to be plated out.) Most commercial metals and some of the more exotic ones can be electroplated; they include, among others, cadmium, zinc, nickel, chromium, copper, gold, silver, tin, lead, rhodium, osmium, selenium, platinum, and germanium. This list is not exclusive; other metals may be pressed into use as design requirements arise from time to time.

Plating most often is used to render the workpiece more corrosion-resistant, but decorative features also are provided or enhanced thereby. On occasion, a soft base metal is plated with a harder metal to impart wear resistance; likewise, the resultant composite material may be designed to exhibit different and/or superior mechanical or other physical properties.

Electroplating implies the addition of metal to a substrate. Accordingly, the process lends itself naturally to buildup of worn surfaces on cutting and forming tools and the like. This buildup must be controlled because of its implications for the finished dimensions of the plated part.

The most familiar example we observe is the universal use of chromium plating on consumer goods and pieces of machinery. Gold plating and silver plating abound. Germanium is electroplated in the production of many solid-state electronic components. Brass, while not a pure metal, can be electroplated by means of a special-composition electrolyte.

Of particular concern in the current working environment is the proper handling and disposition of spent plating solutions. There are restrictions in this regard, and those people engaged in plating operations must exercise stringent precautions in the matter of their ultimate disposal.

Aluminum normally develops a natural thin surface film of aluminum oxide which acts to inhibit corrosion. In **anodizing,** this surface film is thickened by immersion in an electrolytic solution. The protection provided by the thicker layer of oxide is increased, and for many products, the oxide layer permits embellishment with the addition of a dye to impart a desired color. (See Sec. 6.4.)

Vacuum deposition of metal onto metallic or nonmetallic substrates provides special functional or decorative properties. Both the workpiece to be coated and the metal to be deposited are placed in an evacuated chamber, wherein the deposition metal is heated sufficiently to enable atom-size particles to depart from its surface and impinge upon the surface to be coated. The atomic scale of the operation permits deposition of layers less than 1 μm thick; accordingly, it becomes economical to deposit expensive jewelry metals, e.g., gold and silver, on inexpensive substrates. Some major applications include production of highly reflective surfaces, extremely thin conductive surface layers for electronic components, and controlled buildup of deposited material on a workpiece.

A companion process, **sputtering,** operates in a slightly different fashion, but also finds major application in deposition of 1-μm thicknesses on substrates. Other than for deposition of 1-μm thicknesses on solid-state components and circuitry, it enables deposition of hard metals on edges of cutting tools.

Chemical vapor deposition (CVD) is employed widely, but is noted here for its use to produce diamond vapors at ambient temperatures, which are deposited as continuous films from 1 to 1,000 μm thick. The CVD diamond has a dense polycrystalline structure with discrete diamond grains, and in this form it is used to substitute for natural and artificial diamonds in abrasive tools.

VARNISH

Oleoresinous varnishes are classed according to oil length, i.e., the number of gallons of oil per 100 lb of resin. Short oil varnishes contain up to 10 gal of oil per 100 lb of resin; medium, from 15 to 25 gal; long, over 30 gal.

Floor varnish is of medium length and is often made with modified phenolic resins, tung, and linseed oils. It should dry overnight to a tough hard film. Some floor varnishes are rather thin and penetrating so that they leave no surface film. They show scratches less than the orthodox type. Moisture-cured urethane or epoxy ester varnishes are more durable types.

Spar varnish is of long oil length, made usually with phenolic or modified-phenolic resins, tung or dehydrated castor oil, and linseed oil. Other spar varnishes are of the alkyd and urethane types. Spar varnishes dry to a medium-hard glossy film that is resistant to water, actinic rays of the sun, and moderate concentrations of chemicals.

Chemical-resistant varnishes are designed to withstand acid, alkali, and other chemicals. They are usually made of synthetic resins, such as chlorinated rubber, cyclized rubber, phenolic resin, melamine, ureaaldehyde, vinyl, urethane, and epoxy resin. Some of these must be dried by baking. Coal-tar epoxy and coal-tar urethane combinations are also used. **Catalytic varnish** is made with a nonoxidizing film former, and is cured with a catalyst, such as hydrochloric acid or certain amines. **Flat varnish** is made by adding materials, such as finely divided silica or metallic soap, to glossy varnish. Synthetic latex or other emulsion, or aqueous dispersion, such as glue, is sometimes called **water varnish.**

LACQUER

The word *lacquer* has been used for (1) spirit varnishes used especially for coating brass and other metals, (2) Japanese or Chinese lacquer, (3) coatings in which cellulose nitrate (**pyroxylin**), cellulose acetate, or cellulose acetate butyrate is the dominant ingredient, (4) oleoresinous baking varnishes for interior of food cans.

Present-day lacquer primarily refers to cellulosic coatings, clear or pigmented. These lacquers dry by evaporation. By proper choice of solvents, they are made to dry rapidly. Besides cellulosic compounds, they contain resins, plasticizers, and solvent. Cellulose acetate, cellulose acetate butyrate, and cellulose acetate propionate lacquers are nonflammable, and the clear forms have better exterior durability than the cellulose nitrate type. Although cellulosic and acrylic derivatives dominate the lacquer field, compositions containing vinyls, chlorinated hydrocarbons, or other synthetic thermoplastic polymers are of growing importance.

Lac is a resinous material secreted by an insect that lives on the sap of certain trees. Most of it comes from India. After removal of dirt, it is marketed in the form of grains, called seed-lac; cakes, called button lac; or flakes, called shellac. Lac contains up to 7 percent of wax, which is removed to make the refined grade. A bleached, or white, grade is also available.

Shellac varnish is made by "cutting" the resin in alcohol; the cut is designated by the pounds of lac per gallon of alcohol, generally 4 lb. Shellac varnish should always be used within 6 to 12 months of manufacture, as some of the lac combines with the alcohol to form a soft, sticky material.

6.7 WOOD

by Staff, Forest Products Laboratory, USDA Forest Service. Prepared under the direction of David W. Green

(*Note*: This section was written and compiled by U.S. government employees on official time. It is, therefore, in the public domain and not subject to copyright.)

REFERENCES: Freas and Selbo, Fabrication and Design of Glued Laminated Wood Structural Members, *U.S. Dept. Agr. Tech. Bull.*, no. 1069, 1954. Hunt and Garratt, "Wood Preservation," McGraw-Hill. "Wood Handbook: Wood as an Engineering Material," Gen. Tech. Rep. FPL-GTR-113, rev. 1999. "Design Properties of Round, Sawn and Laminated Preservatively Treated Construction Poles as Posts," ASAE Eng. Practice EP 388.2. American Society of Agricultural Engineering Standards, AF&PA, "National Design Specification (NDS) for Wood Construction," American Forest & Paper Association, Washington, 2005. AITC, "Recommended Practice for Protection of Structural Glued Laminated Timber during Transit, Storage, and Erection," AITC III, American Institute of Timber Construction, Englewood, CO, 1994. "Laminated Timber Design Guide," American Institute of Timber Construction, Englewood, CO, 1994. "Timber Construction Manual," Wiley, New York, 1994. "American National Standards for Wood Products—Structural Glued Laminated Timbers," ANSI/AITC A190.1-1992, American National Standards Institute, New York, 1992. ASTM International, "Annual Book of Standards," vol. 04.09, "Wood," American Society for Testing and Materials, Philadelphia, 1993; "Standard Terminology Relating to Wood," ASTM D 9; "Standard Practice for Establishing Structural Grades and Related Allowable Properties for Visually Graded Dimension Lumber," ASTM D 245; "Standard Definitions of Terms Relating to Veneer and Plywood," ASTM D 1038; "Standard Nomenclature of Domestic Hardwoods and Softwoods," ASTM D 1165; "Standard Practice for Establishing Allowable Properties for Visually Graded Dimension Lumber from In-Grade Tests of Full Size Specimens," ASTM D. 1990; "Standard Methods for Establishing Clear Wood Strength Values," ASTM D 2555; "Standard Practice for Evaluating Allowable Properties of Grades of Structural Lumber," ASTM D 2915; "Standard Test Methods for Mechanical Properties of Lumber and Wood Based Structural Material," ASTM D 4761; "Standard Test Method for Surface Burning Characteristics of Building Materials," ASTM E 84; "Standard Test Methods for Fire Tests of Building Construction and Materials," ASTM E 119. ASTM International, "Annual Book of Standards," vol. 04.10. American Society for Testing and Materials, Philadelphia, 1995: "Standard Method for Establishing Design Stresses for Round Timber Piles," ASTM D 2899; "Standard Practice for Establishing Stress Grades for Structural Members Used in Log Buildings," ASTM D 3957; "Standard Practice for Establishing Stresses for Structural Glued Laminated Timber," ASTM D 3737; "Standard Specification for Evaluation for Structural Composite Lumber Products," ASTM D 5456. AWPA, "Book of Standards," American Wood Preserver's Association, Selma, AL, 2005. American Wood Systems, "Span Table for Glulam Timber," APA-EWA, Tacoma, WA, 1994. Cassens and Feist, "Exterior Wood in the South—Selection, Applications, and Finishes," Gen. Tech. Rep. FPL-GTR-69, USDA Forest Service, Forest Products Laboratory, Madison, WI, 1991. Chudnoff, Tropical Timbers of the World, *Agric. Handb.* no. 607, 1984. Forest Products Laboratory, "Wood Handbook: Wood as an Engineering Material," Gen. Tech. Rep. FPL-GTR-113, 1999. Green, "Moisture Content and the Shrinkage of Lumber," Res. Pap. FPL-RP-489, USDA Forest Service, Forest Products Laboratory, Madison WI, 1989. Green and Kretschmann, "Moisture Content and the Properties of Clear Southern Pine," Res. Pap. FPL-RP-531, USDA Forest Service, Forest Products Laboratory, Madison, WI, 1995. Green, Evans, and Craig, Durability of Structural Lumber Products at High Temperatures: Part I, 60°C at 75% RH and 82°C at 30% RH, *Wood Figer Sci.* **35**, no. 4, 2003, pp. 499–523. James, "Dielectric Properties of Wood and Hardboard: Variation with Temperature, Frequency, Moisture Content, and Grain Orientation," Res. Pap. FPL-RP-245, USDA Forest Service, Forest Products Laboratory, Madison, WI, 1975. James, "Electric Moisture Meters for Wood," Gen. Tech. Rep. FPL-GTR-6, USDA Forest Service, Forest Products Laboratory, Madison, WI, 1988. Kohara and Okamoto, Studies of Japanese Old Timbers, *Sci. Rpts. Saikyo. Univ.* no. 7, 1995, pp. 9–20. MacLean, Thermal Conductivity of Wood, *Heat, Piping, Air Cond.* **13**, no. 6, 1941, pp. 380–391. AF&PA, "Design Values for Wood Construction: A Supplement to the National Design Specification for Wood Construction," American Forest & Paper Association, Washington, 2005. Pellerin and Ross, "Nondestructive Evaluation of Wood," Forest Products Society, Madison, WI, 2002. TenWolde et al., "Thermal Properties of Wood and Wood Panel Products for Use in Buildings," ORNL/Sub/87-21697/1, Oak Ridge National Laboratory, Oak Ridge, TN, 1988. Weatherwax and Stamm, The Coefficients of Thermal Expansion of Wood and Wood Products, *Trans. ASME*, **69**, no. 44, 1947, pp. 421–432. Wilkes, Thermo-Physical Properties Data Base Activities at Owens-Corning Fiberglas, in "Thermal Performance of the Exterior Envelopes of Buildings," ASHRAE SP 28, American Society of Heating, Refrigerating, and Air-Conditioning Engineers, Atlanta, 1981.

COMPOSITION, STRUCTURE, AND NOMENCLATURE

by David W. Green

Wood is a naturally formed organic material consisting essentially of elongated tubular elements called cells arranged in a parallel manner for the most part. These cells vary in dimensions and wall thickness with position in the tree, age, conditions of growth, and kind of tree. The walls of the cells are formed principally of chain molecules of cellulose, polymerized from glucose residues and oriented as a partly crystalline material. These chains are aggregated in the cell wall at a variable angle, roughly parallel to the axis of the cell. The cells are cemented by an amorphous material called lignin. The complex structure of the gross wood approximates a rhombic system. The direction parallel to the grain and the axis of the stem is longitudinal (L), the two axes across the grain are radial (R) and tangential (T) with respect to the cylinder of the tree stem. This anisotropy and the molecular orientation account for the major differences in physical and mechanical properties with respect to direction which are present in wood.

Natural variability of any given physical measurement in wood approximates the normal probability curve. It is traceable to the difference in the growth of individual samples and at present cannot be controlled. For engineering purposes, statistical evaluation is employed for determination of safe working limits.

Lumber is classified as **hardwood,** which is produced by the broad-leafed trees (angiosperms), such as oak, maple, and ash; and **softwood,** the product of coniferous trees (gymnosperms), such as pines, larch, spruce, and hemlock. The terms *hard* and *soft* have no relation to the actual hardness of the wood. **Sapwood** is the living wood of pale color on the outside of the stem. **Heartwood** is the inner core of physiologically inactive wood in a tree and is usually darker, somewhat heavier, due to infiltrated material, and more decay-resistant than the sapwood. Other terms relating to wood, veneer, and plywood are defined in ASTM D 9, D 1038, and the "Wood Handbook."

Standard nomenclature of lumber is based on commercial practice which groups woods of similar technical qualities but separate botanical identities under a single name. For listings of domestic hardwoods and softwoods, see ASTM D 1165 and the "Wood Handbook."

The chemical composition of woody cell walls is generally about 40 to 50 percent cellulose, 15 to 35 percent lignin, less than 1 percent mineral, 20 to 35 percent hemicellulose, and the remainder extractable matter of a variety of sorts. Softwoods and hardwoods have about the same cellulose content.

PHYSICAL AND MECHANICAL PROPERTIES OF CLEAR WOOD

by David W. Green, Robert White, Anton TenWolde, William Simpson, Joseph Murphy, and Robert Ross

Moisture Relations

Wood is a hygroscopic material which contains water in varying amounts, depending upon the relative humidity and temperature of the surrounding atmosphere. **Equilibrium conditions** are established as shown in Table 6.7.1. The standard reference condition for wood is **oven-dry** weight, which is determined by drying at 100 to 105°C until there is no significant change in weight.

Moisture content is the amount of water contained in the wood, usually expressed as a percentage of the mass of the oven-dry wood. Moisture can exist in wood as **free water** in the cell cavities as well as water **bound** chemically within the intermolecular regions of the cell wall. The moisture content at which cell walls are completely saturated but at which no water exists in the cell cavities is called the **fiber saturation point.** Below the fiber saturation point, the cell wall shrinks as moisture is removed, and the physical and mechanical properties begin to change as a function of moisture content. **Air-dry wood** has a moisture content of 12 to 15 percent. **Green wood** is wood with moisture content above the fiber saturation point. The moisture content of green wood typically ranges from 40 to 250 percent.

Dimensional Changes

Shrinkage or **swelling** is a result of change in water content within the cell wall. Wood is dimensionally stable when the moisture content is above the fiber saturation point (about 28 percent for shrinkage estimates). Shrinkage is expressed as a percentage of the dimensional change based on the green wood size. Wood is an anisotropic material with respect to shrinkage. **Longitudinal shrinkage** (along the grain) ranges from 0.1 to 0.3 percent as the wood dries from green to oven-dry and is usually neglected. Wood shrinks most in the direction of the annual growth rings (**tangential shrinkage**) and about one-half as much across the rings (**radial shrinkage**). Average shrinkage values for a number of commercially important species are shown in Table 6.7.2. Shrinkage to any moisture condition can be estimated by assuming that the change is linear from green to oven-dry and that about one-half occurs in drying to 12 percent.

Swelling in polar liquids other than water is inversely related to the size of the molecule of the liquid. It has been shown that the tendency

to hydrogen bonding on the dielectric constant is a close, direct indicator of the swelling power of water-free organic liquids. In general, the strength values for wood swollen in any polar liquid are similar when there is equal swelling in the wood.

Swelling in aqueous solutions of sulfuric and phosphoric acids, zinc chloride, and sodium hydroxide above pH 8 may be as much as 25 percent greater in the transverse direction than in water. The transverse swelling may be accompanied by longitudinal shrinkage up to 5 percent. The swelling reflects a chemical change in the cell walls, and the accompanying strength changes are related to the degradation of the cellulose.

Dimensional stabilization of wood cannot be completely attained. Two or three coats of varnish, enamel, or synthetic lacquer may be 50 to 85 percent efficient in preventing short-term dimensional changes. Metal foil embedded in multiple coats of varnish may be 90 to 95 percent efficient in short-term cycling. The best long-term stabilization results from internal bulking of the cell wall by the use of materials such as phenolic resins polymerized in situ or water solutions of polyethylene glycol (PEG) on green wood. The presence of the bulking agents alters the properties of the treated wood. Phenol increases electrical resistance, hardness, compression strength, weight, and decay resistance but lowers the impact strength. Polyethylene glycol maintains strength values at the green wood level, reduces electric resistance, and can be finished only with polyurethane resins.

Mechanical Properties

Average **mechanical properties** determined from tests on clear, straight-grained wood at 12 percent moisture content are given in Table 6.7.2. Approximate standard deviation(s) can be estimated from

$$s = CX$$

where X is average value for species and

$$C = \begin{cases} 0.10 & \text{for specific gravity} \\ 0.22 & \text{for modulus of elasticity} \\ 0.16 & \text{for modulus of rupture} \\ 0.18 & \text{for maximum crushing strength parallel to grain} \\ 0.14 & \text{for compression strength perpendicular to grain} \\ 0.25 & \text{for tensile strength perpendicular to grain} \\ 0.25 & \text{for impact bending strength} \\ 0.10 & \text{for shear strength parallel to grain} \end{cases}$$

Relatively few data are available on tensile strength parallel to the grain. The modulus of rupture is considered to be a conservative estimate for the tensile strength of clear wood.

Mechanical properties remain constant as long as the moisture content is above the fiber saturation point. Below the fiber saturation point, properties generally increase with decreasing moisture content down to about 8 percent. Below about 8 percent moisture content, some properties, principally tensile strength parallel to the grain and shear strength, may decrease with further drying. An approximate adjustment for clear wood properties between about 8 percent moisture and green can be obtained by using an annual compound-interest type of formula

$$P_2 = P_1\left(1 + \frac{C}{100}\right)^{-(M_2 - M_1)}$$

where P_1 is the known property at moisture content M_1, P_2 is the property to be calculated at moisture content M_2, and C is the assumed percentage change in property per percentage change in moisture content. Values of P_1 at 12 percent moisture content are given in Table 6.7.2, and values of C are given in Table 6.7.3. For the purposes of property adjustment, green is assumed to be 23 percent moisture content. The formula should not be used with redwood and cedars. A more accurate adjustment formula is given in "Wood Handbook." Additional data and tests on green wood can be found in "Wood Handbook" and on the Web at www.fpl.fs.fed.us. Data on foreign species are given in "Tropical Timbers of the World."

Table 6.7.1 Moisture Content of Wood in Equilibrium with Stated Dry-Bulb Temperature and Relative Humidity

Temperature (dry-bulb)		Moisture content, % at various relative-humidity levels																			
°F	(°C)	5	10	15	20	25	30	35	40	45	50	55	60	65	70	75	80	85	90	95	98
30	−1.3	1.4	2.6	3.7	4.6	5.5	6.3	7.1	7.9	8.7	9.5	10.4	11.3	12.4	13.5	14.9	16.5	18.5	21.0	24.3	26.9
40	4.2	1.4	2.6	3.7	4.6	5.5	6.3	7.1	7.9	8.7	9.5	10.4	11.3	12.3	13.5	14.9	16.5	18.5	21.0	24.3	26.9
50	9.8	1.4	2.6	3.6	4.6	5.5	6.3	7.1	7.9	8.7	9.5	10.3	11.2	12.3	13.4	14.8	16.4	18.4	20.9	24.3	26.9
60	15	1.3	2.5	3.6	4.6	5.4	6.2	7.0	7.8	8.6	9.4	10.2	11.1	12.1	13.3	14.6	16.2	18.2	20.7	24.1	26.8
70	21	1.3	2.5	3.5	4.5	5.4	6.2	6.9	7.7	8.5	9.2	10.1	11.0	12.0	13.1	14.4	16.0	17.9	20.5	23.9	26.6
80	26	1.3	2.4	3.5	4.4	5.3	6.1	6.8	7.6	8.3	9.1	9.9	10.8	11.7	12.9	14.2	15.7	17.7	20.2	23.6	26.3
90	32	1.2	2.3	3.4	4.3	5.1	5.9	6.7	7.4	8.1	8.9	9.7	10.5	11.5	12.6	13.9	15.4	17.3	19.8	23.3	26.0
100	38	1.2	2.3	3.3	4.2	5.0	5.8	6.5	7.2	7.9	8.7	9.5	10.3	11.2	12.3	13.6	15.1	17.0	19.5	22.9	25.6
110	43	1.1	2.2	3.2	4.0	4.9	5.6	6.3	7.0	7.7	8.4	9.2	10.0	11.0	12.0	13.2	14.7	16.6	19.1	22.4	25.2
120	49	1.1	2.1	3.0	3.9	4.7	5.4	6.1	6.8	7.5	8.2	8.9	9.7	10.6	11.7	12.9	14.4	16.2	18.6	22.0	24.7
130	54	1.0	2.0	2.9	3.7	4.5	5.2	5.9	6.6	7.2	7.9	8.7	9.4	10.3	11.3	12.5	14.0	15.8	18.2	21.5	24.2
140	60	0.9	1.9	2.8	3.6	4.3	5.0	5.7	6.3	7.0	7.7	8.4	9.1	10.0	11.0	12.1	13.6	15.3	17.7	21.0	23.7
150	65	0.9	1.8	2.6	3.4	4.1	4.8	5.5	6.1	6.7	7.4	8.1	8.8	9.7	10.6	11.8	13.1	14.9	17.2	20.4	23.1
160	71	0.8	1.6	2.4	3.2	3.9	4.6	5.2	5.8	6.4	7.1	7.8	8.5	9.3	10.3	11.4	12.7	14.4	16.7	19.9	22.5
170	76	0.7	1.5	2.3	3.0	3.7	4.3	4.9	5.6	6.2	6.8	7.4	8.2	9.0	9.9	11.0	12.3	14.0	16.2	19.3	21.9
180	81	0.7	1.4	2.1	2.8	3.5	4.1	4.7	5.3	5.9	6.5	7.1	7.8	8.6	9.5	10.5	11.8	13.5	15.7	18.7	21.3
190	88	0.6	1.3	1.9	2.6	3.2	3.8	4.4	5.0	5.5	6.1	6.8	7.5	8.2	9.1	10.1	11.4	13.0	15.1	18.1	20.7
200	93	0.5	1.1	1.7	2.4	3.0	3.5	4.1	4.6	5.2	5.8	6.4	7.1	7.8	8.7	9.7	10.9	12.5	14.6	17.5	20.0
210	99	0.5	1.0	1.6	2.1	2.7	3.2	3.8	4.3	4.9	5.4	6.0	6.7	7.4	8.3	9.2	10.4	12.0	14.0	16.9	19.3
220	104	0.4	0.9	1.4	1.9	2.4	2.9	3.4	3.9	4.5	5.0	5.6	6.3	7.0	7.8	8.8	9.9	*	*	*	*
230	110	0.3	0.8	1.2	1.6	2.1	2.6	3.1	3.6	4.2	4.7	5.3	6.0	6.7	*	*	*	*	*	*	*
240	115	0.3	0.6	0.9	1.3	1.7	2.1	2.6	3.1	3.5	4.1	4.6	*	*	*	*	*	*	*	*	*
250	121	0.2	0.4	0.7	1.0	1.3	1.7	2.1	2.5	2.9	*	*	*	*	*	*	*	*	*	*	*
260	126	0.2	0.3	0.5	0.7	0.9	1.1	1.4	*	*	*	*	*	*	*	*	*	*	*	*	*
270	132	0.1	0.1	0.2	0.3	0.4	0.4	*	*	*	*	*	*	*	*	*	*	*	*	*	*

*Conditions not possible at atmospheric pressure.
Source: "Wood Handbook," Forest Products Laboratory, 1999.

Table 6.7.2 Strength and Related Properties of Wood at 12% Moisture Content (Average Values from Tests on Clear Pieces of 2 in × 2 in Cross Section per ASTM D 143)

Kind of wood	Specific gravity, oven-dry volume	Density at 12% m.c., lb/ft³	Shrinkage, % from green to oven-dry condition based on dimension when green — Rad.	Shrinkage, % — Tan.	Static bending: Modulus of rupture, lb/in²	Static bending: Modulus of elasticity, ×10³ lb/in²	Max crushing strength parallel to grain, lb/in²	Compression perpendicular to grain at proportional limit, lb/in²	Tensile strength perpendicular to grain, lb/in²	Impact bending, height of drop in inches for failure with 50-lb hammer	Shear strength parallel to grain, lb/in², avg of R and T	Hardness perpendicular to grain, lb, avg of R and T
Hardwoods												
Ash, white	0.60	42	4.9	7.8	15,400	1,740	7,410	1,160	940	43	1,910	1,320
Basswood	0.37	26	6.6	9.3	8,700	1,460	4,730	370	350	16	990	410
Beech	0.64	45	5.5	11.9	14,900	1,720	7,300	1,010	1,010	41	2,010	1,300
Birch, yellow	0.62	43	7.3	9.5	16,600	2,010	8,170	970	920	55	1,880	1,260
Cherry, black	0.50	35	3.7	7.1	12,300	1,490	7,110	690	560	29	1,700	950
Cottonwood, eastern	0.40	28	3.9	9.2	8,500	1,370	4,910	380	580	20	930	430
Elm, American	0.50	35	4.2	9.5	11,800	1,340	5,520	690	660	39	1,510	830
Elm, rock	0.63	44	4.8	8.1	14,800	1,540	7,050	1,230		56	1,920	1,320
Sweetgum	0.52	36	5.4	10.2	12,500	1,640	6,320	620	760	32	1,600	850
Hickory, shagbark	0.72	50	7.0	10.5	20,200	2,160	9,210	1,760		67	2,430	
Maple, sugar	0.63	44	4.8	9.9	15,800	1,830	7,830	1,470		39	2,330	1,450
Oak, red, northern	0.63	44	4.0	8.6	14,300	1,820	6,760	1,010	800	43	1,780	1,290
Oak, white	0.60	48	5.6	10.5	15,200	1,780	7,440	1,070	800	37	2,000	1,360
Poplar, yellow	0.42	29	4.6	8.2	10,100	1,580	5,540	500	540	24	1,190	540
Tupelo, black	0.50	35	4.2	7.6	9,600	1,200	5,520	930	500	22	1,340	810
Walnut, black	0.35	38	5.5	7.8	14,600	1,680	7,580	1,010	690	34	1,370	1,010
Softwoods												
Cedar, western red	0.32	23	2.4	5.0	7,500	1,110	4,560	460	220	17	990	350
Cypress, bald	0.46	32	3.8	6.2	10,600	1,440	6,360	730	300	24	1,900	510
Douglas-fir, coast	0.48	34	4.8	7.6	12,400	1,950	7,230	800	340	31	1,130	710
Hemlock, eastern	0.40	28	3.0	6.8	8,900	1,200	5,410	650		21	1,060	500
Hemlock, western	0.45	29	4.2	7.8	11,300	1,630	7,200	550	340	26	1,290	540
Larch, western	0.52	38	4.5	9.1	13,000	1,870	7,620	930	430	35	1,360	830
Pine, red	0.46	31	3.8	7.2	11,000	1,630	6,070	600	460	26	1,210	560
Pine, ponderosa	0.40	28	3.9	6.2	9,400	1,290	5,320	580	420	19	1,130	460
Pine, eastern white	0.35	24	2.1	6.1	8,600	1,240	4,800	440	310	18	900	380
Pine, western white	0.38	27	4.1	7.4	9,700	1,460	5,040	470		23	1,040	420
Pine, shortleaf	0.51	36	4.6	7.7	13,100	1,750	7,270	820	470	33	1,390	690
Redwood	0.40	28	2.6	4.4	10,000	1,340	6,150	700	240	19	940	480
Spruce, sitka	0.40	28	4.3	7.5	10,200	1,570	5,610	580	370	25	1,150	510
Spruce, black	0.42	29	4.1	6.8	10,800	1,610	5,960	550		20	1,230	520

SOURCE: Tabulated from "Wood Handbook," "Tropical Woods no. 95, and unpublished data from the USDA Forest Service, Forest Products Laboratory.

Table 6.7.3 Functions Relating Mechanical Properties to Specific Gravity and Moisture Content of Clear, Straight-Grained Wood

	Specific gravity–strength relation*				Change for 1% change in moisture content, %
	Green wood		Wood at 12% moisture content		
Property	Softwood	Hardwood	Softwood	Hardwood	
Static bending					
Modulus of elasticity (10^6 lb/in^2)	$2.331G^{0.76}$	$2.02G^{0.72}$	$2.966G^{0.84}$	$2.39G^{0.70}$	2.0
Modulus of rupture (lb/in^2)	$15,889G^{1.01}$	$17,209G^{1.16}$	$24,763G^{1.01}$	$24,850G^{1.13}$	4.0
Maximum crushing strength parallel to grain (lb/in^2)	$7,207G^{0.94}$	$7,111G^{1.11}$	$13,592G^{0.97}$	$11,033G^{0.89}$	6.0
Shear parallel to grain (lb/in^2)	$1,585G^{0.73}$	$2,576G^{1.24}$	$2,414G^{0.85}$	$3,174G^{1.13}$	3.0
Compression perpendicular to grain at proportional limit (lb/in^2)	$1,360G^{1.60}$	$2,678G^{2.48}$	$2,393G^{1.57}$	$3,128G^{2.09}$	5.5
Hardness perpendicular to grain (lb)	$1,399G^{1.41}$	$3,721G^{2.31}$	$1,931G^{1.50}$	$3,438G^{2.10}$	2.5

*The properties and values should be read as equations; e.g., modulus of rupture for green wood of softwoods $= 15,889G^{1.01}$, where G represents the specific gravity of wood, based on the oven-dry weight and the volume at the moisture condition indicated.

Specific Gravity and Density

Specific gravity G_m of wood at a given moisture condition m is the ratio of the weight of the oven-dry wood W_o to the weight of water displaced by the sample at the given moisture condition w_m:

$$G_m = \frac{W_o}{w_m}$$

This definition is required because volume and weight are constant only under special conditions. The weight density of wood D (unit weight) at any given moisture content is the oven-dry weight plus the contained water divided by the volume of the piece at that same moisture content. Average values for specific gravity oven-dry and weight density at 12 percent moisture content are given in Table 6.7.2. Specific gravity of solid, dry wood substance based on helium displacement is 1.46, or about 91 lb/ft^3.

Conversion of weight density from one moisture condition to another can be accomplished by the following equation ("Standard Handbook for Mechanical Engineers," 9th ed., McGraw-Hill):

$$D_2 = D_1 \frac{100 + M_2}{100 + M_1 + 0.0135D_1(M_2 - M_1)}$$

where D_1 is the weight density, lb/ft^3, which is known for some moisture condition M_1; D_2 is desired weight density at moisture content M_2. Moisture contents M_1 and M_2 are expressed in percentage.

Specific gravity and strength properties vary directly in an exponential relationship $S = KG^N$. Table 6.7.3 gives values for K and the exponent N for various strength properties. The equation is based on more than 160 kinds of wood and yields estimated average values for wood in general. This relationship is the best general index to the quality of defect-free wood.

Load Direction and Relation to Grain of Wood

All strength properties vary with the orthotropic axes of the wood in a manner approximated by Hankinson's formula ("Wood Handbook")

$$N = \frac{PQ}{P \sin^2 \theta + Q \cos^2 \theta}$$

where N is allowable stress induced by a load acting at M angle to the grain direction, lb/in^2; P is allowable stress parallel to the grain, lb/in^2; Q is allowable stress perpendicular to the grain, lb/in^2; and θ is angle between the direction of load and the direction of grain.

The deviation of the grain from the long axis of the member to which the load is applied is known as the slope of grain and is determined by measuring the length of run in inches along the axis for a 1-in deviation of the grain from the axis. The effect of grain slope on the important strength properties is shown in Table 6.7.4.

Rheological Properties

Wood exhibits viscoelastic characteristics. When first loaded, a wood member deforms elastically. If the load is maintained, additional time-dependent deformation occurs. Because of this time-dependent relation, the rate of loading is an important factor to consider in the testing and use of wood. For example, the load required to produce failure in 1 s is approximately 10 percent higher than that obtained in a standard 5-min strength test. Impact and dynamic measures of elasticity of small specimens are about 10 percent higher than those for static measures. Impact strengths are also affected by this relationship. In the impact bending test, a 50-lb (23-kg) hammer is dropped upon a beam from increasing heights until complete rupture occurs. The maximum height, as shown in Table 6.7.2, is for comparative purposes only.

Table 6.7.4 Strength of Wood Members with Various Grain Slopes as Percentages of Straight-Grained Members

Maximum slope of grain in member	Static bending		Impact bending; drop height to failure (50-lb hammer), %	Maximum crushing strength parallel to grain, %
	Modulus of rupture, %	Modulus of elasticity, %		
Straight-grained	100	100	100	100
1 in 25	96	97	95	100
1 in 20	93	96	90	100
1 in 15	89	94	81	100
1 in 10	81	89	62	99
1 in 5	55	67	36	93

SOURCE: "Wood Handbook."

When solid material is strained, some mechanical energy is dissipated as heat. *Internal friction* is the term used to denote the mechanism that causes this energy dissipation. The **internal friction** of wood is a complex function of temperature and moisture content. The value of internal friction, expressed by logarithmic decrement, ranges from 0.1 for hot, moist wood to less than 0.02 for hot, dry wood. Cool wood, regardless of moisture content, has an intermediate value.

The term *fatigue* in engineering is defined as progressive damage that occurs in a material subjected to cyclic loading. *Fatigue life* is a term used to define the number of cycles sustained before failure. Researchers at the USDA Forest Service Forest Products Laboratory have found that small cantilever bending specimens subjected to fully reversed stresses, at 30 Hz with maximum stress equal to 30 percent of estimated static strength and at 12 percent moisture content and 75°F (24°C), have a fatigue life of approximately 30 million cycles.

Thermal Properties

The coefficients of thermal expansion in wood vary with the structural axes. According to Weatherwax and Stamm (*Trans. ASTM* E 69, 1947, p. 421), the longitudinal coefficient for the temperature range +150 to −50°C averages $3.39 \times 10^6/°C$ and is independent of specific gravity. Across the grain, for an average specific gravity oven-dry of 0.46, the radial coefficient α_r is $25.7 \times 10^6/°C$ and the tangential α_t is $34.8 \times 10^6/°C$. Both α_r and α_t vary with specific gravity approximately to the first power. Thermal expansions are usually overshadowed by the larger dimensional changes due to moisture.

Thermal conductivity of **wood** varies principally with the direction of heat with respect to the grain. **Approximate transverse conductivity** can be calculated with a linear equation of the form

$$k = G(B + CM) + A$$

where G is specific gravity, based on oven-dry weight and volume at a given moisture content M percent; for specific gravities above 0.3, temperatures around 75°F (24°C), and moisture contents below 25 percent, the values of constants A, B, and C are $A = 0.129$, $B = 1.34$, and $C = 0.028$ in English units, with k in Btu · in/(h · ft^2 · F) (TenWolde et al., 1988). Conductivity in watts per meter per kelvin is obtained by multiplying the result by 0.144. The effect of temperature on thermal conductivity is relatively minor and increases about 1 to 2 percent per 10°F (2 to 3 percent per 10°C). **Longitudinal conductivity** is considerably greater than transverse conductivity, but reported values vary widely. It has been reported as 1.5 to 2.8 times larger than transverse conductivity, with an average of about 1.8.

Specific heat of wood is virtually independent of specific gravity and varies principally with temperature and moisture content. Wilkes found that the approximate specific heat of dry wood can be calculated with

$$c_{p0} = a_0 + a_1 T$$

where $a_0 = 0.26$ and $a_1 = 0.000513$ for English units (specific heat in Btu per pound per degree Fahrenheit and temperature in degrees Fahrenheit) or $a_0 = 0.103$ and $a_1 = 0.00387$ for SI units [specific heat in kJ/(kg · K) and temperature in kelvins]. The specific heat of moist wood can be derived from

$$c_p = \frac{c_{p0} + 0.01Mc_{p,w}}{1 + 0.01M} + A$$

where $c_{p,w}$ is the specific heat of water [1 Btu/(lb · °F)], or 4.186 kJ/(kg · K), M is the moisture content (percent), and A is a correction factor, given by

$$A = M(b_1 + b_2T + b_3M)$$

with $b_1 = -4.23 \times 10^4$, $b_2 = 3.12 \times 10^5$, and $b_3 = -3.17 \times 10^5$ in English units; and $b_1 = -0.06191$, $b_2 = 2.36 \times 10^4$, and $b_3 = -1.33 \times 10^{-4}$ in SI units. These formulas are valid for wood below fiber saturation at temperatures between 45°F (7°C) and 297°F (147°C). And T is the temperature at which c_{p0} is desired.

The **fuel value** of wood depends primarily upon its dry density, moisture content, and chemical composition. Moisture in wood decreases the fuel value as a result of latent heat absorption of water vaporization. An approximate relation for the fuel value of moist wood (Btu per pound on wet weight basis) (2,326 Btu/lb = 1 J/kg) is

$$H_w = H_D\left(\frac{100 - u/7}{100 + u}\right)$$

where H_D is higher fuel value of dry wood, averaging 8,500 Btu/lb for hardwoods and 9,000 Btu/lb for conifers, and u is the moisture content in percent. The actual fuel value of moist wood in a furnace will be less since water vapor interferes with the combustion process and prevents the combustion of pyrolytic gases. (See Sec. 7 for fuel values and Sec. 4 for combustion.)

Wood undergoes thermal degradation to volatile gases and char when it is exposed to elevated temperature. When wood is directly exposed to the standard fire exposure of ASTM E 119, the **char rate** is generally considered to be 1½ in/h (38 mm/h). The temperature at the base of the char layer is approximately 550°F (300°C). A procedure for calculating the fire resistance rating of an exposed wood member can be found in recent editions of "National Design Specification" (American Forest & Paper Association, 2005 and later). Among other factors, the ignition of wood depends on the intensity and duration of exposure to elevated temperatures. Typical values for rapid ignition are 570 to 750°F (300 to 400°C). In terms of heat flux, a surface exposure to 1.1 Btu/ft^2(13 kW/m^2) per second is considered sufficient to obtain piloted ignition. Recommended "maximum safe working temperatures" for wood exposed for prolonged periods range from 150 to 212°F (65 to 100°C). **Flame spread** values as determined by ASTM E 84 generally range from 65 to 200 for nominal 1-in- (25-mm-) thick lumber. Lists of flame spread index values for different species can be found in "Wood Handbook" and the Website of the American Wood Council. Flame spread can be reduced by impregnating the wood with fire-retardant chemicals or applying a fire-retardant coating.

The **reversible effect of temperature** on the properties of wood is a function of the change in temperature, moisture content of the wood, duration of heating, and property being considered. In general, the mechanical properties of wood decrease when the wood is heated above normal temperatures and increase when it is cooled. The magnitude of the change is greater for green wood than for dry. When wood is frozen, the change in property is reversible; i.e., the property will return to the value at the initial temperature. At constant moisture content and below about 150°F (65°C), mechanical properties are approximately linearly related to temperature. The change in property is also reversible if the wood is heated for a short time at temperatures below about 150°F. Table 6.7.5 lists the changes in properties at −58°F (−50°C) and 122°F (50°C) relative to those at 68°F (20°C).

Permanent loss in properties occurs when wood is exposed to higher temperatures for prolonged periods and then is cooled and tested at normal temperatures. If the wood is tested at a higher temperature after prolonged exposure, the actual strength loss is the sum of the reversible and permanent losses in properties. Permanent losses are higher for heating in steam than in water, and higher when heated in water than when heated in air. Repeated exposure to elevated temperatures is assumed to have a cumulative effect on wood properties. For example, at a given temperature the property loss will be about the same after six exposure of 1-year duration as it would be after a single exposure of 6 years. Figure 6.7.1 illustrates the effect of heating at 150°F (65°C) at 12 percent moisture content on the modulus of rupture relative to the strength at normal temperatures for spruce-pine-fir, Douglas-fir, and southern pine 2 × 4 s. Over the 6-year period, there was little or no change in the modulus of elasticity. Increasing the temperature would be expected to increase the permanent loss in strength; reducing the relative humidity would decrease the loss (see Green et al., 2003).

Electrical Properties

The important electrical properties of wood are **conductivity** (or its reciprocal, **resistivity**), **dielectric constant**, and **dielectric power factor** (see James, 1988).

Table 6.7.5   Approximate Middle-Trends for the Reversible Effect of Temperature on Mechanical Properties of Clear Wood at Various Moisture Conditions

Property	Moisture condition, %	Relative change in mechanical property from 68°F, %	
		At −58°F	At +122°F
Modulus of elasticity parallel to grain	0	+11	−6
	12	+17	−7
	>FSP*	+50	—
Modulus of rupture	≤4	+18	−10
	11–15	+35	−20
	18–20	+60	−25
	>FSP*	+110	−25
Tensile strength parallel to grain	0–12	—	−4
Compressive strength parallel to grain	0	+20	−10
	12–45	+50	−25
Shear strength parallel to grain	>FSP*	—	−25
Compressive strength perpendicular to grain	0–16	—	−20
at proportional limit	≥10	—	−35

*Moisture content higher than the fiber saturation point (FSP).
$T_C = (T_F − 32)(0.55)$.

Fig. 6.7.1 Average residual MOR for solid-sawn 2 × 4 lumber exposed for various times at 66°C (150°F), 75 percent relative humidity and tested at 23°C (73°F), 67 percent relative humidity. DF, Douglas-fir; So. Pine, southern pine; SPF, spruce–pine–fir.

Resistivity approximately doubles for each 10°C decrease in temperature. As moisture content increases from zero to the fiber saturation point (FSP), the resistivity decreases by 10^{10} to 10^{13} times in an approximately linear relationship between the logarithm resistivity and moisture content. The resistivity is about 10^{14} to 10^{16} W · m for oven-dry wood and 10^3 to 10^4 W · m for wood at FSP. As the moisture content increases up to complete saturation, the decrease in resistivity is a factor of only about 50. Wood species also affect resistivity (see James), and the resistivity perpendicular to the grain is about twice that parallel to the grain. Water-soluble salts (some preservatives and fire retardants) reduce resistivity by only a minor amount when the wood has 8 percent moisture content or less, but they have a much larger effect when moisture content exceeds 10 to 12 percent.

The **dielectric constant** of oven-dry wood ranges from about 2 to 5 at room temperature, and it decreases slowly with increasing frequency. The dielectric constant increases as either temperature or moisture content increases. There is a negative interaction between moisture and frequency: At 20 Hz, the dielectric constant may range from 4 for dry wood to 106 for wet wood; at 1 kHz, from 4 dry to 5,000 wet; and at 1 MHz, from about 3 dry to 100 wet. The dielectric constant is about 30 percent greater parallel to the grain than perpendicular to it.

The **power factor** of wood varies from about 0.01 for dry, low-density woods to as great as 0.95 for wet, high-density woods. It is usually greater parallel to the grain than perpendicular. The power factor is affected by complex interactions of frequency, moisture content, and temperature (James, 1975).

The change in electrical properties of wood with moisture content has led to the development of moisture meters for nondestructive estimation of moisture content. Resistance-type meters measure resistance between two pins driven into the wood. Dielectric-type meters depend on the correlation between moisture content and either dielectric constant or power factor, and they require only contact with the wood surface, not penetration.

Wood in Relation to Sound

Transmission of sound and the vibrational properties in wood are functions of a variety of factors. The speed of sound transmission is described by the expression $v = \sqrt{E/\rho}$, in which v is the speed of sound in wood, in/s; E is the dynamic Young's modulus, lb/in²; and ρ is the density of the wood, slugs/in³ (Pellerin and Ross, 2002). Various factors influence the speed of sound transmission; two of the most important factors are grain angle and the presence of degradation from decay. Hankinson's formula, cited previously, adequately describes the relationship between speed of sound transmission and grain angle. The dynamic modulus is about 10 percent higher than the static value and varies inversely with moisture changes by approximately 1.3 percent for each percentage change in moisture content.

Degradation from biological agents can significantly alter the speed at which sound travels in wood. Speed of sound transmission values are greatly reduced in severely degraded wood members. Sound transmission characteristics of wood products are used in one form of nondestructive testing to assess the performance characteristics of wood products. Because speed of sound transmission is a function of the extent of degradation from decay, this technique is used to estimate the extent of severe degradation in large timbers.

PROPERTIES OF LUMBER PRODUCTS
by Roland Hernandez and David W. Green
Visually Graded Structural Lumber

Stress-graded structural lumber is produced under two systems: visual grading and machine grading. Visual structural grading is the oldest stress grading system. It is based on the premise that the mechanical properties of lumber differ from those of clear wood because many growth characteristics of lumber affect its properties; these characteristics can be seen and judged by eye (ASTM D 245). The principal growth feature affecting lumber properties are the size and location of knots, sloping grain, and density.

Grading rules for lumber nominally 2 to 4 in (standard 38 to 89 mm) thick (dimension lumber) are published by grading agencies (listing and addresses are given in "National Design Specification," American Forest & Paper Association, 2005 and later). For most species, allowable properties are based on test results from full-size specimens graded by agency rules, sampled according to ASTM D 2915, and tested according to ASTM D 4761. Procedures for deriving allowable properties from these tests are given in ASTM D 1990. Allowable properties for visually graded hardwoods and a few softwoods are derived from clear-wood data following principles given in ASTM D 2555. Derivation of the allowable strength properties accounts for within-species variability by starting with a nonparametric estimate of the 5th percentile of the data. Thus, 95 of 100 pieces would be expected to be stronger than the assigned property. The allowable strength properties are based on an assumed normal duration of load of 10 years. Tables 6.7.6 and 6.7.7 show the grades and allowable properties for the four most commonly used species groupings sold in the United States. The allowable strength values in bending, tension, shear, and compression parallel to the grain can be multiplied by factors for other load durations. Some commonly used factors are 0.90 for permanent (50-years) loading, 1.15 for snow loads (2 months), and 1.6 for wind/earthquake loading (10 min). The most recent edition of "National Design Specification" should be consulted for updated property values and for property values for other species and size classifications.

Allowable properties are assigned to visually graded lumber at two moisture content levels: green and 19 percent maximum moisture content (assumed 15 percent average moisture content). Because of the influence of knots and other growth characteristics on lumber properties, the effect of moisture content on lumber properties is generally less than its effect on clear wood. The C_M factors of Table 6.7.8 are for adjusting the properties in Tables 6.7.6 and 6.7.7 from 15 percent moisture content to green. The Annex of ASTM D 1990 provides formulas that can be used to adjust lumber properties to any moisture content between green and 10 percent. Below about 8 percent moisture content, some properties may decrease with decreasing values, and care should be exercised in these situations (Green and Kretschmann, 1995).

Shrinkage in commercial lumber differs from that in clear wood primarily because the grain in lumber is seldom oriented in purely radial and tangential directions. Approximate formulas used to estimate shrinkage of lumber for most species are

$$S_w = 6.031 - 0.215M$$
$$S_t = 5.062 - 0.181M$$

where S_w is the shrinkage across the wide [8-in (203-mm)] face of the lumber in a 2 × 8 (standard 38 × 184 mm), S_t is the shrinkage across the narrow [2-in (51-mm)] face of the lumber, and M is moisture content (percent). As with clear wood, shrinkage is assumed to occur below a moisture content of 28 percent. Because extractives make wood less

Table 6.7.6   Base Design Values for Visually Graded Dimension Lumber*
(Tabulated design values are for normal load duration and dry service conditions)

Species and Commercial grade	Size classification, in	Bending F_b	Tension parallel to grain F_t	Shear parallel to grain F_v	Compression perpendicular to grain $F_{c\perp}$	Compression parallel to grain F_c	Modulus of elasticity E	Grading rules agency
			Design values, lb/in²					
		Douglas-fir–Larch						
Select structural		1,500	1,000	180	625	1,700	1,900,000	WCLIB
No. 1 and better	2–4 thick	1,200	800	180	625	1,550	1,800,000	WWPA
No. 1		1,000	675	180	625	1,500	1,700,000	
No. 2	2 and wider	900	575	180	625	1,350	1,600,000	
No. 3		525	325	180	625	775	1,400,000	
Stud		700	450	180	625	850	1,400,000	
Construction	2–4 thick	1,000	650	180	625	1,650	1,500,000	
Standard		575	375	180	625	1,400	1,400,000	
Utility	2–4 wide	275	175	180	625	900	1,300,000	
		Hem–Fir						
Select structural		1,400	925	150	405	1,500	1,600,000	WCLIB
No. 1 and better	2–4 thick	1,100	725	150	405	1,350	1,500,000	WWPA
No. 1		975	625	150	405	1,350	1,500,000	
No. 2	2 and wider	850	525	150	405	1,300	1,300,000	
No. 3		500	300	150	405	725	1,200,000	
Stud		675	400	150	405	800	1,200,000	
Construction	2–4 thick	975	600	150	405	1,550	1,300,000	
Standard		550	325	150	405	1,300	1,200,000	
Utility	2–4 wide	250	150	150	405	850	1,100,000	
		Spruce–Pine–Fir						
Select structural	2–4 thick	1,250	700	135	425	1,400	1,500,000	NLGA
No. 1 – No. 2		875	450	135	425	1,150	1,400,000	
No. 3	2 and wider	500	250	135	425	650	1,200,000	
Stud		675	350	135	425	725	1,200,000	
Construction	2–4 thick	1,000	500	135	425	1,400	1,300,000	
Standard		550	275	135	425	1,150	1,200,000	
Utility	2–4 wide	275	125	135	425	750	1,100,000	

*Lumber dimensions: Tabulated design values are applicable to lumber that will be used under dry conditions such as in most covered structures. For 2- to 4-in-thick lumber, the dry dressed sizes shall be used regardless of the moisture content at the time of manufacture or use. In calculating design values, the natural gain in strength and stiffness that occurs as lumber dries has been taken into consideration as well as the reduction in size that occurs when unseasoned lumber shrinks. The gain in load-carrying capacity due to increased strength and stiffness resulting from drying more than offsets the design effect of size reductions due to shrinkage. Size factor C_F, flat-use factor C_{fu}, and wet-use factor C_M are given in Table 6.7.8.
SOURCE: Table used by permission of the American Forest & Paper Association, Washington, DC.

Table 6.7.7   Design values for Visually Graded Southern Pine Dimension Lumber*
(Tabulated design values are for normal load duration and dry services conditions)

Commercial grade	Size classification, in	Bending F_b	Tension parallel to grain F_t	Shear parallel to grain F_v	Compression perpendicular to grain $F_{c\perp}$	Compression parallel to grain F_c	Modulus of elasticity E	Grading rules agency
Dense select structural	2–4 thick	3,050	1,650	175	660	2,250	1,900,000	SPIB
Select structural		2,850	1,600	175	565	2,100	1,800,000	
Nondense select structural	2–4 wide	2,650	1,350	175	480	1,950	1,700,000	
No. 1 dense		2,000	1,100	175	660	2,000	1,800,000	
No. 1		1,850	1,050	175	565	1,850	1,700,000	
No. 1 nondense		1,700	900	175	480	1,700	1,600,000	
No. 2 dense		1,700	875	175	660	1,850	1,700,000	
No. 2		1,500	825	175	656	1,650	1,600,000	
No. 2 nondense		1,350	775	175	480	1,600	1,400,000	
No. 3 and stud		850	475	175	565	975	1,400,000	
Construction	2–4 thick	1,100	625	175	565	1,800	1,500,000	
Standard		625	350	175	565	1,500	1,300,000	
Utility	4 wide	300	175	175	565	975	1,300,000	
Dense select structural	2–4 thick	2,700	1,500	175	660	2,150	1,900,000	
Select structural		2,550	1,400	175	565	2,000	1,800,000	
Nondense select structural	5–6 wide	2,350	1,200	175	480	1,850	1,700,000	
No. 1 dense		1,750	950	175	660	1,900	1,800,000	
No. 1		1,650	900	175	565	1,750	1,700,000	
No. 1 nondense		1,500	800	175	480	1,600	1,600,000	
No. 2 dense		1,150	775	175	660	1,750	1,700,000	
No. 2		1,250	725	175	565	1,600	1,600,000	
No. 2 nondense		1,150	675	175	480	1,500	1,400,000	
No. 3 and stud		750	425	175	565	925	1,400,000	
Dense select structural	2–4 thick	2,450	1,350	175	660	2,050	1,900,000	
Select structural		2,300	1,300	175	565	1,900	1,800,000	
Nondense select structural	8 wide	2,100	1,100	175	480	1,750	1,700,000	
No. 1 dense		1,650	875	175	660	1,800	1,800,000	
No. 1		1,500	825	175	565	1,650	1,700,000	
No. 1 nondense		1,350	725	175	480	1,550	1,600,000	
No. 2 dense		1,400	675	175	660	1,700	1,700,000	
No. 2		1,200	650	175	565	1,550	1,600,000	
No. 2 nondense		1,100	600	175	480	1,450	1,400,000	
No. 3 and stud		700	400	175	565	875	1,400,000	
Dense select structural	2–4 thick	2,150	1,200	175	660	2,000	1,900,000	
Select structural		2,050	1,100	175	565	1,850	1,800,000	
Nondense select structural	10 wide	1,850	950	175	480	1,750	1,700,000	
No. 1 dense		1,450	775	175	660	1,750	1,800,000	
No. 1		1,300	725	175	565	1,600	1,700,000	
No. 1 nondense		1,200	650	175	480	1,500	1,600,000	
No. 2 dense		1,200	625	175	660	1,650	1,700,000	
No. 2		1,050	575	175	565	1,500	1,600,000	
No. 2 nondense		950	550	175	480	1,400	1,400,000	
No. 3 and stud		600	325	175	565	850	1,400,000	
Dense select structural	2–4 thick	2,050	1,100	175	660	1,950	1,900,000	
Select structural		1,900	1,050	175	565	1,800	1,800,000	
Nondense select structural	12 wide	1,750	900	175	480	1,700	1,700,000	
No. 1 dense		1,350	725	175	660	1,700	1,800,000	
No. 1		1,250	675	175	565	1,600	1,700,000	
No. 1 nondense		1,150	600	90	480	1,500	1,600,000	
No. 2 dense		1,150	575	90	660	1,600	1,700,000	
No. 2		975	550	90	565	1,450	1,600,000	
No. 2 nondense		900	525	90	480	1,350	1,400,000	
No. 3 and stud		575	325	90	565	825	1,400,000	

*For size factor C_F appropriate size adjustment factors have already been incorporated in the tabulated design values for most thicknesses of southern pine dimension lumber. For dimension lumber 4 in thick, 8 in and wider, tabulated bending design values F_b shall be permitted to be multiplied by the size factor $C_F = 1.1$. For dimension lumber wider than 12 in, tabulated bending, tension, and compression parallel-to-grain design values for 12-in-wide lumber shall be multiplied by the size factor $C_F = 0.9$. Repetitive member factor C_r, flat-use factor C_{fu}, and wet-service factor C_M are given in Table 6.7.8.
SOURCE: Table used by permission of the American Forest & Paper Association, Washington, DC.

Table 6.7.8   Adjustment Factors

Size factor C_F for Table 6.7.6 (Douglas-Fir–Larch, Hem-Fir, Spruce-Pine-Fir)

Tabulated bending, tension, and compression parallel to grain design values for dimension lumber 2 to 4 in thick shall be multiplied by the following size factors:

Grades	Width, in	F_b thickness, in 2 & 3	4	F_t	F_c
Select structural	2, 3, and 4	1.5	1.5	1.5	1.15
No. 1 and better	5	1.4	1.4	1.4	1.1
No. 1, no. 2, no. 3	6	1.3	1.3	1.3	1.1
	8	1.2	1.3	1.2	1.05
	10	1.1	1.2	1.1	1.0
	12	1.0	1.1	1.0	1.0
	14 and wider	0.9	1.0	0.9	0.9
Stud	2, 3, and 4	1.1	1.1	1.1	1.05
	5 and 6	1.0	1.0	1.0	1.0
Construction and standard	2, 3, and 4	1.0	1.0	1.0	1.0
Utility	4	1.0	1.0	1.0	1.0
	2 and 3	0.4	—	0.4	0.6

Repetitive-member factor C_r for Tables 6.7.6 and 6.7.7

Bending design values F_b for dimension lumber 2 to 4 in thick shall be multiplied by the repetitive factor $C_r = 1.15$, when such members are used as joists, truss chords, rafters, studs, planks, decking, or similar members which are in contact or spaced not more than 24 in on centers, are not less than 3 in number, and are joined by floor, roof, or other load-distributing elements adequate to support the design load.

Flat-use factor C_{fu} for Tables 6.7.6 and 6.7.7

Bending design values adjusted by size factors are based on edgewise use (load applied to narrow face). When dimension lumber is used flatwise (load applied to wide face), the bending design value F_b shall also be multiplied by the following flat-use factors:

Width, in	Thickness, in 2 and 3	4
2 and 3	1.0	—
4	1.1	1.0
5	1.1	1.05
6	1.15	1.05
8	1.15	1.05
10 and wider	1.2	1.1

Wet-use factor C_M for Tables 6.7.6 and 6.7.7

When dimension lumber is used where moisture content will exceed 19 percent for an extended period, design values shall be multiplied by the appropriate wet-service factors from the following table:

F_b	F_t	F_v	$F_{c\perp}$	F_c	E
0.85*	1.0	0.97	0.67	0.8†	0.9

*When $F_b C_F \leq 1150$ lb/in², $C_M = 1.0$.
†When $F_c C_F \leq 750$ lb/in², $C_M = 1.0$.
SOURCE: Used by permission of the American Forest & Paper Association, Washington, DC.

hygroscopic, less shrinkage is expected in redwood, western redcedar, and northern white cedar (Green, 1989).

The reversible effect of **temperature** on lumber properties appears to be similar to that on clear wood. For simplicity, "National Design Specification" uses conservative factors to account for reversible reductions in properties as a result of heating to 150°F (65°C) or less (Table 6.7.9). No increase in properties is taken for temperatures colder than normal because in practice it is difficult to ensure that the wood temperature remains consistently low. Permanent effects of temperature are not given in "National Design Specification" (see Green et al., 2003).

Table 6.7.9   Temperature Factors C_t for Short-Term Exposure (Reversible Effect)

Design values	In-service moisture conditions	C_t $T \leq 100°F$	$100°F < T \leq 125°F$	$125°F < T \leq 150°F$
F_t, E	Green or dry	1.0	0.9	0.9
F_b, F_v, F_c, and $F_{c\perp}$	$\leq 19\%$ green	1.0	0.8	0.7
		1.0	0.7	0.5

SOURCE: Table used by permission of the American Forest & Paper Association, Washington, DC.

Mechanically Graded Structural Lumber

Machine-stress-rated (MSR) lumber and machine-evaluated lumber (MEL) are two types of mechanically graded lumber. The three basic components of both mechanical grading systems are (1) sorting and prediction of strength through machine-measured nondestructive determination of properties coupled with visual assessment of growth characteristics, (2) assignment of allowable properties based upon strength prediction, and (3) quality control to ensure that assigned properties are being obtained. Grade names for MEL start with an M designation. Grade "names" for MSR lumber are a combination of the allowable bending stress and the average modulus of elasticity; e.g., 1650f-1.4E means an allowable bending stress of 1,650 lb/in^2 (11.4 MPa) and modulus of elasticity of 1.4×10^6 lb/in^2 (9.7 GPa). Selected grades of mechanically graded lumber and their allowable properties are given in Table 6.7.10. Additional grades are available (see AF&PA, 2005).

Structural Composite Lumber

Types of Structural Composite Lumber Structural composite lumber refers to several types of reconstituted products that have been developed to meet the demand for high-quality material for the manufacture of engineered wood products and structures. Two distinct types are commercially available: laminated veneer lumber (LVL) and parallel-strand lumber (PSL).

Laminated veneer lumber is manufactured from layers of veneer with the grain of all the layers parallel. This contrasts with plywood, which consists of adjacent layers with the grain perpendicular. Most manufacturers use sheets of 1/10- to 1/6-in- (2.5- to 4.2-mm-) thick veneer. These veneers are stacked up to the required thickness and may be laid end to end to the desired length with staggered end joints in the veneer. Waterproof adhesives are generally used to bond the veneer under pressure. The resulting product is a billet of lumber that may be up to 1 1/4 in (44 mm) thick, 4 ft (1.2 m) wide, and 80 ft (24.4 m) long. The billets are then ripped to the desired width and cut to the desired length. The common sizes of LVL closely resemble those of sawn dimension lumber.

Parallel-strand lumber is manufactured from strands or elongated flakes of wood. One North American product is made from veneer clipped to 1/2 in (13 mm) wide and up to 8 ft (2.4 m) long. Another product is made from elongated flakes and technology similar to that used to produce oriented strandboard. A third product is made from mats of interconnected strands crushed from small logs that are assembled into the desired configuration. All the products use waterproof adhesive that is cured under pressure. The size of the product is controlled during manufacture through adjustments in the amount of material and pressure applied. Parallel-strand lumber is commonly available in the same sizes as structural timbers or lumber.

Properties of Structural Composite Lumber Standard design values have not been established for either LVL or PSL. Rather, standard procedures are available for developing these design values (ASTM D 5456). Commonly, each manufacturer follows these procedures and submits supporting data to the appropriate regulatory authority to establish design properties for the product. Thus, design information for LVL and PSL varies among manufacturers and is given in their product literature. Generally the engineering design properties compare favorably with or exceed those of high-quality solid dimension lumber. Example design values accepted by U.S. building codes are given in Table 6.7.11.

Glued-Laminated Timber

Structural **glued-laminated timber (glulam)** is an engineered, stress-rated product of a timber laminating plant. It is comprised of assemblies of suitably selected and prepared wood laminations bonded together with adhesives. The grain of all laminations is approximately parallel longitudinally. Individual laminations are typically made from nominal 2-in- (standard 38-mm-) thick lumber when used for straight or slightly cambered members, and nominal 1-in- (standard 19-mm-) thick lumber when used for curved members. Individual lamination pieces may be

joined end to end to produce laminated timbers much longer than the laminating stock itself. Pieces may also be placed or glued edge to edge to make members wider than the input stock. Straight members up to 140 ft (42 m) long and more than 7 ft (2.1 m) deep have been manufactured, with size limitations generally resulting from transportation constraints. Curved members have been used in domed structures spanning over 500 ft (152 m).

The arrangement of lumber grades in the laminated cross section depends on the anticipated loading parallel or perpendicular to the wide faces of the laminations. When loads are applied perpendicular to the wide faces of the laminations, referred to as horizontally laminated, glulam cross sections are typically designed as bending combinations. **Bending combinations** have higher-grade laminations in the outer top and bottom layers to carry the highest stresses and lower-grade laminations in the core layers where stresses are minimal. In situations where all laminations are subjected to the same load, such as compression, tension, or in bending with loads applied parallel to the wide faces of the laminations (vertically laminated), glulam cross sections are typically designed as axial combinations. **Axial combinations** typically have the same lumber grade throughout the cross section.

Manufacture Glulam members used in structural applications in the United States must be manufactured by a laminating plant that meets the requirements of the national standard, ANSI A190.1. This standard contains requirements for production, testing, and certification of the product. Plants meeting these requirements can place their product quality mark on the glulam, which contains vital information regarding the type, species, and properties of the glulam combination. Figure 6.7.2 shows examples of glulam product quality marks and describes the features required in the mark.

Manufacturing standards cover many softwoods and hardwoods; Douglas-fir and southern pine are the most commonly used softwood species. Glulam members can be used in either dry- or wet-use conditions. Dry use, which is defined as a condition resulting in a moisture content of 16 percent or less, permits manufacturing with nonwaterproof adhesives; however, nearly all manufacturers in North America use waterproof adhesives exclusively. For wet-use conditions, these waterproof adhesives are required. For wet-use conditions in which the moisture content is expected to exceed 20 percent, pressure preservative treatment is recommended (AWPA C28). Lumber can be pressure-treated with water-based preservatives prior to gluing, provided that special procedures are followed in the manufacture. For treatment after gluing, oil-based preservatives are generally recommended.

Glulam is generally manufactured at moisture content below 16 percent. For most dry-use applications, it is important to protect the glulam timber from increases in moisture content. End sealers, surface sealers, primer coats, and wrapping may be applied at the manufacturing plant to provide protection from changes in moisture content. Protection will depend upon the final use and finish of the timber.

Special precautions are necessary during handling, storage, and erection to prevent structural damage to glulam members. Padded or non-marring slings are recommended; cable slings or chokers should be avoided unless proper blocking protects the members. Technical manuals on transit, storage, and erection of glulam members are provided by both AITC and APA-EWS.

Design Glulam members are available in standard sizes with standardized design properties. The following standard widths are established to match the widths of standard sizes of lumber, less an allowable amount for finishing the edges of the manufactured beams:

3 or 3 1/8 in (76 or 79 mm)
5 or 5 1/8 in (127 or 130 mm)
6 3/4 in (171 mm)
8 1/2 or 8 3/4 in (216 or 222 mm)
10 1/2 or 10 3/4 in (267 or 273 mm)

Standard beam depths are common multiples of lamination thickness of either 1 3/8 or 1 1/2 in (35 or 38 mm). There are no standard beam lengths, although most uses will be on spans where the length is from 10 to

Table 6.7.10   Design Values for Mechanically Graded Dimension Lumber
(Tabulated design values are for normal load duration and dry service conditions.)

Species and commercial grade	Size classification, in	Design values, lb/in² Bending	Tension parallel to grain	Compression parallel to grain	Modulus of elasticity	Grading rules agency
		Machine-stress-rated lumber				
900f-1.0E		900	350	1,050	1,000,000	WCLIB, WWPA, NELMA, NSLB
1200f-1.2E		1,200	600	1,400	1,200,000	NLGA, WCLIB, WWPA, NELMA, NSLB
1250f-1.4E		1,250	800	1,475	1,400,000	WCLIB
1350f-1.3E		1,350	750	1,600	1,300,000	NLGA, WCLIB, WWPA, NELMA, NSLB
1400f-1.2E		1,400	800	1,600	1,200,000	NLGA
1450f-1.3E		1,450	800	1,625	1,300,000	NLGA, WCLIB, WWPA, NELMA, NSLB
1450f-1.5E		1,450	875	1,625	1,500,000	WCLIB
1500f-1.4E		1,500	900	1,650	1,400,000	NLGA, WCLIB, WWPA, NELMA, NSLB
1600f-1.4E		1,600	950	1,675	1,400,000	NLGA, WWPA
1650f-1.3E		1,650	1,020	1,700	1,300,000	NLGA, SPIB, WCLIB, WWPA, NELMA, NSLB
1650f-1.5E		1,650	1,020	1,700	1,500,000	WCLIB, WWPA
1650f-1.6E-1075f$_t$		1,650	1,075	1,700	1,600,000	WCLIB
1650f-1.6E		1,650	1,175	1,700	1,600,000	WCLIB
1650f-1.8E		1,650	1,020	1,750	1,800,000	WCLIB
1700f-1.6E		1,700	1,175	1,725	1,600,000	WCLIB
1750f-2.0E		1,750	1,125	1,725	2,000,000	NLGA
1800f-1.5E		1,800	1,300	1,750	1,500,000	NLGA, SPIB, WCLIB, WWPA, NELMA, NSLB
1800f-1.6E		1,800	1,175	1,750	1,600,000	WCLIB
1800f-1.8E		1,800	1,200	1,750	1,800,000	SPIB, WWPA
1950f-1.5E	2 & less in thickness	1,950	1,375	1,800	1,500,000	NLGA, SPIB, WCLIB, WWPA, NELMA, NSLB
1950f-1.7E		1,950	1,375	1,800	1,700,000	NLGA
2000f-1.6E		2,000	1,300	1,825	1,600,000	NLGA, SPIB, WCLIB, WWPA, NELMA, NSLB
2100f-1.8E	2 & wider	2,100	1,575	1,875	1,800,000	NLGA, WWPA
2250f-1.7E		2,250	1,750	1,925	1,700,000	NLGA, WCLIB, WWPA
2250f-1.8E		2,250	1,750	1,925	1,800,000	NLGA, SPIB, WCLIB, WWPA, NELMA, NSLB
2250f-1.9E		2,250	1,750	1,925	1,900,000	WCLIB, WWPA
2250f-2.0E-1600f$_t$		2,250	1,600	1,925	2,000,000	WCLIB, WWPA
2250f-2.0E		2,250	1,750	1,925	2,000,000	NLGA, WWPA
2400f-1.8E		2,400	1,925	1,975	1,800,000	NLGA, SPIB
2400f-2.0E		2,400	1,925	1,975	2,000,000	NLGA, SPIB, WCLIB, WWPA, NELMA, NSLB
2500f-2.2E		2,500	1,750	2,000	2,200,000	WCLIB
2500f-2.2E-1925f$_t$		2,500	1,925	2,000	2,200,000	WCLIB
2550f-2.1E		2,550	2,050	2,025	2,100,000	NLGA, SPIB, WCLIB, WWPA, NELMA, NSLB
2700f-2.0E		2,700	1,800	2,100	2,000,000	WCLIB
2700f-2.2E		2,700	2,150	2,100	2,200,000	NLGA, SPIB, WCLIB, WWPA, NELMA, NSLB
2850f-2.3E		2,850	2,300	2,150	2,300,000	NLGA, SPIB, WCLIB, WWPA, NELMA, NSLB
3000f-2.4E		3,000	2,400	2,200	2,400,000	NLGA, SPIB
		Machine-evaluated lumber				
M-10		1,400	800	1,600	1,200,000	NLGA, SPIB
M-11		1,550	850	1,675	1,500,000	NLGA, SPIB
M-12		1,600	850	1,675	1,600,000	NLGA, SPIB
M-13		1,600	950	1,675	1,400,000	NLGA, SPIB
M-14		1,800	1,000	1,750	1,700,000	NLGA, SPIB
M-15		1,800	1,100	1,750	1,500,000	NLGA, SPIB
M-16		1,800	1,300	1,750	1,500,000	SPIB
M-17	2 & less in thickness	1,950	1,300	2,050	1,700,000	SPIB
M-18		2,000	1,200	1,825	1,800,000	SP1B
M-19	2 & wider	2,000	1,300	1,825	1,600,000	NLGA, SPIB
M-20		2,000	1,600	2,100	1,900,000	NLGA, SPIB
M-21		2,300	1,400	1,950	1,900,000	NLGA, SPIB
M-22		2,350	1,500	1,950	1,700,000	NLGA, SPIB
M-23		2,400	1,900	1,975	1,800,000	NLGA, SPIB
M-24		2,700	1,800	2,100	1,900,000	NLGA, SPIB
M-25		2,750	2,000	2,100	2,200,000	NLGA, SPIB
M-26		2,800	1,800	2,150	2,000,000	NLGA, SPIB
M-27		3,000	2,000	2,400	2,100,000	SPIB

Lumber dimensions: Tabulated design values are applicable to lumber that will be used under dry conditions such as in most covered structures. For 2- to 4-in-thick lumber, the *dry* dressed sizes shall be used regardless of the moisture content at the time of manufacture or use. In calculating design values, natural gain in strength and stiffness that occurs as lumber dries has been taken into consideration as well as reduction in size that occurs when unseasoned lumber shrinks. The gain in load-carrying capacity due to increased strength and stiffness resulting from drying more than offsets the design of size reductions due to shrinkage. **Shear parallel to grain F_v and compression perpendicular to grain F_c:** Design values for shear parallel to grain F_{v1} and compression perpendicular to grain $F_{c\perp}$ are identical to the design values given in Tables 6.7.6 and 6.7.7 for No. 2 visually graded lumber of the appropriate species. When the F_v or $F_{c\perp}$ values shown on the grade stamp differ from the values shown in the tables, the values shown on the grade stamp shall be used for design. **Modulus of elasticity E and tension parallel to grain F_t:** For any given bending design value F_b, the average modulus of elasticity E and tension parallel to grain F_t design value may vary depending upon species, timber source, or other variables. The E and F_t values included in the F_b and E grade designations are those usually associated with each F_b level. Grade stamps may show higher or lower values if machine rating indicates the assignment is appropriate. When the E or F_t values shown on a grade stamp differ from the values in Table 6.7.10, the values shown on the grade stamp shall be used for design. The tabulated F_b and F_c values associated with the designated F_b value shall be used for design. **Additional grades are available.**
    SOURCE: Table used by permission of the American Forest & Paper Association, Washington, DC.

Table 6.7.11   Example Design Values for Structural Composite Lumber

Product	Bending stress		Modulus of elasticity		Horizontal shear	
	lb/in²	MPa	×10³ lb/in²	GPa	lb/in²	MPa
LVL	2,800	19.2	2,000	13.8	190	1.31
PSL, type A	2,900	20.0	2,000	13.8	210	1.45
PSL, type B	1,500	10.3	1,200	8.3	150	1.03

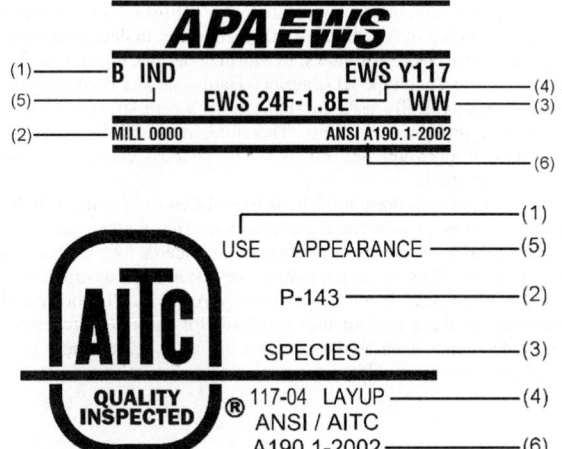

Fig. 6.7.2 Typical product quality stamps showing vital information for glulam combination. 1, structural use of member (B, simple span bending member; C, compression member; T, tension member; CB, continuous or cantilevered span bending member); 2, mill number; 3, species; 4, structural grade designation, indicating species, design bending strength and stiffness; 5, appearance grade designation (framing, industrial, architectural, or premium); 6, identification of conformance to ANSI A190.1 standard.

20 times the depth. Technical publications containing allowable spans for various loadings of standard sizes of beams are available from either AITC or APA-EWS.

The design stresses for beams in bending for dry-use applications are standardized in multiplies of 200 lb/in² (1.4 MPa) with the range of 2,000 to 3,000 lb/in² (13.8 to 20.7 MPa). Modulus of elasticity values associated with these design stresses in bending vary from 1.6 to 2.0 × 10⁶ lb/in². A bending stress of 2,400 lb/in² (16.5 MPa) and a modulus

of elasticity of 1.8×10^6 lb/in² (12.4 GPa) are most commonly specified, and the designer needs to verify the availability of beams with higher values. Design properties must be adjusted for wet-use applications. Detailed information on other design properties for beams as well as design properties and procedures for arches and other uses is given in "National Design Specification" (AF&PA).

Round Timbers

Round timbers in the form of poles, piles, or construction logs represent some of the most efficient uses of forest products because of the minimum of processing required. Poles and piles are generally debarked or peeled, seasoned, graded, and treated with a preservative prior to use. Construction logs are often shaped to facilitate their use. See Table 6.7.12.

Poles   The primary use of wood poles is to support utility and transmission lines. An additional use is for building construction. Each of these uses requires that the poles be pressure-treated with preservatives following the applicable AWPA standard (C1). For utility structures, pole length may vary from 30 to 125 ft (9.1 to 38.1 m). Poles for building construction rarely exceed 30 ft (9.1 m). Southern pines account for the highest percentage of poles used in the United States because of their favorable strength properties, excellent form, ease of treatment, and availability. Douglas-fir and western redcedar are used for longer lengths; other species are also included in the ANSI O5.1 standard (ANSI 1992) that forms the basis for most pole purchases in the United States.

Design procedures for the use of ANSI O5.1 poles in utility structures are described in the "National Electric Safety Code" (NESC). For building construction, design properties developed based on ASTM D 2899 (see ASTM, 1995) are provided in "Timber Construction Manual" (AITC, 1994) or ASAE EP 388.

Piles   Most piles used for foundations in the United States utilize either southern pine or Douglas-fir. Material requirements for timber piles are given in ASTM D 25, and preservative treatment should follow the applicable AWPA standard (C1 or C3). Design stress and procedures are provided in "National Design Specifications."

Construction Logs   Log buildings continue to be a popular form of construction because nearly any available species of wood can be used. Logs are commonly peeled prior to fabrication into a variety of shapes. There are no standardized design properties for construction logs, and when they are required, log home suppliers may develop design properties by following an ASTM standard (ASTM D 3957).

PROPERTIES OF STRUCTURAL PANEL PRODUCTS
by Roland Hernandez

Structural panel products are a family of wood products made by bonding veneer, strands, particles, or fibers of wood into flat sheets. The members of this family are (1) plywood, which consists of products made completely or in part from wood veneer; (2) flakeboard, made

Table 6.7.12   Design Stresses for Selected Species of Round Timbers for Building Construction

Type of timber and species	Design stress					
	Bending		Compression		Modulus of elasticity	
	lb/in²	MPa	lb/in²	MPa	× 10⁶ lb/in²	GPa
Poles*						
Southern pine and Douglas-fir	2,100	14.5	1,000	6.9	1.5	10.3
Western redcedar	1,400	9.6	800	5.5	0.9	6.2
Piles†						
Southern pine	2,400	16.5	1,200	8.3	1.5	10.3
Douglas-fir	2,450	16.9	1,250	8.6	1.5	10.3
Red pine	1,900	13.1	900	6.2	1.3	8.8

*From "Timber Construction Manual" (AITC, 1994).
†From "National Design Specification" (AF&PA, 2005).

from strands, wafers, or flakes; (3) particleboard, made from particles; and (4) fiberboard and hardboard, made from wood fibers. Plywood and flakeboard make up a large percentage of the panels used in structural applications such as roof, wall, and floor sheathing; thus, only those two types will be described here.

Plywood

Plywood is the name given to a wood panel composed of relatively thin layers or plies of veneer with the wood grain of adjacent layers at right angles. The outside plies are called faces or face and back plies, the inner plies with grain parallel to that of the face and back are called cores or centers, and the plies with grain perpendicular to that of the face and back are called crossbands. In four-ply plywood, the two center plies are glued with the grain direction parallel to each ply, making one center layer. Total panel thickness is typically not less than ¹⁄₁₆ in (1.6 mm) or more than 3 in (76 mm). Veneer plies may vary as to number, thickness, species, and grade. Stock plywood sheets usually measure 4 by 8 ft (1.2 by 2.4 m), with the 8-ft (2.4-m) dimension parallel to the grain of the face veneers.

The alternation of grain direction in adjacent plies provides plywood panels with dimensional stability across their width. It also results in fairly similar axial strength and stiffness properties in perpendicular directions within the panel plane. The laminated construction results in a distribution of defects and markedly reduces splitting (compared to solid wood) when the plywood is penetrated by fasteners.

Two general classes of plywood, covered by separate standards, are available: construction and industrial plywood, and hardwood and decorative plywood. Construction and industrial plywood is covered by Product Standard PS 1-95; and hardwood and decorative plywood is covered by ANSI/HPVA HP-1-2004. Each standard recognizes different exposure durability classifications, which are primarily based on the moisture resistance of the glue used, but sometimes also address the grade of veneer used.

The exposure durability classifications for construction and industrial plywood specified in PS-1 are: exterior, exposure 1, intermediate glue (exposure 2), and interior. Exterior plywood is bonded with exterior (waterproof) glue and is composed of C-grade or better veneers throughout. Exposure 1 plywood is bonded with exterior glue, but it may include D-grade veneers. Exposure 2 plywood is made with glue of intermediate resistance to moisture. Interior-type plywood may be bonded with interior, intermediate, or exterior (waterproof) glue. D-grade veneer is allowed on inner and back plies of certain interior-type grades.

The exposure durability classifications for hardwood and decorative plywood specified in ANSI/HPVA HP-1-2004 are, in decreasing order of moisture resistance, as follows: technical (exterior), type I (exterior), type II (interior), and type III (interior). Hardwood and decorative plywood are not typically used in applications where structural performance is a prominent concern. Therefore, most of the remaining discussion of plywood performance will concern construction and industrial plywood.

A very significant portion of the market for construction and industrial plywood is in residential construction. This market reality has resulted in the development of performance standards for sheathing and single-layer subfloor or underlayment for residential construction by APA–The Engineered Wood Association (APA-EWA). Plywood panels conforming to these performance standards for sheathing are marked with grade stamps such as those shown in Fig. 6.7.3 (example grade stamps are shown for different agencies). As seen in this figure, the

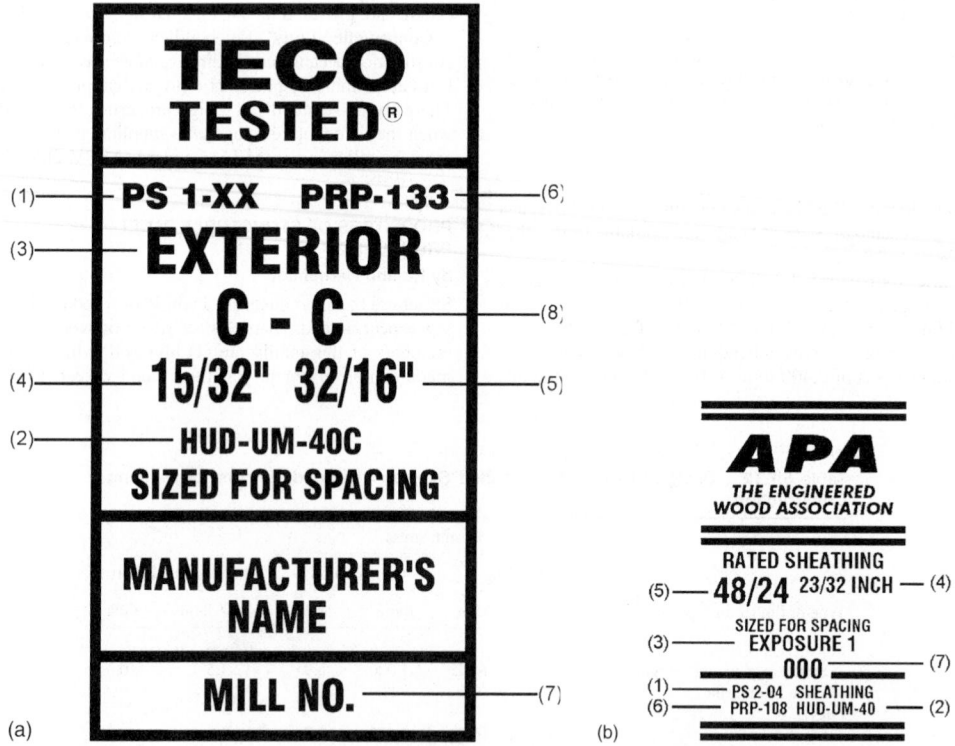

Fig. 6.7.3 Typical grade marks for (*a*) sheathing-grade plywood conforming to Product Standard PS 1-95 and (*b*) sheathing-grade structural-use panel conforming to Product Standard PS 2-95. 1, conformance to indicated product standard; 2, recognition as a quality assurance agency; 3, exposure durability classification; 4, thickness; 5, span rating; 6, conformance to performance-rated product; 7, manufacturer's name or mill number; 8, grade of face and core veneers.

Fig. 6.7.4   Grade stamp for hardwood plywood conforming to ANSI/HPVA HP-1-2004, 1, trademark of Hardwood Plywood and Veneer Association; 2, formaldehyde emission characteristics; 3, structural description; 4, face species or finish type; 5, HPVA mill number; 6, veneer grade of face; 7, flame spread classification; 8, bondline type; 9, standard that governs manufacture.

grade stamps must show (1) conformance to the plywood product standards; (2) recognition as a quality assurance agency by the National Evaluation Services (NES), which is affiliated with the Council of American Building Officials; (3) exposure durability classification; (4) thickness of panel; (5) span rating, 32/16, which refers to the maximum allowable roof support spacing of 32 in (813 mm) and maximum floor joist spacing of 16 in (406 mm); (6) conformance to the performance-rated standard of the agency; (7) manufacturer's name or mill number; and (8) grades of face and core veneers.

All hardwood plywood represented as conforming to American National Standard ANSI/HPVA-HP-1-2004 is identified by one of two methods: either marking each panel with the HPVA plywood grade stamp (Fig. 6.7.4) or including a written statement with this information with the order or shipment. The HPVA grade stamp shows (1) HPVA trademark; (2) formaldehyde emission characteristics; (3) structural description; (4) face species or finish type; (5) HPVA mill number; (6) veneer grade of face; (7) flame spread classification; (8) bondline type; and (9) standard that governs manufacture.

The span-rating system for plywood was established to simplify specification of plywood without resorting to specific structural engineering design. This system indicates performance without the need to refer to species group or panel thickness. It gives the allowable span when the face grain is placed across supports.

If design calculations are desired, a design guide is provided by APA-EWA in "Plywood Design Specifications" (PDS, 2004). The design guide contains tables of grade stamp references, section properties, and allowable stresses for plywood used in construction of buildings and similar related structures. For the design and fabrication of curved panels, plywood-lumber beams, plywood stress-skin panels, plywood sandwich panels, and all-plywood beams, information is available in supplements to the PDS standard from APA-EWA.

If calculations for the actual physical and mechanical properties of plywood are desired, formulas relating the properties of the particular wood species in the component plies to the laminated panel are provided in "Wood Handbook." These formulas could be applied to plywood of any species, provided the basic mechanical properties of the species are known. Note, however, that the formulas yield predicted actual properties (not design values) of plywood made of defect-free veneers.

Structural Flakeboards

Structural flakeboards are wood panels made from specially produced flakes—typically from relatively low-density species, such as aspen or pine—and bonded with an exterior-type water-resistant adhesive. Two major types of flakeboards are recognized, **oriented strandboard** (OSB) and **waferboard**. OSB is a flakeboard product made from wood

strands (long and narrow flakes) that are formed into a mat of three to five layers. The outer layers are aligned in the long panel direction, while the inner layers may be aligned at right angles to the outer layers or may be randomly aligned. In waferboard, a product made almost exclusively from aspen wafers (wide flakes), the flakes are not usually oriented in any direction, and they are bonded with an exterior-type resin. Because flakes are aligned in OSB, the bending properties (in the aligned direction) of this type of flakeboard are generally superior to those of waferboard. For this reason, OSB is the predominant form of structural flakeboard. Panels commonly range from 0.25 to 0.75 in (6 to 19 mm) thick and are 4 by 8 ft (1 by 2 m) in surface dimension. However, thicknesses up to 1.125 in (28.58 mm) and surface dimensions up to 8 by 24 ft (2 by 7 m) are available by special order.

A substantial portion of the market for structural flakeboard is in residential construction. For this reason, structural flakeboards are usually marketed as conforming to a product standard for sheathing or single-layer subfloor or underlayment and are graded as a performance-rated product (PRP-108) similar to that for construction plywood. The Voluntary Product Standard PS 2-95 is the performance standard for wood-based structural-use panels, which includes such products as plywood, composites, OSB, and waferboard. The PS 2-95 is not a replacement for PS 1-95, which contains necessary veneer grade and glue bond requirements as well as prescriptive layup provisions and includes many plywood grades not covered under PS 2-95.

Design capacities of the APA–EWA performance-rated products, which include OSB and waferboard, can be determined by using procedures outlined in the APA–EWA Technical Note N375A. In this reference, allowable design strength and stiffness properties, as well as nominal thicknesses and section properties, are specified based on the span rating of the panel. Additional adjustment factors based on panel grade and construction are provided.

Because of the complex nature of structural flakeboards, formulas for determining actual strength and stiffness properties, as a function of the component material, are not available.

DURABILITY OF WOOD IN CONSTRUCTION
by Stan Lebow and Robert White

Biological Challenge

In the natural ecosystem, wood residues are recycled into the nutrient web through the action of wood-degrading fungi, insects, and other organisms. In some circumstances, these same natural recyclers have the potential to degrade wood used in construction. Termites and decay fungi are the most destructive, but other organisms, such as wood boring beetles and carpenter ants, can be important in some regions. Several types of marine organisms can attack wood used in brackish water and saltwater. If conditions are favorable for the survival of one or more of these wood-destroying organisms, wood that has natural durability or that has been treated with preservatives should be employed to ensure the integrity of the structure.

Conditions Favorable for Biodeterioration

Three environmental components govern the development of wood-degrading organisms within terrestrial wood construction: moisture, oxygen supply, and temperature. Of these, moisture seems most directly influenced by design and construction practices. The oxygen supply is usually adequate except for materials submerged below water or deep within the soil, and wood-degrading organisms have evolved to survive in a wide range of temperatures. For these reasons, most of the following discussion will focus on relationships between wood moisture and potential for wood deterioration.

Decay fungi require free water within the wood cells (above 20 to 25 percent moisture content) to survive. Soil provides a continuing source of moisture. Nondurable wood in contact with the ground will decay most rapidly at the ground line where moisture from soil and the supply of oxygen within the wood support growth of decay fungi. Deep within the soil, a limited supply of oxygen prevents growth of most decay fungi. As the distance above ground increases, the decreasing moisture content limits fungal growth unless the wood is wetted through exposure to rain or from water entrapped as a consequence of design or construction practices. Wood absorbs water through cut ends about 11 times faster than through lateral surfaces. Consequently, decay develops first at the joints in aboveground construction.

Extremely low oxygen content in wood submerged below water will prevent growth of decay fungi, but other microorganisms can slowly colonize submerged wood over decades or centuries of exposure. Thus, properties of such woods need to be reconfirmed when old, submerged structures are retrofitted.

Most wood-degrading insects also prefer moist wood, but their moisture requirements vary somewhat by species. Subterranean termites native to the U.S. mainland establish a continuous connection with soil to maintain adequate moisture in wood that is being attacked above ground. Formosan termites have the capacity to use sources of aboveground moisture without establishing a direct connection with the soil. Dry-wood termites survive only in tropical or neotropical areas with sufficiently high relative humidity to elevate the wood moisture content to levels high enough to provide moisture for insect metabolism. Some types of wood-boring beetles, such as powder post beetles, can colonize wood with low moisture contents. Carpenter ants initially invade wood that is moist or decayed, but may then expand into adjacent dry wood.

Methods for Protecting Wood

Protection with Good Design The most important aspect of protecting wood in construction is use of designs that minimize moisture accumulation. Many wood structures have endured for several hundred years because of sound design and construction practices. In nearly all those old buildings the wood has been kept dry by a barrier over the structure (roof plus overhang), by maintaining a separation between the ground and the wood elements (foundation), and by preventing accumulation of moisture in the structure (ventilation). Where Formosan subterranean termites occur, good designs and construction practices that eliminate sources of aboveground moisture are particularly important. Channeling of water from roof drains, collection of condensate from air conditioners, and proper installation of flashing are examples of important considerations. Other design and construction practices also help to protect wood from insect attack. Chemical or physical barriers are often used to prevent subterranean termites from moving from the soil into a building. Many beetles attack only certain species or groups of woods that may be used in speciality items such as joinery. When these species are used, it is important to use wood that is not pre-infested at the time of construction. Today's engineered wood products will last for centuries if good design practices are used. However, there are many structures where it is impractical to prevent decay and insect attack simply through design. In these situations part of or all the structure may need to be constructed with naturally durable wood species or with wood that has been treated with preservatives.

Naturally Durable Wood The heartwood of old-growth trees of certain species, such as bald cypress, redwood, cedars, and several white oaks, is naturally resistant or very resistant to decay fungi. Heartwood of several other species, such as Douglas-fir, longleaf pine, eastern white pine, and western larch, is moderately resistant to wood decay fungi. Some tropical species, such as ipe, are also imported because of their natural durability. Although decay resistance and resistance to termite attack are correlated, not all species that are decay-resistant are also resistant to termite attack. A more complete listing of naturally durable woods is given in "Wood Handbook." One limitation on the use of these species is that the durability is variable between trees and even between pieces cut from a single tree. In addition, the supply of naturally durable species is limited relative to the demand for durable wood products. Consequently, other forms of protection, such as preservative treatment, are more frequently used in current construction. Naturally durable species are not effective in preventing marine borer attack, and pressure treatment with wood preservatives is required to protect wood used in marine or brackish waters.

Preservative Treatments Wood preservatives have been used for over a hundred years. They are broadly classified as either water type or oil type based on the chemical composition of the preservative and the solvent used during the treating process. Each type of preservative has advantages and disadvantages, depending on the application. The most common oil-type preservatives are creosote, pentachlorophenol, and copper naphthenate. Because wood treated with oil-type preservatives may be oily to the touch or have a noticeable odor, it is not usually used for applications involving frequent human contact or for inhabited structures. Oil-type preservatives may offer improved water repellency and typically do not promote corrosion of fasteners. Water-based preservatives leave a dry, paintable surface and are commonly used to treat wood for residential applications such as decks and fences. They are primarily used to treat softwoods because hardwoods treated with these preservatives may not be well protected from soft-rot attack. Until recently, the most widely used water-type preservative was chromated copper arsenate (CCA). However, CCA has now been voluntarily phased out for most applications around residential areas and where human contact is expected. More recent water-based preservatives, such as alkaline copper quat or copper azole, rely primarily on copper to protect the wood. Water-based wood preservatives can increase susceptibility to corrosion, and metal fasteners used with the treated wood should be hot-dipped galvanized or stainless steel.

Borates are another type of waterborne preservative that may be used to treat interior building components for protection against insect attack. Borates should not be used where they are exposed to soil or rainfall because they are readily leached from the wood. In non-load-bearing applications such as exterior millwork around windows and doors, wood is usually protected with water-repellent preservative treatments that are applied by nonpressure processes. Standards for preservative treatment are published by the American Wood-Preservers' Association, and detailed information on wood preservation is given in "Wood Handbook." It is important to ensure that wood is being treated to standard specifications. The U.S. Department of Commerce American Lumber Standard Committee (ALSC) accredits third-party inspection agencies for treated wood products. Updated lists of accredited agencies can be found on the ALSC Website (www.alsc.org). If the wood has been treated to these specifications, it will have the quality mark or symbol of an ALSC-accredited inspection agency.

Protection from Weathering The combination of sunlight and other weathering agents will slowly remove the surface fibers of wood products. Removal of fibers can be greatly reduced by providing a wood finish; if the finish is properly maintained, the removal of fibers can be nearly eliminated. Information on wood finishes is available in Cassens and Feist, "Exterior Wood in the South."

Protection from Fire   In general, proper design for fire safety allows the use of untreated wood. When there is a need to reduce the potential for heat contribution or flame spread, fire-retardant treatments are available. Although fire-retardant coatings or dip treatments are available, effective treatment often requires that the wood be pressure-impregnated with the fire-retardant chemicals. These chemicals include inorganic salts such as monoammonium and diammonium phosphate, ammonium sulfate, zinc chloride, sodium tetraborate, and boric acid. Resin polymerized after impregnation into wood is used to obtain a leach-resistant treatment. Such amino resin systems are based on urea, melamine, dicyandiamide, and related compounds. An effective treatment can reduce the ASTM E 84 flame spread to less than 25 and show no evidence of significant progressive combustion when the ASTM E 84 test is continued for an additional 20-min period. When the external source of heat is removed, the flames from fire-retardant-treated wood will generally self-extinguish. Many fire-retardant treatments reduce the generation of combustible gases by lowering the thermal degradation temperature. ASTM standards have been developed to address other performance requirements such as initial strength loss due to treatment and kiln drying, strength loss when exposed to elevated temperatures in use, excessive hygroscopicity in areas of high humidity, and loss of fire performance due to outdoor exposure to rainfall.

Effect of Aging

Long exposure of wood to the atmosphere also causes changes in the cellulose. A study by Kohara and Okamoto of sound old timbers of a softwood and hardwood of known ages from temple roof beams shows that the percentage of cellulose decreases steadily over a period up to 1,400 years while the lignin remains almost constant. These changes are reflected in strength losses (Fig. 6.7.5). Impact properties approximate a loss that is nearly linear with the logarithm of time.

Allowable Working Stresses for Preservative-Treated Lumber

Allowable working stresses for preservative-treated lumber usually need not be reduced to account for the effect of the treating process. Tests made by the USDA Forest Service, Forest Products Laboratory, of preservative-treated lumber when undergoing bending, tension, and compression perpendicular to grain show reductions in mean extreme fiber stress from a few percent up to 25 percent, but few reductions in working stresses. Compression parallel to the grain is affected less and modulus of elasticity very little. The effect on horizontal shear can be estimated by inspection for an increase in shakes and checks

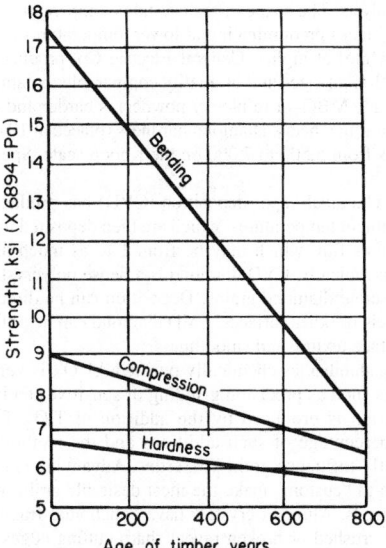

Fig. 6.7.5   Strength loss with age in a hardwood (*Zelkowa serrata*). (*Adapted from information in Kohara and Okamoto, 1955.*)

after treatment. AWPA Standards keep temperatures, heating periods, and pressures to a minimum for required penetration and retention, which precludes the need for adjustment in working stresses.

COMMERCIAL LUMBER STANDARDS

Standard abbreviations for lumber description and size standards for yard lumber are given in "Wood Handbook."

Cross-sectional dimensions and section properties for beams, stringers, joists, and planks are given in the "National Design Specification."

Standard patterns for finish lumber are shown in publications of the grading rules for the various lumber associations.

Information and specifications for construction and industrial plywood are given in Product Standard PS 1-95 and in ANSI/HPVA HP-1-2004 for hardwood and decorative plywood. Information and specifications for structural flakeboard are given in PS 2-95.

6.8   NONMETALLIC MATERIALS
by Ali M. Sadegh

ABRASIVES

REFERENCES: "Abrasives: Their History and Development," The Norton Co. Searle, "Manufacture and Use of Abrasive Materials," Pitman. Heywood, "Grinding Wheels and Their Uses," Penton. "Boron Carbide," The Norton Co. "Abrasive Materials," annual review in "Minerals Yearbook," U.S. Bureau of Mines. "Abrasive Engineering," Hitchcock Publishing Co. Coated Abrasives Manufacturer's Institute, "Coated Abrasives—Modern Tool of Industry," McGraw-Hill. Wick, Abrasives; Where They Stand Today, *Manufacturing Engineering and Management,* **69,** no. 4, Oct. 1972. Burls, "Diamond Grinding; Recent Research and Development," Mills and Boon, Ltd., London. Coes, Jr., "Abrasives," Springer-Verlag, New York. *Proceedings of the American Society for Abrasives Methods.* 1971, ANSI B74.1 to B74.3 1977 to 1993. Washington Mills Abrasive Co., Technical Data, *Mechanical Engineering,* Mar. 1984. "Modern Abrasive Recipes," *Cutting Tool Engineering,* April 1994. "Diamond Wheels Fashion Carbide Tools," *Cutting Tool Engineering,* August 1994. Synthetic Gemstones, *Compressed Air Magazine,* June 1993. Krar and Ratterman, "Superabrasives," McGraw-Hill. Current ASTM and ANSI standards.

Manufactured (Artificial) Abrasives

Manufactured abrasives dominate the scene for commercial and industrial use, because of the greater control over their chemical composition and crystal structure, and their greater uniformity in size, hardness, and cutting qualities, as compared with natural abrasives. Technical advances have resulted in abrasive "machining tools" that are economically competitive with many traditional machining methods, and in some cases replace and surpass them in terms of productivity. Fused electrominerals such as silicon carbide, aluminum oxide, alumina zirconia, alumina chromium oxide, and alumina titania zirconia are the most popular conventional **manufactured abrasives.**

Grains

Manufactured (industrial or synthetic) **diamonds** are produced from graphite at pressures 29.5 to 66.5×10^6 N/m^2 (8 to 18×10^5 lb/in^2) and temperatures from 1,090 to 2,420°C (2,000 to 4,400°F), with the aid of metal catalysts. The shape of the crystal is temperature-controllable, with cubes (black) predominating at lower temperatures and octahedra (yellow to white) at higher. General Electric Co. produces synthetics reaching 0.01 carat sizes and of quality comparable to natural diamond powders. Grade MBG-11 (a blocky powder) is harder and tougher and is used for cutting wheels. Diamond hardness (placed at 10 on the Mohs scale) ranges from 5,500 to 7,000 on the Knoop scale. Specific gravity is 3.521.

Recently, chemical vapor deposition (CVD) has produced diamond vapors at ambient temperatures, which are then deposited on a substrate as a continuous film which may be from 1 to as much as 1,000 μm thick. Microscopically, CVD diamond is a dense polycrystalline structure with discrete diamond grains. Deposition can be directly on tools such as wheels or other surfaces. CVD diamond can be polished to any required surface texture and smoothness.

Crystalline alumina, as chemically purified Al_2O_3, is very adaptable in operations such as precision grinding of sensitive steels. However, a tougher grain is produced by the addition of TiO_2, Fe_2O_3, SiO_2, ZrO_2. The percentage of such additions and the method of cooling the pig greatly influence grain properties. Advantage is taken of this phenomenon to "custom" make the most desirable grain for particular machining needs. Alumina crystals have conchoidal fracture, and the grains when crushed or broken reveal sharp cutting edges and points. Average properties are: density ~3.8, coefficient of expansion ~$0.81 \times 10^{-5}/°C$ ($0.45 \times 10^{-5}/°F$), hardness ~9 (Mohs scale), and melting temperature ~$2,040°C$ ($3,700°F$). Applications cover the grinding of high-tensile-strength materials such as soft and hard steels, and annealed malleable iron.

Silicon carbide (SiC) corresponds to the mineral moissanite, and has a hardness of about 9.5 (Mohs scale) or 2,500 (Knoop), and specific gravity 3.2. It is insoluble in acid and is infusible but decomposes above $2,230°C$ ($4,060°F$). It is manufactured by fusing together coke and sand in an electric furnace of the resistance type. Sawdust is used also in the batch and burns away, leaving passages for the carbon monoxide to escape. The grains are characterized by great brittleness. Abrasives of silicon carbide are best adapted to the grinding of low-tensile-strength materials such as cast iron, brass, bronze, marble, concrete, stone, and glass. It is available under several trade names such as **Carborundum, Carbolon, Crystolon.**

Boron carbide (B₄C), a black crystal, has a hardness of about 9.32 (Mohs scale) or about 2,800 (Knoop), melts at about $2,460°C$ ($4,478°F$) but reacts with oxygen above $983°C$ ($1,800°F$), and is not resistant to fused alkalies. Boron carbide powder for grinding and lapping is obtainable in standard mesh sizes to 240, and to 800 in special finer sizes. It is being used in loose grain form for the lapping of cemented-carbide tools. In the form of molded shapes, it is used for pressure blast nozzles, wire-drawing dies, bearing surfaces for gages, etc.

Boron nitride (BN) when produced at extremely high pressures and temperatures forms tiny reddish to black grains of cubic crystal structure having hardness equal to diamond, and moreover is stable to $1,930°C$ ($3,500°F$). Abrasive powders, such as Borazon, are used extensively for coated-abrasive applications as for grinding tool and die steels and high-alloy steels, particularly where chemical reactivity of diamonds is a problem. Ease of penetration and free-cutting action minimize heat generation, producing superior surface integrity.

Crushed steel is made by heating high-grade crucible steel to white heat and quenching in a bath of cold water. The fragments are then crushed to sizes ranging from fine powder to 1/16 in diam. They are classified as diamond crushed steel, diamond steel, emery, and steelite, used chiefly in the stone, brick, glass, and metal trades.

Rouge and crocus are finely powdered oxide of iron used for buffing and polishing. Rouge is the red oxide; crocus is purple.

Natural Abrasives

Diamond of the **bort** variety, crushed and graded into usable sizes and bonded with synthetic resin, metal powder, or vitrified-type bond, is used extensively for grinding tungsten- and tantalum-carbide cutting tools, and glass, stone, and ceramics.

Corundum is a mineral composed chiefly of crystallized alumina (93 to 97 percent Al_2O_3). It has been largely replaced by the manufactured variety.

Emery, a cheap and impure form of natural corundum which has been used for centuries as an abrasive, has been largely superseded by manufactured aluminum oxide for grinding. It is still used to some extent in the metal- and glass-polishing trades.

Garnet Certain deposits of garnet having a hardness between quartz and corundum are used in the manufacture of abrasive paper. **Quartz** is also used for this purpose. Garnet costs about twice as much as quartz and generally lasts proportionately longer.

Buhrstones and millstones are generally made from cellular quartz. Chasers (or stones running on edge) are also made from the same mineral.

Natural oilstones, the majority being those quarried in Arkansas, are of either the hard or the soft variety. The hard variety is used for tools requiring an extremely fine edge like those of surgeons, engravers, and dentists. The soft variety, more porous and coarser, is used for less exacting applications.

Pumice, of volcanic origin, is extensively used in leather, felt, and woolen industries and in the manufacture of polish for wood, metal, and stone. An artificial pumice is made from sand and clay in five grades of hardness, grain, and fineness.

Infusorial earth or **tripoli** resembles chalk or clay in physical properties. It can be distinguished by absence of effervescence with acid, is generally white or gray in color, but may be brown or even black. Owing to its porosity, it is very absorptive. It is used extensively in polishing powders, scouring soaps, etc., and, on account of its porous structure, in the manufacture of dynamite as a holder of nitroglycerin, also as a nonconductor for steam pipes and as a filtering medium. It is also known as **diatomaceous silica.**

Grinding Wheels For complete coverage and details see current ANSI/ASTM Standards D896 to D3808.

Vitrified Process In wheels, segments, and other abrasive shapes of this type, the abrasive grains are bonded with a glass or porcelain obtained by mixing the grains with such materials as clays and feldspars in various proportions, molding the wheel, drying, and firing at a temperature of $1,370°C$ ($2,500°F$) approx. It is possible to manufacture wheels as large as 60 in diam by this process, and even larger wheels may be obtained by building up with segments. Most of the grinding wheels and shapes (segments, cylinders, bricks, etc.) now manufactured are of the vitrified type and are very satisfactory for general grinding operations.

Silicate Process Wheels and shapes of the silicate type are manufactured by mixing the abrasive grain with sodium silicate (water glass) and fillers that are more or less inert, molding the wheel by tamping, and baking at a moderate temperature. Silicate bonded wheels are considered relatively "mild acting" and, in the form of large wheels, are still used to some extent for grinding-edge tools in place of the old-fashioned sandstone wheels.

Organic Bonded Wheels Organic bonds are used for high-speed wheels, and are equally well adapted to the manufacture of very thin wheels because of their flexibility compared with vitrified wheels. There are three distinct types in the group. The **shellac process** consists of mixing abrasive grains with shellac, heating the mass until the shellac is viscous, stirring, cooling, crushing, forming in molds, and reheating sufficiently to permit the shellac to set firmly upon cooling. Wheels made by this process are used for saw gumming, roll grinding, ball-race and cam grinding, and in the cutlery trade. In the **rubber process** the bond is either natural or synthetic rubber. The initial mixture of grain, rubber, and sulfur (and such special ingredients as accelerators, fillers, and softeners) may be obtained by rolling or other methods. Having formed the wheel, the desired hardness is then developed by vulcanization. Wheels can be made in a wide variety of grain combinations and grades and have a high factor of safety as regards resistance to breakage in service. Wheels made by the rubber process are used for cutoff service on wet-style machines, ball-race punchings, feed wheels and centerless grinders, and for grinding stainless-steel billets and welds, which usually require a high-quality finish. With the **resinoid process,** the practice is to form the wheel by the cold-press process using a synthetic resin. After heating, the resultant bond is an insoluble, infusible product of notable strength and resiliency. Resinoid bond is used for the majority of high-speed wheels in foundries, welding shops, and billet shops, and also for cutoff wheels. The rate of stock removal is generally in direct proportion to the peripheral speed. Resinoid-bonded wheels are capable of being operated at speeds as high as 2,900 m/min (9,500 surface ft/min), as contrasted with 1,980 m/min (6,500 surface ft/min) for most vitrified bonds.

Silicone-coated abrasives, such as Silkote are reported to resist deleterious coolant effects on the bond between resin and grain, because of the silicone's ability to repel entry of coolant.

The **grain size** or **grit** of a wheel is determined by the size or combination of sizes of abrasive grain used. The Grinding Wheel Institute has standardized sizes 8, 10, 12, 14, 16, 20, 24, 30, 36, 46, 54, 60, 70, 80, 90, 100, 120, 150, 180, 200, 220, and 240. The finer sizes, known as flours, are designated as 280, 320, 400, 500, 600, 800, 900, or as F, FF, FFF, and XF.

The **grade** is the hardness or relative strength of bonding of a grinding wheel. The wheel from which grain particles are easily broken away, causing it to wear rapidly, is called soft, and one that is able to retain its particles longer is called hard. The complete range of grade letters used for the order of increasing hardness is

Soft		Hard

A, B, C, D, E, F, G, H, I, J, K, L, M, N, O, P, Q, R, S, T, U, V, W, X, Y, Z

Table 6.8.1 provides a guide to surface finishes attainable with diamond grit.

Table 6.8.1   Guide to Surface Finish Attainable with Diamond Grit

Grit size	Surface finish		Recommended max DOC* per pass	
	μ in (AA)	μm R_a	in	μm
80	26–36			
100	24–32			
105	18–30			
100S	16–26	0.4–0.8	0.001–0.002	25–50
110				
120	16–18			
150	14–16			
180	12–14	0.25–0.50	0.0007–0.0010	17–25
220	10–12			
240	8–10			
320	8	0.12–0.25	0.0004–0.0006	
10–15				
400	7–8	0.08–0.15	0.0003–0.0005	8–12
30–40 μm	7			
20–30	6–7	0.05–0.10	0.0002–0.0004	5–10
	6			
10–20	5–6			
8–16				
6–12	3–5			
4–8	2–4	0.013–0.025	0.00005–0.00007	1–2
3–6	2			
0–2	1			

* DOC = depth of cut.
SOURCE: *Cutting Tool Engineering,* Aug. 1994.

Coated Abrasives

Coated abrasives are "tools" consisting of an abrasive grit, a backing, and an adhesive bond. Grits are generally one of the manufactured variety listed above, and are available in mesh sizes ranging from the coarsest at 12 to the finest at 600. Backings can be cloth, paper, fiber, or combinations and are made in the form of belts and disks for poweroperated tools and in cut sheets for both manual and power usage. Adhesive bonds consist of two layers, a "make coat" and a "size coat." Both natural and synthetic adhesives serve for bond materials. Abrasive coatings can be closed-coat, in which the abrasive grains are adjacent to one another without voids, or open-coat, in which the grains are set at a predetermined distance from one another. The flex of the backing is obtained by a controlled, directional, spaced backing of the adhesive bond.

Abrasive Waterjets

Such tough-to-cut materials as titanium, ceramics, metallic honeycomb structures, glass, graphite, and bonding compounds can be successfully cut with abrasive waterjets. Abrasive waterjets use pressurized water, up to 60 lb/in^2, to capture solid abrasives by flow momentum, after which the mixture is expelled through a sapphire nozzle to form a highly focused, high-velocity cutting "tool." Advantages of this cutting method include: minimal dust, high cutting rates, multidirectional cutting capability, no tool dulling, no deformation or thermal stresses, no fire hazards, ability to cut any material, small power requirements, avoidance of delamination, and reduction of striation. (See also Sec. 13.)

Abrasive waterjet cutting, however, has some limitations including high noise level (80 to 100 dB), safety problems, low material removal rates, inability to machine blind holes or pockets, damage to accidentally exposed machine elements by the particles and/or high-pressure water, and the size of the overall system.

Industrial Sharpening Stones

Sharpening stones come in several forms such as bench stones, files, rubbing bricks, slip stones, and specialties, and are made chiefly from aluminum oxide, such as Alundum, or silicon carbide, such as Crystolon. Grit sizes of fine, medium, and coarse are available.

ADHESIVES

REFERENCES: Cagle (ed.), "Handbook of Adhesive Bonding," McGraw-Hill. Bikerman, "The Science of Adhesive Joints," Academic. Shields, "Adhesives Handbook," CRC Press (Division of The Chemical Rubber Co.). Cook, "Construction Sealants and Adhesives," Wiley. Patrick, "Treatise on Adhesives," Marcel Dekker. NASA SP-5961 (01) Technology Utilization, "Chemistry Technology: Adhesives and Plastics," National Technical Information Services, Virginia. Simonds and Church, "A Concise Guide to Plastics," Reinhold. Lerner, Kotsher, and Sheckman, "Adhesives Red Book," Palmerton Publishing Co., New York. *Machine Design,* June 1976. ISO 6354–1982. "1994 Annual Book of ASTM Standards," vol. 15.06 (Adhesives). ANSI/ASTM Standards D896–D3808. Current ASTM and ANSI Standards.

Adhesives are substances capable of holding materials together in a useful manner by surface attachment. Some of the advantages and disadvantages of adhesive bonding are as follows:

Advantages   Ability to bond similar or dissimilar materials of different thicknesses; fabrication of complex shapes not feasible by other fastening means; smooth external joint surface; economic and rapid assembly; uniform distribution of stresses; weight reduction; vibration damping; prevention or reduction of galvanic corrosion; insulating properties.

Disadvantages   Surface preparation; long cure times; optimum bond strength not realized instantaneously, service-temperature limitations; service deterioration; assembly fire or toxicity; tendency to creep under sustained load.

A broad scheme of classification is given in Table 6.8.2a.

For the vocabulary of adhesives the reader should refer to ISO 6354–1982, and for standard definitions refer to ASTM D907–82. Procedures for the testing of adhesive strength viscosity, storage life, fatigue properties, etc. can be found in ANSI/ASTM D950–D3808.

Thermoplastic adhesives are a general class of adhesives based upon long-chained polymeric structure, and are capable of being softened by the application of heat.

Thermosetting adhesives are a general class of adhesives based upon *cross-linked* polymeric structure, and are incapable of being softened once solidified. A recent development (1994) in cross-linked aromatic polyesters has yielded a very durable adhesive capable of withstanding temperatures of 700°F before failing. It retains full strength through 400°F.

Table 6.8.2a Classification of Adhesives

Origin and basic type		Adhesive material
Natural	Animal	Albumen, animal glue (including fish), casein, shellac, beeswax
	Vegetable	Natural resins (gum arabic, tragacanth, colophony, Canada balsam, etc.); oils and waxes (carnauba wax, linseed oils); proteins (soybean); carbohydrates (starch, dextrins)
	Mineral	Inorganic materials (silicates, magnesia, phosphates, litharge, sulfur, etc.); mineral waxes (paraffin); mineral resins (copal, amber); bitumen (including asphalt)
Synthetic	Elastomers	Natural rubber (and derivatives, chlorinated rubber, cyclized rubber, rubber hydrochloride)
		Synthetic rubbers and derivatives (butyl, polyisobutylene, polybutadiene blends (including styrene and acylonitrile), polyisoprenes, polychloroprene, polyurethane, silicone, polysulfide, polyolefins (ethylene vinyl chloride, ethylene polypropylene)
		Reclaim rubbers
	Thermoplastic	Cellulose derivatives (acetate, acetate-butyrate, caprate, nitrate, methyl cellulose, hydroxy ethyl cellulose, ethyl cellulose, carboxy methyl cellulose)
		Vinyl polymers and copolymers (polyvinyl acetate, alcohol, acetal, chloride, polyvinylidene chloride, polyvinyl alkyl ethers)
		Polyesters (saturated) [polystyrene, polyamides (nylons and modifications)]
		Polyacrylates (methacrylate and acrylate polymers, cyanoacrylates, acrylamide)
		Polyethers (polyhydroxy ether, polyphenolic ethers)
		Polysulfones
	Thermosetting	Amino plastics (urea and melamine formaldehydes and modifications)
		Epoxies and modifications (epoxy polyamide, epoxy bitumen, epoxy polysulfide, epoxy nylon)
		Phenolic resins and modifications (phenol and resorcinol formaldehydes, phenolic-nitrile, phenolic-neoprene, phenolic-epoxy)
		Polyesters (unsaturated)
		Polyaromatics (polyimide, polybenzimidazole, polybenzothiazole, polyphenylene)
		Furanes (phenol furfural)

SOURCE: Shields, "Adhesives Handbook," CRC Press (Division of the Chemical Rubber Co.).

It is likely that widespread applications for it will be found first in the automotive and aircraft industries.

Thermoplastic and thermosetting adhesives are cured (set, polymerized, solidified) by heat, catalysis, chemical reaction, free-radical activity, radiation, loss of solvent, etc., as governed by the particular adhesive's chemical nature.

Elastomers are a special class of thermoplastic adhesive possessing the common quality of substantial flexibility or elasticity.

Anaerobic adhesives are a special class of thermoplastic adhesive (polyacrylates) that set *only* in the *absence* of air (oxygen). The two basic types are: (1) *machinery*—possessing shear strength only, and (2) *structural*—possessing both tensile and shear strength.

Pressure-sensitive adhesives are permanently (and aggressively) tacky (sticky) solids which form immediate bonds when two parts are brought together under pressure. They are available as films and tapes as well as hot-melt solids.

The relative performance of a number of adhesives is given in Table 6.8.2b.

For high-performance adhesive applications (engineering or machine parts) the following grouping is convenient.

Thread locking	Anaerobic acrylic
Hub mounting	Anaerobic acrylic—compatible materials or flow migration unimportant.
	Modified acrylic—large gaps or migration must be avoided.
	Epoxy—maximum strength at high temperatures
Bearing mounting	Anaerobic acrylic—compatible materials necessary and flow into bearing area to be prevented.
	Modified acrylic—for lowest cost
Structural joining	Epoxies and modified epoxies—for maximum strength (highest cost)
	Acrylics—anaerobic or modified cyanacrylates
Gasketing	Silicones—primarily anaerobic

Table 6.8.3 presents a sample of a number of adhesives (with practical information) that are available from various sources. The table is adapted from the rather extensive one found in J. Shields, "Adhesives Handbook," CRC Press (Division of The Chemical Rubber Co.), 1970, by permission of the publisher. Domestic and foreign trade sources are listed there (pages 332–340) and appear coded (in parentheses) in the second column. For other extensive lists of trade sources, the reader is referred to Charles V. Cagle (ed.), "Handbook of Adhesive Bonding," McGraw-Hill, and "Adhesives Red Book," Palmerton Publishing Co., New York.

BRICK, BLOCK, AND TILE

REFERENCES: Plummer and Reardon, "Principles of Brick Engineering, Handbook of Design," Structural Clay Products Institute. Stang, Parsons, and McBurney, Compressive Strength of Clay Brick Walls, *B. of S. Research Paper* 108. Hunting, "Building Construction," Wiley. Amrhein, "Reinforced Masonry Engineering Handbook," Masonry Institute of America. Simpson and Horrbin (eds.), "The Weathering and Performance of Building Materials," Wiley. SVCE and Jeffers (eds.), "Modern Masonry Panel Construction Systems," Cahners Books (Division of Cahners Publishing Co.). ASTM Standards C67-C902. ANSI/ASTM Standards C62-C455.

Brick

In the case of structural and road building material, a small unit, solid or practically so, commonly in the form of a rectangular prism, formed from inorganic, nonmetallic substances and hardened in its finished shape by heat or chemical action. Note that the term is also used collectively for a number of such units, as "a carload of brick." In the present state of the art, the term brick, when used without a qualifying adjective, should be understood to mean such a unit, or a collection of such units, made from clay or shale hardened by heat. When other substances are used, the term brick should be suitably qualified unless specifically indicated by the context.

It is recognized that unless suitably qualified, a brick is a unit of burned clay or shale.

Table 6.8.2b Performance of Adhesive Resins
(Rating 1 = poorest or lowest, 10 = best or highest)

Adhesive resin	Adherence to					Resistance			
	Paper	Wood	Metal	Ceramics	Rubbers	Water	Solvents	Alkali	Acids
Alkyd	6	7	5	6	7	7	2	2	5
Cellulose acetate	4	3	1	3	5	2	3	1	3
Cellulose acetate butyrate	3	3	1	4	5	2	3	I	3
Cellulose nitrate	5	5	1	5	5	3	2	2	4
Ethyl cellulose	3	3	1	3	5	2	3	3	3
Methyl cellulose	5	1	1	3	3	1	6	3	3
Carboxy methyl cellulose	6	1	2	3	2	1	6	1	4
Epoxy resin	10	10	8	8	8	8	9	9	8
Furane resin	8	7	1	8	7	8	9	10	8
Melamine resin	10	10	2	2	2	7	9	5	5
Phenolic resins	9	8	2	6	7	8	10	7	8
Polyester, unsaturated	6	8	2	5	7	7	6	1	6
Polyethylacrylate	3	4	3	5	6	8	2	6	7
Polymethylmethacrylate	2	3	2	3	6	8	3	8	7
Polystyrene	1	3	2	2	5	8	1	10	8
Polyvinylacetate	8	7	7	7	3	3	3	4	6
Polyvinyl alcohol	6	2	2	4	6	I	7	1	3
Polyvinyl acetal	5	7	8	7	7	8	5	3	5
Polyvinyl chloride	5	7	6	7	6	8	6	10	9
Polyvinyl acetate chloride	6	8	6	7	5	8	5	9	9
Polyvinylidene copolymer	4	7	6	7	7	8	7	10	9
Silicons T.S.	4	6	7	7	8	10	7	6	6
Urethane T.S.	8	10	10	9	10	7	8	4	4
Acrylonitrile rubber	3	6	8	6	9	7	5	8	8
Polybutene rubber	3	3	6	2	8	8	3	10	9
Chlorinated rubber	3	5	7	4	7	6	3	10	9
Styrene rubber	5	7	6	5	8	7	3	10	9

SOURCE: Adapted from Herbert R. Simonds and James M. Church, "A Concise Guide to Plastics," 2d ed., Reinhold, 1963, with permission of the publisher.

Brick (Common) Any brick made primarily for building purposes and not especially treated for texture or color, but including clinker and oven-burn brick.

Brick (Facing) A brick made especially for facing purposes, usually treated to produce surface texture or made of selected clays or otherwise treated to produce the desired color.

Brick are manufactured by the dry-press, the stiff-mud, or the soft-mud process. The **dry-press brick** are made in molds under high pressure and from relatively dry clay mixes. Usually all six surfaces are smooth and even, with geometrical uniformity. The **stiff-mud brick** are made from mixes of clay or shale with more moisture than in the dry-press process, but less moisture than used in the soft-mud process. The clay is extruded from an auger machine in a ribbon and cut by wires into the required lengths. These brick may be side-cut or end-cut, depending on the cross section of the ribbon and the length of the section cut off. The two faces cut by wires are rough in texture; the other faces may be smooth or artificially textured. The **soft-mud process** uses a wet mix of clay which is placed in molds under slight pressure.

Brick are highly resistant to freezing and thawing, to attacks of acids and alkalies, and to fire. They furnish good thermal insulation and good insulation against sound transference.

Paving brick are made of clay or shale, usually by the stiff-mud or dry-press process. Brick for use as paving brick are burned to vitrification. The common requirements are as follows: size, $8\frac{1}{2} \times 2\frac{1}{2} \times 4$, $8\frac{1}{2} \times 3 \times 3\frac{1}{2}$, $8\frac{1}{2} \times 3 \times 4$, with permissible variations of $\frac{1}{8}$ in in either transverse dimension, and $\frac{1}{4}$ in in length.

Although brick have always been used in construction, their use has been limited, until recently, to resisting compressive-type loadings. By adding steel in the mortar joints to take care of tensile stresses, **reinforced brick masonry** extends the use of brick masonry to additional types of building construction such as floor slabs.

Sand-lime brick are made from a mixture of sand and lime, molded under pressure and cured under steam at 200°F. They are usually a light gray in color and are used primarily for backing brick and for interior facing.

Cement brick are made from a mixture of cement and sand, manufactured in the same manner as sand-lime brick. In addition to their use as backing brick, they are used where there is no danger of attack from acid or alkaline conditions.

Firebrick (see this section, Refractories).

Specialty Brick A number of types of specialty brick are available for important uses, particularly where refractory characteristics are needed. These include **alumina** brick, **silicon carbide** brick, and **boron carbide** brick.

Other Structural Blocks Important structural units which fall outside the classic definition of brick are as follows: **Concrete blocks** are made with portland cement as the basic binder. Often, these are referred to as **concrete masonry units** (CMUs). The choice of aggregate ranges from relatively dense aggregate such as small crushed stone, small gravel, or coal cinders (the latter, if available) to light aggregate such as sand, limestone tailings, or the like. **Gypsum blocks** are generally used for fire protection in non-load-bearing situations. **Glass blocks** are available where transparency is desired. **Structural clay tile** is widely used in both load-bearing and non-load-bearing situations; a variety of complex shapes are available for hollow-wall construction. For classification of all these products with respect to dimensions and specifications for usage, the references should be consulted.

Brick and block can come with porcelain glazed finish for appearance, ease of cleaning, resistance to weathering, corrosion, etc.

Brick panels (prefabricated brick walls) are economical and find extensive use in modern high-rise construction. Such panels can be constructed on site or in factories.

Table 6.8.3   Properties and Uses of Various Adhesives

Basic type	Curing cycle, time at temp	Service temp range, °C	Adherends	Main uses	Remarks
			Animal		
Animal (hide)	Melted at 70–75°C. Sets on cooling		Paper, wood, textiles	Woodworking, carpet materials, paper, bookbinding	May be thinned with water
Animal (hide) + plasticizers	Applied as a melt at 60°C	<60	Paper, cellulosic materials	Bookbinding, stationery applications	Cures to permanent flexible film
Fish glue	1 h at 20°C	60	Wood, chipboard, paper	General-purpose for porous materials	Rapid setting. Good flexibility Moderate resistance to water. High tack.
Casein	Cold setting after 20-min standing period on mixing		Timber with moisture content	Laminated timber arches and beams, plybox beams, and engineering timber work	Full bond strength developed after seasoning period of 48 h
Casein + 60% latex	Cold setting after 20-min standing period on mixing		Aluminum, wood, phenolic formaldehyde (rigid), leather, rubber	Bonding of dissimilar materials to give flexible, water-resistant bond	Flexible
			Vegetable		
Dextrine	Air drying		Paper, cardboard, leather, wood, pottery	General-purpose glue for absorbent materials	Medium drying period of 2–3h
Dextrine-starch blend	Applied above 15°C air drying	48	Cellulosic materials, cardboard, paper	Labeling, carton sealing, spiral-tube winding	Fast setting. May be diluted with water
Gum arabic	Cold setting		Paper, cardboard	Stationery uses	Fast drying
			Mineral		
Silicate	8 h at 20°C	10–430	Asbestos, magnesia	Lagging asbestos cloth on high-temperature insulation	Unsuitable where moisture: not recommended for glass or painted surfaces
Silicate with chinaclay filler	Dried at 80°C before exposure to heat	−180–1,500	Asbestos, ceramics, brickwork, glass, silver, aluminum, steel (mild)-steel	General-purpose cement for bonding refractory materials and metals. Furnace repairs and gastight jointing of pipe work. Heat-insulating materials	Resistant to oil, gasoline, and weak acids
Sodium silicate	Dried at 20–80°C before exposure to heat	0–850	Aluminum (foil), paper, wood-wood	Fabrication of corrugated fiberboard. Wood bonding, metal foil to paper lamination	Suitable for glass-to-stone bonding
Aluminum phosphate + silica filler	Dried ½ h at 20°C, then ½ h at 70°C + ½ h at 100°C + 1 h at 200°C + 1 h at 250°C. Repeat for 2 overcoatings and finally cure 1 h at 350°C	750	Steels (low-alloy), iron, brass, titanium, copper, aluminum	Strain-gage attachment to heat-resistant metals. Heater-element bonding	Particularly suited to heat-resistant steels where surface oxidation of metal at high temperatures is less detrimental to adhesion
Bitumen/latex emulsion	Dried in air to a tacky state	0–66	Cork, polystyrene (foam), polyvinyl chloride, concrete, asbestos	Lightweight thermal-insulation boards, and preformed sections to porous and non-porous surfaces. Building applications	Not recommended for constructions operated below 0°C
			Elastomers		
Natural rubber	Air-dried 20 min at 20°C and heat-cured 5 min at 140°C		Rubber (styrene butadiene), rubber (latex), aluminum, cardboard, leather, cotton	Vulcanizing cement for rubber bonding to textiles and rubbers	May be thinned with toluene
Natural rubber in hydrocarbon solvent	Air-dried 10 min at 20°C and heat-cured for 20 min at 150°C	100	Hair (keratin), bristle, polyamide fiber	Brush-setting cement for natural- and synthetic-fiber materials	Resistant to solvents employed in oil, paint and varnish industries. Can be nailed without splitting
Rubber latex	Air drying within 15 min		Canvas, paper, fabrics, cellulosic materials	Bonding textiles, papers, packaging materials. Carpet bonding	Resistant to heat. Should be protected from frosts, oils
Chlorinated rabber in hydrocarbon solvents	Air-dried 10 min at 20°C and contact bonded	−20–60	Polyvinyl chloride acrylonitrile butadiene styrene, polystyrene, rubber, wood	General-purpose contact adhesive	Resistant to aging, water, oils, petroleum
Styrene-butadiene rubber lattices	Air drying		Polystyrene (foam), wood, hardboard, asbestos, brickwork	Bonding polystyrene foams to porous surface	

Table 6.8.3   Properties and Uses of Various Adhesives   (Continued)

Basic type	Curing cycle, time at temp	Service temp range, °C	Adherends	Main uses	Remarks
			Elastomers *(Continued)*		
Neoprene/nitrile rubbers in	Dried 30 min in air and bonded under pressure while tacky		Wood, linoleum, leather, paper, metals, nitrile rubbers, glass, fabrics	Cement for bonding synthetic rubbers to metals, woods, fabrics	May be thinned with ketones
Acrylonitrile rubber + phenolic resin	Primer air-dried 60 min at 20°C film cured 60 min at 175°C under pressure. Pressure released on cooling at 50°C	−10–130	Aluminum (alloy)-aluminum to DTD 746	Metal bonding for structural applications at elevated temperatures	Subject to creep at 150°C for sustained loading
Polysulfide rubber in ketone solvent and catalyst	3 days at 25°C	−50–130, withstands higher temps. for short periods	Metals	Sealant for fuel tanks and pressurized cabins in aircraft, where good weatherproof and waterproof properties are required	Resistant to gasoline, oil, hydraulic fluids, ester lubricants. Moderate resistance to acids and alkalies
Silicone rubber	24 h at 20°C (20% R.H.). Full cure in 5 days	−65–260	Aluminum, titanium, steel (stainless), glass, cork, silicone rubber, cured rubber-aluminum, cured rubber-titanium, cured rubber-steel (stainless), aluminum-aluminum (2024 Alclad), cork-cork (phenolic bonded)	General-purpose bonding and sealing applications. Adhesive/sealant for situations where material is expected to support considerable suspended weight. High pressure exposure conditions	Resistant to weathering and moisture
Reclaim rubber	Contact bonded when tacky		Fabric, leather, wood, glass, metals (primed)	General industrial adhesive for rubber, fabric, leather, porous materials	May be thinned with toluene
Polychloroprene	Air-dried 10–20 min at 20°C		Rubber, steel, wood, concrete	Bonding all types of rubber flooring to metals, woods, and masonry	Good heat resistance
Modified polyurethane	3 h at 18°C to 16 h at −15°C	−80–110	Concrete, plaster, ceramics, glass, hardboards, wood, polyurethane (foam), phenol formaldehyde (foam), polystyrene (foam), copper, lead, steel, aluminum	Bonding rigid and semirigid panels to irregular wall surfaces, wall cladding and floor laying. Building industry applications	Foam remains flexible on aging even at elevated temperatures. Will withstand a 12% movement
			Thermoplastic		
Nitrocellulose in ester solvent	Heat set 1 h at 60°C after wet bonding	60	Paper, leather, textiles, silicon carbide, metals	Labeling, general bonding of inorganic materials including metals	Good resistance to mineral oils
Modified methyl cellulose	Dries in air		Vinyl-coated paper, polystyrene foam	Heavy-duty adhesive. Decorating paper and plastics	Contains fungicide to prevent biodeterioration
Ethylene vinyl acetate copolymer + resins	Film transfer at 70–80°C followed by bonding at 150–160°C	60 or 1 h at 90	Cotton (duck)-cotton, resin rubber-leather, melamine laminate—plywood, steel (mild)-steel, acrylic (sheet)-acrylic	Metals, laminated plastics, and textiles. Fabrication of leather goods. Lamination work	Good electrical insulation
Polyvinyl acetate	Rapid setting		Paper, cardboard	Carton sealing in packaging industry	Resistant to water
Synthetic polymer blend	Applied as a melt at 177°C	71	Paper, cardboard, polythene (coated materials)	Carton and paper-bag sealing. Packaging	Resistant to water
Polychloroprene/resin blend in solvent	Air-dried 10 min at 20°C and cured 4 days at 20°C to 7 h at 75°C		Chlorosulfonated polythene, polychloroprene fabrics, polyamide fabrics, leather, wood, textiles	Bonding synthetic rubbers and porous materials. Primer for polyamidecoated fabrics such as nylon, terylene	
Polychloroprene	Air-dried 10–20 min at 20°C		Rubber, steel, wood, concrete	Bonding all types of rubber flooring to metals, woods, and masonry	Good heat resistance
Saturated polyester + isocyanate catalyst in ethyl acetate	Solvent evaporation and press cured at 40–80°C when tacky		Cellulose, cellulose acetate, polyolefins (treated film), polyvinyl chloride (rigid), paper, aluminum (foil), copper (foil)	Lamination of plastic films to themselves and metal foils for packaging industry, printed circuits	Resistant to heat, moisture, and many solvents

Table 6.8.3   Properties and Uses of Various Adhesives   (Continued)

Basic type	Curing cycle, time at temp	Service temp range, °C	Adherends	Main uses	Remarks
			Thermoplastic (*Continued*)		
Cyanoacrylate (*anaerobic*)	15 s to 10 min at 20°C substrate-dependent	Melts at 165	Steel-steel, steel-aluminum, aluminum-aluminum, butyl rubber-phenolic	Rapid assembly of metal, glass, plastics, rubber components	*Anaerobic* adhesive. Curing action is based on the rapid polymerization of the monomer under the influence of basic catalysts. Absorbed water layer on most surfaces suffices to initiate polymerization and brings about bonding
Polyacrylate resin (*anaerobic*)	3 min at 120°C to 45 min at 65°C or 7 days at 20°C	−55–95	Aluminum-aluminum	Assembly requirements requiring high resistance to impact or shock loading. Metals, glass, and thermosetting plastics	*Anaerobic* adhesive
			Thermosetting		
Urea formaldehyde	9 h at 10°C to 1 h at 21°C after mixing powder with water (22%)		Wood, phenolic laminate	Wood gluing and bonding on plastic laminates to wood. Plywood, chipboard manufacture. Boat building and timber engineering	Excess glue may be removed with soapy water
Phenolic formaldehyde + catalyst PX-2Z	Cold setting		Wood	Timber and similar porous materials for outdoor-exposure conditions. Shop fascia panels	Good resistance to weathering and biodeterioration
Resorcinol formaldehyde + catalyst RXS-8	Cured at 16°C to 80°C under pressure		Wood, asbestos, aluminum, phenolic laminate, polystyrene (foam), polyvinyl chloride, polyamide (rigid)	Constructional laminates for marine craft. Building and timber applications. Aluminum-plywood bonding. Laminated plastics	Recommended for severe outdoor-exposure conditions
Epoxy resin + catalyst	24–48 h at 20°C to 20 min at 120°C	100	Steel, glass, polyester-glass fiber composite, aluminum-aluminum	General-purpose structural adhesive	
Epoxy resin + catalyst	8 h at 24°C to 2 h at 66°C to 45 min at 121°C	65	Steel, copper, zinc, silicon carbide, wood, masonry, polyester-glass fiber composite, aluminum-aluminum	Bonding of metals, glass, ceramics, and plastic composites	Cures to strong, durable bond
Epoxy + steel filler (80% w/w)	1–2 h at 21°C	120	Iron, steel, aluminum, wood, concrete, ceramics, aluminum-aluminum	Industrial maintenance repairs. Metallic tanks, pipes, valves, engine casings, castings	Good resistance to chemicals, oils, water
Epoxy + amine catalyst (ancamine LT)	2–7 days at 20°C for 33% w/w catalyst content	−5–60	Concrete, stonework	Repair of concrete roads and stone surfaces	Excellent pigment-wetting properties. Effective under water and suited to applications under adverse wet or cold conditions
Epoxy resin (modified)	4–5 h at 149°C to 20 min at 230°C to 7 min at 280°C	150	Aluminum, steel, ceramics	One-part structural adhesive for high-temperature applications	Good gap-filling properties for poorly fitting joints. Resistant to weather, galvanic action
Epoxy	45 s at 20°C		Gem stones, glass, steel, aluminum-aluminum	Rapid assembly of electronic components, instrument parts, printed circuits. Stone setting in jewelry, and as an alternative to soldering	
Epoxy resin in solvent + catalyst	8 h at 52°C to ½ h at 121°C	−270–371	Aluminum and magnesium alloys for elevated-temperature service	Strain gages for cryogenic and elevated-temperature use. Micro measurement strain gages	Cured material resists outgassing in high vacuum
Epoxy polyamide	8 h at 20°C to 15 min at 100°C	100	Copper, lead, concrete, glass, wood, fiberglass, steel-steel, aluminum-aluminum	Metals, ceramics, and plastics bonding. Building and civil engineering applications	Resists water, acids, oils, greases

Table 6.8.3 Properties and Uses of Various Adhesives (*Continued*)

Basic type	Curing cycle, time at temp	Service temp range, °C	Adherends	Main uses	Remarks
			Thermosetting (Continued)		
Epoxy/polysulfide	24 h at 20°C to 3 h at 60°C to 20 min at 100°C		Asbestos (rigid), ceramics, glass-fiber composites, carbon, polytetrafluoro-ethylene (treated), polyester (film), polystyrene (treated), rubber (treated), copper (treated), tungsten carbide, magnesium alloys, aluminum-aluminum, steel (stainless)-steel	Cold-setting adhesive especially suitable for bonding materials with differing expansion properties	Cures to flexible material. Resistant to water, petroleum, alkalies, and mild acids
Phenol furfural + acid catalyst	2 days at 21°C		Alumina, carbon (graphite)	Formulation of chemically resistant cements. Bedding and joining chemically resistant ceramic tiles	Extremely resistant to abrasion and heat
			Pressure-sensitive		
	Heated by air drying for several hours or 15–30 min at 210°F		Teflon-Teflon, Teflon-metal		Good resistance to acids and alkalies. Excellent electrical properties
			Miscellaneous		
Ceramic-based	Dried for ½ h at 77°C and cured ½ h at 100°C + 1 h at 200°C + 1 h at 250°C. Postcured, 1 h at 350°C	816	Metals	Strain gages, temperature sensors for elevated-temperature work	

SOURCE: Adapted from J. Shields, "Adhesives Handbook," CRC Press (Division of The Chemical Rubber Co., 1970), with the permission of the publisher.

New tile units requiring no mortar joint, only a thin epoxy line, form walls which resemble brick and can be installed three times as fast as conventional masonry.

For further definitions relating to structural clay products, the reader is referred to ANSI/ASTM C43–70(75).

See Table 6.8.4 for brick properties.

CERAMICS

REFERENCES: Kingery, "Introduction to Ceramics," Wiley. Norton, "Elements of Ceramics," Addison-Wesley. "Carbon Encyclopedia of Chemical Technology," Interscience. Humenik, "High-Temperature Inorganic Coatings," Reinhold. Kingery, "Ceramic Fabrication Processes," MIT. McCreight, Rauch, Sr., Sutton, "Ceramic and Graphite Fibers and Whiskers; A Survey of the Technology," Academic. Rauch, Sr., Sutton, McCreight, "Ceramic Fibers and Fibrous Composite Materials," Academic. Hague, Lynch, Rudnick, Holden, Duckworth (compilers and editors), "Refractory Ceramics for Aerospace; A Material Selection Handbook," The American Ceramic Society. Waye, "Introduction to Technical Ceramics," Maclaren and Sons, Ltd., London. Hove and Riley, "Ceramics for Advanced Technologies," Wiley. McMillan, "Glass-Ceramics," Academic. *Machine Design,* May 1983. "1986 Annual Book of ASTM Standards," vols. 15.02, 15.04 (Ceramics). Current ASTM and ANSI Standards.

Ceramic materials are a diverse group of nonmetallic, inorganic solids with a wide range of compositions and properties. Their structure may be either crystalline or glassy. The desired properties are often achieved by high-temperature treatment (firing or burning).

Traditional ceramics are products based on the silicate industries, where the chief raw materials are naturally occurring minerals such as the clays, silica, feldspar, and talc. While silicate ceramics dominate the industry, newer ceramics, sometimes referred to as **electronic** or **technical** ceramics, are playing a major role in many applications. **Glass ceramics** are important for electrical, electronic, and laboratory-ware uses. Glass ceramics are melted and formed as glasses, then converted, by controlled nucleation and crystal growth, to polycrystalline ceramic materials.

Table 6.8.4 Building Brick Made from Clay or Shale
(Standard specifications, ANSI/ASTM C62–81)

Designation*	Min compressive strength (brick flatwise), lb/in² gross area		Max water absorption by 5-h boil, %		Max saturation coefficient†	
	Avg of 5	Indiv.	Avg of 5	Indiv.	Avg of 5	Indiv.
Grade SW	3,000	2,500	17.0	20.0	0.78	0.80
Grade MW	2,500	2,200	22.0	25.0	0.88	0.90
Grade NW	1,500	1,250	No limit		No limit	

* Grade SW includes brick intended for use where a high degree of resistance to frost action is desired and the exposure is such that the brick may be frozen when permeated with water.

Grade MW includes brick intended for use where exposed to temperatures below freezing but unlikely to be permeated with water or where a moderate and somewhat nonuniform degree of resistance to frost action is permissible.

Grade NW includes brick intended for use as backup or interior masonry, or if exposed, for use where no frost action occurs; or if frost action occurs, where the average annual precipitation is less than 20 in.

† The saturation coefficient is the ratio of absorption by 24 h submersion in cold water to that after 5 h submersion in boiling water.

Manufacture Typically, the manufacture of traditional-ceramic products involves blending of the finely divided starting materials with water to form a plastic mass which can be formed into the desired shape. The plasticity of clay constituents in water leads to excellent forming properties. Formation processes include extrusion, pressing, and ramming. Unsymmetrical articles can be formed by "slip-casting" techniques, where much of the water is taken up by a porous mold. After the water content of formed articles has been reduced by drying, the ware is **fired** at high temperature for fusion and/or reaction of the components and for attainment of the desired properties. Firing temperatures can usually be considerably below the fusion point of the pure components through the use of a **flux,** often the mineral feldspar. Following burning, a vitreous ceramic coating, or **glaze,** may be applied to render the surface smooth and impermeable.

Quite different is the **fusion casting** of some refractories and refractory blocks and most glasses, where formation of the shaped article is carried out after fusion of the starting materials.

Properties The physical properties of ceramic materials are strongly dependent on composition, microstructure (phases present and their distribution), and the history of manufacture. Volume pore concentration can vary widely (0 to 30 percent) and can influence shock resistance, strength, and permeability. Most traditional ceramics have a glassy phase, a crystalline phase, and some porosity. The last can be eliminated at the surface by glazing. Most ceramic materials are resistant to large compressive stresses but fail readily in tension. Resistance to abrasion, heat, and stains, chemical stability, rigidity, good weatherability, and brittleness characterize many common ceramic materials.

Products Many traditional-ceramic products are referred to as **whiteware** and include pottery, semivitreous wares, electrical porcelains, sanitary ware, and dental porcelains. Building products which are ceramic include brick and structural tile and conduit, while refractory blocks, as well as many abrasives, are also ceramic in nature. Porcelain **enamels** for metals are opacified, complex glasses which are designed to match the thermal-expansion properties of the substrate. Important **technical ceramics** include magnetic ceramics, with magnetic properties but relatively high electrical resistance; nuclear ceramics, including uranium dioxide fuel elements; barium titanate as a material with very high dielectric constant. Several of the pure oxide ceramics with superior physical properties are being used in electrical and missile applications where high melting and deformation temperatures and stability in oxygen are important. Fibrous ceramic composed of ziconium oxide fibers, such as Zircar, provide optimum combination of strength, low thermal conductivity, and high temperature resistance to about 2,490°C (4,500°F). Partially stabilized zirconia, because of its steel-strong and crack-resisting properties, is finding high-temperature applications, for example, in diesel engines.

Various forms of ceramics are available from manufacturers such as paper, "board," blankets, tapes, and gaskets.

A fair amount of development is under way dealing with **reinforcing ceramics** to yield **ceramic matrix composite** materials. The driving interest here is potential applications in high-temperature, high-stress environments, such as those encountered in aircraft and aerospace structural components. None of the end products are available yet "off the shelf," and applications are likely to be quite circumscribed and relegated to solution of very specialized design problems.

Ceramic inserts suitably shaped and mechanically clamped in tool holders are applied as cutting tools, and they find favor especially for cutting abrasive materials of the type encountered in sand castings or forgings with abrasive oxide crusts. They offer the advantage of resisting abrasion at high tool/chip interface temperatures, and consequently they exhibit relatively long tool life between sharpenings. They tend to be somewhat brittle and see limited service in operations requiring interrupted cuts; when they are so applied, tool life between sharpenings is very short.

CLEANSING MATERIALS

REFERENCES: McCutcheon, "Detergents and Emulsifiers," McCutcheon, Inc. Schwartz, Perry, and Berch, "Surface Active Agents and Detergents," Interscience. *ASTM Special Tech. Pub.* 197. Niven, "Industrial Detergency," Reinhold. McLaughlin, "The Cleaning, Hygiene, and Maintenance Handbook," Prentice-Hall. Hackett, "Maintenance Chemical Specialties," Chemical Publishing Co. Bennett (ed.), "Cold Cleaning with Halogenated Solvents," *ASTM Special Technical Publication* 403, 1966. ANSI/ASTM D459, D534–1979 (definitions and specifications).

Cleansing is the removal of dirt, soil, and impurities from surfaces of all kinds. Means of soil attachment to the surface include simple entrapment in interstices, electrostatically held dirt, wetting of the surface with liquid soils, and soil-surface chemical reaction. A variety of cleansing systems has been developed which are difficult to classify since several soil-removal mechanisms are often involved. A liquid suspending medium, an active cleansing agent, and mechanical action are usually combined. The last may involve mechanical scrubbing, bath agitation, spray impingement, or ultrasonic energy. Cleansing involves detachment from the surface, suspension of solids or emulsification of liquids, or dissolution, either physical or by chemical reaction.

Organic Solvents

Both **petroleum solvents** (mineral spirits or naphtha) and **chlorinated hydrocarbons** (trichloroethylene and perchloroethylene) are used to remove solvent-soluble oils, fats, waxes, and greases as well as to flush away insoluble particles. Petroleum solvents are used in both soak-tank and spray equipment. Chlorinated hydrocarbons are widely used in vapor degreasing (where the metal stock or part is bathed in the condensing vapor), their high vapor density and nonflammability being advantageous. Solvent recovery by distillation can be used if the operation is of sufficient size. Both solvent types are used in garment **dry cleaning** with solvent-soluble detergents.

Fluorocarbons such as UCON solvent 113-LRI (trichlorotrifluoroethane) are nonflammable and are ideal for critical cleaning of mechanical electrical and electronic equipment, especially for whiteroom conditions.

Fluorocarbon solvents are highly proscribed and regulated and are to be used only with utmost care, under close supervision and accountable control. The reader should become familiar with the several references pertaining to health hazards of industrial materials listed under Chlorinated Solvents in this section.

Emulsifiable solvent cleaners contain a penetrating solvent and dissolved emulsifying agent. Following soaking, the surface is flushed with hot water, the resulting emulsion carrying away both soil and solvent. These cleaners are usually extended with kerosinelike solvents.

Alkali Cleansers

Alkali cleansers are water-soluble inorganic compounds, often strong cleansers. Carbonates, phosphates, pyrophosphates, and caustic soda are common, with numerous applications in plant maintenance, material processing, and process-water treatment. Cleansing mechanisms vary from beneficial water softening and suspension of solids to chemical reaction in the solubilization of fats and oils. Certain of these compounds are combined with **detergents** to improve efficiency; these are referred to as **builders.** Boiler and process equipment scales can sometimes be controlled or removed with selected alkali cleansers.

Synthetic Detergents

Detergents concentrate strongly at a solid-liquid or liquid-liquid interface and are thus characterized as **surface active.** In contrast to soaps, they can be tailored to perform over a wide range of conditions of temperature, acidity, and presence of dissolved impurities with little or no foaming. Detergents promote **wetting** of the surface by the suspending medium (usually water), **emulsification** of oils and greases, and **suspension** of solids without redeposition, the last function being the prime criterion of a good detergent. Detergents are classified as **anionic** (negatively charged in solution), **cationic** (positively charged), and **nonionic.** Germicidal properties may influence detergent choice.

For specific applications, the supplier should be consulted since there is a large variety of available formulations and since many detergent systems contain auxiliary compounds which may be diluents, foam promotors, or alkali chemicals. Additives are designed for pH control, water softening, and enhanced suspending power. Formulations have been developed for cleaning food and dairy process equipment, metals processing, metal cleaning prior to electroplating, textile fiber and fabric processing, and industrial building maintenance. Strong, improperly selected detergent systems can cause deterioration of masonry or marble floors, aluminum window frames, water-based paints, and floor tiles, whereas detergents matched to the job at hand can result in increased plant efficiency.

Soaps, the oldest surface-active cleansers, lack the versatility of synthetic detergents but are widely used in the home and in the laundry industry. Properties depend on the fat or oil and alkali used in their preparation and include solvent-soluble soaps for dry cleaning.

Chemical cleaners which attack specific soils include dilute acid for metal oxide removal, for the cleaning of soldered or brazed joints, and for the removal of carbonate scale in process equipment. Oxalic acid (usually with a detergent) is effective on rust.

Chelating agents are organic compounds which complex with several metal ions and can aid in removal of common boiler scales and metal oxides from metal surfaces.

Steam cleaning in conjunction with a detergent is effective on grease-laden machinery.

There are numerous cleansers for the **hands,** including Boraxo, soap jelly and sawdust, lard for loosening oil grime, and linseed oil for paints; repeated use of solvents can be hazardous.

Alcohol ethoxylates are nonionic surfactants suitable for use in the production of maintenance and institutional cleaners. Products such as Tergitol and Neodol 23–6.5 serve in the formulation of liquid detergents for household and industrial uses.

Alcohol ethoxysulfates are anionic surfactants which are suited to the formulation of high-foaming liquid detergents, as for manual dishwashing. They offer the advantages of excellent solubility and biodegradability. Neodol 25–3S40 also exhibits low sensitivity to water hardness.

Enzyme cleaners containing a combination of bacteria culture, enzymes, and nutrients are used to dissolve grease, human waste, and protein stains.

CORDAGE

REFERENCE: Himmelfarb, "The Technology of Cordage Fibers and Rope," Textile Book Publishers, Inc. (division of Interscience Publishers, Inc.).

The term **cordage** denotes any flexible string or line. Usage includes wrapping, baling, hauling, and power transmission in portable equipment. **Twine** and **cord** generally imply lines of $3/8$ in diam or less, with larger sizes referred to as **rope.**

Natural fibers used in cordage are abaca, sisal, hemp, cotton, and jute. For heavy cordage abaca and manila predominate. Hemp is used for small, tarred lines, and henequen for agricultural binder twine.

Rope is made by twisting yarns into strands, with the strands (usually three) **twisted** (laid) into a line. The line twist may be S or Z (see Sec. 10), generally opposite to the twist of the strands, which, in turn, is opposite to the twist of the yarns. The term **lay** designates the number of turns of the strands per unit length of rope but may also characterize the rope properties, a function of the degree of twist of each component. Grades range from **soft lay** (high ultimate strength) to **hard lay** (high abrasion resistance). Cable-laid rope results from twisting together conventional, three-strand rope.

Synthetic fibers are used in cordage because of resistance to rot, high strength, and other special properties. These fibers include nylon for strength, polyester (Dacron) for strength and dimensional stability, vinyls for chemical-plant use, fiberglass for electrical stability, vinyls for chemical-plant use, fiberglass for electrical and chemical properties, and polypropylene for strength and flotation (see Sec. 8).

Braided cordage has been used largely for small diameter lines such as sash cord and clothesline. However, braided lines are now available in larger diameters between 2 and 3 in. The strength of braided rope is slightly superior, and the line has less tendency to elongate in tension and cannot rotate or unlay under load. These properties are balanced against a somewhat higher cost than that of twisted rope.

A no. 1 common-lay rope will conform to the strength and weight table of the Federal Specification TR601A listed below.

For comparison, a **nylon rope** of a given diameter will have about 3 times the breaking strength given above and a **polyester rope** about $2\frac{1}{2}$ times the strength of manila. Both have substantially greater flex and abrasion resistance.

See Table 6.8.5 for manila rope properties.

ELECTRICAL INSULATING MATERIALS

REFERENCES: Plastics Compositions for Dielectrics, *Ind. Eng. Chem.,* **38,** 1946, p. 1090. High Dielectric Ceramics, *Ind. Eng. Chem.,* **38,** 1946, p. 1097. Polystyrene Plastics as High Frequency Dielectrics, *Ind. Eng. Chem.,* **38,** 1946, p. 1121. Paper Capacitors Containing Chlorinated Impregnants, *Ind. Eng. Chem.,* **38,** 1946, p. 1110. "Contributions of the Chemist to Dielectrics," National Research Council, 1947. National Research Council, Conference on Electrical Insulation, (1) Annual Report, (2) Annual Digest of Literature on Dielectrics. Von Hippel, "Di-electric Materials and Applications," Wiley. Birks (ed.), "Modern Dielectric Materials," Academic. Saums and Pendelton, "Materials for Electrical Insulating and Dielectric Functions," Hayden Book, Inc. Licari, "Plastic Coatings for Electronics," McGraw-Hill. Clark, "Insulating Materials for Design and Engineering Practice," Wiley. Mayofis, "Plastic Insulating Materials," Illiffe, Ltd.,

Table 6.8.5   Weight and Strength of Different Sizes of Manila Rope Specification Values

Approx diam, in*	Circumference, in*	Max net weight, lb/ft	Min breaking strength, lb†	Approx diam, in*	Circumference, in*	Max net weight, lb/ft	Min breaking strength, lb	Approx diam, in	Circumference, in	Max net weight, lb/ft	Min breaking strength, lb
$3/16$	$5/8$	0.015	450	$13/16$	$2\frac{1}{2}$	0.195	6,500	$1\frac{13}{16}$	$5\frac{1}{2}$	0.895	26,500
$1/4$	$3/4$	0.020	600	$15/16$	$2\frac{3}{4}$	2.225	7,700	2	6	1.08	31,000
$5/16$	1	0.029	1,000	1	3	0.270	9,000	$2\frac{1}{4}$	7	1.46	41,000
$3/8$	$1\frac{1}{8}$	0.041	1,350	$1\frac{1}{16}$	$3\frac{1}{4}$	0.313	10,500	$2\frac{5}{8}$	8	1.91	52,000
$7/16$	$1\frac{1}{4}$	0.053	1,750	$1\frac{3}{16}$	$3\frac{1}{2}$	0.360	12,000	3	9	2.42	64,000
$1/2$	$1\frac{1}{2}$	0,075	2,650	$1\frac{1}{4}$	$3\frac{3}{4}$	0.418	13,500	$3\frac{1}{4}$	10	2.99	77,000
$9/16$	$1\frac{3}{4}$	0.104	3,450	$1\frac{5}{16}$	4	0.480	15,000	$3\frac{5}{8}$	11	3.67	91,000
$11/16$	2	0.133	4,400	$1\frac{1}{2}$	$4\frac{1}{2}$	0.600	18,500	4	12	4.36	105,000
$3/4$	$2\frac{1}{4}$	0.167	5,400	$1\frac{5}{8}$	5	0.744	22,500				

The approximate length of coil is 1,200 ft for diam $7/16$ in and larger. For smaller sizes it is longer, up to 3,000 ft for $3/16$ in diam.
* 1 in = 0.0254 m; 1 ft = 0.3048 m.
† 1 lbf = 4.448 N.
SOURCE: U.S. government specification TR601A, dated Nov. 26, 1935, formulated jointly by cordage manufacturers and government representatives.

London. Bruins (ed.), "Plastic for Electrical Insulation," Interscience Publishers. Swiss Electrochemical Committee, "Encyclopedia of Electrical Insulating Materials," *Bulletin of the Swiss Association of Electrical Engineers,* vol. 48, 1958. ISO 455–3–1 1981. "1986 Annual Book of ASTM Standards," vols. 10.01–10.03 (Electrical Insulating Materials). Current ASTM and ANSI Standards.

The insulating properties of any material are dependent upon **dielectric strength,** or the ability to withstand high voltages without breakdown; **ohmic resistance,** or the ability to prevent leakage of small currents; and **power loss,** or the absorption of electrical energy that is transformed into heat. Power loss depends upon a number of influences, particularly the molecular symmetry of the insulation and frequency of the voltage, and is the basis of power factor, an important consideration whenever efficient handling of alternating currents is concerned, and a dominating consideration when high frequencies are used, as in radio circuits. Materials may have one of these qualities to a far greater extent than the other; e.g., air has a very high specific resistance but very little dielectric strength and no power loss at any frequency; glass has great dielectric strength yet much lower resistance than air. The **ideal insulator** is one having the maximum dielectric strength and resistance, minimum power loss, and also mechanical strength and chemical stability. Moisture is by far the greatest enemy of insulation; consequently the absence of hygroscopic quality is desirable.

The common insulating materials are described below. For their electrical properties, see Sec. 15.

Rubber See Rubber and Rubberlike Materials.

Mica and Mica Compounds Mica is a natural mineral varying widely in color and composition, and occurs in sheets that can be subdivided down to a thickness of 0.00025 in. White mica is best for electrical purposes. The green shades are the softest varieties, and the white amber from Canada is the most flexible. Mica has high insulating qualities, the best grades having a dielectric strength of 12,000 V per 0.1 mm. Its lack of flexibility, its nonuniformity, and its surface leakage are disadvantages. To offset these, several mica products have been developed, in which small pieces of mica are built up into finished shapes by means of binders such as shellac, gum, and phenolic resins.

Micanite consists of thin sheets of mica built into finished forms with insulating cement. It can be bent when hot and machined when cold, and is obtainable in thicknesses of 0.01 to 0.12 in. Flexible micanite plates, cloth, and paper are also obtainable in various thicknesses. **Megohmit** is similar to micanite except that it is claimed not to contain adhesive matter. It can be obtained in plates, paper, linen, and finished shapes. **Megotalc,** built up from mica and shellac, is similar to the above-named products and is obtainable in similar forms.

Insulating Varnishes Two general types of insulating varnish are used: (1) asphalt, bitumen, or wax, in petroleum solvent, and (2) drying-oil varnishes based on natural oils compounded with resins from natural or synthetic sources. Varnishes have changed greatly in the last few years, since new oils have become available, and particularly since phenolic and alkyd resins have been employed in their manufacture.

Silicone varnishes harden by baking and have electrical properties similar to those of the phenol-aldehyde resins. They are stable at temperatures up to 300°F (138°C). They can be used as wire coatings, and such wire is used in the manufacture of motors that can operate at high temperatures.

Impregnating Compounds Bitumens and waxes are used to impregnate motor and transformer coils, the melted mix being forced into the coil in a vacuum tank, forming a solid insulation when cooled. Brittle compounds, which gradually pulverize owing to vibration in service, and soft compounds, which melt and run out under service temperatures, should be avoided as far as possible.

Oil Refined grades of petroleum oils are extensively used for the insulation of transformers, switches, and lightning arresters. The following specification covers the essential points:

Specific gravity, 0.860; flash test, not less than 335°F; cold test, not more than −10°C (14°F); viscosity (Saybolt) at 37.8°C (100°F), not more than 120 s; loss on evaporation (8 h at 200°F), not more than 0.5 percent; dielectric strength, not less than 35,000 V; freedom

from water, acids, alkalies, saponifiable matter, mineral matter or free sulfur. Moisture is particularly dangerous in oil.

Petroleum oils are used for the impregnation of kraft or manila paper, after wrapping on copper conductors, to form high-voltage power cables for services up to 300,000 V. Oil-impregnated paper insulation is sensitive to moisture, and such cables must be lead-sheathed. The transmission of power at high voltages in underground systems is universally accomplished by such cables.

Chlorinated Hydrocarbons Chlorinated hydrocarbons have the advantage of being nonflammable and are used as filling compounds for transformers and condensers where this property is important. Chlorinated naphthalene and chlorinated diphenyl are typical of this class of material. They vary from viscous oils to solids, with a wide range of melting points. The reader is cautioned to investigate the use of such insulation carefully because of alleged carcinogenic propensities.

Impregnated Fabrics Fabrics serve as a framework to hold a film of insulating material and must therefore be of proper thickness, texture, and mechanical strength, and free from nap and acidity. A wide variety of drying varnishes is used for the impregnation, and the dipping is followed by baking in high-temperature towers. Varnished cambric is used for the wrapping of coils and for the insulation of conductors. These cables have high power factor and must be kept free of moisture, but they are desirable for resisting electrical surges.

Thermosetting substances of the phenol-aldehyde type and of the ureaformaldehyde type first soften and then undergo a chemical reaction which converts them quickly to a strong infusible product. Good properties are available in the phenolics, while numerous special types have been developed for high heat resistance, low-power-factor arc resistance, and other specialized properties. A wide variety of resins is available. The urea plastics are lacking in heat resistance but are suitable for general-purpose molding and have fair arc resistance.

Thermoplastic resins (see Plastics, below) are used for molding and extruding electrical insulations. They differ from the thermosetting resins in that they do not become infusible. **Polyethylene** softens between 99 and 116°C (210 and 240°F). Its dielectric strength and resistivity are high, its power factor is only 0.0003, and its dielectric constant is 2.28. It is used extensively in high-frequency and radar applications and as insulation on some power and communication cables. **Teflon** is a fluorocarbon resin which has electrical properties similar to those of polyethylene. Its softening point is 750°F approx, and it extrudes and molds with difficulty; however, more tractable grades of Teflon are now available. It is resistant to nearly all chemicals and solvents.

Nylon is a synthetic plastic with interesting mechanical and electrical properties. It has only fair water resistance. Its melting point is 198 to 249°C (390 to 480°F), and is used for the molding of coil forms. It can be extruded onto wire in thin layers. Such wires are used in place of the conventional varnish-coated magnet wires in coils and motors. Operation may be at temperatures up to 127°C (260°F).

Paper Except in lead-sheathed telephone cables, the present tendency is to use paper only as a backing or framework for an insulating film or compound, owing to its hygroscopic qualities. **Manila** and **kraft** papers possess the best dielectric and mechanical strength and, when coated with good insulating varnish, are excellent insulators. Various types of paraffined paper are used in condensers. (See also Paper, below.)

Silicone rubber (see Silicones) is a rubberlike material of good physical properties [tensile strength, 400 to 700 lb/in² (2.78 to 4.85 × 10⁶ N/m²) elongation, 200 percent]. It can be operated for long periods of time at temperatures up to 138°C (300°F) or intermittently up to 249°C (480°F).

Ceramics and **glasses** find wide usage as insulating materials where brittleness and lack of flexibility can be tolerated.

Polymeric (plastic) films, particularly the polyester and fluorocarbon types, are being used increasingly where fabrication of the electrical component permits either wrapping or insertion of film chips.

Heat-shrinkable tubing of polyvinylchloride, polyolefin, or polytetrafluoroethylene compositions allow for very tight-fitting insulation around a member, thus affording efficient protection of electrical wires or cables.

FIBERS AND FABRICS

(See also Cordage.)

REFERENCES: Matthews-Mauersberger, "The Textile Fibers," Wiley. Von Bergen and Krauss, "Textile Fiber Atlas," Textile Book Publishers, Inc. Hess, "Textile Fibers and Their Use," Lippincott. Sherman and Sherman, "The New Fibers," Van Nostrand. Kaswell, "Textile Fibers, Yarns and Fabrics," Reinhold. "Harris' Handbook of Textile Fibers," Waverly House. Kaswell, "Wellington Sears Handbook of Industrial Textiles," Wellington Sears Co. Fiber Charts, *Textile World,* McGraw-Hill. ASTM Standards. C. Z. Carroll-Porcynski, "Advanced Materials; Refractory Fibers, Fibrous Metals, Composites," Chemical Publishing Co. Marks, Atlas, and Cernia, "Man-Made Fibers, Fibrous Metals, Composites," Chemical Publishing Co. Marks, Atlas, and Cernia, "Man-Made Fibers, Science and Technology," Interscience Publishers. Frazer, High Temperature Resistant Fibers, *Jour. Polymer Sci.,* Part C, Polymer Symposia, no. 19, 1966. Moncrieff, "Man-Made Fibers," Wiley. Preston and Economy, "High Temperature and Flame Resistant Fibers," Wiley. ANSI/ASTM Standards D1175-D 3940. ANSI/ AATCC Standards 79, 128. "1986–1994 Annual Book of ASTM Standards," vols. 07.01, 07.02 (Textiles). Current ASTM and ANSI Standards.

Fibers are threadlike structural materials, adaptable for spinning, weaving, felting, and similar applications. They may be of natural or synthetic origin and of inorganic or organic composition. **Fabrics** are defined by the ASTM as "planar structures produced by interlacing yarns, fibers, or filaments." A **bonded fabric** (or nonwoven fabric) consists of a web of fibers held together with a cementing medium which does not form a continuous sheet of adhesive material. A **braided fabric** is produced by interlacing several ends of yarns such that the paths of the yarns are not parallel to the fabric axis. A **knitted fabric** is produced by interlooping one or more ends of yarn. A **woven fabric** is produced by interlacing two or more sets of yarns, fibers or filaments such that the elements pass each other essentially at right angles and one set of elements is parallel to the fabric axis. A woven narrow fabric is 12 in or less in width and has a selvage on either side.

Inorganic Fibers

Asbestos is the only mineral fiber of natural origin. Its demonstrated carcinogenic properties, when its particles lodge in the lung, have led to its removal from most consumer products. Formerly ubiquitously applied as a thermal insulator, it has now been proscribed from that use and has been supplanted largely by fiberglass.

Synthetic mineral fibers are **spun glass, rock wool,** and **slag wool.** These fibers can endure high temperatures without substantial loss of strength. Glass fibers possess a higher strength-to-weight ratio at the elastic limit than do other common engineering materials.

Metal filaments (wires) are used as textile fibers where their particular material properties are important.

Carbon (graphite) fibers, such as Thornel, are high-modulus, highly oriented structures characterized by the presence of carbon crystallites (polycrystalline graphite) preferentially aligned parallel to the fiber axis. Depending on the particular grade, tensile strengths can range from 10.3 to 241 $\times$ 10^7 N/m² (15,000 to 350,000 lb/in²). Hybrid composites of carbon and glass fibers find aerospace and industrial applications.

Zirconia fibers (ZrO_2 + HFO_2 + stabilizers), such as Zircar, have excellent temperature resistance to about 2,200°C (4,000°F), and fabrics woven from the fibers serve as thermal insulators.

Natural Organic Fibers

The **animal fibers** include **wool** from sheep, **mohair** from goats, **camel's hair,** and **silk.** Wool is the most important of these, and it may be processed to reduce its susceptibility to moth damage and shrinkage. Silk is no longer of particular economic importance, having been supplanted by one or another of the synthetic fibers for most applications.

The **vegetable fibers** of greatest utility consist mainly of cellulose and may be classified as follows: seed hairs, such as cotton; bast fibers, such as flax, hemp, jute, and ramie; and vascular fibers. Those containing the most cellulose are the most flexible and elastic and may be bleached white most easily. Those which are more lignified tend to be stiff, brittle, and hard to bleach. Vegetable fibers are much less hygroscopic than wool or silk.

Mercerized cotton is cotton fiber that has been treated with strong caustic soda while under tension. The fiber becomes more lustrous, stronger, and more readily dyeable.

Synthetic Organic Fibers

In recent years, a large and increasing portion of commercial fiber production has been of synthetic or semisynthetic origin. These fibers provide generally superior mechanical properties and greater resistance to degradation than do natural organic fibers and are available in a variety of forms and compositions for particular end-use applications. Generic categories of man-made fibers have been established by the Textile Fiber Products Identification Act, 15 USC 70, 72 Stat. 1717.

Among the important fibers are **rayons,** made from regenerated cellulose, plain or acetylated, which has been put into a viscous solution and extruded through the holes of a spinneret into a setting bath. The types most common at present are **viscose** and **acetate.** Cuprammonium rayon, saponified acetate rayon, and high-wet-modulus rayon are also manufactured and have properties which make them suitable for particular applications. Rayon is generally less expensive than other synthetic fibers.

In contrast to these regenerated fibers are a variety of polymer fibers which are chemically synthesized. The most important is **nylon,** including nylon 6.6, a condensation polymer of hexamethylene diamineadipic acid, and nylon 6, a polymer of caprolactam. Nylon possesses outstanding mechanical properties and is widely used in industrial fabrics. Nomex, a high-temperature-resistant nylon retains its most important properties at continuous operating temperatures up to 260°C (500°F). Other polymer fibers possess mechanical or chemical properties which make them a specific material of choice for specialized applications. These include **polyester** fibers, made from polyethylene terephthalate; **acrylic** and **modacrylic** fibers, made from copolymers of acrylonitrile and other chemicals; **Saran,** a polymer composed essentially of vinylidene chloride; and the **olefins,** including polyethylenes and polypropylene.

Teflon, a fluorocarbon resin, has excellent chemical resistance to acids, bases, or solvents. Its useful physical properties range from cryogenic temperatures to 260°C (500°F), high dielectric strength, low dissipation factor, and high resistivity, along with the lowest coefficient of friction of any solid. It is nonflammable and is inert to weather and sunlight.

As part of their manufacture, substantially all synthetic fibers are **drawn** (hot- or cold-stretched) after extrusion to achieve desirable changes in properties. Generally, increases in draw increase breaking strength and modulus and decrease ultimate elongation.

Proximate Identification of Fibers   Fibers are most accurately distinguished under the microscope, with the aid of chemical reagents and stains. A useful rough test is burning, in which the odor of burned meat distinguishes animal fibers from vegetable and synthetic fibers. Animal fibers, cellulose acetate, and nylon melt before burning and fuse to hard rounded beads. Cellulose fibers burn off sharply. Cellulose acetate dissolves in either acetone or chloroform containing some alcohol.

Heat Endurance   Fibers of organic origin lose strength when heated over long periods of time above certain temperatures: cellulose, 149°C (300°F); cellulose acetate, 93.5°C (200°F); nylon, 224°C (435°F); casein, 100°C (212°F); and glass, 316°C (600°F).

Creep   Textile fibers exhibit the phenomenon of creep at relatively low loads. When a textile fiber is subjected to load, it suffers three kinds of distortion: (1) an elastic deformation, closely proportioned to load and fully and instantly recoverable upon load removal; (2) a primary creep, which increases at a decreasing rate with time and which is fully, but not instantaneously, recoverable upon load removal; and (3) a secondary creep, which varies obscurely with time and load and is completely nonrecoverable upon load removal. The relative amounts of these three components, acting to produce the total deformation, vary with the different fibers. The two inelastic components give rise to mechanical hysteresis on loading and unloading.

Felts and Fabrics

A **felt** is a compacted formation of randomly entangled fibers. Wool felt is cohesive because the scaly structure of the wool fibers promotes mechanical interlocking of the tangled fibers. Felts can be made with blends of natural or synthetic fibers, and they may be impregnated with resins, waxes or lubricants for specific mechanical uses. Felts are available in sheets or cut into washers or shaped gaskets in a wide range of thicknesses and densities for packing, for vibration absorption, for heat insulation, or as holders of lubricant for bearings.

Fabrics are woven or knitted from **yarns.** Continuous-filament synthetic fibers can be made into monofilament or multifilament yarns with little or no twist. Natural fibers, of relatively short length, and synthetic **staple** fibers, which are purposely cut into short lengths, must be twisted together to form yarns. The amount of twist in yarns and the tension and arrangement of the weaving determine the appearance and mechanical properties of fabrics. Staple-fiber fabrics retain less than 50 percent of the intrinsic fiber strength, but values approaching 100 percent are retained with continuous-filament yarns. Industrial fabrics can be modified by mechanical and chemical treatments, as well as by coatings and impregnations, to meet special demands for strength and other mechanical, chemical, and electrical properties or to resist insect, fungus, and bacterial action and flammability.

Nomenclature   There are literally dozens of different numbering systems for expressing the relationship between yarn weight and length, all differing and each used in connection with particular fiber types or in different countries. The most common currently used unit is the **denier,** which is the weight in grams of 9,000 m of yarn. A universal

system, based on the **tex** (the weight in grams of 1,000 m of yarn), has been approved by the International Standards Organization and by the ASTM. Relative strengths of fibers and yarns are expressed as the **tenacity,** which reflects the specific gravity and the average cross section of the yarn. Units are grams per denier (gpd) or grams per tex (gpt).

Many textiles can sustain high-energy impact loads because of their considerable elongation before rupture. The total work done per unit length on a fiber or yarn which is extended to the point of rupture can be approximated by multiplying the specific strength by one-half the final extension of that length.

Yarn **twist** direction is expressed as **S** or **Z** twist, with the near-side helical paths of a twisted yarn held in a vertical position comparable in direction of slope to the center portion of one of these letters. Amount of twist is expressed in turns per inch (tpi).

Fabrics are characterized by the composition of the fiber material, the type of weave or construction, the count (the number of yarns per inch in the warp and the filling directions), and the weight of the fabric, usually expressed in ounces per running yard. **Cover factor** is the ratio of fabric surface covered by yarn to the total fabric surface. **Packing factor** is the ratio of fiber volume to total fabric volume.

Tables 6.8.6 to 6.8.8 give physical data and other information about commercial fibers and yarns. The tabulated quantities involving the denier should be regarded as approximate; they are not absolute values such as are used in engineering calculations.

Table 6.8.6   Fiber Properties*

Kind	Source	Length of fiber, in	Width or diam of cells, μm	Specific gravity	Moisture regain,‡ %	Chemical description	Principal uses
Cotton	Plant seed hair	5/8–2	8–27	1.52	8.5	Cellulose	Industrial, household, apparel
Jute†	Plant bast	50–80	15–20	1.48	13.7	Lignocellulose	Bagging, twine, carpet backing
Wool	Animal	2–16	10–50	1.32	17	Protein	Apparel, household, industrial
Viscose	Manufactured	Any	8–43	1.52	11	Regenerated cellulose	Apparel, industrial, household
Cellulose acetate	Manufactured	Any	12–46	1.33	6	Cellulose ester	Apparel, industrial, household
Nylon	Manufactured	Any	8	1.14	4.2	Polyamide	Apparel, industrial, household
Casein	Manufactured	Any	11–28	1.3	4.1	Protein	Apparel
Flax†	Plant bast	12–36	15–17	1.5	12	Cellulose	Household, apparel, industrial
Hemp†	Plant bast	—	18–23	1.48	12	Cellulose	Twine, halyards, rigging
Sisal†	Plant leaf	30–48	10–30	—	—	Lignocellulose	Twine, cordage
Manila†	Plant leaf	60–140	10–30	—	—	Lignocellulose	Rope, twine, cordage
Ramie†	Plant bast	3–10	24–70	1.52	—	Cellulose	Household, apparel, seines
Silk	Silkworm	Any	5–23	1.35	11	Protein	Apparel, household, industrial
Glass	Manufactured	Any	3	2.5	0	Fused metal oxides	Industrial, household
Dacron	Manufactured	Any	8	1.38	0.4	Polyester	Apparel, industrial, household

1 in = 0.025 + m; 1 μ = 10^{-6} m. The more up-to-date term for the micron (μ) is the micrometer (μm).
* Adapted from Smith, Textile Fibers, *proc. ASTM,* 1944; Appel, A Survey of the Synthetic Fibers, *Am. Dyestuff Reporter,* **34,** 1945, pp. 21–26; and other sources.
† These fibers are commercially used as bundles of cells. They vary greatly in width. Width figures given are for the individual cells.
‡ In air at 70°F and 65 percent relative humidity.

With the continuing phasing out of asbestos products, one can substitute, under appropriate conditions, fiberglass woven fabrics. See Fiberglass in this section.

FREEZING PREVENTIVES

Common salt is sometimes used to prevent the freezing of water; it does not, however, lower the freezing point sufficiently to be of use in very cold weather, and in concentrated solution tends to "creep" and to crystallize all over the receptacle. It also actively corrodes metals and has deleterious effects on many other materials, e.g., concrete. For freezing temperatures, see Sec. 18.

Calcium chloride ($CaCl_2$) is a white solid substance widely used for preventing freezing of solutions and (owing to its great hygroscopic power) for keeping sizing materials and other similar substances moist. It does not "creep" as in the case of salt. It does not rust metal but attacks solder.

Calcium chloride solutions are much less corrosive on metal if made alkaline by the use of lime, and also if a trace of sodium chromate is present. They are not suitable for use in automobile radiators, because of corrosive action while hot, and because of tendency of any spray therefrom to ruin the insulation of spark plugs and high-tension cables. For freezing temperatures, see Sec. 18.

Glycerol is a colorless, viscid liquid without odor and miscible with water in all proportions. It should have a specific gravity of approximately 1.25. It has no effect on metals but disintegrates rubber and loosens up iron rust.

Denatured alcohol is free from the disadvantages of calcium chloride, salt, and glycerin solutions, but is volatile from water mixtures which run hot. A solution containing 50 percent alcohol becomes flammable, but it is rarely necessary to use more than 30 percent.

Methyl alcohol solutions were sold widely in the past for use as automotive antifreeze. It is an effective antifreeze, but its fumes are poisonous and it must be used with extreme care. It has been largely

Table 6.8.7   Tensile Properties of Single Fibers

Fiber	Breaking tenacity, gpd	Extension at break, %	Elastic recovery at corresponding strain, %	Elastic modulus,* gpd
Glass	6.0–7.3	3.0–4.0	100 at 2.9	200–300
Fortisan (rayon)	6.0–7.0		100 at 1.2	150–200
			60 at 2.4	
Flax	2.6–7.7	2.7–3.3	65 at 2	
Nylon 6,6	4.6–9.2	16–32	100 at 8	25–50
Nylon 6	4.5–8.6	16–40	100 at 8	25–50
Silk	2.4–5.1	10–25	92 at 2	75–125
Saran	1.1–2.3	15–25	95 at 10	
Cotton	3.0–4.9	3–7	74 at 2	50–100
Steel (90,000 lb/in^2 T.S.)	0.9	28	—	300
Steel (music wire)	3.5	8	—	300
Viscose rayon	1.5–5.0	15–30	82 at 2	50–150
Wool	1.0–1.7	25–35	99 at 2	25–40
Acetate rayon	1.3–1.5	23–34	100 at 1	25–40
Polyester	4.4–7.8	10–25	100 at 2	50–80
Polypropylene	4.0–7.0	15–25	95 at 7	15–50
Polytetrafluoroethylene	1.7	13	—	

* From Kaswell, "Textile Fibers, Yarns, and Fabrics," Reinhold.
SOURCE: Kaswell "Wellington Sears Handbook of Industrial Textiles," Wellington Sears Co., Inc.

supplanted for this use by ethylene glycol solutions, which are applied in automotive practice and for stationary applications where, e.g., it is circulated in subgrade embedded pipes and thus inhibits formation of ice on walking surfaces.

Ethylene glycol (Prestone) is used as a freezing preventive and also permits the use of high jacket temperatures in aircraft and other engines. Sp gr 1.125 (1.098) at 32 (77)°F; boiling point, 387°F; specific heat, 0.575 (0.675) at 68 (212)°F. It mixes with water in all proportions. See Table 6.8.9 for nonfreezing percentages by volume.

Winter-summer concentrates, such as Prestone 11, an ethylene glycol base combined with patented inhibitor ingredients, give maximum freezing protection to about −92°F (−62°C) at 68 percent mixture, and to about −34°F (−37°C) at 50 percent mixture.

Anti-icing of fuels for aircraft is accomplished by adding ethylene glycol monomethyl ether mixtures as Prist.

GLASS

REFERENCES: *Journal of American Ceramic Society,* Columbus, Ohio. *Journal of the Society of Glass Technology,* Sheffield, England. Morey, "Properties of Glass," Reinhold. Scholes, "Handbook of the Glass Industry," Ogden-Watney. "Non-Silica Glasses," *Chem. & Met. Eng.,* Mar. 1946. Phillips, "Glass the

Table 6.8.9   Nonfreezing Percentages by Volume in Solution

	Temperature, °C (°F)			
	−6.7 (20)	−12.2 (10)	−17.8 (0)	−23.4 (−10)
Methyl alcohol	13	20	25	30
Prestone	17	25	32	38
Denatured alcohol	17	26	34	42
Glycerol	22	33	40	47

Miracle-Maker," Pitman. Long, "Propriétés physiques et fusion du verre," Dunod. Eitel-Pirani-Scheel, "Glastechnische Tabellen," Springer. Jebsen-Marwedel, "Glastechnische Fabrikationsfehler," Springer. Shand, "Glass Engineering Handbook," McGraw-Hill. Phillips, "Glass: Its Industrial Applications," Reinhold. Persson, "Flat Glass Technology," Plenum Press, 1969. McMillan, "Glass-Ceramics," Academic. Pye, Stevens, and La Course (eds.), "Introduction to Glass Science," Plenum Press. Jones, "Glass," Chapman & Hall. Doremus, "Glass Science," Wiley. Technical Staffs Division, "Glass," Corning Glass Works. ANSI/ASTM C "1986 Annual Book of ASTM Standards," vol. 15.02 (Glass). Current ASTM and ANSI Standards.

Glass is an inorganic product of fusion which has cooled to a rigid condition without crystallizing. It is obtained by melting together silica, alkali, and stabilizing ingredients, such as lime, alumina, lead, and barium. Bottle, plate, and window glass usually contain SiO_2, Al_2O_3, CaO, and Na_2O. Small amounts of the oxides of manganese and selenium are added to obtain colorless glass.

Special glasses, such as fiberglass, laboratory ware, thermometer glass, and optical glass, require different manufacturing methods and different compositions. The following oxides are either substituted for or added to the above base glass: B_2O_3, ZnO, K_2O, As_2O_3, PbO, etc., to secure the requisite properties. Colored glasses are obtained by adding the oxides of iron, manganese, copper, selenium, cobalt, chromium, etc., or colloidal gold.

Molten glass possesses the ability to be fabricated in a variety of ways and to be cooled down to room temperature rapidly enough to prevent crystallization of the constituents. It is a rigid material at ordinary temperatures but may be remelted and molded any number of times by the application of heat. Ordinary glass is melted at about 1,430°C (2,600°F) and will soften enough to lose its shape at about 594°C (1,100°F).

Window glass is a soda-lime-silica glass, fabricated in continuous sheets up to a width of 6 ft. The sheets are made in two thicknesses, SS and DS, which are, respectively, $\frac{1}{16}$ and $\frac{1}{8}$ in. Both thicknesses are made in A, B, C, and D grades.

Reflective-coated glass, such as Vari-Tran, insulates by reflecting hot sun in summer. Various colors are available, with heat transmissions ranging from 8 to 50 percent.

Table 6.8.8   Temperature and Chemical Effects on Textiles

Fiber	Temperature limit, °F	Resistance to chemicals	Resistance to mildew
Cottom	Yellows 250; decomposes 300	Poor resistance to acids	Attacked
Flax	275	Poor resistance to acids	Attacked
Silk	Decomposes 300	Attacked	Attacked
Glass	Softens 1,350	Resists	Resists
Nylon 6	Sticky 400; melts 420–430	Generally good	Resists
Nylon 6,6	Sticky 455; melts 482	Generally good	Resists
Viscose rayon	Decomposes 350–400	Poor resistance to acids	Attacked
Acetate	Sticky 350–400; melts 500	Poor resistance to acids	Resists
Wool	Decomposes 275	Poor resistance to alkalies	Attacked
Asbestos	1,490	Resists	Resists
Polyester	Sticky 455; melts 480	Generally good	Resists
Polypropylene	Softens 300–310; melts 325–335	Generally good	Resists
Polyethylene	Softens 225–235; melts 230–250	Generally good	Resists
Jute	275	Poor resistance to acids	Attacked

Temperature conversion: $t_c = (\frac{5}{9})(t_F - 32)$.
SOURCE: Fiber Chart, *Textile World,* McGraw-Hill, 1962.

Borosilicate glass is the oldest type of glass to have significant heat resistance. It withstands higher operating temperatures than either limed or lead glasses and is also more resistant to chemical attack. It is used for piping, sight glasses, boiler-gage glasses, sealed-beam lamps, laboratory beakers, and oven cooking ware. This type of glass was the first to carry the Pyrex trademark.

Aluminosilicate glass is similar to borosilicate glass in behavior but is able to withstand higher operating temperatures. It is used for top-of-stove cooking ware, lamp parts, and when coated, as resistors for electronic circuitry.

Plate glass and float glass are similar in composition to window glass. They are fabricated in continuous sheets up to a width of 15 ft and are polished on both sides. They may be obtained in various thicknesses and grades, under names like Parallel-O-Plate and Parallel-O-Float.

Before the introduction of float glass, plate glass sheets were polished mechanically on both sides, which turned out to be the most expensive part of the entire production process. Pilkington, in England, invented the concept of pouring molten glass onto a bed of molten tin and to float it thereon—hence the name *float glass*. In that process, the underside of the glass conforms to the smooth surface of the molten tin surface, and the top surface flows out to a smooth surface. The entire grinding and polishing operations are eliminated, with attendant impact on reducing the cost of the final product. With rare exceptions, all plate glass is currently produced as float glass.

Safety glass consists of two layers of plate glass firmly held together by an intermediate layer of organic material, such as polyvinyl butyral. Safety glass is ordinarily ¼ in thick but can be obtained in various thicknesses. This plate is shatter proof and is used for windshields, bank cashier's windows, etc.

Tempered glass is made from sheet glass in thicknesses up to 1 in. It possesses great mechanical strength, which is obtained by rapidly chilling the surfaces while the glass is still hot. This process sets up a high compression on the glass surfaces, which have the capacity of withstanding very high tensile forces.

Wire glass is a glass having an iron wire screen thoroughly embedded in it. It offers about 1½ times the resistance to bending that plain glass does; even thin sheets may be walked on. If properly made, it does not fall apart when cracked by shocks or heat, and is consequently fire-resistant. It is used for flooring, skylights, fireproof doors, fire walls, etc.

Pressed glass is made by forming heat-softened glass in molds under pressure. Such articles as tableware, lenses, insulators, and glass blocks are made by this process.

Glass blocks find wide application for building purposes, where they form easy-to-clean, attractive, airtight, light-transmitting panels. Glass blocks, such as Vistabrik, are solid throughout, and exceptionally rugged and virtually indestructible. Glass blocks manufactured to have an internal, partially depressurized air gap have energy-conserving, insulating qualities, with "thermal resistance" or R values ranging from 1.1 to almost 2.3 (h · ft² · °F)/Btu. Solar heat transmission can be varied within wide limits by using different colored glasses and by changing the reflection characteristics by means of surface sculpting. Standard blocks are 3⅞ in thick, and can be had in 6 by 6, 8 by 8, 12 by 12, 3 by 6, 4 by 8, and 4 by 12-in sizes. Thinner blocks are also available.

Fiberglass is a term used to designate articles that consist of a multitude of tiny glass filaments ranging in size from 0.0001 to 0.01 in in diam. The larger fibers are used in air filters; those 0.0005 in in diam, for thermal insulation; and the 0.0001- to 0.0002-in-diam fibers, for glass fabrics, which are stronger than ordinary textiles of the same size. Insulating tapes made from glass fabric have found wide application in electrical equipment, such as motors and generators and for mechanical uses. Woven fiberglass has supplanted woven asbestos, which is no longer produced.

Glass is used for many structural purposes, such as store fronts and tabletops, and is available in thicknesses upward of ⁵⁄₁₆ in and in plates up to 72 × 130 in. Plates with customized dimensions to suit specific architectural requirements are available on special order.

Cellular glass is a puffed variety with about 14–15 million cells per cubic foot. It is a good heat insulator and makes a durable marine float. See also Sec. 4.

Vycor is a 96 percent silica glass having extreme heat-shock resistance to temperatures up to 900°C (1,652°F), and is used as furnace sight glasses, drying trays, and space-vehicle outer windows.

Fused silica (silicon dioxide) in the noncrystalline or amorphous state shows the maximum resistance to heat shock and the highest permissible operating temperature, 900°C (1,652°F) for extended periods, or 1,200°C (2,192°F) for short periods. It has maximum ultraviolet transmission and the highest chemical-attack resistance. It is used for astronomical telescopes, ultrasonic delay lines, and crucibles for growing single metal crystals.

Vitreous silica, also called **fused quartz** is made by melting rock crystal or purest quartz sand in the electric furnace. It is unaffected by changes of temperature, is fireproof and acid-resistant, does not conduct electricity, and has practically no expansion under heat. It is used considerably for high-temperature laboratory apparatus. See Plastics, below, for glass substitutes.

Glass ceramics, such as Pyroceram, are melted and formed as glasses, then converted by controlled nucleation and crystal growth to polycrystalline ceramic materials. Most are opaque and stronger than glass, and may also be chemically strengthened.

Leaded glass used to fabricate fine crystal (glasses, decanters, vases, etc.) has a lead content which has been shown to leach out into the liquid contents if the two are in contact for a long time, certainly in the range of months or longer. Often the optical properties of this glass are enhanced by elaborate and delicate designs cut into the surface. The addition of lead to the base glass results in a softer product which lends itself more readily to such detailed artisanry.

Glass beads are used extensively for reflective paints.

Properties of Glass   Glass is a brittle material and can be considered perfectly elastic to the fracture point. The range of Young's modulus is 4 to 14 × 10³ kg/mm² (6 to 20 × 10⁶ lb/in²), with most commercial glasses falling between 5.5 and 9.0 × 10³ kg/mm² (8 and 13 × 10⁶ lb/in²). Theory predicts glass strength as high as 3,500 kg/mm² (5 × 10⁶ lb/in²); fine glass fibers have shown strengths around 700 kg/mm² (10⁶ lb/in²). Glass products realize but a fraction of such properties however, owing to surface-imperfection stress-concentration effects. Often design strengths run from 500 to 1,000 times lower than theoretical. Glass strength also deteriorates when held under stress in atmospheric air (static fatigue), an apparent result of reaction of water with glass; high-strength glasses suffer the greater penalty. Glass also exhibits a timeload effect, or creep, and may even break when subjected to sustained loads, albeit the stresses induced may be extremely low. There exist samples of very old glass (centuries old) which, while not broken, show evidence of plastic flow albeit at ambient temperature all the while. A cylindrical shape, e.g., will slowly degenerate into a teardrop shape, with the thickening at the bottom having been caused by the deadweight alone of the piece of glass. At room temperatures the thermal conductivity of glasses ranges from 1.6 × 10⁻³ to 2.9 × 10⁻³ cal/(s · cm² · °C/cm) [4.65 to 8.43 Btu/(h · ft² · °F/in)].

NATURAL STONES

REFERENCES: Kessler, Insley, and Sligh, *Jour. Res. NBS,* **25,** pp. 161–206. Birch, Schairer, and Spicer, Handbook of Physical Constants, *Geol. Soc. Am. Special Paper* 36. Currier, Geological Appraisal of Dimension Stone Deposits, *USGS Bull.* 1109. ANSI/ASTM Standards C99–C880. "1986 Annual Book of ASTM Standards," vol. 04.08 (Building Stones). Current ASTM and ANSI Standards.

A **stone** or a **rock** is a naturally occurring composite of minerals. Stone has been used for thousands of years as a major construction material because it possesses qualities of strength, durability, architectural adaptability, and aesthetic satisfaction. There are two principal branches of the natural-stone industry—**dimension** stone and **crushed** or broken stone. The uses of the latter vary from aggregate to riprap, in which stones in a broad range of sizes are used as structural support in a matrix or to provide weathering resistance. Dimension stones are blocks or slabs of stone processed to specifications of size, shape, and surface finish. The largest volume today lies in the use of slabs varying from 1 to 4 in in thickness that are mounted on a structure as a protective and aesthetic veneer.

There are two major types of natural stone: **igneous** and **metamorphic** stones, composed of tightly interlocking crystals of one or more minerals, and **sedimentary** rocks, composed of cemented mineral grains in which the cement may or may not be of the same composition as the grains. The major groups of natural stone used commercially are:

Granite, a visibly crystalline rock made of silicate minerals, primarily feldspar and quartz. Commercially, "granite" refers to all stones geologically defined as plutonic, igneous, and gneissic.

Marble, generally a visibly carbonate rock; however, microcrystalline rocks, such as onyx, travertine, and serpentine, are usually included by the trade as long as they can take a polish.

Limestone, a sedimentary rock composed of calcium or magnesium carbonate grains in a carbonate matrix.

Sandstone, a sedimentary rock composed chiefly of cemented, sand-sized quartz grains. In the trade, quartzites are usually grouped with sandstones, although these rocks tend to fracture through, rather than around, the grains. **Conglomerate** is a term used for a sandstone containing aggregate in sizes from the gravel range up.

The above stones can be used almost interchangeably as dimension stone for architectural or structural purposes.

Slate, a fine-grained rock, is characterized by marked cleavages by which the rock can be split easily into relatively thin slabs. Because of this characteristic, slate was at one time widely used for roofing tiles. See Roofing Materials, later in this section.

It is still widely applied for other building uses, such as steps, risers, spandrels, flagstones, and in some outdoor sculptured work.

Formerly, it was used almost universally for blackboards and electrical instrument panels, but it has been supplanted in these applications by plastic materials. Plastic sheets used as a writing surface for chalk are properly called **chalkboards;** the surface on which writing is done with crayons or fluid tip markers is termed **marker board.**

Miscellaneous stones, such as traprock (fine-grained black volcanic rock), greenstone, or argillite, are commonly used as crushed or broken stone but rarely as dimension stone.

Table 6.8.10 lists physical properties and contains the range of values that can be obtained from stones in various orientations relative to their textural and structural anisotropy. For a particular application, where one property must be exactly determined, the value must be obtained along a specified axis.

Selected Terms Applying to the Use of Dimension Stone

Anchor, a metal tie or rod used to fasten stone to backup units.

Arris, the meeting of two surfaces producing an angle, corner, or edge.

Ashlar, a facing of square or rectangular stones having sawed, dressed, or squared beds.

Bond stones, stones projecting a minimum of 4 in laterally into the backup wall; used to tie the wall together.

Cut stone, finished dimension stone—ready to set in place. The finish may be polished, honed, grooved (for foot traffic), or broken face.

Bearing wall, a wall supporting a vertical load in addition to its own weight.

Cavity wall, a wall in which the inner and outer parts are separated by an air space but are tied together with cross members.

Composite wall, a wall in which the facing and backing are of different materials and are united with bond stones to exert a common reaction under load. It is considered preferable, however, not to require the facing to support a load; thus the bond stones merely tie the facing to the supporting wall, as in the case of a veneer.

Veneer or **faced wall,** a wall in which a thin facing and the backing are of different materials but are not so bonded as to exert a common reaction under load.

Definitions may be found in ANSI/ASTM, C119-74 (1980).

PAPER

REFERENCES: Calkin, "Modern Pulp and Papermaking," Reinhold. Griffin and Little, "Manufacture of Pulp and Paper," McGraw-Hill. Guthrie, "The Economics of Pulp and Paper," State College of Washington Press. "Dictionary of Paper," American Pulp and Paper Assoc. Sutermeister, "Chemistry of Pulp and Papermaking," Wiley. Casey, "Pulp and Paper—Chemistry and Technology," Interscience, TAPPI Technical Information Sheets. TAPPI Monograph Series. "Index of Federal Specifications, Standards and Handbooks," GSA. Lockwoods Directory of the Paper and Allied Trades. Casey, "Pulp and Paper; Chemistry and Chemical Technology," 3 vols., Interscience. Libby (ed.), "Pulp and Paper Science and Technology," 2 vols., McGraw-Hill. ASTM Committee D6 on Paper and Paper Products, "Paper and Paperboard Characteristics, Nomenclature, and Significance of Tests," ASTM Special Technical Publication 60-B, 3d ed., 1963. Britt, "Handbook of Pulp and Paper Technology," Van Nostrand Reinhold. Johnson, "Synthetic Paper from Synthetic Fibers," Noyes Data Corp., New Jersey. "1986 Annual Book of ASTM Standards," vol. 15.09 (Paper).

Paper Grades

Specific paper **qualities** are achieved in a number of ways: (1) By selecting the composition of the furnish for the paper machine. Usually more than one pulp (prepared by different pulping conditions or processes) is required. The ratio of long-fibered pulp (softwood) to short-fibered pulp (hardwood or mechanical type), the reused-fiber content, and the use of nonfibrous fillers and chemical additives are important factors. (2) By varying the paper-machine operation. Fourdrinier wire machines are most common, although multicylinder machines and high-speed tissue machines with Yankee driers are also used. (3) By using various finishing operations (e.g., calendering, supercalendering, coating, and laminating).

There is a tremendous number of paper grades, which are, in turn, used in a wide variety of converted products. The following broad classifications are included as a useful guide:

Sanitary papers are tissue products characterized by bulk, opacity, softness, and water absorbency.

Table 6.8.10   Physical and Thermal Properties of Common Stones

Type of stone	Density, lb/ft³	Compressive strength × 10⁻³, lb/in²	Rupture modulus × 10⁻³, lb/in² (ASTM C99-52)	Shearing strength × 10⁻³, lb/in²	Young's modulus × 10⁻⁶, lb/in²	Modulus of rigidity × 10⁻⁶, lb/in²	Poisson's ratio	Abrasion-hardness index (ASTM C241-51)	Porosity, vol %	48-h water absorption (ASTM C97-47)	Thermal conductivity, Btu/(ft · h · °F)	Coefficient of thermal expansion × 10⁻⁶, per °F
Granite	160–190	13–55	1.4–5.5	3.5–6.5	4–16	2–6	0.05–0.2	37–88	0.6–3.8	0.02–0.58	20–35	3.6–4.6
Marble	165–179	8–27	0.6–4.0	1.3–6.5	5–11.5	2–4.5	0.1–0.2	8–42	0.4–2.1	0.02–0.45	8–36	3.0–8.5
Slate	168–180	9–10	6–15	2.0–3.6	6–16	2.5–6	0.1–0.3	6–12	0.1–1.7	0.01–0.6	12–26	3.3–5.6
Sandstone	119–168	5–20	0.7–2.3	0.3–3.0	0.7–10	0.3–4	0.1–0.3	2–26	1.9–27.3	2.0–12.0	4–40	3.9–6.7
Limestone	117–175	2.5–28	0.5–2.0	0.8–3.6	3–9	1–4	0.1–0.3	1–24	1.1–31.0	1.0–10.0	20–32	2.8–4.5

Conversion factors: 1 lbm/ft³ = 16,018 kg/m³. 1 lb/in² = 6.894.8 N/m². 1 Btu/(ft · h · °F) = 623 W/(m · °C).

Glassine, greaseproof, and **waxing papers**—in glassine, high transparency and density, low grease penetration, and uniformity of formation are important requirements.

Food-board products require a good brightness and should be odorless and have good tear, tensile, and fold-endurance properties, opacity, and printability. More stringent sanitary requirements in current practice have caused this product to be supplanted largely by foamed plastics (polyethylene and the like).

Boxboard, misnomered **cardboard,** and a variety of other board products are made on multicylinder machines, where layers of fiber are built up to the desired thickness. Interior plies are often made from wastepaper furnishes, while surface plies are from bleached, virgin fiber. Boards are often coated for high brightness and good printing qualities.

Printing papers include publication papers (magazines), book papers, bond and ledger papers, newsprint, and catalog papers. In all cases, printability, opacity, and dimensional stability are important.

Linerboard and **bag paper** are principally unbleached, long-fiber, kraft products of various weights. Their principal property requirements are high tensile and bursting strengths.

Corrugating medium, in combined fiberboard, serves to hold the two linerboards apart in rigid, parallel separation. Stiffness is the most important property, together with good water absorbency for ease of corrugation.

Corrugated paper, often misnomered **corrugated cardboard,** is an end product of kraft paper, in which a corrugated layer is sandwiched and glued between two layers of cover paper. The whole assembly, when the glue has set, results in a material having a high flexural strength (although it may be directional, depending on whether the material is flexed parallel or crosswise to the corrugations), and high penetration strength. For heavier-duty applications, several sandwiches are themselves glued together and result in a truly superior load-carrying material.

Pulps

The most important source of fiber for paper pulps is wood, although numerous other vegetable substances are used. Reused fiber (wastepaper) constitutes about 30 percent of the total furnish used in paper, principally in boxboard and other packaging or printing papers.

There are three basic processes used to convert wood to papermaking fibers: mechanical, chemical, and "chemimechanical."

Mechanical Pulping (Groundwood) Here, the entire log is reduced to fibers by grinding against a stone cylinder, the simplest route to papermaking fiber. Wood fibers are mingled with extraneous materials which can cause weakening and discoloration in the finished paper. Nonetheless, the resulting pulp imparts good bulk and opacity to a printing sheet. Some long-fibered chemical pulp is usually added. Groundwood is used extensively in low-cost, short-service, and throwaway papers, e.g., newsprint and catalog paper.

Chemical Pulping Fiber separation can also be accomplished by chemical treatment of wood chips to dissolve the lignin that cements the fibers together. From 40 to 50 percent of the log is extracted, resulting in relatively pure cellulosic fiber. There are two major chemical-pulping processes, which differ both in chemical treatment and in the nature of the pulp produced. In **sulfate pulping,** also referred to as the **kraft** or **alkaline** process, the pulping chemicals must be recovered and reused for economic reasons. Sulfate pulps result in papers of high physical strength, bulk, and opacity; low unbleached brightness; slow beating rates; and relatively poor sheet-formation properties. Both bleached and unbleached sulfate pulps are used in packaging papers, container board, and a variety of printing and bond papers. In **sulfite pulping,** the delignifying agents are sulfurous acid and an alkali, with several variations of the exact chemical conditions in commercial use. In general, sulfate pulps have lower physical strength properties and lower bulk and opacity than kraft pulps, with higher unbleached brightness and better sheet-formation properties. The pulps are blended with groundwood for **newsprint** and are used in printing and bond papers, tissues, and glassine.

"Chemimechanical" processes combine both chemical and mechanical methods of defibration, the most important commercial process being the NSSC (Neutral Sulfite Semi Chemical). A wide range of yields and properties can be obtained.

Bleaching

Chemical pulps may be bleached to varying degrees of brightness, depending on the end use. During some bleaching operations, the remaining lignin is removed and residual coloring matter destroyed. Alternatively, a nondelignifying bleach lightens the color of high-lignin pulps.

Refining

The final character of the pulp is developed in the **refining** (beating) operation. Pulp fibers are fibrillated, hydrated, and cut. The fibers are roughened and frayed, and a gelatinous substance is produced. This results in greater fiber coherence in the finished paper.

Sizing, Loading, and Coating

Sizing is used, principally in book papers, to make the paper water-repellant and to enhance interfiber bonding and surface characteristics. Sizing materials may be premixed with the pulp or applied after sheet formation.

Loading materials, or **fillers,** are used by the papermaker to smooth the surface, to provide ink affinity, to brighten color, and to increase opacity. The most widely used fillers are clay and kaolin.

Coatings consisting of pigment and binder are often applied to the base stock to create better printing surfaces. A variety of particulate, inorganic materials are combined with binders such as starch or casein in paper coatings. Coating is generally followed by calendering.

Plastic-Fiber Paper

A tough, durable product, such as Tyvek, can be made from 100 percent high-density polyethylene fibers by spinning very fine fibers and then bonding them together with heat and pressure. Binders, sizes, or fillers are not required. Such sheets combine some of the best properties of the fabrics, films, and papers with excellent puncture resistance. They can readily be printed by conventional processes, dyed to pastel colors, embossed for decorative effects, coated with a range of materials, and can be folded, sheeted, die-cut, sewn, hot-melt-sealed, glued, and pasted. Nylon paper, such as Nomex type 410, is produced from short fibers (floc) and smaller binder particles (fibrids) of a high-temperature-resistant polyamide polymer, formed into a sheet product without additional binders, fillers, or sizes, and calendered with heat and pressure to form a nonporous structure. It possesses excellent electrical, thermal, and mechanical properties, and finds use in the electrical industry. Fiber impregnation of ordinary paper, using glass or plastic fibers, produces a highly tear-resistant product.

Recycling

In the current environmentally conscious climate, almost 50 percent of newsprint itself contains a portion of recycled old newsprint. The remainder, together with almost all other paper products, finds a ready market for recycling into other end products. The small portion that ends up discarded is increasingly used as feedstock for incineration plants which produce power and/or process steam by converting "scrap paper" and other refuse to refuse-derived fuel (RDF). See Sec. 7.4. Only in those geographic areas where it is impractical to collect and recycle paper does the residue finally find its way to landfill dumps.

ROOFING MATERIALS

REFERENCES: Abraham, "Asphalts and Allied Substances," Van Nostrand. Grondal, "Certigrade Handbook of Red Cedar Shingles," Red Cedar Shingle Bureau, Seattle, Wash. ASTM *Special Technical Publication* 409, "Engineering Properties of Roofing Systems," ASTM. Griffin, Jr., "Manual of Built-up Roof Systems," McGraw-Hill. "1986 Annual Book of ASTM Standards," vol. 04.04 (Roofing, Waterproofing and Bituminous Materials). Current ASTM and ANSI Standards.

Asphalt Asphalts are bitumens, and the one most commonly seen in roofing and paving is obtained from petroleum residuals. These are obtained by the refining of petroleum. The qualities of asphalt are affected by the nature of the crude and the process of refining. When the flux asphalts obtained from the oil refineries are treated by blowing air through them while the asphalt is maintained at a high temperature, a material is produced which is very suitable and has good weathering properties.

Coal Tar   Coal tar is more susceptible to temperature change than asphalt; therefore, for roofing purposes its use is usually confined to flat decks.

Asphalt prepared roofing is manufactured by impregnating a dry roofing felt with a hot asphaltic saturant. A coating consisting of a harder asphalt compounded with a fine mineral **filler** is applied to the weather side of the saturated felt. Into this coating is embedded mineral **surfacing** such as mineral granules, powdered talc, mica, or soapstone. The reverse side of the roofing has a very thin coating of the same asphalt which is usually covered with powdered talc or mica to prevent the roofing from sticking in the package. The surfacing used on **smooth-surfaced roll roofing** is usually powdered talc or mica. The surfacing used on **mineral- or slate-surfaced roll roofing** is roofing granules either in natural colors prepared from slate or artificial colors usually made by applying a coating to a rock granule base. **Asphalt shingles** usually have a granular surfacing. They are made in strips and as individual shingles. The different shapes and sizes of these shingles provide single, double, and triple coverage of the roof deck.

Materials Used in Asphalt Prepared Roofing   The **felt** is usually composed of a continuous sheet of felted fibers of selected rag, specially prepared wood, and high-quality waste papers. The constituents may be varied to give a felt with the desired qualities of strength, absorbency, and flexibility. (See above, Fibers and Fabrics.)

The most satisfactory roofing **asphalts** are obtained by air-blowing a steam- or vacuum-refined petroleum residual. Saturating asphalts must possess a low viscosity in order for the felt to become thoroughly impregnated. Coating asphalts must have good weather-resisting qualities and possess a high fusion temperature in order that there will be no flowing of the asphalt after the application to the roof.

Asphalt built-up roof coverings usually consist of several layers of asphalt-saturated felt with a continuous layer of hot-mopped asphalt between the layers of felt. The top layer of such a roof covering may consist of a hot mopping of asphalt only, a top pouring of hot asphalt with slag or gravel embedded therein, or a mineral-surfaced cap sheet embedded in a hot mopping of asphalt.

Wood shingles are usually manufactured in three different lengths: 16, 18, and 24 in. There are three grades in each length: no. 1 is the best, and no. 3 is intended for purposes where the presence of defects is not objectionable. Red-cedar shingles of good quality are obtainable from the Pacific Coast; in the South, red cypress from the Gulf states is preferable. Redwood shingles come 5½ butts to 2 in; lesser thicknesses are more likely to crack and have shorter life. Shingles 8 in wide or over should be split before laying. Dimension shingles of uniform width are obtainable. Various stains are available for improved weathering resistance and altered appearance.

Slate should be hard and tough and should have a well-defined vein that is not too coarse. Roofing slates in place will weigh from 650 to over 1,000 lb per square, depending on the thickness and the degree of overlap. Prudent roofing practice in installing slate roofs requires use of copper (or stainless steel) nails and, as appropriate, the installation of snow guards to prevent snow slides.

Metallic roofings are usually laid in large panels, often strengthened by corrugating, but they are sometimes cut into small sizes bent into interlocking shapes and laid to interlock with adjacent sheets or shingles. Metallic roofing panels of both aluminum and steel are available with a variety of prefinished surface treatments to enhance weatherability. Metal tile and metal shingles are usually made of copper, copper-bearing galvanized steel, tinplate, zinc, or aluminum. The lightest metal shingle is the one made from aluminum, which weighs approximately 40 lb per square. The metal radiates solar heat, resulting in lower temperatures beneath than with most other types of uninsulated roofs. Terne-coated stainless steel is unusually resistant to weathering and corrosion.

Roofing cements and **coatings** are usually made from asphalt, a fibrous filler, and solvents to make the cement workable. Asphalt-base aluminum roof coatings are used to renew old asphalt roofs, and to prolong the service life of smooth-surfaced roofs, new or old.

Tile   Hard-burned clay tiles with overlapping or interlocking edges cost about the same as slate. They should have a durable glaze and be well made. Unvitrified tiles with slip glaze are satisfactory in warm climates, but only vitrified tiles should be used in the colder regions. Tile roofs weigh from 750 to 1,200 lb per square. Properly made, tile does not deteriorate, is a poor conductor of heat and cold, and is not as brittle as slate.

Elastomers

The trend toward irregular roof surfaces—folded plates, hyperbolic paraboloids, domes, barrel shells—has brought the increased use of plastics or synthetic rubber elastomers (applied as fluid or sheets) as roofing membranes. Such membranes offer light weight, adaptability to any roof slope, good heat reflectivity, and high elasticity at moderate temperatures. Negatively, elastomeric membranes have a more limited range of satisfactory substrate materials than conventional ones. Table 6.8.11 presents several such membranes, and the method of use.

Table 6.8.11   Elastomeric Membranes

	Material	Method of application	Number of coats or sheets
Fluid-applied	Neoprene-Hypalon*	Roller, brush, or spray	2 + 2
	Silicone	Roller, brush, or spray	2
	Polyurethane foam, Hypalon* coating	Spray	2
	Clay-type asphalt emulsion reinforced with chopped glass fibers†	Spray	1
Sheet-applied	Chlorinated polyethylene on foam	Adhesive	1
	Hypalon* on asbestos felt	Adhesive	1
	Neoprene-Hypalon*	Adhesive	1 + surface paint
	Tedlar‡ on asbestos felt	Adhesive	1
	Butyl rubber	Adhesive	1
Traffic decks	Silicone plus sand	Trowel	1 + surface coat
	Neoprene with aggregate§	Trowel	1 + surface coat

* Registered trademark of E. I. du Pont de Nemours & Co., for chlorosulfonated polyethylene.
† Frequently used with coated base sheet.
‡ Registered trademark of E. I. du Pont de Nemours & Co. for polyvinyl fluoride.
§ Aggregate may be flint, sand, or crushed walnut shells.
Source: C. W. Griffin, Jr., "Manual of Built-up Roof Systems," McGraw-Hill, 1970, with permission of the publisher.

RUBBER AND RUBBERLIKE MATERIALS (ELASTOMERS)

REFERENCES: Dawson and Porritt, "Rubber: Physical and Chemical Properties," Research Association of British Rubber Manufacturers. Davis and Blake, "The Chemistry and Technology of Rubber," Reinhold. ASTM Standards on Rubber Products. "Rubber Red Book Directory of the Rubber Industry," *The Rubber Age.* Flint, "The Chemistry and Technology of Rubber Latex," Van Nostrand. "The Vanderbilt Handbook," R.T. Vanderbilt Co. Whitby, Davis, and Dunbrook, "Synthetic Rubber," Wiley. "Rubber Bibliography," Rubber Division, American Chemical Society. Noble, "Latex in Industry," *The Rubber Age.* "Annual Report on the Progress of Rubber Technology," Institution of the Rubber Industry. Morton (ed.). "Rubber Technology," Van Nostrand Reinhold. ISO 188–1982. "1986 Annual Book of ASTM Standards," vols. 09.01, 09.02 (Rubber). Current ASTM and ANSI Standards.

To avoid confusion by the use of the word **rubber** for a variety of natural and synthetic products, the term **elastomer** has come into use, particularly in scientific and technical literature, as a name for both natural and synthetic materials which are elastic or resilient and in general resemble natural rubber in feeling and appearance.

The utility of rubber and synthetic elastomers is increased by compounding. In the raw state, elastomers are soft and sticky when hot, and hard or brittle when cold. **Vulcanization** extends the temperature range within which they are flexible and elastic. In addition to vulcanizing agents, ingredients are added to make elastomers stronger, tougher, and harder, to make them age better, to color them, and in general to modify them to meet the needs of service conditions. Few rubber products today are made from rubber or other elastomers alone.

The elastomers of greatest commercial and technical importance today are natural rubber, GR-S, Neoprene, nitrile rubbers, and butyl.

Natural rubber of the best quality is prepared by coagulating the latex of the *Hevea brasilensis* tree, cultivated chiefly in the Far East. This represents nearly all of the natural rubber on the market today.

Unloaded vulcanized rubber will stretch to approximately 10 times its length and at this point will bear a **load** of 13.8×10^6 N/m^2 (10 tons/in^2). It can be compressed to one-third its thickness thousands of times without injury. When most types of vulcanized rubber are stretched, their resistance increases in greater proportion than the extension. Even when stretched almost to the point of rupture, they recover very nearly their original dimensions on being released, and then gradually recover a part of the residual distortion.

Freshly cut or torn raw rubber possesses the power of **self-adhesion** which is practically absent in vulcanized rubber. Cold water preserves rubber, but if exposed to the air, particularly to the sun, rubber goods tend to become hard and brittle. Dry heat up to 49°C (120°F) has little deteriorating effect; at temperatures of 181 to 204°C (360 to 400°F) rubber begins to melt and becomes sticky; at higher temperatures, it becomes entirely carbonized. Unvulcanized rubber is **soluble** in gasoline, naphtha, carbon bisulfide, benzene, petroleum ether, turpentine, and other liquids.

Most rubber is **vulcanized**, i.e., made to combine with sulfur or sulfur-bearing organic compounds or with other chemical cross-linking agents. Vulcanization, if properly carried out, improves mechanical properties, eliminates tackiness, renders the rubber less susceptible to temperature changes, and makes it insoluble in all known solvents. It is impossible to dissolve vulcanized rubber unless it is first decomposed. Other ingredients are added for general effects as follows:

To increase tensile strength and resistance to abrasion: carbon black, precipitated pigments, as well as organic vulcanization accelerators.

To cheapen and stiffen: whiting, barytes, talc, silica, silicates, clays, fibrous materials.

To soften (for purposes of processing or for final properties): bituminous substances, coal tar and its products, vegetable and mineral oils, paraffin, petrolatum, petroleum oils, asphalt.

Vulcanization accessories, dispersion and wetting mediums, etc.: magnesium oxide, zinc oxide, litharge, lime, stearic and other organic acids, degras, pine tar.

Protective agents (natural aging, sunlight, heat, flexing): condensation amines, waxes.

Coloring pigments: iron oxides, especially the red grades, lithopone, titanium oxide, chromium oxide, ultramarine blue, carbon and lampblacks, and organic pigments of various shades.

Specifications should state suitable physical tests. Tensile strength and extensibility tests are of importance and differ widely with different compounds.

GR-S is an outgrowth and improvement of German Buna S. The quantity now produced far exceeds all other synthetic elastomers. It is made from butadiene and styrene, which are produced from petroleum. These two materials are copolymerized directly to GR-S, which is known as a butadiene-styrene copolymer. GR-S has recently been improved, and now gives excellent results in tires.

Neoprene is made from acetylene, which is converted to vinylacetylene, which in turn combines with hydrogen chloride to form chloroprene. The latter is then polymerized to Neoprene.

Nitrile rubbers, an outgrowth of German Buna N or Perbunan, are made by a process similar to that for GR-S, except that acrylonitrile is used instead of styrene. This type of elastomer is a butadiene-acrylonitrile copolymer.

Butyl, one of the most important of the synthetic elastomers, is made from petroleum raw materials, the final process being the copolymerization of isobutylene with a very small proportion of butadiene or isoprene.

Polysulfide rubbers having unique resistance to oxidation and to softening by solvents are commercially available and are sold under the trademark Thiokol.

Ethylene-propylene rubbers are notable in their oxidation resistance. **Polyurethane** elastomers can have a tensile strength up to twice that of conventional rubber, and solid articles as well as foamed shapes can be cast into the desired form using prepolymer shapes as starting materials. **Silicone** rubbers have the advantages of a wide range of service temperatures and room-temperature curing. **Fluorocarbon** elastomers are available for high-temperature service. Polyester elastomers have excellent impact and abrasion resistance.

No one of these elastomers is satisfactory for all kinds of service conditions, but rubber products can be made to meet a large variety of service conditions.

The following examples show some of the important properties required of rubber products and some typical services where these properties are of major importance:

Resistance to abrasive wear: auto-tire treads, conveyor-belt covers, soles and heels, cables, hose covers, V belts.

Resistance to tearing: auto inner tubes, tire treads, footwear, hotwater bags, hose covers, belt covers, V belts.

Resistance to flexing: auto tires, transmission belts, V belts, (see Sec. 8.2 for information concerning types, sizes, strengths, etc., of V belts), mountings, footwear.

Resistance to high temperatures: auto tires, auto inner tubes, belts conveying hot materials, steam hose, steam packing.

Resistance to cold: airplane parts, automotive parts, auto tires, refrigeration hose.

Minimum heat buildup: auto tires, transmission belts, V belts, mountings.

High resilience: auto inner tubes, sponge rubber, mountings, elastic bands, thread, sandblast hose, jar rings, V belts.

High rigidity: packing, soles and heels, valve cups, suction hose, battery boxes.

Long life: fire hose, transmission belts, tubing, V belts.

Electrical resistivity: electricians' tape, switchboard mats, electricians' gloves.

Electrical conductivity: hospital flooring, nonstatic hose, matting.

Impermeability to gases: balloons, life rafts, gasoline hose, special diaphragms.

Resistance to ozone: ignition distributor gaskets, ignition cables, windshield wipers.

Resistance to sunlight: wearing apparel, hose covers, bathing caps.

Resistance to chemicals: tank linings, hose for chemicals.

Resistance to oils: gasoline hose, oil-suction hose, paint hose, creamery hose, packinghouse hose, special belts, tank linings, special footwear.

Table 6.8.12 Comparative Properties of Elastomers

	Natural rubber	GR-S	Neoprene	Nitrile rubbers	Butyl	Thiokol
Tensile properties	Excellent	Good	Very good	Good	Good	Fair
Resistance to abrasive wear	Excellent	Good	Very good	Good	Good	Poor
Resistance to tearing	Very good	Poor	Good	Fair	Very good	Poor
Resilience	Excellent	Good	Good	Fair	Poor	Poor
Resistance to heat	Good	Fair	Good	Excellent	Good	Poor
Resistance to cold	Excellent	Good	Good	Good	Excellent	Poor
Resistance to flexing	Excellent	Good	Very good	Good	Excellent	Poor
Aging properties	Excellent	Excellent	Good	Good	Excellent	Good
Cold flow (creep)	Very low	Low	Low	Very low	Fairly low	High
Resistance to sunlight	Fair	Fair	Excellent	Good	Excellent	Excellent
Resistance to oils and solvents	Poor	Poor	Good	Excellent	Fair	Excellent
Permeability to gases	Fairly low	Fairly low	Low	Fairly low	Very low	Very low
Electrical insulation	Fair	Excellent	Fair	Poor	Excellent	Good
Flame resistance	Poor	Poor	Good	Poor	Poor	Poor
Auto tire tread	Preferred	Alternate				
Inner tube	Alternate				Preferred	
Conveyor-belt cover	Preferred		Alternate			
Tire sidewall	Alternate	Preferred				
Transmission belting	Preferred		Alternate			
Druggist sundries	Preferred					
Gasoline and oil hose				Preferred		
Lacquer and paint hose						Preferred
Oil-resistant footwear			Preferred			
Balloons	Alternate		Preferred			
Jar rings	Alternate	Preferred				
Wire and cable insulation	Alternate	Preferred				

Stickiness: cements, electricians' tapes, adhesive tapes, pressure-sensitive tapes.

Low specific gravity: airplane parts, forestry hose, balloons.

No odor or taste: milk tubing, brewery and wine hose, nipples, jar rings.

Special colors: ponchos, life rafts, welding hose.

Table 6.8.12 gives a comparison of some important characteristics of the most important elastomers when vulcanized. The lower part of the table indicates, for a few representative rubber products, preferences in the use of different elastomers for different service conditions without consideration of cost.

Specifications for rubber goods may cover the chemical, physical, and mechanical properties, such as elongation, tensile strength, permanent set, and oven tests, minimum rubber content, exclusion of reclaimed rubber, maximum free and combined sulfur contents, maximum acetone and chloroform extracts, ash content, and many construction requirements. It is preferable, however, to specify properties such as resilience, hysteresis, static or dynamic shear and compression modulus, flex fatigue and cracking, creep, electrical properties, stiffening, heat generation, compression set, resistance to oils and chemicals, permeability, brittle point, etc., in the temperature range prevailing in service, and to leave the selection of the elastomer to a competent manufacturer.

Latex, imported in stable form from the Far East, is used for various rubber products. In the manufacture of such products, the latex must be compounded for vulcanizing and otherwise modifying properties of the rubber itself. Important products made directly from compounded latex include surgeons' and household gloves, thread, bathing caps, rubberized textiles, balloons, and sponge. A recent important use of latex is for "foam sponge," which may be several inches thick and used for cushions, mattresses, etc.

Gutta-percha and **balata,** also natural products, are akin to rubber chemically but more leathery and thermoplastic, and are used for some special purposes, principally for submarine cables, golf balls, and various minor products.

Rubber Derivatives

Rubber derivatives are chemical compounds and modifications of rubber, some of which have become of commercial importance.

Chlorinated rubber, produced by the action of chlorine on rubber in solution, is nonrubbery, incombustible, and extremely resistant to many chemicals. As commercial **Parlon,** it finds use in corrosion-resistant paints and varnishes, in inks, and in adhesives.

Rubber hydrochloride, produced by the action of hydrogen chloride on rubber in solution, is a strong, extensible, tear-resistant, moisture-resistant, oil-resistant material, marketed as **Pliofilm** in the form of tough transparent films for wrappers, packaging material, etc.

Cyclized rubber is formed by the action of certain agents, e.g., sulfonic acids and chlorostannic acid, on rubber, and is a thermoplastic, nonrubbery, tough or hard product. One form, **Thermoprene,** is used in the **Vulcalock process** for adhering rubber to metal, wood, and concrete, and in chemical-resistant paints. **Pliolite,** which has high resistance to many chemicals and has low permeability, is used in special paints, paper, and fabric coatings. **Marbon-B** has exceptional electrical properties and is valuable for insulation. **Hypalon** (chlorosulphonated polyethylene) is highly resistant to many important chemicals, notably ozone and concentrated sulfuric acid, for which other rubbers are unsuitable.

SOLVENTS

REFERENCES: Sax, "Dangerous Properties of Industrial Materials," Reinhold. Perry, "Chemical Engineers' Handbook," McGraw-Hill. Doolittle, "The Technology of Solvents and Plasticizers," Wiley. Riddick and Bunger, "Organic Solvents," Wiley. Mellan, "Industrial Solvents Handbook," Noges Data Corp., New Jersey. "1986 Annual Book of ASTM Standards," vol: 15.05 (Halogenated Organic Solvents). Current ASTM and ANSI Standards.

The use of solvents is quite widespread throughout industry. The health of personnel and the fire hazards involved should always be considered. Generally, solvents are **organic liquids** which vary greatly in solvent power, flammability, volatility, and toxicity.

Solvents for Polymeric Materials

A wide choice of solvents and solvent combinations is available for use with organic polymers in the manufacture of polymer-coated products and unsupported films. For a given polymer, the choice of solvent system is often critical in terms of solvent power, cost, safety, and evaporation rate. In such instances, the supplier of the base polymer should be consulted.

Alcohols

Methyl alcohol (methanol) is now made synthetically. It is completely miscible with water and most organic liquids. It evaporates rapidly and is a good solvent for dyes, gums, shellac, nitrocellulose, and some vegetable waxes. It is widely used as an antifreeze for automobiles, in shellac solutions, spirit varnishes, stain and paint removers. It is toxic; imbibition or prolonged breathing of the vapors can cause blindness. It should be used only in well-ventilated spaces. Flash point 11°C (52°F).

Ethyl alcohol (ethanol) is produced by fermentation and synthetically. For industrial use it is generally denatured and sold under various trade names. There are numerous formulations of specially denatured alcohols which can legally be used for specified purposes.

This compound is miscible with water and most organic solvents. It evaporates rapidly and, because of its solvent power, low cost, and agreeable odor, finds a wide range of uses. The common uses are antifreeze (see Freezing Preventives, above), shellac solvent, in mixed solvents, spirit varnishes, and solvent for dyes, oils, and animal greases.

Denatured alcohols are toxic when taken orally. Ethyl alcohol vapors when breathed in high concentration can produce the physiological effects of alcoholic liquors. It should be used in well-ventilated areas. Flash point 15.3 to 16.7°C (960 to 62°F).

Isopropyl alcohol (isopropanol) is derived mainly from petroleum gases. It is not as good a solvent as denatured alcohol, although it can be used as a substitute for ethyl alcohol in some instances. It is used as a rubbing alcohol and in lacquer thinners. Flash point is 11°C (52°F).

Butyl alcohol (normal butanol) is used extensively in lacquer and synthetic resin compositions and also in penetrating oils, metal cleaners, insect sprays, and paints for application over asphalt. It is an excellent blending agent for otherwise incompatible materials. Flash point is 29°C (84°F).

Esters

Ethyl acetate dissolves a large variety of materials, such as nitrocellulose oils, fats, gums, and resins. It is used extensively in nitrocellulose lacquers, candy coatings, food flavorings, and in chemical synthesis. On account of its high rate of evaporation, it finds a use in paper, leather, and cloth coatings and cements. Flash point −4.5°C (24°F).

Butyl acetate is the acetic-acid ester of normal butanol. This ester is used extensively for dissolving various cellulose esters, mineral and vegetable oils, and many synthetic resins, such as the vinylites, polystyrene, methyl methacrylate, and chlorinated rubber. It is also a good solvent for natural resins. It is the most important solvent used in lacquer manufacture. It is useful in the preparation of perfumes and synthetic flavors. Flash point 22.3°C (72°F).

Amyl acetate, sometimes known as banana oil, is used mainly in lacquers. Its properties are somewhat like those of butyl acetate. Flash point 21.1°C (70°F).

Hydrocarbons

Some industrial materials can be dangerous, toxic, and carcinogenic. The reader should become familiar with the references listed under Chlorinated Solvents in this section, which discuss the health hazards of industrial materials.

The **aromatic hydrocarbons** are derived from coal-tar distillates, the most common of which are benzene, toluene, xylene (also known as benzol, toluol, and xylol), and high-flash solvent naphtha.

Benzene is an excellent solvent for fats, vegetable and mineral oils, rubber, chlorinated rubber; it is also used as a solvent in paints, lacquers, inks, paint removers, asphalt, coal tar. This substance should be used with caution. Flash point 12°F.

Toluene General uses are about the same as benzol, in paints, lacquers, rubber solutions, and solvent extractions. Flash point 40°F.

Xylene is used in the manufacture of dyestuffs and other synthetic chemicals and as a solvent for paints, rubber, lacquer, and varnishes. Flash point 63 °F.

Hi-flash naphtha or **coal-tar naphtha** is used mainly as a diluent in lacquers, synthetic enamels, paints, and asphaltic coatings. Flash point 100°F.

Petroleum

These hydrocarbons, derived from petroleum, are, next to water, the cheapest and most universally used solvents.

V. M. &. P. naphtha, sometimes called **benzine**, is used by paint and varnish makers as a solvent or diluent. It finds wide use as a solvent for fats, oils, greases and is used as a diluent in paints and lacquers. It is also used as an extractive agent as well as in some specialized fields of cleansing (fat removal). It is used to compound rubber cement, inks, varnish removers. It is relatively nontoxic. Flash point 20 to 45 °F.

Mineral spirits, also called **Stoddard solvent**, is extensively used in dry cleaning because of its high flash point and clean evaporation. It is also widely used in turpentine substitutes for oil paints. Flash point 100 to 110°F.

Kerosine, a No. 1 fuel oil, is a good solvent for petroleum greases, oils, and fats. Flash point 100 to 165°F.

Chlorinated Solvents

Solvents can pose serious health and environmental problems, therefore the reader's attention is called to the following publications: (1) "Industrial Health and Safety; Mutagens and Carcinogens; Solution; Toxicology," vols. 1 to 3, Wiley. (2) "Patty's Industrial Hygiene and Toxicology," vol. 1 (General Principles), vol. 2 (Toxicology), vol. 3 (Industrial Hygiene Practices). (3) Peterson, "Industrial," Prentice-Hall. (4) Sax and Lewis, "Dangerous Properties of Industrial Materials," 7th ed., Van Nostrand Reinhold. Section 18 of this handbook.

The chlorinated hydrocarbons and fluorocarbon solvents are highly proscribed for use in industry and in society in general because of the inherent health and/or environmental problems ascribed to them. These materials are highly regulated and when used should be handled with utmost care and under close supervision and accountable control.

Carbon tetrachloride is a colorless nonflammable liquid with a chloroformlike odor. It is an excellent solvent for fats, oils, greases, waxes, and resins and was used in dry cleaning and in degreasing of wool, cotton waste, and glue. It is also used in rubber cements and adhesives, as an extracting agent, and in fire extinguishers. It should be used only in well-ventilated spaces; prolonged inhalation is extremely dangerous.

Trichlorethylene is somewhat similar to carbon tetrachloride but is slower in evaporation rate. It is the solvent most commonly used for vapor degreasing of metal parts. It is also used in the manufacture of dyestuffs and other chemicals. It is an excellent solvent and is used in some types of paints, varnishes, and leather coatings.

Tetrachlorethylene, also called **perchlorethylene**, is nonflammable and has uses similar to those of carbon tetrachloride and trichlorethylene. Its chief use is in dry cleaning; it is also used in metal degreasing.

Ketones

Acetone is an exceptionally active solvent for a wide variety of organic materials, gases, liquids, and solids. It is completely miscible with water and also with most of the organic liquids. It can also be used as a blending agent for otherwise immiscible liquids. It is used in the manufacture of pharmaceuticals, dyestuffs, lubricating compounds, and pyroxylin compositions. It is a good solvent for cellulose acetate, ethyl cellulose, vinyl and methacrylate resins, chlorinated rubber, asphalt, camphor, and various esters of cellulose, including smokeless powder, cordite, etc. Some of its more common uses are in paint and varnish removers, the storing of acetylene, and the dewaxing of lubricating oils. It is the basic material for the manufacture of iodoform and chloroform. It is also used as a denaturant for ethyl alcohol. Flash point 0°F.

Methylethylketone (MEK) can be used in many cases where acetone is used as a general solvent, e.g., in the formulation of pyroxylin cements and in compositions containing the various esters of cellulose. Flash point 30°F.

Glycol ethers are useful as solvents for cellulose esters, lacquers, varnishes, enamels, wood stains, dyestuffs, and pharmaceuticals.

THERMAL INSULATION

REFERENCES: "Guide and Data Book," ASHRAE. Glaser, "Aerodynamically Heated Structures," Prentice-Hall. Timmerhaus, "Advances in Cryogenic Engineering," Plenum. Wilkes, "Heat Insulation," Wiley. Wilson, "Industrial Thermal Insulation," McGraw-Hill. Technical Documentary Report ML-TDR-64–5, "Thermophysical Properties of Thermal Insulating Materials," Air Force Materials Laboratory, Research and Technology Division, Air Force Systems Command; Prepared by Midwest Research Institute, Kansas City, MO. Probert and Hub (eds.), "Thermal Insulation," Elsevier. Malloy, "Thermal Insulation," Van Nostrand Reinhold. "1986 Annual Book of ASTM Standards," vol. 04.06 (Thermal Insulation). ANSI/ASTM Standards. Current ASTM and ANSI Standards.

Thermal insulation consisting of a single material, a mixture of materials, or a composite structure is chosen to reduce heat flow. **Insulating effectiveness** is judged on the basis of thermal conductivity and depends on the physical and chemical structure of the material. The heat transferred through an insulation results from **solid conduction, gas conduction,** and **radiation.** Solid conduction is reduced by small particles or fibers in loose-fill insulation and by thin cell walls in foams. Gas conduction is reduced by providing large numbers of small pores (either interconnected or closed off from each other) of the order of the mean free paths of the gas molecules, by substituting gases of low thermal conductivity, or by evacuating the pores to a low pressure. Radiation is reduced by adding materials which absorb, reflect, or scatter radiant energy. (See also Sec. 4, Transmission of Heat by Conduction and Convection.)

The **performance** of insulations depends on the temperature of the bounding surfaces and their emittance, the insulation density, the type and pressure of gas within the pores, the moisture content, the thermal shock resistance, and the action of mechanical loads and vibrations. In **transient** applications, the heat capacity of the insulation (affecting the rate of heating or cooling) has to be considered.

The **form** of the **insulations** can be loose fill (bubbles, fibers, flakes, granules, powders), flexible (batting, blanket, felt, multilayer sheets, and tubular), rigid (block, board, brick, custom-molded, sheet and pipe covering), cemented, foamed-in-place, or sprayed.

The choice of **insulations** is dictated by the service-temperature range as well as by design criteria and economic considerations.

Cryogenic Temperatures [below $-102°C$ ($-150°F$)]

(See also Sec. 19.)

At the low temperatures experienced with cryogenic liquids, **evacuated multilayer insulations,** consisting of a series of radiation shields of high reflectivity separated by low-conductivity spacers, are effective materials. Radiation shield materials are aluminum foils or aluminized polyester films used in combination with spacers of thin polyester fiber or glass-fiber papers; radiation shields of crinkled, aluminized polyester film without spacers are also used. To be effective, multilayer insulations require a vacuum of at least 10^{-4} mmHg. **Evacuated powder** and **fiber insulations** can be effective at gas pressures up to 0.1 mmHg over a wide temperature range. **Powders** include colloidal silica (8×10^{-7} in particle diam), silica aerogel (1×10^{-6} in), synthetic calcium silicate (0.001 in), and perlite (an expanded form of glassy volcanic lava particles, 0.05 in diam). Powder insulations can be **opacified** with copper, aluminum, or carbon particles to reduce radiant energy transmission. **Fiber insulations** consist of mats of fibers arranged in ordered parallel layers either without binders or with a minimum of binders. Glass fibers (10^{-5} in diam) are used most frequently. For large process installations and cold boxes, unevacuated perlite powder or mineral fibers are useful.

Cellular glass (see Glass, this section) can be used for temperatures as low as $-450°F$ ($-268°C$) and has found use on liquefied natural gas tank bases.

Foamed organic plastics, using either fluorinated hydrocarbons or other gases as expanding agents, are partially evacuated when gases within the closed cells condense when exposed to low temperatures. Polystyrene and polyurethane foams are used frequently. **Gastight barriers** are required to prevent a rise in thermal conductivity with aging due to diffusion of air and moisture into the foam insulation. Gas barriers

are made of aluminum foil, polyester film, and polyester film laminated with aluminum foil.

Refrigeration, Heating, and Air Conditioning up to 120°C (250°F)

At temperatures associated with commercial refrigeration practice and building insulation, **vapor barriers** resistant to the diffusion of water vapor should be installed on the warm side of most types of insulations if the temperature within the insulation is expected to fall below the dew point (this condition would lead to condensation of water vapor within the insulation and result in a substantial decrease in insulating effectiveness). Vapor barriers include oil- or tar-impregnated paper, paper laminated with aluminum foil, and polyester films. Insulations which have an impervious outer skin or structure require a **vaportight sealant** at exposed joints to prevent collection of moisture or ice underneath the insulation. (See Secs. 12 and 19.)

Loose-fill insulations include powders and granules such as perlite, vermiculite (an expanded form of mica), silica aerogel, calcium silicate, expanded organic plastic beads, granulated cork (bark of the cork tree), granulated charcoal, redwood wool (fiberized bark of the redwood tree), and synthetic fibers. The most widely used fibers are those made of glass, rock, or slag produced by centrifugal attenuation or attenuation by hot gases.

Flexible or **blanket insulations** include those made from organically bonded glass fibers; rock wool, slag wool, macerated paper, or hair felt placed between or bonded to paper laminate (including vapor-barrier material) or burlap; foamed organic plastics in sheet and pad form (polyurethane, polyethylene); and elastomeric closed-cell foam in sheet, pad, or tube form.

Rigid or **board insulations** (obtainable in a wide range of densities and structural properties) include foamed organic plastics such as polystyrene (extended or molded beads); polyurethane, polyvinyl chloride, phenolics, and ureas; balsa wood, cellular glass, and corkboard (compressed mass of baked-cork particles).

Moderate Temperatures [up to 650°C (1200°F)]

The widest use of a large variety of insulations is in the temperature range associated with power plants and industrial equipment. Inorganic insulations are available for this temperature range, with several capable of operating over a wider temperature range.

Loose-fill insulations include diatomaceous silica (fossilized skeletons of microscopic organisms), perlite, vermiculite, and fibers of glass, rock, or slag. Board and blanket insulations of various shapes and degrees of flexibility and density include glass and mineral fibers, expanded perlite bonded with cellulose fiber and asphalt, organic bonded mineral fibers, and cellulose fiberboard.

Sprayed insulation (macerated paper or fibers and adhesive or frothed plastic foam), **insulating concrete** (concrete mixed with expanded perlite or vermiculite), and **foamed-in-place plastic insulation** (prepared by mixing polyurethane components, pouring the liquid mix into the void, and relying on action of generated gas or vaporization of a low boiling fluorocarbon to foam the liquid and fill the space to be insulated) are useful in special applications.

Reflective insulations form air spaces bounded by surfaces of high reflectivity to reduce the flow of radiant energy. Surfaces need not be mirror bright to reflect long-wavelength radiation emitted by objects below 500°F. Materials for reflective insulation include aluminum foil cemented to one or both sides of kraft paper and aluminum particles applied to the paper with adhesive. Where several reflective surfaces are used, they have to be separated during the installation to form airspaces.

Cellular glass can be used up to 900°F (482°C) alone and to 1,200°F (649°C) or higher when combined with other insulating materials and jackets.

High Temperatures [above 820°C (1,500°F)]

At the high temperatures associated with furnaces and process applications, physical and chemical stability of the insulation in an oxidizing, reducing, or neutral atmosphere or vacuum may be required. **Loose-fill insulations** include glass fibers 538°C (1,000°F) useful temperature limit, fibrous potassium titanate 1,040°C (1,900°F), alumina-silica fibers 1,260°C (2,300°F), microquartz fibers, 1,370°C (2,500°F), opacified colloidal alumina 1,310°C (2,400°F in vacuum), zirconia fibers 1,640°C (3,000°F), alumina bubbles 1,810°C (3,300°F), zirconia bubbles 2,360°C (4,300°F), and carbon and graphite fibers 2,480°C (4,500°F) in vacuum or an inert atmosphere.

Rigid insulations include reinforced and bonded colloidal silica 1,090°C (2,000°F), bonded diatomaceous earth brick 1,370°C (2,500°F), insulating firebrick (see Refractories, below), and anisotropic pyrolitic graphite (100:1 ratio of thermal conductivity parallel to surface and across thickness).

Reflective insulations, forming either an airspace or an evacuated chamber between spaced surfaces, include stainless steel, molybdenum, tantalum, or tungsten foils and sheets.

Insulating cements are based on asbestos, mineral, or refractory fibers bonded with mixtures of clay or sodium silicate. Lightweight, castable insulating materials consisting of mineral or refractory fibers in a calcium aluminate cement are useful up to 2,500°F.

Ablators are composite materials capable of withstanding high temperatures and high gas velocities for limited periods with minimum erosion by subliming and charring at controlled rates. Materials include asbestos, carbon, graphite, silica, nylon or glass fibers in a high-temperature resin matrix (epoxy or phenolic resin), and cork compositions.

SILICONES

(See also discussion in Sealants later.)

Silicones are organosilicon oxide polymers characterized by remarkable temperature stability, chemical inertness, waterproofness, and excellent dielectric properties. Among related products are the following:

Water repellents in the form of extremely thin films which can be formed on paper, cloth, ceramics, glass, plastics, leather, powders, or other surfaces. These have great value for the protection of delicate electrical equipment in moist atmospheres.

Oils with high flash points [above 315°C (600°F)], low pour points, −84.3°C (−120°F), and with a constancy of viscosity notably superior to petroleum products in the range from 260 to −73.3°C (500 to −100°F). These oil products are practically incombustible. They are in use for hydraulic servomotor fluids, damping fluids, dielectric liquids for transformers, heat-transfer mediums, etc., and are of special value in aircraft because of the rapid and extreme temperature variations to which aircraft are exposed. At present they are not available as lubricants except under light loads.

Greases and compounds for plug cocks, spark plugs, and ball bearings which must operate at extreme temperatures and speeds.

Varnishes and resins for use in electrical insulation where temperatures are high. Layers of glass cloth impregnated with or bonded by silicone resins withstand prolonged exposure to temperatures up to 260°C (500°F). They form paint finishes of great resistance to chemical agents and to moisture. They have many other industrial uses.

Silicone rubbers which retain their resiliency for −45 to 270°C (−50 to 520°F) but with much lower strength than some of the synthetic rubbers. They are used for shaft seals, oven gaskets, refrigerator gaskets, and vacuum gaskets.

REFRACTORIES

REFERENCES: Buell, "The Open-Hearth Furnace," 3 vols., Penton. *Bull.* R-2-E, The Babcock & Wilcox Co. "Refractories," General Refractories Co. Chesters, "Steel Plant Refractories," United Steel Companies, Ltd. "Modern Refractory Practice," Harbison Walker Refractories Co. Green and Stewart, "Ceramics: A Symposium," British Ceramic Soc. ASTM Standards on Refractory Materials (Committee C-8). Trinks, "Industrial Furnaces," vol. I, Wiley. Campbell, "High Temperature Technology," Wiley. Budnikov, "The Technology of Ceramics and Refractories," MIT Press. Campbell and Sherwood (eds.), "High-Temperature Materials and Technology," Wiley. Norton, "Refractories," 4th ed., McGrawHill. Clauss, "Engineer's Guide to High-Temperature Materials," Addison-Wesley.

Shaw, "Refractories and Their Uses," Wiley. Chesters, "Refractories; Production and Properties," The Iron and Steel Institute, London. "1986 Annual Book of ASTM Standards," vol. 15.01 (Refractories). ANSI Standards B74.10–B74.8. Current ASTM and ANSI Standards.

Types of Refractories

Fire-Clay Refractories Fire-clay brick is made from fire clays, which comprise all refractory clays that are not white burning. Fire clays can be divided into plastic clays and hard flint clays; they may also be classified as to alumina content.

Firebricks are usually made of a blended mixture of flint clays and plastic clays which is then formed, after mixing with water, to the required shape. Some or all of the flint clay may be replaced by highly burned or calcined clay, called **grog.** A large proportion of modern bricks are molded by the dry press or power press process where the forming is carried out under high pressure and with a low water content. Some extruded and hand-molded bricks are still made.

The dried bricks are burned in either periodic or tunnel kilns at temperatures varying between 1,200 and 1,480°C (2,200 and 2,700°F). Tunnel kilns give continuous production and a uniform temperature of burning.

Fire-clay bricks are used for boiler settings, kilns, malleable-iron furnaces, incinerators, and many portions of steel and nonferrous metal furnaces. They are resistant to spalling and stand up well under many slag conditions, but are not generally suitable for use with high lime slags, fluid-coal ash slags, or under severe load conditions.

High-alumina bricks are manufactured from raw materials rich in alumina, such as diaspore and bauxite. They are graded into groups with 50, 60, 70, 80, and 90 percent alumina content. When well fired, these bricks contain a large amount of mullite and less of the glassy phase than is present in firebricks. Corundum is also present in many of these bricks. High-alumina bricks are generally used for unusually severe temperature or load conditions. They are employed extensively in lime kilns and rotary cement kilns, the ports and regenerators of glass tanks, and for slag resistance in some metallurgical furnaces; their price is higher than that for firebrick.

Silica bricks are manufactured from crushed ganister rock containing about 97 to 98 percent silica. A bond consisting of 2 percent lime is used, and the bricks are fired in periodic kilns at temperatures of 1,480 to 1,540°C (2,700 to 2,800°F) for several days until a stable volume is obtained. They are especially valuable where good strength is required at high temperatures. Superduty silica bricks are finding some use in the steel industry. They have a lowered alumina content and often a lowered porosity.

Silica bricks are used extensively in coke ovens, the roofs and walls of open-hearth furnaces, in the roofs and sidewalls of glass tanks, and as linings of acid electric steel furnaces. Although silica brick is readily **spalled** (cracked by a temperature change) below red heat, it is very stable if the temperature is kept above this range, and for this reason stands up well in regenerative furnaces. Any structure of silica brick should be heated up slowly to the working temperature; a large structure often requires 2 weeks or more to bring up.

Magnesite bricks are made from crushed magnesium oxide which is produced by calcining raw magnesite rock to high temperatures. A rock containing several percent of iron oxide is preferable, as this permits the rock to be fired at a lower temperature than if pure materials were used. Magnesite bricks are generally fired at a comparatively high temperature in periodic or tunnel kilns, though large tonnages of unburned bricks are now produced. The latter are made with special grain sizing and a bond such as an oxychloride. A large proportion of magnesite brick made in this country uses raw material extracted from seawater.

Magnesite bricks are basic and are used whenever it is necessary to resist high lime slags, e.g., formerly in the basic open-hearth furnace. They also find use in furnaces for the lead and copper refining industry. The highly pressed unburned bricks find extensive use as linings for cement kilns. Magnesite bricks are not so resistant to spalling as fire-clay bricks.

Dolomite   This rock contains a mixture of $Mg(OH)_2$ and $Ca(OH_2)$, is calcined, and is used in granulated form for furnace bottoms.

Chrome bricks are manufactured in much the same way as magnesite bricks but are made from natural chromite ore. Commercial ores always contain magnesia and alumina. Unburned hydraulically pressed chrome bricks are also made.

Chrome bricks are very resistant to all types of slag. They are used as separators between acid and basic refractories, also in soaking pits and floors of forging furnaces. The unburned hydraulically pressed bricks now find extensive use in the walls of the open-hearth furnace and are often enclosed in a metal case. Chrome bricks are used in sulfite recovery furnaces and to some extent in the refining of nonferrous metals.

The **insulating firebrick** is a class of brick which consists of a highly porous fire clay or kaolin. They are lightweight (about one-half to one-sixth that of fireclay), low in thermal conductivity, and yet sufficiently resistant to temperature to be used successfully on the hot side of the furnace wall, thus permitting thin walls of low thermal conductivity and low heat content. The low heat content is particularly valuable in saving fuel and time on heating up, allows rapid changes in temperature to be made, and permits rapid cooling. These bricks are made in a variety of ways, such as mixing organic matter with the clay and later burning it out to form pores; or a bubble structure can be incorporated in the clay-water mixture which is later preserved in the fired brick. The insulating firebricks are classified into several groups according to the maximum use limit; the ranges are up to 872, 1,090, 1,260, 1,420, and above 1,540°C (1,600, 2,000, 2,300, 2,600, and above 2,800°F).

Insulating refractories are used mainly in the heat-treating industry for furnaces of the periodic type; the low heat content permits noteworthy fuel savings as compared with firebrick. They are also used extensively in stress-relieving furnaces, chemical-process furnaces, oil stills or heaters, and in the combustion chambers of domestic-oilburner furnaces. They usually have a life equal to the heavy bricks that they

replace. They are particularly suitable for constructing experimental or laboratory furnaces because they can be cut or machined readily to any shape. However, they are not resistant to fluid slag.

There are a number of types of **special brick,** obtainable from individual manufactories. High burned kaolin refractories are particularly valuable under conditions of severe temperature and heavy load, or severe spalling conditions, as in the case of high temperature oil-fired boiler settings, or piers under enameling furnaces. Another brick for the same uses is a high-fired brick of Missouri aluminous clay.

Bricks of **silicon carbon,** either nitride or clay bonded, have a high thermal conductivity and find use in muffle walls and as a slag-resisting material.

Other types of refractory that find certain limited use are **forsterite** and **zirconia.** Acid-resisting bricks consisting of a dense body like stoneware are used for lining tanks and conduits in the chemical industry. Carbon blocks are used as linings for the crucibles of blast furnaces.

The **chemical composition** of some of the refractories is given in Table 6.8.13. The **physical properties** are given in Table 6.8.14. Reference should be made to ASTM Standards for details of standard tests, and to ANSI Standards for further specifications and properties.

Standard and Special Shapes

There are a large number of **standard refractory shapes** carried in stock by most manufacturers. Their catalogs should be consulted in selecting these shapes, but the common ones are shown in Table 6.8.15. These shapes have been standardized by the American Refractories Institute and by the Bureau of Simplification of the U.S. Department of Commerce.

Regenerator tile sizes $a \times b \times c$ are: 18×6 or 9×3; 18×9 or 12×4; $22\frac{1}{2} \times 6$ or 9×3; $22\frac{1}{2} \times 9$ or 12×4; $27 \times 9 \times 3$; 27×9 or 12×4; $31\frac{1}{2} \times 12 \times 4$; $36 \times 12 \times 4$.

Table 6.8.13   Chemical Composition of Typical Refractories*

												Resistance to:			
No.	Refractory type	SiO_2	Al_2O_3	Fe_2O_3	TiO_2	CaO	MgO	Cr_2O_3	SiC	Alkalies	Siliceous steel-slag	High-lime steel-slag	Fused mill-scale	Coal-ash slag
1	Alumina (fused)	8–10	85–90	1–1.5	1.5–2.2	...	....	..	......	0.8–1.3‡	E	G	F	G
2	Chrome	6	23	15†	....	...	17	38	......	....	G	E	E	G
3	Chrome (unburned)	5	18	12†	....	...	32	30	......	....	G	E	E	G
4	Fireclay (high-heat duty)	50–57	36–42	1.5–2.5	1.5–2.5	...	....	..	......	1–3.5‡	F	P	P	F
5	Fireclay (super-duty)	52	43	1	2	...	....	..	......	2‡	F	P	F	F
6	Forsterite	34.6	0.9	7.0	....	1.3	55.4							
7	High-alumina	22–26	68–72	1–1.5	3.5	...	....	..	......	1–1.5‡	G	F	F	F
8	Kaolin	52	45.4	0.6	1.7	0.1	0.2	..	......	....	F	P	G§	F
9	Magnesite	3	2	6	....	3	86	..	......	....	P	E	E	E
10	Magnesite (un-burned)	5	7.5	8.5	....	2	64	10	......	....	P	E	E	E
11	Magnesite (fused)	......	....	....	....	...	....	..	......	....	F	E	E	E
12	Refractory porcelain	25–70	25–60	....	....	...	....	..	......	1–5	G	F	F	F
13	Silica	96	1	1	....	2	....	..	......	....	E	P	F	P
14	Silicon carbide (clay-bonded)	7–9	2–4	0.3–1	1	...	....	..	85–90	....	E	G	F	E
15	Sillimanite (mullite)	35	62	0.5	1.5	...	....	..	......	0.5§	G	F	F	F
16	Insulating firebrick (2,600°F)	57.7	36.8	2.4	1.5	0.6	0.5	..	......	....	P	P	G¶	F

E = excellent. G = good. F = fair. P = poor.
* Many of these data have been taken from a table prepared by Trostel, *Chem. Met. Eng.,* Nov. 1938.
† As FeO.
‡ Includes lime and magnesia.
§ Excellent if left above 1,200°F.
¶ Oxidizing atmosphere.

Table 6.8.14   Physical Properties of Typical Refractories*
(Refractory numbers refer to Table 6.8.13)

Refractory no.	Fusion point		Deformation under load, % at °F and lb/in²	Spalling resistance	Reheat shrinkage after 5 h, % at (°F)	Wt. of straight 9-in brick, lb
	°F	Pyrometric cone				
1	3,390+	39+	1 at 2,730 and 50	Good	+0.5(2,910)	9–10.6
2	3,580+	41+	Shears 2,740 and 28	Poor	−0.5 to 1.0 (3,000)	11.0
3	3,580+	41+	Shears 2,955 and 28	Fair	−0.5 to 1.0(3,000)	11.3
4	3,060–3,170	31–33	2.5–10 at 2,460 and 25	Good	±0 to 1.5 (2,550)	7.5
5	3,170–3,200	33–34	2–4 at 2,640 and 25	Excellent	±0 to 1.5 (2,910)	8.5
6	3,430	40	10 at 2,950	Fair	. . . .	9.0
7	3,290	36	1–4 at 2,640 and 25	Excellent	−2 to 4 (2,910)	7.5
8	3,200	34	0.5 at 2,640 and 25	Excellent	−0.7 to 1.0 (2,910)	7.7
9	3,580+	41+	Shears 2,765 and 28	Poor	−1 to 2 (3,000)	10.0
10	3,580+	41+	Shears 2,940 and 28	Fair	−0.5 to 1.5 (3,000)	10.7
11	3,580+	41+		Fair	. . . .	10.5
12	2,640–3,000	16–30		Good		
13	3,060–3,090	31–32	Shears 2,900 and 25	Poor†	+0.5 to 0.8 (2,640)	6.5
14	3,390	39	0–1 at 2,730 and 50	Excellent	+2‡ (2,910)	8–9.3
15	3,310–3,340	37–38	0–0.5 at 2,640 and 25	Excellent	−0 to 0.8 (2,910)	8.5
16	2,980–3,000	29–30	0.3 at 2,200 and 10	Good	−0.2 (2,600)	2.25

Refractory no.	Porosity	Specific heat at 60–1,200°F	Mean coefficient of thermal expansion from 60°F shrinkage point × 10⁵	Mean thermal conductivity, Btu/(ft² · h · °F · in)						
				Mean temperatures between hot and cold face, °F						
				200	400	800	1,200	1,600	2,000	2,400
1	20–26	0.20	0.43	. . .	20	22	24	27	30	32
2	20–26	0.20	0.56	. . .	8	9	10	11	12	12
3	10–12	0.21								
4	15–25	0.23	0.25–0.30	5	6	7	8	10	11	12
5	12–15	0.23	0.25–0.30	6	7	8	9	10	12	13
6	23–26	0.25								
7	28–36	0.23	0.24	6	7	8	9	10	12	13
8	18	0.22	0.23	. . .	. . . .	11	12	13	13	14
9	20–26	0.27	0.56–0.83	. . .	40	35	30	27	26	25
10	10–12	0.26								
11	20–30	0.27	0.56–0.80							
12	. . . .	0.23	0.30	. . .	14	15	17	18	19	20
13	20–30	0.23	0.46§	. . .	8	10	12	13	14	15
14	13–28	0.20	0.24	. . .	. . . .	100	80	65	55	50
15	20–25	0.23	0.30	. . .	10	11	12	13	14	15
16	75	0.22	0.25	. . .	1.6	2.0	2.6	3.2	3.8	

* Many of these data have been taken from a table prepared by Trostel, *Chem. Met. Eng.*, Nov. 1938.
† Excellent if left above 1,200°F
‡ Oxidizing atmosphere.
§ Up to 0.56 at red heat.

The following **arch, wedge,** and **key bricks** have maximum dimensions, $a \times b \times c$ of $9 \times 4\frac{1}{2} \times 2\frac{1}{2}$ in. The minimum dimensions a', b', c', are as noted: no. 1 arch, $c' = 2\frac{1}{8}$; no. 2 arch, $c' = 1\frac{3}{4}$; no. 3 arch, $c' = 1$; no. 1 wedge, $c' = 1\frac{7}{8}$; no. 2 wedge, $c' = 1\frac{1}{2}$ no. 3 wedge, $c' = 2$; no. 1 key, $b' = 4$; no. 2 key, $b' = 3\frac{1}{2}$; no. 3 key, $b' = 3$; no. 4 key, $b' = 2\frac{1}{4}$; edge skew, $b' = 1\frac{1}{2}$; feather edge, $c' = \frac{1}{8}$; no. 1 neck, $a' =$ $3\frac{1}{2}$, $c' = \frac{5}{8}$; no. 2 neck, $a' = 12\frac{1}{2}$; $c = \frac{5}{8}$; no. 3 neck, $a' = 0$, $c' = \frac{5}{8}$; end skew, $a' = 6\frac{3}{4}$; side skew, $b' = 2\frac{1}{4}$; jamb brick, $9 \times 2\frac{1}{2}$; bung arch, $c' = 2\frac{3}{8}$.

Special shapes are more expensive than the standard refractories, and, as they are usually hand-molded, will not be so dense or uniform in structure as the regular brick. When special shapes are necessary, they should be laid out as simply as possible and the maximum size should be kept down below 30 in if possible. It is also desirable to make all special shapes with the vertical dimension as an even multiple of 2½ in plus one joint so that they will bond in with the rest of the brickwork.

Refractory Mortars, Coatings, Plastics, Castables, and Ramming Mixtures

Practically all brickwork is laid up with some type of jointing material to give a more stable structure and to seal the joints. This material may be ground fire clay or a specially prepared mortar containing grog to reduce the shrinkage. The bonding mortars may be divided into three general classes. The first are **air-setting mortars** which often contain chemical or organic binder to give a strong bond when dried or fired at comparatively low temperatures. Many of the air-setting mortars should not be used at extremely high temperatures because the fluxing action of the air-setting ingredient reduces the fusion point. The second class is called **heat-setting mortar** and requires temperatures of over 1,090°C (2,000°F) to produce a good bond. These mortars vary in

Table 6.8.15   Shapes of Firebricks

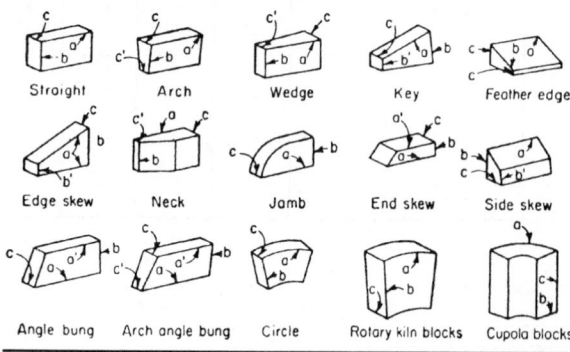

Straight Arch Wedge Key Feather edge

Edge skew Neck Jamb End skew Side skew

Angle bung Arch angle bung Circle Rotary kiln blocks Cupola blocks vitrifying point, some producing a strong bond in the lower temperature ranges, and the others requiring very high temperatures to give good strength. The third classification comprises **special base mortars** such as silica, magnesite, silicon carbide, or chrome, which are specially blended for use with their respective bricks. The chrome-base mortar may be satisfactorily used with fire-clay bricks in many cases.

The refractory bonding mortars should preferably be selected on the advice of the manufacturer of the refractory to obtain good service, although there are a considerable number of independent manufacturers of mortars who supply an excellent product. From 1,330 to 1,778 N (300 to 400 lb) of dry mortar per 1,000 bricks is required for thin joints, which are desirable in most furnace construction. For thicker trowel joints, up to 2,220 N (500 lb) per 1,000 bricks is required. In the case of chrome base mortars, 2,660 N (600 lb) per 1,000 bricks should be allowed, and for magnesite cement 3,550 N (800 lb) per 1,000 bricks.

Coatings are used to protect the hot surface of the refractories, especially when they are exposed to dust-laden gases or slags. These coatings usually consist of ground grog and fireclay of a somewhat coarser texture than the mortar. There are also chrome-base coatings which are quite resistant to slags, and in a few cases natural clays containing silica and feldspar are satisfactory.

The coatings can be applied to the surface of the brickwork with a brush in thin layers about 1/16 in thick, or they may be sprayed on with a cement gun, the latter method generally giving the best results. Some types of coating can be put on in much thicker layers, but care should be taken to assure that the coating selected will fit the particular brick used; otherwise it is apt to peel off in service.

Plastics and ramming mixtures are generally a mixture of fire clay and coarse grog of somewhat the same composition as the original fire-clay brick. They are used in repairing furnace walls which have been damaged by spalling or slag erosion, and also for making complete furnace walls in certain installations such as small boiler furnaces. They are also used to form special or irregular shapes, in temporary wooden forms, in the actual furnace construction.

Some of the plastics and ramming mixtures contain silicate of soda and are air-setting, so that a strong structure is produced as soon as the material is dry. Others have as a base chrome ore or silicon carbide, which make a mixture having a high thermal conductivity and a good resistance to slag erosion. These mixtures are often used in the water walls of large boiler furnaces; they are rammed around the tubes and held in place by small studs welded to the tube walls. The chrome plastic has been used with good success for heating-furnace floors and sub-hearths of open-hearth furnaces.

Castable mixes are a refractory concrete usually containing high-alumina cement to give the setting properties. These find considerable use in forming intricate furnace parts in wooden molds; large structures have been satisfactorily cast by this method. This type of mixture is much used for baffles in boilers where it can be cast in place around the tubes. Lightweight castables with good insulating properties are used to line furnace doors.

Furnace Walls

The modern tendency in furnace construction is to make a comparatively thin wall, anchored and supported at frequent intervals by castings or heat-resisting alloys which, in turn, are held by a structural framework and does not rest on the base. The wall may be made of heavy refractories backed up with insulating material, or of insulating refractory. Table 6.8.16 gives heat losses and heat contents of a number of wall combinations and may enable designers to pick out a wall section to suit their purpose. Solid walls built with standard 9-in brick are made up in various ways, but the hot face has usually four header courses and one stretcher course alternating.

Table 6.8.16   Transmitted Heat Losses and Heat-Storage Capacities of Wall Structures under Equilibrium Conditions
(Based on still air at 80°F)

	Thickness, in	Hot face temperature, °F									
		1,200		1,600		2,000		2,400		2,800	
Wall	Of insulating refractory and firebrick	HL	HS	HL	HS	HL	HS	HL	HS	HL	HS
4½	4½ 20	355	1,600	537	2,300	755	2,900				
	4½ 28	441	2,200	658	3,100	932	4,000	1,241	4,900	1,589	5,900
	4½ FB	1,180	8,400	1,870	11,700	2,660	14,800	3,600	18,100	4,640	21,600
7	4½ 28 + 2½ 20	265	3,500	408	4,900	567	6,500	751	8,100	970	9,800
	4½ FB	423	12,500	660	17,700	917	23,000	1,248	28,200		
9	4½ 28 + 4½ 20	203	4,100	311	5,900	432	7,900	573	9,900	738	12,200
	4½ FB + 4½ 20	285	13,700	437	19,200	615	24,800				
	9 20	181	3,100	280	4,300	395	5,500				
	9 28	233	4,100	349	5,800	480	7,500	642	9,300	818	11,100
	9 FB	658	15,800	1,015	21,600	1,430	27,600	1,900	34,000	2,480	40,300
11½	9 28 + 2½ 20	169	5,700	260	8,000	364	10,500	484	13,100	623	15,800
	9 FB + 2½ 20	335	22,300	514	31,400	718	40,600	962	50,400	1,233	60,300
	9 28 + 4½ 20	143	6,500	217	9,300	305	12,300	404	15,300	514	18,700
	9 FB + 4½ 20	241	24,100	367	34,500	516	44,800	690	55,100		
13½	9 20 + 4½ FB	165	5,300	255	7,300	348	9,900				
	9 28 + 4½ FB	200	6,900	302	9,700	415	12,600	556	15,700	710	19,100
	13½ FB	452	22,300	700	31,000	980	39,900	1,310	49,100	1,683	58,300
16	13½ FB + 2½ 20	275	31,200	423	43,300	588	56,300	780	70,000	994	84,200
18	9 20 + 9 FB	147	8,500	225	11,900	319	15,700				
	9 28 + 9 FB	175	10,700	266	15,100	375	19,700	493	24,600	635	29,800
	13½ FB + 4½ 20	210	34,100	318	48,400	440	62,600	587	77,500	753	92,600
	18 FB	355	28,800	532	40,300	745	52,200	1,000	64,200	1,283	76,500
20½	18 FB + 2½ 20	234	39,000	356	55,400	500	72,000	665	89,200	847	107,000
22½	18 FB + 4½ 20	182	43,200	281	61,000	392	79,200	519	97,700	667	117,600
	22½ FB	287	36,000	435	49,500	612	64,100	814	78,800	1,040	93,400

Conversion factors: $t_c = \frac{5}{9}(t_F - 32)$; 1 in = 0.0254 m.
HL = heat loss in Btu/(ft²)/(h). HS = heat storage capacity in Btu/ft². 20 = 2,000°F insulating refractory. 28 = 2,800°F insulating refractory. FB = fireclay brick.
SOURCE: Condensed from "B & W Insulating Firebrick" Bulletin of The Babcock & Wilcox Co.

Many modern furnaces are constructed with air-cooled walls, with refractory blocks held in place against a casing by alloy steel holders. Sectional walls made up with steel panels having lightweight insulating refractories attached to the inner surface are also used and are especially valuable for use in the upper parts of large boiler furnaces, oil stills, and similar types of construction. The sections can be made up at the plant and shipped as a unit. They have the advantage of low cost because of the light ironwork required to support them.

Many failures in furnace construction result from improper expansion joints. **Expansion joints** should usually be installed at least every 10 ft, although in some low-temperature structures the spacing may be greater. For high-temperature construction, the expansion joint allowance per foot in inches should be as follows: fire clay, $\frac{1}{16}$ to $\frac{3}{32}$; high alumina, $\frac{3}{32}$ to $\frac{1}{8}$; silica, $\frac{1}{8}$ to $\frac{3}{16}$; magnesite, $\frac{1}{4}$; chrome, $\frac{5}{32}$; forsterite, $\frac{1}{4}$. Corrugated paper is often used in the joints.

The **roof** of the furnace is usually either a sprung arch or a suspended arch. A **sprung arch** is generally made of standard shapes using an inside radius equal to the total span. In most cases, it is necessary to build a form on which the arch is sprung.

In the case of arches with a considerable rise, it has been found that an inverted catenary shape is better than a circular shape for stability, and it is possible to run the sidewalls of the furnace right down to the floor in one continuous arch with almost complete elimination of the ironwork. The catenary can be readily laid out by hanging a flexible chain from two points of a vertical wall.

The **suspended arch** is used when it is desirable to have a flat roof (curved suspended arches are also made); it presents certain advantages in construction and repair but is more difficult to insulate than the sprung arch. Special suspended arch shapes are commercially available. The insulating refractory is suited to this type of construction because the steel supports are light and the heat loss is low.

Selection of Refractories

The selection of the most suitable refractory for a given purpose demands experience in furnace construction. A brick that costs twice as much as another brand and gives twice the life is preferable since the total cost includes the laying cost. Furthermore, a brick that gives longer service reduces the shutdown period of the furnace. Where slag or abrasion is severe, brick with a dense structure is desirable. If spalling conditions are important, a brick with a more flexible structure is better, although there are cases where a very dense structure gives better spalling resistance than a more open one.

High-lime slag can be taken care of with magnesite, chrome, or high-alumina brick, but if severe temperature fluctuations are encountered also, no brick will give long life. For coal-ash slag, dense fire-clay bricks give fairly good service if the temperature is not high. At the higher temperatures, a chrome-plastic or silicon carbide refractory often proves successful. When the conditions are unusually severe, air- or water-cooled walls must be resorted to; the water-cooled stud-tube wall has been very successful in boiler furnaces.

With a general freedom from slag, it is often most economical to use an insulating refractory. Although this brick may cost more per unit, it allows thinner walls, so that the total construction cost may be no greater than the regular brick. The substitution of insulating refractory for heavy brick in periodic furnaces has sometimes halved the fuel consumption.

The stability of a refractory installation depends largely on the bricklaying. The total cost, in addition to the bricks, of laying brick varies with the type of construction, locality, and refractory.

Recent Developments in Refractories

Pure-oxide refractories have been developed to permit fabrication of parts such as tubes, crucibles, and special shapes. Alumina (Al_2O_3) is the most readily formed into nonporous pieces and, up to its softening point of 2,040°C (3,690°F), is most useful. Mullite ($2SiO_2 : 3Al_2O_3$), softening at 1,820°C (3,290°F), is used for thermocouple protection tubes, crucibles, and other small pieces. Magnesium oxide (MgO), fusing at 2,800°C (5,070°F), is resistant to metals and slags. Zirconia (ZrO_2), softening at 4,600°F, is very sensitive to temperature changes but can be stabilized with a few percent of lime. Beryllium oxide (BeO), softening at 2,570°C (4,660°F), has a very high thermal conductivity but must be fabricated with great care because of health hazards. Thoria (ThO_2) softens at 3,040°C (5,520°F) and has been used in crucibles for melting active metals and as a potential nuclear fuel.

Refractory carbides, sulfides, borides, silicides, and nitrides have been developed for special uses. Many have high softening points, but all have limited stability in an oxidizing atmosphere. Silicon carbide (SiC) is the most used because of its high thermal and electrical conductivity, its resistance against certain slags, and its relatively good stability in air. Molybdenum silicide ($MoSi_2$) also has considerable resistance to oxidation and, like SiC, can be used for metal-melting crucibles. Cerium sulfide (CeS_2) is a metallic-appearing material of high softening point but no resistance to oxidation. Zirconium nitride (ZrN) and titanium nitride (TiN) are also metalliclike but are not stable when heated in air. Graphite has valuable and well-known refractory properties but is not resistant to oxidation.

Refractory fibers are coming into use quite extensively. Fibers of silica-alumina glass have a use limit of about 1,090°C (2,000°F). They are used for insulating blankets, expansion joints, and other high temperature insulation. Development of higher-temperature fibers is being carried out on a small scale for use as high temperature insulation or mechanical reenforcement.

Nuclear fuels of uranium, thorium, and plutonium oxides or carbides are now extensively used in high temperature reactors.

Space vehicles are using nozzles of refractories of various kinds to withstand the high temperatures and erosion. Nose cones of sintered alumina are now used extensively because of their excellent refractory and electrical properties. Heat shields to protect space vehicles upon reentry are an important use of special refractories.

Physical Properties of High-Purity Refractories

In Table 6.8.17 are shown the properties of some of the more important pure refractory materials. It should be realized that, as purer materials become available and testing methods become more refined, some of these values will change.

SEALANTS

REFERENCES: Damusis (ed.), "Sealants," Reinhold. Flick, "Adhesives and Sealant Compound Formulations," 2d ed., Noyes Publ. *Mech. Eng.*, April 1991. *Engr. News Record*, Nov. 12, 1987; 1972–1993 "Annual Book of ASTM Standards." Baldwin, Selecting High Performance Building Sealants, *Plant Eng.*, Jan. 1976, pp. 58–62. Panek and Cook, "Construction Sealants and Adhesives," 2d ed, Wiley-Interscience. Panek (ed.), Building Seals and Sealants, *ASTM Spec. Tech. Publ.* 606, American Society for Testing Materials, Philadelphia. Klosowski, "Sealants in Construction," Marcel Dekker. Current ASTM and ANSI Standards.

Klosowski's book is an excellent source of information on sealants and serves as an excellent reference. The reader should treat the following material as a guide and should consult the references and manufacturers' catalogs for expanded coverage.

Classifying sealants is somewhat arbitrary, depending as such on a multitude of properties. One of these properties tends to be more important than the rest, and that is *the ability to take cyclic movement*. Low movement ability together with short useful lifetimes tends to be found for the lower-cost sealants, and ability to take cyclic movement along with long useful lifetimes is found for the higher-cost sealants. The remaining key properties are adhesion/cohesion, hardness, modulus, stress relaxation, compression set, and weather resistance. Table 6.8.18 provides a rough comparison of sealant classes.

Key Properties

Adhesion/Cohesion A sealant must stick or adhere to the joint materials to prevent fluid penetration. The sealant must also stick or cohere to itself internally so that it does not split apart and thus allow leakage to occur. The ASTM C-719 test method is recommended for crucial designs and applications.

Table 6.8.17   Physical Properties of Some Dense,* Pure Refractories

Material	Modulus of rupture, 10^3 lb/in^2 at 70°F	Modulus of rupture, 10^3 lb/in^2 at 1,800°F	Modulus of elasticity, 10^6 lb/in^2 at 70°F	Fusion point, °F	Linear coef of expansion, 10^{-6} in/(in · °F) between 65 and 1,800°F	Thermal conductivity, Btu/(ft^2 · h · °F · in), at 212°F	Thermal conductivity, Btu/(ft^2 · h · °F · in), at 1,800°F	Thermal stress resistance
Al$_2$O$_3$	100	60	53	3,690	5.0	210	55	Good
BeO	20	10	45	4,660	4.9	1,450	130	Very good
MgO	14	12	31	5,070	7.5	240	47	Poor
ThO$_2$	12	7	21	5,520	5.0	62	20	Fair
ZrO$_2$	20	15	22	4,600	5.5	15	15	Fair
UO$_2$	12		25	5,070	5.6	58	20	Fair
SiC	24	24	68	5,000†	2.2	390	145	Excel.
BC	50	40	42	4,440	2.5	200	145	Good
BN	7	1	12	5,000	2.6	150	130	Good
MoSi$_2$	100	40	50	3,890	5.1	220	100	Good
C	3	4	2	7,000	2.2	870	290	Good

Conversion factors: 1 lb/in^2 = 6,894.8 N/m$^2 \cdot t_c = \frac{5}{9}(t_F - 32)$. 1 Btu/(ft^2 · h · °F · in) = 225 W/(m^2 · °C · m). 1 Btu/(lbm · °F) = 4,190 kg · °C.
* Porosity, 0 to 5 percent.
† Stabilized.
SOURCES: Norton, "Refractories," 3d ed.; Green and Stewart, "Ceramics: A Symposium"; Ryschkewitsch, "Oxydekramik, der Finstuffstemme," Springer; Campbell, "High Temperature Technology"; Kingery, "Property Measurements at High Temperatures," Wiley.

Hardness   The resistance of a sealant to a blunt probe penetration on a durometer is a measure of its hardness. Any change in this hardness over time will alert the user to check the sealant's useful performance span.

Modulus   The springiness or elastic quality of a sealant is defined by the ratio of force (stress) needed to unit elongation (strain). A high-modulus sealant can exert a rather high force on a joint, so that for weak or marginal substrates it becomes a decided disadvantage. For instance, concrete joints, having rather low tensile strength, will perform poorly with high-modulus sealants.

Stress-Relaxation   Under stress some sealants relax internally over time and stretch; an extreme example is bubble gum. Certain sealants have internal polymer chains that exhibit stress relaxation over time; mastics behave likewise, and low modulus sealants are more likely to do so than are high-modulus sealants.

A small amount of relaxation is useful in continuously tensed joints since it lowers the bond line force. In moving joints, however, a high degree of stress relaxation is problematical because the sealant tends to recover its original shape slowly and incompletely. Such joints tend to pump out their sealants.

Compression Set   If a sealant is unable to reexpand to its original shape after it is compressed, it has experienced **compression set;** i.e., the compressed shape becomes permanent. When the joint reopens, the sealant must sustain high internal and bonding stresses which can cause the sealant to tear off the joint and/or cause internal tearing in the sealant itself as well as possible surface failure of the joint substrate. Polysulfide sealants are prone to such failure. Urethane sealants under combined conditions of joint movement and weathering will exhibit such failure.

Table 6.8.18   Sealant Classes by Movement Capabilities*

Range or movement capability	Sealant types	Comments
Low—near or at 0% of joint movement	1. Oil-based 2. Resin-based 3. Resinous caulks 4. Bituminous-based mastics 5. Polybutene-based 6. PVA (vinyl) latex	Short service life Low cost Major component usually low-cost mineral fillers. For example, putty is mostly finely ground chalk mixed with oil to form a doughlike entity. Some recent ones may contain about 0.2–2% silicone
Medium—0–12½% of joint width	1. Butyls 2. Latex acrylics 3. Neoprenes 4. Solvent-release acrylics	Longer service life Medium cost For protected environments and low movement, service life is perhaps 10–15 years. Typical is 3–10 years Common disadvantage is shrinkage, which may be near 30% in some products. Plasticized types tend to discolor walls
High—greater than 12½% of joint width	1. Polysulfides 2. Urethanes 3. Silicones 4. Proprietary modifications of above	Long service life Higher cost Some advertised as allowing movements of ±25 to ±50%, or + 100 to −50% Generally used in commercial jobs Expanding market into do-it-yourself and over-the-counter trade

SOURCE: Adapted from Klosowski, "Sealants in Construction," Dekker.

Resistance to Weathering   Construction sealants are designed to resist weathering. The nature of the polymer system in silicones provides that resistance, while urethane and polysulfides achieve equivalent resistance only by the use of heavy filler loads. Screening from the sun is important. Heavy filler loads or chemical sunscreens help stop radiation penetration to the internal polymer. Deep joints and opaque joint materials help sealants weather successfully.

Sealant Types

Oil-Based Caulks   Combining a drying oil (such as linseed) with mineral fillers will result in a doughlike material (putty) that can be cooled into a joint. The oils dry and/or oxidize and in about 24 h develop sufficient skin to receive paint. Movement ability is about ±2½ percent of the joint. To prevent porous substrates from wicking in the caulk's oil, a primer should be applied.

Butyl Sealants (Almost Totally Cured Systems Dispersed in a Solvent)   Chemically combining isobutylene and isoprene results in a butyl rubber. End-product variations are gotten by varying both the proportion of starting ingredients and the polymer chain length. Carbon black serves as a reinforcer and stabilizer in the final product. Chlorbutyl rubber is a similar sealant.

Butyl sealants are available as solvent-release caulks, soft deformable types, more rigid gaskets, and hot melts. The better butyls have movement capabilities of ±12½ percent. For applications needing 20 to 30 percent compression, butyl tapes serve well and are used extensively in glazing applications.

Advantages of butyl sealants include moderate cost, good water resistance (not for immersion), and good adhesion without primers.

Disadvantages include poor extension recovery, limited joint movement capability, and odor plus stickiness during application. They also pick up dirt and cause staining. They tend to soften under hot sun conditions such as found in cars with closed windows.

Acrylic Latex (Almost Fully Reacted in a Cartridge)   Acrylic polymers are made by using a surfactant, water, and a catalyst plus appropriate monomers, which results in a high molecular weight polymer, coated with surfactant and dispersed in water. The addition of fillers, plasticizers, and other additives such as silanes for adhesion and ethyl glycol for freeze/thaw stability completes the product. Once extruded, it dries rapidly and can be painted in 30 to 50 min. Movement capability lies between ±7½ and ±12½ percent of joint width.

Advantages include excellent weathering, ease of application, ease of cleanup, low odor, low toxicity, no flammability during cure, and ability to be applied to damp substrates.

Disadvantages include the need for a primer for concrete, most wood, and plastics where joints will move and cause sealant stress. Since they harden when cold and soften when hot, they are not recommended for extremes of temperature cycling. They are not recommended for below-grade, underwater, or chronically damp applications.

Solvent Acrylic Sealants (Solvent Release Sealants) (Almost Totally Cured Systems Dispersed in a Solvent)   This sealant is an acrylic polymer of short chain length, and it is based on an alkyl ester of acrylic and methacrylic acids with various modifications. The system does not quite cure to be a true elastomer, winding up somewhat more like a dry-feeling mastic, and so it exhibits stress relaxation. The best can tolerate ±12½ percent of joint movement, with the more typical between ±8 percent and ±10 percent.

Advantages include excellent adhesion without priming to most substrates (and, by some claims, through dirt and light oils and perhaps even damp surfaces), good weathering, and moderate cost.

Disadvantages include the need to heat the cartridge or dispensing container for easy application in cool weather, offensive odor during cure (thus requiring ventilation in restricted quarters), and slow recovery from either extension or contraction. The cured seal gets very hard in cold weather and loses flexibility, thus compromising movement capability.

Two-Part Polysulfide Sealants   These sealants are the pioneers of high-range products with life expectancies (in certain environments) of up to 20 years and movement capability of up to ±25 percent of joint width. It is widely used in curtain wall construction and in concrete/masonry applications. In glazing applications shielded from direct sunlight, polysulfide sealant is suitable. Since polysulfides also can be produced with excellent solvent resistance, they are used as fuel tank sealants and in other fuel contact applications. Fillers are generally carbon blacks, clays, and mineral materials. Curing agents may be lead peroxide, cyclic amides, diisocyanates, etc. Silanes act as adhesion promoters and stabilizers. Most polysulfides are mixed and dispensed on the job at the time of use.

Advantages include good extensibility, recovery, cure with two-package systems (tack-free in about 36 to 48 h and about 1 week for full cure), adhesion, and resistance to weathering and aging.

Disadvantages include the need for site mixing, requirements for primers on porous surfaces, sun exposure cracking, and compression set effects which emerge in about 5 to 10 years.

One-Part Polysulfide Sealants   These sealants are almost on a par in cured performance to two-part polysulfide sealants. They need no mixing at the time of use and are thus ready to apply. Curing depends upon moisture or oxygen from the air and is relatively slow, requiring days, weeks, or more, after which they become tack-free. Dry and cold weather can lengthen the curing process.

Two-Part Polyurethane Sealants   These sealants are second-generation premium products, and they outperform the polysulfides. They weather well, perform well in expansion joints by tolerating large movement (±25 percent of joint movement), and are tough and resilient. The top of the product line can achieve 85 to 90 percent recovery from compression set. Adhesion is very good. They stay clean for almost the life of the sealant, or about 10 to 20 years. Urethanes tend to be water-sensitive and to bubble if they contact water at the curing surface, and they tend to stiffen with age. Urethane sealants are widely used in construction and for nonglazing joints in walkways and pedestrian traffic areas.

One-Part Urethane Sealants   These sealants are comparable to two-part urethanes, but come in a single package ready to use. Package stability is problematical because of moisture sensitivity. One-part systems take a long time to cure, especially at low temperatures and low humidities.

Silicone Sealants   In some aspects of performance, these sealants represent third-generation products, and others are at least second-generation in quality. Only silicone polymers comprise a true silicone sealant. They generally contain mineral or other inorganic fillers along with functional silane or siloxane cross-crosslinker, with special additives included for specific purposes.

In terms of temperature exposure, silicone polymers are about one-tenth as sensitive as typical hydrocarbons or polyether chains. Thus they extrude easily at both low and high temperatures. Temperature stability ranges from about −40 to 250°F and for some to more than 400°F. Once cured, they maintain elasticity very well at both temperature extremes and weather quite well (lasting 12 to 20 times longer than typical organic sealants, as indicated by weatherometer tests). Warranties often extend from 20 to 50 years. Joint movement performance embraces a wide range from ±12 to +100/−50 percent of joint width. Adhesion qualities are excellent, allowing some common silicones unprimed application to most substrates (including concrete), and some are applicable for service under conditions of total submersion in water.

These performance levels have made silicone sealants the industry standard to which others are compared. Competitive products marketed using terms such as *siliconized, siliconelike, modified with silicone, modified silicone,* and so forth contain a minimal amount of true silicone, usually from 0.1 to 10 percent. Most silicones are fully elastic and exhibit the smallest compression or tension set, ranging from 85 to 99 percent recovery.

Some silicones pick up dirt to varying degrees, but owing to their relatively fast curing times (tack-free in 15 min to 3 h), they are less prone to do so when compared with polyurethane and polysulfides, which have longer cure times. After curing, however, because of their relatively soft surface, silicones tend to accumulate dirt faster. Unfortunately, the dirt cannot be completely washed off. To obtain the advantage of silicone's durability and to avoid dirt pickup, one resorts to overcoating the silicone sealant with a hard silicone resin. Alternately, one can dust the surface of the uncured, tacky silicone with

Table 6.8.19 Summary of Sealant Properties*

Type of sealant	Low performance: Oil-based, One-part	Medium performance: Latex (acrylic), One-part	Medium performance: Butyl skinning, One-part	Medium performance: Acrylic Solvent-release	Polysulfide One-part	Polysulfide Two-part	Urethane One-part	Urethane Two-part	Silicone One-part	Silicone Two-part
Movement ability in percent of joint width (recommended maximum joint movement)	±3	±5, ±2½	±7.5	±10–12½	±25	±25	±25	±25	±25,+100/−50	±12½–±50
Life expectancy, years†	2–10	2–10	5–15	5–20	10–20	10–20	10–20	10–20	10–50	10–50
Service temperature range, °F (°C)	−20 to +150 (−29 to +66)	−20 to +180 (−20 to +82)	−40 to +80 (−40 to +82)	−20 to +180 (−29 to +82)	−40 to +180 (−40 to +82)	−60 to +180 (−51 to +82)	−40 to +180 (−40 to +82)	−25 to +180 (−32 to +82)	−65 to +400 (−54 to +200)	−65 to +400 (−54 to +200)
Recommended application temperature range, °F‡	40–120	40–120	40–120	40–180	40–120	40–120	40–120	40–180	−20 to +160	−20 to +160
Cure time§ to a tack-free condition, h	6	½–1	24	36	24§	36–48§	12–36	24	1–3	½–2
Cure time§ to specified performance, days	Continues	5	Continues	14	30–45	7	8–21	3–5	5–14	¼–3
Shrinkage, %	5	20	20	10–15	8–12	0–10	0–5	0–5	0–5	0–5
Hardness, new (1–6 mo), A scale at 75°F		15–40	10–30	10–25	20–40	20–45	20–45	10–45	15–40	15–40
Hardness, old (5 yr), A scale at 75°F		30–45	30–50	30–55	30–55	20–55	30–55	20–60	15–40	15–50
Resistance to extension at low temperature	Low to moderate	Moderate to high	Moderate to high	High	Low to high	Low to moderate	Low to high	Low to high	Low	Low
Primer required for sealant bond to: Masonry	No	Yes	No	No	Yes	Yes	Yes	Yes	No	No
Metal	No	Sometimes	No	No	Yes	Yes	No	No	?	?
Glass	No	No	No	No	No	No	No	No	No	No
Applicable specifications: United States	TT-C-00593b	ASTM TTS-00230	ASTM TT-S-001657	ASTM C-920 ITS-00230	ASTM C-920 TT-C-00230C	ASTM C-920 TTS-00227E	ASTM C-920 TTS-00230C	ASTM C-920 TTS-00227C	ASTM C-920 TTS-00230C TTS-001543A	ASTM C-920 TTS-001543A TTS-00230C
Canada	CAN 2-19.2M		CGSB 19-GP-14M	CGSB 19-GP-SM	CAN 2-19.13M		CAN 2-19.13M		CAN 2-19.18M	

* Data from manufacturer's data sheets; U.S.-made sealants are generally considered.
† Affected by conditions of exposure.
‡ Some sealants may require heating in low temperatures.
§ Affected by temperature and humidity.
¶ The wide range in performance of the various types of silicone sealants is discussed briefly in the text preceding this table.
SOURCE: Adapted from J. M. Klosowski, "Sealants in Constructions," Marcel Dekker, 1989, by permission.

powdered chalk or some similar material; this can be used as a base upon which to apply paint. Some silicone sealants are formulated to be paintable, but, in general, silicones simply will not accept paint. Silicones abrade easily and are not used in heavily trafficked areas. Silicones which release acetic acid should not be used on marble, galvanized metal, copper, cementaceous substances, or other corrodable materials. Silicones which release amines and the like must not be used on copper and for electrical applications. Neutral-cure systems are available which can be safely used on almost all substrates. The reader is directed to consult the supplier when in doubt about compatibility.

Water-Based Silicones   Water-based silicones combine the durability and longevity of silicones with ease of cleanup, ease of application, and paintability. They are true silicone systems, being dispersions of polymers, crosslinkers, and catalysts in water.

Table 6.8.19 gives a summary of sealant properties.

Miscellaneous

Hot Pours   Several categories of materials are softened by heating and so can be injected or poured into joints (almost exclusively horizontal). The majority of these materials are filled out with bituminous substances such as asphalt tars, coal tars, etc. Such bituminous bases can be blended with urethanes, polysulfides, polyvinyl chlorides, etc. Highway joint sealing constitutes the largest use of hot-pour sealants. They are inexpensive and easy to use, but deteriorate rapidly in typical U.S. climate and suffer surface crazing and lose elasticity (stiffen) in cold weather.

Forever Tacky   Nonhardening sealants, such as Hylomar, remain tacky for years and retain their original sealing qualities very well. They were originally developed for jet-engine joints in the 1950s, but have been applied to resist extreme vibrations and to make emergency repairs. Continued improvements in this product are directed toward broadening its compatibility with a wider range of chemicals and oils. Its stable, tacky condition for long periods enables easy disassembly of parts it has bonded.

Formed-in-Place Gasket   Replacing die-cut gaskets can reduce costs, and for this reason, formed-in-place gaskets have been developed. One such product, Dynafoam, a curable thermoplastic elastomer, forms in-place gaskets by pouring the sealant into the desired shape. Automotive applications include taillight assemblies and sunroof gaskets to seal out moisture, wind, and dust. This type of sealant is designed to fill the niche between traditional hot-melt adhesives and gaskets, such as butyl rubbers or ethyl vinyl acetates, and curable silicone and urethane sealants. Curing time is within a few minutes, and they resist softening up to about 280°F (continuous exposure) and up to 400°F (for periods of 1 h or less).

Foamed-in-Place Sealants   In commercial operations, a sealant such as Dynafoam is heated to 180°F, pumped through a heated hose into a pressurized, dry gas chamber (usually nitrogen), and pumped further through a heated hose and dispensing nozzle. Upon exiting the nozzle, the sealant expands to bead size as the entrained nitrogen expands. The wormlike bead is then applied directly to a part and forms a gasket. The sealant begins curing upon contact with air.

Similar materials in small pressurized canisters contain a room-temperature curing sealant. Upon release, the sealant expands into a foam, fills the joint, and cures in place. It is applied easily to provide local insulation, to fill crevices (e.g., to block wind), and so forth. For extensive operations of this kind, this material is supplied in much larger, yet portable, containers.

6.9   CEMENT, MORTAR, AND CONCRETE
by William L. Gamble

REFERENCES: Neville, "Properties of Concrete," Wiley. Sahlin, "Structural Masonry," Prentice-Hall. Mindless and Young, "Concrete," Prentice-Hall. "BOCA National Building Code," Building Officials and Code Administrators International. "Concrete Manual," U.S. Bureau of Reclamation. ACI 318, "Building Code Requirements for Structural Concrete"; ACI 211.1, "Standard Practice for Selecting Proportions for Normal, Heavyweight, and Mass Concrete"; ACI 304, "Recommended Practice for Measuring, Mixing, Transporting, and Placing Concrete"; American Concrete Institute, Detroit. "Building Code Requirements for Masonry Structures (ACI 530/TMS 402/ASCE 5)," American Concrete Institute, The Masonry Society, and American Society of Civil Engineers.

CEMENT

Normal portland cement is used for concrete, for reinforced concrete, and either with or without lime, for mortar and stucco. It is made from a mixture of about 80 percent carbonate of lime (limestone, chalk, or marl) and about 20 percent clay (in the form of clay, shale, or slag). After being intimately mixed, the materials are finely ground by a wet or dry process and then calcined in kilns to a clinker. When cool, this clinker is ground to a fine powder. During the grinding, a small amount of gypsum is usually added to regulate the setting of the cement. The chemical analysis of 32 American type I cements gives the following average percentage composition: silica (SiO_2), 21.92; alumina (Al_2O_3), 6.91; iron oxide (Fe_2O_3), 2.91; calcium oxide (CaO), 62.92; magnesium oxide (MgO), 2.54; sulfuric oxide (SO_3), 1.72; alkalies (R_2O), 0.82; loss on ignition, 1.50, insoluble residue 0.20.

Types and Kinds of Cements   Five types of portland cements are covered by ASTM specification C150.

Normal portland cement, type I, is used for purposes for which another type having special properties is not required. Most structures, pavements, and reservoirs are built with type I cement.

Modified portland cement, type II, generates less heat from its hydration and is more resistant to sulfate attacks than type I. This cement is used in structures having large cross sections, such as large abutments and heavy retaining walls. It may also be used in drainage where a moderate sulfate concentration exists.

High-early-strength portland cement, type III, is used when high strengths are required in a few days. Use of high-early-strengths will allow earlier removal of forms and shorter periods of curing.

Low-heat portland cement, type IV, generates less heat during hydration than type II and is used for mass concrete construction such as large dams where large temperature increases would create special problems. Type IV cement gains strength more slowly than type I. The tricalcium aluminate content is limited to 7 percent.

Sulfate-resisting portland cement, type V, is a special cement, not readily available, to be used when concrete is exposed to severe sulfate attack. Type V cements gain strength more slowly than type I cement. The tricalcium aluminate content is limited to a maximum of 5 percent.

Air-entraining portland cements purposely cause air, in minute, closely spaced bubbles, to occur in concrete. Entrained air makes the concrete more resistant to the effects of repeated freezing and thawing and of the deicing agents used on pavements. To obtain such cements, air-entraining agents are interground with the cement clinker during manufacture. Types I to III can be obtained as air-entraining cements and are then designated as types IA, IIA, and IIIA, under ASTM C150.

Portland blast-furnace slag cements are made by grinding granulated high-quality slag with portland-cement clinker. Portland blast-furnace slag cement type IS and air-entraining portland blast-furnace slag cement type IS-A are covered by ASTM specification C595. Provisions are also made for moderate-heat-of-hydration cements (MH) and moderate-sulfate-resistance cements (MS), or both (MH-MS).

Type IS cements initially gain strength more slowly but have about the same 28-day strength as type I cements.

White portland cement is used for architectural and ornamental work because of its white color. It is high in alumina and contains less than 0.5 percent iron. The best brands are true portlands in composition.

Portland-pozzolan cement is a blended cement made by intergrinding portland cement and pozzolanic materials. Two types, type IP (portland-pozzolan cement) and type IP-A (air-entraining portland-pozzolan cement), are covered in ASTM specification C595.

Masonry cement, ASTM specification C91, is a blended cement used in place of job cement-lime mixtures to reduce the number of materials handled and to improve the uniformity of the mortar. These cements are made by combining either natural or portland cements with fattening materials such as hydrated lime and, sometimes, with air-entraining admixtures.

Waterproofed cement is sometimes used where a waterproof or water-repellent concrete or mortar is particularly desirable. It is cement ground with certain soaps and oils. The effectiveness is limited to 3 or 4 ft of water pressure.

Shrinkage-compensated cements are special portland cements which expand slightly during the moist curing period, compensating for the shrinkage accompanying later drying. They are used primarily to aid in producing crack-free concrete floor slabs and other members. ASTM C845 covers these cements.

Regulated-set cements are special portland cements formulated to set in very short times, producing usable concrete strengths in regulated times of as little as 1 h or less. Such concretes are obviously well suited to repair work done when it is important to minimize downtime.

Portland Cement Tests   Cement should be tested for all but unimportant work. Tests should be made in accordance with the standard specifications of the ASTM or with the federal specifications where they apply. Samples should be taken at the mill, and tests completed before shipments are made. When this is not possible, samples should be taken at random from sound packages, one from every 10 bbl or 40 bags, and mixed. The total sample should weigh about 6 lb. ASTM requirements for standard portland cements are given in ASTM specification C150.

The **autoclave soundness test** consists of determining the expansion of a 1 in sq neat cement bar 10 in long which, after 24-h storage in 90 percent or greater humidity, is placed in an autoclave, where the pressure is raised to 295 lb/in^2 in about 1 h, maintained for 3 h, and then brought back to normal in $1\frac{1}{2}$ h. Cements that show over 1 percent expansion may show unsoundness after some years of service; the ASTM allows a maximum of 0.80 percent.

Time of Setting   Initial set should not be less than 45 min when Vicat needle is used or 60 min when Gilmore needle is used. Final set should be within 10 h. Cement paste must remain plastic long enough to be properly placed and yet submit to finishing operations in a reasonable time.

Compressive Strength   Minimum requirements for average compressive strength of not less than three 2-in cubes composed of 1 part (by weight) cement and 2.75 parts standard graded mortar sand tested in accordance with ASTM method C109, are shown in Table 6.9.1.

LIME

Common lime, or **quicklime,** when slaked or hydrated, is used for interior plastering and for lime mortar. Mixed with cement, it is used for lime and cement mortar and for stucco. Mortars made with lime alone are not satisfactory for thick walls because of slow-setting qualities. They must never be used under water. Quicklime slakes rapidly with water with much heat evolution, forming calcium hydrate (CaH_2O_2). With proper addition of water, it becomes plastic, and the volume of putty obtained is 2 or 3 times the loose volume of the lime before slaking, and its weight is about $2\frac{1}{2}$ times the weight of the lime. Plastic lime sets by drying, by crystallization of calcic hydrate, and by absorbing carbonic acid from the air. The process of hardening is very slow. Popping is likely to occur in plaster unless the lime is sound, as indicated by an autoclave test at 120 lb/in^2 pressure for 2 h.

Magnesium lime, used for the same purposes as common or high-calcium lime, contains more than 20 percent magnesium oxide. It slakes more slowly, evolves less heat, expands less, sets more rapidly, and produces higher-strength mortars than does high-calcium quicklime.

Pulverized and granulated limes slake completely much more quickly than ordinary lump lime. They are sometimes waterproofed by the addition of stearates and other compounds similar to those used in cement for the same purpose. The waterproofing treatment retards the slaking.

Hydrated lime is a finely divided white powder manufactured by slaking quicklime with the requisite amount of water. It has the advantage over lime slaked on the job of giving a more uniform product, free from unslaked lime. It does not have plasticity or water retention equal to freshly slaked quicklime.

Hydraulic hydrated lime is used for blending with portland cement and as a masonry cement. It is the hydrated product of calcined impure limestone which contains enough silica and alumina to permit the formation of calcium silicates.

Quicklime is covered by ASTM C5 and hydrated limes by ASTM C6, C206, and C821. The testing methods are covered by other ASTM specifications referenced in the above specifications. ASTM C5 contains instructions and cautions for slaking quicklime, as the chemical reactions following the addition of water to quicklime are exothermic and potentially dangerous.

AGGREGATES

Sand

Sand to be used for mortar, plaster, and concrete should consist of clean, hard, uncoated grains free from organic matter, vegetable loam, alkali, or other deleterious substances. Vegetable or organic impurities are particularly harmful. A quantity of vegetable matter so small that it cannot be detected by the eye may render a sand absolutely unfit for use with cement. Stone screenings, slag, or other hard inert material may be substituted for or mixed with sand. Sand for concrete should range in size from fine to coarse, with not less than 95 percent passing a no. 4 sieve, not less than 10 percent retained on a no. 50 sieve, and not more than 5 percent (or 8 percent, if screenings) passing a no. 100 sieve. A straight-line gradation on a graph, with

Table 6.9.1   Minimum Requirements for Average Compressive Strength (without Air Entrainment)

Age of test, days	Storage of test pieces	Compressive strength, lb/in^2				
		Normal	Moderate heat	High early strength	Low heat	Sulfate-resistant
1	1 day moist air	. . . .	. . . .	1,740	. . . .	. . . .
3	1 day moist air, 2 days water	1,740	1,450	3,480	. . . .	1,160
7	1 day moist air, 6 days water	2,760	2,470	. . . .	1,020	2,180
28	1 day moist air, 27 days water	. . . .	. . . .	. . . .	2,470	3,050

1,000 lb/in^2 = 6.895 MPa.
SOURCE: Reprinted from ASTM C150, with permission from ASTM.

Table 6.9.2

Sieve no.	200	100	50	30	16	8	4
Sieve opening, in	0.0029	0.0059	0.0117	0.0234	0.0469	0.0937	0.187
Wire diam, in	0.0021	0.0043	0.0085	0.0154	0.0256	0.0394	0.0606
Sieve size, in		⅜	¾	1	1½	2	3
Sieve opening, in		0.375	0.750	1.00	1.50	2.00	3.00
Wire diam, in		0.0894	0.1299	0.1496	0.1807	0.1988	0.2283

1 in = 25.4 mm

percentages passing plotted as ordinates to normal scale and sieve openings as abscissas to logarithmic scale, gives excellent results.

The grading of sand for mortar depends upon the width of joint, but normally not less than 95 percent should pass a no. 8 sieve, and it should grade uniformly from coarse to fine without more than 8 percent passing a no. 100 sieve.

Sand for plaster should have at least 90 percent passing a no. 8 sieve and not more than 5 percent passing the no. 100 sieve.

Silt or clayey material passing a no. 200 sieve in excess of 2 percent is objectionable.

Test of Sand   Sand for use in important concrete structures should always be tested. The strength of concrete and mortar depends to a large degree upon the quality of the sand and the coarseness and relative coarseness of the grains. Sand or other fine aggregate when made into a mortar of 1 part portland cement to 3 parts fine aggregate by weight should show a tensile strength at least equal to the strength of 1:3 mortar of the same consistency made with the same cement and standard sand. If the aggregate is of poor quality, the proportion of cement in the mortar or concrete should be increased to secure the desired strength. If the strength is less than 90 percent that of Ottawa sand mortar, the aggregate should be rejected unless compression tests of concrete made with selected aggregates pass the requirements. The standard Ottawa sand gradation is described in ASTM specification C109. This sand is supplied by the U.S. Silica Co., Ottawa, IL. The compressive strength of 2-in cubes made from a cement and sand mixture with a 0.9 water-cement ratio and a flow of 100 percent should equal 90 percent of the strength of similar cubes made with graded Ottawa sand.

The ASTM standard test (C40) for the presence of injurious organic compounds in natural sands for cement mortar or concrete is as follows: A 12-oz graduated glass prescription bottle is filled to the 4½-oz mark with the sand to be tested. A 3 percent solution of sodium hydroxide (NaOH) in water is then added until the volume of sand and liquid, after shaking, gives a total volume of 7 liquid oz. The bottle is stoppered, shaken thoroughly, and then allowed to stand for 24 h. A standard-reference-color solution of potassium dichromate in sulfuric acid is prepared as directed in ASTM D154. The color of the clear liquid above the sand is then compared with the standard-color solution; if the liquid is darker than the standard color, further tests of the sand should be made before it is used in mortar or concrete. The standard color is similar to light amber.

Coarse Aggregate

Broken Stone and Gravel   Coarse aggregate for concrete may consist of broken stone, gravel, slag, or other hard inert material with similar characteristics. The particles should be clean, hard, durable, and free from vegetable or organic matter, alkali, or other deleterious matter and should range in size from material retained on the no. 4 sieve to the coarsest size permissible for the structure. For reinforced concrete and small masses of unreinforced concrete, the maximum size should be that which will readily pass around the reinforcement and fill all parts of the forms. Either 1- or 1½-in diam is apt to be the maximum. For heavy mass work, the maximum size may run up to 3 in or larger.

The coarse aggregate selected should have a good performance record, and especially should not have a record of alkali-aggregate reaction which may affect opaline and chert rocks.

Lightweight aggregates are usually pumice, lava, slag, burned clay or shale, or cinders from coal and coke. It is recommended that light-weight

fine aggregate not be used in conjunction with lightweight coarse aggregate unless it can be demonstrated, from either previous performance or suitable tests, that the particular combination of aggregates results in concrete that is free from soundness and durability problems. In case of doubt, the concrete mix should be designed using sand fine aggregate, and lightweight coarse aggregate. Their application is largely for concrete units and floor slabs where saving in weight is important and where special thermal insulation or acoustical properties are desired.

Heavyweight aggregates are generally iron or other metal punchings, ferrophosphate, hematite, magnetite, barite, limenite, and similar heavy stones and rocks. They are used in concrete for counterweights, dry docks, and shielding against rays from nuclear reactions.

Fineness Modulus   The fineness modulus, which is used in the Abrams method as an index of the characteristics of the aggregates, is the sum of the cumulative percentages (divided by 100) which would be retained by all the sieves in a special sieve analysis. The sieves used in this method are nos. 100, 50, 30, 16, 8, and 4 for fine aggregates and these plus the ⅜-, ¾-, 1½-, and 3-in sizes for coarse aggregates. A high fineness modulus indicates a relatively low surface area because the particles are relatively large, which means less water required and, therefore, a higher concrete strength. Aggregates of widely different gradation may have the same fineness modulus.

ASTM standard sieves for analysis of aggregates for concrete have the sizes of opening and wire shown in Table 6.9.2.

WATER

Water for concrete or mortar should be clean and free from oil, acid, alkali, organic matter, or other deleterious substance. Cubes or briquettes made with it should show strength equal to those made with distilled water. Water fit for drinking is normally satisfactory for use with cement. However, many waters not suitable for drinking may be suitable for concrete. Water with less than 2,000 ppm of total dissolved solids can usually be used safely for making concrete. (See Sec. 6.10, Water.)

Seawater can be used as mixing water for plain concrete, although 28-day strength may be lower than for normal concrete. If seawater is used in reinforced concrete, care must be taken to provide adequate cover with a dense air-entrained concrete to minimize risks of corrosion. Seawater should not be used with prestressed concrete.

ADMIXTURES

Admixtures are substances, other than the normal ingredients, added to mortars or concrete for altering the normal properties so as to improve them for a particular purpose. Admixtures are frequently used to entrain air, increase workability, accelerate or retard setting, provide a pozzolanic reaction with lime, reduce shrinkage, and reduce bleeding. However, before using an admixture, consideration must be given to its effect on properties other than the one which is being improved. Most important is consideration of possible changes in the basic mix which might make the admixture unnecessary. Particular care must be used when using two or more admixtures in the same concrete, such as a retarding agent plus an air-entraining agent, to ensure that the materials are compatible with each other when mixed in concrete. The properties of chemical admixtures should meet ASTM C494 and air-entraining agents should meet ASTM C260.

Air-entraining agents constitute one of the most important groups of admixtures. They entrain air in small, closely spaced, separated bubbles in the concrete, greatly improving resistance to freezing and thawing and to deicing agents.

Accelerators are used to decrease the setting time and increase early strength. They permit shorter curing periods, earlier form removal, and placing at lower temperatures. Calcium chloride is the most frequently used accelerator and can be used in amounts up to 2 percent of the weight of the cement, but must never be used with prestressed concrete.

Retarders increase the setting time. They are particularly useful in hot weather and in grouting operations.

Water reducers are used to increase the workability of concrete without an accompanying increase in the water content. The most recent development is the use of high-range water reducers, or superplasticizers, which are polymer liquid materials added to the concrete in the mixer. These materials lead to spectacular temporary increases in slump for a given water content, and may be used to obtain several different results. A normal mix can be transformed into a "flowing" concrete which will practically level itself. Or the addition of the superplasticizer can allow the mix to be redesigned for the same consistency at the time of mixing, but considerably less water will be required and consequently a higher concrete strength can be reached without adding cement, or the same strength can be reached with a lower cement content.

Fly ash from coal-burning power plants can be added to concrete mixes to achieve several effects. The very fine material tends to act as a lubricant to the wet concrete, and thus may be added primarily to increase the workability. Fly ash also enters the chemical reactions involved in the setting of concrete and leads to higher strengths. Fly ash is often used in mass concrete such as dams as a replacement for part of the cement, in order to save some material costs, to reduce the rate at which the hydration produces heat, and to increase the long-term strength gain potential of the concrete. Fly ash is a pozzolan, and should meet requirements of ASTM C618. Free carbon from incomplete combustion must be strictly limited.

Silica Fume Silica fume, also called **condensed silica fume** or **microsilica,** is another pozzolan which may be added to concrete as a supplement or partial replacement to the cement. This material reacts with the lime in the cement and helps lead to very high-strength, low-permeability concrete. The silica-lime reaction can occur very quickly, and retarders are often necessary in concretes containing silica fume. Silica fume should meet the requirements of ASTM C 1240. Concretes containing silica fume tend to be quite dark gray compared with ordinary concrete mixes, and concretes containing fly ash tend to be lighter than the usual concrete gray.

Ground Granulated Blast-Furnace Slag Ground granulated blast-furnace slag has been used in Europe as a supplement to or partial replacement for portland cement for several decades but has become available in North America relatively recently. Concretes made with this material tend to develop very low permeabilities, which improves durability by retarding oxygen and chloride penetration, which in turn protects the reinforcement from corrosion for longer periods. Chloride sources include both deicing chemicals and seawater, and these concretes have been successfully used to resist both these severe environments. They are also sulfate-resistant, which is important in those geographic areas which have sulfate-bearing groundwater. Ground granulated blast-furnace slag should meet the requirements of ASTM C 989.

Other admixtures may be classed as gas-forming agents, pozzolanic materials, curing aids, water-repelling agents, and coloring agents.

MORTARS

Properties desirable in a mortar include (1) good plasticity or workability, (2) low volume change or volume change of the same character as the units bonded, (3) low absorption, (4) low solubility and thus freedom from efflorescence, (5) good strength in bond and ample strength to withstand applied loads, (6) high resistance to weathering.

Mortar Types There are several different types of mortar which are suitable for masonry construction of different kinds, uses, and exposure conditions. The BOCA National Building Code lists several mortar types, and their permitted uses are given in Table 6.9.3. The makeup of these mortars is specified in terms of volumes of materials, as listed in Table 6.9.4, from ASTM C 270. There is considerable leeway given for

Table 6.9.3   Masonry and Mortar Types

Type of masonry	Type of mortar permitted
Masonry in contact with earth	M or S
Grouted and filled cell masonry	M or S
Masonry above grade or interior masonry	
Piers of solid units	M, S, or N
Piers of hollow units	M or S
Walls of solid units	M, S, N, or O
Walls of hollow units	M, S, or N
Cavity walls and masonry bonded hollow walls	
Design wind pressure exceeds 20 lb/ft²	M or S
Design wind pressure 20 lb/ft² or less	M, S, or N
Glass block masonry	S or N
Non-load-bearing partitions and fireproofing	M, S, N, O, or gypsum
Firebrick	Refractory air-setting mortar
Linings of existing masonry, above or below grade	M or S
Masonry other than above	M, S, or N

1 lb/ft² = 47.9 Pa.
SOURCE: BOCA National Building Code.

Table 6.9.4   Mortar Proportions Specification Requirements (Parts by Volume)

Mortar type	Portland cement	Masonry cement	Hydrated lime or lime putty Min	Hydrated lime or lime putty Max	Damp loose aggregate
M	1			¼	
	1	1			
S	1		¼	½	Not less than 2¼ and not more than 3 times the sum of the volumes of the cements and lime used
	½	1			
N	1		½	1¼	
		1			
O		1			
	1		1¼	2½	

SOURCE: ASTM C 270.

the proportions, and the materials are mixed with water to the consistency desired by the mason. The portland cement, masonry cement, and lime may be purchased separately or blended, and for small jobs a dry mix mortar containing everything except the water may be purchased in bags. The hydrated lime tends to give the fresh mortar plasticity, or stickiness, which is necessary for the proper bedding of the masonry or concrete units being laid.

Additional mortar types may be used for reinforced masonry, in which steel bars are used to increase the strength in the same way that concrete is reinforced. Other mortars may be used for filling the cavities in concrete block construction, or concrete masonry unit construction, and this mortar may be referred to as grout.

Design procedures and requirements for concrete and clay masonry are contained in ACI 530/TMS 402/ASCE 5. **Unreinforced masonry** cannot be relied on to resist tension stresses. Compressive stresses permitted under the empirical design rules vary widely, depending on the strength and type of masonry unit. The highest stress permitted is 350 lb/in^2 for high-strength solid bricks, while the lowest is 60 lb/in^2 for the lowest-strength hollow-concrete blocks, where the stresses are computed on the gross area and the units are laid with type M or S mortar. Type N mortar leads to stresses which are 10 to 15 percent lower.

Mortars for Plastering and Stucco   Interior plastering is much less common than it was in the past because of the use of gypsum board products for walls, but is done in various instances. Common lime plaster consists of hydrated lime (usually purchased in prepared, bagged form), clean coarse sand, and hair or fiber. The plastering is normally done in two or three layers, with the base coats containing about equal volumes of lime and sand, plus hair or fiber, and the final layer containing less sand. In three-layer work, the first layer is the scratch coat, the second the brown coat, and the last the white or skim coat. The scratch coat is applied directly to either a masonry wall or the lath in a frame wall.

Skim coat is a finish coat composed of lime putty and fine white sand. It is placed in two layers and troweled to a hard finish. **Gaged skim coat** is skimming mixed with a certain amount of plaster of paris, which makes it a hard finish. **Hard finish** consists of 1 part lime putty to 1 or 2 parts plaster of paris.

Keene's cement, which is an anhydrous calcined gypsum with an accelerator, is much used as a hard-finish plaster.

Gypsum plaster (ASTM C28) lacks the plasticity and sand-carrying capacity of lime plaster but is widely used because of its more rapid hardening and drying and because of the uniformity obtainable as the result of its being put up in bags ready-mixed for use.

Gypsum ready-mixed plaster should contain not more than 3 ft^3 of mineral aggregate per 100 lb of calcined gypsum plaster, to which may be added fiber and material to control setting time and workability. Gypsum neat plaster used in place of sanded plaster for second coat should contain at least 66 percent $CaSO_4 : \frac{1}{2}H_2O$; the remainder may be fiber and retarders. Calcined gypsum for finishing coat may be white or gray. If it contains no retarder, it should set between 20 and 40 min; if retarded, it should set between 40 min and 6 h.

Cement plaster is used where a very hard or strong plaster is required, e.g., for thin metal-lath partitions or as a fire protection. It should contain not more than 2 parts sand, by dry and loose volume, to 1 part portland cement. Lime putty or hydrated lime is added up to 15 percent by volume of the cement.

Under moisture conditions where the plaster will not dry rapidly, **curing** at temperatures above 60 and below 90°F is absolutely necessary if cracking is to be avoided.

Stucco   Stucco is used for exterior plastering and is applied to brick or stone or is plastered onto wood or metal lath. For covering wooden buildings, the stucco is plastered either on wood lath or on metal lath in three coats, using mortar similar to that for brick or stone. Concrete in northern climates exposed to frost should never be plastered but should be finished by rubbing down with carborundum brick or similar tool when the surface is comparatively green. It may also be tooled in various ways. Whenever stucco is used, extreme care must be taken to get a good bond to the supporting surface. While stucco has traditionally been applied with a trowel, it can also be applied pneumatically. Similar

material used for lining metal pipes can be applied with a centrifugal device which "slings" the material onto the pipe wall.

For three-coat work on masonry or wood lath, the first or scratch coat should average ¼ in thick outside the lath or surface of the brick. The thickness of the second coat should be ⅜ to ½ in, while the finish coat should be thin, i.e., ⅛ in or not more than ¼ in. The second coat should generally be applied 24 h after the first or scratch coat. The finish coat should not be applied in less than 1 week after the second coat. Proportions of mix for all coats may be ⅕ part hydrated lime, 1 part portland cement, 3 parts fairly coarse sand, measured by volume. Stucco work should not be put on in freezing weather and must be kept moist for at least 7 days after application of the mortar.

Mineral colors for stucco, if used, should be of such composition that they will not be affected by cement, lime, or the weather. The best method is to use colored sands when possible. The most satisfactory results with colored mortar are obtained by using white portland cement. Prepared patented stuccos which are combinations of cement, sand, plasticizers, waterproofing agents, and pigment are widely used.

CONCRETE

Concrete is made by mixing cement and an aggregate composed of hard inert particles of varying size, such as a combination of sand or brokenstone screenings, with gravel, broken stone, lightweight aggregate, or other material. Portland cement should always be used for reinforced concrete, for mass concrete subjected to stress, and for all concrete laid under water.

Proportioning Concrete   Compressive strength is generally accepted as the principal measure of the quality of concrete, and although this is not entirely true, there is an approximate relation between compressive strength and the other mechanical properties. Methods of proportioning generally aim to give concrete of a predetermined compressive strength.

The concrete mixture is proportioned or designed for a particular condition in various ways: (1) arbitrary selection based on experience and common practice, such as 1 part cement, 2 parts sand, 4 parts stone (written **1 : 2 : 4**); (2) proportioning on the basis of the water/cement ratio, either assumed from experience or determined by trial mixtures with the given materials and conditions; (3) combining materials on the basis of either the voids in the aggregates or mechanical-analysis curves so as to obtain the least voids and thus concrete of the maximum density for a given cement content.

Concrete mixes for small jobs can generally be determined by consultation with a ready-mixed concrete supplier, on the basis of the supplier's experience. If this information is not available, the mixes recommended by ACI Committee 211 and reproduced in Table 6.9.5 can be used as long as the required compressive strength is not higher than perhaps 3,500 lb/in^2. It is important that the mix be kept as dry as can be satisfactorily compacted, by vibration, into the form work since excess water reduces the eventual compressive strength. And the air entrainment is important for concretes which will be exposed to freeze-thaw cycles after curing.

For larger jobs and in all cases where high compressive strength is necessary, the mix must be designed on a more thorough basis. Again, local suppliers of ready-mixed concrete may have records which will be of considerable help. The concrete mix has several contradictory requirements placed on it, so the final product represents a compromise. The hardened concrete must be strong enough for its intended use, it must be durable enough for its expected exposure conditions, and the freshly mixed concrete must be workable enough to be placed and compacted in the forms.

For a given aggregate type and size, the strength is controlled primarily by the water/cement ratio, with a decrease in water content, relative to cement, leading to an increase in strength, as long as the concrete remains workable enough to be placed. The durability is controlled by the water/cement ratio and the air content, assuming a suitable aggregate for the exposure condition. Table 6.9.6, from ACI Committee 211, gives the maximum water/cement ratios which should be used in severe exposure conditions, and Table 6.9.7 gives the water/cement ratios required for various concrete strengths.

Table 6.9.5   Concrete Mixes for Small Jobs

Procedure: Select the proper maximum size of aggregate. Use mix B, adding just enough water to produce a workable consistency. If the concrete appears to be undersanded, change to mix A, and, if it appears over-sanded, change to mix C.

Maximum size of aggregate, in	Mix designation	Approximate weights of solid ingredients per ft^3 of concrete, lb				
			Sand*		Coarse aggregate	
		Cement	Air-entrained concrete†	Concrete without air	Gravel or crushed stone	Iron blast-furnace slag
⅛	A	25	48	51	54	47
	B	25	46	49	56	49
	C	25	44	47	58	51
¾	A	23	45	49	62	54
	B	23	43	47	64	56
	C	23	41	45	66	58
1	A	22	41	45	70	61
	B	22	39	43	72	63
	C	22	37	41	74	65
1½	A	20	41	45	75	65
	B	20	39	43	77	67
	C	20	37	41	79	69
2	A	19	40	45	79	69
	B	19	38	43	81	71
	C	19	36	41	83	72

* Weights are for dry sand. If damp sand is used, increase tabulated weight of sand 2 lb and, if very wet sand is used, 4 lb.
† Air-entrained concrete should be used in all stxuctures which will be exposed to alternate cycles of freezing and thawing. Air-entrainment can be obtained by the use of an air-entraining cement or by adding an air-entraining admixture. If an admixture is used, the amount recommended by the manufacturer will, in most cases, produce the desired air content.
SOURCE: ACI 211.

Table 6.9.6   Maximum Permissible Water/Cement Ratios for Concrete in Severe Exposures*

Type of structure	Structure wet continuously or frequently and exposed to freezing and thawing†	Structure exposed to seawater or sulfates
Thin sections (railings, curbs, sills, ledges, ornamental work) and sections with less than 1-in cover over steel	0.45	0.40‡
All other structures	0.50	0.45‡

* Based on report of ACI Committee 201. Cementitious materials other than cement should conform to ASTM C 618 and C 989.
† Concrete should also be air-entrained.
‡ If sulfate-resisting cement (type II or V of ASTM C 150) is used, the permissible water/cement ratio may be increased by 0.05.
SOURCE: ACI 211.

Table 6.9.7   Relationships between Water/Cement Ratio and Compressive Strength of Concrete

Compressive strength at 28 days, lb/in^2*	Water/cement ratio, by weight	
	Non-air-entrained concrete	Air-entrained concrete
6,000	0.41	—
5,000	0.48	0.40
4,000	0.57	0.48
3,000	0.68	0.59
2,000	0.82	0.74

* Values are estimated average strengths for concrete containing not more than the percentage of air shown in Table 6.9.9. For a constant water/cement ratio, the strength of concrete is reduced as the air content is increased.
Strength is based on 6 × 12 in cylinders moist-cured 28 days at 73.4 ± 3°F (23 ± 1.7°C) in accordance with Section 9(b) of ASTM C31 for Making and Curing Concrete Compression and Flexure Test Specimens in the Field.
Relationship assumes maximum size of aggregate about ¾ to 1 in; for a given source, strength produced for a given water/cement ratio will increase as maximum size of aggregate decreases.
SOURCE: ACI 211.

For the severe-exposure cases, the air content should be between 4.5 and 7.5 percent, with the smaller value being for larger maximum sized aggregate. These tables establish two of the constraints on the mix design. A third constraint is the consistency of the concrete, as determined by the slump test which is described later. Table 6.9.8 gives the same committee's recommendations about the maximum and minimum slump values for various kinds of members. Stiffer mixes, with slumps less than 1 in, can be placed only with very heavy vibration and with great care to achieve the necessary compaction, but the dry mixes can produce very high concrete strengths with only moderate cement contents. Greater slumps are sometimes necessary when the reinforcement is very congested or the members small, such as thin walls or cast-in-place piling. The higher slumps may be produced with the high-range water reducers, or superplasticizers, without the penalty of requiring excessive water content.

The next constraint on the mix is the maximum size of aggregate. Larger aggregate sizes tend to lead to lower cement contents, but the maximum size used is limited by what is available, and in addition usually should be limited to not more than one-fifth the narrowest

Table 6.9.8   Recommended Slumps for Various Types of Construction*

Types of construction	Slump, in	
	Maximum†	Minimum
Reinforced foundation walls and footings	3	1
Plain footings, caissons, and substructure walls	3	1
Beams and reinforced walls	4	1
Building columns	4	1
Pavements and slabs	3	1
Mass concrete	2	1

*Slump may be increased when chemical admixtures are used, provided that the admixture-treated concrete has the same or lower water/cement or water/cementitious material ratio and does not exhibit segregation potential or excessive bleeding.
†May be increased 1 in for methods of consolidation other than vibration.
SOURCE: ACI 211.

Table 6.9.9   Approximate Mixing Water and Air Content Requirements for Different Slumps and Nominal Maximum Sizes of Aggregates

Slump, in	Water, lb/yd^3 of concrete for indicated nominal maximum sizes of aggregate							
	$\frac{3}{8}$ in^a	$\frac{1}{2}$ in^a	$\frac{3}{4}$ in^a	1 in^a	1$\frac{1}{2}$ in^a	2 in^{a,b}	3 in^{b,c}	6 in^{b,c}
Non-air-entrained concrete								
1–2	350	335	315	300	275	260	220	190
3–4	385	365	340	325	300	285	245	210
6–7	410	385	360	340	315	300	270	—
Approximate amount of entrapped air in non-air-entrained concrete, %	3	2.5	2	1.5	1	0.5	0.3	0.2
Air-entrained concrete								
1–2	305	295	280	270	250	240	205	180
3–4	340	325	305	295	275	265	225	200
6–7	365	345	325	310	290	280	260	—
Recommended average^d total air content, percent for level of exposure:								
Mild exposure	4.5	4.0	3.5	3.0	2.5	2.0	1.5^{e,g}	1.0^{e,g}
Moderate exposure	6.0	5.5	5.0	4.5	4.5	4.0	3.5^{e,g}	3.0^{e,g}
Extreme exposure^g	7.5	7.0	6.0	6.0	5.5	5.0	4.5^{e,g}	4.0^{e,g}

^{a} The quantities of mixing water given for air-entrained concrete are based on typical total air content requirements as shown for moderate exposure in the table. These quantities of mixing water are for computing cement contents for trial batches at 68 to 77°F (20 to 25°C). They are maximum for reasonably well-shaped angular aggregates graded within limits of accepted specifications. Rounded aggregates will generally require 30 lb less water for non-air-entrained and 25 lb less for air-entrained concretes. The use of water-reducing chemical admixtures, ASTM C 494, may also reduce mixing water by 5 percent or more. The volume of liquid admixtures is included as part of the total volume of mixing water. Slump values of more than 7 in are only obtained through the use of water-reducing chemical admixtures; they are for concrete containing nominal maximum size aggregate not larger than 1 in.

^{b} The slump values for concrete containing aggregate larger than 1½ in are based on slump tests made after removal of particles larger than 1½ in by wet-screening.

^{c} These quantities of mixing water are for use in computing cement factors for trial batches when 3 in or 6 in nominal maximum size aggregate is used. They are average for reasonably well-shaped coarse aggregates, well-graded from coarse to fine.

^{d} Additional recommendations for air content and necessary tolerances on air content for control in the field are given in a number of ACI documents, including ACI 201, 345, 318, 301, and 302. ASTM C 94 for ready-mixed concrete also gives air-content limits. The requirements in other documents may not always agree exactly so in proportioning concrete consideration must be given to selecting an air content that will meet the needs of the job and also meet the applicable specifications.

^{e} For concrete containing large aggregates which will be wet-screened over the 1½ in sieve prior to testing for air content, the percentage of air expected in the 1½ in minus material should be as tabulated in the 1½ in column. However, initial proportioning calculations should include the air content as a percent of the whole.

^{f} When using large aggregate in low cement factor concrete, air entrainment need not be detrimental to strength. In most cases mixing water requirement is reduced sufficiently to improve the water/cement ratio and to thus compensate for the strength reducing effect of entrained air concrete. Generally, therefore, for these large maximum sizes of aggregate, air contents recommended for extreme exposure should be considered even though there may be little or no exposure to moisture and freezing.

^{g} These values are based on the criteria that 9 percent air is needed in the mortar phase of the concrete. If the mortar volume will be substantially different from that determined in this recommended practice, it may be desirable to calculate the needed air content by taking 9 percent of the actual mortar volume.

SOURCE: ACI 211.

width between forms, one-third the thickness of slabs, three-fourths the minimum clear spacing between reinforcing steel. Extremely strong concretes will require smaller rather than larger maximum aggregate sizes. Once the maximum-size stone has been selected, the water content to produce the desired slump can be estimated from Table 6.9.9, and once the water content has been determined, the cement content is determined from the required water/cement ratio determined earlier. The volume of coarse aggregate is then determined from Table 6.9.10, and when the fraction is multiplied by the dry rodded unit weight, the weight of coarse aggregate per unit volume of concrete can be found.

The weight of sand needed to make a cubic yard of concrete can then be estimated by adding up the weights of materials determined so far, and subtracting from the expected weight of a cubic yard of concrete, which might be estimated at 3,900 lb for an air-entrained concrete and 4,000 lb for a mix without air. This weight should be the surface-dry saturated weight, with a correction made for the moisture content of the sand by increasing the weight of sand and decreasing the amount of mix water. Table 6.9.11 gives the expected ranges of water content of fine and coarse aggregates. Concretes with small-maximum-size aggregates

Table 6.9.10   Volume of Coarse Aggregate per Unit Volume of Concrete

Maximum size of aggregate, in	Volume of dry-rodded coarse aggregate* per unit volume of concrete different fineness moduli of sand			
	2.40	2.60	2.80	3.00
$\frac{3}{8}$	0.50	0.48	0.46	0.44
$\frac{1}{2}$	0.59	0.57	0.55	0.53
$\frac{3}{4}$	0.66	0.64	0.62	0.60
1	0.71	0.69	0.67	0.65
1½	0.75	0.73	0.71	0.69
2	0.78	0.76	0.74	0.72
3	0.82	0.80	0.78	0.76
6	0.87	0.85	0.83	0.81

^{a} Volumes are based on aggregates in oven-dry-rodded condition as described in ASTM C 29 for Unit Weight of Aggregate.

These volumes are selected from empirical relationships to produce concrete with a degree of workability suitable for usual reinforced construction. For less workable concrete such as required for concrete pavement construction they may be increased about 10 percent.

SOURCE: ACI 211.

will be lighter than the values just cited, and concretes with aggregates larger than 1 in will probably be heavier, and the kind of stone making up the coarse aggregate will also make some difference.

Table 6.9.11   Free Moisture Content

	Dry	Damp	Wet
Gravel	0.2	1.0	2.0
Sand	2.0	4.0	7.0

The mix properties just determined represent a starting point, and the final mix will usually be adjusted somewhat after trials have been conducted. Ideally, the trials should be done in a laboratory before the job-site concrete work starts, and the trial mixes should be evaluated for slump, tendency for segregation, ease with which the surface can be finished (especially important for slabs), air content, actual unit weight, actual water requirements for the desired slump, and the compressive strength of the concrete. It must be recognized that the various tables are based on average conditions, and that a particular combination of cement and aggregate may produce a concrete considerably stronger or weaker than expected from these values, and this can be true even when all of the materials meet the appropriate limitations in the ASTM specifications. Admixtures will also change the required mix proportions, and water reducers have the potential of allowing significant reductions in cement content. The tables include values for very large aggregate, up to 6-in maximum size, but aggregate larger than $1\frac{1}{2}$ in will seldom be used in ordinary structures, and such concretes should be left to specialists.

It must also be recognized that consistent quality will be achieved only when care is taken to ensure that the various materials are of consistently good quality. Variations in the size of the sand and coarse aggregate, dirty aggregate, improperly stored cement, very high or low temperatures, and carelessness in any of the batching and mixing operations can reduce the uniformity and quality of the resulting concrete. The addition of extra water "to make it easier to place" is a too-common field error leading to poor concrete, since the extra water increases the water/cement ratio, which reduces the strength and durability. Some of the tests used are described in the following paragraphs.

The **dry rodded weight** is determined by filling a container, of a diameter approximately equal to the depth, with the aggregate in three equal layers, each layer being rodded 25 times using a bullet-pointed rod $\frac{5}{8}$ in diam and 24 in long.

Measurement of materials is usually done by weight. The bulking effect of moisture, particularly on the fine aggregate, makes it difficult to keep proportions uniform when volume measurement is used. Water is batched by volume or weight.

Mixing   In order to get good concrete, the cement and aggregates must be thoroughly mixed so as to obtain a homogeneous mass and coat all particles with the cement paste. Mixing may be done either by hand or by machine, although hand mixing is rare today.

Quality Control of Concrete   Control methods include measuring materials by weight, allowing for the water content of the aggregates; careful limitation of the total water quantity to that designed; frequent tests of the aggregates and changing the proportions as found necessary to maintain yield and workability; constant checks on the consistency by the slump test, careful attention to the placing of steel and the filling of forms; layout of the concrete distribution system so as to eliminate segregation; check on the quality of the concrete as placed by means of specimens made from it as it is placed in the forms; and careful attention to proper curing of the concrete. The field specimens of concrete are usually 6 in diam by 12 in high for aggregates up to $1\frac{1}{2}$ in max size and 8 by 16 in for 3-in aggregate. They are made by rodding the concrete in three layers, each layer being rodded 25 times using a $\frac{5}{8}$-in diam bullet-pointed rod 24 in long.

Machine-mixed concrete is employed almost universally. The mixing time for the usual batch mixer of 1 yd^3 or less capacity should not

be less than 1 min from the time all materials are in the mixer until the time of discharge. Larger mixers require 25 percent increase in mixing time per $\frac{1}{2}$-yd^3 increase in capacity. Increased mixing times up to 5 min increase the workability and the strength of the concrete. Mixing should always continue until the mass is homogeneous.

Concrete mixers of the batch type give more uniform results; few continuous mixers are used. Batch mixers are either (1) rotating mixers, consisting of a revolving drum or a square box revolving about its diagonal axis and usually provided with deflectors and blades to improve the mixing; or (2) paddle mixers, consisting of a stationary box with movable paddles which perform the mixing. Paddle mixers work better with relatively dry, high-sand, small-size aggregate mixtures and mortars and are less widely used than rotating mixers.

Ready-mixed concrete is (1) proportioned and mixed at a central plant (central-mixed concrete) and transported to the job in plain trucks or agitator trucks, or (2) proportioned at a central plant and mixed in a mixer truck (truck-mixed concrete) equipped with water tanks, during transportation. Ready-mixed concrete is largely displacing job-mixed concrete in metropolitan areas. The truck agitators and mixers are essentially rotary mixers mounted on trucks. There is no deleterious effect on the concrete if it is used within 1 h after the cement has been added to the aggregates. Specifications often state the minimum number of revolutions of the mixer drum following the last addition of materials.

Materials for concrete are also **centrally batched,** particularly for road construction, and transported to the site in batcher trucks with compartments to keep aggregates separated from the cement. The truck discharges into the charging hopper of the job mixer. Road mixers sometimes are arranged in series—the first mixer partly mixes the materials and discharges into a second mixer, which completes the mixing.

Consistency of Concrete   The consistency to be used depends upon the character of the structure. The proportion of water in the mix is of vital importance. A very wet mixture of the same cement content is much weaker than a dry or mushy mixture. Dry concrete can be employed in dry locations for mass foundations provided that it is carefully spread in layers not over 6 in thick and is thoroughly rammed. Medium, or quaking, concrete is adapted for ordinary mass-concrete uses, such as foundations, heavy walls, large arches, piers, and abutments. Mushy concrete is suitable as rubble concrete and reinforced concrete, for such applications as thin building walls, columns, floors, conduits, and tanks. A medium, or quaking, mixture has a tenacious, jellylike consistency which shakes on ramming; a mushy mixture will settle to a level surface when dumped in a pile and will flow very sluggishly into the forms or around the reinforcing bars; a dry mixture has the consistency of damp earth.

The two methods in common use for measuring the consistency or workability of concrete are the slump test and the flow test. In the **slump test,** which is the more widely used of the two, a form shaped as a frustum of a cone is filled with the concrete and immediately removed. The slump is the subsidence of the mass below its height when in the cone. The form has a base of 8-in diam, a top of 4-in diam, and a height of 12 in. It is filled in three 4-in layers of concrete, each layer being rodded by 25 strokes of a $\frac{5}{8}$-in rod, 24 in long and bullet-pointed at the lower end.

A test using the penetration of a half sphere, called the **Kelly ball test,** is sometimes used for field control purposes. A 1-in penetration by the Kelly ball corresponds to about 2 in of slump.

Usual limitations on the consistency of concrete as measured by the slump test are given in Table 6.9.8.

The **consistency** of concrete has some relation to its **workability,** but a lean mix may be unworkable with a given slump and a rich mix may be very workable. Certain admixtures tend to lubricate the mix and, therefore, increase the workability at certain slumps. The slump at all times should be as small as possible consistent with the requirements of handling and placing. A slump over 7 in is usually accompanied by segregation and low strength of concrete, unless the high slump was produced with the aid of a superplasticizer.

Forms for concrete should maintain the lines required and prevent leakage of mortar. The pressure on forms is equivalent to that of a liquid with the same density as the concrete, and of the depth placed within 2 h. Dressed lumber or plywood is used for exposed surfaces, and rough lumber for unexposed areas. Wood or steel forms should be oiled before placing concrete.

Placement of concrete can be by dump-bottom buckets, concrete pumps feeding into pipelines and hoses, buggies, chutes, or conveyor belts. Buckets and buggies can handle drier mixes. Pumping will require fairly rich mixes with moderately large slumps and maximum size aggregate not over about ¾ in (20 mm) unless it has been demonstrated that the pump can handle harsher mixes. Pumping has become much more common in recent years, as pumps and mix design procedures have developed. Pumps are often mounted on trucks that have extensive boom arrangements, allowing placement over a large area. Chutes should have a slope not less than 1 vertical to 2 horizontal; the use of flatter slopes encourages the use of excess water, leading to segregation and low strength. Conveyor belts have been used to place concrete in mass concrete dams, where very large quantities are required. Underwater concrete is placed by a dump-bottom bucket or a tremie pipe. Such concrete should have a cement content increase of 15 percent to allow for loss of cement during placement and/or an antiwashout admixture should be used.

Compaction of concrete (working it into place) is accomplished by tampers or vibrators. Vibrators are applied to the outside of the forms and to the surface or interior of the concrete; they should be used with care to avoid producing segregation. The frequency of vibrators is usually between 3,000 and 10,000 pulsations per minute.

Curing of concrete is necessary to ensure proper hydration. Concrete should be kept moist for a period of at least 7 days, and the temperature should not be allowed to fall below 50°F for at least 3 days. Sprayed-on membrane curing compounds may be used to retain moisture. Special precautions must be taken in cold weather and in hot weather.

Weight of Concrete   The following are average weights, lb/ft³, of portland cement concrete:

Sand-cinder concrete	112	Limestone concrete	148
Burned-clay or shale concrete	105	Sandstone concrete	143
Gravel concrete	148	Traprock concrete	155

Watertightness   Concrete can be made practically impervious to water by proper proportioning, mixing, and placing. Leakage through concrete walls is usually due to poor workmanship or occurs at the joints between 2 days' work or through cracks formed by contraction. New concrete can be bonded to old by wetting the old surface, plastering it with neat cement, and then placing the concrete before the neat cement has set. It is almost impossible to prevent contraction cracks entirely, although a sufficient amount of reinforcement may reduce their width so as to permit only seepage of water. For best results, a low-volume-change cement should be used with a concrete of a quaking consistency; the concrete should be placed carefully so as to leave no visible stone pockets, and the entire structure should be made without joints and preferably in one continuous operation. The best waterproofing agent is an additional proportion of cement in the mix. The concrete should contain not less than 6 bags of cement per yd³.

For maximum watertightness, mortar and concrete may require more fine material than would be used for maximum strength. Gravel produces a more watertight concrete than broken stone under similar conditions. Patented compounds are on the market for producing watertight concrete, but under most conditions, equally good results can be obtained for less cost by increasing the percentage of cement in the mix. Membrane waterproofing, consisting of asphalt or tar with layers of felt or tarred paper, or plastic or rubber sheeting in extreme cases, is advisable where it is expected that cracks will occur. Mortar troweled on very hard may produce watertight work.

Concrete to be placed through water should contain at least 7 bags of cement per yd³ and should be of a quaking consistency.

According to Fuller and Thompson (*Trans. ASCE*, **59**, p. 67), watertightness increases (1) as the percentage of cement is increased and in a very much larger ratio; (2) as the maximum size of stone is increased,

provided the mixture is homogeneous; (3) materially with age; and (4) with thickness of the concrete, but in a much larger ratio. It decreases uniformly with increase in pressure and rapidly with increase in the water/cement ratio.

Air-Entrained Concrete   The entrainment of from 3 to 6 percent by volume of air in concrete by means of vinsol resin or other air-bubble-forming compounds has, under certain conditions, improved the resistance of concrete in roads to frost and salt attack. The air entrainment increases the workability and reduces the compressive strength and the weight of the concrete. As the amount of air entrainment produced by a given percentage of vinsol resin varies with the cement, mixture, aggregates, slump, and mixing time, good results are dependent on very careful control.

Concrete for Masonry Units   Mixtures for concrete masonry units, which are widely used for walls and partitions, employ aggregates of a maximum size of ½ in and are proportioned either for casting or for machine manufacture. Cast units are made in steel or wooden forms and employ concrete slumps of from 2 to 4 in. The proportions used are 1 part cement and 3 to 6 parts aggregate by dry and loose volumes. The forms are stripped after 24 h, and the blocks piled for curing. Most **blocks** are made by machine, using very dry mixtures with only enough water present to enable the concrete to hold together when formed into a ball. The proportions used are 1 part cement and 4 to 8 parts aggregate by dry and loose volumes. The utilization of such a lean and dry mixture is possible because the blocks are automatically tamped and vibrated in steel molds. The blocks are stripped from the molds at once, placed on racks on trucks, and cured either in air or in steam. High-pressure-steam (50 to 125 lb/in²) curing develops the needed strength of the blocks in less than 24 h. Most concrete blocks are made with lightweight aggregates such as cinders and burned clay or shale in order to reduce the weight and improve their acoustical and thermal insulating properties. These aggregates not only must satisfy the usual requirements for gradation and soundness but must be limited in the amount of coal, iron, sulfur, and phosphorus present because of their effect on durability, discoloration and staining, fire resistance, and the formation of "pops." Pops form on the surface of concrete blocks using cinders or burned clay and shale as a result of the increase in volume of particles of iron, sulfur, and phosphorus when acted on by water, oxygen, and the alkalies of the cement.

Strength of Concrete

The strength of concrete increases (1) with the quantity of cement in a unit volume, (2) with the decrease in the quantity of mixing water relative to the cement content, and (3) with the density of the concrete. Strength is decreased by an excess of sand over that required to fill the voids in the stone and give sufficient workability. The volume of fine aggregate should not exceed 60 percent of that of coarse aggregate 1½-in max size or larger.

Compressive Strength   Table 6.9.7 gives the results obtainable with first-class materials and under first-class conditions. Growth in strength with age depends in a large measure upon the consistency characteristics of the cement and upon the curing conditions. Table 6.9.12 gives the change in relative strength with age for several water/cement ratios and a wide range of consistencies for a cement with a good age-strength gain relation. Many normal portland cements today show very little gain in strength after 28 days.

Tensile Strength   The tensile strength of concrete is of less importance than the crashing strength, as it is seldom relied upon. The true tensile strength is about 8 percent of the compression strength and must not be confused with the tensile fiber stress in a concrete beam, which is greater.

Table 6.9.12   Variation of Compressive Strength with Age
(Strength at 28 days taken as 100)

Water/cement ratio by weight	3 days	7 days	28 days	3 months	1 year
0.44	40	75	100	125	145
0.62	30	65	100	135	155
0.80	25	50	100	145	165

The tensile strength is not easily measured, and an "indirect tension" test is used, in which a cylinder is loaded along a diametral line. Details of the test are given in ASTM C496. Nearly all the concrete in the plane of the two load strips is in tension. The tensile strength may be computed as $T = 2Pl/(\pi ld)$, where T = splitting tensile strength, lb/in^2; P = maximum load, lb; l = length of cylinder, in; d = diameter of cylinder, in.

Transverse Strength   There is an approximate relationship between the tensile fiber stress of plain concrete beams and their compressive strength. The modulus of rupture is greatly affected by the size of the coarse aggregate and its bond and transverse strength. Quartzite generally gives low-modulus-of-rupture concrete.

The transverse or beam test is generally used for checking the quality of concrete used for roads. The standard beam is 6 by 6 by 20 in, tested on an 18-in span and loaded at the one-third points.

The tensile strength of the concrete as determined by the split cylinder (indirect tension) test is likely to be in the range of 4 to 5 times $\sqrt{f'_c}$, where both f'_c and $\sqrt{f'_c}$ have units of lb/in^2. The modulus of rupture values are likely to be somewhat higher, in the range of 6 to 7.5 times $\sqrt{f'_c}$.

The strength of concrete in direct **shear** is about 20 percent of the compressive strength.

Deformation Properties   Young's modulus for concrete varies with the aggregates used and the concrete strength but will usually be 3 to 5×10^6 lb/in^2 in short-term tests. According to the ACI Code, the modulus may be expressed as

$$E_c = 33w^{3/2}\sqrt{f'_c}$$

where w = unit weight of concrete in lb/ft^3 and both f'_c and $\sqrt{f'_c}$ have lb/in^2 units. For normal-weight concretes, $E_c = 57,000 \sqrt{f'_c}$ lb/in^2.

In addition to the instantaneous deformations, concrete shrinks with drying and is subjected to creep deformations developing with time. Shrinkage strains, which are independent of external stress, may reach 0.05 percent at 50 percent RH, and when restrained may lead to cracking. Creep strains typically reach twice the elastic strains. Both creep and shrinkage depend strongly on the particular aggregates used in the concrete, with limestone usually resulting in the lowest values and river gravel or sandstone in the largest. In all cases, the longer the period of wet curing of the concrete, the lower the final creep and shrinkage values.

Deleterious Actions and Materials

Freezing retards the setting and hardening of portland-cement concrete and is likely to lower its strength permanently. On exposed surfaces such as walls and sidewalks placed in freezing weather, a thin scale may crack from the surface. Natural cement is completely ruined by freezing.

Concrete laid in freezing weather or when the temperature is likely to drop to freezing should have the materials heated and should be protected from the frost, after laying, by suitable covering or artificial heat. The use of calcium chloride, salt, or other ingredients in sufficient quantities to lower the temperature of freezing significantly is not permitted, since the concrete would be adversely affected.

Mica and Clay in Sand   Mica in sand, if over 2 percent, reduces the density of mortar and consequently its strength, sometimes to a very large extent. In crushed-stone screenings, the effect of the same percentage of mica in the natural state is less marked. Black mica, which has a different crystalline form, is not injurious to mortar. Clay in sand may be injurious because it may introduce too much fine material or form balls in the concrete. When not excessive in quantity, it may increase the strength and watertightness of a mortar of proportions $1:3$ or leaner.

Mineral oils which have not been disintegrated by use do not injure concrete when applied externally. Animal fats and vegetable oils tend to disintegrate concrete unless it has thoroughly hardened. Concrete after it has thoroughly hardened resists the attack of diluted organic acids but is disintegrated by even dilute inorganic acids; protective treatments are magnesium fluorosilicate, sodium silicate, or linseed oil. Green concrete is injured by manure but is not affected after it has thoroughly hardened.

Electrolysis injures concrete under certain conditions, and electric current should be prevented from passing through it. (See also Sec. 6.5, Corrosion.)

Seawater attacks cement and may disintegrate concrete. Deleterious action is greatly accelerated by frost. To prevent serious damage, the concrete must be made with a sulfate-resisting cement, a rich mix (not leaner than $1:2:4$), and exceptionally good aggregates, including a coarse sand, and must be allowed to harden thoroughly, at least 7 days, before it is touched by the seawater. Although tests indicate that there is no essential difference in the strengths of mortars gaged with fresh water and with seawater, the latter tends to retard the setting and may increase the tendency of the reinforcement to rust.

6.10   WATER
by Arnold S. Vernick and Amended by Staff

(See also Secs. 4.1, 4.2, and 6.1.)

REFERENCES: "Statistical Abstract of the United States, 1994," 114th ed., U.S. Dept. of Commerce, Economics and Statistics Administration, Bureau of the Census, Bernan Press, Lanham, MD. van der Leeden, Troise, and Todd, "The Water Encyclopedia," Lewis Publishers, Inc., Chelsea, MI. "Hold the Salt," *Compressed Air Magazine*, March 1995. "Water Treatment Plant Design," 2d ed., ASCE and AWWA, McGraw-Hill. Corbitt, "Standard Handbook of Environmental Engineering," McGraw-Hill. AID Desalination Manual, Aug. 1980, U.S. Department of State. Saline Water Conversion Report, *OSW Annual Reports*, U.S. Department of the Interior. "Water Quality and Treatment," AWWA, 1971. "Safe Drinking Water Act," PL-93–523, Dec. 1974 as amended through 1993. "National Primary and Secondary Drinking Water Regulations," Title 40 CFR Parts 141–143, U.S. Environmental Protection Agency. "The A-B-C of Desalting," OSW, U.S. Department of the Interior, 1977. "New Water for You," OSW, U.S. Department of the Interior, 1970. Howe, "Fundamentals of Water Desalination," Marcel Dekker, New York. Ammerlaan, "Seawater Desalting Energy Requirements," *Desalination*, **40**, pp. 317–326, 1982. "National Water Summary 1985," U.S. Geological Survey. "Second National Water Assessment, The Nation's Water Resources 1975–2000," U.S. Water Resources Council, December 1978. Vernick and Walker, "Handbook of Wastewater Treatment Processes," Marcel Dekker, New York.

WATER RESOURCES

Oceans cover 70 percent of the earth's surface and are the basic source of all water. Ocean waters contain about 3½ percent by weight of dissolved materials, generally varying from 32,000 to 36,000 ppm and as high as 42,000 ppm in the Persian Gulf.

About 50 percent of the sun's energy falling on the ocean causes evaporation. The vapors form clouds which precipitate pure water as rain. While most rain falls on the sea, land rainfall returns to the sea in rivers or percolates into the ground and back to the sea or is reevaporated. This is known as the **hydrological cycle**. It is a closed distillation cycle, without additions or losses from outer space or from the interior of the earth.

Water supply to the United States depends on an annual rainfall averaging 30 in (76 cm) and equal to 1,664,800 billion gal (6,301 Gm^3) per year, or 4,560 billion gal (17.3 Gm^3) per day. About 72 percent returns to the atmosphere by direct evaporation and transpiration from trees and plants. The remaining 28 percent, or 1,277 billion gal (4.8 Gm^3) daily, is the maximum supply available. This is commonly called runoff and properly includes both surface and underground flows. Between 33 and 40 percent of runoff appears as groundwater.

Two-thirds of the runoff passes into the ocean as flood flow in onethird of the year. By increased capture, it would be possible to retain about one-half of the 1,277 billion gal (4.8 Gm³), or 638 billion gal (2.4 Gm³), per day as **maximum usable water.**

Withdrawal use is the quantity of water removed from the ground or diverted from a body of surface water. **Consumptive use** is the portion of such water that is discharged to the atmosphere or incorporated in growing vegetation or in industrial or food products.

The estimated withdrawal use of water in the United States in 1990 was 408 billion gal per day, including some saline waters (see Table 6.10.1). This figure increased historically from 140 billion gal per day to a maximum of 440 billion gal per day in 1980. The reduction of 7.3 percent in the decade since then reflects an increased emphasis on water conservation and reuse.

Freshwater withdrawal is about 30 percent of runoff, and consumptive use is about 8 percent of runoff. The consumptive use of water for irrigation is about 55 percent, but another 15 percent is allowed for transmission and distribution losses.

Dividing total water withdrawal by the population of the United States shows a 1990 water use, the **water index,** of 1,620 gal (6,132 L) per capita day (gpcd). It reflects the great industrial and agricultural uses of water, because the withdrawal and consumption of water for domestic purposes is relatively small. The average domestic consumption of water in urban areas of the United States in 1980 was 38 gpcd (144 Lpcd). However, since public water utilities also supply industrial and commercial customers, the average withdrawal by U.S. municipal water systems in 1990 was 195 gpcd (738 Lpcd). The reduction in water withdrawal since 1980 cited above is also reflected in water consumption. Consumption per capita per day peaked at 451 gpcd (1,707 Lpcd) in 1975 and was 370 gpcd (1,400 Lpcd) in 1990, a reduction of 18 percent.

Additions to water resources for the future can be made by (1) increase in storage reservoirs, (2) injection of used water or flood water into underground strata called **aquifers,** (3) covering reservoirs with films to reduce evaporation, (4) rainmaking, (5) saline-water conversion, or (6) wastewater renovation and reuse.

It is equally important to improve the efficient use of water supplies by (1) multiple use of cooling water, (2) use of air cooling instead of water cooling, (3) use of cooling towers, (4) reclamation of waste waters, both industrial and domestic, (5) abatement of pollution by treatment rather than by dilution, which requires additional freshwater.

MEASUREMENTS AND DEFINITIONS

Water **quantities** in this country are measured by U.S. gallons, the larger unit being 1,000 U.S. gal. For agriculture and irrigation, water use is measured in acre-feet, i.e., the amount of water covering 1 acre of surface to a depth of 1 ft. In the British-standard area, the imperial gallon is used. In SI or metric-system areas, water is measured in litres (L) and the larger unit is the cubic metre, expressed as m³, which equals a metric ton of water.

Conversion Table

$$1 \text{ acre} \cdot \text{ft} = 325,850 \text{ U.S. gal or } 326,000 \text{ approx}$$
$$1 \text{ acre} \cdot \text{ft} = 1,233 \text{ t or m}^3$$
$$1 \text{ acre} \cdot \text{ft} = 43,560 \text{ ft}^3$$
$$1 \text{ imperial gal} = 1.20 \text{ U.S. gal}$$
$$1 \text{ m}^3 = 1000 \text{ L}$$
$$1 \text{ L} = 0.2642 \text{ gal}$$
$$1 \text{ metric ton (t)} = 1,000 \text{ kg}$$
$$= 2,204 \text{ lb}$$
$$= (264.2 \text{ U.S. gal})$$
$$= 220 \text{ imperial gal}$$
$$1 \text{ U.S. ton} = 240 \text{ U.S. gal}$$
$$1,000,000 \text{ U.S. gal} = 3.07 \text{ acre} \cdot \text{ft}$$

For stream flow and hydraulic purposes, water is measured in cubic feet per second.

$$1,000,000 \text{ U.S. gal per day} = 1.55 \text{ ft}^3/\text{s} \ (0.44 \text{ m}^3/\text{s})$$
$$= 1,120 \text{ acre} \cdot \text{ft} \ (1,380,000 \text{ m}^3)$$
$$\text{per year}$$

Water **costs** are expressed in terms of price per 1,000 gal, per acre · foot, per cubic foot, or per cubic metre.

$$10\text{¢ per } 1,000 \text{ gal} = \$32.59 \text{ per acre} \cdot \text{ft}$$
$$= 0.075\text{¢ per ft}^3$$
$$= 2.64\text{¢ per m}^3$$

Water **quality** is measured in terms of solids, of any character, which are suspended or dissolved in water. The concentration of solids is usually expressed in parts per million or in grains per gallon. One grain equals 1/7,000 lb (64.8 mg). Therefore, 17.1 ppm = 1 grain per U.S. gal. In the metric system, 1 ppm = 1 g/m^3 = 1 mg/L.

Standards for drinking water, or **potable water,** have been established by the U.S. Environmental Protection Agency. The recommended limit for a specific containment is shown in Table 6.10.2. The limitations indicated in the table apply to public water systems, with the primary regulations being established to protect public health, while the secondary regulations control aesthetic qualities relating to the public acceptance of drinking water.

For **agricultural water,** mineral content of up to 700 mg/L is considered excellent to good. However, certain elements are undesirable, particularly sodium and boron. The California State Water Resources Board limits class I irrigation water to:

	Max mg/L
Sodium, as % of total sodium, potassium, magnesium, and calcium equivalents	60
Boron	0.5
Chloride	177
Sulfate	960

Class II irrigation water may run as high as 2,100 mg/L total dissolved solids, with higher limits on the specific elements, but whether such water is satisfactory or injurious depends on the character of soil, climate, agricultural practice, and type of crop.

Table 6.10.1   U.S. Water Withdrawals and Consumption per Day by End Use, 1990

	Total, 10⁹ gal[a]	Per capita,[b] gal	Irrigation, 10⁹ gal	Public water utilities[c] Total, 10⁹ gal	Public water utilities[c] Per capita,[b] gal	Rural domestic,[d] 10⁹ gal	Industrial and misc.,[e] 10⁹ gal	Steam electric utilities, 10⁹ gal
Withdrawal	408	1,620	137	41	195	7.9	30	195
Consumption	96	370	76	7.1[f]	38[f]	8.9	6.7	4.0

[a] Conversion factor: gal × 3.785 = L.
[b] Based on Bureau of the Census resident population as of July 1.
[c] Includes domestic and commercial water withdrawals.
[d] Rural farm and nonfarm household and garden use, and water for farm stock and dairies.
[e] Includes manufacturing, mining, and mineral processing, ordnance, construction, and miscellaneous.
[f] 1980 data—latest available.
SOURCE: "Statistical Abstract of the United States, 1994," U.S. Dept. of Commerce, Economics and Statistics Administration, Bureau of the Census, 114th ed., Bernan Press, Lanham, MD.

Table 6.10.2   Potable Water Contaminant Limitations

Primary standards			
Organic contaminants		Inorganic contaminants	
Contaminant	MCL, mg/L	Contaminant	MCL, mg/L
Vinyl chloride	0.002	Fluoride	4.0
Benzene	0.005	Asbestos	7 million fibers/L
Carbon tetrachloride	0.005		(longer than 10 μm)
1,2-Dichloroethane	0.005	Barium	2
Trichloroethylene	0.005	Cadmium	0.005
para-Dichlorobenzene	0.075	Chromium	0.1
1,1-Dichloroethylene	0.007	Mercury	0.002
1,1,1-Trichloroethane	0.2	Nitrate	10 (as nitrogen)
cis-1,2-Dichloroethylene	0.07	Nitrite	10 (as nitrogen)
1,2-Dichloropropane	0.005	Total nitrate and nitrite	10 (as nitrogen)
Ethylbenzene	0.7	Selenium	0.05
Monochlorobenzene	0.1	Antimony	0.006
o-Dichlorobenzene	0.6	Beryllium	0.004
Styrene	0.1	Cyanide (as free cyanide)	0.2
Tetrachloroethylene	0.005	Nickel	0.1
Toluene	1	Thallium	0.002
trans-1,2-Dichloroethylene	0.1	Arsenic	0.05
Xylenes (total)	10	Lead	0.015
Dichloromethane	0.005	Copper	1.3
1,2,4-Trichlorobenzene	0.07		
1,1,2-Trichloroethane	0.005	Radionuclides	
Alachlor	0.002	Contaminant	MCL, pCi/L*
Aldicarb	0.003		
Aldicarb sulfoxide	0.004	Gross alpha emitters	15
Aldicarb sulfone	0.003	Radium 226 plus 228	5
Atrazine	0.003	Radon	300
Carbofuran	0.04	Gross beta particle and photon	4m Rem
Chlordane	0.002	Emitters	
Dibromochloropropane	0.0002	Uranium	20 μL (equivalent to
2,4-D	0.07		30 pCi/L)
Ethylene dibromide	0.00005		
Heptachlor	0.0004	Bacteria, Viruses, Turbidity, Disinfectants and By-products—	
Heptachlor epoxide	0.0002	Refer to 40CFR141	
Lindane	0.0002	Secondary standards†	
Methoxychlor	0.04	Contaminant	Level
Polychlorinated biphenyls	0.0005		
Pentachlorophenol	0.001	Aluminum	0.05 to 0.2 mg/L
Toxaphene	0.003	Chloride	250 mg/L
2,4,5-TP	0.05	Color	15 color units
Benzo(a)pyrene	0.0002	Copper	1.0 mg/L
Dalapon	0.2	Corrosivity	Noncorrosive
Di(2-ethylhexyl)adipate	0.4	Fluoride	2.0 mg/L
Di(2-ethylhexyl)phthalate	0.006	Foaming agents	0.5 mg/L
Dinoseb	0.007	Iron	0.3 mg/L
Diquat	0.02	Manganese	0.05 mg/L
Endothall	0.1	Odor	3 threshold odor number
Endrin	0.002	pH	6.5–8.5
Glyphosate	0.7	Silver	0.1 mg/L
Hexachlorobenzene	0.001	Sulfate	250 mg/L
Hexachlorocyclopentadiene	0.05	Total dissolved solids (TDS)	500 mg/L
Oxamyl (Vydate)	0.2	Zinc	5 mg/L
Picloram	0.5		
Simazine	0.004		
2,3,7,8-TCDD (Dioxin)	3×10^{-8}		

* Except as noted.
† Not federally enforceable; intended as guidelines for the states.
SOURCE: EPA Primary Drinking Water Regulations (40CFR141). EPA Secondary Drinking Water Regulations (40CFR143).

Waters containing dissolved salts are called **saline waters,** and the lower concentrations are commonly called **brackish;** these waters are defined, in mg/L, as follows:

Saline	All concentrations up to 42,000
Slightly brackish	1,000–3,000
Brackish	3,000–10,000
Seawaters, average	32,000–36,000
Brine	Over 42,000

Hardness of water refers to the content of calcium and magnesium salts, which may be bicarbonates, carbonates, sulfates, chlorides, or nitrates. Bicarbonate content is called **temporary hardness,** as it may be removed by boiling. The salts in "hard water" increase the amount of soap needed to form a lather and also form deposits or "scale" as water is heated or evaporated.

Hardness is a measure of calcium and magnesium salts expressed as equivalent calcium carbonate content and is usually stated in mg/L (or in grains per gal) as follows: very soft water, less than 15 mg/L; soft

Table 6.10.3   Water Use in U.S. Manufacturing by Industry Group, 1983

Industry	Establishments reporting*	Gross water used — Total — Quantity, 10^9 gal	Gross water used — Total — Average per establishment, 10^6 gal	Water intake, 10^9 gal	Water recycled,† 10^9 gal	Water discharged — Quantity, 10^9 gal	Water discharged — Percent untreated
Food and kindred products	2,656	1,406	529	648	759	552	64.5
Tobacco products	20	34	1,700	5	29	4	N/A
Textile mill products	761	333	438	133	200	116	52.6
Lumber and wood products	223	218	978	86	132	71	63.4
Furniture and fixtures	66	7	106	3	3	3	100.0
Paper and allied products	600	7,436	12,393	1,899	5,537	1,768	27.1
Chemicals and allied products	1,315	9,630	7,323	3,401	6,229	2,980	67.0
Petroleum and coal products	260	6,177	23,758	818	5,359	699	46.2
Rubber, miscellaneous plastic products	375	328	875	76	252	63	63.5
Leather and leather products	69	7	101	6	1	6	N/A
Stone, clay, and glass products	602	337	560	155	182	133	75.2
Primary metal products	776	5,885	7,584	2,363	3,523	2,112	58.1
Fabricated metal products	724	258	356	65	193	61	49.2
Machinery, excluding electrical	523	307	587	120	186	105	67.6
Electrical and electronic equipment	678	335	494	74	261	70	61.4
Transportation equipment	380	1,011	2,661	153	859	139	67.6
Instruments and related products	154	112	727	30	82	28	50.0
Miscellaneous manufacturing	80	15	188	4	11	4	N/A
Total	10,262	33,835	3,297	10,039	23,796	8,914	54.9

* Establishments reporting water intake of 20 million gal or more, representing 96 percent of the total water use in manufacturing industries.
† Refers to water recirculated and water reused.
N/A = not available.
SOURCE: "The Water Encyclopedia," 1990.

water, 15 to 50 mg/L, slightly hard water, 50 to 100 mg/L; hard water, 100 to 200 mg/L; very hard water, over 220 mg/L.

INDUSTRIAL WATER

The use of water within a given industry varies widely because of conditions of price, availability, and process technology (see Table 6.10.3).

When a sufficient water supply of suitable quality is available at low cost, plants tend to use maximum volumes. When water is scarce and costly at an otherwise desirable plant site, improved processes and careful water management can reduce water usage to the minimum. Industrial water may be purchased from local public utilities or self-supplied. Approximately 79 percent of the water used by the chemical industry in the United States comes from company systems, and 66 percent of those self-supplied plants utilize surface water as their source of supply as shown in Table 6.10.4.

The cost of usable water is highly sensitive to a variety of factors including geographic location, proximity to an abundant water source, quality of the water, regulatory limitations, and quantity of water used. Raw surface water not requiring much treatment prior to use can cost as little as 7¢ to 15¢ per 1,000 gal (1.8¢ to 4¢ per m³) in areas of the United States where water is plentiful, or as much as $1.75 to $2.00 per 1,000 gal (46¢ to 53¢ per m³) in water-scarce areas. These costs include collection, pumping, storage tanks, distribution and fire-protection facilities, but treatment, if needed, would add materially to these figures.

The cost of water obtained from public supply systems for industrial use also varies considerably depending upon geographic location and water usage. An EPA survey of over 58,000 public water systems in 1984 indicated water costs as high as $2.30 per 1,000 gal (61¢ per m³) in New England for water use of 3,750 gal, to a low of 88¢ per 1,000 gal (23¢ per m³) in the northwest for water use of 750,000 gal. Water cost in 1983 in Houston, Texas, for the highest volume users, which includes refineries and chemical plants, was $1.14 per 1,000 gal (30¢ per m³).

The use of water by the chemical industry is reflected in Table 6.10.5.

About 80 percent of industrial water is used for cooling, mostly on a once-through system. An open recirculation system with cooling tower

Table 6.10.4   Sources of Water for Chemical Plants

	Quantity, billion gal/year ($\times\ 3.785 \times 10^{-3}$ = billion m³/year)
By source	
Public water systems	727
Company systems	3,399
Surface water	2,251
Groundwater	366
Tidewater	782
Other	200
By type	
Fresh	3,425
Brackish	182
Salt	719

Table 6.10.5   How Water Is Used in Chemical Plants

Use	Quantity, billion gal/year ($\times\ 3.785 \times 10^{-3}$ = billion m³/year)
Process	754
Cooling and condensing	
Electric power generation	480
Air conditioning	71
Other	2,760
Sanitary service	38
Boiler feed	112
Other	81

or spray pond (Fig. 6.10.1) reduces withdrawal use of water by over 90 percent but increases consumptive use by 3 to 8 percent because of evaporation loss. Even more effective reduction in water demand can be achieved by multiple reuse. An example is shown in Fig. 6.10.2.

Quality requirements for general plant use (nonprocess) are that the water be low in suspended solids to prevent clogging, low in total dissolved solids to prevent depositions, free of organic growth and color, and free of iron and manganese salts. Where the water is also used for drinking, quality must meet the Environmental Protection Agency regulations.

Fig. 6.10.1   Open recirculating system.

Fig. 6.10.2   Stepwise or cascade cooling system.

Cooling service requires that water be nonclogging. Reduction in suspended solids is made by settling or by using a coagulating agent such as alum and then settling. For recirculating-type cooling systems, corrosion inhibitors such as polyphosphates and chromates are added; algicides and biocides may be needed to control microorganism growth; for cooling jackets on equipment, hardness may cause scaling and should be reduced by softening.

Process-water quality requirements are often more exacting than potable-water standards; e.g., boiler feedwater must have less than 1 mg/L of dissolved solids (see Sec. 6, Corrosion, and Sec. 9, Steam Boilers). The required quality may be met by the general plant water as available or must be provided by treatment.

Table 6.10.6 shows methods and objectives for industrial-water treatment.

WATER POLLUTION CONTROL

A pollutant is defined in the Federal Clean Water Act as "dredge spoil, solid waste, incinerator residue, sewage, garbage, sewage sludge, munitions, chemical wastes, biological materials, heat, wrecked or discarded equipment, rock, sand, cellar dirt, and industrial, municipal, and agricultural waste discharged into water."

Control of water pollution in the United States is a joint responsibility of the U.S. Environmental Protection Agency and the state environmental protection departments. Promulgation of regulations controlling both industrial and municipal wastewater discharges has been accomplished by the EPA as mandated by the Clean Water Act. Enforcement has been largely delegated to the States, with the EPA serving in an oversight role. Water quality standards for specific streams and lakes have been established by the States, or interstate agencies where appropriate, which frequently mandate tighter control of discharges where local conditions warrant such action.

For the support of fish and aquatic life, water must contain a supply of **dissolved oxygen** (DO). Organic wastes consume oxygen by microbiological action, and this effect is measured by the **biochemical oxygendemand** (BOD) test.

The discharge of industrial wastewater in the United States is controlled by permits issued by the States and/or the EPA to each facility. The permits set specific limits for applicable pollutant parameters such as suspended solids, BOD, COD, color, pH, oil and grease, metals, ammonia, and phenol. In order to achieve these limitations, most industrial facilities must treat their wastewater effluent by either physical, chemical or biological methods.

The investment by industry in water pollution abatement facilities and the annual cost of operating these facilities have increased dramatically over the last two decades. In 1992, manufacturing industries in the United States spent a total of $2.5 billion in capital expenditures and $6.6 billion in operating costs on water pollution abatement facilities (see Table 6.10.7). These amounts represented 32 and 38 percent, respectively, of the total amount spent by industry for all pollution abatement and environmental protection activities. The chemical industry was the major contributor to these totals, spending approximately 40 percent of its capital expenditures and 30 percent of its operating costs for water pollution abatement.

Table 6.10.6   Water Treatment

Treatment	Method	Objective
Clarification	Presedimentation Coagulation Settling Filtering	Remove suspended solids and reduce turbidity, color, organic matter
Disinfection	Add chlorine, 5 to 6 ppm, or continuous-feed to maintain 0.2 to 0.3 ppm residual-free Cl_2	Prevent algae and slime growth
Softening	Cold-lime process	Reduce temporary hardness to 85 ppm; also reduce iron and manganese
	Hot-lime-soda process	Reduce total hardness to 25 ppm
	Zeolite	Reduce total hardness to 5 ppm
Membrane processes	Reverse osmosis	Partial removal of ions; can reduce seawater and brackish water to 550 ppm or less
	Electrodialysis	Partial removal of ions; can reduce 10,000 ppm brackish water to 500 ppm or less
Demineralization	Ion exchange: two-stage or mixed-bed	Remove both positive and negative ions (cations and anions) to provide very pure water
Distillation	Evaporation using steam heat	Produce very pure water—10 ppm or less total solids

Table 6.10.7 Water Pollution Abatement Capital Expenditures and Operating Costs of Manufacturing Industry Groups, 1992

Industry group	Pollution abatement capital expenditures, million $		Pollution abatement gross operating costs, million $*	
	Total†	Water	Total†	Water
Food and kindred products	316.8	202.6	1,312.0	835.7
Lumber and wood products	94.5	18.9	243.0	49.5
Paper and allied products	1,004.6	373.4	1,860.7	822.7
Chemicals and allied products	2,120.9	1,017.3	4,425.1	1,946.8
Petroleum and coal products	2,685.0	492.6	2,585.4	742.8
Stone, clay, and glass products	138.8	20.2	491.2	94.6
Primary metal industries	525.7	123.5	1,993.4	575.0
Fabricated metal products	103.3	42.4	761.2	284.3
Machinery, excluding electrical	150.3	31.7	463.7	160.5
Electrical, electronic equipment	126.6	45.6	657.1	288.8
Transportation equipment	281.0	69.2	1,171.7	347.0
Instruments, related products	89.1	18.8	331.9	89.4
All industries‡	7,866.9	2,509.8	17,466.4	6,576.9

* Includes payments to governmental units.
† Includes air, water, hazardous waste, and solid waste.
‡ Includes industries not shown separately; excludes apparel and other textile products, and establishments with less than 20 employees.
SOURCE: "Statistical Abstract of the United States," 1994.

Physical treatment includes screening, settling, flotation, equalization, centrifuging, filtration, and carbon adsorption. **Chemical treatment** includes coagulation or neutralization of acids with soda ash, caustic soda, or lime. Alkali wastes are treated with sulfuric acid or inexpensive waste acids for neutralization. Chemical oxidation is effective for certain wastes. **Biological treatment** is accomplished by the action of two types of microorganisms: aerobic, which act in the presence of oxygen, and anaerobic, which act in the absence of oxygen. Most organic wastes can be treated by biological processes.

The principal wastewater treatment techniques include the activated sludge process, trickling filters and aerated lagoons, which all employ aerobic microorganisms to degrade the organic waste material. Anaerobic processes are mainly employed for digestion of the sludge produced by biological wastewater treatment.

WATER DESALINATION

An *average seawater* contains 35,000 ppm of dissolved solids, equal to 3½ percent by weight of such solids, or 3.5 lb per 100 lb; in 1,000 gal, there are 300 lb of dissolved chemicals in 8,271 lb of pure water. The principal ingredient is sodium chloride (common salt), which accounts for about 80 percent of the total. Other salts are calcium sulfate (gypsum), calcium bicarbonate, magnesium sulfate, magnesium chloride, potassium chloride, and more complex salts. Because these dissolved chemicals are dissociated in solution, the composition of seawater is best expressed by the concentration of the major ions (see Table 6.10.8). Even when the content of total solids in seawater varies because of dilution or concentration, for instance Arabian Gulf water with total

Table 6.10.8 Ions in Seawater

Ions	ppm	lb/1,000 gal	mg/L
Chloride	19,350	165.6	19,830
Sodium	10,600	91.2	10,863
Sulfate	2,710	23.2	2,777
Magnesium	1,300	11.1	1,332
Calcium	405	3.48	415
Potassium	385	3.30	395
Carbonate and bicarbonate	122	1.05	125
Total principal ingredients	34,872	298.9	35,737
Others	128	1.1	131
Total dissolved solids	35,000	300.0	35,868

SOURCE: Compiled from Spiegler, "Salt Water Purification," and Ellis, "Fresh Water from the Ocean."

dissolved solids concentrations of 45,000 to 50,000 ppm, the proportion of the ions remains almost constant.

The **composition of brackish waters** varies so widely that no average analysis can be given. In addition to sodium chloride and sulfate, brackish well waters often contain substantial amounts of calcium, magnesium, bicarbonate, iron, and silica.

Desalination is the process by which seawater and brackish water are purified. It has been practiced for centuries, primarily aboard sailing vessels, but recently its application has grown significantly in the form of land-based facilities. The International Desalination Association reports that there are more than 9,000 commercial desalination plants in operation worldwide in 1995, whose total capacity exceeds approximately 4 billion gal per day.

Purification of seawater normally requires reduction of 35,000 ppm of total dissolved solids to less than 500 ppm, or a reduction of 70 to 1. For potable water, it is desirable to have a chloride content below 250 mg/L. The oldest method of purification of seawater is **distillation**. This technique has been practiced for over a century on oceangoing steamships. Distillation is used in over 95 percent of all seawater conversion plants and is principally accomplished by multiple-effect distillation, flash distillation, and vapor-compression distillation.

Purification of brackish water requires dissolved solids reductions ranging from a low of 3 to 1, to a high of 30 to 1. In this range, **membrane** desalting processes are more economical. Over 95 percent of all brackish water plants use either the reverse-osmosis process or the electrodialysis process.

In **flash distillation,** the salt water is heated in a tubular brine heater and then passed to a separate chamber where a pressure lower than that in the heating tubes prevails. This causes some of the hot salt water to vaporize, or "flash," such vapor then being condensed by cooler incoming salt water to produce pure distilled water. Flash distillation can be carried on in a number of successive stages (Fig. 6.10.3) wherein the heated salt water flashes to vapor in a series of chambers, each at a lower pressure than the preceding one. The higher the number of stages in such a **multistage flash** (MSF) system, the better the overall yield per heat unit. Today's large-scale MSF plants are designed with 15 to 30 stages and operate at performance ratios of 6 to 10, which means that for every 1 lb (kg) of steam, 6 to 10 lb (kg) of product water is produced.

An advantageous feature of flash distillation is that the separate heating of salt water without boiling causes fewer scale deposits. No scale occurs in the flash chambers as they contain no heated surfaces, the increased concentration of salts remaining in the seawater. Designs of MSF evaporators provide a great number of horizontal stages in one vessel by putting vertical partition walls inside the vessel (Fig. 6.10.3). All large MSF plants are designed this way.

If a saline water is made to boil, by heating it with steam through a heat-transfer surface, and the vapors from the saline water are subsequently condensed, a distillate of practically pure water (< 10 ppm salt) is obtained. This is known as **single-effect** distillation.

Fig. 6.10.3 Multiple-stage (four-stage) flash distillation process.

Multiple-effect distillation is achieved by using the condensation heat of the saline water vapor for boiling additional seawater in a following vessel or **effect** and repeating this step several more times. Each subsequent effect is operating at a somewhat lower temperature and vapor pressure than the previous one. In a single effect, 1 lb (kg) of steam will produce nearly 0.9 lb (kg) of distilled water, a double effect will yield 1.75 lb (kg) of product, and so on. In commercial seawater desalting plants, up to 17 effects have been employed.

Older multieffect plants used tubes submerged in seawater (**submerged-tube evaporators**), but scaling was always a serious problem and this type of design has been phased out in favor of thin-film designs. Only for small, low-temperature waste heat evaporators is a submerged-tube design sometimes applied. In a **thin-film** evaporator, saline water flows as a thin film of water through the inside of a vertical tube (VTE), or over the outside of a bundle of horizontal tubes (HT) after being sprayed on the bundle. Heating of the thin film of seawater is accomplished by condensing steam on the opposite side of the tube wall. See Figs. 6.10.4 and 6.10.5. Horizontal tube designs with vertically stacked effects have been developed.

In **vapor-compression distillation** (Fig. 6.10.6), the energy is supplied by a compressor which takes the vapor from boiling salt water and compresses it to a higher pressure and temperature to furnish the heat for vaporization of more seawater. In so doing, the vapor is condensed to yield distilled water. Vapor compression is theoretically a more efficient method of desalination than other distillation methods. The principle was widely applied during World War II in the form of small, portable units, and still is often used for relatively small plants of up to 200,000 gpd (32 m³/h). The disadvantages of this process lie in the cost, mechanical operation, and maintenance of the compressors.

Reverse osmosis (Figs. 6.10.7 and 6.10.8), in which a membrane permits fresh water to be forced through it but holds back dissolved solids, has now emerged as the most commonly used process for desalting brackish waters containing up to 10,000 ppm of total dissolved solids.

Two commercially available membrane configurations exist: (1) hollow thin-fiber membrane cartridges and (2) spiral-wound sheet membrane modules. Brackish water plants operate at pressures ranging from 250 to 400 lb/in² (17 to 27 bar), which is sufficient to reverse the osmotic flow between brackish water and fresh water.

In the last few years, high-pressure membranes capable of operating at 800 to 1,000 lb/in² (55 to 70 bar) have been commercialized for one-stage desalting of seawater. Several reverse-osmosis seawater plants exceeding 1 mgd (160 m³/h) are presently in operation in Saudi Arabia, Malta, Bahrain, and Florida. Recovery of power from the high-pressure brine discharged from the reverse-osmosis plant is economically attractive for seawater plants because of the high operating pressure and the relatively large quantities of brine (65 to 80 percent of feed) discharged.

Electrodialysis (Fig. 6.10.9) is a proven method of desalination, but where used on ocean water, it is not competitive. The purification of brackish or low-saline waters lends itself most advantageously to the process. The ions of dissolved salts are pulled out of saline water by electric forces and pass from the salt-water compartment into adjacent compartments through membranes which are alternately permeable to positively charged ions or to negatively charged ions.

Electrodialysis equipment typically contains 100 or more compartments between one set of electrodes. The number of cells and the amount of electric current required increase with the amount of purification to be done. A brackish water of 5,000 ppm max dissolved salts can be reduced to potable water with less than 500 ppm. Over 600 electrodialysis plants have been installed since the commercialization of the process 40 years ago. Most of these plants are relatively small: 50,000 gpd (8 m³/h) or less.

Other methods of purifying seawater include freezing and solar evaporation. The freezing process yields ice crystals of pure water, but a certain amount of salt water is trapped in the crystals which then has to be removed by washing, centrifugation, or other means. This process is promising because theoretically it takes less energy to freeze water than to distill it; but it is not yet in major commercial application because in several pilot plants the energy consumption was much higher, and the yield of ice much lower, than anticipated.

Solar evaporation produces about 0.1 gpd of water per ft² of area [0.17 L/(m² · h)], depending on climatic conditions. Capital and maintenance costs for the large areas of glass that are required so far have not made solar evaporation competitive.

A major problem in desalting processes, especially distillation, is the concentration of dissolved salts. Heating seawater above 150°F (65°C) causes **scale**, a deposition of insoluble salts on the heating surfaces, which rapidly reduces the heat transfer. For initial temperatures of 160 to 195°F (71 to 90°C), the addition of polyphosphates results in the formation of a sludge rather than a hard scale. This practice is widely used in ships and present land-based plants but is limited to 195°F (90°C). Some plants use acid treatment that permits boiling of seawater at 240 to 250°F (115 to 121°C). This improves the performance of the process and reduces costs.

In recent years, polymer additives have become available that allow operating temperatures of 230 to 250°F (110 to 121°C) and eliminate the hazardous handling of strong acids and the possibility of severe corrosion of equipment. In reverse osmosis and electrodialysis, scale formation on the membranes is also controlled by adding acid and other additives or by the prior removal of the hardness from the feed by lime softening. Reversing the polarity of the applied dc power can greatly minimize or entirely eliminate the need for pretreatment and additives for the electrodialysis process.

[*Editor's note:* The costs cited in the following paragraphs for equipment and energy reflect the market as of 1995–1996. The reader is advised to make guarded and rational use of the data inasmuch as the totality of current costs for a water treatment plant will include other than those for equipment and energy.]

Energy requirements for the various commercial desalting processes are shown in Fig. 6.10.10 in terms of primary energy sources, such as oil, coal, or gas. The method and arrangement by which primary energy is converted to usable energy, e.g., mechanical shaft power, steam, or electricity, has a profound impact on the energy requirements and cost of every desalting process.

Cost of desalting equipment at the 1-mgd level (160 m³/h), complete and installed, generally ranges from $6 to $8 per gpd ($37,500 to $50,000 per m³/h) installed capacity for both distillation and reverse-osmosis seawater plants.

Electrodialysis and reverse-osmosis brackish water plants of similar capacity usually range from $1.50 to $2.00 per gpd ($9,400 to $12,500 per m³/h). **Total operating costs,** including capital depreciation, power and chemical consumption, maintenance and operating personnel, for a 1-mgd (160-m³/h) facility fall into three categories: (1) seawater single-purpose distillation and reverse osmosis without power recovery = $6 to $8 per 1,000 gal ($1.60 to $2.10 per m³), (2) seawater dual-purpose distillation and reverse osmosis with power recovery = $4 to $6 per 1,000 gal ($1.00 to $1.50 per m³), and (3) brackish water electrodialysis and reverse osmosis = $1 to $2 per 1,000 gal ($0.25 to $0.50 per m³).

Fig 6.10.4 Vertical tube evaporator (VTE). Multieffect distillation process.

Fig 6.10.5 Horizontal tube multieffect evaporator (HTME). Multieffect distillation process.

Fig. 6.10.6 Vapor compression distillation process.

Fig. 6.10.7 Reverse-osmosis cell.

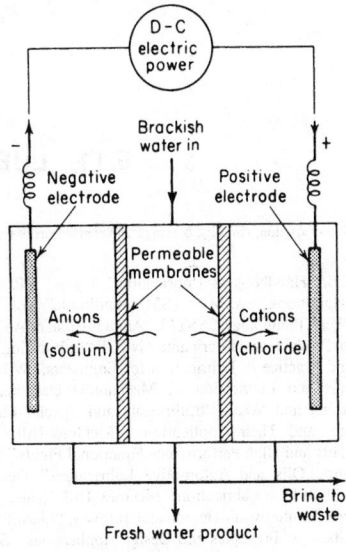

Fig. 6.10.9 Electrodialysis single-compartment process.

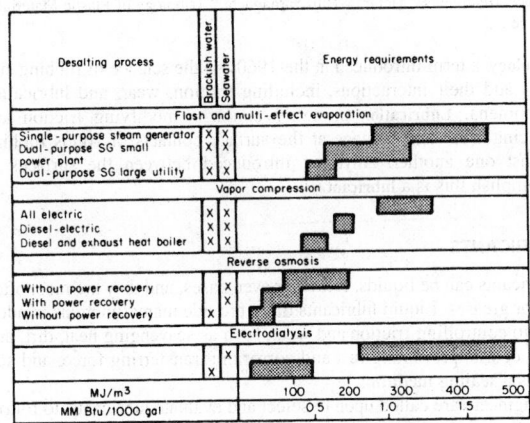

Fig. 6.10.10 Desalting primary energy requirements (*Ammerlaan. Desalination,* **40,** *1982, pp. 317–326.*) Note: 1 metric ton (t) of oil = 43,800 MJ; 1 barrel (bbl) of oil = 6.2 TBtu; 1 kWh = 10.9 MJ = 10,220 Btu (large utility).

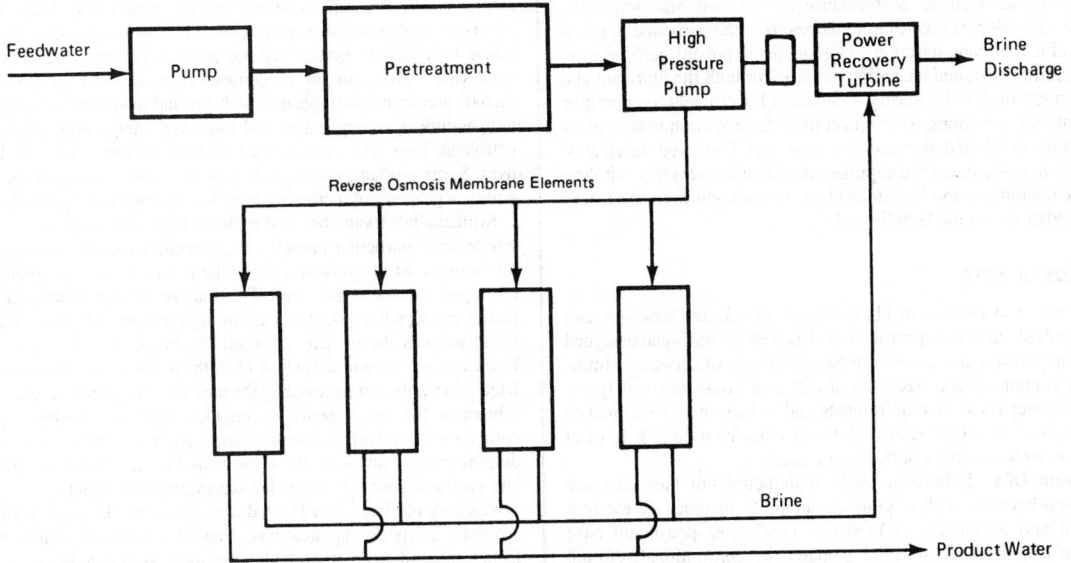

Fig 6.10.8 Reverse-osmosis plant block flow diagram.

6.11 LUBRICANTS AND LUBRICATION
by Glenn E. Asauskas

(For general discussion of friction, viscosity, bearings, and coefficients of friction, see Secs. 3 and 8.)

REFERENCES: STLE, "Handbook of Lubrication," 3 vols., CRC Press. ASME, "Wear Control Handbook." ASM, "ASM Handbook," vol. 18, "Friction, Lubrication, and Wear Technology." ASTM, "Annual Book of ASTM Standards," 3 vols., "Petroleum Products and Lubricants." NLGI, "Lubricating Grease Guide." Fuller, "Theory and Practice of Lubrication for Engineers," Wiley. Shigley and Mischke, "Bearings and Lubrication: A Mechanical Designers' Workbook," McGraw-Hill. Stipina and Vesely, "Lubricants and Special Fluids," Elsevier. Miller, "Lubricants and Their Applications," McGraw-Hill. Shubkin (ed.), "Synthetic Lubricants and High Performance Functional Fluids," Marcel Dekker. Bartz (ed.), "Engine Oils and Automotive Lubrication," Dekker. Hamrock, "Fundamentals of Fluid Film Lubrication," McGraw-Hill. Neale, "Lubrication: A Tribology Handbook," Butterworth-Heinemann. Ramsey, "Elastohydrodynamics," Wiley. Arnell, "Tribology, Principles and Design Applications," Springer-Verlag. Czichos, "Tribology, A Systems Approach to . . . Friction, Lubrication and Wear," Elsevier. Bhushan and Gupta, "Handbook of Tribology: Materials, Coatings, and Surface Treatments," McGraw-Hill. Yamaguchi, "Tribology of Plastic Materials," Elsevier.

Tribology, a term introduced in the 1960s, is the science of rubbing surfaces and their interactions, including friction, wear, and lubrication phenomena. **Lubrication** primarily concerns modifying friction and reducing wear and damage at the surface contacts of solids rubbing against one another. Anything introduced between the surfaces to accomplish this is a **lubricant.**

LUBRICANTS

Lubricants can be liquids, solids, or even gases, and they are most often oils or greases. Liquid lubricants often provide many functions in addition to controlling friction and wear, such as scavenging heat, dirt, and wear debris; preventing rust and corrosion; transferring force; and acting as a sealing medium.

Engineers are called upon to select and evaluate lubricants, to follow their performance in service, and to use them to best advantage in the design of equipment. Lubricant manufacturers and distributors may have hundreds of lubricants in their product line, each described separately in the product literature as to intended applications, properties, and benefits, as well as performance in selected standard tests. Lubricants are selected according to the needs of the particular application. Careful lubricant selection helps obtain improved performance, lower operating cost, and longer service life, for both the lubricant and the equipment involved. Industry's demands for efficient, competitive equipment and operations, which meet the latest environmental regulations, create continued demand for new and improved lubricants. Equipment manufacturers and suppliers specify lubricants that suit their particular equipment and its intended operating conditions, and their recommendations should be followed.

LIQUID LUBRICANTS

A liquid lubricant consists of (1) a mixture of selected base oils and additives, (2) blended to a specific viscosity, with (3) the blend designed to meet the performance needs of a particular type of service. A lubricant may contain several base oils of different viscosities and types, blended to meet viscosity requirements and to help meet performance requirements, or to minimize cost. Additives may be used to help meet viscosity as well as various performance needs.

Petroleum Oils Lubricants made with petroleum base oils are widely used because of their general suitability to much of existing equipment and availability at moderate cost. Most petroleum base oils, often called **mineral oils,** are prepared by conventional refining processes from naturally occurring hydrocarbons in crude oils. The main crude oil types are paraffinic- and naphthenic-based, terms referring to the type of chemical structure of most molecules. Paraffinic oils are most often preferred as lubricants, although naphthenic oils are important in certain applications.

The American Petroleum Institute (API) base stock categories were established in the early 1990s to make it easier for lubricant blenders to interchange one base stock for another (Table 6.11.1). The original five categories were patterned on the base oils which were available at the time. Since that time, the lines between the categories have become less distinct, but these categories are still useful in describing base oils.

Table 6.11.1 API Base Stock Categories

Group	Sulfur, wt %		Saturates, wt %	VI*
I	>0.03	and/or	<90	80–119
II	<0.03	and	>90	80–119
III	<0.03	and	>90	>120
IV		All Polyalphaolefins (PAO)		
V		All Stocks not included in Groups I-IV		
		(Pale Oils and non-PAO Synthetics)		

*All commercial paraffinic base oils have VI > 95.

In general, older-technology base oils, produced by solvent refining and solvent dewaxing technology, fall into group I. More modern base oils, using hydroprocessing technology, fall into groups II and III (group III if the VI is 120 or greater). Polyalphaolefins (PAOs) are group IV, and everything else is group V. Thus, group V can include quite low-quality base stocks such as naphthenic base oils, as well as very high-quality base stocks such as esters.

Synthetic oils are artificially made, as opposed to naturally occurring petroleum fluids. Synthetic oil types include polyalphaolefins (PAOs), diesters, polyol esters, alkylbenzenes, polyalkylene glycols, phosphate esters, silicones, and halogenated hydrocarbons. Synthetic oils include diverse types of chemical compounds, so few generalities apply to all synthetic oils. Generally, synthetic oils are organic chemicals. Synthetic oils cost much more than petroleum oils, and usually they have some lubricant-related properties that are good or excellent, but they are not necessarily strong in all properties. Their stronger properties often include greater high-temperature stability and oxidation resistance, wider temperature operating range, and improved energy efficiency. As with petroleum base oils, some of their weaknesses are correctable by additives. Some synthetic esters, such as polyol esters, have good biodegradability, a property of increasing need for environmental protection.

Synthetic lubricants, because of their high cost, often are used only where some particular property is essential. There are instances where the higher cost of synthetics is justified, even though petroleum lubricants perform reasonably well. This may occur, e.g., where use of synthetics reduces the oxidation rate in high-temperature service and as a result extends the oil life substantially. Properties of some synthetic lubricants are shown in Table 6.11.2, the assessments based on formulated lubricants, not necessarily the base oils. Synthetic oil uses include lubricants for gears, bearings, engines, hydraulic systems, greases, compressors, and refrigeration systems, although petroleum oils tend to dominate these applications. Lubricants blended with both petroleum and synthetic base oils are called **semisynthetic lubricants.**

Vegetable oils have very limited use as base stocks for lubricants, but are used in special applications. Naturally occurring esters, such as from rapeseed or castor plants, are used as the base oils in some

Table 6.11.2   Relative Properties of Synthetic Lubricants*

	Viscosity index	High-temperature stability	Lubricity	Low-temperature properties	Hydrolytic stability	Fire resistance	Volatility
Polyalphaolefins	Good	Good	Good	Good	Excellent	Poor	Good
Diesters	Varies	Excellent	Good	Excellent	Fair	Fair	Average
Polyol esters	Good	Excellent	Good	Good	Good	Poor	Average
Alkylbenzenes	Poor	Fair	Good	Good	Excellent	Poor	Average
Polyalkylene glycols	Excellent	Good	Good	Good	Good	Poor	Good
Phosphate esters	Poor	Excellent	Good	Varies	Fair	Excellent	Average
Silicones	Excellent	Excellent	Poor	Excellent	Fair	—	Good
Fluorinated lubes	Excellent	Excellent	Varies	Fair	Excellent	Excellent	Average

* Comparisons made for typical formulated products with additive packages, not for the base oils alone.
SOURCE: *Courtesy of Texaco's magazine Lubrication,* **78,** no. 4, 1992.

hydraulic fluids, and are used where there are environmental concerns of possibility of an accidental fluid spill or leakage. These oils have a shorter life and lower maximum service temperature, but have a high degree of biodegradability.

Fatty oils, extracted from vegetable, animal, and fish sources, have excellent lubricity because of their glyceride structure, but are seldom used as base oils because they oxidize rapidly. A few percent of fatty oils are sometimes added to petroleum oils to improve lubricity (as in worm-gear applications), or to repel moisture (as in steam engines), and this use is often referred to as **compounding.** They also are used widely in forming soaps which, in combination with other oils, set up grease structures.

Additives are chemical compounds added to base oils to modify or enhance certain lubricant performance characteristics. The amount of additives used varies with the type of lubricant, from a few parts per million in some types of lubricants to over 20 percent in some engine oils. The types of additives used include antioxidants, antiwear agents, extreme-pressure agents, viscosity index improvers, dispersants, detergents, pour point depressants, friction modifiers, corrosion inhibitors, rust inhibitors, metal deactivators, tackiness agents, antifoamants, air-release agents, demulsifiers, emulsifiers, odor control agents, and biocides.

LUBRICATION REGIMES

The viscosity of a lubricant and its additive content are to a large extent related to the lubrication regimes expected in its intended application. It is desirable, but not always possible, to design mechanisms in which the rubbing surfaces are totally separated by lubricant films. **Hydrostatic** lubrication concerns equipment designed such that the lubricant is supplied under sufficient pressure to separate the rubbing surfaces. **Hydrodynamic** lubrication is a more often used regime, and it occurs when the motion of one surface over lubricant on the other surface causes sufficient film pressure and thickness to build up to separate the surfaces. Achieving this requires certain design details, as well as operation within a specific range of speed, load, and lubricant viscosity. Increasing speed or viscosity increases the film thickness, and increasing load reduces film thickness. Plain journal bearings, widely used in automotive engines, operate in this regime. For concentrated nonconforming contacts, such as rolling-element bearings, cams, and certain type of gears, the surface shape and high local load squeeze the oil film toward surface contact. Under such conditions, **elastohydrodynamic** lubrication (EHL) may occur, where the high pressures in the film cause it to become very viscous and to elastically deform and separate the surfaces.

Boundary lubrication occurs when the load is carried almost entirely by the rubbing surfaces, separated by films of only molecular dimensions. **Mixed-film lubrication** occurs when the load is carried partly by the oil film and partly by contact between the surfaces' roughness peaks (asperities) rubbing one another. In mixed-film and boundary lubrication, lubricant viscosity becomes less important and chemical composition of the lubricant film more important. The films, although very thin, are needed to prevent surface damage. At loads up to a certain level, antiwear additives, the most common one being zinc dithiophosphate

(ZDTP), are used in the oil to form films to help support the load and reduce wear. ZDTP films can have thicknesses from molecular to much larger dimensions. In very highly loaded contacts, such as hypoid gears, **extreme-pressure** (EP) additives are used to minimize wear and prevent scuffing. EP additives may contain sulfur, phosphorus, or chlorine, and they react with the high-temperature contact surfaces to form molecular-dimension films. Antiwear, EP, and friction modifier additives function only in the mixed and boundary regimes.

LUBRICANT TESTING

Lubricants are manufactured to have specific characteristics, defined by physical and chemical properties, and performance characteristics. Most of the tests used by the lubricants industry are described in ASTM (American Society for Testing and Materials) Standard Methods. **Physical tests** are frequently used to characterize petroleum oils, since lubricant performance often depends upon or is related to physical properties. Common physical tests include measurements of viscosity, density, pour point, API gravity, flash point, fire point, odor, and color. **Chemical tests** measure such things as the amount of certain elements or compounds in the additives, or the acidity of the oil, or the carbon residue after heating the oil at high temperature. **Performance tests** evaluate particular aspects of in-service behavior, such as oxidation stability, rust protection, ease of separation from water, resistance to foaming, and antiwear and extreme-pressure properties. Many of the tests used are small-size or bench tests, for practical reasons. To more closely simulate service conditions, full-scale standardized performance tests are required for certain types of lubricants, such as oils for gasoline and diesel engines. Beyond these industry standard tests, many equipment builders require tests in specific machines or in the field. The full scope of lubricant testing is rather complex, as simple physical and chemical tests are incapable of fully defining in-service behavior.

VISCOSITY TESTS

Viscosity is perhaps the single most important property of a lubricant. Viscosity needs to be sufficient to maintain oil films thick enough to minimize friction and wear, but not so viscous as to cause excessive efficiency losses. Viscosity reduces as oil temperature increases. It increases at high pressure and, for some fluids, reduces from shear, during movement in small clearances.

Kinematic viscosity is the most common and fundamental viscosity measurement, and it is obtained by oil flow by gravity through capillary-type instruments at low-shear-rate flow conditions. It is measured by ASTM Standard Method D445. A large number of commercially available capillary designs are acceptable. Automated multiple-capillary machines also are sold. Kinematic viscosity is typically measured at 40 and 100°C and is expressed in centistokes, cSt (mm^2/s).

Saybolt viscosity, with units of Saybolt universal seconds (SUS), was once widely used. There is no longer a standard method for direct measurement of SUS. Some industrial consumers continue to use SUS, and it can be determined from kinematic viscosity using ASTM D2161.

Dynamic viscosity is measured for some lubricants at low temperatures (ASTM D4684, D5293, and D2983); it is expressed in centipoise cP (mPa · s). Kinematic and dynamic viscosities are related by the equation cSt = cP/density, where density is in g/cm^3.

Viscosity grades, not specific viscosities, are used to identify viscosities of lubricants in the marketplace, each viscosity grade step in a viscosity grade system having minimum and maximum viscosity limits. There are different viscosity grade systems for automotive engine oils, gear oils, and industrial oils. Grade numbers in any one system are independent of grade numbers in other systems.

ISO Viscosity Grades   The International Standards Organization (ISO) has a viscosity grade system (ASTM D2422) for industrial oils. Each grade is identified by ISO VG followed by a number representing the nominal 40°C (104°F) kinematic viscosity in centistokes for that grade. This system contains 18 viscosity grades, covering the range from 2 to 1,500 cSt at 40°C, each grade being about 50 percent higher in viscosity than the previous one. The other viscosity grade systems are described later.

The **variation of viscosity with temperature** for petroleum oils can be determined from the viscosities at any two temperatures, such as 40 and 100°C kinematic viscosity, using a graphical method or calculation (ASTM D341). The method is accurate only for temperatures above the wax point and below the initial boiling point, and for oils not containing certain polymeric additives.

The **viscosity index** (VI) is an empirical system expressing an oil's rate of change in viscosity with change in temperature. The system, developed some years ago, originally had a 0 to 100 VI scale, based on two oils thought to have the maximum and minimum limits of viscosity-temperature sensitivity. Subsequent experience identified oils far outside the VI scale in both directions. The VI system is still used, and the VI of an oil can be calculated from its 40 and 100°C kinematic viscosity using ASTM D2270, or by easy-to-use tables (ASTM Data Series, DS 39B). Lubricants with high VI have less change in viscosity with temperature change, desirable for a lubricant in service where temperatures vary widely.

Base oils with more than 100 VI can be made from a wide variety of crude oil distillates by solvent refining, by hydrogenation, by selective blending of paraffinic base oils, by adding a few percent of highmolecular-weight polymeric additives called **viscosity index improvers,** or by combinations of these methods. A few highly processed petroleum base oils and many synthetic oils have high VIs.

Effect of Pressure on Viscosity   When lubricating oils are subjected to high pressures, thousands of lb/in^2, their viscosity increases. When oil-film pressures are in this order of magnitude, their influence on viscosity becomes significant. Empirical equations are available to estimate the effect of pressure on viscosity. In rolling-contact bearings, gears, and other machine elements, the high film pressures will influence viscosity with an accompanying increase in frictional forces and load-carrying capacity.

Effect of Shear on Viscosity   The viscosity of lubricating oils not containing polymeric additives is independent of **shear rate,** except at low temperatures where wax is present. Where polymeric additives are used, as in multigrade engine oils, high shear rates in lubricated machine elements cause **temporary viscosity loss,** as the large polymer molecules temporarily align in the direction of flow. Engine oils consider this effect by high-temperature/high-shear-rate measurements (ASTM D4683 or ASTM D4741). Such oils can suffer **permanent viscosity loss** in service, as polymer molecules are split by shear stresses into smaller molecules with less thickening power. The amount of the permanent viscosity loss depends on the type of polymeric additive used, and it is measurable by comparing the viscosity of the new oil to that of the used oil, although other factors need considering, such as any oxidative thickening of the oils.

OTHER PHYSICAL AND CHEMICAL TESTS

Cloud Point   Petroleum oils, when cooled, may become plastic solids, either from wax formation or from the fluid congealing. With some oils, the initial wax crystal formation becomes visible at temperatures slightly above the solidification point. When that temperature is reached at specific test conditions, it is known as the **cloud point** (ASTM D2500). The cloud point cannot be determined for those oils in which wax does not separate prior to solidification or in which the separation is invisible.

The cloud point indicates the temperature below which clogging of filters may be expected in service. Also, it identifies a temperature below which viscosity cannot be predicted by ASTM D2270, as the wax causes a higher viscosity than is predicted.

Pour point is the temperature at which cooled oil will just flow under specific test conditions (ASTM D97). The pour point indicates the lowest temperature at which a lubricant can readily flow from its container. It provides only a rough guide to its flow in machines. Pour point depressant additives are often used to reduce the pour point, but they do not affect the cloud point. Other tests, described later, are used to more accurately estimate oil flow properties at low temperatures in automotive service.

Density and Gravity   ASTM D1298 may be used for determining density (mass), specific gravity, and the API (American Petroleum Institute) gravity of lubricating oils. Of these three, petroleum lubricants tend to be described by using API gravity. The specific gravity of an oil is the ratio of its weight to that of an equal volume of water, both measured at 60°F (16°C), and the API gravity of an oil can be calculated by

$$\text{API gravity, deg} = (141.5/\text{sp. gr. @ } 60°/60°\text{F}) - 131.5$$

The gravity of lubricating oils is of no value in predicting their performance. Low-viscosity oils have higher API gravities than higherviscosity oils of the same crude-oil series. Paraffinic oils have the lowest densities or highest API gravities, naphthenic are intermediate, and animal and vegetable oils have the heaviest densities or lowest API gravities.

Flash and Fire Points   The flash point of an oil is the temperature to which an oil has to be heated until sufficient flammable vapor is driven off to flash when brought into momentary contact with a flame. The fire point is the temperature at which the oil vapors will continue to burn when ignited. ASTM D92, which uses the Cleveland open-cup (COC) tester, is the flash and fire point method used for lubricating oils. In general, for petroleum lubricants, the open flash point is 30°F or (17°C) higher than the closed flash (ASTM D93), and the fire point is some 50 to 70°F or (28 to 39°C) above the open flash point.

Flash and fire points may vary with the nature of the original crude oil, the narrowness of the distillation cut, the viscosity, and the method of refining. For the same viscosities and degree of refinement, paraffinic oils have higher flash and fire points than naphthenic oils. While these values give some indication of fire hazard, they should be taken as only one element in fire risk assessment.

Color   The color of a lubricating oil is obtained by reference to transmitted light; the color by reflected light is referred to as **bloom.** The color of an oil is not a measure of oil quality, but indicates the uniformity of a particular grade or brand. The color of an oil normally will darken with use. ASTM D1500 is for visual determination of the color of lubricating oils, heating oils, diesel fuels, and petroleum waxes, using a standardized colorimeter. Results are reported in terms of the ASTM color scale and can be compared with the former ASTM union color. The color scale ranges from 0.5 to 8; oils darker than color 8 may be diluted with kerosene by a prescribed method and then observed. Very often oils are simply described by visual assessment of color: brown, black, etc. For determining the color of petroleum products lighter than 0.5, the ASTM D156 Test for Saybolt Color of Petroleum Products can be used.

Carbon residue, the material left after heating an oil under specified conditions at high temperature, is useful as a quality control tool in the refining of viscous oils, particularly residual oils. It does not correlate with carbon-forming tendencies of oils in internal-combustion engines. Determination is most often made by the Conradson procedure (ASTM D189). Values obtained by the more complex Ramsbottom procedure (ASTM D524) are sometimes quoted. The correlation of the two values is given in both methods.

Ash Although it is unlikely that well-refined oils that do not contain metallic additives will yield any appreciable ash from impurities or contaminants, measurement can be made by ASTM D482. A more useful determination is **sulfated ash** by ASTM D874, as applied to lubricants containing metallic additives.

Neutralization number and **total acid number** are interchangeable terms indicating a measure of acidic components in oils, as determined by ASTM D664 or D974. The original intent was to indicate the degree of refining in new oils, and to follow the development of oxidation in service, with its effects on deposit formation and corrosion. However, many modern oils contain additives which, in these tests, act similar to undesirable acids and are indistinguishable from them. Caution must be exercised in interpreting results without knowledge of additive behavior. Change in acid number is more significant than the absolute value in assessing the condition of an oil in service. Oxidation of an oil is usually accompanied by an increase in acid number.

Total base number (TBN) is a measure of alkaline components in oils, especially those additives used in engine oils to neutralize acids formed during fuel combustion. Some of these acids get in the crankcase with gases that blow by the rings. Today, TBN is generally measured by ASTM D2896 or D4739, as it provides a better indication of an oil's ability to protect engines from corrosive wear than ASTM D664, which was previously used for TBN measurement.

Antifoam The foaming characteristic of crankcase, turbine, or circulating oils is checked by a foaming test apparatus (ASTM D892), which blows air through a diffuser in the oil. The test measures the amount of foam generated and the rate of settling. Certain additive oils tend to foam excessively in service. Many types of lubricants contain a minute quantity of antifoam inhibitor to prevent excessive foaming in service. With the advent of the HEUI (hydraulically actuated, electronically controlled unit injector) fuel injection system, in diesel trucks, the Navistar 7.3L oil aeration test has become standard in API classification CG-4, CH-4, Cl-4, and Cl-4 plus.

Oxidation Tests Lubricating oils may operate at relatively high temperatures in the presence of air and catalytically active metals or metallic compounds. Oxidation becomes significant when oil is operated above 150°F (66°C). Some lubricating-oil sump temperatures may exceed 250°F (121°C). The oxidation rate doubles for about every 18°F (10°C) rise in oil temperature above 150°F (66°C). The resultant oil oxidation increases viscosity, acids, sludge, and lacquer.

To accelerate oxidation to obtain practical test length, oxidation tests generally use temperatures above normal service, large amounts of catalytic metals, and sufficient oxygen. Two standard oxidation tests, ASTM D943 and D2272, are often applied to steam turbine and similar industrial oils. There are many other standard oxidation tests for other specific lubricants. Because test conditions are somewhat removed from those encountered in service, care should be exercised in extrapolating results to expected performance in service.

Rust Prevention Lubricating oils are often expected to protect ferrous surfaces from rusting when modest amounts of water enter the lubricating system. Many oils contain additives specifically designed for that purpose. Steam turbine and similar industrial oils often are evaluated for rust prevention by ASTM D665 and D3603. Where more protection is required, as when a thin oil film must protect against a moisture-laden atmosphere, ASTM D1748 is a useful method.

Water Separation In many applications, it is desirable for the oil to have good water-separating properties, to enable water removal before excessive amounts accumulate and lead to rust and lubrication problems. The demulsibility of an oil can be measured by ASTM D1401 for light- to moderate-viscosity oils and by ASTM D2711 for heavier-viscosity oils.

Wear and EP Tests A variety of test equipment and methods have been developed for evaluating the wear protection and EP properties **(load-carrying capacity before scuffing)** of lubricant fluids and greases under different types of heavy-duty conditions. The tests are simplified, accelerated tests, and correlations to service conditions are seldom available, but they are important in comparing different manufacturers' products and are sometimes used in lubricant specifications. Each apparatus tends to emphasize a particular characteristic, and a given lubricant will not necessarily show the same EP properties when tested on different machines. ASTM methods D2783 (fluids) and D2509 (grease) use the **Timken EP tester** to characterize the EP properties of industrial-type gear lubricants, in tests with incremented load steps. ASTM D3233 (fluids) uses the **Falex pin** and **vee block tester** for evaluating the EP properties of oils and gear lubricants, using either continuous (method A) or incremented (method B) load increase during the test. Evaluation of wear-prevention properties also can be conducted using this apparatus, by maintaining constant load. ASTM methods D2783 (fluids) and D2596 (grease) use the **four-ball EP tester** for evaluating the EP properties of oils and greases. ASTM methods D4172 (fluids) and D2266 (grease) use the **four-ball wear tester** to evaluate the wear-prevention properties of oils and grease.

GREASES

Lubricating greases consist of lubricating liquid dispersed in a thickening agent. The lubricating liquid is about 70 to 95 wt % of the finished grease and provides the principal lubrication, while the thickener holds the oil in place. Additives are often added to the grease to impart special properties. Grease helps seal the lubricating fluid in and the contaminants out. Grease also can help reduce lubrication frequency, particularly in intermittent operation.

The lubricating liquid is usually a napthenic or paraffinic petroleum oil or a mixture of the two. Synthetic oils are used also, but because of their higher cost, they tend to be used only for specialty greases, such as greases for very low or high temperature use. Synthetic oils used in greases include polyalphaolefins, dialkylbenzenes, dibasic acid esters, polyalkylene glycols, silicones, and fluorinated hydrocarbons. The viscosity of the lubricating liquid in the grease should be the same as would be selected if the lubricant were an oil.

Soaps are the most common thickeners used in greases. Complex soaps, pigments, modified clays, chemicals (such as polyurea), and polymers are also used, alone or in combination. Soaps are formed by reaction of fatty material with strong alkalies, such as calcium hydroxide. In this saponification reaction, water, alcohol, and/or glycerin may be formed as by-products. Soaps of weak alkalies, such as aluminum, are formed indirectly through further reactions. Metallic soaps, such as sodium, calcium, and lithium, are the most widely used thickeners today.

Additives may be incorporated in the grease to provide or enhance tackiness, to provide load-carrying capability, to improve resistance to oxidation and rusting, to provide antiwear or EP properties, and to lessen sensitivity to water. The additives sometimes are solid lubricants such as graphite, molybdenum disulfide, metallic powders, or polymers, which grease can suspend better than oil. Solid fillers may be used to improve grease performance under extremely high loads or shock loads.

The **consistency** or **firmness** is the characteristic which may cause grease to be chosen over oil in some applications. Consistency is determined by the depth to which a cone penetrates a grease sample under specific conditions of weight, time, and temperature. In ASTM D217, a standardized cone is allowed to drop into the grease for 5 s at 77°F (25°C). The resulting depth of fall, or **penetration,** is measured in tenths of millimeters. If the test is made on a sample simply transferred to a standard container, the results are called **unworked penetration.** To obtain a more uniform sample, before the penetration test, the grease is worked for 60 strokes in a mechanical worker, a churnlike device. Variations on unworked and worked penetrations, such as prolonged working, undisturbed penetration, or block penetration, also are described in ASTM D217. The worked 60-stroke penetration is the most widely used indication of grease consistency. The National Lubricating Grease Institute (NLGI) used this penetration method to establish a system of **consistency numbers** to classify differences in grease consistency or firmness. The consistency numbers range from 000 to 6, each representing a range of 30 penetration units, and with a 15 penetration unit gap between each grade, as shown in Table 6.11.3.

As grease is heated, it may change gradually from a semisolid to a liquid state, or its structure may weaken until a significant amount of oil is lost. Grease does not exhibit a true melting point, which implies a sharp change in state. The weakening or softening behavior of grease when heated to a specific temperature can be defined in a carefully controlled heating program with well-defined conditions. This test gives a temperature called the **dropping point,** described in ASTM D566 or

Table 6.11.3   Grease Consistency Ranges

Consistency no.	Appearance	Worked penetration
000	Semifluid	445–475
00	Semifluid	400–430
0	Semifluid	355–385
1	Soft	310–340
2	Medium	265–295
3	Medium hard	220–250
4	Hard	175–205
5	Very hard	130–160
6	Block type	85–115

D2265. Typical dropping points are given in Table 6.11.4. At its **dropping point,** a grease already has become fluid, so the maximum application temperature must be well below the dropping point shown in the table.

The oil constituent of a grease is loosely held. This is necessary for some lubrication requirements, such as those of ball bearings. As a result, on opening a container of grease, free oil is often seen. The tendency of a grease to **bleed** oil may be measured by ASTM D1742. Test results are related to bleeding in storage, not in service or at elevated temperatures.

Texture of a grease is determined by formulation and processing. Typical textures are shown in Table 6.11.4. They are useful in identification of some products, but are of only limited assistance for predicting behavior in service. In general, smooth, buttery, short-fiber greases are preferred for rolling-contact bearings and stringy, fibrous products for sliding service. Stringiness or tackiness imparted by polymeric additives helps control leakage, but this property may be diminished or eliminated by the shearing action encountered in service.

Grease **performance** depends only moderately on physical test characteristics, such as penetration and dropping point. The likelihood of suitable performance may be indicated by these and other tests. However, the final determination of how a lubricating grease will perform in service should be based on observations made in actual service.

SOLID LUBRICANTS

Solid lubricants are materials with low coefficients of friction compared to metals, and they are used to reduce friction and wear in a variety of applications. There are a large number of such materials, and they include graphite, molybdenum disulfide, polytetrafluoroethylene, talc, graphite fluoride, polymers, and certain metal salts. The many diverse types have a variety of different properties, operating ranges, and methods of application. The need for lubricants to operate at extremes of temperature and environment beyond the range of organic fluids, such as in the space programs, helped foster development of solid lubricants.

One method of using solid lubricants is to apply them as thin films on the bearing materials. The film thicknesses used range from 0.0002 to 0.0005 in (0.005 to 0.013 mm). There are many ways to form the films. Surface preparation is very important in all of them. The simplest method is to apply them as **unbonded solid lubricants,** where granular or powdered lubricant is applied by brushing, dipping, or spraying, or in a liquid or gas carrier for ease of application. Burnishing the surfaces is beneficial. The solid lubricant adheres to the surface to some degree by mechanical or molecular action. More durable films can be made by using **bonded solid lubricants,** where the lubricant powders are mixed with binders before being applied to the surface. Organic binders, if used, can be either room-temperature air-cured type or those requiring thermal setting. The latter tend to be more durable. Air-cured films are generally limited to operating temperatures below 300°F (260°C), while some thermoset films may be satisfactory to 700°F (371°C). Inorganic ceramic adhesives combined with certain powdered metallic solid lubricants permit film use at temperatures in excess of 1,200°F (649°C). The performance of solid-lubricant films is influenced by the solid lubricant used, the method of application, the bearing-surface-material finish and its hardness, the binder-solid lubricant mix, the film and surface pretreatment, and the application's operating conditions. Solidlubricant films can be evaluated in certain standard tests, including ASTM D2510 (adhesion), D2511 (thermal shock sensitivity), D2625 (wear endurance and load capacity), D2649 (corrosion characteristics), and D2981 (wear life by oscillating motion).

Many plastics are also solid lubricants compared to the friction coefficients typical of metals. Plastics are used without lubrication in many applications. Strong plastics can be compounded with a variety of solid lubricants to make plastics having both strength and low friction. Solidlubricant powders and plastics can be mixed, compacted, and sintered to form a lubricating solid, such as for a bearing. Such materials also can be made by combining solid-lubricant fibers with other stronger plastic fibers, either woven together or chopped, and used in compressionmolded plastics. Bushings made of low-friction plastic materials may be press-fitted into metal sleeves. The varieties of low-friction plastic materials are too numerous to mention. Bearings made with solid lowfriction plastic materials are commercially available.

Solid lubricants may also be embedded in metal matrices to form solid-lubricant composites of various desirable properties. Graphite is the most common solid lubricant used, and others include molybdenum disulfide and other metal-dichalcogenides. The base metals used include

Table 6.11.4   General Characteristics of Greases

Base	Texture	Typical dropping point, °F(°C)	Max usable temperature,* °F(°C)	Water resistance	Primary uses
Sodium	Fibrous	325–350(163–177)	200 (93)	P–F	Older, slower bearings with no water
Calcium	Smooth	260–290(127–143)	200 (93)	E	Moderate temperature, water present
Lithium 12-hydroxystearate	Smooth	380–395(193–202)	250(121)	G	General bearing lubrication
Polyurea	Smooth	450+(232+)	350(177)	G–E	Sealed-for-life bearings
Calcium complex	Smooth	500+(260+)	350(177)	F–E	
Lithium complex	Smooth	450+(232+)	325(163)	G–E	Used at high temperatures, with frequent relubrication
Aluminum complex	Smooth	500+(260+)	350(177)	G–E	
Organoclay	Smooth	500+(260+)	250(121)	F–E	

* Continuous operation with relatively infrequent lubrication.

copper, aluminum, magnesium, cadmium, and others. These materials are made by a variety of processes, and their uses range from electrical contacts to bearings in heavy equipment.

LUBRICATION SYSTEMS

There are many positive methods of applying lubricating oils and greases to ensure proper lubrication. Bath and circulating systems are common. There are constant-level lubricators, gravity-feed oilers, multiple-sight feeds, grease cups, forced-feed lubricators, centralized lubrication systems, and air-mist lubricators, to name a few. Selecting the method of lubricant application is as important as the lubricant itself. The choice of the device and the complexity of the system depend upon many factors, including the type and quantity of the lubricant, reliability and value of the machine elements, maintenance schedules, accessibility of the lubrication points, labor costs, and other economic considerations, as well as the operating conditions.

The importance of removing foreign particles from circulating oil has been increasingly recognized as a means to minimize wear and avoid formation of potentially harmful deposits. Adequate **filtration** must be provided to accomplish this. In critical systems involving closely fitting parts, monitoring of particles on a regular basis may be undertaken.

LUBRICATION OF SPECIFIC EQUIPMENT

Internal-combustion-engine oils are required to carry out numerous functions to provide adequate lubrication. Crankcase oils, in addition to reducing friction and wear, must keep the engine clean and free from rust and corrosion, must act as a coolant and sealant, and must serve as a hydraulic oil in engines with hydraulic valve lifters. The lubricant may function over a wide temperature range and in the presence of atmospheric dirt and water, as well as with combustion products that blow by the rings into the crankcase. It must be resistant to oxidation, sludge, and varnish formation in a wide range of service conditions.

Engine oils are heavily fortified with additives, such as **detergents,** to prevent or reduce deposits and corrosion by neutralizing combustion by-product acids; **dispersants,** to help keep the engine clean by solubilizing and dispersing sludge, soot, and deposit precursors; **oxidation inhibitors,** to minimize oil oxidation, particularly at high temperatures; **corrosion inhibitors,** to prevent attack on sensitive bearing metals; **rust inhibitors,** to prevent attack on iron and steel surfaces

by condensed moisture and acidic corrosion products, aggravated by low-temperature stop-and-go operation; **pour point depressants,** to prevent wax gelation and improve low-temperature flow properties; **viscosity-index improvers,** to help enable adequate low-temperature flow, along with sufficient viscosity at high temperatures; **antiwear** additives, to minimize wear under boundary lubrication conditions, such as cam and lifter, and cylinder-wall and piston-ring surfaces; **defoamants,** to allow air to break away easily from the oil; and **friction modifiers,** to improve fuel efficiency by reducing friction at rubbing surfaces.

The **SAE viscosity grade system for engine oils** (SAE J300), shown in Table 6.11.5 provides W (winter) viscosity grade steps (for example, 5W) for oils meeting certain primarily low-temperature viscosity limits, and non-W grade steps (for example, 30) for oils meeting high-temperature viscosity limits. The system covers both single-grade oils (for example, SAE 30), not designed for winter use, and multigrade oils (for example, SAE 5 W-30), formulated for year-round service. This system is subject to revision as improvements are made, and the table shows the December 1995 version. Each W grade step has maximum low-temperature viscosity limits for cranking (D5293) and pumping (D4684) bench tests, and a minimum 100°C kinematic viscosity limit. The lower the W grade, the colder the potential service use. Pumping viscosity limits are based on temperatures 10°C below those used for cranking limits, to provide a safety margin so that an oil that will allow engine starting at some low temperature also will pump adequately at that temperature. Each non-W grade step has a specific 100°C kinematic viscosity range and a minimum 150°C high-shear-rate viscosity limit. High-temperature high-shear-rate viscosity limits are a recent addition to J300, added in recognition that critical lubrication in an engine occurs largely at these conditions. Two 150°C high-shear-rate viscosity limits are shown for 40 grade oils, each for a specific set of W grades, the limit being higher for viscosity grades used in heavy-duty diesel engine service.

Gasoline Engines Two systems are used at this time to define gasoline-engine-oil performance levels, not counting systems outside the United States. The API service category system has been used for many years and is the result of joint efforts of many organizations. Since 1970, gasoline engine performance categories in this system have used the letter S (spark ignition) followed by a second letter indicating the specific performance level. Performance category requirements are based primarily on performance in various standard engine tests. The categories available change periodically as higher

Table 6.11.5   SAE Viscosity Grades for Engine Oil[a] (SAE J300 DEC95)

SAE viscosity grade	Low temperature viscosity, cP, max, at temp, °C, max		Viscosity,[d] cSt, at 100°C		High-shear-rate viscosity,[e] cP, min, at 150°C and 10^6 s^{-1}
	Cranking[b]	Pumping[c]	Min	Max	
0W	3,250 at −30	60,000 at −40	3.8	—	—
5W	3,500 at −25	60,000 at −35	3.8	—	—
10W	3,500 at −20	60,000 at −30	4.1	—	—
15W	3,500 at −15	60,000 at −25	5.6	—	—
20W	4,500 at −10	60,000 at −20	5.6	—	—
25W	6,000 at −5	60,000 at −15	9.3	—	—
20	—	—	5.6	< 9.3	2.6
30	—	—	9.3	<12.5	2.9
40	—	—	12.5	<16.3	2.9[f]
40	—	—	12.5	<16.3	3.7[g]
50	—	—	16.3	<21.9	3.7
60	—	—	21.9	<26.1	3.7

[a] All values are critical specifications as defined in ASTM D3244.
[b] ASTM D5293.
[c] ASTM D4684. Note that the presence of any yield stress detectable by this method constitutes a failure, regardless of viscosity.
[d] ASTM D445   Note: 1 cP = 1 mPa · s   1 cSt = 1 mm²/s
[e] ASTM D4683, CEC L-36-A-90 (ASTM D4741)
[f] 0W-40, 5W-40, and 10W-40 grade oils.
[g] 15W-40, 20W-40, 25W-40, and 40 grade oils.

performance categories are needed for newer engines, and as older categories become obsolete when tests on which they are based are no longer available. At this writing, SM is the highest performance level for gasoline engines.

The newer ILSAC system is based primarily on the same tests, but has a few additional requirements. This system was developed by the International Lubricant Standardization and Approval Committee (ILSAC), a joint organization of primarily U.S. and Japanese automobile and engine manufacturers. The first category is in use, ILSAC GF-1.

There are similarities as well as distinct differences between the two systems. Both are displayed on oil containers using API licensed marks. The API service category system uses the API service symbol, and ILSAC system uses the API certification mark, both shown in Fig. 6.11.1. Either or both marks can be displayed on an oil container. In the API service symbol, the performance level (SM), viscosity grade (SAE 5W30), and energy conserving for 5W20, 5W30, and 10W30 viscosity grades

API Service Symbol   API Certification Mark

Fig. 6.11.1 API marks for engine oil containers. *(From API Publication 1509, 12th ed., 1993. Reprinted by courtesy of the American Petroleum Institute.)*

are shown. None of these are shown in the API certification mark of the ILSAC system. The API certification mark remains unchanged as new performance levels are developed and regardless of the viscosity grade of the oil in the container. ILSAC categories require the highest energy-conserving level, and only allow a few viscosity grades, those recommended for newer engines. The viscosity grades allowed are SAE OW-20, 5W-20, 5W-30, and 10W-30. The viscosity grade may be displayed separately. In the ILSAC system, to ensure high-performance-level oils, the API certification mark is licensed annually and only for oils meeting the latest ILSAC performance level.

Diesel Engines API performance-level categories for commercial heavy-duty diesel engine oils, using the API service category system, begin with the letter C (compression ignition), followed by additional letters and numbers indicating the performance level and engine type, the performance letter changing as category improvements are needed. At this writing, Cl-4 plus is the highest performance category for four-cycle turbocharged heavy-duty diesel engines, and CF-2 is the highest for two-cycle turbocharged heavy-duty diesel engines. Viscosity grades used depend on the engine manufacturer's recommendations for the ambient temperature range. SAE 15W-40 oils are very popular for most four-cycle diesel engines, and single-grade oils, such as SAE 40, are used primarily in hotter climates and for two-cycle diesel engines.

The engine oil is increasingly becoming a major part in the total design, affecting performance of critical components as well as engine life. Emission control restrictions are tightening up on diesel engines, and this is affecting the engine oils. Designs of pistons are changing, and rings are being located higher on the pistons and are more subject to carbon deposits. Deposits are being controlled partly by use of proper engine oils. One new engine design with reduced emissions uses the engine oil as hydraulic fluid to operate and improve control of the fuel injectors, and this has created a need for improved oil deaeration performance. Most engine manufacturers maintain lists of oils that perform well in their engines.

Larger diesel engines found in marine, railroad, and stationary service use crankcase oils that are not standardized as to performance

levels, unlike the oils discussed above. They are products developed by reputable oil suppliers working with major engine builders. In general, they are formulated along the same lines as their automotive counterparts. However, particularly in marine service where fuels often contain relatively high levels of sulfur, the detergent additive may be formulated with a high degree of alkalinity to neutralize sulfuric acid resulting from combustion.

Gas engines pose somewhat different lubrication requirements. They burn a clean fuel which gives rise to little soot, but conditions of operation are such that nitrogen oxides formed during combustion can have a detrimental effect on the oil. Suitable lubricants may contain less additive than those which must operate in a sootier environment. However, the quality of the base oil itself assumes greater importance. Special selection of crude source and a high degree of refining must be observed to obtain good performance. The most common viscosity is SAE 40, although SAE 30 is also used. Where cold starting is a factor, SAE 15W-40 oils are available.

Steam Turbines Although they do not impose especially severe lubrication requirements, steam turbines are expected to run for very long periods, often measured in many years, on the same oil charge. Beyond that, the oil must be able to cope with ingress of substantial amounts of water. Satisfactory products consist of highly refined base oils with rust and oxidation (R&O) inhibitors, and they must show good water-separating properties (demulsifiability). There is demand for longer-life turbine oil in the power generation industry, such as 6,000 h or more of life in the ASTM D943 oxidation bench test. This demand is being met by using more severely processed base oils and with improved R&O inhibitors. Most large utility turbines operate with oil viscosity ISO VG 32. Where gearing is involved, ISO VG 100 is preferred. Smaller industrial turbines, which do not have circulating systems typical of large units, may operate with ISO VG 68. Circulating systems should be designed with removal of water in mind, whether by centrifuge, coalescer, settling tank, or a combination of these methods. Where the turbine governor hydraulic system is separate from the central lubrication system, fire-resistant phosphate ester fluids are often used in the governor hydraulic system.

Gas turbines in industrial service are lubricated with oils similar to the ISO VG 32 product recommended for steam turbines; but to obtain acceptable service life under higher-temperature conditions, they may be inhibited against oxidation to a greater extent. There is demand for longer-life oils for gas turbines, and, as for steam turbines, the demand is being met by using more severely processed base oils and improved R&O inhibitors. Gas turbines aboard aircraft operate at still higher bearing temperatures which are beyond the capability of petroleum oils. Synthetic organic esters are used, generally complying with military specifications.

Gears API System of Lubricant Service Designations for Automotive Manual Transmissions and Axles (SAE J308B) defines service levels for automotive gear oils. The service levels include API GL-1 to GL-6 and API PL-1 to PL-2, related to gear types and service conditions. Also, there are SAE Viscosity Grades for Axle and Transmission Lubricants (SAE J306C) for gear oils in automotive service. In this system, there are winter grades from SAE 70 W to 85 W, based on low-temperature viscosity (ASTM D2983) limits and 100°C kinematic viscosity (ASTM D445) limits, and grades from SAE 90 to 250, based on 100°C kinematic viscosity limits. Many multigrade oils are marketed based on it, capable of year-round operation.

The American Gear Manufacturers Association (AGMA) has classification systems for industrial gear lubricants, defined in AGMA 250.04 and 251.02. These systems specify lubricant performance grades by AGMA lubricant numbers, each grade linked to a specific ISO viscosity grade. There are three systems for enclosed gear drives, and they cover (1) rust- and oxidation-inhibited, (2) extreme pressure, and (3) compounded lubricants. Three systems for open gear drives cover (1) rust and oxidation-inhibited, (2) extreme pressure, and (3) residual lubricants.

The various gear types in use differ in severity of lubrication requirements. For spur, bevel, helical, and herringbone gears, elastohydrodynamic lubrication occurs at rated speeds and loads. Oils used for these types of gears need not contain additives that contribute to oil film strength, but may contain antioxidants, rust inhibitors, defoamants, etc., depending on the application. Worm gears need oils with film strength additives, such as compounded lubricants (containing fatty oils in a petroleum base), and use oils of relatively high viscosity, such as ISO VG 460. Other gear types, such as spur, use oils of widely varying viscosities, from ISO VG 46 to 460 or higher. The choice relates to operating conditions, as well as any need to lubricate other mechanisms with the same oil. Hypoid gears, widely used in automobiles, need oils containing effective EP additives. There is increased use of synthetic gear oils where wide operating temperature ranges and need for long life are involved. For example, polyalkylene glycol gear oils, which have high viscosity indices and excellent load capacities, are being used in industrial applications involving heavily loaded or high-temperature worm gears. Synthetic gear oils based on PAO base stocks have performed well in closed gearbox applications. With open gears, the lubricant often is applied sporadically and must adhere to the gear for some time, resisting being scraped off by meshing teeth. Greases, and viscous asphaltic-base lubricants, often containing EP additives, are among the gear lubricants used in this service. Asphaltic products may be diluted with a solvent to ease application, the solvent evaporating relatively quickly from the gear teeth.

Rolling-Contact Bearings   These include ball bearings of various configurations, as well as roller bearings of the cylindrical, spherical, and tapered varieties and needle bearings. They may range in size from a few millimetres to several metres. They are basically designed to carry load under EHL conditions, but considerable sliding motion may exist as well as rolling motion at points of contact between rolling elements and raceways. In addition, sliding occurs between rolling elements and separators and, in roller bearings, between the ends of rollers and raceway flanges. In many instances, it is desirable to incorporate antiwear additives to deal with these conditions.

A choice can frequently be made between oil and grease. Where the lubricant must carry heat away from the bearings, oil is the obvious selection, especially where large quantities can be readily circulated. Also, where contamination by water and solids is difficult to avoid, oil is preferred, provided an adequate purification system is furnished. If circulation is not practical, then grease may be the better choice. Where access to bearings is limited, grease may be indicated. Grease is also preferred in instances where leakage of lubricant might be a problem.

Suitable oils generally contain rust and oxidation inhibitors, and antiwear additives. Viscosity is selected on the basis of EHL considerations or builder recommendations for the particular equipment, and it may range anywhere from ISO VG 32 to 460, typically perhaps ISO VG 68.

Where grease is employed, the selection of type will be governed by operating conditions including temperature, loading, possibility of water contamination, and frequency and method of application. Lithium 12-hydroxystearate grease, particularly of no. 2 consistency grade, is widely used. Where equipment is lubricated for life at the time of manufacture, such as electric motors in household appliances, polyurea greases are often selected.

Air Compressors   **Reciprocating** compressors require lubrication of the cylinder walls, packing, and bearings. Temperatures at the cylinder wall are fairly high, and sufficient viscosity must be provided. Oils of ISO VG from about 68 to 460 may be specified. Since water condensation may be encountered, rust inhibitor is needed in addition to oxidation inhibitor. A principal operating problem concerning the lubricant is development of carbonaceous deposits on the valves and in the piping. This can seriously interfere with valve operation and can lead to disastrous fires and explosions. It is essential to choose an oil with a minimum tendency to form such deposits. Conradson and Ramsbottom carbon residue tests are of little value in predicting this behavior, and builders and reputable oil suppliers should be consulted. Maintenance procedures need to ensure that any deposits are cleaned on a regular basis and not be allowed to accumulate. In units which call for all-loss lubrication to the cylinders, feed rates should be reduced to the minimum recommended levels to minimize deposit-forming tendencies.

Increasingly, certain synthetic oils, such as fully formulated diesters, polyalkylene glycols, and PAOs, are being recommended for longer, trouble-free operation.

Screw compressors present a unique lubrication problem in that large quantities of oil are sprayed into the air during compression. Exposure of large surfaces of oil droplets to hot air is an ideal environment for oxidation to occur. This can cause lacquer deposition to interfere with oil separator operation which is essential to good performance. Although general-purpose rust- and oxidation-inhibited oils, as well as crankcase oils, are often used, they are not the best choice. Specially formulated petroleum oils, in ISO VG 32 to 68, are available for these severe conditions. For enhanced service life, synthetic organic ester fluids of comparable viscosity are often used.

Refrigeration compressors vary in their lubricant requirements depending on the refrigerant gas involved, particularly as some lubricant may be carried downstream of the compressor into the refrigeration system. If any wax from the lubricant deposits on evaporator surfaces, performance is seriously impaired. Ammonia systems can function with petroleum oils, even though ammonia is only poorly miscible with such oils. Many lubricants of ISO VG 15 to 100 are suitable, provided that the pour point is somewhat below evaporator temperature and that it does not contain additives that react with ammonia. Miscibility of refrigerant and lubricant is important to lubrication of some other types of systems. Chlorinated refrigerants (CFCs and HCFCs) are miscible with oil, and highly refined, low-pour, wax-free naphthenic oils of ISO VG 32 to 68 are often used, or certain wax-free synthetic lubricants. CFC refrigerants, which deplete the ozone layer, will cease being produced after 1995. Chlorine-free hydrofluorocarbon (HFC) refrigerants are now widely used, as they are non-ozone-depleting. HFCs are largely immiscible in petroleum-based lubricants, but partial miscibility of refrigerant and lubricant is necessary for adequate lubrication in these systems. Synthetic lubricants based on polyol ester or polyalkylene glycol base oils have miscibility with HFCs and are being used in HFC air conditioning and refrigeration systems.

Hydraulic Systems   Critical lubricated parts include pumps, motors, and valves. When operated at rated load, certain pumps and motors are very sensitive to the lubricating quality of the hydraulic fluid. When the fluid is inadequate in this respect, premature wear occurs, leading to erratic operation of the hydraulic system. For indoor use, it is best to select specially formulated antiwear hydraulic oils. These contain rust and oxidation inhibitors as well as antiwear additives, and they provide good overall performance in ISO VG 32 to 68 for many applications. Newer hydraulic systems have higher operating temperatures and pressures and longer drain intervals. These systems require oils with improved oxidation stability and antiwear protection. Fluids with good filterability are being demanded to allow use of fine filters to protect the critical clearances of the system. Hydraulic fluids may contain VI improver and/or pour point depressant for use in low-temperature outside service, and antifoamants and demulsifiers for rapid release of entrained air and water. Hydraulic equipment on mobile systems has greater access to automotive crankcase oils and automatic transmission fluids and manufacturers may recommend use of these oils in their equipment. Where biodegradability is of concern, fluids made of vegetable oil (rapeseed) are available.

Systems Needing Fire-Resistant Fluids   Where accidental rupture of an oil line may cause fluid to splash on a surface above about 600°F (310°C), a degree of fire resistance above that of petroleum oil is desirable. Four classes of fluids, generally used in hydraulic systems operating in such an environment, are available. **Phosphate esters** offer the advantages of a good lubricant requiring little attention in service, but they require special seals and paints and are quite expensive. **Waterglycol** fluids contain some 40 to 50 percent water, in a uniform solution of diethylene glycol, or glycol and polyglycol, to achieve acceptable fire resistance. They require monitoring in service to ensure proper content of water and rust inhibitor. **Invert emulsions** contain 40 to 50 percent water dispersed in petroleum oil. With oil as the outside phase of the emulsion, lubrication properties are fairly good, but the fluid must be monitored to maintain water content and to ensure that the water remains adequately dispersed. Finally, there are **conventional emulsions** which contain 5 to 10 percent petroleum oil dispersed in water. With the water as the outside phase, the fluid is a rather poor lubricant, and equipment requiring lubrication must be designed and selected to operate with these emulsions. Steps must also be taken to ensure that problems such as rusting, spoilage, and microbial growth are controlled.

Note that *fire-resistant* is a relative term, and it does not mean non-flammable. Approvals of the requisite degree of fire resistance are issued by Factory Mutual Insurance Company and by the U.S. Bureau of Mines where underground mining operations are involved.

Metal Forming   The functions of fluids in machining operation are (1) to cool and (2) to lubricate. Fluids remove the heat generated by the chip/tool rubbing contact and/or the heat resulting from the plastic deformation of the work. Cooling aids tool life, preserves tool hardness, and helps to maintain the dimensions of the machined parts. Fluids lubricate the chip/tool interface to reduce tool wear, frictional heat, and power consumption. Lubricants aid in the reduction of metal welding and adhesion to improve surface finish. The fluids may also serve to carry away chips and debris from the work as well as to protect machined surfaces, tools, and equipment from rust and corrosion.

Many types of fluids are used. Most frequently they are (1) mineral oils, (2) soluble oil emulsions, or (3) chemicals or synthetics. These are often compounded with additives to impart specific properties. Some metalworking operations are conducted in a controlled gaseous atmosphere (air, nitrogen, carbon dioxide).

The choice of a **metalworking fluid** is very complicated. Factors to be considered are (1) the metal to be machined, (2) the tools, and (3) the type of operation. Tools are usually steel, carbide, or ceramic. In operations where chips are formed, the relative motion between the tool and chip is high-speed under high load and often at elevated temperature. In chipless metal forming—drawing, rolling, stamping, extruding, spinning—the function of the fluid is to (1) lubricate and cool the die and work material and (2) reduce adhesion and welding on dies. In addition to fluids, solids such as talc, clay, and soft metals may be used in drawing operations.

In addition to the primary function to lubricate and cool, the cutting fluids should not (1) corrode, discolor, or form deposits on the work; (2) produce undesirable fumes, smoke, or odors; (3) have detrimental physical effects on operators. Fluids should also be stable, resist bacterial growth, and be foam-resistant.

In the machining operations, many combinations of tools, workpieces, and operating conditions may be encountered. In some instances, straight mineral oils or oils with small amounts of additives will suffice, while in more severe conditions, highly compounded oils are required. The effectiveness of the additives depends upon their chemical activity with newly formed, highly reactive surfaces; these combined with the high temperatures and pressures at the contact points are ideal for chemical reactions. Additives include fatty oils, sulfur, sulfurized fatty oils, and sulfurized, chlorinated, and phosphorus additives. These agents react with the metals to form compounds which have a lower shear strength and may possess EP properties. Oils containing additives, either dark or transparent, are most widely used in the industry.

Oils with sulfur- and chlorine-based additives raise some issues of increasing concern, as the more active sulfurized types stain copper and the chlorinated types have environmental concerns and are not suitable for titanium. New cutting oils have been developed which perform well and avoid these concerns.

Water is probably the most effective coolant available but can seldom be used as an effective cutting fluid. It has little value as a lubricant and will promote rusting. One way of combining the cooling properties of water with the lubricating properties of oil is through the use of soluble oils. These oils are compounded so they form a stable emulsion with water. The main component, water, provides effective cooling while the oil and compounds impart desirable lubricating, EP, and corrosion-resistance properties.

The ratio of water to oil will influence the relative lubrication and cooling properties of the emulsion. For these nonmiscible liquids to form a stable emulsion, an emulsifier must be added. Water hardness is an important consideration in the forming of an emulsion, and protection against bacterial growth should be provided. Care in preparing and handling the emulsion will ensure more satisfactory performance and longer life. Overheating, freezing, water evaporation, contamination, and excessive air mixing will adversely affect the emulsion. When one is discarding the used emulsion, it is often necessary to "break" the emulsion into its oil and water components for proper disposal.

Newer-technology soluble oils, using nonionic emulsifiers instead of anionic emulsifiers, have improved properties over earlier-technology oils, and have less tendency to form insoluble soaps with hard water, which results in fewer filter changes, less machine downtime, and lower maintenance costs. They also have longer emulsion life, from greater bacterial and fungal resistance.

Water-soluble chemicals, and synthetics, are essentially solutions or microdispersions of a number of ingredients in water. These materials are used for the same purposes described above for mineral and soluble oils. Water-soluble synthetic fluids are seeing increased use and have greater bioresistance than soluble oils and provide longer coolant life and lower maintenance.

Health Considerations   Increased attention is being focused on hazards associated with manufacture, handling, and use of all types of industrial and consumer materials, and lubricants are no exception. Much progress has been made in removing substances suspected of creating adverse effects on health. Suppliers can provide information on general composition and potential hazards requiring special handling of their products, in the form of Material Safety Data Sheets (MSDSs). It is seldom necessary to go further to ascertain health risks. It is necessary that personnel working with lubricants observe basic hygienic practices. They should avoid wearing oil-soaked clothing, minimize unnecessary exposure of skin, and wash exposed skin frequently with approved soaps.

Used-Lubricant Disposal   The nonpolluting disposal of used lubricants is becoming increasingly important and requires continual attention. More and more legislation and control are being enacted, at local, state, and national levels, to regulate the disposal of wastes. Alternative methods of handling specific used lubricants may be recommended, such as in situ purification and refortifying the oil and returning it to service, often in mobile units provided by the lubricant distributor. Also, waste lubricating oils may be reprocessed into base oils for rerefined lubricant manufacture. The use of waste lubricating oils as fuels is being increasingly regulated because of air pollution dangers. If wastes are dumped on the ground or directly into sewers, they may eventually be washed into streams and water supplies and become water pollutants. They may also interfere with proper operation of sewage plants. Improper burning may contribute to air pollution. Wastes must be handled in such a way as to ensure nonpolluting disposal. Reputable lubricant suppliers are helpful in suggesting general disposal methods, although they cannot be expected to be knowledgeable about all local regulations.

Lubricant management programs are being used increasingly by industry to minimize operating cost, including lubricant, labor, and waste disposal costs. These programs can be managed by lubricant user/supplier teams. Fluid management programs include the following activities: lubrication selection; lubricant monitoring during service; reclamation and refortification; and disposal. The supplier and/or lubricant service firms often have the major role in performing the last three activities. These programs can be beneficial to both lubricant user and supplier.

6.12 PLASTICS

Staff Contribution

REFERENCES: Billmyer, "Textbook of Polymer Science," 3d ed. Brydson, "Plastics Materials," 4th ed. Birley, Heath, and Scott, "Plastics Materials: Properties and Applications." Current publications of and sponsored by the Society of the Plastics Industry (SPI) and similar plastics industry organizations. Manufacturers' specifications, data, and test reports. "Modern Plastics Encyclopedia," McGraw-Hill.

GENERAL OVERVIEW OF PLASTICS

Plastics are ubiquitous engineering materials which are wholly or in part composed of long, chainlike molecules called **high polymers.** While carbon is the element common to all commercial high polymers, hydrogen, oxygen, nitrogen, sulfur, halogens, and silicon can be present in varying proportions. High polymers may be divided into two classes: **thermoplastic** and **thermosetting.** The former reversibly melt to become highly viscous liquids and solidify upon cooling (Fig. 6.12.1). The resultant solids will be elastic, ductile, tough, or brittle, depending on the structure of the solid as evolved from the molten state. Thermosetting polymers are infusible without thermal or mechanical degradation. Thermosetting polymers (often termed **thermosets**) cure by a chainlinking chemical reaction usually initiated at elevated temperature and pressure, although there are types which cure at room temperature through the use of catalysts. The more highly cured the polymer, the higher its heat distortion temperature and the harder and more brittle it becomes.

Fig. 6.12.1 Tensile stress-strain curve for thermoplastic Lucite. *(The Du Pont Co.)*

As a class, plastics possess a combination of physical and mechanical properties which are attractive to the designer. There are certain properties of plastics which cannot be replicated in metals, including light weight and density (specific gravity rarely greater than 2, with the normal value in the range of 1.1 to 1.7—compare this with magnesium, the lightest structural metal of significance, whose specific gravity is 1.75); optical properties which may range from complete clarity to complete opacity, with the ability to be compounded with through colors in an almost limitless range; low thermal conductivity and good electrical resistance; the ability to impart excellent surface finish to parts made by many of the primary production processes. Conversely, when compared to metals, plastics generally have lower elastic modulus—and thus, are inherently more flexible; have lower flexural and impact strength, and inferior toughness; and have lower dimensional stability generally than for most metals.

Some properties of plastics which are most desirable are high strength/weight ratios and the ability to process raw materials through to a finished size and shape in one of several basic operations; this last item has a large impact on the overall costs of finished products by virtue of eliminating secondary operations.

The seemingly endless variety of all types of plastics which are available in the marketplace is continually augmented by new compositions; if an end product basically lends itself to the use of plastics, there is most likely to be some existing composition available to satisfy the design requirements. See Table 6.12.1.

The properties of a given plastic can often be modified by the incorporation of additives into the basic plastic resin and include colorants, stabilizers, lubricants, and fibrous reinforcement. Likewise, where local strength requirements are beyond the capacity of the plastic, ingenious configurations of **metallic inserts** can be incorporated into the manufacture of the plastic part and become structurally integral with the part itself; female threaded inserts and male threaded studs are the most common type of inserts found in plastic part production.

Plastic resins may themselves be used as **adhesives** to join other plastics or other paired materials including wood/wood, wood/plastic, metal/plastic, metal/metal, and wood/metal. Structural **sandwich panels,** with the inner layer of sheet or foamed plastic, are an adaptation of composite construction.

Foamed plastic has found widespread use as thermal insulation, a volume filler, cushioning material, and many lightweight consumer items.

The ability to be recycled is an important attribute of thermoplastics; thermosets are deficient in this regard. In view of the massive amounts of plastic in the consumer stream, **recycling** has become more the norm than the exception; indeed, in some jurisdictions, recycling of spent plastic consumables is mandated by law, while in others, the incentive to recycle manifests itself as a built-in cost of the product—a tax, as it were, imposed at the manufacturing phase of production to spur recycling efforts. (See Recycling on pages following.)

In those cases where secondary operations are required to be performed on plastic parts, the usual chip-producing material-removal processes are employed. In those regards, due diligence must be paid to the nature of the material being cut. Thermoplastics will soften from heat generated during cutting, and their low flexural modulus may require backing to prevent excessive deflection in response to cutting forces; thermosets may prove abrasive (even without fillers) and cause rapid wear of cutting edges. Cutting operations on some soft thermoplastics will experience elastic flow of the material beyond the cutting region, necessitating multiple cuts, each of smaller magnitude.

The raw-materials cost for plastics will vary. The economies of quantity production are self-evident from a study of applicable statistics of plastics production. Consider that in 1993, global production of plastic raw materials was in excess of 100 million metric tons (t). The final overall cost of a finished plastic part is impacted by the cost of raw materials, of course, but further, the use of plastic itself affects the design, manufacture, and shipping components of the final cost; these latter are not inconsequential in arriving at cost comparisons of production with plastics vis-à-vis metals. Suffice it to say that in general, when all costs are accounted for and when design requirements can be met with plastics, the end cost of a plastic item is often decidedly favorable.

RAW MATERIALS

The source of virgin raw materials, or **feedstock,** is generally petroleum or natural gas. The feedstock is converted to monomers, which then become the basis for plastic resins. As recycling efforts increase,

Table 6.12.1 Properties of Plastic Resins and Compounds

	Properties	ASTM test method	Extrusion grade	ABS/Nylon	ABS[a] Injection molding grades Heat-resistant	Medium-impact	High-impact	Platable grade	20% Glass fiber-reinforced
Processing	1a. Melt flow, g/10 min	D1238	0.85–1.0		1.1–1.8	1.1–1.8	1.1–18	1.1	
	1. Melting temperature, °C. T_m (crystalline)								
	T_g (amorphous)		88–120		110–125	102–115	91–110	100–110	100–110
	2. Processing temperature range, °F. (C = compression; T = transfer, I = injection; E = extrusion)		E:350–500	I:460–520	C:325–500 I:475–550	C:325–350 I:390–525	C:325–350 I:380–525	C:325–400 I:350–500	C:350–500 I:350–500
	3. Molding pressure range, 10^3 lb/in²			8–25	8–25	8–25	8–25	8–25	15–30
	4. Compression ratio		2.5–2.7	1.1–2.0	1.1–2.0	1.1–2.0	1.1–2.0	1.1–2.0	
	5. Mold (linear) shrinkage, in/in	D955	0.004–0.007	0.003–0.010	0.004–0.009	0.004–0.009	0.004–0.009	0.005–0.008	0.001–0.002
Mechanical	6. Tensile strength at break, lb/in²	D638[b]	2,500–8,000	4,000–6,000	4,800–7,500	5,500–7,500	4,400–6,300	5,200–6,400	10,500–13,000
	7. Elongation at break, %	D638[b]	20–100	40–300	3–45	5–60	5–75		2–3
	8. Tensile yield strength, lb/in²	D638[b]	4,300–6,400	4,300–6,300	4,300–7,000	5,000–7,200	2,600–5,900	6,700	
	9. Compressive strength (rupture or yield), lb/in²	D695	5,200–10,000		7,200–10,000	1,800–12,500	4,500–8,000		13,000–14,000
	10. Flexural strength (rupture or yield), lb/in²	D790	4,000–14,000	8,800–10,900	9,000–13,000	7,100–13,000	5,400–11,000	10,500–11,500	14,000–17,500
	11. Tensile modulus, 10^3 lb/in²	D638[b]	130–420	260–320	285–360	300–400	150–350	320–380	740–880
	12. Compressive modulus, 10^3 lb/in²	D695	150–390		190–440	200–450	140–300		800
	13. Flexural modulus, 10^3 lb/in² 73°F	D790	130–440	250–310	300–400	310–400	179–375	340–390	650–800
	200°F	D790							
	250°F	D790							
	300°F	D790							
	14. Izod impact, ft · lb/in of notch (⅛-in-thick specimen)	D256A	1.5–12	15–20	2.0–6.5	3.0–6.0	6.0–9.3	4.0–8.3	1.1–1.4
	15. Hardness Rockwell	D785	R75–115	R93–105	R100–115	R102–115	R85–106	R103–109	M85–98, R107
	Shore/Barcol	D2240/ D2583							
Thermal	16. Coef. of linear thermal expansion, 10^6 in/(in · °C)	D696	60–130	90–110	60–93	80–100	95–110	47–53	20–21
	17. Deflection temperature under flexural load, °F 264 lb/in²	D648	170–220	130–150	220–240 annealed 181–193[g]	200–220 annealed	205–215 annealed	190–222 annealed	210–220
	66 lb/in²	D648	170–235	180–195	230–245 annealed	215–225 annealed	210–225 annealed	215–222 annealed	220–230
	18. Thermal conductivity, 10^{-4} cal · cm/(s · cm² · °C)	C177			4.5–8.0				4.8
Physical	19. Specific gravity	D792	1.02–1.08	1.06–1.07	1.05–1.08	1.03–1.06	1.01–1.05	1.04–1.07	1.18–1.22
	20. Water absorption (⅛-in-thick specimen), % 24 h	D570	0.20–0.45		0.20–0.45	0.20–0.45	0.20–0.45		0.18–0.20
	Saturation	D570							
	21. Dielectric strength (⅛-in-thick specimen), short time, V/mil)	D149	350–500		350–500	350–500	350–500	420–550	450–460

NOTE: Footnotes [a] to [f] are at end of table.
SOURCE: Abstracted from "Modern Plastics Encyclopedia," 1995.

			Acetal			
Homopolymer	Copolymer	Extrusion and blow molding grade (terpolymer)	20% Glass-reinforced homopolymer	25% Glass-coupled copolymer	2–20% PTFE-filled copolymer	Chemically lubricated homopolymer
1–20	1–90	1.0	6.0			6
172–184	160–175	160–170	175–181	160–180	160–175	175
I:380–470	C:340–400 I:360–450	E:360–400	I:350–480	I:365–480	I:350–445 I:325–500 E:360–500	I:400–440
10–20	8–20		10–20	8–20	8–20	
3.0–4.5	3.0–4.5	3.0–4.0		3.0–4.5	3.0–4.5	
0.018–0.025	0.020 (Avg.)	0.02	0.009–0.012	0.004 (flow) 0.018 (trans.)	0.018–0.029	
9,700–10,000			8,500–9,000	16,000–18,500	8,300	9,500
10–75	15–75	67	6–12	2–3	30	40
9,500–12,000	8,300–10,400	8,700	7,500–8,250	16,000	8,300	9,500
15,600–18,000 @ 10%	16,000 @ 10%	16,000	18,000 @ 10%	17,000 @ 10%	11,000–12,600	
13,600–16,000	13,000	12,800	10,700–16,000	18,000–28,000	11,500	13,000
400–520	377–464		900–1,000	1,250–1,400	250–280	450
670	450					
380–490	370–450	350	600–730	1,100	310–360	400
120–135			300–360			130
75–90			250–270			80
1.1–2.3	0.8–1.5	1.7	0.5–1.0	1.0–1.8	0.5–1.0	1.4
M92–94, R120	M75–90	M84	M90	M79–90, R110	M79	M90
50–112	61–110		33–81	17–44	52–68	
253–277	185–250	205	315	320–325	198–225	257
324–342	311–330	318	345	327–331	280–325	329
5.5	5.5				4.7	
1.42	1.40	1.41	1.54–1.56	1.58–1.61	1.40	1.42
0.25–1	0.20–0.22	0.22	0.25	0.22–0.29	0.15–0.26	0.27
0.90–1	0.65–0.80	0.8	1.0	0.8–1.0	0.5	1.00
400–500 (90 mil)	500 (90 mil)		490 (125 mil)	480–580	400–410	400 (125 mils)

Table 6.12.1 Properties of Plastic Resins and Compounds *(Continued)*

| | | | Acrylic | | | | Acrylonitrile | |
| | | | Sheet | Molding and extrusion compounds | | | Molding and extrusion | Injection |
Properties	ASTM test method		Cast	PMMA	Impact-modified	Heat-resistant		
1a. Melt flow, g/10 min	D1238			1.4–27	1–11	1.6–8.0		12
1. Melting temperature, °C. T_m (crystalline)							135	
T_g (amorphous)			90–105	85–105	80–103	100–165	95	
2. Processing temperature range, °F. (C = compression; T = transfer, I = injection; E = extrusion)				C:300–425 I:325–500 E:360–500	C:300–400 I:400–500 E:380–480	C:350–500 I:400–625 E:360–550	C:320–345 I:410 E:350–410	380–420
3. Molding pressure range, 10^3 lb/in²				5–20	5–20	5–30	20	20
4. Compression ratio				1.6–3.0		1.2–2.0	2	2–2.5
5. Mold (linear) shrinkage, in/in	D955		1.7	0.001–0.004 (flow) 0.002–0.008 (trans.)	0.002–0.008	0.002–0.008	0.002–0.005	0.002–0.005
6. Tensile strength at break, lb/in²	D638[b]		66–11,000	7,000–10,500	5,000–9,000	9,300–11,500	9,000	
7. Elongation at break, %	D638[b]		2–7	2–5.5	4.6–70	2–10	3–4	3–4
8. Tensile yield strength, lb/in²	D638[b]			7,800–10,600	5,500–8,470	10,000	7,500	9,500
9. Compressive strength (rupture or yield), lb/in²	D695		11,000–19,000	10,500–18,000	4,000–14,000	15,000–17,000	12,000	12,000
10. Flexural strength (rupture or yield), lb/in²	D790		12,000–17,000	10,500–19,000	7,000–14,000	12,000–18,000	14,000	14,000
11. Tensile modulus, 10^3 lb/in²	D638[b]		450–3,100	325–470	200–500	350–650	510–580	500–550
12. Compressive modulus, 10^3 lb/in²	D695		390–475	370–460	240–370	450		
13. Flexural modulus, 10^3 lb/in² 73°F	D790		390–3,210	325–460	200–430	450–620	500–590	480
200°F	D790					150–440		
250°F	D790					350–420		
300°F	D790							
14. Izod impact, ft · lb/in of notch (⅛-in-thick specimen)	D256A		03–0.4	0.2–0.4	0.40–2.5	0.2–0.4	2.5–6.5	2.5
15. Hardness Rockwell	D785		M80–102	M68–105	M35–78	M94–100	M72–78	M60
Shore/Barcol	D2240/ D2583							
16. Coef. of linear thermal expansion, 10^6 in/(in · °C)	D696		50–90	50–90	48–80	40–71	66	66
17. Deflection temperature under flexural load, °F 264 lb/in²	D648		98–215	155–212	165–209	190–310	164	151
66 lb/in²	D648		165–235	165–225	180–205	200–315	172	166
18. Thermal conductivity, 10^{-4} cal · cm/(s · cm² · °C)	C177		4.0–6.0	4.0–6.0	4.0–5.0	2.0–4.5	6.2	6.1
19. Specific gravity	D792		1.17–1.20	1.17–1.20	1.11–1.18	1.16–1.22	1.15	1.15
20. Water absorption (⅛-in-thick specimen), % 24 h	D570		0.2–0.4	0.1–0.4	0.19–0.8	0.2–0.3	0.28	
Saturation	D570							
21. Dielectric strength (⅛-in-thick specimen), short time, V/mil	D149		450–550	400–500	380–500	400–500	220–240	220–240

| Cellulosic | | | | Epoxy | | | | |
| Ethyl cellulose molding compound and sheet | Cellulose acetate | | Cellulose acetate butyrate | Bisphenol molding compounds | | Casting resins and compounds | | |
	Sheet	Molding compound	Molding compound	Glass fiber-reinforced	Mineral-filled	Unfilled	Aluminum-filled	Flexibilized
135	230	230	140	Thermoset	Thermoset	Thermoset	Thermoset	Thermoset
C:250–390 I:350–500		C:260–420 I:335–490	C:265–390 I:335–480	C:300–330 T:280–380	C:250–330 T:250–380			
8–32		8–32	8–32	1–5	0.1–3			
1.8–2.4		1.8–2.6	1.8–2.4	3.0–7.0	2.0–3.0			
0.005–0.009		0.003–0.010	0.003–0.009	0.001–0.008	0.002–0.010	0.001–0.010	0.001–0.005	0.001–0.010
2,000–8,000	4,500–8,000	1,900–9,000	2,600–6,900	5,000–20,000	4,000–10,800	4,000–13,000	7,000–12,000	2,000–10,000
5–40	20–50	6–70	40–88	4		3–6	0.5–3	20–85
		2,500–6,300						
		3,000–8,000	2,100–7,500	18,000–40,000	18,000–40,000	15,000–25,000	15,000–33,000	1,000–14,000
4,000–12,000	6,000–10,000	2,000–16,000	1,800–9,300	8,000–30,000	6,000–18,000	13,000–21,000	8,500–24,000	1,000–13,000
			50–200	3,000	350	350		
					650			1–350
		1,200–4,000	90–300	2,000–4,500	1,400–2,000			
0.4	2.0–8.5	1.0–7.8	1.0–10.9	0.3–10.0	0.3–0.5	0.2–1.0	0.4–1.6	2.3–5.0
R50–115	R85–120	R17–125	R31–116	M100–112	M100–M112	M80–110	M55–85	
								Shore D65–89
100–200	100–150	80–180	110–170	11–50	20–60	45–65	5.5	20–100
115–190		111–195	113–202	225–500	225–500	115–550	190–600	73–250
		120–209	130–227					
3.8–7.0	4–8	4–8	4–8	4.0–10.0	4–35	4.5	15–25	
1.09–1.17	1.28–1.32	1.22–1.34	1.15–1.22	1.6–2.0	1.6–2.1	1.11–1.40	1.4–1.8	0.96–1.35
0.8–1.8	2.0–7.0	1.7–6.5	0.9–2.2	0.04–0.20	0.03–0.20	0.08–0.15	0.1–4.0	0.27–0.5
350–500	250–600	250–600	250–400	250–400	250–420	300–500		235–400

Table 6.12.1 Properties of Plastic Resins and Compounds (*Continued*)

Materials	Properties	ASTM test method	Polychloro-trifluoro-ethylene	Polytetrafluoroethylene Granular	Polytetrafluoroethylene 25% Class fiber-reinforced	Polyvinylidene fluoride Molding and extrusion	Polyvinylidene fluoride EMI shielding (conductive); 30% PAN carbon fiber	Modified PE-TFE 25% Glass fiber-reinforced
Processing	1a. Melt flow, g/10 min	D1238						
	1. Melting temperature, °C. T_m (crystalline)			327	327	141–178		270
	T_g (amorphous)		220			−60 to −20		
	2. Processing temperature range, °F. (C = compression; T = transfer, I = injection; E = extrusion)		C:460–580 I:500–600 E:360–590			C:360–550 I:375–550 E:375–550	I:430–500	C:575–625 I:570–650
	3. Molding pressure range, 10^3 lb/in^2		1–6	2–5	3–8	2–5		2–20
	4. Compression ratio		2.6	2.5–4.5		3		
	5. Mold (linear) shrinkage, in/in	D955	0.010–0.015	0.030–0.060	0.018–0.020	0.020–0.035	0.001	0.002–0.030
Mechanical	6. Tensile strength at break, lb/in^2	D638[b]	4,500–6,000	3,000–5,000	2,000–2,700	3,500–7,250	14,000	12,000
	7. Elongation at break, %	D638[b]	80–250	200–400	200–300	12–600	0.8	8
	8. Tensile yield strength, lb/in^2	D638[b]	5,300			2,900–8,250		
	9. Compressive strength (rupture or yield), lb/in^2	D695	4,600–7,400	1,700	1,000–1,400 @ 1% strain	8,000–16,000		10,000
	10. Flexural strength (rupture, or yield), lb/in^2	D790	7,400–11,000		2,000	9,700–13,650	19,800	10,700
	11. Tensile modulus, 10^3 lb/in^2	D638[b]		58–80	200–240	200–80,000	2,800	1,200
	12. Compressive modulus, 10^3 lb/in^2	D695	150–300	60		304–420		
	13. Flexural modulus, 10^3 lb/in^2 73° F	D790	170–200	80	190–235	170–120,000	2,100	950
	200°F	D790	180–260					450
	250°F	D790						310
	300°F	D790						200
	14. Izod impact, ft · lb/in of notch (⅛-in-thick specimen)	D256A	2.5–5	3	2.7	2.5–80	1.5	9.0
	15. Hardness Rockwell	D785	R75–112			R79–83, 85		R74
	Shore/Barcol	D2240/ D2583	Shore D75–80	Shore D50–65	Shore D60–70	Shore D80, 82 65–70		
Thermal	16. Coef. of linear thermal expansion, 10^6 in/(in · °C)	D696	36–70	70–120	77–100	70–142		10–32
	17. Deflection temperature under flexural load, °F 264 lb/in^2	D648		115		183–244	318	410
	66 lb/in^2	D648	258	160–250		280–284		510
	18. Thermal conductivity, 10^{-4} cal · cm/(s · cm^2 · °C)	C177	4.7–5.3	6.0	8–10	2.4–3.1		
Physical	19. Specific gravity	D792	2.0.8–2.2	2.14–2.20	2.2–2.3	1.77–1.78	1.74	1.8
	20. Water absorption (⅛-in-thick specimen), % 24 h	D570	0	<0.01		0.03–0.06	0.12	0.02
	Saturation	D570						
	21. Dielectric strength (⅛-in-thick specimen), short time, V/mil	D149	500–600	480	320	260–280		425

Fluoroplastics

Furan	Phenolic				
	Molding compounds, phenol-formaldehyde				Casting resins
		High-strength glass fiber-reinforced	Impact-modified		
Asbestos-filled	Wood flour-filled		Fabric- and rag-filled	Cellulose-filled	Unfilled
			0.5–10		
Thermoset	Thermoset	Thermoset	Thermoset	Thermoset	Thermoset
C:275–300	C:290–380 I:330–400	C:300–380 I:330–390 T:300–350	C:290–380 I:330–400 T:300–350	C:290–380 I:330–400	
0.1–0.5	2–20	1–20	2–20	2–20	
	1.0–1.5	2.0–10.0	1.0–1.5	1.0–1.5	
	0.004–0.009	0.001–0.004	0.003–0.009	0.004–0.009	
3,000–4,500	5,000–9,000	7,000–18,000	6,000–8,000	3,500–6,500	5,000–9,000
	0.4–0.8	0.2	1–4	1–2	1.5–2.0
10,000–13,000	25,000–31,000	16,000–70,000	20,000–28,000	22,000–31,000	12,000–15,000
600–9,000	7,000–14,000	12,000–60,000	10,000–14,000	5,500–11,000	11,000–17,000
1,580	800–1,700	1,900–3,300	900–1,100		400–700
		2,740–3,500			
	1,000–1,200	1,150–3,300	700–1,300	900–1,300	
	0.2–0.6	0.5–18.0	0.8–3.5	0.4–1.1	0.24–0.4
R110	M100–115	E54–101	M105–115	M95–115	M93–120
		Barcol 72			
	30–45	8–34	18–24	20–31	68
	300–370	350–600	325–400	300–350	165–175
	4–8	8–14	9–12	6–9	3.5
1.75	1.37–1.46	1.69–2.0	1.37–1.45	1.38–1.42	1.24–1.32
0.01–0.02	0.3–1.2	0.03–1.2	0.6–0.8	05–0.9	0.1–0.36
		0.12–1.5			
	260–400	140–400	200–370	300–380	250–400

Table 6.12.1 Properties of Plastic Resins and Compounds (*Continued*)

			Polyamide				
			Nylon, Type 6				
					Toughened	High-impact	Impact-
Materials / Properties	ASTM test method	Molding and extrusion compound	30–35% Glass fiber-reinforced	33% Glass fiber-reinforced	copolymers and rubber-modified compounds	modified; 30% glass fiber-reinforced	Cast
Processing							
1a. Melt flow, g/10 min	D1238	0.5–10			1.5–5.0		
1. Melting temperature, °C. T_m (crystalline)		210–220	210–220	210–220	210–220	220	227–238
T_g (amorphous)							
2. Processing temperature range, °F. (C = compression; T = transfer; I = injection; E = extrusion)		I:440–550 E:440–525	I:460–550	I:520–550	I:450–580 E:450–550	I:480–550	
3. Molding pressure range, 10^3 lb/in^2		1–20	2–20		1–20	3–20	
4. Compression ratio		3.0–4.0	3.0–4.0		3.0–4.0	3.0–4.0	
5. Mold (linear) shrinkage, in/in	D955	0.003–0.015	0.001–0.005	0.001–0.003	0.008–0.026	0.003–0.005	
Mechanical							
6. Tensile strength at break, lb/in^2	D638[b]	6,000–24,000	24–27,600[c]; 18,900[d]	17,800[c]	6,300–11,000[c]	21,000[c]; 14,500[d]	12,500
7. Elongation at break, %	D638[b]	30–100[c]; 300[d]	2.2–3.6[c]	4.0[c]	150–270[c]	5[c]–8[d]	20–30
8. Tensile yield strength, lb/in^2	D638[b]	13,100[c], 7,400[d]			9,000[c]–9,500		
9. Compressive strength (rupture or yield), lb/in^2	D695	13,000–16,000[c]	19,000–24,000[c]		3,900[c]		17,000
10. Flexural strength (rupture or yield), lb/in^2	D790	15,700[c]; 5,800[d]	34–3,600[c]; 21,000[d]	25,800[c]	5,000–12,000[c]		16,500
11. Tensile modulus, 10^3 lb/in^2	D638[b]	380–464[c]; 100–247[d]	1,250–1,600[c]; 1,090[d]			1,220[c]–754[d]	500
12. Compressive modulus, 10^3 lb/in^2	D695	250[d]					325
13. Flexural modulus, 10^3 lb/in^2 73°F	D790	390–410[c]; 140[d]	1,250–1,400[c]; 800–950[d]	1,110[c]	110–320[c]; 130[d]	1,160[c]–600[d]	430
200° F	D790				60–130[c]		
250°F	D790						
300°F	D790						
14. Izod impact, ft · lb/in of notch (⅛-in-thick specimen)	D256A	0.6–2.2[c]; 30[d]	2.1–3.4[c]; 3.7–5.5[d]	3.5[c]	1.8–No break[c] 1.8–No break[d]	2.2[c]–6[d]	0.7–0.9
15. Hardness Rockwell	D785	R119[c]; M100–105[c]	M93–96c; M78[d]		R81–113[c]; M50		R115–125
Shore/Barcol	D2240/ D2583						D78.83
Thermal							
16. Coef. of linear thermal expansion, 10^6 in/(in · °C)	D696	80–83	16–80		72–120	20–25	50
17. Deflection temperature under flexural load, °F 264 lb/in^2	D648	155–185[c]	392–420[c]	400[c]	113–140[c]	410[c]	330–400
66 lb/in^2	D648	347–375[c]	420–430[c]	430[c]	260–367[c]	428[c]	400–430
18. Thermal conductivity, 10^{-4} cal · cm/(s · cm^2 · °C)	C177	5.8	5.8–11.4				
Physical							
19. Specific gravity	D792	1.12–1.14	1.35–1.42	1.33	1.07–1.17	1.33	1.15–1.17
20. Water absorption (⅛-in-thick specimen), % 24 h	D570	1.3–1.9	0.90–1.2	0.86	1.3–1.7	2.0	0.3–0.4
Saturation	D570	8.5–10.0	6.4–7.0		8.5	6.2	5–6
21. Dielectric strength (⅛-in-thick specimen), short time, V/mil	D149	400[c]	400–450[c]		450–470[c]		500–600

Molding compound	30–33% Glass fiber-reinforced	Toughened 15–33% Glass fiber-reinforced	Antifriction molybdenum disulfide-filled	Lubricated 5% Silicone	30% PTFE	5% Molybdenum disulfide and 30% PTFE
Polyamide (Continued)						
Nylon, Type 66						
255–265	260–265	256–265	249–265	260–265	260–265	260–265
I:500–620	I:510–580	I:530–575	I:500–600	I:530–570	I:530–570	I:530–570
1–25	5–20		5–25			
3.0–4.0	3.0–4.0					
0.007–0.018	0.002–0.006	0.0025–0.0045[c]	0.007–0.018	0.015	0.007	0.01
13,700[c]; 11,000[d]	27,600[c]; 20,300[d]	10,900–20,300[c]; 14,500[d]	10,500–13,700[c]	8,500[c]	5,500[c]	7,500[c]
15–80[c]; 150–300[d]	2.0–3.4[c]; 3–7[d]	4.7[c]; 8[d]	4.4–40[c]			
8,000–12,000[c]; 6,500–8,500[d]	25,000[c]					
12,500–15,000[c] (yld.)	24,000–40,000[c]	15,000[c]–20,000	12,000–12,500[c]			
17,900–1,700[c]; 6,100[d]	40,000[c]; 29,000[d]	17,400–29,900[c]	15,000–20,300[c]	15,000[c]	8,000[c]	1,200[c]
230–550[c]; 230–500[d]	1,380[c]; 1,090[d]	1,230[c]; 943[d]	350–550[c]			
410–470[c]; 185[d]	1,200–1,450[c]; 800[d]; 900	479–1,100[c]	420–495[c]	300[c]	460[c]	400[c]
0.55–1.0[c]; 0.85–2.1[d]	1.6–4.5[c]; 2.6–3.0[d]	>3.2–5.0	0.9–4.5[c]	1.0[c]	0.5[c]	0.6[c]
R120[c]; M83[c]; M95–105[d]	R101–119[c]; M101–102[c]; M96[d]	R107[c]; R115; R116; M86[c]; M70[d]	R119[c]			
80	15–54	43	54	63.0	45.0	
158–212[c]	252–490[c]	446–470	190–260[c]	170	180	185
425–474[c]	260–500[c]	480–495	395–430			
5.8	5.1–11.7					
1.13–1.15	1.15–1.40	1.2–1.34	1.15–1.18	1.16	1.34	1.37
1.0–2.8	0.7–1.1	0.7–1.5	0.8–1.1	1.0	0.55	0.55
8.5	5.5–6.5	5	8.0			
600[c]	360–500		360[c]			

Table 6.12.1 Properties of Plastic Resins and Compounds (*Continued*)

	Properties	ASTM test method	Polycarbonate			
			Unfilled molding and extrusion resins / High viscosity	Glass fiber-reinforced / 10% glass	Impact-modified polycarbonate/polyester blends	Lubricated / 10–15% PTFE, 20% glass fiber-reinforced
Processing	1a. Melt flow, g/10 min	D1238	3–10	7.0		
	1. Melting temperature, °C. T_m (crystalline)					
	T_g (amorphous)		150	150		150
	2. Processing temperature range, °F. (C = compression; T = transfer, I = injection; E = extrusion)		I:560	I:520–650	I:475–560	I:590–650
	3. Molding pressure range, 10^3 lb/in²		10–20	10–20	15–20	
	4. Compression ratio		1.74–5.5		2–2.5	
	5. Mold (linear) shrinkage, in/in	D955	0.005–0.007	0.002–0.005	0.006–0.009	0.002
Mechanical	6. Tensile strength at break, lb/in²	D638[b]	9,100–10,500	7,000–10,000	7,600–8,500	12,000–15,000
	7. Elongation at break, %	D638[b]	110–120	4–10	120–165	2
	8. Tensile yield strength, lb/in²		9,000	8,500–11,600	7,400–8,300	
	9. Compressive strength (rupture or yield), lb/in²	D695	10,000–12,500	12,000–14,000	7,000	11,000
	10. Flexural strength-(rupture or yield), lb/in²	D790	12,500–13,500	13,700–16,000	10,900–12,500	18,000–23,000
	11. Tensile modulus, 10^3 lb/in²	D638[b]	315	450–600		1,200
	12. Compressive modulus, 10^3 lb/in²	D695	350	520		
	13. Flexural modulus, 10^3 lb/in² 73°F	D790	330–340	460–580	280–325	850–900
	200°F	D790	275	440		
	250°F	D790	245	420		
	300°F	D790				
	14. Izod impact, ft · lb/in of notch (⅛-in-thick specimen)	D256A	12–18 @ ⅛ in 2.3 @ ¼ in	2–4	2–18	1.8–3.5
	15. Hardness Rockwell	D785	M70–M75	M62–75; R118–122	R114–122	
	Shore/Barcol	D2240/D2583				
Thermal	16. Coef. of linear thermal expansion, 10^6 in/(in · °C)	D696	68	32–38	80–95	21.6–23.4
	17. Deflection temperature under flexural load, °F 264 lb/in²	D648	250–270	280–288	190–250	280–290
	66 lb/in²	D648	280–287	295	223–265	290
	18. Thermal conductivity, 10^{-4} cal · cm/(s · cm² · °C)	C177	4.7	4.6–5.2	4.3	
Physical	19. Specific gravity	D792	1.2	1.27–1.28	1.20–1.22	1.43–1.5
	20. Water absorption (⅛-in-thick specimen), % 24 h	D570	0.15	0.12–0.15	0.12–0.16	0.11
	Saturation	D570	0.32–0.35	0.25–0.32	0.35–0.60	
	21. Dielectric strength (⅛-in-thick specimen), short time, V/mil	D149	380–>400	470–530	440–500	

| Polyester, thermoplastic | | | | Polyester, thermoset and alkyd | | |
| Polybutylene terephthalate | | Polyethylene terephthalate | | Cast | | Glass fiber-reinforced |
Unfilled	30% Glass fiber-reinforced	Unfilled	30% Glass fiber-reinforced	Rigid	Flexible	Preformed, chopped roving
220–267	220–267	212–265	245–265	Thermoset	Thermoset	Thermoset
		68–80				
I:435–525	I:440–530	I:440–660 E:520–580	I:510–590			C:170–320
4–10	5–15	2–7	4–20			0.25–2
		3.1	2–3			1.0
0.009–0.022	0.002–0.008	0.002–0.030	0.002–0.009			0.0002–0.002
8,200–8,700	14,000–19,000	7,000–10,500	20,000–24,000	600–13,000	500–3,000	15,000–30,000
50–300	2–4	30–300	2–7	<2.6	40–310	1–5
8,200–8,700		8,600	23,000			
8,600–14,500	18,000–23,500	11,000–15,000	25,000	13,000–30,000		15,000–30,000
12,000–16,700	22,000–29,000	12,000–18,000	30,000–36,000	8,500–23,000		10,000–40,000
280–435	1,300–1,450	400–600	1,300–1,440	300–640		800–2,000
375	700					
330–400	850–1,200	350–450	1,200–1,590	490–610		1,000–3,000
			520			
			390			
0.7–1.0	0.9–2.0	0.25–0.7	1.5–2.2	0.2–0.4	>7	2–20
M68–78	M90	M94–101; R111	M90; M100			
				Barcol 35–75	Shore D84–94	Barcol 50–80
60–95	15–25	65×10^{-6}	18–30	55–100		20–50
122–185	385–437	70–150	410–440	140–400		>400
240–375	421–500	167	470–480			
4.2–6.9	7.0	3.3–3.6	6.0–7.6			
1.30–1.38	1.48–1.53	1.29–1.40	1.55–1.70	1.04–1.46	1.01–1.20	1.35–2.30
0.08–0.09	0.06–0.08	0.1–0.2	0.05	0.15–0.6	0.5–2.5	0.01–1.0
0.4–0.5	0.35	0.2–0.3				
420–550	460–560	420–550	430–650	380–500	250–400	350–500

Table 6.12.1 Properties of Plastic Resins and Compounds (Continued)

		Polyethylene and ethylene copolymers					
		Low and medium density		High density			Cross-linked
Properties	ASTM test method	Linear copolymer	LDPE copolymers: Ethylene-vinyl acetate	Poly-ethylene homopolymer	Ultrahigh molecular weight	30% Class fiber-reinforced	Molding grade
1a. Melt flow, g/10 min	D1238		1.4–2.0	5–18			
1. Melting temperature, °C. T_m (crystalline)		122–124	103–110	130–137	125–138	120–140	
T_g (amorphous)							
2. Processing temperature range, °F. (C = compression; T = transfer; I = injection; E = extrusion)		I:350–500 E:450–600	C:200–300 I:350–430 E:300–380	I:350–500 E:350–525	C:400–500	I:350–600	C:240–450 I:250–300
3. Molding pressure range, 10^3 lb/in^2		5–15	1–20	12–15	1–2	10–20	
4. Compression ratio		3		2			
5. Mold (linear) shrinkage, in/in	D955	0.020–0.022	0.007–0.035	0.015–0.040	0.040	0.002–0.006	0.007–0.090
6. Tensile strength at break, lb/in^2	D638[b]	1,900–4,000	2,200–4,000	3,200–4,500	5,600–7,000	7,500–9,000	1,600–4,600
7. Elongation at break, %	D638[b]	100–965	200–750	10–1,200	420–525	1.5–2.5	10–440
8. Tensile yield strength, lb/in^2	D638[b]	1,400–2,800	1,200–6,000	3,800–4,800	3,100–4,000		
9. Compressive strength (rupture or yield), lb/in^2	D695			2,700–3,600		6,000–7,000	2,000–5,500
10. Flexural strength (rupture or yield), lb/in^2	D790					11,000–12,000	2,000–6,500
11. Tensile modulus, 10^3 lb/in^2	D638[b]	38–75	7–29	155–158		700–900	50–500
12. Compressive modulus, 10^3 lb/in^2	D695						50–150
13. Flexural modulus, 10^3 lb/in^2 73°F	D790	40–105	7.7	145–225	130–140	700–800	70–350
200°F	D790						
250°F	D790						
300°F	D790						
14. Izod impact, ft · lb/in of notch (⅛-in-thick specimen)	D256A	1.0–No break	No break	0.4–4.0	No break	1.1–1.5	1–20
15. Hardness Rockwell	D785				R50	R75–90	
Shore/Barcol	D2240/ D2583	Shore D55–56	Shore D17–45	Shore D66–73	Shore D61–63		Shore D55–80
16. Coef. of linear thermal expansion, 10^6 in/(in · °C)	D696		160–200	59–110	130–200	48	100
17. Deflection temperature under flexural load, °F 264 lb/in^2	D648				110–120	250	105–145
66 lb/in^2	D648			175–196	155–180	260–265	130–225
18. Thermal conductivity, 10^{-4} cal · cm/(s · cm^2 · °C)	C177			11–12		8.6–11	
19. Specific gravity	D792	0.918–0.940	0.922–0.943	0.952–0.965	0.94	1.18–1.28	0.95–1.45
20. Water absorption (⅛-in-thick specimen), % 24 h	D570		0.005–0.13	<0.01	<0.01	0.02–0.06	0.01–0.06
Saturation	D570						
21. Dielectric strength (⅛-in-thick specimen), short time, V/mil	D149		620–760	450–500	710	500–550	230–550

| Polyimide | | | | Polypropylene | | | | |
| Thermoplastic | | Thermoset | | Homopolymer | | | Copolymer | |
Unfilled	30% Glass fiber-reinforced	Unfilled	50% Glass fiber-reinforced	Unfilled	10–30% Glass fiber-reinforced	Impact-modified, 40% mica-filled	Unfilled	10–20% Glass fiber-reinforced
4.5–7.5				0.4–38.0	1–20		0.6–44.0	0.1–20
388	388	Thermoset	Thermoset	160–175	168	168	150–175	160–168
250–365	250			−20			−20	
C:625–690 I:734–740 E:734–740	I:734–788	460–485	C:460 I:390 T:390	I:375–550 E:400–500	I:425–475	I:350–470	I:375–550 E:400–500	I:350–480
3–20	10–30	7–29	3–10	10–20			10–20	
1.7–4	1.7–2.3	1–1.2		2.0–2.4			2–2.4	
0.0083	0.0044	0.001–0.01	0.002	0.010–0.025	0.002–0.008	0.007–0.008	0.010–0.025	0.003–0.01
10,500–17,100	24,000	4,300–22,900	6,400	4,500–6,000	6,500–13,000	4,500	4,000–5,500	5,000–8,000
7.5–90	3	1		100–600	1.8–7	4	200–500	3.0–4.0
12,500–13,000		4,300–22,900		4,500–5,400	7,000–10,000		3,000–4,300	
17,500–40,000	27,500	19,300–32,900	34,000	5,500–8,000	6,500–8,400		3,500–8,000	5,500–5,600
10,000–28,800	35,200	6,500–50,000	21,300	6,000–8,000	7,000–20,000	7,000	5,000–7,000	7,000–11,000
300–400	1,720	460–4,650		165–225	700–1,000	700	130–180	
315–350	458	421		150–300				
360–500	1,390	422–3,000	1,980	170–250	310–780	600	130–200	355–510
				50			40	
				35			30	
210	1,175	1,030–2,690						
1.5–1.7	2.2	0.65–15	5.6	0.4–1.4	1.0–2.2	0.7	1.1–14.0	0.95–2.7
E52–99, R129, M95	R128, M104	110M–120M	M118	R80–102	R92–115		R65–96	R100–103
							Shore D70–73	
45–56	17–53	15–50	13	81–100	21–62		68–95	
460–680	469	572–>575	660	120–140	253–288	205	130–140	260–280
				225–250	290–320		185–220	305
2.3–4.2	8.9	5.5–12	8.5	2.8	5.5–6.2		3.5–4.0	
1.33–1.43	1.56	1.41–1.9	1.6–1.7	0.900–0.910	0.97–1.14	1.23	0.890–0.905	0.98–1.04
0.24–0.34	0.23	0.45–125	0.7	0.01–0.03	0.01–0.05		0.03	0.01
1.2								
415–560	528	480–508	450	600			600	

Table 6.12.1 Properties of Plastic Resins and Compounds (*Continued*)

		Polystyrene and styrene copolymers					
		Polystyrene homopolymers			Rubber-modified	Styrene copolymers	High heat-resistant copolymers
						Styrene-acrylonitrile (SAN)	
Properties	ASTM test method	High and medium flow	Heat-resistant	20% Long and short glass fiber-reinforced	High-impact	Molding and extrusion	20% Glass fiber-reinforced
1a. Melt flow, g/10 min	D1238				5.8		
1. Melting temperature, °C. T_m (crystalline)							
T_g (amorphous)		74–105	100–110	115	9.3–105	100–200	
2. Processing temperature range, °F. (C = compression; T = transfer; I = injection; E = extrusion)		C:300–400 I:350–500 E:350–500	C:300–400 I:350–500 E:350–500	I:400–550	I:350–525 E:375–500	C:300–400 I:360–550 E:360–450	I:425–550
3. Molding pressure range, 10^3 lb/in^2		5–20	5–20	10–20	10–20	5–20	
4. Compression ratio		3	3–5		4	3	
5. Mold (linear) shrinkage, in/in	D955	0.004–0.007	0.004–0.007	0.001–0.003	0.004–0.007	0.003–0.005	0.003–0.004
6. Tensile strength at break, lb/in^2	D638[b]	5,200–7,500	6,440–8,200	10,000–12,000	1,900–6,200	10,000–11,900	10,000–14,000
7. Elongation at break, %	D638[b]	1.2–2.5	2.0–3.6	1.0–1.3	20–65	2–3	1.4–3.5
8. Tensile yield strength, lb/in^2	D638[b]		6,440–8,150		2,100–6,000	9,920–12,000	
9. Compressive strength (rapture or yield), lb/in^2	D695	12,000–13,000	13,000–14,000	16,000–17,000		14,000–15,000	
10. Flexural strength (rupture or yield), lb/in^2	D790	10,000–14,600	13,000–14,000	14,000–18,000	3,300–10,000	11,000–19,000	16,300–22,000
11. Tensile modulus, 10^3 lb/in^2	D638[b]	330–475	450–485	900–1,200	160–370	475–560	850–900
12. Compressive modulus, 10^3 lb/in^2	D695	480–490	495–500			530–580	
13. Flexural modulus, 10^3 lb/in^2 73°F	D790	380–490	450–500	950–1,100	160–390	500–610	800–1,050
200°F	D790						
250°F	D790						
300°F	D790						
14. Izod impact, ft · lb/in of notch (⅛-in-thick specimen)	D256A	0.35–0.45	0.4–0.45	0.9–2.5	0.95–7.0	0.4–0.6	2.1–2.6
15. Hardness Rockwell	D785	M60–75	M75–84	M80–95, R119	R50–82; L–60	M80, R83	
Shore/Barcol	D2240/ D2583						
16. Coef. of linear thermal expansion, 10^6 in/(in · °C)	D696	50–83	68–85	39.6–40	44.2	65–68	20
17. Deflection temperature under flexural load, °F 264 lb/in^2	D648	169–202	194–217	200–220	170–205	214–220	231–247
66 lb/in^2	D648	155–204	200–224	220–230	165–200	220–224	
18. Thermal conductivity, 10^{-4} cal · cm/(s · cm^2 · °C)	C177	3.0	3.0	5.9		3.0	
19. Specific gravity	D792	1.04–1.05	1.04–1.05	1.20	1.03–1.06	1.06–1.08	1.20–1.22
20. Water absorption (⅛-in-thick specimen), % 24 h	D570	0.01–0.03	0.01	0.07–0.01	0.05–0.07	0.15–0.25	0.1
Saturation	D570	0.01–0.03	0.01	0.3		0.5	
21. Dielectric strength (⅛-in-thick specimen), short time, V/mil	D149	500–575	500–525	425		425	

Polyurethane					Silicone		
Thermoset			Thermoplastic		Casting resins	Liquid injection molding	Molding and encapsulating compounds
Casting resins		55–65% Mineral-filled potting and casting compounds	Unreinforced molding	10–20% Glass fiber-reinforced molding compounds	Flexible (including RTV)	Liquid silicone rubber	Mineral- and/or glass-filled
Liquid	Unsaturated						
Thermoset	Thermoset	Thermoset	75–137		Thermoset	Thermoset	Thermoset
				120–160			
C:43–250		25 (casting)	I:430–500 E:430–510	I:360–410		I:360–420	C:280–360 I:330–370 T:330–370
0.1–5				8–11		1–2	0.3–6
							2.0–8.0
0.020		0.001–0.002	0.004–0.006	0.004–0.010	0.0–0.006	0.0–0.005	0.0–0.005
175–10,000	10,000–11,000	1,000–7,000	7,200–9,000	4,800–7,500	350–1,000	725–1,305	500–1,500
100–1,000	3–6	5–55	60–180	3–70	20–700	300–1,000	80–800
			7,800–11,000				
20,000				5,000			
700–4,500	19,000		10,200–15,000	1,700–6,200			
10–100			190–300	0.6–1.40			
10–100							
10–100	610		235–310	40–90			
25 to flexible	0.4		1.5–1.8	10–14-No break			
			R > 100; M48	R45–55			
Shore A10–13, D90	Barcol 30–35	Shore A90, D52–85			Shore A10–70	Shore A20–70	Shore A10–80
100–200		71–100		34	10–19	10–20	20–50
Varies over wide range	190–200		158–260	115–130			>500
			176–275	140–145			
5		6.8–10			3.5–7.5		7.18
1.03–1.5	1.05	1.37–2.1	1.2	1.22–1.36	0.97–2.5	1.08–1.14	1.80–2.05
0.2–1.5	0.1–0.2	0.06–0.52	0.17–0.19	0.4–0.55	0.1		0.15
			0.5–0.6	1.5			0.15–0.40
300–500		500–750 @ $\frac{1}{16}$ in.	400	600	400–550		200–550

Table 6.12.1   Properties of Plastic Resins and Compounds   (*Continued*)

Materials	Properties	ASTM test method	Vinyl polymers and copolymers				Molding and extrusion compounds
			PVC molding compound, 20% glass fiber-reinforced	PVC and PVC-acetate MC, sheets, rods, and tubes			Chlorinated polyvinyl chloride
				Rigid	Flexible, unfilled	Flexible, filled	
Processing	1a. Melt flow, g/10 min	D1238					
	1. Melting temperature, °C. T_m (crystalline)						
	T_g (amorphous)		75–105	75–105	75–105	75–105	110
	2. Processing temperature range, °F. (C = compression; T = transfer; I = injection; E = extrusion)		I:380–400 E:390–400	C:285–400 I:300–415	C:285–350 I:320–385	C:285–350 I:320–385	C:350–400 I:395–440 E:360–420
	3. Molding pressure range, 10^3 lb/in²		5–15	10–40	8–25	1–2	15–40
	4. Compression ratio		1.5–2.5	2.0–2.3	2.0–2.3	2.0–2.3	1.5–2.5
	5. Mold (linear) shrinkage, in/in	D955	0.001	0.002–0.006	0.010–0.050	0.008–0.035 0.002–0.008	0.003–0.007
Mechanical	6. Tensile strength at break, lb/in²	D638[b]	8,600–12,800	5,900–7,500	1,500–3,500	1,000–3,500	6,800–9,000
	7. Elongation at break, %	D638[b]	2–5	40–80	200–450	200–400	4–100
	8. Tensile yield strength, lb/in²	D638[b]		5,900–6,500			6,000–8,000
	9. Compressive strength (rupture or yield), lb/in²	D695		8,000–13,000	900–1,700	1,000–1,800	9,000–22,000
	10. Flexural strength (rupture or yield), lb/in²	D790	14,200–22,500	10,000–16,000			14,500–17,000
	11. Tensile modulus, 10^3 lb/in²	D638[b]	680–970	350–600			341–475
	12. Compressive modulus, 10^3 lb/in²	D695					335–600
	13. Flexural modulus, 10^3 lb/in² 73°F	D790	680–970	300–500			380–450
	200°F	D790					
	250°F	D790					
	300°F	D790					
	14. Izod impact, ft · lb/in of notch (⅛-in-thick specimen)	D256A	1.0–1.9	0.4–22	Varies over wide range	Varies over wide range	1.0–5.6
	15. Hardness Rockwell	D785	R108–119				R117–112
	Shore/Barcol	D2240/ D2583	Shore D85–89	Shore D65–85	Shore A50–100	Shore A50–100	
Thermal	16. Coef. of linear thermal expansion, 10^6 in/(in · °C)	D696	24–36	50–100	70–250		62–78
	17. Deflection temperature under flexural load, °F 264 lb/in²	D648	165–174	140–170			202–234
	66 lb/in²	D648		135–180			215–247
	18. Thermal conductivity, 10^{-4} cal · cm/(s · cm² · °C)	C177		3.5–5.0	3–4	3–4	3.3
Physical	19. Specific gravity	D792	1.43–1.50	1.30–1.58	1.16–1.35	1.3–1.7	1.49–1.58
	20. Water absorption (⅛-in-thick specimen), % 24 h	D570	0.01	0.04–0.4	0.15–0.75	0.5–1.0	0.02–0.15
	Saturation	D570					
	21. Dielectric strength (⅛-in-thick specimen), short time, V/mil	D149		350–500	300–400	250–300	600–625

[a] Acrylonitrile-butadiene-styrene.   [b] Tensile test method varies with material: D638 is standard for thermoplastics; D651 for rigid thermoset plastics; D412 for elastomeric plastics; D882 for thin plastics sheeting.   [c] Dry, as molded (approximately 0.2% moisture content).   [d] As conditioned to equilibrium with 50% relative humidity.   [e] Test method is ASTM D4092.   [f] *Pseudo* indicates that the thermoset and thermoplastic components were mixed in the form of pellets or powder prior to fabrication.

conversion of postconsumer plastic items (i.e., scrap) may result in increasing amounts of plastics converted back to feedstock by application of hydrolytic and pyrolitic processes.

The plastic is supplied to primary processers usually in the form of powder, solid pellets, or plastisols (liquid or semisolid dispersions of finely powdered polymer in a nonmigrating liquid).

Compounding, mixing, and blending prepare plastic materials to be fed into any of the various fabrication processes.

PRIMARY FABRICATION PROCESSES

A number of processes are used to achieve finished plastic parts, including some conventional ones usually associated with metalworking and others synonymous with plastics processing. Casting, blow molding, extrusion, forming in metal molds (injection, compression, and transfer molding), expanded-bead molding to make foams, thermoforming of sheet plastic, filament winding over a form, press laminating, vacuum forming, and open molding (hand layup) are widespread plastics fabrication processes.

ADDITIVES

The inherent properties of most plastics can be tailored to impart other desired properties or to enhance existing ones by the introduction of additives. A wide variety of additives allow compounding a specific type of plastic to meet some desired end result; a few additive types are listed below.

Fillers Wood flour, mica, silica, clay, and natural synthetic fibers reduce weight, reduce the volume of bulk resin used, impart some specific strength property (and often, directionality of strength properties), etc.

Blowing or Foaming Agents Compressed gas or a liquid which will evolve gas when heated is the agent which causes the network of interstices in expanded foam. A popular foaming agent, CFC (chlorinated fluorocarbon) is being displaced because of environmental concerns. Non-CFC foaming agents will increasingly replace CFCs.

Mold-Release Compounds These facilitate the removal of molded parts from mold cavities with retention of surface finish on the finished parts and elimination of pickup of molding material on mold surfaces.

Lubricants They ease fabrication and can serve to impart lubricant to plastic parts.

Antistatic Compounds The inherent good electrical insulation property of plastic often leads to buildup of static electric charges. The static charge enhances dust pickup and retention to the plastic surface. At the very least, this may inhibit the action of a mold-release agent; at worst, the leakage of static electric charge may be an explosion or fire hazard.

Antibacterial Agents (Biocides) These act to inhibit the growth of bacteria, especially when the type of filler used can be an attractive host to bacterial action (wood flour, for example).

Colorants Color is imparted to the resin by dyes or pigments. Dyes allow a wider range of color. Colorants enhance the utility of plastic products where cosmetic and/or aesthetic effects are required. Ultraviolet (uv) radiation from sunlight degrades the color unless a uv-inhibiting agent is incorporated into the plastic resin. (See below.)

Flame Retardants These are introduced primarily for safety, and they work by increasing the ignition temperature and lowering the rate of combustion.

Heat Stabilizers Plastics will degrade when subjected to high temperatures; thermoplastics will soften and eventually melt; thermosets will char and burn. Either type of plastic part is subjected to heat during processing and may see service above room temperature during its useful life. The judicious addition of an appropriate heat-stabilizing compound will prevent thermal degradation under those circumstances. Some of the more effective heat-stabilizing agents contain lead, antimony, or cadmium and are subject to increasingly stringent environmental limits.

Impact Modifiers When added to the basic resin, these materials increase impact strength or toughness.

Plasticizers These materials increase the softness and flexibility of the plastic resin, but often there is an accompanying loss of tensile strength, increased flammability, etc.

Ultraviolet Stabilizers These additives retard or prevent the degradation of some strength properties as well as color, which result from exposure to sunlight.

Antioxidants Substances compounded with plastic resin that act to prevent oxidative degeneration which may be brought about by exposure to heat, radiation, impurities, or high shear loads.

ADHESIVES, ASSEMBLY, AND FINISHES

Most adhesives operate to effect solvent-based **bonding,** and they are either one- or two-part materials. While these adhesives cure, the evaporation of volatile organic compounds (VOCs) constitutes an environmental hazard which is subject to increasing regulation. Water-based bonding agents, on the other hand, emit no VOCs and effectively are environmentally benign. Adhesives which are constituted of all solid material contain no solvents or water, hence no evaporants. A small quantity of bonding is performed via **radiation-curable adhesives.** The source of energy is ultraviolet radiation or electron beam radiation.

The adhesives cited above, and others, display a range of properties which must be assessed as to suitability for the application intended.

Welding of plastics most often implies bonding or sealing of plastic sheet through the agent of ultrasonic energy, which, when applied locally, transforms high-frequency ultrasonic energy to heat, which, in turn, fuses the plastic locally to achieve the bond. The design of the joint is critical to the successful application of ultrasonic welding when the pieces are other than thin sheets. **Solvent welding** consists of localized softening of the pieces to be joined and the intimate mixture of the plastic surface material of both parts. The solvent is usually applied with the parts fixed or clamped in place, and sufficient time is allowed for the solvent to react locally with the plastic surfaces and then evaporate.

Most plastic pieces are amenable to the application of **decorative finishes** either during fabrication (e.g., in the mold) or as a secondary operation. Suitable paints are applied as required, often via silk screening. Electrolytic plating, vacuum metallizing, hot stamping, printing, and application of labels via pressure-sensitive adhesives are well-established methods in wide use.

RECYCLING

The feedstock for all plastics is either petroleum or natural gas. These materials constitute a valuable and irreplaceable natural resource. Plastic materials may degrade; plastic parts may break or no longer be of service; certainly, very little plastic per se "wears away." The basic idea behind plastic recycling takes a leaf from the metal industries, where recycling scrap metals has been the modus operandi from ancient times. The waste of a valuable resource implicit in discarding scrap plastic in landfills or the often environmentally hazardous incineration of plastics has led to very strong efforts toward recycling plastic scrap.

Separation of plastic by types, aided by unique identification symbols molded into consumer items, is mandated by law in many localities which operate an active recycling program. Different plastic types present a problem when comingled during recycling, but some postrecycling consumer goods have found their way to the market. Efforts will continue as recycling of plastics becomes universal; the costs associated with the overall recycling effort will become more attractive as economies of large-scale operations come into play.

Ideally, plastic recycling responds best to clean scrap segregated as to type. In reality, this is not the case, and for those reasons, a number of promising scrap processing directions are being investigated. Among those is depolymerization, whereby the stream of plastic scrap is subjected to pyrolitic or hydrolytic processing which results in reversion of the plastic to feedstock components. This regenerated feedstock can then be processed once more into the monomers, then to virgin polymeric resins. Other concepts will come under consideration in the near future.

6.13 FIBER COMPOSITE MATERIALS
by Stephen R. Swanson

REFERENCES: Jones, "Mechanics of Composite Materials," Hemisphere Publishing. Halpin, "Primer on Composite Materials Analysis," 2d ed., Technomic Pub. Co. "Engineered Materials Handbook," vol. 1, "Composites," ASM. Hull, "An Introduction to Composite Materials," Cambridge Univ. Press. Schwartz (ed.), "Composite Materials Handbook," 2d ed., McGraw-Hill.

INTRODUCTION

Composite materials are composed of two or more discrete constituents. Examples of engineering use of composites date back to the use of straw in clay by the Egyptians. Modern composites using fiber-reinforced matrices of various types have created a revolution in high-performance structures in recent years. Advanced composite materials offer significant advantages in strength and stiffness coupled with light weight, relative to conventional metallic materials. Along with this structural performance comes the freedom to select the orientation of the fibers for optimum performance. Modern composites have been described as being revolutionary in the sense that the material can be designed as well as the structure.

The **stiffness** and **strength-to-weight** properties make materials such as carbon fiber composites attractive for applications in aerospace and sporting goods. In addition, composites often have superior resistance to environmental attack, and glass fiber composites are used extensively in the chemical industries and in marine applications because of this advantage. Both glass and carbon fiber composites are being considered for infrastructure applications, such as for bridges and to reinforced concrete, because of this environmental resistance.

The cost competitiveness of composites depends on how important the weight reduction or environmental resistance provided by them is to the overall function of the particular application. While glass fibers usually cost less than aluminum on a weight basis, carbon fibers are still considerably more expensive. Equally important or more important than the material cost is the cost of manufacturing. In some cases composite structures can achieve significant cost savings in manufacturing, often by reducing the number of parts involved in a complex assembly. There is a large variability in cost and labor content between the various methods of composite manufacture, and much attention is currently given to reduce manufacturing costs.

TYPICAL ADVANCED COMPOSITES

Modern composite materials usually, but not always, utilize a **reinforcement** phase and a **binder** phase, in many cases with more rigid and higher-strength fibers embedded and dispersed in a more compliant matrix. Modern applications started with **glass fibers,** although the word **advanced** fibers often means the more **high-performance fibers** that followed, such as carbon, aramid, boron, and silicon carbide. A typical example is carbon fiber that is being widely introduced into aerospace and sporting goods applications. The tensile strength of carbon fiber varies with the specific type of fiber being considered, but a typical range of values is 3.1 to 5.5 GPa (450 to 800 ksi) for fiber tensile strength and stiffness on the order of 240 GPa (35 Msi), combined with a specific gravity of 1.7. Thus the fiber itself is stronger than 7075 T6 aluminum by a factor of 5 to 10, is stiffer by a factor of 3.5, and weighs approximately 60 percent as much. The potential advantages of high-performance fiber structures in mechanical design are obvious. On the other hand, current costs for carbon fibers are on the order of several times to an order of magnitude or more higher than those for aluminum. The cost differential implies that composite materials will be utilized in demanding applications, where increases in performance justify the increased material cost. However, the material cost is only part of the story, for manufacturing costs must also be considered. In many instances it has been possible to form composite parts with significantly fewer individual components compared to metallic structures, thus leading to an overall lower-cost structure. On the other hand, composite components may involve a significant amount of hand labor, resulting in higher manufacturing costs.

FIBERS

Glass Fiber Glass fiber in a **polymeric matrix** has been widely used in commercial products such as boats, sporting goods, piping, and so forth. It has relatively low stiffness, high elongation, moderate strength and weight, and generally lower cost relative to other composites. It has been used extensively where corrosion resistance is important, for piping for the chemical industry, and in marine applications.

Aramid Fiber Aramid fibers (sold under the trade name **Kevlar**) offer higher strength and stiffness relative to glass, coupled with light weight, high tensile strength but lower compressive strength. Both glass fiber and aramid fiber composites exhibit good toughness in impact environments. Aramid fiber is used as a higher-performance replacement for glass fiber in industrial and sporting goods and in protective clothing.

Carbon Fiber Carbon fibers are used widely in aerospace and for some sporting goods, because of their relatively high stiffness and high strength/weight ratios. Carbon fibers vary in strength and stiffness with processing variables, so that different grades are available (high modulus or intermediate modulus), with the tradeoff being between high modulus and high strength. The intermediate-modulus and high-strength grades are almost universally made from a **PAN (polyacrylonitrile) precursor,** which is then heated and stretched to align the structure and remove noncarbon material. Higher-modulus but lower-cost fibers with much lower strength are made from a petroleum pitch precursor.

Other Fibers Boron fibers offer very high stiffness, but at very high cost. These fibers have been used in specialized applications in both aluminum and polymeric matrices. A new fiber being used in textile applications is oriented polyethylene, marketed under the trademark **Spectra.** This fiber combines high strength with extremely light weight. The fiber itself has a specific gravity of 0.97. It is limited to a very low range of temperature, and the difficulty of obtaining adhesion to matrix materials has limited its application in structural composites. A number of other fibers are under development for use with ceramic matrices, to enable use in very high-temperature applications such as engine components. An example is **silicon carbide fiber,** used in **whisker** form.

Table 6.13.1 lists data pertinent to fibers currently available.

MATRICES

Most current applications utilize polymeric matrices. **Thermosetting polymers** (such as **epoxies**) are widely utilized, and a large amount of characterization data are available for these materials. Epoxies provide superior performance but are more costly. Typical cure temperatures are in the range of 121 to 177°C (250 to 350°F). A recent development has been high-toughness epoxies, which offer significantly improved resistance to damage from accidental impact, but at higher cost. Polyester and vinyl ester are often used in less demanding applications, and polyurethane matrices are being considered because of short cure times in high-production-rate environments. **Thermoplastic matrices** are under development and have had limited applications to date. Wider use awaits accumulation of experience in their application and is inhibited in some instances by high material costs. Their use is likely to increase because of the increased toughness they may provide, as well as the

Table 6.13.1   Mechanical Properties of Typical Fibers

Fiber	Fiber diameter, μm	Fiber density		Tensile strength		Tensile modulus	
		lb/in^3	g/cm^3	ksi	GPa	Msi	GPa
E-glass	8–14	0.092	2.54	500	3.45	10.5	72.4
S-glass	8–14	0.090	2.49	665	4.58	12.5	86.2
Polyethylene	10–12	0.035	0.97	392	2.70	12.6	87.0
Aramid (Kevlar 49)	12	0.052	1.44	525	3.62	19.0	130.0
HS carbon, T300	7	0.063	1.74	514	3.54	33.6	230
AS4 carbon	7	0.065	1.80	580	4.00	33.0	228
IM7 carbon	5	0.065	1.80	785	5.41	40.0	276
GY80 carbon	8.4	0.071	1.96	270	1.86	83.0	572
Boron	50–203	0.094	2.60	500	3.44	59.0	407

potential for forming complex shapes. Composites using thermoplastics often are formed to final shape at temperatures of about 315°C (600°F), and they need no cure cycle.

Polymeric matrix materials have a limited temperature range for practical application, with epoxies usually limited to 150°C (300°F) or less, depending on the specific material. Higher-temperature polymers available, such as polyimides, usually display increased brittleness. They are used for cowlings and ducts for jet engines.

Metal matrix composites are utilized for higher-temperature applications than is possible with polymeric matrices. Aluminum matrix has been utilized with boron and carbon fibers. Ceramic matrix materials are being developed for service at still higher temperatures. In this case, the fiber is not necessarily higher in strength and stiffness than the matrix, but is used primarily to toughen the ceramic matrix.

MATERIAL FORMS AND MANUFACTURING

Composite materials come in a wide variety of material forms. The fiber itself may be used in continuous form or as a **chopped fiber.** Chopped glass fibers are used typically to reinforce various polymers, with concomitant lower strength and stiffness relative to continuous fiber composites. Chopped fibers in conjunction with automated fabrication techniques have been utilized to fabricate automotive body parts at high production rates.

Continuous-fiber materials are available in a number of different forms, with the specific form utilized depending on the manufacturing process. Thus it is useful to consider both the material and the manufacturing process at the same time. The fibers themselves have a small diameter, with sizes of 5 to 7 μm (0.0002 to 0.0003 in) typical for carbon fibers. A large number of fibers, from 2,000 to 12,000, are gathered in the manufacturing process to form a **tow** (also called a **roving** or **yarn**). The filament winding process utilizes these tows directly. The tows may be further processed by **prepregging,** the process of coating the individual fibers with the matrix material. This process is widely used with thermosetting polymeric resins. The resin is partially cured, and the resulting **"ply"** is placed on a paper backing. The prepregged material is available in continuous rolls in widths of 75 to 1,000 mm (3 to 40 in). These rolls must be kept refrigerated until they are utilized and the assembled product is cured. Note that the ply consists of a number of fibers through its thickness, and that the fibers are aligned and continuous. Typical volume fractions of fiber are on the order of 60 percent. The material forms discussed here are used in a variety of specific manufacturing techniques. Some of the more popular techniques are described briefly below.

Filament Winding   The process consists of winding the fiber around a mandrel to form the structure. Usually the mandrel rotates while fiber placement is synchronized to proceed in a longitudinal direction. The matrix may be added to the fiber by passing the fiber tow through a matrix bath at the time of placement, a process called **wet winding;** or the tows may be prepregged prior to winding. Filament winding is widely used to make glass fiber pipe, rocket motor cases, and similar products. Filament winding is a highly automated process, with typical

low manufacturing costs. Obviously, it lends itself most readily to axisymmetric shapes, but a number of specialized techniques are being considered for nonaxisymmetric shapes.

Prepreg Layup   This common procedure involves laying together individual plies of prepregged composite into the final laminated structure. A mold may be used to control the part geometry. The plies are laid down in the desired pattern, and then they are wrapped with several additional materials used in the curing process. The objective is to remove volatiles and excess air to facilitate consolidation of the laminate. To this end, the laminate is covered with a **peel ply,** for removal of the other curing materials, and a **breather ply,** which is often a fiberglass net. Optionally, a bleeder may be used to absorb excess resin, although the net resin process omits this step. Finally, the assembly is covered with a vacuum bag and is sealed at the edges. A vacuum is drawn, and after inspection heat is applied. If an autoclave is used, pressure on the order of 0.1 to 0.7 MPa (20 to 100 lb/in^2) is applied to ensure final consolidation. Autoclave processing ensures good lamination but requires a somewhat expensive piece of machinery. Note that the individual plies are relatively thin [on the order of 0.13 mm (0.005 in)], so that a large number of plies will be required for thick parts. The lamination process is often performed by hand, although automated tape-laying machines are available. Although unidirectional plies have been described here, cloth layers can also be used. The bends in the individual fibers that occur while using cloth layers carry a performance penalty, but manufacturing considerations such as drapeability may make cloth layers desirable.

Automated Tape and Tow Placement   Automated machinery is used for tape layup and fiber (tow) placement. These machines can be large enough to construct wing panels or other large structures. Thermosetting matrices have been used extensively, and developments using thermoplastic matrices are underway.

Textile Forms   The individual tows may be combined in a variety of textile processes such as **braiding** and **weaving.** Preforms made in these ways then can be impregnated with resin, often called **resin transfer molding** (RTM). These textile processes can be designed to place fibers in the through-the-thickness direction, to impart higher strength in this direction and to eliminate the possibility of delamination. There is considerable development underway in using braided and/or stitched preforms with RTM because of the potential for automation and high production rates.

DESIGN AND ANALYSIS

The fundamental way in which fiber composites, and in particular continuous-fiber composites, differ from conventional engineering materials such as metals is that their properties are **highly directional.** Stiffness and strength in the fiber direction may be higher than in the direction transverse to the fibers by factors of 20 and 50, respectively. Thus a basic principle is to align the fibers in directions where stiffness and strength are needed. A well-developed theoretical basis called **classical lamination theory** is available to predict stiffness and to calculate stresses within the layers of a laminate. This theory is often applied to filamentwound structures and to textile preform structures, if allowances are made for the undulations in the fiber path. The basic assumption is that fiber composites are orthotropic materials. Thin composite layers require four independent elasticity constants to characterize the stiffness; those are the fiber direction modulus, transverse direction modulus, in-plane shear modulus, and one of the two Poisson ratios, with the other related through symmetry of the orthotropic stress-strain matrix. The material constants are routinely provided by material suppliers. Lamination theory is then used to predict the overall stiffness of the laminate and to calculate stresses within the individual layers under mechanical and thermal loads.

Transverse properties are much lower than those in the fiber direction, so that fibers must usually be oriented in more than one direction, even if only to take care of secondary loads. For example, a laminate may consist of fibers in an axial direction combined with fibers oriented at $\pm 60°$ to this direction, commonly designated as an $[0_m/\pm 60_n]s$ laminate, where m and n refer to the number of plies in the axial and $\pm 60°$ directions and s stands for symmetry with respect to the midplane of the laminate. Although not all laminates are designed to be symmetric, residual stresses will cause flat panels to curve when cooled from elevated-temperature processing unless they are symmetric. Other popular laminates have fibers oriented at $0°$, $\pm 45°$, and $90°$. If the relative amounts of fibers are equal in the $[0/\pm 60]$ or the $[0/\pm 45/90]$ directions, the laminate has in-plane stiffness properties equal in all directions and is thus termed **quasi-isotropic.**

Note that the **stiffness of composite laminates** is less than that of the fibers themselves for two reasons. First, the individual plies contain fibers and often a much less stiff matrix. With high-stiffness fibers and polymeric matrices, the contribution of the matrix can be neglected, so that the stiffness in the fiber direction is essentially the **fiber volume fraction** VF times the fiber stiffness. Fiber volume fractions are often on the order of 60 percent. Second, laminates contain fibers oriented in more than one direction, so that the stiffness in any one direction is less than it would be in the fiber direction of a unidirectional laminate. As an example, the in-plane stiffness of a quasi-isotropic carbon fiber/polymeric matrix laminate is about 45 percent of unidirectional plies, which in turn is approximately 60 percent of the fiber modulus for a 60 percent fiber volume fraction material.

Strength Properties   Fiber composites typically have excellent strength-to-weight properties; see Table 6.13.1. Like stiffness properties, strength properties are reduced by dilution by the weaker matrix,

and because not all the fibers can be oriented in one direction. The overall reduction in apparent strength because of these two factors is similar to that for stiffness, although this is a very rough guide. As an example, the strength of a quasi-isotropic carbon/epoxy laminate under uniaxial tensile load has been reported to be about 35 to 40 percent of that provided by the same material in a unidirectional form.

Fiber composite materials can fail in several different modes. The design goal is usually to have the failure mode be **in-plane failure** due to fiber failure, as this takes advantage of the strong fibers. Unanticipated loads, or poorly designed laminates, can lead to ultimate failure of the matrix, which is much weaker. For example, in-plane shear loads on a [0/90] layup would be resisted by the matrix, with the potential result of failure at low applied load. Adding [± 45] fibers will resist this shear loading effectively.

Fiber composites respond differently to compressive loads than they do to tensile loads. Some fibers, such as aramid and the very high-modulus carbon fibers, are inherently weaker in compression than in tension. It is believed that this is not the case with intermediate-modulus carbon fibers, but even these fibers often display lower compressive strength. The mechanism is generally believed to involve bending or buckling of the very small-diameter fibers against the support provided by the matrix.

Matrix cracking usually occurs in the transverse plies of laminates with polymeric matrices, often at an applied load less than half that for ultimate fiber failure. This is caused by the lower failure strain of the matrix and the stress concentration effect of the fibers, and it can also depend on residual thermal stresses caused by the differences in coefficients of thermal expansion in the fibers and matrix. In some cases these matrix cracks are tolerated, while in other cases they produce undesirable effects. An example of the latter is that matrix cracking increases the permeability in unlined pressure vessels, which can result in pipes weeping.

Another potential failure mode is **delamination.** Delamination is often caused by high interlaminar shear and through-the-thickness tensile stresses, perhaps combined with manufacturing deficiencies such as inadequate cure or matrix porosity. Delamination can start at the edges of laminates due to "edge stresses" which have no counterpart in isotropic materials such as metals.

Fiber composites usually have excellent fatigue properties. However, stress concentrations or accidental damage can result in localized damage areas that can lead to failure under fatigue loading.

Section **7**

Fuels and Furnaces

MARTIN D. SCHLESINGER *Late Consultant, Wallingford Group, Ltd.*
KLEMENS C. BACZEWSKI *Consulting Engineer*
GLENN W. BAGGLEY *Formerly Manager, Regenerative Systems, Bloom Engineering Co., Inc.*
CHARLES O. VELZY *Consultant*
LEONARD M. GRILLO *Principal, Grillo Engineering Co.*
JAMES C. SIMMONS *Senior Vice President, Business Development, Core Furnace Systems Corp.*

7.1 FUELS

by Martin D. Schlesinger and Associates

COAL

by Martin D. Schlesinger

Consultant

REFERENCES: Petrography of American Coals, *U.S. BuMines Bull.* 550. Lowry, "Chemistry of Coal Utilization," Wiley. ASTM, "Standards on Gaseous Fuels, Coal and Coke." Methods of Analyzing and Testing Coal and Coke, *U.S. BuMines Bull.* 638. Karr, "Analytical Methods for Coal and Coal Products," Academic. Preprints, Division of Fuel Chemistry, American Chemical Society.

Coal is a black or brownish-black combustible solid formed by the decomposition of vegetation in the absence of air. Microscopy can identify plant tissues, resins, spores, etc. that existed in the original structure. It is composed principally of carbon, hydrogen, oxygen, and small amounts of sulfur and nitrogen. Associated with the organic matrix are water and as many as 65 other chemical elements. Many trace elements can be determined by spectrometric method D-3683. Coal is used directly as a fuel, a chemical reactant, and a source of organic chemicals. It can also be converted to liquid and gaseous fuels.

Classification and Description

Coal may be classified by rank, by variety, by size and sometimes by use. Rank classification takes into account the degree of metamorphism or progressive alteration in the natural series from lignite to anthracite. Table 7.1.1 shows the classification of coals by rank adopted as standard by the ASTM (method D-388). The basic scheme is according to fixed carbon (FC) and heating value (HV) from a proximate analysis, calculated on the mineral-matter-free (mmf) basis. The higher-rank coals are classified according to the FC on a dry basis and the lower-rank according to HV in Btu on a moist basis. **Agglomerating** character is used to differentiate between certain adjacent groups. Coals are considered agglomerating if, in the test to determine volatile matter, they produce either a coherent button that will support a 500-g weight or a button that shows swelling or cell structure.

For classifying coals according to rank, FC and HV can be calculated to a moisture-free basis by the **Parr formulas,** Eqs. (7.1.1) to (7.1.3) below:

$$FC \text{ (dry, mmf)} = \frac{FC - 0.15S}{100 - (M + 1.08A + 0.55S)} \times 100 \quad (7.1.1)$$

$$VM \text{ (dry, mmf)} = 100 - FC \quad (7.1.2)$$

$$HV \text{ (moist, mmf)} = \frac{Btu - 50S}{100 - (1.08A + 0.55S)} \quad (7.1.3)$$

where FC = percentage of fixed carbon, VM = percentage of volatile matter, M = percentage of moisture, A = percentage of ash, S = percentage of sulfur, all on a moist basis. "Moist" coal refers to the natural bed moisture, but there is no visible moisture on the surface. HV = heating value, Btu/lb (Btu/lb $\times$ 0.5556 = g · cal/g). Because of its complexity, the analysis of coal requires care in sampling, preparation, and selection of the method of analysis.

Figure 7.1.1 shows representative proximate analyses and heating values of various ranks of coal in the United States. The analyses were calculated to an ash-free basis because ash is not a function of rank. Except for anthracite, FC and HV increase from the lowest to the highest rank as the percentages of volatile matter and moisture decrease. The sources and analyses of coals representing various ranks are given in Table 7.1.2.

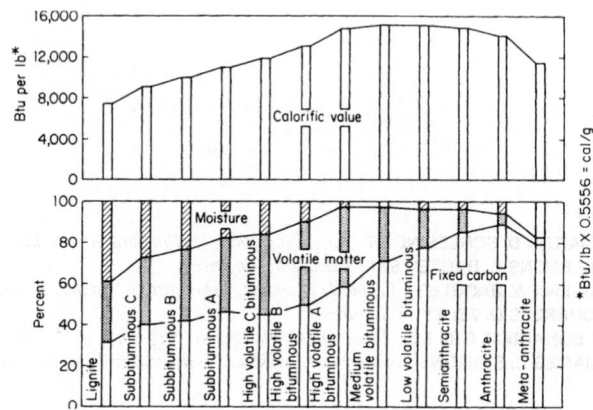

Fig. 7.1.1 Proximate analysis and heating values of various ranks of coal (ash-free basis).

Meta-anthracite is a high-carbon coal that approaches graphite in structure and composition. It usually is slow to ignite and difficult to burn. It has little commercial importance.

Anthracite, sometimes called hard coal, is hard, compact, and shiny black, with a generally conchoidal fracture. It ignites with some difficulty and burns with a short, smokeless, blue flame. Anthracite is used primarily for space heating and as a source of carbon. It is also used in electric power generating plants in or close to the anthracite-producing area. The iron and steel industry uses some anthracite in blends with bituminous coal to make coke, for sintering iron-ore fines, for lining pots and molds, for heating, and as a substitute for coke in foundries.

Semianthracite is dense, but softer than anthracite. It burns with a short, clean, bluish flame and is somewhat more easily ignited than anthracite. The uses are about the same as for anthracite.

Low-volatile bituminous coal is grayish black, granular in structure and friable on handling. It cakes in a fire and burns with a short flame that is usually considered smokeless under all burning conditions. It is used for space heating and steam raising and as a constituent of blends for improving the coke strength of higher-volatile bituminous coals. Low-volatile bituminous coals cannot be carbonized alone in slot-type ovens because they expand on coking and damage the walls of the ovens.

Medium-volatile bituminous coal is an intermediate stage between high-volatile and low-volatile bituminous coal and therefore has some of the characteristics of both. Some are fairly soft and friable, but others are hard and do not disintegrate on handling. They cake in a fuel bed and smoke when improperly fired. These coals make cokes of excellent strength and are either carbonized alone or blended with other bituminous coals. When carbonized alone, only those coals that do not expand appreciably can be used without damaging oven walls.

High-volatile A bituminous coal has distinct bands of varying luster. It is hard and handles well with little breakage. It includes some of the best steam and coking coal. On burning in a fuel bed, it cakes and gives off smoke if improperly fired. The coking property is often improved by blending with more strongly coking medium- and low-volatile bituminous coal.

High-volatile B bituminous coal is similar to high-volatile A bituminous coal but has slightly higher bed moisture and oxygen content and is less strongly coking. It is good coal for steam raising and space heating.

Table 7.1.1 Classification of Coals by Rank (ASTM D388)*

Class	Group	Fixed-carbon limits, percent (dry, mineral-matter-free basis)		Volatile-matter limits, percent (dry, mineral-matter-free basis)		Calorific value limits, Btu/lb (moist,† mineral-matter-free basis)		Agglomerating character
		Equal or greater than	Less than	Greater than	Equal or less than	Equal or greater than	Less than	
I. Anthracitic	1. Meta-anthracite	98	...	...	2	......	......	Nonagglomerating‡
	2. Anthracite	92	98	2	8	......	......	
	3. Semianthracite	86	92	3	14	......	......	
II. Bituminous	1. Low-volatile bituminous coal	78	86	14	22	......	......	Commonly agglomerating¶
	2. Medium-volatile bituminous coal	69	78	22	31	......	......	
	3. High-volatile A bituminous coal	...	69	31	...	14,000§	......	
	4. High-volatile B bituminous coal	...	...	...	...	13,000§	14,000	
	5. High-volatile C bituminous coal	...	...	...	...	11,500	13,000	
						10,500	11,500	Agglomerating
III. Subbituminous	1. Subbituminous A coal	...	...	...	...	10,500	11,500	Nonagglomerating
	2. Subbituminous B coal	...	...	...	...	9,500	10,500	
	3. Subbituminous C coal	...	...	...	...	8,300	9,500	
IV. Lignitic	1. Lignite A	...	...	...	...	6,300	8,300	
	2. Lignite B	...	...	...	...	......	6,300	

* This classification does not include a few coals, principally nonbanded varieties, which have unusual physical and chemical properties and which come within the limits of fixed-carbon or calorific value of the high-volatile bituminous and subbituminous ranks. All of these coals either contain less than 48 percent dry, mineral-matter-free fixed carbon or have more than 15,500 moist, mineral-matter-free British thermal units per pound. Btu/lb × 2.323 = kJ/kg.

† Moist refers to coal containing its natural inherent moisture but not including visible water on the surface of the coal.

‡ If agglomerating, classify in low-volatile group of the bituminous class.

§ Coals having 69 percent or more fixed carbon on the dry, mineral-matter-free basis are classified according to fixed carbon, regardless of calorific value.

¶ It is recognized that there may be nonagglomerating varieties in these groups of the bituminous class, and there are notable exceptions in the high-volatile C bituminous group.

Table 7.1.2 Sources and Analyses of Various Ranks of Coal as Received

Classification by rank	State	County	Bed	Proximate, %				Ultimate, %					Calorific value, Btu/lb*
				Moisture	Volatile matter	Fixed carbon	Ash†	Sulfur	Hydrogen	Carbon	Nitrogen	Oxygen	
Meta-anthracite	Rhode Island	Newport	Middle	13.2	2.6	65.3	18.9	0.3	1.9	64.2	0.2	14.5	9,310
Anthracite	Pennsylvania	Lackawanna	Clark	4.3	5.1	81.0	9.6	0.8	2.9	79.7	0.9	6.1	12,880
Semianthracite	Arkansas	Johnson	Lower Hartshorne	2.6	10.6	79.3	7.5	1.7	3.8	81.4	1.6	4.0	13,880
Low-volatile bituminous coal	West Virginia	Wyoming	Pocahontas no. 3	2.9	17.7	74.0	5.4	0.8	4.6	83.2	1.3	4.7	14,400
Medium-volatile bituminous coal	Pennsylvania	Clearfield	Upper Kittanning	2.1	24.4	67.4	6.1	1.0	5.0	81.6	1.4	4.9	14,310
High-volatile A bituminous coal	West Virginia	Marion	Pittsburgh	2.3	36.5	56.0	5.2	0.8	5.5	78.4	1.6	8.5	14,040
High-volatile B bituminous coal	Kentucky, western field	Muhlenburg	No. 9	8.5	36.4	44.3	10.8	2.8	5.4	65.1	1.3	14.6	11,680
High-volatile C bituminous coal	Illinois	Sangamon	No. 5	14.4	35.4	40.6	9.6	3.8	5.8	59.7	1.0	20.1	10,810
Subbituminous A coal	Wyoming	Sweetwater	No. 3	16.9	34.8	44.7	3.6	1.4	6.0	60.4	1.2	27.4	10,650
Subbituminous B coal	Wyoming	Sheridan	Monarch	22.2	33.2	40.3	4.3	0.5	6.9	53.9	1.0	33.4	9,610
Subbituminous C coal	Colorado	El Paso	Fox Hill	25.1	30.4	37.7	6.8	0.3	6.2	50.5	0.7	35.5	8,560
Lignite	North Dakota	McLean	Unnamed	36.8	27.8	29.5	5.9	0.9	6.9	40.6	0.6	45.1	7,000

* Btu/lb × 2.325 = J/g; Btu/lb × 0.5556 = g · cal/g.
† Ash is part of both the proximate and ultimate analyses.

Some of it is blended with more strongly coking coals for making metallurgical coke.

High-volatile C bituminous coal is a stage lower in rank than the B bituminous coal and therefore has a progressively higher bed moisture and oxygen content. It is used primarily for steam raising and space heating.

Subbituminous coals usually show less evidence of banding than bituminous coals. They have a high moisture content, and on exposure to air, they disintegrate or "slack" because of shrinkage from loss of moisture. They are noncaking and noncoking, and their primary use is for steam raising and space heating.

Lignites are brown to black in color and have a bed moisture content of 30 to 45 percent with a resulting lower heating value than higher-rank coals. Like subbituminous coals, they have a tendency to "slack" or disintegrate during air drying. They are noncaking and noncoking. Lignite can be burned on traveling or spreader stokers and in pulverized form.

The principal ranks of coal mined in the major coal-producing states are shown in Table 7.1.3. Their analyses depend on several factors, e.g., source, size of coal, and method of preparation. Periodic reports are issued by the U.S. Department of Energy, Energy Information Agency. They provide statistics on production, distribution, end use, and analytical data.

Composition and Characteristics

Proximate analysis, sulfur content, and calorific values are the analytical determination most commonly used for industrial characterization of coal. The **proximate analysis** is the simplest means for determining the distribution of products obtained during heating. It separates the products into four groups: (1) water or moisture, (2) volatile matter consisting of gases and vapors, (3) fixed carbon consisting of the carbonized residue less ash, and (4) ash derived from the mineral impurities in the coal. ASTM methods D3712 and D5142 are used; the latter is an instrumental method.

Moisture is the loss in weight obtained by drying the coal at a temperature between 104 and 110°C (220 and 230°F) under prescribed conditions. Further heating at higher temperatures may remove more water, but this moisture usually is considered part of the coal substance. The moisture obtained by the standard method consists of (1) surface or extraneous moisture that may come from external sources such as percolating waters in the mine, rain, condensation from the air, or water from a coal washery; (2) inherent moisture, sometimes called **bed moisture,** which is so closely held by the coal substance that it does not separate these two types of moisture. A coal may be air-dried at room temperature or somewhat above, thereby determining an "air-drying loss," but this result is not the extraneous moisture because part of the inherent moisture also vaporizes during the drying.

Mine samples taken at freshly exposed faces in the mine, which are free from visible surface moisture, give the best information as to inherent or bed moisture content. Such moisture content ranges in value from 2 to 4 percent for anthracite and for bituminous coals of the eastern Appalachian field, such as the Pocahontas, Sewell, Pittsburgh, Freeport, and Kittaning beds. In the western part of this field, especially in Ohio, the inherent moisture ranges from 4 to 10 percent. In the interior fields of Indiana, Illinois, western Kentucky, Iowa, and Missouri, the range is from 8 to 17 percent. In subbituminous coals the inherent moisture ranges from 15 to 30 percent, and in lignites from 30 to 45 percent. The total amount of moisture in commercial coal may be greater or less than that of the coal in the mine. Freshly mined subbituminous coal and lignite lose moisture rapidly when exposed to the air. The extraneous or surface moisture in coal is a function of the surface exposed, each surface being able to hold a film of moisture. Fine sizes hold more moisture than lump. Coal which in the mine does not contain more than 4 percent moisture may in finer sizes hold as much as 15 percent; the same coal in lump sizes, even after underwater storage, may contain little more moisture than originally in the mine.

In the standard method of analysis, the **volatile matter** is taken as the loss in weight, less moisture, obtained by heating the coal for 7 min in a covered crucible at about 950°C (1,742°F) under specified conditions. Volatile matter does not exist in coal as such but is produced by decomposition of the coal when heated. It consists chiefly of the combustible gases, hydrogen, carbon monoxide, methane and other hydrocarbons, tar vapors, volatile sulfur compounds, and some noncombustible gases such as carbon dioxide and water vapor. The composition of the volatile

Table 7.1.3   Principal Ranks of Coal Mined in Various States*

State	Anthracite	Semi-anthracite	Low-volatile bituminous	Medium-vol. bituminous	High-vol. A bituminous	High-vol. B bituminous	High-vol. C bituminous	Subbitumi-nous A	Subbitumi-nous B	Subbitumi-nous C	Lignite
Alabama				x	x						
Alaska						x	x	x	x	x	x
Arkansas		x	x	x	x						x
Colorado					x	x	x	x	x	x	x
Illinois					x	x	x				
Indiana						x	x				
Iowa							x				
Kansas											
Eastern					x	x					
Western					x	x					
Western					x	x	x				
Maryland			x	x	x						
Missouri							x				
Montana						x	x	x	x	x	x
New Mexico					x	x	x	x	x		
North Dakota											x
Ohio					x	x	x				
Oklahoma			x	x	x	x	x				
Pennsylvania	x	x	x	x	x						
South Dakota											x
Tennessee				x	x						
Texas						x	x				x
Utah					x	x	x	x	x		
Virginia		x	x	x	x						
Washington					x	x	x		x		x
West Virginia			x	x	x						
Wyoming						x	x	x	x	x	

* Compiled largely from Typical Analyses of Coals of the United States, *BuMines Bull.* 446, and Coal Reserves of the United States, *Geol Survey Bull.* 1136.

matter varies greatly with different coals: the amount can vary with the rate of heating. The inert or noncombustible gas may range from 4 percent of the total volatile matter in low-volatile coals to 40 percent in subbituminous coals.

The standard method of determining the **fixed carbon** is to subtract from 100 the sum of the percentages of the moisture, volatile matter, and ash of the proximate analysis. It is the carbonaceous residue less ash remaining in the test crucible in the determination of the volatile matter. It does not represent the total carbon in the coal because a considerable part of the carbon is expelled as volatile matter in combination with hydrogen as hydrocarbons and with oxygen as carbon monoxide and carbon dioxide. It also is not pure carbon because it may contain several tenths percent of hydrogen and oxygen, 0.4 to 1.0 percent of nitrogen, and about half of the sulfur that was in the coal.

In the standard method, **ash** is the inorganic residue that remains after burning the coal in a muffle furnace to a final temperature of 700 to 750°C (1,292 to 1,382°F). It is composed largely of compounds of silicon, aluminum, iron, and calcium, with smaller quantities of compounds of magnesium, titanium, sodium, and potassium. The ash as determined is usually less than the inorganic mineral matter originally present in the coal. During incineration, various weight changes take place, such as loss of water of constitution of the silicate minerals, loss of carbon dioxide from carbonate minerals, oxidation of iron pyrites to iron oxide, and fixation of a part of the oxides of sulfur by bases such as calcium and magnesium.

The chemical composition of coal ash varies widely depending on the mineral constituents associated with the coal. Typical limits of ash composition of U.S. bituminous coals are as follows:

Constituent	Percent
Silica, SiO_2	20–40
Alumina, Al_2O_3	10–35
Ferric oxide, Fe_2O_3	5–35
Calcium oxide, CaO	1–20
Magnesium oxide, MgO	0.3–4
Titanium dioxide, TiO_2	0.5–2.5
Alkalies, Na_2O and K_2O	1–4
Sulfur trioxide, SO_3	0.1–12

The ash of subbituminous coals may have more CaO, MgO, and SO_3 than the ash of bituminous coals; the trend may be even more pronounced for lignite ash.

Ultimate analysis expresses the composition of coal as sampled in percentages of carbon, hydrogen, nitrogen, sulfur, oxygen, and ash. The carbon includes that present in the organic coal substance as well as a minor amount that may be present as mineral carbonates. In ASTM practice, the hydrogen and oxygen values include those of the organic coal substance as well as those present in the form of moisture and the water of constitution of the silicate minerals. In certain other countries, the values for hydrogen and oxygen are corrected for the moisture in the coal and are reported separately. The ash is the same as reported in the proximate analysis; the sulfur, carbon, hydrogen, and nitrogen are determined chemically. Oxygen in coal is usually estimated by subtracting the sum of carbon, hydrogen, nitrogen, sulfur, and ash from 100. Many of the analyses discussed can be performed with modern instruments in a short time. ASTM methods are applied when referee data are required.

Sulfur occurs in three forms in coal: (1) **pyritic sulfur,** or sulfur combined with iron as pyrite or marcasite; (2) **organic sulfur,** or sulfur combined with coal substance as a heteroatom or as a bridge atom; (3) **sulfate sulfur,** or sulfur combined mainly with iron or calcium together with oxygen as iron sulfate or calcium sulfate. Pyrite and marcasite are recognized by their metallic luster and pale brass-yellow color, although some marcasite is almost white. Organic sulfur may comprise from about 20 to 85 percent of the total sulfur in the coal. Most freshly mined coal contains only very small quantities of sulfate sulfur; it increases in weathered coal. The total sulfur content of coal mined in the United States varies from about 0.4 to 5.5 percent by weight on a dry coal basis.

The **gross calorific value** of a fuel expressed in Btu/lb of fuel is the heat produced by complete combustion of a unit quantity, at constant volume, in an oxygen bomb calorimeter under standard conditions. It includes the latent heat of the water vapor in the products of combustion. Since the latent heat is not available for making steam in actual operation of boilers, a net caloric value is sometimes determined, although not in usual U.S. practice, by the following formula:

Net calorific value, Btu/lb = gross calorific value, Btu/lb
$$-(92.70 \times \text{total hydrogen, \% in coal})$$

The gross calorific value may also be approximated by Dulong's formula

$$\text{Btu/lb} = 14{,}544\text{C} + 62{,}028\left(\text{H} - \frac{\text{O}}{8}\right) + 4{,}050\text{S}$$

$$(\text{Btu/lb} \times 2.328 = \text{kj/kg})$$

where C, H, O, and S are weight fractions from the ultimate analysis. For anthracites, semianthracites, and bituminous coals, the calculated values are usually within 1½ percent of those determined by the bomb calorimeter. For subbituminous and lignitic coals, the calculated values show deviations often reaching 4 and 5 percent.

Because coal ash is a mixture of various components, it does not have a definite melting point; the gradual softening and fusion of the ash is not merely the successive melting of the various ash constituents but is a more complicated process in which reactions involving the formation of new and more fusible compounds take place.

The **fusibility of coal ash** is determined by heating a triangular pyramide (cone), ¾ in high and ¼ in wide at each side of the base, made up of the ash together with a small amount of organic binder. As the cone is heated, three temperatures are noted: (1) the initial deformation temperature (IDT), or the temperature at which the first rounding of the apex or the edges of the cone occurs; (2) the softening temperature (ST), or the temperature at which the cone has fused down to a spherical lump; and (3) the fluid temperature (FT), or the temperature at which the cone has spread out in a nearly flat layer. The softening interval is the degrees of temperature difference between (2) and (1), the flowing interval the difference between (3) and (2), and the fluidity range the difference between (3) and (1). Of the three, the softening temperature is most widely used. Table 7.1.4 shows ash-fusion data typical of some important U.S. coals.

Data on ash fusion characteristics are useful to the combustion engineer concerned with evaluation of the clinkering tendencies of coals used in combustion furnaces and with corrosion of metal surfaces in boilers due to slag deposits. The kinds of mineral matter occurring in different coals are not well related to rank or geographic location, although

Table 7.1.4   Fusibility of Ash from Some Coals

Seam	Pocahontas no. 3	Ohio no. 9	Pittsburgh	Illinois	Utah	Wyoming	Texas
Type	Low volatile	High volatile	High volatile	High volatile	High volatile	Subbituminous	Lignite
Ash, %	12.3	14.1	10.9	17.4	6.6	6.6	12.8
Temperature, °C*							
Initial deformation	> 1,600	1,325	1,240	1,260	1,160	1,200	1,190
Softening		1,430	1,305	1,330		1,215	1,200
Fluid		1,465	1,395	1,430	1,350	1,260	1,255

* In an oxidizing atmosphere.

there is a tendency for midcontinent coals (Indiana to Oklahoma) to have low ash fusion temperatures. Significance is attached to all of the previously indicated fusion temperatures and the intervals between them. The IDT is sometimes identified with surface stickiness, the ST with plastic distortion or sluggish flow, and the FT with liquid mobility. Long fusion intervals often produce tough, dense, slags; short intervals favor porous, friable structures.

Most bituminous coals, when heated at uniformly increasing temperatures in the absence or partial absence of air, fuse and become plastic. These coals may be designated as either caking or coking in different degrees. **Caking** usually refers to the fusion process in a boiler furnace. **Coking coals** are those that make good coke, suitable for metallurgical purposes where the coke must withstand the burden of the ore and flux above it. Coals that are caking in a fuel bed do not necessarily make good coke in a coke oven. Subbituminous coal, lignite, and anthracite are noncaking.

The **free-swelling index** test measures the free-swelling properties of coal and gives an indication of the caking characteristics of the coal when burned on fuel beds. It is not intended to determine the expansion of coals in coke ovens. The test consists in heating 1 g of pulverized coal in a silica crucible over a gas flame under prescribed conditions to form a coke button, the size and shape of which are then compared with a series of standard profiles numbered 1 to 9 in increasing order of swelling.

The **specific gravity** of coal is the ratio of the weight of solid coal to the weight of an equal volume of water. It is useful in calculating the weight of solid coal as it occurs in the ground for estimating the tonnage of coal per acre of surface. An increase in ash-forming mineral matter increases the specific gravity; e.g., bituminous coals of Alabama, ranging from 2 to 15 percent ash and from 2 to 4.5 percent moisture, vary in specific gravity from 1.26 to 1.37.

Bulk density is the weight per cubic foot of broken coal. It varies according to the specific gravity of the coal, its size distribution, its moisture content, and the amount of orientation when piled. The range of weight from subbituminous coal to anthracite is from 44 to 59 lb/ft³ when loosely piled; when piled in layers and compacted, the weight per cubic foot may increase as much as 25 percent. The weight of fuel in a pile can usually be determined to within 10 to 15 percent by measuring its volume. Typical weights of coal, as determined by shoveling it loosely into a box of 8 ft³ capacity, are as follows: anthracite, 50 to 58 lb/ft³; low- and medium-volatile bituminous coal, 49 to 57 lb/ft³; high-volatile bituminous and subbituminous coal, 42 to 57 lb/ft³.

The **grindability** of coal, or the ease with which it can be ground fine enough for use as a pulverized fuel, is a composite of several specific physical properties such as hardness, tensile strength, and fracture. A laboratory procedure adopted by ASTM (D409) for evaluating grindability, known as the **Hardgrove** machine method, uses a specially designed grinding apparatus to determine the relative grindability or ease of pulverizing coal in comparison with a standard coal, chosen as 100 grindability. Primarily, the ASTM Hardgrove grindability test is used for estimating how various coals affect the capacity of commercial pulverizers. A general relationship exists between grindability of coal and its rank. Coals that are easiest to grind (highest grindability index) are those of about 14 to 30 percent volatile matter on a dry, ash-free basis. Coals of either lower or higher volatile-matter content usually are more difficult to grind. The relationship of grindability and rank, however, is not sufficiently precise for grindability to be estimated from the chemical analysis, partly because of the variation in grindability of the various petrographic and mineral components. Grindability indexes of U.S. coals range from about 20 for an anthracite to 120 for a low-volatile bituminous coal.

Mining

Coal is mined by either underground or surface methods. In underground mining the coal beds are made accessible through shaft, drift, or slope entries (vertical, horizontal, or inclined, respectively), depending on location of the bed relative to the terrain.

The most widely used methods of coal mining in the United States are termed continuous and conventional mining. The former makes use of **continuous miners** which break the coal from the face and load it onto conveyors, shuttle cars, or railcars in one operation. Continuous miners are of ripping, boring, or milling types or hybrid combinations of these. In conventional mining the coal is usually broken from the face by means of blasting agents or by pressurized air or carbon dioxide devices. In preparation for breaking, the coal may be cut horizontally or vertically by cutting machines and holes drilled for charging explosives. The broken coal is then picked up by loaders and discharged to conveyors or cars. A method that is increasing in use is termed **long-wall mining**. It employs shearing or plowing machines to break coal from more extensive faces. Eighty long-wall mines are now in operation. Pillars to support the roof are not needed because the roof is caved under controlled conditions behind the working face. About half the coal presently mined underground is cut by machine and nearly all the mined coal is loaded mechanically.

An important requirement in all mining systems is **roof support.** When the roof rock consists of strong sandstone or limestone, relatively uncommon, little or no support may be required over large areas. Most mine roofs consist of shales and must be reinforced. Permanent supports may consist of arches, crossbars and legs, or single posts made of steel or wood. Screw or hydraulic jacks, with or without crossbars, often serve as temporary supports. Long roof bolts, driven into the roof and anchored in sound strata above, are used widely for support, permitting freedom of movement for machines. Drilling and insertion of bolts is done by continuous miners or separate drilling machines. **Ventilation** is another necessary factor in underground mining to provide a proper atmosphere for personnel and to dilute or remove dangerous concentrations of methane and coal dust. The ventilation system must be well-designed so that adequate air is supplied across the working faces without stirring up more dust.

When coal occurs near the surface, **strip or open-pit mining** is often more economical than underground mining. This is especially true in states west of the Mississippi River where coal seams are many feet thick and relatively low in sulfur. The proportion of coal production from surface mining has been increasing rapidly and now amounts to over 60 percent.

In preparation for surface mining, core drilling is conducted to survey the underlying coal seams, usually with dry-type rotary drills. The overburden must then be removed. It is first loosened by ripping or drilling and blasting. Ripping can be accomplished by bulldozers or scrapers. Overburden and coal are then removed by shovels, draglines, bulldozers, or wheel excavators. The first two may have bucket capacities of 200 yd³ (153 m³). Draglines are most useful for very thick cover or long dumping ranges. Hauling of stripped coal is usually done by trucks or tractor-trailers with capacities up to 240 short tons [218 metric tons (t)]. Reclamation of stripped coal land is becoming increasingly necessary. This involves returning the land to near its original contour, replanting with ground cover or trees, and sometimes providing water basins and arable land.

Preparation

About half the coal presently mined in the United States is cleaned mechanically to remove impurities and supply a marketable product. Mechanical mining has increased the proportion of fine coal and noncoal minerals in the product. At the preparation plant run-of-mine coal is usually given a preliminary size reduction with roll crushers or rotary breakers. Large or heavy impurities are then removed by hand picking or screening. Tramp iron is usually removed by magnets. Before washing, the coal may be given a preliminary size fractionation by screening. Nearly all preparation practices are based on density differences between coal and its associated impurities. **Heavy-medium** separators using magnetite or sand suspensions in water come closest to ideal gravity separation conditions. **Mechanical devices** include jigs, classifiers, washing tables, cyclones, and centrifuges. Fine coal, less than ¼ in (6.3 mm) is usually treated separately, and may be cleaned by froth flotation. Dewatering of the washed and sized coal may be accomplished by screening, centrifuging, or filtering, and finally, the fine coal may be heated to complete the drying. Before shipment the coal may be dust-proofed and freezeproofed with oil or salt.

Removal of sulfur from coal is an important aspect of preparation because of the role of sulfur dioxide in air pollution. Pyrite, the main inorganic sulfur mineral, is partly removed along with other minerals in conventional cleaning. Processes to improve pyrite removal are being developed. These include magnetic and electrostatic separation, chemical leaching, and specialized froth flotation.

Storage

Coal may heat spontaneously, with the likelihood of self-heating greatest among coals of lowest rank. The heating begins when freshly broken coal is exposed to air. The process accelerates with increase in temperature, and active burning will result if the heat from oxidation is not dissipated. The finer sizes of coal, having more surface area per unit weight than the larger sizes, are more susceptible to spontaneous heating.

The **prevention of spontaneous heating** in storage poses a problem of minimizing oxidation and of dissipating any heat produced. Air may carry away heat, but it also brings oxygen to create more heat. Spontaneous heating can be prevented or lessened by (1) storing coal underwater; (2) compressing the pile in layers, as with a road roller, to retard access of air; (3) storing large-size coal; (4) preventing any segregation of sizes in the pile; (5) storing in small piles; (6) keeping the storage pile as low as possible (6 ft is the limit for many coals); (7) keeping storage away from any external sources of heat; (8) avoiding any draft of air through the coal; (9) using older portions of the storage first and avoiding accumulations of old coal in corners. It is desirable to watch the temperature of the pile. A thermometer inserted in an iron tube driven into the coal pile will reveal the temperature. When the coal reaches a temperature of 50°C (120°F), it should be moved. Using water to put out a fire, although effective for the moment, may only delay the necessity of moving the coal. Furthermore, this may be dangerous because steam and coal can react at high temperatures to form carbon monoxide and hydrogen.

Sampling

Because coal is a heterogeneous material, collection and handling of samples that adequately represent the bulk lot of coal are required if the analytical and test data are to be meaningful. Coal is best sampled when in motion, as it is being loaded or unloaded from belt conveyors or other coal-handling equipment, by collecting increments of uniform weight evenly distributed over the entire lot. Each increment should be sufficiently large and so taken to represent properly the various sizes of the coal.

Two procedures are recognized: (1) commercial sampling and (2) special-purpose sampling, such as classification by rank or performance. The **commercial sampling** procedure is intended for an accuracy such that, if a large number of samples were taken from a large lot of coal, the test results in 95 out of 100 cases would fall within ±10 percent of the ash content of these samples. For commercial sampling of lots up to 1,000 tons, it is recommended that one gross sample represent the lot taken. For lots over 1,000 tons, the following alternatives may be used: (1) Separate gross samples may be taken for each 1,000 tons of coal or fraction thereof, and a weighted average of the analytical determinations of these prepared samples may be used to represent the lot. (2) Separate gross samples may be taken for each 1,000 tons or fraction thereof, and the −20 or −60 mesh samples taken from the gross samples may be mixed together in proportion to the tonnage represented by each sample and one analysis carried out on the composite sample. (3) One gross sample may be used to represent the lot, provided that at least four times the usual minimum number of increments are taken. In **special-purpose sampling,** the increment requirements used in the commercial sampling procedure are increased according to prescribed rules.

Specifications

Specifications for the purchase of coal vary widely depending on the intended use, whether for coke or a particular type of combustion unit,

and whether it meets the standards imposed by customers abroad. Attempts at international standardization have met with only limited success. Most of the previously discussed factors in this section must be taken into consideration, for example, the sulfur content and swelling properties of a coal used for metallurgical coke production, the heating value of a coal used for steam generation, and the slagging properties of its ash formed after combustion.

Statistics

Coal production in 2004 was 1.08 billion short tons, 0.5 percent anthracite, 93 percent bituminous, and 6.5 percent lignite. The industry employed about 228,000 miners, 150,000 of them underground. Over 6000 coal mines are in operation in 26 states but over half the production comes from Kentucky, West Virginia, Wyoming, and Pennsylvania. Of the total production, 62 percent is from surface mines. **Transportation** to the point of consumption is primarily by rail (67 percent) followed by barges (11 percent) and trucks (10 percent); about 1 percent moves through pipelines. About 11 percent of the coal consumed is burned at mine-mouth power plants, for it is cheaper and easier to transport electric power than bulk coal.

Several long-distance coal-slurry pipelines are proposed but only one, 273 mi (439 km) long, is in commercial use. The Black Mesa pipeline runs from a coal mine near Kayenta, AZ to the Mohave Power Plant in Nevada. Nominal capacity is 4.8 million short tons (4.35 t) per year, but it usually operates at 3 to 4 million tons per year. The coal concentration is about 47 percent and the particle size distribution is controlled carefully.

Overall energy statistics for the United States show that coal accounts for about 30 percent of the total energy production, with the balance coming from petroleum and natural gas. Below is a distribution of 1 billion tons of coal produced:

Electric power	87.2 percent
Industrial	8.4 percent
Coke	3.8 percent
Commercial and residential	0.6 percent

In 1994 over 100 million tons was exported. The industrial use is primarily for power in the production of food, cement, paper, chemicals, and ceramics.

Reserves of coal in the United States are ample for several hundred years even allowing for the increased production of electric power and the synthesis of fuels and chemicals. The total reserve base is estimated to be almost 4 trillion tons, of which 1.7 trillion tons is identified resources.

BIOMASS FUELS
by Martin D. Schlesinger
Consultant

REFERENCES: Peat, "U.S. Bureau of Mines Mineral Commodities Summaries." Fryling, "Combustion Engineering," Combustion Engineering, Inc., New York. "Standard Classification of Peats, Mosses, Humus and Related Products," ASTM D2607. Lowry, "Chemistry of Coal Utilization," Wiley. United Nations Industrial Development Organization publications.

Biomass conversion to energy continues to be a subject of intensive study for both developed and less developed countries. In the United States, combustion of biomass contributes only a few percent of the total U.S. energy supply, and it is mostly in the form of agricultural wastes and paper. Almost any plant material can be the raw material for gasification from which a variety of products can be created catalytically from the carbon monoxide, carbon dioxide, and hydrogen. Typical commercial products are methanol, ethanol, methyl acetate, acetic anhydride, and hydrocarbons. Alcohols are of particular interest as fuels for transportation and power generation. Producer gas has been introduced into small compression ignition engines, and the diesel oil feed has been reduced. Technical and environmental problems are still not solved in the large-scale use of biomass.

Plants and vegetables are another source of biomass-derived oils. Fatty acid esters have been used in diesel engines alone and in blends with diesel oil, and although the esters are effective, some redesign and changes in operation are required. In developing countries where fossil fuels are costly, deforestation has led to land erosion and its consequences.

Useful biomass materials include most of the precoal organic vegetation such as peat, wood, and food processing wastes like bagasse from sugar production. Digestable food processing wastes can be converted to biogas.

Peat, an early stage in the metamorphosis of vegetable matter into coal, is the product of partial decomposition and disintegration of plant remains in water bogs, swamps or marshlands, and in the absence of air. Like all material of vegetable origin, peat is a complex mixture of carbon, hydrogen, and oxygen in a ratio similar to that of cellulose and lignin. Generally peat is low in essential growing elements (K, S, Na, P) and ash. Trace elements, when found in the peat bed, are usually introduced by leaching from adjacent strata. The Federal Trade Commission specifies that to be so labeled, peat must contain at least 75 percent peat with the rest composed of normally related soil materials. The water content of undrained peat in a bog is 92 to 95 percent but it is reduced to 10 to 50 percent when peat is used as a fuel. Peat is harvested by large earth-moving equipment from a drained bog dried by exposure to wind and sun. The most popular method used in Ireland and the Soviet Union involves harrowing of drum-cut peat and allowing it to field dry before being picked up mechanically or pneumatically.

The **chemical and physical properties of peat** vary considerably, depending on the source and the method of processing. Typical ranges are:

Processed peat	Air-dried	Mulled	Briquettes
Moisture, wt %	25–50	50–55	10–12
Density, lb/ft³	15–25		30–60
Caloric value, Btu/lb	6,200	3,700–5,300	8,000

Proximate analyses of samples, calculated back to a dry basis, follow a similar broad pattern: 55 to 70 percent volatile matter, 30 to 40 percent fixed carbon and 2 to 10 percent ash. The dry, ash-free ultimate analysis ranges from 53 to 63 percent carbon, 5.5. to 7 percent hydrogen, 30 to 40 percent oxygen, 0.3 to 0.5 percent sulfur, and 1.2 to 1.5 percent nitrogen.

World production of peat in 2004 was about 26.5 million tons, mostly from northern Europe. Other significant producers are Ireland, Finland, and Germany. Annual U.S. imports are primarily from Canada, about 1.3 million tons per annum. Several states produce peat, Florida and Michigan being the larger producers. Over a 10-year period, U.S. production declined from 730,000 to about 620,000 short tons (590,000 t) per year. The value of production was $16 million from 67 operations.

By type, peat was about 66 percent reed sedge with the balance distributed between humus, sphagnum, and hypnum. The main uses for peat in the United States are soil improvement, mulch, filler for fertilizers, and litter for domestic animals.

Wood, when used as a fuel, is often a by-product of the sawmill or papermaking industries. The conversion of logs to lumber results in 50 percent waste in the form of bark, shavings, and sawdust. Fresh timber contains 30 to 50 percent moisture, mostly in the cell structure of the wood, and after air drying for a year, the moisture content reduces to 18 to 25 percent. Kiln-dried wood contains about 8 percent moisture. A typical analysis range is given in Table 7.1.5. When additional fuel is required, supplemental firing of coal, oil, or gas is used.

Combustion systems for wood are generally designed specially for the material or mixture of fuels to be burned. When the moisture content is high, 70 to 80 percent, the wood must be mixed with low moisture fuel so that enough energy enters the boiler to support combustion. Dry wood may have a heating value of 8,750 Btu/lb but at 80 percent moisture a pound of wet wood has a heating value of only 1,750 Btu/lb. The heat required just to heat the fuel and evaporate the water is over 900 Btu, and combustion may not occur. Table 7.1.6 shows the

Table 7.1.5   Typical Analysis of Dry Wood Fuels

	Most woods, range
Proximate analysis, %:	
Volatile matter	74–82
Fixed carbon	17–23
Ash	0.5–2.2
Ultimate analysis, %:*	
Carbon	49.6–53.1
Hydrogen	5.8–6.7
Oxygen	39.8–43.8
Heating value	
Btu/lb	8,560–9,130
kJ/kg	19,900–21,250
Ash-fusion temperature, °F:	
Initial	2,650–2,760
Fluid	2,730–2,830

* Typically, wood contains no sulfur and about 0.1 percent nitrogen. Cellulose = 44.5 percent C, 6.2 percent H, 49.3 percent O.

moisture-energy relationship. The usual practice when burning wood is to propel the wood particles into the furnace through injectors along with preheated air with the purpose of inducing high turbulence to the boiler. Furthermore, the wood is injected high enough in the combustion chamber so that it is dried, and all but the largest particles are burned before they reach the grate at the bottom of the furnace. Spreader stokers and cyclone burners work well.

Table 7.1.6   Available Energy in Wood

Moisture, %	Heating value, Btu/lb	Wt water/wt wood
0	8,750	0
20	7,000	0.25
50	4,375	1.00
80	1,750	4.00

Wood for processing or burning is usually sold by the cord, an ordered pile 8 ft long, 4 ft high, and 4 ft wide or 128 ft³ (3.625 m³). Its actual solid content is only about 70 percent, or 90 ft³. Other measures for wood are the cord run, which is measured only by the 8-ft length and 4-ft height; the width may vary. Sixteen-inch-long wood is called **stovewood** or **blockwood.**

Small wood-burning power plants and home heating became popular, but in some areas there was an adverse environmental impact under adverse weather conditions.

Wood charcoal is made by heating wood to a high temperature in the absence of air. Wood loses up to 75 percent of its weight and 50 percent of its volume owing to the elimination of moisture and volatile matter. As a result, charcoal has a higher heating value per cubic foot than the original wood, especially if the final product is compacted in the form of briquettes. Charcoal is marketed in the form of lumps, powder, or briquettes and finds some use as a fuel for curing, restaurant cooking, and a picnic fuel. Its nonfuel uses, particularly in the chemical industry, are as an adsorption medium for purifying gas and liquid streams and as a decolorizing agent.

In addition to peat and wood, several lesser-known fuels are in common use for the generation of industrial steam and power. Aside from their value as a fuel, the burning of wastes minimizes a troublesome disposal problem that could have serious environmental impact. Nearly all these waste fuels are cellulosic in character, and the heating value is a function of the carbon content. Ash content is generally low, but much moisture could be present from processing, handling, and storage. On a moisture- and ash-free basis the heating values can be estimated at 8,000 Btu/lb; more resinous materials about 9,000 Btu/lb. Table 7.1.7 is a list of some typical by-product solid fuels.

Table 7.1.7   By-product Fuels

	Heating value, Btu/lb (dry)	Moisture, % as received	Ash, % moisture-free
Black liquor (sulfate)	6,500	35	40–45
Cattle manure	7,400	50–75	17
Coffee grounds	10,000	65	1.5
Corncobs	9,300	10	1.5
Cottonseed cake	9,500	10	8
Municipal refuse	9,500	43	8
Pine bark	9,500	40–50	5–10
Rice straw or hulls	6,000	7	15
Scrap tires	16,400	0.5	6
Wheat straw	8,500	10	4

Bagasse is the fibrous material left after pressing the juice from sugarcane or harvesting the seeds from sunflowers. The chopped waste usually contains about 50 percent moisture and is burned in much the same manner as wood waste. Spreader stokers or cyclone burners are used. Supplemental fuel is added sometimes to maintain steady combustion and to provide energy for the elimination of moisture. Bagasse can usually supply all the fuel requirements of raw sugar mills. A typical analysis of dry bagasse from Puerto Rico is 44.47 percent C, 6.3 percent H, 49.7 percent O, and 1.4 percent ash. Its heating value is 8,390 Btu/lb.

Furnaces have been developed to burn particular wastes, and some preferences emerge by virtue of particular operating characteristics. Spreader stokers are preferred for wood waste and bagasse. Tangential firing seems to be used for coffee grounds, rice hulls, some wood waste, and chars from coal or lignite. Traveling-grate stokers are used for industrial wastes and coke breeze.

Biogas is readily produced by the anaerobic digestion of wastes. The process is cost-effective in areas remote from natural gas lines. In Asia, for example, there are millions of family biodigesters with a capacity of 8 to 10 m³. Larger-capacity systems of about 2,000 m³ are installed where industrial biodegradable wastes are generated as in communes, feed lots, wineries, food processors, etc. Not only is the gas useful, but also the sludge is a good fertilizer. Remaining parasite eggs and bacteria are destroyed by lime or ammonia treatment. Harvest increases of 10 to 35 percent are reported for rice and corn.

Biogas contains an average of 62 percent methane and 36 percent carbon dioxide; it also has a small amount of nitrogen and hydrogen sulfide. Raw gas heating value is about 600 Btu/ft³ (5,340 cal/m³). The raw gas will burn in an engine, but corrosion can occur unless proper materials of construction are selected. Gas from a garbage site in California is treated to remove acid gases, and the methane is sold to a pipeline system. Most sites, however, do not produce enough gas to use economically and merely flare the collected gas.

Other methods have been demonstrated for converting biomass of variable composition to an energy source of relatively consistent composition. One procedure is to process bulk volume wastes with pressurized carbon monoxide. A yield of about 2 barrels of oil per ton of dry feed is obtained, with a heating value of about 15,000 Btu/lb. Heavy fuel oil from petroleum has a heating value of 18,000 Btu/lb. Pyrolysis is also possible at temperatures up to 1,000°C in the absence of air. The gas produced has a heating value of 400 to 500 Btu/ft³ (about 4,000 kcal/m³). The oil formed has a heating value about 10,500 Btu/lb (24,400 J/g).

Refuse-derived fuels (RDFs) usually refer to waste material that has been converted to a fuel of consistent composition for commercial application. Although several sources exist, the problem lies in gathering the raw material, processing it into a usable form and composition, and delivering it to the point of combustion. Almost any carbonaceous material and a suitable binder can be converted to a form that can be fired into a pulverized fuel or a stoker boiler. Each application should be evaluated to eliminate carryover of low-density material, such as loose paper, and where in the boiler that combustion takes place.

Some installations are cofired with fossil fuels. Most of the by-product fuels in Table 7.1.7 might be used alone or with a binder. An ideal situation would be nearby waste streams from a major production facility. After mixing and extrusion, the pellets would be storable and transported only a short distance. If the pellet density is close to the normal boiler feed, the problems of separation during transportation and firing are reduced or eliminated. The problem of hazardous emissions remains if the waste streams are contaminated; if they are clean-burning, hazardous emissions might reduce the apparent cleanliness of the effluent. (See Sec. 7.4.).

PETROLEUM AND OTHER LIQUID FUELS
by James G. Speight
Western Research Institute

REFERENCES: ASTM, "Annual Book of ASTM Standards," 1993, vols. 05.01, 05.02, 05.03, and 05.04. Bland and Davidson, "Petroleum Processing Handbook," McGraw-Hill, New York. Gary and Handwerk, "Petroleum Refining: Technology and Economics," Marcel Dekker, New York. Hobson and Pohl, "Modern Petroleum Technology," Applied Science Publishers, Barking, England. Mushrush and Speight, "Petroleum Products: Instability and Incompatibility," Taylor & Francis, Washington. Speight, "The Chemistry and Technology of Petroleum," 2d ed., Marcel Dekker, New York.

Petroleum and Petroleum Products

Petroleum accumulates over geological time in porous underground rock formations called reservoirs, where it has been trapped by overlying and adjacent impermeable rock. Oil reservoirs sometimes exist with an overlying gas "cap" in communication with aquifers or with both. The oil resides together with water, and sometimes free gas, in very small holes (pore spaces) and fractures. The size, shape, and degree of interconnection of the pores vary considerably from place to place in an individual reservoir. The anatomy of a reservoir is complex, both microscopically and macroscopically. Because of the various types of accumulations and the existence of wide ranges of both rock and fluid properties, reservoirs respond differently and must be treated individually.

Petroleum occurs throughout the world, and commercial fields have been located on every continent. Reservoir depths vary, but most reservoirs are several thousand feet deep, and the oil is produced through wells that are drilled to penetrate the oil-bearing formations.

Petroleum is an extremely complex mixture and consists predominantly of hydrocarbons as well as compounds containing nitrogen, oxygen, and sulfur. Most petroleums also contain minor amounts of nickel and vanadium.

Petroleum may be qualitatively described as brownish green to black liquids of specific gravity from about 0.810 to 0.985 and having a boiling range from about 20°C (68°F) to above 350°C (660°F), above which active decomposition ensues when distillation is attempted. The oils contain from 0 to 35 percent or more of components boiling in the gasoline range, as well as varying proportions of kerosene hydrocarbons and higher-boiling-point constituents up to the viscous and nonvolatile compounds present in lubricants and the asphalts. The composition of the petroleum obtained from the well is variable and depends on both the original composition of the petroleum in situ and the manner of production and stage reached in the life of the well or reservoir.

The chemical and physical properties of petroleum vary considerably because of the variations in composition. The specific gravity of petroleum ranges from 0.8 (45.3 degrees API) for the lighter crude oils to over 1 (< 10 degrees API) for the near-solid bitumens which are found in many tar (oil) sand deposits. There is also considerable variation in viscosity; lighter crude oils range from 2 to 100 cSt, bitumens have viscosities in excess of 50,000 cSt.

The ultimate analysis (elemental composition) of petroleum is not reported to the same extent as it is for coal since there is a tendency for the ultimate composition of petroleum to vary over narrower limits—carbon: 83.0 to 87.0 percent; hydrogen: 10.0 to 14.0 percent; nitrogen: 0.1 to 1.5 percent; oxygen: 0.1 to 1.5 percent; sulfur: 0.1 to 5.0 percent; metals (nickel plus vanadium): 10 to 500 ppm. The **heat content of petroleum**

Table 7.1.8   Analyses and Heat Values of Petroleum and Petroleum Products

Product	Gravity, deg API	Specific gravity at 60°F	Wt lb/gal	High-heat value, Btu/lb*	Ultimate analysis, %				
					C	H	S	N	O
California crude	22.8	0.917	7.636	18,910	84.00	12.70	0.75	1.70	1.20
Kansas crude	22.1	0.921	7.670	19,130	84.15	13.00	1.90	0.45	
Oklahoma crude (east)	31.3	0.869	7.236	19,502	85.70	13.11	0.40	0.30	
Oklahoma crude (West)	31.0	0.871	7.253	19,486	85.00	12.90	0.76		
Pennsylvania crude	42.6	0.813	6.769	19,050	86.06	13.88	0.06	0.00	0.00
Texas crude	30.2	0.875	7.286	19,460	85.05	12.30	1.75	0.70	0.00
Wyoming crude	31.5	0.868	7.228	19,510					
Mexican crude	13.6	0.975	8.120	18,755	83.70	10.20	4.15		
Gasoline	67.0	0.713	5.935	......	84.3	15.7			
Gasoline	60.0	0.739	6.152	20,750	84.90	14.76	0.08		
Gasoline-benzene blend	46.3	0.796	6.627	......	88.3	11.7			
Kerosine	41.3	0.819	6.819	19,810					
Gas oil	32.5	0.863	7.186	19,200					
Fuel oil (Mex.)	11.9	0.987	8.220	18,510	84.02	10.06	4.93		
Fuel oil (midcontinent)	27.1	0.892	7.428	19,376	85.62	11.98	0.35	0.50	0.60
Fuel oil (Calif.)	16.7	0.9554	7.956	18,835	84.67	12.36	1.16		

* Btu/lb × 2.328 = kJ/kg.

generally varies from 18,000 to 20,000 Btu/lb while the heat content of petroleum products may exceed 20,000 Btu/lb (Table 7.1.8).

Refining Crude Oil   Crude oils are seldom used as fuel because they are more valuable when refined to petroleum products. Distillation separates the crude oil into fractions equivalent in boiling range to gasoline, kerosene, gas oil, lubricating oil, and residual. Thermal or catalytic cracking is used to convert kerosene, gas oil, or residual to gasoline, lower-boiling fractions, and a residual coke. Catalytic reforming, isomerization, alkylation, polymerization, hydrogenation, and combinations of these catalytic processes are used to upgrade the various refinery intermediates into improved gasoline stocks or distillates. The major finished products are usually blends of a number of stocks, plus additives. Distillation curves for these products are shown in Fig. 7.1.2.

Physical Properties of Petroleum Products   Petroleum products are sold in the United States by barrels of 42 gal corrected to 60°F, Table 7.1.9. Their *specific gravity* is expressed on an arbitrary scale termed degrees API.

The **high heat value** (hhv) of petroleum products is determined by combustion in a bomb with oxygen under pressure (ASTM D240). It may also be calculated, in products free from impurities, by the formula

$$Q_v = 22,320 - 3,780d^2$$

in which Q_v is the hhv at constant volume in Btu/lb and d is the specific gravity at 60/60°F.

The low heat value at constant pressure Q_p may be calculated by the relation

$$Q_p = Q_v - 90.8H$$

where H is the weight percentage of hydrogen and can be obtained from the relation

$$H = 26 - 15d$$

Typical heats of combustion of petroleum oils free from water, ash, and sulfur vary (within an estimated accuracy of 1 percent) with the API gravity (i.e., with the "heaviness" or "lightness" of the material) (Table 7.1.10). The heat value should be corrected when the oil contains sulfur by using an hhv of 4,050 Btu/lb for sulfur (ASTM D1405).

The **specific heat** c of petroleum products of specific gravity d and at temperature t (°F) is given by the equation

$$c = (0.388 + 0.00045t)/\sqrt{d}$$

The **heat of vaporization L** (Btu/lb) may be calculated from the equation

$$L = (110.9 - 0.09t)/d$$

The heat of vaporization per gallon (measured at 60°F) is

$$8.34Ld = 925 - 0.75t$$

Fig. 7.1.2   Typical distillation curves.

indicating that the heat of vaporization per gallon depends only on the temperature of vaporization t and varies over the range of 450 for the heavier products to 715 for gasoline (Table 7.1.11). These data have an estimated accuracy within 10 percent when the vaporization is at constant temperature and at pressures below 50 lb/in^2 without chemical change.

Table 7.1.9   Coefficients of Expansion of Petroleum Products at 60°F

Deg API	<15	15–34.9	35–50.9	51–63.9	64–78.9	79–88.9	89–93.9	94–100
Coef of expansion	0.00035	0.0004	0.0005	0.0006	0.0007	0.0008	0.00085	0.0009

Table 7.1.10   Heat Content of Different Petroleums and Petroleum Products

Deg API	Density	High heat value at constant volume Q_v, Btu		Low heat value at constant pressure Q_p, Btu	
at 60°F	lb/gal*	Per lb	Per gal	Per lb	Per gal
10	8.337	18,540	154,600	17,540	146,200
20	7.787	19,020	148,100	17,930	139,600
30	7.305	19,420	141,800	18,250	133,300
40	6.879	19,750	135,800	18,510	127,300
50	6.500	20,020	130,100	18,720	121,700
60	6.160	20,260	124,800	18,900	116,400
70	5.855	20,460	119,800	19,020	112,500
80	5.578	20,630	115,100	19,180	107,000

* Btu/lb $\times$ 2.328 = kJ/kg; Btu/gal $\times$ 279 = kJ/m^3.

Table 7.1.11   Latent Heat of Vaporization of Petroleum Products

Product	Gravity deg API	Average boiling temp, 8F	Heat of vaporization	
			Btu/lb	Btu/gal
Gasoline	60	280	116	715
Naphtha	50	340	103	670
Kerosine	40	440	86	595
Fuel oil	30	580	67	490

Properties and Specifications for Motor Gasoline

Gasoline is a complex mixture of hydrocarbons that distills within the range of 100 to 400°F. Commercial gasolines are blends of straight-run, cracked, reformed, and natural gasolines.

Straight-run gasoline is recovered from crude petroleum by distillation and contains a large proportion of normal hydrocarbons of the paraffin series. Its octane number is too low for use in modern engines, and it is reformed and blended with other products to improve its combustion properties.

Cracked gasoline is manufactured by heating crude-petroleum distillation fractions or residues under pressure, or by heating with or without pressure in the presence of a catalyst. Heavier hydrocarbons are broken into smaller molecules, some of which distill in the gasoline range. The octane number of cracked gasoline is usually above that of straight-run gasoline.

Reformed gasoline is made by passing gasoline fractions over catalysts in such a manner that low-octane-number hydrocarbons are molecularly rearranged to high-octane-number components. Many of the catalysts use platinum and other metals deposited on a silica and/or alumina support.

Natural gasoline is obtained from natural gas by liquefying those constituents which boil in the gasoline range either by compression and cooling or by absorption in oil. Natural gasoline is too volatile for general use, but proper characteristics can be secured by distillation or by blending. It is often blended with gasolines to adjust their volatility characteristics to meet climatic conditions.

Catalytic hydrogenation is used extensively to upgrade gasoline and cracking stocks for blending or further refining. Hydrogenation improves octane number, removes sulfur and nitrogen, and increases storage stability.

The specifications for gasoline (ASTM D439) provide for various volatility classes, varying from low-volatility gasolines to minimize vapor lock to high-volatility gasoline that permits easier starting during cold weather.

Antiknock characteristics of gasolines are very important because engine power output and fuel economy are limited by the antiknock characteristics of the fuel. The antiknock index is currently defined as the average of the research octane number (RON) and motor octane number (MON). The **research octane number** is a measure of antiknock performance under mild operating conditions at low to medium engine speeds. The **motor octane number** is indicative of antiknock performance under more severe conditions, such as those encountered during power acceleration at relatively high engine speeds.

Reformulated motor gasoline is believed to be the answer to many environmental issues that arise from the use of automobiles. In fact, there has been a serious effort to produce reformulated gasoline components from a variety of processes (Table 7.1.12). In addition, methyl-t-butyl ether (MTBE), an additive to maintain the octane ratings of gasoefficient combustion of the hydrocarbons) the emissions of unburned hydrocarbons during gasoline use. However, the ether (MTBE) is believed to have an adverse effect insofar as it appears that aldehyde emissions may be increased.

Table 7.1.12   Production of Reformulated Gasoline Constituents

Process	Objective
Catalytic reformer prefractionation	Reduce benzene
Reformate fractionation	Reduce benzene
Isomerization	Increase octane
Aromatics saturation	Reduce total aromatics
Catalytic reforming	Oxygenate for octane enhancement
MTBE synthesis	Provide oxygenates
Isobutane dehydrogenation	Feedstock for oxygenate synthesis
Catalytic cracker naphtha fractionation	Increase alkylate
	Increase oxygenates
	Reduce olefins and sulfur
Feedstock hydrotreating	Reduce sulfur

Diesel Fuel

Diesel is a liquid product distilling over the range of 150 to 400°C (300 to 750°F). The carbon number ranges on average from about C_{13} to about C_{21}.

The chemical composition of a typical diesel fuel and how it applies to the individual specifications—API gravity, distillation range, freezing point, and flash point—are directly attributable to both the carbon number and the compound classes present in the finished fuel (Tables 7.1.13 and 7.1.14).

Aviation Gasolines Gasolines for aircraft piston engines have a narrower boiling range than motor gasolines; i.e., they have fewer

Table 7.1.13   General ASTM Specifications for Various Types of Diesel Fuels

Specification	Military*	No. 1-D†	No. 2-D‡	No. 1§	No. 2¶
API gravity, deg	40	34.4	40.1	35	30
Total sulfur, percent	0.5	0.5	0.5	0.5	0.5
Boiling point, °C	357	288	338	288	338
Flash point, °C	60	38	52	38	38
Pour point, °C	−6	−18	−6	−18	−6
Hydrogen, wt %	12.5	—	—	—	—
Cetane number	43	40	40	—	—
Acid number	0.3	0.3	0.3	—	—

* MIL-F-16884J (1993) is also NATO F-76.
† High-speed, high-load engines.
‡ Low-speed, high-load engines.
§ Special-purpose burners.
¶ General-purpose heating fuel oil.

low-boiling and fewer high-boiling components. The three grades of aviation gasolines are indicated in Table 7.1.15. Specifications applicable to all three grades of military gasoline are given in Table 7.1.16.

Table 7.1.14 ASTM Methods for Determining Fuel Properties (see also Table 7.1.13)

Specification	ASTM method
API gravity	D1298
Total sulfur, percent	D129
Boiling point, °C	D86
Flash point, °C	D93
Pour point, °C	D97
Hydrogen, wt %	D3701
Cetane number	D613, D976
Acid number	D974

Jet or aviation turbine fuels are not limited by antiknock requirements, and they have wider boiling-point ranges to provide greater availability (ASTM D1655). Fuel JP-4 is a relatively wide-boiling-point range distillate that encompasses the boiling point range of gasoline and kerosene. The average initial boiling point is about 140°F, and the end-point is about 455°F. Fuel JP-5 is a high-flash point kerosine type of fuel with an initial boiling point of 346°F and endpoint of 490°F.

Table 7.1.15 Grades of Aviation Gasoline (ASTM D910)

Grade	Color	Tetraethyl lead content mL/gal, max
80	Red	0.6
100	Green	4.0
100 LL	Blue	2.0

There are a variety of grades (Table 7.1.15) and specifications (Table 7.1.16) for jet fuel because of its use as a commercial and a military fuel (Table 7.1.17). The chemical composition of each jet fuel type, API gravity, distillation range, freezing point, and flash point are directly attributable to both the carbon number and the compound classes present in the finished fuel.

Kerosine is less volatile than gasoline and has a higher flash point, to provide greater safety in handling. Other quality tests are specific gravity, color, odor, distillation range, sulfur content, and burning quality. Most kerosine is used for heating, ranges, and illumination, so it is treated with sulfuric acid to reduce the content of aromatics, which burn

Table 7.1.17 Commercial and Military Specifications for Jet Fuels

Fuel types	Specification
Jet-A	ASTM D1655
JP-4	Mil-T-5624
JP-5	Mil-T-5624
JP-8	Mil-T-83133
JP-10	Mil-P-87107

with a smoky flame. Specification tests for quality control include flash point (minimum 115°F), distillation endpoint (maximum 572°F), sulfur (maximum 0.13 percent), and color (minimum + 16) (ASTM D187).

Diesel fuel for diesel engines requires variability in its performance since the engines range from small, high-speed engines used in trucks and buses to large, low-speed stationary engines for power plants. Thus several grades of diesel fuel are needed (Table 7.1.18) (ASTM D975) for different classes of service:

Grade 1-D: A volatile distillate fuel for engines in service requiring frequent speed and load changes
Grade 2-D: A distillate fuel of lower volatility for engines in industrial and heavy mobile service
Grade 4-D: A fuel for low- and medium-speed engines

An additional guide to fuel selection is the grouping of fuels according to these types of service:

Type C-B: Diesel fuel oils for city bus and similar operations
Type T-T: Fuels for diesel engines in trucks, tractors, and similar service
Type R-R: Fuels for railroad diesel engines
Type S-M: Heavy-distillate and residual fuels for large stationary and marine diesel engines

The combustion characteristics of diesel fuels are expressed in terms of the cetane number, a measure of ignition delay. A short ignition

Table 7.1.16 Specifications for Aviation Gasoline

Test	Test limit	Test method
Distillation:		
Fuel evaporated, 10% min at	167°F (75°C)	D86
Fuel evaporated, 40% max at	167°F (75°C)	
Fuel evaporated, 50% min at	221°F (105°C)	
Fuel evaporated, 90% min at	275°F (135°C)	
Endpoint, max	338°F (170°C)	
Sum of 10% and 50% evaporated temp, min	307°F	
Residue, vol, max %	1.5	
Distillation loss, vol, max %	1.5	
Existent gum, max, mg/100 mL	3.0	D381
Potential gum, 16 h aging, max, mg/100 mL	6.0	D873
Precipitate, max, mg/100 mL	2.0	D873
Sulfur, max, wt %	0.05	D1266 or D2622
Reid vapor pressure at 100°F, min, lb/in^2	5.5	D323
Reid vapor pressure at 100°F, max, lb/in^2	7.0	D323
Freezing point, max	−76°F (−60°C)	D2386
Copper corrosion, max	No. 1	D130
Water reaction:		
Interface rating, max	2	D1094
Vol. change, max, mL	2	D1094
Heating value:		
Net heat of combustion, min, Btu/lb	18,700	D1405
Aniline-gravity product, min	7,500	D611 and D287

Table 7.1.18   Specifications* for Diesel Fuels

Test	ASTM method	1-D	2-D	4-D	U.S. Military spec. Mil-F-16884G
		ASTM grade of diesel fuel			
		Limit			
Flash point, min, °F	D93	100 or legal	125 or legal	130 or legal	140
Water and sediment, vol %, max	D1796	Trace	0.10	0.50	
Viscosity, kinematic, cSt, 100°F	D445				
Min		1.3	1.9	5.5	1.8
Max		2.4	4.1	24.0	4.5
Carbon residue on 10% residuum, % max	D524	015	0.35		0.20
Ash, wt %, max	D482	0.01	0.01	0.10	0.005
Sulfur, wt %, max	D129	0.50	0.50	2.0	1.00
Ignition quality, cetane number, min	D613	40	40	30	45
Distillation temp, °F, 90% evaporated:	D86				
Min			540		
Max		550	640		

* See ASTM D975 and MH-F-16884G specifications for full details.

delay, i.e., the time between injection and ignition, is desirable for a smooth-running engine. Some diesel fuels contain cetane improvers, which usually are alkyl nitrates. The cetane number is determined by an engine test (ASTM D613) or an approximate value, termed the **cetane index** (ASTM D976), can be calculated for fuels that do not contain cetane improvers.

The list of additives used in diesel fuels has grown in recent years because of the increased use of catalytically cracked fuels, rather than exclusively straight-run distillate. In addition to cetane improvers, the list includes antioxidants, corrosion inhibitors, and dispersants. The dispersants are added to prevent agglomeration of gum or sludge deposits so these deposits can pass through filters, injectors, and engine parts without plugging them.

Gas-Turbine Fuels

Five grades of gas-turbine fuels are specified (ASTM D2880) according to the types of service and engine:

Grade O-GT: A naphtha or other low-flash-point hydrocarbon liquid which also includes jet B fuel

Grade 1-GT: A volatile distillate for gas turbines that requires a fuel that burns cleaner than grade 2-GT

Grade 2-GT: A distillate fuel of low ash and medium volatility, suitable for turbines not requiring grade 1-GT

Grade 3-GT: A low-volatility, low-ash fuel that may contain residual components

Grade 4-GT: A low-volatility fuel containing residual components and having higher vanadium content than grade 3-GT

Grade 1-GT corresponds in physical properties to no. 1 burner fuel and grade 1-D diesel fuel. Grade 2-GT corresponds in physical properties to no. 2 burner fuel and grade 2-D diesel fuel. The viscosity ranges of grades 3-GT and 4-GT bracket the viscosity ranges of no. 4, no. 5 light, no. 5 heavy, and no. 6 burner fuels.

Fuel Oils

The characteristics of the grades of fuel oil (ASTM D396) are as follows:

No. 1: A distillate oil intended for vaporizing pot-type burners and other burners requiring this grade of fuel.

No. 2: A distillate oil for general-purpose domestic heating in burners not requiring no. 1 fuel oil.

No. 4 and no. 4 light: Preheating not usually required for handling or burning.

No. 5 light: Preheating may be required depending upon climate and equipment.

No. 5 heavy: Preheating may be required for burning and, in cold climates, may be required for handling.

No. 6: Preheating required for burning and handling.

The sulfur content of no. 1 and no. 2 fuel oil is limited to 0.5 percent (ASTM D396). The sulfur content of fuels heavier than no. 2 must meet the legal requirements of the locality in which they are to be used. The additional refinery processing needed by some residual fuels to meet low-sulfur-content regulations may lower the viscosity enough to cause the fuels to change the grade classification.

Fuel oil used for domestic purposes or for small heating installations will have lower viscosities and lower sulfur content. In large-scale industrial boilers, heavier-grade fuel oil is used with sulfur content (ASTM D129 and D1552) requirements regulated according to the environmental situation of each installation and the local environmental regulations.

The **flash point** (ASTM D93) is usually limited to 60°C (140°F) minimum because of safety considerations. Asphaltene content (ASTM D3279), carbon residue value (ASTM D189 and D524), ash (ASTM D482), water content (ASTM D95), and metal content requirements are included in some specifications.

The **pour point** (ASTM D97), indicating the lowest temperature at which the fuel will retain its fluidity, is limited in the various specifications according to local requirements and fuel-handling facilities. The upper limit is sometimes 10°C (50°F), in warm climates somewhat higher.

Another important specification requirement is the heat of combustion (ASTM D240). Usually, specified values are 10,000 cal/kg (gross) or 9,400 cal/kg (net).

Because of economic considerations residual fuel oil has been replacing diesel fuel for marine purposes. Viscosity specifications had to be adjusted to the particular operational use, and some additional quality requirements had to be allowed for. The main problem encountered during the use of residual fuel oils for marine purposes concerns the stability properties (sludge formation) and even more so the compatibility properties of the fuel.

Blending of fuel oils to obtain lower viscosity values and to mix fuel oils of different chemical characteristics is a source of deposit and sludge formation in the vessel fuel systems. This incompatibility is mainly observed when high-asphaltene fuel oils are blended with diluents or other fuel oils of a paraffinic nature.

Gas oil is a general term applied to distillates boiling between kerosine and lubricating oils. The name was derived from its initial use for making illuminating gas, but it is now used as burner fuel, diesel fuel, and catalytic-cracker charge stock.

GASEOUS FUELS
by James G. Speight
Western Research Institute

REFERENCES: Bland and Davidson, "Petroleum Processing Handbook," McGraw-Hill, New York. Francis and Peters, "Fuels and Fuel Technology," 2d ed., Pergamon, New York, Sec. C, Gaseous Fuels. Goodger, "Alternative Fuels: Chemical Energy Resources," Wiley, New York. Kohl and Riesenfeld, "Gas Purification," Gulf Publishing Co., Houston, TX. Kumar, "Gas Production Engineering," Gulf Publishing Co., TX. Reid, Prausnitz, and Sherwood,

Table 7.1.19   Composition of Natural Gases*

Sample no.	Natural gas from oil or gas wells						Natural, gas from pipelines				
	299	318	393	522	732.	1177	1214	1225	1249	1276	1358
Composition, mole percent:											
Methane	92.1	96.3	67.7	63.2	43.6	96.9	94.3	72.3	88.9	75.4	85.6
Ethane	3.8	0.1	5.6	3.1	18.3	1.7	2.1	5.9	6.3	6.4	7.8
Propane	1.0	0.0	3.1	1.7	14.2	0.3	0.4	2.7	1.8	3.6	1.4
Normal butane	0.3	0.0	1.5	0.5	8.6	0.1	0.2	0.3	0.2	1.0	0.0
Isobutane	0.3	0.0	1.2	0.4	2.3	0.0	0.0	0.2	0.1	0.6	0.1
Normal pentane	0.1	0.0	0.6	0.1	2.7	0.3	Tr	Tr	0.0	0.1	0.0
Isopentane	Tr	0.0	0.4	0.2	3.3	0.0	Tr	0.2	Tr	0.2	0.1
Cyclopentane	Tr	0.0	0.2	Tr.	0.9	Tr	Tr	0.0	Tr	Tr	0.0
Hexanes plus	0.2	0.0	0.7	0.1	2.0	0.1	Tr	Tr	Tr	0.1	Tr
Nitrogen	0.9	1.0	17.4	27.9	3.0	0.6	0.0	17.8	2.2	12.0	4.7
Oxygen	0.2	0.0	Tr	0.1	0.5	Tr	Tr	Tr	Tr	Tr	Tr
Argon	Tr	Tr	0.1	0.1	Tr	0.0	0.0	Tr	0.0	Tr	Tr
Hydrogen	0.0	0.2	0.0	0.0	0.1	0.0	Tr	0.1	0.1	0.0	0.0
Carbon dioxide	1.1	2.3	0.1	0.4	0.5	0.0	2.8	0.1	0.1	0.1	0.2
Helium	Tr	Tr	1.4	2.1	Tr	Tr	Tr	0.4	0.1	0.4	0.1
Heating value	1,062	978	1,044	788	1,899	1,041	1.010	934	1,071	1,044	1,051
Origin of sample	La.	Miss.	N. Mex.	Okla.	Tex.	W.Va.	Colo.	Kan.	Kan.	Okla.	Tex.

* Analyses from *BuMines Bull.* 617 (Tr = trace).
† Calculated total (gross) Btu/ft³, dry, at 60°F and 30 in Hg.

"The Properties of Gases and Liquids," McGraw-Hill. Shekhtman, "Gasdynamic Functions of Real Gases," Hemisphere Publishing Corp., Washington. Speight, "Fuel Science and Technology Handbook," J. G. Speight, ed., Marcel Dekker, New York, pt. 5, pp. 1055 et seq. Speight, "Gas Processing: Environmental Aspects and Methods." Butterworth-Heinemann, Oxford, England. Sychev et al., "Thermodynamic Properties of Propane," Hemisphere Publishing Corp.

Natural gas, which is predominantly methane, occurs in underground reservoirs separately or in association with crude petroleum. But **manufactured gas** is a fuel-gas mixture made from other solid, liquid, or gaseous materials, such as coal, coke, oil, or natural gas. The principal types of manufactured gas are retort coal gas, coke oven gas, water gas, carbureted water gas, producer gas, oil gas, reformed natural gas, and reformed propane or liquefied petroleum gas (LPG). Several processes for making substitute natural gas (SNG) from coal have been developed. **Mixed gas** is a gas prepared by adding natural gas or liquefied petroleum gas to a manufactured gas, giving a product with better utility and higher heat content or Btu value.

Liquefied petroleum gas (LPG) is the term applied to certain specific hydrocarbons and their mixtures, which exist in the gaseous state under atmospheric ambient conditions but can be converted to the liquid state under conditions of moderate pressure at ambient temperature. Thus liquefied petroleum gas is a hydrocarbon mixture usually containing propane ($CH_3.CH_2.CH_3$), n-butane ($CH_3.CH_2.CH_2.CH_3$), isobutane [$CH_3CH(CH_3)CH_3$], and to a lesser extent propylene ($CH_3.CH : CH_2$) or butylene ($CH_3CH_2CH : CH_2$). The most common commercial products are propane, butane, or some mixture of the two, and they are generally extracted from natural gas or crude petroleum. Propylene and butylenes result from cracking other hydrocarbons in a petroleum refinery and are two important chemical feedstocks.

Composition of Gaseous Fuels   The principal constituent of natural gas is methane (CH_4) (Table 7.1.19). Other constituents are paraffinic hydrocarbons such as ethane, propane, and the butanes. Many natural gases contain nitrogen as well as carbon dioxide and hydrogen sulfide. Trace quantities of argon, hydrogen, and helium may also be present. A portion of the heavier hydrocarbons, carbon dioxide, and hydrogen sulfide are removed from natural gas prior to its use as a fuel. Typical natural-gas analyses are given in Table 7.1.19. Manufactured gases contain methane, ethane, ethylene, propylene, hydrogen, carbon monoxide, carbon dioxide, and nitrogen, with low concentrations of water vapor, oxygen, and other gases.

Specifications   Since the composition of natural, manufactured, and mixed gases can vary so widely, no single set of specifications could cover all situations. The requirements are usually based on performances in burners and equipment, on minimum heat content, and on maximum sulfur content. Gas utilities in most states come under the supervision of state commissions or regulatory bodies, and the utilities must provide a gas that is acceptable to all types of consumers and that will give satisfactory performance in all kinds of consuming equipment.

Specifications for liquefied petroleum gases (Table 7.1.20) (ASTM D1835) depend upon the required volatility.

Odorization   Since natural gas as delivered to pipelines has practically no odor, the addition of an odorant is required by most regulations in order that the presence of the gas can be detected readily in case of accidents and leaks. This odorization is provided by the addition of trace

Table 7.1.20   Specifications* for Liquefied Petroleum Gas

	Product designation			Test method
	Propane	Butane	PB mixtures	
Vapor pressure at 100°F, max, psig	210	70		D1267 or D2598
Volatile residue:				
Butane and heavier, %, max	2.5			D2163
Pertane and heavier, %, max		2.0	2.0	D2163
Residual matter:				
Residue on evaporation, + 100 mL, max mL	0.05	0.05	0.05	D2158
Oil-stain observation	Pass	Pass	Pass	D2158
Specific gravity at 60°/60°F		To be reported		D1657 or D2598
Corrosion, copper strip, max	No. 1	No. 1	No. 1	D1838
Sulfur, grains/100 ft³, max	15	15	15	D1266
Moisture content		To be reported		
Free-water content		None	None	D1657

* Refer to ASTM D1835 for full details.

amounts of some organic sulfur compounds to the gas before it reaches the consumer. The standard requirement is that a user will be able to detect the presence of the gas by odor when the concentration reaches 1 percent of gas in air. Since the lower limit of flammability of natural gas is approximately 5 percent, this 1 percent requirement is essentially equivalent to one-fifth the lower limit of flammability. The combustion of these trace amounts of odorant does not create any serious problems of sulfur content or toxicity.

Analysis

The different methods for gas analysis include absorption, distillation, combustion, mass spectroscopy, infrared spectroscopy, and gas chromatography (ASTM D2163, D2650, and D4424). Absorption methods involve absorbing individual constituents one at a time in suitable solvents and recording of contraction in volume measured. Distillation methods depend on the separation of constituents by fractional distillation and measurement of the volumes distilled. In combustion methods, certain combustible elements are caused to burn to carbon dioxide and water, and the volume changes are used to calculate composition. Infrared spectroscopy is useful in particular applications. For the most accurate analyses, mass spectroscopy and gas chromatography are the preferred methods.

ASTM has adopted a number of methods for gas analysis, including ASTM D2650, D2163, and D1717.

Physical Constants The specific gravity of gases, including LP gases, may be determined conveniently by a number of methods and a variety of instruments (ASTM D1070 and D4891).

The heat value of gases is generally determined at constant pressure in a flow calorimeter in which the heat released by the combustion of a definite quantity of gas is absorbed by a measured quantity of water or air. A continuous recording calorimeter is available for measuring heat values of natural gases (ASTM D1826).

Flammability The lower and upper limits of flammability indicate the percentage of combustible gas in air below which and above which flame will not propagate. When flame is initiated in mixtures having compositions within these limits, it will propagate and therefore the mixtures are flammable. A knowledge of flammable limits and their use in establishing safe practices in handling gaseous fuels is important, e.g., when purging equipment used in gas service, in controlling factory or mine atmospheres, or in handling liquefied gases.

Many factors enter into the experimental determination of flammable limits of gas mixtures, including the diameter and length of the tube or vessel used for the test, the temperature and pressure of the gases, and the direction of flame propagation—upward or downward. For these and other reasons, great care must be used in the application of the data. In monitoring closed spaces where small amounts of gases enter the atmosphere, often the maximum concentration of the combustible gas is limited to one fifth of the concentration of the gas at the lower limit of flammability of the gas-air mixture. (See Table 7.1.21.)

The calculation of flammable limits is accomplished by Le Chatelier's modification of the mixture law, which is expressed in its simplest form as

$$L = \frac{100}{p_1/N_1 + p_2/N_2 + \cdots + p_n/N_n}$$

where L is the volume percentage of fuel gas in a limited mixture of air and gas; $p_1, p_2, \ldots, p_n$ are the volume percentages of each combustible gas present in the fuel gas, calculated on an air- and inert-free basis so that $p_1 + p_2 + \cdots + p_n = 100$; and $N_1, N_2, \ldots, N_n$ are the volume percentages of each combustible gas in a limit mixture of the individual gas and air. The foregoing relation may be applied to gases with inert content of 10 percent or less without introducing an absolute error of more than 1 or 2 percent in the calculated limits.

The **rate of flame propagation** or **burning velocity** in gas-air mixtures is of importance in utilization problems, including those dealing with burner design and rate of energy release. There are several methods that have been used for measuring such burning velocities, in both laminar and turbulent flames. Results by the various methods do not agree, but

Table 7.1.21   Flammability Limits of Gases in Air

Gas	Flammable limits in air, vol %	
	Lower	Upper
Methane	5.0	15.0
Ethane	3.0	12.4
Propane	2.1	9.5
Butane	1.8	8.4
Isobutane	1.8	8.4
Pentane	1.4	7.8
Isopentane	1.4	7.6
Hexane	1.2	7.4
Ethylene	2.7	36.0
Propylene	2.4	11.0
Butylene	1.7	9.7
Acetylene	2.5	100.0
Hydrogen	4.0	75.0
Carbon monoxide	12.5	74.2
Ammonia	15.0	28.0
Hydrogen sulfide	4.0	44.0
Natural	4.8	13.5
Producer	20.2	71.8
Blast-furnace	35.0	73.5
Water	6.9	70.5
Carbureted-water	5.3	40.7
Coal	4.8	33.5
Coke-oven	4.4	34.0
High-Btu oil	3.9	20.1

any one method does give relative values of utility. Maximum burning velocities for turbulent flames are greater than those for laminar flames. The bunsen flame method gives results that have significance in gas utilization problems. In this method, the burning velocity is determined by dividing the volume rate of gas flow from the bunsen burner by the area of the inner cone of the flame.

The use of other fuels and equipment to supplement the regular supply of gas during periods of peak demand or in emergencies is known as **peak shaving.** Most gas utilities, particularly natural-gas utilities at the end of long-distance transmission lines, maintain peak-shaving or standby equipment. Also, gas utilities in many cases have established natural-gas storage facilities close to their distribution systems. This allows gas to be stored underground near the point of consumption during periods of low demand, as in summer, and then withdrawn to meet peak or emergency demands, as may occur in winter. Propane-air mixtures are the major supplements to natural gas for peak shaving use. Gas manufactured by cracking various petroleum distillates supplies most of the remaining peak requirements.

SYNTHETIC FUELS
by Martin D. Schlesinger

Consultant
(See also Sec. 7.2.)

REFERENCES: U.S. Department of Energy, Fossil Energy Reports. Whitehurst et al., "Coal Liquefaction," Academic. Speight, "The Chemistry and Technology of Coal," Marcel Dekker. Preprints, Division of Fuel Chemistry, American Chemical Society. Annual International Conferences on Coal Gasification, Liquefaction, and Conversion to Electricity, Department of Chemical and Petroleum Engineering, University of Pittsburgh.

Liquid and gaseous fuels in commercial use can be produced from sources other than petroleum and natural gas. Much new technology has been developed during the twentieth century as the threat of petroleum shortages or isolation loomed periodically. The result was the improvement of known processes and the introduction of new concepts.

A generalized flow sheet Fig. 7.1.3, shows the routes that can be followed from coal to finished products. Liquid and gaseous fuels are formed from coal by increasing its hydrogen to carbon ratio.

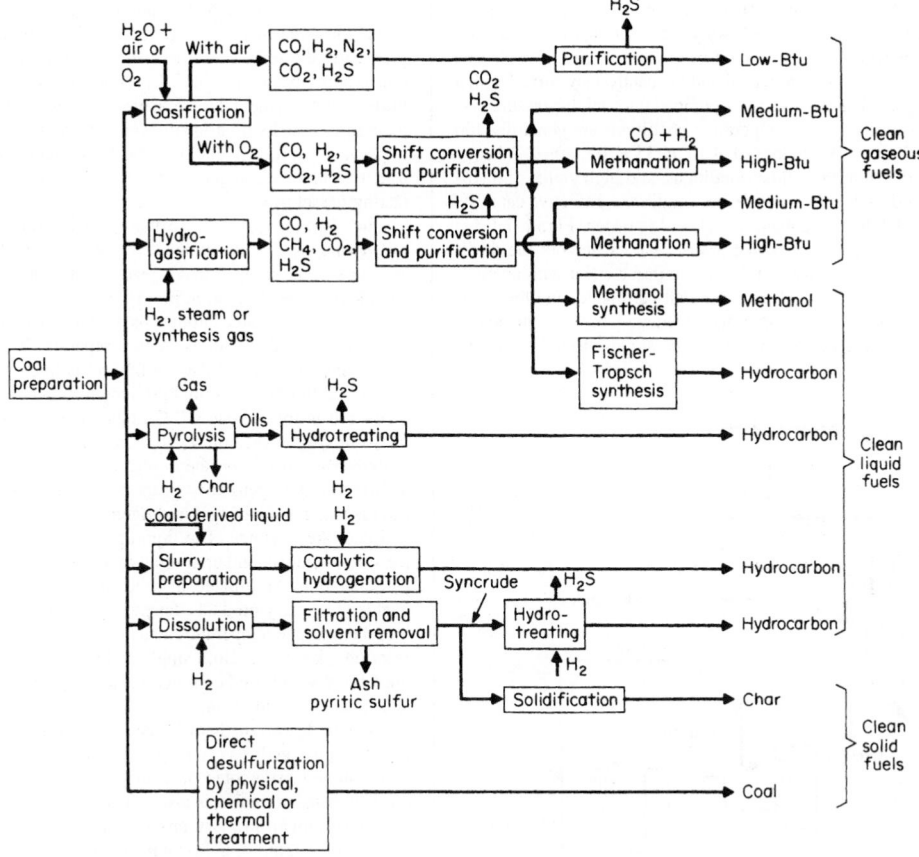

Fig. 7.1.3   Clean fuels from coal.

Primary conversion products are deashed, desulfurized, and further upgraded to a wide range of clean fuels. Gasification can yield clean gases for combustion or synthesis gas with a controlled ratio of hydrogen to carbon monoxide. Catalytic conversion, of synthesis gas to fluids (indirect liquefaction) can be carried out in fixed and fluidized beds and in dilute phase systems. Both gases and liquids are used as the temperature control medium for the exothermic reactions.

Direct liquefaction is accomplished by pyrolysis or hydrogenation and several processes are available for each approach for adding hydrogen to the coal and removing undesirable constituents. The amount of hydrogen consumed is determined by the properties desired in the final product, whether it be a heavy fuel oil, diesel oil, jet fuel, gasoline, or gases. Clean solid fuels, i.e, with low ash and low sulfur contents, can be produced as a product of pyrolysis, from liquefaction residues, or by chemical treatment of the coal.

What must be done with some naturally occurring sources of fuels is exemplified in Fig. 7.1.4. Upgrading involves the addition of hydrogen and more hydrogen is required to upgrade coal than to process petroleum into finished products. The ranges are generalized to indicate the relative need for processing and the range of product distributions. Coal contains very little hydrogen, averaging 0.8 H:C atomic ratio and petroleum has about twice the relative amounts. Premium fuels are in the kerosene/gasoline range, which includes diesel and jet fuels. Commercial transportation fuels contain 15 to 18 wt % hydrogen. At the upper end of the scale is methane with an H : C ratio of 4:1. Not considered in the above is the elimination of mineral matter and constituents such as oxygen, sulfur, and nitrogen, where hydrogen is consumed to form water, hydrogen sulfide, and ammonia. Some oxygen is removed as carbon dioxide.

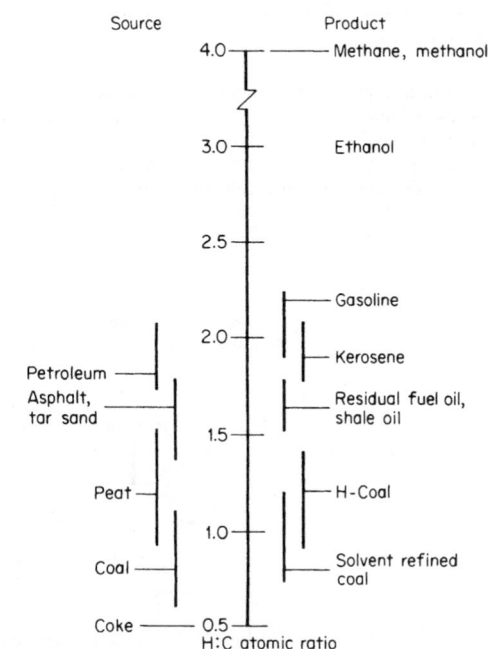

Fig. 7.1.4   Upgrading of carbonaceous sources.

Coal liquefaction is accomplished by four principal methods: (1) direct catalytic hydrogenation, (2) solvent extraction, (3) indirect catalytic by hydrogenation (of carbon monoxide), and (4) pyrolysis. In one approach to direct hydrogenation, coal and the catalyst are mixed with a coalderived recycle oil and the slurry is pumped into a high-pressure system where hydrogen is present. Operating conditions are generally 400 to 480°C and 1,500 to 5,000 lb/in² (10×10^6 to 35×10^6 N/m²). A heavy syncrude is produced at the milder conditions, and high yields of distillable oils are produced at the more severe conditions. Effective catalysts contain iron, molybdenum, cobalt, nickel, and tungsten. Figure 7.1.5 is a schematic flow sheet of the **H-coal process** that carries out the direct hydrogenation by bringing the coal-oil slurry into contact with an ebullating bed of catalyst. The product is generally aromatic, and the gasoline fraction produced after further hydrogenation has a high octane number. Many chemicals of commercial value are also found in the oil.

Fig. 7.1.5   H-coal schematic.

Solvent extraction processing solubilizes and disperses coal in a hydroaromatic solvent that transfers some of its hydrogen to the coal. Ash and insoluble coal are separated from the liquid product to recover recycle oil and product oil. The carbonaceous residue is reacted with steam in a gasifier to produce the hydrogen needed. Figure 7.1.6 is a

schematic flow sheet of the **solvent-refined coal (SRC I)** process. By recycling some of the mineral matter from the coal as a catalyst and increasing the severity of the operating conditions, a lighter hydrocarbon product is formed. SRC I product is a solid at room temperature. Further improvement can be achieved by hydrogenation of the solvent to control the hydroaromatic content of the recycle stream and to improve the product quality. The **Exxon donor solvent (EDS)** process uses this latter technique, and no catalyst is required in the first contacting vessel.

One version of the solvent extraction system is **two-stage hydrogenolysis**. The integrated sequence starts with pulverized coal in a recycle solvent pumped into a reactor where the slurry is hydrogenated for a short time at 2,400 lb/in² (16.5×10^6 N/m³) and 425 and 450°C. Partially hydrogenated product is vacuum distilled to recover solvent and primary product. Solids are next separated by a critical solvent or an antisolvent process and the ash-free oil is hydrogenated at 2,700 lb/in² and 400°C to produce a final product boiling below 350°C. Some oil is recovered in this below 350°C range, and excess higher-boiling oil is recovered in the vacuum distillation step.

Pyrolysis depends on the controlled application of heat without the addition of hydrogen. Most of the carbon is recovered as solid product; liquids and gases having a hydrogen: carbon ratio higher than the original coal are liberated. The liquid product can be hydrotreated further for sulfur removal and upgrading to specification fuels. By-product gas and char must be utilized in order for the process to be economical. Yields of primary products depend on the coal source, the rate of heating, the ultimate temperature reached, and the atmosphere in which the reaction takes place. Both single and multistage processes were developed but few are used commercially except for the production of metallurgical coke and chars.

Catalytic hydrogenation of carbon monoxide is a flexible method of liquefaction. Catalysts can be prepared from Fe, Ni, Co, Ru, Zn, and Th either alone or promoted on a support. Each catalyst gives a different product distribution that is also a function of the method of preparation and pretreatment. Primary products are normally methane and higher-molecular-weight, straight-chain hydrocarbons, alcohols, and organic acids. Operating conditions for the **Fischer-Tropsch-type synthesis** are usually in the range of 300 to 500 lb/in² (2 to 3.5×10^6 N/m³) and 200 to 400°C. Temperature of the exothermic reaction [−39.4 kcal/(g · mol)] is controlled by carrying out the reaction in fixed beds, fluidized beds, slurry, and dilute phase systems with heat removal. Diesel oil from this type of synthesis has a high cetane number; the paraffin waxes can be of high quality. Generally, the gasoline produced by indirect synthesis has a low octane number because of its aliphatic nature. One method of producing a high-octane gasoline is to make methanol from 2:1 H : CO in a first stage and then process the alcohol over a zeolite catalyst. The hydrocarbon product has a narrow boiling range and contains about 30 percent aromatics.

Fig. 7.1.6   Solvent-refined coal.

Fig. 7.1.7 Early schematic flow sheet for SASOL.

In South Africa, coal and natural gas yield about 40 percent of their liquid and gas hydrocarbon fuel requirements. SASOL (South African Synthetic Oil Limited) applies advanced gasification and liquefaction processes to achieve fuels, a spectrum of chemicals and feedstocks for further processing (Fig. 7.1.7). The original fixed bed and circulating bed systems were replaced by more efficient slurry-type reactors. Also, the advanced technology is being introduced into other parts of Africa and Asia.

Shale oil is readily produced by the thermal processing of many shales. The basic technology is available and commercial plants are operated in many parts of the world. The first modern plant in the United States was put on stream in 1983 by the Union Oil Co. in Colorado. About 12,500 tons of raw shale, averaging 35 gal/ton (0.145 m³/1,000 kg), crushed to 5-cm particles, is pushed upward into the retort each day, and at 400 to 500°C crude shale oil is produced from the kerogen in the shale.

The refined product yield was 7,000 bpd (1,115 m³/d) of diesel oil and 3,000 bpd (475 m³/d) of jet fuel. Estimated reserves are equivalent to about 3 billion barrels of shale oil. Figure 7.1.8 is a generalized flow sheet of the process.

Tar sand is a common term for oil-impregnated sediments that can be found in almost every continent. High-grade tar sands have a porosity of 25 to 35 percent and contain about 18 percent by weight of bitumen. The sand grains are wetted by about 2 percent of water, making them hydrophilic and thus more amenable to hot-water extraction. Solvent extraction, thermal retorting, in situ combustion, and steam injection methods have been tested.

Fig. 7.1.8 Shale oil processing.

Reserves of tar sands in the United States are equivalent to about 30 billion barrels of petroleum. Most of the deposits are too deep for surface mining and require in situ treatment before extraction. Some of the surface deposits are worked to produce asphalt for highway application and other minor uses.

Tar sand is a source of hydrocarbon fuels at Syncrude Canada in Alberta. Two commercial plants with combined capacity of 190,000 bpd of synthetic crude oil operate in this region and use the principle of ore mining, hot-water extraction, coking (delayed and fluid), and distillate hydrogenation. Methods for modification of the bitumen in situ and recovery without mining are also under investigation. Properties of conventional petroleum, tar sand bitumen, and synthetic crude oil from the bitumen are given in Table 7.1.22.

Table 7.1.22   Comparison of Tar Sand Bitumen and Synthetic Crude Oil from the Bitumen with Petroleum

	Petroleum	Tar sand bitumen	Synthetic crude oil
API gravity	25–27°	8°	
Viscosity			
cSt at 100°F	3–7	120,000	6
cSt at 210°F		2,000	
Carbon, wt %	86.0	83.1	86.3
Hydrogen, wt %	13.5	10.6	13.4
Nickel, ppm	2–10	100	0
Sulfur, wt %	1–2	4.8	0.15
Nitrogen, wt %	0.2	0.4	0.06
Vanadium, ppm	2–10	250	0
Ash, wt %	0	1.0	0
Carbon residue, wt %	1–5	14.0	0
Pentane insolubles, wt %	< 5	17.0	0

EXPLOSIVES
by J. Edmund Hay

U.S. Department of the Interior

REFERENCES: Meyer, "Explosives," Verlag Chemie. Cook, "The Science of High Explosives," Reinhold. Johansson and Persson, "Detonics of High Explosives," Academic. Davis, "The Chemistry of Powder and Explosives," Wiley. "Manual on Rock Blasting." Aktiebolaget Atlas Diesel, Stockholm. Dick, Fletcher, and D'Andrea, Explosives and Blasting Procedures Manual, *BuMines Inf. Circ.* 8925, 1983.

The term *explosives* refers to any substance or article which is able to function by explosion (i.e., the extremely rapid release of gas and heat) by chemical reaction within itself.

Explosive substances are commonly divided into two types: (1) high or **detonating explosives** are those which normally function by detonation, in which the chemical reaction is propagated by a shock wave that in turn is driven by the energy released; and (2) low or **deflagrating explosives,** normally function by deflagration, in which the chemical reaction is propagated by convective, conductive, and/or radiative heat transfer. High explosives are further divided into two types: primary explosives are those for which detonation is the only mode of reaction, and secondary explosives may either detonate or deflagrate, depending on a variety of conditions. Low explosives are usually pyrotechnics or propellants for guns, rockets, or explosive-actuated devices.

It is important to note the word *normally*. Not only can many high explosives deflagrate rather than detonate under certain conditions, but also some explosives which are normally considered to be "low" explosives can be forced to detonate. Also note that the definition of *explosive* is itself somewhat elastic—the capability to react explosively is strongly dependent on the geometry, density, particle size, confinement, and initiating stimulus, as well as the chemical composition. Historically, immense grief has resulted from ignorance of these facts. In normal use, the term *explosive* refers to those substances or articles which have been classified as explosive by the test procedures recommended by the United Nations (UN) Committee on the Transport of Dangerous Goods.

According to the UN scheme of classification, most high explosives are designated class 1.1 (formerly called *class A*), and most low explosives are designated class 1.3 (formerly called *class B*). Explosive devices of minimal hazard (formerly called *class C*) are designated class 1.4.

However, a very important subclass of secondary high explosives is designated class 1.5 (formerly called *blasting agents* or *nitrocarbonitrates*). The distinction is based primarily on **sensitivity:** In simplified terms, the initiation of detonation of a class 1.5 explosive requires a stronger stimulus than that provided by a detonator with a 0.45-g PETN base charge which exhibits a low tendency to the deflagration-to-detonation transition.

The important physical properties of high explosives include their bulk density, detonation rate, critical and "ideal" diameters (or thickness), "sensitivity," and strength.

The detonation rate is the linear speed at which the detonation propagates through the explosive, and it ranges from about 6,000 ft/s (1.8 km/s) to about 28,000 ft/s (8.4 km/s), depending on the density and other properties of the explosive.

The *critical diameter* (or thickness, in the case of a sheet explosive) is the minimum diameter or thickness at which detonation can propagate through the material. For most class 1.1 explosives, this is between 1 and 30 mm; for class 1.5 explosives, it is usually greater than 50 mm. This value depends to some extent on the confinement provided by the material adjacent to the explosive. For explosives whose diameter (thickness) is only slightly above the critical value, the detonation rate increases with increasing diameter (thickness), finally attaining a value for which no increase in rate is observed for further increases in dimension. The diameter or thickness at which this occurs is called the *ideal diameter* or *thickness*.

Sensitivity and strength are two of the most misunderstood properties of explosives, in that there is a persistent belief that each of these terms refers to a property with a unique value. Each of these properties can be quantitatively determined by a particular test or procedure, but there are far more such procedures than can even be listed in this space, and there are large deviations from correlation between the values determined by these different tests.

Descriptions of some of the more common explosives or types of high explosive are given below.

Ammonium nitrate-fuel oil (ANFO) blasting agents are the most widely used type of explosive product, accounting for about 90 percent of explosives in the United States. They usually contain 5.5 to 6 percent fuel oil, typically no. 2 diesel fuel. If used underground, the oil content must be carefully regulated to minimize the production of toxic fumes. Some ANFO compositions contain aluminum or densifying agents. Premixed ANFO is usually shipped in 50- to 100-lb bags, although bulk shipment and storage are practiced in certain operations. ANFO has a density of 0.85 to 1.0 g/cm^3 and a detonation velocity in the range of 10,000 to 14,000 ft/s (3,000 to 4,300 m/s). ANFO may also incorporate densifying agents and other fuels such as aluminum, and it may be blended with **emulsions** (see below). The density of such compositions may run as high as 1.5 g/cm^3, and the detonation rate as high as 16,000 ft/s (4,800 m/s). ANFO and other blasting agents require the use of high-explosive **primers** to initiate detonation. Commonly used primers are cartridges of ordinary dynamite or specially cast charges of ¼ to ¾ lb of military-type explosives such as **composition B or pentolite**. The efficiency of ANFO lies in the method of loading, which fills the borehole completely and provides good coupling with the burden.

Water-based compositions fall into two types. **Water gels** are gelled solutions of ammonium nitrate containing other oxidizers, fuels, and sensitizers such as amine nitrates or finely milled aluminum, in solution or suspension. **Emulsions** are emulsions of ammonium nitrate solution with oil, and they may contain additional oxidizers, fuels, and sensitizers. These compositions are classified either as explosives or as blasting agents depending on their sensitivity. Both types contain a thickening agent to prevent segregation of suspended solids. For larger operations, mobile mixing trucks capable of high-speed mixing and loading of the composition directly into the large-diameter vertical boreholes have become popular. Another advantage of this on-site mixing and loading is the ability to change composition and strength between bottom and top loads. The density of the explosive-sensitized composition is usually about 1.4 g/cm^3 but may be as high as 1.7 g/cm^3 for the aluminum-sensitized type; detonation velocities vary from 10,000 to 17,000 ft/s (3,000 to 5,200 m/s).

Dynamite is a generic term covering a multitude of nitroglycerine-sensitized mixtures of carbonaceous materials (wood, flour, starch) and oxygen-supplying salts such as ammonium nitrate and sodium nitrate. The nitroglycerin contains ethylene glycol dinitrate or other nitrated compounds to lower its freezing point, and antacids, such as chalk or zinc oxide, are divided into nongelatinous or granular and gelatinous types, the latter containing nitrocellulose. All dynamites are capsensitive.

Straight dynamites are graded by the percentage of explosive oil they contain; this may be as low as 15 percent and as high as 60 percent. A typical percentage formulation for a 40 percent straight dynamite is: nitroglycerin, 40; sodium nitrate, 44; antacid, 2; carbonaceous material, 14. The rate of detonation increases with grade from 9,000 to 19,000 ft/s (2,700 to 5,800 m/s). Straight dynamites now find common use only in ditching where propagation by influence is practiced.

Ammonia dynamites differ from straight dynamites in that some of the sodium nitrate and much of the explosive oil have been replaced by ammonium nitrate. Strength of ammonia dynamites ranges from 15 to 60 percent, each grade having the same weight strength as the corresponding straight dynamite when compared in the ballistic mortar. A typical percentage formula for a 40 percent ammonia dynamite is: explosive oil, 14; ammonium nitrate, 36; sodium nitrate, 33; antacid, 1; carbonaceous material, 16. The rate of detonation, 4,000 to 17,000 ft/s (1,200 to 5,200 m/s), again increases with grade. Low-density, high-weight-strength compositions are popular in many applications, but ANFO has displaced them in numerous operations.

Blasting gelatin is the strongest and highest-velocity explosive used in industrial operations. It consists essentially of explosive oil (nitroglycerin plus ethylene glycol dinitrate) colloided with about 7 percent nitrocellulose. It is completely water-resistant but has a poor fume rating and consequently finds only limited use.

Gelatin dynamites correspond to straight dynamites except that the explosive oil has been gelatinized by nitrocellulose; this results in a cohesive mixture having improved water resistance. Under confinement, the gelatins develop high velocity, ranging from 8,500 to 22,000 ft/s (2,600 to 6,700 m/s) and increasing between the grades of 20 and 90 percent. An approximate percentage composition for a 40 percent grade is: explosive oil, 32; nitrocellulose, 0.7; sulfur, 2; sodium nitrate, 52; antacid, 1.5; and carbonaceous material, 11. In the common grades of 40 and 60 percent, fume characteristics are good, making these types useful for underground hard-rock blasting.

Ammonia gelatin dynamites are similar to the ammonia dynamites except for their nitrocellulose content. These used to be popular in quarrying and hard-rock mining. Their excellent fume characteristics make them suitable for use underground, but again ANFO is widely used. The rates of detonation of 7,000 to 20,000 ft/s (2,000 to 6,000 m/s) are somewhat less than the straight gelatins. A typical percentage composition for the 40 percent grade is: gelatinized explosive oil, 21; ammonium nitrate, 14; sodium nitrate, 49; with antacid and combustible making up the remainder.

The **semigels** are important variants of the ammonia gels; these contain less explosive oil, sodium nitrate, and nitrocellulose and more ammonium nitrate than the corresponding grade of ammonia gel. Rates of detonation fall in the limited range of 10,000 to 13,000 ft/s (3,000 to 4,000 m/s). These powders are cohesive and have good water resistance and good fume characteristics.

Permissible explosives are powders especially designed for use in underground coal mines, which have passed a series of tests established by the Bureau of Mines. The most important of these tests concern incendivity of the explosives—their tendency to ignite methane-air or methane-coal dust-air mixtures. Permissible explosives are either granular or gelatinous; the granular type makes up the bulk of the powders used today. Typically, a granular permissible contains the following, in percent: explosive oil, 9; ammonium nitrate, 65; sodium nitrate, 5; sodium chloride, 10; carbonaceous material, 10; and antacid, 1. Gels contain nitrocellulose for improved water resistance and more explosive oil. Detonation velocities for the granular grades vary from 4,500 to 11,000 ft/s (1,400 to 3,400 m/s), and for the gels from 10,500 to 18,500 ft/s (3,200 to 5,600 m/s). Many water-based permissible formulations with comparable physical and safety properties are now marketed as well.

Liquid oxygen explosives (LOX) once saw considerable use in coal strip mines but have been completely displaced by ANFO or water-based compositions. LOX consisted of bags of pressed carbon black or specially processed char that were saturated with liquid oxygen just before loading into the borehole. The rate of detonation ranged from 12,000 to 18,000 ft/s (3,700 to 5,500 m/s).

Military explosives, originally developed for such uses as bomb, shell, and mine loads and demolition work, have been adapted to many industrial explosive applications. The more common military explosives are listed in Table 7.1.23, with their compositions, ballistic mortar strengths, and detonation velocities.

Table 7.1.23   Physical Characteristics of Military Explosives

Explosive	Composition, %	Ballistic mortar strength (TNT = 100)	Density, g/cm³	Rate of detonation, m/s
80/20 amatol	Ammonium nitrate, 80; TNT, 20	117	Cast	4,500
50/50 amatol	Ammonium nitrate, 50; TNT, 50	122	Cast	5,600
Composition A (pressed)	RDX, 91; wax, 9	134	0.80	4,560
			1.20	6,340
			1.50	7,680
			1.60	8,130
Composition B (cast)	RDX, 59.5; TNT, 39.5; wax, 1	130	1.65	7,660
Composition C-3 (plastic)	RDX, 77; tetryl, 3; mononitrotoluene, 5; dinitrotoluene, 10; TNT, 4; nitrocellulose, 1	145	1.55	8,460
Composition C-4 (plastic)	RDX, 91; dioctyl sebacate, 5.3; polyisobutylene, 2.1; oil, 1.6	—	1.59	8,000
Explosive D	Ammonium picrate	97	0.80	4,000
			1.20	5,520
			1.50	6,660
			1.60	7,040
HBX-1	RDX, 40; TNT, 38; aluminum, 17; desensitizer, 5	130	1.70	7,310
Lead azide*	Lead azide	—	2.0	4,070
			3.0	4,630
			4.0	5,180
PETN	Pentaerythritol tetranitrate	145	0.80	4,760
			1.20	6,340
			1.50	7,520
			1.60	7,920
50/50 Pentolite	PETN, 50; TNT, 50	120	1.20	5,410
			1.50	7,020
			1.60	7,360
			Cast	7,510
Pieric acid	Trinitrophenol	108	1.20	5,840
			1.50	6,800
			Cast	7,350
RDX (cyclonite)	Cyclotrimethylene trinitramine	150	0.80	5,110
			1.20	6,550
			1.50	7,650
			1.60	8,000
			1.65	8,180
Tetryl	Trinitrophenylmethylnitramine	121	0.80	4,730
			1.20	6,110
			1.50	7,160
			1.60	7,510
75/25 Tetrytol	Tetryl, 75; TNT, 25	113	1.60	7,400
TNT	Trinitrotoluene	100	0.80	4,170
			1.20	5,560
			1.50	6,620
			1.60	6,970
			Cast	6,790

* Primary compound for blasting caps.

Amatol was used early in World War II, largely because of the short supply of TNT. Modifications of amatol have been used as industrial blasting agents. **Explosive D,** or ammonium picrate, by virtue of its extreme insensitivity, was used in explosive-filled armor-piercing shells and bombs.

RDX, a very powerful explosive compound, was widely used during World War II in many compositions, of which Compositions A, B, and C were typical. **Composition A** was used as a shell loading; **B** was used as a bomb and shaped charge filling; C, being plastic enough to allow molding to desired shapes, was developed for demolition work. Compositions that also contained aluminum powder were developed for improved underwater performance (**Torpex, HBX**). RDX has found limited commercial application as the base charge in some detonators, the filling for special-purpose detonating fuses or cordeau detonants, and the explosive in small shaped charges used as oil well perforators and tappers for openhearth steel furnaces.

PETN has never found wide military application because of its sensitivity and relative instability. It is used extensively, however, as the core of detonating fuses and in caps and, mixed with **TNT,** in boosters.

Tetryl, once widely used by the military as a booster loading and commercially as a base charge in detonators, has been displaced by other compositions. **Tetrytol** found limited application as a demolition charge.

TNT is a very widely used military explosive. Its stability, insensitivity, convenient melting point (81°C), and relatively low cost have made it the explosive of choice either alone or in admixture with other materials for loadings which are to be cast. A free-flowing pelletized form has found application in certain types of blasting requiring high loading density, where it is used to fill the cavity formed in sprung holes or the free space around the column of other explosives in the borehole.

DUST EXPLOSIONS
by Harry C. Verakis and John Nagy
Mine Safety and Health Administration

REFERENCES: Nagy and Verakis, "Development and Control of Dust Explosions," Dekker. "Classification of Dusts Relative to Electrical Equipment in Class II Hazardous Locations," NMAB 353-4, National Academy of Sciences, Washington. "Fire Protection Handbook," 17th ed., National Fire Protection Assoc. *BuMines Rept. Inv.* 4725, 5624, 5753, 5971, 7132, 7208, 7279, 7507. Eckhoff, "Dust Explosions in the Process Industries," Butterworth-Heinemann.

A dust explosion hazard exists where combustible dusts accumulate or are processed, handled, or stored. The possibility of a dust explosion may often be unrecognized because the material in bulk form presents little or no explosion hazard. However, if the material is dispersed in the atmosphere, the potential for a dust explosion is increased significantly. The first well-recorded dust explosion occurred in a flour mill in Italy in 1785. Dust explosions continue to plague industry and cause serious disasters with loss of life, injuries, and property damage. For example, there were about 100 reportable dust explosions (excluding grain dust) from 1970 to 1980 which caused about 25 deaths and a yearly property loss averaging about $20 million. A complete and accurate record of the number of dust explosions, deaths, injuries, and property damage is unavailable because reporting of each incident is not required unless a fatality occurs or more than five persons are seriously injured. In recent years, most dust explosions have involved wood, grain, resins and plastics, starch, and aluminum. Most of the incidents occurred during crushing or pulverizing, buffing or grinding, conveying, and at dust collectors.

Despite the well-recognized hazards inherent with explosible dusts, the vast amount of technical data accumulated and published, and standards for prevention of and protection from dust explosions, severe property damage and loss of life occur every year. As an example of the severity of dust explosions, there was a series of explosions in four grain elevators during December 1977, which caused 59 deaths, 47 injuries, and nearly $60 million in property damage.

A **dust explosion** is the rapid combustion of a cloud of particulate matter in a confined or partially confined space in which heat is generated at a higher rate than it is dissipated. In a confined space, the explosion is characterized by relatively rapid development of pressure with the evolution of large quantities of heat and reaction products. The condition necessary for a dust explosion to occur is the simultaneous presence of a dust cloud of proper concentration in air or gas that will support combustion and an ignition source.

Dust means particles of materials smaller than 0.016 in in diameter, or those particles passing a no. 40 U.S. standard sieve, 425 μm (this definition relates to the limiting size, not to the average particle size of the material); and *explosible* dust means a dust which, when dispersed, is ignited by spark, flame, heated coil, or in the Godbert-Greenwald furnace at or below 730°C, when tested in accordance with the equipment and procedures described in *BuMines Rept. Inv.* 5624.

Explosibility Factors

Empirical methods and experimental data are the chief guides in evaluating relative dust explosion hazards. A mathematical model correlating some of the numerous interrelated factors affecting dust explosion development in closed vessels has been developed. Details of the model are presented in Nagy and Verakis, "Development and Control of Dust Explosions."

Dust Composition Many industrial dusts are not pure compounds. The severity of a dust explosion varies with the chemical constitution and certain physical properties of the dust. High percentages of noncombustible material, such as mineral matter or moisture, reduce the ease of oxidation, and oxygen requirements influence the explosibility of dusts. Volatile, combustible components in such materials as coals, asphalts, and pitches increase explosibility.

Dust composition also affects the amount and type of products produced in an explosion. Organic materials evolve new gaseous products, whereas most metals form solid oxides during combustion in an air atmosphere.

Particle Size and Surface Area Explosibility of dusts increases with a decrease in particle size. Fine dust particles have greater surface area, more readily disperse into a cloud, mix better with air, remain longer in suspension, and oxidize more rapidly and completely than coarse particles. Decrease of particle size generally results in lower ignition temperature, lower igniting energy, lower minimum explosive concentration, and higher pressure and rates of pressure rise. Some metals, such as chromium, become explosive only at very fine particle sizes (average particle diameter of 3 μm), and almost all metals become pyrophoric if reduced to very fine powder.

Range of Explosibility Most combustible dusts have a well-defined lower limit, but the upper limit is usually indefinite. The upper limit has been determined for only a few dusts, but these data have only limited importance in practice. The range of dust explosibility is normally 0.015 to greater than 10 oz/ft^3 (10 kg/m^3). The optimum concentration producing the strongest dust explosions is about 0.5 to 1.0 oz/ft^3. Table 7.1.24 gives explosion characteristics for a number of dusts at a concentration of 0.5 oz/ft^3. A typical example of the effect of dust concentration on maximum pressure and maximum rate of pressure rise from explosions in closed vessels of various size and shape is shown in Figs. 7.1.9 and 7.1.10.

Ignition Source Ignition sources known to have initiated dust explosions in industry include electric sparks and arcs in fuses, faulty wiring, motors and other appliances, static electrical fuses, faulty wiring, motors and other appliances, static electrical discharges, open flames, frictional, or metallic sparks, glowing particles, overheated bearings and other machine parts; hot electric bulbs, overheated driers, and other hot surfaces; dust layers may also ignite by these sources as well as by spontaneous ignition. Ignition temperatures of many dust clouds are given in Table 7.1.24. Normally the ignition temperature of a dust layer is considerably less than for a dust cloud. The position and intensity of the ignition source affect dust-explosion development; detailed information on these factors is presented in *BuMines Rept. Inv.* 7507.

Table 7.1.24 Explosive Characteristics of Various Dusts*

Type of dust	Ignition temperature of dust cloud, °C	Minimum igniting energy, J	Minimum explosive concentration, oz/ft³	Maximum explosion pressure, lb/in² gage	Maximum rate of pressure rise, lb/(m²)(s)	Terminal oxygen concentration, %†
Agricultural:						
Alfalfa	530	0.320	0.105	92	2,200	
Cereal grass	550	0.800	0.250	52	500	
Cinnamon	440	0.030	0.060	114	3,900	
Citrus peel	730	0.045	0.065	71	2,000	
Cocoa	500	0.120	0.065	55	900	
Coffee	720	0.160	0.085	53	300	
Com	400	0.040	0.055	95	6,000	
Corncob	480	0.080	0.040	110	3,100	
Com dextrine	410	0.040	0.040	105	7,000	
Cornstarch	390	0.030	0.040	115	9,000	
Cotton linters	520	1.920	0.500	48	150	
Cottonseed	530	0.120	0.055	96	3,000	
Egg white	610	0.640	0.140	58	500	
Flax shive	430	0.080	0.080	81	800	
Garlic	360	0.240	0.100	80	2,600	
Grain, mixed	430	0.030	0.055	115	5,500	
Grass seed	490	0.260	0.290	34	400	
Guar seed	500	0.060	0.040	98	2,400	
Gum, Manila (copal)	360	0.030	0.030	88	5,600	
Hemp hard	440	0.035	0.040	103	10,000	
Malt, brewers	400	0.035	0.055	92	4,400	
Milk, skim	490	0.050	0.050	83	2,100	
Pea flour	560	0.040	0.050	95	3,800	
Peanut hull	460	0.050	0.045	82	4,700	
Peat, sphagnum	460	0.050	0.045	84	2,200	
Pecan nutshell	440	0.050	0.030	106	4,400	
Pectin	410	0.035	0.075	112	8,000	
Potato starch	440	0.025	0.045	97	8,000	
Pyrethrum	460	0.080	0.100	82	1,500	
Rauwolfia vomitoria root	420	0.045	0.055	106	7,500	
Rice	440	0.050	0.050	93	2,600	
Safflower	460	0.025	0.055	84	2,900	
Soy flour	550	0.100	0.060	111	1,600	15
Sugar, powdered	370	0.030	0.045	91	1,700	
Walnut shell, black	450	0.050	0.030	97	3,300	
Wheat flour	440	0.060	0.050	104	4,400	
Wheat, untreated	500	0.060	0.065	98	4,400	
Wheat starch	430	0.025	0.045	100	6,500	
Wheat straw	470	0.050	0.055	99	6,000	
Yeast, torula	520	0.050	0.050	105	2,500	
Carbonaceous:						
Asphalt, resin, volatile content 57.5%	510	0.025	0.025	94	4,600	
Charcoal, hardwood mix, volatile content 27.1%	530	0.020	0.140	100	1,800	18
Coal, Colo., Brookside, volatile content, 38.7%	530	0.060	0.045	88	3,200	
Coal, Ill., no. 7, volatile content 48.6%	600	0.050	0.040	84	1,800	15
Coal, Ky., Breek, volatile content 40.6%	610	0.030	0.050	88	4,000	
Coal, Pa., Pittsburgh, volatile content 37.0%	610	0.060	0.055	83	2,300	17
Coal, Pa., Thick Freeport, volatile content, 35.6%	595	0.060	0.060	77	2,200	
Coal, W. Va., no. 2 Gas, volatile content 37.1%	600	0.060	0.060	82	1,600	
Coal, Wyo., Laramie no. 3, volatile content 43.3%	575	0.050	0.050	92	2,000	
Gilsonite, Utah, volatile content 86.5%	580	0.025	0.020	78	3,700	
Lignite, Calif., volatile content 60.4%	450	0.030	0.030	90	8,000	
Pitch, coal tar, volatile content 58.1%	710	0.020	0.035	88	6,000	
Chemical compounds:						
Benzoic acid, C_6H_5COOH	620	0.020	0.030	74	5,500	
Phosphorus pentasulfide, P_2S_5, slowly cooled to give single crystals	280	0.015	0.050	54	10,000+	
Phosphorus pentasulfide, P_2S_5, cooled quickly	290	0.015	0.050	58	7,500	13

Table 7.1.24 Explosive Characteristics of Various Dusts* (Continued)

Type of dust	Ignition temperature of dust cloud, °C	Minimum igniting energy, J	Minimum explosive concentration, oz/ft^3	Maximum explosion pressure, lb/in^2 gage	Maximum rate of pressure rise, lb/(m^2)(s)	Terminal oxygen concentration, %†
Chemical compounds (*Continued*):						
Phthalimide, C6H$_4$(CO)$_2$NH	630	0.050	0.030	79	4,500	
Potassium bitartrate, KHC$_4$H$_4$O$_6$	520					
Salicylanilide, o-HOC$_6$H$_4$CONHC$_6$H$_5$	610	0.020	0.040	61	4,400	
Sodium thiosulfate, anhydrous, Na$_2$S$_2$O$_3$	510					
Sorbic acid, CH$_3$(CH:CH)$_2$ COOH	470	0.015	0.020	88	10,000+	
Sucrose, C$_{12}$H$_{22}$O$_{11}$	420	0.040	0.045	82	4,200	14
Sulfur, S$_8$ 100% finer than 44 μm	210	0.020	0.045	56	3,100	
Sulfur, S$_8$, avg particle size 4 μm	190	0.015	0.035	78	4,700	12
Drugs:						
Aspirin (acetylsalicylic acid), o-CH$_3$COOC$_6$H$_4$COOH, fine	660	0.025	0.050	83	10,000+	
Mannitol (hexahydric alcohol), CH$_2$OH(CHOH)$_4$CH$_2$OH	460	0.040	0.065	82	2,800	
Secobarbital sodium, C$_{12}$H$_{17}$N$_2$O$_3$Na	520	0.960	0.100	54	500	
Vitamin C, ascorbic acid, C$_6$H$_8$O$_6$	460	0.060	0.070	88	4,800	15
Explosives and related compounds:						
Dinitrobenzamide	500	0.045	0.040	94	6,500	
Dinitrobenzoic acid	460	0.045	0.040	92	4,300	
Dinitro-sym-diphenyl-urea (dmitrocarbanilide)	550	0.060	0.095	87	2,500	
Dinitrotoluamide (3,5-dinitro-ortho-toluamide)	500	0.015	0.050	106	10,000−	13
Metals:						
Aluminum	650	0.015	0.045	100	10,000+	2
Antimony	420	1.920	0.420	8	100	16
Boron	470	0.060	0.100	90	2,400	
Cadmium	570	4.000				
Chromium	580	0.140	0.230	56	5,000	14
Cobalt	760					
Copper	900					
Iron	420	0.020	0.100	46	6,000	10
Lead	710					
Magnesium	520	0.020	0.020	94	10,000+	0
Molybdenum	720					
Nickel	950+					
Selenium	950+					
Silicon	780	0.080	0.100	106	10,000+	12
Tantalum	630	0.120	0.200	51	3,700	
Tellurium	550					
Thorium	270	0.005	0.075	48	3,300	0
Tin	630	0.080	0.190	37	1,300	15
Titanium	460	0.010	0.045	80	10,000+	0
Tungsten	950+					
Uranium	20	0.045	0.060	53	3,400	0
Vanadium, 86%	500	0.060	0.220	48	600	13
Zinc	600	0.640	0.480	48	1,800	9
Zirconium	20	0.005	0.045	65	9,000	0
Alloys and compounds:						
Aluminum-cobalt	950	0.100	0.180	78	8,500	
Aluminum-copper	930	1.920	0.280	27	500	
Aluminum-iron	550	0.720	0.500	21	100	
Aluminum-magnesium	430	0.020	0.020	90	10,000	0
Aluminum-nickel	940	0.080	0.190	79	10,000	14
Aluminum-silicon, 12% Si	670	0.060	0.040	74	7,500	
Calcium silicide	540	0.130	0.060	73	10,000+	8
Ferrochromium, high-carbon	790		2.000			19
Ferromanganese, medium-carbon	450	0.080	0.130	47	4,200	
Ferrosilicon, 75% Si	860	0.400	0.420	87	3,600	16
Ferrotitanium, low-carbon	370	0.080	0.140	53	9,500	13
Ferrovanadium	440	0.400	1.300			17
Thorium hydride	260	0.003	0.080	60	6,500	6
Titanium hydride	440	0.060	0.070	96	10,000+	13

Table 7.1.24 Explosive Characteristics of Various Dusts* *(Continued)*

Type of dust	Ignition temperature of dust cloud, °C	Minimum igniting energy, J	Minimum explosive concentration, oz/ft³	Maximum explosion pressure, lb/in² gage	Maximum rate of pressure rise, lb/(m²)(s)	Terminal oxygen concentration, %†
Alloys and compounds *(Continued)*:						
Uranium hydride	20	0.005	0.060	43	6,500	0
Zirconium hydride	350	0.060	0.085	69	9,000	8
Plastics:						
Acetal resin (polyformaldehyde)	440	0.020	0.035	89	4,100	11
Acrylic polymer resin Methyl methacrylate-ethyl acrylate	480	0.010	0.030	85	6,000	11
Alkyd resin Alkyd molding compound	500	0.120	0.155	15	150	15
Allyl resin, allyl alcohol derivative, CR-39	500	0.020	0.035	106	10,000+	13
Amino resin, urea-formaldehyde molding compound	450	0.080	0.075	89	3,600	17
Cellulosic fillers, wood flour	430	0.020	0.035	110	5,500	17
Cellulosic resin, ethyl cellulose molding compound	320	0.010	0.025	102	6,000	11
Chlorinated polyether resin, chlorinated polyether alcohol	460	0.160	0.045	66	1,000	
Cold-molded resin, petroleum resin	510	0.030	0.025	94	4,600	
Coumarone-indene resin	520	0.010	0.015	93	10,000+	14
Epoxy resin	530	0.020	0.020	86	6,000	12
Fluorocarbon resin, fluorethylene polymer	600					
Furane resin, phenol furfural	520	0.010	0.025	90	8,500	14
Ingredients, hexamethylenetetramine	410	0.010	0.015	98	10,000+	14
Miscellaneous resins, petrin acrylate monomer	220	0.020	0.045	104	10,000+	
Natural resin, rosin, DK	390	0.010	0.015	87	10,000+	14
Nylon polymer resin	500	0.020	0.030	89	7,000	13
Phenolic resin, phenol-formaldehyde molding compound	500	0.020	0.030	92	10,000+	14
Polycarbonate resin	710	0.020	0.025	78	4,700	15
Polyester resin, polyethylene terephthalate	500	0.040	0.040	91	5,500	13
Polyethylene resin	410	0.010	0.020	83	5,000	12
Polymethylene resin, carboxypolymethylene	520	0.640	0.115	70	5,500	
Polypropylene resin	420	0.030	0.020	76	5,000	
Polyurethane resin, polyurethane foam	510	0.020	0.025	88	3,700	
Rayon (viscose) flock	520	0.240	0.055	88	1,700	
Rubber, synthetic	320	0.030	0.030	93	3,100	15
Styrene polymer resin, polystyrene latex	500	0.020	0.020	91	7,000	13
Vinyl polymer resin, polyvinyl butyral	390	0.010	0.020	84	2,000	14

* Data taken from the following *Bureau of Mines Reports of Investigations:* RI 5753, "Explosibility of Agricultural Dusts"; RI 5971, "Explosibility of Dusts Used in the Plastics Industry"; RI 6516, "Explosibility of Metal Powders"; RI 7132, "Dust Explosibility of Chemicals, Drugs, Dyes and Pesticides"; RI 7208, "Explosibility of Miscellaneous Dusts." The data were obtained using the equipment described in RI 5624, "Laboratory Equipment and Test Procedures for Evaluating Explosibility of Dusts."

† The terminal oxygen concentration is the limiting oxygen concentration in air-CO_2 atmosphere required to prevent ignition of dust clouds by electric spark.

Turbulence Turbulence has a slight effect on maximum pressure, but a marked effect on the rates of pressure rise for dust explosions. Experiments show the maximum rate of pressure rise in a highly turbulent dust-air mixture can be as much as 8 times higher than in a non-turbulent mixture (*BuMines Repts. Inv.* 5815 and 7507 and Nagy and Verakis).

Moisture and Other Inerts Moisture in a dust absorbs heat and tends to reduce the explosibility of a dust. A high concentration of moisture in the dust also tends to reduce the dispersibility of a dust. An increase in moisture content causes an increase in ignition temperature and a reduction in maximum pressure and rates of pressure rise. However, the amount of moisture required to produce a marked lowering of the explosibility parameters is higher than can ordinarily be tolerated in industrial processes. Most mineral inert dusts admixed with a combustible absorb heat during the combustion reaction and reduce explosibility similar to the action of water. Some chemical compounds, such as sodium and potassium carbonates, act as inhibitors and are more effective than mineral inerts; the limiting inert dust concentration required to prevent ignition and explosion depends on the strength of the igniting source.

Atmospheric Oxygen Concentration The pressure and rate of pressure development decrease as the oxygen concentration in the atmosphere decreases. The ignition sensitivity of dusts decreases with decrease in oxygen concentration and for most dusts, ignition and explosion can be prevented by reducing the oxygen concentration to a safe value. Carbon dioxide, nitrogen, argon, helium, and water vapor are effective diluents. For highly reactive metal powders, only argon and helium are chemically inert. Limiting oxygen concentrations using carbon dioxide as a diluent are given in Table 7.1.24 for many dusts. With carbon dioxide as a diluent, a reduction of oxygen in the atmosphere to 11 percent is sufficient to prevent ignition by sparks for all dusts tested except the metallic powders. With nitrogen as the diluent, ignition of nonmetallic dusts is prevented by diluting the atmosphere to 8 percent oxygen. Some metal dusts, such as magnesium, titanium, and zirconium, ignite by spark in a pure carbon dioxide atmosphere. Freon and halons are sometimes used as diluent gases, but if metal dusts are involved, they can intensify rather than suppress ignition. The limiting oxygen concentration decreases as the dust becomes finer in particle size; limiting oxygen concentration varies slightly with dust concentration and is lowest at concentrations two to five times the stoichiometric mixture.

Fig. 7.1.9 Effect of dust concentration on maximum pressure produced by explosions of cellulose acetate dust in closed vessels.

Relative Dust Explosion Hazards

Table 7.1.24 gives test results of selected dusts whose explosive characteristics have been evaluated in the laboratory by the Bureau of Mines. The data were obtained for dusts passing a no. 200 sieve and represent the most hazardous of the specific materials tested. The values are relative rather than absolute since the test apparatus and experimental procedures affect the results to some degree. The samples were dried before testing only if the moisture content exceeded 5 percent. Detailed description of the equipment and procedures for the small-scale testing are given in *BuMines Rept. Inv.* 5624.

Fig. 7.1.10 Effect of dust concentration on maximum rate of pressure rise by explosions of cellulose acetate dust in closed vessels.

Ignition Temperature   The ignition temperature of a dust cloud was determined by dispersing dust through a heated cylindrical furnace. The ignition temperature is the minimum furnace temperature at which flame appears at the bottom of the furnace in one or more trials in a group of four.

Minimum Energy   The minimum electrical spark energy required to ignite a dust cloud was determined by dispersing the dust in a vertically mounted, 2 ¾-in-diameter, 12-in-long tube. The dust is dispersed by an air blast and then a condenser of known capacitance and voltage is discharged through a spark gap located within the dust dispersion. The top of the tube is enclosed with a paper diaphragm. The minimum energy for ignition of the dust cloud is the least amount of energy required to produce flame propagation 4 in or longer in the tube.

Minimum Concentration   The minimum explosive concentration or the lower explosive limit of a dust cloud was determined in a vertically mounted, 2 ¾-in diameter, 12-in-long tube using a continuous sparkigniting source. A known weight of dust was dispersed within the tube by an air blast. The lowest weight of dust at which sufficient pressure develops to burst a paper diaphragm enclosing the tube or which causes flame to fill the tube is used to calculate the minimum explosive concentration; this calculation is made utilizing the tube volume.

Maximum Pressure and Rates of Pressure Rise   The maximum pressure and rates of pressure rise developed by a dust explosion were determined by dispersing dust in a closed steel tube. A continuous spark is used for ignition. A transcribed pressure-time record is obtained during the test. The maximum rate is the steepest slope of the pressure-time curve. Normally, explosion tests are made at dust concentrations of 0.10 to 2.0 oz/ft³. Maximum pressure is primarily dependent on dust composition and independent of vessel size and shape. The maximum rate of pressure rise increases as vessel size decreases.

Explosibility Index   The overall explosion hazard of a dust is related to the ignition sensitivity and to explosion severity and is characterized by empirical indexes. The ignition sensitivity of a dust cloud depends on the ignition temperature, minimum energy, and minimum concentration. The explosion severity of a dust depends on the maximum pressure and maximum rate of pressure rise. The indexes are not derived from theoretical considerations, but provide a numerical rating consistent with research observations and practical experience. Results obtained for a sample dust are compared with values for a standard Pittsburgh-seam coal dust. The indexes are defined as follows:

Ignition sensitivity =

$$\frac{\text{(ign temp} \times \text{ min energy} \times \text{ min conc) Pittsburgh coal dust}}{\text{(ign temp} \times \text{ min energy} \times \text{ min conc) sample dust}}$$

Explosion severity =

$$\frac{\begin{array}{c}\text{(max explosive pressure}\\ \times \text{ max rate of pressure rise) sample dust}\end{array}}{\begin{array}{c}\text{(max explosive pressure}\\ \times \text{ max rate of pressure rise) Pittsburgh coal dust}\end{array}}$$

Explosibility index = ignition sensitivity × explosion severity

A dust having ignition and explosion characteristics equivalent to the standard Pittsburgh-seam coal has an explosibility index of unity. The relative hazard of dusts is further classified by the following adjective ratings: fire, weak, moderate, strong, or severe. The notation ≪ 0.1 designates a combustible dust presenting primarily a fire hazard as ignition of the dust cloud is not obtained by a spark or flame source, but only by an intense, heated surface source. These ratings are correlated with the empirical indexes as follows:

Type of explosion	Ignition sensitivity	Explosion severity	Index of explosibility
Fire			< 0.1
Weak	< 0.2	< 0.5	< 0.1
Moderate	0.2–1.0	0.50–1.0	0.1–1.0
Strong	1.0–5.0	1.0–2.0	1.0–10
Severe	> 5.0	> 2.0	>10

Prevention of Dust Explosions

(See also Sec. 12.1)

There are codes published by the National Fire Protection Association which contain recommendations for a number of dust-producing industries.

Additional sources of information may be found in the NFPA "Fire Protection Handbook," the Factory Mutual "Handbook of Industrial Loss Prevention," *BuMines Rept. Inv.* 6543, Nagy and Verakis, and Eckhoff.

Safeguards against explosions include, but are not limited to the following:

Good housekeeping: An excellent means to minimize the potential for and extent of an explosion is good housekeeping. Control of dust spillage or leakage and elimination of dust accumulations removes the fuel required for an explosion.

Limited personnel: Wherever a hazardous operation must be performed, the number of persons should be limited to the minimum required for safe operation.

Elimination of Ignition Sources   All sources of ignition should be eliminated from equipment containing combustible dust and from adjacent areas. Open flames or lights and smoking should be prohibited. The use of electric or gas cutting and welding equipment for repairs should be avoided unless dust-producing machinery is shut down and all dust has been removed from the machines and from their vicinity. Proper control methods should be instituted for materials susceptible to spontaneous combustion. Additional safety measures to follow are ground and bonding of all equipment to prevent the accumulation of static electrical charges; strict adherence to the National Electric Code when installing electrical equipment and wiring in hazardous locations; use of magnetic separators to prevent entrance of ferrous materials into dust-grinding mills; use of nonferrous blades in fans through which dust passes; and avoidance of spark-producing tools in certain industries and of highspeed shafting and belts. Safeguards against ignition by lightning should also be considered. (See also NFPA no. 77, Static Electricity and NFPA no. 78, Lightning Protection.)

Building and Equipment Construction   Buildings should be constructed to minimize the collection of dust on beams, ledges, and other surfaces, particularly overhead. Vacuum cleaning is preferable to other methods for dust removal, but soft push brooms may be used without serious hazard. Buildings, including inside partitions, where combustible dusts are handled or stored should be detached units of incombustible construction. Hazardous units within buildings should be separated by substantial fire walls.

Grinders, conveyors, elevators, collectors, and other equipment which may produce dust clouds should be as dust-tight as possible; they should have the smallest practical interior volume and should be constructed to withstand dust explosion pressures. The degree of turbulence within and around an enclosure should be kept to a minimum to prevent dust from being suspended.

Dust collectors should preferably be located outside of buildings or detached rooms and near the dust source. The choice of a suitable dust collector depends on the particle size, dryness, explosibility, dust concentration, gas velocity and temperature, efficiency and space requirements, and economic considerations. (See also Secs. 9 and 18.)

Inerted Atmosphere   Equipment such as grinders, conveyors, pulverizers, mixers, dust collectors, and sacking machines can frequently be protected by using an inerted atmosphere or explosion suppression systems. The inert gas for this purpose may be obtained by dilution of air with flue gases from boilers, internal-combustion engines, or other sources, or by dilution with carbon dioxide, nitrogen, helium from high-pressure cylinders, or gas from inert-gas generators. The amount and rate of application of inert gas required depend upon the permissible oxygen concentration, leakage loss, atmospheric and operating conditions, equipment to be protected, and application method. Addition of inert dusts to the combustible dust may also prevent explosive dust-air mixtures from forming in and around equipment. (See NFPA no. 69, Explosion Prevention Systems.)

Relief Venting   To reduce structural damage and to protect personnel from dust explosions, dust collectors and other equipment and the rooms in which dust-producing machinery is located should be provided with relief vents. Relief vents properly designed and located will sufficiently relieve explosion pressures in most instances and direct explosion gases away from occupied areas. The vents may be unrestricted or free openings; hinged or pivoted sash that swing outward at a low internal pressure; fixed sash with light wall anchorages; scored glass panes; light wall panels; monitors or skylights; paper, metal foil, or other diaphragms that burst at low pressures; poppet-type vent closures; pull-out diaphragms; or other similar arrangements. Vents should be located near potential sources of ignition to keep explosion pressure at a minimum and to prevent a dust explosion from developing into a detonation in long ducts.

Empirical formulas and mathematical methods for calculating the vent area to limit pressure from an explosion have been developed. Unfortunately, because of the many factors involved in determining venting requirements, none of these methods can be considered entirely satisfactory to cover the complete range of situations confronting an equipment or building designer. For example, the maximum pressure that can develop in a vented enclosure during a dust explosion is affected by factors such as the chemical affinity of the combustible material with oxygen, heat of combustion, particle size distribution, degree of turbulence, uniformity of the dust cloud, size and energy of the igniting source, location of the igniting source relative to the vent, area of the vent opening, bursting strength or inertial resistance of the vent closure, initial pressure and initial temperature within an enclosure, and the oxygen concentration of the atmosphere. Because of the complexity of the phenomena during explosion in a vented vessel, information on the required vent area to limit the excess or explosion pressure to be a safe value for a given vessel or structure under specific conditions is still estimated from data obtained by physical tests usually made under severe test conditions. Extremely reactive dusts such as magnesium and aluminum are difficult or nearly impossible to vent successfully if an explosion occurs under optimum conditions. Agricultural dusts, most plastic-type dusts, and other metallic dusts can usually be vented successfully. Materials containing oxygen or a mixture with an oxidant should be subjected to tests before venting is attempted. Information and recommendations on venting are given in NFPA no. 68, "Guide for Venting of Deflagrations" (1994). A mathematical analysis showing the relationship of numerous factors affecting the venting of explosions is presented in "Development and Control of Dust Explosions" (Nagy and Verakis, Dekker).

Information published by others shows that higher values of maximum pressures and rates of pressure rise are normally obtained in vessels larger and differently shaped than the Hartmann apparatus described in *BuMines RI 5624*. Data on maximum pressures and rates obtained from the Hartmann apparatus are shown in Table 7.1.24. A comparison of test data from the Hartmann apparatus and a 1-m³ vessel is shown in "Development and Control of Dust Explosions." NFPA no. 68, "Guide for Venting of Deflagrations" (1994), recommends using the rate of pressure rise data obtained from closed vessels, 1-m³ or larger, in venting calculations.

Combating Dust Fires

The following points should be observed when one is dealing with dust fires, in addition to the usual recommendations for fire prevention and firefighting, including sprinkler protection (see also Secs. 12 and 18).

1. Attention should be directed to the potential hazard of spontaneous heating of dust products, particularly when grinding or pulverizing processes are used.

2. First-aid and firefighting equipment should be installed. Small hoses with spray nozzles or automatic sprinkler systems fitted with spray or fog nozzles are particularly satisfactory. The fine spray wets the dust and is not so likely to raise a dust cloud as with a solid stream. Portable extinguishers used to combat dust fires should be provided with similar devices for safe discharge.

3. Large hose of fire department size giving solid water streams should be used with caution; a dust cloud may be formed with consequent risk of explosion. Plant employees and the fire department should

be advised of this potential hazard in advance. Hose equipped with spray or fog nozzles should be provided and kept ready for an emergency.

4. Fires involving aluminum, magnesium, or some other metal powders are difficult to extinguish. Sand, talc, or other dry inert materials, and special proprietary powders designed for this purpose should be used. These materials should be applied gently to smother the fire. Materials such as hard pitch can completely seal the dust from oxygen and may be used. (See NFPA Code nos. 48, 65, and 651.)

ROCKET FUELS
by Randolph T. Johnson

Indian Head Division, Naval Surface Warfare Center

REFERENCES: Billig, Tactical Missile Design Concepts, Johns Hopkins Applied Physics Laboratory, *Technical Digest*, 1983, **4**, no. 3. Roth and Capener, Propellants, Solid, "Encyclopedia of Explosives and Related Items," Kave, ed., PATR 2700, vol. 8, U.S. Army Armament Research and Development Command, Dover, NJ, 1978. "Solid Propellant Selection and Characterization," NASA Design Criteria Guide, NASA SP 8064, 1971.

Until the development of the solid-fueled Polaris missile in the late 1950s and early 1960s, rocket motors tended to be divided into two distinct categories: rocket motors greater than 3 ft in diameter were liquid-fueled, while smaller rocket motors were solid-fueled. Since that time, a number of quite large rocket motors have been developed which use solid fuel; these include the Minuteman series and, most recently, the solid rocket boosters used on the space shuttle, each of which has over 1,000,000 lb of propellant. There are exceptions to the rule in the other direction, too; e.g., the air-launched Bullpup rocket motor started out as a liquid-fueled unit, changed to a solid fuel, then back to a liquid fuel.

Design Criteria

Before selecting the proper propellant system for a given rocket motor, the designer must consider the parameters by which the design is constrained. The following criteria generally need to be considered when choosing the propellant(s) for a given rocket motor:

Envelope Constraints The designer must consider the volume, mass and shape limits within which the rocket motor is constrained. For example, an air-launched rocket may be limited by the carrying capacity of the aircraft, the size of the launcher, and the size of the payload.

Performance Requirements In general, the requirements imposed on a rocket motor are expressed as the minimum required and/or the maximum allowed to complete the mission and the maximum allowed to prevent damage to the payload, launcher, etc. Such parameters include velocity, range, burn time, and acceleration.

Environmental Conditions The environments seen by various rocket motors differ dramatically, depending on the intended use. For example, a submarine-launched strategic missile lives a pampered life in near ideal conditions while a field-launched barrage rocket or an air-launched missile see near worse-case environments. Some of the parameters to consider in the choice of propellants include temperature limits of storage and operation, vibration and shock spectra to be experienced and survived, and the moisture and corrosion environments the rocket motor (and possibly the propellant) may be expected to encounter. It may be possible to reduce the effects of these environments on the propellant by rocket motor design, but the propellant chemist and the rocket motor designer must be willing to work together to come up with a viable combination.

Safety Requirements Safety considerations include toxic and explosion hazards in the manufacture of the rocket motor and in servicing, handling, and use. The use of the rocket must be considered in the safety margin built into its design. If it is to be used in a "human-rated" system, such as an ejection seat, it must be of more conservative design than if it were to be used in a barrage rocket, for instance.

Service Life The designer must consider the duration over which the rocket motor will be in service. It is, in general, less expensive to

make a number of rocket motors at a time and then store them, than to make the same number in several small lots over a period of time with the accordingly increased start-up and shutdown costs. A very short service life leads to heightened costs for repetitive shipping to and from storage, as well as rework and replacement of overaged units. There is yet another hidden cost of a very short service life: the risk that an overaged unit will be inadvertently used with potentially catastrophic results.

Maintenance The availability of maintenance will affect the choice of propellants for the given unit. The unit with readily available maintenance facilities will have far less severe constraints than the unit which must function after years of storage and/or far from the reach of service facilities.

Smokiness The choice of propellants must be influenced by whether or not the mission will allow the use of a propellant which leaves a smoky trail. Propellants containing a metal fuel, as well as certain other ingredients, leave a large smoky trail which reveals the location of the launch point. This is particularly undesirable in tactical rocket systems for it leaves the user revealed to the enemy. The smoke from metal fuels is termed **primary** smoke as it is a product of the primary combustion, **secondary** smoke comes from the reaction of propellant combustion products with the atmosphere, such as the reaction of hydrogen chloride with moisture in the air to leave a hydrochloric acid cloud, or water vapor condensing in cold air to leave a contrail.

Cost The cost of a rocket motor must be divided into a number of categories including design, test, ingredients, processing, components, surveillance, maintenance, rework, and disposal. The hidden cost of nonfunction (reliability) should also be factored into the equation.

Liquid Propellants versus Solid Propellants

The first choice to be made in the selection of propellants is whether to use a liquid or a solid propellant. Each of these comes with its own set of advantages and disadvantages which must be weighed in the selection process.

Liquid propellants offer the possibility of extremely high impulse per unit weight of propellant. The thrust of a liquid-propellant rocket motor may be easily modulated by controlling the flow rate of the propellant into the combustion chamber.

Liquid propellants, however, tend to be of low density, which leads to large packages for a given total energy. Most liquid propellants have a very limited usable temperature range because of freezing or vaporization. Because they require the use of valves, pipes, pumps, and the like, liquid-propellant rocket motors are relatively complex and require a high level of maintenance.

The use of liquid rocket propellants is predominant in older strategic rocket motors, such as Titan, and in space flight, such as the space shuttle main engines and attitude-control motors.

Solid propellants offer the possibility of use over a wide range of environmental conditions. They offer a significantly higher density than liquid propellants. Solid-propellant rocket motors are much simpler than liquid rocket motors, as the propellant grain forms at least one wall of the combustion chamber. This simplicity of design leads to low maintenance and high reliability. Safety is somewhat higher than with liquid propellants, for there are no volatile and hazardous liquids to spill during handling and storage.

Solid propellants, however, do not attain the impulse levels of the more energetic liquid propellants. Furthermore, once ignited, the thrust-time profile will be as dictated by the propellant surface history and propellant burn rate for the nozzle given; this profile is not easily altered, unlike that of a liquid-propellant rocket motor suitably equipped.

Solid-propellant rocket motors are now used in every size from small thrusters on the Dragon antitank round to the boosters on the space shuttle. They are used when a preprogrammed thrust-time history is appropriate.

Liquid Propellants Once the decision has been made to use liquid propellant, the designer is confronted with the decision of whether to use a monopropellant or a bipropellant system.

A **monopropellant** is a fuel which requires no separate oxidizer, but provides its propulsive energy through its own decomposition. The advantage of a monopropellant is the inherent simplicity of having only one liquid to supply to the combustion chamber. The principal disadvantage of the commonly used monopropellants is their very low impulse compared to most bipropellant or solid-propellant systems. Two of the more commonly used monopropellants are ethylene oxide and hydrogen peroxide.

Bipropellant systems use two liquids, an oxidizer and a fuel, which are merged and burned in a combustion chamber. There is a much wider selection of fuel constituents for bipropellant systems than for monopropellant systems. Bipropellants offer much higher impulse values than do the monopropellants, and higher than those available with solid propellants. The primary disadvantage of bipropellant systems is the need for a far more complex piping and metering system than is required for monopropellants.

Bipropellants may be subdivided further into two categories, **hypergolic** (in which the two constituents ignite on contact) and **nonhypergolic.** Hypergolic systems eliminate the need for a separate igniter, thus decreasing system complexity; however, this increases the fire hazard if the fuel system should leak.

When **cryogenic liquids** such as liquid hydrogen and liquid oxygen are used, the problems of storage become of major significance. A significant penalty in weight and complexity must be paid to store these liquids and hold them at temperature. Furthermore, these liquids charge a significant penalty, because their very low density enlarges the packaging requirements even more. This penalty is in the parasitic weight of the packaging and the increased drag induced by the increased skin area. These problems have limited the use of these propellants to systems where immediate response is not required, such as space launches, as opposed to strategic or tactical systems.

Fuels for bipropellant systems include methyl alcohol, ethyl alcohol, aniline, turpentine, unsymmetrical dimethylhydrazine (UDMH), hydrazine, JP-4, kerosene, hydrogen, and ammonia. **Oxidizers** for bipropellant systems include nitric acid, hydrogen peroxide, oxygen, fluorine, and nitrogen tetroxide. In some cases, mixtures of these will yield superior properties to either ingredient used alone; e.g., 50:50 mixture of UDMH with hydrazine is sometimes used in lieu of either alone.

Solid Propellants   Once the decision has been made to use a solid propellant, one must decide between a case-bonded and cartridgeloaded propellant grain. The decision then must be made among composite, double-base, and composite modified double-base propellants.

Case-Bonded versus Cartridge-Loaded   Case bonding refers to the technique by which the propellant grain is mechanically (adhesively) linked to the motor case. Cartridge-loaded propellant grains are retained in the motor case by purely mechanical means.

Case bonding takes advantage of the strength of the motor case to support the propellant grain radially and longitudinally; this permits the use of low-modulus propellants. This technique also yields high volumetric efficiency.

Since the propellant-to-case bond effectively inhibits the outer surface of the propellant, the quality of this bond becomes critical to the proper function of the rocket motor. This outer inhibition also tends to limit the potential propellant grain surface configurations, thus limiting the interior ballistician's leeway in tailoring the rocket motor ballistics. Since the propellant grain is bonded to the motor case, the grain must be cast into the motor case or secondarily bonded to it. The former method leads to reduced flexibility in scheduling motor manufacture. The second technique can lead to quality control problems if a bare, uninhibited grain is bonded to the motor case, or to a loss in volumetric loading efficiency if a bare grain is inhibited or cast into a premolded form that is bonded to the motor case. Bonding the propellant grain to the motor case also makes it difficult to dispose of the rock motor when it reaches the end of its service life, especially in regard to making the motor case suitable for reuse.

Cartridge loading of the propellant grain offers flexibility of manufacture of the rocket motor, as the motor-case manufacture and propellant-grain manufacture may be pursued independently. This technique

also allows the interior ballistician a free hand in configuring the propellant surface to obtain optimal ballistics. Since the propellant grain is held in the motor case mechanically, it is a simple matter to remove the propellant at the end of its service life and reuse the motor case and associated hardware.

The cartridge-loaded propellant grain must be inhibited in a separate operation, as opposed to the case-bonded grain. The free-standing grain must be supported so that it is not damaged by vibration and shock; correspondingly, the propellant must have substantial strength of its own to withstand the rigors of the support system and the vibrations and shocks that filter through the system. The presence of the mechanical support system and the inhibitor, and the space required to slide the propellant grain into the motor case, prevent the cartridge-loaded propellant grain from having the volumetric loading efficiency of a case bonded propellant grain.

Composite versus Double-Base versus Composite Modified Double-Base   Composite propellants consist primarily of a binder material such as polybutadiene (artificial rubber) and finely ground solid fuels (such as aluminum) and oxidizers (such as ammonium perchlorate). Double-base propellants consist primarily of stabilized nitrocellulose and nitroglycerine. Composite modified double-base propellants use a double-base propellant for a binder, with the solid fillers commonly found in composite propellants.

Composite propellants offer moderately high impulse levels and widely tailorable physical properties. They may be designed to function over a wide range of temperatures and, depending upon the fillers, have high ignition temperatures and consequently very favorable safety features.

Composite propellants at the present time are limited to cast applications. The binders are petroleum-based, and subject to the vagaries of oil availability and price. Many of the formulations for their binders contain toxic ingredients as well as ingredients which are sensitive to moisture. The water sensitivity carries over to the filler materials which tend to dissolve or agglomerate in the presence of moisture in the air. Composite propellants also are notoriously difficult to adequately inhibit once the propellant has fully cured.

Historically, a number of binder materials have been used in the manufacture of composite propellants. Among these are asphalt, polysulfides, polystyrene-polyester, and polyurethanes. Propellants of recent development have been predominantly products of the polybutadiene family: carboxy-terminated polybutadiene (CTPB), hydroxyterminated polybutadiene (HTPB), and carboxy-terminated polybutadiene-acrylonitrile (CTBN). It should be pointed out that the binder material also serves as a fuel, so a satisfactory propellant for many purposes may be manufactured without a separate fuel added to it. While the addition of metallic fuels to the propellant significantly increases the energy of the propellant, it also produces primary smoke in the form of metal oxides. The most commonly added metal is aluminum, but magnesium, beryllium, and other metals have been tried. The most commonly used oxidizer (which makes up the preponderance of the weight of the propellant) is ammonium perchlorate. Other oxidizers which have been used include ammonium nitrate, potassium nitrate, and potassium perchlorate.

Double-base propellants have been used in guns since Alfred Nobel's discovery of ballistite in the nineteenth century. Their first significant use in rocketry came during World War II in barrage rockets. All our modern double-base propellants are descended from JPN, of World War II vintage. These propellants may be cast or extruded, use relatively nontoxic ingredients which are relatively insensitive to water, do not use petroleum derivatives to any appreciable extent, and are easily inhibited by solvent bonding an inert material to the surface of the propellant. It is possible to attain very low temperature coefficients of burn rate with these propellants; it is also possible to attain a double-base propellant which declines in burn rate with pressure over a portion of its usable pressure range, as opposed to the more usual monotonic increase in burn rate. This "mesa" burning, as it is called, allows the interior ballistician to more easily stabilize the pressure and thrust of the rocket motor.

Double-base propellants tend to have low to moderate impulse levels and are limited in temperature range not only by a high glass transition temperature, but also by their tendency to soften, liberate nitroglycerine, and decompose at higher temperatures. These propellants tend to autoignite at low temperatures and tend to be explosive hazards. Their physical properties are essentially fixed by the properties of nitrocellulose and are not easily tailored to changing requirements.

Double-base propellants contain nitrocellulose of various nitration levels (usually 12.6 percent nitrogen), nitroglycerine, and a stabilizer. Various inert plasticizers are added to modify either the flame temperature or the physical properties of the propellant.

Composite modified double-base propellants offer the highest energy levels presently available with solid propellants. This energy comes at an extremely high price, however: the storage and operating temperature limits only differ from minimum to maximum by about 20°F. These propellants are also extremely shock-sensitive and have very low autoignition temperatures. They are limited in use to systems such as submarine-launched ballistic missiles, where the temperature and vibration conditions are closely controlled. Classically, these compositions have contained nitrocellulose, nitroglycerine, aluminum, ammonium perchlorate, and the explosive HMX, which serves both as an oxidizer and gas-producing additive.

Propellant Properties and Interior Ballistics

The thrust and pressure profiles of a rocket motor must be controlled in order to meet the design criteria noted earlier. With liquid and solid propellants, the nominal controls differ dramatically, but in the elimination of spurious pulses and the control of nozzle erosion, the two types of propellant are more similar than different.

Pressure and Thrust Control In design of a liquid-propellant rocket motor, the prime considerations are properly sizing the combustor and nozzle and metering the propellant flow to attain the desired thrust. If variations in thrust are desired, these may be carried out either through preprogramming the flow by orifice size or by pump pressure; similarly, the flow may be varied on command by the operator.

The situation with a solid rocket motor is less clear-cut. The rate of gas generation inside the rocket motor is controlled by the burn rate of the propellant and the amount of burn surface available.

The burn rate of the propellant may be varied by the addition of various chemical catalysts; iron compounds are sometimes added to composite propellants and lead compounds to double-base propellants in order to speed burning. Coolants such as oxamide are added to composite propellants and inert plasticizers are added to double-base propellants to slow burn rates. The burn rate of composite propellants may be changed by changing oxidizers or by modifying the size distribution of the oxidizer. In efforts to achieve ultra-high burn rates, silver or aluminum wires have been added to propellants to increase heat conduction and so speed burning.

The surface area of a propellant grain is controlled by its shape and by the amount and areas in which the propellant grain is inhibited. This surface area tends to change in configuration as the propellant burns. The propellant grain designer must evolve a grain configuration which yields the desired pressure-time and thrust-time history. Most often, the objective in grain design is to achieve a neutral to slightly regressive (constant to slightly decreasing) thrust-time history.

In some cases (as where a high-thrust phase is to be followed by a low- to moderate-thrust phase), it is necessary to develop a grain or grains with varying geometries and/or varying burn rates to achieve the desired thrust-time profile. These results may require such techniques as using tandem or coaxial grains with different geometries and/or different burn rates and perhaps different propellants.

The pressure within the combustion chamber depends not only on the rate of combustion, but also on the size of the nozzle throat. Throat size tends to decrease with heating, but the surface tends to wear away with the hot gases and embedded particles eroding material from the surface of the throat and nozzle exit cone. In some cases, this loss of material may serve to assist the interior ballistician in the quest for the ideal thrust-time profile.

Combustion Instability Rocket motors are sometimes given to sudden, erratic pressure excursions for a number of reasons. Liquid-propellant fuel may surge; solid propellant grains may crack; material may be ejected from the motor and temporarily block the nozzle. Sometimes these excursions cannot be explained by any of the above possibilities, but will be attributed to "unstable combustion." The causes of unstable or "resonant" combustion are still under investigation.

A number of techniques have been developed to reduce or eliminate the pressure excursions brought on by unstable combustion, but there is, as yet, no panacea.

Nozzle Erosion As the propellant gases pass out of the rocket motor through the nozzle, their heat is partially transferred to the nozzle. The heated nozzle material softens and tends to be eroded by the mechanical and chemical action of the propellant gases. The degree of erosion is heightened when a metal fuel is added to the propellant gases, particulate matter is contained in the gases, the gases are corrosive, or the propellant gases are oxidizing.

One technique widely used in reducing nozzle erosion is to add coolant to the propellant formulation. Ablatives are sometimes added to the nozzle or chamber ahead of the nozzle throat; gases generated by the ablating material form a boundary layer to protect the nozzle from the hot propellant gases in the core flow. In liquid propellant rockets, the fuel (where stability permits) may be used as a coolant fluid. Nozzle inserts are probably the most commonly used technique to limit nozzle erosion. The nozzle shell is usually made of aluminum or steel when inserts are used and the nozzle throat is made of heat-resistant material. For a small amount of permissible erosion, carbon inserts are used; when no erosion is acceptable, tungsten or molybdenum inserts are used.

7.2 CARBONIZATION OF COAL AND GAS MAKING
by Klemens C. Baczewski

REFERENCES: Porter, "Coal Carbonization," Reinhold; Morgan, "Manufactured Gas," J. J. Morgan, New York. Wilson and Wells, "Coal, Coke and Coal Chemicals," McGraw-Hill. Elliott, High-Btu Gas from Coal, *Coal Utilization*, Dec. 1961. Kirk-Othmar, "Encyclopedia of Chemical Technology," 2d ed. "1966 Gas Engineers Handbook," The Industrial Press. Elliott, "Chemistry of Coal Utilization," 2d supp. vol., Wiley, 1981. "Marks' Standard Handbook for Mechanical Engineers," 10th ed., McGraw-Hill. "The Making, Shaping and Treating of Steel," 11th ed.," "Ironmaking," AIST (formerly AISE), 1999. DOE/NETL 2004 Gasification Database. ASTM Standards (vol. 05-06).

Carbonization of coal is the process for the production of coke for metallurgical, gas-making, general fuel purposes, and gas for indus-

trial and public-utility use. The process heats coal in the absence of air, yielding solid coke and coke oven gas, containing the organic volatile matter. In by-product plants, the gases contain tars; light oils, mostly benzene and its homologs, used for motor fuels and chemical synthesis; ammonia, usually recovered as ammonium sulfate, used mostly for fertilizer; tar acids (phenol), tar bases (pyridine), and various other chemicals. Residual coke oven gas is used to provide oven heating. In heat recovery systems, the gases are burned inside the ovens and flues, and the hot gas is used to generate steam for power and export. Developments in new designs, pollution control equipment, and the production of formed coke are discussed.

Table 7.2.1   Analyses of Cokes

| | "As-received" basis | | | | | | | | | High heat value, Btu/lb† |
| | Proximate, % | | | | Ultimate,% | | | | | |
Coke type	Moisture	Volatile matter	Fixed carbon	Ash*	Hydrogen	Carbon	Nitrogen	Oxygen	Sulfur	
By-product coke	0.4	1.0	89.6	9.0	0.7	87.7	1.5	0.1	1.0	13,200
Beehive coke	0.5	1.2	88.8	9.5	0.7	87.5	1.1	0.2	1.0	13,100
Low-temperature coke	0.9	9.6	80.3	9.2	3.1	81.0	1.9	2.8	1.0	12,890
Pitch coke	0.3	1.1	97.6	1.0	0.6	96.6	0.7	0.6	0.5	14,100
Petroleum coke	1.1	7.0	90.7	1.2	3.3	90.8	0.8	3.1	0.8	15,050

* Ash is part of both the proximate and ultimate analyses.
† btu/lb × 2.328 = kj/kg.

Gas making is the conversion of coal and biomass, liquid hydrocarbons, and petroleum residues into synthesis gas for chemical production, and for use as fuel for power generation. Systems in use include fixed-bed, fluid-bed, and entrained flow technologies.

CARBONIZATION OF COAL

Coke is the infusible, cellular, coherent, solid material obtained from the thermal processing of coal, pitch, and petroleum residues, and from some carbonaceous materials. This material has a characteristic structure resulting from the decomposition and polymerization of a fused or semiliquid mass. Specific varieties of coke, other than those from coal, are distinguished by prefixing or qualifying word to indicate their source, such as *petroleum coke* and *pitch coke*. A prefix may also be used to indicate the process by which coke is manufactured, e.g., *coke from coal, slot oven coke, beehive coke, gashouse coke,* and *form coke*. See Table 7.2.1.

High-temperature coke (1,650 to 2,000°F; 900 to 1,095°C) for blast-furnace or foundry use is the most common form in the United States. In 1982, almost 100 percent of the production of high-temperature coke was from slot-type ovens, with minimal quantities being produced from nonrecovery beehive and other types of ovens. Blast furnaces utilized 93.1 percent; foundries, 4.3 percent; and other industries, the remainder. With the development of new heat recovery designs, significant quantities from these units are being used. In 2003, about 10 percent of the total U.S. consumption was produced in heat recovery systems.

Low- and medium-temperature cokes (900 to 1,400°F; 480 to 760°C) have limited production in the United States because of a limited market for low-temperature tar and because the coke is not suitable for blast-furnace use.

A high-quality coke provides support for a smooth descent of the blast-furnace burden, while limiting the addition of impurities, providing thermal energy for high metal reduction, and allowing optimal permeability for gas flow up and molten products down. Below the cohesive zone off the furnace, coke is the only solid material, providing strength to support the burden while leaving void spaces. A large size with a narrow range, a high coke strength after reaction (CSR) with CO_2, and ASTM stability are the most important property parameters. These properties impact furnace productivity and coke usage. Average values of coke rates were 0.611 ton per ton of blast-furnace iron in 1975 and 0.444 in 1997. Representative values desired in North America are CSR at 61 or higher, and coke reactivity index (CRI) at 22 or lower. Other tests for determining the physical properties of blast-furnace and foundry cokes that are cited in export specifications are the MICUM, IRSID, ISG, and JIS methods. ("The Making, Shaping and Treating of Steel," (MSTS) 11th ed., "Ironmaking," AIST (formerly AISE), 1999. Curilla et al., Development of Coke Properties for the Improvement of Blast Furnace Operations, *Iron & Steel Technol.,* March 2004, pp. 85–90.)

During the by-product coking process, additional products of commercial value are produced. The products include fuel gas with a heating value of 550 Btu/ft³ (20,500 kJ/m³); tar and light oils that contain benzene, toluene, xylene, and naphthalene; ammonia; and phenols. Heat for the ovens is provided by the fuel gas, supplemented by blast-furnace and natural gas as needed. See Fig. 7.2.1. (MSTS, ibid.)

Foundry Coke   Requirements are somewhat different from those for blast-furnace coke. Chemically, in the cupola the only function of the coke is to furnish heat to melt the iron. Foundry coke should be of large size, generally a 10 : 1 ratio to cupola diameter, typically (1) 6 in by 9 in, or 152 mm by 228 mm; (2) 4 in by 6 in. or 101 mm by 152 mm; and (3) 4 in by 9 in, or 101 mm by 228 mm, It must be strong enough to prevent excessive degradation by impact of the massive iron charged into the cupola shaft. The following characteristics are desired in foundry coke: volatile matter, not over 0.7 percent; fixed carbon, not under 92 percent; ash, not over 7 percent; sulfur, not over 0.65 percent; and moisture, not over 1 percent. The 1½- and 2-in shatter indexes, which are a measure of the ability of coke to withstand breakage by impact, should be 97 or higher. (MSTS, ibid.)

Pitch coke is made from coal tar pitch, whereas **petroleum coke** is made from petroleum refining residues. Both are characterized by high-carbon and low-ash contents and are used primarily for the production of electrode carbon.

Coke consumption in the United States was approximately 28.9 million tons in 1989, 24.7 million tons in 1993, and 19.4 million tons in 2003. The major user is blast-furnace operations, with others, mostly foundry coke, using about 1.5 million tons. (EIA-DOE Annual Energy Review 2003, Sept. 2004.) This trend is expected to continue as electric-arc furnace (EAF) use increases and as direct reduction and coal injection systems are installed. (*Chem. Engrg.,* March 1995, p. 37.)

High-temperature carbonization or coking is carried on in ovens or retorts with flue-wall temperatures of ± 1,800°F (980°C) for the production of foundry coke and up to ±2,550°F (1,400°C) for the production of blast-furnace coke. Typical yields from carbonizing 2,204 lb [1.0 metric ton(t)] of dry coal, containing 30 to 31 percent volatile matter, in a modern oven are coke, 1,590 lb (720 kg); gas, 12,350 ft³ (330 m³); tar, 10 gal (37.85 L); water, 10.5 gal (39.8 L); light oil, 3.3 gal (12.5 L); ammonia, 4.9 lb (2.22 kg).

Coal Characteristics   Despite the vast coal reserves in the United States, most of the coal is not coking coal. Coking coals are only those coals which, according to the ASTM classification by rank, fall into the class of bituminous coal and are in the low-volatile, medium-volatile, high-volatile A or high-volatile B groups and which, when heated in the absence of air, pass through a plastic state and resolidify into a porous mass that is termed *coke*. In determining if an unknown coal is a coking coal, prime importance is placed upon obtaining a freshly mined sample since all coking coals experience oxidation or weathering, which can cause a loss in coking ability.

Coal Blending   The use of a single coal to produce strong metallurgical coke without resultant high-coking pressures and oven wall damage is very rare, and coke producers rely on the blending of coals of varying coking properties to produce strong coke. It is common practice in the coking industry to mix two or more coals to make high-quality coke or to avoid excessive expansion pressures in the oven. High-volatile coals tend to shrink during coking, while low-volatile coals tend to expand. Coal blends are designed to provide a coke of the proper

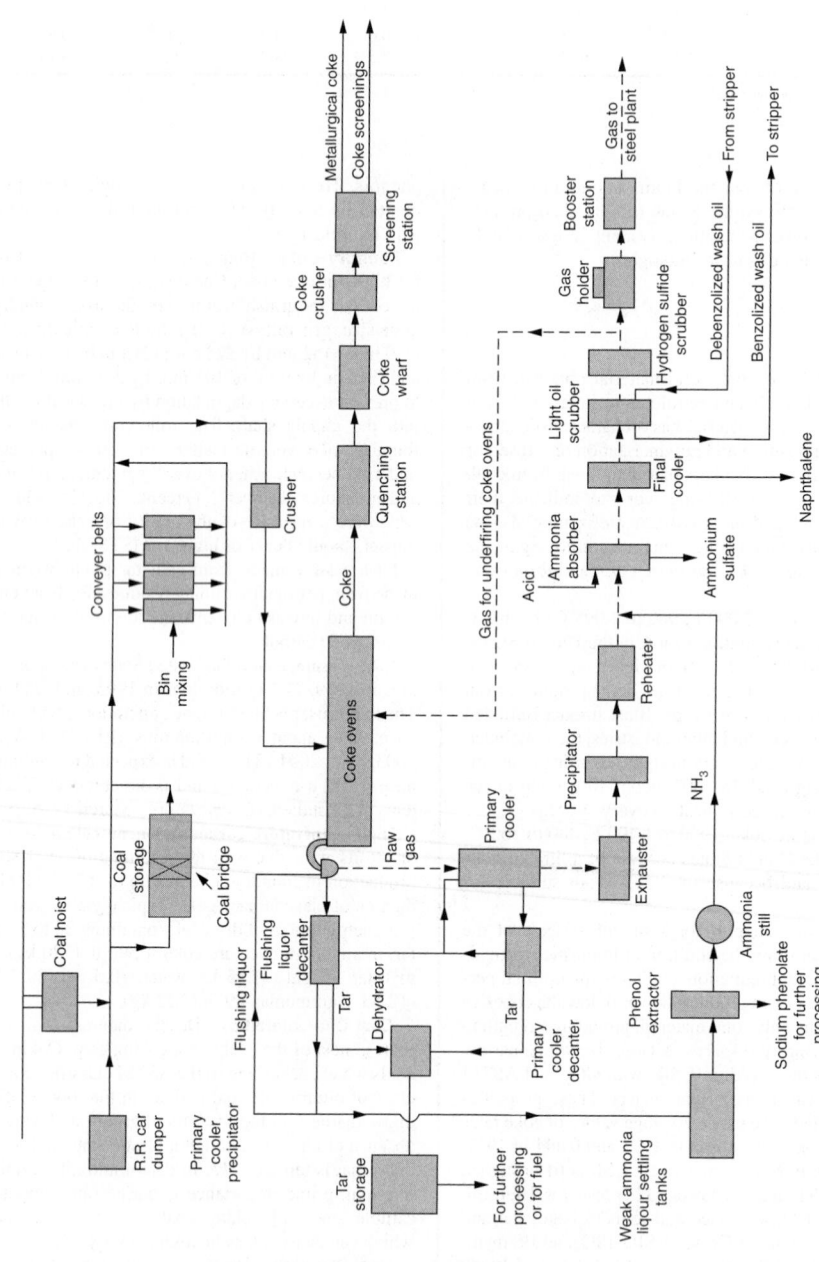

Fig. 7.2.1 By-product coke plant flow diagram.

chemical, physical, and oven pushing properties, and to make use of low-value carbon sources. Selection of candidate coals starts with a survey of vendor data and laboratory tests of the chosen samples. Prediction of coke quality is done from correlations of coal properties developed from pilot and plant experience. Pilot-plant carbonization and plant trials are run next before adoption of the blend. (MSTS, ibid.)

Laboratory tests are used by coal investigators to determine whether particular coals have coking properties and how they can best be used to make coke. Tests are mostly defined in ASTM Standards (vol. 05.06). Coal characteristics are grouped into **chemical** (volatile matter, ash and ash composition, and oxidation), **physical** (sieve size, Hardgrove grindability, ash fusion temperature, bulk density), **rheological** [free-swelling index, Geisler plasticity (fluidity), Arnu dilatometer (swellings)], and **microscopic** [petrographic (maceral analysis, vitrinite reflectance), nonmaceral analysis]. These data are used to predict coal blend properties, using graphical and mathematical correlations.

Important **chemical properties** of coke are moisture, fixed carbon, ash, sulfur, phosphorus, and alkalies. These are dependent on the coals used. **Physical properties** are high stability (ASTM Tumbler test) and coke strength after reaction (CSR) (Japanese reactivity), low coking pressure, and high mass contraction. CSR (ASTM D-5341) is an indicator of coke strength in the blast furnace and is determined by reacting coke with CO_2 at 1,100°C and tumbling the sample. The percent of +9.5-mm sizes surviving is the CSR. The weight loss on reaction is termed the *coke reactivity index* (CRI). Free-swelling index. (ASTM D720) is a fast method to determine if a coke will form a coherent mass. Gieseler plastometer (ASTM D-2639) is the most popular of several dilatometer testers that measure the fluid properties of the coal through the plastic state and into the solidification phase. It is important that temperature intervals of plasticity overlap for coals in a blend. It is generally agreed that the test is useful, although some investigators feel it only indicates that a coal is coking. Others feel that the temperatures at which the coal begins to soften and then resolidify serve as guides to establish which coals are suitable for blending. Still others couple the use of these data with petrographic analysis to predict coke strengths of various blends. While Gieseler is widely used in the United States, two other methods, Audibert-Arnu and the Ruhr test, are frequently used throughout the world. Petrographic composition is determined by the examination of coal under a microscope. Macerals are identified as reactive, semi-inert and inert. Reflectance measurements then are used to determine coal rank. (MSTS, ibid.; Curilla, ibid.).

Pilot-Scale Tests A number of designs of pilot-scale ovens that closely approximate commercial slot coke ovens have been developed. A few have been used solely to produce coke for laboratory testing. Most have been designed with one fixed wall and one movable wall so that data about carbonization pressures in the oven during coking can be collected while coke is made for testing. These ovens usually hold between 400 and 1,000 lb (180 and 450 kg) of coal. Another test oven in use was developed by W. T. Brown (*Proc. ASTM*, **43**, 1943, pp. 314–316) and differs from the movable-wall oven by applying a constant pressure on the charge while heating the coal from only one side.

The European Cokemaking Technology Center (EEZK, Essen, Germany) has operated a minijumbo reactor, which has a chamber width of 34 in (864 mm) and length and height of 40 in (101.6 mm). A demonstration facility, the Jumbo Coking Reactor, is operational at 100 t/d. The chambers of this unit are 34 in (864 mm) wide, 32.8 ft (10 m) high, and 65.6 ft (20 m) long. The concept of such large chambers is supported by full-scale tests conducted on 17-, 24-, and 30-in-wide (450-, 610-, and 760-mm-wide) ovens and is based on preheated coal charging. (Berling et al., *51st Ironmaking Conf. Proc.*, Toronto, Apr. 1992.) A sole heated oven is used to measure contraction and expansion of a coal blend, using a 30-lb (13.6-kg) sample (Curilla, ibid.).

Coking Process Coal produces coke because the particles of coal soften and fuse together when sufficiently heated. Initial softening of the coal, as determined by plastometer tests, occurs at 570 to 820°F (300 to 440°C). At or near the softening temperature, gases of decomposition begin to appear in appreciable quantities, gas evolution

increasing rapidly as the temperature is raised. This evolution of gases within the plastic mass causes the phenomenon that finally results in the cellular structure that is so characteristic of coke. Further temperature increase and decomposition cause hardening into coherent porous coke.

Fig. 7.2.2 Diagrammatic illustration of the progress of carbonization and of composition of the plastic zone.

As a result of the low thermal conductivity of coal (less than one-sixth that of fire clay), and also of semicoke, heat penetrates slowly into the pieces and through the plastic layer, uniform plasticity or complete coalescence does not appear until the temperature (at the heated side of a softening layer) is considerably higher than the softening point of the coal. (See Fig. 7.2.2.)

The **plastic zone** moves slowly from the hot wall of the oven toward the center, at a rate first decreasing with the distance from the wall and then increasing again at the middle of the oven. For several hours after charging a red-hot oven, the center of the charge remains cool. The plastic layer's temperature variation, from one border to the other, is from 700 to 875°F (370 to 468°C), and its thickness is ⅜ to ¾ in depending on the coal, the charge density, and the oven temperature.

In the modem coal-chemical-recovery coke oven, the average rate of travel of the plastic zone is about 0.70 in/h (17.8 mm/per h), and the average **coking rate**, to finished coke in the center, is 0.50 to 0.58 in (12.7 to 14.7 mm); i.e., a 17-in (432-mm) oven may be run on a net coking time of 16 to 17 h. (See Fig. 7.2.3.)

Fig. 7.2.3 Temperature gradients in a cross section of a coal-chemical-recovery coke oven, 17 in wide, at about middepth on a 17-h coking time.

New designs of wide ovens with chamber widths of 21.6 in (550 mm) have been built. The coking time increases slightly, to about 23 h, but not in proportion to the width increase. (Beckmann and Meyer, *52d Ironmaking Conf. Proc.*, Dallas, Mar. 1993.)

The gases and tar vapors travel chiefly outward toward the wall from the plastic layer and from the intermediate partly coked material, finding exit upward through coke and semicoke. Exit through the center core of uncoked coal, except for a very small fraction of the early formed gases, is barred by the relative impermeability of the plastic layer. The final chemical products, including gas, are the result of secondary decompositions and interreactions in the course of this travel.

Average temperatures in various parts of the carbonizing systems, for modern rapid coal-chemical-recovery oven operation, are about as follows: heating flues, at bottom, 2,500 to 2,600°F (1,370 to 1,425°C); heating flues, upper part, 2,150 to 2,450°F (1,175 to 1,345°C); oven wall, inner side (average final), 1,850 to 2,100°F (1,010 to 1,150°C).

Temperature Effects during Carbonization   In industrial carbonizing, higher temperatures at the oven wall and in the outer layers of the charge produce higher gas yield and less tar. The gases and tar vapors change in quality and quantity continuously during the carbonizing period. The percentage content of hydrocarbons and condensibles in the oven gases decreases, and that of hydrogen increases. Passage through the highly heated free space above the charge has the effect of increasing the yield of gas and light oil (reducing, however, the toluene and xylenes) and lowering tar yield, with an increase of naphthalene, anthracene, and lowering tar yield, with increase of naphthalene, anthracene, and free carbon. The rate of decomposition of ammonia increases above 1,450°F (788°C). Modern ovens tend to exercise control of the temperature in this free space.

The progressive change in gas yield and composition during the carbonizing period for a good gas-making coal is about as in Table 7.2.2. Some gas yields and compositions are shown in Table 7.2.3.

Table 7.2.4   Heat Balance as a Function of the Volatile Matter Content*

	% Volatile matter (DAF)			
	23.8	26.5	28.7	33.2
Moisture, %	10.6	10.3	9.7	10.1
Consumption of heat, kcal/kg	492	496	522	559
Waste-gas loss, %	10.9	10.7	108	9.6
Surface loss, %	10.1	9.9	9.5	8.8
Total loss, %	21.0	20.6	20.3	18.4
kcal/kg	103.3	102.2	106.0	102.9
Effective heat (heat of coking), kcal/kg	388.7	393.8	416.0	456.1
Sensible heat in coke, kcal/kg	284.1	278.2	281.0	260.2
Sensible heat in gas, kcal/kg	144.1	152.3	164.9	188.3
Total sensible heat, kcal/kg	428.4	430.5	445.9	448.5
Heat of reaction (exothermic)	+39.7	+36.7	+29.9	−7.6

The heat of reaction is the difference between effective heat and loss by sensible heats.
* W. Weskamp, Influence of the Properties of Coking Coal as a Raw Material on High Temperature Coking in Horizontal Slot Ovens, Glückauf, 103 (5), 1967, 215–225.

The overall thermal efficiency of industrial coal carbonization (useful recovery of heat from total input of heat) is between 86 and 92 percent approximately. External sensible-heat efficiencies are between 65 and 80 percent approximately. Typical heat balances on coal-chemical-recovery ovens are shown in Table 7.2.4.

Heat Used for Carbonization   The total heating value of the gas burned in the flues to heat the ovens varies from 950 to 1,250 Btu/lb (528 to 695 kcal/kg) of wet coal carbonized in efficient installations, depending on the heat required by the individual coals in the blend. Producer and blast-furnace gases are called lean gases. When underfiring with lean gas, both air and gas are regenerated (preheated) in order

Table 7.2.2   Variation in Gas Yield During Carbonization

Period	Volume, m³/Mt	kcal/m³	Volume, ft³/ton	Btu/ft³	Approx composition, %		
					Hydrocarbons	Hydrogen	Oxides of carbon
First quarter	1,130	5,800	3,630	651	41	46	7
Second quarter	1,000	3,400	3,190	610	37	53	7
Third quarter	1,010	5,050	3,250	567	32	59	6
Fourth quarter	585	3,230	1,875	363	8	82	5

Table 7.2.3   Gas Yields and Composition from Various Coal Types with High-Temperature Carbonization

Coal	Temp in inner wall, °F (°C)	Gas yield, ft³/ton (m³/Mt)	Gas composition, %						
			Carbon dioxide	Carbon monoxide	Unsat'd hydro-carbons	Methane	Ethane, etc.	Hydrogen	Nitrogen
Pittsburgh bed, Fayette Co., Pa., V.M.* 33.6	1,950 (1,065)	11,700 (365)	1.3	6.8	3.2	31.1	0	56.5	1.1
Elkhorn bed, Letcher Co., Ky., V.M. 36.6	1,950 (1,065)	11,500 (358)	1.1	7.7	4.0	31.0	0.2	55.0	1.0
Sewell bed, W. Va., V.M. 26.5	1,950 (1,065)	12,000 (375)	0.7	5.5	2.5	26.5	0	64.8	1.0
Pocahontas no. 4, W. Va., V.M. 16.4	1,950 (1,065)	11,900 (372)	0.4	5.0	1.1	18.0	0	75.0	0.5
Illinois, Franklin Co., V.M. 32.1	1,950 (1,065)	12,000 (375)	3.8	14.5	2.8	21.0	0	56.9	1.0
Utah, Sunnyside, V.M. 38.8	1,950 (1,065)	12,600 (394)	3.0	14.5	3.7	26.0	0.5	51.3	1.0

* V.M. = percentage of volatile matter.

to get sufficient flame temperature for the flues. Natural, refinery, and I.P. gases have been used to a limited extent for underfiring. These and coke-oven gas are rich gases and are not preheated, as their flame temperatures are sufficiently high and regeneration would crack their hydrocarbons. Air is preheated in all cases.

CARBONIZING APPARATUS

The current trend in design of coal-chemical-recovery coke ovens is to larger-capacity ovens and improved oven-wall liner brick and wall design to afford increased heat transfer from the heating flues to the oven chamber. Formerly, ovens were usually about 40 ft (12 m) long and from 12 to 16 ft (3.7 to 5.0 m) high. Typically ovens are about 50 ft (15 m) long and 20 to 23 ft (6 to 7 m) high and hold a charge of 35 tons (32 mt) or more. Average oven width is 16 to 19 in (400 to 475 mm), usually 18 in (450 mm), with a taper of 3.0 to 4.5 in (75 to 115 mm) from the pushing end to the coke-discharge end of the oven. New designs, tending toward larger ovens, have been built with oven widths from 21.6 to 24.0 in (550 to 610 mm), 24.75 ft (7.85 m) high, and 51.7 ft (15.75 m) long. Coke production of the 21.6-in oven is about 47.3 tons (43 Mt) per charge. A coke plant in Schwelgern, Germany, started production in March 2003, with the largest ovens, which are 22.6 in (590 mm) wide, 23.2 ft (8.4 m) high, with a volume of 3,280 ft³ (93 m³). The larger oven yields more coke per push, reducing environmental problems while increasing productivity. The larger ovens resist swelling pressure, permitting the use of a wider variety of coals.(Worberg, "Latest Development and Technical Advances in Cokemaking," MET COKE World Summit, Chicago, Oct. 2004.)

Various oven designs are used, distinguished chiefly by their arrangements of the vertical heating flues and waste heat ducts. Two basic designs are used for heating with rich gas (coke-oven gas): (1) the gun-flue design, wherein the fuel gas is introduced via horizontal ducts atop the regenerators and then through nozzles which meter the gas into the vertical flues, and (2) the underjet design, which incorporates a basement, located underneath the regenerators, in the battery structure. Horizontal headers running parallel with the vertical heating walls convey the fuel gas via riser pipes through the regenerators to the vertical heating flues. One design recirculates waste combustion gas from the adjacent heating wall through a duct underneath the oven and regenerator by the jet action of the fuel gas through specially designed nozzles. This provides a leaner gas at the place where combustion occurs and affords a more even vertical heat distribution for tall ovens. Designs for lean gas (blast-furnace gas, producer gas) employ sole flues beneath the regenerators for introducing the fuel gas. Ports control the quantity of gas fed from the sole flues into the regenerators. All modern ovens use regenerators for preheating the combustion air and lean gas. Average waste gas (stack) temperatures range from 450 to 700°F (230 to 370°C).

Coke ovens are built in batteries of as few as 10 to over 100. The Schwelgern plant has two batteries of 70 ovens each. Ovens are arranged so that each row of heating flues, or wall, heats one-half of two adjacent ovens. Modern practice is to build batteries of the maximum number of ovens in a single battery that can be operated by a single work crew to optimize productivity.

Coal is charged from a larry car through openings in the top of the oven. Systems for charging preheated coal into coke ovens have been built to eliminate charging emissions, improve oven productivity, and use larger amounts of weakly coking coals. Two methods are used in drying and preheating coal: (1) two-stage flash drying and preheating (Rosin-Buttner) and (2) single-stage fluidized and entrainment (Cechar) systems. Both approaches use enclosed conveyors to transfer the coal to charging ports on the ovens. Some plants use hot larry units for charging. Heating is by direct firing to coal temperatures of 200 to 250°C (390 to 450°F), requiring flue gas cleaning as scrubbers. Individual systems vary in detail, utilizing features from different approaches of heating and charging. (MSTS, ibid.)

The Japanese have looked at preheating of coal and have decided that it is more effective to briquette a portion of the poorer-quality coals and mix them with the normal coal mixture being charged to the ovens. Briquettes usually make up about 30 percent of the blend, and many Japanese plants have adopted a form of briquette blending. Preheating

is becoming important again in the development of wider ovens and jumbo reactors, which are based on using this technique.

After coking, doors are removed from both ends of the oven, and coke is pushed out of the oven horizontally by a ram operated by a pusher machine, into the hot car. The car is moved during the operation to form a uniform depth of coke on the car floor. When pushed from the oven, the coke has a temperature of about 1,000°C (1,832°F) and has to be cooled. The usual method in the United States is to wet-quench the coke in the hot car by water sprays in a quench station. The water is recirculated, settled to remove solids, and reused. Dry quenching recovers the heat in the coke, to make steam. This is done by lifting the coke into chambers, where it is cooled by recirculating inert gas, which is then cooled in waste heat steam boilers. It has not been practiced in the United States because of technical problems and cost. It is used in Germany, Russia, and Japan. (MSTS, ibid.)

Computerized control of battery heating is being used for new facilities and is applied to existing ones to improve efficiency in the United States, Germany, and Japan. This system integrates control of variables such as charge weight and moisture, excess combustion air, flue temperature, and coke temperature. The performance improvement in one case was the reduction in heating requirements from about 1,500 Btu/lb (832 kcal/kg) dry coal to under 1,200 Btu/lb (666 kcal/kg) dry coal and the stabilization of coking times, ranging from 18 to 24 h (Pfeiffer, *47th Ironmaking Can/. Proc.*, Toronto, Apr. 1988).

A feed-forward system that integrates control of variables such as coal density, heat of carbonization and moisture, excess combustion air, flue temperature, and coke temperature is in use by U.S. Steel Corp. Since the heat of carbonization is difficult to measure accurately, the final coke temperature is used, as feedback, to compensate for changes in carbonization heat and battery efficiency. The use of the computer allows operators time to fine-tune oven temperature profiles which can only be done manually. Distribution of fuel gas flow to each oven is set by graduated fixed orifices, permitting use of central control of gas main pressure and stack draft. Data on gas flow rate and Wobbe index, and pressure, and stack oxygen and temperature are used to control gas main pressure and stack draft. (MSTS, ibid.)

A new development is the PROven® system for individual oven pressure control, by Uhde Gmbh. A water seal is provided on the discharge of the gooseneck to the collection main, the depth of which is controlled by an automated drain valve. During charging and early in the cycle with high gas evolution rates, the drain is open, providing a low-pressure drop for the high flow. As the coking process continues, and the gas rate drops, the drain closes and the water level rises, to maintain the pressure drop at the set point, keeping a uniform pressure at all oven doors. The low pressure levels throughout the cycle almost completely avoid emissions from doors, charging ports, and gooseneck lids. The system is in operation at the Schwelgern coke plant. The arrangement of the water seal and control components is shown in Fig. 7.2.4. (Worberg, ibid.)

Charging position

Fig. 7.2.4 PROven® control valve.

Coal Chemical Recovery The raw coke oven gas is first cooled by direct contact with aqueous flushing liquor as it exists the oven through the gooseneck connection to the collection main. It passes to the primary coolers for a second contact with flushing liquor. These steps condense most of the tar and water from the gas. Some ammonia is absorbed in the water, forming a weak ammonia liquor. Exhausters (usually centrifugal) follow and operate from 6- to 12-in (150- to 300-mm) water column (wc) suction at the inlet to 50- to 80-in (1,250- to 2,000-mm) wc pressure discharge. Electrical precipitators remove the last traces of tar fog. The gas, combined with ammonia vapor stripped from the weak ammonia liquor, then passes through dilute sulfuric acid saturators or scrubbers which recover the ammonia as ammonium sulfate. Phosphoric acid may be used as the absorbent to recover the ammonia as mono- or diammonium phosphate. Low-cost synthetic ammonia produced via reforming of natural gas and hydrocarbon has made recovery of anhydrous coke-oven ammonia uneconomical. Some recent coke plants have been designed with ammonia destruction units.

After direct cooling with water, which removes much of the naphthalene from the gas, the gas is scrubbed of light oil (benzol, toluol, xylol, and solvents) with a petroleum oil. The enriched petroleum oil is stripped of the light oil by steam distillation, and the light oil is usually sold to the local oil companies. Phenols and tar acids are recovered from the ammonia liquor and tar. Pyridine and tar bases are recovered from the ammonia saturator liquor and tar. Distillation of the tar produces cresols, naphthalene, and various grades of road tar and pitches. Acid gases (hydrogen sulfide, hydrogen cyanide) are removed from the gas. The hydrogen sulfide is converted to elemental sulfur. The cyanogen may be recovered as sodium cyanide. The residual gas is used to fire the ovens, for other plant uses, or is sold to local utility. See Fig. 7.2.1.

Pollution Control Enactment of pollution control laws has motivated many developments of design, control devices, and new operating techniques. The continuous emissions caused by leakage from oven doors, charging hold lids, standpipe lids, and oven stacks are being controlled through closer attention to operations and maintenance. Emissions from charging are being contained by the use of a second collecting main, jumper pipes, and U-tube cars. The trend toward larger ovens has decreased emissions by reducing the number of coke pushes, having fewer doors, charge ports, and gas standpipes, and smaller length of door seals. Improved oven pressure control on slot ovens, and operating heat recovery units under a vacuum, minimizes emissions.

Pushing emissions equipment includes hood-duct arrangements with wet scrubbers or baghouses, mobile hoods with wet scrubbers, and coke side enclosures with baghouses. In current practice, a traveling hood is used, integral with the coke transfer car, operating with all coke side functions.

In the coal chemical **recovery** area, emissions are controlled by use of seals on pump and compressor shafts, and valve stems. Piping connections are welded to minimize flanged and screwed fittings. Sulfur and ammonia discharges are controlled by processes that recover them in elemental or chemical forms.

Form coke—the production of shaped coke pieces by extrusion or briquetting of coal fines followed by carbonization—had been practiced for many years in the United Kingdom and Europe to provide a "smokeless" fuel primarily for domestic heating. Form coke for use in low-shaft blast furnaces was produced commercially from brown coal (lignite) at large plants in Lauchhammer and Schwarze-Pumpe, Germany.

A major development was the FMC coke process which was operating commercially at the FMC plant in Kemmerer, WY. Crushed coal is dried, carbonized, and calcined at successively higher temperature, while solids flow continuously through a series of fluid-bed reactors. The mix of coke and pitch is briquetted into pillow shapes up to 2-in sizes and is sent to curing ovens to be devolatilized and hardened into the finished coke product. The process was continuous, and fully contained, so that environmental controls were met with conventional equipment. Coal types suitable for processing include lignites and extend to anthracites. This permits the use of nonmetallurgical coals for coke production. The coke product has been tested in blast furnaces in a 20,000-ton trial at Indian Harbor Works of Inland Steel. Furnace operation was normal up to 50 percent of FMC product in the coke burden. The plant has been shut down.

The **Bergbau-Forschung process** involves devolatilization of low-rank coal to yield a hot char. The hot char is mixed with a fluid coking coal (about 70 percent char and 30 percent coking coal) which becomes plastic at the mixing temperature, and the mixture is briquetted hot in roll presses to produce "green" briquettes containing 7 to 8 percent volatile matter. The **Andt process** is similar to the Bergbau-Faschung process in that it uses about 70 percent noncoking coal with about 30 percent coking coal as binder. The noncoking coal component is conveyed pneumatically from bunkers and introduced at two locations into a horizontal, parallel-flight stream reactor heated by hot products of combustion. The two heated coals are fed by screw feeders into a vertical cylindrical mixer, and the mixture is fed to roll presses and briquetted. The **Consolidation Coal process** employs a heated rotary kiln to produce medium-temperature coke pellets from a mixture of char and coking coal. Pellets are then subjected to final high-temperature carbonization in a vertical-shaft unit. After years of other processes receiving attention there came the **Sumitomo process,** Japan; **Auscoke (BHP),** Australia; the **Sapozhnikov process,** Russia; and **INIEX,** Belgium. After a few years' work on these various processes, it became apparent that production rates, costs, and product quality could not compete with conventional by-product coke ovens, and the test facilities in the United States were shut down.

A program, to use more noncoking coals, increase productivity, save energy, and reduce emissions, was the SCOPE 21 project in Japan. The approach was to upgrade the coal coking quality by rapid preheating and increasing its bulk density. Density can be increased by drying and briquetting, which permits a blend ratio of poor coking coals of 50 percent instead of 20 percent. A pilot plant included fluid-bed drying, followed by a pneumatic preheater and briquetting machine. Prepared coal was transferred by an enclosed conveyor to a test oven, of conventional cross section but shorter in length (7.5 m high by 450 mm wide by 8 m long). Walls were of a super-density, high-conductivity brick as thin as 70 mm. Flue temperatures and coking times were lowered, while producing good-quality coke. Runs were made from March 2002 to March 2003. Commercial application would be continuous, briquetting fines from the fluid-bed dryer, then mixing with heated coarse coal before the pneumatic preheater to be fed to charging machines. (Uebo et al., Development of New Cokemaking Process, SCOPE 21, *AISTech 2004 Proc.* vol. one.)

A form coke project supported by the U.S. Department of Energy and Combustion Resources, Inc. over the past several years has developed a technology and process for making metallurgical-quality coke from alternate feedstocks, including by-product and waste carbonaceous materials. The patent-pending process blends and presses these carbon-containing materials into briquettes of specified size. This product is referred to as **CR Clean Coke** because pollutant emission levels are carefully kept low levels with little or no emissions during processing. A wide range of feedstock materials have been investigated in over 600 tests for run-of-mine coals (coking and noncoking) and waste coal fines of various rank with blend; of coal tars and pitches, coal and biomass chars, met-coke breeze, or petroleum coke. For various coal/pet-coke/tar feedstocks, CR has produced uniform-sized briquettes in commercial-scale briquettes in three nominal sizes: 1, 2, and 3 in (equivalent diameter). Feedstocks are dried if wet and ground to <60 mesh, blended, and briquetted, usually with a tar binder. Test briquettes were coked in an inert atmosphere for 1 to 7 h depending on size. These products were examined according to requirements for conventional met-coke use in a blast furnace (e.g., CRI, CSR, stability, hardness, moisture, volatile percent, fixed carbon, ash, sulfur, alkali, and P). Several formulations have met and frequently exceeded established met-coke specifications, with CSR values ranging from 62 to 79 (average 73.7) and CRI from 22 to 30 (average 26.6). A preliminary process design and cost estimate have been completed for construction and operation of a demonstration plant capable of producing 120,000 tons/yr of **CR Clean Coke.** Blast-furnace test plans are proposed. Tailoring of CR Clean Coke products to other prospective end users, including foundry, sugar, soda ash, and ferrometals industries, presents additional opportunities. Calciner off-gases are treated by using methods typical of chemical recovery plants as appropriate. (Smoot et al., Alternative Coke Production from Unconventional Feedstocks, Met Coke World Summit 2004, Chicago.)

Stamp charging is a process in which the entire coal charge is stamped, or compacted, into a single block and then pushed into the oven for coking. It is an old process and has been in use in Europe and India. The main advantage is the increased bulk density of the charge, permitting the use of relatively poor-quality coals to make good coke. It has never been used in the United States due to the abundance of good coking coals. (MSTS, ibid.)

Heat recovery coke plants were first built in the United States with three test ovens in 1960 in Vansant, VA. Successful operation led to further development by Sun Coke Co., resulting in three U.S. facilities with an installed capacity of 2,500,000 tons/yr in 2005. The plant at Indiana Harbor has a capacity of 1,300,000 tons/yr and exports 94 MW of electric power and 100,000 to 500,000 lb/h of steam. There are four batteries of 67 Jewell-Thompson ovens, charged with 42 to 45 tons on a 48-h cycle. The ovens are 12 ft wide and 47 ft long and horizontal, as in Fig. 7.2.5. Charging and pushing is by a single machine (PCM). Charging is done by a leveling conveyor on the PCM, depositing the coal from the leading edge as it enters the oven. This levels the bed, and the weight of the conveyor compacts the coal, which improves coke quality.

Fig. 7.2.5 Sun coke heat recovery oven.

After charging, the doors are installed and heating begins inside the oven and in the sole flues. Combustion of gas starts above the beds and is completed in the sole flues under the oven floor, with flue gas exiting in uptakes to the waste heat tunnel. Control is accomplished by monitoring crown and sole flue temperatures, adjusting air induced through door ports above the bed and air ports into the sole flues, and controlling the draft by flue gas uptake dampers. Induced air ports are adjusted manually, using settings predicted from repeatable cycle data. Overall operation is controlled by a distributed control system. Flue gas is drawn from the waste heat tunnel through 16 heat recovery steam generators (HRSGs), flue gas desulfurization (FGD), and baghouses to a discharge stack, by exhaust blowers. The HRSGs provide steam to turbogenerators and for export.

On completion of an oven cycle, the doors are removed, the coke guide is set in place at the coke side, and the PCM pushes the hot coke

into a hot car on rails. The car is taken to the baffled wet quench station and transferred to the coke wharf.

Sun Coke has tested and/or used more than 30 blends, with as many as seven coals, for use in blast furnaces. The heat recovery process makes coke which has, in most cases, better CSR and stability values than are predicted from recognized industry models. Actual values of coke stability and CSR were consistently higher than those of slot recovery ovens and movable-wall test ovens. Coke mean size and +2-in size percentages are greater. Actual 2-year mean values at Indiana Harbor are as follows: CSR, 70.0; stability, 62.6; ash, 8.56; sulfur, 0.55; VM, 0.44; mean size, 54.6 mm.

Air emission rates are substantially lower than those of a by-product plant for several reasons:
• The ovens operate under negative pressure (−0.30 to 0.50 in WC), so no door leaks occur.
• The coke discharge system has state-of-the-art emission controls, including a coke side shed with a bag filter evacuation system.
• Charging is done with a closed ventilation and bag filter system. Multiple charge ports are not needed, as the coal side door opening is used.
• The combustion system is an integral part of the process and is operated with time, temperature, and turbulence to destroy all hydrocarbons, and have low CO.
• H_2S and SO_2 are controlled by being combusted and captured in a lime-based FGD system common in coal-fired power plants.

No wastewater is discharged, as all machine cooling water is consumed in the wet quench. Solid wastes consists of the gypsum solids from the FGD system and the discharge from baghouses. (Barkdoll et al., Consistent Coke Quality from Sun Coke Co. Heat Recovery Cokemaking Technology, Coke Summit, Pittsburgh, PA, October 2001; personal communication, Sun Coke Co.)

GASIFICATION

Producer gas and carbureted water gas were common in Europe and the United States for many years and were based on coal and coke. These units gasified the solid fuels by the reaction of oxygen (as air or enriched oxygen) and steam. Oil injection was practiced to improve the heating value. The heating values of the producer gas were approximately 120 Btu/SCF (1,068 kcal/m³) for air-blow units, 250 Btu/SCF (2,225 kcal/m³) or more for oxygen-blown units, and as much as 500 Btu/SCF (4,450 kcal/m³) for the oil-carbureted units. The development of abundant natural-gas supplies and their widespread distribution have supplanted these processes.

Gasification is achieved by partial oxidation of carbon to CO (exothermic reaction). To obtain a mixture of CO and H_2, water is introduced, typically as steam, which reacts endothermically with the coal. The partial oxidation supplies heat to the endotherm. These reactions are described in detail in Elliott, "Chemistry of Coal Utilization" (2d suppl. vol., Wiley Interscience).

In 1974, an oil supply crisis combined with a distribution pinch on natural gas, and interest in converting coal to gaseous and liquid fuels was rekindled in the United States. A demonstration program for synthetic natural gas (SNG) and liquid fuels was established by the U.S. government as the Synthetic Fuels Corporation. As natural-gas distribution and the oil supplies improved, the urgency diminished, and the projects were canceled. A notable exception is the commercial-scale Lurgi plant, which was built as the Great Plains Project in North Dakota, and funded partly by the Department of Energy, to produce SNG. Dakota Gasification Co. operates this plant, which also produces ammonia.

Partial oxidation and gasification of heavy petroleum fractions has long been used to make synthesis gas (syngas) for chemical products. Adapting these systems to coal resulted in an Eastman Chemical Co. facility for acetyl materials from coal. Further clean coal technology studies showed coal gasification having advantages for power plants. Environmental regulations could be met more readily for sulfur emissions by treating a smaller stream than the flue gas from a fossil-fueled plant, while yielding a salable product as opposed to landfill material. The potential efficiency improvement in combined cycles with gas turbines and steam turbines would further reduce the size of the treated stream. Corollary benefits ensue in reducing CO_2 emissions and allowing the use of high-sulfur coals. This has led to the planning and construction of **integrated gasification combined cycle (IGCC)** plants for power in Europe and the United States.

As part of this continuity of interest in coal gasification, research on hot-gas filtration and desulfurization is in progress to reduce thermal losses in gas cooling. Other technologies, such as the demonstration of underground gasification and using biomass wastes, are being pursued.

The DOE/NETL 2004 Gasification Database lists 153 plants worldwide, 117 in operation and under construction and 36 in planning. The United States has 26 units, 20 operating and 6 in planning. Six are on coal or lignite, seven on petroleum coke, and the remainder on natural gas or petroleum. Two coal-fed IGCC units are in operation, and two are in the planning stage. Of the 26, seven are power plants, and the remainder are for chemicals, SNG, and liquid fuels. Twenty-five plants have entrained flow gasifiers, and one is fixed-bed.

Gasifiers come in three types:

1. Fixed-bed with the coal supported by a rotating grate
2. Fluidized-bed in which the fuel is supported by gaseous reactants
3. Entrained flow gasifiers that use very fine particles suspended in a high-velocity gas stream

Of the 153 plants in the Gasification Database, 10 are fixed-bed, 13 are fluid-bed, and 130 are entrained flow.

Fixed-Bed Gasifiers Representative units are **Lurgi, Wellman-Galusha, Koppers-Kerpley, Heurtey,** and **Woodall Duckham.** Early units were air-blown and used coke, with a number built for anthracite. Agitators were added to permit feeding of bituminous coal. Coal feed is a sized lump, with 1½ in (37 mm) being typical. Fines are tolerated only in small amounts. Early units operated at essentially atmospheric pressure, but later units operate at elevated pressure. Temperatures are limited to avoid softening and clinkering of the ash. With air, gas hav-

ing heating values of 120 to 150 Btu/SCF (1,068 to 1,335 kcal/m³) was produced. Use of oxygen allowed gas heating values of 250 to 300 Btu/SCF (2,225 to 2,673 kcal/m³). Coal is fed through lock hoppers to prevent loss of gases and to permit charging into pressurized units. The gasifiers have a distinct upper reduction zone, which dries and preheats the coal. The gasification reactions take place at 1,150 to 1,600°F (620 to 870°C), and the gases leave the unit at 700 to 1,100°F (370 to 595°C). Under these conditions, methane and other light hydrocarbons, naphtha, phenols, tars, oils, and ammonia are generated. The CH_4 is an advantage in SNG production. Tars and oils are removed from the gas stream before further processing to absorb ammonia and acid gases, including carbon dioxide and hydrogen sulfide. The devolatilized coal passes into the lower combustion zone, reaching temperatures of 1,800 to 2,500°F (980 to 1,370°C), depending on the ash softening temperatures. Ash is removed by a rotating grate through lock hoppers.

The **Lurgi process** advanced this concept of a pressurized, oxygen-blown system. Gasifier pressure is 350 to 450 psig (24 to 31 bar). Typical composition of gas from the gasifier with oxygen blowing is as follows:

	Vol % dry basis
C_2H_4	0.42
C_2H_6	0.62
CH_4	11.38
CO	20.24
H_2	37.89
N_2 + A	0.33
CO_2	28.69
H_2S + COS	0.49

After tars and condensibles are removed from the raw gas, the CO shift reactor adjusts the H_2/CO ratio. The H_2/CO ratio is adjusted to values suited for the production of synthetic natural gas, for other chemicals or ammonia if H_2 is suitably optimized. The acid gases (CO_2, H_2S, COS) are removed after the shift reaction prior to synthesis as methanation for SNG. A flow diagram of the process is shown in Fig. 7.2.6. This process has been used in a number of commercial plants, at Sasolburg and

Fig. 7.2.6 Lurgi process.

Secunda in South Africa, in North Dakota, United States, and at Schwarze Pumpe in Germany. The Sasolburg and Secunda plants have 97 gasifiers, utilizing >30 million tons/yr of bituminous coal to produce fuels and chemicals by the Fischer-Tropsch process. The Dakota Gasification Co. plant uses 18,500 tons daily of lignite, in 14 gasifiers, making 130 million SCFD of SNG, and up to 1,200 tons/d of ammonia, plus other products. The Schwarze Pumpe unit has seven gasifiers treating wastes, such as plastics, sewage sludge, rubber, contaminated wood, paint residue, and household waste. Units are also in operation in the Czech Republic and China. (Rudolph, *Oil & Gas Jour.*, Jan. 22, 1973; van Dyk et al., Sasol's Unique Position, Gasification Technologies Conference, Washington, DC, Oct. 2004; Dittus and Johnson, The Hidden Value of Lignite Coal, Gasification Technologies Conference, San Francisco, CA, Oct. 2001.)

A later development is the **British Gas/Lurgi slagging gasifier (BGL)**. Coal is fed with a size distribution of 2 in by 0 with up to 35 percent minus ¼ in (fines). The operation is similar to that of the dry bottom unit except that molten slag is removed through a slag tap, is water-quenched, and is discharged through a lock hopper. Tars, oils, and naphtha can be recycled to the gasifier. Gas composition differs from that of the conventional dry ash unit in that water vapor, CO_2, and CH_4 are lower, and CO is higher, resulting in cold gas efficiencies of 88 percent or more. The unit was tested extensively on a variety of coals including caking types at the British Gas Town Gas Plant in Westfield, Scotland.

A BGL unit has been installed at Schwarze Pump, in Germany, and has been tested on hard coal and lignite. Also gasification of mixtures of feeds with a typical composition in weight percent of 12.5% hard coal, 7.5% lignite, 25% RDF pellets, 45% plastic waste, and 10% combined industrial and sewage sludge has been demonstrated. The gasifier has an internal diameter of 12 ft (3.7 m) and has operated on all coal at 41 t/h. (Sander et al., Operating Results of the BGL Gasifier at Schwarze Pumpe, Gasification Technologies Conference, San Francisco, Oct. 2003.)

Fluid-Bed Gasifiers The **Winkler** and **Kellogg KRW** (formerly Westinghouse) gasifiers are of this design type. Since the bed is truly fluidized, it permits the flexibility of processing solids such as coal and coke. Particle sizes of ⅛ to ⅜ in (2 to 10 mm) are required. Lock hoppers are the method of coal feed. The Winkler has been commercialized in a number of plants (Banchik, Clean Fuels from Coal, *IGT Symp.*, Chicago, Sept. 1973). Caking coals may be preoxidized to avoid agglomeration. The mixing of the bed causes a uniform temperature, so that the distinct regions of oxidation and reduction of the fixed-bed units are absent. The combination of retention time and temperature

cracks tars and oils to CH_4, CO, and H_2. To avoid agglomeration by softening of the ash and loss of fluidization, temperatures are limited to 1,800 and 2,000°F (980 and 1,095°C). Because of the relatively low temperature, these units are primarily useful with reactive coals such as lignite and subbituminous.

In 1986, Rheinbraun started a **high-temperature Winkler (HTW)** demonstration plant in Berrenrath, near Cologne, with a capacity of 25 t/h dried lignite, operating at 150 psig (10 bar). Syngas production was 1.3 million SCFH (34,000 N · m^3/h), having a thermal capacity of 140 MW. The plant used oxygen and steam as the gasifying and fluidizing medium. Both watertube and firetube configurations were tested for gas cooling followed by a hot-gas filter unit. A shift converter and Rectisol acid gas removal prepared the gas for methanol synthesis. Gas cooling by watertube units is suggested for steam pressures above 2,100 psia (150 bar) and firetube units at lower pressures. The plant configuration is shown in Fig. 7.2.7.

From start-up through 1997, operating time was 67,000 h, with an availability of 84 percent in the last 10 years. Further, good experience has been obtained on wastes such as plastics, biomass, municipal solid waste, and sewage sludge.

The hot-gas filter replaced the water scrubber originally installed for particulate removal from the raw gas. It was developed to allow dry dedusting of product gas, and it eliminates handling a wet sludge stream. The filter was successfully tested for 15,500 h.

Another pressurized HTW plant was operated at Wesseling, Germany, from 1989 to 1992, at a coal rate of 7 t/h, at a pressure of 350 psig (25 bar). Operating time was 10,000 h, and tested both air and oxygen as gasification agents.

The HTW process is suited for lignites, subbituminous coals, and reactive bituminous coals, which represent major solid fuels in many countries, and is a viable approach for IGCC facilities. (Renzenbrink et al., High Temperature Winkler Coal Gasification, Gasification Technologies Conference, San Francisco, CA, 1998.)

The Kellogg gasifier is an extension of the air-blown design developed by Westinghouse, which used limestone or dolomite to capture sulfur in the bed, similar to fluid-bed combustion. Operating conditions are 1,900 to 2,000°F (1,040 to 1,050°C), at pressures to 300 psig (21 bar). Product gas is removed through cyclones, where carbon dust and ash are collected and recycled to the gasifier. As the gas has a residual concentration of H_2S and COS, a regenerable hot-gas desulfurization (HGD) system with zinc reagents is included, recovering the sulfur as SO_2, obviating the need for H_2S recovery. This HGD unit is based on petroleum catalytic cracking technology. The SO_2 in the off-gas is captured by limestone in the sulfator-combustor. Remaining fines are

Fig. 7.2.7 Flow sheet of the HTW demonstration plant

removed from the product gas by ceramic filter candles. The gases have a heating value of 138 Btu/SCF (1,230 kcal/m³) and fuel the gas turbines in a combined cycle.

The residual solids in the gasifier contain carbon, sulfided sorbent, and ash. With the alkaline components, the smaller particles tend to agglomerate, and fall, while char particles rise into the reaction zone. The solids, called **lash**, contain sulfided components which are converted to sulfates after leaving the gasifier in a sulfator/combustor.

This design was selected by the Sierra Pacific Power Company for commercial demonstration as a 100-MW IGCC plant as the Piñon Pine Project at the Tracy Station near Reno, NV.

It is not possible to report on the performance of the system under typical operating conditions, as the gasifier and hot-gas cleanup system had not been successfully operated for more than 24 h at a time. The first start-up attempt took place on January 18, 1998, and the eighteenth and last on August 10, 2000. Major problems were not encountered with the gasification itself, but with cooling of the lash, the hot-gas filter, multiple fines removal systems, the fines combustor, and the refractory in the gasifier. While it would be desirable to have an air-blown gasifier, application of the KRW gasifier is dependent on demonstration of the hot-gas solids processing and handling systems. (DOE/NETL-2003-1183, Pinon Pine IGCC Power Project, A DOE Assessment, Dec. 2002.)

Entrained Flow Gasifiers These have been developed as coal gasifiers and as partial oxidation units to produce synthesis gas from liquid hydrocarbons, petroleum residues, or coke. They are characterized by a short residence time (<1 s) with concurrent flow of the feed and gasifying agents. Operating temperatures are high, from 2,200 to 3,500°F (1,200 to 1,927°C), and pressures are as high as 80 bar (1,200 psig). Coal feed systems may be dry with lock hoppers and pneumatic transport, or water slurries of coal. Thus, any feedstock that can be pulverized and dispersed can be processed, including highly caking coals. Oxygen is used as the oxidant reactant to achieve the high temperatures required. This has the additional advantage of not diluting the product gas with nitrogen, particularly if the gas is for synthesis. Product gases are free of tars, condensible hydrocarbons, and phenols. Sulfur compounds such as H_2S and COS must be scrubbed from the gas. The high temperature results in slagging operation of these units. Commercial processes include **General Electric's GE IGCC** (was Texaco), **Shell, ConocoPhillips E-Gas** (was Destec), and **Future Energy.** All operate at elevated pressure. The **Koppers-Totzek** was an atmospheric unit, but it has been developed into a pressurized system, now the **PRENFLO** gasifier.

The **GE IGCC** gasification process is the application of the Texaco partial oxidation process to the use of a slurry feed of coal (60 to 70 wt %)

in water. The unit is operated at temperatures of 1,200 to 1,500°C (2,200 to 2,700°F) at pressures of 27 to 80 bar (400 to 1,200 psig). The slurry is fed with oxygen through a special injection nozzle into the refractory-lined gasifier to produce syngas, while slagging the ash. The syngas goes to a water-quenched unit or a waste heat boiler. The latter is typical for an IGCC facility. Fine ash is removed by a scrubber before conventional removal of H_2S. The molten ash or slag exits the gasifier, is water-quenched, and is removed through lock hoppers for disposal. The slag forms a glassy solid which is nonhazardous. Typical gasifier products are shown in Table 7.2.5. The cleaned gas is used for gas-turbine fuel or for **chemicals manufacture.** The plant configuration can be modified for optimal heat recovery for a power cycle or for maximum H_2 and CO generation for chemical production. A configuration for **power application** is shown in Fig. 7.2.8. The process was demonstrated at a commercial-size (110-MW) combined-cycle (IGCC) unit at the Coolwater project, which tested many coals during its operation from 1984 to 1989. Emissions of SO_2 from the Coolwater demonstration plant were as low as 0.076 lb/10⁶ Btu of coal (0.033 kg/10⁶ kJ) and of NO_x of 0.07 lb/10⁶ Btu of coal (0.03 kg/10⁶ kJ).

Under the Clean Coal Program of DOE, a facility to provide 250 MW at **Tampa Electric Co. Polk Station** began commercial operation in September 1996, with a configuration corresponding to Fig. 7.2.8.

Coals run successfully were from Pittsburgh seam, Indiana, Illinois, Kentucky, and Wyoming mines. Blends of roughly equal portions of coal and petroleum coke ran well. Lower than expected carbon conversion initially reduced thermal efficiency from the design value of 38.6 percent (HHV) to 35.4 percent (HHV). This was caused by the need to operate at a lower temperature to improve refractory life and reduce net operating costs. Recycle of all unconverted carbon has subsequently improved efficiency to 36.7 percent (HHV) and enabled production of slag that is used for cement manufacture, but this requires additional oxygen capacity.

The facility complies with all federal and state requirements for air, water, and solid emissions. Sulfur emissions are lower than those of a coal-fired plant as the sulfur is removed as H_2S more easily from syngas than as a dilute SO_2 from flue gas. Sulfuric acid is produced for sale. Polk's SO_2 emissions typically are 0.12 lb/10⁶ Btu of coal containing 3.5% sulfur. Polk's NO_x emissions are typically 0.05 lb/10⁶/Btu of coal. A hot-gas cleanup sulfur removal system, on a slip stream, was included in the design. Sorbent attrition was excessive during cold testing, precluding further development of hot-gas cleanup at Polk.

The availability of Polk's single gasifier was 50 percent during its first year of commercial operation in 1997. Gasification system availability

Table 7.2.5   Syngas Production from Various Carbonaceous Feeds (GE-IGCC)

Feed type	Pittsb. no. 8	French	Coal Utah	German	S. African	Petroleum coke Delayed	Fluid	Coal liquef. residue Molten	Slurry
Feedstock dry anal, wL%									
Carbon	74.16	78.08	68.21	73.93	65.60	85.50	85.98	68.39	68.39
Hydrogen	5.15	5.26	4.78	4.65	3.51	3.90	2.00	4.75	4.75
Nitrogen	1.18	0.85	1.22	1.50	1.53	1.50	0.98	0.98	0.98
Sulfur	3.27	0.47	0.37	1.08	0.87	5.50	8.31	1.87	1.87
Oxygen	6.70	8.23	15.69	5.85	7.79	0.10	2.27	2.21	2.21
Ash	9.54	7.11	9.73	13.01	20.70	0.50	0.46	21.80	21.80
Higher heating value									
Btu/lb	13,600	14,000	11,800	13,200	11,200	15,400	13,800	12,700	12,700
kcal/kg	7,540	7,780	6,570	7,330	6,220	8,550	7,665	7,060	7,060
Product composition mol %									
Carbon monoxide	39.95	37.36	30.88	39.46	36.53	46.20	47.14	46.31	33.48
Hydrogen	30.78	29.26	26.71	29.33	26.01	28.69	24.33	35.54	28.56
Carbon dioxide	11.43	13.30	15.91	12.59	15.67	10.68	13.16	6.41	13.09
Water	16.43	19.43	25.67	17.47	20.82	12.37	12.67	10.46	23.72
Methane	0.04	0.16	0.22	0.25	0.02	0.17	0.09	0.27	0.23
Nitrogen and argon	0.49	0.37	0.50	0.60	0.68	0.55	0.42	0.45	0.42
Hydrogen sulfide + carbonyl sulfide	0.88	0.12	0.11	0.30	0.27	1.34	2.19	0.56	0.50

Fig. 7.2.8 GE IGCC Gas Cooler Mode.

consistently improved to 80 percent in 2000 and has remained at that level ± 3 percent ever since. Plugging of the convective cooler, a minor heat exchanger in the overall heat recovery scheme, has been the most persistent problem which consistently costs 5 percent availability. Work is in progress to eliminate the pluggage, but if unsuccessful, the system will be reconfigured to eliminate the exchanger.

The plant demonstrated the technical and environmental viability of a large coal gasifier in an IGCC facility, and if continues to operate as a base load unit. (DOE/NETL–2004/1207, Tampa Electric IGCC Project, A DOE Assessment, Aug. 2004.)

Eastman Chemical Company began operation in 1983 of the Chemicals from Coal Facility, using technologies such as GE (Texaco) gasification and Linde AG Rectisol gas cleanup, and Eastman developed processes for chemical production from syngas. Initially, the plant was designed to produce approximately 500 million lb/yr of acetic anhydride and acetic acid to supply a portion of Eastman's acetyl raw material needs. The facility was expanded in 1991, and additional debottlenecking brought the capacity to the current level of approximately 1.14 billion lb/yr. Two gasifiers are campaigned to give gasifier system uptimes in excess of 98 percent with a maximum rate of approximately 1,350 tons/d of coal. The gasifiers are operated at 150 percent of the original design basis. The facility is now the sole source of raw materials for Eastman's acetyl stream and is operationally and economically a proven means of producing acetyl chemicals from coal. (Smith, Eastman Chemical Co., Chemicals from Coal Operations, Gasification Technologies Conference, San Francisco, CA, Oct. 2000.)

Variation in the configuration of other gasifiers are made to improve thermal efficiency. An example is the **ConocoPhillips E-Gas™** (was Destec) unit, which has two stages. The coal slurry is fed to both stages, with that going to the first providing the exothermic heat by partial oxidation by the oxidant. This is absorbed later, in the second stage, by the endothermic gasification of the coal, without oxidant. Operating

conditions of the first stage are approximately 1,450°C (2,600°F) at 27 bar (400 psig), the exit gas being about 1,040°C (1,900°F). The syngas from the gasifier is cooled, generating 1,600-psia (114-bar) steam used in steam turbines. Particulates are removed by a hot candle filter, and the gas is scrubbed to remove H_2S and other sulfur compounds before being fed to the gas turbine–generator sets. The E-Gas gasifier is illustrated in Fig. 7.2.9. The system had operated in a 160-MW facility at Dow Chemical, Plaquemine, LA, for more than 34,000 h on subbituminous coal from April 1987 through November 1995.

The E-Gas technology is being used at the SG Solutions LLC, Wabash River, plant to combine a 100-MW steam-turbine facility with a gas turbine to yield 265 MW on coal or on petroleum coke. Feed rates were 2,450 tons/d of coal and 2,000 tons/d of coke. Shortly after start-up, both gasification and power blocks ran at capacity and within all environmental parameters. Demonstration tests were conducted feeding bituminous coal from Peabody's Hawthorn and other local mines, as well as delayed coke from four refineries.

During the first 2 years of demonstration operation on coal, the net plant heat rate was 8,900 Btu/kWh (9,400 kJ/kWh) HHV, better than the design value of 9,030 Btu/kWh (9,530 kJ/kWh) HHV. Comparable typical values are 10,500 Btu/kWh (11,077 kJ/kWh) for a new coal-fired plant with SO_2 scrubbers. Emissions of SO_2 of 0.1 lb/10^6 Btu, 0.15 lb/10^6 Btu of NO_x, 0.40 lb/10^6 Btu of CO, and no detectible particulates are typical. Solids discharges from the plant are elemental sulfur and ash as a glassy frit, which are both salable products.

During the start-up year, 1996, availability was only 22 percent, primarily due to frequent failure of ceramic elements in the hot-gas filter, but also due to high chloride in the syngas, causing corrosion, degradation of COS hydrolysis catalyst, and excessive wear of gasifier refractory. Corrections, including using metallic filter elements, adding chloride scrubbing, and changing gasifier lining brick, resulted in a second-year availability of 60 percent. The facility is now operated as a base load plant, and at an average availability of about 75 percent from

High-pressure steam

Boiler feed water

Second stage

Product syngas

Coal slurry

First stage

Oxygen

Coal slurry

Oxygen

Coal slurry

Slag quench water

Slag slurry

Fig. 7.2.9   Conoco-Phillips E-Gas™ two-stage gasifier.

2000 to 2003. (DOE/NETL-2002/1164, Wabash River Coal Gasification Repowering Project, A DOE Assessment, July 2002. Personal communication from ConocoPhillips.)

Entrained flow gasifiers in commercial operation as IGCC facilities include the **Shell** and **PRENFLO** units. The **Shell** unit is used in the Boggenum plant in the Netherlands, with a feed of bituminous coal and an output of 253 MW. Puertollano, Spain, is the location of a **PRENFLO** IGCC facility that uses coal and petroleum coke and has an output of 298 MW. Both reactors are single-stage upflow designs. (TSR 008, Gasification of Solid and Liquid Fuels for Power Generation, Dept. of Trade and Industry, United Kingdom, Dec. 1998.)

New Developments   The research in coal conversion includes underground coal gasification (UCG) wherein the coal seam is both reactor and reactant in place. It has a history of development in the United States, United Kingdom, France, Belgium, Spain, Russia, China, and Australia. The technology depends on the evaluation of several characteristics: coal seam characteristics such as dip, thickness, partings, and rock lenses; coal chemistry; its agglomerating and free-swelling properties; water, ash, and sulfur content; boundary strata; overburden depth; faults; bulking factor; and hydrology. Two boreholes are required, one for introduction of oxidant and steam and one to remove the gas make. Connection between them is by one of the methods of (1) air pressurization, (2) artificial galleries, or (3) directional drilling with controlled injection. Method 3 is based on oil and gas technology. Gas processing is managed by the same technologies as aboveground systems. Environmental concerns are the impact on groundwater and surface issues including subsidence. Current operations and tests have provided reliable data.

The program in the United States was begun in Wyoming with a series of test burns, but was abandoned because of environmental problems. A test program in Spain (1992–1998) confirmed the feasibility of using coal seams of depths of 1,800 ft (550 m) with method 3, with movable injection points for oxidants. At Chinchilla, Australia, method one was used in a pilot operation (1999–2003) in a shallow, high-ash coal seam. The test was sufficiently successful for the process to be

offered commercially. The same technology was used in Siberia and Uzbekistan, where two commercial-size power plants are reportedly running (about 200 MWe). China has been running field trials since 1986, using method 2 in abandoned mine shafts, alternating air and steam injection as in earlier Russian practice. Five full-scale tests were underway in 2004, with the gas distributed for residential and industrial heating.

The United Kingdom has studied the feasibility of UCG to utilize deep seams using directional drilling. Of particular interest are undersea coal deposits, supported by offshore oil and gas experience. (Review of the Feasibility of UCG in the UK, Dept. of Trade & Industry (UK) Report, Sept. 2004.)

Gasification of Liquid Hydrocarbons   In the era of manufactured gas in the United States, both base-load and peak-shaving gases were produced by gasifying oils via thermal cracking techniques. Most of the processes produced gases having calorific values compatible with coal gas (coke-oven gas). As natural gas became available, some processes were modified to produce a high-heating-value gas interchangeable with natural gas. These oil-gas units operated on a cyclic (heat-make) basis. To supply the heat for the endothermic thermal cracking of oil, a mass of checker brick was heated to 1,300 to 1,700°F (705 to 925°C) by burning oil and deposited carbon with air. During the "make" cycle, steam and oil were introduced to produce a mixture of hydrogen, methane, saturated and unsaturated hydrocarbons, aromatic oils, tar, and carbon. The supply of natural gas throughout the world displaced the manufactured gas plants by virtue of lower cost, operational simplicity, and reduction of emissions. Some manufactured gas and coke-oven gas were distributed into the early 1980s, but this is no longer current practice. Increased demand for chemicals resulted in the development of processes for production of carbon monoxide, hydrogen, and carbon dioxide. These are for the chemical synthesis of ammonia and methanol, which are feedstocks for many other products. The processes include partial oxidation, catalytic reforming, and hydrogasification.

Partial oxidation processes were developed to produce syngas from liquid feedstocks of any weight, particularly heavy residual oils. These processes produce principally a carbon monoxide-rich gas which is reacted with water in a shift reactor to add hydrogen. Catalytic systems produce methane or other chemicals. Texaco and Shell developed their processes originally for these reasons and later adapted them to coal gasification.

Catalytic reforming of naphtha, natural-gas liquids, and LPG is presently applied commercially for the production of synthetic natural gas. Commercial processes available are the **CRG (catalytic-rich gas)**, Davy Process Technology; **MRG (methane-rich gas)**, Japan Gasoline Co.; and **Gasynthan,** BASF/Lurgi. Processes are similar in that each uses steam reforming of light hydrocarbons over a nickel catalyst. The product gas is a mixture of methane, carbon monoxide, carbon dioxide, and hydrogen. The four basic steps of the process are desulfurization, gasification, methanation, and purification (CO_2 removal and drying).

Up to 2005, **CRG** technology, as licensed by Davy Process Technology, has been used in around 50 applications for the production of substitute natural gas (SNG) from liquid hydrocarbon naphtha or LPG feedstock. Predominantly used throughout the United Kingdom, United States, and Japan, most of these were shut down as natural-gas reserves, and LNG supply chains became well established. A few small units remain operating in Japan for local captive use to meet specific demands. Stream capacities have ranged from less than 0.35 million SCFD (0.01 million N · m³/day) to around 126 million SCFD (3.5 million N · m³/day) for a plant built in the mid-1970s. Following on from this experience, recent increases in energy cost have led to renewed interest in such technology which continues to be developed. Future applications in the bulk methanation of product gas from both coal and biomass gasification are envisaged. (Davy Process Technology, personal communication.)

Hydrogasification The British Gas Corporation has developed the **GRH (gas recycle hydrogenator)** for hydrogenating vaporizable oils to produce synthetic natural gas, and the **FHB (fluid-bed hydrogenation)** of gasifying crudes or heavy oils for synthetic natural-gas production. In the **GRH** process, naphthas, middle distillates, and gas oils that need not be desulfurized are reacted directly with hydrogen-rich gas prepared by steam reforming a rich gas sidestream. In the **FHB** process, crude or heavy oil is preheated and atomized in the presence of coke particles fiuidized by a supply of preheated hydrogen-rich gas. Osaka Gas developed a process, **ARCH,** to hydrogasify coal, producing methane at high temperature (800 to 1,200°C; 1,470 to 2,190°F) and pressure (3 to 7 MPa; 435 to 1,015 psia). A test unit at 5 tons/d was operated, and a 50 tons/d pilot-plant design completed. The work was suspended in 2001, and the technology is available for further development. (Osaka Gas, personal communication.)

7.3 COMBUSTION FURNACES
by Glenn W. Baggley

REFERENCE: Trinks-Mawhinney, "Industrial Furnaces," vols. 1 and 2, Wiley.

FUELS

The selection of the best fuel should be based upon a study of the comparative prepared costs, cleanliness of operation, adaptability to temperature control, labor required, and the effects of each fuel upon the material to be heated and upon the furnace lining. Attention must be paid to the quantity to be burned in each burner, the atmosphere (fuel/air ratio) desired in the furnace, and the uniformity of temperature distribution required, which determines the number and the location of the burners. Common methods of burning furnace fuels are as follows:

Solid Fuels (Almost Entirely Bituminous Coals)

Coal was once a common fuel for industrial furnaces, either hand-fired, stoker-fired, or with powdered coal burners. With the increasing necessity for accurate control of temperature and atmosphere in industrial heating, coal has been almost entirely replaced by liquid and gaseous fuels. It can be expected that methods will be developed for the production of a synthetic gas (natural-gas equivalent) from coal.

Liquid Fuels (Fuel Oil and Tar)

To burn liquid fuels effectively, first it is necessary to atomize the oil into tiny droplets which then vaporize and burn. Atomization can be accomplished mechanically or with the aid of steam or air. With heavy oils and tar, it is important to maintain the proper viscosity of the oil at the atomizer by preheating the fuel.

For larger industrial burners, combustion air is supplied by fans of appropriate capacity and pressure. Combustion air is induced with some smaller burner designs.

Gaseous Fuels

Burners for refined gases (natural gas, synthetic gas, coke-oven gas, clean producer gas, propane, butane):

Two-pipe systems: Include blast burners (open or closed setting), nozzle mixing, luminous flame, excess air (tempered flame), baffle, and radiant-tube burners, all for low-pressure gas and air.

Premix systems: Air and gas mixed in a blower and supplied through one pipe.

Proportioning low-pressure mixers: Air and gas supplied under pressure and proportioned automatically (air aspirating gas or gas inspirating air). The resulting mixture is burned in tunnel burners, radiant-cup, baffle, radiant-tube, ribbon, and line burners.

Pilot flames are generally used to ensure ignition for gas and oil burners. Insurance frequently requires additional safety provision in two main categories: an interconnected pressure system to prevent lighting if any burner in a zone is open, and burner monitors using heat or light to permit ignition.

Burners for crude gas (raw producer-gas, blast-furnace gas, or coke-oven gas):

Simple mixing systems with large orifices and simple mechanisms which cannot become clogged by tar and dirt contained in these gases.

Separate gas and air supplies to the furnace, with all mixture taking place within the furnace.

TYPES OF INDUSTRIAL HEATING FURNACES

Heating furnaces are usually classified according to (1) the purpose for which the material is heated, (2) the nature of the transfer of heat to the material, (3) the method of firing the furnace, or (4) the method of handling material through the furnace.

Purpose Primarily a metallurgical distinction, according as the furnace is intended for tempering, annealing, carburizing, cyaniding, case hardening, forging, heating for forming or rolling, enameling, or for some other purpose.

Transfer of Heat The principal varieties are **oven furnaces,** in which the heat is transferred from the products of combustion of the fuel, in direct contact with the heated material, by convection and direct radiation from the hot gases or by reradiation from the hot walls of the furnace; **muffle furnaces,** in which the heat is conducted through a metal or refractory muffle which protects the heated material from contact with the gases, and is then transferred from the interior of the muffle by radiation to the heated material, which is sometimes surrounded

Fig. 7.3.1 Methods of firing oven furnaces.

by inert gases to exclude air; or **liquid-bath furnaces,** in which a metal pot is heated on the outside or by immersion. This pot contains a liquid heating or processing medium which transfers heat to the material contained in it. This type includes low-temperature tempering furnaces with oil as the heating medium, hardening furnaces using a bath of lead, hardening and cyaniding furnaces with baths of special salts, and galvanizing or tinning furnaces for coating the heating material with zinc or tin. The generally accepted form of muffle is the **radiant-tube** fired furnace, in which the fuel is burned in metal or refractory tubes which radiate heat to the charge. An important form of furnace for temperatures below 1,300°F (700°C) is the **recirculating** type, in which the atmosphere (products of combustion, air, or protective gases) is recirculated rapidly through the heating chamber. **Forced convection** heating is accomplished by a large number of jets of hot gas at high velocity. In **high-speed** heating (or **patterned combustion**), premixed burners are arranged for close application of heat, and with a high-temperature head, very rapid heating is achieved.

Method of Firing   This classification applies principally to the oven type of furnace, and it indicates whether the furnace is direct-fired, overfired, underfired, or heated by radiant tubes. Figure 7.3.1 shows the principles of each of these types. The **direct-fired** method finds increased utilization from constant improvement in the design and control of gas and oil burners, especially for temperatures above 1,200°F (650°C). In **overfired** furnaces, radiant burners fire through the roof and are arranged in patterns to obtain the best temperature distribution. The **underfired** furnace is excellent for temperatures between 800 and 1,800°F (ca. 400 to 1,000°C) because the heated product is protected from the burning fuel. The temperature and atmosphere can be closely controlled, but the temperature is limited by the life of the refractories to about 1,800°F (1,000°C). Many furnaces are now designed for the use of special protective atmospheres and involve the use of **radiant tubes** to avoid any contact with the combustion gases. These fuel-fired tubes of heat-resisting alloy may be horizontal across the furnace above and below the heated material or may be vertical on the sidewalls of the furnace.

Method of Material Handling   In the **batch** type, the heated material charged into the furnace remains in the same position until it is withdrawn after sufficient heating. In a **continuous furnace,** the material is moved through the furnace by mechanical means which include pushers, chain conveyors, reciprocating hearths, rotating circular hearths, cars, walking beams, and roller hearths. Continuous furnaces are principally labor-saving devices and may or may not save fuel.

SIZE AND ECONOMY OF FURNACES

The size of furnace required depends upon the amount of material to be heated per hour, the heating time required, the size of the pieces to be heated, and the amount of heat that can be liberated without excessive damage to the furnace. The efficiency and refractory life obtained depend upon the correctness of furnace size.

Heating Time   For the usual relation of refractory area to stock area, time to heat steel plate from one side for each $\frac{1}{8}$ in (3.18 mm) of thickness varies from 3 min for high-speed heating and 6 to 12 min for heating for forming by usual methods to 20 min for heat treating. Steel cylinders will be heated in one-half these times per $\frac{1}{8}$-in (3.18-mm) diam. Below 800°F (ca. 400°C), the time may be 2 to 3 times these

values. Brass requires about one-half as long as steel to heat, copper 40 percent as long, and aluminum 85 percent as long. The preceding heating times are based on a furnace temperature 50 to 100°F (ca. 25 to 50°C) higher than the final temperature of the heated material. It is assumed that the material is fully exposed to the heat of the furnace. Piling of material in a furnace lengthens the heating time by an amount that must be determined by actual trial. In addition to simple heating, there is frequently additional **time required for soaking** (holding at furnace temperature) to cause metallurgical changes in the material or for some other reason.

The **weight of material in the furnace** at any time is the product of weight of material per hour multiplied by the heating time in hours. If the weight and sizes of pieces involved are known, the **area of the furnace** can then be fixed. The width and length of the furnace to produce this hearth are fixed by the method of firing to be used and by the method of handling material.

The life of a furnace at given temperature depends upon the rate of heating, which may be expressed in pounds per square foot of hearth area per hour. The maximum allowable **rate of heating steel** is about 35 lb/(ft$^2 \cdot$ h) for heat treating, 70 lb for in-and-out rolling-mill furnaces, 100 lb for single-zone continuous furnaces, and 150 lb for multiple-zone furnaces. These are upper limits which should not be used if long life of furnace refractories is expected. These rates are for heating mild steel; they may be about twice as great when heating brass, 2½ times as great for copper, 0.7 as great for alloy steel, and 1.1 times as great when heating aluminum. These maximum allowable rates should be used only for checking the calculation of size, because some shapes and sizes of pieces cannot be properly heated when piled in such a manner as to produce these rates. If the calculated size of the furnace corresponds to a rate of heating that is too great, it should be reduced by making the furnace larger. If the rate is too small, it can sometimes be increased by piling material in a smaller furnace.

EXAMPLE.   To determine furnace size. If a furnace is required to heat 20 pieces per h weighing 30 lb each and requiring a heating time of $\frac{1}{2}$ h, the furnace must be large enough to hold $\frac{1}{2} \times 20 = 10$ pieces. If each piece requires an area of 2 ft^2, the area of the hearth will be $2 \times 10 = 20$ ft^2 for a single layer of pieces in the furnace. If the furnace is of the batch type, a size of 4 ft wide $\times$ 5 ft deep would probably be about right for convenient handling. On checking, the rate of heating is 20 pieces per h $\times$ 30 lb/20 ft^2 = 30 lb/(ft$^3 \cdot$ h). For this rate an under" furnace could be used if the pieces could be more densely piled without seriously interfering with the circulation in the furnace.

The heat released by the fuel in a furnace (heat input) is equal to the sum of the heat required in the heating process (useful heat) plus the heat losses from the furnace. **Heat input** includes the heat of combustion of the fuel, sensible heat in preheated air or fuel, and heat in the material charged. Low-heat values of the fuel are used, and the sensible heat can be calculated from the specific heats of the preheated air, fuel, or material. **Useful heat** includes the heat absorbed by the material in the furnace. Figure 7.3.2 gives heat contents for different metals. In the simple heating of metals, the useful heat applied to the metal includes only the heat absorption, as given in Fig. 7.3.2; but there are many processes that include other requirements, such as drying, where moisture must be heated and evaporated, heating of chemical products where heat is utilized to cause chemical changes, and other special cases. (See also Sec. 4.3.)

Fig. 7.3.2 Heat content of metals.

Fig. 7.3.3 Heat loss from thoroughly heated walls, based on interior area.

Heat losses in a heating furnace include heat lost in waste gases, radiation from and heat absorbed by refractories, heat carried out of the furnace by containers or conveyors, heat lost through openings, and heat in unburned fuel escaping with the products of combustion. The **heat contained in waste gases** depends upon the temperature of these gases as they leave the heating chamber. Table 7.3.1 gives the approximate percentage of heat contained in the flue gases from perfect combustion at different temperatures. These values are about the same for most fuels except producer gas and blast-furnace gas, the losses with which are higher than those given in the table.

Radiation and heat absorption by refractories depend also upon the rate of heating (which determines the interior temperature of the refractories) and upon the refractory area and thickness. Figure 7.3.3 shows the heat radiated through walls of different thickness at various furnace temperatures, for equilibrium conditions, when the wall has reached steady temperatures throughout (see also references at the beginning of this section and Keller, "Flow of Heat through Furnace Hearths," ASME). The **heat carried out by containers and conveyors** is the sensible heat content of these items as they leave the heating chamber. Such losses include the heat in carburizing boxes, pans, chain conveyors, and furnace cars. **Radiation from furnace openings** depend upon the size and shape of the opening and the thickness of the walls in which they are located, as well as upon the temperature of the furnace. Some idea of the magnitude of these losses is given by the values in Table 7.3.2.

The **heat lost in unburned fuel** escaping with the flue gases is small in most furnaces because the fuel can be almost completely consumed.

The **efficiency of an industrial furnace** is the ratio of the heat absorbed by the heated material to the heat of combustion of the fuel burned.

The magnitude of the various heat losses is indicated in Table 7.3.3. Column 1 is for a high-temperature batch-type billet-heating furnace, heating 4,200 lb of billets, per hour, a furnace load at a time, to 2,300°F, at a rate of 25 lb/(ft² · h), averaged over 10 h of operation, and with a fuel consumption of 30 gal of oil per ton of steel heated. Column II is for a large continuous billet-heating furnace of the usual pusher type with a flow of gases opposite to that of the steel, and operating at a rate of 60 lb of steel heated to 2,300°F/ft² of hearth area per hour. Column III is for an underfired batch-type furnace, heating steel to 1,600°F for annealing, at a rate of 30 lb/(ft³ · h).

Table 7.3.1   Average Heat in Waste Products of Combustion at Various Temperatures, Percent of Low Heat Value of the Fuel

Temp of gases, °F	1,000	1,200	1,400	1,600	1,800	2,000	2,200
Temp of gases, °C	540	650	760	870	980	1,090	1,200
% of low heat value in gases	24	28	34	38	45	50	55

Table 7.3.2   Radiation through Openings in Furnace Walls, kBtu/h

	Furnace temp, °F (°C)					
	1,400 (760)			2,200 (1,200)		
	Wall thickness, in*			Wall thickness, in*		
Size of opening, in*	4½	9	18	4½	9	18
4½ × 4½	1.4	1.1	0.8	5.1	4.1	2.8
9 × 9	7.8	6.1	4.5	28.5	22.7	16.8
18 × 18	37	30.5	24.3	137	114	90
24 × 24	71	60	48	264	225	180
36 × 36	173	150	124	650	560	465

* × 25.4 = mm.

Table 7.3.3   Heat Balances for Various Furnace Types, Percent of Heat of Combustion

Disposition of heat	Type I	Type II	Type III
Heat to material, or efficiency	16	49	23
Heat to refractories	20	17	22
Heat lost in flue gases	44	19	40
Heat to water cooling	—	5	—
Heat through openings	20	10	15

Table 7.3.4 gives average requirements in fuel of typical industrial heating furnaces. The values are for furnaces without heat-saving appliances (recuperators, regenerators, or waste-heat boilers) except as noted and show the efficiency and the Btu required in the fuel per net pound of steel heated. To obtain the average amount of any fuel required, this latter figure is divided by the low heat value of the fuel. The values are for average rates of heating. Fuel economy is of small importance as compared with the quality of the product.

FURNACE CONSTRUCTION

When furnace refractories are made up largely of standard bricks and shapes, it is advisable to specify furnace dimensions that can be built with a minimum of cutting. Horizontal flues are made a multiple of 2½ or 3 in (63 or 76 mm) in height, and most other flue dimensions are multiples of 4½ in (114 mm) to correspond to the width and length of standard bricks. The area of furnace flues must be large enough to avoid excessive pressures at maximum fuel rates. Flues should be located so as to promote the circulation of gases in all parts of the furnace. Average allowable **velocities in flues** for furnaces without stacks are:

Furnace temp, °F (°C)	200 (93)	1,000 (538)	1,500 (816)	2,000 (1,093)
Allowable velocity (hot gases), ft/s (m/s)	9 (2.74)	13 (3.97)	15 (4.57)	17 (5.2)

The total **flue areas required** in in²/ft³ of fuel/h (or per gal/h for fuel oil) for furnaces without stacks as temperatures of the products of combustion of 1,000 and 2,000°F are as follows:

Temp, °F (°C)	Fuel oil	Natural gas	Artificial gas	Coke-oven gas	Raw producer gas
1,000(538)	14.0	0.11	0.06	0.05	0.02
2,000 (1,093)	19.0	0.15	0.08	0.06	0.02

The **metal parts of a furnace,** consist of the steel and cast-iron binding, alloy parts exposed to the direct heat of the furnace, and water-cooled steel members. The alloy parts are of nickel or chromium alloys and must be made heavy enough to offset the loss of strength at high temperatures. They are resistant to oxidation at temperatures below 2,000°F (1,093°C). To reduce heat losses, water-cooled members must be insulated.

HEAT-SAVING METHODS

Methods of conserving heat include the use of recuperators or regenerators, waste-heat boilers (see Sec. 9), insulation of refractories, automatic control of temperature and atmosphere, and special attention to the construction and operation of the furnace.

Recuperators and regenerators extract some heat from the escaping flue gases and return it to the furnace by preheating the combustion air or the entering fuel. In **recuperators,** continuous flow of hot gases and cold entering air or gas is maintained through metal or refractory ducts which keep the two gas streams apart but which conduct heat from the hotter stream to the colder. Recuperators are built in the form of self-enclosed units set above the ground or in pits below floor level, and are made of fire-clay tile, silicon carbide, or heat-resisting metal. Overall coefficients of heat transfer in metallic recuperators are between 2.5 and 6.0 Btu/(ft² · h · °F) [14 and 34 W/(m² · °C] and in silicon carbide recuperators about the same; the coefficient for fire-clay recuperators is considerably less than these values. Usual velocities of hot air in recuperators do not exceed 12 ft/s (3.6 m/s) in order to keep pressure drop to a reasonable value.

Regenerators are used where the high temperature of air preheat is required to maintain a high furnace temperature or to obtain high thermal efficiency. When one regenerator serves an entire furnace, it is usually constructed of fire brick and consists of two chambers completely filled with a checkerwork. The flow of flue gases and that of air

Table 7.3.4   Average Net Efficiencies and Fuel Requirements of Various Furnace Types with Good Operation

Type	Temp, °F	Temp, °C	Avg efficiency, %	Avg heat required from fuel, Btu/lb,* of steel
Ingot heating, soaking pits, recuperative	2,000–2,400	1,100–1,300	20	500
Billet heating for forming:				
Batch, in-and-out	2,000–2,400	1,100–1,300	20	1,750
Continuous	2,000–2,400	1,100–1,300	32	1,100
Wire annealing of coils, hood type	1,300–1,500	700–800	16	1,350
Wire annealing of strands, in lead	1,300–1,500	700–800	19	1,100
Wire patenting, strands	1,650	900	21	1,250
Wire baking, coils, continuous	450	230	20	250
Tube annealing, continuous, bright	1,300–1,500	700–800	35	600
Skelp heating, butt weld, continuous	2,900	1,600	25	1,500
Slab heating, continuous, recuperative	2,400	1,300	42	800
Strip coil annealing, hood type	1,250–1,400	680–760	30	600
Hardening, continuous conveyor	1,650	900	21	1,250
Drawing, continuous conveyor	900–1,100	500–600	20	750
Carburizing, gas, continuous	1,750	950	19	1,500

* × 2.326 = kJ/kg.

or gas to be heated are periodically reversed so that the hot gases and cold gases alternately flow through the two sets of chambers. The checkerwork retains the heat of the hot gases and gives it up to the cold gases with each reversal. Another regenerator design employs metal plates. Regenerators are frequently used with glass-melting furnaces and are used almost exclusively where open-hearth furnaces are still employed. Overall coefficient of heat transfer in regenerators is from 1.5 to 2.5 Btu/ft^2 of checkerbrick surface per h per °F temperature difference ([8.5 to 14 W/(m^2 · C)], and the usual mass velocity of hot gas through the openings of the checker is about 0.065 lb/(ft^2 · s) [0.32 kg/(m^2 · s)].

Each burner may also be equipped with its own regenerator. With this design, burners are installed in pairs. When one burner is firing, the products of combustion pass through the second burner and the attached regenerator. The medium in the regenerator recovers and stores heat from the products of combustion. After one cycle, which typically lasts for 20 to 90 s, the functions of the two burners reverse. The burner that was firing now becomes the flue. Combustion air passes through the regenerator of the other burner and is heated by the medium. Typical medium is high-alumina material in ball or grain form. Air is preheated to a temperature within 300°F of the products of combustion. This type of regenerative burner is typically installed in continuous and batch reheating furnaces and aluminum-melting furnaces.

The **savings effected by recuperators or regenerators** depend upon the flue gas temperature and the temperature to which the incoming air or gas is preheated. With a flue gas temperature of 1,600°F, the theoretical savings in fuel with 200°F preheat of combustion air is about 4 percent, with 400°F, 11 percent; with 600°F, 15 percent; and with 800°F, 19 percent.

A recuperator or a regenerator installation, to be a good investment, must show a satisfactory net savings after all costs of investment, repairs, and associated shutdown time lost by such repairs are subtracted from the savings in fuel used or investment savings related to the heat recovery system. For example, it is often possible to reduce the length of continuous furnaces using regenerative burners compared to more conventional designs.

Automatic control prevents the waste of heat by unnecessarily high temperatures, preventable cold periods, and excessive air or unburned fuel from poor combustion. Of even greater importance is the prevention of damage to the heated product from overheating, excessive oxidation, and objectionable chemical reaction between furnace atmosphere and the product (principally decarburization and recarburization). Automatic **temperature** controllers are actuated by thermocouples in the furnace. The thermocouple must not be located in the direct path of the flames, which are not only several hundred degrees hotter than the furnace temperature but are also of extremely variable temperature and not an indication of the average temperature. Automatic control of **atmosphere** for the consistent maintenance of good combustion is accomplished by properly proportioning the fuel and combustion air as they enter the furnace. This is accomplished by the utilization of some characteristic of the flow of one fluid to regulate the flow of the other fluid. Automatic **pressure** control operates the flue

dampers of a furnace to maintain a constant predetermined pressure [usually about 0.01 to 0.05 in (0.25 to 1.25 mm) water] in the heating chamber, which excludes free oxygen from the surrounding atmosphere.

Care in furnace construction, operation, and maintenance is the simplest but often most neglected of all methods of heat savings. A large quantity of fuel can be saved by care in the construction of furnace refractories so that they will remain tight, by attention to the sealing of doors, by taking care that the doors and other openings are closed when not in use, and by maintaining insulation on any water-cooled members in the furnace.

SPECIAL ATMOSPHERES

(See also Sec. 7.5.)

In an increasing number of heat-treating operations, the necessity for improved quality has created a demand for clean- or bright-heating furnaces, in which the heating material is surrounded by a suitable protective gas while it is heated by radiation from electric resistors, radiant tubes, or the walls of a muffle. Table 7.3.5 gives the chemical analysis of common protective gases used in the heat-treating industry.

Type I Purified hydrogen is used for annealing, brazing, and other treatment of low-carbon steel; for the sintering of low-carbon ferrous powders; for the treatment of silicon iron (electrical sheets and strip); for the bright annealing of stainless steels, and the sintering of molybdenum, tungsten, and other metals.

Type II Ammonia is dissociated by steam or electric heat, and is dried by chemical driers. By partial combustion the relative percentages of hydrogen and nitrogen may be varied as shown in Table 7.3.5. The resulting gases from this treatment are cheaper and are used for brazing and sintering copper alloys, and for annealing low-carbon steels. Dissociated ammonia without combustion is used for annealing stainless steels containing nickel, short-cycle heating of all carbon and alloy steels, treatment of silicon iron, and the treatment of cuprous products.

Type III Rich hydrocarbon gas is produced by combustion with about 60 percent of theoretical air (6:1 air/gas ratio when using natural gas) in the presence of a nickel catalyst, followed by cooling to reduce the moisture content. It is used for the annealing of low-carbon steels, for short-cycle hardening of low-carbon steels, for clean annealing of chrome-type stainless steels, for treatment of silicon iron, and for brazing of copper alloys.

Type IV This gas is similar to type III except that about 90 percent of theoretical air is used for combustion. It is used for bright annealing of copper (straight N_2 and CO_2 can also be used for this purpose) and for clean heating of brass and bronze.

Type V This gas is the same as type III but is conditioned by chemical removal of carbon dioxide by monoethanolamine and by drying in chemical driers. It is used for short-cycle treatment of all carbon, alloy, and high-speed steels; for sintering of all ferrous powders; and as a carrier gas for carburizing and carbon restoration with the addition of natural gas or propane.

Table 7.3.5   Protective Gas Atmospheres

Type	Typical analysis					Dew point, °F (°C)
	CO_2	CO	CH_4	H_2	N_2	
I. Hydrogen, purified				100.0		−60 (−51)
II. Dissociated ammonia				75.0–5.0	25.0–95.0	
III. Rich hydrocarbon gas, not conditioned	5.5	9.0	0.8	15.0	69.7	+50 (10)
IV. Lean hydrocarbon gas, not conditioned	11.5	0.7		0.7	87.1	+50 (10)
V. Rich hydrocarbon gas, completely conditioned	0.1	9.5	0.8	15.8	73.8	−60 (−51)
VI. Lean hydrocarbon gas, completely conditioned	0.1	2.8		3.9	93.2	−60 (−51)
VII. Endothermic generator gas	0.5	20.0	1.0	38.0	40.5	+50 (10)

Type VI   This gas is similar to type V except that about 90 percent of theoretical air is used in the combustion. The resulting gas is used for long-cycle treatment of all ferrous materials except stainless steels containing nickel, and is effective in controlling decarburization in all carbon and alloy steels. It is also used for the annealing of brass and bronze.

Type VII   This endothermic gas is made in an externally heated generator with only 25 percent of theoretical air and is cooled to reduce moisture. It is used for short-cycle (under 2 h) heat treating and brazing, usually with small furnace installations. It is also used for dry cyaniding and as a carrier gas for carburizing and carbon restoration.

7.4   MUNICIPAL WASTE COMBUSTION
by Charles O. Velzy and Leonard M. Grillo

REFERENCES: *Proc. Biennial ASME National Waste Processing Conf.*, 1964–2005. "Design Considerations in Heat Recovery from Refuse," International Symposium on Energy Recovery from Refuse, 1975. Velzy et al., eds., "CRC Handbook on Energy Efficiency," chap. 4, Waste-to-Energy Combustion, in press. "Steam, Its Generation and Use," 40th ed., Babcock and Wilcox Co. Kirklin et al., The Variability of Municipal Solid Waste and Its Relationship to the Determination of the Calorific Value of Refuse Derived Fuels, *Resources and Conservation*, **9**, 1982, pp. 281–300.

Waste combustion is a method for processing of solid wastes by the burning of the combustible portions. It reduces the volume of solid wastes and eliminates the possibility of pollution of groundwater from putrescible organic waste, and the residue may serve as a source of mineral constituents and as a fill. With the application of boilers, beneficial use of the energy generated from burning of the waste is possible.

NATURE OF THE FUEL

The refuse which is received at a waste combustor today will contain a high proportion of paper; plastics; some wood; vegetable and animal waste; and varying amounts of cloth, leather, and rubber—together with metal cans, glass, and other noncombustible matter. Collections may also include metal appliances, furniture, tree limbs, waste building material, broken concrete, and other coarse waste matter, commonly classified as rubbish. Yard waste, such as grass clippings, is generally not accepted at a waste combustion facility. There may be a wide variation in moisture content of refuse, depending on the weather. Thus, after a storm, the moisture content may be so high that the combustibility of the waste is adversely impacted. Industrial and hazardous waste should be specifically identified and combustion facilities designed for the particular waste.

TYPES OF FURNACES

The type of furnace for waste combustors is dictated largely by the type of grate around which the furnace is built. Except in small plants, the modern furnace is equipped with a mechanical grate.

The "Controlled Air" Furnace   In the late 1970s, a type of combustion unit, batch fed, utilizing two chambers for staged combustion and followed by a waste heat boiler, was installed in smaller communities in the United States. Such units, while less efficient than larger furnaces, can be factory-assembled in large segments (or modular components) and therefore are less expensive than large, field-erected units. Thus they extended the economics of waste-to-energy plants to the smaller communities. In such plants, the refuse is normally dumped on a receiving floor and is then pushed into a ram for feeding into the combustion units. The smallest units do not have grates, while larger units are fitted with rams to move the material through the furnace as it is burned.

The Rectangular Furnace   Mechanical-grate furnaces are rectangular, with movement provided by travel of the grate or by a reciprocating or rocking action of the grate sections. Refuse is fed by gravity through a vertical chute, by a ram or similar arrangements.

PLANT DESIGN

Capacity   The capacity to be provided is a function of the area and population to be served, and the rate of refuse production for the population served. If records of collections have been kept, the capacity can be determined and forecasts made; lacking records, the quantity of refuse may be estimated as approximately 4 lb (1.8 kg) per capita per day, when there is little or no waste from industry, to 5 lb (2.3 kg) per capita per day where there is some waste from industry. Nearly all facilities, regardless of size, operate 24 h/day, 7 days/week.

Location   An isolated site may be preferred to avoid the possible objections of neighbors to the proximity of a waste disposal plant. However, well-designed and well-operated waste combustors which do not present a nuisance may also be installed in light industrial and commercial areas, thereby avoiding the economic burden of extended truck routes. Since considerable vertical distance is involved in passing refuse through a combustor, there is an advantage in a sloping or hillside site. Collection trucks can then deliver refuse at the higher elevation while the residue trucks operate at the lower elevation with a minimum of site grading.

Refuse-Handling Facilities   Scales should be provided for recording the weight of material delivered by collection trucks. Trucks should then proceed to the tipping floor at the edge of the storage pit. This area, which may be open or enclosed, must be large enough to permit several trucks at a time to maneuver to and from the dumping position. The number of tipping positions is determined by the size of the facility and the type of truck (packer or transfer trailer) delivering the waste.

Since collections usually are limited to one 8-h daily shift (with partial weekend operation) while burning is usually continuous over 24 h, ample **storage** must be provided. Seasonal and cyclic variations must also be considered in establishing the storage requirements.

Refuse storage in larger plants is normally in long, narrow, and deep pits usually extending along the front of the furnaces. If the pit is much over 25 ft in width, it is generally necessary to rehandle refuse dumped from trucks. In some waste combustor plants, floor dumping and storage of refuse are done.

Feeding the Furnaces   In a large waste combustor (pit and crane type), refuse is transferred from storage pit to furnace hopper by a crane equipped with a grapple. (See Sec. 10.) Batch feeding or batch discharge of residue is undesirable because of the resulting variations in furnace temperatures, adverse impact on furnace side walls, and increased air emissions.

In a more modern furnace using mechanized grates, a vertical charging chute, 12 to 14 ft (3.6 to 4.2 m) long, leads from the hopper to the front end of the furnace. This chute is kept full of refuse; feeding is accomplished by the operation of the mechanical grate, or by a ram; the front of the furnace is sealed from cold air; and the fuel is spread over the grate in a relatively thin bed.

In some newer plants, conveyors, and live-bottom bins, shredding and screening of the combustible fraction of the refuse has been utilized to produce a refuse-derived fuel (RDF). See Fig. 7.4.1.

FURNACE DESIGN

The basic design factors which determine furnace capacity are grate area and furnace volume. Both provision for and quantity of underfire

Fig. 7.4.1 Process diagram for a typical RDF system. (*Roy F. Weston, Inc.*)

air, and provision for quantity and method of applying overfire air influence capacity. The required grate area depends upon the selected burning rate, which varies between 60 and 90 lb/(ft² · h) of refuse in practice. Conservative design, with reasonable reserve capacity and reasonable refractory maintenance, calls for a burning rate between 60 and 70 lb/ft² of grate area.

Furnace volume is a function of the rate of heat release from the fuel. A commonly accepted minimum volume is that which results from a heat release of 20,000 Btu/(ft³ · h). Thus, at this rate, if the fuel has a heat content of 5,000 Btu/lb, the burning rate would be 4 lb/(h · ft³) of furnace volume. A conservative design, allowing for some overload and possible quantities of refuse of high heat content, would be from 20 to 30 ft³/ton of rated capacity.

The primary objective of a mechanical **grate** is to convey the refuse automatically from the point of feed through the burning zone to the point of residue discharge with a proper depth of fuel and in a period of time to accomplish complete combustion. The rate of movement of the grate or its parts should be adjustable to meet varying conditions.

A secondary, but important, objective is to stir gently or tumble the refuse to aid in completeness of combustion. In the United States, there are several types of mechanical grates: (1) traveling, (2) rocking, (3) reciprocating, (4) rotating drums, and (5) a proprietary water-cooled rotary combustor. With the traveling grate, stirring is accomplished by building the grate in two or more sections with a drop between sections to tumble the material. Traveling grates are generally used for combusting RDF. The reciprocating and rocking grates tumble the material by movement of the grate elements. The rotating drum grates carry the material over the top of the drum, tumbling the waste as it falls onto each successive drum. The rotary combustor slowly rotates to tumble the material which is inside the cylinder. In Europe, variations of these designs, as well as other types have been developed. The Volund incinerator (Danish) uses a slowly rotating, refractory-lined cylinder or kiln through which the fuel passes as it is burned.

Furnace configuration is largely dictated by the type of grate used. When built with a mechanical grate, the furnace is rectangular in plan and the height is dependent upon the volume required by the limiting rate of heat release.

The total **air capacity** provided in waste combustors must be more than the theoretical amount required for combustion in order to obtain complete combustion and to control temperatures—particularly with dry, high-heat-content refuse. The total combustion-air requirements may range to 10 lb of air/lb of refuse. For the modern mechanical-grate furnace chamber, two blower systems should be provided to supply combustion air to the furnace. Blower capacities can be divided, with half or more from the underfire blower and somewhat less than half from the overfire blower and with dampers on fan inlets and air distribution ducts for control. The pressure on the underfire system for most U.S. grate systems approximates 3 in of water. The pressure on the overfire air should be high enough so that the jet effect on passage through properly proportioned and distributed nozzles in the furnace roof and walls produces sufficient turbulence and retains the gases in the primary furnace chamber long enough to ensure complete combustion.

Heat Recovery   Perhaps the most potentially attractive form of recovery, or extraction, of resources from municipal solid wastes is recovery of energy from the combustion process.

Several options exist when one is considering a waste-to-energy system. These options include mass burning in a refractory-walled furnace with a waste-heat boiler inserted in the flue downstream; mass burning in a water-walled furnace with the convection surface immediately downstream; and refuse preprocessing and separation of the combustible fraction with combustion taking place in a spreader stoker boiler partially in suspension and partially on a grate. This latter option

Table 7.4.1   Energy-from-Waste Plants Larger than 400 Tons/Day Commissioned 1990–1995

Plant location	Plant size tons/d	Start-up year	Type	Energy sold
Broward County, FL, north	2,250	1991	Mass burn water wall	Electricity
Broward County, FL, south	2,250	1991	Mass burn water wall	Electricity
Camden, NJ	1,050	1991	Mass burn water wall	Electricity
Chester, PA	2,688	1991	Rotary kiln water wall	Electricity
Essex County, NJ	2,250	1990	Mass burn water wall	Electricity
Fort Myers, FL (Lee County)	1,200	1994	Mass burn water wall	Electricity
Gloucester County, NJ	575	1990	Mass burn water wall	Electricity
Honolulu, HI	2,165	1990	RDF water wall	Electricity
Hudson Falls, NY	400	1992	Mass burn water wall	Electricity
Huntington, NY	750	1991	Mass burn water wall	Electricity
Huntsville, AL	690	1990	Mass burn water wall	Steam
Kent County, MI	625	1990	Mass burn water wall	Electricity/steam
Lake County, FL	528	1991	Mass burn water wall	Electricity
Lancaster County, PA	1,200	1991	Mass burn water wall	Electricity
Long Beach, CA	1,380	1990	Mass burn water wall	Electricity
Lorton, VA (Fairfax County)	3,000	1990	Mass burn water wall	Electricity
Montgomery County, PA	1,200	1991	Mass burn water wall	Electricity
Pasco County, FL	1,050	1991	Mass burn water wall	Electricity
Rochester, MA	2,700	1988/1993	RDF water wall	Electricity
Southeast CT (Preston, CT)	600	1992	Mass burn water wall	Electricity
Spokane, WA	800	1991	Mass burn water wall	Electricity
Union County, NJ	1,440	1994	Mass burn water wall	Electricity

is generally termed combustion of a refuse-derived fuel. A partial list of waterwall and RDF plants in North America in operation as of the end of 2004 extracting energy from combustion of municipal-type waste is shown in Table 7.4.1. This list excludes plants built for developmental or experimental purposes and plants that utilize specialized industrial wastes.

Figure 7.4.2 illustrates a plant with heat recovery, while Fig. 7.4.3 shows a cross section through a typical RDF facility.

In considering the above options, one should take into account the overall energy balance in the various systems. These systems can be grouped under the following general categories: burning as-received refuse; burning mechanically processed refuse; and gasification. In mechanical processing systems, less heat will be available for use than there was prior to processing.

The tabulation below from published data regarding the production of a fuel gas from refuse will partially illustrate the net energy loss in converting the available energy to another form:

Composition	Percent by volume	Percent by weight	Percent of total carbon
CO	47	62.1	70
H_2	33	3.1	
CO_2	14	29.1	21
CH_4	4	3.0	6
C_2H_2	1	1.4	3
N_2	1	1.3	

Fig. 7.4.2   Typical cross section of a mass-burn facility. (*Roy F. Weston, Inc.*)

Fig. 7.4.3 Typical cross section of an RDF facility. *(Roy F. Weston, Inc.)*

Note that 21 percent of the carbon is in CO_2 which will not burn, while 70 percent of the carbon is in CO where 30 percent of the elemental energy in carbon is no longer available. While this heat is not wasted, it is lost energy not available to do further work.

A tabulation of energy losses and total net available energy, for refuse with an initial heat content of 5,000 Btu/lb, is given in Table 7.4.2.

Of the 5,000 Btu/lb in the refuse as received, the tabulated data indicate the useful energy that may be made available through combustion. While the data are not absolute, the relative magnitudes are meaningful, provided similar degrees of design efficiency and sophistication of control are used for each process.

Other factors to consider in selection and design of heat-recovery facilities include efficiency of boiler facilities, furnace-chamber design, and combustion air supply. While in older plants with waste-heat boilers installed in downstream flues, steam production averaged 1.5 to 1.8 lb/lb of refuse, in newer water-walled furnaces and suspension-fired units, steam production is on the order of 3.0 to 3.5 lb/lb of refuse. The lower efficiency in waste-heat boiler units is due to higher heat losses in the plant stack gases, in turn caused by higher excess air levels required to control combustion temperatures properly in the refractory-lined primary furnace enclosure.

In most water-walled furnaces and furnaces in which shredded combustible refuse fractions are burned, the usual configuration is a tall primary chamber with the gases passing out the top and into the superheater and convection boiler surfaces after completion of combustion. It has been found desirable in mass-burning water-wall plants to coat a substantial height of the primary combustion chamber (where boiler-tube metal temperatures will exceed 500°F) with a refractory material or to overlay metal surfaces with Inconel and to limit average gas velocities to under 15 ft/s. Gas velocity entering the boiler convection bank should be less than 30 ft/s.

In all combustion systems, combustion air is supplied from both under and over the grate. Overfire air is utilized to enhance turbulence and mixing of combustion gases with the combustion air, and for completion of combustion. Accordingly, this air is best introduced through numerous relatively small (1½- to 3-in-diameter) nozzles, at pressures of 20 in of water and higher. Ideally, provision should be made for the introduction of the overfire air at several different elevations in the furnace.

Flues and chambers beyond the furnace convey gases to the air pollution control equipment for removal of fly ash and other pollutants, and to the stack. The draft for a waste combustor is provided by an induced-draft fan. (See Secs. 4 and 14.) Most modern plants include heat recovery equipment and extensive air pollution control equipment.

Air Pollution Control The Federal New Source Performance Standards for air pollution control regulate the 11 pollutants shown in Table 7.4.3. Some jurisdictions have promulgated emission requirements more stringent than the federal standards.

Table 7.4.3 Federal New Source Performance Standards

Pollutant	Emission limit*
Dioxin/furans	13 ng/dscm
Cadmium	0.02 mg/dscm
Lead	0.02 mg/dscm
Mercury	0.08 mg/dscm or 85 percent reduction
Opacity	10 percent
Particulate matter	24 mg/dscm
Hydrogen chloride	25 ppmdv or 95 percent reduction
Nitrogen oxides	150 ppmdv
Sulfur dioxide	30 ppmdv or 80 percent removal
Fugitive ash	5 percent of observation period
Carbon monoxide	50–150 ppmdv (depending on technology)

* dscm = dry standard cubic metres
ppmdv = parts per million dry volume

Table 7.4.2 Energy Losses and Total Net Available Energy for Refuse with Initial Heat Content of 5,000 Btu/lb

Process	Energy loss, %			Total net available energy, Btu/lb
	Processing	Combustion	Total	
As received	1	29	30	3,500
Dry shredding	18	25	43	2,850
Gasification	32	25	57	2,150

To meet current emission requirements, two basic techniques have been utilized on waste combustors: electrostatic precipitation and fabric filters preceded by lime addition in a slurry, so-called "dry scrubbing." Electrostatic precipitation has performed reliably and has given predictable emission test results. The fabric filter/dry scrubber air pollution control system is considered by many to represent the most efficient combination of pollution control systems currently available for particulates and acid gases and is used on most facilities. Selective noncatalytic reduction using ammonia or urea has been applied at some plants for NO_x control. Activated carbon injection for control of mercury and dioxin emissions has been installed on many facilities. Good combustion control is used to limit emission of organics.

Residue Discharge and Disposal   The residue from refuse burning consists of relatively fine, light ash mixed with items such as burned tin cans, partly melted glass, and pieces of metal. Discharge from furnaces may be through a chute into a conveyor trough filled with water for quenching and then carried by flight conveyor to a storage building, or a truck loading facility. Usually there are two conveyor troughs, so arranged that the residue can be discharged to either, one trough being used at a time. Alternatively, a ram discharger, in which ash is submerged in water, effects discharge onto a dry conveyor.

The lower end of the discharge chute leading to the trough is submerged in a water seal to prevent entrance of cold air to the furnace. In design of the conveyor mechanism, the dimensions should be large because of the nature of the material handled, and the metal used should be selected to withstand severe abrasive service. Final disposal of the residue is by dumping at a suitable location, which for modern plants usually means a monofill. Volume required for disposal is 5 to 15 percent of that required for dumping raw refuse.

Miscellaneous Facilities   Good working environment and reasonable comfort for the staff should be provided.

COMBUSTION CALCULATIONS

Among the factors directly affecting design are **moisture** and **combustible content** of refuse as received, heat released by combustion, temperature control, and water requirements. The design of furnaces, chambers, flues, and other plant elements should be based on characteristics which result in large sizes. Controls should provide satisfactory operation for loads below the maximum. The computations which follow are for relatively typical heat releases.

The prime factors in **heat calculations** are the moisture and combustible content of the refuse and heat released by burning the combustible portion of the refuse. The moisture content may vary from 20 to 50 percent by weight, and the combustible content may range from 25 to 70 percent. The combustible portion is composed largely of cellulose and similar materials, mixed with proteins, fats, oils, waxes, rubber, and plastics. The heat released by burning cellulose is approximately 8,000 Btu/lb, while that released by the plastics, fats, oils, etc., is approximately 17,000 Btu/lb. If cellulose, plastics, oil, and fat exist in the refuse in the ratio of 5:1, the heat content of the combustible matter will be 9,500 Btu/lb. The heat content per pound of refuse as received, for varying proportions of moisture and noncombustibles, is given in Table 7.4.4.

Determination of the **air requirement** is illustrated by computation with refuse of 5,000 Btu/lb heat content where the composition is: combustible, 52.5 percent; noncombustible, 22.5 percent.

Carbon and hydrogen are the essential fuel elements in combustion of refuse; sulfur and other elements which oxidize during combustion are present in trace amounts and do not contribute significantly to the heat of combustion. Carbon and hydrogen content can be determined from a complete analysis of the refuse, but such an analysis is of questionable value because of the variable character of refuse and the difficulty of obtaining representative samples. The carbon-to-hydrogen ratio in municipal solid waste is approximately 7.5:1.0. A typical ultimate analysis, which is assumed for the sample calculation, is shown in Table 7.4.5. It is probable that 1 to 3 lb of combustible material per 100 lb of refuse will escape unburned with the residue. For the sake of clarity in the illustrated computations, complete combustion is assumed.

Oxygen requirements and products of combustion can be determined from the reactions as follows:

Carbon	$C + O_2 \rightarrow CO_2$
Atomic wt:	$12 + 32 = 44$
Ratio:	$1 + 2.667 = 3.667$
Hydrogen	$2H_2 + O_2 \rightarrow 2H_2O$
Atomic wt:	$4 + 32 = 36$
Ratio:	$1 + 8 = 9$

The theoretical air required per 100 lb of refuse follows from these figures where air is considered to contain 23.15 percent oxygen.

$$\text{Air required} = (28 \times 2.667 + 3.7 \times 8 - 20.2)/0.2315$$

$$= 363 \text{ lb/100 lb refuse}$$

For waste combustion, furnace temperature must be controlled to minimize refractory maintenance. With no other provision for heat absorption, it is necessary to introduce excess air well beyond the needs for complete combustion, e.g., 140 percent of theoretical so that, in the example cited:

$$\text{Total air} = 1.4 \times 363 = 508 \text{ lb/100 lb refuse}$$

To summarize the quantities for a computation of furnace temperature, a materials balance is given in Table 7.4.6 equating the input to the

Table 7.4.5   Typical Ultimate Analysis

Element	Percentage
Carbon	28.0
Hydrogen	3.7
Oxygen	20.2
Nitrogen	0.4
Sulfur	0.2
Chlorine	0.0
Ash	22.5
Moisture	25.0
Total	**100.0**

Table 7.4.4   Heat Content of Refuse, as Received

Moisture, %	Noncombustible, % 10		Noncombustible, % 15		Noncombustible, % 20		Noncombustible, % 25	
	Comb., %	Heat content*	Comb., %	Heat content*	Comb., %	Heat content*	Comb., %	Heat content*
50	40	3,800	35	3,325	30	2,850	25	2,375
40	50	4,750	45	4,275	40	3,800	35	3,325
30	60	5,700	55	5,225	50	4,750	45	4,275
20	70	6,650	65	6,175	60	5,700	55	5,225

* Btu/lb.

Table 7.4.6   Materials Balance for Furnace lb/100 lb of Refuse

Input:		
Refuse		
Combustible material	52.5	
Moisture	25.0	
Noncombustible	22.5	100.0
Total air, at 140% excess air		
Oxygen	117.7	
Nitrogen	390.7	508.4
Moisture in air		6.7
Residue quench water		5.0
Total		620.1
Output:		
CO_2(28 × 3.3667)		102.7
Air		
Oxygen (117.7 − 84.1)	33.6	
Nitrogen	390.7	424.3
Moisture		
In refuse	25.0	
From burning hydrogen	33.3	
In air	6.7	
In residue quench water	5.0	70.0
Furnace gas subtotal		597.0
Noncombustible material		22.5
Sulfur and Nitrogen		0.6
Total		620.1

furnace and output for 100 lb of refuse. In this tabulation, allowance is made for moisture in the air at a commonly accepted rate of 0.0132 lb/lb of dry air. Some residue quench water will be evaporated, and the moisture added to the flue gases is estimated at 5 lb for each 100 lb of refuse burned. Since the assumed analyses are not precise, an exact balance is not obtained, but the indicated computations are sufficiently accurate for combustor design.

In Table 7.4.6, total air is broken down into oxygen and nitrogen on the basis that 23.15 percent of the air is oxygen. To compute the air in the "output," or flue gas, the nitrogen is the same as the "input." Oxygen

Fig. 7.4.4   Enthalpy of flue gas above 80°F.

is diminished by the amount consumed in combustion. Since carbon and hydrogen unite with oxygen during combustion, the oxygen consumed per 100 lb refuse is

For carbon,   28 × 2.667 = 74.7 lb
For hydrogen,   3.7 × 8 = 29.6 lb
                                      104.3 lb
Less oxygen in fuel   − 20.2 lb
                       Total = 84.1 lb

Abiabatic flame temperature is the maximum theoretical temperature that can be reached by the products of combustion of a specific fuel-air combination. To calculate this temperature, the total heat input in the fuel is adjusted to subtract the heat input required to vaporize moisture in the fuel, moisture produced in combustion of hydrogen, and the residue quench water that is vaporized. A loss is assumed to account for incomplete combustion and other small losses. The remaining heat energy is the sensible heat available in the furnace gas. It can be calculated per 100 lb of refuse from the data of Table 7.4.6 and the enthalpy data of Fig. 7.4.4 as follows:

Input, 100 lb × 5,000 Btu/lb		500,000 Btu
Losses:		
Heat of vaporization deduction		
Moisture		
In refuse	25.0	
From burning hydrogen	33.3	
From residue quench	5.0	
	63.3	
63.3 × 1.050 Btu/lb		66,500
Assumed loss due to incomplete		
combustion and other losses 2%		10.000
Total deduction		−76,500
Sensible heat available in furnace gas		423,500 Btu
Enthalpy of gas, 418.810 Btu ÷ 766.3 lb		553 Btu/lb
% Moisture in furnace gas		
Moisture vaporized	63.3	
Moisture in air	6.7	
	70.0	
70.0 ÷ 766.3 = 9.1%		
Temperature of furnace gas at		1,950°F
553 Btu/lb and 9.1% moisture from		
Fig. 7.4.4.		

RECOVERY

Many modern waste-to-energy (WTE) plants incorporate waste processing/materials recovery facilities. Such facilities are a natural adjunct at plants producing RDF. Relatively active stable markets have developed for newsprint, glass, metals, and certain specific plastics. Recent surveys of waste composition indicate that after recycling, paper and noncombustibles have decreased slightly while high-heat-value plastics have increased significantly as a percentage of municipal solid waste going to disposal. The impact of these changes in waste composition has been to increase slightly the heat content of waste available for processing in WTE plants and to remove some of the more troublesome materials from a materials handling standpoint.

Fly ash has been used to a limited extent as a concrete additive and as a road base. Waste combustor residue has been used for land reclamation in low areas and, in some cases, as a road-base material. Generally, before such use, the fly ash and residue must be tested to ensure that they cannot be classified as hazardous waste.

As this nation's energy needs become more critical in the future, this readily available source of energy should be tapped more frequently. The technology is available now for successful application of these techniques if provision is made for adequate funding and properly trained operating staffs.

7.5 ELECTRIC FURNACES AND OVENS

by James C. Simmons

REFERENCES: Robiette, "Electric Melting and Smelting Practice," Griffin. Campbell, "High-Temperature Technology," Wiley. "Electric-Furnace Steel Proceedings," Annual, AIME. Paschkis, "Industrial Electric Furnaces and Appliances," Interscience. Stansel, "Induction Heating," McGraw-Hill. Ess, The Modern Arc Furnace, *Iron Steel Eng.*, Feb. 1944.

CLASSIFICATION AND SERVICE

The furnaces and ovens addressed in this section generally are those small and medium-size units used in general foundry practice, heat treating, and associated processes. The larger units are generally used for melting large quantities of metal as part of specific production processes such as the production of high-purity alloy steels, processing batches of processed parts receiving vitreous enamel, annealing glass, and so on.

In **resistor furnaces and ovens,** heat is developed by the passage of current through distributed resistors (heating units) mounted apart from the charge. Alternating current of a standard power frequency is used. The furnace service is for heat applications to solids such as heat treatment of metals, annealing glass, and firing of vitreous enamel. Oven service is limited to drying and baking processes usually below 500°F (260°C).

In **induction heaters** heat is developed by currents induced in the charge. The service is heating metals to temperatures below the melting points.

In **induction furnaces** heat is developed by currents induced in the charge. The service is melting metals and alloys.

In **arc furnaces** heat is developed by an arc, or arcs, drawn either to the charge or above the charge. Direct-arc furnaces are those in which the arcs are drawn to the charge itself. In indirect-arc furnaces the arc is drawn between the electrodes and above the charge. A standard power frequency is used in either case. Direct-current (dc) electric power is an alternative source of energy. The general service is melting and refining metals and alloys. The ladle arc furnace is used particularly when a charge of metal is to be processed primarily to refine its chemistry.

In **resistance furnaces** of the submerged-arc type, heat is developed by the passage of current from electrode to electrode through the charge. The manufacture of basic products, such as ferroalloys, graphite, calcium carbide, and silicon carbide, is the general service. Alternating current at a standard power frequency is used. An exception is the use of direct current where the product is obtained by electrolytic action in a molten bath, e.g., in the production of aluminum.

The **characteristics of electric heat** are

1. Precision of the control of the development of heat and of its distribution.

2. The heat development is independent of the nature of the gases surrounding the charge. This atmosphere can be selected at will with reference to the nature of the charge and the chemistry of the heat process. This freedom is often a primary reason for the use of electric heat.

3. The maximum temperature is limited only by the nature of the material of the charge.

The first two characteristics underlie the design of all electric heating apparatus. The third is utilized in thermal processes for the production of certain materials not obtainable in any other way.

RESISTOR FURNACES

Resistor furnaces may be either the batch or the continuous type. Batch furnaces include box furnaces, elevator furnaces, car-bottom furnaces, and bell furnaces. Continuous furnaces include belt-conveyor furnaces, chain-conveyor furnaces, rotary-hearth furnaces, and roller-hearth furnaces.

Standard resistor furnaces are designed to operate at temperatures within the range 1,000 to 2,000°F (550 to 1,200°C). For higher heating chamber temperatures, see Resistors, later in this section.

The **heating chamber** of a standard furnace is an enclosure with a refractory lining, a surrounding layer of heat insulation, and an outer casing of steel plate, or for large furnaces an outer layer of brick or tile, as indicated by Figs. 7.5.1 and 7.5.2. The hearth of a batch furnace often is constructed of a heat-resisting alloy, made in sections to prevent warping. In some continuous furnaces the conveyor forms the hearth; in others a separate hearth is required.

Insulating firebrick—a semirefractory material—is commonly used for the inner lining of the heating chamber. This material has thermal and physical properties intermediate between those of fire-clay brick and heat-insulating materials. A lining of this kind has less heat-storage capacity than a fire-clay brick lining, and its use accordingly decreases the time periods of heating and cooling the chamber and also decreases the stored-heat loss for a given cycle of operation. Other advantages are its high heat-insulating value and light weight.

The maximum temperature of the inner face of the layer of heat insulation determines the character of material required for the insulation. Practically all resistor furnaces have insulation made of diatomite. Composite wall structures with a 4½-in (11-cm) semirefractory lining and a 9- to 13-in (23- to 33-cm) layer of heat insulation represent general practice for standard furnaces. (See also Sec. 4.3.)

Atmospheres A mixture of air and the gases evolved from the charge constitutes a **natural atmosphere** in the heating chamber of a resistor furnace. The composition of such an atmosphere in a batch furnace is variable during a heating cycle. A natural atmosphere in the heating chamber of a continuous furnace is mainly air. Natural atmospheres are used where the extent of the action of oxygen on the charge during the heating cycle is not objectionable and for processes where that chemical action is desired. (See also Sec. 7.3.)

Fig. 7.5.1 Heating chamber with sidewall and hearth resistors.

Fig. 7.5.2 Heating chamber with roof and hearth resistors.

The basis of an **artificial atmosphere** is the exclusion of oxygen (air) from the heating chamber by the substitution of some other gas or mixture of gases. This gas or mixture of gases is selected with reference to the chemical activity of that atmosphere on the charge at the temperature of the heat application. A definite chemical action may be desired, for example, the reduction of any metallic oxide present on the charge, or it may be required that the artificial atmosphere to be chemically inactive. Thus artificial atmospheres are divided into (1) active or process atmospheres and (2) inactive or protective atmospheres. The term "controlled" atmosphere refers generally to a protective atmosphere, but it also includes artificial atmospheres of some degree of chemical activity. An example of a process atmosphere is the use of a hydrocarbon gas to carburize steel. Examples of controlled atmospheres are: the bright annealing of metals, the prevention of decarburization of steel during a heat application, the use of a reducing gas (hydrogen or carbon monoxide) in a copper brazing furnace, etc. In this last example the reducing gas serves to clean the faces of the joint to be made (by removal of any oxide present) and to maintain that cleanliness during the operation. The primary gases for controlled atmospheres are hydrogen and carbon monoxide and nitrogen.

The main uses of controlled atmospheres are (1) the prevention of the formation of oxides on the material of the charge, or conversely the reduction of any oxides present, and (2) the prevention of a change in the carbon content of a steel undergoing a heat treatment. Each of these uses denotes a chemical system in which the reactions are reversible.

The chemical systems relating to metallic oxides are:

A: Oxide + hydrogen $\rightleftarrows$ metal + water vapor
B: Oxide + carbon monoxide $\rightleftarrows$ metal + carbon dioxide

The chemical systems relating to carbon in steel are

E: Methane $\rightleftarrows$ hydrogen + carbon
F: Carbon monoxide $\rightleftarrows$ carbon dioxide + car bon

In artificial atmospheres the volume ratio of the two gases in the heating chamber should be so maintained as to correspond to the desired direction of the chemical activity of the system, or, if no chemical action is desired, to maintain that volume ratio at (or near) its equilibrium value for the temperature of the heat application. The equilibrium volume ratios for each of the four chemical systems A, B, E, and F for carbon steel over the usual range of temperature of heat-treatment processes and for atmospheric pressure are shown in Fig. 7.5.3. There is little tendency toward a change in the carbon content of a steel below the critical range. Oxidation is active down to about 1,100°F (650°C).

Curves E and F of Fig. 7.5.3 show the volume ratios of systems E and F for equilibriums with graphite. The equilibrium volume ratios of these two chemical systems for carbon in solid solution in steel (austenite) depend in each case on the carbon content of the steel. For the methane-hydrogen-carbon system (E) the volume ratio of the two gases at equilibrium with carbon in an unsaturated steel at a given temperature is less than the value shown by curve E. For the carbon monoxide-carbon

dioxide-carbon system (F) the volume ratio of the two gases at equilibrium with the carbon in low- and medium-carbon steel at a given temperature is somewhat greater than the value shown by curve F; for high-carbon steels the equilibrium volume-ratios approach the values of curve F.

In the case of the hydrogen-iron oxide reaction, curve A, the water vapor content of the mixture of gases at equilibrium decreases with decrease of temperature. Hence if a steel is to be cooled in a controlled atmosphere of this kind, the permissible water vapor content of the controlled atmosphere is dictated by the lowest temperature of the operation. The reverse is true of the carbon monoxide-iron oxide reaction, curve B. Thus if at a given temperature the carbon dioxide content of the mixture of carbon monoxide and carbon dioxide is less than the volume for equilibrium at that temperature it will be less than the volume for equilibrium at any lower temperatures and the steel can be cooled in that atmosphere without oxidation.

In the use of mixtures of the gases of the chemical systems noted to form controlled atmospheres for the heat treatment of steel, the interactions of the gases at elevated temperatures must be controlled by removal of all or nearly all the carbon dioxide and water vapor from the heating chamber.

The available data concerning controlled atmospheres for the protection of alloy steels during heat-treatment processes indicate that the technique for alloy steels is much the same as for carbon steels; i.e., a controlled atmosphere suitable for a carbon steel would, in general, be suitable for an alloy steel of the same carbon content.

In the heat treatment of nonferrous metals and alloys the use of either chemical system A or B requires for each oxide a knowledge of the equilibrium volume ratios of the chemical system used over the range of the operating temperature. Individual problems may arise. For example, copper can be bright-annealed in an atmosphere of dry steam—an inactive gas for this application—but the resultant staining of the copper during cooling may be objectionable. Copper usually contains a small percentage of oxide, and when annealing such copper in an atmosphere containing a reducing gas the temperature of the metal must be kept below about 750°F (420°C); otherwise the oxide will be reduced and the copper made brittle.

The foregoing discussion of atmosphere in heating chambers is intended to indicate the principles involved in the use of gases at elevated temperatures. The terms oxidation, reduction, carburization, and decarburization refer here to the chemical condition of a particular atmosphere and not to the extent of its effect on a charge. In all cases the concentration of the active gas or gases, time, temperature, in case of steel the carbon content and the gas pressure, and the catalytic action of hot surfaces within the chamber are important factors in the result obtained.

Bath Heating   Heating for local hardening of edge tools is the most general service. The lead-bath furnace has a working temperature range of 650 to 1,700°F (360 to 950°C). The salt-bath furnace can be adapted to working-temperature ranges within a total range of 300 to 2,350°F (170 to 1,300°C) by the selection of suitable mixture of salts. The two salt baths most generally used are cyanide mixtures and chloride mixtures. The rate of heating by immersion is much faster than obtained by radiation. The rate of heat transfer in a salt bath is about one-half that in the lead bath. An additional use of the salt bath is for cyaniding, in effect a process atmosphere.

Resistors   The resistor of a standard furnace is a sinuous winding mounted on the inner surfaces of the heating chamber as shown in Figs. 7.5.1 and 7.5.2. The resistor winding covers practically the entire surface of the space chosen. Resistors are applied on the basis of 2 to 3 kW/ft² (20 to 30 kW/m²) of wall surface in general practice. The basis of resistor location is radiation to all surfaces of the charge. Hence, the height and width dimensions of the heating chamber indicate the choice between sidewall and roof resistors. In some cases both locations are used. Uniform distribution of heat flow to the charge is obtained by a designed distribution of the surfaces of the resistors supplemented by reradiation from the inner surfaces of the chamber.

Fig. 7.5.3 Equilibrium volume ratios of chemical systems A, B, E, and F for steel. C = carburizing condition; O = oxidizing; D = decarburizing; R = reducing.

Resistors for the majority of standard furnaces are made of 80 Ni, 20 Cr alloy. A nickel-chromium-iron alloy is used in some furnaces for operation only over the lower portion of the furnace temperature range. Both ribbon and cast shapes are in use. The effort in each case is to obtain the maximum surface area per unit length of resistor and at the same time retain sufficient mechanical strength in the resistor winding.

The 80 Ni, 20 Cr alloy is self-protecting against oxidation, but this protection decreases with rise of temperature. The operating temperature of a resistor should be no higher than is needed in each case and should always be at a safe margin below the softening point of the alloy, which is about 2,500°F (1,390°C). This corresponds to a maximum furnace temperature of about 2,100°F (1,170°C). The life of a resistor is also affected by the frequency of heating and cooling. Barring accidents, the resistor of a standard furnace under average conditions of operation has a long life, usually measured in years of service.

The nickel-chromium alloy resistor is used in artificial atmospheres as well as in natural atmospheres. This alloy is not resistant to compounds of sulfur and is affected to some extent by carbon monoxide.

The electric insulation of the resistor circuit is that of its refractory supports at elevated temperatures. This limits the voltage of the circuit to about 600 V. Small furnaces are usually designed for 110 V, medium sizes for 220 V, and the larger units for 440 V. Single phase up to 25 or 30 kW and three phase for higher ratings is general practice.

The resistivity-temperature coefficient of the nickel-chromium alloy permits the operation of resistors of this material on constant-voltage circuits. The rate of heat development in a resistor is proportional to the square of the applied voltage; hence maintenance of normal voltage is desirable. Voltage regulation is not as important as for other types of electrical apparatus because of the heat-storage capacity of the structure of the heating chamber. The power factor of the resistor circuit is practically unity.

High-Temperature Furnaces   Silicon carbide is the basis of a type of nonmetallic resistor for heating-chamber temperatures up to about 2,800°F (1,560°C). The material is formed into rods. Resistors of this material do not require protection against oxidation and are operated on constant-voltage circuits.

Molybdenum resistors are suitable for temperatures up to 3,000°F (1,670°C). Above that temperature the metal begins to vaporize. A molybdenum resistor cannot be operated in a natural atmosphere, and also it must be protected from reactions with silica and carbon. The metal is immune from reactions with sulfur compounds, nitrogen, and water vapor. Hydrogen is the most common artificial atmosphere used with molybdenum resistors. The difference between the cold and hot resistances of the circuit makes a starting device necessary.

Other materials used to some extent for resistors are iron, tungsten, and graphite. These require protection against oxidation.

Temperature Regulation   The temperature of the heating chamber of a resistor furnace is in most cases regulated by a more or less intermittent application of current—the on-and-off method—which is made automatic by instrument control. This method utilizes the heatstorage capacity of the inner lining of the heating chamber as a temperature equalizer. The variation from the normal temperature of the chamber can be kept within less than 7°F (4°C) plus or minus, without undue wear of the temperature-control equipment. Temperature regulation by voltage control is equally applicable to resistor furnaces, and the trend is toward the use of this more accurate method particularly for the more important installations.

Temperature protection for resistor furnaces is obtained by means of a temperature fuse mounted in the heating chamber and connected in the control circuit of the power supply to the furnace.

Multiple-Temperature Control   The resistors of the larger furnaces are divided into two or more circuits. Each circuit can be equipped with individual temperature control. That arrangement provides temperature regulation at more than one location in the heating chamber and is an aid toward maintaining uniform temperature distribution within the chamber.

The subdivision of resistor circuits is used also for zone heating—and zone cooling where needed—in continuous furnaces.

Melting Pots   Resistor heating is applied to melting pots for the soft metals and alloys and for lead baths and for salt baths. The immersion heating unit is used for temperatures up to 950°F (530°C). For higher temperatures the metal pot is heated by resistors mounted outside and around the pot. The assembly in each case includes a heat-insulating wall similar to that of a resistor furnace. Another method of heating applicable only to salt baths is the passage of alternating current (of any frequency) between electrodes immersed in the bath.

Tempering Furnaces   The temperature is comparatively low—below 1,300°F (720°C). Electrically heated oil baths and salt baths are used for tempering many kinds of small parts. Another form of tempering furnace is a vertical resistor furnace with the addition of a removable metal cylinder (or basket) to contain the charge and to provide an annular passageway for the circulation of air (by a fan mounted on the furnace) over the resistors and thence through the charge—an application of forced convection heating.

Sizes   The electrical rating is the general method of expressing the size of a resistor furnace. Sizes up to 100 kW predominate, 100- to 500-kW furnaces are common, and others within the range 500 to 1,000 kW are in service. The data in Table 7.5.1 refer to common sizes of socalled box furnaces for general service.

The **losses** from a resistor furnace for a given heating chamber temperature are as follows: The open-door loss is a variable depending on the area of the door (or doors) and the percentage of the time that the door is open—from a continuous furnace this loss also varies with the type and speed of the conveyor; with artificial atmospheres, the loss of heat in the gases discharged for atmosphere control; the stored-heat loss, a variable that depends on the extent and frequency of the cooling of the furnace within a given period of operation; the heat dissipated from the outer surfaces of the furnace.

The **operating efficiency** is expressed as either pounds of material treated or kWh or kWh per ton. Representative values for average service for the heat treatment of steel range from 7 to 12 lb/kWh. Corresponding values for nonferrous metals and alloys are within the range 12 to 22 lb/kWh.

The general field of the batch furnace is defined by the following conditions: (1) intermittent and varied production; (2) long periods of heating (and in some cases slow cooling); (3) heating service beyond the range of the handling capacity of furnace conveyors; and (4) supplementary heating service. A continuous furnace is indicated where the flow of material to be heated is reasonably uniform and continuous, i.e., mass-production conditions. In some cases, batch furnaces with automatic charging and discharging equipment are essentially continuous furnaces.

Resistor Ovens   The resistor oven is a modification of the resistor furnace to correspond to the low temperatures of drying and baking

Table 7.5.1   Box Resistor Furnaces, 1,850°F (1,030°C) Class

Connected load, kW	Power supply, 200 V	Steel, lb/h at 1,500°F	Time to heat to 1,500°F (830°C) when used previous day, min	Radiation kWh/h, at 1,500°F	Inside			Overall			
					Width	Depth	Height	Width	Depth	Height, door closed	Height, door open
29	1-phase	300	35	4.9	18	36	18	55	89	86	97
45	3-phase	500	35	6.9	24	54	20	61	108	90	101
60	3-phase	650	25	7.8	30	63	23	78	125	90	98
72	3-phase	750	25	9.1	36	72	23	84	135	90	98

processes. The heating chamber is an insulated metal structure with a fresh-air inlet and an exhaust fan for ventilation (the removal of vapors and gases evolved from the charge). A refractory lining is not required. Ovens may be of the batch type with conventional methods of handling the charge of the continuous type, usually with chain conveyors.

The most common type of electric oven is heated by resistors mounted in a separate compartment of the heating-chamber enclosure. The heat transfer is by forced convection which is accomplished by recirculation of the chamber atmosphere by a motor-driven fan.

A resistor oven with filament-type lamps as heating units provides what is generally known as infrared heating. The lamps, usually with self-contained reflectors, surround the charge, and the heat transfer is by radiation, mainly in the infrared portion of the spectrum. This type of oven is best adapted to the continuous heating of charges which present a large surface area in proportion to the mass and which require only surface heating, e.g., baking finishes on sheet products.

Ventilation The vapors and gases evolved from the charge during baking processes are often flammable, and the continuous discharge of these products from the oven chamber is essential for protection against explosions. For detailed recommendations, see *Pamphlet* 74 of the Assoc. Factory Mutual Fire Ins. Cos., Boston.

DIELECTRIC HEATING

The term relates to the heat developed in dielectric materials, such as rubber, glue, textiles, paper, and plastics, when exposed to an alternating electric field. The material to be heated is placed between plate-form electrodes, as indicated in Fig. 7.5.4. It is not necessary that the electrodes be in contact with the charge; hence continuous heating is often practicable.

Fig. 7.5.4   Assembly for dielectric heating.

If the material of the charge is homogeneous and the electric field uniform, heat is developed uniformly and simultaneously throughout the mass of the charge. The thermal conductivity of the material is a negligible factor in the rate of heating. The temperatures and services are within the oven classification.

The frequency and voltage for this class of service depend in each case on the electrical properties of the material of the charge at the temperature specified for the heat application. The frequencies in use range from 2 to 40 MHz; the most common frequencies are from 10 to 30 MHz. It is advisable to select the frequency for heating by trial.

The upper limit of voltage across the electrodes is fixed by the sparkover value and by corona. The permissible voltage gradient depends on the nature of the material of the charge. Values within the range 2,000 to 6,000 V/in (790 to 2,400 V/cm) are found in practice; the voltage across the electrodes should not exceed 15,000 V.

Applications of dielectric heating include setting glue as in plywood manufacture, curing rubber, drying textiles, and the heat treatment of plastics.

INDUCTION HEATING

In induction heating, the lateral surface of the charge is exposed to an alternating magnetic flux. The currents thus induced in the charge flow wholly within its mass. The term **eddy-current heating** is sometimes applied to the method.

A common assembly, if the charge is to be heated to a temperature below its melting point, is to place the charge within a coil as indicated in Fig. 7.5.5. An alternating current in the coil establishes the required alternating magnetic flux around the charge.

A peculiar feature of such assemblies, termed **induction heaters,** is the absence of heat insulation; the coil is water-cooled. Thus, the charge is heated in the open air, or an artificial atmosphere can be used, if the assembly is enclosed. This requires rapid heating with heat cycles measured in minutes or seconds.

The frequency required is a function of the electric and magnetic properties of the charge at the temperature specified for the heat application and of the radius, or one-half the thickness, of the charge. This frequency for a given material increases with decrease of the dimension noted. The frequency in any case is not critical. In practice, 480, 960, 3,000, and 9,600 Hz and around 450 kHz suffice for the entire range of

Fig. 7.5.5   Assembly for induction heating.

induction heating. The highest frequencies needed are those for heating steel charges to temperatures above the Curie point. About ½ in (1¼ cm) diam in this case is the lower limit for 9,600 Hz. This limit dimension is decreased for steel heated to temperatures below the Curie point and for all charges of nonferrous materials.

The operation can be either batch heating or continuous heating as required. Applications include heating for forging, for annealing, for hardening steel, for brazing, soldering, and strain relief. As most of the heat is developed within the annular zone of the charge, the method is particularly well adapted to heating steel parts for surface hardening. A recent application of induction heating is the raising in temperature of billet-size ingots for rolling into merchant bars.

ARC FURNACES

Three types of electric arc furnaces (EAFs) are in common use for melting and refining of ferrous metals: (1) The three-phase ac furnace and (2) the single-phase dc furnace are used primarily for melting; and (3) the ladle furnace (LF) is used for refining of carbon, alloy, and stainless steels. Some EAFs are used for melting and refining of alloy or stainless steels when capital costs prohibit the purchase of ladle furnaces or refining vessels such as the argon-oxygen decarburization (AOD) or vacuum degasser (VD).

Three-Phase Arc Furnaces   The general design of this type of furnace is shown in Fig. 7.5.6. In operation, typically each heat is started by loading the water-cooled and refractory-lined furnace body with scrap and/or alternative iron units (AIUs) such as pig-iron or direct reduced iron in one of two forms, balls or brickets (termed DRI for direct reduced iron balls or HBI for hot briquetted iron). The scrap or AIU is batch-charged into the furnace by using a clamshell charging bucket and dropping the materials when the roof is swung off the furnace, or with the roof remaining on the furnace, continuously charged either through the side of the furnace or through the roof. Arcs are then drawn between the lower ends of the graphite electrodes through the materials; melting proceeds under automatic control until the hearth carries sufficient molten metal to tap the furnace. Many furnaces carry a molten hot heel of liquid metal to aid in the melting of each batch; this

Fig. 7.5.6 Three-phase electric arc furnace (EAF) with water-cooled panels and basic refractory lining.

metal is left on the furnace hearth between heats. The furnace tilts forward for tapping the molten metal and also tilts backward when required to aid in removing slag or to inspect the bottom refractory material. The slag door is opposite the pouring area of the furnace. Melting is aided by burners and injectors mounted in the furnace sidewalls. Burners are typically the oxygen and natural-gas type and are capable of injecting oxygen only when the scrap and AIU are sufficiently melted. Lime, carbon, and other materials can be injected through special nozzles designed into the burners or mounted along with the burners. The injected materials are used to foam the slag for efficient thermal transfer of the arc.

Water-cooled components are used as the majority of the upper sidewalls and roof. Refractory materials are used in the hearth and lower sidewall area and the center section of the roof around the area where the electrodes penetrate the furnace.

Water-Cooled Components Water-cooled elements can be designed as either spray-cooled, tubular or cast sections or panels. The cast sections can be cast iron or copper and have water passages designed into the cast body. The tubular panels are water-flooded and continuously flow water. The spray-cooled panels have spray nozzles arranged to cover the entire surface of the cooled area.

Refractories Furnaces that produce foundry steels operate with acid or basic linings. Acid-lined furnaces have silica brick lining the majority of the walls; the hearth material is ganister or the equivalent. Refractory roofs are typically alumina or silica-alumina materials. The slags of acid-lined furnaces remove no phosphorous or sulfur. Acid lining materials are the least costly refractory materials. To remove phosphorous or sulfur, basic lining materials are used. Magnesite or dolomite materials are the most common basic lining materials used. These materials are well suited for high-lime and FeO content slags that promote the reduction of phosphorus. Sulfur is normally removed during refining in the ladle furnace when a new slag is built, and the old slag is limited during the EAF tapping process. Slag covering the molten bath serves in refining the metal and reduces the heating of wall and roof panels or brick. A foamy slag practice wherein a deep, foamy slag is employed is essential for good thermal transfer of heat from the arc and to prevent the arc from damaging the sidewall and roof panels or refractory materials.

Temperature Arc temperatures approximate 6,300° F (3,500°C); therefore the EAF operation must be carried out so as to protect the

water-cooled elements and refractories as much as possible. During the initial melting phase, the electrodes bore down through the scrap or other materials. The intense heat of the arc is confined to the charge materials which absorb the heat and melt onto the hearth. When the full charge has melted sufficiently, larger quantities of oxygen are supplied to reduce the carbon and silicon in the charge materials to oxides; the oxides are contained in the slag or liberated as gases. The CO generated is combined with oxygen to produce CO_2 either in the slag or in the exhaust gas. The reaction of CO to CO_2 is exothermic and benefits the melting operation when CO_2 is produced (termed *post combustion*) in the EAF. The CO plus oxygen reaction is termed a **chemical energy** reaction. A typical modern EAF heat balance is given in Table 7.5.2.

Table 7.5.2   EAF Heat Balance

Inputs	Old EAF, %	New EAF, %
Electricity	82	65
Burners	0	5
Chemical energy	18	30
Outputs		
Steel	57	57
Slag	11	10
Losses		
Cooling water	6	10
Off-gas	7	20
Miscellaneous	16	3

The EAF was intended initially for alloy and stainless grades of steel; currently, however, the EAF method accounts for approximately 50 percent of all steel produced in North America. In the 1960s and 1970s, the EAF was used to produce long products—rebar and merchant bar. This was followed by production of structural and plate products, and most recently into the flat products area with the development of thin slab casting. EAF producers are attempting to further displace integrated steel producers with a high percentage usage of AIUs which contain low residual alloy levels (copper, nickel, chromium, and molybdenum) in the charge materials. Combining EAF steels with the secondary refining of steel via vacuum processing for the reduction of dissolved gases (hydrogen, oxygen, and nitrogen) promotes a seamless appearance in the steel when compared to steel produced via the integrated steel mill method. Residual elements and dissolved gases have a great effect on the quality of the steel and must be carefully controlled.

The modern electric arc furnace is designed (1) to hold a hot heel of liquid steel at all times to improve arc stability during melting; (2) for a "one-bucket" charge to minimize the heat loss during roof-off operation, or to have a continuous charging system; (3) to have water-cooled panels in the sidewalls and roof; (4) to have a bottom tapping port; and (5) to use a large percentage of chemical energy in burner gas and blowing oxygen. EAFs today typically use between 380 and 650 kWh of energy to produce 1 ton of steel. Theoretically, at 100 percent efficiency, electric energy consumption will be only 320 kWh/ton. The factors that influence the specific consumption of electric energy are the charge weight, tapping weight, weight of fluxes and charge carbon added to the charge materials, tapping temperature, time to produce the heat, number of scrap charges, time between heats, amount of burner gas and distribution points of the gas, amount of blowing oxygen and distribution points of the blowing oxygen, injection of carbon and flux materials into the slag during melting, and type of materials used for melting (including the use of hot metal).

Charging Currently, the majority of furnaces are charged with a bucket of material dropped into the furnace with the roof swung aside. In the past 20 years, continuous charging methods have been developed:

1. The CONSTEEL® process—continuous scrap charging through the furnace sidewall

2. Roof port opening for continuous charging of AIUs

3. Shaft furnace process—continuous feed of scrap through the roof from a specially designed shaft containing the scrap

4. Semicontinuous charging of hot metal through the furnace side-wall or door

The EAF is usually sized by the amount of steel needed for tapping; i.e., the furnace is a 20-ton or 100-ton or 150-ton, etc., furnace, and the rate of production is in tons per hour. Table 7.5.3 summarizes the furnace sizing.

The EAF can be powered by ac, three-phase, or dc. The progress in high-power semiconductor switching technology brought into existence low-cost, efficient dc power supplies. The decision to go with ac or dc technology is usually determined by the power supply available from the local utility company. DC furnaces are more prevalent in areas with small energy-generating capability.

Ladle Furnaces   The technology for today's ladle furnace has its roots in stirring technology. Heating steel in a ladle is a surface reaction;

Table 7.5.3   Three-Phase Arc Furnace Sizes

Shell ID, ft	Charge weight, tons	Nominal power, kVA	Nominal production, tons/h
7	4	2,000	2
9	10	5,000	5
11	29	12,000	10
12.5	40	18,000	22
15	65	40,000	60
20	118	90,000	118
24	195	120,000	215
26	240	150,000	260

the steel must be stirred to achieve temperature homogeneity. Three methods of stirring are (1) induction, (2) porous plug, or (3) top stirring lance. A combination of two or all three of these methods, is possible. The ladle furnace treatment of steel has virtually eliminated the need for refining of plain carbon steels. The uniformity of temperature and chemistry in steel treated in a ladle furnace is remarkable. The preponderance of ladle furnaces in operation today are ac, three-phase furnaces that heat at rates of 5 to 10°F/min. The key to achieving maximum steelmaking production is to use the ladle furnace as a buffer to the melting furnace and continuous casting operation. Ladle furnace designs can be fixed roof or slewing and typically incorporate alloy and flux materials handling systems into their designs. The control systems are PLC-based and can be simple or highly automated. The three basic requirements of a ladle furnace are (1) power source for heating, (2) stirring source for mixing, and (3) a slag system for protection or entrainment.

Arcs   The arc in each phase is maintained between the lower end of the electrode and the top of the charge (or bath, after the molten state is reached). Higher voltages can serve for melting as the size of the furnace increases; thus, where a 7-ft (2.1-m) diam furnace employs 250 V as its highest melting potential, a 15-ft furnace would use 350 V or higher as the top tap. For such a furnace constructed with water-cooled sidewall and roof panels, the application of 500 V would be normal. The furnace transformer is universally of the motor-operated tapchanger type, and in the case of, say, a 10,000 kVA at 55°C rise substation, a secondary voltage variation of more than 150 V is customary. The range of lower voltages used for refining the molten metal is obtained by changing the primary of the main transformer from delta to star connection; this reduces both voltage and capacity to 58 percent of their values with delta primary connection. If, say, 12 tons of steel scrap are to be melted down to fluid in 1 h, then the electric energy needed will approximate 5,000 kWh. With 245 V used as the principal melt-down voltage, the current per phase will have to average close to 12,000 A. A 12-in (30-cm) diam graphite electrode would amply carry this current. Small to large furnaces operate with 600 kVa and even higher powering per ton of charge.

High-Voltage AC Operations

In recent years a number of EAF operations have retrofitted new electric power supplies to supply higher operating voltages. Energy losses

in the secondary circuit are dependent on the secondary circuit reactance and to a greater extent on the secondary circuit current. If power can be supplied at a higher voltage, the current will be lower for the same power input. Operation with a lower secondary circuit current will also result in lower electrode consumption. Thus it is advantageous to operate at as high a secondary voltage as is practical. Of course, this is limited by arc flare to the sidewall and the existing furnace input power. A good foamy slag practice can allow voltage increases of up to 100 percent without adversely affecting flare to the furnace sidewalls. Energy losses can be minimized when reactance is associated with the primary circuit.

Supplementary reactance is not a new technology. In the past, supplementary reactors were used to increase arc stability in small furnaces where there was insufficient secondary reactance. In the past few years, however, this method has been used to increase the operating voltages on the EAF secondary circuit. This is achieved by connecting a reactor in series with the primary windings of the EAF transformer. The insertion of the series reactor drops the secondary voltage to limit the amount of power transferred to the arc. To compensate for this, the furnace transformer secondary voltage is increased, allowing operation at higher arc voltages and lower electrode currents. Typically, new electric equipment must be provided. Some of the benefits attributed to this type of operation are

- More stable arc than for standard operations
- Electrode consumption reduced by 10 percent
- Secondary voltage increased by 60 to 80 percent
- Power savings of 10 to 20 kWh/ton
- System power factor of approximately 0.72
- Furnace power factor of approximately 0.90
- Lower electrical losses due to lower operating current
- Voltage flicker reduced up to 40 percent

Regulation   The characteristics of an arc furnace circuit for a given applied voltage are shown in Fig. 7.5.7. For each voltage there is a value of current that gives maximum power in the furnace. This optimum current is the basis of the regulation of the circuit.

Fig. 7.5.7   Characteristics of an arc furnace circuit.

The control of the power input into direct-arc electric furnaces is effected by the adjustment of the arc length. To accomplish this, the electrode arms are positioned in the "raise" or in the "lower" direction by an automatic regulator. This regulator, which responds within a few cycles, causes the electrode arms to be lowered by gravity when voltage is obtained by closing the circuit breaker. As soon as contact between electrode and scrap charge is established, melting current flows, and this current, whenever excessive, immediately activates a winch motor or hydraulic drive to elevate that particular electrode arm and electrode by the distance corresponding to the diminution in power input needed just at that instant.

The most common type of electrode regulator is an impedance-based device which monitors analog signals of voltage and current. Voltage moves the electrode lower; current raises it. PLC- and PC-based regulators, now also common, convert analog signals to digital form and provide regulation similar to the analog device. An advantage of PLC- and PC-based regulators lies in their ability to capture the signals for use in developing an automatic furnace operation program.

INDUCTION FURNACES

There are two basic types of metal-melting induction furnaces: (1) coreless and (2) core-type. Both types utilize the principle of a transformer. The high-voltage circuit is coupled with that of the low voltage without directly connecting the two circuits. The element responsible for this coupling effect is the magnetic field. Induction heating utilizes the property of the magnetic field, which enables heat to be transferred without direct contact. By correctly disposing the high-voltage winding, which in the case of the induction furnace would be an induction coil or inductor, the magnetic field is directed so that the metal to be heated or melted is made to absorb energy. The temperature attainable is limited solely by the resistance to heat of the surrounding lining material. Induction heating enables any temperature to be achieved while providing for excellent regulation of temperature and metallurgical properties. Any metal which will conduct electric current can be melted in an induction furnace.

Coreless Induction Furnaces   (See Fig. 7.5.8.) This type of furnace consists of a crucible, copper coil, and framework on supports arranged for tilting and pouring. The specially designed induction coil acts as the primary of the transformer. The crucible conforms to conventional refractory practice. A rammed crucible is used for furnaces above 50 kW, and preformed crucibles are used on smaller furnaces such as laboratory units.

Fig. 7.5.8 Coreless induction furnace.

The principle of operation is essentially the same as that of the induction heater previously described. The initial charge in the furnace is cold scrap metal—pieces of assorted dimensions and shapes and a large percentage of voids. As the power is applied and the heat cycle progresses, the charge changes to a body of molten metal; additional cold

metal is added until the molten-metal level is brought to the desired temperature and metallurgical chemistry. The furnace then is tapped.

When the metal in the furnace becomes fluid, depending on whether a line frequency or medium-frequency supply by means of convectors is used, a certain electromagnetic stirring action will occur. This stirring action is peculiar to the induction furnace and aids in the production of certain types of alloys. The stirring action increases as the frequency is reduced.

Line-frequency applications are generally reserved to furnaces having a metal-holding capacity of 800 lb (360 kg) and above. There is always an ideal relationship between the size of a coreless furnace and its operating frequency. As a general rule, a small furnace gives best results at high to medium frequencies and large furnaces work best at the lower frequencies. A frequency is suited to a given furnace when it yields good, fast melting with a gentle stirring action. Too high or too low frequencies are accompanied by undesirable side effects. The tabulation below gives the charge weights and frequencies generally to be used:

Charge weight, lb	Frequency, Hz
2–50	9,600
12–500	3,000
200–15,000	960
800–75,000	60

The coreless induction furnace is usually charged full and tapped empty, although at line frequencies, it may be necessary to retain a certain amount of metal in the furnace to continue the operation, since it is difficult to start the furnace with small metal particles, such as turnings and borings, in a cold crucible. As a result, it is general practice to retain a heel in the furnace of about one-third its molten-metal volume. This problem can be avoided in furnaces of higher frequencies, where start-up can be performed with small-size metal charges without carrying the heel.

Coreless induction furnaces are particularly attractive for melting charges and alloys of known analysis; in essence, the operation becomes one of metal melting with rapidly absorbed electric heat without disturbing the metallurgical properties of the initial charge.

These furnaces are supplied from a single-phase source. In order to obtain a balanced three-phase input, it is necessary specifically to design the electrical equipment for the inclusion of capacitors and suitable reactors, which are generally automatically switched (by inductance changes) during the operation in order to provide a reasonably high power factor. Power factors on such furnaces can be kept at or near unity. In high-frequency coreless induction furnaces, high power factors are necessary to prevent overburdening the motor-generator equipment.

Core-Type Induction Furnaces   (See Fig. 7.5.9.) The transformer is actually wound to conform to a typical transformer design having an iron core and layers of wire acting as a primary circuit. The melting channel acts as a ring short circuit around this transformer in the melting chamber. According to the desired melting capacity, one, two, or three such transformers (or **inductors,** as they are called) may be added to the furnace shell. At all times, the channel must hold sufficient metal to maintain a

Fig. 7.5.9 Core-type induction furnace.

short circuit around the transformer core. Air cooling is used as required to prevent undue heating of the inductor coils and magnetic cores.

The melting output is controlled by varying the voltage supplied to the inductors with the aid of a variable-voltage transformer connected to the primary circuit of the supply. Core-type furnaces always use line frequencies. Voltage or power-input regulation, therefore, can be performed by adjusting the tap setting of the transformer feeding the furnace transformer attached to the furnace shell. These transformers are single-phase units, and by using three such units, a balanced three-phase input can be obtained. The current flowing through the primary inductors by transformation causes a much larger current in the metal loop, whose resistance creates heat for melting.

The core-type furnace is the most efficient type of induction furnace because its iron core concentrates magnetic flux in the area of the magnetic loop, ensuring maximum power transfer from primary to secondary. Efficiency in the use of power can be as high as 95 to 98 percent.

The essential loop of metal must always be maintained in the core-type furnace. If this loop is allowed to freeze by cooling, extreme care is necessary in remelting because the loop may rupture and disrupt the circuit. This could require extensive work in dismantling the coil and restoring the loop. Consequently, core-type furnaces rarely are permitted to cool. This makes alloy changes difficult because a heel of molten metal always is required.

The relatively narrow melting channels must be kept as clean as possible since a high metal temperature exists in this loop. Nonmetallics or tramps in the charge metal tend to accumulate on the walls in the channel area, restricting the free flow of metal and ultimately closing the passage.

This furnace is particularly useful for melting of nonferrous metals such as aluminum, copper, copper alloys, and zinc.

POWER REQUIREMENTS FOR ELECTRIC FURNACES

The **energy required for melting metals** in electric furnaces varies for a given metal or alloy with the size of the furnace, the thickness of the

Table 7.5.4   Energy Consumption of Electric Furnaces

Process	Type of furnace	lb/kWh
Baking finishes on sheet metal	Batch oven	10–18
Baking finishes on sheet metal	Continuous oven	25–30
Baking bread	Continuous oven	10–12
Annealing brass and copper	Batch furnace	10–25
Annealing steel	Batch furnace	5–15
Hardening steel	Batch furnace	7–11
Tempering steel	Batch furnace	15–25
Annealing glass	Continuous furnace	40–100
Vitreous enameling, single coat	Batch furnace	5–8
Vitreous enameling, single coat	Continuous furnace	10–15
Galvanizing	Batch furnace	12–20

Melting metals	Type of furnace	kWh/ton (2,000 lb)
Lead	Resistor	40–50
Solder 50–50	Resistor	40–50
Tin	Resistor	35–50
Zinc	Induction	80–100
Brass	Arc and induction	250–400
Steel, melting only	Arc and induction	450–700
Steel, melting and refining	Arc	600–750
Gray iron	Arc and induction	450–600

Furnace products	Type of furnace	kWh/ton (2,000 lb)
Aluminum	Electrolytic	22,000–27,000
Calcium carbide	Resistance	3,000–6,000
Ferroalloys	Resistance	4,000–8,000
Graphite	Resistance	3,000–8,000
Phosphoric acid	Resistance	5,000–6,000
Silicon carbide	Resistance	8,500–10,000
Smelting iron ore	Resistance	1,650–2,400

refractory lining, the temperature of the molten metal, the rate of melting, and with the degree of the continuity of the operation of the furnace. An estimated efficiency of 50 to 60 percent is often used for preliminary purposes. As is well known, 3- to 6-ton direct-arc furnaces often are used to tap acid foundry steels with the consumption of less than 500 kWh to the ton, and large ingot furnaces of this same type, operating basic-lined on common steels for ingots, give even better results despite the call for several more charges of scrap per heat.

Average values in kWh/ton of molten metal are as follows: yellow brass, 200 to 350; red brass, 250 to 400; copper, 250 to 400; lead, 30 to 50; steel melting, when making high-quality double-slag basic heats, 650 to 800 (Table 7.5.4).

Electrode consumption varies considerably in arc furnaces because of their different constructions and operations. Average values in pounds of electrode per ton of molten metal are: steel melting, with graphite electrodes, 5 to 10; brass melting, with graphite electrodes, 3 to 5. Graphite electrodes have largely superseded carbon electrodes.

SUBMERGED-ARC AND RESISTANCE FURNACES

The resistance furnace is essentially a refractory-lined chamber with electrodes—movable or fixed—buried in the charge. This simplicity permits a wide range of designs and much latitude in dimensions. The general service is heating charges of a refractory nature to bring about chemical reactions or changes in the physical structure of the material of the charge. The energy requirement of each of such processes is a large item in the cost of production. Large units and a favorable power location are the rule. Resistance furnaces also are termed submerged-arc furnaces and/or, in quite a few instances, smelting-type furnaces.

The only limit on the temperature to which a charge can be heated by this method is the temperature at which the materials of the charge are vaporized. For temperatures beyond the limit of refractory linings, the materials of the charge are used to form a protective layer between the core of the charge (through which the current passes) and the walls of the furnace.

Resistance furnaces with movable electrodes may be either single-phase or polyphase. The materials of the charge are fed more or less continuously, and the product is discharged intermittently or continuously as required. In some cases the product is in the molten state; in others the product is a vapor. The usual method of operation is the use of a single operating voltage and a constant power input. The power is regulated by adjustment of the depths of the electrodes in the charge. The load is fairly uniform and, if polyphase, is kept reasonably well balanced.

The **resistance furnace with fixed electrodes** is designed for heating materials in batches and is usually rectangular in shape with an electrode at each end for single-phase operation. The length and cross-sectional area of the path of the current are proportioned to suit the power characteristics of the charge. Refractory materials have negative temperature-resistance coefficients, and hence to maintain constant power in the furnace circuit the applied voltage must be reduced as the temperature of the charge rises in proportion to the square root of the ratio of the initial resistance of the furnace circuit to the resistance of the furnace circuit at the end of the heat cycle. If the materials of the charge are nonconductors at room temperature, a starting circuit is provided by means of a core of carbon—usually coke—placed in the charge. The heat cycles of furnaces of this class generally extend over a period of several days.

Some of the more common **uses of the resistance furnace** are:

Calcium carbide furnaces are charged continuously with lime and coke. These equipments can be either open or closed top. This type of furnace has been built up to 70,000 kVA in electrical powering—covered and sealed for gas collection.

Ferroalloy furnaces for the production of ferrochrome, ferrosilicon, ferromanganese, etc., are usually three-phase furnaces with movable electrodes and are similar in construction to the three-phase arc furnace. The charge is a mixture of the ore (oxide) of the selected metal, scrap iron, and a reducing agent, generally carbon except for very low carbon content alloys, for which some other reducing agent such as aluminum or silicon is required. Six-electrode furnaces often are used for power inputs of 15,000 kVA and more.

The **graphitizing furnace** is of the single-phase batch type. Artificial graphite is made by heating amorphous carbon (coal or coke) while shielded from air to a high temperature—around 4,500°F (2,500°C). The presence of some metallic impurity, such as iron oxide, in the charge appears to be necessary for the conversion of amorphous carbon to graphitic carbon. The raw material for making bulk graphite constitutes both the charge and the protective layer around the core of the charge. Graphite shapes are made from the corresponding shapes of amorphous carbon which are embedded—between the electrodes—in raw material as noted for the manufacture of bulk graphite.

The **silicon carbide furnace** is similar to the graphitizing furnace. The charge is a mixture of sand (silica), coke, sawdust, and a small amount of salt. This mixture is packed around a core of granulated coke to form the initial circuit between the electrodes. The sand and coke are the reacting materials. The sawdust serves to make the charge porous so that the gases formed during the heating of the charge can escape freely. The salt vaporizes and removes impurities, such as iron, in the form of chlorides. The temperature of the process is 2,700 to 3,400°F (1,500 to 1,880°C).

Machine Elements

BY

ALI M. SADEGH *Professor of Mechanical Engineering, The City College of The City University of New York*
GEORGE W. MICHALEC *Late Consulting Engineer*
VITTORIO (RINO) CASTELLI *Senior Research Fellow, Retired, Xerox Corp.; Engineering Consultant*
MICHAEL W. WASHO *Senior Engineer, Retired, Eastman Kodak Company*
JOHN W. WOOD, JR. *Manager of Technical Services, Garlock Klozure*
JAMES DRAGO, P.E. *Manager, Engineering, Garlock Sealing Technologies*
EUGENE A. AVALLONE *Consulting Engineer; Professor of Mechanical Engineering, Emeritus, The City College of The City University of New York*

8.6 PACKINGS, GASKETS, AND SEALS
by John W. Wood, Jr., and James Drago, P.E.

8.7 PIPE, PIPE FITTINGS, AND VALVES
Revised by E. A. Avallone

8.1 MECHANISM
by Ali M. Sadegh

REFERENCES: Beggs, "Mechanism," McGraw-Hill. Hrones and Nelson, "Analysis of the Four Bag Linkage," Wiley. Jones, "Ingenious Mechanisms for Designers and Inventors," 4 vols., Industrial Press. Moliam, "The Design of Cam Mechanisms and Linkages," Elsevier. Chironis, "Gear Design and Application," McGraw-Hill.

NOTE: The reader is referred to the current and near-past professional literature for extensive material on linkage mechanisms. The vast number of combinations thereof has led to the development of computer software programs to aid in the design of specific linkages.

Definition A **mechanism** is that part of a machine which contains two or more pieces so arranged that the motion of one compels the motion of the others, all in a fashion prescribed by the nature of the combination.

LINKAGES

Links may be of any form so long as they do not interfere with the desired motion. The simplest form is four bars *A, B, C,* and *D*, fastened together at their ends by cylindrical pins, and which are all movable in parallel planes. If the links are of different lengths and each is fixed in

Fig. 8.1.1 Beam-and-crank mechanism.

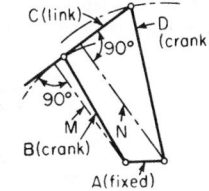

Fig. 8.1.2 Drag-link mechanism.

turn, there will be four possible combinations; but as two of these are similar there will be produced three mechanisms having distinctly different motions. Thus, in Fig. 8.1.1, if *D* is fixed *A* can rotate and *C* oscillate, giving the **beam-and-crank** mechanism, as used on side-wheel steamers. If *B* is fixed, the same motion will result; if *A* is fixed (Fig. 8.1.2), links *B* and *D* can rotate, giving the **drag-link** mechanism used

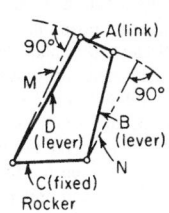

Fig. 8.1.3 Rocker mechanism.

to feather the floats on paddle wheels. Fixing link *C* (Fig. 8.1.3), *D* and *B* can only oscillate, and a **rocker** mechanism sometimes used in straight-line motions is produced. It is customary to call a rotating link a **crank;** an oscillating link a **lever,** or beam; and the connecting link a **connecting rod,** or **coupler.** Discrete points on the coupler, crank, or lever can be pressed into service to provide a desired motion. The fixed link is often enlarged and used as the supporting frame.

If in the linkage (Fig. 8.1.1) the pin joint *F* is replaced by a slotted piece *E* (Fig. 8.1.4), no change will be produced in the resulting motion, and if the length of links *C* and *D* is made infinite, the slotted piece *E* will become straight and the motion of the slide will be that of pure translation, thus obtaining the engine, or **sliding-block, linkage** (Fig. 8.1.5).

If in the sliding-block linkage (Fig. 8.1.5) the long link *B* is fixed (Fig. 8.1.6), *A* will rotate and *E* will oscillate and the infinite links *C* and

D may be indicated as shown. This gives the **swinging-block linkage.** When used as a quick-return motion the slotted piece and slide are usually interchanged (Fig. 8.1.7) which in no way changes the resulting motion. If the short link *A* is fixed (Fig. 8.1.8), *B* and *E* can both rotate,

Figs. 8.1.4 and 8.1.5 Sliding-block linkage.

and the mechanism known as the **turning-block linkage** is obtained. This is better known under the name of the **Whitworth quick-return motion,** and is generally constructed as in Fig. 8.1.9. The **ratio of time of advance to time of return** *H/K* of the two quick-return motions (Figs. 8.1.7 and

Fig. 8.1.6 Swinging-block linkage.

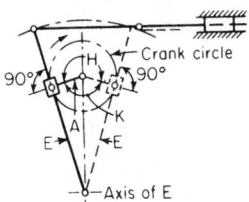

Fig. 8.1.7 Slow-advance, quick-return linkage.

8.1.9) may be found by locating, in the case of the swinging block (Fig. 8.1.7), the two tangent points (*t*) and measuring the angles *H* and *K* made by the two positions of the crank *A*. If *H* and *K* are known, the axis of *E* may be located by laying off the angles *H* and *K* on the crank circle

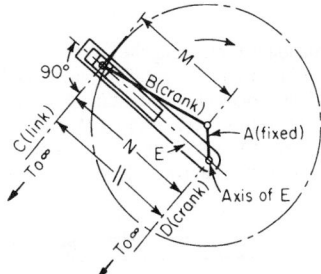

Fig. 8.1.8 Turning-block linkage.

and drawing the tangents *E,* their intersection giving the desired point. For the turning-block linkage (Fig. 8.1.9), determine the angles *H* and *K* made by the crank *B* when *E* is in the horizontal position; or, if the angles are known, the axis of *E* may be determined by drawing a horizontal line through the two crankpin positions (*S*) for the given angle, and the point where a line through the axis of *B* cuts *E* perpendicularly will be the axis of *E*.

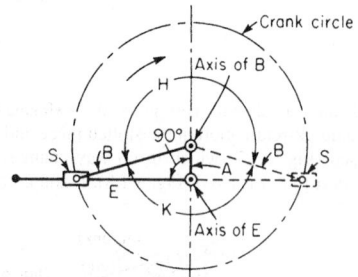

Fig. 8.1.9   Whitworth quick-return motion.

Velocities of any two or more points on a link must fulfill the following conditions (see Sec. 3). (1) Components along the link must be equal and in the same direction (Fig. 8.1.10): $V_a = V_b = V_c$. (2) Perpendiculars to V_A, V_B, V_C from the points *A*, *B*, *C* must intersect at a common point *d*, the **instant center** (or instantaneous axis). (3) The velocities of points *A*, *B*, and *C* are directly proportional to their distances from this center (Fig. 8.1.11): $V_A/a = V_B/b = V_C/c$. For a straight link the tips of

Fig. 8.1.10   **Fig. 8.1.11**

the vectors representing the velocities of any number of points on the link will be on a straight line (Fig. 8.1.12); *abc* = a straight line. To find the velocity of any point when the velocity and direction of any two other points are known, condition 2 may be used, or a combination of conditions 1 and 3. The **linear velocity ratio** of any two points on a

Fig. 8.1.12

linkage may be found by determining the distances *e* and *f* to the instant center (Fig. 8.1.13); then $V_c/V_b = e/f$. This may often be simplified by noting that a line drawn parallel to *e* and cutting *B* forms two similar triangles *efB* and *sAy*, which gives $V_c/V_b = e/f = s/A$. The **angular velocity ratio** for any position of two oscillating or rotating links *A* and *C* (Fig. 8.1.1), connected by a movable link *B*, may be determined by

Fig. 8.1.13

scaling the length of the perpendiculars *M* and *N* from the axes of rotation to the centerline of the movable link. The angular velocity ratio is inversely proportional to these perpendiculars, or $O_C/O_A = M/N$. This method may be applied directly to a linkage having a sliding pair if the two infinite links are redrawn perpendicular to the sliding pair, as indicated in Fig. 8.1.14. *M* and *N* are shown also in Figs. 8.1.1, 8.1.2, 8.1.3, 8.1.5, 8.1.6, and 8.1.8. In Fig. 8.1.5 one of the axes is at infinity; therefore, *N* is infinite, or the slide has pure translation.

Fig. 8.1.14

Forces   A mechanism must deliver as much work as it receives, neglecting friction; therefore, the force at any point *F* multiplied by the velocity V_F in the direction of the force at that point must equal the force at some other point *P* multiplied by the velocity V_P at that point; or the forces are inversely as their velocities and $F/P = V_P/V_F$. It is at times more convenient to equate the moments of the forces acting around each axis of rotation (sometimes using the instant center) to determine the force acting at some other point. In Fig. 8.1.15, $F \times a \times c/(b \times d) = P$.

Fig. 8.1.15

CAMS

Cam Diagram   A cam is usually a plate or cylinder which communicates motion to a follower as dictated by the geometry of its edge or of a groove cut in its surface. In the practical design of cams, the follower (1) must assume a definite series of positions while the driver occupies a corresponding series of positions or (2) must arrive at a definite location by the time the driver arrives at a particular position. The former design may be severely limited in speed because the interrelationship between the follower and cam positions may yield a follower displacement vs. time function that involves large values for the successive time derivatives, indicating large accelerations and forces, with concomitant large impacts and accompanying noise. The second design centers about finding that particular interrelationship between the follower and cam positions that results in the minimum forces and impacts so that the speed may be made quite large. In either case, the desired interrelationship must be put into hardware as discussed below. In the case of high-speed machines, small irregularities in the cam surface or geometry may be severely detrimental.

A stepwise displacement in time for the follower running on a cam driven at constant speed is, of course, impossible because the follower would require infinite velocities. A step in velocity for the follower would result in infinite accelerations; these in turn would bring into being forces that approach infinite magnitudes which would tend to destroy the machine. A step in acceleration causes a large jerk and large shock waves to be transmitted and reflected throughout the parts that generate noise and would tend to limit the life of the machine. A step in jerk, the third derivative of the follower displacement with respect to time, seems altogether acceptable. In those designs requiring or exhibiting clearance between the follower and cam (usually at the bottom of the stroke), as gentle and slow a ramp portion as can be tolerated must be inserted on either side of the clearance region to limit the magnitude of the acceleration and jerk to a minimum. The tolerance on the clearance adjustment must be small enough to assure that the follower will be left behind and picked up gradually by the gentle ramp portions of the cam.

The three most common forms of motion used are uniform motion (Fig. 8.1.16), harmonic motion (Fig. 8.1.17), and uniformly accelerated and retarded motion (Fig. 8.1.18). In plotting the diagrams (Fig. 8.1.18) for this last motion, divide ac into an even number of equal parts and bc

Fig. 8.1.16 Uniform motion.   **Fig. 8.1.17**

into the same number of parts with lengths increasing by a constant increment to a maximum and then decreasing by the same decrement, as, for example, 1, 3, 5, 5, 3, 1, or 1, 3, 5, 7, 9, 9, 7, 5, 3, 1. In order to prevent shock when the direction of motion changes, as at a and b in the uniform motion, the harmonic motion may be used; if the cam is to be operated at high speed, the uniformly accelerated and retarded motion should preferably be employed; in either case there is a very gradual change of velocity.

Fig. 8.1.18   **Fig. 8.1.19**

Pitch Line The actual pitch line of a cam varies with the type of motion and with the position of the follower relative to the cam's axis. Most cams as ordinarily constructed are covered by the following four cases.

FOLLOWER ON LINE OF AXIS (Fig. 8.1.19). To draw the pitch line, subdivide the motion bc of the follower in the manner indicated in Figs. 8.1.16, 8.1.17, and 8.1.18. Draw a circle with a radius equal to the smallest radius of the cam $a0$ and subdivide it into angles $0a1'$, $0a2'$, $0a3'$, etc., corresponding with angular displacements of the cam for positions 1, 2, 3, etc., of the follower. With a as a center and radii $a1$, $a2$, $a3$, etc., strike arcs cutting radial lines at d, e, f, etc. Draw a smooth curve through points d, e, f, etc.

OFFSET FOLLOWER (Fig. 8.1.20). Divide bc as indicated in Figs. 8.1.16, 8.1.17, and 8.1.18. Draw a circle of radius ac (highest point of rise of follower) and one tangent to cb produced. Divide the outer circle into parts $1'$, $2'$, $3'$, etc., corresponding with the angular displacement of

Fig. 8.1.20

the cam for positions 1, 2, 3, etc., of the follower, and draw tangents from points $1'$, $2'$, $3'$, etc., to the small circle. With a as a center and radii $a1$, $a2$, $a3$, etc., strike arcs cutting tangents at d, e, f, etc. Draw a smooth curve through d, e, f, etc.

ROCKER FOLLOWER (Fig. 8.1.21). Divide the stroke of the slide S in the manner indicated in Figs. 8.1.16, 8.1.17, and 8.1.18, and transfer these points to the arc bc as points 1, 2, 3, etc. Draw a circle of radius ak and divide it into parts $1'$, $2'$, $3'$, etc., corresponding with angular

Fig. 8.1.21

displacements of the cam for positions 1, 2, 3, etc., of the follower. With k, $1'$, $2'$, $3'$, etc., as centers and radius bk, strike arcs kb, $1'd$, $2'e$, $3'f$, etc., cutting at $bdef$ arcs struck with a as a center and radii ab, $a1$, $a2$, $a3$, etc. Draw a smooth curve through b, d, e, f, etc.

CYLINDRICAL CAM (Fig. 8.1.22). In this type of cam, more than one complete turn may be obtained, provided in all cases the follower returns to its starting point. Draw rectangle $wxyz$ (Fig. 8.1.22) representing the development of cylindrical surface of the cam. Subdivide the desired motion of the follower bc horizontally in the manner indicated in Figs. 8.1.16, 8.1.17, and 8.1.18, and plot the corresponding angular displacement $1'$, $2'$, $3'$, etc., of the cam vertically; then through the intersection of lines from these points draw a smooth curve. This may best be shown by an example, assuming the following data for the

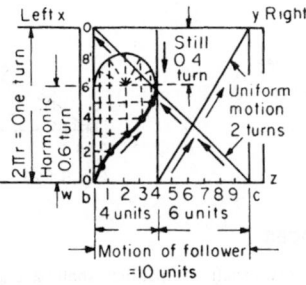

Fig. 8.1.22 Cylindrical cam.

diagram in Fig. 8.1.22: Total motion of follower = *bc*; circumference of cam = $2\pi r$. Follower moves harmonically 4 units to right in 0.6 turn, then rests (or "dwells") 0.4 turn, and finishes with uniform motion 6 units to right and 10 units to left in 2 turns.

Cam Design In the practical design of cams the following points must be noted. If only a small force is to be transmitted, sliding contact may be used, otherwise **rolling contact.** For the latter the pitch line must be corrected in order to get the true slope of the cam. An approximate construction (Fig. 8.1.23) may be employed by using the pitch line as the center of a series of arcs the radii of which are equal to that of the follower roll to be used; then a smooth curve drawn tangent to the arcs will give the slope desired for a roll working on the periphery of the cam

Fig. 8.1.23

(Fig. 8.1.23*a*) or in a groove (Fig. 8.1.23*b*). For plate cams the roll should be a small cylinder, as in Fig. 8.1.24*a*. In cylindrical cams it is usually sufficiently accurate to make the roll conical, as in Fig. 8.1.24*b*, in which case the taper of the roll produced should intersect the axis of the cam. If the pitch line *abc* is made too sharp (Fig. 8.1.25) the follower

Fig. 8.1.24   Plate cam.

will not rise the full amount. In order to prevent this **loss of rise,** the pitch line should have a radius of curvature at all parts of not less than the roll's diameter plus ⅛ in. For the same rise of follower, *a*, the angular motion of the cam, *O*, the slope of the cam changes considerably, as indicated by the heavy lines *A*, *B*, and *C* (Fig. 8.1.26). Care should be taken to keep a moderate slope and thereby keep down the side thrust on the follower, but this should not be carried too far, as the cam would become too large and the friction increase.

Fig. 8.1.25   **Fig. 8.1.26**

ROLLING SURFACES

In order to connect two shafts so that they shall have a definite angular velocity ratio, rolling surfaces are often used; and in order to have no

slipping between the surfaces they must fulfill the following two conditions: the line of centers must pass through the point of contact,

Fig. 8.1.27

and the arcs of contact must be of equal length. The angular velocities, expressed usually in r/min, will be inversely proportional to the radii: $N/n = r/R$. The two surfaces most commonly used in practice, and the only ones having a constant angular velocity ratio, are cylinders where the shafts are parallel, and cones where the shafts (projected) intersect at an angle. In either case there are two possible directions of rotation, depending upon whether the surfaces roll in opposite directions (external contact) or in the same direction (internal contact). In Fig. 8.1.27, $R = nc/(N + n)$ and $r = Nc/(N + n)$; in Fig. 8.1.28, $R = nc/(N - n)$ and $r = Nc/(N - n)$. In Fig. 8.1.29, $\tan B = \sin A/(n/N + \cos A)$ and $\tan C = \sin A/$

Fig. 8.1.28   **Fig. 8.1.29**

$(N/n + \cos A)$; in Fig. 8.1.30, $\tan B = \sin A/(N/n - \cos A)$, and $\tan C = \sin A/(n/N - \cos A)$. With the above values for the angles *B* and *C*, and the length *d* or *e* of one of the cones, *R* and *r* may be calculated.

Fig. 8.1.30

The natural limitations of **rolling without slip,** with the use of pure rolling surfaces limited to the transmission of very small amounts of torque, led historically to the alteration of the geometric surfaces to include teeth and tooth spaces, i.e., **toothed wheels,** or simply **gears.** Modern gear tooth systems are described in greater detail in Sec. 8.3. This brief discussion is limited to the kinematic considerations of some common gear combinations.

EPICYCLIC TRAINS

Epicyclic trains are combinations of gears in which some of or all the gears have a motion compounded of rotation about an axis and a translation or revolution of that axis. The gears are usually connected by a link called an arm, which often rotates about the axis of the first gear. Such trains may be calculated by first considering all gears locked and the arm turned; then the arm locked and the gears rotated. The algebraic sum of the separate motions will give the desired result. The following examples and method of tabulation will illustrate this. The figures on each gear refer to the number of teeth for that gear.

	A	B	C	D
Gear locked, Fig. 8.1.31	+1	+1	+1	+1
Arm locked, Fig. 8.1.31	0	−1	$+1 \times {}^{50}\!/_{20}$	$-1 \times {}^{50}\!/_{20} \times {}^{20}\!/_{40}$
Addition, Fig. 8.1.31	+1	0	$+3\frac{1}{2}$	$-\frac{1}{4}$
Gears locked, Fig. 8.1.32	+1	+1	+1	+1
Arm locked, Fig. 8.1.32	0	−1	$+1 \times {}^{30}\!/_{20}$	$+1 \times {}^{30}\!/_{20} \times {}^{20}\!/_{70}$
Addition, Fig. 8.1.32	+1	0	$+2\frac{1}{2}$	$+1\frac{3}{7}$

In Figs. 8.1.31 and 8.1.32 lock the gears and turn the arm A right-handed through 1 revolution ($+1$); then lock the arm and turn the gear B back to where it started (-1); gears C and D will have rotated the amount indicated in the tabulation. Then the algebraic sum will give the relative turns of each gear. That is, in Fig. 8.1.31, for one turn of the

Figs. 8.1.31 and 8.1.32 Epicyclic trains.

arm, B does not move and C turns in the same direction $3\frac{1}{2}$ r, and D in the opposite direction $\frac{1}{4}$ r; whereas in Fig. 8.1.32, for one turn of the arm, B does not turn, but C and D turn in the same direction as the arm, respectively, $2\frac{1}{2}$ and $1\frac{3}{7}$ r. (*Note:* The arm in the above case was turned $+1$ for convenience, but any other value might be used.)

Figs. 8.1.33 and 8.1.34 Bevel epicyclic trains.

Bevel epicyclic trains are epicyclic trains containing bevel gears and may be calculated by the preceding method, but it is usually simpler to use the general formula which applies to all cases of epicyclic trains:

$$\frac{\text{Turns of } C \text{ relative to arm}}{\text{Turns of } B \text{ relative to arm}} = \frac{\text{absolute turns of } C - \text{turns of arm}}{\text{absolute turns of } B - \text{turns of arm}}$$

The left-hand term gives the value of the train and can always be expressed in terms of the number of teeth (T) on the gears. Care must be used, however, to express it as either plus ($+$) or minus ($-$), depending upon whether the gears turn in the same or opposite directions.

$$\frac{\text{Relative turns of } C}{\text{Relative turns of } B} = \frac{C - A}{B - A} = -1 \qquad \text{(in Fig. 8.1.33)}$$

$$= + \frac{T_E}{T_C} \times \frac{T_B}{T_D} \qquad \text{(Fig. 8.1.34)}$$

OTHER MECHANISMS

Pulley Block (Fig. 8.1.35)   Given the weight W to be raised, the force F necessary is $F = V_W W/V_F = W/n =$ load/number of ropes, V_W and V_F being the respective velocities of W and F.

Differential Chain Block (Fig. 8.1.36)

$$F = V_W W/V_F = W(D - d)/(2D)$$

Fig. 8.1.35 Pulley block.   **Fig. 8.1.36** Differential chain block.

Fig. 8.1.37 Worm and worm wheel.

Fig. 8.1.38 Triplex chain block.

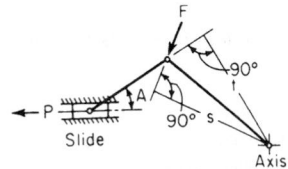

Fig. 8.1.39 Toggle joint.

Worm and Wheel (Fig. 8.1.37) $F = \pi d(n/T)W/(2\pi R) = WP(d/D)/(2\pi R)$, where n = number of threads, single, double, triple, etc.

Triplex Chain Block (Fig. 8.1.38)   This geared hoist makes use of the epicyclic train. $W = FL/\{M[1 + (T_D/T_C) \times (T_B/T_A)]\}$, where T = number of teeth on gears.

Toggle Joint (Fig. 8.1.39) $P = Fs(\cos A)/t$.

Geneva Wheel Mechanism   In this mechanism the driven member makes one-fourth of a revolution for each revolution of the driver. The slots are positioned so that a pin enters and leaves the slots tangentially (Fig. 8.1.40).

Parallel Mechanism   A parallel mechanism that is used to enlarge or to reduce movements is shown in Fig. 8.1.41. Another application of parallel mechanism that is used in drafting is shown in Fig. 8.1.42.

Fig. 8.1.40

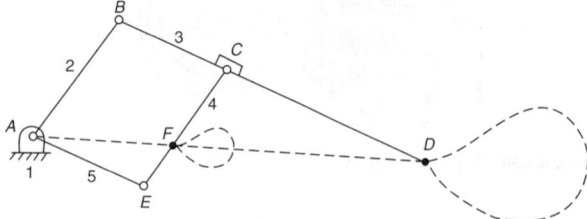

Fig. 8.1.41   Parallel mechanism with magnification.

Harmonic Drive Mechanism

Invented in the 1950s, harmonic drive or a harmonic gear system is a mechanical **speed-changing device**. Harmonic drive transmissions are noted for their high reduction ratios with concentric shafts and their

Fig. 8.1.42   Parallel mechanism.

ability to reduce backlash and vibration in a motion control system. It operates on a different principle from, and has capabilities beyond the scope of, conventional speed changers.

As shown in Figs. 8.1.43 and 8.1.44, a harmonic drive consists of three basic elements: (1) an elastically deformable (flexible) metallic thin-walled ring, having splines or gear teeth on its external surface; (2) a circular thick-walled rigid ring having the same internal splines or gear teeth as in the flexible spline; and (3) an elliptical or similar-shape cam with rotating bearings. The external flexible spline has fewer teeth than the internal circular spline. The flexible spline ring is placed inside the rigid ring. The elliptical cam with rotating bearings is mounted inside the flexible ring and forces the flexible splines (teeth) to push deeply into the rigid ring at two opposite points while rotating. The thin flexible ring deflects elastically as it rolls on the inside of the slightly larger rigid circular ring. The flexible ring rotates at a speed that is the difference in the number of teeth on the rigid ring and on the flexible ring (Fig. 8.1.43). This method basically preloads the teeth, which

Fig. 8.1.43   Harmonic drive mechanisms with two rotating bearings.

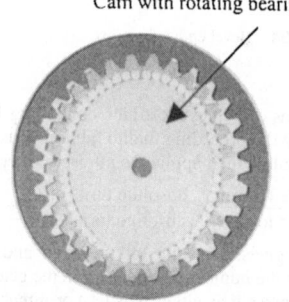

Fig. 8.1.44   Harmonic drive mechanisms with many rotating bearings.

reduces backlash. The flexible spline is made of special steel with a high resistance to fatigue.

The input (motor) shaft is attached to the oval-shaped cam with thin ball bearings placed around the outer circumference of the oval cam. The gear's output shaft is attached to the flexible spline. The circular spline is attached to the gearbox along its outer circumference. If the elliptical cam or the motor shaft rotates in a cw direction, the flexible spline will rotate in a ccw direction and vice versa.

The reduction ratio is calculated by dividing the number of teeth on the rigid ring by the difference between the numbers of teeth on the rigid and flexible rings. For example, if a fixed spline has 80 teeth and its corresponding flexible spline has 78 teeth, then the reduction ratio is $80/(80 - 78) = 40:1$.

Harmonic Drive Limitations   Harmonic drives have single-stage speed reductions of $30:1$ to $350:1$. The lower ratios are limited because of fatigue considerations on the flexible spline ring. The harmonic drive is normally not suitable for high power levels above 6 kW.

Harmonic Drive Applications   Harmonic drives offer many benefits including zero backlash, high positional accuracy, low vibration, small size and light weight, and compact design. The applications are numerous in robotics, machine tools, and medical and aerospace applications.

Slider Crank Mechanism

Kinematics   The slider crank mechanism is widely used in automobile engines, punch presses, feeding mechanisms, etc. For such mechanisms, displacements, velocities, and accelerations of the parts are important design parameters. The basic mechanism is shown in Fig. 8.1.45. The slider displacement x is given by

$$x = r\cos\theta + l\sqrt{1 - [(r/l)\sin\theta]^2}$$

where r = crank length, l = connecting-rod length, θ = crank angle measured from top dead center position. Slider velocity is given by

$$\dot{x} = V = -r\omega\left(\sin\theta + \frac{r\sin 2\theta}{2l\cos\beta}\right)$$

where ω = instantaneous angular velocity of the crank at position θ and

$$\cos\beta = \sqrt{1 - (r/l)^2\sin^2\theta}$$

Slider acceleration is given by

$$a = \ddot{x} = -r\alpha\left(\sin\theta + \frac{r}{l}\frac{\sin 2\theta}{2\cos\beta}\right)$$
$$-r\omega^2\left(\cos\theta + \frac{r}{l}\frac{\cos 2\theta}{\cos\beta} + \frac{r}{l^3}\frac{\sin^2 2\theta}{4\cos^3\beta}\right)$$

where α = instantaneous angular acceleration of the crank at position θ.

Fig. 8.1.45   Basic slider-crank mechanism.

Forces   Neglecting gravity effects, the forces in a mechanism arise from those produced by input and output forces or torques (herewith called static components). Such forces may be produced by driving motors, shaft loads, expanding cylinder gases, etc. The net forces on the various links cause accelerations of the mechanism's masses, and can be thought of as dynamic components. The static components must be borne by the various links, thus giving rise to internal stresses in those parts. The supporting bearings and slide surfaces also feel the effects of these components, as do the support frames. Stresses are also induced by the dynamic components in the links, and such components cause shaking forces and shaking moments in the support frame.

By building onto the basic mechanism appropriately designed countermasses, the support frame and its bearings can be relieved of a significant portion of the dynamic component effects, sometimes called inertia effects. The augmented mechanism is then considered to be "balanced." The static components cannot be relieved by any means, so that the support frame and its bearings must be designed to carry safely the static component forces and not be overstressed. Figure 8.1.46 illustrates a common, simple form of approximate balancing. Sizing of the countermass is somewhat complicated because the total

Fig. 8.1.46   "Balanced" slider-crank mechanism where T = center of mass, S = center of mass of crank A, and Q = center of mass of connecting rod B.

inertia effect is the vector sum of the separate link inertias, which change in magnitude and direction at each position of the crank. The countermass D is sized to contain sufficient mass to completely balance the crank plus an additional mass (effective mass) to "balance" the other links (connecting rod and slider). In simple form, the satisfying condition is approximately

$$M_D e\omega^2 = M_A a\omega^2 + M_{\text{eff}} e\omega^2$$

where a = distance from crankshaft to center of mass of crank (note that most often a is approximately equal to the crank radius r); e = distance from crankshaft to center of mass of countermass; M_{eff} = additional mass of countermass to "balance" connecting rod and slider.

From one-half to two-thirds the slider mass (e.g., $\frac{1}{2}M_c \leq M_{\text{eff}} \leq \frac{2}{3}M_c$) is usually added to the countermass for a single-cylinder engine. For critical field work, single and multiple slider crank mechanisms are dynamically balanced by experimental means.

Forces and Torques   Figure 8.1.47a shows an exploded view of the slider crank mechanism and the various forces and torques on the links (neglecting gravity and weight effects). Inertial effects are shown as broken-line vectors and are manipulated in the same manner as the actual or real forces. The inertial effects are also known as **D'Alembert forces.** The meanings are as follows: $F_1 = -M_D e\omega^2$, parallel to crank A; $F_2 = -M_a e\alpha$, perpendicular to crank A; $F_3 = (z/l)F_9$; F_4 = as found in Fig. 8.1.47b; $F_5 = F_2 - F_7$; $F_6 = F_1 - F_8$; $F_7 = -M_A r\omega^2$, parallel to crank A; $F_8 = -M_A r\alpha$, perpendicular to crank A; $F_9 = -M_B \times$ absolute acceleration of Q, where acceleration of point Q can be found

(a)

(b)

Fig. 8.1.47   (a) Forces and torques; (b) force polygons.

graphically by constructing an acceleration polygon of the mechanism (see, for example, Shigley, "Kinematic Analysis of Mechanisms," McGraw-Hill); $F_{10} = (y/l)F_9$; $F_{11} = -M_c \times$ absolute acceleration of slider $\ddot{x}$; F_{12} = normal wall force (neglecting friction); F = external force on slider's face, where the vector sum $F + F_4 + F_{10} + F_{11} + F_{12} = 0$; T = external crankshaft torque, where the algebraic sum $T + F_3E + F_4W + F_2e + F_7a = 0$ (note that signs must reflect direction of torque); $T_t = F_3E + F_wW$ = transmitted torque; K = the effective

location of force F_9. The distance Δ of force F_9 from the center of mass of connecting rod B (see Fig. 8.1.47a) is given by $\Delta = J_{B(\text{cm})} \times$ angular acceleration of link $B/M_B \times$ absolute acceleration of point Q. Figure 8.1.47b shows the force polygons of the separate links of the mechanism.

8.2 MACHINE ELEMENTS
by Ali M. Sadegh

REFERENCES: American National Standards Institute (ANSI) Standards. International Organization for Standardization (ISO) Standards. Morden, "Industrial Fasteners Handbook," Trade and Technical Press. Parmley, "Standard Handbook of Fastening and Joining," McGraw-Hill. Bickford, "An Introduction to the Design and Behavior of Bolted Joints," Marcel Dekker. Maleev, "Machine Design," International Textbook. Shigley, "Mechanical Engineering Design," McGraw-Hill. *Machine Design* magazine, Penton/IPC. ANSI/Rubber Manufacturers Assn. (ANSI/RMA) Standards. "Handbook of Power Transmission Flat Belting," Goodyear Rubber Products Co. "Industrial V-Belting," Goodyear Rubber Products Co. Carlson, "Spring Designer's Handbook," Marcel Dekker. American Chain Assn., "Chains for Power Transmission and Material Handling—Design and Applications Handbook," Marcel Dekker. "Power Transmission Handbook," DAYCO. "Wire Rope User's Manual," American Iron and Steel Institute. Blake, "Threaded Fasteners—Materials and Design," Marcel Dekker. Young, "Roark's Formulas for Stress and Strain," McGraw-Hill.

NOTE. At this writing, conversion to metric hardware and machine elements continues. SI units are introduced as appropriate, but the bulk of the material is still presented in the form in which the designer or reader will find it available.

SCREW FASTENINGS

At present there exist two major standards for screw threads, namely Unified inch screw threads and metric screw threads. Both systems enjoy a wide application globally, but movement toward a greater use of the metric system continues.

Unified Inch Screw Threads (or Unified Screw Threads)

The Unified Thread Standard originated by an accord of screw thread standardization committees of Canada, the United Kingdom, and the United States in 1984. The Unified Screw-Thread Standard was published by ANSI as American Unified and American Screw Thread Publication B1.1-1974, revised in 1982 and then again in 1989. Revisions did not tamper with the basic 1974 thread forms. In conjunction with Technical Committee No. 1 of the ISO, the Unified Standard was adopted as an ISO Inch Screw Standard (ISO 5864-1978).

Of the numerous and different screw thread forms, those of greatest consequence are

UN—unified (no mandatory radiused root)
UNR—unified (mandatory radiused root; minimum $0.108 = p$)
UNJ—unified (mandatory larger radiused root; recommended $0.150 = p$)
M—metric (inherently designed and manufactured with radiused root; has $0.125 = p$)
MJ—metric (mandatory larger radiused root; recommended $0.150 = p$)

The basic American screw thread profile was standardized in 1974, and it now carries the UN designations (UN = unified). ANSI publishes these standards and all subsequent revisions. At intervals these standards are published with a "reaffirmation date" (that is, R1988). In 1969 an *international* basic thread profile standard was established, and it is designated as M. The ISO publishes these standards with yearly updates. The UN and M profiles are the same, but **UN screws** are

manufactured to **inch** dimensions while **M screws** are manufactured to **metric** dimensions.

The **metric system** has only the two thread forms: **M,** standard for commercial uses, and **MJ,** standard for aerospace use and for aerospace-quality commercial use.

Certain groups of diameter and pitch combinations have evolved over time to become those most used commercially. Such groups are called **thread series.** Currently there are 11 UN series for inch products and 13 **M** series for metric products.

The Unified standard comprises the following two parts:

1. **Diameter-pitch combinations.** (See Tables 8.2.1 to 8.2.5.)
   a. UN inch series:

Coarse	UNC or UNRC
Fine	UNF or UNRF
Extra-fine	UNEF or UNREF
Constant-pitch	UN or UNR

   b. Metric series:

Coarse	M
Fine	M

   NOTE: Radiused roots apply only to external threads. The preponderance of important commercial use leans to UNC, UNF, 8UN (eight-threaded), and metric coarse M. Aerospace and aerospace-quality applications use UNJ and MJ.

2. **Tolerance classes.** The amounts of tolerance and allowance distinguish one thread class from another. Classes are designated by one of three numbers (1, 2, 3), and either letter A for external threads or letter B for internal threads. Tolerance decreases as class number increases. Allowance is specified only for classes 1A and 2A. Tolerances are based on engagement length equal to nominal diameter. 1A/1B—liberal tolerance and allowance required to permit easy assembly even with dirty or nicked threads. 2A/2B—most commonly used for general applications, including production of bolts, screws, nuts, and similar threaded fasteners. Permits external threads to be plated. 3A/3B—for closeness of fit and/or accuracy of thread applications where zero allowance is needed. 2AG—allowance for rapid assembly where high-temperature expansion prevails or where lubrication problems are important.

   Unified screw threads are designated by a set of numbers and letter symbols signifying, in sequence, the nominal size, threads per inch, thread series, tolerance class, hand (only for left hand), and in some instances in parentheses a Thread Acceptability System Requirement of ANSI B1.3.

   EXAMPLE. $\frac{1}{4}$-20 UNC-2A-LH (21), or optionally 0.250-20 UNC-2A-LH (21), where $\frac{1}{4}$ = nominal size (fractional diameter, in, or screw number, with decimal equivalent of either being optional); 20 = number of threads per inch, n; UNC = thread form and series; 2A = tolerance class; LH = left hand (no symbol required for right hand); (21) = thread gaging system per ANSI B1.3.

3. **Load considerations**
   a. *Static loading.* Only a slight increase in tensile strength in a screw fastener is realized with an increase in *root* rounding radius, because minor diameter (hence cross-sectional area at the

root) growth is small. Thus the basic tensile stress area formula is used in stress calculations for all thread forms. See Tables 8.2.2, 8.2.3, and 8.2.4. The designer should take into account such factors as stress concentration as applicable.

b. Dynamic loading. Few mechanical joints can remain absolutely free of some form of fluctuating stress, vibration, stress reversal, or impact. Metal-to-metal joints of very high-modulus materials or non-elastic-gasketed high-modulus joints plus preloading at assembly (preload to be greater than highest peak of the external fluctuating load) can realize absolute static conditions inside the screw fastener. For ordinary-modulus joints and elastic-gasketed joints, a fraction of the external fluctuating load will be transmitted to the interior of the screw fastener. Thus the fastener must be designed for fatigue according to a **static plus fluctuating load** model. See discussion under "Strength" later.

Since **fatigue failures** generally occur at locations of high **stress concentration,** screw fasteners are especially vulnerable because of the abrupt change between head and body, notchlike conditions at the thread roots, surface scratches due to manufacturing, etc. The highest stress concentrations occur at the thread roots. The stress concentration factor can be very large for nonrounded roots, amounting to about 6 for sharp or flat roots, to less than 3 for UNJ and MJ threads which are generously rounded. This can effectively double the fatigue life. UNJ and MJ threads are especially well suited for dynamic loading conditions.

Screw Thread Profile

Basic Profile   The basic profiles of UN and UNR are the same, and these in turn are identical to those of ISO metric threads. Basic thread shape (60° thread angle) and basic dimensions (major, pitch, and minor diameters; thread height; crest, and root flats) are defined. See Fig. 8.2.1.

Design Profile   Design profiles define the maximum material (no allowance) for external and internal threads, and they are derived from the basic profile. UN threads (external) may have either flat or rounded crests and roots. UNR threads (external) must have rounded roots, but may have flat or rounded crests. UN threads (internal) *must* have rounded roots. Any rounding must clear the basic flat roots or crests.

Basic major diameter	Largest diameter of basic screw thread.
Basic minor diameter	Smallest diameter of basic screw thread.
Basic pitch diameter	Diameter to imaginary lines through thread profile and parallel to axis so that thread and groove widths are equal. These three definitions apply to both external and internal threads.
Maximum diameters (external threads)	Basic diameters minus allowance.
Minimum diameters (internal threads)	Basic diameters.
Pitch	$1/n$ (n = number of threads per inch).
Tolerance	Inward variation tolerated on maximum diameters of external threads and outward variation tolerated on minimum diameters of internal threads.

Multiple-threaded screw is a screw that has two or more threads cut beside each other (imagine two or more strings wound side by side around a shaft). **Pitch** is the axial distance between threads. Pitch is equal to the lead in a single-threaded screw. **Lead** is the axial distance the nut advances in one complete revolution of the screw. The lead is equal to the pitch times the number of threads.

Metric Screw Threads

Metric screw thread standardization has been under the aegis of the International Organization for Standardization (ISO). The ISO basic profile is essentially the same as the Unified screw thread basic form, and it is shown in Fig. 8.2.1. The ISO thread series (see Table 8.2.5) are those published in ISO 261-1973. Increased overseas business sparked U.S. interest in metric screw threads, and the ANSI, through its Special Committee to Study Development of an Optimum Metric Fastener System, in joint action with an ISO working group (ISO/TC 1/TC 2), established compromise recommendations regarding metric screw threads. The approved results appear in ANSI B1.13-1979 (Table 8.2.6). This ANSI metric thread series is essentially a selected subset (boxed-in portion of Table 8.2.5) of the larger ISO 261-1973 set. The M profiles of tolerance class 6H/6g are intended for metric applications where inch class 2A/2B has been used.

Metric Tolerance Classes for Threads   Tolerance classes are a selected combination of tolerance grades and tolerance positions applied to length-of-engagement groups.

Tolerance grades are indicated as numbers for crest diameters of nut and bolt and for pitch diameters of nut and bolt. Tolerance is the acceptable variation permitted on any such diameter.

Tolerance positions are indicated as letters, and are allowances (fundamental deviations) as dictated by field usage or conditions. Capital letters are used for internal threads (nut) and lower case for external threads (bolt).

There are three established groups of length of thread engagement, S (short), N (normal), and L (long), for various diameter-pitch combinations. Normal length of thread engagement is calculated from the formula $N = 4.5pd^{0.2}$, where p is pitch and d is the smallest nominal size within each of a series of groupings of nominal sizes.

In conformance with coating (or plating) requirements and demands of ease of assembly, the following tolerance positions have been established:

Bolt	Nut	
e		Large allowance
g	G	Small allowance
h	H	No allowance

See Table 8.2.7 for preferred tolerance classes.

ISO metric screw threads are designated by a set of number and letter symbols signifying, in sequence, metric symbol, nominal size, × (symbol), pitch, tolerance grade (on pitch diameter), tolerance position (for pitch diameter), tolerance grade (on crest diameter), and tolerance position (for crest diameter).

EXAMPLE.   M6 × 0.75-5g6g, where M = metric symbol; 6 = nominal size, × = symbol; 0.75 = pitch-axial distance of adjacent threads measured between corresponding thread points (millimeters); 5 = tolerance grade (on pitch diameter); g = tolerance position (for pitch diameter); 6 = tolerance grade (on crest diameter); g = tolerance position (for crest diameter).

Power Transmission Screw Threads: Forms and Proportions

The **Acme** thread appears in four series [ANSI B1.8-1973 (revised 1988) and B1.5-1977]. Generalized dimensions for the series are given in Table 8.2.8.

The 29° general-purpose thread (Fig. 8.2.2) is used for all Acme thread applications outside of special design cases.

The 29° stub thread (Fig. 8.2.3) is used for heavy-loading designs and where space constraints or economic factors make a shallow thread advantageous.

The 60° stub thread (Fig. 8.2.4) finds special applications in the machine-tool industry.

The 10° modified square thread (Fig. 8.2.5) is, for all practical purposes, equivalent to a "square" thread.

For selected Acme diameter-pitch combinations, see Table 8.2.9.

Table 8.2.1 Standard Series Threads (UN/UNR)*

Nominal size, in		Basic major diameter, in	Series with graded pitches			Series with constant pitches								Nominal size, in
Primary	Secondary		Coarse UNC	Fine UNF	Extra-fine UNEF	4UN	6UN	8UN	12UN	16UN	20UN	28UN	32UN	
0		0.0600	—	80	—	—	—	—	—	—	—	—	—	0
	1	0.0730	64	72	—	—	—	—	—	—	—	—	—	1
2		0.0860	56	64	—	—	—	—	—	—	—	—	—	2
	3	0.0990	48	56	—	—	—	—	—	—	—	—	—	3
4		0.1120	40	48	—	—	—	—	—	—	—	—	—	4
5		0.1250	40	44	—	—	—	—	—	—	—	—	—	5
6		0.1380	32	40	—	—	—	—	—	—	—	—	UNC	6
8		0.1640	32	36	—	—	—	—	—	—	—	—	UNC	8
10		0.1900	24	32	—	—	—	—	—	—	—	—	UNF	10
	12	0.2160	24	28	32	—	—	—	—	—	—	UNF	UNEF	12
¼		0.2500	20	28	32	—	—	—	—	—	UNC	UNF	UNEF	¼
5/16		0.3125	18	24	32	—	—	—	—	—	20	28	UNEF	5/16
3/8		0.3750	16	24	32	—	—	—	—	UNC	20	28	UNEF	3/8
7/16		0.4375	14	20	28	—	—	—	—	16	UNF	UNEF	32	7/16
½		0.5000	13	20	28	—	—	—	—	16	UNF	UNEF	32	½
9/16		0.5625	12	18	24	—	—	—	UNC	16	20	28	32	9/16
5/8		0.6250	11	18	24	—	—	—	12	16	20	28	32	5/8
	11/16	0.6875	—	—	24	—	—	—	12	16	20	28	32	11/16
¾		0.7500	10	16	20	—	—	—	12	UNF	UNEF	28	32	¾
	13/16	0.8125	—	—	20	—	—	—	12	16	UNEF	28	32	13/16
7/8		0.8750	9	14	20	—	—	—	12	16	UNEF	28	32	7/8
	15/16	0.9275	—	—	20	—	—	—	12	16	UNEF	28	32	15/16
1		1.0000	8	12	20	—	—	UNC	UNF	16	UNEF	28	32	1
	1 1/16	1.0625	—	—	18	—	—	8	12	16	20	28	—	1 1/16
1⅛		1.1250	7	12	18	—	—	8	UNF	16	20	28	—	1⅛
	1 3/16	1.1875	—	—	18	—	—	8	12	16	20	28	—	1 3/16
1¼		1.2500	7	12	18	—	—	8	UNF	16	20	28	—	1¼
	1 5/16	1.3125	—	—	18	—	—	8	12	16	20	28	—	1 5/16
1⅜		1.3750	6	12	18	—	UNC	8	UNF	16	20	28	—	1⅜
	1 7/16	1.4375	—	—	18	—	6	8	12	16	20	28	—	1 7/16
1½		1.5000	6	12	18	—	UNC	8	UNF	16	20	28	—	1½
	1 9/16	1.5625	—	—	18	—	6	8	12	16	20	—	—	1 9/16
1⅝		1.6250	—	—	18	—	6	8	12	16	20	—	—	1⅝
	1 11/16	1.6875	—	—	18	—	6	8	12	16	20	—	—	1 11/16
1¾		1.7500	5	—	—	—	6	8	12	16	20	—	—	1¾
	1 13/16	1.8125	—	—	—	—	6	8	12	16	20	—	—	1 13/16
1⅞		1.8750	—	—	—	—	6	8	12	16	20	—	—	1⅞
	1 15/16	1.9375	—	—	—	—	6	8	12	16	20	—	—	1 15/16
2		2.0000	4½	—	—	—	6	8	12	16	20	—	—	2
	2⅛	2.1250	—	—	—	—	6	8	12	16	20	—	—	2⅛
2¼		2.2500	4½	—	—	—	6	8	12	16	20	—	—	2¼
	2⅜	2.3750	—	—	—	—	6	8	12	16	20	—	—	2⅜
2½		2.5000	4	—	—	UNC	6	8	12	16	20	—	—	2½
	2⅝	2.6250	—	—	—	4	6	8	12	16	20	—	—	2⅝
2¾		2.7500	4	—	—	UNC	6	8	12	16	20	—	—	2¾
	2⅞	2.8750	—	—	—	4	6	8	12	16	20	—	—	2⅞
3		3.0000	4	—	—	UNC	6	8	12	16	20	—	—	3
	3⅛	3.1250	—	—	—	4	6	8	12	16	—	—	—	3⅛
3¼		3.2500	4	—	—	UNC	6	8	12	16	—	—	—	3¼
	3⅜	3.3750	—	—	—	4	6	8	12	16	—	—	—	3⅜
3½		3.5000	4	—	—	UNC	6	8	12	16	—	—	—	3½
	3⅝	3.6250	—	—	—	4	6	8	12	16	—	—	—	3⅝
3¾		3.7500	4	—	—	UNC	6	8	12	16	—	—	—	3¾
	3⅞	3.8750	—	—	—	4	6	8	12	16	—	—	—	3⅞
4		4.0000	4	—	—	UNC	6	8	12	16	—	—	—	4
	4⅛	4.1250	—	—	—	4	6	8	12	16	—	—	—	4⅛
4¼		4.2500	—	—	—	4	6	8	12	16	—	—	—	4¼
	4⅜	4.3750	—	—	—	4	6	8	12	16	—	—	—	4⅜
4½		4.5000	—	—	—	4	6	8	12	16	—	—	—	4½
	4⅝	4.6250	—	—	—	4	6	8	12	16	—	—	—	4⅝
4¾		4.7500	—	—	—	4	6	8	12	16	—	—	—	4¾
	4⅞	4.8750	—	—	—	4	6	8	12	16	—	—	—	4⅞
5		5.0000	—	—	—	4	6	8	12	16	—	—	—	5
	5⅛	5.1250	—	—	—	4	6	8	12	16	—	—	—	5⅛
5¼		5.2500	—	—	—	4	6	8	12	16	—	—	—	5¼
	5⅜	5.3750	—	—	—	4	6	8	12	16	—	—	—	5⅜
5½		5.5000	—	—	—	4	6	8	12	16	—	—	—	5½
	5⅝	5.6250	—	—	—	4	6	8	12	16	—	—	—	5⅝
5¾		5.7500	—	—	—	4	6	8	12	16	—	—	—	5¾
	5⅞	5.8750	—	—	—	4	6	8	12	16	—	—	—	5⅞
6		6.0000	—	—	—	4	6	8	12	16	—	—	—	6

* Series designation shown indicates the UN thread form; however, the UNR thread form may be specified by substituting UNR in place of UN in all designations for external use only.
SOURCE: ANSI B1.1-1982; reaffirmed in 1989, reproduced by permission.

Table 8.2.2   Basic Dimensions for Coarse Thread Series (UNC/UNRC)

Nominal size, in	Basic major diameter D, in	Threads per inch n	Basic pitch diameter* E, in	UNR design minor diameter external† K_s, in	Basic minor diameter internal K, in	Section at minor diameter at $D - 2h_b$, in²	Tensile stress area,‡ in²
1 (0.073)§	0.0730	64	0.0629	0.0544	0.0561	0.00218	0.00263
2 (0.086)	0.0860	56	0.0744	0.0648	0.0667	0.00310	0.00370
3 (0.099)§	0.0990	48	0.0855	0.0741	0.0764	0.00406	0.00487
4 (0.112)	0.1120	40	0.0958	0.0822	0.0849	0.00496	0.00604
5 (0.125)	0.1250	40	0.1088	0.0952	0.0979	0.00672	0.00796
6 (0.138)	0.1380	32	0.1177	0.1008	0.1042	0.00745	0.00909
8 (0.164)	0.1640	32	0.1437	0.1268	0.1302	0.01196	0.0140
10 (0.190)	0.1900	24	0.1629	0.1404	0.1449	0.01450	0.0175
12 (0.216)§	0.2160	24	0.1889	0.1664	0.1709	0.0206	0.0242
¼	0.2500	20	0.2175	0.1905	0.1959	0.0269	0.0318
⁵⁄₁₆	0.3125	18	0.2764	0.2464	0.2524	0.0454	0.0524
⅜	0.3750	16	0.3344	0.3005	0.3073	0.0678	0.0775
⁷⁄₁₆	0.4375	14	0.3911	0.3525	0.3602	0.0933	0.1063
½	0.5000	13	0.4500	0.3334	0.4167	0.1257	0.1419
⁹⁄₁₆	0.5625	12	0.5084	0.4633	0.4723	0.162	0.182
⅝	0.6250	11	0.5660	0.5168	0.5266	0.202	0.226
¾	0.7500	10	0.6850	0.6309	0.6417	0.302	0.334
⅞	0.8750	9	0.8028	0.7427	0.7547	0.419	0.462
1	1.0000	8	0.9188	0.8512	0.8647	0.551	0.606
1⅛	1.1250	7	1.0322	0.9549	0.9704	0.693	0.763
1¼	1.2500	7	1.1572	1.0799	1.0954	0.890	0.969
1⅜	1.3750	6	1.2667	1.1766	1.1946	1.054	1.155
1½	1.5000	6	1.3917	1.3016	1.3196	1.294	1.405
1¾	1.7500	5	1.6201	1.5119	1.5335	1.74	1.90
2	2.0000	4½	1.8557	1.7353	1.7594	2.30	2.50
2¼	2.2500	4½	2.1057	1.9853	2.0094	3.02	3.25
2½	2.5000	4	2.3376	2.2023	2.2294	3.72	4.00
2¾	2.7500	4	2.5876	2.4523	2.4794	4.62	4.93
3	3.0000	4	2.8376	2.7023	2.7294	5.62	5.97
3¼	3.2500	4	3.0876	2.9523	2.9794	6.72	7.10
3½	3.5000	4	3.3376	3.2023	3.2294	7.92	8.33
3¾	3.7500	4	3.5876	3.4523	3.4794	9.21	9.66
4	4.0000	4	3.8376	3.7023	3.7294	10.61	11.08

* British: effective diameter.
† See formula under definition of tensile stress area in Appendix B of ANSI B1.1-1987.
‡ Design form. See Fig. 2B in ANSI B1.1-1982 or Fig. 1 in 1989 revision.
§ Secondary sizes.
Source: ANSI B1.1-1982, revised 1989; reproduced by permission.

Name	Symbol		
	Inch bolt or nut	Metric bolt	Metric nut
Major diameter	D	d	D
Pitch diam. (inch); effective diam. (metric)	E	d_2	D_2
Minor diameter	K	d_1	D_1

p = pitch or distance between adjacent threads (inches or mm)

f = flat (inches or mm)

Fig. 8.2.1   Basic thread profile.

Table 8.2.3   Basic Dimensions for Fine Thread Series (UNF/UNRF)

Nominal size, in	Basic major diameter D, in	Threads per inch n	Basic pitch diameter* E, in	UNR design minor diameter external† K_s, in	Basic minor diameter internal K, in	Section at minor diameter at $D - 2h_b$, in^2	Tensile stress area,‡ in^2
0 (0.060)	0.0600	80	0.0519	0.0451	0.0465	0.00151	0.00180
1 (0.073)§	0.0730	72	0.0640	0.0565	0.0580	0.00237	0.00278
2 (0.086)	0.0860	64	0.0759	0.0674	0.0691	0.00339	0.00394
3 (0.099)§	0.0990	56	0.0874	0.0778	0.0797	0.00451	0.00523
4 (0.112)	0.1120	48	0.0985	0.0871	0.0894	0.00566	0.00661
5 (0.125)	0.1250	44	0.1102	0.0979	0.1004	0.00716	0.00830
6 (0.138)	0.1380	40	0.1218	0.1082	0.1109	0.00874	0.01015
8 (0.164)	0.1640	36	0.1460	0.1309	0.1339	0.01285	0.01474
10 (0.190)	0.1900	32	0.1697	0.1528	0.1562	0.0175	0.0200
12 (0.216)§	0.2160	28	0.1928	0.1734	0.1773	0.0226	0.0258
1/4	0.2500	28	0.2268	0.2074	0.2113	0.0326	0.0364
5/16	0.3125	24	0.2854	0.2629	0.2674	0.0524	0.0580
3/8	0.3750	24	0.3479	0.3254	0.3299	0.0809	0.0878
7/16	0.4375	20	0.4050	0.3780	0.3834	0.1090	0.1187
1/2	0.5000	20	0.4675	0.4405	0.4459	0.1486	0.1599
9/16	0.5625	18	0.5264	0.4964	0.5024	0.189	0.203
5/8	0.6250	18	0.5889	0.5589	0.5649	0.240	0.256
3/4	0.7500	16	0.7094	0.6763	0.6823	0.351	0.373
7/8	0.8750	14	0.8286	0.7900	0.7977	0.480	0.509
1	1.0000	12	0.9459	0.9001	0.9098	0.625	0.663
1 1/8	1.1250	12	1.0709	1.0258	1.0348	0.812	0.856
1 1/4	1.2500	12	1.1959	1.1508	1.1598	1.024	1.073
1 3/8	1.3750	12	1.3209	1.2758	1.2848	1.260	1.315
1 1/2	1.5000	12	1.4459	1.4008	1.4098	1.521	1.581

* British: effective diameter.
† See formula under definition of tensile stress area in Appendix B of ANSI B1.1-1982.
‡ Design form. See Fig. 2B of ANSI B1.1-1982, or Fig. 1 in 1989 revision.
§ Secondary sizes.
SOURCE: ANSI B1.1-1982, revised 1989; reproduced by permission.

Table 8.2.4   Basic Dimensions for Extra-Fine Thread Series (UNEF/UNREF)

Nominal size, in — Primary	Nominal size, in — Secondary	Basic major diameter D, in	Threads per inch n	Basic pitch diameter* E, in	UNR design minor diameter external† K_s, in	Basic minor diameter internal K, in	Section at minor diameter at $D - 2h_b$, in^2	Tensile stress area,‡ in^2
	12 (0.216)	0.2160	32	0.1957	0.1788	0.1822	0.0242	0.0270
1/4		0.2500	32	0.2297	0.2128	0.2162	0.0344	0.0379
5/16		0.3125	32	0.2922	0.2753	0.2787	0.0581	0.0625
3/8		0.3750	32	0.3547	0.3378	0.3412	0.0878	0.0932
7/16		0.4375	28	0.4143	0.3949	0.3988	0.1201	0.1274
1/2		0.5000	28	0.4768	0.4573	0.4613	0.162	0.170
9/16		0.5625	24	0.5354	0.5129	0.5174	0.203	0.214
5/8		0.6250	24	0.5979	0.5754	0.5799	0.256	0.268
	11/16	0.6875	24	0.6604	0.6379	0.6424	0.315	0.329
3/4		0.7500	20	0.7175	0.6905	0.6959	0.369	0.386
	13/16	0.8125	20	0.7800	0.7530	0.7584	0.439	0.458
7/8		0.8750	20	0.8425	0.8155	0.8209	0.515	0.536
	15/16	0.9375	20	0.9050	0.8780	0.8834	0.598	0.620
1		1.0000	20	0.9675	0.9405	0.9459	0.687	0.711
	1 1/16	1.0625	18	1.0264	0.9964	1.0024	0.770	0.799
1 1/8		1.1250	18	1.0889	1.0589	1.0649	0.871	1.901
	1 3/16	1.1875	18	1.1514	1.1214	1.1274	0.977	1.009
1 1/4		1.2500	18	1.2139	1.1839	1.1899	1.090	1.123
	1 5/16	1.3125	18	1.2764	1.2464	1.2524	1.208	1.244
1 3/8		1.3750	18	1.3389	1.3089	1.3149	1.333	1.370
	1 7/16	1.4375	18	1.4014	1.3714	1.3774	1.464	1.503
1 1/2		1.5000	18	1.4639	1.4339	1.4399	1.60	1.64
	1 9/16	1.5625	18	1.5264	1.4964	1.5024	1.74	1.79
1 5/8		1.6250	18	1.5889	1.5589	1.5649	1.89	1.94
	1 11/16	1.6875	18	1.6514	1.6214	1.6274	2.05	2.10

* British: effective diameter.
† Design form. See Fig. 2B in ANSI B1.1-1982 or Fig. 1 in 1989 revision.
‡ See formula under definition of tensile stress area in Appendix B in ANSI B1.1-1982.
SOURCE: ANSI B1.1-1982 revised 1989; reproduced by permission.

Table 8.2.5   ISO Metric Screw Thread Standard Series

Nominal size diam, mm — Column* 1	2	3	Series with graded pitches — Coarse	Fine	Pitches, mm — Series with constant pitches — 6	4	3	2	1.5	1.25	1	0.75	0.5	0.35	0.25	0.2	Nominal size diam, mm
0.25			0.075	—	—	—	—	—	—	—	—	—	—	—	—	—	0.25
0.3			0.08	—	—	—	—	—	—	—	—	—	—	—	—	—	0.3
	0.35		0.09	—	—	—	—	—	—	—	—	—	—	—	—	—	0.35
0.4			0.1	—	—	—	—	—	—	—	—	—	—	—	—	—	0.4
	0.45		0.1	—	—	—	—	—	—	—	—	—	—	—	—	—	0.45
0.5			0.125	—	—	—	—	—	—	—	—	—	—	—	—	—	0.5
	0.55		0.125	—	—	—	—	—	—	—	—	—	—	—	—	—	0.55
0.6			0.15	—	—	—	—	—	—	—	—	—	—	—	—	—	0.6
	0.7		0.175	—	—	—	—	—	—	—	—	—	—	—	—	—	0.7
0.8			0.2	—	—	—	—	—	—	—	—	—	—	—	—	—	0.8
	0.9		0.225	—	—	—	—	—	—	—	—	—	—	—	—	—	0.9
1			0.25	—	—	—	—	—	—	—	—	—	—	—	—	0.2	1
	1.1		0.25	—	—	—	—	—	—	—	—	—	—	—	—	0.2	1.1
1.2			0.25	—	—	—	—	—	—	—	—	—	—	—	—	0.2	1.2
	1.4		0.3	—	—	—	—	—	—	—	—	—	—	—	—	0.2	1.4
1.6			0.35	—	—	—	—	—	—	—	—	—	—	—	—	0.2	1.6
	1.8		0.35	—	—	—	—	—	—	—	—	—	—	—	—	0.2	1.8
2			0.4	—	—	—	—	—	—	—	—	—	—	—	0.25	—	2
	2.2		0.45	—	—	—	—	—	—	—	—	—	—	—	0.25	—	2.2
2.5			0.45	—	—	—	—	—	—	—	—	—	—	0.35	—	—	2.5
3			0.5	—	—	—	—	—	—	—	—	—	—	0.35	—	—	3
	3.5		0.6	—	—	—	—	—	—	—	—	—	—	0.35	—	—	3.5
4			0.7	—	—	—	—	—	—	—	—	—	0.5	—	—	—	4
	4.5		0.75	—	—	—	—	—	—	—	—	—	0.5	—	—	—	4.5
5			0.8	—	—	—	—	—	—	—	—	—	0.5	—	—	—	5
		5.5	—	—	—	—	—	—	—	—	—	—	0.5	—	—	—	5.5
6			1	—	—	—	—	—	—	—	—	0.75	—	—	—	—	6
		7	1	—	—	—	—	—	—	—	—	0.75	—	—	—	—	7
8			1.25	1	—	—	—	—	—	—	1	0.75	—	—	—	—	8
		9	1.25	—	—	—	—	—	—	—	1	0.75	—	—	—	—	9
10			1.5	1.25	—	—	—	—	—	1.25	1	0.75	—	—	—	—	10
		11	1.5	—	—	—	—	—	—	—	1	0.75	—	—	—	—	11
12			1.75	1.25	—	—	—	—	1.5	1.25	1	—	—	—	—	—	12
	14		2	1.5	—	—	—	—	1.5	1.25†	1	—	—	—	—	—	14
		15	—	—	—	—	—	—	1.5	—	1	—	—	—	—	—	15
16			2	1.5	—	—	—	—	1.5	—	1	—	—	—	—	—	16
		17	—	—	—	—	—	—	1.5	—	1	—	—	—	—	—	17
	18		2.5	1.5	—	—	—	2	1.5	—	1	—	—	—	—	—	18
20			2.5	1.5	—	—	—	2	1.5	—	1	—	—	—	—	—	20
	22		2.5	1.5	—	—	—	2	1.5	—	1	—	—	—	—	—	22
24			3	2	—	—	—	2	1.5	—	1	—	—	—	—	—	24
		25	—	—	—	—	—	2	1.5	—	1	—	—	—	—	—	25
		26	—	—	—	—	—	—	1.5	—	1	—	—	—	—	—	26
	27		3	2	—	—	—	2	1.5	—	1	—	—	—	—	—	27
		28	—	—	—	—	—	2	1.5	—	1	—	—	—	—	—	28
30			3.5	2	—	—	(3)	2	1.5	—	1	—	—	—	—	—	30
		32	—	—	—	—	—	2	1.5	—	—	—	—	—	—	—	32
	33		3.5	2	—	—	(3)	2	1.5	—	—	—	—	—	—	—	33
		35‡	—	—	—	—	—	—	1.5	—	—	—	—	—	—	—	35‡
36			4	3	—	—	—	2	1.5	—	—	—	—	—	—	—	36
		38	—	—	—	—	—	—	1.5	—	—	—	—	—	—	—	38
	39		4	3	—	—	—	2	1.5	—	—	—	—	—	—	—	39
		40	—	—	—	—	3	2	1.5	—	—	—	—	—	—	—	40
42			4.5	3	—	4	3	2	1.5	—	—	—	—	—	—	—	42
	45		4.5	3	—	4	3	2	1.5	—	—	—	—	—	—	—	45

* Thread diameter should be selected from column 1, 2 or 3; with preference being given in that order.
† Pitch 1.25 mm in combination with diameter 14 mm has been included for spark plug applications.
‡ Diameter 35 mm has been included for bearing locknut applications.
NOTE: The use of pitches shown in parentheses should be avoided wherever possible. The pitches enclosed in the bold frame, together with the corresponding nominal diameters in columns 1 and 2, are those combinations which have been established by ISO Recommendations as a selected "coarse" and "fine" series for commercial fasteners. Sizes 0.25 mm through 1.4 mm are covered in ISO Recommendation R68 and, except for the 0.25-mm size, in ANSI B1.10.
SOURCE: ISO 261-1973, reproduced by permission.

Table 8.2.6 Limiting Dimensions of Standard Series Threads for Commercial Screws, Bolts, and Nuts

Nominal size diam, mm	Pitch P, mm	Basic thread designation	External thread (bolt), mm									Internal thread (nut), mm						Major diam, min
			Tol. class	Allowance	Major diameter Max	Major diameter Min	Pitch diameter Max	Pitch diameter Min	Pitch diameter Tol.	Minor diameter Max*	Minor diameter Min†	Tol. class	Minor diameter Min	Minor diameter Max	Pitch diameter Max	Pitch diameter Min	Pitch diameter Tol.	
1.6	0.35	M1.6	6g	0.019	1.581	1.496	1.354	1.291	0.063	1.151	1.063	6H	1.221	1.321	1.458	1.373	0.085	1.600
1.8	0.35	M1.8	6g	0.019	1.781	1.696	1.554	1.491	0.063	1.351	1.263	6H	1.421	1.521	1.568	1.573	0.085	1.800
2	0.4	M2	6g	0.019	1.981	1.886	1.721	1.654	0.067	1.490	1.394	6H	1.567	1.679	1.830	1.740	0.090	2.000
2.2	0.45	M2.2	6g	0.020	2.180	2.080	1.888	1.817	0.071	1.628	1.525	6H	1.713	1.838	2.000	1.908	0.095	2.200
2.5	0.45	M2.5	6g	0.020	2.480	2.380	2.188	2.117	0.071	1.928	1.825	6H	2.013	2.138	2.303	2.208	0.095	2.500
3	0.5	M3	6g	0.020	2.980	2.874	2.655	2.580	0.075	2.367	2.256	6H	2.459	2.599	2.775	2.675	0.100	3.000
3.5	0.6	M3.5	6g	0.021	3.479	3.354	3.089	3.004	0.085	2.742	2.614	6H	2.850	3.010	3.222	3.110	0.112	3.500
4	0.7	M4	6g	0.022	3.978	3.838	3.523	3.433	0.090	3.119	2.979	6H	3.242	3.422	3.663	3.545	0.118	4.000
4.5	0.75	M4.5	6g	0.022	4.478	4.338	3.991	3.901	0.090	3.558	3.414	6H	3.688	3.878	4.131	4.013	0.118	4.500
5	0.8	M5	6g	0.024	4.976	4.826	4.456	4.361	0.095	3.994	3.841	6H	4.134	4.334	4.605	4.480	0.125	5.000
6	1	M6	6g	0.026	5.974	5.794	5.324	5.212	0.112	4.747	4.563	6H	4.917	5.153	5.500	5.350	0.150	6.000
7	1	M7	6g	0.026	6.974	6.794	6.234	6.212	0.112	5.747	5.563	6H	5.917	6.153	6.500	6.350	0.150	7.000
8	1.25	M8	6g	0.028	7.972	7.760	7.160	7.042	0.118	6.439	6.231	6H	6.647	6.912	7.348	7.188	0.160	8.000
	1	M8 × 1	6g	0.026	7.974	7.794	7.324	7.212	0.112	6.747	6.563	6H	6.918	7.154	7.500	7.350	0.150	8.000
10	1.5	M10	6g	0.032	9.968	9.732	8.994	8.862	0.132	8.127	7.879	6H	8.376	8.676	9.206	9.026	0.180	10.000
	1.25	M10 × 1.25	6g	0.028	9.972	9.760	9.160	9.042	0.118	8.439	8.231	6H	8.646	8.911	9.348	9.188	0.160	10.000
12	1.75	M12	6g	0.034	11.966	11.701	10.829	10.679	0.150	9.819	9.543	6H	10.106	10.441	11.063	10.863	0.200	12.000
	1.25	M12 × 1.25	6g	0.028	11.972	11.760	11.160	11.028	0.118	10.439	10.217	6H	10.646	10.911	11.368	11.188	0.180	12.000
14	2	M14	6g	0.038	13.962	13.682	12.663	12.503	0.160	11.508	11.204	6H	11.835	12.210	12.913	12.701	0.212	14.000
	1.5	M14 × 1.5	6g	0.032	13.968	13.732	12.994	12.854	0.140	12.127	11.879	6H	12.376	12.676	13.216	13.026	0.190	14.000
16	2	M16	6g	0.038	15.962	15.682	14.663	14.503	0.160	13.508	13.204	6H	13.385	14.210	14.913	14.701	0.212	16.000
	1.5	M16 × 1.5	6g	0.032	15.968	15.732	14.994	14.854	0.140	14.127	13.879	6H	14.376	14.676	15.216	15.026	0.190	16.000
18	2.5	M18	6g	0.038	17.958	17.623	16.334	16.164	0.170	14.891	14.541	6H	15.294	15.744	16.600	16.376	0.224	18.000
	1.5	M18 × 1.5	6g	0.032	17.968	17.732	16.994	16.854	0.140	16.127	15.879	6H	16.376	16.676	17.216	17.026	0.190	18.000
20	2.5	M20	6g	0.042	19.958	19.623	18.334	18.164	0.170	16.891	16.541	6H	17.294	17.744	18.600	18.376	0.224	20.000
	1.5	M20 × 1.5	6g	0.032	19.968	19.732	18.994	18.854	0.140	18.127	17.879	6H	18.376	18.676	19.216	19.026	0.190	20.000
22	2.5	M22	6g	0.042	21.958	21.623	20.334	20.164	0.170	18.891	18.541	6H	19.294	19.744	20.600	20.376	0.224	22.000
	1.5	M22 × 1.5	6g	0.032	21.968	21.732	20.994	20.854	0.140	20.127	19.879	6H	20.376	20.676	21.216	21.026	0.190	22.000
24	3	M24	6g	0.048	23.952	23.577	22.003	21.803	0.200	20.271	19.855	6H	20.752	21.252	22.316	22.051	0.265	24.000
	2	M24 × 2	6g	0.038	23.962	23.682	22.663	22.493	0.170	21.508	21.194	6H	21.835	22.210	22.925	22.701	0.224	24.000
27	3	M27	6g	0.048	26.952	26.577	25.003	24.803	0.200	23.271	22.855	6H	23.752	24.252	25.316	25.051	0.265	27.000
	2	M27 × 2	6g	0.038	26.962	26.682	25.663	25.493	0.170	24.508	24.194	6H	24.835	25.210	25.925	25.701	0.224	27.000
30	3.5	M30	6g	0.053	29.947	29.522	27.674	27.462	0.212	25.653	25.189	6H	26.211	26.771	28.007	27.727	0.280	30.000
	2	M30 × 2	6g	0.038	29.962	29.682	28.663	28.493	0.170	27.508	27.194	6H	27.835	28.210	28.925	28.701	0.224	30.000
33	3.5	M33	6g	0.053	32.947	32.522	30.674	30.462	0.212	28.653	28.189	6H	29.211	29.771	31.007	30.727	0.280	33.000
	2	M33 × 2	6g	0.038	32.962	32.682	31.663	31.493	0.170	30.508	30.194	6H	30.835	31.210	31.925	31.701	0.224	33.000
36	4	M36	6g	0.060	35.940	35.465	33.342	33.118	0.224	31.033	30.521	6H	31.670	32.270	33.702	33.402	0.300	36.000
	3	M36 × 3	6g	0.048	35.952	35.577	34.003	33.803	0.200	32.271	31.855	6H	32.752	33.252	34.316	34.051	0.265	36.000
39	4	M39	6g	0.060	38.940	38.465	36.342	36.118	0.224	34.033	33.521	6H	34.670	35.270	36.702	36.402	0.300	39.000
	3	M39 × 3	6g	0.048	38.952	38.577	37.003	36.803	0.200	35.271	34.855	6H	35.752	36.252	37.316	37.051	0.265	39.000

* Design form, see Figs. 2 and 5 of ANSI B1.13M-1979 (or Figs. 1 and 4 in 1983 revision).
† Required for high-strength applications where rounded root is specified.

Source: [Appeared in ASME/SAE Interpretive document Metric Screw Threads, B1.13 (Nov. 3, 1966), pp. 9, 10.] ISO 261-1973, reproduced by permission.

Table 8.2.7   Preferred Tolerance Classes

	Length of engagement														
	External threads (bolts)									Internal threads (nuts)					
	Tolerance position e (large allowance)			Tolerance position g (small allowance)			Tolerance position h (no allowance)			Tolerance position G (small allowance)			Tolerance position H (no allowance)		
Quality	Group S	Group N	Group L	Group S	Group N	Group L	Group S	Group N	Group L	Group S	Group N	Group L	Group S	Group N	Group L
Fine							3h4h	4h	4h5h				4H	5H	6H
Medium		6e	7e6e	5g6g	6g	7g6g	5h6h	6h	7h6h	5G	6G	7G	5H	6H	7H
Coarse					8g	9g8g					7G	8G		7H	8H

NOTE: Fine quality applies to precision threads where little variation in fit character is permissible. Coarse quality applies to those threads which present manufacturing difficulties, such as the threading of hot-rolled bars or tapping deep blind holes.
SOURCE: ISO 261-1973, reproduced by permission.

Table 8.2.8   Acme Thread Series
(D = outside diam, p = pitch. All dimensions in inches.)
(See Figs. 8.2.2 to 8.2.5.)

	Thread dimensions			
Symbols	29° general purpose	29° stub	60° stub	10° modified
t = thickness of thread	$0.5p$	$0.5p$	$0.5p$	$0.5p$
R = basic depth of thread	$0.5p$	$0.3p$	$0.433p$	$0.5p*$
F = basic width of flat	$0.3707p$	$0.4224p$	$0.250p$	$0.4563p†$
G = (see Figs. 8.2.2, 8.2.3, 8.2.4)	$F - (0.52 \times \text{clearance})$	$F - (0.52 \times \text{clearance})$	$0.227p$	$F - (0.17 \times \text{clearance})$
E = basic pitch diam	$D - 0.5p$	$D - 0.3p$	$D - 0.433p$	$D - 0.5p$
K = basic minor diam	$D - p$	$D - 0.6p$	$D - 0.866p$	$D - p$
Range of threads, per inch	1–16	2–16		4–16

* A clearance of at least 0.010 in is added to h on threads of 10-pitch and coarser, and 0.005 in on finer pitches, to produce extra depth, thus avoiding interference with threads of mating parts of a minor or major diameter.
† Measured at crest of screw thread.

Table 8.2.9   Acme Thread Diameter-Pitch Combinations
(See Figs. 8.2.2 to 8.2.5.)

Size	Threads per inch	Size	Threads per inch	Size	Threads per inch	Size	Threads per inch	Size	Threads per inch
¼	16	⅝	8	1¼	5	2¼	3	4	2
5⁄16	14	¾	6	1⅜	4	2½	3	4½	2
⅜	12	⅞	6	1½	4	2¾	3	5	2
7⁄16	12	1	5	1¾	4	3	2		
½	10	1⅛	5	2	4	3½	2		

Fig. 8.2.2   29° Acme general-purpose thread.

Fig. 8.2.4   60° stub Acme thread.

Fig. 8.2.3   29° stub Acme thread.

Three classes (2G, 3G, 4G) of general-purpose threads have clearances on all diameters for free movement. A fourth class (5G) of general-purpose threads has no allowance or clearance on the pitch diameter for purposes of minimum end play or backlash.

Fig. 8.2.5   10° modified square thread.

Raising and Lowering Torques In a single-threaded screw, the torque T_R required for raising load F, or tightening, is

$$T_R = \frac{Fd_m}{2}\left(\frac{l + \pi\mu d_m \sec\alpha}{\pi d_m - \mu l \sec\alpha}\right)$$

where d_m is the mean diameter of the screw, μ is the coefficient of friction between screw and nut threads, l is the lead, and α is one-half of Acme thread angle (2α). Note $\alpha = 0$ for square threads. The torque require to lower the load is

$$T_L = \frac{Fd_m}{2}\left(\frac{\pi\mu d_m \sec\alpha - l}{\pi d_m + \mu l \sec\alpha}\right)$$

When a screw has a collar for lifting a load, additional torque T_c should be added to the lifting torque, $T_c = F\mu_c d_c/2$, where the subscript c is for the collar.

High-Strength Bolting Screw Threads

High-strength bolting applications include pressure vessels, steel pipe flanges, fittings, valves, and other services. They can be used for either hot or cold surfaces where high tensile stresses are produced when the joints are made up. For sizes 1 in and smaller, the ANSI coarse-thread series is used. For larger sizes, the ANSI 8-pitch thread series is used (see Table 8.2.10).

Ball Screws

A ball screw is an improvement over an Acme screw, and it generally consists of a set of bearing balls within the ball nut thread and a ball screw shaft. The thread form in which the bearing balls ride is a semi-circular shape or an oval shape formed from two arcs of the same radius with offset centers; see Figs. 8.2.6 and 8.2.7.

Bearing ball circuit is the closed path that the bearing balls follow through the ball nut. Ball nuts may have one or more circuits.

The primary load path in a ball screw assembly is through the circuit (or circuits) of bearing balls within the ball nut, where a ball enters the

Fig. 8.2.6 Cross section of ball screw and the nut.

Fig. 8.2.7 Ball Screw. (*Courtesy Nook Industries, Inc., Cleveland, Ohio.*)

ball path between the nut and screw carrying the load one or more turns around the screw, as shown in Fig. 8.2.7.

Return guide is a path for returning the bearing balls to the beginning of the circuit. **Pitch** is defined as the axial distance between threads. Pitch is equal to the lead in a single-start screw and is equal to one-half of the lead in double-start screw.

Table 8.2.10   Screw Threads for High-Strength Bolting
(All dimensions in inches)

Size	Threads per inch	Allowance (minus)	Major diam	Major diam tolerance	Max pitch diam*	Max pitch diam tolerance	Minor diam max	Nut max minor diam	Nut max minor diam tolerance	Nut max pitch diam*	Nut max pitch diam tolerance
¼	20	0.0010	0.2490	0.0072	0.2165	0.0026	0.1877	0.2060	0.0101	0.2211	0.0036
⁵⁄₁₆	18	0.0011	0.3114	0.0082	0.2753	0.0030	0.2432	0.2630	0.0106	0.2805	0.0041
⅜	16	0.0013	0.3737	0.0090	0.3331	0.0032	0.2990	0.3184	0.0111	0.3389	0.0045
⁷⁄₁₆	14	0.0013	0.4362	0.0098	0.3898	0.0036	0.3486	0.3721	0.0119	0.3960	0.0049
½	13	0.0015	0.4985	0.0104	0.4485	0.0037	0.4041	0.4290	0.0123	0.4552	0.0052
⁹⁄₁₆	12	0.0016	0.5609	0.0112	0.5068	0.0040	0.4587	0.4850	0.0127	0.5140	0.0056
⅝	11	0.0017	0.6233	0.0118	0.5643	0.0042	0.5118	0.5397	0.0131	0.5719	0.0059
¾	10	0.0019	0.7481	0.0128	0.6831	0.0045	0.6254	0.6553	0.0136	0.6914	0.0064
⅞	9	0.0021	0.8729	0.0140	0.8007	0.0049	0.7366	0.7689	0.0142	0.8098	0.0070
1	8	0.0022	0.9978	0.0152	0.9166	0.0054	0.8444	0.8795	0.0148	0.9264	0.0076
1⅛	8	0.0024	1.1226	0.0152	1.0414	0.0055	0.9692	1.0045	0.0148	1.0517	0.0079
1¼	8	0.0025	1.2475	0.0152	1.1663	0.0058	1.0941	1.1295	0.0148	1.1771	0.0083
1⅜	8	0.0025	1.3725	0.0152	1.2913	0.0061	1.2191	1.2545	0.0148	1.3024	0.0086
1½	8	0.0027	1.4973	0.0152	1.4161	0.0063	1.3439	1.3795	0.0148	1.4278	0.0090
1⅝	8	0.0028	1.6222	0.0152	1.5410	0.0065	1.4688	1.5045	0.0148	1.5531	0.0093
1¾	8	0.0029	1.7471	0.0152	1.6659	0.0068	1.5937	1.6295	0.0148	1.6785	0.0097
1⅞	8	0.0030	1.8720	0.0152	1.7908	0.0070	1.7186	1.7545	0.0148	1.8038	0.0100
2	8	0.0031	1.9969	0.0152	1.9157	0.0073	1.8435	1.8795	0.0148	1.9294	0.0104
2⅛	8	0.0032	2.1218	0.0152	2.0406	0.0075	1.9682	2.0045	0.0148	2.0545	0.0107
2¼	8	0.0033	2.2467	0.0152	2.1655	0.0077	2.0933	2.1295	0.0148	2.1798	0.0110
2½	8	0.0035	2.4965	0.0152	2.4153	0.0082	2.3431	2.3795	0.0148	2.4305	0.0117
2¾	8	0.0037	2.7463	0.0152	2.6651	0.0087	2.5929	2.6295	0.0148	2.6812	0.0124
3	8	0.0038	2.9962	0.0152	2.9150	0.0092	2.8428	2.8795	0.0148	2.9318	0.0130
3¼	8	0.0039	3.2461	0.0152	3.1649	0.0093	3.0927	3.1295	0.0148	3.1820	0.0132
3½	8	0.0040	3.4960	0.0152	3.4148	0.0093	3.3426	3.3795	0.0148	3.4321	0.0133

The Unified form of thread shall be used. Pitch diameter tolerances include errors of lead and angle.
* The maximum pitch diameters of screws are smaller than the minimum pitch diameters of nuts by these amounts.

The advantages of ball screws include a lead accuracy of up to 0.0005 in, quiet and smooth operation, minimal backlash, and high efficiency. Because of these features ball screws have long life and low wear. Ball screws are used in linear motion and linear actuator devices in which high accuracy and axial stiffness are required. Ball screws are available in many types and sizes. For further information the reader is referred to the manufacturers' specifications.

Design Considerations. (1) Flanges may be used with the nut assembly to carry axial and lateral loads. (2) When a ball screw assembly is used in an orientation other than vertical, it is important to orient the return tubes to optimize ball nut operation. As shown in Fig. 8.2.8, return guides down should be avoided.

Preferred   Acceptable   Preferred

Avoid   Avoid

Fig. 8.2.8 Design considerations of ball screw return tubes. (*Courtesy Nook Industries, Inc., Cleveland, Ohio.*)

Screw Threads for Pipes

American National Standard Taper Pipe Thread (ANSI/ASME B1.20.1-1983) This thread is shown in Fig. 8.2.9. It is made to the following specifications: The taper is 1 in 16 or 0.75 in/ft. The basic

Taper of thread 1 in 16 measured on diameter

Imperfect thread due to chamfer on die

Fig. 8.2.9 American National Standard taper pipe threads.

length of the external taper thread is determined by $L_2 = p(0.8D + 6.8)$, where D is the basic outside diameter of the pipe (see Table 8.2.11). Thread designation and notation is written as: nominal size, number of threads per inch, thread series. For example: $\frac{3}{8}$-18 NPT, $\frac{1}{8}$-27 NPSC, $\frac{1}{2}$-14 NPTR, $\frac{1}{8}$-27 NPSM, $\frac{1}{8}$-27 NPSL, 1-11.5 NPSH, where N = National (American) Standard, T = taper, C = coupling, S = straight, M = mechanical, L = locknut, H = hose coupling, and R = rail fittings. Where pressure-tight joints are required, it is intended that taper pipe threads be made up wrench-tight with a sealant. Descriptions of thread series include: NPSM = free-fitting mechanical joints for fixtures, NPSL = loose-fitting mechanical joints with locknuts, NPSH = loose-fitting mechanical joints for hose coupling.

American National Standard Straight Pipe Thread (ANSI/ASME B1.20.1-1983) This thread can be used to advantage for the following: (1) pressure-tight joints with sealer; (2) pressure-tight joints without sealer for drain plugs, filler plugs, etc.; (3) free-fitting mechanical joints for fixtures; (4) loose-fitting mechanical joints with locknuts; and (5) loose-fitting mechanical joints for hose couplings. Dimensions are shown in Table 8.2.12.

American National Standard Dry-Seal Pipe Threads (ANSI B1.20.3-1976 (inch), ANSI B1.20.4-1976 (metric translation) Thread

Table 8.2.11 ANSI Taper Pipe Thread
(All dimensions in inches)
(See Fig. 8.2.9.)

Nominal pipe size	OD of pipe	Threads per inch	Pitch of thread	Hand-tight engagement length L_1	Effective thread external length L_2	Wrench makeup length for internal thread length L_3	Overall length external thread L_4
$\frac{1}{16}$	0.3125	27	0.03704	0.160	0.2611	0.1111	0.3896
$\frac{1}{8}$	0.405	27	0.03704	0.180	0.2639	0.1111	0.3924
$\frac{1}{4}$	0.540	18	0.05556	0.200	0.4018	0.1667	0.5946
$\frac{3}{8}$	0.675	18	0.05556	0.240	0.4078	0.1667	0.6006
$\frac{1}{2}$	0.840	14	0.07143	0.320	0.5337	0.2143	0.7815
$\frac{3}{4}$	1.050	14	0.07143	0.339	0.5457	0.2143	0.7935
1	1.315	11½	0.08696	0.400	0.6828	0.2609	0.9845
1¼	1.660	11½	0.08696	0.420	0.7068	0.2609	1.0085
1½	1.900	11½	0.08696	0.420	0.7235	0.2609	1.0252
2	2.375	11½	0.08696	0.436	0.7565	0.2609	1.0582
2½	2.875	8	0.12500	0.682	1.1375	0.2500	1.5712
3	3.500	8	0.12500	0.766	1.2000	0.2500	1.6337
3½	4.000	8	0.12500	0.821	1.2500	0.2500	1.6837
4	4.500	8	0.12500	0.844	1.3000	0.2500	1.7337
5	5.563	8	0.12500	0.937	1.4063	0.2500	1.8400
6	6.625	8	0.12500	0.958	1.5125	0.2500	1.9462
8	8.625	8	0.12500	1.063	1.7125	0.2500	2.1462
10	10.750	8	0.12500	1.210	1.9250	0.2500	2.3587
12	12.750	8	0.12500	1.360	2.1250	0.2500	2.5587
14 OD	14.000	8	0.12500	1.562	2.2500	0.2500	2.6837
16 OD	16.000	8	0.12500	1.812	2.4500	0.2500	2.8837
18 OD	18.000	8	0.12500	2.000	2.6500	0.2500	3.0837
20 OD	20.000	8	0.12500	2.125	2.8500	0.2500	3.2837
24 OD	24.000	8	0.12500	2.375	3.2500	0.2500	3.6837

Table 8.2.12   ANSI Straight Pipe Threads
(All dimensions in inches)

Nominal pipe size (1)	Threads per inch (2)	Pressure-tight with seals		Pressure-tight without seals		Free-fitting (NPSM)				Loose-fitting (NPSL)			
						External		Internal		External		Internal	
		Pitch diam, max (3)	Minor diam, min (4)	Pitch diam, max (5)	Minor diam, min (6)	Pitch diam, max (7)	Major diam, max (8)	Pitch diam, max (9)	Minor diam, min (10)	Pich diam, max (11)	Major diam, max (12)	Pitch diam, max (13)	Minor diam, min (14)
⅛	27	0.3782	0.342	0.3736	0.3415	0.3748	0.399	0.3783	0.350	0.3840	0.409	0.3989	0.362
¼	18	0.4951	0.440	0.4916	0.4435	0.4899	0.527	0.4951	0.453	0.5038	0.541	0.5125	0.470
⅜	18	0.6322	0.577	0.6270	0.5789	0.6270	0.664	0.6322	0.590	0.6409	0.678	0.6496	0.607
½	14	0.7851	0.715	0.7784	0.7150	0.7784	0.826	0.7851	0.731	0.7963	0.844	0.8075	0.753
¾	14	0.9956	0.925	0.9889	0.9255	0.9889	1.036	0.9956	0.941	1.0067	1.054	1.0179	0.964
1	11½	1.2468	1.161	1.2386	1.1621	1.2386	1.296	1.2468	1.181	1.2604	1.318	1.2739	1.208
1¼	11½	1.5915	1.506	—	—	1.5834	1.641	1.5916	1.526	1.6051	1.663	1.6187	1.553
1½	11½	1.8305	1.745	—	—	1.8223	1.880	1.8305	1.764	1.8441	1.902	1.8576	1.792
2	11½	2.3044	2.219	—	—	2.2963	2.354	2.3044	2.238	2.3180	2.376	2.3315	2.265
2½	8	2.7739	2.650	—	—	2.7622	2.846	2.7739	2.679	2.7934	2.877	2.8129	2.718
3	8	3.4002	3.277	—	—	3.3885	3.472	3.4002	3.305	3.4198	3.503	3.4393	3.344
3½	8	3.9005	3.777	—	—	3.8888	3.972	3.9005	3.806	3.9201	4.003	3.9396	3.845
4	8	4.3988	4.275	—	—	4.3871	4.470	4.3988	4.304	4.4184	4.502	4.4379	4.343
5	8	—	—	—	—	5.4493	5.533	5.4610	5.366	5.4805	5.564	5.5001	5.405
6	8	—	—	—	—	6.5060	6.589	6.5177	6.423	6.5372	6.620	6.5567	6.462
8	8	—	—	—	—	—	—	—	—	8.5313	8.615	8.5508	8.456
10	8	—	—	—	—	—	—	—	—	10.6522	10.735	10.6717	10.577
12	8	—	—	—	—	—	—	—	—	12.6491	12.732	12.6686	12.574

designation and notation include nominal size, number of threads per inch, thread series, class. For example, ⅛-27 NPTF-1, ⅛-27 NPTF-2, ⅛-27 PTF-SAE short, ⅛-27 NPSI, where N = National (American) standard, P = pipe, T = taper, S = straight, F = fuel and oil, I = intermediate. NPTF has two classes: class 1 = specific inspection of root and crest truncation *not* required; class 2 = specific inspection of root and crest truncation *is* required. The series includes: NPTF for all types of service; PTF-SAE short where clearance is not sufficient for full thread length as NPTF; NPSF, nontapered, economical to produce, and used with soft or ductile materials; NPSI nontapered, thick sections with little expansion.

Dry-seal pipe threads resemble tapered pipe threads except the form is truncated (see Fig. 8.2.10), and $L_4 = L_2 + 1$ (see Fig. 8.2.9). Although these threads are designed for nonlubricated joints, as in automobile work, under certain conditions a lubricant is used to prevent galling. Table 8.2.13 lists truncation values.

Tap drill sizes for tapered and straight pipe threads are listed in Table 8.2.14.

Wrench bolt heads, nuts, and wrench openings have been standardized (ANSI 18.2-1972). Bolt head and nut dimensions are given in Table 8.2.15.

Machine Screws

Machine screws are defined according to head types as follows:

Flat Head   This screw has a flat surface for the top of the head with a countersink angle of 82°. It is standard for machine screws, cap screws, and wood screws.

Round Head   This screw has a semielliptical head and is standard for machine screws, cap screws, and wood screws except that for the cap screw it is called *button head*.

Fig. 8.2.10   American National Standard dry-seal pipe thread.

Table 8.2.13   ANSI Dry-Seal Pipe Threads*
(See Fig. 8.2.10.)

Threads per inch n	Truncation, in		Width of flat	
	Min	Max	Min	Max
27 Crest	0.047p	0.094p	0.054p	0.108p
Root	0.094p	0.140p	0.108p	0.162p
18 C	0.047p	0.078p	0.054p	0.090p
R	0.078p	0.109p	0.090p	0.126p
14 C	0.036p	0.060p	0.042p	0.070p
R	0.060p	0.085p	0.070p	0.098p
11½ C	0.040p	0.060p	0.046p	0.069p
R	0.060p	0.090p	0.069p	0.103p
8 C	0.042p	0.055p	0.048p	0.064p
R	0.055p	0.076p	0.064p	0.088p

* The *truncation* and *width-of-flat* proportions listed above are also valid in the metric system.

Fillister Head   This screw has a rounded surface for the top of the head, the remainder being cylindrical. The head is standard for machine screws and cap screws.

Oval Head   This screw has a rounded surface for the top of the head and a countersink angle of 82°. It is standard for machine screws and wood screws.

Hexagon Head   This screw has a hexagonal head for use with external wrenches. It is standard for machine screws.

Socket Head   This screw has an internal hexagonal socket in the head for internal wrenching. It is standard for cap screws.

These screw heads are shown in Fig. 8.2.11; pertinent dimensions are in Table 8.2.16. There are many more machine screw head shapes available to the designer for special purposes, and many are found in the literature. In addition, many different screw head configurations have been developed to render fasteners "tamperproof"; these, too, are found in manufacturers' catalogs or the trade literature.

Eyebolts

Eyebolts are classified as rivet, nut, or screw and can be used on a swivel. See Fig. 8.2.12 and Table 8.2.17. The safe working load may be obtained for each application by applying an appropriate factor of safety.

Table 8.2.14 Suggested Tap Drill Sizes for Internal Pipe Threads

		Taper pipe thread						Straight pipe thread			
	Probable drill oversize cut (mean)	Minor diameter at distance		Drill for use without reamer		Drill for use with reamer		Minor diameter		Theoretical drill size§	Theoretical drill size†
Size		L_1 from large end	$L_1 + L_3$ from large end	Theoretical drill size*	Suggested drill size†	Theoretical drill size‡	Suggested drill size†	NPSF	NPSI		
	1	2	3	4	5	6	7	8	9	10	11
					Inch						
$\frac{1}{16}$–27	0.0038	0.2443	0.2374	0.2405	"C" (0.242)	0.2336	"A" (0.234)	0.2482	0.2505	0.2444	"D" (0.246)
$\frac{1}{32}$–27	0.0044	0.3367	0.3293	0.3323	"Q" (0.332)	0.3254	$\frac{21}{64}$ (0.328)	0.3406	0.3429	0.3362	"R" (0.339)
$\frac{1}{4}$–18	0.0047	0.4362	0.4258	0.4315	$\frac{7}{17}$ (0.438)	0.4211	$\frac{27}{64}$ (0.422)	0.4422	0.4457	0.4375	$\frac{7}{16}$ (0.438)
$\frac{3}{8}$–18	0.0049	0.5708	0.5604	0.5659	$\frac{9}{16}$ (0.562)	0.5555	$\frac{9}{16}$ (0.563)	0.5776	0.5811	0.5727	$\frac{27}{64}$ (0.578)
$\frac{1}{2}$–14	0.0051	0.7034	0.6901	0.6983	$\frac{45}{64}$ (0.703)	0.6850	$\frac{11}{16}$ (0.688)	0.7133	0.7180	0.7082	$\frac{45}{64}$ (0.703)
$\frac{3}{4}$–14	0.0060	0.9127	0.8993	0.9067	$\frac{29}{32}$ (0.906)	0.8933	$\frac{57}{64}$ (0.891)	0.9238	0.9283	0.9178	$\frac{59}{64}$ (0.922)
1–$11\frac{1}{2}$	0.0080	1.1470	1.1307	1.1390	$1\frac{9}{64}$ (1.141)	1.1227	$1\frac{1}{8}$ (1.125)	1.1600	1.1655	1.1520	$1\frac{5}{32}$ (1.156)
$1\frac{1}{4}$–$11\frac{1}{2}$	0.0100	1.4905	1.4742	1.4805	$1\frac{31}{64}$ (1.484)	1.4642	$1\frac{11}{32}$ (1.469)				
$1\frac{1}{2}$–$11\frac{1}{2}$	0.0120	1.7295	1.7132	1.7175	$1\frac{23}{32}$ (1.719)	1.7012	$1\frac{45}{64}$ (1.703)				
2–$11\frac{1}{2}$	0.0160	2.2024	2.1861	2.1864	$2\frac{3}{16}$ (2.188)	2.1701	$2\frac{11}{64}$ (2.172)				
$2\frac{1}{2}$–8	0.0180	2.6234	2.6000	2.6054	$2\frac{39}{64}$ (2.609)	2.5820	$2\frac{27}{64}$ (2.578)				
3–8	0.0200	3.2445	3.2211	3.2245	$3\frac{15}{64}$ (3.234)	3.2011	$3\frac{13}{64}$ (3.203)				
					Metric						
$\frac{1}{16}$–27	0.097	6.206	6.029	6.109	6.1	5.932	6.0	6.304	6.363	6.207	6.2
$\frac{1}{8}$–27	0.112	8.551	8.363	8.438	8.4	8.251	8.2	8.651	8.710	8.539	8.5
$\frac{1}{4}$–18	0.119	11.080	10.816	10.961	11.0	10.697	10.8	11.232	11.321	11.113	11.0
$\frac{3}{8}$–18	0.124	14.499	14.235	14.375	14.5	14.111	14.0	14.671	14.760	14.547	14.5
$\frac{1}{2}$–14	0.130	17.867	17.529	17.737	17.5	17.399	17.5	18.118	18.237	17.988	18.0
$\frac{3}{4}$–14	0.152	23.182	22.842	23.030	23.0	22.690	23.0	23.465	23.579	23.212	23.0
1–$11\frac{1}{2}$	0.203	29.134	28.720	28.931	29.0	28.517	28.5	29.464	29.604	29.261	29.0
$1\frac{1}{4}$–$11\frac{1}{2}$	0.254	37.859	37.444	37.605	37.5	37.190	37.0				
$1\frac{1}{2}$–$11\frac{1}{2}$	0.305	43.929	43.514	43.624	43.5	43.209	43.5				
2–$11\frac{1}{2}$	0.406	55.941	55.527	55.535	56.0	55.121	55.0				
$2\frac{1}{2}$–8	0.457	66.634	66.029	66.177	66.0	65.572	65.0				
3–8	0.508	82.410	81.815	81.902	82.0	81.307	81.0				

* Column 4 values equal column 2 values minus column 1 values.
† Some drill sizes listed may not be standard drills, and in some cases, standard metric drill sizes may be closer to the theoretical inch drill size and standard inch drill sizes may be closer to the theoretical metric drill size.
‡ Column 6 values equal column 3 values minus column 1 values.
§ Column 10 values equal column 8 values minus column 1 values.
SOURCE: ANSI B1.20.3-1976 and ANSI B1.20.4-1976, reproduced by permission.

Fig. 8.2.11 Machine screw heads. (*a*) Flat; (*b*) fillister; (*c*) round; (*d*) oval; (*e*) hexagonal; (*f*) socket.

Driving recesses come in many forms and types and can be found in company catalogs. Figure 8.2.13 shows a representative set.

Setscrews are used for fastening collars, sheaves, gears, etc. to shafts to prevent relative rotation or translation. They are available in a variety of head and point styles, as shown in Fig. 8.2.14. A complete tabulation of dimensions is found in ANSI/ASME B18.3-1982 (R86), ANSI 18.6.2-1977 (R93), and ANSI 18.6.3-1977 (R91). Holding power for various sizes is given in Table 8.2.18.

Hex socket   Phillips   Drilled spanner

Fluted socket   Frearson   Slotted spanner

Slotted   Clutch   Pozidriv   One-way

Fig. 8.2.13 Driving recesses. (*Adapted, with permission, from Machine Design.*)

(a)

(b)

Fig. 8.2.12 Eyebolts.

Table 8.2.15   Width Across Flats of Bolt Heads and Nuts
(All dimensions in inches)

Nominal site or basic major diam of thread	Dimensions of regular bolt heads unfinished, square, and hexagon		Dimensions of heavy bolt heads unfinished, square, and hexagon		Dimensions of cap-screw heads hexagon		Dimensions of setscrew heads		Dimensions of regular nuts and regular jam nuts, unfinished, square, and hexagon (jam nuts, hexagon only)		Dimensions of machine-screw and stove-bolt nuts, square and hexagon		Dimensions of heavy nuts and heavy jam nuts, unfinished, square, and hexagon (jam nuts, hexagon only)	
	Max	Min	Max	Min	Max	Min	Max	Min	Max	Min	Max	Min	Max	Min
No. 0	—	—	—	—	—	—	—	—	—	—	0.1562	0.150		
No. 1	—	—	—	—	—	—	—	—	—	—	0.1562	0.150		
No. 2	—	—	—	—	—	—	—	—	—	—	0.1875	0.180		
No. 3	—	—	—	—	—	—	—	—	—	—	0.1875	0.180		
No. 4	—	—	—	—	—	—	—	—	—	—	0.2500	0.241		
No. 5	—	—	—	—	—	—	—	—	—	—	0.3125	0.302		
No. 6	—	—	—	—	—	—	—	—	—	—	0.3125	0.302		
No. 8	—	—	—	—	—	—	—	—	—	—	0.3438	0.332		
No. 10	—	—	—	—	—	—	—	—	—	—	0.3750	0.362		
No. 12	—	—	—	—	—	—	—	—	—	—	0.4375	0.423		
¼	0.3750	0.362	—	—	0.4375	0.428	0.2500	0.241	0.4375	0.425	0.4375	0.423	0.5000	0.488
⁵⁄₁₆	0.5000	0.484	—	—	0.5000	0.489	0.3125	0.302	0.5625	0.547	0.5625	0.545	0.5938	0.578
⅜	0.5625	0.544	—	—	0.5625	0.551	0.3750	0.362	0.6250	0.606	0.6250	0.607	0.6875	0.669
⁷⁄₁₆	0.6250	0.603	—	—	0.6250	0.612	0.4375	0.423	0.7500	0.728	—	—	0.7812	0.759
½	0.7500	0.725	0.8750	0.850	0.7500	0.736	0.5000	0.484	0.8125	0.788	—	—	0.8750	0.850
⁹⁄₁₆	0.8750	0.847	0.9375	0.909	0.8125	0.798	0.5625	0.545	0.8750	0.847	—	—	0.9375	0.909
⅝	0.9375	0.906	1.0625	1.031	0.8750	0.860	0.6250	0.606	1.0000	0.969	—	—	1.0625	1.031
¾	1.1250	1.088	1.2500	1.212	1.0000	0.983	0.7500	0.729	1.1250	1.088	—	—	1.2500	1.212
⅞	1.3125	1.269	1.4375	1.394	1.1250	1.106	0.8750	0.852	1.3125	1.269	—	—	1.4375	1.394
1	1.5000	1.450	1.6250	1.575	1.3125	1.292	1.0000	0.974	1.5000	1.450	—	—	1.6250	1.575
1⅛	1.6875	1.631	1.8125	1.756	1.5000	1.477	1.1250	1.096	1.6875	1.631	—	—	1.8125	1.756
1¼	1.8750	1.812	2.0000	1.938	1.6875	1.663	1.2500	1.219	1.8750	1.812	—	—	2.0000	1.938
1⅜	2.0625	1.994	2.1857	2.119	—	—	1.3750	1.342	2.0625	1.994	—	—	2.1875	2.119
1½	2.2500	2.175	2.3750	2.300	—	—	1.5000	1.464	2.2500	2.175	—	—	2.3750	2.300
1⅝	2.4375	2.356	2.5625	2.481	—	—	—	—	2.4375	2.356	—	—	2.5625	2.481
1¾	2.6250	2.538	2.7500	2.662	—	—	—	—	2.6250	2.538	—	—	2.7500	2.662
1⅞	2.8125	2.719	2.9375	2.844	—	—	—	—	2.8125	2.719	—	—	2.9375	2.844
2	3.0000	2.900	3.1250	3.025	—	—	—	—	3.0000	2.900	—	—	3.1250	3.025
2¼	3.3750	3.262	3.5000	3.388	—	—	—	—	3.3750	3.262	—	—	3.5000	3.388
2½	3.7500	3.625	3.8750	3.750	—	—	—	—	3.7500	3.625	—	—	3.8750	3.750
2¾	4.1250	3.988	4.2500	4.112	—	—	—	—	4.1250	3.988	—	—	4.2500	4.112
3	4.5000	4.350	4.6250	4.475	—	—	—	—	4.5000	4.350	—	—	4.6250	4.475
3¼	—	—	—	—	—	—	—	—	—	—	—	—	5.0000	4.838
3½	—	—	—	—	—	—	—	—	—	—	—	—	5.3750	5.200
3¾	—	—	—	—	—	—	—	—	—	—	—	—	5.7500	5.562
4	—	—	—	—	—	—	—	—	—	—	—	—	6.1250	5.925

Regular bolt heads are for general use. Unfinished bolt heads are not finished on any surface. Semifinished bolt heads are finished under head.
Regular nuts are for general use. Semifinished nuts are finished on bearing surface and threaded. Unfinished nuts are not finished on any surface but are threaded.

| Hollow oval point | Hollow flat point | Hollow half dog point | Square head cone point | Square head cup point |

Fig. 8.2.14   Setscrews.

Table 8.2.16 Head Diameters (Maximum), in

			Machine screws			
Nominal size	Screw diam	Flat head	Round head	Fillister head	Oval head	Hexagonal head across flats
2	0.086	0.172	0.162	0.140	0.172	0.125
3	0.099	0.199	0.187	0.161	0.199	0.187
4	0.112	0.225	0.211	0.183	0.225	0.187
5	0.125	0.252	0.236	0.205	0.252	0.187
6	0.138	0.279	0.260	0.226	0.279	0.250
8	0.164	0.332	0.309	0.270	0.332	0.250
10	0.190	0.385	0.359	0.313	0.385	0.312
12	0.216	0.438	0.408	0.357	0.438	0.312
$\frac{1}{4}$	0.250	0.507	0.472	0.414	0.507	0.375
$\frac{5}{16}$	0.3125	0.636	0.591	0.519	0.636	0.500
$\frac{3}{8}$	0.375	0.762	0.708	0.622	0.762	0.562

			Cap screws		
Nominal size	Screw diam	Flat head	Button head	Fillister head	Socket head
$\frac{1}{4}$	0.250	$\frac{1}{2}$	$\frac{7}{16}$	$\frac{3}{8}$	$\frac{3}{8}$
$\frac{5}{16}$	0.3125	$\frac{5}{8}$	$\frac{9}{16}$	$\frac{7}{16}$	$\frac{7}{16}$
$\frac{3}{8}$	0.375	$\frac{3}{4}$	$\frac{5}{8}$	$\frac{9}{16}$	$\frac{9}{16}$
$\frac{7}{16}$	0.4375	$\frac{13}{16}$	$\frac{3}{4}$	$\frac{5}{8}$	$\frac{5}{8}$
$\frac{1}{2}$	0.500	$\frac{7}{8}$	$\frac{13}{16}$	$\frac{3}{4}$	$\frac{3}{4}$
$\frac{9}{16}$	0.5625	1	$\frac{15}{16}$	$\frac{13}{16}$	$\frac{13}{16}$
$\frac{5}{8}$	0.625	$1\frac{1}{8}$	1	$\frac{7}{8}$	$\frac{7}{8}$
$\frac{3}{4}$	0.750	$1\frac{3}{8}$	$1\frac{1}{4}$	1	1
$\frac{7}{8}$	—	—	—	$1\frac{1}{8}$	$1\frac{1}{8}$
1	—	—	—	$1\frac{5}{16}$	$1\frac{5}{16}$

Locking Fasteners

Locking fasteners are used to prevent loosening of a threaded fastener in service and are available in a wide variety differing vastly in design, performance, and function. Since each has special features which may make it of particular value in the solution of a given machine problem, it is important that great care be exercised in the selection of a particular design in order that its properties may be fully utilized. These fasteners may be divided into six groups, as follows: seating lock, spring stop nut, interference, wedge, blind, and quick-release. The **seating-lock type** locks only when firmly seated and is therefore free-running on the bolt. The **spring stop-nut type** of fastener functions by a spring action clamping down upon the bolt. The **prevailing torque type** locks by elastic or plastic flow of a portion of the fastener material. A recent development employs an adhesive coating applied to the threads. The **wedge type** locks by relative wedging of either elements or nut and bolt. The **blind type** usually utilizes spring action of the fastener, and the **quick-release**

type utilizes a quarter-turn release device. An example of each is shown in Fig. 8.2.15.

One such specification developed for prevailing torque fasteners by the Industrial Fasteners Institute is based on locking torque and may form a precedent for other types of fasteners as well.

Coach and lag screws find application in wood, or in masonry with an expansion anchor. Figure 8.2.16a shows two types, and Table 8.2.19 lists pertinent dimensions.

Wood screws [ANSI B18.22.1-1975 (R81)] are made in lengths from $\frac{1}{4}$ to 5 in for steel and from $\frac{1}{4}$ to $3\frac{1}{2}$ in for brass screws, increasing by $\frac{1}{8}$ in up to 1 in, by $\frac{1}{4}$ in up to 3 in, and by $\frac{1}{2}$ in up to 5 in. Sizes are given in Table 8.2.20. Screws are made with flat, round, or oval heads. Figure 8.2.16b shows several heads.

Washers [ANSI B18.22.1-1975 (R81)] for bolts and lag screws, either round or square, are made to the dimensions given in Table 8.2.21. For other types of lock washers, see Fig. 8.2.17a and b.

Table 8.2.17 Regular Nut Eyebolts—Selected Sizes
(Thomas Laughlin Co., Portland, Me.)
(All dimensions in inches)
(See Fig. 8.2.12.)

Diam and shank length	Thread length	Eye dimension ID	Eye dimension OD	Approx breaking strength, lb	Diam and shank length	Thread length	Eye dimension ID	Eye dimension OD	Approx breaking strength, lb
$\frac{1}{4} \times 2$	$1\frac{1}{2}$	$\frac{1}{2}$	1	2,200	$\frac{3}{4} \times 6$	3	$1\frac{1}{2}$	3	23,400
$\frac{5}{16} \times 2\frac{1}{4}$	$1\frac{1}{2}$	$\frac{5}{8}$	$1\frac{1}{4}$	3,600	$\frac{3}{4} \times 10$	3	$1\frac{1}{2}$	3	23,400
$\frac{3}{8} \times 4\frac{1}{2}$	$2\frac{1}{2}$	$\frac{3}{4}$	$1\frac{1}{2}$	5,200	$\frac{3}{4} \times 15$	5	$1\frac{1}{2}$	3	23,400
$\frac{1}{2} \times 3\frac{1}{4}$	$1\frac{1}{2}$	1	2	9,800	$\frac{7}{8} \times 8$	4	$1\frac{3}{4}$	$3\frac{1}{2}$	32,400
$\frac{1}{2} \times 6$	3	1	2	9,800	1×6	3	2	4	42,400
$\frac{1}{2} \times 10$	3	1	2	9,800	1×9	4	2	4	42,400
$\frac{5}{8} \times 4$	2	$1\frac{1}{4}$	$2\frac{1}{2}$	15,800	1×18	7	2	4	42,400
$\frac{5}{8} \times 6$	3	$1\frac{1}{4}$	$2\frac{1}{2}$	15,800	$1\frac{1}{4} \times 8$	4	$2\frac{1}{2}$	5	67,800
$\frac{5}{8} \times 10$	3	$1\frac{1}{4}$	$2\frac{1}{2}$	15,800	$1\frac{1}{4} \times 20$	6	$2\frac{1}{2}$	5	67,800

Table 8.2.18   Cup-Point Setscrew Holding Power

Nominal screw size	Seating torque, lb · in	Axial holding power, lb
No. 0	0.5	50
No. 1	1.5	65
No. 2	1.5	85
No. 3	5	120
No. 4	5	160
No. 5	9	200
No. 6	9	250
No. 8	20	385
No. 10	33	540
¼ in	87	1,000
5/16 in	165	1,500
3/8 in	290	2,000
7/16 in	430	2,500
½ in	620	3,000
9/16 in	620	3,500
5/8 in	1,225	4,000
¾ in	2,125	5,000
7/8 in	5,000	6,000
1 in	7,000	7,000

NOTES: 1. Torsional holding power in inch-pounds is equal to one-half of the axial holding power times the shaft diameter in inches.
2. Experimental data were obtained by seating an alloy-steel cup-point setscrew against a steel shaft with a hardness of Rockwell C 15. Screw threads were class 3A, tapped holes were class 2B. Holding power was defined as the minimum load necessary to produce 0.01 in of relative movement between the shaft and the collar.
3. Cone points will develop a slightly greater holding power; flat, dog, and oval points, slightly less.
4. Shaft hardness should be at least 10 Rockwell C points less than the setscrew point.
5. Holding power is proportional to seating torque. Torsional holding power is increased about 6% by use of a flat on the shaft.
6. Data by F. R. Kull, Fasteners Book Issue, *Mach. Des.*, Mar. 11, 1965.

Self-tapping screws are available in three types. **Thread-forming** tapping screws plastically displace material adjacent to the pilot hole. **Thread-cutting** tapping screws have cutting edges and chip cavities (flutes) and form a mating thread by removing material adjacent to the pilot hole. Thread-cutting screws are generally used to join thicker and harder materials and require a lower driving torque than thread-forming screws. **Metallic drive** screws are forced into the material by pressure and are intended for making permanent fastenings. These three types are further classified on the basis of thread and point form as shown in Table 8.2.22. In addition to these body forms, a number of different head types are available. Basic dimensional data are given in Table 8.2.23.

Carriage bolts have been standardized in ANSI B18.5-1971, revised 1990. They come in styles shown in Fig. 8.2.18. The range of bolt diameters is no. 10 (= 0.19 in) to 1 in, no. 10 to ¾ in, no. 10 to ½ in, and no. 10 to ¾ in, respectively.

**Materials, Strength, and Service Adaptability
of Bolts and Screws Materials**

Table 8.2.24 shows the relationship between selected metric bolt classes and SAE and ASTM grades. The first number of a metric bolt class equals the minimum tensile strength (ultimate) in megapascals (MPa) divided by 100, and the second number is the approximate ratio between minimum yield and minimum ultimate strengths.

A variety of specially configured heads is available to prevent vandalism; they go by the generic term **tamper-proof.**

EXAMPLE.   Class 5.8 has a minimum ultimate strength of approximately 500 MPa and a minimum yield strength approximately 80 percent of minimum ultimate strength.

Strength   The fillet between head and body, the thread runout point, and the first thread to engage the nut all create stress concentrations causing local stresses much greater than the average tensile stress in the bolt body. The complexity of the stress patterns renders ineffective the ordinary design calculations based on yield or ultimate stresses. Bolt strengths are therefore determined by laboratory tests on bolt-nut assemblies and published as **proof loads.** Fastener manufacturers are required to periodically repeat such tests to ensure that their products meet the original standards.

In order that a bolted joint remain firmly clamped while carrying its external load P, the bolt must be tightened first with sufficient torque to induce an initial tensile preload F_i. The total load F_B experienced by the bolt is then $F_B = F_i + \varepsilon P$. The fractional multiplier ε is given by $\varepsilon = K_B/(K_B + K_M)$, where K_B = elastic constant of the bolt and $1/K_M = 1/K_N + 1/K_W + 1/K_G + 1/K_J$. K_N = elastic constant of the nut; K_W = elastic constant of the washer; K_G = elastic constant of the gasket; K_J = elastic constant of the clamped surfaces or joint.

By manipulation, the fractional multiplier can be written $\varepsilon = 1/(1 + K_M/K_B)$. When K_M/K_B approaches 0, $\varepsilon \to 1$. When K_M/K_B approaches infinity, $\varepsilon \to 0$. Generally, K_N, K_W, and K_J are much stiffer than K_B, while K_G can vary from very soft to very stiff. In a metal-to-metal joint, K_G is effectively infinity, which causes K_M to approach infinity and ε to approach 0. On the other hand, for a very soft gasketed joint, $K_M \to 0$ and $\varepsilon \to 1$. For a metal-to-metal joint, then, $F_B = F_i + 0 \times P = F_i$; thus no fluctuating load component enters the bolt. In that case, the bolt remains at *static force* F_i at all times, and the static design will suffice. For a very soft gasketed joint, $F_B = F_i + 1 \times P = F_i + P$, which means that if P is a dynamically fluctuating load, it will be superimposed onto the static value of F_i. Accordingly, one must use the *fatigue design* for the bolt. Of course, for conditions between $\varepsilon = 0$ and $\varepsilon = 1$, the load within the bolt body is $F_B = F_i + \varepsilon P$, and again the *fatigue design* must be used.

Fig. 8.2.16a   Coach and lag screws.   **Fig. 8.2.16b**   Wood screws.

Fig. 8.2.15   Locking fasteners.

In general, one wants as much preload as a bolt and joint will tolerate, without damaging the clamped parts, encouraging stress corrosion, or reducing fatigue life. For ungasketed, unpressurized joints under static loads using high-quality bolt materials, such as SAE 3 or better, the preload should be about 90 percent of proof load.

The **proof strength** is the stress obtained by dividing **proof load** by **stress area**. Stress area is somewhat larger than the root area and can be found in thread tables, or calculated approximately from a diameter which is the mean of the root and pitch diameters.

Initial sizing of bolts can be made by calculating area = (% × proof load)/ (proof strength). See Table 8.2.25 for typical physical properties.

NOTE. In European practice, proof stress of a given grade is independent of diameter and is accomplished by varying chemical composition with diameters.

General Notes on the Design of Bolted Joints

Bolts subjected to shock and sudden change in load are found to be more serviceable when the unthreaded portion of the bolt is turned down or drilled to the area of the root of the thread. The drilled bolt is stronger in torsion than the turned-down bolt.

When a **number of bolts** are **employed in fastening** together two parts of a machine, such as a cylinder and cylinder head, the load carried by each bolt depends on its relative tightness, the tighter bolts carrying the greater loads. When the conditions of assembly result in differences in tightness, lower working stresses must be used in designing the bolts than otherwise are necessary. On the other hand, it may be desirable to have the bolts the weakest part of the machine, since their breakage from overload in the machine may result in a minimum replacement cost. In such cases, the breaking load of the bolts may well be equal to the load which causes the weakest member of the machine connected to be stressed up to the elastic limit.

Bolts screwed up tight have an initial stress due to the tightening (preload) before any external load is applied to the machine member. The initial tensile load due to screwing up for a tight joint varies approximately as the diameter of the bolt, and may be estimated at 16,000 lb/in of diameter. The actual value depends upon the applied

Table 8.2.19   Coach and Lag Screws

Diam of screw, in	1/4	5/16	3/8	7/16	1/2	5/8	3/4	7/8	1
No. of threads per inch	10	9	7	7	6	5	4 1/2	4	3 1/2
Across flats of hexagon and square heads, in	3/8	15/32	9/16	21/32	3/4	15/16	1 1/8	1 5/16	1 1/2
Thickness of hexagon and square heads, in	3/16	1/4	5/16	3/8	7/16	17/32	5/8	3/4	7/8

Length of threads for screws of all diameters							
Length of screw, in	1 1/2	2	2 1/2	3	3 1/2	4	4 1/2
Length of thread, in	To head	1 1/2	2	2 1/4	2 1/2	3	3 1/2

Length of screw, in	5	5 1/2	6	7	8	9	10–12
Length of thread, in	4	4	4 1/2	5	6	6	7

Table 8.2.20   American National Standard Wood Screws

Number	0	1	2	3	4	5	6	7	8
Threads per inch	32	28	26	24	22	20	18	16	15
Diameter, in	0.060	0.073	0.086	0.099	0.112	0.125	0.138	0.151	0.164

Number	9	10	11	12	14	16	18	20	24
Threads per inch	14	13	12	11	10	9	8	8	7
Diameter, in	0.177	0.190	0.203	0.216	0.242	0.268	0.294	0.320	0.372

Table 8.2.21   Dimensions of Steel Washers, in

Bolt size	Plain washer Hole diam	OD	Thickness	Lock washer Hole diam	OD	Thickness
3/16	1/4	9/16	3/64	0.194	0.337	0.047
1/4	5/16	3/4	1/16	0.255	0.493	0.062
5/16	3/8	7/8	1/16	0.319	0.591	0.078
3/8	7/16	1	5/64	0.382	0.688	0.094
7/16	1/2	1 1/4	5/64	0.446	0.781	0.109
1/2	9/16	1 3/8	3/32	0.509	0.879	0.125
9/16	5/8	1 1/2	3/32	0.573	0.979	0.141
5/8	11/16	1 3/4	1/8	0.636	1.086	0.156
3/4	13/16	2	1/8	0.763	1.279	0.188
7/8	15/16	2 1/4	5/32	0.890	1.474	0.219
1	1 1/16	2 1/2	5/32	1.017	1.672	0.250
1 1/8	1 1/4	2 3/4	5/32	1.144	1.865	0.281
1 1/4	1 3/8	3	5/32	1.271	2.058	0.312
1 3/8	1 1/2	3 1/4	11/64	1.398	2.253	0.344
1 1/2	1 5/8	3 1/2	11/64	1.525	2.446	0.375
1 5/8	1 3/4	3 3/4	11/64			
1 3/4	1 7/8	4	11/64			
1 7/8	2	4 1/4	11/64			
2	2 1/8	4 1/2	11/64			
2 1/4	2 3/8	4 3/4	3/16			
2 1/2	2 5/8	5	7/32			

Table 8.2.22   Tapping Screw Forms

ASA type and thread form	Description and recommendations
AB	Spaced thread, with gimlet point, designed for use in sheet metal, resin-impregnated plywood, wood, and asbestos compositions. Used in pierced or punched holes where a sharp point for starting is needed.
B	Type B is a blunt-point spaced-thread screw, used in heavy-gage sheet-metal and nonferrous castings.
BP	Same as type B, but has a 45 deg included angle unthreaded cone point. Used for locating and aligning holes or piercing soft materials.
C	Blunt point with threads approximating machine-screw threads. For applications where a machine-screw thread is preferable to the spaced-thread form. Unlike thread-cutting screws, type C makes a chip-free assembly.
U	Multiple-threaded drive screw with steep helix angle and a blunt, unthreaded starting pilot. Intended for making permanent fastenings in metals and plastics. Hammered or mechanically forced into work. Should not be used in materials less than one screw diameter thick.
D	Blunt point with single narrow flute and threads approximating machine-screw threads. Flute is designed to produce a cutting edge which is radial to screw center. For low-strength metals and plastics; for high-strength brittle metals; and for rethreading clogged pretapped holes.
F	Approximate machine-screw thread and blunt point.
G	Approximate machine-screw thread with single through slot which forms two cutting edges. For low-strength metals and plastics.
T	Same as type D with single wide flute for more chip clearance.
BF	Spaced thread with blunt point and five evenly spaced cutting grooves and chip cavities. Wall thickness should be 1½ times major diameter of screw. Reduces stripping in brittle plastics and die castings.
BT	Same as type BF except for single wide flute which provides room for twisted, curly chips.
TT	Thread-rolling screws roll-form clean, screw threads. The plastic movement of the material it is driven into locks it in place. The Taptite form is shown here.

SOURCE: *Mach. Des.*, Mar. 11, 1965.

Table 8.2.23 Self-Tapping Screws

Screw size	Basic major diam, in	Threads per inch					Type U	
		AB	B, BP	C	D, F, G, T	BF, BT	Max outside diam, in	Number of thread starts
00	—	—	—	—	—	—	0.060	6
0	0.060	40	48	—	—	48	0.075	6
1	0.073	32	42	—	—	42		
2	0.086	32	32	56 and 64	56 and 64	32	0.100	8
3	0.099	28	28	48 and 56	48 and 56	28		
4	0.112	24	24	40 and 48	40 and 48	24	0.116	7
5	0.125	20	20	40 and 44	40 and 44	20		
6	0.138	18	20	32 and 40	32 and 40	20	0.140	7
7	0.151	16	19	—	—	19	0.154	8
8	0.164	15	18	32 and 36	32 and 36	18	0.167	8
10	0.190	12	16	24 and 32	24 and 32	16	0.182	8
12	0.216	11	14	24 and 28	24 and 28	14	0.212	8
14	0.242	10	—	—	—	—	0.242	9
$\frac{1}{4}$	0.250	—	14	20 and 28	20 and 28	14		
16	0.268	10						
18	0.294	9						
$\frac{5}{16}$	0.3125	—	12	18 and 24	18 and 24	12	0.315	14
20	0.320	9						
24	0.372	9						
$\frac{3}{8}$	0.375	—	12	16 and 24	16 and 24	12	0.378	12
$\frac{7}{16}$	0.4375	—	10	—	—	10		
$\frac{1}{2}$	0.500	—	10	—	—	10		

torque and the efficiency of the screw threads. Applying this rule to bolts of 1-in diam or less results in excessively high stresses, thus demonstrating why bolts of small diameter frequently fail during assembly. It is advisable to use as large-diameter bolts as possible in pressure-tight joints requiring high tightening loads.

In pressure-tight joints without a gasket the force on the bolt under load is essentially never greater than the initial tightening load. When a gasket is used, the total bolt force is approximately equal to the initial tightening load plus the external load. In the first case, deviations from the rule are a result of elastic behavior of the joint faces without a gasket, and inelastic behavior of the gasket in the latter case. The following generalization will serve as a guide. If the bolt is more yielding than the connecting members, it should be designed simply to resist the initial tension or the external load, whichever is greater. If the probable yielding of the bolt is 50 to 100 percent of that of the connected members, take the resultant bolt load as the initial tension plus one-half the external load. If the yielding of the connected members is probably four to five times that of the bolt (as when certain packings are used), take the resultant bolt load as the initial tension plus three-fourths the external load.

In cases where bolts are subjected to cyclic loading, an increase in the initial tightening load decreases the operating stress range. In certain applications it is customary to fix the tightening load as a fraction of the yield-point load of the bolt.

Table 8.2.24 ISO Metric Fastener Materials

Metric bolt	Metric nut class normally used	Roughly equivalent U.S. bolt materials	
		SAE J429 grades	ASTM grades
4.6	4 or 5	1	A193, B8; A307, grade A
4.8	4 or 5	2	
5.8	5	2	
8.8	8	5	A325, A449
9.9	9	5+	A193, B7 and B16
10.9	10 or 12	8	A490; A354, grade 8D
12.9	10 or 12		A540; B21 through B24

SOURCE: Bickford, "An Introduction to the Design and Behavior of Bolted Joints," Marcel Dekker, 1981; reproduced by permission. See Appendix G for additional metric materials.

A = I. D.
B = O. D.
W = width
T = ave. thickness
$= \dfrac{t_o + t_i}{2}$

Fig. 8.2.17a Regular helical spring lock washer.

External tooth

Internal–external tooth

Internal tooth

Countersunk external tooth

A = I. D.
B = O. D.
C = Thickness

Fig. 8.2.17b Toothed lock washers.

Table 8.2.25a   Specifications and Identification Markings for Bolts, Screws, Studs, Sems,[a] and U Bolts[b]
(Multiply the strengths in kpsi by 6.89 to get the strength in MPa.)

SAE grade	ASTM grade	Metric[c] grade	Nominal diameter, in	Proof strength, kpsi	Tensile strength, kpsi	Yield[d] strength, kpsi	Core hardness, Rockwell min/max	Products[e]
1	A307	4.6	¼ thru 1½	33	60	36	B70/B100	B, Sc, St
2		5.8	¼ thru ¾	55	74	57	B80/B100	B, Sc, St
		4.6	Over ¾ thru 1½	33	60	36	B70/B100	B, Sc, St
4		8.9	¼ thru 1½	65[f]	115	100	C22/C32	St
5	A449 or A325 type 1	8.8	¼ thru 1	85	120	92	C25/C34	B, Sc, St
		7.8	Over 1 thru 1½	74	105	81	C19/C30	B, Sc, St
		8.6	Over 1½ to 3	55	90	58		B, Sc, St
5.1		8.8	No. 6 thru ⅝	85	120		C25/C40	Se
		8.8	No. 6 thru ½	85	120		C25/C40	B, Sc, St
5.2	A325 type 2	8.8	¼ thru 1	85	120	92	C26/C36	B, Sc
7[g]		10.9	¾ thru 1½	105	133	115	C28/C34	B, Sc
8	A354 Grade BD	10.9	¼ thru 1½	120	150	130	C33/C39	B, Sc, St
8.1		10.9	¼ thru 1½	120	150	130	C32/C38	St
8.2		10.9	¼ thru 1	120	150	130	C35/C42	B, Sc
	A574	12.9	0 thru ½	140	180	160	C39/C45	SHCS
		12.9	⅝ thru 1½	135	170	160	C37/C45	SHCS

[a] Sems are screw and washer assemblies.
[b] Compiled from ANSI/SAE J429j; ANSI B18.3.1-1978; and ASTM A307, A325, A354, A449, and A574.
[c] Metric grade is xx.x where xx is approximately 0.01 S_{ut} in MPa and .x is the ratio of the minimum S_y to S_{ut}.
[d] Yield strength is stress at which a permanent set of 0.2% of gage length occurs.
[e] B = bolt, Sc = Screws, St = studs, Se = sems, and SHCS = socket head cap screws.
[f] Entry appears to be in error but conforms to the standard, ANSI/SAE J429j.
[g] Grade 7 bolts and screws are roll-threaded after heat treatment.
NOTE: Company catalogs should be consulted regarding *proof loads*. However, approximate values of proof loads may be calculated from: proof load = proof strength × stress area.
SOURCE: Shigley and Mitchell, "Mechanical Engineering Design," 4th ed., McGraw-Hill, 1983, by permission.

In order to avoid the possibility of bolt failure in pressure-tight joints and to obtain uniformity in bolt loads, some means of determining initial bolt load (**preload**) is desirable. Calibrated torque wrenches are available for this purpose, reading directly in inch-pounds or inch-ounces. Inaccuracies in initial bolt load are possible even when using a torque wrench, owing to variations in the coefficient of friction between the nut and the bolt and, further, between the nut or bolthead and the abutting surface.

An exact method to determine the preload in a bolt requires that the bolt elongation be measured. For a through bolt in which both ends are accessible, the elongation is measured, and the preload force P is obtained from the relationship

$$P = AEe \div l$$

where E = modulus of elasticity, l = original length, A = cross-sectional area, e = elongation. In cases where both ends of the bolt are not accessible, strain-gage techniques may be employed to determine the strain in the bolt, and thence the preload.

High-strength bolts designated ASTM A325 and A490 are almost exclusively employed in the assembly of structural steel members, but they are applied in mechanical assemblies such as flanged joints. The direct tension indicator (DTI) (Fig. 8.2.19), for use with high-strength bolts, allows **bolt preload** to be applied rapidly and simply. The device is a hardened washer with embossed protrusions (Fig. 8.2.19a). Tightening the bolt causes the protrusions to flatten and results in a decrease in the gap between washer and bolthead. The prescribed degree of bolt-tightening load, or preload, is obtained when the gap is reduced to a predetermined amount (Fig. 8.2.19c). A feeler gage of a given thickness is used to determine when the gap has been closed to the prescribed amount (Fig. 8.2.19b and c). With a paired bolt and DTI, the degree of gap closure is proportional to bolt preload. The system is reported to provide bolt-preload force accuracy within +15 percent of that prescribed (Fig. 8.2.19d). The devices are available in both inch and metric series and are covered under ASTM F959 and F959M.

Preload-indicating bolts and nuts provide visual assurance of preload in that tightening to the desired preload causes the wavy flange to flatten flush with the clamped assembly (Fig. 8.2.20).

Fig. 8.2.18 Carriage bolts.

In drilling and tapping cast iron for steel studs, it is necessary to tap to a depth equal to 1½ times the stud diameter so that the strength of the cast-iron threads in shear may equal the tensile strength of the stud. Drill sizes and depths of hole and thread are given in Table 8.2.26.

It is not good practice to drill holes to be tapped through the metal into pressure spaces, for even though the bolt fits tightly, leakage will result that is difficult to eliminate.

Self-sealing Screws and Nuts   Screws and bolts used for vacuum or pressure chambers are sealed by placing elastic, soft metallic, or plastic materials in the shape of an O ring or a washer under the head of a screw or close to the threads of its nut. By tightening the screw or the bolt, the deformable O ring or the soft washer fills the space between the threads of the screw and the member and seals the chamber. For further information, refer to manufacturers.

Screw thread inserts made of high-strength material (Fig. 8.2.21) are useful in many cases to provide increased thread strength and life. Soft or ductile materials tapped to receive thread inserts exhibit improved load-carrying capacity under static and dynamic loading conditions. Holes in which threads have been stripped or otherwise damaged can be restored through the use of thread inserts.

Holes for thread inserts are drilled oversize and specially tapped to receive the insert selected to mate with the threaded fastener used. The standard material for inserts is 18-8 stainless steel, but other materials are available, such as phosphor bronze and Inconel. Recommended insert lengths are given in Table 8.2.27.

Table 8.2.25b ASTM and SAE Grade Head Markings for Steel Bolts and Screws

Grade marking	Specification and material	Grade marking	Specification and material
No mark	SAE grade 1 Low- or medium-carbon steel ASTM A307 Low-carbon steel SAE grade 2 Low- or medium-carbon steel	BC	ASTM A354 grade BC Alloy steel, quenched and tempered
	SAE grade 5 ASTM A449 Medium-carbon steel, quenched and tempered		SAE grade 7 Medium-carbon alloy steel, quenched and tempered, roll-threaded after heat treatment
	SAE grade 5.2 Low-carbon martensite steel, quenched and tempered		SAE grade 8 Medium-carbon alloy steel, quenched and tempered ASTM A 354 grade BD Alloy steel, quenched and tempered
A 325	ASTM A325 type 1 Medium-carbon steel, quenched and tempered; radial dashes optional		SAE grade 8.2 Low-carbon martensite steel, quenched and tempered
A 325	ASTM A325 type 2 Low-carbon martensite steel, quenched and tempered	A 490	ASTM A490 type 1 Alloy steel, quenched and tempered

A 325	ASTM A325 type 3
	Atmospheric corrosion (weathering) steel, quenched
	  and tempered
		A 490	ASTM A490 type 3
			Atmospheric corrosion (weathering) steel, quenched
			  and tempered

SOURCE: ANSI B18.2.1–1981 (R92), Appendix III, p. 41. By permission.

(a)   (b)

(c)

DTI Gaps To Give Required Minimum Bolt Tension

DTI Fitting	325	490
Under bolt head Plain finish DTIs	0.015" (.40 mm)	0.015"
Mechanically galvanized DTIs	0.005" (.125 mm)	—
Epoxy coated on mechanically galvanized DTIs	0.005" (.125 mm)	—
Under turned Element With hardened washers (plain finish)	0.005" (.125 mm)	0.005"

With average gaps equal or less than above, bolt tensions will
be greater than in adjacent listing

Minimum Bolt Tensions
In thousands of pounds (Kips)

Bolt dia.	A325	A490
1/2"	12	—
5/8"	19	—
3/4"	28	35
7/8"	39	49
1"	51	64
1 1/8"	56	80
1 1/4"	71	102
1 3/8"	85	121
1 1/2"	103	—

(d)

Fig. 8.2.19 Direct tension indicators; gap and tension data. *(Source: J & M Turner Inc.)*

Table 8.2.26   Depths to Drill and Tap Cast Iron for Studs

Diam of stud, in	$\frac{1}{4}$	$\frac{5}{16}$	$\frac{3}{8}$	$\frac{7}{16}$	$\frac{1}{2}$	$\frac{9}{16}$	$\frac{5}{8}$	$\frac{3}{4}$	$\frac{7}{8}$	1
Diam of drill, in	$\frac{13}{64}$	$\frac{17}{64}$	$\frac{5}{16}$	$\frac{3}{8}$	$\frac{27}{64}$	$\frac{31}{64}$	$\frac{17}{32}$	$\frac{41}{64}$	$\frac{3}{4}$	$\frac{55}{64}$
Depth of thread, in	$\frac{3}{8}$	$\frac{15}{32}$	$\frac{9}{16}$	$\frac{21}{32}$	$\frac{3}{4}$	$\frac{27}{32}$	$\frac{15}{16}$	$1\frac{1}{8}$	$1\frac{5}{16}$	$1\frac{1}{2}$
Depth to drill, in	$\frac{7}{16}$	$\frac{17}{32}$	$\frac{5}{8}$	$\frac{23}{32}$	$\frac{27}{32}$	$\frac{15}{16}$	$1\frac{1}{32}$	$1\frac{1}{4}$	$1\frac{7}{16}$	$1\frac{5}{8}$

Table 8.2.27   Screw-Thread Insert Lengths
(Heli-Coil Corp.)

Shear strength of parent material, lb/in^2	Bolt material ultimate tensile strenth, lb/in^2				
	60,000	90,000	125,000	170,000	220,000
	Length in terms of nominal insert diameter				
15,000	$1\frac{1}{2}$	2	$2\frac{1}{2}$	3	
20,000	1	$1\frac{1}{2}$	2	$2\frac{1}{2}$	3
25,000	1	$1\frac{1}{2}$	2	2	$2\frac{1}{2}$
30,000	1	1	$1\frac{1}{2}$	2	2
40,000	1	1	$1\frac{1}{2}$	$1\frac{1}{2}$	2
50,000	1	1	1	$1\frac{1}{2}$	$1\frac{1}{2}$

Drill Sizes   Unified thread taps are listed in Table 8.2.28.

RIVET FASTENINGS

Forms and Proportion of Rivets   The forms and proportions of small and large rivets have been standardized and conform to ANSI B18.1.1-1972 (R89) and B18.1.2-1972 (R89) (Fig. 8.2.22a and b).

Materials Specifications for Rivets and Plates   See Secs. 6 and 12.2.

Conventional signs to indicate the form of the head to be used and whether the rivet is to be driven in the shop or the field at the time of erection are given in Fig. 8.2.23. **Rivet lengths** and **grips** are shown in Fig. 8.2.22b.

For **structural riveting**, see Sec. 12.2.

Punched vs. Drilled Plates   Holes in plates forming parts of riveted structures are punched, punched and reamed, or drilled. Punching, while cheaper, is objectionable. The holes in different plates cannot be

Fig. 8.2.20   Load-indicating wavy-flange bolt (or nut).

Fig. 8.2.21   Screw thread insert.

Fig. 8.2.22a   Rivet heads.

Fig. 8.2.22b   Rivet length and grip.

spaced with sufficient accuracy to register perfectly on being assembled. If the hole is punched out, say $\frac{1}{16}$ in smaller than is required and then reamed to size, the metal injury by cold flow during punching will be removed. Drilling, while more expensive, is more accurate and does not injure the metal.

Tubular Rivets

In tubular rivets, the end opposite the head is made with an axial hole (partway) to form a thin-walled, easily upsettable end. As the material at the edge of the rivet hole is rolled over against the surface of the joint, a **clinch** is formed (Fig. 8.2.24a).

Two-part tubular rivets have a thin-walled head with attached thin-walled rivet body and a separate thin-walled expandable plug. The head-body is inserted through a hole in the joint from one side, and the plug from the other. By holding an anvil against the plug bottom and hammering on the head, the plug is caused to expand within the head, thus locking both parts together (Fig. 8.2.24a).

Blind Rivets

Blind rivets are inserted and set all from one side of a structure. This is accomplished by mechanically expanding, through the use of the rivet's built-in mandrel, the back (blind side) of the rivet into a bulb or upset head after insertion. Blind rivets include the *pull type* and *drive-pin type*.

The pull-type rivet is available in two configurations: a self-plugging type and a pull-through type. In the self-plugging type, part of the mandrel remains permanently in the rivet body after setting, contributing additional shear strength to the fastener. In the pull-through type, the entire mandrel is pulled through, leaving the installed rivet empty.

Fig. 8.2.23   Conventional signs for rivets.

Table 8.2.28 Tap-Drill Sizes for American National Standard Screw Threads
(The sizes listed are the commercial tap drills to produce approx 75% full thread)

Size	Coarse-thread series Threads per inch	Coarse-thread series Tap drill size	Fine-thread series Threads per inch	Fine-thread series Tap drill size	Size	Coarse-thread series Threads per inch	Coarse-thread series Tap drill size	Fine-thread series Threads per inch	Fine-thread series Tap drill size
No. 0	—	—	80	$3/64$	$3/4$	10	$21/32$	16	$11/16$
No. 1	64	No. 53	72	No. 53	$7/8$	9	$49/64$	14	$13/16$
No. 2	56	No. 50	64	No. 50	1	8	$7/8$	14	$15/16$
No. 3	48	No. 47	56	No. 45	$1\frac{1}{8}$	7	$63/64$	12	$1\,3/64$
No. 4	40	No. 43	48	No. 42	$1\frac{1}{4}$	7	$1\,7/64$	12	$1\,11/64$
No. 5	40	No. 38	44	No. 37	$1\frac{3}{8}$	6	$1\,7/32$	12	$1\,19/64$
No. 6	32	No. 36	40	No. 33	$1\frac{1}{2}$	6	$1\,21/64$	12	$1\,27/64$
No. 8	32	No. 29	36	No. 29	$1\frac{3}{4}$	5	$1\,35/64$		
No. 10	24	No. 25	32	No. 21	2	$4\frac{1}{2}$	$1\,25/32$		
No. 12	24	No. 16	28	No. 14	$2\frac{1}{4}$	$4\frac{1}{2}$	$2\,1/32$		
$1/4$	20	No. 7	28	No. 3	$2\frac{1}{2}$	4	$2\frac{1}{4}$		
$5/16$	18	F	24	I	$2\frac{3}{4}$	4	$2\frac{1}{2}$		
$3/8$	16	$5/16$	24	Q	3	4	$2\frac{3}{4}$		
$7/16$	14	U	20	$25/64$	$3\frac{1}{4}$	4	3		
$1/2$	13	$27/64$	20	$29/64$	$3\frac{1}{2}$	4	$3\frac{1}{4}$		
$9/16$	12	$31/64$	18	$33/64$	$3\frac{3}{4}$	4	$3\frac{1}{2}$		
$5/8$	11	$17/32$	18	$37/64$	4	4	$3\frac{3}{4}$		

In a drive-pin rivet, the rivet body is slotted. A pin is driven forward into the rivet, causing both flaring of the rivet body and upset of the blind side (Fig. 8.2.24b).

KEYS, PINS, AND COTTERS

Keys and key seats have been standardized and are listed in ANSI B17.1-1967 (R89). Descriptions of the principal key types follow.

Woodruff keys [ANSI B17.2-1967 (R90)] are made to facilitate removal of pulleys from shafts. They should not be used as sliding keys. Cutters for milling out the key seats, as well as special machines for using the cutters, are to be had from the manufacturer. Where the hub of the gear or pulley is relatively long, two keys should be used. Slightly rounding the corners or ends of these keys will obviate any difficulty met with in removing pulleys from shafts. The key is shown in Fig. 8.2.25 and the dimensions in Table 8.2.29.

Square and flat plain taper keys have the same dimensions as gib-head keys (Table 8.2.30) up to the dotted line of Fig. 8.2.26. **Gib-head keys** (Fig. 8.2.26) are necessary when the smaller end is inaccessible for drifting out and the larger end is accessible. It can be used, with care, with all sizes of shafts. Its use is forbidden in certain jobs and places for safety reasons. Proportions are given in Table 8.2.30.

The minimum stock length of keys is 4 times the key width, and maximum stock length of keys is 16 times the key width. The increments of increase of length are 2 times the width.

Sunk keys are made to the form and dimensions given in Fig. 8.2.27 and Table 8.2.31. These keys are adapted particularly to the case of fitting adjacent parts with neither end of the key accessible. **Feather keys** prevent parts from turning on a shaft while allowing them to move in a lengthwise direction. They are of the forms shown in Fig. 8.2.28 with dimensions as given in Table 8.2.31.

In **transmitting large torques,** it is customary to use two or more keys.

Another means for fastening gears, pulleys flanges, etc., to shafts is through the use of mating pairs of tapered sleeves known as **grip springs.** A set of sleeves is shown in Fig. 8.2.29. For further references see data issued by the Ringfeder Corp., Westwood, NJ.

Tapered pins (Fig. 8.2.30) can be used to transmit very small torques or for positioning. They should be fitted so that the parts are drawn together to prevent their working loose when the pin is driven home. Table 8.2.32 gives dimensions of Morse tapered pins.

Fig. 8.2.24a Tubular rivets.

Fig. 8.2.24b Blind rivets.

Table 8.2.29   Woodruff Key Dimensions [ANSI B17.2-1967 (R90)]
(All dimensions in inches)

Key no.	Nominal key size, $A \times B$	Width of key A Max	Width of key A Min	Diam of key B Max	Diam of key B Min	C Max	C Min	D Max	D Min	Distance below E
204	1/16 × 1/2	0.0635	0.0625	0.500	0.490	0.203	0.198	0.194	0.188	3/64
304	3/32 × 1/2	0.0948	0.0928	0.500	0.490	0.203	0.198	0.194	0.188	3/64
305	3/32 × 5/8	0.0948	0.0938	0.625	0.615	0.250	0.245	0.240	0.234	1/16
404	1/8 × 1/2	0.1260	0.1250	0.500	0.490	0.203	0.198	0.194	0.188	3/64
405	1/8 × 5/8	0.1260	0.1250	0.625	0.615	0.250	0.245	0.240	0.234	1/16
406	1/8 × 3/4	0.1260	0.1250	0.750	0.740	0.313	0.308	0.303	0.297	1/16
505	5/32 × 5/8	0.1573	0.1563	0.625	0.615	0.250	0.245	0.240	0.234	1/16
506	5/32 × 3/4	0.1573	0.1563	0.750	0.740	0.313	0.308	0.303	0.297	1/16
507	5/32 × 7/8	0.1573	0.1563	0.875	0.865	0.375	0.370	0.365	0.359	1/16
606	3/16 × 3/4	0.1885	0.1875	0.750	0.740	0.313	0.308	0.303	0.297	1/16
607	3/16 × 7/8	0.1885	0.1875	0.875	0.865	0.375	0.370	0.365	0.359	1/16
608	3/16 × 1	0.1885	0.1875	1.000	0.990	0.438	0.433	0.428	0.422	1/16
609	3/16 × 1 1/8	0.1885	0.1875	1.125	1.115	0.484	0.479	0.475	0.469	5/64
807	1/4 × 7/8	0.2510	0.2500	0.875	0.865	0.375	0.370	0.365	0.359	1/16
808	1/4 × 1	0.2510	0.2500	1.000	0.990	0.438	0.433	0.428	0.422	1/16
809	1/4 × 1 1/8	0.2510	0.2500	1.125	1.115	0.484	0.479	0.475	0.469	5/64
810	1/4 × 1 1/4	0.2510	0.2500	1.250	1.240	0.547	0.542	0.537	0.531	5/64
811	1/4 × 1 3/8	0.2510	0.2500	1.375	1.365	0.594	0.589	0.584	0.578	3/32
812	1/4 × 1 1/2	0.2510	0.2500	1.500	1.490	0.641	0.636	0.631	0.625	7/64
1008	5/16 × 1	0.3135	0.3125	1.000	0.990	0.438	0.433	0.428	0.422	1/16
1009	5/16 × 1 1/8	0.3135	0.3125	1.125	1.115	0.484	0.479	0.475	0.469	5/64
1010	5/16 × 1 1/4	0.3135	0.3125	1.250	1.240	0.547	0.542	0.537	0.531	5/64
1011	5/16 × 1 3/8	0.3135	0.3125	1.375	1.365	0.594	0.589	0.584	0.578	3/32
1012	5/16 × 1 1/2	0.3135	0.3125	1.500	1.490	0.641	0.636	0.631	0.625	7/64
1210	3/8 × 1 1/4	0.3760	0.3750	1.250	1.240	0.547	0.542	0.537	0.531	5/64
1211	3/8 × 1 3/8	0.3760	0.3750	1.375	1.365	0.594	0.589	0.584	0.578	3/32
1212	3/8 × 1 1/2	0.3760	0.3750	1.500	1.490	0.641	0.636	0.631	0.625	7/64

Numbers indicate the nominal key dimensions. The last two digits give the nominal diameter (B) in eighths of an inch, and the digits preceding the last two give the nominal width (A) in thirty-seconds of an inch. Thus, 204 indicates a key 2/32 × 4/8 or 1/16 × 1/2 in; 1210 indicates a key 12/32 × 10/8 or 3/8 × 1 1/4 in.

Fig. 8.2.25 Woodruff key.

Fig. 8.2.26 Gib-head taper stock key.

Table 8.2.30   Dimensions of Square and Flat Gib-Head Taper Stock Keys,* in

Shaft diam	Square type Key Max width W	Square type Key Height at large end,† H	Square type Gib head Height C	Square type Gib head Length D	Square type Gib head Height edge of chamfer E	Flat type Key Max width W	Flat type Key Height at large end,† H	Flat type Gib head Height C	Flat type Gib head Length D	Flat type Gib head Height edge of chamfer E	Tolerance On width (−)	Tolerance On height (+)
1/2–9/16	1/8	1/8	1/4	7/32	5/32	1/8	3/32	3/16	1/8	1/8	0.0020	0.0020
5/8–7/8	3/16	3/16	5/16	9/32	7/32	3/16	1/8	1/4	3/16	5/32	0.0020	0.0020
15/16–1 1/4	1/4	1/4	7/16	11/32	11/32	1/4	3/16	5/16	1/4	3/16	0.0020	0.0020
1 5/16–1 3/8	5/16	5/16	9/16	13/32	13/32	5/16	1/4	3/8	5/16	1/4	0.0020	0.0020
1 7/16–1 3/4	3/8	3/8	11/16	15/32	15/32	3/8	1/4	7/16	3/8	5/16	0.0020	0.0020
1 13/16–2 1/4	1/2	1/2	7/8	19/32	5/8	1/2	3/8	5/8	1/2	7/16	0.0025	0.0025
2 5/16–2 3/4	5/8	5/8	1 1/16	23/32	3/4	5/8	7/16	3/4	5/8	1/2	0.0025	0.0025
2 7/8–3 1/4	3/4	3/4	1 1/4	7/8	7/8	3/4	1/2	7/8	3/4	5/8	0.0025	0.0025
3 3/8–3 3/4	7/8	7/8	1 1/2	1	1	7/8	5/8	1 1/16	7/8	3/4	0.0030	0.0030
3 7/8–4 1/2	1	1	1 3/4	1 3/16	1 3/16	1	3/4	1 1/4	1	13/16	0.0030	0.0030
4 3/4–5 1/2	1 1/4	1 1/4	2	1 7/16	1 7/16	1 1/4	7/8	1 1/2	1 1/4	1	0.0030	0.0030
5 3/4–6	1 1/2	1 1/2	2 1/2	1 3/4	1 3/4	1 1/2	1	1 3/4	1 1/2	1 1/4	0.0030	0.0030

* Stock keys are applicable to the general run of work and the tolerances have been set accordingly. They are not intended to cover the finer applications where a closer fit may be required.
† This height of the key is measured at the distance W equal to the width of the key, from the gib head.

Table 8.2.31   Dimensions of Sunk Keys
(All dimensions in inches. Letters refer to Fig. 8.2.24)

Key no.	L	W	Key no.	L	W	Key no.	L	W	Key no.	L	W
1	1/2	1/16	13	1	3/16	22	1⅜	1/4	54	2¼	1/4
2	1/2	3/32	14	1	7/32	23	1⅜	5/16	55	2¼	5/16
3	1/2	1/8	15	1	1/4	F	1⅜	3/8	56	2¼	3/8
4	5/8	3/32	B	1	5/16	24	1½	1/4	57	2¼	7/16
5	5/8	1/8	16	1⅛	3/16	25	1½	5/16	58	2½	5/16
6	5/8	5/32	17	1⅛	7/32	G	1½	3/8	59	2½	3/8
7	3/4	1/8	18	1⅛	1/4	51	1¾	1/4	60	2½	7/16
8	3/4	5/32	C	1⅛	5/16	52	1¾	5/16	61	2½	1/2
9	3/4	3/16	19	1¼	3/16	53	1¾	3/8	30	3	3/8
10	7/8	5/32	20	1¼	7/32	26	2	3/16	31	3	7/16
11	7/8	3/16	21	1¼	1/4	27	2	1/4	32	3	1/2
12	7/8	7/32	D	1¼	5/16	28	2	5/16	33	3	9/16
A	7/8	1/4	E	1¼	3/8	29	2	3/8	34	3	5/8

The Groov-Pin Corp., New Jersey, has developed a special **grooved pin** (Fig. 8.2.31) which may be used instead of smooth taper pins in certain cases.

Straight pins, likewise, are used for transmission of light torques or for positioning. **Spring pins** have been used widely for many years. Two types shown in Figs. 8.2.32 and 8.2.33 deform elastically in the radial direction when driven; the resiliency of the pin material locks the pin in place. They can replace straight and taper pins and combine the advantages of both, i.e., simple tooling, ease of removal, reusability, ability to be driven from either side.

Cotter pins (Fig. 8.2.34) are used to secure or lock nuts, clevises, etc. Driven into holes in the shaft, the eye prevents complete passage, and the split ends, deformed after insertion, prevent withdrawal.

When two rods are to be joined so as to permit movement at the joint, a round pin is used in place of a cotter. In such cases, the proportions may be as shown in Fig. 8.2.35 (knuckle joint).

Fig. 8.2.27   Sunk key.

Fig. 8.2.28   Feather key.

Fig. 8.2.29   Grip springs.

Fig. 8.2.30   Taper pins.

Fig. 8.2.31   Grooved pins.   **Fig. 8.2.32**   Roll pin.

Fig. 8.2.33   Spiral pins.   **Fig. 8.2.34**   Cotter pin.

Fig. 8.2.35   Knuckle joint.

SPLINES

Involute spline proportions, dimensions, fits, and tolerances are given in detail in ANSI B92.1-1970. External and internal involute splines (Fig. 8.2.36) have the same general form as involute gear teeth, except that

Table 8.2.32   Morse Standard Taper Pins
(Taper, ⅛ in/ft. Lengths increase by ¼ in. Dimensions in inches)

Size no.	0	1	2	3	4	5	6	7	8	9	10
Diam at large end	0.156	0.172	0.193	0.219	0.250	0.289	0.341	0.409	0.492	0.591	0.706
Length	0.5–3	0.5–3	0.75–3.5	0.75–3.5	0.75–4	0.75–4	0.75–5	1–5	1.25–5	1.5–6	1.5–6

the teeth are one-half the depth of standard gear teeth and the pressure angle is 30°. The spline is designated by a fraction in which the numerator is the diametral pitch and the denominator is always twice the numerator.

Fig. 8.2.36   Involute spline.

There are 17 series, as follows: 2.5/5, 3/6, 4/8, 5/10, 6/12, 8/16, 10/20, 12/24, 16/32, 20/40, 24/48, 32/64, 40/80, 48/96, 64/128, 80/160, 128/256. The number of teeth within each series varies from 6 to 50. Both a flat-root and a fillet-root type are provided. There are three **types of fits:** (1) **major diameter**—fit controlled by varying the major diameter of the external spline; (2) **sides of teeth**—fit controlled by varying tooth thickness and customarily used for fillet-root splines; (3) **minor diameter**—fit controlled by varying the minor diameter of the internal spline. Each type of fit is further divided into three classes: (*a*) **sliding**—clearance at all points; (*b*) **close**—close on either major diameter, sides of teeth, or minor diameter; (*c*) **press**—interference on either the major diameter, sides of teeth, or minor diameter. Important basic formulas for tooth proportions are:

$$D = \text{pitch diam}$$
$$N = \text{number of teeth}$$
$$P = \text{diametral pitch}$$
$$p = \text{circular pitch}$$
$$t = \text{circular tooth thickness}$$
$$a = \text{addendum}$$
$$b = \text{dedendum}$$
$$D_O = \text{major diam}$$
$$\text{TIF} = \text{true involute form diam}$$
$$D_R = \text{minor diam}$$

Flat and Fillet Roots

$$D = N/P$$
$$p = \pi/P$$
$$t = p/2$$
$$a = 0.5000/P$$
$$D_O \text{ (external)} = \frac{N + 1}{P}$$

$$\text{TIF (internal)} = \frac{N + 1}{P}$$
$$D_R = \frac{N + 1}{P} \quad \text{(minor-diameter fits only)}$$
$$\text{TIF (external)} = \frac{N - 1}{P}$$

$b = 0.600/P + 0.002$ (For major-diameter fits, the internal spline dedendum is the same as the addendum; for minor-diameter fits, the dedendum of the external spline is the same as the addendum.)

Fillet Root Only

½ through ¹²/₂₄

$$D_O \text{ (internal)} = \frac{N + 1.8}{P}$$
$$D_R \text{ (external)} = \frac{N - 1.8}{P}$$
$$b \text{ (internal)} = 0.900/P$$
$$b \text{ (external)} = 0.900/P$$

¹⁶/₃₂ through ⁴⁸/₉₆

$$D_O \text{ (internal)} = \frac{N + 1.8}{P}$$
$$D_R \text{ (external)} = \frac{N - 2}{P}$$
$$b \text{ (internal)} = 0.900/P$$
$$b \text{ (external)} = 1.000/P$$

Internal spline dimensions are basic while external spline dimensions are varied to control fit.

The advantages of involute splines are: (1) maximum strength at the minor diameter, (2) self-centering equalizes bearing and stresses among all teeth, and (3) ease of manufacture through the use of standard gear-cutting tools and methods.

The design of involute splines is critical in shear. The torque capacity may be determined by the formula $T = LD^2S_s/1.2732$, where L = spline length, D = pitch diam, S_s = allowable shear stress.

Parallel-side splines have been standardized by the SAE for 4, 6, 10, and 16 spline fittings. They are shown in Fig. 8.2.37; pertinent data are in Tables 8.2.33 and 8.2.34.

DRY AND VISCOUS COUPLINGS

A **coupling** makes a semipermanent connection between two shafts. They are of three main types: **rigid, flexible,** and **fluid.**

Rigid Couplings

Rigid couplings are used only on shafts which are perfectly aligned. The **flanged-face coupling** (Fig. 8.2.38) is the simplest of these. The flanges must be keyed to the shafts. The **keyless compression coupling** (Fig. 8.2.39) affords a simple means for connecting abutting shafts without the necessity of key seats on the shafts. When drawn over the slotted tapered sleeve the two flanges automatically center the shafts and provide sufficient contact pressure to transmit medium or light loads. **Ribbed-clamp couplings** (Fig. 8.2.40) are split longitudinally and are bored to the shaft diameter with a shim separating the two halves. It is necessary to key the shafts to the coupling.

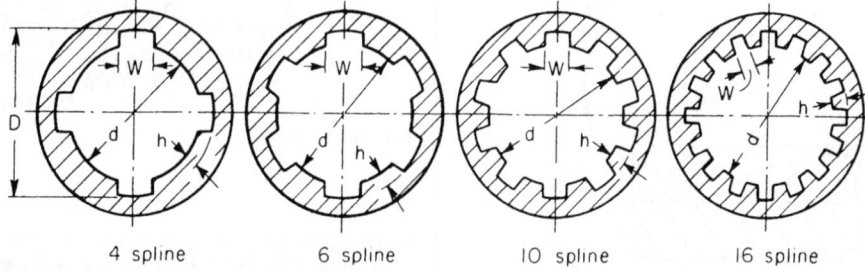

4 spline   6 spline   10 spline   16 spline

Fig. 8.2.37   Parallel-sided splines.

Table 8.2.33   Dimensions of Spline Fittings, in
(SAE Standard)

Nominal diam	4-spline for all fits		6-spline for all fits		10-spline for all fits		16-spline for all fits	
	D max*	W max†	D max*	W max†	D max*	W max†	D max*	W max†
¾	0.750	0.181	0.750	0.188	0.750	0.117		
⅞	0.875	0.211	0.875	0.219	0.875	0.137		
1	1.000	0.241	1.000	0.250	1.000	0.156		
1⅛	1.125	0.271	1.125	0.281	1.125	0.176		
1¼	1.250	0.301	1.250	0.313	1.250	0.195		
1⅜	1.375	0.331	1.375	0.344	1.375	0.215		
1½	1.500	0.361	1.500	0.375	1.500	0.234		
1⅝	1.625	0.391	1.625	0.406	1.625	0.254		
1¾	1.750	0.422	1.750	0.438	1.750	0.273		
2	2.000	0.482	2.000	0.500	2.000	0.312	2.000	0.196
2¼	2.250	0.542	2.250	0.563	2.250	0.351		
2½	2.500	0.602	2.500	0.625	2.500	0.390	2.500	0.245
3	3.000	0.723	3.000	0.750	3.000	0.468	3.000	0.294
3½	—	—	—	—	3.500	0.546	3.500	0.343
4	—	—	—	—	4.000	0.624	4.000	0.392
4½	—	—	—	—	4.500	0.702	4.500	0.441
5	—	—	—	—	5.000	0.780	5.000	0.490
5½	—	—	—	—	5.500	0.858	5.500	0.539
6	—	—	—	—	6.000	0.936	6.000	0.588

* Tolerance allowed of −0.001 in for shafts ¾ to 1¾ in, inclusive; of −0.002 for shafts 2 to 3 in, inclusive; −0.003 in for shafts 3½ to 6 in, inclusive, for 4-, 6-, and 10-spline fittings; tolerance of −0.003 in allowed for all sizes of 16-spline fittings.
† Tolerance allowed of −0.002 in for shafts ¾ in to 1¾ in, inclusive; of −0.003 in for shafts 2 to 6 in, inclusive, for 4-, 6-, and 10-spline fittings; tolerance of −0.003 allowed for all sizes of 16-spline fittings.

Table 8.2.34   Spline Proportions

No. of splines	W for all fits	Permanent fit		To slide when not under load		To slide under load	
		h	d	h	d	h	d
4	0.241D	0.075D	0.850D	0.125D	0.750D		
6	0.250D	0.050D	0.900D	0.075D	0.850D	0.100D	0.800D
10	0.156D	0.045D	0.910D	0.070D	0.860D	0.095D	0.810D
16	0.098D	0.045D	0.910D	0.070D	0.860D	0.095D	0.810D

Flexible Couplings

Flexible couplings are designed to connect shafts which are misaligned either laterally or angularly. A secondary benefit is the absorption of impacts due to fluctuations in shaft torque or angular speed. The **Oldham,** or **double-slider, coupling** (Fig. 8.2.41) may be used to connect shafts which have only lateral misalignment. The **"Fast" flexible coupling** (Fig. 8.2.42) consists of two hubs each keyed to its respective shaft. Each hub has generated splines cut at the maximum possible distance

Fig. 8.2.38   Flanged face coupling.

Fig. 8.2.39   Keyless compression coupling.

from the shaft end. Surrounding the hubs is a casing or sleeve which is split transversely and bolted by means of flanges. Each half of this sleeve has generated internal splines cut on its bore at the end opposite to the flange. These internal splines permit a definite error of alignment between the two shafts.

Another type, the Waldron coupling (Midland-Ross Corp.), is shown in Fig. 8.2.43.

The chain coupling shown in Fig. 8.2.44 uses silent chain, but standard roller chain can be used with the proper mating sprockets. Nylon links enveloping the sprockets are another variation of the chain coupling.

Fig. 8.2.40   Ribbed-clamp coupling.

Steelflex couplings (Fig. 8.2.45) are made with two grooved steel hubs keyed to their respective shafts. Connection between the two halves is secured by a specially tempered alloy-steel member called the "grid."

Fig. 8.2.41   Double-slider coupling.

In the rubber flexible coupling shown in Fig. 8.2.46, the torque is transmitted through a comparatively soft rubber section acting in shear. The type in Fig. 8.2.47 loads the intermediate rubber member in compression. Both types permit reasonable shaft misalignment and are recommended for light loads only.

Fig. 8.2.42   "Fast" flexible coupling.

Universal joints are used to connect shafts with much larger values of misalignment than can be tolerated by the other types of flexible couplings. Shaft angles up to 30° may be used. The Hooke's-type joint (Fig. 8.2.48) suffers a loss in efficiency with increasing angle which

Fig. 8.2.43   Waldron coupling.

may be approximated for angles up to 15° by the following relation: efficiency = $100(1 - 0.003\theta)$, where θ is the angle between the shafts. The velocity ratio between input and output shafts with a single universal joint is equal to

$$\omega_2/\omega_1 = \cos\theta/1 - \sin^2\theta\sin^2(\alpha + 90°)$$

where ω_2 and ω_1 are the angular velocities of the driven and driving shafts, respectively, θ is the angle between the shafts, and α is the angular

Fig. 8.2.44   Chain coupling.

displacement of the driving shaft from the position where the pins on the drive-shaft yoke lie in the plane of the two shafts. A velocity ratio of 1 may be obtained at any angle using two Hooke's-type joints and an intermediate shaft. The intermediate shaft must make equal angles with the main shafts, and the driving pins on the yokes attached to the intermediate shaft must be set parallel to each other.

Fig. 8.2.45   Falk Steelflex coupling.

The Bendix-Weiss "rolling-ball" universal joint provides constant angular velocity. Torque is transmitted between two yokes through a set of four balls such that the centers of all four balls lie in a plane which bisects

the angle between the shafts. Other variations of constant velocity universal joints are found in the Rzeppa, Tracta, and double Cardan types.

Fig. 8.2.46   Rubber flexible coupling, shear type.

Fig. 8.2.47   Rubber flexible coupling, compression type.

Fig. 8.2.48   Hooke's universal joint.

Fluid Couplings

(See also Sec. 11.)
Fluid couplings (Fig. 8.2.49) have two basic parts—the input member, or impeller, and the output member, or runner. There is no mechanical connection between the two shafts, power being transmitted by kinetic energy in the operating fluid. The impeller B is fastened to the flywheel A and turns at engine speed. As this speed increases, fluid within the impeller moves toward the outer periphery because of centrifugal force. The circular shape of the impeller directs the fluid toward the runner C, where its kinetic energy is absorbed as torque delivered by shaft D. The positive pressure behind the fluid causes flow to continue toward the hub and back through the impeller. The toroidal space in both the impeller and runner is divided into compartments by a series of flat radial vanes.

Fig. 8.2.49a   Fluid coupling. (*A*) Flywheel; (*B*) impeller; (*C*) runner; (*D*) output shaft.

Fig. 8.2.49*b*   Schematic of viscous coupling.

The torque capacity of a fluid coupling with a full-load slip of about 2.5 percent is $T = 0.09n^2D^5$, where n is the impeller speed, hundreds of r/min, and D is the outside diameter, ft. The output torque is equal to the input torque over the entire range of input-output speed ratios. Thus the prime mover can be operated at its most effective speed regardless of the speed of the output shaft. Other advantages are that the prime mover cannot be stalled by application of load and that there is no transmission of shock loads or torsional vibration between the connected shafts.

A **hydraulic torque converter** (Fig. 8.2.50) is similar in form to the hydraulic coupling, with the addition of a set of stationary guide vans, the reactor, interposed between the runner and the impeller. All blades in a converter have compound curvature. This curvature is designed to control the direction of fluid flow. Kinetic energy is therefore transferred as a result of both a scalar and vectorial change in fluid velocity. The blades are designed such that the fluid will be moving in a direction parallel to the blade surface at the entrance (Fig. 8.2.51) to each section. With a design having fixed blading, this can be true at only one value of runner and impeller speed, called the design point. Several design modifications are possible to overcome this difficulty. The angle of the blades can be made adjustable, and the elements can be divided into sections operating independently of each other according to the load requirements. Other refinements include the addition of multiple stages in the runner and reactor stages as in steam reaction turbines (see Sec. 9). The advantages of a torque converter are the ability to multiply starting torque 5 to 6 times and to serve as a stepless transmission. As in the coupling, torque varies as the square of speed and the fifth power of diameter.

Fig. 8.2.50   Hydraulic torque converter. (*A*) Flywheel; (*B*) impeller; (*C*) runner; (*D*) output shaft; (*E*) reactor.

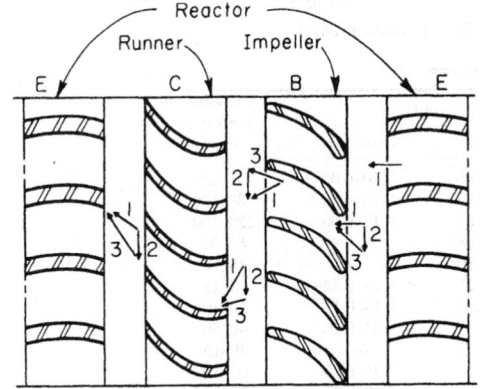

Fig. 8.2.51   Schematic of converter blading. (1) Absolute fluid velocity; (2) velocity vector—converter elements; (3) fluid velocity relative to converter elements.

Optimum efficiency (Fig. 8.2.52) over the range of input-output speed ratios is obtained by a combination converter coupling. When the output speed rises to the point where the torque multiplication factor is 1.0, the clutch point, the torque reaction on the reactor element reverses direction. If the reactor is mounted to freewheel in this opposite direction, the unit will act as a coupling over the higher speed ranges. An automatic friction clutch (see "Clutches," below) set to engage at or near the clutch point will also eliminate the poor efficiency of the converter at high output speeds.

Fig. 8.2.52   Hydraulic coupling characteristic curves. (*Heldt, "Torque Converters and Transmissions," Chilton.*)

Viscous couplings are becoming major players in mainstream front-wheel-drive applications and are already used in four-wheel-drive vehicles.

Torque transmission in a viscous coupling relies on shearing forces in an entrapped fluid between axially positioned disks rotating at different angular velocities (Fig. 8.2.49*b*), all encased in a lifetime leakproof housing. A hub carries the so-called inner disks while the housing carries the so-called outer disks. Silicone is the working fluid.

Operation of the coupling is in normal (slipping) mode when torque is being generated by viscous shear. However, prolonged slipping under severe starting conditions causes heat-up, which in turn causes the fluid, which has a high coefficient of thermal expansion, to expand considerably with increasing temperature. It then fills the entire available space, causing a rapid pressure increase, which in turn forces the disks together into metal-to-metal frictional contact. Torque transmission now increases substantially. This self-induced torque amplification is known as the **hump effect.** The point at which the hump occurs can be set by the design and coupling setup. Under extreme conditions, 100 percent lockup occurs, thus providing a self-protecting relief from overheating as fluid shear vanishes. This effect is especially useful in autos using viscous couplings in their limited-slip differentials, when one wheel is on low-friction surfaces such as ice. The viscous coupling transfers torque to the other gripping wheel. This effect is also useful when one is driving up slopes with uneven surface conditions, such as rain or snow, or on very rough surfaces. Such viscous coupling differentials have allowed a weight and cost reduction of about 60 percent. A fuller account can be found in Barlage, Viscous Couplings Enter Main Stream Vehicles, *Mech. Eng.*, Oct. 1993.

CLUTCHES

Clutches are couplings which permit the disengagement of the coupled shafts during rotation.

Positive clutches are designed to transmit torque without slip. The **jaw clutch** is the most common type of positive clutch. These are made with **square jaws** (Fig. 8.2.53) for driving in both directions or **spiral jaws** (Fig. 8.2.54) for unidirectional drive. Engagement speed should be limited to 10 r/min for square jaws and 150 r/min for spiral jaws. If disengagement under load is required, the jaws should be finish-machined and lubricated.

Friction clutches are designed to reduce coupling shock by slipping during the engagement period. They also serve as safety devices by slipping when the torque exceeds their maximum rating. They may be divided into two main groups, **axial** and **rim clutches,** according to the direction of contact pressure.

The **cone clutch** (Fig. 8.2.55) and the **disk clutch** (Fig. 8.2.56) are examples of axial clutches. The disk clutch may consist of either a

Fig. 8.2.53   Square-jaw clutch.

(a) Right-hand   (b) Left-hand

Fig. 8.2.54   Spiral-jaw clutch.

single plate or multiple disks. Table 8.2.35 lists typical friction materials and important design data. The torque capacity of a disk clutch is given by $T = 0.5ifF_aD_m$, where T is the torque, i the number of pairs of contact surfaces, f the applicable coefficient of friction, F_a the axial

Fig. 8.2.55   Cone clutch.

engaging force, and D_m the mean diameter of the clutch facing. The spring forces holding a disk clutch in engagement are usually of relatively high value, as given by the allowable contact pressures. In order to lower the force required at the operating lever, elaborate linkages are required, usually having lever ratios in the range of 10 to 12. As these linkages must rotate with the clutch, they must be adequately balanced and the effect of centrifugal forces must be considered. Disk clutches

are often run wet, either immersed in oil or in a spray. The advantages are reduced wear, smoother action, and lower operating temperatures. Disk clutches are often operated automatically by either air or hydraulic cylinders as, for examples, in automobile automatic transmissions.

Fig. 8.2.56   Multidisk clutch.

Fig. 8.2.57   Band clutch.

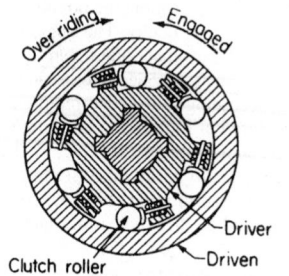

Fig. 8.2.58   Overrunning clutch

Table 8.2.35   Friction Coefficients and Allowable Pressures

Materials in contact	Friction coefficient f			Allowable pressure, lb/in²
	Dry	Greasy	Lubricated	
Cast iron on cast iron	0.2–0.15	0.10–0.06	0.10–0.05	150–250
Bronze on cast iron	—	0.10–0.05	0.10–0.05	80–120
Steel on cast iron	0.30–0.20	0.12–0.07	0.10–0.06	120–200
Wood on cast iron	0.25–0.20	0.12–0.08	—	60–90
Fiber on metal	—	0.20–0.10	—	10–30
Cork on metal	0.35	0.30–0.25	0.25–0.22	8–15
Leather on metal	0.5–0.3	0.20–0.15	0.15–0.12	10–30
Wire asbestos on metal	0.5–0.35	0.30–0.25	0.25–0.20	40–80
Asbestos blocks on metal	0.48–0.40	0.30–0.25	—	40–160
Asbestos on metal, short action	—	—	0.25–0.20	200–300
Metal on cast iron, short action	—	—	0.10–0.05	200–300

SOURCE: Maleev, *Machine Design*, International Textbook, by permission.

Rim clutches may be subdivided into two groups: (1) those employing either a band or block (Fig. 8.2.57) in contact with the rim and (2) overrunning clutches (Fig. 8.2.58) employing the wedging action of a roller or sprag. Clutches in the second category will automatically engage in one direction and freewheel in the other.

HYDRAULIC POWER TRANSMISSION

Hydraulic power transmission systems comprise machinery and auxiliary components which function to generate, transmit, control, and utilize hydraulic power. The **working fluid,** a pressurized incompressible liquid, is usually either a petroleum base or a fire-resistant type. The latter are water and oil emulsions, glycol-water mixtures, or synthetic liquids such as silicones or phosphate esters.

Liquid is pressurized in a **pump** by virtue of its resistance to flow; the pressure difference between pump inlet and outlet results in flow. Most hydraulic applications employ positive-displacement pumps of the gear, vane, screw, or piston type; piston pumps are axial, radial, or reciprocating (see Sec. 14).

Power is transmitted from pump to controls and point of application through a combination of **conduit and fittings** appropriate to the particular application. Flow characteristics of hydraulic circuits take into account fluid properties, pressure drop, flow rate, and pressure-surging tendencies. Conduit systems must be designed to minimize changes in flow velocity, velocity distribution, and random fluid eddies, all of which dissipate energy and result in pressure drops in the circuit (see Sec. 3). Pipe, tubing, and flexible hose are used as hydraulic power conduits; suitable fittings are available for all types and for transition from one type to another.

Controls are generally interposed along the conduit between the pump and point of application (i.e., an actuator or motor), and act to control pressure, volume, or flow direction.

Fig. 8.2.59   Relief valve.

Fig. 8.2.60   Reducing valve.

Pressure control valves, of which an ordinary safety valve is a common type (normally closed), include relief and reducing valves and pressure switches (Figs. 8.2.59 and 8.2.60). Pressure valves, normally closed, can be used to control sequential operations in a hydraulic circuit. **Flow control valves** throttle flow to or bypass flow around the unit being controlled, resulting in pressure drop and temperature increase as pressure energy is dissipated. Figure 8.2.61 shows a simple needle valve with variable orifice usable as a flow control valve. **Directional control valves** serve primarily to direct hydraulic fluid to the point of application. Directional control valves with rotary and sliding spools are shown in Figs. 8.2.62 and 8.2.63.

Fig. 8.2.61   Needle valve.

Fig. 8.2.62   Rotary-spool directional flow valve.

Fig. 8.2.63   Sliding-spool directional flow valve.

A poppet (valve) mechanism is shown in Fig. 8.2.64, a diaphragm valve in Fig. 8.2.65, and a shear valve in Fig. 8.2.66.

Accumulators are effectively "hydraulic flywheels" which store potential energy by accumulating a quantity of pressurized hydraulic fluid

Fig. 8.2.64   Poppet valve.

in a suitable enclosed vessel. The bag type shown in Fig. 8.2.67 uses pressurized gas inside the bag working against the hydraulic fluid outside the bag. Figure 8.2.68 shows a piston accumulator.

Pressurized hydraulic fluid acting against an **actuator** or motor converts fluid pressure energy into mechanical energy. Motors providing

Fig. 8.2.65   Diaphragm valve.   **Fig. 8.2.66**   Shear valve.

continuous rotation have operating characteristics closely related to their pump counterparts. A linear actuator, or cylinder (Fig. 8.2.69), provides straight-line reciprocating motion; a rotary actuator (Fig. 8.2.70) provides arcuate oscillatory motion. Figure 8.2.71 shows a one-shot booster (linear motion) which can be used to deliver sprays through a nozzle.

Hydraulic fluids (liquids and air) are conducted in **pipe, tubing,** or **flexible hose.** Hose is used when the lines must flex or in applications in which fixed, rigid conduit is unsuitable. Table 8.2.36 lists SAE standard hoses. Maximum recommended operating pressure for a broad range of industrial applications is approximately 25 percent of rated bursting pressure. Due consideration must be given to the operating-temperature range; most applications fall in the range from -40 to $200°F$ (-40 to $95°C$). Higher operating temperatures can be accommodated with appropriate materials.

Fig. 8.2.67   Bag accumulator.

Hose fittings are of the screw-type or swaged, depending on the particular application and operating pressure and temperature. A broad variety of hose-end fittings is available from the industry. **Pipe** has the advantage of being relatively cheap, is applied mainly in straight runs, and is usually of steel. Fittings for pipe are either standard pipe fittings for fairly low pressures or more elaborate ones suited to leakproof high-pressure operation.

Fig. 8.2.68   Piston accumulator.

Fig. 8.2.69   Linear actuator or hydraulic cylinder.

Tubing is more easily bent into neat forms to fit between inlet and outlet connections. Steel and stainless-steel tubing is used for the highest-pressure applications; aluminum, plastic, and copper tubing is also used as appropriate for the operating conditions of pressure and temperature. Copper tubing hastens the oxidation of oil-base hydraulic fluids; accordingly, its use should be restricted either to air lines or with liquids which will not be affected by copper in the operating range.

Fig. 8.2.70   Rotary actuator.

Fig. 8.2.71   One-shot booster.

Table 8.2.36   SAE Standard Hoses

100R1A	One-wire-braid reinforcement, synthetic rubber cover
100R1T	Same as R1A except with a thin, nonskive cover
100R2A	Two-wire-braid reinforcement, synthetic rubber cover
100R2B	Two spiral wire plus one wire-braid reinforcement, synthetic rubber cover
100R2AT	Same as R2A except with a thin, nonskive cover
100R2BT	Same as R2B except with a thin, nonskive cover
100R3	Two rayon-braid reinforcement, synthetic rubber cover
100R5	One textile braid plus one wire-braid reinforcement, textile braid cover
100R7	Thermoplastic tube, synthetic fiber reinforcement, thermoplastic cover (thermoplastic equivalent to SAE 100R1A)
100R8	Thermoplastic tube, synthetic fiber reinforcement, thermoplastic cover (thermoplastic equivalent to SAE 100R2A)
100R9	Four-ply, light-spiral-wire reinforcement, synthetic rubber cover
100R9T	Same as R9 except with a thin, nonskive cover
100R10	Four-ply, heavy-spiral-wire reinforcement, synthetic rubber cover
100R11	Six-ply, heavy-spiral-wire reinforcement, synthetic rubber cover

Tube fittings for permanent connections allow for brazed or welded joints. For temporary or separable applications, **flared** or **flareless fittings** are employed (Figs. 8.2.72 and 8.2.73). ANSI B116.1-1974 and B116.2-1974 pertain to tube fittings. The variety of fittings available is vast; the designer is advised to refer to manufacturers' literature for specifics.

Fig. 8.2.72   Flared tube fittings. (*a*) *A 45° flared fitting; (b) triple-lok flared fitting. (Parker-Hannafin Co.)*

Fig. 8.2.73   Ferulok flareless tube fitting. *(Parker-Hannafin Co.)*

Parameters entering into the design of a hydraulic system are volume of flow per unit time, operating pressure and temperature, viscosity characteristics of the fluid within the operating range, and compatibility of the fluid/conduit material.

Flow velocity in suction lines is generally in the range of 1 to 5 ft/s (0.3 to 1.5 m/s); in discharge lines it ranges from 10 to 25 ft/s (3 to 8 m/s).

The pipe or tubing is under internal pressure. Selection of material and wall thickness follows from suitable equations (see Sec. 5). Safety factors range from 6 to 10 or higher, depending on the severity of the application (i.e., vibration, shock, pressure surges, possibility of physical abuse, etc.). JIC specifications provide a guide to the designer of hydraulic systems.

BRAKES

Brakes may be classified as: (1) rim type—internally expanding or externally contracting shoes, (2) band type, (3) cone type, (4) disk or axial type, (5) miscellaneous.

Rim Type—Internal Shoe(s) (Fig. 8.2.74)

$$F = \begin{cases} \dfrac{M_N - M\mu}{d} & \text{clockwise rotation} \\[3mm] \dfrac{M_N + M\mu}{d} & \text{counterclockwise rotation} \end{cases}$$

where

$$M\mu = \frac{\mu P_a Br}{\sin \theta_a} \int_{\theta_1}^{\theta_2} (\sin \theta)(r - d \cos \theta) \, d\theta$$

B = face width of frictional material; P_a = maximum pressure; θ_a = angle to point of maximum pressure (if $\theta_2 > 90°$; then $\theta_a = 90°$; $\theta_2 < 90°$, then $\theta_a = \theta_2$); μ = coefficient of friction; r = radius of drum; d = distance from drum center to brake pivot;

$$M_N = \frac{P_a Brd}{\sin \theta_a} \int_{\theta_1}^{\theta_2} \sin^2 \theta \, d\theta$$

and torque on drum is

$$T = \frac{\mu P_a Br^2 (\cos \theta_1 - \cos \theta_2)}{\sin \theta_a}$$

Self-locking of the brake ($F = 0$) will occur for clockwise rotation when $M_N = M\mu$. This self-energizing phenomenon can be used to advantage without actual locking if μ is replaced by a larger value μ' so that $1.25 \, \mu \le \mu' \le 1.50$, from which the pivot position a can be solved.

Fig. 8.2.74   Rim brake: internal friction shoe.

In automotive use there are two shoes made to pivot in opposition, so that self-energization is present and can be used to great advantage (Fig. 8.2.75).

Fig. 8.2.75   Internal brake.

Rim Type—External Shoe(s) (Fig. 8.2.76)

The equations for M_N and $M\mu$ are the same as above:

$$F = \begin{cases} \dfrac{M_N + M\mu}{d} & \text{clockwise rotation} \\[2ex] \dfrac{M_N - M\mu}{d} & \text{counterclockwise rotation} \end{cases}$$

Self-locking ($F = 0$) can occur for counterclockwise rotation at $M_N = M\mu$.

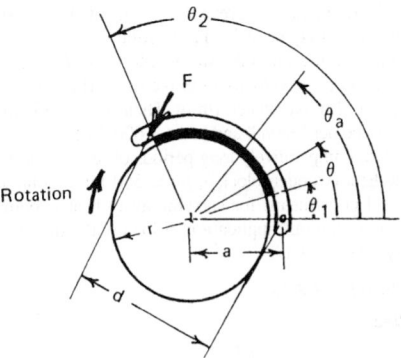

Fig. 8.2.76   Rim brake: external friction shoe.

Band Type (Fig. 8.2.77a, b, and c)

Flexible band brakes are used in power excavators and in hoisting. The bands are usually of an asbestos fabric, sometimes reinforced with copper wire and impregnated with asphalt.

In Fig. 8.2.77a, F = force at end of brake handle; P = tangential force at rim of wheel; f = coefficient of friction of materials in contact;

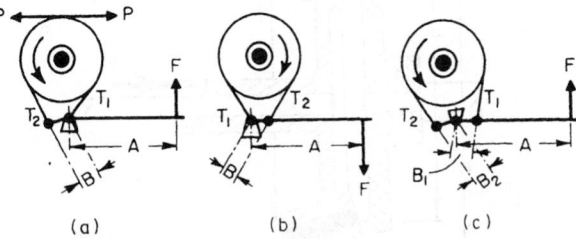

Fig. 8.2.77   Band brakes.

a = angle of wrap of band, deg; T_1 = total tension in band on tight side; T_2 = total tension in band on slack side. Then $T_1 - T_2 = P$ and $T_1/T_2 = 10^{0.0076fa} = 10^b$, where $b = 0.0076fa$. Also, $T_2 = P/(10^b - 1)$ and $T_1 = P \times 10^b/(10^b - 1)$. The values of $10^{0.0076fa}$ are given in Fig. 8.2.93 for a in radians.

For the arrangement shown in Fig. 8.2.77a,

$$FA = T_2 B = PB/(10^b - 1)$$

and

$$F = PB/[A(10^b - 1)]$$

For the construction illustrated in Fig. 8.2.77b,

$$F = PB/\{A[10^b/(10^b - 1)]\}$$

For the differential brake shown in Fig. 8.2.77c,

$$F = (P/A)[(B_2 - 10^b B_1)/(10^b - 1)]$$

In this arrangement, the quantity $10^b B_1$ must always be less than B_2, or the band will grip the wheel and the brake, or part of the mechanism to which it is attached, will rupture.

It is usual in practice to have the leverage ratio A/B for band brakes about 10:1

If f for wood on iron is taken at 0.3 and the angle of wrap for the band is 270°, i.e., subtends three-fourths of the circumference, then $10^b = 4$ approx; the loads required for a given torque will be as follows for the cases just considered and for the leverage ratios stated above:

Band brake, Fig. 8.2.77a	$F = 0.033P$
Band brake, Fig. 8.2.77b	$F = 0.133P$
Band brake, Fig. 8.2.77c	$F = 0.016P$

In the case of Fig. 8.2.77c, the dimension B_2 must be greater than $B_1 \times 10^b$. Accordingly, B_1 is taken as ¼, A as 10, and, since $10^b = 4$, B_2 is taken as 1½.

The principal function of a brake is to absorb energy. This energy appears at the surface of the brake as heat, which must be carried away at a sufficiently rapid rate to prevent burning of the wooden blocks. Suitable proportions may be arrived at as follows:

Let p = unit pressure on brake surface, lb/in^2 = R (reaction against block)/area of block; v = velocity of brake rim surface, ft/s = $2\pi rn/60$, where n = speed of brake wheel, r/min. Then pv = work absorbed per in^2 of brake surface per second, and $pv \le 1,000$ for intermittent applications of load with comparatively long periods of rest and poor means for carrying away heat (wooden blocks); $pv \le 500$ for continuous application of load and poor means for carrying away heat (wooden blocks); $pv \le 1,400$ for continuous application of load with effective means for carrying away heat (oil bath).

Cone Brake (see Fig. 8.2.78)

Uniform Wear

$$F = \frac{\pi P_a d}{2}(D - d)$$

$$T = \frac{\pi \mu P_a d}{8 \sin \alpha}(D^2 - d)$$

where P_a = maximum pressure occurring at $d/2$.

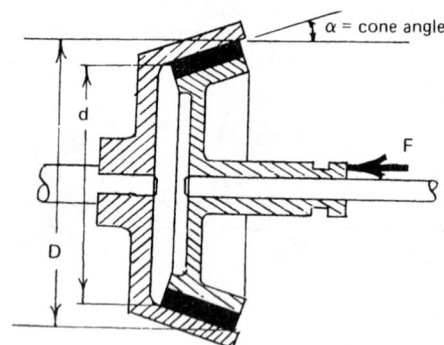

Fig. 8.2.78   Cone brake.

Uniform Pressure

$$F = \frac{\pi P_a}{4}(D^2 - d^2)$$

$$T = \frac{\pi \mu P_a}{12 \sin \alpha}(D^3 - d^3) = \frac{F\mu}{3 \sin \alpha} \times \frac{D^3 - d^3}{D^2 - d^2}$$

Figure 8.2.79 shows a cone brake arrangement used for lowering heavy loads.

Fig. 8.2.79   Cone brake for lowering loads.

Disk Brakes (see Fig. 8.2.80)

Disk brakes are free from "centrifugal" effects, can have large frictional areas contained in small space, have better heat dissipation qualities than the rim type, and enjoy a favorable pressure distribution.

Uniform Wear

$$F = \frac{\pi P_a d}{2}(D - d)$$

$$T = \frac{F\mu}{4}(D + d)$$

Uniform Pressure

$$F = \frac{\pi P_a}{4}(D^2 - d^2)$$

$$T = \frac{F\mu}{3} \times \frac{D^3 - d^3}{D^2 - d^2}$$

These relations apply to a single surface of contact. For caliper disk, or multidisk brakes, the above relations are applied for each surface of contact.

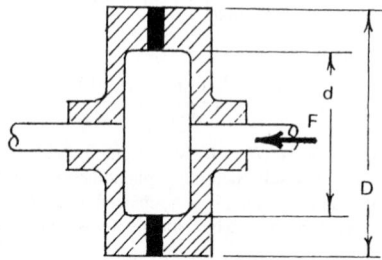

Fig. 8.2.80   Disk brake.

Selected friction materials and properties are listed in Table 8.2.37.

Frequently disk brakes are made as shown in Fig. 8.2.81. The pinion Q engages the gear in the drum (not shown). When the load is to be raised, power is applied through the gear and the connection between B and C is accomplished by the advancing of B along A and the clamping of the friction disks D and d and the ratchet wheel E. The reversal of the motor disconnects B and C. In lowering the load, only as much reversal of rotation of the gear is given as is needed to reduce the force in the friction disks so that the load may be lowered under control.

Fig. 8.2.81   Disk brake.

A **multidisk brake** is shown in Fig. 8.2.82. This type of construction results in an increase in the number of friction faces. The drum shaft is geared to the pinion A, while the motive power for driving comes through the gear G. In raising the load, direct connection is had between G, B, and A. In lowering, B moves relatively to G and forces the friction plates together, those plates fast to E being held stationary by the pawl on E. In the figure, there are three plates fast to E, one fast to G, and one fast to C.

Fig. 8.2.82   Multidisk brake.

Eddy-current brakes (Fig. 8.2.83) are used with flywheels where quick braking is essential, and where large kinetic energy of the rotating

Table 8.2.37   Selected Friction Materials and Properties

Material	Opposing material	Friction coefficient			Max pressure		Max temperature	
		Dry	Wet	In oil	lb/in^2	kPa	°F	°C
Sintered metal	Cast iron or steel	0.1–0.4	0.05–0.01	0.05–0.08	150–250	1,000–1,720	450–1,250	232–677
Wood	Cast iron or steel	0.2–0.35	0.16	0.12–0.16	60–90	400–620	300	149
Leather	Cast iron or steel	0.3–0.5	0.12		10–40	70–280	200	93
Cork	Cast iron or steel	0.3–0.5	0.15–0.25	0.15–0.25	8–14	55–95	180	82
Felt	Cast iron or steel	0.22	0.18		5–10	35–70	280	138
Asbestos—woven	Cast iron or steel	0.3–0.6	0.1–0.2	0.08–0.10	50–100	350–700	400–500	204–260
Asbestos—molded	Cast iron or steel	0.2–0.5	0.08–0.12	0.06–0.09	50–150	350–1,000	400–500	204–260
Asbestos—impregnated	Cast iron or steel	0.32	0.12					
Cast iron	Cast iron	0.15–0.20	0.05	0.03–0.06	150–250	1,000–1,720	500	260
Cast iron	Steel			0.03–0.06	100–250	690–1,720	500	260
Graphite	Steel	0.25	0.05–0.1	0.12 (av)	300	2,100	370–540	188–282

masses precludes the use of block brakes due to excessive heating, as in reversible rolling mills. A number of poles *a* are electrically excited (north and south in turn) and create a magnetic flux which permeates the gap and the iron of the rim, causing eddy current. The flywheel energy

Fig. 8.2.83   Eddy-current brake.

is converted through these currents into heat. The hand brake *b* may be used for quicker stopping when the speed of the wheel is considerably decreased; i.e., when the eddy-current brake is inefficient. Two brakes are provided to avoid bending forces on the shaft.

Electric brakes are often used in cranes, bridges, turntables, and machine tools, where an automatic application of the brake is important as soon as power is cut off. The brake force is supplied by an adjustable spring which is counteracted by the force of a solenoid or a centrifugal thrustor. Interruption of current automatically applies the spring-activated brake shoes. Figures 8.2.84 and 8.2.85 show these types of electric brake.

Fig. 8.2.84   Solenoid-type electric brake.

Fig. 8.2.85   Thrustor-type electric brake.

SHRINK, PRESS, DRIVE, AND RUNNING FITS

Inch Systems   ANSI B4.1-1967 (R87) recommends preferred sizes, allowances, and tolerances for fits between plain cylindrical parts. Such fits include bearing, shrink and drive fits, etc. Terms used in describing fits are defined as follows: **Allowance:** minimum clearance (positive allowance) or maximum interference (negative allowance) between mating parts. **Tolerance:** total permissible variation of size. **Limits of size:** applicable maximum and minimum sizes. **Clearance fit:** one having limits of size so prescribed that a clearance always results when mating parts are assembled. **Interference fit:** In this case, limits are so prescribed that interference always results on assembly. **Transition fit:** This may have either a clearance or an interference on assembly. **Basic size:** one from which limits of size are derived by the application of allowances and tolerances. **Unilateral tolerance:** In this case a variation in size is permitted in only one direction from the basic size.

Fits are divided into the following general classifications: (1) running and sliding fits, (2) locational clearance fits, (3) transition fits, (4) locational interference fits, and (5) force or shrink fits.

1. **Running and sliding fits.** These are intended to provide similar running performance with suitable lubrication allowance throughout the range of sizes. These fits are further subdivided into the following classes:

Class RC1: close-sliding fits. Intended for accurate location of parts which must assemble without perceptible play.

Class RC2: sliding fits. Parts made with this fit move and turn easily but are not intended to run freely; also, in larger sizes they may seize under small temperature changes.

Class RC3: precision-running fits. These are intended for precision work at slow speeds and light journal pressures but are not suitable where appreciable temperature differences are encountered.

Class RC4: close-running fits. For running fits on accurate machinery with moderate surface speeds and journal pressures, where accurate location and minimum play is desired.

Classes RC5 and RC6: medium-running fits. For higher running speeds or heavy journal pressures.

Class RC7: free-running fits. For use where accuracy is not essential, or where large temperature variations are likely to be present, or both.

Classes RC8 and RC9: loose-running fits. For use with materials such as cold-rolled shafting or tubing made to commercial tolerances.

Limits of clearance given in ANSI B4.1-1967 (R87) for each of these classes are given in Table 8.2.38. Hole and shaft tolerances are listed on a unilateral tolerance basis in this reference to give the clearance limits of Table 8.2.38, the hole size being the basic size.

2. **Locational clearance fits.** These are intended for normally stationary parts which can, however, be freely assembled or disassembled. These are subdivided into various classes which run from snug fits for parts requiring accuracy of location, through medium clearance fits (spigots) to the looser fastener fits where freedom of assembly is of prime importance.

Table 8.2.38 Limits of Clearance for Running and Sliding Fits (Basic Hole)
(Limits are in thousandths of an inch on diameter)

Nominal size range, in	Class								
	RC1	RC2	RC3	RC4	RC5	RC6	RC7	RC8	RC9
0–0.12	0.1 0.45	0.1 0.55	0.3 0.95	0.3 1.3	0.6 1.6	0.6 2.2	1.0 2.6	2.5 5.1	4.0 8.1
0.12–0.24	1.5 0.5	0.15 0.65	0.4 1.2	0.4 1.6	0.8 2.0	0.8 2.7	1.2 3.1	2.8 5.8	4.5 9.0
0.24–0.40	0.2 0.6	0.2 0.85	0.5 1.5	0.5 2.0	1.0 2.5	1.0 3.3	1.6 3.9	3.0 6.6	5.0 10.7
0.40–0.71	0.25 0.75	0.25 0.95	0.6 1.7	0.6 2.3	1.2 2.9	1.2 3.8	2.0 4.6	3.5 7.9	6.0 12.8
0.71–1.19	0.3 0.95	0.3 1.2	0.8 2.1	0.8 2.8	1.6 3.6	1.6 4.8	2.5 5.7	4.5 10.0	7.0 15.5
1.19–1.97	0.4 1.1	0.4 1.4	1.0 2.6	1.0 3.6	2.0 4.6	2.0 6.1	3.0 7.1	5.0 11.5	8.0 18.0
1.97–3.15	0.4 1.2	0.4 1.6	1.2 3.1	1.2 4.2	2.5 5.5	2.5 7.3	4.0 8.8	6.0 13.5	9.0 20.5
3.15–4.73	0.5 1.5	0.5 2.0	1.4 3.7	1.4 5.0	3.0 6.6	3.0 8.7	5.0 10.7	7.0 15.5	10.0 24.0

3. **Transition fits.** These are for applications where accuracy of location is important, but a small amount of either clearance or interference is permissible.

4. **Locational interference fits.** Used where accuracy of location is of prime importance and for parts requiring rigidity and alignment with no special requirements for bore pressure.

Data on clearance limits, interference limits, and hole and shaft diameter tolerances for locational clearance fits, transition fits, and locational interference fits are given in ANSI B4.1-1967 (R87).

5. **Force or shrink fits.** These are characterized by approximately constant bore pressures throughout the range of sizes; interference varies almost directly as the diameter, and the differences between maximum and minimum values of interference are small. These are divided into the following classes:

Class FN1: light-drive fits. For applications requiring light assembly pressures (thin sections, long fits, cast-iron external members).

Class FN2: medium-drive fits. Suitable for ordinary steel parts or for shrink fits on light sections. These are about the tightest fits that can be used on high-grade cast-iron external members.

Class FN3: heavy-drive fits. For heavier steel parts or shrink fits in medium sections.

Classes FN4 and FN5: force fits. These are suitable for parts which can be highly stressed. Shrink fits are used instead of press fits in cases where the heavy pressing forces required for mounting are impractical.

In Table 8.2.39 are listed the limits of interference (maximum and minimum values) for the above classes of force or shrink fits for various diameters, as given in ANSI B4.1-1967 (R87). Hole and shaft tolerances to give these interference limits are also listed in this reference.

Metric System   ANSI B4.2-1978 (R94) and ANSI B4.3-1978 (R94) define limits and fits for cylindrical parts, and provide tables listing preferred values.

The standard ANSI B4.2-1978 (R94) is essentially in accord with ISO R286.

ANSI B4.2 provides 22 basic deviations, each for the shaft (a to z plus js), and the hole (A to Z plus Js). International has 18 tolerance grades: IT 01, IT 0, and IT 1 through 16.

IT grades are roughly applied as follows: measuring tools, 01 to 7; fits, 5 to 11; material, 8 to 14; and large manufacturing tolerances, 12 to 16. See Table 8.2.41 for metric preferred fits.

Basic size—The basic size is the same for both members of a fit, and is the size to which limits or deviations are assigned. It is designated by 40 in 40H7.

Table 8.2.39 Limits of Interference for Force and Shrink Fits
(Limits are in thousandths of an inch on diameter)

Nominal size range, in	Class				
	FN 1	FN 2	FN 3	FN 4	FN 5
0.04–0.12	0.05 0.5	0.2 0.85		0.3 0.95	0.3 1.3
0.12–0.24	0.1 0.6	0.2 1.0		0.95 1.2	1.3 1.7
0.24–0.40	0.1 0.75	0.4 1.4		0.6 1.6	0.5 2.0
0.40–0.56	0.1 0.8	0.5 1.6		0.7 1.8	0.6 2.3
0.56–0.71	0.2 0.9	0.5 1.6		0.7 1.8	0.8 2.5
0.71–0.95	0.2 1.1	0.6 1.9		0.8 2.1	1.0 3.0
0.95–1.19	0.3 1.2	0.6 1.9	0.8 2.1	1.0 2.3	1.3 3.3
1.19–1.58	0.3 1.3	0.8 2.4	1.0 2.6	1.5 3.1	1.4 4.0
1.58–1.97	0.4 1.4	0.8 2.4	1.2 2.8	1.8 3.4	2.4 5.0
1.97–2.56	0.6 1.8	0.8 2.7	1.3 3.2	2.3 4.2	3.2 6.2
2.56–3.15	0.7 1.9	1.0 2.9	1.8 3.7	2.8 4.7	4.2 7.2
3.15–3.94	0.9 2.4	1.4 3.7	2.1 4.4	3.6 5.9	4.8 8.4
3.94–4.73	1.1 2.6	1.6 3.9	2.6 4.9	4.6 6.9	5.8 9.4
4.73–5.52	1.2 2.9	1.9 4.5	3.4 6.0	5.4 8.0	7.5 11.6
5.52–6.30	1.5 3.2	2.4 5.0	3.4 6.0	5.4 8.0	9.5 13.6
6.30–7.09	1.8 3.5	2.9 5.5	4.4 7.0	6.4 9.0	9.5 13.6

Deviation—The algebraic difference between a size and the corresponding basic size.

Upper deviation—The algebraic difference between the maximum limit of size and the corresponding basic size.

Lower deviation—The algebraic difference between the minimum limit of size and the corresponding basic size.

Fundamental deviation—That one of the two deviations closest to the basic size. It is designated by the letter H in 40H7.

Tolerance—The difference between the maximum and minimum size limits on a part.

International tolerance grade (IT)—A group of tolerances which vary depending on the basic size, but which provide the same relative level of accuracy within a grade. It is designated by 7 in 40H7 (IT 7).

Tolerance zone—A zone representing the tolerance and its position in relation to the basic size. The symbol consists of the fundamental deviation letter and the tolerance grade number (i.e., H7).

Hole basis—The system of fits where the minimum hole size is basic. The fundamental deviation for a hole basis system is H.

Shaft basis—Maximum shaft size is basic in this system. Fundamental deviation is h. NOTE: Capital letters refer to the hole and lowercase letters to the shaft.

Clearance fit—A fit in which there is clearance in the assembly for all tolerance conditions.

Interference fit—A fit in which there is interference for all tolerance conditions. Table 8.2.40 lists preferred metric sizes. Table 8.2.41 lists preferred tolerance zone combinations for clearance, transition, and interference fits. Table 8.2.42 lists dimensions for the grades corresponding to preferred fits. Table 8.2.43*a* and *b* lists limits (numerical) of preferred hole-basis clearance, transitions, and interference fits.

Stresses Produced by Shrink or Press Fit

STEEL HUB ON STEEL SHAFT.   The maximum equivalent stress, pounds per square inch, set up by a given press-fit allowance (in inches per inch of shaft diameter) is equal to $3x \times 10^7$, where x is the allowance per inch of shaft diameter (Baugher, *Trans. ASME*, 1931, p. 85). The press-fit pressures set up between a steel hub and shaft, for various ratios d/D between shaft and hub outside diameters, are given in Fig. 8.2.86. These

Table 8.2.40   Preferred Sizes (Metric)

First choice	Second choice	First choice	Second choice	First choice	Second choice
1		10		100	
	1.1		11		110
1.2		12		120	
	1.4		14		140
1.6		16		160	
	1.8		18			180
2		20		200	
	2.2		22		220
2.5		25		250	
	2.8		28		280
3		30		300	
	3.5		35			350
4		40		400	
	4.5		45		450
5		50		500	
	5.5		55		550
6		60		600	
	7		70		700
8		80		800	
	9		90		900
				1000	

SOURCE: ANSI B 4.2-1978 (R94), reproduced by permission.

curves are accurate to 5 percent even if the shaft is hollow, provided the inside shaft diameter is not over 25 percent of the outside. The equivalent stress given above is based on the maximum shear theory and is numerically equal to the radial-fit pressure added to the tangential tension in the hub. Where the shaft is hollow, with an inside diameter equal to more than about 25 percent of the outside diameter, the allowance in inches per inch to obtain an equivalent hub stress of 30,000 lb/in² may be determined by using Lamé's thick-cylinder formulas (*Jour. Appl.*

Table 8.2.41   Description of Preferred Fits (Metric)

Hole basis	Shaft basis	Description
ISO symbol		
Clearance fits		
H11/c11	C11/h11	*Loose-running* fit for wide commercial tolerances or allowances on external members.
H9/d9	D9/h9	*Free-running* fit not for use where accuracy is essential, but good for large temperature variations, high running speeds, or heavy journal pressures.
H8/f7	F8/h7	*Close-running* fit for running on accurate machines and for accurate location at moderate speeds and journal pressures.
H7/g6	G7/h6	*Sliding fit* not intended to run freely, but to move and turn freely and locate accurately.
H7/h6	H7/h6	*Locational clearance* fit provides snug fit for locating stationary parts; but can be freely assembled and disassembled.
Transition fits		
H7/k6	K7/h6	*Locational transition* fit for accurate location, a compromise between clearance and interference.
H7/n6	N7/h6	*Locational transition* fit for more accurate location where greater interference is permissible.
Interference fits		
H7/p6*	P7/h6	*Locational interference* fit for parts requiring rigidity and alignment with prime accuracy of location but without special bore pressure requirements.
H7/s6	S7/h6	*Medium-drive* fit for ordinary steel parts or shrink fits on light sections, the tightest fit usable with cast iron.
H7/u6	U7/h6	*Force* fit suitable for parts which can be highly stressed or for shrink fits where the heavy pressing forces required are impractical.

More clearance ↑

More interference ↓

* Transition fit for basic sizes in range from 0 through 3 mm.
SOURCE: ANSI B4.2-1978 (R94), reproduced by permission.

Table 8.2.42   International Tolerance Grades

Basic sizes		Tolerance grades, mm					
Over	Up to and including	IT6	IT7	IT8	IT9	IT10	IT11
0	3	0.006	0.010	0.014	0.025	0.040	0.060
3	6	0.008	0.012	0.018	0.030	0.048	0.075
6	10	0.009	0.015	0.022	0.036	0.058	0.090
10	18	0.011	0.018	0.027	0.043	0.070	0.110
18	30	0.013	0.021	0.033	0.052	0.084	0.130
30	50	0.016	0.025	0.039	0.062	0.100	0.160
50	80	0.019	0.030	0.046	0.074	0.120	0.190
80	120	0.022	0.035	0.054	0.087	0.140	0.220
120	180	0.025	0.040	0.063	0.100	0.160	0.250
180	250	0.029	0.046	0.072	0.115	0.185	0.290
250	315	0.032	0.052	0.081	0.130	0.210	0.320
315	400	0.036	0.057	0.089	0.140	0.230	0.360
400	500	0.040	0.063	0.097	0.155	0.250	0.400
500	630	0.044	0.070	0.110	0.175	0.280	0.440
630	800	0.050	0.080	0.125	0.200	0.320	0.500
800	1,000	0.056	0.090	0.140	0.230	0.360	0.560
1,000	1,250	0.066	0.105	0.165	0.260	0.420	0.660
1,250	1,600	0.078	0.125	0.195	0.310	0.500	0.780
1,600	2,000	0.092	0.150	0.230	0.370	0.600	0.920
2,000	2,500	0.110	0.175	0.280	0.440	0.700	1.100
2,500	3,150	0.135	0.210	0.330	0.540	0.860	1.350

SOURCE: ANSI B4.2-1978 (R94), reproduced by permission.

Mech., 1937, p. A-185). It should be noted that these curves hold only when the maximum equivalent stress is below the yield point; above the yield point, plastic flow occurs and the stresses are less than calculated.

Fig. 8.2.86   Press-fit pressures between steel hub and shaft.

Cast-Iron Hub on Steel Shaft   Where the shaft is solid, or hollow with an inside diameter not over 25 percent of the outside diameter, Fig. 8.2.87 may be used to determine maximum tensile stresses in the cast-iron hub, resulting from the press-fit allowance; for various ratios d/D, Fig. 8.2.88 gives the press-fit pressures. These curves are based on a modulus of elasticity of 30×10^6 lb/in² for steel and 15×10^6 for cast iron. For a hollow shaft with an inside diameter more than about ¼ the outside, the Lamé formulas may be used.

Pressure Required in Making Press Fits   The force required to press a hub on the shaft is given by $\pi f p d l$, where l is length of fit, p the unit press-fit pressure between shaft and hub, f the coefficient of friction,

and d the shaft diameter. Values of f varying from 0.03 to 0.33 have been reported, the lower values being due to yielding of the hub as a consequence of too high a fit allowance; the average is around 0.10 to 0.15. (For additional data see Horger and Nelson, "Design Data and Methods," ASME, 1953, pp. 87–91.)

Fig. 8.2.87   Variation of tensile stress in cast-iron hub in press-fit allowance.

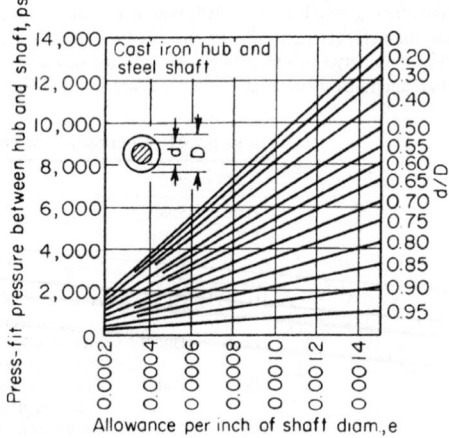

Fig. 8.2.88   Press-fit pressures between cast-iron hub and shaft.

Torsional Holding Ability   The torque required to cause complete slippage of a press fit is given by $T = \frac{1}{2}\pi f p l d^2$. Local slippage will usually occur near the end of the fit at much lower torques. If the torque is alternating, stress concentration and rubbing corrosion will occur at the hub face so that, eventually, fatigue failure may occur at considerably lower torques. Only in cases of static torque application is it justifiable to use ultimate torque as a basis for design.

A designer can often improve shrink-, press-, and slip-fit cylindrical assemblies with adhesives. When applied, adhesives can achieve high frictional force with attendant greater torque transmission without extra bulk, and thus augment or even replace press fits, compensate for differential thermal expansion, make fits with leakproof seals, eliminate backlash and clearance, etc.

The adhesives used are the anaerobic (see Sec. 6.8, "Adhesives") variety, such as Loctite products. Such adhesives destabilize and tend to harden when deprived of oxygen. Design suggestions on the use of such adhesives appear in industrial catalogs.

SHAFTS, AXLES, AND CRANKS

Most shafts are subject to combined bending and torsion, either of which may be steady or variable. Impact conditions, such as sudden starting and stopping, will cause momentary peak stresses greater than those related to the steady or variable portions of operation.

Design of shafts requires a theory of failure to express a stress in terms of loads and shaft dimensions, and an allowable stress as fixed by material strength and safety factor. Maximum shear theory of failure and distortion energy theory of failure are the two most commonly used in shaft design. Material strengths can be estimated from any one of several analytic representations of combined-load fatigue test data, starting from the linear (Soderberg, modified Goodman) which tend to give conservative designs to the nonlinear (Gerber parabolic, quadratic, Kececioglu, Bagci) which tend to give less conservative designs.

When linear representations of material strengths are used, and where both bending and torsion stresses have steady and variable components, the maximum shear theory and the distortion energy theory lead to somewhat similar formulations:

$$d = \left\{ \varepsilon \frac{n}{\pi} \left[\left(\frac{T_a}{S_{se}} + \frac{T_m}{S_{sy}} \right)^2 + \left(\frac{M_a}{S_{se}} + \frac{M_m}{S_{sy}} \right)^2 \right]^{1/2} \right\}^{1/3}$$

where $\varepsilon = 32$ (maximum shear theory) or 48 (distortion energy theory); d = shaft diameter; n = safety factor; T_a = amplitude torque = $(T_{max} - T_{min})/2$ T_m = mean torque = $(T_{max} + T_{min})/2$; M_a = amplitude bending moment = $(M_{max} - M_{min})/2$; M_m = mean bending moment = $(M_{max} + M_{min})/2$; S_{sy} = yield point in shear; S_{se} = completely corrected shear endurance limit = $S'_{se} k_a k_b k_c k_d / K_f$; S'_{se} = statistical average endurance limit of mirror finish, standard size, laboratory test specimen at standard room temperature; k_a = surface factor, a decimal to adjust S'_{se} for other than mirror finish; k_b = size factor, a decimal to adjust S'_{se} for other than standard test size; k_c = reliability factor, a decimal to adjust S'_{se} to other than its implied statistical average of 50 percent safe, 50 percent fail rate (50 percent reliability); k_d = temperature factor, decimal, to adjust S'_{se} to other than room temperature; $K_f = 1 + q(K_t - 1)$ = actual or fatigue stress concentration factor, a number greater than unity to adjust the nominal stress implied by T_a and M_a to a peak stress as induced by stress-raising conditions such as holes, fillets, keyways, press fits, etc. (K_f for T_a need not necessarily be the same as K_f for M_a); q = notch sensitivity; K_t = theoretical or geometric stress concentration factor.

For specific values of endurance limits and various factors the reader is referred to the technical literature (e.g., ASTM, NASA technical reports, ASME technical papers) or various books [e.g., "Machinery's Handbook" (Industrial Press, New York) and machine design textbooks].

If one allows for a variation of at least 15 percent, the following approximation is useful for the endurance limit in bending: $S'_{se} = 0.5 S_{ut}$. This becomes for maximum shear theory $S'_{se} = 0.5(0.5 S_{ut})$ and for distortion energy theory $S'_{se} = 0.577(0.5 S_{ut})$.

Representative or approximate values for the various factors mentioned above were abstracted from Shigley, "Mechanical Engineering Design," McGraw-Hill, and appear below with permission.

Surface Factor k_a

Surface condition	S_{ut}, kips			
	60	120	180	240
Polished	1.00	1.00	1.00	1.00
Ground	0.89	0.89	0.89	0.89
Machined or cold-drawn	0.84	0.71	0.66	0.63
Hot-rolled	0.70	0.50	0.39	0.31
As forged	0.54	0.36	0.27	0.20

Size Factor k_b

$$k_b = \begin{cases} 0.869 d^{-0.097} & 0.3 \text{ in} < d \leq 10 \text{ in} \\ 1 & d \leq 0.3 \text{ in or } d \leq 8 \text{ mm} \\ 1.189 d^{-0.097} & 8 \text{ mm} < d < 250 \text{ mm} \end{cases}$$

Reliability Factor k_c

Reliability, %	k_c
50	1.00
90	0.89
95	0.87
99	0.81

Temperature Factor k_d

Temperature		k_d
°F	°C	
840	450	1.00
940	482	0.71
1,020	550	0.42

Notch Sensitivity q

S_{ut}, kips	Notch radius r, in			
	0.02	0.06	0.10	0.14
60	0.56	0.70	0.74	0.78
100	0.68	0.79	0.83	0.85
150	0.80	0.90	0.91	0.92
200	0.90	0.95	0.96	0.96

See Table 8.2.44 for fatigue stress concentration factors for plain press fits.

A general representation of material strengths (Marin, Design for Fatigue Loading, *Mach. Des.* 29, no. 4, Feb. 21, 1957, pp. 128–131, and series of the same title) is given as

$$\left(\frac{S_a}{S_e} \right)^m + \left(\frac{K S_m}{S_{ut}} \right)^p = 1$$

where S_a = variable portion of material strength; S_m = mean portion of material strength; S_e = adjusted endurance limit; S_{ut} = ultimate strength. Table 8.2.45 lists the constants m, K, and P for various failure criteria.

For purposes of design, safety factors are introduced into the equation resulting in:

$$\left(\frac{\sigma'_{a,p}}{S_e / n_{se}} \right)^m + \left(\frac{K \sigma'_{m,p}}{S_{ut} / n_{ut}} \right)^P = 1$$

where

$$\sigma'_{a,p} = \frac{1}{\pi d^3} \sqrt{(32 n_{Ma} M_a)^2 + 3(16 n_{\tau a} T_a)^2}$$

$$\sigma'_{m,p} = \frac{1}{\pi d^3} \sqrt{(32 n_{Mm} M_m)^2 + 3(16 n_{\tau m} T_m)^2}$$

and n_{ij} = safety factor pertaining to a particular stress (that is, n_a = safety factor for amplitude shear stress).

Stiffness of shafting may become important where critical speeds, vibration, etc., may occur. Also, the lack of sufficient stiffness in shafts may give rise to bearing troubles. Critical speeds of shafts in torsion or bending and shaft deflections may be calculated using the methods of Sec. 5. For shafts of variable diameter see Spotts, "Design of Machine Elements," Prentice-Hall. In order to avoid trouble where sleeve bearings are used, the angular deflections at the bearings in general must be kept within certain limits. One rule is to make the shaft deflection over the bearing width equal to a small fraction of the oil-film thickness. Note that since stiffness is proportional to the modulus of elasticity, alloy-steel shafts are no stiffer than carbon-steel shafts of the same diameter.

Crankshafts For calculating the torsional stiffness of crankshafts, the formulas given in Sec. 5 may be used.

Table 8.2.43a Limits of Fits
(Dimensions in mm)

Preferred hole basis clearance fits

Basic size		Loose-running Hole H11	Shaft c11	Fit	Free-running Hole H9	Shaft d9	Fit	Close-running Hole H8	Shaft f7	Fit	Sliding Hole H7	g6	Fit	Locational clearance Hole H7	Shaft h6	Fit
1	max	1.060	0.940	0.180	1.025	0.980	0.070	1.014	0.994	0.030	1.010	0.998	0.018	1.010	1.000	0.016
	min	1.000	0.880	0.060	1.000	0.955	0.020	1.000	0.984	0.006	1.000	0.992	0.002	1.000	0.994	0.000
1.2	max	1.260	1.140	0.180	1.225	1.180	0.070	1.214	1.194	0.030	1.210	1.198	0.018	1.210	1.200	0.016
	min	1.200	1.080	0.060	1.200	1.155	0.020	1.200	1.184	0.006	1.200	1.192	0.002	1.200	1.194	0.000
1.6	max	1.660	1.540	1.180	1.625	1.580	0.070	1.614	1.594	0.050	1.610	1.598	0.018	1.610	1.600	0.016
	min	1.600	1.480	0.060	1.600	1.555	0.020	1.600	1.584	0.006	1.600	1.592	0.002	1.600	1.594	0.000
2	max	2.060	1.940	0.180	2.025	1.980	0.070	2.014	1.994	0.030	2.010	1.998	0.018	2.010	2.000	0.016
	min	2.000	1.880	0.060	2.000	1.955	0.020	2.000	1.984	0.006	2.000	1.992	0.002	2.000	1.994	0.000
2.5	max	2.560	2.440	0.180	2.525	2.480	0.070	2.514	2.494	0.030	2.510	2.498	0.018	2.510	2.500	0.016
	min	2.500	2.380	0.060	2.500	2.455	0.020	2.500	2.484	0.006	2.500	2.492	0.002	2.500	2.494	0.000
3	max	3.060	2.940	0.180	3.025	2.980	0.070	3.014	2.994	0.030	3.010	2.998	0.018	3.010	3.000	0.016
	min	3.000	2.880	0.060	3.000	2.955	0.020	3.000	2.984	0.006	3.000	2.992	0.002	3.000	2.994	0.000
4	max	4.075	3.930	0.220	4.030	3.970	0.090	4.018	3.990	0.040	4.012	3.996	0.024	4.012	4.000	0.020
	min	4.000	3.855	0.070	4.000	3.940	0.030	4.000	3.978	0.010	4.000	3.988	0.004	4.000	3.992	0.000
5	max	5.075	4.930	0.220	5.030	4.970	0.090	5.018	4.990	0.040	5.012	4.996	0.024	5.012	5.000	0.020
	min	5.000	4.855	0.070	5.000	4.940	0.030	5.000	4.978	0.010	5.000	4.988	0.004	5.000	4.992	0.000
6	max	6.075	5.930	0.220	6.030	5.970	0.090	6.018	5.990	0.040	6.012	5.996	0.024	6.012	6.000	0.020
	min	6.000	5.855	0.070	6.000	5.940	0.030	6.000	5.978	0.010	6.000	5.988	0.004	6.000	5.992	0.000
8	max	8.090	7.920	0.260	8.036	7.960	0.112	8.022	7.987	0.050	8.015	7.995	0.029	8.015	8.000	0.024
	min	8.000	7.830	0.080	8.000	7.924	0.040	8.000	7.972	0.013	8.000	7.986	0.005	8.000	7.991	0.000
10	max	10.090	9.920	0.260	10.036	9.960	0.112	10.022	9.987	0.050	10.015	9.995	0.029	10.015	10.000	0.024
	min	10.000	9.830	0.080	10.000	9.924	0.040	10.000	9.972	0.013	10.000	9.986	0.005	10.000	9.991	0.000
12	max	12.110	11.905	0.315	12.043	11.950	0.136	12.027	11.984	0.061	12.018	11.994	0.035	12.018	12.000	0.029
	min	12.000	11.795	0.095	12.000	11.907	0.050	12.000	11.966	0.016	12.000	11.983	0.006	12.000	11.989	0.000
16	max	16.110	15.905	0.315	16.043	15.950	0.136	16.027	15.984	0.061	16.018	15.994	0.035	16.018	16.000	0.029
	min	16.000	15.795	0.095	16.000	15.907	0.050	16.000	15.966	0.016	16.000	15.983	0.006	16.000	15.989	0.000
20	max	20.130	19.890	0.370	20.052	19.935	0.169	20.033	19.980	0.074	20.021	19.993	0.041	20.021	20.000	0.034
	min	20.000	19.760	0.110	20.000	19.883	0.065	20.000	19.959	0.020	20.000	19.980	0.007	20.000	19.987	0.000
25	max	25.130	24.890	0.370	25.052	24.935	0.169	25.033	24.980	0.074	25.021	24.993	0.041	25.021	25.000	0.034
	min	25.000	24.760	0.110	25.000	24.883	0.065	25.000	24.959	0.020	25.000	24.980	0.007	25.000	24.987	0.000
30	max	30.130	29.890	0.370	30.052	29.935	0.169	30.033	29.980	0.074	30.021	29.993	0.041	30.021	30.000	0.034
	min	30.000	29.760	0.110	30.000	29.883	0.065	30.000	29.959	0.020	30.000	29.980	0.007	30.000	29.987	0.000
40	max	40.160	39.880	0.440	40.062	39.920	0.204	40.039	39.975	0.089	40.025	39.991	0.050	40.025	40.000	0.041
	min	40.000	39.720	0.120	40.000	39.858	0.080	40.000	39.950	0.025	40.000	39.975	0.009	40.000	39.984	0.000
50	max	50.160	49.870	0.450	50.062	49.920	0.204	50.039	49.975	0.089	50.025	49.991	0.050	50.025	50.000	0.041
	min	50.000	49.710	0.130	50.000	49.858	0.080	50.000	49.950	0.025	50.000	49.975	0.009	50.000	49.984	0.000
60	max	60.190	59.860	0.520	60.074	59.900	0.248	60.046	59.970	0.106	60.030	59.990	0.059	60.030	60.000	0.049
	min	60.000	59.670	0.140	60.000	59.826	0.100	60.000	59.940	0.030	60.000	59.971	0.010	60.000	59.981	0.000
80	max	80.190	79.850	0.530	80.074	79.900	0.248	80.046	79.970	0.106	80.030	79.990	0.059	80.030	80.000	0.049
	min	80.000	79.660	0.150	80.000	79.826	0.100	80.000	79.940	0.030	80.000	79.971	0.010	80.000	79.981	0.000
100	max	100.220	99.830	0.610	100.087	99.880	0.294	100.054	99.964	0.125	100.035	99.988	0.069	100.035	100.000	0.057
	min	100.000	99.610	0.170	100.000	99.793	0.120	100.000	99.929	0.036	100.000	99.966	0.012	100.000	99.978	0.000
120	max	120.220	119.820	0.620	120.087	119.880	0.294	120.054	119.964	0.125	120.035	119.988	0.069	120.035	120.000	0.057
	min	120.000	119.600	0.180	120.000	119.793	0.120	120.000	119.929	0.036	120.000	119.966	0.012	120.000	119.978	0.000
160	max	160.250	159.790	0.710	160.100	159.855	0.345	160.063	159.957	0.146	160.040	159.986	0.079	160.040	160.000	0.065
	min	160.000	159.540	0.210	160.000	159.755	0.145	160.000	159.917	0.043	160.000	159.961	0.014	160.000	159.975	0.000

Table 8.2.43b Limits of Fits
(Dimensions in mm)

Preferred hole basis transition and interference fits

Basic size		Locational transition			Locational transition			Locational interference			Medium drive			Force		
		Hole H7	Shaft k6	Fit	Hole H7	Shaft n6	Fit	Hole H7	Shaft p6	Fit	Hole H7	Shaft s6	Fit	Hole H7	Shaft u6	Fit
1	max	1.010	1.006	0.010	1.010	1.010	0.006	1.010	1.012	0.004	1.010	1.020	−0.004	1.010	1.024	−0.008
	min	1.000	1.000	−0.006	1.000	1.004	−0.010	1.000	1.006	−0.012	1.000	1.014	−0.020	1.000	1.018	−0.024
1.2	max	1.210	1.206	0.010	1.210	1.210	0.006	1.210	1.212	0.004	1.210	1.220	−0.004	1.210	1.224	−0.008
	min	1.200	1.200	−0.006	1.200	1.204	−0.010	1.200	1.206	−0.012	1.200	1.214	−0.020	1.200	1.218	−0.024
1.6	max	1.610	1.606	0.010	1.610	1.610	0.006	1.610	1.612	0.004	1.610	1.620	−0.004	1.610	1.624	−0.008
	min	1.610	1.600	−0.006	1.600	1.604	−0.010	1.600	1.606	−0.012	1.600	1.614	−0.020	1.600	1.618	−0.024
2	max	2.010	2.006	0.010	2.010	2.010	0.006	2.010	2.012	0.004	2.010	2.020	−0.004	2.010	2.024	−0.008
	min	2.000	2.000	−0.006	2.000	2.004	−0.010	2.000	2.006	−0.012	2.000	2.014	−0.020	2.000	2.018	−0.024
2.5	max	2.510	2.506	0.010	2.510	2.510	0.006	2.510	2.512	0.004	2.510	2.520	−0.004	2.510	2.524	−0.008
	min	2.500	2.500	−0.006	2.500	2.504	−0.010	2.500	2.506	−0.012	2.500	2.514	−0.020	2.500	2.518	−0.024
3	max	3.010	3.006	0.010	3.010	3.010	0.006	3.010	3.012	0.004	3.010	3.020	−0.004	3.010	3.024	−0.008
	min	3.000	3.000	−0.006	3.000	3.004	−0.010	3.000	3.006	−0.012	3.000	3.014	−0.020	3.000	3.018	−0.024
4	max	4.012	4.009	0.011	4.012	4.016	0.004	4.012	4.020	0.000	4.012	4.027	−0.007	4.012	4.031	−0.011
	min	4.000	4.001	−0.009	4.000	4.008	−0.016	4.000	4.012	−0.020	4.000	4.019	−0.027	4.000	4.023	−0.031
5	max	5.012	5.009	0.011	5.012	5.016	0.004	5.012	5.020	0.000	5.012	5.027	−0.007	5.012	5.031	−0.011
	min	5.000	5.001	−0.009	5.000	5.008	−0.016	5.000	5.012	−0.020	5.000	5.019	−0.027	5.000	5.023	−0.031
6	max	6.012	6.009	0.011	6.012	6.016	0.004	6.012	6.020	0.000	6.012	6.027	−0.007	6.012	6.031	−0.011
	min	6.000	6.001	−0.009	6.000	6.008	−0.016	6.000	6.012	−0.020	6.000	6.019	−0.027	6.000	6.023	−0.031
8	max	8.015	8.010	0.014	8.015	8.019	0.005	8.015	8.024	0.000	8.015	8.032	−0.008	8.015	8.037	−0.013
	min	8.000	3.001	−0.010	8.000	8.010	−0.019	8.000	8.015	−0.024	8.000	8.023	−0.032	8.000	8.028	−0.037
10	max	10.015	10.010	0.014	10.015	10.019	0.005	10.015	10.024	0.000	10.015	10.032	−0.008	10.015	10.037	−0.013
	min	10.000	10.031	−0.010	10.000	10.010	−0.019	10.000	10.015	−0.024	10.000	10.023	−0.032	10.000	10.028	−0.037
12	max	12.018	12.012	0.017	12.018	12.023	0.006	12.018	12.029	0.000	12.018	12.039	−0.010	12.018	12.044	−0.015
	min	12.000	12.001	−0.012	12.000	12.012	−0.023	12.000	12.018	−0.029	12.000	12.028	−0.039	12.000	12.033	−0.044
16	max	16.018	16.012	0.017	16.018	16.023	0.006	16.018	16.029	0.000	16.018	16.039	−0.010	16.018	16.044	−0.015
	min	16.000	16.001	−0.012	16.000	16.012	−0.023	16.000	16.018	−0.029	16.000	16.028	−0.039	16.000	16.033	−0.044
20	max	20.021	20.015	0.019	20.021	20.028	0.006	20.021	20.035	−0.001	20.021	20.048	−0.014	20.021	20.054	−0.020
	min	20.000	20.002	−0.015	20.000	20.015	−0.028	20.000	20.022	−0.035	20.000	20.035	−0.048	20.000	20.041	−0.054
25	max	25.021	25.015	0.019	25.021	25.028	0.006	25.021	25.035	−0.001	25.021	25.048	−0.014	25.021	25.061	−0.027
	min	25.000	25.002	−0.015	25.000	25.015	−0.028	25.000	25.022	−0.035	25.000	25.035	−0.048	25.000	25.048	−0.061
30	max	30.021	30.015	0.019	30.021	30.028	0.006	30.021	30.035	−0.001	30.021	30.048	−0.014	30.021	30.061	−0.027
	min	30.000	30.002	−0.015	30.000	30.015	−0.028	30.000	30.022	−0.035	30.000	30.035	−0.048	30.000	30.048	−0.061
40	max	40.025	40.018	0.023	40.025	40.033	0.008	40.025	40.042	−0.001	40.025	40.059	−0.018	40.025	40.076	−0.035
	min	40.000	40.002	−0.018	40.000	40.017	−0.033	40.000	40.026	−0.042	40.000	40.043	−0.059	40.000	40.060	−0.076
50	max	50.025	50.018	0.023	50.025	50.033	0.008	50.025	50.042	−0.001	50.025	50.059	−0.018	50.025	50.086	−0.045
	min	50.000	50.002	−0.018	50.000	50.017	−0.033	50.000	50.026	−0.042	50.000	50.043	−0.059	50.000	50.070	−0.086
60	max	60.030	60.021	0.028	60.030	60.039	0.010	60.030	60.051	−0.002	60.030	60.072	−0.023	60.030	60.106	−0.057
	min	60.000	60.002	−0.021	60.000	60.020	−0.039	60.000	60.032	−0.051	60.000	60.053	−0.072	60.000	60.087	−0.106
80	max	80.030	80.021	0.028	80.030	80.039	0.010	80.030	80.051	−0.002	80.030	80.078	−0.029	80.030	80.121	−0.072
	min	80.000	80.002	−0.021	80.000	80.020	−0.039	80.000	80.032	−0.051	80.000	80.059	−0.078	80.000	80.102	−0.121
100	max	100.035	100.025	0.032	100.035	100.045	0.012	100.035	100.059	−0.002	100.035	100.093	−0.036	100.035	100.146	−0.089
	min	100.000	100.003	−0.025	100.000	100.023	−0.045	100.000	100.037	−0.059	100.000	100.071	−0.093	100.000	100.124	−0.146
120	max	120.035	120.025	0.032	120.035	120.045	0.012	120.035	120.059	−0.002	120.035	120.101	−0.044	120.035	120.166	−0.109
	min	120.000	120.003	−0.025	120.000	120.023	−0.045	120.000	120.037	−0.059	120.000	120.079	−0.101	120.000	120.144	−0.166
160	max	160.040	160.028	0.037	160.040	160.052	0.013	160.040	160.068	−0.003	160.040	160.125	−0.060	160.040	160.215	−0.150
	min	160.000	160.003	−0.028	160.000	160.027	−0.052	160.000	160.043	−0.068	160.000	160.100	−0.125	160.000	160.190	−0.215

SOURCE: ANSI B4.2-1978 (R94), reprinted by permission.

Table 8.2.44 K_f Values for Plain Press Fits
(Obtained from fatigue tests in bending)

Shaft material	Shaft diam, in	Collar or hub material	K_f	Remarks
0.42% carbon steel	1⅝	0.42% carbon steel	2.0	No external reaction through collar
0.45% carbon axle steel	2	Ni-Cr-Mo steel (case-hardened)	2.3	No external reaction through collar
0.45% carbon axle steel	2	Ni-Cr-Mo steel (case-hardened)	2.9	External reaction taken through collar
Cr-Ni-Mo steel (310 Brinell)	2	Ni-Cr-Mo steel (case-hardened)	3.9	External reaction taken through collar
2.6% Ni steel (57,000 lb/in² fatigue limit)	2	Ni-Cr-Mo steel (case-hardened)	3.3–3.8	External reaction taken through collar
Same, heat-treated to 253 Brinell	2	Ni-Cr-Mo steel (case-hardened)	3.0	External reaction taken through collar

Marine-engine shafts and **diesel-engine crankshafts** should be designed not only for strength but for avoidance of critical speed. (See Applied Mechanics, *Trans. ASME*, **50**, no. 8, for methods of calculating critical speeds of diesel engines.)

Table 8.2.45   Constants for Use in
$(S_a/S_e)^m + (KS_m/S_{ut})^p = 1$

Failure theory	K	P	m
Soderberg	S_{ut}/S_y	1	1
Bagci	S_{ut}/S_y	4	1
Modified Goodman	1	1	1
Gerber parabolic	1	2	1
Kececioglu	1	2	m†
Quadratic (elliptic)	1	2	2

$$^\dagger m = \begin{cases} 0.8914 \text{ UNS G } 10180 \; H_B = 130 \\ 0.9266 \text{ UNS G } 10380 \; H_B = 164 \\ 1.0176 \text{ UNS G } 41300 \; H_B = 207 \\ 0.9685 \text{ UNS G } 43400 \; H_B = 233 \end{cases}$$

PULLEYS, SHEAVES, AND FLYWHEELS

Arms of pulleys, sheaves, and flywheels are subjected to stresses due to condition of founding, to details of construction (such as split or solid), and to conditions of service, which are difficult to analyze. For these reasons, no accurate stress relations can be established, and the following formulas must be understood to be only approximately correct. It has been established experimentally by Benjamin (*Am. Mach.*, Sept. 22, 1898) that thin-rim pulleys do not distribute equal loads to the several pulley arms. For these, it will be safe to assume the tangential force on the pulley rim as acting on half of the number of arms. Pulleys with comparatively thick rims, such as engine band wheels, have all the arms taking the load. Furthermore, while the stress action in the arms is similar to that in a beam fixed at both ends, the amount of restraint at the rim depending on the rim's elasticity, it may nevertheless be assumed for purposes of design that cantilever action is predominant. The bending moment at the hub in arms of thin-rim pulleys will be $M = PL/(\frac{1}{2}N)$, where M = bending moment, in-lb; P = tangential load on the rim, lb; L = length of the arm, in; and N = number of arms. For thick-rim pulleys and flywheels, $M = PL/N$.

For arms of elliptical section having a width of two times the thickness, where E = **width of arm section** at the rim, in, and s_t = intensity of tensile stress, lb/in²

$$E = \sqrt[3]{40PL/(s_tN)} \text{ (thin rim)} = \sqrt[3]{20PL/(s_tN)} \text{ (thick rim)}$$

For single-thickness belts, P may be taken as $50B$ lb and for double-thickness belts $P = 75B$ lb, where B is the width of pulley face, in. Then $E = k \times \sqrt[3]{BL/(s_tN)}$, where k has the following values: for thin rim, single belt, 13; thin rim, double belt, 15; thick rim, single belt, 10; thick rim, double belt, 12. For cast iron of good quality, s_t due to bending may be taken as 1,500 to 2,000. The arm section at the rim may be made from ⅔ to ¾ the dimensions at the hub.

For high-speed pulleys and flywheels, it becomes necessary to check the arm for tension due to rim expansion. It will be safe to assume that each arm is in tension due to one-half the centrifugal force of that portion of the rim which it supports. That is, $T = As_t = Wv^2/(2NgR)$, lb, where T = tension in arm, lb; N = number of arms; v = speed of rim, ft/s; R = radius of pulley, ft; A = area of arm section, in²; W = weight of pulley rim, lb; and s_t = intensity of tensile stress in arm section, lb/in², whence $s_t = WRn^2/(6,800NA)$, where n = r/min of pulley.

Arms of flywheels having heavy rims may be subjected to severe stress action due to the inertia of the rim at sudden load changes. There being no means of predicting the probable maximum to which the inertia may rise, it will be safe to make the arms equal in strength to ¾ of the shaft strength in torsion. Accordingly, for elliptical arm sections,

$$N \times 0.5E^3 s_t = \frac{3}{4} \times 2s_s d^3 \quad \text{or} \quad E = 1.4d\sqrt[3]{s_s/(s_tN)}$$

For steel shafts with s_s = 8,000 and cast-iron arms with s = 1,500,

$$E = 2.4d/\sqrt[3]{N} = 1.3d \text{ (for 6 arms)} = 1.2d \text{ (for 8 arms)}$$

where $2E$ = width of elliptical arm section at hub, in (thickness = E), and d = shaft diameter, in.

Rims of belted pulleys cast whole may have the following proportions (see Fig. 8.2.89):

$$t_2 = \frac{3}{4}h + 0.005D \quad t_1 = 2t_2 + C \quad W = \frac{5}{8}B \text{ to } \frac{5}{4}B$$

where h = belt thickness, $C = \frac{1}{24}W$, and B = belt width, all in inches.

Fig. 8.2.89   Rims for belted pulleys.

Engine band wheels, flywheels, and pulleys run at **high speeds** are subjected to the following stress actions in the rim:

Considering the rim as a free ring, i.e., without arm restraint, and made of cast iron or steel, $s_t = v^2/10$ (approx), where s_t = intensity of tensile stress, lb/in², and v = rim speed, ft/s. For beam action between the arms of a solid rim, $M = Pl/12$ (approx), where M = bending moment in rim, in · lb; P = centrifugal force of that portion of rim between arms, lb, and l = length of rim between arms, in; from which $s_t = WR^2n^2/(450N^2Z)$, where W = weight of entire rim, lb; R = radius of wheel, ft; n = r/min of wheel; and Z = section modulus of rim section, in³. In case the rim section is of the forms shown in Fig. 8.2.89, care must be taken that the flanges do not reduce the section modulus from that of the rectangular section. For **split rims** fastened with bolts the stress analysis is as follows:

Let w = weight of rim portion, lb (with length L, in) lb; w_1 = weight of lug, lb; L_1 = lever arm of lug, in; and s_t = intensity of tensile stress lb/in² in rim section joining arm. Then $s_t = 0.00034n^2R(w_1L_1 + wL/2)/Z$ where n = r/min of wheel; R = wheel radius, ft; and Z = section modulus of rim section, in^3. The above equation gives the value of s_t for bending when the bolts are loose, which is the worst possible condition that may arise. On this basis of analysis, s_t should not be greater than 8,000 lb/in^2. The stress due to bending in addition to the stress due to rim expansion as analyzed previously will be the probable maximum intensity of stress for which the rim should be checked for strength. The flange bolts, because of their position, do not materially relieve the bending action. In case a tie rod leads from the flange to the hub, it will be *safe* to consider it as an additional factor of safety. When the tie rod is kept tight, it very materially strengthens the rim.

A more accurate method for calculating maximum stresses due to centrifugal force in flywheels with arms cast integral with the rim is given by Timoshenko, "Strength of Materials," Pt. II, 1941, p. 98. More exact equations for calculating stresses in the arms of flywheels and pulleys due to a combination of belt pull, centrifugal force, and changes in velocity are given by Heusinger, *Forschung*, 1938, p. 197. In both treatments, shrinkage stresses in the arms due to casting are neglected.

Large flywheels for high rim speeds and severe working conditions (as for rolling-mill service) have been made from flat-rolled steel plates with holes bored for the shaft. A group of such plates may be welded together by circumferential welds to form a large flywheel. By this means, the welds do not carry direct centrifugal loads, but serve merely to hold the parts in position. Flywheels up to 15-ft diam have been constructed in this way.

BELT DRIVES

Flat-Belt Drives

The primary drawback of flat belts is their reliance on belt tension to produce frictional grip over the pulleys. Such high tension can shorten bearing life. Also, tracking may be a problem. However, flat belts, being thin, are not subject to centrifugal loads and so work well over small pulleys at high speeds in ranges exceeding 9,000 ft/min. In light service flat belts can make effective clutching drives. Flat-belt drives have efficiencies of about 90 percent, which compares favorably to geared drives. Flat belts are also quiet and can absorb torsional vibration readily.

Leather belting has an ultimate tensile strength ranging from 3,000 to 5,000 lb/in^2. Average values of breaking strength of good oak-tanned belting (determined by Benjamin) are as follows: single (double) in solid leather 900 (1,400); at riveted joint 600 (1,200); at laced joint 350 lb/in of width. Well-made cemented joints have strengths equal to the belt, leather-laced and riveted joints about one-third to two-thirds as strong, and wire-laced joints about 85 to 95 percent as strong.

Rubber belting is made from fabric or cord impregnated and bound together by vulcanized rubber compounds. The fabric or cord may be of cotton or rayon. Nylon cord and steel cord or cable are also available. Advantages are high tensile strength, strength to hold metal fasteners satisfactorily, and resistance to deterioration by moisture. The best rubber fabric construction for most types of service is made from hard or tight-woven fabric with a "skim coat" or thin layer of rubber between plies. The cord type of construction allows the use of smaller pulley diameters than the fabric type, and also develops less stretch in service. It must be used in the endless form, except in cases where the oil-field type of clamp may be used.

Initial tensions in rubber belts run from 15 to 25 lb/ply/in width. A common rule is to cut belts 1 percent less than the minimum tape-line measurement around the pulleys. For heavy loads, a 1½ percent allowance is usually required, although, because of shrinkage, less initial tension is required for wet or damp conditions. Initial tensions of 25 lb/ply/in may overload shafts or bearings. Maximum safe tight-side tensions for rubber belts are as follows:

Duck weight, oz	28	32	32.66	34.66	36
Tension, lb/ply/in width	25	28	30	32	35

Centrifugal forces at high speeds require higher tight-side tensions to carry rated horsepower.

Rubber belting may be bought in endless form or made endless in the field by means of a vulcanized splice produced by a portable electric vulcanizer. For endless belts the drive should provide take-up of 2 to 4 percent to allow for length variation as received and for stretch in service. The amount of take-up will vary with the type of belt used. For certain drives, it is possible to use endless belts with no provision for take-up, but this involves a heavier belt and a higher initial unit tension than would be the case otherwise. Ultimate tensile strength of rubber belting varies from 280 to 600 lb or more per inch width per ply. The weight varies from 0.02 to 0.03 or more lb/in width/ply. Belts with steel reinforcement are considerably heavier. For horsepower ratings of rubber belts, see Table 8.2.46c.

Arrangements for Belt Drives In belt drives, the centerline of the belt advancing on the pulley should lie in a plane passing through the midsection of the pulley at right angles to the shaft. Shafts inclined to each other require connections as shown in Fig. 8.2.90a. In case guide pulleys are needed their positions can be deteraiined as shown in Fig. 8.2.90a, b, and c. In Fig. 8.2.90d the center circles of the two pulleys to be connected are set in correct relative position in two planes, a being the angle between the planes (= supplement of angle between shafts). If any two points as E and F are assumed on the line of intersection MN of the planes, and tangents EG, EH, FJ, and FK are drawn from them to the circles, the center circles of the guide pulleys must be so arranged that these tangents are also tangents to them, as shown. In other words, the middle planes of the guide pulleys must lie in the planes GEH and JFK.

Table 8.2.46a Service Factors S

Application	Squirrel-cage ac motor		Wound rotor ac motor (slip ring)	Single-phase capacitor motor	DC shunt-wound motor	Diesel engine 4 or more cyl, above 700 r/min
	Normal torque, line start	High torque				
Agitators	1.0–1.2	1.2–1.4	1.2			
Compressors	1.2–1.4		1.4	1.2	1.2	1.2
Belt conveyors (ore, coal, sand)	—	1.4	—	—	1.2	
Screw conveyors	—	1.8	—	—	1.6	
Crushing machinery	—	1.6	1.4	—	—	1.4–1.6
Fans, centrifugal	1.2	—	1.4		1.4	1.4
Fans, propeller	1.4	2.0	1.6	—	1.6	1.6
Generators and exciters	1.2	—	—	—	1.2	2.0
Line shafts	1.4	—	1.4	1.4	1.4	1.6
Machine tools	1.0–1.2	—	1.2–1.4	1.0	1.0–1.2	
Pumps, centrifugal	1.2	1.4	1.4	1.2	1.2	
Pumps, reciprocating	1.2–1.4	—	1.4–1.6	—	—	1.8–2.0

Table 8.2.46b   Arc of Contact Factor K—Rubber Belts

Arc of contact, deg	140	160	180	200	220
Factor K	0.82	0.93	1.00	1.06	1.12

When these conditions are met, the belts will run in either direction on the pulleys.

To avoid the necessity of taking up the **slack** in belts which have become stretched and permanently lengthened, a **belt tightener** such as shown in Fig. 8.2.91 may be employed. It should be placed on the slack side of the belt and nearer the driving pulley than the driven pulley. Pivoted motor drives may also be used to maintain belt tightness with minimum initial tension.

Length of Belt for a Given Drive   The length of an **open belt** for a given drive is equal to $L = 2C + 1.57(D + d) + (D - d)^2/(4C)$, where L = length of belt, in; D = diam of large pulley, in; d = diam of small pulley, in; and C = distance between pulley centers, in. Center

distance C is given by $C = 0.25b + 0.25\sqrt{b^2 - 2(D - d)^2}$, where $b = L - 1.57(D + d)$. When a **crossed belt** is used, the length in $L = 2C + 1.57(D + d) + (D + d)^2/(4C)$.

Step or Cone Pulleys   For belts operating on step pulleys, the pulley diameters must be such that the belt will fit over any pair with equal tightness. With crossed belts, it will be apparent from the equation for length of belt that the sum of the pulley diameters need only be constant in order that the belt may fit with equal tightness on each pair of pulleys. With open belts, the length is a function of both the sum and the difference of the pulley diameters; hence no direct solution of the problem is possible, but a graphical approach can be of use.

A graphical method devised by Smith (*Trans. ASME*, 10) is shown in Fig. 8.2.92. Let A and B be the centers of any pair of pulleys in the set, the diameters of which are known or assumed. Bisect AB in C, and draw CD at right angles to AB. Take $CD = 0.314$ times the center distance L, and draw a circle tangent to the belt line EF. The belt line of any other pair of pulleys in the set will then be tangent to this circle. If the angle

Table 8.2.46c   Horsepower Ratings of Rubber Belts
(hp/in of belt width for 180° wrap)

	Ply						Belt speed, ft/min					
		500	1,000	1,500	2,000	2,500	3,000	4,000	5,000	6,000	7,000	8,000
32-oz fabric	3	0.7	1.4	2.1	2.7	3.3	3.9	4.9	5.6	6.0		
	4	0.9	1.9	2.8	3.6	4.4	5.2	6.5	7.4	7.9		
	5	1.2	2.3	3.4	4.5	5.5	6.5	8.1	9.2	9.8		
	6	1.4	2.8	4.1	5.4	6.6	7.8	9.6	11.0	11.7		
	7	1.6	3.2	4.7	6.2	7.7	9.0	11.2	12.8	13.6		
	8	1.8	3.6	5.3	7.0	8.7	10.2	12.7	14.6	15.5		
32-oz hard fabric	3	0.7	1.5	2.2	2.9	3.5	4.1	5.1	5.8	6.2	6.1	5.5
	4	1.0	2.0	3.0	3.9	4.7	5.5	6.8	7.8	8.3	8.1	7.3
	5	1.3	2.5	3.7	4.9	5.9	6.9	8.5	9.8	10.3	9.1	9.0
	6	1.5	3.0	4.5	5.9	7.1	8.3	10.2	11.7	12.3	12.1	10.7
	7	1.7	3.5	5.2	6.9	8.3	9.7	11.9	13.6	14.3	14.1	12.4
	8	1.9	4.0	5.9	7.9	9.5	11.1	13.6	15.5	16.3	16.0	14.1
	9	2.1	4.5	6.6	8.9	10.6	12.4	15.3	17.4	18.3	17.9	15.8
	10	2.3	5.0	7.3	9.8	11.7	13.7	17.0	19.3	20.3	19.8	17.5
No. 70 rayon cord	3	1.6	3.1	4.6	6.0	7.3	8.6	10.6	12.0	12.7	12.3	10.7
	4	2.1	4.1	6.1	8.0	9.8	11.5	14.5	16.6	17.8	17.8	16.4
	5	2.6	5.1	7.6	10.1	12.3	14.5	18.3	21.1	23.0	23.5	22.2
	6	3.1	6.2	9.2	12.1	14.8	17.5	22.1	25.7	28.1	28.9	27.9
	7	3.6	7.2	10.7	14.1	17.4	20.4	26.0	30.3	33.2	34.5	33.7
	8	4.1	8.2	12.2	16.2	19.9	23.4	29.8	34.8	38.4	40.0	39.4

Table 8.2.46d   Minimum Pulley Diameters—Rubber Belts, in

	Ply						Belt speed, ft/min					
		500	1,000	1,500	2,000	2,500	3,000	4,000	5,000	6,000	7,000	8,000
32-oz fabric	3	4	4	4	4	5	5	5	6	6		
	4	4	5	6	6	7	7	8	9	10		
	5	6	7	9	10	10	11	12	13	14		
	6	9	10	11	13	14	14	16	18	19		
	7	13	14	16	17	18	19	21	22	24		
	8	18	19	21	22	23	24	25	27	29		
32-oz hard fabric	3	3	3	3	4	4	4	4	5	5	6	7
	4	4	4	5	5	6	6	7	7	8	9	12
	5	5	6	7	8	8	9	10	11	12	13	16
	6	6	8	10	11	11	12	13	15	16	18	21
	7	10	12	14	15	15	16	17	19	20	22	26
	8	14	16	17	18	19	20	21	23	24	27	31
	9	18	20	21	22	23	24	25	27	28	31	36
	10	22	24	25	26	27	28	29	31	33	35	41
No. 70 rayon cord	3	5	6	7	7	8	8	9	10	11	12	13
	4	7	8	9	9	10	11	12	12	14	15	17
	5	9	10	11	12	13	13	15	16	17	19	21
	6	13	14	15	16	16	17	18	19	21	23	25
	7	16	17	18	19	20	21	22	23	24	26	29
	8	19	20	22	23	23	24	25	26	28	30	33

EF makes with AB is greater than 18°, draw a tangent to the circle D, making an angle of 18° with AB; and from a center on CD distant 0.298L above C, draw an arc tangent to this 18° line. All belt lines with angles greater than 18° will be tangent to this last drawn arc.

(a)

(b)

(c)   (d)

Fig. 8.2.90 Arrangements for flat-belt drives.

A very slight error in a graphical solution drawn to any scale much under full size will introduce an error seriously affecting the equality of belt tensions on the various pairs of pulleys in the set, and where much power is to be transmitted it is advisable to calculate the pulley diameters from the following **formulas** derived from Burmester's graphical method ("Lehrbuch der Mechanik").

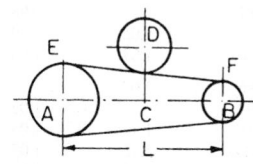

Fig. 8.2.91 Belt tightener.

Fig. 8.2.92 Symbols for cone pulley graphical method.

Let D_1 and D_2 be, respectively, the diameters of the smaller and larger pulleys of a pair, $n = D_2/D_1$, and l = distance between shaft centers, all in inches. Also let $m = 1.58114l - D_0$, where D_0 = diam of both pulleys for a speed ratio $n = 1$. Then $(D_1 + m)^2 + (nD_1 + m)^2 = 5l^2$. First settle on values of D_0, l, and n, and then substitute in the equation and solve for D_1. The diameter D_2 of the other pulley of the pair will then be nD_1. The values are correct to the fourth decimal place.

The speeds given by cone pulleys should increase in a **geometric ratio;** i.e., each speed should be multiplied by a constant a in order to obtain the next higher speed. Let n_1 and n_2 be, respectively, the lowest and highest speeds (r/min) desired and k the number of speed changes. Then $a = \sqrt[k-1]{n_2/n_1}$. In practice, a ranges from 1.25 up to 1.75 and even 2. The ideal value for a in machine-tool practice, according to Carl G. Barth, would be 1.189. In the example below, this would mean the use of 18 speeds instead of 8.

EXAMPLE. Let $n_1 = 16$, $n_2 = 400$, and $k = 8$, to be obtained with four pairs of pulleys and a back gear. From formula, $a = \sqrt[7]{25} = 1.584$, whence speeds will be

16, (16 × 1.584 =) 25.34, (25.34 × 1.584 =) 40.14, and similarly 63.57, 100.7, 159.5, 252.6, and 400. The first four speeds are with the back gear in; hence the back-gear ratio must be 100.7 ÷ 16 = 6.29.

Transmission of Power by Flat Belts The theory of flat-belt drives takes into account changes in belt tension caused by friction forces between belt and pulley, which, in turn, cause belt elongation or contraction, thus inducing relative movement between belt and pulley. The transmission of power is a complex affair. A lengthy mathematical presentation can be found in Firbank, "Mechanics of the Flat Belt Drive," ASME Paper 72-PTG-21. A simpler, more conventional analysis used for many years yields highly serviceable designs.

The turning force (tangential) on the rim of a pulley driven by a flat belt is equal to $T_1 - T_2$, where T_1 and T_2 are, respectively, the tensions in the driving (tight) side and following (slack) side of the belt. (For the relations of T_1 and T_2 at low peripheral speeds, see Sec. 3.) Log $(T_1/T_2) = 0.0076fa$ when the effect of centrifugal force is neglected and $T_1/T_2 = 10^{0.0076fa}$. Figure 8.2.93 gives values of this function. When the speeds are high, however, the relations of T_1 to T_2 are modified by centrifugal stresses in the belt, in which case log $(T_1/T_2) = 0.0076f$ $(1 - x)a$, where f = coefficient of friction between the belt and pulley surface, a = angle of wrap, and $x = 12wv^2/(gt)$ in which w =

Fig. 8.2.93 Values of $10^{0.0076fa}$.

weight of 1 in³ of belt material, lb; v = belt speed, ft/s; g = 32.2 ft/s²; and t = allowable working tension, lb/in². Values of x for leather belting (with $w = 0.035$ and $t = 300$) are as follows:

u	30	40	50	60	70	
x	0.039	0.070	0.118	0.157	0.214	

uv	80	90	100	110	120	130
x	0.279	0.352	0.435	0.526	0.626	0.735

Researches by Barth (*Trans. ASME*, 1909) seem to show that f is a function of the belt velocity, varying according to the formula $f = 0.54 - 140/(500 + V)$ for leather belts on iron pulleys, where V = belt velocity for ft/min. For practical design, however, the following values of f may be used: for leather belts on cast-iron pulleys, $f = 0.30$; on wooden pulleys, $f = 0.45$; on paper pulleys, $f = 0.55$. The treatment of belts with belt dressing, pulleys with cork inserts, and dampness are all factors which greatly modify these values, tending to make them higher.

The **arc of contact** on the smaller of two pulleys connected by an open belt, in degrees, is approximately equal to $180 - 60(D - d)/l$, where D and d are the larger and smaller pulley diameters and l the distance between their shaft centers, all in inches. This formula gives an error not exceeding 0.5 percent.

Selecting a Belt Selecting an appropriate belt involves calculating horsepower per inch of belt width as follows:

$$\text{hp/in} = (\text{demanded hp} \times S)/(K \times W)$$

where demanded hp = horsepower required by the job at hand; S = service factor; K = arc factor; W = proposed belt width (determined from pulley width). One enters a belt manufacturer's catalog with *hp/in*, *belt speed*, and *small pulley diameter*, then selects that belt which has a matching maximum hp/in rating. See Table 8.2.46*a*, *b*, *c*, and *d* for typical values of S, K, hp/in ratings, and minimum pulley diameters.

V-Belt Drives

V-belt drives are widely used in power transmission, despite the fact that they may range in efficiency from about 70 to 96 percent. Such drives consist essentially of endless belts of trapezoidal cross section which ride in V-shaped pulley grooves (see Fig. 8.2.95*a*). The belts are formed of cord and fabric, impregnated with rubber, the cord material being cotton, rayon, synthetic, or steel. V-belt drives are quiet, able to absorb shock and operate at low bearing pressures. A V belt should ride with the top surface approximately flush with the top of the pulley groove; clearance should be present between the belt base and the base of the groove so that the belt rides on the groove walls. The friction between belt and groove walls is greatly enhanced beyond normal values because sheave groove angles are made somewhat less than belt-section angles, causing the belt to wedge itself into the groove. See Table 8.2.55*a* for standard groove dimensions of sheaves.

The cross section and lengths of V belts have been standardized by ANSI in both inch and SI (metric) units, while ANSI and SAE have standardized the special category of automobile belts, again in both units. Standard designations are shown in Table 8.2.47, which also includes minimum sheave diameters. V belts are specified by combining a standard designation (from Table 8.2.47) and a belt length; inside length

for the inch system, and pitch (effective) length for metric system. Sheaves are specified by their pitch diameters, which are used in velocity ratio calculations in which case inside belt lengths must be converted to pitch lengths for computational purposes. Pitch lengths are calculated by adding a conversion factor to inside length (i.e., $L_p = L_s + \Delta$). See Table 8.2.48 for conversion factors. Table 8.2.49 lists standard inside inch lengths L_s and Table 8.2.50 lists standard metric pitch (effective) lengths L_p.

Table 8.2.48   Length Conversion Factors Δ

Belt section	Size interval	Conversion factor	Belt section	Size interval	Conversion factor
A	26–128	1.3	D	120–210	3.3
B	35–210	1.8	D	≥240	0.8
B	≥240	0.3	E	180–210	4.5
C	51–210	2.9	E	≥240	1.0
C	≥240	0.9			

SOURCE: Adapted from ANSI/RMA IP-20-1977 (R88) by permission.

Table 8.2.47   V-Belt Standard Designations—A Selection

Type	Inch standard Section	Inch standard Minimum sheave diameter, in	Metric standard Section	Metric standard Minimum sheave diameter, mm
Heavy-duty	A	3.0	13 C	80
	B	5.4	16 C	140
	C	9.0	22 C	214
	D	13.0	32 C	355
	E	21.04		
Automotive	0.25	2.25	6 A	57
	0.315	2.25	8 A	57
	0.380	2.40	10 A	61
	0.440	2.75	11 A	70
	0.500	3.00	13 A	76
	$^{11}/_{16}$	3.00	15 A	76
	$^3/_4$	3.00	17 A	76
	$^7/_8$	3.50	20 A	89
	1.0	4.00	23 A	102
Heavy-duty narrow	3 V	2.65		
	5 V	7.1		
	8 V	12.3		
Notched narrow	3 VX	2.2		
	5 VX	4.4		
Light-duty	2 L	0.8		
	3 L	1.5		
	4 L	2.5		
	5 L	3.5		
Synchronous belts	MXL			
	XL			
	L			
	H			
	XH			
	XXH			

NOTE: The use of smaller sheaves than minimum will tend to shorten belt life.
SOURCE: Compiled from ANSI/RMA IP-20, 21, 22, 23, 24, 25, 26; ANSI/SAE J636C.

Table 8.2.49   Standard Lengths L_s, in, and Length Correction Factors K_2: Conventional Heavy-Duty V Belts

L_s	Cross section A	B	C	D	E
26	0.78				
31	0.82				
35	0.85	0.80			
38	0.87	0.82			
42	0.89	0.84			
46	0.91	0.86			
51	0.93	0.88	0.80		
55	0.95	0.89			
60	0.97	0.91	0.83		
68	1.00	0.94	0.85		
75	1.02	0.96	0.87		
80	1.04				
81		0.98	0.89		
85	1.05	0.99	0.90		
90	1.07	1.00	0.91		
96	1.08		0.92		
97		1.02			
105	1.10	1.03	0.94		
112	1.12	1.05	0.95		
120	1.13	1.06	0.96	0.88	
128	1.15	1.08	0.98	0.89	
144		1.10	1.00	0.91	
158		1.12	1.02	0.93	
173		1.14	1.04	0.94	
180		1.15	1.05	0.95	0.92
195		1.17	1.06	0.96	0.93
210		1.18	1.07	0.98	0.95
240		1.22	1.10	1.00	0.97
270		1.24	1.13	1.02	0.99
300		1.27	1.15	1.04	1.01
330			1.17	1.06	1.03
360			1.18	1.07	1.04
390			1.20	1.09	1.06
420			1.21	1.10	1.07
480				1.13	1.09
540				1.15	1.11
600				1.17	1.13
660				1.18	1.15

SOURCE: ANSI/RMA IP-20-1977 (R88), reproduced by permission.

Table 8.2.50 Standard Pitch Lengths L_p (Metric Units) and Length Correction Factors K_2

13C		16C		22C		32C	
L_p	K_2	L_p	K_2	L_p	K_2	L_p	K_2
710	0.83	960	0.81	1,400	0.83	3,190	0.89
750	0.84	1,040	0.83	1,500	0.85	3,390	0.90
800	0.86	1,090	0.84	1,630	0.86	3,800	0.92
850	0.88	1,120	0.85	1,830	0.89	4,160	0.94
900	0.89	1,190	0.86	1,900	0.90	4,250	0.94
950	0.90	1,250	0.87	2,000	0.91	4,540	0.95
1,000	0.92	1,320	0.88	2,160	0.92	4,720	0.96
1,075	0.93	1,400	0.90	2,260	0.93	5,100	0.98
1,120	0.94	1,500	0.91	2,390	0.94	5,480	0.99
1,150	0.95	1,600	0.92	2,540	0.96	5,800	1.00
1,230	0.97	1,700	0.94	2,650	0.96	6,180	1.01
1,300	0.98	1,800	0.95	2,800	0.98	6,560	1.02
1,400	1.00	1,900	0.96	3,030	0.99	6,940	1.03
1,500	1.02	1,980	0.97	3,150	1.00	7,330	1.04
1,585	1.03	2,110	0.99	3,350	1.01	8,090	1.06
1,710	1.05	2,240	1.00	3,550	1.02	8,470	1.07
1,790	1.06	2,360	1.01	3,760	1.04	8,850	1.08
1,865	1.07	2,500	1.02	4,120	1.06	9,240	1.09
1,965	1.08	2,620	1.03	4,220	1.06	10,000	1.10
2,120	1.10	2,820	1.05	4,500	1.07	10,760	1.11
2,220	1.11	2,920	1.06	4,680	1.08	11,530	1.13
2,350	1.13	3,130	1.07	5,060	1.10	12,290	1.14
2,500	1.14	3,330	1.09	5,440	1.11		
2,600	1.15	3,530	1.10	5,770	1.13		
2,730	1.17	3,740	1.11	6,150	1.14		
2,910	1.18	4,090	1.13	6,540	1.15		
3,110	1.20	4,200	1.14	6,920	1.16		
3,310	1.21	4,480	1.15	7,300	1.17		
		4,650	1.16	7,680	1.18		
		5,040	1.18	8,060	1.19		
		5,300	1.19	8,440	1.20		
		5,760	1.21	8,820	1.21		
		6,140	1.23	9,200	1.22		
		6,520	1.24				
		6,910	1.25				
		7,290	1.26				
		7,670	1.27				

Source: ANSI/RMA IP-20-1988 revised, reproduced by permission.

For given large and small sheave diameters and center-to-center distance, the needed V-belt length can be computed from

$$L_p = 2C + 1.57(D + d) + \frac{(D - d)^2}{4C}$$

$$C = \frac{K + \sqrt{K^2 - 32(D - d)^2}}{16} \qquad K = 4L_p - 6.28(D + d)$$

where C = center-to-center distance; D = pitch diameter of large sheave; d = pitch diameter of small sheave; L_p = pitch (effective) length.

Arc of contact on the smaller sheave (degrees) is approximately

$$\theta = 180 - \frac{60(D - d)}{C}$$

Transmission of Power by V Belts Unfortunately there is no theory or mathematical analysis that is able to explain all experimental results reliably. Empirical formulations based on experimental results, however, do provide very serviceable design procedures, and together with data published in V-belt manufacturers' catalogs provide the engineer with the necessary V-belt selection tools.

For satisfactory performance under most conditions, ANSI provides the following empirical single V-belt power-rating formulation (inch and metric units) for 180° arc of contact and average belt length:

$$H_r = \left(C_1 - \frac{C_2}{d} - C_3(r - d)^2 - C_4 \log rd\right) rd + C_2 r\left(1 - \frac{1}{K_A}\right)$$

where H_r = rated horsepower in inch units (rated power kW for metric units); C_1, C_2, C_3, C_4 = constants from Table 8.2.52; r = r/min of high-speed shaft times 10^{-3}; K_A = speed ratio factor from Table 8.2.51; d = pitch diameter of small sheave, in (mm).

Selecting a Belt Selecting an appropriate belt involves calculating horsepower per belt as follows:

$$NH_r = (\text{demanded hp} \times K_s)/(K_1 K_2)$$

where H_r = hp/belt rating, either from ANSI formulation above, or from manufacturer's catalog (see Table 8.2.54); demanded hp = horsepower required by the job at hand; K_s = service factor accounting for driver and driven machine characteristics regarding such things as shock, torque level, and torque uniformity (see Table 8.2.53); K_1 = angle of contact correction factor (see Fig. 8.2.94a); K_2 = length correction factor (see Tables 8.2.49 and 8.2.50); N = number of belts.

See Fig. 8.2.94b for selection of V-belt cross section.

V band belts, effectively joined V belts, serve the function of multiple single V belts (see Fig. 8.2.95b).

Long center distances are not recommended for V belts because excess slack-side vibration shortens belt life. In general $D \leq C \leq 3(D + d)$. If longer center distances are needed, then link-type V belts can be used effectively.

Since belt-drive capacity is normally limited by slippage of the smaller sheave, V-belt drives can sometimes be used with a flat, larger pulley rather than with a grooved sheave, with little loss in capacity. For instance the flat surface of the flywheel in a large punch press can serve such a purpose. The practical range of application is when speed ratio is over 3 : 1, and center distance is equal to or slightly less than the diameter of the large pulley.

Table 8.2.51 Approximate Speed-Ratio Factor K_A for Use in Power-Rating Formulation

D/d range	K_A	D/d range	K_A
1.00–1.01	1.0000	1.15–1.20	1.0586
1.02–1.04	1.0112	1.21–1.27	1.0711
1.05–1.07	1.0226	1.28–1.39	1.0840
1.08–1.10	1.0344	1.40–1.64	1.0972
1.11–1.14	1.0463	Over 1.64	1.1106

Source: Adapted from ANSI/RMA IP-20-1977 (R88), by permission.

Table 8.2.52 Constants C_1, C_2, C_3, C_4 for Use in Power-Rating Formulation

Belt section	C_1	C_2	C_3	C_4
		Inch		
A	0.8542	1.342	2.436×10^{-4}	0.1703
B	1.506	3.520	4.193×10^{-4}	0.2931
C	2.786	9.788	7.460×10^{-4}	0.5214
D	5.922	34.72	1.522×10^{-3}	1.064
E	8.642	66.32	2.192×10^{-3}	1.532
		Metric		
13C	0.03316	1.088	1.161×10^{-8}	5.238×10^{-3}
16C	0.05185	2.273	1.759×10^{-8}	7.934×10^{-3}
22C	0.1002	7.040	3.326×10^{-8}	1.500×10^{-2}
32C	0.2205	26.62	7.037×10^{-8}	3.174×10^{-2}

Source: Compiled from ANSI/RMA IP-20-1977 (R88), by permission.

Table 8.2.53 Approximate Service Factor K_s for V-Belt Drives

	Load			
Power source torque	Uniform	Light shock	Medium shock	Heavy shock
Average or normal	1.0–1.2	1.1–1.3	1.2–1.4	1.3–1.5
Nonuniform or heavy	1.1–1.3	1.2–1.4	1.4–1.6	1.5–1.8

Source: Adapted from ANSI/RMA IP-20-1977 (R88), by permission.

Table 8.2.54   Horsepower Ratings of V Belts

Belt section A

Speed of faster shaft, r/min	Rated horsepower per belt for small sheave pitch diameter, in							Additional horsepower per belt for speed ratio				
	2.60	3.00	3.40	3.80	4.20	4.60	5.00	1.02–1.04	1.08–1.10	1.15–1.20	1.28–1.39	1.65–over
200	0.20	0.27	0.33	0.40	0.46	0.52	0.59	0.00	0.01	0.01	0.02	0.03
800	0.59	0.82	1.04	1.27	1.49	1.70	1.92	0.01	0.04	0.06	0.08	0.11
1,400	0.87	1.25	1.61	1.97	2.32	2.67	3.01	0.02	0.06	0.10	0.15	0.19
2,000	1.09	1.59	2.08	2.56	3.02	3.47	3.91	0.03	0.09	0.15	0.21	0.27
2,600	1.25	1.87	2.47	3.04	3.59	4.12	4.61	0.04	0.12	0.19	0.27	0.35
3,200	1.37	2.08	2.76	3.41	4.01	4.57	5.09	0.05	0.14	0.24	0.33	0.43
3,800	1.43	2.23	2.97	3.65	4.27	4.83	5.32	0.06	0.17	0.28	0.40	0.51
4,400	1.44	2.29	3.07	3.76	4.36	4.86*	5.26*	0.07	0.20	0.33	0.46	0.59
5,000	1.39	2.28	3.05	3.71	4.24*	4.48*	4.64*	0.07	0.22	0.37	0.52	0.65
5,600	1.29	2.17	2.92	3.50*				0.08	0.25	0.42	0.58	0.75
6,200	1.11	1.98	2.65*					0.09	0.28	0.46	0.64	0.83
6,800	0.87	1.68*	2.24*					0.10	0.30	0.51	0.71	0.91
7,400	0.56	1.28*						0.11	0.33	0.55	0.77	0.99
7,800	0.31*							0.12	0.35	0.58	0.81	1.04

Belt section B

Speed of faster shaft, r/min	4.60	5.20	5.80	6.40	7.00	7.60	8.00	1.02–1.04	1.08–1.10	1.15–1.20	1.28–1.39	1.65–over
200	0.69	0.86	1.02	1.18	1.34	1.50	1.61	0.01	0.02	0.04	0.05	0.07
600	1.68	2.12	2.56	2.99	3.41	3.83	4.11	0.02	0.07	0.12	0.16	0.21
1,000	2.47	3.16	3.84	4.50	5.14	5.78	6.20	0.04	0.12	0.19	0.27	0.35
1,400	3.13	4.03	4.91	5.76	6.59	7.39	7.91	0.05	0.16	0.27	0.38	0.49
1,800	3.67	4.75	5.79	6.79	7.74	8.64	9.21	0.07	0.21	0.35	0.49	0.63
2,200	4.08	5.31	6.47	7.55	8.56	9.48	10.05	0.09	0.26	0.43	0.60	0.77
2,600	4.36	5.69	6.91	8.01	9.01	9.87	10.36	0.10	0.30	0.51	0.71	0.91
3,000	4.51	5.89	7.11	8.17	9.04	9.73*		0.12	0.35	0.58	0.82	1.05
3,400	4.51	5.88	7.03	7.95*				0.13	0.40	0.66	0.93	1.19
3,800	4.34	5.64	6.65*					0.15	0.44	0.74	1.04	1.33
4,200	4.01	5.17*						0.16	0.49	0.82	1.15	1.47
4,600	3.48							0.18	0.54	0.90	1.25	1.61
5,000	2.76*							0.19	0.59	0.97	1.36	1.75

Belt section C

Speed of faster shaft, r/min	7.00	8.00	9.00	10.00	11.00	12.00	13.00	1.02–1.04	1.08–1.10	1.15–1.20	1.28–1.39	1.65–over
100	1.03	1.29	1.55	1.81	2.06	2.31	2.56	0.01	0.03	0.05	0.08	0.10
400	3.22	4.13	5.04	5.93	6.80	7.67	8.53	0.04	0.13	0.22	0.30	0.39
800	5.46	7.11	8.73	10.31	11.84	13.34	14.79	0.09	0.26	0.43	0.61	0.78
1,000	6.37	8.35	10.26	12.11	13.89	15.60	17.24	0.11	0.33	0.54	0.76	0.97
1,400	7.83	10.32	12.68	14.89	16.94	18.83	20.55	0.15	0.46	0.76	1.06	1.36
1,800	8.76	11.58	14.13	16.40	18.37	20.01	21.31*	0.20	0.59	0.98	1.37	1.75
2,000	9.01	11.90	14.44	16.61	18.37	19.70*		0.22	0.65	1.08	1.52	1.95
2,400	9.04	11.87	14.14					0.26	0.78	1.30	1.82	2.34
2,800	8.38	10.85*						0.30	0.91	1.52	2.12	2.73
3,000	7.76							0.33	0.98	1.63	2.28	2.92
3,200	6.44*							0.36	1.07	1.79	2.50	3.22

* For footnote see end of table on next page.

In general the V-belting of skew shafts is discouraged because of the decrease in life of the belts, but where design demands such arrangements, special deep-groove sheaves are used. In such cases center distances should comply with the following:

For 90° turn $\qquad C_{min} = 5.5(D + W)$
For 45° turn $\qquad C_{min} = 4.0(D + W)$
For 30° turn $\qquad C_{min} = 3.0(D + W)$

where W = width of group of individual belts. Selected values of W are shown in Table 8.2.55d.

Fig. 8.2.94a   Angle-of-contact correction factor, where A = grooved sheave to grooved pulley distance (V to V) and B = grooved sheave to flat-face pulley distance (V to flat).

Fig. 8.2.94b   V-belt section for required horsepower ratings. Letters A, B, C, D, E refer to belt cross section. (See Table 8.2.53 for service factor.)

Figure 8.2.96 shows a 90° turn arrangement, from which it can be seen that the horizontal shaft should lie some distance Z higher than the center of the vertical-shaft sheave. Table 8.2.55c lists the values of Z for various center distances in a 90° turn arrangement.

Cogged V belts have cogs molded integrally on the underside of the belt (Fig. 8.2.97a). Sheaves can be up to 25 percent smaller in diameter with cogged belts because of the greater flexibility inherent in the cogged construction. An extension of the cogged belt mating with a sheave or pulley notched at the same pitch as the cogs leads to a drive particularly useful for timing purposes.

Table 8.2.54   Horsepower Ratings of V Belts   (Continued)

Belt section	Speed of faster shaft, r/min	Rated horsepower per belt for small sheave pitch diameter, in							Additional horsepower per belt for speed ratio				
		12.00	14.00	16.00	18.00	20.00	22.00	24.00	1.02–1.04	1.08–1.10	1.15–1.20	1.28–1.39	1.65–over
D	50	1.96	2.52	3.08	3.64	4.18	4.73	5.27	0.02	0.06	0.10	0.13	0.17
	200	6.28	8.27	10.24	12.17	14.08	15.97	17.83	0.08	0.23	0.38	0.54	0.69
	400	10.89	14.55	18.12	21.61	25.02	28.35	31.58	0.15	0.46	0.77	1.08	1.38
	600	14.67	19.75	24.64	29.33	33.82	38.10	42.15	0.23	0.69	1.15	1.61	2.07
	800	17.70	23.91	29.75	35.21	40.24	44.83	48.94	0.31	0.92	1.54	2.15	2.77
	1,000	19.93	26.94	33.30	38.96	43.86	47.93	51.12	0.38	1.15	1.92	2.69	3.46
	1,200	21.32	28.71	35.05	40.24	44.18*			0.46	1.39	2.31	3.23	4.15
	1,400	21.76	29.05	34.76*					0.54	1.62	2.69	3.77	4.84
	1,600	21.16	27.81*						0.62	1.85	3.08	4.30	5.53
	1,800	19.41							0.69	2.08	3.46	4.84	6.22
	1,950	17.28*							0.75	2.25	3.75	5.25	6.74
		18.00	21.00	24.00	27.00	30.00	33.00	36.00					
E	50	4.52	5.72	6.91	8.08	9.23	10.38	11.52	0.04	0.11	0.18	0.26	0.33
	100	8.21	10.46	12.68	14.87	17.04	19.19	21.31	0.07	0.22	0.37	0.51	0.66
	200	14.68	18.86	22.97	27.00	30.96	34.86	38.68	0.15	0.44	0.73	1.03	1.32
	300	20.37	26.29	32.05	37.67	43.13	48.43	53.58	0.22	0.66	1.10	1.54	1.98
	400	25.42	32.87	40.05	46.95	53.55	59.84	65.82	0.29	0.88	1.47	2.06	2.64
	500	29.86	38.62	46.92	54.74	62.05	68.81	75.00	0.37	1.10	1.84	2.57	3.30
	600	33.68	43.47	52.55	60.87	68.36	74.97	80.63	0.44	1.32	2.20	3.08	3.96
	700	36.84	47.36	56.83	65.15	72.22	77.93*		0.51	1.54	2.57	3.60	4.62
	800	39.29	50.20	59.61	67.36	73.30*			0.59	1.76	2.94	4.11	5.28
	900	40.97	51.89	60.73					0.66	1.98	3.30	4.63	5.94
	1,000	41.84	52.32	60.04 *					0.73	2.21	3.67	5.14	6.60
	1,100	41.81	51.40*						0.81	2.43	4.04	5.65	7.26
	1,200	40.83							0.88	2.65	4.41	6.17	7.93
	1,300	38.84*							0.95	2.87	4.77	6.68	8.59

* Rim speed above 6,000 ft/min. Special sheaves may be necessary.
SOURCE: Compiled from ANSI/RMA IP-20-1998 revised, by permission.

Ribbed V belts are really flat belts molded integrally with longitudinal ribbing on the underside (Fig. 8.2.97b). Traction is provided principally by friction between the ribs and sheave grooves rather than by wedging action between the two, as in conventional V-belt operation. The flat

(a)   **(b)**

Fig. 8.2.95   V- and V-band belt cross section.

upper portion transmits the tensile belt loads. Ribbed belts serve well when substituted for multiple V-belt drives and for all practical purposes eliminate the necessity for belt-matching in multiple V-belt drives.

Fig. 8.2.96   Quarter-turn drive for V belts.

(a)   **(b)**

Fig. 8.2.97   Special V belts. (a) Cogged V belt; (b) ribbed V belt.

Adjustable Motor Bases

To maintain proper belt tensions on short center distances, an adjustable motor base is often used. Figure 8.2.98 shows an embodiment of such a base made by the Automatic Motor Base Co., in which adjustment for proper belt tension is made by turning a screw which opens or closes the center distance between pulleys, as required. The carriage portion of the base is spring loaded so that after the initial adjustment for belt tension

Fig. 8.2.98   Adjustable motor base.

Table 8.2.55a Standard Sheave Groove Dimensions, in

Face width of standard and deep groove sheaves

Face width = $S_g (N_g - 1) + 2 S_e$

where:

N_g = number of grooves

Cross section	Outside diameter range	Groove angle $\alpha \pm 0.33°$	b_g	h_g min	R_B min	$d_B \pm 0.0005$	$S_g \pm 0.025$	S_e	Pitch diameter range	Minimum recommended pitch diameter	$2a$
A	Up through 5.65 Over 5.65	34 38	0.494 ± 0.005 0.504	0.460	0.148 0.149	0.4375 (7/16)	0.625	0.375 +0.090 −0.062	Up through 5.40 Over 5.40	3.0	0.250
B	Up through 7.35 Over 7.35	34 38	0.637 ± 0.006 0.650	0.550	0.189 0.190	0.5625 (9/16)	0.750	0.500 +0.120 −0.065	Up through 7.00 Over 7.00	5.4	0.350
A/B	Up through 7.35 Over 7.35	34 38	0.612 ± 0.006 0.625	0.612	0.230 0.226	0.5625 (9/16)	0.750	0.500 +0.120 +0.065		A = 3.0 B = 5.4	A = 0.620* B = 0.280
C	Up through 8.39 Over 8.39 to and incl. 12.40 Over 12.40	34 36 38	0.879 0.877 ± 0.007 0.895	0.750	0.274 0.276 0.277	0.7812 (25/32)	1.000	0.688 +0.160 −0.70	Up through 7.99 Over 7.99 to and incl. 12.00 Over 12.00	9.0	0.400
D	Up through 13.59 Over 13.59 to and incl. 17.60 Over 17.60	34 36 38	1.256 1.271 ± 0.008 1.283	1.020	0.410 0.410 0.411	1.1250 (1 1/8)	1.438	+0.220 −0.80	Up through 12.99 Over 12.99 to and incl. 17.00 Over 17.00	13.0	0.600
E	Up through 24.80 Over 24.80	36 38	1.527 ± 0.010 1.542	1.270	0.476 0.477	1.3438 (1 11/32)	1.750	1.125 +0.280 −0.090	Up through 24.00 Over 24.00	21.0	0.800

The columns above are grouped under two headings: **Standard groove dimensions** (Groove angle through S_e) and **Drive design factors** (Pitch diameter range, Minimum recommended pitch diameter, $2a$).

* The a values shown for the A/B combination sheaves are the geometrically derived values. These values may be different than those shown in manufacturers' catalogs.

SOURCE: "Dayco Engineering Guide for V-Belt Drives," Dayco Corp., Dayton, OH, 1981, reprinted by permission.

Table 8.2.55*b*   Classical Deep Groove Sheave Dimensions, in

Cross section	Outside diameter range	Groove angle $\alpha \pm 0.33°$	b_g	h_g min	$2a$	$S_g \pm 0.025$	S_e
			Deep groove dimensions				
A	Up through 5.96	34	0.539	0.615	0.560	0.750	$0.438 \begin{array}{l}+0.090\\-0.062\end{array}$
	Over 5.96	38	0.611 ± 0.005				
B	Up through 7.71	34	0.747	0.730	0.710	0.875	$0.562 \begin{array}{l}+0.120\\-0.065\end{array}$
	Over 7.71	38	0.774 ± 0.006				
C	Up through 9.00	34	1.066	1.055	1.010	1.250	$0.812 \begin{array}{l}+0.160\\-0.070\end{array}$
	Over 9.00 to and incl. 13.01	36	$1.085 + 0.007$				
	Over 13.01	38	1.105				
D	Up through 14.42	34	1.513	1.435	1.430	1.750	$1.062 \begin{array}{l}+0.220\\-0.080\end{array}$
	Over 14.42 to and incl. 18.43	36	1.541 ± 0.008				
	Over 18.43	38	1.569				
E	Up through 25.69	36	1.816	1.715	1.690	2.062	$1.312 \begin{array}{l}+0.280\\-0.090\end{array}$
	Over 25.69	38	$1.849 + 0.010$				

SOURCE: "Dayco Engineering Guide for V-Belt Drives," Dayco Corp., Dayton, OH, 1981, reprinted by permission.

Table 8.2.55*c*   *Z* Dimensions for Quarter-Turn Drives, in

Center distance	3V, 5V, 8V Z dimension	A B, C, D Z dimension
20		0.2
30		0.2
40		0.4
50		0.4
60	0.2	0.5
80	0.3	0.5
100	0.4	1.0
120	0.6	1.5
140	0.9	2.0
160	1.2	2.5
180	1.5	3.5
200	1.8	4.0
220	2.2	5.0
240	2.6	6.0

SOURCE: "Dayco Engineering Guide for V-Belt Drives," Dayco Corp., Dayton, OH, 1981, reprinted by permission.

Table 8.2.55*d*   Width *W* of Set of Belts Using Deep-Groove Sheaves, in

No. of belts	V-belt cross section						
	3V	5V	8V	A	B	C	D
1	0.4	0.6	1.0	0.5	0.7	0.9	1.3
2	0.9	1.4	2.3	1.3	1.6	2.2	3.1
3	1.4	2.2	3.6	2.0	2.5	3.4	4.8
4	1.9	3.0	4.0	2.8	3.3	4.7	6.6
5	2.4	3.8	6.2	3.5	4.2	5.9	8.3
6	2.9	4.7	7.6	4.3	5.1	7.2	10.1
7	3.4	5.5	8.9	5.0	6.0	8.4	11.8
8	3.9	6.3	10.2	5.8	6.8	9.7	13.6
9	4.4	7.1	11.5	6.5	7.7	10.9	15.3
10	4.9	7.9	12.8	7.3	8.6	12.2	17.1

SOURCE: "Dayco Engineering Guide for V-Belt Drives," Dayco Corp., Dayton, OH, 1981, reprinted by permission.

has been made by the screw, the spring will compensate for a normal amount of stretch in the belts. When there is more stretch than can be accommodated by the spring, the screw is turned to provide the necessary belt tensions. The carriage can be moved while the unit is in operation, and the motor base is provided for vertical as well as horizontal mounting.

CHAIN DRIVES

Roller-Chain Drives

The advantages of finished steel roller chains are high efficiency (around 98 to 99 percent), no slippage, no initial tension required, and chains may travel in either direction. The basic construction of roller chains is shown in Fig. 8.2.99 and Table 8.2.56.

The shorter the pitch, the higher the permissible operating speed of roller chains. Horsepower capacity in excess of that provided by a single chain may be had by the use of multiple chains, which are essentially parallel single chains assembled on pins common to all strands. Because of its lightness in relation to tensile strength, the effect of centrifugal pull does not need to be considered; even at the unusual speed of 6,000 ft/min, this pull is only 3 percent of the ultimate tensile strength.

Fig. 8.2.99   Roller chain construction.

Sprocket wheels with fewer than 16 teeth may be used for relatively slow speeds, but 18 to 24 teeth are desirable for high-speed service. Sprockets with fewer than 25 teeth, running at speeds above 500 or 600 r/min, should be heat-treated to give a tough wear-resistant surface testing between 35 and 45 on the Rockwell C hardness scale.

If the speed ratio requires the larger sprocket to have as many as 128 teeth, or more than eight times the number on the smaller sprocket, it is usually better, with few exceptions, to make the desired reduction in two or more steps. The ANSI tooth form ASME B29.1 M-1993 allows roller chain to adjust itself to a larger pitch circle as the pitch of the chain elongates owing to natural wear in the pin-bushing joints. The greater the number of teeth, the sooner the chain will ride out too near the ends of the teeth.

Table 8.2.56　Roller-Chain Data and Dimensions, in*

ANSI chain no.	ISO chain no.	Roller			Pin diam	Roller link plate		Dimension			Tensile strength per strand, lb	Recommended max speed r/min		
		Pitch	Width	Diam		Thickness	Height H	A	B	C		12 teeth	18 teeth	24 teeth
25	04C-1	¼	⅛	0.130	0.091	0.030	0.230	0.150	0.190	0.252	925	5,000	7,000	7,000
35	06C-1	⅜	3⁄16	0.200	0.141	0.050	0.344	0.224	0.290	0.399	2,100	2,380	3,780	4,200
41	085	½	¼	0.306	0.141	0.050	0.383	0.256	0.315	—	2,000	1,750	2,725	2,850
40	08A-1	½	5⁄16	0.312	0.156	0.060	0.452	0.313	0.358	0.566	3,700	1,800	2,830	3,000
50	10A-1	⅝	⅜	0.400	0.200	0.080	0.594	0.384	0.462	0.713	6,100	1,300	2,030	2,200
60	12A-1	¾	½	0.469	0.234	0.094	0.679	0.493	0.567	0.897	8,500	1,025	1,615	1,700
80	16A-1	1	⅝	0.625	0.312	0.125	0.903	0.643	0.762	1.153	14,500	650	1,015	1,100
100	20A-1	1¼	¾	0.750	0.375	0.156	1.128	0.780	0.910	1.408	24,000	450	730	850
120	24A-1	1½	1	0.875	0.437	0.187	1.354	0.977	1.123	1.789	34,000	350	565	650
140	28A-1	1¾	1	1.000	0.500	0.218	1.647	1.054	1.219	1.924	46,000	260	415	500
160	32A-1	2	1¼	1.125	0.562	0.250	1.900	1.250	1.433	2.305	58,000	225	360	420
180		2¼	113⁄32	1.406	0.687	0.281	2.140	1.421	1.770	2.592	76,000	180	290	330
200	40A-1	2½	1½	1.562	0.781	0.312	2.275	1.533	1.850	2.817	95,000	170	260	300
240	48A-1	3	1⅞	1.875	0.937	0.375	2.850	1.722	2.200	3.458	135,000	120	190	210

* For conversion to metric units (mm) multiply table values by 25.4.

Table 8.2.57　Selected Values of Horsepower Ratings of Roller Chains

ANSI no. and pitch, in	Number of teeth in small socket	Small sprocket, r/min										
		50	500	1,200	1,800	2,500	3,000	4,000	5,000	6,000	8,000	10,000
25 ¼	11	0.03	0.23	0.50	0.73	0.98	1.15	1.38	0.99	0.75	0.49	0.35
	15	0.04	0.32	0.70	1.01	1.36	1.61	2.08	1.57	1.20	0.78	0.56
	20	0.06	0.44	0.96	1.38	1.86	2.19	2.84	2.42	1.84	1.201	0.86
	25	0.07	0.56	1.22	1.76	2.37	2.79	3.61	3.38	2.57	1.67	1.20
	30	0.08	0.68	1.49	2.15	2.88	3.40	4.40	4.45	3.38	2.20	1.57
	40	0.12	.092	2.03	2.93	3.93	4.64	6.00	6.85	5.21	3.38	2.42
35 ⅜	11	0.10	0.77	1.70	2.45	3.30	2.94	1.91	1.37	1.04	0.67	0.48
	15	0.14	1.08	2.38	3.43	4.61	4.68	3.04	2.17	1.65	1.07	0.77
	20	0.19	1.48	3.25	4.68	6.29	7.20	4.68	3.35	2.55	1.65	1.18
	25	0.24	1.88	4.13	5.95	8.00	9.43	6.54	4.68	3.56	2.31	1.65
	30	0.29	2.29	5.03	7.25	9.74	11.5	8.59	6.15	4.68	3.04	2.17
	40	0.39	3.12	6.87	9.89	13.3	15.7	13.2	9.47	7.20	4.68	—
41 ½	11	0.13	1.01	1.71	0.93	(0.58)	0.43	0.28	0.20	0.15	0.10	
	15	0.18	1.41	2.73	1.49	(0.76)	0.69	0.45	0.32	0.24	0.16	
	20	0.24	1.92	4.20	2.29	(1.41)	1.06	0.69	0.49	0.38		
	25	0.31	2.45	5.38	3.20	(1.97)	1.49	0.96	0.69	0.53		
	30	0.38	2.98	6.55	4.20	(2.58)	1.95	1.27	0.91	0.69		
	40	0.51	4.07	8.94	6.47	(3.97)	3.01	1.95	1.40			
40 ½	11	0.23	1.83	4.03	4.66	(3.56)	2.17	1.41	1.01	0.77	0.50	
	15	0.32	2.56	5.64	7.43	(4.56)	3.45	2.24	1.60	1.22	0.79	
	20	0.44	3.50	7.69	11.1	(7.03)	5.31	3.45	2.47	1.88		
	25	0.56	4.45	9.78	14.1	(9.83)	7.43	4.82	3.45	2.63		
	30	0.68	5.42	11.9	17.2	(12.9)	9.76	6.34	4.54	3.45		
	40	0.93	7.39	16.3	23.4	(19.9)	15.0	9.76	6.99			
50 ⅝	11	0.45	3.57	7.85	5.58	(3.43)	2.59	1.68	1.41	1.20	0.92	
	15	0.63	4.99	11.0	8.88	(5.46)	4.13	2.68	2.25	1.92		
	20	0.86	6.80	15.0	13.7	(8.40)	6.35	4.13	3.46	2.95		
	25	1.09	8.66	19.0	19.1	(11.7)	8.88	5.77	4.83			
	30	1.33	10.5	23.2	25.1	(15.4)	11.7	7.58				
	40	1.81	14.4	31.6	38.7	(23.7)	18.0					

Table 8.2.57   Selected Values of Horsepower Ratings of Roller Chains   (*Continued*)

ANSI no. and pitch, in	Number of teeth in small sprocket	Small sprocket, r/min										
		10	50	100	200	500	700	1,000	1,400	2,000	2,700	4,000
60 ¾	11	0.18	0.77	1.44	2.69	6.13	8.30	11.4	9.41	5.51	(3.75)	1.95
	15	0.25	1.08	2.01	3.76	8.57	11.6	16.0	15.0	8.77	(6.18)	3.10
	20	0.35	1.47	2.75	5.13	11.7	15.8	21.8	23.1	13.5	(9.20)	
	25	0.44	1.87	3.50	6.52	14.9	20.1	27.8	32.2	18.9	(12.9)	
	30	0.54	2.28	4.26	7.94	18.1	24.5	33.8	42.4	24.8	(16.9)	
	40	0.73	3.11	5.81	10.8	24.7	33.5	46.1	62.5	38.2		
80 1	11	0.42	1.80	3.36	6.28	14.3	19.4	19.6	12.8	6.93	4.42	
	15	0.59	2.52	4.70	8.77	20.0	27.1	31.2	18.9	11.0	7.04	
	20	0.81	3.44	6.41	12.0	27.3	37.0	48.1	29.0	17.0		
	25	1.03	4.37	8.16	15.2	34.7	47.0	64.8	40.6	23.8		
	30	1.25	5.33	9.94	18.5	42.3	57.3	78.9	53.3	31.2		
	40	1.71	7.27	13.6	25.3	57.7	78.1	108	82.1	48.1		
100 1¼	11	0.81	3.45	6.44	12.0	27.4	37.1	23.4	14.2	8.29		
	15	1.13	4.83	9.01	16.8	38.3	51.9	37.3	22.5	13.2		
	20	1.55	6.58	12.3	22.9	52.3	70.8	57.5	34.7	20.3		
	25	1.97	8.38	15.5	29.2	66.6	90.1	80.3	48.5	28.4		
	30	2.40	10.2	19.0	35.5	81.0	110	106	63.7	10.0		
	40	3.27	13.9	26.0	48.5	111	150	163	98.1			
120 1½	11	1.37	5.83	10.9	20.3	46.3	46.3	27.1	16.4	9.59		
	15	1.91	8.15	15.2	28.4	64.7	73.8	43.2	26.1			
	20	2.61	11.1	20.7	38.7	88.3	114	66.5	40.1			
	25	3.32	14.1	26.4	49.3	112	152	92.9	56.1			
	30	4.05	17.2	32.1	60.0	137	185	122	73.8			
	40	5.52	23.5	43.9	81.8	187	253	188	59.5			
140 1¾	11	2.12	9.02	16.8	31.4	71.6	52.4	30.7	18.5			
	15	2.96	12.6	23.5	43.9	100	83.4	48.9	29.5			
	20	4.04	17.2	32.1	59.9	137	128	75.2	45.4			
	25	5.14	21.9	40.8	76.2	174	180	105	63.5			
	30	6.26	26.7	49.7	92.8	212	236	138				
	40	8.54	36.4	67.9	127	289	363	213				
160 2	11	3.07	13.1	24.4	45.6	96.6	58.3	34.1				
	15	4.30	18.3	34.1	63.7	145	92.8	54.4				
	20	5.86	25.0	46.4	86.9	198	143	83.7				
	25	7.40	31.8	59.3	111	252	200	117				
	30	9.08	38.7	72.2	125	307	263	154				
	40	12.4	52.8	98.5	184	419	404					
180 2¼	11	4.24	18.1	33.7	62.9	106	64.1	37.5				
	15	5.93	25.3	47.1	88.0	169	102	59.7				
	20	8.10	34.5	64.3	120	260	157	92.0				
	25	10.3	43.9	81.8	153	348	220					
	30	12.5	53.4	99.6	186	424	289					
	40	17.1	72.9	136	254	579	398					
200 2½	11	5.64	24.0	44.8	83.5	115						
	15	7.88	33.5	62.6	117	184						
	20	10.7	45.8	85.4	159	283						
	25	13.7	58.2	109	203	396						
240 3	11	9.08	38.6	72.1	135							
	15	12.7	54.0	101	188							
	20	17.3	73.7	138	257							
	25	22.0	93.8	175	327							

NOTE: The sections separated by heavy lines denote the method of lubrication as follows: type A (left section), manual; type B (middle section), bath or disk; type C (right section), oil stream.
SOURCE: Supplementary action of ANSI B29.1-1975 (R93), adapted by permission.

Idler sprockets may be used on either side of the standard roller chain, to take up slack, to guide the chain around obstructions, to change the direction of rotation of a driven shaft, or to provide more wrap on another sprocket. Idlers should not run faster than the speeds recommended as maximum for other sprockets with the same number of teeth. It is desirable that idlers have at least two teeth in mesh with the chain, and it is advisable, though not necessary, to have an idler contact the idle span of chain.

Horsepower ratings are based upon the number of teeth and the rotative speed of the smaller sprocket, either driver or follower. The pin-bushing bearing area, as it affects allowable working load, is the important factor for medium and higher speeds. For extremely slow speeds, the chain selection may be based upon the ultimate tensile strength of the chain. For chain speeds of 25 ft/min and less, the chain pull should be not more than ⅕ of the ultimate tensile strength; for 50 ft/min, ⅙; for

100 ft/min, $\frac{1}{7}$; for 150 ft/min, $\frac{1}{8}$; for 200 ft/min, $\frac{1}{9}$; and for 250 ft/min, $\frac{1}{10}$ of the ultimate tensile strength.

Ratings for **multiple-strand chains** are proportional to the number of strands. The recommended numbers of strands for multiple chains are 2, 3, 4, 6, 8, 10, 12, 16, 20, and 24, with the maximum overall width in any case limited to 24 in.

The **horsepower ratings** in Table 8.2.57 are modified by the **service factors** in Table 8.2.58. Thus for a drive having a nominal rating of 3 hp, subject to heavy shock, abnormal conditions, 24-h/day operation, the chain rating obtained from Table 8.2.57 should be at least $3 \times 1.7 = 5.1$ hp.

Table 8.2.58   Service Factors for Roller Chains

		Load	
Power source	Smooth	Moderate shock	Heavy shock
Internal combustion engine with hydraulic drive	1.0	1.2	1.4
Electric motor or turbine	1.0	1.3	1.5
Internal combustion engine with mechanical drive	1.2	1.4	1.7

SOURCE: ANSI B29.1-1975, adapted by permission.

Chain-Length Calculations   Referring to Fig. 8.2.100, L = length of chain, in; P = pitch of chain, in; R and r = pitch radii of large and small sprockets, respectively, in; D = center distance, in; A = tangent length, in; a = angle between tangent and centerline; N and n = number

Fig. 8.2.100   Symbols for chain length calculations.

of teeth on larger and smaller sprockets, respectively; $180 + 2a$ and $180 - 2a$ are angles of contact on larger and smaller sprockets, respectively, deg.

$$a = \sin^{-1}[(R - r)/D] \qquad A = D \cos a$$
$$L = NP(180 + 2a)/360 + nP(180 - 2a)/360 + 2D \cos a$$

If L_p = length of chain in pitches and D_p = center distance in pitches,

$$L_p = (N + n)/2 + a(N - n)/180 + 2D_p \cos a$$

Avoiding the use of trigonometric tables,

$$L_p = 2C + (N + n)/2 + K(N - n)^2/C$$

where C is the center distance in pitches and K is a variable depending upon the value of $(N - n)/C$. Values of K are as follows:

$(N - n)/C$	0.1	1.0	2.0	3.0
K	0.02533	0.02538	0.02555	0.02584

$(N - n)/C$	4.0	5.0	6.0
K	0.02631	0.02704	0.02828

Formulas for chain length on multisprocket drives are cumbersome except when all sprockets are the same size and on the same side of the chain. For this condition, the chain length in pitches is equal to the sum of the consecutive center distances in pitches plus the number of teeth on one sprocket.

Actual chain lengths should be in even numbers of pitches. When necessary, an odd number of pitches may be secured by the use of an offset link, but such links should be avoided if possible. An offset link

is one pitch; half roller link at one end and half pin link at the other end. If center distances are to be nonadjustable, they should be selected to give an initial snug fit for an even number of pitches of chain. For the average application, a center distance equal to 40 ± 10 pitches of chain represents good practice.

There should be at least 120° of wrap in the arc of contact on a power sprocket. For ratios of 3 : 1 or less, the wrap will be 120° or more for any center distance or number of teeth. To secure a wrap of 120° or more, for ratios greater than 3 : 1, the center distance must be not less than the difference between the pitch diameters of the two sprockets.

Sprocket Diameters N = number of teeth; P = pitch of chain, in; D = diameter of roller, in. The pitch of a standard roller chain is measured from the center of a pin to the center of an adjacent pin.

$$\text{Pitch diam} = P/\sin \frac{180}{N}$$
$$\text{Bottom diam} = \text{pitch diam} - D$$
$$\text{Outside diam} = P\left(0.6 + \cot \frac{180}{N}\right)$$
$$\text{Caliper diam} = \left(\text{pitch diam} \times \cos \frac{90}{N}\right) - D$$

The exact bottom diameter cannot be measured for an odd number of teeth, but it can be checked by measuring the distance (caliper diameter) between bottoms of the two tooth spaces nearest opposite to each other. Bottom and caliper diameters must not be oversize—all tolerances must be negative. ANSI negative tolerance = $0.003 + 0.001P\sqrt{N}$ in.

Design of Sprocket Teeth for Roller Chains   The **section profile** for the teeth of roller chain sprockets, recommended by ANSI, has the proportions shown in Fig. 8.2.101. Let P = chain pitch; W = chain width (length of roller); n = number of strands of multiple chain; M = overall width of tooth profile section; H = nominal thickness of the link plate, all in inches. Referring to Fig. 8.2.101, $T = 0.93W - 0.006$, for

Fig. 8.2.101   Sprocket tooth sections.

single-strand chain; $= 0.90W - 0.006$, for double- and triple-strand chains; $= 0.88W - 0.006$, for quadruple- or quintuple-strand chains; and $= 0.86W - 0.006$, for sextuple-strand chain and over. $C = 0.5P$. $E = \frac{1}{8}P$. $R(\min) = 1.063P$. $Q = 0.5P$. $A = W + 4.15H + 0.003$. $M = A(n - 1) + T$.

Inverted-tooth (silent) chain drives have a typical tooth form shown in Fig. 8.2.102. Such chains should be operated in an oil-retaining casing with provisions for lubrication. The use of offset links and chains with an uneven number of pitches should be avoided.

Horsepower ratings per inch of silent chain width, given in ANSI B29.2-1957 (R1971), for various chain pitches and speeds, are shown in Tables 8.2.59 and 8.2.60. These ratings are based on ideal drive conditions with relatively little shock or load variation, an average life of 20,000 h being assumed. In utilizing the horsepower ratings of the tables, the nominal horsepower of the drive should be multiplied by a

Fig. 8.2.102   Inverted tooth (silent-chain) drive.

Table 8.2.59 Horsepower Rating per Inch of Chain Width, Silent-Chain Drive (Small Pitch)

Pitch, in	No. of teeth in small sprocket	Small sprocket, r/min											
		500	600	700	800	900	1,200	1,800	2,000	3,500	5,000	7,000	9,000
$\frac{3}{16}$	21	0.41	0.48	0.55	0.62	0.68	0.87	1.22	1.33	2.03	2.58	3.12	3.35
	25	0.49	0.58	0.66	0.74	0.82	1.05	1.47	1.60	2.45	3.13	3.80	4.10
	29	0.57	0.67	0.76	0.86	0.95	1.21	1.70	1.85	2.83	3.61	4.40	4.72
	33	0.64	0.75	0.86	0.97	1.07	1.37	1.90	2.08	3.17	4.02	4.85	—
	37	0.71	0.84	0.96	1.08	1.19	1.52	2.11	2.30	3.48	4.39	5.24	—
	45	0.86	1.02	1.15	1.30	1.43	1.83	2.53	2.75	4.15	5.21	—	—
Type*		I						II			III		

Pitch, in	No. of teeth in small sprocket	Small sprocket, r/min											
		100	500	1,000	1,200	1,500	1,800	2,000	2,500	3,000	3,500	4,000	5,000	6,000
$\frac{3}{8}$	21	0.58	2.8	5.1	6.0	7.3	8.3	9.0	10	11	12	12	12	10
	25	0.69	3.3	6.1	7.3	8.8	10	11	13	14	15	5	15	14
	29	0.80	3.8	7.3	8.5	10	12	13	15	16	18	19	19	18
	33	0.90	4.4	8.3	9.8	12	14	15	18	19	21	21	21	20
	37	1.0	4.9	9.1	11	14	15	16	20	21	24	24	24	—
	45	1.3	6.0	11	13	16	19	20	24	26	28	29	—	—
Type*		I		II				III						

Pitch, in	No. of teeth in small sprocket	Small sprocket, r/min										
		100	500	700	1,000	1,200	1,800	2,000	2,500	3,000	3,500	4,000
$\frac{1}{2}$	21	1.0	5.0	6.3	8.8	10	14	14	15	16	16	—
	25	1.2	5.0	7.5	10	13	16	18	20	21	21	20
	29	1.4	6.3	8.8	13	14	19	21	24	25	25	25
	33	1.6	7.5	10	14	16	23	24	28	29	30	29
	37	1.9	8.8	11	16	19	25	26	30	33	33	—
	45	2.5	10	14	19	23	30	30	36	39	—	—
Type*		I		II				III				

Pitch, in	No. of teeth in small sprocket	Small sprocket, r/min									
		100	500	700	1,000	1,200	1,800	2,000	2,500	3,000	3,500
$\frac{5}{8}$	21	1.6	7.5	10	13	15	19	20	20	20	—
	25	1.9	8.8	11	16	19	24	25	26	26	24
	29	2.1	10	14	19	21	28	30	31	31	29
	33	2.5	11	16	21	25	33	34	36	36	34
	37	2.8	13	18	24	28	36	39	43	41	—
	45	3.4	16	21	29	34	44	46	—	—	—
Type*		I		II			III				

Pitch, in	No. of teeth in small sprocket	Small sprocket, r/min								
		100	500	700	1,000	1,200	1,500	1,800	2,000	2,500
$\frac{3}{4}$	21	2.3	10	14	18	20	23	24	25	24
	25	2.8	13	16	21	25	29	31	31	30
	29	3.1	15	20	26	30	34	36	38	38
	33	3.6	16	23	30	34	39	43	44	44
	37	4.0	19	25	34	39	44	48	49	49
	45	4.9	23	30	40	46	53	56	58	—
Type*		I		II			III			

* Type I: manual, brush, or oil cup. Type II: bath or disk. Type III: circulating pump.
NOTE: For best results, smaller sprocket should have at least 21 teeth.
SOURCE: Adapted from ANSI B29.2M-1982.

service factor depending on the application. Maximum, or worst-case scenario, service factors are listed in Table 8.2.61.

For details on lubrication, sprocket dimensions, etc., see ANSI B29.2M-1957(R71). In utilizing Tables 8.2.59 and 8.2.60 (for a complete set of values, see ANSI B29.2M-1982) the required chain width is obtained by dividing the design horsepower by the horsepower ratings given. For calculating silent-chain lengths, the procedure for roller-chain drives may be used.

Flywheels

In slider crank mechanisms (see Sec. 8.1 Mechanism), one can surmise that both the slider force F and the external crank shaft torque may be functions of crank angle θ. Even if one or the other were deliberately kept constant, the remaining one would still be a function of θ. If a steady-speed crank is desired (ω = constant and α = zero), then external crankshaft torque T must be constantly adjusted to equal transmitted torque T_i. In such a situation a motor at the crankshaft would suffer

Table 8.2.60   Horsepower Rating per Inch of Chain Width, Silent-Chain Drive (Large Pitch)

Pitch, in	No. of teeth in small sprocket	Small sprocket, r/min										
		100	200	300	400	500	700	1,000	1,200	1,500	1,800	2,000
1	21	3.8	7.5	11	15	18	23	29	31	33	33	—
	25	5.0	8.8	14	18	21	28	35	39	41	41	41
	29	5.0	11	16	20	25	33	11	46	50	51	50
	33	6.3	13	18	24	29	38	49	54	59	59	58
	37	6.8	14	20	26	33	43	54	60	65	66	—
	45	8.8	16	25	31	39	51	65	71	76	—	—
Type*		I			II				III			

Pitch, in	No. of teeth in small sprocket	Small sprocket, r/min										
		100	200	300	400	500	600	700	800	1,000	1,200	1,500
$1\frac{1}{4}$	21	6.3	11	18	23	26	30	33	36	40	41	—
	25	7.5	14	20	26	31	36	40	44	50	53	53
	29	8.6	16	24	31	38	43	48	53	59	63	64
	33	9.9	19	28	35	43	49	55	60	69	73	74
	37	11	21	30	40	48	55	63	68	76	81	—
	45	13	26	38	49	59	68	75	81	91	—	—
Type*		I			II			III				

Pitch, in	No. of teeth in small sprocket	Small sprocket, r/min										
		100	200	300	400	500	600	700	800	900	1,000	1,200
$1\frac{1}{2}$	21	8.8	16	24	30	36	40	44	46	49	49	—
	25	10	20	29	38	44	50	55	59	61	65	64
	29	13	24	34	44	51	59	65	70	74	75	76
	33	14	28	39	50	59	68	75	80	85	88	89
	37	16	30	44	59	66	76	84	90	96	99	—
	45	19	38	54	68	81	93	101	108	113	—	—
Type*		I		II			III					

Pitch, in	No. of teeth in small sprocket	Small sprocket, r/min								
		100	200	300	400	500	600	700	800	900
2	21	16	29	40	50	53	63	65	—	—
	25	18	35	49	61	70	78	83	85	85
	29	21	41	58	73	84	93	99	103	103
	33	25	46	66	83	96	106	114	118	118
	37	28	53	75	72	110	124	128	131	—
	45	34	64	90	113	131	144	151	—	—
Type*		I		II			III			

* Type I: manual, brush, or oil cup. Type II: bath or disk. Type III: circulating pump.
NOTE: For best results, small sprocket should have at least 21 teeth.
SOURCE: Adapted from ANSI B29.2-1982.

Table 8.2.61   Service Factors* for Silent-Chain Drives

Application	Fluid-coupled engine or electric motor		Engine with straight mechanical drive		Torque converter drives		Application	Fluid-coupled engine or electric motor		Engine with straight mechanical drive		Torque converter drives	
	10 h	24 h	10 h	24 h	10 h	24 h		10 h	24 h	10 h	24 h	10 h	24 h
Agitators	1.1	1.4	1.3	1.6	1.5	1.8	Line shafts	1.6	1.9	1.8	2.1	2.0	2.3
Brick and clay machinery	1.4	1.7	1.6	1.9	1.8	2.1	Machine tools	1.4	1.7	—	—	—	—
Centrifuges	1.4	1.7	1.6	1.9	1.8	2.1	Mills—ball, hardinge, roller, etc.	1.6	1.9	1.8	2.1	—	—
Compressors	1.6	1.9	1.8	2.1	2.0	2.3	Mixer—concrete, liquid	1.6	1.9	1.8	2.1	2.0	2.3
Conveyors	1.6	1.9	1.8	2.1	2.0	2.3	Oil field machinery	1.6	1.9	1.8	2.1	2.0	2.3
Cranes and hoists	1.4	1.7	1.6	1.9	1.8	2.1	Oil refinery equipment	1.5	1.8	1.7	2.0	1.9	2.2
Crushing machinery	1.6	1.9	1.8	2.1	2.0	2.3	Paper machinery	1.5	1.8	1.7	2.0	1.9	2.2
Dredges	1.6	1.9	1.8	2.1	2.0	2.3	Printing machinery	1.5	2.0	1.4	1.7	1.6	1.9
Elevators	1.4	1.7	1.6	1.9	1.8	2.1	Pumps	1.6	1.9	1.8	2.1	2.0	2.3
Fans and blowers	1.5	1.8	1.7	2.0	1.9	2.2	Rubber plant machinery	1.5	1.8	1.7	2.0	1.9	2.2
Mills—flour, feed, cereal	1.4	1.7	1.6	1.9	1.8	2.1	Rubber mill equipment	1.6	1.9	1.8	2.1	2.0	2.3
							Screens	1.5	1.8	1.7	2.0	1.9	2.2
Generator and excitors	1.2	1.5	1.4	1.7	1.6	1.9	Steel plants	1.3	1.6	1.5	1.8	1.7	2.0
Laundry machinery	1.2	1.5	1.4	1.7	1.6	1.9	Textile machinery	1.1	1.4	—	—	—	—

* The values shown are for the maximum worst-case scenario for each application. The table was assembled from ANSI B29.2M-1982, by permission. Because the listings above are maximum values, overdesign may result when they are used. Consult the ANSI tables for specific design values.

fatigue effects. In a combustion engine the crankshaft would deliver a fluctuating torque to its load.

Inserting a flywheel at the crankshaft allows the peak and valley excursions of ω to be considerably reduced because of the flywheel's ability to absorb energy over periods when $T > T_t$ and to deliver back into the system such excess energy when $T < T_t$. Figure 8.2.103a illustrates the above concepts, also showing that over one cycle of a repeated event the excess $(+)$ energies and the deficient $(-)$ energies are equal. The greatest crank speed change tends to occur across a single large positive loop, as illustrated in Fig. 8.2.103a.

(b)

Fig. 8.2.103 Sizing the flywheel. (a) Variation of torque and crank speed vs. crank angle, showing ΔE_{ab}; (b) graphics for Wittenbauer's analysis.

Sizing the Flywheel For the single largest energy change we can write

$$\Delta E_{ab} \int_a^b (T - T_t)\, d\theta = \frac{J_0}{2} (\omega_2^2 - \omega_1^2)$$
$$= \frac{J_0}{2}(\omega_2 - \omega_1)(\omega_2 + \omega_1)$$

where J_0 = flywheel moment of inertia plus effective mechanism moment of inertia.

Define $\overline{\omega} \doteq (\omega_2 + \omega_1)/2$ and C_s = coefficient of speed fluctuation = $(\omega_2 - \omega_1)/\overline{\omega}$. Hence $\Delta E_{ab} = J_0 C_s \overline{\omega}^2$. Acceptable values of C_s are:

Pumps	0.03–0.05
Machine tools	0.025–0.03
Looms	0.025
Paper mills	0.025
Spinning mills	0.015
Crusher	0.02
Electric generators, ac	0.003
Electric generators, dc	0.002

Evaluating ΔE_{ab} involves finding the integral

$$\int_a^b (T - T_t)\, d\theta$$

which can be done graphically or by a numerical technique such as Simpson's rule.

Wittenbauer's Analysis for Flywheel Performance This method does not involve more computation work than the one described above, but it is more accurate where the reciprocating parts are comparatively heavy. Wittenbauer's method avoids the inaccuracy resulting from the evaluation of the inertia forces on the reciprocating parts on the basis of the uniform nominal speed of rotation for the engine.

Let the crankpin velocity be represented by v_r and the velocity of any moving masses (m_1, m_2, m_3, etc.) at any instant of phase be represented, respectively, by v_1, v_2, v_3, etc. The kinetic energy of the entire engine system of moving masses may then be expressed as

$$E = \frac{1}{2}(m_1 v_1^2 + m_2 v_2^2 + m_3 v_3^2 + \cdots) = \frac{1}{2} M_r v_r^2$$

or, the single reduced mass M_r at the crankpin which possesses the equivalent kinetic energy is

$$M_r = [m_1(v_1/v_r)^2 + m_2(v_2/v_r)^2 + m_3(v_3/v_r)^2 + \cdots]$$

In an engine mechanism, sufficiently accurate values of M_r can be obtained if the weight of the connecting rod is divided between the crankpin and the wrist pin so as to retain the center of gravity of the rod in its true position; usually 0.55 to 0.65 of the weight of the connecting rod should be placed on the crankpin, and 0.45 to 0.35 of the weight on the wrist pin. M_r is a variable in engine mechanisms on account of the reciprocating parts and should be found for a number of crank positions. It should include all moving masses except the flywheel.

The total energy E used in accelerating reciprocating parts from the beginning of the forward stroke up to any crank position can be obtained by finding from the indicator cards the total work done in the cylinder (on both sides of the piston) up to that time and subtracting from it the work done in overcoming the resisting torque, which may usually be assumed constant. The mean energy of the moving masses is $E_0 = \frac{1}{2} M_r v_r^2$.

In Fig. 8.2.103b, the reduced weights of the moving masses $G_F + G_{r5}$ are plotted on the X axis corresponding to different crank positions. $G_F = gM_F$ is the reduced flywheel weight and $G_{r5} = gM_{r5}$ is the sum of the other reduced weights. Against each of these abscissas is plotted the energy E available for acceleration measured from the beginning of the forward stroke. The curve $O123456$ is the locus of these plotted points.

The diagram possesses the following property: Any straight line drawn from the origin O to any point in the curve is a measure of the velocity of the moving masses; tangents bounding the diagram measure the limits of velocity between which the crankpin will operate. The maximum linear velocity of the crankpin in feet per second is $v_2 = \sqrt{2g} \tan a_2$, and the minimum velocity is $v_1 = \sqrt{2g} \tan a_1$. Any desired change in v_1 and v_2 may be accomplished by changing the value of G_F, which means a change in the flywheel weight or a change in the flywheel weight reduced to the crankpin. As G_F is very large compared with G_r and the point 0 cannot be located on the diagram unless a very large drawing is made, the tangents are best formed by direct calculation:

$$\tan \alpha_2 = \frac{v_r^2}{2g}(1 + k) \qquad \tan \alpha_1 = \frac{v_r^2}{2g}(1 - k)$$

where k is the coefficient of velocity fluctuation. The two tangents ss and tt to the curve $O123456$, thus drawn, cut a distance ΔE and on the ordinate E_0. The reduced flywheel weight is then found to be

$$G_F(\Delta E)g/(v_r^2 k)$$

SPRINGS

It is assumed in the following formulas that the springs are in no case stressed beyond the elastic limit (i.e., that they are perfectly elastic) and that they are subject to Hooke's law.

Notation

P = safe load, lb
f = deflection for a given load P, in
l = length of spring, in
V = volume of spring, in^3
S_s = safe stress (due to bending), lb/in^2
S_v = safe shearing stress, lb/in^2
U = resilience, in · lb

For sheet metal and wire gages, ferrous and nonferrous, see Table 8.2.75 and metal suppliers' catalogs.

The **work** in inch-pounds performed in **deflecting a spring** from 0 to f (spring duty) is $U = Pf/2 = s_s^2 V/(CE)$. This is based upon the assumption that the deflection is proportional to the load, and C is a constant dependent upon the shape of the springs.

The **time of vibration** T (in seconds) **of a spring** (weight not considered) is equal to that of a simple circular pendulum whose length l_0 equals the deflection f (in feet that is produced in the spring by the load P). $T = \pi \sqrt{l_0/g}$, where g = acceleration of gravity, ft/s^2.

Springs Subjected to Bending

1. **Rectangular plate spring** (Fig. 8.2.104).

$$P = bh^2 S_s/(6l) \qquad I = bh^3/12 \qquad U = Pf/2 = VS_s^2/(18E)$$
$$f = Pl^3/(3EI) = 4Pl^3/(bh^3 E) = 2l^2 S_s/(3hE)$$

Fig. 8.2.104   Rectangular plate spring.

Fig. 8.2.105   Triangular plate spring.

2. **Triangular plate spring** (Fig. 8.2.105). The elastic curve is a circular arc.

$$P = bh^2 S_s/(6l) \qquad I = bh^3/12 \qquad U = Pf/2 = S_s^2 V/(6E)$$
$$f = Pl^3/(2EI) = 6Pl^3/(bh^3 E) = l^2 S_s/(hE)$$

3. **Rectangular plate spring with end tapered** in the form of a cubic parabola (Fig. 8.2.106). The elastic curve is a circular arc; P, I, and f same as for triangular plate spring (Fig. 8.2.105); $U = Pf/2 = S_s^2 V/(9E)$.

The strength and deflection of **single-leaf flat springs** of various forms are given (Bruce, *Am. Mach.*, July 19, 1900) by the formulas $h = al^2/f$ and $b = cPl/h^2$. The volume of the spring is given by $V = vlbh$. The values of constants a and c and the resilience in inch-pounds per cubic inch are given in Table 8.2.62, in terms of the safe stress S_s. Values of v are given also.

Fig. 8.2.106   Rectangular plate spring: tapered end.

4. **Compound (leaf or laminated) springs.** If several springs of rectangular section are combined, the resulting compound spring should (1) form a beam of uniform strength that (2) does not open between the joints while bending (i.e., elastic curve must be a circular arc). Only the type immediately following meets both requirements, the others meeting only the second requirement.

5. **Laminated triangular plate spring** (Fig. 8.2.107). If the triangular plate spring shown at I is cut into an even number (= $2n$) of strips of equal width (in this case eight strips of width $b/2$), and these strips are combined, a laminated spring will be formed whose carrying capacity will equal that of the original uncut spring; or $P = nbh^2 S_s/(6l)$; $n = 6Pl/(bh^2 S_s)$.

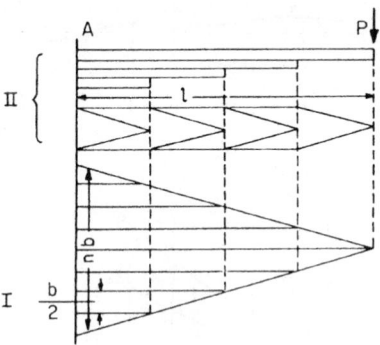

Fig. 8.2.107   Laminated triangular plate spring.

Table 8.2.62   Strength and Deflection of Single-Leaf Flat Springs

Plans and elevations of springs	Load applied at end of spring; $c = 6/S_s$			Plans and elevations of springs	Load applied at center of spring; $c = 6/4S_s$		
	a	U	v		a	U	v
	$\dfrac{S_s}{E}$	$\dfrac{S_s^2}{6E}$	$\dfrac{1}{2}$		$\dfrac{S_s}{4E}$	$\dfrac{S_s^2}{6E}$	$\dfrac{1}{2}$
Parabolic arc	$\dfrac{4S_s}{3E}$	$\dfrac{S_s^2}{6E}$	$\dfrac{2}{3}$		$\dfrac{0.87S_s}{4E}$	$\dfrac{0.70S_s^2}{6E}$	$\dfrac{5}{8}$
	$\dfrac{2S_s}{3E}$	$\dfrac{0.33S_s^2}{6E}$	1	Parabolic arcs	$\dfrac{S_s}{3E}$	$\dfrac{S_s^2}{6E}$	$\dfrac{2}{3}$
	$\dfrac{0.87S_s}{E}$	$\dfrac{0.70S_s^2}{6E}$	$\dfrac{5}{8}$		$\dfrac{1.09S_s}{4E}$	$\dfrac{0.725S_s^2}{6E}$	$\dfrac{3}{4}$
	$\dfrac{1.09S_s}{E}$	$\dfrac{0.725S_s^2}{6E}$	$\dfrac{3}{4}$		$\dfrac{S_s}{6E}$	$\dfrac{0.33S_s^2}{6E}$	I

6. **Laminated rectangular plate spring with lead ends tapered** in the form of a cubical parabola (Fig. 8.2.108); see case 3 above.

Fig. 8.2.108   Laminated rectangular plate spring with leaf end tapered.

7. **Laminated trapezoidal plate spring with leaf ends tapered** (Fig. 8.2.109). The ends of the leaves are trapezoidal and are tapered according to the formula

$$z = \frac{h}{\sqrt[3]{1 + (b_1/b)(a/x - 1)}}$$

8. **Semielliptic springs** (for locomotives, trucks, etc.). Referring to Fig. 8.2.110, the load $2P$ (lb) acting on the spring center band produces a tensional stress $P/\cos a$ in each of the inclined shackle links. This is resolved into the vertical force P and the horizontal force $P \tan a$, which together produce a bending moment $M = P(l + p \tan a)$. The equations

Fig. 8.2.109   Laminated trapezoidal plate spring with leaf ends tapered.

given in (1), (2), and (3) apply to curved as well as straight springs. The bearing force $= 2P = (2nbh^2/6)[S_s/(l + p \tan a)]$, and the deflection $= [6l^2/(nbh^3)]P(l + p \tan a)/E = l^2S_s/(hE)$.

In addition to the bending moment, the leaves are subjected to the tension force $P \tan a$ and the transverse force P, which produce in the upper leaf an additional stress $S = P \tan a/(bh)$, as well as a transverse shearing stress.

Fig. 8.2.110   Semielliptic springs.

In determining the number of leaves n in a given spring, allowance should be made for an excess load on the spring caused by the vibration. This is usually done by decreasing the allowable stress about 15 percent.

The foregoing does not take account of initial stresses caused by the band. For more detailed information, see Wahl, "Mechanical Springs," Penton.

9. **Elliptic springs.** Safe load $P = nbh^2S_s/(6l)$, where $l = \frac{1}{2}$ distance between bolt eyes (less $\frac{1}{2}$ length of center band, where used); deflection $f = 4l^2S_sK/(hE)$, where

$$K = \frac{1}{(1 - r)^3}\left[\frac{1 - r^2}{2} - 2r(1 - r) - r^2 \ln r\right]$$

r being the number of full-length leaves $\div$ total number (n) of leaves in the spring. All dimensions in inches. For semielliptic springs, the deflection is only half as great. Safe load $= nbh^2S_s/(3l)$. (Peddle, *Am. Mach.*, Apr. 17, 1913.)

Coiled Springs   In these, the load is applied as a couple Pr which turns the spring while winding or holds it in place when wound up. If the spindle is not to be subjected to bending moment, P must be replaced by two equal and opposite forces ($P/2$) acting at the circumference of a circle of radius r. The formulas are the same in both cases. The springs are assumed to be fixed at one end and free at the other. The bending moment acting on the section of least resistance is always Pr. The length of the straightened spring $= l$. See Benjamin and French, Experiments on Helical Springs, *Trans. ASME*, **23**, p. 298.

For **heavy closely coiled helical springs** the usual formulas are inaccurate and result in stresses greatly in excess of those assumed. See Wahl, Stresses in Heavy Closely-Coiled Helical Springs, *Trans. ASME*, 1929. In springs 10 to 12 and 15 to 18, the quantity k is unity for lighter springs and has the stated values (supplied by Wahl) for heavy closely coiled springs.

10. **Spiral coiled springs of rectangular cross section** (Fig. 8.2.111).

$$P = bh^2S_s/(6rk) \qquad I = bh^3/12 \qquad U = Pf/2 = S_s^2V/(6Ek^2)$$
$$f = ra = Plr^2/(EI) = 12Plr^2/(Ebh^3) = 2rlS_s/(hEk)$$

For heavy closely coiled springs, $k = (3c - 1)/(3c - 3)$, where $c = 2R/h$ and R is the minimum radius of curvature at the center of the spiral.

Fig. 8.2.111   Spiral coiled spring: rectangular cross section.

Fig. 8.2.112   Cylindrical helical spring: circular cross section.

11. **Cylindrical helical spring of circular cross section** (Fig. 8.2.112).

$$P = \pi d^3S_s/(32rk) \qquad I = \pi d^4/64 \qquad U = Pf/2 = S_s^2V/(8Ek^2)$$
$$f = ra = Plr^2/(EI) = 64Plr^2/(\pi Ed^4) = 2rlS_s/(dEk)$$

For heavy closely coiled springs, $k = (4c - 1)/(4c - 4)$, where $c = 2r/d$.

12. **Cylindrical helical spring of rectangular cross section** (Fig. 8.2.113).

$$P = bh^2S_s/(6rk) \qquad I = bh^3/12 \qquad U = Pf/2 = S_s^2V/(8Ek^2)$$
$$f = ra = Plr^2/(EI) = 12Plr^2/(Ebh^3) = 2rlS_s/(hEk)$$

For heavy closely coiled springs, $k = (3c - 1)/(3c - 3)$, where $c = 2r/h$.

Fig. 8.2.113   Cylindrical helical spring: rectangular cross section.

Springs Subjected to Torsion

The statements made concerning coiled springs subjected to bending apply also to springs 13 and 14.

13. **Straight bar spring of circular cross section** (Fig. 8.2.114).

$$P = \pi d^3 S_v/(16r) = 0.1963 d^3 S_v/r \qquad U = Pf/2 = S_v^2 V/(4G)$$
$$f = ra = 32r^2 lP/(\pi d^4 G) = 2rl S_v/(dG)$$

14. **Straight bar spring of rectangular cross section** (Fig. 8.2.115).

$$P = 2b^2 h S_v/(9r) \qquad K = b/h$$
$$U = Pf/2 = 4S_v^2 V(K^2 + 1)/(45G) \qquad \text{max when } K = 1$$
$$f = ra = 3.6r^2 lP(b^2 + h^2)/(b^3 h^3 G) = 0.8rl S_v(b^2 + h^2)/(bh^2 G)$$

Fig. 8.2.114 Straight bar spring: circular cross section.

Fig. 8.2.115 Straight bar spring: rectangular cross section.

Springs Loaded Axially in Either Tension

NOTE.   For springs 15 to 18, r = mean radius of coil; n = number of coils.

15. **Cylindrical helical spring of circular cross section** (Fig. 8.2.116).

$$P = \pi d^3 S_v/(16rk) = 0.1963 d^3 S_v/(rk)$$
$$U = Pf/2 = S_v^2 V/(4Gk^2)$$
$$f = 64nr^3 P/(d^4 G) = 4\pi nr^2 S_v/(dGk)$$

For heavy closely coiled springs, $k = (4c - 1)/(4c - 4) + 0.615/c$, where $c = 2r/d$.

Fig. 8.2.116   Cylindrical helical spring: circular cross section.

16. **Cylindrical helical spring of rectangular cross section** (Fig. 8.2.117).

$$P = 2b^2 h S_v/(9rk) \qquad K = b/h$$
$$U = Pf/2 = 4S_v^2 V(K^2 + 1)/(45Gk^2) \qquad \text{max when } K = 1$$
$$f = 7.2\pi nr^3 P(b^2 + h^2)/(b^3 h^3 G) = 1.6\pi nr^2 S_v(b^2 + h^2)/(bh^2 Gk)$$

For heavy closely coiled springs, $k = (4c - 1)/(4c - 4) + 0.615/c$, where $c = 2r/b$.

Fig. 8.2.117   Cylindrical helical spring: rectangular cross section.

17. **Conical helical spring of circular cross section** (Fig. 8.2.118a).

l = length of developed spring
d = diameter of wire
r = maximum mean radius of coal
$P = \pi d^3 S_v/(16rk) = 0.1963 d^3 S_v/(rk)$
$U = Pf/2 = S_v^2 V/(8Gk^2)$
$f = 16r^2 lP/(\pi d^4 G) = 16\pi r^3 P/(d^4 G)$
$\quad = rl S_v/(dGk) = \pi nr^2 S_v/(dGk)$

For heavy closely coiled springs, $k = (4c - 1)/(4c - 4) + 0.615/c$, where $c = 2r/d$.

Fig. 8.2.118a Conical helical spring: circular cross section.

Fig. 8.2.118b Conical helical spring: rectangular cross section.

18. **Conical helical spring of rectangular cross section** (Fig. 8.2.118b).

b = small dimension of section
d = large dimension of section
r = maximum mean radius of coil
$K = b/h \ (\leq 1) \qquad P = 2b^2 h S_v/(9rk)$
$U = Pf/2 = 2S_v^2 V(K^2 + 1)/(45Gk^2) \qquad \text{max when } K = 1$
$f = 1.8r^2 lP(b^2 + h^2)/(b^3 h^3 G) = 1.8\pi nr^3 P(b^2 + h^2)/(b^3 h^3 G)$
$\quad = 0.4rl S_v(b^2 + h^2)/(bh^2 Gk) = 0.4\pi nr^2 S_v(b^2 + h^2)(bh^2 Gk)$

For heavy closed coiled springs, $k = (4c - 1)/(4c - 4) + 0.615c$, where $c = 2r/r_o - r_i$.

19. **Truncated conical springs** (17 and 18). The formulas under 17 and 18 apply for truncated springs. In calculating deflection f, however, it is necessary to substitute $r_1^2 + r_2^2$ for r^2, and $\pi n(r_1 + r_2)$ for πnr, r_1 and r_2 being, respectively, the greatest and least mean radii of the coils.

NOTE.   The preceding formulas for various forms of coiled springs are suffi-ciently accurate when the cross-section dimensions are small in comparison with the radius of the coil, and for small pitch. Springs 15 to 19 are for either tension or compression but formulas for springs 17 and 18 are good for compression only until the largest coil flattens out; then r becomes a variable, depending on the load.

Design of Helical Springs

When sizing a new spring, one must consider the spring's available working space and the loads and deflections the spring must experience. Refinements dictated by temperature, corrosion, reliability, cost, etc. may also enter design considerations. The two basic formulas of load and deflection (see item 15, Fig. 8.2.116) contain eight variables (f, P, d, S, r, k, n, G) which prevent one from being able to use a one-step solution. For instance, if f and P are known and S and G are chosen, there still remain d, r, k, and n to be found.

A variety of solution approaches are available, including: (1) slide-rule-like devices available from spring manufacturers, (2) nomographic methods (Chironis, "Spring Design and Application," McGraw-Hill; Tsai, Speedy Design of Helical Compression Springs by Nomography Method, *J. of Eng. for Industry*, Feb. 1975), (3) table methods (Carlson, "Spring Designer's Handbook," Marcel Dekker), (4) formula method (ibid.), and (5) computer programs and subroutines.

Design by Tables

Safe working loads and deflections of cylindrical helical springs of round steel wire in tension or compression are given in Table 8.2.63. The table is based on the formulas given for spring 15. d = diameter of steel wire, in; D = pitch diameter (center to center of wire), in; P = safe working load for given unit stress, lb; f = deflection of 1 coil for safe working load, in.

The table is based on the values of unit stress indicated, and G = 12,500,000. For any other value of unit stress, divide the tabular value by the unit stress used in the table and multiply by the unit stress to be used in the design. For any other value of G, multiply the value of f in the table by 12,500,000 and divide by the value of G chosen. For **square steel wire,** multiply values of P by 1.06, and values of f by 0.75. For **round brass wire,** take S_s=10,000 to 20,000, and multiply values of f by 2 (Howe).

EXAMPLES OF USE OF TABLE 8.2.63.   1. Required the safe load (P) for a spring of ⅜-in round steel with a pitch diameter (D) of 3½ in. In the line headed D, under 3½, is given the value of P, or 678 lb. This is for a unit stress of 115,000 lb/in². The load P for any other unit stress may be found by dividing the 678 by 115,000 and multiplying by the unit stress to be used in the design. To determine the number of coils this spring would need to compress (say) 6 in under a load of (say) 678 lb, take the value of f under 678, or 0.938, which is the deflection of one coil under the given load. Therefore, 6/0.938 = 6.4, say 7, equals the number of coils required. The spring will therefore be 2⅝ in long when closed (7 × ⅜), counting the working coils only, and must be 8⅝ in long when unloaded. Whether there is an extra coil at one end which does not deflect will depend upon the details of the particular design. The deflection in the above example is for a unit stress of 115,000 lb/in². The rule is, divide the deflection by 115,000 and multiply by the unit stress to be used in the design.

2. A ⁷⁄₁₆-in steel spring of 3½-in OD has its coils in close contact. How much can it be extended without exceeding the limit of safety? The maximum safe load for this spring is found to be 1,074 lb, and the deflection of one coil under this load is 0.810 in. This is for a unit stress of 115,000 lb/in². Therefore, 0.810 is the greatest admissible opening between any two coils. In this way, it is possible to ascertain whether or not a spring is overloaded, without knowledge of the load carried.

Design by Formula

A design formula can be constructed by equating calculated stress s_v (from load formula, Fig. 8.2.119) to an allowable working stress in torsion:

$$s_v = \sigma_{\max} = \frac{16rP}{\pi d^3}\left(\frac{4c-1}{4c-4} + \frac{0.615}{c}\right) = \frac{16rPk}{\pi d^3} = \frac{S_v}{K_{sf}}$$

where P = axial load on spring, lb; $r = D/2$ = mean radius of coil, in; D = mean diameter of coil, in (outside diameter minus wire diameter); d = wire diameter, in; $\sigma_{\max}$ = torsional stress, lb/in²; K_{sf} = safety factor; and $c = D/d$. Note that the expression in parentheses [(4c − 1)(4c −4) + 0.615/c] is the **Wahl correction factor** k, which accounts for the added stresses in the coils due to curvature and direct shear. See Fig. 8.2.119. Values of S_v, yield point in shear from standard tests, are strongly dependent on d, hence the availability of S_v in the literature is limited. However, an empirical relationship between S_{uT} and d is available (see Shigley, "Mechanical Engineering Design," McGraw-Hill,

4th ed., p. 452). Using also the approximate relations $S_y = 0.75S_{ut}$ and $S_v = 0.577S_y$ results in the following relationship:

$$S_v = \frac{0.43A}{d^m}$$

where A and m are constants (see Table 8.2.64).

Substituting S_v and rearranging yield the following useful formula:

$$d^{3-m} = \frac{K_{sf}16rPk}{\pi 0.43A}$$

EXAMPLE.   A slow-speed follower is kept in contact with its cam by means of a helical compression spring, in which the minimum spring force desired is 20 lb to assure firm contact, while maximum spring force is not to exceed 60 lb to prevent excessive surface wear on the cam. The follower rod is ¼ inch in diameter, and the rod enclosure where the spring is located is ⅞ inch in diameter. Maximum displacement is 1.5 in.

Choose $r = 0.5 \times 0.75 = 0.375$. From Table 8.2.64 choose $m = 0.167$ and $A = 169,000$. Choose $K_{sf} = 2$. Assume $k = 1$ to start.

$$d^{3-0.167} = \frac{(2)(0.375)(60)(16)}{(\pi)(0.43)(169,000)} = 0.00315$$

$$d = (0.00315)^{0.35298} = 0.131 \text{ in}$$

$$\text{Spring constant} = \frac{\Delta P}{\Delta f} = \frac{40}{1.5} = 26.667$$

and from the deflection formula, the number of active turns

$$n = \frac{(0.131)^4(11,500,000)}{(26.667)(64)(0.375)^3} = 37.6$$

For squared and ground ends add two dead coils, so that

$$n_{\text{total}} = 40$$
$$H = \text{solid height} = (40)(0.131) = 5.24 \text{ in}$$
$$f_0 = \text{displacement from zero to maximum load}$$
$$= 60/26.667 = 2.249$$
$$L_0 = \text{approximate free length} = 5.24 + 2.249 = 7.49 \text{ in}$$

NOTE.   Some clearance should be added between coils so that at maximum load the spring is not closed to its solid height. Also, the nearest commercial stock size should be selected, and recalculations made on this stock size for S_v, remembering to include k at this juncture. If recalculated S_v is satisfactory as compared to published values, the design is retained, otherwise enough iterations are performed to arrive at a satisfactory result. Figure 8.2.120 shows a plot of S_{uT} versus d. To convert S_{uT} to S_v, multiply S_{uT} by 0.43.

$$c = 2(0.375/0.131) = 5.73$$

$$k = \frac{4(5.73)-1}{4(5.73)-4} + \frac{0.615}{5.73} = 1.257$$

$$S_v = \frac{(K_{sf})(16)(0.375)(1.257)(60)}{\pi(0.131)^3} = (K_{sf})(64,072)$$

Now S_{uT} = 235,000 lb/in² (from Fig. 8.2.120) and S_v (tabulated) = (0.43)(235,000) = 101,696 lb/in² so that K_{sf} = 101,696/64,072 = 1.59, a satisfactory value.

NOTE.   The original choice of a generous K_{sf} = 2 was made to hedge against the large statistical variations implied in the empirical formula $S_v = 0.43A/d^m$.

The basis for design of springs in parallel or in series is shown in Fig. 8.2.121.

Belleville Springs   Often called dished, or conical spring, washers, Belleville springs occupy a very small space. They are stressed in a very complex manner, and provide unusual spring rate curves (Fig. 8.2.122a). These springs are nonlinear, but for some proportions, they behave with approximately linear characteristics in a limited range. Likewise, for some proportions they can be used through a spectrum of spring rates, from positive to flat and then through a negative region. The snap-through action, shown at point A in Fig. 8.2.122b, can be

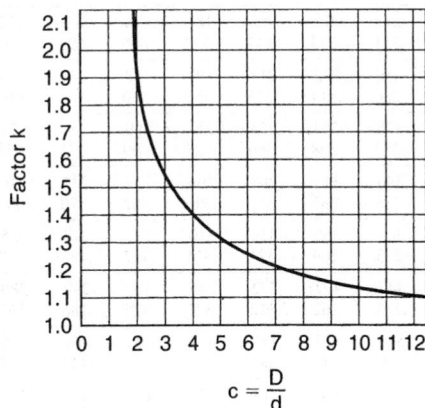

Fig. 8.2.119   Wahl correction factor.

Table 8.2.63   Representative of Safe Working Loads *P* and Deflections *f* of Cylindrical Helical Steel Springs of Circular Cross Section

(For closely coiled springs, divide given load and deflection values by the curvature factor k.)

Allowable unit stress, lb/in²	Diam, in	D	5/32	3/16	1/4	5/16	3/8	7/16	1/2	5/8	3/4	7/8	1
150,000	0.035	P	16.2	13.4	10.0	8.10	6.66	5.75	4.96	4.05	3.39		
		f	.026	.037	.067	.105	.149	.200	.276	.420	.608		
	0.041	P	26.2	21.6	16.2	13.0	10.8	9.27	8.10	6.52	5.35	4.57	
		f	.023	.032	.057	.089	.128	.175	.229	.362	.512	.697	
	0.047	P	39.1	32.6	24.5	19.6	16.4	13.9	12.3	9.80	8.10	6.92	6.14
		f	.019	.028	.049	.078	.112	.153	.200	.311	.449	.610	.800
	0.054	P	59.4	49.6	37.2	29.7	24.6	21.2	18.5	14.7	12.4	10.5	9.25
		f	.016	.024	.043	.067	.098	.133	.174	.273	.390	.532	.695
	0.062	P		74.9	56.1	44.9	37.3	32.0	28.0	22.4	18.6	16.1	13.9
		f		.021	.037	.058	.084	.115	.151	.235	.340	.460	.605
	0.063	P		78.2	58.7	46.9	39.2	33.9	29.4	23.5	19.6	16.8	14.7
		f		.020	.037	.057	.083	.113	.148	.233	.335	.445	.591
	0.072	P		117.	80.7	70.0	58.7	50.2	43.6	35.2	29.0	25.0	21.9
		f		.018	.032	.050	.077	.100	.130	.203	.294	.405	.521
	0.080	P			121	96.6	80.5	69.1	60.4	48.2	48.2	34.6	30.1
		f			.029	.045	.065	.090	.117	.183	.262	.359	.470
140,000	0.092	P			171	136	113	97.6	85.5	68.9	57.3	48.8	42.6
		f			.023	.037	.053	.072	.098	.148	.214	.291	.388
	0.093	P			178	142	118	99.5	89.0	71.2	59.1	50.9	44.3
		f			.023	.036	.052	.071	.093	.146	.211	.286	.376
	0.105	P				204	170	147	127	102	85.4	73.0	63.4
		f				.032	.047	.064	.083	.122	.188	.256	.336
	0.120	P				303	253	217	190	152	126	108	95.2
		f				.028	.041	.055	.073	.114	.174	.223	.296
	0.125	P					286	245	214	171	143	121	107
		f					.039	.053	.069	.109	.169	.213	.278
	0.135	P					359	309	270	217	171	154	135
		f					.036	.049	.064	.106	.145	.198	.260
	0.148	P						408	356	285	237	207	178
		f						.045	.059	.092	.132	.180	.236

Allowable unit stress, lb/in²	Diam, in	D	7/16	1/2	5/8	3/4	7/8	1	1⅛	1¼	1⅜	1½	1⅝	1¾	1⅞	2
140,000	0.156	P	480	418	330	270	234	208	185	167	152	139	128	119	111	104
		f	0.42	.056	.087	.125	.171	.223	.282	.350	.422	.509	.588	.685	.785	.896
	0.162	P		468	376	311	276	234	207	187	170	156	143	134	125	117
		f		.054	.085	.121	.165	.216	.273	.338	.409	.488	.566	.663	.757	.863
	0.177	P		608	487	406	347	305	270	243	223	205	187	174	163	152
		f		.049	.077	.115	.151	.198	.251	.311	.375	.447	.522	.606	.695	.793
125,000	0.187	P		642	522	426	367	320	284	256	233	213	197	183	170	160
		f		.041	.065	.093	.127	.166	.210	.260	.314	.373	.487	.510	.584	.665
	0.192	P		696	556	465	396	348	309	278	254	233	214	199	186	174
		f		.040	.063	.091	.124	.160	.205	.252	.308	.366	.428	.499	.571	.652
	0.207	P			694	579	495	432	385	346	315	288	266	247	232	216
		f			.059	.085	.115	.151	.191	.236	.286	.342	.396	.462	.570	.607
	0.218	P			812	678	580	509	452	408	360	339	310	291	270	255
		f			.055	.080	.109	.142	.180	.223	.269	.321	.374	.437	.488	.570
	0.225	P			895	746	640	560	498	447	407	372	345	320	299	280
		f			.054	.078	.106	.138	.175	.213	.262	.312	.372	.425	.486	.565
	0.244	P			1120	950	811	711	632	570	517	475	438	406	381	356
		f			.049	.071	.098	.138	.161	.200	.240	.287	.336	.391	.449	.537
	0.250	P				1027	880	760	685	617	560	513	476	440	410	385
		f				.070	.095	.131	.157	.191	.236	.281	.328	.385	.439	.524

Allowable unit stress, lb/in²	Diam, in	D,	1¼	1⅜	1½	1⅝	1¾	1⅞	2	2¼	2½	2¾	3
	0.263	P	1195	1125	895	795	717	652	598	551	501	478	448
		f	.066	.089	.118	.149	.183	.224	.266	.312	.363	.416	.475
	0.281	P	1450	1240	1087	969	863	794	724	665	620	580	543
		f	.062	.085	.111	.140	.172	.209	.250	.292	.340	.390	.443
	0.283	P		1264	1110	985	886	805	740	682	634	592	564
		f		.084	.111	.139	.169	.207	.246	.289	.338	.356	.440
115,000	0.312	P		1575	1376	1220	1100	1000	915	845	775	733	687
		f		.070	.092	.116	.144	.174	.207	.242	.283	.322	.368
	0.331	P			1636	1455	1316	1187	1090	1000	932	870	818
		f			.088	.109	.135	.163	.194	.227	.264	.343	.346
	0.341	P			1820	1620	1452	1325	1214	1120	1040	970	910
		f			.082	.105	.127	.156	.186	.218	.256	.293	.330
	0.362	P			2140	1910	1714	1560	1430	1318	1220	1147	1070
		f			.079	.100	.123	.149	.177	.207	.243	.273	.317
	0.375	P				2110	1940	1780	1580	1458	1354	1265	1185
		f				.079	.117	.144	.172	.201	.234	.268	.308
	0.393	P				2430	2180	1984	1820	1680	1560	1458	1365
		f				.092	.114	.137	.164	.195	.223	.256	.292
	0.406	P					2400	2170	2000	1840	1710	1600	1500
		f					.108	.134	.159	.168	.217	.248	.284
	0.430	P					2875	2610	2400	2210	2050	1918	1798
		f					.104	.126	.150	.175	.204	.234	.267
	0.437	P					3000	2730	2500	2310	2140	2000	1800
		f					.100	.124	.148	.173	.201	.231	.264

Allowable unit stress, lb/in²	Diam, in	D,	1¼	1⅜	1½	1⅝	1¾	1⅞	2	2¼	2½	2¾	3
						Pitch diameter D, in							
110,000	0.460	P		3065	2800	2580	2400	2230	2100	1865	1680	1530	1400
		f		.112	.134	.157	.183	.209	.239	.303	.374	.447	.536
	0.468	P		3265	2940	2725	2530	2375	2210	1970	1770	1610	1472
		f		.111	.132	.154	.182	.206	.235	.295	.368	.444	.530
	0.490	P		3675	3270	3115	2890	2710	2535	2245	2025	1840	1690
		f		.106	.126	.148	.172	.196	.225	.284	.351	.424	.506
	0.500	P			3610	3320	3090	2890	2710	2410	2160	1970	1810
		f			.123	.144	.168	.192	.220	.274	.347	.415	.495
	0.562	P				4700	4390	4090	3830	3420	3080	2790	2565
		f				.128	.149	.175	.195	.248	.306	.372	.440
	0.625	P					6100	5600	5260	4660	4210	3825	3505
		f					.134	.154	.176	.218	.275	.328	.397
100,000	0.687	P							6325	5660	5090	4630	4250
		f							.145	.183	.228	.274	.3278
	0.750	P								7400	6640	6030	5540
		f								.178	.218	.252	.299
	0.812	P									8420	7660	7000
		f									.192	.232	.276
	0.875	P									10830	9550	8700
		f									.179	.218	.257
90,0.00	0.937	P										10600	9700
		f										.179	.217
	1.000	P											11780
		f											.206
	1.125	P											
		f											
	1.250	P											
		f											
80,000	1.375	P											
		f											

Table 8.2.64   Constants for Use in $S_v = 0.43A/d^m$

Material	Size range, in	Size range, mm	Exponent m	Constant A kips	Constant A MPa
Music wire*	0.004–0.250	0.10–6.55	0.146	196	2,170
Oil-tempered wire†	0.020–0.500	0.50–12	0.186	149	1,880
Hard-drawn wire‡	0.028–0.500	0.70–12	0.192	136	1,750
Chrome vanadium§	0.032–0.437	0.80–12	0.167	169	2,000
Chrome silicon¶	0.063–0.375	1.6–10	0.112	202	2,000

* Surface is smooth, free from defects, and has a bright, lustrous finish.
† Has a slight heat-treating scale which must be removed before plating.
‡ Surface is smooth and bright, with no visible marks.
§ Aircraft-quality tempered wire; can also be obtained annealed.
¶ Tempered to Rockwell C49 but may also be obtained untempered.
SOURCE: Adapted from "Mechanical Engineering Design," Shigley, McGraw-Hill, 1983 by permission.

Fig. 8.2.120 Minimum tensile strength for the most popular spring materials, spring-quality wire. *(Reproduced from Carlson, "Spring Designer's Handbook," Marcel Dekker, by permission.)*

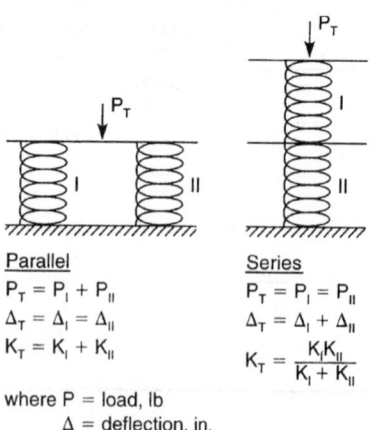

Parallel
$$P_T = P_I + P_{II}$$
$$\Delta_T = \Delta_I = \Delta_{II}$$
$$K_T = K_I + K_{II}$$

Series
$$P_T = P_I = P_{II}$$
$$\Delta_T = \Delta_I + \Delta_{II}$$
$$K_T = \frac{K_I K_{II}}{K_I + K_{II}}$$

where P = load, lb
    Δ = deflection, in.
    K = spring rate, lb/in.

Fig. 8.2.121   Springs in parallel and in series.

Fig. 8.2.122a   Sectional view of Belleville spring.

Fig. 8.2.122b   Load deflection curves for a family of Belleville springs. *(Associated Spring Corp.)*

useful in particular applications requiring reversal of spring rates. These springs are used for very high and special spring rates. They are extremely sensitive to slight variations in their geometry. A wide range is available commercially.

WIRE ROPE

When power source and load are located at extreme distances from one another, or loads are very large, the use of wire rope is suggested. Design and use decisions pertaining to wire ropes rest with the user, but manufacturers generally will help users toward appropriate choices. The following material, based on the Committee of Wire Rope Producers, "Wire Rope User's Manual," current edition, may be used as an initial guide in selecting a rope.

Wire rope is composed of (1) wires to form a strand, (2) strands wound helically around a core, and (3) a core. Classification of wire ropes is made by giving the number of strands, number of minor strands

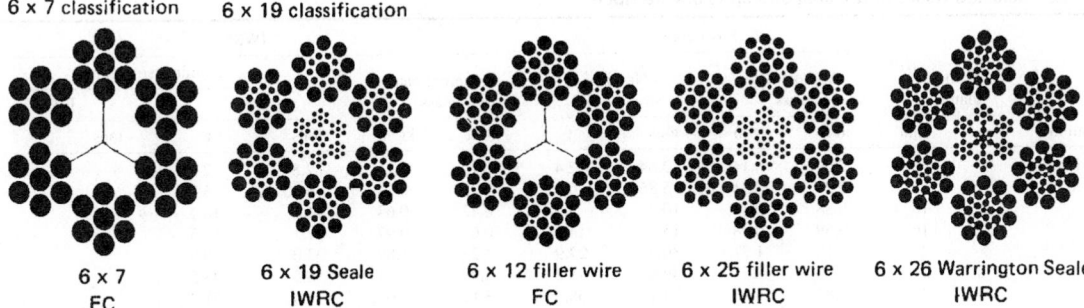

Fig. 8.2.123 Cross sections of some commonly used wire rope construction. (*Reproduced from "Wire Rope User's Manual," AISI, by permission.*)

in a major strand (if any), and nominal number of wires per strand. For example 6 × 7 rope means 6 strands with a nominal 7 wires per strand (in this case no minor strands, hence no middle number). A nominal value simply represents a range. A nominal value of 7 can mean anywhere from 3 to 14, of which no more than 9 are outside wires. A full rope description will also include length, size (diameter), whether wire is preformed or not prior to winding, direction of lay (right or left, indicating the direction in which strands are laid around the core), grade of rope (which reflects wire strength), and core. The most widely used classifications are: 6 × 7, 6 × 19, 6 × 37, 6 × 61, 6 × 91, 6 × 127, 8 × 19, 18 × 7, 19 × 7. Some special constructions are: 3 × 7 (guardrail rope); 3 × 19 (slusher); 6 × 12 (running rope); 6 × 24 and 6 × 30 (hawsers); 6 × 42 and 6 × 6 × 7 (tiller rope); 6 × 3 × 19 (spring lay); 5 × 19 and 6 × 19 (marlin clad); 6 × 25B, 6 × 27H, and 6 × 30G (flattened strand). The diameter of a rope is the circle which just contains the rope. The right-regular lay (in which the wire is twisted in one direction to form the strands and the strands are twisted in the opposite direction to form the rope) is most common. Regular-lay ropes do not kink or untwist and handle easily. Lang-lay ropes (in which wires and strands are twisted in the same direction) are more resistant to abrasive wear and fatigue failure.

Cross sections of some commonly used wire rope are shown in Fig. 8.2.123. Figure 8.2.124 shows rotation-resistant ropes, and Fig. 8.2.125 shows some special-purpose constructions.

The core provides support for the strands under normal bending and loading. Core materials include fibers (hard vegetable or synthetic) or steel (either a strand or an independent wire rope). Most common core designations are: fiber core (FC), independent wire-rope core (IWRC), and wire-strand core (WSC). Lubricated fiber cores can provide lubrication to the wire, but add no real strength and cannot be used in high temperature environments. Wire-strand or wire-rope cores add from 7 to 10 percent to strength, but under nonstationary usage tend to wear from interface friction with the outside strands. Great flexibility can be achieved when wire rope is used as strands. Such construction is

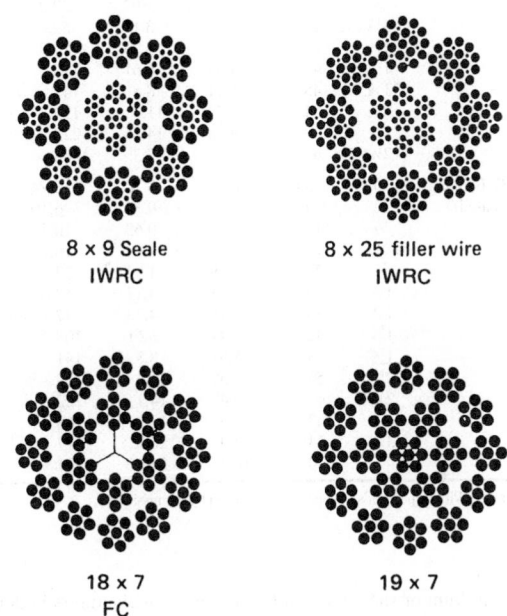

Fig. 8.2.124 Cross section of some rotation-resistant wire ropes. (*Reproduced from "Wire Rope User's Manual," AISI, by permission.*)

very pliable and friction resistant. Some manufacturers will provide plastic coatings (nylon, Teflon, vinyl, etc.) upon request. Such coatings help provide resistance to abrasion, corrosion, and loss of lubricant. *Crushing* refers to rope damage caused by excessive pressures against drum or sheave, improper groove size, and multiple

Fig. 8.2.125 Some special constructions. (*Reproduced from "Wire Rope User's Manual," AISI, by permission.*)

Table 8.2.65   Selected Values of Nominal Strengths of Wire Rope

Classification	Nominal diameter in	Nominal diameter mm	Fiber core Approximate mass lb/ft	Fiber core Approximate mass kg/m	Fiber core Nominal strength IPS tons	Fiber core Nominal strength IPS t	IWRC Approximate mass lb/ft	IWRC Approximate mass kg/m	IWRC Nominal strength IPS tons	IWRC Nominal strength IPS t	IWRC Nominal strength EIP tons	IWRC Nominal strength EIP t
6 × 7 Bright	¼	6.4	0.09	0.14	2.64	2.4	0.10	0.15	2.84	2.58		
(uncoated)	⅜	9.5	0.21	0.31	5.86	5.32	0.23	0.34	6.30	5.72		
	½	13	0.38	0.57	10.3	9.35	0.42	0.63	11.1	10.1		
	⅝	16	0.59	0.88	15.9	14.4	0.65	0.97	17.1	15.5		
	⅞	22	1.15	1.71	30.7	27.9	1.27	1.89	33.0	29.9		
	1⅛	29	1.90	2.83	49.8	45.2	2.09	3.11	53.5	48.5		
	1⅜	35	2.82	4.23	73.1	66.3	3.12	4.64	78.6	71.3		
6 × 19 Bright	¼	6.4	0.11	0.16	2.74	2.49	0.12	0.17	2.94	2.67	3.40	3.08
(uncoated)	⅜	9.5	0.24	0.35	6.10	5.53	0.26	0.39	6.56	5.95	7.55	6.85
	½	13	0.42	0.63	10.7	9.71	0.46	0.68	11.5	10.4	13.3	12.1
	⅝	16	0.66	0.98	16.7	15.1	0.72	1.07	17.7	16.2	20.6	18.7
	⅞	22	1.29	1.92	32.2	29.2	1.42	2.11	34.6	31.4	39.8	36.1
	1⅛	29	2.13	3.17	52.6	47.7	2.34	3.48	56.5	51.3	65.0	59.0
	1⅜	35	3.18	4.73	77.7	70.5	3.5	5.21	83.5	75.7	96.0	87.1
	1⅝	42	4.44	6.61	107	97.1	4.88	7.26	115	104	132	120
	1⅞	48	5.91	8.8	141	128	6.5	9.67	152	138	174	158
	2⅛	54	7.59	11.3	179	162	8.35	12.4	192	174	221	200
	2⅜	60	9.48	14.1	222	201	10.4	15.5	239	217	274	249
	2⅝	67	11.6	17.3	268	243	12.8	19.0	288	261	331	300
6 × 37 Bright	¼	6.4	0.11	0.16	2.74	2.49	0.12	0.17	2.94	2.67	3.4	3.08
(uncoated)	⅜	9.5	0.24	0.35	6.10	5.53	0.26	0.39	6.56	5.95	7.55	6.85
	½	13	0.42	0.63	10.7	9.71	0.46	0.68	11.5	10.4	13.3	12.1
	⅝	16	0.66	0.98	16.7	15.1	0.72	1.07	17.9	16.2	20.6	18.7
	⅞	22	1.29	1.92	32.2	29.2	1.42	2.11	34.6	31.4	39.5	36.1
	1⅛	29	2.13	3.17	52.6	47.7	2.34	3.48	56.5	51.3	65.0	59.0
	1⅜	35	3.18	4.73	77.7	70.5	3.50	5.21	83.5	75.7	96.0	87.1
	1⅝	42	4.44	6.61	107	97.1	4.88	7.26	115	104	132	120
	1⅞	48	5.91	8.8	141	128	6.5	9.67	152	138	174	158
	2⅛	54	7.59	11.3	179	162	8.35	12.4	192	174	221	200
	2⅜	60	9.48	14.1	222	201	10.4	15.5	239	217	274	249
	2⅞	67	11.6	17.3	268	243	12.8	19.0	288	261	331	300
	3⅛	74	13.9	20.7	317	287	15.3	22.8	341	309	392	356
		80	16.4	24.4	371	336	18.0	26.8	399	362	458	415

SOURCE: "Wire Rope User's Manual," AISI, adapted by permission.

layers on drum or sheave. Consult wire rope manufacturers in doubt-ful situations.

Wire-rope materials and their strengths are reflected as grades. These are: traction steel (TS), mild plow steel (MPS), plow steel (PS), improved plow steel (IPS), and extra improved plow (EIP). The plow steel strength curve forms the basis for calculating the strength of all steel rope wires. American manufacturers use color coding on their ropes to identify particular grades.

The grades most commonly available and tabulated are IPS and EIP. Two specialized categories, where selection requires extraordinary attention, are elevator and rotation-resistant ropes.

Elevator rope can be obtained in four principal grades: iron, traction steel, high-strength steel, and extra-high-strength steel.

Bronze rope has limited use; iron rope is used mostly for older existing equipment.

Selection of Wire Rope

Appraisal of the following is the key to choosing the rope best suited to the job: resistance to breaking, resistance to bending fatigue, resistance to vibrational fatigue, resistance to abrasion, resistance to crushing, and reserve strength. Along with these must be an appropriate choice of safety factor, which in turn requires careful consideration of all loads, acceleration-deceleration, shocks, rope speed, rope attachments, sheave arrangements as well as their number and size, corrosive and/or

abrasive environment, length of rope, etc. An approximate selection formula can be written as:

$$DSL = \frac{(NS)K_b}{K_{sf}}$$

where DSL (demanded static load) = known or dead load **plus** additional loads caused by sudden starts or stops, shocks, bearing friction, etc., tons; NS (nominal strength) = published test strengths, tons (see Table 8.2.65); K_b = a factor to account for the reduction in nominal strength due to bending when a rope passes over a curved surface such as a stationary sheave or pin (see Fig. 8.2.126); K_{sf} = safety factor. (For

Fig. 8.2.126   Values of K_{bend} vs. D/d ratios (D = sheave diameter, d = rope diameter), based on standard test data for 6 × 9 and 6 × 17 class ropes. (*Compiled from* "Wire Rope User's Manual," AISI, by permission.)

Table 8.2.66   Suggested Allowable Radial Bearing Pressures of Ropes on Various Sheave Materials

Material	Regular lay rope, lb/in^2				Lang lay rope, lb/in^2			Flattened strand lang lay, lb/in^2	Remarks
	6 × 7	6 × 19	6 × 37	8 × 19	6 × 7	6 × 19	6 × 37		
Wood	150	250	300	350	165	275	330	400	On end grain of beech, hickory, gum.
Cast iron	300	480	585	680	350	550	660	800	Based on minimum Brinell hardness of 125.
Carbon-steel casting	550	900	1,075	1,260	600	1,000	1,180	1,450	30-40 carbon. Based on minimum Brinell hardness of 160.
Chilled cast iron	650	1,100	1,325	1,550	715	1,210	1,450	1,780	Not advised unless surface is uniform in hardness.
Manganese steel	1,470	2,400	3,000	3,500	1,650	2,750	3,300	4,000	Grooves must be ground and sheaves balanced for high-speed service.

SOURCE: "Wire Rope User's Manual," AISI, reproduced by permission.

average operation use $K_{sf} = 5$. If there is danger to human life or other critical situations, use $8 \leq K_{sf} \leq 12$. For instance, for elevators moving at 50 ft/min, $K_{sf} = 8$, while for those moving at 1,500 ft/min, $K_{sf} = 12$.)

Having made a tentative selection of a rope based on the demanded static load, one considers next the wear life of the rope. A loaded rope bent over a sheave stretches elastically and so rubs against the sheave, causing wear of both members. Drum or sheave size is of paramount importance at this point.

Sizing of Drums or Sheaves

Diameters of drums or sheaves in wire rope applications are controlled by two main considerations: (1) the radial pressure between rope and groove and (2) degree of curvature imposed on the rope by the drum or sheave size.

Radial pressures can be calculated from $p = 2T/(Dd)$, where $p =$ unit radial pressure, lb/in^2; $T =$ rope load, lb; $D =$ tread diameter of drum or sheave, in; $d =$ nominal diameter of rope, in. Table 8.2.66 lists suggested allowable radial bearing pressures of ropes on various sheave materials.

All wire ropes operating over drums or sheaves are subjected to cyclical stresses, causing shortened rope life because of fatigue. Fatigue resistance or relative service life is a function of the ratio D/d. Adverse effects also arise out of relative motion between strands during passage around the drum or sheave. Additional adverse effects can be traced to poor match between rope and groove size, and to lack of rope lubrication. Table 8.2.67 lists suggested and minimum sheave and drum ratios for various rope construction. Table 8.2.68 lists relative bending life factors; Figure 8.2.127 shows a plot of relative rope service life versus D/d. Table 8.2.69 lists minimum drum (sheave) groove dimensions. Periodic groove inspection is recommended, and worn or corrugated grooves should be remachined or the drum replaced, depending on severity of damage.

Seizing and Cutting Wire Rope   Before a wire rope is cut, seizings (bindings) must be applied on either side of the cut to prevent rope distortion and flattening or loosened strands. Normally, for preformed ropes, one seizing on each side of the cut is sufficient, but for ropes that are not preformed a minimum of two seizings on each side is recommended, and these should be spaced six rope diameters apart (see Fig. 8.2.128). Seizings should be made of soft or annealed wire or strand, and the width of the seizing should never be less than the diameter of the rope being seized. Table 8.2.70 lists suggested seizing wire diameters.

Wire Rope Fittings or Terminations   End terminations allow forces to be transferred from rope to machine, or load to rope, etc. Figure 8.2.129

Table 8.2.67   Sheave and Drum Ratios

Construction*	Suggested	Minimum
6 × 7	72	42
19 × 7 or 18 × 7 Rotation-resistant	51	34
6 × 19 S	51	34
6 × 25 B flattened strand	45	30
6 × 27 H flattened strand	45	30
6 × 30 G flattened strand	45	30
6 × 21 FW	45	30
6 × 26 WS	45	30
6 × 25 FW	39	26
6 × 31 WS	39	26
6 × 37 SFW	39	26
6 × 36 WS	35	23
6 × 43 FWS	35	23
6 × 41 WS	32	21
6 × 41 SFW	32	21
6 × 49 SWS	32	21
6 × 46 SFW	28	18
6 × 46 WS	28	18
8 × 19 S	41	27
8 × 25 FW	32	21
6 × 42 Tiller	21	14

* WS—Warrington Seale; FWS—Filler Wire Seale; SFW—Seale Filler Wire; SWS—Seale Warrington Seale; S—Seale; FW—Filler Wire.

† $D =$ tread diameter of sheave; $d =$ nominal diameter of rope. To find any tread diameter from this table, the diameter for the rope construction to be used is multiplied by its nominal diameter d. For example, the minimum sheave tread diameter for a ½-in 6 × 21 FW rope would be ½ in (nominal diameter) × 30 (minimum ratio), or 15 in.

NOTE: These values are for reasonable service. Other values are permitted by various standards such as ANSI, API, PCSA, HMI, CMAA, etc. Similar values affect rope life.
SOURCE: "Wire Rope User's Manual," AISI, reproduced by permission.

Table 8.2.68   Relative Bending Life Factors

Rope construction	Factor	Rope construction	Factor
6 × 7	0.61	6 × 36 WS	1.16
19 × 7 or 18 × 7	0.67	6 × 43 FWS	1.16
Rotation-resistant	0.81	6 × 41 WS	1.30
6 × 19 S	0.90	6 × 41 SFW	1.30
6 × 25 B flattened strand	0.90	6 × 49 SWS	1.30
6 × 27 H flattened strand	0.90	6 × 43 FW (2 op)	1.41
6 × 30 G flattened strand	0.89	6 × 46 SFW	1.41
6 × 21 FW	0.89	6 × 46 WS	1.41
6 × 26 WS	1.00	8 × 19 S	1.00
6 × 25 FW	1.00	8 × 25 FW	1.25
6 × 31 WS	1.00	6 × 42 Tiller	2.00
6 × 37 SFW			

SOURCE: "Wire Rope User's Manual," AISI, reproduced by permission.

Table 8.2.69   Minimum Sheave- and Drum-Groove Dimensions*

Nominal rope diameter		Groove radius			
		New		Worn	
in	nm	in	mm	in	mm
¼	6.4	0.135	3.43	.129	3.28
⁵⁄₁₆	8.0	0.167	4.24	.160	4.06
⅜	9.5	0.201	5.11	.190	4.83
⁷⁄₁₆	11	0.234	5.94	.220	5.59
½	13	0.271	6.88	.256	6.50
⁹⁄₁₆	14.5	0.303	7.70	.288	7.32
⅝	16	0.334	8.48	.320	8.13
¾	19	0.401	10.19	.380	9.65
⅞	22	0.468	11.89	.440	11.18
1	26	0.543	13.79	.513	13.03
1⅛	29	0.605	15.37	.577	14.66
1¼	32	0.669	16.99	.639	16.23
1⅜	35	0.736	18.69	.699	17.75
1½	38	0.803	20.40	.759	19.28
1⅝	42	0.876	22.25	.833	21.16
1¾	45	0.939	23.85	.897	22.78
1⅞	48	1.003	25.48	.959	24.36
2	52	1.085	27.56	1.025	26.04
2⅛	54	1.137	28.88	1.079	27.41
2¼	58	1.210	30.73	1.153	29.29
2⅜	60	1.271	32.28	1.199	30.45
2½	64	1.338	33.99	1.279	32.49
2⅝	67	1.404	35.66	1.339	34.01
2¾	71	1.481	37.62	1.409	35.79
2⅞	74	1.544	39.22	1.473	37.41
3	77	1.607	40.82	1.538	39.07
3⅛	80	1.664	42.27	1.598	40.59
3¼	83	1.731	43.97	1.658	42.11
3⅜	87	1.807	45.90	1.730	43.94
3½	90	1.869	47.47	1.794	45.57
3¾	96	1.997	50.72	1.918	48.72
4	103	2.139	54.33	2.050	52.07
4¼	109	2.264	57.51	2.178	55.32
4½	115	2.396	60.86	2.298	58.37
4¾	122	2.534	64.36	2.434	61.82
5	128	2.663	67.64	2.557	64.95
5¼	135	2.804	71.22	2.691	68.35
5½	141	2.929	74.40	2.817	71.55
5¾	148	3.074	78.08	2.947	74.85
6	154	3.198	81.23	3.075	78.11

* Values given are applicable to grooves in sheaves and drums; they are not generally suitable for pitch design since this may involve other factors. Further, the dimensions do not apply to traction-type elevators; in this circumstance, drum- and sheave-groove tolerances should conform to the elevator manufacturer's specifications. Modern drum design embraces extensive considerations beyond the scope of this publication. It should also be noted that dram grooves are now produced with a number of oversize dimensions and pitches applicable to certain service requirements.
SOURCE: "Wire Rope User's Manual," AISI, reproduced by permission.

Fig. 8.2.127 Service life curves for various *D/d* ratios. Note that this curve takes into account only bending and tensile stresses. *(Reproduced from "Wire Rope User's Manual," AISI, by permission.)*

Fig. 8.2.128 Seizings. *(Reproduced from "Wire Rope User's Manual," AISI, by permission.)*

illustrates the most commonly used end fittings or terminations. Not all terminations will develop full strength. In fact, if all of the rope elements are not held securely, the individual strands will sustain unequal loads causing unequal wear among them, thus shortening the effective rope service life. Socketing allows an end fitting which reduces the chances of unequal strand loading.

Wire rope manufacturers have developed a recommended procedure for socketing. A tight wire serving band is placed where the socket base will be, and the wires are unlaid, straightened, and "broomed" out. Fiber core is cut close to the serving band and removed, wires are cleaned with a solvent such as SC-methyl chloroform, and brushed to remove dirt and grease. If additional cleaning is done with muriatic acid this must be followed by a neutralizing rinse (if possible, ultrasonic cleaning is preferred). The wires are dipped in flux, the socket is positioned, zinc (spelter) is poured and allowed to set, the serving band is removed, and the rope lubricated.

A somewhat similar procedure is used in thermoset resin socketing.

Socketed terminations generally are able to develop 100 percent of nominal strength.

Table 8.2.70   Seizing*

Rope diameter		Suggested seizing wire diameter†	
in	mm	in	mm
⅛–⁵⁄₁₆	3.5–8.0	0.032	0.813
⅜–⁹⁄₁₆	9.4–14.5	0.048	1.21
⅝–¹⁵⁄₁₆	16.0–24.0	0.063	1.60
1–1⁵⁄₁₆	26.0–33.0	0.080	2.03
1⅜–1¹¹⁄₁₆	35.0–43.0	0.104	2.64
1¾ and larger	45.0 and larger	0.124	3.15

* Length of the seizing should not be less than the rope diameter.
† The diameter of seizing wire for elevator ropes is usually somewhat smaller than that shown in this table. Consult the wire rope manufacturer for specific size recommendations. Soft annealed seizing strand may also be used.
SOURCE: "Wire Rope User's Manual," AISI, reproduced by permission.

Wire rope socket —poured spelter or resin

Wire rope socket—swaged

Mechanical splice--loop or thimble

Wedge socket

Clips—number of clips varies with rope size and construction

Loop or thimble splice—hand tucked

Fig. 8.2.129 End fittings, or terminations, showing the six most commonly used. (*Reproduced from "Wire Rope User's Manual," AISI, by permission.*)

FIBER LINES

The breaking strength of various fiber lines is given in Table 8.2.71.

Knots, Hitches, and Bends

No two parts of a knot which would move in the same direction if the rope were to slip should lie alongside of and touching each other. The knots shown in Fig. 8.2.130 are known by the following names:

A, bight of a rope; *B*, simple or overhand knot; *C*, figure 8 knot; *D*, double knot; *E*, boat knot; *F*, bowline, first step; *G*, bowline, second step; *H*, bowline, completed; *I*, square or reef knot; *J*, sheet bend or weaver's knot; *K*, sheet bend with a toggle; *L*, carrick bend; *M*, "stevedore" knot completed; *N*, "stevedore" knot commenced; *O*, slip knot; *P*, Flemish loop; *Q*, chain knot with toggle; *R*, half hitch; *S*, timber hitch; *T*, clove hitch; *U*, rolling hitch; *V*, timber hitch and half hitch; *W*, blackwall hitch; *X*, fisherman's bend; *Y*, round turn and half hitch; *Z*, wall knot commenced; *AA*, wall knot completed; *BB*, wall-knot crown commenced; *CC*, wall-knot crown completed.

The bowline *H*, one of the most useful knots, will not slip, and after being strained is easily untied. Knots *H*, *K*, and *M* are easily untied after being under strain. The knot *M* is useful when the rope passes through an eye and is held by the knot, as it will not slip, and is easily untied after being strained. The wall knot is made as follows: Form a bight with strand 1 and pass the strand 2 around the end of it, and the strand 3 around the end of 2, and then through the bight of 1, as shown at *Z* in the figure. Haul the ends taut when the appearance is as shown in *AA*. The end of the strand 1 is now laid over the center of the knot, strand 2 laid over 1, and 3 over 2, when the end of 3 is passed through the bight of 1, as shown at *BB*. Haul all the strands taut, as shown at *CC*. The "stevedore" knot (*M*, *N*) is used to hold the end of a rope from passing through a hole. When the rope is strained, the knot draws up

Table 8.2.71 Breaking Strength of Fiber Lines, lb

Size, in											Poly-propylene (mono-filament)	Esterlon (polyester)
Diam	Cir.	Manila	Composite	Sisal	Sisal mixed	Sisal hemp	Agave or jute	Nylon	Dacron	Poly-ethylene		
$^3/_{16}$	$^5/_8$	450	—	360	340	310	270	1,000	850	700	800	720
$^1/_4$	$^3/_4$	600	—	480	450	420	360	1,500	1,380	1,200	1,200	1,150
$^5/_{16}$	1	1,000	—	800	750	700	600	2,500	2,150	1,750	2,100	1,750
$^3/_8$	$1^1/_8$	1,350	—	1,080	1,010	950	810	3,500	3,000	2,500	3,100	2,450
$^7/_{16}$	$1^1/_4$	1,750	—	1,400	1,310	1,230	1,050	4,800	4,500	3,400	3,700	3,400
$^1/_2$	$1^1/_2$	2,650	—	2,120	1,990	1,850	1,590	6,200	5,500	4,100	4,200	4,400
$^9/_{16}$	$1^3/_4$	3,450	—	2,760	2,590	2,410	2,070	8,300	7,300	4,600	5,100	5,700
$^5/_8$	2	4,400	—	3,520	3,300	3,080	2,640	10,500	9,500	5,200	5,800	7,300
$^3/_4$	$2^1/_4$	5,400	—	4,320	4,050	3,780	3,240	14,000	12,500	7,400	8,200	9,500
$^{13}/_{16}$	$2^1/_2$	6,500	—	5,200	4,880	4,550	3,900	17,000	15,000	8,900	9,800	11,500
$^7/_8$	$2^3/_4$	7,700	—	—	—	—	—	20,000	17,500	10,400	11,500	13,500
1	3	9,000	—	7,200	6,750	6,300	5,400	24,000	20,000	12,600	14,000	16,500
$1^1/_{16}$	$3^1/_4$	10,500	—	8,400	7,870	7,350	6,300	28,000	22,500	14,500	16,100	19,000
$1^1/_8$	$3^1/_2$	12,000	—	9,600	9,000	8,400	7,200	32,000	25,000	16,500	18,300	21,500
$1^1/_4$	$3^3/_4$	13,500	—	10,800	10,120	9,450	8,100	36,500	28,500	18,600	21,000	24,300
$1^5/_{16}$	4	15,000	—	12,000	11,250	10,500	9,000	42,000	32,000	21,200	24,000	28,000
$1^1/_2$	$4^1/_2$	18,500	16,600	14,800	13,900	12,950	11,100	51,000	41,000	26,700	30,000	34,500
$1^5/_8$	5	22,500	20,300	18,000	16,900	15,800	13,500	62,000	50,000	32,700	36,500	41,500
$1^3/_4$	$5^1/_2$	26,500	23,800	21,200	19,900	18,500	15,900	77,500	61,000	39,500	44,000	51,000
2	6	31,000	27,900	24,800	23,200	21,700	18,600	90,000	72,000	47,700	53,000	61,000
$2^1/_8$	$6^1/_2$	36,000	—	—	—	—	—	105,000	81,000	55,800	62,000	70,200
$2^1/_4$	7	41,000	36,900	32,800	30,800	28,700	—	125,000	96,000	63,000	70,000	81,000
$2^1/_2$	$7^1/_2$	46,500	—	—	—	—	—	138,000	110,000	72,500	80,500	92,000
$2^5/_8$	8	52,000	46,800	41,600	39,000	36,400	—	154,000	125,000	81,000	90,000	103,000
$2^7/_8$	$8^1/_2$	58,000	—	—	—	—	—	173,000	140,000	92,000	100,000	116,000
3	9	64,000	57,500	51,200	48,000	44,800	—	195,000	155,000	103,000	116,000	130,000
$3^1/_4$	10	77,000	69,300	61,600	57,800	53,900	—	238,000	190,000	123,000	137,000	160,000
$3^1/_2$	11	91,000	—	—	—	—	—	288,000	230,000	146,000	162,000	195,000
4	12	105,000	94,500	84,000	78,800	73,500	—	342,000	275,000	171,000	190,000	230,000

Breaking strength is the maximum load the line will hold at the time of breaking, The working load of a line is one-fourth to one-fifth of the breaking strength.
SOURCE: Adapted, by permission of the U.S. Naval Institute, Annapolis, MD, and Wall Rope Works, Inc., New York, NY.

Fig. 8.2.130   Knots, hitches, and bends.

tight, but it can be easily untied when the strain is removed. If a knot or hitch of any kind is tied in a rope, its failure under stress is sure to occur at that place. The shorter the bend in the standing rope, the weaker is the knot. The approximate strength of knots compared with the full strength of (dry) rope (= 100), based on Miller's experiments (*Mach.*, 1900, p. 198), is as follows: eye splice over iron thimble, 90; short splice in rope, 80; S and Y, 65; H, O, and T, 60; I and J, 50; B and P, 45.

NAILS AND SPIKES

Nails are either **wire nails** of circular cross section and constant diameter or **cut nails** of rectangular cross section with taper from head to point. The larger sizes are called **spikes**. The length of the nail is expressed in the "penny" system, the equivalents in inches being given in Tables 8.2.72 to 8.2.74. The letter d is the accepted symbol for penny. A keg of nails weighs 100 lb. **Heavy hinge nails** or **track nails** with countersunk heads have chisel points unless diamond points are specified. **Plasterboard nails** are smooth with circumferential grooves and have diamond points. **Spikes** are made either with flat heads and diamond points or with oval heads and chisel points.

WIRE AND SHEET-METAL GAGES

In the metal industries, the word *gage* has been used in various systems, or scales, for expressing the thickness or weight per unit area of thin plates, sheet, and strip, or the diameters of rods and wire. Specific diameters, thicknesses, or weights per square foot have been or are denoted in gage systems by certain numerals followed by the word *gage*, for example, no. 12 gage, or simply 12 gage. Gage numbers for flat rolled products have been used only in connection with thin materials (Table 8.2.75). Heavier and thicker, flat rolled materials are usually designated by thickness in English or metric units.

There is considerable danger of confusion in the use of gage number in both foreign and domestic trade, which can be avoided by specifying thickness or diameter in inches or millimeters.

DRILL SIZES

See Table 8.2.76.

Table 8.2.72   Wire Nails for Special Purposes
(Steel wire gage)

Length, in	Barrel nails		Barbed roofing nails		Barbed dowel nails	
	Gage	No. per lb	Gage	No. per lb	Gage	No. per lb
⅝	15½	1,570	—	—	8	394
¾	15½	1,315	13	729	8	306
⅞	14½	854	12	478	8	250
1	14½	750	12	416	8	212
1⅛	14½	607	12	368	8	183
1¼	14	539	11	250	8	16
1⅜	13	386	11	228	8	145
1½	13	355	10	167	8	131
1¾	—	—	10	143		
2	—	—	9	104		

Length, in	Clout nails		Slating nails		Fine nails		
	Gage	No. per lb	Gage	No. per lb	Length, in	Gage	No. per lb
¾	15	999					
⅞	14	733					
1	14	648	12	425	1	16½	1,280
1⅛	14	580	10½	229			
1¼	13	398	—	—	1	17	1,492
1⅜	13	365					
1½	13	336	10½	190	1⅛	15	757
1¾	—	—	10	144	1⅛	16	984
2	—	—	9	104			

Table 8.2.73   Wire Nails and Spikes

(Steel wire gage)

Size of nail	Length, in	Casing nails		Finishing nails		Clinch nails		Shingle nails	
		Gage	No. per lb	Gage	No. per lb	Gage	No. per lb	Gage	No. per lb
2d	1	15½	940	16½	1.473	14	723	13	434
3d	1¼	14½	588	15½	880	13	432	12	271
4d	1½	14	453	15	634	12	273	12	233
5d	1¾	14	389	15	535	12	234	12	203
6d	2	12½	223	13	288	11	158		
7d	2¼	12½	200	13	254	11	140		
8d	2½	11½	136	12½	196	10	101		
9d	2¾	11½	124	12½	178	10	91.4		
10d	3	10½	90	11½	124	9	70		
12d	3¼	10½	83	11½	113	9	64.1		
16d	3½	10	69	11	93	8	50		
20d	4	9	51	10	65	7	36.4		
30d	4¼	9	45						
40d	5	8	37						

Size of nail	Length, in	Boat nails				Hinge nails				Flooring nails	
		Heavy		Light		Heavy		Light			
		Diam, in	No. per lb	Diam, in	No. per lb	Diam, in	No. per lb	Diam, in	No. per lb	Gage	No. per lb
4d	1½	¼	47	3/16	82	¼	53	3/16	90		
6d	2	¼	36	3/16	62	¼	39	3/16	66	11	168
8d	2½	¼	29	3/16	50	¼	31	3/16	53	10	105
10d	3	⅜	11	¼	24	⅜	12	¼	25	9	72
12d	3¼	⅜	10.4	¼	22	⅜	11	¼	23	8	56
16d	3½	⅜	9.6	¼	20	⅜	10	¼	22	7	44
20d	4	⅜	8	¼	18	⅜	8	¼	19	6	32

Size of nail	Length, in	Common wire nails and brads		Barbed car nails				Spikes			
				Heavy		Light					Approx no. per lb
		Gage	No. per lb	Gage	No. per lb	Gage	No. per lb	Size	Length, in	Gage	
2d	1	15	847	—	—	—	—	10d	3	6	43
3d	1¼	14	548	—	—	—	—	12d	3¼	6	39
4d	1½	12½	294	10	179	12	284	16d	3½	5	31
5d	1¾	12½	254	9	124	10	152	20d	4	4	23
6d	2	11½	167	9	108	10	132	30d	4½	3	18
7d	2¼	11½	150	8	80	9	95	40d	5	2	14
8d	2½	10¼	101	8	72	9	88	50d	5½	1	11
9d	2¾	10¼	92	7	55	8	65	50d	6	1	10
10d	3	9	66	7	50	8	59	—	7	5/16 in	7
12d	3¼	9	61	6	39	7	46	—	8	⅜	4.1
16d	3½	8	47	6	36	7	43	—	9	⅜	3.7
20d	4	6	30	5	27	6	32	—	10	⅜	3.3
30d	4½	5	23	5	24	6	28	—	12	⅜	2.7
40d	5	4	18	4	18	5	22				
50d	5½	3	14	3	14	4	17				
60d	6	2	11	3	13	4	15				

Table 8.2.74 Cut Steel Nails and Spikes
(Sizes, lengths, and approximate number per lb)

Size	Length, in	Common	Clinch	Finishing	Casing and box	Fencing	Spikes	Barrel	Slating	Tobacco	Brads	Shingle
2d	1	740	400	1,100	—	—	—	450	340			
3d	1¼	460	260	880	—	—	—	280	280			
4d	1½	280	180	530	420	—	—	190	220			
5d	1¾	210	125	350	300	100	—	—	180	130		
6d	2	160	100	300	210	80	—	—	—	97	120	
7d	2¼	120	80	210	180	60	—	—	—	85	94	
8d	2½	88	68	168	130	52	—	—	—	68	74	
9d	2¾	73	52	130	107	38	—	—	—	58	62	90
10d	3	60	48	104	88	26	—	—	—	48	50	72
12d	3¼	46	40	96	70	20	—	—	—	—	40	60
16d	3½	33	34	86	52	18	17	—	—	—	27	
20d	4	23	24	76	38	16	14					
25d	4¼	20	—	—	—	—	—					
30d	4½	16½	—	—	30	—	11					
40d	5	12	—	—	26	—	9					
50d	5½	10	—	—	20	—	7½					
60d	6	8	—	—	16	—	6					
—	6½	—	—	—	—	—	5½					
—	7	—	—	—	—	—	5					

Table 8.2.75 Comparison of Standard Gages*
Thickness of diameter, in

Gage no.	BWG; Stubs Iron Wire	AWG; B&S	U.S. Steel Wire; Am. Steel & Wire; Washburn & Moen; Steel Wire	Galv. sheet steel	Manufacturers' standard
0000000	—	—	0.4900	—	—
000000	—	0.580000	0.4615	—	—
00000	—	0.516500	0.4305	—	—
0000	0.454	0.460000	0.3938	—	—
000	0.425	0.409642	0.3625	—	—
00	0.380	0.364796	0.3310	—	—
0	0.340	0.324861	0.3065	—	—
1	0.300	0.289297	0.2830	—	—
2	0.284	0.257627	0.2625	—	—
3	0.259	0.229423	0.2437	—	0.2391
4	0.238	0.204307	0.2253	—	0.2242
5	0.220	0.181940	0.2070	—	0.2092
6	0.203	0.162023	0.1920	—	0.1943
7	0.180	0.144285	0.1770	—	0.1793
8	0.165	0.128490	0.1620	0.1681	0.1644
9	0.148	0.114423	0.1483	0.1532	0.1495
10	0.134	0.101897	0.1350	0.1382	0.1345
11	0.120	0.090742	0.1205	0.1233	0.1196
12	0.109	0.080808	0.1055	0.1084	0.1046
13	0.095	0.071962	0.0915	0.0934	0.0897
14	0.083	0.064084	0.0800	0.0785	0.0747
15	0.072	0.057068	0.0720	0.0710	0.0673
16	0.065	0.050821	0.0625	0.0635	0.0598
17	0.058	0.045257	0.0540	0.0575	0.0538
18	0.049	0.040303	0.0475	0.0516	0.0478
19	0.042	0.035890	0.0410	0.0456	0.0418
20	0.035	0.031961	0.0348	0.0396	0.0359
21	0.032	0.028462	0.03175	0.0366	0.0329
22	0.028	0.025346	0.0286	0.0336	0.0299
23	0.025	0.022572	0.0258	0.0306	0.0269
24	0.022	0.020101	0.0230	0.0276	0.0239
25	0.020	0.017900	0.0204	0.0247	0.0209
26	0.018	0.015941	0.0181	0.0217	0.0179
27	0.016	0.014195	0.0173	0.0202	0.0164
28	0.014	0.012641	0.0162	0.0187	0.0149

Table 8.2.75   Comparison of Standard Gages*   (*Continued*)

Gage no.	BWG; Stubs Iron Wire	AWG; B&S	U.S. Steel Wire; Am. Steel & Wire; Washburn & Moen; Steel Wire	Galv. sheet steel	Manufacturers' standard
29	0.013	0.011257	0.0150	0.0172	0.0135
30	0.012	0.010025	0.0140	0.0157	0.0120
31	0.010	0.008928	0.0132	0.0142	0.0105
32	0.009	0.007950	0.0128	0.0134	0.0097
33	0.008	0.007080	0.0118	—	0.0090
34	0.007	0.006305	0.0104	—	0.0082
35	0.005	0.005615	0.0095	—	0.0075
36	0.004	0.005000	0.0090	—	0.0067
37	—	0.004453	0.0085	—	0.0064
38	—	0.003965	0.0080	—	0.0060
39	—	0.003531	0.0075	—	—
40	—	0.003144	0.0070	—	—

* Principal uses—BWG: strips, bands, hoops, and wire; AWG or B&S: nonferrous sheets, rod, and wire; U.S. Steel Wire: steel wire except music wire; manufacturers' standard: uncoated steel sheets.

Table 8.2.76   Diameters of Small Drills
Number, letter, metric, and fractional drills in order of size (rounded to 4 decimal places)

No.	Ltr	mm	in	Diam, in	No.	Ltr	mm	in	Diam, in	No.	Ltr	mm	in	Diam, in
		0.10		0.0039			1.10		0.0433			2.40		0.0945
		0.15		0.0059			1.15		0.0452	41				0.0960
		0.20		0.0079	56				0.0465			2.45		0.0964
		0.25		0.0098				3/64	0.0468	40				0.0980
		0.30		0.0118			1.20		0.0472			2.50		0.0984
80				0.0135			1.25		0.0492	39				0.0995
		0.35		0.0137			1.30		0.0512	38				0.1015
79				0.0145	55				0.0520			2.60		0.1024
			1/64	0.0156			1.35		0.0531	37				0.1040
		0.40		0.0157	54				0.0550			2.70		0.1063
78				0.0160			1.40		0.0551	36				0.1065
		0.45		0.0177			1.45		0.0570			2.75		0.1082
77				0.0180			1.50		0.0590				7/64	0.1093
		0.50		0.0197	53				0.0595	35				0.1100
76				0.0200			1.55		0.0610			2.80		0.1102
75				0.0210				1/16	0.0625	34				0.1110
		0.55		0.0216			1.60		0.0630	33				0.1130
74				0.0225	52				0.0635			2.90		0.1142
		0.60		0.0236			1.65		0.0649	32				0.1160
73				0.0240			1.70		0.0669			3.00		0.1181
72				0.0250	51				0.0670	31				0.1200
		0.65		0.0255			1.75		0.0688			3.10		0.1220
71				0.0260	50				0.0700				1/8	0.1250
		0.70		0.0275			1.80		0.0709			3.20		0.1260
70				0.0280			1.85		0.0728			3.25		0.1279
69				0.0292	49				0.0730	30				0.1285
		0.75		0.0295			1.90		0.0748			3.30		0.1299
68				0.0310	48				0.0760			3.40		0.1339
			1/32	0.0312			1.95		0.0767	29				0.1360
		0.80		0.0314				5/64	0.0781			3.50		0.1378
67				0.0320	47				0.0785	28				0.1405
66				0.0330			2.00		0.0787				9/64	0.1406
		0.85		0.0334			2.05		0.0807			3.60		0.1417
65				0.0350	46				0.0810	27				0.1440
		0.90		0.0354	45				0.0820			3.70		0.1457
64				0.0360			2.10		0.0827	26				0.1470
63				0.0370			2.15		0.0846			3.75		0.1476
		0.95		0.0374	44				0.0860	25				0.1495
62				0.0380			2.20		0.0866			3.80		0.1496
61				0.0390			2.25		0.0885	24				0.1520
		1.00		0.0393	43				0.0890			3.90		0.1535
60				0.0400			2.30		0.0906	23				0.1540
59				0.0410			2.35		0.0925				5/32	0.1562
		1.05		0.0413	42				0.0935	22				0.1570
58				0.0420				3/32	0.0937			4.00		0.1575
57				0.0430										

Table 8.2.76   Diameters of Small Drills   (Continued)

No.	Ltr	mm	in	Diam, in	No.	Ltr	mm	in	Diam, in	No.	Ltr	mm	in	Diam, in
21				0.1590			6.80		0.2677			11.00		0.4331
20				0.1610			6.90		0.2717				7/16	0.4375
		4.10		0.1614		I			0.2720			11.50		0.4528
		4.2		0.1654			7.00		0.2756				29/64	0.4531
19				0.1660		J			0.2770				15/32	0.4687
		4.25		0.1673			7.10		0.2795			12.00		0.4724
		4.30		0.1693		K			0.2810				31/64	0.4843
18				0.1695				9/32	0.2812			12.50		0.4921
			11/64	0.1718			7.20		0.2835				1/2	0.5000
17				0.1730			7.25		0.2854			13.00		0.5118
		4.40		0.1732			7.30		0.2874				33/64	0.5156
16				0.1770		L			0.2900				17/32	0.5312
		4.50		0.1772			7.40		0.2913			13.50		0.5315
15				0.1800		M			0.2950				35/64	0.5468
		4.60		0.1811			7.50		0.2953			14.00		0.5512
14				0.1820				19/64	0.2968				9/16	0.5625
13				0.1850			7.60		0.2992			14.50		0.5708
		4.70		0.1850		N			0.3020				37/64	0.5781
		4.75		0.1850			7.70		0.3031			15.00		0.5905
			3/16	0.1875			7.75		0.3051				19/32	0.5937
		4.80		0.1890			7.80		0.3071				39/64	0.6093
12				0.1890			7.90		0.3110			15.50		0.6102
11				0.1910				5/16	0.3125				5/8	0.6250
		4.90		0.1920			8.00		0.3150			16.00		0.6299
10				0.1935		O			0.3160				41/64	0.6406
9				0.1960			8.10		0.3189			16.50		0.6496
		5.00		0.1968			8.20		0.3228				21/32	0.6562
8				0.1990		P			0.3230			17.00		0.6693
		5.10		0.2008			8.25		0.3248				43/64	0.6718
7				0.2010			8.30		0.3268				11/16	0.6875
			13/64	0.2031				21/64	0.3281			17.50		0.6890
6				0.2040			8.40		0.3307				45/64	0.7031
		5.20		0.2047		Q			0.3320			18.00		0.7087
5				0.2055			8.50		0.3346				23/32	0.7187
		5.25		0.2066			8.60		0.3386			18.50		0.7283
		5.30		0.2087		R			0.3390				47/64	0.7374
4				0.2090			8.70		0.3425			19.00		0.7480
		5.40		0.2126				11/32	0.3437				3/4	0.7500
3				0.2130			8.75		0.3444				49/64	0.7656
		5.50		0.2165			8.80		0.3464			19.50		0.7677
			7/32	0.2187		S			0.3480				25/32	0.7812
		5.60		0.2205			8.90		0.3504			20.00		0.7874
2				0.2210			9.00		0.3543				51/64	0.7968
		5.70		0.2244		T			0.3580			20.50		0.8070
		5.75		0.2263			9.10		0.3583				13/16	0.8125
1				0.2280				23/64	0.3593			21.00		0.8267
		5.80		0.2283			9.20		0.3622				53/64	0.8281
		5.90		0.2323			9.25		0.3641				27/32	0.8437
	A			0.2340			9.30		0.3661			21.50		0.8464
			15/64	0.2340		U			0.3680				55/64	0.8593
		6.00		0.2362			9.40		0.3701			22.00		0.8661
	B			0.2380			9.50		0.3740				7/8	0.8750
		6.10		0.2402				3/8	0.3750			22.50		0.8858
	C			0.2420		V			0.3770				57/64	0.8906
		6.20		0.2441			9.60		0.3780			23.00		0.9055
	D			0.2460			9.70		0.3819				29/32	0.9062
		6.25		0.2460			9.75		0.3838				59/64	0.9218
		6.30		0.2480			9.80		0.3858			23.50		0.9251
	E		1/4	0.2500		W			0.3860				15/16	0.9375
		6.40		0.2519			9.90		0.3898			24.00		0.9448
		6.50		0.2559				25/64	0.3906				61/64	0.9531
	F			0.2570			10.00		0.3937			24.50		0.9646
		6.60		0.2598		X			0.3970				31/32	0.9687
	G			0.2610		Y			0.4040			25.00		0.9842
		6.70		0.2637				13/32	0.4062				63/64	0.9843
			17/64	0.2656		Z			0.4130				1.0	1.0000
		6.75		0.2657			10.50		0.4134			25.50		1.0039
	H			0.2660				27/64	0.4218					

SOURCE: Adapted from Colvin and Stanley, "American Machinists' Handbook," 8th ed., McGraw-Hill, New York, 1945.

8.3 GEARING
by George W. Michalec

REFERENCES: Buckingham, "Manual of Gear Design," Industrial Press. Cunningham, Noncircular Gears, *Mach. Des.*, Feb. 19, 1957. Cunningham and Cunningham, Rediscovering the Noncircular Gear, *Mach. Des.*, Nov. 1,1973. Dudley, "Gear Handbook," McGraw-Hill. Dudley, "Handbook of Practical Gear Design," McGraw-Hill. Michalec, "Precision Gearing: Theory and Practice," Wiley. Shigely, "Engineering Design," McGraw-Hill. AGMA Standards. "Gleason Bevel and Hypoid Gear Design," Gleason Works, Rochester. "Handbook of Gears: Inch and Metric" and "Elements of Metric Gear Technology," Designatronics, New Hyde Park, NY. Adams, "Plastics Gearing: Selection and Application," Marcel Dekker.

Notation

a = addendum
b = dedendum
B = backlash, linear measure along pitch circle
c = clearance
C = center distance
d = pitch diam of pinion
d_b = base circle diam of pinion
d_o = outside diam of pinion
d_r = root diam of pinion
D = pitch diameter of gear
D_P = pitch diam of pinion
D_G = pitch diam of gear
D_o = outside diam of gear
D_b = base circle diam of gear
D_t = throat diam of wormgear
F = face width
h_k = working depth
h_t = whole depth
inv ϕ = involute function (tan $\phi - \phi$)
l = lead (advance of worm or helical gear in 1 rev)
$l_p(l_G)$ = lead of pinion (gear) in helical gears
L = lead of worm in one revolution
m = module
m_G = gear ratio ($m_G = N_G/N_P$)
m_p = contact ratio (of profiles)
M = measurement of over pins
$n_p(n_G)$ = speed of pinion (gear), r/min
$N_P(N_G)$ = number of teeth in pinion (gear)
n_w = number of threads in worm
p = circular pitch
p_b = base pitch
p_n = normal circular pitch of helical gear
P_d = diametral pitch
P_{dn} = normal diametral pitch
R = pitch radius
R_c = radial distance from center of gear to center of measuring pin
$R_P(R_G)$ = pitch radius of pinion (gear)
R_T = testing radius when rolled on a variable-center-distance inspection fixture
s = stress
t = tooth thickness
t_n = normal circular tooth thickness
$T_P(T_G)$ = formative number of teeth in pinion (gear) (in bevel gears)
v = pitch line velocity
X = correction factor for profile shift
α = addendum angle of bevel gear
γ = pitch angle of bevel pinion
γ_R = face angle at root of bevel pinion tooth
γ_o = face angle at tip of bevel pinion tooth
Γ = pitch angle of bevel gear

Γ_R = face angle at root of bevel gear tooth
Γ_{g_o} = face angle at tip of bevel gear tooth
δ = dedendum angle of bevel gear
$\overline{\Delta C}$ = relatively small change in center distance C
ϕ = pressure angle
ϕ_n = normal pressure angle
ψ = helix or spiral angle
$\psi_P(\psi_G)$ = helix angle of teeth in pinion (gear)
Σ = shaft angle of meshed bevel pair

BASIC GEAR DATA

Gear Types Gears are grouped in accordance with tooth forms, shaft arrangement, pitch, and quality. Tooth forms and shaft arrangements are:

Tooth form	Shaft arrangement
Spur	Parallel
Helical	Parallel or skew
Worm	Skew
Bevel	Intersecting
Hypoid	Skew

Pitch definitions (see Fig. 8.3.1). **Diametral pitch** P_d is the ratio of number of teeth in the gear to the diameter of the pitch circle D measured

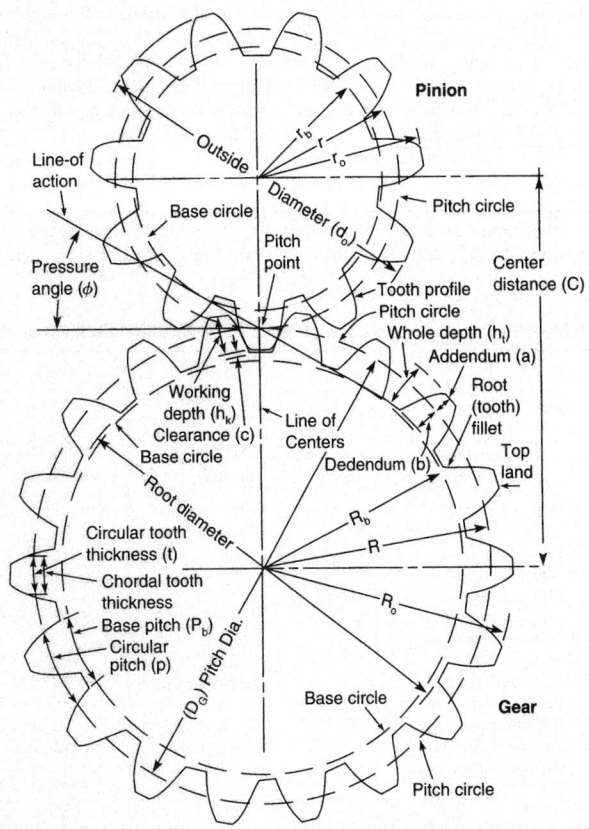

Fig. 8.3.1a Basic gear geometry and nomenclature.

$P_d = 10$
$p = 0.3142$
$P_d = 20$
$p = 0.1571$

Fig. 8.3.1*b*   Comparison of pitch and tooth size.

in inches, $P_d = N/D$. **Circular pitch** p is the linear measure in inches along the pitch circle between corresponding points of adjacent teeth. From these definitions, $P_d p = \pi$. The **base pitch** p_b is the distance along the line of action between successive involute tooth surfaces. The base and circular pitches are related as $p_b = p \cos \phi$, where $\phi =$ the pressure angle.

Pitch circle is the imaginary circle that rolls without slippage with a pitch circle of a mating gear. The pitch (circle) diameter equals $D = N/P_d = Np/\pi$. The basic relation between P_d and p is $P_d p = \pi$.

Tooth size is related to pitch. In terms of diametral pitch P_d, the relationship is inverse; i.e., large P_d implies a small tooth, and small P_d implies a large tooth. Conversely, there is a direct relationship between tooth size and circular pitch p. A small tooth has a small p, but a large tooth has a large p. (See Fig. 8.3.1*b*.) In terms of P_d, coarse teeth comprise P_d less than 20; fine teeth comprise P_d of 20 and higher. (See Fig. 8.3.1*b*.) **Quality** of gear teeth is classified as commercial, precision, and ultraprecision.

Pressure angle ϕ for all gear types is the acute angle between the common normal to the profiles at the contact point and the common pitch plane. For spur gears it is simply the acute angle formed by the common tangent between base circles of mating gears and a normal to the line of centers. For standard gears, pressure angles of 14½°, 20°, and 25° have been adopted by ANSI and the gear industry (see Fig. 8.3.1*a*). The 20° pressure angle is most widely used because of its versatility. The higher pressure angle 25° provides higher strength for highly loaded gears. Although 14½° appears in standards, and in past decades was extensively used, it is used much less than 20°. The 14½° standard is still used for replacement gears in old design equipment, in applications where backlash is critical, and where advantage can be taken of lower backlash with change in center distance.

The **base circle** (or **base cylinder**) is the circle from which the involute tooth profiles are generated. The relationship between the base-circle and pitch-circle diameter is $D_b = D \cos \phi$.

Tooth proportions are established by the addendum, dedendum, working depth, clearance, tooth circular thickness, and pressure angle

(see Fig. 8.3.1). In addition, gear face width F establishes thickness of the gear measured parallel to the gear axis.

For involute teeth, proportions have been standardized by ANSI and AGMA into a limited number of systems using a basic rack for specification (see Fig. 8.3.2 and Table 8.3.1). Dimensions for the basic rack are

Fig. 8.3.2   Basic rack for involute gear systems. $a =$ addendum; $b =$ dedendum; $c =$ clearance; $h_k =$ working depth; $h_t =$ whole depth; $p =$ circular pitch; $r_f =$ fillet radius; $t =$ tooth thickness; $\phi =$ pressure angle.

normalized for diametral pitch = 1. Dimensions for a specific pitch are obtained by dividing by the pitch. Standards for basic involute spur, helical and face gear designs, and noninvolute bevel and wormgear designs are listed in Table 8.3.2.

Table 8.3.2   Gear System Standards

Gear type	ANSI/AGMA no.	Title
Spur and helical	201.02	Tooth Proportions for Coarse-Pitch Involute Spur Gears
Spur and helical	1003-G93	Tooth Proportions for Fine-Pitch Spur and Helical Gearing
Spur and helical	370.01	Design Manual for Fine-Pitch Gearing
Bevel gears	2005-B88	Design Manual for Bevel Gears (Straight, Zerol, Spiral, and Hypoid)
Worm gearing	6022-C93	Design of General Industrial Coarse-Pitch Cylindrical Worm Gearing
Worm gearing	6030-C87	Design of Industrial Double-Enveloping Worm Gearing
Face gears	203.03	Fine-Pitch on Center-Face Gears for 20-Degree Involute Spur Pinions

Table 8.3.1   Tooth Proportions of Basic Rack for Standard Involute Gear Systems

Tooth parameter (of basic rack)	Symbol, Figs. 8.3.1a and 8.3.2	Tooth proportions for various standard systems					
		1	2	3	4	5	6
		Full-depth involute, 14½°	Full-depth involute, 20°	Stub involute, 20°	Coarse-pitch involute spur gears, 20°	Coarse-pitch involute spur gears, 25°	Fine-pitch involute, 20°
1. System sponsors		ANSI and AGMA	ANSI	ANSI and AGMA	AGMA	AGMA	ANSI and AGMA
2. Pressure angle	ϕ	14½°	20°	20°	20°	25°	20°
3. Addendum	a	$1/P_d$	$1/P_d$	$0.8/P_d$	$1.000/P_d$	$1.000/P_d$	$1.000/P_d$
4. Min dedendum	b	$1.157/P_d$	$1.157/P_d$	$1/P_d$	$1.250/P_d$	$1.250/P_d$	$1.200/P_d + 0.002$
5. Min whole depth	h_t	$2.157/P_d$	$2.157/P_d$	$1.8/P_d$	$2.250/P_d$	$2.250/P_d$	$2.2002/P_d\ 0.002$ in
6. Working depth	h_k	$2/P_d$	$2/P_d$	$1.6/P_d$	$2.000/P_d$	$2.000/P_d$	$2.000/P_d$
7. Min clearance	h_c	$0.157/P_d$	$0.157/P_d$	$0.200/P_d$	$0.250/P_d$	$0.250/P_d$	$0.200/P_d + 0.002$ in
8. Basic circular tooth thickness on pitch line	t	$1.5708/P_d$	$1.5708/P_d$	$1.5708/P_d$	$\pi/(2P_d)$	$\pi/(2P_d)$	$1.5708/P_d$
9. Fillet radius in basic rack	r_f	1⅓ × clearance	1½ × clearance	Not standardized	$0.300/P_d$	$0.300/P_d$	Not standardized
10. Diametral pitch range		Not specified	Not specified	Not specified	19.99 and coarser	19.99 and coarser	20 and finer
11. Governing standard:							
ANSI		B6.1	B6.1	B6.1			
AGMA		201.02		201.02	201.02	201.02	1,003-G93

Gear ratio (or **mesh ratio**) m_G is the ratio of number of teeth in a meshed pair, expressed as a number greater than 1; $m_G = N_G/N_P$, where the pinion is the member having the lesser number of teeth. For spur and parallel-shaft helical gears, the base circle ratio must be identical to the gear ratio. The speed ratio of gears is inversely proportionate to their numbers of teeth. Only for standard spur and parallel-shaft helical gears is the pitch diameter ratio equal to the gear ratio and inversely proportionate to the speed ratio. Usually, speed (velocity) ratio $V_r = N_{\text{driver}}/N_{\text{driven}}$. Thus, V_r can be less than or greater than 1.

Metric Gears—Tooth Proportions and Standards

Metric gearing not only is based upon different units of length measure but also involves its own unique design standard. This means that metric gears and American-standard-inch diametral-pitch gears are not interchangeable.

In the metric system the *module m* is analogous to pitch and is defined as

$$m = \frac{D}{N} = \text{mm of pitch diameter per tooth}$$

Note that, for the module to have proper units, the pitch diameter must be in millimeters.

The **metric module** was developed in a number of versions that differ in minor ways. The German DIN standard is widely used in Europe and other parts of the world. The Japanese have their own version defined in JS standards. Deviations among these and other national standards are minor, differing only as to dedendum size and root radii. The differences have been resolved by the new unified module standard promoted by the International Standards Organization (ISO). This unified version (Fig. 8.3.3) conforms to the new SI in all respects. All major industrial

Fig. 8.3.3 The ISO basic rack for metric module gears.

countries on the metric system have shifted to this ISO standard, which also is the basis for American metric gearing. Table 8.3.3 lists pertinent current ISO metric standards.

Table 8.3.3 ISO Metric Gearing Standards

ISO 53:1974	Cylindrical gears for general and heavy engineering—Basic rack
ISO 54:1977	Cylindrical gears for general and heavy engineering—Modules and diametral pitches
ISO 677:1976	Straight bevel gears for general and heavy engineering—Basic rack
ISO 678:1976	Straight bevel gears for general and heavy engineering—Modules and diametral pitches
ISO 701:1976	International gear notation—symbols for geometric data
ISO 1122-1:1983	Glossary of gear terms—Part 1: Geometric definitions
ISO 1328:1975	Parallel involute gears—ISO system of accuracy
ISO 1340:1976	Cylindrical gears—Information to be given to the manufacturer by the purchaser in order to obtain the gear required
ISO 1341:1976	Straight bevel gears—Information to be given to the manufacturer by the purchaser in order to obtain the gear required
ISO 2203:1973	Technical drawings—Conventional representation of gears

Tooth proportions for standard spur and helical gears are given in terms of the basic rack. Dimensions, in millimeters, are normalized for module $m = 1$. Corresponding values for other modules are obtained by multiplying each dimension by the value of the specific module m. Major tooth parameters are described by this standard:

Tooth form: Straight-sided and full-depth, forming the basis of a family of full-depth interchangeable gears.
Pressure angle: 20°, conforming to worldwide acceptance.
Addendum: Equal to module m, which corresponds to the American practice of $1/P_d$ = addendum.
Dedendum: Equal to 12.5m, which corresponds to the American practice of $1.25/P_d$ = dedendum.
Root radius: Slightly greater than American standards specifications.
Tip radius: A maximum is specified, whereas American standards do not specify. In practice, U.S. manufacturers can specify a tip radius as near zero as possible.

Note that the basic racks for metric and American inch gears are essentially identical, but metric and American standard gears are not interchangeable.

The preferred standard gears of the metric system are not interchangeable with the preferred diametral-pitch sizes. Table 8.3.4 lists commonly used pitches and modules of both systems (preferred values are boldface).

Metric gear use in the United States, although expanding, is still a small percentage of total gearing. Continuing industry conversions and imported equipment replacement gearing are building an increasing demand for metric gearing. The reference list cites a domestic source of stock metric gears of relatively small size, in medium and fine pitches. Large diameter coarse pitch metric gears are made to order by many gear fabricators.

FUNDAMENTAL RELATIONSHIPS OF SPUR AND HELICAL GEARS

Center distance is the distance between axes of mating gears and is determined from $C = (n_G + N_p)/(2P_d)$, or $C = (D_G + D_p)/2$. Deviation from ideal center distance of involute gears is not detrimental to proper (conjugate) gear action which is one of the prime superiority features of the involute tooth form.

Contact Ratio   Referring to the top part of Fig. 8.3.4 and assuming no tip relief, the pinion engages in the gear at a, where the outside circle of the gear tooth intersects the line of action ac. For the usual spur gear and pinion combinations there will be two pairs of teeth theoretically in contact at engagement (a gear tooth and its mating pinion tooth considered as a pair). This will continue until the pair ahead (bottom part of Fig. 8.3.4) disengages at c, where the outside circle of the pinion intersects the line of action ac, the movement along the line of action being

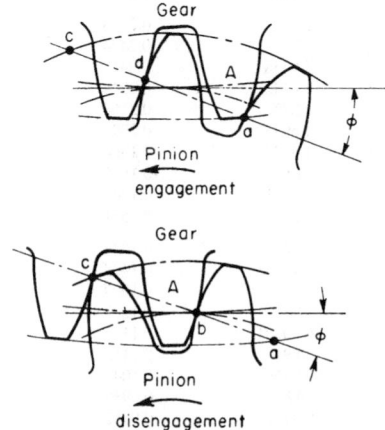

Fig. 8.3.4 Contact conditions at engagement and disengagement.

Table 8.3.4 Metric and American Gear Equivalents

Diametral pitch P_d	Module m	Circular pitch		Circular tooth thickness		Addendum	
		in	mm	in	mm	in	mm
1/2	50.8000	6.2832	159.593	3.1416	79.7965	2.0000	50.8000
0.5080	50	6.1842	157.080	3.0921	78.5398	1.9685	50
0.5644	45	5.5658	141.372	2.7850	70.6858	1.7730	45
0.6048	42	5.1948	131.947	2.5964	65.9734	1.6529	42
0.6513	39	4.8237	122.522	2.4129	61.2610	1.5361	39
0.7056	36	4.4527	113.097	2.2249	56.5487	1.4164	36
3/4	33.8667	4.1888	106.396	2.0943	53.1977	1.3333	33.8667
0.7697	33	4.0816	103.673	2.0400	51.8363	1.2987	33
0.8467	30	3.7105	94.248	1.8545	47.1239	1.1806	30
0.9407	27	3.3395	84.823	1.6693	42.4115	1.0627	27
1	25.4000	3.1416	79.800	1.5708	39.8984	1.0000	25.4001
1.0583	24	2.9685	75.398	1.4847	37.6991	0.9452	24
1.1546	22	2.7210	69.115	1.3600	34.5575	0.8658	22
1.2700	20	2.4737	62.832	1.2368	31.4159	0.7874	20
1.4111	18	2.2263	56.548	1.1132	28.2743	0.7087	18
1.5	16.9333	2.0944	53.198	1.0472	26.5988	0.6667	16.933
1.5875	16	1.9790	50.267	0.9894	25.1327	0.6299	16
1.8143	14	1.7316	43.983	0.8658	21.9911	0.5512	14
2	12.7000	1.5708	39.898	0.7854	19.949	0.5000	12.7000
2.1167	12	1.4842	37.699	0.7420	18.8496	0.4724	12
2.5	10.1600	1.2566	31.918	0.6283	15.9593	0.4000	10.1600
2.5400	10	1.2368	31.415	0.6184	15.7080	0.3937	10
2.8222	9	1.1132	28.275	0.5565	14.1372	0.3543	9
3	8.4667	1.0472	26.599	0.5235	13.2995	0.3333	8.4667
3.1416	8.0851	1.0000	25.400	0.5000	12.7000	0.3183	0.0851
3.1750	8	0.9895	25.133	0.4948	12.5664	0.3150	8.00
3.5	7.2571	0.8976	22.799	0.4488	11.3994	0.2857	7.2571
3.6286	7	0.8658	21.991	0.4329	10.9956	0.2756	7.000
3.9078	6.5	0.8039	20.420	0.4020	10.2101	0.2559	6.5
4	6.3500	0.7854	19.949	0.3927	9.9746	0.2500	6.3500
4.2333	6	0.7421	18.850	0.3710	9.4248	0.2362	6.0000
4.6182	5.5	0.6803	17.279	0.3401	8.6394	0.2165	5.5
5	5.0801	0.6283	15.959	0.3142	7.9794	0.2000	5.080
5.0802	5	0.6184	15.707	0.3092	7.8537	0.1968	5.000
5.3474	4.75	0.5875	14.923	0.2938	7.4612	0.1870	4.750
5.6444	4.5	0.5566	14.138	0.2783	7.0688	0.1772	4.500
6	4.2333	0.5236	13.299	0.2618	6.6497	0.1667	4.233
6.3500	4	0.4947	12.565	0.2473	6.2827	0.1575	4.000
6.7733	3.75	0.4638	11.781	0.2319	5.8903	0.1476	3.750
7	3.6286	0.4488	11.399	0.2244	5.6998	0.1429	3.629
7.2571	3.5	0.4329	10.996	0.2164	5.4979	0.1378	3.500
7.8154	3.25	0.4020	10.211	0.2010	5.1054	0.1279	3.250
8	3.1750	0.3927	9.974	0.1964	4.9886	0.1250	3.175
8.4667	3	0.3711	9.426	0.1855	4.7130	0.1181	3.000
9	2.8222	0.3491	8.867	0.1745	4.4323	0.1111	2.822
9.2364	2.75	0.3401	8.639	0.1700	4.3193	0.1082	2.750
10	2.5400	0.3142	7.981	0.1571	3.9903	0.1000	2.540
10.1600	2.50	0.3092	7.854	0.1546	3.9268	0.0984	2.500
11	2.3091	0.2856	7.254	0.1428	3.6271	0.0909	2.309
11.2889	2.25	0.2783	7.069	0.1391	3.5344	0.0886	2.250
12	2.1167	0.2618	6.646	0.1309	3.3325	0.0833	2.117
12.7000	2	0.2474	6.284	0.1236	3.1420	0.0787	2.000
13	1.9538	0.2417	6.139	0.1208	3.0696	0.0769	1.954
14	1.8143	0.2244	5.700	0.1122	2.8500	0.0714	1.814
14.5143	1.75	0.2164	5.497	0.1082	2.7489	0.0689	1.750
15	1.6933	0.2094	5.319	0.1047	2.6599	0.0667	1.693
16	1.5875	0.1964	4.986	0.0982	2.4936	0.0625	1.587
16.9333	1.5	0.1855	4.712	0.0927	2.3562	0.0591	1.500
18	1.4111	0.1745	4.432	0.0873	2.2166	0.0556	1.411
20	1.2700	0.1571	3.990	0.0785	1.9949	0.0500	1.270
20.3200	1.25	0.1546	3.927	0.0773	1.9635	0.0492	1.250
22	1.1545	0.1428	3.627	0.0714	1.8136	0.0455	1.155
24	1.0583	0.1309	3.325	0.0655	1.6624	0.0417	1.058
25.4000	1	0.1237	3.142	0.0618	1.5708	0.0394	1.000
28	0.90701	0.1122	2.850	0.0561	1.4249	0.0357	0.9071
28.2222	0.9	0.1113	2.827	0.0556	1.4137	0.0354	0.9000
30	0.84667	0.1047	2.659	0.0524	1.3329	0.0333	0.8467
31.7500	0.8	0.0989	2.513	0.04945	1.2566	0.0315	0.8000
32	0.79375	0.0982	2.494	0.04909	1.2468	0.0313	0.7937
33.8667	0.75	0.0928	2.357	0.04638	1.1781	0.0295	0.7500
36	0.70556	0.0873	2.217	0.04363	1.1083	0.0278	0.7056

Table 8.3.4   Metric and American Gear Equivalents   (Continued)

Diametral pitch P_d	Module m	Circular pitch		Circular tooth thickness		Addendum	
		in	mm	in	mm	in	mm
36.2857	**0.7**	0.0865	2.199	0.04325	1.0996	0.0276	0.7000
40	0.63500	0.0785	1.994	0.03927	0.9975	0.0250	0.6350
42.3333	**0.6**	0.0742	1.885	0.03710	0.9423	0.0236	0.6000
44	0.57727	0.0714	1.814	0.03570	0.9068	0.0227	0.5773
48	0.52917	0.0655	1.661	0.03272	0.8311	0.0208	0.5292
50	0.50800	0.0628	1.595	0.03141	0.7976	0.0200	0.5080
50.8000	**0.5**	0.06184	1.5707	0.03092	0.7854	0.0197	0.5000
63.5000	**0.4**	0.04947	1.2565	0.02473	0.6283	0.0157	0.4000
64	0.39688	0.04909	1.2469	0.02454	0.6234	0.0156	0.3969
67.7333	0.375	0.04638	1.1781	0.02319	0.5890	0.0148	0.3750
72	0.35278	0.04363	1.1082	0.02182	0.5541	0.0139	0.3528
72.5714	0.35	0.04329	1.0996	0.02164	0.5498	0.0138	0.3500
78.1538	0.325	0.04020	1.0211	0.02010	0.5105	0.0128	0.3250
80	0.31750	0.03927	0.9975	0.01964	0.4987	0.0125	0.3175
84.6667	**0.3**	0.03711	0.9426	0.01856	0.4713	0.0118	0.3000
92.3636	0.275	0.03401	0.8639	0.01700	0.4319	0.0108	0.2750
96	0.26458	0.03272	0.8311	0.01636	0.4156	0.0104	0.2646
101.6000	0.25	0.03092	0.7854	0.01546	0.3927	0.00984	0.2500
120	0.21167	0.02618	0.6650	0.01309	0.3325	0.00833	0.2117
125	0.20320	0.02513	0.6383	0.01256	0.3192	0.00800	0.2032
127.0000	**0.2**	0.02474	0.6284	0.01237	0.3142	0.00787	0.2000
150	0.16933	0.02094	0.5319	0.01047	0.2659	0.00667	0.1693
169.3333	0.15	0.01855	0.4712	0.00928	0.2356	0.00591	0.1500
180	0.14111	0.01745	0.4432	0.00873	0.2216	0.00555	0.1411
200	0.12700	0.01571	0.3990	0.00786	0.1995	0.00500	0.1270
203.2000	0.125	0.01546	0.3927	0.00773	0.1963	0.00492	0.1250

ab. After disengagement the pair behind will be the only pair in contact until another pair engages, the movement along the line of action for single-pair contact being *bd.* Two pairs are theoretically in contact during the remaining intervals, *ab* + *dc.*

Contact ratio expresses the average number of pairs of teeth theoretically in contact and is obtained numerically by dividing the length of the line of action by the normal pitch. For full-depth teeth, without undercutting, the contact ratio is $m_p = (\sqrt{D_0^2 - D_b^2} + \sqrt{d_0^2 - d_b^2} - 2C \sin \phi)/(2p \cos \phi)$. The result will be a mixed number with the integer portion the number of pairs of teeth always in contact and carrying load, and the decimal portion the amount of time an additional pair of teeth are engaged and share load. As an example, for m_p between 1 and 2:

Load is carried by one pair, $(2 - m_p)/m_p$ of the time.
Load is carried by two pairs, $2(m_p - 1)/m_p$ of the time.
In Figs. 8.3.5 to 8.3.7, contact ratios are given for standard generated gears, the lower part of Figs. 8.3.5 and 8.3.6 representing the effect of undercutting.

These charts are applicable to both standard dimetral pitch gears made in accordance with American standards and also standard metric gears that have an addendum of one module.

Tooth Thickness   For standard gears, the tooth thickness t of mating gears is equal, where $t = p/2 = \pi/(2P_d)$ measured linearly along the arc of the pitch circle. The tooth thickness t_1 at any radial point of the tooth (at diameter D_1) can be calculated from the known thickness t at

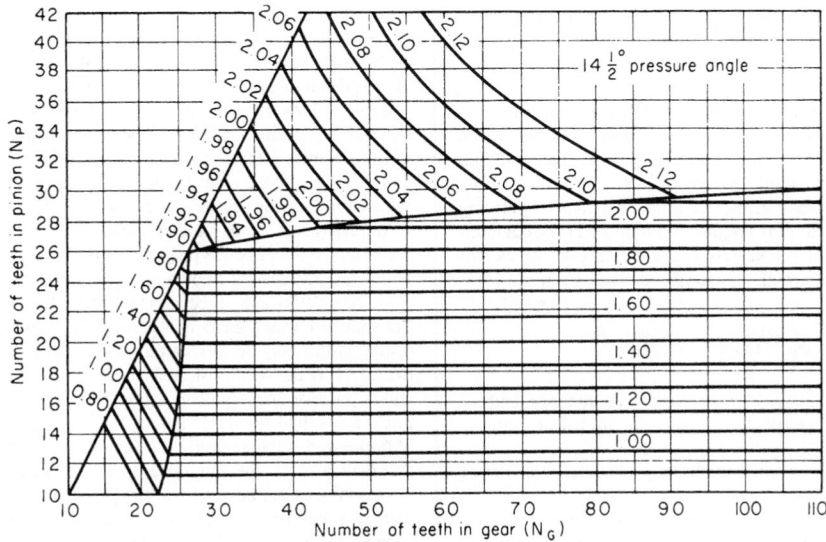

Fig. 8.3.5   Contact ratio, spur gear pairs—full depth, standard generated teeth, $14\frac{1}{2}°$ pressure angle.

Fig. 8.3.6   Contact ratio, spur gear pairs—full-depth standard generated teeth, 20° pressure angle.

Fig. 8.3.7   Contact ratio for large numbers of teeth—spur gear pairs, full-depth standard teeth, 20° pressure angle. (*Data by R. Feeney and T. Wall.*)

Fig. 8.3.8   Geometry of over-pins measurements (*a*) for an even number of teeth and (*b*) for an odd number of teeth.

the pitch radius $D/2$ by the relationship $t_1 = t(D_1/D) - D_1$ (inv $\phi_1 -$ inv ϕ), where inv $\phi = \tan \phi - \phi =$ involute function. Units for ϕ must be radians. Tables of values for inv ϕ from 0 to 45° can be found in the references (Buckingham and Dudley).

Over-plus measurements (spur gears) are another means of deriving tooth thickness. If cylindrical pins are inserted in tooth spaces diametrically opposite one another (or nearest space for an odd number of teeth) (Fig. 8.3.8), the tooth thickness can be derived from the measurement M as follows:

$$t = D(\pi/N + \text{inv } \phi_1 - \text{inn } \phi - d_w/D_b)$$
$$\cos \phi_1 = (D \cos \phi)/2R_c$$
$$R_c = (M - d_w)/2 \qquad \text{for even number of teeth}$$
$$R_c = (M - d_w)/[2 \cos (90/N)] \qquad \text{for odd number of teeth}$$

where $d_w =$ pin diameter, $R_c =$ distance from gear center to center of pin, and $M =$ measurement over pins.

For the reverse situation, the over-pins measurement M can be found for a given tooth thickness t at diameter D and pressure angle ϕ

by the following: inv $\phi_1 = t/D + \text{inv } \phi + d_w/(D \cos \phi) - \pi/N$, $M = D \cos \phi/\cos \phi_1 + d_w$ (for even number of teeth), $M = (D \cos \phi/\cos \phi_1) \cos (90°/N) + d_w$ (for odd number of teeth).

Table values of over-pins measures (see Dudley and Van Keuren) facilitate measurements for all standard gears including those with slight departures from standard. (For correlation with tooth thickness and testing radius, see Michalec, *Product Eng.*, May 1957, and "Precision Gearing: Theory and Practice," Wiley.)

Testing radius R_T is another means of determining tooth thickness and refers to the effective pitch radius of the gear when rolled intimately with a master gear of known size calibration. (See Michalec, *Product Eng.*, Nov. 1956, and "Precision Gearing: Theory and Practice," op. cit.) For standard design gears the testing radius equals the pitch radius. The testing radius may be corrected for small departures Δt from ideal tooth thickness by the relationship, $R_T = R + \Delta t/2 \tan \phi$, where $\Delta t = t_1 - t$ and is positive and negative, respectively, for thicker and thinner tooth thicknesses than standard value t.

Backlash B is the amount by which the width of a tooth space exceeds the thickness of the engaging tooth measured on the pitch circle. Backlash does not adversely affect proper gear function except for lost motion upon reversal of gear rotation. Backlash inevitably occurs because of necessary fabrication tolerances on tooth thickness and center

Table 8.3.5   Metric Spur Gear Design Formulas

To obtain:	From known	Use this formula*
Pitch diameter D	Module; diametral pitch	$D = mN$
Circular pitch p_c	Module; diametral pitch	$p_c = m\pi = \dfrac{D}{N}\pi = \dfrac{\pi}{P}$
Module m	Diametral pitch	$m = \dfrac{25.4}{P}$
No. of teeth N	Module and pitch diameter	$N = \dfrac{D}{m}$
Addendum a	Module	$a = m$
Dedendum b	Module	$b = 1.25m$
Outside diameter D_o	Module and pitch diameter or number of teeth	$D_o = D + 2m = m\,(N + 2)$
Root diameter D_r	Pitch diameter and module	$D_r = D - 2.5m$
Base circle diameter D_b	Pitch diameter and pressure angle ϕ	$D_b = D\cos\phi$
Base pitch p_b	Module and pressure angle	$p_b = m\pi\cos\phi$
Tooth thickness at standard pitch diameter T_{std}	Module	$T_{\text{std}} = \dfrac{\pi}{2}m$
Center distance C	Module and number of teeth	$C = \dfrac{m(N_1 + N_2)}{2}$
Contact ratio m_p	Outside radii, base-circle radii, center distance, pressure angle	$m_p = \dfrac{\sqrt{{}_1R_o^2 - {}_1R_b^2} + \sqrt{{}_2R_o^2 - {}_2R_b^2} - C\sin\phi}{m\,\pi\cos\phi}$
Backlash (linear) B (along pitch circle)	Change in center distance	$B = 2(\Delta C)\tan\phi$
Backlash (linear) B (along pitch circle)	Change in tooth thickness, T	$B = \Delta T$
Backlash (linear) (along line of action) B_{LA}	Linear backlash (along pitch circle)	$B_{LA} = B\cos\phi$
Backlash (angular) B_a	Linear backlash (along pitch circle)	$B_a = 6{,}880\,\dfrac{B}{D}$ (arc minutes)
Min. number teeth for no undercutting, N_c	Pressure angle	$N_c = \dfrac{2}{\sin^2\phi}$

* All linear dimensions in millimeters.

distance plus need for clearance to accommodate lubricant and thermal expansion. Proper backlash can be introduced by a specified amount of tooth thinning or slight increase in center distance. The relationship between small change in center distance $\overline{\Delta C}$ and backlash is $B = 2\,\overline{\Delta C}$ $\tan\phi$ (see Michalec, "Precision Gearing: Theory and Practice").

Total composite error (tolerance) is a measure of gear quality in terms of the net sum of irregularity of its testing radius R_T due to pitch-circle runout and tooth-to-tooth variations (see Michalec, op. cit.).

Tooth-to-tooth composite error (tolerance) is the variation of testing radius R_T between adjacent teeth caused by tooth spacing, thickness, and profile deviations (see Michalec, op. cit.).

Profile shifted gears have tooth thicknesses that are significantly different from nominal standard value; excluded are deviations caused by normal allowances and tolerances. They are also known as **modified gears, long and short addendum gears,** and **enlarged gears.** They are produced by cutting the teeth with standard cutters at enlarged or reduced outside diameters. The result is a relative shift of the two families of involutes forming the tooth profiles, simultaneously with a shift of the tooth radially outward or inward (see Fig. 8.3.9). Calculation of operating conditions and tooth parameters are

$$C_1 = \frac{C\cos\phi}{\cos\phi_1}$$

$$\text{inv}\,\phi_1 = \text{inv}\,\phi + \frac{N_P(t'_G + t'_P) - \pi D_P}{D_P(N_P + N_G)}$$

$$t'_G = t + 2X_G\tan\phi$$
$$t'_P = t + 2X_P\tan\phi$$
$$D'_G = (N_G/P_d) + 2X_G$$
$$D'_P = (N_P/P_d) + 2X_P$$
$$D'_o = D'_G + (2/P_d)$$
$$d'_o = D'_P + (2/P_d)$$

where ϕ = standard pressure angle, ϕ_1 = operating pressure angle, C = standard center distance = $(N_G + N_P)/2P_d$, C_1 = operating center

distance, X_G = profile shift correction of gear, and X_P = profile shift correction of pinion. The quantity X is positive for enlarged gears and negative for thinned gears.

Fig. 8.3.9   Geometry of profile-shifted teeth. (a) Enlarged case; (b) thinned tooth thickness case.

Metric Module Gear Design Equations   Basic design equations for spur gearing utilizing the metric module are listed in Table 8.3.5. (See Designatronics, "Elements of Metric Gear Technology.")

HELICAL GEARS

Helical gears divide into two general applications: for driving parallel shafts and for driving skew shafts (mostly at right angles), the latter often referred to as *crossed-axis* helical gears. The helical tooth form may be imagined as consisting of an infinite number of staggered laminar spur gears, resulting in the curved cylindrical helix.

Pitch of helical gears is definable in two planes. The diametral and circular pitches measured in the plane of rotation (transverse) are defined as for spur gears. However, pitches measured normal to the tooth are related by the cosine of the helix angle; thus normal diametral pitch = $P_{dn} = P_d/\cos\psi$, normal circular pitch = $p_n = p\cos\psi$, and $P_{dn}p_n = \pi$. **Axial pitch** is the distance between corresponding sides of adjacent teeth measured parallel to the gear axis and is calculated as $p_a = p\cot\psi$.

Pressure angle of helical gears is definable in the normal and transverse planes by $\tan \phi_n = \tan \phi \cos \psi$. The transverse pressure angle, which is effectively the real pressure angle, is always greater than the normal pressure angle.

Tooth thickness t of helical gears can be measured in the plane of rotation, as with spur gears, or normal to the tooth surface t_n. The relationship of the two thicknesses is $t_n = t \cos \psi$.

Over-Pins Measurement of Helical Gears Tooth thicknesses t at diameter d can be found from a known over-pins measurement M at known pressure angle ϕ, corresponding to diameter D as follows:

$$R_c = (M - d_w)/2 \qquad \text{for even number of teeth}$$
$$R_c = (M - d_w)/[2 \cos (180/2N)] \qquad \text{for odd number of teeth}$$
$$\cos \phi_1 = (D \cos \phi)/2R_c$$
$$\tan \phi_n = \tan \phi \cos \psi$$
$$\cos \psi_b = \sin \phi_n/\sin \phi$$
$$t = D[\pi/N + \text{inv } \phi_1 - \text{inv } \phi - d_w/(D \cos \phi \cos \psi_b)]$$

Parallel-shaft helical gears must conform to the same conditions and requirements as spur gears with parameters (pressure angle and pitch) consistently defined in the transverse plane. Since standard spur gear cutting tools are usually used, normal plane values are standard, resulting in nonstandard transverse pitches and nonstandard pitch diameters and center distances. For parallel shafts, helical gears must have identical helix angles, but must be of opposite hand (left and right helix directions). The commonly used helix angles range from 15 to 35°. To make most advantage of the helical form, the advance of a tooth should be greater than the circular pitch; recommended ratio is 1.5 to 2 with 1.1 minimum. This overlap provides two or more teeth in continual contact with resulting greater smoothness and quietness than spur gears. Because of the helix, the normal component of the tangential pressure on the teeth produces end thrust of the shafts. To remove this objection, gears are made with helixes of opposite hand on each half of the face and are then known as herringbone gears (see Fig. 8.3.10).

Crossed-axis helical gears, also called *spiral* or *screw gears* (Fig. 8.3.11), are a simple type of involute gear used for connecting nonparallel, nonintersecting shafts. Contact is the point and there is considerably more sliding than with parallel-axis helicals, which limits the load capacity. The individual gear of this mesh is identical in form and specification to a parallel-shaft helical gear. Crossed-axis helicals can connect

Conventional   Staggered   Continuous

Fig. 8.3.10   Herringbone gears.

any shaft angle Σ, although 90° is prevalent. Usually, the helix angles will be of the same hand, although for some extreme cases it is possible to have opposite hands, particularly if the shaft angle is small.

Fig. 8.3.11   Crossed-axis helical gears.

Helical Gear Calculations   For parallel shafts the center distance is a function of the helix angle as well as the number of teeth, that is, $C = (N_G + N_P)/(2P_{dn} \cos \psi)$. This offers a powerful method of gearing shafts at any specified center distance to a specified velocity ratio. For crossed-axis helicals the problem of connecting a pair of shafts for any velocity ratio admits of a number of solutions, since both the pitch radii and the helix angles contribute to establishing the velocity ratio. The formulas given in Tables 8.3.6 and 8.3.7 are of assistance in calculations. The notation used in these tables is as follows:

$N_P(N_G)$ = number of teeth in pinion (gear)
$D_P(D_G)$ = pitch diam of pinion (gear)
$p_P(p_G)$ = circular pitch of pinion (gear)
p = circular pitch in plane of rotation for both gears
P_d = diametral pitch in plane of rotation for both gears
p_n = normal circular pitch for both gears
P_{dn} = normal diametral pitch for both gears
ψ_G = tooth helix angle of gear
ψ_P = tooth helix angle of pinion
$l_P(l_G)$ = lead of pinion (gear)
= lead of tooth helix
$n_P(n_G)$ = r/min of pinion (gear)
Σ = angle between shafts in plan
C = center distance

Table 8.3.6   Helical Gears on Parallel Shafts

To find:	Formula
Center distance C	$\dfrac{N_G + N_P}{2P_{dn} \cos \psi}$
Pitch diameter D	$\dfrac{N}{P_d} = \dfrac{N}{P_{dn} \cos \psi}$
Normal diametral pitch P_{dn}	$\dfrac{P_d}{\cos \psi}$
Normal circular pitch p_n	$p \cos \psi$
Pressure angle ϕ	$\tan^{-1} \dfrac{\tan \phi_n}{\cos \psi}$
Contact ratio m_p	$\dfrac{\sqrt{_GD_o^2 - _GD_b^2} + \sqrt{_PD_o^2 - _PD_b^2} + 2C \sin \phi}{2p \cos \phi} + \dfrac{F \sin \psi}{p_n}$
Gear ratio m_G	$\dfrac{N_G}{N_P} = \dfrac{D_G}{D_P}$

Table 8.3.7   Crossed Helical Gears on Skew Shafts

To find:	Formula
Center distance C	$\dfrac{P_n}{2\pi} \left(\dfrac{N_G}{\cos \psi_G} + \dfrac{N_P}{\cos \psi_P} \right)$
Pitch diameter D_G, D_P	$D_G = \dfrac{N_G}{P_{dG}} = \dfrac{N_G}{P_{dn} \cos \psi_G} = \dfrac{N_G p_n}{\pi \cos \psi_G}$
	$D_P = \dfrac{N_P}{P_{dP}} = \dfrac{N_P}{P_{dn} \cos \psi_P} = \dfrac{n_P p_n}{\pi \cos \psi_P}$
Gear ratio m_G	$\dfrac{N_G}{N_P} = \dfrac{D_G \cos \psi_G}{D_P \cos \psi_P}$
Shaft angle Σ	$\psi_G + \psi_P$

NONSPUR GEAR TYPES*

Bevel gears are used to connect two intersecting shafts in any given speed ratio. The tooth shapes may be designed in any of the shapes shown in Fig. 8.3.12. A special type of gear known as a **hypoid** was

---

* In the following text relating to bevel gearing, all tables and figures have been extracted from Gleason Works publications, with permission.

Fig. 8.3.12 Bevel gear types. (*a*) Old-type straight teeth; (*b*) modern Coniflex straight teeth (exaggerated crowning); (*c*) Zerol teeth; (*d*) spiral teeth; (*e*) hypoid teeth.

developed by Gleason Works for the automotive industry (see *Jour. SAE,* **18,** no. 6). Although similar in appearance to a spiral bevel, it is not a true bevel gear. The basic pitch rolling surfaces are hyperbolas of revolution. Because a "spherical involute" tooth form has a curved crown tooth (the basic tool for generating all bevel gears), Gleason used a straight-sided crown tooth which resulted in bevel gears differing slightly from involute form. Because of the figure 8 shape of the complete theoretical tooth contact path, the tooth form has been called "octoid." Straight-sided bevel gears made by reciprocating cutters are of this type. Later, when curved teeth became widely used (spiral and Zerol), practical limitations of such cutters resulted in introduction of the "spherical" tooth form which is now the basis of all curved tooth bevel gears. (For details see Gleason's publication, "Guide to Bevel Gears.") Gleason Works also developed the generated tooth form Revecycle and the nongenerated tooth forms Formate and Helixform, used principally for mass production of hypoid gears for the automotive industry.

Referring to Fig. 8.3.13, we see that the pitch surfaces of bevel gears are frustums of cones whose vertices are at the intersection of the axes; the essential elements and definitions follow.

Addendum angle α: The angle between elements of the face cone and pitch cone.

Back angle: The angle between an element of the back cone and a plane of rotation. It is equal to the pitch angle.

Back cone: The angle of a cone whose elements are tangent to a sphere containing a trace of the pitch circle.

Back-cone distance: The distance along an element of the back cone from the apex to the pitch circle.

Cone distance A_o: The distance from the end of the tooth (heel) to the pitch apex.

Crown: The sharp corner forming the outside diameter.

Crown-to-back: The distance from the outside diameter edge (crown) to the rear of the gear.

Dedendum angle δ: The angle between elements of the root cone and pitch cone.

Face angle γ_o: The angle between an element of the face cone and its axis.

Face width F: The length of teeth along the cone distance.

Front angle: The angle between an element of the front cone and a plane of rotation.

Generating mounting surface, GMS: The diameter and/or plane of rotation surface or shaft center which is used for locating the gear blank during fabrication of the gear teeth.

Heel: The portion of a bevel gear tooth near the outer end.

Mounting distance, MD: For assembled bevel gears, the distance from the crossing point of the axes to the registering surface, measured along the gear axis. Ideally, it should be identical to the pitch apex to back.

Mounting surface, MS: The diameter and/or plane of rotation surface which is used for locating the gear in the application assembly.

Octoid: The mathematical form of the bevel tooth profile. Closely resembles a spherical involute but is fundamentally different.

Pitch angle Γ: The angle formed between an element of the pitch cone and the bevel gear axis. It is the half angle of the pitch.

Pitch apex to back: The distance along the axis from apex of pitch cone to a locating registering surface on back.

Registering surface, RS: The surface in the plane of rotation which locates the gear blank axially in the generating machine and

the gear in application. These are usually identical surfaces, but not necessarily so.

Root angle γ_R: The angle formed between a tooth root element and the axis of the bevel gear.

Shaft angle Σ: The angle between mating bevel-gear axes; also, the sum of the two pitch angles.

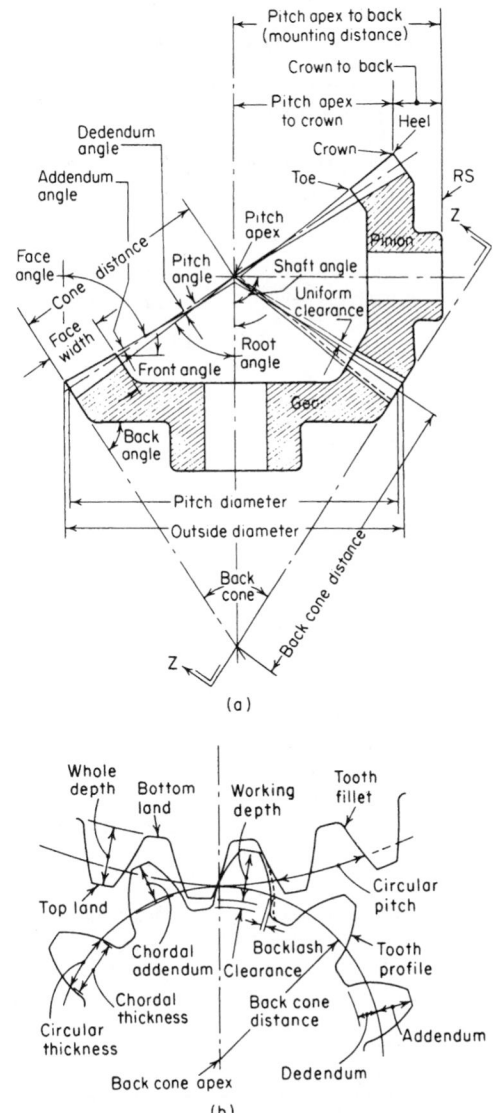

Fig. 8.3.13 Geometry of bevel gear nomenclature. (*a*) Section through axes; (*b*) view along axis Z/Z.

Spiral angle ψ: The angle between the tooth trace and an element of the pitch cone, corresponding to helix angle in helical gears. The spiral angle is understood to be at the mean cone distance.

Toe: The portion of a bevel tooth near the inner end.

Bevel gears are described by the parameter dimensions at the large end (heel) of the teeth. Pitch, pitch diameter, and tooth dimensions, such as addendum are measurements at this point. At the large end of the gear, the tooth profiles will approximate those generated on a spur gear pitch circle of radius equal to the back cone distance. The formative number of teeth is equal to that contained by a complete spur gear. For pinion and gear, respectively, this is $T_P = N_P/\cos \gamma$; $T_G = N_G/\cos \Gamma$, where T_P and T_G = formative number teeth and N_P and N_G = actual number teeth.

Although bevel gears can connect shafts at any angle, most applications are for right angles. When such bevels are in a 1:1 ratio, they are called **mitre gears**. Bevels connecting shafts other than 90° are called **angular bevel gears**. The speeds of the shafts of bevel gears are determined by $n_P/n_G = \sin \Gamma/\sin \gamma$, where $n_P(n_G)$ = r/min of pinion (gear), and $\gamma(\Gamma)$ = pitch angle of pinion (gear).

All standard bevel gear designs in the United States are in accordance with the **Gleason bevel gear system**. This employs a basic pressure angle of 20° with long and short addendums for ratios other than 1:1 to avoid undercut pinions and to increase strength.

20° Straight Bevel Gears for 90° Shaft Angle Since straight bevel gears are the easiest to produce and offer maximum precision, they are frequently a first choice. Modern straight-bevel-gears generators produce a tooth with localized tooth bearing designated by the Gleason registered tradename Coniflex. These gears, produced with a circular cutter, have a slightly crowned tooth form (see Fig. 8.3.12b). Because of the superiority of Coniflex bevel gears over the earlier reciprocating cutter produced straight bevels and because of their faster production, they are the standards for all bevel gears. The design parameters of Fig. 8.3.13 are calculated by the formulas of Table 8.3.8. Backlash data are given in Table 8.3.9.

Angular straight bevel gears connect shaft angles other than 90° (larger or smaller), and the formulas of Table 8.3.8 are not entirely applicable, as shown in the following:

Item 8, shaft angle, is the specified non-90° shaft angle.

Item 10, pitch angles. Shaft angle Σ less than 90°, $\tan \gamma = \sin \Sigma/(N/n + \cos \Sigma)$; shaft angle Σ greater than 90°, $\tan \gamma = \sin (180 - \Sigma)/[N/n - \cos (180° - \Sigma)]$.

Table 8.3.8 Straight Bevel Gear Dimensions*
(All linear dimensions in inches)

1. Number of pinion teeth†	n	5. Working depth	$h_k = \dfrac{2.000}{P_d}$
2. Number of gear teeth†	N	6. Whole depth	$h_t = \dfrac{2.188}{P_d} + 0.002$
3. Diametral pitch	P_d	7. Pressure angle	ϕ
4. Face with	F	8. Shaft angle	Σ

	Pinion	Gear
9. Pitch diameter	$d = \dfrac{n}{P_d}$	$D = \dfrac{N}{P_d}$
10. Pitch angle	$\gamma = \tan^{-1} \dfrac{n}{N}$	$\Gamma = 90° - \gamma$
11. Outer cone distance	$A_O = \dfrac{D}{2 \sin \Gamma}$	
12. Circular pitch	$p = \dfrac{3.1416}{P_d}$	
13. Addendum	$a_{OP} = h_k - a_{OG}$	$a_{OG} = \dfrac{0.540}{P_d} + \dfrac{0.460}{P_d(N/n)^2}$
14. Dedendum‡	$b_{OP} = \dfrac{2.188}{P_d} - a_{OP}$	$b_{OG} = \dfrac{2.188}{P_d} - a_{OG}$
15. Clearance	$c = h_t - h_k$	
16. Dedendum angle	$\delta_P = \tan^{-1} \dfrac{b_{OP}}{A_O}$	$\delta_G = \tan^{-1} \dfrac{b_{OG}}{A_O}$
17. Face angle of blank	$\gamma_O = \gamma + \delta_G$	$\Gamma_O = \Gamma + \delta_P$
18. Root angle	$\gamma_R = \gamma - \delta_P$	$\Gamma_R = \Gamma - \delta_G$
19. Outside diameter	$d_O = d + 2a_{OP} \cos \gamma$	$D_O = D + 2a_{OG} \cos \Gamma$
20. Pitch apex to crown	$x_O = \dfrac{D}{2} - a_{OP} \sin \gamma$	$X_O = \dfrac{d}{2} - a_{OG} \sin \Gamma$
21. Circular thickness	$t = p - T$	$T = \dfrac{p}{2} - (a_{OP} - a_{OG}) \tan \phi - \dfrac{K}{P_d}$ (see Fig. 8.3.14)
22. Backlash	B See Table 8.3.9	
23. Chordal thickness	$t_C = t - \dfrac{t^2}{6d^2} - \dfrac{B}{2}$	$T_C = T - \dfrac{T^2}{6D^2} - \dfrac{B}{2}$
24. Chordal addendum	$a_{CP} = a_{OP} + \dfrac{t^2 \cos \gamma}{4d}$	$a_{CG} = a_{OG} + \dfrac{T^2 \cos \Gamma}{4D}$
25. Tooth angle	$\dfrac{3,438}{A_O}\left(\dfrac{t}{2} + b_{OP} \tan \phi\right)$ minutes	$\dfrac{3,438}{A_O}\left(\dfrac{T}{2} + b_{OG} \tan \phi\right)$ minutes
26. Limit-point width (L.F.)	$W_{LOP} = (T - 2b_{OP} \tan \phi) - 0.0015$	$W_{LOG} = (t - 2b_{OG} \tan \phi) - 0.0015$
27. Limit-point width (S.E.)	$W_{LiP} = \dfrac{A_O - F}{A_O}(T - 2b_{OP} \tan \phi) - 0.0015$	$W_{LiG} = \dfrac{A_O - F}{A_O}(t - 2b_{OG} \tan \phi) - 0.0015$
28. Tool-point width	$W = W_{LiP} -$ stock allowance	$W = W_{LiG} -$ stock allowance

* Abstracted from "Gleason Straight Bevel Gear Design," Tables 8.3.8 and 8.3.9 and Fig. 8.3.4. Gleason Works, Inc.
† Numbers of teeth; ratios with 16 or more teeth in pinion: 15/17 and higher; 14/20 and higher; 13/31 and higher. These can be cut with 20° pressure angle without undercut.
‡ The actual dedendum will be 0.002 in greater than calculated.

Table 8.3.9 Recommended Normal Backlash for Bevel Gear Meshes*

P_d	Backlash range	P_d	Backlash range
1.00–1.25	0.020–0.030	3.50–4.00	0.007–0.009
1.25–1.50	0.018–0.026	4–5	0.006–0.008
1.50–1.75	0.016–0.022	5–6	0.005–0.007
1.75–2.00	0.014–0.018	6–8	0.004–0.006
2.00–2.50	0.012–0.016	8–10	0.003–0.005
2.50–3.00	0.010–0.013	10–12	0.002–0.004
3.00–3.50	0.008–0.011	Finer than 12	0.001–0.003

* The table gives the recommended normal backlash for gears assembled ready to run. Because of manufacturing tolerances and changes resulting from heat treatment, it is frequently necessary to reduce the theoretical tooth thickness by slightly more than the tabulated backlash in order to obtain the correct backlash in assembly. In case of choice, use the smaller backlash tolerances.

For all shaft angles, $\sin \gamma / \sin \Gamma = n/N$; $\Gamma = \Sigma - \gamma$.

Item 13, addendum, requires calculation of the equivalent 90° bevel gear ratio m_{90}, $m_{90} = [N \cos \gamma / (n \cos \Gamma)]^{1/2}$. The value m_{90} is used as the ratio N/n when applying the formula for addendum. The quantity under the radical is always the absolute value and is therefore always positive.

Item 20, pitch apex to crown, $x_o = A_0 \cos \gamma - a_{op} \sin \gamma$, $X_o = A_0 \cos \Gamma - a_{oG} \sin \Gamma$.

Item 21, circular thickness, except for high ratios, K may be zero.

Spiral Bevel Gears for 90° Shaft Angle   The spiral curved teeth produce additional overlapping tooth action which results in smoother gear action, lower noise, and higher load capacity. The spiral angle has been standardized by Gleason at 35°. Design parameters are calculated by formulas of Table 8.3.10.

Angular Spiral Bevel Gears   Several items deviate from the formulas of Table 8.3.10 in the same manner as angular straight bevel gears. Therefore, the same formulas apply for the deviating items with only the following exception:

Item 21, circular thickness, the value of K in Fig. 8.3.15 must be determined from the equivalent 90° bevel ratio (m_{90}) and the equivalent 90° bevel pinion. The latter is computed as $n_{90} = n \sin \Gamma_{90} / \cos \gamma$, where $\tan \Gamma_{90} = m_{90}$.

The zerol bevel gear is a special case of a spiral bevel gear and is limited to special applications. Design and fabrication details can be obtained from Gleason Works.

Fig. 8.3.14   Circular thickness factor for straight bevel gears.

Table 8.3.10 Spiral Bevel Gear Dimensions
(All linear dimensions in inches)

1. Number of pinion teeth	n		5. Working depth	$h_k = \dfrac{1.700}{P_d}$	
2. Number of gear teeth	N		6. Whole depth	$h_i = \dfrac{1.888}{P_d}$	
3. Diametral pitch	P_d		7. Pressure angle	ϕ	
4. Face width	F		8. Shaft angle	Σ	

	Pinion	Gear
9. Pitch diameter	$d = \dfrac{n}{P_d}$	$D = \dfrac{N}{P_d}$
10. Pitch angle	$\gamma = \tan^{-1} \dfrac{n}{N}$	$1 = 90° - \gamma$
11. Outer cone distance	$A_O = \dfrac{D}{2 \sin \Gamma}$	
12. Circular pitch	$p = \dfrac{3.1416}{P_d}$	
13. Addendum	$A_{OP} = h_k - a_{OG}$	$a_{OG} = \dfrac{0.460}{P_d} + \dfrac{0.390}{P_d (N/n)^2}$
14. Dedendum	$b_{OP} = h_t - a_{OP}$	$b_{OG} = h_t - a_{OG}$
15. Clearance	$c = h_t - h_k$	
16. Dedendum angle	$\delta_p = \tan^{-1} \dfrac{b_{OP}}{A_O}$	$\delta_G = \tan^{-1} \dfrac{b_{OG}}{A_O}$
17. Face angle of blank	$\gamma_O = \gamma + \delta_G$	$\Gamma_O = \Gamma + \delta_P$
18. Root angle	$\gamma_R = \gamma - \delta_P$	$\Gamma_R = \Gamma - \delta_G$
19. Outside diameter	$d_o = d + 2a_{OP} \cos \gamma$	$D_o = D + 2a_{OG} \cos \Gamma$
20. Pitch apex to crown	$x_O = \dfrac{D}{2} - a_{OP} \sin \gamma$	$X_O = \dfrac{d}{2} - a_{OG} \sin \Gamma$
21. Circular thickness	$t = p - T$	$T = \dfrac{p}{2} (a_{OP} - a_{OG}) \dfrac{\tan \phi}{\cos \phi} - \dfrac{K}{P_d}$ (see Fig. 8.3.15)
22. Backlash	See Table 8.3.9	
23. Hand of spiral	Left or right	Right or left
24. Spiral angle		35°
25. Driving member		Pinion or gear
26. Direction of rotaion		Clockwise or counterclockwise

SOURCE: Gleason, "Spiral Bevel Gear System."

Hypoid gears are special and are essentially limited to automotive applications.

Fig. 8.3.15 Circular thickness factors for spiral bevel gears with 20° pressure angle and 35° spiral angle. Left-hand pinion driving clockwise or right-hand pinion driving counterclockwise.

WORMGEARS AND WORMS

Worm gearing is used for obtaining large speed reductions between nonintersecting shafts making an angle of 90° with each other. If a wormgear such as shown in Fig. 8.3.16 engages a straight worm, as shown in Fig. 8.3.17, the combination is known as **single enveloping worm gearing.** If a wormgear of the kind shown in Fig. 8.3.16 engages a worm as shown in Fig. 8.3.18, the combination is known as **double enveloping worm gearing.**

Fig. 8.3.16 Single enveloping worm gearing.

Fig. 8.3.17 Straight worm.

With worm gearing, the **velocity ratio** is the ratio between the number of teeth on the wormgear and the number of threads on the worm. Thus, a 30-tooth wormgear meshing with a single threaded worm will have a velocity ratio of 1 : 30; that is, the worm must make 30 rv in order to revolve the wormgear once. For a double threaded worm, there will be 15 rv of the worm to one of the wormgear, etc. High-velocity ratios are thus obtained with relatively small wormgears.

Fig. 8.3.18 Double enveloping worm gearing.

Tooth proportions of the worm in the central section (Fig. 8.3.17) follow standard rack designs, such as 14½, 20, and 25°. The mating wormgear is cut conjugate for a unique worm size and center distance. The geometry and related design equations for a straight-sided cylindrical worm are best seen from a development of the pitch plane (Fig. 8.3.19).

$$D_w = \text{ptich diameter of down} = \frac{n_w p_n}{\pi \sin \lambda}$$

$$p_n = p \cos \lambda = \frac{\pi D_w}{n_w} \sin \lambda$$

$$L = \text{load or worm} = n_w p$$

$$D_g = \text{pitch diameter of wormgear}$$
$$= \frac{N_g}{P_d} = \frac{p N_g}{\pi} = \frac{P_n N_g}{\pi \cos \lambda}$$

$$C = \text{center distance}$$
$$= \frac{D_w + D_g}{2} = \frac{p_n}{2\pi}\left(\frac{N_g}{\cos \lambda} + \frac{n_w}{\sin \lambda}\right)$$

where n_w = number of threads in worm; N_g = number of teeth in wormgear; Z = velocity ratio = N_g/n_w.

The pitch diameter of the wormgear is established by the number of teeth, which in turn comes from the desired gear ratio. The pitch diameter of the worm is somewhat arbitrary. The lead must match the wormgear's circular pitch, which can be satisfied by an infinite number of worm diameters; but for a fixed lead value, each worm diameter has a unique lead angle. AGMA offers a design formula that provides near optimized geometry:

$$D_w = \frac{C^{0.875}}{2.2}$$

where C = center distance. Wormgear face width is also somewhat arbitrary. Generally it will be ⅗ to ⅔ of the worm's outside diameter.

Worm mesh nonreversibility, a unique feature of some designs, occurs because of the large amount of sliding in this type of gearing. For a given coefficient of friction there is a critical value of lead angle below which the mesh is nonreversible. This is generally 10° and lower but is related to the materials and lubricant. Most single thread worm meshes are in this category. This locking feature can be a disadvantage or in some designs can be put to advantage.

Double enveloping worm gearing is special in both design and fabrication. Application is primarily where a high load capacity in small space is desired. Currently, there is only one source of manufacture in the United States: Cone Drive Division of Ex-Cello Corp. For design details and load ratings consult publications of Cone Drive and AGMA Standards.

(a) Cylindrical worm (double-thread example)

(b) Development of worm's pitch cylinder

Fig. 8.3.19 Cylindrical worm geometry and design parameters.

Other Gear Types

Gears for special purposes include the following (details are to be found in the references):

Spiroid (Illinois Tool Works) gears, used to connect skew shafts, resemble a hypoid-type bevel gear but in performance are more like worm meshes. They offer very high ratios and a large contact ratio resulting in high strength. The **Helicon** (Illinois Tool Works) gear is a variation in which the pinion is not tapered, and ratios under 10:1 are feasible.

Beveloid (Vinco Corp.) gears are tapered involute gears which can couple intersecting shafts, skew shafts, and parallel shafts.

Face gears have teeth cut on the rotating face plane of the gear and mate with standard involute spur gears. They can connect intersecting or nonparallel, nonintersecting shafts.

Noncircular gears or **function gears** are used for special motions or as elements of analog computers. They can be made with elliptical, logarithmic, spiral, and other functions. See Cunningham references; also, Cunningham Industries, Inc., Stamford, CT.

DESIGN STANDARDS

In addition to the ANSI and AGMA standards on basic tooth proportions, the AGMA sponsors a large number of national standards dealing with gear design, specification, and inspection. (Consult AGMA, 1500 King St., Arlington, VA 22314, for details.) Helpful general references are AGMA, "Gear Handbook," 390.03 and ANSI/AGMA, "Gear Classification and Inspection Handbook," 2000-A88, which establish a system of quality classes for all gear sizes and pitches, ranging from crude coarse commercial gears to the highest orders of fine and coarse ultra-precision gears.

There are 13 quality classes, numbered from 3 through 15 in ascending quality. Tolerances are given for key functional parameters: runout, pitch, profile, lead, total composite error, tooth-to-tooth composite error, and tooth thickness. Also, tooth thickness tolerances and recommended mesh backlash are included. These are related to diametral pitch and pitch diameter in recognition of fabrication achievability. Data are available for spur, helical, herringbone, bevel, and worm gearing;

and spur and helical racks. Special sections cover gear applications and suggested quality number; gear materials and treatments; and standard procedure for identifying quality, material, and other pertinent parameters. These data are too extensive for inclusion in this handbook, and the reader is referred to the cited AGMA references.

STRENGTH AND DURABILITY

Gear teeth fail in two classical manners: tooth breakage and surface fatigue pitting. Instrument gears and other small, lightly loaded gears are designed primarily for tooth-bending beam strength since minimizing size is the priority. Power gears, usually larger, are designed for both strengths, with surface durability often more critical. Expressions for calculating the beam and surface stresses started with the Lewis-Buckingham formulas and now extend to the latest AGMA formulas.

The Lewis formula for analysis of beam strength, now relegated to historical reference, serves to illustrate the fundamentals that current formulas utilize. In the Lewis formula, a tooth layout shows the load assumed to be at the tip (Fig. 8.3.20). From this Lewis demonstrated that the beam strength $W_b = FSY/P_d$, where F = face width; S = allowable stress; Y = Lewis form factor; P_d = diametral pitch. The form factor Y is derived from the layout as $Y = 2P_d/3$. The value of Y varies with tooth design (form and pressure angle) and number of teeth. In the case of a helical gear tooth, there is a thrust force W_{th} in the axial direction that arises and must be considered as a component of bearing load. See Fig. 8.3.21b. Buckingham modified the Lewis formula to include dynamic effects on beam strength and developed equations for evaluating surface stresses. Further modifications were made by other investigators, and have resulted in the most recent AGMA rating formulas which are the basis of most gear designs in the United States.

Fig. 8.3.20 Layout for beam strength (Lewis formula).

AGMA Strength and Durability Rating Formulas

For many decades the AGMA Gear Rating Committee has developed and provided tooth beam strength and surface durability (pitting resistance) formulas suitable for modern gear design. Over the years, the formulas have gone through a continual evolution of revision and improvement. The intent is to provide a common basis for rating various gear types for differing applications and thus have a uniformity of practice within the gear industry. This has been accomplished via a series of standards, many of which have been adopted by ANSI.

The latest standards for rating bending beam strength and pitting resistance are ANSI/AGMA 2001-C95, "Fundamental Rating Factors and Calculation Methods for Involute, Spur, and Helical Gear Teeth" (available in English and metric units) and AGMA 908-B89, "Geometry Factors for Determining the Pitting Resistance and Bending Strength of Spur, Helical, and Herringbone Gear Teeth." These standards have replaced AGMA 218.01 with improved formulas and details.

The rating formulas in Tables 8.3.11 and 8.3.12 are abstracted from ANSI/AGMA 2001-B88, "Fundamental Rating Factors and Calculation Methods for Involute Spur and Helical Gear Teeth," with permission.

Overload factor K_o is intended to account for an occasional load in excess of the nominal design load W_t. It can be established from experience with the particular application. Otherwise use $K_o = 1$.

Fig. 8.3.21   Forces on spur and helical teeth. (*a*) Spur gear; (*b*) helical gear.

Table 8.3.11   AGMA Pitting Resistance Formula for Spur and Helical Gears
(See Note 1 below.)

$$s_c = C_p \sqrt{W_t K_o K_v K_s \frac{K_m}{dF} \frac{C_f}{I}}$$

where s_c = contact stress number, lb/in^2
   C_p = elastic coefficient,* (lb/in^2)$^{0.5}$ (see text and Table 8.3.13)
   W_t = transmitted tangential load, lb
   K_o = overload factor (see text)
   K_v = dynamic factor (see Fig. 8.3.22)
   K_s = size factor (see text)
   K_m = load distribution factor (see text and Table 8.3.14)
   C_f = surface condition factor for pitting resistance (see text)
   F = net face width of narrowest member, in
   I = geometry factor for pitting resistance (see text and Figs. 8.2.23 and 8.3.24)
   d = operating pitch diameter of pinion, in

$$= \frac{2C}{m_G + 1} \quad \text{for external gears}$$

$$= \frac{2C}{m_G - 1} \quad \text{for internal gears}$$

where C = operating center distance, in
   m_G = gear ratio (never less than 1.0)

Allowable contact stress umber s_{ac}

$$s_c \leq \frac{s_{ac} Z_N C_H}{S_H K_T K_R}$$

where s_{ac} = allowable contact stress number, lb/in^2 (see Tables 8.3.15 and 8.3.16; Fig. 8.3.34)
   Z_N = stress cycle factor for pitting resistance (see Fig. 8.3.35)
   C_H = hardness ratio for pitting resistance (see text and Figs. 8.3.36 and 8.3.37)
   S_H = safety factor for pitting (see text)
   K_T = temperature factor (see text)
   K_R = reliability factor (see Table 8.3.19)

* **Elastic coefficient** C_p can be calculated from the following equation when the paired materials in the pinion-gear set are not listed in Table 8.3.13:

$$C_p = \sqrt{\frac{1}{\pi[(1 - \mu_P^2)/E_P + (1 - \mu_G^2)/E_G]}}$$

where $\mu_P (\mu_G)$ = Poisson's ratio for pinion (gear)
   $E_P (E_G)$ = modulus of elasticity for pinion (gear), lb/in^2

Note 1: If the rating is calculated on the basis of uniform load, the transmitted tangential load is

$$W_t = \frac{33,000P}{v_t} = \frac{2T}{d} = \frac{126,000P}{n_p d}$$

where P = transmitted power, hp
   T = transmitted pinion torque, lb · in
   v_t = pitch line velocity at operating pitch diameter, ft/min = $\dfrac{\pi n_p d}{12}$

Table 8.3.12   AGMA Bending Strength Fundamental Formula for Spur and Helical Gears
(See Note 1 in Table 8.3.11.)

$$s_t = W_t K_o K_v K_s \frac{P_d}{F} \frac{K_m K_B}{J}$$

where s_t = bending stress number, lb/in^2
   K_B = rim thickness factor (see Fig. 8.3.38)
   J = geometry factor for bending strength (see text and Figs. 8.3.25 to 8.3.31)
   P_d = transverse diametral pitch, in^{-1}*;
      P_{dn} for helical gears

$$P_d = \frac{\pi}{p_x \tan \psi_s} = P_{dn} \cos \psi_s \quad \text{for helical gears}$$

where P_{dn} = normal diametral pitch, in^{-1}
   p_x = axial pitch, in
   ψ_s = helix angle at standard pitch diameter

$$\psi_s = \arcsin \frac{\pi}{p_x P_{dn}}$$

Allowable bending stress numbers s_{at}

$$s_t \leq \frac{s_{at} Y_N}{S_F K_T K_R}$$

where s_{at} = allowable bending stress number, lb/in^2 (see Tables 8.3.17 and 8.3.18 and Figs. 8.3.39 to 8.3.41)
   Y_N = stress cycle factor for bending strength (see Fig. 8.3.42)
   S_F = safety factor for bending strength (see text)

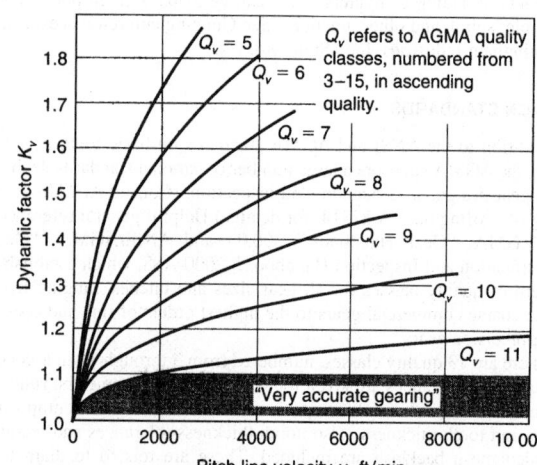

Fig. 8.3.22   Dynamic factor K_v. (*Source: ANSI/AGMA 2001-C95, with permission.*)

Table 8.3.13   Elastic Coefficient C_p

Pinion material	Pinion modulus of elasticity E_P, lb/in² (MPa)	Gear material and modulus of elasticity E_G, lb/in² (MPa)					
		Steel 30×10^6 (2×10^5)	Malleable iron 25×10^6 (1.7×10^5)	Nodular iron 24×10^6 (1.7×10^5)	Cast iron 22×10^6 (1.5×10^5)	Aluminum bronze 17.5×10^6 (1.2×10^5)	Tin bronze 16×10^6 (1.1×10^5)
Steel	30×10^6 (2×10^5)	2,300 (191)	2,180 (181)	2,160 (179)	2,100 (174)	1,950 (162)	1,900 (158)
Malleable iron	25×10^6 (1.7×10^5)	2,180 (181)	2,090 (174)	2,070 (172)	2,020 (168)	1,900 (158)	1,850 (154)
Nodular iron	24×10^6 (1.7×10^5)	2,160 (179)	2,070 (172)	2,050 (170)	2,000 (166)	1,880 (156)	1,830 (152)
Cast iron	22×10^6 (1.5×10^5)	2,100 (174)	2,020 (168)	2,000 (166)	1,960 (163)	1,850 (154)	1,800 (149)
Aluminum bronze	17.5×10^6 (1.2×10^5)	1,950 (162)	1,900 (158)	1,880 (156)	1,850 (154)	1,750 (145)	1,700 (141)
Tin bronze	16×10^6 (1.1×10^5)	1,900 (158)	1,850 (154)	1,830 (152)	1,800 (149)	1,700 (141)	1,650 (137)

Poisson's ratio = 0.30.
SOURCE: ANSI/AGMA 2001-B88; with permission.

Size factor K_s is intended to factor in material nonuniformity due to tooth size, diameter, face width, etc. AGMA has not established factors for general gearing; use $K_s = 1$ unless there is information to warrant using a larger value.

Load distribution factor K_m reflects the nonuniform loading along the lines of contact due to gear errors, installation errors, and deflections. Analytical and empirical methods for evaluating this factor are presented in ANSI/AGMA 2001-C95 but are too extensive to include here. Alternately, if appropriate for the application, K_m can be extrapolated from values given in Table 8.3.14.

Surface condition factor C_f is affected by the manufacturing method (cutting, shaving, grinding, shotpeening, etc.). Standard factors have not been established by AGMA. Use $C_f = 1$ unless experience can establish confidence for a larger value.

Geometry factors I and J relate to the shape of the tooth at the point of contact, the most heavily loaded point. AGMA 908-B89 (Information Sheet, Geometry Factors for Determining the Pitting Resistance and Bending Strength for Spur, Helical and Herringbone Gear Teeth) presents detailed procedures for calculating these factors. The standard also includes a collection of tabular values for a wide range of gear tooth designs, but they are too voluminous to be reproduced here in their entirety. Earlier compact graphs of I and J values in AGMA 218.01 are still valid. They are presented here along with several curves from AGMA 610-E88, which are based upon 218.01. See Figs. 8.3.23 to 8.3.31.

Allowable contact stress s_{ac} and **allowable bending stress S_{at}** are obtainable from Tables 8.3.15 to 8.3.18. Contact stress hardness specification applies to the start of active profile at the center of the face width, and for bending stress at the root diameter in the center of the tooth space and face width. The lower stress values are for general design purposes; the upper values are for high-quality materials and high-quality control. (See ANSI/AGMA 2001-C95, tables 7 through 10, regarding detailed metallurgical specifications; stress grades 1, 2, and 3; and type A and B hardness patterns.)

For **reversing loads,** allowable bending stress values, s_{at} are to be reduced to 70 percent. If the rim thickness cannot adequately support the load, an additional derating factor K_B is to be applied. See Fig. 8.3.38.

Hardness ratio factor C_H applies when the pinion is substantially harder than the gear, and it results in work hardening of the gear and increasing its capacity. Factor C_H applies to only the gear, not the pinion. See Figs. 8.3.36 and 8.3.37.

Safety factors S_H and S_F are defined by AGMA as factors beyond K_O and K_R; they are used in connection with extraordinary risks, human or economic. The values of these factors are left to the designer's judgment as she or he assesses all design inputs and the consequences of possible failure.

Temperature factor $K_T = 1$ when gears operate with oil temperature not exceeding 250°F.

Reliability factor K_R accounts for statistical distribution of material failures. Typically, material strength ratings (Tables 8.3.13 and 8.3.15 to 8.3.18; Figs. 8.3.34 and 8.3.42) are based on probability of one failure in 100 at 10^7 cycles. Table 8.3.19 lists reliability factors that may be used to modify the allowable stresses and the probability of failure.

Strength and Durability of Bevel, Worm, and Other Gear Types

For bevel gears, consult the referenced Gleason publications; for worm gearing refer to AGMA standards; for other special types, refer to Dudley, "Gear Handbook."

Table 8.3.14   Load-Distribution Factor K_m for Spur Gears*

Characteristics of support	Face width, in			
	0–2	6	9	16 up
Accurate mountings, small bearing clearances, minimum deflection, precision gears	1.3	1.4	1.5	1.8
Less rigid mountings, less accurate gears, contact across full face	1.6	1.7	1.8	2.2
Accuracy and mounting such that less than full face contact exists	Over 2.2			

* An approximate guide only. See ANSI/AGMA 2001-C95 for derivation of more exact values.
SOURCE: Darle W. Dudley, "Gear Handbook," McGraw-Hill, New York, 1962.

Fig. 8.3.23 Geometry factor *I* for 20° full-depth standard spur gears. (*Source: ANSI/AGMA 2018-01, with permission.*)

Fig. 8.3.24 Geometry factor *I* for 25° full-depth standard spur gears. (*Source: ANSI/AGMA 2018-01, with permission.*)

Fig. 8.3.25   Geometry factor *J* for 20° standard addendum spur gears. (*Source: ANSI/AGMA 2018-01, with permission.*)

Fig. 8.3.26   Geometry factor *J* for 25° standard addendum spur gears. (*Source: ANSI/AGMA 2018-01, with permission.*)

Fig. 8.3.27 Geometry factor J for 20° normal pressure angle helical gears. (Standard addendum, finishing hob.) (*Source: ANSI/AGMA 2018-01, with permission.*)

Fig. 8.3.28 Geometry factor J for 20° normal pressure angle helical gears. (Standard addendum, full fillet hob.) (*Source: ANSI/AGMA 2018-01, with permission.*)

Fig. 8.3.29 Geometry factor J for 25° normal pressure angle helical gears. (Standard addendum, full fillet hob.) (*Source: ANSI/AGMA 2018-01, with permission.*)

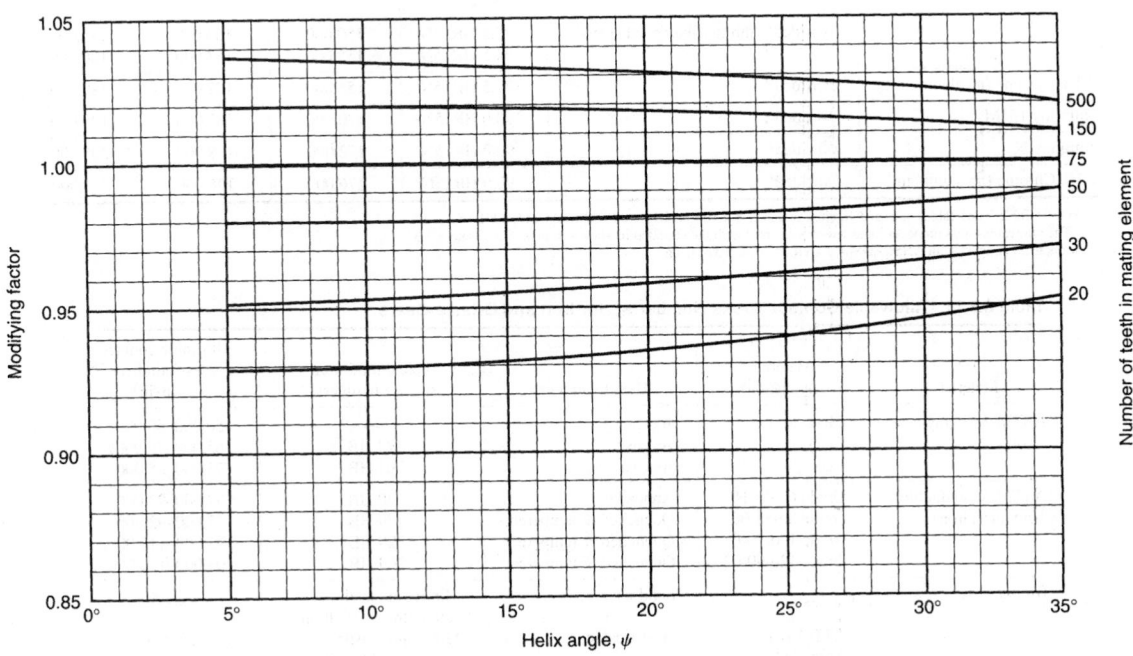

Fig. 8.3.30 Factor J multipliers for 20° normal pressure angle helical gears. The modifying factor can be applied to the J factor when other than 75 teeth are used in the mating element. (*Source: ANSI/AGMA 2018-01, with permission.*)

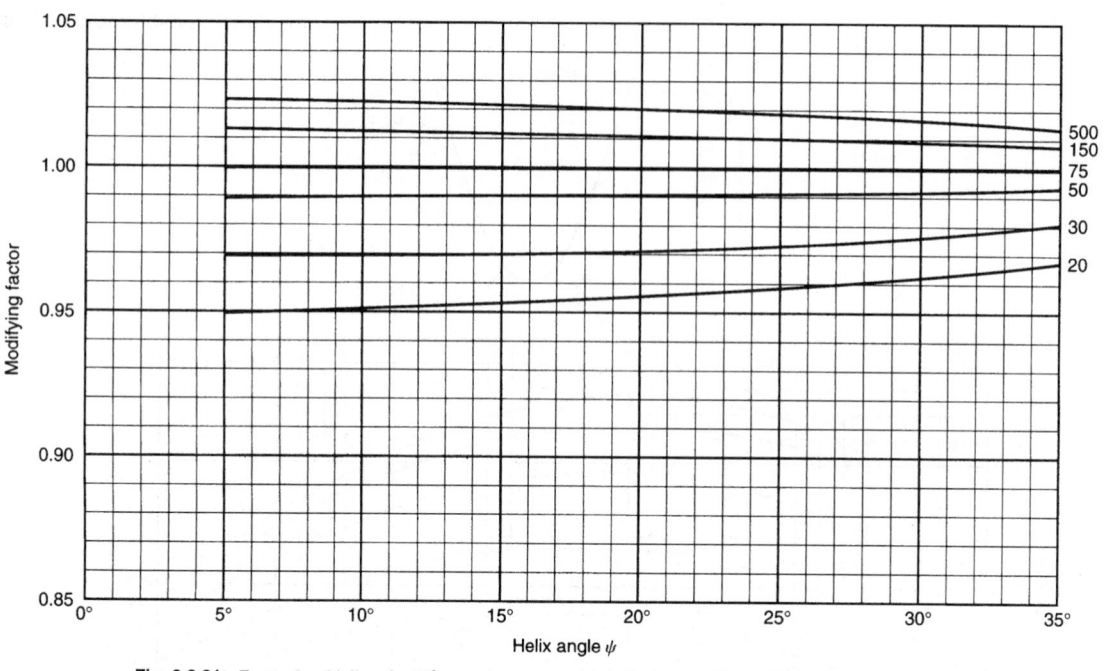

Fig. 8.3.31 Factor J multipliers for 25° normal pressure angle helical gears. The modifying factor can be applied to the J factor when other than 75 teeth are used in the mating element. (*Source: ANSI/AGMA 6010-E88, with permission.*)

Table 8.3.15 Allowable Contact Stress Number s_{ac} for Steel Gears

Material designation	Heart treatment	Minimum surface hardness*	Allowable contact stress number s_{ac}, lb/in²		
			Grade 1	Grade 2	Grade 3
Steel	Through-hardened*	Fig. 8.3.34	Fig. 8.3.34	Fig. 8.3.34	—
	Flame-* or induction-hardened*	50 HRC	170,000	190,000	—
		54 HRC	175,000	195,000	—
	Carburized and hardened*	Table 9 Note 1	180,000	225,000	275,000
	Nitrided* (through-hardened steels)	83.5 HR15N	150,000	163,000	175,000
		84.5 HR15N	155,000	168,000	180,000
2.5% Chrome (no aluminum)	Nitrided*	87.5 HR15N	155,000	172,000	189,000
Nitralloy 135M	Nitrided*	90.0 HR15N	170,000	183,000	195,000
Nitralloy N	Nitrided*	90.0 HR15N	172,000	188,000	205,000
2.5% Chrome (no aluminum)	Nitrided*	90.0 HR15N	176,000	196,000	216,000

Note 1: Table 9 and Tables 7, 8, and 10 cited in Tables 8.3.16 to 8.3.18 are in ANSI/AGMA 2001-95.
* The allowable-stress numbers indicated may be used with the case depths shown in Figs. 8.3.32 and 8.3.33.
SOURCE: Abstracted from ANSI/AGMA 2001-C95, with permission.

Table 8.3.16 Allowable Contact Stress Number s_{ac} for Iron and Bronze Gears

Material	Material designation*	Heat treatment	Typical minimum surface hardness	Allowable contact stress number s_{ac}, lb/in²
ASTM A48 gray cast iron	Class 20	As cast	—	50,000–60,000
	Class 30	As cast	174 HB	65,000–75,000
	Class 40	As cast	201 HB	75,000–85,000
ASTM A536 ductile (nodular) iron	Grade 60-40-18	Annealed	140 HB	77,000–92,000
	Grade 80-55-06	Quenched & tempered	179 HB	77,000–92,000
	Grade 100-70-03	Quenched & tempered	229 HB	92,000–112,000
	Grade 120-90-02	Quenched & tempered	269 HB	103,000–126,000
Bronze	—	Sand-cast	Minimum tensile strength 40,000 lb/in²	30,000
	ASTM B-148 Alloy 954	Heat-treated	Minimum tensile strength 90,000 lb/in²	65,000

* See ANSI/AGMA 2004–B89, "Gear Materials and Heat Treatment Manual."
SOURCE: Abstracted from ANSI/AGMA 2001-C95, with permission.

Table 8.3.17 Allowable Bending Stress Number s_{at} for Steel Gears

Material designation	Heat treatment	Minimum surface hardness	Allowable bending stress number s_{at}, lb/in^2		
			Grade 1	Grade 2	Grade 3
Steel	Through-hardened	Fig. 8.3.39	Fig. 8.3.39	Fig. 8.3.39	—
	Flame* or induction-hardened* with type A pattern	Table 8§	45,000	55,000	—
	Flame* or induction hardened* with type B pattern	Table 8§ Note 1	22,000	22,000	—
	Carburized and hardened*	Table 9§ Note 2	55,000	65,000 or 70,000†	75,000
	Nitrided*‡ (through-hardened steels)	83.5 HR15N	Fig. 8.3.40	Fig. 8.3.40	—
Nitralloy 135M, nitralloy N, and 2.5% Chrome (no aluminum)	Nitrided‡	87.5 HR15N	Fig. 8.3.41	Fig. 8.3.41	Fig. 8.3.41

Note 1: See Table 8 in ANSI/AGMA 2001-C95.
Note 2: See Table 9 in ANSI/AGMA 2001-C95.
* The allowable-stress numbers indicated may be used with the case depths shown in Figs. 8.3.32 and 8.3.33.
† If bainite and microcracks are limited to grade 3 levels, 70,000 lb/in^2 may be used.
‡ The overload capacity of nitrided gears is low. Since the shape of the effective S–N curve is flat, the sensitivity to shock should be investigated before one proceeds with the design.
§ The tabular material is too extensive to record here. Refer to ANSI/AGMA 2001-C95, tables 7 to 10.
Source: Abstracted from ANSI/AGMA 2001-C95, with permission.

Table 8.3.18 Allowable Bending Stress Number s_{at} for Iron and Bronze Gears

Material	Material designation*	Heat treatment	Typical minimum surface hardness	Allowable bending stress number s_{at}, lb/in^2
ASTM A48 gray cast iron	Class 20	As cast	—	5,000
	Class 30	As cast	174 HB	8,500
	Class 40	As cast	201 HB	13,000
ASTM A536 ductile (nodular) iron	Grade 60-40-18	Annealed	140 HB	22,000–33,000
	Grade 80-55-06	Quenched & tempered	179 HB	22,000–33,000
	Grade 100-70-03	Quenched & tempered	229 HB	27,000–40,000
	Grade 120-90-02	Quenched & tempered	269 HB	31,000–44,000
Bronze		Sand-cast	Minimum tensile strength 40,000 lb/in^2	5,700
	ASTM B-148 Alloy 954	Heat-treated	Minimum tensile strength 90,000 lb/in^2	23,600

* See ANSI/AGMA 2004–B89, "Gear Materials and Heat Treatment Manual."
Source: Abstracted from ANSI/AGMA 2001-C95, with permission.

Table 8.3.19 Reliability Factors K_R

Requirements of application	K_R*
Fewer than one failure in 10,000	1.50
Fewer than one failure in 1,000	1.25
Fewer than one failure in 100	1.00
Fewer than one failure in 10	0.85†
Fewer than one failure in 2	0.70†,‡

* Tooth breakage is sometimes considered a greater hazard than pitting. In such cases a greater value of K_R is selected for bending.
† At this value, plastic flow might occur rather than pitting.
‡ From test data extrapolation.
Source: Abstracted from ANSI/AGMA 2001-C95, with permission.

GEAR MATERIALS
(See Tables 8.3.15 to 8.3.18.)

Metals

Plain carbon steels are most widely used as the most economical; similarly, cast iron is used for large units or intricate body shapes. Heat-treated carbon and alloy steels are used for the more severe load- and wear-resistant applications. Pinions are usually made harder to equalize wear. Strongest and most wear-resistant gears are a combination of heat-treated high-alloy steel cores with case-hardened teeth. (See Dudley, "Gear Handbook," chap. 10.) Bronze is particularly recommended for wormgears and crossed helical gears. Stainless steels are limited to special corrosion-resistant environment applications. Aluminum alloys are used for light-duty instrument gears and airborne lightweight requirements.

Sintered powdered metals technology offers commercial high-quality gearing of high strength at very economical production costs. Die-cast gears for light-duty special applications are suitable for many products.

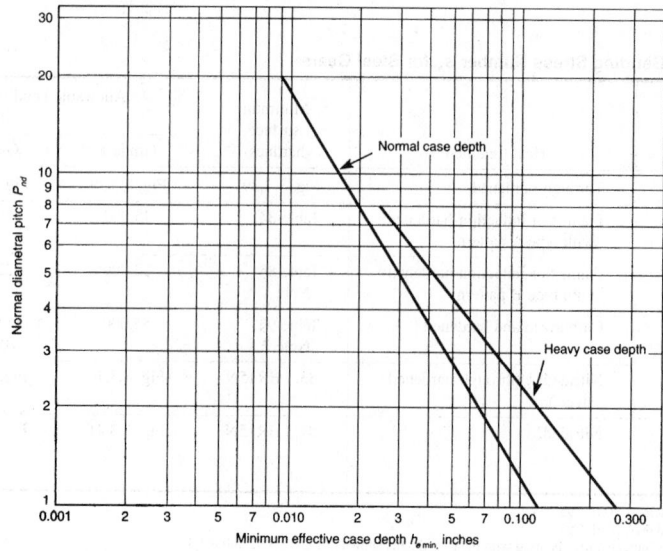

Fig. 8.3.32 Minimum effective case depth for carburized gears $h_{e\min}$. Effective case depth is defined as depth of case with minimun hardness of 50 RC. Total case depth to core carbon is approximately 1.5 times the effective case depth. (*Source: ANSI/AGMA 2001-C95, with permission.*)

Fig. 8.3.33 Minimum total case depth for nitrided gears $h_{c\min}$. (*Source: ANSI/AGMA 2001-C95, with permission.*)

Fig. 8.3.34 Allowable contact stress number for through-hardened steel gears s_{ac}. (*Source: ANSI/AGMA 2001-C95, with permission.*)

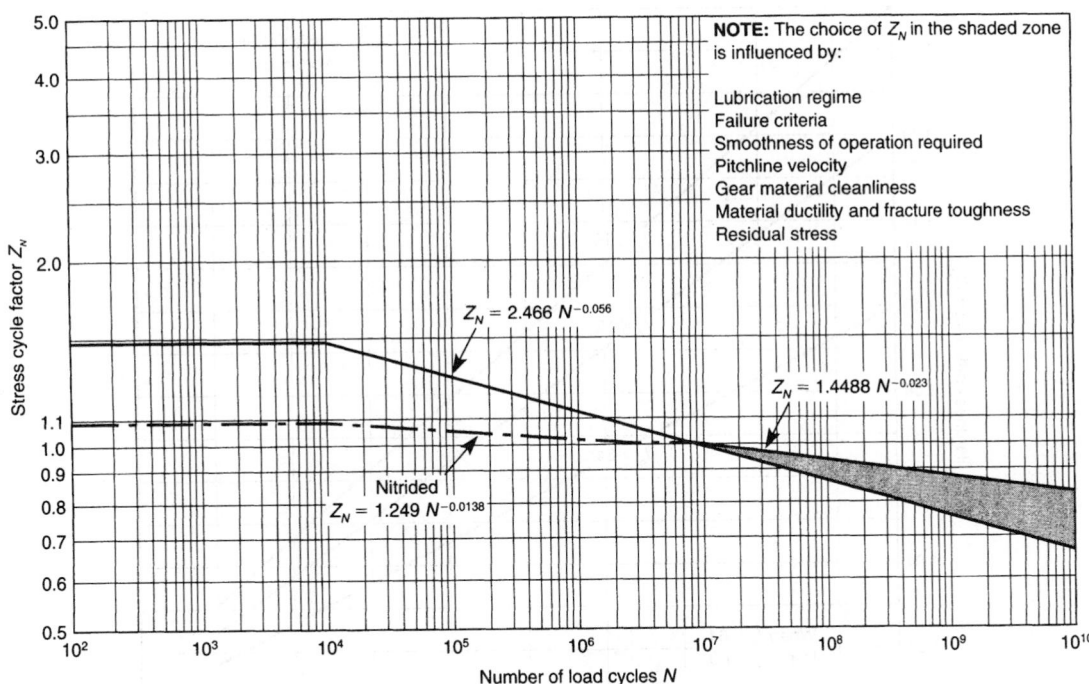

Fig. 8.3.35 Pitting resistance stress cycle factor Z_N. (*Source: ANSI/AGMA 2001-C95, with permission.*)

Fig. 8.3.36 Hardness ratio factor C_H (through-hardened). (*Source: ANSI/AGMA 2001-C95, with permission.*)

Fig. 8.3.37 Hardness ratio factor C_H (surface-hardened pinions). (*Source: ANSI/AGMA 2001-C95, with permission.*)

For power gear applications, heat treatment is an important part of complete and proper design and specification. Heat treatment descriptions and specification tolerances are given in the reference cited below.

Precision gears of the small device and instrument types often require protective coatings, particularly for aircraft, marine, space, and military applications. There is a wide choice of chemical and electroplate coatings offering a variety of properties and protection.

For pertinent properties and details of the above special materials and protective coatings see Michalec, "Precision Gearing," chap. 9, Wiley.

Plastics

In recent decades, various forms of nonmetallic gears have displaced metal gears in particular applications. Most plastics can be hobbed or shaped by the same methods used for metallic gears. However, high-strength composite plastics suitable for good-quality gear molding have become available, along with the development of economical high-speed injection molding machines and improved methods for producing accurate gear molds.

Fig. 8.3.38 Rim thickness factor K_B. (*Source: ANSI/AGMA 2001-C95, with permission.*)

Fig. 8.3.39 Allowable bending stress numbers for through-hardened steel gears s_{at}. (*Source: ANSI/AGMA 2001-C95, with permission.*)

Fig. 8.3.40 Allowable bending stress numbers s_{at} for nitrided through-hardened steel gears (i.e., AISI 4140 and 4340). (*Source: ANSI/AGMA 2001-C95, with permission.*)

Fig. 8.3.41 Allowable bending stress numbers for nitrided steel gears s_{at}. (*Source: ANSI/AGMA 2001-C95, with permission.*)

Fig. 8.3.42 Bending strength stress cycle factor Y_N. (*Source: ANSI/AGMA 2001-C95, with permission.*)

The most significant features and advantages of plastic gear materials are:

Cost-effectiveness of injection molding process
Wide choice of characteristics: mechanical strength, density, friction, corrosion resistance, etc.
One-step production; no preliminary or secondary operations
Uniformity of parts
Ability to integrate special shapes, etc., into gear body
Elimination of machining operations
Capability to mold with metallic insert hubs, if required, for more precise bore diameter or body stability
Capability to mold with internal solid lubricants
Ability to operate without lubrication
Quietness of operation
Consistent with trend toward greater use of plastic housings and components

Plastic gears do have limitations relative to metal gears. The most significant are:

Less load-carrying capacity
Cannot be molded to as high accuracy as machined metal gears
Much larger coefficient of expansion compared to metals
Less environmentally stable with regard to temperature and water absorption
Can be negatively affected by some chemicals and lubricants
Initial high cost in mold manufacture to achieve proper tooth geometry accuracies
Narrower range of temperature operation, generally less than 250°F and not lower than 0°F

For further information about plastic gear materials and achievable precision, consult the cited reference: Michalec, "Precision Gearing: Theory and Practice."

For a comprehensive presentation of gear molding practices, design, plastic materials, and strength and durability of plastic gears, consult the cited reference: Designatronics, "Handbook of Gears: Inch and Metric," pp. T131–T158.

GEAR LUBRICATION

Proper lubrication is important to prevention of premature wear of tooth surfaces. In the basic action of involute tooth profiles there is a significant sliding component along with rolling action. In worm gearing sliding is the predominant consideration. Thus, a lubricant is essential for all gearing subject to measurable loadings, and even for lightly or negligibly loaded instrument gearing it is needed to reduce friction. Excellent oils and greases are available for high unit load, high speed gearing. See Secs. 6.11 and 8.4; also consult lubricant suppliers for recommendations and latest available high-quality lubricants with special-purpose additives.

General and specific information for lubrication of gearing is found in ANSI/AGMA 9005-D94, "Industrial Gear Lubrication," which covers open and enclosed gearing of all types. AGMA lists a family of lubricants in accordance with viscosities, numbered 1 through 13, with a cross-reference to equivalent ISO grades. See Table 8.3.20. For AGMA's lubrication recommendations for open and closed gearing related to pitch line speed and various types of lubrication systems, refer to Tables 8.3.21 to 8.3.23. For worm gearing and other information, see ANSI/AGMA 9005-D94.

Information in Table 8.3.24 may be used as a quick guide to gear lubricants and their sources for general-purpose instrument and medium-size gearing. Lubricant suppliers should be consulted for specific high-demand applications.

Often gear performance can be enhanced by special additives to the oil. For this purpose, colloidal additives of graphite, molybdenum disulfide (MoS_2), and Teflon are very effective. These additives are particularly helpful to reduce friction and prevent wear; they are also very beneficial to reduce the rate of wear once it has begun, and thus they prolong gear life. The colloidal additive combines chemically with the metal surface material, resulting in a tenacious layer of combined material interposed between the base metals of the meshing teeth. The size of the colloidal additives is on the order of 2 μm, sufficiently fine not to interfere with the proper operation of the lubricant system filters. Table 8.3.25 lists some commercial colloidal additives and their sources of supply.

Table 8.3.20 Viscosity Ranges for AGMA Lubricants

Rust- and oxidation-inhibited gear oils, AGMA lubricant no.	Viscosity range[a] mm²/s (cSt) at 40°C	Equivalent ISO grade[a]	Extreme pressure gear lubricants,[b] AGMA lubricant no.	Synthetic gear oils,[c] AGMA lubricant no.
0	28.8–35.2	32		0 S
1	41.4–50.6	46		1 S
2	61.2–74.8	68	2 EP	2 S
3	90–110	100	3 EP	3 S
4	135–165	150	4 EP	4 S
5	198–242	220	5 EP	5 S
6	288–352	320	6 EP	6 S
7, 7 Comp[d]	414–506	460	7 EP	7 S
8, 8 Comp[d]	612–748	680	8 EP	8 S
8A Comp[d]	900–1,100	1,000	8A EP	—
9	1,350–1,650	1,500	9 EP	9 S
10	2,880–3,520	—	10 EP	10 S
11	4,140–5,060	—	11 EP	11 S
12	6,120–7,480	—	12 EP	12 S
13	190–220 cSt at 100°C (212°F)[e]	—	13 EP	13 S

Residual compounds,[f] AGMA lubricant no.	Viscosity ranges[e] cSt at 100°C (212°F)
14R	428.5–857.0
15R	857.0–1,714.0

[a] Per ISO 3448, "Industrial Liquid Lubricants—ISO Viscosity Classification," also ASTM D2422 and British Standards Institution B.S. 4231.
[b] Extreme-pressure lubricants should be used only when recommended by the gear manufacturer.
[c] Synthetic gear oils 9S to 13S are available but not yet in wide use.
[d] Oils marked *Comp* are compounded with 3% to 10% fatty or synthetic fatty oils.
[e] Viscosities of AGMA lubricant no. 13 and above are specified at 100°C (210°F) since measurement of viscosities of these heavy lubricants at 40°C (100°F) would not be practical.
[f] Residual compounds—diluent type, commonly known as solvent cutbacks—are heavy oils containing a volatile, nonflammable diluent for ease of application. The diluent evaporates, leaving a thick film of lubricant on the gear teeth. Viscosities listed are for the base compound without diluent.
CAUTION: These lubricants may require special handling and storage procedures. Diluent can be toxic or irritating to the skin. Do not use these lubricants without proper ventilation. Consult lubricant supplier's instructions.
SOURCE: Abstracted from ANSI/AGMA 9005-D94, with permission.

Table 8.3.21 AGMA Lubricant Number Guidelines for Open Gearing (Continuous Method of Application)[a,b]

Ambient temperature[c] °C (°F)	Character of operation	Pressure lubrication Pitch line velocity Under 5 m/s (1,000 ft/min)	Pressure lubrication Pitch line velocity Over 5 m/s (1,000 ft/min)	Splash lubrication Pitch line velocity Under 5 m/s (1,000 ft/min)	Splash lubrication Pitch line velocity 5–10 m/s (1,000–3,000 ft/min)	Idler immersion Pitch line velocity Up to 1.5 m/s (300 ft/min)
−10 to 15[d] (15–60)	Continuous	5 or 5 EP	4 or 4 EP	5 or 5 EP	4 or 4 EP	8–9 8 EP–9 EP
−10 to 15[d] (15–60)	Reversing or frequent "start/stop"	5 or 5 EP	4 or 4 EP	7 or 7 EP	6 or 6 EP	8–9 8 EP–9 EP
10 to 50[d] (50–125)	Continuous	7 or 7 EP	6 or 6 EP	7 or 7 EP	8–9[f] 8 EP–9 EP	11 or 11 EP
10 to 50[d] (50–125)	Reversing or frequent "start/stop"	7 or 7 EP	6 or 6 EP	9–10[e] 9 EP–10 EP	8–9[f] 8 EP–9 EP	11 or 11 EP

[a] AGMA lubricant numbers listed above refer to gear lubricants shown in Table 8.3.20. Physical and performance specifications are shown in Tables 1 and 2 of ANSI/ASMA 9005-D94. Although both R & O and EP oils are listed, the EP is preferred. Synthetic oils in the corresponding viscosity grades may be substituted where deemed acceptable by the gear manufacturer.
[b] Does not apply to worm gearing.
[c] Temperature in vicinity of the operating gears.
[d] When ambient temperatures approach the lower end of the given range, lubrication systems must be equipped with suitable heating units for proper circulation of lubricant and prevention of channeling. Check with lubricant and pump suppliers.
[e] When ambient temperature remains between 30°C (90°F) and 50°C (125°F) at all times, use 10 or 10 EP.
[f] When ambient temperature remains between 30°C (90°F) and 50°C (125°F) at all times, use 9 or 9 EP.
SOURCE: Abstracted from ANSI/AGMA 9005-D94, with permission.

Table 8.3.22 AGMA Lubricant Number Guidelines for Open Gearing Intermittent Applications[a,b,c]

Gear pitch line velocity does not exceed 7.5 m/s (1,500 ft/min)

Ambient temperature,[d] °C (°F)	Intermittent spray systems[e] R&O or EP lubricant	Intermittent spray systems[e] Synthetic lubricant	Intermittent spray systems[e] Residual compound[f]	Gravity feed or forced-drip method[g] R&O or EP lubricant	Gravity feed or forced-drip method[g] Synthetic lubricant
−10–15 (15–60)	11 or 11 EP	11 S	14 R	11 or 11 EP	11 S
5–40 (40–100)	12 or 12 EP	12 S	15 R	12 or 12 EP	12 S
20–50 (70–125)	13 or 13 EP	13 S	15 R	13 or 13 EP	13 S

[a] AGMA viscosity number guidelines listed above refer to gear oils shown in Table 8.3.20.
[b] Does not apply to worm gearing.
[c] Feeder must be capable of handling lubricant selected.
[d] Ambient temperature is temperature in vicinity of gears.
[e] Special compounds and certain greases are sometimes used in mechanical spray systems to lubricate open gearing. Consult gear manufacturer and spray system manufacturer before proceeding.
[f] Diluents must be used to facilitate flow through applicators.
[g] EP oils are preferred, but may not be available in some grades.
SOURCE: Abstracted from ANSI/AGMA 9005-D94, with permission.

Table 8.3.23   AGMA Lubricant Number Guidelines for Enclosed Helical, Herringbone, Straight Bevel, Spiral Bevel, and Spur Gear Drives[a]

Pitch line velocity [b,c] of final reduction stage	AGMA lubricant numbers,[a,d,e] ambient temperature °C (°F)[f,g]			
	− 40 to − 10 (− 40 to + 14)	− 10 to + 10 (14 to 50)	10 to 35 (50 to 95)	35 to 55 (95 to 131)
Less than 5 m/s (1,000 ft/min)[h]	3 S	4	6	8
5–15 m/s (1,000–3,000 ft/min)	3 S	3	5	7
15–25 m/s (3,000–5,000 ft/min)	2 S	2	4	6
Above 25 m/s (5,000 ft/min)[h]	0 S	0	2	3

[a] AGMA lubricant numbers listed above refer to R&O and synthetic gear oil shown in Table 8.3.20. Physical and performance specifications are shown in Tables 1 and 3 of ANSI/AGMA 9005-D94. EP or synthetic gear lubricants in the corresponding viscosity grades may be substituted where deemed acceptable by the gear drive manufacturer.

[b] Special considerations may be necessary at speeds above 40 m/s (8,000 ft/min). Consult gear drive manufacturer for specific recommendations.

[c] Pitch line velocity replaces previous standards' center distance as the gear drive parameter for lubricant selection.

[d] Variations in operating conditions such as surface roughness, temperature rise, loading, speed, etc., may necessitate use of a lubricant of one grade higher or lower. Contact gear drive manufacturer for specific recommendations.

[e] Drives incorporating wet clutches or overrunning clutches as backstopping devices should be referred to the gear manufacturer, as certain types of lubricants may adversely affect clutch performance.

[f] For ambient temperatures outside the ranges shown, consult the gear manufacturer.

[g] Pour point of lubricant selected should be at least 5°C (9°F) lower than the expected minimum ambient starting temperature. If the ambient starting temperature approaches lubricant pour point, oil sump heaters may be required to facilitate starting and ensure proper lubrication (see 5.1.6 in ANSI/AGMA 9005-D94).

[h] At the extreme upper and lower pitch line velocity ranges, special consideration should be given to all drive components, including bearing and seals, to ensure their proper performance.

SOURCE: Abstracted from ANSI/AGMA 9005-D94, with permission.

Table 8.3.24   Typical Gear Lubricants

Lubricant type	Military specification	Useful temp range, °F	Commercial source and specification (a partial listing)		Remarks and applications
			Source	Identification	
Oils					
Petroleum	MIL-L-644B	− 10 to 250	Exxon Corporation Franklin Oil and Gas Co. Royal Lubricants Co. Texaco	#4035 or Unvis P-48 L-499B Royco 380 1692 Low Temp. Oil	Good general-purpose lubricant for all quality gears having a narrow range of operating temperature
Diester	MIL-L-6085A	− 67 to 350	Anderson Oil Co. Eclipse Pioneer Div., Bendix Shell Oil Co. E. F. Houghton and Co.	Windsor Lube I-245X Pioneer P-10 AeroShell Fluid 12 Cosmolubric 270	General-purpose, low-starting torque, and stable over a wide temperature range. Particularly suited for precision instrument gears and small machinery gears
Diester	MIL-L-7808C	− 67 to 400	Sinclair Refining Co. Socony Mobil Oil Co. Bray Oil Co. Exxon Corporation	Aircraft Turbo S Oil Avrex S Turbo 251 Brayco 880 Exxon Turbine Oil 15	Suitable for oil spray or mist system at high temperature. Particularly suitable for high-speed power gears
Silicone		− 75 to 350	Dow-Corning Corp.	DC200	Rated for low-starting torque and lightly loaded instrument gears
Silicone		− 100 to 600	General Electric Co.	Versilube 81644	Best load carrier of silicone oils with widest temperature range. Applicable to power gears requiring wide temperature ranges
Greases					
Diester oil-lithium soap	MIL-G-7421A	− 100 to 200	Royal Lubricants Co. Texaco	Royco 21 Low Temp. No. 1888	For moderately loaded gears requiring starting torques at low temperatures
Diester oil-lithium soap	MIL-G-3278A	− 67 to 250	Exxon Corporation Shell Oil Co. Sinclair Refining Co. Bray Oil Co.	Beacon 325 AeroShell Grease II Sinclair 3278 Grease Braycote 678	General-purpose light grease for precision instrument gears, and generally lightly loaded gears
Petroleum oil-sodium soap	MIL-L-3545	− 20 to 300	Exxon Corporation Standard Oil Co. of Calif.	Andok 260 RPM Aviation Grease #2	A high-temperature lubricant for high speed and high loads
Mineral oil-sodium soap	—	− 25 to 250	Exxon Corporation	Andok C	Stiff grease that channels readily. Suitable for high speeds and highly loaded gears

Table 8.3.25   Solid Oil Additives

Lubricant type	Temperature range, °F	Source	Identification	Remarks
Colloidal graphite	Up to 1,000	Acheson Colloids Co.	SLA 1275	Good load capacity, excellent temperature resistance
Colloidal MoS$_2$	Up to 750	Acheson Colloids Co.	SLA 1286	Good antiwear
Colloidal Teflon	Up to 575	Acheson Colloids Co.	SLA 1612	Low coefficient of friction

Plastic gears are often operated without any external lubrication, and they will provide long service life if the plastic chosen is correct for the application. Plastics manufacturers and their publications can be consulted for guidance. Alternatively, many plastic gear materials can be molded with internal solid lubricants, such as MoS$_2$, Teflon, and graphite.

GEAR INSPECTION AND QUALITY CONTROL

Gear performance is not only related to the design, but also depends upon obtaining the specified quality. Details of gear inspection and control of subtle problems relating to quality are given in Michalec, "Precision Gearing," Chap. 11.

COMPUTER MODELING AND CALCULATIONS

A feature of the latest AGMA rating standards is that the graphs, including those presented here, are accompanied by equations which allow application of computer-aided design. Gear design equations and strength and durability rating equations have been computer modeled by many gear manufacturers, users, and university researchers. Numerous software programs, including integrated CAD/CAM, are available from these places, and from computer system suppliers and specialty software houses. It is not necessary for gear designers, purchasers, and fabricators to create their own computer programs.

With regard to gear tooth strength and durability ratings, many custom gear house designers and fabricators offer their own computer modeling which incorporates modifications of AGMA formulas based upon experiences from a wide range of applications.

The following organizations offer software programs for design and gear ratings according to methods outlined in AGMA publications: Fairfield Manufacturing Company Gear Software; Geartech Software, Inc.; PC Gears; Universal Technical Systems, Inc. For details and current listings, refer to AGMA's latest "Catalog of Technical Publications."

8.4   FLUID FILM BEARINGS
by Vittorio (Rino) Castelli

REFERENCES: "General Conference on Lubrication and Lubricants," ASME. Fuller, "Theory and Practice of Lubrication for Engineers," 2d ed., Wiley. Booser, "Handbook of Lubrication, Theory and Design," vol. 2, CRC Press. Barwell, "Bearing Systems, Principles and Practice," Oxford Univ. Press. Cameron, "Principles of Lubrication," Longmans Greene. "Proceedings," Second International Symposium on Gas Lubrication, ASME. Gross, "Fluid-Film Lubrication," Wiley. Gunter, "Dynamic Stability of Rotor-Bearing Systems," NASA SP-113, Government Printing Office. Khonsari and Booser, "Applied Tribology," Wiley, 2001. Loeb and Rippel, Determination of Optimum Properties for Hydrostatic Bearings, *ASLE Tribology Trans.* 1958.

Plain bearings, according to their function, may be

Journal bearings, cylindrical, carrying a rotating shaft and a radial load
Thrust bearings, the function of which is to prevent axial motion of a rotating shaft
Guide bearings, to guide a machine element in its translational motion, usually without rotation of the element

In exceptional cases of design, or with a complete **failure of lubrication,** a bearing may run dry. The coefficient of friction is then between 0.25 and 0.40, depending on the materials of the rubbing surfaces. With the **bearing barely greasy,** or when the bearing is well lubricated but the speed of rotation is very slow, boundary lubrication takes place. The coefficient of friction may vary from 0.08 to 0.14. This condition occurs also in any bearing when the shaft is starting from rest if the bearing is not equipped with an oil lift.

Semifluid, or **mixed,** lubrication exists between the journal and bearing when the conditions are not such as to form a load-carrying fluid film and thus separate the surfaces. Semifluid lubrication takes place at comparatively low speed, with intermittent or oscillating motion, heavy load, insufficient oil supply to the bearing (wick or waste-lubrication, drop-feed lubrication). Semifluid lubrication may also exist in thrust bearings with fixed parallel-thrust collars, in guide bearings of machine tools, in bearings with copious lubrication where the shaft is bent or the bearing is misaligned, or where the bearing surface is interrupted by improperly arranged oil grooves. The coefficient of friction in such bearings may range from 0.02 to 0.08 (Fuller, Mixed Friction Conditions in Lubrication, *Lubrication Eng.*, 1954).

Fluid or **complete lubrication,** when the rubbing surfaces are completely separated by a fluid film, provides the lowest friction losses and prevents wear. A certain amount of oil must be fed to the oil film in order to compensate for end leakage and maintain its carrying capacity. Such lubrication can be provided under pressure from a pump or gravity tank, by automatic lubricating devices in self-contained bearings (oil rings or oil disks), or by submersion in an oil bath (thrust bearings for vertical shafts).

Notation

R = radius of bearing, length
r = radius of journal, length
$c = mr = R - r$ = radial clearance, length
W = bearing load, force
μ = viscosity = force × time/length^2
Z = viscosity, centipoise (cP); 1 cP = 1.45×10^{-7} lb · s/in^2
    (0.001 N · s/m^2)
β = angle between load and entering edge of oil film
η = coefficient for side leakage of oil
ν = kinematic viscosity = μ/ρ, length^2/time
R_e = Reynolds number = umr/ν
P_a = absolute ambient pressure, force/area
$P = W/(ld)$ = unit pressure, lb/in^2
N = speed of journal, r/min
m = clearance ratio (diametral clearance/diameter)
F = friction force, force
A = operating characteristic of plain cylindrical bearing

P' = alternate operating characteristic of plain cylindrical bearing
h_0 = minimum film thickness, length
ε = eccentricity ratio, or ratio of eccentricity to radial clearance
e = eccentricity = distance between journal and bearing centers, length
f = coefficient of friction
f' = friction factor = $F/(\pi rlpu^2)$
l = length of bearing, length
d = $2r$ = diameter of journal, length
K_f = friction factor of plain cylindrical bearing
t_w = temperature of bearing wall
t_0 = temperature of air
t_1 = temperature of oil film
u = surface speed, length/time
ω = angular velocity, rad/time
ρ = mass density, mass/length^3
Λ = bearing compressibility parameter = $6\mu\omega r^2/(P_a c^2)$

INCOMPRESSIBLE AND COMPRESSIBLE LUBRICATION

Depending on the fluid employed and the pressure regime, the fluid density may or may not vary appreciably from the ambient value in the load-carrying film. Typically, oils, water, and liquid metals can be considered incompressible, while gases exhibit compressibility effects even at modest loads. The difference comes from the fact that, in incompressible lubricants, fluid flow rates are linearly proportional to pressure differences, whereas for compressible lubricants the mass flow rates are proportional to the difference of some power of the pressure. This is because the pressure affects the fluid density. The bearing behavior is somewhat dissimilar. In incompressible lubrication, gage pressures can be used and the value of the ambient pressure has no effect on the load-carrying capacity, which is linearly related to viscosity and speed. This is not true in compressible lubrication, where the value of ambient pressure has a direct effect on the load-carrying capacity which, in turn, increases with viscosity and speed, but only up to a limit dependent on the bearing geometry. In what follows, incompressible lubrication is treated first and compressible lubrication second.

Incompressible (Plain Cylindrical Journal Bearings)

Fluid lubrication in plain cylindrical bearings depends on the viscosity of the lubricant, the speed of the bearing components, the geometry of the film, and possible external sources of pressurized lubricant. The oil is entrained by the journal into the film by the action of the viscosity which, if the passage is convergent, causes the creation of a pressure field, resulting in a force sufficient to float the journal and carry the load applied to it.

The **minimum film thickness** h_0 determines the closest approach of the journal and bearing surfaces (Fig. 8.4.1). The allowable closest approach depends on the finish of these surfaces and on the rigidity of the journal and bearing structures. In practice, h_0 = 0.00075 in (0.019 mm) is common in electric motors and generators of medium speed, with steel shafts in babbitted bearings; h_0 = 0.003 in (0.076 mm) to 0.005 in (0.127 mm) for large steel shafts running at high speed in babbitted bearings (turbogenerators, fans), with pressure oil-supply for lubrication;

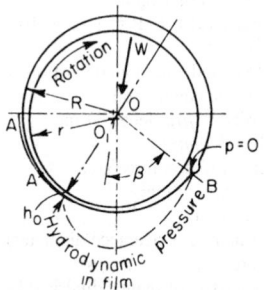

Fig. 8.4.1   Journal bearing with perfect lubrication.

h_0 = 0.0001 in (0.0025 mm) to 0.0002 in (0.005 mm) in automotive and aviation engines, with very fine finish of the surfaces.

Figure 8.4.2 gives the relationship between ε and the load-carrying coefficient A for a plain cylindrical journal. The operating characteristic of the bearing is

$$A = (132/\eta)(1{,}000m)^2[P/(ZN)]$$

In Fig. 8.4.1, β is the angle between the direction of the load W and the entering edge of the load-carrying oil film, in degrees. The entering edge is at the place where the hydrodynamic pressure is equal or nearly equal to the atmospheric pressure and may be at the location of the

Fig. 8.4.2   Eccentricity ratio for a plain cylindrical journal.

oil-distributing groove B, or at the end of the machined recess pocket as at AA. For **complete bearings,** i.e., when the inner surface of the bearing is not interrupted by grooves, β may be taken as 90°. The reason for this assumption is the fact that, where the film diverges, the bearing pumping action tends to generate negative pressure, which liquids cannot sustain. The film **cavitates,** i.e., it breaks up in regions of fluid intermixed with either air or fluid vapor, while the pressure does not deviate substantially from ambient. For a 120° bearing with a central load, β may be taken as 60°.

The coefficient η corrects for side leakage. There is a loss of load-carrying capacity caused by the drop in the hydrodynamic pressure p in the oil film from the midsection of the bearing toward its ends; $p = 0$ at the ends. The value of η depends on the length-diameter ratio l/d and ε, the eccentricity ratio. Values of η are given in Fig. 8.4.3.

Fig. 8.4.3

EXAMPLE 1.   A generator bearing, 6 in diam by 9 in long, carries a vertical downward load of 8,650 lb; N = 720 r/min. The diametral clearance of the bearing is 0.012 in; the bearing is split on its horizontal diameter, and the lower half is relieved 40° down on each side, for oil distribution along journal; the bearing arc is therefore 100°; with the load vertical, β = 50°; bearing temperature 160°F. The absolute viscosity of the oil in the film is 12 centipoises (medium turbine oil). P = W/ld = 160 lb/in^2; μ = 12 × 1.45 × 10^{-7} = 17.4 × 10^{-7} lb · s/in^2. The solution is one of trial and error. By using Fig. 8.4.3 in conjunction with Fig. 8.4.2, only a few trials are necessary to obtain the answer. As a first trial assume ε = 0.85. For an l/d ratio of 1.5 in Fig. 8.4.3, η, the end-leakage factor, will be 0.77. Compute A using this value of η. m = 0.012/6 = 0.002.

$$A = \frac{132}{0.77}(2)^2\frac{160}{12 \times 720} = 12.7$$

Enter Fig. 8.4.2 with this value of a and at $\beta = 50°$, and find that $\varepsilon = 0.9$. This value is larger than the initial assumption for ε. As a second trial, $\varepsilon = 0.88$. Then $\eta = 0.8$, $A = 12.2$, and $\varepsilon = 0.89$. This is a sufficiently close check. The minimum film thickness is $h_0 = mr(1 - \varepsilon) = 0.002 \times 3 \times 0.12 = 0.0007$ in (0.01778 mm).

For severe operating conditions the value of A may exceed 18, the limit of Fig. 8.4.2. For complete journal bearings under extreme operating conditions, Fig. 8.4.4 should be used. The ordinate is P', defined as shown. The curves are drawn for various values of l/d instead of values of β as in Fig. 8.4.2. Values of ε may thus be obtained directly (Dennison, Film-Lubrication Theory and Engine-Bearing Design, *Trans. ASME*, **58**, 1936).

Fig. 8.4.4   Load-carrying parameter in terms of eccentricity.

EXAMPLE 2.   A 360° journal bearing $2\frac{1}{2}$ in diam and $3\frac{7}{8}$ in long carries a steady load of 3,875 lb. Speed $N = 500$ r/min; diametral clearance, 0.0064 in; average viscosity of the oil in the film, 23.4 centipoises (SAE 20 light motor oil at 105°F). $P = 3,875/(2.5 \times 3.875) = 400$ lb/in². Value of $m = 0.0064/2.5 = 0.00256$. Value of $l/d = 1.55$. First, attempt to use Figs. 8.4.2 and 8.4.3 in this solution. Assume eccentricity ratio ε is 0.9. Then, in Fig. 8.4.3, with $l/d = 1.55$, value of η is determined as 0.8. A is calculated as 37. This is completely off scale in Fig. 8.4.2. Consider instead Fig. 8.4.4. Value of P' is computed as

$$P' = 6.9(2.56)^2 \frac{400}{23.4 \times 500} = 1.54$$

In Fig. 8.4.4, enter the curves with $P' = 1.54$, and move left to intersect the curve for $l/d = 1.5$. Drop downward to read a value for $1/(1 - \varepsilon)$ of 16. Then $\frac{1}{16} = 1 - \varepsilon$, or the eccentricity ratio $\varepsilon = \frac{15}{16}$, or 0.94. The minimum film thickness, as in Example $1 = h_0 = mr(1 - \varepsilon)$, or

$$h_0 = 0.00256 \times 1.25 (1 - 0.94) = 0.0002 \text{ in } (0.0051 \text{ mm})$$

Allowable mean bearing pressures in bearings with fluid film lubrication are given in Table 8.4.1. If the load maintains the same magnitude and direction when the journal is at rest (heavily loaded shafts, heavy gears), the mean bearing pressure should be somewhat less than when bearings are loaded only when running.

For internal-combustion-engine bearing design, Etchells and Underwood (*Mach. Des.*, Sept. 1942) list the following maximum design pressures for bearing alloys, pounds per square inch of projected area: lead-base babbitt (75 to 85 percent lead, 4 to 10 percent tin, 9 to 15 percent antimony) 600 to 800; tin-base babbitt (0.35 to 0.6 percent lead, 86 to 90 percent tin, 4 to 9 percent antimony, 4 to 6 percent copper) 800 to 1,000; cadmium-base alloy (0.4 to 0.75 percent copper, 97 percent cadmium, 1 to 1.5 percent nickel, 0.5 to 1.0 percent silver) 1,200 to 1,500; copper-lead alloy (45 percent lead, 55 percent copper) 2,000 to 3,000; copper-lead (25 percent lead, 3 percent tin, 72 percent copper) 3,000 to 4,000; silver (0.5 to 1.0 percent lead on surface, 99 percent silver) 5,000 up. The above pressures are based on fatigue life of 500 h at 300°F bearing temperature, and a bearing metal thickness 0.01 to 0.015 in for lead-, tin-, and cadmium-base metals and 0.25 in for copper, lead, and silver. At lower temperatures the life will be greatly extended.

Much higher pressures are encountered in rolling element bearings, such as ball and roller bearings, and gears. In these situations, the formation of fluid films capable of preventing contact between surface asperities is aided by the increase of viscosity with pressure, as exhibited by most lubricating oils. The relation is typically exponential, $\mu = \mu_0 e^{\alpha p}$, where α is the so-called pressure coefficient of viscosity.

Length-diameter ratios are usually chosen between $l/d = 1$ and $l/d = 2$, although many engine bearings are designed with $l/d = 0.5$, or even less. In shorter bearings, the carrying capacity of the oil film is greatly impaired by the effect of side leakage. Longer bearings are used to restrain the shaft from vibration, as in line shafts, or to position the shaft accurately, as in machine tools. In power machines, the tendency is toward shorter bearings. Typical values are as follows: turbogenerators, 0.8 to 1.5; gasoline and diesel engines for main and crankpin bearings, 0.4 to 1.0, with most values between 0.5 and 0.8; generators and motors, 1.5 to 2.0; ordinary shafting, heavy, with fixed bearings, 2 to 3; light, with self-aligning bearings, 3 to 4; machine-tool bearings, 2 to 4; railroad journal bearings, 1.2 to 1.8.

For the **clearance between journal and bearing** see Fits in Sec. 8. Medium fits may be used for journals running at speeds under 600 r/min, and free fits for speeds over 600 r/min. Kingsbury suggests for these journals a diametral clearance $= 0.002 + 0.001d$ in. In journals running at high speed, diametral clearance $= 0.002d$ should be used in order to lower the friction losses in the bearing. All units are in inches. The most satisfactory clearance should, of course, be based on a complete bearing analysis which includes both load-carrying capacity and heat generation due to friction. For example, a bearing designed to run at the extremely high speed of 50,000 r/min uses a diametral clearance of 0.0025 in for a journal with 0.8-in diameter, giving a clearance ratio, clearance/diameter, of 0.00316.

Table 8.4.1   Current Practice in Mean Bearing Pressures

Type of bearing	Permissible pressure, lb/in², of projected area	Type of bearing	Permissible pressure, lb/in², of projected area
Diesel engines, main bearings	800–1,500	Automotive gasoline engines, main bearings	500–1,000
Crankpin	1,000–2,000	Crankpin	1,500–2,500
Wrist pin	1,800–2,000	Air compressors, main bearings	120– 240
Electric motor bearings	100– 200	Crankpin	240– 400
Marine diesel engines, main bearings	400– 600	Crosshead pin	400– 800
Crankpin	1,000–1,400	Aircraft engine crankpin	700–2,000
Marine line-shaft bearings	25– 35	Centrifugal pumps	80– 100
Steam engines, main bearings	150– 500	Generators, low or medium speed	90– 140
Crankpin	800–1,500	Roll-neck bearings	1,500–2,500
Crosshead pin	1,000–1,800	Locomotive crankpins	1,500–1,900
Flywheel bearings	200– 250	Railway-car axle bearings	300– 350
Marine steam engine, main bearings	275– 500	Miscellaneous ordinary bearings	80– 150
Crankpin	400– 600	Light line shaft	15– 25
Steam turbines and reduction gears	100– 220	Heavy line shaft	100– 150

For high-speed internal-combustion-engine bearings using forced-feed lubrication, medium fits are used. Federal-Mogul recommends the following diametral clearances in inches per inch of shaft diameter for insert-type bearings: tin-base and high-lead babbitts, 0.0005; cadmium-silver-copper, 0.0008; copper-lead, 0.001.

The dependence of the **coefficient of friction** for journal bearings on the bearing clearance, lubricant viscosity, rotational speed, and loading pressure, as reported by McKee and others, is shown in Sec. 3. A plot of the coefficient of friction against the parameter ZN/P is a convenient method for showing this relationship. ZN/P is a parameter based on mixed units. Z is the viscosity in centipoise, N is r/min, P is the mean pressure on the bearing due to the load, pounds per square inch of projected area, and m is the clearance ratio. Values of ZN/P greater than about 30 indicate fluid film conditions in the bearings. If the viscosity of the lubricant becomes lower or if there is a reduction in rotational speed or an increase in load, the value of ZN/P will become smaller until the coefficient of friction reaches a minimum value. Any further reduction in ZN/P will produce breakdown of the oil film, marking the transition from fluid film lubrication with complete separation of the moving surfaces to semifluid or mixed lubrication, where there is partial contact. As soon as semifluid conditions are initiated, there will be a sharp increase in the coefficient of friction. The critical value of ZN/P, where this transition takes place, will be lowest for a rigid bearing and shaft with finely finished surfaces.

Figure 8.4.5 shows a generalization of the relationship between the coefficient of friction for a journal bearing and the parameter ZN/P,

Fig. 8.4.5   Various zones of possible lubrication for a journal bearing.

indicating the various possible lubrication regimes that may be expected. For optimum design, a value of ZN/P somewhere between 30 and 300 would be recommended, but, in any case, the determination of minimum film thickness h_0 should be the deciding parameter. For extremely large values of ZN/P, resulting from high speeds and low loads,

Fig. 8.4.6   Variation of the friction factor of a bearing with eccentricity ratio.

whirl instability may be developed. (See material on gas-lubricated bearings in this section.) With large values of ZN/P and a lubricant having a low kinematic viscosity, turbulent conditions may develop in the bearing clearance.

The friction force in plain journal bearings may be estimated by the use of the expression $F = K_f \mu Nrl/m$, where μ is in lb · s/in^2 units. The value of K_f depends upon the magnitude of ε and the type of bearing. Figure 8.4.6 shows values of K_f for a complete bearing, a 150° partial bearing, and a 120° partial bearing, assuming that the clearance space is at all times filled with lubricant. Note that F is the friction force at the surface of the bearing. Consequently, the friction torque is obtained by multiplying F by the bearing radius.

EXAMPLE 3.   As an illustration of the use of Fig. 8.4.6, determine the friction force in the bearing of Example 2. This is a complete journal bearing 2½-in diam by 3⅞ in. The value of ε was determined as 0.94. From Fig. 8.4.6, $K_f = 2.8$. Then

$$F = \frac{2.8 \times 23.4 \times 1.45 \times 10^{-7} \times 500 \times 1.25 \times 3.875}{0.00256}$$
$$= 8.97 \text{ lb} (4.08 \text{ kg})$$

The coefficient of friction $F/W = 8.97/3,875 = 0.00231$. The mechanical loss in the bearing is $FV/33,000$ hp, where V is the peripheral velocity of the journal, ft/min.

$$\text{Friction hp} = (8.97 \times 500 \times \pi \times 2.5)/(33,000 \times 12)$$
$$= 0.089 \text{ hp} (66.37 \text{ W})$$

Departure from laminarity in the fluid film of a journal bearing will increase the friction loss. Figure 8.4.7 (Smith and Fuller, Journal Bearing Operation at Super-laminar Speeds, *Trans. ASME*, **78**, 1956) shows test results for such bearings, expressed in terms of a Reynolds number for the fluid film, $R_e = umr/v$. Laminar conditions hold up to an R_e of about 1,000. Friction may be calculated for laminar flow by using Fig. 8.4.6 or the left branch of the curve in Fig. 8.4.7, where $f' = 2/R_e$, and which applies to low values of the eccentricity ratio ($K_f = 0.66$). The values from Fig. 8.4.7 may be converted to friction torque T by the use of the expression $T = f' \pi p u^2 r^2 l$, where ρ is the mass density of the lubricant. In Fig. 8.4.7, a transition region spans values of the Reynolds number from 1,000 to 1,600. Here, two types of flow instability can occur. Usually, the first is due to **Taylor vortices** which are wrapped in

Fig. 8.4.7   Friction f' as a function of the Reynolds number for an unloaded journal bearing with $l/d = 1$. *(Smith and Fuller.)*

regular circumferential structures, each of which occupies the entire clearance. The onset of this phenomenon takes place at a value of the Reynolds number exceeding the threshold $R_e = 41.1(r/c)^{1/2}$. The second instability is due to turbulence, occurring at $R_e > 2,000$.

EXAMPLE 4.   A journal bearing is 4.5 in diameter by 4.5 in long. Speed 22,000 r/min. $mr = 0.002$ in. Viscosity μ, 1 cP (water) $= 1.45 \times 10^{-7}$ lb · s/in^2; mass density $p = 62.4/1,728 \times 386 = 9.35 \times 10^{-5}$ lb · s^2/in^4; $v = \mu/\rho = 1.45 \times 10^{-7}/9.35 \times 10^{-5} = 0.155 \times 10^{-2}$ in^2/s; $u = 22,000 \times 2\pi \times 2.25/60 = 5,180$ in/s; $R_e = 5,180 \times 0.002/0.155 \times 10^{-2} = 6,680$. This would indicate turbulence in the film. Value of f' is then $0.078/6,680^{0.43} = 0.078/44.2 = 1.765 \times 10^{-3}$. Friction torque $T = 1.765 \times 10^{-3} \times \pi \times 9.35 \times 10^{-5} \times 5,180^2 \times 2.25^2 \times 4.5$, $T = 317.5$ in · lb. Friction horsepower $= 2\pi TN/12 \times 33,000 = 2\pi \times 317.5 \times 22,000/12 \times 33,000$, FHP $= 111$ (82.77 kW).

In self-contained bearings (electric motor, line shaft, etc.) without external oil or water cooling, the **heat dissipation** is equal to the heat generated by friction in the bearing.

The heat dissipated from the outside bearing wall to the surrounding air is governed by the laws of heat transfer $Q = hS(t_w - t_0)$, where S is the surface area from which the heat is convected, Q is the rate of energy flow; t_w and t_0 are the temperatures of the wall and ambient air, respectively; and h is the heat convection coefficient, which has values from 2.2 Btu/(h · ft² · °F) for still air to 6.5 Btu/(h · ft² · °F) for air moving at 500 ft/min. Calculations of heat loss are extremely important due to the strong temperature dependence of the viscosity of most oils.

The temperature of the oil film will be higher than the temperature of the bearing wall. Typical ranges of values according to Karelitz (*Trans. ASME*, **64**, 1942), Pearce (*Trans. ASME*, **62**, 1940), and Needs (*Trans. ASME*, **68**, 1948) for self-contained bearings with oil bath, oil ring, and waste-packed lubrication are shown in Fig. 8.4.8.

Fig. 8.4.8   Temperature rise of the film.

EXAMPLE 5.   The frictional loss for the generator bearing of Example 1, computed by the method outlined in Example 3, is 0.925 hp with $\varepsilon = 0.88$, $K_f = 1.6$, and $F = 27$ lb. Operating in moving air the heat dissipated by the bearing housing will be $L = 6.5S(t_w - t_0)$. Since this is a self-contained bearing, the heat dissipated is also equal to the heat generated by friction in the oil film, or $L = 0.925 \times 2,545 = 2,355$ Btu/h. With $S = 25 \times 6 \times 9/144 = 9.4$ ft², $t_w = t_0 = 2,355/6.5 \times 9.4 = 38.5°F$. This is the temperature rise of the bearing wall above the ambient room temperature. For an 80°F room, the wall temperature of the bearing would be about 118°F. In Fig. 8.4.8 an oil-ring bearing in moving air with a temperature rise of wall over ambient of 38°F should have a film temperature 50°F higher than that of the wall. The film temperature on the basis of Fig. 8.4.8 will then be 80 +38 + 50, or 168°F. This is close enough to the value of the film temperature of 160°F from Example 1, with which the friction loss in the bearing was computed, to indicate that this bearing can operate without the need for external cooling.

To predict the operating temperature of a self-contained bearing, the cut-and-try method shown above may be used. First, an oil-film temperature is assumed. Viscosity and friction losses are calculated. Then the temperature rise of the wall over ambient is computed so as to dissipate to the atmosphere an amount of heat equal to the friction loss. Lastly from Fig. 8.4.8 the corresponding oil-film temperature is estimated and compared to the value that was originally assumed. A few adjustments of the assumed film temperature will produce satisfactory agreement and indicate the leveling-off temperature of the bearing. Self-contained bearings have been built with diameters of 3, 8, and 24 in (7.62, 20.32, and 60.96 cm) to operate at shaft speeds of 3,600, 1,000, and 200 r/min, respectively. These designs indicate a rough limit for bearings with no external cooling. The highest bearing temperature permissible with normal lubricants is about 210°F (100°C).

The temperature of automotive-type bearings is held within safe limits by using a **pressure-feed oil supply.** Sufficient lubricant is forced through the bearing to act as a coolant and prevent overheating. One widely used practice is to place a circumferential groove at the center of

the bearing to which the oil supply is fed. This is effective as far as cooling is concerned but has the disadvantage of interrupting the active length of the bearing and lowering its l/d ratio (see Fig. 8.4.9). The axial flow through each side of the bearing is given by

$$Q_1 = \frac{\Delta P\, n^3 r^4 \pi}{6\mu b}\left(1 + \frac{3}{2}\varepsilon^2\right)$$

where b is the effective axial length of the half bearing and ΔP is the difference between the oil pressure in the circumferential groove and

Fig. 8.4.9   Bearing with central circumferential groove.

the pressure at the ends of the bearing. The value of the last term in this equation will vary from 1.0 for a concentric shaft and bearing indicated by $\varepsilon = 0$ to a value of 2.5 for the extreme case of the shaft touching the bearing wall, indicated when $\varepsilon = 1$. Most of the heat caused by friction in the bearing is carried away by the circulating oil. Permissible temperature rises for this type of bearing may range from 15 to 50°F (8 to 28°C). In extreme cases a rise of 100°F (55°C) can be tolerated for high-strength bearing materials. The lower values of temperature rise usually indicate needlessly large oil flow. Such a condition will result in an excessive friction loss in the bearing.

EXAMPLE 6.   The bearing of Examples 2 and 3 is lubricated by a circumferential groove with an oil supply pressure of 30 lb/in² and, as before, $\varepsilon = 0.94$, $m = 0.0026$, and $\mu = 23.4 \times 1.45 \times 10^{-7}$ lb·s/in². Length b is about 1.93 in.

$$Q_1 \text{ flow out one side} = \frac{30 \times 0.0026^3 \times 1.25^4 \times \pi}{6 \times 23.4 \times 1.45 \times 10^{-7} \times 1.93}$$
$$\times [1 + 3/2(0.94)^2] = 0.240 \text{ in}^3/\text{s}(3.93 \text{ cm}^3/\text{s})$$

Total flow (two sides) = 0.48 in³/s = 53 lb/h for sp gr = 0.85. The friction loss from Example 3 = 0.089 hp = 226 Btu/h. With a specific heat of 0.5 Btu/(lb · °F) and assuming that all the friction energy is given up to the oil in the form of heat, the temperature rise $\Delta t = 226/0.5 \times 53 = 8.5°F$ (4.72°C).

A definite **minimum rate of oil feed** is required to maintain a fluid film in journal bearings. This makes no allowance for the additional flow that may be needed to cool the bearings. However, many industrial bearings run at relatively low speeds with light loads and, as a consequence, additional oil flow to provide cooling is not necessary. But if a fluid film is desired, a definite minimum amount of lubricant is required. If the volume of lubricant fed to the bearing is less than this minimum requirement, there will not be a complete fluid film in the bearing. Friction will rise, wear will become greater, and the satisfactory service life of such a bearing will be reduced. This minimum lubricant supply can be evaluated by using the equation

$$Q_M = K_M urml$$

where Q_M is the flow rate and K_M is approximately 0.006.

Fig. 8.4.10   Siphon wick.

Fig. 8.4.11   Bottom wick.

EXAMPLE 7. The minimum feed rate for a journal bearing 2⅛-in diam by 2⅛ in long will be determined. Diametral clearance is 0.0045 in; speed, 1,230 r/min; load, 40 lb/in² based on projected area. $u = 1,230 \times \pi \times 2.125 = 10,220$ in/min, $r = 1.062$ in, $m = 0.0045/2.125 = 0.00212$, $l = 2.125$ in. Substituting,

$$Q_M = 0.006 \times 10,220 \times 1.062 \times 0.00212 \times 2.125$$
$$= 0.28 \text{ in}^3/\text{min}$$

(Fuller and Sternlicht, Preliminary Investigation of Minimum Lubricant Requirements of Journal Bearings, *Trans. ASME*, **78**, 1956.)

Many bearings are supplied with oil at low rates of feed by **felts, wicks,** and **drop-feed oilers.** Wicks can supply substantial rates of feed if they are properly designed. The two basic types of wick feed are siphon wicks, as shown in Fig. 8.4.10, and bottom wicks, as shown in Fig. 8.4.11.

Fig. 8.4.12   Oil delivery with siphon wick (Fig. 8.4.10).

Data on oil delivery for these wicks are shown in Figs. 8.4.12 and 8.4.13. The data, from the American Felt Co., are for SAE Fl felts, based on a cross-sectional area of 0.1 in². The flow rate is indicated in drops per minute. One drop equals 0.0026 in³ or 0.043 cm³.

EXAMPLE 8. If it is desired to deliver 12.5 drops/min to a journal bearing, and if the viscosity of the oil is 212 s Saybolt Universal at 70°F, and if *L*, Fig. 8.4.10, is 5 in, what size of round wick would be required? From Fig. 8.4.12, for the stated conditions the delivery rate would be 0.9 drop/min for an area of 0.1 in². If 12.5 drops/min is needed, this would mean an area of 12.5 divided by 0.9 and multiplied by 0.1, or 1.4 in². For a round wick this would mean a diameter of 1⅜ in (3.49 cm).

If a **bottom wick** is considered with L = 4 in, Fig. 8.4.11, then in Fig. 8.4.13 the delivery rate using the same oil would be 1.6 drops/min; and if 12.5 drops/min is required, the area would be 12.5 divided by 1.6 and multiplied by 0.1, or 0.78 in². This would mean a bottom wick of 1 in diam if it is round (2.54 cm).

When journal **bearings** are **started, stopped,** or **reversed,** or whenever conditions are such that the operating value of *ZN/P* falls below the critical value for that bearing, the oil film will be ruptured and metal-to-metal contact will increase friction and cause wear. This condition can be eliminated by using a **hydrostatic oil lift.** High-pressure oil is introduced to the area between the bottom of the journal and the bearing (Fig. 8.4.14). If the pressure and quantity of flow are great enough, the shaft, whether it is rotating or not, will be raised and supported by an oil film. Neglecting axial flow, which is small, the flow up one side is

$$Q_1 = \frac{Wrm^3}{A\mu} \text{ in}^2/\text{s}$$

and the inlet pressure required, $P_o = \mu Q_1 B/(br^2m^3)$, where *b* is the axial length of the high-pressure recess. Values of *A* and *B* are dimensionless factors which represent geometric effects and are given in the following table as a function of ε:

Fig. 8.4.13   Oil delivery with bottom wick (Fig. 8.4.11).

Current practice is to make the total area of the high-pressure recess in a bearing 2½ to 5 percent of the projected area *ld* of the bearing. It is generally desirable to use a check valve in the supply line to the oil lift so that, when the journal builds up a hydrodynamic oil-film pressure, reverse flow of oil in the supply line will be prevented.

Fig. 8.4.14   Diagram of oil lift.

EXAMPLE 9. A 4,000-in-diam journal rests in a bearing of 4.012-in-diam. SAE 30 oil at 100°F (105 cP) is supplied under pressure to a groove at the lowest point in the bearing. Length of bearing, 6 in, length of groove, 3 in, load on bearing, 3,600 lb. What inlet pressure and oil flow are needed to raise the journal 0.004 in?

$$h_0 = mr(1 - \varepsilon)$$
$$0.004 = 0.006(1 - \varepsilon)$$
$$\varepsilon = 0.333$$

From the table, A = 44.5, B = 42.

$$Q_1 = \frac{3,600 \times 2}{44.5 \times 105 \times 1.45 \times 10^{-7}} (0.003)^3$$
$$= 0.287 \text{ in}^3/\text{s, one side } (4.70 \text{ cm}^3/\text{s})$$

Flow from both sides = $(0.287 \times 2) \times {}^{60}/_{231} = 0.149$ gal/min (0.564 1/min).
  Oil supply pressure is

$$P_o = \frac{105 \times 1.45 \times 10^{-7} \times 0.287 \times 42}{3 \times 4} \times \frac{1}{0.003^3} = 566 \text{ lb/in}^2$$

ε	0	0.1	0.2	0.3	0.4	0.5	0.6	0.7	0.75	0.8	0.85	0.9
A	24.0	28.1	33.8	41.6	53.3	72.0	105	173	237	360	613	1,320
B	18.9	23.2	29.0	38.2	52.7	77.9	128	246	344	634	1,260	3,360

ε	0.91	0.92	0.93	0.94	0.95	0.96	0.97	0.98	0.99
A	1,620	2,070	2,620	3,530	5,040	7,800	13,700	30,600	121,000
B	4,340	5,810	8,040	11,800	18,400	32,100	65,300	179,000	348,000

Fig. 8.4.15 Load-carrying capacity and flow for journal bearings *(Loeb)*. Lengths in inches.

An adjustable constant-volume pump or a spur-gear pump with a capacity of about 1,000 lb/in² (6.894 kN/m²) should be used to allow for pressure that may be built up in the line before the journal begins to rise.

Other configurations for hydrostatically lubricated journal bearings are shown in Fig. 8.4.15. These were obtained by means of electric analog solutions (Loeb, Determination of Flow, Film Thickness and Load-Carrying Capacity of Hydrostatic Bearings through the Use of the Electric Analog Field Plotter, *Trans. ASLE*, **1**, 1958). The data from Fig. 8.4.15 are exact for a uniform film thickness corresponding to $\varepsilon = 0$ but may be used with discretion for other values of ε. More recently these solutions are achieved by numerical approximations with the help of digital computers.

Multiple recesses are used in externally pressurized bearings in order to provide local **stiffness.** This term indicates that the bearing resists shaft motions in any direction, and it is achieved by properly arranging the feeding network according to a strategy called **compensation.** Three main types are employed: orifice (and its variant, inherent), capillary, and fixed flow rates. In the first two, the idea is to insert a hydraulic resistance in each of the recess feeding lines and to use a single pump to feed all recesses. The flow rate q through orifices varies with the square root of the pressure drop Δp

$$q \propto \sqrt{\Delta p}$$

while for capillary tubes the relation is linear:

$$q = \frac{\pi \, \Delta p \, d^4}{64 \, l_1 \, \mu}$$

The general rule of thumb in designing orifices or capillary restrictors is to generate a pressure drop approximately equal to that taking place through the bearing, i.e., from the recesses to the ambient. The recess geometry and distribution, on the other hand, are designed so that

$W = 0.5 p_{\text{recess}} \, DL$. Thus, the pump supply pressure is 4 times the average bearing pressure. The bearing stiffness is usually equal to $K = 0.5 p_{\text{recess}} \, DL/c$.

The third method of compensation consists of forcing the same amount of flow to reach each recess regardless of clearance distribution. This can be achieved either by using separate pumps for each recess or by using a hydraulic device called a *flow divider*. With recess distributions as indicated above, the pump pressure need only be double the average bearing pressure; thus, this method of compensation leads to half the power dissipation of the other two. It is commonly used in large machinery, where power consumption must be limited. The polar axis bearings of the 200-in Hale telescope on Mount Palomar were the first large-scale demonstration of this technique. The azimuth axis thrust bearing of the 270-ft-diameter Goldstone radio telescope is probably the largest example of this type of bearing.

ELEMENTS OF JOURNAL BEARINGS

Typical dimensions of solid and split **bronze bushings** are given in Table 8.4.2.

Bronze bushings made from hard-drawn sheets and rolled into cylindrical shape are made with a wall thickness of only $\frac{1}{32}$ in for bearings up to $\frac{1}{2}$ in diam and with a wall thickness of $\frac{1}{16}$ in for bearings from 1 in diam up. The wall thickness of these bearings depends chiefly upon the strength of the material which supports them. Bushings of this type are pressed into place, and the bearing surface is finished by burnishing with a slightly tapered bar to a mirror finish. The allowable bearing pressures may exceed those of cast bronze shown in Table 8.4.1 by 10 to 20 percent.

Babbitt linings in larger bearings are generally employed in thickness of $\frac{1}{8}$ in or over and must be provided with sufficient anchorage in the

Table 8.4.2 Wall Thickness of Bronze Bushings, in

	Diam of journal, in						
	$\frac{1}{4}$	$\frac{1}{4}-\frac{1}{2}$	$\frac{1}{2}-1$	$1-1\frac{1}{2}$	$1\frac{1}{2}-2\frac{1}{2}$	$2\frac{1}{2}-4$	$4-5\frac{1}{2}$
Solid bushing, normal	$\frac{1}{16}$	$\frac{3}{32}$	$\frac{1}{8}$	$\frac{3}{16}$	$\frac{1}{4}$	$\frac{3}{8}$	$\frac{1}{2}$
Split bushing, normal	$\frac{3}{32}$	$\frac{1}{8}$	$\frac{5}{32}$	$\frac{7}{32}$	$\frac{5}{16}$	$\frac{15}{32}$	$\frac{5}{8}$
Solid bushing, thin	$\frac{1}{16}$	$\frac{3}{32}$	$\frac{3}{32}$	$\frac{1}{8}$	$\frac{3}{16}$	$\frac{1}{4}$	$\frac{3}{8}$
Split bushing, thin	$\frac{1}{16}$	$\frac{3}{32}$	$\frac{1}{8}$	$\frac{3}{16}$	$\frac{1}{4}$	$\frac{3}{8}$	$\frac{1}{2}$

supporting shell. The anchors take the form of dovetailed grooves or holes drilled in the shell and counterbored from the outside.

Improved conditions are obtained by sweating or bonding the babbitt to the shell by tinning the latter, using potassium chlorate as flux. Tin-base babbitts and other low-strength materials evidence some yielding when subjected to heavy pressures. This tendency may be alleviated by the use of a thinner layer of the bearing material, fused either to a bronze or to a steel shell. This improves the fatigue life of the bearing material. Standard bearing inserts of this type are available in tin-base babbitts, high-lead babbitts, cadmium alloys, and copper-lead mixtures in diameters up to about 6 in (15.24 cm) (Fig. 8.4.16). A few materials can be obtained in sizes up to 8 in (20.32 cm). Some types are available with flanges or with other special features. The bearing lining may vary from about 0.001 in (0.025 mm) to 0.1 in (2.5 mm) in thickness depending upon the size of the bearing.

Fig. 8.4.16   Bearing insert.

Figure 8.4.17 shows the principal types of bonded babbitt linings. Figure 8.4.17a is for normal operating conditions. Figure 8.4.17b is for more severe operating conditions.

Fig. 8.4.17

General practice for the **thickness of babbitt lining and shells** is as follows: Fig. 8.4.18, $b = \frac{1}{32}d + \frac{1}{8}$ in, $S = 0.18d$ for bronze or steel $= 0.2d$ for cast iron; Fig. 8.4.18a, $t = b/2 + \frac{1}{16}$ in, $W = 1.8t$, $W_1 = 2.2t$.

Solid bronze or steel bushings, when pressed into the bearing housing, must be finished after pressing in. Light press fits and securing by

Fig. 8.4.18

setscrews or keys are preferable to heavy press fits and no keying, since heavy pressure, especially in thin-walled bushings, will set up stresses which will release themselves if bearings should run hot in service and will result in closing in on the journal and scoring when cooling.

Uniform Load Distribution   Misalignment between journal and bearing should never be so great as to cause metallic contact. The

maximum allowable inclination α of the shaft to the bearing is given by $\tan \alpha = md/l$.

Whenever the deflection angle of the bearing installation is greater than α, either the bearing length should be reduced or, if that is not feasible, the bearing should be mounted on a spherical seat to permit self-alignment.

Oil grooves are of two kinds, axial and circumferential; the former distribute the oil lengthwise in the bearing; the latter distribute it around the shaft at the oil hole, and also collect and return oil which would otherwise be forced out at the ends of the bearing. Grooves have often been put into bearings indiscriminatingly, with the result that they scrape off the oil and interrupt the film.

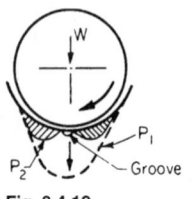

Fig. 8.4.19

In Fig. 8.4.19, W is the resultant force or load, pounds, on the bearing or journal. The radial ordinates P_1, to the dotted curve, show the pressures, lb/in², of the journal on the oil film due to the load when there is no axial groove, while the ordinates P_2, to the solid curve, show the pressures with an incorrectly located groove. Since there is no oil pressure near the groove, the permissible load W must be reduced or the film will be ruptured.

Groove dimensions (Fig. 8.4.20) are given by the following relations: $a = \frac{1}{3}$ wall thickness; $W_o = 2.5a$; $W_d = 3a$; $c = 0.5W_d$; $f = \frac{1}{16}$ in to $0.5W_d$.

In order to maintain the oil film, **the axial distributing groove should be placed in the unloaded sector** of the bearing. The location of grooves in a variety of cases is shown in Figs. 8.4.21 to 8.4.30.

Fig. 8.4.20   Lubrication and drainage grooves.

Horizontal Bearings, Rotational Motion

DIRECTION OF LOAD KNOWN AND CONSTANT

Load downward or inside the lower 60° segment as in the case of ring-oiling bearings (Fig. 8.4.21).

Load at an angle more than 45° to the vertical centerline (Fig. 8.4.22).

In force- or drop-feed oiling, the oil inlet may be anywhere within the no-load sector (Fig. 8.4.23).

Oil can be introduced through the center of the revolving shaft (Fig. 8.4.24).

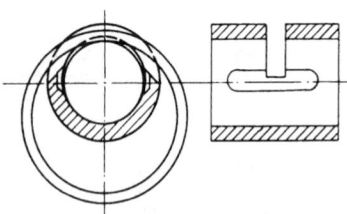

Fig. 8.4.21

Where oil-ring electric-motor bearings will be subjected by the purchaser to belt loads varying from vertical downward to horizontal, a continuous type of oil groove developed by General Electric Co. has proved very successful (Fig. 8.4.25). There are no critical spots with this groove because only a small percentage of the babbitt surface is removed along any axial line.

Fig. 8.4.22

Fig. 8.4.23

Fig. 8.4.24

Fig. 8.4.25   **Fig. 8.4.26**

ROTATING LOAD

For rotating shafts, a circumferential groove at the middle of the bearing and an axial groove on the no-load side (Fig. 8.4.26).

For stationary shafts and rotating bearings, a circumferential groove in the bearing and an axial groove on the no-load side. The oil hole is in the shaft at the midlength of the bearing (Fig. 8.4.27).

Fig. 8.4.27   **Fig. 8.4.28**

LOAD DIRECTION UNCERTAIN

Oil-ring bearings (Figs. 8.4.21 and 8.4.22) may be used, although they have defects under certain load directions. With forced or drop feed, the oil hole enters a circumferential groove at the middle of the bearing and the axial groove is omitted (Fig. 8.4.28). Arrangements for introducing oil through the rotating shaft can be made.

Bearings with Oscillatory Motion

DIRECTION OF LOAD CONSTANT

No oil film can be built up owing to the small sliding velocity, and boundary lubrication will exist. Axial grooves in the loaded sector distribute the lubricant to all parts of the bearing and avoid dry spots (Fig. 8.4.29).

Fig. 8.4.29   **Fig. 8.4.30**

LOAD DIRECTION REVERSED DURING OSCILLATION

Fluid film lubrication is possible, at least during part of the motion, owing to the vacuum caused by shaft moving back and forth. Figure 8.4.30 shows grooving which may be modified to suit local conditions. This arrangement is also advisable for bearings under a load which reverses in direction periodically without any rotation of the bearing. The lubrication may then provide an oil cushion to soften shocks.

Bearing seals are used to prevent oil leakage from the bearing housing and to protect the bearing from outside dust, water, vapors, etc. A drainage groove at the end of the bearing is effective to divert the oil passing through the bearing back into the oil well (Fig. 8.4.31a). The drain holes at the bottom of the groove must be ample for passage of the oil flow.

(a)   (b)   (c)

(d)   (e)   (f)

Fig. 8.4.31 Sealing end grooves.

An oil thrower mounted on the shaft is shown in Fig. 8.4.31b. The bearing housing may be provided with a single (Fig. 8.4.31c) or double collecting groove, or with brass or aluminum strip scrapers (Fig. 8.4.31d), to collect the oil creeping along the shaft.

For protection from dust, etc., felt packing rings are often used (Fig. 8.4.31e). The felt ring is soaked in oil to prevent charring by friction heat. In severe cases, additional protection by a labyrinth runner is very effective (Fig. 8.4.31f).

Standard seals are available for oil and grease retention as shown in Fig. 8.4.32a, b, and c. The seal material that is pressed against the rotating shaft is typically made of synthetic rubber, which is satisfactory for temperatures as high as about 250°F (121°C). Figure 8.4.32a shows the seal material pressed against the shaft by a series of flexible fingers

(a)   (b)   (c)

Fig. 8.4.32   Seals for oil and grease retention.

or leaf springs. In Fig. 8.4.32b a helical garter spring provides the gripping force. In Fig. 8.4.32c the rubber acts as its own spring.

Types of bearings are shown in Figs. 8.4.33 to 8.4.38. They include the principal methods of lubrication and types of construction.

Oiless bearings is the accepted term for self-lubricating bearings containing lubricants in solid or liquid form in their material. Graphite, molybdenum disulfide, and Teflon are used as solid lubricants in one group, and another group consists of porous structures (wood, metal), containing oil, grease, or wax.

Oil grooves: Full lines for load in direction of arrow
Full and dotted lines for indeterminate load

Fig. 8.4.33   Ring-oiled bearing solid bushing.

Fig. 8.4.34   Rigid ring-oiling pillow block. *(Link Belt Co.)*

Section on A–A   Section on B–B   Section on C–C

Fig. 8.4.35   Split bearing with one chain. Main crankshaft bearing; vertical oil engine.

Graphite-lubricated bearings (bridge bearings, sheaves, trolley wheels, high-temperature applications) consist generally of cast bearing bronze as a supporting structure containing various overlapping designs of grooves which are filled with graphite. The graphite is mixed with a binder, and the plastic mass is pressed into the cavities to the hardness of a lead pencil; 45 percent of the bearing area may be graphite.

Porous-metal bearings, compressed from metal powders and sintered, contain up to 35 percent of liquid lubricant. See ASTM B202-45T for sintered bronze and iron bearings, and also Army and Navy Specification AN-B-7G. The porous metal generally consists of a 90-10 coppertin

Section on A-A   Section on B-B

Fig. 8.4.36   Crankshaft main bearing. Horizontal engine with drop-feed lubrication.

bronze with 1½ percent graphite. These bearings do not require oil grooves since capillarity distributes the oil and maintains an oil film. If additional lubrication from an oil well should be provided, oil will be absorbed through the porous wall as required. For high temperatures where oil will carburize, a higher percentage of graphite (6 to 15 percent) is used.

Section through bearing
oil compartment

Fig. 8.4.37

Porous-metal bearings are used where plain metal bearings are impractical because of lack of space, cost, or inaccessibility for lubrication, as in automotive generators and motors, hand power tools, vacuum cleaner motors, and the like.

Fig. 8.4.38

THRUST BEARINGS

At low speeds, shaft shoulders or collars bear against flat bearing rings. The lubrication may be semifluid, and the friction is comparatively high.

For hardened-steel collars on bronze rings, with intermittent service, pressures up to 2,000 lb/in² (13,790 kN/m²) are permissible; for continuous low-speed operation, 1,500 lb/in² (10,341 kN/m²); for steel collars on babbitted rings, 200 lb/in² (1,378.8 kN/m²). In multicollar thrust bearings, the values are reduced considerably because of the difficulty in distributing the load evenly between the several collars.

The performance of the bearing thrust rings is much improved by the introduction of **grooves** with tapered lands as shown in Fig. 8.4.39. The lands extend on either side of the groove. The taper angle of the lands is very slight, so that a pressure oil film is formed between the bearing ring

Fig. 8.4.39 Thrust collar with grooves fitted with tapered lands.

and the collar of the shaft. It is generally known that slightly tapered radial grooves will develop a hydrodynamic load-carrying film, when formed in the manner of Fig. 8.4.39. The taper angle should be on the order of 0.5°. Alternatively, a shallow recessed area that is a couple of film thicknesses deep can be used in place of the taper.

Fig. 8.4.40 Kingsbury thrust bearing with six shoes.

For high speeds or where low friction losses and a low wear rate are essential, **pivoted segmental thrust bearings** are used (Kingsbury thrust bearing, or Michell bearing in Europe). The bearing members in this type are tiltable shoes which rest on hard steel buttons mounted on the bearing housing. The shoes are free to form automatically a wedge-shaped oil film between the shoe surface and the collar of the shaft (Figs. 8.4.40 to 8.4.42).

The **minimum oil-film thickness** h_0, in, between the shoe and the collar, at the trailing edge of the shoe, is approximately

$$h_0 = 0.26 \sqrt{\mu u l / P_{\text{avg}}}$$

where μ is the absolute viscosity; u is the velocity of the collar, on the mean diam; l is the length of a shoe, at the mean diam of the collar, in the direction of sliding motion; P_{avg} is the average load on the shoes. As indicated in Fig. 8.4.40, $b = l$, approximately. The standard thrust bearings have six shoes. Load-carrying capacities of Kingsbury thrust bearings are given in Table 8.4.3.

Fig. 8.4.41 Left half of six-shoe self-aligning equalizing horizontal thrust bearing for load in either axial direction.

The coefficient of friction in Kingsbury thrust bearings, referred to the mean diameter of the shoes, is approximately $f = 11.7 h_0 / l$, where h_0 is computed as shown above. Figures 8.4.41 and 8.4.42 show typical pivoted segmental thrust bearings. They usually embody a system of

Fig. 8.4.42 Half section of mounting for vertical thrust bearing.

rocking levers which are used for alignment and equalization of load on the several shoes (Fig. 8.4.43).

Thrust may be carried on a hydrostatic step bearing as shown schematically in Fig. 8.4.44, where high-pressure oil at P_o is supplied at the

Fig. 8.4.43 Kingsbury thrust bearings. (Developed cylindrical sections.)

center of the bearing from an external pump. The lubricant flows radially outward through the annulus of depth h_0 and escapes at the periphery of the shaft at some pressure P_1 which is usually at atmospheric pressure. An oil film will be present whether the shaft rotates or not. Friction in these bearings can be made to approach zero, depending

Fig. 8.4.44 Hydrostatic step bearing.

Table 8.4.3   Capacities of Six-Shoe Standard-Duty Horizontal and Vertical Thrust Bearings
(Based on viscosity of 150 s Saybolt at operating temperatures. Capacities given may be increased from 10 to 25% if viscosity is increased in same proportion)

Bearing size, in	Area, in²	Speed, r/min						Bearing size, in	Area, in²	Speed, r/min					
		100	200	400	800	1,800	3,600			100	150	200	300	500	700
		Safe load, 10³ lb								Safe load, 10³ lb					
5	12.5	1.44	1.7	2.0	2.4	2.9	3.5	19	180	40.00	44.0	48.0	53.0	60.0	65.0
6	18.0	2.30	2.7	3.2	3.8	4.6	5.5	21	220	51.00	57.0	61.0	68.0	77.0	84.0
7	24.5	3.30	3.9	4.7	5.6	6.8	8.0	23	264	65.00	72.0	77.0	85.0	97.0	105.0
8	32.0	4.60	5.5	6.6	7.8	9.6	11.4	25	312	80.00	88.0	95.0	105.0	119.0	123.0
9	40.5	6.20	7.4	8.8	10.4	13.0	15.0	27	364	97.00	107.0	115.0	127.0	144.0	146.0
10½	55.1	9.20	10.8	13.0	15.4	19.0	22.0	29	420	116.00	128.0	137.0	152.0	168.0	168.0
12	72.0	12.80	15.2	18.0	21.0	26.0	29.0	31	480	137.00	151.0	162.0	180.0	192.0	192.0
13½	91.1	17.20	20.0	24.0	29.0	35.0	36.0	33	544	160.00	177.0	189.0	210.0	220.0	220.0
15	112.5	22.00	26.0	32.0	37.0	45.0		37	684	215.00	235.0	250.0	275.0	275.0	
17	144.5	30.00	36.0	43.0	51.0	58.0		41	840	275.00	305.0	325.0	335.0	335.0	
								45	1012	345.00	385.0	405.0	405.0		

upon the rotational velocity and the viscosity of the lubricant film. Figure 8.4.45 shows the step bearing of a vertical turbogenerator. The load-carrying capacity is

$$W = \frac{P_o \pi}{2} \frac{R^2 - R_o^2}{\ln (R/R_o)}$$

This equation is valid even when the recess is eliminated, in which case R_o becomes the radius of the inlet oil supply pipe. The volume flow rate of lubricant and the clearance are related thus:

$$Q = P_o \pi h_0^3 / [6\mu \ln(R/R_o)]$$

The friction power loss in the bearing is

$$H_f = \frac{\pi\mu\omega^2(R^4 - R_o^4)}{2h_0}$$

The pumping power loss in forcing the lubricant through the bearing is $H_p = Q(P_o - P_1)/\eta$, where η is the efficiency of the pump.

EXAMPLE 10.   A typical 5,000-kW vertical turbogenerator has a thrust load of about 101,000 lb; outside diameter of bearing, 16 in; diam of recess, 10 in; pump efficiency, 0.5; speed, 750 r/min. Substituting these values,

$$101,000 = \frac{P_o \pi}{2} \left[\frac{8^2 - 5^2}{\ln(8/5)} \right]$$

or

$$P_o = 774 \text{ lb/in}^2$$

In practice, about 825 lb/in² is used on this step bearing to provide some margin of safety. Film thickness in the bearing should be from 0.001 to 0.01 in to protect the surfaces from metal-to-metal contact and allow passage of harmful grit that may

find its way into the system. The film thickness determines the oil flow for a given viscosity and pressure. With $h_0 = 0.006$ in (0.1524 mm) and SAE 20 oil at 130°F (29 centipoises), $Q = 8.25 \times \pi \times (0.006)^3/6 \times 29 \times 1.45 \times 10^{-7} \times 0.470 = 46.8$ in³/s (766.91 cm³/s). Flow $= 46.8 \times 60/231 = 12.15$ gal/min (45.99 L/min). The horsepower lost due to friction, $H_f = 750^2 \times 29 \times 1.45(8^4 - 5^4)/383,000 \times 0.006 = 3.58$ hp (2.669 kW). The horsepower lost due to pumping with pump efficiency of 0.5, $H_p = 46.8 \times 825/6,600 \times 0.5 = 11.7$ hp (8.725 kW). The total energy lost $= 11.7 + 3.58 = 15.28$ hp (11.39 kW).

Evaluation of these equations for other film thicknesses will show that the minimum lost energy will occur between $h_0 = 0.004$ and $h_0 = 0.006$ in (0.1016 and 0.1524 mm). The coefficient of friction corresponding to an energy loss of 15.28 hp in the above example is 0.002.

Other configurations for hydrostatically lubricated thrust bearings are shown in Fig. 8.4.46 from Loeb and Rippel. They may be used directly to obtain the value of load-carrying capacity W and flow rate Q.

LINEAR SLIDING BEARINGS

All sliding bearings (Fig. 8.4.47), to wear true, must have the sliding parts of nearly equal lengths. Bearings made in this way will be found not to wear out of true. Oiling is accomplished in several ways, an acceptable method being that shown in Fig. 8.4.48. Short slides in many machine tools are lubricated by oil pads or direct oil application. The weight of the table and work and thrust of the tool cause wear on the bottom and sides of the guides. To compensate for the wear in both directions, bearings are sometimes made V-shaped, as shown in Fig. 8.4.49.

Simpler sliding bearings in machine tools are made with provision for adjustment (as shown in Fig. 8.4.50) of which there are many modifications. Recent applications involving hydrostatic lubrication on machine-tool ways have been very successful.

GAS-LUBRICATED BEARINGS

The fluid-film calculations included in Examples 1 through 10 have assumed that oil (or, in one case, water) was the lubricant. Actually, almost any **process fluid** may be used if proper recognition is given to the viscosity, corrosive action, change in state (where a liquid is close to its boiling point), toxicity, and in the case of a gas, its compressibility. Fluid-film journal and thrust bearings have run successfully, for example, on water, kerosene, gasoline, acid, liquid refrigerants, mercury, molten metals, and a wide variety of gases.

The previous equations for load-carrying capacity, film thickness, friction, and flow may be used for process liquids, but for gases, proper recognition must be made of the compressibility effects.

Because of the great value of gas-lubricated bearings for special applications, and to demonstrate the methods for handling the compressibility action, an introduction to the **design** of **gas-lubricated bearings** follows.

Fig. 8.4.45   Step bearing of a vertical turbogenerator.

Fig. 8.4.46 Load-carrying capacity and flow for several flat thrust bearings *(Loeb).* Lengths in inches.

Naturally, if the change in pressure within the bearing clearance is small compared to ambient pressure, the compressibility effect will be likewise small, and lubrication equations based on liquids may be used. A **compressibility parameter** Λ indicates the extent of this action. For hydrodynamic journal bearings it has the form $\Lambda = 6\mu\omega/(P_a m^2)$. For values of Λ less than one, the previous equations of this section for journal bearings may be used. For values of Λ greater than one, compressibility effects are included through the use of Figs. 8.4.51 to 8.4.54. (Data from Elrod and Burgdorfer, Proceedings First International Symposium on Gas-lubricated Bearings, 1959, and Raimondi, *Trans. ASLE,* vol. IV, 1961.)

Fig. 8.4.47

Fig. 8.4.48

Fig. 8.4.49

Fig. 8.4.50

EXAMPLE 11. Determine the minimum film thickness for a journal bearing 0.5 in (1.27 cm) diameter by 0.5 in long. Ambient pressure 14.7 lb/in^2 abs (101.34 kN/m^2 abs). Speed 12,000 r/min. Load 0.4 lb (0.88 kg). Diametral clearance 0.0005 in (0.0127 mm). Lubricant, air at 100°F and 14.7 lb/in^2 abs (2.68 $\times$ 10^{-9} lb $\cdot$ s/in^2 from Fig. 8.4.55). $m = 0.0005/0.5 = 0.001$ in/in. $\omega = 12,000 \times 2\pi/60 = 1,256$ rad/s, $\Lambda = (6 \times 2.68 \times 10^{-9} \times 1,256)/14.7 \times 0.001^2 = 1.37$, and $W/(dlP_a) = 0.4/0.5 \times 0.5 \times 14.7 = 0.109$. Then, in Fig. 8.4.53 ($l/d = 1$), we find that $\varepsilon = 0.22$, and the minimum film thickness $h_0 = 0.00025(1 - 0.22) = 0.000195$ in (0.00495 mm).

Gas-lubricated journal bearings should be checked for **whirl stability.** Figure 8.4.56 is applicable with sufficient accuracy to bearings where l/d is equal to or greater than one. It is used in conjunction with Fig. 8.4.51 for $l/d = \infty$. The stability parameter is ω_1^* which, for a bearing having only gravity loading, has the value $\omega_1^* = \omega \sqrt{mr/g}$.

EXAMPLE 12. To determine whether the bearing of Example 11 is stable at the running speed of 12,000 r/min, we compute ω_1^* as $1,256 \sqrt{0.00025/386} = 1.015$. The value of eccentricity ratio ε_0 for $l/d = \infty$ is computed from Fig. 8.4.51

Fig. 8.4.51

Fig. 8.4.52

Fig. 8.4.53

Fig. 8.4.54

Figs. 8.4.51–8.4.54 Theoretical load-carrying parameter vs. compressibility parameter for a full journal bearing: $l/d = \infty$. (Fig. 8.4.51), $l/d = 2$ (Fig. 8.4.52), $l/d = 1$ (Fig. 8.4.53), and $l/d = 0.5$ (Fig. 8.4.54). *(Elrod and Raimondi.)*

on the basis of W being the load per inch of bearing length. Thus W(lb/in) = 0.4/0.5 = 0.8 lb/in. For Fig. 8.4.51, $W/(dlP_a) = 0.218$ and in Fig. 8.4.51, we determine $\varepsilon_0 = 0.18$. Then (in Fig. 8.4.56), for $\omega_1^* = 1.015$ and $\Lambda = 1.37$, we find the intersection at about where a curve for $\varepsilon_0 = 0.18$ would be found. The bearing should just be stable. An intersection point on the ε_0 line or to the left should represent a stable condition. An intersection point to the right of the appropriate ε_0 line would predict an unstable condition.

The plain cylindrical journal bearing (360°) is the least stable of possible bearing designs. Control of "half-frequency" whirl has been achieved and the threshold of instability has been raised through modification of the geometry of the plain bearing. The simplest modification is the insertion of axial grooves. Bearings with three or four such grooves have been successful, but lose much stiffness.

A typical three-groove (three-sector) bearing is shown in Fig. 8.4.57.

Half-frequency whirl indicates a dynamic instability in which the journal orbits at approximately one-half of the shaft rotational speed, which coincides with the average speed of the lubricant in the film. If one looked at the bearing geometry from a coordinate system rotating at this whirl speed, one would see the journal attempting to pump lubricant

Fig. 8.4.55 Absolute viscosity of air. *(Iwaski, Sci. Rpts., Research Inst., Tohuku Univ., Ser. A; Kestin and Pliarczyk, Trans. ASME, **56,** 1954.)* Reyns = 1.45×10^{-7} cP.

in one direction and the bearing attempting to do the opposite. The capacity of the film to sustain any load is thus greatly diminished, and failure often occurs. Half-frequency whirl can occur when the dynamic system in which the bearing operates has a natural frequency at approximately one-half the speed of rotation. If the energy dissipation rate is

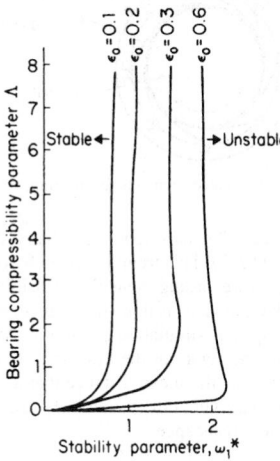

Fig. 8.4.56   Half-frequency translatory whirl threshold for infinite length, 360° journal bearing. *(Castelli and Elrod, Solution for the Stability Problem for 360°, Self Acting Gas Lubricated Bearings, Trans. ASME, 87, Mar. 1965.)*

not sufficiently large, instability occurs. In the dynamic system mentioned above, the stiffness and damping characteristics of the bearing, or bearings, play a major role. Damping arises from squeezing the lubricant in and out of the bearing by the action of the journal vibrations

Fig. 8.4.57   Sector sleeve bearing.

against the viscous resistance. When the journal moves, gas can compress and act as a capacitance rather than flow and act as a damper. Therefore, gas bearings are much more prone to instability than are liquid-lubricated ones.

Aside from avoiding resonance conditions, two often successful techniques for whirl prevention are shown in Figs. 8.4.58 and 8.4.59. The first depicts a complete journal bearing with a rotating geometric artifact causing a synchronous disturbance. This artifact can be a variation in the clearance or an asymmetric mass leading to dynamic unbalance. The second depicts a bearing geometry with shallow, inward-pumping spiral grooves. The depth of the grooves is approximately twice the radial clearance, their angle is from 25° to 30° to the axis, the land-to-groove ratio is 0.5, and the axial extent of the grooved area is one-half the length of the bearing. The grooves can be on either the stator or the rotor, depending on manufacturing convenience. Etching is a common method for production of the grooves. Design data for this excellent type of self-acting gas bearing can be found in Sec. 4 of Gross's book. (See References.)

The tilting-pad journal bearing in Fig. 8.4.60 is considered to be one of the most stable of all possible designs. The shoes, or pads, are supported on rounded pins and are free to pitch and roll through very small angles. Analysis shows that this freedom to move achieves the stability characteristics of these bearings, and also, of course, permits them to be self-aligning. Three-, four-, and five-shoe configurations are often used. Figure 8.4.60 shows an early design used for machine tool grinding spindles.

When applied to gas lubrication, pressurized gas may be supplied through a drilled pivot (hydrostatic lubrication), as an aid in starting and

to provide a reserve of load-carrying capacity (Fig. 8.4.61). See Gunter, Hinkle, and Fuller, Design Guide for Gas-Lubricated, Tilting-Pad Journal and Thrust Bearings, NYO-2512-1, U.S. AEC, I-A 2392-3-1, Nov. 1964, Contract AT 30-1-2512.

Fig. 8.4.58   Stabilizing rotating geometry. (Clearance greatly exaggerated.)

Fig. 8.4.59   Spiral groove bearing shown as outward-pumping. (Clearance is greatly exaggerated.)

A bearing design development that is simple, inexpensive, and very stable is the foil bearing. It is also very tolerant of thermal distortions and possible loss of clearance resulting from elevated temperatures. Probably the most widely used of several designs is shown in Fig. 8.4.62. It consists of overlapping metal shims, anchored at the base end like a cantilever beam. The "free" ends deflect and are able to automatically form their own operating clearance. These bearings are widely used in aircraft cabin cooling and for auxiliary power supply systems (Suriano, Dayton, and Woessner, Test Experience with Turbine-end Foil Bearing Equipped Gas Turbine Engines, ASME Paper 83-GT-73, 1983).

Thrust bearings of the tilting-pad variety are less susceptible to compressibility effects and may be considered as liquid-lubricated for values of Λ (suitable for thrust bearings) less than about 30. $\Lambda = 6u\mu l/(h_0^2 P_a)$ where l is the length of the shoe in the direction of sliding and U is the linear velocity at the mean radius. However, the shoes

Fig. 8.4.60   Filmatic bearing. *(Courtesy Cincinnati Milacron Corp.)*

should not be made flat for gas operation but should have a crowned contour (see Fig. 8.4.63). (Gross, "Gas Film Lubrication," Wiley.) An approximate value for the crown is to make $\delta = \frac{3}{4}h_0$. The tilting-pad bearing design is probably the most common gas bearing presently in existence. Every hard-disk computer memory since the early 1960s has had its read-write heads supported by self-acting tilting-pad sliders. Hundreds of millions of such units, called *flying heads*, have been manufactured to date. Some designs employ the crown geometry while,

Fig. 8.4.61   Cross-sectional view, spring-mounted pivot assembly. *(Courtesy of The Franklin Institute Research Labs.)*

most commonly, heads with flat multiple sliders with straight ramps in their forward sections are used. The reason for the multiple thin sliders is the achievement of maximum damping possible. The typical minimum film heights have decreased steadily through the years from 1 μm

Fig. 8.4.62   Bending-dominated segments foil bearing.

(40 millionths of an inch) 25 years ago to less than 0.2 μm (8 millionths of an inch) currently (1995). This trend is driven by the achievement of the higher and higher recording densities possible at lower flying heights. Design of these devices is done rather precisely from first principles by means of special simulation programs. At these low clearances, allowance must be made for the finiteness of the **molecular mean free path,** which represents the mean distance that a gas molecule must travel between collisions. This effect manifests itself in a lowering of viscosity and wall shear resistance.

Fig. 8.4.63   Schematic of tilting-pad shoe, showing crown height δ.

Gas-lubricated hydrostatic bearings, unlike liquid-lubricated bearings, cannot be designed on the basis of fixed flow rate. They are designed instead to have a pressure loss produced by an **orifice restrictor** in the supply line. Such throttling enables the bearing to have load-carrying capacity and stiffness. For maximum stiffness the pressure drop in the orifice may be about one-half of the manifold supply pressure. For a circular thrust bearing with a single circular orifice, the load-carrying capacity is given with sufficient accuracy by the equation previously used for liquids (see Fig. 8.4.44). $W = (P_R - P_a/2)[R^2 - R_0^2/\ln(R/R_0)]$, where P_R is the recess pressure, lb/in^2 abs. The flow volume, however, is given by $Q_0 = \pi h_0^3/[6\mu \ln(R/R_0)](P_0^2 - P_1^2)/2P_0$. Q_0 and P_0 refer to recess conditions, and Q_1 and P_1 refer to ambient conditions. Pressures are absolute.

EXAMPLE 13.   A circular thrust bearing 6 in (15.24 cm) diameter with a recess 2 in (5.08 cm) diameter has a film thickness of $h_0 = 0.0015$ in (0.0381 mm). $P_0 = 30$ lb/in^2 gage or 44.7 lb/in^2 abs (308.16 kN/m^2). P_1 is room pressure, 14.7 lb/in^2 abs (101.34 kN/m^2 abs). Depth of recess is 0.02 in. Applied load is 375 lb. $Q_0 = (\pi \times 0.0015^3)/(6 \times 2.68 \times 10^{-9}\ln 3)(44.7^2 - 14.7^2)/(2 \times 44.7)$, $Q_0 = 12.3$ in^3/s (201.6 cm^3/s) at recess pressure. Converted to free air, $Q_1 = Q_0(P_0/P_1)$ with isothermal expansion, $Q_1 = 12.3(44.7/14.7) = 37.4$ in^3/s (612.87 cm^3/s), or $Q_1 = 37.4 \times 60 = 2,244$ in^3/min (36.77 L/min). Actual measured flow = 2,440 in^3/min (39.98 L/min).

Externally pressurized gas bearings are not as easily designed as liquid-lubricated ones. Whenever a volume larger than approximately that of the film is present between the restrictor and the film, a phenomenon known as **air hammer** or **pneumatic instability** can take place. Therefore, in practical terms, recesses cannot be used and orifice restrictors must be obtained by the smallest flow cross-section at the very entrance to the film; this area is equal to the perimeter of the inlet holes multiplied by the local height of the film. This technique is called **inherent compensation.** Unfortunately, as one can readily see, the area of the restrictors is smaller where the film is smaller; thus, the stiffness is lower than that obtainable by incompressible lubrication. Design data are available in Sec. 5 of Gross's book (see References).

8.5 BEARINGS WITH ROLLING CONTACT

by Michael W. Washo

REFERENCES: Anti-Friction Bearing Manufacturers Association, Inc. (AFBMA), Method of Evaluating Load Ratings. American National Standards Institute (ANSI), Load Ratings for Ball and Roller Bearings. AFBMA, "Mounting Ball and Roller Bearings." Guyer, "Rolling Bearing Handbook and Troubleshooting Guide."

COMPONENTS AND SPECIFICATIONS

Rolling-contact bearings are designed to support and locate rotating shafts or parts in machines. They transfer loads between rotating and stationary members and permit relatively free rotation with a minimum of friction. They consist of **rolling elements (balls or rollers)** between an **outer** and **inner ring. Cages** are used to space the rolling elements from each other. Figure 8.5.1 illustrates the common terminology used in describing rolling-contact bearings.

Fig. 8.5.1   Radial contact bearing terminology.

Rings   The inner and outer rings of a rolling-contact bearing are normally made of 52100 AISI steel, hardened to Rockwell C 60 to 67. The rolling-element raceways are accurately ground in the rings to a very fine finish (8 μin or less).

Rings are available for special purposes in such materials as stainless steel, ceramics, and plastic. These materials are used in applications where extreme conditions exist.

Rolling Elements   Normally the rolling elements, balls or rollers, are made of the same material and finished like the rings. Other rolling-element materials, such as stainless steel, ceramics, Monel, and plastics, are used in conjunction with various ring materials. Bearings that use ceramic balls and steel rings are referred to as **hybrid** bearings. They last longer in extreme conditions.

Cages   Cages, sometimes called separators or retainers, are used to space the rolling elements from each other. Cages are furnished in a wide variety of materials and construction. Pressed-steel cages, riveted or clinched and filled nylon, are most common. Solid machined cages are used where greater strength or higher speeds are required. They are fabricated from bronze or phenolic-type materials. At high speeds, the phenolic type operates more quietly with a minimum amount of friction. Bearings without cages are referred to as full-complement.

A wide variety of rolling-contact bearings are normally manufactured to standard boundary dimensions (bore, outside diameter, width) and tolerances which have been standardized by the AFBMA. All bearing manufacturers conform to these standards, thereby permitting interchangeability. ANSI has for the most part adopted these and published them jointly as AFBMA/ANSI standards as follows:

Title	Standard	Title	Standard
Terminology	1	Ball Standards	10
Gaging Practice	4	Roller Load Ratings	11
Mounting Dimensions	7	Instrument Bearings	12
Mounting Accessories	8.2	Vibration and Noise	13
Ball Load Ratings	9	Basic Boundary Dimensions	20

The Annular Bearing Engineers Committee (ABEC) of the AFBMA has established progressive levels of precision for ball bearings. Designated as ABEC-1, ABEC-5, ABEC-7, and ABEC-9, these standards specify tolerances for bore, outside diameter, width, and radial runout. Similarly, roller bearings have established precision levels as RBEC-1 and RBEC-5.

PRINCIPAL STANDARD BEARING TYPES

The selection of the type of rolling-contact bearing depends upon many considerations, as evidenced by the numerous types available. Furthermore, each basic type of bearing is furnished in several **standard "series"** as illustrated in Fig. 8.5.2. Although the bore is the same, the outside diameter, width, and ball size are progressively larger. The result is that a wide range of load-carrying capacity is available for a given size shaft, thus giving designers considerable flexibility in selecting standard-size interchangeable bearings. Some of the more common bearings are illustrated below and their characteristics described briefly.

Fig. 8.5.2   Bearing standard series.

Ball Bearings

Single-Row Radial (Fig. 8.5.3)   This bearing is often referred to as the **deep groove** or conrad bearing. Available in many variations—single or double shields or seals. Normally used for radial and thrust loads (maximum two-thirds of radial).

Maximum Capacity (Fig. 8.5.4)   The geometry is similar to that of a deep-groove bearing except for a **filling slot.** This slot allows more balls in the complement and thus will carry heavier radial loads. However, because of the filling slot, the thrust capacity in both directions is reduced drastically.

Double-Row (Fig. 8.5.5)   This bearing provides for heavy radial and light thrust loads without increasing the OD of the bearing. It is approximately 60 to 80 percent wider than a comparable single-row bearing. Because of the filling slot, thrust loads must be light.

Internal Self-Aligning Double-Row (Fig. 8.5.6)   This bearing may be used for primarily radial loads where **self-alignment** ($\pm$ 4°) is required. The self-aligning feature should not be abused, as excessive misalignment or thrust load (10 percent of radial) causes early failure.

Angular-Contact Bearings (Fig. 8.5.7)   These bearings are designed to support **combined radial and thrust** loads or heavy thrust loads depending on the contact-angle magnitude. They may be mounted in pairs (Fig. 8.5.8) which are referred to as **duplex bearings:** back-to-back, tandem, or face-to-face. These bearings may be preloaded to minimize axial movement and deflection of the shaft. They are commonly manufactured as hybrid bearings using silicon nitride balls and steel races. They are meant for service at high speeds and temperatures.

Fig. 8.5.3

Fig. 8.5.4

Fig. 8.5.5

Fig. 8.5.6

Fig. 8.5.7

Fig. 8.5.8

Fig. 8.5.11

Fig. 8.5.12

Fig. 8.5.9

Fig. 8.5.10

Fig. 8.5.13

Fig. 8.5.14

Fig. 8.5.15

Ball Bushings (Fig. 8.5.9) This type of bearing is used for linear motions on hardened shafts (Rockwell C 58 to 64). Some types can be used for linear and rotary motion.

Split-Type Ball Bearing (Fig. 8.5.10) This type of ball or roller bearing has split inner ring, outer ring, and cage. They are assembled by screws. This feature is expensive but useful where it is difficult to install or remove a solid bearing.

Roller Bearings

Cylindrical Roller (Fig. 8.5.11) These bearings utilize cylinders with approximate length/diameter ratio ranging from 1:1 to 1:3 as rolling elements. Normally used for heavy radial loads. Especially useful for free axial movement of the shaft. Highest speed limits for roller bearings.

Needle Bearings (Fig. 8.5.12) These bearings have rollers whose length is at least 4 times their diameter. They are most useful where space is a factor and are available with or without inner race. If shaft is used as inner race, it must be hardened and ground. Full-complement type is used for high loads, oscillating, or slow speeds. Cage type should be used for rotational motion. They cannot support thrust loads.

Tapered-Roller (Fig. 8.5.13) These bearings are used for heavy radial and thrust loads. The bearing is designed so that all elements in the rolling surface and the raceways intersect at a common point on the axis: thus **true rolling** is obtained. Where maximum system rigidity is required, the bearings can be adjusted for a preload. They are available in double row.

Spherical-Roller (Fig. 8.5.14) These bearings are excellent for heavy radial loads and moderate thrust. Their internal self-aligning feature is useful in many applications such as HVAC fans.

Thrust Bearings

Ball Thrust Bearing (Fig. 8.5.15) It may be used for low-speed applications where other bearings carry the radial load. These bearings are made with shields, as well as the open type.

Straight-Roller Thrust Bearing (Fig. 8.5.16) These bearings are made of a series of short rollers to minimize the skidding, which causes twisting, of the rollers. They may be used for moderate speeds and loads.

Tapered-Roller Thrust (Fig. 8.5.17) It eliminates the skidding that takes place with straight rollers but causes a thrust load between the

Fig. 8.5.16

Fig. 8.5.17

ends of the rollers and the shoulder on the race. Thus speeds are limited because the roller end and race flange are in sliding contact.

Selection of Ball or Roller Bearing

Selection of the type of rolling-element bearing is a function of many factors, such as load, speed, misalignment sensitivity, space limitations, and desire for precise shaft positioning. However, to determine if a ball or roller bearing should be selected, the following **general rules** apply:

1. Ball bearings function on theoretical point contact. Thus they are suited for higher speeds and lighter loads than roller bearings.

2. Roller bearings are generally more expensive except in larger sizes. Since they function theoretically on line contact, they will carry heavy loads, including shock, more satisfactorily, but are limited in speed.

Use Fig. 8.5.18 as a general guide to determine if a ball or roller bearing should be selected. This figure is based on a rated life of 30,000 h.

Fig. 8.5.18 Guide to selection of ball or roller bearings.

ROLLING-CONTACT BEARINGS' LIFE, LOAD, AND SPEED RELATIONSHIPS

An accurate knowledge of the load-carrying capacity and expected life is essential in the proper selection of ball and roller bearings. Bearings that are subject to millions of different stress applications fail owing to fatigue. In fact, **fatigue** is the only cause of **failure** if the bearing is properly lubricated, mounted, and sealed against the entrance of dust or dirt and is maintained in this condition. For this reason, the **life** of an individual bearing is defined as the total number of revolutions or hours at a given constant speed at which a bearing runs before the first evidence of fatigue develops.

Definitions

Rated Life L_{10} The number of revolutions or hours at a given constant speed that 90 percent of an apparently identical group of bearings will complete or exceed before the first evidence of fatigue develops; i.e., 10 out of 100 bearings will fail before rated life. The names **Minimum life** and L_{10} **life** are also used to mean rated life.

Basic Load Rating C The radial load that a ball bearing can withstand for one million revolutions of the inner ring. Its value depends on bearing type, bearing geometry, accuracy of fabrication, and bearing material. The basic load rating is also called the **specific dynamic capacity**, the **basic dynamic capacity**, or the **dynamic load rating.**

Equivalent Radial Load P Constant stationary radial load which, if applied to a bearing with rotating inner ring and stationary outer ring, would give the same life as that which the bearing will attain under the actual conditions of load and rotation.

Static Load Rating C_0 Static radial load which produces a maximum contact stress of 580,000 lb/in^2 (4,000 MPa).

Static Equivalent Load P_0 Static radial load, if applied, which produces a maximum contact stress equal in magnitude to the maximum contact stress in the actual condition of loading.

Bearing Rated Life

Standard formulas have been developed to predict the statistical rated life of a bearing under any given set of conditions. These formulas are based on an exponential relationship of load to life which has been established from extensive research and testing.

$$L_{10} = \left(\frac{C}{P}\right)^K \times 10^6 \qquad (8.5.1)$$

where L_{10} = rated life, r; C = basic load rating, lb; P = equivalent radial load, lb; K = constant, 3 for ball bearings, 10/3 for roller bearings.

To convert to hours of life L_{10}, this formula becomes

$$L_{10} = \frac{16{,}700}{N}\left(\frac{C}{P}\right)^K \qquad (8.5.2)$$

where N = rotational speed, r/min. Table 8.5.1 lists some common design lives vs. the type of application. These may be altered to suit unusual circumstances.

Load Rating

The **load rating** is a function of many parameters, such as number of balls, ball diameter, and contact angle. Two load ratings are associated with a rolling-contact bearing: basic and static load rating.

Basic Load Rating C This rating is *always* used in determining bearing life for all speeds and load conditions [see Eqs. (8.5.1) and (8.5.2)].

Static Load Rating C_0 This rating is used only as a check to determine if the maximum allowable stress of the rolling elements will be exceeded. It is *never* used to calculate bearing life.

Values for C and C_0 are readily attainable in any bearing manufacturer's catalog as a function of size and bearing type. Table 8.5.2 lists the basic and static load ratings for some common sizes and types of bearings.

Equivalent Load

There are two **equivalent-load** formulas. Bearings operating with some finite speed use the equivalent radial load P in conjunction with C [Eq. (8.5.1)] to calculate bearing life. The static equivalent load is used in comparison with C_0 in applications when a bearing is highly loaded in a static mode.

Table 8.5.1 Design-Life Guide

Application	Design life, h, L_{10}	Application	Design life, h, L_{10}
Agricultural equipment	3,000–6,000	Domestic appliances	1,000–2,000
Aircraft engines	1,000–3,000	Electric motors:	
Aircraft jet engines	1,500–4,000	Domestic	1,000–2,000
Automotive:		Industrial	20,000–30,000
Bus, car	2,000–5,000	Elevator	8,000–15,000
Trucks	1,500–2,500	Fans:	
Blowers:	20,000–30,000	Industrial	8,000–15,000
Continuous 8-h service	20,000–40,000	Mine ventilation	40,000–50,000
Continuous 24-h service	40,000–60,000	Gearing units (multipurpose)	8,000–15,000
Continuous 24-h service (extreme reliability)	100,000–200,000	Intermittent service	8,000–15,000
Compressors	40,000–60,000	Paper machines	50,000–60,000
Conveyors	20,000–40,000	Pumps	40,000–60,000

Table 8.5.2   Approximate Basic and Static Load Ratings vs. Types and Sizes
(Ratings are in pounds) 1 lb = 4,448 N

Bearing bore, mm	Ball single-row 200 series		Ball single-row 300 series		Ball double-row 200 series		Roller cylindrical 300 series		Roller spherical 22200 series	
	C_0	C	C_0	C	C_0	C	C_0	C	C_0	C
10	600	1,040	850	1,430	800	1,210	1,020	1,960		
12	680	1,180	1,040	1,650	1,250	1,820	1,350	2,540		
15	780	1,330	1,220	1,970	1,430	2,030	1,520	2,820		
17	1,000	1,660	1,470	2,340	1,840	2,510	2,070	3,700		
20	1,390	2,220	1,760	2,730	2,540	3,480	2,560	4,490		
25	1,560	2,420	2,350	3,550	2,858	3,780	3,720	6,360		
30	2,250	3,360	3,120	4,600	4,110	5,140	5,070	8,460		
35	3,070	4,430	4,020	5,770	5,600	6,700	6,400	10,400		
40	3,520	5,040	5,020	7,060	6,430	7,680	7,930	12,500	11,800	15,200
45	4,000	5,660	6,130	8,430	7,320	8,620	9,310	14,700	12,600	15,900
50	4,450	6,070	8,010	10,750	8,130	9,220	11,600	17,900	13,600	16,800
55	5,630	7,500	9,400	12,410	10,300	11,400	12,600	19,100	16,500	20,300
60	6,950	9,070	10,902	14,179	12,700	13,800	15,200	22,800	20,800	25,200
65	7,660	9,900	12,516	16,051	14,000	15,000	19,900	29,000	25,500	30,200
70	8,410	10,714	14,240	18,030	15,400	16,300	21,400	30,800	27,500	31,900
75	9,190	11,610	16,080	19,600	16,900	17,300	23,200	32,900	29,100	33,100
80	10,010	12,550	18,020	21,230	18,300	19,100	27,000	38,100	32,100	36,800
85	11,750	14,490	20,080	22,880	19,500	19,700	30,900	43,300	38,200	43,200
90	13,630	16,540	22,250	24,580	22,100	22,600	35,200	48,800	44,500	49,800
95	15,650	18,740	24,530	26,300	28,600	28,600	39,500	54,200	48,800	54,700
100	17,800	21,130	29,430	29,940	32,500	32,100	44,700	60,800	55,700	61,900
110	20,100	23,000	32,040	31,800	30,500	30,700	53,200	70,500	72,000	78,400

Equivalent Radial Load P   All bearing loads are converted to an equivalent radial load. Equation (8.5.3) is the general formula used for both ball and roller bearings.

$$P = XR + YT \qquad (8.5.3)$$

where P = equivalent radial loads, lb; R = radial load, lb; T = thrust (axial) load, lb; X and Y = radial and thrust factors (Table 8.5.3). The

Table 8.5.3   Radial and Thrust Factors

Bearing type	X_1	Y_1	X_2	Y_2
Single-row ball	1.0	0.0	0.56	1.40
Double-row ball	1.0	0.75	0.63	1.25
Cylindrical roller	1.0	0.0	1.0	0.0
Spherical roller	1.0	2.5	0.67	3.7

empirical X and Y factors in Eq. (8.5.3) depend upon the geometry, loads, and bearing type. Average X and Y factors can be obtained from Table 8.5.3. Two values of X and Y are listed. The set $X_1\,Y_1$ or $X_2\,Y_2$ giving the largest equivalent load should always be used.

Static Equivalent Load P_0   The static equivalent load may be compared directly to the static load rating C_0. If P_0 is greater than the C_0 rating, permanent deformation of the rolling element will occur. Calculate P_0 as follows:

$$P_0 = X_0 R = Y_0 T \qquad (8.5.4)$$

where P_0 = static equivalent load, lb; X_0 = radial factor (see Table 8.5.4); Y_0 = thrust factor (see Table 8.5.4); R = radial load, lb T = thrust (axial) load, lb. If a load higher than the basic static load rating is imposed while rotating, the deformation is distributed evenly and no practical impairment occurs until the deformation becomes quite large. Some equipment operates with loads greatly exceeding the static capacity, such as bearings supporting artillery (twice static capacity), or aircraft control pulleys (four times static capacity). The load which will fracture a bearing is approximately eight times the static load rating.

Table 8.5.4   Radial and Thrust Factors

Type of bearing	X_0	Y_0
Single-row ball	0.6	0.5
Double-row ball	0.6	0.5
Spherical roller, 22200 series	1.0	2.9
Cylindrical roller	1.0	0.0

Oscillating loads, where the motion is such that the rolling element rotates less than half a revolution, approach static load conditions. This type of load is conducive to rapid **false brinelling** and requires special lubrication techniques.

Required Capacity

The basic load rating C is very useful in the selection of the type and size of bearing. By calculating the required capacity needed for a bearing in a certain application and comparing this with known capacities, a bearing can be selected. To calculate the required capacity, the following formula can be used:

$$C_r = \frac{P(L_{10}N)^{1/K}}{Z} \qquad (8.5.5)$$

where C = required capacity, lb; L_{10} = rated life, h; P = equivalent radial load, lb; K = constant, 3 for ball bearings, 10/3 for roller bearings; Z = constant, 25.6 for ball bearings, 18.5 for roller bearings; N = rotation speed, r/min.

LIFE ADJUSTMENT FACTORS

Modifications to Eq. (8.5.2) can be made, based on a better understanding of causes of fatigue. Influencing factors include

1. Reliability factors for survival rates greater than 90 percent
2. Improved raw materials and manufacturing processes for ball bearing rings and balls
3. The beneficial effects of elastrohydrodynamic lubricant films

Equation (8.5.2) can be rewritten to reflect these influencing factors:

$$L_{10}\,\text{modified} = A_1 A_2 A_3 \frac{16{,}700}{N}\left(\frac{C}{P}\right)^K \qquad (8.5.6)$$

where A_1 = statistical life reliability factor for a chosen survival rate, A_2 = life-modifying factor reflecting bearing material type and condition, and A_3 = elastohydrodynamic lubricant film factor.

Factor A_1

Reliability factors listed in Table 8.5.5 represent rates from 90 to 99 percent. Using rates other than 90 percent will yield larger bearings for equivalent loads and, therefore, will increase costs. Rates higher than 90 percent should be used only when absolutely necessary.

Table 8.5.5   Reliability Factor A_1 for Various Survival Rates

Survival rate, %	Bearing life notation	Reliability factor A_1
90	L_{10}	1.00
95	L_5	0.62
96	L_4	0.53
97	L_3	0.44
98	L_2	0.33
99	L_1	0.21

Factor A_2

While not formally recognized by AFBMA, estimated A_2 factors are commonly used as represented by the values in Table 8.5.6. The main considerations in establishing A_2 values are the material type, melting procedure, mechanical working and grain orientation, and hardness.

Table 8.5.6   Life-Modifying Factor A_2

Material	Factor A_2
AISI 440C, Air melted	0.025
SAE 52100, Vacuum processed	1.0
M50 VIM-VAR melted	2.0

Factor A_3

This factor is based on elastohydrodynamic lubricant film calculations which relate film thickness and surface finish to fatigue life. A factor of 1 to 3 indicates adequate lubrication, with 1 being the minimum value for which the fatigue formula can still be applied. As A_3 goes from 1 to 3, the life expectancy will increase proportionately, with 3 being the largest value for A_3 that is meaningful. If A_3 is less than 1, poor lubrication conditions are presumed. Calculations for A_3 are beyond the scope of this section.

Speed Limits

Many factors combine to determine the limiting speeds of ball and roller bearings. It depends on several factors, like bearing size, inner- or outer-ring rotation, contacting seals, radial clearance and tolerances, operating loads, type of cage and cage material, temperature, and type of lubrication. A convenient check on speed limits can be made from a *dn* value. The *dn* value is a direct function of size and speed and is dependent on type of lubrication. It is calculated by multiplying the bore in millimeters (mm) by the speed in r/min.

$$dn = \text{bore (mm)} \times \text{speed (r/min)} \qquad (8.5.7)$$

A guide for *dn* values is listed in Table 8.5.7. When these values are exceeded, bearing life is shortened. The values are only a guide for

approaching difficulties and can be exceeded by special bearings, lubrication, and application.

Table 8.5.7   *dn* Values vs. Bearing Types

Bearing type	Series	Max *dn* value Grease	Max *dn* value Oil
Single-row ball	100, 200, 300, 400, 30, in	200,000	300,000
Double-row ball	200, 300	160,000	220,000
Cylindrical roller	200, 300	150,000	200,000
Spherical roller	22200	120,000	170,000

Friction

One of the assets of rolling-contact bearings is their low friction. The **coefficient of friction** varies appreciably with the type of bearing, load, speed, lubrication, and sealing element. For rough calculations the following coefficients can be used for normal operating conditions and favorable lubrication:

Single-row ball bearings	0.0015
Roller bearings	0.0018

Excess grease, contact seals, etc., will increase these values, and allowances should be made.

PROCEDURE FOR DETERMINING SIZE, LIFE, AND BEARING TYPE

Basically, three common situations may be encountered in the analysis of a bearing system; bearing-size selection, bearing-type selection, and bearing-life determination. Each of these problems requires the following conditions to be known; radial load, thrust load, and speed. The static load capacity is not considered in the following procedures but should be analyzed if the bearing rotational speed is slow or if the bearing is idle for a period of time.

Bearing Size Selection

Known type and series:
1. Select desired design life (Table 8.5.1).
2. Calculate equivalent radial load P [Eq. (8.5.3)].
3. Calculate required capacity C_r [Eq. (8.5.5)].
4. Compare C_r with capacities C in Table 8.5.2. Select first bore size having a capacity C greater than C_r.
5. Check bearing speed limit [Eq. (8.5.7)].

Bearing-Type Selection

Known bore size and life:
1. Select ball or roller bearing (Fig. 8.5.18).
2. Calculate equivalent load P [Eq. (8.5.3)] for various bearing types (conrad, spherical, etc.).
3. Calculate C_r [(Eq. (8.5.5)].
4. Compare C_r with capacities C in Table 8.5.3, and select the type that has a capacity equal to or greater than C_r.
5. Check bearing speed limit [Eq. (8.5.7)].

Bearing-Life Determination

Known bearing size:
1. Select ball or roller bearing (Fig. 8.5.18).
2. Calculate equivalent radial load P [Eq. (8.5.3)].
3. Select basic load rating C from Table 8.5.3.
4. Calculate rated life L_{10} [Eq. (8.5.1) or (8.5.2)].
5. Check calculated life with design life.

BEARING CLOSURES

Rolling-element bearings are made with a wide variety of **closures.** Basically, they are open, shielded, or sealed (Figs. 8.5.19 and 8.5.20).

Shielded bearings have a small clearance between the stationary shield and rotating ring. This provides reasonable exclusion of dirt without an

Fig. 8.5.19

Fig. 8.5.20

increase in friction. **Sealed bearings** have a flexible lip in contact with the inner ring. Friction is increased, but more effective retention of lubricant and exclusion of dirt is obtained. Bearings are also available with solid fillers (graphite, polymeric oils) that fill the open spaces and seal the bearings.

BEARING MOUNTING

Correct **mounting** of a rolling-contact bearing is essential to obtain its rated life. Many types of mounting methods are available. The selection of the proper method is a function of the accuracy, speed, load, and cost of the application. The most common and best method of bearing retention is a press fit against a shaft shoulder secured with a locknut. End caps are used to secure the bearing against the housing shoulder (Fig. 8.5.21). Retaining rings are also used to fix a bearing on a shaft or in a housing (Fig. 8.5.22). Each shaft assembly normally must provide for expansion by allowing one end to float. This can be accomplished by

Fig. 8.5.21

Fig. 8.5.22

allowing the bearing to expand linearly in the housing or by using a straight roller bearing on one end. Care must be exercised when designing a **floating installation** because it requires a slip fit. An excessively loose fit will cause the bearing to spin on the shaft or in the housing.

Table 8.5.8 lists shaft and housing tolerances for press fits with ABEC 1 precision applications (pumps, gear reducers, electric motors, etc.) and ABEC 7 precision applications (grinding spindles, etc.).

Mounted units such as **ball-bearing pillow blocks** (Fig. 8.5.23) are frequently used for fans and conveyors. Three common methods are used to attach the bearing to the shaft; setscrew, eccentric locking collar, and taper-sleeve adapter.

Table 8.5.8   Shaft and Housing Tolerances for Press Fit

Bearing bore, mm	Shaft tolerances, in, ABEC 1 precision	Bearing bore, mm	Shaft tolerances, in, ABEC 7 precision
4–6	+ 0.0000 − 0.0002	4–30	+ 0.00000 − 0.00015
7–17	+ 0.0000 − 0.0003	35–50	+ 0.0000 − 0.0002
20–50	+ 0.0000 − 0.0004	55–80	+ 0.0000 − 0.0003
55–80	+ 0.0000 − 0.0005	85–120	+ 0.00000 − 0.00035
85–120	+ 0.0000 − 0.0006		

Bearing OD, mm	Housing tolerances, in, ABEC 1 precision	Bearing OD, mm	Housing tolerances, in, ABEC 7 precision
16–30	+ 0.0008 − 0.0000	16–80	+ 0.0002 − 0.0000
32–47	+ 0.0010 − 0.0000	85–120	+ 0.0003 − 0.0000
52–80	+ 0.0012 − 0.0000	125–225	+ 0.0004
85–120	+ 0.0014 − 0.0000		
125–180	+ 0.0016 − 0.0000		

Fig. 8.5.23

Setscrew   Figure 8.5.24 illustrates the use of an extended inner-ring bearing held to the shaft with a setscrew. This is a simple method and is suitable only for lightly loaded bearings.

Fig. 8.5.24   Fig. 8.5.25

Eccentric Locking Collar   Figure 8.5.25 illustrates the use of an extended inner-ring bearing held to the shaft with an eccentric collar. This method tends to keep the shaft centered in the bearing more concentrically than the setscrew method. It is suitable for light to moderate loads.

Taper-Sleeve Adapter Figure 8.5.26 illustrates the use of a taper-sleeve adapter to mount the bearing on the shaft. It provides uniform

concentric contact between the shaft and bearing bore. However, skill is required to tighten the locking nut enough to keep the sleeve from spinning on the shaft and yet not so tight where the clearance is removed from the bearing. To minimize this problem, pillow block bearings called **pop release** are available. Their use eliminates measuring clearances.

LUBRICATION

Rolling-contact bearings need a **fluid lubricant** to obtain or exceed their rated life. In the absence of high-temperature environment, only a small amount of lubricant is required for excellent perfor-

Fig. 8.5.26

mance. Excess lubricant will cause heating of the bearing and accelerate the deterioration of the lubricant. Optimum lubrication of rolling-contact bearings can be predicted by elastohydrodynamic theory (EHD). It has been shown that film thickness is sensitive to bearing speed of operation and lubricant viscosity properties and, moreover, that the film thickness is virtually insensitive to load.

Grease is commonly used for lubrication of rolling-contact bearings because of its convenience and minimum maintenance. A high-quality lithium-based NLGI 2 grease should be used for temperatures up to 180°F (82°C), or polyurea-based grease for temperatures up to 300°F (150°C). In applications involving high speed, oil lubrication is often necessary. Table 8.5.9 can be used as a general guide in selecting oil of the proper viscosity for rolling-contact bearings.

In applications using grease, it is necessary to replenish the lubricant. Relubrication intervals in hours of operation are dependent on

Table 8.5.9   Oil-Lubrication Viscosity
(Viscosity in ISO identification numbers*)

Bearing bore, mm	Bearing speed, r/min				
	10,000	3,600	1,800	600	50
4–7	68	150	220		
10–20	32	68	150	220	460
25–45	10	32	68	150	320
50–70	7	22	68	150	320
75–90	3	10	22	68	220
100	3	7	22	68	220

* ISO identification number = midpoint viscosity in centistokes at 40°C.

Table 8.5.10   Ball-Bearing Grease Relubrication Intervals
(Hours of operation)

Bearing bore, mm	Bearing speed, r/min				
	5,000	3,600	1,750	1,000	200
10	8,700	12,000	25,000	44,000	220,000
20	5,500	8,000	17,000	30,000	150,000
30	4,000	6,000	13,000	24,000	127,000
40	2,800	4,500	11,000	20,000	111,000
50		3,500	9,300	18,000	97,000
60		2,600	8,000	16,000	88,000
70			6,700	14,000	81,000
80			5,700	12,000	75,000
90			4,800	11,000	70,000
100			4,000	10,000	66,000

temperature, speed, and bearing size. Table 8.5.10 is a general guide which represents the time after which it is advisable to add a small amount of grease in order to safeguard the bearings. The intervals are valid up to 160°F (71°C) and should be divided by 2 for cylindrical roller bearings and by 10 for spherical roller bearings.

8.6   PACKINGS, GASKETS, AND SEALS
by John W. Wood, Jr., and James Drago, P.E.

REFERENCES: Staniar, "Plant Engineering Handbook," McGraw-Hill. Thorn, Rubber and Plastic Packings, *Rubber Age*, Jan. 1956. Roberts, Gaskets and Bolted Joints, *Jour. Applied Mechanics*, June 1950. Nonmetallic Gaskets, *Mach. Des.*, Nov. 1954. Elonka, Basic Data on Seals, a *Power* reprint, McGraw-Hill. Fluidtec Engineered Products, Training Manuals. Brink, et al., "Handbook of Fluid Sealing," McGraw-Hill.

Packings, gaskets, and **seals** are materials used to control or stop leakage of fluids (liquids and/or gases) or of solid dry products through mechanical clearances when the contained material is under pressure or vacuum. The mechanical clearances to be sealed may be between either fixed or moving components.

Gaskets are compressible materials installed in static clearances, which normally exist between parallel flanges or concentric cylinders. Sealing of flat flange gaskets is effected by compressive loading achieved through bolting or other mechanical means. The full-face gasket (Fig. 8.6.1) is not normally recommended because the material outside the bolt holes does not improve the seal. However, it prevents bending moments that may damage brittle flange materials such as cast iron or plastic. The simple ring gasket (Fig. 8.6.2) is more efficient and economical. With irregularly shaped flanges, bolt holes may serve to locate the gasket. Bolt holes can be placed in lobes on the outer perimeter or simply cut into a full-face template of the flange. **Metal-to-metal** fits require a recess whose depth and volume are engineered for the gasket to be used. A gasket, such as an O-ring (Fig. 8.6.13), extends above

the groove sufficiently to provide the minimum compression required for initial seating. Depending on the seal configuration, the internal fluid pressure can automatically provide additional sealing force. **Warped, wavy,** or irregular flanges (often resulting from corrosion, wear and tear, welding, other fabrication, or as found in glass-lined equipment) require gaskets that are softer and/or thicker than normal, to compensate for surface imperfections. Excessive thickness or volume of gasket material, even if the gasket is installed in a groove, must be avoided to prevent distortion or "mushrooming," which can result in inadequate loading. Tongue-and-groove joints (Fig. 8.6.4) confine the gasket material and may adapt to the extra thickness, within limits. The use of thicker gaskets also means an increase in gasket creep, which increases **bolt relaxation** and loss of compressive load on the gasket. A decrease in bolt load can compromise the integrity of the joint's seal.

In addition to the types (Figs. 8.6.5 to 8.6.7) shown, there are the machined metal profile gasket (Fig. 8.6.8) and solid metal designs in flat, round, kammprofile and either octagonal or oval API ring joint gaskets for extreme pressures and temperatures to seal against steam, oil, and gases. These types have very low compressibilities, and their behavior depends on the geometry of their cross sections. The envelope gasket (Fig. 8.6.3), usually polytetrafluoroethylene (PTFE) with a variety of cores, is particularly useful for corrosive, noncontaminating service, average pressure in wavy, glass-lined equipment. Envelope gasket thickness can be adjusted or "shimmed" to fill extreme flange waviness.

Fig. 8.6.1–8.6.36 Packings, gaskets, and diaphragms.

Cylindrical or **concentric** gasketing uses a retaining gland follower and is mechanically loaded, e.g., the standard mechanical joint for cast-iron pipe (Fig. 8.6.10) or the condenser tube-sheet ferrule (Fig. 8.6.11). Cup-shaped gaskets are designed to be self-tightening under pressure (Fig. 8.6.12). The O-ring (Fig. 8.6.13) located in an annular groove and pre-compressed as in the grooved flange, is a self-energized gasket. A cylindrical ring with internal single lip or double lips, also automatic in action, is quite common in pipe joints.

A variety of gasketing materials are listed in Table 8.6.1 together with common applications. Beyond rubber are many elastomeric mate-rials generally similar in mechanical behavior but varying as to temperature limits and fluid compatibility (see also Sec. 6).

The **proper design** of a gasketed joint requires flange rigidity to avoid distortion, flange surface finish and texture commensurate with gasket type, adequate compressive stress on the gasket, and adequate bolt loading. The load must seat the gasket, i.e., cause the material to flow into and fill flange irregularities. Once the pressure of the fluid being contained by the joint is applied, the residual compressive stress on the gasket must be sufficient to maintain the seal. These values, known, respectively, as the **seating stress factor y,** expressed in force per unit area,

Table 8.6.1   Common Usage of Gasketing Materials*

Type	Service principally for
Sheet rubber	Water
Cloth-inserted rubber sheet	Water
Cork composition, cellulose fiber sheet	Oil, low-pressure
Rubberized cloth (Fig. 8.6.9)	Hot water (boiler manholes, etc.)
Compressed fiber sheet	All services up to 750°F (400°C)
Corrugated sheet metal with filling (Fig. 8.6.5)	Steam, oil at high temperatures
Metal jacket over compressible fiber center (Fig. 8.6.6)	Steam, oil at high temperatures
Spirally wound steel strip with intervening graphite layers (Fig. 8.6.7)	Steam, oil at high temperatures

*Asbestos material may be found in old equipment; current, new, and/or replacement parts are made of other materials suitable to the service.

and the **gasket factor *m*,** which is dimensionless. The factors vary with gasket material, gasket thickness, and tightness required by the joint designer. The ASME Boiler and Pressure Vessel Code, Section VIII, Division I, Appendix 2, gives the details for typical joint design and tabulates values for *y* and *m* for various generic gasket materials. The *m* and *y* factors are to be used for flange design, not for developing torque values for existing flanges. The *m* and *y* values are provided by gasket manufacturers for their products. The leak rate allowed during tests used to arrive at the *m* and *y* values greatly affects the magnitude of these values. Manufacturers should provide that leak rate to designers.

The concept of *m* and *y* gasket factors was introduced to the ASME Code pertaining to flange design in the 1940s. In the 1980s efforts by the Materials Technology Institute of the Chemical Process Industries (MTI), the Pressure Vessel Research Council (PVRC), and the University of Montreal's Ecole Polytechnique embarked on new methods of characterizing gasket materials. The creation of gasket factors G_b, *a*, and G_s in conjunction with the tightness parameter T_p gives flange designers an alternative to *m* and *y* values. While the new methods of flange design based on the new factors are used, they have not been accepted as part of the ASME Boiler and Pressure Vessel Code.

High bolt loading resulting in high bolt stretch is desirable for tight and enduring gasket joints, but it must not crush the gasket material or permanently deform the bolt material, i.e., cause bolt yield. Crushing-strength values, which vary with thickness and temperature, can be obtained from gasket manufacturers. Consistent with the condition of the flanges, the thinner the gasket, the more efficient the joint. The use of thinner sheet gaskets minimizes leakage and decreases gasket creep, which maximizes gasket performance.

O-rings are used as static seals (Fig. 8.6.13) between flanges and as dynamic or static seals between concentric surfaces (Figs. 8.6.20 and 8.6.32). An O-ring must be either stretched and/or compressed to function. If the O-ring is not compressed or stretched while in service, it is the wrong size. O-ring type seals are available in various profiles (e.g., round, delta, D, T, lobe, cored, hollow ring, and K, to name a few). See Fig. 8.6.37. They can be molded, extruded, or lathe-cut. Each profile has its application advantages, and users should work with their suppliers to apply the proper geometry and material for each application. Aerospace Materials Specifications (AMS), Air Force/Navy Specifications (AN), Military Specifications (M; MIL; MS) and National Aeronautical Specifications (NAS) are some of the various standards organizations which address O-rings. Following their specifications ensures conformity and uniform quality. O-rings are available in a wide range of elastomeric and thermoplastic compounds as well as metal. Chemical compatibility, temperature limits, size (ID and c/s), durometer, and profile of the selected compound are the main concerns in making O-ring choices. Data on the design of O-ring joints are available from suppliers. The nominal pressure limit for O-rings based on typical mechanical clearances is 1,500 psig (10 MPa) without backup rings and 3,000 psig (20 MPa) with backup rings. If clearances can be eliminated, as in a flanged joint with close metal-to-metal contact, no limit can be set.

Other self-energizing joints, such as the boiler hand-hole plate (Fig. 8.6.9), need only sufficient load to effect an initial seal.

Valve disks are specialized gaskets designed for joints that are frequently broken and reseated. Disks for globe valves (Fig. 8.6.14) are

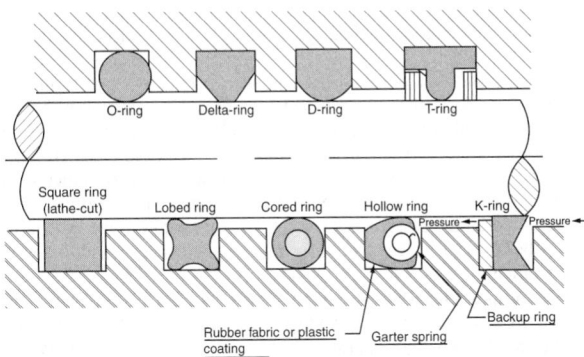

Fig. 8.6.37   O-rings in various cross sections.

usually encased in a disk holder with a swivel mounting, which ensures precise reseating without abrasion during the closing and opening cycles. They are made of firm rubber for bib washers, hard rubber and phenolics for more severe service, and plastics such as nylon and PFTE. Pump valves (Fig. 8.6.15) are described in Sec. 14. Rubber **valve seats** are used with metal valve disks on some pumps, e.g., the rotary drilling pump valve (Fig. 8.6.16). Plastics are also used for seats, notably in ball valves.

Dynamic packings include all packings that operate on moving surfaces. To retain fluid under pressure, they are subjected to the hydraulic load. When no pressure exists, as in many oil-seal applications, the packing is mechanically loaded by a spring (Fig. 8.6.28) or by its own resiliency. Dynamic packings operate like bearings, wherein the lubricant serves as both a separating film and a coolant. The film is vital for satisfactory service life, but some leakage will occur. Low-viscosity fluids and high pressures add to leakage problems, as both require thin films to minimize leakage. This causes higher friction and generates heat, which is the single most detrimental factor in packing life. Deep packings (packing sets with many rings) reduce leakage but increase frictional heat, particularly at high speeds. Five rings of packing are considered optimum. Normally the fluid being sealed serves as the lubricant. Maximum efficiency can be attained when oil is the fluid being sealed; in decreasing order of efficiency are clean water, solvents, and fluids containing solids. These are progressively more unsatisfactory unless supplemental cooling and lubrication in the form of a flush are provided. Lubrication may be provided by using a lantern ring in the center of the packing set through which fluid is fed to the packings (Fig. 8.6.27). The pressure of the fluid supply should be slightly higher [5 to 10 psig (0.35 to 0.7 bar · g)] than the pressure in the packing sealing chamber. The process media govern the choice of flushing fluid. The two need to be compatible. In cases of extreme contamination, the lantern ring is moved to the bottom of the packing set to introduce clean fluid, dilute the contaminated media, and prevent abrasives from migrating to the dynamic sealing interface. Use of a lantern ring also seals against air being drawn into the system when equipment is operating at negative head or starts under vacuum. Centrifugal pumps equipped in this manner are said to have a **water seal.**

Dynamic packings are classified in three ways:

1. On the basis of shape of the surfaces: cylindrical, conical, spherical, or flat. Cylindrical packings are in turn classified according to whether they pack on the outside perimeter, as in piston packings (Figs. 8.6.17 to 8.6.20) or inside perimeter, as on rods or shafts (Figs. 8.6.21 to 8.6.28). Other examples are: conical, the plug cock lining (Fig. 8.6.29);

spherical, the ball joint (Fig. 8.6.30); and flat, the mechanical seals (Figs. 8.6.31 and 8.6.32).

2. On the basis of the type of motion: rotary, oscillating, reciprocating, or helical (as in a rising-stem valve packing).

3. On the basis of how the seal is energized: nonautomatic soft or braided compression packing tightened by external means, usually a gland follower; automatic preformed, molded shapes whose geometry facilitates the service's internal pressure (what the seal is sealing) to tighten the packing.

The **selection of a compression packing** type is a matter of performance expectations, service conditions, and economics. In most cases several types are available, some of which, although expensive, yield exceptional service. A cheaper packing type could yield unsatisfactory service. Service requirements most often dictate the final choice of a packing material. The packing must be compatible with the fluid being sealed, operating pressure in the system, temperature, type of equipment being sealed, ease of adjustment/maintenance, and ease of replacement.

For **reciprocating elements,** the **O-ring** (Fig. 8.6.20) is extremely simple. It is a precision part manufactured to close tolerances, as must be the seat into which it is placed. As an elastomeric material completely exposed to the operating fluid, it is subject to chemical degradation. The O-ring material must be carefully chosen to ensure compatibility with the fluid being sealed; the wrong choice will lead to either shrinkage or swelling, with premature failure of the O-ring. It is best suited to medium-pressure service from 1,500 to 3,000 lb/in^2 (100 to 200 bar·g) with backup rings and intermittent movement, as in hydraulic cylinder or valve stem service. It is not recommended for pump service. Backup rings are preferably of blocklike cross section in either tetrafluoroethylene (PTFE) or similar material, avoiding the thin spiral type. The split **piston ring** (Fig. 8.6.17), usually cast iron, is widely used in gas, oil, and steam engines and compressors. Large pistons frequently employ segmental rings similar to floating metal rod packings (Fig. 8.6.24) but facing outward. **Floating metal packing rings** are made of numerous radial or tangential segments, making it possible for them to contract on the shaft; they are assembled in sets of two to break the joints and are held together with garter springs. They are used for steam, gas, or air, in either engines or compressors under the most severe operating conditions and at pressures up to 35,000 lb/in^2 (2410 bar·g). Normally oil lubrication is provided; for less severe service, filled PTFE rings perform very well in dry gases without auxiliary lubrication. Step-, scarf-, or butt-cut rings of laminated cotton fabric, bonded with an elastomer or phenolic resin, are employed in water pumps, gasoline pumps, etc. They may float similar to cast iron piston rings, or be retained by a gland follower, as in Fig. 8.6.18. **Cups** (Fig. 8.6.19) are fully automatic and very tight; cups in their inverted form, with the lip on the ID, are known as **flange packings** and are also fully automatic and very tight. They are used principally for slow-speed applications. **Nested V and conical rings** (Figs. 8.6.22 and 8.6.23) are automatic, though often provided with a gland follower to effect initial fit. They are made of a wide range of materials including homogeneous elastomers and polymers, and reinforced woven fibers (cotton, aramid, or fiberglass) for severe duty. They range in hardness from soft and flexible to semirigid. Use of multiple rings allows them to be of the cut or split type for ease of installation and replacement. **Braided, soft,** or **jamb packings** are normally formed in square or rectangular sections with a butt joint staggered from ring to ring at installation. Many materials are employed to make braided packing. Fibers such as synthetic blends, aramid, PTFE graphite, carbon, fiberglass, or flax are saturated with wax, custom lubricating compounds, PTFE suspensoid solutions, or rubber compounds. Metal foil and tinsel are used for high-temperature and high-pressure conditions. Packings containing woven or braided fibers also may contain wire for strength and pressure resistance. For pipe expansion joints, see Sec. 8.

Rotary shafts are generally packed with adjustable soft packings, with the notable exception of the mechanical seals (Figs. 8.6.31 and 8.6.32); where pressures are low, nested V or conical styles may be used. At zero or negligible pressures, the oil seal, a spring-loaded flange packing

(Fig. 8.6.28), is very widely used. Where some leakage can be tolerated, the labyrinth (Fig. 8.6.25) and controlled-gap seals are used, particularly on high-speed equipment such as steam and gas turbines. **Soft packings** are of the same general type as those used for reciprocating service, with the fiber braid lubricated with grease and graphite or with polytetrafluoroethylene fibers and suspensoid. Aramid, carbon, and graphite fibers filled with various lubricants and reinforcements are used at higher speeds and fluid pressures. Fiber braid with PTFE suspensoid is widely applied on valve stems operating below 500°F (260°C) and on centrifugal pumps. This material is an insulator, however, and results in high heat buildup on the dynamic surface; a better choice lies in use of a packing with better heat-transfer characteristics, such as one containing carbon or graphite. For continuous rotary service, **automatic packings** energized by internal pressure are best restricted to low pressure because their tightness under high pressure results in overheating. For intermittent service, as on valve stems, they are excellent.

Oil seals (Fig. 8.6.28) are unique flange packings having an elastomer lip generally bonded to a metal cup which is press-fitted into a smooth cylindrical bore. Basically, an oil seal is a flange packing with a flexible lip and a narrow contact area about 1⁄16 in (1.6 mm) wide which, under pressure, causes extreme local heating and wear. They are recommended only for nonpressure service and perform best in good lubricating media. To accommodate shaft runout up to 0.020 in (0.5 mm) depending on the rotating speed, the lip is spring-loaded with a coil or finger spring. Finger springs are safer inasmuch as they are molded into the elastomer and are less likely to become dislodged and cause shaft damage. Since the lip is completely exposed to the sealed fluid, particular care should be taken to ensure compatibility between the elastomer and the fluid. Temperature, speed at the surface contacting the seal, equipment type, internal pressure (positive or negative) if any, space available to insert the seal, dynamic runout, and misalignment are all parameters that must be considered when one is selecting an oil seal. For applications in which there is no internal pressure or elevated fluid level (such as a gearbox with an oil sump), bearing isolators are a viable choice to protect bearings and their lubricants from contamination. Bearing isolators work on principles similar to those in labyrinth seals, shown in Figs. 8.6.25 and 8.6.26. An isolator seals due to the tortuous path the fluid must take, and there are no contacting/rubbing surfaces to wear. Isolators are very effective in keeping solid contaminants and water spray (used for equipment wash-down) out of the bearing area and lubricating fluid.

Mechanical, Rotary, or End Face Seals

The greatest advancements in the design of end face mechanical seals have come about in response to environmental regulations; requirements to minimize energy consumption and operating costs; safety; and concerns over loss of the product which is being sealed. The application of seals to replace packing in rotary equipment has increased dramatically and continues.

All end face mechanical seals (Figs. 8.6.31 and 8.6.32) consist of four parts: a stationary flat face, a rotating flat face, secondary sealing elements (usually elastomeric), and a flexible loading device. The assembled seal is placed and effects proper leak control. The two flat-face seal rings (one stationary, one rotating) rub and create the primary seal. Normally, the flat seal rings have different hardness values, and the soft one is narrower than the hard one. Secondary sealing elements prevent leakage between the rotating shaft and the rotating seal ring, and they block the leakage path around the outside of the stationary seal face. They also serve as gaskets between the assembled parts (i.e., gland plate and housing). The flexible loading device usually consists of one or more springs which press the flat seal rings together. Spring loading ensures a seal when there is little or no hydraulic pressure available to press the faces together and helps maintain constant pressure between the faces as the soft (sacrificial) face wears down. The springs also act as vibration dampers to mitigate against the intrusion of transmitted vibrations, which may affect the efficient operation of the seal assembly.

Types of End Face Mechanical Seals

1. *Inside-mounted.* The seal head is mounted inside the stuffing box (Fig. 8.6.38a).

Fig. 8.6.38   Rotary end face seal. (*a*) Inside the seal chamber/stuffing box; (*b*) outside the seal chamber/stuffing box.

2. *Outside-mounted.* The seal head is mounted outside the stuffing box (Fig. 8.6.38b).

3. *Unbalanced seal.* The full hydraulic pressure in the seal chamber is transmitted to the seal faces (Fig. 8.6.39a).

Fig. 8.6.39   Rotary end face seals showing (*a*) unbalanced and (*b*) balanced configurations.

4. *Balanced.* The seal elements are designed to reduce the hydraulic forces transmitted to the seal faces (Fig. 8.6.39b). A complete balance is not practical.

5. *Rotary seal.* In this design, the springs rotate with the shaft.

6. *Stationary seal.* In this design, the springs do not rotate with the shaft.

7. *Metal bellows.* Welded or formed metal bellows exert a spring load; there is no dynamic secondary seal element (Fig. 8.6.40).

Fig. 8.6.40   Rotary end face seal with metal bellows and "t" clamp stationary.

8. *Double seal.* Two mechanical seals are mounted back to back, face to face, or in tandem, between which a barrier fluid (liquid or gas) can be introduced for environmental control (Fig. 8.6.41).

Fig. 8.6.41   Back-to-back double seal. Barrier fluid must have an inlet and an outlet.

End face mechanical seal materials must satisfy a number of design requirements, including chemical compatibility between the sealed fluid and the seal materials, ability of the seal materials to remain serviceable under the worst operating conditions, and ability to provide a reasonably long life in service at the operating conditions. Mating faces of the seals can be made from ordinary materials like bronze and PTFE

Fig. 8.6.42   High-performance lip seal with modified PTFE elastomer. Illustration shows single and staged elements.

for mild service on up to carbon, carbides, stainless steels, and other exotic alloys as service conditions become more severe. Hard faces can utilize ceramics, tungsten and silicon carbides, and hard coatings over base metals (chromium oxide over stainless steel 316SS, subsequently lapped flat).

Secondary seal materials are usually elastomeric and include these:

Elastomer	Temperature range
Nitrile	−75 to 250°F (−59 to 121°C)
Neoprene®	−65 to 250°F (−54 to 121°C)
Viton®	−40 to 450°F (−40 to 232°C)
Ethylene propylene	−65 to 300°F (−59 to 149°C)
Kalrez®	32 to 550°F (−0 to 288°C)
Aflas®	−23 to 400°F (−31 to 205°C)

Environmental controls as applied to seals are special techniques used to control the environment in which the seal operates. Some examples include discharge return recirculation, suction return, flush from an outside source, flush, and quench and drain. The controlling techniques often require supplementary equipment such as heaters, coolers, pumping rings, external circulation devices, various special solenoids and valves compatible with the working environment, and so on. Sealed products that solidify, vaporize, abrade, or carry contaminants are successfully sealed with end face seals augmented by environmental controls.

High Performance Lip Seals The nature of some sealed products is such that end face mechanical seals are not applicable. In many difficult instances of that type, sealing can be achieved with high-performance modified PTFE lip seals (Fig. 8.6.42). Gylon® is such a material which can serve in seals operating over a wide range of pressures, temperatures, and rotating speeds. It is particularly useful to seal against some dry products (air purge may be required), viscous resins, salting solutions, and products which tend to solidify on seal faces. Dry running is possible under some circumstances. Unlike conventional lip seal material, restructured PTFE lip seals in multiples can operate from high vacuums (10^{-3} inHg) up to 10 bar (150 lb/in^2), within a temperature range of -130 to $+500°F$ (-90 to $+260°C$), and exhibit excellent compatibility with a wide range of sealed fluids. Manufacturers' literature will provide data showing the effect of temperature and rotating speed on the permissible operating pressure.

For extremely high speeds, where it is desirable to eliminate all rubbing contact, the **labyrinth** seal (Fig. 8.6.25) is chosen. This seal is not fluid-tight but restricts serious flow by means of a torturous path and induced turbulence. It is widely used on steam turbines (Sec. 9.4). Where no leakage is permissible, a liquid seal based on the U-tube principle (Fig. 8.6.26) may be used. The natural weight of the liquid is amplified by centrifugal force so that under high rotating speed a fair pressure differential can be sealed. Another noncontacting seal is the **controlled gap seal** which is being used on gas turbines where pressure differentials are not excessive and a small amount of leakage can be tolerated. The seal consists of a ring with a shaft clearance in the range of 0.0005 to 0.0015 in (0.013 to 0.038 mm) and is made of exotic heat-resisting materials capable of maintaining that clearance at all operating temperatures. Usually one end of the ring is faced to form an axial seal against the inside of its housing.

Diaphragms are a form of dynamic packing but include the requirements of a gasket where they are gripped or held in position. In service they are leakless, although generally limited in travel. By literally rolling one cylinder inside another, considerable increase in travel is possible. This type is often called a bellows, and a simple application is the mechanical seal suspension shown in Fig. 8.6.31. In the diaphragm valve (Fig. 8.6.33) the diaphragm replaces both the conventional stem packing and valve disk. **Diaphragms** of fabric such as cotton or nylon (except friable materials such as glass) covered with an elastomer suitable for the fluids and temperatures involved are used in pumps (fuel pump, Fig. 8.6.35) and in motors (Fig. 8.6.34) to operate valves, switches, and other controls. Correctly designed diaphragms are made with slack to permit a natural rolling action. Flat sheet stock should be used only where limited travel is desired. An unusual application is shown in Fig. 8.6.36, where the diaphragm is under balanced fluid pressure on both sides and is unstressed. Thin sheet metal, usually with concentric corrugations, is used where movement is limited and long life is desired. Where considerable movement is involved, the possibility of fatigue must be considered.

PTFE and Gylon diaphragms are used with chemically aggressive fluids. Experience has demonstrated that PTFE has a tendency toward cold flow, which leads to leaking at the clamp areas; Gylon has proved more dimensionally stable and serviceable.

8.7 PIPE, PIPE FITTINGS, AND VALVES
Revised by E. A. Avallone

REFERENCES: M. L. Nayyar, "Piping Handbook," McGraw-Hill. ANSI Code for Power Piping. ASTM Specifications. Tube Turns Division, Natural Cylinder Gas Co., catalogs. Crane Co., catalogs and bulletins. Grinnell Co., Inc., "Piping Design and Engineering." M. W. Kellogg Co., "Design of Piping Systems," Wiley. United States Steel Co., catalogs and bulletins.

EDITOR'S NOTE: The several piping standards listed in this section are subject to continuing periodic review and/or modification. It is suggested that the reader make inquiry to the issuing organizations (see Table 8.7.1) as to the currency of a given standard as listed.

PIPING STANDARDS

Codes for various piping services have been developed by nationally recognized engineering societies, standardization bodies, and trade associations. The sound engineering practices incorporated in these codes generally cover minimum safety requirements for the selection of materials, dimensions, design, fabrication, erection, and testing of piping systems. By means of interpretation and revision these codes continually reflect the knowledge gained through experience, testing, and research.

Generally, piping codes form the basis for many state and municipal safety laws. Compliance with a code which has attained this status is mandatory for all systems included within the jurisdiction. Although some of today's piping installations are not within the scope of any mandatory code, it is advisable to comply with the applicable code in the interests of safety and as a basis for contract negotiations. Contracts with various agencies of the federal government are regulated by federal specifications or rules. These often do not have a direct connection with the codes enumerated below.

The reader is cautioned that the **piping standards** are changing more often than in previous years. Although the formulas and other data provided are in accordance with the code rules in effect at the time of publication, it must be recognized that code rules may change, and piping engineering and design work performed in accordance with information contained herein does not provide complete assurance that all extant code requirements have been met. The reader is urged to become familiar with the specific code edition and addenda applicable in a particular project, for they may contain mandatory requirements applicable to the particular project.

The **ASME Boiler and Pressure Vessel Code** is mandatory in many cities, states, and provinces in the United States and Canada. Local application of this code into law is not uniform, making it necessary to investigate the city or state laws which have jurisdiction over the installation in question. Compliance with this code is required in all locations to qualify for insurance approval.

Section I: "Power Boilers" concerns all piping connections to power boilers or superheaters including the first stop valve on single boilers, or including the second stop valve for cross-connected multiple-boiler installations. Section I refers to ASME B31.1 which contains rules for design and construction of "boiler external piping." "Boiler external piping" is under the jurisdiction of Section I and requires inspection and code stamping in accordance with Section I even though the rules for its design and construction are contained in the ASME Code for Pressure Piping, section B31.1.

Section II: "Material Specifications" provides detailed specifications of the materials which are acceptable under this code. (These specifications generally are identical to the corresponding ASTM Standards.)

Section III: "Nuclear Components" includes all nuclear piping. It is the responsibility of the designer to determine whether or not a particular piping system is "nuclear" piping, since Section III makes this determination the responsibility of the designer. In general, piping whose failure could result in the release of radiation which would endanger the public or plant personnel is considered "nuclear" piping.

Section VIII: "Unfired Pressure Vessels" concerns piping only to the extent of the flanged or threaded connections to the pressure vessel, except that the entire section will apply in those special cases where unfired pressure vessels are made from pipe and fittings.

Section IX: "Welding and Brazing Qualifications" establishes the minimum requirements for ASME Code welding.

Section XI: "Rules for Inservice Inspection of Nuclear Power Plant Components" contains rules for the examination and repair of components throughout the life of the plant.

The ASME Code for Pressure Piping B31 is, at present, a nonmandatory code in the United States except where U.S. state legislative bodies and Canadian provinces have adopted this code as a legal requirement. The minimum safety requirements of these codes have been accepted by the industry as a standard for all piping outside the jurisdiction of other codes. The piping systems covered by the separate sections of this code are listed below:

Power Piping	B31.1
Fuel Gas Piping	B31.2
Chemical Plant and Petroleum Refinery Piping	B31.3
Liquid Petroleum Transportation Piping Systems	B31.4
Refrigeration Piping	B31.5
Gas Transmission and Distribution Piping Systems	B31.8
Building Service Piping	B31.9

Several other engineering societies and trade associations have also issued standards covering piping. Foremost among these is the American Society for Testing and Materials (ASTM), the American National Standards Institute (ANSI), the American Water Works Association (AWWA), the American Petroleum Institute (API), and the Manufacturers Standardization Society of the Valve and Fitting Industry (MSS).

Additional piping specifications have been issued by the American Welding Society (AWS), the Pipe Fabrication Institute (PFI), the National Fire Protection Association (NFPA), the Copper Development Association (CDA), the Plastics Pipe Institute (PPI), and several others.

The piping standards issued by the ASTM are most commonly referred to in specifications covering piping for power plants, chemical plants, refineries, pulp and paper mills, and other industrial plants. The large majority of ASTM Standards has also been issued by the ASME in Section II of the ASME Boiler and Pressure Vessel Code. The same specification numbers are applied by the ASME as were originally assigned by the ASTM.

The ANSI formerly prepared the various standards of the B31 Code for Pressure Piping. These standards are now issued by the ASME. The ANSI, however, continues to prepare and issue various standards covering pipe fittings, flanges, and other piping components. Note that ASME B16 prepares and issues standards for fittings, flanges, etc.

The AWWA has issued various standards for waterworks applications. The majority of these involve ductile iron pipe, ductile iron and cast iron pipe fittings, etc.

The MSS has prepared various standards for valves, hangers, and fittings, generally involving the lower range of pressures and temperatures.

Table 8.7.1 gives the most **commonly used piping standards** and the organizations from which the standards are available.

PIPING, PIPE, AND TUBING

The term **piping** generally is broadly applied to pipe, fittings, valves, and other components that convey liquids, gases, slurries, etc.

The term **pipe** is applied to tubular products of dimensions and materials commonly used for pipelines and connections, formerly designated as iron pipe size (IPS). The outside diameter of all weights and kinds of IPS pipe is of necessity the same for a given pipe size on account of threading. Nevertheless, most pipe is furnished unthreaded with ends prepared for butt welding.

The word **tube** (or **tubing**) is generally applied to tubular products as utilized in boilers, heat exchangers, instrumentation, and in the machine, aircraft, automotive, and related industries.

Pipe and Tube Products—General

Commercial pipe and tube products are grouped into various classifications generally based on the application or use and not on the manufacturing method. Most tubular products fall into one of three very broad classifications: (1) pipe, (2) pressure tubes, and (3) mechanical tubes. Each classification falls into various subgroupings, which may have been defined and standardized differently by the different trade or user groups. The same standard materials specifications may apply to several of the (user) classifications. For example, ASTM A120 or A53 pipe may be used for applications representing refrigeration, pressure, and nipple service.

Cost considerations enter into the selection of specific piping materials. In some sizes, prices of pipe made to different materials specifications may vary, whereas in other sizes, they may be identical.

Within the broad **use classifications** listed above, the **production method classifications** are also recognized. These are primarily (1) seamless wrought pipe, (2) seamless cast pipe, and (3) seam-welded pipe or tubes. The large variety of single and combination pipe- or tube-forming methods can produce different characteristics and properties in essentially identical pipe materials. In addition, the final finishing can result in hot-finished or cold-finished products. Cold-finishing may be accomplished by reducing or by expanding. Heat treatments may also affect the properties of the finished product.

Piping

On the basis of user classification, the more commonly used types of pipe are tabulated in Table 8.7.2. This listing ignores method of manufacture, size range, wall thickness, and finish, for which the different user groups may have developed different standard requirements.

Standard Pipe Mechanical service pipe is produced in three classes of wall thickness—standard weight, extra strong, and double extra strong. It is available as welded or seamless pipe of ordinary finish and dimensional tolerances, produced in sizes up to 12-in nominal OD. This pipe is used for structural and mechanical purposes. Certain applications have other requirements for size, surface finish, or straightness.

Refrigeration Pipe This pipe is also known as ice-machine pipe or ammonia pipe. It may be butt-welded, lap-welded, electric-resistance-welded, or seamless and is intended for use as a conveyor of refrigerants. This pipe is suitable for coiling, bending, and welding. The sizes commonly used range from ¾ to 2 in. The piping is produced in random and double random lengths in standard line pipe sizes and weights. Double random lengths are used as ice-rink pipe. It can be produced with plain ends, with threaded ends only, or with threaded ends and line pipe couplings, as desired.

Dry-Kiln Pipe This pipe is butt-welded, electric-resistance-welded, or seamless pipe for use in the lumber industry. It is produced in standard-weight pipe sizes of ¾, 1, and 1¼ in. Joints are designed to permit subsequent "makeup" after expansion has occurred. Dry-kiln pipe is commonly produced with threaded ends and couplings and in random lengths.

Pressure Pipe Pressure pipe is used for conveying fluids or gases at normal, subzero, or elevated temperatures and/or pressures. It generally is not subjected to external heat application. The range of sizes is ⅛-in nominal size to 36-in actual OD. It is produced in various wall thicknesses. Pressure piping is furnished in random lengths, with threaded or plain ends, as required. Pressure pipe generally is hydrostatically tested at the mill.

Line Pipe Line pipe is seamless or welded pipe produced in sizes from ⅛-in nominal OD to 48-in actual OD. It is used principally for conveying gas, oil, or water. Line pipe is produced with ends which are

Table 8.7.1   Commonly Used Piping Standards (Latest Current Standards)

ASTM Specifications	ASTM Specifications (*Cont.*)	ASTM Specifications (*Cont.*)	ASTM Specifications (*Cont.*)
*A 20	*A 516	*C 361	*F 492
*A 36	*A 522	*C 582	*F 493
*A 47	*A 524	*C 599	*F 546
*A 48	*A 530	*D 1503	*F 599
*A 53	*A 537	*D 1527	
*A 105	*A 553	D 1600	SAE Specifications
*A 106	*A 579	D 1694	
*A 120	*A 571	*D 1785	*J 513f
*A 126	*A 587	*D 2104	*J 514j
*A 134	*A 645	*D 2235	*J 518c
*A 135	*A 658	*D 2239	
*A 139	*A 671	*D 2241	ASNT Standard
*A 167	*A 672	*D 2282	
*A 179	*A 675	*D 2310	SNT TC-1A
*A 181	*A 691	*D 2412	*A3.0
*A 182	*B 12	*D 2446	*A5.1
*A 193	*B 21	*D 2447	*A5.4
*A 194	*B 26	*D 2464	*A5.5
*A 197	*B 42	*D 2465	
*A 202	*B 43	*D 2466	Copper Development Assn.
*A 203	*B 61	*D 2467	
*A 204	*B 62	*D 2468	Copper Tube Handbook
*A 211	*B 68	*D 2469	
*A 216	*B 75	*D 2513	EJMA Standards, Current
*A 217	*B 88	D 2517	
*A 225	*B 96	*D 2560	*American National Standards
*A 234	*B 98	*D 2564	
*A 240	*B 127	*D 2609	A21.14
*A 263	*B 148	*D 2657	A21.52
*A 264	*B 150	*D 2662	A58.1
*A 265	*B 152	*D 2666	B1.1
*A 268	*B 160	*D 2672	B1.20.3
*A 269	*B 161	*D 2683	B1.20.1
*A 276	*B 162	D 2737	B1.20.7
*A 278	*B 164	*D 2740	B16.1
*A 283	*B 165	*D 2837	B16.3
*A 285	*B 166	*D 2846	B16.4
*A 299	*B 167	*D 2855	B16.5
*A 302	*B 168	*D 2992	B16.9
*A 307	*B 169	*D 2996	B16.9a
*A 312	*B 209	*D 2997	B16.10
*A 320	*B 210	D 3000	B16.11
*A 325	*B 211	D 3035	B16.14
*A 333	*B 221	D 3036	B16.15
*A 334	*B 241	D 3139	B16.18
*A 335	*B 247	*D 3140	B16.20
*A 338	*B 283	D 3197	B16.21
*A 350	*B 333	*D 3261	B16.22
*A 351	*B 335	D 3287	B16.24
*A 352	*B 337	*D 3309	B16.25
*A 353	*B 345	*E 112	B16.26
*A 354	*B 361	*E 114	B16.28
*A 358	*B 366	*E 125	B16.33
*A 369	*B 402	*E 142	B16.34
*A 370	*B 407	*E 155	B16.36
*A 376	*B 443	*E 165	B16.38
A 377	*B 444	*E 186	B16.39
*A 381	*B 446	*E 272	B16.42
*A 387	*B 466	*E 280	B18.2.1
*A 395	*B 467	*E 310	B18.2.2
*A 403	*B 574	*E 446	B36.10
*A 409	*B 575	*E 709	B36.19
*A 420	*B 581	*F 336	B46.1
*A 426	*B 582	F 423	B93.11
*A 430	*B 584	*F 437	
*A 437	*B 619	*F 438	PPI Technical Reports
*A 451	*B 620	*F 439	
*A 452	*B 621	*F 441	TR-21
*A 453	*C 14	*F 442	
*A 494	*C 296	*F 443	Pipe Fabrication Institute
*A 515	*C 301	*F 491	ES-7-1962

Federal Specification

WW-P-421D

NOTE: Footnotes appear at the end of the table.

Table 8.7.1   Commonly Used Piping Standards (Latest Current Standards)   (Continued)

API Standards†	API Standards (*Cont.*)	MSS Standard Practices (*Cont.*)	MSS Standard Practices (*Cont.*)
5B, 10th ed.	*C151	SP-9	SP-82
5L, 32d ed.	*C200	SP-25	SP-83
5LE, 2d ed.	*C207	SP-42	SP-85
5LP, 4th ed.	C208	SP-43	SP-86
5LR, 4th ed.	C300	SP-44	SP-87
5LS, 12th ed.	*C301	SP-45	SP-88
5LX, 24th ed.	C302	SP-51	SP-89
*526, 2d ed.	C400	SP-53	SP-90
593, 2d ed.	*C402	SP-55	SP-91
594, 2d ed.	*C500	*SP-58	SP-92
595, 2d ed.	*C504	SP-60	SP-93
597, 3d ed.	*C600	SP-61	SP-94
599, 2d ed.	C900	SP-65	
600, 8th ed.		SP-67	**CGA**
601, 5th ed.	**ASME Codes**	SP-68	G-4.1
602, 4th ed.		SP-69	
603, 3d ed.	*ASME, Boiler and Pressure	SP-70	**NACE**
604, 4th ed.	Vessel Code	SP-71	Corrosion Data Survey
*605, 3d ed.	*Section V, incl. addenda through	SP-72	
606, 1st ed.	W82	SP-73	**NBS**
609, 2d ed.	*Section VIII, Division 1	SP-75	PS 15
*C101-1967	*Section VIII, Division 2	SP-77	
*C110	*Section IX, incl. addenda through	SP-78	NFPA Specifications, current
*C111	W82	SP-79	
*C115		SP-80	Uniform Building Code, current
*C150	**MSS Standard Practices**	SP-81	Aluminum Assn.
	SP-6		

The referenced standards are available from the listed organizations:

Standards sources			
Alum. Assn.	Aluminum Association 900 19th St., NW, Washington, DC 20006 202 862-5100	MSS	Manufacturers Standardization Society of the Valve and Fittings Industry 127 Park Street, NE, Vieona, VA 22180 703 281-6613
ANSI	American National Standards Institute, Inc. 11 West 42d St., New York, NY 10036 212 642-4900	NACE	National Association of Corrosion Engineers P.O. Box 986 Katy, TX 77450 713 492-0535
API	American Petroleum Institute 1220 L Street, NW, Washington, DC 20005-8029 202 682-8000	NIST	National Institute of Standards and Technology (U.S. Dept. of Commerce): Publications available from Superintendent of Documents
ASME	The American Society of Mechanical Engineers 345 East 47th Street, New York, NY 10017 212 705-7722		United States Government Printing Office Washington, DC 20402 202 541-3000
ASNT	American Society for Nondestructive Testing 3200 Riverside Drive, Columbus, OH 43221 614 488-7921	NFPA	National Fire Protection Association P.O. Box 9101 1 Batterymarch Park, Quincy, MA 02269-9101 617 770-3000
ASTM	ASTM International (formerly American Society for Testing and Materials) 100 Barr Harbor Drive, P.O. Box C700 West Conshohocken, PA 19428-2959 610 832-9585	PFI	Pipe Fabrication Institute Box 173, Lenore Avenue, Springdale, PA 15144-1518 412 274-4722
AWWA	American Water Works Association 6666 W. Quincy Avenue, Denver, CO 80235 303 754-7711	PPI	Plastics Pipe Institute 65 Madison Avenue, Morristown, NJ 07960-6078 No telephone listed
AWS	American Welding Society 2501 N.W. 7th Street, Miami, FL 33125 305 642-7090	SAE	Society of Automotive Engineers 400 Commonwealth Drive Warrendale, PA 15096 412 776-4841
CDA	Copper Development Association 260 Madison Avenue, New York, NY 10016 212 251-7234	UBC	Uniform Building Code International Conference of Building Officials 5360 South Workman Mill Road
(a) CGA	Compressed Gas Association 1235 Jefferson Davis Highway Arlington, VA 22202		Whittier, CA 90601 213 699-0541
(a) EJMA	Expansion Joint Manufacturers Association 25 North Broadway, North Tarrytown, NY 10591 914 382-0040		
Fed. Spec.	Federal Specification: Superintendent of Documents United States Government Printing Office Washington, DC 20402 202 541-3000		

* Indicates that the standard has been approved as an American National Standard by the American National Standards Institute.
† Including supplements to these API Standards through spring 1981.

Table 8.7.2   Major Pipe Classification and Examples of Applications

Identification of pipe	Uses
Standard	Mechanical (structural) service pipe, low-pressure service pipe, refrigeration (ice-machine) pipe, ice-rink pipe, dry-kiln pipe
Pressure	Liquid, gas, or vapor service pipe, service for elevated temperature or pressure, or both
Line	Threaded or plain end, gas, oil, and steam pipe
Water well	Reamed and drifted, water-well casing, drive pipe, driven well pipe, pump pipe, turbine-pump pipe
Oil country tubular goods	Casing, well tubing, drill pipe
Other pipe	Conduit, piles, nipple pipe, sprinkler pipe, bed-stead tubing

plain, threaded, beveled, grooved, flanged, or expanded, as required for various types of mechanical couplers, or for welded joints. When threaded ends and couplings are required, recessed couplings are normally supplied.

Water-Well Pipe   Water-well pipe is welded or seamless steel pipe used for conveying water for municipal and industrial applications. Pipelines for such purposes involve flow mains, transmission mains, force mains, water mains, or distribution mains. The mains are generally laid underground. Sizes range from ⅛- to 106-in OD in a variety of wall thicknesses. Pipe is produced with ends suitably prepared for mechanical couplers, with plain ends beveled for welding, with ends fitted with butt straps for field welding, or with bell-and-spigot joints with rubber gaskets for field joining. Pipe is produced in double random lengths of about 40 ft, single random lengths of about 20 ft, or in definite cut lengths, as specified. Wall thicknesses vary from 0.068 in for ⅛-in nominal OD to 1.00 in for 106-in actual OD.

When required, water-well pipe is produced with a specified coating or lining or both. For example, cement-mortar lining and coatings are extensively used.

Oil Country Goods   Casing is used as a structural retainer for the walls of oil or gas wells. It is also used to exclude undesirable fluids, and to confine and conduct oil or gas from productive subsurface strata to the ground level. Casing is produced in sizes 4½- to 20-in OD. Size designations refer to actual outside diameter and weight per foot. Ends are commonly threaded and furnished with couplings. When required, the ends are prepared to accommodate other types of joints.

Drill Pipe   Drill pipe is used to transmit power by rotary motion from ground level to a rotary drilling tool below the surface and also to convey flushing media to the cutting face of the tool. Drill pipe is produced in sizes 2⅜- to 6⅝-in OD. Size designations refer to actual outside diameter and weight per foot. Drill pipe is generally upset, either internally or externally, or both, and is furnished with threaded ends and couplings, threaded only, or prepared to accommodate other types of joints.

Tubing is used within the casing of oil wells to conduct oil to ground level. It is produced in sizes 1.050- to 4.500-in OD in several weights per foot. Ends are threaded and fitted with couplings and may or may not be upset externally.

Other Pipe Classifications   Rigid conduit pipe is welded or seamless pipe intended especially for the protection of electrical wiring systems. Conduit pipe is not subjected to hydrostatic tests unless so specified. It is furnished in standard-weight pipe sizes from ¼- to 6-in OD in 10-ft lengths, with plain ends or with threaded ends and couplings, as specified. Although some specifications of rigid conduit pipe list lengths to 20 ft, the National Electric Code, 1965, limits lengths to 10 ft.

Piling pipe is welded or seamless pipe for use as piles, where the cylinder section acts as a permanent load-carrying member or where it acts as a shell to form cast-in-place concrete piles. Specifications provide for the choice of three grades by minimum tensile strength, in which the sizes listed are 8⅝- to 24-in OD in a variety of wall thicknesses and in two length ranges. Ends are plain or beveled for welding.

Nipple pipe is standard-weight, extra-strong, or double-extra-strong welded or seamless pipe produced for the manufacture of pipe nipples. Standard-weight pipe with threaded ends is also used in sprinkler systems. Nipple pipe is commonly produced in random lengths with plain ends in nominal sizes ⅛- to 12-in OD. Close OD tolerances, sound welds, good threading properties, and surface cleanliness are essential in this product. It is commonly coated with oil or zinc and well protected in shipment. When reference is made to ASTM Specifications for this application, Specification A120 is generally used for diameters to 5-in OD and A53 for diameters of 5 in and over.

Standard Pipe Sizes   Standard pressure, line, and other pipe with plain ends for welding or with threaded ends is standardized in two ranges. Diameters of 12 in and less have a nominal size which represents approximately that of the inside diameter of standard-weight pipe. The nominal outside diameter is standard, regardless of weight. Increase in wall thickness results in a decrease of the inside diameter.

The standardization of pipe sizes over 12 in is based on the actual outside diameter, the wall thickness, and the weight per foot.

The principal dimensions, weights, and characteristics of commercial piping materials are summarized in Table 8.7.3.

The weights of butt-welding elbows, tees, and laterals and flanges are given in Table 8.7.4 for 3-in standard pipe. The data shown are illustrative; complete data for all standard pipe sizes are readily available in pipe manufacturers' catalogs.

The weights of welding reducers are for one size reduction and are thus only approximately correct for other reductions.

Hot-finished or cold-drawn seamless low-alloy steel tubes generally are process-annealed at temperatures between 1,200 and 1,350°F.

Austenitic stainless-steel tubes are usually annealed at temperatures between 1,800 and 2,100°F, with specific temperatures varying somewhat with each grade. This is generally followed by pickling, unless bright-annealing was done.

Mechanical Tubing

Unlike pipe and pressure tubes, mechanical tubing is generally classified by the method of manufacture and the degree of finish. Examples of classifications are "seamless hot-finished," "cold-drawn welded," "flash-in-grade," etc.

Seamless Tubes   Seamless tubes are available as either hot- or cold-finished. They are normally made in sizes from 0.187-in OD to 10.750-in OD.

Dimensions for hot-finished mechanical tubes are provided in Table 8.7.5. Dimensions for cold-finished tubes are listed in Table 8.7.6.

Welded Tubes   Welded tubes generally are produced by electric resistance methods. Where required, the welding flash is removed with a cutting tool. Industry practice normally recognizes a number of finish conditions which are summarized in Table 8.7.7.

Flash-in Type Tubing   This tubing is generally limited to applications where nothing is inserted in the tube.

Flash-Controlled Tubing   This tubing is used where moderate control of the inside diameter is required. Generally, the outside and inside diameters are specified.

For special materials, the equations listed below for weights of tubes and weights of contents of tubes are helpful.

$$\text{Weight of tube, lb/ft} = F \times 10.68 \times T \times D - T$$

where T = wall thickness, in; D = outside diameter, in; F = relative weight factor.

The weight of tube calculation is based on low-carbon steel weighing 0.2833 lb/in³ and is extended to other materials through the factor F.

Relative weight factor F

Aluminum	0.35
Brass	1.12
Cast-iron	0.91
Copper	1.14
Ferritic stainless steel	1.02
Steel	1.00
Wrought iron	0.98

Table 8.7.3 Properties of Commercial Steel Pipe

Nominal pipe size, outside diameter, in	Schedule number* a	b	c	Wall thickness, in	Inside diameter, in	Inside area, in²	Metal area, in²	Outside surface, ft²/ft	Inside surface, ft²/ft	Weight, lb/ft†	Weight of water, lb/ft	Moment of inertia, in⁴	Section modulus, in³	Radius of gyration, in
⅛ 0.405	40	Std	10S	0.049	0.307	0.0740	0.0548	0.106	0.0804	0.186	0.0321	0.00088	0.00437	0.1271
	80	XS	40S	0.068	0.269	0.0568	0.0720	0.106	0.0705	0.245	0.0246	0.00106	0.00525	0.1215
			80S	0.095	0.215	0.0364	0.0925	0.106	0.0563	0.315	0.0157	0.00122	0.00600	0.1146
¼ 0.540	40	Std	10S	0.065	0.410	0.1320	0.0970	0.141	0.1073	0.330	0.0572	0.00279	0.01032	0.1694
	80	XS	40S	0.088	0.364	0.1041	0.1250	0.141	0.0955	0.425	0.0451	0.00331	0.01230	0.1628
			80S	0.119	0.302	0.0716	0.1574	0.141	0.0794	0.535	0.0310	0.00378	0.01395	0.1547
⅜ 0.675			5S	0.065	0.710	0.396	0.1582	0.220	0.1859	0.538	0.1716	0.01197	0.0285	0.2750
			10S	0.065	0.545	0.2333	0.1246	0.177	0.1427	0.423	0.1011	0.00586	0.01737	0.2169
	40	Std	40S	0.091	0.493	0.1910	0.1670	0.177	0.1295	0.568	0.0827	0.00730	0.02160	0.2090
	80	XS	80S	0.126	0.423	0.1405	0.2173	0.177	0.1106	0.739	0.0609	0.00862	0.02554	0.1991
½ 0.840			5S	0.065	0.710	0.3959	0.1583	0.220	0.1859	0.538	0.171	0.0120	0.0285	0.2750
			10S	0.083	0.674	0.357	0.1974	0.220	0.1765	0.671	0.1547	0.01431	0.0341	0.2692
	40	Std	40S	0.109	0.622	0.304	0.2503	0.220	0.1628	0.851	0.1316	0.01710	0.0407	0.2613
	80	XS	80S	0.147	0.546	0.2340	0.320	0.220	0.1433	1.088	0.1013	0.02010	0.0478	0.2505
	160			0.187	0.466	0.1706	0.383	0.220	0.1220	1.304	0.0740	0.02213	0.0527	0.2402
		XXS		0.294	0.252	0.0499	0.504	0.220	0.0660	1.714	0.0216	0.02425	0.0577	0.2192
¾ 1.050			5S	0.065	0.920	0.655	0.2011	0.275	0.2409	0.684	0.2882	0.02451	0.0467	0.349
			10S	0.083	0.884	0.614	0.2521	0.275	0.2314	0.857	0.2661	0.02970	0.0566	0.343
	40	Std	40S	0.113	0.824	0.533	0.333	0.275	0.2157	1.131	0.2301	0.0370	0.0706	0.334
	80	XS	80S	0.154	0.742	0.432	0.435	0.275	0.1943	1.474	0.1875	0.0448	0.0853	0.321
	160			0.218	0.614	0.2961	0.570	0.275	0.1607	1.937	0.1284	0.0527	0.1004	0.304
		XXS		0.308	0.434	0.1479	0.718	0.275	0.1137	2.441	0.0641	0.0579	0.1104	0.2840
1 1.315			5S	0.065	1.185	1.103	0.2553	0.344	0.310	0.868	0.478	0.0500	0.0760	0.443
			10S	0.109	1.097	0.945	0.413	0.344	0.2872	1.404	0.409	0.0757	0.1151	0.428
	40	Std	40S	0.133	1.049	0.864	0.494	0.344	0.2746	1.679	0.374	0.0874	0.1329	0.421
	80	XS	80S	0.179	0.957	0.719	0.639	0.344	0.2520	2.172	0.311	0.1056	0.1606	0.407
	160			0.250	0.815	0.522	0.836	0.344	0.2134	2.844	0.2261	0.1252	0.1903	0.387
		XXS		0.358	0.599	0.2818	1.076	0.344	0.1570	3.659	0.1221	0.1405	0.2137	0.361
1¼ 1.660			5S	0.065	1.530	1.839	0.326	0.434	0.401	1.107	0.797	0.1038	0.1250	0.564
			10S	0.109	1.442	1.633	0.531	0.434	0.378	1.805	0.707	0.1605	0.1934	0.550
	40	Std	40S	0.140	1.380	1.496	0.669	0.434	0.361	2.273	0.648	0.1948	0.2346	0.540
	80	XS	80S	0.191	1.278	1.283	0.881	0.434	0.335	2.997	0.555	0.2418	0.2913	0.524
	160			0.250	1.160	1.057	1.107	0.434	0.304	3.765	0.458	0.2839	0.342	0.506
		XXS		0.382	0.896	0.631	1.534	0.434	0.2346	5.214	0.2732	0.341	0.411	0.472
1½ 1.900			5S	0.065	1.770	2.461	0.375	0.497	0.463	1.274	1.067	0.1580	0.1663	0.649
			10S	0.109	1.682	2.222	0.613	0.497	0.440	2.085	0.962	0.2469	0.2599	0.634
	40	Std	40S	0.145	1.610	2.036	0.799	0.497	0.421	2.718	0.882	0.320	0.326	0.623
	80	XS	80S	0.200	1.500	1.767	1.068	0.497	0.393	3.631	0.765	0.391	0.412	0.605
	160			0.281	1.338	1.406	1.429	0.497	0.350	4.859	0.608	0.483	0.508	0.581
		XXS		0.400	1.100	0.950	1.855	0.497	0.288	6.408	0.412	0.568	0.598	0.549
				0.525	0.850	0.567	2.267	0.497	0.223	7.710	0.246	0.6140	0.6470	0.5200
				0.650	0.600	0.283	2.551	0.497	0.157	8.678	0.123	0.6340	0.6670	0.4980

NOTE: See footnotes at end of table.

Table 8.7.3 Properties of Commercial Steel Pipe (Continued)

Nominal pipe size, outside diameter, in	Schedule number* a	b	c	Wall thickness, in	Inside diameter, in	Inside area, in²	Metal area, in²	Outside surface, ft²/ft	Inside surface, ft²/ft	Weight, lb/ft†	Weight of water, lb/ft	Moment of inertia, in⁴	Section modulus, in³	Radius of gyration, in
2 2.375			5S	0.065	2.245	3.96	0.472	0.622	0.588	1.604	1.716	0.315	0.2652	0.817
			10S	0.109	2.157	3.65	0.776	0.622	0.565	2.638	1.582	0.499	0.420	0.802
	40	Std	40S	0.154	2.067	3.36	1.075	0.622	0.541	3.653	1.455	0.666	0.561	0.787
	80	XS	80S	0.218	1.939	2.953	1.477	0.622	0.508	5.022	1.280	0.868	0.731	0.766
	160			0.343	1.689	2.240	2.190	0.622	0.442	7.444	0.971	1.163	0.979	0.729
		XXS		0.436	1.503	1.774	2.656	0.622	0.393	9.029	0.769	1.312	1.104	0.703
				0.562	1.251	1.229	3.199	0.622	0.328	10.882	0.533	1.442	1.2140	0.6710
				0.687	1.001	0.787	3.641	0.622	0.262	12.385	0.341	1.5130	1.2740	0.6440
2½ 2.875			5S	0.083	2.709	5.76	0.728	0.753	0.709	2.475	2.499	0.710	0.494	0.988
			10S	0.120	2.635	5.45	1.039	0.753	0.690	3.531	2.361	0.988	0.687	0.975
	40	Std	40S	0.203	2.469	4.79	1.704	0.753	0.646	5.793	2.076	1.530	1.064	0.947
	80	XS	80S	0.276	2.323	4.24	2.254	0.753	0.608	7.661	1.837	1.925	1.339	0.924
	160			0.375	2.125	3.55	2.945	0.753	0.556	10.01	1.535	2.353	1.637	0.894
		XXS		0.552	1.771	2.464	4.03	0.753	0.464	13.70	1.067	2.872	1.998	0.844
				0.675	1.525	1.826	4.663	0.753	0.399	15.860	0.792	3.0890	2.1490	0.8140
				0.800	1.275	1.276	5.212	0.753	0.334	17.729	0.554	3.2250	2.2430	0.7860
3 3.500			5S	0.083	3.334	8.73	0.891	0.916	0.873	3.03	3.78	1.301	0.744	1.208
			10S	0.120	3.260	8.35	1.274	0.916	0.853	4.33	3.61	1.822	1.041	1.196
	40	Std	40S	0.216	3.068	7.39	2.228	0.916	0.803	7.58	3.20	3.02	1.724	1.164
	80	XS	80S	0.300	2.900	6.61	3.02	0.916	0.759	10.25	2.864	3.90	2.226	1.136
	160			0.437	2.626	5.42	4.21	0.916	0.687	14.32	2.348	5.03	2.876	1.094
		XXS		0.600	2.300	4.15	5.47	0.916	0.602	18.58	1.801	5.99	3.43	1.047
				0.725	2.050	3.299	6.317	0.916	0.537	21.487	1.431	6.5010	3.7150	1.0140
				0.850	1.800	2.543	7.073	0.916	0.471	24.057	1.103	6.8530	3.9160	0.9840
3½ 4.000			5S	0.083	3.834	11.55	1.021	1.047	1.004	3.47	5.01	1.960	0.980	1.385
			10S	0.120	3.760	11.10	1.463	1.047	0.984	4.97	4.81	2.756	1.378	1.372
	40	Std	40S	0.226	3.548	9.89	2.680	1.047	0.929	9.11	4.28	4.79	2.394	1.337
	80	XS	80S	0.318	3.364	8.89	3.68	1.047	0.881	12.51	3.85	6.28	3.14	1.307
		XXS		0.636	2.728	5.845	6.721	1.047	0.716	22.850	2.550	9.8480	4.9240	1.2100
4 4.500			5S	0.083	4.334	14.75	1.152	1.178	1.135	3.92	6.40	2.811	1.249	1.562
			10S	0.120	4.260	14.25	1.651	1.178	1.115	5.61	6.17	3.96	1.762	1.549
				0.188	4.124	13.357	2.547	1.178	1.082	8.560	5.800	5.8500	2.6000	1.5250
	40	Std	40S	0.237	4.026	12.73	3.17	1.178	1.054	10.79	5.51	7.23	3.21	1.510
	80	XS	80S	0.337	3.826	11.50	4.41	1.178	1.002	14.98	4.98	9.61	4.27	1.477
	120			0.437	3.626	10.33	5.58	1.178	0.949	18.96	4.48	11.65	5.18	1.445
				0.500	3.500	9.621	6.283	1.178	0.916	21.360	4.160	12.7710	5.6760	1.4250
	160			0.531	3.438	9.28	6.62	1.178	0.900	22.51	4.02	13.27	5.90	1.416
		XXS		0.674	3.152	7.80	8.10	1.178	0.825	27.54	3.38	15.29	6.79	1.374
				0.800	2.900	6.602	9.294	1.178	0.759	31.613	2.864	16.6610	7.4050	1.3380
				0.925	2.650	5.513	10.384	1.178	0.694	35.318	2.391	17.7130	7.8720	1.3060
5 5.563			5S	0.109	5.345	22.44	1.868	1.456	1.399	6.35	9.73	6.95	2.498	1.929
			10S	0.134	5.295	22.02	2.285	1.456	1.386	7.77	9.53	8.43	3.03	1.920
	40	Std	40S	0.258	5.047	20.01	4.30	1.456	1.321	14.62	8.66	15.17	5.45	1.878
	80	XS	80S	0.375	4.813	18.19	6.11	1.456	1.260	20.78	7.89	20.68	7.43	1.839

Steel pipe dimension and property table (continuation page; columns read left-to-right: Nominal size / OD, Schedule No., Wall designation, Stainless schedule (S), Wall thickness, Inside diameter, Inside area, Metal area, Sq ft outside surface per ft, Sq ft inside surface per ft, Weight of pipe (lb/ft), Weight of water (lb/ft), Moment of inertia I, Section modulus Z, Radius of gyration r).

Nom. size / OD	Sched. No.	Desig.	Sched. (S)	Wall t	Inside dia.	Inside area	Metal area	Sq ft out./ft	Sq ft in./ft	Wt. pipe	Wt. water	I	Z	r
	120			0.500	4.563	16.35	7.95	1.456	1.195	27.04	7.09	25.74	9.25	1.799
	160			0.625	4.313	14.61	9.70	1.456	1.129	32.96	6.33	30.0	10.80	1.760
		XXS		0.750	4.063	12.97	11.34	1.456	1.064	38.55	5.62	33.6	12.10	1.722
				0.875	3.813	11.413	12.880	1.456	0.998	43.810	4.951	36.6450	13.1750	1.6860
				1.000	3.563	9.966	14.328	1.456	0.933	47.734	4.232	39.1110	14.0610	1.6520
6 / 6.625			5S	0.109	6.407	32.2	2.231	1.734	1.677	5.37	13.98	11.85	3.58	2.304
			10S	0.134	6.357	31.7	2.733	1.734	1.664	9.29	13.74	14.40	4.35	2.295
				0.219	6.187	30.100	4.410	1.734	1.620	15.020	13.100	22.6600	6.8400	2.2700
	40	Std	40S	0.280	6.065	28.89	5.58	1.734	1.588	18.97	12.51	28.14	8.50	2.245
	80	XS	80S	0.432	5.761	26.07	8.40	1.734	1.508	28.57	11.29	40.5	12.23	2.195
	120			0.562	5.501	23.77	10.70	1.734	1.440	36.39	10.30	49.6	14.98	2.153
	160			0.718	5.189	21.15	13.33	1.734	1.358	45.30	9.16	59.0	17.81	2.104
		XXS		0.864	4.897	18.83	15.64	1.734	1.282	53.16	8.17	66.3	20.03	2.060
				1.000	4.625	16.792	17.662	1.734	1.211	60.076	7.284	72.1190	21.7720	2.0200
				1.125	4.375	15.025	19.429	1.734	1.145	66.084	6.517	76.5970	23.1240	1.9850
8 / 8.625			5S	0.109	8.407	55.5	2.916	2.258	2.201	9.91	24.07	26.45	6.13	3.01
			10S	0.148	8.329	54.5	3.94	2.258	2.180	13.40	23.59	35.4	8.21	3.00
				0.219	8.187	52.630	5.800	2.258	2.150	19.640	22.900	51.3200	11.9000	2.9700
	20			0.250	8.125	51.8	6.58	2.258	2.127	22.36	22.48	57.7	13.39	2.962
	30			0.277	8.071	51.2	7.26	2.258	2.113	24.70	22.18	63.4	14.69	2.953
	40	Std	40S	0.322	7.981	50.0	8.40	2.258	2.089	28.55	21.69	72.5	16.81	2.938
	60			0.406	7.813	47.9	10.48	2.258	2.045	35.64	20.79	88.8	20.58	2.909
	80	XS	80S	0.500	7.625	45.7	12.76	2.258	1.996	43.39	19.80	105.7	24.52	2.878
	100			0.593	7.439	43.5	14.96	2.258	1.948	50.87	18.84	121.4	28.14	2.847
	120			0.718	7.189	40.6	17.84	2.258	1.882	60.63	17.60	140.6	32.6	2.807
	140			0.812	7.001	38.5	19.93	2.258	1.833	67.76	16.69	153.8	35.7	2.777
		XXS		0.875	6.875	37.1	21.30	2.258	1.800	72.42	16.09	162.0	37.6	2.757
				0.906	6.813	36.5	21.97	2.258	1.784	74.69	15.80	165.9	38.5	2.748
	160			1.000	6.625	34.454	23.942	2.258	1.734	81.437	14.945	177.1320	41.0740	2.7190
				1.125	6.375	31.903	26.494	2.258	1.669	90.114	13.838	190.6210	44.2020	2.6810
10 / 10.750			5S	0.134	10.482	86.3	4.52	2.815	2.744	15.15	37.4	63.7	11.85	3.75
			10S	0.165	10.420	85.3	5.49	2.815	2.728	18.70	36.9	76.9	14.30	3.74
				0.219	10.312	83.52	7.24	2.815	2.70	24.63	36.2	100.46	18.69	3.72
	20			0.250	10.250	82.5	8.26	2.815	2.683	28.04	35.8	113.7	21.16	3.71
	30			0.307	10.136	80.7	10.07	2.815	2.654	34.24	35.0	137.5	25.57	3.69
	40	Std	40S	0.365	10.020	78.9	11.91	2.185	2.623	40.48	34.1	160.8	29.90	3.67
	60	XS	80S	0.500	9.750	74.7	16.10	2.815	2.553	54.74	32.3	212.0	39.4	3.63
	80			0.593	9.564	71.8	18.92	2.815	2.504	64.33	31.1	244.9	45.6	3.60
	100			0.718	9.314	68.1	22.63	2.815	2.438	76.93	29.5	286.2	53.2	3.56
	120			0.843	9.064	64.5	26.24	2.815	2.373	89.20	28.0	324	60.3	3.52
				0.875	9.000	63.62	27.14	2.815	2.36	92.28	27.6	333.46	62.04	3.50
	140			1.000	8.750	60.1	30.6	2.815	2.091	104.13	26.1	368	68.4	3.47
	160			1.125	8.500	56.7	34.0	2.815	2.225	115.65	24.6	399	74.3	3.43
				1.250	8.250	53.45	37.31	2.815	2.16	126.832	23.2	428.17	79.66	3.39
				1.500	7.750	47.15	43.57	2.815	2.03	148.19	20.5	478.59	89.04	3.31

Table 8.7.3 Properties of Commercial Steel Pipe (Continued)

Nominal pipe size, outside diameter, in	Schedule number* a	b	c	Wall thickness, in	Inside diameter, in	Inside area, in²	Metal area, in²	Outside surface, ft²/ft	Inside surface, ft²/ft	Weight, lb/ft†	Weight of water, lb/ft	Moment of inertia, in⁴	Section modulus, in³	Radius of gyration, in
12 12.750	20		5S	0.156	12.438	121.4	6.17	3.34	3.26	20.99	52.7	122.2	19.20	4.45
	30		10S	0.180	12.390	120.6	7.11	3.34	3.24	24.20	52.2	140.5	22.03	4.44
				0.250	12.250	117.9	9.84	3.34	3.21	33.38	51.1	191.9	30.1	4.42
				0.330	12.090	114.8	12.88	3.34	3.17	43.77	49.7	248.5	39.0	4.39
		Std	40S	0.375	12.000	113.1	14.58	3.34	3.14	49.56	49.0	279.3	43.8	4.38
	40			0.406	11.938	111.9	15.74	3.34	3.13	53.53	48.5	300	47.1	4.37
		XS	80S	0.500	11.750	108.4	19.24	3.34	3.08	65.42	47.0	362	56.7	4.33
	60			0.562	11.626	106.2	21.52	3.34	3.04	73.16	46.0	401	62.8	4.31
	80			0.687	11.376	101.6	26.04	3.34	2.978	88.51	44.0	475	74.5	4.27
				0.750	11.250	99.40	28.27	3.34	2.94	96.2	43.1	510.7	80.1	4.25
	100			0.843	11.064	96.1	31.5	3.34	2.897	107.20	41.6	562	88.1	4.22
				0.875	11.000	95.00	32.64	3.34	2.88	110.9	41.1	578.5	90.7	4.21
	120			1.000	10.750	90.8	36.9	3.34	2.814	125.49	39.3	642	100.7	4.17
				1.125	10.500	86.6	41.1	3.34	2.749	139.68	37.5	701	109.9	4.13
	140			1.250	10.250	82.50	45.16	3.34	2.68	153.6	35.8	755.5	118.5	4.09
	160			1.312	10.126	80.5	47.1	3.34	2.651	160.27	34.9	781	122.6	4.07
14 14.000			5S	0.156	13.688	147.20	6.78	3.67	3.58	23.0	63.7	162.6	23.2	4.90
			10S	0.188	13.624	145.80	8.16	3.67	3.57	27.7	63.1	194.6	27.8	4.88
				0.210	13.580	144.80	9.10	3.67	3.55	30.9	62.8	216.2	30.9	4.87
				0.219	13.562	144.50	9.48	3.67	3.55	32.2	62.6	225.1	32.2	4.87
	10			0.250	13.500	143.1	10.80	3.67	3.53	36.71	62.1	255.4	36.5	4.86
				0.281	13.438	141.80	12.11	3.67	3.52	41.2	61.5	285.2	40.7	4.85
	20			0.312	13.376	140.5	13.42	3.67	3.50	45.68	60.9	314	44.9	4.84
				0.344	13.312	139.20	14.76	3.67	3.48	50.2	60.3	344.3	49.2	4.83
	30	Std		0.375	13.250	137.9	16.05	3.67	3.47	54.57	59.7	373	53.3	4.82
	40			0.437	13.126	135.3	18.62	3.67	3.44	63.37	58.7	429	61.2	4.80
				0.469	13.062	134.00	19.94	3.67	3.42	67.8	58.0	456.8	64.3	4.79
		XS		0.500	13.000	132.7	21.21	3.67	3.40	72.09	57.5	484	69.1	4.78
	60			0.593	12.814	129.0	24.98	3.67	3.35	84.91	55.9	562	80.3	4.74
		XXS		0.625	12.750	127.7	26.26	3.67	3.34	89.28	55.3	589	84.1	4.73
	80			0.750	12.500	122.7	31.2	3.67	3.27	106.13	53.2	687	98.2	4.69
	100			0.937	12.126	115.5	38.5	3.67	3.17	130.73	50.0	825	117.8	4.63
	120			1.093	11.814	109.6	44.3	3.67	3.09	150.67	47.5	930	132.8	4.58
	140			1.250	11.500	103.9	50.1	3.67	3.01	170.22	45.0	1017	146.8	4.53
	160			1.406	11.188	98.3	55.6	3.67	2.929	189.12	42.6	1127	159.6	4.48
16 16.000			5S	0.165	15.670	192.90	8.21	4.19	4.10	28	83.5	257	32.2	5.60
			10S	0.188	15.624	191.70	9.34	4.19	4.09	32	83.0	292	36.5	5.59
	10			0.250	15.500	188.7	12.37	4.19	4.06	42.05	81.8	384	48.0	5.57
	20	Std		0.312	15.376	185.7	15.38	4.19	4.03	52.36	80.5	473	59.2	5.55
	30	XS		0.375	15.250	182.6	18.41	4.19	3.99	62.58	79.1	562	70.3	5.53
	40			0.500	15.000	176.7	24.35	4.19	3.93	82.77	76.5	732	91.5	5.48
	60			0.656	14.688	169.4	31.6	4.19	3.85	107.50	73.4	933	116.6	5.43
	80			0.843	14.314	160.9	40.1	4.19	3.75	136.46	69.7	1157	144.6	5.37

Nom. size (OD)	Sched. No.	Desig.	Wall thickness	Inside dia.	Inside area	Metal area	Sq ft outside surface per ft	Sq ft inside surface per ft	Weight pipe (lb/ft)	Weight water (lb/ft)	I (in⁴)	Section modulus (in³)	Radius gyration (in)
16 (16.000) *(cont.)*	100		1.031	13.938	152.6	48.5	4.19	3.65	164.83	66.1	1365	170.6	5.30
	120		1.218	13.564	144.5	56.6	4.19	3.55	192.29	62.6	1556	194.5	5.24
	140		1.437	13.126	135.3	65.7	4.19	3.44	223.64	58.6	1760	220.0	5.17
	160		1.593	12.814	129.0	72.1	4.19	3.35	245.11	55.9	1894	236.7	5.12
18 (18.000)		5S	0.165	17.670	245.20	9.24	4.71	4.63	31	106.2	368	40.8	6.31
		10S	0.188	17.624	243.90	10.52	4.71	4.61	36	105.7	417	46.4	6.30
	10		0.250	17.500	240.5	13.94	4.71	4.58	47.39	104.3	549	61.0	6.28
	20		0.312	17.376	237.1	17.34	4.71	4.55	59.03	102.8	678	75.5	6.25
		Std	0.375	17.250	233.7	20.76	4.71	4.52	70.59	101.2	807	89.6	6.23
	30		0.437	17.126	230.4	24.11	4.71	4.48	82.06	99.9	931	103.4	6.21
		XS	0.500	17.00	227.0	27.49	4.71	4.45	93.45	98.4	1053	117.0	6.19
	40		0.562	16.876	223.7	30.8	4.71	4.42	104.75	97.0	1172	130.2	6.17
	60		0.750	16.500	213.8	40.6	4.71	4.32	138.17	92.7	1515	168.3	6.10
	80		0.937	16.126	204.2	50.2	4.71	4.22	170.75	88.5	1834	203.8	6.04
	100		1.156	15.688	193.3	61.2	4.71	4.11	207.96	83.7	2180	242.2	5.97
	120		1.375	15.250	182.6	71.8	4.71	3.99	244.14	79.2	2499	277.6	5.90
	140		1.562	14.876	173.8	80.7	4.71	3.89	274.23	75.3	2750	306	5.84
	160		1.781	14.438	163.7	90.7	4.71	3.78	308.51	71.0	3020	336	5.77
20 (20.000)		5S	0.188	19.634	302.40	11.70	5.24	5.14	40	131.0	574	57.4	7.00
		10S	0.218	19.564	300.60	13.55	5.24	5.12	46	130.2	663	66.3	6.99
	10		0.250	19.500	298.6	15.51	5.24	5.11	52.73	129.5	757	75.7	6.98
	20	Std	0.375	19.250	291.0	23.12	5.24	5.04	78.60	126.0	1114	111.4	6.94
	30	XS	0.500	19.000	283.5	30.6	5.24	4.97	104.13	122.8	1457	145.7	6.90
	40		0.593	18.814	278.0	36.2	5.24	4.93	122.91	120.4	1704	170.4	6.86
	60		0.812	18.376	265.2	48.9	5.24	4.81	166.40	115.0	2257	225.7	6.79
			0.875	18.250	261.6	52.6	5.24	4.78	178.73	113.4	2409	240.9	6.77
	80		1.031	17.938	252.7	61.4	5.24	4.70	208.87	109.4	2772	277.2	6.72
	100		1.281	17.438	238.8	75.3	5.24	4.57	256.10	103.4	3320	332	6.63
	120		1.500	17.00	227.0	87.2	5.24	4.45	296.37	98.3	3760	376	6.56
	140		1.750	16.500	213.8	100.3	5.24	4.32	341	92.6	4200	422	6.48
	160		1.968	16.064	202.7	111.5	5.24	4.21	379.01	87.9	4590	459	6.41
22 (22.000)		5S	0.188	21.624	367.3	12.88	5.76	5.66	44	159.1	766	69.7	7.71
		10S	0.218	21.564	365.2	14.92	5.76	5.65	51	158.2	885	80.4	7.70
	10		0.250	21.500	363.1	17.18	5.76	5.63	58	157.4	1010	91.8	7.69
	20	Std	0.375	21.250	354.7	25.48	5.76	5.56	87	153.7	1490	135.4	7.65
	30	XS	0.500	21.000	346.4	33.77	5.76	5.50	115	150.2	1953	177.5	7.61
			0.625	20.750	338.2	41.97	5.76	5.43	143	146.6	2400	218.2	7.56
			0.750	20.500	330.1	50.07	5.76	5.37	170	143.1	2829	257.2	7.52
	60		0.875	20.250	322.1	58.07	5.76	5.30	197	139.6	3245	295.0	7.47
	80		1.125	19.750	306.4	73.78	5.76	5.17	251	132.8	4029	366.3	7.39
	100		1.375	19.250	291.0	89.09	5.76	5.04	303	126.2	4758	432.6	7.31
	120		1.625	18.750	276.1	104.02	5.76	4.91	354	119.6	5432	493.8	7.23
	140		1.875	18.250	261.6	118.55	5.76	4.78	403	113.3	6054	550.3	7.15
	160		2.125	17.750	247.4	132.68	5.76	4.65	451	107.2	6626	602.4	7.07

Table 8.7.3 Properties of Commercial Steel Pipe (Continued)

Nominal pipe size, outside diameter, in	Schedule number* a	b	c	Wall thickness, in	Inside diameter, in	Inside area, in²	Metal area, in²	Outside surface, ft²/ft	Inside surface, ft²/ft	Weight, lb/ft†	Weight of water, lb/ft	Moment of inertia, in⁴	Section modulus, in³	Radius of gyration, in
24 24.000			5S	0.218	23.564	436.1	16.29	6.28	6.17	55	188.9	1152	96.0	8.41
	10			0.250	23.500	434	18.65	6.28	6.15	63.41	188.0	1316	109.6	8.40
	20	Std		0.375	23.250	425	27.83	6.28	6.09	94.62	183.8	1943	161.9	8.35
		XS		0.500	23.000	415	36.9	6.28	6.02	125.49	180.1	2550	212.5	8.31
	30			0.562	22.876	411	41.4	6.28	5.99	140.80	178.1	2840	237.0	8.29
				0.625	22.750	406	45.9	6.28	5.96	156.03	176.2	3140	261.4	8.27
	40			0.687	22.626	402	50.3	6.28	5.92	171.17	174.3	3420	285.2	8.25
				0.750	22.500	398	54.8	6.28	5.89	186.24	172.4	3710	309	8.22
				0.875	22.250	388.6	63.54	6.28	5.83	216	168.6	4256	354.7	8.18
	60			0.968	22.064	382	70.0	6.28	5.78	238.11	165.8	4650	388	8.15
	80			1.218	21.564	365	87.2	6.28	5.65	296.36	158.3	5670	473	8.07
	100			1.531	20.938	344	108.1	6.28	5.48	367.40	149.3	6850	571	7.96
	120			1.812	20.376	326	126.3	6.28	5.33	429.39	141.4	7830	652	7.87
	140			2.062	19.876	310	142.1	6.28	5.20	483.13	134.5	8630	719	7.79
	160			2.343	19.314	293	159.4	6.28	5.06	541.94	127.0	9460	788	7.70
26 26.000	10			0.250	25.500	510.7	19.85	6.81	6.68	67	221.4	1646	126.6	9.10
				0.312	25.376	505.8	25.18	6.81	6.64	86	219.2	2076	159.7	9.08
		Std		0.375	25.250	500.7	30.19	6.81	6.61	103	217.1	2478	190.6	9.06
	20	XS		0.500	25.000	490.9	40.06	6.81	6.54	136	212.8	3259	250.7	9.02
				0.625	24.750	481.1	49.82	6.81	6.48	169	208.6	4013	308.7	8.98
				0.750	24.500	471.4	59.49	6.81	6.41	202	204.4	4744	364.9	8.93
				0.875	24.250	461.9	69.07	6.81	6.35	235	200.2	5458	419.9	8.89
				1.000	24.000	452.4	78.54	6.81	6.28	267	196.1	6149	473.0	8.85
				1.125	23.750	443.0	87.91	6.81	6.22	299	192.1	6813	524.1	8.80
28 28.000	10			0.250	27.500	594.0	21.80	7.33	7.20	74	257.3	2098	149.8	9.81
				0.312	27.376	588.6	27.14	7.33	7.17	92	255.0	2601	185.8	9.79
		Std		0.375	27.250	583.2	32.54	7.33	7.13	111	252.6	3105	221.8	9.77
	20	XS		0.500	27.000	572.6	43.20	7.33	7.07	147	248.0	4085	291.8	9.72
	30			0.625	26.750	562.0	53.75	7.33	7.00	183	243.4	5038	359.8	9.68
				0.750	26.500	551.6	64.21	7.33	6.94	218	238.9	5964	426.0	9.64
				0.875	26.250	541.2	74.56	7.33	6.87	253	234.4	6865	490.3	9.60
				1.000	26.000	530.9	84.82	7.33	6.81	288	230.0	7740	552.8	9.55
				1.125	25.750	520.8	94.98	7.33	6.74	323	225.6	8590	613.6	9.51
30 30.000	10		5S	0.250	29.500	683.4	23.37	7.85	7.72	79	296.3	2585	172.3	10.52
			10S	0.312	29.376	677.8	29.19	7.85	7.69	99	293.7	3201	213.4	10.50
		Std		0.375	29.250	672.0	34.90	7.85	7.66	119	291.2	3823	254.8	10.48
	20	XS		0.500	29.000	660.5	46.34	7.85	7.59	158	286.2	5033	335.5	10.43
	30			0.625	28.750	649.2	57.68	7.85	7.53	196	281.3	6213	414.2	10.39
	40			0.750	28.500	637.9	68.92	7.85	7.46	234	276.6	7371	491.4	10.34
				0.875	28.250	620.7	80.06	7.85	7.39	272	271.8	8494	466.2	10.30
				1.000	28.000	615.7	91.11	7.85	7.33	310	267.0	9591	639.4	40.30
				1.125	27.750	604.7	102.05	7.85	7.26	347	262.2	10653	710.2	10.22

Nominal pipe size, in	Outside diameter, in	Schedule	Wall thickness, in	Inside diameter, in	Inside area, in²	Area of metal, in²	Sq ft outside surface per ft	Sq ft inside surface per ft	Weight per ft pipe, lb	Weight per ft water, lb	Moment of inertia, in⁴	Section modulus, in³	Radius of gyration, in
32	32.000		0.250	31.500	779.2	24.93	8.38	8.25	85	337.8	3141	196.3	11.22
		10	0.312	31.376	773.2	31.02	8.38	8.21	106	335.2	3891	243.2	11.20
		Std	0.375	31.250	766.9	37.25	8.38	8.18	127	332.5	4656	291.0	11.18
		20 XS	0.500	31.000	754.7	49.48	8.38	8.11	168	327.2	6140	383.8	11.14
		30	0.625	30.750	742.5	61.59	8.38	8.05	209	321.9	7578	473.6	11.09
		40	0.688	30.624	736.6	67.68	8.38	8.02	230	319.0	8298	518.6	11.07
			0.750	30.500	730.5	73.63	8.38	7.98	250	316.7	8990	561.9	11.05
			0.875	30.250	718.3	85.52	8.38	7.92	291	311.6	10372	648.2	11.01
			1.000	30.000	706.8	97.38	8.38	7.85	331	306.4	11680	730.0	10.95
			1.125	29.750	694.7	109.0	8.38	7.79	371	301.3	13023	814.0	10.92
34	34.000		0.250	33.500	881.2	26.50	8.90	8.77	90	382.0	3773	221.9	11.93
		10	0.312	33.376	874.9	32.99	8.90	8.74	112	379.3	4680	275.3	11.91
		Std	0.375	33.250	867.8	39.61	8.90	8.70	135	376.2	5597	329.2	11.89
		20 XS	0.500	33.000	855.3	52.62	8.90	8.64	179	370.8	7385	434.4	11.85
		30	0.625	32.750	841.9	65.53	8.90	8.57	223	365.0	9124	536.7	11.80
		40	0.688	32.624	835.9	72.00	8.90	8.54	245	362.1	9992	587.8	11.78
			0.750	32.500	829.3	78.34	8.90	8.51	266	359.5	10829	637.0	11.76
			0.875	32.250	816.4	91.01	8.90	8.44	310	354.1	12501	735.4	11.72
			1.000	32.000	804.2	103.67	8.90	8.38	353	348.6	14114	830.2	11.67
			1.125	31.750	791.3	116.13	8.90	8.31	395	343.2	15719	924.7	11.63
36	36.000		0.250	35.500	989.7	28.11	9.42	9.29	96	429.1	4491	249.5	12.64
		10	0.312	35.376	982.2	34.95	9.42	9.26	119	426.1	5565	309.1	12.62
		Std	0.375	35.250	975.8	42.01	9.42	9.23	143	423.1	6664	370.2	12.59
		20 XS	0.500	35.000	962.1	55.76	9.42	9.16	190	417.1	8785	488.1	12.55
		30	0.625	34.750	948.3	69.50	9.42	9.10	236	411.1	10872	604.0	12.51
		40	0.750	34.500	934.7	83.01	9.42	9.03	282	405.3	12898	716.6	12.46
			0.875	34.250	920.6	96.50	9.42	8.97	328	399.4	14903	827.9	12.42
			1.000	34.000	907.9	109.96	9.42	8.90	374	393.6	16851	936.2	12.38
			1.125	33.750	894.2	123.19	9.42	8.83	419	387.9	18763	1042.4	12.34
42	42.000		0.250	41.500	1352.6	32.82	10.99	10.86	112	586.4	7126	339.3	14.73
		Std	0.375	41.250	1336.3	49.08	10.99	10.80	167	579.3	10627	506.1	14.71
		XS	0.500	41.000	1320.2	65.18	10.99	10.73	222	572.3	14037	668.4	14.67
		20	0.625	40.750	1304.1	81.28	10.99	10.67	276	565.4	17373	827.3	14.62
		30	0.750	40.500	1288.2	97.23	10.99	10.60	330	558.4	20689	985.2	14.59
		40	1.000	40.000	1256.6	128.81	10.99	10.47	438	544.8	27080	1289.5	14.50
			1.250	39.500	1225.3	160.03	10.99	10.34	544	531.2	33233	1582.5	14.41
			1.500	39.000	1194.5	190.85	10.99	10.21	649	517.9	39181	1865.7	14.33

*Schedule numbers: Standard-weight pipe and schedule 40 are the same in all sizes through 10 in; from 12 in through 24 in, standard weight pipe has a wall thickness of ⅜ in. Extra-strong-weight pipe and schedule 80 are the same in all sizes through 8 in; from 8 in through 24 in, extra-strong-weight pipe has a wall thickness of ½ in. Double-extra-strong-weight pipe has no corresponding schedule number.

a: ANSI B36.10 steel pipe schedule numbers
b: ANSI B36.10 steel pipe nominal wall thickness designation
c: ANSI B36.19 stainless steel pipe schedule numbers
† The ferritic steels may be about 5% less and the austenitic stainless steels about 2% greater than the values shown in this table, which are based on weights for carbon steel.

NOTE: The following formulas are used in the computation of the values shown in the table:

$$\text{Weight of pipe per foot (pounds)} = 10.6802t(D - t)$$
$$\text{Weight of water per foot (pounds)} = 0.3405d^2$$
$$\text{Square feet outside surface per foot} = 0.2618D$$
$$\text{Square feet inside surface per foot} = 0.2618d$$
$$\text{Inside area (square inches)} = 0.785d^2$$
$$\text{Area of metal (square inches)} = 0.785(D^2 - d^2)$$
$$\text{Moment of inertia (inches}^4) = 0.0491(D^4 - d^4) = A_m R_g^2$$
$$\text{Section modulus (inches}^3) = \frac{0.0982(D^4 - d^4)}{D}$$
$$\text{Radius of gyration (inches)} = 0.25\sqrt{D^2 + d^2}$$

where A_m = area of metal, in²; d = inside diameter, in; D = outside diameter, in; R_g = radius of gyration, in; t = pipe wall thickness, in.

Table 8.7.4 Weight of Standard Pipe Fittings and Materials, 3-in Size (3.500-in OD)*

Schedule no.	40	80	160	XXS
Wall designation	Std	XS		XXS
Thickness, in	0.216	0.300	0.438	0.600
Pipe, lb/ft	7.58	10.25	14.32	18.58
Water, lb/ft	3.20	2.86	2.35	1.80

Welding fittings

	40	80	160	XXS
L.R. 90° elbow	**4.6** 0.8	**6.1** 0.8	**8.4** 0.8	**10.7** 0.8
S.R. 90° elbow	**3** 0.5	**4** 0.5		
L.R. 45° elbow	**2.4** 0.3	**3.2** 0.3	**4.4** 0.3	**5.4** 0.3
Tee	**7.4** 0.8	**9.5** 0.8	**12.2** 0.8	**14.8** 0.8
Lateral	**13** 1.8	**19** 1.8		
Reducer	**2.2** 0.3	**2.9** 0.3	**3.7** 0.3	**4.7** 0.3
Cap	**1.4** 0.5	**1.8** 0.5	**3.5** 0.5	**3.7** 0.5

Insulation

Temperature range, °F		100–199	200–299	300–399	400–499	500–599	600–699	700–799	800–899	900–999	1,000–1,099	1,100–1,200
85% magnesia calcium silicate	Nom. thick., in	1	1	1½	2	2	2½	3	3	3	3½	3½
	lb/ft	1.25	1.25	2.08	3.01	3.01	4.07	5.24	5.24	5.24	6.65	6.65
Combination	Nom. thick., in						2½	3	3	3	3	3½
	lb/ft						5.07	6.94	6.94	6.94	9.17	9.17
†Asbestos fiber-sodium silicate	Nom. thick., in	1	1	1	1½	1½	2	2	3	3	3½	3½
	lb/ft	1.61	1.61	1.61	2.74	2.74	3.98	3.98	6.99	6.99	8.99	8.99

Pressure rating, lb/in²

	Cast iron		Steel						
	125	250	150	300	400	600	900	1,500	2,500
Flanges									
Screwed or slip-on	9 / 1.5	17 / 1.5	9 / 1.5	17 / 1.5	20 / 1.5	20 / 1.5	37 / 1.5	61 / 1.5	102 / 1.5
Welding neck			11 / 1.5	19 / 1.5	27 / 1.5	27 / 1.5	38 / 1.5	61 / 1.5	113 / 1.5
Lap joint			9 / 1.5	17 / 1.5	19 / 1.5	19 / 1.5	36 / 1.5	60 / 1.5	99 / 1.5
Blind	10 / 1.5	19 / 1.5	10 / 1.5	20 / 1.5	24 / 1.5	24 / 1.5	38 / 1.5	61 / 1.5	105 / 1.5
Flanged fittings									
S.R. 90° elbow	26 / 3.9	46 / 4	32 / 3.9	53 / 4		67 / 4.1	98 / 4.3	150 / 4.6	
L.R. 90° elbow	30 / 4.3	50 / 4.3	40 / 4.3	63 / 4.3					
45° elbow	22 / 3.5	41 / 3.6	28 / 3.5	46 / 3.6		60 / 3.8	93 / 3.9	135 / 4	
Tee	39 / 5.9	67 / 6	52 / 5.9	81 / 6		102 / 6.2	151 / 6.5	238 / 6.9	
Valves									
Flanged bonnet gate	66 / 7	112 / 7.4	70 / 4	125 / 4.4		155 / 4.8	260 / 5	410 / 5.5	
Flanged bonnet globe or angle	56 / 7.2	121 / 7.6	60 / 4.3	95 / 4.5		155 / 4.8	225 / 5	495 / 5.5	
Flanged bonnet check	46 / 7.2	100 / 7.6	60 / 4.3	70 / 4.4		120 / 4.8	150 / 4.9	440 / 5.8	
Pressure seal bonnet–gate							208 / 3	235 / 3.2	
Pressure seal bonnet–globe							135 / 2.5	180 / 3	

* The data shown are illustrative; complete data for all standard pipe sizes are readily available in manufacturers' catalogs.

† 16 lb/ft³ density. (Existing installations only.)

NOTES: Boldface type is weight in pounds. Lightface type beneath weight is weight factor for insulation.

Insulation thicknesses and weights are based on average conditions and do not constitute a recommendation for specific thicknesses of materials. Insulation weights are based on 85% magnesia and hydrous calcium silicate at 11 lb/ft³. The listed thicknesses and weights of combination covering are the sums of the inner layer of diatomaceous earth at 21 lb/ft³ and the outer layer at 11 lb/ft³.

Insulation weights include allowances for wire, cement, canvas, bands and paint, but no special surface finishes.

To find the weight of covering on flanges, valves, or fittings, multiply the weight factor by the weight per foot of covering used on straight pipe.

Valve weights are approximate. When possible, obtain weights from the manufacturer.

Cast-iron valve weights are for flanged end valves; steel weights for welding end valves.

All flanged fitting, flanged valve and flange weights include the proportional weight of bolts or studs to make up all joints.

Table 8.7.5   Diameter and Wall-Thickness Tolerances for Seamless Hot-finished Mechanical Tubing of Carbon and Alloy Steel (AISI)*

Specified size, OD, in	Ratio of wall thickness to OD	OD tolerance		Wall thicknesses tolerance, %							
				0.109 in and under		Over 0.109 to 0.172 in, incl.		Over 0.172 to 0.203 in, incl.		Over 0.203 in	
		Over	Under	Over	Under	Over	Under	Over	Under	Over	Under
Under 3	All wall thicknesses	0.023	0.023	16.5	16.5	15	15	14	14	12.5	12.5
3–5½, excl.	All wall thicknesses	0.031	0.031	16.5	16.5	15	15	14	14	12.5	12.5
5½–8 excl.	All wall thicknesses	0.047	0.047					14	14	12.5	12.5
8–10¾, incl.	5% and over	0.047	0.047							12.5	12.5
8–10¾, incl.	Under 5%	0.063	0.063							12.5	12.5

* The common range of sizes of hot-finished tubes is 1½ to and including 10¾ in outside diameter with wall thickness not less than 0.095 in (no. 13 BWG) or 3% or more of the outside diameter. For sizes under 1½ or over 10¾ in outside diameter, the tolerances are commonly negotiated between the purchaser and producer.
SOURCE: AISI, "Steel Products Manual."

Weight of contents of tube, lb/ft $= G \times 0.3405 \times (D - 2T)^2$

where G = specific gravity of contents; T = tube wall thickness, in; D = tube outside diameter, in.

The weight per foot of steel pipe is subject to the tolerances listed in Table 8.7.8.

The designation *sink-draw tubes* is specified where close control over the outer diameter is required with normal tolerance applying to the wall thickness. Smoothness of the inside surface is not controlled, except that the flash is generally controlled to a height of 0.005 or 0.010 in maximum.

Mandrel-drawn tubes usually are normalized after welding by passing the tubes through a continuous atmosphere-controlled furnace. After descaling, the tubes are cold-drawn through a die with a mandrel on the inside of the tube. These tubes provide maximum control over surface finish, outside or inside diameters, and wall thickness. The normalizing heat treatment removes the effects of welding and provides a uniform microstructure around the tube circumference.

The different finish classifications may result in substantial differences in the mechanical properties of the steel material.

Typical examples for low-carbon steel material are given in Table 8.7.9. Differences in carbon content and other chemistry, heat treatment, etc., may significantly change these typical values.

Other Tubing Types   Among other tube classifications are sanitary tubing usually made of 18% Cr-8% Ni stainless steel and available as seamless or welded tubing. This tubing is used extensively in the dairy, beverage, and food industries. Sanitary tubing is generally available in sizes from 1- to 4-in OD. It may be furnished either hot- or cold-finished. The tubes are normally annealed at temperatures above 1,900°F.

Some welded tube is also produced by fusion-welding methods utilizing either the inert-gas tungsten-arc-welding or gas-shielded consumable metal-arc-welding process. This tubing is generally more expensive than the resistance-welded types.

The butt-welded cold-finished tubes are made from hot-rolled or cold-rolled strip and fusion-welded. This tubing is usually furnished as sink-drawn or mandrel-drawn.

Butt-welded tubing is made in heavier wall thicknesses than the resistance-welded tube.

Several tubing materials used in the automobile industry are covered by specifications of the Society of Automotive Engineers, "SAE Handbook."

Pressure Tubing

Pressure-tube applications commonly involve external heat applications, as in boilers or superheaters.

Pressure tubing is produced to the actual outside diameter and minimum wall or average wall thickness specified by the purchaser. Pressure tubing may be hot- or cold-finished.

The wall thickness is normally given in decimal parts of an inch rather than as a fraction or gage number. When gage numbers are given without reference to a gage system, Birmingham wire gage (BWG) is implied.

Pressure tubing is usually made from steel produced by the basic oxygen or electric-furnace processes.

Seamless pressure tubing may be either hot-finished or cold-drawn. Cold-drawn steel tubing is frequently process-annealed at temperatures

above 1,200°F. To ensure quality, maximum hardness values are frequently specified. For example, in ASME Specification SA192, Specification for Seamless Carbon Steel Boiler Tubes for High-Pressure Service, the following maximum hardness values are given. Specifications for boiler tubes generally reference the ASME Standards such as SA178, SA192, SA210, etc.

Boiler tubes	Brinell hardness no. (tubes 0.200 in and over in wall thickness)	Rockwell hardness no. (tubes less than 0.200 in in wall thickness)
Hot-finished tubes	137	B77
Cold-drawn (normalized) tubes	125	B72

Piping Fabrication Methods   Commercial steel pipe is furnished most commonly as seamless pipe. Seamless pipe can be produced by (1) piercing, (2) hollow forging, and (3) forging, turning, and boring. Commercial pipe is also produced by welding involving (1) electric resistance welding, (2) electric fusion welding, or (3) submerged arc welding. Some mills are prepared to extrude small-diameter pipe and tubing in a variety of geometric shapes or to cast large-diameter steel pipe.

Electric-Fusion Welding   Flat plate, known as **skelp,** is prepared in proper width and thickness for the desired pipe inside and outside diameters. It is then charged into an electric furnace and, when the proper welding temperature has been reached, is drawn through a funnel-shaped die so shaped that the plate is gradually formed into the shape of a tube, with the edges of the plate being forced squarely together and fused. The formed pipe then passes through a series of rolls in which it is sized or drawn to final dimensions.

Electric-Resistance Welding   For pipes or tubes sized 4-in (10.2-cm) OD and under, strip is fed onto a set of forming rolls which consists of horizontal and vertical rollers so placed as to gradually form the flat strip into a tube. The tube form then passes to the welding electrodes. The electrodes are copper disks connected to the secondary of a revolving transformer assembly. The copper-disk electrodes make contact on each side of the seam of the tube form; a flow of current takes place across the seam, and temperature is raised to the welding point. Outside flash is removed by a cutting tool as the tube leaves the electrodes; inside flash is removed either by an air hammer or by passing a mandrel through the welded tube after the tube has been cooled.

Submerged-Arc Electric Welding   This process is used for pipes from 24- to 36-in (61.0- to 9.1.4-cm) OD. Flat plate is first pressed into a U and later into an O shape. The O shape is placed in an automatic welder and backed up on the inside by a water-cooled copper shoe. Two electrodes in close proximity are used. The electrodes are not in actual contact with the pipe. The current passes from one electrode through a granular flux and across the gap in the pipe to the second electrode. The high temperature of the arc heats the edges of the plate;

Table 8.7.6 Diameter and Wall-Thickness Tolerances for Seamless Cold-Worked Mechanical Tubing of Carbon and Alloy Steel (AISI)*

Size, OD, in	Unannealed of finish-annealed				Soft annealed or normalized				Quenched and tempered				Wall thickness all conditions, %	
	OD, in		ID, in		OD, in		ID, in		OD, in		ID, in			
	Over	Under	Over	Under	Over	Under	Over	Under	Over	Under	Over	Under	Over	Under
3/16–1/2, excl.†‡	0.004	0		0.005	0.006	0.002			0.010	0.010			15	15
1/2–1½, excl.†‡§¶	0.005	0		0.010	0.008	0.002	0.002	0.008	0.015	0.015	0.015	0.015	10	10
1½–3½, excl.†‡§¶	0.010	0	0.005	0.015	0.015	0.005	0.005	0.015	0.030	0.030	0.030	0.030	10	10
3½–5½, excl.§¶	0.015	0	0.035	0.035	0.023	0.007	0.015	0.025	0.045	0.045	0.045	0.045	10	10
5½–8, excl.¶ wall less than 5% OD	0.030	0.030	0.025	0.025	0.060	0.060	0.070	0.070					10	10
5½–8, excl., wall from 5 to 7.5% OD	0.020	0.020	0.015	0.030	0.040	0.040	0.050	0.050					10	10
5½–8, excl.§ wall over 7.5% OD	0.030	0	0.050	0.050	0.045	0.015	0.037	0.053					10	10
8–10¾, incl.¶ wall less than 5% OD	0.045	0.045	0.040	0.040									10	10
8–10¾, incl., wall from 5 to 75% OD	0.035	0.035	0.015	0.040									10	10
8–10¾, incl.§ wall over 7.5% OD	0.045	0											10	10

* For tolerances closer than those indicated, availability, and applicable tolerances for tubing less than 3/16 in OD or larger than 10¾-in OD, the producer should be consulted.

† For those tubes with inside diameter less than ½ in (or less than ⅝ in when the wall thickness is more than 20% of the outside diameter), which are not commonly drawn over a mandrel. Note § is not applicable. Unless otherwise agreed upon by the purchaser and producer, the wall thickness may vary 15% over and under that specified, and the inside diameter is governed by the outside diameter and wall-thickness tolerances shown.

‡ For tubes with inside diameter less than ½ in (or less than ⅝ in when the wall thickness is more than 20% of the outside diameter), which can be produced by the rod or bar mandrel process, the tolerances are as shown in the above table except that the wall-thickness tolerances are 10% over and under the specified wall thickness.

§ Many tubes with inside diameter less than 50% of outside diameter, or with wall thickness more than 25% of outside diameter, or with wall thickness over 1¼ in, or weighing more than 90 lb/ft are difficult to draw over a mandrel. Unless otherwise agreed upon by the purchaser and producer the inside diameter may vary over or under by an amount equal to 10% of the wall thickness and the wall thickness may vary 12½% over and under that specified.

¶ Tubing having a wall thickness less than 3% of the outside diameter cannot be straightened properly without a certain amount of distortion. Consequently, such tubes, while having an average outside diameter and inside diameter within the tolerances shown in the above table, require an ovality tolerance of 0.5% over and under nominal outside diameter, this being in addition to the tolerances indicated in the above table.

Source: AISI, "Steel Products Manual."

Table 8.7.7   Finish Classifications Normally Recognized as Welded Mechanical Tubing

	Type	Condition
A	Flash-in, hot-rolled	Made by longitudinally forming and welding hot-rolled strip; the interior welding flash remains in place
B	Flash-in, cold-rolled	Made by longitudinally forming and welding cold-rolled strip; the interior flash remains in place
C	Flash controlled–0.010 in max, hot-rolled	Same as A, except that the interior flash is partially removed so that the height remaining does not exceed 0.010 in
D	Flash controlled–0.010 in max, cold-rolled	Same as C, except that cold-rolled strip has been used
E	Flash controlled–0.005 in max, hot-rolled	Same as C, except that flash height on tube inside does not exceed 0.005 in
F	Flash controlled–0.005 in max, cold-rolled	Same as E, except that cold-rolled strip has been used
G	Sink-drawn, hot-rolled	Tubing with flash controlled to 0.010 in max, which has been descaled and cold-drawn through a die to cold-finish the exterior surface
H	Sink-drawn, cold-rolled	Tubing with flash controlled to 0.010 in max, which has been cold-drawn through a die to cold-finish the exterior surface
I	Mandrel-drawn	Cold-rolled tubing with the interior flash removed and drawn through a die and over a mandrel to cold-finish the exterior and interior surfaces

Table 8.7.8   Weight Tolerances of Steel Piping

Specification	Size	Tolerance, %	
ASTM A53	Std wt	+5	−5
	XS wt	+5	−5
ASTM A120	XXS wt	+10	−10
ASTM A106	Sch 10–120	+6.5	−3.5
	Sch 140–160	+10	−3.5
ASTM A335	12 in and under	+6.5	−3.5
	Over 12 in	+10	−5
ASTM A312 ASTM A376	12 in and under	+6.5	−3.5
API 5L	Std. wt Reg wt XS wt XXS wt	+10	−3.5
	Spec. plain end pipe	+10	−5

a welding rod placed just over the seam is thereby melted and metal is deposited in the groove. After the outside weld has been made, the pipe is conveyed to an inside welder where a similar operation is carried on, except that no backup shoe is needed.

Seamless Tubing and Pipe   A heated billet is brought into contact with tapered revolving rolls in such a way that the billet is pulled into the space allowed between the rolls. A piercing mandrel is placed in this space; the soft center of the billet makes it possible for the rolls to draw the billet over the mandrel, producing a hollow shell. When the billet has entirely passed over the mandrel, it is in the form of a thick-walled seamless tube. The heavy-walled tube is then passed to a rolling mill which reduces the tube to pipe of proper outside diameter and wall thickness.

The method of fabrication described above is limited as to diameter and thickness. For seamless alloy tubes and for heavy-wall carbon-steel tubes or pipe, a process known as **cupping and drawing** is frequently used. A circular flat plate of proper diameter and thickness is heated, placed in a hydraulic press, and pressed by a ram through a die. The cup so formed is reheated and pressed through a smaller die, thus elongating the cup so that it becomes a short cylinder with one closed end. This short cylinder is then placed in a horizontal drawbench, and with reheating as necessary, is pushed by a ram through dies of successively smaller diameters until the desired outer diameter is reached.

Forged, Turned, and Bored Tubing   In this process, the ingot is heated and forged to a rough cylindrical shape, oversize in both diameter and length. The forging is then placed in a lathe and the outside turned down to the desired outer diameter. Rough ends are then removed so that the finally desired length is obtained. The cylinder is then placed in a boring mill, and the inside bored out until the desired wall thickness is secured. Because of the relatively high cost, this process is now rarely used.

Hollow-Forged Pipe and Tubing   In this process, ingots are cast and their ends cropped; then they are placed in a furnace and heated to a specified temperature. The heated ingot is placed in a press where it is pierced. This hollow cylinder, open at one end, is then descaled and drawn over a mandrel on a horizontal drawbench. The closed end is then burned off, and the hollow forging is chemically descaled. Following this, the forging is straightened, placed in a lathe, and the outer diameter machined to a true dimension. The inside is dressed to remove scale, but no machining is done on the inside.

Carbon-steel piping is most frequently used as manufactured in accordance with ASTM Specifications A106 and A53 (or ASME Specifications SA106 and SA53). The chemical compositions of these two materials are identical except for the deoxidation practice which applies to the A106 pipe. Both are subjected to physical tests, but those for A106 are more rigorous. A53 and A106 are made in grades A and B; grade B has higher strength properties but is less ductile and, for this reason, grade A is permitted only for cold bending or close coiling. When carbon steel is intended for use in welded construction at temperatures in excess of 775°F (413°C), consideration should be given to the possibility of graphite formation.

Chromium-molybdenum steel has been used for temperatures up to 1,100°F (593°C). In the small diameters, the material is usually available in the seamless construction; because of the inability of the seamless mills to fabricate large-diameter and heavy-walled pipe, it may be necessary to resort to the more expensive hollow-forged or forged-and-bored piping for higher pressures and temperatures. The material for a high-temperature piping system should be selected after a careful review of technical and economic considerations; the following is intended only as being indicative of recent and current practice.

For temperatures up to 1,000°F, 1¼% Cr-½% Mo (A335, grade P11) is used. For temperatures from 950 to 1050°F, 2¼% Cr-1% Mo (A335, P22) generally is used. Where there is a combination of high temperatures

Table 8.7.9   Typical Mechanical Properties of Resistance-Welded Mechanical Carbon-Steel Tubing

Type	Yield strength, lb/in²	Tensile strength, lb/in²	Elongation, %	Hardness, Rockwell B
Flash-in or flash-controlled, hot-rolled	48,000	58,000	29	68
Flash-in or flash-controlled, cold-rolled	68,000	76,000	17	84
Normalized	34,000	52,000	39	61
Sink-drawn	73,000	76,000	20	84
Mandrel-drawn	80,000	83,000	15	86

and erosive action, 5% Cr-½% Mo (A335, grade 5) or other more highly chromium-molybdenum or chromium stainless steels have been used.

Stainless-steel piping is available in a variety of compositions, most popular of which are ASTM A213, grade TP304 (16% Cr–8% Ni), and ASTM A213, grade TP316 (18% Cr–12% Ni and 3% Mo). For high-temperature service, type 34 stainless steels are used (18% Cr–8% Ni and stabilized with columbium). This material may be used up to 1,200°F (649°C); particular care must be given to choice of welding filler metal to avoid brittleness in the welds.

The permissible stress values for a large variety of piping materials at low and elevated temperatures are provided in a basic reference: ASME B31.3.

PIPE FITTINGS

The various major piping materials are also produced in the form of standard fittings. Among the more widely used are wrought-steel fittings, welded-steel fittings, cast-steel fittings, cast-iron fittings, ductile-iron fittings, malleable-iron fittings, brass and copper fittings, aluminum fittings, etc. Other major nonferrous piping materials are also produced in the form of cast and wrought fittings.

Cast-iron, ductile-iron, and malleable-iron fittings are made by conventional founding methods for a variety of joints including bell-and-spigot, flanged, and mechanical (gland-type), or other proprietary joint designs.

Schedule Designations Over 100 years ago piping was designated as standard, extra-strong, and double extra-strong. There was no provision for thin-walled pipe, and no intervening standard thicknesses between the three schedules, which covered too great a spread to be economical without intermediate weights. Table 8.7.3 lists piping as a function of the schedule number which is given, approximately, by the following relationship: Schedule no. $= 1,000 \, P/(SE)$, where P is operating pressure, lb/in² gage, S is allowable stress, lb/in², and E is the quality factor (Tables 8.7.10 and 8.7.11).

Table 8.7.10 Basic Casting Quality Factors E_c
(Abstracted in part from ASME B31.3 with permission)

Spec. no.	Description	E_c*
Iron		
FS-WW-P421c	Centrifugally cast pipe	1.00
A 377	Centrifugally cast pipe	1.00
A 47	Malleable iron castings	1.00
A 48	Gray iron castings	1.00
A 126	Gray iron castings	1.00
A 197	Cupola malleable iron castings	1.00
A 278	Gray iron castings	1.00
A 338	Malleable iron castings	1.00
A 395	Ductile and ferritic ductile iron castings	0.80
A 571	Austenitic ductile iron castings	0.80
Carbon steel		
A 216	Carbon steel castings	0.80
A 352	Ferritic steel castings	0.80
Low and intermediate alloy steel		
A 426	Centrifugally cast pipe	1.00
A 217	Martensitic stainless and alloy castings	0.80
A 352	Ferritic steel castings	0.80
Stainless steel		
A 451	Centrifugally cast pipe	0.90
A 452	Centrifugally cast pipe	0.85
A 351	Austenitic steel castings	0.80
Copper and copper alloy		
B 61	Steam bronze castings	0.80
B 62	Composition bronze castings	0.80
B 148	Al–bronze and Si–Al–bronze castings	0.80
B 584	Copper alloy castings	0.80
Nickel and nickel alloy		
A 494	Nickel and nickel alloy castings	0.80
Aluminum alloy		
B 26, temper F	Aluminum alloy castings	1.00
B 26, temper T6, T71	Aluminum alloy castings	0.80

* Refer to ASME B31.3 for details.

Table 8.7.11 Basic Quality Factors for Longitudinal Weld Joints in Pipes, Tubes, and Fittings, E_j
(Abstracted in part from ASME B31.3 with permission)

Spec. no.	Class material	Description	E_j*
Carbon steel			
API 5L		Seamless	1.00
		Electric resistance welded	0.85
		Electric fusion welded, double butt, straight or spiral	0.85
		Furnace butt welded	0.60
A 53	Type S	Seamless	1.00
	Type E	Electric resistance welded	0.85
	Type F	Furnace butt welded	0.60
A 105		Forgings and fittings	1.00
A 106		Seamless	1.00
A 120		Seamless	1.00
		Electric resistance welded	0.85
		Furnace butt welded	0.60
A 134		Electric fusion welded, single butt, straight or spiral	0.80
A 135		Electric resistance welded	0.85
A 139		Electric fusion welded, straight or spiral	0.80
A 179		Seamless	1.00
A 181		Forgings and fittings	1.00
A 211		Spiral welded	0.75
A 234		Seamless and welded fittings	1.00
A 333		Seamless	1.00
		Electric resistance welded	0.85
A 334		Seamless	1.00
A 350		Forgings and fittings	1.00
A 369		Seamless	1.00
A 381		Electric fusion welded, 100% radiograph	1.00
		Electric fusion welded, spot radiograph	0.90
		Electric fusion welded, as manufactured	0.85
A 420		Welded fittings, 100% radiograph	1.00
A 524		Seamless	1.00
A 587		Electric resistance welded	0.85
A 671	12, 22	Electric fusion welded, 100% radiograph	1.00
	13, 23	Electric fusion welded, double butt	0.85
A 672	12, 22	Electric fusion welded, 100% radiograph	1.00
	13, 23	Electric fusion welded, double butt	0.85
A 691	12, 22	Electric fusion welded, 100% radiograph	1.00
	13, 23	Electric fusion welded, double butt	0.85
Low and intermediate alloy steel			
A 182		Forgings and fittings	1.00
A 234		Seamless and welded fittings	1.00
A 333		Seamless	1.00
		Electric resistance welded	0.85
A 334		Seamless	1.00

Table 8.7.11   Basic Quality Factors for Longitudinal Weld Joints in Pipes, Tubes, and Fittings, E_j (Continued)

Spec. no.	Class material	Description	E_j*
Low and intermediate alloy steel (Cont.)			
A 335		Seamless	1.00
A 350		Forgings and fittings	1.00
A 369		Seamless	1.00
A 420		Welded fittings, 100% radiograph	1.00
A 671	12, 22	Electric fusion welded, 100% radiograph	1.00
	13, 23	Electric fusion welded, double butt	0.85
A 672	12, 22	Electric fusion welded, 100% radiograph	1.00
	13, 23	Electric fusion welded, double butt	0.85
A 691	12, 22	Electric fusion welded, 100% radiograph	1.00
	13, 23	Electric fusion welded, double butt	0.85
Stainless steel			
A 182		Forgings and fittings	1.00
A 268		Seamless	1.00
		Electric fusion welded, document butt	0.85
		Electric fusion welded, single butt	0.80
A 269		Seamless	1.00
		Electric fusion welded, double butt	0.85
		Electric fusion welded, single butt	0.80
A 312		Seamless	1.00
		Electric fusion welded, double butt	0.85
		Electric fusion welded, single butt	0.80
A 358	1, 3, 4	Electric fusion welded, 100% radiograph	1.00
	5	Electric fusion welded, spot radiograph	0.90
	2	Electric fusion welded, double butt	0.85
A 376		Seamless	1.00
A 403		Seamless fitting	1.00
		Welded fitting, 100% radiograph	1.00
		Welded fitting, double butt	0.85
		Welded fitting, single butt	0.80
A 409		Electric fusion welded, double butt	0.85
		Electric fusion welded, single butt	0.80
A 430		Seamless	1.00
Copper and copper alloy			
B 42		Seamless	1.00
B 43		Seamless	1.00
B 68		Seamless	1.00
B 75		Seamless	1.00
B 88		Seamless	1.00
B 466		Seamless	1.00
B 467		Electric resistance welded	0.85
		Electric fusion welded, double butt	0.85
		Electric fusion welded, single butt	0.80
Nickel and nickel alloy			
B 160		Forgings and fittings	1.00
B 161		Seamless	1.00
B 164		Forgings and fittings	1.00
B 165		Seamless	1.00
B 166		Forgings and fittings	1.00
B 167		Seamless	1.00

Spec. no.	Class material	Description	E_j*
Nickel and nickel alloy (Cont.)			
B 366		Seamless and welded fittings	1.00
B 407		Seamless	1.00
B 444		Seamless	1.00
B 619		Electric resistance welded	0.85
		Electric fusion welded, double butt	0.85
		Electric fusion welded, single butt	0.80
Unalloyed titanium			
B 337		Seamless	1.00
		Electric fusion welded, double butt	0.85
Aluminum alloy			
B 210		Seamless	1.00
B 241		Seamless	1.00
B 247		Forgings and fittings	1.00
B 345		Seamless	1.00
B 361		Seamless fittings	1.00

* Refer to ASME B31.3 for details.

Commercial sizes of steel pipe are known by their nominal inside diameter (ID) from ⅛ in (0.3175 cm) to 12 in (30.5 cm). Above 12-in ID, pipe is usually known by its outside diameter (OD). All classes of pipe of a given nominal size have the same OD, the extra thickness for different weights being on the inside.

Thickness of Pipe   The following notes, covering power piping systems, have been abstracted from Part 2 of the Code for Power Piping (ASME B31.1).

For inspection purposes, the minimum thickness of pipe wall to be used for piping at different pressures and for temperatures not exceeding those for the various materials listed in ASME B31.3 shall be determined by the formula

$$t_m = \frac{PD}{2(SE + Py)} + A \qquad (8.7.1)$$

where t_m = minimum pipe-wall thickness, in, allowable on inspection; P = maximum internal service pressure, lb/in² gage (plus water-hammer allowance in case of cast-iron conveying liquids); D = OD of pipe, in; S = maximum allowable stress in material due to internal pressure, lb/in²; E = quality factor, y = a coefficient, values for which are listed in Table 8.7.12; A = allowance for threading, mechanical strength, and corrosion, in, with values of A listed in Table 8.7.13.

Table 8.7.12   Values of y
(Interpolate for intermediate values) (ASME B31.1)

	Temp, °F (°C)					
	900 (482) and below	950 (510)	1,000 (538)	1,050 (566)	1,100 (593)	1,150 (621) and above
Ferritic steels	0.4	0.5	0.7	0.7	0.7	0.7
Austenitic steels	0.4	0.4	0.4	0.4	0.5	0.7

Table 8.7.13   Values of A
(ASME B31.1)

Type of pipe	A, in
Cast-iron pipe, centrifugally cast	0.14
Cast-iron, pit-cast	0.18
Threaded-steel, wrought-iron, or nonferrous pipe:	
⅜ in and smaller	0.05
½ in and larger	Depth of thread
Grooved-steel, wrought-iron or nonferrous pipe	Depth of groove
Plain-end steel, wrought-iron or tube:	
1 in and smaller	0.05
1¼ in and larger	0.065
Plain-end nonferrous pipe or tube	0.000

The thickness of ductile-iron pipe conveying liquid may be taken from Table 8.7.18, using the pressure class next higher than the maximum internal service pressure in pounds per square inch. Where ductile-iron pipe is used for steam service, the thickness should be calculated by Eq. (8.7.1).

Plain-end pipe includes pipe joined by flared compression couplings, lapped joints, and by welding, i.e., by any method that does not reduce the wall thickness of the pipe at the joint.

Physical and Chemical Properties of Pipes, Tubes, etc.   The design of piping for operation above 750°F (399°C) presents many problems not encountered at lower temperatures. For the properties of steel applicable to high-temperature service (as well as to ordinary service) for pipes, tubes, fittings, bolting material, etc., see Sec. 6. For a discussion of creep properties, see Sec. 5.

Piping of thickness designed in accordance with Eq. (8.7.1) may be used for any combination of pressure and temperature for which S and E values are listed in Table 8.7.11 and ASME B31.3. The following summarizes piping industry practice.

Steam Pressures above 250 lb/in^2 (1,724 kPa), and Not above 2,500 lb/in^2 (17,238 kPa), Temperatures Not above 1,100°F (593°C)   For pressures in excess of 100 lb/in^2 (690 kPa), the pipe may be seamless steel (A106), (A312), (A335), or (A376); or electric-fusion-welded steel (A691); or forged-and-bored steel (A369); or automatic-welded steel (A312). For pressures between 250 and 600 lb/in^2 (9,224 and 22,137 N/m^2) the pipe may be seamless steel (A106) or (A53); electric-fusion-welded steel (A155); electric-resistance-welded steel (A135) or (A53). For pressures of 250 lb/in^2 (1,724 kPa) and lower and for service up to 750°F (399°C), any of the following may be used: electric-fusion-welded steel (A134) or (A139); electric-resistance-welded steel (A135); seamless or welded steel (A53). Grade A seamless pipe (A106) or (A53); or grade A electric-welded pipe (A53), (A135), or (A139) is used for close coiling or cold bending. Pipe permissible for services specified may be used for temperatures higher than 750°F (399°C), unless otherwise prohibited, if the S and E values of Table 8.7.11 and ASME B31.3 are used when calculating the required wall thickness.

Because of several failures in seam-welded 1¼% Cr–½% Mo or 2¼% Cr–1% Mo piping produced to ASTM Specification A155 operating at temperatures above 950°F, a preference has developed for seamless piping in these applications. Nevertheless, in general, seam-welded piping has provided entirely satisfactory service in high-temperature applications above 950°F.

Valves and fittings must have flange openings or welded ends, and valves must have external stem threads. Valves must be of cast or forged steel or may be fabricated from plate and pipe. Valves of nonferrous materials are generally cast or forged. Forged and cast-steel threaded valves and fittings may be used up to 300 lb/in^2 and 500°F for 3 (2) [1½] in pipe, and pressure from 250 to 400 (400 to 600) [600 to 2,500] lb/in^2. Malleable-iron threaded fittings (300 lb/in^2) may be used for pressures not greater than 300 lb/in^2 and temperatures not over 500°F. Valves 8 in and larger should have the bypass of at least ¾ in, commercial size. See Manufacturers Standardization Society SP-45 for recommended size of bypass valves. Welded fittings may be used of the same material and thickness as the pipe to which they are to be connected.

Steam Pressures from 125 to 250 lb/in^2 (862 to 1,724 kPa), Temperature Not above 450°F (232°C)   Pipe may be electric-fusion-welded steel (A134 or A139). Copper and brass may be used if the temperature does not exceed 406°F. Cast iron may also be used. For close coiling or cold bending, grade A seamless steel (A53); or grade A electric-welded steel (A53), (A135), or (A139) is suitable. Pipe permissible for this service may be used for temperatures above 450°F (232°C) if the proper S and E are used in calculating the pipe-wall thickness.

Valves below 3 in may have inside stem screws. Stop valves 8 in and over must be bypassed. Bodies, bonnets, and yokes are of cast iron, malleable iron, steel, bronze, brass, or Monel. Flanged-steel fittings must conform to the class 300 ANSI Standard B16.5; if of cast iron, to the class 250 ANSI Standard B16.1; or, for threaded fittings, to the ANSI Standard B16.4. Malleable-iron threaded fittings must conform to the class 300 ANSI B16.3 standard, except that the class 150 ANSI

Standard B16.3 may be used for pressures not greater than 150 lb/in^2. Welded fittings may be used.

Steam Pressures from 25 to 125 lb/in^2 (172 to 802 kPa) Temperatures Not above 450°F   Pipe may be of steel, ductile iron, copper, or brass; valve bodies of cast iron, malleable iron, ductile iron, steel, or brass. Fittings are of class 125 or class 150 American Standard cast iron with screwed or flanged ends, or of ductile or malleable iron with screwed ends.

Steam Pressures 25 lb/in^2 (172 kPa) and Less, Temperature up to 450°F   Pipe may be of steel, brass, copper, or cast iron. Flanged fittings conform to the class 25 ANSI Standard B16.1. Screwed fittings are of the class 125 ANSI Standard B16.4 or of the class 150 ANSI Standard B16.3 for malleable iron, or conform to B16.15 for cast bronze. Welded-steel fittings are extensively used.

Pipe coils are made from any of the commercial sizes of iron, steel, brass, and copper pipe and tubing. Limiting center-to-center dimensions, to which pipe coils can be fabricated in sizes ¾ to 2 in, are given in Table 8.7.14. Steel tubing cannot be bent to the absolute limits of brass or copper.

Table 8.7.14   Center-to-Center Dimensions of Pipe Coils

Nominal pipe size, in	Recommended and advisable minimum, in	
	Schedule 40	Schedule 80
¾	3½	2½
1	4	3
1¼	5	4
1½	6	5
2	8	6

Seamless mechanical tubing is obtainable in outside diameters ranging from ¼ to 10¾ in and in wall thickness from 20 gage to 2 in (0.091 to 5.08 cm). Oval, square, rectangular, and other special shapes can be obtained in various sizes and wall thicknesses. Mechanical tubing is available either hot-finished or cold-drawn, but is furnished principally cold-drawn. It is readily adaptable to varied treatment by expansion, cupping, tapering, swaging, flanging, coiling, welding, and similar manipulations. Typical of the many uses are aircraft tubing, automobile axle housings, driveshafts, drive-shaft housings, tie rods, steering columns, piston rods and pins, gear rings, roller-bearing cases and cones, cylinders for various purposes, machine parts, sleeves, bushings, spacers, surgical instruments, and hypodermic needles. Table 8.7.15 lists weights and dimensions of round seamless-steel tubes for sizes that have by common usage become standard. Detailed information on mechanical tubing for any particular applications can be obtained from manufacturers.

Dimensions and weights of **condenser** and **heat-exchanger tubes** are given in Table 8.7.16 and of **boiler** tubes in Table 8.7.17.

Spiral Pipe   Spiral pipe is strong lightweight steel pipe with a single continuous welded helical seam from end to end stiffening it throughout. It is listed in sizes 6- to 42-in ID (15.24- to 106.7-cm), in various thicknesses, and in lengths up to 40 ft (12.19 m). It is used for high- and low-pressure water lines, vacuum lines, exhaust-steam lines, low-pressure air lines, sand and gravel slurry conveying and similar services. It is also used extensively by the petroleum industry, for oil and gas lines, for low-pressure steam lines, etc.

Spiral pipe may be asphalt-coated or galvanized. The pipe is designed for special joints, flanges, and lightweight fittings, but the ANSI flanges and fittings can be furnished, if desired.

The **sleeve-type coupling** illustrated in Fig. 8.7.1 is particularly suitable for plain-end pipe and is widely used. A gasket is used to make a tight joint. Advantages of this coupling are low cost, the use of unskilled labor in making the connections, and the fact that small changes in alignment and grade can be made with regular straight lengths of pipe by a movement in the coupling. This type of coupling is used extensively in long oil lines.

Table 8.7.15 Approx Weight of Round Seamless Cold-Finished Carbon-Steel Mechanical Tubing, lb/ft*

(Carbon 0.25% max. Standard sizes for warehouse stocks random lengths. United States Steel Corporation)

Wall thickness		OD, in																		
in	g or in	⅜	½	⅝	¾	⅞	1	1⅛	1¼	1⅜	1½	1⅝	1¾	1⅞	2	2⅛	2¼	2⅜	2½
0.035	20 g	0.127	0.174	0.221	0.267	0.314	0.361	0.407	0.454	—	0.548								
0.049	18 g	0.171	0.236	0.301	0.367	0.432	0.498	0.563	0.629	0.694	0.759	0.825	0.890	—	1.02				
0.058	17 g	—	0.274	0.351	0.429		0.584												
0.065	16 g	0.215	0.302	0.389	0.476	0.562	0.649	0.736	0.823	0.909	0.996	1.08	1.17	1.26	1.34	1.43	1.52	1.60	1.69
0.083	14 g	—	0.370	0.480	0.591	0.702	0.813	0.924	1.03	1.15	1.26	—	—	—	1.70				
0.095	13 g	—	0.411	0.538	0.665	0.791	0.918	1.05	1.17	1.30	1.43	1.55	1.68	1.81	1.93	2.06	2.19	2.31	2.44
0.109	12 g	—	0.455	0.601	0.746	0.892	1.04	1.18	1.33	—	1.62								
0.120	11 g	—	0.487	0.647	0.807	0.968	1.13	1.29	1.45	1.61	1.77	1.93	2.09	2.25	2.41	2.57	2.73	2.89	3.05
0.134	10 g	—	—	0.703	0.882	—	1.24	1.42	1.60	1.78	1.96	2.13	2.31	—	2.67				
0.156	5/32	—	—	0.781	0.990	1.20	1.41	1.61	1.82	2.03	2.24	2.45	2.66	2.86	3.07	3.28	3.49	3.70	3.91
0.188	3/16	—	—	—	1.13	1.38	1.63	1.88	2.13	2.38	2.63	2.89	3.14	3.39	3.64	3.89	4.14	4.39	4.64
0.219	7/32	—	—	—	—	1.53	1.83	2.12	2.41	2.70	3.00	3.29	3.58	3.87	4.17	4.46	4.75	5.04	5.34
0.250	¼	—	—	—	1.34	1.67	2.00	2.34	2.67	3.00	3.34	3.67	4.01	4.34	4.67	5.01	5.34	5.67	6.01
0.281	9/32	—	—	—	—	—	—	—	2.91	—	3.66	4.03	4.41	4.78	5.16	—	5.91	6.28	6.66
0.313	5/16	—	—	—	—	—	—	—	3.13	3.55	3.97	4.39	4.80	5.22	5.64	6.06	6.48	6.89	7.31
0.375	⅜	—	—	—	—	—	—	—	3.50	4.01	4.51	5.01	5.51	6.01	6.51	7.01	7.51	8.01	8.51
0.438	7/16	—	—	—	—	—	—	—	—	—	4.97	5.53	6.14	6.72	7.31	7.89	8.48	9.06	9.65
0.500	½	—	—	—	—	—	—	—	—	—	5.34	6.01	6.68	7.34	8.01	8.68	9.35	10.0	10.7
0.625	⅝	—	—	—	—	—	—	—	—	—	—	—	—	—	9.18	—	10.8	11.7	12.5

To obtain weights in kg/m, multiply tabular values shown by 1.42.

* Other standard sizes, in certain standard wall thicknesses, vary by ⅛-in increments for 2½ to 3½ in; by ¼-in increments from 3½ to 7½ in; ½-in increments from 7½ to 10½ in OD. There are also standard sizes for every ¹⁄₁₆ in from ⅜ to 1⅝ in OD.

Table 8.7.16 Steel Condenser and Heat-Exchanger Tubes

(Dimensions and weights. United States Steel Corporation)

OD, in	Thickness, in	Avg wall			Min wall		
		ID, in	Area of metal, in²*	Weight per ft, lb†	ID, in	Area of metal, in²*	Weight per ft, lb†
½	0.035	0.430	0.0511	0.1738	0.423	0.0558	0.1898
	0.050	0.400	0.0707	0.2403	0.390	0.0769	0.2614
	0.065	0.370	0.0888	0.3020	0.357	0.0963	0.3272
⅝	0.035	0.555	0.0649	0.2205	0.548	0.0709	0.2412
	0.050	0.525	0.0903	0.3071	0.515	0.0985	0.3348
	0.065	0.495	0.1144	0.3888	0.482	0.1243	0.4227
	0.085	0.455	0.1442	0.4902	0.438	0.1561	0.5308
¾	0.050	0.650	0.1100	0.3738	0.640	0.1201	0.4082
	0.065	0.620	0.1399	0.4755	0.607	0.1524	0.5181
	0.085	0.580	0.1776	0.6037	0.563	0.1928	0.6556
	0.095	0.560	0.1955	0.6646	0.541	0.2119	0.7204
⅞	0.050	0.775	0.1296	0.4406	0.765	0.1417	0.4817
	0.065	0.745	0.1654	0.5623	0.732	0.1805	0.6136
	0.085	0.705	0.2110	0.7172	0.688	0.2296	0.7804
	0.095	0.685	0.2328	0.7914	0.666	0.2530	0.8599
1	0.050	0.900	0.1492	0.5073	0.890	0.1633	0.5551
	0.065	0.870	0.1909	0.6491	0.857	0.2086	0.7090
	0.085	0.830	0.2443	0.8306	0.813	0.2663	0.9052
	0.095	0.810	0.2701	0.9182	0.791	0.2940	0.9994
1¼	0.050	1.150	0.1885	0.6408	1.140	0.2065	0.7020
	0.065	1.120	0.2420	0.8226	1.107	0.2647	0.8999
	0.085	1.080	0.3111	1.058	1.163	0.3397	1.155
	0.095	1.060	0.3447	1.172	1.041	0.3761	1.278
	0.105	1.040	0.3777	1.284	1.019	0.4117	1.399
1½	0.050	1.400	0.2278	0.7743	1.390	0.2497	0.8488
	0.065	1.370	0.2930	0.9962	1.357	0.3209	1.091
	0.085	1.330	0.3779	1.285	1.313	0.4053	1.378
	0.095	1.310	0.4193	1.426	1.291	0.4581	1.557
	0.105	1.290	0.4602	1.564	1.269	0.5024	1.708
1¾	0.065	1.620	0.3441	1.170	1.606	0.3803	1.293
	0.085	1.580	0.4446	1.512	1.561	0.4907	1.668
	0.095	1.560	0.4939	1.679	1.539	0.5448	1.852
	0.105	1.540	0.5426	1.845	1.517	0.5902	2.007
	0.120	1.510	0.6145	2.089	1.484	0.6766	2.300
2	0.065	1.870	0.3951	1.343	1.856	0.4370	1.486
	0.085	1.830	0.5114	1.738	1.811	0.5649	1.920
	0.095	1.810	0.5685	1.933	1.789	0.6275	2.133
	0.105	1.790	0.6251	2.125	1.767	0.6896	2.344
	0.120	1.760	0.7087	2.409	1.734	0.7812	2.656

* Multiply values shown by 0.0645 to obtain areas in cm².
† Multiply values shown by 1.42 to obtain weights in kg/m.

Table 8.7.17 Seamless-Steel Boiler Tubes

Outside diam, in	Thickness BWG	Thickness in	Mfg.* wt, lb/ft	Outside diam, in	Thickness BWG	Thickness in	Mfg.* wt, lb/ft	Outside diam, in	Thickness BWG	Thickness in	Mfg.* wt, lb/ft
1	13	0.095	1.037	2½	12	0.109	3.171	4½	10	0.134	7.103
	12	0.109	1.168		11	0.120	3.457		9	0.148	7.817
	11	0.120	1.263		10	0.134	3.835		8	0.165	8.702
	10	0.134	1.384		9	0.148	4.207		7	0.180	9.447
1¼	13	0.095	1.323	2¾	12	0.109	3.504	5	9	0.148	8.720
	12	0.109	1.502		11	0.120	3.823		8	0.165	9.711
	11	0.120	1.628		10	0.134	4.244		7	0.180	10.550
	10	0.134	1.793		9	0.148	4.658		6	0.203	11.810
1½	13	0.095	1.619	3	12	0.109	3.838	5½	9	0.148	9.622
	12	0.109	1.836		11	0.120	4.189		8	0.165	10.720
	11	0.120	1.994		10	0.134	4.652		7	0.180	11.650
	10	0.134	2.201		9	0.148	5.110		6	0.203	13.050
1¾	13	0.095	1.910	3¼	11	0.120	4.555	6	7	0.180	12.750
	12	0.109	2.169		10	0.134	5.061		6	0.203	14.290
	11	0.120	2.360		9	0.148	5.061		5	0.220	15.410
	10	0.134	2.610		8	0.165	6.179		4	0.238	16.640
2	13	0.095	2.201	3½	11	0.120	4.921				
	12	0.109	2.503		10	0.134	5.469				
	11	0.120	2.726		9	0.148	6.012				
	10	0.034	3.018		8	0.065	6.683				
2¼	13	0.095	2.492	4	10	0.134	6.286				
	12	0.109	2.837		9	0.148	6.915				
	11	0.120	3.092		8	0.165	7.693				
	10	0.134	3.427		7	0.180	8.347				

* Multiply values shown by 1.42 to obtain weights in kg/m.
SOURCE: United States Steel Corporation.

Fig. 8.7.1 Sleeve type, plain end coupling.

CAST-IRON AND DUCTILE-IRON PIPE

Cast-iron pipe was extensively produced and used in certain applications until about 1960, but now shares the field with **ductile-iron pipe**. Both provide excellent service when appropriately applied, although ductile-iron pipe has largely supplanted cast-iron pipe when employed because of its superior strength and ductility, i.e., when adapted to underground and submerged service because of its superior corrosion resistance compared to steel pipe. Ductile-iron pipe is now extensively utilized for water, gas, sewage, culverts, drains, etc. It is produced in a wide range of sizes for varying pressures. Nevertheless, steel pipe, when properly coated and wrapped, can also provide adequate resistance to corrosion when placed in certain soils.

Pipe fittings are available as cast-iron pipe fittings and ductile-iron pipe fittings. Both types of fittings are used with ductile-iron pipe.

Ductile-iron pipe may be obtained in various thicknesses and weights with (1) flanges cast on, (2) ends threaded for screwed-on flanges, (3) ends prepared for mechanical joint, (4) ends grooved or shouldered for patented coupling, (5) one end bell, other end spigot, and (6) one end hub, other end spigot. **Bell-and-spigot** ends are most popular for underground work; hub-and-spigot ends are most frequently used for sewage systems in enclosed spaces. Spigot-end joints are prepared by tightly tamping in hemp or jute at the bottom of the recess with a yarning iron and then pouring in molten lead; the lead, when cooled, is caulked in tightly with a caulking iron and makes a gastight joint. Alternatively, an elastomeric **gasket** (usually Neoprene) may be assembled in the bell-and-spigot joint. For exposed piping, flanged ends are used, the joints being made up with gaskets. Flanged pipe has superior strength and tightness of the joint and is used where pipelines can be well supported. The bell-and-spigot joint possesses greater flexibility and provides for expansion and contraction. It is therefore suitable for water pipe and is largely used for that purpose. Figure 8.7.2 shows a typical form of this joint for ordinary pressures. Figure 8.7.3 shows one form of this joint for ordinary pressures. Figure 8.7.3 shows one form of **mechanical joint** suitable for water, gas, or oil. Other forms of joint, plain-end pipe with couplings, and threaded pipe also are manufactured. Cast-iron and ductile-iron pipe, fittings, and valves have been found unsuitable for superheated steam service. The Code for Pressure Piping, B31.1 (Power Piping), states that cast-iron or ductile-iron pipe may be used for steam service not over 250 lb/in^2 or 406°F (1,724 kPa or 208°C) provided that it meets the requirements as dictated by Eq. (8.7.1).

Fig. 8.7.2 Standard bell-and-spigot joint.

Fig. 8.7.3 Mechanical joint.

Wall thicknesses for the various conditions which ductile-iron pipe is designed to meet are determined in accordance with the requirements of ANSI 21.50 (AWWA C150).

Hubless pipe (often termed **no-hub**), cast or ductile iron, is rapidly replacing the bell-and-spigot form of cast pipe. The ends of the cast pipe are thickened to accommodate an elastomeric gasket and a mechanical compression coupling; see Fig. 8.7.4b. In such nonpressure installations, the outer metallic sleeve of the coupling is configured to allow the gasket to be compressed at assembly and render the joint leak-free. This type of joint is utilized primarily aboveground; underground, the metallic sleeve and bolts, albeit of select noncorroding material, ultimately may fail from corrosion in the embedding soil or backfill. A complete line of fittings are available for hubless pipe.

Ductile-iron pipe is made by **centrifugal casting,** in which molten iron is admitted to the interior of a sand-lined or metal-lined mold, the mold being rotated at high speeds so that the molten metal is thrown by centrifugal force against the lining. ANSI specifications have been prepared for the various combinations of fabrication procedure and intended end use.

Table 8.7.18 lists thicknesses and weight data for centrifugally cast ductile-iron pipe intended for use with water or other liquids.

Fig. 8.7.4   (*a*) Universal ductile-iron pipe and joint. (*b*) Hubless cast pipe: (1) Cast pipe end; (2) assembled compression fitting.

Table 8.7.18   Standard Weights and Thicknesses of Ductile-Iron Bell-and-Spigot Pipe for Water

Nominal size, in	Class 50, 50 lb/in² (345 kPa) 115-ft (35.05-m) head			Class 100, 100 lb/in² (690 kPa) 231-ft (70.41-m) head			Class 150, 150 lb/in² (1,034 kPa) 346-ft (105.5-m) head			Approx lb lead per joint 2 in thick	Approx lb hemp or jute per joint
	Thickness, in	OD, in	Wt, lb per avg ft	Thickness, in	OD, in	Wt, lb per avg ft	Thickness, in	OD, in	Wt, lb per avg ft		
3	0.32	3.96	12.4	0.32	3.96	12.4	0.32	3.96	12.4	6.2	0.17
4	0.35	4.80	16.5	0.35	4.80	16.5	0.35	4.80	16.5	7.5	0.21
6	0.38	6.90	25.9	0.38	6.90	25.9	0.38	6.90	25.9	10.3	0.31
8	0.41	9.05	37.0	0.41	9.05	37.0	0.41	9.05	37.0	13.3	0.44
10	0.44	11.10	49.1	0.44	11.10	49.1	0.44	11.10	49.1	16.0	0.53
12	0.48	13.20	63.7	0.48	13.20	63.7	0.48	13.20	63.7	19.0	0.61
14	0.48	15.30	74.6	0.51	15.30	78.8	0.51	15.65	80.7	22.0	0.81
16	0.54	17.40	95.2	0.54	17.40	95.2	0.54	17.80	97.5	30.0	0.94
18	0.54	19.50	107.6	0.58	19.50	114.8	0.58	19.92	117.2	33.8	1.00
20	0.57	21.60	125.9	0.62	21.60	135.9	0.62	22.06	138.9	37.0	1.25
24	0.63	25.80	166.0	0.68	25.80	178.1	0.73	26.32	194.0	44.0	1.50
30	0.79	32.00	257.6	0.79	32.00	257.6	0.85	32.00	275.4	54.3	2.06
36	0.87	38.30	340.9	0.87	38.30	340.9	0.94	38.30	365.9	64.8	3.00
42	0.97	44.50	442.0	0.97	44.50	442.0	1.05	44.50	475.3	75.3	3.62
48	1.06	50.80	551.6	1.06	50.80	551.6	1.14	50.80	589.6	85.5	4.37

Pipe weights indicated are approximate and include allowance for bell based on a 16-ft laying length. Calculations are for pipe laid without blocks, on flat-bottom trench, with tamped backfill under 5 ft of cover. Thicknesses given above include allowance for water hammer and factory tolerance.
  To obtain weights in kg/m, multiply values shown in lb per avg ft by 1.42.
  SOURCE: Condensed from Table 8.2 of Specification AWWA C 108.

The employment of ductile-iron pipe for gas supply and distribution is second in importance only to its use for carrying water. Bell-and-spigot gas pipe is similar in design to bell-and-spigot water pipe (Fig. 8.7.2). For flanged gas pipe, the class 25 ANSI B16.1 Standard flanges are approved for maximum gas pressures of 25 lb/in² (172 kPa). The class 125 ANSI B16.1 Standard flanges are approved for gas pressures of 125 lb/in² (862 kPa), up to 4 in nominal pipe size; 100 lb/in² (689 kPa), 6 to 12 in; and 80 lb/in² (552 kPa), 16 to 48 in. The type of joint shown in Fig. 8.7.2 is also widely used for gas.

Flexible-Joint Pipe   The necessity for crossing streams and other waterways and of laying pipelines into them has led to the development of various forms of flexible-joint pipe adapted to laying under water, which are capable of motion through several degrees without leakage. Figure 8.7.5 shows one style of such joint which has an adjustment of about 15° in standard sizes.

In selecting the thickness of a pipe for a submerged line, the internal-service pressure is seldom the determining factor, as ample allowance should be made to minimize the risk of breakage in laying and to withstand external shocks from floating ice or other objects. The dimensions and weights given in Table 8.7.19 are typical of those listed by several manufacturers.

"Universal" pipe (Fig. 8.7.4) is ductile-iron pipe with hub-and-spigot ends, the contact surfaces of which are machined on a taper, giving an iron-to-iron joint, By making the tapers of slightly different pitch, the joint provides for flexibility while remaining tight. Two bolts to the joint are sufficient, except for pressures above 175 lb/in² (1,207 kPa). Universal ductile-iron pipe is used largely for carrying gas and water and is suitable for all pressures and services. The pipe is tested with hydrostatic pressure of 300 to 500 lb/in² (2,067 to 3,448 kPa). All universal pipe and special castings of a given diameter and of any class are interchangeable with those of a different class. Standard laying lengths are 6 ft (1.83 m). Thicknesses and weights of standard types up to 16 in are given in Table 8.7.20. Information on other types and sizes and on fittings may be obtained from pipe producers.

Fittings for Ductile-Iron Water Pipe   Flanged fittings of the dimensions of the ANSI standard for steam are not often used with ductile-iron water pipe. The longer fittings of the AWWA are generally preferred because of low friction loss. The dimensions of the flanged fittings of this class conform very closely to the dimensions of the bell-and-spigot fittings of the AWWA. The flange thicknesses and drillings conform to those of the ANSI standards. These fittings, both flange and bell-and-spigot type, are made in a great variety of forms known as "standard special fittings." For dimensions and weights, see manufacturers' catalogs or standard specifications of the AWWA.

Fig. 8.7.5   Flexible joint. (See Table 8.7.19 for dimensions.)

Table 8.7.19 Dimensions and Weights of Flexible-Joint Pipe*
(Dimensions refer to Fig. 8.7.5.)

| Nominal diam, in | Class | Dimensions, in | | | Bolts | | | Average metal thickness, in | Weight† of pipe, incl. bell, lb per 12-ft length |
		A	B	C	No. required	Size, in	Length, in		
4	B	12.13	9.75	5.00	8	0.75	4.50	0.45	290
4	C	12.13	9.75	5.00	8	0.75	4.50	0.48	305
4	D	12.13	9.75	5.00	8	0.75	4.50	0.52	325
6	B	14.25	11.75	7.10	12	0.75	4.50	0.48	440
6	C	14.25	11.75	7.10	12	0.75	4.50	0.51	460
6	D	14.25	11.75	7.10	12	0.75	4.50	0.55	490
8	B	17.25	14.75	9.30	12	0.75	5.25	0.51	635
8	C	17.25	14.75	9.30	12	0.75	5.25	0.56	680
8	D	17.25	14.75	9.30	12	0.75	5.25	0.60	720
10	B	20.56	18.00	11.40	16	0.75	5.25	0.57	905
10	C	20.56	18.00	11.40	16	0.75	5.25	0.62	965
10	D	20.56	18.00	11.40	16	0.75	5.25	0.68	1,035
12	B	23.75	21.00	13.50	16	0.75	6.25	0.62	1,200
12	C	23.75	21.00	13.50	16	0.75	6.25	0.68	1,280
12	D	23.75	21.00	13.50	16	0.75	6.25	0.75	1,375

* United States Pipe and Foundry Co.
† Weights do not include follower rings, bolts, or gaskets. For sizes above 12 in, see manufacturers' catalogs. Multiply weights shown by 38.7 to obtain weight in kg per m or by 0.4536 to obtain weight in kg per 12-ft length.

Table 8.7.20 Standard Weights and Thicknesses of Universal Ductile-Iron Pipe
(Central Foundry Co.)

| Nominal ID, in | Class 150, 150 lb/in² (1,034 kPa) | | | Class 250, 250 lb/in² (1,724 kPa) | | |
| | Approx thickness, in | Estimated wt, lb per* | | Approx thickness, in | Estimated wt, lb per* | |
		ft	6-ft length		ft	6-ft length
2	0.25	7	42	0.31	8	48
3	0.30	11¼	67½	0.34	12½	75
4	0.32	16½	99	0.37	18	108
6	0.36	26.6	160	0.43	30	180
8	0.39	38½	231	0.47	44¼	265½
10	0.43	53½	321	0.50	60½	363
12	0.47	69	414	0.53	77½	465
14	0.50	87	522	0.565	98½	591
16	0.53	106	636	0.60	121	726

ID, in	2	3	4	6	8	10	12	14	16
Bolt sizes, in	½ × 4	½ × 4¼	⅝ × 5	¾ × 6	⅞ × 6¾	1 × 7½	1 × 8	1⅛ × 9	1¼ × 9½

* Multiply wts. by 0.4536 to obtain wt. in kilograms.

Cast-iron soil pipe and fittings are of the hub-and-spigot form, similar in design to the pipe shown in Fig. 8.7.2. Tapped openings and pipe plugs are threaded in accordance with the taper pipe thread requirements of ANSI B1.20.1.

The ANSI standard, Threaded Cast-Iron Pipe for Drainage, Vent, and Waste Services, ANSI A40.5, covers two types of pipes having threaded joints in nominal pipe sizes 1¼ to 12 in and in lengths 5 to 27 ft. One type has external threads on both ends; the other type has external threads on one end and an internal threaded drainage hub on the other end.

A large volume of piping put in place forms part of the drainage, waste, and vent, or **DWV**, system. The pipe material most commonly includes cast and ductile iron, thin-wall copper, and a wide range of plastic compositions (ABS, PE, PVC, etc.)

PIPES AND TUBES OF NONFERROUS MATERIALS

Brass tubing is commercially available in the form known as yellow brass, an alloy consisting of approximately 65 percent copper and 35 percent zinc, and is used principally for ornamental work and hand railings, It has a density of approximately 0.3 lb/in³ (8.31 g/cm³), the exact density being dependent upon the specific chemical composition. **Brass piping** is most frequently furnished as red brass, an alloy consisting of

approximately 85 percent copper and 15 percent zinc. This alloy, having a density of about 0.32 lb/in³ (8.86 g/cm³), has been found to be structurally superior to the yellow brasses and is used where the fluid being conveyed has corrosive properties.

Copper is available either as pipe or as tubing. In the form of piping, it has the same outer diameter as that of standard steel pipe (Tables 8.7.21 and 8.7.22). As tubing, it is used for a variety of purposes, such as for compressed-air instrumentation lines, hydraulic control lines around machinery, domestic oil-burner and heating systems, and for general plumbing purposes. **Copper tubing** (Table 8.7.23) is furnished in 12-ft (3.7-m) and 20-ft (6.1-m) straight lengths or in coils of 100-ft (30.5-m) length. **Type K** tubing, in coils, is used for underground work where the minimum number of joints, combined with greater thickness of type K tubing, is of distinct advantage. **Type L** tubing, usually in straight lengths, is used as a principal piping material for plumbing systems in homes and buildings; this is largely due to the economy of installation made possible by the use of soldered fittings. Copper deteriorates rapidly under high temperatures and repeated stresses. At a temperature of 360°F (182°C) its strength is reduced 15 percent, and on this account it should never be used for high steam pressures and temperatures.

Rigid, thin-wall copper tubing (99.9 percent pure copper) is widespread in **DWV** systems. The plain ends are meant to be soldered to a

Table 8.7.21   Sizes and Weights of SPS Copper and 85 Red Brass Pipe*

Standard size, in	Nominal dimensions, in			Theoretical areas based on nominal dimensions			Theoretical weight, lb/ft	Allowable internal pressure, lb/in^2			
	Outside diameter	Inside diameter	Wall thickness	Cross-sectional area of bore, in^2	External surface, ft^2/lin. ft	Internal surface ft^2/lin. ft		100°F, S = 6,000	200°F, S = 4,800	300°F, S = 4,700	400°F, S = 3,000
¼	0.540	0.410	0.065	0.132	0.141	0.107	0.376	1,500	1,200	1,180	740
⅜	0.675	0.545	0.065	0.233	0.177	0.143	0.483	1,170	940	910	580
½	0.840	0.710	0.065	0.396	0.220	0.186	0.613	920	740	720	460
¾	1.050	0.920	0.065	0.665	0.275	0.241	0.780	730	580	560	360
1	1.315	1.185	0.065	1.10	0.344	0.310	0.989	580	460	450	290
1¼	1.660	1.530	0.065	1.84	0.435	0.401	1.26	450	360	350	220
1½	1.900	1.770	0.065	2.46	0.497	0.464	1.45	400	310	300	190
2	2.375	2.245	0.065	3.96	0.622	0.588	1.83	300	240	240	140
2½	2.875	2.745	0.065	5.92	0.753	0.719	2.22	250	200	190	120
3	3.500	3.334	0.083	8.73	0.916	0.874	3.45	260	210	210	130
3½	4.000	3.810	0.095	11.4	1.05	0.998	4.52	270	210	210	130
4	4.500	4.286	0.107	14.4	1.18	1.12	5.72	270	210	210	130
5	5.562	5.298	0.132	22.0	1.46	1.39	8.73	270	210	210	130
6	6.625	6.309	0.158	31.3	1.73	1.65	12.4	270	210	210	130
8	8.625	8.215	0.205	53.0	2.26	2.15	21.0	270	210	210	130
10	10.750	10.238	0.256	82.3	2.81	2.68	32.7	270	210	210	130
12	12.750	12.124	0.313	115.	3.34	3.18	47.4	280	220	220	130

* The values in the table above are based on the formula in The Code for Pressure Piping, ANSI B31:

$$t_m = \frac{PD}{2S + 0.8P} + C \quad \text{or when } C \text{ is } O \quad P = \frac{2St_m}{D - 0.8t_m}$$

where t_m = minimum pipe wall thickness, in; P = maximum rated internal working pressure, lb/in^2; D = outside diameter of pipe, in; S = allowable stress in material due to internal pressure, at operating temperature, lb/in^2; C = allowance for threading, mechanical strength, and/or corrosion, in. The allowable internal pressures apply to the pipe itself after brazing.

SOURCE: Copper Development Association.

wide range of available fittings. Tubing from 1½- to 4-in dia ($\approx$0.050- to 0.060-in wall) is most common; sizes up to 8-in diam are available for extreme conditions, but the cost then becomes quasi-prohibitive.

Commercial sizes of **aluminum tubing** are listed by the manufacturers in even outside diameters and in wall thicknesses conforming to Stubs gage. Aluminum pipe is available as listed in Table 8.7.24. To obtain the approximate weight per foot of aluminum pipe or tubing, a weight of 0.098 lb/in^3 (2.71 g/cm^3) may be used.

Lead pipe is supplied in straight lengths, in coils, or in reels.

Block tin is a term used in the metal trade to refer to products made wholly from strictly pure high-grade tin. While tin pipe has many and varied uses, its most important applications are in types of equipment handling liquids intended for human consumption. Tin pipe does not corrode, and therefore does not contaminate most of the liquids passing through it.

Zirconium tubing has applications primarily in nuclear power.

Titanium tubing, as pure metal or alloyed, is used in the chemical processing industries and in some heat-exchange equipment such as power plant surface condensers. Titanium alloys are stronger, but suffer reduced corrosion resistance and weldability versus the unalloyed metal. A range of sizes, wall thicknesses, and associated fittings (generally cast) are available.

Glass pipe has unique applications in the chemical and food processing industries. Borosilicate glass imparts its chemical resistance. A small number of fittings are available; by and large, however, fittings are custom-made.

Plastic pipes and **tubes** are available in a wide range of diameters and thicknesses, with Table 8.7.25 generally representative. The plastic used is resistant to attack by many chemicals, light in weight, flexible, and available in coiled form so that installation time is low. It is used for a variety of purposes including drainage, irrigation, sewage, and for conveying chemical solutions or waters that would attack metal piping. Plastic pipe used in gas service is listed in Table 8.7.26. Caution should be used in selection of plastic piping insofar as service temperature is concerned: e.g., polyethylene is suitable for a maximum temperature of 120°F (49°C). Table 8.7.27 lists corrosion-resistance data for polyethylene plastic piping.

Rigid **plastic tubing** can serve in DWV systems. Many plastic compositions are available, although ABS, PE, and PVC predominate. Connections are made with solvents compatible with the plastic materials bonded. A wide range of fittings are available.

Pipes with Special Linings   For use in lines through which are passed solutions containing more or less free acid or other corrosive agents, standard pipe, valves, and fittings may be **lead-lined, tin-lined, or rubber-lined,** to resist corrosive action. This lining prolongs the life of the pipe and also gives it additional strength. For mine service in coal districts where the drainage water is more or less impregnated with sulfur or free sulfuric acid, **wood-lined pipe** and fittings are sometimes used. For special service, **seamless-copper-lined pipe** is also used. The **cement lining** of ductile-iron and steel pipe for water and other services is advantageous because of its protection against unusual destructive agencies and its ability to prevent tuberculation. Standard **hard-rubber pipe** and fittings have been developed for working pressures of 50 lb/in^2 (345 kPa) at normal temperatures. Standard sizes run from ¼- to 4-in diam (0.635- to 10.2-cm), in 10-ft (3.05-m) lengths. For temperature above 120°F (49°C), the use of hard-rubber-lined steel pipe is recommended. This pipe is suitable for conveying strong acids and chemicals.

Glass-lined metallic pipe, likewise, finds special applications. The fittings are subject to specialty design in view of the delicate nature of the material. **Rubber-lined pipe** is a specialty item, but is available for discrete end uses. **Plastic-lined pipe** can be obtained with a variety of such linings. Saran and polypropylene are among the more common such plastics used.

VITRIFIED, WOODEN-STAVE, AND CONCRETE PIPE

Vitrified pipe is used extensively for drains and sewerage systems. Burnt-clay tile, being rendered impervious to water by glazing, is by far the best material for sewage purposes as it is not attacked by acids. Dimensions are given in Table 8.7.28 (see also Fig. 8.7.6). For sizes larger than 36 in (91.4 cm) and other data, refer to the publications of the Clay Products Assoc., Chicago (see ASTM Standard C-700).

Wood-stave pipe (Fig. 8.7.7) is used to a large extent for municipal water supply, outfall sewers, mining, irrigation, and various other uses

Table 8.7.22 Allowable Internal Pressures, lb/in², for Temperatures up to 400°F for Red Brass and Copper Pipe*

	Pipe with threaded ends								Pipe with plain ends for use with welded, brazed, or soldered fittings							
	Red brass				Copper				Red brass				Copper			
Standard size, in	100°F S=8,000	200°F S=8,000	300°F S=8,000	400°F S=5,000	100°F S=6,000	200°F S=4,800	300°F S=4,700	400°F S=3,000	100°F S=8,000	200°F S=8,000	300°F S=8,000	400°F S=5,000	100°F S=6,000	200°F S=4,800	300°F S=4,700	400°F S=3,000
								Regular								
⅛	370	370	370	230	280	210	210	130	2,640	2,640	2,640	1,650	1,980	1,570	1,540	980
¼	870	870	870	550	650	510	510	320	2,610	2,610	2,610	1,640	1,960	1,560	1,530	970
⅜	890	890	890	570	670	530	510	320	2,280	2,280	2,280	1,440	1,710	1,350	1,330	840
½	900	900	900	570	670	530	520	320	2,160	2,160	2,160	1,350	1,620	1,280	1,260	800
¾	810	810	810	520	610	480	470	300	1,800	1,800	1,800	1,130	1,350	1,070	1,050	670
1	630	630	630	400	480	370	370	230	1,580	1,580	1,580	1,000	1,190	940	920	590
1¼	690	690	690	430	520	410	400	250	1,440	1,440	1,440	900	1,080	890	840	530
1½	630	630	630	400	480	370	370	230	1,280	1,280	1,280	800	960	750	740	470
2	540	540	540	350	410	320	310	190	1,050	1,050	1,050	670	790	620	610	380
2½	450	450	450	280	340	260	250	160	1,040	1,040	1,040	650	780	620	610	380
3	510	510	510	320	380	300	290	180	1,000	1,000	1,000	630	750	590	580	370
3½	570	570	570	370	430	330	330	200	1,000	1,000	1,000	630	750	590	580	370
4	510	510	510	320	380	300	290	180	880	880	880	550	660	530	520	320
5	410	410	410	270	310	230	240	140	710	710	710	450	540	420	410	260
6	340	340	340	220	260	190	200	120	600	600	600	380	450	350	340	220
8	360	360	360	236	270	270	210	130	560	560	560	350	420	320	320	200
10	360	360	360	230	270	210	210	130	520	520	520	330	390	300	300	200
12	320	320	320	200	240	190	200	120	450	450	450	280	340	260	250	160
								Extra-Strong								
⅛	1,960	1,960	1,960	1,240	1,470	1,160	1,140	720	4,630	4,630	4,630	2,900	3,470	2,750	2,710	1,730
¼	2,210	2,210	2,210	1,340	1,660	1,310	1,290	820	4,200	4,200	4,200	2,640	3,150	2,500	2,460	1,570
⅜	1,840	1,840	1,840	1,150	1,380	1,090	1,070	680	3,360	3,360	3,360	2,100	2,520	2,000	1,960	1,250
½	1,760	1,760	1,760	1,100	1,320	1,040	1,030	660	3,130	3,130	3,130	1,970	2,350	1,860	1,830	1,160
¾	1,510	1,510	1,510	950	1,130	900	880	560	2,560	2,560	2,560	1,600	1,920	1,520	1,500	960
1	1,340	1,340	1,340	850	1,010	790	780	490	2,360	2,360	2,360	1,490	1,770	1,400	1,370	880
1¼	1,160	1,160	1,160	730	880	730	680	430	1,950	1,950	1,950	1,220	1,460	1,160	1,140	720
1½	1,090	1,090	1,090	680	820	660	640	410	1,770	1,770	1,770	1,120	1,330	1,050	1,030	660
2	1,000	1,000	1,000	630	750	590	570	360	1,520	1,520	1,520	950	1,140	910	890	560
2½	970	970	970	620	730	580	560	360	1,600	1,600	1,600	1,000	1,200	950	930	590
3	910	910	910	570	680	530	530	340	1,420	1,420	1,420	900	1,070	840	830	530
3½	860	860	860	550	650	510	500	310	1,300	1,300	1,300	820	980	790	760	480
4	840	840	840	530	630	490	480	300	1,230	1,230	1,230	770	920	730	710	460
5	770	770	770	480	580	450	440	340	1,080	1,080	1,080	680	810	640	630	400
6	800	800	800	500	600	470	460	290	1,060	1,060	1,060	670	800	620	610	380
8	710	710	710	450	530	430	410	260	910	910	910	570	680	540	540	340
10	550	550	550	350	420	340	330	200	710	710	710	450	540	440	420	240

* The values above are based on the formula in The Code for Pressure Piping, ANSI B31:

$$t_m = \frac{PD}{2S + 0.8P} + C \qquad \text{or} \qquad P = \frac{2S(t_m - C)}{D - 0.8(t_m - C)} \qquad \text{or when } C \text{ is O} \qquad P = \frac{2St_m}{D - 0.8t_m}$$

where t_m = minimum pipe wall thickness, in; P = maximum rated internal working pressure, lb/in²; D = outside diameter of pipe, in; S = allowable stress in annealed material due to internal pressure, at operating temperature, lb/in²; C = allowance for threading, mechanical strength and/or corrosion, in. C equals 0.05 for sizes ⅜ in and smaller and the depth of the thread for all other sizes, and equals O for pipe with plain ends. These allowable internal pressures apply to the pipe itself and do not take into account the limitations which may be imposed by the type of joint and the joining material.

SOURCE: Copper Development Association.

Table 8.7.23   Sizes and Weights of Copper Tubes

Nominal size, in	OD, in, types* K, L	ID, in		Wall thickness, in		Permissible variation of mean OD, in		Weight, † lb/ft	
						Types K, L			
		Type K	Type L	Type K	Type L	Annealed	Hard-drawn	Type K	Type L
⅜	0.500	0.402	0.430	0.049	0.035	0.0025	0.001	0.269	0.198
½	0.625	0.527	0.545	0.049	0.040	0.0025	0.001	0.344	0.285
⅝	0.750	0.652	0.666	0.049	0.042	0.0025	0.001	0.418	0.362
¾	0.875	0.745	0.785	0.065	0.045	0.003	0.001	0.641	0.455
1	1.125	0.995	1.025	0.065	0.050	0.0035	0.0015	0.839	0.655
1¼	1.375	1.245	1.265	0.065	0.055	0.004	0.0015	1.04	0.884
1½	1.625	1.481	1.505	0.072	0.060	0.0045	0.002	1.36	1.14
2	2.125	1.959	1.985	0.083	0.070	0.005	0.002	2.06	1.75
2½	2.625	2.435	2.465	0.095	0.080	0.005	0.002	2.93	2.48
3	3.125	2.907	2.945	0.109	0.090	0.005	0.002	4.00	3.33
3½	3.625	3.385	3.425	0.120	0.100	0.005	0.002	5.12	4.29
4	4.125	3.857	3.905	0.134	0.110	0.005	0.002	6.51	5.38
5	5.125	4.805	4.875	0.160	0.125	0.005	0.002	9.67	7.61
6	6.125	5.741	5.845	0.192	0.140	0.005	0.002	13.9	10.2
8	8.125	7.583	7.725	0.271	0.200	0.006	0.003	25.9	19.3
10	10.125	9.449	9.625	0.338	0.250	0.008	0.004	40.3	30.1
12	12.125	11.315	11.565	0.405	0.280	0.008	0.004	57.8	40.4

* Type K recommended for underground service and general plumbing. Type L suitable for interior plumbing and other services.
† Multiply these values by 1.48 to obtain weight in kg/m.
SOURCE: American Brass Co.

Fig. 8.7.6   Vitrified pipe.

Elevation of pipe and coupling

Section A-A

Cross section

Side elevation

Fig. 8.7.7   Wood-stave pipes.

providing for the transportation of water. The water carried may be hot, cold, or acid. It is made either untreated or creosoted by a vacuum and pressure process. This process uses 8 lb of creosote per cubic foot of wood treated. The untreated pipe is most used where the pipe is constantly full of water, and the wood therefore completely saturated, although in many such instances the creosoted wood is used to give assurance of permanence. (See also Sec. 6.)

Wood-stave pipe is made in two types: machine-banded pipe and continuous-stave pipe. **Machine-banded pipe** is banded with wire and is made with wood or metal collars, or with inserted joints. **Continuous-stave pipe** is manufactured in units consisting of staves, bands, and shoes, shipped in knocked-down form, and constructed in the trench. In building this type of pipe, the staves are laid so as to break joints and the completed pipe is without joints. Continuous-stave pipe is banded with individual bands, ranging in size from ⅜ to 1 in (0.95 to 2.54 cm), depending upon the size of the pipe. A factor of safety of 4 is maintained in the band, based on an ultimate strength of 60,000 lb/in² (414 MPa) of cross section. The maximum pressure to which a continuous-stave pipe may be subjected depends upon the size of the pipe. The head for small pipes may run as high as 400 ft (121.9 m) while in the largest sizes the head would be less than 200 ft (61 m).

Machine-banded pipe is made for pressures of 50 to 400 ft (15.2 to 121.9 m). The staves are made from redwood or Douglas-fir lumber, dried and carefully selected. The inside and outside of the staves are

Fig. 8.7.8   Reinforced-concrete pipe.

Table 8.7.24 Aluminum Piping

Nominal pipe size, in	Schedule no.	OD, in	ID, in	Wall thickness, in	Weight per linear foot, lb, plain ends†	Cross-sectional wall area, in²	Inside cross-sectional area, in²	Moment of inertia, in⁴	Section modulus, in³	Radius of gyration, in
⅛	40‡	0.405	0.269	0.068	0.085	0.0720	0.0568	0.0011	0.0053	0.1215
	80§	0.405	0.215	0.095	0.109	0.0925	0.0363	0.0012	0.0060	0.1146
¼	40‡	0.540	0.364	0.088	0.147	0.1250	0.1041	0.0033	0.0123	0.1628
	80§	0.540	0.302	0.119	0.185	0.1574	0.0716	0.0038	0.0139	0.1547
⅜	40‡	0.675	0.493	0.091	0.196	0.1670	0.1909	0.0073	0.0216	0.2090
	80§	0.675	0.423	0.126	0.256	0.2173	0.1405	0.0086	0.0255	0.1991
½	40‡	0.840	0.622	0.109	0.294	0.2503	0.3039	0.0171	0.0407	0.2613
	80§	0.840	0.546	0.147	0.376	0.3200	0.2341	0.0201	0.0478	0.2505
¾	10	1.050	0.884	0.083	0.297	0.2521	0.6138	0.0297	0.0566	0.3432
	40‡	1.050	0.824	0.113	0.391	0.3326	0.5333	0.0370	0.0705	0.3337
	80§	1.050	0.742	0.154	0.510	0.4335	0.4324	0.0448	0.0853	0.3214
1	5	1.315	1.185	0.065	0.300	0.2553	1.103	0.0500	0.0760	0.4425
	10	1.315	1.097	0.109	0.486	0.4130	0.9452	0.0757	0.1151	0.4382
	40‡	1.315	1.049	0.133	0.581	0.4939	0.8643	0.0873	0.1328	0.4205
	80§	1.315	0.957	0.179	0.751	0.6388	0.7193	0.1056	0.1606	0.4066
1¼	5	1.660	1.530	0.065	0.383	0.3257	1.839	0.1037	0.1250	0.5644
	10	1.660	1.442	0.109	0.625	0.5311	1.633	0.1605	0.1934	0.5497
	40‡	1.660	1.380	0.140	0.786	0.6685	1.496	0.1947	0.2346	0.5397
	80§	1.660	1.278	0.191	1.037	0.8815	1.283	0.2418	0.2913	0.5238
1½	5	1.900	1.770	0.065	0.441	0.3747	2.461	0.1579	0.1662	0.6492
	10	1.900	1.682	0.109	0.721	0.6133	2.222	0.2468	0.2598	0.6344
	40‡	1.900	1.610	0.145	0.940	0.7995	2.036	0.3099	0.3262	0.6226
	80§	1.900	1.500	0.200	1.256	1.068	1.767	0.3912	0.4118	0.6052
2	5	2.375	2.245	0.065	0.555	0.4717	3.958	0.3149	0.2652	0.8170
	10	2.375	2.157	0.109	0.913	0.7760	3.654	0.4992	0.4204	0.8021
	40‡	2.375	2.067	0.154	1.264	1.074	3.356	0.6657	0.5606	0.7871
	80§	2.375	1.939	0.218	1.737	1.477	2.953	0.8679	0.7309	0.7665
2½	5	2.875	2.709	0.083	0.856	0.7280	5.764	0.7100	0.4939	0.9876
	10	2.875	2.635	0.120	1.221	1.039	5.453	0.9873	0.6868	0.9750
	40‡	2.875	2.469	0.203	2.004	1.704	4.788	1.530	1.064	0.9474
	80§	2.875	2.323	0.276	2.650	2.254	4.238	1.924	1.339	0.9241
3	5	3.500	3.334	0.083	1.048	0.8910	8.730	1.301	0.7435	1.208
	10	3.500	3.260	0.120	1.498	1.274	8.346	1.822	1.041	1.196
	40‡	3.500	3.068	0.216	2.621	2.228	7.393	3.017	1.724	1.164
	80§	3.500	2.900	0.300	3.547	3.016	6.605	3.894	2.225	1.136
3½	5	4.000	3.834	0.083	1.201	1.021	11.55	1.960	0.9799	1.385
	10	4.000	3.760	0.120	1.720	1.463	11.10	2.755	1.378	1.372
	40‡	4.000	3.548	0.226	3.151	2.680	9.887	4.788	2.394	1.337
	80§	4.000	3.364	0.318	4.326	3.678	8.888	6.281	3.140	1.307
4	5	4.500	4.334	0.083	1.354	1.152	14.75	2.810	1.249	1.562
	10	4.500	4.260	0.120	1.942	1.651	14.25	3.963	1.761	1.549
	40‡	4.500	4.026	0.237	3.733	3.174	12.73	7.232	3.214	1.510
	80§	4.500	3.826	0.337	5.183	4.407	11.50	9.611	4.272	1.477
5	40‡	5.563	5.047	0.258	5.057	4.300	20.01	15.16	5.451	1.878
	80§	5.563	4.813	0.375	7.188	6.112	18.19	20.67	7.432	1.839
6	40‡	6.625	6.065	0.280	6.564	5.581	28.89	28.14	8.496	2.246
	80§	6.625	5.761	0.432	9.884	8.405	26.07	40.49	12.22	2.195
8	30	8.625	8.071	0.277	8.543	7.265	51.16	63.35	14.69	2.953
	40‡	8.625	7.981	0.322	9.878	8.399	50.03	72.49	16.81	2.938
	80§	8.625	7.625	0.500	15.01	12.76	45.66	105.7	24.51	2.878
10		10.750	10.192	0.279	10.79	9.178	81.59	125.9	23.42	3.704
	30	10.750	10.136	0.307	11.84	10.07	80.69	137.4	25.57	3.694
	40‡	10.750	10.020	0.365	14.00	11.91	78.85	160.7	29.90	3.674
	80§	10.750	9.750	0.500	18.93	16.10	74.66	211.9	39.43	3.628
12	30	12.750	12.090	0.330	15.14	12.88	114.8	248.5	38.97	4.393
	‡	12.750	12.000	0.375	17.14	14.58	113.1	279.3	43.81	4.377
	§	12.750	11.750	0.500	22.63	19.24	108.4	361.5	56.71	4.335
Construction pipe										
		2.00	1.900	0.050	0.3602	0.3063	2.835	0.1457	0.1457	0.6897
		3.00	2.900	0.050	0.5449	0.4634	6.605	0.5042	0.3361	1.043
		4.00	3.900	0.050	0.7297	0.6205	11.95	1.210	0.6051	1.397
		5.00	4.896	0.052	0.9506	0.8083	18.83	2.474	0.9896	1.749
		6.00	5.876	0.062	1.360	1.157	2.712	5.098	1.699	2.100
		7.00	6.856	0.072	1.843	1.567	36.92	9.403	2.687	2.450
		8.00	7.812	0.094	2.745	2.335	47.93	18.24	4.561	2.795

* Aluminium Co. of America.
† Weights calculated for 6061 and 6063. For 3003 multiply by 1.010.
‡ Also designated as standard pipe.
§ Also designated as extra-heavy or extra-strong pipe. All calculations based on nominal dimensions.

Table 8.7.25   Commerical Sizes (IPS) and Weights of Polyvinyl Chloride (PVC) Pipe*

Nominal size, in	Schedule	Wall thickness,* in	OD, in	ID, in	Theoretical weight,† lb/ft	Calculated min bursting pressure, lb/in² Note 1	Calculated min bursting pressure, lb/in² Note 2
¼	40	0.088	0.540	0.364	0.076	2,490	1,950
	80	0.119	0.540	0.302	0.096	3,620	2,830
½	40	0.109	0.840	0.622	0.153	1,910	1,490
	80	0.147	0.840	0.546	0.185	2,720	2,120
¾	40	0.113	1.050	0.824	0.203	1,540	1,210
	80	0.154	1.050	0.742	0.265	2,200	1,720
1	40	0.133	1.315	1.049	0.305	1,440	1,130
	80	0.179	1.315	0.957	0.385	2,020	1,580
1¼	40	0.140	1.660	1.380	0.409	1,180	920
	80	0.191	1.660	1.278	0.550	1,660	1,300
1½	40	0.145	1.900	1.610	0.489	1,060	830
	80	0.200	1.900	1.500	0.653	1,510	1,180
2	40	0.154	2.375	2.067	0.640	890	690
	80	0.218	2.375	1.939	0.910	1,290	1,010
3	40	0.216	3.500	3.068	1.380	840	660
	80	0.300	3.500	2.900	1.845	1,200	940
4	40	0.237	4.500	4.026	1.965	710	560
	80	0.337	4.500	3.826	2.710	1,040	810

* Thicknesses listed are minimum values. Tolerance is generally – 0 + 10%.
† These representative values are not specified in ASTM D1785.
NOTES:
1. Materials are PVC 1120, 1220, and 4116. A fiber stress of 6,400 lb/in² was used in bursting pressure calculations.
2. Materials are PVC 2112, 2116, and 2120. A fiber stress of 5,000 lb/in² was used in bursting pressure calculations.
SOURCE: Abstracted from ASTM Specification D1785.

Table 8.7.26   Dimensions of Plastic Pipe for Gas Service (AGA Requirements)

Nominal size, in	Nominal OD, in	Nominal ID, in	Sleeved weight, lb	Max working pressure at 73°F, lb/in²
½	0.625	0.500	0.054	250
¾	0.875	0.750	0.079	175
1	1.125	1.000	0.103	125
1¼	1.375	1.250	0.127	110
1½	1.625	1.500	0.152	90
1¾	1.875	1.750	0.177	80
2	2.125	2.000	0.200	75
2½	2.660	2.500	0.320	75
3	3.190	3.000	0.457	75
4	4.250	4.000	0.800	75

dressed to conform to the circumferential lines, and the edges of the staves dressed to conform to the radial lines.

Wooden pipe is largely built in western sections of the United States, close to the natural lumber market. The sizes of machine-banded pipe range from 2 to 24 in (5.08 to 61 cm), and of the continuous-stave pipe from 6- to 20-ft (15.2-cm to 6.1-m) inside diameter.

Pipe made from **plywood** is molded in lengths up to 11 ft (3.35 m) in diameters 3 in (7.62 cm) and up, and in wall thicknesses to specifications. Tubes made from fiber, by a molding process, are obtainable in a variety of sizes and lengths.

Concrete pipe is an important factor in sewer, conduit, railroad, culvert, and water-pipe construction. The pipe, as usually made, is constructed of concrete reinforced longitudinally with bars and transversely with wire mesh or steel bands. It is made in sections of definite length, with the longitudinal reinforcement so disposed as to provide for the interlocking of one section with another, and so formed that when these are locked together and cemented they form a continuous line of pipe free from leakage or seepage. Various forms of joints are used, all capable of taking care of expansion. Figure 8.7.8 shows one type of construction. Concrete pipe is manufactured in a great variety of

diameters, thicknesses, and lengths to suit almost any requirement arising in practice. (See also Sec. 6.)

Asbestos-cement pipe, known by the trade name **Transite** pipe in this country, was developed initially in Europe. It was widely used in the United States for many years in a large variety of services. It was made of a mixture of portland cement and asbestos fiber, was highly resistant to corrosion, and provided outstanding service in mine drainage systems, waterworks systems, gas lines, and sewerage systems. It was manufactured in diameters from 3 to 36 in (7.6 to 91 cm), in stock lengths of 13 ft (4 m), and in pressure classes of 50, 100, 150, and 200 lb/in² (349, 689, 1,034, and 1,379 kPa). While it is no longer manufactured, it is still in place in some of the above-cited applications. Repairs to existing Transite piping generally devolves into replacing it with pipe made of a material compatible with its service, often with fiberglass-reinforced pipe, plastic pipe, or some form of metal pipe.

FITTINGS FOR STEEL PIPE

American National Standard Cast-Iron Pipe Flanges and Flanged Fittings for Maximum Working Saturated Steam Pressure of 25, 125, and 250 lb/in² (172, 862, and 1,724 kPa)

INTRODUCTORY NOTES

ANSI B16.1 lists complete dimensional data and drilling template details for these classes of pipe.

Sizes   The sizes of the fittings in the following tables are nominal pipe sizes. In the class 25 standard, the nominal pipe size is the same as the port diameter of the fittings, for *all* sizes. In the class 125 and class 250 standards the nominal pipe size is the same as the port diameter of fittings for pipe having inside diameters of 12 in (30.5 cm) and smaller. For pipe 14 in (35.6 cm) and larger, the corresponding outside diameter of the pipe is given, and consequently the fittings will have a smaller port diameter.

Pressure Rating   In the class 25 standard the sizes 36 in (91.4 cm) and smaller may also be used for maximum nonshock working hydraulic pressures of 43 lb/in² gage (296 kPa) or a maximum gas pressure of 25 lb/in² gage (172 kPa), at or near the ordinary range of air

Table 8.7.27   Corrosion-Resistance Data, Polyethylene Pipe

Reagent	Performance at: 75°F (24°C)	120°F (49°C)	Reagent	Performance at: 75°F (24°C)	120°F (49°C)	Reagent	Performance at: 75°F (24°C)	120°F (49°C)
Acetic acid, glacial*	F	NG	Ferrous sulfate, 15 percent aq.	E	E	Potassium chloride, saturated	E	E
Acetic acid, 10 percent*	E	E	Fluorine	E	NG	Potassium dichromate	E	E
Acetone	NG	NG	Fluosilicic acid, concentrated	E	F	Potassium hydroxide	E	E
Ammonia, dry gas	E	E	Formaldehyde, 40 percent	E	E	Potassium nitrate	E	E
Ammonium hydroxide, 10 percent	E	E	Formic acid, 50 percent	E	E	Potassium permanganate	E	E
Ammonium hydroxide, 28 percent	E	E	Furfuryl alcohol	NG	NG	Silicic acid	E	E
Amyl acetate	NG	NG	Gasoline	NG	NG	Silver nitrate	E	E
Aniline	E	F	Hydrobromic acid	E	E	Sodium benzoate	E	E
Benzene	NG	NG	Hydrochloric acid, 10 percent	E	E	Sodium bisulfite	E	E
Bromine	NG	NG	Hydrochloric acid, 37 percent	E	E	Sodium carbonate, concentrated	E	E
Butyraldehyde	E	G	Hydrofluoric acid, 48 percent	E	E	Sodium chloride, saturated solution	E	E
Calcium chloride, saturated	E	E	Hydrofluoric acid, 75 percent	E	F	Sodium hydroxide, 10 percent	E	E
Calcium hydroxide	E	E	Hydrogen peroxide, 30 percent	E	G	Sodium hydroxide, 50 percent	E	E
Calcium hypochlorite	E	E	Hydrogen peroxide, 90 percent	G	NG	Sodium sulfate	E	E
Carbon disulfide	NG	NG	Lactic acid, 90 percent	E	E	Stannic chloride, saturated	E	E
Carbon tetrachloride	NG	NG	Linseed oil	E	E	Stearic acid, 100 percent	E	E
Carbonic acid	E	E	Lubricating oil	NG	NG	Sulfuric acid, 10 percent	E	E
Chlorine, dry gas	F	NG	Magnesium chloride	E	E	Sulfuric acid, 30 percent	E	E
Chlorine, liquid	NG	NG	Magnesium sulfate	E	E	Sulfuric acid, 60 percent	E	F
Chlorosulfonic acid	NG	NG	Methyl bromide	NG		Sulfuric acid, 98 percent	F	NG
Citric acid, saturated	E	E	Methyl isobutyl ketone	F	NG	Tannic acid	E	E
Copper sulfate	E	E	Nitric acid, 10 percent	E	E	Toluene	NG	NG
Cyclohexanone	NG	NG	Nitric acid, 30–50 percent	E	E	Trichlorobenzene	F	NF
Diethylene glycol	E	E	Nitric acid, 70 percent	E	E	Trichloroethylene	NG	NG
Dioxane	E	G	Oleic acid	F	NG	Vinegar	E	E
Ethyl acetate	F	NG	Phosphoric acid, 30 percent	E	E	Xylene	NG	NG
Ethyl alcohol, 35 percent	NG	NG	Phosphoric acid, 90 percent	E	NG	Zinc chloride	E	E
Ethyl butyrate	F	NG	Photographic developer	E	E	Zinc sulfate	E	E
Ethylene dichloride	NG	NG	Potassium borate	E	E			
Ferric chloride	E	E	Potassium carbonate	E	E			

Corrosion-resistance data given in this table based on laboratory tests conducted by the manufacturers of the materials covered, and are indicative only of the conditions under which the tests were made. This information may be considered as a basis for recommendation, but not as a guarantee. Materials should be tested under actual service to determine suitability for a particular purpose.
E = excellent, G = good, F = fair, NG = not good.
* Polyethylene is permeable to acetic acid.

temperatures. In the class 125 standard, the sizes 12 in (30.5 cm) and smaller may also be used for maximum nonshock working hydraulic pressure of 175 lb/in² gage (1,207 kPa) at or near the ordinary range of air temperatures. In the class 250 standard, the sizes 12 in and smaller may be used for maximum nonshock working hydraulic pressures of 400 lb/in² gage (2,758 kPa) at or near the ordinary range of air temperatures.

Facing   All class 25 and class 125 cast-iron flanges and flanged fittings are plain faced, i.e., without projection or raised face. All class 250 cast-iron flanges and flanged fittings have a raised face $\frac{1}{16}$ (0.16 cm) high. The raised face is included in the minimum flange thickness and center-to-face dimensions.

An **inspection limit** of $\pm \frac{1}{32}$ in (0.08 cm) is allowed on all center-to-contact-surface dimensions for sizes up to and including 10 in (25.4 cm) and $\pm \frac{1}{8}$ in on sizes larger than 10 in.

Dimensions   In the class 25 standard, the flange diameters, bolt circles, and number of bolts are the same as in the class 125 ANSI Standard B16.1 with a reduction in the thickness of flanges and bolt

Table 8.7.28   Standard-Strength Vitrified Clay Pipe
(Dimensions refer to Fig. 8.7.6)

Size D, in	Laying length L Nominal, ft	Limit of minus variation,* in/ft length	Max difference in length of two opposite sides, in	OD of barrel, in Min	Max	ID of socket $\frac{1}{2}$ in above base D_s, in Min	Max	Depth of socket L_s, in Nominal	Min	Thickness of barrel T, in Nominal	Min	Thickness of socket at $\frac{1}{2}$ in from outer end T_s, in Nominal	Min
4	2, 2½ 3	¼	5/16	4⅞	5⅛	5¾	6⅛	1¾	1½	½	7/16	7/16	⅜
6	2, 2½, 3	¼	⅜	7 1/16	7 7/16	8 3/16	8⅝	2½	2	⅝	9/16	½	7/16
8	2, 2½, 3	¼	7/16	9¼	9¾	10½	11	2½	2¼	¾	11/16	9/16	½
10	2, 2½, 3	¼	7/16	11½	12	12¾	13¼	2⅝	2⅜	⅞	13/16	⅝	9/16
12	2, 2½, 3	¼	7/16	13¾	14 5/16	15⅛	15¾	2¾	2½	1	15/16	¾	11/16
15	3, 4	¼	½	17 3/16	17 13/16	18⅝	19¼	2⅞	2⅝	1¼	1⅛	15/16	⅞
18	3, 4	¼	½	20⅝	21 7/16	22¼	23	3	2¾	1½	1⅜	1⅛	1 1/16
21	3, 4	¼	9/16	24⅛	25	25⅞	26¾	3¼	3	1¾	1⅝	1 5/16	1 3/16
24	3, 4	⅜	9/16	27½	28½	29⅜	30⅜	3⅜	3⅛	2	1⅞	1½	1⅜
27	3, 4	⅜	9/16	31	32⅛	33	34⅛	3½	3¼	2¼	2⅛	1 11/16	1 9/16
30	3, 4	⅜	⅝	34⅜	35⅝	36½	37¾	3⅝	3⅜	2½	2⅜	1⅞	1¾
33	3, 4	⅜	⅝	37⅝	38 15/16	39⅞	41¼	3¾	3½	2⅝	2½	2	1 13/16
36	3, 4	⅜	11/16	40¾	42¼	43¼	44¾	4	3¾	2¾	2⅝	2 1/16	1⅞

When ordering standard-strength vitrified-clay pipe, give the size of pipe (ID) and the laying strength wanted, and refer to ASTM Specification C-700. Standard lengths of pipe shown meet normal practice in various sections of the country. Manufacturers' stocks include those lengths conforming to local practice.
* There is no limit for plus variation.

diameters, thereby maintaining interchangeability between the two standards.

The center-to-face and face-to-face dimensions of class 25 standard fittings are the same as for the class 125 standard.

Bolting   Drilling templates are in multiples of four, so that fittings may be made to face in any quarter. **Bolt holes** straddle the centerline. For bolts smaller than $1\frac{3}{4}$ in (4.45 cm) the bolt holes are drilled $\frac{1}{8}$ in larger in diameter than the nominal size of the bolt. Holes for bolts $1\frac{3}{4}$ in and larger are drilled $\frac{1}{4}$ in (0.64 cm) larger than nominal diameter of bolts. **Bolts** of steel are with standard "rough square heads" and the nuts are of steel with standard "rough hexagonal" dimensions; all as given in the American Standard on Wrench Head Bolts and Nuts and Wrench Openings of the National Screw Thread Commission. For bolts, $1\frac{3}{4}$-in (4.45-cm) diam and larger, bolt studs with a nut on each end are recommended.

Hexagonal nuts for pipe sizes 1 to 48 in (2.54 to 122 cm) in the class 125 standard and 1 to 16 in (40.6 cm) in the class 250 standard can be conveniently pulled up with open wrenches of minimum design of heads. Hexagonal nuts for pipe sizes 48 to 96 in (244 cm) in the class 125 standard and 18 to 48 in (45.7 to 122 cm) in the class 250 standard can be conveniently pulled up with box wrenches.

Spot Facing   The bolt-holes of classes 25, 125, and 250 cast-iron flanges and flanged fittings are not spot-faced for ordinary service. When required, the flanges and fittings in sizes 30 in (76.2 cm) and larger may be spot-faced or back-faced to the minimum thickness of flange with a plus tolerance of $\frac{1}{8}$ in (0.32 cm).

Reducing Fittings   Reducing elbows and side-outlet elbows carry same dimensions center to face as straight-size elbows corresponding to the size of the larger opening.

Tees, side-outlet tees, crosses, and laterals sizes 16 in (40.6 cm) and smaller, reducing on the outlet or branch, have the same dimension center to face and face to face as straight-size fittings corresponding to the size of the larger opening. Sizes 18 in (45.7 cm) and larger, reducing on the outlet or branch, are made in two lengths depending on the size of the outlet as given in the tables of dimensions.

Tees, crosses, and laterals, reducing on the run only, have the same dimensions center to face and face to face as straight-size fittings corresponding to the size of the larger opening.

Reducers and eccentric reducers for all reduction have the same face-to-face dimensions for the larger opening.

Special double-branch elbows whether straight or reducing have the same dimension center to face as straight-size elbows corresponding to the size of the larger opening.

Side-outlet elbows and side-outlet tees have all openings on intersecting centerlines.

Elbows   Special degree elbows ranging from 1 to 45° have the same center-to-face dimensions given for 45° elbows, and those over 45° and up to 90° have the same center-to-face dimensions given for 90° elbows. The angle designation of an elbow is its deflection from straight-line flow and is the angle between its flange faces.

Threaded Companion Flanges   Threaded companion flanges in the class 25 standard should not be thinner than those in the class 125 standard on sizes 24 in (61 cm) and smaller. Other types of flanges may vary in thickness.

Laterals   Laterals (Y branches) both straight and reducing sizes 8 in and larger are reinforced to compensate for the inherent weakness in the casting design.

The American National Standard B16.1 covers also dimensions of base elbows and base tees and anchorage bases for straight tees and reducing tees.

American National Standard cast-iron pipe flanges and flanged fittings are available for maximum nonshock working hydraulic pressure of 800 lb/in² gage (5,516 kPa) at ordinary air temperatures.

Assembly of Flanged Joints   The optimum degree of tightening occurs when a stress of 30,000 lb/in² (207 MPa) is uniformly reached in each flange stud or bolt. For a modulus of elasticity of 30,000,000 lb/in² (207 GPa) a stress of 30,000 lb/in² occurs when the elongation, determined with a dial indicator or micrometer, is 0.001 in/in of stud length

measured between centers of nuts. Uniform tension in flange bolts may also be obtained by use of a torque wrench; bearing surfaces of nuts must have a good machine finish, and threads must be properly lubricated for reliable results with a torque wrench. The following torque values have been found to give 30,000 lb/in² stress in studs:

Study diam, in	Threads per inch	Torque, lb · ft*
$\frac{5}{8}$	11	89
$\frac{3}{4}$	10	107
$\frac{7}{8}$	9	162
1	8	244
$1\frac{1}{8}$	8	322
$1\frac{1}{4}$	8	410
$1\frac{3}{8}$	8	510
$1\frac{1}{2}$	8	615

* Multiply by 0.149 to obtain torque in kilogram-metres.

American National Standard Steel Pipe Flanges and Flanged Fittings

INTRODUCTORY NOTES

Pressure Ratings and Tests   These standards are known as "American Class 150, 300, 400, 600, 900, 1,500, and 2,500 Steel Flange Standards" (ANSI B16.5).

Sizes   The size of the fittings and companion flanges are listed in the standard.

Materials   The flanged fittings and flanges should be either steel castings or steel forgings of the grade complying with the ASTM specifications recommended under these standards for the various pressure-temperature ratings for which these standards are designed.

Bolting material including nuts and washers are based on a high-grade product equal to that given in ASTM Standard Specifications for Alloy-Steel Bolting Material for High Temperature Service (Table 8.7.29) and with physical and chemical requirements in accordance with the tables given under ANSI B16.5. **Commercial steel bolts should not be used at steam pressures over 250** lb/in² (1,724 kPa) and temperatures over 450°F (232°C). Nuts should be of carbon or alloy steel. Washers when used under nuts should be of forged or rolled carbon steel.

Bolting   Drilling templates are in multiples of four, so that fittings may be made to face in any quarter. Bolt holes straddle the centerlines. Bolt holes are drilled $\frac{1}{8}$ in (0.32 cm) larger in diameter than the nominal size of bolt. Bolts or bolt studs threaded at both ends may be used and should be equipped with cold-punched or cold-pressed semifinished nuts of American National Standard rough dimensions, chamfered and trimmed.

All bolts and bolt studs having diameters 1 in and smaller, and the corresponding nuts are threaded with the American National Standard screw thread, coarse thread series, medium fit, class 3, while those bolts and bolt studs whose diameters are $1\frac{1}{8}$ in (2.86 cm) and larger have special threads of the American National form whose pitch is $\frac{1}{8}$ in (8 threads per inch). It is recommended that these special threads be allowed a pitch-diameter tolerance of -0.006 in and a lead tolerance of ± 0.002 in.

Bolt studs with a nut at each end are recommended for high-temperature service.

The allowable working fiber stress, considering internal allowable working pressure only, in bolting material for valve bonnet flanges, cleanout flanges, etc., is not to exceed 9,000 lb/in² (62 MPa) assuming the pressure to act upon an area circumscribed by the periphery of the outside of the contact surface.

Metal Thickness   Minimum metal thicknesses specified in the tables are based on an allowable fiber stress of 7,000 lb/in² (48 MPa), using the modified Barlow formula of the ASME Boiler Construction Code for cylindrical sections and adding 50 percent to the thickness thus determined to compensate for the shape of the fittings. The minimum commercial casting thickness is considered to be $\frac{1}{4}$ in; therefore, the standards do not show thicknesses less than this. The minimum thickness in these standards means the minimum thickness in any part of the finished casting.

Table 8.7.29 Mechanical Properties of Alloy Steel Bolting Material for High-Temperature Service

Ferritic steels

Grade	Diameter, in (mm)	Minimum tempering temperature,* °F (°C)	Tensile strength, min ksi (MPa)	Yield strength, min 0.2% offset, ksi (MPa)	Elongation in 2 in (50.8 mm), min. %	Reduction of area, min %	Hardness, max
B5, 4 to 6% chromium	Up to 4 (101.6), incl.	1,100 (593)	100 (690)	80 (550)	16	50	
B6, 13% chromium	Up to 4 (101.6), incl.	1,100 (593)	110 (760)	85 (585)	15	50	
B6X, 13% chromium	Up to 4 (101.6), incl.	1,100 (593)	90 (620)	70 (485)	16	50	26 HRC
B7, chromium-molybdenum	2½ (63.5) and under	1,100 (593)	125 (860)	105 (720)	16	50	
	Over 2½ to 4 (63.5 to 101.6)	1,100 (593)	115 (790)	95 (655)	16	50	
	Over 4 to 7 (101.6 to 117.8)	1,100 (593)	100 (690)	75 (515)	18	50	
B7M,† chromium-molybdenum	2½ (63.5) and under	1,150 (620)	100 (690)	80 (550)	18	50	235 HB or 99 HRB
B16, chromium-molybdenum-vanadium	2½ (63.5) and under	1,200 (650)	125 (860)	105 (720)	18	50	
	Over 2½ to 4 (63.5 to 101.6)	1,200 (650)	110 (760)	95 (655)	17	45	
	Over 4 to 7 (101.6 to 177.8)	1,200 (650)	100 (690)	85 (585)	16	45	

Austenitic steels

Class and grade, diameter, in (mm)	Heat treatment‡	Tensile strength, min ksi (MPa)	Yield strength, min 0.2% offset, ksi (MPa)	Elongation in 2 in (50.8 mm), min %	Reduction of area, min %	Hardness, max
Class 1: B8, B8C, B8M, B8P, B8T, B8LN, B8MLN, all diameters	Carbide solution treated	75 (515)	30 (205)	30	50	223 HB or 96 HRB
Class 1A: B8A, B8CA, B8MA, B8PA, B8TA, B8LNA, B8MLNA, B8NA, B8MNA, all diameters	Carbide solution treated in finished condition	75 (515)	30 (205)	30	50	192 HB or 90 HRB
Class 1B: B8N, B8MN, all diameters	Carbide solution treated	80 (550)	35 (240)	30	40	223 HB or 96 HRB
Class 1C: B8R, all diameters	Carbide solution treated	100 (690)	55 (380)	35	55	271 HB or 28 HRC
B8RA, all diameters	Carbide solution treated in finished condition	100 (690)	55 (380)	35	55	271 HB or 28 HRC
B8S, all diameters	Carbide solution treated	95 (655)	50 (345)	35	55	271 HB or 28 HRC
B8SA, all diameters	Carbide solution treated in finished condition	95 (655)	50 (345)	35	55	271 HB or 28 HRC
Class 2: B8, B8C, B8P, B8T, B8N, ¾ (19.05) and under	Carbide solution treated and strain-hardened	125 (860)	100 (690)	12	35	321 HB or 35 HRC
Over ¾ to 1 (19.05 to 25.4) incl.		115 (790)	80 (550)	15	35	321 HB or 35 HRC
Over 1 to 1¼ (25.4 to 31.6) incl.		105 (720)	65 (450)	20	35	321 HB or 35 HRC
Over 1¼ to 1½ (31.6 to 37.9) incl.		100 (690)	50 (345)	28	45	321 HB or 35 HRC
Class 2: B8M, B8MN, ¾ (19.05) and under	Carbide solution treated and strain-hardened	110 (760)	95 (655)	15	45	321 HB or 35 HRC
Over ¾ to 1 (19.05 to 25.4) incl.		100 (690)	80 (550)	20	45	321 HB or 35 HRC
Over 1 to 1¼ (25.4 to 31.6) incl.		95 (655)	65 (450)	25	45	321 HB or 35 HRC
Over 1¼ to 1½ (31.6 to 37.9) incl.		90 (620)	50 (345)	30	45	321 HB or 35 HRC

* For sizes ¾ in. in diameter and smaller, a maximum hardness of 241 HB (100 HRB) is permitted.
† To meet the tensile requirements, the Brinell hardness shall be over 201 HB (94 HRB).
‡ Class 1 is solution treated. Class 1A is solution treated in the finished condition for corrosion resistance: heat treatment is critical due to physical property requirement. Class 2 is solution-treated and strain-hardened. Austenitic steels in the strain-hardened condition may not show uniform properties throughout the section particularly in sizes over ¾ in (19.05 mm) in diameter.

The modified Barlow formula is as follows: For pipes having nominal diameters of ¼ to 5 in, $P + 125 = 2S(t - 0.065)/D$. For pipes of nominal diameters over 5 in, $P = 2S(t - 0.1)/D$, where P is the working pressure, lb/in^2, t is the thickness of wall of pipe, in; D is the actual outside diameter of pipe, in; and S is 7,000 lb/in^2 (48 MPa).

Ring Joints   The dimensions used for ring and groove joint facings were developed by a committee of the API. The corresponding dimensions and ring numbers incorporated in the ANSI Standard are identical with those given in API Standard 5G. The dimension for the depth of groove is added to the basic flange thickness, which makes it necessary to include separate tables of dimensions for fittings having the ring joint facing.

Fitting Dimensions   An inspection limit of ±¹⁄₃₂ in is allowed on all center-to-contact surface dimensions for sizes up to and including 10 in, and ±¹⁄₁₆ in on sizes larger than NPS 10. An inspection limit of ±¹⁄₁₆ in is allowed on all contact-surface to contact-surface dimensions for sizes up to and including NPS 10, and ±¹⁄₈ in, on sizes larger than NPS 10.

When elbows having longer radii than specified in the standards are required, the use of pipe bends is recommended.

Laterals   The 45° laterals of the larger sizes may require additional reinforcement to compensate for the inherent weakness in this shape of casting.

Valve Dimensions   Face-to-face and end-to-end dimensions of ferrous valves for the various pressures are in accordance with the requirements of ANSI B16.10 for ferrous flanged valves.

Reducing Fittings   Reducing fittings have the same center-to-flange edge dimensions as those of straight-size fittings of the largest opening.

Side-Outlet Fittings   All side-outlet fittings have all openings on the intersecting centerlines.

Welding Neck Flanges   The materials, facings, spot facings, etc., conform to the requirements given for other flanges, with the additional provision that the carbon content of the steel shall not exceed 0.35 percent.

Templates for drilling and center to contact-surface dimensions of the American Standard class flanges and flanged fittings are listed in the Standard.

Flanged Pipe Joints

The usual form of pipe joint is that made up by bolting together flanges cast or forged integral with the pipe or fitting, threaded flanges, loose flanges on pipes with lapped ends, and flanges arranged for welding. These forms are illustrated in Fig. 8.7.9. The threaded joint is satisfactory for low and medium steam pressures. The lapped joint is permitted in the same sizes and service ratings as for joints with integral flanges. It is extensively used in high-class work. With the ring joint a higher pressure can be maintained with the same total bolt stress than is possible with the flat gasket type of joint. The welded joint eliminates possibility of leakage between flange and pipe. It is very successful in lines subject to high temperatures and pressures and heavy expansion strains. The welding-neck flange is available in the various pipe sizes. Specific requirements covering the application of all the types of joints in common use are outlined in the Code for Power Piping (ASME B31.1 and B31.3).

Facing of Flanges   Various styles of finish are used on the faces of flanges, for the purpose of the retention of the gasket used to make a tight joint. Those in general use are as follows: plain straight face, plain face corrugated or scored, male and female, tongue and groove, and raised face.

The **plain straight face** has the entire face of the flange faced straight across and may be used with either a full face or ring gasket. The **plain face, serrated or V-grooved,** is a plain face upon which concentric grooves have been cut with either a round-nose or V-shaped tool. This finish is sometimes of advantage when the service demands an exceptionally thick, loosely woven fibrous or soft metallic gasket, because the roughening of the faces of the flanges tends to keep the gasket from blowing out. The **male-and-female** facing consists of a recess in one flange and a corresponding raised face or projection on the other mating flange extending from the inside of the pipe nearly to the inside

Fig. 8.7.9   Welded flange joints and ring joint. (*a*) Forged steel, screwed flange, back-welded and refaced; (*b*) forged steel, slip-on welding flange, welded front and back, refaced; (*c*) forged steel, welding neck flange, butt-welded to pipe; (*d*) lap-welding nipple, butt-welded to pipe; (*e*) ring joint.

of the bolt holes. In the **tongue-and-groove** facing, the tongue or raised face and the groove or recess are narrow rings located between the bolt holes and the port. The male-and-female and the tongue-and-groove facing have been extensively used, particularly on hydraulic lines. To a more limited extent they have been used also on high-pressure steam lines. Both these types, however, have in common several objectionable features from the standpoint of manufacture, erection, and maintenance. These objections are removed by the use of the **raised-face** facing, which consists of a high narrow raised ring on each of the mating flanges, whose inside diameters is the same as that of the pipe or port. It is particularly recommended for high-pressure steam and hydraulic lines. **Gaskets** used in this type of joint are either soft fibrous material or soft metal and extend from the inside of the pipe to the bolt holes. Only the small portion in contact with the narrow raised face is subjected to the compressive effect of the bolts.

The following advantages are claimed for the raised-face type of facing: all mating of flanges has been eliminated; any valve or fitting may be removed from the line without springing the line apart; the gasket is automatically centered by its outer edge coming in contact with the bolts; the outside edges of the flanges are far enough apart to make it possible to determine whether the joint has been properly made.

Unions may be classified as **screw** and **flange.** Typical designs are shown in Fig. 8.7.10, where at the top left is represented a female screw union of the gasket type, at the top right a female screw union having a brass to iron seat that is noncorrosive and a ground joint that eliminates the need for a gasket, and at the bottom a flange union of the gasket type. As in the case of other pipe fittings, unions and union fittings are available in the various pipe sizes and in materials and designs suitable for any service conditions. Very large flange unions can be made by bolting together two screwed companion flanges.

Threaded Fittings

Threaded fittings are made of cast iron, malleable iron, cast steel, forged steel, or brass. Plain standard fittings are generally used for low-pressure gas and water, as in house plumbing and railing work, while the beaded fitting is the standard steam, air, gas, or oil fitting. Screwed fittings are supplied with a large factor of safety. The questions of strength involve much more than the pressure from within the pipe which induces a comparatively low stress in the material. The greater strains come from expansion, contraction, weight of piping, settling, water hammer, etc.

Fig. 8.7.10 Types of pipe unions.

The dimensions of ferrous plugs, bushings, locknuts, and caps with pipe threads are covered by ANSI B16.14.

The normal **amount of thread engagement** necessary to make a tight joint for ANSI Standard pipe thread joints as recommended by Crane Co. is as follows:

Size of pipe, in	1/8	1/4	3/8	1/2	3/4	1	1 1/4	1 1/2	2
Length of thread, in	1/4	3/8	3/8	1/2	9/16	11/16	11/16	11/16	3/4

Size of pipe, in	2 1/2	3	3 1/2	4	5	6	8	10	12
Length of thread, in	15/16	1	1 1/16	1 1/8	1 1/4	1 5/16	1 7/16	1 5/8	1 3/4

The Manufacturers' Standardization Society of Valve and Fitting Industry (MSS) has standardized malleable-iron and brass threaded fittings for several pressures.

Cast-bronze threaded fittings are made in both the class 125 and 250 standards. They are used for any water pipe where bad water makes steel pipe undesirable. Bronze fittings may be had in iron pipe sizes. Forged-steel threaded fittings are made for cold water or oil-working pressures up to 6,000 lb/in^2 (41.4 MPa) hydrostatic. The ANSI has approved standard B16.26 for cast copper alloy fittings for flared copper tubes for maximum cold-water service pressure of 175 lb/in^2 (1,207 kPa).

Railing Fittings Fittings of special construction and of tighter weight than standard steam, gas, and water pipe fittings are widely used for hand railings around areaways, on stairs, for office enclosures with gates, and for permanent ladders. Railing fittings are made in various styles, generally globe-shaped in body, with ends reduced to take thread and recessed to cover all threads. They are furnished in malleable iron, black and galvanized, and in brass.

Special railing-fitting joints are available, such as the slip-and-screwed joint, where the post connection is screwed and the rim of the fitting is so made that the rail will slip into the fitting and allow for an angular variation of several degrees, being fastened by pins which are riveted over and filed smooth. The flush-joint stair-rail fitting is another special style of fitting which provides a hand rail with even surfaces at the joints.

Drainage fittings, have no pockets for the lodgment of solids, and the length of the thread chamber is such that when the pipe is threaded to the American National Standard dimensions, the end of the pipe will practically touch the shoulder when screwed in. They are especially adapted to plumbing work and vacuum-cleaning pipe installations. Dimensions conform to ANSI Standard B16.12, Cast-Iron Threaded Drainage Fittings.

The development of standards for **cast-iron long-turn sprinkler fittings** was begun by the National Fire Protection Assoc. in 1914 with a study of the peculiar needs of fittings intended for fire-protection purposes. These fittings (screwed and flanged) are rated at 175 and 250 lb/in^2 (1,207 and 1,724 kPa).

American National Standard Air Gaps and Backflow Preventers in Plumbing Systems, ANSI A40.4 and A40.6, were prepared to establish minimum requirements for plumbing, including water-supply distributing systems, drainage and venting systems, fixtures, apparatus, and devices, and the standardization of plumbing equipment in general.

Ammonia valves and fittings must provide a high margin of safety against accidents. Flanged valves and fittings have tongue-and-groove faces to assure tightness at the joints and against blowing out gaskets. Gaskets are compressed asbestos sheet. Threaded valves and fittings have long threads and are recessed so that the joints may be soldered. These valves and fittings are made of malleable iron, ductile iron, ferrosteel, or forged steel; depending on the size and style. Valves are all iron, with steel stems, and have special lead disk faces or steel disks. Copper or brass must not be used in their construction. Flanged valves are generally interchangeable with flanged fittings. All valves and fittings for ammonia are tested to 300 lb/in^2 (2,069 kPa) air pressure under water. For dimensions of valves, fittings, and specialties for ammonia, refer to manufacturers' catalogs.

Soldered-Joint Fittings The American standard for these fittings (ANSI B16.18) covers certain dimensions of soldered-joint wrought metal and cast brass fittings for copper water tubing including (1) detailed dimensions of the bore, (2) minimum specifications for materials, (3) minimum inside diameter of the fitting, (4) metal thickness for both wrought metal and cast brass fittings, and (5) general dimensions for cast brass fittings including center-to-shoulder dimensions for both straight and reducing cast fittings. Pressure and temperature ratings are also given. Sizes of the fittings are identified by the nominal tubing size as covered by the Specifications for Copper Water Tube (ASTM B88).

Valves

The face-to-face dimensions of ferrous flanged and welding end valves are given in ANSI B16.10. The types covered are:

Wedge-Gate Valves Cast iron, for 125-, 175-, and 250-lb/in^2 (862, 1,207, and 1,724 kPa) steam service pressure and 800-lb/in^2 (5,516 kPa) hydraulic pressure, and steel, for 150-, 300-, 400-, 600-, 900-, and 1,500-lb/in^2 (1,034-, 2,068-, 2,758-, 4,137-, 6,206-, and 10,343-kPa) steam service pressures (see Fig. 8.7.11).

Fig. 8.7.11 Wedge gate valves.

Double-Disk Gate Valves Cast iron, for 125- and 250-lb/in^2 (862- and 1,724-kPa) steam service pressure and 800-lb/in^2 (5,516-kPa) hydraulic pressure.

Globe and Angle Valves Cast iron, for 125- and 250-lb/in^2 (862- and 1,724-kPa) steam service pressure, and steel, for 150-, 300-, 400-, 600-, 900-, 1,500-, and 2,500-lb/in^2 (862-, 2,068-, 2,758-, 4,137-, 6,206-, 10,343-, and 17,238-kPa) steam service pressures.

Swing-Check Valves Cast iron, for 125- and 250-lb/in^2 (862- and 1,724-kPa) steam service pressure and 800-lb/in^2 (5,516-kPa) hydraulic pressure, and steel, for 150-, 300-, 400-, and 600-lb/in^2 (1,034-, 2,068-, 2,758-, and 4,137-kPa) steam service pressures.

Except for ring-joint facings to the face-to-face dimension for flanged valves is the distance between the faces of the connecting end flanges upon which the gaskets are actually compressed, i.e., the "contact surfaces."

All flanges for class 125 cast-iron valves are plain-faced. The facings of the class 250 cast-iron, and the class 150 and 300 steel valves have a

Fig. 8.7.12   Globe valve and angle valve.

$1/16$-in raised face which is included in the contact-surface to contact-surface dimensions. The contact-surface to contact-surface dimensions of steel valves for class 400 and higher pressures and for cast-iron valves for class 800 hydraulic pressure include a $1/4$-in raised face.

The end-to-end dimensions for **welding-end** valves for sizes NPS 1 to 8 are the same as the contact-surface to contact-surface dimensions given in the tables for steel valves. For details of welding bevel see ANSI B16.10 and Fig. 8.7.15.

A plus or minus **tolerance** of $1/16$ in is allowed on all face-to-face dimensions of valves NPS 10 and smaller, and a tolerance $1/8$ on sizes NPS 12 and larger.

Cocks   The ordinary plug cock operated by a handle or wrench is a form of valve in comparatively small sizes suitable for ordinary service only. The ASME Code for Pressure Piping requires that where cocks are used for high-temperature service they shall be so designed as to prevent galling, either by making the plugs of different material from the body of the cock or by treating the plugs to ensure different physical properties. By means of special design features that eliminate the tendency to leak and stick, the plug-cock type of valve has become available in large sizes and for severe service conditions. Sizes are listed as high as 30 in and are gear-operated in the larger sizes. For further details, refer to manufacturers' catalogs.

Expansion and Flexibility

Piping systems must be designed so that they (1) will not fail because of excessive stresses, (2) will not produce excessive thrusts or moments at connected equipment, or (3) will not leak at joints because of expansion of the pipe. Flexibility is provided by changes of direction in the piping through the use of bends or loops, or provision may be made to absorb thermal strains by use of expansion joints. All, or portions, of the pipe may be corrugated to improve flexibility; in many systems, however, sufficient change is provided by the geometry of the layout to make unnecessary the use of either expansion joints or corrugated sections of piping. Proper cold springing is beneficial in assisting the piping system to attain its most favorable condition. Because of plastic flow of the piping material, hot stresses tend to decrease with time while cold stresses tend to increase with time; their sum, called the stress range, remains substantially constant. For this reason no credit is warranted with regard to stresses; for calculation of forces and moments, the effect of cold spring is recognized by use of a cold-spring factor varying from 0 to 1 for cold spring varying from 0 to 100 percent.

The allowable stress range S_A is calculated by

$$S_A = f(1.25S_c + 0.25S_h)$$

where S_c and S_h are the S values for the minimum cold and maximum hot conditions, respectively. The stress-reduction factor f is a function of the number of hot-to-cold-to-hot (full) temperature cycles anticipated over the life of the plant, as follows:

Total no. of full temp cycles over expected life	Stress-reduction factor
7,000 and less	1.0
14,000 and less	0.9
22,000 and less	0.8
45,000 and less	0.7
100,000 and less	0.6
Over 100,000	0.5

The bending and torsional stresses calculated (see paragraph 119.6.4 of ANSI B31.1.0-1983) are used to determine the maximum computed expansion stress $S_E = \sqrt{S_b^2 + 4S_t^2}$, where S_b and S_t are bending and torsional stresses, respectively. S_E must not exceed the allowable stress range S_A.

In recent years, many principal high-temperature steam lines have either been analyzed, tested in a model-testing machine, or both. No rigid rule is stipulated for the requirement of analysis or model test; however, the Code for Pressure Piping suggests that when the following criterion is not satisfied, need for an analysis is indicated: $DY/(L - U)^2 \leq 0.03$, where D is the nominal pipe size, in; Y is the resultant of movements to be absorbed by pipeline, in; U is the length of straight line joining the anchor points, ft; and L is the length of the developed line axis, ft. A wide assortment of software is available which makes computer analysis attractive, rapid, and widely applicable.

Expansion Joints for Steam Pipelines   In many instances it may be economical to care for thermal expansion by use of expansion joints. For low-pressure steam lines, the use of packed expansion joints may be feasible; experience has indicated that packed joints are difficult to maintain when used on high-pressure lines. Figure 8.7.13 shows a type of joint that has been successfully used for high-pressure, high-

Fig. 8.7.13   Expansion joint for steam line. (*Croll-Reynolds, Inc.*)

temperature service. The bellows is designed to take either axial, lateral, or combined axial and lateral deflections. The internal sleeve guides movement of the joint and also protects the flexible bellows from direct contact with the fluid being handled. Face-to-face dimensions and permissible axial and lateral deflections are available from manufacturers' specifications and catalogs.

Where large lateral deflections are to be absorbed, two expansion joints separated by a length of pipe as shown in Fig. 8.7.14 may be used. With such an arrangement, the lateral deflection permissible with one joint only may be increased many times. Tie rods, as shown, should always be installed to protect the joint against overtravel and externally to guide movement of the joint.

Fig. 8.7.14   Arrangement of expansion joints for large lateral deflection. (*Croll-Reynolds, Inc.*)

Table 8.7.30 Thermal Expansion Data

Material	Coefficient	Temp range: 70°F (21°C) to:													
		70 (21)	200 (93)	300 (149)	400 (205)	500 (260)	600 (316)	700 (371)	800 (427)	900 (482)	1,000 (538)	1,100 (593)	1,200 (649)	1,300 (705)	1,400 (760)
Carbon steel: carbon-moly steel low-chrome	A		6.38	6.60	6.82	7.02	7.23	7.44	7.65	7.84	7.97	8.12	8.19	8.28	8.36
steels (through 3% Cr)	B	0	0.99	1.82	2.70	3.62	4.60	5.63	6.70	7.81	8.89	10.04	11.10	12.22	13.34
Intermediate alloy steals: 5 Cr Mo-9 Cr Mo	A		6.04	6.19	6.34	6.50	6.66	6.80	6.96	7.10	7.22	7.32	7.41	7.49	7.55
	B	0	0.94	1.71	2.50	3.35	4.24	5.14	6.10	7.07	8.06	9.05	10.00	11.06	12.05
Austenitic stainless steels	A		9.34	9.47	9.59	9.70	9.82	9.92	10.05	10.16	10.29	10.39	10.48	10.54	10.60
	B	0	1.46	2.61	3.80	5.01	6.24	7.50	8.80	10.12	11.48	12.84	14.20	15.56	16.92
Straight chromium stainless steels: 12 Cr,	A		5.50	5.66	5.81	5.96	6.13	6.26	6.39	6.52	6.63	6.72	6.78	6.85	6.90
17 Cr, and 27 Cr	B	0	0.86	1.56	2.30	3.08	3.90	4.73	5.60	6.49	7.40	8.31	9.20	10.11	11.01
25 Cr-20 Ni	A		7.76	7.92	8.08	8.22	8.38	8.52	8.68	8.81	8.02	9.00	9.08	9.12	9.18
	B	0	1.21	2.18	3.20	4.24	5.33	6.44	7.60	8.78	9.95	11.12	12.31	13.46	14.65
Monel 67: Ni-30 Cu	A		7.84	8.02	8.20	8.40	8.58	8.78	8.96	9.16	9.34	9.52	9.70	9.88	10.04
	B	0	1.22	2.21	3.25	4.33	5.46	6.64	7.85	9.12	10.42	11.77	13.15	14.58	16.02
Monel 66: Ni-29 CuAl	A		7.48	7.68	7.90	8.09	8.30	8.50	8.70	8.90	9.10	9.30	9.50	9.70	9.89
	B	0	1.17	2.12	3.13	4.17	5.28	6.43	7.62	8.86	10.16	11.50	13.00	14.32	15.78
Aluminum	A		12.95	13.28	13.60	13.90	14.20								
	B	0	2.00	3.66	5.39	7.17	9.03								
Gray cast iron	A		5.75	5.93	6.10	6.28	6.47	6.65	6.83	7.00	7.19				
	B	0	0.90	1.64	2.42	3.24	4.11	5.03	5.98	6.97	8.02				
Bronze	A		10.03	10.12	10.23	10.32	10.44	10.52	10.62	10.72	10.80	10.90	11.00		
	B	0	1.56	2.79	4.05	5.33	6.64	7.95	9.30	10.68	12.05	13.47	14.92		
Brass	A		9.76	10.00	10.23	10.47	10.69	10.92	11.16	11.40	11.63	11.85	12.09		
	B	0	1.52	2.76	4.05	5.40	6.80	8.26	9.78	11.35	12.98	14.65	16.39		
Wrought iron	A		7.32	7.48	7.61	7.73	7.88	8.01	8.13	8.29	8.39				
	B	0	1.14	2.06	3.01	3.99	5.01	6.06	7.12	8.26	9.36				
Copper-nickel (70/30)	A		8.54	8.71	8.90										
	B	0	1.33	2.40	3.52										

A = mean coefficient of thermal expansion $\times\ 10^6$, in/(in · °F) in going from 70°F (21°C) to indicated temperature.
B = linear thermal expansion, in/100 ft in going from 70°F (21°C) to indicate temperature.
Multiply values of A shown by 1.8 to obtain coefficient of expansions in cm/(cm · °C).
Multiply values of B shown by 8.33 to obtain linear expansion in cm per 100 m.
SOURCE: ANSI B31.1.

Table 8.7.30, extracted from the Code for Pressure Piping, lists thermal-expansion data for both ferrous and nonferrous piping. For expansion at temperatures intermediate between those shown, straight-line interpolation is permitted.

The **rubber expansion joint** has become an established part of pipeline equipment. Its special field of application is on low-pressure and vacuum lines in condenser applications, etc., and it is recommended for pressures up to 25 lb/in² (172 kPa) gage where the maximum temperature does not exceed 250°F. Standard joints for pressure installations are reinforced to withstand working pressures up to 125 lb/in² (862 kPa) gage and temperatures up to 200°F. Joints are available in all standard pipe sizes.

Welding in Power-Plant Piping

The majority of main-cycle and service steel piping in modern steam power plants is of welded construction. Steel pipe of NPS 2 and smaller is generally socket-welded; larger-size piping is usually butt-welded. Frequently, depending on location and scheduling, piping larger than NPS 2 is prefabricated; smaller piping is shipped to the construction site in random lengths and is fabricated concurrently with installation. Small-sized chromium-molybdenum piping requiring bending is frequently also shop-fabricated so as to avoid high field preheat, welding, and stress-relieving costs. It is desirable to schedule shipment of hangers so that they will be available at the job site upon arrival of the prefabricated piping; this avoids the expense of providing, installing, and later removing temporary hangers and supports. Aside from the economy of welded construction, it is a virtual necessity in high-pressure, high-temperature work because of danger of leakage if joints are flanged.

Shop welds are frequently made by automatic or semiautomatic submerged-arc or inert-gas shielded-arc processes; field welds are generally of the manual type and are usually done by the shielded metal-arc and/or inert-gas metal-arc processes. Welding in power piping systems, whether in the shop or at the job site, must be done by welders who have qualified under provisions of the Code for Pressure Piping or the ASME Boiler and Pressure Vessel Code.

End Preparation for Butt Welds Figure 8.7.15 shows the end preparation recommended (not required) for piping whose wall thickness is ¾ in or less, and Fig. 8.7.16 shows that required for piping

Fig. 8.7.15 Recommended end preparation for pipe wall thickness of ¾ in or less.

Fig. 8.7.16 Recommended end preparation for pipe wall thickness greater than ¾ in.

with wall thickness above ¾ in. During the welding process, to avoid entrance of welding material into the pipe, backing rings may be used as shown in Fig. 8.7.17a, b, and c. Consumable inserts are also available. They are recommended for installation in piping systems which require a smooth, unobstructed interior surface. Note that thick-walled pipes (over ¾ in) are taper-bored on the inside in order that they may receive a tapered, machined backing ring.

Preheating Prior to start of welding, many materials require preheat to a specified temperature: preheat may be done by electrical-resistance or induction heating or by ring-type gas burners placed concentrically with the pipe. The preheat temperature is measured by indicating crayons or by thermocouple pyrometers and must be maintained during the welding operation. Table 1, Appendix D, of the Code for Pressure Piping lists materials used in piping systems and the appropriate temperatures for preheat. In general, the following is indicative of the intent only; for specific instances, the Code must be consulted.

Carbon steel and **wrought iron** should be preheated to a "hand-hot" condition if the ambient temperature at time of field installation is 32°F (0°C) or less: carbon steels which have minimum tensile properties of 70,000 lb/in² (483 MPa) or higher should be preheated to 250°F

Fig. 8.7.17 Recommended backing ring types. (*a*) Butt joint with split backing ring; (*b*) butt joint with bored pipe ends and solid machined or split backing ring; (*c*) butt joint with taper-bore ends and machined backing ring.

(121°C); under other conditions, preheat is not mandatory, but some purchasers insist that the contractor preheat heavy-walled piping such as boiler feed.

Low-alloy steels with a chromium content not exceeding ¾ percent and low-alloy steels with a total alloy content not exceeding 2 percent are required to be preheated to a minimum temperature of 300°F (149°C).

Alloy steels with a chromium content between ¾ and 2 percent and low-alloy steels with a total alloy content not exceeding 2¾ percent require preheating to 375°F (191°C) minimum. Those with a total alloy content greater than 2¾ percent but not exceeding 10 percent require preheating to a temperature of 450°F (232°C) minimum.

High-alloy steels containing the martensitic phase require preheating to 450°F (232°C) minimum; preheating is a matter of agreement between the purchaser and contractor in the case of welding high-alloy ferritic steels (ASTM A240 and A268). The possible advantages of preheat have not been established in the case of welding high-alloy austenitic steels, and for this reason the Code for Pressure Piping states that preheat is optional for these materials.

Welding procedure varies with material and welding process. In general, the pipe ends must be cleaned of oil or grease, and excessive amounts of scale or rust should be removed. The size and type of welding rod must be stated; the number of layers or passes is determined by the thickness of the pieces being joined. All slag or flux remaining on any bead of welding must be removed before laying down the next successive bead; any cracks or blowholes that appear on the surface of any bead must be chipped or ground away before the next bead of weld material is deposited. Throughout the welding process, it is essential that the minimum specified preheat temperature be maintained.

Stress Relieving Welded joints in all carbon-steel material whose thickness is ¾ in (1.91 cm) or greater must be stress-relieved at a temperature of 1,100°F (593°C) or over for a period of time proportioned on the basis of at least 1 h/in of pipe-wall thickness (but in no case less than ½ h) and then allowed to cool slowly (generally under a blanket)

and uniformly. No stress relief is required for joints in carbon-steel piping whose wall thickness is less than ¾ in.

Welded joints in alloy steels with a wall thickness of ½ in (1.27 cm) or greater, having a chromium content not exceeding ¾ percent, and low-alloy steels with a total alloy content not exceeding 2 percent require stress-relieving at a temperature of 1,200°F (649°C) or over for a period of time proportioned on the basis of at least 1 h/in (0.4 h/cm) of wall thickness, but in no case less than ½ h.

Welded joints in alloy steels having a chromium content exceeding ¾ percent, or a total alloy content exceeding 2 percent, except high-alloy ferritic (ASTM A240, A268) and austenitic steels, regardless of wall thickness, require stress relief at a temperature of 1,200°F or over for a period of time proportioned on the basis of at least 1 h/in of wall thickness, but in no case less than ½ h. Stress relief of high-alloy ferritic steel (A240, A268) and austenitic steels is not required but may be performed as agreed upon by purchaser and contractor. In welds between austenitic and ferritic materials, stress relieving is optional and, if used, shall be a matter of agreement between the purchaser and contractor. Because of the difference between the coefficients of thermal expansion of the two dissimilar materials, careful consideration should be given to the selection of a heat treatment, if any, that will be beneficial to the welded joint.

Graphitization is precipitation of carbon at the grain boundaries in the heat-affected zone during the welding process. Such a phenomenon occurs when some metals operate at high temperatures for extended periods. It has been observed particularly in carbon-molybdenum steels that operate at 900°F (482°C) or higher.

Graphitization does not generally occur in carbon-molybdenum steels with over 1 percent molybdenum. It also has generally not occurred in the chromium-molybdenum low-alloy steels operated at temperatures between 900°F (482°C) and 1,050°F (566°C). Where graphitization has occurred, the two most commonly used methods for rehabilitation of the pipe are (1) gouging out the heat-affected zone of the weld deposit and rewelding the area with electrodes depositing carbon-molybdenum weld metal, followed by a stabilization heat treatment at 1,300°F (704°C) for 4 h, or (2) solution annealing the weld joints at 1,800°F (982°C), followed by a stabilization heat treatment.

Pipe Supports

The Code for Pressure Piping includes many types of supports and gives directions for their application. A proper pipe support must have a strong rigid base properly supported, and an adjustable roll construction which will maintain the alignment in any direction. It is important to avoid friction caused by the movement of the pipe in the support and to have all parts of sufficient strength to maintain alignment at all times. Wire hangers, band iron hangers, wooden hangers, hangers made from small pipe, and hangers having one vertical pipe support do not maintain alignment.

The direction of expansion in a pipe run can be predetermined by anchoring one end, both ends, or the middle. **Anchors** must be firmly fastened to a rigid and heavy part of the power-plant structure, and must also be securely fastened to the pipe; otherwise the equipment for absorbing expansion is useless, and severe stresses may be thrown on parts of the piping system. Some methods of support are shown in Figs. 8.7.18 and 8.7.19. Welded steel brackets (Fig. 8.7.18*a*) are available in light, medium, and heavy weights. Many types of supports can

(a)   (b)   (c)   (d)

Fig. 8.7.18 Methods of supporting pipes.

Fig. 8.7.19   Pipe supports on cast-iron rolls.

be mounted on these brackets, such as the anchor chair shown on the bracket at (*a*), pipe roller supports of the type at (*c*), pipe roll stands of various types such as shown in Fig. 8.7.19, pipe seats, etc. Figure 8.7.18*b* illustrates one of the many types of adjustable ring hangers in use. The split ring hanger can be applied after the pipeline is in place. At (*c*) in Fig. 8.7.18 is shown a spring cushion pipe roll hanger recommended for service where constant support is required and compensation must be made for movement of the piping. The support afforded by the hanger of Fig. 8.7.18*c* is constant only in the sense that some degree of support is always present. It might be more appropriately termed a variable-support device. The springs provide an efficient means of absorbing the vibration. Figure 8.7.18*d* shows one of the many types of pipe saddle supports available. Figure 8.7.19 shows a cast-iron pipe roll stand designed for cases where vertical adjustment is not necessary but where provision must be made for expansion and contraction of the pipeline. Several designs of such stands with provision for vertical adjustment and of the same general dimensions are also available. One type of cast-iron roll and plate, illustrated in Fig. 8.7.19, provides for expansion and contraction where vertical adjustment is not necessary. If necessary, the baseplate can be raised or lowered by use of shims. Detailed information and dimensions of a great variety of pipe supports can be found in manufacturers' catalogs.

In supporting a high-temperature piping system, it is necessary to provide for expansion and contraction due to cyclic changes. It is often possible to find a point of zero movement along the run of a long line and to support a considerable portion of the total load by a rigid hanger or support of the type shown in Figs. 8.7.18 and 8.7.19. However, for other portions of the run, some form of spring support is often indicated. For relatively light lines, which are not subjected to excessive movements from hot to cold positions, a variable spring hanger will frequently suffice; for heavy lines, or those in which expansion movements are great, it is advisable to use constant support of counterweighted hangers so that transfer of weight to other hangers or equipment connections is prevented. Parts (*a*) and (*b*) of Fig. 8.7.20 indicate, respectively, a horizontal and vertical run of piping supported by a **constant-support hanger**. Figure 8.7.20*c* and 8.7.21*a* indicate horizontal runs supported by **variable-spring hangers**. Figure 8.7.21*b* shows a riser supported by a variable spring beneath a base elbow. Figure 8.7.21*c* indicates a **sway brace** that is used to control vibration and undesirable movement in a piping system.

The principal supports utilized for the support of critical piping involve constant-support hangers, variable-spring hangers, rigid hangers, and restraints.

Constant-Support Hangers   This type of hanger provides a constant supporting force for the piping system throughout its full range of vertical pipe movement. This is accomplished through use of a spring coil working in conjunction with a lever in such a way that the spring force times its distance to the lever pivot is always equal to the pipe load times its distance to the lever pivot. This type of support is "thermally invisible," as the supporting force equals the pipe weight throughout its entire expansion or contraction cycles.

These hangers are used on systems or at locations where stresses are considered critical. Pipe weight reactions or transfer of loads are not imposed in the system or connections with this type of device.

As the load is considered constant with the unit in travel, readings from inspections are based on travel. The readings are taken from the position of an indicator and its relation to the numbers on the travel scale. The scale is divided into 10 divisions. H or high on the scale is equal to (0.0), M or midway on the scale is equal to (5.0), and L or low on the scale is equal to 10.

The design settings were obtained from the position of the factory-installed buttons that are placed adjacent to the travel scale.

"Perfect" readings would be if the indicator were to line up with the white button (cold) and the red (hot); however, this is rarely the case.

Fig. 8.7.20   Constant support and variable-spring hangers.

Fig. 8.7.21   Spring hangers and sway brace.

Generally, readings are considered acceptable and not noteworthy as long as they reflect movement consistent with design in both direction and length. A general rule is that when the hot setting is higher than the cold setting, then movement is down from cold to hot. If the cold setting is higher, movement is up.

The following terms are normally applied to these devices:

Actual travel: Anticipated movement of the pipe from design. The hot and cold position stickers are a function of this movement.

Total travel: The maximum movement a support can accept without danger of topping or bottoming out. The scale from H (0.0) to L (10.0) is a function of this.

Topped out: The indicator is above the high point and in contact with the end of the slot. This condition means the support is unloaded or is in the process of unloading.

Bottomed out: The indicator is below the (10.0) point and in contact with the end of the travel slot; the support is overloaded.

Variable-Spring Hangers These devices are installed at locations where stresses are not considered to be critical or where movement and economics permit their use.

The inherent characteristics of a variable are such that the supporting force varies with the spring's deflection. Movement of the pipe causes the spring to extend or compress. This results in a change or variance in the supporting force. Since the weight of the pipe is the same in either condition, hot or cold, the variation results in pipe weight transfer to equipment and adjacent hangers and consequently additional stresses in the piping system. The effects of this variation are usually considered during the original design.

In addition, as it is desirable to support the actual weight of the pipe when the system is hot, when the stresses tend to become most critical, the hot load is the dead weight of the pipe. The cold load is actually under- or oversupporting the pipe, depending on the movement from cold to hot.

The general rule for determining movement is similar to that of constant supports. If the hot load is higher than the cold, then pipe movement is down from cold to hot. If the cold load is higher, then movement is up.

Unlike constant supports, the readings from variables are measured in pounds. The readings are taken by noting the position of the indicator relative to a load scale that is adjacent to the travel slot.

The **distance between supports** will vary with the kind of piping and the number of valves and fittings. Supports should be provided near changes in direction, branch lines, and particularly near valves. The weight of piping must not be carried through valve bodies. In establishing the location of pipe supports, the designer should be guided by two requirements: (1) the horizontal span must not be so long that sag in the pipe will impose an excessive stress in the pipe wall and (2) the pipeline must be pitched downward so that the outlet of each span is lower than maximum sag in the span. Otherwise entrapped water can result in severe water hammer and pipe swings, particularly during plant start-up of steam piping.

Fabrication and installation practices are provided in MSS Standard Practice SP-89. Table 8.7.31 lists spacing for standard-weight pipe supports.

Pipe Insulation

(See Secs. 4 and 6 for heat-transmission data.)

The value of a steam-pipe covering is measured by its ability to reduce heat losses. This might range from 50 percent for small, low-temperature lines to 90 percent for large, high-temperature lines. Many **pipe-insulating materials** are available: 85 percent magnesia, foam glass, calcium silicate, and various forms of diatomaceous earths. Some of these materials are suited for relatively low temperatures only, others are best suited for high temperatures, and still others are suitable over a considerable temperature range.

Pipe insulation is applied in molded sections 3 ft long. For high-temperature work, the insulation is applied in at least two layers with the joints staggered so as to prevent a direct channel for heat loss. Because of its maximum-temperature limitation of about 600°F (316°C), 85 percent magnesia is used as the second layer with a high-temperature-resistant material placed in direct contact with the pipe. The molded insulation is fastened securely in place with copper or galvanized wire and is then given a surface finish; indoor pipes are first sheathed with resin paper and covered with canvas, either pasted or sewed; outdoor pipes may be weather-protected by a coating of asphaltic-type waterproofing compound, they may be sheathed and canvased and then given a weatherproof surface, so they may be encased in metallic (steel or aluminum) jackets.

The heat loss from an insulated pipe appears in three phases: heat passes by *conduction* through the metallic pipe walls and through the insulating material; it then is dissipated from the outdoor surface of the insulation by *convection* and by *radiation*. Extremely accurate calculations must also take into account the temperature drop by convection through the film on the inside surface of the pipe. The task of accurately calculating heat losses is somewhat tedious, since the convection and radiation losses are related to the surface temperature (outside of insulation), which is unknown until conduction losses are balanced against surface losses.

For combined convection and radiation coefficients for bare pipes, and all necessary formulas to permit trial-and-error calculations, see Sec. 4. Insulation manufacturers publish data which give heat losses for wide ranges of pipe size and temperature.

Identification of Piping

The American National Standards Institute has approved a **Scheme for the Identification of Piping Systems** (ANSI A13.1). This scheme is limited to the identification of piping systems in industrial plants, not including pipes buried in the ground, and electric conduits. Fittings, valves, and pipe coverings are included, but not supports, brackets, or other accessories.

Classification by Color All piping systems are classified by the nature of the material carried. Each piping system is placed, by the nature of its contents, in the following classifications:

Class	Color
F—Fire-protection equipment	Red
D—Dangerous materials	Yellow (or orange)
S—Safe materials	Green (or the achromatic colors, white, black, gray, or aluminum)
P—Protective materials	Bright blue
V—Extra valuable materials	Deep purple

Method of Identification At conspicuous places throughout a piping system, color bands should be painted on the pipes to designate to which one of the five main classes it belongs. If desired, the entire length of the piping system may be painted the main classification color.

Further, the actual contents of a piping system may be indicated by, preferably, a stenciled legend of standard size giving the name of the contents in full or abbreviated form. These legends should be placed on the color bands. The identification scheme may be extended by the use of colored stripes placed at the edges of the colored bands.

Table 8.7.31 Maximum Spacing of Pipe Supports at 750°F (399°C)*

Nominal pipe size, in	1	1½	2	3	4	6	8	10	12	14	16	18	20	24
Maximum span, ft	7	9	10	12	14	17	19	22	23	25	27	28	30	32
Maximum span, m	2.13	2.74	3.05	3.66	4.27	5.18	5.79	6.71	7.01	7.62	8.23	8.53	9.14	9.75

* This tabulation assumes that concentrated loads, such as valves and flanges, are separately supported. Spacing is based on a combined bending and shear stress of 1,500 lb/in² when pipe is filled with water; under this condition, sag in pipeline between supports will be approximately 0.1 in.

The bands, legends, and stripes should be placed at intervals throughout the piping system, preferably adjacent to valves and fittings to ensure ready recognition during operation, repairs, and at times of emergency.

A recommended classification, under this color scheme of materials carried in pipes, includes, as dangerous, combustible gases and oils, hot water and steam above atmospheric pressure; as safe, compressed air, cold water, and steam under vacuum.

Pressure Hose

Hose with durable rubber lining may be obtained to withstand any needed pressure. If the rubber compound is properly made, the life of a hose will be 7 to 10 years or more.

American National Fire-Hose Coupling Screw Thread (ANSI B1.20-7)   This standard is intended to cover the threaded part of fire-hose couplings, hydrant outlets, standpipe connections, and at other special fittings on fire lines, where fittings of the nominal diameters given in Table 8.7.32 are used. It also includes the limiting dimensions of the field inspection gages. The American National Standard form of thread must be used.

Table 8.7.32   Dimensions of Standard Fire-Hose Couplings
(All dimensions in inches. Letters refer to Fig. 8.7.22)

Inside diam, C	Diam of thread, D	No. of threads per inch	L	I	H	J	T
$2\frac{1}{2}$	$3\frac{1}{16}$	$7\frac{1}{2}$	1	$\frac{1}{4}$	$\frac{15}{16}$	$\frac{3}{16}$	$\frac{11}{16}$
3	$3\frac{5}{8}$	6	$1\frac{1}{8}$	$\frac{5}{16}$	$1\frac{1}{16}$	$\frac{1}{4}$	$\frac{13}{16}$
$3\frac{1}{2}$	$4\frac{1}{4}$	6	$1\frac{1}{8}$	$\frac{5}{16}$	$1\frac{1}{16}$	$\frac{1}{4}$	$\frac{13}{16}$
$4\frac{1}{2}$	$5\frac{3}{4}$	4	$1\frac{1}{4}$	$\frac{7}{16}$	$1\frac{3}{16}$	$\frac{3}{8}$	$\frac{15}{16}$

SOURCE: ANSI B1.20.7

American National Standard Hose-Coupling Screw Threads (ANSI B 1.20-7)   These standards apply to the threaded parts of hose couplings, valves, nozzles, and all other fittings used in direct connection with hose intended for fire protection or for domestic, industrial, or general service in nominal sizes given in Table 8.7.33. The American National

Standard thread form is used. This coupling is similar in design to the fire-hose couplings illustrated in Fig. 8.7.22.

Fig. 8.7.22 Typical form of standard coupling.

Flexible metal hose and tubing are available for a wide range of conditions of temperature, pressure, vibration, and corrosion, and are made in two basic constructions, corrugated or interlocked, and in either bronze or steel.

The corrugated type (Fig. 8.7.23) may have either annular or helical corrugated formations, usually covered with metal braid, and is adapted to high-pressure high-temperature leak-proof service. Some typical applications include diesel-engine exhaust hose, reciprocating flexible connections, loading and unloading hose, saturated and superheated steam lines, lubricating lines, gas and oil lines, vibration connections, etc.

The interlocked type is made in several ways; the fully interlocked type is illustrated in Fig. 8.7.24. Typical applications include wiring conduit, cable armor, decorative wiring covering, dust-collective tubing, grease and oil connections, flexible spouts, and moderate-pressure oil lines.

Fig. 8.7.23 Flexible metal hose.

Fig. 8.7.24 Interlocked flexible metal hose.

Standard couplings and fittings can be attached to flexible metal hose or tubing by various methods such as brazing or welding. Each type of hose construction has limits of service use and proved application usages. Information and recommendations as to the type and size to use under any given conditions should be obtained from the manufacturers.

Table 8.7.33   Dimensions of Standard Hose Couplings
(All dimensions in inches. Letters refer to Fig. 8.7.22)

Service and nominal size	Inside diam, C	Diam of thread, D	No. of threads per inch	L	I	H	T
Garden:							
$\frac{1}{2}$, $\frac{5}{8}$, $\frac{3}{4}$	$\frac{25}{32}$	$1\frac{1}{16}$	$11\frac{1}{2}$	$\frac{9}{16}$	$\frac{1}{8}$	$\frac{17}{32}$	$\frac{3}{8}$
Chemical:							
$\frac{3}{4}$, 1	$1\frac{1}{32}$	$1\frac{3}{8}$	8	$\frac{5}{8}$	$\frac{5}{32}$	$\frac{19}{32}$	$\frac{15}{32}$
Fire:							
$1\frac{1}{2}$	$1\frac{17}{32}$	2	9	$\frac{5}{8}$	$\frac{5}{32}$	$\frac{19}{32}$	$\frac{15}{32}$
Other connections:							
$\frac{1}{2}$	$\frac{17}{32}$	$\frac{13}{16}$	14	$\frac{1}{2}$	$\frac{1}{8}$	$\frac{15}{32}$	$\frac{5}{32}$
$\frac{3}{4}$	$\frac{25}{32}$	$1\frac{1}{32}$	14	$\frac{9}{16}$	$\frac{1}{8}$	$\frac{17}{32}$	$\frac{3}{8}$
1	$1\frac{1}{32}$	$1\frac{9}{32}$	$11\frac{1}{2}$	$\frac{9}{16}$	$\frac{5}{32}$	$\frac{17}{32}$	$\frac{3}{8}$
$1\frac{1}{4}$	$1\frac{9}{32}$	$1\frac{5}{8}$	$11\frac{1}{2}$	$\frac{5}{8}$	$\frac{5}{32}$	$\frac{19}{32}$	$\frac{15}{32}$
$1\frac{1}{2}$	$1\frac{17}{32}$	$1\frac{7}{8}$	$11\frac{1}{2}$	$\frac{5}{8}$	$\frac{5}{32}$	$\frac{19}{32}$	$\frac{15}{32}$
2	$2\frac{1}{32}$	$2\frac{11}{32}$	$11\frac{1}{2}$	$\frac{3}{4}$	$\frac{3}{16}$	$\frac{23}{32}$	$\frac{19}{32}$

SOURCE: ANSI B1.20.7.

Section **9**

Power Generation

BY

EUGENE A. AVALLONE *Consulting Engineer; Professor of Mechanical Engineering, Emeritus, The City College of The City University of New York*
EZRA S. KRENDEL *Emeritus Professor of Operations Research and Statistics, University of Pennsylvania*
RAMA RAMAKUMAR *PSO/Albrecht Naeter Professor and Director, Engineering Energy Laboratory, Oklahoma State University*
CHARLES P. BUTTERFIELD *Chief Engineer, National Wind Technology Center, National Renewable Energy Laboratory*
EDUARD MULJADI *Senior Engineer, National Wind Technology Center, National Renewable Energy Laboratory*
ERICH A. FARBER *Distinguished Service Professor Emeritus and Director Emeritus of Solar Energy and Energy Conversion Laboratory, University of Florida*
KENNETH A. PHAIR *Senior Project Engineer, Shaw Stone & Webster*
JOHN B. KITTO, JR., P.E. *Babcock & Wilcox Co.*
FREDERICK G. BAILY *Consulting Engineer, Steam Turbines, General Electric Co.*
WILLIAM J. BOW. *Director (Retired), Heat Transfer Products Dept., Foster Wheeler Energy Corp.*
DONALD E. BOLT *Engineering Manager, Heat Transfer Products Dept., Foster Wheeler Energy Corp.*
DENNIS N. ASSANIS *Professor of Mechanical Engineering, University of Michigan*
DAVID E. COLE *Director, Office for the Study of Automotive Transportation, Transportation Research Institute, University of Michigan*
TIMOTHY J. JACOBS *Research Fellow, Department of Mechanical Engineering, University of Michigan*
D. J. PATTERSON *Professor of Mechanical Engineering, Emeritus, University of Michigan*
JOHN H. LEWIS *Engineering Consultant; formerly Engineering Staff, Pratt & Whitney Division, United Technologies Corp.; Adjunct Associate Professor, Hartford Graduate Center, Rensselaer Polytechnic Institute*
WILLIAM H. DAY *President, Longview Energy Associates, LLC; formerly Founder and Board Chairman, The Gas Turbine Association*
ANDREW C. KLEIN *Professor of Nuclear Engineering, Oregon State University; Director of Training, Education and Research Partnership, Idaho National Laboratories*
ROBERT D. STEELE *Project Manager, Voith Siemens Hydro Power Generation, Inc.*

9.1 SOURCES OF ENERGY

Contributors are shown at the head of each category.

REFERENCES: Latest available published data from the following: "Reserves of Crude Oil, Natural Gas Liquids, and Natural Gas in the U.S. and Canada," American Petroleum Institute. "Annual Statistical Review—Petroleum Industry Statistics," American Petroleum Institute. Worldwide Issue, *Oil & Gas Jour.* annually. "Potential Supply of Natural Gas in the U.S.," Mineral Resources Institute. Colorado School of Mines. Coal resources in the United States, *U.S. Geol. Surv. Bull.* 1412. Geological Estimates of Undiscovered Recoverable Oil and Gas Resources in the U.S., *U.S. Geol Surv. Circ.* 725, United Nations Statistical Office, "Statistical Yearbook," New York, U.N. Department of Economic and Social Affairs. Bureau of Mines, Metals, Minerals, and Fuels, vol. I of "Minerals Yearbook" published annually. "Coal Data," National Coal Association. "International Coal," National Coal Association. Annual Technical Literature Data Base, *Power,* McGraw-Hill.

INTRODUCTION
Revised by E. A. Avallone

Global energy requirements are supplied primarily by fossil fuels, nuclear fuels, and hydroelectric sources; about 1 to 2 percent of global requirements are supplied from other miscellaneous sources. In the United States in 1994, total domestic power requirements were supplied approximately as follows: 70 percent fossil fuel (of which coal accounted for 58 percent), 20 percent nuclear fuel, 10 percent from hydroelectric sources, and less than 1 percent from all other sources. In spite of the large increase in nuclear generated power, both in the United States and globally, coal continues to be the major fuel consumed.

In the United States, new power plants constructed at this time are designed to consume fossil fuels—primarily coal, many with gas, and a few with petroleum. The situation with regard to nuclear power is complicated by a number of circumstances; see Sec. 9.8. Energy statistics and accompanying data stay current for a short time. Bear in mind that when quantities of known reserves of fuel of all types are stated, there is implied the significant matter of whether they are, indeed, producible in a given economic climate. Estimates for additional reserves remaining to be discovered are available, by and large, only for the United States. At any given time, the situation with regard to estimates of recoverable fuel sources is subject to wide swings whose source is manifold: national and international politics, environmental concerns, significant progress in energy conservation, unsettled political and social conditions in locations within which reside much of the world reserves of fossil fuels, economic impact of financing, effects of inflation, and so on. The references cited, in their most current form, will provide the reader with realistic and authoritative compilations of data.

Fossil Fuels

Petroleum Proved reserves of crude oil and natural gas liquids in the United States, based upon estimated discovered quantities which geological and engineering data demonstrate with reasonable certainty to be recoverable in future years from presently known reservoirs under existing economic and operating conditions, are published annually by the American Petroleum Institute. Estimates of additional remaining producible reserves which will be discovered, proved, and produced in the future from the total original oil in place, are derived by *U.S. Geol. Surv. Circ.* 725 from present and projected conditions in the industry.

Estimates of proved crude oil reserves in all countries of the world are published by *Oil and Gas Journal.* New discoveries are continually adding to and changing proved reserves in many parts of the world, and these estimates are indicative of producible quantities.

Natural Gas Proved reserves of natural gas in the United States, based upon the same definition as for crude oil and natural gas liquids, are estimated annually by the American Gas Association. The estimated

total additional potential supply remaining to be discovered is prepared by the Potential Gas Committee, sponsored by the Potential Gas Agency, Colorado School of Mines Foundation, Inc.

Estimates of proved reserves of natural gas in all countries of the world are published by *Oil and Gas Journal.* As with crude oil, large additional natural gas reserves are currently being discovered and developed in Alaska, the arctic regions, offshore areas, northern Africa, and other locations remote from consuming markets. Valid estimates of additional probable remaining reserves in the world are not available.

Coal (See also Secs. 7.1 and 7.2.) Authoritative information about reserves of coal is presented in *Geol. Surv. Bull.* 1412, Coal Resources of the United States. Remaining U.S. proved reserves (1974) of bituminous, subbituminous, lignite, and anthracite have been estimated by mapping and exploration of areas with 0 to 3,000-ft overburden. The U.S. Geological Survey (USGS) estimates probable additional resources in unmapped and unexplored areas with 0 to 3,000-ft overburden and in areas with 3,000- to 6,000-ft overburden. Slightly more than one-half of the proved reserves are considered producible (at this time) because of favorable depth of overburden and thickness of coal strata. Approximately 30 percent of all ranks of coal are commerically available in beds less than 1,000 ft deep. The USGS estimates that about 65 percent contains less than 1 percent sulfur; most of the low-sulfur coals are located west of the Mississippi. *USGS Bull* 1412 also estimates global coal resources, but in view of the questionable validity of much of the global data, it can but offer gross approximations. (See Sec. 7.1.)

Shale Oil The portion of total U.S. reserves of oil from oil shale, measured or proved, considered minable and amenable to processing is estimated to be over 50 billion bbl (30 billion m^3), based upon grades averaging 30 gal/ton in beds at least 100 ft thick (*USGS Bull* 1412). Most oil shale occurs in Colorado. No commercial production is expected for many years. World reserves occur largely in the United States and Brazil, with small quantities elsewhere.

Tar Sands Large deposits are in the Athabasca area of northern Alberta, Canada, estimated (2004) capable of producing 100 to 300 billion bbl (15.9 to 47.7 billion m^3) of oil. About 6.3 billion bbl (1.0 billion m^3) has been proved economically recoverable within the radius of the present large mining and recovery plant in Athabasca. Commercial quantities of oil have been produced there since the 1960s. Sizable

---

Table 9.1.1 Major U.S. Coal-Producing Locations

Anthracite and semianthracite
Pennsylvania

Bituminous coal
Illinois
West Virginia
Kentucky
Colorado
Pennsylvania
Ohio
Indiana
Missouri

Subbituminous coal
Montana
Alaska
Wyoming
New Mexico

Lignite
North Dakota
Montana

deposits are located elsewhere; they have not been exploited to date, meaningful data for them are not available, and there is no report of those other deposits having been worked. (See Sec. 7.1.)

Nuclear Fuels

Uranium Reserves of uranium in the United States are reported by the Department of Energy (DOE). The proved reserves, usually presented in terms of quantity of U_3O_8, refer to ore deposits (concentrations of 0.01 percent, or 0.0016 oz/lb ore, are viable) of grade, quantity, and geological configuration that can be mined and processed profitably with existing technology. *Estimated additional resources* refer to uranium surmised to occur in unexplored extensions of known deposits or in undiscovered deposits in known uranium districts, and which are expected to be discovered and economically exploitable in the given price range. The total of these uranium reserves would yield about 3,000 tons of U_3O_8. United States uranium resources are located mainly in New Mexico, Wyoming, and Colorado.

Thorium Total known resources of thorium, the availability of which is considered reasonably assured, are estimated in the millions of tons of thorium oxide. Additional actual reserves will increase in response to the demand and concomitant market price. Most of the larger known resources are in India and Brazil. There seems to be little prospect of significant requirements for thorium as a nuclear fuel in the near future.

Hydroelectric Power

Hydroelectric and Pumped Storage for Electric Generation Although most available sites for economical production of hydroelectric energy have been developed, some additional hydroelectric capacity will be provided at new sites or by additions at existing plants. Increased pumped storage capacity will be limited by the availability of suitable sites and a dependable supply of economical pumping energy. The flexibility of operation of a pumped storage plant in meeting sudden load changes and its ability to provide high inertia spinning reserve at low operating cost are additional benefits that can weigh heavily in favor of this type of installation, particularly in the future if (when) the proportion of nuclear capacity in service increases. At this time, hydro and pumped storage account for about 8 to 12 percent of electricity generated from all sources of energy in the United States.

World installed hydropower capacity presently is located about 40 percent in North America and 40 percent in Europe.

ALTERNATIVE ENERGY, RENEWABLE ENERGY, AND ENERGY CONVERSION: AN INTRODUCTION
Revised by E. A. Avallone

REFERENCES: AAAS, *Science*. Hottell and Howard, "New Energy Technology—Some Facts and Assessments," MIT Press. Fisher, "Energy Crises in Perspective," Wiley-Interscience. Hammond, Metz, and Maugh, "Energy and the Future," AAAS.

Many sources of raw energy have been proposed or used for the generation of power. Only a few sources—fossil fuels, nuclear fission, and elevated water—are dominant in practical applications today.

A more complete list of sources would include fossil fuels (coal, petroleum, natural gas); nuclear (fission); wood and vegetation; elevated water supply; solar; winds; tides; waves; geothermal; muscles (human, animal); industrial, agricultural, and domestic wastes; oceanic thermal gradients; oceanic currents. There are others.

Historically, wood, muscles, elevated water, and wind were prominent. These sources were superseded in the industrial era by fossil fuels, with nuclear energy the most recent addition. This dominance rests in the suitability of the thermal sources for practical stationary and transportation power plants. Features of acceptability include reliability, flexibility, portability, maneuverability, size, bulk, weight, efficiency, economy, maintenance, and costs. The plant for transportation service must be self-contained. For stationary service there is wider latitude for choice.

The dominant end product, especially for stationary applications, is electricity, because of its favorable distribution and control features. However, there is no practical way of storing electric energy. Electricity must be generated at the instant of its use. Reliability and continuity of service consequently dictate the need for reserve, alternate, and interconnection supports. Pumped storage, coal piles, and tanks of liquid and gaseous fuels, e.g., offer the necessary continuity, flexibility, and reliability.

Raw energy sources, other than fuels (fossil and nuclear) and elevated water, are particularly deficient in this storage aspect. For example, wind power is best for jobs that can wait for the wind, e.g., pumping water or grinding grain. Solar power, to avoid foul weather and the darkness of night, could call for desert locations or extraterrestrial satellites.

Despite such limitations an energy-intensive society can expect to see increasing efforts to harness many of the raw energy sources cited. Several of these topics are treated in the following pages to show the factual and technical progress that has been made to adapt sources to practicality.

MUSCLE-GENERATED POWER
by Ezra S. Krendel, Amended by Staff

REFERENCES: Whitt and Wilson, "Bicycling Science," 2d ed., MIT Press. Harrison, Maximizing Human Power Output by Suitable Selection of Motion Cycle and Load, *Human Factors,* **12**, 1970. Krendel, Design Requirements for Man-Generated Power, *Ergonomics,* **3**, 1960. Wilkie, Man as a Source of Mechanical Power, *Ergonomics,* **3**, 1960. Brody, "Bioenergetics and Growth," Reinhold.

The use of human muscles to generate work will be examined from two points of view. The first is that of measuring the energy expended in gross, long duration physical activities such as marching, forestry work, freight handling, and factory work. The second is that of determining the useful mechanical work which can be performed by specified muscle groups for brief or extended periods of time in well defined work situations, such as pedaling or cranking.

Labor

Over an 8-h day for a 48-h week, a useful norm for a 35-year-old laborer for total power expenditure, including basal metabolism energy, is 0.49 hp (366 W). Of this total expenditure, approximately 0.1 hp (75 W) is available for useful work. A 20-year-old man can generate about 15 percent more power than this norm, and a 60-year-old man about 20 percent less. The total energy or power expenditure is needed for determining nutritional requirements for classes of labor. A rule of thumb for power developed by European males can be expressed as a function of age and duration of effort in minutes for work lasting from 4 to about 480 min, assuming that 20 percent of the total output is useful power.

Age, years	Useful horsepower (t in min)
20	hp = 0.40 − 0.10 log t
35	hp = 0.35 − 0.09 log t
60	hp = 0.30 − 0.08 log t

For a well-trained man, useful power production by pedaling, hand cranking, or a combination of the two for working durations of from 20 to 120 s may be summarized as follows (t is in seconds):

Arms and legs	hp = $4.4t^{-0.40}$
Legs only	hp = $2.8t^{-0.40}$
Arms only	hp = $1.5t^{-0.40}$

There are examples of well-trained athletes generating between 1.5 and 2 hp for efforts of 5 to 20 s, using both arms and legs to generate power.

For pedaling efforts of from 1 to about 100 min, the useful power generated may be expressed as hp = 0.53 − 0.13 log t (t is in minutes).

Work scheduling, either as rhythmic work activity or with rest stops for recuperation, the temperature and humidity of the environment, and the detailed nature of the laborer's diet are factors which influence ability to generate and maintain the above nominal power values. These considerations should be factored in for specific work situations.

Steady State and Transient

When a human and a passive mechanism are working together to generate power, the following conditions obtain: Energy is available both from stores residing in the muscles [a total usefully available energy of about 0.6 hp · min (27 kJ), usually applied in transient bursts of activity] and from the oxidation of foods (for producing steady state power). For an aerobic transient activity, energy production depends on the mass of muscle which can be brought into effective contact with the power transmission mechanism. For example, bicycle pedaling is an effective use of a large muscle mass. For steady-state activity, assuming adequate food for fuel energy, power generated depends on the oxygen supply and the efficiency with which oxygenated blood can be transported to the muscles as well as on the muscle mass.

The **physiological limit,** determined by oxygen-respiration capacity, for steady-state useful mechanical power generation is between 0.4 and 0.54 hp (300 and 400 W), depending on the man's physical condition.

Useful power production may be achieved by such methods as rowing, cranking, or pedaling. The **highest values** of human-generated horsepower using robust subjects have been achieved using a rowing assembly which restrained nonuseful motions of the torso and major limbs. Under these conditions up to 2 hp (1,500 W) was generated over intervals of 0.6 s, and averages of about 1 hp (750 W) were generated over 2 min.

In order to approach an optimal **conversion efficiency** (mechanical work/food energy) of 25 percent, a mechanism would be required to store and to transmit energy from the body muscle masses when they were operating at optimal efficiency. This condition occurs when the force exerted by the muscle is about one-half of its maximum and the speed of muscle movement one-quarter of its maximum. Data on both force and speed for a given set of muscles are best measured in situ. Optimal conversion efficiency and maximum output power do not occur together.

Examples of High Output

Data for human-generated power come from measurements of subjects with different kinds of training, skill, body builds, diets, and motivation using a variety of mechanical devices such as bicycles, ergonometers, and variations on rowing machines. For strong, healthy young men, aggregations of such data for power produced in an interval t of 10 to 120 s can be approximated as follows:

$$hp = 2.5t^{-0.40}$$

For world-class athletes this becomes hp $= 0.25 + 2.5t^{-40}$. These values can be exceeded for bursts of power of less than 10 s. For longterm efforts of from 2 to about 200 min, the aggregated data for useful power generated by strong, healthy young men can be approximated as follows (t in minutes):

$$hp = 0.50 - 0.13 \log t$$

For world-class athletes this becomes hp $= 0.65 - 0.13 \log t$. The pilot of the Gossamer Albatross, who flew 22 mi from England to France in 2 h 55 min on August 12, 1979 entirely by pedal-generated power, sustained an output of about $\frac{1}{3}$ hp (250 W) during the flight.

Maximum power output occurs at a load impedance of 5 to 10 times the size of the human being's source impedance.

Brody has developed detailed nomograms for determining the energetic cost of muscular work by farm animals; these nomograms are useful for precise cost-effectiveness comparisons between animal and mechanical power generation methods. A 1,500- to 1,900-lb horse can work continuously for up to 10 h/day at a rate of 1 hp, or equivalently pull 10 percent of its body weight for a total of 20 mi/day, and retain its vigor to an advanced age. Brody's work allows the following approximations for estimating the useful power output of work animals of varying sizes: The ratio of the power exerted in maximal energy production for a few

seconds to the maximum steady-state power maintained for 5 to 30 min to the power produced in sustained heavy work over a 6- to 10-h day is approximately 25:4:1. For any one of these conditions, it has been found that, for healthy, mature specimens,

$$hp_{animal} = hp_{man}(mass \ of \ animal/mass \ of \ man)^{0.73}$$

Thus, from the previously given horsepower magnitudes for men, one can compute the power generated by ponies, horses, bullocks, or elephants under the specified working conditions.

WIND POWER
by Rama Ramakumar, Charles P. Butterfield, and Eduard Muljadi

REFERENCES: NREL technical information at Internet address: http://gopher.nrel.gov.70. AWEA information at Internet address: awea.wind-net@notes.igc.apc.org. Hansen and Butterfield, Aerodynamics of Horizontal-Axis Wind Turbines, *Ann. Rev. Fluid Mech.*, **25**, 1993, pp. 115–149. Touryan, Strickland, and Berg, "Electric Power from Vertical-Axis Wind Turbines," *J. Propulsion*, **3**, no. 6,1987. Betz, "Introduction to the Theory of Flow Machines," Pergamon, New York. Eldridge, "Wind Machines," 2d ed., Van Nostrand Reinhold, New York. Glauert, "Aerodynamic Theory," Durand, ed., 6, div. L, p. 324, Springer, Berlin, 1935. Richardson and McNerney, Wind Energy Systems, *Proc. IEEE*, **81**, no. 3, Mar. 1993, pp. 378–389. Elliott et al., "Wind Energy Resource Atlas," Wilson and Lissaman, Applied Aerodynamics of Wind Power Machines, Oregon State University Report, 1974. Eggleston and Stoddard, *Wind Turbine Engineering Design*, New York, Van Nostrand Reinhold. Spera, *Wind Turbine Technology*, ASME Press, New York. Gipe, "Wind Power for Home and Business," Chelsea Green Publishing Company. Ramakumar et al., Economic Aspects of Advanced Energy Technologies, *Proc. IEEE*, **81**, no. 3, Mar. 1993, pp. 318–332.

Wind is one of the oldest widely used sources of energy. Although its use is many centuries old, it has not been a dominant factor in the energy picture of developed countries for the past 50 years because of the abundance of fossil fuels. Recently, the realization that fossil fuels are in limited supply has awakened the need to develop wind power with modern technology on a large scale. Consequently, there has been a tremendous resurgence of effort in wind power in just the past few years. The state of knowledge is rapidly increasing, and the reader is referred to the current literature and the NREL Internet address cited above for information on the latest technology. Wind energy is one of the lowest-cost forms of renewable energy. Since the collection area is perpendicular to the ground surface, wind energy systems pose minimal burden on the land area and can coexist with many farming and other activities. By the end of 2005, total global installed wind power capacity has exceeded 50 GW, with Europe contributing slightly higher than 60 percent and North America approaching 20 percent of this total. For the latest status on worldwide use of wind energy, the reader is referred to the American Wind Energy Association (AWEA) at the Internet address cited above. The fundamental principles of wind power technology do not change and are discussed here.

Wind Turbines The essential ingredient in a **wind energy conversion system (WECS)** is the wind turbine, traditionally called the windmill. The predominant configurations are **horizontal-axis propeller turbines (HAWTs)** and **vertical-axis wind turbines (VAWTs),** the latter most often termed **Darrieus rotors.** In the performance analysis of wind turbines, the propeller devices were studied first, and their analysis set the current conventions for the evaluation of all turbines. HAWT sizes and ratings have been steadily increasing. From 15 to 30-m-diam systems rated at 50 to 300 kW in the 1980s, the dawn of the new millennium saw the popular emergence of 1.5-MW units with 66-m rotor diameter. Larger units rated at 3.5 MW (100-m diam) and 5 MW (120-m diam) are in various stages of development.

General Momentum Theory for Horizontal-Axis Turbines Conventional analysis of horizontal-axis turbines begins with an axial momentum balance originated by Rankine using the control volume depicted in Fig. 9.1.1. The turbine is represented by a porous disk of area A which extracts energy from the air passing through it by reducing its pressure: on the upstream side the pressure has been raised above atmospheric by the slowing airstream; on the downstream side pressure is lower, and atmospheric pressure will be recovered by further slowing of the stream. V is original wind speed, decelerated to $V(1 - a)$ at the turbine disk, and to $V(1 - 2a)$ in the wake of the turbine (a is called the interference factor). Momentum analysis predicts the axial thrust on the turbine of radius R to be

$$T = 2\pi R^2 \rho V^2 a(1 - a) \qquad (9.1.1)$$

where air density, ρ, equals 0.00237 lbf · s²/ft⁴ (or 1.221 kg/m³) at sea-level standard-atmosphere conditions.

Fig. 9.1.1   Control volume.

Application of the mechanical energy equation to the control volume depicted in Fig. 9.1.1 yields the prediction of power to the turbine of

$$P = 2\pi R^2 \rho V^3 a(1 - a)^2 \qquad (9.1.2)$$

This power can be nondimensionalized with the energy flux E in the upstream wind covering an area equal to the rotor disk, i.e.,

$$E = \tfrac{1}{2}\rho V^3 \pi R^2 \qquad (9.1.3)$$

The resulting power coefficient is

$$C_p = \frac{P}{E} = 4a(1 - a)^2 \qquad (9.1.4)$$

This power coefficient has a theoretical maximum at $a = \frac{1}{3}$ of $C_p = 0.593$. This result was first predicted by Betz and shows that the load placed on a windmill must be optimized to obtain the best power output: If the load is too small (small a), too much of the power is carried off with the wake; if the load is too large (large a), the flow is excessively obstructed and most of the approaching wind passes around the turbine.

This derivation includes some important assumptions which limit its accuracy and applicability. In particular, the portion of the kinetic energy in the swirl component of the wake is neglected. Partial accounting for the rotation in the wake has been included in the analysis of Glauert with the resulting prediction of ideal power coefficient as a function of turbine tip speed ratio $X = \Omega R/V$ (where Ω is the angular velocity of the turbine) shown in Fig. 9.1.2. Clearly, the swirl is made up of wasted kinetic energy and is largest for a high-torque, low-speed turbine. Actual farm, multiblade, and two- or three-bladed turbines show somewhat lower than ideal performance because of drag effects neglected in ideal flow analysis, but the high-speed two- or three-bladed turbines do tend to yield higher efficiency than low-speed multiblade windmills.

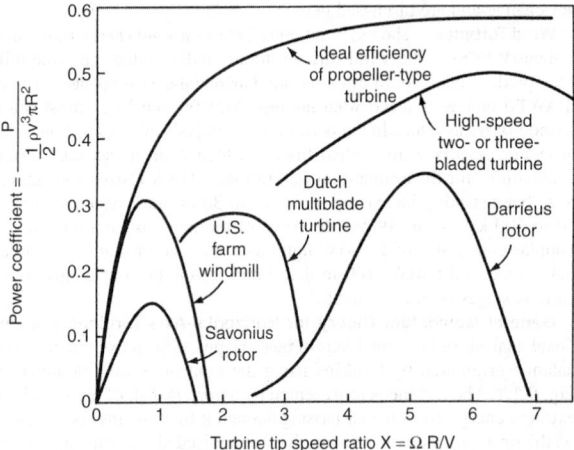

Fig. 9.1.2   Performance curves for wind turbines.

Blade Element Theory for Horizontal-Axis Turbines   Wilson and Eggleston describe blade element theory as a mechanism for analyzing the relationship between the individual airfoil properties and the interference factor a, the power produced P, and the axial thrust T of the turbine. Rather than the stream tube of Fig. 9.1.1, the control volume consists of the annular ring bounded by streamlines depicted in Fig. 9.1.3. It is assumed that the flow in each annular ring is independent of the flow in all other rings.

Fig. 9.1.3   Annular ring control volume.

A schematic of the velocity and force vector diagrams is given in Fig. 9.1.4. The turbine is defined by the number B of its blades, by the variation of chord c, by the variation in blade angle θ, and by the shape of blade sections $a' = \omega/(2\Omega)$, where ω is the angular velocity of the air just behind the turbine and Ω is the turbine angular velocity. Also W is the velocity of the wind relative to the airfoil. Note that the angle ϕ will be different for each blade element, since the velocity of the blade is a function of the radius. In order to keep the local flow angle of attack $\alpha = \theta - \phi$ at a suitable value, it will generally be necessary to construct twisted blades, varying θ with the radius. The sectional lift and drag coefficients C_L and C_D are obtained from empirical airfoil data and are unique functions of the local flow angle of attack $\alpha = \theta - \phi$ and the local Reynolds number of the flow. The entire calculation requires trial-and-error procedures to obtain the axial interference factor a and the angular velocity fraction a'. It can, however, be reduced to programs for small computers.

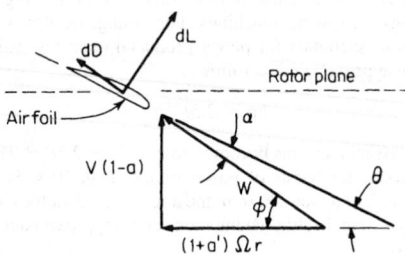

Fig. 9.1.4   Velocity and force vector diagrams.

A typical solution for steady-state operation of a two- or three-bladed wind axis turbine is shown in Fig. 9.1.2. When optimized, these turbines run at high tip speed ratios. The curve shown in Fig. 9.1.2 for the two- or three-bladed wind turbine is for constant blade pitch angle. These turbines typically have pitch change mechanisms which are used to feather the blades in extreme wind conditions. In some instances the blade pitch is continuously controlled to assist the turbine to maintain constant speed and appropriate output. Turbines with continuous pitch control typically have flatter, more desirable operating curves than the one depicted in Fig. 9.1.2.

The traditional U.S. farm **windmill** has a large number of blades with a high solidity ratio σ. (σ is the ratio of area of the blades to swept area of the turbine πR^2.) It operates at slower speed with a lower power coefficient than high-speed turbines and is primarily designed for good starting torque.

The curves depicted in Fig. 9.1.2 representing the performance of high- and low-speed wind axis turbines are theoretically predicted performance curves which have been experimentally confirmed.

Vertical-Axis Turbines   The Darrieus rotor looks somewhat like an eggbeater (Fig. 9.1.5). The blades are high-performance symmetric airfoils formed into a gentle curve to minimize the bending stresses in the blades. There are usually two or three blades in a turbine, and as shown in Fig. 9.1.2, the turbines operate efficiently at high speed. Wilson shows that VAWT performance analysis also takes advantage of the same momentum principles as the horizontal axis wind turbines. However, the blade element momentum analysis becomes much more complicated (see Touryan et al.).

Fig. 9.1.5   Darrieus rotor.

Care must be taken not to overemphasize the aerodynamic efficiency of wind turbine configurations. The most important criterion in evaluating WECSs is the power produced on a per-unit-cost basis.

Drag Devices   Rotors utilizing drag rather than lift have been constructed since antiquity, even though they are bulky and limited to low coefficients of performance. The Savonius rotor is a modern variation of these devices; in practice it is limited to small sizes. Eldridge describes the history and theory of this type of windmill.

Augmentation   Occasionally, the use of structures designed to concentrate and equalize the wind at the turbine is proposed. For its size, the most effective of these has been a short diffuser (hollow cone) placed around and downwind of a wind turbine. The disadvantage of such augmentation devices is the cost of the bulky static structures required.

Rotor Configuration Trends   Hansen and Butterfield describe some trends in turbine configurations which have developed from 1975 to 1995. Although no single configuration has emerged which is clearly superior, HAWTs have been more widely used than VAWTs. Only about 3 percent of turbines installed to date are VAWTs.

HAWT rotors are generally classified according to rotor orientation (upwind or downwind of the tower), blade articulation (rigid or teetering), and number of blades (generally two or three). **Downwind turbines** were favored initially in the United States, but the trend has been toward greater use of **upwind turbines**.

Downwind orientation allows blades to deflect away from the tower when thrust loading increases. Coning can also be easily introduced to decrease mean blade loads by balancing aerodynamic loads with centrifugal loads. Figure 9.1.6 shows typical upwind and downwind configurations along with definitions for blade coning and yaw orientation.

Free yaw, or passive orientation with the wind direction, is also possible with downwind configurations, but yaw must be actively controlled with upwind configurations. Free-yaw systems rely on rotor thrust loads and blade moments to orient the turbine. Net yaw moments for rigid rotors are sensitive to inflow asymmetry caused by turbulence, wind shear, and vertical wind. These are in addition to the moments caused by changes in wind direction which are commonly, though often incorrectly, considered the dominant cause of yaw loads.

Fig. 9.1.6   HAWT configurations. *(Courtesy of Atlantic Orient Corp.)*

Some early downwind turbine designs developed a reputation for generating subaudible noise as the blades passed through the tower shadow (tower wake). Most downwind turbines operating today have greater tower clearances and lower tip speeds, which result in negligible infrasound emissions (Kelley and McKenna, 1985).

Blade Articulation   Several different rotor blade articulations have been tested. Only two have survived—the three-blade, rigid rotor and the two-blade, teetered rotor. The rigid, three-blade rotor attaches the blade to a hub by using a stiff cantilevered joint. The first bending natural frequency of such a blade is typically greater than twice the rotor rotation speed 2*p*. Cyclic loads on rigid blades are generally higher than on teetering blades of the same diameter. Richardson and McNerney describe a 33-m, 300-kW turbine currently under development which reflects a mature version of this configuration.

Teetered, two-blade rotors use relatively stiff blades rigidly connected to a hub, but the hub is attached to the main drive shaft through a **teeter hinge.** This type rotor is commonly used in tail rotors and some main rotors on helicopters. Two-blade rotors usually require teeter hinges or flexible root connections to reduce dynamic loading resulting from non-axisymmetric mass moments of inertia. In normal operation, the cyclic loads on the teetering rotor are low, but there is risk of teeterstop bumping ("mast bumping" in helicopter terminology) that can greatly increase dynamic loads in unusual situations.

Number of Blades   Most two-blade rotors operating today use teetering hinges, but all three-blade rotors use rigid root connections. For small turbines (smaller than 50-ft diameter) rigid, three-blade rotors are inexpensive and simple and have the lowest system cost. As the turbines become larger, blade weight (and hence cost) increases in proportion to the third power of the rotor diameter, whereas power output increases only as the square of the diameter. This makes it cost-effective to reduce the number of blades to two and to add the complexity of a teeter hinge or flex beams to reduce blade loads. In the midscale rotor size (15 to 30 m), it is difficult to determine whether three rigidly mounted blades or two teetered blades are more cost-effective. In many cases, the choice between two- and three-blade rotors has been driven by designers' lack of experience and the potential risk of high development cost rather than by technical and economic merit. Currently 10 percent of the turbines installed are two-bladed, yet approximately 60 percent of all new designs being considered in the United States are two-blade, teetered rotors.

Design Problems   A key design consideration is survival in severe storms. Various systems for furling the rotor, feathering the blades, or braking the shaft have been employed; failure of these systems in a high wind has been known to cause severe damage to the turbine.

A different, but related, consideration is the control of the turbine after a loss of electrical load, which also could cause severe overspeeding and catastrophic failure.

The other major cause of mechanical failure is the high level of vibration and alternating stresses. Loosening of inappropriately chosen fasteners is common. Fatigue considerations must be taken into account, especially at the rotor blade root.

Resonant oscillations are also possible if exciting frequencies and structural frequencies coincide. The dominant exciting frequency tends to be the blade passage frequency, which is equal to the number of blades times the revolutions per second. An important structural frequency in HAWTs is the natural frequency of the tower. One design approach is to make the tower so stiff that the exciting frequency is always below the lowest natural frequency of the tower. Another is to permit the tower to be more flexible, but manage the speed of the turbine such that the exciting frequency is never at a structural frequency for any significant length of time.

Use of Wind Energy Conversion Systems   Historically, wind energy conversion systems were first used for milling grain and for pumping water. These tasks were ideally suited for wind power sources, since the intermittent nature of the wind did not adversely affect the operation.

The largest impact of wind power on the energy picture in the developed countries of the world is expected to be in the generation of electric power. In most cases, this involves feeding power into the power grid, and requires induction or synchronous generators. These generators require that the rotor turn at a constant speed. Wind turbines operate more efficiently (aerodynamically speaking) if they turn at an optimum ratio of tip speed to wind speed. Thus the use of variable-speed operation, using power electronics to obtain constant-frequency utility-grade ac power, has become attractive. Richardson describes the modern use of variable speed in wind turbines.

Gipe explains that in remote locations, where the power grid is not accessible and the first few units of electric energy may be very valuable, dc generation with storage and/or wind and diesel "village power systems" have been used. These systems are now being optimized to supply stable, constant-frequency ac electric energy.

Power in the Wind   Since wind is air in motion, the power in wind can be expressed as

$$P_w = \frac{1}{2}\rho V^3 A \qquad (9.1.5)$$

where P_w = power, W; A = reference area, m^2; V = wind speed, m/s; ρ = air density, kg/m^3. Since V appears to the third power, the wind speed is clearly very important. Figure 9.1.7 is a map of the United States showing regions of annual average available wind power.

The wind speed at a location is random; thus it can be modeled as a continuous random variable in terms of a density function $f(v)$ or a distribution function $F(v)$. The Weibull distribution is commonly used to model wind:

$$F(v) = 1 - \exp\left[-(v/\alpha)^\beta\right] \qquad (9.1.6)$$

$$f(v) = \beta(v^{\beta-1}/\alpha^\beta)\exp\left[-(v/\alpha)^\beta\right] \qquad (9.1.7)$$

In Eqs. (9.1.6) and (9.1.7), α and β are two parameters which can be adjusted to fit available data over the study period, typically one month. They can be calculated from the sample mean m_v and the sample variance σ_v^2 using the following equations:

$$m_v = \alpha\Gamma(1 + 1/\beta) \qquad (9.1.8)$$

$$(\sigma_v/m_v)^2 = \Gamma(1 + 2/\beta)\Gamma^2(1 + 1/\beta) - 1 \qquad (9.1.9)$$

Typically, the sample mean is the only piece of information readily available for many potential sites. In such cases, a knowledge of the variability of the wind speed can be used to select an appropriate value for β, which can be used in (9.1.8) to obtain α. A good compromise value for β is about 4 for wind regimes with low variances.

In addition to α and β, several other parameters are used to characterize wind regimes. Some of the important ones are listed below.

$$\text{Mean cubed wind speed} = \langle v^3 \rangle$$

$$= \int_0^\infty v^3 f(v)\, dv \qquad (9.1.10)$$

$$\text{Cube factor } K_c = (\langle v^3 \rangle)^{1/3}/m_v \qquad (9.1.11)$$

$$= \alpha^3\Gamma(1 + 3/\beta)$$

$$\text{for Weibull model}$$

$$\text{Average power density} = P_{av} = \frac{1}{2}\rho\langle v^3 \rangle \quad \text{W/m}^2 \quad (9.1.12)$$

$$\text{Energy pattern factor} = K_{ep} = \langle v^3 \rangle/m_v^3 = K_c^3 \qquad (9.1.13)$$

Values of K_{ep} range from 1.5 to 3 for typical wind regimes.

The annual average available wind power for the contiguous United States is shown in Fig. 9.1.7. The values shown must be regarded as averages over large areas. The possibility of finding small pockets of sites with excellent wind regimes because of special terrain any where in the country should not be overlooked. The variability of the wind can also be shown in terms of a speed duration curve. Figure 9.1.8 shows the wind speed duration curve for Plum Brook, OH, for 1972.

Wind speed varies with the height above ground level (Fig. 9.1.9a). Anemometers are usually located at a height of 10 m above ground level. The long-term average wind speed at height h above ground can be expressed in terms of the average wind speed at 10-m height using a one-seventh power law:

$$(v/v_{10\,m}) = (h/10)^{1/7} \qquad (9.1.14)$$

The power $\frac{1}{7}$ in the power law equation above depends on surface roughness and other terrain-related factors and can range from 0.1 to 0.3. The value of $\frac{1}{7}$ should be regarded as a compromise value in the absence of other information regarding the terrain. Clearly, it is advantageous to construct an adequately high support tower for a wind energy conversion system.

Table 9.1.2 shows average and peak wind velocities at locations within the continental United States.

Wind to Electric Power Conversion   The ease with which wind energy can be converted to rotary mechanical energy and the maturity of electromechanical energy converters and solid-state power conditioning equipment clearly point to wind-to-electric conversion as the most promising approach to harnessing wind power in usable form.

The electric power output of a wind-to-electric conversion system can be expressed as

$$P_e = \frac{1}{2}\rho\eta_g\eta_m\eta_p A C_p V^3 \qquad (9.1.15)$$

where P_e = electric power output, W; η_g and η_m = efficiencies of the electric generator and the mechanical interface, respectively; η_p = efficiency of the power conditioning equipment (if employed). The product of these efficiencies and the coefficient of performance (Fig. 9.1.2) usually will be in the range of 20 to 35 percent.

The electrical equipment needed for wind-to-electric conversion depends, above all, on whether the aeroturbine is operated in the constant-speed, nearly constant-speed, or variable-speed mode. With constant-speed and nearly constant-speed operation, the power coefficient C_p in Eq. (9.1.15) becomes a function of wind speed. If variable-speed mode is used, it is possible to operate the turbine at a constant optimum C_p over a range of wind speeds, thus extracting a larger fraction of the energy in the wind.

Most of the small wind turbines use permanent magnet alternators and are operated in isolation for battery charging, water pumping, ice making, etc. Sometimes they are connected to the grid or operated in conjunction with a local diesel generator. Utility-scale systems are equipped with different types of generators, often with power electronic subsystems, and are always connected to inject power into a conventional grid.

Generator Types   Conventional induction generators operate at nearly constant speed with a slip varying from 0 to 2 percent. The stator is directly connected to the grid without any power converter.

Fig. 9.1.7 Gridded map of annual average wind energy resource estimates in the United States and Paerto Rico.

9-9

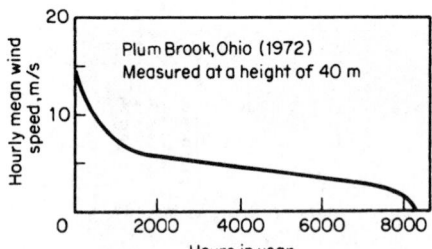

Fig. 9.1.8   Wind variability at Plum Brook, OH (1972).

Fig. 9.1.9a   Typical variation of mean wind velocity with height.

Typically, capacitor banks are connected in parallel to compensate for the reactive power needed by the generator. This type of generator limits the operation of the system to almost constant speed and the associated increased stresses in the structural elements.

Wound rotor induction generators with external rotor resistances placed in the rotating rotors and adjusted electronically allow operating speeds to vary from 0 to 10 percent above synchronous speed. They also require reactive power compensation using switched capacitors.

Doubly fed induction generators with slip rings employ rotors fed by three-phase power electronic converters operating at variable frequency.

This scheme allows operating speeds to vary between +30 and −30 percent of synchronous speed, needing a power converter rated only about 30 percent of the full rated output of the system. This flexibility dramatically reduces the stresses on the mechanical and rotating elements and opens the possibility to have fairly large sizes and ratings.

If a power converter rated at the full capacity of the system is used, any type of generator can be employed and the power converter can provide the reactive power compensation. This will isolate the turbine generator from the grid.

Control   Figure 9.1.9b shows a simplified block diagram of control systems implemented in a typical variable-speed wind turbine. Variables with an asterisk indicate reference (or command) quantities, and the others indicate measured quantities. There are two main subsystems, one controlling the generator and the other, the turbine.

Fig. 9.1.9b   Control schemes for wind electric conversion systems. (a) Generator control; (b) Turbine control.

The turbine control is accomplished by pitching the blade. At low and medium wind speeds, the pitch angle is controlled to operate at maximum energy capture. At high wind speeds, after the rotor reaches its rated speed, the pitch angle is controlled to shave the aerodynamic power and limit the rotor speed of the wind turbine. Employing separate generator and turbine controls works very harmoniously. Having the turbine control capability is a big plus because the ratings of the generator and power

Table 9.1.2   Wind Velocities in the United States*

Station	Avg velocity, mi/h	Prevailing direction	Peak velocity, mi/h	Station	Avg velocity, mi/h	Prevailing direction	Peak velocity, mi/h
Albany, N.Y.	9.0	S	71	Louisville, Ky.	8.7	S	68
Albuquerque, N.M.	8.8	SE	90	Memphis, Tenn.	9.9	S	57
Atlanta, Ga.	9.8	NW	70	Miami, Fla.	12.6	—	132
Boise, Idaho	9.6	SE	61	Minneapolis, Minn.	11.2	SE	92
Boston, Mass.	11.8	SW	87	Mt. Washington, N.H.	36.9	W	150
Bismarck, N. Dak.	10.8	NW	72	New Orleans, La.	7.7	—	98
Buffalo, N.Y.	14.6	SW	91	New York, N.Y.	14.6	NW	113
Burlington, Vt.	10.1	S	72	Oklahoma City, Okla.	14.6	SSE	87
Chattanooga, Tenn.	6.7	—	82	Omaha, Neb.	9.5	SSE	109
Cheyenne, Wyo.	11.5	W	75	Pensacola, Fla.	10.1	NE	114
Chicago, Ill.	10.7	SSW	87	Philadelphia, Pa.	10.1	NW	88
Cincinnati, Ohio	7.5	SW	49	Pittsburgh, Pa.	10.4	WSW	73
Cleveland, Ohio	12.7	S	78	Portland, Maine	8.4	N	76
Denver, Colo.	7.5	S	65	Portland, Ore.	6.8	NW	57
Des Moines, Iowa	10.1	NW	76	Rochester, N.Y.	9.1	SW	73
Detroit, Mich.	10.6	NW	95	St. Louis, Mo.	11.0	S	91
Duluth, Minn.	12.4	NW	75	Salt Lake City, Utah	8.8	SE	71
El Paso, Tex.	9.3	N	70	San Diego, Calif.	6.4	WNW	53
Galveston, Tex.	10.8	—	91	San Francisco, Calif.	10.5	WNW	62
Helena, Mont.	7.9	W	73	Savannah, Ga.	9.0	NNE	90
Kansas City, Mo.	10.0	SSW	72	Spokane, Wash.	6.7	SSW	56
Knoxville, Tenn.	6.7	NE	71	Washington, D.C.	7.1	NW	62

U.S. Weather Bureau records of the average wind velocity, and fastest mile, at selected stations. The period of record ranges from 6 to 84 years, ending 1954. No correction for height of station above ground.

* See also ASCE Standard (ASCE/SEI 7-05, with Supplement No. 1) "Minimum Design Loads for Buildings and Other Structures," for updated basic wind velocities (ASCE, 2005).

converter can be minimized. Thus the generator and power converter do not have to be used to control turbine aerodynamics.

System Integration   Typically, wind turbines generate electrical energy at low voltages (480 to 690 V). Pad-mounted transformers at the base of the towers step up the voltage to distribution voltage level (11 to 66 kV). Outputs of pad-mounted transformers are connected to a substation at the point of interconnection where the output power is stepped up to transmission line level (115 to 345 kV) to be transmitted over long distances.

Wind farms consist of several wind turbines, with total ratings approaching 30 to 300 MW. As the level of penetration into conventional power systems increases, the impact of wind farms on the overall power system in terms of stability, reliability, availability, security, power quality, economics, and protection against faults should be studied thoroughly, and the results incorporated in the planning process.

Economics   Costs of wind energy systems are often divided into two categories: annualized fixed costs and operation and maintenance (O&M) costs. Annualized fixed costs are comprised primarily of the cost of capital required to purchase and install the turbines. In addition, they include certain fixed costs such as taxes and insurance. O&M costs include scheduled and unscheduled maintenance and the levelized cost for major equipment overhauls.

The initial capital cost of a wind turbine system includes the cost of the turbine, installation, and balance of plant. Turbine costs are often expressed in terms of nameplate rating ($/kW). In 1995, utility-grade turbines cost on the order of $800 per kilowatt. Installation and balance of plant costs add approximately 20 percent. The cost of capital varies, but (in 1995) was often estimated as 8 percent per annum for wind energy projects. Other fixed costs were estimated at around 3 percent of the installed turbine cost. The **fixed charge rate (FCR),** combined capital and other fixed costs, was approximately 11 percent per annum. O&M costs in modern wind farms are around $0.01 per kilowatthour (1995).

In addition to capital and O&M costs, an economic assessment of wind energy systems must account for system performance. A commonly used parameter that describes the production of useful energy by wind and other energy systems is the **capacity factor C,** also called the **plant factor** or **load factor.** It is the ratio of the annual energy produced (AEP) to the energy that would be produced if the turbine operated at full-rated output throughout the year:

$$C_f = \frac{\text{AEP}}{8{,}760 P_R} \qquad (9.1.16)$$

where AEP is in kWh, 8,760 is the number of hours in 1 year, and P_R is the unit's nameplate rating in kW.

In order of decreasing importance, C_f is affected by the average power available in the wind, speed vs. duration curve of the wind regime, efficiency of the turbine, and reliability of the turbine. Variable-speed turbines which tend to have low cut-in speeds and high efficiency in low winds exhibit better capacity factors than constant-speed turbines. Modern utility-grade turbines at good sites (class 4) can achieve capacity factors in the range of 30 to 50 percent.

The combination of cost and performance can be used to calculate the cost of energy (COE) as follows:

$$\text{COE} = \frac{\text{FCR} \times \text{ICC}}{8{,}760 C_f} + (\text{O\&M}) \qquad (9.1.17)$$

where FCR is the fixed charge rate for the cost of capital and for other fixed charges such as taxes and insurance, ICC is the installed capital cost of the turbine and balance of plant in dollars per kilowatt, and (O&M) is the pro-rated operation and maintenance cost per kilowatthour of electric energy generated. This method is useful to estimate the cost of energy for different technologies or sites. However, for investment decisions, more detailed analyses that include the effects of various investment strategies, tax incentives, and environmental factors should be performed. Ramakumar et al. (see Reference) discuss the economic aspects of advanced energy technologies, including wind energy systems.

Levelized cost of wind-generated electricity has come down from U.S. $ 0.38/kWh in 1980 to about U.S. $ 0.025/kWh to U.S. $ 03.5/kWh in 2005.

A photograph of a typical, modern large on-shore wind farm consisting of 1.5-MW units is shown in Fig. 9.1.9c. A view of the world's first commercial application of offshore wind turbines over 3 MW in size is shown in Fig. 9.1.9d.

Figure 9.1.9c   Wind turbines rated at 1.5 MW with 72-m-diam rotors at TVIG's Kimball wind farm installation in Nebraska. (Photo courtesy of Tennessee Valley Infrastructure Group Inc.)

Figure 9.1.9d   Located about 10 km off the coast of Arklow, Ireland, the Arklow Bank offshore wind park consists of seven GE Wind 3.6-MW wind turbines. Construction was completed in 2003. (*Photo: Robert Thresher.*)

POWER FROM BIOMASS
Revised by E. A. Avallone

Vegetation offers, by photosynthesis, a natural process for the storage of solar energy. The efficiency of the photosynthetic process for the conversion of the sun's rays into a usable fuel form is low (less than 2 percent is probably realistic). Wood, wood waste, sawdust, hogged fuel, bagasse, straw, and tanbark have heating values ranging to 10,000 ± Btu/lb (see Sec. 7.1). They may be incinerated for disposal as waste material or burned directly for the subsequent production of steam or hot water, most often used in the processing activities of the plant, e.g., hot water soak of logs for plywood peeling and steam for drying in paper mills. In food processing, fruit pits and nut shells have been used to generate a portion of the in-house requirements for steam.

The alternative to direct burning of the so-called biofuels lies in their possible conversion to gaseous fuel by gasification at high temperature in the presence of air. Pyrolitic treatment can render biofuels to fractions of liquids and gases that have thermal value. In both cases, the solid residue remaining also has some thermal value which can be utilized in normal combustion.

Tree farming, with controlled growth and cutting, proposes to balance harvesting plans to load demands; e.g., Szego and Kemp (*Chem. Tech.,* May 1973) project a 400-mi^2 "energy plantation" to serve a 400-MW steam electric plant. Such proposals would utilize proved steam power plant cycles and equipment for novel breeding, growing, harvesting, preparation, and combustion of vegetation. (See also Sec. 7.1.)

The **photosynthesis process** is basic to all agricultural practice. The human animal has long known how to convert grain to alcohol. It can be said that as long as we can grow green stuff we should be able to harness some of the sun's energy. The prohibition era in the United States saw many efforts to use the alcohol production capacity of the nation to offer alcohol as a substitute or supplementary fuel for internal combustion engines. Ethanol (C_2H_5OH) and methanol (CH_3OH) have properties that are basically attractive for internal combustion engines, to wit, smokeless combustion, high volatility, high octane ratings, high compression ratios ($R_v > 10$). Heating values are 9,600 Btu/lb for methanol and 12,800 Btu/lb for ethanol. On a volume basis these translate, respectively, to 63,000 and 85,000 Btu/gal for methanol and ethanol. Gasoline, by comparison, has 126,000 Btu/gal (20,700 Btu/lb). (See Sec. 7 for values.) The blending of ethanol and methanol with gasolines (9 ± gasoline to 1 ± alcohol) has been used particularly in Europe since the 1930s as a suitable internal combustion engine fuel. The miscibility of the lighter alcohols with water and gasoline introduces corrosion problems for engine parts and lowers the octane number. Higher-carbon alcohols (e.g., butyl) which are immiscible with water are possible blending substitutes, but their availability and cost are not presently attractive. Properties such as flash point would introduce further problems. While these constitute some of the unsolved technical problems, the basic principle of harnessing the sun's energy through vegetation will continue as a provocative challenge not only in the field of power generation but also as a solution for the perennial problem of waste from farms and wood product manufacturers.

SOLAR ENERGY
by Erich A. Farber

REFERENCES: Daniels, "Direct Use of the Sun's Energy," Yale. ASHRAE, "Solar Energy Use for Heating and Cooling of Buildings." Duffie and Beckman, "Solar Engineering of Therma Processes," Wiley. Lunde, "Solar Thermal Engineering," McGraw-Hill. Kreider and Kreith, "Solar Heating and Cooling," McGraw-Hill. ASHRAE, "Handbook of Fundamentals, 1993," "Handbook of Applications, 1982." Yellott, Selective Reflectance, *Trans. ASHRAE,* **69**, 1963. Hay, Natural Air Conditioning with Roof Ponds and Movable Insulation, *Trans. ASHRAE,* **75**, part I, p. 165, 1969. Backus, "Solar Cells," IEEE. Bliss, Atmospheric Radiation near the Surface of the Ground, *Solar Energy,* **5**, no. 3, 1961. Yellott, Passive and Hybrid Cooling Research, "Advances in Solar Energy—1982," American Solar Energy Soc., Boulder. Haddock, Solar Energy Collection, Concentration, and Thermal Conversion—A Review, *Energy Sources,* **13**, 1991, pp. 461–482. "Proceedings of the Intersociety Energy Conversion Conferences," held annually, The University of Florida, "Erich A. Farber Archives." "European Directory of Renewable Energy, Suppliers and Services," James & James Ltd., London, England.

Notation

A, R, T = subscripts denoting absorbed, reflected, and transmitted solar radiation
C = concentration ratio
c = subscript denoting collector cover
c_p = specific heat of fluid, Btu/(lb · °F)
I_{DN} = direct normal solar intensity, Btu/(ft^2 · h)
I_d = diffuse radiation, Btu/(ft^2 · h)
I_0 = radiation intensity beyond earth's atmosphere, Btu/(ft^2 · h)
I_r = reflected solar radiation, Btu/(ft^2 · h)
I_{SC} = solar constant; normal incidence intensity at average earth-sun distance, Btu/(ft^2 · h)
I_t = total solar radiation, Btu/(ft^2 · h)

L = latitude, deg
m = air mass
o, i = subscripts denoting outgoing and incoming fluid conditions
q = rate of heat flux, Btu/(ft^2 · h)
q_I = heat flow through insulation, Btu/(ft^2 · h)
T_p = temperature of absorbing surface, °R
U = overall coefficient of heat transfer, Btu/(ft^2 · h · °F)
w_f = flow rate of collecting fluid, lb/(h · ft^2)
ϕ = solar azimuth, deg from south
α, ρ, τ = absorptance, reflectance, and transmittance for solar radiation
β = solar altitude, deg
δ = solar declination, deg
ε = emittance for long-wave radiation
γ = wall-solar azimuth, deg
λ = unit of wavelength, μm
Σ = angle of tilt from horizontal, deg
θ = incident angle, deg, from perpendicular to surface

Introduction and Scope

The sun exerts forces upon the earth and radiates solar energy produced within the sun by nuclear fusion. A small fraction of that energy is intercepted by the earth and is converted by nature to heat, winds, ocean currents, waves, tides; makes plants grow, some of which over millions of years produced fossil fuels (oil, coal, and gas); and creates biomass which can be burned to generate heat and/or power. Solar energy is implicit in many subject areas treated elsewhere in the Handbook; only the more direct uses such as water heating, space heating and cooling, swimming pool heating, solar distillation, solar drying and cooking, solar furnaces, solar engines, solar electricity generation, and solar assisted transportation will be treated here. The total field is widely termed *alternative or renewable sources of energy and their conversion.*

Solar Energy Utilization   Solar energy reaches the earth's surface as shortwave electromagnetic radiation in the wavelength band between 0.3 and 3.0 μm; its peak spectral sensitivity occurs at 0.48 μm (Fig. 9.1.10). Total solar radiation intensity on a horizontal surface at sea level varies from zero at sunrise and sunset to a noon maximum which can reach 340 Btu/(ft^2 · h) (1,070 W/m2) on clear summer days. This inexhaustible source of energy, despite its variability in magnitude and

Fig. 9.1.10 Spectral distribution of solar radiation and radiation emitted by blackbody at 95°F (35°C).

direction, can be used in three major processes (Daniels, "Direct Use of the Sun's Energy," Yale; Zarem and Erway, "Introduction to the Utilization of Solar Energy," McGraw-Hill): (1) **Heliothermal,** in which the sun's radiation is absorbed and converted into heat which can then be used for many purposes, such as evaporating seawater to produce salt or distilling it into potable water; heating domestic hot water supplies; house heating by warm air or hot water; cooling by absorption refrigeration; cooking; generating electricity by vapor cycles and thermoelectric processes; attaining temperatures as high as 6,500°F (3,600°C) in solar furnaces. (2) **Heliochemical,** in which the shorter wavelengths can cause chemical reactions, can sustain growth of plants and animals, can convert carbon dioxide to oxygen by photosynthesis, can cause degradation and fading of fabrics, plastics, and paint, can be used to detoxify toxic waste, and can increase the rate of

Table 9.1.3 Annual Variation in Solar Declination and Solar Radiation Intensity beyond the Earth's Atmosphere

Date	Jan. 1	Feb. 1 Nov. 10	Mar. 1 Oct. 13	Apr. 1 Sept. 12	May 1 Aug. 12	June 1 July 12	July 1
Declination, deg	−23.0	−17.1	−7.7	+4.4	+15.0	+22.0	+23.1
Ratio, I_0/I_{sc}	1.033	1.029	1.017	1.000	0.983	0.971	0.967
Intensity I_0,	449	447	442	435	427	422	420
Btu/(ft^2 · h) (W/m^2)	1,416	1,409	1,393	1,370	1,347	1,331	1,324

chemical reactions. (3) **Helioelectrical,** in which part of the energy between 0.33 and 1.3 μm can be converted directly to electricity by photovoltaic cells. Silicon solar batteries have become the standard power sources for communication satellites, orbiting laboratories, and space probes. Their use for terrestrial power generation is currently under intensive study, with primary emphasis upon cost reduction. Other methods include thermoelectric, thermionic, and photoelectromagnetic processes and the use of very small antennas in arrays for the conversion of solar energy to electricity (Antenna Solar Energy to Electricity Converter/ASETEC, Air Force Report, AF C FO 8635-83-C-0136, Task 85-6, Nov. 1988).

Solar Radiation Intensity In space at the average earth-sun distance, 92.957 million mi (150 million km), solar radiation intensity on a surface normal to the sun's rays is 434.6 ± 1 Btu/(ft^2 · h) (1,370 ± 3 W/m^2). This quantity, called the solar constant I_{SC}, undergoes small (±1 percent) periodic variations which affect primarily the shortwave portion of the spectrum (Abbott, in Moon, Standard Polar Radiation Curves, *Jour. Franklin Inst.*, Nov. 1940). Recent measurements using satellites give essentially the same results. Since the earth-sun distance varies throughout the year, the intensity beyond the earth's atmosphere I_0 also varies by ±3.3 percent (Table 9.1.3). The great seasonal variations in terrestrial solar radiation intensity are due to the earth's tilted axis, which causes the solar declination δ (the angle between the earth's equatorial plane and earth-sun line) to change from 0° on Mar. 21 and Sept. 21 to −23.5° on Dec. 21 and + 23.5° on June 21.

In passing through the earth's atmosphere, the sun's radiation is partially and selectively absorbed, scattered, and reflected by water vapor and ozone, air molecules, natural dust, clouds, and artificial pollutants. Some of the scattered and reflected energy reaches the earth as diffuse or sky radiation I_d.

The intensity of the direct normal radiation I_{DN} depends upon the clarity and the amount of precipitable moisture in the atmosphere and the length of the atmospheric path, which is determined by the solar altitude β and expressed in terms of the air mass m, which is the ratio of the existing path length to the path length when the sun is at the zenith. Except at very low solar altitudes, $m = 1.0/\sin \beta$.

Figure 9.1.10 shows relative values of the spectral intensity of solar radiation in space for $m = 0$ (Thekaekara, Solar Energy outside the Earth's Atmosphere, *Solar Energy*, 14, no. 2, 1973) and at sea level (Moon, Standard Solar Radiation Curves, *Jour. Franklin Inst.*, Nov. 1940) for a solar altitude of 30° ($m = 2.0$). Table 9.1.4 shows the variation at 40° north latitude throughout typical clear summer (June 21) and winter (Dec. 21) days of solar altitude and azimuth (measured from the south), direct normal radiation, total solar irradiation of horizontal and vertical south-facing surfaces.

The total solar irradiation reaching a terrestrial surface is the sum of the direct, diffuse, and reflected components: $I_t = I_{DN} \cos \theta + I_d + I_r$, where θ is the incident angle between the sun's rays and a line perpendicular to the receiving surface and, I_r is the shortwave radiation reflected from adjacent surfaces.

Direct beam solar radiation intensity is measured by **pyroheliometers** with collimating tubes to exclude all but the direct rays from their sensors, which may use calorimetric, thermoelectric, or photovoltaic means to produce a response proportional to the irradiation rate. Similar but uncollimated instruments called **pyranometers** are used to measure the total radiation from sun and sky; when their sensors are shaded from the sun's direct rays, they also can measure the diffuse component.

Incident Angle Determination The incident angle θ affects both the direct solar intensity and the solar optical properties of the irradiated surface. For a flat surface tilted at an angle Σ from the horizontal, $\cos \theta = \cos \beta \cos \gamma \sin \Sigma + \sin \beta \cos \Sigma$. For vertical surfaces, $\Sigma = 90°$; so $\cos \theta = \cos \beta \cos \gamma$; for horizontal surfaces, $\Sigma = 0°$ and $\theta = 90° - \beta$ (See ASHRAE, "Handbook of Fundamentals," for values of solar altitude, azimuth, and direct normal radiation throughout the year for 0 to 56° north latitude.)

Solar Optical Properties of Transparent Materials When solar radiation with total intensity I_t falls on a transparent material, part of the energy is reflected, part is absorbed, and the remainder is transmitted. At any instant,

$$I_t = q_\tau + q_A + q_R = I_t(\tau + \alpha + \rho)$$

The sum of the solar optical properties τ, α, and ρ must equal unity, but the individual values depend upon the incident angle and wavelength

Table 9.1.4 Solar Altitude and Azimuth, Direct Normal Radiation, and Total Solar Radiation on Horizontal and Vertical South-Facing Surfaces, June 21 and Dec. 21, for 40° North Latitude

June 21, declination = +23.45°							
Time: A.M.; P.M.	6; 6	7; 5	8; 4	9; 3	10; 2	11; 1	12; 12
Solar altitude, deg	14.8	26.0	37.4	48.8	59.8	69.2	73.5
Solar azimuth, deg	108.4	99.7	90.7	80.2	65.8	41.9	0.0
Direct normal irradiation, Btu/(ft^2 · h)	154	215	246	262	272	276	278
Total irradiation, Btu/(ft^2 · h)							
On horizontal surface	60	123	182	233	273	296	304
On vertical south surface	10	14	16	47	74	92	98
Dec. 21, declination = −23.45°							
Time: A.M.; P.M.	6; 6	7; 5	8; 4	9; 3	10; 2	11; 1	12; 12
Solar altitude, deg			5.5	14.0	20.7	25.0	26.6
Solar azimuth, deg			53.0	41.9	29.4	15.2	0.0
Direct normal irradiation, Btu/(ft^2 · h)			88	217	261	279	284
Total irradiation, Btu/(ft^2 · h)							
On horizontal surface			14	65	107	119	143
On vertical south surface			56	163	221	252	263

Values adapted from ASHRAE, "Handbook of Applications," 1982.

Fig. 9.1.11 Variation with incident angle of solar optical properties of ³⁄₃₂-in (2.4-mm) clear glass and ¼-in (6.3-mm) heat-absorbing glass.

of the radiation, the composition of the material, and the nature of any coatings which may be applied to the surfaces.

For uncoated single-strength (³⁄₃₂-in or 2.4-mm) clear window glass (Fig. 9.1.11), solar transmittance at normal incidence (θ = 0°) is approximately 0.90, but the transmittance for longwave thermal radiation (5 μm) is virtually zero. Thus glass acts as a "heat trap" by admitting solar radiation readily but retaining most of the heat produced by the absorbed sunshine. This "greenhouse effect," which is also exhibited but to a lesser degree by some plastic films (see Whillier, Plastic Covers for Solar Collectors, *Solar Energy*, 7, no. 3, 1964), is the basis for most heliothermal processes. Heat absorbing glass [¼ in (6.3 mm) thick (Fig. 9.1.11)], which absorbs more than 50 percent of the incident solar radiation, is widely used by architects to reduce the heat and glare

admitted through unshaded windows. Reflective coatings (Yellott, Selective Reflectance, *Trans. ASHRAE*, **69**, 1963) have been developed to serve similar purposes.

For all types of glass, transmittance falls and reflectance rises as θ exceeds about 30°. Absorptance increases somewhat owing to the increased path length and then drops off sharply toward zero as θ exceeds 60°.

Absorptance and Emittance of Opaque Surfaces Opaque materials absorb or reflect all the incident sunshine. The absorptance α for solar radiation and the emittance ε for longwave radiation at the temperature of the receiving surface are particularly important in heliotechnology. For a true blackbody, the absorptance and emittance are equal and do not change with wavelength. Most real surfaces have reflectances and absorptances which vary with wavelength (Fig. 9.1.12). Aluminum foil has a consistently low absorptance and high reflectance over the entire spectrum from 0.25 to 25 μm, while black paint has a high absorptance and low reflectance. White paint, however, has low shortwave (solar) absorptance, but beyond 3 μm its absorptance and reflectance are virtually the same as for black paint.

Solar collectors require a high α/ε ratio, while surfaces which should remain cool, such as rooftops or space vehicles, should have low ratios since their objective usually is to absorb as little solar radiation and emit as much longwave radiation as possible. Special surface treatments have been developed (see ASHRAE, "Solar Energy Use for Heating and Cooling of Buildings," 1977) for which the ratio α/ε is above 7.0, making them suitable for solar collectors; others with ratios as low as 0.15 are useful as heat rejectors for space applications (see Table 9.1.5). In addition, the absorptance can be changed and controlled by paint composition (grain material and size, binder, thickness etc.) and surface configuration, both large and small.

Equilibrium Temperatures for Concentrating Collectors When a surface is irradiated, its temperature rises until the rate of solar radiation absorption equals the rate at which heat is removed from the surface. If no heat is intentionally removed, the maximum temperature which can be attained by a blackbody (α = ε) is found from $I_{DN}C\alpha = 0.1713\varepsilon(T_p/100)^4$, where C is the concentration ratio. Figure 9.1.13 shows the variation of blackbody equilibrium temperatures for earth and near space where I_{DN} = 320 and I_0 = 435 Btu/ft² · h (1,000 and 1,370W/m²).

For flat-plate collectors, C = 1.0; so their maximum attainable temperatures are below 212°F (100°C) unless a selective surface is used with $\alpha/\varepsilon > 1.0$, or both radiation and convection loss are suppressed by the use of multiple glass cover plates. Only the direct component of the total solar radiation can be concentrated, and concentrating collectors must follow the sun's apparent motion across the sky or use heliostats which serve the same function. Diffuse radiation cannot be concentrated effectively.

Fig. 9.1.12 Variation with wavelength of reflectance and absorptance for opaque surfaces.

Table 9.1.5 Solar Absorptance, Longwave Emittance, and Radiation Ratio for Typical Surfaces

Surface or material	Shortwave (solar) absorptance α	Longwave emittance ε	Radiation ratio, α/ε
Flat, oil-based paints:			
Black	0.90	0.90	1.00
Red	0.74	0.90	0.82
Green	0.50	0.90	0.55
Aluminum	0.45	0.90	0.50
White	0.25	0.90	0.28
Whitewash on galvanized iron	0.22	0.90	0.25
Building materials:			
Asbestos slate	0.81	0.96	0.84
Tar paper, black	0.93	0.93	1.00
Brick, red	0.55	0.92	0.59
Concrete	0.60	0.88	0.68
Sand, dry	0.82	0.90	0.92
Glass	0.04–0.70	0.84	
Metals:			
Copper, polished	0.18	0.04	4.50
Copper, oxidized	0.64	0.60–0.90	1.03–0.71
Aluminum, polished	0.30	0.05	6.00
Selective surfaces:			
Tabor, electrolytic	0.90	0.12	7.50
Silicon cell, uncoated	0.94	0.30	3.13
Black cupric oxide on copper	0.91	0.16	5.67

Fig. 9.1.14   Typical flat-plate solar radiation collectors.

Flat-Plate Collectors   Direct, diffuse, and reflected solar radiation can be collected and converted into heat by flat-plate collectors (Fig. 9.1.14). These generally use blackened metal plates which are finned, tubed, or otherwise provided with passages through which water, air, or other fluids may flow and be heated to temperatures as much as 100 to 150°F (55 to 86°C) above the ambient air. The actual temperature rise may be estimated from the heat balance for a unit area of collector surface:

$$q_A = I_t \tau_c \alpha_p = w_f c_p(t_o - T_i) \div q_l + U(t_p - t_a)$$

The loss from the back of the collector plate q_l can be minimized by the use of adequate insulation. The radiation component of the loss from the upper surface can be reduced (Zarem and Erway, "Introduction to the Utilization of Solar Energy," McGraw-Hill; ASHRAE, "Solar Energy Use for Heating and Cooling of Buildings") by using selective reflectance coatings with high α/ε ratios and by using covers which are transparent to solar radiation but opaque to longwave emissions (see Table 9.1.6). Both convection and radiation can be reduced by the use of honeycomb structures in the airspace between the cover and the collector plate (see Hollands, Honeycomb Devices in Flat-Plate Solar Collectors, *Solar Energy*, **9**, no. 3, 1965). For the transparent covers, glass is best since it lets through the shortwave solar radiation but stops the longwave radiation given off by the collector plate. Plastics do not have this trapping characteristic, do have a shorter life, may lose their transparency, and outgas when heated. The fumes may condense on other surfaces, forming a film to reduce the collector performance.

Table 9.1.6 Transmittance and Overall Heat-Transfer Coefficients for Collectors with Glass and Plastic Covers

Type and number of covers	None	One glass	One plastic	Two glass	Two plastic
Solar transmittance	1.00	0.90	0.92	0.81	0.85
Overall coefficient U	3.90	1.12	1.30	0.71	0.87

Fig. 9.1.13   Variation with concentration ratio of equilibrium temperatures for earth and space.

Applications of Heliotechnology

Solar Drying   Probably the largest use of solar energy over the centuries—the drying of agricultural crops, evaporation of ocean and salt lake ponds for salt production, etc.—has led more recently to more efficient dryers. The newer dryers prevent rain and dew from rewetting the materials. Simple inexpensive solar dryers—essentially transparent covers—sometimes are supplemented by air heaters providing hot air to dry crops, fish, etc., especially in tropical regions where daily rains prevent efficient natural drying. They are used for curing wood and to remove moisture from mining ores (especially if they have been washed) to reduce shipping costs. Some uneconomical mining operations have been made profitable by the use of solar drying.

Swimming Pool Heating   Probably the widest U.S. commercial application of solar energy today is in swimming pool heating. To extend the swimming season, a transparent cover floating on the surface of the pool, with as few air bubbles underneath as possible, will raise the temperature of the water by up to 20°F (11°C). Flat-plate type of heaters on roofs, often made of rubber or plastics, can be used instead of or in addition to the pool cover. Approximately 1,000 Btu/(ft² · day) [3.1 kWh/(m² · day)] can be expected on average from a reasonably good collector. Since in many places a fence is required around a pool, the fence can incorporate collectors. They are not as efficient because of less favorable orientation, but they serve a dual purpose. The developing countries are interested in solar swimming pool heating since they lack fossil fuels and the currency to buy them, but want to attract tourists with modern conveniences.

Solar Ponds   If water in ponds or reservoirs contains salts in solution, the warmer layers will have higher concentrations and, being heavier, will sink to the bottom. The hot water on the bottom is insulated against heat losses by the cooler layers above. Heat can be extracted from these ponds for power generation. Large solar ponds have been used in the Mideast along the Dead and Red seas, along the Salton Sea; a number of artificial ponds have been established elsewhere. The inexpensive extraction of heat at over 200°F (94°C) still requires improvement. Solar ponds are, effectively, large inexpensive solar collectors. Their heat can be used to power vapor engines and turbines; these, in turn, drive electric generators.

Solar Stills   Covering swimming pools, ponds, or basins with airtight covers (glass or plastic) will condense the water vapors on the underside of the covers. The condensate produced by the solar energy can be collected in troughs as distilled water. Deep-basin stills have a water depth of several feet (between approximately 0.5 and 1.5 m) and require renewal only every few months. Shallow-basin stills have a water depth of about 0.5 to 2.0 in (approximately 1 to 5 cm) and have to be fed and flushed out frequently. Solar stills can be designed to also collect rainwater (in Florida this can double the freshwater production). If the supply water is not contaminated and only too high in solids content, it can be mixed with the distilled water to increase the actual output. This is often done for farm animal water and water used for irrigation.

The glass-covered roof-type solar still (Fig. 9.1.15a) is in wide use in arid areas for the production of drinking water from salty or brackish sources. The sun's rays enter through the cover glasses, warm the water, and thus produce vapor which condenses on the inner surface of the cover. The water droplets coalesce and flow downward into the discharge troughs, while the remaining brine is periodically replaced with a new supply of nonpotable water. Daily yield ranges from 0.4 lb/ft² (2 kg/m²) of water surface in winter to 1.0 lb/ft² (5 kg/m²) in summer (Daniels, "Direct Use of the Sun's Energy," Yale, p. 174).

Inflated plastic films (Fig. 9.1.15b) have also been used to cover solar stills, but their greatest success has been achieved in controlled-environment greenhouses where the vapor which transpires from plant leaves is condensed and reused at the plant roots. Stills made of inflatable plastic also are equipment in survival kits, on lifeboats, etc. Wicks in some of the solar stills can improve their performance. Most plastics have to be surface-treated for this application to produce film condensation (for good solar transmission) rather than dropwise condensation, which reflects a considerable fraction of the impinging solar radiation.

Fig. 9.1.15   Shallow-basin horizontal-surface solar stills. (*a*) Glass-covered roof type; (*b*) inflated plastic type.

Solar Water Heaters   These can be the pan (batch) type—a tank or basin with transparent cover—or tube collector type, described previously. The simple thermosiphon solar water heater (Fig. 9.1.16) with a glass-covered flat-plate collector is used in thousands of homes all over the world, but mainly in Australia, Japan, Israel, North Africa, and Central and South America. Under favorable climatic conditions

Fig. 9.1.16   Thermosiphon type of water heater. (*The collector angle *A* depends on the latitude. More favorable orientation toward the winter sun when days are shorter can produce the same amount of hot water all year round. In Florida, *A* is 10°.)

(abundant sunshine and moderate winter temperatures) they can produce 30 to 50 gal (110 to 190 L) of water at temperatures up to 160°F (70°C) in summer and 120°F (50°C) in winter. Auxiliary electric heaters are often used to produce higher temperatures during unfavorable winter weather.

In the United States and Europe, solar collectors are usually placed on the roof, with the hot water storage tank lower. This requires a small circulating pump. The pump is controlled by a timer, a temperature sensor in the collector, or a differential temperature controller. Ideally, the pump runs when heat can be added to the water in the tank.

To protect the system from freezing, the collectors are drained, manually or automatically. The system can also be designed with a primary circuit containing antifreeze and effecting heat transfer with a heat exchanger in the tank, or by use of a double-walled tank. Another method for protection is a dual system, in which the water drains from the collectors when the pump stops. This is the preferred method for large systems.

Auxiliary heaters, usually electric, are used often in solar water heaters to handle overloads and unusually bad weather conditions.

Solar House Heating and Cooling   House heating can be accomplished in temperate climates by collecting solar radiation with flat-plate devices (Fig. 9.1.14) mounted on south-facing roofs or walls (in the northern hemisphere). Water or air, warmed by solar radiation, can be used in conventional heating systems, with small auxiliary fuel-burning apparatus available for use during protracted cloudy periods. Excess heat collected during the day can be stored for use at night in insulated tanks of hot water or beds of heated gravel. Heat-of-fusion storage systems, which use salts that melt and freeze at moderate temperatures, may also be used to improve heat storage capacity per unit volume.

Solar air conditioning and refrigeration can be done with absorption systems supplied with moderately high-temperature (200°F or 93°C) working fluids from high-performance good flat-plate collectors. The economics of solar energy utilization for domestic purposes become much more favorable when the same collection and storage apparatus can be used for both summer cooling and winter heating. One such system (see Hay, Natural Air Conditioning with Roof Ponds and Movable Insulation, *Trans. ASHRAE*, **75**, part I, 1969, p. 165) uses a combination of shallow ponds of water on horizontal rooftops with panels of insulation which may be moved readily to cover or uncover the water surfaces. During the winter, the ponds are uncovered during the day to absorb solar radiation and covered at night to retain the absorbed heat. The house is warmed by radiation from metallic ceiling panels which are in thermal contact with the roof ponds. During the summer, the ponds are covered at sunrise to shield them from the daytime sun, and uncovered at sunset to enable them to dissipate heat by radiation, convection, and evaporation to the sky.

Figure 9.1.17 is a schematic of a more typical active solar house heating, cooling, and hot water system. Flat-plate collectors, roughly 1,000 ft² (93 m²) to provide both 3 tons of air conditioning and hot water, properly oriented on the roof (they can actually be the roof), supply hot water to a storage tank (usually buried) of about 2,000 gal (about 7,570-L) capacity. A heat-exchanger coil, submerged in and near the top of the tank, provides domestic hot water. When heat is needed, the thermostat orders the following: Valves V_1 and V_2 close, V_3 and V_4 open, and pump P_C circulates hot water through the house as long as required. When cooling is required, the thermostat orders the following: Valves V_1 and V_2 open, V_3 and V_4 close, and pumps P_B and P_C feed hot water to the air conditioner, which, in turn, produces chilled water for circulation through the house as long as needed. This system can be used over a wide range of latitudes, and only the heating and cooling duty cycles will change (i.e., more heating in the north and more cooling in the south). The proper orientation of the collectors will depend upon the duty cycles. In the northern hemisphere, the collectors will face south. When used mostly for heating, they are inclined to the horizontal by latitude plus up to 20°. When they are used mostly for cooling, the inclination will be latitude minus up to 10°. The basic idea is to orient the collectors more favorably toward the winter sun when mostly heating and more favorably toward the summer sun when mostly cooling. The collectors can be made adjustable at extra cost. Air can be used instead of water with rock bin storage and blowers instead of pumps. Blowers require more energy to run.

The air conditioner can be a conventional type, driven by solar engines or solar electricity, or preferably an absorption system (e.g., low-temperature NH_3/H_2O), a jet air conditioning system, a liquid or solid dessicant system, or other direct energy conversion systems described later (in "Direct Energy Conversion").

Solar cooking utilizes (1) a sun-following broiler-type device with a metallized parabolic reflector and a grid in the focal area where cooking pots can be placed; (2) an oven-type cooker comprising an insulated box with glass covers over an open end which is pointed toward the sun. When reflecting wings are used to increase the solar input, temperatures as high as 400°F (204°C) are reached at midday. Large solar cookers for community cooking in third world country villages can be floated on water and thus easily adjusted to point at the sun. For cooking when the sun does not shine, oils or other fluids can be heated to a high temperature, 800°F (around 425°C), with a solar concentrator when the sun shines, and then stored. A range similar to an electric range but with the hot oil flowing through the coils, at adjustable rates, is then used to cook with solar energy 24 h/day.

Solar Furnaces Precise paraboloid concentrators can focus the sun's rays upon small areas, and if suitable receivers are used, temperatures up to 6,500°F (3,600°C) can be attained. The concentrator must be able to follow the sun, either through movement of the paraboloidal reflector itself (Fig. 9.1.13) or by the use of a heliostat which tracks the sun and reflects the rays along a horizontal or vertical axis into the concentrator. This pure, noncontaminating heat can be used to produce highly purified materials through zone refining, in a vacuum or controlled atmosphere. This allows us to grow crystals of high-temperature materials, crystals not existing in nature, or to do simple things such as determining the melting points of exotic materials. Other methods of heating contaminate these materials before they melt. With solar energy the materials can be sealed in a glass or plastic bulb, and the solar energy can be concentrated through the glass or plastic onto the target. The glass or plastic is not heated appreciably since the energy is not highly concentrated when it passes through it.

Solar furnaces are used in high-temperature research, can simulate the effects of nuclear blasts on materials, and at the largest solar furnace in the world (France) produce considerable quantities of highly purified materials for industry.

Power from Solar Energy During the past century (Zarem and Erway, "Introduction to the Utilization of Solar Energy," McGraw-Hill) many attempts have been made to generate power from solar radiation through the use of both flat-plate and concentrating collectors. Hot air and steam engines have operated briefly, primarily for pumping irrigation water, but none of these attempts have succeeded commercially because of high cost, intermittent operation, and lack of a suitable means for storing energy in large quantities. With the rapid rise in the cost of conventional fuels and the increasing interest in finding pollution-free sources of power, attention has again turned to parabolic trough concentrators and selective surfaces (high α/ε ratios) for producing high-temperature working fluids for Rankine and Brayton cycles. Because of the cost of Rankine engines, often operating on other than water-based working fluids, Stirling engines (discussed later) and other small engines (phase shift), turbines, gravity machines, etc., have been

Fig. 9.1.17   Schematic of typical solar house heating, cooling, and hot water system.

developed, their application and use being directly related to the cost and availability of fossil fuels. The price of crude oil rose from $2.50 to over $80.00 per barrel during the period from 1973 to 2006. When fossil fuel costs are high, solar energy conversion methods become competitive and attractive.

Flat-plate collectors utilizing both direct and diffuse solar radiation can be used to drive small, low-temperature Stirling engines, which operate off the available hot water; in turn, the engines can circulate water from the buried storage tank through the collectors on the roof. These engines have low efficiency, but that is not quite so important since solar energy is free. Low efficiency generally implies larger, and thus more expensive, equipment. The application cited above is ideal for small circulating electric pumps powered by solar cells.

Concentrating collector systems, having higher conversion efficiencies, can only utilize the direct portion of the solar radiation and in most cases need tracking mechanisms, adding to the cost. A 10-MW plant was built and had been operating in Barstow, CA until recently. That plant was used both for feasibility studies and to gather valuable operating experience and data. Although intrinsically attractive by virtue of zero fuel cost, solar-powered central plants of this type are not quite yet state of the art. Proposed plants of this and competitive types require abundant sunshine most of the time; obviously, they are not very effective on cloudy days, and their siting would appear to be circumscribed to desertlike areas.

Direct Conversion of Solar Radiation to Electricity   Photovoltaic cells made from silicon, cadmium sulfide, gallium arsenide, and other semiconductors (see "Solar Cells," National Academy of Science, Washington, 1972) can convert solar radiation directly to electricity without the intervention of thermal cycles. Of primary importance today are the silicon solar batteries which are used in large numbers to provide power for space probes, orbiting laboratories, and communication satellites. Their extremely high cost and relatively low efficiency have thus far made them noncompetitive with conventional power sources for large-scale terrestrial applications, but intensive research is currently underway to reduce their production cost and to improve their efficiency. Generation of power from solar radiation on the earth's surface encounters the inherent problems of intermittent availability and relatively low intensity. At the maximum noon intensity of 340 Btu/(ft$^2 \cdot$ h) (1,080 W/m^2) and 100 percent energy conversion, 10ft^2 (1.1 m^2) of collection area would produce 1 thermal kilowatt, but with a conversion efficiency of 10 percent, the area required for an electrical kilowatt approaches 100 ft^2 (9.3 m^2). Thus very large collection areas are essential, regardless of what method of conversion may be employed. However, the total amount of solar radiation falling on the arid southwestern section of the United States is great enough to supply all the nation's electrical needs, provided that the necessary advances are made in collection, conversion, and storage of the unending supply of energy from the sun.

Solar Transportation   Presently the best application of solar energy to transportation seems to be electric vehicles, although solar-produced hydrogen could be used in hydrogen-propelled cars. Since there is not enough surface area on these vehicles to collect the solar energy needed for effective propulsion, storage is needed. Vehicles have been designed and built so that batteries are charged by solar energy. Charging is by Rankine engine-, Stirling engine-, or other engine-driven generators or by photovoltaic panels. A number of utilities, government agencies, municipalities, and universities have electric vehicles, cars, trucks, or buses, the batteries in which are charged by solar energy. For general use a nationwide pollution-free system is proposed, with solar battery charging stations replacing filling stations. They would provide, for a fee, charged batteries in exchange for discharged ones. A design objective to help implement this concept would require that the change of batteries be effected quickly and safely. Electric cars with top speeds of over 65 mi/h (88 km/h) and a range of 200 mi (320 km) have been built. Regenerative braking (the motor becomes a generator when slowing down, charging the batteries), especially in urban driving, can increase the range by up to 25 percent.

Closure   Our inherited energy savings, in the form of fossil fuels, cannot last forever; indeed, an energy income must be part of the overall picture. That income, in the form of solar energy, will have to assume a larger role in the future. Fossil fuels will definitely fade from the picture at some time; it behooves us to plan now for the benefit of future generations.

For the successful application of solar energy, as with any other source of energy, each potential use must be analyzed carefully and the following criteria must be met:

1. Use the minimum amount of energy to do the task (efficiency).
2. Use the best overall energy source available.
3. The end result must be feasible and workable.
4. Cost must be reasonable.
5. The end result must fit the lifestyle and habits of the user.

Schemes which have failed in the past have violated one or more of these criteria.

In utilizing conventional fossil fuels, the energy conversion equipment is a capital cost to which must be added the periodic cost of fuel. Solar energy conversion systems, likewise, represent a capital cost, but there is no periodic cost for fuel. To be competitive, solar capital costs must be less than the total cost of a conventional plant (capital cost plus fuel). There exist circumstances where this is, indeed, the case. Financing will continue to look favorably on such investments.

There are several reasons why solar energy conversion has not had a wider impact, especially in the fossil-fuel-rich countries: lack of awareness of the long-term problems associated with fossil fuel consumption; the fact that solar energy conversion equipment is not as available as would be desirable; the current continuing supply of fossil fuels at very competitive prices; and so forth. There will come a time, however, when the bank of fossil fuels will have been exhausted; solar energy conversion looms large in the future.

A significant consideration with regard to fossil fuels is the realization that they constitute an irreplaceable source of raw materials which ought really to be husbanded for their greater utility as feedstocks for medicines, fertilizers, petrochemicals, etc. Their utility in serving these purposes overrides their convenient use as cheap fossil fuels burned for their energy content alone.

Environmental considerations (pollution, global warming, etc.) and the ever-increasing costs of fossil fuels have become extremely powerful driving forces in encouraging the use of renewable, nonpolluting energy sources (solar, wind, etc.).

During the last decade many countries and many states of the United States have introduced legislation setting goals and requiring that by certain dates a fixed percentage of the energy used (in some cases as much as 30 percent) come from renewable energy sources. Many of those jurisdictions have attached penalties if these goals are not met by certain dates (2008, 2010, etc.). Some of the legislation is very specific; e.g., a certain number of cars have to be operated by renewable energy sources by prescribed dates. More than 600 utilities offer "Green Power" options to their customers. In 2006, Spain legislated that any new building must include some solar energy system.

GEOTHERMAL POWER
by Kenneth A. Phair

REFERENCES: Assessment of Geothermal Resources of the United States—1978, *U.S. Geol. Surv. Circ. 790,* 1979, "Geothermal Resources Council Transactions," vol. 23, Geothermal Resources Council, Davis, CA. Getting the Most out of Geothermal Power, *Mech. Eng.,* publication of ASME, Sept. 1994. "Geothermal Program Review XII," U.S. Department of Energy, DOE/GO 10094-005, 1994.

Geothermal energy is a naturally occurring, renewable source of thermal energy. Thermal energy within the earth approaches the surface in many different geologic formations. Volcanic eruptions, geysers, fumaroles, hot springs, and mud pots are visual indications of geothermal energy.

Significant **geothermal reserves** exist in many parts of the world. The U.S. Geological Survey, in *Circ. 790,* has estimated that in the United States alone there is the potential for 23,000 megawatts (MW) of electric power generation for 30 years from recoverable hydrothermal (liquid- or steam-dominated) geothermal energy. Undiscovered reserves may add significantly to this total. Many of the known resources can be developed using current technology to generate electric power and for various direct uses. For other reserves, technical breakthroughs are necessary before this energy source can be fully developed.

Power generation from geothermal energy is cost-competitive with most combustion-based power generation technologies. In a broader picture, geothermal power generation offers additional benefits to society by producing significantly less carbon dioxide and other pollutants per kilowatt-hour than combustion-based technologies.

Electric power was first generated from geothermal energy in 1904. Active worldwide development of geothermal resources began in earnest in 1960 and continues. In 2004, the capacity of geothermal power plants worldwide exceeded 9,000 MW. Table 9.1.7 lists the installed capacity by country.

Table 9.1.7 Worldwide Geothermal Capacity, 2004

Country	Installed capacity, MW
United States	2,898
Philippines	1,909
Mexico	953
Indonesia	807
Italy	785
Japan	535
New Zealand	437
Iceland	170
Costa Rica	162
Kenya	121
El Salvador	119
Others	263
Total	**9,159**

Geothermal power plants are typically found in areas with "recently" active volcanoes, crustal stretching, and continuing seismic activity. Since the mid-1990s, significant additional geothermal power generation facilities were installed in Indonesia, Italy, Mexico, and the Philippines. Many countries not included in Table 9.1.7 also have significant geothermal resources that are not yet developed.

Although geothermal energy is a renewable resource, economic development of geothermal resources usually extracts energy from the reservoir at a much higher rate than natural recharge can replenish it. Therefore, facilities that use geothermal energy should be designed for high efficiency to obtain maximum benefit from the resource.

Geothermal Resources Geothermal resources may be described as hydrothermal, hot dry rock, geopressured, or magma.

Hydrothermal resources contain hot water, steam, or a mixture of water and steam. These fluids transport thermal energy from the reservoir to the surface. Reservoir pressures are usually sufficient to deliver the fluids to the surface at useful pressures, although some liquid-dominated resources require downhole pumps for fluid production. Hydrothermal resources may be geologically closed or open systems. In a closed system, the reservoir fluids are contained within an essentially impermeable boundary. Communication and fluid transport within the reservoir occur through fractures in the reservoir rock. There is little, if any, natural replenishment of fluids from outside the reservoir boundary. An open system allows influx of cold subsurface fluids into the reservoir as the reservoir pressure decreases. Hydrothermal reservoirs have been found at depths ranging from 400 ft (122 m) to over 10,000 ft (3,050 m).

Hot dry rock resources are geologic formations that have high heat content but do not contain meteoric or magmatic waters to transport thermal energy. Thus water must be injected to carry the energy to the surface. The difficulty in recovering a sufficient percentage of the injected water and the limited thermal conductivity of rock have hindered development of hot dry rock resources. Because of the vast amount of energy in these resources, additional research and development is justified to evaluate whether it is technically exploitable. Research and development of hot dry rock resources is proceeding in Australia, France, Japan, and the United States.

Geopressured resources are liquid-dominated resources at unusually high pressure. They occur between 5,000 and 20,000 ft (1,500 and 6,100 m), contain water that varies widely in salinity and dissolved minerals, and usually contain a significant amount of dissolved gas. Pressures in such reservoirs vary from about 3,000 to 14,000 lb/in^2 gage (21 to 96 MPa) with temperatures between 140°F (60°C) and 360°F (182°C). The largest geopressured zones in the United States exist beneath the continental shelf in the Gulf of Mexico, near the Texas, Louisiana, and Mississippi coasts. Other zones of lesser extent are scattered throughout the United States.

Magma resources occur as formations of molten rock that have temperatures as high as 1,300°F (700°C). In most regions in the continental United States, such resources occur at depths of 100,000 ft (30,500 m) or more. However, in the vicinity of current or recent volcanic activity, magma chambers are believed to be within 20,000 ft (6,100 m) of the surface. The Department of Energy initiated a magma energy research program in 1975, and an exploratory well was drilled in the Long Valley Caldera in California in 1989. The drilling program was planned in four phases to reach a final depth of 20,000 ft (6,100 m) or a temperature of 900°F (500°C), whichever is reached first. The first phases have been completed, but the deep, and most significant, drilling remains to be done.

Exploration Technology Geothermal sites historically have been identified from obvious surface manifestations such as hot springs, fumaroles, and geysers. Some discoveries have been made accidentally during exploring or drilling for other natural resources. This approach has been replaced by more scientific prospecting methods that appraise the extent, as well as the physical and thermodynamic properties, of the reservoir. Modern methods include geological studies involving aerial, surface, and subsurface investigations (including remote infrared sensing) and geochemical analyses which provide a guide for selecting specific drilling sites. Geophysical methods include drilling, measuring the temperature gradient in the drill hole, and measuring the thermal conductivity of rock samples taken at various depths.

Resource Development Extraction of fluids from a geothermal resource entails drilling large-diameter production wells into the reservoir formation. Bottom hole temperatures in hydrothermal wells and hot dry rock formations can exceed 500°F (260°C). Geopressured resources have lower temperatures but offer energy in the form of fluids at unusually high pressures that frequently contain significant amounts of dissolved combustible gases. Although research and development projects continue to seek ways to efficiently extract and use the energy contained in hot dry rock, geopressured, and magma resources, virtually all current geothermal power plants operate on hydrothermal resources. Advances in technology are required to extract the thermal energy in many known geothermal resources. The U.S. Department of Energy's R&D initiative for Enhanced Geothermal Systems (EGS) intends to develop the technology to extract the energy in resources with high heat flow but with low natural permeability and/or low fluid content.

Production Facilities For most projects, a number of wells drilled into different regions of the reservoir are connected to an aboveground piping system. This system delivers the geothermal fluid to the power plant. As with any fluid flow system, the geothermal reservoir, wells, and production facilities operate with a specific flow vs. pressure relationship. Figure 9.1.18 shows a typical steam deliverability curve for a 110-MW geothermal power plant.

Resource permeability; the number, depth, and size of the wells; and the surface equipment and piping arrangement all contribute to make the deliverability curve different for each power plant. Production of geothermal fluids over time results in declining deliverability. For one-half or more of the operating life of a reservoir, the deliverability can usually be held constant by drilling additional production wells into other regions of the resource. As the resource matures, this technique

Fig. 9.1.18 Typical deliverability curve; steam flow to power plant [1,000 lb/h = (453 kg/h)] vs. turbine inlet pressure [100 lb/in² gage (689 kPa gage)].

ceases to provide additional production. The deliverability curve begins to change shape and slope as deliverability declines. The power plant design must be matched to the deliverability curve if maximum generation from the resource is to be achieved.

Geothermal Power Plants A steam-cycle geothermal power plant is very much like a conventional fossil-fueled power plant, but without a boiler. There are, however, significant differences. The turbines, condensers, noncondensable gas removal systems, and materials used to fabricate the equipment are designed for the specific geothermal application. With geothermal steam delivered to the power plant at approximately 100 lb/in² gage (689 kPa), only the low-pressure sections of a conventional turbine generator are used. Additionally, the geothermal turbine must operate with steam that is far from pure. Chemicals and compounds in solid, liquid, and gaseous phases are transported with the steam to the power plant. At the power plant, the steam passes through a separator that removes water droplets and particulates before it is delivered to the turbine. Geothermal turbines are of conventional design with special materials and design enhancements to improve reliability in geothermal service. Turbine rotors, blades, and diaphragms operate in a wet, corrosive, and erosive environment. High-alloy steels, stainless steels, and titanium provide improved durability and reliability. Still, frequent overhauls are necessary to maintain reliability and performance. The high moisture content and the corrosive nature of the condensed steam require effective moisture removal techniques in the later (low-pressure) stages of the turbine. Scale formation on rotating and stationary parts of the turbine occurs frequently. Water washing of the turbine at low-load operation is sometimes used between major overhauls to remove scale.

Most geothermal power plants use direct-contact condensers. Only when control of hydrogen sulfide emissions has been required or anticipated have surface condensers been used. Surface condensers in geothermal service are subject to fouling on both sides of the tubes. Power plants in The Geysers in northern California use conservative cleanliness factors to account for the expected tube-side and shell-side fouling. Some plants have installed on-line tube-cleaning systems to combat tube-side fouling on a continuous basis, whereas other plants mechanically clean the condenser tubes to restore lost performance.

Noncondensable gas is transported with the steam from the geothermal resource. The gas is primarily carbon dioxide but contains lesser amounts of hydrogen sulfide, ammonia, methane, nitrogen, and other gases. Noncondensable gas content can range from 0.1 percent to more than 5 percent of the steam. The makeup and quantity of noncondensable gas vary not only from resource to resource but also from well to well within a resource. The noncondensable gas removal system for a geothermal power plant is substantially larger than the same system for a conventional power plant. The equipment that removes and compresses the noncondensable gas from the condenser is one of the largest consumers of auxiliary power in the facility, requiring up to 15 percent

of the thermal energy delivered to the power plant. A typical system uses two stages of compression. The first stage is a steam jet ejector. The second stage may be another steam jet ejector, a liquid ring vacuum pump, or a centrifugal compressor. The choice of equipment selected for the second stage is influenced by project economics and the amount of gas to be compressed.

The chemicals and compounds in geothermal fluids are highly corrosive to the materials normally used for power plant equipment and facilities. The chemical content of geothermal fluids is unique to each resource; therefore, each resource must be evaluated separately to determine suitable materials for system components. Carbon steel usually will degrade at alarmingly high rates when exposed to geothermal fluids. Corrosion-resistant materials such as stainless steel may perform satisfactorily, but may experience rapid, unpredictable local failures depending upon the composition of the geothermal fluid. Based on experience with a number of geothermal resources:

Carbon steel with a corrosion allowance is usually suitable for transporting dry geothermal steam.

Geothermal condensate and cooling water usually require corrosion-resistant piping and equipment.

Because noncondensable gas is also corrosive, special materials are usually required.

Copper is extremely vulnerable to attack from the atmosphere surrounding a geothermal power plant. Therefore, copper wire and electrical components should be protected with tin plating and isolated from the corrosive atmosphere.

Within the context of these generalities, the fluids at each resource must be evaluated before construction materials are chosen.

The steam Rankine cycle used in fossil-fueled power plants is also used in geothermal power plants. In addition, a number of plants operate with binary cycles. Combined cycles also find application in geothermal power plants. The basic cycles are shown in Fig. 9.1.19.

The direct steam cycle shown in Fig. 9.1.19a is typical of power plants at The Geysers in northern California, the world's largest geothermal field. Steam from geothermal production wells is delivered to power plants through steam-gathering pipelines. The wells are up to 1 mi or more from the power plant. The number of wells required to supply steam to the power plant varies with the geothermal resource as well as the size of the power plant. Each 55-MW power plant in The Geysers receives steam from between 8 and 23 production wells.

A flash steam cycle for a liquid-dominated resource is shown in Fig. 9.1.19b. Geothermal brine or a mixture of brine and steam is delivered to a flash vessel at the power plant by either natural circulation or pumps in the production wells. At the entrance to the flash vessel, the pressure is reduced to produce flash steam, which then is delivered to the turbine. This cycle has been used at power plants in California, Nevada, Utah, and many other locations around the world. Increased thermal efficiency is available from the use of a second, lower-pressure flash to extract more energy from the geothermal fluid. However, this technique must be approached carefully because dissolved solids in the geothermal fluids will concentrate and may precipitate as more steam is flashed from the fluid. The solids also tend to form scale at lower temperatures, resulting in clogged turbine nozzles and rapid buildup in equipment and piping to unacceptable levels.

A binary cycle is the economic choice for hydrothermal resources with temperatures below approximately 330°F (166°C). A binary cycle uses a secondary heat-transfer fluid instead of steam in the power generation equipment. A typical binary cycle is shown in Fig. 9.1.19c. Binary cycles can be used to generate electric power from resources with temperatures as low as 250°F (121°C). The binary cycle shown in Fig. 9.1.19c uses isobutane as the heat-transfer fluid. It is representative of units of about 10-MW capacity. Many small modular units of 1- or 2-MW capacity use pentane as the binary fluid. Heat from geothermal brine vaporizes the binary fluid in the brine heat exchanger. The binary fluid vapor drives a turbine generator. The turbine exhaust vapor is delivered to an air-cooled condenser where the vapor is condensed. Liquid binary fluid drains to an accumulator vessel before being pumped back to the brine heat exchangers to repeat the cycle.

Fig. 9.1.19 Geothermal power cycles. T = turbine, G = generator, C = condenser, S = separator, E = heat exchanger.

Binary-cycle geothermal plants are in operation in several countries. In the United States, they are located in California, Nevada, Utah, and Hawaii.

A geothermal combined cycle is shown in Fig. 9.1.19d. Just as combustion-based power plants have achieved improved efficiencies by using combined cycles, geothermal combined cycles also show improved efficiencies. Some new power plants in the Philippines use a combination of steam and binary cycles to extract more useful energy from the geothermal resource. Existing steam-cycle plants can be modified with a binary bottoming cycle to improve efficiency.

Cycle optimization is critically important to maximize the power generation potential of a geothermal resource. Selecting optimum cycle design parameters for a geothermal power plant does not follow the practices used for fossil-fueled power plants. While a higher turbine inlet pressure will improve the efficiency of the power plant, a lower turbine inlet pressure may result in increased generation over the life of the resource. The resource deliverability curve (Fig. 9.1.18) is used with turbine and cycle performance predictions to determine the flow and turbine inlet pressure that will yield maximum generation. The technical optimum must then be subjected to an economic analysis to identify the best parameters for the power plant design. Because the shape and slope of the deliverability curve vary from resource to resource, the optimum turbine inlet conditions will likewise vary.

Direct Use   There are substantial geothermal resources with temperatures less than 250°F (121°C). While these resources cannot currently be used to generate electric power economically, they can be used for various low-temperature direct uses. Services such as district heating, industrial process heating, greenhouse heating, food processing, and aquaculture farming have been provided by geothermal fluids. For these applications, corrosion and fouling of surface equipment must be addressed in the system design.

The **geothermal heat pump (GHP)** is another direct use of the earth's thermal energy. The GHP, however, does not require a high-temperature geothermal reservoir. The GHP uses essentially constant-temperature groundwater as a heat source or heat sink in a conventional, reversible, water-to-air heat pump cycle for building heating or cooling. The ground, groundwater, and local climatic conditions must be included in the design of a GHP for a specific location. Systems are currently available for residential (single- and multifamily) dwellings, offices, and small industrial buildings.

Environmental Considerations   Geothermal fluids contain many chemicals and compounds in solid, liquid, and gaseous phases. For both environmental protection and resource conservation, spent geothermal liquids are returned to the reservoir in injection wells. This limits the release of compounds to the environment to a small amount of liquid lost as drift from the cooling tower and noncondensable gases. Problems with arsenic, boron, and mercury contamination have been encountered in the immediate vicinity of cooling towers at geothermal power plants. The noncondensable gases, composed primarily of carbon dioxide, usually also contain hydrogen sulfide. Along with its noxious odor, hydrogen sulfide is hazardous to human and animal life. Although many geothermal power plants do not currently control the release of hydrogen sulfide, others use process systems to oxidize the hydrogen sulfide to less toxic compounds. A number of the process systems produce 99.9 percent pure sulfur that can be sold as a by-product. Using geothermal energy for power generation and other direct applications provides environmental benefits. Carbon dioxide released from a geothermal power plant is approximately 90 percent less than the amount released from a combustion-based power plant of equal size, and they create little, if any, liquid or solid waste.

STIRLING (HOT AIR) ENGINES
by Erich A. Farber

Hot air engines, frequently referred to as Stirling engines, are heat engines with regenerative features in which air; other gases such as H_2, He, N_2; or even vapors are used as working fluids, operating, theoretically at least, on the Stirling or Ericsson cycle (see Sec. 4.1) or modifications of them. While the earlier engines of this type were bulky, slow, and low in efficiency, a number of new developments have addressed these deficiencies. Stirling engines are multifuel engines and have been driven by solid, liquid, or gaseous fuels, and in some cases with solar energy. They can be reciprocating or rotary, include special features, run quietly, are relatively simple in construction (no valves, no electrical systems), and if used with solar energy, produce no waste products. (See Walker, "Stirling Engines," Clarendon Press, Oxford; and the *Proceedings of the Intersociety Energy Conversion Engineering Conferences*, held annually.)

The Philips Stirling Engines The Philips Laboratory (in Holland) seems to have developed the first efficient, compact hot air or Stirling engine. It operates at 3,000 r/min, with a hot chamber temperature of 1,200°F (650°C), maximum pressure of 50 atm, and mep of 14 atm (14.1 bar). The regenerator consists of a porous coil of thin wires having 95 percent efficiency, saving about three-fourths of the heat required by the working fluid. The exhaust gases preheat the air, saving about 70 percent of this loss.

Single-cylinder engines, up to 90 hp (67 kW), and multicylinder engines of several hundred horsepower have been constructed with mechanical efficiencies of 90 percent and thermal efficiencies of 40 percent. Heat pipes incorporated in the designs improve the heat transfer characteristics. Philips Stirling engines have been installed in clean-air buses on an experimental basis. Exhaust estimates for an 1,800-kg car are C_xH_y, 0.02 g/mi (0.012 g/km); CO, 1.00 g/mi (0.62 g/km); NO (25 percent recirculated), 0.16 g/mi (0.099 g/km).

Much of the efforts at Philips in recent years have gone into Stirling engine component development, special design features, and even special fuel sources. Engines with rhombic drive were replaced by double-acting machines with "wobble-plate" or, as later referred to, as "swash plate" drive, reducing the weight and complexity of the design. Work with high temperature, efficient hydrogen storage in metallic hydrides offers the possibility of using hydrogen as fuel for transportation applications.

GMR Stirling Thermal Engines A cooperative program between the Philips Research Laboratory and General Motors Corporation resulted in the development of several engines. One, weighing 450 lb (200 kg) and operating at mean pressure of 1,500 lb/in² (103.4 bar), produces 30 hp (22 kW) at 1,500 r/min with a 39 percent efficiency and 40 hp (30 kW) at 2,500 r/min with a 33.3 percent efficiency. Another weighing 127 lb (57 kg) and operating at a mean pressure of 1,000 lb/in² (6.9 MN/m²), produces 6 hp (4.5 kW) at 2,400 r/min with a 29.6 percent efficiency and 8.63 hp (6.4 kW) at 3,600 r/min with a 26.4 percent efficiency.

One such engine was used for a portable Stirling engine electric generator set; another was installed in a Stirling engine electric hybrid car. A 360 hp (265 kW) marine engine was delivered to the U.S. Navy. Another 400-hp (295-kW) engine with special control features was built and tested, and could reverse its direction of rotation almost instantaneously.

Ford-Philips Stirling Engine Development In 1972, Ford Motor Company and Philips entered into a joint development program and developed Stirling engines which were installed experimentally in then-current automobile models.

MAN/MWM Stirling Engines The German company Entwicklungsgruppe Stirlingmotor MAN/MWM, in cooperation with Philips, developed a single acting engine with rhombic drive which developed 30 hp (22 kW) at 1500 r/min and formed the basic test unit for a four-cylinder 120-hp (88-kW) engine. Some double acting engines have been developed. In cooperation with the Battelle Institut, Frankfurt, a 15-kVA Stirling engine hydroelectric generator was developed. It operated at 3,000 r/min, pressurized with helium, with an efficiency of 25 percent.

United Stirling Engines United Stirling AB (Sweden) in cooperation with Philips developed Stirling engines for boats, including those of the Swedish Royal Navy, and engines for buses. One generated 200 hp (145 kW) at 3,000 r/min and a mean helium pressure of 220 atm (22.3 MN/m²).

Internally Focusing Regenerative Gas Engines These engines, conceived at the Solar Energy Laboratory of the University of Wisconsin, use solar energy, concentrated by a parabolic reflector and directed through a quartz dome upon an internal absorber. This reduces the heat losses, since the engine has no external high-temperature heat transfer surfaces. A small working model of this engine has been built at Battelle Memorial Institute and was demonstrated driving a small fan.

Fractional-Horsepower Solar Hot Air Engines The Solar Energy and Energy Conversion Laboratory of the University of Florida has developed small (¼ to ⅓ hp; 0.186 to 0.25 kW) solar hot air engines (some of them converted lawnmower engines). The actual power output of the engines is determined by the size of the solar concentrator rather than by the engine. Some of these engines are self-supercharging to increase power. Water injection, self-acting, increased power by 19 percent. The average speed of the closed-cycle engines is about 500 r/min; average conversion efficiency is about 9 percent. Open-cycle engines separate the heating process from the working cycle, allowing the design of high-speed or low-speed engines as desired. Any heat source can be used with these engines, such as solar, wood, farm wastes, etc. They are simple, rugged, and designed for possible use in developing countries.

The Stirling Engine for Space Power General Motors Corporation, under contract to the U.S. Air Force Aeronautical Systems Command, adapted the GMR Stirling engine to possible space applications. A 3-kW engine was built utilizing NaK heated to 1,250°F (677°C) as a heat source and water at 150°F (66°C) as the cooling medium. The engine is pressurized to a mean pressure of 1,500 lb/in² (103.4 bar), giving an efficiency of 27 percent at 2,500 r/min. The weight of this solar energy conversion system is 550 lb (249 kg). Chemical, nuclear, or other energy sources can also be used.

Free-Piston Stirling Engines The free-piston Stirling engines, pioneered principally by William Beale, consist of displacer and power pistons, coupled by springs, inside one cylinder. They are relatively simple, self-starting, and if pressurized, can be hermetically sealed. The power piston can be coupled to a pump piston since the motion is reciprocating. Single- and double-acting engines have been designed, built, and demonstrated for water pumping and electricity generation. Some of the engines are presently under evaluation by the U.S. Agency for International Development for possible use in developing countries. They can use alternative energy sources, principally solar.

Closed-Environment Stirling Engines A number of Stirling engines have been developed to utilize energy sources which do not require coupling to the external environment. Such engines can be powered by specially prepared fuel sources or by stored energy.

Artificial-Heart Stirling Engines Considerable interest has been shown in the possible use of Stirling engines either to assist weakened hearts or to replace them if they have been damaged beyond repair. The program is supported by the National Heart Institute and has involved many organizations (e.g., Philips, Westinghouse, Aerojet-General, McDonnell-Douglas, University of California, Washington State University). Most of the engines are powered by nuclear fuel sources.

Low-Temperature Stirling Engines In many applications, low-temperature sources such as exhaust gases from combustion, waste steam, and hot water from solar collectors are available. Several groups (University of Florida, University of Wisconsin, Zagreb University in former Yugoslavia, etc.) are working on the development of low-temperature Stirling engines. Models for demonstration have been built and their performance has been evaluated.

Heat Pump and Cryocooler Stirling Engines A Stirling engine can be driven by any mechanical source or by another Stirling engine, and when so motored becomes a heat pump or cooler, depending upon the effects desired and utilized. Special duplex designs for Stirling engines lend themselves especially well for these applications. A number of private companies and public laboratories are involved in this development. Philips manufactured small cryocoolers in the past and sold them throughout the world.

Liquid Piston Stirling Engines Liquid piston engines are extremely simple. The basic liquid piston Stirling engine consists of two U tubes. A pipe connects the two ends of one of the U tubes with one end of the other. The unconnected end of the second U tube is left open. Both U tubes are filled with liquid, thus forming the liquid pistons. The closed U tube liquid acts as the displacer and the other as the power unit. The section of the connecting pipe between the displacer U tube ends contains the regenerator. This engine is referred to as the basic **Fluidyne**.

Even though these engines have been around for a long time much development work is still needed. Their efficiencies are still extremely low.

Closure During their history, Stirling engines have experienced periods of high interest and rapid development. Stirling Engines for Energy Conversion in Solar Energy Units (Trukhov and Tursunbaev, *Geliotekhnika,* **29,** no. 2, 1993, pp. 27–31) summarizes the performance of 15 Stirling engines. With supply temperatures of about 600°C, their output varies from 0.55 to 43.2 kW, their speed varies from 833 to 4,000 r/min, and their conversion efficiencies vary from 12.5 to 30.3 percent. Development work continues on some problem areas (seals, hydrogen embrittlement, weight, etc.). Interest in the potential of these engines remains high, as indicated by the many Stirling engine papers presented and published yearly in each of the "Proceedings of the Intersociety Energy Conversion Conferences."

POWER FROM THE TIDES
by E. A. Avallone

REFERENCES: The Rance Estuary Tidal Power Project, *Pub. Util. Ftly.,* Dec. 3, 1964. *Mech. Eng.,* Ap. 1984. ASCE Symposium, 1987: Tidal Power. Gray and Gashus, "Tidal Power," Department of Commerce, NOAA, Water for Energy, *Proc. 3d Intl. Symp.,* 1986.

The tides are a renewable source of energy originating in the gravitational pull of the moon and sun, coupled with the rotation of the earth. The consequent portion of the earth's rotation is a mean ocean tide of 2 ± ft (0.6 m). The seashore periodic variation of the tides averages 12 h 25 min.

Tidal power is derivable from the large periodic variations in tidal flows and water levels in certain oceanic coastal basins. Suitable configurations of the continental shelves and of the coastal profiles result in reflection and resonance that amplify normally small bulges to ranges as high as 50 ± ft (15 ± m).

Principal tidal-power sites include the North Sea [12 ft (3.6 m) average tidal range]; the Irish Sea [22 ft (6.7 m)]; the west coast of India [23 ft (7 m)]; the Kimberly coast of western Australia [40 ft (12 m)]; San Jose Bay on the east coast of the Argentine [23 ft (7 m)]; the Kislaya Guba (Kisgalobskaia Bay) near the White Sea (no data); St. Michel (including the Rance estuary) on the Brittany coast of France [26 ft (8 m)]; the Bristol Channel (Severn) in England [32 ft (9.8 m)]; the Bay of Fundy (including the Chignecto Bay between New Brunswick and Nova Scotia and the Minas Basin in Nova Scotia) [40 ft (12 m)]; Passamaquoddy Bay between Maine and New Brunswick [18 ft (5.5 m)].

The great attraction for a cost-free source of power continues. From 2000 to 2005, a number of tidal power projects have been funded and are either in experimental use or in the real planning stage. An installation in Norway has delivered power to the local electric distribution grid. An installation in New York City—in the East River—is underway and is scheduled to begin experimental runs in the near future (2007). A large installation is under construction in the vicinity of Inchon, South Korea, which will exploit the high tidal flows there. China has an ambitious proposal to install a 300-MW plant to exploit large flows in tidal lagoons on its coast; to date, this project is only in the planning stage, and it awaits funding.

The harnessing of the tides reaches back into ancient history. Tidal mills, typically with undershot water wheels, were used in New England raceway estuaries, with reversible features for ebb and flood conditions. These power applications were suitable for purposes such as grinding grain, but their number and size were small. In recent times the unique tidal ranges to 50 ± ft (15 ± m) have prompted many studies, proposals, and projects for most of the regions cited above. The attraction for the utilization of tides to generate electric power lies in the facts that there results no air or thermal pollution, the source is effectively inexhaustible, and the construction work related to the tidal power plant is relatively benign in its environmental impact. Despite these efforts for the generation of electricity, there are only four tidal power developments in actual service (1995)—the Rance estuary in France

(240,000 kW), the Kislaya Guba in Russia (400 kW), the Bay of Fundy in Canada (20,000 kW), and a small pilot plant in Kiangshia, China.

Developments take one of two general forms: **single-basin** or **multiple-basin.** A single-basin project, such as the Rance, has a dam, sluices, locks, and generating units in a structure separating a tidal basin from the sea. Water is trapped in the basin after a high tide. As the water level outside the basin falls with the tide, flow from the basin through turbines generates power. Power also may be generated when a basin emptied during a low tide is refilled on a rising tide. Numerous variations in operation are possible, depending on tide conditions and the relationship between the tide cycles and the load cycles. Pumping into or out of the basin increases the availability of the installed capacity for peak load service. A multiple-basin development, such as projected for Chignecto or Passamaquoddy, generally has the power house between two basins. Sluices between the sea and the basins are so arranged that one basin is filled twice a day on high tide and the other emptied twice a day on low tide. Power output can be made continuous.

The amount of **energy available** from a tidal development is proportional to the basin area and to the square of the tidal range. Head variations are large in tidal projects during generating cycles and on a daily, monthly, and annual basis owing to various cosmic factors. Intermittent power, as from all single-basin plans and from two-basin plans with low-capacity factor, implies that the output can best be utilized as peaking capacity. Because of low heads, particularly toward the end of any generating cycle when pools have been drawn down, the cost of adding generating units only to tidal projects is well over the total installation cost of alternative peaking capacity. To be economically competitive with alternative capacity, the tidal projects must produce enough energy to pay the power-plant costs and also to pay for the dams and other costs, such as general site development, transmission, operation, and replacements.

The **risks and uncertainties** involved in designing, pricing, building, and operating capital-intensive tidal works, and the technological developments in alternative types of generating capacity, have tended to defeat tidal developments. Civil works are too extensive; transmission distances to load centers are too great; the required scale of development is too large for existing loads; the coordination of system demands and tidal generation requires interconnections for economic loading; the ultimate capacity of all the world's tidal potential is practically insignificant to meet the world's demands for electricity. In addition, the matter of interrupting tidal action and storage of tidewaters in large basins for extended periods of time raises the possibility of saltwater infiltration of adjacent underground fresh water supplies which, in many cases, are the source of drinking water for the contiguous areas.

UTILIZATION OF ENERGY OF OCEAN WAVES
Revised by E. A. Avallone

According to Albert W. Stahl, USN (*Trans, ASME,* **13,** p. 438), the total energy of a series of **trochoidal deep-sea waves** may be expressed as follows: hp per ft of breadth of wave = $0.0329 \times H^2\sqrt{L}(1 - 4.935H^2/L^2)$, where H = height of wave, ft, and L = length of wave between successive crests, ft. For example, with $L = 25$ ft and $L/H = 50$, hp = 0.04; with $L = 100$ ft and $L/H = 10$, hp = 31.3. Not much more than a quarter of the total energy of such waves would probably be available after reaching shallow water, and apparatus rugged enough for this purpose would doubtless be unable to utilize more than a third of this amount. **Wave motors** brought out from time to time have depended for their operation largely on the lifting power of the waves.

Gravity waves may be only a few feet high yet develop as much as 50 kW/ft of wavefront. Historical wave motors utilize (1) the kinetic energy of the waves by a device such as a paddle wheel or turbine or (2) the potential energy from devices such as a series of floats or by impoundment of water above sea level. Few devices proposed utilize both forms of energy. Jacobs (*Power Eng.,* Sept. 1956) has analyzed the periodic fluctuation of "seiching" of the water level of harbors or basins where, with a resonant port, a 1,000-ft wavefront might be used to achieve a liquid piston effect for the compression of air, the air to be subsequently used in an air turbine.

The principle of using an oscillating column of displaced air has been employed for many years in buoys and at lighthouses, where the wave-actuated rise and fall of the column of air actuated sound horns. A wave-actuated air turbine and electric generator have been operating on the Norwegian coast to study feasibility and to gather operating data. Another similar unit has been emplaced recently off the Scottish coast, and while serving to provide operational data, it also feeds about 2 MW of power into the local grid. If the results are favorable, this particular type of unit may be expanded at this site or emplaced at other similar sites.

A variation of capture of sea wave energy is to cause waves to spill over a low dam into a reservoir, whence water is conducted through water turbines as it flows back to sea level. Any attempt to channel significant quantities of water by this method would require either a natural location or one in which large concrete structures (much of them under water) are emplaced in the form of guides and dams. The capital expenses implicit in this scheme would be enormous.

The extraction of sea wave energy is attractive because not only is the source of energy free, but also it is nonpolluting. Most probably, the capture of wave energy for beneficial transformation to electric power may be economically effected in isolated parts of the world where there are no viable alternatives. Remote island locations are candidates for such installations.

An installation is proposed off Washington state's Olympic Peninsula. While relatively small, it may serve as an in situ experimental installation as well as a source of electric power to the nearby community.

UTILIZATION OF HEAT ENERGY OF THE SEA
Revised by E. A. Avallone

REFERENCES: Claude and Boucherot, *Compt. rend.*, 183, 1926, pp. 929–933. *The Engineer*, 1926, p. 584. Anderson and Anderson, *Mech. Eng.*, Apr. 1966. Othmer and Roels, Power, Fresh Water, and Food from Cold, Deep Sea Water, *Science*, Oct. 12, 1973. Roe and Othmer, *Mech. Eng.*, May 1971. Veziroglu, "Alternate Energy Sources: An International Compendium." Department of Commerce. NOAA, Water for Energy, *Proc. 3d Intl. Symp.*, 1986. Current *Proc. of Third World Water Forum.*

Deep seawater, e.g., at 1-mi (1.6-km) depth in some tropical regions, may be as much as 50°F (28°C) colder than the surface water. This difference in temperature is a fundamental challenge to the power engineer, as it offers a potential for the conversion of heat into work. The Carnot cycle (see Sec. 4.1) specifies the limits of conversion efficiency. Typically, with a heat source surface temperature $T_1 = 85°F$ (545°R, 29.4°C, 303 K), and a heat sink temperature T_2 50° lower, or $T_2 = 35°F$ (495°R, 1.7°C, 275 K), the ideal Carnot cycle thermal efficiency = $(T_1 - T_2)/T_1 = [(545 - 495)/545]100 = 9.2$ percent.

Some units in experimental or pilot operation have demonstrated actual thermal efficiency in the range of 2 to 3 percent. These efficiencies, both ideal and actual, are far lower than those obtained with fossil or nuclear fuel-burning plants. The fundamental attraction of **ocean thermal energy conversion (OTEC)** is the vast quantity of seawater exhibiting sufficient difference in temperature between shallow and deep layers. In reality, the sea essentially represents a limitless store of solar energy which manifests itself in the warmth of seawater, especially in the top, shallow layers. Water temperatures fluctuate very little over time; thus the thermodynamic properties are relatively constant. In addition, the thermal energy is available on a 24-h basis, can be harnessed to serve a plant on land or offshore, provides "free" fuel, and results in a nonpolluting recycled effluent.

Consider the extent of the ocean between latitudes of 30°S and 30°N, and ascribe a temperature difference between shallow and very deep waters of 20°F. The theoretical energy content comes to about 8×10^{21} Btu. Of this enormous amount of raw energy, the actual amount posited for eventual recovery in the best of circumstances via an OTEC system is a miniscule percentage of that total. The challenge is to develop practical machinery to harness thermal energy of the sea in a competitive way.

The concept was put forward first in the nineteenth century, and it has been reduced to practice in several experimental or pilot plants in the past decades. There are three basic conversion schemes: closed Rankine cycle, open Rankine cycle, and mist cycle. In the closed cycle, warm surface water is pumped through a heat exchanger (boiler) which transfers heat to a low-temperature, high-vapor-pressure working fluid (e.g., ammonia). The working fluid vaporizes and expands, drives a turbine, and is subsequently cooled by cold deep water in another heat exchanger (condenser). The heat exchangers are large; the turbine, likewise, is large, by virtue of the low pressure of the working fluid flowing through; enormous quantities of water are pumped through the system. In the open cycle, seawater itself is the working fluid. Steam is generated by flash evaporation of warm surface water in an evacuated chamber (boiler), flows through a turbine, and is cooled by pumped cold deep water in a direct-contact condenser. The reduction of heat-transfer barriers between working fluid and seawater increases the overall system efficiency and requires smaller volumes of seawater than are used in the closed cycle. Introduction of a closed-cycle heat exchanger into the open-cycle scheme results in a slightly reduced thermal efficiency, but provides a valuable by-product in the form of freshwater, which is suitable for human and animal consumption and may be used to irrigate vegetation. A plant of this last type operates in Hawaii and generates 210 kW of electric power, with a 50-kW net surplus power available after supplying all pumping and other house power. This plant continues to provide operational and feasibility data for further development.

An experimental installation is proposed offshore in the island nation of Palau, in the western Pacific. The project is dual-purpose, in that it will provide electric power as well as potable drinking water which results from the processing of cold seawater through low-pressure distilling equipment. Favorable results there may be extended to other island enclaves which depend on rainwater as their sole source of drinking water. In that case, effectively, potable water could be the main product; the generation of power is secondary.

The mist cycle mimics the natural cycle which converts evaporated seawater to rain which is collected and impounded and ultimately flows through a hydraulic turbine to generate power.

Problems encountered in the implementation of any OTEC system revolve on material selection, corrosion, maintenance, and significant fouling of equipment and heat-transfer surfaces by marine flora and fauna. Although development of OTEC systems will continue, with a view to their application in fairly restricted locales, there is no prospect that the systems will make any significant impact on power generation in the foreseeable future.

POWER FROM HYDROGEN
Revised by E. A. Avallone

REFERENCES: Stewart and Edeskuty, Alternate Fuels for Transportation, *Mech. Eng.*, June 1974. Winshe et al., Hydrogen: Its Future Role in the Nation's Energy Economy, *Science*, **180**, 1973.

Hydrogen offers many attractive properties for use as fuel in a power plant. Fundamentally it is a "clean" fuel, smokeless in combustion with no particulate products, and if burned with oxygen, water vapor is the sole end product. If, however, it is burned with air, some of the nitrogen may combine at elevated temperature to form NO_x, a troublesome contaminant. If carbon is present as a fuel constituent, or if it can be picked up from a source such as a lubricant, the carbon introduces further contaminant potentials, e.g., carbon monoxide and cyanogens.

Basically the potential cleanliness of combustion is supported by other properties that make hydrogen a significant fuel, to wit, prevalence as a chemical element, calorific value, ignition temperature, explosibility limits, diffusivity, flame emissivity, flame velocity, ignition energy, and quenching distance.

Hydrogen offers a unique calorific value of 61,000 Btu/lb (140,000 kJ/kg). With a specific volume of 190 ft³/lb (12 m³/kg) this translates to 319 Btu/ft³ (12,000 kJ/m³) at normal pressure and temperature, 14.7 lb/in² absolute and 32°F (1 bar at 0°C). These figures, particularly on the volume basis, introduce many practical problems because hydrogen, with a critical point of −400°F (33 K) at 12.8 atm, is a gas at all normal, reasonable temperatures. When compared with alternative fuels, results are as shown in Table 9.1.8.

Table 9.1.8   Bulk and Calorific Power of Selected Fuels (Approximate and Comparative)

Fuel	State	Sp. wt., lb/ft³	Sp. gr.	Btu/lb	Btu/ft³	Btu/gal
Hydrogen	Gas (NTP)	0.0052	0.07	61,000	320	(40)
Natural gas	Gas (NTP)	0.042	0.67	24,000	1,000	(130)
Gasoline (reg., 90 oct.)	Liquid	46	0.72	20,500	950,000	125,000
Ethanol (99 oct.)	Liquid	49	0.79	12,800	620,000	82,000
Methanol (98 oct.)	Liquid	49	0.79	9,600	480,000	64,000
Hydrogen	Liquid (36°R, 14.7 lb/in² abs)	4.4	0.07	56,000	240,000	32,000
Coal	Piled	50	0.8	12,000	600,000	80,000

These figures demonstrate the volumetric deficiency of gaseous hydrogen. High-pressure storage (50 to 100 atm) is a dubious substitute for the gasoline tank of an automobile. Liquefaction calls for cryogenic elements (Secs. 11 and 19). Chemical compounds, metallic hydrides, hydrazene, and alcohols are potential alternates, but practicality and cost are presently disadvantageous.

Hydrogen has been used to power a number of different vehicles. Its use as a rocket fuel is well documented; in that application, cost is no concern. Experimental use in automotive and other commercial vehicles with slightly altered internal combustion engines is still in its very early stages. Hydrogen as a source of energy as applied to power generation has advanced in experimental automotive vehicles to demonstrate its feasibility. Since liquid hydrogen fuel must be carried in special pressurized tanks, the design of cars with hydrogen-burning engines will present difficult, but not intractable, problems. At the very least, the whole new infrastructure for dispensing liquid hydrogen fuel at "filling stations" will require massive changes. The most likely manifestation of hydrogen-powered vehicles will take the form of hydrogen fuel cells. (See, in this section, "Direct Energy Conversion.") Aircraft jet engines have been powered successfully for short flight times on an experimental basis, and it is conjectured that among the first successful commercial applications of hydrogen as a source of power will be as fuel for jet aircraft sometime in the mid twenty-first century.

In spite of the demonstrated thermodynamic advantages inherent in hydrogen as a source of power, in the current competitive economic market for fuels, hydrogen still faces daunting problems because of its high cost and difficulties related to its efficient storage and transportation.

DIRECT ENERGY CONVERSION
by Erich A. Farber

REFERENCES: Kaye and Welsh, "Direct Conversion of Heat to Electricity," Wiley. Chang, "Energy Conversion," Prentice-Hall. Shive, "Properties, Physics and Design of Semiconductor Devices," Van Nostrand. Bredt, Thermoelectric Power Generation, *Power Eng.*, Feb.–Apr. 1963. Wilson, Conversion of Heat to Electricity by Thermionic Conversion, *Jour. Appl. Phys.*, Apr. 1959. Angrist, "Direct Energy Conversion," Allyn & Bacon, Harris and Moore, Combustion—MHD Power Generation for Central Stations, *IEEE Trans. Power Apparatus and Systems*, **90**, 1971. Roberts, Energy Sources and Conversion Techniques, *Am. Scientist*, Jan.–Feb. 1973. Poule, Fuel Cells: Today and Tomorrow, *Heating, Piping, and Air Conditioning*, Sept. 1970. Fraas, "Engineering Evaluation of Energy Systems," McGraw-Hill. Kattani, "Direct Energy Conversion, Addison-Wesley. Commercialization of Fuel Cell Technology, *Mech. Eng.*, Sept. 1992, p. 82. The Power of Thermionic Energy Conversion, *Mech. Eng.*, Sept. 1993, p. 78. Fuel Cells Turn Up the Heat, *Mech. Eng.*, Dec. 1994, p. 62. "Proceedings of the Intersociety Energy Conversion Conferences," published yearly.

In contrast to the conventional thermal cycle for the conversion of heat into electricity are several more direct methods of converting thermal and chemical energy into electric power. The methods which seem to have the greatest potential possibilities are thermoelectric, thermionic, magnetohydrodynamic (MHD), fuel cell, and photovoltaic. The principles of operation of these processes have long been known, but technological and economic obstacles have limited their use. New applications, materials, and technology now provide increased impetus to the development of these processes.

Thermoelectric Generation

Thermoelectric generation is based on the phenomenon, discovered by **Seebeck** in 1821, that current is produced in a closed circuit of two dissimilar metals if the two junctions are maintained at different temperatures, as in thermocouples for measuring temperature. A thermoelectric generator is a low-voltage, dc device. To obtain higher voltages, the elements must be stacked. Typical thermocouples produce potentials on the order of 50 to 70 μV/°C and power at efficiencies on the order of 1 percent.

Certain semiconductors have thermoelectric properties superior to conductor materials, with resultant improved efficiency. The criterion for evaluating material characteristics for thermoelectric generation is the **figure of merit Z**, measured in (°C)$^{-1}$ and defined as $Z = S^2/(\rho K)$, where S = Seebeck coefficient, V/°C; ρ = electrical resistivity, $\Omega \cdot$ cm; K = thermal conductivity, W/(°C $\cdot$ cm).

An ideal thermoelectric material would have a high Seebeck coefficient, low electrical resistivity, and low thermal conductivity. Unfortunately, materials with low electrical resistivity have a high thermal conductivity since both properties are dependent, to some extent, on the number of free electrons in the material. The maximum conversion efficiency of a thermoelectric generator is a function of the figure of merit, the hot junction temperature, and the temperature difference between the hot and cold junctions.

In some types of thermoelectric materials, the voltage difference between the hot and cold junctions results from the flow of negatively charged electrons (n type, hot junction positive), whereas in other types, the voltage difference between the cold and hot junctions results from the flow of positively charged voids vacated by electrons (p type, cold junction positive). Since the voltage output of a typical semiconductor thermoelectric couple is low (about 100 to 300 μV/°C temperature difference between the hot and cold junctions), it is advantageous to use both p- and n-type materials in constructing a thermoelectric generator. The two types of materials make it possible to connect the thermojunctions in series electrically and in parallel thermally (Fig. 9.1.20).

Fig. 9.1.20

Typical semiconductor thermoelectric materials are compounds and alloys of lead, selenium, tellurium, antimony, bismuth, germanium, tin, manganese, cobalt, and silicon. To these materials, minute quantities of "dopants" such as boron, phosphorus, sodium, and iodine are sometimes added to improve properties. Typical Z values for the more commonly used thermoelectric materials are in the range of 0.5×10^{-3} to 3.0×10^{-3} $(°C)^{-1}$. The onset of deleterious thermochemical effects at elevated temperatures, such as sublimation or reaction, limit the materials' application. Bismuth telluride alloys, which have the highest Z values, cannot be used beyond a hot-side temperature of about 300°C without encountering undue degradation. Silicon-germanium alloys have high-temperature capability up to 1,000°C that can take advantage of higher Carnot efficiencies. However, these alloys possess low Z values. Optimized designs of thermoelectric junctions using semiconductor materials have resulted in experimental conversion efficiencies as high as 13 percent; however, the efficiency of practical thermoelectric generators is lower, e.g., 4 to 9 percent. Materials which have higher figures of merit (2 or 3×10^{-3}) and which are capable of operating at higher temperatures (800 to 1,000°C) are required for an appreciable improvement in efficiency.

Thermoelectric-generation technology has matured considerably through its application to nuclear power systems for space vehicles where modules as large as 500 W have been used. It is also used in terrestrial applications such as gas pipeline cathodic protection and power for microwave repeater stations. Development work continues, but the use of this technology is expected to be limited to special cases where power source selection criteria other than efficiency and first cost will dominate.

Thermoelectric Cooling

The **Peltier effect,** discovered in 1834, is the inverse of the Seebeck effect. It involves the heating or cooling of the junction of two thermoelectric materials by passing current through the junction. The effectiveness of the thermojunction as a cooling device has been greatly increased by the application of semiconductor thermoelectric materials. Typical applications of thermoelectric coolers include electronic circuit cooling, small-capacity ice makers, and dew-point hygrometers, small refrigerators, freezers, portable coolers or heaters, etc. These devices make it possible to preserve vaccines, medicines, etc., in remote areas and in third world countries during disasters or military conflicts.

Thermionic Generation

Thermionic generation, proposed by **Schlicter** in 1915, uses a thermionic converter (Fig. 9.1.21), which is a vacuum or gas-filled device with a hot electron "emitter" (cathode) and a cold electron "collector" (anode) in or as part of a suitable gastight enclosure, with electrical connections to the anode and cathode, and with means for heating the cathode and cooling the anode. A thermionic generator is a low-voltage dc device.

Fig. 9.1.21

Figure 9.1.22 is a plot of the electron energy at various places in the converter. The abscissa is cathode-anode spacing, and the ordinate is electron energy. The base line corresponds to the energy of the electrons in the cathode before heating. Heating the cathode imparts sufficient energy to some of the electrons to lift them over the **work function barrier** (retaining force) at the surface of the cathode into the interelectrode space. (The lower the work function, the easier it is for an electron to

Fig. 9.1.22

escape from the surface of the cathode.) If it is assumed that the electrons can follow path *a* to the anode with only a small loss of energy, they will "drop down" the work function barrier as they join the electrons in the anode still retaining some of their potential energy (**Fermi level**), which is available to cause an electric current to flow in the external circuit. The work function of the anode should be as small as possible. The anode should be maintained at a lower temperature to prevent anode emission or back current. This pattern presumes that the electrons could follow path *a* from the cathode to the anode with little interference. Since, however, electrons are charged particles, those in the space between the cathode and anode form a space charge barrier, as shown by *b*. This space charge barrier limits the electrons emitted from the cathode. Space charge formation can be reduced by close spacing of the cathode and anode surfaces or by the introduction of a suitable gas atmosphere that can be ionized by heating and thus neutralize the space charge. In vacuum-type thermionic converters, the spacing between cathode and anode must be less than 0.02 mm to get as many as 10 percent of the electrons over to the cathode and to achieve an efficiency of 4 to 5 percent. In gas-filled converters, the negative electron space charge is neutralized by positive ions. Cesium vapor is used for this purpose. At low pressure, it will also lower the work function of the anode, and at high pressure, it can, in addition, be used to adjust the work function of the cathode. Efficiencies as high as 17 percent have been obtained with gas-filled converters operating at a cathode temperature of 1,900°C (2,173 K). The output voltage is 1 to 2 V, so the units must be connected in series for reasonable utilization voltages.

Thermionic development results have been encouraging, but major technical challenges remain to be resolved before reliable, long-life converters become available. Effort has been focused on problem areas such as the limited life of emitter materials, leaktightness of the converter, and dimensional stability of the converter gap. Studies of thermionic converters incorporated into nuclear reactors and as a topping cycle for fossil fuel fired steam generators as well as for space and solar applications have received the greatest attention.

Fuel Cells

The **fuel cell** is an electrochemical device in which electric energy is generated by chemical reaction without altering the basic components (electrodes and electrolyte) of the cell itself. It is a low-voltage, dc device. To obtain higher voltages, the elements must be stacked. The fact that electrode and electrolyte are invariant distinguishes the fuel cell from the primary cell and storage battery. The fuel cell dates back to 1839, when **Grove** demonstrated that the electrolysis of water could be reversed using platinum electrodes. The fuel cell is unique in that it converts chemical energy to electric energy without an intermediate conversion to heat energy; its efficiency is therefore independent of the thermodynamic limitation of the Carnot cycle. In practical units, however, its efficiency is comparable with the efficiency of Carnot limited engines.

Figure 9.1.23 is a simplified version of a hydrogen or hydrocarbon fuel cell with air or oxygen as the other reactant. The fuel is supplied to the anode, where it is ionized, freeing electrons, which flow in the external circuit, and hydrogen ions, which pass through the electrolyte to the cathode, which is supplied with oxygen. The oxygen is ionized by electrons flowing into the cathode from the external circuit. The ionized oxygen and hydrogen ions react to form water. Electrodes for this

Fig. 9.1.23

Fig. 9.1.24

type of cell are usually porous and impregnated with a catalyst. In a simple cell of this type, chemical and catalytic action take place only at the line (**notable surface of action**) where the electrolyte, gas, and electrode meet. One of the objectives in designing a practical fuel cell is to increase the notable surface of action. This has been accomplished in a number of ways, but usually by the creation of porous electrodes within which, in the case of gas diffusion electrodes, the fuel and oxidant in gaseous state can come in contact with the electrolyte at many sites. If the electrolyte is a liquid, a delicate balance must be achieved in which surface tension and density of the liquid must be considered and gas pressure and electrode pore size must be chosen to hold their interface inside the electrode. If the gas pressure is too high, the electrolyte is excluded from the electrode, gas leaks into the electrolyte, and ion flow stops; if the gas pressure is too low, drowning of the electrode occurs and electron flow stops.

Fuel cells may be classified broadly by operating temperature level, type of electrolyte, and type of fuel. Low-temperature (less than 150°C) fuel cells are characterized by the need for good and expensive catalysts, such as platinum and relatively simple fuel, such as hydrogen. High temperatures (500 to 1,000°C) offer the potential for use of hydrocarbon fuels and lower-cost catalysts. Electrolytes may be either acidic or alkaline in liquid, solid, or solid-liquid composite form. In one type of fuel cell, the electrolyte is a solid polymer.

Low-temperature fuel cells of the hydrogen-oxygen type, one a solid-polymer electrolyte type, and the other using free KOH as an electrolyte, have been successfully applied in generating systems for U.S. space vehicles. High-temperature fuel cell development has been primarily in molten carbonate cells (500 to 700°C) and solid-electrolyte (zirconia) cells (1,100°C), but no significant practical applications have resulted.

Considerable study and development work has been done toward the application of fuel cell generating systems to bulk utility power systems. Low-temperature cells of the phosphoric acid matrix and solid-polymer electrolyte types using petroleum fuels and air have been considered. Cell efficiencies of about 50 percent have been achieved; but with losses in the fuel reformers and electrical inverters, the overall system efficiency becomes of the order of 37 percent. In this application fuel cells have environmental advantages, such as low noise, low atmospheric emissions, and low heat rejection requirements. Additional development work is necessary to overcome the disadvantage of high catalyst costs and requirements for expensive fuels. More than 50 phosphoric acid fuel cell units, having a capacity of 200 kW, are in use. Companies in Canada, Germany, and the United States have demonstrated fuel cells in passenger bus propulsion systems. They are cooperating now on the development of proton exchange fuel cells. Other U.S. and Japanese companies are developing power plants for transportation.

Summary   Fuel cells require further development before they will be practical for widespread use, particularly for automotive applications. Certainly, fuel cells are in the forefront in the arena of public discussion, and their great advantages (high conversion efficiency, low—or no—pollution) are extolled. Most fuel cells which have been thoroughly studied and developed use very expensive fuels, fuels that, in many cases, are derived from or made from and with fossil fuels. Many fuel cells must operate at high temperature. An example is the hydrogen fuel cell, utilizing derived or produced hydrogen. At this time, hydrogen is produced by the consumption of other fossil fuel, via processes operating at moderate efficiency and which give rise to considerable pollution products.

When one considers the overall process, from source to hydrogen to fuel cell to release of energy, the net result is not as attractive as claimed. The pollution that ensues from the process of delivering hydrogen must rightfully be entered into the accounting for the hydrogen fuel cell advantages and disadvantages.

Magnetohydrodynamic Generation

Magnetohydrodynamic generation utilizes the movement of electrically conducting gas through a magnetic field. Normally, it results in a high-voltage, dc output, but it can be designed to provide alternating current. In the simple open-cycle MHD generator (Fig. 9.1.24), hot, partially ionized, compressed gas, which is the product of combustion, is expanded in a duct and forced through a strong magnetic field. Electrodes in the sides of the duct pick up the potential generated in the gas, so that current flows through the gas, electrodes, and external load. Temperature in excess of 3,000 K is necessary for the required ionization of gas, but this can be reduced by the addition of a seeding material such as potassium or cesium. With seeding, the gas temperature may be reduced to the order of 2,750 K. The temperature of the gas leaving the generator is about 2,250 K. Although the efficiency of the basic MHD channel is on the order of 70 percent, only a portion of the available thermal energy can be removed in the channel. The remainder of the energy contained in the hot exhaust gas must be removed by a more conventional steam cycle. In this combined-cycle plant, the exhaust gas from the MHD generator is passed sequentially through an air preheater, the steam superheater and boiler, and an economizer and stack gas cooler. The air preheater is necessary to raise the temperature of incoming combustion air to some 1,900 K in order to obtain the initial gas temperature of 2,750 K.

The potential improvement in efficiency from the use of MHD generator in a combined-cycle plant is in the order of 15 to 30 percent. An overall steam plant efficiency of 38 percent could be raised to some 45 to 54 percent. Contrasted to other methods for direct conversion, MHD generation appears best suited to large blocks of power. For example, an MHD generator 75 m long with an average magnetic field of 5 T (attained by means of a superconductive magnet) would have a net output of about 1,000 MW dc at 5 to 10 kV. Typically, this would provide topping energy for a steam plant of about 500 MW.

Although the MHD topping cycle offers the highest peak cycle temperature and thermodynamic cycle efficiency of any system that has been studied, none of the generators tested have yielded enough efficiency to account for even half of the power required to supply oxidant to the combustor. Serious materials problems have also been experienced, with severe erosion, corrosion, and thermal stresses in the electrodes and insulators. Slow progress in the solution of these and other difficulties diminishes the prospect for a viable MHD system in the foreseeable future.

Closed-cycle MHD generators are also under study for bulk power generation. They are of two types: first, one in which the working fluid is an inert gas such as argon seeded with cesium; and second, the liquidmetal type in which the working fluid is a helium-sodium mixture. Closed-cycle MHD generation offers the potential for high efficiency with considerably lower peak cycle temperatures, lower pressure ratios, and lower average magnetic-flux density.

Photovoltaic Generation

Photovoltaic generation utilizes the direct conversion of light energy to electric energy and stems from the discovery by **Becquerel** in 1839 that a voltage is generated when light is directed on one of the electrodes in an electrolyte solution. Subsequent work using selenium led to the development of the photoelectric cell and the exposure meter. It results in low-voltage direct current. To obtain higher voltages, the elements must be stacked.

Photovoltaic effect is the generation of electric potential by the ionization by light energy (photons) of the area at or near the p-n junction of a semiconductor. The p-n junction constitutes a one-way potential barrier which permits the passage of photon-generated (−) electrons from the p to the n material and (+) "holes" from the n to the p material. The resulting excess of (−) electrons in the n material and (+) holes in the p material produces a voltage at the terminals comparable to the junction potential.

Solar cells have been a useful source of electricity since about 1960 and have enjoyed widespread use for small amounts of electric energy in remote locations. They have proved particularly well suited for use in spacecraft; in fact, much of the effort in photovoltaic R&D has been funded by the space program. Solar cells have been applied also to remote weather monitoring and recording stations (some of them equipped with transmitters to send the data to collecting stations), traffic control devices, buoys, channel markers, navigational beacons, etc. They can also be used for applications such as battery chargers for cars, boats, flashlights, tools, watches, calculators, and emergency radios. Cost reduction through the use of polycrystalline or thin-film techniques constitutes a major development effort.

A commercially available solar cell is constructed of a 0.3-mm-thick silicon wafer (2×2 cm or 2×6 cm) that is doped with boron to give it a p-type characteristic. It is then diffused with phosphorus to a depth of about 10^{-4} cm (n-type layer), and subsequently electrically contacted with titanium-silver or gold-nickel. Contacting on the light exposure side is limited to maximize transmission of light into the cell. The cell is coated with antireflection material to reduce losses due to light reflection. The cell is then electrically coupled to the intercell circuit by soldering. The method of fabricating solar cells is complex and expensive.

The efficiency of a photovoltaic cell varies with the spectrum of the light. The maximum theoretical efficiency of a single-junction, single-transition silicon cell with solar illumination is about 22 percent. Actual cell performance has been realized at 12 to 15 percent efficiency at 0.6 V open circuit and 0.02 W/cm². Advanced cells using materials such as gallium arsenide and cadmium telluride offer maximum theoretical efficiencies above 25 percent. Further cell efficiency improvement is being investigated by concentrating the sunlight with a Fresnel lens or parabolic mirror and by selecting the light spectrum.

Factors reducing the efficiency of conversion of light energy into electricity using solar cells include: (1) the fact that only a certain bandwidth of the solar spectrum can be effective, (2) structural defects and chemical impurities within the materials, (3) reflection of incident light, and (4) cell internal resistance. These factors lower the overall efficiency of solar arrays to about 6 to 8 percent. Another drawback is the intermittent nature of the solar source, which necessitates the use of an energy storage facility.

It is commonly agreed that substantial investments will be necessary to make solar cell energy conversion economically competitive with terrestrial fossil fuel fired or nuclear power plants. Space applications remain a practical use of this technology, since long life and reliability override cost considerations.

Other Energy Converters

Each of the converters described in the following section has characteristics which makes it suitable for specific tasks. Some produce low-voltage direct current and must be stacked for higher voltage output. Other converters produce high-voltage direct or alternating current depending upon their design. High voltages can be used to drive Klystron or X-ray tubes, or similar equipment, and can be transmitted without step-up transformers. Some medium- or low-temperature converters can be used in low-grade energy (heat) applications; in medical practice, e.g., body heat can drive heart pacemakers, artificial heart pumps, internal medicine dispensing devices, and organ monitoring equipment.

Electrohydrodynamic Converters   When positive ions are transported by neutral hot gases against an electric field, high potential differences result. The charges produced by the ions do work when allowed to flow through a load. These devices are also called **electrogasdynamic (EGD)** converters. If the gases are allowed to condense, producing small liquid droplets, the devices are often referred to as **aerosol EGD** converters. A number of different basic designs exist.

Van der Graaff Converter   This device operates on the same principle as the EHD converter, except that a belt is used instead of hot gases to transport the ions against an electric field. Very high potential differences are produced, which may be utilized in high-energy particle accelerators, atom smashers, artificial lightning generators, and the like.

Ferroelectric Converters   Certain materials exhibit a rapid change in their dielectric constant k around their Curie temperature. The voltage produced by a charge is the charge divided by the capacitance, or $V = Q/C$. The variation between capacitance and dielectric constant is expressed by $C_h = (k_h/k_c)C_c$, where c = cold and h = hot.

A capacitor containing ferroelectric material is charged when the capacitance is high. When the temperature is changed to lower the dielectric constant, the capacitance is lowered, with an accompanying increase in voltage. The charge is dissipated through a load when the voltage is high, resulting in the performance of work. Barium titenate, e.g., can produce a fivefold voltage swing between temperatures of 100 and 120°C. Thermocycling could be produced by the sun on a spinning satellite.

Ferromagnetic Converters   Ferromagnetic material is used to complete the magnetic circuit of a permanent magnet. The ferromagnetic material is heat-cycled through its Curie point, producing flux changes in a coil wrapped around the magnet. The operating conditions can be selected by the Curie temperature of the ferromagnetic material. Gadolinium, e.g., has a Curie point near room temperature.

Piezoelectric Converters   When axisymmetric crystals are compressed parallel to their polar axes, they become polarized; i.e., positive charges are generated on one side, and negative charges are generated on the other side of the crystal. The induced compressive stresses can be produced mechanically or by heating the crystal. The resulting potential difference will do work when allowed to flow through a load. In a reverse process, imposing a potential difference between the ends of the crystal will result in a compressive stress within the crystal. The imposition of alternating current will result in controlled oscillations useful in sonar, ultrasound equipment, and the like.

Pyroelectric Converters   Some materials become electrically polarized when heated, and the conversion of heat to electricity can be utilized as it is in piezoelectric converters.

Bioenergetic Converters   Energy requirements for medical devices used to monitor or control the performance of human organs (heart, brain, etc.) range from a few microwatts to a few watts. In many cases, body heat is sufficient to operate an energy converter which, in turn, will power the device.

Nernst Effect Converters   When heat flows through certain semiconductors exposed to a magnetic field perpendicular to the direction of heat flow, an electric potential difference will be induced along a third orthogonal axis. This conversion of heat to electricity can be utilized to do work. In a reverse fashion, crossing a magnetic and an electric field will produce a temperature difference (Ettinghausen effect, the reverse of the Nernst effect). This reverse conversion is useful in electric heating, cooling, and refrigeration.

Thermophotovoltaic Converters   A radiant heat source surrounded by photovoltaic cells will result in the radiant energy being converted to electricity. Source radiation and photovoltaic cell characteristics can be controlled to operate anywhere in the spectrum.

Photoelectromagnetic Converters   When certain semiconductors (Cu_2O, for example) are placed in a tangential magnetic field and illuminated by visible light, there will result an electric potential difference along an axis orthogonal to the other two axes. The resulting flow of electric current can be used to do work.

Magnetothermoelectric Converters　A magnetic field applied to certain thermoelectric semiconductors produces electric potential differences, useful in power generation.

Superconducting Converters　The phase transition in a superconductor can be utilized similarly to a ferroelectric converter. Thermal cycling of the superconductor material will produce alternating current in the coil surrounding it. An idealized analysis for niobium at 8 K yields a conversion efficiency of about 44 percent.

Magnetostrictive Converters　Changes in dimensions of materials in a magnetic field produce electric potential differences, thus converting mechanical energy to electricity. The effects can be reversed by combining an electric field with a magnetic field to produce dimensional changes.

Electron Convection Converters　When a liquid is heated (sometimes to the boiling point), electrons and neutral atoms are emitted from the liquid surface. The flow of vapor transports the electrons upward, where they are collected on screens. The vapor condenses and recycles into the liquid pool. High electric potential differences can be produced in this manner between the screens and the liquid pool, allowing the subsequent flow of current to do work. The process is similar to that for EGD converters.

Electrokinetic Converters　Certain fluids flowing through capillary tubes due to pressure gradients produce an electric potential difference between the ends of the capillaries, converting flow (kinetic) energy to electricity.

Particle-Collecting Converters　When an alpha, beta, or gamma particle emitter is surrounded by a collector surface, an electric potential difference is produced between the emitter and the collector. Biased screen grids can improve the performance.

EHD Water Drop Converter　Two separate streams of water coming from the same reservoir, in falling, are allowed to break up into droplets. At the breakup points, each stream is surrounded by a short metal cylinder. Each cylinder is connected electrically to a screen at the bottom of the opposite stream. High potential differences are produced between the two metal cylinders.

Photogalvanic Converters　Photochemical reactions often produced by solar radiation (especially at the shorter wavelengths) can be used to generate electricity. Concentration of solar energy can increase the power of the converters considerably. The actual processes are similar to those in fuel cells.

The field of instrumentation provides other techniques which could become useful as energy conversion devices. While many of the methods cited and described above are not economically competitive with conventional conversion methods in current use, some are adapted to unique situations where the matter of cost becomes inconsequential. Certainly, it is expected that as progress is made in the field of energy conversion, certain techniques will be refined to greater practicality, and others will be developed.

9.2 STEAM BOILERS

by John B. Kitto, Jr., P.E.

REFERENCES: The Babcock & Wilcox Company, "Steam—Its Generation and Use." Combustion Engineering, Inc. (now Alstom), "Combustion—Fossil Power Systems." Elliott, "Standard Handbook of Powerplant Engineering," McGraw-Hill. Cohen, "The ASME Handbook on Water Technology in Thermal Power Systems," American Society of Mechanical Engineers. Gunn, "Industrial Boilers," Longman Scientific. Clapp, "Modern Power Station Practice: Boilers and Ancillary Plant," Pergamon Press. "Boiler and Pressure Vessel Code," "Performance Test Code for Fired Steam Generators," American Society of Mechanical Engineers. Also, *Proc. ASME* and the Joint Power Generation Conference.

Steam boilers or steam generators will convert the thermal energy from the combustion of one or more solid, liquid, or gaseous fuels into steam for a variety of applications. Primary among these are electric power generation and industrial process heating. Steam flow rates and operating conditions vary dramatically from one boiler to another: 1,000 lb/h (0.1 kg/s) to more than 10 million lb/h (1,260 kg/s); 14.7 psi (0.10 MPa) and 212°F (100°C) to 4,500 psi (31.0 MPa); and 1,100°F (593°C) or more in advanced cycles. Since the combustion of fuels generates air pollutants, a key element in boiler and power system design is the control and reduction of regulated emissions.

Steam boilers are fabricated from pressure parts (tubes, tube sheets, and cylindrical vessels) which separate the products of combustion (or the **flue gas**) from the water while permitting heat transfer to take place and generate steam. Two basic boiler design configurations are used: the **fire-tube** boiler with the **flue gas** passing through the tubes and **water-tube** boilers with the water and steam passing through the tubes. Cost, size, technical, and safety considerations dictate that water-tube boilers, the subject of this section, supply the vast majority of steam used today, especially in thermal power stations.

Water-tube boilers generally consist of a large open **furnace** volume where the fuel combustion takes place, and heat is absorbed primarily by thermal radiation from the flue gas to the water-cooled furnace enclosure walls and surfaces. The furnace is followed by a **convection** section or pass where heat transfer from the flue gas takes place primarily by convective heat transfer as the flue gas passes over tube bundles containing steam and/or water.

A specialized steam generator is used in nuclear power applications where pressurized hot water from a nuclear reactor core is used to generate steam in a lower-pressure water-steam loop for power production.

In all cases, the function of boiler design is to produce the desired quantity of steam at the specified temperature and pressure conditions while providing the most economical design that is safe and meets environmental control standards.

FUELS AVAILABLE FOR STEAM GENERATION

(See also Secs. 7, 9.1, and 9.8.)

A large variety of materials and heat sources can be used for steam generation. In the absence of other considerations, boilers are designed to use the most economical fuel or combination of fuels available. These include natural gas, residual oil, and coal. Some typical solid fuels are: anthracite coal, bituminous and subbituminous coal, lignite, coke breeze, fluid petroleum coke (4 to 5 percent volatile), petroleum coke (9 to 14 percent volatile; with auxiliary fuel), wood, bark, bagasse, and other agricultural wastes. Other less common energy sources, many suitable for steam generation, are described in Secs. 7.4 and 9.1.

In the past fuel sources nearest the plant generally supplied the boiler, but now complex economics of fuel cost, storage, transportation, emissions control, and operating requirements determine fuel use. In many cases, boilers must be designed to efficiently burn a variety of coals or other fuels which can change over time. Many industries use their by-products for steam generation. Paper mills burn the black liquor from the cooking of wood pulp. Steel mill boilers may be fired with blast furnace and coke oven gases. Oil refineries burn lean CO gas, a waste product, in combination with a richer gas. The increased cost of basic fuels has caused many other industries to consider the use of their own combustible or heat-bearing by-products. The use of municipal wastes for steam generation is also widespread (Sec. 7.4).

EFFECT OF FUEL ON BOILER DESIGN

Fuel is the governing factor in boiler design. Clean natural gas leads to the simplest design. If it is the only fuel, the boiler can be relatively small and compact. When solid or liquid fuels are to be used, the boilers will be larger because of the need to provide the required furnace volume for combustion and to accommodate slag and ash. Combustion of fossil fuels also releases air pollutants such as nitrogen oxides (NO_x), sulfur dioxide (SO_2), particulate, and selected other emissions which must be reduced to regulated levels before the combustion products are released. NO_x is formed from fuel-bound nitrogen and nitrogen in the combustion air. NO_x formation can be limited through effective combustion system and furnace design, but emissions must frequently be further controlled through the use of postcombustion equipment. Where significant amounts of sulfur are present in the fuel, postcombustion control equipment is usually required to control SO_2 release. In some applications, fluidized-bed combustion technology where the fuel is burned in an air fluidized bed of limestone can be effective for SO_2 emission control. Where the fuel contains a significant amount of inert solids, fly ash collectors, electrostatic precipitators, and fabric filters are required to meet government regulations. These emission control technologies are discussed later.

Slag and Ash

All modern pulverized coal-fired boilers are designed for **dry-ash** or **dry-bottom** furnace operation. Most of the ash, typically 70 to 80 percent, is entrained in the flue gas and carried out of the furnace as **fly ash**. Some of the coarser fly ash particles are collected in hoppers arranged under the economizer and air heater. Finer fly ash particles remain in suspension and are carried through the unit and collected in particulate control equipment downstream of the boiler. The relatively small portion of the ash collecting on the furnace walls and other heat-transfer surfaces is removed by effective sootblower and other cleaning equipment operation. The remaining 20 to 30 percent of the ash (**bottom ash**) is collected in the hopper formed by the front and rear wall tube panels at the bottom of the furnace. A limited number of existing boilers have been designed with **slag-tap** or **wet-bottom** furnaces which burn coals having low ash fusion temperatures and which are designed to have high temperatures near the furnace floor so that the ash is kept molten for tapping.

Ash passing through the boiler is subject to various chemical reactions and physical forces which lead to deposition on heat-transfer surfaces. Such ash deposition is defined as slagging or fouling due primarily to the difference in deposition mechanisms. **Slagging** is the formation of molten, partially fused or resolidified deposits on furnace surfaces exposed to radiant heat from combustion. Slagging can also extend into the convection surfaces if the flue gas temperatures are not reduced sufficiently. **Fouling** is defined as the formation of high-temperature bonded deposits on convection surfaces, such as superheaters and reheaters, which are not exposed to radiant heat. In general, fouling is caused by the vaporization of volatile inorganic elements in the coal during combustion. As heat is absorbed from the flue gas and the temperatures fall in the convection section of the boiler, components formed by these volatile elements condense on ash particles and heating surfaces, forming a "glue" which drives the deposition process. Both slagging and fouling can reduce the effectiveness of the heat-absorbing surfaces and in the case of extreme conditions can result in uncontrolled ash deposits, which can increase pressure drop and ultimately block portions of the flow passages, requiring the unit to be shut down for manual deposit removal.

Slagging and fouling characteristics are influenced by the chemical composition and physical characteristics of the coal ash. The amount of ash, as well as its chemical and physical characteristics, varies over a wide range, not only from coal mine to coal mine, but also from different parts of the same mine. Thus, in the design of new boilers or selection of new coal sources for existing units, it is essential to have a thorough knowledge of the coal and compare this to the considerable operating data that have been accumulated over time.

Coal ash is classified into two categories based upon chemical composition. **Bituminous** ash is defined as having more Fe_2O_3 than CaO plus MgO. **Lignitic** ash is defined as having more CaO plus MgO than Fe_2O_3. Bituminous ash is generally a characteristic of higher-rank coals (Sec. 7.1 definition) found in the eastern United States. Lower-rank western U.S. coals tend to have lignitic ash. Because the quantitative and qualitative evaluations of mineral forms are extremely difficult, the relatively simple **elemental ash analyses** are performed on coal ash samples prepared in accordance with the ASTM International Standard D-3174 ashing procedure (**ASTM ash**). Elements in the ash are quantitatively measured by using a combination of emission spectroscopy and flame photometry, and they are reported as weight percentages of their oxides. Coal ash is consistently found to be composed mainly of silicon, aluminum, iron, and calcium and smaller quantities of magnesium, titanium, sodium, and potassium. The elemental analysis also identifies phosphorous as P_2O_5 and sulfur as sulfur trioxide, SO_3.

Unfortunately, elemental ash analyses alone do not directly identify the compounds that cause deposition, or directly identify the mechanisms of deposit formation. As a result, coal ash research has developed methods to correlate elemental analysis data and other characteristics (such as ash fusibility, sintering strength, and viscosity) of ASTM ash with the slagging and fouling behavior observed in full-scale boilers and in test facilities. The measurement of ash fusibility temperatures is the most widely used method for predicting ash behavior at elevated temperatures. The preferred procedure in the United States is ASTM D-1857 which defines and provides procedures to determine **initial deformation temperature (IT or IDT)**, **softening temperature (ST)**, **hemispherical temperature (HT)**, and **fluid temperature (FT)**. These ash fusion temperatures are determined under both oxidizing and reducing furnace conditions, although reducing conditions are assumed if not specified. In practical terms for dry-bottom furnaces, fusion temperatures provide an indication of the temperature range over which portions of the ash will be a molten fluid or semimolten plastic state. High fusion temperatures indicate that the ash released will cool quickly to a nonsticky state, resulting in minimum potential for slagging. Conversely, low fusion temperatures indicate that ash will remain in a molten state longer, exposing more of the furnace surface or convection surface to potential deposition.

Iron has a dominating influence on the behavior of bituminous-type ash and the ash fusion temperatures, as shown in Fig. 9.2.1. In the

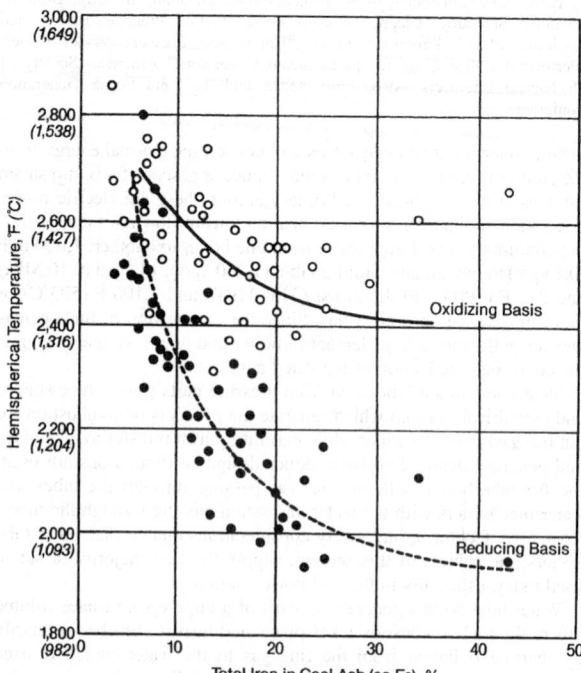

Fig. 9.2.1   Influence of iron on coal ash hemispherical fusion temperature.

completely oxidized form, iron tends to raise fusion temperatures while in the lesser oxidized form it tends to lower them. The iron content and its degree of oxidation also have a great influence on the ash viscosity, which increases with ferric percentage. Based upon the weight percentage elemental ash analysis, some of the parameters used to evaluate slagging and fouling potential include

$$\text{Ferric percentage} = \frac{Fe_2O_3 \times 100}{Fe_2O_3 + 1.11FeO + 1.43Fe}$$

$$\text{Base/acid ratio} = B/A = \frac{Fe_2O_3 + CaO + MgO + Na_2O + K_2O}{SiO_2 + Al_2O_3 + TiO_2}$$

$$\text{Dolomite percentage} = \frac{(CaO + MgO) \times 100}{Fe_2O_3 + CaO + MgO + Na_2O + K_2O}$$

The most accurate indicators of coal ash slagging potential, such as ash viscosity, and fouling potential, such as sintering strength, can be complex, time-consuming, and costly to acquire. As a result, indices based upon the ash elemental analysis and ash fusion temperatures have been developed to provide criteria for various aspects of boiler design with slagging indices establishing furnace design criteria and fouling indices establishing criteria for convection pass design. These indices are generally used to establish the **slagging classification** and **fouling classification** as low, medium, high, or severe. Because of their significantly different characteristics, bituminous ash and lignitic ash are treated separately:

Bituminous Ash Indices

$$\text{Slagging index: } R_s = (B/A) \times S$$

where S = wt % sulfur, on dry coal basis
B/A = base/acid ratio defined above

Slagging classification: low for R_s less than 0.6; medium for 0.6 to 2.0; high for 2.0 to 2.6; severe for greater than 2.6.

$$\text{Fouling index: } R_f = (B/A) \times Na_2O$$

where Na_2O = wt % from coal ash analysis.

Fouling classification: low for R_f less than 0.2; medium for 0.2 to 0.5; high for 0.5 to 1.0; severe for greater than 1.0.

Lignitic Ash Indices

$$\text{Slagging index: } R_s^* = [(\text{Max } HT) + 4(\text{Min } IT)]/5$$

where Max HT equals the higher of the reducing or oxidizing hemispherical softening temperatures and Min IT equals the lower of the reducing or oxidizing initial deformation temperature.

Slagging classification: low for R_s^* greater than 2,450°F; medium for 2,250 to 2,450°F; high for 2,100 to 2,250°F; severe for less than 2,100°F.

Fouling index: Na_2O percentage in the coal ash

For $(CaO + MgO + Fe_2O_3) > 20$ percent by weight of coal ash, fouling classification: low to medium for $Na_2O < 3$; high for Na_2O of 3 to 6; severe for Na_2O greater than 6.
For $(CaO + MgO + Fe_2O_3) < 20$ percent by weight of coal ash, fouling classification: low to medium for $Na_2O < 1.2$; high for Na_2O of 1.2 to 3; severe for Na_2O greater than 3.

Ash Reflectivity   Ashes from certain coals produce furnace deposits that reduce heat transfer by high reflectivity as well as by thermal insulation. This is particularly true of low-sulfur, low-sodium coals found in the western United States, Powder River Basin in Wyoming and Montana. Reflective deposits can significantly reduce furnace heat absorption and increase the furnace exit gas temperature, even when only a very thin deposit is present. This can result in excessive slagging and fouling in the upper furnace and convection pass, respectively. These deposits can be difficult to remove and require special attention in the selection of ash cleaning equipment and cleaning media. Field experience and laboratory tests are used to evaluate the potential impact and boiler design factors.

Effect on Boiler Design   The boiler furnace must have sufficient volume to completely burn the fuel, provide sufficient surface to cool the flue gas and ash particles to a temperature suitable for admission to the convection pass, and minimize the formation of NO_x emissions. In general for coal-fired units, it is the second criterion that determines the minimum furnace size. The **slagging classification** establishes the upper limit on the furnace exit gas temperature (FEGT) required to minimize the potential for slagging in the superheater and closely spaced convection surfaces. This, in turn, establishes the total furnace volume and heat-transfer surface requirements. In addition, the furnace must be correctly proportioned with respect to width, depth, and height to minimize slagging with ample clearance provided between the burners and the furnace walls and furnace hoppers. The slagging classification also determines the locations, quantity, and spacing of furnace wall sootblowers and long, retractable sootblowers. Fig. 9.2.2 shows the effect of

Fig. 9.2.2   Influence of slagging potential on furnace size.

different slagging characteristics on a 660-MW large utility boiler. The boilers are assumed to have the same width for the purposes of illustration with the boiler setting height and furnace depth varied to accommodate the slagging classification and characteristics of the different fuels.

In the convection pass design, the fouling classification determines the specific side spacing (perpendicular to flow) dimension at the temperature entering the tube bank with severe fouling coals requiring the widest spacing. The fouling classification and temperature entering the bank also establish the tube bank depth (parallel to flue gas flow). Cavities are provided between the tube banks for long retractable sootblowers.

Sootblower Systems

While slagging and fouling in boilers can be minimized by proper design and operation, auxiliary equipment for online cleaning of furnace walls and removing deposits from convection surfaces must be supplied to maintain capacity and efficiency. Superheated steam is the most common cleaning medium, but saturated steam, compressed air, and water are also used. The cleaning media jets from the sootblower nozzles dislodge the dry or sintered ash and slag, which then fall into hoppers or travel along with the flue gas to the particulate removal equipment.

Types of sootblowers vary with their location in the boiler unit, the severity of ash or slag conditions, and the arrangement of the heat-absorbing surfaces.

Furnace walls are generally cleaned with **wall blowers** (Fig. 9.2.3) which project a nozzle assembly into the furnace for blowing and then retract it behind the wall tubes for protection after operation.

A water lance sootblower may be used for dense slag layers which are resistant to mechanical impact pressure alone. The rapid quenching and mechanical impact energy create cracks and fissures, dislodging the slag in pieces. A more recent development is the use of a water jet through a sidewall opening to clean surfaces on the opposite side of the furnace.

Fig. 9.2.3 Retracting sootblowers. (*a*) Furnace wall blower; (*b*) long-lance blower.

Fig. 9.2.4 Rotating sootblower with multiple nozzles.

Tube banks in high-gas-temperature zones, such as slag screens, superheaters, and reheaters, where slag or sintered ash may accumulate, are generally cleaned by **long-lance retracting-type blowers** (Fig. 9.2.3). The lance, which rotates or oscillates as it advances into the boiler, is fitted with nozzles to supply a powerful cleaning action and is retracted from the boiler for protection when it is not operating.

Tube banks located in low-gas-temperature zones, including the economizer and boiler sections, where uncooled metals have satisfactory life and ash removal is easier, usually can be cleaned by **multiple-nozzle rotating-type sootblowers** (Fig. 9.2.4). However, long-lance retracting-type blowers may be necessary for very wide boilers, for extended cleaning ranges, or where the ash tends to accumulate.

Sootblowers for air heaters generally are arranged to blow through the plate or tube assemblies with single- or multiple-nozzle elements moving in an arc or straight-line motion. These blowers may also supply water for washing the air-heater surface.

Advancements have occurred in automated sootblower control systems. Automatic control systems have historically been arranged to operate the blowers in a prescribed sequence at time intervals adjusted to blower-cleaning requirements or to receive signals from the blower unit's instruments and controls so as to operate the sootblowers selectively in the various heat-absorbing sections in order to maintain the required cleanliness and heat absorption. Modern boiler cleaning systems can incorporate on-line diagnostic sensors and advanced control algorithms to increase cleaning effectiveness and boiler efficiency while reducing the consumption of sootblower media and reducing potential damage to the boiler. In addition, flue gas temperature can be used with heat balance and heat-transfer calculations for a particular tube bank to determine a real-time **fouling factor** which can be used to detect deposit accumulation over time. Advanced "intelligent" sootblower control systems then select the individual sootblowers and cleaning cycle to address the specific deposition area.

COMBUSTION AND FUEL PREPARATION SYSTEMS

Stoker Firing Systems

Mechanical stoker/grate system designs are available to burn a wide range of fuels including all forms of coal, sludge, wood, wood waste, other biomass and agricultural wastes, as well as residential, commercial, and industrial refuse. Modern mechanical stoker firing systems consist of

1. A **stoker** or fuel feeding system
2. A stationary or moving **grate** assembly to support the burning mass of fuel and admit air from under the grate
3. An **overfire air** (OFA) system above the fuel bed to complete the combustion and limit air pollutant emissions
4. An ash discharge system

The term *stoker* often refers to not only the fuel feeder but also to the full feed/grate/air system.

There are two general types of systems; **underfeed,** in which fuel and air are both supplied from below the grate, and **overfeed,** in which fuel is supplied from above and air from below. Overfeed systems are further divided into two types: **mass feed,** in which the fuel is continuously fed to one end

Fig. 9.2.5 Spreader stoker with traveling grate.

of the grate and travels horizontally across the grate as it burns, and **spreader,** in which the fuel is thrown or spread across the grate area. Underfeed and small mass feed stokers are uncommon today.

Spreader stokers, in combination with either a traveling or vibrating grate, are the most commonly used systems for a steaming capacity range from 75,000 to 900,000 lb/h (9.5 to 113 kg/s). They respond quickly to changes in steam demand, have good **turndown capability** (maximum to minimum load), and can be used with a variety of fuels. Figure 9.2.5 shows a typical spreader stoker with continuous ash discharge traveling grate. The rotating mechanical feeder on the left side spreads the fuel evenly across the grate area. Fine particles burn in suspension with the coarser particles landing on the grate to complete the combustion. The traveling grate moves the fuel bed back toward the feeder, below which is the ash discharge. The traveling grate is an endless moving belt which consists of a series of chains attached to grate bars with air metering holes. To achieve high combustion efficiency, fine particles carried over from the combustion must be captured in a mechanical dust collector, collected in hoppers, and then returned to the furnace by a **carbon reinjection** system. The split of the air supplied under the grate and injected above the fuel bed is carefully controlled to maximize burnout of the fuel and to minimize the formation of NO_x.

Water- and air-cooled vibrating grates are also frequently used in spreader stoker firing systems. As shown in Fig. 9.2.6, the vibrating

grate consists of water-cooled grate panels that create a furnace floor which is mounted on leaf-spring supports. As the fuel bed burns, a vibrating unit creates a back-and-forth motion which moves the fuel bed toward the ash discharge. Air-swept spouts (fuel spreaders) are combined with vibrating grates in many wood- and bark-fired boilers.

Pulverizers

Pulverized coal (PC) is the dominant method of firing coal in modern utility boilers. Pulverizers are used to grind coal into a fine powder with 70 percent of the ground coal passing through a 200-mesh ($74\text{-}\mu\text{m}$) screen for typical PC applications (sample collected using ASTM procedure D-197). Utility systems in operation today are **direct-fired** with the coal ground in the pulverizer directly transported to the burner with a portion of the combustion air (**primary air**). Pulverizer models range in total capacity from 1.5 to 115 tons/h of coal (0.4 to 29 kg/s) with a typical pulverizer shown in Fig. 9.2.7. The primary air entering the pulverizer is preheated to temperatures as high as 700°F (371°C) to dry the moisture in the coal and provide stable, efficient combustion. The inlet temperature is controlled to keep the pulverizer outlet temperature to a specified range which is largely determined by the volatile matter content in the coal. Coal with low volatile matter content may require higher air-coal temperatures to ensure stable combustion. The typical pulverizer outlet temperature is 150°F (66°C). Table 9.2.1 provides outlet temperature ranges for a variety of coal types.

The three principal types of pulverizers may be classified by their speed: low (< 75 r/min), medium (75 to 250 r/min), and high (> 250 r/min). Figure 9.2.7 shows a slow-speed, **roller-and-race** type of pulverizer.

Fig. 9.2.7 Slow-speed pulverizer, roller-and-race type. (*Courtesy The Babcock & Wilcox Company.*)

Fig. 9.2.6 Water-cooled vibrating grate stoker.

Table 9.2.1 Typical Pulverizer Outlet Temperatures

Fuel type	Volatile content,%*	Exit temperature, °F (°C)†
Lignite	All values	120–140 (49–60)
Subbituminous	All values	130–150 (54–66)
High-volatile bituminous	Above 31	140–175 (60–79)
Medium- and low-volatile bituminous	14–31	160–200 (71–93)
Anthracite and coal waste	0–14	200–210 (93–99)
Petroleum coke	0–8	200–250 (93–121)

* Volatile content is on a dry, mineral-matter-free basis.
† The capacity of the pulverizer is adversely affected for exit temperatures below 125°F (52°C) when grinding high-moisture lignites.

Medium-speed pulverizers are generally either the **ball-and-race** type (Fig. 9.2.8) or the **bowl-and-roller** type (Fig. 9.2.9). All three of these pulverizers are referred to as vertical spindle pulverizers. High-speed pulverizers are usually of the attrition type, shown in Fig. 9.2.10. Some older boilers are equipped with **ball-and-tube** coal pulverization equipment, in which the coal is ground in a horizontal rotating cylinder that is partially filled with small-diameter tumbling balls. In all pulverizers, provision is made to remove and dispose of **pyrite** (iron sulfide, FeS_2) and other mineral matter from the coal which is hard to grind and which has no material heating value.

The capacity of a specific pulverizer is a function of the grindability of the coal, the desired coal fineness (percent passing through 200-mesh), and the moisture. **Grindability** defines the ease with which a specific coal can be ground and is quantified by the Hardgrove Grindability Index (HGI), as described in ASTM D-409. Higher HGI indicates easier to grind coals. The required fineness is established based upon the coal and firing system. Figure 9.2.11 correlates the HGI and the fineness with a pulverizer **capacity factor** for roller-and-race and ball-and-race types of pulverizers. The expected pulverizer capacity is then the capacity factor times the pulverizer **base capacity** for the specific size and model, plus a margin. An additional factor must be included where high moisture is present (moisture determined according to ASTM D-3302).

Fig. 9.2.8 Medium-speed pulverizer, ball-and-race type. (*Courtesy The Babcock & Wilcox Company.*)

Fig. 9.2.9 Medium-speed pulverizer, bowl-and-roller type. (*Courtesy ALSTOM Power Inc.*)

Fig. 9.2.10 High-speed pulverizer, attrition type.

Fig. 9.2.11 Pulverizer capacity factors for varying fineness and grindability: roller-and-race and ball-and-race types.

Burners and Combustion Systems

Advanced burners mix air and fuel to ensure rapid ignition and effective combustion (> 99 percent fuel utilization) while minimizing the formation of nitrogen oxides (NO_x). In pulverized coal firing, a portion of the total air flow (typically 15 to 25 percent at full load, but as low as 11 percent for ball-and-tube pulverizers and as high as 33 percent for some lignite fuels), called **primary air**, is mixed with the pulverized coal to convey it to the burners. The remaining portion, or **secondary air**, is introduced at the burner from a **wind box**.

Multiple low-NO_x circular burners (See Fig. 9.2.12) located in the walls of the lower boiler furnace are the centerpiece of the overall combustion system which includes the fuel preparation (including pulverizers), air-flow distribution, igniters, furnace design, and integrated combustion control system including monitors, diagnostic sensors, and flame safety

Fig. 9.2.12 Low-NO_x circular burner for pulverized coal firing.

Fig. 9.2.13 Air staging for NO_x reduction in pulverized coal wall-fired boilers.

equipment. Coal and primary are introduced through the central coal pipe while the secondary air is introduced through several concentric zones with controlled flow rates and swirl rates induced by the spin vanes. Approximately 75 percent of the NO_x formed from pulverized coal combustion is from fuel-bound nitrogen (**fuel NO_x**) while the balance is from nitrogen in the combustion air (referred to as **thermal NO_x** because the NO_x formation is strongly dependent upon high combustion temperatures). By controlling the local stoichiometry within the flame and the secondary air-fuel mixing rates, NO_x formation can be minimized. NO_x emissions can be further reduced by using **air staging** in the furnace (Fig. 9.2.13)—a portion of the secondary air (referred to as **overfire air**, or **OFA**) is removed from the burners and introduced later in the combustion process through air injection **NO_x ports.** Modern pulverized-coal-fired boilers typically operate with 15 to 20 percent **excess air** above the theoretical amount needed at full load and with higher amounts as boiler load is reduced. Individual pulverized coal circular burners have maximum heat input rates of 50 to 300 million Btu/h (15 to 88 MW). For modern applications with high volatile bituminous or subbituminous coal, combustion should be stable without the use of igniters from about 30 to 100 percent load. The actual minimum is dependent on the coal, burner design, and pulverizer load. Operation below this load is performed with igniters in service. Specialized burners are available for difficult to combust coals (typically low-volatile fuels and high-moisture and/or ash fuels).

Oil and natural gas are also usually burned in specially designed low-NO_x circular burners where all the combustion air is supplied from the wind box and specialized combustion components are used. Fuel oil is atomized into a fine mist which is dispersed into the airstream for combustion, using an individual atomizer located in the centerline of the burner. Atomizers are classified as one of two types and are selected based upon economics and the individual application. **Dual fluid atomizers** use steam or compressed air to supply most of the energy to create the fine fuel-oil mist. Steam atomization is generally preferred for its safety characteristics and the creation of a finer mist. Several designs are available with up to 300 million Btu/h (88-MW) heat input. Depending upon the design, steam pressures as high as 200 psig (1.4 MPa gage) are used. Dry steam with 20 to 40°F (11 to 22°C) superheat is recommended. **Mechanical atomizers** use relatively high fuel pressure to create the droplets, using one of several designs. Depending upon the design, oil pressure requirements range from 300 to 1,000 psig (2.1 to 6.9 MPa gage) and maximum firing rates range from 70 to 200 million Btu/h (21 to 59 MW). A typical turndown ratio for oil firing is on the order of 6:1 depending upon the fuel characteristics, flexibility of the delivery system, and atomization technique. The primary source of NO_x with heavy fuel oil firing is fuel NO_x, while thermal and fuel NO_x are important for distillate oils.

Natural gas and other commercially high-heating-value gaseous fuels can be burned with circular burners by admitting the gas through a series of radial spuds or through a centrally located high-capacity gas element, which is sited at the burner centerline. The latter is usually employed where multifuel capacity including the use of coal is needed. The maximum practical limit for gas input per burner is 200 million Btu/h (59 MW). Gas pressure at the burner is generally in the range of 8 to 12 psig (55 to 83 kPa gage) for the multiradial spud design and as high as 20 psig (138 kPa gage) for the high-capacity gas element. The design of the gas spuds is critical to minimizing the formation of NO_x from natural-gas-fired burners.

Thermal NO_x is the primary source of NO_x emissions from natural gas and distillate fuel oil firing. Fuel NO_x is the primary source when firing heavy fuel oil. A variety of techniques have been used to reduce NO_x on boilers using older oil and gas firing equipment. Low excess air (LEA) and burners out of service (BOOS) have been used as has air staging. An additional technique referred to as **flue gas recirculation (FGR)** has been successfully used where fuel NO_x is a relatively small contributor to total NO_x emissions. Flue gas from the economizer is returned to the combustion airstream where it helps reduce the peak flame temperatures and thereby the thermal NO_x formed.

When **corner-fired burners** (Fig. 9.2.14) are used, the mixing of the fuel and combustion air takes place in the furnace (Fig. 9.2.14a). Coal

Fig. 9.2.14   Corner-fired tangential tilting burners for pulverized coal, oil, or gas. (*a*) Plan section; (*b*) burners tilted down; (*c*) burners tilted up.

Fig. 9.2.15   Cyclone furnace.

and air are introduced into the furnace at distinct elevations in the corners of the furnace. Nozzles direct the coal and airstreams at slight tangents to an imaginary firing circle in the vertical center of the furnace, to create a rotating fireball that fills the furnace volume. For steam temperature control at lower loads, the nozzles can be tilted down (Fig. 9.2.14b) or up (Fig. 9.2.14c) in unison to lower or raise the fireball and lower or raise the furnace exit gas temperatures. Air staging can also be used in this design to reduce NO_x emissions.

Cyclone Furnaces

The cyclone furnace was originally developed to burn low-ash-fusion coals and to retain most of the coal ash in the slag, which is then tapped from the furnace, thus preventing the passage of the ash through the heat-absorbing surfaces. Cyclone furnaces have also been designed for and successfully operated with a wider range of coals including subbituminous and lignite. The coal, crushed so that approximately 95 percent passes through a 4-mesh (4.75-mm) screen, is admitted with primary air in a tangential manner to the primary burner (Fig. 9.2.15). The finer particles burn in suspension while the coarser particles are thrown by centrifugal force to the outer wall of the cyclone furnace. The wall surface with its sticky coating of molten slag retains most of the particles of coal until they burn and leave their molten ash on the wall. The molten ash drains into the boiler furnace and then, through an opening in the boiler furnace floor, into the slag-collecting tank.

The secondary air, which is admitted tangentially at the top of the cyclone, vigorously scrubs the coal particles on the wall, and combustion is completed at heat release rates of 450,000 to 800,000 Btu/(h · ft³) (4.7 to 8.3 MW/m³). The primary-furnace walls (consisting of fully studded tubes) also are wetted with molten ash and help to catch ash particles that are not retained in the cyclone furnaces. Gas, when available as a backup fuel, can be burned by injection through openings at the bottom of the secondary-air ports. Oil as a backup fuel is burned by spraying it axially into the cyclone through the primary burner or by firing it tangentially through an oil element located in the secondary-air port.

Historically, cyclone-equipped boilers have produced relatively high uncontrolled levels of NO_x emissions because of the high combustion temperatures. Various techniques have been developed to control NO_x emissions from existing units.

Fluidized-Bed Combustion

Fluidized-bed boilers can be designed to burn a wide variety of fuels, including all coals, in an air-suspended mass (bed) of solid particles while simultaneously reducing air pollutant formation and emission. **Fluidized-bed combustion (FBC)** is particularly attractive for difficult to burn fuels with high ash or moisture. Where the fuel contains sulfur, the bed material feed can be limestone (**reagent**) which ultimately absorbs the sulfur dioxide (SO_2) during the combustion process. Where no sulfur is present, an inert material such as sand and/or the ash from the fuel can form the bed of solids. Nitrogen oxides (NO_x) formation and emissions are controlled by utilizing relatively low combustion temperatures and by staging the air addition to the bed. By carefully controlling the

Fig. 9.2.16 150-MW utility reheat internal recirculation, fluidized circulating bed (FCB) boiler, coal-fired.

Fig. 9.2.17 Combustible efficiency loss, spreader stoker.

bed temperature, emission of nitrogen oxides (NO_x) can be minimized while still optimizing the SO_2 capture. Depending upon the reagent bed stoichiometry, the limestone reactivity, and the fluidized-bed type (see below), SO_2 reductions of 90 to 95 percent are typically achieved. Where tighter NO_x emission limits are required, NO_x can be further reduced through selective noncatalytic reduction (SNCR) of NO_x by injecting ammonia into the furnace at the proper locations.

Fluidized-bed boilers are generally classified into two types: **circulating fluidized bed (CFB).** or **bubbling fluidized bed (BFB).** Figure 9.2.16 shows the main features of a large CFB boiler which is the dominant technology used for utility applications. Air is supplied through a wind box and air distributor in the furnace floor, creating an initial bubbling fluidized bed in the bottom of the furnace. The addition of the overfire air then provides sufficient gas velocity to carry the bed solids vertically upward through the full furnace shaft before the bulk of the solids are collected in the primary particle collectors at the furnace exit and returned to the furnace. Secondary solids collection is provided by multicyclone dust collectors downstream of the superheat and reheat convection surfaces with the solids also being returned to the furnace to maintain total solids inventory. The furnace temperature is controlled by heat absorption to the furnace enclosure water walls which in turn is primarily controlled by the furnace local solids density. CFBs are operated to maintain bed temperatures between 1,470 and 1,650°F (799 and 899°C) for applications burning coal and other sulfur-containing fuels to maximize SO_2 absorption and minimize NO_x emissions.

BFB boilers have been found to be particularly attractive and economical in retrofit applications and in cases burning biomass and waste fuels containing a high noncombustible fraction. BFBs operate at lower superficial flue gas velocities than CFBs. This creates a distinct stationary bubbling or churning bed of solids in the lower furnace where the combustion takes place. BFB boilers operate with bed temperatures between 1,350 and 1,650°F (732 and 899°C) depending upon the fuel moisture, ash content, and alkali content in the ash. Bed temperature is maintained for biomass and other high-volatility fuels by adjusting the split between primary and overfire air and sometimes by the use for flue gas recirculation. For coal, bed temperature is maintained by the use of in-bed heat-transfer surface and compartmentalized wind box designs.

Unburned Combustible Loss

The unburned combustible loss in the fly ash from pulverized-coal firing varies with the furnace heat liberation, type of furnace cooling, use

of slag-tap or dry-ash removal, volatility and fineness of the coal, excess air, and type of burner. There is practically no combustible in the fluid slag from slag-tap furnaces. Although the hopper refuse from dry-ash furnaces usually is low in combustible, the combustible may be appreciable in some cases especially where very low NO_x emissions are achieved. The fly ash from cyclone-furnace boilers also has a very low combustible content.

Fly ash combustible from spreader stokers varies widely with the rating, size, and type of coal burned. The combustible carryover at high rates of firing is relatively large, but reinjection of the fly ash is common, and the loss in efficiency can be reduced as shown in Fig. 9.2.17.

The combustible loss in the fly ash from solid fuels is determined by withdrawing a representative sample of fly ash and flue gas from the boiler outlet flue, or stack, at the same velocity in the sampling tip as the gas velocity in the flue (see ASME Test Code). The rates of flue-gas flow and fly ash collection are measured, and the dust loading in pounds per 1,000 lb of flue gas is then calculated. The combustible in fly ash also can be measured. Combustion efficiency loss from unburned combustible in ash can be calculated by using the following simplified relationship:

Combustion efficiency loss

$$= \frac{14{,}500 \times (\%\,\text{carbon in residue})/[100 - (\%\,\text{carbon in residue})]}{(\text{coal HHV})/(\%\,\text{ash})}$$

where coal HHV = higher heating value of coal as fired, Btu/lb; % ash = percent of inerts in coal as-fired; % carbon in residue = weighted average of measured % carbon in fly ash leaving the steam generator, economizer hopper ash, and boiler bottom ash. For a typical dry-bottom opposed-wall fired boiler:

Electrostatic precipitator = 75%

Economizer = 5%

Boiler bottom = 20%

This assumes that the pyrite loss through the pulverizer rejects is small and/or the percent carbon loss in the pyrite is about the same as the weighted average. The equation accounts for efficiency losses from total combustibles which are primarily in the form of carbon.

BOILER TYPES

Water-tube boilers are available in a wide range of steaming capacities— from as low as 1,000 lb/h (0.1 kg/s) to as high as 10,000,000 lb/h (1,260 kg/s). By integrating the components—pressure parts, furnaces, fuel burners, fans, and controls—boiler manufacturers have produced a broad series of standardized and economical steam generating units, with steam capacities up to 600,000 lb/h (76 kg/s), burning oil or natural gas. For capacities up to 230,000 lb/h (29 kg/s) most units can be shipped by rail or truck as a completely shop-assembled package (Fig. 9.2.18). For larger units, barges or ocean vessels may be used if the site is suitably located.

For higher steam capacity needs typical of industrial and small utility applications, larger field-erected boilers are used, although the erection

Fig. 9.2.18 Water-tube package boiler for natural gas and light oil firing; 60,000 lb/h (7.6 kg/s) steam flow.

of large shop-assembled modules facilitates field erection and reduces cost. Such field-erected units can be designed to burn most solid, liquid, and gaseous fuels alone or in combination. Figure 9.2.19 shows a boiler capable of multifuel firing using biomass (bark), pulverized coal, and natural gas. Such two-drum boilers are typically designed for pressures

Fig. 9.2.19 Two-drum type of boiler designed for multifuel firing. (*Courtesy The Babcock & Wilcox Company.*)

up to 1,800 psi (12.4 MPa), although pressures as high as 2,200 psi (15.2 MPa) have been used. For pulverized coal, oil, and natural gas firing, steam capacities range up to 1,200,000 lb/h (151 kg/s). For stoker coal firing, steam capacities range up to 400,000 lb/h (50 kg/s), and for stoker wood and biomass, steam capacities can reach 600,000 lb/h (76 kg/s.) The **boiler bank** consisting of a large number of tubes connecting the lower water drum to the steam drum (shown in Fig. 9.2.19) is typically used in lower-pressure boilers to aid in steam generation where the water-cooled furnace enclosure provides insufficient boiling surface.

Designs of high-pressure, high-temperature, utility-type boilers ranging from steam capacities of 700,000 to 10,000,000 lb/h (88 to 1,260 kg/s) are classed as radiant-type boilers. In radiant boilers, virtually all the steam is generated in the tubes forming the furnace enclosure walls from thermal radiation to these tubes from the hot combustion gases. Figure 9.2.20 illustrates a modern large, pulverized coal wall-fired steam-water natural circulation radiant boiler while Fig. 9.2.21 illustrates a smaller natural circulation unit which has been designed for oil and natural gas wall firing. A typical boiler with assisted circulation utilizing tangential firing techniques is shown in Fig. 9.2.22. Such drum-type natural or assisted circulation boilers are restricted to operating pressures below the critical pressure of water [3,200 psia (22.1 MPa)] because of circulation and steam separation characteristics, and generally have superheater outlet pressures below 2,685 psia (18.51 MPa). Main steam and reheat steam superheat temperatures of 1,005°F (541°C) are common.

To increase steam cycle and power plant efficiency, higher steam pressures and temperatures are often used with once-through forced circulation boilers where the steam-water circuitry is designed so that water is continuously converted to steam without the use of a large steam drum. Such units can be designed for constant-pressure operation at all loads or variable-pressure operation where the operating pressure can be reduced as the load is reduced. Figure 9.2.23 shows a modern universal pressure radiant boiler which is designed for full once-through forced circulation and a main superheat outlet pressure of 3,500 psi (24.1 MPa) at full load. The unit is designed with advanced (**ultrasupercritical**) steam conditions with a main superheat outlet temperature of 1,100°F (593°C) and reheat superheat temperature of 1,125°F (607°C). This boiler employs a spiral circuitry furnace waterwall construction where the tubes spiral around the furnace perimeter (versus vertical tube orientation in the other boilers discussed so far), which permits the unit to operate in variable-pressure mode with subcritical pressure pump-assisted circulation at low loads.

A variety of other boiler designs are used to meet the specialized needs for steam generation from waste energy or process by-product fuels. Figure 9.2.24 shows a specialized waste heat recovery boiler [or **heat recovery steam generator (HRSG)**] which is used to recover heat from the exhaust of a gas turbine in a combined cycle power system. The turbine exhaust gas enters the unit at moderate temperatures [typically less than 1,200°F (649°C)], and heat is recovered as the gas passes cross-flow over a series of modular tube bundles where water is heated to saturation temperature, steam is generated, and steam is superheated for power cycle application. The designs have become quite integrated with multiple pressure loops to maximize the energy recovered at a minimum equipment cost. Steam from the HRSG supplies a steam-turbine power cycle which adds to the power produced by the gas turbine.

The exhaust gas from oil refinery fluid catalytic-cracking units is used as a fuel in specialized **carbon monoxide (CO) fired boilers.** Historically these have used refractory-lined circular boiler enclosures, but newer designs are replacing these original units with rectangular waterwall furnace enclosures. Supplemental fuel firing is required to raise the temperature of the CO to the ignition point and to ensure complete burning of the combustibles in the CO gas stream.

Process recovery (PR) boilers used in the pulp and paper industry are specially designed for the recovery of chemicals in the spent cooking liquors from Kraft, sulfite, soda, and other papermaking processes. The unique recovery boiler furnace design permits the combustion of the

Fig. 9.2.20 A 600-MW natural-circulation radiant boiler, coal-fired, 4,410,000-lb/h (556 kg/s) steam, 2,507 psi (17.3 MPa), 1,006°F (541°C) steam temperature; 1,002°F (539°C) reheat steam temperature. (*Courtesy The Babcock & Wilcox Company.*)

Fig. 9.2.21 A 320-MW natural-circulation radiant boiler, oil- and gas-fired; 2,275,000 lb/h (287 kg/s) steam 2,610-psi (18.0-MPa) pressure; 1,006°F (541°C) steam temperature; 1,006°F (541°C) reheat steam temperature (*Courtesy The Babcock & Wilcox Company.*)

Fig. 9.2.22 A 500-MW controlled or assisted-circulation radiant boiler with tangential coal firing. (*Courtesy ALSTOM Power Inc.*)

Fig. 9.2.23 A 750-MW universal pressure radiant boiler; full sliding pressure; supercritical pressure once-through flow at full load; coal-fired 5,500,000 lb/h (693 kg/s) steam, 3,500-psi (24.1-MPa) pressure, 1,100°F (593°C) steam temperature; 1,125°F (607°C) reheat steam temperature. (*Courtesy The Babcock & Wilcox Company.*)

organic elements in the spent liquor while reducing the oxidized inorganic material in a pile or bed supported on the furnace floor. In the Kraft process, the molten inorganic chemicals or smelt in the bed is discharged to a tank for further processing and ultimate reuse in the papermaking process.

BOILER COMPONENTS

Furnaces

A furnace is an enclosure provided for the combustion of fuel. The enclosure confines the products of combustion and is capable of withstanding the high temperatures developed and the pressures involved with operation. Its dimension and geometry are adapted to the rate of heat release, the type of fuel, and the method of firing so as to promote complete burning of combustible and provide suitable disposal of the ash.

Water-cooled furnaces are used with most boiler units and for all types of fuel and methods of firing. Water cooling of the furnace walls reduces the transfer of heat to the structural members and, consequently, their temperature can be limited to that which will meet the requirements of strength. Water-cooled tube construction facilitates large furnace dimensions and optimum arrangements of roofs, hoppers, arches, and burners; and the use of tubular screens, platens, or division walls to increase the amount of heat-absorbing surface in the combustion zone. The use of water-cooled furnaces also reduces the external heat losses.

Heat-absorbing surfaces in the furnace receive heat from the products of combustion and contribute directly to steam generation while lowering the furnace exit gas temperature (FEGT). In the boiler furnace, all the principal heat-transfer mechanisms take place simultaneously. These include intersolid radiation between suspended solid particles, tubes, and refractory materials; nonluminous gas radiation from the

Fig. 9.2.24 Triple pressure heat recovery steam generator (HRSG).

products of combustion; convection heat transfer from the gases to the furnace walls; and heat conduction through deposits and tube metals. (See also Sec. 4.) The absorption effectiveness of the furnace surfaces is influenced by the deposits of ash or slag.

Furnaces vary in shape and size, in the location and spacing of burners, in the disposition of heat-absorbing surfaces, and the general overall arrangement of arches and hoppers. Flame shape and length affect the distribution of radiation and heat absorption by water-cooled surfaces.

Analytical solution of the transfer of heat in the furnaces of steam-generating units is extremely complex, and it is not possible to calculate furnace outlet gas temperature by theoretical methods alone. Nevertheless, the furnace outlet gas temperature must be accurately predicted, since this temperature determines the design of the remainder of the boiler unit, particularly that of the superheater and reheater. The design calculations are based upon test results supplemented by data accumulated from operating experience and application of the principles of heat transfer and the characteristics of fuels and slags. Today this is supplemented by numerical computational methods.

The curve in Fig. 9.2.25 shows the gas temperatures at the furnace outlet of typical boiler units when firing coal, and by reference oil and gas. The furnace exit gas temperatures vary considerably with coal firing because of the insulating effect of ash and slag deposits on the heat-absorbing surfaces. The amount of surface is the major factor in overall furnace heat absorption, and the heat released and available for absorption per unit time per unit area of effective heat-absorbing surface is therefore used as the basis for correlation. The heat released and available for absorption is the sum of the heat content of the fuel fired and the sensible heat of the combustion air, less the sum of the heat unavailable owing to the unburned portion of the fuel and latent heat of the water vapor formed from moisture in the fuel and the combustion of hydrogen.

Empirical furnace design methods are being supplemented with numerical computational methods as the level of detail increases and confidence improves. Fundamental relationships for thermal radiation, combustion, energy, and turbulent flow are combined with thermophysical properties and other experimentally determined factors to provide useful engineering information. Detailed results include three-dimensional distributions of gas temperatures and combustion species such as air pollutants as well as heat flux on the furnace walls (see Fig. 9.2.26).

The common forms of furnace water-cooled wall construction found in operation today are shown in Fig. 9.2.27. However, most boilers now

Fig. 9.2.25 Approximate relationship of furnace exit gas temperature (FEGT) for a modern water-cooled dry ash furnace with pulverized bituminous coal firing. Values for oil firing fall in the bottom half of the range. Values for gas firing generally fall at the upper boundary.

Fig. 9.2.26 Numerical model-predicted furnace wall, flat projected heat flux distribution for a modern low-NO$_x$ pulverized bituminous coal-fired dry-bottom furnace [1W/m^2 = 0.317 Btu (h · ft^2)].

Fig. 9.2.27   Water-cooled furnace construction types. (*a*) Tangent tube wall; (*b*) welded membrane tube wall; (*c*) flat studs welded to sides of tubes; and (*d*) full stud tube wall, refractory covered.

use **membrane tube walls** in which a steel bar, or membrane, is welded between adjacent tubes. This construction facilitates the fabrication of water-cooled walls in large shop-assembled tube panels and facilitates ash removal.

Additional furnace cooling in the form of tubular platens, division walls, or wide-spaced tubular screens may be used. In high heat input zones, the tubes can be protected by refractory coverings anchored to the tubes by studs. Peak heat-absorption rates of furnace-wall tubes in the combustion zone may be well in excess of 130,000 Btu/(h · ft^2) (0.4 MW/ m^2) of projected surface, but the average heat absorption rate for the furnace is considerably lower (Fig. 9.2.28).

Fig. 9.2.28   General range of average heat absorption rates in water-cooled coal-fired furnaces.

Furnace walls must be adequately supported with provision for thermal expansion and with reinforcing buckstays to withstand the lateral forces caused by the difference between the furnace pressure and the surrounding atmosphere. The furnace enclosure must prevent air infiltration when the furnace is operated under negative pressure and gas leakage when the furnace is operated at positive pressure.

Superheaters and Reheaters

The addition of heat to steam after evaporation increases the temperature and the enthalpy of the fluid. The heat is added to the steam in boiler components called superheaters and reheaters, which are comprised of tubular elements exposed to the high-temperature gaseous products of combustion.

The advantages of superheat and reheat in power generation result from thermodynamic gain in the Rankine cycle (see Sec. 4.1) and from the reduction of heat losses due to moisture in the low-pressure stages of the turbine. With high steam pressures and temperatures, more useful energy is available, but the advances to high steam temperature often are restricted by the strength and the oxidation resistance of the steel and the metallic alloys currently available and economically practical for use in boiler pressure-part and turbine-blade construction.

The term superheating is applied to the higher-pressure steam and the term reheating to the lower-pressure steam which has given up some of its energy during expansion in the high-pressure turbine. With high initial steam pressure, one or more stages of reheating may be employed to improve the cycle thermal efficiency.

Separately fired superheaters have been used, but superheaters usually are installed as an integral part of the steam-generating unit and broadly classified as **radiant** or **convection** types, depending upon the predominant method of heat transfer to the heat-absorbing surfaces.

The quantity of heat absorbed and the amount of superheat attained are dependent upon the size, location, and arrangement of the heat-absorbing surfaces; the temperature differentials between the gas, the tube metal, and the steam; and the heat-transfer coefficients. Steam-temperature characteristics of radiant- and convection-type superheaters are shown in Fig. 9.2.29, as well as the effect of using a combination of these types to produce a more uniform steam temperature over a wide operating range.

Fig. 9.2.29 Comparative radiant and convection superheater characteristics.

Superheaters of the predominantly **radiant type** usually are arranged for direct exposure to the very high-temperature furnace gases and, in some designs, form a part of the furnace enclosure. In other designs, the surface is arranged in the form of tubular loops or platens, on wide lateral spacing, extending into the furnace (see Fig. 9.2.23). Such surface is exposed to high-temperature furnace gases traveling at relatively low velocities, and the transfer of heat is principally by thermal radiation.

Convection-type superheaters are installed beyond the furnace exit where the gas temperatures are lower than those in the zones where radiant-type superheaters are used. The tubes are usually arranged in the form of parallel elements on close lateral spacing and in tube banks extending partially or completely across the width of the gas stream, with the gas flowing through the relatively narrow spaces between the tubes. High gas flow rates and, thus, high convection heat-transfer rates are obtained at the expense of gas-pressure drop through the tube bank.

Superheaters, shielded from the furnace combustion zone by arches or wide-spaced screens of steam-generating tubes, which receive heat by thermal radiation from the high-temperature gases in cavities or intertube spaces and also by convection due to the relatively high velocity gas flow through the tube banks, have both radiant and convection characteristics.

Superheaters may utilize tubes arranged in the form of hairpin loops connected in parallel to inlet and outlet headers; or they may be of the continuous-tube type, where each element consists of a number of tube loops in series between the inlet and outlet headers. The latter arrangement permits the use of large tube banks, thus increasing the amount of heat-absorbing surface that can be installed and providing economy of space and reduction of cost. Either type may be designed for the drainage of the condensate which forms within the tubes during outages of the unit, or they may be used in pendant arrangements which are not drainable but, usually, have simpler and better supports. Nondrainable superheaters require additional care during start-up to remove the condensate by evaporation. Both types require that every tube have sufficient steam flow to prevent overheating during operation. In start-up, when flow is insufficient, gas temperatures entering the superheater must be controlled to limit tube metal temperatures to safe values for the material used.

The heat transferred from high-temperature gases by thermal radiation and convection is conducted through the metal tube wall and imparted by convection to the high-velocity steam in the tubes. The removal of heat by the steam is necessary to keep the tube metals within in a safe temperature range consistent with the temperature limits of oxidation and the creep or rupture strength of the materials (see Sec. 5). Allowable design stresses for various steels and alloys, as established by the ASME Code, are listed in Table 9.2.2 (see also Sec. 6). The practical use limit for each material is also indicated. Current editions of the code should be checked for latest revisions. For economic reasons, it is customary to use low-carbon steel in the steam inlet sections of the superheater and, progressively, more costly alloys as the metal temperatures increase.

The rate of steam flow through superheater tubes must be sufficiently high to keep the metal temperatures within a safe operating range and to ensure good distribution of flow through all the elements connected in parallel circuits. This can be accomplished by arrangements which provide for multiple passes of the steam flow through the superheater tube banks. Excessive steam-flow rates, while providing lower tube-metal temperatures, should be avoided, since they result in high pressure drop with consequent loss of thermodynamic cycle efficiency.

The spacing of the tubular elements in the tube bank and, consequently, the gas flow rate and convection heat transfer are governed primarily by the types of fuel fired, draft-loss considerations, and the fouling and erosive characteristics of fuel ash carried in the gas stream. Gas flow velocity limits for coal-fired units are typically set to control fly ash erosion in the superheater and reheater surfaces. The velocity limits are determined based upon the ash quantity (mass per unit heat input basis) and the relative proportion of abrasive constituents in the ash (primarily Al_2O_3 and SiO_2). Typical limits range from 65 ft/s (20 m/s) for relatively nonabrasive low-ash coals to 45 ft/s (14 m/s) or less for coals with high ash quantities and/or abrasive ash. In the higher-gas-temperature zones, the adherence and the accumulation of ash deposits can reduce the gas-flow area and, in some cases, may completely bridge the space between tubes. Thus, as gas temperatures increase, it is customary to increase the tube spacings in the tube banks to avoid excessive draft loss and to facilitate ash removal.

Economizers

Economizers remove heat from the moderately low-temperature combustion gases after the gases leave the steam-generating and superheating/reheating sections of the boiler. Economizers are, in effect, feedwater heaters which receive water from the boiler feed pumps and deliver it at a higher temperature to the steam generator. Economizers are used instead of additional steam-generating surface, since the feedwater and, consequently, the heat-receiving surface are at temperatures below the saturated-steam temperature. Thus, the gases can be cooled to lower temperature levels for greater heat recovery and economy.

Table 9.2.2 Typical Superheater and Reheater Tube Materials—Maximum Allowable Design Stress,1,000 lb/in²

SI conversions: $C = (F - 32) \times 5/9$; kPa $= 6.895 \times$ lb/in²

Material	Allowable stresses, 1,000 lb/in² Temperature, °F											
	700	750	800	850	900	950	1,000	1,050	1,100	1,150	1,200	1,250
SA-210C	18.3	14.8	12.0	9.3	6.7	4.0						
SA-209T1a	16.8	16.4	15.9	15.4	13.7	8.2						
SA-213T2	15.7	15.4	14.9	14.5	13.9	9.2	5.9					
SA-213-T12	15.8	15.5	15.3	14.9	14.5	11.3	7.2	4.5				
SA-213-T11	15.1	14.8	14.4	14.0	13.6	9.3	6.3	4.2				
SA-213-T22	16.6	16.6	16.6	16.6	13.6	10.8	8.0	5.7	3.8	2.4		
SA-213-T23	20.5	20.3	19.9	19.5	18.9	17.8	14.3	11.2	8.4	5.5		
SA-213-T91	22.9	22.2	21.3	20.3	19.1	17.8	16.3	14.0	10.3	7.0	4.3	
SA-213 304H	15.8	15.5	15.2	14.9	14.6	14.3	14.0	12.4	9.8	7.7	6.1	4.7
SA-213 347H	16.8	16.8	16.8	16.8	16.7	16.6	16.4	16.2	14.1	10.5	7.9	5.9

SOURCE: Table 1A, Section IID, "Properties of the ASME Boiler and Pressure Vessel Code," 2004 edition.

Economizers are forced-flow, once-through, convection heat-transfer devices, usually consisting of steel tubes, to which feedwater is supplied at a pressure above that in the steam-generating section and at a rate corresponding to the steam output of the boiler unit plus any blowdown to remove boiler water contaminants. They are classed as horizontal- or vertical-tube types, according to geometrical arrangement; as longitudinal or crossflow, depending upon the direction of gas flow with respect to the tubes; as parallel or counterflow, with respect to the relative direction of gas and water flow; as steaming or nonsteaming, depending on the thermal performance; and as bare-tube or extended-surface, according to the type of heat-absorbing surface. Staggered or in-line tube arrangements may be used. The arrangement of tubes affects the gas flow through the tube bank, the draft loss, the heat-transfer characteristics, and the ease of cleaning.

The size of an economizer is governed by economic considerations involving the cost of fuel, the comparative cost and thermal performance of alternate steam-generating or air-heater surface, the feedwater temperature, and the desired exit gas temperature. In many cases, it is more economical to use both an economizer and an air heater.

The temperatures of the economizer tube metals generally approximate those of the water flowing within the tubes, and thus with low feedwater temperatures, condensation and external corrosion are encountered in those locations where the tube-metal temperature is below that of the acid or water dew point of the gas (see Fig. 9.2.30). Internal corrosion and pitting also may be experienced if the feedwater contains significant levels of dissolved oxygen. Therefore, it is imperative to maintain feedwater temperatures above the dew-point temperature of the gas and to provide suitable deaeration of the feedwater for the removal of oxygen.

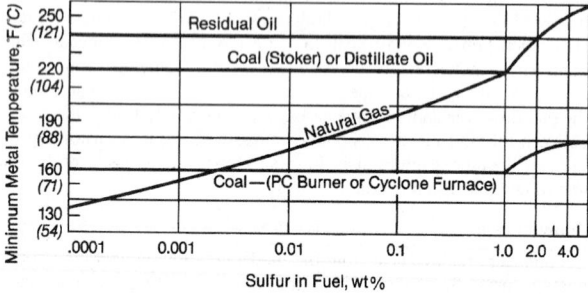

Fig. 9.2.30 Limiting metal temperatures to minimize external corrosion on economizers or air heaters when fuel containing sulfur is burned.

Air Heaters

Air heaters, like economizers, remove heat from the relatively low-temperature combustion gases. The temperature of the inlet air is less than that of the water to the economizer, and thus it is possible to reduce the temperature of the gaseous products of combustion further before they are discharged to the stack.

The heat recovered from the combustion gases is recycled to the furnace with the combustion air and, when added to the thermal energy released from the fuel, is available for absorption by the steam-generating unit, with a resultant gain in overall thermal efficiency. The use of preheated combustion air accelerates ignition and promotes rapid and complete burning of the fuel.

Air heaters are usually classed as **recuperative** or **regenerative**. Both types utilize the convection transfer of heat from the gas stream to a metal or other solid surface and the convection transfer of heat from this surface to the air. In recuperative air heaters, exemplified by the tubular or plate types (Fig. 9.2.31), the stationary metal parts form a separating boundary between the heating and cooling fluids, and the heat passes by conduction through the metal wall. There are two commonly used types of regenerative air heaters (Fig. 9.2.32). In Fig. 9.2.32a, the heat transfer members are moved alternately through the gas and air streams

Fig. 9.2.31 Recuperative-type tubular air heaters, two gas passes, single counterflow air pass.

undergoing successive heating and cooling cycles and transferring heat by the thermal-storage capacity of the members. The other type of regenerative air heater (Fig. 9.2.32b) has stationary elements, and the alternate flow of gas and air is controlled by rotating the inlet and outlet connections.

Recuperative and regenerative air heaters may be arranged either vertically or horizontally and for either parallel or counterflow of the gas and air. The gases are usually passed through the tubes of tubular air heaters to facilitate cleaning, although in some designs, particularly for marine installations, the air flows through the tubes.

Improved heat transfer and better utilization of the heat-absorbing surfaces are obtained with a counterflow of the gases and the use of small flow channels. Regenerative type air heaters readily lend themselves to these two principles and thus offer high capability in minimum space. However, regenerative air heaters have the disadvantages of air leakage into the gas stream and the transport of fly ash into the combustion air system.

The products of combustion from most fuels contain a high percentage of water vapor, and thus condensation will be experienced in air heaters if the exposed metal surfaces are cooled below the dew point of the gas. Minute concentrations of sulfur trioxide in the gases, originating from the combustion of sulfur and varying with the sulfur content of the fuel and the method of firing, combine with the water vapor in the combustion gases to form sulfuric acid, which may condense on the metal surfaces at acid-dew-point temperatures as high as 250 to 300°F (121 to 149°C), well above the water dew point. Such condensation leads to corrosion and/or the fouling of the gas-flow area. It is most likely to occur during the winter when the entering-air temperature is low, and at low operating loads or in localized sections at the cold-air inlet if there is poor distribution of the air or the gas flowing through the air heater. Corrosion and fouling can be prevented by the use of auxiliary steam-heated air heaters located ahead of the air inlet, by recirculating heated air from the outlet duct, or by bypassing a portion of the cold air to reduce the airflow through the air heater. Both recuperative and regenerative air heaters often are designed with separate corrosion sections arranged to facilitate the replacement of the vulnerable cold-end portions.

Steam Temperature, Adjustment, and Control

The control of steam temperatures is vital to the life of high-temperature equipment and to the economy of power generation. Actual, or operating, steam temperatures below the design temperature reduce the thermodynamic efficiency and increase fuel cost, and temperatures above the design temperature reduce the margins of reserve in the strength of tubes, headers, piping, valves, and turbine elements. Further, sudden or extreme temperature variations may cause destructive stresses, particularly in rotating equipment.

It is sometimes necessary, because of the complexities involved in the design evaluation of heat-transfer rates and variability of fuel characteristics, to modify installed equipment to obtain the required steam temperatures. Such changes might involve the installation of baffles for absorbed in the furnace, in the superheater, and/or in the reheater; (3) selective use of burners at different elevations in the furnace or the use of tilting burners to change the location of the combustion zone with respect to the furnace heat-absorbing surface; (4) attemperation or controlled cooling of the steam at superheater inlet, at superheater outlet, or between the primary and secondary stages of the superheater; (5) control of the firing rate in divided furnaces; and (6) control of the firing rate relative to the pumping rate of water in forced-flow once-through boilers.

The speed of response differs for the various methods, and the control of steam temperature by gas bypass or flame position is slower than that by spray-water attemperation. The operating controls for these methods can be arranged for manual, automatic, or combination adjustment, and the use of more than one method often facilitates the maintaining of constant steam temperature over a wider range of boiler load.

The attemperation of superheated steam by direct-contact water spray (Fig. 9.2.33) results in an equivalent increase in high-pressure steam generation without thermal loss. Spray attemperation requires the use of essentially pure water, such as condensate, to avoid impurities in the steam.

Ash and slag deposits on superheater and reheater surfaces reduce heat transfer and lower steam temperatures. Similar deposits on the furnace walls and steam-generating surface ahead of the superheater and/or reheater also reduce heat transfer to those surfaces, resulting in higher-temperature gas to the superheater and reheater and, consequently,

Fig. 9.2.32 Two rotary regenerative air heater designs. (*a*) Ljungstrom (rotating heating surface); (*b*) Rothemuhle (rotating hoods).

Fig. 9.2.33 Spray-type direct-contact attemperator.

increased steam temperatures. Thus the control of surface cleanliness is an important factor in the control of steam temperature.

Increased excess air results in higher steam temperatures because of reduction in furnace radiant-heat absorption, the greater amount of gas, and the increased convection heat transfer in the superheater and/or reheater. Operation with feedwater temperatures below that anticipated also results in increased steam temperatures because of the greater firing rate required to maintain steam generation.

OPERATING CONTROLS
(See also Sec. 16.)

The need for operating instruments and manual or automatic controls varies with the size and type of equipment, method of firing, and proficiency of operating personnel.

Safe operation and efficient performance require information relative to the (1) water level in the boiler drum; (2) burner performance; (3) steam and feedwater pressures; (4) superheated and reheated steam temperatures; (5) pressures of the gas and air entering and leaving principal components; (6) feedwater and boiler-water chemical conditions and particle carryover; (7) operation of feed pumps, fans, and fuel-burning and fuel-preparation equipment; (8) relationship of the actual combustion air passing through the furnace to that theoretically required for the fuel fired; (9) temperatures of the fuel, water, gas, and air entering and leaving the principal components of the boiler unit; and (10) fuel, feedwater, steam, and air flows so as to monitor operating conditions continuously and to make such adjustments as might be necessary.

the distribution of gas through the superheater and the removal, or addition, of tubular elements in the superheater or in those components preceding the superheater which affect the temperature of the gas to the superheater. For routine operation, it is essential to provide a suitable means of controlling steam temperature to compensate for the likely variations in fuel, heat transfer, and surface cleanliness conditions. These may include (1) damper control of the gases to the superheater and/or reheater; (2) recirculation of the low-temperature gaseous products of combustion to the furnace to change the relative amounts of heat

Control of the various functions to maintain the desired operating conditions may be accomplished on small-capacity boilers by the manual adjustment of valves, dampers, and motor speeds. Most oil- and gas-fired package boilers are equipped with automatic controls to purge the furnace, to start and stop the burners, and to maintain the required steam pressure and water level. The operating requirements of utility and large industrial boilers dictate the use of automatic controls for the major variables, such as feedwater flow, firing rate, and steam temperature. The type of boiler and its components generally establish the basic mode of control. Analog controls of either the pneumatic or the electric type are still in use but are largely being replaced with advanced integrated digital control systems which manage all the normal boiler functions. These systems are designed with the maximum amount of monitoring and control at the lowest possible control system level.

Sequence controls often are applied in the start-up of utility boilers to program the furnace purge, burner light-off, and burner control. Interlocks are essential to ensure the proper starting and firing sequence and to alarm or automatically shut down the unit in the event of the failure of essential auxiliaries.

BOILER PERFORMANCE

Boiler Circulation

The water-steam pressure part components are arranged for the most economical system to provide the continuous supply of steam at the specified operating conditions. The full circulation system (excluding reheater) for a natural-circulation subcritical pressure drum type coal-fired steam generator is shown in Fig. 9.2.34.

Adequate circulation in the steam-generating section of a boiler (Circuit D.E.F.G.H in Fig. 9.2.34) is required to prevent overheating of the heat-absorbing surfaces, and it may be provided naturally by gravitational forces, mechanically by pumps, or by a combination of both methods. Figure 9.2.35 shows four common boiler circulation systems.

Natural circulation is produced by the difference in the densities of the water in the unheated downcomers and the steam-water mixture in the heated steam generating tubes. This density differential provides a large circulating force. The downcomers and the heated circuits are so designed that the resistance to flow from friction, local component pressure losses, and acceleration of the two-phase steam water flow through the system balances the circulating force at the desired total circulating flow.

Fig. 9.2.34 Simplified diagram of a coal-fired stream-water natural-circulation boiler system.

Water Circuit   A-B-C-D
Two-Phase Circuit   D-E-F-G-H
Steam Circuits   H-I-L-M-N-O-P-Q-R
                       H-J-K-M-N-O-P-Q-R

Fig. 9.2.35 Common fossil-fuel boiler circulation systems.

The subcritical pressure forced-recirculation or assisted-circulation type of boiler (Fig. 9.2.35b) uses a steam drum similar to that used with natural-circulation boilers. The supply of water to the furnace walls and the boiler surfaces flows from this drum to a circulating pump, which supplies the pressure necessary to force the water through the water-steam mixture circuits and then back to the drum, where the steam and water are separated. The recirculating pump power required is equivalent to about 0.5 percent of the heat input to the boiler. Resistor orifices are required at the entrance to each tube, or circuit, to control flow distribution.

In assisted-circulation designs, the velocities or flows are independent of boiler rating, thus facilitating the use of smaller connecting piping and, sometimes, smaller-diameter furnace-wall tubing than that used with natural-circulation units. Both drum-type natural- and assisted-circulation boilers operate with essentially saturated steam temperatures in all parts of the steam-evaporating sections, and they are used for drum operating pressures up to approximately 3,000 psi (20.7 MPa).

In natural- and assisted-circulation boilers, it is essential to wet inside surfaces continually with water of the two-phase water-steam mixture to prevent dryout and overheating heat-absorbing surfaces. This dryout phenomenon is commonly referred to as departure from nucleate boiling (DNB) or critical heat flux (CHF). Satisfactory cooling of the heat-absorbing steam-generating surfaces is dependent upon the pressure, heat flux, water-steam mass flux, percent steam by weight (quality), tube diameter, and the tube's internal geometry. The heat flux is one of the most predominant of these parameters, and the rate of furnace heat absorption at the maximum firing rate generally dictates design considerations. Satisfactory circulation design also includes the evaluation of two-phase flow stability/instability limits, steam separator and drum flow limits, minimum tube flow velocity limits, and sensitivity (flow increases for heat flux increases at all expected operating conditions.)

In forced-flow once-through boilers (Fig. 9.2.35c), the water from the feed supply is pumped to the inlet of the heat-absorbing circuits. Evaporation, or change of state, takes place along the length of the circuit, and when evaporation is completed the steam is superheated. These units do not require steam or water drums and, in most cases, use relatively small-diameter tubes. The water flow to the unit is the same as the steam output, and fluid velocities greater than those needed for natural- or assisted-circulation units must be used at full load so as to maintain adequate velocities at the low loads and satisfactory tube-metal temperatures at all loads.

The transition from a liquid to a vapor at, or above, the critical steam pressure of 3,200 psi (221 MPa) is dependent upon temperature and takes place without a change in density. Thus, separation of steam and water is impossible and forced-flow once-through boilers must be used.

Traditional forced-flow once-through boilers must be operated above a specified minimum flow in order to maintain adequate water velocities in the furnace-wall tubes. However, the turbogenerator can be operated at any load by the use of a bypass system that diverts the excess flow to a flash tank for heat recovery. The bypass system also can be used as a pressure relieving system, as the source of low-pressure steam to the turbine during start-up, and as a means of controlling steam temperature to the turbine during hot restarts.

Combined circulation units utilize forced once-through flow with flow recirculation in the furnace walls to provide satisfactory water velocities during start-up and low-load operations. In this design, some of the water at the exit of the furnace circuits is mixed with the incoming feedwater, flows to and through a circulating pump, and then passes to the furnace-wall inlet headers. The use of combined circulation increases the water velocities in the furnace tubes at low loads, and since recirculation is not used at the higher loads, there is no increase in velocity or in the resistance to flow at the higher loads.

Boilers are also designed for variable-pressure operation with operating pressure varying as the boiler load changes. Near full load, supercritical pressure once-through operation occurs. At reduced loads, operating pressure is reduced below the critical pressure, and a part-load recirculation system (Fig. 9.2.35d) is used to separate the steam-water flow and return the water with the feedwater to the economizer inlet.

Steam Separation

In drum-type boilers, using either natural or assisted circulation, a mixture of steam and water is delivered to the upper or steam drum where separation of the steam and water takes place and a water level is maintained. The water is recycled through downcomers to the heat-absorbing circuits, and the steam is discharged from the top of the drum for use as saturated steam or the supply to the superheater.

Separation of the steam and water by buoyant or gravitational force requires a relatively large cross-sectional area and, consequently, a low fluid velocity within the drum as well as an effective difference in the fluid densities, which decreases as pressure is increased. Steam entrainment in the recycled water impedes circulation; and water entrainment in the outlet steam transports dissolved or suspended particulate matter into the superheater, steam piping, and turbine, where particle deposition can cause overheating of the tubes or flow obstructions in the turbine blading with subsequent loss of capacity, efficiency, and dynamic balance.

Gravity separation of steam and water may be satisfactory in low-pressure, low-duty boilers. This type of separation can be augmented by the use of baffles which utilize a change in direction to throw out the water droplets, or by dry pipes.

In high-pressure boiler units, particularly those employing high evaporative ratings, a part of the circulating head can be utilized to provide a separating force, many times greater than that of gravity, in centrifugal separating devices such as the cyclone steam separators shown in Fig. 9.2.36. These separators deliver steam-free water to the drum and downcomers, and discharge steam with a minimum of water entrainment. Secondary steam-and-water separation is accomplished by passing the steam at a low velocity through sinuously shaped passages between closely spaced corrugated plates, which provide a large surface area for intercepting entrained boiler-water particles. In modern

Fig. 9.2.36   Steam drum internal arrangement with cyclone steam separators and scrubber elements.

high-capacity boilers, the steam leaving the steam drum contains less than 0.1 ppm of total solids. This purity is generally acceptable for commercial practice. However, where further refinement in steam purity is required to remove boiler contaminants such as silica, the steam can be washed with condensate or feedwater of sufficient purity.

Flow of Gas Through Boiler Unit

During the combustion of fuel and the transfer of heat to the heat-absorbing surfaces, it is necessary to maintain sufficient pressure to overcome the resistance to flow imposed by the burning equipment, tube banks, directional turns, and the flues and dampers in the system. The resistance to air and gas flows depends upon the arrangement of the equipment and varies with the rates of flow and the temperatures of the air and gas. Figure 9.2.37 provides a simplified air-gas flow schematic for a coal-fired boiler.

Fig. 9.2.37 Simplified boiler air-gas flow path diagram: primary air (A1-B-D-E); secondary air (A2-B-C-E); and products of combustion (E-F-G-H-I).

The term draft denotes the difference between atmospheric pressure and some lower pressure existing in the furnace or gas passages of a steam-generating unit. Draft loss is defined as the difference in static pressure of a gas between two points in a system and is the result of the resistance to flow. These terms originated with the use of the so-called natural-draft units in which the pressure differentials are obtained from a chimney or stack which produces static pressures throughout the boiler

setting that are below that of the atmosphere. The terms are rather loosely applied to modern boilers using induced draft and/or forced draft, mechanically produced by fans, in which the pressures throughout the boiler unit may be well above atmospheric pressure.

Forced-draft fans, handling cold, clean air, provide the most economical source of energy to produce flow through high-capacity units (see Sec. 14.5). Induced-draft fans, handling hot flue gases, require more power and are subject to fly ash erosion. However, they facilitate operation by providing a draft in the boiler setting and thus prevent the outward leakage of gas through joints or crevices in the boiler enclosure. Boilers built for positive-pressure operation are generally referred to as pressure-fired units, while those using induced-draft fans are classed as suction units. When both forced- and induced-draft fans are used, the boilers are designated as balanced-draft units, the most common design today.

The pressure drop throughout the unit is caused by the fluid friction in the gas stream and the shock losses at the turns or the contractions and enlargements of sections. It can be calculated as a function of the fluid mass flow and fluid properties in accordance with the principles of fluid flow (see Sec. 3.3). It is essential in the design of a boiler to determine the sum of all component resistances in the flow system at the maximum load in order to establish the fan requirements. It is customary to specify test-block static head, temperature, and capacity requirements of the fan in excess of those calculated so as to allow for departure from ideal flow conditions and to provide a satisfactory margin of reserve.

Stack effect is caused by the difference in densities resulting from the difference in the temperatures of two vertical columns of gas. In a chimney, or stack, the stack effect is due to the difference between the confined hot gas and the cooler surrounding air and the equal static pressure at the top or free outlet of the stack. The stack effect, which varies with the height and the mean temperature of the columns, can be calculated from the data in Table 9.2.3. The effect is the static draft produced by a stack, at sea level, with no gas flow. When flow occurs, a portion of the stack effect is used to establish gas velocity and the remainder is used to overcome the resistance of the connected system, including the dampers and the stack. The limit of natural-draft capacity is reached when these forces are in balance with the dampers in a wide-open position. Stack performance may be favorably or adversely affected by external factors such as the wind and the atmospheric conditions. The available draft varies directly with the barometric pressure for altitudes above sea level.

Stack effects also exist within the boiler setting and are most pronounced in tall units with vertical gas passes. The individual gas columns within the setting may aid the head produced by the fan or chimney if the flow is upward or may reduce it if the flow is downward. The net stack effect, and its overall influence on the performance of the fan, may be calculated from the data in Table 9.2.3, taking into account the positive or negative effects. The relationship between local static pressures and the atmospheric pressure is important, since gas may blow into the room through an open inspection door at the top of a furnace, even though a strong draft, or negative pressure, exists at some other elevation.

Performance

Specific guarantees relating to the boiler or steam generator scope include the maximum continuous rating (MCR) capacity (steam flows, temperatures, and pressures), boiler efficiency, auxiliary power

Table 9.2.3   Stack Effect of Pressure Difference, in of Water for 1 ft of Vertical Height (mm of Water for Each 1 m of Vertical Height)
Barometer = 29.92 inHg (759.97 mmHg)

Avg temp in flue, °F (°C)	Air temp outside flue, °F (°C)			
	40°F (5°C)	60°F (15°C)	80°F (25°C)	100°F (35°C)
250 (125)	0.0041 inH$_2$ O/ft (0.346 mmH$_2$O/m)	0.0035 (0.303)	0.0030 (0.262)	0.0025 (0.224)
500 (250)	0.0070 (0.563)	0.0064 (0.520)	0.0059 (0.480)	0.0054 (0.442)
1,000 (500)	0.0098 (0.788)	0.0092 (0.774)	0.0087 (0.708)	0.0082 (0.665)
1,500 (750)	0.0112 (0.902)	0.0106 (0.858)	0.0100 (0.818)	0.0095 (0.780)
2,000 (1,000)	0.0120 (0.972)	0.0114 (0.925)	0.0108 (0.887)	0.0103 (0.850)
2,500 (1,250)	0.0125 (1.018)	0.0119 (0.975)	0.0114 (0.934)	0.0109 (0.896)

No. of Cooled Furnace Walls

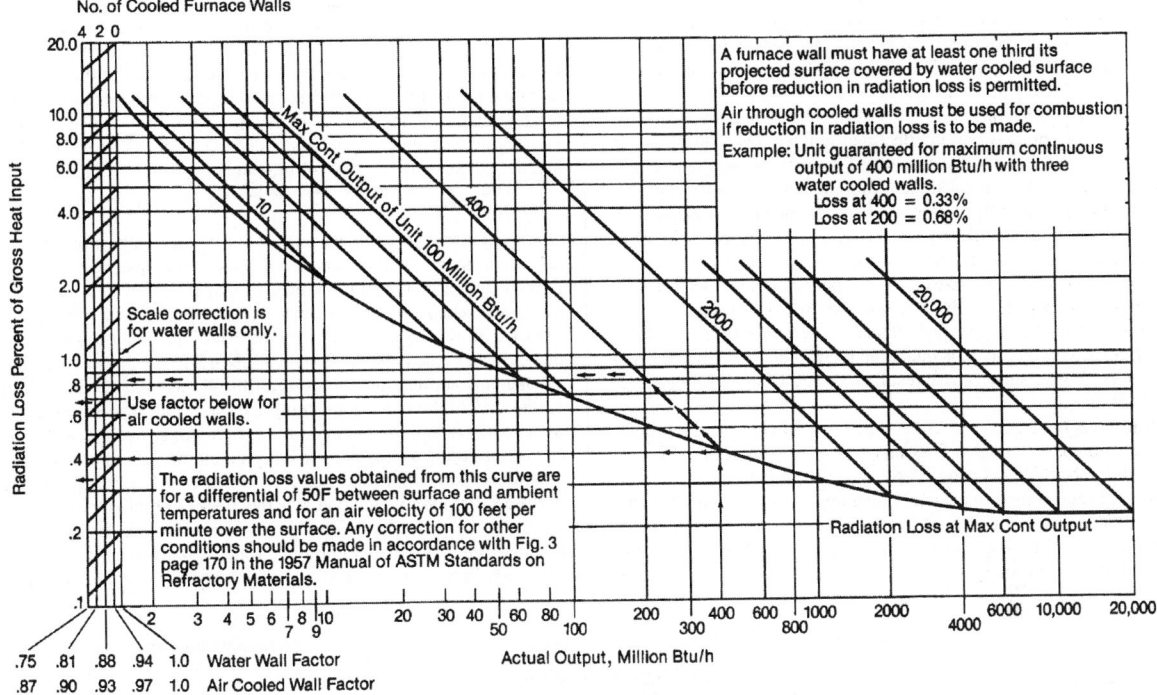

Fig. 9.2.38 External heat loss from boiler setting based upon gross heat input (ABMA).

consumption, water- and steam-side pressure drop, superheater and reheater steam temperature control range, spray water quantities, and emissions of NO_x, CO, and volatile organic compounds (VOCs). If the spray water source is downstream of the feedwater heaters, then the superheater spray is not considered. For subcritical-pressure, drum-type boilers, steam purity is commonly included.

Anticipated-performance data for several rates of operation may be given to the purchaser in addition to the guaranteed-performance data. Guarantees may be demonstrated by acceptance tests, conducted in accordance with established codes, as agreed upon by the parties in the contract. However, acceptance tests are more difficult to perform as unit size and capacity increase, and overall performance usually is determined from the operating data. Warranties of materials and the quality of manufacture and erection are usually considered separately from those pertaining to operating performance.

Heat balances account for the thermal energy entering the system in terms of its ultimate useful heat absorption or thermal loss. Methods of measuring and calculating the quantities involved in heat balances are presented in the ASME Performance Test Code PTC4 for Fired Steam Generators.

Both the heat input and the heat absorption may be very large quantities. Therefore, unless elaborate precautions are taken in the sampling and the measurement of fuel and steam quantities, it is difficult to obtain test data having the degree of accuracy required to determine the actual efficiency of the boiler unit. For this reason, boiler efficiency usually is established from the heat losses, since each of the thermal losses is a relatively small percentage of the heat entering the system and reasonable errors in measurement will not appreciably affect the final result.

The principal thermal losses are those due to the sensible heat in the gases leaving the unit, latent-heat losses associated with the evaporation of fuel moisture and formation of water vapor resulting from burning of hydrogen in the fuel, unburned-combustible loss, loss from the boiler setting or enclosure due to external convection and radiation, and wet ashpit loss. The first two losses can be derived from the fuel analysis, the exit gas temperature, and the analysis of the flue gas (see Sec. 4). The unburned-combustible loss can be

established by a qualitative and quantitative sampling of refuse and fly ash. The radiation from the boiler setting can be estimated in detail, but it also can be approximated from Fig. 9.2.38. The heat loss to the ashpit of large units can be determined by measuring the quantity of quenching water evaporated from the furnace ash hopper or the slag tank; and for small units the ashpit loss is included in the heat loss from the boiler setting. The sum of these heat losses, expressed as a percentage of the total heat input, is the total measurable loss. The anticipated or guaranteed performance data include a tolerance for the so-called manufacturer's margin (unaccountable losses) in the order of 0.5 percent, depending upon the type of fuel fired. The thermal efficiency of the unit is established by subtracting the sum of all these losses from 100 percent.

ENVIRONMENTAL CONTROL SYSTEMS

(See also Sec. 18.)

As of this edition, the primary-air pollutant emissions controlled on utility and most industrial boilers are particulate, nitrogen oxides (NO_x), and sulfur dioxide (SO_2). The strategies for control are formulated by considering the range of specific fuels to be burned and their sources, specific kind and extent of emission levels mandated, and economic factors such as the specific boiler design, location, new or existing equipment, plant age, and remaining life. In most cases, overall control of individual pollutants for a specific fuel involves some combination of precombustion, combustion, and postcombustion technologies. This section provides a brief overview of the major postcombustion technologies. The control of NO_x emissions by combustion modifications is covered under "Burners and Combustion Systems" and "Stokers." The control of SO_2 during combustion is covered under "Fluidized-Bed Combustion."

Control of Particulate

Particulate emissions from boilers arise from noncombustible, ash-forming mineral matter in the fuel that is released during the combustion process and carried with the products of combustion. Three general types of equipment are used to reduce particulate emissions.

Mechanical collectors are generally cyclone collectors and have been widely used on small boilers with less stringent emission limits. They are low-cost, simple, compact, and rugged, but conventional cyclones are limited in collection efficiencies of about 90 percent and are poor at collecting the finest particulate.

Electrostatic precipitators (ESPs) electrically charge the ash particles in the products of combustion to collect and remove them. As shown in Fig. 9.2.39, an ESP normally consists of a series of parallel, vertical metallic plates (collecting electrodes) forming lanes through which the products of combustion flow. Centered between the plates are discharge electrodes which provide the particle charging and electric field. ESPs are available for most utility and industrial applications, and collection efficiencies can be expected to exceed 99.9 percent of the inlet ash loading. ESPs are considered to be less sensitive to plant upsets, and they have low pressure drop. ESP performance is sensitive to fly ash loading, ash resistivity (ability to acquire and retain a charge), and coal sulfur content. Lower-sulfur coals result in lower ESP performance or large ESP sizes.

Fabric filters (or baghouses) collect the ash particles as the products of combustion pass through a filter material. As shown in Fig. 9.2.40, a large utility fabric filter is comprised of a series of individual compartments, with each compartment containing up to several thousand long, vertically supported, small-diameter fabric bags. The ash cake which builds up on the filter bag surface is dislodged and removed by one of three techniques: reversing the gas flow temporarily, shaking the bags,

or using a pulse jet of air down the bag. Collection efficiencies can be expected to exceed 99.9 percent of the inlet ash loading. When combined with dry SO_2 scrubbing, they can enhance SO_2 removal as the gas passes through the filter cake on the bag surface. Fabric filters are more sensitive to plant temperature excursions and upsets because of the filter material, and they are a higher pressure-drop device by nature. Fabric filters are more capable of collecting the finest fraction of the ash and are not sensitive to fuel sulfur content.

Control of NO$_x$

Oxides of nitrogen (NO_x) are formed during the combustion process, as discussed under "Burners and Combustion Systems," and are initially limited by the combustion system design. If further NO_x reduction is necessary. **selective catalytic reduction (SCR)** NO_x control systems can be installed, most frequently between the exit of the boiler economizer and the air heater (see Figs. 9.2.20 and 9.2.23). Such designs are referred to as **high-dust** applications because all the fly ash passes through them on the way to the particulate control system. **Low-dust** systems can be installed after a particulate control system but typically involve higher capital costs. A reagent, typically ammonia, is uniformly injected and mixed into the products of combustion (see Fig. 9.2.23) upstream of the SCR reactor which contains the blocks of high-surface-area catalyst. As the products of combustion pass through the catalyst, the ammonia and NO_x react to produce water vapor and gaseous nitrogen. The most common catalyst types are coated metal **plate** catalyst,

Fig. 9.2.39   Electrostatic precipitator.

Fig. 9.2.40   Pulse jet fabric filter (or baghouse) for fly ash particulate removal.

homogeneous ceramic **honeycomb** catalyst, and coated **corrugated** glass fiber substrate catalyst.

The catalyst performance is temperature-dependent with an optimal range of 700 to 750°F (371 to 399°C) although specialized catalyst can perform somewhere in the wider temperature window of 625 to 800°F (329 to 427°C) depending upon the gas composition and modified catalyst formulation. Some catalysts are also specially formulated to be resistant to certain constituents of the products of combustion which can poison and deactivate the catalyst surface. SCR units can effectively reduce NO_x levels leaving the boiler by 90 percent or greater depending upon the site- and boiler-specific conditions. Commercial projects are usually designed for 70 to 90 percent reduction. Ammonia is supplied in anhydrous or aqueous forms and from systems which convert urea to gaseous ammonia.

Control of SO_2

Sulfur dioxide (SO_2) comes from sulfur in the fuel. The dominant method of SO_2 reduction for coal-fired units is through the installation of a **flue gas desulfurization (FGD)** system downstream of the boiler. There are two basic types of FGD systems in wide use today, wet FGD scrubbers and dry FGD scrubbers. The selection of the most appropriate system depends upon the site- and fuel-specific economics.

Wet scrubbing (see Fig. 9.2.41) is centered on a large-diameter absorber module in which a sorbent slurry consisting of water mixed with a reagent (usually limestone, lime, magnesium-enhanced lime, or sodium carbonate) is contacted with the products of combustion. The SO_2 is first absorbed by the aqueous solution and then reacted with the reagent in the slurry. Limestone ($CaCO_3$) based systems with forced oxidation of the residual slurry by-product to produce wallboard-quality gypsum ($CaSO_4 \cdot 2H_2O$) are the dominant systems being installed. Wet scrubbing is highly efficient with > 97 percent removal of the SO_2 at calcium-to-sulfur molar ratios of close to 1.0. The full scrubber system consists of reagent receiving and preparation equipment, the absorber module with recirculation pumps, and the spent slurry dewatering system. The absorbers are designed with corrosion-resistant linings or materials to prevent chemical attack. One or more absorber modules (more commonly one on new systems) are located between the forced-draft fan and the stack. Performance-enhancing additives can be used to reduce power consumption or increase removal efficiency.

Fig. 9.2.41 Wet flue gas desulfurization system (absorber module).

Dry scrubbing (see Fig. 9.2.42), which is more correctly referred to as **spray dryer absorption (SDA)**, involves the spraying of an aqueous sorbent slurry into a reaction vessel under conditions which permit the slurry droplets to dry as they contact the hot gas [~300°F(~149°F)]. The SO_2 reaction occurs during the drying process and results in a dry particulate containing reaction products and unreacted sorbent or reagent entrained

in the products of combustion flow along with the fly ash. These materials are then captured in the downstream particulate control equipment. The reagent is usually lime. Scrubber modules are located between the air heater and particulate control equipment, which often is a fabric filter to take advantage of the increased removal possible, since the unreacted lime collected in the bag filter cake reacts with the remaining SO_2. Such

Fig. 9.2.42 Dry flue gas desulfurization system (spray dryer module).

SDA plus fabric filter systems can be designed to remove 90 percent or more of the inlet SO_2. The SDA reactor is made of carbon steel. The full SDA system consists of reagent receiving and preparation equipment, the SDA, a fabric filter, and spent reagent/ash collection and disposal. A key attribute of the system is the dry waste residue.

WATER TREATMENT

The objectives of water treatment in boiler systems include (1) minimizing deposits of scale or sludge on internal tube surfaces, (2) reducing and managing corrosion of the internal boiler surfaces, and (3) aiding in the delivery of steam of required purity to the steam turbine or application. Scale and sludge deposits tend to collect preferentially in the high-heat-flux zones of the boiler where they retard heat flow, raise metal temperatures, and potentially lead to overheating and tube failure. Sludge and solid particles can settle out in some areas and cause flow restrictions. Corrosion (see also Sec. 6.5) due to chemical imbalance, excessive concentration, or dissolved gases can cause general metal loss and ultimately lead to tube failures. The presence of internal deposits can accelerate underdeposit corrosion because of chemical concentration and elevated temperatures. Certain chemical reactions produce intergranular attack of metals and may lead to embrittlement and ultimate fracture.

A comprehensive discussion of water and treatment practices for power boilers is provided in "The ASME Handbook on Water Technology in Thermal Power Systems," American Society of Mechanical Engineers, 1989. A brief overview discussion is provided here.

Raw Water

Raw water treatment is necessary for all natural water sources to remove impurities and prepare water for steam generation. Concentrations are typically expressed as parts weight fraction of the impurity per million parts water or **parts per million (ppm)**. For very low concentration **parts per billion (ppb)** is used. Treatment of raw water for use as boiler feedwater and makeup water involves one or more of the following processes:

1. *Removal of suspended solids*
2. *Chemical treatment for the removal of hardness.* Calcium and magnesium are the principal hardness impurities in water which have retrograde solubility with increasing temperature and form scale deposits on heated surfaces. They must be removed to very low levels.
3. *Cation exchange for the removal of hardness*
4. *Ion-exchange demineralization for complete removal of dissolved solids.* Several types of inorganic resins have the ability to selectively remove undesirable cations and anions of impurities from water by exchange with hydrogen or hydroxyl ions. The resins are later regenerated. **Mixed bed ion-exchange demineralizers** use a mixture of the anion and cation resins.
5. *Evaporation*
6. *Reverse osmosis*

Feedwater and Makeup Water

Boiler **feedwater** is generally a mix of returned steam condensate and fresh makeup water from the treated raw water supply. For utility boilers, most of the steam is usually returned as condensate, and only 1 to 2 percent makeup is necessary. However, for some industrial cycles there is little or no return condensate; so up to 100 percent makeup may be necessary. Overall feedwater parameters which must be controlled include dissolved solids, pH, dissolved oxygen, hardness, suspended solids, total organic carbon (TOC), oil, chlorides, sulfides, alkalinity, and acid- or base-forming tendencies. At a minimum, boiler feedwater must be softened water for low-pressure boilers and demineralized water for high-pressure boilers. Removal of oxygen and other dissolved gases adequate for boiler feedwater applications is generally accomplished by thermal deaeration of the water in open-type feedwater heaters as well as tray or spray-type deaerating heaters. Thermal deaeration can reduce feedwater oxygen concentrations to less than 7 ppb. It also essentially eliminates dissolved carbon dioxide, nitrogen, and argon.

Chemical agents are generally used to scavenge residual oxygen. Traditional oxygen scavengers have been sodium sulfite for low-pressure boilers [< 900 psig (6.2 MPa gage)] and hydrazine for high-pressure boilers. Hydrazine has been identified as a carcinogen, and thus the use of alternate high-temperature oxygen scavengers (erythorbic acid, diethylhdroxylamine, hydroquinone, and carbohydrazide) has increased. Scavengers are generally fed at the exit of the condensate polishing system and/or the boiler feed pump suction.

When the boiler feedwater pH falls below a minimum value, ammonium hydroxide or an alternate alkalizer is added. Common alternative pH control agents include neutralizing amines, such as cyclohexylamine and morpholine. For high-pressure boilers with superheaters, the more complex amines are thermally-unstable and decomposition products can be problematic.

For many boilers, a large fraction of the feedwater is returned as condensate. Condensate has been purified by prior evaporation, and in many cases makeup water can be directly mixed with the condensate to supply feedwater to the boiler. However, steam condensate can be contaminated by corrosion products or by in-leakage of condenser cooling water. Where the returning condensate is contaminated to such an extent that it does not meet the feedwater quality requirements, mixed bed ion-exchange demineralization purifiers are used to remove the dissolved impurities and filter out the suspended solids. This is referred to as **condensate polishing** and is essential for satisfactory operation of once-through utility boilers where the feedwater requirements are especially stringent. Condensate polishing is also recommended for all boilers operating with all-volatile treatment (AVT) (see below).

Boiler Water

In boilers, water is converted to steam, and the steam leaves the boiler in a relatively pure state. In drum-type recirculating boilers, impurities other than gases are retained and concentrate in the boiler water. High concentrations of foam-producing solids in the boiler water contribute to particle and water carryover and contamination of steam. To limit the dissolved solids to specified levels, a portion of the boiler water is intermittently or continuously removed (referred to as **blowdown**), disposed of, and replaced with fresh feedwater. Blowdown is normally expressed as a percent relative to the total steam flow. In once-through boilers, basically all the material entering with the feedwater is carried out with the steam or deposited on the internal boiler surfaces. Therefore, more stringent feedwater quality limits are required, and any boiler water treatment must take this into account to ensure that the quality requirements are met. Boiler water quality limits by the American Boiler manufacturers Association are shown in Table 9.2.4.

Customized chemistry limits and treatment practices must be established for each boiler. These limits depend upon the steam purity requirements, feedwater chemistry, and boiler design. They also depend upon boiler owner/operator preferences in the development of an overall plant water treatment program and system. Because corrosion in boilers is usually minimized by maintaining alkaline boiler water, the pH or total alkalinity (a measure of all reactives of all reactives that have the ability to neutralize acids, expressed in ppm) is closely monitored and controlled. Direct boiler water treatment (**internal treatment**) practices commonly used to control boiler water chemistry include

1. **Traditional all-volatile treatment:** No solid chemicals are added, and boiler water chemistry is controlled by feedwater treatment alone. Feedwater pH is controlled with ammonia and hydrazine, or a suitable alternative is added to scavenge residual oxygen.

2. **Oxygen treatment:** For boilers with copper-free feedwater trains and in the absence of impurities [cation conductivity $< 0.15 \mu S/cm$ at 77°F (25°C)], a controlled low level of oxygen is maintained in the boiler which helps establish and maintain an especially protective Fe^{3+} iron oxide coating. Usually used in high-pressure once-through boilers, it has also been used successfully in drum boilers. The target oxygen concentration is 0.050 to 0.150 ppm for once-through boilers and 0.040 ppm for drum-type boilers.

3. **Coordinated phosphate treatment:** This treatment controls the boiler water alkalinity with mixtures of disodium and trisodium phosphate added to the drum through a chemical feed pipe. The objective is to maintain the pH of the boiler water and underdeposit boiler water concentration within an acceptable range.

4. **High-alkalinity phosphate treatment (low-pressure boilers only):** Used in industrial boilers less than 1,000 psig (6.9 MPa gage) where full demineralized makeup water is not cost-effective, trisodium phosphate and sodium hydroxide are added to the drum to maintain the pH at 10.8 to 11.4 and phosphate at acceptable levels.

5. **High-alkalinity dispersant, polymer, and chelant treatment (low-pressure boilers only):** In industrial boilers with substantial hardness (e.g., 0.1 ppm as $CaCO_3$) and pressures below 1,000 psig (6.9 MPa gage), the additives promote the formation of stable calcium and magnesium complexes which remain in solution or suspension and can be removed by blowdown. Boiler water is typically maintained at a pH of 10.0 and 11.4.

CARE OF BOILERS

The care of power boilers is delineated in Section VII of the ASME Boiler and Pressure Vessel Code. Principal considerations include the initial preparation of new equipment for service; normal operation, including routine start-up and shutdown; emergency operations; inspection and maintenance; and idle storage. In all these phases, the handling of equipment is the responsibility of the operator, but recommendations and operating instructions supplied by the manufacturer should be thoroughly understood and followed.

The initial preparation of new boiler units for service, or of old equipment after completing major alterations or repairs, involves the removal of construction or foreign material from the setting and from the interior of pressure parts, hydrostatic testing and inspection for leaks, and the boiling out of the unit with an alkaline solution for the removal of grease and other deposits in the steam-generating pressure parts. The boiler unit is fired at a low rate during boilout. This procedure also facilitates the desired slow drying of any refractories used in the setting. After boilout and flushing, corrosion products remain in the feedwater system and boiler in the form of iron oxide and mill scale. Chemical cleaning to remove these should be delayed until after full-load operation in most cases, so that the scale and oxides are carried into the boiler from the feedwater system. Exceptions include boilers with full-flow condensate polishing systems and boilers in which the preboiler system has been chemically cleaned. These units can be chemically cleaned immediately.

After the boiler is filled with high-purity water and brought back into operation, the pressure is raised to test and set the safety valves to code requirements; the superheater and the steam piping are blown out to remove any foreign material; and the boiler is placed on the line for a period of low-load operation, during which the auxiliary equipment, controls, and interlocks are test-operated. After these operations, it is advisable to shut down, cool, and drain the boiler unit prior to a thorough internal and external inspection and any adjustments or modifications required to the equipment.

Normal operation involves the orderly start-up and shut-down of equipment and operation, under controlled conditions, to meet plant requirements. Statistics show that about 80 percent of the recorded furnace explosions occur during start-up and low-load operation, and particular care must be taken during such operations to prevent such explosions. The National Fire Protection Association "NFPA 85: Boiler and Combustion Systems Hazard Code" delineates the preferred sequence of starting fuel-burning equipment, the recommended minimum flame-monitoring equipment and safety interlocks, the recommended fuel-transport piping systems, the recommended purging procedures, and the procedures to be taken in the event of burner or furnace flame-out.

Normal operation also entails the maintenance of specified feedwater and boiler water conditioning, design steam and metal temperatures, clean gas passages and heat-absorbing surfaces, and ash removal.

In most cases the rate of firing during start-up is limited by the allowable metal temperatures in the superheater and reheater until steam flow through the turbine is established. In some cases, the time for both start-up and shutdown may be determined by the time required to limit the thermal stresses in the drum and headers. Then, after the steam flow is

Table 9.2.4   Recommended Limits of Boiler Water (BW) Concentration

Drum pressure, psig	Maximum boiler water solids, ppm	Steam TDS corresponding to max. BW TDS	Maximum total alkalinity, ppm, as $CaCO_3$	Maximum suspended solids, ppm
0–300	3,500	1.0	*	15
301–450	3,000	1.0	*	10
451–600	2,500	1.0	*	8
601–750	1,000	0.5	*	3
751–900	750	0.5	*	2
901–1,000	625	0.5	*	1
1,001–1,800	100	0.1[†]		1
1,801–2,350	50	0.1[†]	Not applicable[†]	1
2,351–2,600	25	0.05[†]		1
2,601–2,900	15	0.05[†]		1

* 20% of actual boiler water solids. For TDS $\leq$ 100 ppm, the total alkalinity is dictated by the boiler water treatment.
† Does not include vaporous silica carryover.
Source: ABMA, 1995.

established, the temperatures and the temperature differentials in the various parts of the boiler and turbine control the permissible rate of firing.

Emergency operations are usually the direct result of abnormal conditions such as the failure of the feedwater supply, the rupture of a pressure part, the interruption of the fuel supply, the loss of air, or a burner flame-out. Automatic safety interlocks usually are installed which trip the fuel supply and shut down the unit if these or other hazardous conditions are experienced. Abnormal operating conditions, which might become hazardous if allowed to persist, such as low (fuel-rich) or high (air-rich) air-fuel ratios or the failure of essential auxiliaries, require appropriate action to correct operating conditions. An operator who cannot correct an abnormal condition must determine whether operation can continue and, if not, must shut down the unit in the proper manner or activate the emergency trip through the automatic interlock system.

Inspection and maintenance should be performed during regularly scheduled outages. A list of the known items requiring repair or maintenance should be prepared before the outage and should be supplemented by any additional items noted in thorough inspection of the boiler and auxiliary equipment during the outage. A major item on the work list should be the maintenance of internal and external cleanliness. External cleaning is usually accomplished by water washing or air lancing. The internal surfaces of large-capacity units are generally chemically cleaned under competent supervision.

Boilers removed from service for more than 7 days should be laid up using one of several accepted practices. Alternatives for fluid-side layup include nitrogen blanketing, wet layup with treated demineralized water, and dry dehumidified layup. For extended outages it is important that the preboiler feedwater train, balance of the water-steam cycle, auxiliary equipment, and fire side of the boiler be adequately preserved. For idle periods longer than 6 months, dry dehumidified storage is usually recommended. For boilers with nondrainable components, water removal issues must be weighed against the advantages of dry layup. Reheaters are generally stored dry because they cannot easily be isolated from the turbine. The use of nitrogen blanketing to maintain a positive pressure in the system for either dry or wet storage can be very effective in preventing air in-leakage and the associated corrosion. However, the boiler must be well sealed to prevent excessive leakage.

CODES

The ASME Boiler and Pressure Vessel Code, initiated in 1914 and supplemented by continuing revisions, contains the basic rules for the safe design, the construction, and the materials for steam generating units. Its legal status depends upon its adoption by state or municipal authority. The Code is administered by the National Board of Boiler and Pressure Vessel Inspectors. This organization also has established the "National Board Inspection Code," or NBIC, which provides rules and guidelines for in-service inspection, repair, and alterations of pressure-retaining items including steam generators. The adoption of both codes has been widespread, and they form the basis for the pertinent legal requirements in all but a few localities throughout the United States and Canada.

NUCLEAR STEAM GENERATORS

Commercial nuclear power plants provide a specialized application for steam generators, or boilers. As discussed in Sec. 9.8, each pressurized water reactor (PWR) combines two to four steam generators with a reactor, pressurizer, and multiple coolant pumps to form the nuclear steam supply system (NSSS). The 69 operating PWR plants in the United States use a total of 209 steam generators. In addition, the 22 Canada Deuterium Uranium (CANDU) pressurized heavy-water reactor (PHWR) systems installed in Canada use an additional 184 with as many as 12 steam generators per reactor, although 4 is more typical.

These nuclear steam generators provide the principal isolation barrier between the **primary** high-pressure water coolant system in contact with the reactor and the **secondary** lower-pressure water power cycle, as discussed in Sec. 9.8. Nuclear steam generators also serve two other functional requirements: (1) they provide a heat sink for the reactor core, and (2) they supply the specified flow rate of steam from the feedwater

supply at the temperature and pressure necessary to efficiently drive the steam turbine/electric generator.

There are two fundamental steam generator designs in operation. The dominant design is the **recirculating steam generator (RCG)** which recirculates water on the secondary side in much the same way as a drum-type fossil-fuel boiler works. The other design is a **once-through steam generator (OTSG)** which converts the feedwater to superheated steam without water recirculation.

As shown in Fig. 9.2.43, each RSG is a vertical shell, inverted U-tube heat exchange with steam water separation equipment located above the tube bundle inside the upper shell (or drum). A cylindrical shroud or bundle wrapper surrounds the tube bundle, separating it from the lower shell. This creates an annular region which serves as the downcomer to return the recirculated water from the steam separators to the tube bundle inlet at the bottom of the unit. The U-tubes are sealed to the tube sheet at the bottom of the bundle which separates the primary and secondary flow loops. The hot primary reactor water enters half of the

Fig. 9.2.43 Nuclear power recirculating steam generator. (*Courtesy The Babcock Wilcox Company*).

lower steam generator head through the primary nozzle and passes through the tubes, giving up heat, before exiting through the other half of the lower steam generator head and returning to the reactor. The secondary water flow enters the bottom of the tube bundle and passes upward over the outside of the tubes, generating a steam-water mixture (10 to 40 percent steam by weight) before passing through the steam-water separators. The effectively moisture-free steam (generally < 0.25 percent water and as low as < 0.10 percent) is sent to the power turbine while the recirculated water is returned to the bottom of the tube bundle through the downcomer annulus. Feedwater is most often injected into the downcomer near the top of the annulus to be mixed with the recirculating flow from the separators.

Horizontal tube supports are provided at various elevations along the vertical straight tube bundle to provide lateral support. Additional tube supports or restraining systems are required in the upper U-shaped part of the bundle to position the tubes and prevent excessive flow-induced vibration.

The primary operating pressure is typically 2,235 psig (15.4 MPa gage) with the average temperature around 580°F (304°C) depending on the system supplier design. The secondary operating pressure is usually between 740 and 1,070 psig (5.1 and 7.4 MPa gage), again depending upon the system supplier design.

The steam generator is designed in accordance with Section III of the ASME Boiler and Pressure Vessel Code.

9.3 STEAM ENGINES
Revised by E. A. Avallone

REFERENCES: Ewing, "The Steam Engine and Other Engines," Cambridge. Ripper, "Steam Engine Theory and Practice," Longmans. Heck, "The Steam Engine and Turbine," Van Nostrand. Allen, "Uniflow, Back Pressure, and Steam Extraction Engines," Pitman. Peabody, "Valve Gears for Steam Engines," Wiley. Zeuner, "Treatise on Valve Gears," Spon. Spangler, "Valve Gears," Wiley. Dalby, "Valves and Valve Gear Mechanisms," Longmans.

EDITOR'S NOTE: Although largely relegated to history and nostalgia, small steam engines are still manufactured in the United States, primarily as replacement items. Steam locomotives in the United States cater only to the tourist trade; they are kept in repair and in service mainly with parts manufactured locally. The remaining steam locomotives in India are in the last stage of being phased out of service; all are scrapped and replaced by diesel, electric, or diesel-electric locomotives. China is still engaged in the manufacture of steam locomotives, and has based its railroad system on their continued use.

The reciprocating steam engine is a major, if not the major, hallmark in the pantheon of machinery most intimately associated with the profession and practice of mechanical engineering, and it was at the heart of the early industrial era. It dominated power generation for stationary and transportation service for more than a century until the development of the steam turbine and the internal-combustion engine. The mechanisms were numerous (see Patent Office listings), but practicality essentially standardized the positive-displacement, double-acting, piston and cylinder design in a vertical or horizontal configuration. These engines were heavy, cast-iron structures, e.g., 50 to 100 lb/hp; they had low piston speed (600 to 1,200 ft/min); long stroke (up to 6 ft); low turning speeds (50 to 500 r/min); steam conditions less than 300 lb/in² (dry saturated, or 100 to 200°F superheat); noncondensing or condensing (25 ± inHg vacuum); sized from children's toys to 25,000 hp. Diversity of valve gear was an inventor's paradise with many suitable for reversible operation, as in locomotive, rolling-mill, and ship applications. Most of the machine elements known today, such as cylinders, piston rods, crossheads, connecting rods, crankshafts, flywheels, and governors, were developed in steam engines.

These engines utilize the expansive power of steam. Theoretically, the more the steam can be expanded in the engine cylinder, the better will be the economy. Practical losses, which occur in every steam engine, limit the expansion ratio and result in a minimum steam rate for a definite degree of expansion. Cylinder dimensions preclude the practical utilization of high-vacuum conditions because of the high specific volume (e.g., 339 ft³/lb dry and saturated at 2 inHg abs). The steam turbine can effectively utilize maximum vacuum.

WORK AND DIMENSIONS OF THE STEAM ENGINE

Thermodynamic principles define the limits of the conversion efficiency of heat into work (see Rankine and Carnot cycles, Sec. 4). In fact, the historical development of thermodynamic principles was primarily aimed in the nineteenth century at defining the thermal performance of steam engines and steam power plants. It can be said that the steam engine did more for thermodynamics than thermodynamics did for the steam engine. Steam tables and steam charts (e.g., temperature-entropy; enthalpy-entropy; and pressure-volume) are essential to evaluate theoretical and actual equipment performance.

The pressure-volume diagram, or indicator card, is of primary significance in both the design and operation of the reciprocating piston and cylinder mechanism—not only steam engines, but also internal-combustion engines, air compressors, and pumps. The utility of the p-v diagram is often lost in attempts to improve fluid-dynamic reciprocating mechanisms. The work and power are the consequence of the difference in pressure on the two sides of the piston, expressed as mean effective pressure (mep or p_m). This is applied to the cylinder dimensions and rotating speed in the **"plan" equation**

$$\text{hp} = \frac{p_m L a n}{33,000} \tag{9.3.1}$$

where p_m = mep, lb/in²; L = stroke, ft; a = effective piston area, in²; n = number of cycles completed per min; 33,000 = mechanical equivalent of horsepower, ft · lb/min, by definition.

A typical steam-engine indicator card is shown in Fig. 9.3.1, identifying the established nomenclature for the events of the cycle. Superimposed is a theoretical diagram, with expansion, but without clearance or compression. The terminal pressure controls the steam, or water, rate. The term "back pressure" is used for pressures above the

Fig. 9.3.1 Typical steam engine indicator diagram.

atmosphere, while "condenser pressure" is used when engines operate with negative exhaust pressure. The volume ratio $v_j/v_h = R$ is called the **ratio of expansion.**

The average ordinate, or **mean effective pressure** p_m [applicable in Eq. (9.3.1)] for the ideal cycle diagram $ghjkl$ is

$$p_m = p[(1 + \log_e R)/R] - p \qquad (9.3.2)$$

The expansion phase h to j of the cycle is *logarithmic* where $pv = $ constant. It is not isothermal but reasonably approximates expansion (and compression) in actual engines where condensation and evaporation on cylinder walls, heads, pistons, and valves are cyclically recurrent.

The value of R is commonly around 4 in simple engines. It should increase as P increases and as p decreases, usually between 3 and 5. It should be higher in jacketed than in unjacketed engines. The efficiency of the engine depends largely on the value chosen for R. This is dictated by service requirements, e.g., load variation, load fluctuation, engine governing type (cutoff vs. throttle), overload capacity, steam cost, and economics. Efficiency, however, is not the sole requirement for all installations. The high starting torque of a steam locomotive dictates a valve gear that allows full-stroke admission of steam with zero expansion, a rectangular indicator card with consequent maximum p_m.

Figure 9.3.1 shows that the theoretical card has a larger area than the actual card. The value of p_m obtained from Eq. (9.3.2) must be multiplied by the **diagram factor** to obtain the actual p_m under the assumed conditions. This factor may have a value between 0.7 and 0.95. The actual p_m is obtained from the card drawn on the indicator drum under running conditions of the engine. The card area, graphically measured by a planimeter (see Sec. 16), is divided by the card length to get the average equivalent height of the card. The spring scale of the instrument, applied to this average height, gives the mean effective pressure actually obtaining within the engine cylinder. This value, when introduced in Eq. (9.3.1), gives the indicated horsepower (ihp) of the engine.

Losses in steam-engine cylinders are (1) incomplete expansion; (2) initial condensation; (3) throttling, affected by valve and port area; and (4) radiation, which can be considered constant. The point of best steam economy occurs with the p_m at which the total of all losses is a minimum.

Mechanical Efficiency and Shaft Output

$$\text{Brake horsepower (bhp)} = \text{ihp} \times \text{mechanical efficiency} \qquad (9.3.3)$$

$$\text{Friction horsepower} = \text{ihp} - \text{bhp} \qquad (9.3.4)$$

bhp	25	50	75	100
Friction hp	6	6	6	6
hp	31	56	81	106
Mechanical efficiency, %	81	89.5	92.5	94

Figure 9.3.2 shows the effect of mechanical efficiency and generator efficiency on the steam rate of an engine-generator set, both as to relative magnitude and as to location of the minimum values.

Engine economy may be improved by a number of methods: (1) **separation of inlet and outlet ports;** (2) **steam jackets,** applied to cylinders and heads to keep surfaces hot and dry; (3) **multiple expansion.** Condensation

Fig. 9.3.2 Engine steam rate curves (*a*) at the steam cylinder, lb/(ihp · h); (*b*) at the engine shaft, lb/(bhp · h); (*c*) at the generator, lb/kWh.

losses are related to the temperature difference existing in the cylinder. Cylinders in series reduce this temperature difference and allow more complete overall expansion of the steam. Two, three, and four cylinders in series, as in **compound-, triple-,** and **quadruple-expansion** engines were common constructions, but the successive improvement is smaller for each additional stage. (4) **Superheating** gives the vapor the properties of a gas, reduces cylinder condensation, and necessitates decisive changes in engine design. The overall improvement in performance and water rate is so substantial that superheat is prevalent in practice. (5) The **uniflow** arrangement was the last great improvement in design (Figs. 9.3.3 and 9.3.4). Its high economy results from the high temperature of the residual steam at the end of compression. This temperature, aided by jackets, is higher than the live steam temperature, so that initial condensation is reduced to a negligible amount. For **condensing operation** the engines are built with steam valves only (two-valve type), Fig. 9.3.3. Clearance pockets are provided, with either hand-operated or automatic valves to permit operation with atmospheric exhaust or against back pressure. **Noncondensing uniflow** engines have, in addition, auxiliary exhaust valves to reduce the otherwise large clearance required (four-valve type). If back pressure is variable, exhaust-valve *gears* are designed to change the length of compression while the engine is in operation (Fig. 9.3.4).

Fig. 9.3.3 Condensing uniflow engine cylinder, with clearance pockets.

Fig. 9.3.4 Noncondensing uniflow engine cylinder, with auxiliary exhaust valves.

Engine Steam (Water) Rates The efficiency of steam engines has generally been expressed in terms of pounds of steam per horsepower-hour. The term water rate was frequently used because of the convenient accuracy in weighing liquid water in the condensing plant. Figure 9.3.5 shows, on a percentage basis, two types of performance curves, one in lb/hph and the other in lb/h. The latter is identified as the "total steam" or **Willans line** and is straight for an engine with fixed cutoff and variable initial pressure. Figure 9.3.6 reflects the impact of steam pressure and superheat on the steam consumption of a condensing uniflow engine.

The **Rankine cycle** (see Sec. 4) is the accepted thermodynamic standard for comparing the performance of steam prime movers (engines and turbines). It is predicated on complete isentropic (reversible adiabatic) expansion of the steam from initial to back pressure. It is shown on the *p-v* basis in Fig. 9.3.7. There is no compression or clearance. The water rate of this cycle is most conveniently calculated by use of the Mollier chart (Sec. 4) where

$$\text{Rankine steam rate} = 2{,}545/(h_1 - h_2) \quad \text{lb/(hp · h)} \qquad (9.3.5)$$

Fig. 9.3.5   Actual steam (water) rate, lb/(hp · h), and Willans line, lb/h, for a noncondensing uniflow engine, percentage basis.

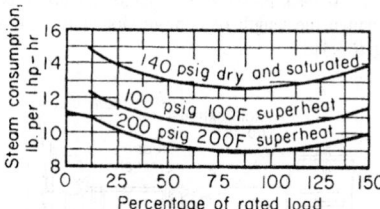

Fig. 9.3.6   Steam consumption of uniflow engine with 27.5-in vacuum.

and h_1 and h_2 are the initial and exhaust enthalpies, Btu/lb (constant entropy). The lowest point of the actual steam-rate curve (Fig. 9.3.6) is usually referred to as the Rankine rate, and the ratio of Rankine rate to actual rate is the **Rankine efficiency ratio (RER).** This ratio may vary from 0.5 to 0.9, depending on the type of engine.

When clearance, compression, and incomplete expansion are introduced, the methods of Sec. 4 should be used to evaluate steam rate. This involves steam tables and charts to find the net work of the indicator card from the positive and negative areas of the several component phases. Figure 9.3.8 shows a theoretical steam-rate curve, computed by such methods, together with the actual test curve for the engine.

Engine Details   Valves and valve-gear types range from the simplest D slide to gridiron, double-slide (Meyer or Rider), rocking, piston, releasing and nonreleasing Corliss, and poppet.

Fig. 9.3.7   Rankine cycle, *p-v* diagram.

Fig. 9.3.8   Theoretical and test steam rate curves. Uniflow engine 20×24 at 200 r/min; throttle conditions 150 lb/in² saturated steam; exhaust to atmosphere.

Volumetric clearance should be made as small as possible. Slide and piston valves bring clearance volumes to 12 to 15 percent, Corliss valves 6 to 10 percent, and poppet valves 4 to 8 percent.

Valve and Port Sizes   Flow area of valves and cross section of ports are usually determined by port area $= AS/C$ in², where A is net piston area, in², S is mean piston speed, ft/min, and C is a constant, ft/min. Values of C are approximately 9,000 to 15,000 ft/min for inlet and 6,000 to 7,000 ft/min for outlet. Small valves and ports represent lost work.

Superheated Steam   With high-temperature steam, the cylinder and parts design must allow for free expansion. Poppet and piston-valve cylinders can easily meet these requirements. The orthodox type of Corliss cylinder, however, is not suitable for highly superheated steam.

9.4   STEAM TURBINES
by Frederick G. Baily

REFERENCES: Stodola, "Steam and Gas Turbines," trans. by L. C. Loewenstein, McGraw-Hill. Bartlett, "Steam Turbine Performance and Economics," McGraw-Hill. Warren, Development of Steam Turbines for Main Propulsion of High-Powered Combatant Ships, *Trans. Soc. Naval Architects Marine Engrs.,* 1946. Newman, Modern Extraction Turbines, *Power Plant Eng.,* Jan.–Apr., 1945. Salisbury, "Steam Turbines and Their Cycles," Wiley. Campbell and Heckman, Tangential Vibration of Steam Turbine Buckets, *Trans. ASME,* **47,** 1925, pp. 643–671. Campbell, Protection of Steam Turbine Disk Wheels from Axial Vibration, *Trans. ASME,* **46,** 1924, pp. 31–139. Deak and Baird, A Procedure for Calculating the Packet Frequencies of Steam Turbine Exhaust Blades, *Trans. ASME,* **85,** series A, Oct. 1963.

Steam turbines have established a wide usefulness as prime movers, and are manufactured in many different forms and arrangements. They are used to drive many different types of apparatus, e.g., electric generators, pumps, compressors, and for driving ship propellers, through suitable gears. When designed for variable-speed operation, a turbine may be operated over a considerable speed range, which may be of advantage in many applications. Steam turbines range in output capacity from a few horsepower to over 1,500 MW. The largest ones are used for generator drive in central power stations.

Turbines are classified descriptively in various ways.

1. **By steam supply and exhaust conditions,** e.g., condensing, nondensing, automatic extraction, mixed pressure (in which steam is supplied from more than one source at more than one pressure), regenerative extractions, reheat.

2. **By casing or shaft arrangement,** e.g., single casing, tandem compound (two or more casings with the shaft coupled together in line), cross-compound (two or more shafts not in line, often at different speeds).

3. **By number of exhaust stages in parallel** as regards steam flow, e.g., two-flow, four-flow, six-flow.

4. **By details of stage design,** e.g., **impulse** or **reaction.**

5. **By direction of steam flow** in the turbine, e.g., axial flow, radial flow, tangential flow. In this country, radial-flow steam turbines have not been used; there are quite a few such machines abroad. Axial-flow units predominate; some small turbines in this country operate on the tangential-flow principle.

6. **Whether single-stage or multistage.** Small turbines, or those designed for small energy drop, may have only one stage; larger units are always multistage.

7. **By type of driven apparatus,** e.g., generator, mechanical, or ship drive.

8. **By nature of steam supply,** e.g., fossil-fuel-fired boiler, or light-water nuclear reactor.

Any particular turbine unit may be described under one or more of these classifications, e.g., a single-casing, condensing, regenerative extraction fossil unit, or a tandem-compound, three-casing, four-flow steam-reheat nuclear unit.

Turbine-Stage Design A turbine stage consists of a **stationary set of blades,** often called **nozzles,** and a moving set adjacent thereto, called **buckets,** or **rotor blades.** These stationary and rotating blades act together to allow the steam flow to do work on the rotor, which can be transmitted to the **load** through the **shaft** on which the rotor assembly is carried. Classical turbine-stage design recognized two distinct designs of turbine stage, "impulse" and "reaction" (see classification 4 above). In the **impulse** stage, the total pressure drop for the stage is taken across the nozzles or stationary element, the flow through the buckets or rotor blades then being substantially at constant static pressure. This may be extended to include flow through an additional set of stationary "intermediate" blades and another row of buckets, or rotor blades (Curtis or two row stages). See velocity diagrams, Figs. 9.4.1 and 9.4.2.

Fig. 9.4.1   Impulse turbine with single-velocity stages.

In the **reaction** stage, the total pressure drop assigned to the stage is divided equally between the stationary blades and the rotor blades, giving rise to a velocity diagram, as shown in Fig. 9.4.3a. As can be seen, there arises a marked difference in the shapes of the rotor blades in the two classical designs; the impulse buckets do much more turning of the stream; the reaction-bucket shape is more nearly the same as the nozzle-blade shape.

Fluid flow theory recognizes that only in rare cases can an axial-flow turbine stage be either pure impulse or pure reaction. The annulus following the nozzle exit is filled with steam flowing with a high tangential

Fig. 9.4.2   Impulse turbine with multivelocity stages.

velocity, i.e., a vortex, confined between inner and outer boundaries, and for equilibrium to exist, there must be a gradient in static pressure from a lower-than-average value at the inner boundary to a higher-than-average value at the outer boundary. The amount of this depends upon the boundary **radius ratio** R_{outer}/R_{inner}. Thus only for a radius ratio near 1.0 (small-height blades) can it be said that any one pressure condition exists for the stage. All axial-flow turbine stages of larger radius tend to be more nearly impulse at the inner diameter and more nearly reaction at the outer diameter.

Fig. 9.4.3a   Reaction turbine.

The designers of multistage steam turbines have continued to refine the efficiency of their steam paths over the years. For example, the 1950s saw the broad introduction of **twisted free-vortex staging.** Vane profiles have been improved since the 1950s by drawing on developments made in their aerodynamic laboratories and those of others. Complex computational fluid-dynamics computer codes based on true three-dimensional formulations of the inviscid Euler and viscous Navier-Stokes equations, and the supercomputers on which to run them, have recently become available. Manufacturing technology advancement permits the creation of steam-path components of almost any desired three-dimensional configuration. The result has been the introduction of **controlled-vortex staging** in the 1990s, in which the flow is biased toward the more efficient midsection of the blade, and secondary flow and profile losses are minimized by refinement in nozzle and bucket profiles, as illustrated in Fig. 9.4.3b.

Differences in basic mechanical construction of axial-flow turbine stages exist. Generally, the reaction type of turbine has continued as in the past with a "drum rotor" and stationary blades fixed in the casing, while the impulse type continues as a "diaphragm-and-wheel" construction. However, the mechanical construction bears no fixed relation to the degree of impulse or reaction adopted in the blading design. Designers employ the mechanical construction which they deem suitable for best reliability and efficiency.

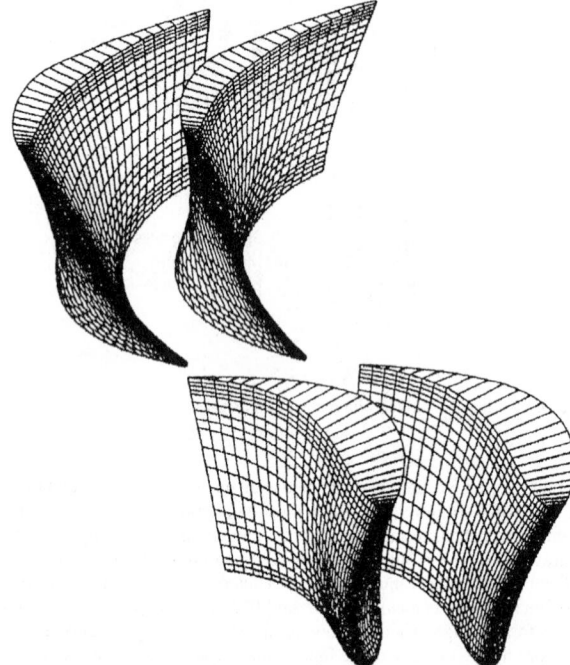

Fig. 9.4.3b Controlled vortex stage. (*© General Electric Company, 1994. All rights reserved. Used with permission.*)

An impulse stage, when employed for the first expansion, permits nozzle group control, i.e., steam admission to each group, opening and closing, successively, in response to load changes. This improves the efficiency at low loads through the reduction of the loss due to throttling.

A greater enthalpy drop can be employed per stage with the impulse element, particularly in the case of multivelocity elements, thus reducing the number of stages in a turbine. This is of special importance in the first expansion if the nozzle chamber is not integral with the turbine casing, so that the casing is not subjected to the initial steam conditions. If turbine elements of equal blade speed could have equal efficiency, one 2 row impulse element would equal four 1 row impulse elements or 16 rows (8 pairs) of reaction elements.

General Advantages of Steam Turbines Compared with reciprocating engines steam turbines require less floor space, lighter foundations, and less attendance; have a lower lubricating-oil consumption, with no internal lubrication, the exhaust steam being free from oil; have no reciprocating masses with their resulting vibrations; have uniform torque; have no rubbing parts excepting the bearings; have great overload capacity, great reliability, low maintenance cost, and excellent regulation; are capable of operating with higher steam temperature and of expanding to lower exhaust pressure than the reciprocating steam engine. Their efficiencies may be as good as steam engines for small powers, and much better at large capacities. Single units can be built of greater capacity than can any other type of prime mover. Small turbines cost about the same as reciprocating engines; larger turbines cost much less than corresponding sizes of reciprocating engines, and they can be built in capacities never reached by reciprocating engines. Combustion-gas turbines possess many of the advantages of steam turbines but are not available in ratings much exceeding 180 and 265 MW for 60- and 50-Hz service, respectively.

STEAM FLOW THROUGH NOZZLES AND BUCKETS IN IMPULSE TURBINES

Nozzles For the general treatment of the flow of steam and for maximum weight of flow of saturated steam, see Sec. 4.

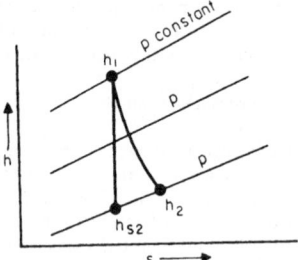

Fig. 9.4.4 Enthalpy-entropy diagram.

The **theoretical work** obtainable from the expansion of 1 lb of steam is equal to the enthalpy drop in isentropic expansion $h_1 - h_{s2}$, in Btu/lb, and the spouting velocity in $223.8 \sqrt{h - h_{s2}}$ ft/s (m/s $= 44.7 \sqrt{h'_1 - h'_{s2}}$, where h' is in kJ/kg). The actual expansion is not isentropic but follows a path such as $h_1 h_2$ on the enthalpy-entropy diagram (Fig. 9.4.4), and the available work becomes $h_1 - h_2$.

The **nozzle efficiency** is $(h_1 - h_2)/(h_1 - h_{s2})$.

The required **throat area** of the nozzle is $A_t = Wv_t/V_t$, and the **mouth area** is $A_m = Wv_m/V_m$, where v is specific volume, V is velocity, and the subscripts relate to throat and mouth, respectively

In the case of a subcritical pressure ratio, throat and mouth conditions will coincide; with a supercritical pressure ratio, the mouth area will be larger than the throat area. For **nozzle velocity coefficients** based on tests, see Keenan and Kraft, *Trans. ASME*, **71**, 1949, pp. 773–787. The **bucket-velocity coefficient** is the ratio of the average exit velocity from the bucket divided by the velocity equivalent of the total energy available to the bucket, i.e., the sum of inlet-velocity energy and pressure-drop energy. Typical values of tests on impulse buckets are shown in Fig. 9.4.5, when the bucket inlet angle is 3 to 5° larger than the exit angle. These are for subsonic flow. Reaction buckets can have the same coefficients as nozzles. For intermediate cases between pure impulse and pure reaction, interpolate between the values for these limiting cases.

Fig. 9.4.5 Impulse-bucket velocity coefficients.

If, in the velocity diagram for a usual turbine stage, V_1 = actual nozzle exit velocity, V_2 = bucket relative entrance velocity, V_3 = bucket relative exit velocity, V_4 = absolute leaving velocity, V_0 = velocity corresponding to total stage available energy, then the **diagram efficiency** is $(V_1^2 - V_2^2 + V_3^2 - V_4^2)/V_0^2$.

Diagram efficiencies calculated with nozzle and bucket coefficients given above will be higher than the efficiencies derived by turbine tests, because of losses not existing in stationary tests of nozzles and buckets.

For supersonic bucket velocities, the impulse bucket-velocity coefficient is lower by the factors in Table 9.4.1.

Table 9.4.1 Impulse Bucket-Velocity Coefficient Factors

Mach no.	<	1.0	1.2	1.3	1.5	1.75	2.0
Factor		1.0	0.997	0.995	0.978	0.928	0.816

Turbine Stage Efficiency Single-row stages of short blade length have relatively lower efficiency, owing to inner and outer sidewall losses; stages with longer blades are therefore higher in efficiency. Figure 9.4.6 shows typical values of stage efficiency for single-row stages, plotted against the wheel-speed steam-speed ratio, with pitch diam, inches/nozzle area, square inches, as a parameter. These curves reflect the net total of losses: (1) friction losses in nozzles (stationary blades); (2) friction losses in buckets (rotor blades); (3) rotation loss of rotor; (4) leakage loss between inner circumference of stationary element and rotor; (5) leakage loss between tip of rotor blades and casing; (6) moisture and supersaturation losses, if steam is wet (not included in curves of Fig. 9.4.6).

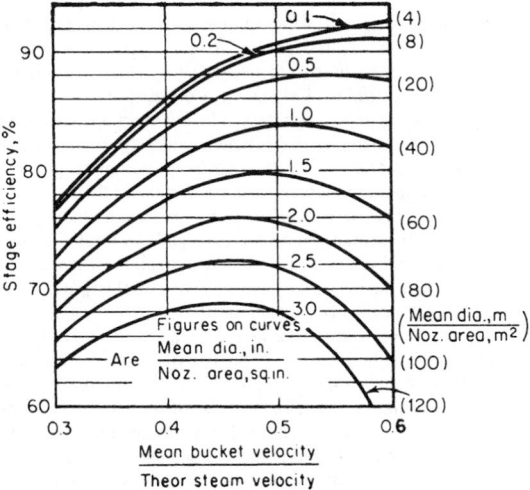

Fig. 9.4.6 Turbine single-row stage efficiency.

Nozzles and bucket friction losses are minimized by good aerodynamic design and by increasing **aspect ratio** (blade length/steam passage width). Rotation losses depend upon disk or rotor dimensions and surrounding stationary parts. Exact values for rotation loss depend on several factors; the following formulas may be relied upon for all usual purposes:

$$L_d = 0.042D^2w(U/100)^{2.9} = (1.4 \times 10^{-12})(D')^2w'(U')^{2.9}$$
$$L_b = 0.187Dwh^{1.25}(U/100)^{2.9}$$
$$= 3.34 \times 10^{-11}D'w'(h')^{1.25}(U')^{2.9}$$

where L_d = rotation loss of disk carrying buckets, kW; L_b = rotation loss of one row of buckets, kW; U = wheel speed at pitch diameter, ft/s; U', m/s; D = pitch diameter (at center line of nozzle), ft; D', mm; w = density of steam, lb/ft³; w', kg/m³; h = mean bucket height, in; h', mm. L_b must be figured for each row of buckets. L_d plus the sum of the values of L_b gives the total rotation loss in kilowatts for dry saturated steam. The formula for L_b is approximate.

Leakage loss of steam between inner circumference of stationary element and rotor is minimized by maintaining minimum practical clearance and by use of labyrinth packings (see Figs. 9.4.14 and 9.4.15 and accompanying text). Leakage loss of steam between tip of rotor blades and casing is similar to that through labyrinths between shaft and stationary parts; the magnitude depends upon the clearance area and the amount of reaction; other things being equal, the larger the percentage of reaction, the larger the leakage (see Fig. 9.4.9). Thus designers often employ considerable amounts of reaction in stages with long blades where such losses are small; this improves the net efficiency. In stages with short blades, the best net efficiency obtains with near impulse design. The curves in Fig. 9.4.6 reflect the effects of these practices.

The presence of moisture in the steam causes extra losses. These are probably mainly due to three factors:

1. Effect of supersaturation; i.e., the steam in expanding rapidly does not remain in equilibrium but tends to be more or less supercooled; thus less than theoretical equilibrium energy is available.

2. The presence of water drops increases friction losses in the steam itself.

3. Water drops tend to move more slowly than the vapor; they strike the rotor blades at unfavorable velocities and exert a braking effect.

Figure 9.4.7 gives correction factors which may be applied to the values from Fig. 9.4.6 to arrive at stage efficiencies in the "wet" region. Curves are identified by initial superheat, °F, or by initial quality, percent.

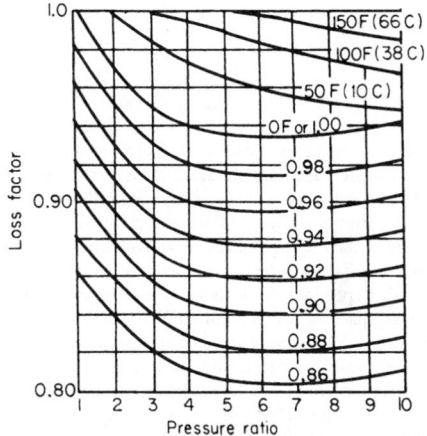

Fig. 9.4.7 Supersaturation and moisture loss.

Two-row stages, with one set of nozzles, have lower basic efficiency than single-row stages; they are useful for the first, or governing, stage in small to medium units. They are no longer employed in large central station designs. The approximate relative efficiency level of these stages is shown in Fig. 9.4.8.

Fig. 9.4.8 Variation of the efficiency of turbine elements with velocity ratio.

The losses occurring in a turbine stage are partially recoverable in succeeding stages in a multistage turbine because the energy available to succeeding stages is increased above that resulting from isentropic expansion. The amount of this factor depends upon the temperature at which the loss takes place, and the turbine exhaust temperature. Values for the reheat or energy regain factor are shown in Fig. 9.4.9.

Turbine Steam-Flow Requirements The steam flow required by a turbine is related to its power output and steam conditions by the following expressions:

Flow, lb/h = (TSR/efficiency) × power output, hp or kW

Fig. 9.4.9 Energy regain chart.

where TSR = theoretical steam rate, lb/(hp · h) = 2,545/available energy, Btu/lb, or TSR, lb/kWh = 3,412.14/available energy, Btu/lb.

Values of TSR are given in tables or may be calculated. In case of mixed-pressure or extraction turbines, the various sections of the turbine where flows are added or subtracted must be treated separately.

Turbine Steam-Path Design   The basic quantities required for this are the steam conditions (i.e., inlet pressure and temperature and exhaust pressure), the required flow (see above), and the turbine speed. The latter is often fixed by the requirements of the driven machine; if not, the choice of speed by the designer is based upon experience, and factors such as space or weight limitations, efficiency requirements, stress limits, or required exhaust area. Often several preliminary layouts are needed to arrive at the best design. If it appears that a single stage will suffice, the problem is simple, since the entire available energy is allotted to the one stage. If a multistage machine is required, the total available energy must be divided properly between the various stages.

For a given wheel-speed/steam-speed ratio, the energy to be allotted to the stage is directly proportional to the square of the product (r/min × D), D being the rotor mean diameter. If available energy AE is in Btu/lb, D, in, and velocity ratio, 0.50, then

$$AE = (r/min \times D)^2/(6.57 \times 10^8)$$

The sum of the AE values per stage for all the stages must equal the total energy on the turbine, which is greater than the isentropic available energy by the amount of the reheat factor.

Having determined the energy to be assigned to each stage, the steam pressure in each stage is fixed, and this, together with the enthalpy determined from the efficiency of the machine from inlet, determines the steam specific volume. The velocity ratio and the degree of reaction decided upon fix the velocity diagram, from which the velocities through the stationary and rotating blades are determined. With the flow Q, lb/h, the velocity V, ft/s, and the volume v, ft³/lb, determined, theoretical areas required are given by

$$A = Qv/(25V)   \text{in}^2$$

The actual area required will be larger than theoretical because of friction losses; a value of 0.98 for nozzle flow coefficient is reasonable.

There is no fixed criterion for the number of stages to be used in a steam-turbine design, and experienced designers differ in the number they will choose for any particular design. The stage velocity ratio, the mean diameter, and often the r/min are subject to judgment, bearing in mind the general relationships shown in Figs. 9.4.6, 9.4.7, and 9.4.9. Cross-compound arrangements are possible, allowing higher speed for the high-pressure section, where steam volume flow is smaller, and a lower speed for the low-pressure section, where the final exhaust area desired may require long blades on a large diameter.

A detailed calculation of the efficiencies and energy outputs of each stage can be summed up to the total "internal used energy" of the turbine, which, when divided by the isentropic available energy, results in an **"internal efficiency."** Then, account must be taken of other losses to arrive at the turbine overall efficiency. These losses are

1. Exhaust loss, i.e., the kinetic energy corresponding to the absolute velocity of the steam leaving the last stage, plus the pressure drop

through the exhaust connection to the turbine outlet flange, where exhaust pressure is by custom measured as a static pressure.

2. Pressure drops through interconnecting piping if turbine has more than one casing
3. Shaft end-packing leakages
4. Valve-stem leakages, if any
5. Inlet-valve and intermediate-valve pressure-drop losses
6. Bearing, oil-pump, and coupling power losses
7. If the turbine drives a generator or gear, the losses of these elements

LOW-PRESSURE ELEMENTS OF TURBINES

In condensing turbines expanding to high vacuum, the steam increases in specific volume as it passes through the stages, exhausting at about 1,000 times the inlet volume. The increase in specific volume from stage to stage is greatest in the latter few stages, which are commonly designed for pressure ratios of about 2:1. Considerations of efficiency and economics dictate providing reasonably low interstage steam velocity in these stages, and reasonable leaving loss from the last-stage blades. These requirements are satisfied by providing a steam path whose cross-sectional area increases in proportion to the specific volume. The area increase may be achieved by increasing blading length, by increasing the mean diameter of the steam path, by arranging the last expansions in multiple flow, or by some combination of two or more. The relatively small single casing unit of Fig. 9.4.10 is provided with increasing area by the first two techniques.

The ratio of the last-stage blade height to the mean diameter of the steam path may be as high as 0.35 in the last stage of large central station units. With such a ratio, there is a variation in blade velocity between the inner and the outer radii of the steam path that cannot be properly satisfied by any one steam velocity. Losses incidental to this may be minimized or eliminated by providing blades with warped surfaces, the sections at the inner radius partaking of impulse form with relatively small inlet angles, and those at the outer radius of reaction form with large inlet angles. The longest last-stage blades in service today have tips whose speed exceeds sonic velocity. Many are of transonic design employing subsonic profiles toward the inner radius, blending into supersonic profiles at the tip.

Among the methods of obtaining increased low-pressure blade areas through multiple flowing are:

1. A single-casing turbine with the last elements arranged for double flow.
2. The steam expansion divided between two casings, coupled in tandem, and driving a single main generator, the low-pressure casing being arranged for double flow. This arrangement is commonly extended to provide tandem-compound turbines with four and six exhaust ends. Figure 9.4.11 illustrates the application of a double-flow low-pressure casing to a unit of medium rating.
3. Cross-compound turbines, in which the steam expansion is divided between two or more separate casings driving separate generators, electrically synchronized. The low-pressure casings are usually double-flow and can be arranged so as to provide two, four, or six exhaust ends. This system permits the turbine elements to be operated at different speeds, selected as appropriate to the respective steam volumes. It lends itself to geared applications, such as marine propulsion machinery, when two or more pinions of different diameters and speeds drive a single-output gear wheel.
4. Divided-flow turbines in which steam expands in a series of elements in a single flow to a point where the flow is divided, with, perhaps, one-third continuing expansion within the same casing to condenser pressure, the remaining two-thirds expanding to condenser pressure in a separate double-flow casing. This construction has been used to provide triple-flow exhaust ends.

The **leaving loss** at the exit from the last row of blades is $V_{c2}^2/50,100$ Btu/lb of steam, where V_{c2} is the absolute terminal velocity in feet per second [or $(V_{c2}')^2/2,000$ kJ/kg, where V_{c2}' is in m/s]. The presence of moisture in the steam causes **moisture loss.** The acceleration of

Fig. 9.4.10   Cross section of a multistage impulse condensing turbine rated at 30,000 kW. *(General Electric Co.)*

moisture particles is less as the density of the steam decreases; hence the difference between the velocities of the particles and the steam increases. As indicated in Fig. 9.4.12, with steam velocity, V_{s1} and moisture velocity V_m leaving the stationary blades and with bucket velocity V_b, the velocities relative to the moving blades of the steam are V_{c1} and of the moisture V_{m1}. The component V_{m2} of the moisture relative to the moving blades is opposite to the direction of their motion and is proportional to the force acting on the back of the blades, needed to accelerate the water to blade speed, and results in negative work. This negative work can be calculated when the weight of moisture per pound of steam and V_m are known. The results of many tests indicate that the *efficiency of a stage is reduced about 1 percent for each 1 percent moisture present in the steam.*

The presence of moisture particles will result in erosion of the blades along their inlet edges if V_{m1} is too large—unless the moisture content is very small. The rate of erosion is reduced by using materials of inherently high erosion resistance, or protective shielding of hardened materials along the inlet edge. Attached Stellite shields and thermally hardened edges are commonly employed. Protective materials and processes are carefully chosen to minimize the possibility of corrosion.

Experience with the usual 12-chrome alloy bucket steels indicates that a threshold velocity exists at about $V_b = 900$ ft/s (270 m/s), below which impact erosion does not normally occur. With alloys specifically selected for their erosion resistance and with Stellite shields, satisfactory service has been obtained with values of V_b in excess of 1,900 ft/s (580 m/s). Centrifugal stress limits the application of steel blades to a tip velocity of about 2,000 ft/s (610 m/s). Titanium alloys provide high strength, low density (and hence low centrifugal stress), combined with excellent inherent moisture erosion resistance. Titanium bucket designs are available with a tip speed of 2,200 ft/s (670 m/s), without the need for separate erosion protection shielding. The severity of erosion penetration is dependent upon the thermodynamic properties of the stage and the effectiveness of reducing moisture content by means of interstage collection and drainage as well.

Rotative Speed   The speed selected greatly influences weight and cost. With two geometrically similar turbines, one having twice the linear dimensions of the other, the steampath areas of the larger, and hence its capacity, would be 4 times and its weight 8 times as great as those of the smaller. The weight per unit of capacity with similar machines increases inversely as the speed with strict geometrical similarity. For this reason, the highest possible low-pressure blade speeds and r/min are selected. Machines of different speeds are not usually made strictly geometrically similar, and the reduction of specific weight is not so rapid as the above rule would indicate. With large-capacity turbines, blade speeds, at the outer radius, can reach 2,200 ft/s (670 m/s). With high speeds and small dimensions, the turbine can operate with higher steam temperature and greater temperature fluctuations because of lighter casing walls and less mass of rotor; the amount of distortion is less with more uniform heating; the turbine can be heated and put in service more quickly; space requirements are less; and dynamic loadings on foundations are less.

Balancing (See Secs. 3 and 5) With a rotor, or a component of a rotor, of relatively short axial length (such as a disk), static balancing may suffice. Single bodies of more than half the diameter in axial length are usually dynamically balanced by the use of balancing machines. The balancing may be done at less than the running speed, since a bladed turbine rotor cannot be rotated in air at a speed approaching the running speed, or at high speed in an evacuated spin facility, or with a combination of both. Balance at full speed may not be satisfactory unless the balance weights are applied at points diametrically opposite the errors in balance, so that balance corrections must frequently be provided along the length of the rotor. Five balance planes are commonly used for the rotors of large central-station turbines, of which three are in the rotor body between bearings and two are in the overhung couplings.

Unsatisfactory operation of turbine rotors in service, resembling a simple imbalance, may be caused by nonuniform material or nonuniform heating of the rotor. The latter may be caused by permitting a rotor to remain stationary in a hot casing, or by packing rubbing which may apply frictional heating to the "high" side of the rotor, leading to further bowing into the rub. Care must be taken to see that nonuniformity of material which could cause rotor distortion with heat is avoided, and to avoid rubbing of the packings on the rotor. Turning gears are generally

Fig. 9.4.11 Cross Section of a 160,000-kW tandem-compound, double-flow reheat turbine. (*Westinghouse Electric Corp.*)

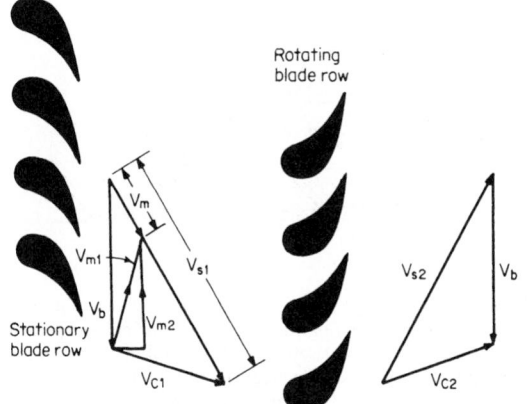

Fig. 9.4.12   Velocity of steam and of the moisture in steam.

used to keep the rotor turning at low speed to maintain uniform temperature when the turbine is shut down and cooling, and for a prolonged period before starting.

TURBINE BUCKETS, BLADING, AND PARTS

Figure 9.4.13 shows various blade fastenings. Blades are subject to vibration and possible fatigue fracture if their natural frequency is resonant with some applied vibration force. Forced vibrations may arise from the following causes (see also Sec. 5):

Fig. 9.4.13   Steam-turbine blade fastenings.

1. Variations in steam forces. The blade frequencies should not be even multiples of the running speed, nor should they be resonant with the frequency of passing nozzle partitions or exhaust-hood struts.

2. Shock, the result of blades being subjected to discontinuous steam flow, such as may be caused by incomplete peripheral steam admission or extraction.

3. Torsional vibrations of the rotor.

High-speed, low-pressure blades of condensing turbines are usually of tapering section and have a warped surface in order to provide appropriate blade angles throughout their length. Long blades of this type frequently have their natural frequency between 3 and 4 times the running speed or even lower. Such blades should always be specially tuned to have a margin in frequency away from running-speed stimuli.

Margins from running speed to assure freedom from fatigue due to resonant vibration and transverse to the plane of the wheel are as follows:

Frequency, cycles per revolution	2	3	4	5
Margin between critical and running speeds, %:				
Within wheel plane (tangential)	15	10	5	5
Transverse to wheel plane (axial)	20	10	10	5

Higher-frequency buckets whose frequencies cannot assuredly be made nonresonant should be designed with adequate strength to resist such stimuli as may occur under service conditions.

Blade Materials   The material in most general use is a low-carbon stainless steel of the following composition: Cr, 12 to 14 percent; C,

0.10 to 0.12; Mn, 0.08 max; P, 0.03 max; S, 0.05 max; Si, 0.25 max. Its physical characteristics in the heat-treated condition at room temperature may be tensile strength, 100,000 lb/in^2 (690 MPa); yield point, 80,000 lb/in^2 (550 MPa); elongation, 21 percent; reduction of area, 60 percent. For the higher-temperature blades, particularly on large machines, it is practice to use alloyed chrome steel to achieve the required strength and oxidation-erosion resistance (see also Sec. 6).

Rotor Materials   Since steam turbines operate at high speeds, rotor materials must be of very high integrity and of basically high strength. In addition, the material should be "tough" at the temperatures at which it is to be highly stressed. A measure of this toughness may be obtained by running Charpy notch impact tests at various temperatures (see Sec. 5). In large modern machines the rotor forgings are almost exclusively made of steel melted in basic electric furnaces and vacuum poured to achieve freedom from internal defects. Turbine rotor forgings are usually made with small amounts of alloying elements such as Ni, Cr, V, or Mo.

Casing and Bolting Materials   High-temperature and -pressure casings are almost always made of castings in order to achieve the complicated shapes required by these components. The alloy compositions used are selected so as to provide good weldability and castability as well as good physical properties. Low-temperature and -pressure casings are usually fabricated from steel plate. Bolts are made of forged or rolled materials.

The practical use of higher steam pressures and temperatures is limited by the strength and cost of available materials.

Leakage   Metallic labyrinth packings are employed to (1) reduce internal steam leakage from stage to stage, (2) prevent steam from escaping the turbine from elevated-pressure ends, and (3) prevent air from leaking into the turbine at subatmospheric-pressure shaft ends. Interstage packings usually employ single rings with multiple teeth. End packings are arranged in multiple rings. At the high-pressure end, the leakage of steam past the first few rings may be carried to a lower-pressure stage of the turbine so that the outer rings need only prevent the leakage of low-pressure steam to atmosphere. The annulus above the last ring, or group of rings, is connected to a packing exhauster which maintains a pressure slightly below atmospheric. In consequence, no steam leaks out along the shaft, while a small amount of air is drawn past the final packing to the exhauster. Vacuum packings are provided with an inner annulus which is supplied with steam above atmospheric pressure. Steam flows inward from that annulus to supply the leakage toward the vacuum end, and outward toward a packing-exhauster connection. Thus steam is prevented from escaping along the shaft, while air is prevented from being drawn into the turbine.

Some turbines use carbon end packings or water seals. Carbon packings consist of one or more rings of pure carbon made in sections of 90 or 120° and held toward the shaft with small clearances by means of springs. The springs should have an axial component of force to hold the rings against the side of the box.

Labyrinths dependent upon radial clearances are shown in Fig. 9.4.14. In the design with "high and low" teeth, heavy teeth are cut on the turbine shaft, and thin teeth are part of a renewable packing ring which is made in segments backed and held inward by flat springs. The key indicated in the figure prevents turning of the segments. These types require that the rotor remain sensibly concentric with the stator but do not require a close axial adjustment.

Labyrinths dependent upon axial clearances are shown in Fig. 9.4.15a. These require the maintenance of a close axial adjustment of the rotor.

The flow through a labyrinth may be approximately determined by the formula

$$W = 25KA\sqrt{\frac{(P_1/V_1)[1 - (P_2/P_1)^2]}{N - \ln(P_2/P_1)}}$$

where W = mass flow of steam, lb/h; K = experimentally determined coefficient; A = area through packing clearance space, in^2; P_1 = initial pressure, lb/in^2 abs; V_1 = initial specific volume of steam, ft^3/lb; P_2 = final pressure, lb/in^2 abs; N = number of throttlings.

Fig. 9.4.14   Labyrinths with radial clearance.

The value of K for interlocking labyrinths where the flow velocity is effectively destroyed between throttlings is approximately 50 and is independent of clearance for usual clearance values.

Fig. 9.4.15a   Labyrinths with axial clearance.

For noninterlocking labyrinths, i.e., stationary teeth against a straight cylindrical shaft, the value of K varies with the ratio of tooth spacing to radial clearance, being about 120 for tooth spacing of five times the radial clearance, reducing to approximately 50 for a tooth spacing fifty times the clearance.

Brush Seals   The 1990s and 2000s have seen the application of brush seals to steam turbines. They consist of a multiplicity of fine wires inclined at an angle of approximately 45° to the surface of the shaft or to the bucket covers. The use of resilient bristles permits closer clearances than do the rigid teeth of a labyrinth seal, and the resulting leakage flow is typically one-third that of the labyrinth. Fig. 9.4.15b. shows brush seals incorporated into the bodies of two labyrinth packing rings in series, as is frequently retrofitted to existing high-pressure units. Flow and pressure drop from left to right. The bristles are shown supported by a backing plate on the downstream side and protected by a forward plate on the upstream side. Both plates and the bristles are welded into segments for the same arc length as the labyrinth segments.

Fig. 9.4.15b   Brush seal installation.

Turning Gears   Large turbines are equipped with turning gears to rotate the rotors slowly during warming up, cooling off, and particularly during shutdown periods of several days when it may be necessary to start the turbine again on short notice. The object is to maintain the shaft or rotor at an approximately uniform temperature circumferentially, so as to maintain straightness and preserve the balance. Turning gears permit an appreciable reduction in starting time, particularly following a relatively short shutdown.

It is seldom necessary to use high-pressure oil to lift the journals off their bearings when using a turning gear. A low-pressure motor-driven oil pump is used which floods the bearings with about half their usual flow of oil. The turning gear is made powerful enough to start the rotor and rotate it at low speed.

High-Temperature Bolting   The bolting of high-pressure, high-temperature joints, particularly turbine-shell or valve-bonnet joints, is very exacting. It is worthwhile to taper the threads of either the male or the female element so that the engagement of the threads throughout the length of the engaged thread portion will give approximately uniform bearing. The reliability of taper-threaded bolts is superior to that of parallel-threaded bolts.

Thrust Bearings   They must be designed to carry axial rotor thrust in either direction, with sufficient margin to take care of unusual operating conditions. Thrust runners may be machined solid on the shaft or can be a separate piece shrunk on and secured from endwise motion. The stationary bearing surfaces may be of the pivoted-shoe type, or made in solid plates with babbitt or other bearing-material facing, with grooves for oil supply and "lands" to carry the thrust load.

Axial thrust on the turbine rotor is caused by pressure and velocity differences across the rotor blades, pressure differences from one side to the other on wheels or rotor bodies, and pressure differences across shaft labyrinths which have steps in diameter. The net thrust is the sum of all these effects, some of which may be in one direction, and some in the opposite direction. Rotor-blade and wheel or body thrust are usually in the direction of steam flow. It is usual to balance this thrust either partially or completely by proper choice of shaft packing diameters and pressure differences so that the net thrust is not too large. The thrust bearing must be made large enough so that it is not overloaded by the net thrust. In this respect it is necessary to foresee all operating conditions which may influence the net thrust and allow for these; in addition some margin must be allowed for abnormal or unforeseen circumstances which may occur in service. (See also Sec. 8.)

Controls   Steam turbines are nearly always equipped with speed-control governors, and with separate overspeed governors. The only exceptions are special cases where it is judged that the possibility of overspeed due to loss of load is exceedingly remote. The speed-control governor may be arranged for a wide range of speed setting in the case of a variable-speed turbine. The steam flow-control valve or valves are operated by this governor, usually through a hydraulic relay mechanism. The overspeed governor is usually of an overisochronous type, arranged to trip at 10 percent over normal full speed (on some small turbines, 15 percent), actuating quick-closing stop valves to shut off the steam supply to the turbine. Speed-control governing systems are usually designed so that the overspeed-governing system is not brought into action on sudden loss of full load.

On automatic extraction machines, the speed governor must be correlated with the extraction-pressure controlling-valve system.

On fossil-reheat turbines, because of the large stored steam volume in the reheater and piping, and on nuclear turbines, because of the moisture separator/steam reheater and piping volumes, it is necessary to protect the turbine from overspeed on sudden loss of load by shutting off this stored steam ahead of the lower-pressure stages. It is done by intermediate intercept valves, actuated by a governor set slightly higher than the speed-control governor. An intermediate stop valve actuated by the overspeed trip is usually added in series for additional protection.

Speed-governing systems can be supplied of various sensitivities and speed ranges, to suit the requirements of the driven apparatus. Both mechanical-hydraulic and electrohydraulic systems are employed. Analog electrohydraulic systems were introduced during the 1960s as electronic components of sufficient reliability for turbine service became available. Digital systems were introduced in the 1980s and have become the standard control technology. Modern systems control transient thermal stress and the expenditure of low-cycle fatigue life of high-temperature components, in addition to controlling speed and load. (See "Steam Temperature—Starting and Loading.")

Steam turbines are usually provided with a supervisory instrumentation system. That for a large central station unit may process several hundred channels of information, providing alarms and/or trips for abnormal parameters. Data may be stored on paper charts, on magnetic media, or in computer memory depending upon the design of the system. Displays frequently employ color cathode-ray tubes.

INDUSTRIAL AND AUXILIARY TURBINES

Low-capacity turbines are employed for services such as engine-room auxiliaries and small generating sets. Usually they comprise a single turbine element. Their efficiency may be less than that of a corresponding reciprocating engine, but they are employed because of their compactness and because they require no internal lubrication. The exhaust steam is free from oil and grease and is available for heating purposes. They are frequently coupled to the driven machine by means of speed-reducing gears. Turbines of this type are usually of the axial-flow type, but the **tangential helical-flow** turbine, in which the steam is directed tangentially and radially inward by nozzles against buckets milled in the wheel rim and made to flow in a helical path reentering the buckets one or more times, is also used. Such machines have generally been limited to small single-stage designs, and are very simple and rugged.

In **back-pressure turbines,** the exhaust steam is employed for some heating process, and the turbine work may be a by-product. If all the exhaust steam is condensed in heat-absorbing apparatus and returned to the system, the thermal efficiency of the system may be over 90 percent. One application is the **superposition** of a high-pressure system on lower-pressure power units, with the exhaust from the high-pressure turbine power units, with the exhaust from the high-pressure turbine going to the low-pressure steam mains. By this device, an old power station can be rehabilitated and its capacity increased. Two methods of operation are in use: (1) with constant intermediate pressure as when the lower-pressure power units operate also with steam from existing lower-pressure boilers and (2) with variable intermediate pressure as when the low-pressure units receive steam only from the back-pressure turbine.

Boiler-feed-pump-drive turbines have been used extensively as part of the power-plant system, especially for large, high-pressure plants where the required feed-pump power may amount to 4 percent of the gross plant output, and for large nuclear units. The turbine and pump can be matched as to rotative speed. These turbines are variously integrated into the main cycle. The most common present practice is to use condensing, nonextracting turbines supplied with steam in the range of 150 to 200 lb/in² abs (1,000 to 1,400 kPa) taken from the exhaust of the intermediate sections of fossil turbines, or from the inlet of the low-pressure sections of nuclear units. These turbines normally have a connection to the main steam supply for starting and low-load operation. Similar auxiliary turbines are frequently used to drive the forced- or induced-draft fans of large fossil-fuel-fired boilers.

With **extraction turbines,** partly expanded steam is extracted for external process use at one or more points. The turbines may be either condensing or noncondensing. Extraction turbines are usually designed to sustain full rated output, with or without extraction, and are provided with automatic regulating mechanisms to deliver steam from the extraction points at constant pressure, as long as there is sufficient power load to permit the necessary flow. The use of such extraction turbines, particularly with high initial pressures in connection with many industrial processes requiring moderate- or low-pressure steam, results frequently in a high efficiency of power production, i.e., the only heat required in such a plant over and above that to provide the required process steam is the heat equivalent of the power generated by the steam before extraction. This means that such power can be produced at nearly 100 percent thermal efficiency.

Figure 9.4.16 illustrates a typical double-automatic, condensing extraction turbine, providing two controlled extraction pressures. In this case, the unit is equipped with internal spool valves at both extraction points. Grid and poppet-type valves have been used for this purpose.

Fig. 9.4.16   Double automatic condensing extracting turbine, 25,000 kW. *(General Electric Co.)*

Fig. 9.4.17a A 16,000-hp cross-compound marine turbine designed for steam conditions of 600 lb/in2 gage, 850°F, 1½ inHg abs. High-pressure section for 6,550-r/min normal speed. Low-pressure section in Fig. 9.4.17b. *(General Electric Co.)*

The extraction-stage valves are under the control of an extraction-pressure-actuated governor; they determine the flow to the subsequent stages of the turbine and maintain the pressure in the extraction stage. The operation of the valves is by means of a pilot valve controlling the admission of high-pressure fluid to an actuating cylinder, which, in turn, opens or closes the valves to the nozzle ports to the succeeding stage. Extractions of this kind are called pressure-controlled extractions, and the pressure is maintained practically constant over a wide range if the load is sufficient to permit the required steam flow.

Mechanical-drive turbines are commonly applied where moderate to high power and/or precise speed control of the driven machine are needed. Typical applications include the powering of papermaking machines and the driving of fluid compressors in petrochemical plants. So many sizes and types are available from the manufacturers, and they have been adapted to so many applications, that it is impossible to give here more than a general description. These turbines are commonly built in sizes from a few to several thousand horsepower. If the speed of the driven machine is low, a reduction gear may be used in order to reduce the size and cost of the driving turbine and to improve its efficiency. Mechanical-drive turbines have a wide range of application, being adaptable for any steam conditions and a wide variety of speeds. They can be equipped with speed governors suited to the requirements, i.e., of very good stability and accuracy, if this is desirable, and arranged for various constant speed settings over a speed range as wide as 10:1.

Main Propulsion Marine Turbines (See also Sec. 11.3.) Steam turbines were commonly employed for the propulsion of naval and merchant ships into the 1970s. The availability of aero-derivative gas turbines of light weight, high capacity, and acceptable efficiency has led to their use for the propulsion of oil-fired naval vessels. The rapid rise in the price of fuel oil in the 1970s led merchant ship operators away from steam propulsion to the use of the more efficient diesel engine. Steam turbines continue to be used for the propulsion of all nuclear-powered vessels. Marine turbines are basically the same as central-station or industrial turbines except that usually the turbine is divided into a high-pressure and a low-pressure element, each geared through a common low-speed gear to the propeller shaft. The advantages of this compound arrangement are that two high-speed pinions divide the load on a common low-speed gear, thus reducing gear weight when compared with a single turbine. The high-pressure turbine can be made higher-speed than the low-pressure turbine and so be better adapted to the low volume flow. Each turbine can have a short rugged shaft, and either turbine can be used to propel the ship in an emergency.

Geared marine turbines require a reversing element for operating the vessel astern. This is typically a two-stage impulse turbine with two 2-row, or one 2-row and one 1-row, velocity stages arranged in the exhaust space of the low-pressure ahead turbine, so as to operate under vacuum under normal ahead conditions. The rotation loss of such an astern turbine is about ½ percent under normal ahead conditions.

Being directly geared to the propeller shaft, marine turbines must work at variable speeds. Overspeed governors are not required but are sometimes applied as a precautionary measure. Control is effected in most cases by sequentially operated nozzle valves. A typical marine turbine rated 16,000 hp is shown in Fig. 9.4.17. Ratings of 70,000 hp have been built, and larger sizes are realistic.

LARGE CENTRAL-STATION TURBINES

Figures 9.4.18 and 9.4.19 show examples of large central-station turbines as built by two manufacturers.

Figure 9.4.18 illustrates a 3,600-r/min tandem-compound four-flow unit rated 500 MW. It is a single-reheat fossil unit for nominal

Fig. 9.4.17b Same marine turbine as in Fig. 9.4.17a, but showing low-pressure section for 3,750-r/min normal speed. Low-pressure section contains two-stage reversing element. *(General Electric Co.)*

inlet-steam conditions of 2,400 lb/in^2 gage (16,650 kPa) 1,000°F (538°C), reheat to 1,000°F (538°C). The right-hand casing is a combined high-pressure and reheat section. Steam flows from the left center to the right through the impulse-type governing stage, then reverses, flowing to the left through the nine reaction-type high-pressure stages, and exhausts from the casing to the reheat section of the boiler. Reheated steam reenters that casing at its right center, flowing to the right through five reheat stages, turning once more to flow to the left between the inner and outer casings, finally exhausting up and to the left to the two double-flow low-pressure casings on the left end of the unit. The inlet stop and control valves, the reheat stop and intercepter valves, and the generator are not shown.

Four-flow units of this general type employ last-stage rotor blades from 25 to 40 in. (600 to 900 mm) long and in ratings up to about 700 MW. Significantly higher ratings require dividing the functions of the combined casing into a separate single-flow high-pressure casing and a separate two-flow reheat casing, for a total of four casings. The use

of the long titanium last-stage buckets makes this configuration suitable for ratings up to 1,200 MW. In cases requiring additional exhaust area, a third double-flow low-pressure element can be employed for a total of five casings. Tandem-compound 3,600- and 3,000-r/min units are commonly offered for ratings up to about 1,200 MW. Somewhat larger ratings can be accommodated by 3,600/3,600- or 3,600/ 1,800-r/min cross-compound units, but economic considerations makes their application infrequent.

Figure 9.4.19 illustrates an 1,800-r/min tandem-compound six-flow turbine designed for steam from light-water nuclear reactors. Reactors of both the boiling and pressurized-water types raise steam at 1,000 lb/in^2 abs (7,000 kPa) approximately with little or no initial superheat, so that the initial temperature is about 545°F (285°C), with a fraction of 1 percent of moisture frequently present. The poorer steam conditions result in higher steam rates than seen by fossil turbines. The lower initial pressure causes larger initial specific volume. In consequence, a typical nuclear turbine must accommodate 2.5 to 4 times the initial volume flow,

Fig. 9.4.18 Cross section of a 500-MW tandem-compound, quadruple-flow 3,600-r/min reheat turbine. *(Westinghouse Electric Corp.)*

Fig. 9.4.19   Cross section of a 1,300-MW class tandem-compound, six-flow 1,800-r/min turbine for steam from light-water nuclear reactors, showing one of the three low-pressure sections. *(General Electric Co.)*

and about 1½ times the exhaust volume flow of a fossil unit of the same rating. These considerations and the fact that the low temperature of the steam results in high moisture content in the expansion lead to the choice of 1,800 r/min. In halving the speed, diameters are less than doubled, balancing the advantages of larger steam-path area to accommodate large flow, while reducing velocities to minimize the occurrence of impact moisture erosion. The shortened energy range due to the lower initial conditions requires only two kinds of casings, high pressure and low pressure, compared with the three needed by fossil-reheat units.

Referring to Fig. 9.4.19, the steam enters the double-flow nozzle boxes of the high-pressure section, to the left, through stop and control valves which are not shown. It flows in both directions through the impulse-type stages, exhausting through four connections on each end of the shell. At the exhaust, the pressure is reduced to 200 lb/in² abs (1,400 kPa) approximately, and the moisture content is increased to 12 percent. That moisture poses an erosion risk and a performance loss to the low-pressure section following. It is current practice to dry the steam in an external moisture separator, frequently combined with one or two stages of steam-to-steam reheating, before admission to the low-pressure casings. Figure 9.4.20 is a cross section through a combination moisture separator and two-stage steam reheater. The exhaust from the high-pressure turbine enters the shell at the bottom, flowing upward through the inclined corrugated-plate moisture-separating elements, which remove essentially all the entrained water. It continues upward,

flowing over the tubes of the first-stage bundle, which are supplied with steam extracted from the high-pressure turbine, approaching to within 20 to 50°F (11 to 28°C) of that temperature. Next, it flows over the tubes of the second stage bundle, which are supplied with initial steam, approaching to within 20 to 50°F (11 to 28 °C) of that temperature, or to 495 to 525°F (257 to 274°C). The steam leaves the vessel at the top and is admitted to the low-pressure sections of Fig. 9.4.19, through stop and intercept valves, not shown. The last-stage blade length is in the range of 38 to 52 in (960 to 1,320 mm).

Units of the type described are built to ratings of approximately 1,300 MW, with larger sizes available. Similar four-flow units are employed for ratings up to approximately 1,000 MW.

Other types of nuclear reactors, such as the high-temperature gas-cooled reactor and the liquid-metal-cooled fast-breeder reactor, produce steam conditions at temperature and pressure levels comparable with fossil-fuel-fired boilers, leading to the use of 3,600-r/min units similar to fossil practice.

STEAM TURBINES FOR COMBINED CYCLES
(See also Sec. 9.7.)

Modern combustion **gas turbines (GTs)** have ratings approaching 250 MW. Considerable sensible heat is available in the gas-turbine exhaust-gas flow which ranges from 1,000 to 1,100°F (530 to 600°C) in temperature. In the **combined cycle,** the gas-turbine exhaust is directed to an unfired steam boiler called a **heat-recovery steam generator (HRSG).** The steam generated in the HRSG is admitted to a steam turbine, whose electrical output is approximately one-half that of the gas turbine. The resulting **combined thermal efficiency** can range from 50 to approaching 60 percent, better than that of any other available power cycle. High efficiency combined with the clean exhaust of gas turbines burning natural gas has created a broad field of application for combined cycles and the associated steam turbines.

Combined cycles can be arranged in several ways. In the *single-shaft* configuration, the gas and steam turbines are coupled together, driving a single generator. Units rated 413 MW have been manufactured for 50-Hz power systems, in which the gas turbine contributes 263 MW while the steam turbine generates 150 MW. The *multiple-shaft* arrangement uses a separate generator for each steam turbine and gas turbine. Each gas turbine has its own HRSG. The steam output from two or more GT/HRSG pairs can be manifolded to a single steam turbine. For example, a typical application for 60-Hz power systems combines two 181 MW gas turbines with one 201-MW steam turbine for a combined output of 563 MW. Typical steam conditions are 1,800 lb/in² gage (12,510 kPa), 1,050°F (566°C) with reheat in the HRSGs back to 1,050°F (566°C). Some cycles use inlet pressure as high as 2,200 lb/in² gage (15,270 kPa).

Steam turbines in combined-cycle service operate following the gas turbine(s), with their admission valves wide open accepting the full instantaneous output of the HRSG(s). The valves are located away from the turbine casings and are used only for speed control upon starting and for overspeed protection in the event of loss of load. The result is a very simple symmetric casing arrangement which reduces transient thermal stress and helps the steam turbine follow the potentially rapid changes in gas-turbine output. These features can be seen in Fig. 9.4.21. A two-casing reheat unit with single-flow exhaust is illustrated.

Fig. 9.4.20   Cross section of a combination moisture separator and two-stage steam reheater for use with nuclear reactor turbines.

Fig. 9.4.21 Two-casing reheat combined-cycle turbine with single-flow down exhaust. (*General Electric Co.*)

STEAM-TURBINE PERFORMANCE

The ideal **steam rate** (steam consumption, lb/kWh) of a simple turbine cycle is $3,412.14/(h_1 - h_{s2})$, where h is in Btu/lb [or kg/kWs = $1/(h_1' - h_{s2}')$, h' in kJ/kg]. (See Fig. 9.4.4.)

The actual steam rate is $3,412.14/[\eta_t(h_1 - h_{s2})]$, where η_t is the **engine efficiency** of the turbine only, inclusive of mechanical losses {or kg/kWs = $1/[\eta_t(h_1' - h_{s2}')]$}.

The actual steam rate of turbine and generator is $3,412.14/\eta_e(h_1 - h_{s2})]$, where η_e is the engine efficiency including all mechanical and electrical losses {or kg/kWs = $1/[\eta_e(h_1' - h_{s2}')]$}.

The enthalpy of the steam leaving the turbine elements h_2 is

$$h_2 = h_1 - \eta_s(h_1 - h_{s2}) = (Wh_1 - 3,412.14P_g)/W$$

where η_s is the engine efficiency of the steam path, inclusive of leakages and losses but exclusive of mechanical and electrical losses; W is the steam flow, lb/h; and the gross output P_g is the net output plus mechanical and electrical losses, kW [or $h_2' = h_1' - \eta_s(h_1' - h_{s2}') = (W'h_1' - P_g)/W'$, where W' is flow in kg/s].

Table 9.4.2 gives steam rates for the ideal simple turbine cycle through a wide range of operating conditions. The performance of a turbine is usually expressed as a steam rate in the case of machines having no extraction or admission of steam between inlet and exhaust, which is generally true of small units and most noncondensing turbines.

Turbines having automatic pressure-controlled extractions or admissions of steam between inlet and exhaust usually have their performance expressed by a chart showing required throttle flow vs. load for varying amounts of steam extracted or admitted at specified conditions.

Turbines working on regenerative and/or reheat cycles, condensing, usually have performance expressed as a heat rate, based upon a carefully specified heat cycle arrangement. This is usually illustrated by a diagram that defines all the surrounding conditions. See, for example, Fig. 9.4.26, actual cycle.

The above methods for expressing turbine performance are more satisfactory for most application than the use of turbine "engine efficiencies." However, it is useful to know the general range of turbine efficiency realized in practice. The engine efficiency of a turbine depends mainly upon the flow areas and diameter of stages, the average velocity ratio, as can be deduced from Fig. 9.4.6, 9.4.7, and 9.4.8, the number of turbine stages, and the steam conditions. With so many variables, it is not possible to do more than show a general picture of efficiency as a function of rating, as in Fig. 9.4.22, for multistage condensing turbines. Noncondensing turbines will usually have similar efficiency levels; automatic extraction turbines will generally be slightly lower because of extra losses in the control-stage sections.

Approximate steam rates for turbines operating without auxiliary admissions or extractions of steam between inlet and exhaust may be estimated for any turbine rating by dividing theoretical steam rate, corresponding to inlet steam pressure and temperature and exhaust pressure, by the appropriate turbine efficiency from Fig. 9.4.22.

Table 9.4.2 Theoretical Steam Rates for Typical Steam Conditions, lb/kWh*

	Initial pressure, lb/in² gage															
	150	250	400	600	600	850	850	900	900	1,200	1,250	1,250	1,450	1,450	1,800	2,400
	Initial temp, °F															
	365.9	500	650	750	825	825	900	825	900	825	900	950	825	950	1000	1000
	Initial Superheat, °F															
	0	94.0	201.9	261.2	336.2	297.8	372.8	291.1	366.1	256.3	326.1	376.1	232.0	357.0	377.9	337.0
	Initial enthalpy, Btu/lb															
Exhaust pressure	1,195.5	1,261.8	1,334.9	1,379.6	1,421.4	1,410.6	1,453.5	1,408.4	1,451.6	1,394.7	1,438.4	1,468.1	1,382.7	1,461.2	1,480.1	1,460.4
in Hg abs																
2.0	10.52	9.070	7.831	7.083	6.761	6.580	6.282	6.555	6.256	6.451	6.133	5.944	6.408	5.900	5.668	5.633
2.5	10.88	9.343	8.037	7.251	6.916	6.723	6.415	6.388	6.584	6.256	6.061	6.536	6.014	5.773	5.733
3.0	11.20	9.582	8.217	7.396	7.052	6.847	6.530	6.819	6.502	6.699	6.362	6.162	6.648	6.112	5.862	5.819
4.0	11.76	9.996	8.524	7.644	7.282	7.058	6.726	7.026	6.694	6.894	6.541	6.332	6.835	6.277	6.013	5.963
lb/in² gage																
5	21.69	16.57	13.01	11.05	10.42	9.838	9.288	9.755	9.209	9.397	8.820	8.491	9.218	8.351	7.874	7.713
10	23.97	17.90	13.83	11.64	10.95	10.30	9.705	10.202	9.617	9.797	9.180	8.830	9.593	8.673	8.158	7.975
20	28.63	20.44	15.33	12.68	11.90	11.10	10.43	10.982	10.327	10.490	9.801	9.415	10.240	9.227	8.642	8.421
30	33.69	22.95	16.73	13.63	12.75	11.80	11.08	11.67	10.952	11.095	10.341	9.922	10.801	9.704	9.057	8.799
40	39.39	25.52	18.08	14.51	13.54	12.46	11.66	12.304	11.52	11.646	10.831	10.380	11.309	10.134	9.427	9.136
50	46.00	28.21	19.42	15.36	14.30	13.07	12.22	12.90	12.06	12.16	11.284	10.804	11.779	10.531	9.767	9.442
60	53.90	31.07	20.76	16.18	15.05	13.66	12.74	13.47	12.57	12.64	11.71	11.20	12.22	10.90	10.08	9.727
75	69.4	35.77	22.81	17.40	16.16	14.50	13.51	14.28	13.30	13.34	12.32	11.77	12.85	11.43	10.53	10.12
80	75.9	37.47	23.51	17.80	16.54	14.78	13.77	14.55	13.55	13.56	12.52	11.95	13.05	11.60	10.67	10.25
100		45.21	26.46	19.43	18.05	15.86	14.77	15.59	14.50	14.42	13.27	12.65	13.83	12.24	11.21	10.73
125		57.88	30.59	21.56	20.03	17.22	16.04	16.87	15.70	15.46	14.17	13.51	14.76	13.01	11.84	11.28
150		76.5	35.40	23.83	22.14	18.61	17.33	18.18	16.91	16.47	15.06	14.35	15.65	13.75	12.44	11.80
160		86.8	37.57	24.79	23.03	19.17	17.85	18.71	17.41	16.88	15.41	14.69	16.00	14.05	12.68	12.00
175			41.16	26.29	24.43	20.04	18.66	19.52	18.16	17.48	15.94	15.20	16.52	14.49	13.03	12.29
200			48.24	29.00	26.95	21.53	20.05	20.91	19.45	18.48	16.84	16.05	17.39	15.23	13.62	12.77
250			69.1	35.40	32.89	24.78	23.08	23.90	22.24	20.57	18.68	17.81	19.11	16.73	14.78	13.69
300				43.72	40.62	28.50	26.53	27.27	25.37	22.79	20.62	19.66	20.89	18.28	15.95	14.59
400				72.2	67.0	38.05	35.43	35.71	33.22	27.82	24.99	23.82	24.74	21.64	18.39	16.41
425				84.2	78.3	41.08	38.26	38.33	35.65	29.24	26.21	24.98	25.78	22.55	19.03	16.87
600						78.5	73.1	68.11	63.4	42.10	37.03	35.30	34.50	30.16	24.06	20.29

* From Theoretical Steam Rate Tables—compatible with the 1967 ASME Steam Tables, ASME 1969.

Fig. 9.4.22 Turbine efficiencies vs. capacity.

Fig. 9.4.23 Correction factor for condensing mechanical-drive turbines.

A short method for calculating extraction-turbine performance is illustrated by the following example:

EXAMPLE. Assume a 12,500-kW automatic-extraction-condensing unit operating at 10,000 kW, with 175,000 lb/h extraction for process at 150 psig with no extraction for feedwater heating, and throttle steam conditions of 850 lb/in², 825°F, exhaust at 2 inHg abs.

PROCEDURE. Find theoretical steam rates (TSR) from Table 9.4.2 or steam charts; TSR_1 for 850 lb/in², 825°F, 2 inHg is 6.58 lb/kWh; TSR_2 for 850 lb/in², 825°F to 150 lb/in² is 18.61 lb/kWh.

Turbine-generator efficiency from Table 9.4.3, single autoextraction at 80 percent rating (10,000 kW on a 12,500-kW unit), is 78 percent. Efficiency correction for autoextraction (see Table 9.4.3) is 0.92. Then actual steam rate (ASR) is TSR/(efficiency × correction); $ASR_1 = 6.58/78\% \times 0.92 = 9.17$ lb/kWh; $ASR_2 = 18.61/78\% \times 0.92 = 25.9$ lb/kWh; kW generation from extraction flow = extraction flow/$ASR_2 = 175,000/25.9 = 6,760$ kW.

kW to be generated by condenser flow = $10,000 - 6,760 = 3,240$.

Condenser steam flow required is $3,240 \times 9.17 = 29,700$ lb/h, or (say) 30,000 lb/h. Total steam flow to throttle then is $175,000 + 30,000 = 205,000$ lb/h.

Figure 9.4.23 gives correction factors for speed which differ from 3,600 r/min and is representative of units designed for about 4 inHg abs exhaust pressure.

Mechanical-Drive Turbines Table 9.4.3 and Figs. 9.4.22 and 9.4.23 provide efficiency-estimating data for typical condensing turbines, primarily for 3600-r/min generator drive. Figure 9.4.24 shows approximate values for turbine efficiency to be expected for noncondensing mechanical-drive units designed for a broad range of horsepower rating, speed, and inlet steam conditions.

Large Central-Station Turbine-Generators

REFERENCES: Spencer, Cotton, and Cannon, A Method for Predicting the Performance of Steam Turbine-Generators · · · 16,500 kW and Larger, *Trans. ASME,* ser. A, Oct. 1963. Baily, Cotton, and Spencer, Predicting the Performance of Large Steam Turbine-Generators Operating with Saturated and Low Superheat Steam Conditions, *Proc. Amer. Power Conf.,* 1967; discussion of foregoing, *Combustion,* Sept. 1967. Spencer and Booth, Heat Rate Performance of Nuclear Steam Turbine-Generators, *Proc. Amer. Power Conf.,* 1968. Baily, Booth, Cotton, and Miller, "Predicting the Performance of 1800-rpm Large Steam Turbine-Generators Operating with Light Water-Cooled Reactors," General Electric publication GET-6020, 1973. "Heat Rates for Fossil Reheat Cycles Using General Electric Steam Turbine-Generators 150,000 kW and Larger," General Electric publication GET-2050C, 1974.

The performance of central-station turbine-generators is generally expressed as **heat rate**, Btu/kWh, the ratio of the heat added to the cycle in Btu/h, to generation, in kW. Heat rate may be converted to **thermal efficiency** using the relationship, Efficiency = 3,412.14/heat rate (or 1/heat rate expressed in kJ/kWs). Heat rates are calculated by the preparation of a **heat balance**, which considers steam conditions, steam flow, turbine-expansion efficiency, packing leaking losses, exhaust loss at the end of the low-pressure expansion (perhaps other casings as well), mechanical losses, electrical losses associated with the generator, moisture separation and reheat if present, and extraction for feedwater heating. **Gross heat rate** is calculated without consideration of the power consumed by the boiler feed pump. **Net heat rate** does consider pump power and is higher (poorer) than gross by a factor related to the initial steam pressure. If the pump is driven by an auxiliary turbine, as is the

Table 9.4.3   Basic Efficiency for Steam Turbines, Straight Condensing at Rated Load*

kW capacity	Equivalent mechanical drive, hp	Initial steam conditions (gage pressure and temp.)				
		250 lb/in² 500°F	400 lb/in² 650°F	600 lb/in² 750°F	800 lb/in² 825°F	1,250 lb/in² 900°F
875	1,200	63	63	62		
1,875	2,600	76	67	66		
2,500	3,500	69	69	68		
5,000	6,900		74	73	73	
7,500	10,300		76	75	75	
12,500	17,200		78	78	78	77
15,625	21,500		79	79	79	77
20,000	27,100		79	80	79	79

* Efficiency correction factors, mechanical drive and auto extraction-condensing turbines—multiply basic efficiencies by:

	At 80% rating	At 100% rating
Single autoextraction-condensing	0.92	0.96
Double autoextraction-condensing	0.88	0.92

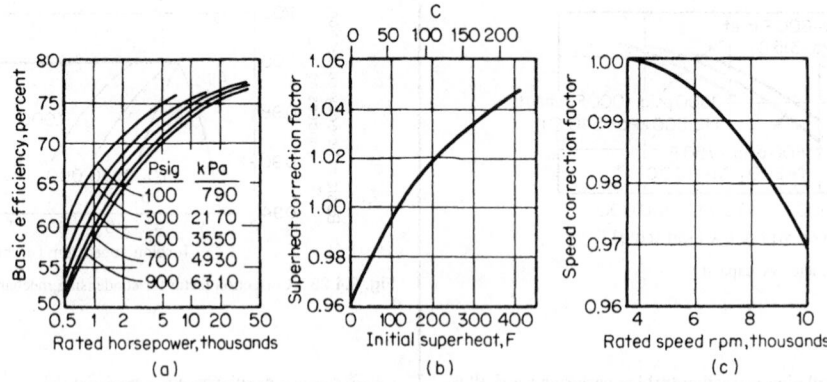

Fig. 9.4.24 Mechanical-drive turbine efficiencies. (*a*) Basic efficiency, 3,600 r/min. Figures on curve are inlet steam pressure in lb/in^2 gage. (*b*) Superheat correction factor. (*c*) Rated-load speed-correction factor.

present usual practice, net heat rate is the natural result of the heat-balance calculation, and gross heat rate has little meaning. **Net station heat rate** considers the auxiliary power required by the rest of the power-plant equipment, and the boiler efficiency of fossil plants. It is generally about 3 percent higher than net heat rate in nuclear plants (3 percent auxiliary power, 100 percent "boiler" efficiency), and about 16 percent higher than net heat in the case of a coal-fired plant (4 percent auxiliary power, 90 percent boiler efficiency).

A typical current value of net station heat rate for the fossil steam conditions is 9,000 Btu/kWh (2.64 kJ/kWs), equivalent to a thermal efficiency of 38 percent. A typical net station heat rate for a large light-water nuclear-reactor plant is about 10,100 Btu/kWh (2.96 kJ/ kWs), or about 34 percent thermal efficiency.

Table 9.4.4 lists representative net heat rates for large fossil turbines of various types and steam conditions. Steam pressures in excess of 3,500 lb/in^2 gage (24,200 kPa) and initial and reheat temperatures in excess of 1,000°F (538°C) were frequently employed in the past. However, a number of operating and economic considerations have led to near standardization on the single reheat cycle with initial pressure of 2,400 or 3,500 lb/in^2 gage (16,650 or 24,240 kPa), with initial and reheat temperature of 1,000°F (538°C), although there is some renewed interest in "ultra-supercritical" units in the Far East.

Table 9.4.5 lists some representative net heat rates for large nuclear turbines for service with steam from boiling-water reactors (BWR), at 950 lb/in^2 gage (6,650 kPa), ½ percent initial moisture. Values for other light-water reactors may be approximated by reducing heat rate by

1 percent for each 100 lb/in^2 (690 kPa) pressure increase, reducing heat rate by 0.15 percent for reducing initial moisture to 0 percent, reducing heat rate by 0.3 percent for each 10°F (6°C) of initial superheat provided.

Reheating with Regenerative Cycle

REFERENCES: Reynolds, Reheating in Steam Turbines, *Trans. ASME*, **71**, 1949, p. 701. Harris and White, Development in Resuperheating in Steam Power Plants, *Trans. ASME*, **71**, 1949, p. 685.

Reheating is commonly used on large fossil central-station turbines. It is accomplished by taking the steam from the turbine after partial expansion, reheating it in a separate section of the boiler, and returning it to the next lower-pressure section of the turbine. Reheating results in lowering of the turbine heat rate by approximately 5 percent; the exact improvement is dependent on several factors. Roughly speaking, 40 percent of the improvement comes from having added heat to the cycle at a higher-than-average temperature (thermodynamic gain), and the remaining 60 percent comes from improvement in turbine efficiency due to reduced moisture loss and increased reheat factor.

Reheating can theoretically be done any number of times, but because of extra cost of apparatus and piping, and the steam pressure drops required in practice (8 to 10 percent of the reheat pressure), the economic gains diminish rapidly with more than one reheating (see Fig. 9.4.25). In a few cases, two reheatings are employed. (See also Sec. 4.)

Table 9.4.4 Representative Net Heat Rates for Large Fossil Central-Station Turbine-Generators

Nominal rating MW, at 1.5 inHg abs	Steam conditions			Tandem compound, 3,600 r/min. last-stage buckets				Boiler feed-pump drive	Net heat rate, Btu/kWh, at rated load and steam conditions, and at exhaust pressure, inHg abs				
	Throttle pressure lb/in2 gage	Temp, °F	Reheat temp, °F	No. of rows	Length in	Exhaust area, ft^2	Approx kW/ft^2		1.5	2	3	4	5
150	1,800	1,000	1,000	2	26	82	1,820	Motor	8,010	8,060.	8,230	8,440	8,630
235	1,800	1,000	1,000	2	26	82	2,860	Motor	8,240	8,240	8,290	8,380	8,500
250	1,800	1,000	1,000	2	30	111	2,250	Motor	8,080	8,100	8,220	8,400	8,620
250	1,800	1,000	1,000	2	30	111	2,250	Turbine	8,030	8,060	8,200	8,390	8,610
250	2,400	1,000	1,000	2	30	111	2,250	Turbine	7,850	7,890	8,030	8,240	8,450
500	2,400	1,000	1,000	4	30	222	2,250	Turbine	7,790	7,830	7,970	8,170	8,370
700	2,400	1,000	1,000	4	33.5	264	2,650	Turbine	7,860	7,870	7,970	8,130	8,320
1,000	2,400	1,000	1,000	6	30	334	3,000	Turbine	7,920	7,930	8,000	8,100	8,250
500	3,500	1,000	1,000	4	30	222	2,250	Turbine	7,620	7,660	7,820	8,030	8,220
700	3,500	1,000	1,000	4	33.5	264	2,650	Turbine	7,670	7,690	7,810	7,980	8,170
1,000	3,500	1,000	1,000	6	30	334	3,000	Turbine	7,710	7,730	7,810	7,940	8,090
1,100	3,500	1,000	1,000	6	33.5	397	2,770	Turbine	7,680	7,700	7,310	7,960	8,140

Table 9.4.5   Representative Net Heat Rates for Large Nuclear Central-Station Turbine-Generators

Warranted reactor thermal power, MWt	Nominal turbine rating, MWe at 2 inHg abs	Tandem compound, 1,800 r/min, last stage buckets				Net heat rate, Btu/k/Wh, at warranted reactor thermal power, at rated steam-conditions, and at exhaust pressure, inHg abs				
		No. of rows	Length, in	Exhaust, area, ft²	Approx kW/ft²	1.5	2	3	4	5
2,440	840	4	38	423	1,980	9,950	9,950	10,090	10,190	10,410
2,440	850	4	43	495	1,720	9,810	9,820	9,950	10,170	10,440
2,890	1,010	6	38	634	1,590	9,750	9,780	9,950	10,200	10,480
2,890	990	4	43	495	2,000	9,980	9,980	10,050	10,200	10,410
3,580	1,230	6	38	634	1,540	9,910	9,920	10,000	10,170	10,380
3,580	1,250	6	43	743	1,680	9,780	9,790	9,930	10,160	10,430
3,830	1,310	6	38	634	2,070	9,990	9,990	10,050	10,190	10,390
3,830	1,330	6	43	743	1,790	9,840	9,850	9,960	10,170	10,420

All units boiling-water reactor steam conditions of 965 lb/in² abs, 1,190.8 Btu/lb, and two-stage steam reheat with 25°F approach to reheating steam temperature.

The throttle and condenser steam-flow rates for a given turbine output are reduced approximately 17 and 13 percent, respectively, by reheating once to the initial temperature, as compared with no reheat with the same initial steam conditions.

Fig. 9.4.25   Approximate gains due to reheating.

The maximum gain in heat rate from one reheating with a fixed-percentage pressure drop through the reheating system occurs when the reheat pressure is about 0.15 of the initial pressure. In practice, however, the reheat pressure is higher, 0.20 to 0.30 times initial pressure, because of the extra cost of larger piping, valves, etc., required for lower reheater pressures owing to the larger steam volume.

Regenerative Feedwater Heating
(See also Sec. 4.)

The heat consumption of a turbine may be reduced by heating the condensate (feedwater) in stages by the condensation of steam extracted at various points from the turbine. This is shown diagrammatically for an ideal cycle and a more practical cycle in Fig. 9.4.26. The difference between the two is in the use of mixing heaters in the ideal cycle with each discharge pumped back while the practical cycle has closed heaters with cascaded drains in the upper and pumped drains in the lowest heater, together with some pressure drop between turbine and heaters and a terminal temperature difference between saturated-steam temperature in the heater and feedwater temperature coming out. Usually the difference between such an ideal and a practical cycle is about 1 ½ percent. A deaerating type of contact heater with no terminal difference may be substituted for one of the closed heaters as is shown in Fig. 9.4.26. Other variations are the use of (1) all open-contact heaters, or (2) drain coolers to reduce the loss due to cascading the drips, or (3) a desuperheating section on the top heater to get a higher

final feed temperature, thereby approaching most closely to the ideal cycle.

Figures 9.4.27 and 9.4.28 and Table 9.4.6 supply data on the results of regenerative heating based on the ideal cycle of Fig. 9.4.26. Figure 9.4.27 shows the reduction in heat rate for various initial pressures and temperatures at 1 inHg abs (3.4 kPa) exhaust pressure, for various feed-water temperatures and number of heaters. The increase in

Fig. 9.4.26   Comparison of ideal and actual cycles for regenerative feedwater heating.

Fig. 9.4.27 Reduction in heat rate by use of ideal regenerative cycle, with 1-inHg back pressure.

Fig. 9.4.28 Increase in steam flow necessary to maintain the same power output when using the ideal regenerative cycle, with 1-inHg back pressure.

throttle flow necessary to maintain the same power output when extracting steam for feedwater heating is shown in Fig. 9.4.28.

INSTALLATION, OPERATION, AND MAINTENANCE CONSIDERATIONS

Steam turbines are capable of long life and high reliability with relatively little maintenance, if proper attention is paid to their installation, operation, and preventive maintenance. This section considers four areas proved to be of particular importance by operating experience.

Steam Temperature—Starting and Loading

REFERENCES: Mora et al., "Design and Operation of Large Fossil-Fueled Steam Turbines Engaged in Cyclic Duty," ASME/IEEE Joint Power Generation Conference, Oct. 1979. Spencer and Timo, Starting and Loading of Large Steam Turbines, *Proc. Amer. Power Conf.,* 1974. Ipsen and Timo, The Design of

Table 9.4.6 Total Steam Bled, Percent of Throttle Flow

Final feed temp		Steam pressure and temperature							
		400 lb/in² gage (2,900 kPa)		600 lb/in² gage 825°F (4,200 kPa, 440°C)		1,250 lb/in² gage 950°F (8,700 kPa, 510°C)		1,500 lb/in² gage 1,050°F 10,400 kPa, 565°C)	
		Stages of feedwater heating							
°F	°C	2	10	2	10	2	10	2	10
150	65	7.0	7.1	6.9	7.0				
200	93	11.4	11.8	11.3	11.6	11.2	11.5	10.7	11.0
250	121	15.6	16.2	15.5	16.0	15.5	15.9	12.6	13.1
300	149	19.6	20.6	19.5	20.4	19.5	20.3	18.8	19.5
350	177	23.6	24.8	23.5	24.6	23.6	24.6	20.7	21.6
400	204	27.1	29.0	27.1	28.8	27.4	28.9	26.4	27.8
450	232			30.2	32.5	31.1	33.2	30.0	32.0
500	260					35.0	37.4	33.9	36.1

Turbines for Frequent Starting, *Proc. Amer. Power Conf.,* 1969. Timo and Sarney, "The Operation of Large Steam Turbines to Limit Cyclic Shell Cracking," ASME Paper 67-WA/PWR-4, 1967.

Changing steam temperature at constant load or changing load at constant temperature subjects rotors and shells to thermal transients. Whereas temperature and load may be changed in seconds, it may take heavy metal sections hours to reach equilibrium with the new temperatures imposed on their surfaces. Parts are subjected to transient thermal stresses which may deplete their low-cycle thermal fatigue life. Repeated thermal cycles may lead to full expenditure of the available fatigue life of the part, followed by surface cracking. Further cycles tend to drive the cracks deeper into the affected part, leading to steam leakage through shells or vibration problems with rotors. Cracks tend to be driven deeper by downward steam-temperature changes, since the surface chills faster than the underlying material and is stressed in tension.

Steam-turbine manufacturers publish specific starting and loading instructions for their units. Data are provided such that the operator may select loading rates that stay within an acceptable expenditure of total low-cycle fatigue life per starting or loading cycle. For example, if a unit is expected to be started and loaded daily for a 30-year life, total cycles will be about 10,000, and it would be desirable to avoid exceeding 0.01 percent life expenditure per daily cycle.

Water-Induction Damage

REFERENCES: Recommended Practices for the Prevention of Water Damage to Steam Turbines Used for Electric Power Generation, ANSI/ASME TDP-1-1985 Fossil Fueled Plants. ASME Standard No. TWPDS-1, Part 2—Nuclear Fueled Plants, Apr. 1973.

Any connection to a steam turbine is a potential source of water either by induction from external equipment or by accumulation of condensed steam. The sources include the following along with their piping and drains: main and reheat steam systems; reheat attemperating system (fossil units); bypass systems, crossaround piping, moisture separator/reheater system (nuclear units); extraction system and feedwater heaters; steam-seal system; turbine-drain system. Water induction may lead to steam-path damage such as broken buckets, thrust-bearing failure, rotor bowing, and shell distortion, which may be indicated by abnormal vibration or differential expansion, or inability to turn the rotor on turning gear.

Water induction may be prevented by proper system installation, provision of protective and indicating devices, and periodic testing, inspection, and maintenance. Detailed recommendations are given in the ANSI/ASME and ASME standards and in manufacturers' instructions.

Lubricating-Oil and Hydraulic-Fluid Purity

Most steam turbines are provided with a lubricating-oil system consisting of reservoir, pumps, coolers, and piping to provide the thrust and journal bearings with a generous supply of oil at the proper temperature and viscosity. Some units also use the lube-oil system as a source of fluid power for control devices and steam-valve actuation. It is most important to assure the cleanliness and purity of the lube oil at all times to avoid bearing and journal damage, or control-system malfunction. Bearings have failed because of oil starvation caused by clogged lines. At least one overspeed failure has resulted from the silting of control devices with rust caused by the entry of water into the oil system.

Units employing an electrohydraulic control system frequently use a synthetic fire-resistant fluid for the high-pressure control hydraulics, separate from the petroleum oil used in the bearing lubrication system. High-pressure hydraulic systems employing synthetic fluid offer advantages in size reduction, speed of response, and fire safety. However, because of the need for very close clearances between small parts at high pressures, and because of the poorer rust-preventing properties of the fluid, cleanliness is of even greater importance than with oil-based systems.

Turbine manufacturers provide equipment such as conditioners to maintain bulk lubricating oil purity and full-flow filters to remove contaminants before they enter bearings. Further, they provide instructions for cleaning oil systems by oil flushing between installation and first operation, and for maintaining the required oil and hydraulic-fluid purity. It is important that these be followed carefully.

Steam Purity

REFERENCES: Lindinger and Curren, "Corrosion Experience in Large Steam Turbines," ASME/IEEE Joint Power Generation Conference, Oct. 1981. Bussert et al., "The Effect of Water Chemistry on the Reliability of Modern Large Steam Turbines," ASME/IEEE Joint Power Generation Conference, Sept. 1978. McCord et al., "Stress Corrosion Cracking of Steam Turbine Materials," National Association of Corrosion Engineers, Apr. 1975.

Boiler feedwater treatments have traditionally been designed to remove solids that might clog steam passages in the boiler, remove salts that could cause scaling of tube surfaces and interfere with heat transfer, prevent corrosion of tube surfaces by reducing oxygen content and by maintaining pH at a high level, and provide "clean" steam to the turbine.

At one time the main concern with steam quality in the turbine was the level of silica present, since that chemical tends to deposit in the steam passages and causes reduction in capacity and efficiency. In general, the monitoring and control of silica has been well developed, and as steam turbines have increased in rating, the passages have increased in area, so that the net effect has been a reduction in the extent of losses in efficiency and capacity from deposits.

On the other hand, the growth in unit ratings has been accomplished without a proportional growth in physical size, and has resulted in greater power densities per casing, per stage, and per pound of material. Such increased duty has required the use of higher-strength alloys operating at higher stresses. As a result, modern turbine components are more susceptible to stress-corrosion cracking than those in older, smaller units, and therefore require better control of contaminants in the steam.

The **feedwater treatment** in most fossil fuel stations is designed to provide a sufficiently low level of contaminants so that stress-corrosion cracking of turbine components should not be a problem. Both the "zero solids" and the "coordinated phosphate" treatments can be controlled to provide steam of acceptable chemistry. Unfortunately, there are a number of situations in which undesirable chemicals can be introduced in the steam in spite of well-intentioned "normal" water-control practices. For example, the composition of coatings put on turbine components for corrosion protection during shipment, storage, and installation must be controlled. The chemistry of solutions used for the removal of coatings during installation, and the methods used, must be carefully regulated. The turbine must be protected during the chemical cleaning of related components such as the boiler, condenser, and feedwater heaters. Critical components have been damaged by fumes from cleaning. The feedwater system must be designed so that only water of high purity is used for boiler desuperheater sprays, and for the turbine exhaust-hood sprays used to limit temperature during light-load operation. Condensate demineralizers must be operated and regenerated so as to ensure that they do not introduce the harmful chemical they are intended to remove. In the event of a leak into the condenser of impure cooling water, it is important to avoid changing the feedwater treatment in such a way that the turbine is subjected to harmful contaminants introduced to protect other station components.

In each of these undesirable instances, the average concentration of chemicals in the steam can be quite low, but high local concentrations can be developed through several mechanisms. For example, dilute solutions can enter crevices not washed by flowing steam; as water evaporates on heating, the concentration of the solution wetting the surfaces tends to increase. In the case of expansion-joint bellows, chemicals contained in steam condensing on shutdown or cold start-up tend to be trapped and concentrated in the bottom of the bellows convolutions.

Succeeding cycles can lead to dry residue or to concentrated solutions during operating conditions which provide moisture. In the case of the steam path, as the expansion crosses into the moisture region, the first droplets of water condensed from the steam will tend to contain most of the contaminants. Concentration-enhancement factors of 100 to 1,000 can be achieved. In modern reheat turbines the early-moisture region occurs in one of the later few stages of the low-pressure section, and at light loads can occur on the most highly stressed last-stage buckets.

Extreme care must be exercised in protecting turbines from chemical contamination during installation, operation, and maintenance. Any deviation from sound feedwater-treatment practice during condenser leaks should be done with the full realization that damage to the turbine may result.

9.5   POWER PLANT HEAT EXCHANGERS
by William J. Bow, assisted by Donald E. Bolt

NOTE: Standards for this industry retain USCS units except as indicated in the text.

SURFACE CONDENSERS

REFERENCE: Heat Exchange Institute Standards for Steam Surface Condensers.

The power plant surface condenser is attached to the low-pressure exhaust of a steam turbine (see Figs. 9.5.1 and 9.5.2). Its purposes are (1) to produce a vacuum or desired back pressure at the turbine exhaust for the improvement of plant heat rate, (2) to condense turbine exhaust steam for reuse in the closed cycle, (3) to deaerate the condensate, and (4) to accept heater drains, makeup water, steam drains, and start-up and emergency drains.

Fig. 9.5.1   Equipment arrangement, schematic.

An economical turbine back pressure is from 1.0 to 3.5 inHg abs. The factors involved in establishing this pressure are involved and will not be discussed here.

An equipment diagram of a closed power plant cycle is shown in Fig. 9.5.1.

For a condenser to deaerate the condensate, it must remove oxygen and other noncondensable gases to an acceptable level compatible with material selection and/or chemical treatment of the feedwater (condensate). Depending on materials and treatment, the dissolved O_2 level must normally be kept below 0.005 cm³/L for turbine units operating with high-pressure and -temperature steam.

Deaeration in a condenser is accomplished by applying Henry's law, which states that the concentration of the dissolved gas in a solution is directly proportional to the partial pressure of that gas in the free space above the condensate level in the hot well, with the exception of those gases (e.g., $CO_2 + NH_3$) which unite chemically with the solvent. In a condenser droplets of condensate are continually scrubbed with steam, liberating the O_2 and permitting it to flow to the low-pressure air-removal section, where it is discharged to the atmosphere by the air-removal equipment.

To remove the last traces of O_2 from the condensate, an ammonia compound such as hydrazine is normally added. Free ammonia is liberated in the cycle and is either removed with the noncondensables as a gas or is condensed and retained in the condensate, depending on the detailed design of the condenser air-removal section. If the ammonia is concentrated as a liquid, it can be very corrosive to certain copper-base materials.

Most condenser manufacturers have **tube-bundle configurations** unique to their design philosophy. Basically, pressure losses from turbine exhaust to the air offtake are kept to a minimum and tubes are arranged to promote good heat-transfer rates. Small condensers are usually cylindrical, whereas large ones are rectangular for better utilization of space. Most turbines exhaust downward into the condenser, but condensers are also built to accommodate side as well as axial exhaust turbines.

Because of the inherent strength of cylindrical shapes as opposed to flat plates, condenser **water boxes** are generally made with curved surfaces. This has come about because of the increased pressure resulting from cooling towers, which in turn, are the result of environmental influences. With a cooling tower, pressures are in the 60 to 80 lb/in² range, whereas with water from lakes, rivers, etc., where a siphon system can be employed, water-box design pressures are in the 20 to 30 lb/in² range.

As a general rule, **tube selection** is based on economics; 18 BWG admiralty metal has been satisfactory for freshwater service and 90-10 copper-nickel material likewise for seawater. The current trend is to use 22 BWG titanium or one of the new specially formulated stainless-steel tube materials for this application. Titanium has found favor as condenser tubing material. Its long life offsets the higher initial cost. Titanium in alloyed form suffers from reduced corrosion resistance and weldability but generally exhibits higher tensile strength when compared to the unalloyed form. Material prices fluctuate greatly, and selection can be influenced by first cost. Lost revenue due to downtime caused by tube leaks or other causes, particularly in larger units, can usually justify the use of more exotic and expensive materials.

Low-pressure feedwater heaters are frequently located in the steam-inlet neck of a condenser. This is done to minimize pressure drop in the extraction steam piping and to utilize floor space surrounding the condenser better.

A sufficient number of **tube supports** must be provided within the condenser so that the tubes will not vibrate excessively, which will cause tubes to rub or crack circumferentially. During low-water-temperature operation, the steam entering the condenser will often reach sonic velocities, causing severe **flow-induced vibration** and ultimate tube failures if the tube support system is inadaquate.

Where once-through boiler or nuclear steam generators are used, it is imperative to dispose of large quantities of steam during starting and stopping of a turbine unit. The condenser, because of its large volume, has been used as a convenient dumping place for this steam. Means must be provided within the condenser to accommodate the high-energy steam without damage to condenser tubing, structural members, or the low-pressure end of the turbine.

Fig. 9.5.2 One-pass rectangular surface condenser.

A major factor in condenser performance is **tube cleanliness.** Tubes can be plugged with leaves, marine life, and other debris deposited on the face of the tube sheets. By valving, various arrangements for back washing, or flushing, are effected to remove these materials.

The inside surfaces of the tubes can also become fouled by scaling, corrosion products, silt, products of biofouling such as slimes, or a combination of these. To clean the inside surfaces of the tubes, it is necessary to force brushes or plugs through them individually by the use of water or air jets. To do this, the affected portion of the condenser must be shut down. As an alternative, recirculating sponge rubber ball or brush systems are available that can be used during normal operation. To remove scale formations, sponge rubber balls with abrasive coatings are effective, but their continued use will shorten tube life.

Movement due to **temperature differences** between turbine and condenser is usually accommodated by a stainless-steel bellow or rubber-belt-type **expansion joint.** For small units the condenser may be supported on springs and rigidly connected to the turbine. To accommodate differential expansion between condenser shell and tubes, a flexible diaphragm or other expansion element may be installed.

The **tube-to-tube sheet joint** is usually rolled but has been welded in certain installations. Proper material selection must be made regardless of whether the joint is rolled or welded. Currently, considerable emphasis

is placed on inward leakage of circulating water into the condenser steam space. Double tube sheets have been supplied for some large central-station condensers for nuclear units. They are standard in the Navy's nuclear-powered ships and submarines to minimize this source of contamination.

Condensers for some large turbine units having two or three low-pressure ends are designed for **multipressure application.** Usually there will be a gain, in the form of either heat-rate improvement or a reduction in capital investment, with a multipressure condenser. A rather complex analysis must be made for each application before a decision can be made.

The current version of "Standards for Steam Surface Condensers" is published by the Heat Exchange Institute in Cleveland, OH. These are available to the public and provide guidance in solving most structural and sizing problems related to surface condensers.

Performance Calculations and Sizing

Notaion

S = condenser tube surface area, ft^2
C_c = cleanliness factor
C_1 = heat-transfer-rate constant
C_m = material and gage factor

C_t = temperature correction factor
c_p = specific heat, Btu/lb, °F
G = circulating-water quantity, gal/min
h = enthalpy, Btu/lb
h_r = heat rejected by steam, Btu/lb
L = length of water travel, ft
NPSH = net positive suction head, ft
OD = tube outside diameter, in
Q = heat transferred, Btu/h
R = temperature rise $(t_o - t_i)$, °F
TDH = total dynamic head, ft
t = tube thickness, in
TTD = terminal temperature difference = $t_s - t_o$
t_i = inlet-water temperature, °F
t_o = outlet-water temperature, °F
t_s = saturation temperature in condenser, °F
U_o = overall heat-transfer rate, Btu/(ft² · h · °F)
V = water velocity, ft/s
W_s = steam to be condensed, lb/h
Δt_m = log arithmic mean temperature difference, °F

In **sizing** a condenser, the steam flow and heat rejected to the condenser are obtained from the turbine heat balance. Table 9.5.1 gives representative steam flows; heat rejected to the condenser is approximately 950 Btu/lb of steam for nonreheat turbines and 975 Btu/lb for reheat

Table 9.5.1   Steam Flow to Condensers*

Turbine throttle conditions	Weight flow, lb/kWh
600 lb/in², 825°F	7.08
850 lb/in², 900°F	6.45
1,250 lb/in², 950°F	5.80
1,450 lb/in², 1,000°F	5.44
1,450 lb/in², 1,000°F, 1,000°F	4.70

* Approximate values for unit sizes to 100,000 kW and exhaust at 1.0 inHg abs.
NOTE: For exhaust pressures other than 1.0 inHg abs, multiply tabular values by 1.02 (1.04)[1.06] for 1.5 (2.0)[2.5] in.

machines. Figure 9.5.3 shows approximate average water temperatures for the United States; local water temperature should be used, when known. The number of passes is usually dictated by the plant arrangement, with total length of water travel and tube diameter dictated by economic considerations. Normally, small-diameter tubes, single-pass condensers are used where water is plentiful, and larger-diameter tubes, two-pass condensers when water is scarce. The vacuum, or back pressure, is determined by economic evaluation, but Table 9.5.2 gives normal recommended values for average water temperatures. Table 9.5.3 is a pressure-temperature conversion table.

Table 9.5.2   Normal Condenser Pressures and Circulating-Water Temperatures

Water inlet temp t_i, °F	Normal back pressure, inHg abs
55	1.0
70	1.5
80	2.0
85	2.5
90	3.0
95	3.5

A **cleanliness factor** is applied to the heat-transfer rate of new, clean tubes to allow for gradual decrease by fouling. A standarized cleanliness factor of 85 percent is frequently used, but this can often be misleading and even erroneous. The fouling is attributable to (1) sedimentation, (2) scaling, (3) steam-side deposits, (4) corrosion, and (5) biological growth. The fouling is correctly determined by use; in some cases the cleanliness factor may be 90 percent, whereas in other cases it may never rise above 75 percent.

Velocities normally used are: for clean water, 7.0 ft/s; for very clean water with cooling towers, 8.0 ft/s; and for seawater, with entrained sand, as low as 6.0 ft/s to minimize erosion. Prevalent velocities are

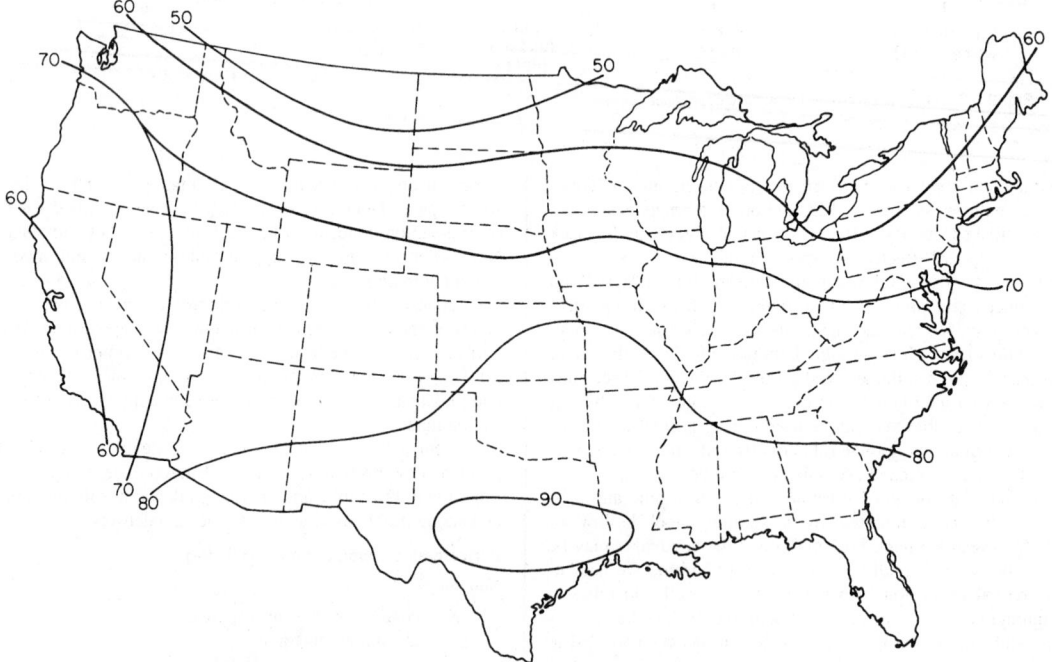

Fig. 9.5.3   Average inlet temperature of circulating water, °F, United States.

Table 9.5.3   Pressure-Temperature Conversion Table for Water*

Abs press, inHg	Sat temp, °F	Abs press, inHg	Sat temp, °F	Abs press, inHg	Sat temp, °F	Abs press, inHg	Sat temp, °F
0.20	34.57	1.35	88.36	2.50	108.71	8.00	152.24
0.25	40.23	1.40	89.51	2.55	109.38	9.00	157.09
0.30	44.96	1.45	90.64	2.60	110.06	10.00	161.49
0.35	49.06	1.50	91.72	2.65	110.72	11.00	165.54
0.40	52.64	1.55	92.77	2.70	111.37	12.00	169.28
0.45	55.89	1.60	93.81	2.75	112.01	13.00	172.78
0.50	58.80	1.65	94.80	2.80	112.63	14.00	176.05
0.55	61.48	1.70	95.78	2.85	113.25	15.00	179.14
0.60	63.96	1.75	96.73	2.90	113.86	16.00	182.05
0.65	66.26	1.80	97.65	2.95	114.45	17.00	184.82
0.70	68.41	1.85	98.56	3.00	115.06	18.00	187.45
0.75	70.43	1.90	99.43	3.20	117.35	19.00	189.96
0.80	72.32	1.95	100.30	3.40	119.51	20.00	192.37
0.85	74.13	2.00	101.14	3.60	121.57	21.00	194.68
0.90	75.84	2.05	101.96	3.80	123.53	22.00	196.90
0.95	77.48	2.10	102.77	4.00	125.43	23.00	199.03
1.00	79.03	2.15	103.56	4.20	127.21	24.00	201.09
1.05	80.53	2.20	104.33	4.40	128.94	25.00	203.08
1.10	81.96	2.25	105.09	4.60	130.61	26.00	205.00
1.15	83.33	2.30	105.85	4.80	132.20	27.00	206.87
1.20	84.64	2.35	106.58	5.00	133.76	28.00	208.67
1.25	85.93	2.40	107.30	6.00	140.78	29.00	210.43
1.30	87.17	2.45	108.01	7.00	146.86	29.921	212.00

* See also Sec. 4 for further data.

6.5 ft/s with aluminum-brass tubes, 7.0 ft/s with admiralty metal, and 8 + ft/s with stainless steel or titanium. **Water-temperature rise** is about 10°F for single-pass condensers and 15°F for two-pass condensers, with a minimum 5° terminal temperature difference (TTD).

Approximate general rules for condensers serving turbines rated up to 100 MW. The **surface area,** ft^2, is equal to the steam flow, lb/h, divided by 10 for single-pass condensers and by 7.5 for two-pass condensers. **Circulating-water quantity,** gal/min, is equal to the area, ft^2, for a two-pass condenser and is twice the area for a single-pass condenser. Condenser **proportions** are given in Table 9.5.4. Empty **weight** of an installed condenser is 5 to 6 lb/ft^2 of surface.

Table 9.5.4   Typical Small Condenser Proportions

Surface area, ft^2	Effective tube lengths, ft		
	¾-in OD	⅞-in OD	1-in OD
1,000–1,750	8, 10, 12, 14		
2,000–2,500	10, 12, 14, 16		
2,750–4,750	12, 14, 16, 18	12, 14, 16, 18	
5,000–7,000	14, 16, 18, 20	14, 16, 18, 20	14, 16, 18, 20
7,250–14,000	16, 18, 20, 22	16, 18, 20, 22	16, 18, 20, 22
15,000–19,000		18, 20, 22, 24	18, 20, 22, 24
20,000–27,500		20, 22, 24, 26	20, 22, 24, 26
30,000–47,500		22, 24, 26, 28	22, 24, 26, 28
50,000–75,000		24, 26, 28, 30	24, 26, 28, 30

Condenser Calculations

It is normally a tedious process to calculate the size and performance of a condenser directly by use of the logarithmic mean temperature difference. But using the following formula, the sizing can readily be determined by iteration, and performance can be found directly.

$$t_s = \frac{R}{\ln^{-1}[(S/G)(U_o/500)] - 1} + t_o \qquad (9.5.1)$$

Note that the specific heat c_p is normally neglected in condenser work, it being considered insignificant.

The basic diagram is shown in Fig. 9.5.4, and the applicable equations are

$$Q = U_o S \Delta t_m \qquad (9.5.2)$$

$$Q = 500 G c_p (t_o - t_i) \qquad (9.5.3)$$

$$R = W_s h_r / (500 G) \qquad (9.5.4)$$

$$U_o = C_1 \sqrt{V} C_t C_m C_c \qquad (9.5.5)$$

EXAMPLE.   Calculate the surface and circulating-water requirements to condense 445,000 lb of steam per h to an absolute pressure of 2.00 inHg abs. The heat rejected is 980 Btu/lb of steam and is absorbed by circulating seawater at an average inlet temperature of 75°F. The condenser is to be two-pass with ⅞-in 18-gage 26-ft-active-length aluminum-brass tubes; cleanliness factor = 85 percent.

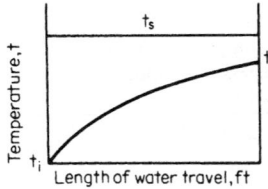

Fig. 9.5.4   Temperature vs. water travel, surface condenser.

SOLUTION.   The heat transfer rate U_o is a function of the tube diameter, material, gage, water velocity, and cleanliness factors within the velocity limits of 3 and 10 ft/s. For ⅝- and ¾-tubes $C_1 = 267$; for 7/8- and 1-in tubes $C_1 = 263$, and for 1⅛- and 1¼-in tubes $C_1 = 259$. The inlet-water-temperature correction factor is obtained from Table 9.5.7. The material and gage correction factor C_m is obtained from Table 9.5.6, and the cleanliness factor C_c is given in the conditions of the example.

Assume a surface of 59,000 ft^2 based on the typical steam loading for such a unit and a water velocity of 6.5 ft/s. For the selected surface and water velocity, by using the information in Table 9.5.5, the number of gallons per minute required by this selection is determined. It is now possible to determine the steam temperature that this selection would produce, by using Eq. (9.5.1).

The results show a steam temperature of 99°F, indicating that a smaller surface is required to attain a 2-inHg abs condenser pressure. Try a smaller unit until a surface with no more than three significant figures will yield a steam temperature of no more than 101.14°F. The surface arrived at is 54,400 ft² with 43,900 gal/min of cooling water.

Table 9.5.5   Tube Characteristics

OD and ext. surface, ft²/ft	BWG no.	ID, in	Water, gal/min @ 1 ft/s velocity
⅝-in OD 0.1636	16	0.495	0.600
	18	0.527	0.680
	20	0.555	0.754
	22	0.569	0.793
¾-in OD 0.1963	16	0.620	0.941
	18	0.652	1.041
	20	0.680	1.132
	22	0.694	1.179
⅞-in OD 0.2291	16	0.745	1.359
	18	0.777	1.478
	20	0.805	1.586
	22	0.819	1.642
1-in OD 0.2618	16	0.870	1.853
	18	0.902	1.992
	20	0.930	2.117
	22	0.944	2.182

Table 9.5.6   Material and Gage Factor

Tube material	C_m BWG no.						
	24	22	20	18	16	14	12
Admiralty metal	1.03	1.02	1.01	1.00	0.98	0.96	0.93
Arsenical copper	1.04	1.03	1.03	1.02	1.01	1.00	0.98
Copper iron 194	1.04	1.04	1.03	1.03	1.02	1.01	1.00
Aluminum brass	1.02	1.02	1.01	0.99	0.97	0.95	0.92
Aluminum bronze	1.02	1.01	1.00	0.98	0.96	0.93	0.89
90–10 Cu-Ni	0.99	0.98	0.96	0.93	0.89	0.85	0.80
70–30 Cu-Ni	0.97	0.95	0.92	0.88	0.83	0.78	0.71
Cold-rolled carbon steel	1.00	0.98	0.97	0.93	0.89	0.85	0.80
Stainless steel type 304/316	0.90	0.86	0.82	0.75	0.69	0.62	0.54
Titanium	0.94	0.91	0.88	0.82	0.77	0.71	0.63

Table 9.5.7   Inlet-Water-Temperature Correction Factors

Inlet temp t_p °F	Correction factor C_t
40	0.740
50	0.830
60	0.920
70	1.000
80	1.045
90	1.075
100	1.100

If this condenser is single-shell with one main turbine exhaust inlet and has no auxiliary turbine exhausts entering it (such as from a turbine-driven boiler feed pump), then the air-removal equipment must have a minimum capacity of 10 scfm of air at 1-inHg abs suction pressure. (See Table 9.5.8.)

Performance curves (Fig. 9.5.5) are drawn for a condenser to show the back pressure for various condenser steam loads and inlet-water temperatures maintaining a minimum TTD of 5°. The zero-load back pressure and the cutoff pressure are shown in Fig. 9.5.6, where the cutoff

Table 9.5.8   Venting-Equipment Capacities*

Effective steam flow each main exhaust opening, lb/h	One condenser shell		
		Total number of exhaust openings	
		1	2
Up to 25,000	scfm†	3.0	4.0
	Dry air, lb/h	13.5	18.0
	Water vapor, lb/h	29.7	39.6
	Total mixture, lb/h	43.2	57.6
25,001–50,000	scfm†	4.0	5.0
	Dry air, lb/h	18.0	22.5
	Water vapor, lb/h	39.6	49.5
	Total mixture, lb/h	57.6	72.0
50,001–100,000	scfm†	5.0	7.5
	Dry air, lb/h	22.5	33.8
	Water vapor, lb/h	49.5	74.4
	Total mixture, lb/h	72.0	108.2
100,001–250,000	scfm†	7.5	12.5
	Dry air, lb/h	33.8	56.2
	Water vapor, lb/h	74.4	123.6
	Total mixture, lb/h	108.2	179.8
250,001–500,000	scfm†	10.0	15.0
	Dry air, lb/h	45.0	67.5
	Water vapor, lb/h	99.0	148.5
	Total mixture, lb/h	144.0	216.0
500,001–1,000,000	scfm†	12.5	20.0
	Dry air, lb/h	56.2	90.0
	Water vapor, lb/h	123.6	198.0
	Total mixture, lb/h	179.8	288.0
1,000,001–2,000,000	scfm†	15.0	25.0
	Dry air, lb/h	67.5	112.5
	Water vapor, lb/h	148.5	247.5
	Total mixture, lb/h	216.0	360.0

* Heat Exchange Institute.
† 14.7 lb/in² abs at 70°F.
NOTE: These tables are based on air leakage only and the air-vapor mixture at 1 inHg abs and 71.5°F.

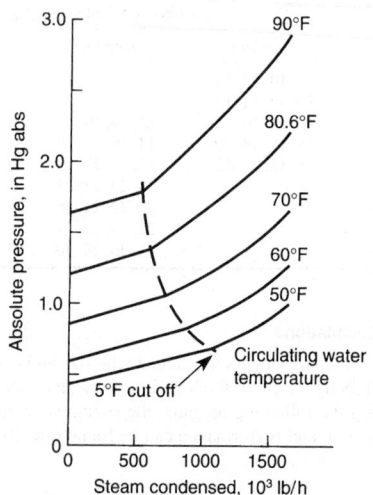

Fig. 9.5.5   Representative performance curves for a 157,500-ft² one-pass surface condenser, 1-in, 18-gage, 37.8-ft-active-length aluminum-brass tubes. Performance is based on 85 percent clean tubes, 221,400 gal/min cooling water at a velocity of 7 ft/s, and 968 Btu/lb rejected to the cooling water.

Fig. 9.5.6 Cutoff and zero load vacuums. Curve *A*, cutoff (except where the absolute pressure is limited to 5°F TTD); curve *B*, zero load.

limitation is set by the air-removal equipment. A combined turbine-condenser performance curve is sometimes drawn in which heat rejected vs. back pressure for a given turbine load is superimposed on the condenser performance.

Fig. 9.5.7 Friction loss for water flowing in 18 BWG tubes, low velocity. *(Heat Exchange Institute.)*

Circulating-water pumps are usually half capacity, with 3 or 4 percent additional allowance on each pump for miscellaneous heat exchangers. The total dynamic head (TDH) is the sum of the condenser and water-box friction (Figs. 9.5.7 to 9.5.11), the pipe loss, and any unrecovered static head. Normal heads are about 20 ft; for cooling-tower installations, 60 ft. (See also Sec. 14 and Fig. 9.5.12.)

Condensate pumps are usually full capacity, determined by condenser flow plus any heater drains dumped into the condenser and

Fig. 9.5.8 Friction loss for water flowing in 18 BWG tubes, high velocity. *(Heat Exchange Institute.)*

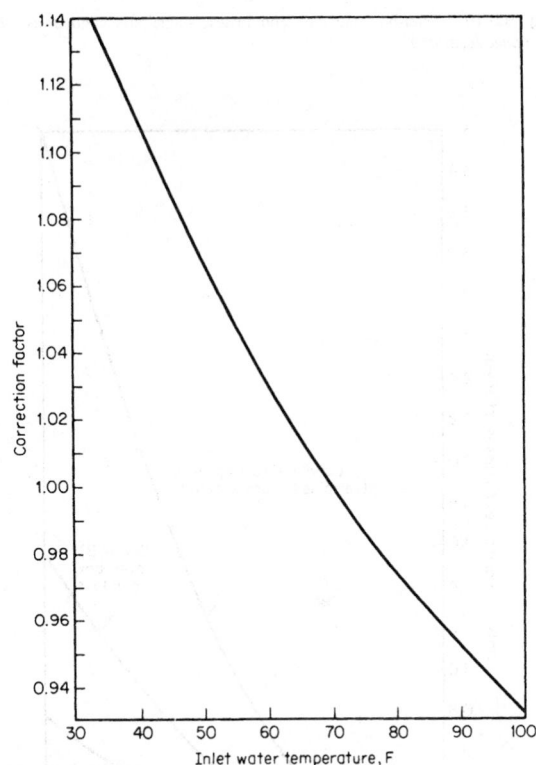

Fig. 9.5.9 Temperature correction for friction losses in tubes (see Figs. 9.5.7 and 9.5.8). *(Heat Exchange Institute.)*

5 percent additional margin. Vertical-pit-type designs are commonly used because of low net-positive-suction-head (NPSH) requirements with water at the boiling point. TDH ranges from a normal value of 100 ft on small plants to 800 ft on larger plants. (See Sec. 14.)

Fig. 9.5.10 Water-box and tube end losses, single-pass condensers. *(Heat Exchange Institute.)*

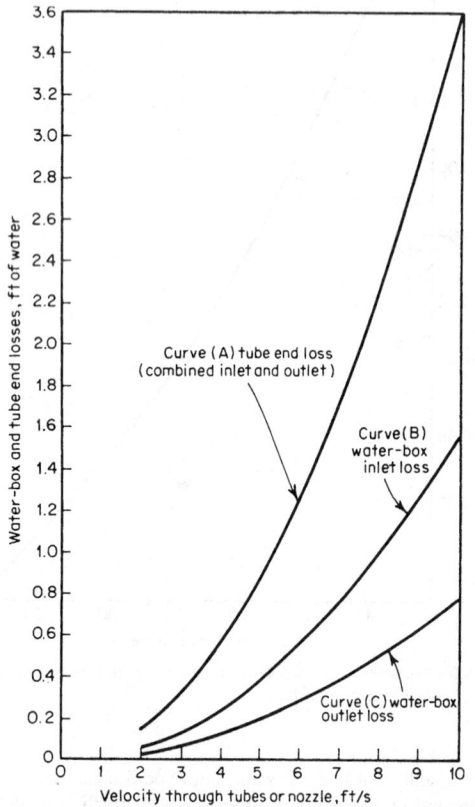

Fig. 9.5.11 Water-box and tube end losses, two-pass condensers. *(Heat Exchange Institute.)*

Fig. 9.5.12 Condenser configuration selection for a given heat load, $W_s = 477,355$ lb/h; $h = 975$ Btu/lb; $t_i = 70.0°F$; 1.5 inHg abs; 18-gage admiralty metal tubes; 7.0-ft/s water velocity; 85 percent cleanliness factor.

AIR-COOLED CONDENSERS

The air-cooled condenser is used where adaquate water is not available in sufficient quantities to permit the use of conventional cooling towers. The water lost by evaporation and cooling tower blow-down is approximately equal to the steam condensed from the turbine exhaust.

In an air-cooled condenser, the steam is condensed inside tubes while cooling air flows over fins on the external surfaces. These tubes are arranged in the area where packing or fill would be arranged in a conventional cooling tower, and propeller fans (see Sec. 14) supply air across the tube surfaces.

Aluminum is generally the material of choice for tubing because of its excellent heat transfer properties and availability. Items to be considered in design are the overall heat-transfer coefficient, steam-side pressure drop, uniformity of steam distribution, noncondensible gas concentration for air removal, and, of course, provision for freeze protection. Turbine design back pressures are normally in the range of 5 in Hga when served by air-cooled condensers.

DIRECT-CONTACT CONDENSERS

REFERENCES: Heat Exchange Institute Standards for Direct Contact and Low Level Condensers.

When (1) low investment is desired and (2) condensate recovery is not a factor, direct-contact condensers are effective. They are relatively simple to build and operate, are limited to sizes less than 250,000 lb of steam per h, and are built in three types: (1) barometric, (2) low-level, and (3) jet.

Figure 9.5.13 shows a self-supported counterflow **barometric condenser** with tail pipe, hot well, and air ejector. Steam and cooling water flow in opposite directions, with the coldest water for final condensation and cooling of noncondensables. The air pump must handle that part of the air disengaged from the cooling water as well as air leakage. The head required to pump water is pipe friction plus static head, minus 75 percent approx of the design vacuum. The barometric condenser is usually placed outdoors, and the water leg must be more than 34 ft high.

The **low-level condenser** substitutes a pump for the water leg of the barometric condenser to remove the liquid from the vacuum space. The **jet condenser** utilizes the aspirating effect of a jet for the entrainment of noncondensables and the consequent elimination of a separate air pump.

In the usual direct-contact condenser, where steam and raw circulating water are mixed, the recovery of pure condensate is precluded; greater feedwater makeup is necessary, and poorer vacuums are attained than with surface condensers. Direct-contact-condenser installations are not found in large plants, but there is some recent interest in their use with dry cooling towers.

Fig. 9.5.13 Counterflow barometric condenser with two-stage air ejector. *(Ingersoll-Rand Co.)*

Table 9.5.9 Air in Cooling Water

Water temp, °F	Air, ft³/min per 1,000 gal/min
35	4.0
40	3.78
50	3.35
60	2.97
70	2.68
80	2.41
90	2.21
100	2.0

Fig. 9.5.15 Direct-contact condenser performance terms.

Table 9.5.9 gives values of expected air content of cooling waters; Fig. 9.5.14 gives typical performance curves; Fig. 9.5.15 defines flows W, temperatures t, and enthalpies h needed for the basic heat-balance equation $W_s(h_s - h_2) = W_w(t_2 - t_1)$. The latent heat $(h_s - h_2)$ is frequently taken as 950 Btu/lb of steam.

Shell materials are usually steel plate; with dirty or corrosive water, bronze, stainless steel, or linings of ceramic, plastic, or rubber may be used.

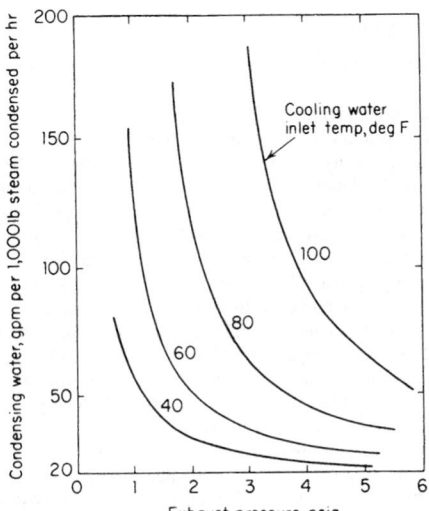

Fig. 9.5.14 Representative performance curves for a direct-contact condenser.

AIR EJECTORS

REFERENCE: Heat Exchange Institute, Standards for Steam Jet Ejectors.

The steam-jet ejector is used to remove noncondensable air and gases from condensers. It consists of a suction chamber, diffuser, and steam nozzle. The high-velocity jet of steam issuing from the nozzle entrains the gas, and the kinetic energy of the mixture is converted to pressure energy in the diffuser. Ratings are in lb/h at a given suction pressure. Ejectors are operated in series or are staged for absolute suction pressures of $1 \pm$ inHg abs. They will handle wet or dry mixtures, they are easy to operate, installation costs are low, and there are no moving parts—with consequent long life, high sustained efficiency, and low maintenance.

Two-stage ejectors, with surface-type inter- and aftercondensers (Fig. 9.5.16), are common in steam-surface-condenser application. The main condensate serves as a coolant, returning heat regeneratively to the boiler feed system. Two-stage ejectors have a shutoff pressure of 0.5 inHg abs approx. Two elements are usually provided, one serving as a spare.

Computation of **ejector performance** requires application of Dalton's law of mixtures (see Sec. 4), where basically the total pressure is the sum of the condensable vapor (steam) pressure and the noncondensable gas (air) pressure. In power plant condensers, the mixture is saturated with steam so that the temperature of the mixture fixes the partial pressure of the vapor.

Air ejectors are usually rated with suction conditions of 1 inHg abs and 7.5°F subcooling. Since the saturation temperature of steam at 1 inHg abs is 79°F, the mixture entering the ejector is at 71.5°F. The partial pressure of the vapor at 71.5°F is 0.78 inHg abs, and the partial pressure of the air is $1 - 0.78 = 0.22$ inHg abs. Therefore, an ejector with a standard rating (see Table 9.5.8) of 12.5 ft³/min of free dry air at 30 inHg abs handles, by the gas laws, 12.5(30/0.22) = 1,704 ft³/min at the ejector suction.

Motive steam to the ejector is usually supplied from turbine throttle conditions, e.g., 600 lb/in², 750°F; 1,500 lb/in², 1,000°F; 2,500 lb/in², 1,050°F, through a reducing valve to a pressure not higher than 600 lb/in².

The **inter- and aftercondenser,** one- or two-pass, is a small heat exchanger which condenses the motive steam and allows the noncondensable gases, such as O_2 and CO_2 to be expelled to the atmosphere.

However, the highly soluble ammonia gas may be returned to the feed-water system. The water-box pressure of tube-side pressure must be designed for the condensate-pump shutoff head. The pressure drop through the inter- and aftercondensers varies from 1 to 10 ft of water, with 3 ft as a reasonable average.

Fig. 9.5.16 Two-stage air ejector with surface-type inter- and aftercondensers. *(Westinghouse Electric Corp.)*

Fig. 9.5.17 Single-stage air ejector.

Normal **materials** used are: nozzle, stainless steel type 303; steam chest, steel; inlet-air chamber, cast steel; diffuser, aluminum bronze; inter- and aftercondenser shell, steel; inter- and aftercondenser tubes, stainless steel type 304; inter- and after-condenser tube sheet, carbon steel.

Hogging and Priming Ejectors   Single-stage steam-jet ejectors (Fig. 9.5.17) are used for evacuating, or **hogging**, the air from the steam side of a surface condenser, in 15 or 20 min time, so that steam flow can be established and the condenser brought on the line. These ejectors are usually noncondensing and exhaust to the atmosphere. Ratings (Table 9.5.10) are typically in lb/h of free dry air at 70°F, suction conditions of 15 inHg abs, shutoff pressure about 2 inHg abs, motive steam pressures and temperatures as with multistage ejectors.

Priming ejectors are also single-stage noncondensing units and are used to withdraw air from the water side and to fill the condenser with circulating water during start-up periods. The hogging ejector may also serve this priming function by a suitable system of piping and valves.

Materials normally employed are: for the nozzle, stainless steel, 18-8 chrome nickel; steam chest, steel; air-inlet chamber and diffuser, cast steel.

Table 9.5.10   Hogger Capacities*

Total steam condensed, lb/h	scfm†—dry air at 10 inHg abs design suction pressure
Up to 100,000	50
100,001–250,000	100
250,001–500,000	200
500,001–1,000,000	350
1,000,001–2,000,000	700
2,000,001–3,000,000	1,050
3,000,001–4,000,000	1,400
4,000,001–5,000,000	1,750
5,000,001–6,000,000	2,100
6,000,001–7,000,000	2,450
7,000,001–8,000,000	2,800
8,000,001–9,000,000	3,150
9,000,001–10,000,000	3,500

* HEAT EXCHANGE INSTITUTE.
† 14.7 lb/in² abs at 70°F.
NOTE: In the range of 500,000 lb/h steam condensed and above, the table provides evacuation of the air in the condenser and low-pressure turbine from atmospheric pressure to 10 inHg abs in about 30 min if the volume of condenser and low-pressure turbine is assumed to be 26 ft³/(1,000 lb/h) of steam condensed.

VACUUM PUMPS

REFERENCE: Woodman, Rotary or Steam-Jet Air Pumps, *Power,* Aug. 1948.

Mechanical vacuum pumps, used for hogging, holding, and priming, are of several types for power-plant service: (1) reciprocating—piston, diaphragm; (2) rotary—sliding-vane, oval-water- seal, eccentric-rotor. (See Sec. 14 for details.) Multistage pumps usually need intercooling. A silencer is generally provided on the air discharge. Normally, two half-capacity pumps are installed, both operating during hogging and one adequate for holding service. The relative capabilities of a vacuum pump and hogging and holding ejectors are illustrated in Fig. 9.5.18 (note the considerable hogging capacity of the vacuum pump on start-up).

Advantages of vacuum pumps compared to steam-jet air ejectors include (1) system independent of a steam supply; (2) start-up when steam is not available, as on a once-through boiler system; (3) system capable of completely automatic operation; (4) high-pressure steam piping eliminated; (5) recycling of noncondensables and ammonia eliminated; (6) quieter operation. Disadvantages include (1) damage by water entering the intake (except when pumps having a liquid rotating seal); (2) higher initial cost; (3) higher maintenance cost. Operating costs are approximately equal. Various combination arrangements are found in practice, e.g., an air-operated air-ejector first stage followed by

Fig. 9.5.18   Relative capabilities of air-removal pumps.

a vacuum-pump second stage, with consequent savings in motor-horse-power and space requirements.

COOLING TOWERS

REFERENCES: "Evaluated Weather Data for Cooling Equipment Design," Fluor Products Co. Cooling Towers, *Power,* Mar. 1963. Baker and Shryock, A Comprehensive Approach to the Analysis of Cooling Tower Performance, *Trans. ASME,* 1961.

The prevalent type of cooling tower (see Figs. 9.5.19 and 9.5.22) dissipates heat by the evaporation of some of the water sprayed into the air circulated through the tower. It is used where water is in limited supply, where temperature pollution of natural water bodies is to be avoided, where water conservation is to be effected, or where otherwise polluted sources must be avoided. Figure 9.5.20 illustrates the functional cycle and the significant basic terms (see also Sec. 4).

Wet-bulb temperature, for design purposes, should not exceed the maximum expected value more than 5 percent of the time during summer (June to September). (See Fig. 9.5.21.)

Approach is the difference in temperature between the cold water leaving the tower and the ambient wet bulb.

Cooling range is the difference in temperature between the hot water entering and the cold water leaving the tower.

Drift is the water lost as mist or droplets entrained by the circulating air and discharged to the atmosphere. It is in addition to the evaporative loss and is minimized by good design.

Fig. 9.5.19   Schematic for induced-draft, counterflow, cooling tower installation.

Makeup is the water required to replace total losses by evaporation, drift, blowdown, and small leaks.

The early, simple atmospheric towers have, because of reliance on natural air circulation, high pumping heads, excessive spray losses, and

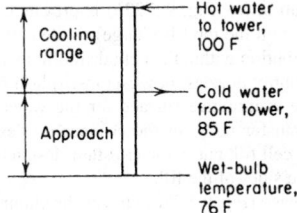

Fig. 9.5.20   Cooling tower terms.

Fig. 9.5.21   Wet-bulb temperature, °F, isoclines at the 5 percent level, United States. *(Adapted from Fluor Products Co., by permission.)*

Fig. 9.5.22   Cooling tower types. (*a*) Forced draft; (*b*) crossflow-induced draft; (*c*) hyperbolic.

makeup, been largely superseded by three important types: (1) forced-draft, (2) induced-draft, and (3) hyperbolic (Fig. 9.5.22).

The **mechanical forced-draft tower** (Fig. 9.5.22*a*) gives (1) a controllable air supply from fans conveniently located for inspection and maintenance at ground level, and (2) reduced water- pumping head. Nonuniform distribution of air over the ground area of the tower cell, recirculation of vapor from the tower discharge to the tower inlet, with its deleterious effects and fan icing in cold weather, and limitations on the physical diameter of fans are all problems with forced-draft towers.

The **induced-draft tower** (Fig. 9.5.22*b*) is prevalent in U.S. practice. The fan is mounted on the top (discharge) of the cell, with consequent improved air distribution within the cell; drift eliminators reduce makeup requirements; spray nozzles, downspouts, splash plates, and splash bars ensure ample evaporative surface for the water, with maximum volumetric heat-transfer rates. In the **counterflow** design, air is introduced beneath the cell fill, but in the **crossflow** design (Fig. 9.5.22*b*), air is introduced at the sides of the fill.

The **hyperbolic tower** (Fig. 9.5.22*c*) utilizes the chimney effect (height, $300 \pm$ ft) for natural circulation. It has been favored in large European installations where the prevalent lower dry-bulb ambient temperature gives a greater difference in density between entrance to and exit from the tower. The wide approach keeps the unit size within practical bounds, and the savings in fan power support the higher investment. Recent installations in the United States attest to its attractiveness.

Precautions are necessary to avoid freezing troubles in cold weather, fire hazards with intermittent operations, corrosion, scale, and microbiological growth problems. **Location** must avoid recirculation: for tower lengths to 250 ft, the long-axis orientation should be parallel to the prevailing wind; towers longer than 250 ft should be arranged broadside to the prevailing wind. The tower should be isolated as much as possible; adjacent heat sources compel the specification of higher design wet-bulb temperatures.

Performance Calculations

(See Fig. 9.5.19.)

Applicable equations are

$$W_{a1}h_{a1} + W_{v1}h_{v1} + W_{wA}h_{fA} = W_{a2}h_{a2} + W_{v2}h_{v2} + W_{wB}h_{fB} \quad (9.5.6)$$

$$W_{wB} = W_{wA} - (W_{v2} - W_{v1}) \qquad W_{a1} - W_{a2} \quad (9.5.7)$$

$$W_{wA}(h_{fA} - h_{fB}) = W_{a2}h_{a2} + W_{v2}h_{v2}$$
$$- (W_{a1}h_{a1} + W_{v1}h_{v1}) - (W_{v2} - W_{v1})h_{fB} \quad (9.5.8)$$

$$h_{fA} - h_{fB} = t_{wA} - t_{wB} \quad (9.5.9)$$

and

$$W_a h_a + W_v h_v$$

= total heat, from psychrometric chart   (See Sec. 4.)   (9.5.10)

$$W_{wA}(t_{wA} - t_{wB}) = \text{total heat at 2}$$
$$- \text{total heat at 1} - (W_{v2} - W_{v1})h_{fB} \quad (9.5.11)$$

where W = flow, lb/h; a = air; w = water; v = vapor; h = enthalpy, Btu/lb; f = fluid; t = temperature, °F; 1, 2, A, B = locations.

EXAMPLE.   Given: flow = 80,000; wet bulb = 76°F; dry bulb = 86°F; range (also condenser rise) = $100 - 85 = 15°F$; approach = 9°F. Find: economical tower size, circulating-pump power, fan capacity and power, and makeup water.

SOLUTION.   (1) *Economical tower size:* A study must be made to tie in with condenser. Heat to be dissipated is constant. Vary rise (range), approach, and then compare capital costs of çooling tower, circulating pump, and condenser with operating costs for fan and pump power. (2) *Circulating-water-pump power;* bhp = gal/min × TDH/3,960 × eff. TDH = condenser friction + pipe friction + static head. (3) *Fan capacity and power:* If air leaves tower saturated at 95°F, then from psychrometric charts (Sec. 4),

$$\text{Total heat} = 1 \text{ lb dry air } C_p(t_a - 0°) + W_v h_g \qquad \text{at } t_a \quad (9.5.12)$$

For air and vapor at 1, vapor = 120 grains/lb dry air. By Eqs. (9.5.10) and (9.5.12), total heat = $1 \times 0.24(86 - 0) + (120/7,000)1,099.2 = 39.5$ Btu/lb dry air. For air and vapor at 2, vapor = 256 grains/lb dry air. Total heat = 63.1 per lb dry air. Sp vol (air and vapor) = 14.80 ft³/lb dry air. By Eq. (9.5.11), $W_{wA}(100 - 85) = 63.1 - 39.5 - (256 - 120) \frac{63}{7,000} = 1.50$ lb water/lb dry air. 80,000 gal/min/(1.50 × 500) = 26.6×10^6 lb dry air/h, or $26.6 \times 10^6 \times 14.80/60 = 6.58 \times 10^6$ ft³/min. With test data on static-pressure requirements for the tower and with the fan characteristics, the horsepower to drive the fan can be calculated. (4) *Makeup:* $W_{v2} - W_{v1} = (256 - 120) 26.6 \times 10^6/7,000 = 515,000$ lb/h.

Normally, a cooling tower is purchased for only one guarantee point. It is well, however, to have **performance** curves (Fig. 9.5.23) showing operation for various wet-bulb temperatures and cooling ranges. The **investment** for a cooling tower is essentially a matter of water flow and is influenced by approach, range, and wet bulb (Fig. 9.5.24). The **cost evaluation** should include consideration of tower frame and fill, fans,

Fig. 9.5.23   Cooling tower performance. Design for 85°F cold water, 95°F hot water, 78°F wet-bulb temperature, 10° range; 1,000-gal/min cell.

motors, basin and pump pit, pump head, fan horsepower, freight, labor, and erection. In the choice of a tower for a power plant, there should be coordinated study and evaluation of the turbine and condenser for best overall economy.

The **height** of a field-erected induced-draft tower, from basin curb to fan deck, ranges from 8 to 50 ft; widths vary from 6 to 60 ft; lengths from 8 to 500 ft; fan-stack height between 2 and 15 ft.

Materials used are: frame—redwood (treated or untreated), Douglas fir (treated), steel (galvanized), concrete; fill—red wood, plastics (polyethylene, polypropylene), cement asbestos* board (extruded); casing—cement asbestos* board (corrugated, slat), redwood, fiber glass, aluminum, concrete; fan blades—aluminum, glass-reinforced polyester, stainless steel, monel; fan hubs—cast iron, galvanized steel, stainless steel; fan stack—redwood, steel, masonite, drift eliminators—redwood, cement asbestos* board; spray nozzles—ceramic, bakelite; louvers—redwood, cement asbestos* board.

Fig. 9.5.24 Comparative cooling tower costs. (*Foster Wheeler Corp.*)

DRY COOLING TOWERS, WITH DIRECT-CONTACT CONDENSERS

REFERENCE: Ritchings and Lotz, Economics of Closed vs. Open Cooling Water Cycles, *Power Eng.*, May and June 1963.

The thermodynamic requirement for a heat sink can generally best be met with a natural body of water such as a river, lake, or ocean. When water supplies are limited, a **"wet" cooling tower** (water losses less than 2 or 3 percent) may be used for reclamation. A **"dry" cooling tower** will further reduce this loss. Figure 9.5.25 shows an application with a direct-contact condenser. The finned surfaces of the dry cooling tower

Fig. 9.5.25 Dry cooling tower system.

are substituted for the tubes of a surface condenser. Heat-transfer rates (see Sec. 4) are appreciably lower, but the extended surface, relatively low in cost, may be installed in an induced-draft tower or in a hyperbolic tower, as in England for a 125,000-kW unit. Problems of recirculation, freezing, and air leakage must be solved. If a minor evaporative loss is permissible, peaking capacity is obtained by wetting the heat-exchange surface in the tower.

SPRAY PONDS

If land area is available, a spray pond may serve as an alternative reclamation system. Pipelines are typically arranged on 50-ft centers, with five nozzles (50 ± gal/min each) every 12 ft of pipeline. Cooling is limited by the relatively short period of contact between spray and air. Drift, especially with adverse wind conditions, can make losses higher than on towers. Generally, if cooling efficiencies above 50 percent are required, a tower is favored, where efficiency = $(t_h - t_c) \times 100/(t_h - t_{wb})$, where the temperature t subscripts are h = hot water, c = cold water, wb = wet bulb of the ambient air. (See Wright and Kirsopp,

* Existing plants only.

Pond Surface Cooling of a Chemical Plant Cooling Water, *Proc. Am. Power Conf.*, 1960.)

CLOSED FEEDWATER HEATERS

REFERENCES: Heat Exchange Institute Standards for Closed Feedwater Heaters. ASME Boiler and Pressure Vessel Code, Sec. VIII.

Feedwater heaters are used (1) to effect the thermodynamic gains of the regenerative steam cycle (see Secs. 4 and 9), and (2) to raise water temperatures to a sufficient value for the avoidance of thermal shock to the boiler metal. The **number and types of heaters** usually employed are: (1) plant sizes of up to 70 mW, two closed low-pressure heaters, one open heater, one closed high-pressure heater; (2) plant sizes of 75 to 300 mW, two or three closed low-pressure heaters, one open heater, two closed high-pressure heaters; (3) fossil-fueled plant sizes above 300 mW, three or four closed low-pressure heaters, one open heater, two or three high-pressure heaters. Nuclear units require very large feedwater flows, necessitating double or triple parallel strings of heaters. There are generally five or six low-pressure and one high-pressure heater (no open heaters) in each string.

Minimum heater cost prevails with minimum restrictive specifications, e.g., horizontal, two-pass, high water velocity (10 ft/s at 60°F), no length limits. Overall heater length is limited by maximum available tube lengths of 100 ft for copper alloys, admiralty metal, and copper-nickel and 85 ft for Monel. With U-tube construction, this results in heater lengths of about 48 and 40 ft, respectively. A general rule, to ensure good steam distribution, is that the length in feet shall not exceed the shell diameter in inches plus 2; i.e., with a 30-in diam shell, the length should not exceed 32 ft. **Pressure drop** through the tubes must be economically evaluated as it varies approximately with the square of the water velocity.

If a **length restriction** is imposed, the designer may have to substitute a four-pass arrangement for the two-pass design, with consequent large-diameter shell and water chamber, heavier walls, more tube holes to be drilled, more tubes to be installed, and a cost increase. If a **pressure-drop restriction** is imposed, a lower water velocity results, with more tubes, larger shell and chamber diameter, and more surface because of the lower heat-transfer rate. Vertical heaters, with appropriate construction details, are also higher in cost.

The **condensing heater** (Fig. 9.5.26), like a surface condenser, performs at constant saturation temperature outside the tubes with no condensate subcooling. Feedwater is generally heated to within 5°F of the

Fig. 9.5.26 Basic heater section for condensing heater.

saturation temperature. The addition of a **drain-cooling section** (Fig. 9.5.27), with a cover or enveloping baffle around the tubes of the inlet pass, can reduce the drip temperature to within 10 or 15°F of the incoming water. When the steam is at sufficient pressure (above 125 lb/in² abs) and contains enough superheat (more than 250°), a **desuperheating section**

Fig. 9.5.27 Basic heater section with drain cooling and desuperheater sections added.

(Fig. 9.5.27), using an enveloping baffle around the tubes at feedwater outlet, can raise the water temperature above the saturation temperature of the entering steam. Drain coolers and desuperheating sections improve overall plant heat rate by reduction of energy degradation. Subcooling of drains reduces flashing, erosion, vibration, and noise when drains are dropped to a lower-pressure heater. Special attention should be given to drains flashed to the main condenser, as this is a direct thermodynamic loss. As an alternate, these drains can be pumped forward into the feedwater line.

Heat-transfer rates for a condensing feedwater heater range, overall, between 500 and 900 Btu/(h · ft² · °F). (See Figs. 9.5.28 and 9.5.29.)

Fig. 9.5.28   Temperature vs. water travel, closed-feed heater.

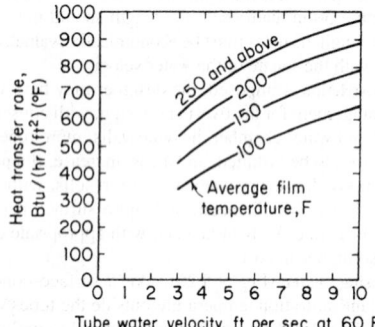

Fig. 9.5.29   Approximate condenser heat-transfer rates with 18 BWG admiralty metal tubes. Average film temperature = $t_s - 0.8\Delta t_m$.

Heat-transfer rates for drain-cooling and desuperheating sections must be separately evaluated, with overall ranges from 300 to 500 on the former and 80 to 140 on the latter. (See Figs. 9.5.30 and 9.5.31.) Normally, the three sections are included in the same shell (Fig. 9.5.27); when the cooler becomes too large, as with the lowest-pressure heater, a separate shell-and-tube exchanger may be justified.

Low-pressure heaters (upstream of the boiler feed pump) are usually designed for tube-side pressure of less than 90 lb/in² gage. This is based on the shutoff head of the condensate pump plus 10 percent. For pressures up to 300 to 400 lb/in² gage, a bolted cover with flange and

Fig. 9.5.30   Approximate drain-cooler heat-transfer rates.

gasket (Fig. 9.5.32a) is used, while higher pressures usually justify a welded joint to eliminate the gasket (Fig. 9.5.32b and c). A manway is utilized for access to the tube sheet. Tubes are generally rolled into the tube sheet.

High-pressure heaters (downstream of the boiler feed pump) are designed for shutoff head of the pump plus 10 percent, resulting in

Fig. 9.5.31   Approximate desuperheater section heat-transfer rates.

about 1,500 lb/in² gage for a nuclear plant and up to 5,000 lb/in² gage for fossil-fuel plants. Hemispherical head construction (Fig. 9.5.32c) and welded tube-to-tube sheet joints are almost exclusively used. Since the cost of the tubes represents the largest portion of the heater cost, material selection is important (see Table 9.5.11). Depending on the desired water-chemistry requirements, low-pressure heaters usually use admiralty metal, 90-10 Cu-Ni, stainless or carbon steel. High-pressure heaters require the use of the higher-strength alloys of 70-30 Cu-Ni, monel, stainless, or carbon steel. The thinnest wall possible [$t = Pd/(2s + 0.8p)$] or the industry minimum standard of 18 BWG (20 BWG for stainless steel) is utilized.

Fig. 9.5.32   Feedwater heater channels. (a) Bolted; (b) dished; (c) hemispherical.

Table 9.5.11 Material and Gage Correction Factor C_m
(For ⅝- to 1-in tubes)

BWG no.	Tube material							
	Ars-Cu	Admiralty	90-10 Cu-Ni	80-20 Cu-Ni	70-30 Cu-Ni	Monel	18-8 SS	Carbon steel
18	1.00	1.00	0.97	0.95	0.92	0.89	0.85	0.96
17	1.00	1.00	0.94	0.91	0.87	0.85	0.80	0.92
16	1.00	1.00	0.91	0.88	0.84	0.82	0.77	0.89
15	1.00	0.99	0.89	0.86	0.82	0.79	0.74	0.87
14	1.00	0.96	0.85	0.82	0.77	0.75	0.70	0.83
13	0.98	0.93	0.81	0.78	0.73	0.70	0.65	0.79

Performance Calculations

Notation

A = heat-transfer surface, ft^2
C_m = material and gage correction factor, Table 9.5.11
D = tube ID, in, Table 9.5.12
d = OD of tube, in (⅝ and ¾ in are most common)
F_1 = friction factor, Table 9.5.13
F_2 = water-temperature correction factor, Table 9.5.14
h = enthalpy, Btu/lb
k = tube diameter and gage factor, Table 9.5.15
L = active length of tubes per pass, ft
N = number of passes
n = number of tubes
P = design pressure, lb/in^2
p = pressure, lb/in^2
Q = total heat transferred, Btu/h
S = allowable design stress at tube design temperature, lb/in^2, from ASME Code
t = wall thickness, in (see Table 9.5.16); temperature, °F
U_o = overall heat-transfer rate, Btu/(h · ft^2 · °F) (the material correction factor, Table 9.5.11, must be applied to the overall heat-transfer rate)
V = water velocity, ft/s
W = steam flow, lb/h
W_w = feedwater flow, lb/h
Δp = pressure drop, lb/in^2
Δt_m = logarithmic mean temperature difference, °F (Fig. 9.5.28)

Table 9.5.12 Feedwater-Heater Tube Constants

Tube BWG	⅝-in OD		¾-in OD		⅞-in OD	
	D	$D^{1.24}$	D	$D^{1.24}$	D	$D^{1.24}$
18	0.527	0.452	0.652	0.589	0.777	0.731
17	0.509	0.433	0.634	0.567	0.759	0.710
16	0.495	0.418	0.620	0.554	0.745	0.695
15	0.481	0.404	0.606	0.538	0.731	0.678
14	0.458	0.380	0.584	0.514	0.709	0.652
13	—	—	0.560	0.488	0.685	0.625

Table 9.5.15 Tube Diameter and Gage Factor k

Diam, in	BWG no.				
	20	18	17	16	15
⅝	0.217	0.2407	0.2580	0.2728	0.289
¾	0.1732	0.1887	0.1995	0.2089	0.218
⅞	0.1445	0.1550	0.1624	0.1686	0.1748
1	0.1238	0.1314	0.1369	0.1413	0.1458

Table 9.5.13 Friction Loss, lb/in^2 per ft of Straight Travel

Water velocity (at 60°F), ft/s	F_1
3.0	0.018
4.0	0.031
5.0	0.047
6.0	0.065
7.0	0.085
8.0	0.108
9.0	0.134
10.0	0.162

Table 9.5.16 Tube-Wall Thickness

Gage, BWG	Thickness, in
15	0.072
16	0.065
17	0.058
18	0.049
19	0.042
20	0.035
22	0.028

Table 9.5.14 Water-Temperature Correction Factor*

Average water temp, °F	F_2
100	0.905
150	0.840
200	0.795
250	0.770
300	0.755
350	0.750
400	0.755
450	0.757
500	0.795
550	0.830
600	0.885

* Average water temperature = $t_s - \Delta t_m$.

Heat-balance data will provide the major parameters of steam pressure, feedwater flows, and temperatures. The choice of tube diameter and material must be made first. Tube thickness is calculated by

$$t = Pd/(2S + 0.8P)$$

Test pressure is 1.5 times design pressure.

Condensing-heater-size equations are

$$Q = UA\,\Delta t_m = W_w(h_3 - h_2) = W_s(\Delta h_s)$$

(See Figs. 9.5.28 and 9.5.29.) The minimum terminal temperature difference is 2°F.

The active length of tubes is calculated from $L = 500AV/(NW_w k)$, where maximum velocity is 10 ft/s at 60° F; the number of tubes, from $n = 3.82A/(Ld)$; the pressure drop, from $\Delta p = (L + 5.5D)F_1 F_2 N/D^{1.24}$.

The shell design pressure is 20 percent greater than the maximum operating pressure, rounded to the higher 25 lb/in^2 as a minimum.

A **drain-cooling section,** when added, is treated as a separate heat exchanger where the quantities t_v, t_o, t_i, W_s, and W_w are known; t_2 is the only unknown and is to be calculated from a heat balance on the section or $Q = W_s(h_v - h_o) = W_w(h_2 - h_1)$. Compute Δt_m; find the area of the drain-cooler section A_{dc} from $Q = U_o A_{dc}\ \Delta t_m$, with overall U_o from Fig. 9.5.30.

A **desuperheating section** is similarly calculated with t_v, t_6, t_4, p_6, h_6, h_4, W_s known. Allow approximately 1 percent pressure drop through the section to get p_v, and 80 to 90 percent desuperheating (30°F superheat entering condensing section) to obtain t_5 and h_5. By a heat balance on section, calculate h_3 from $Q = W_s(h_6 - h_5) = W_w(h_4 - h_3)$. With equivalent t_3, calculate Δt_m and substitute in $Q = U_o A\ \Delta t_m$ for area of desuperheating section (overall U_o from Fig. 9.5.31).

Materials

Heaters are designed in accordance with the Heat Exchange Institute Standards for Closed Feedwater Heaters and the ASME Boiler and Pressure Vessel Code, Sec. VIII, using the following materials:

Tubes:

Material		Max temp, °F	
		Rolled joint	Welded joint
Admiralty metal		350	
90-10 Cu-Ni	ASME	400	450
80-20 Cu-Ni	SB-395	400	475
70-30 Cu-Ni		400	525
70-30 Cu-Ni, stress-relieved		400	525
Monel, ASME SB-163		400	600
Carbon steel, ASME SA-556, grades A, B, and C		350	650
Stainless steel, ASME SA-249; TP 304 (welded)		350	650

Shell, nozzles: carbon-steel pipe and plant, ASME SA-515 grade 70; SA-106 grade B.

Channels, or heads, channel covers: carbon-steel plate, forgings, and pipe, ASME SA-515 grade 70; SA-105 grade II; SA-106 grade B.

Baffles and tube supports: steel plate, ASME SA-283 grade C.

Tube sheets: carbon-steel plate and forgings, ASME SA-515 grade 70; SA-105 grade II.

Carbon steel is used for temperatures up to 800°F. Chromium-molybdenum steel (ASME SA-387 grade C) is used for higher temperatures.

OPEN, DEAERATING, AND DIRECT-CONTACT HEATERS

REFERENCE: Heat Exchange Institute Standards for Deaerators and Deaerating Heaters.

Deaerators or deaerating heaters serve (1) to degasify feedwater and thus reduce equipment corrosion (see Sec. 6), (2) to heat feedwater regeneratively and improve thermodynamic efficiency (see Sec. 9), and (3) to provide storage, positive submergence, and surge protection on the boiler feed-pump suction (see Sec. 14).

Removal of oxygen and carbon dioxide from boiler feedwater and process water at elevated temperature is essential for adequate conditioning. In some power-plant applications, a well-designed surface condenser gives adequate deaeration and the accompanying exclusive use of closed feedwater heaters.

A modern **deaerator** will, by mechanical action, reduce O_2 content of effluent to less than 0.005 cm^3/L and CO_2 content to a negligible amount. Water must (1) be heated to and kept at saturation temperature, as the gas solubility is zero at the boiling point of the liquid, and (2) be mechanically agitated by spraying or cascading over trays for effective scrubbing, release, and removal of gases. Gases must be swept away by an adequate supply of steam. Since the water is heated to saturation conditions, the terminal temperature difference is zero with maximum improvement in associated turbine heat rate. Extremely low partial gas pressures, dictated by Henry's law, call for large volumes of scrubbing steam. **Vent condensers** of the shell-and-tube or direct-contact types, cooled by incoming feed, serve to recover heat and water before release of gases to the atmosphere.

The **tray-type deaerator** (Fig. 9.5.33) is prevalent. While it has some tendency to scale, it will operate at wide load conditions and is practically independent of water-inlet temperature. Trays can be loaded to some 10,000 lb/(ft$^2 \cdot$ h), and the deaerator seldom exceeds 8 ft in height.

Fig. 9.5.33 Tray-type deaerating heater section.

The **spray type** uses a high-velocity steam jet to atomize and scrub the preheated water. Applications are (1) marine service, where it is unaffected by ship roll and pitch, and (2) industrial plants where operating pressures are stable. It requires a temperature gradient, e.g., 50°F minimum, to produce the fine sprays and vacuums with the cold water required.

Materials

Open-feed heaters are designed to the ASME Boiler and Pressure Vessel Code. Materials used are: shell—steel, ASME SA-285 grade C and SA-515 grade 70; trays—stainless steel, type 304; baffles in vent condenser—stainless steel, type 304; spray valves—stainless steel, type 304.

EVAPORATORS

For general treatment of evaporators used in the preparation of makeup water, see Sec. 6.

9.6 INTERNAL COMBUSTION ENGINES

by D. N. Assanis, D. E. Cole, T. J. Jacobs, and D. J. Patterson

REFERENCES: Lichty "Internal-Combustion Engines," McGraw-Hill. Obert, "Internal-Combustion Engines and Air Pollution," Harper & Row. Taylor and Taylor, "The Internal-Combustion Engine," International Textbook. Vincent, "Supercharging the Internal-Combustion Engine," McGraw-Hill. Crouse, "Automotive Engine Design," McGraw-Hill. Patterson and Henein, "Emissions from Combustion Engines," Ann Arbor Science. Starkman, "Combustion Generated Air Pollution," Plenum. *Trans. ASME Trans. SAE Proc. I Mech E, Automobile Division, Automotive Inds.* Ferguson, "Internal Combustion Engines," Wiley. Heywood, "Internal Combustion Engine Fundamentals," McGraw-Hill. Blair, "The Basic Design of Two-Stroke Engines," SAE. Watson and Janota, "Turbo-charging the Internal Combustion Engine," Wiley. Benson and Whitehouse, "Internal Combustion Engines," Pergamon. Eastwood, "Critical Topics in Exhaust Gas Aftertreatment," Research Studies Press.

GENERAL FEATURES

Cycles

(See also Sec. 4.1)

In piston-type **internal combustion engines,** the combustion process is assumed to occur at constant volume (Fig. 9.6.1), at constant pressure (Fig. 9.6.2), or by some combination of these (Fig. 9.6.3). The constant-volume process is characteristic of the **spark ignition** or **Otto cycle;** the

Fig. 9.6.1 Otto cycle (constant-volume combustion, no excess air). Process of the ideal cycle: *a-b,* suction stroke, admission of the charge; *b-c,* compression stroke; at *c,* ignition of the compressed charge; *c-d,* combustion; *d-e,* expansion; at *e,* exhaust valve opens; *b-a,* exhaust stroke.

Fig. 9.6.2 Diesel cycle (constant-pressure combustion, 50 percent excess air). Processes of the ideal cycle: *a-b,* suction stroke, admission of air; *b-c,* compression stroke; *c-d,* injection and combustion of the fuel; *d-e,* expansion; at *e,* exhaust valve opens; *b-a,* exhaust stroke.

constant pressure is found only in the slow-speed **compression ignition** or **diesel cycle;** with both processes, the cycle is sometimes called a **mixed, combination,** or **limited pressure cycle.** Actually, the indicator card obtained from a high-speed, mixed-cycle, compression ignition engine may be similar to that obtained from a spark ignition engine. The fundamental differences are the methods of mixing the air and fuel (before compression in the Otto cycle, and usually near the end of compression in the diesel cycle) and the methods of ignition.

The nominal **compression ratio** (usually specified) is the displacement plus clearance volume divided by the clearance volume. The actual compression ratio is appreciably less than the nominal value because of late intake valve or port closing.

Fig. 9.6.3 Mixed cycle (constant-volume and constant-pressure combustion, 50 percent excess air). Processes of the ideal cycle: *a-b,* suction stroke, admission of air; *b-c,* compression of air; *c-d,* ignition and constant-volume combustion; *e-f,* expansion; at *f,* exhaust begins; *b-a,* exhaust stroke. Broken lines represent ideal cycle; solid lines show characteristics of actual cycle. Primes attached to letters show actual locations of events.

The **compression pressure** may be estimated from the relation $p = r_a^{1.33}$, where p_m is the intake manifold pressure and r_a is the actual compression ratio.

The actual compression pressure can be determined with a compression pressure gage that traps the gases by means of a check valve, thus indicating the maximum pressure under motoring conditions.

Spark ignition engines use volatile liquids or gases as fuel; have compression ratios between 6:1 and 12:1 (limited by combustion knock) and compression pressures from below 150 to above 300 lb/in² (1,030 to 2,060 kPa); use carburetors, gas mixing valves, or fuel injection systems; and operate on the Otto cycle. Gasoline is the fuel commonly used in airplane, automobile, small marine, small stationary, and tractor engines. Commercial gases such as blast furnace gas, coal gas, coke oven gas, carbureted water gas, producer gas, and natural gas are used in large stationary engines. The mixtures used generally are near chemically correct. Combustion pressures are usually 3.5 to 5 times the compression pressures. Load and speed are usually controlled by throttling the charge; piston speeds above 3,000 ft/min (15 m/s) are permissible.

Advantages are low first cost, low specific weight, low cranking effort required, wide variation obtainable in speed and load, high mechanical efficiency, and fairly low specific fuel consumption at high compression ratios and loads.

Compression ignition engines use liquid fuels of low volatility varying from fuel oil and distillates to crude oil (see Sec. 7.1), have compression ratios between 11.5:1 and 22:1 and compression pressures 400 to 700 lb/in² (2,760 to 4,830 kPa), and operate on the diesel or mixed cycle. Generally no ignition devices are used, although some lower-compression-ratio and multiple-chamber engines may require starting aids. Load and speed are controlled by varying the fuel quantity injected.

The **dual-fuel engine** is a diesel engine with a compression ratio which is too low to result in ignition, at the desired time. A pilot (small) injection of liquid fuel with good ignition quality is used to initiate the combustion process.

Advantages are low specific fuel consumption, high thermal efficiency at part loads, possibly lower fuel cost, no preignition, low CO and moderate hydrocarbon emission at low and moderate loads without exhaust treatment, suitability for two-stroke operation, and excellent durability (by design). The lower-compression engines are of simpler construction (usually valveless two-stroke cycle), are more lightweight, and have lower first cost, lower operating expense, and higher mechanical efficiency than the higher-compression engines.

The **four-stroke cycle** (four-cycle) engine requires four piston strokes or two crankshaft revolutions per cycle (Figs. 9.6.1 and 9.6.2). This engine cycle is used almost exclusively in automobile, tractor, and aircraft engines of all types and sizes and in engines of other classifications with the exception of larger outboard and handheld engines. Small four-cycle engines are always single-acting (combustion on only one side of the piston). The pumping loss, indicated by the area between the exhaust and the induction curves (Fig. 9.6.12), depends on valve openings and speed and amounts to 3 to 7 percent of the engine indicated power. Throttling increases the pumping loss and reduces the positive indicator card area. Fuel is introduced either in the intake charge or by direct cylinder injection.

Advantages compared with two-cycle crankcase compression spark ignition engines include wider variation in speed and load, cooler pistons, common crankcase in multicylinder engines, good lubrication without oil addition to the fuel, no fuel loss during exhaust stroke, lower specific fuel consumption, lower pumping losses, less exhaust dilution, lower hydrocarbon emissions, positive-displacement inlet and exhaust processes, and easier power regulation.

The **two-stroke cycle** (two-cycle) engine requires only one crankshaft revolution for each cycle. Exhaust ports in the cylinder wall are uncovered by the piston, or exhaust valves in the cylinder head are opened near the end (at 60 to 88 percent) of the expansion stroke, permitting the escape of exhaust gases and reducing the pressure in the cylinder (Fig. 9.6.4). The charge of air or combustible mixture flows into and is compressed in a separate crankcase compartment for each cylinder or by a compressor or a blower to slightly above atmospheric pressure. Intake ports are uncovered by the piston or intake valves and

Fig. 9.6.4 Indicator card for two-cycle diesel engine (16:1 compression ratio).

are opened soon after the opening of the exhaust, and the compressed charge flows into the cylinder, expelling most of the exhaust products, some charge escaping with the exhaust. Blowers or compressors, driven either mechanically or by an exhaust turbine, are often used to supply sufficient air to scavenge the cylinder and clearance space and, in some cases, to supercharge the engine. Although small engines use only ports and crankcase compression (Fig. 9.6.5), large engines usually have separate compressors or blowers and either ports or both valves and ports. With crankcase compression, low scavenging of the cylinder occurs, and the volumetric efficiency is low (30 to 70 percent). The crankcase compression work amounts to 7 to 12 percent of the total work, but scavenging blowers with displacements 20 to 80 percent greater than the piston displacement may require as high as 30 percent of the indicated work in high-speed engines. To eliminate hydrocarbon emissions from scavenging, fuel is added by direct cylinder injection or timed intake port injection (gasoline).

Advantages compared with four-cycle engines include the 50 to 80 percent greater, power output per crank revolution, smoother operation, low cost for valveless designs employing crankcase compression, low NO_x emissions (spark ignited), and light weight.

Fig. 9.6.5 Small two-cycle gasoline engine with crankcase compression.

ANALYSIS OF ENGINE PROCESS

Cycle Analysis The theoretical thermal efficiency η_a of the **air standard Otto cycle** depends only on the volumetric compression ratio r_c and working fluid heat capacity, which for pure air at room temperature (Sec. 4.1) is given by

$$\eta_a = 1 - \left(\frac{1}{r_c}\right)^{0.4} \tag{9.6.1}$$

For the **air standard diesel cycle** (constant-pressure combustion),

$$\eta_a = 1 - \frac{r_d^{1.4} - 1}{1.4 r_c^{0.4}(r_d - 1)} \tag{9.6.2}$$

where r_d is the volumetric expansion ratio of the constant-pressure combustion process.

However, the diesel engine is more closely approximated by the limited pressure combustion (dual) cycle for which the thermal efficiency is given by

$$\eta_a = 1 - \frac{r_p r_d^{1.4} - 1}{r_c^{0.4}[r_p - 1 + 1.4 r_p(r_d - 1)]} \tag{9.6.3}$$

where r_p is the pressure ratio during the constant-volume heat addition.

The efficiencies indicated by the air standard analysis are much higher than can be attained in practice, principally because of variable specific heats, dissociation, and heat and time losses. The improved fuel-air cycle analysis takes into additional consideration variable fuel/air mixture properties and dissociation of combustion products and results in lower, more realistic values for efficiencies (Figs. 9.6.6 and 9.6.7). It more closely approximates a fast-burn, low-heat-rejection engine. Real cycle analyses also include combustion, heat losses, and leakages, resulting in thermal efficiencies approximately 80 percent of those predicted by the fuel-air cycle analyses, but requiring complex and time-intensive computer computations.

The **mean effective pressure (mep)** is equal to net work divided by volumetric displacement, and is proportional to torque. Horsepower follows from

$$\text{hp} = P_{\text{mep}} SA \text{ rpm}/K \tag{9.6.4}$$

where P_{mep} = mean effective pressure, lb/in² (kPa); S = stroke, ft (m); A = total piston area, in² (m²); rpm = number of cycles completed per min; K = 33,000 (0.4566 for SI units).

Increasing the compression ratio and the air-fuel ratio increases the thermal efficiency of both Otto (Fig. 9.6.6) and diesel cycles (Fig. 9.6.7). Mean effective pressures depend on charge properties and thermal efficiency, both of which depend on compression ratio, air and fuel supplies (Figs. 9.6.8 and 9.6.9), and boost pressure.

Fig. 9.6.6 Indicated thermal efficiencies of various Otto cycle analyses and engine test results.

Fig. 9.6.7 Indicated thermal efficiency of the air standard diesel cycle with constant-pressure combustion.

Deviations from Ideal Processes

Changes in atmospheric pressure and temperature change power output although they do not appreciably affect thermal efficiency. The correction formulas adopted by SAE for brake power output are, for spark ignition engines,

$$\text{bhp}_s = \text{bhp}_o \left(\frac{P_s}{P_o}\right)\sqrt{\frac{T_o}{T_s}}$$

and for diesel engines,

$$\text{bhp}_s = \text{bhp}_o \left(\frac{P_s}{P_o}\right)\left(\frac{T_o}{T_s}\right)^{0.7}$$

where subscripts o and s indicate observed and standard conditions, respectively, P_o being the dry barometric pressure.

Pumping work (area EFGAE, Fig. 9.6.10) for the Otto cycle is required to overcome the resistance to flow of fresh charge into and exhaust products out of the cylinder. It varies from almost 0 for an engine running at slow speed with wide-open throttle to about 3 lb/in² (21 kPa) mep at high speed, and to 10 lb/in² (69 kPa) with idle throttle position. The negative loop (area AXFGA, Fig. 9.6.10) in a normally aspirated engine represents about 60 percent of the pumping work.

Volumetric efficiency is the ratio of the volume of air (for a liquid-fuel engine) and charge (for a gas engine) actually admitted, measured at atmospheric pressure and temperature, to the displacement volume. High manifold and cylinder temperatures and resistance to flow reduce the

Fig. 9.6.8 MEP for Otto cycle engine [air and liquid iso-octane supplied at 60°F (16°C)].

Fig. 9.6.9 MEP for diesel cycle engine [air and liquid dodecane supplied at 60°F (16°C)].

Fig. 9.6.10 Throttled Otto cycle.

volumetric efficiency. High-output engines have maximum volumetric efficiencies of 85 to 90 percent at rated speeds. Supercharged engines can have volumetric efficiencies well above 100 percent or can be designed to maintain high volumetric efficiencies with an increase in speed.

With fixed valve timing, volumetric efficiencies of automobile engines (Fig. 9.6.11) diminish appreciably as piston speeds exceed 1,200 to 1,600 ft/min (6 to m/s), decreasing to approximately 60 percent at top speeds of 2,500 to 3,200 ft/min (13 to 16 m/s). Volumetric efficiency varies appreciably with valve timing and the number of valves, especially intake valves. Variable timing is used to increase high-and low-speed volumetric efficiency.

Two-cycle crankcase compression engines have volumetric efficiencies of about 60 percent at low speeds and 50 percent or below at high speeds.

Fuels with high latent heats and low boiling temperatures reduce the charge temperature and result in high volumetric efficiencies (Table 9.6.1).

Air cleaners, governors, and carburetors increase the resistance to flow and decrease the volumetric efficiency.

The **compression stroke** begins with heat transfer from the cylinder walls to the charge but ends with heat transfer from the charge to the walls. The net result is a heat loss of about 0.5 to 1.5 percent of the heat value of the charge. Heat loss during compression lowers the

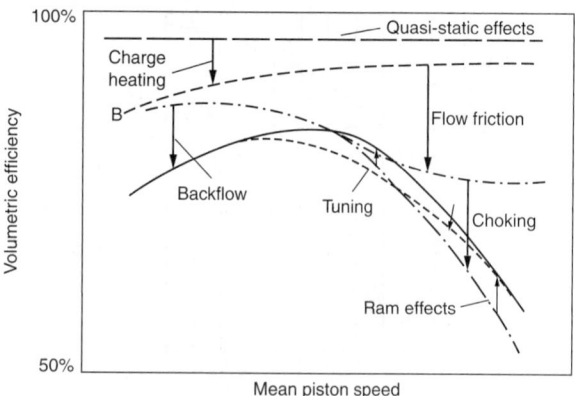

Fig. 9.6.11   The effect of engine factors on volumetric efficiency as a function of mean piston speed at a fixed valve timing. Quasi-static effects include exhaust residual and fuel evaporation factors.

compression curve and work, but lowers the expansion work a greater amount, thus slightly reducing the net work output.

The mean value of the exponent n in the approximate equation $pv^n = $ constant for the polytropic compression of a correct mixture of octane and air is about 1.33. For the diesel cycle in which the gas under compression is mainly air, the exponent is about 1.38 for a 15:1 compression ratio. Heat transfer into the charge increases the exponent, but leakage lowers its value.

The **combustion process** in the spark-ignition engine begins before the end of the compression stroke and ends after the beginning of the expansion stroke. Rassweiler, Withrow, and Cornelius (*Trans. SAE,* **34**, 1939) found a maximum variation of 30 percent in combustion time for a single-cylinder engine while variations of both maximum cylinder pressure and rate of pressure rise of about 30 percent were found by Patterson (SAE paper 660129) in a V-8 engine. These effects are thought to be due primarily to cyclic variations in swirl and turbulence at the spark plug, but can arise also from variations in air/fuel ratio and homogeneity of the mixture. Poor distribution in multicylinder engines also causes variations between cylinders. It has been shown that the loss due to the actual finite combustion time is 3 to 6 percent of the ideal efficiency with constant-volume combustion. Any energy released before or after top-center piston position has an availability corresponding to the expansion ratio for the piston position at which the energy is released. Heat loss during the combustion process amounts to 5 to 7 percent of the heat supply of the charge at rated outputs and speeds for automotive engines. Combustion in the diesel engine begins almost simultaneously at several locations depending on pressures, temperature, and mixture composition.

Flame Travel   Flame begins at the spark plug in the spark ignition engine or at various points in the combustion chamber of the compression ignition engine and travels in all directions through the mixture. At the end of combustion, the first part of the charge to burn has reached a higher temperature than the last portion to burn, except in the constant-pressure process. Rassweiler and Withrow (Trans. *SAE,* **30**, 1935, p. 125) observed temperature differences of over 400°F (200°C) (non-knocking) and over 600°F (315°C) (knocking) in a nonturbulent combustion chamber.

The combustion of part of the charge compresses the remainder. Thus with 30 percent of the mass burned, the volume of the unburned portion will be only about 35 percent of the total-volume. This increases the temperature of the last fraction to burn, which in a constant-volume process is an increase of about 500°F (280°C) above its temperature at the beginning of combustion. In some cases, this may produce almost instantaneous combustion of remaining unburned parcels of charge, which could produce the undesirable abnormal combustion condition called **combustion knock.**

The mean adiabatic exponent for the **expansion process** is about 1.22, varying from about 1.20 at the beginning to 1.25 at the end of the process. Heat transfer and leakage increase the exponent. The heat transfer during the expansion stroke amounts to 8 to 12 percent of the heat value of the charge.

Blowdown begins at 80 to 90 percent of the expansion stroke. This reduces the work about 1 to 2 percent. High gas velocities [1,200 to 1,500 ft/s (360 to 460 m/s)] are attained, and heat transfer, including that through exhaust port walls, amounts to 10 to 20 percent of the heat value of the charge, but represents practically no loss of work or availability.

The **exhaust process** is the discharge of some of the gases from the cylinder by the piston. The exhaust pressure is usually above atmospheric (Fig. 9.6.12) but may run below atmospheric for part of the stroke. Heat loss during exhaust amounts to 3 to 5 percent of the heat value of the charge but does not represent any loss of work or availability. At full load about 80 percent of the gases in the cylinder escape during blowdown, and about 15 percent are pushed out of the cylinder during exhaust, the remainder being left in the clearance space at the

Fig. 9.6.12   Pumping loop of Otto cycle engine.

Fig. 9.6.13   Exhaust temperatures for Otto cycle engines.

Table 9.6.1   Relative Ideal Volumetric Efficiencies with Liquid Fuels
(Air and liquid fuel at 60°F)

Fuel	Chemically correct air-fuel ratio	Heat required per lb of fuel for complete evaporation, Btu/lb (J/g)	Percent evap with 200 Btu/lb fuel	Suction temperature, °F (°C)		Relative volumetric efficiency	
				Complete evaporation	200 Btu/lb fuel	Complete evaporation	200 Btu/lb fuel
Gasoline	14.2–14.8	150 (349)	81	106 (41)	81 (27)	100	100
Benzene	13.26	169 (392)	100+	50 (10)	68 (20)	111	103
Ethyl alcohol	8.99	425 (989)	55	71 (22)	55 (13)	107	105
Methyl alcohol	6.46	519 (1,207)	49	68 (20)	45 (7)	108	107

end of the exhaust stroke. At partial load, the blowdown pulse is smaller and the residual fraction higher. Variable valve timing and/or lift can change the residual fraction considerably.

Exhaust gas temperatures vary with speed and load (Figs. 9.6.13 and 9.6.14), high loads and speeds resulting in the highest temperatures.

Fig. 9.6.14 Exhaust temperatures for diesel cycle engines.

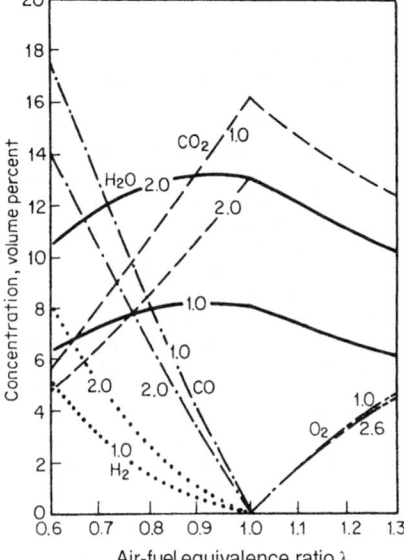

Fig. 9.6.15 Theoretical mole percent of principal combustion products of hydrocarbon fuels for fuel hydrogen/carbon ratios. Air/fuel equivalence ratio, i.e., the inverse of the fuel equivalence ratio ϕ, is the actual mixture air/fuel ratio divided by the chemically correct air/fuel ratio.

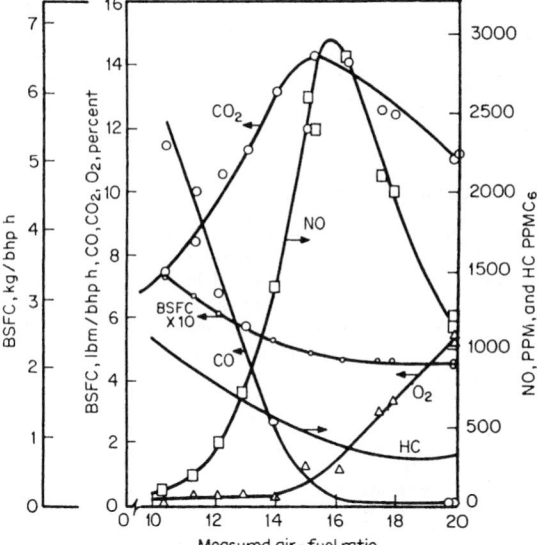

Fig. 9.6.16 Gasoline engine basic specific fuel consumption and emissions vs. air/fuel ratio without exhaust after treatment.

The first gases escaping during release are at the highest temperatures. The low values for diesels at light loads result from the lean-burning nature of conventional diesel engines.

Exhaust Gas Analysis The composition of exhaust varies depending upon the equivalence ratio and the hydrogen/carbon ratio of the fuel. Figure 9.6.15 shows the calculated composition including water (wet basis) for complete combustion. Charts of theoretical exhaust composition for various fuels have been prepared by D'Alleva (*General Motors Res. Rept.,* **372,** 1960) and Eltinge (*SAE Trans.,* 1968). In practice, non-equilibrium products, namely, unburned hydrocarbons and nitrogen oxide, appear in trace quantities. Figure 9.6.16 shows measured exhaust composition on a dry basis for a fuel with a hydrogen/carbon ratio of 1.85, typical of gasoline used for emission certification.

Equations for calculating the air/fuel ratio from exhaust CO, CO_2, HC, and O_2 have been developed by Spindt (*SAE Trans.,* 1966) and others. Such determinations and those for calculation of remaining but unmeasured exhaust species depend upon carbon, hydrogen, oxygen, and nitrogen balances employing measured values for some exhaust constituents (Obert, "Internal Combustion Engines").

Table 9.6.2 summarizes the principal exhaust gas constituents and combustion efficiency of an older gasoline engine without exhaust after

Table 9.6.2   Exhaust Gas Constituents and Combustion Efficiency of Older Gasoline Engines*

Air	Percent by volume						$\frac{H_2O}{CO_2}$	$\frac{H_2O}{Fuel}$	Combustion[†]
Fuel	CO_2	O_2	CO	H_2	N_2	H_2O			eff., percent
11	8.76	0.15	9.14	4.66	77.08	13.78	1.57	0.972	66.7
12	10.18	0.44	6.65	3.39	79.13	13.93	1.37	1.043	73.8
13	11.60	0.59	4.31	2.20	81.09	14.16	1.22	1.122	81.5
14	13.02	0.63	2.09	1.07	82.99	14.46	1.11	1.205	89.6
15	13.23	1.35	0.99	0.50	83.72	14.09	1.06	1.247	93.8
16	12.62	2.49	0.68	0.35	83.65	13.30	1.05	1.256	94.8
17	12.00	3.55	0.48	0.25	83.51	12.54	1.05	1.261	95.5
18	11.45	4.49	0.30	0.16	83.39	11.88	1.04	1.267	96.2
19	10.90	5.36	0.20	0.10	83.23	11.25	1.03	1.269	96.5
20	10.40	6.15	0.11	0.06	83.07	10.68	1.03	1.272	96.9
21	9.92	6.86	0.08	0.04	82.90	10.16	1.03	1.271	96.9
22	9.44	7.55	0.06	0.03	82.71	9.65	1.02	1.268	96.8
23	9.00	8.18	0.05	0.03	82.53	9.19	1.02	1.266	96.7
24	8.60	8.74	0.06	0.03	82.37	8.78	1.02	1.264	96.6

* Gerrish and Voss, *NACA Rept.* 616, 1937.
† Low-heat-value basis.

treatment. HC and NO were not measured. Modern engines, both gasoline and diesel, have combustion efficiencies approaching 99 percent and, with exhaust aftertreatment and the excellent fuel distribution made possible with fuel injection systems, have exhaust constituents that approach the theoretical levels of Fig. 9.6.15. Consequently, there is little fuel economy gain with better combustion efficiency per se.

U.S. AUTOMOBILE ENGINES

The size (displacement) of U.S. automobile engines generally varies from 61 in³ (1.0 L) to 488 in³ (8 L). Rated horsepower and displacement vary greatly depending on application with horsepower per unit of displacement generally increasing during the period from 1994 to 2005 (Fig. 9.6.17). Most U.S. automobiles are spark-ignited and are fueled with gasoline, or a blend with ethanol (minor).

Fig. 9.6.17   Trend of average U.S. automobile power density. (*Source: Center for Automotive Research, Ann Arbor, MI.*)

Emission control and the need for better fuel economy brought about (1) variable valve timing and lift, (2) cylinder cutout, (3) more precise and accurate fuel metering (carburetion and fuel injection), (4) controlled inlet air temperatures, (5) electronically controlled fuel management (most with closed-loop feed-back to maintain a chemically correct mixture), (6) exhaust gas recirculation, and (7) more stable ignition timing. Current advertised horsepower is based on net rather than gross horsepower. Net horsepower is measured with a fully equipped engine, including all parts needed to perform its intended function unaided, such as fuel pump, oil pump, intake air system, and exhaust system.

With a fixed valve timing system, the closure of intake valves (approximately 50° crankshaft rotation after bottom dead center) is selected to give a rising volumetric efficiency and torque as the rotation speed is increased. At high speeds, restriction in the intake system, developed primarily by the intake valves, begins to throttle the engine. Automotive engines typically develop maximum torque at approximately 3000 r/min. Despite reduced torque at higher speeds, the power peaks at 5000 to 6000 r/min. Higher powers are obtained with electronic fuel injection, variable valve timing, and other modifications, such as variable intake tuning. Hybrid vehicles combining smaller conventional gasoline or diesel engines, augmented by battery-powered electric motors, are in the marketplace in increasing numbers.

Engines have been produced with various numbers of cylinders over the years (from 2 to 16), but the most common engines in modern cars

Fig. 9.6.18   GM Northstar V8, engine (2006). (*Courtesy: General Motors.*)

are in-line four-cylinder designs for smaller cars and V-6 or V-8 designs (Fig. 9.6.18) for larger cars. All V-8 engines have an angle of 90° between the two banks of four cylinders, whereas the V-6 engines have an angle of either 60° or 90° between two banks of three cylinders. Smaller engines, in terms of both cylinder number and displacement, provide improved fuel economy, which is particularly important because of rigid federal corporate average fuel economy (CAFE) standards. Neither four- nor six-cylinder engines operate as smoothly as the V-8 because of less effective dynamic balancing and greater torque fluctuations. With the rapid growth in the use of transverse engines with front drive, and smaller and lighter vehicle structures with sloping aerodynamic front hoods, packaging constraints on automobile engines are severe. For this reason, push-rod engines remain popular in some applications, as they require less physical vehicle space.

All modern engines employ compact, rigid structures. Cylinder blocks and heads are made of cast iron or aluminum. Aluminum use is increasing in order to reduce weight. Other materials, such as magnesium have been used for the engine block. Four-cylinder engines use three or five crankshaft bearings; the V-6 uses four, and the V-8 uses five. Cylinder bore is usually greater than piston stroke.

Liquid cooling passages are provided around the bore of each cylinder as well as at the hot regions of the combustion chambers and exhaust ports in the head. Modern pistons are almost always made of aluminum. Crankshafts and connecting rods are usually made of cast iron or forged steel. Four-cylinder engines and some six-cylinder engines use a separate crank throw (eccentric portion of the crankshaft) for each connecting rod, whereas in V-8s, connecting rods for corresponding cylinders on each side of the V are side by side on the same crank throw.

Detachable cylinder heads contain the inlet and exhaust valves (overhead valves). Often engines employ a single exhaust and inlet valve, although there is a growing trend toward designs with three or four valves per cylinder (Table 9.6.3). Most automotive engines use belt or chain-driven overhead camshafts. Exhaust gas is directed through short passages to the exhaust collector or manifold, thereby minimizing heat loss and improving catalytic converter performance.

The charge is typically delivered through a compact inlet manifold, although intake tuning is often used, some with variable geometry. Fuel management is provided by an electronic fuel injection system with one injector for each inlet port and injection timing for each cylinder. In most cases, feedback from an exhaust gas oxygen sensor is used to provide precise control of air/fuel ratio and optimize performance of the emission control system. Electronic ignition virtually eliminates the need for periodic ignition system maintenance. Much attention has been directed to reducing friction losses in pistons, rings, valve train,

Table 9.6.3 Selected U.S. Automobile Engines

Make and model	Type and no.*	Bore	B/S ratio	Displacement, in³ (L)	Valves per cyl.	Comp. ratio	bhp	r/min	bhp/L	Vehicle weight, lb
General Motors										
Saturn L-Series	DOHC I-4	86.00 mm	0.91	134 (2.2)	4	10	140	6,500	63.6	3,237
								7,000(auto)		
Cadillac CTS	DOHC V-6	89 mm	1.02	170 (2.8)	4	10	210	7,200(man)	75	3,960
Buick Century	OHV V-6	89.00 mm	1.06	191 (3.1)	2	9.6	175	6,000	54.5	3,342
Chevrolet Impala	OHV V-6	92.00 mm	1.1	204 (3.4)	2	9.5	180	6,000	52.9	3,389
Pontiac Grand Prix	OHV V-6	96.52 mm	1.12	231 (3.8)	2	9.4	200	6,000	56.6	3,583
Chevrolet Silverado	OHV V-6	101.6 mm	1.15	272 (4.3)	2	9.2	195	5,600	45.3	4,360
Buick Rainier	OHV V-8	96.01 mm	1.04	325 (5.3)	2	9.95	300	5,750	56.6	4,417
Ford										
Ford Focus	I-4	3.44 in	1.05	121 (2.0)	4	10	136		68	
Ford Mustang GT	OHV V-8	3.6 in	1	281 (4.6)	3	9.8	300		65.2	
Ford Taurus	V-6	3.50 in	1.13	181 (3.0)	2?	10	201		67	
Ford Five Hundred	V-6	3.5 in	1.12	183 (3.0)	4	10	203		67.7	
Ford GT	DOHC V-8 supercharged	90.2 mm	0.85	330 (5.4)	4	8.4	550		101.9	
Daimler Chrysler										
Dodge Neon	SOHC I-4	87.5 mm	1.05	2.0 L	4	9.3	132	5,600	66	2,581
Dodge Charger	SOHC V-6	96 mm	1.19	5.7 L	4	10	250	6,400	53.2	3,800
Chrysler 300	DOHC V-6	86 mm	1.1	2.7 L	4	9.7	190	6,400	70.4	3,721
Chrysler Sebring	DOHC I-4	87.4 mm	0.87	2.4 L	4	9.4	150	5,500	62.5	3,172
Chrysler Pacifica	SOHC V-6	96 mm	1.19	3.5 L	4	10	250	6,400	71.4	4,383
Jeep Grand Cherokee	SOHC V-6	93 mm	1.02	3.7 L	2	9.6	210	5,200	56.76	4,195
Jeep Wrangler	DOHC I-4	88 mm	0.87	2.4 L	4	9.5	147	5,200	61.25	3,235

* I = in-line, V = vee type.
SOURCE: 2005 Manufacturers' websites.

bearings, belts, and engine accessories. This especially improves part-load fuel economy.

Quiet operation is essential in automobile engines and requires close manufacturing control of clearances. Most engines have hydraulic valve lifters and means of rotating valves to prolong life. Exhaust valve seats are hardened or incorporate inserts to prevent recession .

It is standard practice to provide pistons with three rings above the wrist pin—two narrow compression rings above one oil scraper ring, with several drain holes from the back of the ring groove to return oil to the crankcase. All compression rings are narrow, are made of cast iron from 0.063 to 0.0787 in (0.160 to 0.200 cm) wide, and have a wear-resistant coating, such as chromium (particularly for the top ring) and tin, or compounds (iron oxide, phosphates), which minimizes the running-in period without danger of scuffing or scoring the cylinders. Oil scraper rings are of steel or cast iron, generally are a little less than 3⁄16 in (0.476 cm) wide, and are provided with internal expander springs to increase pressure against the cylinder walls. Many oil scraper rings, especially those of steel, also use wear-resistant coatings. Low friction bore or piston skirt coatings are sometimes used.

Liquid coolant under a pressure of 15 lb/in² (103 kPa) is standard practice. Coolant jacket temperatures are controlled by thermostats which open at approximately 195°F (90.6°C), permitting coolant from the cylinder heads to flow to the radiator. All engines either are designed with coolant all around each cylinder or are "siamesed" and with jackets the full length of the piston travel. The water jacket cools the oil also.

FOREIGN AUTOMOBILE ENGINES

Foreign automobile engines sold in the United States are increasingly similar to engines produced in the United States, reflecting the growth of the global economy. (See Table 9.6.4.) These similarities include size and basic design features. They are mostly four-stroke, Otto cycle engines; the rest are diesels. All engines are liquid-cooled. Engines produced for areas of the world other than North America exhibit a broader range of characteristics. Very small cars with engines of less than 40 in³ (0.66 L) and 35 hp are found in some areas. Engines produced for the U.S. market are slightly larger, on average, in terms of piston displacement, than those used elsewhere. Non-North American engines often have very high specific output in terms of power per unit displacement because of taxation based on engine displacement, higher vehicle speed limits, and different pollution requirements. Many countries tax larger engines at a higher rate than smaller engines.

Most engines are normally aspirated, although some are turbocharged. Spark-ignition engines are often sized about 0.5 to 0.6 L/cyl, with some outside this range. Vehicle size and power requirements along with smoothness, efficiency, cost, size, and weight limitations determine the number of cylinders.

The maximum horsepower of foreign automobile engines ranges from approximately 50 to over 400 hp, equivalent to the range for U.S. engines. For a number of years, power has decreased because of emphasis on fuel economy and emission controls, but power levels have once more increased in response to consumer demand.

A number of foreign engines were considered "high tech" engines, with such features as three and four valves per cylinder, aluminum cylinder heads and blocks, and lightweight precision components. These features are now common in all cars produced worldwide. See Tables 9.6.3 and 9.6.4. Passenger car diesel engines are most popular in non-North American areas because of high fuel cost there. The newly developed hybrid systems are most popular in North America, although higher fuel cost worldwide is homogenizing engine offerings.

Table 9.6.4   Selected Foreign Automobile Engines

Make and model	Type and no.	Bore	B/S ratio	Displacement, L	Valves per cyl.	Comp. ratio	bhp	Rated rpm	bhp/L	Curb weight, lb
VW Golf		82.55 mm		2.0						
	SOHC I-4		0.89		2	10	115	5,200	57.5	2,972
	DOHC I-4			1.8						
VW Jetta 1.8t	Turbo	86.4 mm	0.94		5	9.5	180	5,500	100	3,197
	SOHC I-4			1.9						
VW Jetta Diesel	Turbo	79.5 mm	0.83		2	18.5	100	4,000	52.6	3,197
BMW 325i	DOHC I-6	3.35 in	0.97	3.0	4	10.7	215	6,250	71.7	3,285
BMW 750i	DOHC V-8	3.66 in	1.05	4.8	4	10.5	360	6,300	75	4,486
Toyota Camry	DOHC I-4	88.4 mm	0.92	2.4	4	9.6	160	5,700	66.7	3,109
Toyota Corolla	DOHC I-4	79.0 mm	0.86	1.8	4	10	130	6,000	72.2	2,502
	DOHC I-4 hybrid									
Toyota Prius	Electric	75.0 mm	0.89	1.5	4	13	76 (+67 electric)[†]	5,000	70.7 (95.3)[†]	2,890
Subaru Outback	SOHC Flat-4	99.5 mm	1.26	2.5	4	10	168	5,600	67.2	3,309
Mercedes CL-Class	SOHC V-8	97 mm	1.15	5.0	3	10	302	5,600	60.4	4,085
Mercedes E-Class	DOHC V-6	92.9 mm	1.08	3.5	4	10.7	268	6,000	76.6	3,691
Lexus GS300	DOHC I-6	87.4 mm	1.06	3.0	4	11.5	245	6,200	81.7	3,536
Nissan Maxima	DOHC V-6	95.5 mm	1.17	3.5	4	10.3	265	5,800	2.7	3,432
Lotus Elise	DOHC I-4	80 mm	0.9	1.8	4	11.5	190	7,800	105.6	1,777
Acura TL	SOHC V-6	89 mm	1.03	3.2	4	11	270	6,200	84.4	3,482
Audi A6	DOHC V-6	84.5 mm	0.91	3.1	5	12.5	255	6,500	82.3	3,957
Honda Civic	SOHC I-4	75.0 mm	0.79	1.7	4	9.5	115	6,100	67.65	2,449

* I = in-line, V = vee type.
[†] Numbers in parentheses are with the addition of an electric motor.
SOURCE: 2005 Manufacturers' websites.

TRUCK AND BUS ENGINES

Truck and bus engines (Figs. 9.6.19 and 9.6.20) are similar to automobile engines but, in general, are larger, having 300- to 1,000-in³ (4.6- to 16-L) displacement, are more rugged, and run at lower speeds. Both Otto and diesel cycles are used, with the diesel predominant in the large and medium sizes. The two-cycle blower scavenged diesel engine is being replaced by four-stroke designs (Fig. 9.6.20). Water cooling is universally used. The gasoline engines are naturally aspirated and have compression ratios slightly lower than automobile engines but are valved similarly, have similar performance curves, and attain maximum brake horsepower (bhp) (55 to above 300) at a mean piston speed aver-

Fig. 9.6.20 GM Duramax medium-duty four-stroke diesel engine with bowl-in-piston combustion chamber, direct fuel injection, and turbocharging. (*Courtesy of GM.*)

aging about 2,600 ft/min (13 m/s). Maximum torque is obtained at about 1,500 r/min, appreciably lower than for automobile engines. Diesel engines, which are often supercharged and/or turbocharged, have piston speeds ranging from 1,500 to 2,100 ft/min (7.6 to 11 m/s) and operate at somewhat lower speeds than gasoline engines. Truck and bus gasoline engines can operate at speeds slightly higher than those at which maximum power is attained. Some diesel engines can be operated near the maximum power speed. Most are operated below (Fig. 9.6.21).

Truck and bus engines are usually built with four or six cylinders in line or with eight cylinders in V-type construction. The bore/stroke ratio

Fig. 9.6.19 Cummins model KT 450-hp turbocharged, four-stroke diesel engine.

Table 9.6.5 Selected Diesel Engines for Truck, Bus, Tractor, Marine, and Industrial Uses

Make and model	Designed for*	No.	Bore, in	B/S ratio	Displacement, in³	Compression ratio	Max continuous bhp @ r/min	Max torque lb · ft	(N · m)	At r/min
Case 445/M2	Tr	4	4.09	0.79	272		76 @ 2,200	236	(320)	1,400
Caterpillar C7	B, T	6	4.33	0.87	441	16.5	190–210 @ 2,500	520	(705)	1,440
Deere 4024T	Ind, Tr	4	3.4	0.83	149	18.2	41 @ 2,800	129	(175)	1,500
Detroit Diesel Series 60	T	6	5.12	0.81	778	17.25	390 @ 1,800	1,350		1,200
Deutz BF4M1013	Ind, T	4	4.25	0.83	290	17.6	125 @ 2,300		(464)	1,400
International DT 570	Ind, T	6	4.59	0.8	570	17.5	310 @ 2,200	950		2,200
Mack AI-350	T	6	4.875	0.75	728	16	350 @ 1,950	1,260	(1,708)	1,300
Perkins 1104C-44T	Ind	4	4.09	0.83	269	18.25	90 @ 2,400	274	(371)	1,400
Allis-Chalmers 2800	Ind, T	6	3.88	0.91	301	16.25	68 @ 2,200	225	(305)	1,400
Cummins B-785-C	Ind	8	5.50	1.33	785	15.9	220 @ 2,300	560	(759)	1,600
MAN DO826	B, T	6	4.25	0.86	419	16	270 @ 2,400	710⁻	(960)	1,500

* B = bus; T = truck; Ind = industrial; Tr = tractor.

Fig. 9.6.21 Torque and power profiles for 2006 GM 6600 6.6-L V-8 Turbo Duramax medium-duty four-stroke diesel engine. (See Fig. 9.6.20.)

ranges from about 0.75 to 1.4, the latter engines having the larger values (see Tables 9.6.5 and 9.6.6).

TRACTOR ENGINES

Both spark ignition (gasoline or liquefied petroleum gas) and compression ignition (diesel) engines are used in tractors. Almost all engines above 100 hp are diesels. Most engines are four-cycle, water-cooled, with valve-in-head arrangement. Integral bore, dry-sleeve, or wetsleeve cylinder liner arrangements are utilized. One-, two-, three-, four-, five-, six-, and eight-cylinder engines are used. All engines except the six-cylinder are usually in-line in order to minimize the width of the hood in the interest of driver visibility. Tractor applications have a high load factor, are relatively low-speed, and deliver high low-speed torque (Fig. 9.6.22). Off-road engine equipment now must meet emission standards in the United States.

STATIONARY ENGINES

Stationary engines may be either Otto or diesel cycle, use either liquid or gaseous fuel, and use the two- or four-stroke cycle. The power output ranges up to about 48,000 bhp per engine. Piston rods, which permit the crankcase to be isolated from blowby gases and the bottom face

Table 9.6.6 Selected Gasoline Engines for Truck, Bus, Tractor, Marine, and Industrial Uses

Make and model	Designed for*	No.	Bore, in (cm)	B/S ratio	Displacement, in³ (L)	Compression ratio	Rating† bhp at r/min		Max torque, lb · ft (N · m), at r/min	
Allis-Chalmers G-153	Ind	4	3.44 (8.74)	0.83	153 (2.5)	8	52	2,400	132 (179)	1,460
Jeep Cherokee	T, Tr, Gp, Ind	I-6	3.88 (9.53)	1.13	242 (4.0)	8.8	190	4,750	215 (292)	1,400
Case 201 G	Tr, Gp, Ind	4	4.0 (10.16)	—	251 (4.1)	7.5	87	2,400	221 (300)	1,400
Chevrolet 350	Gp	V-8	4.0 (10.16)	1.15	350 (5.7)	9.1	210	4,000	—	
Chevrolet 454	T	8	4.25 (10.79)	1.06	454 (7.4)	8.3	250	4,000	365 (495)	2,800
Chrysler H-225	M, Ind	6	3.41 (8.66)	0.83	225 (3.7)	8.4	132	4,000	203 (275)	2,000
Continental R 821-46	Tr, Ind	4	3.03 (7.70)	0.92	95.5 (1.6)	8.6	67.2	5,000	84 (114)	2,400
Dodge CT 900	T	8	4.19 (10.64)	1.12	413 (6.8)	7.5	238	3,600	355 (481)	2,000
Ford 158-G	Tr	3	4.20 (10.67)	1.10	158 (2.6)	7.8	45.7	2,100	120 (163)	1,350
Ford Aerostar	Gp	V-6	3.5 (8.89)	1.11	182 (3.0)	9.2	135	4,600	—	
Ford F150	T	I-6	4.0 (10.16)	1.00	300 (4.9)	8.8	145	3,400	—	
Hercules G-3400	Tr, Ind	6	4.0 (10.16)	0.89	339 (5.6)	7.5	143	2,800	311 (422)	1,400
International V5-401	T	8	4.13 (10.49)	1.10	400 (6.6)	7.7	226	3,600	355 (481)	2,000
Minneapolis-Moline HD 425	Tr	6	4.25 (10.79)	0.85	425 (7.0)	8.1	133	2,000	378 (512)	800
Oliver 1655	Tr	6	3.75 (9.53)	0.94	265 (4.4)	8.5	84	2,200	220 (298)	1,300
Volkswagen 124A	Ind	4	3.38 (8.59)	1.24	96.7 (1.6)	7.7	53	3,600	75.5 (102)	2,500

* B = bus; T = truck; Tr = tractor; M = marine; Gp = general purpose; Ind = industrial.
† hp without accessories.

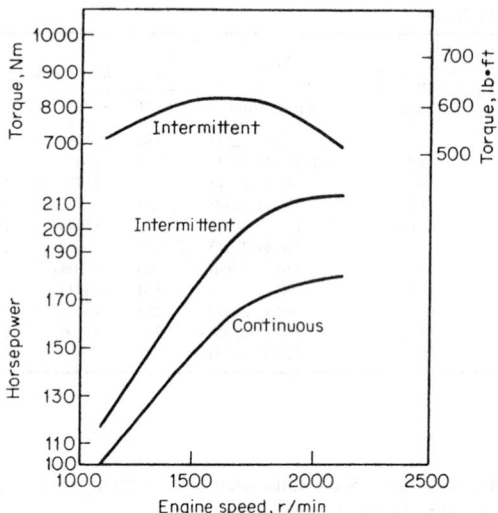

Fig. 9.6.22   Typical performance of a tractor diesel engine.

Fig. 9.6.23   Cross section of Sulzer V-type ZV-40 engine. Bore = 400 mm; stroke = 480 mm.

of the piston to function as an air pump, are commonly used. Nearly all engines are now single-acting. In the larger sizes, cylinders are built up of several parts, cylinder liners are used, and long pistons for single-acting diesel engines usually have about five piston rings. The valves are usually massive and in the larger sizes may be cooled. The exhaust pipe is sometimes water-cooled, and the pistons for large-bore engines are always liquid-cooled with oil or water. In general, there is much less integral construction (Fig. 9.6.23) in the large stationary engine than in the automobile type.

An interesting medium-speed, four-stroke diesel engine, V-type design with rotating pistons, is illustrated in Fig. 9.6.23. The symmetric piston is cooled with lubricating oil, and besides the reciprocating motion, it performs a slow rotating motion around its longitudinal axis. At each stroke, a new oil-wetted portion of the piston is always in contact with the pressure side of the liner, thus minimizing wear, oil consumption, and local overheating of the liner resulting from blowby.

Stationary engines usually operate at constant speed and are governed by throttling the charge of the Otto engine and by varying the amount of fuel injected into the diesel engine. Compression ratios as low as 6:1 are used with spark ignition engines and as low as 12:1 with

compression ignition engines. Stationary engines are usually built with 1 to 12 cylinders, with cylinders in a vertical line, or with a V-type arrangement having up to 16 or more cylinders.

MARINE ENGINES
(See also Sec. 11.3)

Marine engines may be either Otto or diesel cycle and may be normally aspirated or supercharged. Outboard engines range in power from less than 1 to more than 200 bhp and are generally of the two-cycle gasoline type. Modified automotive gasoline engines are also used in larger boats. Diesel engines range in power from below 20 to 48,000 bhp and may be either two- or four-cycle. Both automotive and stationary types are used in marine work; consequently, wide variations in specific weights are found. (See Tables 9.6.5, 9.6.6, and 9.6.7).

Table 9.6.7   Selected Outboard Engines

	Cylinders				Rating,		
Make and model	Type and no.*	Bore, in (cm)	B/S ratio	Displacement, in³ (cm³)	bhp	at r/min	Weight, lb (kg)
Mercury							
Verado 135	I-4	3.23 (8.2)	1	105.7 (1,732)	135	5,200	510 (231)
OptiMax 150	V-6	3.50 (8.9)	1.32	153 (2,507)	153	5,250	431 (195)
OptiMax 90	I-3	3.63 (9.2)	1.21	92.96 (1,526)	90	5,000	375 (170)
40 EFI	I-3	2.56 (6.5)	0.9	45.6 (747)	40	5,500	216 (98)
Johnson							
J25PL4	I-3	2.56 (6.5)	1.08	36.4 (597)	25	5,000	212 (96)
J115PL	V-4	3.6 (9.1)	1.4	105.4 (1,727)	115	4,500	335 (152)
J175PL	V-6	3.6 (9.1)	1.4	158 (2,589)	175	4,500	383 (174)
Evinrude							
E60DPL	I-2	3.6 (9.1)	1.4	52.7 (863)	60	5,750	240 (109)
E200DPL	V-6	3.6 (9.1)	1.4	158 (2,589)	200	5,000	419 (190)

* I = in-line, V = vee type.
SOURCE: 2005 Manufacturers' websites.

Fig. 9.6.24 Johnson two-cycle, 30-hp outboard motor.

Fig. 9.6.26 Specific fuel consumption curves.

Fig. 9.6.25 Performance of Johnson outboard motor.

Fig. 9.6.27 Performance of six-cylinder $3\frac{7}{16} \times 4\frac{3}{4}$ in (8.73 × 12.07 cm) Chrysler marine engine with 7:1 compression ratio.

The outboard two-cycle engines (see Table 9.6.7) have crankcase compression (Fig. 9.6.24) and may have two ports, three ports, or two ports with a rotary or reed crankcase inlet valve for the highest outputs (Fig. 9.6.25). Otto four-cycle engines have either L-head or valve-in-head construction, and diesel four-cycle engines are invariably of valve-in-head construction.

Regular gasoline is required for Otto cycle engines, the fuel and lubricating oil being mixed for outboard two-cycle engines. A comparison of a number of fuel-consumption curves for various types of engines shows the large low-speed marine diesel engine as having the lowest specific fuel consumption (Fig. 9.6.26).

Low-speed engines are directly connected to the propellers, and high-speed engines are connected through reduction gearing. Outboard engines have maximum speeds of 4,500 to 6,000 r/min with the piston speeds ranging from 1,000 to about 2,500 ft/min (5.0 to 12.7 m/s).

Maximum engine speed is obtained at the intersection of the propeller/horsepower and the maximum horsepower curves (Fig. 9.6.27). This may occur well below the speed for maximum horsepower or beyond this speed. The effect of throttling to reduce speed is to increase the specific fuel consumption considerably.

SMALL INDUSTRIAL, UTILITY, AND RECREATIONAL VEHICLE GASOLINE ENGINES

The Otto cycle engine with two or fewer cylinders is predominant in this class (Table 9.6.8). Fuel economy is generally not a major concern, whereas simplicity and performance are important. For reciprocating engines, the L-head design is predominant. Both high- and low-speed engines are used, depending on the application. Engines of this type generally employ air cooling of the cylinder and charge cooling of the interior. Water or liquid cooling is used on a few high-output small engines. Higher-mep air-cooled engines used forced fan cooling. Two-stroke engines are often used in applications where minimum weight and cost are important.

Fuel and ignition systems and other auxiliary equipment are usually less sophisticated than those on larger engines. A given engine is often used in a wide range of applications.

Engines used in motorcycles and in all-terrain vehicles are generally very sophisticated and constructed with great precision. Features such as overhead valves and overhead cams are common. The specific output of these engines is often high as measured by power per unit of piston displacement.

Table 9.6.8 Selected Small Gasoline Engines

| | | | Cylinders | | | | | Continuous rating, bhp at r/min | Max torque, lb·ft (N·m), at r/min | Weight, lb (kg) |
Make and model	Cycles	Type	Bore, in (cm)	B/S ratio	Displacement, in³ (cm³)	Compression ratio	Valve location*			
Briggs & Stratton 100900	4	V	2.5 (6.50)	1.17	10.43 (171)	6.2	L	3.40 3,600	5.9 (8.0) 3,100	30.5 (13.9)
Briggs & Stratton 80400	4	H	2.38 (6.04)	1.36	7.75 (127)	6.2	L	2.55 3,600	4.6 (6.2) 3,100	25.25 (11.9)
Chrysler 500	2	H, V	2.00 (6.45)	1.26	5.00 (82)	6.5		3.33 7,000	3 (4.1) 5,000	8.75 (3.9)
Homelite XL-123	2	H	1.81 (4.60)	1.31	3.60 (59)	8.3				13.0 (5.9)
Kohler MV18	4	H, V	3.12 (7.92)	1.13	42.2 (692)	6	L	18 3,600	26 (35) 2,600	130 (59)
O & R 13A	2	V	1.25 (3.18)	1.15	1.34 (22)	9		1.0 6,300	0.88 (1.2) 5,200	3.75 (1.70)
Onan AI	4	V	2.75 (6.99)	1.10	14.90 (244)	6.25	L	3.86 3,600	8 (10.8) 2,100	150 (68.2)
Rockwell Zf-400	2	V	2.56 (6.50)	1.00	24.29 (398)	7.5		24.0 6,250	28 (38) 6,250	62 (28.2)
Tecumseh LAV 35-1A	4	V	2.50 (6.35)	1.36	9.06 (149)	—	L	2.27 2,400	5.25 (7.12) 3,100	21.5 (9.77)
Tecumseh HH160	4	H	3.5 (8.89)	1.22	27.66 (453)	8.7	I	16.0 3,600	25 (33.9) 2,400	84 (38.2)
WE400 Wisconsin TRA-10D	4	V	3.3 (8.4)	1.20	23.7 (388)	6.5	I	10.0 3,400	16.5 (22.4) 2,800	80 (36.4)
Fichtel & Sachs KM 914	4	H	—	—	18.5 (303)	8	RC	20 5,000	—	56 (25.5)

* I = valve in head; L = L head; RC = rotary combustion.
SOURCE: *Auto. Ind.*, **48**, no. 7, Apr. 1, 1973, pp. 104–106.

Table 9.6.9 Selected Medium-Size Diesel Engines*

| | | | Cylinders | | | | Max rating, bhp at r/min | Weight, lb (kg) |
Make and model	Cylinder arrangement†	Cycles	No.	Bore, in (cm)	B/S ratio	Displacement, in³ (L)	Compression ratio	
Alco 251	V	4	12	9.0 (22.86)	0.86	8,016 (131.4)	11.5	3,150 1,000	35,581 (16,152)
DeLaval GVB-16	V	4	16	13.0 (33.0)	0.87	31,856 (522.0)	—	2,900 1,000	141,000 (64,005)
Electro-Motive 12-645	V	2	12	9.125 (22.0)	0.92	7,740 (128.4)	16	2,950 950	28,600 (11,350)
Fairbanks-Morse 12-38D 1/8	OP	2	12	8.13 (20.65)	0.81	12,443 (203.9)	16.1	3,840 900	45,000 (20,430)
White-Superior 40-VX-12	V	4	12	10.0 (25.4)	0.95	9,896 (162.2)	—	1,800 1,000	

* "Diesel and Gas Turbine Catalog," 1994. All engines are turbocharged.
† V = V type; OP = opposed.

LOCOMOTIVE ENGINES

Diesel engines are used for locomotive service because of high thermal efficiency (overall diesel brake thermal efficiency of about 38 to 40 percent with the net locomotive traction efficiency of about 32 percent for a diesel locomotive, compared with 6 percent for a steam locomotive), high availability for service, elimination of destructive pounding of rails, and elimination of shutdown fuel losses. The locomotive diesels are usually lighter, have lower displacement per cylinder, but higher outputs per cubic inch of displacement than the corresponding stationary diesel engine. These engines are built with 4 to 20 cylinders. Usually one engine is used per locomotive, and one or more locomotives are used per train. On an experimental basis, locomotive diesel engines have used coal-water slurries and natural gas as fuel. The convenience and low cost of conventional liquid diesel fuel will cause liquid fuels to predominate into the foreseeable future. (See Table 9.6.9.)

AIRCRAFT ENGINES

Small aircraft piston engines are air-cooled and invariably four-stroke. All types are naturally aspirated or supercharged. These engines have the lowest specific weight of piston engines (Table 9.6.10). Aircraft engines always have valve-in-head construction; they also have low specific fuel consumption, and the guaranteed minimum may be as low as 0.40 lb (0.18 kg)/(bhp · h). Maximum cruising bmep ranges from 100 to 150 lb/in^2 (690 to 1,030 kPa), while takeoff bmep ranges from 120 to 200 lb/in^2 (830 to 1,380 kPa), depending principally on the compression ratio, fuel, and supercharge.

Aircraft engines are valved to obtain high mean effective pressure at rated speed. High-output engines usually operate with compression ratios and supercharging levels that prohibit full-throttle operation at or near sea level with the specified fuel. Such engines can be operated at full throttle at or above a critical altitude, beyond which the power decreases with an increase in altitude. Below the critical altitude, the power decreases with a decrease in altitude, at constant intake-manifold pressure, because of the increase in exhaust back pressure and ambient temperature. Turbocharged engines, however, maintain essentially constant-power with constant manifold pressure from sea level to a critical altitude, which can be as high as 20,000 ft (6,100 m).

The performance of an aircraft engine varies with speed, manifold pressure, and altitude. At any given speed and with a fixed supercharger or blower ratio (supercharger r/min to engine r/min) and wide-open throttle, the bhp of an engine varies almost linearly with the density of the atmosphere. This characteristic makes it possible to estimate the power curves at altitude if the sea-level output is known. The power at 20,000-ft (6,100-m) altitude is about 50 percent of the sea-level wide-open-throttle output for any given speed.

Low specific fuel consumption is obtained by operating with reduced r/min for the desired power with a lean mixture setting, the optimum output at this point being about one-half to two-thirds of the normal rated output of the engine.

Aircraft engine cylinders are limited to a maximum diameter of about 6 in (0.15 m) by piston cooling and knocking difficulties that arise with large cylinders, high compression ratios, and high intake manifold pressures. The bore/stroke ratio is commonly above 1.0. The larger piston engine sizes are being replaced by turbine-type engines.

Piston aircraft engines are mostly of the horizontally opposed design (desirable profile) with an even number of cylinders, usually four, six, or eight. They are equipped with a dual ignition system, resulting in more reliable combustion.

WANKEL (ROTARY) ENGINES

The Wankel (rotary) SI engine is used in the Mazda RX series. A triangular rotor rotates on an eccentric shaft inside an epitrochoidal housing. The rotor tips are in constant contact with the housing and form three working chambers. The operating cycle (Fig. 9.6.28) follows the four-stroke cycle principle, although the output shaft rotates at 3 times rotor speed, providing one power stroke per crankshaft revolution (as in the single-cylinder two-cycle). A fixed gear meshes with an internal gear in the rotor to maintain the proper relationship between the rotor and housing. Housing cooling is by either water or air and rotor cooling by oil or fuel/air charge. Auxiliary equipment (fuel system, ignition system, etc.) is similar to that used on piston engines. A major problem is the sealing grid (90° intersection of seals and single-line seal at apexes of rotor), and this tends to produce high hydrocarbon emissions.

Intake   Compression   Ignition   Expansion   Exhaust

Fig. 9.6.28 Operating cycle of Wankel rotating combustion engine.

Table 9.6.10 Selected Piston-Type Aircraft Engines

			Cylinders					Ratings (METO)†			Weight§		
										At sea level (SL) or altitude, ft	bmep at max bhp, lb/in² (kPa)	Engine, dry, lb (kg)	Fuel octane no. required
Make and model	No.	Arrange-ments*	Bore, in (cm)	Stroke, in (cm)	Displacement, in³ (cm³)	Comp. ratio	bhp‡	r/min					
Avco Lycoming													
0-320-E	4	H	5.13 (13.03)	3.88 (9.86)	319.8 (5,240.6)	7.0	150	2,700	SL		138 (952)	244 (110.4)	80/87
I0-360-A	4	H	5.13 (13.03)	4.38 (9.86)	361.0 (5,915.7)	8.7	200	2,700	SL		163 (1,120)	293 (132.6)	100/130
TIGO-541-E	6	H	5.13 (13.03)	4.38 (9.86)	541.5 (8,873.6)	7.3	425	3,200	15,000		195 (1,340)	700 (316.8)	100/130
Teledyne Continental													
0-200	4	H	4.06 (10.31)	3.88 (9.86)	201 (3,293.8)	7.1	100	2,750	SL		143 (936)	190 (86.0)	80/87
10-470	6	H	5.0 (12.7)	4.0 (10.16)	471 (7,718.3)	8.6	260	2,600	SL		166 (1,140)	465 (210.5)	100/130
GTSIO-520-F	6	H	5.25 (13.34)	4.0 (10.16)	520 (8,521.3)	7.5	435	3,400	19,000		195 (1,340)	647 (292.8)	100/130

* H = horizontal (production of radial and vertical cylinder arrangements has been discontinued).
† METO = maximum except during takeoff.
‡ Takeoff power is usually higher than the recommended maximum continuous power.
§ Weights are for dry engines without hub or starter.

Table 9.6.11 Data on Fuel Properties

Fuel	Formula (phase)	Molecular weight	Specific gravity (density* kg/m³)	Heat of vaporization, kJ/kg†	Specific heat Liquid, kJ/(kg·K)	Specific heat Vapor c_p, kJ/(kg·K)	Higher heating value, MJ/kg	Lower heating value, MJ/kg	LHV of stoich. mixture, MJ/kg	$(A/F)_s$	$(F/A)_s$	Fuel octane rating RON	Fuel octane rating MON
Practical fuels													
Gasoline	$C_nH_{1.87n}(l)$	~110	0.72–0.78	305	2.4	~1.7	47.3	44.0	2.83	14.6	0.0685	92–98	80–90
Light diesel	$C_nH_{1.8n}(l)$	~170	0.84–0.88	270	2.2	~1.7	44.8	42.5	2.74	14.5	0.0690	—	—
Heavy diesel	$C_nH_{1.7n}(l)$	~200	0.82–0.95	230	1.9	~1.7	43.8	41.4	2.76	14.4	0.0697	—	—
Natural gas‡	$C_nH_{3.8n}N_{0.1n}(g)$	~18	(−0.79†)	—	—	~2	50	45	2.9	14.5	0.069	—	—
Pure hydrocarbons													
Methane	$CH_4(g)$	16.04	(0.72*)	509	0.63	2.2	55.5	50.0	2.72	17.23	0.0580	120	120
Propane	$C_3H_8(g)$	44.10	0.51 (2.0*)	426	2.5	1.6	50.4	46.4	2.75	15.67	0.0638	112	97
Isooctane	$C_8H_{18}(l)$	114.23	0.692	308	2.1	1.63	47.8	44.3	2.75	15.13	0.0661	100	100
Cetane	$C_{16}H_{34}(l)$	226.44	0.773	358		1.6	47.3	44.0	2.78	14.82	0.0675		
Benzene	$C_6H_6(l)$	78.11	0.879	433	1.72	1.1	41.9	40.2	2.82	13.27	0.0753	—	115
Toluene	$C_7H_8(l)$	92.14	0.867	412	1.68	1.1	42.5	40.6	2.79	13.50	0.0741	120	109
Alcohols													
Methanol	$CH_4O(l)$	32.04	0.792	1,103	2.6	1.72	22.7	20.0	2.68	6.47	0.155	106	92
Ethanol	$C_2H_6O(l)$	46.07	0.785	840	2.5	1.93	29.7	26.9	2.69	9.00	0.111	107	89
Other fuels													
Carbon	$C(s)$	12.01	~2‡	—	—	—	33.8	33.8	2.70	11.51	0.0869		
Carbon monoxide	$CO(g)$	28.01	(1.25*)	—	—	1.05	10.1	10.1	2.91	2.467	0.405		
Hydrogen	$H_2(g)$	2.015	(0.090*)	—	—	1.44	142.0	120.0	3.40	34.3	0.0292		

(l) = liquid phase; (g) = gaseous phase; (s) = solid phase; RON = research octane number; MON = motor octane number.
* Density in kg/m³ at 0°C and 1 atm.
† At 1 atm and 25°C for liquid fuels; at 1 atm and boiling temperature for gaseous fuels.
‡ Typical values.

SOURCES: E. M. Goodger, "Hydrocarbon Fuels; Production, Properties and Performance of Liquids and Gases," Macmillan, London, 1975. E. F. Obert, "Internal Combustion Engines and Air Pollution," Intext Educational Publishers, 1973. C. F. Taylor, "The Internal Combustion Engine in Theory and Practice," vol. 1, MIT Press, 1966. J. W. Rose and J. R. Cooper (eds.), "Technical Data on Fuel," 7th ed., British National Committee, World Energy Conference, London, 1977.

Advantages include small size, low weight, simple and fewer components (compared with piston-type four-cycle engines), superior breathing, no valves (rotor and its seals serve as valves), lower friction, low-octane fuel requirement, low NO emissions, and no reciprocating unbalance.

FUELS

Fuels for internal combustion engines are predominantly petroleum products. Natural and manufactured gases are also used in limited quantities. Data on properties of fuels which are commonly used for internal combustion engines are summarized in Table 9.6.11. Additional details on fuels are given in Sec. 7.1.

Gasoline

Gasoline consists of various amounts of many hydrocarbons, each having its vapor, pressure, and temperature characteristics. Also included are small amounts of additives such as deposit modifiers, antioxidants, metal deactivators, antirust agents, anti-icing agents, detergents, upper cylinder lubricants, and dyes. Different grades of gasoline are marketed on the basis of octane number. The overall volatility of a gasoline sample is specified in U.S. industry by the ASTM Distillation Test (D86) and the ASTM Vapor Pressure Test (D323, Reid Method). The distillation test records the increasing temperature of the vapor in the neck of a flask vs. percent evaporated as 100 mL of fuel is gradually heated and distilled. The vapor pressure test gives the pressure in a split-chamber bomb which contains the fuel sample and room air in thermal equilibrium with a 100°F (311 K) temperature bath. The pressure, corrected to standard initial bomb air conditions, gives the **Reid vapor pressure (RVP)** of the gasoline. Typical results for summer- and winter-grade gasolines are given in Table 9.6.12.

In general, volatility is specified by giving the initial boiling point (IBP); the 10, 50, and 90 percent evaporation points; and the endpoint (EP) temperatures as well as the Reid vapor pressure. These data indicate the effects of fuel volatility on engine performance, e.g., ease of starting, warm-up, and fuel economy (Figs. 9.6.29 and 9.6.30).

Reformulated Gasoline (RFG) The composition of gasoline is known to affect the exhaust and evaporative emissions. The C_4 and C_5 paraffins and olefins (photochemically active) are the main contributors to evaporative losses. Reformulated gasolines have much reduced amounts, and thus lower vapor pressures (Reid vapor pressure maximum of 48 kPa) and a maximum of 6 percent by volume olefins. Benzene, a toxic compound, is limited to a maximum of 0.85 percent by volume, while aromatics (photochemically active, deposits) as a class are limited to 25 percent by volume. Oxygenated fuel components (methanol, ethanol, etc.) are sometimes added at a level of 1.8 to 2.2 percent by weight. This reduces CO somewhat, especially from older vehicles without feedback fuel controls. For the federal RFG standards, sulfur is limited to 130 ppm; this lowers acid emissions and lengthens engine life. Further, the ASTM 90 percent distillation temperature is limited to 165°C. This minimizes deposits and leads to less enrichment required upon cold starting. Such a fuel may give fewer miles per gallon. The California Air Resources Board (CARB) applies even more stringent limits, with sulfur held to 30 ppm and the 90 percent distillation temperature limited to 143°C.

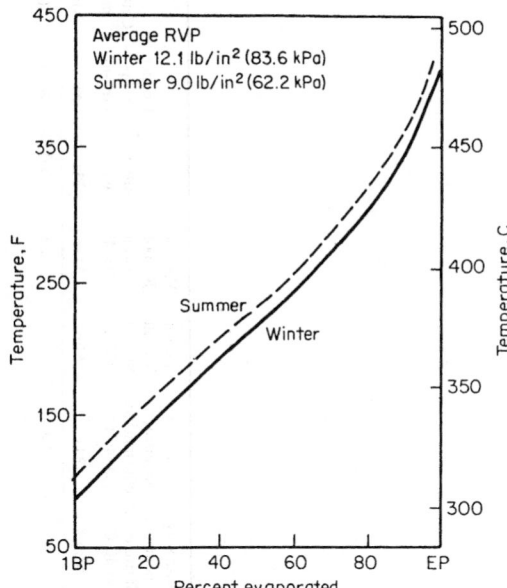

Fig. 9.6.29   Distillation curves for typical winter and summer gasolines. *(Ethyl Corp.)*

Fig. 9.6.30   Effects of volatility characteristics on engine performance.

Diesel Fuels   Automotive and railroad diesel fuels are derived from petroleum refinery products which are referred to as **middle distillates**. They have a higher distillation temperature range than gasoline. The properties of commercial distillate diesel fuels depend on the refinery practices employed and the nature of the crude oil from which they are derived, and thus vary substantially. Such fuels generally boil over a range between

Table 9.6.12   Average Volatility Characteristics of U.S. Motor Gasolines

| | Summer | | Winter | |
	Premium	Regular	Premium	Regular
IBP temp, °F (°C)	93 (34)	94 (34)	82 (28)	82 (28)
10% temp, °F (°C)	126 (52)	125 (52)	107 (42)	107 (42)
50% temp, °F (°C)	216 (102)	207 (97)	207 (97)	198 (92)
90% temp, °F (°C)	321 (160)	338 (170)	316 (157)	333 (167)
Endpoint temp, °F (°C)	402 (205)	414 (212)	397 (203)	407 (209)
RVP, lb/in² (kPa)	8.9 (61.3)	8.8 (60.6)	12.3 (84.8)	12.1 (83.4)

Table 9.6.13 Detailed Requirements for Diesel Fuel Oils[a,h,i] (ASTM D975)

Grade of diesel fuel oil	Flash point, °C (°F) Min	Cloud point °C (°F) Max	Water and sediment, vol % Max	Carbon residue on, 10 percent residuum, % Max	Ash, wt % Max	Distillation temperatures, °C (°F), at 90% point Min	Distillation temperatures, °C (°F), at 90% point Max	Viscosity kinematic, cSt (mm²/s), at 40°C Min	Viscosity kinematic, cSt (mm²/s), at 40°C Max	Viscosity Saybolt, SUS at 100°F Min	Viscosity Saybolt, SUS at 100°F Max	Sulfur,[d] wt % Max	Copper strip corrosion Max	Cetane no. Min
No. 1-D—a volatile distillate fuel oil for engines in service requiring frequent speed and load changes	38 (100)	b	0.05	0.15	0.01	—	288 (550)	1.3	2.4	—	34.4	0.0015	No. 3	40[f]
No. 2-D—a distillate fuel oil of lower volatility for engines in industrial and heavy mobile service	52 (125)	b	0.05	0.35	0.01	282[c] (540)	338 (640)	1.9	4.1	32.6	40.1	0.0015	No. 3	40[f]
No. 4-D—a fuel oil for low- and medium-speed engines	55 (130)	b	0.05	—	0.10	—	—	5.5	24.0	45.0	125.0	2.0	—	30[f]

[a] To meet special operating conditions, modifications of individual limiting requirements may be agreed upon between purchaser, seller, and manufacturer.

[b] It is unrealistic to specify low-temperature properties that will ensure satisfactory operation on a broad basis. Satisfactory operation should be achieved in most cases if the cloud point (or wax appearance point) is specified at 6°C (10°F) above the tenth percentile minimum ambient temperature for the area in which the fuel will be used. The tenth percentile minimum ambient temperatures for the months of October, November, December, January, and March are given in ASTM D 975 in maps for the 48 contiguous states and Alaska. This guidance is of a general nature; some equipment designs, use of flow improver additives, fuel properties, and/or operations may allow higher or require lower cloud point fuels. Appropriate low-temperature operability properties should be agreed on between the fuel supplier and purchaser for the intended use and expected ambient temperatures.

[c] When cloud point less than −12°C (10°F) is specified, the minimum viscosity shall be 1.7 cSt and the 90% point shall be waived.

[d] In countries outside the U.S., other sulfur limits may apply.

[e] Where cetane number by method D 613 is not available, method D 976 may be used as an approximation. Where there is disagreement, method D 613 shall be preferred method.

[f] Low atmospheric temperature as well as engine operation at high altitudes may require system use of fuels with higher cetane ratings.

[g] Millimetre squared per second (official SI unit) (Note: 1 cSt = 1 mm²/s).

[h] The values in SI units are to be regarded as the standard. The values in U.S. Customary system units are for information only.

[i] Nothing in this specification shall preclude observance of federal, state, or local regulations which may be more restrictive.

163 and 371°C (325 and 700°F). Their makeup can represent various combinations of volatility, ignition quality, viscosity, sulfur level, gravity, and other characteristics. Additives may be used to enhance specific properties. The SAE recommended practices are described in SAE J313 (July 2004). Permissible limits of significant diesel fuel properties are summarized in ASTM's classification chart D975, reproduced as Table 9.6.13.

Alternative Fuels   Other than gasoline and diesel fuels, alternative fuels used in various parts of the world include propane, frequently referred to as LPG; hydrous ethyl alcohol (ethanol), particularly in Brazil; gasohol, a blend of 10 percent by volume of gasoline and 90 percent by volume of denatured anhydrous ethanol. Other possible fuels include compressed or liquefied natural gas (CNG or LNG)—primarily methane, other alcohols—particularly methanol, which is also used in racing, and biomass fuels. Details can be found in SAE J1297.

GAS EXCHANGE PROCESSES

Intake Manifolds   The intake manifold distributes the air (diesel) or the air and fuel (Otto) to various cylinders of a multicylinder engine. Intake manifolds are **streamlined** (always with diesel), the **rake type,** or a combination of both. In spark-ignited engines with port fuel injection, the issue of manifold wetting is virtually eliminated. Runners may be long for tuning at low engine speeds or short to minimize flow losses (and improve tuning at high engine speeds). In diesel engines the runners are normally short, especially in medium-speed engines.

The overlapping of the intake valve periods produces complicated disturbances in the intake manifold. This has resulted in dual-level manifolds for six- and eight-cylinder engines with carburetors. With port fuel injection, single-level manifolds are often used, but dynamic air charging may lead to cylinder-to-cylinder air maldistribution.

Port fuel-injected engines have reduced volumetric efficiency due to the presence of vaporized fuel in the mixture (see Fig. 9.6.11). However, the process of vaporizing the fuel in the air helps to improve volumetric efficiency, as fuel vaporization lowers the mixture temperature, thereby increasing its density. The dew points or condensation temperatures for fuel in an air/fuel mixture depend on the fuel, air/fuel ratio, and total pressure (Table 9.6.14), and they may be determined from equilibrium air distillation data.

Exhaust Manifolds   In addition to adequate flow area, the most important consideration in exhaust manifold design is expansion. Small multicylinder engines employ a one-piece exhaust manifold. This is anchored at the center of the engine, and elongated holes are provided at other points for expansion. Large engines employ multipiece manifolds with expansion joints. They are usually water-cooled in marine applications to prevent strain and for safety reasons. Exhaust manifolds can be insulated to conserve energy for turbocharging and/or exhaust aftertreatment.

Manifold flow area should be large enough to avoid local back pressure buildup during exhaust and in connection with the total exhaust system should not raise back pressure excessively at maximum flow. A flow area of 0.7 times the intake manifold area may be used for four-cycle

engines. For two-cycle engines, the exhaust flow area usually exceeds the intake flow area. Overlapping of the exhaust valve periods in multicylinder engines makes multiple manifolds desirable.

Mufflers (see Sec. 12.6)   In order to be efficient as sound silencers, exhaust mufflers must decrease the exhaust gas velocity and either absorb the sound waves or cancel them by interference with other waves from the same source. Mufflers should have volumes 6 to 8 times the piston displacement and may contain baffles with or without holes. Mufflers that cancel sound waves by interference usually break the waves into two parts which follow different paths and meet again, out of phase, before leaving the muffler (Davis et al., NACA Rept. 1192).

Exhaust **muffle pits** are used with large engines, are usually made of concrete, have a volume about 20 times the piston displacement, and may be open or provided with baffling or partly filled with loose stone through which the gases must pass. A stack is usually provided for the escape of the gases, and a drain for the discharge of the condensed water from the exhaust products.

Exhaust **back pressure** should be kept to a minimum since an increase of 1 lb/in^2 (6.9 kPa) in back pressure decreases the maximum power output about 2 percent, about 1 percent being due to more exhaust work and the balance of the effect of increased clearance gas pressure on volumetric efficiency.

Supercharging

Supercharging increases the amount of charge per cycle (above that of the normally aspirated engine) by increasing its pressure and hence its density. Supercharging permits more fuel to be burned and is a practical means to increase engine power, but it increases mechanical and thermal stresses in the engine. For aircraft piston engines, it is a means to high-power output for takeoff and to offset the rare atmosphere at high altitude. For diesel engines supercharging results in smaller and lighter power plants. It is especially practical for diesel engines because supercharging does not add to the fuel quality required, and because diesel engines do not use combustion air as effectively as spark ignited engines.

Three alternative methods can be used to achieve supercharging. In **mechanical supercharging,** a positive-displacement type of Roots blower or compressor is geared directly to the engine shaft and thus is driven by engine power. This type is desirable for variable-speed engines (diesel railcar) where high torque is required at various speeds, since the pressure and capacity characteristics of this type do not decrease with speed. In **turbocharging,** exhaust energy drives a turbine, which, in turn, provides the necessary work to drive the turbocharger's compressor, mounted on the same shaft as the turbine (Fig. 9.6.31). The centrifugal compressor is particularly adaptable to aircraft engines because of high capacity, small size, and low weight. It is also used with truck diesel engines to improve the volumetric efficiency and power at high engine speeds. Centrifugal blowers may have one or two stages with cooling between stages or after the single or last stage. The rotors run at maximum speeds of 15,000 to 70,000 r/min. Finally, in **pressure wave supercharging,** wave action and resonant tuning in the intake and exhaust systems are used to compress the charge and thereby boost volumetric efficiency.

Table 9.6.14   Condensation Temperatures of Liquid Fuels, °F (°C)

| Fuel | Deg API | ASTM distillation | | | | Condensation temp. | | | |
| | | Initial point | 50 percent | 90 percent | Endpoint | Atmospheric pressure Air/fuel ratio | | Pressure, ⅔ atmospheric Air/fuel ratio | |
						12:1	15:1	12:1	15:1
Gasoline	60	100 (38)	249 (121)	354 (179)	400 (204)	118 (48)	114 (46)	111 (44)	107 (42)
Kerosene	43	350 (177)	450 (232)	510 (266)	550 (288)	204 (96)	200 (93)	201 (94)	197 (92)

Fig. 9.6.31 Schematic of exhaust turbosupercharger installation on diesel engine cylinder.

Supercharging a spark ignition, premixed-charge engine which has the optimum compression ratio for the available fuel causes combustion knock and necessitates lowering the compression ratio, enriching the mixture, or increasing the octane number of the fuel. Supercharging a diesel engine increases the maximum pressure and permits lowering the compression ratio or changing the fuel injection timing, if the limiting pressure is already attained without supercharging. The use of variable engine compression ratio together with supercharging is very desirable but difficult and costly.

Turbocharging of diesel engines is usual for most applications, including truck, locomotive, and marine uses. The exhaust gas turbine provides more complete expansion of the combustion gases than is practical in the engine cylinder and contributes to improved thermal efficiency. Higher output also increases the engine mechanical efficiency (Buchi, *ASME Trans.*, **59**, pp. 85–96). Supercharging assists with NO_x control without a fuel economy penalty. The fuel economy and other performance characteristics of a supercharged Sulzer marine diesel engine of 30-in (760-mm) cylinder diameter are shown in Fig. 9.6.32.

Fig. 9.6.32 Performance curves of a turbocharged diesel engine. *(Sulzer.)*

Fig. 9.6.33 Common scavenging systems.

Scavenging Two-Stroke Cycle Engines

Crankcase compression for each cylinder or an engine driven blower may be used for scavenging and charging the cylinders of a two-stroke cycle engine (Fig. 9.6.33). With crankcase compression, used on some gasoline outboard marine engines, a third port uncovered by the piston near the top of the stroke, a rotary valve, or an automatic poppet valve may be used. The top of the piston is shaped to deflect the gases entering the cylinder and to prevent short-circuiting to the exhaust port. A "dead" spot of gases may remain in the lower center of the cylinder or in the upper corners.

The intake may be inclined and tangential, giving the entering charge an upward helical motion (Fig. 9.6.34). This motion keeps the entering charge near the cylinder walls, forcing the exhaust gases to the center of the cylinder and down to the exhaust port. Some engines employ loop scavenging (Fig. 9.6.35).

Fig. 9.6.34 Hesselman helical-loop system.

Fig. 9.6.35 M-A-N loop scavenging.

Some large diesel engines, such as the Detroit Diesel engine specified in Table 9.6.15, have poppet exhaust valves in the cylinder head, eliminating the hot exhaust ports from the cylinder walls. Intake ports can then completely surround the cylinder. The resulting flow pattern is referred to as **uniflow scavenging.** Typical blower pressures and related conditions are also shown in Table 9.6.15.

Opposed-piston engines have intake and exhaust ports located in the cylinder walls at opposite ends of the cylinder. A blower forces the entering charge through the intake ports which extend around the entire circumference of the cylinder.

The two-stroke cycle Sulzer diesel engine (Fig. 9.6.36) makes use of a crosshead and piston rod design. The air in the space between the piston and the diaphragm carrying the piston rod stuffing box is displaced by the descending piston and is forced into the cylinder through the intake ports as they are uncovered by the piston. This aids

Table 9.6.15   Detroit Diesel Two-Stroke Cycle Diesel Engine

Engine type	Air-box pressure at 500 to 2,300 r/min		Max bmep*		Max imep*	
	inHg	cmHg	lb/in²	kPa	lb/in²	kPa
Naturally aspirated	1–9	2.54–22.86	101	697	129	890
Turbocharged	1.5–40	3.81–101.62	136	938	163	1,125

* At 2,300 r/min.

the scavenging of the cylinder. Supercharging is provided by exhaust gas turbine-driven blowers. Air from these blowers passes through intercoolers and is admitted through automatic one-way valves to the scavenging chambers. Electric motor-driven air blowers are sometimes used on large engines to assist scavenging at idle and low power.

Fig. 9.6.36   A Sulzer series RL two-stroke crosshead diesel engine.

The crosshead design, with its stuffing box, also prevents combustion products from contaminating the crankcase oil system and is commonly used on large engines. This RL design (Fig. 9.6.36) is used for cylinder diameters of 560, 660, 760, and 1,900 mm. The largest cylinder, with a stroke of 1,900 mm, delivers a normal output of 4,000 bhp at 102 r/min corresponding to a bmep of 1,430 kPa (207 lb/in²) and a piston speed of 6,480 m/s (1,300 ft/min). Figure 9.6.32 shows the performance curves of this type up to an overload of 5,000 bhp per cylinder. Smaller cylinders produce correspondingly lower outputs with roughly the same mep and piston speed.

FUEL-AIR MIXTURE PREPARATION

Mixture Distribution   Perfect distribution of fuel-air mixture consists of equal distribution of both air and fuel to the various cylinders of a multicylinder engine for each cycle. The compression pressure of each cylinder is an indication of the air distribution if all cylinders have identical valve timing and compression ratios and no leakage. Cylinders receiving lean mixtures develop less power than those receiving correspondingly rich mixtures. Poor distribution can be masked by enriching the mixture, and high-power output attained at the cost of high fuel consumption rates and HC and CO emission increases. The CO, CO_2, and O_2 analysis of the exhaust gases from the individual cylinders and spark plug temperatures (thermocouple in the center electrode) may be used to determine the mixture distribution.

Distribution of load (and fuel) and variation in injection timing between cylinders of a diesel engine are similarly indicated by the

exhaust gas composition and temperature of the individual cylinders, the mixtures in the various cylinders being lean under all conditions.

Air lines, extending from the engine to the outside air, should be designed for air velocities of 50 to 100 ft/s (15 to 30 m/s). Air cleaners remove dirt particles and reduce piston, ring, and cylinder wear. Air silencers, used at the entrance to the air lines, are combined with cleaners for automotive purposes.

Gasoline and **diesel fuel lines** should be designed for maximum velocities under 1 ft/s (0.3 m/s) for gravity feed. Gasoline lines should be in a cool location, because heating above 100°F (38°C) may vaporize the lighter fractions and cause **vapor lock.** Sediment and water traps as well as strainers or filters should be located in the fuel line ahead of the fuel injector or carburetor, fuel pump, or injection pump.

Gas lines should be designed for gas velocities of 30 to 60 ft/s (9 to 18 m/s). A pressure drop of 0.07 to 0.12 in (1.8 to 3 mm) of water should be allowed per 100 ft (30 m) of pipe. Close calculation of the **size of gas pipe** for industrial gas installations is undesirable, because of the reduction of the normal pipe cross section by tar and dust deposits. A large-diameter storage chamber should be located in the pipe near the engine. The gas line should be free of traps and as free as possible from changes in direction of flow, and it should be pitched slightly from both ends toward a low point in the center, where a sealed drain should be located. All water seals in the gas line should be as cool as possible, to minimize evaporation and the breaking of the seal.

Mixture Preparation in Spark Ignition Engines

Most automotive engines use stoichiometric mixtures at idle and part-load operation, but transition to slightly rich mixtures near and at full power. A stoichiometric mixture, relative to slightly rich mixtures, provides good fuel economy and low engine-out emissions of CO and HC. Exhaust gas is recirculated (EGR) into the fresh mixture to control nitrogen oxides. The dilution reduces the need for throttling and thus improves part-load fuel economy. In smaller engines, a slightly rich mixture is required for idling and small throttle openings, transitioning to near stoichiometric at medium loads, and may be stoichiometric or rich at full load to improve power but is limited by emissions considerations (Fig. 9.6.37).

For many years, carburetion has been used to meter, atomize (if liquid), and mix the fuel with the air flowing to the engine. This method is still in use for mixing gaseous fuels with air, for small engines such as

Fig. 9.6.37   Mixture ratios for spark ignition engines.

those used in household and off-road applications, for piston-type air-cooled engines, and for older passenger cars. However, except for parts of the world with less stringent emission standards, carburetion has been replaced by fuel injection for automotive applications. The two methods of fuel-air mixture preparation are described below.

Carburetion

The basic principle of carburetion is illustrated in Fig. 9.6.38. A float maintains a constant fuel level in a float chamber which is vented to the atmosphere or air entrance of the carburetor. The float chamber fuel level is slightly below the outlet of the fuel jet. The engine air flows through a venturi tube which has a throat diameter of 0.75 to 0.85 of the diameter of the carburetor bore. The reduction in pressure at the venturi throat causes fuel to flow from the float chamber through the main fuel jet into the airstream. Fuel atomization is accomplished by the velocity difference between the air and fuel. Vaporization of the fuel continues while flowing to the engine cylinder but is usually not completed in the intake manifold at wide-open throttle.

Fig. 9.6.38 Principle of carburetion.

The metering characteristic of a carburetor is the weight ratio of fuel to air supplied over the operating range of airflow. These carburetor flow characteristics are commonly shown on a graph, similar to Fig. 9.6.37, which gives the fuel/air ratios at idle, along the "road-load curve" for an automobile engine, and at wide-open throttle. Although computations can be made to establish the fuel/air ratio under specified conditions, estimates are approximate because air is usually bled into the fuel channels to help control the fuel flow. Also, airflow through the venturi is pulsating in nature and makes both airflow and fuel flow somewhat unpredictable. Actual carburetor characteristics approach the desirable characteristics indicated in Fig. 9.6.37.

When the throttle is opened quickly, there is a momentary lag of the fuel flow in relation to the airflow due to the formation of a liquid film on the intake manifold walls. This necessitates the use of some device (accelerating pump or well) to enrich the mixture during sudden accelerations and thereby avoid engine backfire. At idle, the depression of the throat of the venturi is insufficient to overcome the viscosity of the fuel, thus requiring special idle circuits to overcome this problem and to enrich the otherwise very dilute, high-residual cylinder contents. Provisions are also made for mixture enrichment at high loads. In the 1980s, with the enforcement of more stringent emissions standards and the introduction of closed-loop control of the fuel/air ratio to ensure operation in a narrow window about the stoichiometric ratio, the electronic carburetor was introduced (Fig. 9.6.39). In this device, the flow area for fuel metering is computer-controlled by a pulse-width-modulated solenoid.

Carburetors for piston-type air-cooled aircraft engines provide fuel/air ratios that vary from 0.11 at idle to 0.085 for cruising rich (cruising lean being about 0.072). As the full-power condition is approached, the mixture richness is increased to aid cylinder cooling,

Fig. 9.6.39 Electronic carburetor. *(Amann, "The Automotive Spark-Ignition Engine—An Historical Perspective," ASME-ICE vol. 8, 1989, pp. 33–45.)*

becoming about 0.108 at maximum output. Aircraft carburetors must compensate for the effect of altitude on the air density and the resulting change in the mixture ratio. Compensation for the enrichment that occurs with increasing altitude is usually provided by reducing the area of the main fuel-metering orifice. This is accomplished by moving a tapered needle in the orifice, which may be done manually or automatically by means of an air-pressure-sensitive bellows. A carburetor providing a 0.067 fuel/air ratio at sea level would provide a fuel/air ratio of about 0.091 at an altitude of 20,000 ft (6,100 m).

Fuel Injection in Spark Ignition Engines A majority of new automobile and truck engines and some small marine and utility engines employ fuel injection. Fuel is injected into the supercharger, intake manifold, intake valve ports, or combustion chambers of spark ignition engines. Fuel injection consists of an in-tank pump, common-rail recirculating supply, and injection nozzles. With single-point, central fuel injection above the throttle, as typified by the throttle body injector of Fig. 9.6.40, one or two injectors with injection pressure of about 70 kPa are used for multicylinder engines. This system provides an

Fig. 9.6.40 Throttle body injector. *(Amann, "The Automotive Spark-Ignition Engine—An Historical Perspective," ASME-ICE vol. 8, 1989, pp. 33–45.)*

electronically controlled injector at the intake manifold entrance where the carburetor was formerly positioned. With multipoint fuel injection, an injector is positioned in each port, and pressures from 400 to 700 kPa are used to avoid vapor formation in the tip located in the hot spit-back region just upstream of the intake valve. In certain cases, an additional injector is used to supplement the fuel flow during starting.

The advantages of fuel injection over carburetion are the more uniform distribution of fuel between the engine cylinders, the reduction of combustion knock caused by nonuniform fuel distribution, the elimination of heat to the intake manifold to obtain the desirable fuel vaporization (thus obtaining higher-power outputs because of higher charge density), more rapid throttle response, and the use of less volatile fuel. Carburetor and induction system icing are eliminated. With an injection system, it is easier to cut off the fuel flow during deceleration (coasting) than is the case with the carburetor with its fuel-wetted intake manifold. Note that throttle body injection encounters problems associated with the slower transport of fuel than air during transients, as with carburetors.

Fuel injection systems may have mechanical or electronic control of fuel flow, with the electronic type being predominant. In mechanical systems, such as the Bosch K-Jetronic, fuel is continuously injected into the port where fuel quantity is mechanically determined by movement of an air valve. In electronic systems, an electromagnetic injector is pulsed on and off by the computer to supply the desired fuel quantity in proportion to the air. The **injector pulse** width may be trimmed by an exhaust sensor signal to achieve the desired mixture strength. The airflow must be sensed to establish the correct fuel/air ratio for various engine conditions. This can be done by sensing the airflow with a venturi or equivalent, a hot-film, hot-wire, or other air anemometer (such as in the L-Jetronic system), or by sensing the principal engine variables which control engine air consumption. The latter is referred to as a **speed-density** control system. A system of this type is shown schematically in Fig. 9.6.41. Engine speed and manifold pressure and air temperature are used to calculate the engine airflow from prior dynamometer data, a method which has proved less accurate in practice.

For central fuel injection designs, the injector opening angle may be independent of engine cylinder events (asynchronous). For port injection designs, the opening is normally synchronized with cylinder events to open before the intake valve opens. This minimizes axial

Fig. 9.6.41 Schematic of electronic gasoline injection system for an automobile. (*a*) Gasoline tank; (*b*) fuel filter; (*c*) fuel supply pump; (*d*) auxiliary idling air valve; (*e*) fuel pressure regulator; (*f*) air filter; (*g*) throttle plate; (*h*) fuel injection valve; (*i*) throttle position switch; (*k*) full load pressure switch; (*l*) manifold pressure sensor; (*m*) ignition distributor with trigger contacts; (*n*) electronic control unit; (*o*) temperature sensor; (*p*) battery. (*Robert Bosch.*)

mixture stratification which occurs when the injector supplies fuel when the intake valve is open. Such axial stratification may be used advantageously to create a lean-burning engine (Toyota). Injection of fuel into the intake port usually occurs early in the intake stroke, to provide the maximum time for fuel evaporation and mixing to occur. Injection into the intake port provides good mixing as the fuel and air are drawn at high velocity through the intake valve passages, and it isolates the injector from combustion gases at high pressure and temperature. Direct injection into the cylinder occurs during intake or early compression, and it usually results in charge stratification.

Spray targeting in port fuel injection is especially important for fuel evaporation, wall wetting, and engine emissions. Typically when the spray is directed at the center of the intake valve, toward its valve stem base, hydrocarbon and CO emissions are minimized (Fig. 9.6.42). Injector design, spray angle, and droplet size distribution are important. Quader (1989) also showed that targeting the fuel spray at the intake valve centerline toward the valve head resulted in the lowest smoke and HC emissions under high-speed and heavy-load conditions. Some designs are less affected by deposits. Air-assisted fuel sprays provide smaller droplets and may be an advantage in both port and direct cylinder injection. Air-assisted direct cylinder injection has shown promise

Fig. 9.6.42 Effects of spray angle and injection targeting on engine emissions. (*a*) Effect of dual stream separation angle; (*b*) effect of injection direction. (*Iwata et al, SAE paper 870125, 1987.*)

for improved economy and lower emissions in two-stroke, spark ignition, orbital combustion process (OCP) engines.

Fuel Injection in Diesel Engines   The fuel injection system for diesel engines usually consists of a pump, fuel line, and nozzle. This system provides control of the beginning, rate and duration of injection, and desired fuel quantity per injection, and atomizes and partially distributes the fuel in the combustion chamber. Methods of fuel injection include pump or unit injectors and common-rail systems. In the **pump injection system** (Fig. 9.6.43), a pump forces a desired quantity of fuel through a high-pressure fuel line and an atomizing nozzle into the combustion chamber. Unit injectors combine the plunger pump and spray nozzle, thus eliminating the high-pressure fuel line. Injection pressures usually range from 1,500 to 7,000 lb/in² (10,300 to 48,000 kPa) but may run above 30,000 lb/in² (206,000 kPa) at high injection rates. Fuel quantity control with this type of injection is usually affected by controlling a bypass valve in the pump or the pump discharge line or by varying the time of closure or opening of the fuel-pump inlet port. A separate fuel-pump is commonly required for each engine cylinder, although rotating plunger pumps serving from two to eight cylinders

Fig. 9.6.43 Pump injection.	Fig. 9.6.44 Common rail.

per plunger are in common use, especially on small engines. In the **common-rail system** (Fig. 9.6.44), a constant pressure is maintained in the fuel discharge line, and the discharge of fuel to the engine cylinder is controlled by the lift, timing, and duration of opening of the fuel valve. Considerable capacity of the discharge line is necessary in this case, a triplex pump being considered desirable to supply fuel at a constant rate. A relatively new injection system, used in Caterpillar and International engines, is the **hydraulic electronic unit injector (HEUI)** system. Oil is pressurized up to 3,000 to 3,500 psi (21,000 to 24,000 kPa). The pressurized oil acts on a stepped piston which pressurizes fuel up to 25,000 psi (170,000 kPa). Injection control via a small shuttle piston is done electronically.

Pumps Fuel pumps for pump injection systems require rugged construction, are usually actuated by a cam, and have constant mechanical stroke. The pump plunger and barrel (Fig. 9.6.45) are a very close lapped fit to minimize leakage. The inlet port is uncovered near the bottom of the plunger stroke, the low vapor pressure in the pump cylinder causing fuel flow into the barrel. On the upward plunger stroke, fuel flows past the discharge valve and out of the spring-loaded nozzle. Fuel spray from the nozzle continues until the pump bypass port is uncovered. The pressure then subsides, the nozzle closes, and the pump delivery valve closes to maintain a desired residual line pressure until the next injection. Fuel quantity control is obtained by turning the pump

Fig. 9.6.45 Bosch plunger, barrel, and delivery valve assembly.

plunger in the barrel, which varies the time of uncovering the bypass port by the plunger helix. This in turn varies the duration of the effective plunger stroke and the amount of fuel injected. The high pressures which occur in these injection systems, together with the compressibility of the fuel oil, give rise to transient pressure wave phenomena in the connecting lines which change injection characteristics.

Fuel Lines The tubing connecting the injection pump with the spray nozzle must be thick-walled, of smooth and uniform bore, and made of ductile material to permit bending. Tubes range in size from ¼-in (0.64-cm) OD and ¹⁄₁₆-in (0.16-cm) ID to ½-in (1.3-cm) OD and ³⁄₁₆-in (0.48-cm) ID. Fuel lines should be the same length for each cylinder, to obtain the same transient hydraulic characteristics. Pressure waves due to plunger motion and nozzle efflux characteristics are developed, travel back and forth, and create disturbances during the injection process (De Juhasz, *Trans. ASME, Oil and Gas Power,* **60**, p. 2; Bolt et al., *SAE Trans.,* 1971, paper 710569), making short lines desirable.

Fuel Injection Nozzles Injection nozzles are usually hydraulically operated, having a differential-diameter valve (Fig. 9.4.46), held on its seat by a spring, which opens when the fuel line pressure is increased by the injection pump. The valve closing pressure is always lower than the opening pressure, depending on the relative effective valve areas exposed to fuel pressure. The valves may be either the pintle or multipleorifice type (Fig. 9.6.46). The former has a single orifice through which the end of the valve protrudes, and it is used mostly with precombustion chambers. A pintle may be designed to restrict the fuel flow as it opens; this results in a throttling nozzle. The pintle nozzle provides a hollow conical spray with an included angle which varies from 4° to 60° with various pintle sizes. The multiple-orifice type of nozzle (Fig. 9.6.46c) is used principally with the open combustion chamber. The resulting fuel spray consists of a core and surrounding envelope of atomized fuel.

Fig. 9.6.46 Solid injection closed nozzle.

Diesel Fuel Injection System Characteristics At any given setting, most pump and nozzle fuel injection systems deliver more fuel per cycle with a decrease in speed, except under the idling condition of very low speed and small fuel delivery. The rising fuel delivery curve with decrease in speed at full load contributes to the good torque rise and lugging ability of the high-speed diesel. At the idle condition, decrease in speed reduces fuel delivery and causes the engine to run still more slowly, etc. The injection quantity and timing of electronically controlled injection systems is determined by a variety of driver inputs, sensors, and software algorithms.

COMBUSTION CHAMBERS

Spark Ignition Engines

Most combustion chambers for four-cycle engines, especially automotive, have an open design and employ overhead valves actuated by either pushrods or overhead cams. This design provides fast burning of the charge because of the concentrated combustion volume and minimizes surface area and thus heat losses and wall quenching. Figure 9.6.47 shows classical, two-valve chamber designs. In each case, the spark plug location and chamber shape yield a relation between flame front area and flame travel similar to that in Fig. 9.6.48. For a given engine, the maximum rate of pressure development is proportional to the peak flame-front area viewed at TDC. For the chambers of Fig. 9.6.47a, b, and d, virtually all the volume is in the head. This design allows large valves and minimizes piston inertia. The hemispherical or domed head of Fig. 9.6.41b allows very large valves and with a domed piston provides a combustion chamber of normal compression ratio. Aircraft engines usually have this chamber shape. The chamber of Fig. 9.6.47c, used by Jaguar in its V-12 design, is in the piston bowl. The bathtub chamber of Fig. 9.6.47d is very compact for good fuel economy. To suppress knock, each chamber has the spark close to the exhaust valve and a well-cooled, quench region in the end gas far from the spark plug. While all these chambers have a squish area to generate turbulence, compression-induced turbulence is low and mixture motion depends on intake-induced swirl, which is generally quite considerable.

More recently, pent-roof heads with four valves per cylinder and a centrally located spark plug have become very popular (Fig. 9.6.47).

This arrangement gives minimum flame travel distance and hence fast burn, as well as large valve flow area and thus high volumetric efficiency. In such designs, turbulence has been generated through the complex interaction of the two inlet flows which produce a tumbling motion. Other fast-burn chambers have been designed by Nissan and Ricardo. In the Nissan NAPS-Z combustion system, there are twin spark plugs and an induction system producing a comparatively high level of axial swirl (Fig. 9.6.49). In Ricardo's high-ratio compact chamber (HRCC), the combustion chamber is centered on the exhaust valve. Chamber compactness, high turbulence, and lean burning allow this design to operate at compression ratios exceeding 11 without knock.

Fig. 9.6.48 Flame-front area curves for fast- and normal-burn combustion chambers.

Where cost or engine height is a principal concern, a design such as the L-head (Fig. 9.6.50), in which the valves are actuated from below directly by the cam, is commonly used. Spark plug location can compensate to some extent for the slow burn rate of the elongated chamber. The L-head has a relatively high surface area, and thus greater heat loss and wall quenching, and a small space for valves.

Combustion Roughness The high rate of pressure rise during combustion at relatively low engine speeds (1,000 r/min) and the maximum combustion pressure near peak indicated torque engine speeds

Fig. 9.6.50 L-head engine.

Fig. 9.6.47 Classical combustion chambers for spark-ignition engines; (a) Wedge; (b) hemispherical head; (c) bowl in piston; (d) bathtub head. (*Courtesy: R. Stone, SAE, 1995.*)

Fig. 9.6.49 Modern fast-burn combustion chambers. (a) Four-valve pent-roof; (b) NAPS-Z with twin spark plugs and high axial swirl; (c) high-ratio compact chamber (HRCC). (*Collins and Stokes, SAE, 1983.*)

subject the engine mechanism to severe stresses which may result in combustion roughness noise unless the pressure rate is controlled. Andon and Marks (*SAE Trans.* **72**, 1964, p. 636) found that spark plug location, combustion chamber shape, and degree of turbulence are the principal factors that can be varied to control combustion roughness and the resulting noise. In a given engine, the maximum combustion rate of pressure rise can be reduced by retarding the spark timing or decreasing the volumetric efficiency.

A rate of pressure rise of 30 lb/in² (207 kPa) per degree of crank travel results in near-maximum efficiency, and higher rates of pressure rise usually result in combustion roughness noise. Increasing the rigidity of the crankshaft increases the natural frequency of vibration and reduces the deflection of the structure caused by combustion and inertial forces and thereby reduces roughness noise.

Die-cast aluminum transmits roughness noise more readily than cast iron. Structurally stiff compact engines such as the V-block are better than the in-line engines in this regard.

Compression Ignition Engines

Compression ignition engines have valve-in-head construction if operating with four cycles. Two types of combustion chamber arrangements are used: a single chamber, referred to as the *open* or *direct* injection type, and a two-chamber, referred to as a *prechamber* or *divided-chamber* type.

Open Combustion Chambers (Direct Injection) Fuel is injected under high pressure, usually through a multiple-orifice nozzle, directly into the clearance space or chamber between the piston and the cylinder head. The piston head is usually conformed (Fig. 9.6.51) to fit the fuel spray, and swirl moves the air into the fuel spray. Air swirl is accomplished by intake port design, by shrouding the intake valve, or, in the two-stroke cycle engine, by using tangential intake ports. High turbulence is accomplished by having the piston closely approach part of the cylinder head. This forces the gases out of the small clearances

(a)   (b)

Fig. 9.6.51 (*a*) Open combustion chamber with swirl produced by intake ports (left-hand image); (*b*) open combustion chamber with high turbulence and some swirl (right-hand image).

and agitates the mixture. Open chambers are used for small automotive as well as heavy-duty diesel engines.

Precombustion chambers are divided into two parts, the major volume being between the piston and the cylinder head and connected by a small passageway to the minor volume located in the cylinder head (Fig. 9.6.52). Fuel is injected into only the smaller chamber, and except under light loads, partial combustion occurs and discharges the burning mixture into the large chamber in which combustion is completed. This type of combustion chamber produces smooth combustion but has fairly high fluid friction and heat transfer losses.

Turbulence or **divided combustion chambers** are modifications of the precombustion chamber, having the major chamber in the cylinder head

Fig. 9.6.52 Precombustion chamber.

and usually only a small clearance space between piston and cylinder head (Fig. 9.6.53). The passage between the two chambers is considerably larger than that in the precombustion chamber. Close piston clearance produces high turbulence in the prechamber and promotes rapid combustion. Part of the prechamber containing the transfer passage commonly is thermally insulated. An electrically heated hot-wire glow plug is used to assist cold starting, together with compression ratios up to 24:1.

Fig. 9.6.53 Ricardo divided combustion chamber.

Stratified-Charge Engines

A **stratified-charge spark-ignited engine** has potential for burning lean mixtures for improved fuel economy and reduced emissions. In theory, an overall lean mixture is formed in the combustion chamber, while controlled to be slightly richer than stoichiometric in the vicinity of the spark plug at the time of ignition. The richer portion is thus easily ignited, and this in turn ignites the remaining lean mixture. Some of the benefits claimed are (1) lower flame temperature, (2) improved cycle efficiency, (3) less heat loss, (4) less dissociation, and (5) less pumping loss. In addition, less NO_x would be expected because of the lower combustion temperatures; likewise, less unburned hydrocarbon and CO emissions would result because of the overall excess of oxygen.

For effective stratification, very careful development has proved to be necessary to obtain complete combustion of fuel under the wide range of speed and load condition required of an automotive type of engine. Distinct means for accomplishing the stratified charge condition have been explored, including the following.

1. A single combustion chamber with a well-controlled rotating air motion is used. This arrangement is illustrated (Fig. 9.6.54) by the Texaco combustion process (TCP), patented in 1949. Ford Motor Company's programmed combustion (PROCO) is a similar, single-chamber concept, but has only been produced in very limited quantities.

Fig. 9.6.54 Texaco combustion process (TCP).

Fig. 9.6.55   Mitsubishi gasoline direct injection (GDI) engine concept.

2. In the wall-guided concept (Figs. 9.6.55 and 9.6.56), the preparation of the stratified fuel-air charge relies on a wall-guided system. The fuel spray is injected from the side and directed toward the piston surface, so as to impinge on the spherical piston cavity and be reflected toward the spark plug by the tumble.

3. In the air-guided concept (Fig. 9.6.56), the spray does not impinge on the piston. The injector is mounted half way between the side and the center. Preparation of the stratified charge is accomplished via a combination of swirl and tumble in the charge.

4. In the spray/jet guided concept, the injector is mounted at the center and is the primary means for providing the sprayed fuel prominent swirl motion to enable fuel-air mixing. Note that the spark plug is off to the side, but placed nearer the injector than the other two concepts.

5. There is axially stratified charge. By turning on the port fuel injector toward the end of the filling process, a rich mixture is produced near the spark plug, while the overall mixture may be lean. This method is known as the *Toyota lean-burn system.*

Some production vehicles employing stratified-charge concepts have appeared in other global markets such as Japan, but are not sold in North America because of current emission regulations.

Homogeneous Charge Compression Ignition

Recent developments in engine control technology have enabled the prospects of a fairly novel form of engine combustion: **homogeneous charge compression ignition (HCCI).** HCCI operates by inducting a homogeneous mixture of fuel and air (similar to conventional spark-ignited engines), but using compression ignition to ignite the fuel (similar to conventional diesel engines). Major challenges, such as precise control over the start of combustion, continue to keep HCCI under rigorous development without any use in production vehicles yet.

One practical implementation of an HCCI-type combustion mode is **low-temperature premixed compression ignition (PCI)** combustion. Used with direct-injection configurations, PCI operates on the principle of utilizing an extremely long ignition delay (the time between start of injection and start of combustion) via strategic injection timings and large rates of EGR to create a combustion process that burns premixed fuel and air at ultralow temperatures. The major benefit of PCI combustion is dramatically reduced emissions of nitrogen oxides and particular matter (Jacobs et al., *SAE Trans. Jour. Eng.*, **114**, 2005). Nissan has demonstrated its PCI combustion concept in production use.

SPARK IGNITION COMBUSTION

Ignition System   The ignition system usually consists of a 12-V storage battery, induction coil, and high-tension distributor, or a high-tension magneto with a distributor. Recent automotive practice employs electronic means to interrupt primary current in place of a contact breaker, and the trend is toward individual coils for each plug. For aircraft, the low-tension distribution system to transformer spark plugs, or other apparatus for obtaining a high-tension spark at the plug, appears to eliminate some of the troubles of the conventional ignition system in aircraft engines at high altitude. For details on ignition systems, see Sec. 15.1.

The usual firing order for in-line engines is 1, 3, 4, 2, or 1, 2, 4, 3 for four cylinders; 1, 5, 3, 6, 2, 4 (U.S. automobile engines) or 1, 4, 2, 6, 3, 5 for six cylinders in line; and 1, 6, 5, 4, 3, 2 for V-6 engines. A common procedure for V-8 engines is to number the cylinders from front to rear, with the odd numbers in the left bank, as viewed from the driver's seat. For this arrangement a typical firing order is 1, 8, 4, 3, 6, 5, 7, 2. Radial engines have firing orders that skip every other cylinder. For a nine-cylinder, single-row engine the order is 1, 3, 5, 7, 9, 2, 4, 6, 8.

Spark plug gaps vary from 0.020 to 0.080 in (0.5 to 2.0 mm). The smaller gaps require 4,000 to 8,000 V while the larger gaps require 10,000 to 34,000 V. In wedge-shaped, two-valve combustion chambers, spark plugs were typically located near the exhaust valve (hottest spot) so that the flame progressed toward the cooler part of the combustion chamber; this arrangement permitted higher compression ratios without combustion knock. In modern, four-valve, pent-roof combustion chambers, a central spark plug location results in minimum combustion time and good tolerance for dilution. Dual ignition (two spark plugs located opposite each other) also reduces combustion time and results in a slight gain in efficiency (usually less than 1 percent) compared with single ignition with optimum spark advance. Rotary (Wankel) engines employ one or two spark plugs per rotor. Single plugs are centrally located either above or below the trochoid waist. In modern, direct injection, stratified-charge, lean-burn engines, optimum placement of the spark plug is crucial to provide a combustible mixture near its vicinity.

Flame Speed   Flame speeds in spark ignition engines are low immediately after ignition, attain maximum values when about one-half the combustion chamber has been inflamed, and decrease toward the end of the piston. Mean flame speeds are a maximum for mixtures usually 10 to 20 percent richer than the chemically correct air/fuel ratio, and they vary with fuel, engine speed, and turbulence.

At 900-ft/min (4.5-m/s) mean piston speed, mean flame speeds with maximum-power gasoline-air mixtures and optimum spark advance range from about 50 to 100 ft/s (15 to 30 m/s). Mean flame speeds under the same mixture, spark, and speed conditions in the CFR Waukesha engine are estimated at 55 (70) [80] ft/s (17, 21, 24 m/s) for a 4.6:1 (6.5:1)

Fig. 9.6.56   Various designs implementing the stratified-charge combustion concept.

Fig. 9.6.57 Effect of spark advance on traction force and octane requirement.

and power. Spark advance is measured by the number of degrees the crankshaft rotates between the time of the spark and the end of the compression stroke. **Optimum spark advance** is the minimum advance which develops the **maximum brake torque (MBT)**. Usually this causes (1) the maximum pressure to peak at about 16° after top center and (2) 50 percent of the charge to be burned by 10° after top center (Heywood, 1988). The torque lost by a spark timing earlier or later than optimum is at first only slight but increases rapidly as the timing further deviates from optimum (see Fig. 9.6.57).

Optimum spark advance depends principally on air-fuel mixture, amount of residual gas, emission control requirements, combustion chamber design, turbulence, engine speed, number of spark plugs, and spark plug location. The optimum spark advance is also affected by knock considerations, as described later. Maximum power air/fuel ratios require the minimum spark advance. Low-speed engines require 10° to 20° spark advance, and high-speed automotive engines require 30° to 40° spark advance. Full load requires less spark advance than part throttle. Spark advance in contemporary automotive engines is based on computer-stored values and is a function of speed, load, and other engine operating variables.

Autoignition The autoignition temperature of an air-fuel mixture is the lowest temperature at which chemical reaction proceeds at a rate sufficient to result eventually (long time lag) in inflammation. This temperature depends principally on the air-fuel mixture, fuel properties, mixture pressure, and test apparatus. Subjecting a mixture to a temperature higher than the autoignition temperature results in inflammation after a shorter time lag, there being a temperature for each mixture which results in practically instantaneous ignition (Table 9.6.16). An increase in pressure decreases the ignition temperature of an air-fuel mixture (Table 9.6.17). See CRC Report on Project CF-37-53.

[8:1] compression ratio, while with ethyl alcohol–air mixtures the corresponding mean flame speeds are 50 (65) [75] ft/s (15, 20, 23 m/s).

The mean flame speed in an automotive engine under the foregoing conditions is estimated from the optimum spark advance (approximately five-ninths of combustion time) to be 90 ft/s (27 m/s) for a 6.5:1 compression ratio. In the same engine, mean flame speeds with optimum conditions vary from about 60 ft/s (18 m/s) at 500 ft/min (2.5 m/s) to about 170 ft/s (0.9 m/s) at 3,000 ft/min (15 m/s) mean piston speed.

Spark Advance Since combustion requires finite time, it is necessary to ignite the mixture in the cylinder before the piston reaches the end of its compression stroke. This should distribute the combustion process before and after top center in order to obtain maximum torque

Table 9.6.16   Ignition Temperatures of Air-Gas Mixtures at Atmospheric Pressure

Gas	Chemical symbol	With ignition lag*		Instantaneous† ignition	
		% gas in mixture	Avg temp, °F (K)	% gas in mixture	°F (K)
Hydrogen	H_2	8–24	1,130 (893)	10	1,377 (1,020)
Carbon monoxide	CO	13–47	1,204 (924)	50	1,708 (1,204)
Methane	CH_4	5–38	1,202‡ (923)	25	>1,832 (1,273)
Ethane	C_2H_6	§	968‡ (793)		
Propane	C_3H_8	§	914‡ (763)		
Acetylene	C_2H_2	4–22	804 (702)		
Ethylene	C_2H_4	6–19	1,004 (813)	10	1,832 (1,273)
Benzene	C_6H_6			50	1,943 (1,335)
Ether	$C_4H_{10}O$			50	1,892 (1,306)
Gasoline:					
92 octane					734 (663¶)
100 octane					804 (1,702¶)

* Dixon and Coward, *Trans. Chem. Soc.*, **95**, 1909, p. 514.
† David, *Trans. Chem. Soc.*, **111**, 1917, p. 1003.
‡ Minimum value, since spread in values was appreciable.
§ Composition not reported.
¶ Jones, *U.S. Bur. Mines Bull*, 1946.

Table 9.6.17   Ignition Temperatures at Various Air Pressure, °F (K)

Gas	Air pressure, atm								
	1	3	5	7	10	15	20	25	30
Hydrogen*	1,134 (885)	1,132 (884)	1,112 (873)	1,100 (866)					
Methane*	1,343 (1,001)	1,269 (960)	1,215 (930)	1,175 (908)					
Gasoline†			590 (583)		480 (522)	420 (489)			
Kerosine†			670 (628)		490 (528)	430 (494)	400 (478)	385 (469)	
Gas oil†			580 (578)		500 (533)	450 (505)	435 (497)	415 (486)	
Machine oil†			710 (650)		610 (594)	550 (561)	520 (544)		
Benzene†			685 (636)		585 (580)	545 (558)	530 (550)	515 (541)	500 (533)
Cylinder oil†			825 (714)		710 (650)	645 (614)	620 (600)	595 (586)	572 (573)
Benzol†					1,095 (864)	970 (794)	930 (772)	905 (758)	880 (744)

* Bone and Townsend, "Flame and Combustion in Gases, p. 69.
† Tausz and Schulte, *Zeit. V.D.I.*, 1924, p. 574.

Compression ignition engines, such as diesel engines, rely on autoignition for initiation of combustion. However, in spark ignition engines, autoignition of the gas ahead of the flame front is undesirable as it leads to combustion knock. This abnormal combustion phenomenon is described in greater detail in the following section.

COMBUSTION KNOCK

Combustion knock in the internal combustion engine is produced by the spontaneous combustion or autoignition of an appreciable portion of the charge (Rassweiler and Withrow, *Trans, SAE,* **31**, 1936, p. 297), which results in an extremely rapid local pressure rise and sharp metallic knock. In the spark ignition engine, this occurs when the last fraction of the charge to burn is compressed appreciably above its autoignition temperature by the fraction that has burned. However, if the lag in time (ignition lag) between the attainment of this temperature and the spontaneous local appearance of the flame is increased by the use of a knock-suppressor blending agent or additive, then the flame from the spark plug will be permitted to travel through the unburned fraction before it can knock. Thus knocking tendency depends on the autoignition temperature, ignition lag, and flame speed of the air-fuel mixture, which are all partly influenced by the combustion chamber shape.

The tendency of a spark ignition engine to knock increases almost directly with spark advance (Fig. 9.6.57). Prior to the introduction of exhaust emission control requirements, it was general practice in automobile engine design to use a spark advance a few degrees less than optimum at wide-open throttle. This materially reduces the tendency to knock with only a negligible loss of torque. At part-throttle operation, where combustion knock may not be a problem, the spark is advanced closer to optimum. However, a compromise may be made between spark advance for good torque and spark retard for emission control.

In the compression ignition engine, combustion knock occurs when the ignition lag is comparatively long, thus permitting the evaporation of a considerable portion of the injected fuel, which suddenly inflames and produces the characteristic knock at the initiation of combustion. Thus knock suppression requires long ignition lags (high octane, low cetane) in spark ignition engines and short ignition lags (high cetane, low octane) in compression ignition engines. Most automotive engines now employ knock sensors, providing feedback to the engine control computer to alter spark advance.

Knock Rating of Fuels

For detailed information regarding methods, apparatus, reference fuels, and suppliers of apparatus and reference fuels, see ASTM Manual of Engine Test Methods for Rating Fuels.

Octane Number

The octane number (ON) of a fuel is the percentage by volume of iso-octane (2,2,4 trimethylpentane) in a mixture of iso-octane and normal heptane which matches the unknown fuel in knock tendency when compared by a specific procedure in the ASTM CFR knock-testing engine. A fuel of high octane will have a low cetane number and vice versa (Fig. 9.6.58).

The **performance number** scale for aviation gasolines (Fig. 9.6.58) is designed to relate fuel rating to average knock-limited performance (imep). A gasoline of the 100/130 grade indicates that in the aviation test (lean mixture, normally aspirated) the fuel is equivalent to isooctane in imep while in the supercharge test (rich mixture) it permits an imep of 130 percent of that of iso-octane.

There are four methods—**research, motor, aviation** (lean mixture), and **supercharge** (rich mixture) (ASTM D 908, 357, 614, and 909, respectively)—in general use in the rating of gasolines according to knocking tendency. These methods vary principally in the engine operating conditions and the means for indicating knock intensity (see ASTM manual).

Engine conditions, as indicated by the air or mixture temperature and by the coolant temperature, are less severe in the research method than in the other methods. Thus, some fuels rate lower by the motor method than by the research method. The sensitivity of a fuel is defined quantitatively as the difference between the research and motor octane

Fig. 9.6.58 Relationship between antiknock scales (octane number and performance number) and cetane number. *(Brewster and Kesley, SAE, 1963.)*

numbers. Although most gasolines rate lower by the motor method than by the research method, straight-run gasolines have about the same rating by both methods. With present fuels and engines, the average of the research and motor method numbers provides the best correlation with road ratings. This value is posted on the fuel station pump, with the vehicle's recommended fuel published in the owner's manual.

The use of the aviation and supercharge methods for aviation gasolines results in two performance numbers (such as 100/130). This grading indicates that the fuel is less sensitive than iso-octane to the difference in engine severity in the two test methods.

Road ratings of gasoline are determined by means of borderline knock curves for the test fuel and various blends of the reference fuels at various spark advances. The car is accelerated with wide-open throttle, and the speed at which steady knock ceases is recorded as the knock die-out speed (CRC procedure E-2). A plot of these data on a sparkadvance vs. knock die-out speed chart (Fig. 9.6.59) makes possible the determination of the octane number rating of the test fuel over the desired speed range.

Mechanical octane number is the resulting decrease in fuel octane requirements brought about by mechanical changes in engine design or

Fig. 9.6.59 Road octane rating.

Fig. 9.6.60   Effect of fuel composition on knock-limited compression ratio.

control. Design variables include ignition control, combustion chamber design (including spark plug location cooling), valve timing, fuel system, and engine-transmission combinations. (See Table 9.6.18.)

Knock Suppression   The antiknock characteristics of a fuel can be improved by blending in either another fuel of better antiknock characteristics or a knock-suppressor fuel additive (Fig. 9.6.60). Tetraethyllead (TEL) and tetramethyllead (TML) are the most effective knock suppressors (Table 9.6.19) and have been used extensively in small amounts (up to 3 mL/gal of automotive fuels) in many gasolines. Their addition has been eliminated because of adverse effects on health and exhaust emissions. Other compounds can raise antiknock quality, but the much larger amount needed typically changes volatility and stoichiometric point significantly and may affect engine life, emissions, and lubrication.

Cetane Number   The cetane number (CN) of a diesel fuel is determined by matching its ignition quality with that of a blend of two reference fuels, normal cetane (CN = 100) and heptamethylnonane (CN = 15). The comparison is made in an ASTM CFR diesel engine (Waukesha Motor Co.) using a specified procedure (ASTM D 613 or CRC-F-S). Cetane number for the reference blend is calculated by

$$CN = (\text{vol } \% \ n\text{-cetane}) + 0.15(\text{vol } \% \text{ heptamethylnonane})$$

Table 9.6.18   Antiknock Indices (ASTM D4814)

Unleaded fuel* (for vehicles that can or must use unleaded fuel)	
Antiknock index (RON + MON)/2	Application
87	Designed to meet antiknock requirements of most 1971 and later model vehicles
89	Satisfies vehicles with somewhat higher antiknock requirements
91+	Satisfies vehicles with high antiknock requirements

* Unleaded fuel having an antiknock index of at least 87 should also have a minimum motor octane number of 82 in order to adequately protect those vehicles that are sensitive to motor octane quality.
SOURCE: Abstracted from ASTM D4814, with permission.

Table 9.6.19   Relative Effect of Antiknock Compounds

Compound	Chemical symbol	Weight for a given effect, g
Tetraethyllead (TEL)	$Pb(C_2H_6)_4$	0.0295
Aniline	$C_6H_5NH_2$	1.0000
Ethyl iodide	C_2H_5I	1.55
Ethyl alcohol	C_2H_5OH	4.75
Xylene	$C_6H_4(CH_3)_2$	8.0
Toluene	$C_6H_5CH_3$	8.8
Benzene	C_6H_6	9.8

SOURCE: "International Critical Tables," vol. 2, pp. 162–163.

The **ignition quality** of a fuel for a compression ignition engine is indicated by the time delay between beginning of injection and the start of rapid pressure rise caused by combustion. The compression ratio in the test engine is adjusted to result in a 13° crank-angle delay for the unknown fuel when injection begins at 13° before top center. The mixture of reference fuels which has the same lag under the same conditions indicates the cetane number of the fuel.

The factors that tend to aggravate knocking in a spark ignition gasoline engine tend to suppress it in a diesel engine. Thus, octane has a low, while heptane has a high, cetane number. High-CN fuels burn more smoothly and start cold engines more readily than low-CN fuels.

Cetane number has been correlated with the mid-boiling point and the API gravity of the fuel, the relation for computing the approximate cetane number being (*Jour. Inst. Petroleum*, **30**, 1944, p. 193)

$$CN = 175.4(\log \text{ mid-boiling pt, °F}) + 1.98(\text{API gr}) - 496$$

The deviation from engine test CN for 579 fuels was less than ±5 CN for 94 percent of the fuels (see also Sec. 7.1).

Ignition accelerators increase the cetane number of diesel fuels, but the susceptibility of the fuel to the accelerator decreases with an increase in the amount of the accelerator (Fig. 9.6.61). Commercial amyl nitrate is slightly less effective than the pure isoamyl nitrate (Bogen and Wilson, *Petroleum Refiner*, July 1944).

Fig. 9.6.61   Improvement in CN with ignition accelerators. *(Bogen and Wilson.)*

Permissible Compression Ratios   Data obtained on single-cylinder test engines and multicylinder automotive engines show wide variations in the compression ratios producing incipient knocking of fuels of various octane numbers. High compression ratios in engines using fuels of a given octane number do not necessarily indicate excellent design and high performance, since valve timing, manifolding, and porting may act as throttling devices, or spark timing may be retarded. Average compression ratios vary considerably depending on the design and fuel used. An increase in cylinder size usually necessitates a lower compression ratio (Fig. 9.6.62). The larger diesel engines require less compression to

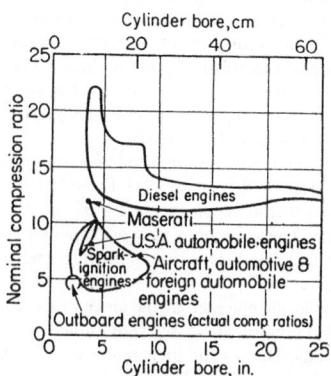

Fig. 9.6.62   Range of compression ratios as a function of cylinder bore.

obtain the desired temperatures since the heat transfer effect is less than in the smaller engine.

OUTPUT CONTROL

Speed and **load regulation** are usually obtained in four-stroke cycle gasoline engines by throttling the charge or by quantitative governing. The air/fuel ratio may vary some with load, depending on the engine requirements and emission control system. Intake throttling reduces the pressure in the cylinder at the end of the intake stroke and thereby reduces the mass of charge trapped per cycle. While throttling is the simplest means for regulation and provides a stable idling condition, it increases the pumping work and consequently decreases the efficiency of the engine. Presently, there is a trend toward using variable valve timings and variable valve lift as a means of controlling load without increasing pumping losses. In direct injection, stratified-charge engines, load control may be accomplished through control of the amount of fuel injected per cylinder.

Power control of spark-ignited, two-stroke cycle engines is difficult because intake throttling reduces or prevents cylinder scavenging. Qualitative governing is used in some cases with constant-speed gas engines. The fuel is throttled while the air is unrestricted. Qualitative governing is limited by the lean limit of flammability. In addition, lean mixtures burn more slowly, causing a decrease in fuel conversion efficiency. Advancing the ignition timing, as the mixture is made leaner, reduces this loss of efficiency.

Governing by variation of ignition timing, as used on some small two-stroke cycle gasoline engines, is undesirable since this is regulation by variation of thermal efficiency and may result in overheating the engine. Variation of ignition timing is required for each change in speed and load to obtain optimum results.

Power control or regulation of diesel engines is obtained by varying the amount of fuel injected per cycle, which provides increasing richness of the fuel-air mixture as the engine load is increased. Maximum specific fuel consumption usually occurs at about 85 percent of the maximum load.

Diesel engines require a speed governor to provide a stable idle. Mechanical, hydraulic, and electronic speed-control governors are used on diesel engines. These control the fuel pump injection quantity and provide speed regulations at all operating conditions.

COOLING SYSTEMS

Engine cylinders must be cooled to maintain a lubricant film on the cylinder walls and other sliding surfaces. The cylinder heads, pistons, and exhaust valves are cooled to prevent combustion knock or destruction of these parts due to overheating. The lubricant must be cooled to maintain the desirable viscosity under operating conditions. Liquid or air cooling systems are used. Pistons, exhaust valves, and lubricants in comparatively small or low-duty engines are sufficiently cooled by contact with other engine parts or the lubricant between them and do not require separate cooling systems.

Liquid Cooling   The heat removed by the cooling liquid from the cylinders and cylinder heads ranges from 15 to 20 percent of the energy input for large diesel engines, 20 to 35 percent for automotive engines, and may run as high as 40 percent for automotive engines at one-third load. An increase in speed decreases the heat loss to the coolant. Heat loss can range from 40 to 50 percent of the brake output for large diesel engines and 100 to 150 percent of the brake output for automotive engines.

Pistons in small engines are cooled by heat transfer to the cylinder walls and lubricant. In some cases, an appreciable quantity of oil is directed against the piston head to maintain desirable head temperatures.

Some large engines have even used liquid-cooled pistons. Liquid is usually fed to the piston by swing joints or telescopic connections. The coolant jacket surrounding the high-temperature areas of the engine is maintained at a pressure considerably above atmospheric—10 to 15 lb/in^2 (70 to 100 kPa). This permits operation at higher coolant temperatures and/or decreases the tendency for local boiling in the region of "hot spots," which could lead to component failures. The average percentage distribution of the coolant to the various jackets is approximately 60 percent to the cylinder jacket and 40 to the cylinder-head jacket with uncooled piston; 50 to the cylinder jacket, 25 to the cylinder-head jacket, and 25 to the piston and piston rod with liquid-cooled piston.

The cooling system may be noncirculating, in which case cold water is supplied, flows through the water jacket, and is wasted. With a temperature rise in large engines of 90°F (32°C), from 2 to 4 gal/(bhp · h) [7.6 to 15 L/(bhp · h)] of coolant must be supplied. In some engines, coolant outlet temperatures as low as 120°F (49°C) must be maintained, which, with high inlet temperatures, may increase the required quantity of coolant to as much as 10 gal/(bhp · h) (38 L)/(bhp · h). Coolant is circulated at a rate of 25 to 50 gal/(bhp · h) [95 to 190 L/(bhp · h)] with a temperature rise of 10 to 20°F (5.6 to 11°C) while flowing through the water jacket.

Natural circulation (**thermosiphon**), with low velocities, requiring large connections, may be used if the coolant forms a complete circuit. In this case, hot coolant rises in the engine jacket, flows to the radiator, is cooled, descends, and flows back to the engine jacket.

Small stationary engines are often **hopper-cooled.** This is an **evaporative cooling system** requiring about 4 to 6 lb/(bhp · h) [1.8 to 2.7 kg/(bhp · h)] of makeup water. The ASTM CFR knock-testing engines are cooled by evaporation, the vapor being condensed by a **reflex condenser** (copper coil with cold water flowing through) and returned to the cylinder jacket.

In high-output engines, the coolant is directed at the hottest spot, usually the exhaust valve seat, with either an external or internal inlet manifold from a common cylinder block jacket; otherwise vapor bubbles will form, cling to the surface, and cause overheating. In vertical engines, the coolant usually flows upward, around the cylinder barrel into the cylinder head jacket, and to the outlet. Higher output and reduced knocking tendency have resulted from directing the entering coolant at the hot spots in the cylinder head and then letting the coolant flow downward around the cylinders.

The **size of piping** required for the inlet to the jacket may be calculated for 3-ft/s (1-m/s) coolant velocity, and for the discharge 2 ft/s (0.6 m/s) for large engines. In engines with recirculation, these values are increased up to 10 to 15 ft/s (3 or 4.5 m/s), and the size of the inlet line to the circulating pump is usually larger than the outlet from the coolant jacket. It is desirable to have an open or visible outlet from each separately cooled part, showing the flow of coolant. To ensure totally filled systems which are as air-free as possible, overflow bottles may be used. Water is an outstanding liquid coolant. However, where freeze protection is necessary or where higher jacket temperatures are desired, other cooling media are generally used. For example, light-duty engines typically use a mixture of water and ethylene glycol. (See Sec. 6.8.)

Air Cooling   Many small industrial engines and most motorcycle and aircraft engines are air-cooled. Even here, however, higher specific output engines are increasingly liquid-cooled. Air cooling eliminates the necessity for water or other liquid cooling media, coolant jackets, pumps, radiators, and coolant connections, but necessitates single-cylinder construction, finning, baffles, and, in some cases, blowers. Air-cooled engines also tend to be more noisy since the combustion chamber walls radiate sound and the engine structure is less rigid. The permissible compression ratio and output of air-cooled aircraft engines depend on the efficiency of the cooling of the cylinder head, exhaust valves, and seats. Long fins [1 to 2 in (25 to 50 mm) depending on cylinder size], closely spaced [0.10 to 0.20 in (2.5. to 5 mm)], with baffles directing the air at high velocity at the hot spots, have made high-output aircraft engines possible at the expense of considerable air resistance or drag. Properly designed cooling ducts may make use of the heat added to the cooling air to obtain a jet effect, thereby reducing drag.

Oil Cooling   Shearing of various oil films by moving parts and oil contact with hot parts of an engine cause a rise in oil temperature until energy input to the oil equals the energy transferred by the oil to the cooler parts that it contacts. Oil coolers are required for high-duty engines to maintain oil temperatures of 200°F (94°C) or under. Temperatures of 250°F (120°C) are not uncommon in automobile engines on hot days: 300°F (150°C) is considered too high, particularly for oils which decompose rapidly under conditions causing the high oil temperatures. The desired oil temperature is maintained by circulating the oil through a radiator or cooler to an oil sump and then through the engine. Transmission oil cooling is also accomplished by placing plate coolers in the end tanks of radiators or with separate radiators.

LUBRICATION

(See also Secs. 6.10 and 8.4)

General   A suitable liquid lubricant is vital to the satisfactory operation of an internal combustion engine. It prevents excessive wear and deposit accumulation and removes heat from the areas of relatively high temperature within the engine.

Most engine oils are composed of base oils and additives. The base oils are usually mineral oils, but some are synthetic oils. Chemical additives are incorporated in engine oil formulations to impart desirable performance characteristics which are not provided by base oils alone.

Physical Properties   Both the physical and chemical properties of an engine oil affect its performance. The principal physical property involved is viscosity. It must be low enough at low temperatures to permit cranking and starting and high enough at high temperatures to provide an adequate oil film between rabbing surfaces. Sufficiently high viscosity is also required to help prevent excessive oil consumption at high temperatures. Another physical property affecting consumption is volatility. Consumption can be undesirably high with an oil of excessive volatility. Both viscosity and volatility are influenced by base oil selection. Viscosity can also be modified by means of additives.

Chemical additives perform a number of other functions in an engine oil. For example, oxidation inhibitors reduce oxidative and thermal degradation, which can lead to varnish and sludge deposition and to excessive thickening. Detergent-dispersant additives suspend insoluble products formed during engine operation, so they can be removed from the engine when the used oil is drained. Antiwear agents help protect rubbing surfaces, especially those subject to boundary lubrication. Antifoam agents prevent foam and air entrainment and their adverse effects on oil pressure and heat transfer. Rust inhibitors eliminate the formation of corrosion products on ferrous surfaces and reduce corrosive wear. Friction-reducing additives lower boundary friction.

SAE Classifications   Engine oils are classified by the Society of Automotive Engineers in two general groups, viscosity (SAE J300 standard) and performance (SAE J183 standard). Viscosity is measured at both 0°F (−18°C) and 210°F (98.9°C). SAE 5W, 10W, 15W, 20W, and 25W viscosity numbers are related to 0°F (−18°C) measurements; SAE 20, 30, 40, 50, and 60 to 210°F (98.9°C) measurements. Multigrade oils, such as SAE 5W-20, SAE 10W-30, and SAE20W-40, are those which have viscosity characteristics meeting requirements at both temperatures. Multigrade oils can be used throughout the year, reduce friction and wear, and provide better lubrication and fuel economy during a cold start; e.g., the viscosity of an SAE 20W-40 oil will be less than that of an SAE 40 oil at ambient conditions.

The SAE performance classification includes engine oils ranging in performance quality from that of a straight mineral oil plus a small amount of pour-point and/or foam depressants, to those required in severe passenger car service and heavy-duty truck and off-highway operation. Letter designations such as SG apply to active test techniques for oils for gasoline engines existing since 1988; designations CD and CE apply to oils for naturally aspirated and turbocharged diesel engine service; and designation CD-II applies to oils for two-stroke cycle diesel engines. Performance of these oils is evaluated in both single- and multicylinder engine tests. Factors measured include rust, wear, deposit formation, oil consumption, oil thickening, and fuel economy.

REFERENCES: Most of the foregoing information on engine lubrication is covered in greater detail, and references are provided in the "SAE Handbook." See also the list of references in Sec. 6.11.

AIR POLLUTION

(See also Sec. 18.1)

General Relationship

Automobile engine exhaust was first detected to contribute to air pollution in 1942, where smog created serious visibility and health problems in Los Angeles, CA (Haagen-Smit, *Sci. Amer.,* **210,** no. 1, 1964, p. 25). Smog itself is a combination of smoke (particulates) and ground-level ozone. Ground-level ozone, unlike the beneficial stratospheric ozone, arises from sunlight-induced photochemical reactions between nitrogen dioxide and hydrocarbons (Caplan, SAE Pt. 12, p. 20) and can cause severe eye, lung, and plant life irritation. Engines are a contributor of both nitrogen dioxide and hydrocarbons to the atmosphere.

In addition to nitrogen dioxide and hydrocarbons, engines emit the toxic gas carbon monoxide, particulates (unburned carbon particles), and sulfur dioxide. Table 9.6.20 identifies the historical trend of mobile engine emissions and the relative contribution they make to total atmospheric emissions. In spite of increased driven mileage from 1 trillion mi in 1970 to a projected 4 trillion mi in 2000, nearly all mobile engine source emissions decreased over the span of the three decades. This trend has resulted from improved engine design, improved fuel technology, and the universal use of catalytic converters in new automobiles. Furthermore, the trend is expected to continue as more stringent government regulations become fully enforceable in 2007.

Locally, human-made pollutants can exceed those from natural sources and can become dominant. However, emitted mass does not reflect pollution severity. For example, 1 unit mass of oxides of sulfur may equal 30 units of carbon monoxide or more in terms of health effects.

Emissions from internal combustion engines include those from blowby, evaporation, and exhaust. These can vary considerably in amount and composition depending on engine type, design, and condition; fuel-system type; fuel volatility; and engine operating point. For an automobile without emission control it is estimated that of the organic emissions, 20 to 25 percent arise from blowby, 60 percent from the exhaust, and the balance from evaporative losses primarily from the fuel tank and to a lesser extent from the carburetor (if used). All other nonhydrocarbon emissions emanate from the exhaust.

Table 9.6.20 Annual U.S. Emissions from Mobile Engine Sources, Million Tons and Percent of Total Emissions*

	1970		1980		1990		2000	
	(Million tons)	%	(Million tons)	%	(Million tons)	%	(Million tons)	%
PM10	0.644	5.3	0.689	11.2	0.715	22.2	0.552	17.8
CO	174.6	88.5	160.5	90.3	131.7	91.7	92.2	90
NO$_x$	15.3	56.8	14.85	54.8	13.4	53.2	12.6	56.2
SO$_2$	0.551	1.8	0.717	2.8	0.874	3.8	0.697	4.3
VOC	18.5	54.9	16.1	53.4	12.1	52.1	8.0	47.2

* Stated values are calculated.
SOURCE: EPA.

At least 200 hydrocarbon (HC) and other organic compounds have been identified in exhaust. Some of those compounds, such as the olefins, are referred to as **reactive** since they react rapidly and to a great extent in the atmosphere to form ozone. Others such as the paraffins are virtually unreactive. Methane is deemed unreactive. Of several reactivity scales, the Carter scale (Table 9.6.21) has become most widely used. It ranks nonmethane organic gases (NMOGs) on a scale from 0 to 11 g ozone/g NMOG. The reactivity of a mixture is computed as the sum of the grams of ozone from individual species, with lower levels typical of the larger molecules. A value of 3.42 g NMOG/mi is representative of non-catalyst-equipped cars.

Table 9.6.21   Carter Scale of Reactivity

Nonmethane organic gas	g ozone/g NMOG (approx. range)
Methane	0
C_2-C_5 paraffins	0–1.5
Acetylene	0
Benzene	0.42
C_6 + C_8 paraffins	2–0.2
Toluene (and higher aromatics)	2–10
Alcohols	0.5–2.5
I-alkenes	9–1
Diolefins	11–9
Aldehydes	3–7
Internally double-bonded olefins	10–3

SOURCE: Carter and Atkinson, *Emu. Sc. Tech.* **23**, 1989, p. 864.

Atmospheric oxides of nitrogen (commonly termed NO_x) are a mixture of nitrogen oxide (NO) and nitrogen dioxide (NO_2). Oxides of sulfur (SO_Z) arising from fuel sulfur are a mixture of SO_2, SO_3, and SO_4. Sulfuric acid may be a product in some oxidizing catalytic exhaust converters.

Emission quantities are most often reported as mole percent for the principal exhaust constituents (CO, H_2O, CO_2, H_2, O_2, N_2) and as moles per million moles (ppm) for trace constituents (HC, NO). Hydrocarbons are reported commonly as either ppm equivalent hexane (C_6H_{14}) or as ppm carbon (C_1). One ppm C_6 is equivalent to 6 ppm C. Other reporting parameters are (1) specific emission rate, g pollutant/kWh or g pollutant/vehicle mi (km), and (2) emission index, g pollutant/1,000 g fuel.

Emission Sources

Homogenous Combustion and Spark Ignition Engines   ORGANIC GAS EMISSIONS. Emissions of unburned hydrocarbon and other organic gases can be generated from several sources, as shown in Fig. 9.6.63. The largest source is thought to be flow during compression of un-burned fuel-air mixture into the crevice volume between the piston and liner above the top ring; the crevice mixture returns to the combustion during expansion at a time when it is too cool to burn. A similar mechanism is caused by the absorption of fuel vapor into oil layers and combustion chamber deposits during compression, and subsequent desorption during expansion and exhaust.

Emissions also are generated from flame extinction or wall quenching due to the inability of the flame to propagate totally to a surface or into a crevice. Daniel ("Sixth Symposium on Combustion," Reinhold, 1957, p. 886) described the result of wall quenching to be a visibly dark layer of unburned and partially burned fuel-air mixture left adjacent to all engine combustion chamber surfaces. Subsequent studies showed that the layer adjacent to single walls is largely burned prior to being exhausted. Thus the organic emission from the engine is mainly due to crevice quenching, of which the top piston land clearance is the major source.

Under light-load engine operating conditions where cylinder exhaust residual is high, incomplete flame propagation may leave a portion of the combustion chamber contents unburned. This can produce organic emissions of several thousand ppm C_6. When residual exceeds 15 percent by weight, some incomplete combustion may occur. In some

Fig. 9.6.63   Emission sources of unburned hydrocarbon and other organic gases. *(Amann, "The Automotive Spark-Ignition Engine—An Historical Perspective," ASME-ICE vol. 8, 1989, pp. 33–45.)*

stratified-charge, direct-injection designs, bulk quenching as the flame comes in contact with cooler or leaner gases may also be present.

Wankel engine organic emissions arise from the above sources, but significant additional quantities may arise from mixture leakage through the leading rotor apex seal directly into the exhaust.

Two-stroke engines take in a mixture of fuel and air, and where intake pressure exceeds exhaust during the scavenging or overlap period, raw-mixture loss to the exhaust creates very high organic emissions. Emissions from such two-cycle engines may be 10 times higher than those from four-cycle engines. Note also that most two-stroke designs utilize once-through lubrication; i.e., the oil is mixed with the fuel, typically in a fuel/oil ratio of 50:1. Hence, increased deposits and oil films generate organic emissions by physical and chemical storage and release of fuel components. Fuel injection with proper timing can reduce engine-out organic emissions to near four-stroke engine levels.

Any unburned hydrocarbons and other organic gases escaping the cylinder during exhaust can still oxidize in the exhaust temperature and manifold if the exhaust temperature is high enough for the available residence time and if there is sufficient oxygen present. In addition, where combustion is close to stoichiometric, oxidation of unburned gases is accomplished in catalysts (see later).

CARBON MONOXIDE EMISSIONS. The principal source of exhaust carbon monoxide (CO) is rich-mixture combustion. Figure 9.6.16 shows exhaust composition as mixture strength is varied. Cycle-to-cycle and cylinder-to-cylinder maldistribution of the fuel-air mixture, especially during transients, increases CO. Additional CO may arise from incomplete combustion of organics in an exhaust treatment device.

OXIDES OF NITROGEN EMISSIONS. The principal oxide of nitrogen produced inside the engine cylinder is nitrogen oxide (NO). However, since the residence time of gases within the reaction zone is very short, NO forms primarily in postflame combustion gases and does not attain chemical equilibrium. The formation depends primarily on peak combustion gas temperature and to a lesser extent on oxygen content. Once formed, NO tends to be frozen and persists in high, nonequilibrium concentrations during expansion and exhaust. Nitrogen oxide variation with mixture strength is shown in Fig. 9.6.16. Other important variables that affect NO emissions include the amount of exhaust gas recirculation (EGR) and spark timing. When mixtures are more than 30 percent lean or where air is injected into the exhaust, about 5 percent of NO may be oxidized to NO_2 prior to leaving the exhaust pipe. Typically NO and NO_2 emissions are measured together with a chemiluminescent analyzer, and their sum is reported as NO_x emissions.

ALDEHYDE EMISSIONS. Aldehydes are incomplete combustion products of the oxidation of organics. They arise from the same wall quenching and incomplete combustion mechanisms as the unburned organic emissions themselves, but in addition may be formed in low-temperature exhaust system oxidation reactions. Typically gasoline engine exhaust contains 100 ppm C aldehyde, less for rich and more for lean mixtures. Methanol fuel tends to produce high emission levels of formaldehyde.

PARTICULATE EMISSIONS. Particle emissions from engines consist primarily of carbon, liquid hydrocarbons, sulfates, metals, and metal oxides. Metals arise from fuel and lubricant additives such as lead, phosphorus, zinc, and barium, as well as engine wear and corrosion. Sulfuric acid particles may be emitted as a result of reactions involving SO_2 over platinum metal catalysts. SO_2 is formed during combustion from fuel sulfur. Liquid aerosols of heavy hydrocarbons are emitted not only during starting, but also during running. Of these, polynuclear aromatics (PNAs) are health concerns. Catalytic converters virtually eliminate most hydrocarbons, especially those with long molecular chains which tend to be the most reactive.

Heterogeneous Combustion and Diesel Engines The hydrocarbon, carbon monoxide, and nitrogen oxide emission formation in heterogeneous combustion processes in diesel engines derives from the same basic mechanisms as in the spark ignition engine. However, the relative importance of each is different. In the diesel engine, the emission formation in combustion can be considered to arise from a broad distribution of individual fuel-air elements, some of which burn rich and others lean. Patterson and Henein (Emissions from Combustion Engines and Their Control, *Ann Arbor Sci.*, 1972, p. 260) have suggested the various emission sources in the combustion of a fuel spray injected into swirling air, as shown in Fig. 9.6.64. In addition, fuel impingement on chamber surfaces contributes to unburned hydrocarbons and carbon smoke particles. Hydrocarbon emissions from diesels contain many heavy fuel types. Normally, heated sampling lines are required to avoid HC condensation during sample analysis.

Fig. 9.6.64   Emission formation in a fuel spray injected in swirling air.

The amount of carbon, carbon monoxide, hydrocarbon, and nitrogen oxide emissions from diesel engines varies considerably with engine and fuel system design as well as with the operating point. Moreover, because of the high air mass flow rate at partial load, emissions from diesel and gasoline engines cannot be compared directly except on a mass rate emission basis. Diesel emissions as mole percent or ppm may be corrected for dilution to an equivalence ratio of unity for comparison with spark ignition engine data. On a corrected basis, diesel engines emit in the range of 50 to 500 ppm C_6 and NO_x up to 3,000 ppm. Diesel engines, since they run lean, tend to show higher NO_2/NO ratios (between 10 and 30 percent), especially at light loads when cooler regions quench any NO_2 formed in the reaction zone (Hiliard and Wheeler, SAE 790691, *SAE Trans.*, **88**, 1979). Carbon monoxide emission from the diesel engine is very low, typically under 1,000 ppm (0.1 percent).

Diesel engines emit considerable amounts of carbon smoke and odor constituents. Most carbon arises from pyrolysis in rich combustion regions. Spindt, Barnes, and Somers (SAE paper 710605, 1971) have identified low concentrations of many potential oxidation products and certain fuel fractions as being the odor-producing compounds. Human-panel techniques are commonly used to measure odor.

Strategies for Controlling Spark Ignition Engine Emissions: Effects of Engine Design Variables

Combustion Chamber Shape The shape of the combustion chamber affects organic emissions through a change in surface area and thus heat loss. Surface area relative to the entire clearance volume is an indicator of the volume fraction of organic emissions. This may be stated as

$$\text{Organic, ppm} \propto \frac{A_s}{V_c}$$

where A_s is the chamber surface area and V_c the clearance volume at TDC. According to Sheffler (SAE, Pt. 12, p. 60), the following change the surface/volume ratio, thereby changing hydrocarbon emissions: chamber shape, bore/stroke ratio, compression ratio, and cylinder displacement. In automotive reciprocating engines, surface/volume ratios range from 5 to 8 in^2/in^3 (2 to 3 cm^2/cm^3). Wankel rotary engines have surface/volume ratios about 50 percent higher. In well-designed engines, combustion chamber shape has little, if any, effect on CO. A fast-burning chamber increases NO because of an increase in peak temperature.

Ignition System Under most running conditions, improvements in spark energy, rise time, and spark duration produce little improvement in emissions. However, a high-energy, long-duration spark in conjunction with extended electrodes and wider spark gap has been found to minimize the increase in organic emissions due to misfire, as mixtures are leaned well beyond what is chemically correct. (See Quader, *SAE Trans.*, **85**, 1976.)

Exhaust back pressure and valve overlap affect organics and NO through changes in the exhaust residual. Increasing overlap or back pressure lowers NO_x and may lower organic emissions. At very light loads, excessive residual leads to incomplete combustion, thereby increasing organic emissions.

Induction Systems and Fuel-Air Mixture Preparation Optimum mixture preparation and distribution are a key to minimum emissions. With carburetors, high-velocity manifolds, increased manifold heating, and staged throttles have shown emission reductions (see Bartholomew, SAE Pt. 12, 1971). Port fuel injection should be well-targeted on the intake valve to minimize hydrocarbon and soot emission. (See Kosowski, SAE paper 850294, 1985.)

Effects of Operating Variables

Mixture ratio and ignition timing are the most important operating variables affecting emissions. Figure 9.6.16 shows the effect of the mixture ratio on major emission amounts. Retarding ignition timing from the best-efficiency setting reduces HC and NO_x emissions significantly. Excessive retard increases HC and CO emissions. Increasing engine speed reduces HC emissions as ppm. NO_x emissions increase as load increases. Increasing coolant temperature normally reduces HC emissions and increases NO_x emissions.

Limiting fuel enrichment during warm-up is critical to low emissions, especially without exhaust treatment devices. With exhaust treatment, emissions may be reduced to virtually zero once the system is warmed up. Thus the only significant emissions from such systems are those generated during starting and warm-up.

Exhaust Treatment Devices

Catalytic or noncatalytic exhaust treatment devices may be used to clean up remaining exhaust emissions. Thermal reactors are noncatalytic devices which rely on homogeneous bulk gas reactions to oxidize CO and HC. Commonly they appear to be enlarged exhaust manifolds, but they may have internal baffling. In thermal reactors, NO_x is unaffected.

Fig. 9.6.65 Conversion efficiency of three-way catalyst (TWC) as a function of air/fuel ratio. *(Kummer, Prog. Energy Combust. Sci.,* **6,** *1981, pp. 177–199. Reprinted with permission.)*

Reactions are enhanced by increased exhaust temperature (reduced compression ratio, retarded timing) or by increasing exhaust combustibles (rich mixtures). Typically, temperatures of 1,500°F (800°C) or more are required for peak efficiency. Usually the engine is run rich to give 1 percent CO, and air is injected into the exhaust. A typical thermal reactor residence time is 100 ms. Required reactor volume is commonly 1.5 times engine displacement. Maximum allowable reactor temperature is governed by materials and is typically 1,750°F (950°C). Thermal reactors are seldom used because the required engine calibration produces low fuel economy.

Catalytic systems are capable of reducing NO_x as well as oxidizing CO and HC. However, a reducing environment for NO_x treatment is required which necessitates a richer than chemically correct engine mixture ratio. A two-bed converter may be used in which air is injected into the second stage to oxidize CO and HC. It is highly efficient but results in lower fuel economy. Single-stage, three-way catalysts (TWCs) are widely used, but they require extremely precise fuel control to be effective. As illustrated in Fig. 9.6.65, only in the vicinity of stoichiometric ratio is the efficiency high for all three pollutants. Such TWC systems employ either a zirconia or a titanium oxide exhaust oxygen sensor (EGO, HEGO, or UEGO) and a feedback fuel system to maintain the required mixture strength near stoichiometric. Figure 9.6.66 identifies the typical placement of various emission control technologies on an automotive engine. Figure 9.6.18 also shows the close-coupled three-way catalyst with oxygen sensor. Conversion efficiencies for typical automotive three-way catalyst emissions (HC, CO, and NO_x) range between 70 and 99 percent over the U.S. transient emissions drive cycle for a hot, stabilized engine/catalyst. Conversion efficiencies range between 40 and 99 percent for vehicles operating in high-speed accelerations.

Catalyst support beds may be of the pellet or honeycomb (monolith) type. Materials for reducing catalysts include rhodium, Monel, and ruthenium. Materials for oxidizing catalysts include platinum and palladium. The amount of active catalyst material required may be as low as 0.05 troy oz (1.4 g) per vehicle system. Catalyst volume is about 50 to 100 percent of the engine displacement. The weight of the catalyst bed typically is about 2.2 to 5 lb (1 to 2.3 kg) excluding the metal container. Many aged catalysts can oxidize at 85 percent or higher efficiency at temperatures as low as 800 to 1,000°F (430 to 540°C). Above 1,500°F (820°C), they deteriorate rapidly. Compared with thermal reactors, catalytic converters warm up slowly because of the larger mass of material. Catalytic converters may be located farther from the engine than thermal reactors, since converters operate efficiently at lower temperatures. However, this slows warm-up. Monolithic converters, which can normally be operated hotter, can be smaller and can be placed closer to the engine. Catalysts with metal substrates (monoliths) warm up even faster and are more resistant to high-temperature deterioration. A small cold-start catalyst located near the engine, especially

Fig. 9.6.66 Typical placement of engine emission control technology.

one with supplementary heating, is very effective. (See Laing, *Auto. Eng.*, Apr. 1994, p. 51.)

Strategies for Controlling Diesel Emissions: Engine Design and Operating Variables

Horrocks (*Proc. Inst. Mech. Engr.*, **208**, 1994, pp. 289–297) has summarized the strategies for controlling particulate and HC emissions as follows: (1) optimization of fuel injection and air motion; (2) good atomization of the fuel spray, particularly at idle, at partial load, and during transients; (3) precise control of fuel injection timing, particularly during transients; (4) minimization of parasitic losses in the combustion chamber; (5) low sac volume or valve cover orifice nozzles for direct injection; (6) minimization of contribution of lubricating oil; (7) central injector and combustion chamber for high-speed, direct injection engines; (8) modulated EGR; (9) electronic controls; (10) rapid warm-up of engine.

Control of NO_x emissions from diesels presents additional challenges since suppression of NO formation tends to cause an increase in particulates. Nevertheless, the following strategies are frequently used (Horrocks, op. cit.): (1) injection retard; (2) injection rate shaping; (3) intercooling of turbocharged engines; (4) centralized combustion for high-speed, direct injection; (5) modulated EGR; (6) electronic controls.

Exhaust Treatment Devices

Since diesel engines burn lean overall, the exhaust will always contain excess oxygen. Hence the reduction of NO_x with conventional three-way catalysts is not feasible. Currently, NO_x is removed from the diesel exhaust by either selective catalytic reduction (e.g., Feeley et al., SAE paper 950747) or catalyzed thermal decomposition of NO into O_2 and N_2. In addition, lean NO_x catalysts are under development.

Diesel particulate traps have been developed which employ ceramic or metal filters (see Wade et al., SAE 830083, 1983). Thermal and catalytic regeneration can burn out the material stored. Particulate standards of 0.2 g/mi may necessitate such traps. Both fuel sulfur and aromatic content contribute strongly to particulates. Catalysts have been developed for diesels which are very effective in oxidizing the organic portion of the particulate (see Ball and Stack, SAE paper 902110, 1990).

Emission Regulations

Passenger Vehicles (Cars, Trucks, and SUVs) Federal regulations on automotive engine emissions have been enforced in the United States since the implementation of the Clean Air Act in 1970. The state of California has had its own regulations in place since before 1970 and has often applied more stringent standards on automobiles driven there. In 1990, the Clean Air Act Amendments gave authority to the EPA to enforce further restrictions on the allowable emissions from engine-drive automobiles. Additionally, standards were stiffened for heavy-duty on

highway truck engines, off-road equipment engines, handheld engines, and marine engines.

The most recent U.S. EPA standards are collectively known as Tier II, and they will be fully implemented by 2009. Tier I standards were the first set of reductions enforced under the 1990 Clean Air Act Amendments. Under Tier I and older standards, allowable vehicle emissions were a function of both vehicle weight and engine (or fuel) type. Thus heavier vehicles were allowed to emit higher per-mile pollution due to the larger engine. Similarly, emission control technology was required to function for at least 50,000 mi. Under Tier II regulations, vehicles are no longer differentiated based on vehicle weight or engine type. Instead, a vehicle is certified to one of eight emission categories, or bins (see Table 9.6.22). Any vehicle is allowed to be certified in any one of the eight bins; however, a manufacturer's passenger vehicle fleet average must meet bin 5 standards. Additionally, Tier II requires emission control technology to function for 120,000 mi. Table 9.6.22 lists the standards associated with each bin after full implementation in 2009 that a vehicle must meet for 120,000 mi. Tighter standards are required for most emissions within the vehicle's first 50,000 mi. During the phase-in period from 2004 to 2009, additional transition bins (bins 9 and 10) as well as relaxed standards for certain vehicle classes in other bins are in place to smooth the implementation.

A vehicle is certified by measuring the emissions, with diluted bag samples, produced as it is driven through a predetermined drive cycle, and then calculating the average emission on a gram per mile basis. Traditionally, passenger vehicles only underwent the **Federal Test Procedure (FTP)** 75 drive cycle, which involves a 20-min test of accelerations, decelerations, and steady cruises on a chassis dynamometer. The vehicle is started after a 12-h soak in a 60 to 86°F (16 to 30°C) ambient temperature room. Top cycle speed is 56.7 mi/h (91.3 km/h). To reflect the assumed typical behavior of a U.S. driver, vehicles are now required to also pass **supplemental FTP (SFTP)** drive cycles, one (SC03) which includes more aggressive accelerations and decelerations to higher speeds and another (US06) which includes high-speed steady cruises reaching greater than 75 mi/h (120.7 km/h).

Heavy-Duty Engines Similar to passenger vehicles, medium- and heavy-duty trucks must meet stringent emission standards. One major difference between passenger vehicle engines and heavy-duty truck engines is the burden of certification: passenger vehicles are certified by the vehicle manufacturer (on a gram per mile basis) whereas heavy-duty engines are certified by the engine manufacturer (on a gram per horsepower per hour basis). Similar to passenger vehicles, truck engines are "driven" through an FTP cycle on an engine dynamometer. Diluted emissions are recorded and normalized by the engine cycle's average horsepower. With the advent of 2004 regulations, truck engines must also certify to a steady-state emissions cycle. The engine is operated at several predetermined steady-state points, each having a different weighting factor. The weighted average of emissions from the

Table 9.6.22 U.S. Passenger Vehicle Emission Standards for Cars and Light Trucks (EPA Tier II Standards)*

Bin no.	NOMG,† g/mi	CO, g/mi	NO_x, g/mi	PM, g/mi	HCHO,‡ g/mi
8	0.125	4.2	0.20	0.02	0.018
7	0.090	4.2	0.15	0.02	0.018
6	0.090	4.2	0.10	0.01	0.018
5	0.090	4.2	0.07	0.01	0.018
4	0.070	2.1	0.04	0.01	0.011
3	0.055	2.1	0.03	0.01	0.011
2	0.010	2.1	0.02	0.01	0.004
1	0	0	0	0	0

*Standards apply to all vehicle classes, regardless of weight or engine type. Values shown are for 2009 full implementation and apply to the vehicle for 120,000 mi.
† NMOG = nonmethane organic gases, which include unburned hydrocarbons.
‡ HCHO = formaldehyde.
SOURCE: EPA (www.epa.gov).

several steady-state points must not exceed the same standard applied to the FTP cycle. Furthermore, the engine cannot emit more than 1.5 times the standard, a "not-to-exceed" specification, at any operating point.

Heavy-duty truck engines were first regulated in 1980, and the standards have since undergone several reductions. The next phase-in of emission reductions begins in 2007, with full fleet compliance by 2010. Table 9.6.23 summarizes the standards that truck engines must meet by 2010. The engines must comply with the listed standards for the full useful "life" which is 435,000 mi.

Other Regulations In addition to the standards applied to road vehicles, such as passenger vehicles and on-highway trucks, emission regulations are now applied to off-road vehicles (such as heavy construction equipment), utility and handheld engines, outboard and

Table 9.6.23   U.S. Heavy-Duty Truck Engine Regulations*

NO, g/(hp · h)	PM, g/(hp · h)	NMHC, g/(hp · h)	CO, g/(hp · h)
0.2	0.01	0.14	15.5

*Full compliance required by 2010. Full useful life of heavy-duty truck engines is 435,000 mi.
SOURCE: EPA (www.epa.gov) and DieselNet (www.dieselnet.com).

personal watercraft engines, and large marine engines. Similarly, engine regulations are enforced in other countries or political alignments such as Japan, Argentina, Australia, Brazil, Canada, China, Hong Kong, India, Korea, Mexico, Russia, and the European Union. The standards vary considerably relative to the regulations imposed by the United States, which has the most stringent regulations.

9.7   GAS TURBINES
by John H. Lewis and William H. Day

REFERENCES: Kerrebrock, "Aircraft Engines and Gas Turbines," MIT Press. Koff, Spanning the Globe with Jet Propulsion, 1991, AIAA paper 2987. "Turbine Engine Directory," Flight Int., June 8–14, 1994. Hill and Peterson, "Mechanics and Thermodynamics of Propulsion," Addison-Wesley. Bejan, "Advanced Engineering Thermodynamics," Wiley. Horlock, "Combined Cycle Powerplants," Pergamon. "Gas Turbine World Handbook," published annually, Pequot Publishing, Inc.

INTRODUCTION

The **gas turbine** is a turbine-type engine which operates with gas (usually air) as distinguished from steam or water as the working substance. In its most basic configuration, the gas-turbine engine consists of a compressor to draw in and compress the gas, a combustor or heat source to add energy to the compressed gas, and a turbine to extract power from the heated gas flow. Practical applications of gas turbines first occurred from 1939 to 1941. In 1939, the firm of Brown Boveri of Switzerland used a gas turbine to generate electricity. Also in 1939, the first flight of an aircraft powered by a gas turbine developed by Hans von Ohain took place in Germany. Another aircraft gas turbine was developed by Frank Whittle, who powered an aircraft in 1941 in England. From these early applications, the gas turbine has been developed to the point where today it is the most important

aircraft power plant in use. In addition to its aviation applications, the gas turbine is a significant power plant for electric power generation, pipeline pumping power, and marine propulsion. It is a key component of combined-cycle power plants and cogeneration plants (see Fig. 9.7.18 and Sec. 9.4).

Depending upon its development roots and applications, the gas turbine is referred to by many names. In aviation, it is called a jet engine, a turbojet, a jet turbine, a turboprop, a turbofan, and a fanjet. In land and marine applications it is called a gas turbine, a turboshaft, a combustion turbine, and an industrial turbine. In the industrial world, engines generally fall into two categories. **Heavy-duty** or **frame machines** refer to gas turbines designed and built strictly for ground applications. **Aeroderivative machines,** as the name implies, are gas turbines originally designed for aircraft use and adapted for industrial applications. A typical aeroderivative is shown in Fig. 9.7.1. Future trends to marry heavy industrial and aerospace technologies will blur this distinction between the two types.

Gas turbines offer significant advantages compared with other types of engines because they are compact, lightweight, reliable, and efficient. They are capable of rapid start-up, follow transient loading well, and can be operated remotely or left unattended. They have a long service life, long service intervals, and low maintenance costs. Cooling

Fig. 9.7.1   Cross section of Pratt and Whitney FT8 gas turbine derived from the JT8D commercial aircraft engine.

fluids are usually not required. Their compact size high power-to-weight ratio, and high reliability make gas turbines the ideal power plant for aircraft and for portable power plants. In industrial applications, the engine's small footprint allows relatively small foundations and enclosure buildings.

FUELS

Gas turbines use a wide variety of **gaseous and liquid fuels.** For ground and marine applications, natural gas and light distillate oils are most commonly used, but gasified coal and heavy residual oils can also be used. When fuels that contain significant amounts of sulfur, salt, vanadium, and other metals are used, the hot-section components such as the combustor and turbine must be designed to withstand the corrosive effects of these impurities. For aircraft applications, specific kerosine-type fuels have been developed that control important qualities of the fuel, to obtain maximum performance.

THERMODYNAMIC CYCLE BASIS

The gas turbine operates on the principle of the **Brayton** (or **Joule) cycle.** The working substance, usually ambient air, is compressed adiabatically to high pressure where heat is added at constant pressure to elevate the temperature, at which point it is expanded, also adiabatically, back to the original pressure. Because of the expansion from a higher temperature, the work extracted in the expansion process exceeds the work required for compression by an amount that equals the net output work of the cycle. The cycle is completed at the original pressure with heat rejection in the exhaust. See Fig. 9.7.2. At the end of the expansion process, the working substance temperature is higher than its initial temperature by an amount that corresponds to the waste energy to be dissipated at constant pressure to the surroundings. Unique to the Bray-ton cycle are the heat addition and heat rejection processes at (ideally) constant pressure. The constant-pressure process lends itself to continuous flow of the working substance. When coupled with compressors and turbines which operate on the rotating turbomachinery

1-2 Adiabatic compression

2-3 Constant pressure heat addition

3-4 Adiabatic expansion

4-1 Constant pressure heat rejection

Fig. 9.7.2   Brayton cycle temperature vs. entropy diagram. Ideal cycle.

principle (see Sec. 14), the entire gas turbine acts as a **continuous flow engine.** The continuously flowing working substance (air) results in very high power density. The gas turbine's intrinsic compactness favors its use over other engine types in many power generation applications.

Temperature-entropy diagrams describe the Brayton cycle. Figure 9.7.2 shows the ideal case where all component efficiencies are 100 percent. Figure 9.7.3 shows the real cycle with nonideal components,

Fig. 9.7.3   Brayton cycle temperature vs. entropy diagram. Real cycle reflects component inefficiencies and pressure losses.

including friction losses in the ducting between the compressor and turbine.

The **ideal Brayton cycle efficiency** η is related to the cycle pressure ratio by the following expression, where P_{max}/P_{min} = pressure ratio, R = gas constant, and c_p (the latter being specific heat at constant pressure properties of the working substance):

$$\eta \text{ Brayton, ideal} = 1 - 1/(P_{max}/P_{min})^{R/C_p} \qquad (9.7.1)$$

Thus with ideal components, the gas-turbine efficiency will be increasingly high as its characteristics maximum/minimum pressure ratio continues to increase approaching 100 percent when the pressure ratio becomes infinitely large. With real components, however, when the pressure ratio increases, the component inefficiencies become larger contributors to exhaust waste heat. The net result is that efficiency reaches a peak at some level and then falls off. The point where this peaking occurs is directly related to the quantity of cycle heat addition, which, in turn, is related to the maximum cycle temperature. As illustrated in Fig. 9.7.4, the peak shifts to a higher efficiency level at high pressure ratio as the turbine temperature increases. At the same time, the specific power output has a similar trend of increasing up to a peak and then falling off as the pressure ratio increases, while the amount of specific output power increases continuously with heat addition as determined by the maximum cycle temperature. These trends are illustrated in Fig. 9.7.5.

Among manufacturers and users, the **maximum cycle temperature** is variously referred to as turbine inlet temperature, combustor exit temperature,

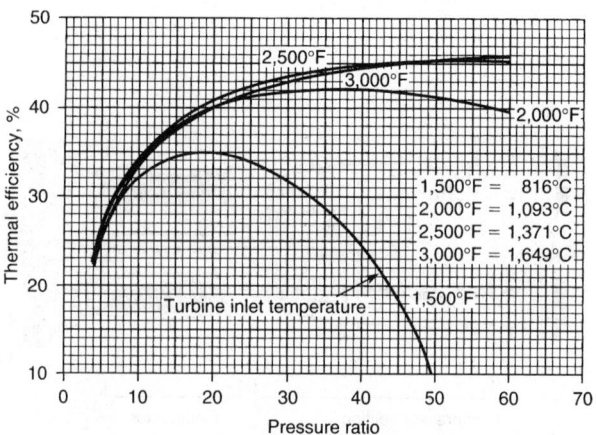

Fig. 9.7.4   Representative Brayton cycle variation of thermal efficiency with pressure ratio at different turbine inlet temperatures.

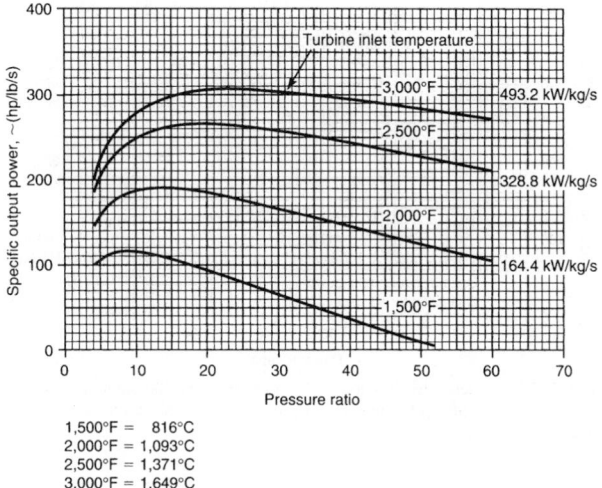

1,500°F = 816°C
2,000°F = 1,093°C
2,500°F = 1,371°C
3,000°F = 1,649°C

Fig. 9.7.5 Representative Brayton cycle variation of specific output power with different pressure ratios and turbine inlet temperatures.

turbine rotor inlet temperature, or firing temperature. Precise definition depends on the design details for a specific engine.

Specific power refers to output per unit mass flow of the working substance, expressed in hp/(lb · s) or its electrical equivalent kW/(kg · s). As shown in Fig. 9.7.5, a given cycle fixes specific power. When specific power is constant, absolute power then varies directly with the mass flow rate of the working substance. The mass flow quantity, in turn, sets the overall size of the machine

$$\text{Power} = \text{specific power} \times \text{mass flow rate}$$
$$= \text{hp/(lb} \cdot \text{s)} \times \text{(lb/s)}$$
$$= \text{kW/(kg} \cdot \text{s)} \times \text{(kg/s)}$$

Generally, the gas turbine's working substance is atmospheric air which is continuously drawn into the front of the compressor through inlet ducting and is expelled back to atmosphere through exhaust ducting. In this case, the minimum cycle pressure is identical to the surrounding ambient pressure while the initial working substance temperature is the ambient temperature of the surrounding atmosphere. The schematic in Fig. 9.7.6 shows the typical component arrangement of a continuous flow simple-cycle gas turbine. In this typical arrangement, the heat addition section consists of an internal combustor where fuel is added directly into the through-flowing air and is continuously burned. The rotating compressor is attached to the turbine by a shaft which transmits the power needed for compression. The excess power available from expansion through the turbine at high temperature is fed by a rotating power output shaft to an external load.

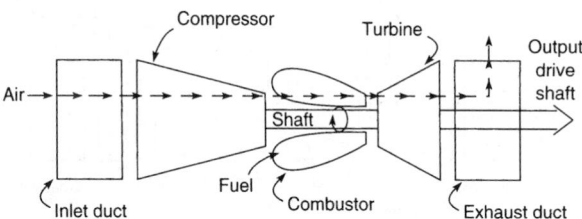

Fig. 9.7.6 Simple Brayton cycle gas-turbine component arrangement.

BRAYTON CYCLE VARIATIONS

The following paragraphs describe variations to the Brayton cycle. In some cases these variations are represented by practical machines that have been or are now seeing commercial use. The combined cycle (Fig. 9.7.18) has come to be most widely used in the electric power generation market. There are three main reasons why combined-cycle (CC) plants have become so popular: (1) The gas-turbine manufacturer need not make any special modifications to the basic simple cycle to adapt it to the CC application; (2) combined cycles are by far the most efficient of all the various possible variants; and (3) as the temperature of the gas-turbine exhaust gas reduces, while it passes through and gives up heat to the heat recovery steam generator, a **selective catalytic reactor (SCR)** can be placed at an appropriate location in the boiler system. The SCR system is designed to reduce NO_x emissions to extremely low levels. As a consequence, combined-cycle plants can gain local permits to operate in environmentally sensitive areas.

Regenerative Cycle (Fig. 9.7.7). In this cycle, air exiting the compressor is directed through a heat exchanger, or **regenerator** before being introduced to the combustion section. Then the heated air exhausting from the turbine passes through the other side of the regenerator. By this means, some of the exhaust heat is recovered and used as an additional source of energy to heat the high-pressure air, instead of being dissipated as waste heat in the low-pressure exhaust. The effect of

Fig. 9.7.7 Regenerative Brayton cycle gas-turbine component arrangement.

regeneration is to increase cycle efficiency, but it is only effective at low levels of pressure ratio. This is because as the pressure ratio increases, the compressor exit air temperature increases and the turbine exhaust temperature is lowered, thereby reducing the regenerator temperature difference necessary for heat transfer. At some point, depending on the maximum cycle temperature, the compressor exit and exhaust temperatures become equal and there can be no further regenerative heat recovery. Figure 9.7.8 compares regenerative cycle trends with those of the simple cycle.

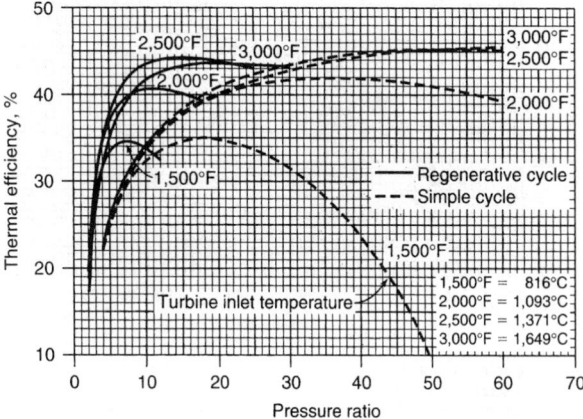

Fig. 9.7.8 Regenerative Brayton cycle variation of thermal efficiency with pressure ratio at different turbine inlet temperatures.

Fig. 9.7.9   Intercooled Brayton cycle gas-turbine component arrangement.

Fig. 9.7.11   Intercooled Brayton cycle variation of specific output power with pressure ratio at different turbine inlet temperatures, optimized for output power and for efficiency.

Intercooled Cycle (Fig. 9.7.9) In this cycle, the compression process is split between two compressors. An **intercooling heat exchanger** is placed between the front and rear compressors to reduce the temperature of the working substance. This cycle takes advantage of the thermodynamic principle that the work of compression is directly proportional to the level of the entering temperature. With a lowered inlet temperature, the work input needed to drive the rear compressor is reduced for the same pressure ratio. The reduced compressor work is directly available as increased net cycle output. This is true for both ideal and real cycles. In the ideal cycle, because the intercooler is a source of added heat rejection, the intercooling cycle efficiency is always lower than that of a simple cycle having the same pressure ratio. However, in real cycles, the intercooling effect can help offset the effect of real component efficiencies. In some circumstances, intercooling improves both specific power output and efficiency over simple cycle levels. Theoretically, the maximum power output occurs when the total pressure ratio is split evenly between front and rear compressors. However, maximum cycle efficiency occurs at a split which depends on the turbine inlet temperature as well as the efficiency levels of compressors and turbines. Typically, peak efficiency occurs when intercooling takes place well in front of the midpoint of the compression process. Figures 9.7.10 and 9.7.11 compare intercooled thermal efficiency and specific power output trends for pressure-ratio splits optimized for power output and for pressure-ratio splits optimized for efficiency.

Reheat Cycle (Fig. 9.7.12) In this cycle, the turbine expansion process is interrupted at some intermediate point before it reaches minimum (or

ambient) pressure in the exhaust. At this point, additional heat is added to the working substance by combustion at constant pressure in a second combustor. This takes advantage of the basic thermodynamic principle that the work of any expansion process increases in direct proportion to the temperature entering the turbine. (This is the converse of the principle that applies to compression, as cited above for the intercooled process.) The elevated temperature entering the second turbine

Fig. 9.7.12   Reheat Brayton cycle gas-turbine component arrangement.

increases its work output; this additional work appears directly as greater power on the output shaft. Since the extra work is more than offset by the added heat energy of the second combustor, the result is also reduced cycle efficiency.

Associated with the **reheat cycle** is an increase in exhaust temperature which ordinarily represents higher exhaust waste heat. However, in the case of a **combined cycle,** the higher exhaust temperature is a source of waste heat recovery that can be used to generate steam. The higher exhaust temperature can improve cycle efficiency of the steam system to the extent that the **overall system efficiency** can be improved. Reheat thus lends itself to use with exhaust waste heat recovery systems.

Intercooled and Regenerative Cycle This cycle (Fig. 9.7.13) has both an intercooler in the middle of the compression process and a regenerator

Fig. 9.7.10   Intercooled Brayton cycle variation of thermal efficiency with pressure ratios at different turbine inlet temperatures, optimized for output power and for efficiency.

1,500°F = 816°C
2,000°F = 1,093°C
2,500°F = 1,371°C
3,000°F = 1,649°C

Fig. 9.7.13   Intercooled regenerative Brayton cycle gas-turbine component arrangement.

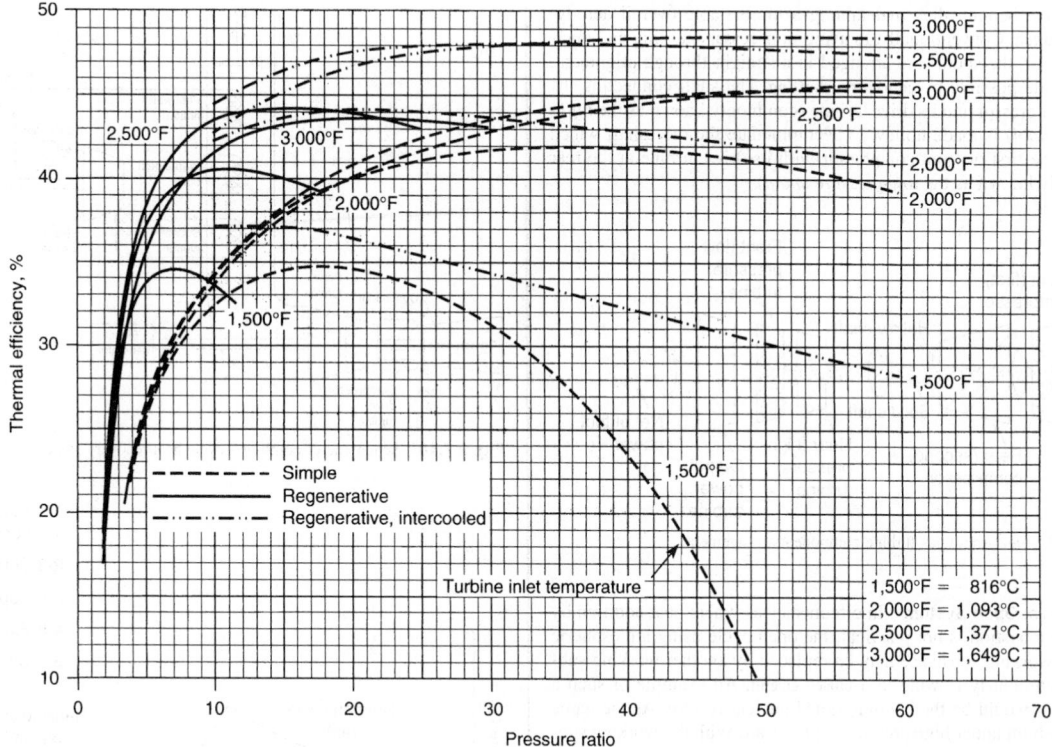

Fig. 9.7.14 Variation of Brayton cycle thermal efficiency with pressure ratio and at different turbine inlet temperatures, for regenerative cycle and for simple cycle.

to capture exhaust waste heat. The intercooling results in a lower temperature leaving the rear compressor. This circumvents the temperature difference limitation reached in regeneration. That is, at increasingly high pressure-ratio levels, the turbine exhaust temperature remains sufficiently higher than the compressor exit temperature to sustain heat exchange and thus waste heat recovery. Overall, the cycle efficiency of the intercooled regenerative cycle is superior to that of the simple Brayton cycle and all its common variants over a wide range of cycle pressure-ratio levels. Figure 9.7.14 shows variation of efficiency for an intercooled regenerative cycle compared to simple and pure regenerative cycles.

In actual practice, the choice of a simple cycle or one of its variants depends on overall economics applied to a specific use. The extra cycle benefits must trade favorably against the additional cost and complexity of the added heat exchangers or combustors and the penalties of the added ducting in between.

CONFIGURATION VARIATIONS

Since the gas turbine is made up of rotating compressors and turbines connected by driveshafts, there are various ways to arrange these elements. The simplest, Fig. 9.7.15, is the single shaft or "spool" arrangement. In this case, the compressor, turbine, and output driveshaft all turn at the same rotational speed.

Free Turbine (Fig. 9.7.16) To provide more flexibility by allowing the load to turn independently at a speed different from that of the compressor and its drive turbine, a **free turbine** configuration connects the compressor and its turbine on one driveshaft, while hot air continues to expand through a separate turbine which delivers net output power. The free turbine finds use either in electric generator power applications, where the generator must operate at a fixed speed (either 3,600 r/min or 3,000 r/min for 60- or 50-Hz installations, respectively), or in pumping applications,

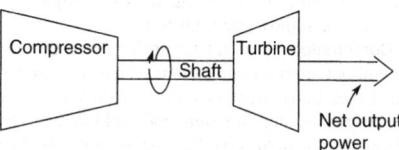

Fig. 9.7.15 Single-spool gas-turbine component arrangement.

Fig. 9.7.16 Free turbine gas-turbine component arrangement.

where speed can vary. In either case, the compressor at partial load can operate at its most favorable speed independent of the driveshaft load.

Dual Spool (Fig. 9.7.17) At very high compressor pressure ratios, the part load operation of a compressor on a single spool tends to become unstable. When a compressor designed for high pressure ratio is forced to operate at low pressure ratio (as load is reduced), the front and rear portions of the single-spool compressor become aerodynamically unbalanced. This is alleviated by dividing the compression process into two or more spools on separate shafts driven by separate turbines. With this arrangement, the front compressor can slow down more rapidly than the one behind, thus maintaining a more stable overall aerodynamic balance. Conventionally, to avoid unnecessary turning and ducting of the air, the spools are arranged in a straight-through alignment. This is made possible by **concentric shafting.** The shaft connecting the second compressor and its drive turbine (commonly known as the *high-pressure spool*) is made hollow to accommodate the driveshaft which connects the first or front compressor with its drive turbine at the rear (commonly known as the *low-pressure spool*).

Fig. 9.7.17   Dual-spool gas-turbine component arrangement.

Closed vs. Open System   In circumstances where heat addition by internal combustion is not practical, the open cycle using atmospheric air as the working substance may be replaced by another working substance continuously flowing in a **closed circuit.** An example of such a power plant would be the helium-cooled nuclear reactor. As the name implies, helium under high pressure replaces air. With the working substance completely contained, the minimum cycle pressure can be made much higher than ambient. Even such a low-molecular-weight gas as helium can be compressed into a low volume and thus keep the overall power plant relatively compact. Helium lends itself to the nuclear reactor application by virtue of its being almost completely chemically inert, which prevents radioactive transport.

External Combustion   The gas turbine can maintain its continuous flow feature without internal combustion; such is the case with the direct use of coal as a fuel. Early attempts to use coal directly in internally fired gas turbines always resulted in severe and rapid damage to the turbine materials. Associated with the coal burning are not only the erosive particulates but also highly corrosive impurities such as sulfur. As an alternative, gas-turbine power plants using indirect firing of coal were also considered, and at least one was built in Germany. It, too, was a failure due to the corrosive attack on the heat exchanger, largely by the alkali metals in the coal. There is now no significant development taking place on external combustion with coal. The alternative for use of coal in gas turbines now receiving considerable attention is coal gasification. See the description in the "Applications" section of this chapter and in Sec. 7.2.

WASTE HEAT RECOVERY SYSTEMS

A characteristic of gas-turbine engines is the incentive to operate at as high a "firing" (turbine inlet) temperature as the prevailing technology will allow. This incentive comes from the direct benefit to both specific output power and cycle efficiency, as illustrated in Figs. 9.7.4 and 9.7.5. Associated with the high maximum temperature is a high exhaust temperature which, if not utilized, represents waste heat dissipated to the atmosphere. Schemes to capture this high-temperature waste heat are prevalent in industrial applications of the gas turbine.

Combined Cycle   This term commonly refers to the scheme that places a boiler and superheater directly behind the gas turbine to generate steam and produce additional power in a steam turbine (Rankine cycle). A simplified version of a combined cycle is shown in Fig. 9.7.18. Figure 9.7.19 illustrates the combined-cycle process on a temperature vs. entropy diagram. The steam portion is sometimes called the **bottoming** cycle. Early combined cycles had **fired** as well as **unfired** versions. The supplementary firing was required because of the high

Fig. 9.7.18   Schematic of combined-cycle heat recovery system.

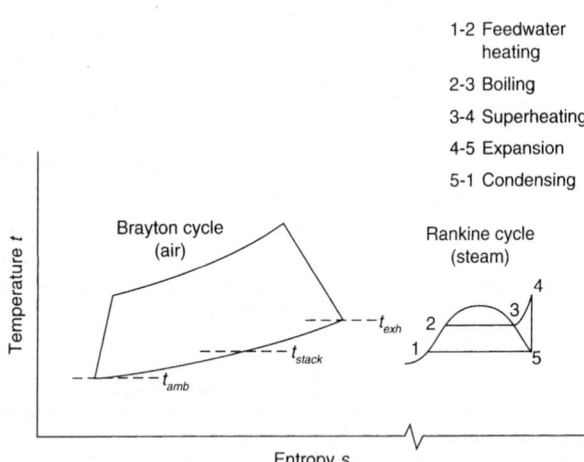

1-2 Feedwater heating

2-3 Boiling

3-4 Superheating

4-5 Expansion

5-1 Condensing

Fig. 9.7.19   Combined-cycle temperature vs. entropy diagram.

sensitivity of the Rankine cycle to maximum superheat temperature, as shown in Fig. 9.7.20. When the gas-turbine exhaust temperature is too low to sustain a good steam cycle efficiency, the extra combustion in the gas-turbine exhaust could result in an overall system efficiency improvement. Modern gas turbines, with their increasingly high turbine inlet and exhaust temperatures, can approach or exceed 60 percent in **overall combined-cycle plant efficiency** without supplementary firing.

Fig. 9.7.20   Variation of typical steam system heat recovery with turbine exhaust temperature.

If η_{gt} is the gas-turbine (Brayton) cycle efficiency and η_{st} is the steam bottoming cycle (Rankine) efficiency, then the overall combined-cycle plant efficiency η_{cc} is:

$$\eta_{cc} = \eta_{gt} + (1 - \eta_{gt})\,\eta_{st}\!\left(\frac{t_{exh} - t_{stack}}{t_{exh} - t_{amb}}\right) \qquad (9.7.2)$$

where t_{stack} is the minimum practical exhaust temperature to be released into the atmosphere after heat recovery, t_{exh} is the temperature of the gas turbine exhaust gases leaving the turbine, and t_{amb} is the outside ambient temperature. (Typically, stack exhaust temperatures are limited to 200 to 300°F (93 to 150°C) to avoid condensation of the potentially corrosive combustion products.)

Steam Injection (Fig. 9.7.21) As a variation of the conventional combined cycle described above, some gas-turbine installations generate steam by exhaust waste heat recovery and, instead of using that steam in a separate steam turbine, inject the steam directly back into the turbine section of the gas turbine. In this system, the turbine acts not just on air (plus combustion products) as the working substance, but instead on a mixture of air products and steam. Such a scheme eliminates the need for additional equipment and is thus less expensive than the normal combined cycle. This system has a drawback in that direct steam injection consumes water at the plant site and thus raises environmental concerns.

Fig. 9.7.21 Arrangement of components for gas turbine with steam injection.

Cogeneration There are many power plants used to provide the total energy needs of industries which use process heat. In these instances the exhaust heat of the gas turbine provides a source of high energy that can be used directly in the plant process rather than for producing more power. This is particularly true in the chemical industry, where many operational **total energy plants** can be found. In addition, some of or all the exhaust waste heat can be used for **district heating** where, especially in colder climates, steam or warm water generated from the gas-turbine exhaust waste heat can be circulated around the community to heat homes and buildings. Use of gas-turbine exhaust for heating is especially suited to provide the total energy needs of moderate-size to large institutions such as schools, hospitals, or commercial complexes like shopping malls.

Fig. 9.7.22 Typical partial-load performance characteristics for fixed output speed.

OPERATING CHARACTERISTICS

Partial Load Gas turbines respond rapidly to throttle variations. They are capable of accelerating from idle to full power in minutes or, in the case of the lightweight flight engine designs, seconds. They can be started in a matter of minutes and, via automatic controls, can operate for long periods of time unattended. Partial-load fuel efficiency remains high. Typical partial-load performance characteristics are shown in Fig. 9.7.22 for fixed output rotating speed (r/min) and in Fig. 9.7.23 for a variable-speed (free-turbine) drive.

Fig. 9.7.23 Typical partial-load performance characteristics for variable output speed in a free-turbine drive.

Ambient Conditions Gas-turbine operation is sensitive to ambient conditions. Being directly dependent on mass flow for a fixed cycle, power output varies directly with air density and thus ambient pressure. Variations in ambient temperature affect the thermodynamic cycle and engine operating characteristics. Normally, any particular gas-turbine model will have an associated **power rating curve** which varies as shown in Fig. 9.7.24.

Fig. 9.7.24 Typical variation of output power with ambient temperature.

GAS-TURBINE COMPONENTS

The primary components of a gas turbine consist of a **compressor,** a **combustor** and a **turbine.** Other elements of a gas turbine, depending on the installation, are the electronic control, inlet to the compressor, exhaust nozzle, power turbine, regenerators, and intercoolers. The component that poses the greatest aerodynamic and thermodynamic design challenge, as well as having the greatest influence on engine performance, is the compressor.

Compressors   Compressors can be centrifugal flow, axial flow, or a combination centrifugal and axial flow. In a **centrifugal flow compressor,** the airflow enters at the hub and is then turned from axial to radial by the compressor rotor. Early gas turbines and some modern small engines employ centrifugal compressors. In an **axial flow compressor,** air flows in an axial direction through a series of rotating blades and stationary vanes which are concentric with the axis of rotation. All large gas turbines employ multistage axial flow compressors because of their higher efficiency and capacity. Figure 9.7.25 illustrates the rotor, stators, and assembly of a multistage axial flow compressor. To achieve high pressure ratios and maintain stability at all operating conditions, **dual-spool** or three-spool arrangements are employed, which permit each spool

to operate at optimum speed. Interstage bleeds, discharge bleeds, and variable-angle vanes are also employed to avoid surge. Compression ratios in any one spool can exceed 20:1 with spools of 10 or more stages, and polytropic efficiency can be as high as 92 percent.

Combustors   A number of arrangements are used for the **combustion chamber** including side-by-side can annular combustors or a single annular combustion chamber (Fig. 9.7.26). Fuel is introduced through a manifold and an arrangement of multiple fuel nozzles. The **combustor** must maintain high combustion efficiency, and low levels of pollutant emissions and smoke, have a minimum pressure loss, maintain stable combustion over a wide range of operating conditions, and produce a uniform temperature in the hot gases distributed to the turbine. The combustor must be designed to utilize some of the airflow to atomize the fuel to achieve efficient combustion, generate turbulent mixing needed to produce a uniform exit temperature, and cool the metal parts of the combustor for durability at high operating temperature. Burner efficiency usually exceeds 99 percent except at low power in the idle region.

Turbines   The **turbine** consists of one or more stages of blades and interstage guide vanes located immediately to the rear of the combustor section. Figure 9.7.27 illustrates the rotors or wheels, sets of stationary

Rotor   Stator   Assembly

Fig. 9.7.25   Components and assembly of axial flow compressor.

Can annular combustion chamber

Annular combustion chamber

Fig. 9.7.26   Combustion chamber (combustor) configurations.

Turbine rotors (wheels)   Nozzles (vanes)   Assembly

Fig. 9.7.27   Turbine elements and assembly.

vanes or nozzles, and their assembly. The turbine extracts kinetic energy from the heated gas stream and converts it to shaft power to drive the compressors and accessories. In the case of a turboprop, turbofan, or turboshaft, power is also extracted to drive a propeller, a fan, or an output shaft (see Sec. 11.5). The arrangement of the turbine is similar to that of the compressor with two major differences. First, since the expansion process is aerodynamically easier than the compression process, turbines have fewer stages than compressors. Second, the gases are much hotter and, therefore, require materials and cooling schemes to maintain the life of the turbine parts. High-temperature alloys, internal cooling passages (for air usually supplied from the compressor), and ceramic thermal barrier coatings are used in the turbine. The normal range of turbine efficiency is 90 to 95 percent.

APPLICATIONS

A worldwide directory of currently available **gas-turbine aircraft engines** published in *Flight International* (June 8–14, 1994) lists 37 manufacturers and 166 models. A much larger number of companies are manufacturers or suppliers of parts or components to the original equipment manufacturers. More than 700 model designations are now or have been in service.

For **industrial gas turbines** above 500 kW, the "2004–2005 Gas Turbine World Handbook" lists 131 gas-turbine models on the market, from 16 original equipment manufactures (OEMs). In addition there are a number of companies that manufacture or package gas turbines under license from the OEMs. Frame-type gas turbines account for more than three-fourths of the gas turbines shipped annually, measured in megawatt capacity of the shipments. Aeroderivatives, from four OEMs, are limited in power output to a maximum of about 50 MW due to the size of the aircraft engines from which they are derived, while frame-type gas turbines are available up over 300 MW. This limits the applicability of the aeroderivatives to small to medium-size power plants for electric power generation, while much of the power generation market is in large combined-cycle power plants. Figures 9.28 and 9.29 show cutaway views of an aeroderivative engine (see also Figure 9.7.1) and a large, heavy-duty industrial design respectively.

Military Aviation   Military aircraft, including fixed-wing combat and transport as well as helicopters, are almost exclusively gas-turbine-powered. Fighter aircraft are powered with afterburning turbofan engines. Modern heavy-lift cargo and troop transports are powered by high-bypass-ratio turbofans often derived from commercial engines. Many older cargo, troop, tanker, and utility aircraft still in service are powered by turboprop engines. Most helicopters are powered by turboshaft engines. New developments in military engine technology aim at increasing overall performance by increasing the power of the core engine components through use of higher-temperature materials and turbine technology. The military strives to achieve a balanced tradeoff of performance, weight, and stealth against reliability, maintainability, and cost.

Commercial Aviation   Commercial aircraft are primarily powered by turbofan and turboprop engines. Design of large commercial wide-body transports tends toward long-range twin-engine turbofan power, but three-and four-engine aircraft are still being manufactured. With twin-engine aircraft approaching the size and range of four-engine aircraft, thrust requirements are ever increasing. Turbofan engines in the 60,000- to 90,000-lb (267- to 400-kN) thrust range have been certified, and versions to 100,000 lb (444.8 kN) are planned. Medium-size commercial transports are also turbofan-powered, with engines in the 20,000- to 40,000-lb (89- to 178-kN) thrust range. Small regional or commuter aircraft are primarily powered by turboprop engines from 500 to 3,000 hp (372 to 2,237 kW). Commercial engine technology is aimed at higher bypass ratios for increased thrust and reduced emissions and noise; higher component efficiency for better fuel consumption; and increased reliability. Future developments include advanced ducted propulsors (ADPs) which couple a gear-driven prop fan to the gas turbine, geared fan engines, and engines for the second generation of supersonic aircraft.

Electric Power   Of the three types of industrial gas-turbine applications—electric power generation, mechanical drive, and marine propulsion—electric power generation typically amounts to over 90 percent of the annual shipments of new capacity, as measured in megawatts. Gas turbines, including combined cycles, are now the fastest-growing type of power plant in the world. According to the Energy Information Agency, part of the U.S. Department of Energy, in 2001 natural gas fuel (virtually all used in gas turbines) accounted for 18.4 percent of the world's electric power generation. That proportion is projected to grow to 25.2 percent by 2025, with a growth rate about 3 times that of coal, the next-fastest-growing fuel. The reasons are that gas turbines and combined cycles, compared to other types of power plants, are low in capital cost, easy to install, and highly efficient, with combined-cycle efficiencies approaching 60 percent compared to at best about 40 percent for coal-fired steam-turbine power plants. Additionally the emissions from gas-turbine power plants are far less than those from coal-fired steam-turbine plants, which adds to their popularity—helped by the low carbon/hydrogen ratio of natural gas compared to coal, which, in addition to high efficiency, further reduces CO_2 emissions per kilowatthour produced. Some gas-turbine installations use oil fuel, but natural gas is becoming more widely available worldwide and is usually the preferred fuel where it is available, because it results in lower maintenance cost for the gas turbine and produces lower emissions. When gas turbines began to be used in the late 1950s to early 1960s, they were mostly simple cycle in "peaking" service, i.e., operated only at times of high demand for electricity. But by the 1990s very large, efficient, and low-capital-cost combined cycles became available, and these products now dominate the market for new central station power plants.

Fig. 9.7.28 Cutaway view of FT8 aeroderivative industrial gas turbine. (*Courtesy Pratt & Whitney Aircraft.*)

Fig. 9.7.29 Cutaway view of M701 large heavy-duty industrial design. (*Courtesy Mitsubishi Heavy Industries.*)

Fig. 9.7.30   Integrated gasification combined cycle (IGCC) system diagram. (*U.S. Department of Energy.*)

Integrated Gasification Combined Cycles (IGCCs)   The advent of large natural-gas-fired combined cycles, coupled with the worldwide desire to reduce emissions from power generation plants, has given new life to a relatively old concept—coal gasification. The concept of gasifying coal started in the 1940s but was not considered seriously as a power plant option until the late 1990s to early 2000s, when the price for natural gas escalated substantially at the same time as the Kyoto Treaty focused attention on reducing emissions of CO_2. The concept is shown in Fig. 9.7.30.

A major incentive for IGCCs is that by gasifying the coal it is possible to remove, before combustion, contaminants such as mercury, lead, and sulfur to much lower levels than is possible in a conventional coal-fired plant, where the contaminants are substantially diluted in the exhaust gas.

Development of IGCCs started in the 1970s following the energy crisis of 1973 and was accelerated by the second energy crisis in 1979. The resulting incentive in the United States and in Europe to find ways to produce electric power from coal with increased efficiency and reduced emissions prompted government funding for IGCC plants. Two coal-based IGCCs were built in the United States, in Florida (Tampa Electric) and Indiana (Cinergy), and in Europe, Nuon in the Netherlands and Elcogas in Spain. All the plants were relatively small, 250- to 320-MW plant output, but large enough for one large gas-turbine combined cycle. None of them would have been competitive economically without government subsidies, but the plants provided valuable demonstration and experience that the technology held promise. Also, IGCCs fed with petroleum coke or with tars or asphalt have been built without government subsidies in the United States, Europe, and Asia. Additionally, in the United States there was a DOE-funded coal gasification plant in

Kingsport, TN, which produced product gases and final chemical products clean enough for human consumption in pharmaceuticals, although at high capital cost. The plant demonstrated 98 percent reliability and helped convince many utilities that IGCC is a viable technology on a technical basis, the question being whether it could be competitive economically vs. conventional coal-fired steam-turbine plants. As of the early 2000s, the capital cost of an IGCC was about 10 to 15 percent higher per kilowatt than that of a conventional coal-fired plant.

In 2004 the U.S. Department of Energy announced the funding of two larger-scale IGCC plants based on the latest existing gas-turbine and gasifier technology and in 2005 launched a solicitation for a major research and development program for the next level of technology in IGCCs to be built in about 2010. The intent is to enable IGCCs to be fully competitive with conventional coal-fired plants and produce substantially lower emissions, including the possibility of stripping off the CO_2 and sequestering it (see the cycle diagram in Fig. 9.7.30).

In the long term, IGCCs provide the possibility that gas turbines may become the prime movers for coal-based as well as natural-gas-based electric power generation systems.

Marine gas turbines are almost exclusively aeroderivatives because of their light weight and compact size. They range in size up to 57,000 hp, as listed in the "2004–2005 Gas Turbine World Handbook." They provide power for most of the surface warships (e.g., cruisers, destroyers, frigates, corvettes) recently added to the world's fleets. Two new market opportunities are emerging. The first is the growing market for commercial lightweight, very fast (40 + kn) novel ferries and cargo ships—catamaran, slim monohull, etc.

The second is being created by the increasingly severe emissions standards for ships. Most of the world's cargo and cruise ships are diesel-powered due to their ability to use poor-quality, less expensive fuels. However, as emissions regulations are tightened for SO_x and NO_x, the fuels must be lower in sulfur, and the power plants must reduce NO_x. Gas turbines are more capable than diesels of reducing NO_x levels, so they are gaining in market share.

Mechanical Drive Gas Turbines   The increased popularity of natural gas as a preferred fuel for power generation and home heating has stimulated the market for gas turbines in mechanical drive applications, largely for natural gas pipeline pumping but also for driving large compressors in process plants such as oil refineries and liquefied natural gas (LNG) installations. The "2004–2005 Gas Turbine World Handbook" lists mechanical drive gas turbines from 900 to 70,000 hp, in a mix of aeroderivative and frame-type units. Most applications are simple cycle, although a number that are in process plants are combined-cycle configurations. The aeroderivatives have higher efficiency, but frame-type units are also popular for remote of critical applications due to their durability and low maintenance requirements. A characteristic of these applications is sometimes unstaffed stations and often the high economic cost consequences of failure. Consequently many owners are willing to sacrifice some efficiency for lower risk of downtime.

Microturbines   Gas turbines in smaller sizes than 500 kW are used in several applications. The gas turbines usually have centrifugal compressors and, in some cases, radial inflow turbines. There are three manufacturers who specialize in this type of gas turbine, although some of the bigger companies are beginning to develop microturbines to add to their existing product lines. The most common applications are for **distributed generation,** often for small commercial or industrial companies that have a need for the exhaust heat for a process, for heating or for cooling via absorption refrigeration. It is generally not economical for a user to buy a microturbine for peaking power, although it can be a

good economic choice for backup power to protect against the loss of power from the grid. Due to the economy-of-scale effects, the cost per kilowatt is high and the efficiency is low for this type of gas turbine; but if the user's requirement for exhaust heat is a good match for the gas turbine, and if the user's process runs 24/7, a microturbine can be a worthwhile investment. Some users have a varying electric or heat load and buy a microturbine plus exhaust heat recovery system strictly for baseload operation. A current issue is the **interconnect requirement**—the requirements imposed by the local utility on the way that users can connect to the grid, and the price for backup power in case in microturbine has downtime when the power need is present. This is as much a regulatory and political issue as a technical one.

Small Engines for Aerospace Applications   Gas turbines in smaller sizes [less than 1,000 hp (746 kW)] are used in several fields of application. Such engines usually feature centrifugal compressors and are all manufactured by specialized companies. In the ground transportation field, automobile and truck manufacturers have experimented since 1950 with gas turbines, primarily of the low-pressure-ratio (less than 10:1) regenerative type. While several prototypes have been tried over the years, displacement of the widespread, well entrenched automobile piston engine has not taken place, with one exception: The U.S. Army has found the gas turbine to be well suited as a tank engine. As with the automobile engine, gas turbines have been experimentally tried in railroad trains, but have yet to make inroads to replace the diesel engine.

Small gas turbines are used as a compact, simple, and reliable power source on board aircraft for start-ups and auxiliary power. They are known as auxiliary power units or APUs. Similar engines find use as quick-start standby units to drive generators for emergency power. Finally, very small jet engines provide propulsion for remotely controlled aircraft with surveillance or cruise missile military missions.

9.8   NUCLEAR POWER
by Andrew C. Klein

REFERENCES: Etherington, "Nuclear Engineering Handbook," McGraw-Hill. Hogerton, "The Atomic Energy Deskbook," Reinhold. Kramer, "Nuclear Energy—What It Is and How It Acts," Technical Publishing Co. Blatz, "Radiation Hygiene Handbook," McGraw-Hill. Ellis, "Nuclear Technology for Engineers," McGraw-Hill. Glasstone, "Source Book on Atomic Engineering," Van Nostrand. Murray, "Nuclear Reactor Physics," Prentice-Hall. Annual reports of the International Atomic Energy Agency, Vienna. "Nuclear News," LaGrange Park, IL. Cole, "Understanding Nuclear Power," Gower. Glasstone and Sesonake, "Nuclear Reactor Engineering," 3d ed.; Murray and Powell, "Understanding Radioactive Waste," 3d ed.; Nero, "Guidebook to Nuclear Reactors," 1979; Rahn et al., "A Guide to Nuclear Power Technology," 1984.

NOTE: The material related to nuclear power plant safety and licensing was revised and amended by personnel of the Nuclear Regulatory Commission (NRC) on official time. That material, therefore, is in the public domain and not covered by copyright.

FISSION AND FUSION ENERGY

Nuclear power is the energy derived from the fission (or splitting) of the nuclei of heavy elements, such as uranium or thorium, or from the fusion (or combining) of the nuclei of light elements, such as deuterium or tritium. Particles set in motion by these processes yield heat energy virtually instantaneously. The amount of energy released per atom exceeds by a factor of several million the amount of energy obtainable per atom in a chemical reaction, such as the burning of fossil fuels. While control of the fusion process is still surrounded by tremendous technical problems, considerable success has been achieved in utilizing heat energy produced by fission for power generation, propulsion, industrial production, and scientific experiments.

NUCLEAR PHYSICS

A nuclear reaction occurs when the particles making up the nucleus of an atom are rearranged—as distinct from a chemical reaction, which is caused by changes in the electron structure surrounding the nucleus.

Fundamental (elementary) particles are, theoretically, the irreducible constituents of the material world. New experimental techniques lead to discovery of more particles; advances in theory indicate that some are compounds of particles. Except for the electron and proton, all the fundamental particles are unstable.

An **electron** carries one unit of negatively charged electricity [defined as 1.6×10^{-19} coulomb (C)] and has a mass at rest equal to 1/1,836 times the mass of the hydrogen atom. Electrons emitted by radioactive atoms are called **beta (β) rays** and have energies up to several MeV (million electron volts).

A **proton** carries one unit of positively charged electricity equal in magnitude to that of the electron and has a mass, at rest, of 1.0072764 atomic mass units [u when 1 u (1.66057×10^{-27} kg) is defined as 1/12 the mass of a neutral atom of the most abundant isotope of carbon]. It is identical to the nucleus of the hydrogen atom. Protons are not emitted spontaneously from atomic nuclei.

A **neutron** has a mass approximately equal to that of the proton but lacks an electric charge. It cannot be detected by subjecting it to electric or magnetic fields, nor can its presence be shown by its passage through cloud chambers. The presence of neutrons can be shown only by their interaction with other particles.

An **atom,** which is the basic unit of any chemical element, consists of a central nucleus surrounded by planetary electrons. The number of its electrons determines its chemical characteristics and in electrically neutral atoms equals the positive charge on its nucleus. The nuclei of all atoms except the hydrogen atom are composed of protons and neutrons. The nucleus of the hydrogen atom consists of a single proton, which gives it an atomic number of 1. This number is designated by the letter Z. The number of neutrons in the nucleus is represented by the letter N. From this it follows that the mass number A, which represents the atomic weight, is equal to $N + Z$. All with atomic numbers above 92 (uranium) must be produced artificially and have a short half-life.

Isotopes are elements which have identical chemical characteristics but different atomic weights, i.e., certain atoms of the same element have different numbers of neutrons in their nuclei. There are considerably more than a thousand isotopes of the known elements. Only 320 of these exist in nature, of which approximately 40 are unstable and radioactive.

For example, hydrogen has three isotopes. One is ordinary hydrogen with a proton nucleus. Another is deuterium, whose nucleus consists of a proton and a neutron. The third is tritium, which consists of one proton and two neutrons. At the heavy end of the natural periodic table is uranium, with 92 protons. When it has 146 neutrons, it is $_{92}U^{238}$, which comprises 99.28 percent of natural uranium. The remainder of natural uranium consists of 0.006 percent U^{234} and 0.71 percent U^{235}. The latter isotope, with 92 protons and 143 neutrons, is the only one of the three which is readily fissionable and is the one utilized in most of the nuclear power reactors operating in the United States. It is separated from the natural element in the enrichment desired by gaseous diffusion or centrifuge separation processes in government facilities.

Energy and mass are used interchangeably in nuclear physics, and the total energy of a nucleus can be determined by measurement of its exact mass m, which is related to its energy E by the Einstein equation $E = mc^2$, where c is the velocity of light, 3×10^{10} cm/s. This law of equivalence of mass and energy unifies two long-accepted laws: the conservation of energy and the conservation of matter. Singly, the two older laws must be regarded as high-order approximations, adequate in all engineering fields except atomic energy. The exact mass m of a nucleus differs by only a few hundredths of a percent from an integral number A of proton masses, but this small deviation is of significance in nuclear theory, since it measures the difference in energy between the nucleus and its separate components, i.e., its binding energy. The energies of nuclides are recorded in the tables in terms of their masses.

Physical atomic weights do not quite equal chemical atomic weights because the oxygen used in the chemical determinations is a mixture of isotopes, whereas in mass spectrographic measurements the single isotope C^{12} is used for reference. The relationship is: Physical atomic weight = 1.000280 × chemical atomic weight.

Planck's Constant Since radiation has the properties of both waves and particles, it is theorized that in the emission or absorption of energy by atoms or molecules, the process takes place by steps, each step being the emission or absorption of an individual quantity of electromagnetic energy, which is considered the elementary quantum of action. It is known as Planck's constant and has the value $h = 6.624 \times 10^{-27}$ erg · s. It is used in virtually all quantum relationships, including Schrödinger's equation and Heisenberg's uncertainty principle, and in the formula for the energy levels of atomic hydrogen. The standard notion is $\hbar = h/2\pi$, where $\hbar$ signifies the angular momentum of an orbiting electron.

Relativistic Mass When moving particles approach the speed of light, the approximation of kinetic energy, $\Sigma \frac{1}{2}m_0v^2$, fails; it then becomes necessary to use the equation $E = mc^2$ and to use the relativistic mass, a function that increases with velocity according to the law $m - m_0/\sqrt{1 - (v/c)^2}$, where m_0 is the rest mass (i.e., the mass of newtonian mechanics) and v is the coordinate velocity of the particle.

Binding energy is the work required to disintegrate an atom completely into Z protons and $A - Z$ neutrons. It is a measure of the total kinetic and potential energy of the nucleons in the nucleus and may be calculated from the equations

$$B = C^2\Delta \quad \text{and} \quad \Delta = Zm_H + (A - Z)m_n - m$$

where m is the mass of the nuclide, m_H the mass of the hydrogen atom, and m_n the mass of the neutron, all on the physical atomic-mass scale (u). The quantity of Δ is known as the mass defect. The energy involved is usually stated in MeV.

Fusion Figure 9.8.1 shows that if two deuterium (H^2) nuclei react to produce one He^3 nucleus plus one neutron, the total binding energy of the four particles (two neutrons plus two protons) involved in the reaction is increased from approximately 4.4 to 7.6 MeV, releasing 3.2 MeV for the reaction. Such a process is called fusion. The deuterium-deuterium fusion reaction may also produce an H^3 plus an H^1 nucleus, with a release of 4.0 MeV.

Fig. 9.8.1 Binding energy per nuclear particle for stable nuclei.

For fusion to occur, the two nuclei must approach each other with exceptionally high kinetic energy in order to overcome their electrostatic repulsion; and since kinetic energy is proportional to absolute temperature, effective fusion can occur only at extremely high temperatures. For fusion to continue as a chain reaction, it is further necessary that nuclear collisions be sufficiently frequent to maintain the high temperature in spite of heat radiation. The condition necessary for frequent collisions is high pressure.

A light-element chain reaction is the principal source of heat in the sun, hydrogen being converted to helium as the net result of a multistage process. The temperature and pressure at which this reaction is sustained at the center of the sun are 35 million °F and 1.5×10^{11} lb/in² respectively.

A deuterium-deuterium reaction can be sustained at temperatures and pressures which are considerably lower than those in the solar reactions but still fantastically high by terrestrial standards. Achievement of a controlled light-element chain reaction for power generation by fusion depends on producing high particle velocity and high density locally by electromagnetic means, without development of correspondingly high temperatures and pressures at the container walls.

Fission is the division of an atomic nucleus into parts of comparable mass, either naturally (spontaneously) or under bombardment (induced) with neutrons, α particles, γ rays, deuterons, or protons. In elements with $Z > 90$, fission can be produced by neutrons of low or moderate energy.

Fission is important because in the process neutrons are emitted that may, under certain conditions, be utilized to produce further fissions, thus leading to a self-sustaining, or "chain," reaction. It is thus possible to "burn" uranium to obtain an energy release per atom approximately 10^8 greater than from a chemical fuel. The energy release in the fission of U^{235} is

	MeV
Kinetic energy of fission products	168
Kinetic energy of fission neutrons	5
Energy of γ rays	10
Energy of β rays	5
Energy of neutrons	11
	199

The total of 199 MeV is equivalent to 3.2×10^{-11} W · s, which indicates that it requires the fissioning of 1.12×10^{17} atoms to release a kilowatthour of energy. This number of atoms forms only a speck of matter six thousands of an inch in diameter.

Radioactivity occurs when an unstable nucleus undergoes atomic disintegration by emitting α, β, or $\beta+$ particles, γ or X-ray electromagnetic radiation.

The **alpha** (α) **particle** is identical with the nucleus of a helium atom. The emission of an α particle creates a new nucleus, with Z reduced by 2 and A decreased by 4 mass units.

Beta (β) **decay** involves emission of an electron from the nucleus of an atom, thereby increasing Z by unity.

The **gamma** (γ) **ray** is electromagnetic radiation originating in the nucleus—as differentiated from X-rays, which usually are less energetic and arise from energy adjustments as electrons move between orbital shells outside the nucleus. Gamma rays are emitted instantaneously when a neutron is captured in a nucleus and are frequently emitted following ejection of a particle from a radioactive nucleus.

The **positron** $(\beta+)$ is a particle whose mass is the same as that of an electron and whose charge is equal in magnitude but opposite in sign. Positron emission decreases Z by unity.

Electron capture (EC) results when an unstable atom decays by capturing an orbital electron in the nucleus, resulting also in a decrease of Z by unity. This capture produces a vacancy in the orbital shell which is filled by an electron moving from an outer shell, giving the rise to X-ray characteristics.

Half-life is the time required for half the original nuclei in a sample of an isotope to decay; it is given as $0.693/\lambda$ s. Half-lives of the radioactive nuclides vary from 10^{-7} s to 10^{10} years.

UTILIZATION OF FISSION ENERGY

Nonpower Types

Nuclear reactors having a purpose other than the generation of usable power may be classified into two general types: research and test reactors and weapons material production reactors.

The latter category of **production reactors** includes both light- and heavy-water and gas-cooled reactors of high thermal power (comparable in thermal output to large power reactors). Their fuel cycle is adjusted to produce plutonium and tritium of weapons quality for nuclear weapons programs. Most production reactors operate with outlet temperatures near or below the boiling point of water, so the thermal heat is wasted to the environment. A few **"dual-purpose"** production reactors operate at temperatures which allow the coupling of useful electric power output facilities.

Teaching reactors have the characteristics of very little excess reactivity, very low power capability (a few to a few thousand watts of heat energy), and relatively easy access to the core. Teaching reactors are used mainly for the purpose of instruction in the behavior of a nuclear reactor. With a teaching reactor novice reactor operators may experience the response of reactor instrumentation as reactivity is inserted and criticality is approached, "just critical" behavior, and power increases or decreases.

Teaching reactors are generally owned and operated by universities. In addition to being used for operator training, they are used in a variety of nuclear-engineering instructional experiments. Typical experiments include measurements of neutron-flux distribution; measurement of neutronic characteristics such as generation time, leakage fraction, thermal-utilization factor, resonance escape probability, and other characteristics affecting criticality and kinetics; and measurement of parameters affecting control such as control-rod worth and reactivity feedback coefficients. Although the neutron flux available from a teaching reactor is relatively small, it is nevertheless significant and can be used for limited production of radioactive isotopes.

Research reactors generally have a power capability of more than 1.0 MW. They are for the purpose of providing an intense radiation source, both neutrons and gamma rays. A research reactor is usually designed to have a high leakage flux in order to facilitate its use as a radiation source.

The neutrons from a research reactor have a variety of uses including neutron activation analysis, neutron radiography, neutron diffractometry, and production of radioactive isotopes. The gamma rays are applied in materials testing and in studies using the interaction of gamma rays with the nuclei of various elements.

In **neutron activation analysis,** an unknown sample is irradiated in a calibrated neutron flux, and the constituents of the sample become radioactive through neutron absorption. The radiation from the sample is then analyzed. Since the radiation from a specific radioactive isotope is unique to that isotope, a qualitative and quantitative determination of the constituents of the sample is possible. Extremely small quantities of radioactivity can be detected; therefore, neutron activation analysis is a very sensitive analytical tool.

Neutron radiography is a technique similar to X-radiography. The perturbation of a collimated beam of neutrons that has passed through the radiographed item is determined by film exposure or other means. Unlike x-rays that serve to outline dense, highly absorptive materials, neutron radiography serves to outline light materials that preferentially scatter neutrons from the incident beam.

Neutron diffractometry is a technique used to determine the molecular and crystalline structure of materials. The analysis is accomplished through a precise determination of the diffraction pattern from a collimated monoenergetic beam of neutrons that has interacted with the sample.

Certain radioactive isotopes are useful and effective in the **treatment of some diseases.** Isotopes that do not occur naturally can be produced in large quantities by exposure of the precursor element to the neutron fluxes available from research or test reactors.

Test reactors generally have a power capability ranging from 25 to 100 thermal MW. They are used largely for the purpose of providing an experimental radiation environment similar to that which exists in the core of a power reactor. They are also used for large scale radioactive isotope production.

Power Cycles

Although the energy of fission appears as kinetic energy (85 percent) and photon energy of γ rays, there is no practical method of converting this energy directly into useful work on a large scale. Therefore, a nuclear reactor must be treated as a heat source which differs from a chemical heat source in that no oxygen is required and the heat does not have to be removed from gaseous combustion products which possess poor heat-transfer properties.

In the large power capacities associated with nuclear power plants, the **Rankine vapor cycle** is the one in use. The Rankine cycle appears to be best suited for nuclear power because (1) the maximum practicable operating temperature for a reactor corresponds to or is lower than the temperature used in the conventional Rankine vapor cycle; (2) the problem of containment of radioactivity is better solved by this cycle than by other cycles; (3) reactors are most economical in large sizes, as are Rankine engines.

Three variations of the Rankine cycle (Fig. 9.8.2) are being used. The **direct cycle** is the least complicated and thermodynamically the most desirable. Its disadvantages are the facts that the boiling process does not permit the high power densities which are attainable with liquid cooling and that radioactive steam is carried into the turbine and condenser. The **indirect cycle** gives greater power density and eliminates radioactivity in the turbine, although it too produces steam at lower pressures and temperatures than those considered most efficient for modern turbines. Both concepts are highly developed and commercially available. Plants utilizing sodium or other liquid metal as the heat-exchange medium use an intermediate-link system to isolate the water system from radioactive sodium, but the ability to operate at high temperature with liquid-metal coolants permits the generation of high-temperature steam to utilize the present state of steam turbine technology.

The only other cycle which has been given consideration is the **Brayton cycle.** This cycle is being examined with renewed interest

Fig. 9.8.2 Plant cycles—coupling with a working fluid. *(Amorosi, Selection of Reactors, "Nuclear Engineering Handbook," McGraw-Hill, 1957.)*

because of the rapid advances in gas-cooled reactor technology and the availability of large gas turbines suitable for reliable utility industry service. The use of an open cycle using air has been discarded because of the spread of activation products from the turbine exhaust and the corrosion problems in the reactor core and fuel elements. The use of a closed Brayton cycle with the latest high-temperature gas-cooled reactors (HTGRs) has potential advantages in that the gas temperatures can match the latest gas-turbine designs and a very compact plant can be built.

The ability of a nuclear reactor to produce high temperatures not limited by chemical reaction equilibrium has generated interest in advanced concepts such as magnetohydrodynamics and thermionic generators. These systems are limited by the availability of suitable materials of construction and appear destined to remain in the research phase until suitable materials are developed.

Power Reactor Types

During the early years in the development of power reactors, a large number of possible reactor designs were investigated. In the present state of the industry, the designs which have gained the greatest acceptance all are solid fuel in the form of uranium dioxide. The moderator may be either graphite or water and the coolant may be water, gas, or liquid sodium. Experimental power reactors using a solution of uranium compounds in water or molten salt have been built but have not assumed industrial importance because of many unsolved corrosion problems. Light-water (moderated and cooled) reactors (LWRs), comprised of the boiling water reactor (BWR) and pressurized water reactor (PWR), represent about 99 percent of the operating reactors in the United States. A high-temperature gas-cooled reactor (HTGR) is also a feasible concept in which there is significant interest. The liquid-metal-cooled fast reactor (LMFR) concept is employed in operating test reactors and has been successfully implemented in full-scale operating plants in France and Russia. For the efficient utilization of the energy represented by the world's uranium reserves, a successful breeder reactor is desirable since the present generation of reactors are all consumers of only the naturally occurring fissionable isotope of uranium, U^{235}.

For utilization of natural uranium as a fuel, D_2O is substituted for H_2O as the coolant moderator. The reactor must be refueled more frequently than enriched reactors to maintain an adequate operating reactivity margin. However, D_2O reactors such as the Canadian designs (CANDU) utilize on-line refueling to minimize shutdown time.

Boiling water reactors have a simple design and can utilize relatively thin-walled vessels and pipes because they operate at moderate pressures compared with pressurized water reactors. Fuel cladding temperatures are only slightly higher than steam temperatures, and there is an inherent safety factor because the steam-void volume increases on a transient power increase. Some of the problems caused by radioactivity carryover, such as maintenance on the turbine and condenser and the prevention of radioactive leakage, are offset by the cost of the boiler in pressurized water reactors. However, boiling water reactor vessels, despite their lower design pressure, are larger and heavier than pressurized water vessels of similar rating and extensive water purification equipment is required to remove corrosion products and other impurities from the feedwater before introduction into the reactor.

Limitations on the **power density** imposed by exit voids (i.e., the vapor volume of the exiting steam-water mixture from the reactor core) are a disadvantage in that low-power density contributes to high fuel inventory charges. Load changes are accomplished by steam bypass control or rods since adjustment of the turbine throttle and consequent reduction in steam flow will cause an increase in pressure in the reactor, which in turn will collapse the steam bubbles and increase reactivity. The control system functions to maintain constant reactor pressure, and reactivity control is achieved in part by varying the recirculation rate in the reactor. The net decomposition of the water into O and H is much greater than in a pressurized water reactor.

Pressurized Water Reactors   Properties of water and steam which led to their predominance as general-purpose heat-transfer media have also caused their widespread application as a reactor coolant. A major disadvantage of water results from its relatively high vapor pressure. However, this can be partially overcome by allowing boiling in the reactor. Thermal efficiencies up to 36 percent are possible.

The use of H_2O as a coolant and moderator is based on well-developed technology which indicates that the ultimate size (or capacity) will be dictated primarily by heat transfer requirements. The average heat flux q/A is around 200,000 Btu/(h · ft^2) (630,000 W/m^2), with a maximum of 600,000 Btu/(h · ft^2) approximately (1,890,000 W/m^2).

The **light-water** (H_2O) **reactors** require slightly enriched fuel. However, the cost of enrichment is economically justified because the increased power density reduces inventory charges. To provide sufficient neutron moderation, the H_2O/U volume ratio is kept slightly above 2:1.

The use of oxide fuel minimizes corrosion. Design problems are reasonably well understood, although costly structural provisions must be made to load and unload fuel because of the high pressures. Fuel enrichment usually runs between 1.5 and 4.5 percent.

Control rods, burnable neutron absorbers, and soluble boron in the coolant are used to control excess reactivity, achieve high fuel burnup, and maintain power distribution within the design limits.

Gas-cooled reactors offer low fuel and operating costs because of their ability to utilize natural uranium as fuel. This advantage is offset by higher capital costs caused by the large system size, in turn resulting from the lower power density usually used. More advanced designs such as the high-temperature gas-cooled reactor and its variants can use low enriched uranium and are able to use higher temperatures and the required structural material for such temperatures. These newer gas-cooled designs are much more compact than the natural-uranium reactors which they are beginning to supplant.

Carbon dioxide is the coolant usually used in natural uranium reactors, as it is nontoxic, only mildly radioactive, nonflammable, noncontaminating in case of leakage, and relatively inexpensive. Helium makes an attractive coolant, and its superior performance at high temperatures can be utilized. Helium, being inert, does not chemically react with graphite and the reactor structure at high temperatures as does carbon dioxide.

The latest gas-cooled reactor designs are modular designs which use direct Brayton cycles to achieve high efficiency. These advanced modular designs use either solid blocks of graphite-fuel mixtures with coolant channels, or small balls (~8-cm diam) of fuel-graphite to achieve nuclear criticality. Modifications of these high-temperature reactors can be used with thermochemical reactions to split water, thus providing a means for producing hydrogen for proposed transportation fuel needs.

Breeder reactors operate at extremely high power densities (as they are unmoderated) and are liquid-metal-cooled. Their attractiveness stems from an ability to "breed," i.e., produce more fuel than they consume.

Aside from savings in fuel cost, this characteristic is believed necessary to conserve the world's supply of energy.

Two conflicting definitions of a breeder reactor are in common use: (1) a nuclear reactor that produces more fissionable material than it consumes, regardless of type of fuel used; (2) a nuclear reactor that produces the same species of material as it consumes, regardless of the net gain or loss. The first definition has wider acceptance in the United States.

The selection of a coolant for a breeder reactor is restricted by the requirement that it not act as a moderator. This requirement rules out the use of water, although the attainment of some breeding in a water-cooled reactor has been investigated. The choice of coolant is limited to helium or liquid metal.

Choice of a liquid-metal coolant is based mainly on nuclear properties and on the engineering difficulties associated with melting temperatures and corrosion effects. Those of prime interest are sodium, sodium-potassium alloy, bismuth, lead, and lead-bismuth alloy. Sodium has been selected as the preferred coolant because of its availability, good nuclear properties, and the short half-life of its induced radioactivity. However, in the presence of oxygen, it burns and can react violently with water.

Advantages of operating in the fast neutron spectrum include (1) structural material for the reactor core can be selected without consideration of neutron absorption; (2) very little excess reactivity is needed to compensate for fission product buildup; and (3) reprocessed fuel in which fission product removal may not be complete can be used.

The disadvantages are the increased damage to structural materials by the high fast flux and the control problems resulting from the short neutron lifetime.

Currently (2005) efforts are underway to develop next-generation (so-called generation IV) reactor designs for future deployment around 2020. Currently second- and third-generation nuclear reactor power plants are used effectively around the world. Generation-III+ reactors have been designed, and some have received design certification in the United States from the NRC, and are ready for worldwide deployment. Many countries are in the process of developing and expanding their nuclear power programs, aimed at generating reasonably priced electric power, clean water, and other useful by-products, all the while reducing discharge of greenhouse gases.

Fourth-generation reactors are under design to meet exacting requirements related to low cost capital construction, short construction times, easier fuel cycle management, minimization of radioactive waste, enhanced safety, and better control against proliferation of radioactive waste. These concepts hold great promise for safe and reliable fission nuclear power plant deployment in the twenty-first century.

Fuel and Waste Cycle

The steps and processes in a typical fuel cycle for light-water reactors are described below. Figure 9.8.3 illustrates the relationship of these steps and processes.

Natural Uranium   Uranium occurs naturally in ores in very low concentrations, less than ½ percent by weight. In certain very unusual cases, the uranium content in the ore, called ore grade, may be as high as 10 percent, but these occurrences are extremely rare.

Uranium ores are mined by conventional techniques and then crushed and ground in a mill, which has special equipment for the recovery and purification of the ore into uranium concentrates. The product is in the form of a powder called yellow cake, which has the chemical form of either ammonium or sodium diuranate, both of which are yellow, hence the name.

Conversion of U_3O_8 to UF_6   The process of enriching the U^{235} isotope (from 0.711 percent to the 2 or 3 percent required by water reactors) requires that the uranium be in gaseous form. The gas UF_6 is used; so the natural U_3O_8 must be converted to natural UF_6.

Enrichment   alters the relative amounts of U^{235} and U^{238} in uranium, raising the U^{235} concentration from 0.711 percent to the percentage required for use in a reactor. By so doing, the input (feed) stream of natural UF_6 is broken into two output (product plus tails) streams, one enriched and one depleted.

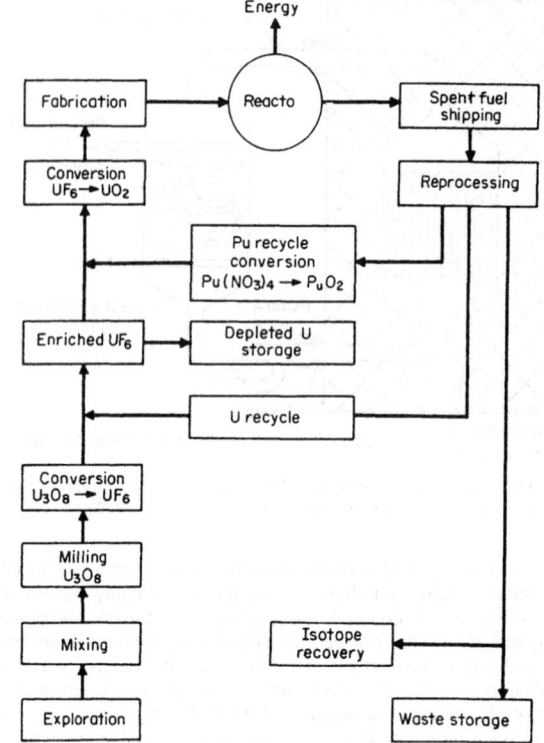

Fig. 9.8.3   Nuclear fuel cycle.

The process of enrichment makes use of the mass difference between U^{235} and U^{238} isotopes. In a gaseous form, the average energies of all UF_6 molecules are equal whether the U atom is U^{235} or U^{238}. Since the energies are equal, the average velocity of the lighter $U^{235}F_6$ molecule is greater than that of the $U^{238}F_6$. Because of this greater velocity and, hence, a greater momentum, the $U^{235}F_6$ molecules can diffuse through a barrier more easily. Gas diffusion plants work on this principle.

In the **gas centrifuge** process, gaseous UF_6 is fed into centrifuges which rotate at very high speed. The heavier U^{238}-bearing molecules drift to the outside wall, while those containing the U^{235} atoms accumulate close to the axis of the machine. The separation process is enhanced by longitudinal gas circulation along the wall and near the axis, in currents counter to each other. This also allows end scoops to collect the enriched and depleted streams. Centrifuges are connected in cascade to obtain specific enrichment assay levels and the cascades are operated in parallel to obtain adequate amounts of enriched product. A plant will contain several thousand stages. Electric power demands for the centrifuge enrichment process are typically 5 to 10 percent those of the diffusion process for similar enrichment and amount. The term **separative work unit (SWU)** is used to express the quantity. Centrifuge plants are in use in European countries, but the United States continues to use the more energy-intensive diffusion process, after abandoning a major centrifuge plant half-built.

There are a number of other possible separation processes for uranium (or for that matter, other isotopes of any material). These include various thermal separation processes and plasma processes, activated either by electronic or laser energy, which work on atomic or molecular ions with separation by magnetic field. The lighter isotope will have a slightly smaller radius of curvature and can be caught in a pocket separate from the heavier. Several of these processes have been proven at the experimental level, and some have been used to produce significant amounts of material in the past.

Fabrication The gaseous enriched UF_6 that comes out of the enrichment plant is delivered in standard cylindrical gas bottles. The UF_6 is then converted to UO_2, the fuel form is which it is used. The UO_2 is in the form of a powder which is then milled to provide suitable sintering properties. The powder is formed into pellets by cold pressing, sintered, and ground to final dimensions. The pellets are loaded into a zirconium alloy tube (cladding) which is backfilled with helium and sealed at each end with a welded-end plug. The fuel rods are assembled into complete fuel assemblies. The number of fuel rods in an assembly varies from 50 to 200 or more, depending on the reactor type and specific design.

Spent Fuel Storage and Transportation Several options are available for managing spent fuel after its discharge from the reactor. It may be stored at the reactor site for a minimum time of 3 months or longer and then shipped in a shielded cask to a reprocessing plant for recovery of residual fuel valves. This procedure is used in most power reactors, except in the United States. Alternatively, it may be stored at reactor water pools, or in an away-from-reactor storage facility for an extended period of time, and then shipped to an interim dry storage site, possibly after fuel mechanical compaction, or sent to a geological repository for long-term disposal. This latter procedure is mandated by current U.S. law.

Reprocessing and Reconversion Spent nuclear fuel has substantial residual value in its uranium and plutonium, and possibly in other fission products or transuranic elements. The plutonium and other transuranics are the longer-lived radioactive components. The spent fuel is mechanically chopped and dissolved in acid, and the solution processed through subsequent steps to purify the plutonium, uranium, and any other desired product. The remaining fission wastes remain in the high-level waste streams. The plutonium product is usually in the form of a nitrate, and the uranium, either directly or in a subsequent process as UF_6.

Uranium and Plutonium Recycle As described above, the components of value in the spent fuel are the remaining uranium, now of lower enrichment, and the plutonium generated through neutron capture in U^{238}. The uranium can either be reenriched or blended with new enriched uranium and used to make new fuel elements. The plutonium can also be used either alone or mixed with uranium in making new fuel elements. A breeder reactor, by definition, will reuse the plutonium as fuel. Water reactors are nearly all designed to reuse fuel, but this is currently not general practice in the United States. By law the United States has mandated the use of a once-through cycle, without reprocessing the spent fuel. The economics of the two approaches depend on the relative cost of newly mined uranium.

Waste Disposal Disposal of **low-level radioactive wastes,** from power plants or other radioactive material uses, such as medical is usually done by dry burial in a dedicated site.

High-level waste in the form of solid spent fuel containing transuranics as well as high-level fission products, or reprocessing plant high-level waste converted to a solid form as a glass or ceramic, can be disposed of in a mined geologic repository to ensure long-term isolation of the waste from the environment. Current U.S. policy calls for acceptance of commercial high-level waste by the Department of Energy for disposal in mined repositories. At this time (2005) it seems unlikely that the U.S. government will be able to use those repositories for long-time storage of spent nuclear fuel. The electric power producing utilities pay a fee on the basis of kilowatthours generated. Isolation will be by the use of both engineered (waste package and container) and geologic (geologic media, such as salt, basalt, or granite) barriers. Other countries, after separating the uranium and plutonium, generally plan to store the waste for extended periods of time before final disposal. Key technical areas in implementation include the development of highly stable materials for encapsulating the waste (e.g., glass, ceramic, corrosion-resistant metals), methodologies for predicting repository performance over extended time periods, and repository closure techniques. Demonstration of the ability to safely dispose of waste has become critical to public acceptance of nuclear power in many countries.

PROPERTIES OF MATERIALS
(See also Table 9.8.1.)

Materials used in nuclear reactors can be divided into six basic categories: fuel, radiation, control, coolant, structural, and shielding. While the nature of the materials may be different, they must be physically and chemically stable and compatible with their environment for a sufficient period to enable the reactor to operate predictably, stably, and economically under the particular set of parameters imposed by the design. (See also Sec. 6.)

In addition to the environmental conditions to which materials may ordinarily be exposed, reactor materials must also sustain the extraordinary bombardment of X-rays (photons) and other atomic particles which are products of nuclear fission. In their passage through matter, these photons and atomic particles may inflict severe damage, since the energy of these particles is spent mainly in ejecting or pulling electrons from their orbits, i.e., producing **ionization.** Gamma photons and neutrons are highly penetrating, whereas fission fragments have a path under 0.001 in the solids and liquids (but cause intense ionization over that path).

The **behavior of materials** following breakage of a bond by ionization may differ markedly. The fragments of simple, highly stable molecules such as water (H_2O) or carbon dioxide (CO_2) will most probably recombine. However, it is highly probable that the parts of complex **organic compounds** such as lubricants and electrical insulation will suffer permanent fragmentation and/or recombination in a different manner to produce new molecules.

Metals, on the other hand, are essentially unaffected by ionization since the probability of displacement of excited atoms in a metal lattice is small and the atoms will merely revert to their original status on ridding themselves of their excess energy. The atoms of **nonconducting crystals** and glasses are similarly immobile, and most of their properties are insensitive to the ionization. However, since these crystals are nonconducting, some of the electrons displaced by ionization become trapped in lattice vacancies, producing changes of optical properties (such as darkening of glass by gamma irradiation) and change of electrical resistivity.

On direct collision, **fast neutrons** may displace atoms from a chemical compound without affording opportunity for immediate recombination. Similarly, **energetic neutrons** may displace atoms from their normal lattice positions in a metal or other crystalline material and push them into abnormal interstitial positions in the lattice, generally producing an effect on metals similar to work hardening. The hardness and tensile strength of the metal will increase, but the metal becomes more brittle. **Thermal neutrons** affect the properties of metals only in those cases where the probability of absorption is high enough to cause significant transmutation to other elements.

Most severe damage occurs when the atoms of a fuel material fission and produce two or more new atoms from each of the original fissioned atoms. Volumetric changes occur by the substitution of these additional atoms in the lattice for the original atom, and the lattice is further damaged by the intrusion of the high-velocity fission fragments.

Fuel Materials

Fuel materials are generally divided into two categories: fissionable and fertile. The **fissionable materials** are those isotopes of uranium and plutonium which fission upon interaction with thermal neutrons. These are the isotopes U^{233} U^{235}, Pu^{239}, and Pu^{241}, which have odd atomic weights. The **fertile materials** are those isotopes of uranium and plutonium which have even atomic weights. These absorb neutrons under proper conditions and undergo a series of nuclear reactions, with the eventual formation of fissionable isotopes. Thus U^{238} forms Pu^{239}, and Pu^{240} forms Pu^{241} In addition, thorium 232 forms the fissionable isotope U^{233}.

Both fissionable and fertile materials fall short of possessing the **properties considered ideal** for a nuclear fuel, namely, corrosion resistance, good thermal conductivity, strength, and ductility. To overcome these problems as well as to prevent fission products from entering the

Table 9.8.1 General Properties of Reactor Materials before Irradiation*†

Materials	Density, g/cm³	Melting point, °C	Specific heat, cal/(mol·°C)	Thermal conductivity,† cal/(s·cm·°C)	Coefficient of thermal expansion, 10⁻⁶ per °C	Ultimate tensile strength,† 10³ lb/in²	Modulus of elasticity,† 10⁴ lb/in²	Capture cross sections, 0.025 eV (σ_c, barns)
Fuels and fertile materials								
Uranium, normal	19.3	1,133 ± 1	6.649	100°C, 0.063	25–125°C, 45.8	56	24	7.68
Thorium, normal	11.71	1,690 ± 10	100°C, 6.59	100°C, 0.090	30–100°C, 11.5	30.6	10.6	7.56
Plutonium[239]	19.60	623 ± 7		100°C, 0.061		125	13–16	1.026
Uranium oxide (UO₂)	10.96	2,500–2,600	15.38	100°C, 0.022	22–363°C, 9.3–10.8		21	24.3
Al-16 weight % U alloy		~650		200°C, 0.42	20–100°C, 20	17.7	10.6	
Zr-4 weight % U alloy	6.72	1,840 ± 25		100°C, 0.033	190–300°C, 6.66	63	14.4	
Diluents and cladding materials								
Zirconium	6.50	1,845 ± 25	6.31	50°C, 0.05	5.82	30–38	13.8	0.18
Zircaloy 2	6.55	1,820 ± 25		<0.04	20–100°C, 5.2 ± 0.6	65	14	
Aluminum (28)	2.694	600.2	100°C, 6.07	0.53	20–100°C, 23.8	13	10.3	0.26
Magnesium	1.74	650	25°C, 6.08	18°C, 0.376	49°C, 26	Annealed, 27; 27°C, 22	Annealed, 6.5	0.069
MoSi₂	6.24	1,870		370°C, 0.088	0–1,500°C, 5.1			
Solid moderators								
Graphite	2.27	Sublimes at 3,656 ± 25°C	2.066	100°C, 037–0.48	20–250°C, 1.9–4.0	0.500–2.4	0.5–1.2	0.0032
Be₂C	6.24	1,870						
Beryllium	1.847	1,315	100°C, 13.69	30°C, 0.02	25–200°C, 7.7		400°C, 23	
Sintered BeO†	2.2–2.8	2,550 ± 25	100°C, 7.7	100°C, 0.34	25–100°C, 11.54	29	44	0.01
SiC	3.21	Decomposes at 2,500°C	327°C, 10.06	200°C, 0.19; 400°C, 0.060	25–100°C, 5.5; 0–1,700°C, 4.4	400°C, 15; 1,700°C, 50	400°C, 39	0.00072
Structural materials								
Stainless steel (347)	8.027	1,427	100°C, 0.12‡	100°C, 0.037	20–100°C, 16.5	85	29	3
Inconel X	8.51	1,395–1,425	25–100°C, 0.109‡	0–100°C, 0.036	20–100°C, 11.5	Annealed, 100–120		4.1
Molybdenum	10.2	2,622 ± 10	100°C, 6.24	0°C, 0.32	5.1	67	48	2.7
Titanium (commercial purity)	4.507	1,690	0–500°C, 6.61	25°C, 0.41	25°C, 8.5	85–100	15.5	5.8
Niobium	8.57	2,415	0°C, 6.01			1,200°C, 14.7		1.1
Control materials								
Cadmium	8.694	321	25–32°C, 6.19	0.22	20–100°C, 31.8	10.3	7.1–10	2,450
Hafnium	13.36	2,130 ± 15	25–2,227°C, 6.16		0–100°C, 5.9	67.5	14	105

* All properties change with irradiation at room temperature except when specified.
† Thermal conductivity, strength, and the modulus of elasticity vary directly with the density; however, the coefficient of thermal expansion appears unaffected.
‡ Cal/(g·°C).

Sources: H. A. Seller, Properties of Reactor Materials, "Nuclear Engineering Handbook," McGraw-Hill 1958; "Reactor Physics Constants," ANL 5800, 1963.

coolant, various techniques are used, including plating, cladding, and alloying.

Materials for this use must be corrosion-resistant, be compatible with the fuel materials, be physically stable under irradiation, have good heat transfer characteristics, have good nuclear properties, and be reasonably fabricable. From a consideration of cross section alone, only aluminum, beryllium, magnesium, and zirconium are attractive for use in thermal reactors. Magnesium usually is ruled out on the basis of strength and corrosion resistance but has been widely used in some gas-cooled systems. In fast reactors, a number of other materials become attractive, the most important being stainless steel, which also is used under special conditions in thermal reactors. Consideration also is being given to ceramic materials, BeO, Be_2C, SiC, and $MoSi_2$, which are very attractive in all phases except resistance to thermal shock.

Moderating material must be capable of reducing neutron energy very rapidly and should have good strength at high temperatures, corrosion resistance, thermal and radiation stability, and reasonable cost. Such properties are not available in a single material. Graphite is the most widely used solid moderator. Attention also is being given to beryllium and its compounds, deuterium, oxygen, and hydrogen.

The nuclear and physical properties considered desirable for **fuel diluents and cladding** are, in general, desirable for structural materials. For the latter, however, more emphasis is placed on strength and corrosion resistance than on nuclear properties. Stainless steel, aluminum, zirconium, molybdenum, titanium, niobium, and their alloys are used. As operating temperatures continue to rise, ceramics and cermets will have to be developed for this purpose.

Materials used for **control purposes** must have a high cross section for absorption of neutrons, adequate strength, low mass to permit movement, good corrosion resistance, chemical and dimensional stability under heat and irradiation, and low cost. Boron, cadmium, and hafnium

are given the most consideration in various metallic and ceramic forms and as compounds dissolved in the coolant.

Coolant Properties

The choice of a coolant also is a compromise, since no known material possesses all the **desirable characteristics.** These are good heat transfer coefficient and good heat capacity on a volume basis, low absorption cross section, low vapor pressure at the operating temperature, minimum of chemical reaction with the material contacted by the coolant, good resistance to forming γ emitters with long half-lives, stability under irradiation and high temperatures, and workable melting points.

Coolants are divided into three broad **classes:** aqueous and organic liquids, liquid metals, and gases. **Water** has found the widest application, largely because it is readily available and economical, has fairly good heat transfer properties, and offers no corrosion problems which cannot be handled. Its most serious drawback is its high vapor pressure. **Heavy water** has superior qualifications except for cost, which is high. Diphenyl and other **organic liquids** have demonstrated an ability to stand up under irradiation but are handicapped by a tendency to deposit decomposition products on the fuel element plates. See Table 9.8.2.

Liquid metals are attractive because they provide high heat-transfer rates and do not require pressurizing to operate at high temperatures. Sodium is most attractive because of its high boiling point and high heat conductivity, but it reacts violently with water and has poor lubricating qualities. Lithium 7 better meets qualifications but is harder to obtain and is costly.

Gas coolants offset their advantage of being inert at elevated temperatures with poor heat transfer qualities. They must be operated at high pressures to be useful in power reactors. Sodium hydroxide and yellow

Table 9.8.2   Properties of Nuclear Coolants*†

			Aqueous and organic liquids		
	Water	Heavy water	Diphenyl	Dowtherm-A	Santowax R
Density, lb/ft³	42.4 (600°F)	46.2 (600°F)	46.4 (675°F)	46.6 (675°F)	52.29 (675°F)
Viscosity, lb/(h · ft)	0.20 (600°F)	0.220 (600°F)	0.353 (675°F)	0.7986 (675°F)	0.630 (675°F)
Melting point, °F	32	39	156	53.2	293
Boiling point, °F	212	214	49	496	687–784
Heat capacity, Btu/(lb · °F)	1.54(600°F)	1.68 (600°F)	0.637 (675°F)	0.675 (675°F)	0.539 (675°F)
Thermal conductivity,	0.294 (600°F)	0.294 (600°F)	0.0763 (675°F)	0.098 (675°F)	0.064 (675°F)
Capture cross section σ_a, 0.025 eV (barns)	0.66	0.0011	3.3	3.3	4.6

				Gases			
	Hydrogen	Helium	CO_2	Air	Nitrogen	Steam	Neon
Density, lb/ft³	0.00185	0.00375	0.041	0.0275	0.026	0.0167	0.0190
Viscosity, lb/(h · ft)	0.042	0.094	0.081	0.089	0.086	0.069	0.1445
Heat capacity, Btu/(lb · °F)	3.625	1.245	0.283	0.263	0.268	0.518	0.246
Thermal conductivity, Btu/(h · ft · °F)	0.224	0.159	0.0328	0.0337	0.0327	0.0341	0.0536
Prandtl no.	0.660	0.735	0.6988	0.6898	0.7097	1.048	0.6632

				Liquid metals				
	Gallium	Lithium	Potassium	Rubidium	Sodium	Tin	Na (56%) K(44%)	Pb (44.5%) Bi (55.5%)
Density, lb/ft³	359	29.9	44.6	84.4	51.2	421	48.8	625.5
Viscosity, lb/(h · ft)	1.836	1.188	0.3816	0.415	0.5040	2.736	0.435	2.88
Melting point, °F	85.86	354	147	102	208	449	66.2	257
Boiling point, °F	3,601	2,403	1,400	1,270	1,621	4,118	1,518	3,038
Heat capacity, Btu/(lb · ft · °F)	0.082	1.0	0.180	0.0877	0.3005	0.0639	0.2484	0.035
Thermal conductivity, Btu/(h · ft · °F)	13.0	18.0	21.15	13.2	37.7	19.0	16.35	8.05
Capture cross section σ_a, 0.025 eV (barns)	2.77	71.0†	1.97	0.70	0.505	0.60		0.094

* At 1,000°F unless otherwise specified. Adapted from L. Green, Reactor Coolant Properties, *Nucleonics,* 19, no. 11, which contains a detailed bibliography.
† Thermodynamic properties of a variety of other specific materials are listed also in Secs. 4.1, 4.2, and 6.1.

phosphorus are among other materials considered as coolants, but their acceptability has not been demonstrated.

FISSION REACTOR DESIGN

Neutron Balance   The main objective of reactor design is to achieve neutron balance during reactor operation. At steady state, the number of neutrons produced during fission is equal to the neutrons lost by absorption and leakage processes. The net rate of change of neutron density n (neutrons per cubic centimetre) is given by the neutron conservation equation:

$$\frac{\partial n}{\partial t} = \text{production} - \text{absorption} - \text{leakage}$$

If $\partial n/\partial t$ is positive, the reactor is said to be supercritical; if it is zero, the reactor is at steady state and is said to be critical; if it is negative, the reactor is subcritical.

Cross Section   The microscopic cross section σ for a particular nuclear process is the effective target area of the nucleus with which a neutron must interact to produce the given reaction. The unit of σ barn (1 barn = 10^{-24} cm^2). Neutron cross sections (absorption, scattering, and fission) constitute the basic nuclear data for reactor design. Absorption cross sections for thermal neutrons range from 0.00046 barn for deuterium to 3.3×10^6 barns for xenon^{135}. Moderating atoms for the slowing down of fast neutrons have low atomic mass, small absorption, and large scattering cross sections. Examples are hydrogen, deuterium beryllium, and carbon. Table 9.8.3 gives thermal neutron cross sections in barns for selected elements (based upon the naturally occurring combinations of isotopes).

Multiplication Factor   The number of neutrons produced in any one generation in a reactor for each neutron produced in a previous generation is called the multiplication factor k. If $k = 1$, the reactor is critical; if $k < 1$, it is subcritical; and if $k > 1$, it is supercritical. For an infinite thermal reactor, multiplication factor $k_\infty = \eta f p \varepsilon \cdot \eta$ is the number of fission neutrons produced per neutron absorption in fuel, f is the thermal utilization factor (neutron absorption rate in fuel/total absorption rate), ε is the fast-fission factor, and p is the resonance capture escape probability. The leakage of neutrons from the reactor core is accounted for by multiplying k_∞ by the factor $1/(1 + M^2 B^2)$, where M^2 is the migration area (sum of the squares of the diffusion length L and the slowing-down length τ) and B^2 is the geometric buckling ($B^2 = 0$ for an infinite reactor).

Breeding Ratio   If the ratio of a number of fissile nuclides produced by the capture of neutrons by the fertile nuclides to the number of fissile nuclides consumed is greater than unity, it is called the **breeding ratio;** if it is less than unity, it is called the conversion ratio. For breeding, the value of η (defined above) must be greater than 2, as one neutron is required to maintain the chain reaction by fission, another neutron maintains breeding by its capture by the fertile nuclide, and the remaining to compensate for the loss of neutrons due to absorption and leakage. Neutron economy favors the breeding of Pu^{239} fissile nuclides from fast neutron capture by U^{238} fertile nuclides. This breeding process is the basis of LMFBR (liquid-metal fast breeder reactor). On the other

hand, thermal breeders are based upon the breeding of uranium 233 fissile nuclides from thorium 232 fertile nuclides by the capture of a thermal neutron.

Diffusion Theory   Multienergy group diffusion theory is presently the basis of reactor design. For a critical reactor the neutron balance equation reduces to

$$D_g \nabla^2 \phi_g - \sum R_g \phi_g + S_g = 0$$

where ϕ_g and S_g are the gth energy group flux and the slowing-down neutron source; D_g and R_g are the diffusion coefficient and removal cross section. The latter are called group constants, which are obtained by the analysis of the $\sum R_g \phi_g$ basic fuel lattice. The solution of the above equations provides the reactor multiplication factor, reactor critical size, neutron fluxes, power distributions, isotopics of elements, and fuel-cycle length.

Control   Extra fuel is required for depletion to compensate fuel lost in producing energy, to account for the loss of reactivity due to the buildup of fission products, and to account for the change in reactivity due to changes in fuel and moderator temperatures. The control of excess reactivity is undertaken by the control rods of strongly absorbing materials having large absorption cross sections (such as boron, hafnium, or silver-indium-cadmium) and the burnable poison rods containing boron or gadolinium. In PWR, soluble boron is also employed. Control rods are inserted to shut down the reactor and are withdrawn to make the reactor critical.

A fraction of prompt neutrons (0.0065) are delayed neutrons. The presence of the delayed neutrons makes control of the reactor easier.

Thermal

Sources of Heat   The energy produced by fissioning appears in several forms, each of which eventually degrades to heat. To provide for removal of this heat, the core, or active portion of the reactor, usually consists of uranium bearing fuel elements with passages around these elements for flow of coolant. The rate of coolant flow and the temperature of the coolant must be such as to permit removal of all the heat generated in the elements without exceeding allowable materials properties.

Hot Spot and Hot Channel Factors   The problem of heat removal is complicated by the fact that heat is not generated uniformly throughout the core. The geometric configuration of most power reactors is approximately represented by a finite cylinder of height H and radius R. Diffusion theory predicts the following thermal neutron flux pattern for an unreflected, unrodded, and homogenized cylindrical reactor as a function of axial Z and radial r variables:

$$\phi(z, r) = \phi_0 \left(\cos \frac{\pi Z}{H} \right) J_0 \frac{2.405r}{R}$$

where ϕ_0 is peak flux and J_0 is the Bessel function.

In addition, the power generation is influenced by mechanical factors such as variables in fuel element dimensions, fuel concentrations, and flow. These nuclear and mechanical factors are generally called hot spot

Table 9.8.3   Absorption (σ_α) and Scattering (σ_s) Cross Section for Thermal Neutrons, in barns

Element	σ_α	σ_s	Element	σ_α	σ_s	Element	σ_α	σ_s
Aluminum	0.23	1.4	Helium	0.007	0.8	Potassium	1.97	1.5
Beryllium	0.01	7	Hydrogen	0.33	38	Plutonium^{239}	1.026	9.6
Bismuth	0.032	9	Iron	2.53	11	Sodium	0.505	4.0
Boron	755	4	Lead	0.17	11	Thorium	7.56	12.6
Cadmium	2,450	7	Magnesium	0.063	3.6	Tin	0.6	4
Carbon	0.0032	4.8	Molybdenum	2.7	7	Titanium	5.6	4
Chromium	2.9	3.0	Nickel	4.6	17.5	Uranium	7.68	8.3
Copper	3.7	7.2	Niobium	1.1	5	Zirconium	0.18	8
Hafnium	105	8	Oxygen	<0.0002	4.2			

NOTE: Individual isotopes have markedly different cross section. For example, U^{235} has total σ_α = 682 barns and fission cross section σ_f = 580 barns.

factors. Thus the peak power density of heat flux in the core is greater than the average by an amount indicated by the hot spot factor. In addition, hot channel factors are identified which indicate how much greater the temperature rise of coolant is in the hot channel than in the average channel. As an example of how such factors are used, consider a reactor in which the axial heat generation pattern is a cosine and the total flow passes uniformly once through all channels of the core; the maximum fuel element surface temperature T_{sm} depends on the hot channel and hot spot factors $F_{\Delta T}$ and F_θ and on the three temperatures—the coolant temperature at the core inlet T_1, the coolant temperature rise in the core ΔT, and the average film temperature drop in the core θ_a. (F_θ includes the axial peak-to-average of $\pi/2$.)

$$T_{sm} = T_1 + \tfrac{1}{2}F_{\Delta T}\Delta T + \tfrac{1}{2}\sqrt{F_{\Delta T}^2\Delta T^2 + 4F_\theta^2\theta_a^2}$$

Design Criteria   The maximum fuel element surface temperature T_{sm} is of interest because it may be limited by corrosion effects or by a desire to avoid film boiling, a condition which lowers the heat transfer from the fuel to the coolant. However, other limiting conditions also exist. In addition the internal fuel element temperature during all anticipated reactor transients is generally limited to a maximum value (generally below the melting point of the fuel) or to a maximum average value.

Coolant temperature rise generally is limited indirectly by fuel temperature limits. Occasionally, however, thermal stresses within the fuel element may impose a direct limit. In pressurized water reactors, the coolant temperature is limited not to exceed the saturation temperature of the fluid; thus it is more significant to speak of the maximum enthalpy rise, which is related directly to the steam quality of the fluid leaving the hot channel.

Density changes within the coolant may be limited to provide control stability in reactors that use boiling heat transfer.

Coolant velocity usually is limited to a maximum value by erosion or by vibration resonances, and sometimes is required to exceed a minimum value because of crud deposition. The burnout condition is determined in a given fuel element geometry for a particular combination of fuel element heat flux, primary coolant flow, and primary coolant enthalpy. Steady-state, local, or bulk boiling may have to be limited, even when burnout is not a danger, because of pitting of fuel element surfaces or crud deposition on the fuel surfaces. The burnout problem is most acute during reactor transients; hence steady state burnout generally is not limiting.

Mechanical

The mechanical design of nuclear reactors is influenced by three factors not encountered with other apparatus: the need to control criticality, the effects of irradiation, and the high power density produced in the core.

Control of Criticality   The following requirements are introduced by the need to control criticality:

1. Not only must components perform reliably, but also if they fail to function as designed, the reactor will shut down. Small parts located elsewhere in the core should be so designed that if they fail they do not interfere with the action of control rods in shutting down the reactor.

2. Emergency shutdown mechanisms must be fast-acting devices to reduce the time delay in initiating shutdown.

3. During normal operation, components of the core must be rigid enough so that deflections due to mechanical or thermal stresses do not introduce unpredicted criticality variations.

Effects of Irradiation   Neutron and gamma irradiations cause the following effects, which must be considered in the design of reactor systems and reactor components:

1. Radiation effects on properties of materials—particularly on the brittle fracture characteristics of the reactor vessel, on the relaxation characteristics of bolts, and on crevice corrosion.

2. Heat generation brought about by attenuation of radiation within components materials—probably insofar as such heat influences heat removal requirements, thermal stresses, and thermal distortions of parts.

3. Induced radioactivity in corrosion and wear products that could be carried by the coolant and deposited in such regions as pumps and valves.

High Power Density   A nuclear reactor requires a large amount of heat transfer surface for removal of the high power per unit volume which can be produced. Thus the reactor consists of many small, closely packed fuel elements. Mechanical tolerances are often appreciable fractions of the dimensions and spacing of these elements and therefore influence thermal performance. The influence of such tolerances on performance must be evaluated. Parts also must be rigid and vibration-free in the face of the high fluid flow rates required for power removal. The high power density may also lead to large thermal stresses in fuel elements.

Furthermore, moving parts may have to operate without benefit of organic lubricants because such lubricants tend to sludge under irradiation, and if deposited on heat surfaces could interfere with removal of heat from the core.

Residual radioactivity also influences programs and facilities for refueling of the core and for periodic maintenance of components in the reactor system. In addition, personnel exposure by the above radiation requires adequate shielding.

Shielding   To protect reactor operating personnel against damaging biological effects of neutrons and gamma rays, shielding is required around a nuclear reactor. Neutron and gamma ray fluxes in the range of 10^{13} to 10^{14} must be attenuated to 10^3 particles/(cm^2 · s) to meet the tolerance radiation level.

To attenuate gamma rays, which interact primarily with the orbital electrons of atoms, a material with a high atomic number containing a high density of these electrons is required. Examples are lead, tungsten, depleted uranium, or concrete containing high-Z elements in the form of scrap or heavy ore.

To attenuate neutrons, they must be slowed down and then absorbed. Hydrogenous materials such as water, concrete, or polyethylene are excellent moderators. The slowed neutrons must be absorbed without producing high-energy-capture gamma rays by using boron 10. Therefore, some gamma shielding outside the neutron shield is generally required.

Electrical

Neutron level instruments, located adjacent to but outside the core or core vessel, are used for the basic control of nuclear reactors. Thermocouples, flowmeters, pressure taps, and fission chambers are used inside the core for power calibration of nuclear instruments and for determining the distribution of power and/or temperature within the core; in turn, the power distribution is used to compute heat fluxes and fuel temperatures for comparison with limiting values.

Neutron level instruments utilize the fact that neutrons can produce ionizing particles to indicate their presence. The two most common processes used for this purpose are the (n,α) reaction with boron and the fission reaction. The resulting ion pair is used to produce ionization of a gas in a tube containing a central anode and outer electrode. The electric discharge produced by the ionization is used for measurement purposes. The detailed design features vary widely with the range of the instrument and the type of reactor in which it is to be used.

Different neutron level instruments are employed during the reactor operation at various power levels, because of the wide range of sensitivity required to monitor neutron fluxes from start-up to full-power condition. In power reactors at least three instrument ranges are used with the overlap between the adjacent ranges. These are (1) source range, (2) intermediate range, and (3) power range. During the reactor start-up (source range) the BF$_3$ proportional counters are used to measure neutron flux in the range of 10^{-1} to 5×10^4 neutrons/(cm^2 · s). The compensated ionization chambers are employed to monitor neutron flux in the intermediate power range of 2.5×10^2 to 2.5×10^{10} neutrons/(cm^2 · s). In the power range, where neutron flux ranges from 2.5×10^7 to 2.5×10^{10} neutrons/(cm^2 · s), uncompensated ionization chambers are employed.

Compensated ionization chambers are designed to count neutrons against a background of gamma radiation. Uncompensated chambers are useful for providing axial power distribution.

All the above instruments are located outside the core and monitor neutron flux, hence reactor power. The latter is also obtained from the measure of ΔT (change in temperature) across the reactor core from the temperature measuring instraments located in the primary loop of the reactor. However, these instruments do not provide local flux peaking and local anomalous coolant temperature situations. In-core instrumentation must be used for this purpose.

In-core instrumentation is employed to provide information on neutron flux and fuel-assembly-coolant outlet temperature at a few selected locations in the reactor core. Thermocouples measure the coolant outlet temperature and thus provide the rise in local coolant temperature, a measure of the increase in local reactor power. Miniature, fixed, and movable fission chambers containing U_3O_8 which is 90 to 95 percent enriched U^{235} constitute neutron flux detectors. The experimental data obtained during the reactor operation from the in-core instrumentation provide the check against the reactor design parameters such as the assembly distributions and hot channel factors.

In addition to thermocouples and fission chambers, other in-core instruments such as pressure, displacement, and strain transducers are also desirable. None of these instruments except thermocouples have operated satisfactorily over a long period. The installation of in-core instrumentation is vital to the safe and economic operation of the reactor.

NUCLEAR POWER PLANT ECONOMICS

NOTE: The example below is illustrative only. The reality and currency of all factors which enter into consideration of the overall economics of nuclear power plants must be reckoned with for any real case under study and/or analysis.

Decisions by electric utilities as to what type of generating plants to order to meet future base load requirements are based largely, though by no means wholly, on a comparison of the costs for which the alternative types can produce electricity over their lifetimes. Other factors, which have become relatively more important in recent years, include adequacy and reliability of fuel supply, including dependence on foreign sources; environmental and safety considerations; availability of suitable sites; public acceptance as influenced particularly by environmental opposition; the costs, uncertainties, and delays involved in government regulation; lead times required from initial decision to get plants on line; and plant reliability and availability.

As a practical matter in the United States, utilities seeking to add to their base load capacity have the choice only of plants utilizing nuclear fuel or one of the fossil fuels; coal, gas, or oil. Most usable hydrosites have already been developed.

The discussion of nuclear generating costs here is directed to light water reactors of either the pressurized water reactor (PWR) or boiling water reactor (BWR) types. Although other types are in use (e.g., high-temperature gas-cooled) or under development (e.g., fast breeder), the overwhelming preponderance of reactors in commercial use are, and will for some time continue to be, light-water types.

It is customary to compare generating costs for nuclear and fossil fueled plants in three broad categories: (1) fixed costs on capital investment, (2) fuel costs, and (3) operating, maintenance, and insurance costs.

Fixed Costs on Capital Investment

In standard accounting practice, the capital investment in a nuclear power plant is considered to be the total cost of building the plant and placing it in commercial operation. Accounting classification systems employed by both NRC and FPC provide separately for "direct" and "indirect" costs. Direct costs are associated on an item-by-item basis with land and land rights, the equipment and structures which comprise the complete power plant, and coolant and moderator materials. Equipment and structures are customarily subdivided into reactor plant and equipment, turbine plant and equipment, other electrical equipment, equipment and facilities of a general nature, and transmission plant. Indirect costs consist mainly of expenses which apply to all portions of the physical plant, including engineering and design costs, construction facilities, taxes, interest during construction, and the owner's administrative and overhead costs.

The total of these items tends to be higher for nuclear than for fossil fueled plants largely because of (1) the need for containment shells, shielding, instrumentation, and other measures to contain radioactivity and ensure safety; (2) the greater complexity, hence greater costs, of reactor equipment compared to conventional boiler equipment; and (3) the lower turbine steam temperatures and pressures, and the lower thermal efficiencies of nuclear plants compared to conventional plants, requiring the use of larger, hence costlier, turbines. The discrepancy is wider in warm climates, which permit fossil fuel plants to utilize outdoor (unenclosed) construction.

Mitigating factors which prevent the construction cost discrepancy between nuclear and coal plants, in particular, from being still larger are (1) the lower cost of fuel handling equipment in nuclear plants and (2) the need in coal plants but not in nuclear plants, for smoke control and ash removal equipment.

Unit construction costs (expressed in dollars per kilowatt) for both nuclear and fossil fueled plants are lower at any given time as plants increase in size because of various economies of scale. The decreases are sharper in the case of nuclear plants, largely because of the prominence of certain minimum and relatively fixed costs to ensure safety. The existence of these inescapable costs places small nuclear plants at a particularly great economic disadvantage.

Adding a nuclear unit to a generating station which already has one or more nuclear units is less costly than constructing a first nuclear unit of otherwise identical characteristics. The principal savings in multipleunit plants result from use of a developed site; reduced engineering and design efforts; and use of joint facilities and equipment such as cooling water intake and discharge facilities, control rooms, warehouses, shops, offices, roads, fuel handling and storage facilities, and temporary construction facilities.

The unit capital costs of nuclear plants often can be reduced substantially over their lifetimes by using improved fuel cores to achieve outputs higher than the original rating, which is usually conservative. To take advantage of this opportunity, allowance for higher output must be made at the outset in the capacity of turbine generators. Conversely, should it be necessary to derate plants to take account of safety uncertainties such as the effectiveness of emergency core-cooling systems, the effect would be to increase unit capital costs.

Total construction costs of nuclear plants have risen rapidly since about 1963, when it was believed that a large light-water plant could be built for $125 per kilowatt or less. Plants now are being estimated to cost in excess of $2,000 per kilowatt. The principal factors making for such increases in cost have been inflation, stretch-outs of plant schedules (adding to interest during construction), lower labor productivity, increases in the scope of projects to add safety and environmental features, and more stringent quality control.

Some of these factors may continue to push unit construction costs upward in the future. Counterbalancing factors tending toward a moderation in unit costs include continued increases in plant size, the likelihood of greater standardization of systems and components, a larger sales volume over which to spread manufacturing overhead costs, and improved field assembly and construction methods. One should take note of the fact that the historical tendency in the industry has been to underestimate construction costs by a wide margin.

The fixed cost component of generating costs (mills per kilowatt-hour) can be obtained from construction costs (dollars per kilowatt) by applying an annual fixed charge rate and an annual plant capacity factor.

Annual fixed charges cover all costs that are in direct proportion to the initial capital investment and are independent of the extent to which the facilities represented by this investment are used. A simple procedure for evaluation of a plant, which approximates the levelized annual charges obtained through utility accounting procedures, is to use constant annual fixed charges, i.e., to apply a constant fixed charge rate each year to the initial investment.

The fixed charge rate for a utility will vary because of a number of factors including amount of state and local taxes, plant life and accounting method for depreciation purposes, and debt-equity composition of the utility's financial structure. A representative range of fixed charge rates for depreciating capital of an investor-owned utility illustrate the components and relative importance of these items, it being understood that the actual components and range will vary, depending on the time of implementation and the then-current financial climate.

Component	Fixed charge range, %
Return on investment	7.0–9.0
Depreciation	1.0–1.3
Interim replacements of short-lived assets	0.3–0.5
Property insurance	0.3–0.4
Federal income taxes	2.0–2.8
State and local taxes	2.9–3.5
	13.5–17.5

A lower fixed charge rate would be applied to the nondepreciating capital (land and land rights and working capital) because depreciation, interim replacements, and property insurance are eliminated and associated adjustments are made in taxes. Not included in annual fixed charges are relatively constant fixed costs, such as nuclear liability insurance or plant staffing expense, which are only indirectly related to plant investment.

The annual capacity factor is calculated by dividing the kilowatthours generated by the plant in a year by the kilowatthours that would have been generated had the plant operated at full capacity throughout all the hours in the year (8.760 except in leap years).

The conversion from capital to generating costs may be illustrated as follows: Assume a plant of 1,000,000-kW capacity costing $2 billion, with annual fixed charges of 14 percent and operating at 80 percent capacity factor. The construction-cost contribution to generating costs would then be 40 mills/kWh, calculated as follows:

$$\frac{\$2,000,000,000 \times 14\%}{1,000,000 \text{ kW} \times 8,760 \text{ h} \times 80\%} = 40 \text{ mills/kWh}$$

Fuel Costs

The economic advantage of nuclear power plants lies essentially in fuel costs which are substantially lower than those of fossil fueled plants.

The fuel cost advantage is implicit in the compactness of nuclear fuel: 685 lb of uranium, if fully consumed, contains the energy equivalent of 1.7 million tons of coal, 7.2 million bbl of oil, or 32 billion ft^3 of natural gas. It has not yet proved possible in any water cooled reactor to consume more than a small fraction of the uranium used in nuclear fuel; yet this fraction is high enough to yield the fuel cost advantage noted above.

Determining nuclear fuel costs is a far more complicated matter than determining fossil fuel costs because of the need to take account of a complex sequence of events which begins long before the uranium fuel is inserted in the reactor and ends long after the spent fuel has been removed from the reactor. The entire sequence, which is called the nuclear fuel cycle, includes mining and milling of uranium, refining of uranium and conversion to uranium hexafluoride, enrichment of the uranium in the isotope uranium 235, conversion of enriched uranium to fuel material, fabrication of reactor fuel elements, use of fuel elements in nuclear power plants, recovery and marketing of by-product plutonium (not in the United States), chemical reprocessing of spent fuel to obtain reusable fuel material (not in the United States), and disposal of radioactive wastes.

Some principal items of nuclear fuel cost and possibilities of change include the following:

Fabrication The cost of fuel element fabrication is taken to include all the steps necessary to change a starting material (UF$_6$ in the case of enriched uranium elements) into a usable element. These steps involve chemical conversion to a powder or metal, metallurgical and mechanical processing to form and clad the elements, inspection, testing, and scrap recovery. Fabrication costs are expected to continue declining as

greater quantities are handled, processing becomes more efficient, and fuel design is improved.

Recycling of Plutonium Uranium-fueled light-water reactors in normal operation produce a certain amount of fissionable plutonium 239 through capture of neutrons in the nonfissionable isotope uranium 238. The separation is done by well-established chemical processes, the end products being uranium dioxide and plutonium dioxide. The plan to recycle spent nuclear fuel resulted in several reprocessing plants being built in the United States, but only one was ever operated. Widespread public concern over proliferation of nuclear weapons and the possible use of plutonium by terrorists and others resulted in a ban on reprocessing in the late 1970s. The ban was lifted in 1980, but since that time there has been no economic incentive to embark on an extended program of industrial reprocessing of nuclear fuel. This situation is entirely different elsewhere; reprocessing is currently a going operation in France, Russia, Japan, Great Britain, and elsewhere. The confluence of many factors will foster increased reprocessing of spent nuclear fuels, among which will be the matter of depletion of fossil fuels, accumulated pollution products from the combustion thereof, growth of power requirements worldwide, and so on. Certainly, at such future time as breeder reactors may become viable, they will provide an ideal place to utilize plutonium derived from reprocessed spent nuclear fuel produced in light-water reactors.

Shipping Charge The costs of transporting irradiated fuel elements from a reactor site to a chemical processing plant include charges for freight, shipping casks, handling, and insurance. The shape of the fuel element (having an effect on the size of cask required), the irradiation level, the cooling time required, the shipping distance, the route and type of transport, and the regulatory requirements are all factors which affect the cost.

Enrichment Enrichment in the fissionable isotope uranium 235 is accomplished at present (in the United States) only in DOE-owned plants.

Burnup The amount of heat that can be obtained from a given weight of nuclear fuel before it is discharged from the reactor is known as the fuel's burnup. Higher burnups are expected as a result of technological improvements. This would have the consequences of reducing the amount of fuel that must be fabricated, reprocessed, and shipped per unit of electrical output. Cost reductions should be achieved as a result, but they may be limited by the fact that fuel elements capable of achieving the higher burnups may be more expensive to fabricate and process and may require higher enrichment.

Chemical Reprocessing and Waste Disposal The cost of chemical reprocessing, less any income from the plutonium or uranium recovered, is usually a net credit for foreign reactors. For U.S. and foreign reactors, the cost of waste disposal is an appreciable cost factor. Under U.S. law, that cost is assessed against the user utility as the power is generated, even though the disposal will take place decades later.

Decommissioning The cost of decommissioning the nuclear power plant when it is eventually closed down is treated differently in different countries. In the United States, these costs are required to be guaranteed by various accounting practices, which in effect accumulate a reserve for costs to be accrued in the future. Reactors decommissioned and dismantled are utilized as case studies to accumulate data on the costs incurred thereby. Other reactors that have been shut down have been defueled, and either stored under guard, or entombed in concrete, with ultimate decommissioning postponed to the future.

Operating and Maintenance Costs

The main cost items included in this category are station labor, training of replacement personnel, expendable materials, and supplies other than fuel, maintenance operations, and insurance.

The novelty and complexity of nuclear plants, a conservative approach to health and safety problems, and the need for specialized knowledge are all factors requiring a more skilled staff in nuclear plants and a longer period of training. Problems of coal and ash handling and maintenance of tubes and related boiler components require a larger working force in coal plants. While insurance costs have been higher for nuclear plants, it is expected that a continuance of safe operating experience will eventually equalize this item.

Numbers of operating personnel per generating unit are less for a plant having two or more nuclear units than for a single-unit plant. Increasing the size of a nuclear unit does not materially affect the number of people required to operate it. For these two reasons, one can expect declining personnel requirements per kilowatt of capacity at such time as the nuclear power industry expands present plant capacity and/or adds new plants to the overall inventory.

Total Generating Costs

Although many uncertainties becloud cost estimates for both nuclear and fossil fuel plants, comparisons tended to favor nuclear power in many parts of the United States on either economic or non-economic bases, or both. The continued and expanded use of fossil fuels in power plants is driven by several considerations, not the least of which is the current apparant plentiful supply at reasonable prices. In looking to the future, however, there are certain problems which are recognized and which must be addressed in the matter of fossil fuels. Domestic supplies of low-sulfur oil are diminishing, and dependence on imports makes for uncertain supply and is always subject to sharp price increases. Domestic coal is plentiful, but most of it, especially in the eastern United States, has a relatively high sulfur content which requires extensive and expensive steps to be taken to desulfurize it or to treat the effluent gases to meet local air pollution standards. Vast quantities of natural gas have come on the market in the last decade, but it is recognized that this source, too, is finite. Oil and natural gas have required deeper and deeper wells, in locations which add to the difficulties of extraction (i.e., offshore platforms in deep waters) and shipment.

The large-scale plans for expansion of the nuclear power plant base in the United States were dramatically scaled back in the 1980s, for a number of reasons. Nationwide, actual and forecasted growth of demand was markedly lower; the regulatory climate and length of time required from inception to the plant's going on-line were such that construction financing presented onerous economic problems which largely offset the erstwhile advantages to be gained from the use of nuclear fuel; and there was denial of retraction of plant certification by local authorities.

In the past several years, efforts have been made to reduce the time required from planning to final operation of nuclear plants. To that end, the industry has designed a number of standard size and type nuclear reactors which, when approved by the NRC, do not have to undergo the long-time, one-of-a-kind review for each proposed installation. Although no plants are on order in the United States, the major U.S. fabricators and suppliers of nuclear reactors are engaged, either solely or in partnership, with the growing nuclear energy programs in many other countries. It is expected that the store of expertise gained thereby will be available for application in the United States at such future time when, as expected, nuclear power plants once more will serve to meet growing demands. At this time (2005), approximately 20 percent of the electric power generated in the United States originates in nuclear reactors. Outside the United States, there is great activity in the emplacement and operation of an expanded nuclear power base in the electric power generating industries. In addition to consolidation of new reactor designs and the construction of new power plants worldwide, existing reactors are being approved by the NRC for license extensions beyond their original 30- to 40-year license. Typical license extensions are for 20 years.

Effect on Fuel Resources

The emergence of nuclear power as a viable energy source comes at a time when additional environmentally acceptable sources of energy are sorely needed, both in the United States and in other countries, to meet continued rapid increases in demand. As indicated above, domestic production of both natural gas and oil appears to be approaching its peak and may decline in the future. Dependence on foreign sources entails risks with regard to dependability of supply and the U.S. balance of payments. New technology is needed in order to be able to mine and use this country's extensive coal reserves in an environmentally and economically acceptable manner.

Uranium supplies are limited and, if utilized only in light-water receptors, would rapidly diminish. Breeder reactors, can utilize most (60 percent or more) of our uranium (and/or thorium) resources rather than just a minute percentage.

A practically unlimited source of energy is available from the deuterium in the world's oceans should development work now underway in several countries lead to production of useful power from controlled fusion reactions. The technical problems are formidable, however, and the harnessing of fusion as an energy source in large power generating plants is not expected until the latter half of the twenty-first century.

NUCLEAR POWER PLANT SAFETY

Through the end of 2005, over 3,000 reactor-years of operation had been accumulated by commercial nuclear power plants in the United States, and about 8,000 worldwide, not including military or shipboard reactors (International Atomic Energy Agency, "Nuclear Power Reactors of the World"). This operational history has been accompanied by an impressive safety record with respect to health and safety impacts on the public, yet accidents have occurred. The most disastrous occurred in 1986 at the Chernobyl nuclear power plant complex located in the then U.S.S.R., resulting in loss of life and long-term radioactive contamination over a large area in the vicinity of the plant as well as transport of short-lived radioactive contamination dictated by the prevailing local weather patterns. Ultimate disposition of the destroyed reactor, now encased in an added containment structure ("sarcophagus"), remains a concern. In the United States, two of the worst commercial accidents occurred at the Brown's Ferry facility in 1975 and the Three Mile Island (TMI) facility in 1979. In the Brown's Ferry accident, a fire burned through hundreds of electrical power and control cables, threatening the plant's capability to be shut down and maintained in a safe condition. There was ultimately no damage to the reactor and no off-site radioactive releases occurred. At Three Mile Island significant reactor damage occurred through a sequence of events that began with a stuck-open primary system relief valve. Radioactivity was released to the environment, mostly in the form of noble gas fission products. However, estimated exposures to members of the public were found by a Federal interagency working group to be exceedingly small, even for the maximally exposed individual. The TMI reactor vessel did not fail despite relocation of approximately 19 tons of molten core materials. Although these accidents had minimal public impact, safety systems and plant operators were severely challenged during the events.

Commercial reactors have inherent safety features that prevent them from exploding like bombs. In a fission bomb, the fissioning of the fuel is caused by neutrons whose energy is about the same as the level at which they are produced by fission of the material in the previous generations. As a result, the average neutron lifetime is very small, and this allows the supercritical chain reaction to increase the power level to incredibly high levels before the device explosively destructs itself. In a power reactor, light-water or gas-cooled, enrichments of the fuel is very small (a few percent). As a result, essentially all fissioning of the fuel is caused by neutrons that have been reduced greatly in energy by the moderator (water or carbon). This results in a long average neutron lifetime compared to that in a bomb. If the chain reaction is allowed to increase in an uncontrolled fashion, this long lifetime will prevent the power of the reactor from rising to very high levels before the inherent design characteristics of negative temperature and void coefficients cause the reaction to shut down. While the Chernobyl reactor had design deficiencies in this regard, the violence of the accident was very far removed from what would result from detonation of a nuclear bomb.

There are two fundamental inherent safety features that contribute to the controllability of commercial power reactors. The first of these is the **Doppler effect.** As the power level of a reactor increases rapidly, the temperature of the fuel increases very rapidly. This rise in temperature causes the low-energy resonant peaks in the microscopic cross section to broaden, effectively creating a larger probability that neutrons will suffer nonfission captures by the U^{238} in the fuel. The second inherent safety feature is caused by the physical densities of materials, particularly the moderator, that decrease as the temperature increases. This results in higher probabilities that neutrons will escape from the reactor rather than cause fission in the fuel. Both of these inherent safety features feed back to the chain reaction in a negative way such that an uncontrolled increase in power level tends to shut the reactor down. This is called a **negative temperature coefficient.** Steam voids in a water moderator have a similar effect.

In spite of the inherent safety features, a reactor still provides a potential hazard to plant workers and members of the public in the vicinity of the plant. This is because of the generation of highly radioactive fission products within the fuel. The essential safety challenge is to assure that these fission products are kept confined at all times.

Nuclear power plant safety is based on a defense-in-depth philosophy. The defense-in-depth relies not only upon the inherent safety features, but also on high quality standards that provide a high degree of assurance that accidents are not likely to occur, and that if they do occur, the radioactivity will be contained locally.

High quality begins with the selection of highly qualified personnel to design, construct, and operate a plant. Each applicant for a license to build a nuclear plant must institute a quality assurance program to be applied to design, fabrication, construction, and testing of the structures, systems, and components of the facility. An applicant for a license to operate a plant must also provide for appropriate managerial and administrative controls to be applied to assure safe operation. The function of the quality assurance program is to provide a high degree of confidence that the structures, systems, and components in the plant will perform as intended during the service life of the plant.

Nevertheless, accidents can occur, and other safety features are designed to keep fission products confined in the event of a broad range of accidents that could conceivably occur. Two categories of anticipated accidents are undercooling events and positive reactivity addition events. An undercooling event can be caused by such things as ruptures in the reactor cooling system, loss of flow through the reactor core, or loss of feedwater flow. Positive reactivity addition events are the result of such things as inadvertent control-rod withdrawals or dissolved boron dilution, main steam line breaks (pressurized water reactor), or overfeeding of a steam generator.

There are various types of engineered safety features found in today's commercial reactors. They can be grouped into four major areas: (1) reactor protection system, (2) emergency core-cooling systems, (3) containment, and (4) emergency power supplies. The following discussion of these features is generally applicable to light-water-cooled reactors, pressurized water and boiling water reactors, but the same general features are applicable to other types.

The purpose of the **reactor protection system** is to rapidly insert the control rods into the reactor upon receipt of a signal indicative of a malfunction that potentially threatens the integrity of the core. For example, a low reactor coolant system pressure is indicative of a loss-of-coolant accident. High temperatures in the reactor coolant system or the core indicate the possibility of a loss of coolant or loss of flow through the core. The reactor protection system processes the various input signals and compares them to predetermined set points. If a set point is reached or exceeded, the system is designed to automatically insert the control rods into the reactor. The control rods absorb neutrons and quickly reduce the fission rate in the core. This rapidly drops the power level in the reactor and aids in prevention of damage to the fuel.

In the event of a loss-of-coolant accident, it is not sufficient to shut the reactor down in order to prevent damage to the fuel. Decay heat produced from the radioactive decay of fission products (initially some 4 percent of the previous operating power, but decreasing rapidly with time after stopping the chain reaction) would cause damage to the fuel if the coolant inventory losses were not made up by an injection system. **Emergency core-cooling systems** are designed to permit injection of coolant at high reactor coolant system pressures that exist during small ruptures as well as at low reactor coolant system pressures resulting from large breaks. Redundancy is provided to assure that equipment failures do not result in the loss of capability to keep the core covered with water. Large sources of water are stored to meet the long-term

cooling requirements to remove the decay heat. In the event that a pipe rupture results in the depletion of the stored injection water, provisions are made to allow recirculation of water from the emergency sumps in the containment building back through the injection pumps. In time the rate of heat generation drops to the point where forced-circulation cooling is no longer required. After the Three Mile Island core damage event the system was cooling by natural convection to the surroundings at atmospheric pressure after about 3 years.

Containment features provide an important function essential to the defense-in-depth philosophy. There are actually three levels in the containment scheme. The first level is the fuel cladding, which not only mechanically holds the fuel pellets in the core, but also provides a boundary to prevent migration of fission products to the coolant. The second level is the reactor cooling system pressure boundary. This consists of the reactor vessel system piping, and other components designed to withstand pressures in the order of hundreds of atmospheres without failure. Parts of these systems that are outside containment in normal operation, as in a boiling water reactor, are isolated at the containment building boundary in case of an accident. Thus even if the fuel cladding should fail, the pressure boundary is likely to keep the fission products contained. In the event that the accident would cause both of these boundaries to fail, a third level is provided in the form of a containment building. The typical containment building consists of a steel shell surrounded by a concrete structure subjected to precompression by steel cables. Thick steel only, reinforced concrete, and combinations have all been used in containment schemes, most of which are designed to withstand 5- to 10-atm pressure, based on containing the total energy available during an accident. The destroyed Chernobyl reactor did not have a containment building; it had only confinement provisions for small local pipe breaks.

Emergency power supplies also exhibit redundancy and defense-in-depth features. The power requirements for auxiliary systems within a power plant are normally supplied from the reactor output itself through the main electrical generator and auxiliary station transformer. In the event of an accident, the reactor would not be available to produce useful power; thus the preferred power source for supplying emergency electrical loads is from the transmission lines from other power plants. At least two off-site power supplies are required. In case of loss of these sources, redundant on-site sources are also required, usually powered by quick-start diesel generators. Battery systems are also needed to provide certain direct-current loads and some alternating-current loads through inverters. Control, computer, and instrument supplies are typically so served.

Protection is provided not only for events originating in the plant but also for external events. Plants are designed to withstand safely the worst anticipated earthquakes, floods, and winds. Containment and other safety features of nuclear power plants generally offer significant protection against consequences of accidents exceeding the severity of design-basis accidents. In recent years, experimental work and probabilistic studies have been carried out to determine the extent of such margins and discover residual vulnerabilities that would then become the basis for possible further safety improvements, generally on a plant-specific basis.

As the operating years of plants have accumulated, there has been increasing attention to detecting and counteracting age-related degradation of plant materials and equipment. A notable example is the varying degrees of radiation-induced embrittlement of reactor vessel materials, which if unchecked should increase vulnerability to overcooling accidents (pressurized thermal shock).

Another development in recent years has been the increasing use of probabilistic safety assessments as a partial guide to plant safety actions as well as safety regulation. In such assessments, **fault trees** are used to trace malfunctions through all antecedent events to the initiating causes and **event trees** to trace undesired events through all their consequences to ultimate outcomes. Known or estimated probabilities are assigned at each step in the chain of causation, to derive probabilities of sequences of safety (or risk) interest.

The various inherent and engineered safety features, along with the high quality standards which are required in the design, construction, and operation of a nuclear power plant, make accidental releases of radioactivity to the environment extremely unlikely. Nonetheless, additional defense-in-depth is provided in the form of trained operators and procedural adherence. As a final feature of the defense-in-depth philosophy, emergency planning is required that establishes the protective measures that would be taken to protect workers and members of the public against possible radioactive releases. Such emergency measures are required to be practiced at least annually in the United States.

Further assurance of nuclear plant safety is provided by the regulatory requirements summarized in the next section.

NUCLEAR POWER PLANT LICENSING

NOTE. The following describes the process in the United States. Nearly all countries have similar processes, but the degree of public participation and appeal opportunities may differ in significant degree, generally resulting in less complex and time-consuming processes in other countries, especially those with nationally owned industries and utilities.

Licensing of commercial power reactors is conducted in accordance with the Atomic Energy Act of 1954, as amended. The applicable regulations appear in Title 10 of the Code of Federal Regulations (10 CFR). Licensing is currently conducted by the NRC in a two-step process involving the issuance of a construction permit (CP) and then an operating license (OL) when the plant is completed. The process begins with the filing by a utility of an application for the construction permit and is followed by safety, environmental, security, and antitrust reviews by the NRC staff. These reviews are conducted in parallel. Following the staff reviews, and report to the NRC, another review is conducted by the Advisory Committee on Reactor Safeguards (ACRS), a body independent of the NRC, which gives a formal opinion letter to the NRC. Finally a mandatory public hearing is held by a three-member Atomic Safety and Licensing Board (ASLB), which reports to the NRC. If all issues are resolved satisfactorily, the staff is allowed by the NRC to issue the construction permit. The applicant may then begin construction, provided other local, state, and federal requirements are met for a variety of nonnuclear matters such as water discharge, building construction, worker safety, etc. The total number of permits required may exceed one hundred.

Several years before the plant construction is completed the applicant files an application to the NRC for an operating license. The process that follows is similar to that for the construction permit except that the public hearing is not mandatory during this step. However, a hearing may be requested by affected members of the public or by the commission (NRC). In practice this usually occurs on request from some member of the public.

Major features of the licensing process, are summarized below.

Preliminary Safety Analysis Report (PSAR) The PSAR is the primary component of the application for a CP. It generally comprises 10 or more volumes of information concerning site characteristics, plant design features, operating features, and documentation of compliance with NRC regulations.

Final Safety Analysis Report (FSAR) The FSAR accompanies the application for an operating license. This document provides more detailed information than the PSAR about changes in design, requests for information from the NRC staff during the CP phase, changes in regulations since the issuance of the CP, and conformance with the additional regulations that are only applicable at the OL stage.

NRC Staff and ACRS Safety Reviews The applicant's safety analysis reports provide the basis for the safety reviews. The staff reviews the reports against standard NRC acceptance criteria derived from the regulations and documents the results of these reviews in **Safety Evaluation Reports (SERs)** at both steps in the licensing process. Clarifications are usually required from the applicant both in writing and in meetings. Following completion of the SERs, the ACRS independently completes its reviews. These ACRS reviews typically include many meetings with the NRC staff and applicant. At the conclusion of the ACRS review, a letter is sent to the chairman of the NRC providing the ACRS recommendation as to whether the CP or OL should be issued. Supplements to the SERs are issued by the staff that incorporate appropriate changes as a result of the ACRS recommendations.

Environmental Review (ER) This review is held at both the CP and OL steps pursuant to the National Environmental Policy Act of 1969. The staff review is based on a Safety Evaluation Report (SER) prepared by the applicant. Upon completion of the staff s analysis, a draft Environmental Statement is prepared and distributed for comment to the various federal, state, and local agencies and to members of the public. Comments are taken into account in the preparation of a final Environmental Statement.

Antitrust Reviews This review is conducted by the NRC and the attorney general (Department of Justice) to ensure that operation of the facility will not be in violation of the antitrust laws.

Public Hearings Separate hearings may be held on safety and environmental matters, or the issues may be consolidated into a single hearing. If an antitrust hearing is required, it is always held separately. Hearings are normally held close to the reactor site to allow local public participation.

License Issuance and Commission Reviews Since 1981 it has been NRC practice for the staff to initially issue an operating license restricting operations of the plant to power levels not exceeding 5 percent of the full power rating. This low power license is issued after the completion of the safety, environmental, and antitrust reviews, and after completion of the public hearings, and resolution of any appeals. The exception to this is that public hearings on off-site emergency planning need not be completed prior to the issuance of a low-power license. The low-power license permits plant heat-up and reactor system testing, which typically takes 2 or 3 months. After the major remaining safety issues are resolved and the staff has confidence that the applicant can operate the plant safely, the staff brings the matter before the five-member commission (NRC) for review. Following this review and a favorable finding by the commission, the staff is allowed to issue a license amendment authorizing full-power operation of the plant. Federal courts may review the commission decision on appeal from interested parties, with or without a stay of execution of the license as the court may decide. Following the term of the full-power license, the license can be renewed for a period up to 20 years in accordance with the requirements established by the NRC.

The NRC has issued a safety goal policy statement that established goals that broadly define an acceptable level of radiological risk that might be imposed on the public as a result of nuclear power plant operation. To implement the goals, numerical objectives on core damage frequency and containment performance have been established. The safety goals are among the considerations underlying NRC's generic regulatory actions, but do not supplant NRC regulations and are not applied to individual plants.

Continuing Review A licensed reactor is subject to continual oversight by the NRC throughout its operating life to assure that regulations and license conditions are observed. The NRC maintains an active program for evaluating and resolving generic safety issues, i.e., issues relevant to all plants or to a class of plants. Plant or procedural modifications are required if needed for adequate safety or justified on a safety-cost tradeoff basis. Licensees are required to monitor the performance or condition of certain structures, systems, and components, and the effectiveness of maintenance, to provide reasonable assurance that capability to perform their intended safety functions will be retained.

Reactor licensees are also required to perform a probabilistic risk assessment, called an **individual plant examination,** for their facilities. These risk assessment analyses are a result of the NRC policy regarding assessment of beyond-design-basis accidents (severe accidents) and are intended to identify any plant vulnerabilities needing attention.

On-site resident inspectors carry out a continuous inspection program and are augmented by specialized inspectors based in the NRC regional offices or headquarters. Significant events are subjected to thorough investigations to determine causes and to assure corrective actions. Should a safety hazard ever be presented at a facility, the NRC is empowered to take any necessary enforcement action, including the shutdown of the plant.

Standardized Design Standard plant designs can lead to greater assurance of safety, reduced costs through quality production, and less licensing review time by NRC. Under NRC's procedures, a standard design is reviewed using bounding site conditions for the range of sites at which the plant might be located. For safety considerations which do not involve site-specific issues, the licensing review is therefore conducted only once, even though many applicants may reference the design. The NRC has certified several standard nuclear power plant designs, and it strongly encourages future applicants for licenses to reference them in their applications. The operation of multiple plants having essentially the same design features will provide a substantial safety benefit because experience gained at one facility may be applied to other like facilities. This should lead to more effective maintenance of key safety components and improved reliability of plant operations. Standardization would also stimulate programs of quality control and lead to better and faster training of operators and workers. It appears likely that when U.S. utilities do begin ordering nuclear plants again, the standard plant concept will be used, as has successfully been done in other countries.

Operator Licenses In addition to issuing operating licenses to the owners of nuclear power plants, the NRC also licenses persons who actually manipulate the controls of a reactor, at two responsibility levels, reactor operator and senior reactor operator. The licensing process involves training, experience requirements, and examinations, both written and practical, to verify an acceptable degree of knowledge on the part of the candidate before the license is granted or renewed. Because of the high cost of using an operating reactor for training and demonstration of practical facility in operation, many utilities use elaborate replica simulators of the control-room panels and systems of the power plant for operator training.

Decommissioning Five years before an operating license is to expire, licensees are required to inform the NRC for approval of their program for funding the management of spent fuel until disposal in a permanent repository. No later than 1 year before the license is to expire, the licensee must file an application to terminate its operating license, together with a detailed plan for decommissioning. With license renewal, decommissioning follows the period of extended operation.

Radioactive Waste The nuclear waste generated at nuclear power plants is generally classified as low-level waste, mixed waste, and spent fuel, although these plants are not significant generators of mixed waste. The low-level and mixed wastes can be shipped off-site to regional storage facilitates, if available. If not available, the low-level and mixed waste can be safely stored on-site until a facility is available. There is currently no permanent repository for storage of spent fuel, although the Department of Energy is working to establish one. Spent fuel is stored in spent fuel storage pools at each facility, and when they are filled, additional storage can be provided on-site by dry storage in independent spent fuel storage installations (ISFSIs). The spent fuel has normally been allowed to cool for at least 5 years before it is placed in an ISFSI; these facilities are certified for 20 years, can be recertified, if needed, and have a design life of 50 years.

OTHER POWER APPLICATIONS

Nuclear Ships The earliest and most successful of the efforts to develop nuclear power for ship propulsion were those by the U.S. AEC and U.S. Navy. These initial efforts toward this end predated and contributed to the development of large electric power generators. The U.S. Navy launched the *USS Nautilus* in 1954 and followed it with about 125 other operational nuclear submarines and 13 operational surface vessels. Additional submarines and surface vessels have been committed by the U.S. Navy. Nuclear submarines and ships are also being operated by the British, Russian, and French and Chinese navies.

Although a ship can be propelled by diesel engines, steam engines, gas turbines, and other devices, the steam turbine has emerged as the principal modern marine propulsion unit. When a nuclear reactor of a type developed for generating stations is used to produce steam, the combustion process is eliminated. Such a propulsion plant is especially advantageous for a submarine, since it is then completely independent of the earth's atmosphere. The U.S. Navy's nuclear ships and submarines all employ the pressurized water type of nuclear reactor.

On commercial ships, the advantages of nuclear power are related to higher-speed operation over long run routes. Experience indicates that an increase in speed usually produces substantial increases in cargoes, thereby increasing a ship's load factor. In conventional ships, the power required for higher speed is increased by the third power of the speed ratio. This means larger engines and greater fuel storage capacity. On the other hand, a nuclear ship can operate continuously at the maximum level of its hull form without penalty in payload. Such advantages are especially desirable for icebreakers, large bulk cargo vessels, and high-speed long-distance passenger service. Efforts in the 1960s and 1970s to build commercial nuclear-powered ships resulted in the N.S. *Savannah* in the United States, the N.S. *Otto Hahn* in West Germany, the N.S. *Mutsu* in Japan, and the icebreaker *Lenin* in the Soviet Union. All proved to be failures. The first two were technical successes, but economic failures. The *Savannah* is a museum, and the *Otto Hahn* had her power plant removed and replaced by diesel propulsion units. The *Mutsu* had a radiation shield leak, never operated, and was eventually broken up. The *Lenin* had her nuclear power plant replaced by a more modern one, and with three larger nuclear-powered icebreakers and a large supply ship, also nuclear-powered, operates in the Russian Arctic. Economics of a state-owned ship are difficult to assess, but these are the only unarmed nuclear-powered surface ships in operation.

Other Vehicles The use of nuclear power in airplanes, automobiles, and railroad trains has been deterred by the requirements for heavy shielding to protect against radioactivity and collisions.

Remote Power Sources Small nuclear power plants have been built and operated by the U.S. Corps of Engineers in such remote regions as the Greenland ice cap and the Antarctic ice sheet. A barge-mounted plant was also built and operated in the Panama Canal Zone, but was scrapped.

Isotopic Power Generators Isotopic power generators are direct conversion devices that convert the energy of decaying radioisotopes to electricity, using direct thermoelectric or thermionic devices, or thermal conversion systems. (See Sec. 9.1.) The first isotopic power generator, completed in 1959, produced 2.5 W of power and used polonium 210 as fuel. The first commercial isotopic power generator, fueled by strontium 90, became available in 1966, and produced 25 W of power. There are now eight radioisotopes available in this area (see Table 9.8.4).

Table 9.8.4 Radioisotopic Materials

Radioisotope	Half-life, yr	Typical power density,* W/cm^3	Radiation present†	Shielding required
Strontium^{90}	28	1.4	Beta and bremsstrahlung	Heavy
Cesium^{137}	30	0.21	Beta and gamma	Heavy
Cerium^{144}	0.78	24.5	Beta and gamma	Heavy
Promethium^{147}	2.7	1.8	Beta	Minor
Polonium^{210}	0.38	1,210	Alpha	Minor
Plutonium^{238}	89	3.5	Alpha	Minor
Curium^{242}	0.45	1,150	Alpha	Minor
Curium^{244}	18	26.4	Alpha and neutron	Moderate

* Reduced below theoretical maxima in order to account for expected isotopic impurities.
† Of dominant importance in heat generation and for consideration of radiation shielding.
Source: AEC TID-20079.

Plutonium 238 was used in the long-range space probes Voyager 1 and 2.

Space Power Sources Most satellites and space vehicles use solar and battery power sources. However for very high powers, into the multikilowatt region, and for very deep space probes where solar energy is not available, nuclear sources (either radioisotopes or reactor heat systems) are required. Several dozen such systems have been deployed since 1970. Most are isotopic, but the United States orbited one reactor system in 1965. The Soviet Union has put several reactor-powered systems into orbit. On at least two occasions such U.S.S.R. systems have decayed back into the earth's atmosphere, and in one instance radioactive parts landed in Canada.

NUCLEAR FUSION

Development of nuclear fusion reactors for the production of useful energy will require significant advances from the present state of the art in both plasma physics and engineering. A large fusion effort, especially in physics research, has been carried on since 1956 in many industrial nations, notably the Soviet Union, Japan, the United States, and several in western Europe; considerable progress has been made.

There are a number of possible **fuel cycles** involving the fusion of light ions into heavier atoms, the most promising of which are shown in Table 9.8.5. The most accessible of these is the deuterium-tritium (DT) reaction, which involves fusion of two hydrogen isotopes. Even this lowest-temperature reaction of the possible fusion cycles has a threshold

Table 9.8.5   Possible Fusion Fuel Cycles

Reactor equation	Approx threshold plasma temp*	Approx avg energy gain per fusion
$D + T \rightarrow {}^4He\ (3.5\ MeV) + n(14.1\ MeV)$	10 keV	1,800
$D + D \begin{array}{l} \nearrow {}^3He\ (0.82\ MeV) + n(2.45\ MeV) \\ \searrow T\ (1.01\ MeV) + p(3.02\ MeV) \end{array}$	50 keV	70
$D + {}^3He \rightarrow {}^4He\ (3.6\ MeV) + p(14.7\ MeV)$	100 keV	180

* 1 keV = 11,000,000 K approximately.

temperature of about 50 million K, and a practical reactor would require operating temperatures of about 100 million K. At these temperatures the ionized gas or plasma cannot be contained in vessels of conventional materials.

Fortunately, ionized plasmas can be contained by magnetic or inertial forces on earth. (The level of gravitation forces which produce fusion in the sun and stars cannot be achieved on earth.) Research seeking how this can be done at high enough combinations of temperatures, pressures, and densities to obtain practical net energy output is the main effort of fusion programs worldwide. This research is concentrated in three major approaches:

1. **Toroidal sytems,** especially of the tokamak variety, in which very high currents are circulated through an endless plasma confined by toroidal (doughnut-shaped) magnetic systems of various physical cross sections and magnetic shaped fields.

2. **Magnetic mirror systems,** in which the ends of a solenoidal (linear) magnetic field are "plugged" by a combination of electrostatic and magnetic forces to make the plasma reflect back and forth.

3. **Inertial systems,** in which small pellets of fusionable material are rapidly compressed by beams of laser-produced light, electrons, heavy ions, or other compressive forces. This is the principle employed in nuclear fusion weapons.

For practical energy production, a DT plasma should have a temperature of about 100 million K. This is usually expressed as equivalent ion temperature in electron volts (1 eV = 11,000 K approx.). Coincidentally with this temperature the product of density (in ions per cubic centimetre) and confinement time (in seconds) should reach 10^{14}. These conditions have been achieved experimentally.

Three large tokamak installations (TFTR in the United States, JET in Europe, and JT-60 in Japan) have operated. Each of these machines created plasma conditions approximating those required in practical power producing reactors, although they are not in themselves net power producing devices. At the Lawrence Livermore National Laboratory in the United States a very large laser device is being used to attempt to establish similar conditions by inertial means, using very tiny pellets filled with DT mixtures. All these devices are experimental.

After practical reactor conditions have been achieved in a confined plasma, much engineering development associated with removing the energy and converting it into useful form will be required. Developmental work on fusion engineering continues, but formidable obstacles lie ahead, especially in the requirements for materials capable of surviving the environment of burning plasmas for long periods of time.

Although scientific feasibility seems ensured and was demonstrated in the late 1990s, the very considerable engineering problems yet unresolved put the operation of a demonstration fusion power reactor well into the twenty-first century, with any significant effect on world energy supplies occurring at some future time. Since uranium resources appear adequate for many decades, the essentially unlimited fuel supplies for fusion (from the oceans) may not be needed until then.

9.9 HYDRAULIC TURBINES
by Robert D. Steele

REFERENCES: "Water Power Engineering," McGraw-Hill. Creager and Justin, "Hydroelectric Handbook," Wiley. Daugherty and Ingersoll, "Fluid Mechanics," McGraw-Hill. Hammitt, "Cavitation and Multiphase Flow Phenomena," McGraw-Hill. Kovalev, "Hydroturbines, Design and Construction," Russian translated by U.S. Department of the Interior and National Science Foundation. Mosonyi, "Water Power Development," Hungarian Academy of Science.

GENERAL

Notation

B = width of distributor or height of wicket gates, in (mm)
D = diameter of runner, in (mm)
D_l = diameter of runner, at centerline of guide case, in (mm)
D_{th} = throat diameter of runner, in (mm)

D_r = relative diameter of runner, in (mm) (= entrance diameter for low-speed Francis turbines; throat diameter for high-speed Francis turbines; bore of throat ring at centerline of blade for propeller turbines)

D_d = diameter top of draft tube, in (mm)
D_p = pitch diameter of impulse-turbine runner, in (mm)
d = jet diameter of impulse turbine, in (mm)
e = overall efficiency of turbine $= e_h e_m$
e_h = hydraulic efficiency (including draft tube)
e_m = mechanical efficiency
g = acceleration of gravity, ft/s^2 (m/s^2)
H = net effective head, ft (m)
b = head change due to load change, ft (m)
n = r/min

n_1 = r/min at 1 ft (m) head = $n/\sqrt{H}$
n_s = specific speed = $n\sqrt{P}/H^{5/4}$
P = horsepower = kW/0.746
P_1 = horsepower (kW) at 1 ft (m) head = $P/H^{3/2}$
Q = discharge, ft³/s (m³/s)
Q_1 = discharge at 1 ft (m) head = $Q/\sqrt{H}$, ft³/s (m³/s)
t = thickness, in (mm), or time, s
u = circumferential velocity of a point on runner, ft/s (m/s)
V = absolute velocity of water, ft/s (m/s)
V_u = tangential components of absolute velocity = $V\cos\alpha$
V_r = component of V in radial plane
v = velocity of water relative to runner, ft/s (m/s)
w = weight of water lb/ft³ (kg/m³)
z = number of runner buckets (blades for propeller turbines)
α = angle between V and u (measured between positive directions)
β = angle between v and u (measured between positive directions)
θ = angle of water deflection relative to bucket
ϕ = peripheral coefficient $[\pi Dn/(720\sqrt{2gH})][\pi Dn/(60,000\sqrt{2gH})]$
σ = cavitation coefficient

Fundamental Formulas The actual power of a hydraulic turbine is the theoretical power P_t multiplied by the turbine efficiency e.
Power in U.S. Customary System (USCS) Units

$$P = eP_t = \frac{eHQw}{550}$$

In the power equation above, **power** is calculated in horsepower. In the following text, when power P is used in stated equations, it is in horsepower.

Power in Metric Units

$$P = eP_t = \frac{eHQw \times .746}{76.04} = \frac{eHQw}{101.93}$$

The metric equation stated above for power calculates power in kilowatts. In the following text, where power P is used in metric equations (shown in parentheses), the power stated in the equations is in kilowatts. [The customary unit of horsepower in the United States should not be confused with the term *metric horsepower* (see the conversion and equivalency tables of this handbook).]

The laws of proportionality for homologous turbines are:

For constant runner diam	For constant head	For variable diam and head
$p \propto H^{3/2}$	$P \propto D^2$	$P \propto D^2 H^{3/2}$
$n \propto H^{1/2}$	$n \propto 1/D$	$n \propto H^{1/2}/D$
$Q \propto H^{1/2}$	$Q \propto D^2$	$Q \propto D^2 H^{1/2}$

Fig. 9.9.1   Medium-head Francis turbine.

Nomenclature   The nomenclature used throughout this section is based on NEMA's standards publication HT1-1957, "Hydraulic Turbines, Governors and Accessory Equipment," which also contains definitions.

Types of Turbines   Three characteristic types of hydraulic turbines are now in general use: the **Francis reaction** type (Figs. 9.9.1 and 9.9.2); the **propeller reaction** type (Fig. 9.9.3) and the **impulse** type (Fig. 9.9.12). The propeller type may be further divided into fixed and adjustable blade types. All three types have in common a stationary guide case (or nozzle in the case of the impulse type) in which the static head is transformed partly, or wholly, into velocity, and a revolving part, the runner.

Fig. 9.9.2   Typical profile of Francis runners. (*a*) Low specific speed; (*b*) high specific speed.

Fig. 9.9.3   Adjustable-blade Kaplan turbine.

In the guide case of the impulse turbine, the static head is completely transformed into velocity, so that air surrounds both the jet issuing from the nozzle and the runner.

In the guide case of the reaction types, the static head is only partly transformed into velocity, leaving an overpressure between the guide case and the runner. This overpressure causes an acceleration of the relative velocity of the water passing through the runner, the discharge area of which is smaller than the entrance area. Except when operating vented at low loads, the water passages are completely filled with water from the intake to the end of the draft tube.

Selection of Turbine Type and Casing   Impulse turbines receive their water supply directly from a pipeline. Francis and propeller reaction types are set in a case of concrete or metal. While the limits of head to which the impulse and reaction types are adopted may be fairly well defined in practice, as roughly outlined in Table 9.9.1, there is no definite line where the application of one type ends and the other begins. The choice between impulse and reaction types depends upon the size of the unit as well as upon the head and other considerations.

Specific Speed   The common basis of comparison between turbine runners of different types and between runners of the same type but different design and characteristics is termed *specific speed* n_s. This is the

Table 9.9.1   General Arrangements of Turbine Installations and Usual Head Limits Employed

Type	Setting	Construction	No. of runners	Usual head limits for direct-connected units	
				ft	m
Reaction turbines, 5 to 1,000ft (1.5 to 300m) head	Axial flow	Vertical or horizontal or slanted	1	5–60	1.5–20
	Encased	Concrete vertical	1	15–130	5–40
		Cast or welded plate steel. Stainless steel. Vertical or horizontal	1 or 2	30–1,600	10–500
Impulse turbines, 500 to 6,000 ft (150 to 1,800 m) head	Encased	Horizontal (1–2 nozzles)	2	500–6,000	150–1,800
		Vertical (1–6 nozzles)	1	500–6,000	150–1,800

relationship between the speed of a runner at the point of highest efficiency and the maximum power output at this speed regardless of size. However, since both power and speed vary with head, specific speed is defined as the relationship between the speed n_1 and power P_1 at 1 ft (m) head. Subscript 1 denotes that the value is reduced by the similarity law to a 1 ft (m) head basis. Since $n \propto 1/D$ and $P \propto D^2$, the product $n_1 \sqrt{P_1}$ remains a constant regardless of the size of the runner and is designated the specific speed of the runner. The term *specific speed* for this relationship stems from the fact that $n_1 \sqrt{P_1}$ also is the value of the speed in r/min at best efficiency which the runner would have if operated under 1 ft (m) head, the runner being of such size as to develop 1 hp (kW) ($P_1 = 1$).

Since $n \propto \sqrt{H}$ and $P \propto H^{3/2}$, the specific speed of any runner operating under head H will be $n_s = n \sqrt{P}/H^{5/4}$ where n is the best efficiency speed and P the maximum power output at this speed, all at head H.

Selecting the Speed   Hydraulic turbines are usually connected to ac generators. The turbine speed must agree with one of the synchronous speeds required for the system frequency. The prevailing frequency for most systems in the United States is 60 Hz. Synchronous speeds are determined by the formula $n = 120 \times$ frequency/number of poles in generator. The number of poles must be even.

The speed should be as high as practicable, as the higher the speed the less expensive will be the turbine and generator and the more efficient will be the generator.

A convenient way of determining the highest practicable speed is by the relation of the specific speed to the head. For Francis turbines this may be taken as $n_s = 900/\sqrt{H}(1,900/\sqrt{H})$; for propeller-type turbines as $n_s = 1,000/\sqrt{H}(2,100/\sqrt{H})$. The adoption of these values of specific speed is predicated on the use of a reasonable setting of the unit with reference to tailwater level, i.e., selection of proper cavitation coefficient.

Number of Units   From the standpoint of reducing the number of auxiliaries and the amount of associated equipment and also reducing initial and maintenance costs for the entire plant, the number of units should be kept to a minimum. Also the larger the unit, the higher the efficiency and generally the lower the cost per unit power output. However, other considerations, such as flexibility of operation, higher efficiency operation during low-load demands, and minimum loss of capacity during shutdown for repair or maintenance, might dictate the use of multiple units where one unit would be feasible in terms of physical size. For some projects, the physical size of the unit has been limited to the maximum size runner that could be shipped in one piece; this is largely due to the extra manufacturing costs involved in furnishing split runners. However, since split runners present no serious mechanical difficulties, the tendency in recent years has been to disregard this limitation.

REACTION TURBINES

Francis   Figure 9.9.1 shows a Francis-type reaction turbine for medium head. The runner consists of a relatively large number of shrouded buckets. Movable wicket gates with axes parallel to the turbine shaft control the flow. This type of turbine is normally used for heads ranging from 75 to 1,600 ft (25 to 500 m). Specific speeds vary from 15 to 100 (50 to 400).

Propeller Turbines, Fixed and Adjustable Blade   The propeller turbine has a runner which is normally provided with from 3 to 10 unshrouded blades, either fixed or adjustable. This type of turbine usually is used for heads from 5 ft (1.5 m) up to 120 ft (40 m), although in a few cases they have been used for heads up to 200 ft (60 m). The higher the head, the greater the number of blades used. Specific speeds vary from 80 to 250 (300 to 1,000). Propellers have very steep efficiency vs. power curves (4 and 5, Fig. 9.9.5). Adjustable-blade propeller runners are used to produce a flat efficiency curve over a wide range of power (6, Fig. 9.9.5) and to produce considerably more power beyond the maximum efficiency point than can be obtained with a fixed-blade runner of equal diameter. For fixed-blade runners, the blade angle is usually set between 16 and 28°, where maximum efficiency occurs. For adjustable blade runners, the blade angle may vary from $-10°$ min to $+45°$ max. Normally, automatic oil-pressure-operated blades are used in a **Kaplan turbine** (Fig. 9.9.3). The blades are adjusted by means of an oil-operated piston located either within the main shaft or in the runner hub. The oil is admitted to and discharged from the piston by means of a distributor head either on top of the generator shaft or surrounding the main shaft. The oil pressure is supplied from the governor oil pressure system. The controls are so arranged that the blade tilt varies automatically with the wicket gate opening to produce a maximum efficiency envelope curve (Fig. 9.9.9). A large operating-cylinder capacity is required. The capacity S, ft · lb (J) may be approximated from the formula $S = 20Pn_s^{1/4}/\sqrt{H}(14n_s^{1/4}/\sqrt{H}$ If antifriction bearings are used in the hub instead of bronze bearings for the blade trunnions, this formula becomes $S = 10Pn_s^{1/4}/\sqrt{H}(7n_s^{1/4}/\sqrt{H}$. However, since the hydraulic unbalance on the blades varies with speed, gate opening, and the location of the blade pivot axis, the oil-operated cylinder size should be calculated from model test data.

Axial-Flow Turbines   Axial-flow turbines use the propeller type runner with either fixed or adjustable blades. Their characteristic feature is the straight-through, or nearly straight-through, water passageway from intake to discharge. The shaft is, therefore, either horizontal, vertical, or inclined (Fig. 9.9.4). The spiral or semispiral case and elbow draft tube, which require substantial widths and depth of excavation, are

Fig. 9.9.4   Axial-flow bulb turbine.

eliminated. Therefore, with a reduction in height and area of the powerhouse and the turbine's suitability for location directly within the dam, an overall construction-cost savings may be obtained for the power plant part of the project compared with conventional vertical units. This may make it possible to build or redevelop power plants for low heads or in small sizes that have previously been considered uneconomical.

In recent years, axial-flow turbines have been installed for use with tidal power since they can be arranged to operate with water flowing in either direction and, when required, to pump with water flowing in either direction.

There are four general types of axial-flow turbines. The first type has the generator rotor mounted around the periphery of the turbine runner. This type of turbine and generator arrangement has been called a **rim** turbine. The second type is the **bulb** turbine, where the generator is directly connected to the turbine at a submerged elevation. The bulb turbine has the generator enclosed in a streamlined, watertight housing located in the water passageway on either the upstream or downstream side of the runner (Fig. 9.9.4). The third type of axial-flow turbine is the **pit** turbine. In the pit turbine arrangement, a high-speed generator is driven from the high-speed output shaft of a speed increaser, and the low-speed input torque to the speed increaser is supplied from the lowspeed shaft of the turbine. The generator of an axial-flow pit turbine can be mounted in line with the turbine, or in smaller units the generator may be mounted inclined or vertical.

The fourth type of axial-flow turbine is the **tube** type, which has the generator located outside the water passages. The construction characteristic of this turbine is the slight bend of the water passageway that permits the extension of the turbine shaft to the outside of the water passageway. The tube turbine is typically arranged so that the generator is on the downstream side of the turbine, but it could be constructed to have the generator on the upstream side. The shaft from the turbine to the generator of a tube turbine can be inclined to reduce excavation, thereby raising the generator higher above tailwater elevations. In some cases, a gear-type speed increaser is used to reduce the combined equipment cost and generator size and weight.

Axial-flow turbines are suitable for heads up to at least 90 ft (30 m) with basically the same limitations that apply to the conventional Kaplan or other propeller-type turbines. Maximum unit capacity is limited, however, by maximum practical speed increaser torques and maximum practical horizontal generator capacities. This appears to be approximately 100 MW at present. Either fixed-position or movable wicket gates can be used. Power and efficiency performance is comparable to conventional vertical shaft propeller turbines.

Design The water entering the runner is given a whirl or tangential component by the guide vanes. This whirl is taken out by the runner so that, at the design point (point of best efficiency), the water leaves the runner without appreciable whirl. According to Euler's theorem, the power of a turbine in steady motion equals the angular velocity multiplied by the change in angular momentum experienced by the mass of water flowing in a unit time in its passage through the turbine. This is expressed in the following equations:

$$P = 62.4Q(u_1V_{u1} - u_2V_{u2})/(550g)$$
$$= QHe_n/8.81$$
(9.9.1)

and

$$u_1V_{u1} - u_2V_{u2} = e_hgH$$
$$P = 1{,}000Q(u_1v_{ul} - u_2v_{u2})/(101.93g)$$
$$= QHe_hw/101.93$$
(9.9.2)

(See notation at beginning of this section.) Subscript 1 refers to the inlet edge of the bucket and 2 to the discharge edge.

In practice, e_h may be taken as 0.92 to 0.94, and the term u_2V_{u2} may be omitted as it becomes zero for discharge in a radial plane, $\alpha_2 = 90°$.

The right hand member of Eq. (9.9.2) may be regarded as a fixed value for a given head. Consequently, in order to have a high-speed-type runner (u_1, large) the whirl V_{u1} placed in the water by the guide case

must be small. This means that for the best load, low-speed low-capacity (small n_s) runners will have relatively small gate openings and short wicket gates, whereas high-speed high-capacity (large n_s) runners will have relatively large gate openings and long wicket gates.

Equations (9.9.1) and (9.9.2) are used in designing the runner and guide case. The P and Q used in Eq. (9.9.1) are for the point of best efficiency and not for the rated values. For specific speeds up to about 60 (250) the power at best efficiency may be taken as 85 percent of the rated power. Above that specific speed the point of best efficiency occurs at higher loads as indicated in Fig. 9.9.5. Adjustable-blade propeller turbine-runner blades are laid out for an intermediate load of 75 to 80 percent of the rated load. The actual design of a runner should be worked out by an experienced designer on the basis of coordinated test data.

Fig. 9.9.5 Variation of efficiency vs. load for reaction turbines.

The general proportions of a normal line of turbine runners may be determined as follows for any combination of net head H and rated power P. (Rated power is usually considered to be 95 percent of full gate power.)

First select the specific speed appropriate to the head (see "Specific Speed" above). Next calculate n from the formula $n_s = n\sqrt{P}/H^{5/4}$. If $n/H^{5/4}$ is not a synchronous speed, select the nearest synchronous speed and calculate the corresponding n_s. This value of n_s may be used in Fig. 9.9.6 to determine the peripheral coefficient ϕ for D_r and D_1. Then the diameter may be determined from the formula

$$\phi = \pi Dn/720\sqrt{2gH}\ [\ = \pi Dn/(60{,}000\sqrt{2gH})]$$

Fig. 9.9.6 Turbine characteristics and specific speed.

Other proportions of the runner may be determined from Fig. 9.9.7, which shows the ratio to D_r of width of distributor B; diameter at top of draft tube D_d; and centerline of distributor to top of draft tube A. See Fig. 9.9.1 for D_r, B, and A.

Fig. 9.9.7 Turbine characteristics and specific speed.

The values of unit power P_1 and unit speed n_1 for $D_r = 100$ in (2,540 mm), which correspond to the specific speed of the runner may be determined from Fig. 9.9.6. The following formulas may then be used: $P = P_1 H^{3/2}(D_r/100)^2 [= P_1 H^{3/2}(D_r/(2,540)^2]$ and $n = n_1 \sqrt{H}$ $(100/D_r) [n_1 \sqrt{H}(2,540/D_r)]$. These formulas may be used to verify the diameter calculated from the peripheral coefficient, and will prove useful in selecting runner sizes for units which must meet power requirements other than rated power at design head. In such cases the values of n_1 and P_1 from Fig. 9.9.6 should be used in conjunction with Fig. 9.9.8.

For preliminary powerhouse layouts these ratios of diameter to D_r may be used for all specific speeds: gate circle, 1.20; stay ring, 1.65; pit liner, 1.50; unit spacing, 3.5 to 4.5.

The clearance space between the discharge tips of the wicket gates and the entrance edge of the runner buckets is treated as a free vortex space in which the tangential component of the absolute velocity of discharge from the gates (V_u) increases in inverse proportion to the radius while the component in a radial plane (V_r) increases in inverse proportion to the area.

To obtain the best efficiency, the water must enter the runner without shock and leave it with as little velocity as possible; i.e., the entrance angles of the buckets must be approximately in line with the relative direction of the entering flow and the direction of the absolute discharge

Fig. 9.9.8 Variation of unit power with unit speed and specific speed.

velocity should be approximately in a radial plane (axial in the case of propeller runners). This condition prevails only at the most efficient load. Above that load the water enters the draft tube with reverse whirl and below that load with a forward whirl.

The shape of the buckets should be kept as simple as possible, consistent with meeting the proper angles. The curvature of the bucket along a flow line is made greatest near the entrance and reduces as the discharge edge is approached, somewhat the shape of a portion of a parabola.

Turbine Characteristics Speed, power, and efficiency characteristics vary widely with specific speed as shown in Figs. 9.9.5 and 9.9.6. The variation of unit power P_1 with the unit speed (as a percent of normal basis) is shown in Fig. 9.9.8 for typical specific speeds ranging from 20 to 160 (70 to 600). For low specific speed Francis-type turbines, P_1 decreases as n_1 increases, as when operating at heads below normal. This is due to the centrifugal effect, which tends to decrease the flow as n_1 increases. This characteristic of low specific speed Francis runners makes them less suitable for operation under extreme variations in head. With the higher specific speeds and particularly with the propeller type, P_1 increases as n_1 increases above normal. This makes the higher specific speeds particularly suitable for operation at widely varying heads.

Fig. 9.9.9 Variation of efficiency vs. power for fixed- and adjustable-blade propeller turbines.

Figure 9.9.9 shows the advantage of blade adjustment for propeller-type turbines. By angular adjustment of the blades as the load changes, the peaked power-efficiency curve of the fixed blade propeller is transformed into a flat power-efficiency curve, and the maximum power output is considerably increased.

Runaway Speed If a turbine runner is allowed to revolve freely without load and with the wicket gates wide open, it will overspeed to a value called the *runaway speed*. The runaway speed, at normal head, varies with the specific speed and for Francis turbines ranges from 170 percent (normal speed = 100 percent) at low specific speed [$n_s = 20$ (76)] to some 195 percent at high specific speed [$n_s = 100$ (383)]. For propeller turbines, the runaway speed varies with blade angle—the steeper the blade angle, the lower the runaway speed. For fixed-blade propellers with the blades set from 16 to 28°, where maximum efficiency is usually obtained, the runaway speed ranges approx from 255 to 180 percent, respectively. For adjustable blade turbines, where the minimum blade angle is sometimes as low as 6 to 16° in order to obtain high efficiency at part load, the maximum possible runaway speed will be about 290 to 270 percent, respectively. However, with adjustable-blade propeller turbines, there is from the standpoint of efficiency an optimum relationship between runner-blade angle and wicket gate opening, usually controlled by a cam in the operating mechanism; the higher the gate opening, the steeper the blade angle. Thus the combination of wide-open gate and minimum design blade angle can only occur in the so-called "off-cam" position, which is an extremely rare possibility. In some units, this maximum possible off-cam runaway speed is reduced by limiting the minimum design blade angle to 16°. Another method is the use of a runaway speed limiter. There are several designs in use. One of the most frequently used is a valve designed to open by centrifugal force at speeds of 120 percent (approximately). The open valve bypasses the runner blade servomotor piston, thus equalizing the oil pressure on both sides of this piston. If the runner blades are designed to open because of hydraulic unbalance at overspeeds, they will then go to a higher blade angle under off-cam overspeed conditions. The off-cam runaway speed thus can be limited to some 185 percent if the runner is designed for a maximum blade angle of 28 to 40°.

For all turbines, if the maximum head is higher than the normal head, the runaway speed will be increased in proportion to the square root of the head. Therefore, runaway speeds should be based on the maximum operating head rather than on the normal head. Any runaway speed above 180 percent can increase appreciably the cost of the generator.

Turbine Thrust The water passing through the turbine creates a thrust load, that is carried by a thrust bearing, typically supplied with the generator. For vertical units, the thrust T consists of the hydraulic thrust T_h plus the weight of the turbine runner and shaft T_w, lb (kg). Units have been constructed with the thrust bearing supporting the total weight of rotating components of the turbine and generator, plus the hydraulic thrust, located on the turbine head cover.

For horizontal or inclined units, the thrust T consists of the hydraulic thrust T_h and a component of the weight T_w, of the rotating components, if the unit is inclined. The thrust bearing can be located in the generator, in a speed increaser, in the hydraulic turbine, or in a separate thrust bearing housing.

The hydraulic thrust may be obtained by using a **thrust coefficient K_t** multiplied by the weight, lb (kg), of a circular column of water whose diameter is the diameter of the runner D_r, in (mm), and whose height is equal to the maximum head H_m, ft (m), under which the turbine must operate, or

$$T_h = \frac{K_t D_r^2 H_m}{2.94} \quad \text{lb} \qquad \left(T_h = \frac{K_t D_r^2 H_m}{1{,}273} \quad \text{kg} \right)$$

The thrust coefficient for Francis turbines varies with the specific speed and may be taken as approximately $K_t = n_s/250$ ($K_t = n_s/954$). In using this value of K_t, it is assumed that the runner seals are properly arranged and that the crown of the runner inside the seals is properly drained; otherwise, the thrust may be much higher.

For propeller turbines, with either fixed or adjustable blades, the thrust coefficient K_t may be taken as 0.90 for all specific speeds.

Reaction Turbine Elements

Runner and Wearing Rings The number of buckets z for Francis runners ranges from about 21 for low specific speed to 12 for high specific speed. The approximate number may be taken as $z = 55/n_s^{1/3}$ ($86/n_s^{1/3}$).

Materials that can readily be repaired by welding with either carbon-steel or stainless-steel electrodes are selected for the construction of runners. Most runners are made of cast carbon steel, cast stainless steel, formed carbon steel plate, formed stainless-steel plates, or a combination of carbon steel and stainless steel. The majority of new equipment is supplied manufactured of stainless steel, using a combination of martensitic and austenitic stainless-steel casting, forgings, and plates. The use of high-strength stainless steel for runners is increasing for high heads and for conditions where cavitation pitting may be a design consideration.

The number of blades for propeller-type turbines ranges from 10 for low specific speeds to 3 for high specific speeds. Propeller runner blades are usually made of cast carbon steel, formed carbon steel, formed stainless steel, or cast stainless steel. If they are made of cast steel or formed carbon plate steel, the surfaces over which pitting may be expected are overlay-welded with stainless steel before finishing to final contour. In place of overlay welding, solid stainless-steel inserts are sometimes welded into propeller blades, and stainless trim is added to the periphery of the blades.

The functions of the runner seals for Francis turbines are to prevent excessive leakage loss and thus improve the efficiency and control the hydraulic thrust.

Rolled, forged, and cast steels make excellent wearing rings for low- and medium-head units. To prevent seizure in case of contact the rotating ring material should differ from that of the stationary rings. Stainless steel, bronze, or steel with bronze inserts make excellent wearing rings. For extremely high heads, stainless steel should be used to prevent undue wear and erosion.

Seal clearances are made as small as practicable to reduce leakage, particularly on high-head units. Larger clearances are required with water-lubricated main bearings, subject to considerable wear before being readjusted.

Main Shaft and Bearings The main shaft, for a typical vertical hydraulic turbine unit, must be rigid and should be made of a medium-grade carbon-steel forging. The size of the shaft is determined by limiting the torsional stress from 3,000 to 6,000 lb/in² (20 to 41 N/mm²).

The shafting system of a horizontal or an inclined hydraulic turbine unit is subject to the torsional stress produced by the torque transmitted and the bending stress produced by the weight of the runner, the shaft, and the components that are attached to the shafting system. The design of a shaft for a horizontal unit must be subjected to review by experienced mechanical engineering designers to ensure acceptable stress levels, to account for points of stress concentration, and to determine the overall safety margin of a design. The allowable stress levels in a horizontal shaft must be evaluated after the fatigue properties of the shafting material and the points of stress concentration are taken into consideration. Should the bearing be designed to be water-lubricated, a stainlesssteel sleeve is installed on the shaft to provide a bearing surface for the guide bearing. A stainless-steel sleeve is typically placed on the shaft in the area of a packing box or stuffing box.

Vertical hydraulic turbines are usually provided with one main bearing located in the head cover, as near the runner as practicable. The main turbine bearing is generally a babbitted oil bearing, supplied in halves, or as independently adjustable segmented babbitted shoes. The oil is supplied to the bearing by an independent low-pressure oiling system. Self-lubricated babbitted shell or shoe bearings, containing an oil reservoir, and pumping grooves in the babbitt are sometimes used.

Water-lubricated bearings of wood, lignum vitae, rubber, or special composition are sometimes preferred, particularly for small and medium-size propeller-type turbines where the bearing is located at the bottom of the head cover cone, where the packing box and the oil-lubricated bearing would be inaccessible, and where it would be difficult to avoid water contamination. With the water-lubricated bearing, the packing box is placed above the bearing.

Spiral Case The spiral case must be proportioned so as to cause relatively low friction losses, as well as to prevent eddying which would travel into the runner and affect its efficiency. The cases generally are made of (1) metal: cast steel, cast iron, or steel plate, and (2) concrete. Metal cases are made as complete spirals, customarily with uniform or slightly increasing velocity from the throat to the small end. When a valve is used at the case entrance, the net area at the centerline of the valve should be at least equal to the case entrance area.

Concrete cases may be complete spirals or semispirals, rectangular or oval in cross section. There should be no piers close to the turbine.

Rated load velocities for general practice for cases of various types are shown in Fig. 9.9.10. Higher velocities are sometimes used with the larger units to reduce size and cost.

Fig. 9.9.10 Case velocities.

Stay Ring   That part of the guide apparatus between the spiral case and the wicket gates, containing stationary stay vanes, is called the stay ring. The water is accelerated within this space as it approaches the gates. The number of stay vanes employed is usually made equal to or one-half the number of gates. They are placed at the angle which will cause the least obstruction to the flow.

The stay ring is cast integral with cast cases and is made separately of welded or cast steel for concrete or steel plate spiral cases. It should be a continuous ring to facilitate erection, and very rigid, because it serves as a foundation for the rest of the turbine and generator.

Wicket Gates and Operating Mechanism   Wicket gates control the power and speed of the turbine. The number of gates m ranges from 12 for small units to 28 for large units. The overall dimensions of the turbine decrease as the number of gates increases.

To prevent interference between the gates and the runner buckets, which may cause noise and vibration, the discharge tips of the fully open gates should be kept well away from the inlet edges of the runner buckets of Francis turbines, the radial clearance being large enough to prevent the gate tips from overhanging the curved part of the discharge ring. Gate tip diameter = 1.1 (maximum inlet diameter D_r).

The height and angular movement of the gates increase with the specific speed. The angular movement varies from 15° for low specific speed to 80° for high specific speed.

Most wicket gates are made of cast steel; a few, for higher heads, are made of stainless steel; some, for lower heads, are weldments built from rolled materials and castings.

Each gate connection to the operating ring should be provided with a breaking element to protect the gate and other mechanism in case of an obstruction. Because of the inherent instability of a free gate and its tendency to flutter, it is advisable to install gate-restraining devices to hold the gate if a breaking element fails. This device also avoids potential resonant flutter with the natural frequencies in the system. Each gate should also be provided with stops to prevent it from striking the runner or reversing after the breaking element fails.

The capacity G, ft · lb (J), of the oil pressure cylinder or cylinders (sometimes known as governor capacity) may be computed from the formula

$$G = 18Pn_s^{1/4}/\sqrt{H}   (13Pn_s^{1/4}/\sqrt{H})$$

Some units now provide individual servomotors on each wicket gate. This synchronized system allows the hydraulic control system to dampen a tendency to gate flutter.

One or more vacuum breakers or air valves are installed in the head cover to admit air to the runner or draft tube, to improve efficiency at low gate openings or to alleviate draft tube vortex cavitation. An air valve is also necessary on propeller-type units to break the vacuum under the head cover and to help prevent the runner from "screwing up" in the water when the gates are suddenly closed. The air valve is piped to the outside of the powerhouse above the floodwater level.

Draft Tubes   The draft tube may serve the double purpose of (1) allowing the turbine to be set above tailwater level, without loss of head, to facilitate inspection and maintenance, and (2) regaining, by diffuser action, the major portion of the kinetic energy delivered to it from the runner.

At rated load the velocity at the upstream end of the tube for modern units ranges from 24 to 40 ft/s (7 to 12 m/s), representing from 9 to 25 ft (3 to 8 m) head. As the specific speed is increased and the head reduced, it becomes increasingly more important to have an efficient draft tube. Good practice limits the velocity at the discharge end of the tube to 5 to 10 ft/s (2 to 3 m/s), representing less than 1 ft (0.3 m) velocity head loss.

The elbow type of tube is now used with most vertical turbine installations. With this type the vertical portion begins with a conical section which gradually flattens in the elbow section and then discharges horizontally through substantially rectangular sections to the tailrace. Most of the regain of energy takes place in the vertical portion, very little in the elbow section which is shaped to deliver the water to the horizontal portion so that the regain may be efficiently completed. Figure 9.9.11 shows proportions of a good elbow tube. One or two vertical piers are placed in the horizontal portion of the tube for structural reasons.

Fig. 9.9.11   Elbow draft tube layout.

Most tubes are made of concrete with a steel plate lining extending from the upper end to a point where the velocity has been sufficiently reduced [say 20 ft/s (7 m/s)] to prevent erosion of the concrete. Pier noses are also lined where necessary to prevent erosion and for structural reasons.

IMPULSE TURBINES

Impulse turbines are utilized when the head is too high for the practical use of Francis turbines, which is normally a head exceeding 1,600 ft (500 m). Impulse turbines are also sometimes used for heads below 1,600 ft (500 m) where excessive erosion due to foreign materials in the water presents a problem. The main disadvantage of impulse turbines, especially for low heads, is their low specific speed. In the past, this was overcome on conventional horizontal shaft units by the use of two runners or two jets per runner. In recent years, the vertical shaft multijet impulse turbine (Fig. 9.9.12) has become popular.

The efficiency obtainable from a horizontal shaft impulse turbine is about 90 percent. Field tests on multijet vertical units have shown efficiencies as high as 91.5 percent. The use of multiple jets on the vertical units reduces the percentage loss due to windage of the runner. The unit can be operated with a reduced number of jets at part load, thus increasing part load efficiency. Six jets are about the maximum number that can be used on one runner without jet interference.

Selection of Speed   The first consideration is the selection of suitable n_s. As with reaction turbines, there is a relation which determines approximately the limiting value of specific speed n_s (per jet) for any head. For heads around 1,000 ft (300 m), n_s = 5.0 to 5.5 (19 to 21) gives high efficiency. For heads around 2,000 ft (600 m), maximum efficiency is attained near n_s = 4.0 to 5.0 (15 to 19). Care has to be exercised in selecting an n_s (per jet), especially for the higher heads, so that the proper number of buckets can be used on the wheel disk. The higher the n_s, the fewer the number of buckets required (Fig. 9.9.13) but the smaller the pitch circle. The latter can create stress problems in the attachment of the buckets to the disk. If the resulting speed n is quite low for the power to be developed, the n_s of the unit and consequently the speed n can be increased by increasing the number of runners or the number of jets per runner. The n_s of the unit will then be n_s (per jet) times the square root of the number of jets.

Fig. 9.9.12 Vertical-shaft multijet impulse turbine installation.

Basic Dimensions The pitch diameter D_p of the runner is twice the distance from the center of the runner to the axial centerline of the jet and is determined by the value of the coefficient ϕ selected, $D_p = 1,840\phi\sqrt{H}/n\,(84{,}578\phi\sqrt{H}/n)$. The variation of ϕ with n_s is established by experience and model tests (Fig. 9.9.13).

The quantity of water Q discharged per jet is $550P/(wHe)$ $[101.93P/(wHe)]$, where P is the horsepower (kilowatts) developed by each jet.

The diameter of the jet is $4.85\sqrt{Q}/H^{1/4}$ in $(0.55\sqrt{Q}/H^{1/4}$ m), using a velocity coefficient of 0.97 for the jet.

The **velocity** in the inlet should not exceed $0.12\sqrt{2gH}$ (approximately), and it is considered good practice not to exceed an absolute value of 40 ft/s (12 m/s).

The **valve** employed in the inlet pipe should be of a type which, when open, permits a smooth, uninterrupted flow, free of eddies or obstructions. Modern preference is for rotary sphere valves for this purpose.

The **housing** serves primarily to carry off the discharged water to the tail pit below. On horizontal shaft units, at the place where the runner receives the discharge from the jet, the housing should be about 10 to 12

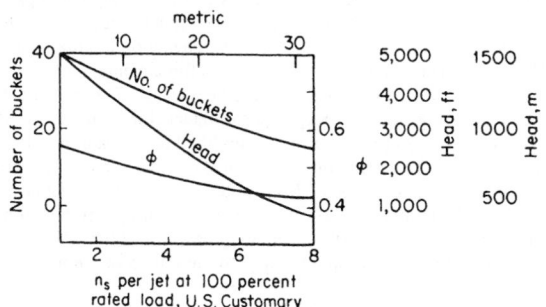

Fig. 9.9.13 Impulse turbine characteristics and proportions.

times the jet diameter. At the place where the runner has been cleared of the discharge, the housing should be as narrow as possible to decrease windage. Suitable baffling should be used to carry the discharge away from the buckets. The housing should be vented adequately near the center of the runner.

For vertical-shaft units, the housing should be of ample size to prevent discharge water from interfering with the buckets. The housing should also be vented adequately near the center of the runner.

The **setting** of the horizontal-shaft unit should be such that the lower edges of the buckets are at least 3 ft (1 m) above maximum tailwater elevation. For vertical shaft units, this distance should be at least 5 ft (1.5 m). Impulse turbine discharge entrains large amounts of air, and unless this is completely replaced, a vacuum will be produced in the housing which will draw the tailwater up and drown out the runner. Thus the roof of the discharge tunnel or passageway, for both horizontal- and vertical-shaft units, should be at least 3 ft (1m) above the maximum operating tailwater elevation to permit the free circulation of air to the runner (Fig. 9.9.12).

In cases where extremely high tailwater occurs for a short period, compressed air can be used to depress the tailwater.

Experience has shown that the ft³/s (m³/s) of air (at the discharge pressure of the compressor) required to maintain the depressed tailwater is about 15 percent of the ft/3/s (m³/s) of water (Q) discharged by the turbine.

Runner The force F exerted by a stream upon a stationary bucket (Fig. 9.9.14) is

$$F = MV(1 - \cos\theta) = wQV(1 - \cos\theta)/g$$

where M = mass per s and θ = angle in deg through which the water is turned relative to the bucket (see section Y-Y, Fig. 9.9.14). If the bucket is moving in the direction of the stream with a velocity u, $F = wQ\,(V - u)$ $(1 - \cos\theta)/g$. The work done by the stream $= Fu = wQn\,(V - u)$ $(1 - \cos\theta)/g$. When θ is greater than 90°, the cosine becomes negative. Thus the closer the angle θ is to 180°, the greater the amount of work done with the same Q and V. However, in order to prevent the water from striking the succeeding bucket, θ should never be 180° (see section Y-Y,

Fig. 9.9.14   Impulse turbine bucket design.

Fig. 9.9.14). Also, since the water discharged from the bucket must have some velocity, the value of u is made somewhat less than $V/2$.

The angle β_1 at the centerline of the bucket should be calculated from the velocity triangle (Fig. 9.9.14). The value of β_1 is greatest as the entrance lip E enters the jet. The angle of the underside of the bucket should be greater than β_1 to avoid pitting.

In terms of the jet diameter which will give maximum guaranteed or rated capacity, the approximate proportions of the buckets are $B = 3d$, $L = 2.6d$, $D = 0.85d$ (Fig. 9.9.14). The contour of the entrance lip E is important, since the manner in which this lip comes in contact with the jet affects the path of the water passing to the preceding bucket. The size of the bucket with relation to the jet diameter also determines the percentage of full load at which maximum efficiency occurs. The larger the bucket, the higher this percentage of load.

The number of buckets on a runner should be such that no water can pass through the buckets without being deflected by the buckets. Figure 9.9.13 indicated approximately the most efficient number.

The **needle nozzle** should be placed as close to the buckets as possible, as the jet tends to lose its compactness of form shortly after emerging from the nozzle. The needle and the seat of the nozzle should be designed for easy replacement and should be of a material highly resistant to erosion. The needle should terminate in a cone of 30 to 45°. The nozzle tip should have a subtended angle of 45 to 60° and a diameter at discharge about 20 percent greater than the calculated diameter of the jet. The maximum diameter of the needle should be about 15 percent greater than the discharge diameter of the nozzle tip. The diameter of the upstream portion of the nozzle should be such that the velocity does not exceed $0.10\sqrt{2gH}$. (See Lowy, Efficiency Analysis of Pelton Wheels, *Trans. ASME,* Aug. 1944, for additional information on the design of impulse turbines.)

Regulation   The inertia of the water flowing through the long penstocks usually employed with impulse turbines prohibits rapid reduction in velocity because of the pressure rise which would occur. Therefore, to minimize the speed rise following a sudden load rejection, it is necessary to reduce the hydraulic power delivered to the runner without changing the flow in the penstock too rapidly. This is usually accomplished by placing a governor controlled jet deflector between the needle nozzle and the runner. The governor moves this deflector rapidly into the jet, cutting off the load. It is not unusual for the deflector to cut off the entire jet in 1½ s. Since the deflector acts on the jet after it leaves the nozzle, there is no change of flow in the penstock;

hence, there is no pressure rise. The governor then moves the needle at a permissible rate (in terms of pressure rise), with simultaneous automatic withdrawal of the deflector. The jet is finally reduced the necessary amount to correspond to the reduced load. The needle must also move slowly in the opening direction for oncoming loads to avoid penstock collapse because of large pressure drops.

Runaway Speed   The runaway speed for impulse turbines ranges from 160 to 190 percent of normal speed, depending upon the specific speed of the runner; the higher the specific speed, the higher the runaway speed.

REVERSIBLE PUMP/TURBINES

In this type of project, surplus low-value energy from either hydro or thermal plants is used to pump water during off-peak periods to an elevated reservoir, where it becomes available for generating high-value peaking energy. While separate pumps and hydraulic turbines of conventional design can and have been used for this purpose, the development of single reversible pump/turbines has made many pumped storage projects economically feasible.

Figure 9.9.15 is a cross section of a reversible pump/turbine, showing that, with the exception of the runner, it is essentially a conventional turbine. The wicket gates must be designed for flow in both directions. A few units have been built without movable wicket gates. The runner is essentially a pump runner modified for optimum performance while generating power. (See Sec. 14 for the design of centrifugal pumps and their characteristics.) Conventional turbine runners, because of their short blades, are not suited for the pumping cavitation requirements.

The reversible pump/turbines have certain fundamental performance characteristics which are inherent in the design. The relationship between pumping and generating performance for a given specific speed is more or less fixed and can be modified only to a minor degree by alterations in the design. Figure 9.9.16 shows the expected generating performance, and Fig. 9.9.17 shows the expected pumping performance based on model tests of a reversible pump/turbine with a specific speed $n_s = 47.3$ (180) generating and $n_s = 2,600$ (50) pumping (see "Centrifugal and Axial Pumps" for definition of pump specific speed).

The best efficiency, with reversible pump/turbines, occurs at a lower speed when generating than when pumping. This can be compensated for by using a generator/motor capable of operating at two speeds with

Fig. 9.9.15 Reversible pump turbine.

a constant frequency. Several units of this type have been built, but since two-speed generator/motors cost considerably more than those built for single speed, a careful study should be made of the advantages to be obtained in operating at two speeds.

Single reversible pump/turbines can be built for any heads up to 2,000 ft (600 m). Beyond this, either multiple-stage reversible units or separate pumps and turbines should be used. Some of the more notable reversible pump/turbine installations are shown in Table 9.9.2.

Runaway Speed The runaway speed for pump/turbines is considerably lower than that for conventional turbines. It ranges from 150 percent of normal for low specific speed runners to 175 percent for high specific speed runners.

Figs. 9.9.16 Generating

Fig. 9.9.17 Pumping.

Figs. 9.9.16 and 9.9.17 Typical performance curves for a pump turbine.

Table 9.9.2   Reversible Pump/Turbine Installations

| | Generating | | | | Pumping | | | | |
| | Power | | Head | | Capacity | | Head | | Year in |
Plant	1,000 hp	MW	ft	m	ft³/s	m³/s	ft	m	operation
Pedreira "A"	5.2	3.9	49	15	690	19.5	49	15	1939
Hiwassee	62	46	190	58	3,900	111	205	62.5	1956
Taum Sauk	220	165	790	241	2,650	75.0	765	233	1963
Cruachan	100	74.6	1,130	343	1,010	28.6	1,175	358	1966
Cabin Creek	166	125	1,190	363	1,375	39.0	1,060	323	1967
Villarino	135	100	1,250	382	1,020	29	1,325	404	1969
Kisenyami	240	180	755	220	3,880	110	645	197	1969
Drakensburg	360	269	1,380	422	1,685	42	1,443	440	1977
Dinorwic	405	302	1,682	513	1,647	46.7	1,771	540	1975
Bajina Basta	422	385	1,968	600	1,520	43	2,132	650	1976
Raccoon Mt.	465	350	940	287	3,850	110	1,000	305	1974
Bath Co.	510	380	1,080	329	4,100	116	1,100	335	1983
Palmiet	335	250	988	301	2,012	57	1,000	305	1988
La Muela	284	212	1,725	525	1,165	33	1,637	499	1990
BadCreek	482	360	1,200	367	3,531	100	1,240	378	1991
Mingtan	368	275	1,322	403	2,896	82	1,345	410	1992
Shisanling	273	204	1,411	430	1,335	37.8	1,588	484	1996
Guangzhou II	410	306	1,673	509	2,030	57.5	1,742	531	2000
Goldisthal	360	269	990	302	2,825	80	1,004	306	2003
Taian	335	250	1,344	410	3,397	96.2	820	250	2005

MODEL TESTS

Model tests serve several purposes. They are primarily used to check turbine runner, wicket gate, draft tube, casing, and (sometimes) inlet works designs for optimum performance. Correctly interpreted, they may also be used as a reliable indication of the performance of the units in the field. In many cases, purchasers specify the performance of a homologous model test, which is used as an acceptance test of the unit in lieu of field tests. In such cases, the field conditions, particularly the casing and draft tube, must be reproduced faithfully. Model tests should be run in accordance with the International Code for Model Acceptance Tests of Hydraulic Turbines, Publication 193; International Code for Model Acceptance Tests of Storage Pumps, Publication 497 of the International Electrotechnical Commission (IEC); and International Standard Publication, IEC 995, that supplements IEC 193 and 497.

Figure 9.9.18 shows typical model test results of a reaction-type turbine, in which the power P at 1 ft (0.3 m) head and the efficiency are each plotted against the speed n at 1 ft (0.3 m) head, for each of several gate openings.

Fig. 9.9.18 Model test curves for a Francis runner ($D_r = D_{th}$ = 16.2 in (0.41 m)) and n_s = about 65 (250).

Laws of Proportionality for Homologous Turbines The laws of proportionality shown previously are used to calculate from model tests the power, speed, and discharges of a homologous turbine of different diameter under a different head. The field unit will have a somewhat higher efficiency and power owing to proportionally smaller frictional and bearing losses. Expected field efficiency is customarily computed from the model efficiency by the Moody formula

$$c' = 1 - \frac{2}{3}\left[(1 - e)\left(\frac{D}{D'}\right)^{1/5}\right]$$

in which the prime letters refer to the field installation. The step-up efficiency is usually computed for the point of best efficiency only, and the corresponding differential is applied as a constant value from, say, half load to full load.

CAVITATION

Cavitation occurs when the pressure at any point in the flowing water drops below the vapor pressure of the water.

The relationship which produces cavitation is between vapor pressure, barometric pressure, setting of the runner with respect to tailwater, and net effective head on the turbine and is expressed by the **Thoma cavitation coefficient** $\sigma = (H_b - H_v - H_S)/H$, where H_b = barometric head, ft (m) of water; H_v = vapor pressure of water, abs; H_s = elevation, ft (m) of the runner above tailwater, measured at the centerline of the distributor of a Francis runner and at the centerline of the blades of a propeller runner (if the runner is submerged, H_s becomes negative); and H = total or net effective head, ft (m), on the turbine. The setting of the runner depends upon the value of σ, which varies with specific speed n_s of the runner and the individual characteristics of a particular runner design. In practice, the model of the proposed runner is first tested with relatively high back pressure (H, small or negative). Then the back pressure is reduced in increments until the breaking point, as indicated by a drop in power, efficiency, and discharge, is reached. This breaking point is designated as the critical σ and will vary with gate opening and speed and, on propeller turbines, with blade angle. Consequently, σ must be determined for a range of limiting conditions.

In the absence of cavitation tests, the value of σ should not be lower than $\sigma = n_s^{3/2}/2{,}000\,(n_s^{3/2}/15{,}000)$ for Francis and propeller runners and $\sigma = n_s^2/25{,}000\,(n_s^2/350{,}000)$ for adjustable-blade propeller runners.

The value of σ at which a plant operates, depending largely upon the setting of the runner with respect to tailwater, is called the **plant** σ. To avoid excessive cavitation, the plant σ should exceed the critical σ. The greater this margin, the less possibility of cavitation during operation. For a general discussion of cavitation phenomena, see Knapp, Recent Investigations of the Mechanics of Cavitation and Cavitation Damage, *Trans. ASME*, Oct. 1955. Laboratory tests and experience have shown that materials having a high resistance to cavitation erosion (pitting) and suitable for use in hydraulic turbines are the stainless steels and aluminum bronzes, especially when used as welding overlays. (See Rheingans, "Resistance of Various Materials to Cavitation Damage," ASME Report of 1956 Cavitation Symposium.)

SPEED REGULATION

(See also Sec. 16.)

Regulation is accomplished by changing the flow of water to the turbine. The flow is controlled by the wicket gates of reaction turbines and by the needle valve or jet deflector of impulse turbines. The governor moves the gates or needle in response to speed changes resulting from load or head changes.

A schematic diagram of a classic governor is shown in Fig. 9.9.19. The parts consist of a speed-responsive device, a power element that

Fig. 9.9.19 Schematic diagram of governor.

changes the gates or needle position, and a follow-up or compensating device that prevents hunting.

The **speed-responsive device** originally was a pair of spring-loaded fly-balls mounted directly on the turbine shaft, or driven from the shaft by belt or gears, or driven by an electric motor that receives its power either from the bus line or from an independent generator driven from the main turbine generator shaft.

The **power element** consists of oil-operated power cylinders or servo-motors which operate the turbine gates or needle. Oil pumps and a pressure tank or accumulator maintain a supply of oil. A valve operated by the speed sensor controls the flow of oil to the servomotors or acts as a pilot valve controlling a larger relay valve, which in turn controls the oil to the servomotors. The pump capacity is usually 3 servomotor volumes per minute. The capacity of the pressure tank is generally made 20 times the servomotor volume, allowing for 8 volumes of oil and 12 volumes of air. The velocity of oil in the pipelines is kept below 15 ft/s (5 m/s). When using individual servomotors on each gate, these values are usually increased 50 percent to balance the usual damping effect of the common massive two-servomotor system.

The **follow-up** or **compensating device** connects the power piston of the servomotor to the control valve, usually through a dashpot, and causes the motion of gates or needle to stop when they have moved sufficiently to compensate for the load change.

The time for a full stroke or traverse of the governor is controlled by the rate of flow of oil to the servomotors; most governors have provisions for varying this time. The gate opening changes at a uniform rate over the major portion of the stroke and at a somewhat slower rate at the ends of the stroke. The governor *dead time,* or the elapsed time from the initial speed change to the first movement of the gates, is usually less than 0.2 s.

Electric Governor   In recent years, the functional requirements placed upon the hydraulic-turbine governor have increased to the point where electrical control of the hydraulic turbine is attractive in view of the simplicity with which electrical signals can be manipulated. The basic elements of an electric-hydraulic governing system are (1) permanent magnet generator (PMG) or equivalent for measurement of turbine speed and the means for transmittal of such speed signals to the electrical portion of the governor; (2) an electric circuit sensitive to speed variations about some adjustable reference point; (3) amplifying circuits to convert speed reference changes, speed error signals, and auxiliary signals into a useful electric current; (4) an electrohydraulic transducer to transform the electric current into a hydraulic output signal; (5) hydraulic amplifying equipment to deliver suitable power and the desired signal to the gate servomotors as a function of the output of the electro-hydraulic transducer; (6) power supplies for the electric and hydraulic portions of the control. For further particulars of the electric governor, see Leum, Electric Governors for Hydro Turbines, ASME Paper 62-WA165.

Electronic Governors   With the advent of minicomputers, microprocessors, and programmable controllers, many of the traditional mechanical functions in the hydraulic turbine governor have been replaced electronically. While the basic elements of the governor remain the same, the use of electronics in the feedback sensing and control circuits has greatly reduced the size and mechanical complexity of these units. The basic elements of the modern electronic governoring system are: (1) Inductive pickups or permanent magnetic generators or equivalent to measure the turbine speed and to transmit the information to the processor. (2) Electronic circuitry to establish the speed-control function of the governor (integrated circuits or digital programming). Basically the circuitry consists of a proportional, an integrator, and a derivative function. (3) A servo valve or equivalent to translate the electronic signal from the processor to a hydraulic signal. (4) Where required, a means to hydraulically amplify the output from the servo valve so that a suitable power level can be delivered to the gate and blade servomotors. (5) Auxiliary equipment as required in the form of power supplies, etc. to support the system.

Speed Regulation Requirements   Usually, a sufficient measure of the regulation provided is the maximum speed rise resulting from sudden rejection of full load, as from the breaker tripping. A maximum speed rise of 30 percent of normal speed for this condition is a common limitation.

Speed Rise Following Load Reduction   For sudden load reductions, the approximate speed rise is

$$n_t/n = [1 + 1,620,000T_t P_t (1 + h/H)^{3/2}/WR^2 n^2]^{1/2} \quad \text{(USCS)}$$
$$= [1 + 365,000T_t P_t (1 + h/H)^{3/2}/GD^2 n^2]^{1/2} \quad \text{(SI)}$$

where n_t is the speed, rpm, at the end of time T_t; n is the speed, rpm, before the load decrease; T_t is the time interval, seconds, for the governor to adjust the flow to the new load; P_t is the reduction in load, hp (kW); h is the head rise caused by the retardation of the flow, ft (m); H is the net effective head before the load change, ft (m); WR^2 is the product of the revolving parts weight, lb, and the square of their radius of gyration, ft; GD^2 is the product of the revolving parts weight, kg, and the square of their diameter of gyration, m. For approximate values of h, see below. Very rapid gate closure produces a reduction of pressure in the draft tube and the possibility of breaking the water column, with subsequent violent resurge which may damage the turbines.

Speed Drop Following Load Increase   For sudden load increases, the approximate speed drop is

$$n_t/n = [1 - 1,620,000T_t P_t/WR^2 n^2(1 - h/H)^{3/2}]^{1/2} \quad \text{(USCS)}$$
$$= [1 - 365,000T_t P_t/GD^2 n^2(1 - h/H)^{3/2}]^{1/2} \quad \text{(SI)}$$

where P_t is the actual load increase and h is the head drop caused by the increase of the flow. If the speed drop is to be determined for a given increase in gate opening, the governor time T_t for making this increase and the normal change in load for the change in gate opening, under constant head H, can be used in the following formula:

$$n_t/n = [1 - 1,620,000T_t P_t (1 - h/H)^{3/2}/WR^2 n^2]^{1/2} \quad \text{(USES)}$$
$$= [1 - 365,000T_t P_t (1 - h/H)^{3/2}/GD^2 n^2]^{1/2} \quad \text{(SI)}$$

The actual change in load, however, will be $P_t (1 - h/H)^{3/2}$.

For derivation of the above speed variation formulas and for a more accurate determination, see Strowger and Kerr, Speed Changes of Hydraulic Turbines for Sudden Changes of Load, *Trans. ASME,* 1926; and Rich, "Hydraulic Transients," McGraw-Hill.

Water Hammer in Penstocks   If a gate movement is considered as a series of instantaneous movements with a very small interval between each movement, the pressure variation in the penstock following the gate movement will be the effect of a series of pressure waves, each caused by one of the instantaneous small gate movements. For a steel penstock, the velocity of the pressure wave $a = 4,660/\sqrt{1 + d/100t}$ $(1,420/\sqrt{1 + d/100t})$, where d is the penstock diameter, in (m), and t is the penstock wall thickness, in (m). The pressure change at any point along the penstock at any time after the start of the gate movement may be calculated by summing up the effect of the individual pressure waves. See "Symposium on Water Hammer," ASME, 1933; and Parmakian, "Water Hammer Analysis," Dover.

Approximate formulas (De Sparre) for the increase in pressure h, ft (m), following gate closure, are given below. They are quite accurate for pressure rises not exceeding 50 percent of the initial pressure, which includes most practical cases.

$$h = aV/g \qquad\qquad \text{for } K < 1, N < 1$$
$$h = aV/\{g[N + K(N - 1)]\} \qquad \text{for } K < 1, N > 1$$
$$h = aV/[g(2N - K)] \qquad\qquad \text{for } K > 1, N > 1$$

where $K = aV/(2gH)$; $N = aT/(2L)$. V and H are the penstock velocity, ft/s (m/s), and head, ft (m), prior to closure; L is the penstock length, ft (m), and T is the time of gate closure. For full load rejection, T may be taken as 85 percent of the total gate traversing time to allow for nonuniform gate motion.

For pressure drop following a complete gate opening, the following formula (S. Logan Kerr) may be used with T not less than $2L/a$:

$$h = \frac{aV}{g}\left(\frac{-K + \sqrt{K^2 + N^2}}{N^2}\right) = \text{pressure, drop, ft (m)}$$

Pressure variations exceeding 40 percent rise and about 25 percent drop should be avoided. When the control, directly by the governor, causes undesirable pressure variations, a surge tank, a pressure regulator, or a jet deflector may be used. A **surge tank** is a standpipe with an atmospheric tank, attached to the penstock as close as possible to the casing inlet. The tank provides a reservoir and expansion chamber for the water demand or the water rejection following sudden gate movements, so that sudden accelerations or decelerations of the flow in the penstock are avoided.

Pressure regulators may be of either the water-wasting or water-saving type. The **water-wasting type** is a synchronous bypass, generally attached to the turbine casing. It is operated directly from the governor, or the gate mechanism of the turbine, and wastes such an amount as to keep the total water discharge equal at all times to the full-load discharge of the turbine. The bypass is a needle nozzle or a mushroom-shaped disk valve or a cone valve which opens and is partly balanced hydraulically by a piston under pipeline pressure. The **water-saving type** permits the regulator to open upon rapid closure of the turbine gates, and then close slowly, so that the total water discharge is gradually reduced and finally limited to that through the turbine, adjusted for the new load.

AUXILIARIES

Valves and head gates are provided for shutting off the water to each turbine for safety, for ease of maintenance, and to reduce water leakage losses. Motor-operated steel head gates are generally used for low- and medium-head plants with concrete scroll cases, although a few still use stop logs. The butterfly type of valve placed close to the turbine casing is suitable for medium- and high-head units with metal casings of circular inlet diameter. Butterfly valves of 8-ft (2.5 m) diameter for 1,000-ft (305 m) head and 27-ft (8.2 m) diameter for 100-ft (30 m) head have been built. The new flow through valve or biplane valve is becoming more popular for this application. In recent years, rotary sphere valves have replaced gate valves for high heads and where the loss through the butterfly valve is excessive because of the obstruction to flow by the valve disk.

COMPUTER-AIDED DESIGN

The continued development, refinement, and use of proven high-technology computer-aided design and analysis tools help create new generations of designs. Design objectives today are to produce equipment which has characteristics of reliability, easy servicing, operational flexibility, and high performance. The success of this approach is made possible by integrating design history and field experience into hydraulic, mechanical, and operational computer design and analysis systems.

TURBINE TESTS

Field testing of hydraulic turbines to determine the absolute efficiency and output involves careful and accurate measurement of the power available in the water supplied to the turbine [water hp (kW)] and the turbine output [developed hp (kW)] e = developed hp (kW)/water hp (kW) = $550P/(wQH)$ [$101.93P/(QHw)$]. The tests should be conducted in accordance with the International Code for Field Acceptance Tests of Hydraulic Turbines and/or Storage Pumps, IEC Publications 41 and 198.

Because of the difficulties and costs involved in making accurate measurements of horsepower, net head, and discharge in the field, there has been a trend in recent years to dispense with the field test, especially where a laboratory test on a homologous model turbine is available. Instead, an index test is made on the unit in the field, which measures the turbine output and relative discharge under various conditions. Index tests should be conducted in accordance with the International Code for Field Acceptance Tests of Hydraulic Turbines, IEC Publication 41.

Materials Handling

BY

VINCENT M. ALTAMURO *President, VMA Inc., Toms River, NJ*
ERNST K. H. MARBURG *Vice President, Total Quality and Standards, Retired, Columbus*
McKinnon Corp.

10.1 MATERIAL HOLDING, FEEDING, AND METERING
by Vincent M. Altamuro

FACTORS AND CONSIDERATIONS

The movement of material from the place where it is to the place where it is needed can be time consuming, expensive, and troublesome. The material can be damaged or lost in transit. It is important, therefore, that it be done smoothly, directly, with the proper equipment and so that it is under control at all times. The several factors that must be known when a material handling system is designed include:

1. Form of material at point of origin, e.g., liquid, granular, sheets, etc.

2. Characteristics of the material, e.g., fragile, radioactive, oily, etc.

3. Original position of the material, e.g., under the earth, in cartons, etc.

4. Flow demands, e.g., amount needed, continuous or intermittent, timing, etc.

5. Final position, where material is needed, e.g., distance, elevation differences, etc.

6. In-transit conditions, e.g., transocean, jungle, city traffic, inplant, etc., and any hazards, perils, special events or situations that could occur during transit

7. Handling equipment available, e.g., devices, prices, reliability, maintenance needs, etc.

8. Form and position needed at destination

9. Integration with other equipment and systems

10. Degree of control required

Other factors to be considered include:
1. Labor skills available
2. Degree of mechanization desired
3. Capital available
4. Return on investment
5. Expected life of installation

Since material handling adds expense, but not value, it should be reduced as much as possible with respect to time, distance, frequency, and overall cost. A straight steady flow of material is usually most efficient. The use of mechanical equipment rather than humans is usually, but not always, desirable—depending upon the duration of the job, frequency of trips, load factors, and characteristics of the material. When equipment is used, maximizing its utilization, using the correct equipment, proper maintenance, and safety are important considerations.

The proper material handling equipment can be selected by analyzing the material, the route it must take from point of origin to destination, and knowing what equipment is available.

FLOW ANALYSIS

The path materials take through a plant or process can be shown on a flowchart and floor plan. See Figs. 10.1.1 and 10.1.2. These show not only distances and destinations, but also time required, loads, and other information to suggest which type of handling equipment will be needed.

Description of operation	Hdlg.	Move	Insp.	Temp. stor.	Perm. stor.	People req.	Dist. moved	Time req.	Equipment used
Unload from truck to receiving dock	○	○	□	△	△	1	10 ft	2 min. 10s	Hand truck
Inspect and tag	○	○	□	△	△	1		30s	
Move to raw material storage	○	○	□	△	△	1	40 ft	1 min. 10s	Hand truck
Raw material storage	○	○	□	△	△				
Move to machine	○	○	□	△	△	1	120 ft.	1 min. 30s	Hand truck
Set in machine	○	○	□	△	△	1		10 min.	Manual handling
Place finished stock on pallet	○	○	□	△	△	1		12 min 10s	Manual handling
Move to inspection and test	○	○	□	△	△	1	110 ft	1 min.	Pallet truck
Move to packing	○	○	□	△	△	1	60 ft	1 min. 25s	Pallet truck
Move to finished goods storage	○	○	□	△	△	1	70 ft	1 min. 10s	Pallet truck
Finished goods storage	○	○	□	△	△				
Move to shipping dock	○	○	□	△	△	1	40 ft	1 min. 10s	Pallet truck
Load on trucks	○	○	□	△	△	1	10 ft	2 min. 30s	Manual handling
Total manpower travel and time						11	460 ft	34 min. 45s	

Fig. 10.1.1 Material flowchart.

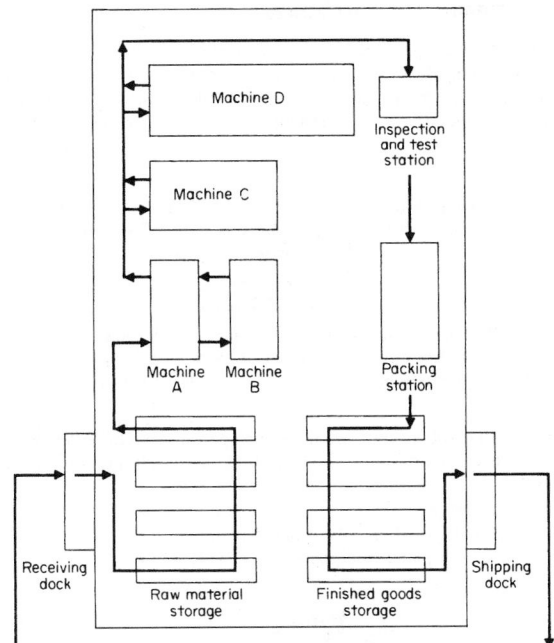

Fig. 10.1.2   Material flow floor plan.

CLASSIFICATIONS

Material handling may be divided into classifications or actions related to the stage of the process. For some materials the first stage is its existence in its natural state in nature, e.g., in the earth or ocean. In other cases the first stage would be its receipt at the factory's receiving dock. Subsequent material handling actions might be:

1. Holding, feeding, metering
2. Transferring, positioning
3. Lifting, hoisting, elevating
4. Dragging, pulling, pushing
5. Loading, carrying, excavating
6. Conveyor moving and handling
7. Automatic guided vehicle transporting
8. Robot manipulating
9. Identifying, sorting, controlling
10. Storing, warehousing
11. Order picking, packing
12. Loading, shipping

FORMS OF MATERIAL

Material may be in solid, liquid, or gaseous form. Solid material may be in end-product shapes or in intermediate forms of slabs, sheets, bars, wire, etc., which may be rigid, soft, or amorphous. Further, each item may be unique or all may be the same, allowing for indiscriminate selection and handling. Items may be segregated according to some characteristic or all may be commingled in a mixed container. They might have to be handled as discrete items, or portions of a bulk supply may be handled. If bulk, as coal or powder, their size, size distribution, flowability, angle of repose, abrasiveness, contaminants, weight, etc., must be known. In short, before any material can be held, fed, sorted, and transported, all important characteristics of its form must be known.

HOLDING DEVICES

Devices used to hold material should be selected on the basis of the form of the material, what operations, if any, are to be done to it while

being held (e.g., heated, mixed, macerated, dyed, etc.), and how it is to be fed out of the device. Some such devices are tanks, bins, hoppers, reels, spools, bobbins, trays, racks, magazines, tubes, and totes. Such devices may be stationary—either with or without moving parts—or moving, e.g., vibrating, rotating, oscillating, or jogging.

The material holding system is the first stage of a material handling system. Efforts should be made to assure that the holding devices never become empty, lest the entire system's flow stops. They can be refilled manually or automatically and have signals to call for refills at carefully calculated trigger points.

Fig. 10.1.3   Reciprocating feeder. Either hopper or tube reciprocates to obtain, orient, and feed items.

MATERIAL FEEDING AND METERING MODES

Material can be fed from its holding devices to processing areas in several modes:

1. Randomly or by selection of individual items
2. Pretested or untested
3. Linked together or separated
4. Oriented or in any orientation
5. Continuous flow or interruptible flow
6. Specified feed rate or any rate of flow
7. Live (powered) or unpowered action
8. With specific spaces between items or not

Fig. 10.1.4   Centerboard feeder. Board, with edge shaped to match part, dips and then rises through parts to select part, which then slides into feed groove.

FEEDING AND METERING DEVICES

The feeding and metering devices selected depend on the above factors and the properties of the material. Some, but not all, possible devices include:

1. Pumps, dispensers, applicators
2. Feedscrews, reciprocating rams, pistons
3. Oscillating blades, sweeps, arms
4. Belts, chains, rotaries, turntables
5. Vibratory bowl feeders, shakers
6. Escapements, mechanisms, pick-and-place devices

Fig. 10.1.5 Rotary centerboard feeder. Blade, with edge shaped to match part, rotates through parts, catching those oriented correctly. Parts slide on contour of blade to exit onto feeder rail.

Figures 10.1.3 to 10.1.8 illustrate some possible feeding and metering devices.

Fig. 10.1.6 Escapement feeder. One of several variations. Escapement is weighted or spring-loaded to stop part from leaving track until it is snared by the work carrier. Slope of track feeds next part into position.

TRANSFERRING AND POSITIONING

There is frequently a need to transfer material from one machine to another, to reposition it within a machine, or to move it a short distance. This can be accomplished through the use of ceiling-, wall-, column-, or floor-mounted hoists, mechanical-advantage linkages, powered

Fig. 10.1.7 Spring blade escapement. Advancing work carrier strips off one part. Remainder of parts slide down feed track. Work carrier may be another part of the product being assembled, resulting in an automated assembly.

manipulators, robots, airflow units, or special devices. Also turrets, dials, indexers, carousels, or conveyors can be used; their motion and action can be:

1. In-line (straight line)
   a. Plain (without pallets or platens)
   b. Pallet or platen equipped

Fig. 10.1.8 Lead screw feeder. Parts stack up behind the screw. Rotation of screw feeds parts with desired separation and timing.

2. Rotary (dial, turret, or table)
   a. Horizontal
   b. Vertical
3. Shuttle table
4. Trunnion
5. Indexing or continuous motion
6. Synchronous or nonsynchronous movement
7. Single- or multistation

10.2　LIFTING, HOISTING, AND ELEVATING
by Ernst K. H. Marburg and Associates

CHAIN
by Joseph S. Dorson and Ernst K. H. Marburg
Columbus McKinnon Corporation

Chain, because of its energy absorption properties, flexibility, and ability to follow contours, is a versatile medium for lifting, towing, pulling, and securing. The American Society for Testing and Materials (ASTM) and National Association of Chain Manufacturers (NACM) have divided chain into two groups: welded and weldless. Welded chain is further categorized into grades by strength and used for many high-strength industrial applications, while weldless chain is used for light-duty low-strength industrial and commercial applications. The grade designations recognized by ASTM and NACM for welded steel chain are 30, 43, 70, 80, and 100. These grade numbers relate the mean stress in newtons per square millimeter (N/mm²) at the minimum breaking load. As an example, the mean stress of Grade 30 chain at its minimum breaking load is 300 N/mm² while that for Grades 100, 80, 70, and 43 is 1000, 800, 700 and 430 N/mm², respectively.

Graded welded steel chain is further divided by material and use. Grades 30, 43, and 70 are welded carbon-steel chains intended for a variety of uses in the trucking, construction, agricultural, and lumber industries, while Grade 80 is welded alloy-steel chain used primarily for overhead lifting (sling) applications.

As a matter of convenience when discussing graded steel chains, chain designers have developed an expression that contains a constant for each given grade of chain regardless of the material size. This expression is $M = N \times$ (diam.)$^2 \times 2000$ where M is the chain minimum ultimate strength in pounds; N is a constant: 114 for Grade 100, 91 for Grade 80, 80 for Grade 70, 49 for Grade 43, and 34 for Grade 30. Note that N has units of lb/in² and diam. is the actual material stock diameter in inches.

Note that the foregoing expression can also be stated as chain ultimate strength $= N \times$ (diam.)$^2 \times$ (short tons) where short tons $= 2,000$. Therefore, knowing the value of the constant N and the stock diameter of the chain, the minimum ultimate strength can be determined.

As an example, the ultimate strength of ½-in. Grade 80 chain can be verified as follows by using the information contained in Table 10.2.1:

$$M = (91 \text{ lb/in}^2)(0.512 \text{ in})^2 \times 2000 = 47,700 \text{ lb (214 kN)}$$

Table 10.2.1   Grade 80 and Grade 100 Alloy Chain Specifications
Note: For overhead lifting applications, only alloy chain should be used.

Nominal chain size		Material diameter		Working load limit (max.)		Proof test* (min.)		Minimum breaking force*		Inside length (max.)		Inside width range	
in	mm	in	mm	lb	kg	lb	kN	lb	kN	in	mm	in	mm
Grade 80													
7/32	5.5	0.217	5.5	2,100	970	4,200	19.0	8,400	38.0	0.69	17.6	0.281–0.325	7.14–8.25
9/32	7.0	0.276	7.0	3,500	1,570	7,000	30.8	14,000	61.6	0.90	22.9	0.375–0.430	9.53–10.92
5/16	8.0	0.315	8.0	4,500	2,000	9,000	40.3	18,000	80.6	1.04	26.4	0.430–0.500	10.92–12.70
3/8	10.0	0.394	10.0	7,100	3,200	14,200	63.0	28,400	126.0	1.26	32.0	0.512–0.600	13.00–15.20
1/2	13.0	0.512	13.0	12,000	5,400	24,000	107.0	48,000	214.0	1.64	41.6	0.688–0.768	17.48–19.50
5/8	16.0	0.630	16.0	18,100	8,200	36,200	161.0	72,400	322.0	2.02	51.2	0.812–0.945	20.63–24.00
3/4	20.0	0.787	20.0	28,300	12,800	56,600	252.0	113,200	504.0	2.52	64.0	0.984–1.180	25.00–30.00
7/8	22.0	0.866	22.0	34,200	15,500	68,400	305.0	136,800	610.0	2.77	70.4	1.080–1.300	27.50–33.00
1	26.0	1.020	26.0	47,700	21,600	95,400	425.0	190,800	850.0	3.28	83.2	1.280–1.540	32.50–39.00
1¼	32.0	1.260	32.0	72,300	32,800	144,600	644.0	289,200	1,288.0	4.03	102.4	1.580–1.890	40.00–48.00
Grade 100													
7/32	5.5	0.217	5.5	2,700	1,220	5,400	23.8	10,800	47.6	0.69	17.6	0.281–0.325	7.14–8.25
9/32	7.0	0.276	7.0	4,300	1,950	8,600	38.5	17,200	77.0	0.90	22.9	0.375–0.430	9.53–10.92
5/16	8.0	0.315	8.0	5,700	2,600	11,400	51.0	22,800	102.0	1.04	26.4	0.430–0.500	10.92–12.70
3/8	10.0	0.394	10.0	8,800	4,000	17,600	79.0	35,200	158.0	1.26	32.0	0.512–0.600	13.00–15.20
1/2	13.0	0.512	13.0	15,000	6,800	30,000	134.0	60,000	268.0	1.64	41.6	0.688–0.768	17.48–19.50
5/8	16.0	0.630	16.0	22,600	10,300	45,200	201.0	90,400	402.0	2.02	51.2	0.812–0.945	20.63–24.00
3/4	20.0	0.787	20.0	35,300	16,000	70,600	315.0	141,200	630.0	2.52	64.0	0.984–1.180	25.00–30.00
7/8	22.0	0.866	22.0	42,700	19,400	85,400	381.0	170,800	762.0	2.77	70.4	1.080–1.300	27.50–33.00
1	26.0	1.020	26.0	59,700	27,100	119,400	531.0	238,800	1,062.0	3.28	83.2	1.280–1.540	32.50–39.00
1¼	32.0	1.260	32.0	90,400	41,000	180,800	805.0	361,600	1,610.0	4.03	102.4	1.580–1.890	40.00–48.00

*The proof test and minimum breaking force loads *shall not* be used as criteria for service or design purposes.
SOURCE: National Association of Chain Manufacturers; copyright material reprinted with permission.

While graded steel chains are not classified as calibrated chain, since links are manufactured to accepted commercial tolerances, they are sometimes used for power transmission in pocket wheel applications. In most such applications, however, power transmission chain is special in that link dimensions are held to very close tolerances and the chain is processed to achieve special wear and strength properties. More in-depth discussions of chain, chain fittings, power transmission, and special chain follow in this section.

Grade 80 and Grade 100 Welded Alloy-Steel Chain (Sling Chain)

The ASTM and NACM specifications for alloy-steel chain (sling chain) have evolved over a period of many years dating back to the mid-1950s. The load and strength requirements for slings and sling chain have unified and are currently specified at Grade 80 and Grade 100 levels in the ASME standard for slings, and the NACM and ASTM specifications for alloy-steel chain. Table 10.2.1 relates the ASTM and NACM Grade 80 and Grade 100 specifications. As noted, the working load limit or maximum load to be applied during use is approximately 25 percent of the minimum breaking force. Because the endurance limit is approximately 18 percent of the minimum breaking strength for these grades, applications that call for a high number of load cycles near the working load limit should be reconsidered to maintain the load within the endurance limit. As an alternative, a larger size chain with a higher working load limit could be applied.

Alloy-steel chain manufactured to the Grade 80 specification varies in hardness among producers from about 360 to 400 Brinell. Similarly, Grade 100 chain varies among producers from about 400 to 440 Brinell. The hardness range for Grade 80 chain has narrowed, on average, in recent years from a wider range of 250 to 450 Brinell that was common in the 1960s. The current values for Grade 80 equate to a material tensile strength of about 175,000 lb/in^2 or an average chain tensile strength of approximately 116,000 lb/in^2 (800 N/mm^2). Thus, the tensile strength expressed for Grade 80 chain provides a breaking strength of $91 \times (\text{diam})^2$ short tons. Similarly, the material tensile strength for Grade 100 is about 215,000 lb/in^2 for an average chain tensile strength of approximately 145,000 lb/in^2 (1000 N/mm^2). Grade 100 chain for the tensile strength expressed thus has a breaking strength of $114 \times (\text{diam})^2$ short tons. Note that the approximate conversion from USCS units to SI units is $1.0 \ (\text{diam})^2$ short tons $= 8.78$ N/mm^2. As an example, 91 $(\text{diam})^2$ short tons $= 800$ N/mm^2.

As is the case with chain strength, chain dimensions have become standardized in the interest of interchangeability of chain and chain fittings. Thus, the dimensions presented in Table 10.2.1 are intended to assure interchangeability of products between manufacturers.

Present ASTM and NACM chain specifications call for a minimum elongation of 20 percent at rupture. An original value of 15 percent was established in 1924 as an attempt to guarantee against brittle fracture and to give visible warning of overloading. For low-hardness carbon steel chains (under 200 Brinell) in use prior to the advent of hardened alloy steel, the requirement for sling chains to have a 15 percent minimum elongation at rupture provided some usefulness as an overload warning signal. However, since high-quality Grade 80 and Grade 100 alloy-steel chain of 360 Brinell and higher has replaced carbon steel chain in high-strength applications, elongation as an indicator of overload is not so apparent. Hardened steel chain achieves about 50 percent of its total elongation during the final 10 percent of loading just prior to rupture. For this reason, only about 50 percent of the total elongation is useful as a visual indicator of overloading. Thus, it is increasingly important that sling chain and slings of optimum strength, toughness, and durability be properly selected, applied, and maintained as advised in applicable standards.

Total deformation (plastic plus elastic) at fracture is important as one of the determinants of energy absorption capability and therefore of impact resistance. However, its companion determinant—breaking strength—is equally important. The only way to measure their composite effect is by impact tests carried out on actual chain samples—not by tests on Charpy or other prepared specimens of material. Special impact testers equipped with high-speed measuring and integration devices have been developed for this purpose.

General-Purpose Welded Carbon-Steel Chains

Grade 30 (proof coil) chain is made from low-carbon steel containing about 0.08 percent carbon. It has a minimum breaking strength of approximately $34 \times (\text{diam})^2$ short tons at 125 Brinell, while its working

Table 10.2.2   Grade 30 Proof Coil Chain Specifications

Note: Not to be used in overhead lifting applications.

Nominal chain size		Material diameter		Working load limit (max.)		Proof test* (min.)		Minimum breaking force*		Inside length (max.)		Inside width (min.)	
in	mm	in	mm	lb	kg	lb	kN	lb	kN	in	mm	in	mm
⅛	4.0	0.156	4.0	400	180	800	3.6	1,600	7.2	0.94	23.9	0.25	6.4
³⁄₁₆	5.5	0.217	5.5	800	365	1,600	7.2	3,200	14.4	0.98	24.8	0.30	7.7
¼	7.0	0.276	7.0	1,300	580	2,600	11.6	5,200	23.2	1.24	31.5	0.38	9.8
⁵⁄₁₆	8.0	0.331	8.4	1,900	860	3,800	16.9	7,600	33.8	1.29	32.8	0.44	11.2
⅜	10.0	0.394	10.0	2,650	1,200	5,300	23.6	10,600	47.2	1.38	35.0	0.55	14.0
⁷⁄₁₆	11.9	0.488	11.9	3,700	1,680	7,400	32.9	14,800	65.8	1.64	41.6	0.65	16.6
½	13.0	0.512	13.0	4,500	2,030	9,000	40.0	18,000	80.0	1.79	45.5	0.72	18.2
⅝	16.0	0.630	16.0	6,900	3,130	13,800	61.3	27,600	122.6	2.20	56.0	0.79	20.0
¾	20.0	0.787	20.0	10,600	4,800	21,200	94.3	42,400	188.6	2.76	70.0	0.98	25.0
⅞	22.0	0.866	22.0	12,800	5,810	25,600	114.1	51,200	228.2	3.03	77.0	1.08	27.5
1	26.0	1.020	26.0	17,900	8,140	35,800	159.1	71,600	318.2	3.58	90.9	1.25	31.7

*The proof test and minimum breaking force loads *shall not* be used as criteria for service or design purposes.
SOURCE: National Association of Chain Manufacturers; copyright material reprinted with permission.

Table 10.2.3   Grade 43 High-Test Chain Specifications

Note: Not to be used in overhead lifting applications.

Nominal chain size		Material diameter		Working load limit (max.)		Proof test* (min.)		Minimum breaking force*		Inside length (max.)		Inside width (min.)	
in	mm	in	mm	lb	kg	lb	kN	lb	kN	in	mm	in	mm
¼	7.0	0.276	7.0	2,600	1,180	3,900	17.3	7,800	34.6	1.24	31.5	0.38	9.8
⁵⁄₁₆	8.7	0.343	8.7	3,900	1,770	5,850	26.0	11,700	52.0	1.29	32.8	0.44	11.2
⅜	10.0	0.406	10.3	5,400	2,450	8,100	36.0	16,200	72.0	1.38	35.0	0.55	14.0
⁷⁄₁₆	11.9	0.468	11.9	7,200	3,270	10,800	48.0	21,600	96.0	1.64	41.6	0.65	16.6
½	13.0	0.531	13.5	9,200	4,170	13,800	61.3	27,600	122.6	1.79	45.5	0.72	18.2
⅝	16.0	0.630	16.0	13,000	5,910	19,500	86.5	39,000	173.0	2.20	56.0	0.79	20.0
¾	20.0	0.787	20.0	20,200	9,180	30,300	134.7	60,600	269.4	2.76	70.0	0.98	25.0
⅞	22.0	0.866	22.0	24,500	11,140	36,750	163.3	73,500	326.6	3.03	77.0	1.08	27.5

* The proof test and minimum breaking force loads *shall not* be used as criteria for service or design purposes.
SOURCE: National Association of Chain Manufacturers; copyright material reprinted with permission.

Table 10.2.4   Grade 70 Transport Chain Specifications

Note: Not to be used in overhead lifting applications.

Nominal chain size		Material diameter		Working load limit (max.)		Proof test* (min.)		Minimum breaking force*		Inside length (max.)		Inside width (min.)	
in	mm	in	mm	lb	kg	lb	kN	lb	kN	in	mm	in	mm
¼	7.0	0.281	7.0	3,150	1,430	6,300	28.0	12,600	56.0	1.24	31.5	0.38	9.8
⁵⁄₁₆	8.7	0.343	8.7	4,700	2,130	9,400	41.8	18,800	83.6	1.29	32.8	0.44	11.2
⅜	10.0	0.406	10.3	6,600	2,990	13,200	58.7	26,400	117.4	1.38	35.0	0.55	14.0
⁷⁄₁₆	11.9	0.468	11.9	8,750	3,970	17,500	77.8	35,000	155.4	1.64	41.6	0.65	16.6
½	13.0	0.531	13.5	11,300	5,130	22,600	100.4	45,200	200.8	1.79	45.5	0.72	18.2
⅝	16.0	0.630	16.0	15,800	7,170	31,600	140.4	63,200	280.8	2.20	56.0	0.79	20.0
¾	20.0	0.787	20.0	24,700	11,200	49,400	219.6	98,800	439.2	2.76	70.0	0.98	25.0

* The proof test and minimum breaking force loads *shall not* be used as criteria for service or design purposes.
SOURCE: National Association of Chain Manufacturers; copyright material reprinted with permission.

load limit is 25 percent of its minimum breaking force. Applications include load securement, guard rail chain, and boat mooring. It is not for use in critical or lifting applications. Table 10.2.2 gives pertinent ASTM-NACM size and load data.

Grade 43 (high test) chain is made from medium carbon steel having a carbon content of 0.15 to 0.22 percent. It is typically heat-treated to a minimum breaking strength of $49 \times (\text{diam})^2$ short tons at a hardness of 200 Brinell. Table 10.2.3 presents the ASTM-NACM specifications for Grade 43 chain. As shown, the working load limit is approximately 35 percent of the minimum breaking force. The major use of this chain grade is load securement, although it also finds use for towing and dragging in the logging industry.

Grade 70 (transport) chain was previously designated *binding chain* by NACM. This is a high-strength chain which finds use as a load securement medium on log and cargo transport trucks. It may be made from carbon or low-alloy steel with boron or manganese added in some cases to provide a uniformly hardened cross section. Carbon content is in the range of 0.22 to 0.30 percent. The chain is heat-treated to a minimum breaking force of $80 \times (\text{diam})^2$ short tons with the working load limit at 25 percent of the minimum breaking force. Although this chain approaches the strength of Grade 80, it is not recommended for overhead lifting because of the greater energy absorption capability and toughness demanded of sling chain. Table 10.2.4 presents the ASTM-NACM specifications for Grade 70 chain.

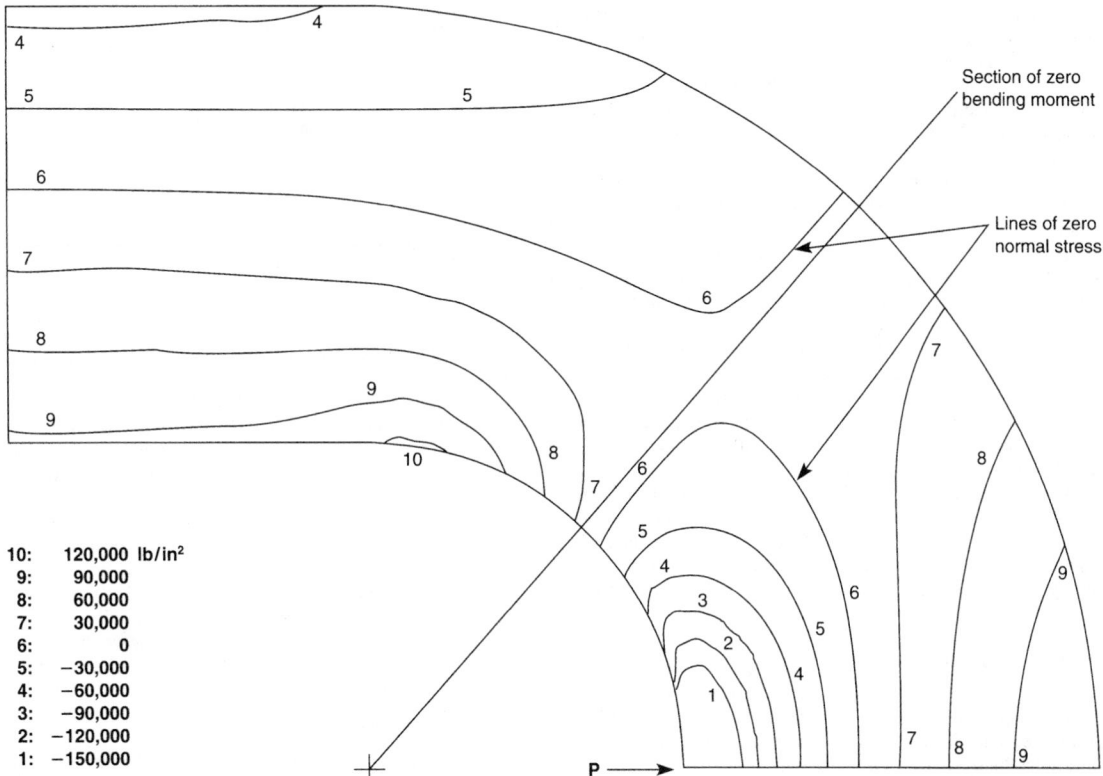

Fig. 10.2.1   Stress distribution in a ⅜-in (10-mm) Grade 80 chain link at rated load.

Chain Strength

Welded chain links are complex, statically indeterminate structures subjected to combinations of bending, shear, and tension under normal axial load. When a link is loaded, the maximum tensile stress normal to the section occurs in the outside fiber at the end of the link (see Fig. 10.2.1), which shows the stress distribution in a ⅜-in (10-mm) Grade 80 chain link with rated load, determined by finite element analysis (FEA). In Fig. 10.2.1, stress distribution is obtained from FEA by assuming that peak stress is below the elastic limit of the material. If the material is ductile and the peak stress from linear static FEA exceeds the elastic limit, plastic deformation will occur, resulting in a distribution of stresses over a larger area and a reduction of peak stress.

It is interesting to note that the highest stress in the chain link is compressive and occurs at the point of loading on the inside end of the link. Conversely, traveling around the curved portion of the link, the compressive and tensile stresses reverse and the outside of the link barrels are in compression. Since the outside of the barrels of the link, unlike the link ends, are unprotected, they are vulnerable to damage such as wear from abrasion, nicks, and gouges. Fortunately, since the damage occurs in an area of compressive stress, their potentially harmful effects are reduced.

Maximum shear stress, as indicated in Fig. 10.2.1, occurs on a plane approximately 45° to the x axis of the link where the bending moment is zero. Chains such as Grade 80 with low to medium hardness (BHN 400 or less) normally break when overloaded in this region. As hardness increases, there is a tendency for the fracture mode to shift from one of shear as described above to one of tension bending through the end of the link where the maximum tensile stress occurs. A member made of ductile material and subjected to uniaxial stress fails only after extensive plastic deformation of the material. The yielding results in deformation, causing a reduction in the link's cross section until final failure occurs in shear.

Combined stresses referred to initially in this section result in higher stresses than those computed by considering the load applied uniformly in simple tension across both barrels. For example, a ⅜-in (10-mm) Grade 80 alloy chain with a BHN of 360 has a material strength of 180,000 lb/in² (1240.9 MN/m²). Based on a specified ultimate strength of the chain of 91 × (diam)² tons, which is 28,300 lb (12,800 kg), the average stress in each barrel of the chain would be 116,000 lb/in² (799.7 MN/m²). The actual breaking strength, however, because of the effects of the combined stresses, is magnified by a factor of 180,000/116,000 = 1.5 over the average stress in each barrel.

Most chain manufacturers now manufacture a through-hardened commercial alloy sling chain, discussed in this section. Although designed primarily for sling service, it is used in many original equipment applications. Typical strength of this Grade 80 chain is 91 × (diam)² tons with a BHN of 360 and a minimum elongation of 15 percent. This combination of strength, wear resistance, and energy absorption characteristics makes it ideal for applications where the utmost dependability is required under rugged conditions. Typical inside width and pitch dimensions are 1.25 × diam and 3.20 × diam.

Chain End Fittings

Most industrial chains must be equipped with some type of end fittings. These usually consist of oblong links, pear-shaped links, or rings (called masters) on one end to fit over a crane hook and some variety of hook or enlarged link at the other end to engage the load. The hooks are normally drop-forged from carbon or alloy steel and subsequently heat-treated. They are designed (using curve-beam theories) to be compatible in strength with the chain for which they are recommended. Master links or rings must fit over the rather large section thicknesses of crane hooks, and they therefore must have large inside dimensions. For this reason they are subjected to higher bending moments, requiring larger section diameters than those on links with a narrower configuration.

CM Chain Division of Columbus McKinnon recommends that oblong master links be designed with an inside width of 3.5 × diam and an inside length of 7.0 × diam, where diam is the section diameter in inches. Using a material with a yield strength of 100,000 lb/in², the diameter can be calculated from the relationship diam = (WLL)/20,000¹ᐟ², where WLL is the working load limit in pounds.

Rings require a somewhat greater section diameter than oblong links to withstand the same load without deforming because of the higher bending moment. **Master rings** with an inside diameter of 4 × diam can be sized from the relationship diam = (WLL)/15,000¹ᐟ² in. Rings require a 15 percent larger section diameter and a 33 percent greater inside width than oblong links for the same load-carrying capacity, and the use of the less bulky oblong links is therefore preferred.

Pear-shaped master links are not nearly as widely in demand as oblong master links. Some users, however, prefer them for special applications. They are less versatile than oblong links and can be inadvertently reversed, leading to undesirable bending stresses in the narrow end due to jamming around the thick saddle of the crane hook.

Hooks and **end links** are attached to sling chains by means of coupling links, either welded or mechanical. Welded couplers require special equipment and skill to produce quality and reliability compatible with the other components of a sling and, consequently, must be installed by sling chain manufacturers in their plants. The advent of reliable mechanical couplers such as **Hammerloks** (CM Chain Division), made from high-strength alloy forgings, have enabled users to repair and alter the sling in the field. With such attachments, customized slings can be assembled by users from component parts carried in local distributor stocks.

Welded-Link Wheel Chains

These differ from sling chains and general-purpose chains in two principal aspects: (1) they are precisely calibrated to function with pocket or sprocket wheels and (2) they are usually provided with considerably higher surface hardness to provide adequate wear life.

The most widely used variety is hoist load wheel chain, a short link style used as the lifting chain in hand, electric, and air-powered hoists. Some roller chain is still used for this purpose, but its use is steadily declining because of the higher strength per weight ratio and three-dimensional flexibility of welded link chain. Wire rope is also sometimes used in hoist applications. However, it is much less flexible than either welded link or roller chain and consequently requires a drum of 20 to 50 times the rope diameter to maintain bending stresses within safe limits. Welded link chain is used most frequently in hoists with three- or four-pocket wheels; however, it is also used in hoists with multipocket wheels. Pocket wheels, as opposed to cable drums, permit the use of much smaller gear reductions and thus reduce the hoist weight, bulk, and cost.

Welded load wheel chain also has 3 times as much impact absorption capability as a wire rope of equal static tensile strength. As a result, wire rope of equal impact strength costs much more than load wheel chain. Since average wire rope life in a hoist is only about 5 percent of the life of chain, overall economics are greatly in favor of chain for hoisting purposes.

To achieve maximum flexibility, hoist load wheel chain is made with link inside dimensions of pitch = 3 × diam and width = 1.25 × diam. Breaking strength is 90 × (diam)² tons, and endurance limit is about 18 × (diam)² tons. In hand-operated hoists, it can be used safely at working loads up to one-fourth the ultimate strength (a design factor of 4). In powered hoists, the higher operating speeds and expectancy of more lifts during the life of the hoist require a somewhat higher design factor of safety. CM Chain Division recommends a factor of 7 for such units where starting, stopping, and resonance effects do not cause the dynamic load to exceed the static chain load by more than 25 percent. For this condition, chain fatigue life will exceed 500,000 lifts, the normal maximum requirement for a powered hoist. Standard hoist load wheel chains, now produced in sizes from 0.125-in (3-mm) through 0.562-in (14-mm) diameter, meet requirements for working load limits up to approximately 5 or 6 short tons (5 long tons) for hand-operated hoists and 3 short tons (2.5 long tons) for power-operated hoists. More complete technical information on this special chain product and on chain-hoist design considerations is available from CM Chain Division.

Table 10.2.5   Conveyor Chain Specifications

Chain size, mm	Link pitch, mm	Minimum breaking force, 1000 lb (454 kg)	
		Case-hardened	Through-hardened
14	50	36.5	50.0
18	64	65.0	92.2
22	86	75.0	121.5
26	92	90.0	169.8

Source: Columbus McKinnon Corporation.

Another widespread use of welded link chain is in conveyor applications where high loads and moderate speeds are encountered. As contrasted to link-wheel chain used in hoists, conveyor chain requires a longer pitch. For example, the pitch of hoist chain is approximately 3.0 × diam and conveyor chain is 3.5 × diam. The longer-pitch chain, in addition to being more economical, accommodates the use of shackle connectors between sections of the chain to which flight bars are attached.

CM Corporation, for example, manufactures conveyor chains in four metric sizes and in two different thermal treatments (Table 10.2.5). These chains are precisely calibrated to operate in pocket wheels. They may have a high surface hardness (500 Brinell) to provide adequate wear life for ash conveyors in power generating plants and to provide high strength and good wear properties for drive chain applications. Also, the long-pitch chain is used extensively in longwall mining operations to move coal in the mines.

Power Transmission

Power transmission, in the sense used to describe the transfer of power in machines from one shaft to another one close by, is a relatively new field of application for welded-link chain. Roller chain and other pin-link chains have been widely used for this purpose for many years. However, recent studies have shown that welded-link chain can be operated at speeds to 3,000 ft/min. Its three-dimensional flexibility, which can sometimes eliminate the need for direction-change components, and its high strength-weight ratio suggest the possibility for significant cost savings in some power-transmission applications.

Miscellaneous Special Chains

Special requirements calling for corrosion or heat resistance, nonmagnetic properties, noncontamination of dyes and foodstuffs, and spark resistance have resulted in the development of many special chains. They have been produced from beryllium copper, bronze, monel, Inconel, Hadfield's manganese, and aluminum and from a wide variety of AISI analyses of stainless and nonstainless alloys. However, the need for such special chains, other than those of stainless steel, is so infrequent that they are seldom carried as stock items.

WIRE ROPE
by Larry D. Means
Means Engineering and Consulting

Wire rope is a machine composed of many moving parts (Fig. 10.2.2). It has three main segments, the wires that make up the individual strands, the multiwire strands helically laid around the central core, and the core. The combination of these three components varies in both complexity and configuration to produce many different wire ropes designed for specific applications and having special characteristics. A typical 6 by 25 filler wire construction has 150 wires in its outer strands alone. These wires and the strands they compose act together and independently around the core as the wire rope operates. The allowed tolerances for the individual wires and the clearances between the wires in the strands are in the same range as those found associated with engine bearings.

The basic component of a wire rope is the wires. These are normally round, but in some ropes may be shaped. These wires are drawn in a

Fig. 10.2.2 Components of a wire rope.

process called "cold forming" or "cold wire drawing" subsequent to the base material having been subjected to special heat treating called "patenting." Uncoated high-carbon steel is by far the most predominant base material. Wire ropes manufactured utilizing these wires are normally referred to as "bright" ropes. High-carbon steel wires are available in various strength levels defined by ASTM A-1007-00, *Standard Specification for Carbon Steel Wire for Wire Rope*. The wire levels are selected by the wire rope manufacturer to optimize the performance and operating characteristics of the finished product.

Other types of wire are available. Galvanized wire is used to enhance the wire rope's corrosion resistance. Galvanized wire is produced by two different procedures. Wires produced by the "galvanized to finished size" wire process is first cold drawn to a predetermined bright wire diameter then coated in a galvanizing line to the required finished wire diameter. Wires produced with wire manufactured in this manner have a 10 percent lower tensile strength than the comparable diameter bright wire. Therefore, wire ropes made from this type wire have a 10 percent lower minimum breaking force than the same size and grade of bright rope. Wires produced by the drawn galvanized process are galvanized prior to being drawn to the required final diameter and, consequently, have the same tensile strength as the same diameter and grade bright wire. Therefore, wire ropes produced by utilizing drawn galvanized wires have the same minimum breaking force as the same diameter and grade bright wire rope.

In addition to galvanized wire ropes, various stainless steel wire ropes are available. To meet the requirements of numerous corrosive-type environments, several grades are typically available. Most of the common grades are typically the 18 percent chromium and 8 percent nickel alloys. The wire rope manufacturer should be consulted for the specific stainless steel wire ropes available to meet the specific requirements of any situation. It should be noted that the majority of the stainless steel wire ropes are not normally available from stock and will require ample times for production and minimum production lengths.

The first thing usually identified with a wire rope is its diameter. This characteristic defines the strength and the ancillary equipment, i.e., sheaves, drums, and fittings, that can be used with the rope. The diameter normally refers to a rope's *nominal diameter*, which is the diameter that will be found in tables, specifications, etc. The actual diameter of a new, unused wire rope will normally be larger than the nominal diameter. Most specifications allow a tolerance of $-0/+5\%$ of nominal diameter.

This is an important factor when it comes to inspection and evaluation for continued service. Methods for measuring a wire rope's diameter are available in industry publications.

Wire ropes are available with strand constructions numbering in the hundreds. The number of strands and the construction of the individual strands play a major role in determining the wire rope's operating characteristics. Standard wire ropes are available with 6 to 8 outer strands varying. Special rotation resistant wire ropes are available with from 12 to 17 or more outer strands and total strands numbering 40 or more. The applications for wire rope are so varied it is impractical to attempt to list specifics in this section. Particulars on these ropes can be obtained from wire rope distributors and manufacturers. An excellent source of information on this subject is the *Wire Rope User's Manual*, published by the Wire Rope Technical Board. Specifications for most of the typical wire ropes are covered in Federal Specification RR-W-410, *Wire Rope and Strand*, and in ASTM A 1023, *Standard Specification for Carbon Steel Wire Ropes for General Purposes*.

The wire rope manufacturers have devised a system to group their products into segments of similar products. These are called rope classifications. The more common wire ropes fall into classifications shown in Table 10.2.6

Table 10.2.6   Classifications of Wire Rope

Standard wire rope classification	Number of outer strands	Wires per strand	Maximum number of outer wires
6 × 7	6	3–14	9
6 × 19	6	15–26	12
6 × 36	6	27–49	18
7 × 19	7	15–26	12
7 × 36	7	27–49	18
8 × 19	8	15–26	12
8 × 36	8	27–49	18
Rotation resistant wire rope classification	Number of outer strands	Wires per strand	Maximum number of outer wires
8 × 19	9 max.	15–26	12
19 × 7	10 min.	6–9	8
19 × 19	10 min.	15–26	12
35 × 7	15 min.	6–9	8
35 × 19	15 min.	15–26	12

There are numerous specialty wire ropes with added features to enhance certain operating characteristics. These include compacted ropes, ropes with compacted strands, plastic-filled ropes, plastic-coated ropes, plastic-coated-core ropes, flattened-strand ropes, and others. Additionally, many rope manufacturers combine more than one of these operating enhancing production methods to produce highly specialized wire ropes for specific applications.

Wire rope "lay" defines three different properties of a specific wire rope. Each of them has important bearing on the rope's operation and use.

The first is the direction the strands lie in the rope. When you look at a wire rope, the strands will either have a right-hand spiral, **right lay**, or a left-hand spiral, **left lay**. This characteristic can have importance with respect to proper spooling on a drum and in balancing torque in multiple rope lifts or suspension. While left lay is available in some rope constructions, right lay is much more common.

The second meaning of lay indicates the relationship between the direction the wires are laid in the strands compared to the direction the strands are laid in the rope. In **regular lay** ropes, the wires in the strands are laid in the opposite direction as those strands are laid in the rope. The wires in a regular lay rope appear to run straight down the length of the rope almost parallel to the longitudinal axis of the rope. Regular lay rope is the rope commonly used in most applications. In **lang lay** ropes, the wires in the strands lie in the same direction as those strands lie in the rope. The wires in a lang lay rope appear to angle across the axis of the rope. Lang lay ropes are available in many rope constructions but are normally associated with applications where abrasion is the predominant mode of degradation.

The third meaning of lay means the distance along the length of a wire rope that it takes any strand to make one complete spiral around the rope (Fig. 10.2.3). This measurement is one that is commonly used in inspection and removal criteria relating to the number of broken wires allowable for continued utilization. It has other significance in specialized uses, i.e., mine hoisting, that are particular to those industries.

ONE ROPE LAY

Fig. 10.2.3 Wire rope lay.

The **grade** of a wire rope is used to define the wire rope **minimum breaking force**. In standard wire ropes, several grades are normally available. The most common grade for standard wire ropes is **extra improved plow steel, or EIPS.** Another grade available is **extra extra improved plow steel,** or **EEIPS,** which normally has a 10 percent higher minimum breaking force than the same diameter and construction EIPS wire rope. See Table 10.2.7 for minimum breaking forces of a few typical wire rope diameters. Other grades may be available from the manufacturers. The minimum breaking forces for different wire ropes are listed in manufacturer's brochures, RR-W-410, ASTM A 1023, and numerous other sources. This section does not address the grades associated with the elevator industry, as they have special grades and specifications for wire ropes utilized in elevators.

Table 10.2.7   Minimum Breaking Force, 6 × 19 and 6 × 36 Classification Uncoated IWRC Wire Rope

Nominal diameter, in	Approx. weight, lb/ft	EIPS, tons	EEIPS, tons
½	0.42	13.3	14.6
¾	0.95	29.4	32.4
1	1.68	51.7	56.9
1¼	2.63	79.9	87.9
1½	3.78	114	125
2	6.72	198	217
3	15.1	425	468

The **core** of a wire rope is the central element of the rope that provides the foundation and support for the finished wire rope. There are common cores used in the majority of wire ropes. The first is a **fiber core.** While it may be natural fiber for special application, synthetic fiber cores are the most prevalent in wire ropes produced today. In most ropes, the fiber core is made up of filaments stranded into a fiber rope of the synthetic material, but, in some special ropes, the core can be a solid synthetic core. The second core type is the **independent wire rope core** commonly referred to as **IWRC.** The name identifies what this is. In this case, the core is actually a small-diameter wire rope used as the core of a larger-diameter finished rope. IWRC ropes make up the majority of the wire ropes in use today. The third core type is a **wire strand core** or WSC. Again, the name describes the process in which a single strand is used as the core of a larger diameter rope.

The terms and descriptions listed above are the language of wire rope. While they are used in specific ways to describe certain characteristics of the wire rope, they are important in communicating with others about the product. Once the rope has been identified, the next thing to consider is the equipment on which it is going to be used. One of the first things to consider is how and with what the wire rope is going to be terminated or attached to other equipment. **Fittings** are normally used for this purpose. There are a multitude of these attachment devices available. They include thimbles to minimize the damage to the rope at termination points; wedge-type sockets to attach the rope to

other equipment or to drums; mechanical, pressed, or swaged end fittings for attachment or joining; wire rope clips for forming attachment points; and many others. One of the most important factors to consider in making an attachment with any fitting or end termination is that the connection may not develop the full strength of the wire rope. The efficiency of fittings is readily available from the manufacturers of these items or from most wire rope manufacturers' catalogs.

A **drum** is a cylindrical-shaped member with plates, called flanges, attached to both ends used as the means to transmit power to the rope and allow that power to move objects. Drums can either be smooth-faced or grooved. They can be welded, cast, or machined of various materials, but are almost always metallic. The method of wire rope attachment varies by drum manufacturer and by application. Again, it is important for the designer to determine the efficiency of the drum attachment. Another thing the designer should consider is that most industry standards specify a minimum number of wraps that must remain on the drum during normal operation. Hoisting drums are illustrated in Figs. 10.2.4 and 10.2.5

Fig. 10.2.4 Hoisting drum.

Fig. 10.2.5 Hoisting drum.

Where there is side draft on the rope, movable **idlers** are provided to align the rope and groove. The idlers may be moved parallel to the face of the drum by the side pressure of the rope or may be driven positively sideways, thus eliminating friction and increasing the life of rope. In Fig. 10.2.4, idlers sheave c revolves between fixed collars on shaft a, which is connected to the drum shaft by sprocket and chain. The shaft is prevented from rotating by a feather key. On sprocket b, a nut is held from moving sideways by flanges; the shaft a with sheave c moves in the direction of its axis. In an alternative construction, the idler shaft is threaded but held stationary, and the sheave hub is a nut. The sheave is turned by the friction of the rope, which causes it to travel back and forth. Figure 10.2.5 shows a construction used when the side draft is excessive. The upright rollers a are moved sideways by screw b, which is driven by sprocket and chain c from the drum shaft. For winding the rope on the drum in several layers, a clutch is provided that reverses the direction at the end of travel.

A **sheave** is a grooved wheel or pulley used with either wire rope or a fiber rope to change the direction and the point of application of a pulling force. Sheaves utilized with wire ropes must be maintained in good condition to minimize the wear on the wire rope and the sheave. When bent over a sheave, the rope's load-induced stretch causes wear to both the rope and the groove. This wear is also influenced by the sheave material and the radial pressure between the wire rope and the groove. It should be noted that these factors are also relative to drums and rollers on which the rope operates. Values for radial pressures with various materials are available in the *Wire Rope User's Manual*, published by the Wire Rope Technical Board.

Sheaves, drums, and rollers must be of the correct design if optimum service is to be obtained. All wire ropes operating over sheaves are subjected to contact cyclic bending stresses, resulting in fatigue-type deterioration to the rope. The relationship between the rope diameter d and the sheave diameter D, expressed as D/d, is a critical factor for this ratio that will determine the rope's service life. Values for this ratio vary by rope construction and should be maximized when possible. It should be noted that, in addition to causing accelerated wear, a D/d smaller than recommended can result in a loss of efficiency of the wire rope itself. Values of D/d and loss of efficiency reductions are available in the *Wire Rope User's Manual*, published by the Wire Rope Technical Board.

Since most wire ropes are used in conjunction with drums and sheaves, the grooves in the drums or sheaves used with wire ropes must be properly sized to facilitate the optimum service life of the wire rope. In addition to the sheave overall diameter, the groove configuration is also critical for optimum service life. For general-purpose wire ropes, grooves in new drums and sheaves should nominally be 6 percent larger than the nominal diameter of the rope to be used in them. The diameter of a groove should never exceed 10 percent larger than the nominal wire rope diameter to be used in it. Any time a groove has worn to a diameter 2½ percent greater than the rope's nominal diameter, it should be regrooved or replaced. Sheave and drum grooves should be checked periodically with groove gauges. Gauges for measuring the worn condition are available and should be used during the normal inspection of the equipment utilizing wire rope.

Another factor of rope wear is directly related to the wire rope alignment with the sheaves and/or the drum, referred to as **fleet angle**. Fleet angles should be in the range of ½° to 1½° for smooth faced drums and ½° to 2° for grooved drums. Angles smaller than ½° should be avoided, as this will cause the rope to pile up. Angles larger than recommended result in poor spooling, accelerated rope wear in both the sheave and on the drum, plus accelerated wear to the sheave itself.

Blocks consisting of one or more sheaves are used in many wire rope applications. When the wire rope is used in these blocks, there is a resultant loss in efficiency due to the friction in the sheave bearings. This loss is dependent upon the bearing material being used. When multipart reeving is used, these factors should be considered. Values for the efficiency of various bearing materials are available from the bearing manufacturer.

Wire Rope Hoist Brakes

Small wire rope hoists are provided with hand-operated band brakes or electrically operated disk brakes; larger wire rope hoists have mechanically operated post brakes. On electric hoists, it is usual to apply the brake with a weight or spring and to remove it by a solenoid. Where controlled rate of application of a brake is desirable, a **thrustor** is frequently used. This is a self-contained unit consisting of a vertical motor, a centrifugal pump in a piston, and a cylinder filled with oil. Starting the motor raises the piston and connected counterweight. Stopping the motor permits the weight to fall. An adjustable range of several seconds in falling sets the brake slowly. Thrustors are built with considerably greater load capacities and stroke lengths than solenoids. Should the current fail, the brakes are automatically applied. On steam hoists, the brake is taken off by a steam piston and cylinder instead of by a solenoid. On overhead cranes, load brakes are used to sustain the load automatically at any point and occasionally to regulate the speed when lowering.

In one type of **load brake** (Fig. 10.2.6), the motor A drives drum B through the load brake on intermediate shaft D. A spider C is keyed to shaft D, its inner end supporting one end of a coiled bronze spring of square section E. The opposite end of the spring is fixed in flange F, which is loosely fitted to shaft D and directly attached to pinion G. Any relative angular motion of flange F and spider C alters the closeness of

the coiling of spring E, consequently altering its outside diameter (considered as a drum). This outer surface is one of the friction surfaces of the brake, the other being provided by the internal face of drum H, which revolves loosely on shaft D at one end and on flange F at the

Fig. 10.2.6 Load brake.

other. Drum H is restrained from moving in one direction by ratchet I and pawls J. The exterior of drum H is grooved for heat dissipation.

The action of the **load brake** in Fig. 10.2.6 is as follows: When hoisting, the brake revolves as shown by the arrow. Pawls J permit drum H to revolve; consequently, the whole mechanism is locked and revolves as one piece. When stopping the load, the downward pull of the load reacts to drive drum H against pawl J. Flange F therefore moves slightly in an angular direction relative to spider C, and spring E consequently untwists until it grips the interior of drum H, thus locking the load. The action is such that the grip is slightly more than necessary to hold the load. Reversing the motor for lowering the load drives the interior of the brake surface against drum H so that the power consumed is the amount necessary to overcome the excess holding power of the brake over the load reaction.

Load brakes of the **disk** and **cone** types are also used and embody the same principle of pawl locks and differential action. The choice of brake type should be based on considerations of smoothness of working and lack of chatter, as well as on the power requirements for lowering at different values of load within the range of the crane.

In **regenerative braking,** the motor, when overhauled, acts as a generator to pump current back into the line.

HOLDING MECHANISMS

Lifting Tongs

Figure 10.2.7 shows the type of clamps used for lifting plates. The cams a grip the plate when the chains tighten, preventing slipping. Self-closing

Fig. 10.2.7 Lifting clamps.

tongs (Fig. 10.2.8) are used for handling logs, manure, straw, etc. The rope *a* is attached to and makes several turns around drum *b*; chains *c* are attached to the bucket head *e* and to drum *b*. When power is applied to rope *a*, drum *b* is revolved, winding itself upon chains *c* and closing the tongs. To open, slack off on rope *a*, holding tongs on rope *d*, attached to head *e*.

Fig. 10.2.8   Lifting tongs.

Special Mechanical and Vacuum Holding Mechanisms

There are many specially fabricated holding mechanisms used in general industrial applications. They may be mechanical or vacuum and are used for many differing applications. Mechanical devices include clamps, grabs, grips, beams, tongs, and lifters used to hold such objects as sheets, coils, ingots, drums, and many other objects including those supported by slings suspended from a beam. A grip force may be applied to the object via mechanical or electrical means. Design considerations include grip force, grip ratio, and mechanical design factors.

Vacuum devices usually consist of one or more suction pads attached to a framework capable of supporting the object to be held. A vacuum is applied to attach and then shut off to release the object. Vacuum devices are used to lift sheets, bags, boxes, and many other objects that have a sufficiently smooth surface upon which the suction pad can be attached. Design considerations include those for the mechanical framework plus the suction pad's capacity, which is a function of the area of the pad and the vacuum applied.

Lifting Magnets

By Richard A. Dahlin, *Walker Magnetics*

Lifting magnets are used in many different industries for handling all types of magnetic materials, and they are available in many different sizes and shapes to accommodate the various lifting applications. In the scrap handling and reclamation industries, magnets lift all types of scrap iron, castings, and steel shredding. The steel industry uses magnets to handle hot billets, tubes, rails, plates, and pipes, and to load and unload large coils of steel into annealing ovens. In material warehouses they lift structural steel materials singly, in bundles or layers, and load and unload flame and plasma tables. Lifting magnets are used throughout industrial plants, machine shops, steel supply, and die shops in the receiving and shipping areas, storeroom, near cutoff saws, buring and welding tables, and for loading and unloading various machine tools. They are often the fastest and the only way for a single operator to load, handle, and unload magnetic material.

To help in understanding the lifting power of magnets, magnetic power is often illustrated as lines of magnetic force flowing from the magnet's north pole to south pole. Anything that limits the flow of these magnetic lines of force reduces the magnet's lifting capacity. Many factors limit the flow of these lines of force. For instance, magnetic lines of force do not flow easily through air. They need iron in order to flow freely. Anything that creates a space or an air gap between a magnet and

the load limits the flow of magnetic lines of force and reduces the lifting capacity of a magnet. The greater the number of lines of magnetic force flowing from a magnet into the load, the greater the effectiveness of the magnet. The thicker the load, the more lines of magnetic force are able to flow. However, after a certain thickness of load, no additional lines of force will flow because the magnet has reached its full capacity. The thinner the load, the less iron is available, and fewer lines of magnetic force flow from the magnet into the load. Therefore, the lifting capcacity of the magnet is reduced. Low-carbon steels, such as AISI 1020 steel, are almost as good conductors of magnetic lines of force as pure iron. However, many other alloys contain nonmagnetic materials, which reduce the ability of magnetic lines of force to flow into the load. An alloy such as AISI 300 series of stainless steel is almost as poor a conductor of magnetic lines of force as air. AISI 416 stainless steel is considered magnetic, but it contains enough chromium that a magnet can develop only one-half as much force on an AISI 416 stainless steel load as it can on an AISI 1020 steel load. Elevated temperatures can reduce the ability of magnetic lines of force to flow into the load. When low-carbon steel is heated above its curie temperature of 700°C (1390°F), it will become austenitic and lose its magnetic properties.

There are basically three types of lifting magnets: electromagnets, permanent magnets, and electro-permanent lifting magnets.

Electromagnets (Figs. 10.2.9 and 10.2.10) become magnetized when electrical direct current (dc) is passed through an insulated electrical

Fig. 10.2.9   Circular lifting electromagnet. (*Walker.*)

wire or strip conductor that is wound around the low-carbon or cast steel core of the magnet. The strength of the magnetic field developed is dependent on the amount of the current and the number of turns of the conductor wrapped around the magnet's core. Electromagnets can be designed to overcome significant air gaps that allow them to gather and lift scrap material and bundles of steel that often provide minimum contact with the poles of the lifting magnet. Because of the amount of heat that can be generated within these magnets by the electrical current, electromagnets are not usually designed for continuous operation in order to protect their electrical insulation. Exceeding the typical duty cycles of 50 to 75 percent that match the lifting application will result

Fig. 10.2.10   DC battery-powered lifting electromagnet. (*Walker.*)

magnet material neodymium iron boron to produce the magnetic hold-ing force. These lifting magnets are manually turned on and off by mov-ing a handle that in turn internally redirects the magnetic alignment of the magnetic lines of force. To turn on the magnet, the permanent magnets within the lifting magnet are positioned to direct the magnetic lines of force to the lifting magnets' lifting surfaces (poles) in order to complete the magnetic circuit through the load to be lifted. To turn off the magnet, the permanent magnets are positioned to direct the mag-netic lines of force to remain within the lifting magnet. Turning these magnets on and off requires the interruption of internal magnetic forces that can become quite large. Thus the maximum rated lifting capacity of permanent lifting magnets is limited to acceptable actuation forces. The strength of the magnetic field developed is dependent on the mag-net design and the properties of the permanent-magnet material used in the magnet. With the introduction of the newer higher-energy magnet material neodymium iron boron, it has been possible to reduce the over-all size and weight of permanent lifting magnets. Permanent lifting magnets can be operated with a 100 percent duty cycle without any con-cerns for an electrical power interruption or the generation of internal heat. However, the permanent-magnet material is affected by higher temperatures. Its magnetic properties are reduced as the temperature is increased, thus permanent lifting magnets are not recommended for use on hot material. These lifting magnets are primarily used as self-contained close-proximity lifting magnets.

Electro-permanent lifting magnets (Fig. 10.2.12) are permanent lifting magnets that are energized and de-energized electrically without moving any internal parts. These lifting magnets use two types of per-manent magnets, fixed (or static) magnets and reversible magnets, along with an electrical coil that surrounds the reversible magnets to change their polarity orientation. In the past, fixed magnets were

in reduced lifting capacity because of the excessive heat that will be built up in the magnet. Duty cycle rating (D.C.%) is defined as (time on $\times$ 100)/(time off + time on) = D.C.%. It is expressed as a percent, with a maximum of 10 minutes time on to avoid overheating the mag-net. When these magnets are used on high-temperature loads it is important to keep the magnet-coil temperature within the design limits of the electrical insulation. While energized, an electromagnet is a high-ly inductive device that requires a special magnet controller that allows the current to be reversed in order to effectively release the load and also absorb the large amounts of energy stored within the magnet. Also, to prevent the load from being dropped in the event of a power inter-ruption, many electromagnets are supplied with a battery backup sys-tem that provides instantaneous power transfer to battery power. Battery-powered electromagnets and other compact electromagnets are self-contained magnets that are used in many industrial applications. These magnets are considered close-proximity lifting magnets because the operator is required to manually position the lifting magnet on the load, and/or manually guide the load during its movement. Magnets that are remotely controlled and operated close to people are also considered close-proximity magnets.

Permanent lifting magnets (Fig. 10.2.11) utilize permanent-magnet materials such as ferrite (ceramic), Alnico, and the newer rare-earth

Fig. 10.2.11   Permanent lifting magnet. (*Walker.*)

Fig. 10.2.12   Electro-permanent lifting magnet. (*Walker.*)

usually made of ferrite, but today the newer higher-energy neodymium iron boron is becoming more common. The reversible magnets are usually made of Alnico. The fixed and reversible magnets are oriented so they can combine their magnetic lines of force into the load to be lifted or direct the magnetic lines of force to remain within the lifting magnet to release the load when the polarity is reversed. Turning the magnet on is achieved by placing the electro-permanent magnet on the load and then applying a short dc pulse to the coil to reverse the polarity of the reversible magnet to direct the magnetic lines of force from within the magnet body to the load. Electro-permanent lifting magnets will remain on, without the need of any battery backup systems, until another electrical dc pulse is applied to turn it off.

Electro-permanent lifting magnets, like permanent lifting magnets, have been significantly larger and heavier than electromagnets of equivalent lifting capacity. However, this difference has been greatly reduced with the development of the newer, higher-energy rare-earth permanent magnetic materials. Electro-permanent magnets perform well in lifting large solid loads such as plates, slabs, coils, and rounds. They are also used as self-contained close-proximity lifting magnets.

Lifting magnets can lift and move steel very efficiently without the need for the attachment of slings, hooks, or cables. They are often the fastest and the only way for a single operator to load, handle, and unload magnetic material. Electromagnets, permanent lifting magnets, and electro-permanent lifting magnets have different benefits and limitations. Therefore, all aspects of the lifting application must be understood when choosing the appropriate lifting magnet.

Buckets and Scrapers

One type of **Williams self-filling dragline bucket** (Fig. 10.2.13) consists of a bowl with the top and digging end open. The shackles at c are so attached that the cutting edge will penetrate the material. When filled,

Fig. 10.2.13   Self-filling dragline bucket.

the bucket is raised by keeping the line a taut and lifting on the fall line d. In this position, the bucket is carried to the dumping position, where by slacking off on line a, it is dumped. Line b holds up line a and the bridle chains while dumping. Such buckets are built in sizes from ¾ yd³ to as large as 85 yd³ (0.6 to 65 m³), weighing 188,000 lb (85,000 kg).

Figure 10.2.14 shows an **open-bottom-type scraper.** Inhauling on cable a causes the sloping bottom plate to dig and load until upward pressure

Fig. 10.2.14   Scraper bucket. (*Sauerman.*)

of the material against the top prevents further loading. The scraper continues its forward haul to the dumping point. Pulling on cable b deposits the load and returns the scraper to the excavation point. Scrapers are made in sizes from ⅓ to 20 yd³ capacity (0.25 to 15 m³).

The **Hayward clamshell** type (Fig. 10.2.15) is used for handling coal, sand, gravel, etc., and for other flowable materials. The holding rope a is made fast to the head of the bucket. The closing rope b makes several wraps around and is made fast to drum d, mounted on the shaft to which the scoops are pivoted. The chains c are made fast to the head of the bucket and to the smaller diameter of drum d. When power is applied to rope b, it causes the drum to wind itself up on chain c, raising the drum and closing the bucket. To dump, hold rope a and slack off rope b. The digging power of the bucket is determined by its weight and by the ratio of the diameters of the large and small parts of the drum. Hayward also has an **orange-peel** bucket that operates like the clamshell type (Fig. 10.2.15) but has four blades pivoted to close.

Fig. 10.2.15   Hayward grab bucket.

The **Hayward electrohydraulic** single-rope type (Fig. 10.2.16) is used not only for handling ore and sand but also where it is desirable to hook the bucket on a derrick or crane hook. The bucket requires only to be hung from the crane or derrick hook by the eye and an electric line plugged in. No ropes are fed into the bucket and no time-consuming

Fig. 10.2.16   Hayward electrohydraulic single-rope-type grab bucket.

shifting of lines is required—thus facilitating changing from magnet to bucket. The bowls are suspended from the head, which contains a complete electrohydraulic power unit consisting of an ac or dc motor, hydraulic pump, directional valve, sump, filters, and cylinders. When current is turned on, the motor drives the pump which provides system pressure. Energizing the solenoids in the directional valve directs the fluid flow to either end of the cylinders, which are attached to the blade arms, thus opening or closing the bucket. The system uses a pressure-compensated variable-displacement pump which enters a no-delivery mode when system pressure is reached during the closing or opening action, thus avoiding any heat generation and reducing the load on the motor while maintaining system pressure. The 1 yd³ (0.76 m³) bucket is provided with a 15-hp (11.2-kW) motor and 15 gal/min (0.95 1/s) pump. The 2 and 3 yd³ (1.5 to 2.3 m³) buckets have motors of 20 to 25 hp (15 to 19 kW), according to duty.

HOISTS

Hand-Chain Hoists

Hand chain hoists are portable lifting devices suspended from a hook and operated by pulling on a hand chain. There are two types currently in common use, the high-speed, high-efficiency hand hoist (Fig. 10.2.17) and the economy hand hoist. The economy hand hoist is similar in appearance to the high-efficiency hand hoist except the handwheel diameter is smaller. This smaller diameter accounts for a smaller headroom dimension for the economy hoist. High-speed hoists have mechanical efficiency as high as 80 percent, employ Weston self-energizing brakes for load control, and can incorporate load-limiting devices which prevent the lifting of excessive overloads. They require less hand chain pull and less chain overhaul for movement of a given load at the expense of the larger head size when compared to economy hoists. As noted, this larger head size also creates a minimum headroom dimension somewhat greater than that for the economy hoists. High-speed hoists find application in production operations where low operating effort and long life are important. Economy hoists find application in construction and shop use where more compact size and lower unit cost compensate for the higher operating effort and lower efficiency.

Fig. 10.2.17   High-speed hand hoist. (*CM Hoist Div., Columbus McKinnon.*)

Both hoist types function in a similar manner. The hand chain operates over a handwheel which is connected via the brake mechanism and gear train to a pocket wheel over which the load chain travels. The brake is disengaged during hoisting by a one-way ratchet mechanism. In lowering, the hand chain is pulled continuously in a reverse direction to overcome brake torque, thus allowing the load to descend.

Although a majority of hand chain hoist applications are fixed hook-suspended applications, sometimes a hand hoist is mounted to a trolley to permit horizontal movement of the load, as in a forklift battery-changing operation. This is accomplished by directly attaching the hoist hook to the trolley load bar, or by using an integrated trolley hoist which combines a high-speed hoist with a trolley to save headroom (Fig. 10.2.18).

Hand chain hoists are available in capacities to 50 tons (45 tonnes). Tables 10.2.8 and 10.2.9 provide data for high-speed and economy hoists in readily available capacities to 10 tons (9 tonnes), respectively.

Fig. 10.2.18   Trolley hoist. Shown with hand chain removed for clarity. (*CM Hoist Div., Columbus McKinnon.*)

Pullers or Come-Alongs

Pullers, or come-alongs, are lever-operated chain or wire rope hoists (Fig. 10.2.19). Because they are smaller in size and lower in weight than equivalent capacity hand chain hoists, they find use in applications where the travel (take-up) distance is short and the lever is within easy reach of the operator. They may be used for lifting or pulling at any angle as long as the load is applied in a straight line between hooks. There are two types of lever-operated chain or wire rope hoists currently in use: long handle and short handle. The long-handle variety, because of direct drive or numerically lower gear ratios, allows a greater load chain take-up per handle stroke than the short-handle version. Conversely, the short-handle variety requires less handle force at the expense of chain take-up distance because of numerically higher gear ratios. Both types find use in construction, electrical utilities and industrial maintenance for wire tensioning, machinery skidding, and lifting applications.

Fig. 10.2.19   Puller or come-along hoist. (*CM Hoist Div., Columbus McKinnon.*)

A reversible ratchet mechanism in the lever permits operation for both tensioning and relaxing. The load is held in place by a Weston-type brake or a releasable ratchet. Lever tools incorporating load-limiting devices via slip clutch, spring deflection, or handle-collapsing means are also available. Table 10.2.10 provides data for lever operated chain hoists.

Electric Hoists

Electric hoists are used for repetitive or high-speed lifting. Two types are available: chain (see Fig. 10.2.20) (both link and roller), in capacities to 15 tons (13.6 tonnes); wire rope (see Fig. 10.2.21), in ratings to 25 tons (22.7 tonnes). The typical hoist has a drum or sprocket centered in the frame, with the motor and gearing at opposite ends, the motor shaft passing through or alongside the drum or sprocket. Many have an integral load-limiting device to prevent the lifting of gross overloads.

Electric hoists are equipped with at least two independent braking means. An electrically released brake causes spring-loaded disk brake

Table 10.2.8   Typical Data for High-Speed Hand Chain Hoists with 8-ft Chain Lift

| Capacity | | Hand chain pull to lift rated capacity load | | Retracted distance between hooks (headroom) | | Net weight | | Hand chain overhaul to lift load 1 ft | |
short tons	tonnes	lb	kg	in	mm	lb	kg	ft	m
¼	0.23	23	10.4	13	330.2	33	15.0	22.5	6.9
½	0.45	46	20.9	13	330.2	33	15.0	22.5	6.9
1	0.91	69	31.3	14	355.6	36	16.3	30	9.1
1½	1.36	80	36.3	17.5	444.5	59	26.8	40.5	12.3
2	1.81	83	37.6	17.5	444.5	60	27.2	52	15.8
3	2.7	85	38.6	21.5	546.1	84	38.1	81	24.7
4	3.6	88	39.9	21.5	546.1	91	41.3	104	31.7
5	4.5	75	34.0	24.5	622.3	122	55.3	156	47.5
6	5.4	90	40.8	25.5	647.7	127	57.6	156	47.5
8	7.3	89	40.3	35.5	901.7	207	93.9	208	63.4
10	9.1	95	43.1	35.5	901.7	219	99.3	260	79.2

SOURCE: CM Hoist Division, Columbus McKinnon.

Table 10.2.9   Typical Data for Economy Hand Chain Hoist with 8-ft Chain Lift

| Capacity | | Hand chain pull to lift rated capacity load | | Retracted distance between hooks (headroom) | | Net weight | | Hand chain overhaul to lift load 1 ft | |
short tons	tonnes	lb	kg	in	mm	lb	kg	ft	m
½	0.45	45	20	10	254	15	7	32	9.8
1	0.91	74	34	12	305	22	10	39	11.9
2	1.81	70	32	17	432	53	24	77	23.5
3	2.7	54	25	22	559	69	31	154	46.9
5	4.5	88	40	24	610	74	34	154	46.9
10	9.1	90	41	32	813	139	63	308	93.9

SOURCE: CM Hoist Division, Columbus McKinnon.

Table 10.2.10   Typical Data for Short- and Long-Handle Lever-Operated Chain Hoists

| Capacity | | Retracted distance between hooks (headroom) | | Handle pull to lift rated capacity | | | | Net weight* | | | |
| | | | | Short | | Long | | Short | | Long | |
short tons	tonnes	in	mm	lb	kg	lb	kg	lb	kg	lb	kg
¾	0.68	11	279	35	15.8	58	26.3	18	8.2	14	6.4
1½	1.4	14	356	40	18.1	83	37.6	27	12.2	24	10.9
3	2.7	18	457	73	33.1	95	43.1	45	20.4	44	20.0
6	5.4	23	584	77	35.0	96	43.5	66	30.0	65	29.5

* With chain to allow 52 in (1322 mm) of hook travel.
SOURCE: CM Hoist Division, Columbus McKinnon.

Fig. 10.2.20   Electric chain hoist. (*CM Hoist Div., Columbus McKinnon.*)

Fig. 10.2.21   Electric wire-rope hoist. (*CM Hoist Div., Columbus McKinnon.*)

plates to engage when current is off. When the hoist motor is activated, a solenoid overcomes the springs to release the brake. In the lowering direction, the motor acts as a generator, putting current back into the line and controlling the lowering speed. Some electric hoists use the same type of Weston brake as is used in hand hoists, but with this type, the motor must drive the load downward so as to try to release the brake. This type of brake generates considerable heat that must be dissipated—usually through an oil bath. The heat generated may also lower the useful-duty cycle of the hoist. If the Weston brake is used, an additional auxiliary hand-released or electrically released friction brake must be provided since the Weston brake will not act in the raising direction. All

electric hoists have upper-limit devices; lower-limit devices are standard on chain hoists, optional on wire-rope hoists. Control is usually by push button: pendant ropes from the controller are obsolescent. Control is "deadman" type, the hoist stopping instantly upon release. Modern multiple-speed and variable-speed ac controls have made dc hoists obsolete. Single-phase ac hoists are available to 1 hp, polyphase in all sizes.

The hoist may be suspended by an integral hook or bolt-type lug or may be attached to a trolley rolling on an I beam or monorail. The trolley may be plain (push type), geared (operated by a hand chain), or motor-driven; the latter types are essential for heavier loads. Table 10.2.11 gives data for electric **chain hoists,** and Table 10.2.12 for **wire-rope hoists.**

Table 10.2.11 Typical Data for Electric Chain Hoists*

Capacity		Lifting speed		Retracted distance between hooks (headroom)		Net weight		Motor	
short tons	tonnes	ft/min	m/min	in	mm	lb	kg	hp	kW
1/8	0.11	32	10	15	381	62	28	1/4	0.19
1/8	0.11	60	18	15	381	68	31	1/2	0.37
1/4	0.23	16	5	15	381	62	28	1/4	0.19
1/4	0.23	32	10	15	381	68	31	1/2	0.37
1/2	0.45	8	2.5	18	457	73	33	1/4	0.19
1/2	0.45	16	5	15	381	68	31	1/2	0.37
1/2	0.45	32	10	16	406	114	52	1	0.75
1/2	0.45	64	20	16	406	121	55	2	1.49
1	.91	8	2.5	18	457	79	36	1/2	0.37
1	.91	16	5	16	406	114	52	1	0.75
1	.91	32	10	16	406	121	55	2	1.49
2	1.81	8	2.5	23	584	139	63	1	0.75
2	1.81	16	5	23	584	146	66	2	1.49
3	2.7	5.5	1.5	25	635	163	74	1	0.75
3	2.7	11	3.3	25	635	170	77	2	1.49
5	4.5	10	3	37	940	668	303	5	3.73
5	4.5	24	7	37	940	684	310	7 1/2	5.59
7 1/2	6.8	7	2	40	1016	929	421	5	3.73
7 1/2	6.8	16	5	40	1016	957	434	7 1/2	5.59
10	9.1	7	2	42	1067	945	429	5	3.73
10	9.1	13	4	42	1067	961	436	7 1/2	5.59
15	13.6	4	1.2	48	1219	1155	524	5	3.73
15	13.6	8	2.5	48	1219	1167	529	7 1/2	5.59

*Up to and including 3-ton hook suspended with 10-ft lift. 5- through 15-ton plain trolley suspended with 20-ft lift. Headroom distance is beam to high hook.
SOURCE: CM Hoist Division, Columbus McKinnon Corporation.

Table 10.2.12 Typical Data for Electric Wire Rope Hoists*

Capacity		Lifting speed		Beam to high hook distance		Net weight		Motor	
short tons	tonnes	ft/min	m/min	in	mm	lb	kg	hp	kW
1/2	0.45	60	18	27	686	430	195	4.5	3.36
3/4	0.68	37	11	27	686	430	195	4.5	3.36
1	0.91	30	9	25	635	465	211	4.5	3.36
1	0.91	37	11	27	686	435	197	4.5	3.36
1	0.91	60	18	27	686	445	202	4.5	3.36
1 1/2	1.36	18	5.5	25	635	465	211	4.5	3.36
1 1/2	1.36	30	9	25	635	465	211	4.5	3.36
1 1/2	1.36	37	11	27	686	445	202	4.5	3.36
2	1.81	18	5.5	25	635	465	211	4.5	3.36
2	1.81	30	9	25	635	480	218	4.5	3.36
3	2.7	18	5.5	25	635	560	254	4.5	3.36
5	4.5	13	4	30	762	710	322	4.5	3.36
7 1/2	6.8	15	4.6	36	914	1100	499	7.5	5.59
10	9.1	13.5	4.1	40	1016	1785	810	10	7.46
15	13.6	13	4	49	1245	2510	1139	15	11.19

*Up to and including 5-ton with plain trolley. 7 1/2- to 15-ton with motor-driven trolley.
SOURCE: CM Hoist Division, Columbus McKinnon.

Air Hoists

Air hoists are similar to electric hoists except that air motors are used. Hoists with roller chain are available to 1 ton capacity, with link chain to 3 short tons (2.7 tonnes), and with wire rope to 15 short tons (13.6 tonnes) capacity. The motor may be of the rotary-vane or piston type. The piston motor is more costly but provides the best starting and low-speed performance and is preferred for larger-capacity hoists. A brake, interlocked with the controls, automatically holds the load in neutral; control movement releases the brake, either mechanically or by air pressure. Some air hoists also include a Weston-type load brake. The hoist may be suspended by a hook, lug, or trolley; the latter may be plain, geared, or air-motor-driven. Horizontal movement is limited to about 25 ft (7.6 m) because of the air hose, although a runway system is available with a series of normally closed ports that are opened by a special trolley to supply air to the hoist.

Air hoists provide **infinitely variable speed,** according to the movement of the control valve. Very high speeds are possible with light loads. When severely overloaded, the air motor stalls without damage. Air hoists are smaller and lighter than electric hoists of equal capacity and can be operated in explosive atmospheres. They are more expensive than electric hoists, require mufflers for reasonably quiet operation, and normally are fitted with automatic lubricators in the air supply.

Jacks and Power Screw Actuators

Jacks are portable, hand-operated devices for moving heavy loads through short distances. There are three types in common use: screw jacks, ratchet jacks, and hydraulic jacks. Bell-bottom screw jacks (Fig. 10.2.22) are available in capacities to 24 tons and lifting ranges to 14 in. The screw is rotated by a bar inserted in holes in the screw head. Geared journal jacks will lift up to 75 short tons (68 tonnes). A lever ratchet mechanism turns a pinion which drives a gear affixed to a lifting screw causing an internally threaded nut affixed to the jack head to travel. Ductile iron jacks (Fig. 10.2.23) consist of a cast-steel housing in which

Fig. 10.2.22   Bell-bottom screw jack.

Fig. 10.2.23   Ratchet jack.

the lever pivots. The rack-toothed bar passes through the hollow housing; the load may be lifted either on the top or on a toe extending from the bottom of the bar. The lever pawl may be biased either to raise or to lower the bar, the housing pawl holding the load on the return lever stroke. Track jacks are ratchet jacks that may be tripped to release the load. They are used for railroad-track work but not for industrial service where the tripping features might be hazardous. **Hydraulic jacks** (Fig. 10.2.24) consist of a cylinder, a piston, and a lever-operated pump. Capacities to 100 short tons (91 tonnes) and lifting heights to 22 in. (0.56 m) are available. Jacks 25 short tons (22.7 tonnes) and larger may be provided with two pumps, the second pump being a high-speed unit for rapid travel at partial load.

Power screw actuators are integral gear-driven power screw devices that convert rotary motion and power into linear movement of a load. Power screw actuators are available in many variations. Configurations

Fig. 10.2.24   Hydraulic jack.

are: translating screw, rotating screw (translating nut), and translating tube. Gearing types are: worm gearing, bevel gearing, and spur/helical gearing. Mounting methods are: base mounting, trunnion mounting and clevis mounting. Available options include: stop nuts, bellows boots, limit switches, motor mounts, and integral speed reducers and motors. Perhaps the greatest single advantage of power screw actuators is that multiple units can be tied together mechanically for synchronized travel. Power screw actuators are of three basic types.

Machine screw actuators (Fig. 10.2.25) feature a sliding-contact power thread (ACME, STUB ACME, modified square, or metric trapezoidal) mechanism. Advantages are: lowest cost, self-locking operation (most machine screw actuators will not backdrive), quietest operation, availability of widest range of lifting capacities, availability of internal keying of translating screw models, availability of an antibacklash feature (to control power screw axial free movement), and availability of stainless steel construction. Disadvantages are: lowest efficiency, lowest duty cycle capability, highest power usage, requirement for largest drives, and increasing backlash from wear during life. Machine screw actuators are available in lifting capacities to 250 tons (228 tonnes).

Fig. 10.2.25   Upright translating machine screw actuator. (*Duff Norton.*)

Ball screw actuators (Fig. 10.2.26) feature a rolling contact recirculating ball power thread mechanism. Advantages are: highest efficiency, faster speed capability, highest duty cycle capability, lowest power usage, requirement for smallest drives, near-constant backlash during life, and predictable power thread life. Disadvantages are higher cost

Fig. 10.2.26   Inverted rotating ball screw actuator. (*Duff Norton.*)

and self-lowering operation (most ball screw actuators require brakes to prevent backdriving). Ball screw actuators are available in lifting capacities to 50 tons (45 tonnes).

Roller screw actuators feature a rolling contact planetary roller power thread mechanism. Advantages are: higher efficiency, highest load capacity, longest life, highest speed and acceleration capability, high stiffness, high positional accuracy, near-constant backlash during life, predictable power thread life, and quieter operation. Disadvantages are highest cost and self-lowering operation (most roller screw actuators require brakes to prevent backdriving). Roller screw actuators are available with lifting capacities to 90 tons (82 tonnes).

MINE HOISTS AND SKIPS
by Burt Garofab
Pittston Corporation

There are two types of hoists, the drum hoist and the friction hoist. On a **drum hoist,** the rope is attached to the drum and is wound around and stored on the drum. A drum hoist may be single-drum or double-drum. There are also several different configurations of both the single-drum and the double-drum hoist. A drum hoist may be further divided as either unbalanced, partially balanced, or fully balanced.

On a **friction hoist,** which is often referred to as a **Koepe hoist,** the rope is not wound around the drum, but rather passes over the drum. A friction hoist may utilize a single rope or multiple ropes.

The operation of hoists may be controlled manually, automatically, or semiautomatically. There are also arrangements that use combinations of the different control types.

There are currently three types of grooving being used on new drums for drum hoists. They are helical grooving, counterbalance (Lebus) grooving, and antisynchronous grooving. Helical grooving is used primarily for applications where only a single layer of rope is wound around the drum. Counterbalance and antisynchronous grooving are used for applications where multiple layers of rope are wound around the drum.

Mine hoists are generally divided into (1) **metal-mine hoists** (e.g., iron, copper, zinc, salt, gypsum, silver, gold, ores) and (2) **coal-mine hoists.** These classes subdivide into main hoists (for handling ores or coal) and hoists for men, timbers, and supplies. They are designed for operating

(1) mine shafts, vertical and inclined, balanced and unbalanced; and (2) slopes, balanced and unbalanced. When an empty cage or platform descends while the loaded cage or platform ascends, as when both cables are wound on a single drum, the machine is referred to as a **balanced hoist.** Most medium- and large-sized hoists normally operate in balance, as the tonnage obtained for a given load and rope speed is about double that for an unbalanced hoist and the power consumption per ton hoisted is lower. Balanced operation can also be obtained by a counterweight. The counterweight (approximately equal to all the dead loads plus one-half the live load) is usually installed in guides within a single shaft compartment. The average depth of ore mines is about 2,000 ft (610 m), and that of coal mines is close to 500 ft (152 m). Most ore-mine hoists are of the double-drum type, normally hosting in balance, each drum being provided with a friction clutch for changing the relative positions of the two skips when operating from various levels.

Hoists for coal mines are principally of the keyed-drum type, for operating in balance from one level. For high rope speeds in shallow shafts, it is generally advantageous to use **combined cylindrical and conical drums.** The cylindroconical drum places the maximum rope pull (weight of rope and loaded skip) on the small diameter so that during the acceleration period of the cycle, the weight of the opposing skip is offering the greatest counterbalance torque, reducing motor peak loads and slightly reducing the power consumption. The peak reduction obtained becomes greater when shafts are shallower and hoisting speeds are higher, provided that the proportion between the smaller and the large diameters is increased as these conditions increase. By varying the ratio of diameters and the distribution of rope on the drum profiles with respect to the periods of acceleration, retardation, and constant speed, static and dynamic torques can be modified to produce the most economical power consumption and the minimum size of motor. The conical drum is not applicable for multiple-level operation, and except in very special cases, only a single layer of rope can be used.

Skip hoists for industrial purposes such as power-plant fuel handling and blast-furnace charging are similar to shallow-lift slow-speed coal-mine hoists in that they operate from a single level. Speeds of 100 to 400 ft/min (0.5 to 2 m/s) are usual. For blast-furnace charging with combined bucket and load weights up to 31,000 lb (14,000 kg) and a speed of 500 ft/min (2.5 m/s), modern plants consist of straight-drum geared engines, frequently with Ward Leonard control.

Industrial skip hoists may be specified where the lift is too high for a bucket elevator, where the lumps are too large for elevator buckets, or where the material is pulverized and extremely abrasive or actively corrosive. For high lifts having a vertical or nearly vertical path, the skip with supporting structure usually costs less than a bucket elevator or an inclined-belt conveyor with bridge. Typical paths are shown in Fig. 10.2.27, paths C and D being suitable when the load is received through a track hopper.

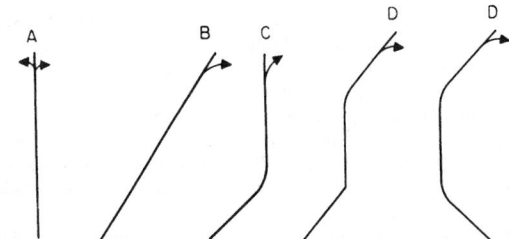

Fig. 10.2.27   Typical paths for skip hoists. (*a*) Vertical with discharge to either side; (*b*) straight inclined run; (*c*) incline and vertical; (*d*) incline, vertical, incline.

The skip may be **manually loaded** direct from a dump car or **automatically loaded** by a pivoted chute, which is actuated by the bucket and which, when upturned, serves as a cutoff gate (Fig. 10.2.28).

For small capacities, the skip can be manually loaded with semiautomatic control. When the bucket has been filled, the operator pushes the start button and the bucket ascends, dumps, and returns to loading position. With automatic loading and larger capacity, the skip may have full automatic control. For economy, the bucket is counterbalanced by a weight, usually equaling the weight of the empty bucket plus half the load. For large capacity, a balanced skip in which one bucket rises as the other descends may be used. High-speed skips usually have automatic slowdown (two-speed motor) as the bucket nears the loading and discharge points.

Fig. 10.2.28   Automatic loader (in loading position).

There are various types of wire ropes used in hoisting. Hoist ropes can be categorized into three main types: round strand, flattened strand, and locked coil. **Round strand rope** is used on drum hoists in applications when a single layer of rope is wound on the drum. **Flattened strand rope** is used on drum hoists when multiple layers of rope are wound on the drum. Flattened strand rope can also be used on friction hoists. **Locked coil rope** is used on friction hoists.

Plow steel and improved plow steel are the most commonly used grades; the latter is used where the service is severe. Some state mining regulations require higher factors of safety than the usual hoisting requirements. The working capacity of new ropes is usually computed by using the minimum breaking strength given in the manufacturer's tables and the following factors of safety: rope lengths of 500 ft (152 m) or less, minimum factor 8; 500 to 1,000 ft (152 to 305 m), 7; 1,000 to 2,000 ft (305 to 610 m), 6; 2,000 to 3,000 ft (610 to 914 m), 5; 3,000 ft (914 m) or over, 4½. These are gross factors between the rated minimum breaking strength of the rope and the maximum static pull due to suspended load plus rope weight. The net factor, which should be used in dealing with large capacities and great depth, must take into consideration stresses due to acceleration and bending around the drum, together with suitable allowances for shock. With 6-by-19 wire construction, the pitch diameter of the drums is generally not less than sixty times the diameter of the rope. Drums are made of either cast iron or cast steel machine-grooved to suit the size of rope (see above). In large hoists, a lifting device is installed at the free rope end of the drum to assist the rope in doubling back over the first layer.

Brakes   There are three main types of brakes used on hoists: the jaw brake, the parallel motion brake, and disk brakes. Brake control is accomplished through air or hydraulics. Brake shoes are steel with attached friction material surfaces.

Hoist Motors   Determining the proper size of motor for driving a hoist calls for setting up a definite cycle of duty based upon the required daily or hourly tonnage.

The **permissible hoisting speed** for mine hoists largely depends upon the depth of the shaft; the greater the depth, the higher the allowable speed.

Conservative maximum hoisting speeds, as recommended by *Bu. Mines Bull.* 75, are as follows:

Depth of shaft		Hoisting speed	
ft	m	ft/min	m/s
500 or less	150 or less	1,200	6
500–1,000	150–300	1,600	8
1,000–2,000	300–600	2,000	10
2,000–3,000	600–900	3,000	15

High hoisting speeds call for rapid **acceleration** and **retardation**. For small hoists, the rate of acceleration may be made as low as 0.5 ft/s² (0.15 m/s²). An average value of 3 ft/s² is adopted for large hoists with fairly high speeds. Exceptional cases may require up to 6 ft/s² (1.9 m/s²). The speed should also be considered with regard to the weight of the material to be hoisted per trip. The question of whether the load should be increased and the speed reduced or vice versa is controlled by local conditions, mining laws, and practical experience. The rest period assigned to the duty cycle, i.e., the requisite time for loading at the bottom and unloading at the top, is dependent upon the equipment employed. With skips loaded from underground ore-storage hoppers, 5 to 6 s is the minimum that can be assumed. Unless special or automatic provision is made, the loading time should be taken as 8 to 10 s minimum. When the (1) hoisting speed, (2) weight of skip or cage, (3) weight of load, (4) periods of acceleration, (5) retardation, and (6) time for loading have been decided upon, the next step is to ascertain the "root-mean-square" equivalent continuous load-heating effect on the motor, taking into account rope and load weights, acceleration and deceleration of all hanging and rotating masses, and friction of sheaves, machines, etc. The friction load is usually taken as constant throughout the running period of the cycle. The overall efficiency of the mechanical parts of a single-reduction-geared hoist averages 80 percent; that of a first-motion hoist is closer to 85 percent. The motor selected must have sufficient starting torque to meet the temporary peaks of any cycle, including, in the case of balanced hoists, the requirements of trips out of balance.

Electrical equipment for driving mine hoists is of four classes:

1. *Direct-current motors* with resistance control for small hoists, usually series-wound but occasionally compound-wound in conjunction with dynamic braking control.

2. *Alternating-current slip-ring-type motor* with secondary resistance.

3. *Ward Leonard system of control* for higher efficiency, particularly on short lifts at high rope speeds, where the rheostatic losses during acceleration and retardation represent a large proportion of the net work done during the cycle; for accuracy of speeds, with high-speed hoists; and for equalization of power demands. Complete control of the speed from standstill to maximum is obtained for all values of load from maximum positive to maximum negative. The lowering of unbalanced loads without the use of brakes is as readily accomplished as hoisting.

4. The *Ilgner Ward Leonard system* consists of a flywheel directly connected to a Ward Leonard motor-generator set and a device for automatically varying the speed through the secondary rheostatic control of the slip-ring induction motor driving the set. This form of equipment is used under conditions that prohibit the carrying of heavy loads or where power is purchased under heavy reservation charges for peak loads. It limits the power taken from the supply circuit to a certain predetermined value; whatever is required in excess of this value is produced by the energy given up by the flywheel as its speed is reduced.

ELEVATORS, DUMBWAITERS, ESCALATORS
by Louis Bialy
Otis Elevator Corporation

The advent of the safety elevator changed the concept of the city by making high-rise buildings possible. Elevators are widely used to transport passengers and freight vertically or at an incline in buildings and structures. Elevators are broadly classified as low-rise, medium-rise, and high-rise units. Low-rise elevators typically serve buildings with between 2 and 7 floors, medium-rise elevators serve buildings with between 5 and 20 floors, while high-rise elevators serve buildings with more than 15 floors. The speed of the elevator is indexed to the rise of the building so that the overall flight time from bottom to top, or vice versa, is approximately the same. A typical **flight time** is about one minute. Typical speeds for low-rise elevators are up to 200 ft/min (1.0 m/s). Typical speeds for medium rise elevators are up to 400 ft/min (2.0 m/s). High rise elevators typically travel at speeds of up to 1,800 ft/min (9.0 m/s).

Low rise elevators are usually oil-power hydraulic devices. The simplest version consists of a hydraulic jack buried in the ground beneath the elevator car. The jack is approximately centrally located beneath the car and the ram or plunger is connected to the platform or structure which supports the car. The car is guided by guiderails which cover the full rise of the elevator hoistway. Guideshoes or guiderollers typically guide the elevator as it ascends and descends the hoistway. The hydraulic cylinder is equipped with a cylinder head which houses the seals and bearing rings which locate the ram. The ram typically runs clear of the inner cylinder wall. Buried cylinders are typically protected against corrosion.

Hydraulic power is usually supplied by a screw-type positive-displacement pump driven by an induction motor. It is common for the motor and pump to be coaxially mounted and submersed in the hydraulic reservoir. Operating pressures are typically 300 to 600 lb/in² (2 to 4 MPa). A hydraulic control valve controls the flow of oil to the hydraulic jack and hence the speed of the elevator. In the down direction, the pump is not powered, and the elevator speed is controlled by bleeding fluid through the control valve.

Another manifestation of the **hydraulic elevator** is called the *holeless hydraulic* elevator. These typically have one or more hydraulic jacks mounted vertically alongside the elevator car, the plunger being either directly attached to the car or connected to the car by steel wire cables (also known as *wire ropes*). Elevators of the latter type are known as *roped hydraulic elevators*. These elevators are often roped 1 : 2 so that the elevator moves at twice the speed of the hydraulic ram. The rise of the elevator is also twice the stroke of the ram. Holeless hydraulic elevators with the ram directly connected to the elevator car may have single-stage jacks or multistage telescopic jacks, depending on the required rise.

Medium- and high-rise elevators are typically traction-driven units, i.e., the rotary motion of the drive sheave is transmitted to the suspension system via friction. The suspension system usually comprises steel wire cables or ropes, although other systems are emerging. Example of these include: ropes manufactured from synthetic fibers (such as aramid) encased in an elastomeric jacket and flat belts comprising several steel wire cords (each consisting of multiple wires) encased in an elastomeric coating. The elevators are typically counterweighted so that the motor and drive need overcome only the unbalanced load. The **counterweight** mass is typically the car mass plus approximately half the duty load (load of passengers or freight). With high-rise elevators, the weight of the rope is neutralized by compensating chains hung from the underside of the counterweight and looped to the bottom of the car. If the compensation weight matches the suspension rope weight, then irrespective of the position of the car, there will be no imbalance due to rope weight. The drive sheave is usually grooved to guide the rope and enhance the traction. Roping ratios may be 1 : 1 or 2 : 1. With 1 : 1 ratios the car speed is the same as the rope speed. With 2 : 1 ratios the elevator speed is half of the rope speed.

Medium-rise elevators are typically driven by geared machines that transmit the motor power to the drive sheave. Gear reduction ratios are typically in the range from 12 : 1 to 30 : 1.

Right-angle worm reduction gear sets are most common; however, helical gears are also used. These have higher operating efficiency and low wear characteristics.

High-rise elevators are typically driven by gearless machines which provide the smoothest and most precise performance of all elevators (see Fig. 10.2.29).

Fig. 10.2.29   Electric traction elevator.

Motors   AC motors have virtually completely replaced dc motors to power the driving sheave on new elevators. A large population of dc motors still exists on old elevators, and some new elevators still use this technology. Direct current motors have the advantages of good starting torque and ease of speed control. Elevator motors are obliged to develop double-rated torques at 125 percent rated current and have frequent starts, stops, reversals, and runs at constant speed. With sparkless commutation required under all conditions, commercial motors cannot as a rule meet these requirements. For constant voltage, cumulative-compound motors with heavy series fields are used. The series fields are gradually short-circuited as the motor comes up to speed, after which the motor operates shunt. The shunt field is excited permanently, the current being reduced to a low value with increased resistance instead of the circuit opening when the motor is not in operation.

Drive Control for DC Motors   The relatively few new elevators being supplied with dc drive motors are equipped with thyristor [silicon controlled rectifier (SCR)] drive controls. Older elevators still operational in the field are equipped with Ward Leonard (voltage control) systems using motor/generator sets. With voltage control, an individual dc generator is used for each behavior. The generator may be driven by either a dc or an ac motor (see Ward Leonard and Ilgner systems, above).

AC Motor and Control   Squirrel-cage induction motors are extensively used for elevators because of their ruggedness and simplicity. The absence of brushes is considered a distinct advantage. The motors are typically three-phase machines driven by variable-voltage, variable-frequency (VVVF) drives. The VVVF current is provided by an inverter. High-current-capacity power transistors modulate the power to the motor. Motion of the rotor or the elevator is often monitored by devices such as optical encoders that provide feedback to the control system for closed loop control. Permanent-magnet-rotor synchronous motors with wound field coils are becoming more prevalent for elevator use, because of their high efficiency, compact design, and reliability. These are invariably driven by VVVF drives.

Elevators installed in the United States are usually required to comply with the ANSI/ASME A17.1 Safety Code or other applicable local codes. The A17.1 code imposes specific safety requirements pertaining to the mechanical, structural, and electrical integrity of the elevator. Examples of **code requirements** include: factors of safety for mechanical and structural elements; the number and type of wire ropes to be used; the means of determining the load capacity of the elevator; the requirements for speed governors and independent safety devices to arrest the car should the car descend at excessive speeds; the requirements for brakes to hold the stationary car with loads of up to 125 percent of rated load; the requirements for buffers at the bottom of the elevator shaft to decelerate the elevator or counterweight should they descend beyond the lowest landing; the type and motion of elevator and hoistway doors; the requirement that doors not open if the elevator is not at a landing; the requirement that door motion cease or reverse if a passenger or obstruction is in their path and that the kinetic energy of the door motion be limited so as to minimize the impact with a passenger or obstruction should they fail to stop in time. The code also requires electrical protective devices which stop the car should it transcend either upper or lower terminal landings, remove power to the motor, and cause the brake to apply should the speed exceed governor settings, etc. The A17.1 code also provides special requirements for elevators located in high seismic risk zones.

Loads Table 10.2.13 shows the rated load of passenger elevators.

Table 10.2.13 Rated Load of Passenger Elevators

Rated Capacity,	lb	2,500	3,500	4,500	6,000	9,000	12,000
	kg	1,135	1,590	2,045	2,725	4,085	5,450
Net inside area,	ft²	29.1	38.0	46.2	57.7	80.5	103.0
Platform area,	m²	2.7	3.5	4.3	5.4	7.5	9.6

Efficiencies and Energy Dissipation per Car Mile The overall efficiency and hence the energy dissipation varies greatly from elevator to elevator, depending on the design, hoistway conditions, loading conditions, acceleration and deceleration profiles, number of stops, etc.

For a load of 2,500 lb (1,130 kg) for **geared traction elevators** driven at 350 ft/min (1.75 m/s) by SCR-controlled dc motor or VVVF-controlled ac motor, typical values are as follows: efficiency 60 percent, energy dissipation 4.5 kWh (16 MJ).

Under same conditions for **gearless traction elevators** traveling at 700 ft/min (3.5 m/s), the efficiency may be 70 percent and the energy dissipation 4 kWh (14.5 MJ).

Note that efficiencies are based on net load, i.e., full load minus overbalance.

Car Mileage and Stops (per elevator, 8-h day)

Office buildings, local elevators: intensive service, 12 to 20 car miles (19 to 32 car kilometers), making about 150 regular stops per mile

Express elevators: 20 to 40 car miles (32 to 64 car kilometers), making about 75 to 100 regular stops per mile

Department store elevators: 4 to 8 car miles (6 to 13 car kilometers), making about 350 regular stops per mile.

Operational Control of Elevators Most modern passenger elevators are on fully automatic group collective control. Each group of elevators is controlled by a dispatching system which assigns specific elevators to answer specific hall calls. Modern dispatch systems are microprocessor-based or have some equivalent means of processing information so that each call registered for an elevator is answered in a timely manner. The determination as to which elevator is assigned to answer a specific call is a complex process which in the case of microprocessor-controlled dispatch systems requires the use of sophisticated algorithms which emulate the building dynamics. Some dispatch systems use advanced decision-making processes based on *artificial intelligence* such as *expert systems* and *fuzzy logic* to optimize group collective service in buildings. With better dispatching systems it is possible to achieve excellent elevator service in buildings with the minimum number of elevators, or the efficient service of taller buildings without devoting more space to elevators.

The operational control systems also provide signals for door opening and closing.

Dumbwaiters follow the general design philosophy of elevators except that code requirements are somewhat more relaxed. For example, roller-link chain can be used for support instead of wire rope. Moreover, a single steel wire rope can be used instead of the mandatory minimum of three for traction-type elevators and two for drum-type elevators. Dumbwaiters are not intended for the transportation of people.

Escalators have the advantages of continuity of motion, great capacity, and small amounts of space occupied and current consumed for each passenger carried. Escalators are built with step widths of 24, 32, and 40 in (610, 810, and 1015 mm). The angle of incline is 30° from the horizontal, and the speed is 90 to 120 ft/min (0.45 to 0.6 m/s).

Approximate average handling capacities for escalators traveling at 90 ft/min are 2,000, 2,300, and 4,000 passengers per hour for 24-in, 32-in, and 40-in (610-, 810-, and 1015-mm) escalator step sizes, respectively. Higher-speed escalators have proportionally higher passenger-carrying capacity.

Escalators are equipped with a handrail on each side, mounted on the balustrade. The handrail moves at the same speed as the escalator. Escalator steps are so arranged that as they approach the upper and lower landings they recede so that they are substantially level with the floor for safe embarkation and disembarkation of the escalator.

10.3 DRAGGING, PULLING, AND PUSHING
by Harold V. Hawkins
revised by Ernst K. H. Marburg and Associates

HOISTS, PULLERS, AND WINCHES

Many of the fundamental portable lifting mechanisms such as hoists or pullers (see above) can also be used forcefully to drag or pull materials. In addition, nonmobile versions, called **winches**, utilizing hoisting drums can also be used.

LOCOMOTIVE HAULAGE, COAL MINES
by Burt Garofab
Pittston Corporation

The haulage system of an underground mine is used to transport personnel and material between the face and the portal. It can be subdivided into face haulage and main haulage. The main haulage system extends from the end of the face haulage system to the outside.

Rubber-tired haulage at the coal face was introduced in 1935 and received further impetus with the introduction of the rubber-tired shuttle car in 1938. Crawler-mounted loaders and rubber-tired universal coal cutters completed the equipment needed for complete off-track mining. This off-track mining caused a revolution in face haulage, since it eliminated the expense of laying track in the rooms and advancing the track as the face of the coal advanced. It also made gathering locomotives of the cable-reel, crab-reel, and battery types obsolete. Practically all the gathering duty is now performed by rubber-tired shuttle cars, chain conveyors, extensible belt conveyors, and other methods involving offtrack equipment. Some mines eliminated track completely by having belt conveyors carry the coal to the outside.

Most coal mines today utilize a combination of haulage systems. Personnel and supplies are transported by either rail or rubber-tired haulage. Coal is transported by belt or rail, predominantly belt.

Locomotives used in rail haulage may be trolley wire powered, battery powered, or diesel powered. Battery and diesel powered locomotives are commonly used to transport supplies in mines which utilize belt as the main coal haulage system. Trolley-wire-powered locomotives are used when rail is the main coal haulage system. The sleek, streamlined, fast, and easy-riding **Jeffrey** eight-wheel four-motor locomotives are available in 27-, 37-, and 50-short ton (24-, 34-, and 45-tonne) sizes. This type of locomotive has two four-wheel trucks, each having two motors. The trucks, having Pullman or longitudinal-type equalizers with snubbers, will go around a 50 ft (15 m) radius curve, have short overhang, and provide a very easy ride. This construction is easy on the track, with consequent low track-maintenance cost. Speed at full load ranges from 10 to 12.5 mi/h (4.5 to 5.6 m/s), and the maximum safe speed is approximately 30 mi/h (13.4 m/s).

The **eight-wheel locomotive** usually has a box frame; series-wound motors with single-reduction spur gearing; 10 steps of straight parallel, full electropneumatic contactor control; dynamic braking; 32-V battery-operated control and headlights, with the battery charged automatically from the trolley; straight air brakes; air-operated sanders; air horn; automatic couplers; one trolley pole; two headlights at each end; and blowers to ventilate the traction motors. The equipment is located so that it is easily accessible for repair and maintenance.

The eight-wheel type of locomotive has, to a great extent, superseded the tandem type, consisting of two four-wheel two-motor locomotives coupled together and controlled from the cab of one of the units of the tandem.

The older Jeffrey **four-wheel-type locomotive** is available in 11-, 15-, 20-, and 27-short ton (10-, 13.6-, 18-, 24.5-tonne) nominal weights. The 20- and 27-short ton (18- and 24.5-tonne) sizes have electrical equipment very much like that of the eight-wheel-type locomotive. Speeds are also comparable. The 15-short ton (13.6-tonne) locomotive usually has semielectropneumatic contactor control, rather than full electropneumatic control, and dynamic braking but usually does not have the 32-V battery-operated control. The 11-short ton (10-tonne) locomotive has manual control, with manual brakes and sanders, although contactor control, air brakes, blowers, etc., are optional.

Locomotive motors have a horsepower rating on the basis of 1 h at 75°C above an ambient temperature of 40°C. Sizes range from a total of 100 hp (75 kW) for the 11-ton (10-tonne) to a total of 720 hp for the 50-ton.

The following formulas are recommended by the Jeffrey Mining Machinery Co. to **determine the weight of a locomotive** required to haul a load. Table 10.3.1 gives haulage capacities of various weights of locomotives on grades up to 5 percent. The tabulation of haulage capacities shows how drastically the tons of trailing load decrease as the grade increases. For example, a 50-ton locomotive can haul 1,250 tons trailing load on the level but only 167 tons up a 5 percent grade.

The following formulas are based on the use of steel-tired or rolled-steel wheels on clean, dry rail.

Weight of locomotive required on level track:

$$W = L(R + A)/(0.3 \times 2{,}000 - A)$$

Weight of locomotive required to haul train up the grade:

$$W = L(R + G)/(0.25 \times 2{,}000 - G)$$

Weight of locomotive necessary to start train on the grade:

$$W = L(R + G + A)/(0.30 \times 2{,}000 - G - A)$$

where W is the weight in tons of locomotive required; R is the frictional resistance of the cars in pounds per ton and is taken as 20 lb for cars with antifriction bearings and 30 lb for plain-bearing cars; L is the weight of the load in tons; A is the acceleration resistance [this is 100 for 1 mi/(h · s) and is usually taken at 20 for less than 10 mi/h or at 30 from 10 to 12 mi/h, corresponding to an acceleration of 0.2 or 0.3 mi/(h · s)]; G is the grade resistance in pounds per ton or 20 lb/ton for each percent of grade (25 percent is the running adhesion of the locomotive, 30 percent is the starting adhesion using sand); 2,000 is the factor to give adhesion in pounds per ton.

Where the grade is in favor of the load:

$$W = L(G - R)/(0.20 \times 2{,}000 - G)$$

To brake the train to a stop on grade:

$$W = L(G + B - R)/(0.20 \times 2{,}000 - G - B)$$

where B is the braking (or decelerating) effort in pounds per ton and equals 100 lb/ton for a braking rate of 1 mi/(h · s) or 20 lb/ton for a braking rate of 0.2 mi/(h · s) or 30 lb for a braking rate of 0.3 mi/(h · s). The adhesion is taken from a safety standpoint as 20 percent. It is not advisable to rely on using sand to increase the adhesion, since the sandboxes may be empty when sand is needed.

Time in seconds to brake the train to stop:

$$s = \frac{\text{mi/h (start)} - \text{mi/h (finish)}}{\text{deceleration in mi/(h · s)}}$$

Distance in feet to brake the train to a stop:

$$\text{ft} = [\text{mi/h (start)} - \text{mi/h (finish)}] \times s \times 1.46/2$$

Storage-battery locomotives are used for hauling muck cars in tunnel construction where it is inconvenient to install trolley wires and bond the track as the tunnel advances. They are also used to some extent in metal mines and in mines of countries where trolley locomotives are not permitted. They are often used in coal mines in the United States for hauling supplies. Their first cost is frequently less than that for a trolley installation. They also possess many of the advantages of the trolley locomotive and eliminate the danger and obstruction of the trolley wire. Storage-battery locomotives are limited by the energy that is stored in the battery and should not be used on steep grades or where large, continuous overloads are required. Best results are obtained where light

Table 10.3.1   Haulage Capacities of Locomotives with Steel-Tired or Rolled-Steel Wheels*

Grade		Weight of locomotive, tons†					
		11	15	20	27	37	50
Level	Drawbar pull, lb	5,500	7,500	10,000	13,500	18,500	25,000
	Haulage capacity, gross tons	275	375	500	675	925	1,250
1%	Drawbar pull, lb	5,280	7,200	9,600	12,960	17,760	24,000
	Haulage capacity, gross tons	132	180	240	324	444	600
2%	Drawbar pull, lb	5,260	6,900	9,200	12,420	17,020	23,000
	Haulage capacity, gross tons	88	115	153	207	284	384
3%	Drawbar pull, lb	4,840	6,600	8,800	11,880	16,280	22,000
	Haulage capacity, gross tons	67	82	110	149	204	275
4%	Drawbar pull, lb	4,620	6,300	8,400	11,340	15,540	21,000
	Haulage capacity, gross tons	46	63	84	113	155	210
5%	Drawbar pull, lb	4,400	6,000	8,000	10,800	14,800	20,000
	Haulage capacity, gross tons	31	50	67	90	123	167

* Jeffrey Mining Machinery Co.
† Haulage capacities are based on 20 lb/ton rolling friction, which is conservative for roller-bearing cars. Multiply lb by 0.45 to get kg and tons by 0.91 to get tonnes.

and medium loads are to be handled intermittently over short distances with a grade of not over 3 percent against the load.

The general construction and mechanical features are similar to those of the four-wheel trolley type, with battery boxes located either on top of the locomotive or between the side frames, according to the height available. The motors are rugged, with high efficiency. Storage-battery locomotives for coal mines are generally of the explosion-tested type approved by the Bureau of Mines for use in gaseous mines.

The **battery** usually has sufficient capacity to last a single shift. For two- or three-shift operation, an extra battery box with battery is required so that one battery can be charging while the other is working on the locomotive. Motor-generator sets or rectifiers are used for charging the batteries. The overall efficiency of the battery, motor, and gearing is approximately 63 percent. The speed varies from 3 to 7 mi/h, the average being 3½ to 4½ mi/h. Battery locomotives are available in sizes from 2 to 50 tons. They are usually manufactured to suit individual requirements, since the sizes of motors and battery are determined by the amount of work that the locomotive has to do in an 8-h shift.

INDUSTRIAL CARS

Various types of narrow-gage industrial cars are used for handling bulk and package materials inside and outside of buildings. Those used for bulk material are usually of the dumping type, the form of the car being determined by the duty. They are either pushed by workers or drawn by mules, locomotives, or cable. The **rocker side-dump car** (Fig. 10.3.1) consists of a truck on which is mounted a V-shaped steel body supported on rockers so that it can be tipped to either side, discharging material. This type is mainly used on construction work. Capacities vary from ⅔ to 5 tons for track gages of 18, 20, 24, 30, 36, and 56½ in. In the **gable-bottom car** (Fig. 10.3.2), the side doors *a* are hinged at the top and controlled by levers *b* and *c*, which lock the doors when closed. Since this type of car discharges material close to the ground on both sides of

Fig. 10.3.1   Rocker side-dump car.

Fig. 10.3.2   Gable-bottom car.

the track simultaneously, it is used mainly on trestles. Capacities vary from 29 to 270 ft³ for track gages of 24, 36, 40, and 50 in. The **scoop dumping car** (Fig. 10.3.3) consists of a scoop-shaped steel body pivoted at *a* on turntable *b*, which is carried by the truck. The latch *c* holds the body in a horizontal position, being released by chain *d* attached to handle *e*. Since the body is mounted on a turntable, the car is used for service where it is desirable to discharge material at any point in the

circle. This car is made with capacities from 12 to 27 ft³ to suit local requirements. The **hopper-bottom car** (Fig. 10.3.4) consists of a hopper on wheels, the bottom opening being controlled by door *a*, which is operated by chain *b* winding on shaft *c*. The shaft is provided with handwheel and ratchet and pawl. The type of door or gate controlling the bottom opening varies with different materials.

Fig. 10.3.3   Scoop dumping car.

Fig. 10.3.4   Hopper-bottom car.

The **box-body dump car** (Fig. 10.3.5) consists of a rectangular body pivoted on the trucks at *a* and held in horizontal position by chains *b*. The side doors of the car are attached to levers so that the door is automatically raised when the body of the car is tilted to its dumping position. The cars can be dumped to either side. On the large sizes, where rapid dumping is required, dumping is accomplished by compressed air. This type of car is primarily used in excavation and quarry work, being loaded by power shovels. The greater load is placed on the side on which the car will dump, so that dumping is automatic when the operator releases the chain or latch. The car bodies may be steel or steel-lined wood. **Mine cars** are usually of the four-wheel type, with low bodies, the doors being at one end, and pivoted at the top with latch at

Fig. 10.3.5   Box-body dump car.

the bottom. Industrial **tracks** are made with rails from 12 to 45 lb/yd (6.0 to 22 kg/m) and gages from 24 in to 4 ft 8½ in (0.6 to 1.44 m). Either steel or wooden ties are used. Owing to its lighter weight, the steel tie is preferred where tracks are frequently moved, the track being made up in sections. Industrial cars are frequently built with one wheel attached to the axle and the other wheel loose to enable the car to turn on short-radius tracks. Capacities vary from 4 to 50 yd³ (3.1 to 38 m³) for track gages of 36 to 56½ in (0.9 to 1.44 m), with cars having weights from 6,900 to 80,300 lb (3,100 to 36,000 kg). The frictional resistance per ton (2,000 lb) (8,900 N) for different types of mine-car bearings are given in Table 10.3.2.

Table 10.3.2   Frictional Resistance of Mine Car Bearings

	Drawbar pull					
	Level track		2% grade		4% grade	
Types of bearings	lb/short ton	N/tonne	lb/short ton	N/tonne	lb/short ton	N/tonne
Spiral roller	13	58	15	67	46	205
Solid roller	14	62	18	80	53	236
Self-oiling	22	98	31	138		
Babbitted, old style	24	107	40	178		

DOZERS, DRAGLINES

The dual capability of some equipment, such as **dozers** and **draglines,** suggests that it should be mentioned as prime machinery in the area of materials handling by dragging, pulling, or pushing. Dozers are described in the discussion on earthmoving equipment since their basic frames are also used for power shovels and backhoes. In addition, dozers perform the auxiliary function of pushing carryall earthmovers to assist them in scraping up their load. Dragline equipment is discussed with below-surface handling or excavation. The same type of equipment that would drag or scrape may also have a lifting function.

MOVING SIDEWALKS

Moving horizontal belts with synchronized balustrading have been introduced to expedite the movement of passengers to or from railroad trains in depots or planes at airports (see belt conveyors). A necessary feature is the need to prevent the clothing of anyone (e.g., a child) sitting on the moving walk from being caught in the mechanism at the end of the walk. Use of a comblike stationary end fence protruding down into longitudinal slots in the belt is an effective preventive.

CAR-UNLOADING MACHINERY

Four types of devices are in common use for unloading material from all types of open-top cars: crossover and horn dumps, used to unload mine cars with swinging end doors; rotary car dumps, for mine cars without doors; and tipping car dumps, for unloading standard-gage cars where large unloading capacity is required.

Crossover Dump   Figure 10.3.6 shows a car in the act of dumping. Figure 10.3.7 shows a loaded car pushing an empty car off the dump. A section of track is carried by a platform supported on rockers a. An extension bar b carries the weight c and the brake friction bar d. A hand lever controls the brake, acting on the friction bar and placing the dumping under the control of the operator. A section of track e in front

Fig. 10.3.6   Crossover dump: car unloading.

Fig. 10.3.7   Crossover dump: empty car being pushed away.

of the dump is pivoted on a parallel motion and counterbalanced so that it is normally raised. The loaded car depresses the rails e and, through levers, pivots the horns f around the shafts g, releasing the empty car. The loaded car strikes the empty car, starting it down the inclined track. After the loaded car has passed the rails e, the springs return the horns f so that they stop the loaded car in the position to dump. Buffer springs on the shaft g absorb the shock of stopping the car. Since the center of gravity of the loaded car is forward of the rockers, the car will dump automatically under control of the brake. No power is required for this dump, and one operator can dump three or four cars per minute.

Rotary Gravity Dump   (Fig. 10.3.8) This consists of a steel cylinder supported by a shaft a, its three compartments carrying three tracks.

The loaded car b is to one side of the center and causes the cylinder to rotate, the material rolling to the chute beneath. The band brake c, with counterweight d, is operated by lever e, putting the dumping under control of the operator. No power is required; one operator can dump two or three cars per minute.

Fig. 10.3.8   Rotary gravity dump.

Rotary-power dumpers are also built to take any size of open-top railroad car and are frequently used in power plants, coke plants, ports, and ore mines to dump coal, coke, ore, bauxite, and other bulk material. They are mainly of two types: (1) single barrel, and (2) tandem.

The McDowell-Wellman Engineering Co. dumper consists of a revolving cradle supporting a platen (with rails in line with the approach and runoff car tracks in the upright position), which carries the car to be dumped. A blocking on the dumping side supports the side of the car as the cradle starts rotating. Normally, the platen is movable and the blocking is fixed, but in some cases the platen is fixed and the blocking movable. Where there is no variation in the car size, the platen and blocking are both fixed. The cradle is supported on two end rings, which are bound with a rail and a gear rack. The rail makes contact with rollers mounted in sills resting on the foundation. Power through the motor rotates the cradle by means of two pinions meshing with the gear racks. The angle of rotation for a complete dump is 155° for a normal operation, but occasionally, a dumper is designed for 180° rotation. The clamps, supported by the cradle, start moving down as the dumper starts to rotate. These clamps are lowered, locked, released, and raised either by a gravity-powered mechanism or by hydraulic cylinders.

With the advent of the **unit-train system,** the investment and operating costs for a dumper have been reduced considerably. The design of dumpers for unit train has improved and results in fewer maintenance problems. The use of rotary couplers on unit train eliminates uncoupling of cars while dumping because the center of the rotary coupler is in line with the center of rotation of the dumper.

Car Shakers   As alternatives to rotating or tilting the car, several types of car shakers are used to hasten the discharge of the load. Usually the shaker is a heavy yoke equipped with an unbalanced pulley rotated at 2,000 r/min by a 20-hp (15-kW) motor. The yoke rests upon the car top sides, and the load is actively vibrated and rapidly discharged. While a car shaker provides a discharge rate about half that of a rotary dumper, the smaller investment is advantageous.

Car Positioner   As the popularity of unit-train systems consisting of rail cars connected by rotary couplers has increased, more rotary dumping stations have been equipped with an automatic train positioner developed by McDowell-Wellman Engineering Company. This device consists of a carriage moving parallel to the railroad track actuated by either hydraulic cylinders or wire rope driven by a winch, which carries an arm that rotates in a vertical plane to engage the coupling between the cars. The machines are available in many sizes, the largest of which are capable of indexing 200-car trains in one- or two-car increments through the rotary dumper. These machines or similar ones are also available from FMC/Materials Handling System Division, Heyl and Patterson, Inc., and Whiting Corp.

10.4 LOADING, CARRYING, AND EXCAVATING

by Ernst K. H. Marburg and Associates

CONTAINERIZATION

The proper packaging of material to assist in handling can significantly minimize the handling cost and can also have a marked influence on the type of handling equipment employed. For example, partial carload lots of liquid or granular material may be shipped in rigid or nonrigid containers equipped with proper lugs to facilitate in-transit handling. Heavy-duty rubberized containers that are inert to most cargo are available for repeated use in shipping partial carloads. The nonrigid container reduces return shipping costs, since it can be collapsed to reduce space. Disposable lightweight corrugated-cardboard shipping containers for small and medium-sized packages both protect the cargo and permit stacking to economize on space requirements. The type of container to be used should be planned or considered when the handling mechanism is selected.

SURFACE HANDLING
by Colin K. Larsen

Blue Giant Equipment Co.

Lift Trucks and Palletized Loads

The basis of all efficient handling, storage, and movement of unitized goods is the cube concept. Building a cube enables a large quantity of unit goods to be handled and stored at one time. This provides greater efficiency by increasing the volume of goods movement for a given amount of work. Examples of cube-facilitating devices include pallets, skids, slip sheets, bins, drums, and crates.

The most widely applied cube device is the pallet. A **pallet** is a low platform, usually constructed of wood, incorporating openings for the forks of a lift truck to enter. Such openings are designed to enable a lift truck to pick up and transport the pallet containing the cubed goods.

Lift truck is a loose term for a family of pallet handling/transporting machines. Such machines range from manually propelled low-lift devices (Fig. 10.4.1) to internal combustion and electric powered ride-on high-lift devices (Fig. 10.4.2). While some machines are substitutes in terms of function, each serves its own niche when viewed in terms of individual budgets and applications.

Pallet trucks are low-lift machines designed to raise loaded pallets sufficiently off the ground to enable the truck to transport the pallet horizontally. Pallet trucks are available as manually operated and propelled models that incorporate a hydraulic pump and handle assembly (Fig. 10.4.1). This pump and handle assembly enables the operator to raise the truck forks, and push/pull the load. Standard manual pallet trucks are available in lifting capacities from 4,500 to 5,500 lb

(2,045 to 2,500 kg), with customer manufactured models to 8,000 lb (3,636 kg). While available in a variety of fork sizes, by far the most common is 27 in wide × 48 in long (686 mm × 1,219 mm). This size accommodates the most common pallet sizes of 40 in × 48 in (1,016 mm × 1,219 mm) and 48 in × 48 in (1,219 mm × 1,219 mm). Pallet trucks are also available in motorized versions, equipped with dc electric motors to electrically raise and transport. The power supply for these

Fig. 10.4.2 Lift truck powered by an internal-combustion engine.

trucks is an on-board lead-acid traction battery that is rechargeable when the truck is not in use. Control of these trucks is through a set of lift, lower, speed, and direction controls fitted into the steering handle assembly. Powered pallet trucks are available in walk and ride models. Capacities range from 4,000 to 10,000 lb (1,818 to 4,545 kg), with forks up to 96 in (2,438 mm) long. The longer fork models are designed to allow the truck to transport two pallets, lined up end to end.

Stackers, as the name implies, are high lift machines designed to raise and stack loaded pallets in addition to providing horizontal transportation.

Fig. 10.4.3 Counterbalanced electric-battery-powered pallet truck. (*Blue Giant.*)

Fig. 10.4.1 Manually operated and propelled pallet truck.

Stackers are separated into two classes: straddle and counterbalanced. **Straddle stackers** are equipped with legs which straddle the pallet and provide load and truck stability. The use of the straddle leg system results in a very compact chassis which requires minimal aisle space for turning. This design, however, does have its trade-offs insomuch as the straddles limit the truck's usage to smooth level floors. The limited leg underclearance inherent in these machines prohibits their use on dock levelers for loading/unloading transport trucks. Straddle stackers are available from 1,000 to 4,000 lb (455 to 1,818 kg) capacity with lift heights to 16 ft (4,877 mm). **Counterbalanced stackers** utilize a counter-weight system in lieu of straddle legs for load and vehicle stability (Fig. 10.4.3). The absence of straddle legs results in a chassis with increased underclearance which can be used on ramps, including dock levelers. The counterbalanced chassis, however, is longer than its straddle counterpart, and this requires greater aisle space for maneuvering. For materials handling operations that require one machine to perform a multitude of tasks, and are flexible in floor layout of storage areas, the counterbalanced stacker is the recommended machine.

Off-Highway Vehicles and Earthmoving Equipment

by B. Douglas Bode, John Deere and Co.

The movement of large quantities of bulk materials, earth, gravel, and broken rock in road building, mining, construction, quarrying, and land clearing may be handled by **off-highway vehicles.** Such vehicles are mounted on large pneumatic tires or on crawler tracks if heavy pulling and pushing are required on poor or steep terrain. Width and weight of the rubber-tired equipment often exceed highway legal limits, and use of grouser tracks on highways is prohibited. A wide range of working tool attachments, which can be added (and removed) without modification to the basic machine, are available to enhance the efficiency and versatility of the equipment.

Proper selection of size and type of equipment depends on the amount, kind, and density of the material to be moved in a specified time and on the distances, direction, and steepness of grades, footing for traction, and altitude above sea level. Time cycles and pay loads for production per hour can then be estimated from manufacturers' performance data and job experience. This production per hour, together with the corresponding owning, operating, and labor costs per hour, enables selection by favorable cost per cubic yard, ton, or other pay unit.

Current rapid progress in the development of off-highway equipment will soon make any description of size, power, and productivity obsolete. However, the following brief description of major off-highway vehicles will serve as a guide to their applications.

Crawler Tractors   These are track-type prime movers for use with mounted bulldozers, rippers, winches, cranes, cable layers, and side booms rated by net engine horsepower in sizes from 40 to over 500 hp; maximum traveling speeds, 5 to 7 mi/h (8 to 11 km/h). Crawler tractors develop drawbar pulls up to 90 percent or more of their weight with mounted equipment.

Wheel Tractors   Sizes range from rubber-tired industrial tractors for small scoops, loaders, and backhoes to large, diesel-powered, two- and four-wheel drive pneumatic-tired prime movers for propelling scrapers and wagons. Large, four-wheel-drive, articulated-steering types also power bulldozers.

Bulldozer—Crawler Type   (Fig. 10.4.4) This is a crawler tractor with a front-mounted blade, which is lifted by hydraulic or cable power control. There are four basic types of moldboards: straight, semi-U and U (named by top-view shape), and angling. The angling type, often called **bullgrader** or **angledozer,** can be set for side casting 25° to the right or left of perpendicular to the tractor centerline, while the other blades can be tipped forward or back through about 10° for different digging conditions. All blades can be tilted for ditching, with hydraulic-power tilt available for all blades.

APPLICATION.   This is the best machine for pioneering access roads, for boulder and tree removal, and for short-haul earthmoving in rough terrain. It push-loads self-propelled scrapers and is often used with a rear-mounted ripper to loosen firm or hard materials, including rock, for

Fig. 10.4.4   Crawler tractor with dozer blade. (*John Deere.*)

scraper loading. U blades drift 15 to 20 percent more loose material than straight blades but have poor digging ability. Angling blades expedite sidehill benching and backfilling of trenches. Loose-material capacity of straight blades varies approximately as the blade height squared, multiplied by length. Average capacity of digging blades is about 1 yd^3 loose measure per 30 *net* hp rating of the crawler tractor. Payload is 60 to 90 percent of loose measure, depending on material swell variations.

Bulldozer—Wheel Type   This is a four-wheel-drive, rubber-tired tractor, generally of the hydraulic articulated-steering type, with front-mounted blade that can be hydraulically raised, lowered, tipped, and tilted. Its operating weights range to 150,000 lb, with up to 700 hp, and its traveling speeds range from stall to about 20 mi/h for pushing and mobility.

APPLICATION.   It is excellent for push-loading self-propelled scrapers, for grading the cut, spreading and compacting the fill, and for drifting loose materials on firm or sandy ground for distances up to 500 ft. Useful tractive effort on firm earth surfaces is limited to about 60 percent of weight, as compared with 90 percent for crawler dozers.

Compaction Equipment   Compactors are machines, either self-propelled or pull-type, consisting of drums, rollers, or wheels and combinations of wheels and drums. Drums can be smooth or sheepsfoot (drums with protrusions to provide a blunt point of contact with the soil), standard or vibrating. They contact the surface, causing pressure to compact the soil. Sizes range from small walk-behind units to land-fill compactors weighing 100,000 lb (45.4 tonnes) or more.

APPLICATION.   Sheepsfoot compactors are used to compact soil in road building, dams, and site preparation. Smooth drums and rubber-tired wheels are used to finish compacting granular surfaces and asphalt pavement. Compaction is needed to increase soil and surface density to support heavy loads.

Loader—Crawler Type   (Fig. 10.4.5) This is a track-type prime mover with front-mounted bucket that can be raised, dumped, lowered, and tipped by power control. Capacities range from 0.7 to 5.0 yd^3 (0.5 to 3.8 m^3), SAE rated. It is also available with grapples for pulpwood, logs, and lumber.

Fig. 10.4.5   Crawler tractor with loader bucket. (*John Deere.*)

APPLICATION.   It is used for digging basements, pools, ponds, and ditches; for loading trucks and hoppers; for placing, spreading, and compacting earth over garbage in sanitary fills; for stripping sod; for removing steel-mill slag; and for carrying and loading pulpwood and logs.

Loader—Wheel Type   (Fig. 10.4.6) This is a four-wheel, rubber-tired, articulated-steer machine equipped with a front-mounted, hydraulic-powered bucket and loader linkage that loads material into the bucket through forward motion of the machine and which lifts, transports, and discharges the material. The machine is commonly referred to as a *four-wheel-drive loader tractor*. Bucket sizes range from ½ yd^3 (0.4 m^3) to more than 20 yd^3 (15 m^3), SAE rated capacity. The addition

Fig. 10.4.6   Four-wheel-drive loader. (*John Deere.*)

of a quick coupler to the loader linkage permits convenient interchange of buckets and other working tool attachments, adding versatility to the loader. Rigid-frame machines with variations and combinations of front/rear/skid steer, front/rear drive, and front/rear engine are also used in various applications.

APPLICATION.   Four-wheel-drive loaders are used primarily in construction, aggregate, and utility industries. Typical operations include truck loading, filling hoppers, trenching and backfilling, land clearing, and snow removal.

Loader—Skid Steer   (Fig. 10.4.7) This is a four-wheel or rubber-tracked skid-steer unit equipped with a hydraulically operated loader linkage and bucket that loads material into the bucket by forward movement of the unit and which lifts, transports loads, and dumps material. Bucket sizes range from ½ yd^3 (0.38 m^3) to 1¾ yd^3 (1.34 m^3), with the weight of these units ranging from 4000 lb (1.8 tonnes) to 10,000 lb (4.5 tonnes). These units use a quick coupler on the front to accommodate a large variety of other attachments such as forks, blades, rakes, and a host of hydraulically driven attachments such as brooms, hammers, planers, augers, and trenchers. The small size and the versatility of the unit contribute to its increased popularity.

APPLICATION. Skid-steer loaders are used in all types of job sites where small power units are required, such as landscaping and nurseries, home

Fig. 10.4.7   Skid-steer loader. (*John Deere.*)

and commercial construction, bridge and road repair, and agriculture. They are used to lift, transport, load materials with buckets or forks, break concrete with hammers, plane high spots, and grade and smooth a landscape. They are a true multipurpose utility unit.

Backhoe Loader   (Fig. 10.4.8) This is a self-propelled, highly mobile machine with a main frame to support and accommodate both the rear-mounted backhoe and front-mounted loader. The machine was designed with the intention that the backhoe will normally remain in place when the machine is being used as a loader and vice versa. The backhoe digs, lifts, swings, and discharges material while the machine is stationary. When used in the loader mode, the machine loads material into the bucket through forward motion of the machine and lifts, transports, and discharges the material. Backhoe loaders are categorized according to digging depth of the backhoe. Backhoe loader types include variations of front/rear/articulated and all-wheel steer and rear/four-wheel drive.

Fig. 10.4.8   Backhoe loader. (*John Deere.*)

APPLICATION.   Backhoe loaders are used primarily for trenching and backfilling operations in the construction and utility industries. Quick couplers for the loader and backhoe are available which quickly interchange the working tool attachments, thus expanding machine capabilities. Backhoe loader mobility allows the unit to be driven to nearby job sites, thus minimizing the need to load and haul the machine.

Scrapers   (Fig. 10.4.9) This is a self-propelled machine, having a cutting edge positioned between the front and rear axles which loads, transports, discharges, and spreads material. Tractor scrapers include open-bowl and self-loading types, with multiple steer and drive axle variations. Scraper rear wheels may also be driven by a separate rear-mounted engine which minimizes the need for a push tractor. Scraper ratings are provided in cubic yard struck/heaped capacities. Payload capacities depend on loadability and swell of materials but approximate the struck capacity. Crawler tractor-drawn, four-wheel rubber-tired scrapers have traditionally been used in a similar manner—normally in situations with shorter haul distances or under tractive and terrain conditions that are unsuitable for faster, self-propelled scrapers.

APPLICATION. Scrapers are used for high-speed earth moving, primarily in road building and other construction work where there is a need to move larger volumes of material relatively short distances. The convenient

Fig. 10.4.9   Two-axle articulated self-propelled elevating scraper. (*John Deere.*)

control of the cutting edge height allows for accurate control of the grade in either a cut or fill mode. The loaded weight of the scraper can contribute to compaction of fill material. All-wheel-drive units can also load each other through a push-pull type of attachment. Two-axle, four-wheel types have the best maneuverability; however, the three-axle type is sometimes preferred for operator comfort on longer, higher-speed hauls.

Motor Grader (Fig. 10.4.10) This is a six-wheel, articulated-frame self-propelled machine characterized by a long wheelbase and mid-mounted blade. The blade can be hydraulically positioned by rotation about a vertical axis—pitching fore/aft, shifting laterally, and independently raising each end—in the work process of cutting, moving, and spreading material, to rough- or finish-grade requirements. Motor graders range in size to 60,000 lb (27,000 kg) and 275 hp (205 kW) with typical transport speeds in the 25-mi/h (40-km/h) range. Rigid-frame machines with various combinations of four/six wheels, two/four-wheel drive, front/rear-wheel steer are used as dictated by the operating requirements.

Fig. 10.4.10 Six-wheel articulated-frame motor grader. *(John Deere.)*

APPLICATION. Motor graders are the machine of choice for building paved and unpaved roads. The long wheelbase in conjunction with the midmounted blade and precise hydraulic controls allows the unit to finish-grade road beds within 0.25 in (6 mm) prior to paving. The weight, power, and blade maneuverability enable the unit to perform all the necessary work, including creating the initial road shape, cutting the ditches, finishing the bank slopes, and spreading the gravel. The motor grader is also a cost-effective and vital part of any road maintenance fleet.

Excavators (Fig. 10.4.11) This is a mobile machine that is propelled by either a crawler-track or rubber-tired undercarriage, with the unique feature being an upper structure that is capable of continuous rotation and a wide working range. The unit digs, lifts, swings, and dumps material by action of the boom, arm, or telescoping boom and bucket. Excavators include the hoe type (bucket cuts toward the unit and

Fig. 10.4.11 Excavator with tracked undercarriage. *(John Deere.)*

generally downward) and the shovel type (bucket cuts toward the unit and generally upward). Weight of the machines range from mini [2000 lb (0.9 tonnes)] to giant [1,378,000 lb (626 tonnes)] with power ratings from 8 to 3644 hp (5.9 to 2717 kW).

APPLICATION. The typical attachment for the unit is the bucket, which is used for trenching for the placement of pipe and other underground utilities, digging basements and footings for building foundations, loading trucks in mass excavation sites, and maintaining and grading steep slopes of retention ponds. Other specialized attachments include hydraulic hammers and compactors, thumbs, clamshells, grapples, and long-reach fronts that expand the capabilities of the excavator.

Dump Trucks (Fig. 10.4.12) A dump truck is a self-propelled machine, having an open body or bin, that is designed to transport and dump or spread material. Loading is accomplished by means external to the dumper with another machine. Types are generally categorized into rigid-frame or articulated-steer. Most rigid-frame trucks are rear dump with two axles and front axle steer while articulated trucks most commonly are rear dump with three axles and front frame steer. Some bottom dumps are configured as a tractor trailer with up to five axles.

Fig. 10.4.12 Articulated dump truck. *(John Deere.)*

APPLICATION. Dump trucks are used for hauling and dumping blasted materials in mines and quarries and earth, sand and gravel for building roads and dams, and materials for site development. The units are capable of 30 to 40 mi/h (50 to 65 km/h) when loaded, depending on terrain and slope.

Owning and Operating Costs These include depreciation; interest, insurance, taxes; parts, labor, repairs, and tires; fuel, lubricant, filters, hydraulic-system oil, and other operating supplies. This is reduced to cost per hour over a service life of 4 to 6 years of 2,000 h each—average 5 years, 10,000 h. Owning and operating costs of diesel-powered bulldozers and scrapers, excluding operator's wages, average 3 to 4 times the delivered price in 10,000 h.

ABOVE-SURFACE HANDLING

Monorails

by David T. Holmes, Lift-Tech International

Materials can be moved from point to point on an overhead fixed track by wheeled carriers or trolleys that roll on the top surface of the track lower flange. Track members supporting the trolleys can be structural I beams, wide-flange or H beams, and specially fabricated rails having an I shape and special lower flat flanges for improved rolling characteristics. Track sections may be straight, curved, or a combination thereof. The wheel tread diameter and surface finish determine the ease with which it rolls on a track. Figure 10.4.13 illustrates a typical trolley that will negotiate both straight and curved track sections. Typical dimensions are given in Table 10.4.1. The trolley may be plain hand push, geared hand wheel with hand chain, or motor driven. To raise and

Table 10.4.1   Typical Monorail Trolley Dimensions

Capacity, short tons	I-beam range (depth), in	Wheel-tread diam, in	Net weight, lb	B,* in	C, in	H, in	M, in	N, in	Min beam radius, * in
½	5—10	3½	32	3¼	4⅛	9⅞	2⅜	6¼	21
1	5—10	3½	32	3¼	4⅛	9⅞	2⅜	6¼	21
1½	6—10	4	52	3⅞	4⅝	11⅜	2¹⁵⁄₁₆	7⁷⁄₁₆	30
2	6—10	4	52	3⅞	4⅝	11⅜	2¹⁵⁄₁₆	7⁷⁄₁₆	30
3	8—15	5	88	4⁷⁄₁₆	5⅜	13½	2¹³⁄₁₆	7¹⁵⁄₁₆	42
4	8—15	5	88	4⁷⁄₁₆	5⅜	13½	2¹³⁄₁₆	7¹⁵⁄₁₆	42
5	10—18	6	137	5³⁄₁₆	6³⁄₁₆	15⅜	3⁵⁄₁₆	10⅛	48
6	10—18	6	137	5³⁄₁₆	6³⁄₁₆	15⅜	3⁵⁄₁₆	10⅛	48
8	12—24	8	279	5½	7⁷⁄₁₆	21⅜	4³⁄₁₆	13¾	60
10	12—24	8	279	5½	7⁷⁄₁₆	21⅜	4³⁄₁₆	13¾	60

Metric values, multiply tons by 907 for kg, inches by 25.4 for mm, and lb by 0.45 for kg.
* These dimensions are given for minimum beam.
SOURCE: CM Hoist Division, Columbus McKinnon.

Fig. 10.4.13   Monorail trolley. (*CM Hoist Div., Columbus McKinnon.*)

lower a load, a hoist must be attached to the trolley. For low headroom, the trolley can be built into the hoist and is known as a monorail trolley hoist.

Overhead Traveling Cranes

by David T. Holmes, Lift-Tech International

An overhead crane is a mechanism used to lift, transport, and lower loads. It consists of a traveling bridge that supports a movable trolley or a stationary hoist mechanism. The bridge structure has single or multiple horizontal girders supported at each end by a structural end truck assembly. This assembly incorporates wheels, axles, and bearings and rides on an overhead fixed runway structure installed at right angles to the bridge girders. The movable trolley travels the full length of the bridge girders and incorporates a hoist mechanism for lifting and lowering loads. When the bridge is equipped with a stationary hoist it may be mounted at any position in the length of the bridge girders.

Overhead cranes are provided as top-running or under-running. This refers to the structural section on which the wheel of the structural end truck assembly travels. Top-running cranes utilize single- or double-flange wheels that ride on the top surface of a railroad-type rail or square bar. Under-running cranes utilize single-flange wheels that ride on the top surface of the lower flange of a structural I beam, wide-flange beam, or H beam like the trolley on the monorail illustrated in Figure 10.4.13.

Overhead cranes provide three directions of motion at right angles to each other. This permits access to any point in the cube of space over which the crane operates. Light-capacity bridge and trolley horizontal traverse motions may be propelled by hand or geared wheel with hand chain, while electric or pneumatic motors may drive all capacities. The hoist motion, which is entirely independent of the trolley motion, may be propelled by a geared wheel with hand chain or a motor like the bridge or trolley. Overhead cranes are most commonly powered by electricity. In hazardous atmospheres, as identified in the National Electrical Code, cranes are frequently powered by compressed air and are provided with spark-resistant features. Occasionally overhead cranes are powered by hydraulics.

Specifications 70 and 74 of the Crane Manufacturers Association of America (CMAA) have established crane service classes for overhead cranes so that the most economical crane can be obtained for the installation. Service classifications A through F are defined as follows:

Class A (Standby or Infrequent Service)   This service class covers cranes that may be used in installations such as powerhouses, public utilities, turbine rooms, and transformer stations where precise handling of equipment at slow speeds with long idle periods between lifts is required. Capacity loads may be handled for initial installation of equipment and for infrequent maintenance.

Class B (Light Service)   This service covers cranes that may be used in repair shops, light assembly operations, service buildings, light warehousing, etc., where service requirements are light and the speed is slow. Loads may vary from no load to occasional full-rated loads with 2 to 5 lifts per hour, averaging 10 feet per lift.

Class C (Moderate Service)   This service covers cranes that may be used in machine shops or papermill machine rooms, etc., where service requirements are moderate. In this type of service the crane will handle loads that average 50 percent of the rated capacity with 5 to 10 lifts per hour, averaging 15 feet, not over 50 percent of the lift at rated capacity.

Class D (Heavy Service)   This service covers cranes that may be used in heavy machine shops, foundries, fabricating plants, steel warehouses, container yards, lumber mills, etc., and standard-duty bucket and magnet operations where heavy-duty production is required. In this type of service, loads approaching 50 percent of the rated capacity will be handled constantly during the working period. High speeds are desirable for this type of service, with 10 to 20 lifts per hour averaging 15 feet, not over 65 percent of the lifts at rated capacity.

Class E (Severe Service)   This type of service requires a crane capable of handling loads approaching rated capacity throughout its life. Applications may include magnet, bucket, magnet/bucket combination cranes for scrap yards, cement mills, lumber mills, fertilizer plants, container handling, etc., with 20 or more lifts per hour at or near the rated capacity.

Class F (Continuous Severe Service)   This type of service requires a crane capable of handling loads approaching rated capacity continuously under severe service conditions throughout its life. Applications may include custom-designed specialty cranes essential to performing the critical work tasks affecting the total production facility. These cranes must provide the highest reliability with special attention to ease of maintenance features.

Single-girder cranes have capacities from ½ to 30 tons with spans up to 60 feet in length. Typical speeds for motions of motor-driven single-girder cranes as suggested by the Crane Manufactures Association of America are given in Table 10.4.2. Single-girder cranes are provided for service classes A through D.

A basic top-running single-girder crane is illustrated in Figure 10.4.14. It consists of an I-beam girder *a* supported at each end by a two-wheeled structural end truck assembly *b*. The trolley *c* traverses the length of the girder on the top surface of the lower flange of the I beam.

Table 10.4.2   Single-girder crane speeds

| Capacity, tons | Suggested operating speeds, ft/min | | | | | | | | |
| | Hoist | | | Trolley | | | Bridge | | |
	Slow	Medium	Fast	Slow	Medium	Fast	Slow	Medium	Fast
3	14	35	45	50	80	125	50	115	175
5	14	27	40	50	80	125	50	115	175
7.5	13	27	38	50	80	125	50	115	175
10	13	21	35	50	80	125	50	115	175
15	13	19	31	50	80	125	50	115	175
20	10	17	30	50	80	125	50	115	175
25	8	14	29	50	80	125	50	115	175
30	7	14	28	50	80	125	50	115	150

A hoist is attached to the trolley to provide the lifting mechanism. The trolley and hoist may be an integral unit. The bridge travels on the runway by pulling on the hand chain d, turning sprocket wheel that is attached to shaft f. The pinions at each end of the crane attached to shaft f mesh with the gears g which are attached to the wheel axles. An underrunning crane is similar except for having two wheels at each wheel location that travel on the top surface of the lower flange of an I beam, wide-flange beam, or H beam. In lieu of the hand-geared drive, the bridge may be driven by a motor through a gear reduction to the shaft f. When the bridge is motor-driven, a pendant push-button station suspended from the crane or hoist operates the motions.

Fig. 10.4.14   Hand-powered crane.

Double-girder cranes have capacities up to 200 tons, with spans of 20 to 200 feet. Larger capacities and spans are built for special applications. Typical speeds for motions of motor-driven double-girder cranes as suggested by the Crane Manufacturers Association of American are given in Tables 10.4.3 for floor-controlled cranes and in Table 10.4.4 for cab-controlled cranes. Double-girder cranes are provided for service classes A through F.

A basic top-running double-girder crane is illustrated in Figure 10.4.15. Two bridge girders a are supported at each end by structural end trucks having two or more wheels depending upon the capacity of the crane. The bridge travels on the runway by motor c driving shaft d through a gear reduction. Shaft d is coupled to the wheel axle directly or through a gear mesh. The top-running integral trolley hoist b is set on the rails mounted to the top of the girders a traversing their length. Both trolley and hoist are motor-driven. An operator's cab e may be suspended at any position along the length of the girders. It contains the motor controls, master switches or push-button station, hydraulic brake, and warning device.

The two girders are standard structural shapes for low-capacity cranes up to 60 feet in span. High-capacity and long-span cranes will use box girders fabricated of plates or plates and channels to provide strength and stiffness. Provisions are made in the structural end truck to prevent it dropping more than 1 inch if there is an axle failure. A hydraulic brake mounted to the bridge motor shaft is used on cab-operated cranes to stop the crane. Floor-operated cranes use a spring-set, electrically-released brake.

The top-running trolley hoist has the following basic parts: structural frame, hoist mechanism, and trolley traverse drive. The structural frame supports the hoist mechanism and imparts the load to the trolley drive.

Table 10.4.3   Double-girder crane speeds—floor-controlled

| Capacity, tons | Suggested operating speeds, ft/min | | | | | | | | |
| | Hoist | | | Trolley | | | Bridge | | |
	Slow	Medium	Fast	Slow	Medium	Fast	Slow	Medium	Fast
3	14	35	45	50	80	125	50	115	175
5	14	27	40	50	80	125	50	115	175
7.5	13	27	38	50	80	125	50	115	175
10	13	21	35	50	80	125	50	115	175
15	13	19	31	50	80	125	50	115	175
20	10	17	30	50	80	125	50	115	175
25	8	14	29	50	80	125	50	115	175
30	7	14	28	50	80	125	50	115	150
35	7	12	25	50	80	125	50	115	150
40	7	12	25	40	70	100	40	100	150
50	5	11	20	40	70	100	40	100	150
60	5	9	18	40	70	100	40	75	125
75	4	9	15	40	70	100	30	75	125
100	4	8	13	30	60	80	25	50	100
150	3	6	11	25	60	80	25	50	100

NOTE: Consideration must be given to length of runway for the bridge speed, span of bridge for the trolley speed, average travel, distance and spotting characteristics required.

Table 10.4.4 Double-girder crane speeds—cab-controlled

Capacity, tons	Suggested operating speeds, ft/min								
	Hoist			Trolley			Bridge		
	Slow	Medium	Fast	Slow	Medium	Fast	Slow	Medium	Fast
3	14	35	45	125	150	200	200	300	400
5	14	27	40	125	150	200	200	300	400
7.5	13	27	38	125	150	200	200	300	400
10	13	21	35	125	150	200	200	300	400
15	13	19	31	125	150	200	200	300	400
20	10	17	30	125	150	200	200	300	400
25	8	14	29	100	150	175	200	300	400
30	7	14	28	100	125	175	150	250	350
35	7	12	25	100	125	150	150	250	350
40	7	12	25	100	125	150	150	250	350
50	5	11	20	75	125	150	100	200	300
60	5	9	18	75	100	150	100	200	300
75	4	9	15	50	100	125	75	150	200
100	4	8	13	50	100	125	50	100	150
150	3	6	11	30	75	100	50	75	100

NOTE: Consideration must be given to length of runway for the bridge speed, span of bridge for the trolley speed, average travel, distance and spotting characteristics required.

Fig. 10.4.15 Electric traveling crane.

The hoist mechanism includes the motor, motor brake, load brake, gearing, helically grooved rope drum, upper sheave assembly, lower block with hook, and wire rope. The wire rope is reeved from the drum over the sheaves of the upper sheave assembly and the lower block. To prevent over wrapping of the wire rope on the drum and the engagement of the frame by the lower block, limit switches are provided to stop the motors at the upper limit of travel. Frequently, limit switches are also provided to limit the lower travel of the lower block to prevent the rope from totally unwinding and reverse winding onto the rope drum. The trolley traverse drive includes the motor, motor brake, gearing, axles, and wheels. On large-capacity trolleys, an auxiliary hoist of lower capacity and higher lifting speed is provided. For low headroom applications, an under-running trolley is used, placing the hoisting mechanism up between the girders. Table 10.4.5 provides capacities and dimensions for standard industrial cranes.

Electrically powered cranes use either alternating current or direct current power sources. Alternating current is predominately used. Power is delivered to the crane bridge by rolling or sliding collectors in contact with the electrical conductors mounted on the runway or by festooned multiconductor cables. In a similar manner, power is delivered to the trolley hoist with the conductor system being supported from the bridge girders. The motors and controls are designed for crane and hoist service to operate specifically on one of the power sources. Alternating current motors are generally of the single or two-speed squirrel cage or wound-rotor type designed for operation on a 230 or 460-V, three-phase, 60-Hz power source. Standard crane and hoist motor control uses contactors, relays, and resistors. Variable-frequency drives provide precise load control with a wide range of speeds when used with squirrel cage motors. Direct current motors can be either series-wound or shunt-wound type designed for operation on a 250-V power source.

Gantry Cranes

A gantry crane is an adaptation of an overhead crane. Most gantry crane installations are outdoors where an elevated runway is long and not cost-effective to erect. Figure 10.4.16 illustrates a double leg gantry crane. A structural leg a is attached to each end of the bridge and is supported by structural end trucks having two or more wheels, depending upon the capacity of the crane. The crane rides on railroad-type rails mounted on the ground and is driven by motor b through a gear reduction

Fig. 10.4.16 Gantry crane.

Table 10.4.5   Dimensions, Loads, and Speeds for Industrial-Type Cranes [a, d, e]

Capacity main hoist, tons, 2,000 lb	Span, ft	Std. lift, main hoist, ft [b]	Dimensions, refer to Fig. 10.4.13							Max load per wheel, lb [c]	Runway rail, lb/yd	X, in	No. of bridge wheels
			A	B min	C	D	E [b]	F	H				
6	40	36	5⅝"	3'9"	3¼"	29⅞"	31⅞"		8'0"	9,470	25	12	4
	60	53	5⅝"	3'9"	3¼"	29⅞"	31⅞"		9'6"	12,410	25	12	4
	80	86	5⅝"	3'9"	3¼"	29⅞"	31⅞"		11'6"	15,440	25	12	4
	100	118	5⅝"	3'9"	3¼"	29⅞"	31⅞"		14'6"	19,590	40	12	4
10	40	36	5⅝"	3'9"	3¼"	29⅞"	31⅞"		8'0"	15,280	25	12	4
	60	53	5⅝"	3'9"	3¼"	29⅞"	31⅞"		9'6"	18,080	40	12	4
	80	86	5⅝"	3'9"	3¼"	29⅞"	31⅞"		11'6"	20,540	40	12	4
	100	118	5⅝"	3'9"	3¼"	29⅞"	31⅞"		14'6"	24,660	60	12	4
16	40	28	5⅝"	3'9"	4'0"	28⅛"	38½"		8'0"	20,950	60	12	4
	60	40	5⅝"	3'9"	4'0"	28⅛"	38½"		9'6"	23,680	60	12	4
	80	62	5⅝"	3'9"	4'0"	28⅛"	38½"		11'6"	26,140	60	12	4
	100	84	5⅝"	3'10"	4'0"	28⅛"	38½"		14'6"	30,560	80	12	4
20 5 ton aux	40	22	5⅝"	3'9"	4'0"	40⅜"	28⅜"	2'2"	8'0"	26,350	60	18	4
	60	30	5⅝"	3'9"	4'0"	40⅜"	28⅜"	2'2"	9'6"	30,000	80	18	4
	80	46	6⅜"	3'11"	4'0"	40⅜"	28⅜"	2'2"	11'6"	33,370	60	18	4
	100	63	6⅜"	3'11"	4'0"	40⅜"	28⅜"	2'2"	14'6"	38,070	80	18	4
30 5 ton aux	40	22	6⅜"	3'11"	4'6"	41⅞"	28⅞"	1'11½"	9'6"	37,300	80	24	4
	60	22	6⅜"	3'11"	4'6"	41⅞"	28⅞"	1'11½"	9'6"	40,050	135	24	4
	80	34	7⅞"	3'11"	4'6"	40⅜"	27⅞"	1'11½"	11'6"	44,680	80	24	4
	100	48	7⅞"	3'11"	4'6"	40⅜"	27⅞"	1'11½"	14'6"	51,000	135	24	4
40 5 ton aux	40	25	7⅞"	5'5"	6'6"	46⅞"	17⅝"	2'6¾"	9'6"	48,630	135	24	4
	60	25	7⅞"	5'5"	6'6"	46⅞"	17⅝"	2'6¾"	9'6"	52,860	135	24	4
	80	40	7⅞"	5'5"	6'6"	46⅞"	17⅝"	2'6¾"	11'6"	59,040	135	24	4
	100	54	7⅞"	5'5"	6'6"	46⅞"	17⅝"	2'6¾"	14'6"	65,200	135	24	4
50 10 ton aux	40	25	7⅞"	6'10"	7'6"	56⅜"	29¼"	3'5¼"	10'6"	55,810	135	24	4
	60	25	7⅞"	6'7"	7'6"	56⅜"	29¼"	3'5¼"	10'6"	62,050	135	24	4
	80	32	9"	6'7"	7'6"	55¼"	28⅛"	3'5¼"	11'6"	68,790	135	24	4
	100	46	9"	6'7"	7'6"	55¼"	28⅛"	3'5¼"	14'6"	76,160	176	24	4
60 10 ton aux	40	25	9"	6'8"	7'6"	55¼"	28⅛"	3'5¼"	10'6"	65,970	135	24	4
	60	25	9"	6'8"	7'6"	55¼"	28⅛"	3'5¼"	10'6"	72,330	135	24	4
	80	32	9"	6'8"	7'6"	55¼"	28⅛"	3'5¼"	11'6"	79,230	175	24	4
	100	46	9"	6'10"	7'6"	55¼"	28⅛"	3'5¼"	14'6"	89,250	175	24	4

[a] Lift-Tech International, Inc.
[b] For each 10 ft extra lift, increase H by X.
[c] Direct loads, no impact.
[d] These figures should be used for preliminary work only as the data varies among manufacturers.
[e] Multiply ft by 0.30 for m, ft/min by 0.0051 for m/s, in by 25.4 for mm, lb by 0.45 for kg, lb/yd by 0.50 for kg/m.

to shaft *c*, which drives the vertical shafts *d* through a bevel gear set. Shaft *d* is coupled to the wheel axles by bevel and spur gear reductions. Gantry cranes are also driven with separate motors, brakes, and gear reductions mounted adjacent to the drive wheels of the structural end truck in lieu of the bridge-mounted cross-shaft drive. Gantry cranes are available with similar capacities and spans as overhead cranes.

Special-Purpose Overhead Traveling Cranes

Overhead and gantry cranes can be specially designed to meet special application requirements: furnace charging cranes that charge scrap metal into electric furnaces, wall traveling cranes that use a runway system on one wall of a building, lumber handling cranes that travel at extremely high speeds and are equipped with special reeving systems to provide stability of the loads, polar cranes that operate on a single rail circular runway, and stacker cranes that move materials in and out of racks are examples of these applications.

Jib Cranes

Jib cranes find common use in manufacturing. Figure 10.4.17 shows a **column jib crane**, consisting of pivoted post *a* and carrying boom *b*, on which travels either an electric or a hand hoist *c*. The post *a* is attached to building column *d* so that it can swing through approximately 270°. Cranes of this type have been replaced by such other methods of handling materials as the mobile lift truck and other types of cranes. Column jib cranes are built with radii up to 20 ft (6 m) and for loads up

Fig. 10.4.17   Column jib crane.

to 5 tons (4.5 tonnes). Yard jib cranes are generally designed to meet special conditions.

Derricks

by Ronald M. Kohner, Landmark Engineering Services, Ltd.

Derricks are stationary lifting devices used to raise and reposition heavy loads. A diagonal latticed strut, or boom, from which loads are suspended is typically pinned near the base of a second vertical latticed strut, or mast. The object to be lifted is raised or lowered vertically by means of a pulley and rope system suspended from the tip of the boom. When one end of the rope is spooled onto a drum, the object is raised vertically toward the boom tip. When the rope is unspoiled from the drum, the object is lowered. The bottom end of the boom is pinned near the base of the mast to allow it to be pivoted through a vertical arc usually slightly less than 90°. A second rope and pulley system connects the top of the boom to the top of the vertical mast. Spooling or unspooling this second rope from another drum moves the boom through its vertical arc. As the boom pivots in the vertical plane, any load that is suspended from the tip of the boom will be moved horizontally—either closer to or farther away from the base of the mast and boom. The top of the derrick mast is typically held in place either by two rigid tension legs fastened to anchor sills, as in a stiff-leg derrick (Fig. 10.4.18), or by a system of guy lines anchored some distance from the derrick, as in a guy derrick (Fig. 10.4.19). Both the boom and mast can be rotated horizontally about the mast's vertical axis to provide lateral movement of any load suspended from the boom tip. A derrick can therefore be used to move a load up or down (hoisting), closer or farther away from the derrick base (luffing), and side to side (swinging).

The term *derrick* is originally derived from a London hangman who achieved such notoriety that his name became synonymous with the tall gallows on which he plied his gruesome trade. In early days derricks consisted of booms and masts that were simply large timbers fitted with steel end caps to which the base pins or pulley systems could be attached. Drum power was supplied by manually operated cranks. Steel soon replaced timber as the primary material for booms and masts. Power was provided first by steam boiler then by electric motors or internal combustion engines. Lifting capacities grew from a few hundred pounds for the earliest manually powered derricks to over 600 tons. Some guy derricks have even been adapted to mount on mobile

Fig. 10.4.18   Steel stiff-leg derrick. (*HydraLift AmClyde.*)

Fig. 10.4.19   Self-slewing steel guy derrick. (*HydraLift AmClyde.*)

crawler and truck type cranes to increase the inherent lifting capacity of these machines (Fig. 10.4.20). Due to their fixed location and the effort required to assemble/disassemble them, the use of derricks in recent years has been limited to a relatively few special-purpose applications. They have largely been overshadowed by mobile-type cranes in the lifting industry.

Mobile Cranes

Mobile cranes are characterized by their ability to move from lifting job to lifting job with relative ease. This is accomplished by mounting the crane structure and hoist machinery on some sort of mobile base. Another distinguishing feature of this type of crane is the strut, or boom, that protrudes diagonally upward from the crane body to provide a lifting point from which heavy loads can be suspended. A rope and pulley system is hung from the tip of this boom to provide a means for raising and lowering loads. Internal combustion engines provide power to spool the pulley rope onto drums to raise heavy loads.

The diagonal position of the boom is controlled by one of two basic systems. With the first arrangement, the tip of the boom is tied back to the rear of the crane with a second rope and pulley system. Spooling this rope on or off a drum raises or lowers the tip of the boom, thereby changing its angle in a vertical arc about hinge pins at the bottom of the

boom. Booms suspended by ropes in this fashion are typically lattice-type structures built in sections that can be fastened together to lengthen or shorten the boom (Fig. 10.4.21). Boom systems of this type reach lengths of over 500 ft.

Fig. 10.4.21   Lattice boom crane. (*Construction Safety Association of Ontario.*)

The second type of boom system uses hydraulic cylinders pushing up on the bottom of the boom to support it at the desired angle. Extending or retracting the hydraulic boom support cylinder raises or lowers the tip of the boom in a vertical arc about its bottom hinge pins. Booms supported in this fashion are typically composed of several nested rectangular box sections that can be telescoped in or out with another internal hydraulic cylinder to change the length of the boom (Fig. 10.4.22). Boom systems of this type reach lengths of several hundred feet. With either type of boom system, sideways movement of a suspended load is achieved by rotating the boom horizontally on a swing roller system.

The first mobile cranes were associated with the railroads and moved from location to location on their tracks. As time progressed the need to travel beyond the confines of the rail bed gave rise to crane mountings that could move over virtually any reasonably solid and level terrain.

Fig. 10.4.20   Guy derrick.

Fig. 10.4.22   Telescoping boom crane. (*Crane Institute of America, Inc.*)

Today, mobile cranes are grouped into one of two general categories based on the travel system on which the crane is mounted.

One category of cranes is mounted on crawler tracks much like those on a military tank (Fig. 10.4.23). The second category is mounted on various types of rubber-tired vehicles. Crawler mountings offer the advantage of a more rigid base for the boom system and allow the entire crane to move forward or backward while a load remains suspended from the boom. The disadvantage of crawlers is their very slow travel speed, which imposes a practical limit on the distance that can be traveled. Crawler-mounted cranes are typically loaded on trucks or railcars for extended travel distances between jobs. Crawler cranes range in capacity from a few tons to over 1,000 tons.

Fig. 10.4.23 Crawler-mounted lattice boom crane. (*Crane Institute of America. Inc.*)

The second category of mobile cranes is mounted on rubber-tired vehicles and can be further subdivided by the configuration of that vehicle. Boom-and-hoist systems that are mounted on relatively standard, commercially available flatbed trucks are commonly termed **boom trucks** (Fig. 10.4.24). They are generally shorter in boom length and lower in lifting capacity than other rubber-tired cranes. Booming and hoisting functions are powered by the truck's own engine. These cranes are equipped with hydraulic telescopic booms and are light enough to travel easily over most roads for rapid deployment. They range in capacity from a few tons to near 35 tons.

Fig. 10.4.24 Boom truck. (*Construction Safety Association of Ontario.*)

Larger rubber-tired cranes are mounted on truck carriers specially built to support the swing roller system on which bigger cranes must be mounted (Fig. 10.4.25). These truck carriers are built stronger than conventional trucks to withstand the heavier loads imposed by crane service. They are driven from a second control location remote from the position where the operator of the crane is seated. Like all wheel-mounted cranes, they are equipped with self-contained outriggers to increase the crane's stability beyond that provided by the relatively narrow tire system. Deployment of the outriggers for maximum lifting stability generally raises the crane off its wheels, thereby fixing the crane's location when loads are being lifted. Truck-mounted cranes were

Fig. 10.4.25 Carrier-mounted lattice boom crane. (*Crane Institute of America, Inc.*)

originally geared for travel only on improved roads. Today, with both off-road and highway gearing, truck-mounted cranes are available in an all-terrain (AT) configuration. They are available with either rope-supported lattice booms or hydraulic telescopic-type booms. They range in capacity from a few tons to over 1,000 tons.

A third category of rubber-tire-mounted mobile crane is the rough terrain (RT) crane (Fig. 10.4.26). It is distinguished by higher ground clearance with much larger tires than other wheel-mounted cranes. Unlike conventional truck cranes, the steering and travel controls are typically located at the same operator's station as the hoist, thereby placing all crane movements under the control of a single individual. Like most other wheel-mounted cranes, an integral outrigger system provides the necessary stability for lifting heavy loads but confines the crane to a fixed location when on outriggers. Rough terrain cranes are typically configured only with hydraulic telescopic-type booms. These cranes in the 30- to 60-ton capacity range have become the general-purpose workhorses for most job sites where cranes are required. Capacities range from a few tons to over 100 tons.

Fig. 10.4.26 Rough terrain crane. (*Construction Safety Association of Ontario.*)

Numerous attachments other than simple booms have been developed for rope-supported lattice-boom-type mobile cranes to address specific problems associated with lifting work. Several of these attachments are as follows:

1. *Jib.* A smaller, lighter strut mounted on top of a standard crane boom. It is typically offset at some slight angle forward from the main boom. It allows lighter loads to be lifted to higher elevations than is possible on a standard boom alone (Fig. 10.4.21). Jibs are also used on top of hydraulic telescopic booms.

2. *Luffing jib.* A smaller, lighter strut mounted on top of a standard crane boom. With a luffing jib, the angle that the jib is offset forward of the main boom can be varied with a rope-and-pulley system. This offers a wider selection of boom angle/jib angle combinations than are possible with a fixed jib. Luffing-type jibs are available on either lattice or hydraulic telescopic-type booms (Fig. 10.4.27).

Fig. 10.4.27 Telescoping boom crane with luffing jib. (*Crane Institute of America, Inc.*)

3. *Tower attachment.* The normal diagonally oriented working boom is mounted on the top of a second strut, or tower, that rises vertically from the crane body. Protruding up and out from the vertical tower allows the working boom to maintain sufficient clearance to reach up over the corner of tall structures and handle loads while the crane base stands close to the front of the structure. Tower attachments are typically available only on some lattice boom cranes (Fig. 10.4.28).

Fig. 10.4.28 Mobile tower crane. (*Construction Safety Association of Ontario.*)

4. *Ringer attachment.* A rail-like structural ring is placed completely around the tracks of a conventional crawler-mounted crane. A large supplemental counterweight is placed on top of a roller bed on the rear of the ring to increase the crane's stability. The crane's boom is mounted on a similar roller bed on top of the ring in front of the crane. A second strut, or mast, is mounted behind the ringer boom and protrudes up diagonally to the rear. The tip of the boom is attached by a rope and pulley system to the tip of the mast. The purpose of this attachment is to increase the capacity of a standard crawler-mounted crane by adding extra counterweight to the crane along with a mast to provide more advantageous geometry for the boom suspension system. When equipped with a Ringer attachment, the crane can no longer move with a load suspended. The patented Ringer attachment increases a standard crawler crane's capacity from 300 tons to over 1,000 tons (Fig. 10.4.29).

Fig. 10.4.29 Ringer crane. (*Construction Safety Association of Ontario.*)

5. *Sky Horse attachment.* The first of several similar attachments offered by different crane manufacturers that place an auxiliary counterweight on a wheeled trailer system directly behind the crane. A second strut, or mast, is mounted behind the Sky Horse boom and protrudes up diagonally to the rear (Fig. 10.4.30). The purpose of this attachment is to increase the capacity of a standard crawler-mounted crane by adding extra counterweight while maintaining mobility with a suspended load as the auxiliary counterweight follows behind on its own wheels. The mast provides improved geometry for the boom suspension system. The patented Sky Horse attachment approximately doubles the capacity of a standard crawler crane. Other similar attachments are known by names such as Maxer and Superlift (Fig. 10.4.31).

Fig. 10.4.31 Superlift crane. (*Construction Safety Association of Ontario.*)

Fig. 10.4.30 Sky Horse crane.

Mobile cranes have grown from capacities of 20 or 30 tons several decades ago to well over 1,000 tons today. Boom lengths can reach elevations over 500 ft. Sizes are largely limited by the weight and external dimensions of crane components when moving the machines between job sites. Current equipment can move rapidly between work sites and deploy for lifting in a matter of minutes. They have become one of the most popular and necessary tools in the industry's toolbox.

Cableways

Cableways are **aerial hoisting and conveying devices** using suspended steel cable for their tracks, the loads being suspended from carriages and moved by gravity or power. The most common uses are transporting material from open pits and quarries to the surface; handling construction material in the building of dams, docks, and other structures where the construction of tracks across rivers or valleys would be uneconomical; and loading logs on cars. The maximum clear **span** is 2,000 to 3,000 ft (610 to 914 m); the usual spans, 300 to 1,500 ft (91 to 457 m). The gravity type is limited to conditions where a grade of at least 20 percent is obtainable on the track cable. Transporting cableways move the load from one point to another. Hoisting transporting cableways hoist the load as well as transport it.

A transporting cableway may have one or two fixed track cables, inclined or horizontal, on which the carriage operates by gravity or power. The gravity transporting type (Fig. 10.4.32-I) will either raise or lower material. It consists of one track cable *a* on which travels the wheeled carriage *b* carrying the bucket. The traction rope *c* attached to the carriage is made fast to power drum *d*. The inclination must be sufficient for the carriage to coast down and pull the traction rope after it. The carriage is hauled up by traction rope *c*. Drum *d* is provided with a brake to control the lowering speed, and material may be either raised or lowered. When it is not possible to obtain sufficient fall to operate the load by gravity, traction rope *c* (Fig. 10.4.32-II) is made endless so that carriage *b* is drawn in either direction by power drum *d*. Another type of inclined cableway, shown in Fig. 10.4.32-III, consists of two track cables *aa*, with an endless traction rope *c*, driven and controlled by drum *d*. When material is being lowered, the loaded bucket *b* raises the empty carriage *bb*, the speed being controlled by the brake on the drum.

When material is being raised, the drum is driven by power, the descending empty carriage assisting the engine in raising the loaded carriage. This type has twice the capacity of that shown in Fig. 10.4.32-I.

A **hoisting and conveying cableway** (Fig. 10.4.32-IV) hoists the material at any point under the track cable and transports it to any other point. It consists of a track cable *a* and carriage *b*, moved by the endless traction rope *c* and by power drum *d*. The hoisting of the load is accomplished by power drum *e* through fall rope *f*, which raises the fall block *g* suspended from the carriage. The fall-rope carriers *h* support the fall rope; otherwise, the weight of this sagging rope would prevent fall block *g* from lowering when without load. Where it is possible to obtain a minimum inclination of 20° on the track cable, the traction-rope drum *d* is provided with a brake and is not power-driven. The carriage then

Fig. 10.4.32   Cableways.

descends by gravity, pulling the fall and traction ropes to the desired point. Brakes are applied to drum *d*, stopping the carrier. The fall block is lowered, loaded, and raised. If the load is to be carried up the incline, the carriage is hauled up by the fall rope. With this type, the friction of the carriage must be greater than that of the fall block or the load will run down. A novel development is the use of self-filling grab buckets operated from the carriages of cableways, which are lowered, automatically filled, hoisted, carried to dumping position, and discharged.

The **carriage speed** is 300 to 1,400 ft/min (1.5 to 7.1 m/s) [in special cases, up to 1,800 ft/min (9.1 m/s)]; average hoisting speed is 100 to 700 ft/min. The **average loads** for coal and earth are 1 to 5 tons (0.9 to 4.5 tonnes); for rock from quarries, 5 to 20 tons; for concrete, to 12 yd^3 (9.1 m^3) at 50 tons.

The **deflection of track cables** with their maximum gross loads at midspan is usually taken as 5½ to 6 percent of the span. Let S = span between supports, ft; L = one-half the span, ft; w = weight of rope, lb/ft; P = total concentrated load on rope, lb; h = deflection, ft; H = horizontal tension in rope, lb. Then $h = (wL + P)L/2H$; $P = (2h - wL^2)/L = (8hH - wS^2)/2S$.

For track cables, a **factor of safety** of at least 4 is advised, though this may be as low as 3 for locked smooth-coil strands that use outer wires of high ultimate strength. For traction and fall ropes, the sum of the load and bending stress should be well within the elastic limit of the rope or, for general hoisting, about two-thirds the elastic limit (which is taken at 65 percent of the breaking strength). Let P = load on the rope, lb; A = area of metal in rope section, in^2; E = 29,500,000; R = radius of curvature of hoisting drum or sheave, whichever is smaller, in; d = diameter of individual wires in rope, in (for six-strand 19-wire rope, d = $^1/_{15}$ rope diam; for six-strand 7-wire rope, d = $^1/_9$ rope diam). Then load stress per in^2 = T_1 = P/A, and bending stress per in^2 = T_b = $Ed/2R$. The radius of curvature of saddles, sheaves, and driving drums is thus important to fatigue life of the cable. In determining the horsepower required, the load on the traction ropes or on the fall ropes will govern, depending upon the degree of inclination.

Cable Tramways

Cable tramways are aerial conveying devices using suspended cables, carriages, and buckets for transporting material over level or mountainous country or across rivers, valleys, or hills (they transport but do not hoist). They are used for handling small quantities over long distances, and their construction cost is insignificant compared with the construction costs of railroads and bridges. Five types are in use:

Monocable, or Single-Rope, Saddle-Clip Tramway   Operates on grades to 50 percent gravity grip or on higher grades with spring grip and has capacity of 250 tons/h (63 kg/s) in each direction and speeds to 500 ft/min. Single section lengths to 16 miles without intermediate stations or tension points. Can operate in multiple sections without transshipment to any desired length [monocables to 170 miles (274 km) over jungle terrain are practical]. Loads automatically leave the carrying moving rope and travel by overhead rail at angle stations and transfer points between sections with no detaching or attaching device required. Main rope constantly passes through stations for inspection and oiling. Cars are light and safe for passenger transportation.

Single-Rope Fixed-Clip Tramway   Endless rope traveling at low speed, having buckets or carriers fixed to the rope at intervals. Rope passes around horizontal sheaves at each terminal and is provided with a driving gear and constant tension device.

Bicable, or Double-Rope, Tramway   Standing track cable and a moving endless hauling or traction rope traveling up to 500 ft/min (2.5 m/s). Used on excessively steep grades. A detacher and attacher is required to open and close the car grip on the traction rope at stations. Track cable is usually in sections of 6,000 to 7,000 ft (1,830 to 2,130 m) and counterweighted because of friction of stiff cable over tower saddles.

Jigback, or Two-Bucket, Reversing Tramway   Usually applied to hillside operations for mine workings so that on steep slopes loaded bucket will pull unloaded one up as loaded one descends under control of a brake. Loads to 10 tons (9 tonnes) are carried using a pair of track cables and an endless traction rope fixed to the buckets.

To-and-Fro, or Single-Bucket, Reversing Tramway   A single track rope and a single traction rope operated on a winding or hoist drum. Suitable for light loads to 3 tons (2.7 tonnes) for intermittent working on a hillside, similar to a hoisting and conveying cableway without hoisting feature.

The **monocable tramway** (Fig. 10.4.33) consists of an endless cable *a* passing over horizontal sheaves *d* and *e* at the ends and supported at intervals by towers. This cable is moved continuously, and it both supports and propels carriages *b* and *c*. The carriages either are attached permanently to the cable (as in the **single-rope fixed clip tramway**), in

Fig. 10.4.33   Single-rope cable tramway.

which case they must be loaded and dumped while in motion, or are attached by friction grips so that they may be connected automatically or by hand at the loading and dumping points. When the tramway is lowering material from a higher to a lower level, the grade is frequently sufficient for the loaded buckets *b* to raise the empty buckets *c*, operating the tramway by gravity, the speed being controlled by a brake on grip wheel *d*.

The **bicable tramway** (Fig. 10.4.34) consists of two stationary track cables *a*, on which the wheeled carriages *c* and *d* travel. The endless traction rope *b* propels the carriages, being attached by friction grips. Figure 10.4.35 shows the arrangement of the **overhead type**. The track cable *a* is supported at intervals by towers *b*, which carry the saddles *c*

Fig. 10.4.34   Double-rope cable tramway.

in which the track cable rests. Each tower also carries the sheave *d* for supporting traction rope *e*. The self-dumping bucket *f* is suspended from carriage *g*. The grip *h*, which attaches the carriage to traction rope *e*, is controlled by lever *k*. In the **underhung type**, shown in Fig. 10.4.36, track cable *a* is carried above traction rope *e*. Saddle *c* on top of the tower supports the track cable, and sheave *d* supports the traction cable. The sheave is provided with a rope guard *m*. The lever *h*, with a roller on the end, automatically attaches and detaches the grip by coming in contact with guides at the loading and dumping points. The carriages move in

Fig. 10.4.35   Overhead-type double-rope cable tramway.

only one direction on each track. On steep downgrades, special hydraulic speed controllers are used to fix the speed of the carriages.

The **track cables** are of the special locked-joint smooth-coil, or tramway, type. Nearly all wire rope is made of plow steel, with the old cast-steel type no longer being in use. The track cable is usually provided with a smooth outer surface of *Z*-shaped wires for full lock type

Fig. 10.4.36   Underhung-type double-rope cable tramway.

or with a surface with half the wires *H*-shaped and the rest round. Special tramway couplers are attached in the shops with zinc or are attached in the field by driving little wedges into the strand end after inserting the end into the coupler. The second type of coupling is known as a **dry socket** and, though convenient for field installation, is not held in as high regard for developing full cable strength. The usual spans for level ground are 200 to 300 ft (61 to 91 m). One end of the track cable is anchored; the other end is counterweighted to one-quarter the breaking strength of the rope so that the horizontal tension is a known quantity. The **traction ropes** are made six-strand 7-wire or six-strand 19-wire, of cast or plow steel on hemp core. The maximum diameter is 1 in, which limits the length of the sections. The traction rope is endless and is driven by a drum at one end, passing over a counterweighted sheave at the other end.

Figure 10.4.37 shows a **loading terminal**. The track cables *a* are anchored at *b*. The carriage runs off the cable to the fixed track *c*, which makes a 180° bend at *d*. The empty buckets are loaded by chute *e* from the loading bin, continue around track *c*, are automatically gripped to

Fig. 10.4.37   Cable-tramway loading terminal.

traction cable *f*, and pass on the track cable *a*. Traction cable *f* passes around and is driven by drum *g*. When the carriages are permanently attached to the traction cable, they are loaded by a moving hopper, which is automatically picked up by the carriage and carried with it a short distance while the bucket is being filled. Figure 10.4.38 shows a **discharge terminal**. The carriage rolls off from the track cable *a* to the fixed track *c*, being automatically ungripped. It is pushed around the 180° bend of track *c*, discharging into the bin underneath and continuing on track *c*

Fig. 10.4.38   Cable-tramway discharge terminal.

until it is automatically gripped to traction cable *f*. The counterweights *h* are attached to track cables *a*, and the counterweight *k* is attached to the carriage of the traction-rope sheave *m*. The **supporting towers** are A frames of steel or wood. At abrupt vertical angles the supports are placed close together and steel tracks installed in place of the cable. Spacing of towers will depend upon the capacity of the track cables and sheaves and upon the terrain as well as the bucket spacing.

Stress In Ropes   (Roebling) The deflection for track cables of tramways is taken as one-fortieth to one-fiftieth of the span to reduce the grade at the towers. Let S = span between supports, ft; h = deflection, ft; P = gross weight of buckets and carriages, lb; Z = distance between buckets, ft; W_1 = total load per ft of rope, lb; H = horizontal tension of rope, lb. The formulas given for cableways then apply. When several buckets come in the span at the same time, special treatment is required for each span. For large capacities, the buckets are spaced close together, the load may be assumed to be uniformly distributed, and the live load per linear foot of span = P/Z. Then $H = W_1 S^2/8h$, where W_1 = (weight of rope per ft) + (P/Z). When the buckets are not spaced closely, the equilibrium curve can be plotted with known horizontal tension and vertical reactions at points of support.

For figuring the traction rope, t_0 = tension on counterweight rope, lb; t_1, t_2, t_3, t_4 = tensions, lb, at points shown in Fig. 10.4.39; n = number of carriers in motion; a = angle subtended between the line connecting the tower supports and the horizontal; W_1 = weight of each loaded carrier, lb; W_2 = weight of each empty carrier, lb; w = weight of traction

Fig. 10.4.39   Diagram showing traction rope tensions.

rope, lb/ft; L = length of tramway of each grade a, ft; D = diameter of end sheave, ft; d = diameter of shaft of sheave, ft; $f_1 = 0.015$ = coefficient of friction of shaft; $f_2 = 0.025$ = rolling friction of carriage wheels. Then, if the loads descend, the maximum stress on the loaded side of the traction rope is

$$t_2 = t_1 + \sum (Lw \sin a + \tfrac{1}{2}nW_1 \sin a)$$
$$- f_2 \sum (Lw \cos a + \tfrac{1}{2}nW_1 \cos a)$$

where $t_1 = \tfrac{1}{2}t_0[1 - f_1(d/D)]$. If the load ascends, there are two cases: (1) driving power located at the lower terminal, (2) driving power at the upper terminal. If the line has no reverse grades, it will operate by gravity at a 10 percent incline to 10 tons/h capacity and at a 4 percent grade for 80 tons/h. The preceding formula will determine whether it will operate by gravity.

The **power required** or developed by tramways is as follows: Let V = velocity of traction rope, ft/min; P = gross weight of loaded carriage, lb; p = weight of empty carriage, lb; N = number of carriages on one track cable; $P/50$ = friction of loaded carriage; $p/50$ = friction of empty carriage; W = weight of moving parts, lb; E = length of tramway divided by difference in levels between terminals, ft. Then, power required is

$$ hp = \frac{NV}{33,000}\left(\frac{P-p}{E} \pm \frac{P+p}{50}\right) \pm 0.0000001\,WV $$

Where power is developed by tramways, use 80 instead of 50 under $P + p$.

BELOW-SURFACE HANDLING (EXCAVATION)

Power Shovels

Power shovels stand upon the bottom of the pit being dug and dig above this level. Small machines are used for road grading, basement excavation, clay mining, and trench digging; larger sizes are used in quarries, mines, and heavy construction; and the largest are used for removing overburden in opencut mining of coal and ore. The uses for these machines may be divided into two groups: (1) **loading,** where sturdy machines with comparatively short working ranges are used to excavate material and load it for transportation; (2) **stripping,** where a machine of very great dumping and digging reaches is used to both excavate the material and transport it to the dump or wastepile. The **full-revolving shovel,** which is the only type built at the present (having entirely displaced the old railroad shovel), is usually composed of a crawler-mounted truck frame with a center pintle and roller track upon which the revolving frame can rotate. The revolving frame carries the swing and hoisting machinery and supports, by means of a socket at the lower end and cable guys at the upper end, a boom carrying guides for the dipper handles and machinery to thrust the dipper into the material being dug.

Figure 10.4.40 shows a full-revolving shovel. The dipper a, of cast or plate steel, is provided with special wear-resisting teeth. It is pulled through the material by a steel cable b wrapped on a main drum c. Gasoline engines are used almost exclusively in the small sizes, and diesel, diesel-electric, or electric power units, with Ward Leonard control, in the large machines. The commonly used sizes are from ½ to 5 yd³ (0.4 to 3.8 m³) capacity, but special machines for coal-mine stripping are built with buckets holding up to 33 yd³ (25 m³) or even more. The very large machines are not suited for quarry or heavy rock work. Sizes up to 5 yd³ (3.8 m³) are known as **quarry machines.** Stripping shovels are crawler-mounted, with double-tread crawlers under each of the four corners and with power means for keeping the turntable level when traveling over uneven ground. The crowd motion consists of a chain

Fig. 10.4.40   Revolving power shovel.

which, through the rack-and-pinion mechanism, forces the dipper into the material as the dipper is hoisted and withdraws it on its downward swing. On the larger sizes, a separate engine or motor is mounted on the boom for crowding. A separate engine working through a pinion and horizontal gear g swings the entire frame and machinery to bring the dipper into position for dumping and to return it to a new digging position. Dumping is accomplished by releasing the hinged dipper bottom, which drops upon the pulling of a latch. With gasoline-engine or diesel-engine drives, there is only one prime mover, the power for all operations being taken off by means of clutches.

Practically all **power shovels** are readily **converted** for operation as **dragline excavators, or cranes.** The changes necessary are very simple in the case of the small machines; in the case of the larger machines, the installation of extra drums, shafts, and gears is required, in addition to the boom and bucket change.

The **telescoping boom,** hydraulically operated excavator shown in Fig. 10.4.41 is a versatile machine that can be quickly converted from the rotating-boom power shovel shown in Fig. 10.4.41a to one with a crane boom (Fig. 10.4.41b) or backhoe shovel boom (Fig. 10.4.41c). It can dig ditches reaching to 22 ft (6.7 m) horizontally and 9 ft 6 in (2.9 m) below grade; it can cut slopes, rip, scrape, dig to a depth of 12 ft 6 in (3.8 m), and load to a height of 11 ft 2 in (3.4 m). It is completely hydraulic in all powerized functions.

Dredges

Placer dredges are used for the mining of gold, platinum, and tin from placer deposits. The usual maximum digging depth of most existing dredges is 65 to 70 ft (20 to 21 m), but one dredge is digging to 125 ft (38 m). The dredge usually works with a bank above the water of 8 to 20 ft (2.4 to 6 m). Sometimes hydraulic jets are employed to break down these banks ahead of the dredge. The excavated material is deposited astern, and as the dredge advances, the pond in which the dredge floats is carried along with it.

The digging element consists of a chain of closely connected buckets passing over an idler tumbler and an upper or driving tumbler. The chain is mounted on a structural-steel ladder which carries a series of

(c) Backhoe shovel boom

(b) Crane boom

(a) Shovel with rotating boom

Fig. 10.4.41   Hydraulically operated excavator. (*Link-Belt.*)

rollers to provide a bearing track for the chain of buckets. The upper tumbler is placed 10 to 40 ft (3 to 13 m) above the deck, depending upon the size of the dredge. Its fore-and-aft location is about 65 percent of the length of the ladder from the bow of the dredge. The ladder operates through a well in the hull, which extends from the bow practically to the upper-tumbler center. The material excavated by the buckets is dumped by the inversion of the buckets at the upper tumbler into a hopper, which feeds it to a revolving screen.

Placer dredges are made with buckets ranging in capacity from 2 to 20 ft³ (0.06 to 0.6 m³). The usual speed of operation is 15 to 30 buckets per min, in the inverse order of size.

The digging reaction is taken by stern spuds, which act as pivots upon which the dredge, while digging, is swung from side to side of the cut by swinging lines which lead off the dredge near the bow and are anchored ashore or pass over shore sheaves and are dead-ended near the lower tumbler on the digging ladder. By using each spud alternately as a pivot, the dredge is fed forward into the bank.

Elevator dredges, of which dredges are a special classification, are used principally for the excavation of sand and gravel beds from rivers, lakes, or ocean deposits. Since this type of dredge is not as a rule required to cut its own flotation, the bow corners of the hull may be made square and the digging ladder need not extend beyond the bow. The bucket chain may be of the close-connected placer-dredge type or of the open-connected type with one or more links between the buckets. The dredge is more of an elevator than a digging type, and for this reason the buckets may be flatter across the front and much lighter than the placer-dredge bucket.

The excavated material is usually fed to one or more revolving screens for classification and grading to the various commercial sizes of sand and gravel. Sometimes it is delivered to sumps or settling tanks in the hull, where the silt or mud is washed off by an overflow. Secondary elevators raise the material to a sufficient height to spout it by gravity or to load it by belt conveyors to the scows.

Hydraulic dredges are used most extensively in river and harbor work, where extremely heavy digging is not encountered and spoil areas are available within a reasonable radius of the dredge. The radius may vary from a few hundred feet to a mile or more, and with the aid of booster pumps in the pipeline, hydraulic dredges have pumped material through distances in excess of 2 mi (3.2 km), at the same time elevating it more than 100 ft (30 m). This type of dredge is also used for sand-and-gravel-plant operations and for land-reclamation work. Levees and dams can be built with hydraulic dredges. The usual maximum digging depth is about 50 ft (16 m). Hydraulic dredges are reclaiming copper stamp-mill tailings from a depth of 115 ft (35 m) below the water, and a depth of 165 ft (50 m) has been reached in a land-reclamation job.

The usual type of hydraulic dredge has a **digging ladder** suspended from the bow at an angle of 45° for the maximum digging depth. This ladder carries the suction pipe and cutter, with its driving machinery, and the swinging-line sheaves. The cutter head may have applied to it 25 to 1,000 hp (3.7 to 746 kW). The 20-in (0.5-m) dredge, which is the standard, general-purpose machine, has a cutter drive of about 300 hp. The usual operating speed of the cutter is 5 to 20 r/min.

The material excavated by the cutter enters the mouth of the suction pipe, which is located within and at the lower side of the cutter head. The material is sucked up by a centrifugal pump, which discharges it to the dump through a pipeline. The shore discharge pipe is usually of the telescopic type, made of No. 10 to 3/10-in (3- to 7.5-mm) plates in lengths of 16 ft (5 m) so that it can be readily handled by the shore crew. Floating pipelines are usually made of plates from ¼ to ½ in (6 to 13 mm) thick and in lengths of 40 to 100 ft (12 to 30 m), which are floated on pontoons and connected together through rubber sleeves or, preferably, ball joints. The floating discharge line is flexibly connected to the hull in order to permit the dredge to swing back and forth across the cut while working without disturbing the pipeline.

Pump efficiency is usually sacrificed to make an economical unit for the handling of material, which may run from 2 to 25 percent of the total volume of the mixture pumped. Most designs have generous clearances and will permit the passage of stone which is 70 percent of the pipeline

diameter. The pump efficiencies vary widely but in general may run from 50 to 70 percent.

Commercial dredges vary in size and discharge-pipe diameters from 12 to 30 in (0.3 to 0.8 m). Smaller or larger dredges are usually special-purpose machines. A number of 36-in (0.9-m) dredges are used to maintain the channel of the Mississippi River. The power applied to pumps varies from 100 to 3,000 hp (75 to 2,200 kW). The modern 20-in (0.5-m) commercial dredge has about 1,350 bhp (1,007 kW) applied to the pump.

Diesel dredges are built for direct-connected or electric drives, and modern steam dredges have direct-turbine or turboelectric drives. The steam turbine and the dc electric motor have the advantage that they are capable of developing full rating at reduced speeds.

Within its scope, the hydraulic dredge can work more economically than any other excavating machine or combination of machines.

Dragline Excavators

Dragline excavators are typically used for digging open cuts, drainage ditches, canals, sand, and gravel pits, where the material is to be moved 20 to 1,000 ft (6 to 305 m) before dumping. They cannot handle rock unless the rock is blasted. Since they are provided with long booms and mounted on turntables, permitting them to swing through a full circle, these excavators can deposit material directly on the spoil bank farther from the point of excavation than any other type of machine. Whereas a shovel stands below the level of the material it is digging, a dragline excavator stands above and can be used to excavate material under water.

Figure 10.4.42 shows a self-contained dragline mounted on crawler treads. The drive is almost exclusively gasoline in the small sizes and diesel, diesel-electric, or electric, frequently with Ward Leonard control, in the large sizes. The boom a is pivoted at its lower end to the turntable, the outer end being supported by cables b, so that it can be raised or lowered to the desired angle. The scraper bucket c is supported

Fig. 10.4.42 Dragline excavator.

by cable d, which is attached to a bail on the bucket, passes over a sheave at the head of the boom, and is made fast to the engine. A second cable e is attached to the front of the bucket and made fast to the second drum of the engine. The bucket is dropped and dragged along the surface of the material by cable e until filled. It is then hoisted by cable d, drawn back to its dumping position, e being kept tight until the dumping point is reached, when e is slacked, allowing the bucket to dump by gravity. After the bucket is filled, the boom is swung to the dumping position while the bucket is being hauled out. A good operator can throw the bucket 10 to 40 ft (3 to 12 m) beyond the end of the boom, depending on the size of machine and the working conditions. The depth of the cut varies from 12 to 75 ft (3.7 to 23 m), again depending on the size of machine and the working conditions. With the smaller machines and under favorable conditions, two or even three trips per minute are possible; but with the largest machines, even one trip per minute may not be attained. The more common sizes are for handing ¾- to 4-yd³ (0.6- to 3-m³) buckets with boom lengths up to 100 or 125 ft (30 to 38 m), but machines have been built to handle an 8-yd³ (6-m³) bucket with a boom length of 200 ft (60 m). The same machine can handle a 12-yd³ (9-m³) bucket with the boom shortened to 165 ft (50 m).

Slackline Cableways

Used widely in sand-and-gravel plants, the **slackline cableway** employs an open-ended dragline bucket suspended from a carrier (Fig. 10.4.43) which runs upon a track cable. It will dig, elevate, and convey materials in one continuous operation.

Fig. 10.4.43   Slackline-cable bucket and trolley.

Figure 10.4.44 shows a typical slackline-cableway operation. The bucket and carrier is *a*; *b* is the track cable, inclined to return the bucket and carrier by gravity; *c* is a tension cable for raising or lowering the track cable; *d* is the load cable; and *e* is a power unit with two friction drums having variable speeds. A mast or tower *f* is used to support guide and tension blocks at the high end of the track cable; a movable tail

tower *g* supports the lower end of the track cable. The bucket is raised and lowered by tensioning or slacking off the track cable. The bucket is loaded, after lowering, by pulling on the load cable. The loaded bucket, after raising, is conveyed at high speed to the dumping point and is

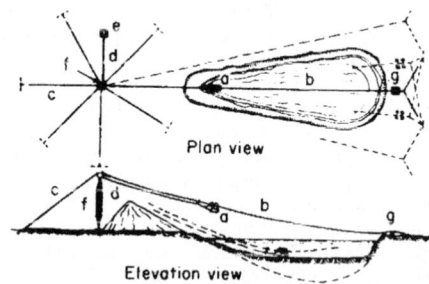

Fig. 10.4.44   Slackline-cable plant. (*Sauerman.*)

returned at a still higher speed by gravity to the digging point. The cableway can be operated in radial lines from a mast or in parallel lines between two moving towers. It will not dig rock unless the rock is blasted. The depth of digging may vary from 5 to 100 ft.

10.5   CONVEYOR MOVING AND HANDLING
by Vincent M. Altamuro

OVERHEAD CONVEYORS
By Ivan L. Ross
Acco Chain Conveyor Division

Conveyors are primarily horizontal-movement, fixed-path, constant-speed material handling systems. However, they often contain inclined sections to change the elevation of the material as it is moving, switches to permit alternate paths and "power-and-free" capabilities to allow the temporary slowing, stopping, or accumulating of material.

Conveyors are used, not only for transporting material, but also for in-process storage. To the extent possible, materials should be processed while being moved, so that value is added to it as it is continuously transported. They may be straight, curved, closed-loop, irreversible, or reversible. Some types of conveyors are:

air blower	monorail
apron	pneumatic tube
belt	power-and-free
bucket	roller
car-on-track	screw
carousel	skate wheel
chain	slat
flight	spiral
hydraulic	towline
magnetic	trolley

Tables 10.5.1 and 10.5.2 describe and compare some of these.

Conveyors are often used as integral components of assembly systems. They bring the correct material, at the required rate, to each worker and then to the next operator in the assembly sequence. Figure 10.5.1 shows ways conveyors can be used in assembly.

Overhead conveyor systems are defined in two general classifications: the basic trolley conveyor and the power-and-free conveyor, each of which serves a definite purpose.

Trolley conveyors, often referred to as overhead power conveyors, consist of a series of trolleys or wheels supported from or within an

overhead track and connected by an endless propelling means, such as chain, cable, or other linkages. Individual loads are usually suspended from the trolleys or wheels (Fig. 10.5.2). Trolley conveyors are utilized for transportation or storage of loads suspended from one conveyor which follows a single fixed path. They are normally used in applications where a balanced, continuous production is required. Track sections range from lightweight "tee" members or tubular sections, to medium- and heavy-duty I-beam sections. The combinations and sizes of trolley-propelling means and track sections are numerous. Normally this type of conveyor is continually in motion at a selected speed to suit its function.

Power-and-free conveyor systems consist of at least one power conveyor, but usually more, where the individual loads are suspended from one or more free trolleys (not permanently connected to the propelling means) which are conveyor-propelled through all or part of the system. Additional portions of the system may have manual or gravity means of propelling the trolleys.

Worldwide industrial, institutional, and warehousing requirements of in-process and finished products have affected the considerable growth and development of power-and-free conveyors. Endless varieties of size, style, and color, in all imaginable product combinations, have extended the use of power-and-free conveyors. The power-and-free system combines the advantages of continuously driven chains with the versatile traffic system exemplified by traditional monorail unpowered systems. Thus, high-density load-transportation capabilities are coupled each with complex traffic patterns and in-process or workstation requirements to enable production requirements to be met with a minimum of manual handling or transferring. Automatic dispatch systems for coding and programming of the load routing are generally used.

Trolley Conveyors

The load-carrying member of a trolley conveyor is the trolley or series of wheels. The wheels are sized and spaced as a function of the imposed

Table 10.5.1   Types of Conveyors Used in Factories

Type of conveyor	Description	Features/limitations
Overhead		
Power-and-free	Carriers hold parts and move on overhead track. With inverted designs, track is attached to factory floor. Carriers can be transferred from powered to "free" track.	Accumulation; flexible routing; live storage; can hold parts while production steps are performed; can travel around corners and up inclines.
Trolley	Similar to power and free, but carriers cannot move off powered track.	Same as power and free, but flexible routing is not possible without additional equipment
Above-floor		
Roller, powered	Load-carrying rollers are driven by a chain or belt.	Handles only flat-bottomed packages, boxes, pallets or parts; can accumulate loads; can also be suspended from ceiling for overhead handling.
Roller, gravity	Free-turning rollers; loads are moved either by gravity or manual force.	When inclined, loads advance automatically.
Skate wheel	Free-turning wheels spaced on parallel shafts.	For lightweight packages and boxes; less expensive than gravity rollers.
Spiral tower	Helix-shaped track which supports parts or small "pallets" that move down track.	Buffer storage; provides surge of parts to machine tools when needed.
Magnetic	Metal belt conveyor with magnetized bed.	Handles ferromagnetic parts, or separates ferromagnetic parts from nonferrous scrap.
Pneumatic	Air pressure propels cylindrical containers through metal tubes.	Moves loads quickly; can be used overhead to free floor space.
Car-on-track	Platforms powered by rotating shaft move along track.	Good for flexible manufacturing systems; precise positioning of platforms; flexible routing.
In-floor		
Towline	Carts are advanced by a chain in the floor.	High throughput; flexible routing possible by incorporating spurs in towline; tow carts can be manually removed from track; carts can travel around corners.

SOURCE: *Modern Materials Handling*, Aug. 5, 1983, p. 55.

Table 10.5.2   Transportation Equipment Features

Equipment	Most practical travel distance	Automatic load or unload	Typical load	Throughput rate	Travel path	Typical application
Conveyors						
Belt	Short to medium	No	Cases, small parts	High	Fixed	Take-away from picking, sorting
Chain	Short to medium	Yes	Unit	High	Fixed	Deliver to and from automatic storage and retrieval systems
Roller	Short to long	Yes	Unit, case	High	Fixed	Unit: same as chain Case: sorting, delivery between pick station
Towline	Medium to long	Yes	Carts	High	Fixed	Delivery to and from shipping or receiving
Wire-guided vehicles						
Cart	Short to long	Yes	Unit	Low	Fixed	Delivery between two drop points
Pallet truck	Short to long	Yes	Unit	Low	Fixed	Delivery between two drop points
Tractor train	Short to long	Yes	Unit	Medium	Fixed	Delivery to multiple drop points
Operator-guided vehicles						
Walkie pallet	Short	No	Unit	Low	Flexible	Short hauls at shipping and receiving
Rider pallet	Medium	No	Unit	Low	Flexible	Dock-to-warehouse deliveries
Tractor train	Short to long	No	Unit	Medium	Flexible	Delivery to multiple drop points

SOURCE: *Modern Materials Handling*, Feb. 22, 1980, p. 106.

load, the propelling means, and the track capability. The load hanger (carrier) is attached to the conveyor and generally remains attached unless manually removed. However, in a few installations, the load hanger is transferred to and from the conveyor automatically.

Fig. 10.5.1 Conveyors used in assembly operations. (*a*) Belt conveyor, with diverters at each station; (*b*) carousel circulates assembly materials to workers; (*c*) multiple-path conveyors allow several products to be built at once; (*d*) gravity roller conveyors; (*e*) towline conveyor, moving assemblies from one group of assemblers to another. (*Modern Materials Handling, Nov. 1979, page 116.*)

The trolley conveyor can employ any chain length consistent with allowable propelling means and drive(s) capability. The track layout always involves horizontal turns and commonly has vertical inclines and declines.

When a dimensional layout, load spacing, weights of moving loads, function, and load and unload points are determined, the chain or cable pull can be calculated. Manufacturer's data should be used for frictional values. A classical point-to-point analysis should be made, using the most unfavorable loading condition.

In the absence of precise data, the following formulas can be used to find the approximate drive effort:

$$\text{Max drive effort, lb} = A + B + C$$
$$\text{Net drive effort, lb} = A + B + C - D$$

$$\text{Max drive effort, } N = (A + B + C)\,9.81$$
$$\text{Net drive effort, } N = (A + B + C - D)\,9.81$$

where $A = fw$ [where w = total weight of chain, carriers, and live load, lb (kg) and f = coefficient of friction]; $B = wS$ [where w = average carrier load per ft (m), lb/ft (kg/m) and S = total vertical rise, ft (m)]; $C = 0.017f(A + B)N$ (where N = sum of all horizontal and vertical curves, deg); and $D = w'S'$ [where w' = average carrier load per ft (m), lb/ft (kg/m) and s' = total vertical drop, ft (m)].

For conveyors with antifriction wheels, with clean operating condition, the coefficient of friction f may be 0.13.

Fig. 10.5.2   Typical conveyor chains or cable.

Where drive calculations indicate that the allowable chain or cable tension may be exceeded, multiple-drive units are used. When multiple drives are used for constant speed, high-slip motors or fluid couplings are commonly used. If variable speed is required with multiple drives, it is common to use direct-current motors with direct-current supply and controls. For both constant and variable speed, drives are balanced to share drive effort. Other than those mentioned, many various methods of balancing are available for use.

Complexities of overhead conveyors, particularly with varying loads on inclines and declines, as well as other influencing factors (e.g., conveyor length, environment, lubrication, and each manufacturer's design recommendations), usually require detail engineering analysis to select the proper number of drives and their location. Particular care is also required to locate the take-up properly.

The following components or devices are used on trolley-conveyor applications:

Trolley Assembly   Wheels and their attachment portion to the propelling means (chain or cable) are adapted to particular applications, depending upon loading, duty cycle, environment, and manufacturer's design.

Carrier Attachments   These are made in three main styles: (1) enclosed tubular type, where the wheels and propelling means are carried inside; (2) semienclosed tubular type, where the wheels are enclosed and the propelling means is external; and (3) open-tee or I-beam type, where the wheels and propelling means are carried externally.

Sprocket or Traction Wheel Turn   Any arc of horizontal turn is available. Standards usually vary in increments of 15° from 15 to 180°.

Roller Turns   Any arc of horizontal turn is available. Standards usually vary in increments of 15° from 15 to 180°.

Track Turns   These are horizontal track bends without sprockets, traction wheels, or rollers; they are normally used on enclosed-track conveyors where the propelling means is fitted with horizontal guide wheels.

Track Hangers, Brackets, and Bracing   These conform to track size and shape, spaced at intervals consistent with allowable track stress and deflection applied by loading and chain or cable tensions.

Track Expansion Joints   For use in variable ambient conditions, such as ovens, these are also applied in many instances where conveyor track crosses building expansion joints.

Chain Take-Up Unit   Required to compensate for chain wear and/or variable ambient conditions, this unit may be traction-wheel, sprocket, roller, or track-turn type. Adjustment is maintained by screw, screw spring, counterweight, or air cylinder.

Incline and Decline Safety Devices   An "anti-back-up" device will ratchet into a trolley or the propelling means in case of unexpected reversal of a conveyor on an incline. An "anti-runaway" device will sense abnormal conveyor velocity on a decline and engage a ratchet into a trolley or the propelling means. Either device will arrest the uncontrolled movement of the conveyor. The anti-runaway is commonly connected electrically to cut the power to the drive unit.

Drive Unit   Usually sprocket or caterpillar type, these units are available for constant-speed or manual variable-speed control. Common speed variation is 1 : 3; e.g., 5 to 15 ft/min (1.5 to 4.5 m/min). Drive motors commonly range from fractional to 15 hp (11 kW).

Drive Unit Overload   Overload is detected by electrical, torque, or pull detection by any one of many available means. Usually overload will disconnect power to the drive, stopping the conveyor. When the reason for overload is determined and corrected, the conveyor may be restarted.

Equipment Guards   Often it is desirable or necessary to guard the conveyor from hostile environment and contaminants. Also employees must be protected from accidental engagement with the conveyor components.

Transfer Devices   Usually unique to each application, automatic part or carrier loading, unloading, and transfer devices are available. With growth in the use of power-and-free, carrier transfer devices have become rare.

Power-and-Free Conveyor

The power-and-free conveyor has the highest potential application wherever there is a requirement for other than a single fixed-path flow (trolley conveyor). Power-and-free conveyors may have any number of automatic or manual switch points. A system will permit scheduled transit and delivery of work to the next assigned station automatically. Accumulation (storage) areas are designed to accommodate in-process inventory between operations.

The components and chain-pull calculation discussed for powered overhead conveyors are basically applicable to the power-and-free conveyor. Addition of a secondary free track surface is provided for the work carrier to traverse. This free track is usually disposed directly below the power rail but is sometimes found alongside the power rail. (This arrangement is often referred to as a "side pusher" or "drop finger.") The power-and-free rails are joined by brackets for rail (free-track)

(a) Open type

(b) Enclosed type

Fig. 10.5.3   Conveyor tracks.

continuity. The power chain is fitted with pushers to engage the work-carrier trolley. Track sections are available in numerous configurations for both the power portion and the free portion. Sections will be enclosed, semienclosed, or open in any combinations. Two of the most common types are shown in Fig. 10.5.3.

As an example of one configuration of power-and-free, Fig. 10.5.4 shows the ACCO Chain Conveyor Division enclosed-track power-and-free rail. In the cutaway portion, the pushers are shown engaged with the work-carrier trolley.

The pushers are pivoted on an axis parallel to the chain path and swing aside to engage the pusher trolley. The pusher trolley remains engaged on level and sloped sections. At automatic or manual switching points, the leading dispatch trolley head which is not engaged with the

Fig. 10.5.4   Power-and-free conveyor rail and trolley heads. (*ACCO Chain Conveyor Division.*)

chain is propelled through the switch to the branch line. As the chain passes the switching point, the pusher trolley departs to the right or left from pusher engagement and arrives on a free line, where it is subject to manual or controlled gravity flow.

The distance between pushers on the chain for power-and-free use is established in accordance with conventional practice, except that the minimum allowable pusher spacing must take into account the wheelbase of the trolley, the bumper length, the load size, the chain velocity, and the action of the carrier at automatic switching and reentry points. A switching headway must be allowed between work carriers. An approximation of the minimum allowable pusher spacing is that the pusher spacing will equal twice the work-carrier bumper length. Therefore, a 4-ft (1.2-m) work carrier would indicate a minimum pusher spacing of 8 ft (2.4 m).

The load-transmission capabilities are a function of velocity and pusher spacing. At the 8-ft (2.4-m) pusher spacing and a velocity of 40 ft/min (0.22 m/s), five pushers per minute are made available. The load-transmission capability is five loads per minute, or 300 loads per hour.

Method of Automatic Switching from a Powered Line to a Free Line   Power-and-free work carriers are usually switched automatically. To do this, it is necessary to have a code device on the work carrier and a decoding (reading) device along the track in advance of the track switch. Figure 10.5.5 shows the equipment relationship. On each carrier, the free trolley carries the code selection, manually or automatically introduced, which identifies it for a particular destination or routing. As the free trolley passes the reading station, the trolley intelligence is decoded and compared with a preset station code and its current knowledge of the switch position and branch-line condition; a decision is then reached which results in correct positioning of the rail switch.

In Fig. 10.5.5, the equipment illustrated includes a transistorized readout station *a*, which supplies 12 V direct current at stainless-steel code brushes *b*. The code brushes are "matched" by contacts on the encoded trolley head *c*. When a trolley is in register and matches the code-brush positioning, it will be allowed to enter branch line *d* if the line is not full. If carrier *e* is to be entered, the input signal is rectified and amplified so as to drive a power relay at junction box f, which in turn actuates solenoid g to operate track switch h to the branch-line position. A memory circuit is established in the station, indicating that a full-line condition exists. This condition is maintained until the pusher pin of the switched carrier clears reset switch j.

Fig. 10.5.5   Switch mechanism exiting trolley from power rail to free rail. (*ACCO Chain Conveyor Division.*)

The situations which can be handled by the decoding stations are as follows: (1) A trolley with a matching code is in register; there is space in the branch line. The carrier is allowed to enter the branch line. (2) A trolley with a matching code is in register; there is no space in the branch line. The carrier is automatically recirculated on the powered system and will continue to test its assigned destination until it can be accommodated. (3) A trolley with a matching code is in register; there is no space in the branch line. The powered conveyor is automatically stopped, and visual and/or audible signals are started. The conveyor can be automatically restarted when the full-line condition is cleared, and the waiting carrier will be allowed to enter. (4) A trolley with a non-matching code is in register; in this case, the decoding station always returns the track switch to the main-line position, if necessary, and bypasses the carrier.

Use and Control of Carriers on Free and Gravity Lines   Free and gravity lines are used as follows: (1) To connect multiple power-and-free conveyors, thus making systems easily extensible and permitting different conveyor designs for particular use. (2) To connect two auxiliary devices such as vertical conveyors and drop-lift stations. (3) As manned or automatic workstations. In this case, the size of the station depends on the number of carriers processed at one time, with additional

space provision for arriving and departing carriers. (Considerable knowledge of work-rate standards and production-schedule requirements is needed for accurate sizing.) (4) As manned or automatic storage lines, especially for handling production imbalance for later consumption.

Nonmanned automatic free lines require that the carrier be controlled in conformance with desired conveyor function and with regard to the commodity being handled. The two principal ways to control carriers in nonmanned free rails are (1) to slope the free rail so that all carriers will start from rest and use incremental spot retarders to check velocity, and (2) to install horizontal or sloped rails and use auxiliary power conveyor(s) to accumulate carriers arriving in the line.

In manned free lines, it is usual to have slope at the automatic arrival and automatic departure sections only. These sections are designed for automatic accumulation of a finite number of carriers, and retarding or feeding devices may be used. Throughout the remainder of the manned station, the carrier is propelled by hand.

Method of Automatic Switching from a Free Line to a Powered Line   Power-and-free work carriers can be reentered into the powered lines either manually or automatically. The carrier must be integrated with traffic already on the powered line and must be entered so that it will engage with a pusher on the powered chain. Figure 10.5.6 illustrates a typical method of automatic reentry. The carrier a is held at a rest on a slope in the demand position by the electric trolley stop b. The demand to enter enables the sensing switches c and d mounted on the powered rail to test for the availability of a pusher. When a pusher that is not propelling a load is sensed, all conditions are met and the carrier is released by the electric trolley stop such that it arrives in the pickup position in advance of the pusher. A retarding or choke device can be used to keep the entering carrier from overrunning the next switch position or another carrier in transit. The chain pusher engages the pusher trolley and departs the carrier. The track switch f can remain in the branch-line position until a carrier on the main line g would cause the track switch to reset. Automatic reentry control ensures that no opportunity to use a pusher is overlooked and does not require the time of an operator.

Power-and-Free Conveyor Components   All components used on trolley conveyors are applied to power-and-free conveyors. Listed below are a few of the various components unique to power-and-free systems.

TRACK SWITCH.   This is used for diverting work carriers either automatically or manually from one line or path to another. Any one system may have both. Switching may be either to the right or to the left. Automatically, stops are usually operated pneumatically or electromechanically. Track switches are also used to merge two lines into one.

Fig. 10.5.6   Switch mechanism reentering trolley from free rail to power rail. (*ACCO Chain Conveyor Division.*)

TROLLEY STOPS.   Used to stop work carriers, these operate either automatically or manually on a free track section or on a powered section. Automatically, stops are usually operated pneumatically or electromechanically.

STORAGE.   Portions or spurs of power-and-free conveyors are usually dedicated to the storage (accumulation) of work carriers. Unique to the design, type, or application, storage may be accomplished on (1) level hand-pushed lines, (2) gravity sloped lines (usually with overspeed-control retarders), (3) power lines with spring-loaded pusher dogs, or (4) powered lines with automatic accumulating free trolleys.

INCLINES AND DECLINES.   As in the case of trolley conveyors, vertical inclines and declines are common to power-and-free. In addition to safety devices used on trolley conveyors, similar devices may be applied to free trolleys.

LOAD BARS AND CARRIERS.   The design of load bars, bumpers, swivel devices, index devices, hooks, or carriers is developed at the time of the initial power-and-free investigation. The system will see only the carrier, and all details of system design are a function of its design. How the commodity is being handled on or in the carrier is carefully considered to facilitate its use throughout the system, manually, automatically, or both.

VERTICAL CONVEYOR SECTIONS.   Vertical conveyor sections are often used as an accessory to power-and-free. For practical purposes, the vertical conveyor can be divided into two classes of devices:

DROP (LIFT) SECTION.   This device is used to drop (or lift) the work carrier vertically to a predetermined level in lieu of vertical inclines or declines. One common reason for its use is to conserve space. The unit may be powered by a cylinder or hoist, depending on the travel distance, cycle time, and load. One example of the use of a drop (lift) section is to receive a carrier on a high level and lower it to an operations level. The lower level may be a load-unload station or processing station. Automatic safety stops are used to close open rail ends.

INTERFLOOR VERTICAL CONVEYOR.   When used for interfloor service and long lifts, the vertical conveyor may be powered by high-speed hoists or elevating machines. In any case, the carriers are automatically transferred to and from the lift, and the dispatch control on the carrier can instruct the machine as to the destination of the carrier. Machines can be equipped with a variety of speeds and operating characteristics. Multiple carriers may be handled, and priority-call control systems can be fitted to suit individual requirements.

NONCARRYING CONVEYORS

Flight Conveyors   Flight conveyors are used for moving granular, lumpy, or pulverized materials along a horizontal path or on an incline seldom greater than about 40°. Their principal application is in handling coal. The flight conveyor of usual construction should not be specified for a material that is actively abrasive, such as damp sand and ashes. The **drag-chain conveyor** (Fig. 10.5.7) has an open-link chain, which serves, instead of flights, to push the material along. With a hard-faced concrete or cast-iron trough, it serves well for handling ashes. The return run is, if possible, above the carrying run, so that the dribble will be back into the loaded run. A feeder must be provided unless the feed is

otherwise controlled, as from a tandem elevator or conveyor. As the feeder is interlocked, either mechanically or electrically, the feed stops if the conveyor stops. Flight conveyors may be classified as **scraper type** (Fig. 10.5.8), in which the element (chain and flights) rests on the trough; **suspended-flight type** (Fig. 10.5.9), in which the flights are carried clear of the trough by shoes resting on guides; and **suspended-chain type** (Fig. 10.5.10), in which the chain rests on guides, again carrying the flights clear of the trough. These types are further differentiated as **single-strand** (Figs. 10.5.8 and 10.5.9) and **double-strand** (Fig. 10.5.10). For lumpy material, the latter has the advantage since the lumps will enter the trough without interference. For heavy duty also, the double strand has the advantage, in that the pull is divided between

Fig. 10.5.9 Single-strand suspended-flight conveyor.

Fig. 10.5.10 Double-strand roller-chain flight conveyor.

two chains. A special type for simultaneous handling of several materials may have the trough divided by longitudinal partitions. The material having the greatest coefficient of friction is then carried, if possible, in the central zone to equalize chain wear and stretch.

Improvements in the welding and carburizing of welded-link chain have made possible its use in flight conveyors, offering several significant advantages, including economy and flexibility in all directions. Figure 10.5.11 shows a typical scraper-type flight cast from malleable iron incorporated onto a slotted conveyor bed. The small amount of fines that fall through the slot are returned to the top of the bed by the returning flights. Figure 10.5.12 shows a double-chain scraper conveyor in which the ends of the flights ride in a restrictive channel. These types of flight conveyors are driven by pocket wheels.

Fig. 10.5.11 Scraper flight with welded chain.

Flight conveyors of small capacity operate usually at 100 to 150 ft/min (0.51 to 0.76 m/s). Large-capacity conveyors operate at 100 ft/min (0.51 m/s) or slower; their long-pitch chains hammer heavily against the drive-sprocket teeth or pocket wheels at higher speeds.

Fig. 10.5.7 Drag chain.

Fig. 10.5.8 Single-strand scraper-flight conveyor.

A conveyor steeply inclined should have closely spaced flights so that the material will not avalanche over the tops of the flights. The capacity of a given conveyor diminishes as the angle of slope increases. For the heaviest duty, hardened-face rollers at the articulations are essential.

Fig. 10.5.12 Scraper flight using parallel welded chains.

Cautions in Flight-Conveyor Selections With abrasive material, the trough design should provide for renewal of the bottom plates without disturbing the side plates. If the conveyor is inclined and will reverse when halted under load, a solenoid brake or other automatic backstop should be provided. Chains may not wear or stretch equally. In a double-strand conveyor, it may be necessary to shift sections of chain from one side to the other to even up the lengths.

Intermediate slide gates should be set to open in the opposite direction to the movement of material in the conveyor.

The **continuous-flow conveyor** serves as a conveyor, as an elevator, or as a combination of the two. It is a slow-speed machine in which the material moves as a continuous core within a duct. Except with the **Redler** conveyor, the element is formed by a single strand of chain with closely spaced impellers, somewhat resembling the flights of a flight conveyor.

The **Bulk flo** (Fig. 10.5.13) has peaked flights designed to facilitate the outflow of the load at the point of discharge. The load, moved by a positive push of the flights, tends to provide self-clearing action at the end of a run, leaving only a slight residue.

The **Redler** (Fig. 10.5.14) has skeletonized or U-shaped impellers which move the material in which they are submerged because the resistance to slip through the element is greater than the drag against the walls of the duct.

Fig. 10.5.13 Bulk-flow continuous-flow elevator.

Materials for which the continuous-flow conveyor is suited are listed below in groups of increasing difficulty. The constant C is used in the power equations below, and in Fig. 10.5.16.

$C = 1$: clean coal, flaxseed, graphite, soybeans, copra, soap flakes
$C = 1.2$: beans, slack coal, sawdust, wheat, wood chips (dry), flour
$C = 1.5$: salt, wood chips (wet), starch
$C = 2$: clays, fly ash, lime (pebble), sugar (granular), soda ash, zinc oxide
$C = 2.5$: alum, borax, cork (ground), limestone (pulverized)

Among the materials for which special construction is advised are bauxite, brown sugar, hog fuel, wet coal, shelled corn, foundry dust, cement, bug dust, and hot brewers' grains. The machine should not be specified for ashes, bagasse, carbon-black pellets, sand and gravel,

sewage sludge, and crushed stone. Fabrication from corrosion-resistant materials such as brass, monel, or stainless steel may be necessary for use with some corrosive materials.

Where a single **runaround conveyor** is required with multiple feed points and some recirculation of excess load, the Redler serves. The U-frame flights do not squeeze the loads, as they resume parallelism after separating when rounding the terminal wheels. As an elevator, this machine will also handle sluggish materials that do not flow out readily. A pusher plate opposite the discharge chute can be employed to enter

Fig. 10.5.14 Redler U-type continuous-flow conveyor.

between the legs of the U flights to push out such material. When horizontal or inclined, continuous-flow elevators such as the Redler are normally self-cleaning. When vertical, the Redler type can be made self-cleaning (except with sticky materials) by use of special flights.

Continuous-flow conveyors and elevators do not require a feeder (Fig. 10.5.15). They are self-loading to capacity and will not overload, even though there are several open- or uncontrolled-feed openings, since the duct fills at the first opening and automatically prevents the entrance of additional material at subsequent openings. Some special care may be required with free-flowing material.

Fig. 10.5.15 Shallow-track hopper for continuous-flow conveyor with feed to return run.

The duct is easily insulated by sheets of asbestos cement or similar material to reduce cooling in transit. As the duct is completely sealed, there is no updraft where the lift is high. The material is protected from exposure and contamination or contact with lubricants. The handling capacity for horizontal or inclined lengths (nearly to the angle of repose of the material) approximates 100 percent of the volume swept through by the movement of the element. For steeper inclines or elevators, it is between 50 and 90 percent.

If the material is somewhat abrasive, as with wet bituminous coal, the duct should be of corrosion-resistant steel, of extra thickness, and the chain pins should be both extremely hard and of corrosion-resistant material.

A long horizontal run followed by an upturn is inadvisable because of radial thrust. Lumpy material is difficult to feed from a track hopper. An automatic brake is unnecessary, as an elevator will reverse only a few inches when released.

The motor horsepower P required by continuous-flow conveyors for the five arrangements shown in Fig. 10.5.16 is given in the accompanying formulas in terms of the capacity T, in tons per h; the horizontal run H, ft; the vertical lift V, ft; and the constant C, values for which are given above. If loading from a track hopper, add 10 percent.

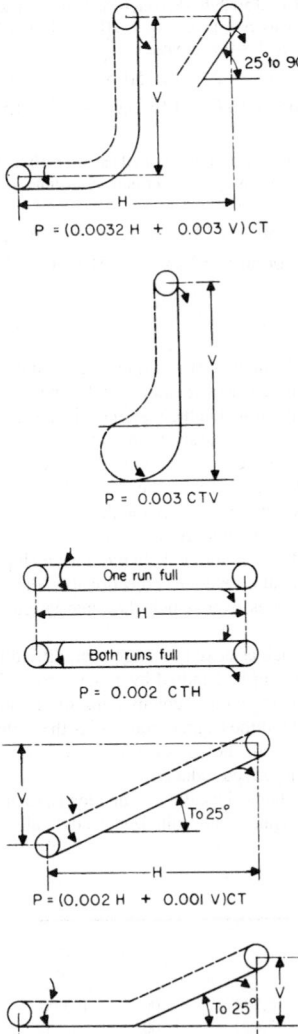

Fig. 10.5.16 Continuous-flow-conveyor arrangements.

Screw Conveyors

The screw, or spiral, conveyor is used quite widely for pulverized or granular, noncorrosive, nonabrasive materials when the required capacity is moderate, when the distance is not more than about 200 ft (61 m), and when the path is not too steep. It usually costs substantially less than any other type of conveyor and is readily made dusttight by a simple cover plate.

The conveyor will handle lumpy material if the lumps are not large in proportion to the diameter of the helix. If the length exceeds that advisable for a single conveyor, separate or tandem units are readily arranged. Screw conveyors may be inclined. A standard-pitch helix will handle material on inclines up to 35°. The reduction in capacity as compared with the capacity when horizontal is indicated in the following table:

Inclination, deg	10	15	20	25	30	35
Reduction in capacity, percent	10	26	45	58	70	78

Abrasive or corrosive materials can be handled with suitable construction of the helix and trough.

The standard screw-conveyor helix (Fig. 10.5.17) has a pitch approximately equal to its outside diameter. Other forms are used for special cases.

Fig. 10.5.17 Spiral conveyor.

Short-pitch screws are advisable for inclines above 29°.

Variable-pitch screws, with short pitch at the feed end, automatically control the flow to the conveyor so that the load is correctly proportioned for the length beyond the feed point. With a short section either of shorter pitch or of smaller diameter, the conveyor is self-loading to capacity and does not require a feeder.

Fig. 10.5.18 Cut-flight conveyor.

Cut flights (Fig. 10.5.18) are used for conveying and mixing cereals, grains, and other light materials.

Ribbon screws (Fig. 10.5.19) are used for wet and sticky materials, such as molasses, hot tar, and asphalt, which might otherwise build up on the spindle.

Paddle screws are used primarily for mixing such materials as mortar and bitulithic paving mixtures. One typical application is to churn ashes and water to eliminate dust.

Fig. 10.5.19 Ribbon conveyor.

Standard constructions have a plain or galvanized-steel helix and trough. For abrasives and corrosives such as wet ashes, both helix and trough may be of hard-faced cast iron. For simple abrasives, the outer edge of the helix may be faced with a renewable strip of Stellite or

Table 10.5.3 Capacities and Speed of Spiral Conveyors

Group	Max percent of cross section occupied by the material	Max density of material, lb/ft^3 (kg/m^3)	Max r/min for diameters	
			6 in (152 mm)	20 in (508 mm)
1	45	50 (800)	170	110
2	38	50 (800)	120	75
3	31	75 (1,200)	90	60
4	25	100 (1,600)	70	50
5	12½		30	25

Group 1 includes light materials such as barley, beans, brewers grains (dry), coal (pulv.), corn meal, cottonseed meal, flaxseed, flour, malt, oats, rice, wheat. The value of the factor F is 0.5.

Group 2 includes fines and granular materials. The values of F are alum (pulv.), 0.6, coal (slack or fines), 0.9; coffee beans, 0.4; sawdust, 0.7; soda ash (light), 0.7; soybeans, 0.5; fly ash, 0.4.

Group 3 includes materials with small lumps mixed with fines. Values of F are alum, 1.4; ashes (dry), 4.0; borax, 0.7; brewers grains (wet), 0.6; cottonseed, 0.9; salt, course or fine, 1.2; soda ash (heavy), 0.7.

Group 4 includes semiabrasive materials, fines, granular and small lumps. Values of F are acid phosphate (dry), 1.4; bauxite (dry), 1.8; cement (dry), 1.4; clay, 2.0; fuller's earth, 2.0; lead salts, 1.0; limestone screenings, 2.0; sugar (raw), 1.0; white lead, 1.0; sulfur (lumpy), 0.8; zinc oxide, 1.0.

Group 5 includes abrasive lumpy materials which must be kept from contact with hanger bearings. Values of F are wet ashes, 5.0; flue dirt, 4.0; quartz (pulv.), 2.5; silica sand, 2.0; sewage sludge (wet and sandy), 6.0.

similar extremely hard material. For food products, aluminum, bronze, monel metal, or stainless steel is suitable but expensive.

Table 10.5.3 gives the capacities, allowable speeds, percentages of helix loading for five groups of materials, and the factor F used in estimating the power requirement.

Table 10.5.4 gives the handling capacities for standard-pitch screw conveyors in each of the five groups of materials when the conveyors are operating at the maximum advised speeds and in the horizontal position. The capacity at any lower speed is in the ratio of the speeds.

Power Requirements The power requirements for horizontal screw conveyors of standard design and pitch are determined by the Link-Belt Co. by the formula that follows. Additional allowances should be made for inclined conveyors, for starting under load, and for materials that tend to stick or pack in the trough, as with cement.

$$H = \text{hp at conveyor head shaft} = (ALN + CWLF) \times 10^{-6}$$

where A = factor for size of conveyor (see Table 10.5.5); C = quantity of material, ft^3/h; L = length of conveyor, ft; F = factor for material (see Table 10.5.3); N = r/min of conveyor; W = density of material, lb/ft^3.

The motor size depends on the efficiency E of the drive (usually close to 90 percent); a further allowance G, depending on the horsepower, is made:

H	1	1–2	2–4	4–5	5
G	2	1.5	1.25	1.1	1.0

$$\text{Motor hp} = HG/E$$

When the material is distributed into a bunker, the conveyor has an open-bottom trough to discharge progressively over the crest of the pile so formed. This trough reduces the capacity and increases the required power, since the material drags over the material instead of over a polished trough.

If the material contains unbreakable lumps, the helix should clear the trough by at least the diameter of the average lump. For a given capacity, a conveyor of larger size and slower speed is preferable to a conveyor of minimum size and maximum speed. For large capacities

and lengths, the alternatives—a flight conveyor or a belt conveyor—should receive consideration.

EXAMPLES. 1. Slack coal 50 lb/ft^3 (800 kg/m^3); desired capacity 50 tons/h (2,000 ft^3/h) (45 tonnes/h); conveyor length, 60 ft (18 m); 14-in (0.36-m) conveyor at 80 r/min. F for slack coal = 0.9 (group 2).

$$H = (255 \times 60 \times 80 + 2,000 \times 50 \times 60 \times 0.9)/1,000,000 = 6.6$$
$$\text{Motor hp} = (6.6 \times 1.0)/0.90 = 7.3 \quad (5.4 \text{ kW})$$

Use 7½-hp motor.

2. Limestone screenings, 90 lb/ft^3; desired capacity, 10 tons/h (222 ft^3/h); conveyor length, 50 ft; 9-in conveyor at 50 r/min. F for limestone screenings = 2.0 (group 4).

$$H = (96 \times 50 \times 50 + 222 \times 90 \times 50 \times 2.0)/1,000,000 = 2.24$$
$$\text{Motor hp} = (2.24 \times 1.25)/0.90 = 2.8$$

Use 3-hp motor.

Chutes

Bulk Material If the material is fragile and cannot be set through a simple **vertical chute, a retarding chute** may be specified. Figure 10.5.20 shows a ladder chute in which the material trickles over shelves instead of falling freely. If it is necessary to minimize breakage when material is fed from a bin, a vertical box chute with flap doors opening inward, as shown in Fig. 10.5.21, permits the material to flow downward only from the top surface and eliminates the degradation that results from a converging flow from the bottom of the mass.

Straight inclined chutes for coal should have a slope of 40 to 45°. If it is found that the coal accelerates objectionably, the chute may be provided with cross angles over which the material cascades at reduced speed (Fig. 10.5.22).

Lumpy material such as coke and large coal, difficult to control when flowing from a bin, can be handled by a chain-controlled feeder chute with a screen of heavy endless chains hung on a sprocket shaft (Fig. 10.5.23). The weight of the chain curtain holds the material in the chute. When a feed is desired, the sprocket shaft is revolved slowly, either manually or by a motorized reducer.

Unit Loads Mechanical handling of unit loads, such as boxes, barrels, packages, castings, crates, and palletized loads, calls for methods and

Table 10.5.4 Screw-Conveyor Capacities
(ft^3/h)

Group	Conveyor size, in*							
	6	9	10	12	14	16	18	20
1	350	1,100	1,600	2,500	4,000	5,500	7,600	10,000
2	220	700	950	1,600	2,400	3,400	4,500	6,000
3	150	460	620	1,100	1,600	2,200	3,200	4,000
4	90	300	400	650	1,000	1,500	2,000	2,600
5	20	68	90	160	240	350	500	650

*Multiply by 25.4 to obtain mm.

Table 10.5.5 Factor *A*
(Self-lubricating bronze bearings assumed)

Diam of conveyor, in	6	9	10	12	14	16	18	20	24
mm	152	229	254	305	356	406	457	508	610
Factor *A*	54	96	114	171	255	336	414	510	690

mechanisms entirely different from those adapted to the movement of bulk materials.

Fig. 10.5.20 Ladder chute.

Fig. 10.5.21 Box chute with flap doors. Chute is always full up to discharging point.

Spiral chutes are adapted for the direct lowering of unit loads of various shapes, sizes, and weights, so long as their slide characteristics do not vary widely. If they do vary, care must be exercised to see that items accelerating on the selected helix pitch do not crush or damage those ahead.

Fig. 10.5.22 Inclined chute with cross angles.

Fig. 10.5.23 Chain-controlled feeder chute.

A spiral chute may extend through several floors, e.g., for lowering parcels in department stores to a basement shipping department. The opening at each floor must be provided with automatic closure doors, and the design must be approved by the Board of Fire Underwriters.

At the discharge end, it is usual to extend the chute plate horizontally to a length in which the loads can come to rest. A tandem gravity roll conveyor may be advisable for distribution of the loads.

The **sheet-metal spiral** (Fig. 10.5.24) has a fixed blade and can be furnished in varying diameters and pitches, both of which determine the maximum size of package that can be handled. These chutes may have receive and discharge points at any desired floors. There are certain kinds of commodities, such as those made of metal or bound with wire or metal bands, that cannot be handled satisfactorily unless the spiral chute is designed to handle only that particular commodity. Sheet-metal spirals can be built with double or triple blades, all mounted on the same standpipe. Another form of sheet-metal spiral is the open-core type, which is especially adaptable for handling long and narrow articles or for use where the spiral must wind around an existing column or pass through floors in locations limited by beams or girders that cannot be conveniently cut or moved. For handling bread or other food products, it is customary to have the spiral tread made from monel metal or aluminum.

Fig. 10.5.24 Metal spiral chute.

CARRYING CONVEYORS

Apron Conveyors

Apron conveyors are specified for granular or lumpy materials. Since the load is carried and not dragged, less power is required than for screw or scraper conveyors. Apron conveyors may have stationary skirt or side plates to permit increased depth of material on the apron, e.g., when used as a feeder for taking material from a track hopper (Fig. 10.5.25) with

Fig. 10.5.25 Track hopper and apron feeder supplying a gravity-discharge bucket-elevator boot.

controlled rate of feed. They are not often specified if the length is great, since other types of conveyor are substantially lower in cost. Sizes of lumps are limited by the width of the pans and the ability of the conveyor to withstand the impact of loading. Only end discharge is possible. The apron conveyor (Fig. 10.5.26) consists of two strands of roller chain separated by overlapping apron plates, which form the carrying surface, with sides 2 to 6 in (51 to 152 mm) high. The chains are driven by sprockets at one end, take-ups being provided at the other end. The conveyors always pull the material toward the driving end. For light duty, flangeless rollers on flat rails are used; for heavy duty, single-flanged rollers and T rails are used. Apron conveyors may be run without feeders, provided that the opening of the feeding hopper is made sufficiently narrow to prevent material from spilling over the sides of the conveyor after passing from the opening. When used as a conveyor, the **speed** should not exceed 60 ft/min (0.30 m/s); when used as a feeder, 30 ft/min (0.15 m/s). Table 10.5.6 gives the capacities of apron feeders with material weighing 50 lb/ft³ (800 kg/m³) at a speed of 10 ft/min (0.05 m/s).

Chain pull for horizontal-apron conveyor:

$$2LF(W + W_1)$$

Chain pull for inclined-apron conveyor:

$$L(W + W_1)(F \cos \theta + \sin \theta) + WL(F \cos \theta - \sin \theta)$$

where L = conveyor length, ft; W = weight of chain and pans per ft, lb; W_1 = weight of material per ft of conveyor, lb; θ = angle of inclination, deg; F = coefficient of rolling friction, usually 0.1 for plain roller bearings or 0.05 for antifriction bearings.

Fig. 10.5.26   Apron conveyor.   **Fig. 10.5.27**   Open-top carrier.

Bucket Conveyors and Elevators

Open-top bucket carriers (Fig. 10.5.27) are similar to apron conveyors, except that dished or bucket-shaped receptacles take the place of the flat or corrugated apron plates used on the apron conveyor. The carriers will operate on steeper inclines than apron conveyors (up to 70°), as the buckets prevent material from sliding back. Neither sides extending above the tops of buckets nor skirtboards are necessary. **Speed,** when loaded by a feeder, = 60 ft/min (max)(0.30 m/s) and when dragging the load from a hopper or bin, ≤30 ft/min. The **capacity** should be calculated on the basis of the buckets being three-fourths full, the angle of inclination of the conveyor determining the loading condition of the bucket.

V-bucket carriers are used for elevating and conveying nonabrasive materials, principally coal when it must be elevated and conveyed with one piece of apparatus. The length and height lifted are limited by the strength of the chains and seldom exceed 75 ft (22.9 m). These carriers can operate on any incline and can discharge at any point on the horizontal run. The size of lumps carried is limited by the size and spacing of the buckets. The carrier consists of two strands of roller chain

separated by V-shaped steel buckets. Figure 10.5.28 shows the most common form, where material is received on the lower horizontal run, elevated, and discharged through openings in the bottom of the trough of the upper horizontal run. The material is scraped along the horizontal trough of the conveyor, as in a flight conveyor. The steel guard plates *a*

Fig. 10.5.28   V-bucket carrier.

at the right prevent spillage at the bends. Figure 10.5.29 shows a different form, where material is dug by the elevator from a boot, elevated vertically, scraped along the horizontal run, and discharged through gates in the bottom of the trough. Figure 10.5.30 shows a variation of the type shown in Fig. 10.5.29, requiring one less bend in the conveyor. The troughs are of steel or steel-lined wood. When feeding material to the horizontal run, it is advisable to use an automatic feeder driven by

Fig. 10.5.29 and 10.5.30   V-bucket carriers.

power from one of the bend shafts to prevent overloading. Should the buckets of this type of conveyor be overloaded, they will spill on the vertical section. The drive is located at *b*, with take-up at *c*. The speed should not exceed 100 ft/min (0.51 m/s) when large material is being handled, but when material is small, speed may be increased to 125 ft/min (0.64 m/s). The best results are obtained when speeds are kept low. Table 10.5.7 gives the **capacities** and **weights** based on an even and continuous feed.

Pivoted-bucket carriers are used primarily where the path is a runaround in a vertical plane. Their chief application has been for the dual

Table 10.5.6   Capacities of Apron Conveyors

Width between skirt plates		Capacity, 50 lb/ft³ (800 kg/m³) material at 10 ft/min (0.05 m/s) speed			
		Depth of load, in (mm)			
in	mm	12 (305)	16 (406)	20 (508)	24 (610)
24	610	22 (559)	30 (762)		
30	762	26 (660)	37 (940)	47 (1,194)	56 (1,422)
36	914	34 (864)	45 (1,143)	56 (1,422)	67 (1,702)
42	1,067	39 (991)	52 (1,321)	65 (1,651)	79 (2,007)

Table 10.5.7   Capacities and Weights of V-Bucket Carriers*

Buckets				Capacity, tons of coal per hour at 100 ft/min	Weight per ft of chains and buckets, lb
Length, in	Width, in	Depth, in	Spacing, in		
12	12	6	18	29	36
16	12	6	18	32	40
20	15	8	24	43	55
24	20	10	24	100	65
30	20	10	24	126	70
36	24	12	30	172	94
42	24	12	30	200	105
48	24	12	36	192	150

* Multiply in by 25.4 for mm, ton/h by 0.25 for kg/s or by 0.91 for tonnes/h.

duty of handling coal and ashes in boiler plants. They require less power than V-bucket carriers, as the material is carried and not dragged on the horizontal run. The length and height lifted are limited by the strength of the chains. The length seldom exceeds 500 ft (152 m) and the height lifted 100 ft (30 m). They can be operated on any incline and can discharge at any point on the horizontal run. The size of lumps is limited by the size of buckets. The maintenance cost is extremely low. Many carrier installations are still in operation after 40 years of service. Other applications are for hot clinker, granulated and pulverized chemicals, cement, and stone.

The carrier consists of two strands of roller chain, with flanged rollers, between which are pivoted buckets, usually of malleable iron. The drive (Fig. 10.5.31) is located at *a* or *a'*, the take-up at *b*. The material is fed to the buckets by a feeder at any point along the lower horizontal run, is elevated, and is discharged on the upper horizontal run.

Fig. 10.5.31   Pivoted-bucket carrier.

The tripper *c*, mounted on wheels so that it can be moved to the desired dumping position, engages the cams on the buckets and tips them until the material runs out. The buckets always remain vertical except when tripped. The chain rollers run on T rails on the horizontal sections and between guides on the vertical runs. Speeds range from 30 to 60 ft/min (0.15 to 0.30 m/s).

After dumping, the overlapping bucket lips are in the wrong position to round the far corner; after rounding the take-up wheels, the lap is wrong for making the upturn. The Link-Belt **Peck** carrier eliminates this by suspending the buckets from trunnions attached to rearward cantilever extensions of the inner links (Fig. 10.5.32). As the chain rounds

the turns, the buckets swing in a larger-radius curve, automatically unlatch, and then lap correctly as they enter the straight run.

The pivoted-bucket carrier requires little attention beyond periodic lubrication and adjustment of take-ups. For the dual service of coal and ash handling, its only competitor is the skip hoist.

Fig. 10.5.32   Link-Belt Peck carrier buckets.

Table 10.5.8 shows the capacities of pivoted-bucket carriers with materials weighing 50 lb/ft³ (800 kg/m³), with carriers operating at 40 to 50 ft/min (0.20 to 0.25 m/s), and with buckets loaded to 80 percent capacity.

Bucket elevators are of two types: (1) chain-and-bucket, where the buckets are attached to one or two chains; and (2) belt-and-bucket, where the buckets are attached to canvas or rubber belts. Either type may be vertical or inclined and may have continuous or noncontinuous buckets. Bucket elevators are used to elevate any bulk material that will not adhere to the bucket. Belt-and-bucket elevators are particularly well adapted to handling abrasive materials which would produce excessive wear on chains. Chain-and-bucket elevators are frequently used with perforated buckets when handling wet material, to drain off surplus water. The length of elevators is limited by the strength of the chains or belts. They may be built up to 100 ft (30 m) long, but they average 25 to 75 ft (7.6 to 23 m). Inclined-belt elevators operate best on an angle of about 30° to the vertical. At greater angles, the sag of the return belt is excessive, as it cannot be supported by rollers between the head and foot pulleys. This applies also to single-strand chain elevators. Double-strand chain elevators, however, if provided with roller chain, can run on an angle, as both the upper and return chains are supported by rails.

Table 10.5.8   Capacities of Pivoted-Bucket Carriers with Coal or Similar Materials Weighing 50 lb/ft³ (800 kg/m³) at Speeds Noted

Bucket pitch × width		Capacity of coal		Speed	
in	mm	Short ton/h	tonne/h	ft/min	m/s
24 × 18	610 × 457	35–45	32–41	40–50	0.20–0.25
24 × 24	610 × 610	50–60	45–54	40–50	0.20–0.25
24 × 30	610 × 762	60–75	54–68	40–50	0.20–0.25
24 × 36	610 × 914	70–90	63–82	40–50	0.20–0.25

The size of lumps is limited by the size and spacing of the buckets and by the speed of the elevator.

Fig. 10.5.33 Continuous bucket.

Continuous-bucket elevators (Fig. 10.5.33 and Table 10.5.9) usually operate at 100 ft/min (0.51 m/s) or less and are single- or double-strand. The contents of each bucket discharge over the back of the preceding bucket. For maximum capacity and a large proportion of lumps, the buckets extend rearward behind the chain runs. The elevator is then called a supercapacity elevator (Fig. 10.5.34 and Table 10.5.10).

Gravity-discharge elevators operate at 100 ft/min (0.51 m/s) or less and are double-strand, with spaced V buckets. The path may be an L, an inverted L, or a runaround in a vertical plane (Fig. 10.5.28). Along the horizontal run, the buckets function as pushers within a trough. An elevator with a tandem flight conveyor costs less. For a runaround path, the pivoted-bucket carrier requires less power and has lower maintenance costs.

Table 10.5.9   Continuous Bucket Elevators*

Bucket size, in	Max lump size, in		Capacity with 50 lb material at 100 ft/min tons/h
	All lumps	10% lumps	
10 × 5 × 8	¾	2½	17
10 × 7 × 12	1	3	21
12 × 7 × 12	1	3	25
14 × 7 × 12	1	3	30
14 × 8 × 12	1¼	4	36
16 × 8 × 12	1½	4½	42
18 × 8 × 12	1½	4½	46

* Multiply in by 25.4 for mm, lb by 0.45 for kg, tons by 0.91 for tonnes.

As bucket elevators have no **feed control**, an interlocked feeder is desirable for a gravity flow. Some types scoop up the load as the buckets round the foot end and can take care of momentary surges by spilling the excess back into the boot. The continuous-bucket elevator, however, must be loaded after the buckets line up for the lift, i.e., when the gaps between buckets have closed.

Fig. 10.5.34 Supercapacity bucket.

Belt-and-bucket elevators are advantageous for grain, cereals, glass batch, clay, coke breeze, sand, and other abrasives if the temperature is not high enough to scorch the belt [below 250°F (121°C) for natural rubber].

Elevator casings usually are sectional and dusttight, either of ³⁄₁₆-in (4.8-mm) sheet steel or, better, of aluminum. If the elevator has considerable height, its cross section must be sufficiently large to prevent sway contact between buckets and casing. Chain guides extending the length of both runs may be provided to control sway and to prevent piling up of the element, at the boot, should the chain break. **Caution:** Indoor high elevators may develop considerable updraft tending to sweep up light, pulverized material. Provision to neutralize the pressure differential at the top and bottom may be essential. Figure 10.5.35 shows the **cast-iron boot** used with centrifugal-discharge and V-bucket chain elevators and belt elevators. Figure 10.5.36 shows the general form of a belt-and-bucket elevator

Table 10.5.10   Supercapacity Elevators*
(Link-Belt Co.)

Bucket, in length × width × depth	Max lump size, large lumps not more than 20%, in	Capacity with 50 lb material tons/h
16 × 12 × 18	8	115
20 × 12 × 18	8	145
24 × 12 × 18	8	175
30 × 12 × 18	8	215
24 × 17 × 24	10	230
36 × 17 × 24	10	345

* Multiply in by 25.4 for mm, lb by 0.45 for kg, tons by 0.91 for tonnes.

with **structural-steel boot and casing.** Elevators of this type must be run at sufficient speed to throw the discharging material clear of the bucket ahead.

Capacity   Elevators are rated for capacity with the buckets 75 percent loaded. The buckets must be large enough to accommodate the lumps, even though the capacity is small.

Fig. 10.5.35 Cast-iron boot.

Power Requirements   The motor horsepower for the continuous-bucket and supercapacity elevators can be approximated as

$$\text{Motor hp} = (2 \times \text{tons/h} \times \text{lift, ft})/1{,}000$$

The motor horsepower of gravity-discharge elevators can be approximated by using the same formula for the lift and adding for the horsepower of the horizontal run the power as estimated for a flight conveyor. For a vertical runaround path, add a similar allowance for the lower horizontal run.

Fig. 10.5.36 Structural-steel boot and casing.

Belt Conveyors

The belt conveyor is a heavy-duty conveyor available for transporting large tonnages over paths beyond the range of any other type of mechanical conveyor. The capacity may be several thousand tons per hour, and the distance several miles. It may be horizontal or inclined upward or downward, or it may be a combination of these, as outlined in Fig. 10.5.37. The limit of incline is reached when the material tends to slip on the belt surface. There are special belts with molded designs to assist in keeping material from slipping on inclines. They will handle pulverized, granular, or lumpy material. Special compounds are available if material is hot or oily.

In its simplest form, the conveyor consists of a head or drive pulley, a take-up pulley, an endless belt, and carrying and return idlers. The spacing of the carrying idlers varies with the width and loading of the belt and usually is 5 ft (1.5 m) or less. Return idlers are spaced on 10-ft (3.0-m) centers or slightly less with wide belts. Sealed antifriction idler bearings are used almost exclusively, with pressure-lubrication fittings requiring attention about once a year.

Fig. 10.5.37   Typical arrangements of belt conveyors.

Belts   Belt width is governed by the desired conveyor capacity and maximum lump size. The standard rubber belt construction (Fig. 10.5.38) has several plies of square woven cotton duck or synthetic fabric such as rayon, nylon, or polyester cemented together with a rubber compound called "friction" and covered both top and bottom with rubber to resist abrasion and keep out moisture. Top cover thickness is

Fig. 10.5.38   Rubber-covered conveyor belt.

determined by the severity of the job and varies from $\frac{1}{16}$ to $\frac{3}{4}$ in (1.6 to 19 mm). The bottom cover is usually $\frac{1}{16}$ in (1.6 mm). By placing a layer of loosely woven fabric, called the breaker strip, between the cover and outside fabric ply, it is often possible to double the adhesion of the cover to the carcass. The belt is rated according to the tension to which it may safely be subjected, and this is a function of the length and lift of the conveyor. The standard Rubber Manufacturers Association (RMA) multiple ply ratings in lb/in (kg/mm) of width per ply are as follows:

RMA multiple ply No.	MP35	MP43	MP50	MP60	MP70
Permissible pull, lb/in (kg/mm) of belt width per ply, vulcanized splice	35 (0.63)	43 (0.77)	50 (0.89)	60 (1.07)	70 (1.25)
Permissible pull, lb/in (kg/mm) of belt width per ply, mechanical splice	27 (0.48)	33 (0.60)	40 (0.71)	45 (0.80)	55 (0.98)

Thus, for a pull of 4,200 lb (1,905 kg) and 24-in (0.61-m) belt width, a five-ply MP35 could be used with a vulcanized splice or a five-ply MP50 could be used with a mechanical splice.

High-Strength Belts   For belt conveyors of extremely great length, a greater strength per inch of belt width is available now through the use of improved weaving techniques that provide straight-warp synthetic fabric to support the tensile forces (Uniflex by Uniroyal, Inc., or Flexseal by B. F. Goodrich). Strengths go to 1,500 lb/in width tension rating. They are available in most cover and bottom combinations and have good bonding to carcass. The number of plies is reduced to two instead of as many as eight so as to give excellent flexibility. Widths to 60 in are available. Other conventional high-strength fabric belts are available to

somewhat lower tensile ratings of 90 (1.61), 120 (2.14), 155 (2.77), 195 (3.48), and 240 (4.29) lb/in (kg/mm) per ply ratings.

The B. F. Goodrich Company has developed a steel-cable-reinforced belt rated 700 to 4,400 lb/in (12.5 to 78.6 kg/mm) of belt width. The belt has parallel brass-plated 7 by 19 steel airplane cables ranging from $\frac{5}{32}$ to $\frac{3}{8}$ in (4.0 to 9.5 mm) diameter placed on $\frac{1}{2}$- to $\frac{3}{4}$-in (12.7- to 19.0-mm) centers.

These belts are used for long single-length conveyors and for high-lift, extremely heavy duty service, e.g., for taking ore from deep open pits, thus providing an alternative to a spiraling railway or truck route.

Synthetic rubber is in use for belts. Combinations of synthetic and natural rubbers have been found satisfactory. Synthetics are superior under special circumstances, e.g., neoprene for flame resistance and resistance to petroleum-based oils, Buna N for resistance to vegetable, animal, and petroleum oils, and butyl for resistance to heat (per RMA).

Belt Life   With lumpy material, the impact at the loading point may be destructive. Heavy lumps, such as ore and rock, cut through the protective cover and expose the carcass. The impact shock is reduced by making the belt supports flexible. This can be done by the use of idlers with cushion or pneumatic tires (Fig. 10.5.39) or by supporting the idlers on rubber mountings. Chuting the load vertically against the belt should be avoided. Where possible, the load should be given a movement in the direction of belt travel. When the material is a mixture of lumps and fines, the fines should be screened through to form a cushion for the lumps. Other destructive factors are belt overstressing, belts

Fig. 10.5.39   Pneumatic-tired idlers applied to belt feeder at loading point of belt conveyor.

running out of line and rubbing against supports, broken idlers, and failure to clean the belt surface thoroughly before it comes in contact with the snub and tripper pulleys. Introduction of a 180° twist in the return belt (B. F. Goodrich Co.) at both head and tail ends can be used to keep the clean side of the belt against the return idlers and to prevent buildup. Using one 180° twist causes both sides of the belt to wear evenly. For each twist, 1 ft of length/in of belt width is required.

Idler Pulleys   Troughing idlers are usually of the three-pulley type (Fig. 10.5.40), with the troughing pulleys at 20°. There is a growing tendency toward the use of 35 and 45° idlers to increase the volume capacity of a belt; 35° idlers will increase the volume capacity of a given belt 25 to 35 percent over 20° idlers, and 45° idlers, 35 to 40 percent.

Fig. 10.5.40   Standard assembly of three-pulley troughing idler and return idler.

The bearings, either roller or ball type, are protected by felt or labyrinth grease seals against the infiltration of abrasive dust. A belt running out of line may be brought into alignment by shifting slightly forward one end or the other with a few idler sets. **Self-aligning idlers** (Fig. 10.5.41) should be spaced not more than 75 ft (23 m) apart. These call attention to the necessity of lining up the belt and should not serve as continuing correctives.

Drive Belt slip on the conveyor-drive pulley is destructive. There is little difference in tendency to slip between a bare pulley and a rubber-lagged pulley when the belt is clean and dry. A wet belt will adhere to a lagged pulley much better, especially if the lagging is grooved. Heavy-duty conveyors exposed to the possibility of wetting the belt are

Fig. 10.5.41 Self-aligning idler.

generally driven by a head pulley lagged with a ½-in (12.7-mm) rubber belt and with ¼- by ¼-in (6.4- by 6.4-mm) grooves spaced ½ in (12.7 mm) apart and, preferably, diagonally as a herringbone gear. A snub pulley can be employed to increase the arc of contact on the head pulley, and since the pulley is in contact with the dirty side of the belt, a belt cleaner is essential. The belt cleaner may be a high-speed bristle brush, a spiral rubber wiper (resembling an elongated worm pinion), circular disks mounted slantwise on a shaft to give a wiping effect when rotated, or a scraper. Damp deposits such as clay or semifrozen coal dirt are best removed by multiple diagonal scrapers of stainless steel.

A belt conveyor should be emptied after each run to avoid a heavy starting strain. The motor should have high starting torque, moderate starting-current inrush, and good characteristics when operating under full load. The double-squirrel-cage ac motor fulfills these requirements.

Heavy-Duty Belt-Conveyor Drives For extremely heavy duty, it is essential that the drive torque be built up slowly, or serious belt damage will occur. The hydraulic clutch, derived from the hydraulic automobile clutch, serves nicely. The best drive developed to date is the **dynamatic clutch** (Fig. 10.5.42). This has a magnetized rotor on the extended motor shaft, revolving within an iron ring keyed to the reduction gearing of the conveyor. The energizing current is automatically built up over a period that may extend to 2 min, and the increasing magnetic pull on the ring builds up the belt speed.

Fig. 10.5.42 Operating principle of the dynamic clutch.

Take-ups For short conveyors, a screw take-up is satisfactory. For long conveyors, the expansion and contraction of the belt with temperature changes and the necessity of occasional cutting and resplicing make a weighted gravity take-up preferable, especially if a vulcanized splice is used. The take-up should, if possible, be located where the slack first occurs, usually just back of the drive except in a conveyor inclined

downward (retarding conveyor), when the take-up is located at the downhill end.

Trippers The load may be removed from the belt by a diagonal or V plow, but a tripper that snubs the belt backward is standard equipment. Trippers may be (1) stationary, (2) manually propelled by crank, or (3) propelled by power from one of the snubbing pulleys or by an independent motor. The discharge may be chuted to either side or back to the belt by a deflector plate. When the tripper must move back to the load-receiving end of the conveyor, it is usual to incline the belt for about 15 ft (4.6 m) to match the slope up to the tripper top pulley. As the lower tripper snub pulleys are in contact with the dirty side of the belt, a cleaner must be provided between the pulleys. A scraper in light contact with the face of the pulley may be advisable.

Belt Slope The slopes (in degrees) given in Table 10.5.11 are the maximum permissible angles for various materials.

Table 10.5.11 Maximum Belt Slopes for Various Materials, Degrees

Coal: anthracite, sized; mined, 50 mesh and under; or mined and sized	16
Coal, bituminous, mined, run of mine	18
Coal: bituminous, stripping, not cleaned; or lignite	22
Earth, as excavated, dry	20
Earth, wet, containing clay	23
Gravel, bank run	20
Gravel, dry, sharp	15–17
Gravel, pebbles	12
Grain, distillery, spent, dry	15
Sand, bank, damp	20–22
Sand, bank, dry	16–18
Wood chips	27

SOURCE: Uniroyal, Inc.

Determination of Motor Horsepower The power required to drive a belt conveyor is the sum of the powers required (1) to move the empty belt, (2) to move the load horizontally, and (3) to lift the load if the conveyor is inclined upward. If (3) is larger than the other two, an automatic brake must be provided to hold the conveyor if current fails. A solenoid brake is usual. The power required to move the empty belt is given by Fig. 10.5.43. The power to move 100 tons/h horizontally is given by the formula hp = 0.4 + 0.00325L, where L is the distance between centers, ft. For other capacities the horsepower is proportional.

Fig. 10.5.43 Horsepower required to move belt conveyor empty at 100 ft/ min (0.15 m/s).

Table 10.5.12 Troughed Conveyor-Belt Capacities*

Tons per hour (TPH) with belt speed of 100 ft/min (0.51 m/s).

Belt width, in	Equal length, 3 roll										Long center, 3 roll						
	20°			35°			45°				35°			45°			
SCA†	0°	10°	30°	0°	10°	30°	0°	10°	30°	CED‡	0°	10°	30°	0°	10°	30°	CED‡
12	10	14	24	16	20	28	19	29	35	.770							
24	52	74	120	83	102	143	98	115	150	1.050	82	101	141	96	113	149	1.05
30	86	121	195	137	167	232	161	188	244	1.095	111	144	212	133	163	225	1.12
42	177	249	400	282	345	476	332	386	500	1.130	179	248	394	216	281	417	1.22
60	375	526	843	598	729	1,003	702	815	1,053	1.187	266	417	734	324	468	772	1.35
72	548	768	1,232	874	1,064	1,464	1,026	1,190	1,535	1.205	291	516	987	356	573	1,030	1.44

*Tons per hour (TPH) = value from table × $\dfrac{\text{(actual material wt., lb/ft}^3)}{100}$ × $\dfrac{\text{(actual belt speed, ft/min)}}{100}$

† Surcharge angle (see Fig 10.5.44).

‡ Capacity calculated for standard distance of load from belt edge: $(0.55\,b + 0.9)$, where b = belt width, inches. For constant 2-in edge distance (CED) multiply by CED constant as given in this table. For slumping materials (very free flowing), use capacities based upon 2-in CED. This includes dry silica sand, dry aerated portland cement, wet concrete, etc., with surcharge angle 5° or less. For metric units multiply in by 25.4 for mm; tons per hour by 0.91 for tonne per hour; ft/min by 0.0051 for m/s.

Table 10.5.13 Minimum Belt Width for Lumps

Belt width,	in	12	18	24	30	42	60	72
	mm	305	457	610	762	1,067	1,524	1,829
Sized material,	in	2	4	5	6	8	12	14
	mm	51	102	127	152	203	305	356
Unsized material,	in	4	6	8	10	14	20	24
	mm	102	152	203	254	356	508	610

SOURCE: Uniroyal, Inc.

Table 10.5.14 Maximum Belt Speeds, ft/min, for Various Materials*

Width of belt, in	Light or free-flowing materials, grains, dry sand, etc.	Moderately free-flowing sand, gravel, fine stone, etc.	Lump coal, coarse stone, crushed ore	Heavy sharp lumpy materials, heavy ores, lump coke
12–14	400	250		
16–18	500	300	250	
20–24	600	400	350	250
30–36	750	500	400	300
42–60	850	550	450	350

* Multiply in by 25.4 for mm, ft/min by 0.005 for m/s.

The capacity in tons per hour for materials of various weights per cubic foot is given by Table 10.5.12. Table 10.5.13 gives minimum belt widths for lumps of various sizes. Table 10.5.14 gives advisable maximum belt speeds for various belt widths and materials. The speed should ensure full-capacity loading so that the entire belt width is utilized.

Drive Calculations From the standpoint of the application of power to the belt, a conveyor is identical with a power belt. The determining factors are the coefficient of friction between the drive pulley and the belt, the tension in the belt, and the arc of contact between the pulley and the belt. The arc of the contact is increased up to about 240° by using a

Fig. 10.5.44 (a) Troughed belt; (b) flat belt. (Uniroyal, Inc.)

snub pulley and up to 410° by using two pulleys geared together or driven by separate motors and having the belt wrapped around them in the form of a letter S. The resistance to be overcome is the sum of all the frictional resistances throughout the length of the conveyor plus, in the case of a rising conveyor, the resistance due to lifting the load. The sum of the conveyor and load resistances determines the working pull that has to be transmitted to the belt at the drive pulley. The total pull is increased by the slack-side tension necessary to keep the belt from slipping on the pulley. Other factors adding to the belt pull are the component of the weight of the belt if the conveyor is inclined and a take-up pull to keep the belt from sagging between the idlers at the loading point. These, however, do not add to the working pull. The maximum belt pull determines the length of conveyor that can be used. If part of the conveyor runs downgrade, the load on it will reduce the working pull. In moderate-length conveyors, stresses due to acceleration or deceleration are safely carried by the factor of safety used for belt-life calculations.

The total or maximum tension T_{max} must be known to specify a suitable belt. The effective tension T_e is the difference between tight-side tension and slack-side tension, or $T_e = T_1 - T_2$. The coefficient of friction between rubber and steel is 0.25; with a lagged pulley, between rubber and rubber, it is 0.55 for ideal conditions but should be taken as 0.35 to allow for loss due to dirty conditions.

Except for extremely heavy belt pulls, the tandem drive is seldom used since it is costly; the lagged-and-grooved drive pulley is used for most industrial installations.

For a belt with 6,000-lb (26,700-N) max tension running on a bare pulley drive with 180° wrap (Table 10.5.15), $T = 1.857T_e = 6,000$ lb; $T_e = 3,200$ lb (14,200 N). Such a belt, 30 in wide, might be a five-ply MP50, a reduced-ply belt rated at 200 lb/in, or a steel-cable belt with $\frac{5}{32}$-in (4-mm) cables spaced on 0.650-in (16.5-mm) centers. The last is the most costly.

Table 10.5.15 Ratio T_1 to T_e for Various Arcs of Contact with Bare Pulleys and Lagged Pulleys
(Coefficients of friction 0.25 and 0.35, respectively)

Belt wrap, deg	180	200	210	215	220	240
Bare pulley	1.85	1.72	1.67	1.64	1.62	1.54
Lagged pulley	1.50	1.42	1.38	1.36	1.35	1.30

In an inclined belt with single pulley drive, the T_{max} is lowest if the drive is at the head end and increases as the drive shifts toward the foot end.

Allowance for Tripper The belt lifts about 5 ft to pass over the top snub pulley of the tripper. Allowance should be made for this lift in determining the power requirement of the conveyor. If the tripper is propelled by the belt, an allowance of 1 hp (0.75 kW) for a 16-in (406-mm) belt, 3 hp (2.2 kW) for a 36-in (914-mm) belt, or 7 hp (5.2 kW) for a 60-in (1,524-mm) belt is ample. If a rotary cleaning brush is driven from one of the snub shafts, an allowance should be made which is approximately the same as that for the propulsion of the tripper.

Magnetic pulleys are frequently used as head pulleys on belt conveyors to remove tramp iron, such as stray nuts or bolts, before crushing; to concentrate magnetic ores, such as magnetic or nickeliferous pyrrhotite, from nonmagnetic material; and to reclaim iron from foundry refuse. A chute or hopper automatically receives the extracted material as it is drawn down through the other nonmagnetic material, drawn around the end pulley on the belt, and finally released as the belt leaves the pulley. Light-duty permanent-magnet types [for pulleys 12 to 24 in (305 to 610 mm) in diameter] will separate material through a 2-in (51-mm) layer on the belt. Heavy-duty permanent-magnet units (12 to 24 in in diameter) will separate material if the belt carries over 2 in of material or if the magnetic content is very fine. If larger units are available for special applications. So effective and powerful are the permanent-magnet types that **electromagnetic pulleys** are available only in the larger sizes, from 18 to 48 in in diameter. The permanent-magnet type requires no slip rings, external power, or upkeep.

Overhead magnetic separators (Dings), both electromagnetic and Ceramox permanent-magnet types, for suspension above a belt conveyor are also available for all commercial belt widths to pull magnetic material from burden as thick as 40 in and at belt speeds to 750 ft/min. These may or may not be equipped with a separately encompassing belt to be self-cleaning. Wattages vary from 1,600 to 17,000. The permanent-magnet type requires no electric power and has nonvarying magnet strength. An alternate type of belt protection is to use a Ferro Guard Detector (Dings) to stop belt motion if iron is detected.

Trippers of the fixed or movable type are used for discharging material between the ends of a belt conveyor. A **self-propelling tripper** consists of two pulleys, over which the belt passes, the material being discharged into the chute as the belt bends around the upper pulley. The pulleys are mounted on a frame carried by four wheels and power-driven. A lever on the frame and stops alongside the rails enable the tripper, taking power from the conveyor belt, to move automatically between the stops, thus distributing the material. Rail clamps are provided to hold the tripper in a fixed position when discharging. Motor-driven trippers are used when it is desirable to move the tripper independently of the conveyor belt. Fixed trippers have their pulleys mounted on the conveyor framework instead of on a movable carriage.

Shuttle conveyors are frequently used in place of trippers for distributing materials. They consist of a reversible belt conveyor mounted upon a movable frame and discharging over either end.

Belt-Conveyor Arrangements Figure 10.5.37 shows **typical arrangements of belt conveyors.** *a* is a level conveyor receiving material at one end and discharging at the other. *b* shows a level conveyor with traveling tripper. The receiving end of the conveyor is depressed so that the belt will not be lifted against the chute when the tripper is at its extreme loading end. *c* is a level conveyor with fixed trippers. *d* shows an inclined end combined with a level section. *e* is a combination of level conveyor, vertical curve, and horizontal section. The radius of the vertical curve depends upon the weight of the belt and the tension in the belt at the point of tangency. This must be figured in each case and is found by the formula: min radius, ft = belt tension at lowest point of curve divided by weight per ft of belt. The belt weight should be for the worn belt with not over $\frac{1}{16}$-in (1.6-mm) top cover. *f* is a combination of level conveyor receiving material from a bin, a fixed dump, and inclined section, and a series of fixed trippers.

Portable conveyors are widely used around retail coal yards, small power plants, and at coal mines for storing coal and reclaiming it for loading into trucks or cars. They are also used for handling other bulk materials. They consist of a short section of chain or belt conveyors mounted on large wheels or crawler treads and powered with a gasoline engine or electric motor. They vary in length from 20 to 90 ft and can handle up to 250 tons/h (63 kg/s) of coal. For capacities greater than what two people can shovel onto a belt, some form of power loader is necessary.

Sectional-belt conveyors have come into wide use in coal mines for bringing the coal from the face and loading it into cars in the entry. They consist of short sections (6 ft or more) of light frame of special low-type construction. The sections are designed for ease of connecting and disconnecting for transfer from one part of the mine to another. They are built in lengths up to 1,000 ft (305 m) or more under favorable conditions and can handle 125 tons/h (32 kg/s) of coal.

Sliding-belt conveyors use belts sliding on decks instead of troughed belts carried on rollers. Sliding belts are used in the shipping rooms of department stores for handling miscellaneous parcels, in post offices for handling mail bags, in chemical plants for miscellaneous light waste, etc. The decking preferably is of maple strips. If of steel, the deck should be perforated at intervals to relieve the vacuum effect between the bottom of the belt and the deck. Cotton or balata belts are best. The speed should be low. The return run may be carried on 4-in straight-face idlers. The power requirement is greater than with idler rollers.

The **oscillating conveyor** is a horizontal trough varying in width from 12 to 48 in (305 to 1,219 mm), mounted on rearward-inclined cantilever supports, and driven from an eccentric at 400 to 500 r/min. The effect is to "bounce" the material along at about 50 ft/min (0.25 m/s) with minimum wear on the trough. The conveyor is adapted to abrasive or hot fragmentary materials, such as scrap metals, castings, or metal chips. The trough bottom may be a screen plate to cull fine material, as when cleaning sand from castings, or the trough may have louvers and a ventilating hood to cool the moving material. These oscillating

conveyors may have unit lengths up to 100 ft (30 m). Capacities range from a few tons to 100 tons/h (25 kg/s) with high efficiency and low maintenance.

Feeders

When material is drawn from a hopper or bin to a conveyor, an **automatic feeder** should be used (unless the material is dry and free-running, e.g., grain). The satisfactory operation of any conveyor depends on the material being fed to it in an even and continuous stream. The automatic feeder not only ensures a constant and controlled feed, irrespective of the size of material, but saves the expense of an operator who would otherwise be required at the feeding point. Figure 10.5.45 shows a **reciprocating-plate feeder,** consisting of a plate mounted on four wheels and forming the bottom of the hopper. When the plate is moved forward, it carries the material with it; when moved back, the plate is withdrawn from under the material, allowing it to fall into the chute. The plate is moved by connecting rods from cranks or eccentrics. The capacity of this feeder is determined by the length and number of strokes, width of plate, and location of the adjustable gate. The number of strokes should not exceed 60 to 70 per min. When used under a track hopper, the material remaining on the plate may freeze in winter, as this type of feeder is not self-clearing.

Fig. 10.5.45   Reciprocating-plate feeder.

Vibrating Feeders   The vibrating feeder consists of a plate inclined downward slightly and vibrated (1) by a high-speed unbalanced pulley, (2) by electromagnetic vibrations from one or more solenoids, as in the Jeffrey Manufacturing Co. feeder, or (3) by the slower pulsations secured by mounting the plate on rearward-inclined leaf springs.

The **electric vibrating feeder** (Fig. 10.5.46) operates magnetically with a large number of short strokes (7,200 per min from an alternating current in the small sizes and 3,600 from a pulsating direct current in the larger sizes). It is built to feed from a few pounds per minute to 1,250 tons/h (315 kg/s) and will handle any material that does not adhere to the pan. It is self-cleaning, instantaneously adjustable for capacity, and controllable from any point near or remote. It is usually supported from above with spring shock absorbers *a* in each hanger, but it may be supported from below with similar springs in the supports. A modified form can be set to feed a weighed **constant amount** hourly for process control.

Fig. 10.5.46   Electric vibrating feeder.

Roller Conveyors

The principle involved in **gravity roller conveyors** is the control of motion due to gravity by interposing an antifriction trackage set at a definite grade. Roller conveyors are used in the movement of all sorts of package goods with smooth surfaces which are sufficiently rigid to prevent sagging between rollers—in warehouses, brickyards, building-supply yards, department stores, post offices, and the manufacturing and shipping departments of industrial manufacturers.

The **rollers vary in diameter** and strength from 1 in, with a capacity of 5 lb (2.3 kg) per roller, up to 4 in (102 mm), with a capacity of 1,800 lb (816 kg) per roller. The heavier rollers are generally used in foundries and steel mills for moving large molds, castings, or stacks of sheet steel. The small roller is used for handling small, light objects. The spacing of the rollers in the frames varies with the size and weight of the objects to be moved. Three rollers should be in contact with the package to prevent hobbling. The grade of fall required to move the object varies from 1½ to 7 percent, depending on the weight and character of the material in contact with the rollers.

Figure 10.5.47 shows a typical cross section of a roller conveyor. Curved sections are similar in construction to straight sections, except that in the majority of cases multiple rollers (Fig. 10.5.48) are used to keep the package properly lined up and in the center of the conveyor.

Fig. 10.5.47   Gravity roller conveyor.

Fig. 10.5.48   Multiple-roller conveyor.

Figure 10.5.49 illustrates a **wheel conveyor,** used for handling bundles of shingles, fruit boxes, bundles of fiber cartons, and large, light cases. The wheels are of ball-bearing type, bolted to flat-bar or angle-frame rails.

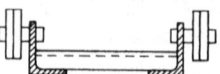

Fig. 10.5.49   Wheel-type conveyor.

When an installation involves a **trunk line with several tributary runs,** a simple two-arm deflector at each junction point holds back the item on one run until the item on the other has cleared. **Power-operated** roller conveyors permit handling up an incline. Usually the rolls are driven by sprockets on the spindle ends. An alternative of a smooth deck and pusher flights should be considered as costing less and permitting steeper inclines.

Platform conveyors are single- or double-strand conveyors (Fig. 10.5.50) with plates of steel or hardwood forming a continuous platform on which the loads are placed. They are adapted to handling heavy drums or barrels and miscellaneous freight.

Fig. 10.5.50   Double-strand platform conveyor.

Pneumatic Conveyors

The pneumatic conveyor transports dry, free-flowing, granular material in suspension within a pipe or duct by means of a high-velocity airstream or by the energy of expanding compressed air within a comparatively dense column of fluidized or aerated material. Principal uses are (1) dust collection; (2) conveying soft materials, such as grain, dry foodstuff (flour and feeds), chemicals (soda ash, lime, salt cake), wood chips, carbon black, and sawdust; (3) conveying hard materials, such as fly ash, cement, silica metallic ores, and phosphate. The need in processing of bulk-transporting plastic pellets, powders, and flour under

contamination-free conditions has increased the use of pneumatic conveying.

Dust Collection   All pipes should be as straight and short as possible, and bends, if necessary, should have a radius of at least three diameters of the pipe. Pipes should be proportioned to keep down friction losses and yet maintain the air velocities that will prevent settling of the material. Frequent cleanout openings must be provided. Branch connections should go into the sides of the main and deliver the incoming stream as nearly as possible in the direction of flow of the main stream. Sudden changes in diameter should be avoided to prevent eddy losses.

When **vertical runs** are short in proportion to the horizontal runs, the size of the riser is locally restricted, thereby increasing the air velocity and producing sufficient lifting power to elevate the material. If the vertical pipes are comparatively long, they are not restricted, but the necessary lifting power is secured by increased velocity and suction throughout the entire system.

The area of the main at any point should be 20 to 25 percent in excess of the sums of the branches entering it between the point in question and the dead end of the main. Floor sweepers, if equipped with efficient blast gates, need not be included in computing the main area. The diameter of the connecting pipe from machine to main and the section required at each hood are determined by experience. The sum of the volumes of each branch gives the total volume to be handled by the fan.

Fan Suction   The maintained resistance at the fan is composed of (1) suctions of the various hoods, which must be chosen from experience, (2) collector loss, and (3) loss due to pipe friction.

The **pipe loss** for any machine is the sum of the losses in the corresponding branch and in the main from that branch to the fan. For each elbow, add a length equal to 10 diameters of straight pipe. The total loss in the system, or static pressure required at the fan, is equal to the sum of (1), (2), and (3).

For conveying soft materials, a fan is used to create a suction. The suspended material is collected at the terminal point by a separator upstream from the fan. The material may be moved from one location to another or may be unloaded from barge or rail car. Required conveying velocity ranges from 2,000 ft/min (10.2 m/s) for light materials, such as sawdust, to 3,000 to 4,000 ft/min (15.2 to 20.3 m/s) for medium-weight materials, such as grain. Since abrasion is no problem, steel pipe or galvanized-metal ducts are satisfactory. Unnecessary bends and fittings should be avoided to minimize power consumption.

For conveying hard materials, a water-jet exhauster or steam exhauster is used on suction systems, and a positive-displacement blower on pressure systems. A mechanical exhauster may also be used on suction systems if there is a bag filter or air washer ahead of the exhauster. The largest tonnage of hard material handled is fly ash. A single coal-fired, steam-electric plant may collect more than 1,000 tons (907 tonnes) of fly ash per day. Fly ash can be conveyed several miles pneumatically at 30 tons (27 tonnes) or more per hour using a pressure conveyor. Another high-tonnage material conveyed pneumatically is cement. Individual transfer conveyors may handle several hundred tons per hour. Hard materials are usually also heavy and abrasive. Required conveying velocities vary from 4,000 to 5,000 ft/min (20.3 to 25.4 m/s). Heavy cast-iron or alloy pipe and fittings are required to prevent excessive wear.

Vacuum pneumatic ash-handling systems have the airflow induced by steam- or water-jet exhausters, or by mechanical blowers. Cyclone-type Nuveyor receivers collect the ash for storage in a dry silo. A typical Nuveyor system is shown in Fig. 10.5.51. The conveying velocity is dependent upon material handled in the system. Fly ash is handled at approximately 3,800 ft/min (19.3 m/s) and capacity up to 60 tons/h (15.1 kg/s). Positive-pressure pneumatic ash systems are becoming more common because of higher capacities required. These systems can convey fly ash up to 1½ mi (2.4 km) and capacities of 100 tons/h (25.2 kg/s) for shorter systems.

The **power requirement** for pneumatic conveyors is much greater than for a mechanical conveyor of equal capacity, but the duct can be led along practically any path. There are no moving parts and no risk of injury to the attendant. The vacuum-cleaner action provides dustless

Fig. 10.5.51   Nuveyor pneumatic ash-handling system. (*United Conveyor Corp.*)

operation, sometimes important when pulverized material is unloaded from boxcars through flexible hose and nozzle. A few materials build up a state-electric charge which may introduce an explosion risk. Sulfur is an outstanding example. Sticky materials tend to pack at the elbows and are unsuitable for pneumatic handling.

The performance of a pneumatic conveyor cannot be predicted with the accuracy usual with the various types of mechanical conveyors and elevators. It is necessary to rely on the advice of experienced engineers.

The **Fuller-Kinyon system** for transporting dry pulverized material consists of a motor- or engine-driven pump, a source of compressed air for fluidizing the material, a conduit or pipe-line, distributing valves (operated manually, electropneumatically, or by motor), and electric bin-level indicators (high-level, low-level, or both). The impeller is a specially designed differential-pitch screw normally turning at 1,200 r/min. The material enters the feed hopper and is compressed in the decreasing pitch of the screw flights. At the discharge end of the screw, the mass is introduced through a check valve to a mixing chamber, where it is aerated by the introduction of compressed air. The fluidized material is conveyed in the transport line by the continuing action of the impeller screw and the energy of expanding air. Practical distance of transportation by the system depends upon the material to be handled. Cement has been handled in this manner for distances up to a mile. The most important field of application is the handling of portland cement. For this material, the Fuller-Kinyon pump is used for such operations as moving both raw material and finished cement within the cement-manufacturing process; loading out; unloading ships, barges, and railway cars; and transferring from storage to mixer plant on large construction jobs.

The **Airslide** (registered trademark of the Fuller Company) **conveyor system** is an air-activated gravity-type conveyor using low-pressure air to aerate or fluidize pulverized material to a degree which will permit it to flow on a slight incline by the force of gravity. The conveyor comprises an enclosed trough, the bottom of which has an inclined air-permeable surface. Beneath this surface is a plenum chamber through which air is introduced at low pressure, depending upon the application. Various control devices for controlling and diverting material flow and for controlling air supply may be provided as part of complete systems. For normal conveying applications, the air is supplied by an appropriate fan; for operation under a head of material (as in a storage bin), the air is supplied by a rotary positive blower. The Airslide conveyor is widely used for horizontal conveying, discharge of storage bins, and special railway-car and truck-trailer transport, as well as in stationary blow-tank-type conveying systems. An important feature of this conveyor is low power requirement.

Hydraulic Conveyors

Hydraulic conveyors are used for handling boiler-plant ash or slag from an ash hopper or slag tank located under the furnace. The material is flushed from the hopper to a grinder, which discharges to a jet pump or a mechanical pump for conveying to a disposal area or a dewatering bin (Fig. 10.5.52). Water requirements average 1 gal/lb of ash.

Fig. 10.5.52   Ash-sluicing system with jet pump. (*United Conveyor Corp.*)

CHANGING DIRECTION OF MATERIALS ON A CONVEYOR

Some material in transit can be made to change direction by being bent around a curve. Metal being rolled, wire, strip, material on webs, and the like, can be guided to new directions by channels, rollers, wheels, pulleys, etc.

Some conveyors can be given curvatures, such as in overhead monorails, tow car grooves, tracks, pneumatic tube bends, etc. Other, straight sections of conveyors can be joined by curved sections to accomplish directional changes. Figures 10.5.53 to 10.5.56 show some examples of these. In some cases, the simple intersection, or overlapping, at an angle of two conveyors can result in a direction change. See Figs. 10.5.57 and 10.5.58. When not all of the material on the first section of conveyor is to be diverted, sweeps or switches may be used. Figure 10.5.59 shows how fixed diverters set at decreasing heights can direct boxes of certain sizes onto other conveyors. Switches allow the material to be sent in one direction at one time and in another at other times. In Fig. 10.5.60 the diverter can swing to one side or the other. In Fig. 10.5.61, the switching section has a dual set of wheels, with the set being made higher at the time carrying the material in their angled direction. Pushers, as shown in Fig. 10.5.62 can selectively divert material, as can air blasts for lightweight items (Fig. 10.5.63) and tipping sections (Fig. 10.5.64).

Fig. 10.5.59   Fixed diverters. Diverters do not move. They are preset at heights to catch and divert items of given heights.

Fig. 10.5.53   Fixed curve turn section. Joins two conveyors and changes direction and level of material flow.

Fig. 10.5.54   Fixed curve turn section. Joins two conveyors and changes direction of material flow. Turn section has no power. Incoming items push those ahead of them around curve. Wheels, balls, rollers, etc. may be added to turn section to reduce friction of dead plate section shown.

Fig. 10.5.60   Moving diverter. Material on incoming conveyor section can be sent to either outgoing section by pivoting the diverter to one side or the other.

Fig. 10.5.55   Fixed curve turn section. Uses tapered rollers, skate wheels, balls, or belt. May be level or inclined, "dead" or powered.

Fig. 10.5.61   Wheel switch. Fixed wheels carry material on incoming conveyor to outgoing conveyor until movable set of wheels is raised above the fixed set. Then material is carried in the other direction.

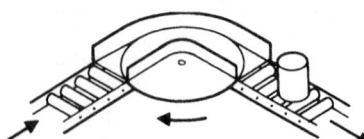

Fig. 10.5.56   Fixed curve turn section. Disk receives item from incoming conveyor section, rotates it, and directs it onto outgoing conveyor section.

Fig. 10.5.62   Pusher diverter. When triggered, the pusher advances and moves item onto another conveyor. A movable sweep can also be used.

Fig. 10.5.57   Simple intersection. Incoming conveyor section overlaps outgoing section and merely dumps material onto lower conveyor.

Fig. 10.5.63   Air diverter. When activated, a jet of air pushes a lightweight item onto another conveyor.

Fig. 10.5.58   Simple intersection. Both incoming and outgoing sections are powered. Rotating post serves as a guide.

Fig. 10.5.64   Tilting section. Sections of main conveyor move along, level, until activated to tilt at chosen location to dump material onto another conveyor.

10.6 AUTOMATIC GUIDED VEHICLES AND ROBOTS

by Vincent M. Altamuro

AUTOMATIC GUIDED VEHICLES

Driverless towing tractors guided by wires embedded into or affixed onto the floor have been available since the early 1950s. Currently, the addition of computer controls, sensors that can monitor remote conditions, real-time feedback, switching capabilities, and a whole new family of vehicles have created automatic guided vehicle systems (AGVs) that compete with industrial trucks and conveyors as material handling devices.

Most AGVs have only horizontal motion capabilities. Any vertical motion is limited. Fork lift trucks usually have more vertical motion capabilities than do standard AGVs. Automatic storage and retrieval systems (AS/RS) usually have very high rise vertical capabilities, with horizontal motions limited only to moving to and down the proper aisle. Power for the automatic guided vehicle is usually a battery, like that of the electric industrial truck. Guidance may be provided several ways. In *electrical* or *magnetic* guidance, a wire network is usually embedded in a narrow, shallow slot cut in the floor along the possible paths. The electromagnetic field emitted by the conductor wire is sensed by devices on-board the vehicle which activate a steering mechanism, causing the vehicle to follow the wire. An *optical* guidance system has a similar steering mechanism, but the path is detected by a photosensor picking out the bright path (paint, tape, or fluorescent chemicals) previously laid down. The embedded wire system seems more popular in factories, as it is in a protective groove, whereas the optical tape or painted line can get dirty and/or damaged. In offices, where the AGV may be used to pick up and deliver mail, the optical system may be preferred, as it is less expensive to install and less likely to deteriorate in such an environment. However, in carpeted areas, a wire can easily be laid under the carpet and operate invisibly. That wire can be relocated if the desired path of the AGV changes.

In a **laser beam** guidance system, a laser scanner on the vehicle transmits and receives back an infrared laser beam that reads passive retroreflective targets that are placed at strategic points on x and y coordinates in the facility. The vehicle's on-board computers take the locations and distances of the targets and calculate the vehicle's position by triangulation. The locations of the loads to be picked up, the destinations at which they are to be dropped off, and the paths the vehicle is to travel are transmitted from the system's base station. Instructions are converted to inputs to the vehicle's steering and driving motors. A variation of the system is for the laser scanner to read bar codes or radio-frequency identification (RFID) targets to get more information regarding their mission than merely their current location. Other, less used, guidance methods are **infrared,** whereby line-of-sight signals are sent to the vehicle, and **dead reckoning,** whereby a vehicle is programmed to traverse a certain path and then turned loose.

The directions that an AGV can travel may be classified as unidirectional (one way), bidirectional (forward or backward along its path), or omnidirectional (all directions). Omnidirectional AGVs with five or more on-board microprocessors and a multitude of sensors are sometimes called **self-guided vehicles** (SGVs).

Automatic guided vehicles require smooth and level floors in order to operate properly. They can be weatherized so as to be able to run outdoors between buildings but there, too, the surface they travel on must be smooth and level.

AGV equipment can be categorized as:
1. Driverless tractors
2. Guided pallet trucks
3. Unit load transporters and platform carriers
4. Assembly or tool bed robot transporters

Driverless tractors can be used to tow a series of powerless material handling carts, like a locomotive pulls a train. They can be routed, stopped, coupled, uncoupled, and restarted either manually, by a programmed sequence, or from a central control computer. They are suited to low-volume, heavy or irregularly shaped loads which have to be moved over longer distances than would be economical for a conveyor.

Guided pallet trucks, like the conventional manually operated trucks they replace, are available in a wide range of sizes and configurations. In operation, they usually are loaded under the control of a person, who then drives them over the guide wire, programs in their desired destination, then switches them to automatic. At their destination, they can turn off the main guidepath onto a siding, automatically drop off their load, then continue back onto the main guidepath. The use of guided pallet trucks reduces the need for conventional manually operated trucks and their operators.

Unit load transporters and platform carriers are designed so as to carry their loads directly on their flat or specially contoured surface, rather than on forks or on carts towed behind. They can either carry material or work-in-process from workstation to workstation or they can be workstations themselves and process the material while they transport it.

The assembly or tool bed type of AGV is used to carry either work-in-process or tooling to machines. It may also be used to carry equipment for an entire process step—a machine plus its tooling—to large, heavy, or immovable products.

Robot transporters are used to make robots mobile. A robot can be fitted atop the transporter and carried to the work. Further, the robot can process the work as the transporter carries it along to the next station, thereby combining productive work with material handling and transportation.

Most AGVs and SGVs have several safety devices, including flashing in-motion lights, infrared scanners to slow them down when approaching an obstacle, sound warnings and alarms, stop-on-impact bumpers, speed regulators, and the like.

ROBOTS

A robot is a machine constructed as an assemblage of links joined so that they can be articulated into desired positions by a reprogrammable controller and precision actuators to perform a variety of tasks. Robots range from simple devices to very complex and "intelligent" systems by virtue of added sensors, computers, and special features. See Figure 10.6.1 for the possible components of a robotic system.

Robots, being programmable multijointed machines, fall between humans and fixed-purpose machines in their utility. In general, they are most useful in replacing humans in jobs that are dangerous, dirty, dull, or demeaning, and for doing things beyond human capabilities. They are usually better than conventional "hard" automation in short-run jobs and those requiring different tasks on a variety of products. They are ideal for operations calling for high consistency, cycle after cycle, yet are changeable when necessary. In contrast to "fixed" machines, they are "flexible," meaning that they are reprogrammable.

There are several hundred types and models of robots. They are available in a wide range of shapes, sizes, speeds, load capacities, and other characteristics. Care must be taken to select a robot to match the requirements of the tasks to be done. One way to classify them is by their intended application. In general, there are industrial, commercial, laboratory, mobile, military, security, medical, service, hobby, home, and personal robots. While they are programmable to do a wide variety of tasks, most are limited to one or a few categories of capabilities, based on their construction and specifications. Thus, within the classification of industrial robots, there are those that can paint but not assemble products, and those that can assemble products, some specialize in assembling very small electronic components while others make automobiles.

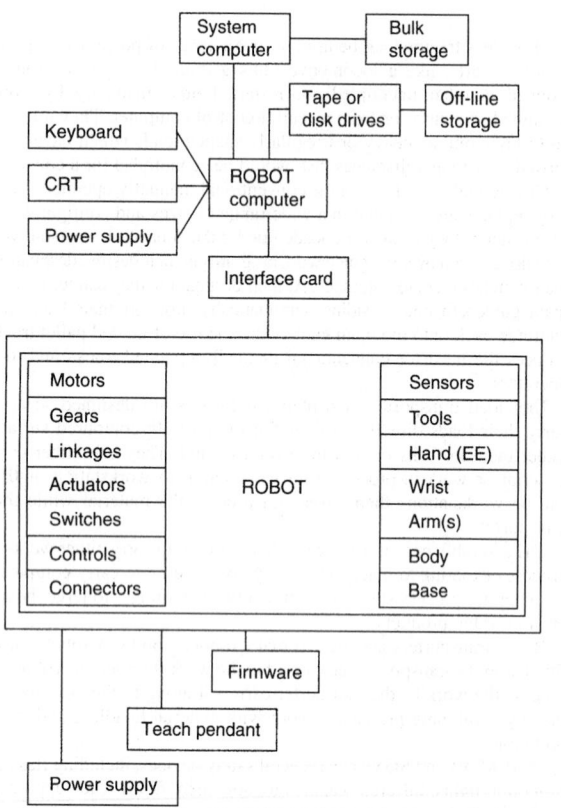

Fig. 10.6.1   Robotic system schematic. *(Robotics Research Division, VMA, Inc.)*

Some of the common uses of industrial robots include: loading and unloading machines, transferring work from one machine to another, assembling, cutting, drilling, grinding, deburring, welding, gluing, painting, inspecting, testing, packing, palletizing, and many others.

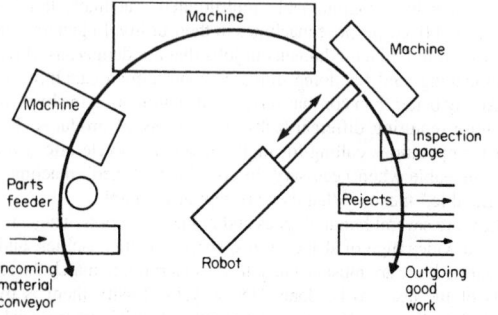

Fig. 10.6.2   Robot work cell. Robot is positioned to service a number of machines clustered about it.

Rather than have a robot perform only one task, it is sometimes possible to select a model and its tooling so that it can be positioned to do several jobs. Figure 10.6.2 shows a robot set in a work cell so that it can service the incoming material, part feeders, several machines, inspection gages, reject bins, and outgoing product conveyors arranged in an economical cluster around it. Robots are also used in chemical mixing and measuring; bomb disassembly; agriculture; logging; fisheries; department stores; amusement parks; schools; prisons; hospitals; nursing homes; nuclear power plants; warfare; spaces; underwater; surveillance; police, fire, and sanitation departments; inside the body; and an increasing number of innovative places.

Robot Anatomy

All robots have certain basic sections, and some have additional parts that give them added capabilities. All robots must have a power source, either mechanical, electrical, electromechanical, pneumatic, hydraulic, nuclear, solar, or a combination of these. They all must also have a means of converting the source of power to usable power and transmitting it: motors, pumps, and other actuators, and gears, cams, drives, bars, belts, chains, and the like. To do useful work, most robots have an assemblage of links arranged in a configuration best-suited to the tasks it is to do and the space within which it is to do them. In some robots this is called the *manipulator,* and the links and the joints connecting them are sometimes referred to as the robot's *body, shoulder, arm,* and *wrist.* A robot may have no, one, or multiple arms. Multiarmed robots must have control coordination protocols to prevent collisions, as must robots that work very close to other robots. Some robots have an immovable base that connects the manipulator to the floor, wall, column, or ceiling. In others the base is attached to, or part of, an additional section that gives it mobility by virtue of components such as wheels, tracks, rollers, "legs and feet," or other devices.

All robots need an intelligence and control section that can range from a very simple and relatively limited type up to a very complex and sophisticated system that has the capability of continuously interacting with varying conditions and changes in its environment. Such advanced control systems would include computers of various types, a multitude of microprocessors, memory media, many input/output ports, absolute and/or incremental encoders, and whatever type and number of sensors may be required to make it able to accomplish its mission. The robot may be required to sense the positions of its several links, how well it is doing its mission, the state of its environment, and other events and conditions. In some cases, the sensors are built into the robots; in others they are handled as an adjunct capability—such as machine vision, voice recognition, and sonic ranging—attached and integrated into the overall robotic system.

Other items that may be built into the robots or handled as attachments are the **tooling**—the things that interface the robot's arm or wrist and the workpiece or object to be manipulated and which accomplish the intended task. These are called the robot's *end effectors* or *end-of-arm tooling* (EOAT) and may be very simple or so complex that they become a major part of the total cost of the robotic system. A gripper may be more than a crude open or closed clamp (see Fig. 10.6.3), jaws equipped with sensors (see Figure 10.6.4), or a multifingered hand complete with several types of miniature sensors and capable of articulations that even the human hand cannot do. They may be binary or servoed. They may be purchased from stock or custom-designed to do specific tasks. They may be powered by the same type of energy as the basic robot or they may have a different source of power. A hydraulically powered robot may have pneumatically powered grippers and tooling, for example, making it a hybrid system. Most end effectors and EOAT are easily changeable so as to enable the robot to do more tasks.

Another part of the anatomy of many robots is a safety feature. Many robots are fast, heavy, and powerful, and therefore a source of danger. Safety devices can be both built into the basic robot and added as an adjunct to the installation so as to reduce the chances that the robot will do harm to people, other equipment, the tooling, the product, and itself. The American National Standards Institute, New York City, issued

ANSI/RIA R 15.06-1986 that is supported as a robot safety standard by the Robotic Industries Association, Dearborn, MI.

Finally, ways of programming robots, communicating with them, and monitoring their performance are needed. Some of the devices used for these purposes are teach boxes on pendants, keyboards, panels, voice recognition equipment, and speech synthesis modules.

Fig. 10.6.3   A simple nonservo, no-sensor robot gripper.

Linear position →
Angular position →

Strain gages for torque and force exerted and weight lifted

Gripping force
Tactile and slip
Light beam and photodetector for presence sensing

Also possible:
• Velocity and acceleration sensors
• Vision
• Sniffer
• Voice recognition
• Speech synthesis or other sounds

Also possible:
• Inductance or capacitance proximity sensor
• Radar or sonar distance ranging sensors

Fig. 10.6.4   A complex servoed, multisensor robot gripper.

Robot Specifications

While most robots are made "to stock" and sold from inventory, many are made "to order" to the specifications required by the intended application.

Configuration   The several links of a robot's manipulator may be joined to move in various combinations of ways relative to one another. A joint may turn about its axis (rotational axis) or it may translate along its axis (linear axis). Each particular arrangement permits the robot's control point (usually located at the center of its wrist flange, at the center of its gripper jaws, or at the tip of its tool) to move in a different way and to reach points in a different area. When a robot's three movements are along translating joints, the configuration is called *cartesian* or *rectangular* (Fig. 10.6.5). When one of the three joints is made a rotating joint, the configuration is called *cylindrical* (Fig. 10.6.6). When two of the three joints are rotational, the configuration is *called polar* or *spherical* (Fig. 10.6.7). And if all three joints are rotational, the robot is

said to be of the *revolute* or *jointed-arm* configuration (Fig. 10.6.8). The selective compliance assembly robot arm (SCARA) is a special type of revolute robot in which the joint axes lie in the vertical plane rather than in the horizontal (Fig. 10.6.9).

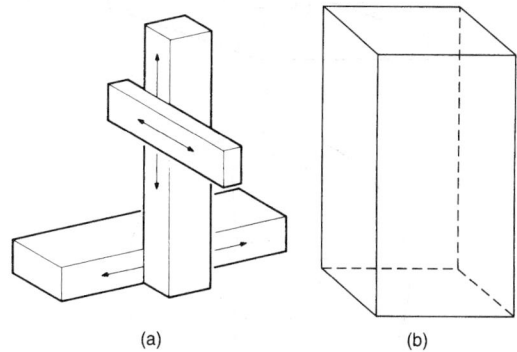

(a)   (b)

Fig. 10.6.5   Cartesian- or rectangular-coordinate robot configuration. (*a*) Movements; (*b*) work envelope.

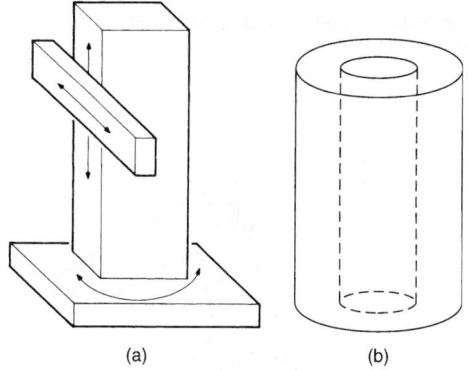

(a)   (b)

Fig. 10.6.6   Cylindrical-coordinate robot configuration. (*a*) Movements; (*b*) work envelope.

Articulations   A robot may be described by the number of articulations, or jointed movements, it is capable of doing. The joints described above that establish its configuration give the typical robot three degrees of freedom (DOF) and at least three more may be obtained from the motions of its wrist—these being called *roll, pitch,* and *yaw* (Fig. 10.6.10). The mere opening and closing of a gripper is usually not regarded as a degree of freedom, but the many positions assumable by a servoed gripper may be called a DOF and a multifingered mechanical hand has several additional DOFs. The added abilities of mobile robots to traverse land, water, air, or outer space are, of course, additional degrees of freedom. Figure 10.6.11 shows a revolute robot with a three-DOF wrist and a seventh DOF, the ability to move in a track along the floor.

Size   The physical size of a robot influences its capacity and its capabilities. There are robots as large as gantry cranes and as small as grains of salt—the latter being made by micromachining, which is the same process used to make integrated circuits and computer chips. "Gnat" robots, or "microbots," can be made using nanotechnology and can be small enough to travel inside a person's body to detect, report out, and possibly treat a problem. Some robots intended for light assembly work are designed to be approximately the size of a human so as to make the installation of the robot into the space vacated by the replaced human as easy and least disruptive as possible.

(a)

(b)

Fig. 10.6.7 Polar- or spherical-coordinate robot configuration. (*a*) Movements; (*b*) work envelope.

(a)

(b)

Fig. 10.6.8 Revolute– or jointed-arm–coordinate robot configuration. (*a*) Movements; (*b*) work envelope.

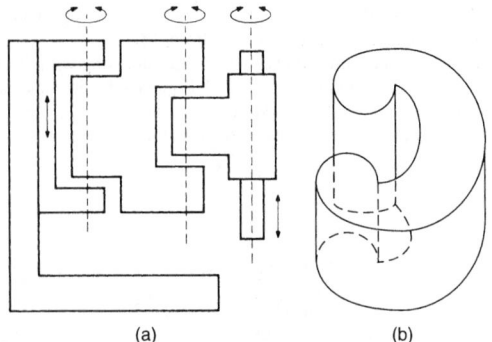

(a)   (b)

Fig. 10.6.9 SCARA coordinate robot configuration. (*a*) Movements; (*b*) work envelope.

Yaw

Pitch

Roll

Fig. 10.6.10 A robot wrist able to move in three ways: roll, pitch, and yaw.

Elbow extension

Wrist

Shoulder swivel

Pitch   Yaw   Roll

Arm sweep
or
body rotation

Body
traverse

Fig. 10.6.11 A typical industrial robot with three basic degrees of freedom, plus three more in its wrist, and a seventh in its ability to move back and forth along the floor.

Workspace   The extent of a robot's reach in each direction depends on its configuration, articulations, and the sizes of its component links and other members. Figure 10.6.12 shows the side and top views of the points that the industrial robot in Fig. 10.6.11 can reach. The solid geometric space created by subtracting the inner (fully contracted) from the outer (full extended) possible positions of a defined point (e.g., its control point, the center of its gripper, or the tip of its tool) is called the robot's *workspace* or *work envelope*. For the rectangular-coordinate

Fig. 10.6.12   Side and top views of the points that the robot in Fig. 10.6.11 can reach. The three-dimensional figure generated by these accessible end points is shown in Fig. 10.6.8*b*.

configuration robot, this space is a rectangular parallelepiped; for the cylindrical-configuration robot, it is a cylinder with an inaccessible space hollowed out of its center; for the polar and revolute robots, it consists of spheres with inaccessible spots at their centers; and for the SCARA configuration, it is the unique solid shown in the illustration. See Figs. 10.6.5 to 10.6.9 for illustrations of the work envelope of robots of the various configurations. For a given application, one would select a robot with a work envelope slightly larger than the space required for the job to be done. Selecting a robot too much larger than is needed could increase the costs, control problems, and safety concerns. Many robots have what is called a *sweet spot:* that smaller space where the performance of most of the specifications (payload, speed, accuracy, resolution, etc.) peaks.

Payload   The payload is the weight that the robot is designed to lift, hold, and position repeatedly with the accuracy advertised. People reading robot performance claims should be aware of the conditions under which they are promised, e.g., whether the gripper is empty or at maximum payload, or the manipulator is fully extended or fully retracted. Payload specifications usually include the weight of the gripper and other end-of-arm tooling, therefore the heavier those devices become as they are made more complex with added sensors and actuators, the less workpiece payload the robot can lift.

Speed   A robot's speed and cycle time are related, but different, specifications. There are two components of its speed: its acceleration and deceleration rate and its slew rate. The acceleration/deceleration rate is the time it takes to go from rest to full speed and the time it takes to go from full speed to a complete stop. The slew rate is the velocity once the robot is at full speed. Clearly, if the distance to go is short, the robot may not reach full speed before it must begin to slow to stop, thus making its slew rate an unused capability. The cycle time is the time it takes for the robot to complete one cycle of picking a given object up at given height, moving it a given distance, lowering it, releasing it, and returning to the starting point. This is an important measure for those concerned with productivity, or how many parts a robot can process in an hour. Speed, too, is a specification that depends on the conditions under which it is measured—payload, extension, etc.

Accuracy   The accuracy of a robot is the difference between where its control point goes and where it is instructed or programmed to go.

Repeatability   The repeatability or precision of a robot is the variance in successive positions each time its control point returns to a taught position.

Resolution   The resolution of a robot is the smallest incremental change in position that it can make or its control system can measure.

Robot Motion Control

While a robot's number of degrees of freedom and the dimensions of its work envelope are established by its configuration and size, the paths along which it moves and the points to which it goes are established by its control system and the instructions it is given. The motion paths generated by a robot's controller are designated as point-to-point or continuous. In point-to-point motion, the controller moves the robot from starting point *A* to end point *B* without regard for the path it takes or of any points in between. In a continuous-path robot controller, the path in going from *A* to *B* is controlled by the establishment of *via* or *way points* through which the control point passes on its way to each end point. If there is a conditional branch in the program at an end point such that the next action will be based on the value of some variable at that point, then the point is called a *reference point*.

Some very simple robots move only until their links hit preset stops. These are called *pick-and-place* robots and are usually powered by pneumatics with no servomechanism or feedback capabilities.

The more sophisticated robots are usually powered by hydraulics or electricity. The hydraulic systems require pumps, reservoirs, and cylinders and are good for lifting large loads and holding them steady. The electric robots use stepper motors or servomotors and are suited for quick movements and high-precision work. Both types of robots can use harmonic drives (Fig. 10.6.13), gears, or other mechanisms to reduce the speed of the actuators.

The harmonic drive is based on a principle patented by the Harmonic Drive unit of the Emhart Division of the Black & Decker Corp. It permits dramatic speed reductions, facilitating the use of high-speed actuators in robots. It is composed of three concentric components: an

Fig. 10.6.13   A harmonic drive, based on a principle patented by the Harmonic Drive unit of the Emhart Division of the Black & Decker Corp. As the elliptical center wave generator revolves, it deforms the inner toothed Flexspline into contact with the fixed outer circular spline which has two more teeth. This results in a high speed-reduction ratio.

elliptical center (called the *wave generator),* a rigid outer ring (called the *circular spline*), and a compliant inner ring (called the *Flexspline*) between the two. A speed-reduction ratio equal to one-half of the number of teeth on the Flexspline is achieved by virtue of the wave generator rotating and pushing the teeth of the Flexspline into contact with the stationary circular spline, which has two fewer teeth than the Flexspline. The result is that the Flexspline rotates (in the opposite direction) the distance of two of its teeth (its circular pitch × 2) for every full rotation of the wave generator.

Robots also use various devices to tell them where their links are. A robot cannot very well be expected to go to a new position if it does not know where it is now. Some such devices are potentiometers, resolvers, and encoders attached to the joints and in communication with the controller. The operation of a rotary-joint resolver is shown in Fig. 10.6.14. With one component on each part of the joint, an excitation input in one part induces sine and cosine output signals in coils (set at right angles in the other part) whose differences are functions of the amount of angular offset between the two. Those angular displacement waveforms are sent to an analog-to-digital converter for output to the robot's motion controller.

Encoders may be of the contact or noncontact type. The contact type has a metal brush for each band of data, and the noncontact type either reflects light to a photodetector or lets the light pass or not pass to a photodetector on its opposite side, depending on its pattern of light/dark

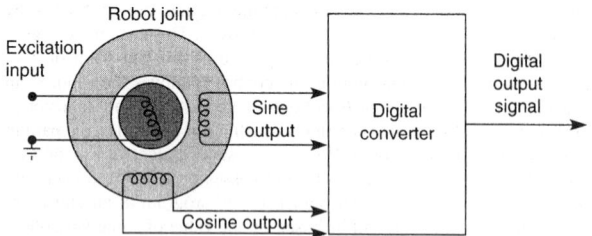

Fig. 10.6.14 Rotary-joint resolver. Resolver components on the robot's joint send analog angular displacement signals to a digital converter for interpretation and output to the robot's controller.

or clear/solid marks. Both types of encoders may be linear or rotary—that is, shaped like a bar or a disk—to match the type of joint to which they are attached. The four basic parts of a rotary optical encoder (the light sources, the coding disk, the photodetectors, and the signal processing electronics) are all contained in a small, compact, sealed package.

Encoders are also classified by whether they are incremental or absolute. Figure 10.6.15 shows two types of incremental optical encoder disks. They have equally spaced divisions so that the output of their photodetectors is a series of square waves that can be counted to tell the amount of rotational movement of the joint to which they are attached. Their resolution depends on how many distinct marks can be painted or etched on a disk of a given size. Disks with over 2,500 segments are available, such that the counter recording 25 impulses would know that the joint rotated 1/100 of a circle, reported as either degrees or radians, depending on the type of controller. It can also know the speed of that movement if it times the intervals between incoming pulses. It will not be able to tell the direction of the rotation if it uses a single-track, or tachometer-type, disk, as shown in Fig. 10.6.15*a*. A quadrature device, as shown in Fig. 10.6.15*b*, has two output channels, A and B, which are offset 90°, thus allowing the direction of movement to be determined by seeing whether the signals from A lead or lag those from B.

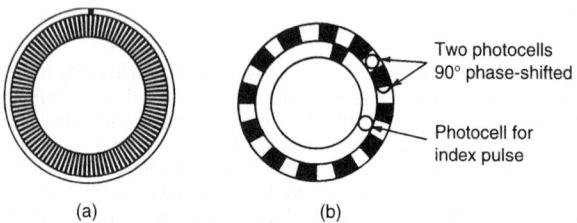

Fig. 10.6.15 Two types of incremental optical encoder disks. Type *a*, a single track incremental disk, has a spoke pattern that shutters the light onto a photodetector. The resulting triangular waveform is electronically converted to a square-wave output, and the pulses are counted to determine the amount of angular movement of the joint to which the encoder components are attached. Type *b*, a quadrature incremental disk, uses two light sources and photodetectors to determine the direction of movement.

The major limitation of incremental encoders is that they do not tell the controller where they are to begin with. Therefore, robots using them must be sent "home," to a known or initial position of each of its links, as the first action of their programs and after each switch-off or interruption of power. Absolute encoders solve this problem, but are

more complex and more expensive. They have a light source, a photodetector, and two electrical wires for each track or concentric ring of marks. Each mark or clear spot in each ring represents a bit of data, with the whole word being made up by the number of rings on the disk. Figure 10.6.16 shows 4-bit-word and 8-bit-word disks. Each position of the joint is identified by a specific unique word. All of the tracks in a segment of the disk are read together to input a complete coded word (address) for that position. The codes mostly used are natural binary, binary-coded decimal (BCD), and Gray. Gray code has the advantage that the state (ON/OFF or high/low) of only one track changes for each position change of the joint, making error detection easier.

Fig. 10.6.16 Two types of absolute optical encoder disks. Type *a* is for 4-bit-word systems, and type *b* is for 8-bit-word systems. Absolute encoders require a light source and photodetector set for each bit of word length.

All robot controllers require a number of input and output channels for communication and control. The number of I/O ports a robotic system has is one measure of its capabilities. Another measure of a robot's sophistication is the number and type of microprocessors and microcomputers in its system. Some have none; some have a microprocessor at each joint, as well as one dedicated to mathematical computations, one to safety, several to vision and other sensory subsystems, several to internal mapping and navigation (if they are mobile robots), and a master controller micro- or minicomputer.

Programming Robots

The methods of programming robots range from very elementary to advanced techniques. The method used depends on the robot, its controller, and the task to be performed. At the most basic level, some robots can be programmed by setting end stops, switches, pegs in holes, cams, wires, and so forth, on rotating drums, patch boards, control panels, or the robot's links themselves. Such robots are usually the pick-and-place type, where the robot can stop only at preset end points and nowhere in between, the number of accessible points being 2^N, where N is the robot's number of degrees of freedom. The more advanced robots can be programmed to go to any point within their workspace and to execute other commands by using a teach box on a tether (called a *teach pendant*) or a remote control unit. Such devices have buttons, switches, and sometimes a joystick to instruct the robot to do the desired things in the desired sequence. Errors can be corrected by simply overwriting the mistake in the controller's memory. The program can be tested at a slow speed, then, when perfected, can be run at full speed by merely turning a switch on the teach box. This is called the *walk-through* method of programming.

Another way some robots can be programmed is to physically take hold of it (or a remote representation of it) and lead it through the desired sequence of motions. The controller records the motions when in *teach mode* and then repeats them when switched to *run mode*. This is called the *lead-through* method of programming. The foregoing are all *teach-and-repeat* programming methods and are done on-line, that is, on a specific robot as it is in place in an installation but not doing other work while it is being programmed.

More advanced programming involves the use of a keyboard to type in textual language instructions and data. This method can be done on line or off line (away from and detached from the robot, such that it can be working on another task while the program is being created). There are many such programming languages available, but most are limited to use on one or only a few robots. Most of these languages are explicit, meaning that each of their instructions gives the robot a specific command to do a small step (e.g., open gripper) as part of a larger task. The more advanced higher-level implicit languages, however, give instructions at the task level (e.g., assemble 200 of product XYZ), and the robot's computer knows what sequence of basic steps it must execute to do that. These are also called *object-level* or *world-modeling languages* and *model-based systems*. Other ways of instructing a robot include spoken commands and feedback from a wide array of sensors, including vision systems. Telecherics, teleoperators, and other mechanisms under the continuous control of a remotely located operator are not true robots, as they are not autonomous.

Using Robots

The successful utilization of robots involves more than selecting and installing the right robot and programming it correctly. For an assembly robot, for example, care must be taken to design the product so as to facilitate assembly by a robot, presenting the component piece parts and other material to it in the best way, laying out the workplace (work cell) efficiently, installing all required safety measures, and training programming, operating, and maintenance employees properly.

10.7 MATERIAL STORAGE AND WAREHOUSING
by Vincent M. Altamuro

The warehouse handling of material is often more expensive than in-process handling, as it frequently requires large amounts of space, expensive equipment, labor, and computers for control. Warehousing activities, facilities, equipment, and personnel are needed at both ends of the process—at the beginning as receiving and raw materials and purchased parts storage, and at the end as finished goods storage, picking, packing, and shipping. These functions are aided by various subsystems and equipment—some simple and inexpensive, some elaborate and expensive.

The rapid and accurate identification of materials is essential. This can be done using only human senses, humans aided by devices, or entirely automated. Bar codes have become an accepted and reliable means of identifying material and inputting that data into an information and control system.

Material can be held, stacked, and transported in simple devices, such as shelves, racks, bins, boxes, baskets, tote pans, pallets and skids, or in complex and expensive computer-controlled systems, such as automated storage and retrieval systems.

IDENTIFICATION AND CONTROL OF MATERIALS

Materials must be identified, either by humans or by automated sensing devices, so as to:
1. Measure presence or movement
2. Qualify and quantify characteristics of interest
3. Monitor ongoing conditions so as to feed back corrective actions
4. Trigger proper marking devices
5. Actuate sortation or classification mechanisms
6. Input computing and control systems, update data bases, and prepare analyses and summary reports

To accomplish the above, the material must have, or be given, a unique code symbol, mark, or special feature that can be sensed and identified. If such a coded symbol is to be added to the material, it must be:
1. Easily and economically produced
2. Easily and economically read
3. Able to have many unique permutations—flexible
4. Compact—sized to the package or product
5. Error resistant—low chance of misreading, reliable
6. Durable
7. Digital—compatible with computer data formats

The codes can be read by contact or noncontact means, by moving or stationary sensors, and by humans or devices.

There are several technologies used in the identification and control of materials. The most important of these are:
1. Bar codes
2. Radio-frequency identification (RFID)
3. Smart cards
4. Machine vision
5. Voice recognition
6. Magnetic stripes
7. Optical character recognition (OCR)

All can be used for automatic data collection (ADC) purposes and as part of a larger electronic data interchange (EDI), which is the paperless, computer-to-computer, intercompany, and international communications and information exchange using a common data language. The dominant data format standard in the world is Edifact (Electronic Data Interchange for Administration, Commerce, and Transportation).

Bar codes are one simple and effective means of identifying and controlling material. Bar codes are machine-readable patterns of alternating, parallel, rectangular bars and spaces of various widths whose combinations represent numbers, letters, or other information as determined by the particular symbology, or language, employed. Several types of bar codes are available. While multicolor bar codes are possible, simple black-and-white codes dominate because a great number of permutations are available by altering their widths, presence, and sequences. Some codes are limited to numeric information, but others can encode complete alphanumeric plus special symbols character sets. Most codes are binary digital codes, some with an extra parity bit to catch errors. In each bar code, there is a unique sequence set of bars and spaces to represent each number, letter, or symbol. One-dimensional, or linear, codes print all of the bars and spaces in one row. Two-dimensional stacked codes arrange sections above one another in several rows so as to condense more data into a smaller space. To read stacked codes, the scanner must sweep back and forth so as to read all of the rows. The most recent development is the two-dimensional matrix codes that contain much more information and are read with special scanners or video cameras. Many codes are of the n of m type, i.e., 2 of 5, 3 of 9, etc. For example, in the **2-of-5 code**, a narrow bar may represent a 0 bit or OFF and a wide bar a 1 or ON bit. For each set of 5 bits, 2 must be ON—hence, 2 of 5. See Table 10.7.1 for the key to this code. A variation to the basic 2 of 5 is the **Interleaved 2-of-5**, in which the spaces between the bars also can be either narrow or wide, permitting the reading of a set of bars, a set of spaces, a set of bars, a set of spaces, etc., and yielding more information in less space. See Fig. 10.7.1. Both the 2-of-5 and the Interleaved 2-of-5 codes have ten numbers in their character sets.

Code 39, another symbology, permits alphanumeric information to be encoded by allowing each symbol to have 9 modules (locations) along the bar, 3 of which must be ON. Each character comprises five bars and four spaces. Code 39 has 43 characters in its set (1-10, A-Z, six symbols, and one start/stop signal). Each bar or space in a bar code is called an element. In Code 39, each element is one of two widths, referred to as *wide* and *narrow*. As in other bar codes, the narrowest element is referred to as the X dimension and the ratio of the widest to narrowest element widths is referred to as the N of the code. For each code, X and

Table 10.7.1   Bar Code 2-of-5 and I 2/5 Key

Character	Code				
	1	2	4	7	P
0	0	0	1	1	0
1	1	0	0	0	1
2	0	1	0	0	1
3	1	1	0	0	0
4	0	0	1	0	1
5	1	0	1	0	0
6	0	1	1	0	0
7	0	0	0	1	1
8	1	0	0	1	0
9	0	1	0	1	0

Wide bars and spaces = 1
Narrow bars and spaces = 0

N are constants. These values are used to calculate length of the bar code labels. The *Y* dimension of a bar code is the length (or height) of its bars and spaces. It influences the permissible angle at which a label may be scanned without missing any of the pattern. See Table 10.7.2 for the Code 39 key. The patterns for the Code 39 bars and spaces are designed in such a way that changing or misreading a single bit in any of them

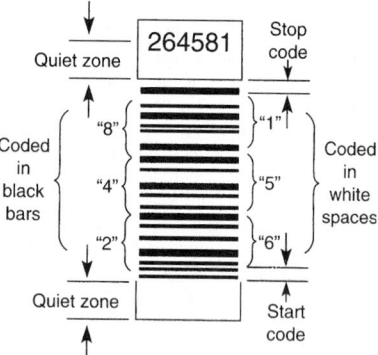

Fig. 10.7.1   A sample of a Uniform Symbology Specification Interleaved 2-of-5 bar code (also called I 2/5 or ITF).

results in an illegal code word. The bars have only an even number of wide elements and the spaces have only an odd number of wide elements. The code is called *self-checking* because that design provides for the immediate detection of single printing or reading errors. See Fig. 10.7.2. Code 39 was developed by Intermec (Everett, WA), as was **Code 93,** which requires less space by permitting more sizes of bars and spaces. Code 93 has 128 characters in its set (the full ASCII set) and permits a maximum of about 14 characters per inch, versus Code 39's maximum of about 9. Code 39 is *discrete,* meaning that each character is printed independently of the other characters and is separated from the characters on both sides of it by a space (called an intercharacter gap) that is not part of the data. Code 93 is a *continuous* code, meaning that there are no intercharacter gaps in it and all spaces are parts of the data symbols.

Code 128, a high-density alphanumeric symbology, has 128 characters and a maximum density of about 18 characters per inch (cpi) for numeric data and about 9 cpi for alphanumeric data. A Code 128 character comprises six elements: three bars and three spaces, each of four possible widths, or modules. Each character consists of eleven *1X* wide modules, where *X* is the width of the narrowest bar or space. The sum of the number of bar modules in any character is always even (even parity), while the sum of the space modules is always odd (odd parity), thus permitting self-checking of the characters. See Table 10.7.3 for the

Table 10.7.2   Uniform Symbology Specification Code 39 Symbol Character Set

Char.	Encodation Pattern	bsbsbsbsb	ASCII
0		000110100	48
1		100100001	49
2		001100001	50
3		101100000	51
4		000110001	52
5		100110000	53
6		001110000	54
7		000100101	55
8		100100100	56
9		001100100	57
A		100001001	65
B		001001001	66
C		101001000	67
D		000011001	68
E		100011000	69
F		001011000	70
G		000001101	71
H		100001100	72
I		001001100	73
J		000011100	74
K		100000011	75
L		001000011	76
M		101000010	77
N		000010011	78
O		100010010	79
P		001010010	80
Q		000000111	81
R		100000110	82
S		001000110	83
T		000010110	84
U		110000001	85
V		011000001	86
W		111000000	87
X		010010001	88
Y		110010000	89
Z		011010000	90
−		010000101	45
.		110000100	46
SPACE		011000100	32
$		010101000	36
/		010100010	47
+		010001010	43
%		000101010	37
S/S		010010100	none

Note: "S/S" denotes special Code 39 Start and Stop Character.
Note: In the columns headed "b" and "s" 1 is used to represent a wide element and 0 is used to represent a narrow element.

Code 128 key. Code 128 is a continuous, variable-length, bidirectional code.

The **Universal Product Code** (U.P.C.) is a 12-digit bar code adopted by the U.S. grocery industry in 1973. See Figure 10.7.3 for a sample of the U.P.C. standard symbol. Five of the six digits in front of the two long bars in the middle [called the *center guard pattern* (CGP)] identify the item's manufacturer and five of the six digits after the CGP identifies the specific product. There are several variations of the U.P.C., all those in the United States are controlled by the Uniform Code Council, located in Dayton, OH. One variation is an 8-digit code that identifies both the manufacturer and the specific product with one number. The U.P.C. uses prefix and suffix digits separated from the main number. The prefix digit is the *number system character.* See Table 10.7.4 for the key to these prefix numbers. The suffix is a "modulo-10" check digit. It serves to confirm that the prior 11 digits were read correctly. See Figure 10.7.4 for the method by

Table 10.7.3 Uniform Symbology Specification Code 128 Symbol Character Set

Value	Code set A	Code set B	Code set C	Encoding Pattern	bsbsbs
0	SP	SP	00		212222
1	!	!	01		222122
2	"	"	02		222221
3	#	#	03		121223
4	$	$	04		121322
5	%	%	05		131222
6	&	&	06		122213
7	'	'	07		122312
8	(	(	08		132212
9	)	)	09		221213
10	*	*	10		221312
11	+	+	11		231212
12	,	,	12		112232
13	.	.	13		122132
14	-	-	14		122231
15	/	/	15		113222
16	0	0	16		123122
17	1	1	17		123221
18	2	2	18		223211
19	3	3	19		221132
20	4	4	20		221231
21	5	5	21		213212
22	6	6	22		223112
23	7	7	23		312131
24	8	8	24		311222
25	9	9	25		321122
26	:	:	26		321221
27	;	;	27		312212
28	<	<	28		322112
29	=	=	29		322211
30	>	>	30		212123
31	?	?	31		212321
32	@	@	32		232121
33	A	A	33		111323
34	B	B	34		131123
35	C	C	35		131321
36	D	D	36		112313
37	E	E	37		132113
38	F	F	38		132311
39	G	G	39		211313
40	H	H	40		231113
41	I	I	41		231311
42	J	J	42		112133
43	K	K	43		112331
44	L	L	44		132131
45	M	M	45		113123
46	N	N	46		113321
47	O	O	47		133121
48	P	P	48		313121
49	Q	Q	49		211331
50	R	R	50		231131
51	S	S	51		213113
52	T	T	52		213311
53	U	U	53		213131
54	V	V	54		311123
55	W	W	55		311321

Value	Code set A	Code set B	Code set C	Encoding Pattern	bsbsbs
56	X	X	56		331121
57	Y	Y	57		312113
58	Z	Z	58		312311
59	[	[	59		332111
60	\	\	60		314111
61	]	]	61		221411
62	^	^	62		431111
63	_	_	63		111224
64	NUL	`	64		111422
65	SOH	a	65		121124
66	STX	b	66		121421
67	ETX	c	67		141122
68	EOT	d	68		141221
69	ENQ	e	69		112214
70	ACK	f	70		112412
71	BEL	g	71		122114
72	BS	h	72		122411
73	HT	i	73		142112
74	LF	j	74		142211
75	VT	k	75		241211
76	FF	l	76		221114
77	CR	m	77		413111
78	SO	n	78		241112
79	SI	o	79		134111
80	DLE	p	80		111242
81	DC1	q	81		121142
82	DC2	r	82		121241
83	DC3	s	83		114212
84	DC4	t	84		124112
85	NAK	u	85		124211
86	SYN	v	86		411212
87	ETB	w	87		421112
88	CAN	x	88		421211
89	EM	y	89		212141
90	SUB	z	90		214121
91	ESC	{	91		412121
92	FS	|	92		111143
93	GS	}	93		111341
94	RS	~	94		131141
95	US	DEL	95		114113
96	FNC 3	FNC 3	96		114311
97	FNC 2	FNC 2	97		411113
98	SHIFT	SHIFT	98		411311
99	CODE C	CODE C	99		113141
100	CODE B	FNC 4	CODE B		114131
101	FNC 4	CODE A	CODE A		311141
102	FNC 1	FNC 1	FNC 1		411131

Value				Encoding Pattern	bsbsbs
103	START A				211412
104	START B				211214
105	START C				211232

				Encoding Pattern	bsbsbsb
	STOP				2331112

Note: The numeric values in the "b" and "s" columns represent the number of modules in each of the symbol characters' bars and spaces.

Note: Dashed line indicates leading edge of adjacent character.

Code 39

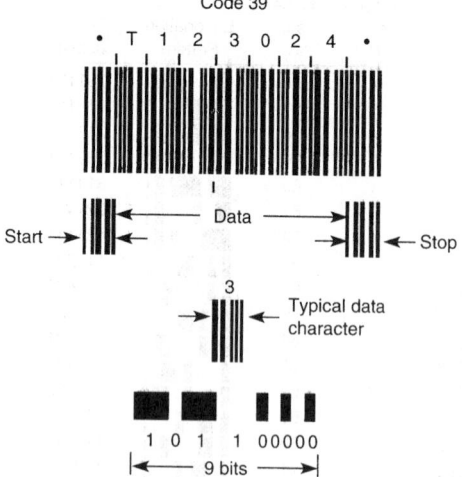

Fig. 10.7.2 A sample of Uniform Symbology Specification Code 39 bar code.

which the check digit is calculated. Each U.P.C. character comprises two bars and two spaces, each of which may be one, two, three, or four modules wide, such that the entire character consumes a total of seven modules of space. See Table 10.7.5 for the key to the code. It will be noted that the codes to the left of the CGP all start with a zero (a space)

(Not to scale)

Fig. 10.7.3 A sample of the Universal Product Code Standard Symbol. (*U.P.C. Symbol Specification Manual, January 1986, Copyright © 1986 Uniform Code Council, Inc. The Uniform Code Council prohibits the commercial use of their copyrighted material without prior written permission.*)

and end with a one (a bar) while those to the right of the CGP all start with a one and end with a zero. This permits the code to be scanned from either left to right or right to left (that is, it is omnidirectional) and allows the computer to determine the direction and whether it needs to be reversed before use by the system. The U.P.C. label also has the number printed in human-readable form along the bottom.

When integrated into a complete system, the U.P.C. can serve as the input data point for supermarket pricings, checkouts, inventory control, reordering stock, employee productivity checks, cash control, special promotions, purchasing habits and history of customers holding store "club" cards, sales analyses, and management reports. Its use increases checkout speed and reduces errors. In some stores, it is combined with

Table 10.7.4   U.P.C. Prefix Number System Character Key

The human-readable character identifying the encoded number system will be shown in the left-hand margin of the symbol as per Fig. 10.7.3.

Number system character	Specified use
0	Regular U.P.C. codes (Versions A and E)
2	Random-weight items, such as meat and produce, symbol-marked at store level (Version A)
3	National Drug Code and National Health Related Items Code in current 10-digit code length (Version A). Note that the symbol is not affected by the various internal structures possible with the NDC or HRI codes.
4	For use without code format restriction with check digit protection for in-store marking of nonfood items (Version A).
5	For use on coupons (Version A).
6, 7	Regular U.P.C. codes (Version A).
1, 8, 9	Reserved for uses unidentified at this time.

SOURCE: U.P.C. Symbol Specification Manual, January 1986, © Copyright 1986 Uniform Code Council, Inc. The Uniform Code Council prohibits the commercial use of their copyrighted material without prior written permission.

The following example will illustrate the calculation of the check character for the symbol shown in Fig. 10.7.3. Note that the code shown in Fig. 10.7.3 is in concurrent number system 0.

Step 1: Starting at the left, sum all the characters in the *odd* positions (that is, first from the left, third from the left, and so on), *starting with the number system character.*

(For the example, 0 + 2 + 4 + 6 + 8 + 0 = 20.)

Step 2: Multiply the sum obtained in Step 1 by 3.

(The product for the example is 60.)

Step 3: Again starting at the left, sum all the characters in the even positions.

(For the example, 1 + 3 + 5 + 7 + 9 = 25.)

Step 4: Add the product of step 2 to the sum of step 3.

(For the example, the sum is 85.)

Step 5: The modulo-10 check character value is the smallest number which when added to the sum of step 4 produces a multiple of 10.

(In the example, the check character value is 5.)

The human-readable character identifying the encoded check character will be shown in the right-hand margin of the symbol as in Fig. 10.7.3.

Fig. 10.7.4 U.P.C. check character calculation method. (*U.P.C. Symbol Specification Manual, January 1986, Copyright © 1986 Uniform Code Council, Inc. The Uniform Code Council prohibits the commercial use of their copyrighted material without prior written permission.*)

a speech synthesizer that voices the description and price of each item scanned. It is a vital component in a fully automated store.

As with many new technologies, different users use different protocols. New users must choose to use an existing protocol or create one of their own better suited to their particular needs. Soon there arise several camps of conflicting "standards," where one universally accepted standard is essential for the full application of the technology in all products, systems, companies, and countries. The longer users of each different design go separate ways, the more difficult it is for them to change, as their investment in hardware and software increases. Such was the case with bar codes. Hence, in January 1, 2005, in order to have their bar code compatible with the rest of the world, and as directed by the Uniform Code Council, the U.S. and Canada added a digit to their 12-digit UPC to make a 13-digit pattern that would match the European Article Numbering system pattern and become the global standard.

Table 10.7.5   U.P.C. Encodation Key

Encodation for U.P.C. characters, number system character, and modulo check character.

Decimal value	Left characters (Odd parity—O)	Right characters (Even parity—E)
0	0001101	1110010
1	0011001	1100110
2	0010011	1101100
3	0111101	1000010
4	0100011	1011100
5	0110001	1001110
6	0101111	1010000
7	0111011	1000100
8	0110111	1001000
9	0001011	1110100

The encodation for the left and right halves of the regular symbol, including UPC characters, number system character and modulo check character, is given in the following chart, which is applicable to version A. Note that the left-hand characters always use an odd number (3 or 5) of modules to make up the dark bars, whereas the right-hand characters always use an even number (2 or 4). This provides an "odd" and "even" parity encodation for each character and is important in creating, scanning, and decoding a symbol.

SOURCE: U.P.C. Symbol Specification Manual, January 1986, © Copyright 1986 Uniform Code Council, Inc. The Uniform Code Council prohibits the commercial use of their copyrighted material without prior written permission.

Also, the Uniform Code Council and EAN International combined to form a global code-controlling organization, called GSI, headquartered in Brussels.

Over the years, nearly 50 different one-dimensional bar code symbologies have been introduced. Around 20 found their way into common use at one time or another, with six or seven being most important. These are the Interleaved 2-of-5, Code 39, Code 93, Code 128, Codabar, Code 11, and the U.P.C. Table 10.7.6 compares some of the characteristics of these codes. The choice depends on the use by others in the same industry, the need for intercompany uniformity, the type and amount of data to be coded, whether mere numeric or alphanumeric plus special symbols will be needed, the space available on the item, and intracompany compatibility requirements.

Some bar codes are limited to a fixed amount of data. Others can be extended to accommodate additional data and the length of their label is variable. Figure 10.7.5 shows a sample calculation of the length of a Code 39 label.

Example used:

  Symbology: Code 39

  System: Each bar or space = one element
                Each character = 9 elements (5 bars and 4 spaces)
                Each character = 3 wide and 6 narrow elements
                (That is, 3 of 9 elements must be wide)

Width of narrow element = X dimension = 0.015 in (0.0381 cm)
Wide to narrow ratio = N value = 3 : 1 = 3
Length segments per data character = 3(3) + 6 = 15
Number of data characters in example = 6
Length of data section = 6 × 5   = 90.0 segments of length
Start character   = 15.0
Stop character   = 15.0
Intercharacter gaps (at 1 segment per gap):
No. of gaps = (data characters − 1)
      + start + stop
      = (6 − 1) + 1 + 1 = 7   =   7.0
Quiet zones
  (2, each at least 10 segments)   = 20.0
                Total segments = 147.0
Times X, length of each segment   ×0.015
                   = 2.205 in (5.6007 cm)

Fig. 10.7.5   Calculation of a bar code label length.

The foregoing one-dimensional, linear bar codes are sometimes called *license plates* because all they can do is contain enough information to identify an item, thereby permitting locating more information stored in a host computer. Two-dimensional stacked codes contain more information. The stacked codes, such as **Code 49,** developed by Dr. David Allais at Intermec, **Code 16K,** invented by Ted Williams, of Laserlight Systems, Inc., Dedham, MA, and others are really one-dimensional codes in which miniaturized linear codes are printed in multiple rows, or tiers. They are useful on small products, such as electronic components, single-dose medicine packages, and jewelry, where there is not enough space for a one-dimensional code. To read them, a scanner must traverse each row in sequence. Code 49 (see Fig. 10.7.6) can arrange all 128 ASCII encoded characters on up to eight stacked rows. Figure 10.7.7 shows a comparison of the space required for the same amount (30 characters) of alphanumeric information in three bar code symbologies: Code 39, Code 93, and Code 49. Code 16K (Fig. 10.7.8), which stands for 128 squared, shrinks and stacks linear codes in from 2

Fig. 10.7.6   Code 49. (*Uniform Symbology Specification Code 49, Copyright © 1993 Automatic Identification Manufacturers, Inc.*)

to 16 rows. It offers high-data-density encoding of the full 128-character ASCII set and double-density encoding of numerical data strings. It can fit 40 characters of data in the same space as eight would take in Code 128, and it can be printed and read with standard printing and scanning equipment.

Fig. 10.7.7   A comparison of the space required by three bar codes. Each bar code contains the same 30 alphanumeric characters.

Two-dimensional codes allow extra information regarding an item's description, date and place of manufacture, expiration date, lot number, package size, contents, tax code, price, etc. to be encoded, rather than merely its identity. Two-dimensional matrix codes permit the encoding of very large amounts of information. Read by video cameras or special laser scanners and decoders, they can contain a complete data file about

Fig. 10.7.8   Code 16K with human-readable interpretation. (*Uniform Symbology Specification Code 16K, Copyright © 1993 Automatic Identification Manufacturers, Inc.*)

Table 10.7.6 Comparison of Bar Code Symbologies

	Interleaved	Code 39	Codabar	Code 11	U.P.C./EAN	Code 128	Code 93
Date of inception	1972	1974	1972	1977	1973	1981	1982
Industry-standard specification	AIM ANSI UPCC AIAG	AIM ANSI AIAG HIBC	CCBBA ANSI	AIM	UPCC IAN	AIM	AIM
Government support		DOD					
Corporate sponsors	Computer Identics	Intermec	Welch Allyn	Intermec		Computer Identics	Intermec
Most-prominent application area	Industry	Industry	Medical	Industry	Retail	New	New
Variable length	No[a]	Yes	Yes	Yes	No	Yes	Yes
Alphanumeric	No	Yes	No	No	No	Yes	Yes
Discrete	No	Yes	Yes	Yes	No	No	No
Self-checking	Yes	Yes	Yes	No	Yes	No	No
Constant character width	Yes	Yes	Yes[b]	Yes	Yes	Yes	Yes
Simple structure (two element widths)	Yes	Yes	No	No	No	No	No
Number of data characters in set	10	43/128	16	11	10	103/128	47/128
Density[c]:							
Units per character	7–9	13–16	12	8–10	7	11	9
Smaller nominal bar, in	0.0075	0.0075	0.0065	0.0075	0.0104	0.010	0.008
Maximum characters/inch	17.8	9.4	10	15	13.7	9.1	13.9
Specified print tolerance at maximum density, in							
Bar width	0.0018	0.0017	0.0015	0.0017	0.0014	0.0010	0.0022
Edge to edge					0.0015	0.0014	0.0013
Pitch					0.0030	0.0029	0.0013
Does print tolerance leave more than half of the total tolerance for the scanner?	Yes	Yes	No	Yes	No	No	Yes
Data security[d]	High	High	High	High	Moderate	High	High

[a] Interleaved 2 of 5 is fundamentally a fixed-length code.

[b] Using the standard dimensions Codabar has constant character width. With a variant set of dimensions, width is not constant for all characters.

[c] Density calculations for interleaved 2 of 5, Code 39, and Code 11 are based on a wide-to-narrow ratio of 2.25 : 1. Units per character for these symbols are shown as a range corresponding to wide/narrow ratios from 2 : 1 to 3 : 1. A unit in Codabar is taken to be the average of narrow bars and narrow spaces, giving about 12 units per character.

[d] High data security corresponds to less than 1 substitution error per several million characters scanned using reasonably good-quality printed symbols. Moderate data security corresponds to 1 substitution error per few hundred thousand characters scanned. These values assume no external check digits other than those specified as part of the symbology and no file-lookup protection or other system safeguards.

[e] AIM = Automatic Identification Mfrs Inc (Pittsburgh)
ANSI = American National Standards Institute (New York)
AIAG = Automative Industry Action Group (Southfield, MI)
IAN = Intl Article Numbering Assn (Belgium)
DOD = Dept of Defense
CCBBA = Committee for Commonality in Blood Banking (Arlington, VA)
UPCC = Uniform Product Code Council Inc (Dayton, OH)

SOURCE: Intermec—A Litton Company, as published in the March, 1987 issue of *American Machinist* magazine, Penton Publishing Co., Inc.

the product to which they are affixed. In many cases, this eliminates the need to access the memory of a host computer. That the data file travels with (on) the item is also an advantage. Their complex patterns and interpretation algorithms permit the reading of complete files even when part of the pattern is damaged or missing.

Code PDF417, developed by Symbol Technologies Inc., Bohemia, NY (and put in the public domain, as are most codes), is a two-dimensional bar code symbology that can record large amounts of data, as well as digitized images such a fingerprints and photographs. To read the high-density PDF417 code, Symbol Technologies Inc. has developed a rastering laser scanner. This scanner is available in both a hand-held and fixed-mount version. By design, PDF417 is also scannable using a wide range of technologies, as long as specialized decoding algorithms are used. See Fig. 10.7.9 for an illustration of PDF417's ability to compress Abraham Lincoln's Gettysburg Address into a small space. The minimal unit that can represent data in PDF417 is called a *codeword*. It comprises 17 equal-length modules. These are grouped to form 4 bars and 4 spaces, each from one to six modules wide, so that they total 17 modules. Each unique codeword pattern is given one of 929 values. Further, to utilize different data compression algorithms based on the data type (i.e., ASCII versus binary) the system is multimodal, so that a codeword value may have different meanings depending on which of the various modes is being used. The modes are switched automatically

Fig. 10.7.9   A sample of PDF417 Encoding: Abraham Lincoln's Gettysburg Address. *(Symbol Technologies, Inc.)*

during the encoding process to create the PDF417 code and then again during the decoding process. The mode-switching instructions are carried in special codewords as part of the total pattern. Thus, the value of a scanned codeword is first determined (by low-level decoding), then its meaning is determined by virtue of which mode is in effect (by high-level decoding). Code PDF417 is able to detect and correct errors because of its sophisticated error-correcting algorithms. Scanning at angles is made possible by giving successive rows of the stacked code different identifiers, thereby permitting the electronic "stitching" of the individual codewords scanned. In addition, representing the data by using codewords with numerical values that must be interpreted by a decoder permits the mathematical correction of reading errors. At its simplest level, if two successive codewords have values of, say, 20 and 52, another codeword could be made their sum, a 72. If the first codeword were read, but the second missed or damaged, the system would subtract 20 from 72 and get the value of the second codeword. More complex high-order simultaneous polynomial equations and algorithms are also used. The PDF417 pattern serves as a portable data file that provides local access to large amounts of information about an item (or person) when access to a host computer is not possible or economically feasible. PDF417 provides a low-cost paper-based method of transmitting machine-readable data between systems. It also has use in WORM (written once and read many times) applications. The size of the label and its aspect ratio (the shape, or relationship of height to width) are variable so as to suit the user's needs.

Data Matrix, a code developed by International Data Matrix Inc., Clearwater, FL, is a true matrix code. It may be square or rectangular, and can be enlarged or shrunk to whatever size [from a 0.001-in (0.254-mm), or smaller, to a 14-in (36-cm), or larger, square] suits the user and application. Figure 10.7.10 shows the code in three sizes, but with the same contents. Its structure is uniform—square cells all the same size for a given pattern. To fit more user data into a matrix of a specific size, all the internal square cells are reduced to a size that will accommodate the

01234567890123456789

01234567890123456789

01234567890123456789

Fig. 10.7.10 Three sizes of Data Matrix patterns, all with the same contents. *(International Data Matrix, Inc.)*

amount of data to be encoded. Figure 10.7.11 shows five patterns, all the same size, containing different amounts of data. As a user selects additional data, the density of the matrix naturally increases. A pattern can contain from 1 to 2,000 characters of data. Two adjacent sides are always solid, while the remaining two sides are always composed of

0123456789
0123456789

0123456789
ABCDEFGHIJKLMN
OPQRSTUVWXYZ

0123456789ABCD
EFGHIJKLMNOPQ
RSTUVWXYZabcd
efghijklmnopqrstuv
wxyz

0123456789ABCD
EFGHIJKLMNOPQ
RSTUVWXYZabcd
efghijklmnopqrstuv
wxyz!@#$%^&*()_
+[]:"<>?~',./;'[]-=

01234567890123456789012345678901234567890123456789
01234567890123456789012345678901234567890123456789
01234567890123456789012345678901234567890123456789
01234567890123456789012345678901234567890123456789
01234567890123456789012345678901234567890123456789
01234567890123456789012345678901234567890123456789
01234567890123456789012345678901234567890123456789
01234567890123456789012345678901234567890123456789
01234567890123456789012345678901234567890123456789
01234567890123456789012345678901234567890123456789

Fig. 10.7.11 Five Data Matrix patterns, all the same size but containing different amounts of information. *(International Data Matrix, Inc.)*

alternating light and dark cells. This signature serves to have the video [charge-coupled device (CCD)] camera locate and identify the symbol. It also provides for the determination of its angular orientation. That capability permits a robot to rotate the part to which the code is affixed to the proper orientation for assembly, test, or packaging. Further, the information it contains can be program instructions to the robot. The labels can be printed in any combination of colors, as long as their difference provides a 20 percent or more contrast ratio. For aesthetic or security reasons, it can be printed in invisible ink, visible only when illuminated with ultraviolet light. The labels can be created by any printing method (laser etch, chemical etch, dot matrix, dot peen, thermal transfer, ink jet, pad printer, photolithography, and others) and on any type of computer printer through the use of a software interface driver provided by the company. The pattern's binary code symbology is read with standard CCD video cameras attached to a company-made controller linked to the user's computer system with standard interfaces and a Data Matrix Command Set. The algorithms for error correction that it uses create data values that are cumulative and predictable so that a missing or damaged character can be deduced or verified from the data values immediately preceding and following it.

Code One (see Fig. 10.7.12), invented by Ted Williams, who also created Code 128 and Code 16K, is also a two-dimensional checker-board or matrix type code read by a two-dimensional imaging device, such as a CCD video camera, rather than a conventional scanner. It can be read at any angle. It can encode the full ASCII 256-character set in addition to four function characters and a pad character. It can encode from 6 to over 2,000 characters of data in a very small space. If more data is needed, additional patterns can be linked together.

123456789012345678901 2

Fig. 10.7.12   Code One. *(Laserlight Systems, Inc.)*

Figure 10.7.13 shows a comparison of the sizes of four codes—Code 128, Code 16K, PDF-417, and Code One—to illustrate how much smaller Code One is than the others. Code One can fit 1,000 characters into a 1-in (2.54-cm) square of its checkerboard-like pattern. All Code One patterns have a finder design at their center. These patterns serve several purposes. They permit the video camera to find the label and determine its position and angular orientation. They eliminate the need for space-consuming quiet zones. And, because the internal patterns are different for each of the versions (sizes, capacities, and configurations) of the code, they tell the reader what version it is seeing. There are 10 different versions and 14 sizes of the symbol. The smallest contains 40 checkerboard-square type bits and the largest has 16,320. Code 1H, the

Code 128

Code 16K

PDF-417

Code One

Fig. 10.7.13   Comparative sizes of four symbols encoding the same data with the same *x* dimension. *(Laserlight Systems, Inc.)*

largest version, can encode 3,550 numeric or 2,218 alphanumeric characters, while correcting up to 2,240 bit errors. The data is encoded in rectangular tile groups, each composed of 8-bit squares arranged in a four-wide by two-high array. Within each tile, each square encodes one bit of data, with a white square representing a zero and a black square representing a one. The squares are used to create the 8-bit bytes that are the symbol's characters. The most significant bit in the character pattern is in the top left corner of the rectangle, proceeding left to right through the upper row and then the lower row, to the least significant bit in the bottom right corner of the pattern. The symbol characters are ordered left to right and then top to bottom, so that the first character is in the top left corner of the symbol and the last character is in the bottom right corner.

The fact that the patterns for its characters are arrays of spots whose images are captured by a camera, rather than the row of bars that one-dimensional and stacked two-dimensional codes have, frees it from the *Y*-dimension scanning angle restraint of those codes and makes it less susceptible to misreads and no-reads. The size of each square data bit is controlled only by the smallest optically resolvable or printable spot.

Bar codes may be printed directly on the item or its packaging or on a label that is affixed thereon. Some are scribed on with a laser. In many cases, they are printed on the product or package at the same time other printing or production operations are being performed, making the cost of adding them negligible. Labels may be produced in house or ordered from an outside supplier.

A recent application (made compulsory in 2006) of bar codes is on hospital medications to ensure that patients get the correct dose of the correct drug at the correct time so as to reduce the thousands of hospitalized patients and nursing home residents who die annually from drug errors. Matching the bar code on the patient's wristband with the drugs prescribed can also catch the administration of drugs which might cause an allergic reaction, drugs in conflict with other drugs, and drugs that should be given in liquid rather than pill form and vice versa.

In addition to the bar code symbology, an automatic identification system also requires a scanner or a video camera. A scanner is an electrooptical sensor which comprises a light source [typically a light-emitting diode (LED), laser diode, or helium-neon-laser tube], a light detector [typically a photodiode, photo-integrated circuit (photo-IC), phototransistor, or charge-coupled device (CCD)], an analog signal amplifier, and (for digital scanners) an analog-to-digital converter. The system also needs a microprocessor-based decoder, a data communication link to a computer, and one or more output or actuation devices. See Fig. 10.7.14.

The scanner may be hand-held or mounted in a fixed position, but something must move relative to the other, either the entire scanner, the scanner's oscillating beam of light, or the package or item containing the bar code. For moving targets, strobe lights are sometimes used to freeze the image while it is being read.

The four basic categories of scanners are: hand-held fixed beam, hand-held moving beam, stationary fixed beam, and stationary moving beam. Some scanners cast crosshatched patterns on the object so as to catch the bar code regardless of its position. Others create holographic laser patterns that wrap a field of light beams around irregularly shaped articles. Some systems require noncontact between the scanner and the bar code, but most are of the noncontact type. A read-and-beep system emits a sound when a successful reading has been made. A specialized type of scanner, a slot scanner, reads bar codes on badges, cards, envelopes, documents, and the like as they are inserted in the slot and swiped.

In operation, the spot or aperture of the scanner beam is directed by optics onto the bar code and reflected back onto a photodetector. The relative motion of the scanner beam across the bar code results in an analog response pattern by the photodetector as a function of the relative reflectivities of the dark bars and light spaces. A sufficient contrast between the bars, spaces, and background color is required. Some scanners see red bars as white. Others see aluminum as black, requiring the spaces against such backgrounds to be made white.

The system anticipates an incoming legitimate signal by virtue of first receiving a zero input from the "quiet zone"—a space of a certain

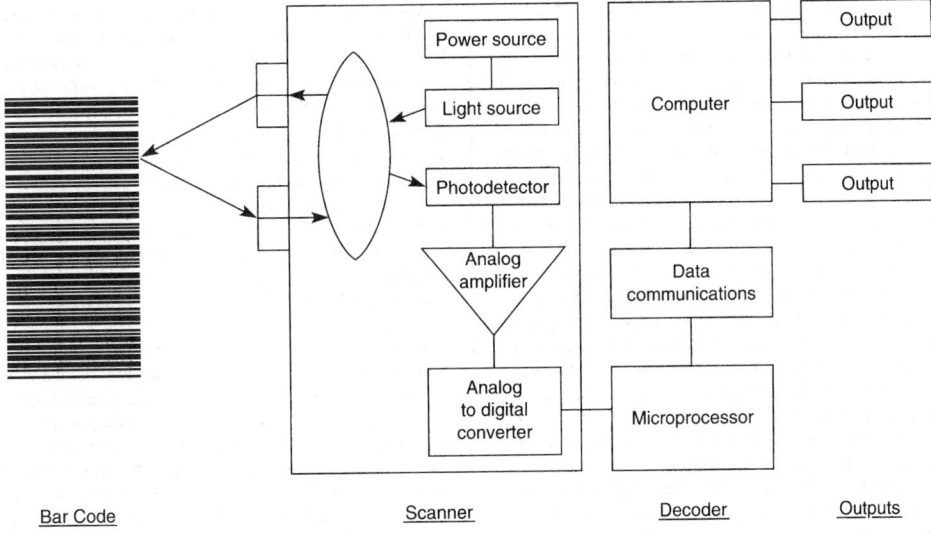

Bar Code   Scanner   Decoder   Outputs

Fig. 10.7.14 Components of a bar code scanner.

width (generally at least 10 times the code's X dimension) that is clear of all marks and which precedes the start character and follows the stop character. The first pattern of bars and spaces (a character) that the scanner sees is the start/stop code. These not only tell the system that data codes are coming next and then that the data section has ended, but in bidirectional systems (those that can be scanned from front to back or from back to front) they also indicate that direction. In such bidirectional systems, the data is read in and then reversed electronically as necessary before use. The analog signals pass through the scanner's signal conditioning circuitry, were they are amplified, converted into digital form, tested to see whether they are acceptable, and, if so, passed to the decoder section microprocessor for algorithmic interpretation.

The decoder determines which elements are bars and which are spaces, determines the width (number of modules) of each element, determines whether the code has been read frontward or backward and reverses the signal if needed, makes a parity check, verifies other criteria, converts the signal into a computer language character—the most common of which is asynchronous American Standard Code for Information Interchange (ASCII)—and sends it, via a data communication link, to the computer. Virtually all automatic identification systems are computer-based. The bar codes read become input data to the computer, which then updates its stored information and/or triggers an output.

Other types of automatic identification systems include radio-frequency identification, smart cards, machine vision, voice recognition, magnetic stripes, and optical character recognition.

Radio-frequency identification has many advantages over the other identification, control, and data-collection methods. It is immune to factory noise, heat, cold, and harsh environments. Its tags can be covered with dirt, grease, paint, etc. and still operate. It does not require line of sight from the reader, as do bar codes, machine vision, and OCR. It can be read from a distance, typically 15 ft (4.6 m). Its tags can both receive and send signals. Many types have higher data capacities than one-dimensional bar codes, magnetic stripes, and OCR. Some tags can read/write and process data, much like smart cards. However, RFID also has some disadvantages, compared to other methods. The tags are not human-readable and their cost for a given amount of data capacity is higher than that of some other methods. The RFID codes are currently closed—that is, a user's code is not readable by suppliers or customers—and there are no uniform data identifiers (specific numeric or alphanumeric codes that precede the data code and that identify the data's industry, category, or use) like the U.P.C. has.

Since 2004, the U.S. military, faced with frequent separations of wounded personnel and their medical records, has recorded medical data on passive RFID chips sewn into wristbands, embedded in dog tags, and otherwise attached to the person, so that the records and the wounded travel together. The attached RFID chips also serve to identify wounded or killed personnel. The chips, which can hold 2 kbits of information, are also used by the military to monitor and warn individuals going into and out of chemical, biological, or radioactive "hot" areas and also to keep track of the movements, levels, and ages of munitions and supplies.

An RFID system consists of a transceiver and a tag, with a host computer and peripheral devices possible. The transceiver has a coil or antenna, a microprocessor, and electronic circuitry, and is called a *reader* or *scanner*. The tag can also have an antenna coil, a capacitor, a nonvolatile type of memory, transponder electronic circuitry, and control logic—all encapsulated for protection.

The system's operations are very simple. The reader transmits an RF signal to the tag, which is usually on a product, tote bin, person, or other convenient carrier and can be made small and concealed. This signal tells the tag that a reader is present and wants a response. The activated tag responds by broadcasting a coded signal on a different frequency. The reader receives that signal and processes it. Its output may be as simple as emitting a sound for the unauthorized removal of an item containing the tag (retail theft, library books, and the like), flashing a green light and/or opening a door (employee ID badges), displaying information on a human-readable display, directing a robot or automatic guided vehicle (AGV), sending a signal to a host computer, etc. The signals from the reader to the tags can also provide energy to power the tag's circuitry and supply the digital clock pulses for the tag's logic section.

The tags may be passive or active. Passive tags use power contained in the signal from the reader. The normal range for these is 18 in (0.46 m) or less. Active tags contain a battery for independent power and usually operate in the 3- to 15-ft (0.9- to 4.6-m) range. Tags programmed when made are said to be *factory programmed* and are read-only devices. Tags programmed by plugging them into, or placing them very near, a programmer are called *field-programmed* devices. Tags whose program or data can be changed while they are in their operating locations are called *in-use programmed* devices. The simplest tags contain only one bit of code. They are typically used in retail stores to detect shoplifting and are called *electronic article surveillance* (EAS), *presence sensing*, or *Level I* tags. The next level up are the tags that, like bar codes, identify an object and serve as an input code to a database in a host computer. These "electronic license plates" are called *Level II* tags and typically have a capacity of between 8 and 128 bits. The third level devices are called *Level III*, or *transaction/routing* tags. They hold up to 512 bits so that the item to which they are attached may be described as well as identified. The next level tags, *Level IV,* are called *portable databases,* and, like matrix bar codes, carry large amounts of information in text form, coded in ASCII. The highest current level tags incorporate microprocessors, making them capable of data processing and decision making, and are the RF equivalent of smart cards.

Smart cards, also called chipcards, or memory cards, are the size of credit cards but twice as thick. They are composed of self-contained circuit boards, stacked memory chips, a microprocessor, a battery, and input/output means, all laminated between two sheets of plastic which usually contain identifiers and instructions. They are usually read by inserting them into slots and onto pins of readers. The readers can be free-standing devices or components of production and inventory control systems, computer systems, instruments, machines, products, maintenance records, calibration instruments, store checkout registers, vending machines, telephones, automatic teller machines, road toll booths, doors and gates, hospital and patient records, etc. One smart card can hold the complete data file of a person or thing and that data can be used as read and/or updated.

The cards may be described as being in one of four classifications: **contactless cards,** those which contain a tiny antenna, such that energy as well as data are exchanged via electromagnetic fields rather than direct physical contact and are used mostly for general access, such as transportation; **intelligent memory cards,** those having an EEPROM (electrically erasable programmable read-only memory), a logic circuit, and security feature and which are used mainly as telephone, shopping, and health cards; **microcontroller cards,** which have an EEPROM, RAM, ROM, I/O, and a CPU, making them a complete computer system with a security feature and are typically used as network access cards and banking cards; and **cryptocards,** which contain microcontrollers plus a mathematics coprocessor for cryptographic applications and high-security situations.

Cards with up to 64-Mbytes of flash memory are available. They are made by tape-automated bonding of 20 or more blocks of two-layered memory chips onto both sides of a printed-circuit board.

Machine vision can be used to:

1. Detect the presence or absence of an entire object or a feature of it, and sort, count, and measure.
2. Find, identify, and determine the orientation of an object so that a robot can go to it and pick it up.
3. Assure that the robot or any other machine performed the task it was supposed to do, and inspect the object.
4. Track the path of a seam so that a robot can weld it.
5. Capture the image of a matrix-type bar code for data input or a scene for robotic navigation (obstacle avoidance and path finding).

In operation, the machine-vision system's processor digitizes the image that the camera sees into an array of small square cells (called picture elements, or pixels). It classifies each pixel as light or dark or a relative value between those two extremes. The system's processor then employs methods such as pixel counting, edge detection, and connectivity analysis to get a pattern it can match (a process called *windowing and feature analysis*) to a library of images in its memory, triggering a specific output for each particular match. Rather than match the entire image to a complete stored pattern, some systems use only common feature tests, such as total area, number of holes, perimeter length, minimum and maximum radii, and centroid position. The resolution of a machine vision system is its field of view (the area of the image it captures) divided by its number of pixels (e.g., $256 \times 256 = 65,536, 600 \times 800 = 480,000$). The system can evaluate each pixel on either a binary or a gray scale. Under the binary method, a threshold of darkness is set such that pixels lighter than it are assigned a value of 1 and those darker than it are given a value of 0 for computer processing. Under the gray scale method, several intensity levels are established, the number depending on the size of the computer's words. Four bits per pixel permit 16 levels of gray, 6 bits allow 64 classifications, and 8-bit systems can classify each pixel into any of 256 values, but at the cost of more computer memory, processing time, and expense. As the cost of vision systems drops, their use in automatic identification increases. Their value in reading matrix-type bar codes and in robotics can justify their cost.

Voice recognition systems process patterns received and match them to a computer's library of templates much like machine vision does. Memory maps of sound pattern templates are created either by the system's manufacturer or user, or both. A *speaker-dependent* system recognizes the voice of only one or a few people. It is taught to recognize their words by having the people speak them several times while the system is in the teach mode. The system creates a speech template for each word and stores it in memory. The system's other mode is the recognition mode, during which it matches new input templates to those in its memory. *Speaker-independent* systems can recognize almost anyone's voice, but can handle many fewer words than speaker-dependent systems because they must be able to recognize many template variations for each word. The number of words that each system can recognize is constantly being increased.

In operation, a person speaks into a microphone. The sound is amplified and filtered and fed into an analyzer, thence to a digitizer. Then a synchronizer encodes the sound by separating its pattern into equal slices of time and sends its frequency components to a classifier where it is compared to the speech templates stored in memory. If there is a match, the computer produces an output to whatever device is attached to the system. Some systems can respond with machine-generated speech. The combination of voice recognition and machine speech is called *speech processing* or *voice technology.* Products that can "hear" and "talk" are called *conversant products.* Actual systems are more complicated and sophisticated than the foregoing simplified description. The templates, for example, instead of having a series of single data points define them, have numerical fields based on statistical probabilities. This gives the system more tolerant template-to-template variations. Further, and to speed response, neural networks, fuzzy logic, and associative memory techniques are used.

Some software uses expert systems that apply grammatical and context rules to the inputs to anticipate that a word is a verb, noun, number, etc. so as to limit memory search and thereby save time. Some systems are adaptive, in that they constantly retrain themselves by modifying the templates in memory to conform to permanent differences in the speakers' templates. Other systems permit the insertion by each user of an individual module which loads the memory with that person's particular speech templates. This increases the capacity, speed, and accuracy of the system and adds a new dimension: security against use by an unauthorized person. Like fingerprints, a person's voice prints are unique, making voice recognition a security gate to entry to a computer, facility, or other entity. Some systems can recognize thousands of words with almost perfect accuracy. Some new products, such as the VCP 200 chip, of Voice Control Products, Inc., New York City, are priced low enough, by eliminating the digital signal processor (DSP) and limiting them to the recognition of only 8 or 10 words (commands), to permit their economic inclusion in products such as toys, cameras, and appliances.

A major use of voice recognition equipment is in industry where it permits assemblers, inspectors, testers, sorters, and packers to work with both hands while simultaneously inputting data into a computer directly from the source. They are particularly useful where keyboards are not suitable data-entry devices. They can eliminate paperwork, duplication, and copying errors. They can be made to cause the printing of a bar code, shipping label, or the like. They can also be used to instruct robots—by surgeons describing and directing an operation, for example—and in many other applications.

Magnetic stripes encode data on magnetic material in the form of a strip or stripe on a card, label, or the item itself. An advantage to the method is that the encoded data can be changed as required.

Optical character recognition uses a video camera to read numbers, letters, words, signs, and symbols on packages, labels, or the item itself. It is simpler and less expensive than machine vision but more limited.

The information collected from the foregoing can be used at the location where it is obtained, observed by a human, stored in the sensing device for later input to a computerized system, or transmitted in real time to a central location via either fixed wires or by a wireless means. Users should be aware that wireless signals travel through the air, making them more vulnerable than wired connections to reception by unauthorized persons, unless security features are included. Devices equipped with Bluetooth components for shorter-range transmissions or WiFi (wireless fidelity) technology for longer ranges can communicate wirelessly with other devices so equipped. WiFi requires the use of locations called access points or "hot spots" to connect to the Internet, networks of portable computers, or other WLANs (wireless local area networks).

The industry's trade group is the Automatic Identification Manufacturers, Inc. (AIM), located in Pittsburgh.

STORAGE EQUIPMENT

The functions of storage and handling devices are to permit:

The greatest use of the space available, stacking high, using the "cube" of the room, rather than just floor area

Multiple layers of stacked items, regardless of their sizes, shapes, and fragility

"Unit load" handling (the movement of many items of material each time one container is moved)

Protection and control of the material

Shelves and Racks Multilevel, compartmentalized storage is possible by using shelves, racks, and related equipment. These can be either prefabricated standard size and shape designs or modular units or components that can be assembled to suit the needs and space available. Variations to conventional equipment include units that slide on floor rails (for denser storage until needed), units mounted on carousels to give access to only the material wanted, and units with inclined shelves in which the material rolls or slides forward to where it is needed.

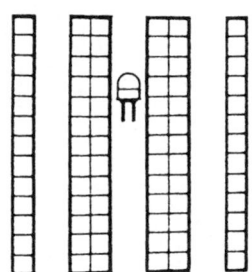

Fig. 10.7.15 Pallet racks. (*Modern Materials Handling, Feb. 22, 1980, p. 85.*)

Figure 10.7.15 shows the top view of a typical arrangement of pallet racks providing access to every load in storage. Storage density is low because of the large amount of aisle space. Figure 10.7.16 shows multiple-depth pallet racks with proportionally smaller aisle requirement. Figure 10.7.17 shows cantilever racks for holding long items. Figure 10.7.18 shows flow-through racks. The racks are loaded on one side and emptied from the other. The lanes are pitched so that the loads advance as they are

Fig. 10.7.16 Multiple-depth pallet racks. (*Modern Materials Handling, Feb. 22, 1980, p. 85.*)

removed. Figure 10.7.19 shows mobile racks. Here the racks are stored next to one another. When material is required, the racks are separated and become accessible. Figure 10.7.20 shows racks where the picker passes between the racks. The loads are supported by arms attached to

Fig. 10.7.17 Cantilever racks. (*Modern Materials Handling, Feb. 22, 1980, p. 85.*)

uprights. Figure 10.7.21 shows block storage requiring no rack. Density is very high. The product must be self-supporting or in stacking frames. The product should be stored only a short time unless it has a very long shelf life.

Fig. 10.7.18 Flow-through racks. (*Modern Materials Handling, Feb. 22, 1980, p. 85.*)

Bins, Boxes, Baskets, and Totes Small, bulk, odd-shaped, or fragile material is often placed in containers to facilitate unit handling. These bins, boxes, baskets, and tote pans are available in many sizes, strength grades, and configurations. They may be in one piece or with hinged flaps for ease of loading and unloading. They may sit flat on the

Fig. 10.7.19 Mobile racks and shelving. (*Modern Materials Handling, Feb. 22, 1980, p. 85.*)

Fig. 10.7.20 Drive-in drive-through racks. (*Modern Materials Handling, Feb. 22, 1980, p. 85.*)

floor or be raised to permit a forklift or pallet truck to get under them. Most are designed to be stackable in a stable manner, such that they nest or interlock with those stacked above and below them. Many are sized to fit (in combination) securely and with little wasted space in trucks, ships, between building columns, etc.

Fig. 10.7.21 Block storage.
*(Modern Materials Handling,
Feb. 22, 1980, p. 85.)*

Fig. 10.7.22 Single-faced pallet.

Pallets and Skids Pallets are flat, horizontal structures, usually made of wood, used as platforms on which material is placed so that it is unitized, off the ground, stackable, and ready to be picked up and moved by a forklift truck or the like. They can be single-faced (Fig. 10.7.22), double-faced, nonreversible (Fig. 10.7.23) double-faced, reversible (Fig. 10.7.24), or solid (slip pallets). Pallets with overhanging stringers are called *wing pallets,* single or double (Figs. 10.7.25 and 10.7.26).

Fig. 10.7.23 Double-faced pallet.

Fig. 10.7.24 Double-faced reversible pallet.

Fig. 10.7.25 Single-wing pallet.

Fig. 10.7.26 Double-wing pallet.

Skids typically stand higher off the ground than do pallets and stand on legs or stringers. They can have metal legs and framing for extra strength and life (Fig. 10.7.27), be all steel and have boxes, stacking alignment tabs, hoisting eyelets, etc. (Fig. 10.7.28).

Fig. 10.7.27 Wood skid with metal legs.

Fig. 10.7.28 Steel skid box.

AUTOMATED STORAGE/RETRIEVAL SYSTEMS

An automated storage/retrieval system (AS/RS) is a high-rise, high-density material handling system for storing, transferring, and controlling inventory in raw materials, work-in-process, and finished goods stages. It comprises:

1. Structure
2. Aisle stacker cranes or storage-retrieval machines and their associated transfer devices
3. Controls

Fig. 10.7.29 Schematic diagram of an automated storage and retrieval system (AS/RS) structure. *(Harnischfeger Corp.)*

The AS/RS structure (see Fig. 10.7.29) is a network of steel members assembled to form an array of storage spaces, arranged in bays, rows, and aisles. These structures typically can be 60 to 100 ft (18 to 30 m) high, 40 to 60 ft (12 to 18 m) wide and 300 to 400 ft (91 to 121 m) long. Usually, they are erected before their protective building, which might be merely a "skin" and roof wrapped around them.

The AS/RS may contain only one aisle stacker crane which is transferred between aisles or it may have an individual crane for each aisle. The function of the stacker crane storage-retrieval machine is to move down the proper aisle to the proper bay, then elevate to the proper storage space and then, via its shuttle table, move laterally into the space to deposit or fetch a load of material.

AS/RS controls are usually computer-based. Loads to be stored need not be given a storage address. The computer can find a suitable location, direct the S/R machine to it, and remember it for retrieving the load from storage.

ORDER PICKING

Material is often processed or manufactured in large lot sizes for more economical production. It is then stored until needed. If the entire batch is not needed out of inventory at one time, order picking is required. Usually, to fill a complete order, a few items or cartons of several different materials must be located, picked, packed in one container, address labeled, loaded, and delivered intact and on time.

In picking material, a person can go to the material, the material can be brought to the person doing the packing, or the process can be automated, such that it is released and routed to a central point without a human order picker.

When the order picking is to be done by a human, it is important to have the material stored so that it can be located and picked rapidly, safely, accurately, and with the least effort. The most frequently picked items should be placed in the most readily accessible positions. Gravity racks, with their shelves sloped forward so that as one item is picked, those behind it slide or roll forward into position for ease of the next picking, are used when possible.

LOADING DOCK DESIGN

In addition to major systems and equipment such as automatic guided vehicles and automated storage/retrieval systems, a material handling capability also requires many auxiliary or support devices, such as the various types of forklift truck attachments, slings, hooks, shipping cartons, and packing equipment and supplies. Dock boards to bridge the gap between the building and railroad cars or trucks may be one-piece portable plates of steel or mechanically or hydraulically operated leveling mechanisms. Such devices are required because the trucks that deliver and take away material come in a wide variety of sizes. The shipping-receiving area of the building must be designed and equipped to handle these truck size variations. The docks may be flush, recessed, open, closed, side loading, saw tooth staggered, straight-in, or turn-around.

Section **11**

Transportation

BY

CHARLES A. AMANN *Principal Engineer, KAB Engineering*
V. TERREY HAWTHORNE *late Senior Engineer, LTK Engineering Services*
KEITH L. HAWTHORNE *Vice President–Technology, Transportation Technology Center, Inc.*
MICHAEL C. TRACY *Rear Admiral, U.S. Navy*
MICHAEL W. M. JENKINS *Professor, Aerospace Design, Georgia Institute of Technology*
SANFORD FLEETER *McAllister Distinguished Professor, School of Mechanical Engineering,*
  Purdue University
WILLIAM C. SCHNEIDER *Retired Assistant Director Engineering/Senior Engineer, NASA*
  Johnson Space Center and Visiting Professor, Texas A&M University
G. DAVID BOUNDS *Senior Engineer, Duke Energy Corp.*

11.7 PIPELINE TRANSMISSION
by G. David Bounds

11.8 CONTAINERIZATION
(Staff Contribution)

11.1 AUTOMOTIVE ENGINEERING
by Charles A. Amann

REFERENCES: Sovran and Blazer, "A Contribution to Understanding Automotive Fuel Economy and Its Limits," SAE Paper 2003–01–2070. Toboldt, Johnson, and Gauthier, "Automotive Encyclopedia," Goodheart-Willcox Co., Inc. "Bosch Automotive Handbook," Society of Automotive Engineers. Gillespie, "Fundamentals of Vehicle Dynamics," Society of Automotive Engineers. Hellman and Heavenrich, "Light-Duty Automotive Technology Trends: 1975 Through 2003," Environmental Protection Agency. Davis et al., "Transportation Energy Book," U.S. Department of Energy. Various publications of the Society of Automotive Engineers, Inc. (SAE International).

AUTOMOTIVE VEHICLES

The dominant mode of personal transportation in the United States is the light-duty automotive vehicle. This category is composed not only of passenger cars, but of light-duty trucks as well. Light-duty trucks include passenger vans, SUVs (suburban utility vehicles), and pickup trucks. Among light-duty trucks, the distinction between personal and commercial vehicles is blurred because they may be used for either purpose. Beyond the light-duty truck, medium- and heavy-duty trucks are normally engaged solely in commercial activities.

In 2003, there were 135.7 million passenger cars registered in the United States, and 87.0 million light trucks. Purchasing preferences of the citizenry have caused a gradual shift from passenger cars to light trucks in recent decades. In 1970, light trucks comprised a 14.8 percent share of light-vehicle sales. By 2003, that share had swollen to nearly half of new vehicle sales.

Among heavier vehicles, there were 7.9 million heavy trucks (more than two axles or more than four tires) registered in 2003, of which 26 percent were trailer-towing truck tractors. In addition, there were 0.7 million buses in use, 90 percent of which were employed in transporting school students.

U.S. highway vehicles traveled a total of 2.89 trillion miles in 2003. Of this mileage, 57 percent was accumulated by passenger cars, 34 percent by light trucks, and 8 percent by heavy trucks and buses.

In 2000, 34 percent of U.S. households operated one vehicle, 39 percent had two, and 18 percent had three or more, leaving only 9 percent of households without a vehicle. The 2001 National Household Travel Survey found that the most common use of privately owned vehicles was in trips to or from work, which accounted for 27 percent of vehicle miles. Other major uses were for shopping (15 percent), other family or personal business (19 percent), visiting friends and relatives (9 percent), and other social or recreational outlets (13 percent).

According to a 2002 survey by the U.S. Department of Transportation, slightly over 11 billion tons of freight were transported in the United States. Of this, 67 percent was carried by truck, 16 percent by rail, 4 percent over water, and 6 percent by pipeline. Much of the balance was transported multimodally, with truck transport often being one of those modes.

Trucks are often grouped by **gross vehicle weight (GVW)** rating into light duty (0 to 14,000 lb), medium-duty (14,001 to 33,000 lb), and heavy-duty (over 33,000 lb) categories. They are further subdivided into classes 1 through 8, as outlined in Table 11.1.1. Examples of Classes 6 through 8 are illustrated by the silhouettes of Fig. 11.1.1.

TRACTIVE FORCE

The **tractive force** (F_{TR}) required to propel an automobile, measured at the tire-road interface, is composed of four components—the **rolling resistance** (F_R), the **aerodynamic drag** (F_D), the **inertia force** (F_I), and the **grade requirement** (F_G). That is, $F_{TR} = F_R + F_D + F_I + F_G$.

The force of rolling resistance is $C_R W$, where C_R is the rolling resistance coefficient of the tires and W is vehicle weight. For a given tire,

Table 11.1.1 Gross Vehicle Weight Classification

GVW group	Class	Weight kg (max)
6000 lb or less	1	2722
6001–10,000 lb	2	4536
10,001–14,000 lb	3	6350
14,001–16,000 lb	4	7258
16,001–19,500 lb	5	8845
19,501–26,000 lb	6	11,794
26,001–33,000 lb	7	14,469
33,001 lb and over	8	14,970

Class 8 (33,001 lb or over)

COE sleeper   Cement   Extra-heavy tandem conventional

Class 7 (26,001 to 33,000 lbs.)

High-profile COE   Trash   Medium conventional   Fuel

Class 6 (19,500 to 26,000 lbs.)

Single-axle van   Stake   COE van

Fig. 11.1.1 Medium and heavy-duty truck classification by gross weight and vehicle usage. (*Reprinted from SAE, SP-868, © 1991, Society of Automotive Engineers, Inc.*)

C_R is normally considered constant with speed, although it does rise slowly at very high speeds. In the days of bias-ply tires, it had a typical value of 0.015, but with modern radial tires, coefficients as low as 0.006 have been reported. The rolling resistance of cold tires is greater than these warmed-up values, however. On a 70°F day, it may take 20 minutes of driving for a tire to reach its equilibrium running temperature. Under-inflated tires also have higher rolling resistance. Figure 11.1.2

Fig. 11.1.2 Dependence of rolling resistance on inflation pressure for FR78-14 tire, 1280-lb load and 60-mi/h speed. (*"Tire Rolling Losses," SAE Proceedings P-74.*)

illustrates how the rolling resistance of a particular tire falls off with increasing inflation pressure.

Aerodynamic drag force is given by $C_D A \rho V^2/2g$, where C_D is the aerodynamic drag coefficient, A is the projected vehicle frontal area, ρ is the ambient air density, V is the vehicle velocity, and g is the gravitational constant. The frontal area of most cars falls within the range of 20 to 30 ft² (1.9 to 2.8 m²). The frontal areas of light trucks generally range from 27 to 38 ft² (2.5 to 3.5 m²). For heavy-duty trucks and truck-trailer combinations, frontal area can range from 50 to 100 ft² (4.6 to 9.3 m²). The drag coefficient of a car is highly dependent on its shape, as illustrated in Fig. 11.1.3. The power required to overcome aerodynamic drag is the product of the aerodynamic component of tractive force and vehicle velocity. That power is listed at various speeds in Figure 11.1.3 for a car with a frontal area of 2 m². For most contemporary cars, C_D falls between 0.30 and 0.35, although values around 0.20 have been reported for certain advanced concepts. For light trucks, C_D generally falls between 0.40 and 0.50. Drag coefficients for heavy-duty trucks and truck-trailers range from 0.6 to over 1.0, with 0.7 being a typical value for a Class 8 tractor-trailer.

	Drag coefficient	Drag power in kW, average values for $A = 2$ m² at various speeds		
		40 km/h	120 km/h	160 km/h
Open convertible	0.5 ... 0.7	1	27	63
Station wagon (2-box)	0.5 ... 0.6	0.91	24	58
Conventional form (3-box)	0.4 ... 0.55	0.78	21	50
Wedge shape, headlights & bumpers integrated in body	0.3 ... 0.4	0.58	16	37
Optimum streamlining	0.15 ... 0.20	0.29	7.8	18

Fig. 11.1.3 Drag coefficient and aerodynamic power requirements for various body shapes. *(Bosch, "Automotive Handbook," SAE.)*

The tractive force at the tire patch that is associated with changes in vehicle speed is given by $F_I = (W/g)\,a$, where a is vehicle acceleration. If the vehicle is decelerating, then a (hence F_I) is negative.

If the vehicle is operated on a grade, an additional tractive force must be supplied. Within the limits of grades normally encountered, that force is $F_G = W \tan \theta$, where θ is the angle of the grade, measured from the horizontal.

For constant-speed driving over a horizontal roadbed, F_I and F_G are both zero, and the tractive force is simply $F_{TR} = F_R + F_D$. This is termed the **road load** force. In Fig. 11.1.4, curve T shows this road-load force requirement for a car with a test weight of 4000 lb (1814 kg). Curves A and R represent the aerodynamic and rolling resistance components of the total resistance, respectively, on a level road with no wind. Curves T′ paralleling curve T show the displacement of curve T for operation on 5, 10, and 15 percent grades. Curve E shows the tractive force delivered from the engine at full throttle. The differences between the E curve and the T curves represent the tractive force available for acceleration. The intersections of the E curve with the T curves indicate maximum speed capabilities of 98, 98, and 70 mi/h on grades of 0, 5, and 10 percent. Operation on a 15 percent grade is seen to exceed the capability of the engine in this transmission gear ratio. Therefore, operation on a 15 percent grade requires a transmission downshift. Tractive forces on the ordinate of Fig. 11.1.4 can be converted to power by multiplying by the corresponding vehicle velocities on the abscissa.

Fig. 11.1.4   Traction available and traction required for a typical large automobile.

PERFORMANCE

In the United States, a commonly used performance metric is acceleration time from a standstill to 60 mi/h. A zero to 60 mi/h acceleration at wide-open throttle is seldom executed. However, a vehicle with inferior acceleration to 60 mi/h is likely to do poorly in more important performance characteristics, such as the ability to accelerate uphill at highway speeds, to carry a heavy load or pull a trailer, to pass a slow-moving truck, or to merge into freeway traffic.

The most important determinant of zero-to-60 mi/h acceleration time is vehicle weight-to-power ratio. To calculate this acceleration time rigorously, the appropriate power available from the engine at each instant, diminished for the inefficiencies of the transmission and driveline. Furthermore, the actual weight of the vehicle is replaced by its effective weight (W_{eff}), which includes the rotational inertia of the wheels, driveline, transmission, and engine. In the highest (drive) gear, W_{eff} may exceed W by about 10 percent. In lower gears (higher input/output speed ratios), the engine and parts of the transmission rotate at higher speeds for a given vehicle velocity. Because rotational energy varies as the square of angular velocity, operation in lower gears increases the effective mass of the vehicle more significantly.

The maximum acceleration capability calculated in this manner may be limited by loss of traction between the driving wheels and the road surface. This is dependent upon the coefficient of friction between tire and road. For a new tire, this coefficient can range from 0.85 on dry concrete to 0.5 if water is puddled on the road. Under the same conditions, the coefficient for a moderately worn tire can range from 1.0 down to 0.25. Using special rubber compounds, coefficients as high as 1.8 have been attained with special racing tires.

To avoid all these variables when estimating the acceleration performance of a typical light-duty vehicle under normal driving conditions on a level road, various empirical correlations have been developed by fitting test data from large fleets of production cars. Such correlations for the time from zero to 60 mi/h (t_{60}) have generally taken the form, $C(P/W)^n$. In one simple correlation, t_{60} is approximated in seconds when P is the rated engine horsepower, W is the test weight in pounds, $C = 0.07$ and $n = 1.0$. According to the U.S. Environmental Protection Agency, the fleet-average t_{60} for new passenger cars has decreased rather consistently from 14.2 s in 1975 to 10.3 s in 2001. Over that same time span, it has decreased from 13.6 to 10.6 s for light-duty trucks.

FUEL CONSUMPTION

Automotive fuels are refined from petroleum. In 1950, less than 10 percent of U.S. petroleum consumed was imported, but that fraction has grown to 52.8 percent in 2002. Much of this imported petroleum comes from politically unstable parts of the world. Therefore, it is in the national interest to reduce automotive fuel consumption, or conversely, to increase automotive fuel economy.

In the United States, passenger-car fuel-economy standards were promulgated by Congress in 1975 and took effect in 1978. Standards for light-duty trucks took effect in 1982. The fuel economies of new vehicles are monitored by the U.S. Environmental Protection Agency (EPA). Each vehicle model is evaluated by running that model over prescribed transient driving schedules of velocity versus time on a chassis dynamometer. Those driving schedules are plotted in Figs. 11.1.5 and 11.1.6. The **urban driving schedule** (Fig. 11.1.5) simulates 7.5 mi of driving at an average speed of 19.5 mi/h. It is the same schedule used in measuring emissions compliance. It requires a 12-h soak of the vehicle at room temperature, with driving initiated 20 s after the cold start. The 18 start-stop cycles of the schedule are then followed by engine shutdown and a 10-min hot soak, after which the vehicle is restarted and the first five cycles repeated. The **highway driving schedule** (Fig. 11.1.6) is free of intermediate stops, covering 10.2 mi at an average speed of 48.2 mi/h.

The **combined fuel economy** of each vehicle model is calculated by assuming that the 55 percent of the distance driven is accumulated in the urban setting and 45 percent on the highway. Consequently, the combined value is sometimes known as the **55/45 fuel economy**. It is derived from MPG$_{55/45}$ = $1/[(0.55$ MPG$_U$ + 0.45 MPG$_H)]$, where MPG represents fuel economy in mi/gal, and subscripts U and H signify the urban and highway schedules, respectively.

Each manufacturer's **CAFE (corporate average fuel economy)** is the sales-weighted average of the 55/45 fuel economies for its vehicle sales in a given year. CAFE standards are set by the U.S. Congress. From 1990–2004, the CAFE standard for passenger cars was 27.5 mi/gal. For light-duty trucks, it was 20.7 mi/gal from 1976–2004. Manufacturers failing to meet the CAFE standard in any given year are fined for their shortfall.

The 55/45 fuel economy of a vehicle depends on its tractive-force requirements, the power consumption of its accessories, the losses in its transmission and driveline, the vehicle-to-engine speed ratio, and the fuel-consumption characteristics of the engine itself. For a given level of technology, of all these variables, the most significant factor is the EPA test weight. This is the curb weight of the vehicle plus 300 lb. In a given model year, test weight correlates with vehicle size, which is loosely related for passenger cars to interior volume. The various EPA classification of vehicles are listed in Table 11.1.2, along with their fleet-average test weights and fuel economies, for the 2003 model year.

After the CAFE regulation had been put in place, the driving public complained of its inability to achieve the fuel economy on the road that had been measured in the laboratory for regulatory purposes. Over-the-road fuel consumption is affected by ambient temperature, wind, hills, and driver behavior. Jack-rabbit starts, prolonged engine idling, short trips, excessive highway speed, carrying extra loads, not anticipating stops at traffic lights and stop signs, tire underinflation and inadequate

Fig. 11.1.5   Urban driving schedule.

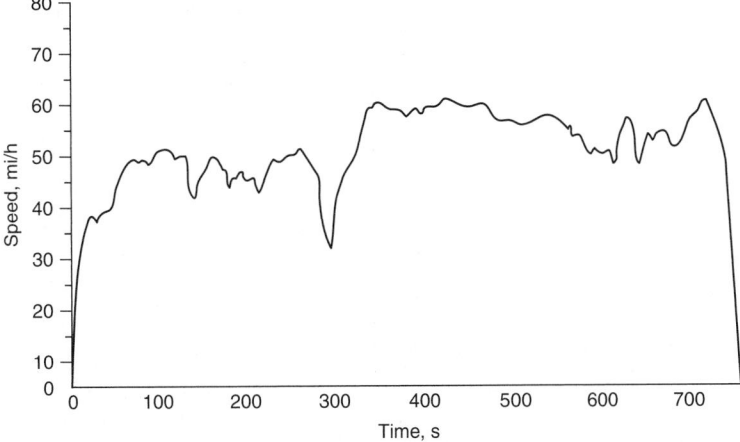

Fig. 11.1.6   Highway driving schedule.

Table 11.1.2   Characteristics of U.S. New-Car Fleet (2003 model year)

	Passenger Cars			
	Sales fraction	Volume, ft³	Test weight, lb	Adj. 55/45 mpg
Two-seater	0.010	—	3444	21.5
Minicompact	0.006	81.5	3527	22.7
Subcompact	0.033	95.5	3379	23.5
Compact	0.198	103.9	3088	27.6
Midsize	0.156	114.4	3582	23.8
Large	0.081	125.1	3857	22.1
Station wagons				
Small	0.025	116.0	3296	25.0
Midsize	0.015	133.2	3687	28.2
Large	0.001	170.2	4500	19.9

	Light trucks (gross vehicle weight rating < 8500 lb)			
	Sales fraction	Wheel base, in.	Test weight, lb	Adj. 55/45 mpg
Vans				
Small	0.000	<105	—	—
Midsize	0.073	105–115	4345	20.3
Large	0.008	>115	5527	15.4
SUVs				
Small	0.017	<109	3508	21.6
Midsize	0.128	109–124	4140	19.0
Large	0.089	>124	5383	15.8
Pickups				
Small	0.013	<100	3750	20.1
Midsize	0.028	100–110	3966	18.7
Large	0.119	>110	4978	16.1

SOURCE: "Light-Duty Automotive Technology and Fuel Economy Trends, 1975 through 2003," U.S. EPA, April 2003.

maintenance all contribute to higher fuel consumption. To compensate for the difference between the EPA 55/45 fuel economy and the over-the-road fuel economy achieved by the average driver, EPA tested a fleet of cars in typical driving. It determined that on average, the public should expect to achieve 90 percent of MPG_U in city driving and 78 percent of MPG_H on the highway. Applying these factors to the urban and highway economies measured in the laboratory results in **adjusted fuel economy** for each of the two driving schedules. These can be combined to derive an adjusted 55/45 fuel economy, as listed for 2003 in Table 11.1.2. Typically, the adjusted 55/45 fuel economy is about 15 percent less than the measured laboratory fuel economy on which CAFE regulation is based. The historical trend in adjusted 55/45 fuel economy for new cars, light-duty trucks, and the combined fleet of both are represented for each year in Fig. 11.1.7. The decline in fuel economy for the combined fleet since 1988 reflects the influence of an increasing consumer preference for less fuel-efficient vans, SUVs, and pickups instead of passenger cars.

TRANSMISSIONS

The torque developed by the engine is fed into a **transmission.** One function of the transmission is to control the ratio of engine speed to vehicle velocity, enabling improved fuel economy and/or performance. This may be done either with a **manual transmission,** which requires the driver to select that ratio with a gear-shift lever, or with **automatic transmission,** in which the transmission controls select that ratio unless overridden by the driver. Another function of the transmission is to provide a means for operating the vehicle in reverse.

Because an internal-combustion engine cannot sustain operation at zero rotational speed, a means must be provided that allows the engine to run at its idle speed while the driving wheels of the vehicle are stationary. With the manual transmission, this is accomplished with a **dry friction clutch** (Fig. 11.1.8). During normal driving, the pressure plate is held tightly against the engine flywheel by compression springs, forcing the flywheel and plate to rotate as one. At idle, depression of the clutch pedal by the driver counters the force of the springs, freeing the engine to idle while

the pressure plate and driving wheels remain stationary. The transmission gearbox on the output side of the clutch changes its output/input speed ratio, typically in from three to six discrete gear steps. As the driver shifts the transmission from one gear ratio to the next, he disengages the clutch and then slips it back into engagement as the shift is completed.

With an automatic transmission, such a dry clutch is unnecessary at idle because a fluid element between the engine and the gearbox accommodates the required speed disparity between the engine and the driving wheels. However, the **wet clutch,** cooled by transmission fluid, is used in the gearbox to ease the transition from one gear set to another during shifting. Typically, the hydraulically operated **multiple-disk clutch** is used (Fig. 11.1.9). The **band clutch** is also used in automatic transmissions.

The fluid element used between the engine and gearbox in early automatic transmissions was the **fluid coupling** (Fig. 11.1.10). It assumes the shape of a split torus, filled with transmission fluid. Each semitorus is lined with radial blades. As the engine rotates the driving member, fluid is forced outward by centrifugal force. The angular momentum of this swirling flow leaving the periphery to the driving member is then removed by the slower rotating blades of the driven member as the fluid flows inward to reenter the driving member at its inner diameter. This transmits the torque on the driving shaft to the driven shaft with an input/output torque ratio of unity, independent of input/output speed ratio. That gives the fluid coupling the torque delivery characteristic of a slipping clutch. Transmitted torque being proportional to the square of input speed, very little output torque is developed when the engine is idling. This minimizes vehicle creep at stoplights. Because torque is transmitted from the fastest to the slowest member in either direction, the engine may be used as a brake during vehicle declaration. This characteristic also permits starting the engine by pushing the car.

In modern automatic transmissions, the usual fluid element is the three-component **torque converter,** equipped with a one-way clutch (Fig. 11.1.11a). As with the fluid coupling, transmission fluid is centrifuged outward by the engine-driven pump and returned inward through a turbine connected to the output shaft delivering torque to the gearbox. However, a stator is interposed between the turbine exit and the pump inlet. The shaft supporting the stator is equipped with **sprag clutch.**

Fig. 11.1.7   Historical trends in average adjusted combined fuel economy for new-vehicle fleet. *(U.S. Environmental Protection Agency, April 2003.)*

Fig. 11.1.8   Single-plate dry-disk friction clutch.

Fig. 11.1.9   Schematic of two hydraulically operated multiple-disk clutches in an automatic transmission. *(Ford Motor Co.)*

The blading in the torque converter is nonradial, being oriented to provide torque multiplication. When the output shaft is stalled, the sprag locks the stator in place, multiplying the output torque to from 2.0 to 2.7 times the input torque (Fig. 11.1.11*b*). As the vehicle accelerates and the output/input speed ratio rises, the torque ratio falls toward 1.0. Beyond that speed ratio, known as the **coupling point,** the direction of the force imposed on the stator by the fluid reverses, the sprag allows the stator to free wheel, and the torque converter assumes the characteristic of a fluid coupling.

It is clear from Fig. 11.1.11*b* that multiplication of torque by the torque converter comes at the expense of efficiency. Therefore, most modern torque-converter transmissions incorporate a hydraulically operated **lockup clutch** capable of eliminating the speed differential between pump and turbine above a preset cruising speed, say 40 mi/h in high gear. In the lower gears, lockup is less likely to be used. When a significant increase in engine power is requested, the clutch is also disengaged to capitalize on the torque multiplication of the torque

Fig. 11.1.10   Fluid coupling.

Fig. 11.1.11   Torque converter coupling: (*a*) section; (*b*) characteristics.

converter. During its engagement, the efficiency-enhancing elimination of torque-converter slip comes with a loss of the torsional damping quality of the fluid connection. Sometimes the loss of damping is countered by allowing modest slippage in the lockup clutch.

A cross section through the gearbox of a three-speed manual transmission is shown in Fig. 11.1.12. The input and output shafts share a common axis. Below, a parallel countershaft carries several gears of differing diameter, one of which is permanently meshed with a gear on the input shaft. Additional gears of differing diameter are splined to the output shaft. Driver operation of the gear-shift lever positions these splined gears along the output shaft, determining which pair of gears on the output shaft and the countershaft is engaged. In direct drive (high gear), the output shaft is coupled directly to the input shaft. Reverse is secured by interposing an idler gear between gears on the output shaft and the countershaft. Helical gears are favored to minimize noise production. During gear changes, a **synchromesh** device, acting as a friction clutch, brings the gears to be meshed approximately to the correct speed just before their engagement. Typically, transmission gear ratios follow an approximately geometric pattern. For example, in a four-speed gearbox, the input/output speed ratios might be 2.67 in first, 1.93 in second, 1.45 in third, and 1.0 in fourth, or direct, drive.

The gearbox for a typical automatic transmission uses planetary gear sets, any one set of which can supply one or two gear-reduction ratios, and reverse, by actuating multiple-disk or band clutches that lock various elements of the planetary system in Fig. 11.1.13. These clutches are actuated hydraulically according to a built-in control schedule that utilizes such input signals as vehicle velocity and engine throttle position.

Fig. 11.1.12   Three-speed synchromesh transmission. (*Buick.*)

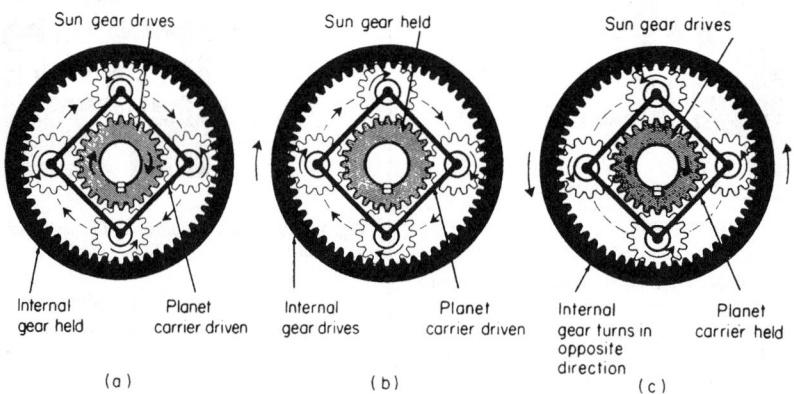

Fig. 11.1.13   Planetary gear action: (*a*) Large speed reduction: ratio = 1 + (internal gear diam.)/(sun gear diam.) = 3.33 for example shown. (*b*) Small speed reduction: ratio = 1 + (sun gear diam.)/(internal gear diam.) = 1.428. (*c*) Reverse gear ratio = (internal gear diam.)/(sun gear diam.) = −2.33.

Fig. 11.1.14   Three-element torque converter and planetary gear. (*General Motors Corp.*)

A schematic of such an automatic transmission, combining a torque converter and a compound planetary gearbox, is shown in Fig. 11.1.14.

Increasingly, **electronic transmission control** is taking over from totally hydraulic control. Although hydraulic actuation is retained, electronic modules control gear selection and modulate hydraulic pressure in accordance with torque flow, providing smoother shifts.

The influence of the transmission on the interaction between a vehicle and its roadbed is illustrated by Fig. 11.1.15, where the tractive force required to propel the vehicle is plotted against its road speed. The hyperbolas of constant horsepower superimposed on these coordinates are independent of engine characteristics. The family of curves rising from the left axis at increasing rates trace the traction requirements of the vehicles as it travels at constant speed on grades of increasing steepness. Two curves representing the engine operating at full throttle in high gear are overlaid on this basic representation of vehicle characteristics. One is for a manual transmission followed by a differential with an input/output speed ratio of 3.6. The other is for the same engine with a torque-converter transmission using a differential with a speed ratio of 3.2. With either transmission, partial closing of the engine throttle allows operation below these two curves, but operation above them is not possible. The difference in tractive force between the engine curve and the road-load curve indicates the performance potential—the excess power available for acceleration or for hill climbing on a level road. The high-speed intersection of these two curves indicates that with either transmission, the vehicle can sustain a speed of 87 mi/h on a grade of nearly 5 percent. From 87 mi/h down to 37 mi/h, the manual transmission shows slightly greater performance potential. However, at 49 mi/h the torque converter reaches its coupling point, torque

multiplication begins to occur, and below 37 mi/h its performance potential in high gear exceeds that of the manual transmission by an increasing margin. This performance disadvantage of the manual transmission may be overcome by downshifting. This transmission comparison illustrates why the driver is usually satisfied with an automatic transmission that has one fewer forward gear ratio than an equivalent manual transmission.

An evolving transmission development is the **continuously variable transmission** (CVT), which changes the engine-to-vehicle speed ratio without employing finite steps. This greater flexibility promises improved fuel economy. Its most successful implementation in passenger cars to date varies transmission input/output speed ratio by using a steel V-belt composed of individual links held together by steel bands. This belt runs between variable-geometry sheaves mounted on parallel input and output shafts. Each of these two sheaves is split in the plane of rotation such that moving its two halves apart decreased its effective diameter, and vice versa. The control system uses hydraulic pressure to move the driving sheaves apart while concurrently moving the driven sheaves together in such a way that the constant length of the steel belt is accommodated. Thus the effective input/output diameter ratio, hence speed ratio, of the sheaves can be varied over a range of about 6:1. To accommodate the infinite speed ratio required when the vehicle is stopped and the engine is idling, and to manage vehicle acceleration from a standstill, this type of CVT typically includes a torque converter or a wet multidisk clutch.

DRIVELINE

The **driveline** delivers power from the transmission output to the driving wheels of the vehicle, whether they be the rear wheels or the front. An important element of the driveline is the **differential,** which in its basic form delivers equal torque to both left and right driving wheels while allowing one wheel to rotate faster than the other, as is essential during turning. Fig. 11.1.16 is a cross section through the differential

Fig. 11.1.15   Comparative traction available in the performance of a fluid torque-converter coupling and a friction clutch.

Fig. 11.1.16   Rear-axle hypoid gearing.

Fig. 11.1.17   Rear axle. (*Oldsmobile.*)

of a rear-wheel driven vehicle as viewed in the longitudinal plane. It employs hypoid gearing, which allows the axis of the driving pinion to lie below the centerline of the differential gear. This facilitates lowering the intruding hump in the floor of a passenger car, which is needed to accommodate the longitudinal drive shaft. Figure 11.1.17 is a cross-sectional view of the differential from above, showing its input connection to the drive shaft, and its output connection to one of the driving wheels. The driving pinion delivers transmission-output torque to a ring gear, which is integral with a differential carrier and rotates around a transverse axis. A pair of beveled pinion gears is mounted on the pinion pin, which serves as a pinion shaft and is fixed to the carrier. The pinion gears mesh with a pair of beveled side gears splined to the two rear axles. In straight-ahead driving, the pinion gears do not rotate on the pinion gear shaft. In a turn, they rotate in opposite directions, thus accommodating the difference in speeds between the two rear wheels.

If one of the driving wheels slips on ice, the torque transmitted to that wheel falls essentially to zero. In the basic differential, the torque on the opposite wheel then falls to the same level and the tractive force is insufficient to move the vehicle. Various **limited-slip differentials** have been devised to counter this problem. In one type, disk clutches are introduced into the differential. When one wheel begins to lose traction, energizing the clutches transmits power to the wheel with the best traction.

Figure 11.1.17 illustrates a **semifloating rear axle.** This axle is supported by bearings at each of its ends. The vehicle weight supported by the rear wheel is imposed on the axle shaft and transmitted to the axle housing through the outboard wheel bearing between the shaft and that housing. In contrast, most commercial vehicles use a **full-floating rear axle.** With this arrangement the load on the rear wheel is transmitted directly to the axle housing through a bearing between it and the wheel itself. Thus freed of vehicle support, the full-floating axle can be withdrawn from the axle housing without disturbing the wheel. In both types of rear axle, of course, the axle shaft transmits the driving torque to the wheels.

With **rear-wheel drive,** a driveshaft (or **propeller shaft**) connects the transmission output to the differential input. Misalignment between the latter two shafts is accommodated through a pair of **universal joints** (U joints) of the Cardan (or Hooke) type, one located at either end of the driveshaft. With this type of universal joint, the driving and the driven shaft each terminate in a slingshot-like fork. The planes of the forks are oriented perpendicular to one another. Each fork is pinned to the perpendicular arms of a cruciform connector in such a way that rotary motion can be transmitted from the driving to the driven shaft, even when they are misaligned. Vehicles with a long wheel base may use a two-piece driveshaft. Then a third U joint is required, and an additional bearing is placed near that third U joint to support the driveshaft.

With **front-wheel drive,** the traditional driveshaft becomes unnecessary as the engine is coupled directly to a **transaxle,** which combines the transmission and the differential in a single unit that is typically housed in a single die-cast aluminum housing. A cutaway view of a transaxle with an automatic transmission appears in Fig. 11.1.18. Output from a transversely mounted engine is delivered to the torque converter, which shares a common centerline with the gearbox and its various gears and clutches. The gearbox output is delivered to the transfer-shaft gear, which rotates about a parallel axis. Its shaft contains the parking sprag. The gear at the opposite end of this shaft feeds torque into the differential, which operates around a third rotational axis, as seen at the bottom of Fig. 11.1.19. The axle shafts driving the front wheels may be of unequal length. A **constant-velocity (CV) joint** is located at each end of each front-drive axle. During driving, the front axles may be required to flex through angles up to 20° while transmitting substantial torque—a duty too severe for the type of U joint commonly used on the driveshaft of a rear-wheel-drive vehicle. Two types of CV joints are in common use. In the **Rzeppa joint,** the input and output shafts are connected through a balls-in-grooves arrangement. In the **tripod joint,** the connection is through a rollers-in-grooves arrangement.

To enhance operation on roads with a low-friction-coefficient surface, various means for delivering torque to all four wheels have been devised. In the **four-wheel drive** system, two-wheel drive is converted to four-wheel drive by making manual adjustments to the driveline,

Fig. 11.1.18   Cross-section view of a typical transaxle. (*Chrysler Corp.*)

Fig. 11.1.19 Automatic transaxle used on a front-wheel-drive car. (*Chrysler Corp.*)

normally with the vehicle at rest. In **all-wheel drive** systems, torque is always delivered to all four wheels, by a system capable of distributing power to the front and rear wheels at varying optimal ratios according to driving conditions and the critical limits of vehicle dynamics. In vehicles capable of driving through all four wheels, a **transfer case** receives the transmission output torque and distributes it through drive shafts and differentials to the front and rear axles.

SUSPENSIONS

The suspension system improves ride quality and handling in driving on irregular roadbeds and during maneuvers like turning and braking. The vehicle body is supported on the front and rear **springs,** which are important elements in the suspension system. Three types used are the **coil spring,** the **leaf spring,** and the **torsion bar.**

Also important are the **shock absorbers.** These piston-in-tube devices dampen spring oscillation, road shocks, and vibration. As the hydraulic shock-absorber strut telescopes, it forces fluid through a restrictive orifice, thus damping its response to deflections imposed by the springs.

Especially in the front suspension, most vehicles incorporate an **antiroll bar** (sway bar, stabilizer). This is a U-shaped bar mounted transversely across the vehicle, each end attached to an axle. During high-speed turns, torsion developed in the bar reduces the tendency of the vehicle to lean to one side, thus improving handling.

A wide variety of suspension systems have seen service.

Front Suspension All passenger cars use independent front-wheel suspension, which allows one wheel to move vertically without disturbing the opposite wheel. Rear-wheel drive cars typically use the **short-and-long-arm (SLA) system** illustrated in Fig. 11.1.20. The **steering knuckle** is held between the outboard ends of the upper and lower **control arms** (wishbones) by spherical joints. The upper control arm is always shorter than the lower. The load on these control arms is generally taken by coil springs acting on the lower arm, or by torsion-bar springs mounted longitudinally.

Figure 11.1.21 shows a representative application of the **spring strut (MacPherson) system** for the front suspension of front-wheel drive cars. The lower end of the shock absorber (strut) is mounted within the coil spring and connected to the steering knuckle. The upper end is anchored to the body structure.

Rear Suspensions The trailing-arm rear suspension illustrated in Fig. 11.1.22 is used on many passenger cars. Each longitudinal trailing arm pivots at its forward end. Motion of its opposite end is constrained by a coil spring. One or more links may be added to enhance stability.

Using leaf springs in place of coil springs is common practice on trucks. The Hotchkiss drive, used with a solid rear axle, employs longitudinal leaf springs. They are connected to the chassis at their front and rear ends, with the rear axle attached near their midpoints.

Fig. 11.1.20 Front-wheel suspension for rear-wheel-drive car.

Fig. 11.1.21 Typical MacPherson strut, coil spring front suspension for front-wheel-drive car. (*Chrysler Corp.*)

Fig. 11.1.22 Trailing-arm type rear suspension with coil springs, used on some front-wheel-drive cars. (*Chrysler Corp.*)

Active Suspension A variety of improvements to the traditional passive suspension system have been introduced. Generally they require some form of input energy. The air suspension widely used in heavy-duty trucks, for example, requires an air compressor. To an increasing degree, advanced systems employ sensors and electronic controls. Some of the improvements available with active suspension are improved ride quality (adaptation to road roughness), control of vehicle height (low for reduced aerodynamic drag on highways; high for off-road driving), and vehicle leveling (to correct change in pitch when the trunk is heavily loaded; avoid pitching during braking and heavy accelerations).

STEERING

When turning a corner, each wheel traces an arc of different radius, as illustrated in Fig. 11.1.23. To avoid tire scrubbing when turning a corner of radius R, the point of intersection M for the projected front-wheel axes must fall in a vertical plane through the center of the rear axle. This requires each front wheel to turn through a different angle. For small turning angles of the vehicle, α and β can be shown to be simple functions of W, L, and R. A steering geometry conforming to Fig. 11.1.23 is known as **Ackerman geometry,** and L/R defines the **Ackerman angle.** The difference in front-wheel angles, $\beta - \alpha$, varies with turning radius R. Having the α and β corresponding to Ackerman geometry improves handling in low-speed driving, as in parking lots.

At higher speeds, dynamic effects become significant. In a turn, the vehicle may encounter either **understeer** or **oversteer.** With understeer, the vehicle fails to turn as sharply as the position of the steering wheel suggests it should. With oversteer, the vehicle turns more sharply than intended, and in extreme cases may spin around. These tendencies are amplified by increased speed and are affected by many variables, e.g., tire pressure, weight distribution, design of the suspension system, and the application of brakes.

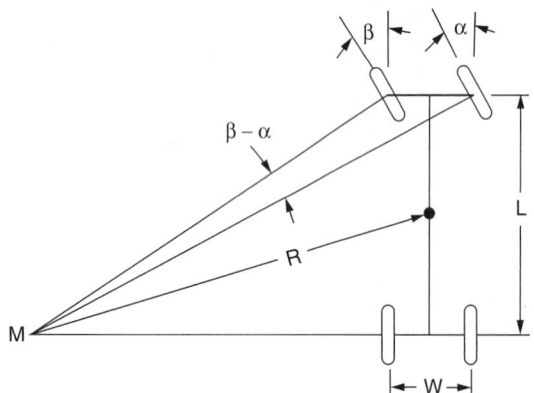

Fig. 11.1.23 Ackerman steering geometry.

The physical implementation of a front-wheel steering system is shown schematically from above in Fig. 11.1.24. The front wheels are steered around the axes of their respective **kingpins.** For improved handling, the axes of the two kingpins are not vertical. In the transverse plane they lean inward, and in the longitudinal plane they lean rearward. The **steering arms** are the lever arms for turning the front wheels. Each **tie rod** is pinned at one end to a steering arm and at its opposite end to some type of motion transformer, the purpose of which is to convert the rotational input of the **steering column** into translational motion at the inner ends of the tie rods.

Fig. 11.1.24 Schematic of front-wheel steering system.

A common motion transformer is the **rack and pinion,** in which a **pinion gear** attached to the steering column engages a geared bar, or **rack,** causing it to slide to the left or right as the **steering wheel** is turned. The ends of the rack are attached to the inner ends of the tie rods. In the variation of the rack-and-pinion illustrated in Fig. 11.1.25a, the rotation of the pinion on the end of the steering shaft causes a rotatable **worm gear** to control the tie-rod positions as it moves left and right like the rack.

With the **rotatable ball** system of Fig. 11.1.25b, a worm is mounted on the end of the steering column. Its spiral grooves contain ball bearings. The outer surfaces of the balls rotate in spiral grooves lining the inside of a **ball nut.** The outer surface of the ball nut contains gear teeth that mesh with a **gear sector** on the **pitman arm.** The end of pitman arm is connected to the tie rods through a series of links.

(a)   (b)

Fig. 11.1.25 Two commonly used types of steering gear. (*a*) Variation of rack-and-pinion using rotating worm in place of translating rack; (*b*) recirculating ball.

With **variable-ratio steering,** the response of the front wheels to steering-wheel rotation is increased as the steering wheel approaches the extremes of its travel. This is especially helpful in curbside parallel parking. Variable-ratio steering can be achieved by nonuniformly spacing the worm grooves in Fig. 11.1.25a, or by varying the lengths of the teeth on the sector of the pitman arm in Fig. 11.1.25b.

Most modern U.S. cars have **power-assisted steering.** A pressurized hydraulic fluid is used to augment the steering effort input by the driver, but if the engine is off or if there is a failure in the power-steering system, the driver retains steering control, albeit with greater effort.

The power-steering circuit contains an engine-driven hydraulic **pump,** a fluid reservoir, a hose assembly, return lines, and a **control valve.** One of a variety of pumps that has been used is the rotary-vane pump of Fig. 11.1.26. As the rotor is turned by a belt drive from the engine, the radially slidable vanes are held against the inner surface of the eccentric case by centrifugal force, springs, and/or hydraulic pressure. Not shown are the inlet and exit ports, and a pressure-relief valve.

A balanced-spool control valve is represented in Fig. 11.1.27. In straight-ahead driving, centering springs hold the spool in the neutral position. In the position illustrated, however, the driver has turned the steering wheel to position the spool valve at the right extremity of its travel.

Fig. 11.1.26 Rotary power-steering pump. (*General Motors, Saginaw.*)

Fig. 11.1.27 Control valve positioned for full-turn power steering assistance using maximum pump pressure.

That directs all of the pump discharge flow to the left side of a piston, moving the piston rod to the right. The piston illustrated can, for example, be incorporated into the rack of a rack-and-pinion steering system to add to the turning effort put into the steering wheel by the driver.

Some cars now use electric power steering. Both the need for hydraulic fluid and the equipment pictured in Fig. 11.1.27 are eliminated. They are replaced by a variable-speed electric motor, and electronic sensors that match the appropriate degree of assistance to driver demand and road resistance. Compared to hydraulic power steering, electric power steering effects a small improvement in fuel economy by reducing the parasitic load on the engine.

BRAKING SYSTEMS

Brakes The moving automobile is stopped with hydraulic **service brakes** on all four wheels. The service brakes are engaged by depressing and holding down a self-returning foot pedal. The mechanical **parking brake**, used to prevent the vehicle from rolling when parked, acts on the rear wheels. It is engaged with an auxiliary foot pedal or a hand pull and is held by a ratchet until released. **Drum brakes**, once used extensively on all four wheels, are now used primarily on rear wheels. Typically, **disk brakes** are used on the front wheels, but sometimes on the rear wheels.

A **Bendix dual-servo** drum brake is illustrated in Fig. 11.1.28c. It is **self-energizing** because the rotating drum tries to drag the brake lining and shoe counter-clockwise, along with it. Because the anchor pin prevents this for the secondary shoe, and the adjusting screw prevents it for the primary shoe, both are called **leading shoes**. This causes both shoes to push more strongly against the drum, increasing the friction force. This self-energization is lost when the wheel turns in the reverse direction. With this brake configuration, typically a linkage turns a notched

wheel on the adjusting screw when the brake is applied with the car moving in reverse. This automatically adjusts for lining wear.

In the arrangement of Fig. 11.1.28a, the anchor pin has been moved 180°. The leading shoe continues to be self-energized, but that quality is lost on the **trailing (following) shoe**. In this arrangement the leading shoe provides most of the braking torque and experiences most of the wear when traveling forward. In reverse, however, the trailing shoe in forward travel becomes the leading shoe and provides most of the braking torque. This arrangement is used on the rear wheels of some front-wheel drive cars.

The **Caliper disk brake** is illustrated in Fig. 11.1.29. The disk rotates with the wheel. The fixed caliper straddles the disk. Within the caliper, the pistons in opposing hydraulic cylinders press the attached pads against opposite sides of the disk when actuated.

Fig. 11.1.29 Schematic of caliper disk brake.

As friction heats the brake lining and the surface on which it rubs during stopping, the coefficient of friction often falls. This undesirable characteristic causes what is known as **brake fade**. Because brakes are normally applied only intermittently, the time available for cool down between successive brake applications keeps brake fade at a tolerable level. However, closely spaced panic stops from high speeds can cause objectionable fade. The "disk" illustrated in Fig. 11.1.29 often consists of two parallel disks that are joined together by radial vanes. The ability of air to access the nonfriction side of each of these disks, and the air pumped between the disks by the vanes when the wheel rotates, help to avoid brake overheating. Brake fade is more probable with drum brakes because the more enclosed nature of their construction makes brake cooling more difficult.

Many modern disk brakes use a single-cylinder caliper. One cylinder and its actuating piston and attached pad are retained on the inboard side. However, the outboard cylinder is eliminated, the pad on that side instead being fastened directly to the inner surface of the caliper. Instead of being fixed, the caliper is allowed to slide inboard on mounting bolts when the single piston is actuated. Transverse movement of the caliper on its mounting bolts then causes each pad to rub on its adjacent disk surface.

Because disk brakes are not self-energizing, and the rubbing area of the pads is less than the rubbing area of the shoes on an equivalent drum

Fig. 11.1.28 Three types of internal-expanding brakes.

brake, a higher hydraulic pressure is required with disk brakes than with drums. If disk brakes are used on the front wheels and drums on the rear, then provisions are made in the upstream hydraulic system to provide that difference.

Cast iron is the preferred material for the rubbing surface on both rotors and drums. The **brake lining** on pad or shoe may contain high-temperature resins, metal powder, solid lubricant, abrasives, and fibers of metal, plastic, and ceramic. Friction coefficients from 0.35 to 0.50 are typical with these combination. The asbestos once common in brake linings has been abandoned because of the adverse health effect of asbestos wear particles.

Hydraulics   When the brake pedal is depressed, it moves pistons in the master cylinder, as illustrated in Fig. 11.1.30. This forces pressurized brake fluid through lines to the wheel cylinders for brake application. The magnitude of the fluid pressure is proportional to the magnitude of the force applied to the brake pedal. As is typical, this dual master cylinder contains two pistons in tandem (Fig. 11.1.31), providing a split system. One operates the rear brakes and the other the front brakes, providing the ability to brake the vehicle even if one of the two systems should fail. Late model cars often use a diagonal-split arrangement, where each piston in the dual master cylinder serves diagonally opposite wheels.

Fig. 11.1.30   "Split" hydraulic brake system.

Fig. 11.1.31   Typical design of a dual master cylinder for a split brake system.

When the brake pedal is released, a spring in the master cylinder returns the pistons to their original positions. This uncovers ports in the master cylinder, allowing communication with the brake-fluid reservoirs on the master cylinder and relieving the pressure in the wheel cylinders. A check valve maintains a minimum pressure in the system to prevent the entrance of air. Air in the system makes the brakes feel spongy, and, if it gains entrance, must be bled from the system.

A combination valve is usually included in vehicles using disks in the front and drums in the rear. Among its functions, the **proportioning valve** regulates pressure individually to the front and rear wheels to compensate for the different pressure requirements of disks and drums, and the **pressure differential valve** checks to ensure that both sides of a diagonally split system experience the same pressures.

The conventional cast-iron master cylinder with integral reservoir has been largely replaced by the aluminum master cylinder with translucent plastic reservoirs for visual inspection of brake-fluid level. The specially formulated brake fluid must be free of petroleum-based oil, which would cause swelling of rubber parts in the braking system. Double-walled steel lines are used to distribute the brake fluid. Near the wheels, these rigid lines are replaced by flexible hydraulic hoses to accommodate wheel movement.

Power Brakes   Power brakes relieve the driver of much physical effort in retarding or stopping a car. Several types have been used, but on passenger cars, the **vacuum-suspended type** illustrated in Fig. 11.1.32 is the most common. The power section contains a large diaphragm. In the unapplied condition, the chambers on each side of the diaphragm are exposed to intake-manifold vacuum. When the brake pedal is depressed, first the vacuum passage is closed and then, with further depression, air is admitted to one side of the diaphragm. The pressure difference across the diaphragm, acting over its area, creates an axial force that is applied to the piston in the master cylinder. This augments the force provided directly by the driver's foot. Without requiring an additional reservoir, the vacuum captured allows at least one stop after the engine operation has terminated.

Fig. 11.1.32   Power-assisted brake installation.

Another type of power brake used a **hydraulic booster**. The accumulator, a reservoir of pressurized fluid provided by the power-steering pump, provides the force that moves the piston in the master cylinder.

Heavy-duty tractor-trailer combinations require braking forces far in excess of what can be supplied from a simple foot pedal. **Air brakes** are common in such applications. Air supplied from an engine-driven air compressor is stored in a high-pressure tank, from which the air is withdrawn for brake augmentation as needed.

Retarders   Using traditional wheel brakes to control the road speed of a heavily laden tractor-trailer rig on long downhill runs promotes brake fade and accelerates brake-lining wear. Several types of retarders are available that decelerate the vehicle independent of wheel brakes.

With the **compression-release braking system** (often called the **Jake brake**), fuel injection in the diesel engine is terminated and engine valve timing is temporarily changed so that the exhaust valve opens near the end of the compression stroke. This releases the air compressed during compression directly into the exhaust system and imposes the negative work of compression on the crankshaft. A disadvantage of this system is that the sudden intermittent release of high-pressure air from the cylinder creates considerable noise.

The **hydrodynamic retarder** consists essentially of a centrifugal fluid pump that is run by the driveshaft. The high tangential velocity of the discharge fluids is removed by stator vanes as the fluid is returned to the pump inlet. Because such a pump consumes little torque at low rotational speeds, a booster retarder that is geared to rotate at a higher speed may be added in parallel. Since the braking energy generated appears as heat in the recirculating fluid, adequate cooling must be provided for it.

The **electrodynamic retarder** resembles an eddy-current dynamometer. Voltage applied from the battery or the alternator to the field coils controls braking torque. The electrodynamic retarder offers greater braking torque at low speeds than the hydrodynamic retarder. However, its braking ability falls off significantly as it is heated by the energy absorbed, and heat rejection is problematic.

Antilock Brakes   At a steady forward speed, vehicle velocity equals product of the rolling radius of the tire and its angular velocity. When the brakes are applied, the vehicle velocity exceeds that product by a difference known as **slip**. As slip in any wheel increases, its braking force rises to a peak, then slowly decreases until the wheel locks up and the tire merely slides over the road surface. At lockup, the brakes have not only lost much of their effectiveness, but steering control is also sacrificed. The **antilock braking system** (ABS) incorporates wheel-speed sensors, an electronic control unit (ECU), and solenoids to release and reapply pressure to each brake. When the ECU determines from a wheel-speed signal that lockup is imminent, it signals the corresponding solenoid to release hydraulic pressure in that brake line. Once that wheel has accelerated again, the solenoid reapplies its brake-line pressure. This sequence is repeated in each wheel at a high frequency, thus avoiding lockup and generally shortening stopping distance during severe braking or on slippery roads.

Traction Control   The **traction control system** (TCS) helps maintain traction and directional stability, especially on slippery roads. It uses many of the same components as ABS. If one driven wheel spins, the ECU applies the brake on that wheel until traction is regained. If both driven wheels spin, the ECU temporarily overrides the command signal from the accelerator pedal to reduce engine output torque.

Electronic Stability Control   The step beyond ABS and TCS is electronic stability control, which helps the driver to maintain steering during severe vehicle maneuvers. By adding sensor measurements of steering-wheel angle and vehicle yaw rate to the ECU input, it is possible to reduce the incidence of inadvertent departure of the vehicle from the road, or of vehicle rollover. This is accomplished by controlling the wheel brakes as with TCS, and in addition, by temporarily reducing engine torque.

Stopping Distance   The service brakes are the most important element in arresting a moving vehicle, but they derive some help from other sources. Aerodynamic drag provides a retarding force. Because it varies with the square of speed, it is most influential at highway speeds. The engine, when it is driven by the inertia of the vehicle during braking, and transmission and driveline losses, all contribute resistance to forward motion. On the other hand, during forward motion the inertia of these components, as well as that of the wheels, all oppose the action of the brakes. The braking force from the motoring engine can be especially significant if it is joined to the driven wheels by a manual transmission. The force can be amplified by downshifting that transmission to increase the speed of the crankshaft. With the conventional automatic transmission, however, engine braking is minimal because the torque converter is relatively ineffective in transmitting torque in the reverse direction. A small retarding force is also derived from the rolling resistance of the tires.

In braking, the sequence of events involves, successively, (1) the driver reaction time t_r, during which the driver responds to an external signal by engaging the brake pedal, (2) a response time from pedal engagement to full pedal depression, (3) a buildup time, during which brake line pressure increases to develop a substantial braking force, and (4) a prolonged active braking time. In estimating minimum stopping distance from initial vehicle velocity V, the vehicle may be assumed to move at its original speed during the reaction time, response time, and half of the buildup time. The sum of the latter two is the delay time t_d. During the second half of the buildup time and the active braking time, the braking force depends on μ, the average coefficient of friction between tire and roadbed. Then the stopping distance is $S = V(t_r + t_d) + V^2/2\mu g$, where g is the gravitational constant. The reaction time typically consumes 1 ± 0.7 s. For brakes in good condition, the delay time may range from 0.35 s for passenger cars to 0.55 s for large vehicles. The static friction coefficient depends on speed and the condition of tires and road. For new tires on dry paved roads, it is in the range of 0.75 to 0.85, increasing about 20 percent with wear. On moderately wet pavement the coefficient for new tires may fall to the 0.55 to 0.65 range. On wet pavement, the coefficient falls even more, especially at high speeds. The sliding friction coefficient, applicable when the wheels are locked and skid over the roadbed, is less than the static coefficient.

WHEELS AND TIRES

The total weight of a vehicle is the sum of the **sprung weight** supported by the tires and suspension system, and the **unsprung weight**. Wheels and tires are primary contributors to unsprung weight. For a given vehicle, the lower the unsprung weight, the better the acceleration performance, and the ride and handling qualities on rough roads. Because of cost considerations, most automotive wheels are fabricated from sheet steel. Special lower weight wheels have been forged or cast in aluminum, however, and even made of magnesium.

Maintenance of proper front-wheel alignment is important for optimum vehicle performance and handling. Normally, proper **camber, caster**, and **toe-in** are specified by the manufacturer.

Camber is the tilt angle of the wheel plane with respect to the longitudinal plane of the vehicle. When the plane of the wheel is perpendicular to the ground, the camber angle is 0°. During operation, that angle is affected by the changing geometry of the suspension system as the vehicle is loaded or traverses a bump in the road, or as the body rolls during cornering. Taking such factors into account, the manufacturer specifies the proper camber angle for each vehicle design when the vehicle is at rest. With positive camber, the distance between wheels is greater at their tops than between their bottoms, where they contact the roadbed. Normally, a zero to slightly negative camber is specified.

The **caster** angle is the angle between the kingpin axis (axis around which the wheel is steered) and a radial line from the center of the wheel to the ground, when both are projected onto a longitudinal plane. When the caster is positive, that radial line intersects the ground behind the intersection of the kingpin axis with the ground. Increasing the caster angle increases the tendency of a turned wheel to restore itself to the straight-ahead position as the vehicle moves forward. This self-restoring tendency can also be influenced somewhat by the inclination of the kingpin in the transverse plane.

When viewing an unloaded vehicle at rest from above, the front wheels may be set either to **toe in** or **toe out** slightly to compensate for the effects of driving forces and kingpin inclination. This adjustment influences the tendency of the wheel to shimmy. Toe-in is typical for rear-wheel-drive cars, while the front wheels of front-wheel-drive cars may be set to toe out.

The automotive pneumatic **tire** performs four **main functions**: supporting a moving load; generating steering forces; transmitting driving and braking forces; and providing isolation from road irregularities by acting as a spring in the total suspension system. A tire made up of two basic parts: the tread, or road-contacting part, which must provide traction and resist wear and abrasion; and the body, consisting of rubberized fabric that gives the tire strength and flexibility.

The three **principal types** of automobile and truck tires are the cross-ply or bias-ply, the radial-ply, and the bias-belted (Fig. 11.1.33). In the **bias-ply** design, the cords in each layer of fabric run at an angle from

Fig. 11.1.33   Three types of tire construction.

one bead (rim edge) to the opposite bead. To balance the tire strength symmetrically across the tread center, another layer is added at an opposing angle of 30 to 38°, producing a two-ply tire. The addition of two criss-crossed plies makes a four-ply tire.

In the **radial tire** (used on most automobiles), the cords run straight across from bead to bead. The second layer of cords runs the same way; the plies do not cross over one another. Reinforcing belts of two or more layers are laid circumferentially around the tire between the radial body piles and the tread. The **bias-belted** tire construction is similar to that of the conventional bias-ply tire, but it has circumferential belts like those in the radial tire.

Of the three types, the radial-ply offers the longest tread life, the best traction, the coolest running, the highest gasoline mileage, and the greatest resistance to road hazards. The bias-ply provides a softer, quieter ride and is least expensive. The bias-belted design is intermediate between the good quality bias-ply and the radial tire. It has a longer tread life, and it gives a smoother low-speed ride than the radial tire.

Tire specifications are molded into the sidewall. Consider, for example, the specification "P225/60TR16." The initial P designates "passenger car" (LT = light truck, ST = trailer). Next, 225 is the tire width in mm. Following the slash, 60 is the aspect ratio, i.e., the ratio of sidewall height, from bead to tread, to tire width, expressed as a percentage. T indicates a maximum speed rating of 190 km/h. Speed designators progress from S = 180 km/h through T (190 km/h), U (200 km/h), H (210 km/h), V (240 km/h) to Z (>240 km/h). On some tires the speed rating is omitted. R indicates radial construction (D = bias ply; B = bias belted). Finally, 16 indicates a wheel with a 16-in. rim diameter. Alternatively, such a tire could be labeled "P225/60R 89T," where the speed rating comes last and is preceded by a numerical load index indicating carrying capacity.

For passenger cars, the recommended **tire pressure** is posted on the rear edge of the driver's door. **Underinflation** causes excessive tire flexing and heat generation in the tire, more rapid wear, poor handling, and poor fuel economy. **Overinflation** causes a rough ride, unusual wear, poor handling, and increased susceptibility to tire damage from road hazards.

As a safety measure, a dashboard signal of tire under-inflation can be provided for the driver. Early versions of such a system made use of the wheel-speed signal necessary for traction control. Any wheel that consistently rotates at a higher speed than the rest suggests a reduced running radius, as would occur with an underinflated tire. Later versions measure the pressure of individual tires with wheel-mounted sensors, electronically transmitting their signals to an on-board receiver for dashboard display.

An out-of-balance wheel causes uneven tire wear and adversely affects ride quality. Wheel-tire assemblies are balanced individually by attaching balance weights to the rim of the wheel. For **static balance**, a proper weight on the outside rim of the wheel suffices. For the **dynamic balance** required at high speeds, weights are necessary on both the outside and inside rims. Dynamic balancing necessitates spinning the wheel to determine correct placement of the balance weights.

The trunk space occupied by a fifth wheel on which a **spare tire** is mounted to replace a flat tire has long bothered consumers and manufacturers alike. To regain part of that space, some cars have used a collapsible tire that must be inflated when pressed into service, using either a compressed-air storage reservoir or an onboard air compressor. Alternatively, some cars store a smaller-than-standard compact spare tire that has been preinflated to a high pressure. The run-flat tire is another option. It eliminates the need to carry a jack. It may rely upon an extra stiff sidewall to carry the load if the tire is punctured, or upon a plastic donut built into the tire. Such run-flat tires eliminate not only the storage-space requirement for carrying any spare, but also the need to carry a jack. When used, all of these alternatives to carrying a fifth wheel/tire combination entail some sacrifice in ride quality, and a limitation on allowable travel distance and speed.

HEATING, VENTILATING, AND AIR CONDITIONING (HVAC)

The **heating, ventilation, and air conditioning** (HVAC) system is installed to provide a comfortable environment within the passenger compartment. Fresh air for HVAC usually enters through an opening in front of the windshield and into a plenum that separates rain from the ambient air. The airflow rate is augmented by a variable-speed electrically driven blower.

In cold weather, heater air is passed through a finned heat-exchanger core, where it is heated by the engine coolant. Compartment temperature is controlled by (1) mixing ambient air with heated air, (2) mixing heated air with recirculated air, or (3) changing blower speed. Provision is made to direct heated air against the interior surface of the windshield to prevent the formation of ice or fog. Figure 11.1.34 schematically illustrates a system in which a multispeed blower drives fresh air through (1) a radiator core or (2) a bypass. The flow rate and direction of the air entering the passenger compartment are controlled with manually operated doors and vanes in the discharge ducts.

Fig. 11.1.34 Heater airflow diagram.

For cooling in hot weather, an electric clutch is engaged to operate the air-conditioning compressor, which is belt-driven from the engine (Fig. 11.1.35). The pressurized, hot gaseous refrigerant exiting the

Fig. 11.1.35 Schematic of air conditioning system using HFC-134a refrigerant. *(Reprinted with permission from SAE, SP-916, © 1992, Society of Automotive Engineers, Inc.)*

compressor enters the condenser, a heat exchanger typically located in front of the engine radiator. Ambient air passed through this heat exchanger by a fan cools the refrigerant, liquefying it. The high-pressure liquid then passes through a liquid reservoir and on to an expansion valve that drops the refrigerant pressure as it sprays the refrigerant into the evaporator. The evaporator is a heat exchanger normally mounted on the firewall. Compartment air circulated through the evaporator rejects its heat to the colder refrigerant, thus cooling as it vaporizes the refrigerant in preparation for reentry of the refrigerant into the compressor. Moisture condensed from the ambient air during cooling is leaked onto the roadbed.

Figure 11.1.36 shows schematically a **combined air-heating and air-cooling system**. Various dampers control the proportions of fresh and recirculated air to the heater and evaporator cores. Compartment-air temperature is controlled electronically by a thermostat that modulates the mixing of cooled and heated air, and cycles the compressor on and off through a magnetic clutch. During defrosting, mixing cooled air in with heated air helps to remove humidity from the air directed to the interior of the windshield.

Fig. 11.1.37   Separate body and frame design.

Fig. 11.1.36   Combined heater and air conditioner. *(Chevrolet.)*

Fig. 11.1.38   Unit-construction design for automobiles.

For many years the refrigerant in automotive air conditioners was CFC-12, a chlorinated fluorocarbon also known as F-12. About 1990 the depletion of stratospheric ozone, which shields humankind from harmful solar ultraviolet radiation, became an environmental concern. The chlorine in CFC-12 having been identified as a contributor to the problem, the use of CFC-12 was banned worldwide. About the same time, concern mounted about the threat of global warming from the anthropogenic production of greenhouse gases. Carbon dioxide from the combustion of fossil fuels is a major contributor. It was estimated that on a mass basis, CFC-12 was about 10,000 times as serious a greenhouse gas as carbon dioxide. In the United States, an immediate response to the ozone depletion concern was to require mechanics to capture spent CFC-12 for recycling, rather than venting it to the atmosphere, every time an air conditioner was serviced. Beyond that, within a few years, CFC-12 was replaced by chlorine-free HFC-134a, which has a global warming effect about one-eighth that of CFC-12. Because HFC-134a is still 1300 times as serious a greenhouse gas as carbon dioxide, a replacement for HFC-134a is being sought.

AUTOMOBILE CONSTRUCTION

Light-duty vehicles include passenger cars and light-duty trucks. Light-duty trucks include passenger vans, sports utility vehicles (SUVs), and pickup trucks. The **layout of light-duty vehicles** generally falls into one of four categories: (1) front engine, rear-wheel drive—used mostly in large cars and light-duty trucks, (2) front engine, front-wheel drive—used mostly in small cars and passenger vans, (3) rear or midengine, rear-wheel drive—sometimes used in sports cars, and (4) four-wheel drive—used mostly in light-duty trucks and premium large cars.

The two most common basic body constructions, illustrated in Fig. 11.1.37 and Fig. 11.1.38, are **body-and-frame** and **unit construction**. The body-and-frame design uses a body that is bolted to a separate frame. Most of the suspension, bumper, and brake loads are transmitted to the frame. Body-and-frame construction dominates in pickups and SUVs. With unit construction, most common in passenger cars, the frame is eliminated. Its function is assumed by a framework consisting of stamped sheet-metal sections joined together by welding. Outer panels are then fastened to this framework, primarily by bolting.

A third construction option that has been borrowed from aircraft practice and used to a limited extent is the **space frame** and **monocoque** combination. The space frame is a skeleton to which the body panels are attached, but the body assembly is designed in a manner that causes it to contribute significantly to the strength of the vehicle.

An important aspect of auto design is passenger protection. Crumple zones are designed into the structure, front and rear, to absorb collision impacts. Beams may be built into doors to protect against side impact.

The 1990s drive for improved passenger-car fuel economy, hence smaller and lighter cars, and for improved corrosion resistance has led to a gradual shift in material usage. This trend is evident in Table 11.1.3. The weight percentage of ferrous content has declined consistently, but remains over half the total weight of the average car. The shift from conventional to high-strength steel is evident. In areas likely to experience rust, the steel may be galvanized. The percentage use of lighter aluminum has doubled. The use of plastics and composites, while still less than 10 percent of the car weight, has increased by 50 percent on a mass basis.

The interest in recycling materials from scrapped vehicles has also increased. In the mid-1990s, 94 percent of scrapped vehicles were processed for recycling. Approximately 75 percent of vehicle content was recycled. Table 11.1.4 indicates the disposition of materials from recycled vehicles.

Table 11.1.3 Average Materials Consumption for a Domestic Automobile

Material	1978 Pounds	1978 Percent	1985 Pounds	1985 Percent	2001 Pounds	2001 Percent
Ferrous materials						
Conventional steel	1880.0	53.8%	1481.5	46.5%	1349.0	40.8%
High-strength steel	127.5	3.6	217.5	6.8	351.5	10.6
Stainless steel	25.0	0.7	29.0	0.9	54.5	1.6
Other steels	56.0	1.6	54.5	1.7	25.5	0.8
Iron	503.0	14.4	468.0	14.7	345.0	10.4
Total ferrous	2591.5	74.1%	2250.5	70.6%	2125.5	64.2%
Nonferrous materials						
Powdered metal[1]	16.0	0.5	19.0	0.6	37.5	1.1
Aluminum	112.0	3.2	138.0	4.3	256.5	7.8
Copper	39.5	1.1	44.0	1.4	46.0	1.4
Zinc die castings	28.0	0.8	18.0	0.5	11.0	0.3
Plastics/composites	176.0	5.0	211.5	6.6	253.0	7.6
Rubber	141.5	4.1	136.0	4.3	145.5	4.4
Glass	88.0	2.5	85.0	2.7	98.5	3.0
Fluids & lubricants	189.0	5.4	184.0	5.8	196.0	5.9
Other materials[2]	112.5	3.2	101.5	3.2	139.5	4.2
Total nonferrous	902.5	25.8%	937.02	29.4%	1183.5	35.7%
Total	3494.0 lb		3187.5 lb		3309.0 lb	

[1] Powdered metals may include ferrous materials.
[2] Other materials include sound deadeners, paint and anti-corrosion dip, ceramics, cotton, cardboard, battery lead, magnesium castings, etc.
SOURCE: Reproduced with permission from American Metal Market. Copyright 2004 American Metal Market LLC, a division of Metal Bulletin plc. All rights reserved.

Table 11.1.4 Materials Disposition from Recycled Vehicles

Recycled materials

Ferrous metals	Nonferrous metals	Fluids	Shredder residue
• Steel	• Aluminum	• Engine oil	• Plastic
• Iron	• Copper	• Coolant	• Rubber
	• Lead	• Refrigerant	• Glass
	• Zinc		• Fluids
			• Other

11.2 RAILWAY ENGINEERING
by V. Terrey Hawthorne and Keith L. Hawthorne
(in collaboration with E. Thomas Harley, Charles M. Smith, Robert B. Watson, and C. A. Woodbury)

REFERENCES: *Proceedings,* Association of American Railroads (AAR), Mechanical Division. *Proceedings,* American Railway Engineering Association (AREA). *Proceedings,* American Society of Mechanical Engineers (ASME). "The Car and Locomotive Cyclopedia," Simmons-Boardman. J. K. Tuthill, "High Speed Freight Train Resistance," *University of Illinois Engr. Bull., 376,* 1948. A. I. Totten, "Resistance of Light Weight Passenger Trains," *Railway Age,* **103,** Jul. 17, 1937. J. H. Armstrong, "The Railroad—What It Is, What It Does," Simmons-Boardman, 1978. Publications of the International Government-Industry Research Program on Track Train Dynamics, Chicago. Publications of AAR Research and Test Department, Chicago. Max Ephraim, Jr., "The AEM-7—A New High Speed, Light Weight Electric Passenger Locomotive," ASME RT Division, 82-RT-7, 1982. W. J. Davis, Jr., *General Electric Review,* October 1926. R. A. Allen, Conference on the Economics and Performance of Freight Car Trucks, October 1983. "Engineering and Design of Railway Brake Systems," The Air Brake Association, Sept. 1975. Code of Federal Regulations, Title 40, Protection of Environment. Code of Federal Regulations, Title 49, Transportation. *Standards and Recommended Practices and Proceedings,* Association of American Railroads (AAR) Mechanical Division, www.aar.com. "Standard for the Design and Construction of Passenger Railroad Rolling Stock," American Public Transportation Association (APTA), www.apta.com. "Manual for Railway Engineering," American Railway Engineering and Maintenance-of-Way Association (AREMA), www.arema.com. *Proceedings,* American Society of Mechanical Engineers (ASME), www.asme.org. "The Car and Locomotive Cyclopedia of American Practices," Simmons-Boardman Books, Inc., www.transalert.com. Publications of AAR Research and Test Department, www.aar.com. Code of Federal Regulations, Title 40, Protection of Environment and Title 49, Transportation, published by the Office of the Federal Register, National Archives and Records Administration. "The Official Railway Equipment Register—Freight Connections and Freight Cars Operated by Railroads and Private Car Companies of North America," R.E.R. Publishing Corp., East Windsor, NJ, www.cbizmedia.com.

DIESEL-ELECTRIC LOCOMOTIVES

Diesel-electric locomotives and electric locomotives are classified by wheel arrangement; letters represent the number of adjacent driving axles in a rigid truck (A for one axle, B for two axles, C for three axles, etc.). Idler axles between drivers are designated by numerals. A plus sign indicates articulated trucks or motive power units. A minus sign indicates separate nonarticulated trucks. This nomenclature is fully explained in RP-5523, issued by the Association of American Railroads (AAR). Virtually all modern locomotives are of either B-B or C-C configuration.

The high efficiency of the diesel engine is an important factor in its selection as a prime mover for traction. This efficiency at full or partial load makes it ideally suited to the variable service requirements of routine railroad operations. The diesel engine is a constant-torque machine that cannot be started under load and hence requires a variably coupled transmission arrangement. The electric transmission system allows it to make use of its full rated power output at low track speeds for starting as well as for efficient hauling of heavy trains at all speeds. Examples of the most common diesel-electric locomotive types in service are shown in Table 11.2.1. A typical diesel-electric locomotive is shown in Fig. 11.2.1.

Diesel-electric locomotives have a dc generator or rectified alternator coupled directly to the diesel engine crankshaft. The generator/alternator is electrically connected to dc series traction motors having nose suspension mountings. Many recent locomotives utilize gate turn-off inverters and ac traction motors to obtain the benefits of increased adhesion and higher tractive effort. The gear ratio for the axle-mounted bull gears to the motor pinions that they engage is determined by the locomotive speed range, which is related to the type of service. A high ratio is used for freight service where high tractive effort and low speeds are common, whereas high-speed passenger locomotives have a lower ratio.

Diesel Engines   Most new diesel-electric locomotives are equipped with either V-type, two strokes per cycle, or V-type, four strokes per cycle, engines. Engines range from 8 to 20 cylinders each. Output power ranges from 1,000 hp to over 6,000 hp for a single engine application. These medium-speed diesel engines have displacements from 560 to over 1,000 in^3 per cylinder.

Two-cycle engines are aspirated by either a gear-driven blower or a turbocharger. Because these engines lack an intake stroke for natural aspiration, the turbocharger is gear-driven at low engine speeds. At higher engine speeds, when the exhaust gases contain enough energy to drive the turbocharger, an overriding clutch disengages the gear train. Free-running turbochargers are used on four-cycle engines, as at lower speeds the engines are aspirated by the intake stroke.

The engine control governor is an electrohydraulic or electronic device used to regulate the speed and power of the diesel engine.

Electrohydraulic governors are self-contained units mounted on the engine and driven from one of the engine camshafts. They have integral oil supplies and pressure pumps. They utilize four solenoids that are actuated individually or in combination from the 74-V auxiliary generator/battery supply by a series of switches actuated by the engineer's throttle. There are eight power positions of the throttle, each corresponding to a specific value of engine speed and horsepower. The governor maintains the predetermined engine speed through a mechanical linkage to the engine fuel racks, which control the amount of fuel metered to the cylinders.

Computer engine control systems utilize electronic sensors to monitor the engine's vital functions, providing both electronic engine speed control and fuel management. The locomotive throttle is interfaced with an on-board computer that sends corresponding commands to the engine speed control system. These commands are compared to input from timing and engine speed sensors. The pulse width and timing of the engine's fuel injectors are adjusted to attain the desired engine speed and to optimize engine performance.

One or more centrifugal pumps, gear driven from the engine crankshaft, force water through passages in the cylinder heads and liners to provide cooling for the engine. The water temperature is automatically controlled by regulating shutter and fan operation, which in turn controls the passage of air through the cooling radiators. The fans are driven by electric motors.

The lubricating oil system supplies clean oil at the proper temperature and pressure to the various bearing surfaces of the engine, such as the crankshaft, camshaft, wrist pins, and cylinder walls. It also provides

Table 11.2.1   Locomotives in Service in North America

Locomotive* Builder	Model	Service	Arrangement	Weight min./max. (1,000 lb)	Tractive effort Starting† for min./max. weight (1,000 lb)	At continuous‡ speed (mi/h)	Number of cylinders	Horsepower rating (r/min)
Bombardier	HR-412	General purpose	B-B	240/280	60/70	60,400 (10.5)	12	2,700 (1,050)
Bombardier	HR-616	General purpose	C-C	380/420	95/105	90,600 (10.0)	16	3,450 (1,000)
Bombardier	LRC	Passenger	B-B	252 nominal	63	19,200 (42.5)	16	3,725 (1,050)
EMD	SW-1001	Switching	B-B	230/240	58/60	41,700 (6.7)	8	1,000 (900)
EMD	MP15	Multipurpose	B-B	248/278	62/69	46,822 (9.3)	12	1,500 (900)
EMD	GP40-2	General purpose	B-B	256/278	64/69	54,700 (11.1)	16	3,000 (900)
EMD	SD40-2	General purpose	C-C	368/420	92/105	82,100 (11.0)	16	3,000 (900)
EMD	GP50	General purpose	B-B	260/278	65/69	64,200 (9.8)	16	3,500 (950)
EMD	SD50	General purpose	C-C	368/420	92/105	96,300 (9.8)	16	3,500 (950)
EMD	F40PH-2	Passenger	B-B	260 nominal	65	38,240 (16.1)	16	3,000 (900)
GE	B18-7	General purpose	B-B	231/268	58/67	61,000 (8.4)	8	1,800 (1,050)
GE	B30-7A	General purpose	B-B	253/280	63/70	64,600 (12.0)	12	3,000 (1,050)
GE	C30-7A	General purpose	C-C	359/420	90/105	96,900 (8.8)	12	3,000 (1,050)
GE	B36-7	General purpose	B-B	260/280	65/70	64,600 (12.0)	16	3,600 (1,050)
GE	C36-7	General purpose	C-C	367/420	92/105	96,900 (11.0)	16	3,600 (1,050)
EMD	AEM-7	Passenger	B-B	201 nominal	50	33,500 (10.0)	NA¶	7,000 §
EMD	GM6C	Freight	C-C	365 nominal	91	88,000 (11.0)	NA	6,000 §
EMD	GM10B	Freight	B-B-B	390 nominal	97	100,000 (12.0)	NA	10,000 §
GE	E60C	General purpose	C-C	364 nominal	91	82,000 (22.0)	NA	6,000 §
GE	E60CP	Passenger	C-C	366 nominal	91	34,000 (55.0)	NA	6,000 §
GE	E25B	Freight	B-B	280 nominal	70	55,000 (15.0)	NA	2,500 §

* Engines: Bombardier—model 251, 4-cycle, V-type, 9 × 10½ in cylinders.
   EMD—model 645E, 2-cycle, V-type, 9¹⁄₁₆ × 10 in cylinders.
   GE—model 7FDL, 4-cycle, V-type, 9 × 10½ in cylinders.
† Starting tractive effort at 25% adhesion.
‡ Continuous tractive effort for smallest pinion (maximum).
§ Electric locomotive horsepower expressed as diesel-electric equivalent (input to generator).
¶ Not applicable.

Fig. 11.2.1   A typical diesel-electric locomotive. (*Electro-Motive Division, General Motors Corp.*)

Fig. 11.2.2   Axle-hung traction motors for diesel-electric and electric locomotives.

oil internally to the heads of the pistons to remove excess heat. One or more gear-type pumps driven from the crankshaft are used to move the oil from the crankcase through filters and strainers to the bearings and the piston cooling passages, after which the oil flows by gravity back to the crankcase. A heat exchanger is part of the system as all of the oil passes through it at some time during a complete cycle. The oil is cooled by engine cooling water on the other side of the exchanger. Paper-element cartridge filters in series with the oil flow remove fine impurities in the oil before it enters the engine.

Electric Transmission Equipment   A dc main generator or three-phase ac alternator is directly coupled to the diesel engine crankshaft. Alternator output is rectified through a full-wave bridge to keep ripple to a level acceptable for operation of series field dc motors or for input to a solid-state inverter system for ac motors.

Generator power output is controlled by (1) varying engine speed through movement of the engineer's controller and (2) controlling the flow of current in its battery field or in the field of a separate exciter generator. Shunt and differential fields (if used) are designed to maintain constant generator power output for a given engine speed as the load and voltage vary. The fields do not completely accomplish this, thus the battery field or separately excited field must be controlled by a load regulator to provide the final adjustment in excitation to load the engine properly. This field can also be automatically

deenergized to reduce or remove the load to prevent damage to the power plant or other traction equipment when certain undesirable conditions occur.

The engine main generator or alternator power plant functions at any throttle setting as a constant-horsepower source of energy. Therefore, the main generator voltage must be controlled to provide constant power output for each specific throttle position under the varying conditions of train speed, train resistance, atmospheric pressure, and quality of fuel. The load regulator, which is an integral part of the governor, accomplishes this within the maximum safe values of main generator voltage and current. For example, when the locomotive experiences an increase in track gradient with a consequent reduction in speed, traction motor counter emf decreases, causing a change in traction motor and main generator current. Because this alters the load demand on the engine, the speed of the engine tends to change to compensate. As the speed changes, the governor begins to reposition the fuel racks, but at the same time a pilot valve in the governor directs hydraulic pressure into a load regulator vane motor, which changes the resistance value of the load regulator rheostat in series with the main generator excitation circuit. This alters the main generator excitation current and consequently, main generator voltage and returns the value of main generator power output to normal. Engine fuel racks return to normal, consistent with constant values of engine speed.

The load regulator is effective within maximum and minimum limit values of its rheostat. Beyond these limits the power output of the engine is reduced. However, protective devices in the main generator excitation circuit limit its voltage to ensure that values of current and voltage in the traction motor circuits are within safe limits.

Auxiliary Generating Apparatus   DC power for battery charging, lighting, control, and cab heaters is provided by a separate generator, geared to the main engine. Voltage output is regulated within 1 percent of 74 V over the full range of engine speeds. Auxiliary alternators with full-wave bridge rectifiers are also utilized for this application. Modern locomotives with on-board computer systems also have power supplies to provide "clean" dc for sensors and computer systems. These are typically 24-V systems.

Traction motor blowers are mounted above the locomotive underframe. Air from the centrifugal blower housings is carried through the underframe and into the motor housings through flexible ducts. Other designs have been developed to vary the air output with cooling requirements to conserve parasitic energy demands. The main generator/alternators are cooled in a similar manner.

Traction motors are nose-suspended from the truck frame and bearing-suspended from the axle (Fig. 11.2.2). The traction motors employ series exciting (main) and commutating field poles. The current in the series field is reversed to change locomotive direction and may be partially shunted through resistors to reduce counter emf as locomotive speed is increased. Newer locomotives dispense with field shunting to improve commutation. Early locomotive designs also required motor connection changes (series, series/parallel, parallel) referred to as transition, to maintain motor current as speed increased.

AC traction motors employ a variable-frequency supply derived from a computer controlled solid-state inverter system, fed from the rectified alternator output. Locomotive direction is controlled by reversing the sequence of the three-phase supply.

The development of the modern traction alternator with its high output current has resulted in a trend toward dc traction motors that are permanently connected in parallel. DC motor armature shafts are equipped with grease-lubricated roller bearings, while ac traction motor shafts have grease-lubricated roller bearings at the free end and oil-lubricated bearings at the pinion drive end. Traction motor support bearings are usually of the plain sleeve type with lubricant wells and spring-loaded felt wicks, which maintain constant contact with the axle surface; however, many new passenger locomotives have roller support bearings.

Electrical Controls   In the conventional dc propulsion system, electropneumatic or electromagnetic contactors are employed to make and break the circuits between the traction motors and the main generator. They are equipped with interlocks for various control circuit functions. Similar contactors are used for other power and excitation circuits of lower power (current). An electropneumatic or electric motor-operated cam switch, consisting of a two-position drum with copper segments moving between spring loaded fingers, is generally used to reverse traction motor field current ("reverser") or to set up the circuits for dynamic braking. This switch is not designed to operate under load. On some dc locomotives these functions have been accomplished with a system of contactors. In the ac propulsion systems, the power control contactors are totally eliminated since their function is performed by the solid-state switching devices in the inverters.

Locomotives with ac traction motors utilize inverters to provide phase-controlled electrical power to their traction motors. Two basic arrangements of inverter control for ac traction motors have emerged. One arrangement uses an individual inverter for each axle. The other uses one inverter per truck. Inverters are typically GTO (gate turn-off) or IGBT (insulated-gate bipolar transistor) devices.

Cabs   In order to promote uniformity and safety, the AAR has issued standards for many locomotive cab features, RP-5104. Locomotive cab noise standards are prescribed by 49 CFR § 229.121.

Propulsion control circuits transmit the engineer's movements of the throttle lever, reverse lever, and transition or dynamic brake control lever in the controlling unit to the power-producing equipment of each unit operating in multiple in the locomotive consist. Before power is applied, all reversers must move to provide the proper motor connections for the direction of movement desired. Power contractors complete the circuits between generators and traction motors. For dc propulsion systems, excitation circuits then function to provide the proper main generator field current while the engine speed increases to correspond to the engineer's throttle position. In ac propulsion systems, all power circuits are controlled by computerized switching of the inverter.

To provide for multiple unit operation, the control circuits of each locomotive unit are connected by jumper cables. The AAR has issued Standard S-512 covering standard dimensions and contact identification for 27-point control jumpers used between diesel-electric locomotive units.

Wheel slip is detected by sensing equipment connected either electrically to the motor circuits or mechanically to the axles. When slipping occurs on some units, relays automatically reduce main generator excitation until slipping ceases, whereupon power is gradually reapplied. On newer units, an electronic system senses small changes in motor current and reduces motor current before a slip occurs. An advanced system recently introduced adjusts wheel creep to maximize wheel-to-rail adhesion. Wheel speed is compared to ground speed, which is accurately measured by radar. A warning light and/or buzzer in the operating cab alerts the engineer, who must notch back on the throttle if the slip condition persists. With ac traction the frequency selected prevents wheel slip if the wheel tries to exceed a speed dictated by the input frequency.

Batteries   Lead-acid storage batteries of 280 or 420 ampere-hour capacity are usually used for starting the diesel engine. Thirty-two cells on each locomotive unit are used to provide 64 V to the system. (See also Sec. 15.) The batteries are charged from the 74-V power supply.

Air Brake System   The "independent" brake valve handle at the engineer's position controls air pressure supplied from the locomotive reservoirs to the brake cylinders on only the locomotive itself. The "automatic" brake valve handle controls the air pressure in the brake pipe to the train (Fig. 11.2.3). On more recent locomotives, purely pneumatic braking systems have been supplanted by electropneumatic systems. These systems allow for more uniform brake applications, enhancing brake pipe pressure control and enabling better train handling. The AAR has issued Standard S-5529, "Multiple Unit Pneumatic Brake Equipment for Locomotives."

Fig. 11.2.3   Automatic-brake valve-handle positions for 26-L brake equipment.

Compressed air for braking and for various pneumatic controls on the locomotive is usually supplied by a two-stage, three-cylinder compressor connected directly or through a clutch to the engine crankshaft, or with electric motor drive. A compressor control switch is activated to maintain a pressure of approximately 130 to 140 lb/in^2 (896 to 965 kPa) in the main reservoirs.

Dynamic Braking   On most locomotives, dynamic brakes supplement the air brake system. The traction motors are used as generators to convert the kinetic energy of the locomotive and train into electrical energy, which is dissipated through resistance grids located near the locomotive roof. Motor-driven blowers, designed to utilize some of this braking energy, force cooler outside air over the grids and out through roof hatches. By directing a generous and evenly distributed air stream over the grids, their physical size is reduced, in keeping with the relatively small space available in the locomotive. On some locomotives, resistor grid cooling is accomplished by an engine-driven radiator/braking fan, but energy conservation is causing this arrangement to be replaced by motor-driven fans that can be energized in response to need, using the parasitic power generated by dynamic braking itself.

By means of a cam-switch reverser, the traction motors are connected to the resistance grids. The motor fields are usually connected in series across the main generator to supply the necessary high excitation current. The magnitude of the braking force is set by controlling the traction motor excitation and the resistance of the grids. Conventional dynamic braking is not usually effective below 10 mi/h (16 km/h), but it is very useful at 20 to 30 mi/h (32 to 48 km/h). Some locomotives are equipped with "extended range" dynamic braking which enables dynamic braking to be used at speeds as low as 3 mi/h (5 km/h) by shunting out grid resistance (both conventional and extended range are shown on Fig. 11.2.4). Dynamic braking is now controlled according to the "tapered" system, although the "flat" system has been used in the past. Dynamic braking control requirements are specified by AAR Standard S-5018. Dynamic braking is especially advantageous on long grades where accelerated brake shoe wear and the potential for thermal damage to wheels could otherwise be problems. The other advantages are smoother control of train speed and less concern for keeping the pneumatic trainline charged. Dynamic brake grids can also be used for a self-contained load-test feature, which permits a standing locomotive to be tested for power output. On locomotives equipped with ac traction motors, a constant dynamic braking (flat-top) force can be achieved from the horsepower limit down to 2 mi/h (3 km/h).

Fig. 11.2.4 Dynamic-braking effort versus speed. (*Electro-Motive Division, General Motors Corp.*)

Performance

Engine Indicated Horsepower The power delivered at the diesel locomotive drawbar is the end result of a series of subtractions from the original indicated horsepower of the engine, which takes into account the efficiency of transmission equipment and the losses due to the power requirements of various auxiliaries. The formula for the engine's indicated horsepower is:

$$\text{ihp} = \frac{PLAN}{33,000} \qquad (11.2.1)$$

where P = mean effective pressure in the cylinder, lb/in^2; L = length of piston stroke, ft; A = piston area, in^2; and N = total number of cycles completed per min. Factor P is governed by the overall condition of the engine, quality of fuel, rate of fuel injection, completeness of combustion, compression ratio, etc. Factors L and A are fixed with design of engine. Factor N is a function of engine speed, number of working chambers, and strokes needed to complete a cycle.

Engine Brake Horsepower In order to calculate the horsepower delivered by the crankshaft coupling to the main-generator, frictional losses in bearings and gears must be subtracted from the **indicated horsepower** (ihp). Some power is also used to drive lubricating oil pumps, governor, water pumps, scavenging blower, and other auxiliary devices. The resultant horsepower at the coupling is **brake horsepower** (bhp).

Rail Horsepower A portion of the engine bhp is transmitted mechanically via couplings or gears to operate the traction motor blowers, air compressor, auxiliary generator, and radiator cooling fan generator or alternator. Part of the auxiliary generator electrical output is used to run some of the auxiliaries. The remainder of the engine bhp transmitted to the main generator or main alternator for traction purposes must be multiplied by generator efficiency (usually about 91 percent), and the result again multiplied by the efficiency of the traction motors (including power

circuits) and gearing to develop rail horsepower. Power output of the main generator for traction may be expressed as

$$\text{Watts}_{\text{traction}} = E_g \times I_m \qquad (11.2.2)$$

where E_g is the main-generator voltage and I_m is the traction motor current in amperes, multiplied by the number of parallel paths or the dc link current in the case of an ac traction system.

Rail horsepower may be expressed as

$$\text{hp}_{\text{rail}} = V \times \text{TE}/375 \qquad (11.2.3)$$

where V is the velocity, mi/h, and TE is tractive effort at the rail, lb.

Thermal Efficiency The thermal efficiency of the diesel engine at the crankshaft, or the ratio of bhp output to the rate at which energy of the fuel is delivered to the engine, is about 33 percent. Thermal efficiency at the rail is about 26 percent.

Drawbar Horsepower The drawbar horsepower represents power available at the rear of the locomotive to move the cars and may be expressed as

$$\text{hp}_{\text{drawbar}} = \text{hp}_{\text{rail}} - \text{locomotive running resistance} \times \frac{V}{375} \qquad (11.2.4)$$

where V is the speed in mi/h. Train resistance calculations are discussed later under "Vehicle/Track Interaction." Theoretically, therefore, drawbar horsepower available is the power output of the diesel engine less the parasitic loads and losses described above.

Speed-Tractive Effort At full throttle the losses vary somewhat at different values of speed and tractive effort, but a curve of tractive effort plotted against speed is nearly hyperbolic. Figure 11.2.5 is a typical speed-tractive effort curve for a 3,500-hp (2,600-kW) freight locomotive. The diesel-electric locomotive has full horsepower available over the entire speed range (within the limits of adhesion described below). The reduction in power as continuous speed is approached is known as *power matching*. This allows multiple operation of locomotives of different ratings at the same continuous speed.

Adhesion In Fig. 11.2.6 the maximum value of tractive effort represents the level usually achievable just before the wheels slip under average rail conditions. Adhesion is usually expressed as a percentage of vehicle weight on drivers, with the nominal level being 25 percent. This means that a force equal to 25 percent of the total locomotive weight on drivers is available as tractive effort. Actually, adhesion will vary widely with rail conditions, from as low as 5 percent to as high as 35 percent or

Fig. 11.2.5 Tractive effort versus speed.

Fig. 11.2.6 Typical tractive effort versus speed characteristics. (*Electro-Motive Division, General Motors Corp.*)

more. Adhesion is severely reduced by lubricants which spread as thin films in the presence of moisture on running surfaces. Adhesion can be increased with sand applied to the rails from the locomotive sanding system. More recent wheel slip systems permit wheel creep (very slow controlled slip) to achieve greater levels of tractive effort. Even higher adhesion levels are available from ac traction motors; for example, 45 percent at start-up and low speed, with a nominal value of 35 percent.

Traction Motor Characteristics Motor torque is a function of armature current and field flux (which is a function of field current). Since the traction motors are series-connected, armature and field current are the same (except when field shunting circuits are introduced), and therefore tractive effort is solely a function of motor current. Figure 11.2.7 presents a group of traction motor characteristic curves with tractive effort, speed and efficiency plotted against motor current for full field (FF) and at 35 (FS1) and 55 (FS2) percent field shunting. Wheel diameter and gear ratio must be specified when plotting torque in terms of tractive effort. (See also Sec. 15.)

Traction motors are usually rated in terms of their maximum continuous current. This represents the current at which the heating due to electrical losses in the armature and field windings is sufficient to raise

the temperature of the motor to its maximum safe limit when cooling air at maximum expected ambient temperature is forced through it at the prescribed rate by the blowers. Continuous operation at this current level ideally allows the motor to operate at its maximum safe power level, with waste heat generated equal to heat dissipated. The tractive effort corresponding to this current is usually somewhat lower than that allowed by adhesion at very low speeds. Higher current values may be permitted for short periods of time (as when starting). These ratings are specified in intervals of time (minutes) and are posted on or near the load meter (ammeter) in the cab.

Maximum Speed Traction motors are also rated in terms of their maximum safe speed in r/min, which in turn limits locomotive speed. The gear ratio and wheel diameter are directly related to speed as well as the maximum tractive effort and the minimum speed at which full horsepower can be developed at the continuous rating of the motors. Maximum locomotive speed may be expressed as follows:

$$(\text{mi/h})_{\max} = \frac{\text{wheel diameter (in)} \times \text{maximum motor r/min}}{\text{gear ratio} \times 336} \quad (11.2.5)$$

where the gear ratio is the number of teeth on the gear mounted on the axle divided by the number of teeth on the pinion mounted on the armature shaft.

Locomotive Compatibility The AAR has developed two standards in an effort to improve compatibility between locomotives of different model, manufacture, and ownership: a standard 27-point control system (Standard S-512, Table 11.2.2) and a standard control stand (RP-5132). The control stand has been supplanted by a control console in many road locomotives (Fig. 11.2.8).

Energy Conservation Efforts to improve efficiency and fuel economy have resulted in major changes in the prime movers, including

Fig. 11.2.7 Traction motor characteristics. (*Electro-Motive Division, General Motors Corp.*)

Table 11.2.2 Standard Dimensions and Contact Identification of 27-Point Control Plug and Receptacle for Diesel-Electric Locomotives

Receptacle point	Function	Code	Wire size, AWG
1	Power reduction setup, if used	(PRS)	14
2	Alarm signal	SG	14
3	Engine speed	DV	14
4*	Negative	N	14 or 10
5	Emergency sanding	ES	14
6	Generator field	GF	12
7	Engine speed	CV	14
8	Forward	FO	12
9	Reverse	RE	12
10	Wheel slip	WS	14
11	Spare		14
12	Engine speed	BV	14
13	Positive control	PC	12
14	Spare		14
15	Engine speed	AV	14
16	Engine run	ER	14
17	Dynamic brake	B	14
18	Unit selector circuit	US	12
19	2d negative, if used	(NN)	12
20	Brake warning light	BW	14
21	Dynamic brake	BG	14
22	Compressor	CC	14
23	Sanding	SA	14
24	Brake control/power reduction control	BC/PRC	14
25	Headlight	HL	12
26	Separator blowdown/ remote reset	SV/RR	14
27	Boiler shutdown	BS	14

*Receptacle point 4—AWG wire size 12 is "standard" and AWG wire size 10 is "Alternate standard" at customer's request. A dab of white paint in the cover latch cavity must be added for ready identification of a no. 10 wire present in a no. 4 cavity.

Fig. 11.2.8 Locomotive control console. (*Electro-Motive Division, General Motors Corp.*)

more efficient turbocharging, fuel injection, and combustion. The auxiliary (parasitic) power demands have also been reduced by improvements, which include fans and blowers that move only the air required by the immediate demand, air compressors that declutch when unloaded, and a selective low-speed engine idle. Fuel-saver switches permit dropping trailing locomotive units off the line when less than maximum power is required, while allowing the remaining units to operate at maximum efficiency.

Emissions The U.S. Environmental Protection Agency has promulgated regulations aimed at reducing diesel locomotive emissions, especially oxides of nitrogen (NO_x). These standards also include emissions reductions for hydrocarbons (HC), carbon monoxide (CO), particulate matter (PM), and smoke. These standards are executed in three "tiers." Tier 0 going into effect in 2000, Tier 1 in 2002, and Tier 3 in 2005. EPA locomotive emissions standards are published under 40 CFR § 85, § 89, and § 92.

ELECTRIC LOCOMOTIVES

Electric locomotives are presently in very limited use in North America. Freight locomotives are in dedicated service primarily for hauling coal or mineral products. Electric passenger locomotives are used in high-density service in the northeastern United States.

Electric locomotives draw power from overhead catenary or third-rail systems. While earlier systems used either direct current up to 3,000 V or single-phase alternating current at 11,000 V, 25 Hz, the newer systems in North America use 25,000 or 50,000 V at 60 Hz. The higher voltage levels can be used only where clearances permit. A three-phase power supply was tried briefly in this country and overseas many years ago, but was abandoned because of the complexity of the required double catenary.

While the older dc locomotives used resistance control, ac locomotives have used a variety of systems including Scott-connected transformers; series ac motors; motor generators and dc motors; ignitrons and dc motors; silicon thyristors and dc motors; and, more recently, chopper control and inverter with ac motors. Examples of the various electric locomotives in service are shown in Table 11.2.1.

An electric locomotive used in high-speed passenger service is shown in Fig. 11.2.9. High short-time ratings (Figs. 11.2.10 and 11.2.11)

Fig. 11.2.9 Modern electric high-speed locomotive.(*Courtesy of Bombardier Transportation.*)

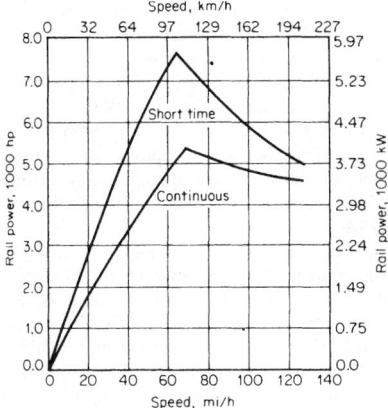

Fig. 11.2.10 Speed-horsepower characteristics (half-worn wheels). [Gear ratio = 85/36; wheel diameter = 50 in (1270 mm); ambient temperature = 60°F (15.5°C).] (*Courtesy of M. Ephraim, ASME, Rail Transportation Division.*)

Fig. 11.2.11 Tractive effort versus speed (half-worn wheels). [Voltage = 11.0 kV at 25 Hz; gear ratio = 85/36; wheel diameter = 50 in (1270 mm).] (*Courtesy of M. Ephraim, ASME, Rail Transportation Division.*)

render electric locomotives suitable for passenger service where high acceleration rates and high speeds are combined to meet demanding schedules.

The modern electric locomotive in Fig. 11.2.9 obtains power for the main circuit and motor control from the catenary through a pantograph. A motor-operated switch provides for the transformer change from series to parallel connection of the primary windings to give a constant secondary voltage with either 25-kV or 12.5-kV primary supply. The converters for armature current consist of two asymmetric-type bridges for each motor. The traction-motor fields are each separately fed from a one-way connected field converter.

Identical control modules separate the control of motors on each truck. Motor sets are therefore connected to the same transformer winding. Wheel slip correction is also modularized, utilizing one module for each two-motor truck set. This correction is made with a complementary wheel-slip detection and correction system. A magnetic pickup speed signal is used for the basic wheel slip. Correction is enhanced by a magnetoelastic transducer used to measure force swings in the traction motor reaction rods. This system provides the final limit correction.

To optimize the utilization of available adhesion, the wheel-slip control modules operate independently to allow the motor modules to

receive different current references, depending on their respective adhesion conditions.

All auxiliary machines—air compressor, traction motor blower, cooling fans, etc.—are driven by three-phase, 400-V, 60-Hz induction motors powered by a static inverter that has a rating of 175 kVA at a 0.8 power factor. When cooling requirements are reduced, the control system automatically reduces the voltage and frequency supplied to the blower motors to the required level. As a backup, the system can be powered by the static converter used for the head-end power requirements of the passenger cars. This converter has a 500-kW, 480-V, three-phase, 60-Hz output capacity and has a built-in overload capacity of 10 percent for half of any 1-hour period.

Dynamic brake resistors are roof-mounted and are cooled by ambient airflow induced by locomotive motion. The dynamic brake capacity relative to train speed is shown in Fig. 11.2.12. Regenerative braking can be utilized with electric locomotives by returning braking energy to the distribution system.

Fig. 11.2.12 Dynamic-brake performance. (*Courtesy of M. Ephraim, ASME, Rail Transportation Division.*)

Many electric locomotives utilize traction motors identical to those used on diesel-electric locomotives. Some, however, use frame-suspended motors with quill-drive systems (Fig. 11.2.13). In these systems, torque is transmitted from the traction motor by a splined coupling to a quill shaft and rubber coupling to the gear unit. This transmission allows for greater relative movement between the traction motor (truck frame) and gear (wheel axle) and reduces unsprung weight.

Fig. 11.2.13 Frame-suspended locomotive transmission. (*Courtesy of M. Ephraim, ASME, Rail Transportation Division.*)

FREIGHT CARS

Freight Car Types

Freight cars are designed and constructed for either general service or specific ladings. The Association of American Railroads (AAR) has established design and maintenance criteria to assure the interchangeability of freight cars on all North American railroads. Freight cars that do not conform to AAR standards have been placed in service by special agreement with the railroads over which the cars operate. The AAR "Manual of Standards and Recommended Practices" specifies dimensional limits, weights, and other design criteria for cars, which may be freely interchanged between North American railroads. This manual is revised annually by the AAR. Many of the standards are reproduced in the "Car and Locomotive Cyclopedia," which is revised periodically. Safety appliances (such as end ladders, sill steps, and uncoupling levers), braking equipment, and certain car design features and maintenance practices must comply with the Safety Appliances Act, the Power Brake Law, and the Freight Car Safety Standards covered by the Federal Railroad Administration's (FRA) Code of Federal Regulations (Title 49 CFR). Maintenance practices are set forth in the "Field Manual of the AAR Interchange Rules" and repair pricing information is provided in the companion "AAR Office Manual."

The AAR identifies most cars by nominal capacity and type. In some cases there are restrictions as to type of load (i.e., food, automobile parts, coil steel, etc.). Most modern cars have nominal capacities of either 70 or 100 tons.* A 100-ton car is limited to a maximum gross rail load of 263,000 lb (119.3 t*) with 6½ by 12 in (165 by 305 mm) axle journals, four pairs of 36-in (915-mm) wheels and a 70-in (1,778-mm) rigid truck wheel base. Some 100-ton cars are rated at 286,000 lb and are handled by individual railroads on specific routes or by mutual agreement between the handing railroads. A 70-ton car is limited to a maximum gross rail load of 220,000 lb (99.8 t) with 6- by 11-in (152- by 280-mm) axle journals, four pairs of 33-in (838-mm) wheels with trucks having a 66-in (1,676-mm) rigid wheel base.

On some special cars where height limitations are critical, 28-in (711-mm) wheels are used with 70-ton axles and bearings. In these cases wheel loads are restricted to 22,400 lb (10.2 t). Some special cars are equipped with two 125-ton two-axle trucks with 38-in (965-mm) wheels in four-wheel trucks having 7 by 14 in (178 by 356 mm) axle journals and 72-in (1,829-mm) truck wheel base. This application is prevalent in articulated double-stack (two-high) container cars. Interchange of these very heavy cars is by mutual agreement between the operating railroads involved.

The following are the most common car types in service in North America. Dimensions given are for typical cars; actual measurements may vary.

Box cars (Fig. 11.2.14a) come in six popular types:

1. Standard box cars may have either sliding or plug doors. Plug doors provide a tight seal from weather and a smooth interior. Unequipped box cars are usually of 70-ton capacity. These cars have tongue-and-groove or plywood lining on the interior sides and ends with nailable floors (either wood or steel with special grooves for locking nails). The cars carry typical general merchandise lading: packaged, canned, or bottled foodstuffs; finished lumber; bagged or boxed bulk commodities; or, in the past when equipped with temporary door fillers, bulk commodities such as grain [70-ton: $L = 50$ ft 6 in (15.4 m), $H = 11$ ft 0 in (3.4 m), $W = 9$ ft 6 in (2.9 m), truck centers = 40 ft 10 in (12.4 m)].

2. Specially equipped box cars usually have the same dimensions as standard cars but include special interior devices to protect lading from impacts and over-the-road vibrations. Specially equipped cars may have hydraulic cushion units to dampen longitudinal shock at the couplers.

3. Insulated box cars have plug doors and special insulation. These cars can carry foodstuffs such as unpasteurized beer, produce, and dairy products. These cars may be precooled by the shipper and maintain a heat loss rate equivalent to 1°F (0.55°C) per day. They also can protect loads from freezing when operating at low ambient temperatures.

4. Refrigerated box cars are used where transit times are longer. These cars are equipped with diesel-powered refrigeration units and are primarily used to carry fresh produce and meat. They are often 100-ton cars [$L = 52$ ft 6 in (16.0 m), $H = 10$ ft 6 in (3.2 m), truck centers = 42 ft 11 in (13.1 m)].

5. "All door" box cars have doors which open the full length of the car for loading package lumber products such as plywood and gypsum board.

6. High cubic capacity box cars with an inside volume of 10,000 ft^3 (283 m^3) have been designed for light density lading, such as some automobile parts and low-density paper products.

Box car door widths vary from 6 to 10 ft for single-door cars and 16 to 20 ft (4.9 to 6.1 m) for double-door cars. "All door" cars have clear doorway openings in excess of 25 ft (7.6 m). The floor height above rail for an empty uninsulated box car is approximately 44 in (1,120 mm) and for an empty insulated box car approximately 48 in (1,220 mm). The floor height of a loaded car can be as much as 3 in (76 mm) lower than the empty car.

Covered hopper cars (Fig. 11.2.14b) are used to haul bulk commodities that must be protected from the environment. Modern covered hopper cars are typically 100-ton cars with roof hatches for loading and from two to six bottom outlets for discharge. Cars used for dense commodities, such as fertilizer or cement, have two bottom outlets, round roof hatches, and volumes of 3,000 to 4,000 ft^3 (84.9 to 113.2 m^3) [100-ton: $L = 39$ ft 3 in (12.0 m). $H = 14$ ft 10 in (4.5 m), truck centers = 26 ft 2 in (8.0 m)]. Cars used for grain service (corn, wheat, rye, etc.) have three or four bottom outlets, longitudinal trough roof hatches, and volumes of from 4,000 to 5,000 ft^3 (113 to 142 m^3). Cars used for hauling plastic pellets have four to six bottom outlets (for pneumatic unloading with a vacuum system), round roof hatches, and volumes of from 5,000 to 6,000 ft^3 (142 to 170 m^3) [100-ton: $L = 65$ ft 7 in (20.0 m), $H = 15$ ft 5 in (4.7 m), truck centers = 54 ft 0 in (16.5 m)].

Open-top hopper cars (Fig. 11.2.14c) are used for hauling bulk commodities such as coal, ore, or wood chips. A typical 100-ton coal hopper car will vary in volume depending on the light weight of the car and density of the coal to be hauled. Volumes range from 3,900 to 4,800 ft^3 (110 to 136 m^3). Cars may have three or four manually operated hopper doors or a group of automatically operated doors. Some cars are equipped with rotating couplers on one end of allow rotary dumping without uncoupling [100-ton: $L = 53$ ft 1½ in (16.2 m), $H = 12$ ft 8½ in (3.9 m), truck centers = 40 ft 6 in (12.3 m)]. Cars intended for aggregate or ore service have smaller volumes for the more dense commodity. These cars typically have two manual bottom outlets [100-ton: $L = 40$ ft 8 in (12.4 m), $H = 11$ ft 10 in (3.6 m), truck centers = 29 ft 9 in (9.1 m)]. Hopper cars used for wood chip service are configured for low-density loads. Volumes range from 6,500 to 7,500 ft^3 (184 to 212 m^3).

High-side gondola cars (Fig. 11.2.14d) are open-top cars typically used to haul coal or wood chips. These cars are similar to open-top hopper cars in volume but require a rotary coupler on one end for rotary dumping to discharge lading since they do not have bottom outlets. Rotary-dump coal gondolas are usually used in dedicated, unit-train service between a coal mine and an electric power plant. The length over coupler pulling faces is approximately 53 ft 1 in (16.2 m) to suit the standard coal dumper [100-ton: $L = 50$ ft 5½ in (15.4 m), $H = 11$ ft 9 in (3.6 m), $W = 10$ ft 5 in (3.2 m), truck centers = 50 ft 4 in (15.3 m)]. Wood chip cars are used to haul chips from sawmills to paper mills or particle board manufacturers. These high-volume cars are either rotary-dumped or, when equipped with end doors, end-dumped [100-ton: $L = 62$ ft 3 in (19.0 m), $H = 12$ ft 7 in (3.8 m), $W = 10$ ft 5 in (3.2 m), truck centers = 50 ft 4 in (15.3 m)]. Rotary dump aggregate or ore cars, called "ore jennies," have smaller volumes for the high-density load.

Bulkhead flat cars (Fig. 11.2.14e) are used for hauling such commodities as packaged finished lumber, pipe, or, with special inward canted floors, pulpwood. Both 70- and 100-ton bulkhead flats are used. Typical deck heights are approximately 45 to 50 in (1,143 to 1,270 mm) [100-ton: $L = 66$ ft 2 in (20.2 m), $W = 10$ ft 5 in (3,175 m), $H = 14$ ft 2⁹/₁₆ in (4.3 m), truck centers = 48 ft 2 in (14.7 m)]. Special "center beam" bulkhead flat cars designed for pulpwood and lumber service have a full-height, longitudinal beam from bulkhead to bulkhead.

Tank cars (Fig 11.2.14f) are used for liquids, compressed gases, and other ladings, such as sulfur, which can be loaded and unloaded in a molten state. Nonhazardous liquids such as corn syrup, crude oil, and mineral spring water are carried in nonpressure cars. Cars used to haul hazardous substances such as liquefied petroleum gas (LPG), vinyl chloride, and anhydrous ammonia are regulated by the U.S. Department of Transportation. Newer and earlier-built retrofitted cars equipped for hazardous commodities have safety features including safety valves;

---

* 1 ton = 1 short ton = 2,000 lb; 1 t = 1 metric ton = 1,000 kg = 2205 lb.

Fig. 11.2.14   Typical freight cars.

especially designed top and bottom "shelf" couplers, which increase the interlocking effect between couplers and decrease the danger of disengagement due to derailment; head shields on the ends of the tank to prevent puncturing; bottom outlet protection if bottom outlets are used; and thermal insulation and jackets to reduce the risk of rupturing in a fire. These features resulted from industry-government studies in the RPI-AAR Tank Car Safety Research and Test Program.

Cars for asphalt, sulfur, and other viscous liquid service have heating coils on the shell so that steam may be used to liquefy the lading for discharge [pressure cars, 100-ton: volume = 20,000 gal (75.7 m³), L = 59 ft 11¾ in (18.3 m), truck centers = 49 ft ¼ in (14.9 m)] [nonpressure, 100-ton: volume = 21,000 gal (79.5 m³), L = 51 ft 3¼ in (15.6 m), truck centers = 38 ft 11¼ in (11.9 m)].

Conventional intermodal cars are 89 ft 4 in (27.2-m) intermodal flat cars equipped to haul one 45-ft (13.7-m) and one 40-ft (12.2-m) trailer with or without end-mounted refrigeration units, two 45-ft (13.7-m) trailers (dry vans), or combinations of containers from 20 to 40 ft (6.1 to 12.2 m) in length. Hitches to support trailer fifth wheels may be fixed for trailer-only cars or retractable for conversion to haul containers or to facilitate driving

trailers onto the cars in the rare event where "circus" loading is still required. In some cars a drawbar connects two units so three 57-ft trailers can be loaded, with one trailer over the drawbar. Trailer hauling service (Fig. 11.2.14g) is called TOFC (trailer on flat car); container service (Fig. 11.2.14h) is called COFC (container on flat car) [70-ton: L = 89 ft 4 in (27.1 m), W = 10 ft 3 in (3.1 m), truck centers = 64 ft 0 in (19.5 m)].

Introduction of larger trailers in highway service has led to the development of alternative TOFC and COFC cars. Other techniques involve articulation of well-type car bodies or skeletonized spine car bodies, assembled into multiunit cars for lift-on loading and lift-off unloading (stand-alone well car, Fig. 11.2.14n and articulated well car, Fig. 11.2.14o). These cars are typically composed of from three to ten units. Well cars consist of a center well for double-stacked containers [10-unit spine car; L = 46 ft 6 in (14.2 m) per end unit, L = 465 ft 3½ in (141.8 m)].

In the 1980s another approach was to haul larger trailers on two axle spine cars. These cars, used either singly or in multiple combinations, can haul a single trailer from 40 to 48 ft long (12.2 to 14.6 m) with end-mounted refrigeration unit and 36 or 42-in (914 to 1070 mm) spacing between the kingpin and the front of the trailer.

Bilevel and trilevel auto rack cars (Fig. 11.2.14*i*) are used to haul finished automobiles and other vehicles. Most recent designs of these cars feature fully enclosed racks to provide security against theft and vandalism [70-ton: $L = 89$ ft 4 in (27.2 m), $H = 18$ ft 11 in (5.8 m), $W = 10$ ft 7 in (3.2 m), truck centers $= 64$ ft 0 in (19.5 m)]. Because of increased clearances provided by some railroads for double-stack cars, auto rack cars with a height of 20 ft 2 in (6.1 m) are now being produced.

Mill gondolas (Fig. 11.2.14*j*) are 70- or 100-ton open-top cars principally used to haul pipe, structural steel, scarp metal, and when specially equipped, coils of aluminum or tinplate and other steel materials [100-ton: $L = 52$ ft 6 in (16.0 m), $W = 9$ ft 6 in (2.9 m), $H = 4$ ft 6 in (1.4 m), truck centers $= 43$ ft 6 in (13.3 m)].

General-purpose or **machinery flat cars** (Fig. 11.2.14*k*) are 70- or 100-ton cars used to haul machinery such as farm equipment and highway tractors. These cars usually have wood decks for nailing lading restraint dunnage. Some heavy-duty six-axle cars are used for hauling off highway vehicles such as army tanks and mining machinery [100-ton, four-axle: $L = 60$ ft 0 in (18.3 m), $H = 3$ ft 9 in (1.1 m), truck centers $= 42$ ft 6 in (13.0 m)] [200-ton, 8-axle: $L = 44$ ft 4 in (13.5 m), $H = 4$ ft 0 in (1.2 m), truck centers $= 33$ ft 9 in (10.3 m)].

Depressed-center flat cars (Fig. 11.2.14*l*) are used for hauling transformers and other heavy, large materials that require special clearance considerations. Depressed-center flat cars may have four- or six-axle or dual four-axle, trucks with span bolsters, depending on weight requirements.

Schnabel cars (Fig. 11.2.14*m*) are special cars for transformers and power plant components. With these cars the load itself provides the center section of the car structure during shipment. Some Schnabel cars are equipped with hydraulic controls to lower the load for height restrictions and shift the load laterally for wayside restrictions [472-ton: $L = 22$ ft 10 in to 37 ft 10 in (7.0 to 11.5 m), truck centers $= 55$ ft 6 in to 70 ft 6 in (16.9 to 21.5 m)]. Schnabel cars must be operated in special trains.

Freight Car Design

The AAR provides specifications to cover minimum requirements for design and construction of new freight cars. Experience has demonstrated that the AAR specifications alone do not ensure an adequate design for all service conditions. The designer must be familiar with the specific service and increase the design criteria for the particular car above the minimum criteria provided by the AAR. The AAR requirements include stress calculations for the load-carrying members of the car and physical tests that may be required at the option of the AAR Equipment Engineering Committee (EEC) approving the car design. In the application for approval of a new or untired type of car, the EEC may require either additional calculations or tests to assess the design's ability to meet the AAR minimum requirements. These tests might consist of a static compression test of 1,000,000 lb (4.4 MN), a static vertical test applied at the coupler, and impact tests simulating yard impact conditions.

In some cases, it is advisable to operate an instrumented prototype car in service to detect problems which might result from unexpected track or train-handling input forces. The car design must comply with width and height restrictions shown in AAR clearance plates furnished in the specifications, Fig. 11.2.15. In addition, there are limitations on

Fig. 11.2.15 AAR Equipment diagrams. (*a*) Plate B unrestricted interchange service; (*b*) Plate C limited interchange service; (*c*) Plate H controlled interchange service.

LIGHT CAR CONDITIONS

CARS MAY BE CONSTRUCTED TO AN EXTREME WIDTH
OF 9' 11" AND TO THE OTHER LIMITS OF THIS DIAGRAM
WHEN TRUCK CENTERS DO NOT EXCEED 58' 11" AND
WHEN, WITH TRUCK CENTERS OF 58' 11", THE
SWING-OUT AT ENDS OF CAR DOES NOT EXCEED THE
SWING-OUT AT CENTER OF CAR ON A 13° CURVE; A CAR
TO THESE DIMENSIONS IS DEFINED AS THE BASE CAR.

WHEN TRUCK CENTERS EXCEED 58' 11", CAR AND
CONTAINER WIDTH SHALL BE REDUCED TO COMPENSATE
FOR THE INCREASED SWING-OUT AT CENTER AND/OR
ENDS OF CAR ON A 13° CURVE SO THAT THE WIDTH OF
CAR SHALL NOT PROJECT BEYOND THE CENTER OF
TRACK MORE THAN THE BASE CAR.

MAXIMUM CAR WIDTH FOR VARIOUS TRUCK CENTERS,
AT CENTER OF CAR, ARE SHOWN IN PLATE H-1.
MAXIMUM CAR WIDTH AT LOCATIONS OTHER THAN
CENTER OF CAR ARE SHOWN ON PLATE D.

WHEN TRUCK CENTERS ARE LESS THAN 58' 11",
THE FULL WIDTH PERMITTED BY PLATES C AND C-1
MAY BE UTILIZED.

CARS WITH RAIL LOADS IN EXCESS OF 65,760 LB
PER AXLE CANNOT BE OPERATED IN UNRESTRICTED
INTERCHANGE. HOWEVER, THEY MAY BE PERMITTED
UNDER CONTROLLED CONDITIONS WHERE SPECIAL
AGREEMENT HAS BEEN REACHED BETWEEN
PARTICIPATING RAILROADS TO SO HANDLE.

NOTE:
RESTRICTED TO ROADS ONLY
FROM WHICH SPECIFIC CLEARANCE
HAS BEEN OBTAINED FROM THE
HANDLING LINES.

(c)

Fig. 11.2.15 (*Continued*)

the height of the center of gravity of the loaded car and on the vertical and horizontal curving capability allowed by the clearance provided at the coupler. The AAR provides a method of calculating the minimum radius curve that the car design can negotiate when coupled to another car of the same type or to a standard AAR "base" car. In the case of horizontal curves, the requirements are based on the length of the car over the pulling faces of the couplers.

Freight cars are designed to withstand single-end impact or coupling loads on the basis of the type of energy absorption, cushioning, provided in the car design. Conventional friction, elastomer, or combination draft gears or short-travel hydraulic cushion units that provide less than 6 in (152 mm) of travel require a structure capable of withstanding a 1,250,000-lb (5.56-MN) impact load. For cars with hydraulic units that provide greater than 14 in (356 mm) of travel, the required design impact load is 600,000 lb (2.7 MN). In all cases, the structural connections to the car must be capable of withstanding a static compressive (squeeze) end load of 1,000,000 lb (4.44 MN) or a dynamic (impact) compressive load of 1,250,000 lb (5.56 MN).

The AAR has adopted requirements for unit trains of high-utilization cars to be designed for 3,000,000 mi (4.8 Gm) of service based upon fatigue life estimates. General-interchange cars that accumulate less mileage in their life should be designed for 1,000,000 mi (1.6 Gm) of service. Road environment spectra for various locations within the car are being developed for different car designs for use in this analysis. The fatigue strengths of various welded connections are provided in the AAR "Manual of Standards and Recommended Practices," Section C, Part II.

Many of the design equations and procedures are available from the AAR. Important information on car design and approval testing is contained in the AAR "Manual of Standards and Recommended Practice," Section C-II M-1001, Chapter XI.

Freight-Car Suspension

Most freight cars are equipped with standard three-piece trucks (Fig. 11.2.16) consisting of two side-frame castings and one bolster casting. Side-frame and bolster designs are subjected to both static and fatigue test requirements specified by the AAR. The bolster casting is equipped with a female centerplate bowl upon which the car body rests and with side bearings generally located 25 in (635 mm) on each side of the centerline. In most cases, the side bearings have clearance to the car body and are equipped with either flat sliding plates or rollers. In some cases, constant-contact side bearings provide a resilient material between the car body and the truck bolster.

Unit beam roller bearing truck components

70. Wheel	82. Truck side frame
71. Axle	83. Truck springs
72. Truck dead lever	84. Truck side bearing
73. Dead lever fulcrum	85. Side bearing roller
74. Dead lever fulcrum bracket	86. Truck bolster
75. Brake beam	87. Truck center plate cast
76. Bottom rod	88. Truck live lever
77. Roller bearing adapter	89. Center pin
78. Roller bearing assembly	90 Horizontal wear plate
79. End cap	91. Vertical wear plate
80. End cap retaining bolt	92. Brake shoe key
81. Locking plate	93. Brake shoe

Fig. 11.2.16 Three-piece freight car truck. (*Courtesy AAR Research and Test Department.*)

The centerplate arrangement may have a variety of styles of wear plates or friction materials and a vertical loose or locked pin between the truck centerplate and the car body.

Truck springs nested into the bottom of the side-frame opening support the end of the truck bolster. Requirements for spring designs and the grouping of springs are generally specified by the AAR. Historically, the damping provided within the spring group has utilized a combination of springs and friction wedges. In addition to friction wedges, in more recent years, some cars have been equipped with hydraulic damping devices that parallel the spring group.

A few trucks have a "steering" feature that includes an interconnection between axles to increase the lateral interaxle stiffness and decrease the interaxle yaw stiffness. Increased lateral stiffness improves the lateral stability and decreased yaw stiffness improves the curving characteristics.

Freight-Car Wheel-Set Design

A freight-car wheel set consists of two wheels, one axle, and two bearings. Cast- and wrought-steel wheels are used on freight cars in North America (AAR "Manual of Standards and Recommended Practice," Sec. G). Freight-car wheels are subjected to thermal loads from braking, as well as mechanical loads at the wheel-rail interface. Experience with thermal damage to wheels has led to the requirement for "low stress" or curved-plate wheels (Fig. 11.2.17). These wheels are less susceptible to the development of circumferential residual tensile stresses which render the wheel vulnerable to sudden failure if a flange or rim crack occurs. New wheel designs introduced for interchange service must be evaluated by using a finite-element technique employing both thermal and mechanical loads (AAR S-660).

Fig. 11.2.17 Wheel-plate designs. (*a*) Flat plate; (*b*) parabolic plate; (*c*) S-plate curved wheel.

Freight-car wheels range in diameter from 28 to 38 in (711 to 965 mm), depending on car weight (Table 11.2.3). The old AAR standard tread profile (Fig. 11.2.18*a*) has been replaced with the AAR-1B (Fig. 11.2.18*c*) profile which represents a "worn" profile to minimize early tread loss due to wear and provide a more stable profile over the life of the tread. Several variant tread profiles, including the AAR-1B, were developed from the basic Heumann design, Fig. 11.2.18*b*. One of these, for application in Canada, provided increasing conicity into the throat of the flange, similar to the Heumann profile. This reduces curving resistance and extends wheel life.

Wheels are also specified by chemistry and heat treatment. Low-stress wheel designs of Classes B and C are required for freight cars. Class B wheels have a carbon content of 0.57 to 0.67 percent and

are rim-quenched. Class C wheels have a carbon content of 0.67 to 0.77 percent and are also rim-quenched. Rim quenching provides a hardened running surface for a long wear life. Lower carbon levels than those in Class B may be used where thermal cracking is experienced, but freight-car equipment generally does not require their use.

Axles used in interchange service are solid steel forgings with raised wheel seats. Axles are specified by journal size for different car capacities, Table 11.2.3.

Most freight car journal bearings are grease-lubricated, tapered-roller bearings (see Sec. 8). Current bearing designs eliminate the need for periodic field lubrication.

Wheels are mounted and secured on axles with an interference fit. Bearings are mounted with an interference fit and retained by an end cap bolted to the end of the axle. Wheels and bearings for cars in interchange service must be mounted by an AAR-inspected and -approved facility.

Special Features

Many components are available to enhance the usefulness of freight cars. In most cases, the design or performance of the component is specified by the AAR.

Coupler Cushioning Switching of cars in a classification yard can result in relatively high coupler forces at the time of the impact between the moving and standing cars. Nominal coupling speeds of 4 mi/h (6.4 km/h) or less are sometimes exceeded, with lading damage a possible result. Conventional cars are equipped with an AAR-approved draft gear, usually a friction-spring energy-absorbing device, mounted between the coupler and the car body. The rated capacity of draft gears ranges between 20,000 ft-lb (27.1 kJ) for earlier units to over 65,000 ft · lb (88.1 kJ) for later designs. Impact forces of 1,250,000 lb (5.56 MN) can be expected when a moving 100-ton car strikes standing cars at 8 to 10 mi/h (12.8 to 16 km/h). Hydraulic cushioning devices are available to reduce the impact force to 500,000 lb (2.22 MN) at coupling speeds of 12 to 14 mi/h (19 to 22 km/h). These devices may be mounted either at each end of the car (end-of-car devices) or in a long beam which extends from coupler to coupler (sliding centersill devices).

Lading Restraint Many forms of lading restraint are available, from tie-down chains for automobiles on rack cars to movable bulkheads for boxcars. Most load-restraining devices are specified by the AAR "Manual of Standards and Recommended Practices" and approved car loading arrangements are specified in AAR "Loading Rules," a multivolume publication for enclosed, open-top, and TOFC and COFC cars.

Covered Hopper Car Discharge Gates The majority of covered hopper cars are equipped with rack-and-pinion–operated sliding gates, which allow the lading to discharge by gravity between the rails. These gates can be operated manually, with a simple bar or a torque-multiplying wrench, or mechanically with an impact or hydraulic wrench. Many special covered hopper cars have discharge gates with nozzles and metering devices for vacuum or pneumatic unloading.

Coupling Systems The majority of freight cars are connected with AAR standard couplers. A specification has been developed to permit the use of alternative coupling systems such as articulated connectors, drawbars, and rotary-dump couplers; see AAR M-215.

Table 11.2.3   Wheel and Journal Sizes of Eight-Wheel Cars

Nominal car capacity, ton	Maximum gross weight, lb	Journal (bearing) size, in	Bearing AAR class	Wheel diameter, in
50	177,000	5½ × 10		33
*	179,200	6 × 11	E	28
70	220,000	6 × 11	E	33
100	263,000	6½ × 12	F	36
110†	286,000	6½ × 12	F	36
110	286,000	6½ × 9	K	36
125†	315,000	7 × 12	G	38

* Load limited by wheel rating.
† Not approved for free interchange.

Fig. 11.2.18 Wheel-tread designs. (*a*) Obsolete standard AAR; (*b*) Heumann; (*c*) new AAR-1B.

Freight-Train Braking

The retarding forces acting on a railway train are rolling and mechanical resistance, aerodynamic drag, curvature, and grade, plus that force resulting from friction of brake shoes rubbing the wheel treads. On locomotives so equipped, dynamic or rheostatic brake using the traction motors as generators can provide all or a portion of the retarding force to control train speed.

Quick-action automatic air brakes of the type specified by the AAR are the common standard in North America. With the automatic air brake system, the brake pipe extends through every vehicle in the train, connected by hoses between each locomotive unit and car. The front and rear end brake pipe angle cocks are closed.

Air pressure is provided by compressors on the locomotive units to the main reservoirs, usually at 130 to 150 lb/in^2 (900 to 965 kPa). (Pressure values are gage pressures.) The engineer's automatic brake valve, in "release" position, provides air to the brake pipe on freight trains at reduced pressure, usually at 75, 80, 85, or 90 lb/in^2 (520, 550, 585, or 620 kPa), depending on the type of service, train weight, grades, and speeds at which a train will operate. In passenger service, brake pipe pressure is usually 90 or 110 lb/in^2 (620 to 836 kPa).

When brake pipe pressure is increased, the control valve allows the reservoir capacity on each car and locomotive to be charged and at the same time connects the brake cylinders to exhaust. Brake pipe pressure is reduced when the engineer's brake valve is placed in a "service" position, and the control valve cuts off the charging function and allows the reservoir air on each car to flow into the brake cylinder. This moves the piston and, through a system of levers and rods, pushes the brake shoes against the wheel treads.

When the engineer's automatic brake valve is placed in the emergency position, the brake pipe pressure (BP) is reduced very rapidly. The control valves on each car move to the emergency-application position and rapidly open a large vent valve, exhausting brake pipe pressure to atmosphere. This will serially propagate the emergency application through the train at from 900 to 950 ft/s (280 to 290 m/s.) With the control valve in the emergency position both auxiliary- and emergency-reservoir volumes (pressures) equalize with the brake cylinder, and higher brake cylinder pressure (BCP) results, building up at a faster rate than in service applications.

The foregoing briefly describes the functions of the fundamental automatic air brake based on the functions of the control valve. AAR-approved brake equipment is required on all freight cars used in interchange service. The functions of the control valve have been refined to permit the handling of longer trains by more uniform brake performance. Important improvements in this design have been (1) reduction of the time required to apply the brakes on the last car of a train,

(2) more uniform and faster release of the brakes, and (3) availability of emergency application with brake pipe pressure greater than 40 lb/in^2 (275 kPa). Faster brake response has been achieved by a number of railroads using "distributed power," where radio-controlled locomotives are spaced within the train or placed at both ends of the train.

The braking ratio of a car is defined as the ratio of brake shoe (normal) force to the car's rated gross weight. Two types of brake shoes, high-friction composition and high-phosphorus cast iron, are used in interchange service. As these shoes have very different friction characteristics, different braking ratios are required to assure uniform train braking performance (Table 11.2.4). Actual or net shoe forces are measured with calibrated devices. The calculated braking ratio R (nominal) is determined from the equation

$$R = PLANE \times 100/W \qquad (11.2.6)$$

where P = brake cylinder pressure, 50 lb/in^2 gage; L = mechanical ratio of brake levers; A = brake cylinder area, in^2; N = number of brake cylinders; E = brake rigging efficiency = $E_r \times E_b \times E_c$; and W = car weight, lb.

To estimate rigging efficiency consider each pinned joint and horizontal sliding joint as a 0.01 loss of efficiency; i.e., in a system with 20 pinned and horizontal sliding joints, $E_r = 0.80$. For unit-type (hangerless) brake beams $E_b = 0.90$ and for the brake cylinder $E_c = 0.95$, giving an overall efficiency of 0.684 or 68.4 percent.

The total retarding fore in pounds per ton may be taken as:

$$F = (PLef/W) = F_g G \qquad (11.2.7)$$

where P = total brake-cylinder piston force, lbf; L = multiplying ratio of the leverage between cylinder pistons and wheel treads; ef = product of the coefficient of brake shoe friction and brake rigging efficiency; W = loaded weight of vehicle, tons; F_g = force of gravity, 20 lb/ton/percent grade; G = ascending grade, percent.

Stopping distance can be found by adding the distance covered during the time the brakes are fully applied to the distance covered during the equivalent instantaneous application time.

$$S = \frac{0.0334 V_2^2}{\left[\dfrac{W_n B_n (p_a/p_n) ef}{W_a} \right] + \left(\dfrac{R}{2000} \right) \pm (G)} + 1.467\, t_1 \left[V_1 - \left(\frac{R + 2000G}{91.1} \right) \frac{t_1}{2} \right] \qquad (11.2.8)$$

where S = stopping distance, ft; V_1 = initial speed when brake applied, mi/h; V_2 = speed, mi/h, at time t_1; W_n = weight on which

Table 11.2.4   Braking Ratios, AAR Standard S-401

Type of brake rigging and shoes	With 50 lb/in^2 brake cylinder pressure			Hand brake*
	Percent of gross rail load		Maximum percent of light weight	Minimum percent of gross rail load
	Min.	Max.		
Conventional body-mounted brake rigging or truck-mounted brake rigging using levers to transmit brake cylinder force to the brake shoes				
Cars equipped with cast iron brake shoes	13	20	53	13
Cars equipped with high-friction composition brake shoes	6.5	10	30	11
Direct-acting brake cylinders not using levers to transmit brake cylinder force to the brake shoes				
Cars equipped with cast iron brake shoes				
Cars equipped with high-friction composition brake shoes	6.5	10	33	11
Cabooses†				
Cabooses equipped with cast iron brake shoes			35–45	
Cabooses equipped with high-friction composition brake shoes			18–23	

* Hand brake force applied at the horizontal hand brake chain with AAR certified or AAR approved hand brake.
† Effective for cabooses ordered new after July 1, 1982, hand brake ratios for cabooses to the same as lightweight ratios for cabooses.
NOTE: Above braking ratios also apply to cars equipped with empty and load brake equipment.

braking ratio B_n is based, lb (see Table 11.2.5 for values of W_n for freight cars; for passenger cars and locomotives: W_n is based on

Table 11.2.5   Capacity versus Load Limit for Railroad Freight Cars

Capacity, ton	W_n, 1,000 lb
50	177
70	220
100	263
110	286
125	315

empty or ready-to-run weight); B_n = braking ratio (total brake shoe force at stated brake cylinder, lb/in², divided by W_n); P_n = brake cylinder pressure on which B_n is based, usually 50 lb/in²; P_a = full brake cylinder pressure; e = overall rigging and cylinder efficiency, decimal; f = typical friction of brake shoes, see below; R = total resistance, mechanical plus aerodynamic and curve resistance, lb/ton; G = grade in decimal, + upgrade, − downgrade; t_1 = equivalent instantaneous application time, s.

Equivalent instantaneous application time is that time on a curve of average brake cylinder buildup versus time for a train or car where the area above the buildup curve is equal to the area below the curve. A straight-line buildup curve starting at zero time would have a t_1 of half the total buildup time.

The friction coefficient f varies with the speed; it is usually lower at high speed. To a lesser extent, it varies with brake shoe force and with the material of the wheel and shoe. For stops below 60 mi/h (97 km/h), a conservative figure for a high-friction composition brake shoe on steel wheels is approximately

$$ef = 0.30 \qquad (11.2.9)$$

In the case of high-phosphorus iron shoes, this figure must be reduced by approximately 50 percent.

P_n is based on 50 lb/in² (345 kPa) air pressure in the cylinder; 80 lb/in² is a typical value for the brake pipe pressure of a fully charged freight train. This will give a 50 lb/in² (345 kPa) brake cylinder pressure during a full service application on AB equipment, and a 60 lb/in² (415 kPa) brake cylinder pressure with an emergency application.

To prevent wheel sliding, $F_R \leq \phi W$, where F_R = retarding force at wheel rims resisting rotation of any pair of connected wheels, lb; ϕ = coefficient of wheel-rail adhesion or friction (a decimal); and W = weight upon a pair of wheels, lb. Actual or adhesive weight on wheels when the vehicle is in motion is affected by weight transfer (force transmitted to the trucks and axles by the inertia of the car body through the truck center plates), center of gravity, and vertical oscillation of body weight upon truck springs. The value of ϕ varies with speed as shown in Fig. 11.2.19.

The relationship between the required coefficient ϕ of wheel-rail adhesion to prevent wheel sliding and rate of retardation A is miles per hour per second may be expressed by $A = 21.95\phi_1$.

There has been some encouraging work to develop an electric brake that may eventually obsolete the present pneumatic brake systems.

Test Devices   Special devices have been developed testing brake components and cars on a repair track.

End-of-Train Devices   To eliminate the requirement for a caboose crew car at the end of the train, special electronic devices have been developed to transmit the end-of-train brake-pipe pressure to the locomotive operator by telemetry.

PASSENGER EQUIPMENT

During the past two decades most mainline or long-haul passenger service in North America has become a function of government agencies, i.e., Amtrak in the United States and Via Rail in Canada. Equipment for

Fig. 11.2.19   Typical wheel-rail adhesion. Track has jointed rails. *(Air Brake Association.)*

intraurban service is divided into three major categories: commuter rail, heavy rail rapid transit, and light rail transit, depending upon the characteristics of the service. Commuter rail equipment operates on conventional railroad rights-of-way, usually intermixed with other long-haul passenger and freight traffic. Heavy-rail rapid transit (HRT), often referred to as "metro," operates on a dedicated right-of-way, which is commonly in subways or on elevated structures. Light-rail transit (LRT) utilizing light rail vehicles (LRV) has evolved from the older trolley or streetcar concepts and may operate in any combination of surface, subway, and elevated dedicated rights-of-way, semireserved surface rights-of-way with grade crossing, and shared rights-of-way with other traffic on surface streets. In a few cases LRT shares the trackage with freight operations, which are time-separated to comply with FRA regulations.

Since mainline and commuter rail equipment operates over conventional railroad rights-of-way, the structural design is heavy to provide the FRA-required "crashworthiness" of vehicles in the event of collisions with other trains or at grade crossings with automotive vehicles. Although HRT vehicles are designed to stringent structural criteria, the requirements are somewhat different, since the operation is separate from freight equipment and there are usually no highway grade crossings. Minimum weight is particularly important for transit vehicles so that demanding schedules over lines with close station spacing can be met with minimum energy consumption.

Main-Line Passenger Equipment   There are four primary passenger train sets in operation across the world: diesel locomotive hauled, diesel multiple unit (DMU), electric locomotive hauled, and electric MU (EMU). In recent years the design of mainline passenger equipment has been controlled by specifications provided by APTA and the operating authority. Most of the newer cars provided for Amtrak have had stainless steel structural components. These cars have been designed to be locomotive-hauled and to use a separate 480-V three-phase power supply for heating, ventilation, air conditioning, food car services, and other control and auxiliary power requirements. Figure 11.2.20a shows an Amtrak coach car of the Superliner class; Fig. 11.2.20b shows an intercity high-speed coach.

Trucks for passenger equipment are designed to provide a superior ride to freight-car trucks. As a result, passenger trucks include a form of "primary" suspension to isolate the wheel set from the frame of the truck. A softer "secondary" suspension is provided to isolate the truck from the car body. In most cases, the primary suspension uses either coil springs, elliptical springs, or elastomeric components. The secondary suspension generally utilizes either large coil ring springs or pneumatic springs with special leveling valves to control the height of the car body. Hydraulic dampers are also applied to improve the vertical and lateral ride quality.

Fig. 11.2.20 Mainline passenger cars. (*a*) Double-deck coach; (*b*) Single-deck coach. (*Courtesy of Bombardier Transportation.*)

Fig. 11.2.21   Commuter rail car. *(Courtesy of Bombardier Transportation.)*

Commuter Rail Passenger Equipment   Commuter rail equipment can be either locomotive-hauled or self-propelled (Fig. 11.2.21). Some locomotive-hauled equipment is arranged for "push-pull" service. This configuration permits the train to be operated with the locomotive either pushing or pulling the train. For push-pull service, some of the passenger cars must be equipped with cabs to allow the engineer to operate the train from the end opposite the locomotive during the push operation. All cars must have control trainlines to connect the lead (cab) car to the trailing locomotive. Locomotive-hauled commuter rail cars use AAR-type H couplers. Most self-propelled (single- or multiple-unit) cars use other automatic coupler designs, which can be mechanically coupled to an AAR coupler with an adaptor.

Heavy Rail Rapid Transit Equipment   This equipment is used on traditional subway/elevated properties in such cities as Boston, New York, Philadelphia (Fig. 11.2.22), and Chicago in a semiautomatic mode of operation over dedicated rights-of-way which are constrained by limiting civil features. State-of-the-art subway-elevated properties include such cities as Washington, Atlanta, Miami, and San Francisco,

Fig. 11.2.22   Heavy rail transit car. *(Courtesy of Bombardier Transportation.)*

where the equipment provides highly automated modes of operation on rights-of-way with generous civil alignments.

The cars can operate bidirectionally in multiple with as many as 12 or more cars controlled from the leading cab. They are electrically propelled, usually from a dc third rail, which makes contact with a shoe insulated from and supported by the frame of the truck. Occasionally, roof-mounted pantographs are used. Voltages range from 600 to 1,500 V dc.

The cars range from 48 to 75 ft (14.6 to 22.9 m) over the anticlimbers, the longer cars being used on the newer properties. Passenger seating varies from 40 to 80 seats per car, depending upon length and the local policy (or preference) regarding seated to standee ratio. Older properties require negotiation of curves as sharp as 50 ft (15.2 m) minimum radius with speeds up to only 50 mi/h (80 km/h) on tangent track, while newer properties usually have no less than 125 ft (38.1 m) minimum radius curves with speeds up to 75 mi/h (120 km/h) on tangent track.

All North American properties operate on standard-gage track with the exception of the 5 ft 6 in (1.7 m) San Francisco Bay Area Rapid Transit (BART), the 4 ft 10 in (1.5 m) Toronto Transit Subways, and the 5 ft 2½ in (1.6 m) Philadelphia Southeastern Pennsylvania Transportation Authority (SEPTA) Market-Frankford line. Grades seldom exceed 3 percent, and 1.5 to 2.0 percent is the desired maximum.

Typically, newer properties require maximum acceleration rates of between 2.5 and 3.0 mi/(h · s) [4.0 to 4.8 km/(h · s)] as nearly independent of passenger loads as possible from 0 to approximately 20 mi/h (32 km/h). Depending upon the selection of motors and gearing, this rate falls off as speed is increased, but rates can generally be controlled at a variety of levels between zero and maximum.

Deceleration is typically accomplished by a blended dynamic and electropneumatic friction tread or disk brake, although a few properties use an electrically controlled hydraulic friction brake. In some cases energy is regenerated and sent back into the power system. All of these systems usually provide a maximum braking rate of between 3.0 and 3.5 mi/(h · s) [4.8 and 5.6 km/(h · s)] and are made as independent of passenger loads as possible by a load-weighing system that adjusts braking effort to suit passenger loads. Some employ regenerative braking to supplement the other brake systems.

Dynamic braking is generally used as the primary stopping mode and is effective from maximum speed down to 10 mi/h (16 km/h) with friction braking supplementation as the characteristic dynamic fade occurs. The friction brakes provide the final stopping forces. Emergency braking rates depend upon line constraints, car subsystems, and other factors, but generally rely on the maximum retardation force that can be provided by the dynamic and friction brakes within the limits of available wheel-to-rail adhesion.

Acceleration and braking on modern properties are usually controlled by a single master controller handle that has power positions in one direction and a coasting or neutral center position and braking positions in the opposite direction. A few properties use foot-pedal control with a "deadman" pedal operated by the left foot, a brake pedal operated by the right foot, and an accelerator (power) pedal also operated by the right foot. In either case, the number of positions depends upon property policy and control subsystems on the car. Control elements include motor-current sensors and some or all of the following: speed sensors, rate sensors, and load-weighing sensors. Signals from these sensors are processed by an electronic control unit (ECU), which provides control functions to the propulsion and braking systems. The propulsion systems currently include pilot-motor-operated cams which actuate switches or electronically controlled unit switches to control resistance steps, or chopper or inverter systems that electronically provide the desired voltages and currents at the motors.

In some applications, dynamic braking utilizes the traction motors as generators to dissipate energy through the onboard resistors also used in acceleration. It is expected that regenerative braking will become more common since energy can be returned to the line for use by other cars. Theoretically, 35 to 50 percent of the energy can be returned, but the present state-of-the-art is limited to a practical 20 percent on properties with large numbers of cars on close headways.

Car bodies are made of welded stainless steel or low-alloy high-tensile (LAHT) steel of a design that carries structural loads. Earlier problems with aluminum, primarily electrolytic action among dissimilar metals and welding techniques, have been resolved and aluminum is also used on a significant number of new cars.

Trucks may be cast or fabricated steel with frames and journal bearings either inside or outside the wheels. Axles are carried in roller-bearing journals connected to the frames so as to be able to move vertically against a variety of types of primary spring restraint. Metal springs or air bags are used as the secondary suspension between the trucks and the car body. Most wheels are solid cast or wrought steel. Resilient wheels have been tested but are not in general service on heavy rail transit equipment.

All heavy rail rapid transit systems use high-level loading platforms to speed passenger flow. This adds to the civil construction costs, but is necessary to achieve the required level of service.

Light Rail Transit Equipment   The cars are called light rail vehicles (Fig. 11.2.23 shows typical **LRVs**) and are used on a few remaining

(a)

(b)

Fig. 11.2.23 Light-rail vehicle (LRV). (*a*) Articulated; (*b*) nonarticulated streetcar systems such as those in Boston, Philadelphia, and Toronto on city streets and on state-of-the-art subway, surface, and elevated systems such as those in Calgary, Edmonton, Los Angeles, San Diego, and Portland (Oregon) and in semiautomated modes over partially or wholly reserved rights-of-way.

As a practical matter, the LRV's track, signal systems, and power systems utilize the same subsystems as heavy rail rapid transit and are not "lighter" in a physical sense.

Most LRVs are designed to operate bidirectionally in multiple, with up to four cars controlled from the leading cab. The LRVs are electrically propelled from an overhead contact wire, often of catenary design, that makes contact with a pantograph or pole on the roof. For reasons of wayside safety, third-rail pickup is not used. Voltages range from 550 to 750 V dc.

The cars range from 60 to 65 ft (18.3 to 19.8 m) over the anticlimbers for single cars and from 70 to 90 ft (21.3 to 27.4 m) for articulated cars. The choice is determined by passenger volumes and civil constraints.

Articulated cars have been used in railroad and transit applications since the 1920s, but have found favor in light rail applications because of their ability to increase passenger loads in a longer car that can negotiate relatively tight-radius curves. These cars require the additional mechanical complexity of the articulated connection and a third truck between two car-body sections.

Passenger seating varies from 50 to 80 or more seats per car, depending upon length, placement of seats in the articulation, and the policy regarding seated/standee ratio. Older systems require negotiation of curves down to 30 ft (9.1 m) minimum radius with speeds up to only 40 mi/h (65 km/h) on tangent track, while newer systems usually have no less than 75 ft (22.9 m) minimum radius curves with speeds up to 65 mi/h (104 km/h) on tangent track.

The newer properties all use standard-gage track; however, existing older systems include 4 ft 10 in (1,495 mm) and 5 ft 2½ in (1,587 mm) gages. Grades have reached 12 percent but 6 percent is now considered the maximum and 5 percent is preferred.

Typically, newer properties require maximum acceleration rates of between 3.0 to 3.5 mi/(h · s) [4.8 to 5.6 km/(h · s)] as nearly independent of passenger load as possible from 0 to approximately 20 mi/h (32 km/h). Depending upon the selection of motors and gearing, this rate falls off as speed is increased, but the rates can generally be controlled at a variety of levels.

Unlike most heavy rail rapid transit cars, LRVs incorporate three braking modes: dynamic, friction, and track brake, which typically provide maximum service braking at between 3.0 and 3.5 mi/(h · s) [4.8 to 5.6 km/(h · s)] and 6.0 mi/(h · s) [9.6 km/(h · s)] maximum emergency braking rates. The dynamic and friction brakes are usually blended, but a variety of techniques exist. The track brake is intended to be used primarily for emergency conditions and may or may not be controlled with the other braking systems.

The friction brakes are almost exclusively disk brakes, since LRVs use resilient wheels that can be damaged by tread-brake heat buildup. No single consistent pattern exists for the actuation mechanism. All-electric, all-pneumatic, electropneumatic, electrohydraulic, and electropneumatic over hydraulic are in common use.

Dynamic braking is generally used as the primary braking mode and is effective from maximum speed down to about 5 mi/h (8 km/h) with friction braking supplementation as the characteristic dynamic fade occurs.

As with heavy rail rapid transit, the emergency braking rates depend upon line constraints, car-control subsystems selected, and other factors, but the use of track brakes means that higher braking rates can be achieved because the wheel-to-rail adhesion is not the limiting factor.

Acceleration and braking on modern properties are usually controlled by a single master controller handle that has power positions in one direction and a coasting or neutral center position and braking positions in the opposite direction. A few properties use foot-pedal control with a "deadman" pedal operated by the left foot, a brake pedal operated by the right foot, and an accelerator (power) pedal also operated by the right foot. In either case, the number of positions depends upon property policy and control subsystems on the car.

Control elements include motor-current sensors and some or all of the following: speed sensors, rate sensors, and load-weighing sensors. Signals from these sensors are processed by an electronic control unit (ECU), which provides control functions to the propulsion and braking systems.

The propulsion systems currently include pilot-motor-operated cams that actuate switches or electronically controlled unit switches to control resistance steps. Chopper and inverter systems that electronically provide the desired voltages and currents at the motors are also used.

Most modern LRVs are equipped with two powered trucks. In two-section articulated designs (Fig. 11.2.23a), the third (center) truck may be left unpowered but usually has friction and track brake capability. Some European designs use three powered trucks but the additional cost and complexity have not been found necessary in North America.

Unlike heavy rail rapid transit, there are three major dc-motor configurations in use: the traditional series-wound motors used in bimotor trucks, the European-derived monomotor, and a hybrid monomotor with a separately excited field—the last in chopper-control version only.

The bimotor designs are rated between 100 and 125 shaft hp per motor at between 300 and 750 V dc, depending upon line voltage and series or series-parallel control schemes (electronic or electromechanical control). The monomotor designs are rated between 225 and 250 shaft hp per motor at between 300 and 750 V dc (electronic or electromechanical control).

The motors, gear units (right angle or parallel), and axles are joined variously through flexible couplings. In the case of the monomotor, it is supported in the center of the truck, and right-angle gearboxes are mounted on either end of the motor. Commonly, the axle goes through the gearbox and connection is made with a flexible coupling arrangement. Electronic inverter control drives with ac motors have been applied in recent conversions and new equipment.

Dynamic braking is achieved in the same manner as in heavy rail rapid transit.

Unlike heavy rail rapid transit, LRV bodies are usually made only of welded LAHT steel and are of a load-bearing design. Because of the semireserved right-of-way, the risk of collision damage with automotive vehicles is greater than with heavy rail rapid transit and the LAHT steel has been found to be easier to repair than stainless steel or aluminum. Although LAHT steel requires painting, this can be an asset since the painting can be performed in a highly decorative manner pleasing to the public and appropriate to themes desired by the cities.

Trucks may be cast or fabricated steel with either inside or outside frames. Axles are carried in roller-bearing journals, which are usually resiliently coupled to the frames with elastomeric springs as a primary suspension. Both vertical and a limited amount of horizontal movement occur. Since tight curve radii are common, the frames are usually connected to concentric circular ball-bearing rings, which, in turn, are connected to the car body. Air bags, solid elastomeric springs, or metal springs are used as a secondary suspension. Resilient wheels are used on virtually all LRVs.

Newer LRVs have low-level loading doors and steps, which minimize station platform costs.

TRACK

Gage   The gage of railway track is the distance between the inner sides of the rail heads measured 0.626 in (15.9 mm) below the top running surface of the rail. North American railways are laid with a nominal gage of 4 ft 8½ in (1,435 mm), which is known as *standard gage*. Rail wear causes an increase in gage. On sharp curves [over 8°, see Eq. (11.2.10)], it has been the practice to widen the gage to reduce rail wear.

Track Structure   In basic track structure is composed of six major elements: rail, tie plates, fasteners, cross ties, ballast, and subgrade (Fig. 11.2.24).

Rail   The American Railway Engineering and Maintenance of Way Association had defined **standards** for the cross section of rails of

Fig. 11.2.24 Track structure.

various weights, which are identified by their weight (in pounds) per yard. The AREMA standards provide for rail weights varying from 90 to 141 lb/yd. However, recommendations for new rail purchases are limited to weights of 115 lb/yd and above. Many railroads historically used special sections of their own design; however, the freight industry is converging on the use of the 115-, 136-, and 141-lb/yd sections. The standard length in use is 80 ft (31.5 m), although some mills are now considering making longer lengths available.

The prevailing weight of rails used in mainline track is from 112 to 141 lb/yd. Secondary and branch lines are generally laid with 90- to 115-lb/yd or heavier rail that has been previously used and partly worn in mainline service. In a very limited application, rail weights as low as 60 lb/yd exist in some tracks having very light use. In the past, rail sections were connected by using bolted joint bars. Current practice is the use of continuous welded rail (CWR). In general, rails are first welded into lengths averaging 1,440 feet (440 m) or more. After these strings, or "ribbons," are installed, they are later field welded to eliminate conventional joints. At one time the gas-fusion welding process was common, but today almost all rail is welded by the electric-flash or aluminothermic processes. CWR requires the use of more rail anchors than jointed rail in order to clamp the base of the rail and position it on the cross ties to prevent rail movement. This is necessary to restrain longitudinal forces that result from thermal expansion and contraction with ambient temperature changes. Failure to adequately restrain the rail can result in track buckling in extremely hot weather or in rail "pull aparts" in extremely cold weather. Care must be exercised to install the rail at temperatures that do not approach high or low extremes.

A variety of heat-treating and alloying processes have been used to produce rail that is more resistant to wear and less susceptible to fatigue failure. Comprehensive studies indicate that both gage face and top-of-rail lubrication can dramatically reduce the wear rate of rail in curves (see publications of AAR Research and Test Department). Friction modification using top-of-rail lubrication must be carefully managed so that it does not interfere with traction required to move and brake trains. Reduced wear and higher permissible axle loads have led to fatigue becoming the dominant cause for rail replacement at extended life. Grinding the running surface of rail has been greatly increased to reduce the damage caused by fatigue.

Tie Plates Rail is supported on tie plates that restrain the rail laterally, transmit the vehicle loads to the cross tie, and position (cant) the rail for optimum vehicle performance. Tie plates cant the rail toward the gage side. Cants vary from 1:14 to 1:40, depending on service conditions and operating speeds, with 1:40 being most common.

Fasteners Rail and tie plates are fixed to the cross ties by fasteners. Conventional construction with wood cross ties utilizes steel spikes driven into the tie through holes in the tie plate. In some cases, screw-type fasteners are used to provide more resistance to vertical forces. Elastic, clip-type fasteners are used on concrete and, at times, on wooden cross ties to provide a uniform, resilient attachment and longitudinal restraint in lieu of anchors. This type of fastening system also reduces the likelihood of rail overturns.

Cross Ties To retain gage and provide a further distribution of vehicle loads to the ballast, lateral cross ties are used. The majority of cross ties used in North America are wood. However, concrete cross ties are being used more frequently in areas where track surface and lateral

stability are difficult to maintain and in high-speed passenger corridors. Slab track designs are now being given scrutiny for high-utilization areas.

Ballast Ballast serves to further distribute vehicle loads to the subgrade, restrain vertical and lateral displacement of the track structure, and provide for drainage. Typical mainline ballast materials are limestone, trap rock, granite, or slag. Lightly built secondary lines have used gravel, cinders, or sand in the past.

Subgrade The subgrade serves as the interface between the ballast and the native soil. Subgrade material is typically compacted or stabilized soil. Instability of the subgrade due to groundwater permeation is a concern in some areas. Subgrade instability must be analyzed and treated on a case-by-case basis.

Track Geometry Railroad maintenance practices, state standards, AREMA standards, and, more recently, FRA safety standards (49 CFR Title § 113) specify limitations on geometric deviations for track structure. Among the characteristics considered are:

Gage
Alignment (the lateral position of the rails)
Profile (the elevation of the rail)
Curvature
Easement (the rate of change from tangent to curve)
Superelevation (the elevation of the outer or "high" rail in a curve compared to the inner or "low" rail)
Runoff (the rate of change in superelevation)
Cross level (the relative height of opposite rails)
Twist (the change in cross level over a specified distance)

FRA standards specify maximum operating speeds by class of track ranging from Class 1 (lowest acceptable) to Class 9. Track geometry is a significant consideration in vehicle suspension design. For operating speeds greater than 110 mi/h, FRA has promulgated regulations (track Classes 6 through 9) that require qualification of locomotives and cars through a test program over each operating route. These tests call for comprehensive measurement of wheel/rail forces and vehicle accelerations (49 CFT § 213).

Curvature The curvature of track is designated in terms of "degrees of curvature." The degree of curve is the number of degrees of central angle subtended by a chord of 100-ft length (measured on the track centerline). Equation (11.2.10) gives the approximate radius R in feet for ordinary railway curves.

$$R = \frac{5,730}{\text{degrees per 100-ft chord}} \qquad (11.2.10)$$

On important main lines where trains are operated at relatively high speed, the curves are ordinarily not sharper than 6 to 8°. In mountainous territory, and in other rare instances, curves as sharp as 18° occur. Most diesel and electric locomotives are designed to traverse curves up to 21°. Most uncoupled cars will pass considerably sharper curves, the limiting factor in this case being the clearance of the truck components to car body structural members or equipment or the flexibility of the brake connections to the trucks (AAR "Manual of Standards and Recommended Practices," Sec. C).

Clearances For new construction, the AREMA standard clearance diagrams provide for a clear height of 23 ft (7.0 m) above the tops of the railheads and for a width of 18 ft (5.5 m). Where conditions require, as in tunnels and bridges, these dimensions are reduced at the top and bottom of the diagram. For tracks entering buildings, an opening 16 ft (4.9 m) wide and 18 ft (5.5 m) high is recommended and will ordinarily suffice to pass the largest locomotives and most freight cars. A standard car clearance diagram for railway bridges (Fig. 11.2.25) represents the AAR recommendation for new construction. Minimum clearances are dictated by state authorities, and are summarized in Chapter 28, Table 3.3, of the AREMA Manual. Individual railroads have their own standards for existing lines and new construction, which are generally more stringent than the legal requirements. The actual clearance limitations on a particular railroad can be found in the official publication "Railway Line Clearances," Railway Equipment and Publishing Co.

Fig. 11.2.25 Wayside clearance diagram. *(AREA "Manual for Railway Engineering—Fixed Properties.")*

Track Spacing The distance between centers of mainline track commonly varies between 12 and 15 ft (3.6 and 4.6 m). Twelve-foot spacing exists, however, only on older construction. Most states require a minimum of 14 ft 0 in (4.3 m) between main tracks; however, many railroads have adopted 15 ft 0 in (3.6 m) as their standard. In a few cases, where space permits, 20 ft (6.1 m) is being used.

Turnouts Railway turnouts are defined by frog number. The frog number is the cotangent of the frog angle, which is the ratio of the lateral displacement of the diverging track per unit of longitudinal distance along the nondiverging track. Typical turnouts in yard and industry trackage range from number 7 to number 10. For mainline service, typical turnouts range from number 10 to number 20. The other important part of a turnout is the switch. The switch has movable points that direct the wheel down the straight or diverging side of the turnout. The higher the turnout number, the longer the switch, and the higher the allowable speed for trains traversing the diverging track. Allowable speeds are governed by the curvature of the diverging tack, and the angle of the switch. As a general rule, the allowable speed (in mi/h) is slightly less than twice the turnout number. In recent track construction for high-speed service special turnouts with movable point frogs have been applied.

Noise Allowable noise levels from railroad operation are prescribed by 40 CFR § 201.

VEHICLE-TRACK INTERACTION

Train Resistance The resistance to a train in motion along the track is of prime interest, as it is reflected directly in locomotive energy requirement. This resistance is expressed in terms of pounds per ton of train weight. **Gross train resistance** is that force that must be overcome by the locomotives at the driving-wheel–rail interface. **Trailing train resistance** must be overcome at the rear drawbar of the locomotive.

There are two classes of resistance that must be overcome: inherent and incidental. **Inherent resistance** includes the rolling resistance of bearings and wheels and aerodynamic resistance due to motion through still air. It may be considered equal to the force necessary to maintain motion at constant speed on level tangent track in still air. **Incidental resistance** includes resistance due to grade, curvature, wind, and vehicle dynamics.

Inherent Resistance Of the elements of inherent resistance, at low speeds rolling resistance is dominant, but at high speeds aerodynamic resistance is the predominant factor. Attempts to differentiate and evaluate the various elements through the speed range are a continuing part of

industry research programs to reduce train resistance. At very high speeds, the effect of air resistance can be approximated as an aid to studies in its reduction by means of cowling and fairing. The resistance of a car moving in still air on straight, level track increases parabolically with speed. Because the aerodynamic resistance is independent of car weight, the resistance in pounds per ton decreases as the weight of the car increases. The total resistance in pounds per ton of a 100-ton car is much less than twice as great as that of a 50-ton car under similar conditions. With known conditions of speed and car weight, inherent resistance can be predicted with reasonable accuracy. Knowledge of track conditions will permit further refining of the estimate, but for very rough track or extremely cold ambient temperatures, generous allowances must be made. Under such conditions normal resistance may be doubled. A formula proposed by Davis (General Electric Review, Oct. 1926) and revised by Tuthill ("High Speed Freight Train Resistance." *Univ. of Illinois Engr. Bull.*, **376**, 1948) has been used extensively for inherent freight-train resistances at speeds up to 40 mi/h:

$$R = 1.3W + 29n + 0.045WV + 0.0005\,AV^2 \quad (11.2.11)$$

where R = train resistance, lb/car; W = weight per car, tons; V = speed, mi/h; n = total number of axles; A = cross-sectional area, ft^2.

With freight train speeds of 50 to 70 mi/h (80 to 112 km/h), it has been found that actual resistance values fall considerably below calculations based on the above formula. Several modifications of the Davis equation have been developed for more specific applications. All of these equations apply to cars trailing locomotives.

1. Davis equation as modified by I. K. Tuthill (*University of Illinois Engr. Bull*, **376**, 1948):

$$R = 1.3W + 29n + 0.045WV + 0.045V^2 \quad (11.2.12)$$

Note: In the Totten modification, the equation is augmented by a matrix of coefficients when the velocity exceeds 40 mi/h.

2. Davis equation as modified by the Canadian National Railway:

$$R = 0.6W + 20n + 0.01WV + 0.07V^2 \quad (11.2.13)$$

3. Davis equation as modified by the Canadian National Railway and former Erie-Lackawanna Railroad for trailers and containers on flat cars:

$$R = 0.6W + 20n + 0.01WV + 0.2V^2 \quad (11.2.14)$$

Other modifications of the Davis equation have been developed for passenger cars by Totten ("Resistance of Light Weight Passenger Trains," *Railway Age*, **103**, July 17, 1937). These formulas are for passenger cars pulled by a locomotive and do not include head-end air resistance.

1. Davis equations as modified by Totten for streamlined passenger cars:

$$R = 1.3W + 29n + 0.045WV + [0.00005 + 0.060725(L/100)^{0.88}]V^2$$
$$(11.2.15)$$

2. Davis equations as modified by Totten for nonstreamlined passenger cars

$$R = 1.3W + 29n + 0.045WV + [0.00005 + 0.1085(L/100)^{0.7}]V^2$$
$$(11.2.16)$$

where L = car length in feet.

Aerodynamic and Wind Resistance Wind-tunnel testing has indicated a significant effect on freight train resistance resulting from vehicle spacing, open tops of hopper and gondola cars, open boxcar doors, vertical side reinforcements on railway cars and intermodal trailers, and protruding appurtenances on cars. These effects can cause significant increases in train resistance at higher speeds. For example, the spacing of intermodal trailers or containers greater than approximately 6 ft can result in a new frontal area to be considered in determining train resistance. Frontal or cornering ambient wind conditions can also have an adverse effect on train resistance, which is increased with discontinuities along the length of the train.

Curve Resistance Train resistance due to track curvature varies with speed and degree of curvature. The behavior of rail vehicles in curve negotiation is the subject of several ongoing AAR studies.

Lubrication of the rail gage face or wheel flanges has become common practice for reducing friction and the resulting wheel and rail wear. Recent studies indicate that flange and/or gage face lubrication can significantly reduce train resistance on tangent track as well (Allen, Conference on the Economics and Performance of Freight Car Trucks, October 1983). In addition, a variety of special trucks (wheel assemblies) that reduce curve resistance by allowing axles to steer toward a radial position in curves have been developed. For general estimates of car resistance and locomotive hauling capacity on dry (unlubricated) rail with conventional trucks, speed and gage relief may be ignored and a figure of 0.8 lb/ton per degree of curvature used.

Grade resistance depends only on the angle of ascent or descent and relates only to the gravitational forces acting on the vehicle. It equates to 20 lb/ton for each percent of grade, or 0.379 lb/ton for each foot per mile rise.

Acceleration   The force (tractive effort) required to accelerate the train is the sum of the forces required for linear acceleration and that required for rotational acceleration of the wheels about their axle centers. A linear acceleration of 1 mi/h · s (1.6 km/h · s) is produced by a force of 91.1 lb/ton. The rotary acceleration requirement adds 6 to 12 percent, so that the total is nearly 100 lb/ton (the figure commonly used) for each mile per hour per second. If greater accuracy is required, the following expression is used:

$$R_a = A(91.05W + 36.36n) \qquad (11.2.17)$$

where R_a = the total accelerating force, lb; A = acceleration, mi/h · s; W = weight of train, tons; n = number of axles.

Acceleration and Distance   If a distance of S ft the speed of a car or train changes from V_1 to V_2 mi/h, the force required to produce acceleration (or deceleration if the speed is reduced) is

$$R_a = 74(V_2^2 - V_1^2)/S \qquad (11.2.18)$$

The coefficient, 74, corresponds to the use of 100 lb/ton. This formula is useful in the calculation of the energy required to climb a grade with the assistance of stored energy. In any train-resistance calculation or analysis, assumptions with regard to acceleration will generally submerge all other variables; e.g., an acceleration of 0.1 mi/h · s (0.16 km/h · s) requires more tractive force than that required to overcome inherent resistance for any car at moderate speeds.

Starting Resistance   Most railway cars are equipped with roller bearings requiring a starting force of 5 or 6 lb/ton.

Vehicle Suspension Design   The primary consideration in the design of the vehicle suspension system is to isolate track input forces from the vehicle car body and lading. In addition, there are a few specific areas of instability that railway suspension systems must address; see AAR "Manual of Standards and Recommended Practice," Sec. C-II-M-1001, Chap. XI.

Harmonic roll is the tendency of a freight car with a high center of gravity to rotate about its longitudinal axis (parallel to the track). This instability is excited by passing over staggered low rail joints at a speed that causes the frequency of the input for each joint to match the natural roll frequency of the car. Unfortunately, in many car designs, this occurs for loaded cars at 12 to 18 mi/h (19.2 to 28.8 km/h), a common speed for trains moving in yards or on branch lines where tracks are not well maintained. Many freight operations avoid continuous operation in this speed range.

This adverse behavior is more noticeable in cars with truck centers approximately the same as the rail length. The effect of harmonic roll can be mitigated by improved track surface and by damping in the truck suspension.

Pitch and bounce are the tendencies of the vehicle to either translate vertically up and down (bounce), or rotate (pitch) about a horizontal axis perpendicular to the centerline of track. This response is also excited by low track joints and can be relieved by increased truck damping.

Yaw is the tendency of the car to rotate about its axis vertical to the centerline of track. Yaw responses are usually related to truck hunting.

Truck hunting is an instability inherent in the design of the truck and dependent on the stiffness parameters of the truck and on wheel conicity (tread profile). The instability is observed as "parallelogramming" of the truck components at a frequency of 2 to 3 Hz, causing the car body to yaw or translate laterally. This response is excited by the effect of the natural frequency of the gravitational stiffness of the wheel set when the speed of the vehicle approaches the kinematic velocity of the wheel set. This problem is discussed in analytic work available from the AAR Research and Test Department.

Superelevation   As a train passes around a curve, there is a tendency for the cars to roll toward the outside of the curve in response to centrifugal force acting on the center of gravity of the car body (Fig. 11.2.26a). To compensate for this effect, the outside rail is superelevated, or raised, relative to the inside rail (Fig. 11.2.26b). The amount of superelevation for a particular curve is based upon the radius of the curve and the operating speed of the train. The "balance" or equilibrium speed for a given curve is that speed at which the weight of the vehicle is equally distributed on each rail. The FRA allows a railroad to operate with 3 in of unbalance, or at the speed at which equilibrium would exist if the superelevation were 3 in greater than that which exists. The maximum superelevation is usually 6 in but maybe lower if freight operation is used exclusively.

Longitudinal Train Action   Longitudinal train (slack) action is associated with the dynamic action between individual cars in a train. An example would be the effect of starting a long train in which the couplers between each car had been compressed (i.e., bunched up). As the locomotive begins to pull the train, the slack between the locomotive and the first car must be traversed before the first car begins to move. Next the slack between the first and second car must traversed before the second car beings to move, and so on. Before the last car in a long train begins to move, the locomotive and the moving cars may be traveling at a rate of several miles per hour. This effect can result in coupler forces sufficient to cause the train to break in two.

Longitudinal train (slack) action is also induced by serial braking, undulating grades, or by braking on varying grades. The AAR Track Train Dynamics Program has published guidelines titled "Track Train Dynamics to Improve Freight Train Performance" that explain the causes of undesirable train action and how to minimize its effects. Analysis of the forces developed by longitudinal train action requires the application of the Davis equation to represent the resistance of each vehicle based upon its velocity and location on a grade or curve. Also, the longitudinal stiffness of each car and the tractive effort of the locomotive must be considered in equations that model the kinematic response of each vehicle in the train. Computer programs are available from the AAR to assist in the analysis of longitudinal train action.

Fig. 11.2.26   Effect of superelevation on center of gravity of car body.

11.3 MARINE ENGINEERING
by Michael C. Tracy

REFERENCES: Harrington, "Marine Engineering," SNAME, 1992. Lewis, "Principles of Naval Architecture," SNAME, 1988/89. Myers, "Handbook of Ocean and Underwater Engineering," McGraw-Hill, 1969. "Rules for Building and Classing Steel Vessels," American Bureau of Shipping. Rawson and Tupper, "Basic Ship Theory," Oxford: Butterworth-Heinemann, 2001. Gillmer and Johnson, "Introduction to Naval Architecture," Naval Institute, 1982. Barnaby, "Basic Naval Architecture," Hutchinson, London, 1967. *Jour. Inst. Environ, Sci.* Taggert, "Ship Design and Construction," SNAME, 1980. *Trans. Soc. Naval Architects and Marine Engrs.,* SNAME. *Naval Engrs. Jour.,* Am. Soc. of Naval Engrs., ASNE. *Trans. Royal Inst. of Naval Architects,* RINA. Figures and examples herein credited to SNAME have been included by permission of The Society of Naval Architects and Marine Engineers.

Marine engineering is an integration of many engineering disciplines directed to the development and design of systems of transport, warfare, exploration, and natural-resource retrieval which have one thing in common: operation in or on a body of water. Marine engineers are responsible for the engineering systems required to propel, work, or fight ships. They are responsible for the main propulsion plant; the powering and mechanization aspects of ship functions such as steering, anchoring, cargo handling, heating, ventilation, air conditioning, etc.; and other related requirements. They usually have joint responsibility with naval architects in areas of propulsor design; hull vibration excited by the propeller or main propulsion plant; noise reduction and shock hardening, in fact, dynamic response of structures or machinery in general; and environmental control and habitability. Marine engineering is a distinct multidiscipline and characteristically a dynamic, continuously advancing technology.

THE MARINE ENVIRONMENT

Marine engineers must be familiar with their environment so that they may understand fuel and power requirements, vibration effects, and propulsion-plant strength considerations. The outstanding characteristic of the open ocean is its irregularity in storm winds as well as under relatively calm conditions. The **irregular sea** can be described by statistical mathematics based on the superposition of a large number of regular waves having different lengths, directions, and amplitudes. The characteristics of idealized regular waves are fundamental for the description and understanding of realistic, irregular seas. Actual sea states consist of a combination of many sizes of waves often running in different directions, and sometimes momentarily superimposing into an exceptionally large wave. Air temperature and wind also affect ship and machinery operation.

The effects of the marine environment also vary with water depth and temperature. As a ship passes from deep to shallow water, there is an appreciable change in the potential flow around the hull and a change in the wave pattern produced. Additionally, silt, sea life, and bottom growth may affect seawater systems or foul heat exchangers.

MARINE VEHICLES

The platform is additionally a part of marine engineers' environment. Ships are supported by a buoyant force equal to the weight of the volume of water displaced (Archimedes' principle). For surface ships, this weight is equal to the total weight of the structure, machinery, outfit, fuel, stores, crew, and useful load. The principal sources of resistance to propulsion are skin friction and the energy lost to surface waves generated by moving in the interface between air and water. Minimization of one or both of these sources of resistance has generally been a primary objective in the design of marine vehicles.

Displacement Hull Forms

Displacement hull forms are the familiar monohull, the catamaran, and the submarine. The moderate-to-full-displacement **monohull** form provides the best possible combinations of high-payload-carrying ability, economical powering characteristics, and good seakeeping qualities. A more slender hull form achieves a significant reduction in wave-making resistance, hence increased speed; however, it is limited in its ability to carry topside weight because of the low transverse stability of its narrow beam. **Multihull ships** provide a solution to the problem of low transverse stability. They are increasingly popular in sailing yachts, high-speed passenger ferries, and research and small support ships. Sailing catamarans, with their superior transverse stability permitting large sail-plane area, gain a speed advantage over monohull craft of comparable size. A powered catamaran has the advantage of increased deck space and relatively low roll angles over a monohull ship. The **submarine,** operating at depths which preclude the formation of surface waves, experiences significant reductions in resistance compared to a well-designed surface ship of equal displacement.

Planing Hull Forms

The planing hull form, although most commonly used for yachts and racing craft, is used increasingly in small, fast commercial craft and in coastal patrol craft. The weight of the planing hull ship is partially borne by the dynamic lift of the water streaming against a relatively flat or V-shaped bottom. The effective displacement is hence reduced below that of a ship supported statically, with significant reduction in wave-making resistance at higher speeds.

High-Performance Ships

In a search for high performance and higher speeds in rougher seas, several advanced concepts to minimize wave-making resistance have been developed, such as hydrofoil craft, surface-effect vehicles, and small-waterplane-area twin hull (SWATH) forms (Fig. 11.3.1). The **hydrofoil craft** has a planing hull that is raised clear of the water through dynamic lift generated by an underwater foil system. The **surface-effect vehicles** ride on a cushion of compressed air generated and maintained in the space between the vehicle and the surface over which it hovers or moves. The most practical vehicles employ a peripheral-jet principle, with flexible skirts for obstacle or wave clearance. A rigid sidewall craft, achieving some lift from hydrodynamic effects, is more adaptable to marine construction techniques. The **SWATH** gains the advantages of the catamaran, twin displacement hulls, with the further advantage of minimized wave-making resistance and wave-induced motions achieved by submarine-shaped hulls beneath the surface and the small water-plane area of the supporting struts at the air-water interface.

SEAWORTHINESS

Seaworthiness is the quality of a marine vehicle being fit to accomplish its intended mission. In meeting their responsibilities to produce seaworthy vehicles, marine engineers must have a basic understanding of the effects of the marine environment with regard to the vehicle's (1) **structure,** (2) **stability** (3) **seakeeping**, and (4) **resistance** and **powering** requirements.

Units and Definitions

The introduction of the International System of units (SI) to the marine engineering field presented somewhat of a revolutionary change. Reference will generally be made to both USCS and SI systems in subsequent examples.

Fig. 11.3.1   Family of high-performance ships.

The **displacement** Δ is the weight of the water displaced by the immersed part of the vehicle. It is equal (1) to the buoyant force exerted on the vehicle and (2) to the weight of the vehicle (in equilibrium) and everything on board. Displacement is expressed in long tons (1.01605 metric tons). Since the specific weight of seawater averages about 64 lb/ft³, displacement in saltwater is measured by the displaced volume ∇ in ft³ divided by 35. In freshwater, divide by 35.9.

The standard maritime industry metric weight unit is the metric ton (tonne) of 1,000 kilograms. Under the SI system, the mass displacement of a ship is commonly represented in metric tons, which under gravity, leads to a force displacement represented in metric tonneforce or newtons.

Two measurements of a merchant ship's earning capacity that are of significant importance to its design and operation are deadweight and tonnage. The **deadweight** of a ship is the weight of cargo, stores, fuel, water, personnel, and effects that the ship can carry when loaded to a specific load draft. Deadweight is the difference between the load displacement, at the minimum permitted freeboard, and the light displacement, which comprises hull weight and machinery. Deadweight is expressed in long tons (2,240 lb each) or MN. The volume of a ship is expressed in tons of 100 ft³ (2.83 m³) each and is referred to as its **tonnage.** Charges for berthing, docking, passage through canals and locks, and for many other facilities are based on tonnage. **Gross tonnage** was based on cubic capacity below the tonnage deck, generally the uppermost complete deck, plus allowances for certain compartments above, which are used for cargo, passengers, crew, and navigating equipment. Deduction of spaces for propulsion machinery, crew quarters, and other prescribed volumes from the gross tonnage leaves the **net tonnage**.

The dimensions of a ship may refer to the molded body (or form defined by the outside of the frames), to general outside or overall dimensions, or to dimensions on which the determination of tonnage or of classification is based. There are thus (1) molded dimensions, (2) overall dimensions, (3) tonnage dimensions, and (4) classification dimensions. The published rules and regulations of the classification societies and the U.S. Coast Guard should be consulted for detailed information.

The **designed load waterline** (DWL) is the waterline at which a ship would float freely, at rest in still water, in its normally loaded or designed condition. The keel line of most ships is parallel to DWL. Some keel lines are designed to slope downward toward the stern, termed **designed drag**.

A vertical line through the intersection of DWL and the foreside of the stem is called the **forward perpendicular, FP**. A vertical line through the intersection of DWL with the afterside of the straight portion of the rudder post, with the afterside of the stern contour, or with the centerline of the rudder stock (depending upon stern configuration), is called the **after perpendicular, AP**.

The **length** on the designed load waterline, L_{WL}, is the length measured at the DWL, which, because of the stern configuration, may be

equal to the **length between perpendiculars,** L_{pp}. This is generally the same as classification-society-defined length, with certain stipulations. The extreme length of the ship is the **length overall,** L_{OA}.

The **molded beam** B is the extreme breadth of the molded form. The extreme or maximum breadth is generally used, referring to the extreme transverse dimension taken to the outside of the plating.

The **draft** T (molded) is the distance from the top of the keel plate or bar keel to the load waterline. It may refer to draft amidships, forward, or aft.

Trim is the longitudinal inclination of the ship usually expressed as the difference between the draft forward, T_F, and the draft aft, T_A.

Coefficients of Form   Assume the following notation: L = length on waterline; B = beam; T = draft; ∇ = volume of displacement; A_{WP} = area of water plane; A_M = area of midship section, up to draft T; v = speed in ft/s (m/s); and V = speed in knots.

Block coefficient, $C_B = \nabla/LBT$, may vary from about 0.38, for high-powered yachts and destroyers, to greater than 0.90 for slow-speed seagoing cargo ships and is a measure of the fullness of the underwater body.

Midship section coefficient, $C_M = A_M/BT$, varies from about 0.75 for tugs or trawlers to about 0.99 for cargo ships and is a measure of the fullness of the maximum section.

Prismatic coefficient, $C_P = \nabla/LA_M = C_B/C_M$, is a measure of the fullness of the ends of the hull, and is an important parameter in powering estimates.

Water-plane coefficient, $C_{WP} = A_{WP}/LB$, ranges from about 0.67 to 0.95, is a measure of the fullness of the water plane, and may be estimated by $C_{WP} \approx \frac{2}{3}C_B + \frac{1}{3}$.

Displacement/length ratio, $\frac{\Delta}{L} = \Delta/(L/100)^3$, is a measure of the slenderness of the hull, and is used in calculating the power of ships and in recording the resistance data of models.

A similar coefficient is the **volumetric coefficient,** $C_V = \nabla/(L/10)^3$, which is commonly used as this measure.

Table 11.3.1 presents typical values of the coefficients with representative values for Froude number $(v/\sqrt{gL})$, discussed under "Resistance and Powering."

Structure

The structure of a ship is a complex assembly of small pieces of material. Common hull structural materials for small boats are wood, aluminum, and fiberglass-reinforced plastic. Large ships are nearly always constructed of steel, with common application of alloyed aluminum in superstructures of some ship types, particularly for KG-critical designs.

The methods for design and analysis of a ship's structure have undergone significant improvement over the years, with research from the ocean environment, shipboard instrumentation, model testing, computer programming and probabilistic approaches. Early methods consisted of

Table 11.3.1   Coefficients of Form

Type of vessel	$\dfrac{v}{\sqrt{gL}}$	C_B	C_M	C_P	C_{WP}	$\dfrac{\Delta}{(L/100)^3}$
Great Lakes ore ships	0.116–0.128	0.85–0.87	0.99–0.995	0.86–0.88	0.89–0.92	70–95
Slow ocean freighters	0.134–0.149	0.77–0.82	0.99–0.995	0.78–0.83	0.85–0.88	180–200
Moderate-speed freighters	0.164–0.223	0.67–0.76	0.98–0.99	0.68–0.78	0.78–0.84	165–195
Fast passenger liners	0.208–0.313	0.56–0.65	0.94–0.985	0.59–0.67	0.71–0.76	75–105
Battleships	0.173–0.380	0.59–0.62	0.99–0.996	0.60–0.62	0.69–0.71	86–144
Destroyer, cruisers	0.536–0.744	0.44–0.53	0.72–0.83	0.62–0.71	0.67–0.73	40–65
Tugs	0.268–0.357	0.45–0.53	0.71–0.83	0.61–0.66	0.71–0.77	200–420

standard strength calculations involving still water and quasi-static analysis along with conventions and experience from proven designs. These methods were extremely useful and remain the basis, as well as a starting tool, for comparative analysis with the newer applications and rules in use today. For detailed discussion of these methods, see the SNAME publications referenced earlier.

Weight, buoyancy, and load curves (Figs. 11.3.2 and 11.3.3) are developed for the ship for the determination of shear force and bending moment. Several extreme conditions of loading may be analyzed. The weight curves include the weights of the hull, superstructure, rudder, and castings, forgings, masts, booms, all machinery and accessories, solid ballast, anchors, chains, cargo, fuel, supplies, passengers, and baggage. Each individual weight is distributed over a length equal to the distance between frames at the location of that particular weight.

The traditional method of calculating the **maximum design bending moment** involved the determination of weight and buoyancy distributions with the ship poised on a wave of trochoidal form whose length L_w is equal to the length of the ship, L_{pp}, and whose height $H_w = L_w/20$. To more accurately model waves whose lengths exceed 300 ft, $H_w = 1.1\sqrt{L}$ has been widely used. With the wave crest amidships, the ship is in a **hogging** condition (Fig. 11.3.4a); the deck is in tension and the bottom shell in compression. With the trough amidships, the ship is in a **sagging** condition (Fig. 11.3.4b); the deck is in compression and the

Fig. 11.3.2   Representative buoyancy and weight curves for a ship in still water.

Fig. 11.3.3   Representative moment and net-load curves for a ship in still water.

Fig. 11.3.4   Hogging (a) and sagging (b) conditions of a vessel. *(From "Principles of Naval Architecture," SNAME, 1967.)*

bottom in tension. For cargo ships with the machinery amidships, the hogging condition produces the highest bending moment, whereas for tankers and ore carriers with the machinery aft, the sagging condition gives the highest bending moment.

The limitations of the above approach should be apparent; nevertheless, it has been an extremely useful one for many decades, and a great deal of ship data have been accumulated upon which to base refinements. Many advances have been made in the theoretical prediction of the actual loads a ship is likely to experience in a realistic, confused seaway over its life span. Today it is possible to predict reliably the bending moments and shear forces a ship will experience over a short term in irregular seas. Long-term prediction techniques use a probabilistic approach which associates load severity and expected periodicity to establish the design load.

The maximum longitudinal bending stress normally occurs in the deck or bottom plating in vicinity of the midship section. In determining the **design section modulus** Z which the continuous longitudinal material in the midship section must meet, an **allowable bending stress** σ_{all} must be introduced into the bending stress equation. Based on past experience, an appropriate choice of such an allowable stress is $1.19\,L^{1/3}$ tons/in^2 ($27.31\,L^{1/3}$ MN/m^2) with L in ft (m).

For ships under 195 ft (90 m) in length, strength requirements are dictated more on the basis of locally induced stresses than longitudinal bending stresses. This may be accounted for by reducing the allowable stress based on the above equation or by reducing the K values indicated by the trend in Table 11.3.2 for the calculation of the ship's bending moment. For aluminum construction, Z must be twice that obtained for steel construction. Minimum statutory values of section modulus are published by the U.S. Coast Guard in "Load Line Regulations." Shipbuilding classification societies such as the American Bureau of Shipping have section modulus standards somewhat greater. Rules for section modulus have a permissible (allowable) bending stress based on the hull material.

The **maximum bending stress** at each section can be computed by the equation

$$\sigma = M/Z$$

where σ = maximum bending stress, lb/in^2 (N/m^2) or tons/in^2 (MN/m^2); M = maximum bending moment, ft · lb (N · m) or ft · tons; Z = section modulus, in^2 · ft (m^3); $Z = I/y$, where I = minimum vertical moment

Table 11.3.2   Constants for Bending Moment Approximations

Type	Δ tons (MN)	L ft (m)	K^*
Tankers	35,000–150,000 (349–1495)	600–900 (183–274)	35–41
High-speed cargo passenger	20,000–40,000 (199–399)	500–700 (152–213)	29–36
Great Lakes ore carrier	28,000–32,000 (279–319)	500–600 (152–183)	54–67
Trawlers	180–1,600 (1.79–16)	100–200 (30–61)	12–18
Crew boats	65–275 (0.65–2.74)	80–130 (24–40)	10–16

*Based on allowable stress of $\sigma_{\text{all}} = 1.19\sqrt[3]{L}$ tons/in^2($27.31\sqrt[3]{L}$ MN/m^2) with L in ft (m).

of inertia of section, in$^2 \cdot$ ft^2 (m^4), and y = maximum distances from neutral axis to bottom and strength deck, ft (m).

The **maximum shear stress** can be computed by the equation

$$\tau = Vac/It$$

where τ = horizontal shear stress, lb/in^2 (N/m^2); V = shear force, lb (N); ac = moment of area above shear plane under consideration taken about neutral axis, ft^3 (m^3); I = vertical moment of inertia of section, in$^2 \cdot$ ft^2 (m^4); t = thickness of material at shear plane, ft (m).

For an approximation, the **bending moment** of a ship may be computed by the equation

$$M = \Delta L/K \quad \text{ft} \cdot \text{tons (MN} \cdot \text{m)}$$

where Δ = displacement, tons (MN); L = length, ft (m); and K is as listed in Table 11.3.2.

The above formulas have been successfully applied in the past as a convenient tool for the initial assessment of strength requirements for various designs. At present, computer programs are routinely used in ship structural design. Empirical formulas for the individual calculation of both still water and wave-induced bending moments have been derived by the classification societies. They are based on ship length, breadth, block coefficient, and effective wave height (for the wave-induced bending moment). Similarly, permissible bending stresses are calculated from formulas as a function of ship length and its service environment. Finite element techniques in the form of commercially available computer software packages combined with statistical reliability methods are being applied with increasing frequency.

For ships, the maximum longitudinal bending stresses occur in the vicinity of the midship section at the deck edge and in the bottom plating. Maximum shear stresses occur in the shell plating in the vicinity of the quarter points at the neutral axis. For long, slender girders, such as ships, the maximum shear stress is small compared with the maximum bending stress.

The structure of a ship consists of a grillage of stiffened plating supported by longitudinals, longitudinal girders, transverse beams, transverse frames, and web frames. Since the primary stress system in the hull arises from longitudinal bending, it follows that the longitudinally continuous structural elements are the most effective in carrying and distributing this stress system. The strength deck, particularly at the side, and the keel and turn of bilge are highly stressed regions. The shell plating, particularly deck and bottom plating, carry the major part of the stress. Other **key longitudinal elements,** as shown in Fig. 11.3.5, are longitudinal deck girders, main deck stringer plate, gunwale angle, sheer strake, bilge strake, inner bottom margin plate, garboard strake, flat plate keel, center vertical keel, and rider plate.

Transverse elements include deck beams and transverse frames and web frames which serve to support and transmit vertical and transverse loads and resist hydrostatic pressure.

Good structural design minimizes the structural weight while providing adequate strength, minimizes interference with ship function, includes access for maintenance and inspection, provides for effective continuity of the structure, facilitates stress flow around deck openings and other stress obstacles, and avoids square corner discontinuities in the plating and other stress concentration "hot spots."

A longitudinally framed ship is one which has closely spaced longitudinal structural elements and widely spaced transverse elements. A transversely framed ship has closely spaced transverse elements and widely spaced longitudinal elements. Longitudinal framing systems are generally more efficient structurally, but because of the deep web frames supporting the longitudinals, it is less efficient in the use of internal space than the transverse framing system. Where interruptions of open internal spaces are unimportant, as in tankers and bulk carriers, longitudinal framing is universally used. However, modern practice tends increasingly toward longitudinal framing in other types of ships also.

Stability

A ship afloat is in vertical equilibrium when the force of **gravity,** acting at the ship's **center of gravity** G, is equal, opposite, and in the same

Fig. 11.3.5   Key structural elements.

vertical line as the force of **buoyancy,** acting upward at the **center of buoyancy** B. Figure 11.3.6 shows that an upsetting, transverse couple acting on the ship causes it to rotate about a longitudinal axis, taking a list ϕ. G does not change; however, B moves to B_1, the centroid of the new underwater volume. The resulting couple, created by the transverse separation of the two forces ($\overline{GZ}$) opposes the upsetting couple, thereby righting the ship. A static stability curve (Fig. 11.3.7), consisting of

Fig. 11.3.6   Ship stability.

Fig. 11.3.7   Static stability curve.

values for the righting arm $\overline{GZ}$ plotted against angles of inclination (ϕ), gives a graphic representation of the static stability of the ship. The primary indicator for the safety of a ship is the maximum righting arm it develops and the angle heel at which this righting arm occurs.

For small angles of inclination ($\phi < 10°$), centers of buoyancy follow a locus whose instantaneous center of curvature M is known as the transverse **metacenter**. When M lies above G ($\overline{GM} > 0$), the resulting gravity-buoyancy couple will right the ship; the ship has positive stability. When G and M are coincident, the ship has neutral stability. When G is above $M(\overline{GM} < 0)$, negative stability results. Hence $\overline{GM}$, known as the **transverse metacentric height**, is an indication of **initial stability** of a ship.

The transverse **metacentric radius** $\overline{BM}$ and the vertical location of the center of buoyancy are determined by the design of the ship and can be calculated. Once the vertical location of the ship's center of gravity is known, then $\overline{GM}$ can be found. The vertical center of gravity of practically all ships varies with the condition of loading and must be determined either by a careful calculation or by an inclining experiment.

Minimum values of $\overline{GM}$ ranging from 1.5 to 3.5 ft (0.46 to 1.07 m), corresponding to small and large seagoing ships, respectively, have been accepted in the past. The $\overline{GM}$ of passenger ships should be under 6 percent of the beam to ensure a reasonable (comfortable) rolling period. This relatively low value is offset by the generally large range of stability due to the high freeboard of passenger ships.

The question of **longitudinal stability** affects the trim of a ship. As in the transverse case, a longitudinal metacenter exists, and a longitudinal metacentric height $\overline{GM}_L$ can be determined. The moment to alter trim 1 in, $\overline{MT1}$, is computed by $\overline{MT1} = \Delta\overline{GM}_L/12L$, ft · tons/in and moment to alter trim 1 m is $\overline{MT1} = 10^6\Delta\overline{GM}_L/L$,N · m/m. Displacement Δ is in tons and MN, respectively.

The **location of a ship's center of gravity** changes as small weights are shifted within the system. The vertical, transverse, or longitudinal component of movement of the center of gravity is computed by $\overline{GG}_1 = w \times d/\Delta$, where w is the small weight, d is the distance the weight is shifted in a component direction, and Δ is the displacement of the ship, which includes w.

Vertical changes in the location of G caused by **weight addition** or **removal** can be calculated by

$$KG_1 = \frac{\Delta\,\overline{KG} \pm w\overline{K}g}{\Delta_1}$$

where $\Delta_1 = \Delta \pm w$ and $\overline{K}g$ is the height of the center of gravity of w above the keel.

The free surface of the liquid in fuel oil, lubricating oil, and water storage and service tanks is deleterious to ship stability. The weight shift of the liquid as the ship heels can be represented as a virtual rise in G, hence a reduction in $\overline{GM}$ and the ship's initial stability. This virtual rise, called the **free surface effect,** is calculated by the expression

$$GG_v = \gamma_i i/\gamma_w \nabla$$

where γ_i, γ_w = specific gravities of liquid in tank and sea, respectively; i = moment of inertia of free surface area about its longitudinal centroidal axis; and ∇ = volume of displacement of ship. The ship designer can minimize this effect by designing long, narrow, deep tanks or by using baffles.

Seakeeping

Seakeeping is how a ship moves in response to the ocean's wave motion and determines how successfully the ship will function in a seaway. These motions in turn directly affect the practices of marine engineers. The motion of a floating object has six degrees of freedom. Figure 11.3.8 shows the conventional ship coordinate axes and ship motions. Oscillatory movement along the x axis is called **surge**; along the y axis, **sway**; and along the z axis, **heave**. Rotation about x is called **roll**; about y, **pitch**; and about z, **yaw**.

Rolling has a major effect on crew comfort and on the structural and bearing requirements for machinery and its foundations. The natural period of roll of a ship is $T = 2\pi K/\sqrt{gGM}$, where K is the radius of gyration of virtual ship mass about a longitudinal axis through G. K varies from 0.4 to 0.5 of the beam, depending on the ship depth and transverse distribution of weights.

Angular acceleration of roll, if large, has a very distressing effect on crew, passengers, machinery, and structure. This can be minimized by increasing the period or by decreasing the roll amplitude. Maximum angular acceleration of roll is

$$d^2\phi/dt^2 = -4\pi^2(\phi_{max}/T^2)$$

where ϕ_{max} is the maximum roll amplitude. The period of the roll can be increased effectively by decreasing $\overline{GM}$; hence the lowest value of $\overline{GM}$ compatible with all stability criteria should be sought.

Pitching is in many respects analogous to rolling. The natural period of pitching, bow up to bow down, can be found by using the same expression used for the rolling period, with the longitudinal radius of gyration K_L substituted for K. A good approximation is $K_L = L/4$, where L is the length of the ship. The natural period of pitching is usually between one-third and one-half the natural period of roll. A by-product of pitching is **slamming,** the reentry of bow sections into the sea with heavy force.

Heaving and **yawing** are the other two principal rigid-body motions caused by the sea. The amplitude of heave associated with head seas, which generate pitching, may be as much as 15 ft (4.57 m); that arising

Fig. 11.3.8   Conventional ship coordinate axes and ship motions.

from beam seas, which induce rolling, can be even greater. Yawing is started by unequal forces acting on a ship as it quarters into a sea. Once a ship begins to yaw, it behaves as it would at the start of a rudder-actuated turn. As the ship travels in a direction oblique to its plane of symmetry, forces are generated which force it into heel angles independent of sea-induced rolling.

Analysis of ship movements in a seaway is performed by using probabilistic and statistical techniques. The seaway is defined by a mathematically modeled wave energy density spectrum based on data gathered by oceanographers. This wave spectrum is statistically calculated for various sea routes and weather conditions. One way of determining a ship response amplitude operator or transfer function is by linearly superimposing a ship's responses to varying regular waves, both experimentally and theoretically. The wave spectrum multiplied by the ship response transfer function yields the ship response spectrum. The ship response spectrum is an energy density spectrum from which the statistical character of ship motions can be predicted. (See Edward V. Lewis, Motions in Waves, "Principles of Naval Architecture," SNAME, 1989.)

Resistance and Powering

Resistance to ship motion through the water is the aggregate of (1) wave-making, (2) frictional, (3) pressure or form, and (4) air resistances.

Wave-making resistance is primarily a function of Froude number, $\mathrm{Fr} = v/\sqrt{gL}$, where v = ship speed, ft/s (m/s); g = acceleration of gravity, ft/s^2 (m/s^2); and L = ship length, ft (m). In many instances, the dimensional speed-length ratio $V/\sqrt{L}$ is used for convenience, where V is given in knots. A ship makes at least two distinct wave patterns, one from the bow and the other at the stern. There also may be other patterns caused by abrupt changes in section. These patterns combine to form the total wave system for the ship. At various speeds there is mutual cancellation and reinforcement of these patterns. Thus, a plot of total resistance of the ship versus Fr or $V/\sqrt{L}$ is not smooth but shows humps and hollows corresponding to the wave cancellation or reinforcement. Normal procedure is to design the operating speed of a ship to fall at one of the low points in the resistance curve.

Frictional resistance is a function of Reynolds number (see Sec. 3). Because of the size of a ship, the Reynolds number is large and the flow is always turbulent.

Pressure or **form resistance** is a viscosity effect but is different from frictional resistance. The principal observed effects are boundary-layer separation and eddying near the stern.

It is usual practice to combine the wave-making and pressure resistances into one term, called the **residuary resistance,** assumed to be a function of Froude number. The combination, although not strictly legitimate, is practical because the pressure resistance is usually only 2 to 3 percent of the total resistance. The frictional resistance is then the only term which is considered to be a function of Reynolds number and can be calculated. Based on an analysis of the water resistance of flat, smooth planes, Schoenherr gives the formula

$$R_f = 0.5\rho S v^2 C_f$$

where R_f = frictional resistance, lb (N); ρ = mass density, lb/s$^2 \cdot$ ft^4 (kg/m^3); S = wetted surface area, ft^2 (m^2); v = velocity, ft/s (m/s); and C_f is the frictional coefficient computed from the ITTC formula

$$C_f = \frac{0.075}{(\log_{10} \mathrm{Re} - 2)^2}$$

and Re = Reynolds number = vL/v.

Through towing tests of ship models at a series of speeds for which Froude numbers are equal between the model and the ship, **total model resistance** (R_{tm}) is determined. Residuary resistance (R_{rm}) for the model is obtained by subtracting the frictional resistance (R_{fm}). By **Froude's law of comparison,** the residuary resistance of the ship (R_{rs}) is equal to R_{rm} multiplied by the ratio of ship displacement to model displacement. Total ship resistance (R_{ts}) then is equal to the sum of R_{rs}, the calculated ship

frictional resistance (R_{fs}), and a correlation allowance that allows for the roughness of ship's hull opposed to the smooth hull of a model.

$$
\begin{aligned}
R_{tm} &= \text{measured resistance} \\
R_{rm} &= R_{tm} - R_{fm} \\
R_{rs} &= R_{tm}(\Delta s/\Delta m) \\
R_{ts} &= R_{rs} + R_{fs} + R_a
\end{aligned}
$$

A nominal value of 0.0004 is generally used as the correlation allowance coefficient C_a. $R_a = 0.5\rho S v^2 C_a$.

From R_{ts}, the **total effective power** P_E required to propel the ship can be determined:

$$P_E = \frac{R_{ts}v}{550} \quad \text{ehp} \qquad \left(P_E = \frac{R_{ts}v}{1{,}000} \quad \text{kW}_E \right)$$

where R_{ts} = total ship resistance, lb (N); v = velocity, ft/s (m/s); and $P_E = P_s \times$ P.C., where P_s is the shaft power (see Propulsion Systems) and P.C. is the propulsive coefficient, a factor which takes into consideration mechanical losses, propeller efficiency, and the flow interaction with the hull. P.C.= 0.45 to 0.53 for high-speed craft; 0.50 to 0.60, for tugs and trawlers; 0.55 to 0.65, for destroyers; and 0.63 to 0.72, for merchant ships.

Figure 11.3.9 illustrates the specific effective power for various displacement hull forms and planing craft over their appropriate speed regimes. Figure 11.3.10 shows the general trend of specific resistance versus Froude number and may be used for coarse powering estimates.

Fig. 11.3.9 Specific effective power for various speed-length regimes. (*From "Handbook of Ocean and Underwater Engineering," McGraw-Hill, 1969.*)

Fig. 11.3.10 General trends of specific resistance versus Froude number.

In early design stages, the hull form is not yet defined, and model testing is not feasible. Reasonably accurate power calculations are made by using preplotted series model test data of similar hull form, such as from Taylor's Standard Series and the Series 60. The Standard Series data were originally compiled by Admiral David W. Taylor and were based on model tests of a series of uniformly varied twin screw, cruiser hull forms of similar geometry. Revised Taylor's Series data are available in "A Reanalysis of the Original Test Data for the Taylor Standard Series," Taylor Model Basin, Rept. no. 806.

Air Resistance   The air resistance of ships in a windless sea is only a few percent of the water resistance. However, the effect of wind can have a significant impact on drag. The wind resistance parallel to the ship's axis is roughly 30 percent greater when the wind direction is about 30° ($\pi/6$ rad) off the bow vice dead ahead, since the projected above-water area is greater. Wind resistance is approximated by $R_A = 0.002B^2V_R^2$ lb ($0.36B^2V_R^2$ N), where B = ship's beam, ft (m) and V_R = ship speed relative to the wind, in knots (m/s).

Powering of Small Craft   The American Boat and Yacht Council, Inc. (ABYC) provides guidance for determining the maximum power for propulsion of outboard boats, to evaluate the suitability of power installed in inboard boats, and to determine maneuvering speed. Figures 11.3.11 and 11.3.12 demonstrate such data. The latest standard review dates are available at www.abycinc.org.

The development of high-power lightweight engines facilitated the evolution of planing hulls, which develop dynamic lift.

Small high-speed craft primarily use gasoline engines for powering because of weight and cost considerations. However, small high-speed diesel engines are commonly used in pleasure boats.

Fig. 11.3.11   Boat brake power capacity for length-width factor under 52. *(From "Safety Standards for Small Craft," ABYC, 1974.)*

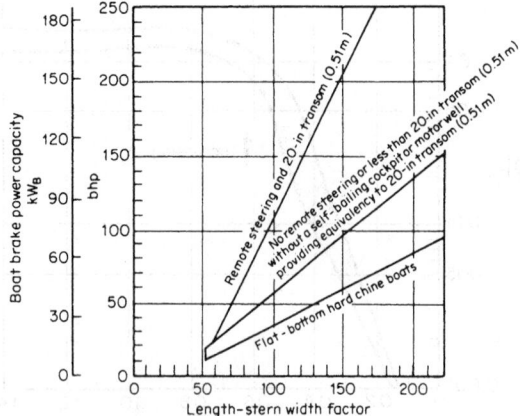

Fig. 11.3.12   Boat brake power capacity for length-width factor over 52. *(From "Safety Standards for Small Craft," ABYC, 1974.)*

ENGINEERING CONSTRAINTS

The constraints affecting marine engineering design are too numerous, and some too obvious, to include in this section. Three significant categories, however, are discussed. The geometry of the hull forms immediately suggests **physical constraints.** The interaction of the vehicle with the marine environment suggests **dynamic constraints,** particularly vibration, noise, and shock. The broad topic of **environmental protection** is one of the foremost engineering constraints of today, having a very pronounced effect on the operating systems of a marine vehicle.

Physical Constraints

Formerly, tonnage laws in effect made it economically desirable for the propulsion machinery spaces of a merchant ship to exceed 13 percent of the gross tonnage of the ship so that 32 percent of the gross tonnage could be deducted in computing net tonnage. In most design configurations, however, a great effort is made to minimize the space required for the propulsion plant in order to maximize that available to the mission or the money-making aspects of the ship.

Specifically, **space** is of extreme importance as each component of support equipment is selected. Each component must fit into the master compact arrangement scheme to provide the most efficient operation and maintenance by engineering personnel.

Weight constraints for a main propulsion plant vary with the application. In a tanker where cargo capacity is limited by draft restrictions, the weight of machinery represents lost cargo. Cargo ships, on the other hand, rarely operate at full load draft. Additionally, the low weight of propulsion machinery somewhat improves inherent cargo ship stability deficiencies. The weight of each component of equipment is constrained by structural support and shock resistance considerations. Naval shipboard equipment, in general, is carefully analyzed to effect weight reduction.

Dynamic Constraints

Dynamic effects, principally mechanical vibration, noise, shock, and ship motions, are considered in determining the dynamic characteristics of a ship and the dynamic requirements for equipment.

Vibration (See Sec. 9.) Vibration analyses are especially important in the design of the propulsion shafting system and its relation to the excitation forces resulting from the propeller. The main propulsion **shafting** can vibrate in longitudinal, torsional, and lateral modes. Modes of **hull vibration** may be vertical, horizontal, torsional, or longitudinal; may occur separately or coupled; may be excited by synchronization with periodic harmonics of the propeller forces acting either through the shafting, by the propeller force field interacting with the hull afterbody, or both; and may also be set up by unbalanced harmonic forces from the main machinery, or, in some cases, by impact excitation from slamming or periodic-wave encounter.

It is most important to reduce the excitation forces at the source. Very objectionable and serious vibrations may occur when the frequency of the exciting force coincides with one of the hull or shafting-system natural frequencies.

Vibratory forces generated by the propeller are (1) alternating pressure forces on the hull due to the alternating hydrodynamic pressure fields of the propeller blades; (2) alternating propeller-shaft bearing forces, primarily caused by wake irregularities; and (3) alternating forces transmitted throughout the shafting system, also primarily caused by wake irregularities.

The most effective means to ensure a satisfactory level of vibration is to provide adequate clearance between the propeller and the hull surface and to select the propeller revolutions or number of blades to avoid synchronism. Replacement of a four-blade propeller, for instance, by a three- or five-blade, or vice versa, may bring about a reduction in vibration. Singing propellers are due to the vibration of the propeller blade edge about the blade body, producing a disagreeable noise or hum. Chamfering the edge is sometimes helpful.

Vibration due to variations in **engine torque reaction** is hard to overcome. In ships with large diesel engines, the torque reaction tends to produce an athwartship motion of the upper ends of the engines. The motion is increased if the engine foundation strength is inadequate.

Foundations must be designed to take both the thrust and torque of the shaft and be sufficiently rigid to maintain alignment when the ship's hull is working in heavy seas. Engines designed for foundations with three or four points of support are relatively insensitive to minor working of foundations. Considerations should be given to using flexible couplings in cases when alignment cannot be assured.

Torsional vibration frequency of the prime mover–shafting-propeller system should be carefully computed, and necessary steps taken to ensure that its natural frequency is clear of the frequency of the main-unit or propeller heavy-torque variations; serious failures have occurred. Problems due to resonance of engine torsional vibration (unbalanced forces) and foundations or hull structure are usually found after ship trials, and correction consists of local stiffening of the hull structure. If possible, engines should be located at the nodes of hull vibrations.

Although more rare, a **longitudinal vibration** of the propulsion shafting has occurred when the natural frequency of the shafting agreed with that of a pulsating axial force, such as propeller thrust variation.

Other forces inducing vibrations may be vertical inertia forces due to the acceleration of reciprocating parts of an unbalanced reciprocating engine or pump; longitudinal inertia forces due to reciprocating parts or an unbalanced crankshaft creating unbalanced rocking moments; and horizontal and vertical components of centrifugal forces developed by unbalanced rotating parts. Rotating parts can be balanced.

Noise   The noise characteristics of shipboard systems are increasingly important, particularly in naval combatant submarines where remaining undetected is essential, and also from a human-factors point of view on all ships. Achieving significant reduction in machinery noise level can be costly. Therefore desired noise levels should be analyzed. Each operating system and each piece of rotating or reciprocating machinery installed aboard a submarine are subjected to intensive airborne (noise) and structureborne (mechanical) vibration analyses and tests. Depending on the noise attenuation required, similar tests and analyses are also conducted for all surface ships, military and merchant.

Shock   In naval combatant ships, shock loading due to noncontact underwater explosions is a major design parameter. Methods of qualifying equipment as "shock-resistant" might include "static" shock analysis, "dynamic" shock analysis, physical shock tests, or a combination.

Motions   Marine lubricating systems are specifically distinguished by the necessity of including list, trim, roll, and pitch as design criteria. The American Bureau of Shipping requires satisfactory functioning of lubricating systems when the vessel is permanently inclined to certain angles athwartship and fore and aft. In addition, electric generator bearings must not spill oil under momentary roll specifications. Military specifications are similar, but generally add greater requirements for roll and pitch.

Bearing loads are also affected by ship roll and pitch accelerations, and thus rotating equipment often must be arranged longitudinally.

Environmental Constraints

The **Refuse Act of 1899** (33 U.S.C. 407) prohibits discharge of any refuse material from marine vehicles into navigable waters of the United States or into their tributaries or onto their banks if the refuse is likely to wash into navigable waters. The term **refuse** includes nearly any substance, but the law contains provisions for sewage. The Environmental Protection Agency (EPA) may grant permits for the discharge of refuse into navigable waters.

The **Oil Pollution Act of 1961** (33 U.S.C. 1001-1015) prohibits the discharge of oil or oily mixtures (over 100 mg/L) from vehicles generally within 50 mi of land; some exclusions, however, are granted. (The Oil Pollution Act of 1924 was repealed in 1970 because of supersession by subsequent legislation.)

The **Port and Waterways Safety Act of 1972** (PL92-340) grants the U.S. Coast Guard authority to establish and operate mandatory vehicle traffic control systems. The control system must consist of a VHF radio for ship-to-shore communications, as a minimum. The Act, in effect, also extends the provisions of the Tank Vessel Act, which protects against hazards to life and property, to include **protection of the marine environment**. Regulations stemming from this Act govern standards of tanker design, construction, alteration, repair, maintenance, and operation.

The most substantive marine environmental protection legislation is the **Federal Water Pollution Control Act (FWPCA)**, 1948, as amended (33 U.S.C. 466 et seq.). The 1972 and 1978 Amendments contain provisions which greatly expand federal authority to deal with **marine pollution**, providing authority for control of pollution by oil and hazardous substances other than oil, and for the assessment of penalties. **Navigable waters** are now defined as ". . . the waters of the United States, including territorial seas."

"Hazardous substances other than oil" and harmful quantities of these substances are defined by the EPA in regulations. The Coast Guard and EPA must ensure removal of such discharges, costs borne by the responsible vehicle owner or operator if liable. **Penalties** now may be assessed by the Coast Guard for any discharge. The person in charge of a vehicle is to notify the appropriate U.S. government agency of any discharge upon knowledge of it.

The Coast Guard regulations establish **removal procedures** in coastal areas, contain regulations to **prevent discharges** from vehicles and transfer facilities, and regulations governing the **inspection of cargo vessels** to prevent hazardous discharges. The EPA regulations govern inland areas and non-transportation-related facilities.

Section 312 of the FWPCA as amended in 1972 deals directly with **discharge of sewage**. The EPA must issue standards for marine sanitation devices, and the Coast Guard must issue regulations for implementing those standards. In June, 1972, EPA published standards that now prohibit discharge of any sewage from vessels, whether it is treated or not. Federal law now prohibits, on all inland waters, operation with marine sanitation devices which have not been securely sealed or otherwise made visually inoperative so as to prevent overboard discharge.

The **Marine Plastic Pollution Control Act of 1987** (PL100-220) prohibits the disposal of plastics at sea within U.S. waters.

The **Oil Pollution Act of 1990** (PL101-380) provides extensive legislation on liability and includes measures that impact the future use of single-hull tankers.

PROPULSION SYSTEMS

(See Sec. 9 for component details.)

The basic operating requirement for the main propulsion system is to propel the vehicle at the required sustained speed for the range or endurance required and to provide suitable maneuvering capabilities. In meeting this basic requirement, the marine propulsion system integrates the power generator/prime mover, the transmission system, the propulsor, and other shipboard systems with the ship's hull or vehicle platform. Figure 11.3.13 shows propulsion system alternatives with the most popular drives for fixed-pitch and controllable-pitch propellers.

Definitions for Propulsion Systems

Brake power P_B, bhp (kW$_B$), is the power delivered by the output coupling of a prime mover before passing through speed-reducing and transmission devices and with all continuously operating engine-driven auxiliaries in use.

Shaft power P_s, shp (kW$_s$), is the net power supplied to the propeller shafting after passing through all reduction gears or other transmission devices and after power for all attached auxiliaries has been taken off. Shaft power is measured in the shafting within the ship by a torsionmeter as close to the propeller or stern tube as possible.

Delivered power P_D, dhp (kW$_D$), is the power actually delivered to the propeller, somewhat less than P_s because of the power losses in the stern tube and other bearings between the point of measurement and the propeller. Sometimes called **propeller power,** it is also equal to the effective power P_E, ehp (kW$_E$), plus the power losses in the propeller and the losses in the interaction between the propeller and the ship.

Fig. 11.3.13 Alternatives in the selection of a main propulsion plant. *(From "Marine Engineering" SNAME, 1992.)*

Normal shaft power or **normal power** is the power at which a marine vehicle is designed to run most of its service life.

Maximum shaft power is the highest power for which propulsion machinery is designed to operate continuously.

Service speed is the actual speed maintained by a vehicle at load draft on its normal route and under average weather and sea conditions typical of that route and with average fouling of bottom.

Designed service speed is the speed expected on trials in fair weather at load draft, with clean bottom, when machinery is developing a specified fraction of maximum shaft power. This fraction may vary but is of order of 0.8.

MAIN PROPULSION PLANTS

Steam Plants

The basic steam propulsion plant contains main boilers, steam turbines, a condensate system, a feedwater system, and numerous auxiliary components necessary for the plant to function. A heat balance calculation, the basic tool for determining the effect of various configurations on plant thermal efficiency, is demonstrated for the basic steam cycle. Both fossil-fuel and nuclear energy sources are successfully employed for marine applications.

Main Boilers (See Sec. 9.) The pressures and temperatures achieved in steam-generating equipment have increased steadily over recent years, permitting either a higher-power installation for a given space or a reduction in the size and weight of a given propulsion plant.

The **trend in steam pressures** and **temperatures** has been an increase from 600 psig (4.14 MN/m²) and 850°F (454°C) during World War II to 1,200 psig (8.27 MN/m²) and 950°F (510°C) for naval combatants in and since the postwar era. For merchant ships, the progression has been from 400 psig (2.76 MN/m²) and 750°F (399°C) gradually up to 850 psig (5.86 MN/m²) and 850°F (454°C) in the 1960s, with some boilers at 1,500 psig (10.34 MN/m²) and 1,000°F (538°C) in the 1970s.

The **quantity of steam** produced by a marine boiler ranges from approximately 1,500 lb/h (680 kg/h) in small auxiliary boilers to over 400,000 lb/h (181,500 kg/h) in large main propulsion boilers. Outputs of 750,000 lb/h (340,200 kg/h) or more per boiler are practical for high-power installations.

Most marine boilers are **oil-fired.** Compared with other fuels, oil is easily loaded aboard ship, stored, and introduced into the furnace, and does not require the ash-handling facilities required for coal firing.

Gas-fired boilers are used primarily on power or drill barges which are fixed in location and can be supplied from shore (normally classed in the Ocean Engineering category). At sea, tankers designed to carry liquefied natural gas (LNG) may use the natural boil-off from their cargo gas tanks as a supplemental fuel **(dual fuel system).** The cargo gas boil-off is collected and pumped to the boilers where it is burned in conjunction with oil. The quantity of boil-off available is a function of the ambient sea and air temperatures, the ship's motion, and the cargo loading; thus, it may vary from day to day.

For economy of space, weight, and cost and for ease of operation, the trend in **boiler installations** is for fewer boilers of high capacity rather than a large number of boilers of lower capacity. The minimum installation is usually two boilers, to ensure propulsion if one boiler is lost; one boiler per shaft for twin-screw ships. Some large ocean-going ships operate on single boilers, requiring exceptional reliability in boiler design and operation.

Combustion systems include forced-draft fans or blowers, the fuel oil service system, burners, and combustion controls. Operation and maintenance of the combustion system are extremely important to the efficiency and reliability of the plant. The best combustion with the least possible excess air should be attained.

Main Turbines (See Sec. 9.) Single-expansion (i.e., single casing) marine steam turbines are fairly common at lower powers, usually not exceeding 4,000 to 6,000 shp (2,983 to 4,474 kW$_S$). Above that power range, the turbines are usually double-expansion (cross-compound) with high- and low-pressure turbines each driving the main reduction gear through its own pinion. The low-pressure turbine normally contains the reversing turbine in its exhaust end. The main condenser is either underslung and supported from the turbine, or the condenser is carried on the foundations, with the low-pressure turbine supported on the condenser.

The inherent **advantages** of the steam turbine have favored its use over the reciprocating steam engine for all large, modern marine steam propulsion plants. Turbines are not size-limited, and their high temperatures and pressures are accommodated safely. Rotary motion is simpler than reciprocating motion; hence unbalanced forces can be eliminated in the turbine. The turbine can efficiently utilize a low exhaust pressure; it is lightweight and requires minimum space and low maintenance.

The **reheat cycle** is the best and most economical means available to improve turbine efficiencies and fuel rates in marine steam propulsion plants. In the reheat cycle, steam is withdrawn from the turbine after partial expansion and is passed through a heat exchanger where its temperature is raised; it is then readmitted to the turbine for expansion to condenser pressure. Marine reheat plants have more modest steam conditions than land-based applications because of lower power ratings and a greater reliability requirement for ship safety.

The reheat cycle is applied mostly in high-powered units above 25,000 shp (18,642 kW$_s$) and offers the maximum economical thermal efficiency that can be provided by a steam plant. Reheat cycles are not used in naval vessels because the improvement in efficiency does not warrant the additional complexity; hence a trade-off is made.

Turbine foundations must have adequate rigidity to avoid vibration conditions. This is particularly important with respect to periodic variations in propeller thrust which may excite longitudinal vibrations in the propulsion system.

Condensate System   (See Sec. 9.)   The condensate system provides the means by which feedwater for the boilers is recovered and returned to the feedwater system. The major components of the condensate system of a marine propulsion plant are the **main condenser,** the **condensate pumps,** and the **deaerating feed tank** or **heater.** Both singlepass and two-pass condenser designs are used; the single-pass design, however, allows somewhat simpler construction and lower water velocities. The single-pass condenser is also adaptable to scoop circulation, as opposed to pump circulation, which is practical for higher-speed ships. The deaerating feedwater heater is supplied from the condensate pumps, which take suction from the condenser hot well, together with condensate drains from steam piping and various heaters.

The deaerating feed heater is normally maintained at about 35 psig (0.241 MN/m^2) and 280°F (138°C) by auxiliary exhaust and turbine extraction steam. The condensate is sprayed into the steam atmosphere at the top of the heater, and the heated feedwater is pumped from the bottom by the feed booster or main feed pumps. In addition to removing oxygen or air, the heater also acts as a surge tank to meet varying demands during maneuvering conditions. Since the feed pumps take suction where the water is almost saturated, the heater must be located 30 to 50 ft (9.14 to 15.24 m) above the pumps in order to provide enough positive suction head to prevent flashing from pressure fluctuations during sudden plant load change.

Feedwater System   The feedwater system comprises the pumps, piping, and controls necessary to transport feedwater to the boiler or steam generator, to raise water pressure above boiler pressure, and to control flow of feedwater to the boiler. **Main feed pumps** are so vital that they are usually installed in duplicate, providing a standby pump capable of feeding the boilers at full load. Auxiliary steam-turbine-driven centrifugal pumps are usually selected. A typical naval installation consists of three main feed booster pumps and three main feed pumps for each propulsion plant. Two of the booster pumps are turbine-driven and one is electric. Additionally, an electric-motor-driven emergency booster pump is usually provided. The total capacity of the main feed pumps must be 150 percent of the boiler requirement at full power plus the required recirculation capacity. Reliable feedwater regulators are important.

Steam Plant—Nuclear   (See Sec. 9.)   The compact nature of the energy source and not requiring air for combustion are the most significant characteristics of nuclear power for marine applications. The fission of one gram of uranium per day produces about one megawatt of power. (One pound produces about 600,000 horsepower.) In other terms, the fission of 1 lb of uranium is equivalent to the combustion of about 86 tons (87,380 kg) of 18,500 Btu/lb (43.03 MJ/kg) fuel oil. This characteristic permits utilization of large power plants on board ship without the necessity for frequent refueling or large bunker storage. Economic studies, however, show that nuclear power, as presently developed, is best suited for military purposes, where the advantages of high power and endurance override the pure economic considerations. As the physical size of nuclear propulsion plants is reduced, their economic attractiveness for commercial marine application will increase.

Major differences between the nuclear power and fossil-fuel plants are: (1) The **safety aspect** of the nuclear reactor system—operating personnel must be shielded from fission product radiation, hence the size and weight of the reactor are increased and maintenance and reliability become more complicated; (2) the steam produced by a pressurized-water reactor plant is saturated and, because of the **high moisture content** in the turbine steam path, the turbine design requires more careful attention; (3) the steam pressure provided by a pressurized-water reactor plant varies with output, the maximum pressure occurring at no load

and decreasing approximately linearly with load to a minimum at full power; hence, **blade stresses** in a nuclear turbine increase more rapidly with a decrease in power than in a conventional turbine. Special attention must be given to the design of the control stage of a nuclear turbine. The *N.S. Savannah* was designed for 700 psig (4.83 MN/m^2) at no load and 400 psig (2.76 MN/m^2) at full power.

Steam Plant—Heat Balance   The heat-balance calculation is the basic analysis tool for determining the effect of various steam cycles on the thermal efficiency of the plant and for determining the quantities of steam and feedwater flow.

The thermal arrangement of a **simple steam cycle,** illustrated in Fig. 11.3.14, and the following simplified analysis are taken from "Marine Engineering," SNAME, 1992.

Fig. 11.3.14   Simple steam cycle. *(From "Marine Engineering," SNAME, 1992.)*

The unit is assumed to develop 30,000 shp (22,371 kW$_s$). The steam rate of the main propulsion turbines is 5.46 lb/(shp · h) [3.32 kg/ (kW$_S$ · h)] with throttle conditions of 850 psig (5.86 MN/m^2) and 950°F (510°C) and with the turbine exhausting to the condenser at 1.5 inHg abs (5,065 N/m^2). To develop 30,000 shp (22,371 kW$_S$), the throttle flow must be 163,800 lb/h (74,298 kg/h).

The generator load is estimated to be 1,200 kW, and the turbogenerator is thus rated at 1,500 kW with a steam flow of 10,920 lb/h (4,953 kg/h) for steam conditions of 850 psig (5.86 MN/m^2) and 950°F (510°C) and a 1.5 inHg abs (5,065 N/m^2) back pressure. The total steam flow is therefore 174,720 lb/h (79,251 kg/h).

Tracing the steam and water flow through the cycle, the flow exhausting from the main turbine is 163,800 − 250 = 163,550 lb/h (74,298 − 113 = 74,185 kg/h), 250 lb/h (113 kg/h) to the gland condenser, and from the auxiliary turbine is 10,870 lb/h (4,930 kg/h), 50 lb/h (23 kg/h) to the gland condenser. The two gland leak-off flows return from the gland leak-off condenser to the main condenser. The condensate flow leaving the main condenser totals 174,720 lb/h (79,251 kg/h).

It is customary to allow a 1°F (0.556°C) hot-well depression in the condensate temperature. Thus, at 1.5 inHg abs (5,065 N/m^2) the condenser saturation temperature is 91.7°F (33.2°C) and the condensate is 90.7°F (32.6°C). Entering the gland condenser there is a total energy flow of 174,720 × 59.7 = 10,430,784 Btu/h (11,004,894,349 J/h). The gland condenser receives gland steam at 1,281 Btu/lb (2,979,561 J/kg) and drains at a 10°F (5.6°C) terminal difference or 101.7°F (38.6°C). This adds a total of 300 × (1,281 − 69.7) or 363,390 Btu/h (383,390,985 J/h) to the condensate, making a total of 10,794,174 Btu/h (11,388,285,334 J/h) entering the surge tank. The feed leaves the surge tank at the same enthalpy with which it enters.

The feed pump puts an amount of heat into the feedwater equal to the total power of the pump, less any friction in the drive system. This friction work can be neglected but the heat from the power input is a significant quantity. The power input is the total pump head in feet of feedwater, times the quantity pumped in pounds per hour, divided by the mechanical equivalent of heat and the efficiency. Thus,

Heat equivalent of feed pump work

$$= \frac{144 \Delta P v_f Q}{778 E} \quad \text{Btu/h} \quad \left(\frac{\Delta P v_f Q}{E} \quad \text{J/h} \right)$$

where ΔP = pressure change, lb/in^2 (N/m^2); v_f = specific volume of fluid, ft^3/lb (m^3/kg); Q = mass rate of flow, lb/h (kg/h); and E = efficiency of pump.

Assuming the feed pump raises the pressure from 15 to 1,015 lb/in^2 abs (103,421 to 6,998,178 N/m^2), the specific volume of the water is 0.0161 ft^3/lb (0.001005 m^3/kg), and the pump efficiency is 50 percent, the heat equivalent of the feed pump work is 1,041,313 Btu/h (1,098,626,868 J/h). This addition of heat gives a total of 11,835,487 Btu/h (12,486,912,202 J/h) entering the boiler. Assuming no leakage of steam, the steam leaves the boiler at 1,481.2 Btu/lb (3,445,219 J/kg), with a total thermal energy flow of 1,481.2 × 174,720 = 258,795,264 Btu/h (273,039,355,300 J/h). The difference between this total heat and that entering [258,795,264 − 11,835,487 = 246,959,777 Btu/h (260,552,443,098 J/h)] is the net heat added in the boiler by the fuel. With a boiler efficiency of 88 percent and a fuel having a higher heating value of 18,500 Btu/lb (43,030,353 J/kg), the quantity of fuel burned is

Fuel flow rate = 246,959,777/(18,500)(0.88) = 15,170 lb/h
or   260,552,443,098/(43,030,353)(0.88) = 6,881 kg/h

The specific fuel rate is the fuel flow rate divided by the net shaft power [15,170/30,000 = 0.506 lb/(shp · h) or 6881/22,371 = 0.308 kg/(kW$_S$ · h)]. The heat rate is the quantity of heat expended to produce one horsepower per hour (one kilowatt per hour) and is calculated by dividing the net heat added to the plant, per hour, by the power produced.

Heat rate
= (15,170 lb/h)(18,500 Btu/lb)/30,000 shp = 9,335 Btu/(ship · h)
or   296,091,858,933/22,371 = 13,235,522 J/(kW$_s$ · h)

This simple cycle omits many details that must necessarily be included in an actual steam plant. A continuation of the example, developing the details of a complete analysis, is given in "Marine Engineering," SNAME, 1992. A heat balance is usually carried through several iterations until the desired level of accuracy is achieved. The first heat balance may be done from approximate data given in the SNAME Technical and Research Publication No. 3-11, "Recommended Practices for Preparing Marine Steam Power Plant Heat Balances," and then updated as equipment data are known.

Diesel Engines

(See Sec. 9.)

While steam plants are custom-designed, diesel engines and gas turbines are selected from commercial sources at discrete powers. Diesel engines are referred to as being high, medium, or low speed. Low-speed diesels are generally categorized as those with engine speeds less than about 300 r/min, high-speed in excess of 1,200 r/min. Low-speed marine diesel engines are directly coupled to the propeller shaft. Unlike steam turbines and gas turbines, which require special reversing provisions, most direct-drive diesel engines are readily reversible.

Slow-speed diesel engines are well-suited to marine propulsion. Although larger, heavier, and initially more expensive than higher-speed engines, they generally have lower fuel, operating, and maintenance costs. Slow-speed engine parts take longer to wear to the same percentage of their original dimension than high-speed engine parts. Largebore, low-speed diesel engines have inherently better combustion performance with lower-grade diesel fuels. However, a well-designed high-speed engine which is not overloaded can give equally good service as a slow-speed engine.

Medium- and High-Speed Diesels The number of medium- and high-speed diesel engines used in marine applications is relatively small compared to the total power of such engines produced. The medium- and high-speed marine engine of today is almost universally an adaptation of engines built in quantities for service in automotive and stationary applications. The automotive field contributes high-speed engine applications in the 400-hp (298-kW) range, with engine speeds commonly varying from 1,800 to 3,000 r/min, depending on whether use is continuous or intermittent. Diesels in the 1,200- to 1,800-r/min range are typical of off-highway equipment engines at powers from 500 to upward of 1,200 hp (373 to 895 kW). Medium-speed diesel engines in units from 6 to 20 cylinders are available at ratings up to and exceeding 8,000 hp (5,966 kW) from both locomotive and stationary engine manufacturers. These applications are not all-inclusive, but only examples of the wide variety of diesel ratings commercially available today.

Some of the engines were designed with **marine applications** in mind, and others require some modification in external engine hardware. The changes are those needed to suit the engine to the marine environment, meaning salt-laden air, high humidity, use of corrosive seawater for cooling, and operating from a pitching and rolling platform. It also may mean an installation made in confined spaces. In order to adapt to this environment, the prime requisite of the marine diesel engine is the ability to resist corrosion. Because marine engines may be installed with their crankshafts at an angle to the horizontal and because they are subjected to more motion than in many other applications, changes are also necessary in the lubricating oil system. The air intake to a marine engine may not be dust-free and dirt-free when operating in harbors, inland waters, or close offshore; therefore, it is as important to provide a good air cleaner as in any automotive or stationary installation.

Diesel engines are utilized in all types of marine vehicles, both in the merchant marine and in the navies of the world. The power range in which diesel engines have been used has increased directly with the availability of higher-power engines. The line of demarcation in horsepower between what is normally assigned to diesel and to steam has continually moved upward, as has the power installed in ships in general.

Small boats in use in the Navy are usually powered by diesel engines, although the gas turbine is being used in special boats where high speed for short periods of time is the prime requirement. Going a little higher in power, diesels are used in many kinds of workboats such as fishing boats, tugs, ferries, dredges, river towboats and pushers, and smaller types of cargo ships and tankers. They are used in the naval counterparts of these ships and, in addition, for military craft such as minesweepers, landing ships, patrol and escort ships, amphibious vehicles, tenders, submarines, and special ships such as salvage and rescue ships and icebreakers. In the nuclear-powered submarine, the diesel is relegated to emergency generator-set use; however, it is still the best way to power a nonnuclear submarine when not operating on the batteries. This will likely change, giving way to the new technology of air-independent propulsion systems for nonnuclear submarines.

Diesel engines are used either singly or in multiple to drive propeller shafts. For all but high-speed boats, the rpm of the modern diesel is too fast to drive the propeller directly with efficiency, and some means of speed reduction, either mechanical or electrical, is necessary. If a single engine of the power required for a given application is available, then a decision must be made whether it or several smaller engines should be used. The decision may be dictated by the available space. The diesel power plant is flexible in adapting to specific space requirements. When more than one engine is geared to the propeller shaft, the gear serves as both a speed reducer and combining gear. The same series of engines could be used in an electric-drive propulsion system, with even greater flexibility. Each engine drives its own generator and may be located independently of other engines and the propeller shaft.

Low-Speed, Direct-Coupled Diesels Of the more usual prime mover selections, only low-speed diesel engines are directly coupled to the propeller shaft. This is due to the low rpm required for efficient propeller operation and the high rpm inherent with other types of prime movers.

A rigid hull foundation, with a high resistance to vertical, athwartship, and fore-and-aft deflections, is required for the low-speed diesel. The engine room must be designed with sufficient overhaul space and with access for large and heavy replacement parts to be lifted by cranes.

Because of the low-frequency noise generated by low-speed engines, the operating platform often can be located at the engine itself. Special control rooms are often preferred as the noise level in the control room can be significantly less than that in the engine room.

Electric power may be produced by a generator mounted directly on the line shafting. Operation of the entire plant may be automatic and remotely controlled from the bridge. The engine room is often completely unattended for 16 h a day.

Gas Turbines

(See Sec. 9.)

The gas turbine has developed since World War II to join the steam turbine and the diesel engine as alternative prime movers for various shipboard applications.

In gas turbines, the efficiency of the components is extremely important since the compressor power is very high compared with its counterpart in competitive thermodynamic cycles. For example, a typical marine propulsion gas turbine rated at 20,000 bhp (14,914 kW_B) might require a 30,000 hp (22,371 kW) compressor and, therefore, 50,000 hp (37,285 kW) in turbine power to balance the cycle.

The basic **advantages** of the gas turbine for marine applications are its simplicity, small size, and light weight. As an internal-combustion engine, it is a self-contained power plant in one package with a minimum number of large supporting auxiliaries. It has the ability to start and go on line very quickly. Having no large masses that require slow heating, the time required for a gas turbine to reach full speed and accept the load is limited almost entirely by the rate at which energy can be supplied to accelerate the rotating components to speed. A further feature of the gas turbine is its low personnel requirement and ready adaptability to automation.

The relative simplicity of the gas turbine has enabled it to attain outstanding records of reliability and maintainability when used for aircraft propulsion and in industrial service. The same level of reliability and maintainability is being achieved in marine service if the unit is properly applied and installed.

Marine units derived from aircraft engines usually have the gas generator section, comprising the compressor and its turbine, arranged to be removed and replaced as a unit. Maintenance on the power turbine, which usually has the smallest part of the total maintenance requirements, is performed aboard ship. Because of their light weight, small gas turbines used for auxiliary power or the propulsion of small boats, can also be readily removed for maintenance.

Units designed specifically for marine use and those derived from industrial gas turbines are usually maintained and overhauled in place. Since they are somewhat larger and heavier than the aviation-type units, removal and replacement are not as readily accomplished. For this reason, they usually have split casings and other provisions for easy access and maintenance. The work can be performed by the usual ship repair forces.

Both **single-shaft** and **split-shaft** gas turbines can be used in marine service. Single-shaft units are most commonly used for generator drives. When used for main propulsion, where the propeller must operate over a very wide speed range, the single-shaft unit must have a controllable-pitch propeller or some equivalent variable-speed transmission, such as an electric drive, because of its limited speed range and poor acceleration characteristics. A multishaft unit is normally used for main propulsion, with the usual arrangement being a split-shaft unit with an independent variable-speed power turbine; the power turbine and propeller can be stopped, if necessary, and the gas producer kept in operation for rapid load pickup. The use of variable-area nozzles on the power turbine increases flexibility by enabling the compressor to be maintained at or near full speed and the airflow at low-power turbine speeds. Nearly full power is available by adding fuel, without waiting for the compressor to accelerate and increase the airflow. Where low-load economy is of importance, the controls can be arranged to reduce the compressor speed at low loads and maintain the maximum turbine inlet and/or exhaust temperature for best efficiency. Since a gas turbine inherently has a poor part-load fuel rate performance, this variable-area nozzle feature can be very advantageous.

The **physical arrangement** of the various components, i.e., compressors, combustion systems, and turbines, that make up the gas turbine is influenced by the thermodynamic factors (i.e., the turbine connected to a compressor must develop enough power to drive it), by mechanical considerations (i.e., shafts must have adequate bearings, seals, etc.), and also by the necessity to conduct the very high air and gas flows to and from the various components with minimum pressure losses.

In marine applications, the gas turbines usually **cannot be mounted rigidly** to the ship's structure. Normal movement and distortions of the hull when under way would cause distortions and misalignment in the turbine and cause internal rubs or bearing and/or structural failure. The turbine components can be mounted on a subbase built up of structural sections of sufficient rigidity to maintain the gas-turbine alignment when properly supported by the ship's hull.

Since the gas turbine is a high-speed machine with output shaft speeds ranging from about 3,600 r/min for large machines up to 100,000 r/min for very small machines (approximately 25,000 r/min is an upper limit for units suitable for the propulsion of small boats), a **reduction gear is necessary** to reduce the speed to the range suitable for a propeller. Smaller units suitable for boats or driving auxiliary units, such as generators in larger vessels, frequently have a reduction gear integral with the unit. Larger units normally require a separate reduction gear, usually of the double- or triple-reduction type.

A gas turbine, in common with all turbine machinery, is **not inherently reversible** and must be reversed by external means. Electric drives offer ready reversing but are usually ruled out on the basis of weight, cost, and to some extent efficiency. From a practical standpoint there are two alternatives, a reversing gear or a controllable reversible-pitch (CRP) propeller. Both have been used successfully in gas-turbine-driven ships, with the CRP favored.

Combined Propulsion Plants

In some shipboard applications, diesel engines, gas turbines, and steam turbines can be employed effectively in various combinations. The prime movers may be combined either mechanically, thermodynamically, or both. The leveling out of specific fuel consumption over the operating speed range is the goal of most combined systems.

The gas turbine is a very flexible power plant and consequently figures in most possible combinations which include combined diesel and gas-turbine plants (**CODAG**); combined gas- and steam-turbine plants (**COGAS**); and combined gas-turbine and gas-turbine plants (**COGAG**). In these cycles, gas turbines and other engines or gas turbines of two different sizes or types are combined in one plant to give optimum performance over a very wide range of power and speed. In addition, combinations of diesel or gas-turbine plants (**CODOG**), or gas-turbine or gas-turbine plants (**COGOG**), where one plant is a diesel or small gas turbine for use at low or cruising powers and the other a large gas turbine which operates alone at high powers, are also possibilities. Even the combination of a small nuclear plant and a gas turbine plant (**CONAG**) has undergone feasibility studies.

COGAS   Gas and steam turbines are connected to a common reduction gear, but thermodynamically are either independent or combined. Both gas- and steam-turbine drives are required to develop full power. In a thermodynamically independent plant, the boost power is furnished by a lightweight gas turbine. This combination produces a significant reduction in machinery weight.

In the thermodynamically combined cycle, commonly called STAG (steam and gas turbine) or COGES (used with electric-drive systems), energy is recovered from the exhaust of the gas turbine and is used to augment the main propulsion system through a steam turbine. The principal advantage gained by the thermodynamic interconnection is the potential for improved overall efficiency and resulting fuel savings. In this arrangement, the gas turbine discharges to a heat-recovery boiler where a large quantity of heat in the exhaust gases is used to generate

Fig. 11.3.15 COGAS system schematic.

steam. The boiler supplies the steam turbine that is geared to the propeller. The steam turbine may be coupled to provide part of the power for the gas-turbine compressor. The gas-turbine may be used to provide additional propulsion power, or its exhaust may be recovered to supply heat for various ship's services (Fig. 11.3.15).

Combined propulsion plants have been used in several applications in the past. **CODOG** plants have serviced some Navy patrol gunboats and the Coast Guard's Hamilton class cutters. Both **COGOG** and **COGAS** plants have been used in foreign navies, primarily for destroyer-type vessels.

PROPULSORS

The force to propel a marine vehicle arises from the rate of change of momentum induced in either the water or air. Since the force produced is directly proportional to the mass density of the fluid used, the reasonable choice is to induce the momentum change in water. If air were used, either the cross-sectional area of the jet must be large or the velocity must be high. A variety of propulsors are used to generate this stream of water aft relative to the vehicle: screw propellers, controllable-pitch propellers, water jets, vertical-axis propellers, and other thrust devices. Figure 11.3.16 indicates the type of propulsor which provides the best efficiency for a given vehicle type.

Screw Propellers

The screw propeller may be regarded as part of a helicoidal surface which, as it rotates, appears to "screw" its way through the water, driving water aft and the vehicle forward. A propeller is termed **"right-handed"** if it turns clockwise (viewed from aft) when producing ahead thrust; if

Fig. 11.3.16 Comparison of optimum efficiency values for different types of propulsors. *(From "Marine Engineering," SNAME, 1992.)*

counterclockwise, **"left-handed."** In a twin-screw installation, the starboard propeller is normally right-handed and the port propeller left-handed. The surface of the propeller blade facing aft, which experiences the increase in pressure, producing thrust, is the **face** of the blade; the forward surface is the **back.** The face is commonly constructed as a true helical surface of constant pitch; the back is not a helical surface. A **true helical surface** is generated by a line rotated about an axis normal to itself and advancing in the direction of this axis at constant speed. The distance the line advances in one revolution is the **pitch.** For simple propellers, the pitch is constant on the face; but in practice, it is common for large propellers to have a reduced pitch toward the hub and less usually toward the tip. The pitch at 0.7 times the maximum radius is usually a representative mean pitch; maximum lift is generated at that approximate point. Pitch may be expressed as a dimensionless ratio, P/D.

The **shapes** of blade outlines and sections vary greatly according to the type of ship for which the propeller is intended and to the designer's ideas. Figure 11.3.17 shows a typical design and defines many of the terms in common use. The **projected area** is the area of the projection of the propeller contour on a plane normal to the shaft, and the **developed area** is the total face area of all the blades. If the variation of helical cord length is known, then the true blade area, called **expanded area,** can be obtained graphically or analytically by integration.

Fig. 11.3.17 Typical propeller drawing. *(From "Principles of Naval Architecture," SNAME, 1988.)*

Diameter = D Pitch ratio = P/D Pitch = P
Blade thickness ratio = t/D No. of blades = 4
Pitch angle = ϕ
Disk area = area of tip circle = $\pi D^2/4 = A_O$
Developed area of blades, outside hub = A_D
Developed area ratio = DAR = A_D/A_O
Projected area of blades (on transverse plane) outside hub = A_P
Projected area ratio = PAR = A_P/A_O
Blade width ratio = BWR = (max. blade width)/D
Mean width ratio = MWR = $[A_D/\text{length of blades (outside hub)}]/D$

Consider a section of the propeller blade at a radius r with a pitch angle ϕ and pitch P working in an unyielding medium; in one revolution of the propeller it will advance a distance P. Turning n revolutions in unit time it will advance a distance $P \times n$ in that time. In a real fluid, there will be a certain amount of yielding when the propeller is developing thrust and the screw will not advance $P \times n$, but some smaller distance. The difference between $P \times n$ and that smaller distance is called the **slip velocity. Real slip ratio** is defined in Fig. 11.3.18.

A wake or a frictional belt of water accompanies every hull in motion; its velocity varies as the ship's speed, hull shape, the distance from the ship's side and from the bow, and the condition of the hull surface. For ordinary propeller design the **wake velocity** is a fraction w of the ship's speed. Wake velocity = wV. The **wake fraction** w for a ship may be obtained from Fig. 11.3.19. The velocity of the ship relative to the ship's wake at the stern is $V_A = (1 - w)V$.

The **apparent slip ratio** S_A is given by

$$S_A = (Pn - V)/Pn = 1 - V/Pn$$

Although real slip ratio, which requires knowledge of the wake fraction, is a real guide to ship performance, the apparent slip ratio requires

Fig. 11.3.18 Definition of slip. *(From "Principles of Naval Architecture," SNAME, 1988.)*

$$\tan \varphi = Pn/2\pi nr = P/2\pi r$$

$$\text{Real slip ratio} = S_R = MS/ML = (Pn - V_A)/Pn = 1 - V_A/Pn$$

Fig. 11.3.19 Wake fractions for single- and twin-screw models. *(From "Principles of Naval Architecture," SNAME, 1988.)*

only ship speed, revolutions, and pitch to calculate and is therefore often recorded in ships' logs.

For P in ft (m), n in r/min, and V in knots, $S_A = (Pn - 101.3\ V)/Pn$; $[S_A = (Pn - 30.9V)/Pn]$.

Propeller Design The design of a marine propeller is usually carried out by one of two methods. In the first, the design is based upon charts giving the results of open-water tests on a series of model propellers. These cover variations in a number of the design parameters such as pitch ratio, blade area, number of blades, and section shapes. A propeller which conforms with the characteristics of any particular series can be rapidly designed and drawn to suit the required ship conditions.

The second method is used in cases where a propeller is heavily loaded and liable to cavitation or has to work in a very uneven wake pattern; it is based on purely theoretical calculations. Basically this involves finding the chord width, section shape, pitch, and efficiency at a number of radii to suit the average circumferential wake values and give optimum efficiency and protection from cavitation. By integration of the resulting thrust and torque-loading curves over the blades, the thrust, torque, and efficiency for the whole propeller can be found.

Using the first method and **Taylor's propeller** and **advance coefficients** B_p and δ, a convenient practical design and an initial estimate of propeller size can be obtained

$$B_P = \frac{n(P_D)^{0.5}}{(V_A)^{2.5}} \qquad \left[\frac{1.158n(P_D)^{0.5}}{(V_A)^{2.5}}\right]$$

$$\delta = \frac{nD}{V_A} \qquad \left(\frac{3.281\,nD}{V_A}\right)$$

$$\eta_O = \frac{TV_A}{325.7P_D\eta_R} \qquad \left(\frac{TV_A}{1,942.5P_D\eta_R}\right)$$

where B_p = Taylor's propeller coefficient; δ = Taylor's advance coefficient; n = r/min; P_D = delivered power, dhp (kW$_D$); V_A = speed of

advance, knots; D = propeller diameter, ft (m); T = thrust, lb (N); η_O = open-water propeller efficiency; η_R = relative rotative efficiency ($0.95 < \eta_R < 1.0$, twin-screw; $1.0 < \eta_R < 1.05$, single-screw), a factor which corrects η_O to the efficiency in the actual flow conditions behind the ship.

Assume a reasonable value for n and, using known values for P_D and V_A (for a useful approximation, $P_D = 0.98P_S$), calculate B_p. Then enter Fig. 11.3.20, or an appropriate series of Taylor or Troost propeller charts, to determine δ, η_O, and P/D for optimum efficiency. The charts and parameters should be varied, and the results plotted to recognize the most suitable propeller.

Fig. 11.3.20 Typical Taylor propeller characteristic curves. *(From "Handbook of Ocean and Underwater Engineering," McGraw-Hill, 1969.)*

Propeller cavitation, when severe, may result in marked increase in rpm, slip, and shaft power with little increase in ship speed or effective power. As cavitation develops, noise, vibration, and erosion of the propeller blades, struts, and rudders are experienced. It may occur either on the face or on the back of the propeller. Although cavitation of the face has little effect on thrust and torque, extensive cavitation of the back can materially affect thrust and, in general, requires either an increase in blade area or a decrease in propeller rpm to avoid. The erosion of the backs is caused by the collapse of cavitation bubbles as they move into higher pressure regions toward the trailing edge.

Avoidance of cavitation is an important requirement in propeller design and selection. The Maritime Research Institute Netherlands (MARIN) suggests the following **criterion** for the minimum blade area required **to avoid cavitation:**

$$A_P^2 = \frac{T^2}{1,360(p_0 - p_v)^{1.5}V_A} \qquad \text{or} \qquad \left[\frac{T^2}{5.44(p_0 - p_v)^{1.5}V_A}\right]$$

where A_p = projected area of blades, ft^2 (m^2); T = thrust, lb (N); p_0 = pressure at screw centerline, due to water head plus atmosphere, lb/in^2 (N/m^2); p_v = water-vapor pressure, lb/in^2 (N/m^2); and V_A = speed of advance, knots (1 knot = 0.515 m/s); and

$$p_0 - p_v = 14.45 + 0.45h \qquad \text{lb/in}^2$$
$$\text{or} \qquad (p_0 - p_v = 99,629 + 10,179h \qquad \text{N/m}^2)$$

where h = head of water at screw centerline, ft (m).

A four-blade propeller of 0.97 times the diameter of the three-blade, the same pitch ratio, and four-thirds the area will absorb the **same power** at the same rpm as the three-blade propeller. Similarly, a two-blade propeller of 5 percent greater diameter is approximately equivalent to a three-blade unit. Figure 11.3.20 shows that propeller efficiency increases with decrease in value of the propeller coefficient B_p. A low value of B_p in a slow-speed ship calls for a large-diameter (optimum) propeller. It is frequently necessary to limit the diameter of the propeller and accept the accompanying loss in efficiency.

The number of propeller blades is usually three or four. Four blades are commonly used with single-screw merchant ships. Recently five, and even seven, blades have been used to reduce vibration. Alternatively, highly skewed blades reduce vibration by creating a smoother interaction between the propeller and the ship's wake. Highly loaded propellers of fast ships and naval vessels call for large blade area, preferably three to five blades, to reduce blade interference and vibration.

Propeller fore-and-aft clearance from the stern frame of large single-screw ships at 70 percent of propeller radius should be greater than 18 in. The stern frame should be streamlined. The clearance from the propeller tips to the shell plating of twin-screw vessels ranges from about 20 in for low-powered ships to 4 or 5 ft for high-powered ships. U.S. Navy practice generally calls for a propeller tip clearance of $0.25D$. Many high-speed motorboats have a tip clearance of only several inches. The immersion of the propeller tips should be sufficient to prevent the drawing in of air.

EXAMPLE. Consider a three-blade propeller for a single-screw installation with $P_s = 2,950$ shp (2,200 kW$_s$), $V = 17$ knots, and $C_B = 0.52$. Find the propeller diameter D and efficiency η_O for $n = 160$ r/min.

From Fig. 11.3.19, the wake fraction $w = 0.165$. $V_A = V(1 - 0.165) = 14.19$ knots. $P_D = 0.98P_s = 2,891$ dhp (2,156 kW$_D$). At 160 r/min, $B_p = 160(2,891)^{0.5}/(14.19)^{2.5} = 11.34$. [$1.158 \times 160 (2,156)^{0.5}/(14.19)^{2.5} = 11.34$.] From Fig. 11.3.20, the following is obtained or calculated:

r/min	B_p	P/d	η_O	δ	D, ft (m)	P, ft (m)
160	11.34	0.95	0.69	139	12.33 (3.76)	11.71 (3.57)

For a screw centerline submergence of 9 ft (2.74 m) and a selected projected area ratio PAR = 0.3, investigate the cavitation criterion. Assume $\eta_R = 1.0$. Minimum projected area:

$$T = 325.7 \, \eta_O \eta_R P_D/V_A = 325.7 \times 0.69 \times 1.0 \times 2,891/14.19$$
$$= 45,786 \text{ lb}$$
$$(T = 1,942.5 \times 0.69 \times 2,156/14.19 = 203,646 \text{ N})$$
$$p_0 - p_v = 14.45 + 0.45(9) = 18.5 \text{ lb/in}^2$$
$$(p_0 - p_v = 99,629 + 10,179(2.74 = 127,519 \text{ N/m}^2)$$
$$A_p^2 = (45,786)^2/1,360(18.5)^{1.5}(14.19) = 1,365, A_p$$
$$= 36.95 \text{ ft}^2 \text{ minimum}$$
$$A_p^2 = (203,646)^2/5.44(127,519)^{1.5}(14.19) = 11.79, A_p$$
$$= 3.43 \text{ m}^2 \text{ minimum}$$

Actual projected area: PAR $= A_p/A_0 = 0.3$, $A_p = 0.3A_0 = 0.37\pi (12.33)^2/4 = 35.82$ ft^2 (3.33 m^2).

Since the selected PAR does not meet the minimum cavitation criterion, either the blade width can be increased or a four-blade propeller can be adopted. To absorb the same power at the same rpm, a four-blade propeller of about 0.97 (12.33) = 11.96 ft (3.65 m) diameter with the same blade shape would have an $A_p = 35.82 \times 4/3 \times (11.96/12.33)^2 = 44.94$ ft^2 (4.17 m^2), or about 25 percent increase. The pitch = 0.95 (11.96) = 11.36 ft (3.46 m). The real and apparent slip ratios for the four-blade propeller are:

$$S_R = 1 - (101.3)(14.19)/(11.36)(160) = 0.209$$
$$S_A = 1 - (30.9)(17.0)/(3.46)(160) = 0.051$$

that is, 20.9 and 5.1 percent (S_A calculated using SI values).

Since the projected area in the three-blade propeller of this example is only slightly under the minimum, a three-blade propeller with increased blade width would be the more appropriate choice.

Controllable-Pitch Propellers

Controllable-pitch propellers are screw propellers in which the blades are separately mounted on the hub, each on an axis, and in which the pitch of the blades can be changed, and even reversed, while the propeller is turning. The pitch is changed by means of an internal mechanism consisting essentially of hydraulic pistons in the hub acting on crossheads. Controllable-pitch propellers are most suitable for vehicles which must meet different operating conditions, such as tugs, trawlers, ferries, minesweepers, and landing craft. As the propeller pitch is changed, the engine can still run at its most efficient speed. Maneuvering is more rapid since the pitch can be changed more rapidly than could the shaft revolutions. By use of controllable-pitch propellers, neither reversing mechanisms nor astern turbines in turbine-powered vehicles are necessary in reciprocating engines, nor astern turbines in turbine-powered vehicles, especially important in gas-turbine installations. Except for the larger hub needed

to house the pitch-changing mechanism, the controllable-pitch propeller can be made almost as efficient as the solid, fixed-blade propeller. Gas turbines and controllable reversible-pitch propellers (CRPs) are generally used in all recent U.S. Navy surface ship designs (except carriers).

Water Jet

This method usually consists of an impeller or pump inside the hull, which draws water from outside, accelerates it, and discharges it astern as a jet at a higher velocity. It is a reaction device like the propeller but in which the moving parts are contained inside the hull, desirable for shallow-water operating conditions and maneuverability. The overall efficiency is lower than that of the screw propeller of diameter equal to the jet orifice diameter, principally because of inlet and ducting losses. Other disadvantages include the loss of volume to the ducting and impeller and the danger of fouling of the impeller. Water jets have been used in several of the U.S. Navy's hydrofoil and surface-effect vehicles.

Vertical-Axis Propellers

There are two types of vertical-axis propulsor systems consisting of one or two vertical-axis rotors located underwater at the stern. Rotor disks are flush with the shell plating and have five to eight streamline, spade-like, vertical impeller blades fitted near the periphery of the disks. The blades feather during rotation of the disk to produce a maximum thrust effect in any direction desired. In the **Kirsten-Boeing** system the blades are interlocked by gears so that each blade makes a half revolution about its axis for each revolution of the disk. The blades of the **Voith-Schneider** system make a complete revolution about their own axis for each revolution of the disk. A bevel gear must be used to transmit power from the conventional horizontal drive shaft to the horizontal disk; therefore, limitations exist on the maximum power that can be transmitted. Although the propulsor is 30 to 40 percent less efficient than the screw propeller, it has obvious maneuverability advantages. Propulsors of this type have also been used at the bow to assist in maneuvering.

Other Thrust Devices

Pump Jet In a pump-jet arrangement, the rotating impeller is external to the hull with fixed guide vanes either ahead and/or astern of it; the whole unit is enclosed in a duct or long shroud ring. The duct diameter increases from the entrance to the impeller so that the velocity falls and the pressure increases. Thus the impeller diameter is larger, thrust loading less, and the efficiency higher; the incidence of cavitation and noise is delayed. A penalty is paid, however, for the resistance of the duct.

Kort Nozzles In the Kort nozzle system, the screw propeller operates in a ring or nozzle attached to the hull at the top. The longitudinal sections are of airfoil shape, and the length of the nozzle is generally about one-half its diameter. Unlike the pump-jet shroud ring, the Kort nozzle entrance is much bigger than the propeller, drawing in more water than the open propeller and achieving greater thrust. Because of the acceleration of the water into the nozzle, the pressure inside is less; hence a forward thrust is exerted on the nozzle and the hull. The greatest advantage is in a tug, pulling at rest. The free-running speed is usually less with the nozzle than without. In some tugs and rivercraft, the whole nozzle is pivoted and becomes a very efficient steering mechanism.

Tandem and Contrarotating Propellers Two or more propellers arranged on the same shaft are used to divide the increased loading factor when the diameter of a propeller is restricted. Propellers turning in the same direction are termed **tandem**, and in opposite directions, **contrarotating**. In tandem, the rotational energy in the race from the forward propeller is augmented by the after one. Contrarotating propellers work on coaxial, contrary-turning shafts so that the after propeller may regain the rotational energy from the forward one. The after propeller is of smaller diameter to fit the contracting race and has a pitch designed for proper power absorption. Such propellers have been used for years on torpedoes to prevent the torpedo body from rotating. Hydrodynamically, the advantages of contrarotating propellers are increased propulsive efficiency, improved vibration characteristics, and higher blade frequency. Disadvantages are the complicated gearing, coaxial shafting, and sealing problems.

Supercavitating Propellers When cavitation covers the entire back of a propeller blade, an increase of rpm cannot further reduce the pressure at the back but the pressure on the face continues to increase as does the total thrust, though at a slower rate than before cavitation began. Advantages of such fully cavitating propellers are an absence of back erosion and less vibration. Although the characteristics of such propellers were determined by trial and error, they have long been used on high-speed racing motor boats. The blade section design must ensure clean separation of flow at the leading and trailing edges and provide good lift-drag ratios for high efficiency. Introducing air to the back of the blades (**ventilated propeller**) will ensure full cavitation and also enables use at lower speeds.

Partially Submerged Propellers The appendage drag presented by a propeller supported below high-speed craft, such as planing boats, hydrofoil craft, and surface-effect ships, led to the development of partially submerged propellers. Although vibration and strength problems, arising from the cyclic loading and unloading of the blades as they enter and emerge from the air-water interface, remain to be solved, it has been demonstrated that efficiencies in partially submerged operation, comparable to fully submerged noncavitating operation, can be achieved. The performance of these propellers must be considered over a wide range of submergences.

Outboard Gasoline Engine Outboard gasoline engines of 1 to 300 bhp (0.75 to 225 kW$_B$), combining steering and propulsion, are popular for small pleasure craft. Maximum speeds of 3,000 to 6,000 r/min are typical, as driven by the propeller horsepower characteristics (see Sec. 9).

PROPULSION TRANSMISSION

In modern ships, only large-bore, slow-speed diesel engines are directly connected to the propeller shaft. Transmission devices such as mechanical speed-reducing gears or electric drives (generator/motor transmissions) are required to convert the relatively high rpm of a compact, economical prime mover to the relatively low propeller rpm necessary for a high propulsive efficiency. In the case of steam turbines, medium and high-speed diesel engines, and gas turbines, speed reduction gears are used. Gear ratios vary from relatively low values for medium-speed diesels up to approximately 50 : 1 for a compact turbine design. Where the prime mover is unidirectional, the drive mechanism must also include a reversing mechanism, unless a CRP is used.

Reduction Gears

Speed reduction is usually obtained with reduction gears. The simplest arrangement of a marine reduction gear is the **single-reduction,** single-input type, i.e., one pinion meshing with a gear (Fig. 11.3.21). This arrangement is used for connecting a propeller to a diesel engine or to an electric motor but is not used for propelling equipment with a turbine drive.

Fig. 11.3.21 Single-reduction, single-input gear. *(From "Marine Engineering," SNAME, 1992.)*

The usual arrangement for turbine-driven ships is the **double-reduction,** double-input, articulated type of reduction gear (Fig. 11.3.22). The two input pinions are driven by the two elements of a cross-compound turbine. The term *articulated* applies because a flexible coupling is generally provided between the first reduction or primary gear wheel and the second reduction or secondary pinion.

The **locked-train** type of double-reduction gear has become standard for high-powered naval ships and, because it minimizes the total weight and size of the assembly, is gaining increased popularity for higher-powered merchant ships (Fig. 11.3.23).

Fig. 11.3.22 Double-reduction, double-input, articulated gear. *(From "Marine Engineering," SNAME, 1992.)*

Fig. 11.3.23 Double-reduction, double-input, locked-train gear. *(From "Marine Engineering," SNAME, 1992.)*

Electric Drive

Electric propulsion drive is the principal alternative to direct- or geared-drive systems. The prime mover drives a generator or alternator which in turn drives a propulsion motor which is either coupled directly to the propeller or drives the propeller through a low-ratio reduction gear.

Among its **advantages** are the ease and convenience by which propeller speed and direction are controllable; the freedom of installation arrangement offered by the electrical connection between the generator and the propulsion motor; the flexibility of power use when not used for propulsion; the convenience of coupling several prime movers to the propeller without mechanical clutches or couplings; the relative simplicity of controls required to provide reverse propeller rotation when the prime mover is unidirectional; and the speed reduction that can be provided between the generator and the motor, hence between the prime mover and the propeller, without mechanical speed-reducing means.

The disadvantages are the inherently higher first cost, increased weight and space, and the higher transmission losses of the system. The advent of **superconductive electrical machinery** suitable for ship propulsion, however, has indicated order-of-magnitude savings in weight, greatly reduced volume, and the distinct possibility of lower costs. Superconductivity, a phenomenon occurring in some materials at low temperatures, is characterized by the almost complete disappearance of electrical resistance. Research and development studies have shown that size and weight reductions by factors of 5 or more are possible. With successful development of superconductive electrical machinery and resolution of associated engineering problems, monohulled craft, such as destroyers, can benefit greatly from the location flexibility and maneuverability capabilities of superconductive electric propulsion, but the advantages of such a system will be realized to the greatest extent in the new high-performance ships where mechanical-drive arrangement is extremely complex.

In general, electric propulsion drives are employed in marine vehicles requiring a high degree of maneuverability such as ferries, icebreakers, and tugs; in those requiring large amounts of special-purpose power such as fireboats, and large tankers; in those utilizing nonreversing, high-speed, and multiple prime movers; and in deep-submergence vehicles. Diesel-electric drive is ideally adapted to bridge control.

Reversing

Reversing may be accomplished by stopping and reversing a reversible engine, as in the case of many reciprocating engines, or by adding reversing elements in the prime mover, as in the case of steam turbines. It is impracticable to provide reversing elements in gas turbines; hence a reversing capability must be provided in the transmission system or in the propulsor itself. Reversing reduction gears for such transmissions are available up to quite substantial powers, and CRP propellers are also used with diesel or gas-turbine drives. Electrical drives provide reversing by dynamic braking and energizing the electric motor in the reverse direction. Waterjets reverse inserting reversing thrust buckets into the discharge stream.

Marine Shafting

A main propulsion shafting system consists of the equipment necessary to transmit the rotative power from the main propulsion engines to the

Fig. 11.3.24 Shafting arrangement with strut bearings. *(From "Marine Engineering," SNAME, 1992.)*

Fig. 11.3.25 Shafting arrangement without strut bearings. *(From "Marine Engineering," SNAME, 1992.)*

propulsor; support the propulsor; transmit the thrust developed by the propulsor to the ship's hull; safely withstand transient operating loads (e.g., high-speed maneuvers, quick reversals); be free of deleterious modes of vibration; and provide reliable operation.

Figure 11.3.24 is a shafting arrangement typical of multishaft ships and single-shaft ships with transom sterns. The shafting must extend outboard a sufficient distance for adequate clearance between the propeller and the hull. Figure 11.3.25 is typical of single-screw merchant ships.

The shafting located inside the ship is **line shafting.** The outboard section to which the propeller is secured is the **propeller shaft** or **tail shaft.** The section passing through the stern tube is the **stern tube shaft** unless the propeller is supported by it. If there is a section of shafting between the propeller and stern tube shafts, it is an **intermediate shaft.**

The principal dimensions, kind of material, and material tests are specified by the classification societies for marine propulsion shafting. The American Bureau of Shipping Rules for Steel Vessels specifies the minimum diameter of propulsion shafting, tail shaft bearing lengths, bearing liner and flange coupling thicknesses, and minimum fitted bolt diameters.

Tail or propeller shafts are of larger diameter than the line or tunnel shafting because of the propeller-weight-induced bending moment; also, inspection during service is possible only when the vessel is drydocked.

Propellers are fitted to shafts over 6 in (0.1524 m) diameter by a taper fit. Aft of the propeller, the shaft is reduced to 0.58 to 0.68 of its specified diameter under the liner and is fitted with a relatively fine thread, nut, and keeper. Propeller torque is taken by a long flat key, the width and total depth of which are, respectively, about 0.21 and 0.11 times the shaft diameter under the liner. The forward end of the key slot in the shaft should be tapered off in depth and terminate well aft of the large end of the taper to avoid a high concentrated stress at the end of the keyway. Every effort is made to keep saltwater out of contact with the steel tail shaft so as to prevent failure from corrosion fatigue.

A cast-iron, cast-steel, or wrought-steel-weldment **stern tube** is rigidly fastened to the hull. In single-screw ships it supports the propeller shaft and is normally provided with a packing gland at the forward end as seawater lubricates the stern-tube bearings. The aft stern-tube bearing is made four diameters in length; the forward stern-tube bearing is

much shorter. A bronze liner, ¾ to 1 in (0.0191 to 0.0254 m) thick, is shrunk on the propeller shaft to protect it from corrosion and to give a good journal surface. Lignum-vitae inserts, epoxy applications, or rubber linings have been used for stern-tube bearings; the average bearing pressure is about 30 lb/in² (206,842 N/m²). Rubber and phenolic compound bearings, having longitudinal grooves, are also extensively used. White-metal oil-lubricated bearings and, in Germany, roller-type stern bearings have been fitted in a few installations.

The line or tunnel shafting is laid out, in long equal lengths, so that, in the case of a single-screw ship, withdrawal of the tail shaft inboard for inspection every several years will require only the removal of the next inboard length of shafting. For large outboard or water-exposed shafts of twin-screw ships, a bearing spacing up to 30 shaft diameters has been used. A greater ratio prevails for steel shafts of power boats. Bearings inside the hull are spaced closer—normally under 15 diameters if the shaft is more than 6 in (0.1524 m) diameter. Usually, only the bottom half of the bearing is completely white-metaled; oil-wick lubrication is common, but oil-ring lubrication is used in high-class installations.

Bearings are used to support the shafting in essentially a straight line between the main propulsion engine and the desired location of the propeller. Bearings inside the ship are most popularly known as lineshaft bearings, steady bearings, or spring bearings. Bearings which support outboard sections of shafting are stern-tube bearings if they are located in the stern tube and strut bearings when located in struts.

The propeller thrust is transmitted to the hull by means of a main thrust bearing. The main thrust bearing may be located either forward or aft of the slow-speed gear.

HIGH-PERFORMANCE SHIP SYSTEMS

High-performance ships have spawned new hull forms and propulsion systems which were relatively unknown just a generation ago. Because of their unique characteristics and requirements, conventional propulsors are rarely adequate. The prime thrust devices range from those using water, to water-air mixtures, to large air propellers.

The total propulsion power required by these vehicles is presented in general terms in the Gabrielli and Von Karman plot of Fig. 11.3.26. Basically, there are those vehicles obtaining lift by displacement and those obtaining lift from a wing or foil moving through the fluid. In

Fig. 11.3.26 Lift-drag ratio versus calm weather speed for various vehicles.

Fig. 11.3.27 Some foil arrangements and sections.

order for marine vehicles to maintain a reasonable lift-to-drag ratio over a moderate range of speed, the propulsion system must provide increased thrust at lower speeds to overcome low-speed wave drag. The weight of any vehicle is converted into total drag and, therefore, total required propulsion and lift power.

Lightweight, compact, and efficient propulsion systems to satisfy extreme weight sensitivity are achieved over a wide range of power by the marinized aircraft gas-turbine engine. Marinization involves addition of a power turbine unit, incorporation of air inlet filtration to provide salt-free air, addition of silencing equipment on both inlet and exhaust ducts, and modification of some components for resistance to salt corrosion. The shaft power delivered is then distributed to the propulsion and lift devices by lightweight shaft and gear power transmission systems.

A comparison of high-performance ship propulsion system characteristics is given in Table 11.3.3.

Hydrofoil Craft

Under foilborne conditions, the support of the hydrofoil craft is derived from the dynamic lift generated by the foil system, and the craft hull is lifted clear of the water. Special problems relating to operation and directly related to the conditions presented by waves, high fluid density, surface piercing of foils and struts, stability, control, cavitation, and

ventilation of foils have given rise to foil system configurations defined in Fig. 11.3.27.

The hydrofoil craft has three modes of operation: **hullborne, takeoff,** and **foilborne.** The thrust and drag forces involved in hydrofoil operation are shown in Fig. 11.3.28. Design must be a compromise between the varied requirements of these operating conditions. This may be difficult, as with propeller design, where maximum thrust is often required at takeoff speed (normally about one-half the operating speed).

Fig. 11.3.28 Thrust and drag forces for a typical hydrofoil.

Table 11.3.3 U.S. Navy High-Performance Ship Propulsion System Characteristics

Ship	Engine	Propulsor/lift
Hydrofoil, PHM, 215 tons (2.1 MN), 50 knots	Foilborne: one LM-2500 GT, 22,000 bhp (15,405 kW$_B$), 3,600 r/min	Foilborne: one 2-stage axial-flow waterjet, twin aft strut inlets
Surface-effect ship, SES-200, 195 tons (1.94 MN), 30 + knots	Propulsion: two 16V-149TI diesel, 1,600 bhp (1,193 kW$_B$) Lift: two 8V-92TI diesel, 800 bhp (597 kW$_B$)	Two waterjet units Two double-inlet centrifugal lift fans
Air-cushion vehicle, LCAC, 150 tons (1.49 MN), 50 knots	Propulsion and lift: four TF-40B GT, 3,350 bhp (2,498 kW$_B$), 15,400 r/min	Two ducted-air propellers, 12 ft (3.66 m) dia., 1,250 r/min; four double-entry centrifugal lift fans, 1,900 r/min
SWATH, SSP, 190 tons (1.89 MN), 25 knots	Propulsion: two T-64 GT, 3,200 bhp (2,386 kW$_B$), 1,000 r/min at output of attached gearbox	Two subcavitating controllable-pitch propellers, 450 r/min

A propeller designed for maximum efficiency at takeoff will have a smaller efficiency at design speeds, and vice versa. This is complicated by the need for supercavitating propellers at operating speeds over 50 knots. Two sets of propellers may be needed in such a case, one for hull-borne operation and the other for foilborne operation. The selection of a surface-piercing or a fully submerged foil design is determined from specified operating conditions and the permissible degree of complexity. A **surface-piercing system** is inherently stable, as a deviation from the equilibrium position of the craft results in corrective lift forces. **Fully submerged foil** designs, on the other hand, provide the best lift-drag ratio but are not stable by themselves and require a depth-control system for satisfactory operation. Such a control system, however, permits operation of the submerged-foil design in higher waves.

Surface-Effect Vehicles (SEV)

The air-cushion-supported **surface-effect ship (SES)** and **air-cushion vehicle (ACV)** show great potential for high-speed and amphibious operation. Basically, the SEV rides on a cushion of air generated and maintained in the space between the vehicle and the surface over which it moves or hovers. In the SES, the cushion is captured by rigid sidewalls and flexible skirts fore and aft. The ACV has a flexible seal or skirt all around. Figure 11.3.29 illustrates two configurations of the surface-effect principle.

Skegs or side walls   Flexible Skirt

Fig. 11.3.29   Two configurations of the surface-effect principle.

Since the ACV must rise above either land or water, the propulsors use only air for thrust, requiring a greater portion of the total power allocated to lift than in the SES design as shown in Fig. 11.3.30.

While SES technology has greatly improved the capability of amphibious landing vehicles, it is rapidly being developed for high-speed large cargo vessels.

Small Water-Plane Area Twin Hull

The **SWATH,** configured to be relatively insensitive to seaway motions, gains its buoyant lift from two parallel submerged, submarinelike hulls which support the main hull platform through relatively small vertical struts. The small hull area at the waterline lowers the response of the hull to waves. The same small waterplane area makes the SWATH hull form sensitive to changes in weight and load distribution. In this configuration, a propulsion arrangement choice must be with regard to the location of the main engines. In the U.S. Navy's SSP, power from two independently operated gas turbines is transmitted down each aft

Fig. 11.3.30   Comparison of power requirements: SES versus ACV.

strut by a group of three wide-belt chain drives with speed reduction to the propellers. By contrast, the T-AGOS 19 series SWATH vessels employ diesel-electric drive, with both motors and propulsors in the lower hull.

CARGO SHIPS

The development of containerization in the 1950s sparked a growth in ship size. More efficient cargo-handling methods and the desire to reduce transit time led to the emergence of the high-speed container vessel.

The Container Vessel

Because of low fuel costs in the 1950s and 1960s, early container vessel design focused on higher speed and container capacity. Rising fuel costs shifted emphasis toward ship characteristics which enhanced reduced operating costs, particularly as affected by fuel usage and trade route capacity. By the early 1980s, ships were generally sized for TEU (twenty-foot-equivalent-unit) capacity less than 2,500 and speeds of about 20 knots, lower if fuel was significant in the overall operating costs. Largely a function of the scale of trade route capacity, container ships carrying in excess of 5,000 TEUs are common today. The larger the trade capacity, the larger the containerships operating the route. However, many trade route scales do not support even moderately sized containership service.

The **on deck** and **under deck** container capacity varies with the type of vessel, whether a **lift on/lift off** vessel, a bulk carrier equipped for containers, or a **roll on/roll off** vessel. Extensive studies have been conducted to optimize the hull form and improve the operating efficiency for the modern containership.

11.4 AERONAUTICS
by M. W. M. Jenkins

REFERENCES: National Advisory Committee for Aeronautics, *Technical Reports* (designated *NACA-TR* with number), *Technical Notes* (*NACA-TN* with number), and *Technical Memoranda* (*NACA-TM* with number). British Aeronautical Research Committee *Reports and Memoranda* (designated *Br. ARC-R & M* with number). *Ergebnisse der Aerodynamischen Versuchsanstalt zu Göttingen.* Diehl, "Engineering Aerodynamics," Ronald. Reid, "Applied Wing Theory," McGraw-Hill. Durand, "Aerodynamic Theory," Springer. Prandtl-Tietjens, "Fundamentals of Hydro- and Aeromechanics," and "Applied Hydro- and Aeromechanics," McGraw-Hill. Goldstein, "Modern Developments in Fluid Dynamics," Oxford. Millikan, "Aerodynamics of the Airplane," Wiley. Von Mises, "Theory of Flight," McGraw-Hill. Hoerner, "Aerodynamic Drag," Hoerner. Glauert, "The Elements of Airfoil and Airscrew Theory," Cambridge. Milne-Thompson, "Theoretical

Hydrodynamics," Macmillan. Munk, "Fluid Dynamics for Aircraft Designers," and "Principles of Aerodynamics," Ronald. Abraham, "Structural Design of Missiles and Spacecraft," McGraw-Hill. "U.S. Standard Atmosphere, 1976," U.S. Government Printing Office.

DEFINITIONS

Aeronautics is the science and art of flight and includes the operation of heavier-than-air aircraft. An **aircraft** is any weight-carrying device designed to be supported by the air, either by buoyancy or by dynamic action. An **airplane** is a mechanically driven fixed-wing aircraft, heavier than air, which is supported by the dynamic reaction of the air against its wings. An aircraft supported by buoyancy is called an **airship**, which can be either a balloon, a hot-air type, or of semirigid or rigid construction. These airships can be free floating or, if of semirigid or rigid design, propelled by fixed or vectoring propeller or jet propulsion. Ascent or descent is achieved primarily through helium gas or hot air, venting, and by the use of disposable ballast. A **helicopter** is a kind of aircraft lifted and moved by a large propeller mounted horizontally above the fuselage. It differs from the **autogiro** in that this propeller is turned by motor power and there is no auxiliary propeller for forward motion. A **ground-effect machine (GEM)** is a heavier-than-air surface vehicle which operates in close proximity to the earth's surface (over land or water), never touching except at rest, being separated from the surface by a cushion or film of air, however thin, and depending entirely upon aerodynamic forces for propulsion and control.

Aerodynamics is the branch of aeronautics that treats of the motion of air and other gaseous fluids and of the forces acting on solids in motion relative to such fluids. Aerodynamics falls into velocity ranges, depending upon whether the velocity is below or above the local speed of sound in the fluid. The velocity range below the local speed of sound is called the **subsonic** regime. Where the velocity is above the local speed of sound, the flow is said to be **supersonic**. The term **transonic** refers to flows in which both subsonic and supersonic regions are present. The **hypersonic** regime is that speed range usually in excess of five times the speed of sound.

STANDARD ATMOSPHERE

The U.S. Standard Atmosphere, 1976 of Table 11.4.1a is the work of the U.S. Committee on Extension to the Standard Atmosphere (COESA), established in 1953, which led to the 1958, 1962, 1966, and 1976 versions of this standard. It is an idealized, steady-state representation of the atmosphere and extends to approximately 621.4 miles (1000 kilometres) above the surface. It is assumed to rotate with the earth.

It is the same as COESA's "U.S. Standard Atmosphere—1962" below approximately 32 miles (51 km) but replaces this standard above this altitude. Figure 11.4.1 shows the variation of base temperature K with altitude Z up to approximately 76 miles (122 km). Above 281,952 feet, our distinct altitude zones are identified (Table 11.4.1b) in which

Table 11.4.1a U.S. Standard Atmosphere, 1976

Altitude h, ft*	Temp T, °F†	Pressure ratio, p/p_0	Density ratio, ρ/ρ_0	$(\rho_0/\rho)^{0.5}$	Speed of sound V_s, ft/s‡
0	59.00	1.0000	1.0000	1.000	1,116
5,000	41.17	0.8320	0.8617	1.077	1,097
10,000	23.34	0.6877	0.7385	1.164	1,077
15,000	5.51	0.5643	0.6292	1.261	1,057
20,000	−12.62	0.4595	0.5328	1.370	1,036
25,000	−30.15	0.3711	0.4481	1.494	1,015
30,000	−47.99	0.2970	0.3741	1.635	995
35,000	−65.82	0.2353	0.3099	1.796	973
36,089	−69.70	0.2234	0.2971	1.835	968
40,000	−69.70	0.1851	0.2462	2.016	968
45,000	−69.70	0.1455	0.1936	2.273	968
50,000	−69.70	0.1145	0.1522	2.563	968
55,000	−69.70	0.09001	0.1197	2.890	968
60,000	−69.70	0.07078	0.09414	3.259	968
65,000	−69.70	0.05566	0.07403	3.675	968
65,800	−69.70	0.05356	0.07123	3.747	968
70,000	−67.30	0.04380	0.05789	4.156	971
75,000	−64.55	0.03452	0.04532	4.697	974
80,000	−61.81	0.02725	0.03553	5.305	977
85,000	−59.07	0.02155	0.02790	5.986	981
90,000	−56.32	0.01707	0.02195	6.970	984
95,000	−53.58	0.01354	0.01730	7.600	988
100,000	−50.84	0.01076	0.01365	8.559	991

* × 0.3048 = metres.
† × (°F − 32)/1.8 = °C.
‡ × 0.3048 = m/s.

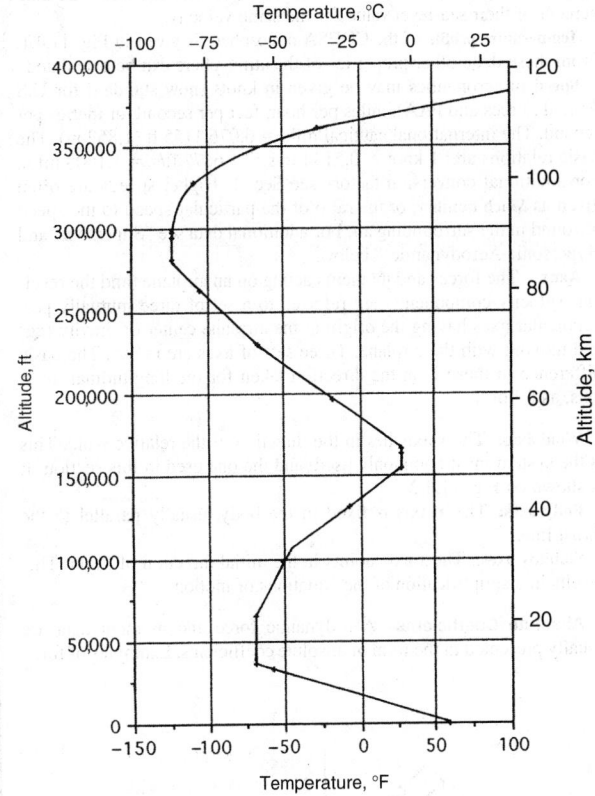

Fig. 11.4.1   Temperature as a function of altitude. *(COESA-1976.)*

Table 11.4.1b Reference Levels and Function Designations for Each of Four Zones of the Upper Atmosphere

Base Altitude (Z), km	Base altitude, ft	Base temperature K	Base temperature °F	Form of function relating K to Z (four zones)	Base values ρ/ρ_0	Base values p/p_0
86	281,952	186.9	−123.3	Linear-constant	5.680×10^{-6}	3.685×10^{-6}
91	298,320	186.9	−123.3	Elliptical	2.335×10^{-6}	1.518×10^{-6}
110	361,152	240.0	−27.7	Linear-increasing	0.0793×10^{-6}	0.0701×10^{-6}
120	393,888	360.0	188.3	Exponential	0.0181×10^{-6}	0.0251×10^{-6}
1000	3,280,992	1000.0	1340.3	—	2.907×10^{-15}	7.416×10^{-14}

temperature altitude relationships are defined relative to the base altitude of each zone. These four zones constitute the current standard atmosphere in terms of base temperature K from 53.4 miles to approximately 621.4 miles (1000 km). For further definition of the standard atmosphere in this four-zone region the reader is referred to U.S. Department of Commerce document "U.S. Standard Atmosphere, 1976," ADA035728.

The assumed sea-level conditions are: pressure, p_0 = 29.91 in (760 mm Hg) = 2,116.22 lb/ft²; mass density, ρ_0 = 0.002378 slugs/ft³ (0.001225 g/cm³); T_0 = 59°F (15°C).

UPPER ATMOSPHERE

High-altitude atmospheric data have been obtained directly from balloons, sounding rockets, and satellites and indirectly from observations of meteors, aurora, radio waves, light absorption, and sound effects. At relatively low altitudes, the earth's atmosphere is, for aerodynamic purposes, a uniform gas. Above 250,000 ft, day and night standards differ because of dissociation of oxygen by solar radiation. This difference in density is as high as 35 percent, but it is usually aerodynamically negligible above 250,000 ft since forces here will become less than 0.05 percent of their sea-level value for the same velocity.

Temperature profile of the COESA atmosphere is given in Fig. 11.4.1. From these data, other properties of the atmosphere can be calculated.

Speed in aeronautics may be given in knots (now standard for U.S Armed Forces and FAA), miles per hour, feet per second, or metres per second. The international nautical mile = 6,076.1155 ft (1,852 m). The basic relations are: 1 knot = 0.5144 m/s = 1.6877 ft/s = 1.1508 mi/h. For additional conversion factors see Sec. 1. Higher speeds are often given as **Mach number,** or the ratio of the particular speed to the speed of sound in the surrounding air. For additional data see "Supersonic and Hypersonic Aerodynamics" below.

Axes The forces and moments acting on an airplane (and the resultant velocity components) are referred to a set of three mutually perpendicular axes having the origin at the airplane center of gravity (cg) and moving with the airplane. Three sets of axes are in use. The basic difference in these is in the direction taken for the longitudinal, or x, axis, as follows:

Wind axes: The x axis lies in the direction of the relative wind. This is the system most commonly used, and the one used in this section. It is shown on Fig. 11.4.2.

Body axes: The x axis is fixed in the body, usually parallel to the thrust line.

Stability axes: The x axis points in the initial direction of flight. This results in a simplification of the equations of motion.

Absolute Coefficients Aerodynamic force and moment data are usually presented in the form of absolute coefficients. Examples of force

coefficients are: lift $C_L = L/qS$; drag $C_D = D/qS$; side force $C_Y = Y/qS$; where q is the dynamic pressure $\frac{1}{2}\rho V^2$, ρ = air mass-density, and V = air speed. Examples of moment coefficients are $C_m = M/qSc$; roll $C_l = L/qSb$; and yaw $C_n = N/qSb$; where c = mean wing chord and b = wing span.

Section Coefficients Basic test data on wing sections are usually given in the form of section lift coefficient C_l and section drag coefficient C_d. These apply directly to an infinite aspect ratio or to two-dimensional flow, but aspect-ratio and lift-distribution corrections are necessary in applying to a finite wing. For a wing having an elliptical lift distribution, $C_L = (\pi/4)C_l$, where C_l is the centerline value of the local lift coefficient.

SUBSONIC AERODYNAMIC FORCES

When an airfoil is moved through the air, the motion produces a pressure at every point of the airfoil which acts normal to the surface. In addition, a frictional force tangential to the surface opposes the motion. The sum of these pressure and frictional forces gives the **resultant force** R acting on the body. The point at which the resultant force acts is defined as the **center of pressure,** c.p. The resultant force R will, in general, be inclined to the airfoil and the relative wind velocity V. It is resolved (Fig. 11.4.3) either along wind axes into

L = **lift** = component normal to V
D = **drag** or resistance = component along V, or along body axes into
N = **normal force** = component perpendicular to airfoil chord
T = **tangential force** = component along airfoil chord

Instead of specifying the center of pressure, it is convenient to specify the moment of the air forces about the so-called *aerodynamic center,* a.c. This point lies at a distance a (about a quarter-chord length) back of the leading edge of the airfoil and is defined as the point about which the moment of the air forces remains constant when the angle of attack α is changed. Such a point exists for every airfoil. The force R acting at the c.p. is equivalent to the same force acting at the a.c. plus a moment equal to that force times the distance between the c.p. and the a.c. (see Fig. 11.4.3). The location of the aerodynamic center, in terms of the chord c and the section thickness t, is given for NACA 4- and 5-digit series airfoils approximately by

$$a/c = 0.25 - 0.40(t/c)^2$$

Fig. 11.4.3   Forces acting on an airfoil. (*a*) Actual forces; (*b*) equivalent forces through the aerodynamic center plus a moment.

Positive moments: X→Y, Y→Z, Z→X

Fig. 11.4.2   Wind axes.

The distance $x_{c.p.}$ from the leading edge to the center of pressure expressed as a fraction of the chord is, in terms of the moment $M_{c/4}$ about the quarter-chord point,

$$M_{c/4} = (\tfrac{1}{4} - x_{c.p.}) \cdot cN$$

$$x_{c.p.} = \frac{1}{4} - \frac{M_{c/4}}{cN}$$

From dimensional analysis it can be seen that the **air force on a body** of length dimension l moving with velocity V through air of density ρ can be expressed as

$$F = \varphi \cdot \rho V^2 l^2$$

where φ is a coefficient that depends upon all the dimensionless factors of the problem. In the case of a wing these are:

1. **Angle of attack** α, the inclination between the chord line and the velocity V
2. **Aspect ratio** $A = b/c$, where b is the span and c the mean chord of the wing
3. **Reynolds number** $\mathrm{Re} = \rho V l/\mu$, where μ is the coefficient of dynamic viscosity of air
4. **Mach number** V/V_s, where V_s is the velocity of sound
5. Relative surface roughness
6. Relative turbulence

The dependence of the force coefficient φ upon α and A can be theoretically determined; the variation of φ with the other parameters must be established experimentally, i.e., by model tests.

AIRFOILS

Applying Bernoulli's equation to the flow around a body, if p represents the **static pressure**, i.e., the atmospheric pressure in the undisturbed air, and if V is superposed as in Fig. 11.4.4 and if p_1, V_1 represent the pressure and velocity at any point 1 at the surface of the body,

$$p + \tfrac{1}{2}\rho V^2 = p_1 + \tfrac{1}{2}\rho V_1^2$$

The maximum pressure occurs at a point s on the body at which the velocity is zero. Such a point is defined as the **stagnation point**. The maximum pressure increase occurs at this point and is

$$p_s - p = \tfrac{1}{2}\rho V^2$$

This is called the stagnation pressure, or **dynamic pressure**, and is denoted by q. It is customary to express all aerodynamic forces in terms of $\frac{1}{2}\rho V^2$, hence

$$F = \tfrac{1}{2}\varphi\rho V^2 l^2 = q\varphi l^2$$

For the case of a wing, the forces and moments are expressed as

$$\text{Lift} = L = C_L \tfrac{1}{2}\rho V^2 S = C_L qS$$
$$\text{Drag} = D = C_D \tfrac{1}{2}\rho V^2 S = C_D qS$$
$$\text{Moment} = M = C_M \tfrac{1}{2}\rho V^2 Sc = C_M qSc$$

where S = wing area and c = wing chord.

The lift produced can be determined from the intensity of the circulatory flow or **circulation** Γ by the relation $L' = \rho V \Gamma$, where L' is the lift per unit width of wing. In a **wing of infinite aspect ratio**, the flow is two-dimensional and the lift reaction is at right angles to the line of relative motion. The **lift coefficient** is a function of the angle of attack α and by mathematical analysis $C_L = 2\pi \sin \alpha$, which for small angles becomes $C_L = 2\pi\alpha$. Experiments show that $C_L = 2\pi\eta(\alpha - \alpha_0)$, where α_0 is the

angle of attack corresponding to zero lift. $\eta \approx 1 - 0.64(t/c)$, where t is airfoil thickness.

In **a wing of finite aspect ratio**, the circulation flow around the wing creates a strong vortex trailing downstream from each wing tip (Fig. 11.4.5). The effect of this is that the direction of the resultant velocity at the wing is tilted downward by an **induced angle of attack** $\alpha_i = w_i/V$ (Fig. 11.4.6). If friction is neglected, the resultant force R is now tilted to the rear by this same angle α_i. The lift force L is approximately the same as R. In addition, there is also a drag component D_i, called the **induced drag**, given by

$$D_i = L \tan \alpha_i = L w_i/V$$

For a given geometrical angle of attack a the **effective angle of attack** has been reduced by α_i, and thus

$$C_L = 2\pi\eta(\alpha - \alpha_0 - \alpha_i)$$

According to the Lanchester-Prandtl theory, for **wings having an elliptical lift distribution**

$$\alpha_i = w_i/V = C_L/\pi A \qquad \text{rad}$$
$$D_i = L^2/\pi b^2 q \quad \text{and} \quad C_{Di} = D_i/qS = C_L^2/\pi A$$

where b = span of wing and $A = b^2/S$ = aspect ratio of wing. These results also apply fairly well to wings not differing much from the elliptical shape. For **square tips**, correction factors are required:

$$\left. \begin{aligned} \alpha_i &= \frac{C_L}{\pi A}(1 + \tau) \quad \text{where } \tau = 0.05 + 0.02A \\[2mm] C_{Di} &= \frac{C_L^2}{\pi A}(1 + \sigma) \quad \text{where } \sigma = 0.01A - 0.01 \end{aligned} \right\} \quad \text{for } A < 12$$

These formulas are the basis for transforming the characteristics of rectangular wings from an **aspect ratio** A_1 to an aspect ratio A_2:

$$\alpha_2 = \alpha_1 + \frac{57.3 C_L}{\pi}\left(\frac{1 + \tau_2}{A_2} - \frac{1 + \tau_1}{A_1}\right) \quad \text{deg}$$

$$C_{D2} = C_{D1} + \frac{C_L^2}{\pi}\left(\frac{1 + \sigma_2}{A_2} - \frac{1 + \sigma_1}{A_1}\right)$$

For elliptical wings the values of τ and σ are zero. Most wing-section data are given in terms of aspect ratios 6 and ∞. For other values, the preceding formulas must be used.

Fig. 11.4.5   Vortex formation at wing tips.

Fig. 11.4.6   Induced angle of attack.

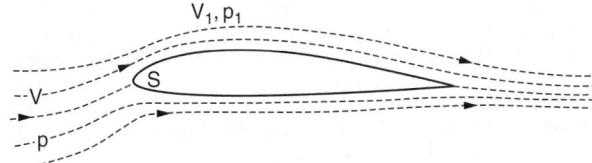

Fig. 11.4.4   Fluid flow around an airfoil.

Characteristics of Wings Wings have **airfoil sections** that can be constant from wing root to tip, or sections that vary across the wing to tailor the aerodynamic properties of the wing. Wing aerodynamic properties are expressed in terms of the dimensionless coefficients C_L, C_D, and C_M. **Airfoil** coefficients, termed section characteristics, are usually denoted as C_l, C_d, and C_m, and refer to wings of infinite aspect ratio. Figure 11.4.10 presents tunnel test results from early NACA studies on 6 percent and 12 percent thick two-dimensional sections.

The lift coefficient C_L is a linear function of the angle of attack up to a critical angle called the **stalling angle** (Fig. 11.4.7). The **maximum lift coefficient** C_{Lmax} which can be reached is one of the important characteristics of a wing because it determines the landing speed of the airplane.

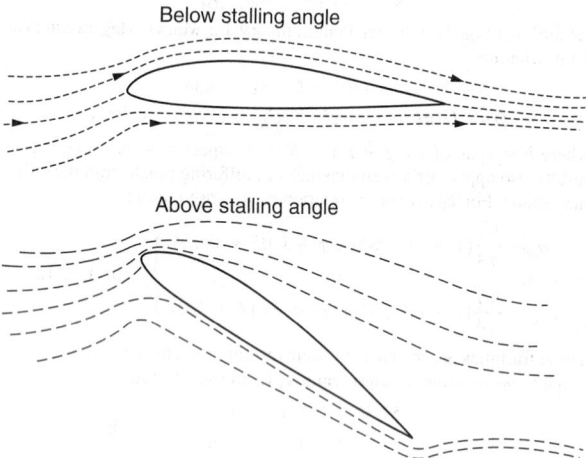

Fig. 11.4.7   Stalling angle of an airfoil.

The **drag of a wing** is made up of three components: the **profile drag** D_0, the **induced drag** D_i, and the **wave drag** D_W, near the speed of sound. The profile drag is due principally to surface friction. At aspect ratio ∞ or at zero lift the induced drag is zero, and thus the entire drag is profile drag. In Fig. 11.4.8 the coefficient of induced drag $C_{Di} = C_L^2/\pi A$ is also

Fig. 11.4.8   Polar-diagram plot of airfoil data.

plotted for a wing. The difference $C_D - C_{Di} = C_{DO}$, the profile-drag coefficient. Among the desirable characteristics of a wing is a small value of the minimum profile-drag coefficient and a large value of C_L/C_D (L/D).

The moment characteristics of a wing are obtainable from the curve of the center of pressure as a function of α or by the **moment coefficient taken about the aerodynamic center** $C_{Ma.c.}$ as a function of C_L. A forward motion of the c.p. as α is increased corresponds to an unstable wing section. This instability is undesirable because it requires a large download on the tail to counteract it.

The characteristics of a wing section (Fig. 11.4.9) are determined principally by its **mean camber line**, i.e., the curvature of the median line of the profile, and secondly by the thickness distribution along the chord. In the NACA system of designation, when a four-digit number is used such as 2412, the significance is always first digit = maximum camber in percent of chord, second digit = location of position of maximum camber in tenths of chord measured from the leading edge (that is, 4 stands for 40 percent), and the last two figures indicate the maximum thickness in percent of chord. NACA also developed a 5-digit and 6-digit series of airfoils, the latter referred to as *laminar flow sections*. More recently, NASA work has centered on developing low-speed sections for general aviation aircraft. Today most major aerospace companies no longer use the early NACA section data and prefer to develop their own geometric and aerodynamic section properties, through either computational or experimental means. Extensive reshaping of airfoil leading edge geometries has yielded significant increases in maximum lift coefficients, and high-speed sections, termed "peaky" or "supercritical" sections have permitted cruise Mach numbers to increase significantly. These high-speed sections retain a large region of high pressure in front of the airfoil crest, resulting in a peak in pressure coefficient there at high speeds.

Fig. 11.4.9   Characteristics of a wing section.

Selection of Wing Section

In selecting a wing section for a particular airplane the following factors are generally considered:

1. Maximum lift coefficient C_{Lmax}
2. Minimum drag coefficient C_{Dmin}
3. Moment coefficient at zero lift C_{m0}
4. Maximum value of the ratio C_L/C_D

For certain special cases it is necessary to consider one or more factors from the following group:

5. Value of C_L for maximum C_L/C_D
6. Value of C_L for minimum profile drag
7. Maximum value of $C_L^{3/2}/C_D$
8. Maximum value of $C_L^{1/2}/C_D$
9. Type of lift curve peak (stall characteristics)
10. Drag divergence Mach number M_{DD}

Characteristics of airfoil sections are given in *NACA-TR* 586, 647, 669, 708, and 824. Figure 11.4.10 gives data on two typical sections: 0006 is often used for tail surfaces, and $NACA_I$ 65-212 is especially suitable for the wings of subsonic airplanes. For Mach numbers greater than 0.6 thin wings must be used.

The dimensionless coefficients C_L and C_D are functions of the **Reynolds number** Re $= \rho Vl/\mu$. For wings, the characteristic length l is taken to be the chord. In standard air at sea level

$$\text{Re} = 6{,}378 V_{\text{ft/s}} \cdot l_{\text{ft}} = 9{,}354 V_{\text{mi/h}} \cdot l_{\text{ft}}$$

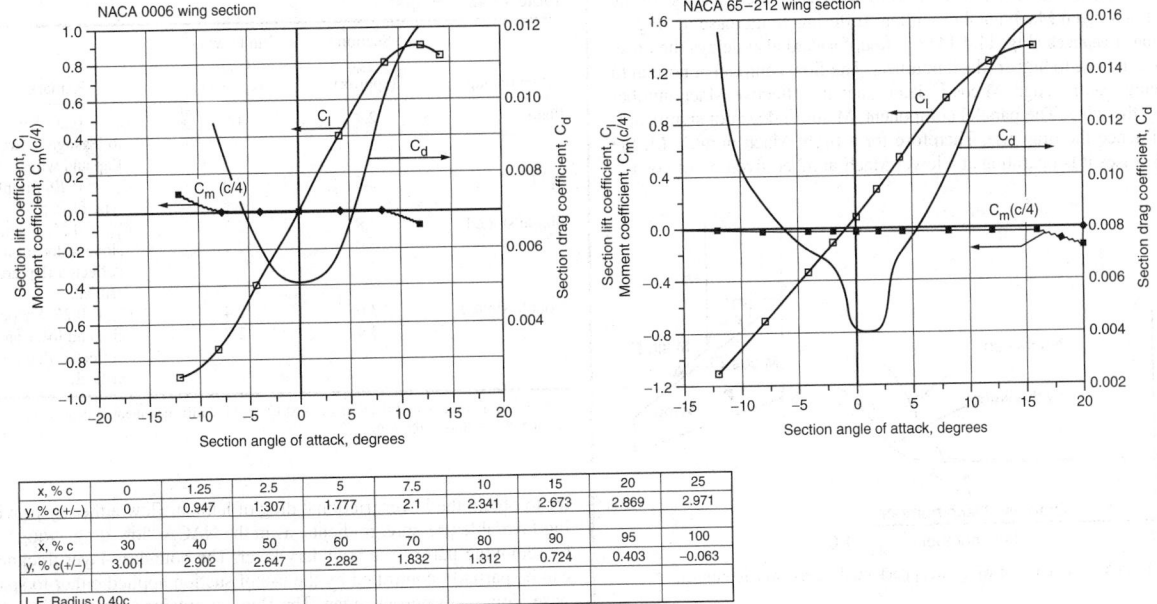

x, % c	0	1.25	2.5	5	7.5	10	15	20	25
y, % c(+/−)	0	0.947	1.307	1.777	2.1	2.341	2.673	2.869	2.971

x, % c	30	40	50	60	70	80	90	95	100
y, % c(+/−)	3.001	2.902	2.647	2.282	1.832	1.312	0.724	0.403	−0.063

L.E. Radius: 0.40c

Fig. 11.4.10 Properties of two airfoil sections. *(Courtesy of NACA.)*

For other heights, multiply by the coefficient $K = (\rho/\mu)/(\rho_0/\mu_0)$. Values of K are as follows:

Transonic Airfoils At airplane Mach numbers of about M = 0.75, normal airfoil sections begin to show a greater drag, which increases

Altitude, ft*	0	5,000	10,000	15,000	20,000	25,000	30,000	40,000	50,000	60,000
K	1.000	0.884	0.779	0.683	0.595	0.515	0.443	0.300	0.186	0.116

* × 0.305 = m.

The variation of C_L and C_D with the Reynolds number is known as **scale effect.** These are shown in Figs. 11.4.11 and 11.4.12 for some typical airfoils. Figure 11.4.12 shows in dashed lines also the theoretical variation of the drag coefficient for a smooth flat plate for laminar and turbulent flow, respectively, and also for the transition region.

In addition to the Reynolds number, airfoil characteristics also depend upon the **Mach number** (see "Supersonic and Hypersonic Aerodynamics" below).

Laminar-flow Wings NACA developed a series of **low-drag wings** in which the distribution of thickness along the chord is so selected as to maintain **laminar flow** over as much of the wing surface as possible.

The low-drag wing under controlled conditions of surface smoothness may have drag coefficients about 30 percent lower than those obtained on normal conventional wings. Low-drag airfoils are so sensitive to roughness in any form that the full advantage of laminar flow is unobtainable.

Fig. 11.4.12 Scale effect on minimum profile-drag coefficient.

sharply as the speed of sound is approached. Airfoil sections known as **transonic airfoils** have been developed which significantly delay this **drag rise** to Mach numbers of 0.9 or greater (Fig. 11.4.13). Advantage may

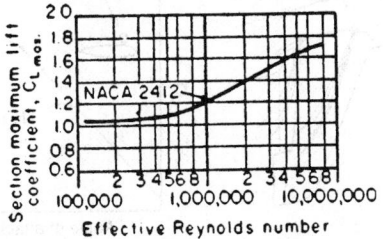

Fig. 11.4.11 Scale effect on section maximum lift coefficient.

Fig. 11.4.13 Drag rise delay due to transonic section.

also be taken of these airfoil sections by increasing the thickness of the airfoil at a given Mach number without suffering an increase in drag.

Wing sweepback (Fig. 11.4.14) or sweepforward also delays the onset of the drag rise to higher Mach numbers. The flow component normal to the wing leading edge, $M \cos \Gamma$ determines the effective Mach number felt by the wing. The parallel component, $M \sin \Gamma$, does not significantly influence the drag rise. Therefore for a flight Mach number M, the wing senses it is operating at a lower Mach number given by $M \cos \Gamma$.

Fig. 11.4.14 Influence of wing sweepback on drag rise Mach number.

Flaps and Slots

The maximum lift of a wing can be increased by the use of **slots**, retractable or fixed slats on the leading edge, or **flaps** on the trailing edge. **Fixed slots** are formed by rigidly attaching a curved sheet of metal or a small auxiliary airfoil to the leading edge of the wing. The trailing-edge flap is used to give increased lift at moderate angles of attack and to increase C_{Lmax}. Theoretical analyses of the effects are given in *NACA-TR* 938 and *Br. ARCR&M* 1095. Flight experience indicates that the four types shown in Fig. 11.4.15 have special advantages over other known types. Values of C_{Lmax} for these types are shown in Table 11.4.2. All flaps cause a diving moment ($-C_m$) which must be trimmed out by a download on the tail, and this download reduces C_{Lmax}. The correction is ΔC_{Lmax} (trim) = $\Delta C_{mo}(l/c)$, where l/c is the tail length in mean chords.

Boundary-Layer Control (BLC) This includes numerous schemes for (1) maintaining laminar flow in the boundary layer in the flow over a wing or (2) preventing flow separation. Schemes in the first category

Fig. 11.4.15 Trailing-edge flaps: (1) plain flap; (2) split flap; (3) single-slotted flap; (4) double-slotted flap (retracted); (5) double-slotted flap (extended).

Table 11.4.2 Flaps*

Type of flap	Section C_{Lmax} (approx)	Finite wing C_{Lmax} (approx)	Remarks
Plain	2.4	1.9	$C_{f/c} = 0.20$. Sensitive to leakage between flap and wing.
Split	2.6	2.0	$C_{f/c} = 0.20$. Simplest type of flap.
Single slotted	2.8 3.0	2.2 2.4	$C_{f/c} = 0.25$. Shape of slot and location of deflected flap are critical.
Double slotted	3.0 3.4	2.4 2.7	$C_{f/c} = 0.25$. Shape of slot and location of deflected flap are critical.

* The data is for wing thickness $t/c = 0.12$ or 0.15. Lift increment is dependent on the leading-edge radius of the wing.

try to obtain the lower frictional drag of laminar flow either by providing favorable pressure gradients, as in the NACA "low-drag" wings, or by removing part of the boundary layer. The boundary-layer thickness can be partially controlled by the use of suction applied either to spanwise slots or to porous areas. The flow so obtained approximates the laminar skin-friction drag coefficients (see Fig. 11.4.12). Schemes in the second category try to delay or improve the stall. Examples are the leading-edge slot, slotted flaps, and various forms of suction or blowing applied through transverse slots. The slotted flap is a highly effective form of boundary-layer control that delays flow separation on the flap.

Spanwise Blowing Another technique for increasing the lift from an aerodynamic surface is spanwise blowing. A high-velocity jet of air is blown out along the wing. The location of the jet may be parallel to and near the leading edge of the wing to prevent or delay leading-edge flow separation, or it may be slightly behind the juncture of the trailing-edge flap to increase the lift of the flap. Power for spanwise blowing is sensitive to vehicle configuration and works most efficiently on swept wings. Other applications of this technique may include local blowing on ailerons or empennage surfaces (Fig. 11.4.16).

The **pressure distributions** on a typical airfoil are shown in Fig. 11.4.17. In this figure the ratio p/q is given as a function of distance along the chord. In Fig. 11.4.18 the effect of addition of an external airfoil flap is shown.

Airplane Performance

In **horizontal flight** the lift of the wings must be equal to the weight of the airplane, or $L = nW$, where W is the weight (lb or kg) of the airplane and n is 1.00. When the airplane is in a turn or is performing a pull-up, $n > 1.00$. The following equation determines the minimum or stall speed of the airplane at the appropriate value of n:

$$nW = C_{Lmax} \frac{1}{2} \rho V_{min}^2 S \quad \text{or} \quad V_{min} = \sqrt{\frac{2nW}{C_{Lmax} \rho S}} \quad (11.4.1)$$

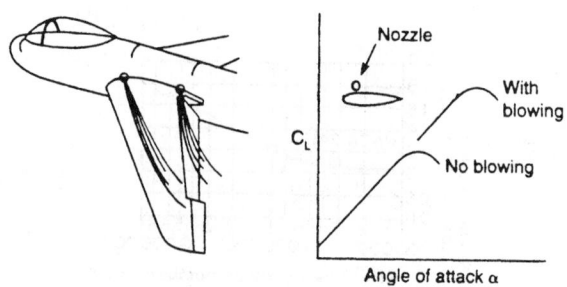

Fig. 11.4.16 Influence of spanwise blowing.

Fig. 11.4.17 Pressure distribution over a typical airfoil.

Fig. 11.4.18 Pressure distribution on an airfoil-flap combination.

Fig. 11.4.19 Thrust-drag relationships for airplane performance at sea level and 30,000 ft.

Fig. 11.4.20 Summary of airplane speed variations with altitude.

This is called the *stalling speed* and when multiplied by a suitable safety factor, such as 1.2, can be termed the **minimum safe flight speed.**

The interplay of airplane **drag** D and **thrust** T dictates the airplane performance. Usually graphical means are used to display performance capability. Here the airplane **drag polar** and thrust available, at sea level, are superimposed as shown in Fig. 11.4.19. In level flight at constant speed $T = D$, where T can be generated either by propeller, rocket, or jet engine. For propeller-driven airplanes, T becomes TV and D becomes DV in performance analyses. Our discussion will center on **jet propulsion** with a summary showing significant propeller-induced equation changes.

Figure 11.4.19 shows that at maximum B and minimum A speeds, $T = D$. The minimum thrust-dictated speed is usually less than the stall speed; the latter then dictates the lower performance boundary speed. Between the maximum and minimum speeds $T > D$, and this excess in T available can be used to climb (dh/dt), or to accelerate in level flight. Similar curves can be constructed for a range of altitudes (a 30,000-foot set is shown), resulting in the performance boundaries as shown in Fig. 11.4.20. Here are minimum and maximum speeds, best climb speed, stall speed, and best cruise speed. Maximum and minimum speeds are equal at the absolute ceiling where $dh/dt = 0$.

Rate of climb, or airplane **vertical velocity**, is given by

$$\frac{dh}{dt} = V\left(\frac{T - D}{W}\right) \qquad \text{ft/s} \qquad (11.4.2)$$

where V is in ft/s.

Climb angle θ is given by

$$\theta = \sin^{-1}\left(\frac{T - D}{W}\right) \qquad \text{deg} \qquad (11.4.3)$$

Jet engine thrust reduces with increase in altitude approximately as density reduces. At some altitude $(T - D)$ becomes zero (therefore $dh/dt = 0$) and the **absolute ceiling** of the airplane is reached, beyond

which the aircraft cannot fly in steady flight. Climb angle becomes zero also and the airplane can fly at one speed only. A practical maximum altitude, termed **service ceiling,** is assumed to occur when $dh/dt =$ 100 ft/min. The absolute ceiling can be determined from the curve of maximum rate of climb as a function of altitude when this is approximated by a straight line, usually accurate at the higher altitudes. See Fig. 11.4.21. For a linear decrease in the rate of climb the following approximations hold:

$$\text{Absolute ceiling } H = (r_0\, h)/(r_0 - r) \qquad \text{ft} \qquad (11.4.4)$$

where subscript 0 refers to sea level rate of climb and r is the rate of climb, in ft/min, at altitude h.

$$\text{Service ceiling } h_s = H\,(r_0 - 100)/r_0 \qquad \text{ft} \qquad (11.4.5)$$

Altitude climbed in t minutes is

$$h = H\,(1 - e^k) \qquad \text{ft} \qquad (11.4.6)$$

where $k = (-r_0\, t)/H$.

Fig. 11.4.21   Characteristic rate of climb curves.

Time to climb to altitude h is

$$t = 2.303\left(\frac{H}{r_0}\right)\log\left(\frac{H}{H-h}\right) \quad \text{min} \qquad (11.4.7)$$

Power off **minimum glide angle** γ is given by

$$\gamma = \tan^{-1}\left(\frac{2}{(L/D)_{\max}}\right) \quad \text{deg} \qquad (11.4.8)$$

The minimum glide angle corresponds to maximum straight line **glide distance measured along the ground**, s_g, given by

$$s_g = h_g(L/D)_{\max} \quad \text{ft} \qquad (11.4.9)$$

where h_s is the start of glide altitude in feet.

For performance purpose, an airplane **drag polar** is required, which will vary with Mach number. This polar will also change when flaps and landing gears are extended, and is given by the generic equation,

$$C_D qS = (C_{D0} + C_L^2/Ae\pi)qS \qquad (11.4.10)$$

where C_{D0} = profile drag coefficient, f(Mach number) = C_{D0}(clean aircraft) + ΔC_{D0} (flaps) + ΔC_{D0} (gear). A = wing aspect ratio, e = Oswald's span efficiency factor, $\pi = 3.14159$, $q = \frac{1}{2}\rho V^2$, S = wing reference area.

A typical clean aircraft drag variation with V, complete with the corresponding L/D is shown in Fig. 11.4.22 which shows that a maximum value of L/D occurs at a given velocity which is termed the best **endurance speed**. The **best range speed** occurs at 0.866 $(L/D)_{\max}$ as shown. Table 11.4.3 shows typical values of C_{D0} for various types of jet aircraft.

The ratio L/D, (C_L/C_D), is sometimes called the **aerodynamic efficiency** of the airplane. Mission requirements dictate the magnitude for competitive values of this ratio which is primarily controlled by the value of the **wing aspect ratio** and C_{D0}. Airliners usually have high aspect ratio values and fighter aircraft low values. See Fig. 11.4.23. As seen in the following equations, maximizing L/D will maximize range and endurance. Range, more accurately termed **maximum range**, is defined as the maximum **distance** flown on a given quantity of fuel. **Maximum endurance** is defined as the maximum **time** spent in the air on a given quantity of fuel.

Range R at constant altitude is

$$R = 2\sqrt{\frac{2}{\rho S}}\frac{1}{c}\frac{C_L^{1/2}}{C_D}\left(W_0^{1/2} - W_1^{1/2}\right) \quad \text{ft} \qquad (11.4.11)$$

where ρ = air density in slugs/ft³, c = specific fuel consumption in lb of fuel/lb thrust · s, V = aircraft velocity in ft/s. Subscript 0 refers to the

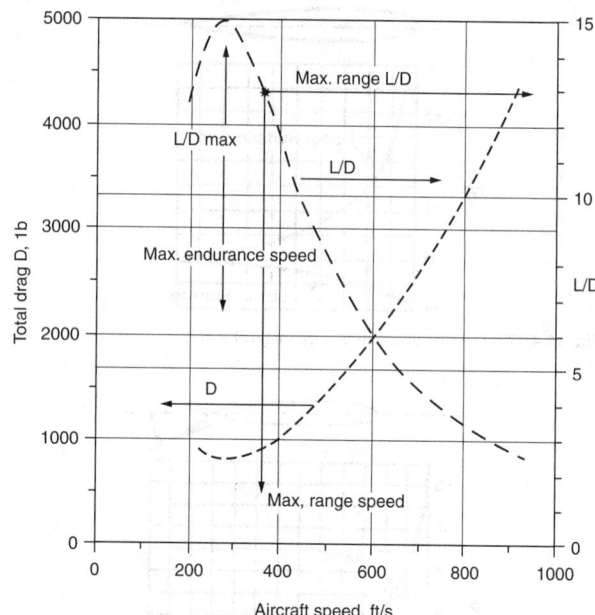

Fig. 11.4.22   Aircraft aerodynamic properties: drag polar and L/D ratio.

Table 11.4.3   Total C_{D0} Values for Typical Aircraft Types (Clean Aircraft)

Aircraft type	Subsonic C_{D0}
Supersonic twin-jet fighter	0.0237
Narrow-body twin-jet transport	0.0166
Wide-body four-jet transport	0.0170

segment weight at start of mission and subscript 1 to segment weight at end of mission.

Endurance E, in hours, is given by

$$E = \frac{1}{c}\left(\frac{C_L}{C_D}\right)\ln\left(\frac{W_0}{W_1}\right) \quad \text{h} \qquad (11.4.12)$$

In each instance, c and (L/D) correspond to the appropriate altitude and speed, and $(W_0 - W_1)$ represents fuel used.

For **propeller-powered airplanes**, the above equations, starting at Eq. (11.4.2) become

$$\frac{dh}{dt} = \left(\frac{TV - DV}{W}\right) \quad \text{ft/s} \qquad (11.4.13)$$

where V is in ft/s. If thrust horsepower available $P_{Ta} = \eta \cdot P_a$, where P_a = **available engine horsepower**, η is the propeller efficiency, and P_r = horsepower required, then

$$\frac{dh}{dt} = 550\left(\frac{P_{Ta} - P_a}{W}\right) \quad \text{ft/s} \qquad (11.4.14)$$

$$\text{Range } R = \frac{\eta}{c}\left(\frac{C_L}{C_D}\right)\ln\left(\frac{W_0}{W_1}\right) \quad \text{ft} \qquad (11.4.15)$$

where the units of c are (lb fuel)/ft · lb/s) · (s).

$$\text{Endurance } E = \left(\frac{\eta}{c}\right)\frac{C_L^{3/2}}{C_D}(2\rho S)^{1/2}\left(W_1^{-1/2} - W_0^{-1/2}\right)\left(\frac{1}{3600}\right) \quad \text{h}$$
$$(11.4.16)$$

See Table 11.4.4 for dimensions and performance of selected airplanes.

F-117A

b = 43.3 ft
S = 912.7 ft²

Aspect ratio = 2.05

Aircraft type	Wing aspect ratio
Fighters	**2.00 - 6.00**
Transports	**7.00 - 9.00**

L-1011-1

b = 155.3 ft
S = 3456 ft²

Aspect ratio = 7.00

Fig. 11.4.23 Typical aircraft configurations. (*Drawings courtesy of Lockeed-Martin.*)

Power Available The maximum efficiency η_m and the diameter of a propeller to absorb a given power at a given speed and rpm are found from a propeller-performance curve. The thrust horsepower at maximum speed P_{Tm} is found from $P_{Tm} = \eta_m p_m$. The **thrust horsepower at any speed** can be approximately determined from the ratios given in the following table:

$V/V_{\max}$		0.20	0.30	0.40	0.50	0.60
P_T/P_{TM}	Fixed-pitch propeller	0.29	0.44	0.57	0.68	0.77
	constant-rpm propeller	0.47	0.62	0.74	0.82	0.88
$V/V_{\max}$		0.70	0.80	0.90	1.00	1.10
P_T/P_{Tm}	Fixed-pitch propeller	0.84	0.90	0.96	1.00	1.03
	constant-rpm propeller	0.93	0.97	0.99	1.00	1.00

The **brake horsepower** of an engine decreases with increase of altitude. The variation of P_T with altitude depends on engine and propeller characteristics with average values as follows:

h, ft	0	5,000	10,000	15,000	20,000	25,000
P_T/P_{T0} (fixed pitch)	1.00	0.82	0.66	0.52	0.41	0.30
P_T/P_{T0} (controllable pitch)	1.00	0.85	0.71	0.59	0.48	0.38

Parasite Drag

The drag of the nonlifting parts of an airplane is called the **parasite drag.** It consists of two components: the frictional and the eddy-making drag.

Frictional drag, or **skin friction,** is due to the viscosity of the fluid. It is the force produced by the viscous shear in the layers of fluid immediately

Table 11.4.4 Principle Dimensions and Performance of Typical Air Vehicles

	General Dynamics	Lockheed Martin	Raytheon Aircraft	Bell Helicopter Textron	Boeing	Boeing	Airbus	Airbus	Northrop-Grumman	Agusta-Westland
Designation model no.	Gulfstream G 500	F-35C	Beech King Air C-90B	427	747-400	737-800	A380-800	A310-200	B-2A	Super Lynx 300
Type	Executive transport	Fighter	General aviation	Utility helicopter	Wide-body turbofan	Narrow-body turbofan	Wide-body turbofan	Wide-body turbofan	Bomber	Naval helicopter
No passengers	14-19	Pilot only	12	6	416	162	555	220-280	2 crew	9
Cargo capacity, lb	N.A.	15,000	N.A.	N.A.	6,025 ft^3	1,555 ft^3	N.A.	N.A.	40,000 payload	N.A.
Span, ft	93.5	35	50.3	37*	211.4	112.6	261.8	144	172	42*
Overall length, ft	96.4	51.4	35.5	42.9	231.8	129.5	239.5	153.1	69	50
Overall height, ft	25.8	15	14.3	11.4	63.7	41.2	79.1	51.1	17	12
Wing area, ft^2	1136.5	620	294	N.A.	5,650	1,341	9,095	2,360	5,140	N.A.
Weight empty, lb	47,800	30,000	6,810	3,485	398,800	91,660	611,000	174,700	N.A.	N.A.
Weight gross, lb	85,500	60,000	10,100	6,000	875,000	174,200	1,234,600	313,100	336,500	11,750
Power plant	2 × RR Tay-BR710	1 × PWF 135-PW-400	2 × PWC-PT6A-21	2 × PWC PW-206D	4 × PW4056	2 × CFM56-7B	4 × RR Trent 970	2 × GE CF6	4 × GE-F118	2 × LHTEC CTS800-4N
High speed, mi/h (Mach)	(0.89)	(1.6+)	283	160	(0.92)	(0.82)	(0.89)	(0.84)	N.A.	173
Cruise speed, mi/h (Mach)	(0.85)	N.A.	283	N.A.	(0.85)	(0.788)	(0.85)	N.A.	N.A.	N.A.
Range, mil	7,200	N.A.	1,640	411	8,356	3,522	9,200	N.A.	N.A.	550

NOTES: *rotor diameter.
N.A. = not available.
SOURCE: *Aviation Week, 2003 Source Book*, Copyright 2003 by The McGraw-Hill Companies.

adjacent to the body. It is always proportional to the wetted area, i.e., the total surface exposed to the air.

Eddy-making drag, sometimes called **form drag,** is due to the disturbance or wake created by the body. It is a function of the shape of the body.

The total drag of a body may be composed of the two components in any proportion, varying from almost pure skin friction for a plate edgewise or a good streamline form to 100 percent form drag for a flat plate normal to the wind. A **steamline form** is a shape having very low form drag. Such a form creates little disturbance in moving through a fluid.

When air flows past a surface, the layer immediately adjacent to the surface adheres to it, or the tangential velocity at the surface is zero. In the transition region near the surface, which is called the **boundary layer,** the velocity increases from zero to the velocity of the stream. When the flow in the boundary layer proceeds as if it were made up of laminae sliding smoothly over each other, it is called a **laminar boundary layer.** If there are also irregular motions in the layers normal to the surface, it is a **turbulent boundary layer.** Under normal conditions the flow is laminar at low Reynolds numbers and turbulent at high Re with a transition range of values of Re extending between 5×10^5 and 5×10^7. The profile-drag coefficients corresponding to these conditions are shown in Fig. 11.4.12.

For laminar flow the friction-drag coefficient is practically independent of surface roughness and for a flat plate is given by the Blasius equation,

$$C_{DF} = 2.656/\sqrt{\text{Re}}$$

The turbulent boundary layer is thicker and produces a greater frictional drag. For a smooth flat plate with a turbulent boundary layer

$$C_{DF} = 0.91/\log \text{Re}^{2.58}$$

For rough surfaces the drag coefficients are increased. The drag coefficients as given above are based on *projected area* of a double-surfaced plane. If *wetted* area is used, the coefficients must be divided by 2. The frictional drag is $D_F = C_{DF}\, qS$, where S is the *projected area,* or $D_F = \frac{1}{2}C_{DF}\, qA$, where A is the wetted area.

Values of C_{DF} for double-surfaced planes may be estimated from the following tabulation:

Laminar flow (Blasius equation):

Re	10	10^2	10^3	10^4	10^5	10^6
C_{DF}	0.838	0.265	0.0838	0.0265	0.0084	0.0265

Turbulent flow:

Re	10^5	10^6	10^7	10^8	10^9	10^{10}
C_{DF}	0.0148	0.0089	0.0060	0.0043	0.00315	0.0024

These are double-surface values which facilitate direct comparison with wing-drag coefficients based on projected area. For calculations involving wetted area, use one-half of double-surface coefficients. Interpolations in the foregoing tables must allow for the logarithmic functions; i.e., the variation in C_{DF} is not linear with Re.

Drag Coefficients of Various Bodies

For **bodies with sharp edges** the drag coefficients are almost independent of the Reynolds number, for most of the resistance is due to the difference in pressure on the front and rear surfaces. Table 11.4.5 gives $C_D = D/qS$, where S is the maximum cross section perpendicular to the wind.

Table 11.4.5 Drag Coefficients

Object	Proportions	Attitude	C_D
Rectangular plate, sides a and b	$\dfrac{a}{b} = \begin{matrix}1\\4\\8\\12.5\\25\\50\\\infty\end{matrix}$		1.16 1.17 1.23 1.34 1.57 1.76 2.00
Two disks, spaced a distance l apart	$\dfrac{l}{d} = \begin{matrix}1\\1.5\\2\\3\end{matrix}$		0.93 0.78 1.04 1.52
Cylinder	$\dfrac{l}{d} = \begin{matrix}1\\2\\4\\7\end{matrix}$		0.91 0.85 0.87 0.99
Circular disk			1.11
Hemispherical cup, open back			0.41
Hemispherical cup, open front, parachute			1.35
Cone, closed base			$\alpha = 60°, 0.51$ $\alpha = 30°, 0.34$

For **rounded bodies** such as **spheres, cylinders,** and **ellipsoids** the drag coefficient depends markedly upon the Reynolds number, the surface roughness, and the degree of turbulence in the airstream. A sphere and a cylinder, for instance, experience a sudden reduction in C_D as the Reynolds number exceeds a certain critical value. The reason is that at low speeds (small Re) the flow in the boundary layer adjacent to the body is laminar and the flow separates at about 83° from the front (Fig. 11.4.24). A wide wake thus gives a large drag. At higher speeds (large Re) the boundary layer becomes turbulent, gets additional energy from the outside flow, and does not separate on the front side of the sphere. The drag coefficient is reduced from about 0.47 to about 0.08 at a critical Reynolds number of about 400,000 in free air. Turbulence in the airstream reduces the value of the critical Reynolds number (Fig. 11.4.25). The Reynolds number at which the sphere drag $C_D = 0.3$ is taken as a criterion of the amount of turbulence in the airstream of wind tunnels.

Cylinders   The drag coefficient of a cylinder with its axis normal to the wind is given as a function of Reynolds number in Fig. 11.4.26. Cylinder drag is sensitive to both Reynolds and Mach number. Figure 11.4.26 gives the Reynolds number effect for M = 0.35. The increase in C_D because of Mach number is approximately:

M	0.35	0.4	0.6	0.8	1.0
C_D increase, percent	0	2	20	50	70

(See *NACA-TN* 2960.)

Laminor boundary layer   Turbulent boundary layer
(early separation—   (later separation—
wide woke)   narrow woke)

Fig. 11.4.24   Boundary layer of a sphere.

Fig. 11.4.25   Drag coefficients of a sphere as a function of Reynolds number and of turbulence.

Fig. 11.4.26   Drag coefficients of cylinders and spheres.

Streamline Forms   The drag of a streamline body of revolution depends to a very marked extent on the Reynolds number. The difference between extreme types at a given Reynolds number is of the same order as the change in C_D for a given form for values of Re from 10^6 to 10^7.

Tests reported in *NACA-TN* 614 indicate that the shape for minimum drag should have a fairly sharp nose and tail. At Re = 6.6×10^6 the best forms for a fineness ratio (ratio of length to diameter) 5 have a drag

$$D = 0.040qA = 0.0175qV^{2/3}$$

where A = max cross-sectional area and V = volume. This value is equivalent to about 1.0 lb/ft^2 at 100 mi/h. For Re > 5×10^6, the drag coefficients vary approximately as Re$^{0.15}$.

Minimum drag on the basis of cross-sectional or frontal area is obtained with a fineness ratio of the order of 2 to 3. Minimum drag on the basis of contained volume is obtained with a fineness ratio of the order of 4 to 6 (see *NACA-TR* 291).

The following table gives the ordinates for good streamline shapes: the **Navy strut,** a two-dimensional shape; and the Class C **airships,** a three-dimensional shape. Streamline shapes for high Mach numbers have a high fineness ratio.

Percent length		1.25	2.50	4.00	7.50	10.0	12.50
Percent of max ordinate	Navy strut	26	37.1	52.50	63.00	72.0	78.50
	Class C airship	20	33.5	52.60	65.80	75.8	83.50
Percent length		20	40	60	80	90	
Percent of max ordinate	Navy strut	91.1	99.5	86.1	56.2	33.8	
	Class C airship	94.7	99.0	88.5	60.5	49.3	

Test data on the *RM*-10 shape are given in *NACA-TR* 1160. This shape is a parabolic-arc type for which the coordinates are given by the equation $r_x = X/7.5(1.0 - X/L)$.

Nonintrusive flow field measurements about airfoils can be made by laser Doppler velocimeter (LDV) techniques. A typical LDV measurement system setup is shown in Fig. 11.4.27a. A seeded airflow generates the necessary reflective particles to permit two-dimensional flow directions and velocities to be accurately recorded.

Fig. 11.4.27a   Typical laser Doppler velocimeter measurement system.

CONTROL, STABILITY, AND FLYING QUALITIES

Control   An airplane is controlled in flight by imposing yawing, pitching, and rolling moments by use of rudders, elevators, and ailerons. In some instances special airplane configurations will use mixtures or combinations of these control methods. It is possible to control aircraft through the use of two controls by suitable blending of aileron and rudder inputs.

Stability   An airplane is **statically stable** if the airframe moments, due to a disturbance, tend to return the airplane to its original attitude. It is **dynamically stable** if the oscillations produced by the static stability are rapidly damped out. A **statically unstable** aircraft will produce upsetting moments instead of restoring moments. A **dynamically unstable** aircraft will generate oscillations increasing in amplitude with time. A statically unstable or a dynamically unstable aircraft can be made stable by the use of suitable automatic flight control systems.

Longitudinal static and dynamic stability, and longitudinal trim, are commonly achieved through the use of a suitable sized horizontal tail. Under these conditions the pitching moment equation, representing forces and moments on the airframe, is conditioned as follows: $C_m = 0$ for trim, and $(\partial C_m/\partial \alpha) < 0$ for stability, where C_m is the overall pitching moment coefficient about the aircraft center of gravity and α is the aircraft angle of attack. When a definitive value is chosen for $(\partial C_m/\partial \alpha)$ in the design process (such as -0.03), this is known as the design **static margin.** In a similar manner the vertical tail size is determined from lateral/directional stability and lateral trim requirements. The trim requirement is usually overriding in fixing vertical tail size for multiengine aircraft when critical engine-out design requirements are evaluated.

It can be shown that the forces and moments, inertial and aerodynamic, acting on an aircraft can be represented by six simultaneous, linear, small perturbation differential equations. These are called *the equations of motion of an aircraft,* and they represent the three forces and three moments assumed to act on the aircraft center of gravity (Fig. 11.4.27b). For conventional aircraft configurations, it is common practice to separate these two sets of three each: one set representing longitudinal motions and one set the lateral-directional motions of the aircraft.

Fig. 11.4.27b Six equations of motion.

In this separation, the assumption is that these motions do not normally influence each other, therefore they can be solved separately to establish the separate dynamic behaviors. The longitudinal set is characterized by two distinct oscillations called *short period* and *long period.* Short period motions usually influence angle of attack and pitch angle θ, whereas the long period, called "phugoid," primarily influences the height excursions and small speed changes. The lateral-directional set characterizes the motions in spiral ψ, the so-called Dutch roll mode, a mixture of bank angle ϕ and sideslip β motions, and in the bank angle response. All response behavior is characterized by the oscillatory parameters, relative damping ratio ζ, and undamped natural frequency ω_n, as well as time to one-half amplitude $(t_{1/2})$, or time to double amplitude (t_2). It is the magnitudes of these parameters that establish the **flying** or **handling qualities** of an aircraft.

Flying Qualities Requirements These are for military aircraft and civil aircraft and are in the form of Federal Aviation Requirements (FARs) and Military Standards (Mil. STD). These imply certain levels for the response parameters to ensure that the aircraft will achieve satisfactory pilot acceptability in the operation of the aircraft. The response parameter requirements vary with mission segments. Typical longitudinal response values for two aircraft, in cruise, are:

Large Jet Transport, W = 630,000 lb, V = 460 knots, altitude = 40,000 ft

	$t_{1/2}$, s	ω_n, rad/s	ζ
Short period	1.86	1.00	0.39
Phugoid	211.00	0.07	0.05

Small Civil Propeller Aircraft, W = 2,800 lb, V = 104 knots, altitude = sea level

	$t_{1/2}$, s	ω_n, rad/s	ζ
Short period	0.28	3.6	0.69
Phugoid	40.3	0.22	0.08

Rolling (or **banking**) does not produce any lateral shift in the center of the lift, so that there is no restoring moment as in pitch. However, when banked, the airplane sideslips toward the low wing. A fin placed above the center of gravity gives a lateral restoring moment that can correct the roll and stop the slip. The same effect can be obtained by **dihedral,** i.e., by raising the wing tips to give a transverse V. An effective dihedral of 1 to 3° on each side is generally required to obtain stability in roll. In low-wing monoplanes, 2° effective dihedral may require 8° or more of geometrical dihedral owing to interference between the wing and fuselage.

In a **yaw** or **slip** the line of action of the lateral force depends on the size of the effective vertical fin area. Insufficient fin surface aft will allow the skid or slip to increase. Too much fin surface aft will swing the nose of the plane around into a tight spiral. Sound design demands sufficient vertical tail surface for adequate directional control and then enough dihedral to provide lateral stability.

When moderate positive effective dihedral is present, the airplane will possess **static lateral stability,** and a low wing will come up automatically with very little yaw. If the dihedral is too great, the airplane may roll considerably in gusts, but there is little danger of the amplitude ever becoming excessive. **Dynamic lateral stability** is not assured by static stability in roll and yaw but requires that these be properly proportioned to the damping in roll and yaw.

Spiral instability is the result of too much fin surface and insufficient dihedral.

HELICOPTERS

REFERENCES: *Br. ARC-R & M* 1111, 1127, 1132, 1157, 1730, and 1859. *NACA-TM* 827, 836, 858. *NACA-TN* 626, 835, 1192, 3323, 3236. *NACA-TR* 434, 515, 905, 1078. *NACA Wartime Reports* L-97, L-101, L-110. NACA, "Conference on Helicopters," May, 1954. Gessow and Myers, "Aerodynamics of the Helicopter," Macmillan.

Helicopters derive lift, propulsion force, and control effect from adjustments in the blade angles of the rotor system. At least two rotors are required, and these may be arranged in any form that permits control over the reaction torque. The common arrangements are main lift rotor; auxiliary torque-control or tail rotor at 90°; two main rotors side by side; two main rotors fore and aft; and two main rotors, coaxial and oppositely rotating.

The helicopter rotor is an actuator disk or momentum device that follows the same general laws as a propeller. In calculating the rotor performance, the major variables concerned are diameter D, ft (radius R); tip speed, $Vt = \Omega R$, ft/s; angular velocity of rotor $\Omega = 2\pi n$, rad/s; and rotor solidity q (= ratio of blade area/disk area). The rotor performance is usually stated in terms of coefficients similar to propeller coefficients. The rotor coefficients are:

$$\text{Thrust coefficient } C_T = T/\rho(\Omega R)^2 \pi R^2$$
$$\text{Torque coefficient } C_Q = Q/\rho(\Omega R)^2 \pi R^3$$
$$\text{Torque } Q = 5,250 \text{ bhp/rpm}$$

Hovering The **hovering flight** condition may be calculated from basic rotor data given in Fig. 11.4.28. These data are taken from full-scale rotor tests. Surface-contour accuracy can reduce the total torque coefficient 6 to 7 percent. The power required for hovering flight is greatly reduced near the ground. Observed flight-test data from various sources are plotted in Fig. 11.4.29. The ordinates are heights above the ground measured in rotor diameters.

Effect of Gross Weight on Rate of Climb Figure 11.4.30 from Talkin (*NACA-TN* 1192) shows the rate of climb that can be obtained by reducing the load on a helicopter that will just hover. Conversely, given

Fig. 11.4.28 Static thrust performance of NACA 8-H-12 blades.

Fig. 11.4.29 Observed ground effect on Sikorsky-type helicopters.

Fig. 11.4.30 Effect of gross weight on the rate of climb of a helicopter.

the rate of climb with a given load, the curves determine the increase in load that will reduce the rate of climb to zero; i.e., they determine the maximum load for which hovering is possible.

Performance with Forward Speed The performance of a typical single-main-rotor-type helicopter may be shown by a curve of C_{TR}/C_{QR} plotted against $\mu = v/\pi nD$, where v is the speed of the helicopter, as in Fig. 11.4.31. This curve includes the parasite-drag effects which are appreciable at the higher values of μ. It is only a rough approximation

Fig. 11.4.31 Performance curve for a single-main-rotor-type helicopter.

to the experimentally determined values. C_{TR} may be calculated to determine C_{QR}, from which Q is obtained.

The performance of rotors at forward speeds involves a number of variables. For more complete treatment, see *NACA-TN* 1192 and *NACA Wartime Report* L-110.

GROUND-EFFECT MACHINES (GEM)

For data on air-cushion vehicles and hydrofoil craft, see Sec. 11.3 "Marine Engineering."

SUPERSONIC AND HYPERSONIC AERODYNAMICS

REFERENCES: Liepmann and Roshko, "Elements of Gas Dynamics," Wiley. "High Speed Aerodynamics and Jet Propulsion" 12 vols., Princeton. Shapiro, "The Dynamics and Thermodynamics of Compressible Fluid Flow," Vols. I, II, Ronald. Howarth (ed.), "Modern Developments in Fluid Mechanics—High Speed Flow," Vols. I, II, Oxford. Kuethe and Schetzer, "Foundations of Aerodynamics," Wiley. Ferri, "Elements of Aerodynamics of Supersonic Flows," Macmillan. Bonney, "Engineering Supersonic Aerodynamics," McGraw-Hill.

The effect of the compressibility of a fluid upon its motion is determined primarily by the **Mach number M.**

$$M = V/V_s \qquad (11.4.17)$$

where V = speed of the fluid and V_s = speed of sound in the fluid which for air is 49.02 $\sqrt{T}$, where T is in °R and V_s in ft/s. The Mach number varies with position in the fluid, and the compressibility effect likewise varies from point to point.

If a *body moves through the atmosphere,* the overall compressibility effects are a function of the Mach number $\overline{M}$ of the body, defined as

$$\overline{M} = \text{velocity of body/speed of sound in the atmosphere}$$

Table 11.4.6 lists the useful **gas dynamics relations** between velocity, Mach number, and various fluid properties for isentropic flow. (See also Sec. 4, "Heat.")

$$V^2 = 2(h_0 - h) = 2C_P(T_0 - T)$$

$$\frac{V^2}{V_{s0}^2} = \frac{M^2}{1 + \dfrac{\gamma - 1}{2}M^2}$$

$$\left(\frac{p}{p_0}\right)^{(\gamma-1)/\gamma} = \frac{T}{T_0} = \left(\frac{\rho}{\rho_0}\right)^{\gamma-1} = \left(\frac{V_s}{V_{s0}}\right)^2 = \frac{1}{1 + \dfrac{\gamma - 1}{2}M^2}$$

$$\frac{\frac{1}{2}\rho V^2}{P_0} = \frac{(\gamma/2)M^2}{\left(1 + \dfrac{\gamma - 1}{2}M^2\right)^{\gamma/(\gamma-1)}} \qquad (11.4.18)$$

$$\left(\frac{A}{A^*}\right)^2 = \frac{1}{M^2}\left[\frac{2}{\gamma + 1}\left(1 + \frac{\gamma - 1}{2}M^2\right)\right]^{(\gamma+1)/(\gamma-1)}$$

where h = enthalpy of the fluid, A = stream-tube cross section normal to the velocity, and the subscript $_0$ denotes the isentropic stagnation condition reached by the stream when stopped frictionlessly and adiabatically (hence isentropically). V_{s0} denotes the speed of sound in that medium.

The superscript * denotes the conditions occurring when the speed of the fluid equals the speed of sound in the fluid. For M = 1, Eqs. (11.4.18) become

$$\frac{T^*}{T_0} = \left(\frac{V_s^*}{V_{s0}}\right)^2 = \frac{2}{\gamma + 1}$$

$$\frac{p^*}{p_0} = \left(\frac{2}{\gamma + 1}\right)^{\gamma/(\gamma-1)} \qquad (11.4.19)$$

$$\frac{\rho^*}{\rho_0} = \left(\frac{2}{\gamma + 1}\right)^{1/(\gamma-1)}$$

For air $p^* = 0.52828p_0$; $\rho^* = 0.63394\rho_0$; $T^* = 0.83333T_0$.

Table 11.4.6 Isentropic Gas Dynamics Relations

M	$\dfrac{p}{p_0}$	$\dfrac{V}{V_{s0}}$	$\dfrac{A}{A^*}$	$\dfrac{\frac{1}{2}\rho V^2}{p_0}$	$\dfrac{\rho V}{\rho_0 V_{s0}}$	$\dfrac{\rho}{\rho_0}$	$\dfrac{T}{T_0}$	$\dfrac{V_s}{V_{s0}}$
0.00	1.000	0.000	∞	0.000	0.000	1.000	1.000	1.000
0.05	0.998	0.050	11.59	0.002	0.050	0.999	0.999	1.000
0.10	0.993	0.100	5.82	0.007	0.099	0.995	0.998	0.999
0.15	0.984	0.150	3.91	0.016	0.148	0.989	0.996	0.998
0.20	0.972	0.199	2.964	0.027	0.195	0.980	0.992	0.996
0.25	0.957	0.248	2.403	0.042	0.241	0.969	0.988	0.994
0.30	0.939	0.297	2.035	0.059	0.284	0.956	0.982	0.991
0.35	0.919	0.346	1.778	0.079	0.325	0.941	0.976	0.988
0.40	0.896	0.394	1.590	0.100	0.364	0.924	0.969	0.984
0.45	0.870	0.441	1.449	0.123	0.399	0.906	0.961	0.980
0.50	0.843	0.488	1.340	0.148	0.432	0.885	0.952	0.976
0.55	0.814	0.534	1.255	0.172	0.461	0.863	0.943	0.971
0.60	0.784	0.580	1.188	0.198	0.487	0.840	0.933	0.966
0.65	0.753	0.624	1.136	0.223	0.510	0.816	0.922	0.960
0.70	0.721	0.668	1.094	0.247	0.529	0.792	0.911	0.954
0.75	0.689	0.711	1.062	0.271	0.545	0.766	0.899	0.948
0.80	0.656	0.753	1.038	0.294	0.557	0.740	0.887	0.942
0.85	0.624	0.795	1.021	0.315	0.567	0.714	0.874	0.935
0.90	0.591	0.835	1.009	0.335	0.574	0.687	0.861	0.928
0.95	0.559	0.874	1.002	0.353	0.577	0.660	0.847	0.920
1.00	0.528	0.913	1.000	0.370	0.579	0.634	0.833	0.913
1.05	0.498	0.950	1.002	0.384	0.578	0.608	0.819	0.905
1.10	0.468	0.987	1.008	0.397	0.574	0.582	0.805	0.897
1.15	0.440	1.023	1.018	0.407	0.569	0.556	0.791	0.889
1.20	0.412	1.057	1.030	0.416	0.562	0.531	0.776	0.881
1.25	0.386	1.091	1.047	0.422	0.553	0.507	0.762	0.873
1.30	0.361	1.124	1.066	0.427	0.543	0.483	0.747	0.865
1.35	0.337	1.156	1.089	0.430	0.531	0.460	0.733	0.856
1.40	0.314	1.187	1.115	0.431	0.519	0.437	0.718	0.848
1.45	0.293	1.217	1.144	0.431	0.506	0.416	0.704	0.839
1.50	0.272	1.246	1.176	0.429	0.492	0.395	0.690	0.830
1.55	0.253	1.274	1.212	0.426	0.478	0.375	0.675	0.822
1.60	0.235	1.301	1.250	0.422	0.463	0.356	0.661	0.813
1.65	0.218	1.328	1.292	0.416	0.443	0.337	0.647	0.805
1.70	0.203	1.353	1.338	0.410	0.433	0.320	0.634	0.796
1.75	0.188	1.378	1.387	0.403	0.417	0.303	0.620	0.788
1.80	0.174	1.402	1.439	0.395	0.402	0.287	0.607	0.779
1.85	1.161	1.425	1.495	0.386	0.387	0.272	0.594	0.770
1.90	0.149	1.448	1.555	0.377	0.372	0.257	0.581	0.762
1.95	0.138	1.470	1.619	0.368	0.357	0.243	0.568	0.754
2.00	0.128	1.491	1.688	0.358	0.343	0.230	0.556	0.745
2.50	0.059	1.667	2.637	0.256	0.219	0.132	0.444	0.667
3.00	0.027	1.793	4.235	0.172	0.137	0.076	0.357	0.598
3.50	0.013	1.884	6.790	0.112	0.085	0.045	0.290	0.538
4.00	0.007	1.952	10.72	0.074	0.054	0.028	0.238	0.488
4.50	0.003	2.003	16.56	0.049	0.035	0.017	0.198	0.445
5.00	0.002	2.041	25.00	0.033	0.023	0.011	0.167	0.408
10.00	0.00002	2.182	536.00	0.002	0.001	0.005	0.048	0.218

SOURCE: Emmons, "Gas Dynamics Tables for Air," Dover Publications, Inc., 1947.

For subsonic regions flow behaves similarly to the familiar hydraulics or incompressible aerodynamics: In particular, an increase of velocity is associated with a decrease of stream-tube area, and friction causes a pressure drop in a tube. There are no regions where the local flow velocity exceeds sonic speed. For **supersonic regions,** an increase of velocity is associated with an increase of stream-tube area, and friction causes a pressure rise in a tube. M = 1 is the dividing line between these two regions. For **transonic flows,** there are regions where the local flow exceeds sonic velocity, and this mixed-flow region requires special analysis. For **supersonic flows,** the entire flow field, with the exception of the regime near stagnation areas, has a velocity higher than the speed of sound.

The hypersonic regime is that range of very high supersonic speeds (usually taken as M > 5) where even a very streamlined body causes disturbance velocities comparable to the speed of sound, and stagnation temperatures can become so high that the gas molecules dissociate and become ionized.

Shock waves may occur at locally supersonic speeds. At low velocities the fluctuations that occur in the motion of a body (at the start of the motion or during flight) propagate away from the body at essentially the speed of sound. At higher subsonic speeds, fluctuations still propagate at the speed of sound in the fluid away from the body in all directions, but now the waves cannot get so far ahead; therefore, the force coefficients increase with an increase in the Mach number. When the Mach number of a body exceeds unity, the fluctuations instead of traveling away from the body in all directions are actually left behind by the body. These cases are illustrated in Fig. 11.4.32. As the supersonic speed is increased, the fluctuations are left farther behind and the force coefficients again decrease.

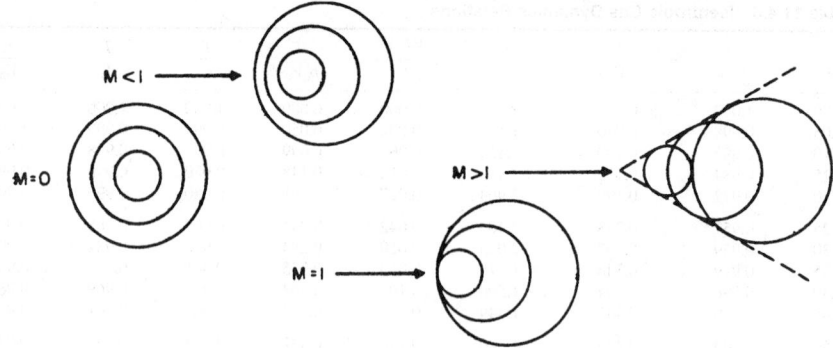

Fig. 11.4.32   Propagation of sound waves in moving streams.

Where, for any reason, waves form an envelope, as at M > 1 in Fig. 11.4.32, a wave of finite pressure jump may result.

The speed of sound is higher in a higher-temperature region. For a compression wave, disturbances in the compressed (hence high-temperature) fluid will propagate faster than and overtake disturbances in the lower-temperature region. In this way shock waves are formed.

For a **stationary normal shock wave** the following properties exist, where subscripts 1 and 2 refer to conditions in front of and behind the shock, respectively,

$$\left(\frac{V_{S2}}{V_{S1}}\right) = \sqrt{\left(\frac{T_2}{T_1}\right)} = \left(\frac{a_2}{a_1}\right) \tag{11.4.20a}$$

$$\left(\frac{V_2}{V_1}\right) = \left(\frac{M_2}{M_1}\right) \cdot \left(\frac{a_2}{a_1}\right) \tag{11.4.20b}$$

In the following equations, also for a normal shock, the second form shown is for air with $\gamma = 1.4$.

$$\left(\frac{p_2}{p_1}\right) = \frac{2\gamma M_1^2 - (\gamma - 1)}{\gamma + 1} = \left(\frac{7M_1^2 - 1}{6}\right) \tag{11.4.20c}$$

$$\left(\frac{\rho_2}{\rho_1}\right) = \frac{(\gamma + 1)M_1^2}{(\gamma - 1)M_1^2 + 2} = \left(\frac{6M_1^2}{M_1^2 + 5}\right) \tag{11.4.20d}$$

$$M_2 = \left[\frac{(\gamma - 1)M_1^2 + 2}{2\gamma M_1^2 - (\gamma - 1)}\right]^{1/2} = \left(\frac{M_1^2 + 5}{7M_1^2 - 1}\right)^{1/2} \tag{11.4.20e}$$

$$\left(\frac{p_{20}}{p_{10}}\right) = \left[\frac{(\gamma + 1)M_1^2}{(\gamma - 1)M_1^2 + 2}\right]^{\gamma/(\gamma-1)} \cdot \left[\frac{\gamma + 1}{2\gamma M_1^2 - (\gamma - 1)}\right]^{1/(\gamma-1)}$$

$$= \left(\frac{6M_1^2}{M_1^2 + 5}\right)^{3.5} \cdot \left(\frac{6}{7M_1^2 - 1}\right)^{2.5} \tag{11.4.20f}$$

$$\left(\frac{T_2}{T_1}\right) = \frac{1 + \left(\frac{\gamma - 1}{2}\right)M_1^2}{1 + \left(\frac{\gamma - 1}{2}\right)M_2^2} = \left(\frac{1 + 0.2M_1^2}{1 + 0.2M_2^2}\right) \tag{11.4.20g}$$

The subscript 20 refers to the stagnation condition after the shock, and the subscript 10 refers to the stagnation condition ahead of the shock.

In a normal shock M, V, and p_0 decrease; p, ρ, T, and s increase, whereas T_0 remains unchanged. These relations are given numerically in Table 11.4.7.

If the shock is moving, these same relations apply *relative to the shock*. In particular, if the shock advances into stationary fluid, it does so at the speed V_1 which is always greater than the speed of sound in the stationary fluid by an amount dependent upon the shock strength.

When an air vehicle is traveling at high subsonic, transonic, or supersonic speeds, shock waves will form on it. A shock wave is a thin region of air through which air physical properties change. Shock waves can be normal or oblique to the air vehicle direction of flight. If oblique, it behaves exactly like a normal shock to the normal component of the stream. The tangential component is left unchanged. Thus the resultant velocity not only drops abruptly in magnitude but also changes discontinuously in direction. Figure 11.4.33 gives the relations between M_1, M_2, θ_W, and δ.

In **supersonic flow past a two-dimensional wedge** with semiangle δ, one **strong** and one **weak** oblique shock wave will theoretically form attached to the leading edge. See Fig. 11.4.34a. In practice only the weak one will be present and downstream of this oblique shock the resultant flow will be parallel to the wedge surface. It is possible to use Eqs. (11.4.20) to generate the flow characteristics about wedges with various wedge angles or more easily with the help of Fig. 11.4.33.

The relationship between δ, M_1, and θ_W is shown in Fig. 11.4.34b. These are related through the equation

$$\tan \delta = \left[\frac{2 \cot \theta_W (M_1^2 \sin^2 \theta_W - 1)}{M_1^2 (\gamma + \cos 2\theta_W) + 2)}\right]$$

The results of a numerical evaluation of the above equation for $1.5 < M_1 < 20$, $\gamma = 1.4$ (for air) are shown graphically in Fig. 11.4.35. The locus of the maximum wedge angle line differentiates between weak and strong shocks. For small wedge angles, the shock angle differs only slightly from $\sin^{-1}(1/M_1)$, the **Mach angle**, and the velocity component normal to this wave is the speed of sound. The pressure jump is small and is given approximately by

$$p - p_1 = \left[\frac{\gamma p_1 \overline{M}_1^2}{\sqrt{\overline{M}_1^2 - 1}}\right] \delta$$

where δ is the wedge semiangle.

For wedge semiangles that exceed the corresponding Mach number lines, a detached shock will form at the leading edge (Fig. 11.4.36). For example, if $\delta = 40°$ and $M_1 = 3.00$, then a detached bow wave will form in front of the wedge leading edge (see Fig. 11.4.35). Behind this detached shock there will be a region of subsonic flow reaching the wedge leading edge.

Exact solutions also exist for **supersonic flow past a cone.** Above a certain supersonic Mach number, a conical shock wave is attached to the apex of the cone. Figures 11.4.37 and 11.4.38 show these exact relations; they are so accurate that they are often used to determine the Mach number of a stream by measuring the shock-wave angle on a cone of known angle. For small cone angles the shock differs only slightly from the **Mach cone**, i.e., a cone whose semiapex angle is the Mach angle; the pressure on the cone is then given approximately by

$$p - p_1 = \gamma p_1 \overline{M}_1^2 \delta^2 (\ln 2/\delta \sqrt{\overline{M}_1^2 - 1}) \tag{11.4.21}$$

where δ is the cone semiangle.

Table 11.4.7   Normal Shock Relations

M	p_2/p_1	p_{20}/p_1	p_{20}/p_{10}	M_2	V_{s2}/V_{s1}	V_2/V_1	T_2/T_1	ρ_2/ρ_1
1.00	1.000	1.893	1.000	1.000	1.000	1.000	1.000	1.000
1.05	1.120	2.008	1.000	0.953	1.016	0.923	1.033	1.084
1.10	1.245	2.133	0.999	0.912	1.032	0.855	1.065	1.169
1.15	1.376	2.266	0.997	0.875	1.047	0.797	1.097	1.255
1.20	1.513	2.408	0.993	0.842	1.062	0.745	1.128	1.342
1.25	1.656	2.557	0.987	0.813	1.077	0.700	1.159	1.429
1.30	1.805	2.714	0.979	0.786	1.091	0.660	1.191	1.516
1.35	1.960	2.878	0.970	0.762	1.106	0.624	1.223	1.603
1.40	2.120	3.049	0.958	0.740	1.120	0.592	1.255	1.690
1.45	2.286	3.228	0.945	0.720	1.135	0.563	1.287	1.776
1.50	2.458	3.413	0.930	0.701	1.149	0.537	1.320	1.862
1.55	2.636	3.607	0.913	0.684	1.164	0.514	1.354	1.947
1.60	2.820	3.805	0.895	0.668	1.178	0.492	1.388	2.032
1.65	3.010	4.011	0.876	0.654	1.193	0.473	1.423	2.115
1.70	3.205	4.224	0.856	0.641	1.208	0.455	1.458	2.198
1.75	3.406	4.443	0.835	0.628	1.223	0.439	1.495	2.279
1.80	3.613	4.670	0.813	0.617	1.238	0.424	1.532	2.359
1.85	3.826	4.902	0.790	0.606	1.253	0.410	1.573	2.438
1.90	4.045	5.142	0.767	0.596	1.268	0.398	1.608	2.516
1.95	4.270	5.389	0.744	0.586	1.284	0.386	1.647	2.592
2.00	4.500	5.640	0.721	0.577	1.299	0.375	1.688	2.667
2.50	7.125	8.526	0.499	0.513	1.462	0.300	2.138	3.333
3.00	10.333	12.061	0.328	0.475	1.637	0.259	2.679	3.857
3.50	14.125	16.242	0.213	0.451	1.821	0.235	3.315	4.261
4.00	18.500	21.068	0.139	0.435	2.012	0.219	4.047	4.571
4.50	23.458	26.539	0.092	0.424	2.208	0.208	4.875	4.812
5.00	29.000	32.653	0.062	0.415	2.408	0.200	5.800	5.000
10.00	116.500	129.220	0.003	0.388	4.515	0.175	20.388	5.714

SOURCE: Emmons, "Gas Dynamics Tables for Air," Dover Publications, Inc., 1947.

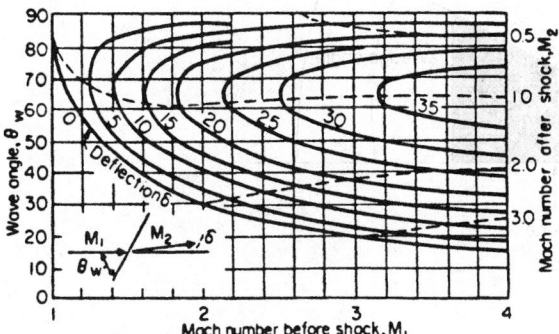

Fig. 11.4.33 Oblique shock wave relations.

A **nozzle** consisting of a single contraction will produce at its exit a jet of any velocity from M = 0 to M = 1 by a proper adjustment of the pressure ratio. For use as a subsonic wind-tunnel nozzle where a uniform parallel gas stream is desired, it is only necessary to connect the supply section to the parallel-walled or open-jet test section by a smooth gently curving wall. If the radius of curvature of the wall is nowhere less than the largest test-section cross-sectional dimension, no flow separation will occur and a good test gas stream will result.

When a **converging nozzle** connects two chambers with the pressure drop beyond the critical [$(p/p_0) < (p^*/p_0)$], the Mach number at the exit of the nozzle will be 1; the pressure ratio from the supply section to the nozzle exit will be critical; and all additional expansion will take place outside the nozzle.

A nozzle designed to supply a supersonic jet at its exit must converge to a minimum section and diverge again. The area ratio from the minimum section to the exit is given in the column headed A/A^* in Table 11.4.6. A converging-diverging nozzle with a pressure ratio p/p_0 = (gas pressure)/(stagnation pressure), Table 11.4.6, gradually falling from unity to zero will produce shock-free flow for all exit Mach numbers from zero to the subsonic Mach number corresponding to its area ratio. From this Mach number to the supersonic Mach number corresponding to the given area ratio, there will be shock waves in the nozzle. For all smaller pressure ratios, the Mach number at the nozzle exit will not change, but additional expansion to higher velocities will occur outside of the nozzle.

To obtain a uniform parallel shock-free supersonic stream, the converging section of the nozzle can be designed as for a simple converging nozzle. The diverging or supersonic portion must be designed to produce and then cancel the expansion waves. These nozzles would perform as designed if it were not for the growth of the boundary layer. Experience indicates that these nozzles give a good first approximation to a uniform parallel supersonic stream but at a somewhat lower Mach number.

For **rocket nozzles** and other thrust devices the gain in thrust obtained by making the jet uniform and parallel at complete expansion must be balanced against the loss of thrust caused by the friction on the wall of the greater length of nozzle required. A simple conical diverging section, cut off experimentally for maximum thrust, is generally used (see also Sec. 11.5 "Jet Propulsion and Aircraft Propellers").

For transonic and supersonic flow, **diffusers** are used for the recovery of kinetic energy. They follow the test sections of supersonic wind tunnels and are used as inlets on high-speed planes and missiles for ram recovery. For the first use, the diffuser is fed a nonuniform stream from the test section (the nonuniformities depending on the particular body under test) and should yield the maximum possible pressure-rise ratio. In missile use the inlet diffuser is fed by a uniform (but perhaps slightly yawed) airstream. The maximum possible pressure-rise ratio is important but must provide a sufficiently uniform flow at the exit to assure good performance of the compressor or combustion chamber that follows.

In simplest form, a **subsonic diffuser** is a diverging channel, a nozzle in reverse. Since boundary layers grow rapidly with a pressure rise, subsonic diffusers must be diverged slowly, 6 to 8° equivalent cone angle, i.e., the apex angle of a cone with the same length and area ratio.

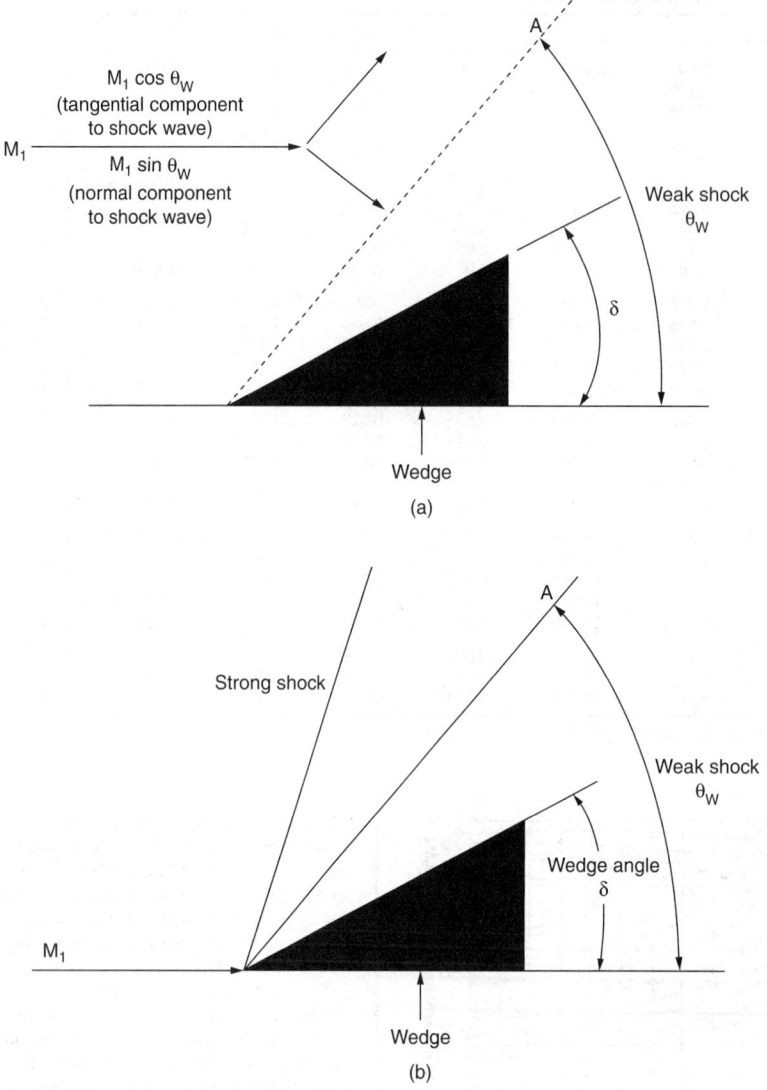

Fig. 11.4.34 Shock waves on a wedge.

Similarly, a **supersonic diffuser** in its simplest form is a supersonic nozzle in reverse. Both the convergent and divergent portions must change cross section gradually. In principle it is possible to design a shock-free diffuser. In practice shock-free flow is not attained, and the design is based upon minimizing the shock losses. Oblique shocks should be produced at the inlet and reflected a sufficient number of times to get compression nearly to M = 1. A short parallel section and a divergent section can now be added with the expectation that a weak normal shock will be formed near the throat of the diffuser.

For **a supersonic wind tunnel,** the best way to attain the maximum pressure recovery at a wide range of operating conditions is to make the diffuser throat variable. The ratio of diffuser-exit static pressure to the diffuser-inlet (test-section outlet) total pressure is given in Table 11.4.8. These pressure recoveries are attained by the proper adjustment of the throat section of a variable diffuser on a supersonic wind-tunnel nozzle.

A **supersonic wind tunnel** consists of a compressor or compressor system including precoolers or aftercoolers, a supply section, a supersonic nozzle, a test section with balance and other measuring equipment, a diffuser, and sufficient ducting to connect the parts. The **minimum pressure ratio** required from supply section to diffuser exit is given in Table 11.4.8. Any pressure ratio greater than this is satisfactory. The extra pressure ratio is automatically wasted by additional shock waves that appear in the diffuser. The compression ratio required of the compressor system must be greater than that of Table 11.4.8 by at least an amount sufficient to take care of the pressure drop in the ducting and valves. The latter losses are estimated by the usual hydraulic formulas.

After selecting a compressor system capable of supplying the required maximum pressure ratio, the test section area is computed from

$$A = 1.73 \frac{Q}{V_{s0}} \frac{A}{A^*} \frac{p_e}{p_0} \qquad \text{ft}^2 \qquad (11.4.22)$$

where Q is the inlet volume capacity of the compressors, ft^3/s; V_{s0} is the speed of sound in the supply section, ft/s; A/A^* is the area ratio given in Table 11.4.6 as a function of M; p_e/p_0 is the pressure ratio given in Table 11.4.8 as a function of M.

Fig. 11.4.35 Oblique shock properties for air: strong and weak shock variables.

Fig. 11.4.37 Wave angles for supersonic flow around cones.

The nozzle itself is designed for uniform parallel airflow in the test section. Since for each Mach number a different nozzle is required, the nozzle must be flexible or the tunnel so arranged that fixed nozzles can be readily interchanged.

The Mach number of a test is set by the nozzle selection, and the Reynolds number is set by the inlet conditions and size of model. The Reynolds number is computed from

$$\text{Re} = \text{Re}_0 D(p_0/14.7)(540T_0)^{1.268} \qquad (11.4.23)$$

where Re_0 is the Reynolds number per inch of model size for atmospheric temperature and pressure, as given by Fig. 11.4.39, D is model diam, in; p_0 is the stagnation pressure, lb/in^2; T_0 is the stagnation temperature, °R.

For a closed-circuit tunnel, the Reynolds number can be varied independently of the Mach number by adjusting the mass of air in the system, thus changing p_0.

Intermittent-wind tunnels for testing at high speeds do not require the large and expensive compressors associated with continuous-flow tunnels. They use either a large vacuum tank or a large pressure tank (often in the form of a sphere) to produce a pressure differential across the test

section. Such tunnels may have steady flow for only a few seconds, but by careful instrumentation, sufficient data may be obtained in this time. A **shock tube** may be used as an intermittent-wind tunnel as well as to study shock waves and their interactions; it is essentially a long tube of constant or varying cross section separated into two parts by a frangible diaphragm. High pressure exists on one side; by rupturing the diaphragm, a shock wave moves into the gas with the lower pressure. After the shock wave a region of steady flow exists for a few milliseconds. Very high stagnation temperatures can be created in a shock tube, which is not the case in a wind tunnel, so that it is useful for studying hypersonic flow phenomena.

Wind-tunnel force measurements are subject to errors caused by the model support strut. Wall interference is small at high Mach numbers for which the reflected model head wave returns well behind the model. Near $M = 1$, the wall interference becomes very large. In fact, the tunnel chokes at Mach numbers given in Table 11.4.5, at

$$\frac{A}{A^*} = \cfrac{1}{1 - \cfrac{\text{area of model projected on test-section cross section}}{\text{area of test-section cross section}}}$$

$$(11.4.24)$$

There are two choking points: one subsonic and one supersonic. Between these two Mach numbers, it is impossible to test in the tunnel. As these Mach numbers are approached, the tunnel wall interference becomes very large.

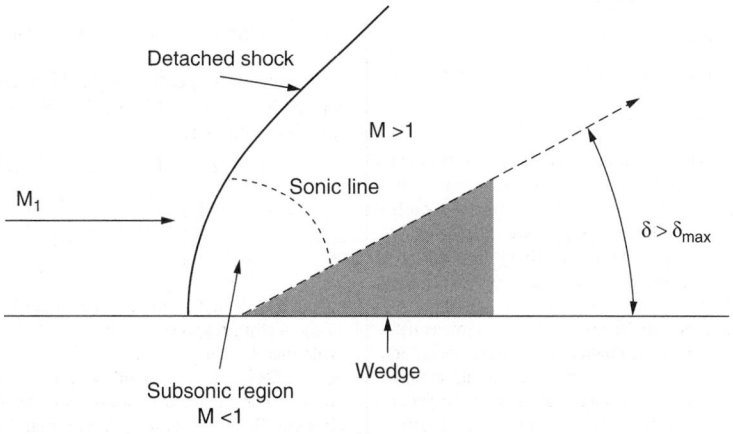

Fig. 11.4.36 Detached shock relations on a wedge.

Fig. 11.4.38 Pressure calculations for supersonic flow around cones.

Table 11.4.8

M	1	1.5	2	3	4
P_e/P_0	0.83	0.69	0.50	0.23	0.10

Fig. 11.4.39 Reynolds number–Mach number relation (for a fixed model size and stagnation condition).

For tests in this range of Mach number, specially constructed **transonic wind tunnels** with perforated or slotted walls have been built. The object here is to produce a mean flow velocity through the walls that comes close to that which would have existed there had the body been moving at that speed in the open. Often auxiliary blowers are needed to produce the necessary suction on the walls and to reinject this air into the tunnel circuit in or after the diffuser section.

Drag is difficult to predict precisely from wind-tunnel measurements, especially in the transonic regime. **Flight tests** of rocket-boosted models which coast through the range of Mach numbers of interest are used to obtain better drag estimates; from telemetered data and radar or optical sighting the deceleration can be determined, which in turn yields the drag.

LINEARIZED SMALL-DISTURBANCE THEORY

As the speed of a body is increased from a low subsonic value (Fig. 11.4.40a), the local Mach number becomes unity somewhere in the fluid along the surface of the body, i.e., the lower critical Mach number is reached. Above this there is a small range of transonic speed in which a supersonic region exists (Fig. 11.4.40b). Shock waves appear in this region attached to the sides of the body (Fig. 11.4.40c) and grow with increasing speed. At still higher transonic speeds a detached shock wave appears ahead of the body, and the earlier side shocks either disappear or move to the rear (Fig. 11.4.40d). Finally, for a sharp-nosed body, the head wave moves back and becomes attached (Fig. 11.4.40e). The flow is now generally supersonic everywhere, and the transonic regime is replaced by the supersonic regime. With the appearance of shock waves there occurs a considerable alteration of the pressure distribution, and the center of pressure on airfoil sections moves from the one-fourth chord point back toward the one-half chord point. There is an associated increase of drag and, often, flow separation at the base of the shock.

The redistribution of pressures and the motion of shock waves over the wing surfaces through the transonic regime demands special consideration in the design of control surfaces so that they do not become ineffective by separation or inoperative by excessive loading.

If a body is slender (i.e., planes tangent to its surface at any point make small angles with the flight direction), the disturbance velocities caused by this body will be small compared with the flight speed and, excluding the hypersonic regime, small compared with the speed of sound. This permits the use of linearized small-disturbance theory to predict the approximate flow past the body. This theory relates the steady flow past the body at subsonic speeds to the incompressible flow past a distorted version of this body (the "generalized Prandtl-Glauert rule"). For **thin airfoil sections** at subsonic speeds, this theory predicts

$$C_L = 2\pi\alpha/\sqrt{1 - M^2} \qquad \alpha \text{ in radians} \qquad (11.4.25)$$

where $L = C_L qS$ and $q = (\gamma/2)p_0 M^2$.

An approximation to the **subsonic** lift curve slope for wings is given by:

$$C_{L\alpha} = \left(\frac{2\pi A}{\left[2 + \left\{ \left(\frac{A^2\beta^2}{0.90} \right) \left(1 + \frac{\tan^2\Lambda_{c/2}}{\beta^2} \right) + 4 \right\}^{1/2} \right]} \right) \text{ per radian}$$

where A = aspect ratio, $\beta = (1 - M^2)^{0.5}$, $\Lambda_{c/2}$ = semichord sweep angle. This can be applied to wings of all aspect ratios, taper ratios, and for quarter chord sweep angles of up to 60°.

The corresponding **induced drag,** or **drag due to lift,** is given by

$$C_{Di} = \frac{C_L^2}{\pi A e}$$

where e is the span efficiency factor, which is usually between 0.85 and 0.95.

At supersonic speeds, the Prandtl-Glauert rule is also applicable if we replace β by $\lambda = \sqrt{M^2 - 1}$. For **thin airflow sections** (see Fig. 11.4.41) this theory predicts

$$C_L = 4\alpha/\sqrt{M^2 - 1} \qquad \alpha \text{ in radians} \qquad (11.4.26)$$

$$C_D = \frac{2}{\lambda} \int_0^1 \left[\left(\frac{dy_u}{dx} \right)^2 + \left(\frac{dy_l}{dx} \right)^2 \right] d\left(\frac{x}{c} \right) + \frac{4a^2}{\lambda} + C_{DF}$$

$$(11.4.27)$$

where dy_u/dx and dy_l/dx are the slopes of the upper and lower surfaces of the airfoil, respectively, and C_{DF} is the skin-friction drag coefficient. Note that there is a drag due to thickness and a drag due to lift at supersonic speeds for an airfoil section where there is none at subsonic speeds. For symmetric double-wedge airfoil sections, the thickness drag coefficient becomes $(4/\lambda)(t/c)^2$, and for symmetric biconvex airfoil sections it is $(16/3\lambda)(t/c)^2$, where t is the maximum thickness.

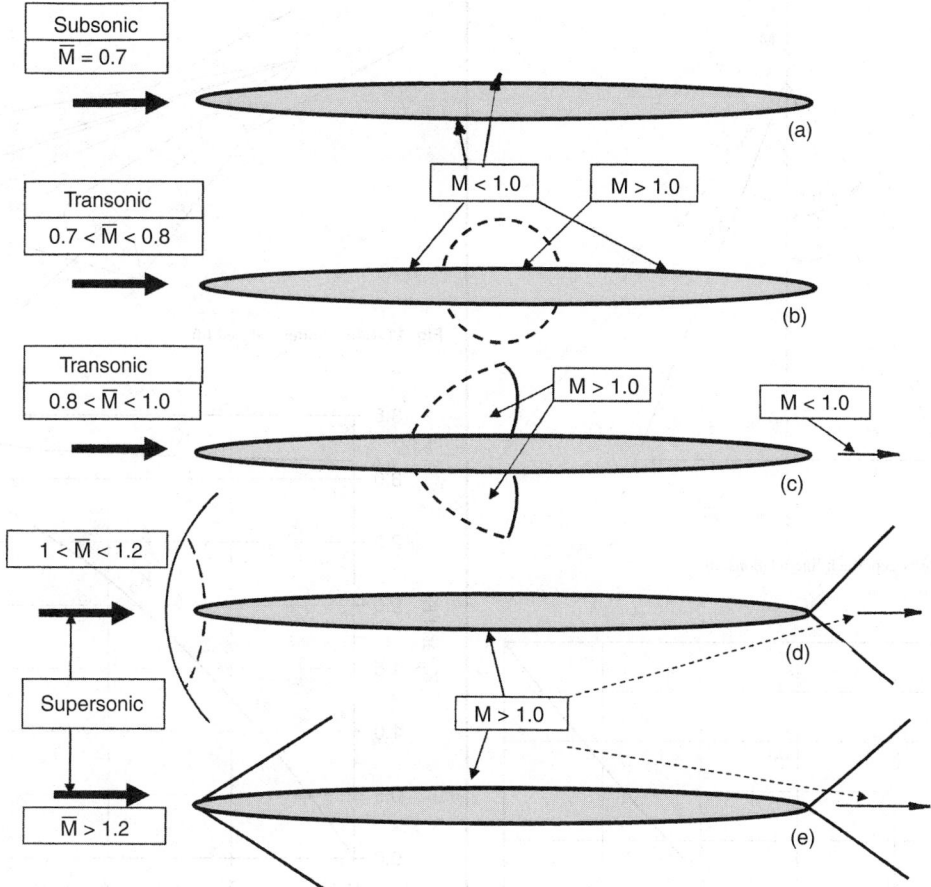

Fig. 11.4.40   Regions of flow about an airfoil (Mach numbers are approximate).

Fig. 11.4.41   Supersonic airfoil section.

Within the small disturbance approximation, a disturbance at a point in supersonic flow can affect only the points in the downstream **Mach cone**. Thus a rectangular wing with constant airfoil section has two-dimensional flow on all parts of the wing except the points within the Mach cones. At Mach numbers near unity these tip Mach cones cover nearly the entire wing, whereas at high Mach numbers they cover only a small part.

Research at NACA (NACA TR 839, NACA TR 900) established that for **supersonic** wings, swept and unswept, the relative location of the Mach lines and the wing planform (Fig. 11.4.42) is important in the determination of the magnitudes of the wing lift curve slopes and the drag due to lift. For swept wings, the wing possesses a **subsonic** leading edge when $\mu > \varepsilon$, (Fig. 11.4.43), and a **supersonic** leading edge when $\varepsilon > \mu$. When $\varepsilon = \mu$ the leading edge is termed **sonic**.

For **thin airfoil section supersonic delta wings** with **subsonic leading edges**, lift can be approximated by

$$\beta\left(\frac{C_L}{\alpha}\right) = \left(\frac{2\pi^2}{\pi + \lambda}\right)\left(\frac{\tan \varepsilon}{\tan \mu}\right)$$

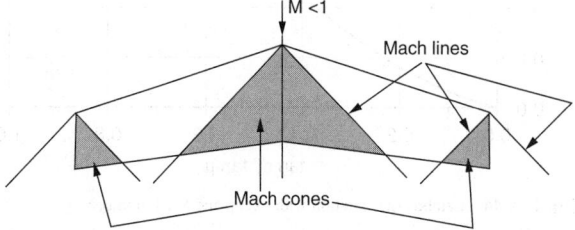

Fig. 11.4.42   Mach cones and Mach lines for wing with supersonic leading edge.

where λ is given in Fig. 11.4.44 and $\beta = (M^2 - 1)^{0.5}$, and the corresponding drag due to lift approximated by

$$C_{Di} = \left(\frac{C_L^2}{\pi A}\right)\left[2\left(1 + \frac{\lambda}{\pi}\right) - \sqrt{1 - \beta^2 \tan^2 \varepsilon}\right]$$

For **delta wings with supersonic leading edges**, the corresponding first approximations are

$$\left(\frac{C_L}{\alpha}\right) = \frac{4}{\sqrt{M^2 - 1}} \text{ per radian} \qquad \text{and} \qquad C_{Di} = \frac{4\alpha^2}{\sqrt{M^2 - 1}}$$

For delta wings with sharp leading edges, research at NASA (NASA TN-D 4739) established the presence of a **vortex lift** component at low

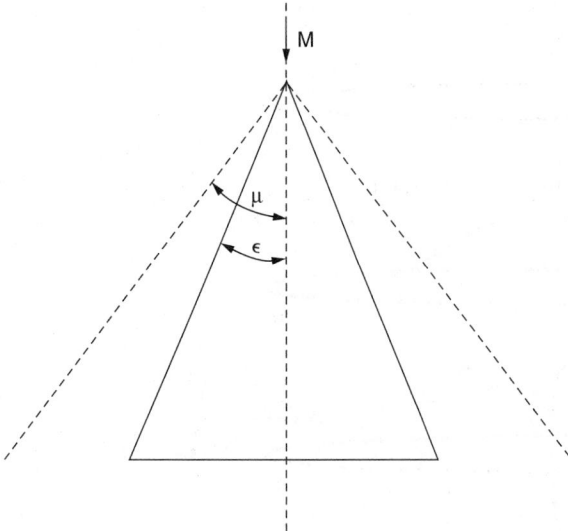

Fig. 11.4.43 Wing sweep/Mach line relationship.

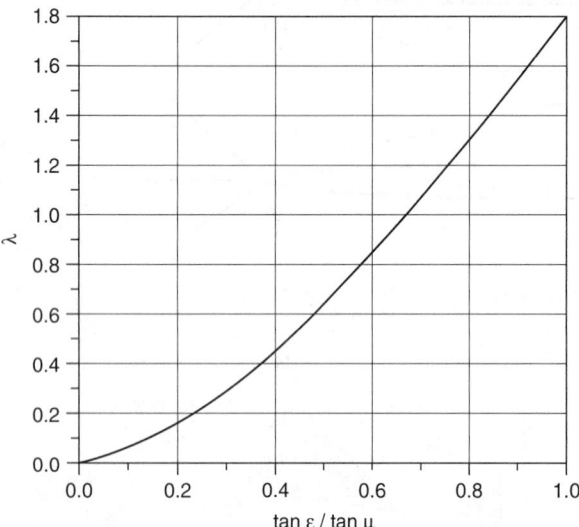

Fig. 11.4.44 Lambda (λ) variation with sweep and Mach number.

speed. Figure 11.4.45 illustrates the formation of these vortices which generates a total lift coefficient given by

$$C_L = K_p \sin \alpha \cos^2 \alpha + K_v \cos \alpha \sin^2 \alpha$$

Drag due to this lift is given by

$$\Delta C_{Di} = K_p \sin^2 \alpha \cos \alpha + K_v \sin^3 \alpha$$

where K_p and K_v are given in Fig. 11.4.46.

Rectangular wings in supersonic flight are not amenable to simple closed form solutions for either lift or drag due to lift and the reader is referred to NACA TN 2114 for lift and drag estimations.

The lift of a slender **axially symmetric body** at small angle of attack is nearly independent of Mach number and is given approximately by

$$C_L = 2\alpha \qquad (11.4.28)$$

where the lift coefficient is based on the cross-sectional area of the base. The drag for $\alpha = 0$ at subsonic speeds is made up of skin friction and

Fig. 11.4.45 Vortex-induced lift.

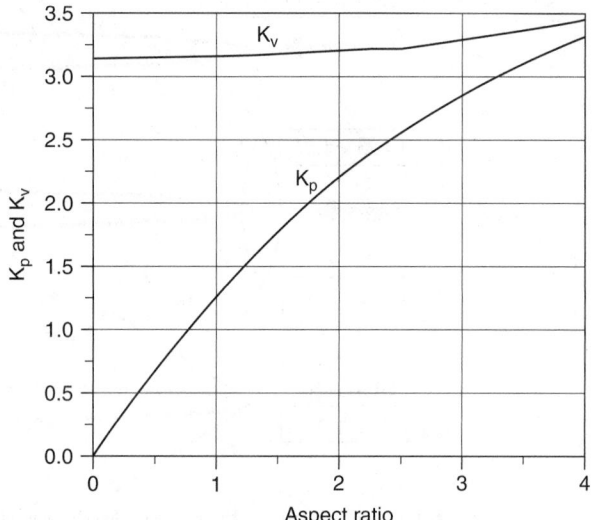

Fig. 11.4.46 Variations of K_p and K_v with wing aspect ratio.

base drag; a dead-air region exists just behind the blunt base, and the pressure here is below ambient, causing a rearward suction which is the base drag. At supersonic speeds a wave drag is added, which represents the energy dissipated in shock waves from the nose. Figure 11.4.47 shows a typical drag-coefficient curve for a body of revolution where the **fineness ratio** (ratio of length to diameter) is 12.2. The skin-friction drag coefficient based on wetted area is the same as that of a flat plate, within experimental error.

Fig. 11.4.47 Drag coefficients for parabolic-arc (*NACA-RM* 10) body (calculations from experimental data for a 30,000-ft altitude, $D = 12$ in, from *NACA-TR* 1160 and 1161, 1954).

At transonic and moderate supersonic speeds, the wave drag of a **wing body combination** can be effectively reduced by making the cross-sectional area distribution (including wings) a smooth curve when plotted versus fuselage station. This is the **area rule**; its application results in a decided indentation in the fuselage contour at the wing juncture. **Lift interference** effects also occur, especially when the body diameter is not small compared with the wing span. If the wing is attached to a cylindrical body, and if the wing-alone lift coefficient would have been C_L, then the lift carried on the body is $K_B C_L$, and the lift carried on the wing is $K_W C_L$. Figure 11.4.48 shows the variation of K_B and K_W with body diameter to wing-span ratio.

The **effect of compressibility on skin-friction drag** is slight at subsonic speeds, but at supersonic speeds, a significant reduction in skin-friction coefficient occurs. Figure 11.4.49 shows the turbulent-boundary-layer mean skin-friction coefficient for a cone as a function of wall temperature and Mach number. An important effect at high speed is **aerodynamic heating.**

Figure 11.4.50 illustrates the velocity profile behind the shock wave of a body traveling at hypersonic speed. The shock-wave front represents an area of high-temperature gas which radiates energy to the body, but boundary-layer convective heating is usually the major contributor. Behind this front is shown the velocity gradient in the boundary layer. The decrease in velocity in the boundary layer is brought about by the forces of interaction between fluid particles and the body (viscosity). This change in velocity is accompanied by a change in temperature and is dependent on the characteristics of the boundary layer. For example, heat transfer from a turbulent boundary layer may be of an order of magnitude greater than for laminar flow.

If the gas is brought to rest instantaneously, the total energy in the flow is converted to heat and the temperature of the air will rise. The

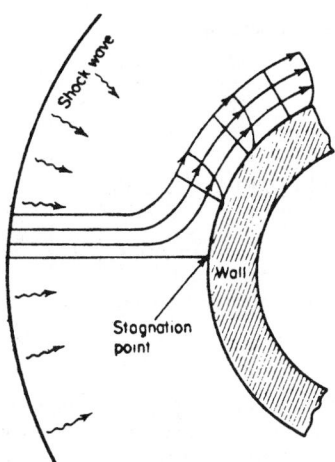

Fig. 11.4.50 Velocity profile behind a shock wave.

resulting temperature is known as the **stagnation temperature** (see stagnation point, Fig. 11.4.50).

$$T_S = T_\infty\left(1 + \frac{\gamma - 1}{2}M^2\right) \quad (11.4.29)$$

T_∞ is the ambient temperature of the gas at infinity, and γ is the specific-heat ratio of the gas. For undissociated air, $\gamma = 1.4$.

In general, this simple, one-dimensional relationship between velocity and temperature does not hold for temperature in the boundary layer. The laminae of the boundary layer are not insulated from each other, and there is cross conduction. This is associated with the **Prandtl number** Pr, which is defined as

$$\text{Pr} = C_p\mu/k_r \quad (11.4.30)$$

where k_r = thermal conductivity of fluid, C_p = specific heat of fluid at constant pressure, and μ = absolute viscosity of fluid.

Defining the **recovery factor** r as the ratio of the rise in the idealistic wall and stagnation temperature over the free-stream temperature,

$$r = (T_{aw} - T_\infty)/(T_s - T_\infty) = (\text{Pr})^n \quad (11.4.31)$$

For laminar flow, $r = (\text{Pr})^{1/2}$; and for turbulent flow, $r = (\text{Pr})^{1/3}$.

Prandtl number Pr greatly complicates the thermal computations, but since it varies only over a small range of values, a recovery factor of 0.85 is generally used for laminar flow and 0.90 for turbulent flow.

For a thermally thin wall, the rate of change of the surface temperature is a function of the rate of total heat input and the surface's ability to absorb the heat.

$$dT_w/dt = q_T/wcb \quad (11.4.32)$$

where t = time, q_T = forced convective heating + radiation heating − heat radiation from the skin, w = density of skin material, c = specific heat of skin material, b = skin thickness, and T_w = skin temperature.

The heat balance may then be written

$$wcb\left(\frac{dT_w}{dt}\right) = k_c\left[T_0\left(1 + r\frac{\gamma_B - 1}{2}M_0^2\right) - T_w\right] + \alpha G_s A_p - \epsilon\sigma T_w^4 \quad (11.4.33)$$

where A_p = correction factor to account for area normal to radiation source, γ_B = specific heat ratio of boundary layer, ϵ = radiative emissivity of surface, σ = Stefan-Boltzmann constant [17.3 × 10⁻¹⁰ Btu/(h · ft² · °R⁴)], α = surface absorptivity, G_s = solar irradiation Btu/(h·ft²).

Fig. 11.4.48 Wing-body lift-interface factors.

Fig. 11.4.49 Laminar-skin-friction coefficient for a cone.

For small time increments Δt,

$$\frac{dT_w}{dt} = \frac{T_{w2} - T_{w1}}{\Delta t}$$

Then

$$T_{w2} = T_{w1} + \frac{\Delta t}{wcb}\left\{ h_c\left[T_0\left(1 + \frac{\gamma_B - 1}{2}M_0^2 \right) - T_w \right] \right.$$
$$\left. + \alpha G_s A_p - T_w^4 \right\} \qquad (11.4.34)$$

The local heat-transfer coefficient h_c is defined as

$$h_c = \frac{k_r}{r^{4/3}}C_p C_{f\rho 0} V g \qquad (11.4.35)$$

where C_p = specific heat of air; C_f = local skin-friction coefficient; ρ_0 = density outside of boundary layer; V = velocity outside of boundary layer; g = acceleration of gravity.

For a cone, $k_r = 1,800$ (laminar Re $< 2 \times 10^6$) and $k_r = 1,800 r^{4/3}$ (turbulent).

For a flat-plate transition, Reynolds number is 1×10^6.

Figure 11.4.49 gives laminar-skin-friction coefficient for a cone. For a flat plate, multiply C_f by $\sqrt{3/2}$. For turbulent skin-friction coefficient for a cone,

$$\frac{0.242}{\sqrt{A^2 C_f T_w / T_0}}(\sin^{-1}\psi + \sin^{-1}\theta)$$
$$= 0.41 \log \frac{Re}{2} C_f - 1.26 \log \frac{T_w}{T_0}$$

where

$$\psi = \frac{2A^2 - B}{\sqrt{B^2 + 4A^2}} \qquad A^2 = \frac{(\gamma_0 - \frac{1}{2})M_0^2}{T_w / T_0}$$

$$\theta = \frac{B}{\sqrt{B^2 + 4A^2}} \qquad B = \frac{1 + (\gamma_0 - \frac{1}{2})M_0^2}{T_w / T_0} - 1$$

For a flat plate, use Re instead of Re/2 in the above equation.

Figure 11.4.51 shows data on stagnation and adiabatic wall temperatures.

Fig. 11.4.51   Variation of stagnation and adiabatic wall temperature with Mach number.

The primary **measurements in aerodynamics** are pressure measurements. A well-aligned pitot tube with an impact-pressure hole at its nose and static-pressure holes 10 diameters or more back from the nose will accurately measure the **impact pressure** and the **static pressure** of a uniform gas stream. Up to sonic speed the impact pressure is identical with stagnation pressure, but at supersonic speeds, a detached shock wave forms ahead of the probe, through which there is drop in stagnation pressure; the portion of the shock wave just ahead of the probe is normal, so that the equation for water flow in layer form over horizontal tubes (see "Transmission of Heat by Conduction and Convection" in Sec. 4) gives the relation of the measured impact pressure to the isentropic stagnation pressure.

Force measurements of total lift, drag, side force, pitching moment, yawing moment, and rolling moment on models are made on wind-tunnel balances just as at low speed. Small internal-strain-gage balances are often used to minimize strut interference.

An open thermocouple is unreliable for determining **stagnation temperature** if it is in a stream of high velocity. Figure 11.4.52 shows a simple temperature probe in which the fluid is decelerated adiabatically before its temperature is measured. The recovery factor used in equations, for this probe, accurately aligned to the stream, is 0.98.

Fig. 11.4.52   Stagnation-temperature probe (recovery factor = 0.98).

Optical measurements in high-speed flow depend on the variation of index of refraction with gas density. This variation is given by

$$n = 1 + k\rho \qquad (11.4.36)$$

where $k = 0.116$ ft³/slug for air. A Mach-Zehnder **interferometer** (Fig. 11.4.53a), is capable of giving accurate density information for two-dimensional and axially symmetric flows. The **Schlieren** optical system (Fig. 11.4.53b) is sensitive to density gradients and is the most commonly used system to determine location of shock waves and regions of compression or expansion. The **shadowgraph** optical system is the simplest system (Fig. 11.4.53c) and is sensitive to the second space derivative of the density.

Fig. 11.4.53   Optical systems for observing high-speed flow phenomena. (a) Interferometer; (b) Schlieren (two-mirror) system; (c) shadowgraph.

FURTHER READING: "Aerospace Source Book 2003," Aviation Week & Space Technology, McGraw-Hill, New York. "Air International," 1989–2001, Vol. 36–60, Key Publishing Ltd, Stamford, Lincolnshire, U.K. Anderson, J. D., Jr., 1999, "Aircraft Performance and Design," WCB/McGraw-Hill, Boston. Jenkinson, D., Simpson, P., and Rhodes, D. "Civil Jet Aircraft Design," AIAA Education Series, Washington, DC. Loftin, L. K., Jr., 1980. "Subsonic Aircraft: Evolution and the Matching of Size to Performance," NASA, Reference Publication 1060. Nicolai, L. M., 1984, "Fundamentals of Aircraft Design," 1st ed., METS, San Jose, CA. Raymer, D. P., 1989, "Aircraft Design—A Conceptual Approach," AIAA Educational Series, Washington, DC. Rolls-Royce, plc, 1986, "The Jet Engine," Rolls-Royce, plc, Derby, U.K. Roskam, J., 1980, "Aircraft Design, Parts I through VIII," DAR Corporation, Lawrence, KS. Gessow, A., and Myers, G. C., Jr., 1952, "Aerodynamics of the Helicopter," Macmillan, New York. Huenecke, K., 1987, "Modern Combat Aircraft Design," Naval Institute Press, Annapolis, MD. Khoury, G. A., and Gilbert, J. D., 1999, "Airship Technology," Cambridge Aerospace Series 10, Cambridge University Press, Cambridge, U.K. Mair, W. A., and Birdshall, D. J., 1992, "Aircraft Performance," Cambridge Aerospace Series 5, Cambridge University Press, Cambridge, U.K. McCormick, B. W., Jr., 1967, "Aerodynamics of V/STOL Flight," Academic Press, New York. Schaufele, R., 2000, "The Elements of Aircraft Preliminary Design," Aires, Santa Ana, CA. Shapiro, J., 1955, "Principles of Helicopter Engineering," McGraw-Hill, New York. Smetana, F. O., 2001, "Flight Vehicle Performance and Aerodynamic Control," AIAA Educational Series, Washington, DC. Stinton, D., 1988, "The Anatomy of the Airplane," 2nd ed., Blackwell Science, Oxford, and AIAA, Washington, DC. Stinton, D., 1983, "The Design of the Airplane," Van Nostrand, New York, Stinton, D., 1996, "Flying Qualities and Flight Testing of the Airplane," AIAA Educational Series, Washington, DC.

11.5   JET PROPULSION AND AIRCRAFT PROPELLERS
by Sanford Fleeter

REFERENCES: Zucrow, "Aircraft and Missile Propulsion," Vol. 2, Wiley. Hill and Peterson, "Mechanics and Thermodynamics of Propulsion," Addison-Wesley. Sutton, "Rocket Propulsion Elements," Wiley. Shorr and Zaehringer, "Solid Propellant Technology," Wiley. Forestor and Kuskevics, "Ion Propulsion," AIAA Selected Reprints. Theodorsen, "Theory of Propellers," McGraw-Hill. Smith, "Propellers for High Speed Flight," Princeton. Zucrow, "Principles of Jet Propulsion," Wiley. "Aircraft Propeller Handbook," Depts. of the Air Force, Navy, and Commerce, U.S. Government Printing Office, 1956. Kerrebrock, "Aircraft Engines and Gas Turbines," MIT Press. Oates, ed., "The Aerothermodynamics of Aircraft Gas Turbine Engines," AFAPL-TR-78–52, 1978. Fleeter, "Aeroelasticity for Turbomachine Applications," *AIAA J. of Aircraft,* **16,** no. 5, 1979. Bathie, "Fundamentals of Gas Turbines," Wiley. "Aeronautical Technologies for the Twenty-First Century," National Academy Press. Mattingly, Heiser, and Daley, "Aircraft Engine Design," AIAA.

Fig. 11.5.1   Ideal propeller in the relative coordinate system.

Newton's reaction principle, based on the second and third laws of motion, is the theoretical basis for all methods of propelling a body either in (or on) a fluid medium or in space. Thus, the aircraft propeller, the ships screw, and the jet propulsion of aircraft, missiles, and boats are all examples of the application of the reaction principle to the propulsion of vehicles. Furthermore, all jet engines belong to that class of power plants known as **reaction engines.**

Newton's second law states that a change in motion is proportional to the force applied; i.e. a force proportional to the rate of change of the velocity results whenever a mass is accelerated: $F = ma$. This can be expressed in a more convenient form as $F = d(mv)/dt$, which makes it clear that the application of the reaction principle involves the time rate of increase of the momentum (mv) of the body. It should be noted that the same force results from either providing a small acceleration to a large mass or a large acceleration to a small mass.

Newton's third law defines the fundamental principle underlying all means of propulsion. It states that for every force acting on a body there is an equal and opposite reaction force.

The application of this reaction principle for propulsion involves the indirect effect of increasing the momentum of a mass of fluid in one direction and then utilizing the reaction force for propulsion of the vehicle in the opposite direction. Thus, the reaction to the time rate of increase of the momentum of that fluid, called the **propulsive fluid,** creates a force, termed the **thrust,** which acts in the direction of motion desired for the propelled vehicle. The known devices for achieving the propulsion of bodies differ only in the methods and mechanisms for achieving the time rate of increase in the momentum of the propulsive fluid or matter.

The function of a propeller is to convert the power output of an engine into useful thrust. To do this it accelerates a mass of air in the direction opposite to the direction of flight, generating thrust via Newton's reaction principle. Figure 11.5.1 illustrates schematically, in the relative coordinate system for steady flow, the operating principle of the **ideal aircraft propeller.** Power is supplied to the propeller, assumed to be equivalent to an **actuator disk,** which imparts only an axial acceleration to the air flowing through it. The rotation of the actuator disk produces a **slipstream** composed of the entire mass of air flowing in the axial direction through the area of the actuator disk, i.e., the circle swept by the propeller. Atmospheric air enters the slipstream with the flight speed V_0 and mass flow rate m_a. It leaves the slipstream with the **wake velocity** w. The thrust is given in Eq. (11.5.5). Propulsion systems utilizing propellers will be termed **propeller propulsion.** (For detailed discussions of the airplane propeller, see the section "Aircraft Propellers.")

Jet propulsion differs from propeller propulsion in that the propulsive fluid (or matter), instead of being caused to flow around the propelled body, is ejected from within the propelled body in the form of one or more high-speed jets of fluid or particles. In fact, the jet propulsion engine is basically designed to accelerate a large stream of fluid (or matter) and to expel it at a high velocity. There are a number of ways of accomplishing this, but in all instances the resultant reaction or thrust exerted on the engine is proportional to the time rate of increase of the momentum of the fluid. Thus, jet engines produce thrust in an analogous manner to the propeller and engine combination. While the propeller gives a small acceleration to a large mass of air, the jet engine gives a larger acceleration to a smaller mass of air.

Theoretically, there are no restrictions on either the type of matter, called the **propellant,** for forming a high-velocity jet or the means for producing the **propulsive jet.** The selection of the most suitable propellant and the most appropriate jet propulsion engine is dictated by the specific mission for the propelled vehicle. For example, in the case of the jet propulsion of a boat, termed **hydraulic jet propulsion,** the propulsive jet is formed from the water on which the boat moves. For the practical propulsion of bodies through either the atmosphere or in space, however, only two types of propulsive jets are suitable:

1. For propulsion within the atmosphere, there is the jet formed by expanding a highly heated, compressed gas containing atmospheric air as either a major or a sole constituent. Such an engine is called either an **airbreathing** or a **thermal-jet engine**. If the heating is accomplished by burning a fuel in the air, the engine is a **chemical thermal-jet engine**. If the air is heated by direct or indirect heat exchange with a nuclear-energy source, the engine is termed a **nuclear thermal-jet engine**.

2. For propulsion both within and beyond the atmosphere of earth, there is the exhaust jet containing no atmospheric air. Such an exhaust jet is termed a **rocket jet**, and any matter used for creating the jet is called a **propellant**. A rocket may consist of a stream of gases, solids, liquids, ions, electrons, or a plasma. The assembly of all the equipment required for producing the rocket jet constitutes a **rocket engine**.

Modern airbreathing engines may be segregated into two principal types: (1) **ramjet engines**, and (2) **turbojet engines**. The turbojet engines are of two types: (1) the **simple turbojet engine**, and (2) the **turbofan**, or **bypass, engine**.

Rocket engines can be classified by the form of energy used for achieving the desired jet velocity. The three principal types of rocket engines are (1) **chemical rocket engines**, (2) **nuclear heat-transfer rocket engines**, and (3) **electric rocket engines**.

The propulsive element of a jet-propulsion engine, irrespective of type, is the **exhaust nozzle** or orifice. If the exhaust jet is gaseous, the assembly comprising all the other components of the jet-propulsion engine constitutes a gas generator for supplying highly heated, high-pressure gases to the exhaust nozzle.

These classifications of jet-propulsion engines apply to the basic types of engines. It is possible to have combinations of the different types of thermal-jet engines and also combinations of thermal-jet engines with rocket engines; only the principal types are discussed here.

ESSENTIAL FEATURES OF AIRBREATHING OR THERMAL-JET ENGINES

In the subsequent discussions a *relative coordinate system* is employed wherein the atmospheric air flows toward the propulsion system with the flight speed V_0, and the gases leave the propulsion system with the velocity w, *relative to the walls of the propulsion system*. Furthermore, steady-state operating conditions are assumed.

Ramjet Engine Figure 11.5.2 illustrates schematically the essential features of the simple ramjet engine. It comprises three major components: a diffusion or inlet system, (0–2); a combustion chamber (2–7); and an exhaust nozzle (7–9). In this simple ramjet engine, apart from the necessary control devices, there are no moving parts. However, for accelerating the vehicle or for operating at different Mach numbers, a variable geometry diffuser and exhaust may be required.

The operating principle of the ramjet engine is as follows. The freestream air flowing toward the engine with a velocity V_0 and Mach number M_0 is decelerated by the inlet diffuser system so that it arrives at the entrance to the combustion chamber with a low Mach number, on the order of $M_2 = 0.2$. When the inlet-flow Mach number is supersonic,

this diffuser system consists of a supersonic diffuser followed by a subsonic diffuser. The supersonic diffuser decelerates the inlet flow to approximately unity Mach number at the entrance to the subsonic diffuser; this deceleration is accompanied by the formation of shock waves and by an increase in the pressure of the air (diffusion). In the subsonic diffuser the air is further diffused so that it arrives at the entrance to the combustion chamber with the required low Mach number. If P_0 is the total pressure of the free-stream air having the Mach number M_0 and P_2 that for the air entering the combustion system with the Mach number M_2, then it is desirable that P_2/P_0 be a maximum.

In the combustion chamber a fuel is burned in the air, thereby raising the total temperature of the gases entering the exhaust nozzle (see Sec. 7) to approximately $T_7 = 4,260°R$ (2,366 K). Most generally a liquid hydrocarbon fuel is used, but experiments have been conducted with solid fuels, liquid hydrogen, liquid methane, and "slurries" of metallic fuels in a liquid fuel. The combustion process is not quite isobaric because of the pressure drops in the combustion chamber, because of the increase in the momentum of the working fluid due to heat addition, and because of friction. The hot gases are discharged to the atmosphere, after expanding in the exhaust nozzle, with the relative velocity $w = V_9$.

Since the ramjet engine can function only if there is a ram pressure rise at the entrance to the combustion chamber, it is not self-operating at zero flight speed. It must, therefore, be accelerated to a flight speed which permits the engine to develop sufficient thrust for accelerating the vehicle it propels to the design flight Mach number. Consequently, a ramjet-propelled missile, for example, must either be launched by dropping it from an airplane or be **boosted** to the required flight speed by means of **launching**, or **bootster**, rockets. It appears that the most appropriate flight regime for the ramjet is at low to moderate supersonic Mach numbers, between approximately $M_0 = 2$ and 5. The lower limit is associated with the necessary ram pressure rise while the upper limit is set by the problem of either cooling or protecting the outer skin of the engine body.

Scramjet Low-cost access to space is one of the major reasons for the interest in **hypersonic propulsion** (speeds above Mach 5). A vehicle to provide access to space must be capable of reaching very high Mach numbers, since orbital velocity (17,000 mi/h) is about Mach 24. The ramjet engine, which must slow the entering air to a low subsonic velocity, cannot be used much above Mach 5 or 6. The solution is to decelerate the flow in the engine to a supersonic Mach number lower than the flight Mach number but greater than the local speed of sound. This propulsion device is the **supersonic combustion ramjet**, or **scramjet**, which is capable, in principle, of operation to Mach 24. As shown schematically in Fig. 11.5.3, in contrast to the ramjet, the scramjet engine requires that fuel be added to the air, mixed, and burned, all at supersonic velocities, which is a significant technical challenge.

The advantage derived from supersonic combustion of the liquid fuel (usually liquid hydrogen) is that the diffuser of the scramjet engine is required to decelerate the air entering the engine from M_0 to only approximately $M_2 = 0.35M_0$ instead of the $M_2 = 0.2$, which is essential with subsonic combustion. This elimination of the subsonic diffusion increases the diffuser efficiency, reduces the static pressure in the combustor (and, therefore, the engine weight and heat transfer rate), and increases the velocity in the combustor (thereby decreasing engine frontal area). It is the combination of the above effects that makes ramjet propulsion at high flight Mach numbers (>7) feasible.

Pulsejets may be started and operated at considerably lower speeds than ramjets. A pulsejet is a ramjet with an air inlet which is provided with a set of shutters loaded to remain in the closed position. After the pulsejet engine is launched, ram air pressure forces the shutters to open, fuel is injected into the combustion chamber, and is burned. As soon as the pressure in the combustion chamber equals the ram air pressure, the shutters close. The gases produced by combustion are forced out the jet nozzle by the pressure that has built up within the chamber. When the pressure in the combustion chamber falls off, the shutters open again, admitting more air, and the cycle repeats at a high rate.

Fig. 11.5.2 Ramjet engine.

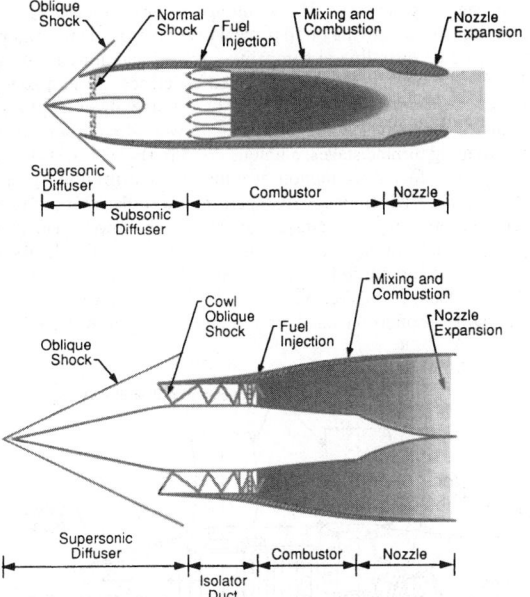

Fig. 11.5.3   Ramjet and scramjet flow configurations.

Simple Turbojet Engine   Figure 11.5.4 illustrates schematically the principal features of a simple turbojet engine, which is basically a gas-turbine engine equipped with a propulsive nozzle and diffuser. Atmospheric air enters the engine and is partially compressed in the diffusion system, and further compressed to a much higher pressure by the air compressor, which may be of either the axial-flow or centrifugal type. The highly compressed air then flows to a combustion chamber wherein sufficient fuel is burned to raise the total temperature of the gases entering the turbine to approximately $T_4 = 2,160°R$ (1,200 K) for an uncooled turbine. The maximum allowable value for T_4 is limited by metallurgical and stress considerations; it is desirable, however, that T_4 be as high as possible. The combustion process is approximately isobaric. The highly heated air, containing approximately 25 percent of combustion products, expands in the turbine, which is directly connected to the air compressor, and in so doing furnishes the power for driving the air compressor. From the turbine the gases pass through a tailpipe which may be equipped with an **afterburner**. The gases are expanded in a suitably shaped exhaust nozzle and ejected to the atmosphere in the form of a high-speed jet.

Like the ramjet engine, the turbojet engine is a continuous-flow engine. It has an advantage over the ramjet engine in that its functioning does not depend upon the ram pressure of the entering air, although the amount of ram pressure recovered does affect its overall economy and performance. The turbojet is the only airbreathing jet engine that has been utilized as the sole propulsion means for piloted aircraft. It appears to be eminently suited for propelling aircraft at speeds above 500 mi/h (805 km/h). As the design flight speed is increased, the ram

Fig. 11.5.4   Simple turbojet engine.

pressure increases rapidly, and the characteristics of the turbojet engine tend to change over to those of the ramjet engine. Consequently, its top speed appears to be limited to that flight speed where it becomes more advantageous to employ the ramjet engine. It appears that for speeds above approximately 2,000 mi/h (3,219 km/h) it is advantageous to use some form of a ramjet engine.

The thrust of the simple turbojet engine increases rapidly with T_4, because increasing T_4 increases the jet velocity V_j. Actually, V_j increases faster than the corresponding increase in T_4. It is also an inherent characteristic of the gas-turbine engine which produces shaft power, called a **turboshaft** engine, that its useful power increases proportionally faster than a corresponding increase in its turbine inlet temperature T_4. Because of the decrease in strength of turbine materials with increase in temperature, the turbine blades, stators, and disks require cooling at $T_4 > 2,160°R$ (1,200 K) approximately. A typical advanced turbine environment has **gas temperatures** of 3,600°R (2,000 K), with cooling provided to maintain a blade **metal temperature** of 2,250°R (1,250 K). Note that gas temperatures are being pushed toward the 4,600°R (2,556 K) stoichiometric limit of JP4 fuel. The cooling air is bled from the compressor at the appropriate stage (or stages) and used to cool the stator blades or rotor blades by convective, film, or transpiration heat transfer. Up to 10 percent of the compressor air may be bled for turbine cooling, and this air is "lost" for turbine work for that blade row where it is used for cooling. Consequently "trade" studies must be made to "weigh" the increased complexity of the engine and the turbine work loss due to air bleed against the increased engine performance associated with increased T_4.

As in the case of any gas-turbine power plant, the efficiencies of the components of the turbojet engine have an influence on its performance characteristics, but its performance is not nearly as sensitive to changes in the efficiency of its component machines as is a gas turbine which delivers shaft power. (See Sec. 9.)

As noted previously, two types of compressors are currently employed, the axial-flow compressor and the centrifugal compressor. Irrespective of the type, the objectives are similar. The compressor must be reliable, compact, easy to manufacture, and have a small frontal area. Because of the limited air induction capacity of the centrifugal compressor, also called the radial compressor, engines for developing thrusts above 7,000 lb (31 kN) at static sea level employ axial-flow compressors (see Sec. 14).

The turbojet engine exhibits a rather flat thrust-versus-speed curve. Because the ratio of the takeoff thrust to thrust in flight is small, certain operational problems exist at takeoff. Since the exhaust gases from the turbine contain considerable excess air, the jet velocity, and consequently the thrust, can be increased by burning additional fuel in the tailpipe upstream from the exhaust nozzle. By employing **"tailpipe burning,"** or **"afterburning,"** as it is called, the thrust can be increased by 35 percent and at 500 mi/h, in a tactical emergency, by approximately 60 percent. With afterburning, the temperature of the gases entering the nozzle T_7, is of the order of 3,800°R (2,110 K).

Turbofan, or Bypass, Engine   For a fixed turbine inlet temperature, the jet velocity from a simple turbojet engine propelling an airplane at subsonic speed is relatively constant. The propulsive efficiency depends on the ratio of the flight speed to the jet velocity and increases as the ratio increases. On the other hand, the thrust depends on the difference between the jet velocity and the flight speed; the larger the difference, the larger the thrust per unit mass of air induced into the engine. By reducing the jet velocity and simultaneously increasing the mass rate of airflow through the engine, the **propulsive efficiency** can be increased without decreasing the thrust. To accomplish this, a **turbofan or bypass engine** is required.

Figure 11.5.5 illustrates schematically the components of a turbofan engine. There are two turbines, a low-pressure turbine (LPT) and a high-pressure turbine (HPT); one drives the air compressor of the hot-gas generator and the other drives the fan. Air enters the fan at the rate of $\dot{m}_{aF}$ and is ejected through the nozzle of area A_{7F} with the jet velocity $V_{jF} < V_j$, where V_j is the jet velocity attained by the air flowing through the hot-gas generator with the mass flow rate $\dot{m}_{a1}$; the hot-gas generator is basically a turbojet engine.

Fig. 11.5.5 Schematic arrangement of components of turbofan engines (subsonic flight).

A key parameter is the **bypass ratio**, defined as the ratio of the air mass flow rate through the fan bypass duct to that through the hot-gas generator ($\dot{m}_{aF}/\dot{m}_{a1}$). Bypass ratios of 5 to 10 are in use, with values of about 10 utilized in modern large commercial transport engines, for example, the Trent 800, the GE-90, and the PW-4000. The hot-gas generator is basically a turbojet engine. The fuel is added to $\dot{m}_{a1}$ at the rate $\dot{m}_f$. The hot-gas stream ($\dot{m}_{a1} + \dot{m}_f$) is discharged to the atmosphere through A_7 with the velocity V_j. Turbofan engines produce an "overall" jet velocity V_{jTF} which is smaller than that for a turbojet engine operating with the same P_3/P_2 and T_4. The arrangement shown in Fig. 11.5.5 is for subsonic propulsion. Fuel can, of course, be burned in the fan air m_{aF} for increasing the thrusts of the engines. Practically all the newer commercial passenger aircraft, including regional jets, are propelled by turbofan engines. The advantage of the lower effective jet velocity V_{jTF} is twofold: (1) It increases the propulsive efficiency η_p by reducing $\nu = V_{jTF}/V_0$ and, consequently, raises the value of η_0. (2) The reduced jet velocity reduces the jet noise; the latter increases with approximately the eighth power of the jet speed.

The next generation of large transport engines will be conventional high-bypass engines, but will have a larger thrust capability and will offer further improvements in specific fuel consumption, achieved by using even higher overall pressure ratios (over 50:1) and turbine inlet temperature (over 1,850 K or 2,870°F) combined with even more efficient components (Fig. 11.5.6).

A highly loaded, multiblade swept variable-pitch propeller, the **propfan** or **advanced turboprop**, could be combined with the latest turbine engine technology. The resulting advanced turboprop would offer a potential fuel saving of 50 percent over an equivalent-technology turbofan engine operating at competitive speeds and altitudes because of the turboprop's much higher installed efficiency. To eliminate the gearbox development for a 20,000 shaft horsepower engine, the propfans of the **unducted fan (UDF)** engine were directly driven with counterrotating turbine stages, a unique concept (Fig. 11.5.7). The projected specific fuel consumption at cruise for the resulting gearless, counterrotating UDF engine was 30 percent lower than that of the most modern turbofan engines, and about 50 percent lower than that of engines presently in use on 150-passenger airplanes. The fuel-saving potential of the turboprop has already been demonstrated. If economic conditions change, this concept could replace conventional high-bypass-ratio turbofans on small aircraft and military transports.

Fig. 11.5.7 UDF counterrotating turbine configuration.

The work of traditional gas turbines is based on the constant-pressure **Brayton cycle**. However, the **Humphrey cycle**'s constant-volume detonation combustion, or **pulse detonation**, has potential of increased efficiency and fuel consumption as compared to constant-pressure-combustion turbofan and turbojet engines, while having fewer moving parts and being able to operate at both subsonic and supersonic speeds. The **pulse detonation engine** (PDE) is, thus, a new engine technology currently being developed.

PDEs are essentially tubes with fuel and air admitted into one end, with **detonation combustion** occurring rather than the deflagration combustion utilized in current gas turbines. Note that there are two types of combustion: deflagration and detonation. Consider a tube containing a fuel-air mixture. The fuel is ignited by a spark at one end and the combustion reaction then propagates down the tube. With deflagration, the burning is relatively slow. Detonation combustion, in contrast, results in a high-pressure detonation or shock wave that travels down the tube at supersonic speeds, resulting in a nearly constant-volume heat addition process that produces a high pressure in the combustor and provides the thrust. This wave compresses and ignites the fuel and air in a narrow high-pressure zone, thereby resulting in burning at thousands of metres per second. Exhaust gases are expelled at the open end of the tube and, as the exhaust gases clear out, a new cycle begins. A schematic of PDE operation is shown in Fig. 11.5.8. The operation of multitube configurations at high frequencies (100+ Hz) produces near-constant thrust.

To generate the critical high-pressure combustion zone, the fuel, air, and the ignition spark must be precisely coordinated, thereby generating the **deflagration-to-detonation** transition (DDT). The DDT process, whereby an ordinary flame suddenly accelerates into a detonation, has four phases. In the first phase, deflagration is initiated by an electric spark. The second phase is characterized by flame acceleration and shock wave formation. This is followed by the formation and amplification of explosion centers from pockets of reactants that create small blast waves. Finally, in the last phase, a strong detonation wave is formed by coupling between the amplified blast waves and the reaction zone.

Fig. 11.5.6 Efficiency trends.

Fuel mixed with air

Detonation wave exits
Air drawn in by reduced pressure

Detonation inititated

High pressure gas
fills chamber

Detonation moves through
fuel-air mixture

Fig. 11.5.8 Pulse detonation engine schematic.

There are three categories of pulse-detonation-based propulsion systems being considered. **Pure PDEs** are the simplest and comprise an array of detonation tubes, an inlet, and a nozzle. The upper speed limit of a pure PDE is about Mach 4 with a hydrocarbon fuel, but could be higher with hydrogen. **Combined-cycle PDEs** consist of a pure PDE combined with a ramjet/scramjet flow path or other propulsion cycle, where each cycle operates in a different speed range to optimize overall engine performance. **Hybrid PDEs** involve pure PDEs in combination with turbomachinery. For example, a hybrid turbofan-PDE would combine a turbofan with a PDE: the turbofan core engine would still power the large fan, but the bypass air would flow into a ring of PDEs, thereby producing significantly more thrust.

ESSENTIAL FEATURES OF ROCKET ENGINES

Figure 11.5.9 is a block-type diagram illustrating the essential features of a rocket engine, which comprises three main components: (A) a supply of propellant material contained in the rocket-propelled vehicle, (B) a propellant feed and metering system, and (C) a **thrust chamber,** also called a **rocket motor, thrustor,** or **accelerator.** In any rocket engine, energy must be added to the propellant as it flows through the thrustor. In Fig. 11.5.9 propellant material from the supply is metered and fed to the thrust chamber, where energy is added to it. As a consequence of the energy addition, the propellant is discharged from the thrustor with the jet velocity V_j. The thrustor F acts in the direction opposite to that for the jet velocity V_j.

Fig. 11.5.9 Essential features of a rocket engine.

Chemical Rocket Engines

All chemical rocket engines have two common characteristics: (1) They utilize chemical reactions in a thrust chamber to produce a high-pressure, high-temperature gas at the entrance to a converging-diverging exhaust nozzle; (2) the hot propellant gas expands in flowing through the exhaust nozzle, and the expansion process converts a portion of the thermal energy, released by the chemical reaction, into the kinetic energy associated with a high-velocity gaseous-exhaust jet. Chemical rocket engines may be grouped into (1) liquid-bipropellant rocket engines, (2) liquid-monopropellant rocket engines, and (3) solid-propellant rocket engines.

Liquid-Bipropellant Rocket Engine Figure 11.5.10 illustrates the essential features of a liquid-bipropellant rocket engine employing **turbopumps** for feeding two propellants, an **oxidizer** and a **fuel,** to a rocket motor. The motor comprises (1) an injector, (2) a combustion chamber, and (3) a converging-diverging exhaust nozzle. The liquid propellants are fed under pressure, through the injector, into the combustion chamber, where they react chemically to produce large volumes of high-temperature, high-pressure gases. For a given propellant combination, the **combustion temperature** T_c depends primarily on the oxidizer-fuel ratio (by weight), termed the **mixture ratio,** and to a lesser extent upon the combustion static pressure p_c. When the mass rate of flow of the liquid propellants equals that of the exhaust gases, the combustion pressure remains constant—the mode of operation that is usually desired.

Fig. 11.5.10 Essential features of a liquid-bipropellant rocket engine.

If the bipropellants react chemically when their liquid streams come in contact with each other, they are said to be **hypergolic.** Propellants which are not hypergolic are said to be **diergolic,** and some form of ignition system is required to initiate combustion.

Except in those cases where the operating duration of the rocket motor is very short or where the combustion temperature is low, means must be provided for protecting the interior walls of the motor. The two most common methods are ablative cooling and regenerative cooling. In **ablative cooling** the inner surfaces of the thrust chamber are covered with an ablative material which vaporizes, thereby providing some cooling. Ordinarily, the material leaves a "char" which acts as a high-temperature insulating material. For high-performance engines which are to operate for relatively long periods, regenerative cooling is employed. In **regenerative cooling,** one of the propellants is circulated around the walls before injection into the motor. In some engines the regenerative cooling is combined with local liquid film cooling at critical areas of the thrust chamber. See Table 11.5.1 for bipropellant combinations.

Refer to Figure 11.5.10. By removing the oxidizer tank, the oxidizer pump, and the plumbing associated with the liquid oxidizer, one obtains the essential elements of a liquid-monopropellant rocket engine. Such an engine does not require a liquid oxidizer to cause the monopropellant to decompose and release its thermochemical energy.

There are basically three groups of monopropellants: (1) liquids which contain the fuel and oxidizer in the same molecule, e.g., hydrogen peroxide (H_2O_2) or nitromethane (CH_3NO_2); (2) liquids which contain either the oxidizer or the fuel constituent in an unstable molecular arrangement, e.g., hydrazine (N_2H_4); and (3) synthetic mixtures of liquid fuels and oxidizers. The most important liquid monopropellants are hydrazine and hydrogen peroxide (up to 98% H_2O_2). Hydrazine can be decomposed by a suitable metal catalyst, which is ordinarily packed in a portion of the thrust chamber at the injector end. The decomposition of hydrazine yields gases at a temperature of 2,260°R (1,265 K); the gases are a mixture of hydrogen and ammonia. Hydrogen peroxide can be readily decomposed either thermally, chemically, or catalytically. The most favored method is catalytically; a series of silver screens coated with samarium oxide is tightly packed in a decomposition chamber located at the injector end of the thrust chamber. At $P_c = 300$ psia, the

Table 11.5.1 Calculated Values of Specific Impulse for Liquid Bipropellant Systems[a]

[$P_c = 1,000$ lbf/in^2 ($6,896 \times 10^3$ N/m^2) to $P_e = 1$ atm]

Shifting equilibrium; isentropic expansion; adiabatic combustion; one-dimensional flow

| Oxidizer | Fuel | $\dot{m}_o/\dot{m}_f$ | ρ | Temperature T_c | | c^* | I (s) |
				°R	K		
Chlorine trifluoride (CTF) (ClF$_3$)	Hydrazine	2.80	1.50	6553	3640	5961	293.1
	MMH[b]	2.70	1.41	5858	3254	5670	286.0
	Pentaborane (B$_5$H$_9$)	7.05	1.47	7466	4148	5724	289.0
Fluorine (F$_2$)	Ammonia	3.30	1.12	7797	4332	7183	359.5
	Hydrazine	2.30	1.31	8004	4447	7257	364.0
	Hydrogen	7.70	0.45	6902	3834	8380	411.1
	Methane	4.32	1.02	7000	3889	6652	343.8
	RP-1	2.62	1.21	6839	3799	6153	318.0
Hydrogen peroxide (100% H$_2$O$_2$)	Hydrazine	2.00	1.26	4814	2674	5765	287.4
	MMH	3.44	1.26	4928	2738	5665	284.8
IRFNA[c]	Hydine[d]	3.17	1.26	5198	2888	5367	270.3
	MMH	2.57	1.24	5192	2885	5749	275.5
Nitrogen tetroxide (N$_2$O$_4$)	Hydrazine	1.30	1.22	5406	3002	5871	292.2
	Aerozine-50[e]	2.00	1.21	5610	3117	5740	289.2
	MMH	2.15	1.20	5653	3141	5730	288.7
Oxygen (LOX) (O$_2$)	Hydrazine	0.91	1.07	5667	3148	6208	312.8
	Hydrogen	4.00	0.28	4910	2728	7892	291.2
	Methane	3.35	0.82	6002	3333	6080	310.8
	Pentaborane	2.21	0.91	7136	3964	6194	318.1
	RP-1	2.60	1.02	6164	3424	5895	300.0

[a] Based on "Theoretical Performance of Rocket Propellant Combinations," Rocketdyne Corporation, Canoga Park, CA.
[b] MMH: Monomethyl hydrazine (N$_2$H$_3$CH$_3$).
[c] IRFNA: Inhibited red fuming nitric acid; 84.4% (HNO$_3$), 14% (N$_2$O$_4$), 1% (H$_2$O), 0.6% (HF).
[d] Hydine: 60% UDMH[f], 40% Deta[g].
[e] Aerozine-50: 50% UDMH[f], 50% hydrazine (N$_2$H$_4$).
[f] UDMII: Unsymmetrical dimethylhydrazine (CH$_3$)$_2$N$_2$H$_4$.
[g] DETA: Diethylenetriamine (C$_2$H$_{13}$N$_3$).

decomposition of hydrogen peroxide (98%) yields gases having a temperature of 2,260°R (1,256 K).

The specific impulse obtainable for a liquid monopropellant is considerably smaller than that obtainable from liquid-bipropellant systems. Consequently, monopropellants are used for such auxiliary purposes as thrust vernier control, attitude reaction controls for space vehicles and missiles, and for gas generation.

Solid-propellant Rocket Engine Figure 11.5.11 illustrates schematically a solid-propellant rocket engine employing an **internal-burning case-bonded grain;** the latter burns radially outward at a substantially constant rate. A solid propellant contains both its fuel and the requisite oxidizer. If the fuel and oxidizer are contained in the molecules forming the solid propellant, the propellant is termed a **double-base propellant.** Those solid propellants wherein a solid fuel and a solid oxidizer form an intimate mechanical mixture are termed either

composite or heterogeneous solid propellants. The chemical reaction of a solid propellant is initiated by an igniter. In general, the configuration of a propellant grain can be designed so that the area of the burning surface of the propellant varies to give a prescribed thrust-versus-time curve.

Nuclear Heat-transfer Rocket Engine

Figure 11.5.12 illustrates schematically a nuclear heat-transfer rocket engine employing a **solid-core reactor.** The heat generated by the fissions of the uranium nucleus is utilized for heating a gaseous propellant, such as hydrogen, to a high temperature of 4,000°R (2,200 K) approx at the entrance section of the exhaust nozzle. The hot gas is ejected to the surroundings after expansion in a converging-diverging exhaust nozzle. The basic difference between the operating principles of a nuclear heat-transfer and a chemical rocket engine is the substitution of nuclear fission for chemical reaction as the source of heat for the propellant gas. Since the exhaust nozzle is the propulsive element, the

Fig. 11.5.11 Essential features of an internal-burning case-bonded solid-propellant rocket engine.

Fig. 11.5.12 Schematic arrangement of a nuclear heat-transfer rocket engine.

remaining components of the nuclear heat-transfer rocket engine constitute the hot-gas generator.

The propellant in a nuclear heat-transfer rocket engine functions basically as a fluid for cooling the solid-core nuclear reactor. Its reaction is not based on energy considerations alone, but on such properties as specific heat, latent heat, molecular weight, and liquid density and on certain practical considerations.

Electric Rocket Engines

Several different types of electric rocket engines have been conceived; the principal types are described below. All of them require some form of power plant for generating electricity. Thus the power plant may be nuclear, solar-cell batteries, thermoelectric, or other. The choice of the type of power plant depends upon the characteristics of both the electric rocket engine and the space-flight mission. Figure 11.5.13 illustrates diagrammatically an electric rocket engine, comprising (1) a **nuclear power source,** (2) an energy-conversion unit for obtaining the desired form of electric energy, (3) a propellant feed and metering system, (4) an electrically operated thrustor, and (5) the requisite control devices. Electric rocket engines may be classified as (1) electrothermal, (2) electromagnetic, and (3) electrostatic.

Electrothermal Rocket Engine   Figure 11.5.14a illustrates the type of engine which uses electric power for heating gaseous propellant to a high temperature before ejecting it through a converging-diverging exhaust nozzle. If an electric arc is employed for heating the propellant, the engine is called a **thermal arc-jet rocket engine.**

Fig. 11.5.13   Essential features of an electric rocket engine.

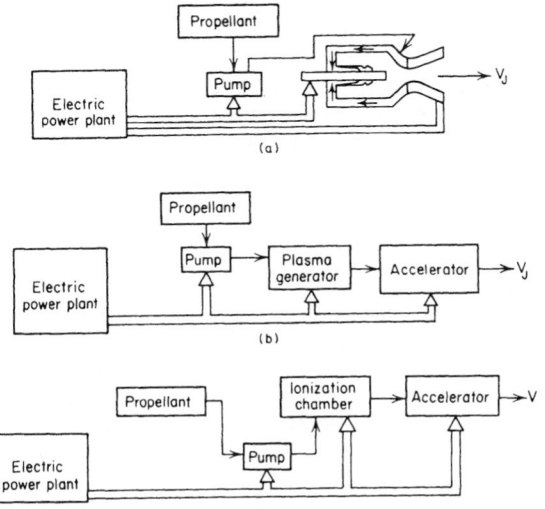

Fig. 11.5.14   Electric rocket engine. (a) Electrothermal rocket engine; (b) electromagnetic rocket engine; (c) electrostatic (ion) engine.

Electromagnetic Rocket Engine   Figure 11.5.14b illustrates an electromagnetic, or plasma, rocket engine. There is a wide variety of such engines, but all of them utilize the same operating principle. A plasma (a neutral ionized conducting gas) is accelerated by means of its interaction with either a stationary or a varying magnetic field. Basically, a plasma engine differs from a conventional electric motor by the substitution of a conducting plasma for a moving armature.

Electrostatic Rocket Engine   Figure 11.5.14c illustrates schematically the electrostatic, or ion, rocket engine, comprising (1) an electric power plant; (2) a propellant supply; (3) ionization apparatus; (4) an ion accelerator; and (5) an electron emitter for **neutralizing** the ion beam ejected from the accelerator. Its operating principle is based on utilizing electrostatic fields for accelerating and ejecting electrically charged particles with extremely large velocities. The overall objective is to transform thermal energy into the kinetic energy associated with an extremely high-velocity stream of electrically neutral particles ejected from one or more thrustors. **Mercury ion engines** are currently being tested. The engine provides thrust by a rapid, controlled discharge of ions created by the ionization of atoms of the mercury fuel supply. Solar cells convert sunlight into electricity, which is then used to bombard the mercury with free electrons. This causes one electron to be stripped from each mercury atom, resulting in ionized mercury atoms. The ions mix with other electrons to form a plasma. A magnetic field set up between two metal screens at the aft end of the engine draws the positively charged ions from the plasma and accelerates them at very high exit speeds out the back of the engine. This creates a high-velocity exit beam producing low-level but extremely long-term thrust.

All electric rocket engines are low-thrust devices. The interest in such engines stems from the fact that chemical-rocket technology, for example, the space shuttle, is now so advanced that it has become feasible to place heavy *payloads,* such as a satellite equipped with an electric rocket engine, into an earth orbit. An electric rocket engine which is virtually inoperable terrestrially can be positioned in space and can then operate effectively because of the absence of aerodynamic drag and strong gravitational fields. Consequently, if a vehicle equipped with an electric rocket engine is placed in an earth orbit, the low-thrust electric rocket engine, under the conditions in space, will accelerate it to a large vehicle velocity if the operating time for the engine is sufficiently long.

In a chemical rocket engine, the energy for propulsion, as well as the mass ejected through the exhaust nozzle, is provided by the propellants, but the energy which can be added to a unit mass of the propellant gas is a fixed quantity, limited by the nature of the chemical bonds of the reacting materials. For that reason, chemical rocket engines are said to be **energy-limited rocket engines**. Although a nuclear heat-transfer rocket engine is also energy-limited, the limitation is imposed by the amount of energy that can be added per unit mass of propellant without exceeding the maximum allowable temperature for the materials employed for the solid-core reactor.

In an electric rocket engine, on the other hand, the energy added to the propellant is furnished by an electric power plant. The power available for heating the propellant is limited by the maximum power output of the electric power plant, accordingly, electric rocket engines are **power-limited**.

NOTATION

a = acoustic speed
a_0 = acoustic speed in free-stream air
a_2 = burning rate constant for solid propellant
A_e = cross-sectional area of exit section of exhaust nozzle
A_p = area of burning surface for solid propellant
A_t = cross-sectional area of throat of exhaust nozzle
c_p = specific heat at constant pressure
c_v = specific heat at constant volume
$c^* = P_c A_t / \dot{m}$ = characteristic velocity for rocket motor
C_d = discharge coefficient for exhaust nozzle
C_F = thrust coefficient

C_{Fg} = gross-thrust coefficient for ramjet engine
C_{Fn} = net thrust coefficient
D or $\mathcal{D}$ = drag
$D_i = \dot{m}_a V_0$ = ram drag for airbreathing or thermal-jet engine
d = diameter
d_p = diameter of propeller
E_{in} = rate at which energy is supplied to propulsion system
E_f = calorific value of fuel
E_p = calorific value of rocket propellants
$f = \dot{m}_f/\dot{m}_a$ = fuel-air ratio
$f' = f/\eta_B$ = fuel-air ratio for ideal combustion chamber
F = force or thrust
$F_i = m_e w - m_i V_0 + (p_e - p_0)A_e$ = thrust due to internal flow
$F_j = m_e V_j$ = jet thrust
$F_p = (p_e - p_0)A_e$ = pressure thrust
$F_g = F_i$ = gross thrust for ramjet engine
$F_g SFC$ = gross-thrust specific-fuel consumption for ramjet engine
g_c = correction factor defined by Newton's second law of motion
h = static specific enthalpy
Δh_n = enthalpy change for exhaust nozzle
H = total (stagnation) specific enthalpy
ΔH_c = lower heating value of fuel
I = specific impulse
$I_a = F/gm_a$ = air specific impulse
J = mechanical equivalent of heat; advance ratio, propeller
$k = c_p/c_v$ = specific heat ratio
L or $\mathscr{L}$ = lift
$\mathfrak{M}$ = mass
m = mass rate of flow
$\dot{m}_a$ = mass rate of air consumption for thermal-jet engine
$\dot{m}_e$ = mass rate of flow of gas leaving propulsion system
$\dot{m}_f$ = mass rate of fuel consumption
$\dot{m}_i$ = mass rate of flow of gas into propulsion system
$\dot{m}_o$ = mass rate of oxidizer consumption for rocket engine
$\dot{m}_p = \dot{m}_o + \dot{m}_f$ = mass rate of propellant consumption for rocket engine
$\overline{m}$ = molecular weight
M_e = momentum of gases leaving propulsion system in unit time
M_i = momentum of gases entering propulsion system in unit time
$M_0 = V_0/a_0$ = Mach number of free-stream air (flight speed)
n = revolutions per unit time
p = absolute static pressure; pitch of propeller blade
p_c = absolute static pressure of gases in combustion chamber
p_e = absolute static pressure in exit section of exhaust nozzle
p_0 = absolute static pressure of free-stream ambient air
P = absolute total (stagnation) pressure
P_c = absolute total pressure at entrance to exhaust nozzle
$P_0 = p_0\left(1 + \dfrac{k-1}{2}M_0^2\right)^{k/(k-1)}$ = absolute total pressure of free-stream air
$\mathscr{P}$ = propulsion power
$\mathscr{P}_L$ = leaving loss
$\mathscr{P}_T$ = thrust power
$q = \rho V^2/2$ = dynamic pressure
$q_0 = \rho_0 V_0^2/2 = k_0 p_0 M_0^2/2$ = dynamic pressure of free-stream air
Q = torque
Q_i = heat supplied to actual combustor
Q_i' = heat supplied to ideal combustor (no losses)
r_0 = linear burning rate for solid propellant
$R = R_u/\overline{m}$ = gas constant
R_u = universal gas constant $\overline{m}R$
t = absolute static temperature
t_p = temperature of solid propellant prior to ignition
T = absolute total (stagnation) temperature
T_c = absolute total temperature of gas entering exhaust nozzle of rocket motor

T_2 = absolute total temperature at entrance to air compressor (see Fig. 11.5.4)
T_3 = absolute total temperature at exit section of air compressor (see Fig. 11.5.4)
T_3' = absolute total temperature at exit section of ideal compressor (see Fig. 11.5.4) operating between same pressure limits as actual compressor
T_4 = absolute total temperature at entrance to turbine (see Fig. 11.5.4)
T_5 = absolute total temperature at exit from turbine (see Fig. 11.5.4)
T_5' = absolute total temperature at exit from ideal turbine
$u = \pi n d_p$ = propeller tip speed
V = velocity
V_F = forward speed of propeller
V_j = effective jet velocity
V_0 = velocity of free-stream air (flight speed)
w = velocity of exit gases relative to walls of exhaust nozzle or velocity of air in ultimate wake of propeller
$\dot{W}$ = weight rate of flow
$\dot{W}_o$ = weight rate of flow of oxidizer
$\dot{W}_p$ = weight rate of flow of propellants

Greek

$\alpha = T_4/t_0 = \alpha_d \alpha_1$ = cycle temperature ratio; angle of attack
$\alpha_d = T_2/t_0$ = diffusion temperature ratio
$\alpha_1 = T_4/T_2$
β = bypass ratio for turbofans; helix angle for propeller blade
$\delta = P/P_{std}$ corrected pressure
η = efficiency
$\eta_B = f'/f$ = efficiency of combustion for thermal-jet engine
$\eta_c = (T_3' - T_2)/(T_3 - T_2)$ = isentropic efficiency of compressor
η_d = isentropic efficiency of diffuser
$\eta_n = \varphi^2$ = isentropic efficiency of exhaust nozzle
$\eta_o = \mathscr{P}_T/E_{in}$ = overall efficiency of propulsion system
$\eta_p = \mathscr{P}_T/(\mathscr{P}_T + \mathscr{P}_L)$ = ideal propulsive efficiency
$\eta_t = (T_4 - T_5)/(T_4 - T_5')$ = isentropic efficiency of turbine
$\eta_{th} = \mathscr{P}_T/E_{in}$ = thermal efficiency of propulsion engine
$\lambda = gI_a/\sqrt{2gJc_p t_0}$ = thrust parameter for turbojet engine; $\frac{1}{2} + \frac{1}{2}\cos\phi$ = divergence coefficient for exhaust nozzle for rocket engines
$v = V_0/w$ = speed ratio
ρ = density
$\overline{\rho}$ = mean density
$\Omega = \sqrt{k\left(\dfrac{2}{k+1}\right)^{(k-1)(2k-1)}}$
ω = angular velocity of propeller shaft
ω' = rate of rotation of slipstream at propeller
ω'' = rate of rotation of slipstream in ultimate slipstream
Φ = fan velocity coefficient = V_F/u
ϕ = semiangle of exhaust-nozzle divergence; effective helix angle
$\varphi = \sqrt{\eta_n}$ = velocity coefficient for exhaust nozzle
ρ = density
σ = solidity of propeller
$\Theta = (P_3/p_0)^{(k-1)/k} = \Theta_d \Theta_c$ = cycle pressure ratio parameter
$\Theta_c = (P_3/P_2)^{(k-1)/k}$ = compressor pressure-ratio parameter
$\Theta_d = (P_2/p_0)^{(k-1)/k}$ = diffuser pressure-ratio parameter
$\Theta_n = (P_7/p_0)^{(k-1)/k} = (P_7/p_9)^{(k-1)/k}$ = nozzle pressure-ratio parameter
$\Theta_t = (P_4/P_5)^{(k-1)/k}$ = turbine pressure-ratio parameter
$\theta = T/T_{std}$ = corrected temperature

Subscripts

(a) numbered
    0 = free stream
    1 = entrance to subsonic diffuser

2 = exit from subsonic diffuser
3 = entrance to combustion chamber
4 = entrance to turbine of turbojet engine
5 = exit from turbine of turbojet engine
6 = tail-pipe entrance
7 = entrance to exhaust nozzle
8 = throat section of exhaust nozzle
9 = exit section of exhaust nozzle

(b) lettered

a = air
B = burner or combustion chamber
b = blade
c = compressor
d = diffuser
e = exit section; effective
F = fan
f = fuel
h = hydraulic
n = nozzle
o = overall or oxidizer
p = propellant
P = propulsive
std = standard
t = turbine or throat, as specified in text

Statement on Units   The dynamic equations employed in the following sections are written for *consistent sets of units,* i.e., for sets in which 1 *unit of force* = 1 *unit of mass* × 1 *unit of acceleration.* Consequently, the gravitational correction factor g_c in Newton's equation $F = (1/g_c)ma$ (see notation) has the numerical value unity and is omitted from the dynamic equations. When the equations are used for calculation purposes, g_c should be included and its appropriate value employed.

THRUST EQUATIONS FOR JET-PROPULSION ENGINES

Refer to Fig. 11.5.15, which illustrates schematically a rotationally symmetrical arbitrary propulsion engine immersed in a uniform flow field. Because of the reactions between the fluid flowing through the engine, called the **internal flow,** and the interior surfaces wetted by the internal flow a resultant **axial force** is produced, that is, a force collinear with the longitudinal axis of the engine. If an axial force acts in the **forward direction,** employing a relative coordinate system, it is called a **thrust;** if it acts in the **backward direction** it is called a **drag.** Similarly, the resultant axial force due to the **external flow,** the flow passing over the external surfaces of the propulsion system, is a thrust or drag depending upon whether it acts in the forward direction or the backward direction.

Application of the momentum equation of fluid mechanics to the generalized propulsion system illustrated in Fig. 11.5.15 gives the following equation for the thrust F_i due to the **internal flow.** Thus, if it is assumed that the external flow is frictionless, there is no change in the rate of momentum for the external flow between S_i and S_e. Hence, the thrust $F = F_i$ is due entirely to the internal flow, and

$$F = F_i = m_e w - \dot{m}_i V_0 + (p_e - p_0)A_e \quad (11.5.1)$$

Fig. 11.5.15   Generalized jet-propulsion system.

In Eq. (11.5.1), $\dot{m}_e w = F_j =$ **jet thrust,** $\dot{m}_i V_0 = D_i =$ **ram drag,** and $(p_e - p_0)A_e = F_p =$ **pressure thrust.** Hence,

$$F_i = F_j - D_i + F_p \quad (11.5.2)$$

It is convenient to introduce a fictitious **effective jet velocity** V_j, which is defined by

$$F = \dot{m}_e V_j - \dot{m}_i V_0 = \dot{m}_e w - \dot{m}_i V_0 + (p_e - p_0)A_e \quad (11.5.3)$$

If the gases are expanded completely, in the exhaust nozzle, then $w = V_j$ and $F_p = 0$. No appreciable error is introduced, in general, if it is assumed that $F_p = 0$. For **thermal jet engines,** $\dot{m}_i = \dot{m}_a$ (see notation), and $\dot{m}_e = \dot{m}_a + \dot{m}_f$. Let $f = \dot{m}_f/\dot{m}_a$, $v = V_0/V_j$; then

$$F_i = \left(\frac{1 + f}{v} - 1\right)\dot{m}_a V_0 + (p_e - p_0)A_e \quad (11.5.4)$$

In the case of uncooled turbojet engines, $\dot{m}_f$ is not significantly different from the fraction of m_a utilized for cooling the bearings and turbine disk. Consequently, no significant error is introduced by assuming that $\dot{m}_e = \dot{m}_a + \dot{m}_f \approx \dot{m}_a = \dot{m}$. Hence, for a simple **turbojet engine,** one may write

$$F_e = \dot{m}_a(V_j - V_0) = \dot{m}V_0(1/v - 1) \quad (11.5.5)$$

Equation (11.5.5) is also the thrust equation for an **ideal propeller;** in that case V_j is the **wake velocity** for the air leaving its slipstream. The thrust equation for a hydraulic jet propulsion system has the same form as Eq. (11.5.5). Let V_0 denote the speed of a boat, and $\dot{m}_w$ the mass rate of flow of the water entering the hydraulic pump and discharged by the exit nozzle with the velocity $w = V_j$ relative to the boat. Hence, for hydraulic jet propulsion

$$F = \dot{m}_s(V_j - V_0) = \dot{m}_w V_0(1/v - 1) \quad (11.5.6)$$

where $v = V_0/V_j$.

In the case of **rocket engines,** since they do not consume atmospheric air, $\dot{m}_a = 0$, and the flow of gas out of the rocket motor, under steady-state conditions, is equal to $\dot{m}_p - \dot{m}_o + \dot{m}_f$ (see notation).

The effective jet velocity V_j is larger than $V_e =$ **exit velocity** if $p_e > p_0$; that is, if the gases are **underexpanded.** The effective jet velocity is a useful criterion because it can be determined accurately from the measured values of F_i and $\dot{m}_p$ obtained from a static firing test of the rocket motor.
The ratio $F_i/\dot{m}_p$ is denoted by I and called either the **specific thrust** or the **specific impulse.** Hence

$$I = F/\dot{m}_p = V_j/g_c \quad (11.5.7)$$

Although the dimensions of specific impulse are force/(mass)(s), it is conventional to state its units as *seconds.*

Equation (11.5.5) for the thrust of a **turbojet engine,** when expressed in terms of the effective jet velocity V_j, becomes

$$F = \dot{m}_a(V_j - V_0) \quad (11.5.8)$$

POWER AND EFFICIENCY RELATIONSHIPS

In a jet-propulsion engine the **propulsion element** is the **exhaust nozzle,** and the rate at which energy is supplied to it is called the **propulsion power,** denoted by $\mathcal{P}$. The rate at which the propulsion system does useful work is termed the **thrust power** $\mathcal{P}_T$ and is given by

$$\mathcal{P}_T = FV_0 \quad (11.5.9)$$

Assume that $p_e = p_0$, and that the only energy loss in a propulsion system is the **leaving loss** $\mathcal{P}_L = \dot{m}(V_j - V_0)^2/2$, that is, the kinetic energy associated with the jet gases discharged from the system; then the propulsive power is given by

$$\mathcal{P} = \mathcal{P}_L + \mathcal{P}_T \quad (11.5.10)$$

The **ideal propulsive efficiency** is defined, in general, by

$$\eta_P = \mathcal{P}_T/(\mathcal{P}_T + \mathcal{P}_L) = \text{thrust power/propulsion power} \quad (11.5.11)$$

For a turbojet engine and hydraulic jet propulsion,

$$\eta_P = 2v/(1 + v) \qquad (11.5.12)$$

For a chemical rocket engine,

$$\eta_P = 2v/(2 + v^2) \qquad (11.5.13)$$

The propulsive efficiency η_P is of more or less academic interest. Of more importance is the overall efficiency η_o, which for airbreathing and rocket engines is defined by

$$\eta_o = \eta_{th}\eta_P \qquad (11.5.14)$$

where

$$\eta_{th} = \mathcal{P}/E_{in} = \text{thermal efficiency of system} \qquad (11.5.15)$$

and E_{in} is the rate at which energy is supplied to the propulsion system.

The overall efficiency of a hydraulic jet-propulsion system is given by

$$\eta_o = \eta_{th}\eta_h\eta_p = \eta_h\eta_{th}[2v/(1 + v)] \qquad (11.5.16)$$

where η_{th} is the thermal efficiency of the power plant which drives the water pump, and η_h is the hydraulic efficiency of the water pump. To achieve a reasonable fuel consumption rate, η_h must have a larger value. For an airbreathing jet engine, $E_{in} = \dot{m}_f(E_f + V_0^2/2)$, and

$$\eta_o = \frac{2v}{1 + v} \frac{\mathcal{P}}{\dot{m}_f(E_f + V_0^2/2)} \qquad (11.5.17)$$

The ratio $\dot{m}_f/F$ is called the **thrust specific-fuel consumption** (TSFC) and is measured in mass of fuel per hour per unit of thrust. Hence,

$$\text{TSFC} = \dot{m}_f/F = f\dot{m}_a/F \qquad (11.5.18)$$

For a rocket engine, $E_{in} = \dot{m}_p(E_p + V_0^2/2)$, so that

$$\eta_o = \frac{2v}{1 + v^2} \frac{\mathcal{P}}{\dot{m}_p(E_p + V_0^2/2)} \qquad (11.5.19)$$

PERFORMANCE CHARACTERISTICS OF AIRBREATHING JET ENGINES

The Ramjet Engine   In ramjet technology the thrust due to the internal flow F_i is called the **gross thrust** and denoted by F_g. Refer to Fig. 11.5.2 and assume $\dot{m}_0 \approx \dot{m}_9 \approx \dot{m}_a$. Then

$$F_g = F_i \approx \dot{m}_a(V_j - V_0) \qquad (11.5.20)$$

In level unaccelerated flight F_g is equal to the external drag of the propelled vehicle and the ramjet body.

It is customary to express the thrust capabilities of the engine in terms of the **gross-thrust coefficient** C_{Fg}. If A_m is the maximum cross-sectional area of the ramjet engine, $q_0 = \rho_0 V_0^2/2 = k_0 p_0 M_0^2/2 =$ the **dynamic pressure** of the free-stream air, then

$$C_{Fg} = \frac{F_g}{q_0 A_m} = \frac{2F_g}{A_m k_0 p_0 M_0^2} \qquad (11.5.21)$$

In terms of the Mach numbers M_0 and M_9,

$$C_{Fg} = \frac{2A_9/A_m}{k_0 M_0^2}\left(\frac{P_9}{P_0}\frac{1 + k_9 M_9^2}{P_9/p_9} - 1\right) - 2\frac{A_0}{A_m} \qquad (11.5.22)$$

In a fixed-geometry engine, M_9 depends upon the total temperature T_7, the total pressure P_7, the fuel-air ratio f, the nozzle area ratio A_e/A_t, and the efficiency of the nozzle η_n. For estimating purposes, $k_0 = 1.4$ and $k_9 = 1.28$, when the engine burns a liquid-hydrocarbon fuel. The manner in which C_{Fg} varies with altitude M_0 and fuel-air ratio f is shown schematically in Fig. 11.5.16.

If $\alpha = T_7/t_0 =$ cycle temperature ratio, and k_B is the mean value of k for the combustion gases, the rate at which heat is supplied to the engine is

$$Q_i = Q_i'/\eta_B = \dot{m}_a c_{pB} t_0(\alpha - T_2)/\eta_B \qquad (11.5.23)$$

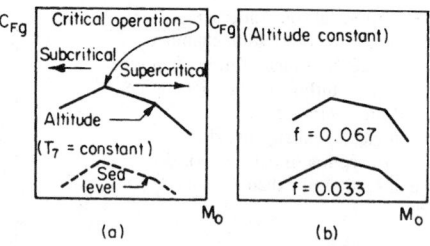

Fig. 11.5.16   Effect of altitude and fuel-air ratio on the gross-thrust coefficient of a fixed-geometry ramjet engine. (*a*) Effect of altitude; (*b*) effect of fuel-air ratio.

It is readily shown that

$$Q_i = \frac{A_0 p_0 V_0}{\eta_B J}\frac{k_B}{k_B - 1}\left[\alpha - \left(1 + \frac{k_0 - 1}{2}M_0^2\right)\right] \qquad (11.5.24)$$

The **gross-thrust specific-fuel consumption** (F_gSFC) is, by definition,

$$F_g\text{SFC} = \dot{m}_f/F_g \qquad (11.5.25)$$

The principal sources of loss are aerothermodynamic in nature and cause a decrease in total pressure between stations 0 and 7. For estimating purposes, assuming $M_0 = 2.0$ and $T_6 = 3,800°R$ (2,111 K), the total pressures across different sections of the engine may be assumed to be approximately those tabulated below.

Part of engine	Total pressure ratio
Supersonic diffuser (0–1)	$P_1/P_0 = 0.92$
Subsonic diffuser (1–2)	$P_2/P_1 = 0.90$
Flameholders (2–7)	$P_7/P_2 = 0.97$
Combustion chamber (2–7)	$P_7/P_2 = 0.92$
Exhaust nozzle (7–9)	$P_9/P_7 = 0.97$

The Simple Turbojet Engine   A good insight into the design performance characteristics of the turbojet engine is obtained conveniently by making the following assumptions: (1) The mass rate of flow of working fluid is identical at all stations in the engine; (2) the thermodynamic properties of the working fluid are those for air; (3) the air is a perfect gas and its specific heats are constants; (4) there are no pressure losses due to friction or heat addition; (5) the exhaust nozzle expands the working fluid completely so that $p_9 = p_0$; and (6) the auxiliary power requirements can be neglected. Refer to Fig. 11.5.4 and let

$$\Theta = \left(\frac{P_3}{p_0}\right)^{(k-1)/k} \quad \Theta_d = \left(\frac{P_2}{p_0}\right)^{(k-1)/k} \quad \Theta_c = \left(\frac{P_3}{P_2}\right)^{(k-1)/k}$$

$$\Theta_t = \left(\frac{P_4}{P_5}\right)^{(k-1)/k} \quad \Theta_n = \left(\frac{P_7}{p_0}\right)^{(k-1)/k} = \left(\frac{P_7}{p_9}\right)^{(k-1)/k}$$

$$\alpha = T_4/t_0 \qquad \alpha_d = T_2/t_0 \qquad \alpha_1 = T_4/T_2$$

In view of the assumptions,

$$\Theta = \Theta_d\Theta_c = \Theta_t\Theta_n \qquad (11.5.26)$$

and

$$\alpha = \alpha_d\alpha_1 \qquad (11.5.27)$$

The diffuser pressure-ratio parameter Θ_d is given by

$$\Theta_d = 1 + \eta_d[(k - 1)/2]M_0^2 \qquad (11.5.28)$$

where $\eta_d =$ isentropic efficiency of diffuser (0.75 to 0.90 for well-designed systems).

The turbine pressure-ratio parameter Θ_t is given by

$$1/\Theta_t = 1 - [(\Theta_c - 1)/\alpha_1\eta_c\eta_t] \qquad (11.5.29)$$

where $\eta_t =$ the isentropic efficiency of the turbine (0.90 to 0.95).

Heat supplied, per unit mass of air, is

$$Q_1 = (c_p/\eta_B)(T_4 - T_3)$$

or

$$Q_i = \frac{c_p t_0}{\eta_B \eta_c} \alpha_d \left(\frac{\alpha}{\alpha_d} \eta_c - \eta_c - \Theta_c + 1 \right) \quad (11.5.30)$$

where η_c is the isentropic efficiency of the air compressor (0.85 to 0.90), and η_B is the efficiency of the burner (0.95 to 0.99).

The enthalpy change for the exhaust nozzle Δh_n is given by

$$\frac{\Delta h_n}{c_p t_0} = \frac{\eta_n}{\eta_c} [\alpha \eta_c - \alpha_d (\Theta_c - 1)]$$

$$\times \left\{ 1 - \frac{\alpha_1 \eta_c \eta_t}{\Theta_c [1 + \eta_d (\alpha_d - 1)][\alpha_1 \eta_c \eta_t - (\Theta_c - 1)]} \right\} \quad (11.5.31)$$

The **specific thrust,** also called the **air specific impulse,** is

$$I_a = F_i / \dot{m}_a g = (\sqrt{2 \Delta h_n} - V_0)/\dot{m}_a \quad (11.5.32)$$

The overall efficiency η_o is given by

$$\eta_o = \eta_{th} \eta_p = \frac{I_a V_0}{J Q_i}$$

$$= 2 \eta_B \eta_c \frac{\lambda M_0 \left(\dfrac{k-1}{2} \right)^{1/2}}{\alpha \eta_c - \left(1 + \dfrac{k-1}{2} M_0^2 \right)(\eta_c + \Theta_c - 1)} \quad (11.5.33)$$

where η_B ranges from 0.95 to 0.99 and λ is the dimensionless turbojet specific thrust parameter.

The TSFC for a turbojet engine is given by

$$\text{TSFC} = \frac{\dot{m}_f}{F} \quad (11.5.34)$$

It can be shown by dimensional analysis, if the effects of Reynolds numbers are neglected, that the variables entering into the performance of a given turbojet engine may be grouped as indicated in Table 11.5.2.

Table 11.5.2

Nondimensional group	Uncorrected form	Corrected form
Flight speed	$V_0/\sqrt{t_0}$	$V_0/\sqrt{\theta}$
Rotational speed	$N/\sqrt{T}$	$N/\sqrt{\theta}$
Air flow rate	$\dot{W}_a \sqrt{T}/D^2 P$	$\dot{W}_a \sqrt{\theta}/\delta$
Thrust	$F/D^2 P$	F/δ
Fuel flow rate	$\dot{W}_f J \Delta H_c / D^2 P \sqrt{T}$	$\dot{W}_f/\delta \sqrt{\theta}$

$\theta = T/T_{std} = T/519 \ (T/288) = $ corrected temperature (exact value for T_{std} is 518.699°R).
$\delta = P/p_{std} = P/14.7 \ (P/1.013 \times 10^5) = $ corrected pressure.

Figure 11.5.17 is a design-point chart presenting λ and η_o as functions of α, P_3/P_2, and Θ_c for the subsonic performance of simple turbojet engines. The curves apply to propulsion at 30,000 ft (9.14 km) altitude for engines having characteristic data indicated in the figure. Curve A illustrates the effect of pressure ratio P_3/P_2 or Θ_c for engines operating with $\alpha = 5.0$. It is seen that increasing the compressor pressure ratio increases η_o but it approaches a maximum value at P_3/P_2 approximately equal to 19. Curve B applies to the design point of engines having a fixed compressor pressure ratio of 10 : 1 but having different values of α, that is, turbine inlet temperature. It shows that increasing α much above $\alpha = 5$ reduces the overall efficiency and consequently TSFC. Curve C presents the design-point characteristics for engines operating with $\alpha = 5$ but having different compressor pressure ratios. The maximum value of λ is obtained at a relatively low pressure ratio: approximately $P_3/P_2 = 5$. Curve D presents the design-point characteristics of engines having a fixed compressor pressure ratio of

$\eta_d = 0.91$ $\gamma = 1.4$
$\eta_c = 0.85$ $z = 30{,}000$ ft (9.14 km)
$\eta_t = 0.96$ $M_0 = 0.82$
$\eta_n = 0.93$

Fig. 11.5.17 Dimensionless thrust parameter λ and overall efficiency η_0 as functions of the compressor pressure ratio (simple turbojet engine).

10 : 1 but different cycle temperature ratios. Increasing α gives significant increases in λ. Hence, large values of turbine inlet temperature T_4 offer the potential for large thrusts per unit of frontal area for such turbojet engines.

Figure 11.5.18 presents the design-point characteristics for engines having $\eta_c = 0.85$, $\eta_t = 0.90$, $\eta_r = 0.93$, and $\eta_B = 0.99$. Different values of diffuser efficiency are presented for flight Mach numbers $M_0 = 0.8$ ($\eta_d = 0.91$), $M_0 = 1.60$ ($\eta_d = 0.90$), and $M_0 = 2.4$ ($\eta_d = 0.88$). Two different classes of engines are considered, those with $P_3/P_2 = 4.0$ and those with $P_3/P_2 = 12.0$. The dimensionless thrust λ is plotted as a function of α with M_0 as a parameter. It is evident that increasing λ, that is, the turbine inlet temperature T_4, offers the advantage of a large increase in thrust per unit area of the engine under all the conditions considered in Fig. 11.5.18.

Figure 11.5.19 presents the effect of flight speed on the design-point performance, at 30,000 ft (9.14 km) altitude, for engines having $\eta_c = 0.85$, $\eta_t = 0.90$, $\eta_B = 0.95$, $T_4 = 2{,}000°R$ (1,111 K) and burning a fuel having a heating value $\Delta H_c = 18{,}700$ Btu/lbm (10,380 kcal/kgm). The magnitude of the compressor pressure ratio P_3/P_2 approaches unity when V_0 is approximately 1,500 mi/h, indicating that at higher speeds the compressor and turbine are superfluous; i.e., the propulsion requirements would be met more adequately by a ramjet engine. Raising T_4 tends to delay the aforementioned condition to a high value of V_0.

It is a characteristic of turbojet engines, since $\eta_o = \eta_{th} \eta_p$, that for a given M_0 increasing η_{th} causes Δh_n to increase, and hence the jet velocity V_j. As a consequence, η_p is decreased. It is this characteristic which causes $\eta_o = f(\alpha)$, curve B of Fig. 11.5.17, to be rather flat for a wide range of values of $\alpha = T_4/t_0$.

The curves of Figs. 11.5.17 to 11.5.19 are design-point performance curves; each point on a curve is the design point for a different turbojet engine. The performance curves for a specific engine are either

Fig. 11.5.18 Dimensionless thrust parameter λ for a simple turbojet engine and its overall efficiency η_0 as functions of the cycle temperature ratio α, with the flight Mach number as a parameter.

Fig. 11.5.19 *TSFC* and specific thrust I_a functions of the compressor pressure ratio, with flight speed as a parameter (simple turbojet engine).

computed from experimental data pertaining to its components operating over a wide range of conditions or obtained from testing the complete turbojet engine. It is customary to present the data in standardized form.

The Turbofan or the Ducted-Fan Engine   Refer to Fig. 11.5.5b. Assume that the hot-gas flow ($\dot{m}_{a1} + \dot{m}_f$) is ejected through the converging nozzle, area A_j, with the effective jet velocity V_j, and that the airflow $\dot{m}_{aF}$ is ejected through the converging annular nozzle, area A_{7F}, with the jet velocity V_{jF}. Let $\beta = \dot{m}_{aF}/\dot{m}_{a1}$ denote the bypass ratio, approximately 5 to 10 for modern transport engines.

The thrust equation for the turbofan engine is

$$F = \dot{m}_{a1}(1 + f)V_j + \beta\dot{m}_{a1}V_{jF} - \dot{m}_{a1}(1 + \beta)V_0 \quad (11.5.35)$$

where $V_{jF} < V_j$. The thrust per unit mass of airflow, or air specific impulse, is

$$I_a = F/\dot{m}_{a1}(1 + \beta) \quad (11.5.36)$$

The TSFC is given by

$$\text{TSFC} = \dot{m}_f/F \quad (11.5.37)$$

One of the most important applications of turbofans is for **high-subsonic transport aircraft** where the fuel consumption in cruise is a major consideration. In general, as the bypass ratio increases, the TSFC decreases but the engine weight and external drag increase. The **optimum bypass ratio** for any particular application is determined by a trade-off between fuel weight, which is reduced by increasing the bypass ratio, and engine weight and external drag, which increase with bypass ratio for a given thrust. In general, the optimum value of β is that which makes the difference $(F - D_e)$ a maximum, where D_e is the external drag of the engine. Up to a certain point, the more sophisticated and efficient the engine, the larger the bypass ratio.

To achieve the maximum potential for the turbofan engine, high values of turbine inlet temperature T_4 are essential; i.e., cooled turbine stator and rotor blades must be developed. There is an upper limit to T_4; above it the TSFC begins to increase. As T_4 is increased, however, the optimum value of β must also be increased to avoid excessive leaving losses in the propulsive jet.

Mechanical Design

Blades   New technologies being developed to achieve higher thrust-to-weight ratios in advanced gas turbine engines also cause higher vibratory blade row responses and stresses that delay engine development and may adversely affect engine reliability. In particular, the design trend toward higher stage loadings and higher specific flow are being attained through increased tip speeds, lower radius ratios, lower aspect ratios, and fewer stages. The resultant axial-flow compressor blade designs utilize thin, low-aspect-ratio blading with corresponding high steady-state stresses. Also, the mechanical damping is considerably reduced in newer rotor designs, particularly those with integral blade-disk configurations (**blisks**) and in those without shrouds. These are similar trends in turbine blading, but they are further complicated by increased inlet and cooling temperatures that have resulted in thin-walled, complex cooling passage designs. As a result, advanced axial-flow blade designs, both compressors and turbines, feature low-aspect-ratio blading which affect the **structural integrity** of the blading. These problems can be classified into two general categories: (1) **flutter** and (2) **forced response**. A fundamental parameter common to both categories is the **reduced frequency**, $k = \omega b/V$, where b is the blade semichord, ω the frequency of vibration, and V the relative free-stream velocity. For engine applications, the frequency corresponds generally to one of the lower-blade or coupled-blade disk **natural frequencies**.

Under certain conditions, a blade row operating in a completely uniform flow can enter into a self-excited vibrational **instability** termed **flutter**. The vibration is sustained by the extraction of energy from the uniform flow during each vibratory cycle. The outstanding feature of flutter is that **high stresses** exist in the blading, leading to short-term, high-cycle, **fatigue failures**. Its importance is evidenced by the fact that if a portion of a single blade fails due to flutter, the result may be instantaneous and total loss of engine power.

Figure 11.5.20 is a typical compressor map showing schematically the more common types of flutter. At subsonic Mach numbers, up to 0.8 to 0.9, either positive or negative **stall flutter** in the **bending** or **torsion** mode may occur. It is caused by operating beyond some critical airfoil angle of attack. A **critical reduced frequency** value of 0.8 has been cited for stall flutter; i.e., if $k < 0.8$ stall flutter is possible. **Choking flutter** usually occurs at negative incidence angles at a part-speed condition with the blade operating either subsonically or transonically. The physical mechanism of choke flutter is not fully understood; both high negative

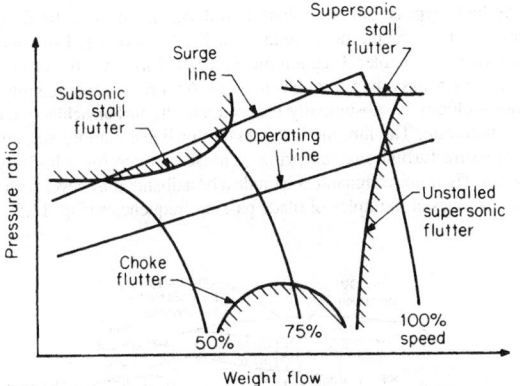

Fig. 11.5.20 Common types of compressor flutter.

incidence angles and choked flow are viable candidates. **Supersonic unstalled flutter** is largely associated with thin **fan blades** operating at supersonic relative flow velocities at the tip section. It is caused by a lagging of the unsteady aerodynamic forces relative to the blade motion and imposes a limit on the **high-speed operation** of the engine. The stress encountered during this type of flutter can be catastrophically large, decreasing rapidly as a constant speed line is traversed toward higher pressure ratios. High-speed operation near the surge line can lead to **supersonic stall flutter.** Because this flutter is associated with **high pressure ratios,** there can be strong shocks present within the blade passages, and in some circumstances, ahead of the blade row.

Destructive aerodynamic **forced responses** have been noted in all engine blading. These failure-level vibratory responses occur when a periodic forcing function, with frequency equal to a natural blade **resonant frequency,** acts on a blade row. Responses of sufficient magnitude to fail blades have been generated by a variety of sources including upstream and downstream blade rows, flow distortion, bleeds, and mechanical sources.

The importance of forced response lies in the **multiple sources** of excitation for any rotating system. The rotor speeds at which forced responses occur are predicted with **"Campbell"** or **speed-frequency** diagrams. These display the **natural frequency** of each blade mode versus rotor speed and, at the same time, the forcing function frequency (or **engine order** E lines) versus rotor speed, as indicated schematically in Fig. 11.5.21. These E lines represent the loci of available **excitation energy** at any rotational speed for 1, 2, 3, etc. excitations per revolution.

Wherever these curves intersect, a potential source of destructive forced response exists. Not all intersections can be avoided. Design practice is to eliminate the lower-order excitations from the operating range whenever possible. This is because the sources of most forced response energy usually result from these lower-order resonances, particularly the 2E line.

High-cycle fatigue (HCF) of turbomachine blading resulting from flow-induced vibrations is a significant problem throughout the gas turbine industry. To address this problem, various approaches have been developed. In these, the response of a tuned airfoil row, i.e., a rotor with all airfoils having the same structural properties and thus identical natural frequencies, is analyzed. In fact, there are small airfoil-to-airfoil structural property variations that result, for example, from the manufacturing process or as a consequence of in-service wear. These are collectively referred to as **mistuning** and are known to lead to significant increases in airfoil resonant response amplitudes as compared to that of the tuned airfoil row, with mistuning thus often cited as a source of HCF in gas turbine engines. Hence, the key metric that characterizes the resonant response of mistuned bladed disks is the **amplification factor,** defined as the ratio of a blade's response amplitude on a mistuned bladed disk to its response amplitude on a tuned bladed disk.

Turbomachinery rotors typically have been bladed disks, with individual airfoils inserted into a slotted disk and retained by means of a dovetail or fir-tree attachment. However, advances in manufacturing techniques have resulted in bladed-disk assemblies with increased uniformity; i.e., the blade-to-blade natural frequency variation, termed mistuning, is small. In addition, new design and manufacturing techniques have enabled the development and implementation of **integrally bladed rotors** (IBRs) wherein the blades and disk are machined from a single piece of material. IBRs have even less blade-to-blade mistuning than do bladed disks. Unfortunately, smaller mistuning does not translate into smaller amplification factors. On the contrary, as mistuning decreases, the maximum amplification factor on a disk tends to increase until some threshold is reached, at which point it decreases rapidly toward 1. Figure 11.5.22 shows this trend; it is based on studies wherein many hundreds of bladed disks are simulated for each percent mistuning, with the maximum resulting amplification factor taken. The decrease in mistuning due to new manufacturing methods is on the negative slope portion of Fig. 11.5.22, thus rendering the new manufacturing methods a strong impetus for improving models which predict mistuned bladed disk response. In contrast to the maximum amplification factor of Fig. 11.5.22, there is a **theoretical maximum amplification factor** that depends only on the number of blades N; specifically, amplification factor $= 0.5(1 + \sqrt{N})$. Taking these theoretical maxima into account during design is an important first step in building blade rows resistant to HCF failure.

Environmental Issues

The increasing worldwide concern for the environment has led to two technical issues becoming significant: (1) airport noise and (2) the Earth's ozone layer.

Fig. 11.5.21 Schematic speed-frequency diagram.

Fig. 11.5.22 Amplification factor versus mistuning level.

Aeroacoustics is an increasingly important issue in the design of advanced gas turbine engines for commercial aircraft. This is because engine certification requires meeting prevailing **International Civil Aviation Organization (ICAO) Chapter 3 noise regulations** established in 1978. Aircraft are certified to a standard that is actually a combination of three measures: the noise generated as the plane approaches the runway radiating to the community under its flight path; the noise it generates during takeoff radiating to the areas surrounding the runway; and the noise under the flight path further out in the community during takeoff (Fig. 11.5.23).

Fig. 11.5.23   Aircraft certification noise measurement locations.

The ICAO Chapter 3 regulations specify approach, sideline, and takeoff perceived noise levels (EPNL) as a function of the aircraft takeoff gross weight (Fig. 11.5.24). The total or cumulative margin for a given aircraft defined as the sum of the margin at all three conditions. Note that all current-technology aircraft meet ICAO Chapter 3 regulations.

Air travel has increased dramatically since 1978. To meet the demands of community groups for stricter aircraft noise requirements, ICAO has established a new set of limits which are to become effective in 2006—the **Chapter 4 rule**. These new limits require aircraft to be 10 dB (cumulative) quieter than current Chapter 3 levels. The use of trades to meet the new rules was also eliminated. With Chapter 3, an aircraft could exceed the limits on one measure of noise, takeoff, as long as it was lower than the limits at the other conditions by an offsetting amount. With Chapter 4, an aircraft must operate below the Chapter 3 curve at all three conditions: approach, sideline, and takeoff.

Fig. 11.5.24   Chapter 3 regulations specify EPNL at approach, sideline, and takeoff.

The high-bypass-ratio turbofans introduced in the late 1960s had a major effect on engine noise reduction. ICAO recognized this when it established the Chapter 3 regulations in 1978. Turbofans direct more air through the bypass than through the core. As a result, the maximum jet exhaust velocity is substantially reduced and the **turbomachinery noise** is more important. The fan, the low-pressure or boost compressor, and the low-pressure turbine are the primary noise sources for a high-bypass turbofan. Their noise signatures include a broadband noise level with large spikes or tones at multiples of blade passing frequencies (Fig. 11.5.25).

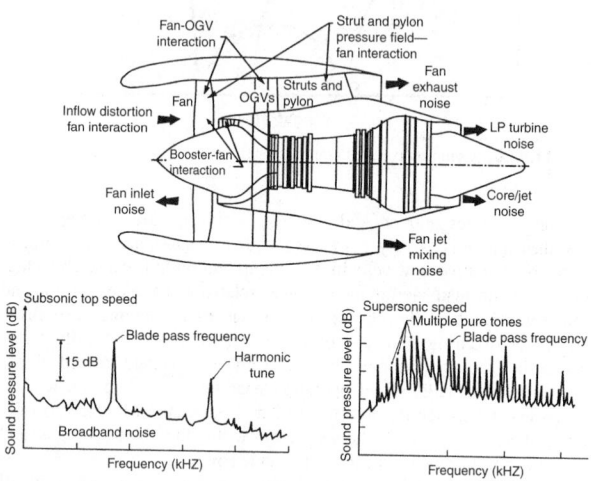

Fig. 11.5.25   High-bypass-ratio turbofan noise sources and noise spectrum.

Current noise control and reduction methods for high-bypass turbofans are a combination of turbomachinery noise source control and suppression. Source control is accomplished by increasing the axial spacing between adjacent blade rows and by selecting blade and vane number combinations to produce *cutoff*, whereby the highest-energy acoustic modes decay exponentially with distance along the ducting. Source suppression is achieved with acoustic liners in the fan inlet and exhaust ducts and the core exhaust duct. As a result, future advanced turbofans will have much lower levels of turbomachinery noise. Thus, the dominant engine sources in these future engines are the fan at takeoff and approach and the turbulent mixing in the jet exhaust stream at takeoff.

The **emissions** generated by the combustion of aviation fuel include carbon dioxide (CO_2), carbon monoxide (CO), oxides of nitrogen (NO_x), and particulates such as soot and oxides of sulfur (SO_x). The effect that each has on the atmosphere depends on the altitude at which it is emitted and its concentration. Most commercial aircraft cruise in the **troposphere**, which ranges from an altitude of 11 mi (58,000 ft) at the equator to 6 mi (33,000 ft) at the poles. In this region, the temperature decreases with increasing altitude, resulting in large air motions and mixing. Increased NO_x concentration results in increased ozone concentration through complex photochemical reactions with CO and hydroxyls (HO_x). Ozone also acts as a global warming gas in the troposphere. Methane is another global warming gas, although NO_x reduces the concentration of methane.

It has been projected that, by 2050, aircraft emissions will increase by a factor of 4 because of an increase in air traffic and that the total contribution to **global warming** by aircraft will be a factor of 2 greater than from CO_2 alone. In addition, aircraft exhaust plumes contribute to global warming. Contrails are formed by water in the exhaust freezing in cold atmospheric conditions, with their formation aided by the particulates present in exhaust plumes. These contrails may turn into cirrus clouds, which lead to a net increase in warming.

In summary, pollutants generated by aircraft engines contribute to global warming because of the altitude at which they operate and the projected growth of the aviation industry. Hence, there is a distinct need to develop ultralow-emission combustor designs for aircraft engines. The two primary approaches to ultralow-emission combustors are rich and lean front-end designs. Each has demonstrated the potential to achieve low emissions.

The International Civil Aviation Organization (ICAO) standards limit aircraft engine emissions in the vicinity of an airport, up to 3,000 ft altitude. A standardized **landing-takeoff cycle (LTO)** is specified, which includes time at each of four engine power settings. Typical levels of engine emissions as a function of engine thrust are presented in Fig. 11.5.26. Emissions of carbon monoxide (CO) and unburned hydrocarbons (UHC) are dominant at low power conditions such as idle, and rapidly decrease to negligible levels as engine power increases. Most modern combustor designs result in very low levels of CO and UHC at all cruise power settings through efficient combustion that minimizes fuel burn. Nitrogen oxides (NO_x) and soot particle emissions or smoke both exhibit the opposite trend, increasing with engine power.

Fig. 11.5.26   Typical aircraft combustor emission characteristics.

Aircraft spend much more time at takeoff, climb, and cruise settings than at idle. Thus, NO_x is the dominant pollutant by mass in a flight. In fact, the NO_x **produced** at cruise is significantly higher than that produced during the LTO cycle. NO_x emissions are a result of the high temperatures in the combustor, with the rate of NO_x production highly nonlinear with temperature. This is shown in Fig. 11.5.27, with the NO_x production rate presented as a function of the combustion reaction **equivalence ratio,** which is the ratio of fuel to air in the actual mixture divided by the fuel-to-air ratio for stoichiometric conditions. The peak is in the region of stoichiometric combustion, i.e., an equivalence ratio of 1.0.

Rich-front-end combustors are found in most current engines. Fuel is introduced at the front and mixed with some of the total combustor airflow to create a stable rich reaction zone (Fig. 11.5.28), with the remainder of the airflow introduced in a mixing zone. The combination of rich oxygen-depleted products from the front end quicky react with the mixing air and are then diluted to exit conditions. This mixing zone

Fig. 11.5.27 NO_x production rate versus equivalence ratio.

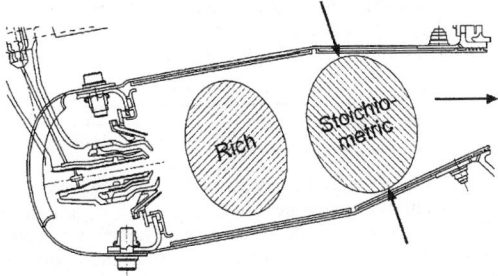

Fig. 11.5.28   Schematic of rich-front-end combustor.

can be highly nonuniform, resulting in local stoichiometric combustion with high NO_x production rates.

The **lean-front-end combustor** focuses on mixing the air and fuel in the front end of the combustor exactly to create a combustion zone that is, on average, below the peak stoichiometric equivalence ratio of 1.0. The lower flame temperature at these lean conditions significantly reduces the NO_x production rate. Lean-front-end combustors require multiple fuel stages to achieve the operability requirements for aircraft application. Typical modern engine designs require the combustor to operate with a factor of 3 higher exit fuel-air ratio at maximum power than at minimum power. A lean combustion system must be optimized to a much narrower range of fuel-air ratios to avoid high CO production, if too lean, and high NO_x production, if too rich. This issue has been addressed by combustors with many small air and fuel injectors and sequentially fueling a subset of these as engine power increases.

CRITERIA OF ROCKET-MOTOR PERFORMANCE

In most applications of rocket motors the objective is to produce a large thrust for a limited period. The criteria of rocket-motor performance are, therefore, related to thrust-producing capabilities and operating duration. If $\dot{W}_p$ denotes the rate at which the rocket motor consumes propellants, P_c the total pressure at the entrance to the exhaust nozzle, and C_w the **weight flow coefficient,** then

$$\dot{W}_p = C_w P_c A_t \qquad (11.5.38)$$

Ordinarily, the values of C_w as a function of P_c and **mixture ratio** $\dot{W}_o/\dot{W}_f$ are determined experimentally. If no experimental data are available, the values of C_w can be calculated with good accuracy from the calculated thermodynamic properties of the combustion gases. If T_c is the total temperature, and P_c the total pressure of the combustion gases at the entrance section of the nozzle, then, for steady operating conditions, $\dot{W}_p = \dot{W}_g$,

$$\dot{W}_g = C_d A_t P_c \Omega \sqrt{\frac{g}{RT_c}} \qquad (11.5.39)$$

where

$$\Omega = \left(\frac{2}{k+1}\right)^{(k-1)/[2(k-1)]} \sqrt{k} \qquad (11.5.40)$$

An equation similar to Eq. (11.5.38) can be written for the thrust developed by the rocket motor. Thus

$$F = C_F P_c A_t \qquad (11.5.41)$$

The values of the **thrust coefficient** C_F for specific propellant combinations are determined experimentally as functions of P_c and the ratio $\dot{W}_o/\dot{W}_f$. When no experimental values are available, C_F can be calculated from the following equation. Therefore

$$C_F = C_d \lambda \phi \Omega \sqrt{\frac{2k}{k-1} Z_t} + \left(\frac{p_e}{P_c} - \frac{p_0}{P_c}\right)\frac{A_e}{A_t} \qquad (11.5.42)$$

where

$$Z_t = 1 - \left(\frac{p_e}{P_c}\right)^{(k-1)/k}$$

Equation (11.5.42) shows that C_F is independent of the combustion temperature T_c and the molecular weight of the combustion gases. The nozzle area ratio A_e/A_t is given by the formula

$$\frac{A_e}{A_t} = \frac{(P_c/p_e)^{1/k}}{\left(\dfrac{k+1}{2}\right)^{1/(k-1)} \sqrt{[(k+1)/(k-1)]Z_t}} \quad (11.5.43)$$

Figure 11.5.29 presents C_F calculated by means of the preceding equations, as a function of the area ratio A_e/A_t, with P_c/p_0 as a parameter, for gases having $k = 1.28$; the calculations assumed $\lambda = 0.983$ and $C_d = \varphi = 1.0$). The curves do not take into account the separation phenomena which occur when A_e/A_t is larger than that required to expand the gases to p_0.

Fig. 11.5.29 Calculated thrust coefficient versus expansion ratio for different pressure ratios (based on $k = 1.28$, $\lambda = 0.9830$, $C_d = \varphi = 1.0$).

The specific impulse I is defined by Eq. (11.5.8) and can be shown to be given by

$$I = F/\dot{m}_p = C_F/C_w = V_j/g_c \quad (11.5.44)$$

A criterion which is employed for stating the performance of rocket motor is the characteristic velocity c^*. By definition

$$c^* = V_j/C_F = P_cA_t/\dot{m} = I/C_F \quad (11.5.45)$$

The values of c^* for a given propellant combination are determined experimentally. It should be noted that the value of c^* is independent of the thrust. Basically, c^* measures the effectiveness with which the thermochemical energy of the propellants is converted into effective jet kinetic energy. When no experimental data are available the value of c^* can be estimated quite closely from

$$c^* = \sqrt{RT_c}/\Omega \quad (11.5.46)$$

Table 11.5.1 presents values of specific impulse for some possible rocket-propellant combinations.

Solid-propellant rocket motors may be segregated into (1) **double-base powders** and (2) **composite, or heterogeneous, propellants**. Double-base powders are gelatinized colloidal mixtures of nitroglycerin and cellulose to which certain stabilizers have been added. Heterogeneous, or composite, propellants are physical mixtures of a solid oxidizer in powder form and some form of solid fuel, such as a plastic or rubberlike material.

Solid-propellant rocket motor technology has made tremendous strides. It is possible to produce solid propellants with a wide range of linear burning rates and with good physical properties. It has been demonstrated experimentally that large thrust rocket motors with chamber diameters of approximately 22 ft (6.7 m) are feasible. As a result, the solid-propellant rocket motor has displaced the liquid rocket engine from a large number of applications, which heretofore were considered

the province of the liquid rocket engine. In military applications, there is a strong tendency to supplant liquid rocket engines with solid rocket engines. Only in such areas as the large thrust launching rockets for boosting manned vehicles into space have the liquid rocket engines reigned supreme; the space shuttle is equipped with liquid-bipropellant rocket engines, but it is boosted with auxiliary solid-propellant rocket motors.

The following considerations are what make solid-propellant rocket motors attractive: (1) They are simpler in construction than liquid rocket engines; (2) they are easier to handle in the field; (3) they have instant readiness for use; (4) they have very good storage properties; (5) they have a larger average density $\bar{\rho}_p$ than most liquid-propellant combinations; (6) they have considerably fewer parts; and (7) their reliability in practice has been very good. Furthermore, once the performance characteristics and physical properties of a solid propellant have been established, it is a relatively straightforward engineering problem to design and develop solid-propellant rocket motors of widely different thrust levels and burning times t_R.

Rocket motors which burn double-base propellants find the greatest use in weaponry, e.g., artillery rockets, antitank weapons, etc.

Rocket motors burning heterogeneous, also called **composite**, solid propellants are used for propelling all kinds of military missiles and for sounding rockets. Modern composite propellants have three basic types of ingredients: (1) a fuel which is an organic polymer, called the **binder**; (2) a finely powdered oxidizer, ordinarily ammonium perchlorate (NH_4NO_3); and (3) various additives, for catalyzing the combustion process, for increasing the propellant density, for increasing the specific impulse (e.g., powdered aluminum), for improving the physical properties of the propellant, and the like. After the ingredients have been thoroughly mixed, the resulting viscous fluid is poured, usually under vacuum to prevent the formation of voids, directly into the chamber of the rocket motor which contains a suitable mandril for producing the desired configuration for the propellant grain. The propellant is cured by polymerization to form an elastomeric material. Some of the more common binders are butadiene copolymers and polyurethane.

Modern composite solid propellants have densities ranging from 1.60 to 1.70, depending upon the formulation, and specific impulse values of approximately 245 s, based on a combustion pressure of 1000 psia and expansion to 14.7 psia.

In the case of solid-propellant rockets the rate of propellant consumption $\dot{m}_p$ is related to the **linear burning rate** for the propellant r_0, that is, the rate at which the burning surface of the propellant recedes normal to itself as it burns. For practical purposes it can be assumed that the linear burning rate r_0 is given by

$$r_0 = a_2 p_c^n \quad (11.5.47)$$

where a_2 and n are constants which are determined by experiment.

If ρ_p denotes the density of the solid propellant and A_p the area of the burning surface, then the propellant consumption rate $\dot{m}_p$ is given by

$$\dot{m}_p = A_p\rho_p r_0 = a_2 A_p\rho_p p_c^n \quad (11.5.48)$$

Figure 11.5.30 presents the linear burning rate as a function of the combustion pressure for several propellants.

The linear burning rate r_0 is influenced by the temperature of the solid propellant t_p prior to its ignition. Low values of t_p reduce the burning rate, and vice versa. Consequently, the temperature of the propellant must be given in presenting data on linear burning rates. Figure 11.5.31 presents r_0 as a function of p_c for $t_p = -40, 60$, and $140°F$ for a composite propellant manufactured by the Aerojet-General Corp., Azusa, Calif.

AIRCRAFT PROPELLERS

The function of the aircraft propeller (or airscrew) is to convert the torque delivered to it by an engine into the thrust for propelling an airplane. If the airplane is in steady, level flight, the propeller thrust and

Fig. 11.5.30 (a) Linear burning rate as a function of combustion pressure for double-base propellants; (b) burning-rate characteristics of several heterogeneous propellants at 60°F (15.5°C). (Aerojet-General Corp.)

Fig. 11.5.31 Effect of propellant temperature upon the linear burning rate of a heterogeneous propellant. (Aerojet-General Corp.)

the airplane drag equal each other. If the airplane is climbing, the thrust must overcome the drag plus the weight component of the airplane.

The propeller may be the sole thrust-producing element, but it can also be employed in conjunction with another thrust-producing device, such as the exhaust jet of a **turboprop** or **gas-turbine engine.** It may also be employed as a **windmilling** device for producing power.

The conventional propeller consists of two or more equally spaced radial blades, which are rotated at a substantially uniform angular velocity. At any arbitrary radius, the section of a blade has the shape of an airfoil, but as the hub is approached, the sections become more nearly circular because of considerations of strength rather than aerodynamic performance. The portions of the blades in the vicinity of the hub

contribute, at best, only a small portion of the thrust developed by the propeller blades.

Each section of the propeller blade experiences the aerodynamic reactions of an airfoil of like shape moving through the air in a similar manner. Furthermore, at each radius, the section of one blade forms with the corresponding sections on the other blades a series of similar airfoils, which follow one another as the propeller rotates.

The torque of the rotating propeller imparts a rotational motion to the air flowing through it and, furthermore, causes the pressure immediately behind the propeller to increase while that in front of it is reduced; i.e., the air is sucked toward the front of the propeller and pushed away behind it. A slipstream is created as illustrated in Fig. 11.5.1. The ratio of the cross-sectional area of the slipstream at any station to that of the actuator disk is termed the **slipstream contraction.**

The thrust produced by the propeller follows the relationships developed from the momentum theorem of fluid mechanics for jet engines as presented in Fig. 11.5.15 and Eq. (11.5.5). The thrust equation for the airbreathing engine (simplified) and the propeller is

$$F = \dot{m}(w - V_0) = \rho A V_0 (w - V_0) \qquad (11.5.49)$$

In the case of a propeller the mass flow m is large and the velocity rise across the propeller is small, as compared with a jet engine of the same thrust. However, the thrust of propellers falls off rapidly as air compressibility effects on the propeller become large (flight speeds of 300 to 500 mi/h).

Propeller efficiency is defined as the useful work divided by the energy supplied to the propeller:

$$\eta_p = \frac{P_T}{P} = \frac{FV_0}{P} = \frac{m(v - V_0)V_0}{m/2(w^2 - V_0^2)} = \frac{2v}{1 + v} \qquad (11.5.50)$$

where $v = V_0/w$.

Since w is only slightly larger than V_0, the ideal efficiency is of the order of 0.95, and actual efficiencies are in the range of 0.5 to 0.9.

Propeller Theory

Axial-Momentum Theory The axial-momentum theory of the propeller is basically a special case of the generalized jet-propulsion system presented in Fig. 11.5.15. It is only a first approximation of the action of the propeller and cannot be employed for design purposes. It neglects such factors as the drag of the blades, energy losses due to slipstream rotation, blade interference, and air compressibility effects. Because of those losses, an actual propeller requires power to rotate at zero thrust, and at zero thrust its propulsive efficiency is, of course, zero.

Blade-Element Theory This theory, called **strip theory,** takes into account the profile losses of the blade sections. It is the theory most commonly employed in designing a propeller blade and for assessing the off-design performance characteristics of the propeller.

Each section of the propeller is considered to be a rotating airfoil. It is assumed that the radial flow of air may be neglected, so that the flow over a blade section is two-dimensional. Figure 11.5.32 illustrates diagrammatically the velocity vectors pertinent to a blade section of length dr, located at an arbitrary radius r. The projection of the axis of rotation is OO', the plane of rotation is OC, the blade angle is β, the angle of attack for the air flowing toward the blade (in the relative coordinate system) with the velocity V_0 is α, the tangential velocity of the blade

Fig. 11.5.32 Vector diagram for a blade element of a propeller.

Fig. 11.5.33 Aerodynamic forces acting on a blade element.

element is $u = 2\pi nr$, and the **pitch** or **advance angle** is ϕ. From Fig. 11.5.32

$$\alpha = \beta - \phi \qquad (11.5.51)$$

and

$$\phi = \tan^{-1}(V_0/\pi nd) \qquad (11.5.52)$$

Figure 11.5.33 illustrates the forces acting on the blade element. If b denotes the width of the blade under consideration and C_L and C_D are the lift and drag coefficients of the blade (airfoil) section, then the thrust force dF acting on the blade element is given by

$$dF = \tfrac{1}{2}\rho V_0^2 b \, dr C_L \frac{\cos(\phi + \gamma)}{\cos \gamma \sin^2 \phi} \qquad (11.5.53)$$

The corresponding torque force dQ is given by

$$dQ = \tfrac{1}{2}\rho V_0^2 b \, dr C_L \frac{\sin(\phi + \gamma)}{\cos \gamma \sin^2 \phi} \qquad (11.5.54)$$

where

$$\tan \gamma = d\mathcal{D}/d\mathcal{L} = C_D/C_L = \mathcal{D}/\mathcal{L} \qquad (11.5.55)$$

where $\mathcal{D}$ and $\mathcal{L}$ denote drag and lift, respectively.

The propulsion efficiency of the blade element, termed the **blading efficiency,** is defined by

$$\eta_b = \frac{V_0 \, dF}{u \, dQ} = \frac{\tan \phi}{\tan(\phi + \gamma)} = \frac{\mathcal{L}/\mathcal{D} - \tan \phi}{\mathcal{L}/\mathcal{D} + \cot \phi} \qquad (11.5.56)$$

The value of ϕ which makes η_b a maximum is termed the **optimum advance angle** ϕ_{opt}. Thus

$$\phi_{\text{opt}} = \frac{\pi}{4} - \frac{\gamma}{2} = 45° - \frac{57.3}{2(\mathcal{L}/\mathcal{D})} \qquad (11.5.57)$$

The maximum blade efficiency is accordingly

$$(\eta_b)_{\max} = \frac{2\gamma - 1}{2\gamma + 1} = \frac{2(\mathcal{L}/\mathcal{D}) - 1}{2(\mathcal{L}/\mathcal{D}) + 1} \qquad (11.5.58)$$

A force diagram similar to Fig. 11.5.33 can be constructed for each element of the propeller blade, taking into account the variation in ϕ with the radius. The resultant forces acting on the propeller are obtained by summing (integrating) those forces acting on each blade element.

The integrated lift coefficient is

$$C_{Li} = \int_{r_i/R}^{r/R = 1.0} C_{Lr} \left(\frac{r}{R}\right)^3 d\left(\frac{r}{R}\right) \qquad (11.5.59)$$

where C_{Lr} = propeller-blade-section lift coefficient and r/R = fraction of propeller-tip radius.

The integrated lift coefficient is in the range 0.15 to 0.70. There is a dilemma in the design and selection of blade sections since high static thrust requires high C_{Li} which gives relatively low η_p at cruise. Cruise conditions require low C_{Li} for good η_p. Variable camber propellers permit optimization of C_{Li}, but at the expense of weight and complexity.

Vortex Theory The simple two-dimensional blade-element theory can be further improved by taking account of the actual velocity distribution in the slipstream and thus determining the actual loading on the blading by the so-called *vortex theory*. The blades are represented by bound vortex filaments sweeping out vortex sheets to infinity. The vortex sheets are distorted both by the slipstream contraction and by the profile drag of the blades. If such distortions are neglected, the induced velocity in the wake can be calculated by assuming the vortex sheets to be tubular. Thus, in Fig. 11.5.32, the induced axial velocity becomes the arithmetic mean of the axial velocities at far distances upstream and downstream of the rotor disk; and the induced rotational velocity at the disk becomes $\omega - 0.5\omega'$, where ω' is the rotation of the slipstream in the plane of the propeller. Hence, the angle of attack α for a rotor element is given by

$$\alpha = \beta - \tan^{-1}[V_F/r(\omega - 0.5\,\omega')] \qquad (11.5.60)$$

where V_F is the induced axial velocity in the plane of the propeller.

The addition of the vortex effects improves the representation of the propeller by the blade sections. There are, however, three-dimensional flow effects which are not included, and the following coefficients have been suggested to include this effect (which reduces the lift coefficient).

Lift Curve Slope Correction Factors for Three-Dimensional Airfoils

Radial station (ratio r/R)	Correction factor
(0.2)(0.3)(0.45)(0.6)(0.7)	0.85
0.8	0.80
0.9	0.70
0.95	0.65

SOURCE: Departments of the Air Force, Navy, and Commerce, "Propeller Handbook."

Performance Characteristics

The pitch and the angle ϕ (Fig. 11.5.32) have different values at different radii along a propeller blade. It is customary, therefore, to refer to all parameters determining the overall characteristics of a propeller to their values at either $0.7r$ or $0.75r$.

For a given blade angle β, the angle of attack α at any blade section is a function of the **velocity coefficient** V_0/u, where $u = \pi nd_p$; it is customary, however, to replace V_0/u by its equivalent, the **advance ratio** $J = V_0/nd_p$.

For given values of β and n, the angle of attack α attains its maximum value when $V_0 = 0$, with $\phi = 0$. In other words, the propeller thrust F is maximum at takeoff. As V_0 increases, α decreases and so does the thrust. Finally, a forward speed is reached for which the thrust is zero.

Increasing $J = V_0/nd_p$, for a constant blade angle, causes the propeller to operate successively as a fan, propeller, brake, and windmill. Most of the operation, however, is conducted in the propeller state. At takeoff, the propeller is in the fan state. Windmilling must be avoided because it may overspeed the engine and damage it.

The lift coefficient C_L is a linear function of α up to the stalling angle, and C_D is a quadratic function of α. Furthermore, the lift $\mathcal{L}$ is zero at a negative value for α. Figure 11.5.34 presents the $\mathcal{L}/\mathcal{D}$ ratios of typical propeller sections. Figure 11.5.35 presents the blade (section) efficiency as a function of the pitch angle ϕ for different values of $\mathcal{L}/\mathcal{D}$. Propellers normally operate at $\mathcal{L}/\mathcal{D}$ ratios of 20 or more. Supersonic propellers operate with $\mathcal{L}/\mathcal{D}$ ratios of approximately 10, and from Fig. 11.5.34 it is apparent that they must operate close to the optimum pitch angle ϕ to obtain usable blade efficiencies.

Propeller Coefficients It can be shown, neglecting the compressibility of the air, that

$$f(V_0, n, d_p, \rho, F) = 0 \qquad (11.5.61)$$

Fig. 11.5.34   Maximum lift-drag ratios for representative sections.

Fig. 11.5.35   Blade-element efficiency versus pitch angle.

By dimensional analysis, the following coefficients are obtained for expressing the performances of propellers having the same geometry:

$$F = \rho n^2 d_p^4 C_F \qquad Q = \rho n^2 d_p^5 C_Q \qquad \mathcal{P} = \rho n^3 d_p^5 C_P \qquad (11.5.62)$$

C_F, C_Q, and C_P are termed the **thrust, torque,** and **power coefficients,** respectively. They are independent of the size of the propeller. Consequently, tests of small-scale models can be employed for obtaining the values of F, Q, and $\mathcal{P}$ for geometrically similar full-scale propellers. Hence, the ideal propulsive efficiency of a propeller is given by

$$\eta_p = \frac{FV_0}{\mathcal{P}} = \frac{C_F}{C_P}\frac{V_0}{d_p n} = J\frac{C_F}{C_P} \qquad (11.5.63)$$

Other useful coefficients are the **speed-power coefficient** C_S and the **torque-speed coefficient** C_{QS}. Thus,

$$C_S = V_0 5\sqrt{\rho/\mathcal{P}n^2} \qquad C_{QS} = V_0\sqrt{\rho d_p^3/Q} \qquad (11.5.64)$$

A commonly employed factor related to the geometry of a propeller is the **solidity** σ. By definition,

$$\sigma = c/p \qquad (11.5.65)$$

The solidity varies from radius to radius; at a given radius, σ is proportional to the power-absorption capacity of the annulus of blade elements.

The **activity factor** of a blade AF is an arbitrary measure of the blade solidity and hence of its ability to absorb power. It takes the form

$$AF = \frac{100,000}{16}\int_{r_h/R}^{r/R=1.0} \frac{c}{d_p}\left(\frac{r}{R}\right)^3 d\left(\frac{r}{R}\right) \qquad (11.5.66)$$

where $R = d_p/2$, r = radius at any section, and r_h = spinner radius.

Performance

Static Thrust   From the simple axial-momentum theory, the static thrust F_0 of an actuator disk of diameter d_p is given by

$$F_0 = (\pi\rho/2)^{1/3}(\mathcal{P}d_p)^{2/3} = 10.4\text{bhp} \times d_p^{2/3} \qquad \text{(at sea level)} \quad (11.5.67)$$

The **thrust horsepower** (thp) is accordingly

$$\text{thp} = \frac{F_0}{\text{bhp}} = (38,400/Nd_p)(1/C_p^{1/3}) \qquad (11.5.68)$$

Equations (11.5.67) and (11.5.68) assume uniform axial velocity, no rotation of the air, and no profile losses. Consequently, they may be in error by as much as 50 percent. Hesse suggests the use of the equations

$$\eta_{p\,\text{static}} = \sqrt{\frac{2}{\pi}}\frac{C_F^{3/2}}{C_P} = 0.798\frac{C_F^{3/2}}{C_P} \qquad (11.5.69)$$

and

$$F = 10.42[d_p(\eta_{p\,\text{static}})\text{shp}]^{2/3}(\rho/\rho_0)^{1/3} \qquad (11.5.70)$$

together with data computed for a specific propeller, as illustrated in Fig. 11.5.36. The static thrust for a given power input increases with the diameter and the number of blades. Controllable pitch propellers selected for good overall performance will develop 3 to 4 lb of thrust per shp.

Fig. 11.5.36   Static-thrust performance of a propeller; four-bladed, 100 activity-factor propellers of various integrated design C_L's. *(From Hamilton Standard Division, United Aircraft Corporation.)*

At flight conditions the performance may be obtained from generalized charts presenting C_P as a function of the advance ratio J, with the propeller efficiency and blade pitch angle at 75 percent radius, $\theta_{3/4}$, as parameters. Figure 11.5.37 is such a chart.

Fig. 11.5.37   Propeller performance—efficiency in forward flight; four-bladed, 100 activity factor, 0.500 integrated design C_L propeller. *(From Hamilton Standard Division, United Aircraft Corporation.)*

Reverse Thrust By operating the propeller blades at large negative angles of attack, reversed thrust can be developed. In that condition, the blades are stalled, and as in the case of static thrust, the forces acting on the blades cannot be calculated. In practice, the reverse-pitch stop is set by trial and error so that the propeller can absorb the rated power at zero speed, i.e., at $V_0 = 0$. With fixed-pitch reversing, the power absorbed when there is forward speed will be smaller than the rated value, but the reversed thrust is usually substantially larger than the static forward thrust corresponding to the rated power. The landing roll of an aircraft with reversed thrust plus brakes is approximately 45 percent of that with brakes alone.

Compressibility Effects As the relative velocity between an airfoil and the free-stream air approaches the sonic speed, the lift coefficient C_L decreases rapidly, and simultaneously there is a large increase in the drag coefficient C_D. The propeller is subject to those effects. Since the acoustic speed decreases from sea level to the isothermal altitude, the effects appear at lower tip speeds at high altitude. Furthermore, the effect is more pronounced at higher tip speeds. Shock waves are formed at the leading edge of a propeller blade when its pitch-line velocity approaches the local sonic velocity of the air. As a consequence, the coefficient C_F is reduced and C_P is increased, thereby adversely affecting the propulsive efficiency of the propeller.

Because of the high propulsive efficiency of propellers, the application of advanced technologies may provide **turboprop**-powered transports with cruise speeds and other desirable characteristics comparable to advanced turbofan-powered transports, but with a 15 to 20 percent reduction in fuel consumption.

Advanced turboprop propeller blades operate at high power loading, are extremely thin, and have swept leading edges to minimize compressibility losses. Shaping the spinner and nacelle minimizes choking and compressibility losses. Efficiencies of about 70 percent at Mach 0.8 and design power loadings have been achieved in research propellers. Such propellers may be relatively noisy because the tips may be slightly supersonic at Mach 0.8 cruise. Attenuating the noise within the cabin wall will likely add airframe weight and reduce the fuel savings.

Mechanical Design

Blades The principal blade loadings are (1) steady tensile, due to centrifugal forces; (2) steady bending, due to the aerodynamic thrust and torque forces; and (3) vibratory bending, due to cyclic variations in airloads and other excitations originating in the engine. The most serious and limiting stresses usually result from the vibratory loadings. The principal vibratory loading results from the cyclic variation in angle of attack of the blades when the axis of rotation is pitched or yawed relative to airstream. When the axis is pitched up, such as when the aircraft is in a high-angle-of-attack climb, the blades are at a higher angle of attack on the downstroke than when on the upstroke. This results in a once per revolution, $1E$, variation in aerodynamic loading on the blades, usually referred to as $1E$ **aerodynamic excitation,** and a steady **side force** and **yawing moment,** which are transmitted through the shaft to the aircraft.

The degree of pitch or yaw of the propeller axis is usually measured in terms of the **excitation factor,** defined as

$$\text{Excitation factor} = \psi(V_i/400)^2 \qquad (11.5.71)$$

where ψ = angle of pitch or yaw, deg; V_i = indicated flight velocity, mi/h. Another factor sometimes used is Aq, where $A = \psi^\circ$, $q = 12\rho V_i^2$, Aq = excitation factor × 410.

Because of the restoring moment of the **centrifugal loads** on the blade elements when the blade deflects in bending, centrifugal forces have the effect of apparently stiffening the blade in **bending.** In Fig. 11.5.38, bending natural frequency increase with rpm. At some rpm, the bending natural frequency will come into resonance with the $1E$ excitation, at which point small loadings will be greatly magnified (limited only by the damping present in the blade).

Propellers are designed so that the $1E$ resonant speed is above any expected operating speed. However, there will always be some magnification of the vibratory loads (due to proximity to resonant speed),

Fig. 11.5.38 Relation between blade natural frequency and major excitations.

resulting in disproportionately high vibratory stresses. As indicated in Fig. 11.5.38, blades normally pass through a $2E$ **resonance** in coming up to speed (usually not critical because it is a transient condition) and may pass through or be close to resonance with other modes.

The **natural frequencies** of the rotating propeller blade can be computed with good accuracy, but since the excitation varies from aircraft to aircraft and is usually not well defined, vibration surveys must be made of all new installations. These often result in the definition of ranges at which continuous operation should be avoided. **Allowable vibratory** stresses are limited by the fatigue strength of the material. Unlike the steady-stress limits, **fatigue strength** is not a unique function of the material but is a function of surface finish as well. A sharp notch in the surface reduces the fatigue strength to a very low value. If propellers were designed to hold vibratory stresses below the lowest possible fatigue strength, the blades would be unacceptably heavy. Thus propellers are designed on the maintenance of a reasonably smooth surface finish. Since the blade surfaces are constantly subject to **store damage,** blades must be regularly inspected to make sure that the finish does not deteriorate below design standards.

Propeller blades are stalled when operating at low forward speeds and when reversing (see Performance, above). Under these conditions torsion-mode **stall flutter** can occur.

In general, $1E$ **considerations** control the design of the **inboard** portions of a blade, and **stall flutter** the **outboard** sections.

There are five types of **blade construction** in common use today: **solid aluminum; fabricated steel; one-piece steel; monocoque, fiberglass construction,** supported on a steel shank; and a construction with a **hollow steel central spar and a lightweight fiberglass cover** that forms the aerodynamic contour of the blade. As in any beam designed on the basis of bending loads, it is desirable to concentrate the structural material at the maximum radii, as in the flanges of a simple I beam, and omit it from the center, where it carries no load. This is accomplished, with steel blades, by making them hollow. The structural material is concentrated in the outer fibers, or surface. In general, steel blades weigh about 80 to 85 percent of the equivalent solid aluminum blade. The advantage of steel increases with size. The blades normally constitute from one-third to one-half of the total propeller weight.

The **hub** retains the blades and contains the pitch-change motor. A **pitch-change mechanism** is provided so that the blades can be positioned at the proper angle of attack α to absorb the desired power at the desired rpm, regardless of flight speed. If the pitch is unchanged, as in Fig. 11.5.39, the angle of attack is excessive at low speed. This prevents the engine from reaching rated speed and delivering rated power. The centrifugal loads on the blade elements tend to rotate the blades toward flat pitch (see Fig. 11.5.40). The pitch-change motor acts against this moment by applying a pitch-increasing moment. The addition of **counterweights,** shown dotted in Fig. 11.5.40, reduces the moment and the size of the pitch-change motor. Counterweights, however, add to the radial centrifugal load of the blades, necessitating a heavier retention, and consequently do not necessarily reduce the net weight of the propeller.

Most pitch-change mechanisms employ a **hydraulic piston,** mounted on the hub, with feed through the propeller shaft and with rotation of the blades by means of gears or links. **Electric motors** and **direct mechanical drives** have been used successfully. In the past, the weight of all systems, when fully developed, has turned out to be about the same.

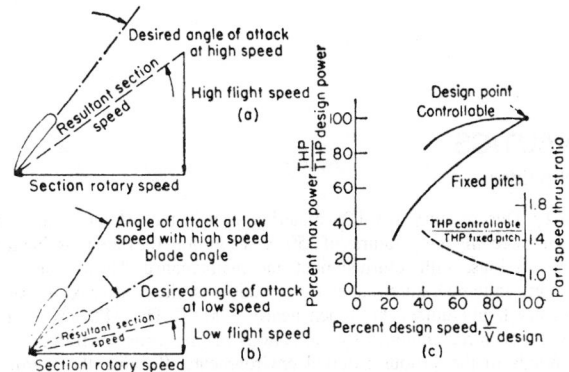

Fig. 11.5.39   Effect of flight speed on the desired blade angle.

Fig. 11.5.40   Twisting effect of centrifugal loads on blades.

Action of the pitch-change motor is limited by **low-** and **high-pitch stops**. These prevent the blades from assuming negative angles with reverse thrust. On reciprocating engines, the low-pitch stop is set to allow the propeller to absorb approximately rated power at zero forward speed. A lower setting results in higher windmilling drags in gliding flight, which could be dangerous if carried too far. In reversing propellers the low-pitch stop is automatically removed when the controls are set for reverse pitch. The high-pitch stop prevents the propeller from reversing through the positive range. In feathering propellers it is set at the angle which gives zero aerodynamic moment about the shaft.

Propeller feathering is provided in order to reduce the drag of a dead engine. To accomplish it, the blade-angle range is extended to 90°, approximately. Auxiliary pitch-change power is provided to complete the feathering as the engine slows down and to unfeather when the engine is stationary.

Reversing is provided in order to give propeller braking on the landing roll. It is accomplished by removing the flight low-pitch stop and driving the blades to a high negative **reverse-pitch angle**. The principal mechanical problem is to provide a mechanism which has the least possible chance of going into reverse inadvertently.

The **propeller control** is that portion of the system which regulates the blade angle. In light aircraft, this may consist merely of a mechanism which sets a given blade angle corresponding to a given position of the cockpit control. In such cases the propeller acts as a fixed-pitch propeller except when activated by the pilot. In most cases, the propeller control regulates blade angle so as to maintain a preset rotary rpm, irrespective of flight speed or shaft power. The **basic elements** of this system

Fig. 11.5.41   Propeller-speed governor.

normally consist of a spring-balanced, engine-drive flyweight delivering an error signal which directs the servo, or pitch-change motor, to increase or decrease blade angle (Fig. 11.5.41). The control loop is closed by rpm feedback through the main engine. This basic system may be embellished with anticipating and delay devices to maintain speed in rapid throttle movements, synchronizing to coordinate two or more engines, and with overriding features to activate the reversing and feathering sequences. (See also Sec. 16.)

Propeller **controls** are fundamentally the same, with several added secondary features. First, because the engine is capable of a very rapid increase or decrease in torque and because of the high gear ratio between the propeller and engine shaft, anticipating features are necessary to prevent overspeeds in the propeller from producing large overspeeds and possible failure in the turbine. Second, because of the high idle rpm needed to maintain the pressure ratio in the compressor, the low-pitch stop of the turboprop engine propeller must be set substantially lower than for a piston engine. This is not true of a free-turbine drive and is only partially true of a split-compressor engine. In the event of an **engine failure**, the propeller, sensing the loss in rpm, will govern down to the low-pitch stop. This may lead to large, often uncontrollable, negative thrust. To avoid this possibility, an engine-shaft torque signal is put into the control, a negative torque automatically activating the feathering system. This portion of the system is known as the **ENT (emergency negative-thrust)** system. Third, because of the high idle rpm characteristic of the turboprop engine, and the resulting low blade angles at low power in the ground-taxi regime, provision must be made to cut out the governing system and substitute direct blade-angle regulation (known as **beta control**). This is required because the propeller will not govern under these conditions as the torque-to-blade-angle relationship goes to zero. The propeller control of a turboprop engine is normally mechanically coordinated directly to the engine throttle, leaving only a single control, with so-called power lever, in the aircraft.

The shaft on which the propeller is mounted is subjected to the steady-thrust load, vibratory loads due to the propeller side, and **gyroscopic forces** (the $1E$ moment, acting on the blades, is reactionless in propellers with three or more blades). In large aircraft, the shaft size is usually determined by the propeller side force. In aircraft used for aerobatics, gyroscopic loads may become limiting. The **gyroscopic moment** on the propeller shaft is given by $2\omega_1 \cdot \omega_2(I_1 - I_2)$, where ω_1 = angular velocity of maneuver, ω_2 = angular velocity of shaft, I_1 = polar moment of inertia of propeller about shaft, I_2 = moment of inertia of propeller about diameter in plane of rotation. (See also Sec. 3.)

11.6 ASTRONAUTICS

by William C. Schneider

REFERENCES: "Progress in Astronautics and Aeronautics," AIAA series, Academic Press. "Advances in Astronautical Sciences," American Astronautical Society series. "Handbook of the British Astronautical Association." Koelle, "Handbook of Astronautical Engineering," McGraw-Hill. Purser, Faget, and Smith, "Manned Spacecraft: Engineering Design and Operation," Fairchild Publications. Herrick, "Astrodynamics," Vol. 1, Van Nostrand Reinhold. Cohen, "Space Shuttle," Yearbook of Science and Technology, 1979, McGraw-Hill.

SPACE FLIGHT
by Aaron Cohen

The science of astronautics deals with the design, construction, and operation of craft capable of flight through interplanetary or interstellar space. In addition to orbits around, and trajectories between, such bodies as stars, planets, and planetoids, the ascents from and descents to the surface of such bodies are considered to be part of the field.

The purposes of such craft are numerous and include the transportation of people and cargo, the transmission or relaying of signals, and the carrying of instruments for the measurement of the characteristics of (1) the environment through which the craft flies, (2) the surface of the celestial body over which the craft flies, or (3) astronomical objects or phenomena. The commercial applications of space are now being realized. Many communication, weather, and earth-sensing satellites have been deployed. The space shuttle which provides a method to go to and from low earth orbits offers an expanded use of space.

Processing of materials in a microgravity environment is proving to be very reliable. In microgravity the effects of thermal convection and gravity are eliminated, hence purer materials can be manufactured. Containerless processing can also be accomplished which will lead to additional applications in microgravity materials technology. The space shuttle allows payloads up to 32,000 lb to be placed in earth orbit at altitudes of 150 to 200 nautical miles (nmi) and inclinations of 28.5° to 90°. A description of the space shuttle is contained in Table 11.6.1. Using expendable launch vehicles or the space shuttle and propulsion-assistance modules, satellites can be placed in various orbits. A polar orbit is one whose orbital plane includes the earth's axis and permits line-of-sight contact between the spacecraft and every point on earth on a periodic basis. Synchronous orbits have angular rates equal to the rotational rate of the earth about its axis, providing a spacecraft locus of positions at nearly constant longitude. Orbital-parameter variation provides a wide variety of other orbits, permitting the selection of characteristics most suitable for the specific flight objective, such as communication relays, meteorological observations, earth surface monitoring, or interplanetary trajectories.

SHUTTLE THERMAL PROTECTION SYSTEM TILES
by William C. Schneider
Texas A & M University

REFERENCES: "National Space Transportation System Reference Document," NASA, June 1988. Moser and Schneider, "Strength Integrity of the Space Shuttle Orbiter Tiles," AIAA First Flight Testing Conference, Nov. 11–12, 1981, Las Vegas, paper AIAA-81-2469. Schneider and G. J. Miller, "The Challenging Scales of the Bird (Shuttle Tile Structural Integrity)," Space Shuttle Technical Conference, June 28–30, Houston. O'Hearn, "Application of Probabilistic Analysis to Aerospace Structural Design," AIAA 81-0231.

The thermal protection system (TPS) consists of various materials applied externally to the outer structural skin of the shuttle orbiter to passively maintain the skin within acceptable temperatures, primarily during the entry phase of the mission. The orbiter's outer structural skin is constructed primarily of aluminum and a limited amount of graphite epoxy. During entry, the TPS materials protect the orbiter outer skin from exceeding temperatures of 350°F. In addition, they are reusable for 100 missions, with refurbishment and maintenance. These materials perform in temperature ranges from −250°F in the cold soak of space to entry temperatures that reach nearly 3,000°F. The TPS also withstands the forces induced by deflections of the orbiter airframe as it responds to the various external environments. Because the thermal protection system is installed on the outside of the orbiter skin, it establishes the aerodynamics over the vehicle. The orbiter interior temperatures also are controlled by internal insulation, heaters, and purging techniques in the various phases of the mission.

The TPS is a passive thermal insulation system consisting of materials selected for stability at high temperatures and weight efficiency. Additionally, the TPS has to be reusable flight after flight. These highly constraining requirements drove the design to the highly reliable, passive, lightweight, reusable system that, flight after flight, protects the Shuttle Orbiter. However, the TPS was never designed to withstand significant impacts of any kind any more than an automobile windshield was designed to withstand significant impacts. The region of the orbiter vehicle where each material is used is depicted in Fig. 11.6.1. These materials are as follows:

1. **Reinforced carbon-carbon** (RCC) protects areas where the surface temperatures exceed 2,300°F during entry, such as the wing leading edges and the nose cap.

2. Black **high-temperature reusable surface insulation** (HRSI) tiles protect areas where the surface temperatures are below 2,300°F but above 1,200°F. These tiles have a black surface coating, necessary for entry emittance. HRSI tiles are used in areas on the upper forward fuselage, including around the forward fuselage windows; the entire underside of the vehicle where RCC is not used; portions of the orbital maneuvering system and reaction control system pods; the leading and trailing edges of the vertical stabilizer; wing glove areas; elevon trailing edges; adjacent to the RCC on the upper wing surface; the base heat shield; the interface with wing leading edge RCC; and the upper body flap surface.

3. Black tiles, called **fibrous refractory composite insulation** (FRCI), with higher strength and weight were developed later in the thermal protection system program. FRCI tiles replace some of the HRSI tiles in selected areas of the orbiter.

4. **Low-temperature reusable surface insulation** (LRSI) tiles protect areas where surface temperatures are below 1,200°F but above 700°F and have a white surface coating to provide better thermal characteristics on orbit. These white tiles are used on the upper wing and vertical tail and in selected areas of the forward, mid-, and aft fuselages, etc.

5. After the initial delivery of Columbia to the assembly facility, an **advanced flexible reusable surface insulation** (AFRSI) was developed. This material consists of composite quilted fabric insulation batting sewn between two layers of white fabric. AFRSI was used on Discovery and Atlantis to replace the vast majority of the LRSI tiles. Following its seventh flight, Columbia also was modified to replace most of the LRSI tiles with portions on the upper wing. The AFRSI blankets provide improved producibility and durability, reduced fabrication and installation time and costs, and a weight reduction over that of the LRSI tiles.

6. White blankets made of coated Nomex **felt reusable surface insulation** (FRSI) protect areas where surface temperatures are below 700°F and are used on the upper payload bay doors, portions of the mid-fuselage and aft fuselage sides.

7. Additional materials are used in other special areas such as the thermal panes for the windows; metal for the forward reaction control system fairings and elevon seal panels on the upper wing-to-elevon interface; and a combination of white- and black-pigmented silica cloth

Table 11.6.1   Specifications of the Space Shuttle System and Its Components

Component	Characteristic	Specifications
Overall system	Length	184.2 ft (56.1 m)
	Height	76.6 ft (23.3 m)
	System weight	
	Due-east launch	4,490,800 lbm (2037.0 Mg)
	104° launch azimuth	4,449,000 lbm (2018.0 Mg)
	Payload weight	
	Due-east launch	65,000 lbm (29.5 Mg)
	104° launch azimuth	32,000 lbm (14.5 Mg)
External tank	Diameter	27.8 ft (8.5 m)*
	Length	154.4 ft (47.1 m)
	Weight	
	Launch	1,648,000 lbm (747.6 Mg)
	Inert	70,990 lbm (32.2 Mg)
Solid rocket booster (SRB)	Diameter	12.2 ft (3.7 m)
	Length	149.1 ft (45.4 m)
	Weight (each)	
	Launch	1,285,100 lbm (582.9 Mg)
	Inert	176,300 lbm (80.0 Mg)
	Launch thrust (each)	2,700,000 lbf (12.0 MN)
Separation motors (each SRB), four aft, four forward	Thrust (each)	22,000 lbf (97.9 kN)
Orbiter	Length	121.5 ft (37.0 m)
	Wingspan	78.1 ft (23.8 m)
	Taxi height	57 ft (17.4 m)
	Weight	
	Inert	161,300 lbm (73.2 Mg)
	Landing	
	With payload	203,000 lbm (92.1 Mg)
	Without payload	173,000 lbm (78.5 Mg)
	Cross range	1,100 nmi (2037.2 km)
Payload bay	Length	60 ft (18.3 m)
	Diameter	15 ft (4.6 m)
Main engine (three)	Vacuum thrust (each)	470,000 lbf (2090.7 kN)
Orbital maneuvering system engines (two)	Vacuum thrust (each)	6,000 lbf (26.7 kN)
Reaction control system engines		
38 engines	Vacuum thrust (each)	870 lbf (3870 N)
6 vernier rockets	Vacuum thrust (each)	24 lbf (111.2 N)

* Includes spray-on foam insulation.

Fig. 11.6.1   Thermal protection system regions.

for thermal barriers and gap fillers around operable penetrations, such as main and nose landing gear doors, egress and ingress flight crew side hatch, umbilical doors, elevon cove, forward RCS, RCS thrusters, mid-fuselage vent doors, payload bay doors, rudder/speed brake, and gaps between TPS tiles in high differential pressure areas.

The area of the shuttle orbiter vehicle protected by the TPS tiles is shown in Fig. 11.6.1 and is designated as reusable surface insulation. As can be seen in Fig. 11.6.2, the tiles protect the vehicle from surface temperatures in the range of 700°F to 2,300°F.

The RSI tiles, shown conceptually in Fig. 11.6.3, are designed to radiate 80 to 90 percent of the reentry heat back to space before it can be conducted to the Orbiter's structure. Since the structure deforms in-plane during a mission, continuous covering for large areas would be prohibitive because of the weak RSI material, therefore the nominal 6-in-square tile with a 0.05-in gap between was used. To accommodate the flight-induced out-of-plane structural deformation, the individual tiles are individually mounted on a compliant strain isolation pad (SIP) made of Nomex felt, the same material as FRSI. This design concept of using individual tiles on SIP becomes fairly obvious with a limited amount of stress analysis using the RSI physical properties shown in Table 11.6.2.

As the LI-900 RSI 13-psi ultimate flatwise tension strength allowable was being established, the strength properties of the tile/SIP system were also being established. The effective joint allowable of this interface was approximately 50 percent that of the LI-900 material. This reduction in strength was caused by stress concentrations in the RSI because of the "stiff spots" introduced into the SIP by the needling felting process. The challenge of accommodating these stiff spots for the more highly loaded tiles, as opposed to using the heavier, higher-strength LI-2200, was met by locally densifying (strengthening) the underside of the tile. A solution of colloidal silica particles is applied to the noncoated tile underside and baked in an oven at 3,500°F for 3 hours. The densified layer produced is

0.1 in thick and increases the weight of a typical 6- by 6-in tile by only 0.06 lb. For load distribution, the densified layer serves as a structural plate that distributes the concentrated SIP loads evenly into the weaker unmodified RSI material. As shown in Fig. 11.6.4, this forces the failure surface to occur above the densified layer. This enhancement in itself offers somewhat of a fail-safe feature for thermal protection, but the major enhancement is in the improved tile/SIP system strength shown in Fig. 11.6.5.

There were many tiles that had standard 6- by 6-in square planforms but the numerous special tiles created a greater structural design challenge. It was necessary to design thousands of unique tiles that had compound curves, which interfaced with thermal barriers and hatches, which had penetrations for instrumentation and structural access, etc. The overriding challenge then was to assure the strength integrity of the tiles with a confidence that there was no greater chance than 1/10,000 of losing any one of the more than 10,000 more critical tiles (i.e., a probability of tile failure of no greater than $1/10^8$). The statistical treatment of the tile strength integrity is addressed in the first reference.

To accomplish this magnitude of system reliability and still minimize the weight it was necessary to define the detail loads and environments on each tile. Each tile experienced stresses induced by the following combined sources:
- Substrate or structure out-of-plane displacement
- Aerodynamic loads on the tile
- Tile accelerations due to vibration and acoustics
- Mismatch between tile and structure at installation
- Thermal gradients in the tile
- Residual stress due to tile manufacture
- Substrate in-plane displacement

Literally years of complex structural analysis, hundreds of tile specimens on high-speed-aircraft flow tests, tens of multitile wind tunnel tests, approximately 40 full-scale structural vibroacoustic tests

Fig. 11.6.2 Orbiter isotherms. (*From Moser et al.; reprinted by permission of the American Institute of Aeronautics and Astronautics, Inc.*)

Fig. 11.6.3 Tile design. (*From Moser et al.; reprinted by permission of the American Institute of Aeronautics and Astronautics, Inc.*)

Table 11.6.2 Properties of Reusable Surface Insulation

	Low density	High density
Material	High-purity silica fiber, LI-900	High-purity silica fiber, LI-2200
Density	9 lb/ft^3	22 lb/ft^3
Strength		
Flatwise tension	13 psi	35 psi
In-plane tension	41 psi	100 psi
Modulus of elasticity	6000 psi	36,000 psi
Thermal coefficient of expansion	2.7×10^{-7} in/in/°F	2.7×10^{-7} in/in/°F

SOURCE: AIAA Paper 81–2469. From Moser et al., reprinted by permission of the American Institute of Aeronautics and Astronautics, Inc.

• High strength at SIP interface and better stress distribution

• Failure always occurs above SIP interface

Fig. 11.6.4 Salient features of densified tiles. (*From Moser et al.; reprinted by permission of the American Institute of Aeronautics and Astronautics, Inc.*)

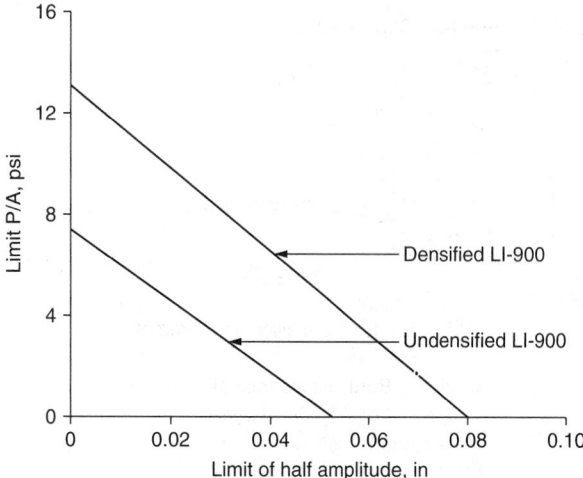

Fig. 11.6.5 Allowable flatwise tension in presence of substrate deflection. *(From Moser et al.; reprinted by permission of the American Institute of Aeronautics and Astronautics, Inc.)*

(containing hundreds of tiles each), and approximately 10,000 on-the-vehicle structural tile proof tests served in concert to design and verify the integrity of the TPS tile design.

DYNAMIC ENVIRONMENTS
by Michael B. Duke
NASA

REFERENCES: Harris and Crede, "Shock and Vibration Handbook," McGrawHill. Beranek, "Noise Reduction," McGraw-Hill. Crandall, "Random Vibration," Technology Press, MIT. Neugebauer, The Space Environment, *Tech. Rep. 34-229*, Jet Propulsion Lab., Pasadena, 1960. Hart, Effects of Outer Space Environment Important to Simulation of Space Vehicles, *AST Tech Rep. 61-201.* Cornell Aeronautical Lab. Barret, Techniques for Predicting Localized Vibratory Environments of Rocket Vehicles, *NASA TN D-1836.* Wilhold, Guest, and Jones, A Technique for Predicting Far Field Acoustic Environments Due to a Moving Rocket Sound Source, *NASA TN D-1832.* Eldred, Roberts, and White, Structural Vibration in Space Vehicles, *WADD Tech. Rep. 61-62.* Bolt, Beranek, and Newman, Exterior Sound and Vibration Fields of a Saturn Vehicle during Static Firing and during Launching, *U.S.A. Ord. Rep. 764.*

The forces required to propel a space vehicle from the launch pad are tremendous. The dynamic pressures generated in the atmosphere by large rocket engines are exceeded only by nuclear blasts. Slow initial ascent from the launch pad is followed by rapid acceleration and high g loading, and acoustic and aerodynamic forces drive every point of the vehicle surface. The space-vehicle velocity quickly becomes supersonic and continues to hypersonic velocities. Gradually, as the space vehicle leaves the atmosphere, the aerodynamic forces recede. Then suddenly the rocket thrust decays and is followed by the ignition of another rocket, which quickly develops thrust and continues to accelerate the space vehicle. Finally, the space vehicle becomes weightless. In completing a mission, a space vehicle is subjected to the additional environmental extremes of planetary landings, escape, and earth reentry.

These environments represent basic criteria for space-vehicle design. For operational reliability, the most important are the shock, vibration, and acoustic environments which are present to varying magnitudes in all operational phases and which constitute major engineering problems.

It is not possible to have original environmental data prior to fabrication and testing of a particular structure. Vibration, shock, and acoustic noise adversely affect structural integrity and vehicle reliability. These environments must be considered prior to design and fabrication and

then again for design verification through ground testing. Precise vibration and acoustic environmental predictions are complicated because of structural nonlinearities, random forcing functions, and multiple degree-of-freedom systems. Current environmental-prediction techniques consist mainly of extrapolating measured data obtained from existing vehicles. Acoustic environments produced during ground test are important factors in evaluating ground-facility locations and design, personnel safety, and public relations.

A source of sound common to all jet and rocket-engine propulsion systems is the turbulent exhaust flow. This high-velocity flow produces pressure fluctuations, referred to as noise, and has adverse effects on the vehicle and its operations. Also, far-field uncontrolled areas may be subjected to intense acoustic-sound-pressure levels which may require personnel protection or sound-suppression devices. (See also Sec. 12.)

The sound-generation mechanism is presented in empirical and analytical form, utilizing measured sound pressures and the known engine operational parameters. This combination of empirical terms and analytical reasoning provides a generalized relationship to predict sound-pressure levels for the vehicle structure and also for far-field areas. It has been shown by Lighthill (*Proc. Royal Soc.,* 1961) that acoustic-power generation of a subsonic jet is proportional to $\rho V^8 D^2/a_0^3$, where ρ is the density of the ambient medium, a_0 is the speed of sound in the ambient medium, V is the jet or rocket exit exhaust velocity, and D is a characteristic dimension of the engine. Attempts have been made to extend this relationship to allow supersonic-jet acoustic-power predictions to be made, but no one scheme is accepted for broad use.

From acoustic-data-gathering programs utilizing rocket engines, various trends have been noted which facilitate acoustic predictions. For rocket engines of up to 5×10^9 W mechanical power, the acoustic efficiency shows a nonlinear rate of increase. For greater mechanical power, the acoustic efficiency approaches a value slightly higher than 0.5 percent (Fig. 11.6.6).

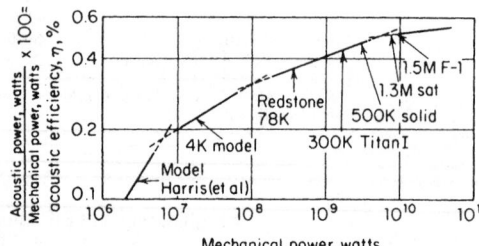

Fig. 11.6.6 Acoustic-efficiency trends.

Turbulent-exhaust-flow sound-source mechanisms also exhibit directional characteristics. Figure 11.6.7 shows data from a Saturn rocket vehicle launch with the microphone flush-mounted on the vehicle skin at approximately 100 ft above the nozzle plane.

During the period between on-pad firing and post-lifting-off or main-stage flight, the exhaust-flow direction changes. The directivity has also shifted 90° to retain its inherent relationship to the exhaust stream. The sound-pressure-level variation during a small time interval—between on-pad condition and shortly after liftoff—depends on the changes which have occurred in the directivity and the jet-exhaust-velocity vector (Fig. 11.6.7). The change in sound-pressure level is 20 dB

Fig. 11.6.7 Vehicle acoustic environment.

approximately in this case and is indicative of the significance of a sound source's directional properties, which depend on velocity and temperature of the jet.

The acoustic environments produced by a given conventional engine can be empirically described by that engine's flow parameters and geometry. It is difficult to predict the acoustic environments for non-conventional engines such as the plug-nozzle and expansion-deflection engines.

The turbulent-boundary-layer problem of high-velocity vehicle flights is difficult to evaluate in all but the simplest cases. Only for attached, homogeneous, subsonic flows can the boundary-layer noise be estimated with confidence.

The vibratory environment on rocket vehicles consists of total-vehicle bending vibrations, in which the vehicle is considered a nonuniform beam, with low-frequency responses (0 to 20 Hz) and localized vibrations with frequencies up to thousands of hertz. (See also Sec. 5.)

The response of a simple structure to a dynamic forcing function may be described by the universal equation of motion given as $M\ddot{x} + c\dot{x} + Kx = F(t)$ or $F(t)x(t) = (M - K\omega^2) - j(c/\omega)$ and is a complex function containing real and imaginary quantities. In the case of rocket-vehicle vibrations, $F(t)$ is a random forcing phenomenon. Thus, the motion responses must also be random. Random vibration may be described as vibration whose instantaneous magnitudes can be specified only by probability-distribution functions giving the probable fraction of the total time that the magnitude, or some sequence of magnitudes, lies within a specified range. In this type of vibration, many frequencies are present simultaneously in the waveform (Fig. 11.6.8). Power spectral density (PSD; in units of g^2/Hz) is defined as the limiting mean-square value of a random variable, in this case acceleration per unit bandwidth; i.e.,

$$PSD = \lim_{\Delta f \to 0} \bar{g}^2/\Delta f = d\bar{g}^2/df$$

Fig. 11.6.8   Random-vibration spectrum of a space-vehicle structure.

The sources of vibration excitation are (1) acoustic pressures generated by rocket-engine operation; (2) aerodynamic pressures created by boundary-layer fluctuations; (3) mechanically induced vibration from rocket-engine pulsation, which is transmitted throughout the vehicle structure; (4) vibration transmitted from adjacent structure; and (5) localized machinery. Since the mean square acceleration is proportional to power, the total vibrational power at any point on the vehicle may be expressed as

$$Pv \doteq Pv(1) + Pv(2) + Pv(3) + Pv(4) + Pv(5)$$

This is illustrated in Fig. 11.6.9 for an arbitrary vehicle. In many instances, the structure is principally excited by only one of these sources and the remaining can be considered negligible (Fig. 11.6.10). The vibration characteristics of structure B, taken from a skin panel, closely correspond to the trends of the acoustic pressures during flight. However, structure A, taken from a rocket-engine component, exhibits stationary trends, indicating no susceptibility to the impinging acoustic field. Thus, for this structure, it can be assumed that only $Pv(3)$ is significant, whereas for structure B, only $Pv(1)$ is the principal exciting source during the on-pad phase and only $Pv(2)$ is significant during the maximum dynamic-pressure phase. This also holds true for captive firings of the vehicle, in which only $Pv(1)$ is significant for structure B and $Pv(3)$ for structure A.

Fig. 11.6.9   Sources of vibration excitement.

Fig. 11.6.10   Typical structural responses to a random-vibration environment.

SPACE-VEHICLE TRAJECTORIES, FLIGHT MECHANICS, AND PERFORMANCE
by O. Elnan, W. R. Perry, J. W. Russell, A. G. Kromis, and D. W. Fellenz
University of Cincinnati

REFERENCES: Russell, Dugan, and Stewart, "Astronomy," Ginn. Sutton, "Rocket Propulsion Elements," Wiley. White, "Flight Performance Handbook for Powered Flight Operations," Wiley. "U.S. Standard Atmosphere, 1962," NASA, USAF, U.S. Weather Bureau. Space Flight Handbooks, *NASA SP*-33, *SP*-34, and *SP*-35. Gazley, Deceleration and Heating of a Body Entering a Planetary Atmosphere from Space, "Visits in Astronautics," Vol. I, Pergamon. Chapman, An Approximate Analytical Method for Studying Entry into Planetary Atmospheres, *NACA-TN* 4276, May 1958. Chapman, An Analysis of the Corridor and Guidance Requirements for Super-circular Entry into Planetary Atmospheres, *NASA-TN* D-136, Sept 1959. Loh, "Dynamics and Thermodynamics of Planetary Entry," Prentice-Hall. Grant, Importance of the Variation of Drag with Lift in Minimization of Satellite Entry Acceleration, *NASA-TN* D-120, Oct. 1959. Lees, Hartwig, and Cohen, Use of Aerodynamic Lift during Entry into the Earth's Atmosphere, *Jour. ARS,* **29,** Sept 1959. Gervais and Johnson, Abort during Manned Ascent into Space, *AAS,* Jan. 1962.

Notation

A = reference area
A_z = launch azimuth
B = ballistic coefficient
C_L = lift coefficient
e = eccentricity
f = stage-mass fraction (WP/W_A)
H = total energy
H_∞ = hyperbolic excess energy

I = specific impulse
Δ_i = plane-change angle
L = lift force
m = mass
n = mean motion $(2\pi/p_{sid})$
p = ambient atmospheric pressure
q = dynamic pressure
R = earth's equatorial radius
S = wing area
T = transfer time, time, absolute temperature
t_{pp} = time after perigee passage
V_c = injection velocity
V = velocity
V_g = velocity loss to gravity
W = weight
W_L = payload weight
W_P = propellant weight
Y = distance
α_{Ω} = right ascension of the ascending node
$\gamma = \theta - 90°$ = flight-path angle used in reentry analysis
v = true anomaly
ω = earth's rotational velocity
ψ = central angle
$\sigma = \rho/\rho_0$ = relative atmospheric density
Υ = vernal equinox
A_e = nozzle exit area
a = semimajor axis
C_D = drag coefficient
D = drag force
E = eccentric anomaly
F = thrust
g = acceleration due to gravity
H_{esc} = escape energy
h = altitude
i = inclination
J_2 = oblateness coefficient of the earth's potential (second-zonal harmonic)
M = molecular weight, mean anomaly
nm = nautical miles
P = period of revolution
P_e = nozzle exit pressure
q_s = stagnation-point heat-transfer rate
r = radius
s = range
t = time
V_∞ = hyperbolic excess velocity
V_{id} = ideal velocity
ΔV = impulsive velocity
W_A = stage weight $(W_O - W_L)$
W_O = gross weight
X = distance
α = angle of attack, right ascension
β = exponent of density-altitude function
λ = longitude
τ = hour angle
ω = argument of perigee
ϕ' = latitude
ρ = atmospheric density
θ = flight-path angle

Subscripts

a = apogee
e = entry
f = final
o = sea level conditions at 45° latitude
s = space fixed
vac = vacuum
circ = circular orbital condition

esc = escape
i = initial
p = perigee
M = mean
SL = sea level
sid = sidereal

ORBITAL MECHANICS
by O. Elnan and W. R. Perry
University of Cincinnati

The motion of the planets about the sun, as well as that of a satellite in its orbit about a planet, is governed by the inverse-square force law for attracting bodies. In those cases where the mass of the orbiting body is small relative to the central attracting body, it can be neglected. This simplifies the analysis of the orbital motion. The shape of the orbit is always a conic section, i.e., ellipse, parabola, or hyperbola, with the central attracting body at one of the foci. Parabolic and hyperbolic orbits are open, terminate at infinity, and represent cases where the orbiting body escapes the central force field of the attracting body.

The **laws of Kepler** for satellite orbits are (1) the radius vector to the satellite from the central body sweeps over equal areas in equal times; (2) the orbit is an ellipse, with the central attracting body at one of its foci; and (3) the square of the period of the satellite is proportional to the cube of the semimajor axis of the orbit.

Six **orbital elements** are required to describe the orbit in the orbit plane and the orientation of the plane in inertial space. The three elements that define the orbit are the semimajor axis a, eccentricity e, and period of revolution P. The orientation of the orbit (Fig. 11.6.11) is defined by the right ascension of the ascending node α_{Ω}, inclination of the orbit plane to the earth's equatorial plane i, and the argument of perigee ω, which is the central angle measured in the orbit plane from ascending node to perigee. The following equations compute these orbital elements and other orbital parameters.

$$a = (r_a + r_p)/2 = \mu/V_aV_p = r_a/(1 + e) = r_p/(1 - e)$$
$$e = (r_a/a) - 1 = (V_p - V_a)/(V_p + V_a) = (r_pV_p^2/\mu) - 1$$
$$r = a(1 - e^2)/(1 + e \cos v)$$
$$r_a = a(1 + e) = r_p(1 + e)/(1 - e)$$
$$r_p = a(1 - e) = r_a(1 - e)/(1 + e)$$
$$V = [\mu(2/r - 1/a)]^{0.5}$$
$$P = 2\pi(a^3\mu)^{0.5} \quad \text{(for circular orbits, } r = a\text{)}$$
$$V_a = V_p(1 - e)/(1 + e) = [\mu(1 - e)/a(1 + e)]^{0.5}$$
$$V_p = V_a(1 + e)/(1 - e) = [\mu(1 + e)/a(1 - e)]^{0.5}$$
$$v = \cos^{-1}\{[2r_ar_p - r(r_a + r_p)]/r(r_a - r_p)\}$$
$$t_{pp} = (a^3/\mu)^{0.5}\{\cos^{-1}[(a - r)/ae] - e[1 - (a - r)^2/(ae)^2]^{0.5}\}$$
$$M = nt_{pp} = E - e \sin E$$
$$r = a(1 - e \cos E)$$
$$V_{esc} = (2\mu/r)^{0.5}$$

Fig. 11.6.11 Elements of satellite orbit around a planet.

The elements may be determined from known injection conditions. For example,

$$1/a = V_i^2 \mu - 2/r_i$$
$$e^2 = \cos^2 V_i (r_i V_i^2/\mu - 1)^2 + \sin^2 V_i$$

General precession in longitude of vernal equinox:

$$x = 50''.2575 + 0''.0222T$$

where T is in Julian centuries of 36,525 mean solar days reckoned from 1900 Jan. 0.5 UT.

Measurement of Time

The concept of time is based on the position of celestial bodies with respect to the observer's meridian as measured along the geocentric celestial equator. The mean sun and the vernal equinox (Y) are used to define mean solar time and sidereal time. The mean sun, rather than the apparent (true) sun, is used since the latter's irregular motion from month to month gives a variation in the length of a day. The mean sun moves with uniform speed along the equator at a rate equal to the average of the true sun's angular motion during the year.

One sidereal day is the interval between two successive transits of the vernal equinox across the observer's meridian. A mean solar day is the interval between two successive transits of the mean sun. Local civil time (LCT) is defined to be the hour angle of the mean sun plus 12 h. Local sidereal time is the hour angle of the vernal equinox plus 12 h. For an observer at the Greenwich meridian, the local civil time is Greenwich mean time (GMT) or universal time (UT). The earth is divided into 24 time zones, each 15° of longitude wide. Each zone keeps the time of the standard meridian in the middle of the zone. (See Fig. 11.6.12.)

Time $T = \tau + 12$ h
Local sidereal time LST $= \tau_Y + 12$ h
Greenwich sidereal time GST $=$ LST $+ \lambda(w)$
Local civil time LCT $= \tau_M + 12$ h $=$ mean solar time
Equation of time ET $= \tau - \tau_M = \alpha_M - \alpha$
Greenwich mean time GMT $=$ UT $= \tau_M$(Greenwich) $+ 12$ h $=$ universal time

$$LCT = UT - \lambda(w)$$
$$GMT = ZT \pm \lambda°/15$$

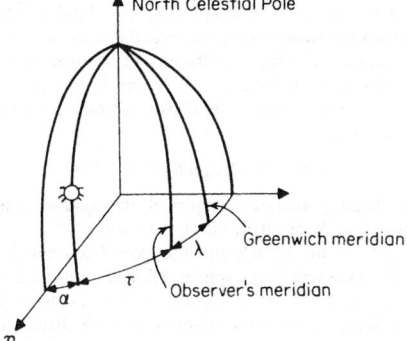

Fig. 11.6.12   Time-zone geometry.

The "American Ephemeris and Nautical Almanac" gives the equation of time for every day of the year and conversions of mean solar time to and from sidereal time.

Hohmann Transfer   This is a maneuver between two coplanar orbits where the elliptical transfer orbit is tangent at its perigee to the lower orbit and tangent at its apogee to the higher orbit. To transfer from a low circular orbit to a higher one, the impulsive velocity required is the difference between the velocities in the circular orbits and the velocities in the corresponding points on the transfer orbit. Using the general velocity

Fig. 11.6.13   Satellite lifetimes in elliptic orbits. $\rho = $ ARDC, 1959; $B = C_D A/2m = 1.0$ ft^2/slug; $e = $ initial eccentricity; --- $= $ decay histories.

equation above, the transfer velocities are $\Delta V_i = \{\mu[2/r_i - 2/(r_i + r_f)]\}^{0.5} - (\mu/r_i)^{0.5}$ and $\Delta V_f = (\mu/r_f)^{0.5} - \{\mu[2/r_f - 2/(r_i + r_f)]\}^{0.5}$, where V_i is the orbiting velocity in low orbit and V_f is the orbiting velocity in the high orbit.

Orbital Lifetime   The satellite orbit will gradually decay to lower altitudes because of the drag effects of the atmosphere. This drag force has the form $D = \frac{1}{2} C_D A_\rho V^2$, where values of the drag coefficient C_D may range from 2.0 to 2.5. The atmospheric density, as a function of altitude, can be used for computing an estimated orbital lifetime of a satellite (Fig. 11.6.13). (See NASA, "U.S. Standard Atmosphere" and "Space Flight Handbooks.")

Perturbations of Satellite Orbits   There are many secular and periodic perturbations of satellite orbits due to the effects of the sun, the moon, and some planets. The most significant perturbation on close-earth satellite orbits is caused by the oblateness of the earth. Its greatest effects are the precession of the orbit along the equator (nodal regression) and the rotation of the orbit in the orbit plane (advance of perigee). The nodal-regression rate is given to first-order approximation by

$$\dot\alpha_\Omega = -\tfrac{3}{2}(\mu/a^3)^{0.5} J_2 [R/a(1 - e^2)]^2 \cos i$$

The mean motion of the perigee is

$$\dot\omega = J_2 n(2 - (\tfrac{5}{2}) \sin^2 i)/a^2(1 - e^2)^2$$

LUNAR- AND INTERPLANETARY-FLIGHT MECHANICS
by J. W. Russell
University of Cincinnati

The extension of flight mechanics to the areas of lunar and interplanetary flight must include the effects of the sun, moon, and planets on the transfer trajectories. For most preliminary performance calculations, the **"sphere-of-influence,"** or **"patched-conic,"** method—whereby the multibody force field is treated as a series of central force fields—provides sufficient accuracy. By this method, the trajectories in the central force field are calculated to the "sphere of influence" of each body, and the velocity and position of the extremals are then matched to give a continuous trajectory. After the missions have been finalized, the precision trajectories are obtained by numerically integrating the equations of motion which include all the perturbative elements. It is necessary to know the exact positions of the sun, moon, and planets in their respective orbits.

Lunar-Flight Mechanics The moon moves about the earth in an orbit having an eccentricity of 0.055 and an inclination to the ecliptic of about 5.145°. The sun causes a precession of the lunar orbit about the ecliptic, making the inclination of the lunar plane to the earth's equatorial plane oscillate between 18.5 and 28.5° over a period of 18.5 years approx. To compute precision earth-moon trajectories, it is necessary to know the precise launch time to be able to include the perturbative effects of the sun, moon, and planets. The data presented are based on the moon being at its mean distance from earth and neglect the perturbation of the sun and planets; they are therefore only to be considered as representative data. The injection velocity (see Gazley) at 100 nmi for earth-moon trajectories and the impulsive velocity for braking into a 100-nmi orbit about the moon are shown in Fig. 11.6.14 as a function of transfer time. Additional energy is required to offset losses due to gravity and aerodynamic drag as well as to provide plane-change and launch-window capabilities.

Fig. 11.6.14 Lunar-mission velocity requirements.

Interplanetary-Flight Mechanics The fact that planetary orbits about the sun are not coplanar greatly restricts the number of feasible interplanetary trajectories. The plane of the interplanetary trajectory must include the position of the departure planet at departure, the sun, and the position of the target planet at arrival. The necessary plane change can be prohibitive even though the relative inclinations of the planetary orbits are small. The impulsive velocity required to effect a plane change for a spacecraft departing earth can be approximated by $\Delta V = 195{,}350 \sin{(\Delta_i/2)}$, where Δ_i is the amount of plane change required and ΔV is in ft/s.

For interplanetary flight, the "ideal" total energy that must be imparted to the spacecraft is the energy required to escape the gravitational field of the departure planet plus the energy required to change path about the sun so as to arrive at the target planet at the desired position and time. The energy required to escape the gravitational attraction of a planet can be determined by Keplerian mechanics to be $H_{esc} = \mu/r$, where μ is the gaussian gravitational constant and r is the distance from the center of the planet. After escaping the departure planet, the velocity must be altered in both magnitude and direction—in order to arrive at the target planet at the chosen time—by supplying additional energy, $H_\infty = \frac{1}{2} V_\infty^2$.

For determination of vehicle size necessary to inject the spacecraft into the interplanetary trajectory, it is convenient to express the total energy $H = H_{esc} + H_\infty$ in terms of the required injection velocity

$$V_c = \sqrt{2(H_{esc} + H_\infty)} \qquad \text{or} \qquad V_c = \sqrt{(2\mu/r) + V_\infty^2}$$

Hyperbolic excess velocities V_∞ for earth-to-Mars missions are shown in Table 11.6.3 as a function of date. These velocities are near optimum for the trip time and year given but do not represent absolute-optimum trajectories. Mars has a cyclic period with respect to earth of 17 years approx; hence the energy requirements for Mars missions from earth also follow approximate 17-year cycles.

Figure 11.6.15 shows how the hyperbolic excess velocities of Table 11.6.3 can be converted to stage characteristic velocities for a vehicle leaving a 100-nmi earth parking orbit. These are simple cases.

Table 11.6.3   Hyperbolic-Excess-Velocity Requirements (ft/s) for Typical Mars Missions

Trip time, days	Opposition year				
	1982	1984	1986	1988	1990
100	27,241	23,012	18,070	15,569	19,955
150	15,940	13,088	11,437	10,520	13,792
200	12,268	10,021	11,486	9,308	12,463
250	10,510	12,688	13,410	10,480	15,784

SOURCE: Gammal, "Space Flight Handbooks."

Fig. 11.6.15 Stage-velocity for earth departure from a 100-nmi parking orbit (specific impulse = 450 s).

In many recent planetary missions, use has been made of planetary flybys to gain velocity and aim the space probe. These "gravity assists" significantly increase the capability of interplanetary flight.

ATMOSPHERIC ENTRY
by D. W. Fellenz
University of Cincinnati

A vehicle approaching a planet possesses a considerable amount of energy. The entry vehicle must be designed to dissipate this energy without exceeding its limits with respect to maximum decelerations or heating.

The trajectory parameters of an entering vehicle are determined largely by its initial trajectory conditions (suborbital, orbital, superorbital), by the ratio of gas-dynamic forces acting upon it, and by its mass (ballistic factor, lift-drag ratio) and the type of atmosphere it is entering.

Planetary atmospheres (Table 11.6.4) can be assumed, as a first approximation, to have exponential density-altitude distributions: $\sigma = \rho/\rho_{SL} = e^{-\beta h}$, where

$$\beta = -(1/\rho)(d\rho/dh) = Mg/RT$$

For more exacting analyses, empirical atmospheric characteristics (e.g., U.S. Standard Atmosphere) have to be used.

The trajectory of the vehicle in flight-path fixed notation with the assumption of a nonrotating atmosphere is described as $dV/dt = g \sin \gamma - (C_D A/m)(\rho/2)V^2$ and $(V/\cos \gamma)(d\gamma/dt) = g - (V^2/r) - (C_L A/m)(\rho/2)(V^2/\cos \gamma)$. Solutions exist for direct ($\gamma > 5°$) ballistic entry and for equilibrium glide-lifting entry for $L/D > 1$. General solutions of the equations of motion have been obtained for shallow entry of both ballistic and lifting bodies. For a survey of analytical methods available, see Chapman, Loh, Grant, Lees, and Gervais.

The **energy of an earth satellite** at 200 mi altitude is about 13,000 Btu/lb, and a vehicle entering at escape velocity possesses twice this energy. This energy is transformed, through the mechanism of gas-dynamic drag, into thermal energy in the air around the vehicle, of which only a fraction enters the vehicle surface as heat. This fraction depends on the characteristics of the boundary-layer flow, which is determined by shape, surface condition, and Reynolds number. Figure 11.6.16 illustrates the **energy conversion** where it is necessary to manage a given

Table 11.6.4   Planetary Atmospheres

Planet	V_{esc}, ft/s	Gases	M, gmol^{-1}	T, K	β^{-1}, ft	$\beta\tau$	ρ_{SL}, lb/ft^3
Venus	34,300	2% N_2 98% CO_2	40	250–350	2×10^4	1,006	1.0
Earth	36,800	78% N_2 21% O_2	29	240	2.35×10^4	880	0.0765
Mars	16,900	95% N_2 3% CO_2	28	130–300	6×10^4	132	0.0062
Jupiter	195,000	89% H_2 11% He	2.2	100–200	6×10^4	3,600	

Fig. 11.6.16 Energy conversion during reentry. (*After Gazley in Knoelle, "Handbook of Astronautical Engineering."*)

amount of energy in a way that minimizes structural and heat-protection weights, operational restraints, and cost. It would be most desirable if this energy could be dissipated at a constant rate, but with constant vehicle parameters, decelerations vary proportionally to ρV^2 and heating rates proportionally to ρV^3. The selection of a heat-protection system depends on the type of entry flown. Lifting entry from orbit results in relatively long flight times (in the order of ½ h) as compared with 10 min in the case of a steep ballistic entry. **Radiative-heat-transfer systems** favor low heating rates over long time periods. **Ablative systems** favor short heat pulses. In fact, longer soaking periods may melt the ablation coating without getting the benefit of heat absorption through multiple-phase changes. A more uniform dissipation of energy can be achieved by modulation of vehicle parameters.

Determination of an **entry-vehicle configuration** is a process of iteration. The entry-flight profile and the entry and recovery procedures are mainly determined by whether experiments or passengers are carried. The external shape is determined by requirements for hypersonic glide capability and subsonic handling and landing characteristics and also by the relations between aerodynamic shape, heat input, and structural-materials characteristics. Intermediate results are fed back into the evaluation of performance and operational effectiveness of the total transportation system.

If reentry capability from any point of the ascent trajectory is desired for a manned vehicle, vehicle constraints and trajectory requirements must be compatible. **Performance-optimized trajectories** encompass combinations of relatively small velocities and large flight-path angles, which would result in considerable decelerations. In such cases, either reshaping the ascent trajectory or adding velocity at the time of abort, resulting in lower entry-flight-path angles, is effective (see Fig. 11.6.17). Lower flight-path angles during ascent depress the trajectory and increase drag losses. **Atmospheric entry is initiated** by changing the vehicle-velocity vector so that the virtual perigee of the descent ellipse comes to lie inside the atmosphere. The retrovelocity requirement for entry from low orbit is between 250 and 500 ft/s, depending on the range desired from deorbit to landing. For nonlifting entry, maximum deceleration and heat input into the vehicle are largely a function of entry angle, with the ballistic factor

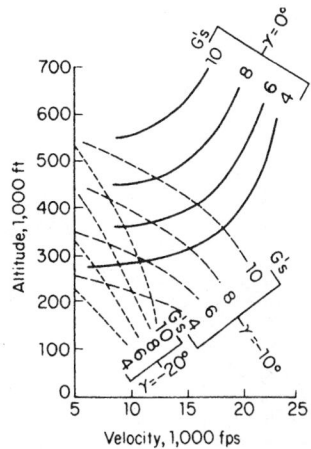

Fig. 11.6.17 Suborbital entry of a lifting vehicle. $W/C_D A = 28$ lb/ft; $L/D = 0.7$.

$W/C_D A$ determining the altitude at which the maximum decelerations and heating rates occur. Deceleration can be readily determined through the approximate relation $-(1/g)(dv/dt) \approx q/(W/C_D A)$. Decelerations and temperatures are drastically reduced with increasing lift-drag ratio (Fig. 11.6.18) and decreasing $W/C_D A$ (Figs. 11.6.19 and 11.6.20). This effect is particularly beneficial at steeper entry angles. The combination of longer flight times with lower heating rates, however, may actually result in a larger total heat input into the vehicle.

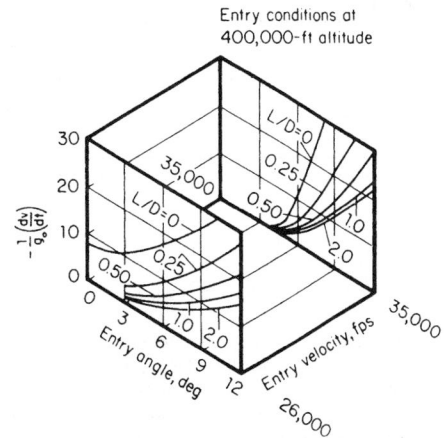

Fig. 11.6.18 Peak decelerations for entry at constant L/D.

Fig. 11.6.19 Glide reentry, dynamic pressure. $W/C_D A = 1$ lb/ft^2; $\gamma = 2°$; $\frac{1}{2}\rho_\infty V = (W/C_D A)(\frac{1}{2}\rho_\infty V^2)$; $h = h_1 - 23{,}500 \ln (W/C_D A)$; $V = V_1$.

Fig. 11.6.20 Glide reentry, stagnation-heating rate. $W/C_D A = 1$ lb/ft^2; $\gamma = 2°$; $q_s = q_{s1} W/C_D A$; $h = h_1 - 23{,}500 \ln (W/C_D A)$; $V = V_1$.

The **influence of lift** on the reduction of decelerations is greatest for the step up to $L/D \approx 1$. The influence of $W/C_D A$ on decelerations disappears beyond $L/D \approx 0.6$. Higher, hypersonic L/D ratios serve to

Table 11.6.5

L/D	Lateral, nmi	Longitudinal, nmi
0.5	$\approx$ 200	$\approx$ 2,000
1	$\approx$ 600	$\approx$ 4,000
1.5	$\approx$ 1,200	$\approx$ 6,000
2	$\approx$ 2,000	$\approx$ 9,000

* Assumptions: $\gamma_e = 1°$; $b_e = 400{,}000$ ft; $V_e = 26{,}000$ ft/s.

improve maneuverability. For entry from low orbit, the landing area selection (footprint) can be increased (see Table 11.6.5).

For entry at parabolic or hyperbolic speeds, it is necessary to dissipate sufficient energy so that the planet can capture the vehicle. Vehicle mass and aerodynamic characteristics determine the minimum allowable entry angle, the skip limit. The vehicle must be steered between the skip limit and the angle for maximum tolerable deceleration and heating. (See Figs. 11.6.21 and 11.6.22.)

Fig. 11.6.21 Comparison of decelerations and duration for entry into various atmospheres from decaying orbits. *(After Chapman.)*

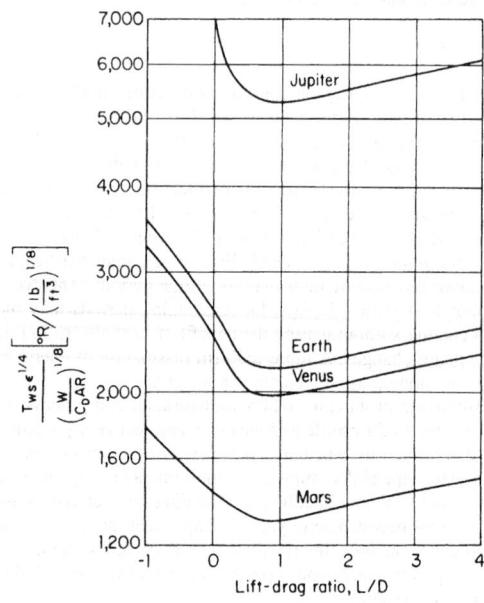

Fig. 11.6.22 Maximum surface temperature for entry into various planets from decaying orbits. *(After Chapman.)*

ATTITUDE DYNAMICS, STABILIZATION, AND CONTROL OF SPACECRAFT
by M. R. M. Crespo da Silva
University of Cincinnati

REFERENCES: Crespo da Silva, Attitude Stability of a Gravity-stabilized Gyrostat Satellite, *Celestial Mech.,* 2, 1970; Non-linear Resonant Attitude Motions in Gravity-stabilized Gyrostat Satellites, *Int. Jour. Non-linear Mech.,* **7,** 1972; On the Equivalence Between Two Types of Vehicles with Rotos, *Jour. Brit. Int. Soc.,* **25,**1972. Thomson, Attitude Dynamics of Satellites, in Huang and Johnson (eds.), "Developments in Mechanics," Vol. 3, 1965. Kane, Attitude Stability of Earth

Pointing Satellites, *AIAA Jour.*, **3**, 1965. Kane and Mingori, Effect of a Rotor on the Attitude Stability of a Satellite in a Circular Orbit, *AIAA Jour.*, **3**, 1965. Lange, The Drag-free Satellite, *AIAA Jour.*, **2**, 1964. Fleming and DeBra, Stability of Gravity-stabilized Drag-free Satellites, *AIAA Jour.*, **9**, 1971. Kendrick (ed.), TWR Space Book, 3d ed., 1967. Greensite, "Analysis and Design of Space Vehicle Flight Control Systems," Spartan Books, 1970.

A spacecraft is required to maintain a certain angular orientation, or attitude, in space in order to perform its mission adequately. For example, within a certain tolerance, it may be required to point a face toward the earth for communications and observation purposes, as well as a solar panel toward the sun for power generation. A vehicle in space is subject to several external forces which produce a moment about its center of mass tending to change its attitude. The environmental moments that can act on a spacecraft can be due to solar radiation pressure, aerodynamic forces, magnetic forces, and the gravity-gradient effect. The relative importance of each of the above moments depends on the spacecraft design and on how close it is to a central attracting body. Most often, the effect on the vehicle's attitude of the first three moments mentioned above is undesirable, although they have been used occasionally for attitude control (e.g., Mariner IV, Tiros, OAO). Micrometeorite impacts can also produce a deleterious effect on the vehicle's attitude.

The most common ways to stabilize a vehicle's attitude in space are the gravity-gradient and the spin-stabilization methods. When a spacecraft is subject to the gravitational force of a central attracting body E, its mass element near the center of the gravity field will be subject to a greater force than that acting on the mass elements farther from E. This creates a moment about the center of mass, C, of the spacecraft, which depends on the inertia properties of the vehicle and also on the distance r from E to C. This gravity-gradient moment is given by

$$\mathbf{M} = (3GM_e/r^3)\hat{\mathbf{r}} \times (\mathbf{I} \cdot \hat{\mathbf{r}})$$

where $\mathbf{I}$ is the inertia dyadic of the spacecraft, referred to its mass center, and $\hat{\mathbf{r}}$ is the unit vector in the direction from E to C. As a specific example, consider the elongated orbiting spacecraft shown in Fig. 11.6.23. The body-fixed x axis (yaw) departs an angle θ_3 from the local vertical, and the body-fixed z axis (pitch) remains normal to the orbital plane. The gravitational moment about C is readily found to be as given in the equation below (it is assumed that the maximum dimension of the vehicle is much smaller than the distance r).

$$\mathbf{M} = (3GM_e/2r^3)(I_x - I_y)(\sin 2\theta_3)\hat{\mathbf{z}}$$

(I_x and I_y are the spacecraft's moments of inertia about the x and y axes, respectively.) It is seen that unless the spacecraft's x axis is vertical or horizontal, this moment is nonzero, and for $I_x < I_y$ it tends to force the x axis of the vehicle to oscillate about the local vertical. Thus, the gravity-gradient moment provides a passive, and therefore very reliable, means for stabilizing the vehicle's attitude.

The orientation of a rigid body with respect to a reference frame is described by a set of three Euler angles. Figures 11.6.24 and 11.6.25 show a rigid body in orbit around a central attracting point E and the coordinate systems used to describe its attitude. The unit vectors $\hat{\mathbf{a}}_1$, $\hat{\mathbf{a}}_2$, and $\hat{\mathbf{a}}_3$, with origin at the spacecraft's center of mass, define the orbital reference frame. The vector $\hat{\mathbf{a}}_1$ points in the direction of the vector from

E to C; $\hat{\mathbf{a}}_3$ in the direction of the orbital angular velocity (whose magnitude is denoted by n); and $\hat{\mathbf{a}}_2$ is such that $\hat{\mathbf{a}}_2 = \hat{\mathbf{a}}_3 \times \hat{\mathbf{a}}_1$. The unit vectors $\hat{\mathbf{x}}$, $\hat{\mathbf{y}}$, and $\hat{\mathbf{z}}$ in Figs. 11.6.24 and 11.6.25 are directed along the three principal axes of inertia of the spacecraft. The attitude of the vehicle with respect to the orbiting frame is defined by the three successive rotations θ_1 (yaw), θ_2 (roll), and θ_3 (pitch) as shown in Fig. 11.6.25.

Using the above Euler angles, the gravity-gradient moment (about C) acting on the vehicle is given as

$$\mathbf{M} = \tfrac{3}{2}n^2[\hat{\mathbf{x}}(I_y - I_z)\sin 2\theta_2 \sin \theta_3$$
$$+ \hat{\mathbf{y}}(I_x - I_z)\sin 2\theta_2 \cos \theta_3 + \hat{\mathbf{z}}(I_x - I_y)\cos^2 \theta_2 \sin 2\theta_3]$$

If the spacecraft is subject to an external (other than the gravitational attraction of E) torque $\mathbf{T} = T_x\hat{\mathbf{x}} + T_y\hat{\mathbf{y}} + T_z\hat{\mathbf{z}}$ about its center of mass C, and if its attitude deviations θ_1, θ_2, and θ_3 remain small, its linearized equations of motion are obtained as given in the following equations.

$$I_x\ddot{\theta}_1 + n\dot{\theta}_2(I_z - I_y - I_x) + n^2\theta_1(I_z - I_y) = T_x$$
$$I_y\ddot{\theta}_2 + n\dot{\theta}_1(I_x - I_z + I_y) - 4n^2\theta_2(I_x - I_z) = T_y$$
$$I_z\ddot{\theta}_3 + 3n^2(I_y - I_x)\theta_3 = T_z$$

Care must be taken in the design of a gravity-gradient stabilized satellite in order to guarantee that $I_x/I_z < I_y/I_z < 1$ for the attitude motions to be stable in the presence of the least amount of damping. For augmenting the gravity-gradient moments, booms are added to the satellite in order to make its inertia ellipsoid thinly shaped. It is interesting to note that theoretically stable motions can also be obtained when the axis of minimum moment of inertia is normal to the orbital plane. However, this orientation is unstable if damping is present, and this is what occurs in practice since all actual vehicles have parts that can move relative to each other, such as an antenna.

It is seen from the equations above that for infinitesimal oscillations the pitch motion of the satellite is uncoupled from its coupled roll-yaw motions. The natural frequency ω_p of the infinitesimal pitch oscillations is given by

$$\omega_p = n\sqrt{3(I_y - I_x)/I_z}$$

and those of the coupled roll-yaw oscillations are given by the roots of the polynomial $\omega^4 - a_2 n^2 \omega^2 - a_0 n^4 = 0$, where

$$a_2 = 1 - [3 + (I_z - I_y)/I_x](I_x - I_z)/I_y$$
$$a_0 = 4(I_z - I_y)(I_x - I_z)/(I_x I_y)$$

For noninfinitesimal oscillations the pitch motion is coupled to the roll-yaw motion through nonlinear terms in the equations of motion. This coupling may give rise to an internal undesirable energy interchange between the modes of the oscillation, causing the roll-yaw motion, if uncontrolled, to oscillate slowly between two bounds. The upper bound of this nonlinear resonant roll-yaw motion may be much greater than (and independent of) the roll-yaw initial condition. For given values of I_x, I_y, and I_z, it can be decreased only by reducing the initial conditions of the pitch motion. This phenomenon can be excited if $\omega_p \approx 2\omega_1$ or $\omega_p \approx 2\omega_2$, where ω_1 and ω_2 are the two natural frequencies of the roll-yaw oscillations. The same phenomenon can also be excited if $\omega_p \approx \omega_1$ (or $\omega_p \approx \omega_2$). However, the observation of this latter

Fig. 11.6.23   Dumbell satellite.

Fig. 11.6.24   Orbiting satellite.

Fig. 11.6.25   Coordinate frames and Euler angles.

resonance requires a much finer "tuning" between the modes of oscillation, and therefore it is of lesser importance.

As the altitude of the orbit increases, the gravity-gradient moment decreases rapidly. For precise pointing systems, spin stabilization is used. Historically, spin stabilization was the first method used in space and is still the most commonly employed today. Simplicity and reliability are the advantages of this method when only two-axis stabilization is needed. By increasing the spin rate, the vehicle can be made very stiff in resisting disturbance moments.

A rigid body spins in a stable manner about either its axis of maximum or of minimum moment of inertia. However, in the presence of energy dissipation, even if it is infinitesimal, a spin about the axis of minimum moment of inertia leads to an unstable roll-yaw motion. Spinning satellites are built in such a way as to achieve symmetry about the spin axis which, in practice, is the axis of maximum moment of inertia. A spin of a rigid body about its intermediate axis of inertia leads to an unstable roll-yaw motion. Since damping can destroy roll-yaw stability, an individual analysis of the equations of motion must be performed to guarantee the stability of the motions when the spacecraft houses a specific damper.

For planet-orbiting spinning satellites (for which the gravitational moment can now be viewed as a disturbance) the stability of the roll-yaw motion depends not only on the magnitude of the spin rate but also on the direction of its angular velocity (relative to the orbital reference frame) due to spin. Denoting the satellite's angular velocity due to spin by $\mathbf{s} = s\hat{\mathbf{z}}$, the following inequalities must be satisfied for the roll-yaw motions of a spinning symmetric ($I_x = I_y$) satellite in a circular orbit to be stable in the presence of an infinitesimal amount of damping:

$$\left(1 + \frac{s}{n}\right)\frac{I_z}{I_x} - 1 > 0$$

$$\left(4 + \frac{s}{n}\right)\frac{I_z}{I_x} - 4 > 0$$

Very often, a spinning pitch wheel is added to a satellite to provide additional stiffness for resisting motions out of the orbital plane. Also, the inclusion of such a wheel provides more freedom to the spacecraft designer when specifying the range of the inertia ratios to guarantee that the attitude motions of the vehicle are stable.

Let us assume that a pitch wheel with axial moment of inertia I_w is connected to the spacecraft shown in Fig. 11.6.24. It is assumed that the wheel is driven by a motor at a constant angular velocity $\mathbf{s} = s\hat{\mathbf{z}}$ relative to the main body of the spacecraft. If the attitude deviations of the main body with respect to the orbiting reference frame remain small, the linearized equations for the attitude motion of the vehicle are now given as

$$I_x\ddot{\theta}_1 + n\theta_2(I_z - I_y - I_x + \beta I_z) + n^2\theta_1(I_z - I_y + \beta I_z) = T_x$$
$$I_y\ddot{\theta}_2 + n\dot{\theta}_1(I_x - I_z + I_y - \beta I_z) - n^2\theta_2[4(I_x - I_z) - \beta I_z] = T_y$$
$$I_z\ddot{\theta}_3 + 3n^2(I_y - I_x)\theta_3 = T_z$$

In these equations, the parameter β is defined as $\beta = I_w s/I_z n$, and I_x, I_y, I_z refer to the principal moments of inertia of the entire vehicle (rotor included).

Figure 11.6.26 shows a "stability chart" for this spacecraft when its center of mass, C, is describing a circular orbit around the central attracting point E. In this figure, a cut along the plane $\beta = 0$ produces the stability conditions for a gravity-gradient satellite, whereas a cut along the plane $K_1 + K_2 = 0$ (that is, $I_x = I_y$) gives rise to the inequalities for a spin-stabilized vehicle. Note that a rotor's spin rate with opposite sign to the orbital angular speed ($s < 0$) may destabilize the roll-yaw motions. Nonlinear resonances, as previously discussed, can also be excited in this vehicle. However, since the natural frequencies of the roll-yaw motions now also depend on the magnitude and sign of the internal angular momentum due to the rotor, the resonance lines are displaced in the region shown in Fig. 11.6.26 when the spin rate s of the rotor is changed to a different constant. Therefore, this phenomenon, when excited, can be avoided simply by changing the parameter β.

Fig. 11.6.26 Stable region of the vehicle parameter space.

For a spacecraft to perform its mission adequately during a long period of time, an attitude control system to maintain the vehicle's attitude within specified limits is necessary. Typically, this is done by means of a set of mass expulsion jets mounted on the vehicle, which are actuated by the output of a controller that receives information from the attitude sensors.

Drag-Free Satellites   For scientific applications that require the use of a satellite that follows a purely gravitational orbit, a translational control system is incorporated into the vehicle. Typical applications include satellite geodesy and navigation where accurate ephemeris prediction is needed. A drag-free satellite consists of a vehicle that contains a spherical cavity which houses an unsupported spherical proof mass. Sensors in the satellite detect the relative position of the proof mass inside the cavity and activate a set of thrustors placed in the vehicle, forcing it to follow the proof mass without touching it. Since the proof mass is shielded by the satellite from nongravitational forces, it follows a purely gravitational orbit.

METALLIC MATERIALS FOR AEROSPACE APPLICATIONS
by Stephen D. Antolovich
Revised by Robert L. Johnston
NASA

REFERENCES: MIL-HDBK-5C, "Metallic Materials and Elements for Aerospace Vehicle Structures," Sept. 1976. Wolf and Brown (eds.), "Aerospace Structural Metals Handbook," 1978, Mechanical Properties Data Center, Traverse City, Michigan. (This handbook gives information on mechanical, physical, and chemical properties; fabrication; availability; etc. in five published volumes encompassing ferrous, nonferrous, light metal alloys, and nonferrous heat-resistant alloys.) "Damage Tolerant Design Handbook," MCIC-HB-01, Metals and Ceramics Information Center, Battelle Columbus Laboratories, 1975, parts 1 and 2. (These two documents provide the most up-to-date and complete compilation of fracture-mechanics data for alloy steels, stainless steels, aluminum, and titanium alloys.) "Metals Handbook," 8th ed., 11 vols., American Society for Metals: Vol. I, Properties and Selection of Metals, 1961; Vol. II, Heat Treating, Cleaning and Finishing, 1964; Vol. III, Machinery, 1967; Vol. IV, Forming, 1969; Vol. V, Forging and Casting, 1970; Vol. VI, Welding and Brazing, 1971; Vol. VII, Atlas of Microstructures, 1972, Vol. VIII, Metallography, Structure and Phase Diagrams, 1973; Vol. IX, Fractography and Atlas of Fractographs, 1974; Vol. X, Failure Analysis and Prevention, 1975; Vol. XI, Nondestructive Testing and Quality Control, 1976. "Metals Handbook," 9th ed., 5 vols., American Society for Metals: Vol. I, Properties and Selection: Nonferrous Alloys and Pure Metals, 1979; Vol. III, Properties and Selection: Stainless Steels, Tool Materials and Special Purpose Metals, 1980; Vol. IV, Heat Treating; Vol. V, Surface Cleaning, Finishing and Coating. "Structural Alloys Handbook," Mechanical Properties Data Center, Belfour Stulen, Inc., Traverse City, Michigan, 1978. "Titanium Alloys Handbook," MCIC-HB-05, Metals and Ceramics Information Center, Battelle Columbus Laboratories, 1972.

A wide variety of materials are utilized in aerospace applications where performance requirements place extreme demands on the materials.

Some of the more important engineering properties to be considered in the selection of materials are (1) strength-to-weight ratio, (2) density, (3) modulus of elasticity, (4) strength and toughness at operating temperature, (5) resistance to fatigue damage, and (6) environmental effects on strength, toughness, and fatigue properties. Factors such as weldability, formability, castability, quality control, and cost must be considered.

Although it is desirable to maximize the strength-to-weight ratio to achieve design goals, materials with high strength-to-weight ratios generally are sensitive to the presence of small cracks and may exhibit catastrophic brittle failure at stresses below the nominal design stress. The problem of premature brittle fracture is accentuated as the temperature decreases and the yield and ultimate strength increase. For these reasons it is imperative that the tendency toward brittle failure be given major consideration for low-temperature applications. For example, liquid oxygen, a common oxidizer, boils at −297°F (−183°C) whereas liquid hydrogen, an important fuel, boils at −423°F (−253°C), and the problem of brittle failure is predominant. The fracture-mechanics approach has gained wide acceptance, and nondestructive inspection (NDI) characterization of flaws allows safe operating stresses to be calculated. A considerable amount of fracture-mechanics data has been collected and evaluated ("Damage Tolerant Design Handbook"). Stress corrosion cracking and corrosion fatigue become important for repetitive-use components such as aircraft airframes, landing gear, fan shafts, fan and compressor blades, and disks, in addition to brittle fracture problems such as fatigue crack formation and propagation.

The presence of flaws must be assumed (they may occur either during processing or fabrication) and design must be based on **fatigue crack propagation** (FCP) properties (Sec. 5). FCP of many materials follows an equation of the form

$$da/dN = R(\Delta K)^n \qquad (11.6.1)$$

where a refers to crack length, N is the number of cycles, K is the fluctuation in the stress-intensity parameter, and R and n are material constants. Equation (11.6.1) can be integrated to determine the cyclic life of a component provided that the stress-intensity parameter is known (Sec. 5). Although fatigue-crack initiation is generally retarded by having smooth surfaces, shot-peening, and other surface treatments, FCP of commercial aerospace alloys is relatively insensitive to metallurgical treatment and, to a reasonable approximation, whole classes of materials (i.e., aluminum alloys, high-strength steels, etc.) can be represented within the same scatter band.

It should be noted that increases in the plane-strain fracture toughness K_{IC} result in increased stress, through its linear relationship to the stress-intensity parameter, which in turn causes an exponential increase in the FCP rate and may introduce a fatigue problem where previously there was none.

The effect of environment on materials used for aerospace problems is usually to aggravate the FCP rate. It should be emphasized that fatigue loads have a synergistic effect on the environmental component of cracking and that the crack growth rate during corrosion fatigue is rarely the sum of the FCP and stress-corrosion cracking rates. In weldments of castings or wrought material, tests are needed to simulate anticipated use conditions and to prove weld soundness.

Materials in the turbine section of jet engines are subjected to deleterious gases such as O_2 and SO_2, high temperatures and stresses, and cyclic loads. Consequently the most significant metallurgical factors to consider are microstructural stability, resistance to oxidation and sulfidization, creep resistance, and resistance to low cycle fatigue. Nickelbase superalloys are generally used in the turbine section of the engine for both disks and blades. (See "Aerospace Structural Metals Handbook," Vols. II and IIA, for specific properties.) Although the Nibase superalloys generally exhibit excellent strength and oxidation resistance at temperatures up to 1,800°F (982.22°C), the interactive effects of sustained and cyclic loads are difficult to handle because, at high temperature, plastic deformation of these materials is time-dependent. An approach based on strain range partitioning is described in ASTM STP 520, 1973, pp. 744–782. The utility of this approach is that frequency and temperature dependence are incorporated into the

analysis by the way in which the strain is partitioned and reasonable envelopes of the expected life can be calculated.

STRUCTURAL COMPOSITES
by Terry S. Creasy
Texas A & M University

REFERENCES: MIL-HDBK-17, "Plastics for Aerospace Vehicles," Sept. 1973. Advanced Composites Design Guide, Vols. I–V, 3d ed., Sept. 1976. Herakovich, "Mechanics of Fibrous Composites," 1998.

High-performance composite materials are fiber-filled matrix materials with specific strength (strength/mass) and specific stiffness (modulus/mass) greater than that found in monolithic materials such as high-performance alloys or polymers (Fig. 11.6.27). The fibers contribute high strength and stiffness to the composite. The fibers and matrix must be chemically compatible and structurally sound at the application temperature.

Fig. 11.6.27 Specific properties of nonmetallic reinforcement materials compared with most metals. *(Hercules Aerospace Co.)*

A first-order estimate of the composite's effective Young's modulus in the fiber (primary) direction can be found using the rule of mixtures relation:

$$E_1 = (E_f - E_m)V_f + E_m$$

where the Es indicate the Young's modulus of the composite (1), the fiber (f), and the matrix (m), and the volume fraction of the fiber in the composite is V_f. The fiber volume fraction of fiber composites can vary between 0.5 and 0.7, i.e., 50 to 70 percent. Inspection of this relation shows that the effective Young's modulus E_1 is less than the fiber's modulus but greater than the modulus of the matrix material. Fiber composites are anisotropic—their properties are strong functions of the material orientation. The composite's stiffness perpendicular to the fiber direction (transverse direction) comes from this equation:

$$E_2 = \frac{E_m}{V_f\left(\dfrac{E_m}{E_f} - 1\right) + 1}$$

In the range of typical fiber volume fraction, the transverse modulus E_2 is only moderately larger than that of the matrix alone. Combined layers of composite produce a laminated structural material. As the primary and secondary modulus equations suggest, the properties will vary with the orientation of the material. The fiber placement must address the state of stress in the design.

Material selection must address two issues: (1) the fibers must provide the specific strength and stiffness needed for the design and (2) the matrix must be compatible with the fibers and they must survive the maximum use temperature for the application.

Table 11.6.6 Properties of Fibers Used in Composites

Fiber	Modulus, 10^6 lb/in^2	Tensile strength, ksi	Tensile failure strain, %	Specific gravity	Density, lb/in^3
Boron (4 mil)	60	460	0.8	2.60	0.094
Carbon (Thornel 300)	34	470	1.1	1.74	0.063
Carbon (IM8)	45	750	1.9	1.79	0.065
Carbon (pitch)	120	350	0.4	2.18	0.079
Kevlar 49	19	400	1.8	1.45	0.052
Kevlar 29	12	400	3.8	1.45	0.052
Glass (E)	10.5	250–300	2.4	2.54	0.092
Glass (S)	12.5	450	3.6	2.48	0.090

Tables 11.6.6 and 11.6.7 list typical properties of fiber and matrix materials.

The fatigue behavior of several fiber/matrix systems appears in Fig. 11.6.28. The response of these composites can be compared with 2024-T3 aluminum alloy.

Table 11.6.7 Properties of Matrix Materials Used in Composites

Material	Maximum use temperature, °F
Polyester	250
Polysulfone	250
Epoxy	350
Polyarylsulfone	400
Phenolic	450
Silicone	450
Polyimide	650
Aluminum	700
Magnesium	700
Glass	2,500
Carbon* (W/C)	3,200

* With oxidation resistance.

Fig. 11.6.28 Fatigue behavior of unidirectional composites and aluminum. *(Hercules Aerospace Co.)*

Finally, composite materials provide two unique capabilities for astronautic structures. Some carbon fibers have a negative coefficient of thermal expansion of approximately -0.2×10^{-6} in/in. With careful design a composite structure can compensate for expansion and contraction of other components. In addition, carbon fibers provide high thermal conductivity in the fiber direction. Thermal management of astronautic systems can take advantage of this capability.

MATERIALS FOR USE IN HIGH-PRESSURE OXYGEN SYSTEMS
by Robert L. Johnston
NASA

Oxygen is relatively reactive at ambient conditions and extremely reactive at high pressures. The design of high-pressure oxygen systems requires special consideration of materials and designs. The types of potential ignition sources that could be present in even a simple component are quite varied. Sources of ignition include electrical failures, particle impact in high-flow regions, pneumatic shock, adiabatic compression, fretting or galling, and Helmholtz resonance in blind passages. Other sources of ignition energy are also possible. A comprehensive review of potential design problems is available in NASA Reference Publication 1113, A.C. Bond et al., "Design Guide for High Pressure Oxygen Systems."

Materials currently used in high-pressure oxygen systems range from ignition-resistant materials like Monel 400 to materials of widely varying ignitability like butyl rubber and the silicones. The range of ignitability of various materials is shown in Figs. 11.6.29 and 11.6.30.

While material selection alone cannot preclude ignition, proper choices can markedly reduce the probability of ignition and limit propagation. Selecting materials with small exothermic heats of combustion will reduce the possibility of propagation. Materials with high heats of

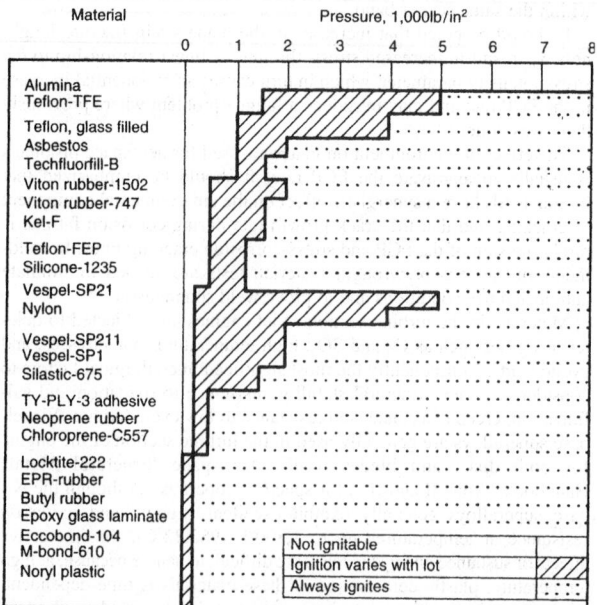

Fig. 11.6.29 Ignition variability of current oxygen-system nonmetallic materials.

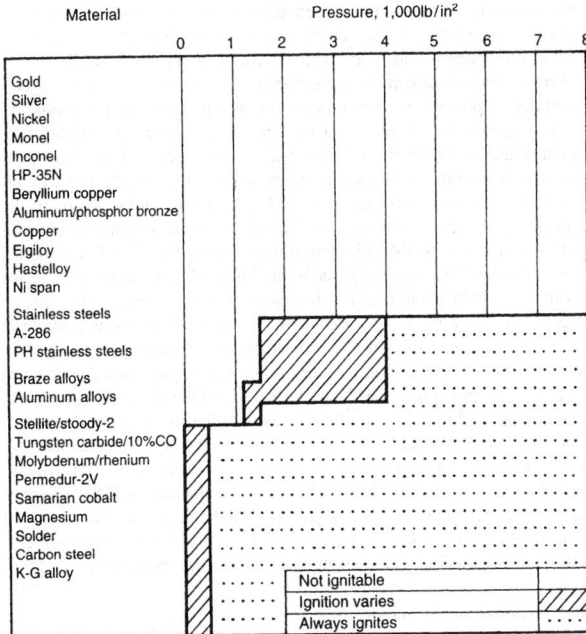

Fig. 11.6.30   Ignition variability of current oxygen-system metallic materials.

Table 11.6.9   Recommended Materials

Application	Material
Component bodies	Monel
	Inconel 718
Tubing and fittings	Monel
	Inconel 718
Internal parts	Monel
	Inconel 718
	Beryllium copper
Springs	Beryllium copper
	Elgiloy
Valve seats	Gold or silver over Monel or Inconel 718
Valve balls	Sapphire
Lubricants	Batch- or lot-tested Braycote 3L-38RP
	Batch- or lot-tested Everlube 812
O-seals and backup rings	Batch- or lot-tested Viton
	Batch- or lot-tested Teflon
Pressure vessels	Inconel

combustion (stainless steels) or very high heats of combustion (aluminum) should be avoided. A summary of heats of combustion, as well as other design properties of a few currently used materials, is shown in Table 11.6.8.

The materials listed in Table 11.6.9 have superior resistance to ignition and fire propagation in high-pressure oxygen systems. Monel alloys are available in the necessary range of hardnesses. Springs can be made of Elgiloy. Sapphire poppet balls should replace tungsten carbide or steel balls because sapphire has a lower level of reactivity (and therefore is less combustible) than either tungsten carbide or steel and is more resistant than tungsten carbide to breakup under mechanical impact in an oxygen environment.

Titanium and its alloys, normally attractive as materials for pressure vessels, cannot be used for oxygen vessels because they are impact-sensitive in oxygen. Inconel is a good choice for vessels for high-pressure oxygen.

In no case should an alloy be used at an oxygen pressure above which it can be ignited by particle impact. This criterion would limit the use of aluminum alloys and stainless steels to pressures below 800 lb/in^2 to allow some margin for error in test results and impact predictions.

METEROID/ORBITAL DEBRIS SHIELDING
by Eric L. Christiansen
NASA

REFERENCES: Anderson, (ed.) Smith (compiler), "Natural Orbital Environment Guidelines for Use in Aerospace Vehicle Development," NASA TM 4527, 1994. Liou et al., The New NASA Orbital Debris Engineering Model ORDEM2000, NASA TP-2002-210780, 2002. Christiansen, Meteroid/Debris Shielding, NASA TP-2003-210788, 2003.

Spacecraft shielding is designed to prevent a majority of the meteoroid/orbital debris particles that can impact the spacecraft during its lifetime from causing serious damage that would endanger the continuing operation of the spacecraft and/or endanger crew survivability. Because of spacecraft size and mass constraints, it is not possible with current shielding technology to eliminate completely the risk from meteoroid/orbital debris (M/OD) impact. Current practice calls for designing shields to meet or exceed protection requirements, which are generally expressed in terms of a minimum acceptable reliability level or success criteria (i.e., a probability of *not* being struck by a M/OD particle that will completely penetrate through the spacecraft shielding). Shielding is an important component of the overall strategy used to reduce the risk from meteoroid/orbital debris impact, which can also include other operational means, such as collision warning and avoidance to reduce the risk from orbital debris impact.

Spacecraft M/OD Environment

NASA standard meteoroid and debris environment models are used in the risk assessments to determine the cumulative flux of particles with a diameter that exceeds the ballistic limits. The environment models indicate that there are many more smaller particles than larger particles. Thus, raising the shielding performance in terms of the M/OD particles sizes the shielding can "stop" decreases the flux of penetrating particles and improves spacecraft reliability.

Meteoroid Environment Model   Meteoroids are the natural particles in orbit about the sun, which have quite high impact velocities relative to spacecraft in orbit about earth. Meteoroid velocities range from 11 km/s to 72 km/s, with an average for earth orbiting spacecraft of 19 km/s. A majority of meteoroids impacting a spacecraft are thought to originate from comets, with relatively low particle densities, ranging from 2 g/cm^3 (for particles 1 μg and less) to 0.5 g/cm^3 (particle mass 0.01 g and greater). This is contrasted with the meteorites that survive atmospheric entry and are found on earth's surface, which are higher

Table 11.6.8   Physical Properties of Typical Oxygen System Metallic Materials

Property	Aluminum alloys	Stainless steel	Inconel 718	Monel 400
Density, lb/in^3	0.10	0.28	0.30	0.31
Tensile, ultimate, lb/in^2	40,000	140,000	180,000	145,000
Maximum use temperature, °F	350	950	1,200	1,000
Heat of combustion, Btu/lb	130,000	33,500	15,100	14,200
Maximum use pressure, lb/in^2	1,250	1,250	8,000	8,000

11-120   ASTRONAUTICS

density and thought to originate mainly from asteroids. The meteoroid environment model to be used for shielding design is defined in the NASA documentation.

Orbital Debris Environment Model   Orbital debris is human-created pieces of spacecraft that are in orbit about earth (Fig. 11.6.31). Orbital debris impact velocities are lower than meteoroids, generally impacting spacecraft in low earth orbit at relative speeds of less than 1 km/s to just over 15 km/s, with an average velocity of about 9 km/s for a 400-km altitude. The debris environment threat is composed of metallic fragments, paint, aluminum oxide, and other components of spacecraft and solid-rocket motor exhaust. Typically, for debris risk assessments, orbital debris particle density is assumed to be 2.8 g/cm³, corresponding to aluminum metal. Because the orbital debris environment is tied to human activities in space, it is much more dynamic then the meteoroid environment, and the orbital debris environment definitions are subject to periodic updates and revisions as more data is collected on the amount and evolution of orbital debris in earth orbit. The current debris environment model for purposes of shielding design is defined by NASA TP-2002-210780 and referred to as ORDEM2000.

Fig. 11.6.31   Orbital debris "snapshot."

Design Requirements

Vehicles from the early years of space exploration have used the probabilistic approach to design meteoroid shielding. Table 11.6.10 provides a listing of historical M/OD protection design requirements. Generally, each program defines "critical" penetrations as those that would endanger the survivability of the vehicle and/or crew, although requirements for mission success and functionality have been applied as well.

Space Station Design Requirements   The Space Station has meteoroid/debris protection requirements consistent with past programs. As such, it carries by far the most capable meteoroid/debris shields ever flown. This is because ISS is larger and exposed longer than other space vehicles. It operates at higher altitudes in general than other spacecraft and its operation will extend into the future. These factors increase the expected number of meteoroid/debris impacts. To meet comparable protection requirements, ISS shielding must be more effective. For instance, most ISS critical hardware exposed to the M/OD flux in the velocity vector (front) or port/starboard (side) directions will be protected by shields effective at stopping 1- to 1.3-cm-diameter aluminum debris particle at typical impact velocity and angle (9 km/s, 45°). In comparison, the Mir space station is able to stop 0.3-cm particles, the space shuttle orbiter is capable of stopping 0.2- to 0.5-cm particles, and Apollo and Skylab were able to stop 0.15 to 0.2-cm particles under similar impact conditions. ISS will also have the ability to maneuver from ground-trackable debris particles (typically >10 cm diameter). By virtue of the large internal volume of ISS, its crews will have time to locate and isolate leaks, if they were to occur, by closing hatches. Hole repair kits will be available and crews will be trained to repair a leak in a module if it occurs. Crew escape vehicles will be docked to ISS in the event of a major event requiring evacuation.

Risk Assessment Process

The approach to evaluating and designing protection systems to meet requirements is illustrated in Fig. 11.6.32. By using this methodology, the analyst will be able to accurately evaluate spacecraft risks from M/OD impact, identify zones or areas of the spacecraft that are the "risk drivers" that control the M/OD risk, and evaluate options to reduce risk. Each major step in the risk assessment process is described in the following sections.

Space Geometry   Spacecraft size is directly proportional to the number of impacts as given in Eq. (11.6.1). Risk of M/OD impact failure increases as the area and time exposed to the M/OD flux increases.

$$N = \sum_{i=1}^{n} N_i = \sum_{i=1}^{n} (FAt)_i \qquad (11.6.1)$$

where N is the number of impacts causing failure and is equal to the sum of the impact failures in each region (N_i) over all regions ($i = 1$ to n) of the spacecraft. N_i is found from the product of flux (F, number/m²-year) of meteoroid and debris impacts that exceed the failure ballistic limits of a region, exposed area (A, m²), and time (t, year). It is important to break down the spacecraft geometry into different regions, each with a

Table 11.6.10 Historical Meteoroid/Debris Shielding Requirements		
Vehicle	Environments	Required probability of no penetration (PNP)
Apollo command module	Meteoroid	0.996 per 8.3-day mission
Skylab module	Meteoroid	0.995 for 8-month mission
Shuttle orbiter	Meteoroid	0.95 for 500 missions (design requirement)
	Meteoroid and debris	0.995 per mission (for critical penetration)
	Meteoroid and debris	0.984 per mission (for radiator leak— mission abort)
Spacelab module	Meteoroid	0.999 for 7-day mission
Hubble space telescope	Meteoroid and debris	0.95 for 2 years
ISS	Meteoroid and debris	0.98 to 0.998 per critical element over 10 years

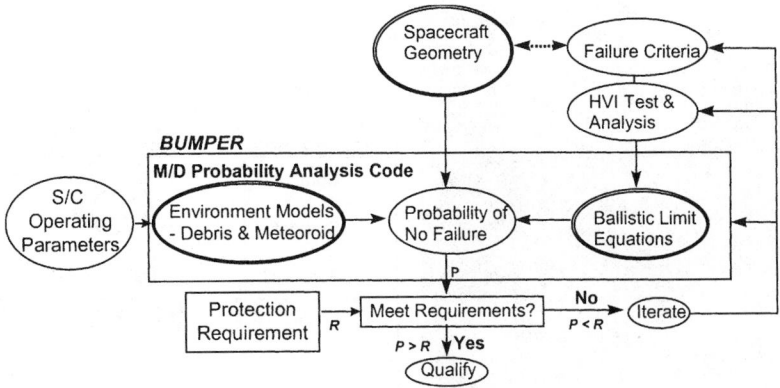

Fig. 11.6.32 Meteoroid/orbital debris risk assessment methodology.

different exposed area, materials of construction, shielding configurations, and thickness. For instance, there are over 400 different shields protecting International Space Station (ISS) habitable modules and external critical items (pressure vessels, control moment gyros) which have been optimized according to their location and exposure to the M/OD threat to provide the necessary M/OD protection.

Impact Directionality The front and sides of a spacecraft are more exposed to debris impact, while the front, sides, and top (zenith) are more exposed to meteoroid impact. The Long Duration Exposure Facility (LDEF) had 20 times more craters observed on the forward face compared to the aft, and 200 times more craters on the forward than the earth-facing side (Fig. 11.6.33). Heavier shielding is typically placed on front and sides of spacecraft with fixed orientation relative to the velocity direction to protect from higher impact rates on these surfaces.

Spacecraft Failure Criteria A key step in the risk assessment process is to precisely define what is meant by "failure," that is, to define potential damage modes from M/OD impact and, for each, to quantify the minimum amount of damage that can lead to failure of spacecraft hardware. This definition establishes the failure criterion or criteria that are assigned to each region of the spacecraft geometry model. For instance, M/OD damage modes and failure criteria for elements of the Space Station are given in Table 11.6.11.

Hypervelocity Impact Tests Hypervelocity impact (HVI) tests are an integral part of the analyses conducted to ensure adequate design of spacecraft M/OD shielding. Two-stage light-gas guns (LGGs) are used to accelerate projectiles up to 7 km/s (Fig. 11.6.34). LGGs are capable

Fig. 11.6.33 Long duration exposure facility.

of launching a variety of different and well-controlled projectile shapes. A disadvantage is that LGGs are capable of velocities that cover only a fraction of the orbital debris threat. Since average orbital debris velocity in low-earth orbit is on the order of 9 km/s, LGG can directly simulate only 40 percent of the orbital debris threat.

Other techniques exist to launch projectiles at speeds over 10 km/s. For instance, an inhibited shaped-charge launcher (ISCL) at Southwest Research Institute (SwRI) is capable of launching aluminum projectiles, 0.25 g to 2 g mass, in the shape of a cylinder to 11.5 km/s (Fig. 11.6.35). Another high-speed launcher that has provided useful information on shield capabilities in excess of 10 km/s is the three-stage hypervelocity launcher (HVL) developed at Sandia National Laboratories.

Table 11.6.11   Damage Modes and Failure Criteria for Space Station Elements
(Note: Critical failure criteria marked with*.)

Critical elements	Thermal protection system (TPS) damage	Spall/perforation of shield exterior of pressure shell	Perforation of pressure shell	Uncontrolled decompression	Catastrophic rupture	Detonation
Crew modules			X*	X	X	
Pressure vessels (liquids and gases)		X*1	X*1		X	X
Control moment gyros (rotating equipment)		X*			X	
Crew return vehicles	X		X*	X	X	

[1] Failure criterion for Russian PVs is perforation of pressure shell. Failure criterion of other designs is perforation or detached spall of shielding surrounding PV.

Functional elements	Surface degradation	Leak	Short or open circuit
Radiator panels	X	X	
Radiator flex/hard lines		X	
Thermal loop lines and exchangers		X	
Power cables			X
Batteries		X	X
Solar arrays	X		X
Data lines/cables			X
Window outer panes	X		

Fig. 11.6.34   NASA JSC-WSTF two-stage light-gas gun.

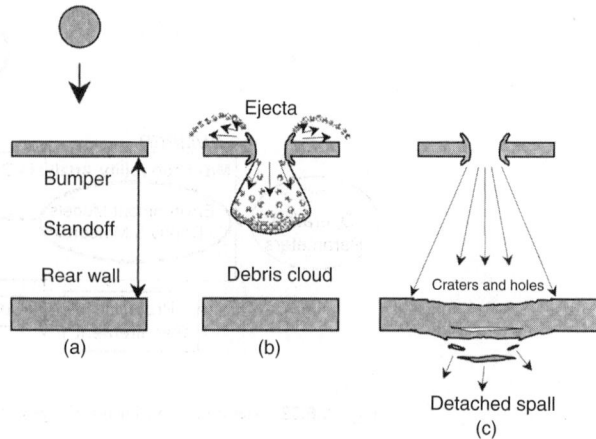

Fig. 11.6.36   Whipple shield. (*a*) Whipple shields consist of a bumper, standoff (gap or spacing), and real wall. (*b*) Hypervelocity impacts will generate a cloud of bumper and projectile debris that can contain solid fragments, liquid, and vapor particles. (*c*) The rear wall must survive the fragments and debris cloud impulsive loading. It could fail by perforation from solid fragments, spall, or tear and petal from the impulsive loading.

This launches thin disks (*L/d* = 0.1 to 0.2, mass = 0.2 to 1 g) of aluminum and titanium from 10 to 15 km/s with some bowing and tilting of the projectile. Hydrocode simulations are used to assess projectile shape effects on shield response.

Ballistic Limit Equations

Ballistic limit equations (BLEs) or "penetration" equations are based on the HVI test results, numerical simulations, and analytical assessments (Christiansen, 2003). BLEs describe the particle sizes that are on the failure threshold of a particular spacecraft component. The ability of a particular shield to withstand hypervelocity impact is predicted by a BLE. BLEs are a function of the target "failure criteria"; target configuration, materials, and thickness; and projectile parameters such as velocity, angle, shape, and density. M/OD risk assessments require a BLE defined for each surface of a spacecraft, for each failure criterion assessed. Shield design equations have also been developed that provide a means to estimate shielding parameters needed to defend against specific projectile threats.

Because of the complexity of spacecraft and their HVI failure modes, many different ballistic limit equations exist for a wide variety of shielding types, spacecraft components and hardware, and differing damage/failure modes. A description of ballistic limit equations for one particular type of shield, the basic Whipple shield, is provided in the following section. This example illustrates the considerations involved in developing ballistic limit equations.

Whipple Shield Ballistic Limit Equations

Fred Whipple proposed in the 1940s a meteoroid shield for spacecraft consisting of a thin "sacrificial" bumper at a distance from a rear wall.

The Whipple shield is shown in Fig. 11.6.36. The function of the first sheet, or bumper, is to break up the projectile into a cloud of material containing both projectile and bumper debris. This cloud expands while moving across the standoff, resulting in the impactor momentum being distributed over a wide area of the rear wall (Figs. 11.6.37 to 11.6.39). The back sheet must be thick enough to withstand the blast loading from the debris cloud and any solid fragments that remain. For most conditions, a Whipple shield results in a significant weight reduction over a single plate, which must contend with deposition of the projectile kinetic energy in a very localized area.

Whipple shields were used to provide protection to the Apollo command module and lunar lander. Apollo shields were capable of stopping 0.16-cm-diameter aluminum particles at 7 km/s, normal impact angle (0°).

Whipple Shield Design Equations   For design purposes, Eq. (11.6.2) to (11.6.4) provide bumper and rear wall thickness to defeat a given threat particle (assuming $V_n \geq 7$ km/s).

$$t_b = c_b m_p / \rho_b = c_b d \rho_p / \rho_b \qquad (11.6.2)$$

$$t_w = c_w d^{0.5} M^{1/3} (\rho_p \rho_b)^{1/6} \rho_w^{-1} (V \cos \theta) S^{-3/4} \sigma_h'^{-1/2} \qquad (11.6.3)$$

where $c_b = 0.25$ when $S/d < 30$, $c_b = 0.20$ when $S/d \geq 30$, and $c_w = 0.79 \, \text{cm}^{-3/4} \cdot \text{s} \cdot \text{g}^{1/3} \cdot \text{km}^{-1}$. Normalized yield stress (unitless) $\sigma_h' = (\sigma/70 \, \text{ksi})$.

The ballistic limit criterion is no perforation or spall of the rear wall. Bumper fragments become the primary source of rear wall damage at impact angles greater than 65°. Therefore, for oblique angles over 65°,

Fig. 11.6.35   SwRI's inhibited shaped-charge launcher. Various size charges are available that are capable of launching 0.25 to 2 g aluminum projectiles up to 11.5 km/s (left view). Projectiles are typically in the shape of a hollow cylinder (right view.)

Fig. 11.6.37 A 0.32-cm projectile impacts bumper at 6.8 km/s, left to right, with 1 μs between frames. [High-speed camera film (*photograph by Cordin*).]

Fig. 11.6.38 Ejecta and debris cloud.

the calculated rear wall thickness should be constrained to 65°. Coefficient c_w should be adjusted by the factor $[(S/d)/(S/d)_0]$ where $(S/d)_0 = 15$, when $S/d < 15$, as follows.

$$c_w = 0.79 \, [(S/d)/(S/d)_0]^{-0.185} \qquad (11.6.4)$$

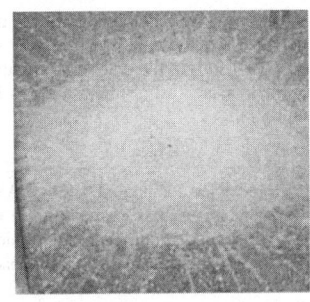

Fig. 11.6.39 Rear wall after impact.

Whipple Shield Ballistic Limits for All Impact Velocities Typical ballistic limits for a Whipple shield at all impact velocities are illustrated in Fig. 11.6.40 for normal impact by an aluminum sphere. The Whipple shield ballistic limit is compared to a monolithic, single aluminum plate of same mass as the combined Whipple shield's bumper and rear wall. Protection capabilities are defined for three penetration regimes based on normal component velocity (V_n) for an aluminum impact on a Whipple shield: ballistic regime ($V_n < 3$ km/s), fragmentation regime (3 km/s $< V_n < 7$ km/s), and melt/vaporization regime ($V_n > 7$ km/s). Important shield physical and material properties that influence ballistic limits of the shield vary as a function of impact velocity. Equations used to quantify ballistic limits for various types of shields are provided in the literature (Christiansen, 2003).

A key factor governing the performance of spaced shields is the "state" of the debris cloud projected from the bumper toward the rear wall. The debris cloud may contain solid, liquid, or vaporized projectile and bumper materials, or a combination of the three states, depending

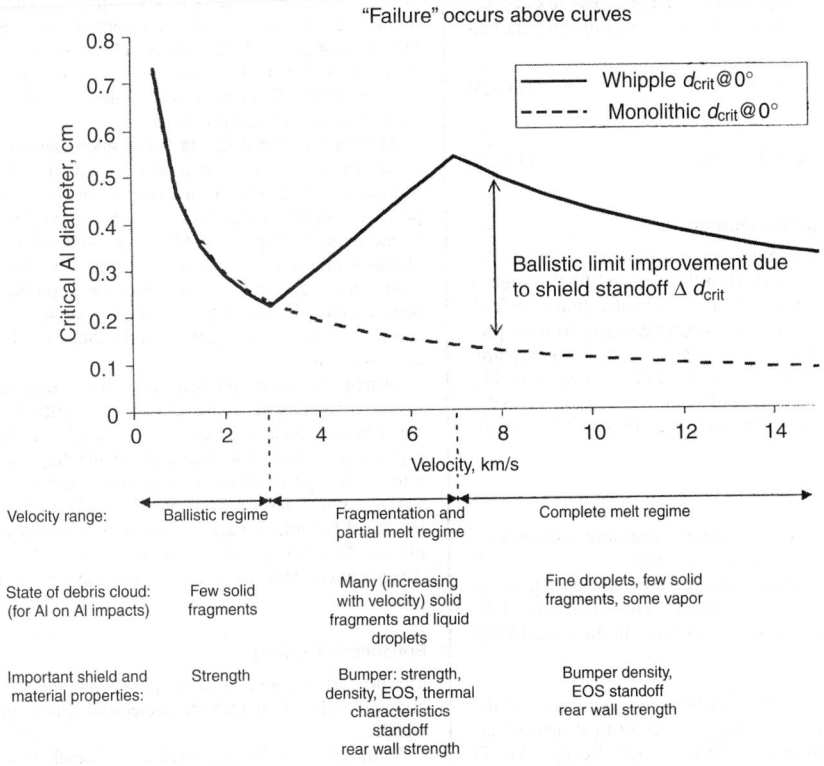

Fig. 11.6.40 Ballistic limits for equal mass monolithic target and Whipple shield. Failure criterion is threshold perforation or detached spall from rear wall. Monolithic target is 0.44-cm-thick Al 6061T6. Whipple shield consists of 0.12-cm-thick Al 6061T6 bumper followed at 10 cm by 0.32-cm-thick Al 6061T6 rear wall.

on the initial impact pressure. Solid fragments in the debris cloud are generally more penetrating when they contact the rear wall than liquid or vapor particles. Blast loading in the rear wall is a function of shield standoff, which becomes of increasing importance as impact velocity increases above about 5 km/s.

Bumper Code

The BUMPER code has been the standard in use by NASA and contractors to perform meteoroid/debris risk assessments since the early 1960s. Over that period of time, it has undergone extensive revisions and updates. NASA has applied BUMPER to risk assessments for space station, shuttle, Mir, extravehicular mobility units (i.e., "space suits"), and other satellites and spacecraft. Significant effort has been expended to validate BUMPER and "benchmark" it to other M/OD risk assessment codes used by some ISS international partners. Figure 11.6.31 illustrates where BUMPER fits in the risk assessment process. The BLEs and M/OD environment models are embedded into BUMPER. A finite-element model (FEM) that describes the spacecraft geometry is created in and input into BUMPER. BUMPER calculates the number of failures by determining the number of M/OD particles that exceed the ballistic limits for each element of the FEM, and calculates the total number of failures by summing the individual elements. BUMPER results are used to determine the risk drivers, that is, the areas on the spacecraft that control the risk. Emphasis on risk reduction is placed on lowering the risk for the drivers.

Probability of No Penetration (PNP)

Probability of no penetration (PNP) failure is assessed on the basis of Poisson statistics [Eq. (11.6.5)], where N is defined by Eq. (11.6.1). This is NASA's standard approach used to assess meteoroid shielding since Apollo. The same probabilistic approach has been extended to designing spacecraft protection systems for the combined meteoroid and debris environments, assessing risks of functional failure, or calculating risks of impact damage exceeding a user-defined size (diameter or depth). The same statistical approach is used to assess risks of any kind of damage extent that can be expressed in terms of different ballistic limit equations.

$$PNP = \exp(-N) \qquad (11.6.5)$$

The risk of penetration is:

$$Risk = 1 - PNP \qquad (11.6.6)$$

Analysis Iteration and Design Qualification

Assessed PNP is compared to requirements for M/OD protection. The shielding design effort is successful when the assessed PNP is greater than the required PNP. As illustrated in Fig. 11.6.31, iteration of the risk assessment and risk reduction process is often necessary to meet protection requirements and optimize the design, i.e., meet the requirements with less weight, lower volume (less standoff), less cost, etc. The last step in the process is to conduct final shield verification by analysis (supported by HVI test as appropriate), prior to final design certification and acceptance.

Risk Reduction Practices

Engineering practices to decrease risk, improve shielding performance, and decrease shielding mass are summarized below.

Find M/OD Risk Drivers   Perform detailed assessment of penetration risks for the overall vehicle to determine the zones that control the risks. Selectively improve the protection capability in the areas identified as risk drivers.

Shielding Enhancement

1. Optimize the shielding weight distribution to account for the directional M/OD distribution. Shielding on each critical item is tailored for the environment and its location on station. Because M/OD impact rate is highest on forward and side surfaces, more capable shielding (heavier or with greater standoff) is applied to these surfaces and less on the earth-facing surface. Shielding is also reduced in areas where shadowing from neighboring structures will reduce impacts. The goal of this effort is to equalize the ratio of risk to area for the various vehicle zones. The risk/area ratio for each zone is assessed and optimized by repeated BUMPER code runs.

2. Incorporate enhancements such as increasing the spacing between bumper and rear wall, or using higher-performance alloys and materials for the rear wall.

3. Implement more-fundamental changes such as incorporating more efficient, multibumper shielding concepts.

4. Incorporate toughening materials, such as ceramic fabric and high-strength fabric into the multilayer insulation (MLI) thermal blanket that is often integral to the M/OD shielding.

Impact Reduction

1. *Maximize shadowing.* Take advantage of shielding/shadowing from neighboring items. Locate external critical equipment to trailing or earth-facing surfaces to reduce M/OD impact rates, or put them in areas highly shadowed by other hardware.

2. *Select low frontal area orientations.* It has been shown that the best orientation for a relatively short cylinder (length/diameter of 1.8) in terms of lowest meteoroid/debris impact risk is with cylindrical length axis oriented perpendicular to the orbital plane. A vertical (gravity-gradient) orientation (with length axis parallel to earth radial) has a 30 percent higher M/OD impact risk. An orientation with length axis parallel to velocity vector is in the middle.

3. *Flight attitude.* Selecting the best flight attitude by pointing the most vulnerable surfaces aft or toward earth is standard procedure for the shuttle. Other spacecraft can take a similar approach.

Insert Stored Energy Equipment   After use, stored energy equipment should be made inert if possible. For instance, completely depressurize any storage tank after it is emptied. The risk of catastrophic rupture is eliminated when stress levels in the pressure wall are made negligible by depressurizing to a small value. This would require design modifications to implement for propellant tanks and other fluid storage tanks. Another example is to keep spare flywheels, gyros, and other momentum storage devices in an inactive state until required. A corollary idea for gaseous pressure vessels is to use them in series (not parallel) and deplete first the pressure vessels in the locations that are most exposed to M/OD impact (those in the forward or side positions), followed by less exposed positions.

Reduce Hazards if Shield Penetration Occurs   Design and operational options may exist to reduce hazards if a penetration occurs. For instance, some hatches to unoccupied modules can be kept closed to prevent a depressurization of an entire station if a penetration occurred to the module. A perforation into an unoccupied module with hatch closed would not result in loss of crew from the fragments/shrapnel, light flash, acoustic overpressure, or depressurization. Vent lines between modules could be left open to allow for some air circulation and to keep pressures equalized to facilitate hatch opening during normal operations.

Critical Damage Detection, Repair, and Replacement (R&R)   Inspection and repair of impact damage to critical areas of reentry vehicles (such as the crew return vehicle attached to ISS) can be used as a supplement to M/OD shielding for maintaining flight worthiness. Some impact damage to thermal protection systems (TPS) on earth-return vehicles is not a hazard while on orbit (and may therefore be undetected) but could become hazardous later during reentry aerodynamic heating phases. Properly placed instrumentation with correct sensitivity to detect critical TPS damage can help support the inspection/detection process.

Enhanced Shielding

The primary purpose of developing improved shielding is to provide higher levels of spacecraft meteoroid/debris protection with less weight.

To illustrate the issue, consider the shielding required to stop 1-cm-diameter aluminum projectiles at 7 km/s, 0° impact angle. Four shield concepts to meet the requirement are given in Fig. 11.6.41: a conventional aluminum Whipple shield with a 10.2-cm standoff, a

Areal density (kg/m²)				
	Whipple $S = 10$ cm	Stuffed Whipple $S = 10$ cm	Whipple $S = 30$ cm	Multishock $S = 30$ cm
Bumper:	7.0	10.6	5.6	5.2
Rear wall:	17.2	6.6	7.5	3.8
Total:	24.2	17.3	13.1	9.0
Surface area (m²)				
Bumper:	152	152	175	175
Rear wall:	141	141	141	141
Mass (kg) including 30% of bumper for support mass				
Bumper:	1060	1620	980	910
Support:	320	490	300	270
Rear wall:	2420	940	1060	540
Total:	3800	3050	2340	1720
(include support)				

Fig. 11.6.41 Shielding concepts comparison. (Projectile in all cases 1-cm-diameter aluminum, 1.5 g, 7 km/s, normal impact.)

Nextel/Kevlar stuffed Whipple shield with the same standoff, a Whipple shield with a 30-cm standoff, and a Nextel multishock shield concept. The stuffed Whipple shield incorporates a blanket between the outer aluminum bumper and inner pressure wall that combines two materials: Nextel™ ceramic fabric and Kevlar™ high-strength fabric. The shielding mass estimates are made assuming the shielding encloses a cylinder, with 4.2-m inside diameter by 8.5 m long. Stuffed Whipple and multishock shields are described in more detail later in this section. But it is clear for this example that there are significant mass savings by using advanced shielding concepts (i.e., up to 50 percent reduction).

Also, it is possible to trade weight for protection capability (i.e., capability and shield PNP are related), so it can be shown that lower weight and more effective protection in terms of higher PNP are possible by using Nextel/Kevlar stuffed Whipple and multishock shields, compared to conventional Whipple shields.

SPACE-VEHICLE STRUCTURES
by Thomas L. Moser and Orvis E. Pigg
NASA

REFERENCES: MIL-HDBK-5D, "Metallic Materials and Elements for Aerospace Vehicle Structures." Advanced Composites Design Guide, Sept. 1976, Library Accession no. AD916679 through AD916683. Ashton, Halpin, Pertit, "Primer on Composite Materials Analysis," Technomic Pur. Co. Aeronautical Structures Manual (Vols. I–III), NASA TMX-73305. Bruhn, "Analysis and Design of Flight Vehicle Structures," Tri-State Offset Co., Cincinnati. Barton, "Fundamentals of Aircraft Structures," Prentice-Hall, 1948. Roark, "Formulas for Stress and Strain," McGraw-Hill. "Nastran Theoretical Manual," Computer Software Management and Information Control (COSMIC), University of Georgia. Perry, "Aircraft Structures," McGraw-Hill. Gathwood, "Thermal Stress," McGraw-Hill. Zienkiewicz, "The Finite Filement," McGraw-Hill. MIL-STD-810D, "Environmental Test Methods and Engineering Guidelines." Chapman, "Heat Transfer," Macmillan.

Space vehicles include a large variety of vehicle types, like their predecessors, earth vehicles. Airplanes were the first generation of space vehicles, and our definition of space vehicle has changed as the airplane approached the limits of the earth's atmosphere. Space, in this discussion, is the region beyond the earth's atmosphere.

Many space vehicles are hybrids since they are required to operate in the earth's atmosphere and beyond. The space shuttle orbiter is such a hybrid space vehicle. The Apollo command and service module is also a hybrid, but the Apollo lunar module, which primarily functioned outside the earth's atmosphere, was a space vehicle.

The design procedure for any vehicle structure is basically the same:

1. Establish the design requirements and criteria
2. Define the loads and environments
3. Perform the design and analysis
4. Iterate the previous two steps (as required) because of the interrelation between configuration and loads
5. Verify the design

There are conditions and requirements which are unique for hybrid and space-vehicle structures.

Minimum weight and high reliability are significant but opposing requirements for space-vehicle structures which must be balanced to achieve an efficient, reliable structure design. Depending on the mission of the particular space vehicle, the ratio between the weight of the earth launch vehicle and the payload (e.g., interplanetary probe) can be as high as 400 to 1. The large weight leverages require that the design be as efficient as possible. Structural materials, therefore, must be stressed (or worked) as close to the capabilities as possible. Like aircraft structures, space-vehicle structures should be designed for a zero margin of safety (MS):

$$MS = \frac{\text{material ultimate allowable}}{\text{FS (limit stress)}} - 1$$

1. *Material ultimate allowable.*   The stress at which the material will fail.

2. *FS (factor of safety).*   This factor usually ranges between 1.25 and 1.50, for space structures, depending on the maturity of the design, the behavior of the material, the confidence in the loads, etc.

3. *Limit stress.*   The maximum (reasonable) stress that the structural component would be expected to encounter.

Failure can be based on rupture, collapse, or yield as dictated by the functional requirements of the component. High reliability is necessary for most space vehicle structures since a failure is usually catastrophic or nonrepairable. Total structural reliability is composed of many of the design elements—loads, materials, allowables, load paths, factor of safety, analysis techniques, etc. If the reliability is specified and the statistical parameters are known for each design element, then the process is straightforward. A total structural reliability is seldom specified and good engineering practice is used for the solution of the reliability of each design element. Commonly used to assure structural reliability are (1) material allowables equivalent to A values from MIL-Handbook-5A for safe-life structure and B values for fail-safe structures; (2) limit loads which correspond to events and conditions which would not be exceeded more often than 3 times in 1,000 occurrences; (3) factor of safety equal to 1.4 for crew-carrying space vehicles and 1.25 for those without crews; and (4) verification of the structural design by test demonstration. A good structural design is one which has the correct balance between minimum weight and high reliability.

Space vehicles are usually manufactured and assembled on earth and transported to space. The launch environments of acceleration, vibration, aerodynamic pressure, rapid pressure change, and aerodynamic heating coupled with the space environments of extremely low pressure, wide ranges in temperature, meteroids, and radiation dictate that the design engineers accurately define the magnitudes of these environments and the correct time-compatible combinations of these environments if realistic loads are to be determined. Many natural environments have been defined and are documented in the references. The induced environments, which are a function of the system and/or structural characteristics (e.g., dynamic pressure, vibration, etc.), must be determined by analysis, test, or extrapolation from similar space vehicles. Quantifying the induced loads for the structural designers is one of the major consumers of time and money in any aerospace program and one of the most important. Load determination can be a very complex task. For further guidance the reader is directed to the references.

For a space vehicle which is not exposed to the launch loads, the load determination is greatly simplified. In the absence of an appreciable atmosphere and gravitational accelerations the predominant external force on a space-vehicle structure is that of the propulsion and attitude control systems and the resulting vibration loads. Even though not induced by an external force, thermal stresses and/or deflections can be a major design driver. Space vehicles commonly experience larger thermal gradients because of the time- or attitude-related exposure to solar radiation.

The thermal effects of stress and deflections of the structure require that temperature distributions be defined consistent with the strength and deflection requirements. From a strength integrity standpoint, temperature distribution on a space vehicle can be as important as the pressure distribution on an aircraft. The significance and sensitivity of thermal effects are greatly reduced by the use of composite materials for the structure. The materials can be layered to achieve, within limits,

the desired thermal coefficient of expansion and thereby reduce the thermal effects.

Space vehicles which return to the earth must withstand the extremely high temperatures of atmospheric reentry. This produces dynamic temperature distributions which must be combined with the aerodynamic pressures and vehicle dynamic loads.

Innovative, clever, and efficient structural designs minimize the effects of loads associated with launch for a vehicle which is to remain in space. This is accomplished by attaching the space vehicle to the booster rocket to minimize scar weight and to reduce the aerodynamic loads by the use of shrouds, fairings, or controlling the attitude during atmospheric flight.

Efficiencies of weight-critical structures have been greatly improved by more accurately quantifying the loads and stress distributions in complex structural configurations. **Finite element methods** (FEM) of analysis and large capacity, high-speed digital computers have enabled the increased fidelity of analyses. In this process, the structure is modified with finite elements (beams, bars, plates, or solids) by breaking the structure into a group of nodal points and connecting the nodal points with the finite elements which best represent the actual structure. Loads are applied at the nodal points or on the elements which then distribute the loads to the nodal points. Constraints are also applied to the nodal points to support or fix the structure being analyzed. The finite element program then assembles all the data into a number of linear simultaneous equations and solves for the unknown displacements and then for internal loads and stresses, as required. Many finite element programs have been developed to perform a variety of structural analyses including static, buckling, vibration, and response.

The loads in a structure depend on the characteristics of the structure and the characteristics of the structure are largely dictated by the loads. This functional dependency between loads and structures requires an iterative design process.

Classically, when structures are worked to stresses near the capability of the materials, e.g., airplane structures, the design is verified by test demonstration. This method of verification was common before high-fidelity structural mathematical models were employed and is still used when practical to apply the design loads and environments to a dedicated test article to verify the design. Design deficiencies are then corrected depending on the importance and the manufacturing schedule.

Space-vehicle structures often do not lend themselves to the classical test approach because of the multitude of loads and environments which must be applied, because testing in earth's gravity precludes realistic loadings, and because of the size and configuration. Verification by analyses with complementary component tests and tests of selected load conditions are common.

INFLATABLE SPACE STRUCTURES FOR HUMAN HABITATION
by William C. Schneider
Texas A & M University

Inflatable space structures have been considered for many years and even some have been put into practice for applications such as large reflecting satellites, shock attenuation, and reentry vehicles. Their clear advantage over rigid structures from the standpoint of weight, compactness, and large volume potential have been recognized for years, but only recently, as a result of the development of very high strength fabrics, has serious development of inflatable structures for human application taken place. NASA, as well as some private aerospace companies, recognize the increasing demand for human-inhabited inflatable space structures in astronautics. As newer, even higher strength, fabrics are developed, the full utilization of inflatable structures for human space will be realized. In the following subsections the TransHab and the portable hyperbaric chamber are discussed, and they represent only two examples of the emerging inflatable space structure technology for human habitation.

Fig. 11.6.42   TransHab inflatable habitation module.

TransHab
by Horacio de la Fuente
NASA

One example of the significant advantages derived from the use of inflatable technology for human space applications was demonstrated by the TransHab inflatable habitation module. Figure 11.6.42 shows a unit developed at the NASA Johnson Space Center. The TransHab is an inflatable human habitation module (30 ft long × 25 ft in diameter) originally conceived as an interplanetary module for use in spaceflight missions beyond low earth orbit. The long transit times required for these types of missions create several significant design challenges. Because multiple crewmembers will be living in a confined space for long periods of time, psychological health required the spacecraft to have a large-volume living space yet be lightweight. The ability of inflatable structures to be folded and packaged during the launch environment is a key feature that allows this requirement to be met. The advantages of inflatable modules, however, go beyond the obvious volumetric packing efficiency. The inflatable TransHab is folded and packaged tightly during the launch phase of the mission and is inflated only after it is in the quiescent space environment. Huge weight savings are realized when compared to a large rigid shell structure, since the rigid shell must be designed to withstand the severe steady-state, transient, and vibration launch loads as a large, rigid shell. There is, however, no requirement for the inflated TransHab to react to the launch loads. It must only withstand the differential pressure experienced in space.

The TransHab module design was later adapted to serve as a habitation module for the International Space Station. This change primarily resulted in modifications to some of the inflatable shell layers. As described later, the shell layers most affected by this change were the

orbital debris protection layer, the atomic oxygen protection layer, and the restraint layer, which was strengthened to be compatible with the 14.7-lb/in^2 internal pressure of the International Space Station.

Central Structure Core   Turning the large volume into a functional living environment creates additional challenges. Once inflated, the space habitation module requires several internal integrated subsystems such as environmental control and life support, power distribution, communication, and data handling, as well as crew mobility and translations aids. In addition, to maintain psychological health during the long stay times, dedicated living spaces for exercise, personal hygiene, food preparation and consumption, and sleeping are required. All of the above require internal structure as well as a significant amount of preintegration. For this reason the TransHab is designed as a preintegrated system. The integrated systems approach for TransHab used a hard central structural core, shown in Fig. 11.6.43, to provide the necessary backbone structure to allow for the preintegration and checkout of all the subsystems before launch, to support the subsystems during launch, and to provide the necessary internal structure for the module after inflation.

The inflatable shell of the TransHab is folded and wrapped around the central structural core during launch, then released and inflated on orbit, leaving a central structural core running down the central axis of the inflated module. All of the necessary structural elements needed to support the internal distributed systems, crew translation aids, and dedicated living spaces are obtained from the central structural core, which is already inside the module. Many of the structural elements of the central structural core are repositioned after inflation to provide a new function within the module. A good example is the use of removable shelf elements in the core. These shelves, which contain preintegrated subsystem components, are used to strengthen the core during launch and are

Launch support for subsystems

Central backbone on orbit

Fabric Floor

Removable Shelves

Fig. 11.6.43   Central structural core.

then repositioned adjacent to the core after inflation. Another example is a stretched fabric flooring system that partitions the internal volume into three separate "floors" after inflation. These floor system elements were also used to provide a protective barrier between the central structural core and the inflatable shell during launch.

The approach of reusing elements for multiple purposes is a common feature used throughout the TransHab design.

Inflatable Shell   The design of the inflatable shell is driven by several requirements. These requirements include the need for flexure resistance, high tensile strength, low air permeability, internal puncture resistance, protection from meteoroid and orbital debris, thermal insulation, and protection against atomic oxygen. Finding a single material construction that meets all these requirements is difficult. As shown in Fig. 11.6.44, the TransHab design uses a multilayer construction to satisfy all of these requirements. This allows each layer to be optimized for its specific function. In addition, the use of separate layers loosely connected to each other allows the layers to slip relative to each other during the folding process. The different layers used in the TransHab design are an inner protective liner, a redundant bladder system, a strong restraint layer, a multilayer meteoroid and orbital debris layers, a multilayer insulation system, and an atomic oxygen protection layer. These layers are described briefly below.

Inner protective liner. The primary function of the inner liner is to protect the bladder layers from internal puncture and from flammability

hazards. It uses a Nomex fabric backed by Kevlar felt. This layer is directly attached to the innermost bladder layer.

Bladder system. The primary function of the bladder system is to contain the internal gas, in this case air, from escaping. Because this layer is so critical, the TransHab design uses more than one bladder for redundancy in the event of a puncture. The bladder layers are made from a polyurethane-based laminated film, which provides extremely low permeability as well as excellent flexure characteristics at low temperature. The bladder system is oversized so that it transfers the entire pressure load to the restraint layer.

Restraint layer. The primary function of the restraint layer is to react the shell loads created by internal pressure. The restraint layer is composed of interwoven Kevlar straps running in a circumferential and a longitudinal direction.

Orbital debris protection layer. The primary function of the orbital debris protection layer is to act as a shield to protect the inboard layers from space debris impacts. To shield against hypervelocity impacts, the debris protection layer consists of multiple layers of Nextel and Kevlar™ cloth separated by foam spacers. While the spacing of the debris protection layers is needed for the shield to function, it also makes the inflatable shell difficult to fold and package. The use of open cell foam inside individual bags allows the foam spacers to be vacuum packed and collapsed for easy folding. After deployment of the inflatable shell, the lack of external atmospheric pressure allows the foam to expand back to its original state. As a result, the debris protection layer contributes approximately 16 inches to the overall thickness of the inflatable shell after deployment.

Thermal insulation layer. The primary function of the thermal insulation layer is to protect the inboard layers from the extreme on-orbit temperatures. The thermal insulation layer for TransHab uses standard multilayer insulation blankets with some modifications to account for having the shell constrained during launch and then released on orbit.

Atomic oxygen layer. The primary function of the atomic oxygen layer is to protect the inboard layers from atomic oxygen degradation in low earth orbit. The TransHab module uses a standard Beta-cloth layer for this purpose.

Integrity Tests   Two of the key TransHab features, the ability to protect against space debris and the structural integrity of the TransHab restraint layer design, were verified by test. Sample cross sections of the debris protection layer were shot with a hypervelocity gun. The debris protection layer successfully stopped an approximately 2-cm-diameter aluminum projectile traveling at 7 km/s from penetrating into the bladder layers. A 1-g particle traveling at 11.5 km/s was also tested successfully. The structural integrity of the

Thermal Insulation Layer   *Redundant Bladders*

SPACE ← → **Pressurized Volume**

Atomic Oxygen Cover   *Inner Liner*

Debris Shielding   *Restraint Layer*

Fig. 11.6.44   Multilayer inflatable shell.

Fig. 11.6.45 Structural integrity test article.

restraint layer was test-verified through hydrostatic pressure testing. A full-scale-diameter, yet shorter length, test article, shown in Fig. 11.6.45, was fabricated and hydrostatically pressurized. While space habitation modules are designed to withstand a structural factor of safety equal to 2 (2× the operating pressure) and are typically 14 ft in diameter, the TransHab test article was 23 ft in diameter and was tested to a factor of safety equal to 4 (4× the operating pressure). Because pressure vessel stresses increase linearly with diameter, the test showed that the test article had much greater structural capability than its metallic counterpart. Since the structural weight of the TransHab module was comparable to that of existing metallic habitation modules, the test showed the significant advantages of inflatable habitation modules.

Portable Hyperbaric Chamber
by Chris Hansen and Jim Locke
NASA

The assembly and maintenance of the International Space Station requires that astronauts perform many extravehicular activities (EVAs), or spacewalks. The interior habitable compartment contains a nitrogen/oxygen atmosphere maintained at sea level atmospheric pressure (14.7 lb/in² or 760 mm Hg). Current spacesuit designs must be operated at significantly lower pressures, so that they are flexible enough to allow the performance of meaningful work. The NASA extravehicular mobility unit (EMU) spacesuit operates in space at a pressure of 4.3 lb/in² (220 mm Hg), or an earth altitude equivalent of approximately 30,000 ft. Depressurizing from sea level to 30,000 ft can cause the nitrogen gas dissolved in body tissues to leave solution and form gas bubbles, which can lead to decompression sickness (DSC), or "the bends." The risk of such altitude DCS can be mitigated by breathing pure oxygen prior to depressurization. This safely decreases tissue nitrogen content and decreases bubble formation. However, even with these procedures, the risk of serious, or even fatal, DCS remains. DCS can be effectively treated by using a hyperbaric chamber, a room containing air at high pressure (typically equivalent to a depth of 60 ft of seawater). Breathing pure oxygen under such high atmospheric pressure forces the nitrogen bubbles back into solution. In general, patients can be easily cured of even severe DCS if the hyperbaric chamber treatment is initiated quickly. Since it takes more than 24 hours to leave the space station, reenter the earth's atmosphere, land, and reach a terrestrial hyperbaric chamber, placing such a chamber on the station would be highly desirable. Unfortunately, hyperbaric chambers are typically very large and heavy, making their use on the space station impractical.

However, with the development of lightweight, high-strength inflatable construction techniques developed for the TransHab project, it was possible to design and build a portable hyperbaric chamber that was light enough to put on the space station, and could be stowed away in a relatively small space. In the late 1990s, a small NASA group was funded to develop a prototype for a portable hyperbaric chamber (PHC) that could be tested and eventually used on the space station. In the first phase of the project, the team concentrated on the structural components of the chamber, since it was considered to be the most difficult technical challenge. While TransHab had revolutionized inflatable space structures, the PHC had to be greatly scaled down from TransHab while still utilizing, as much as possible, the materials research and construction techniques previously developed. The PHC had to be just big enough to accommodate two astronauts: one (a patient) with DCS and a second to tend to the possibly incapacitated patient.

The structure of the PHC is shown in Fig. 11.6.46. It is composed of an inner, airtight polyurethane bladder surrounded by a woven Kevlar restraint layer. The restraint layer is actually composed of two sets of orthogonal interwoven Kevlar straps. One end of the chamber has a high-strength 7050 aluminum interface ring that acts as a doorway to the chamber and provides the sealing surface for the hatch and feedthroughs for air, pure oxygen, communication equipment, etc. The hatch, also high-strength aluminum, is machined in an oval shape, allowing the hatch to be internally sealing, yet allowing it to be removed by rotating it sideways and sliding it through the oval doorway. By having an internally sealing hatch, the hatch structure is greatly reduced in size and the need for numerous heavy latches is eliminated. The interface ring has series of steel brackets with Delrin rollers that connect the restraint layer to the interface ring. The Delrin rollers reduce the stress concentration in the straps at the interface ring

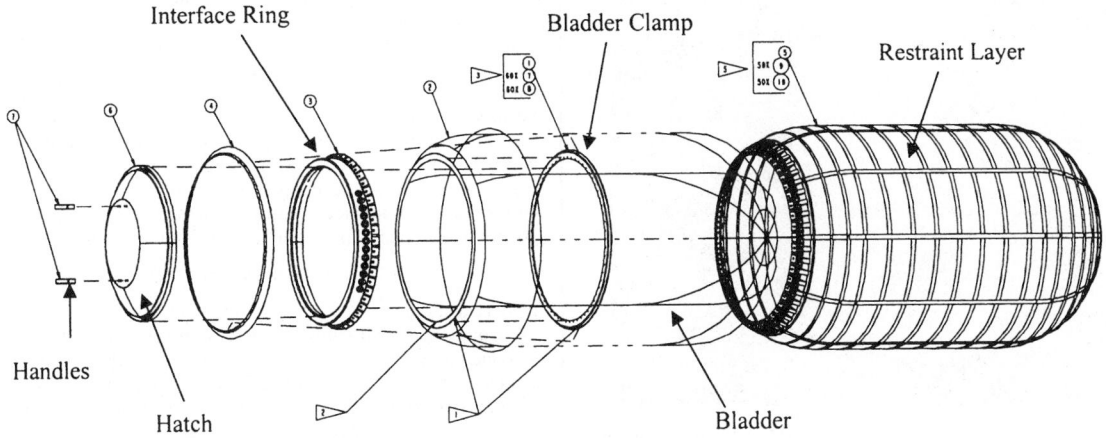

Fig. 11.6.46 Portable hyperbaric chamber structure (shown in an exploded view).

Fig. 11.6.47 Cross section of the hatch, interface ring, and restraint layer interface.

enough to allow the full pressure of the chamber to be developed without failure of the Kevlar straps. The seal between the hatch and the interface ring is formed with an O ring. The O-ring interface and the Delrin roller are shown in Fig. 11.6.47.

One of the most challenging parts of the structural design was developing a method of manufacturing the restraint layer. Made from 1-in Kevlar straps, the restraint layer had to be woven into the cylindrical shape of the chamber. Unlike TransHab, the restraint layer shape could not be formed into a flat pattern because it did not have an interface ring at each end. The second interface ring was eliminated

Fig. 11.6.48 Portable hyperbaric chamber structural prototype during low-pressure leak test.

to save weight, although it complicated the restraint layer design. However, this decision led to an estimated 30 percent reduction in weight.

After completion of the structural prototype, a pressurized hydrostatic test was performed. The working pressure of the chamber is 1.8 atm, and the chamber is designed to withstand 4.0 atm without failure. The final prototype is shown in Fig. 11.6.48 during a low-pressure leak test. All testing of the PHC was successful and demonstrated the flexibility of the inflatable structure technology developed during the TransHab project.

VIBRATION OF STRUCTURES
by Lawrence H. Sobel
University of Cincinnati

REFERENCES: Crandall and McCalley, Numerical Methods of Analysis, chap. 28, in Harris and Crede (eds.), "Shock and Vibration Handbook," Vol. 2, McGraw-Hill. Hurty and Rubinstein, "Dynamics of Structures," Prentice-Hall. Meirovitch, "Analytical Methods in Vibrations," Macmillan. Przemieniecki, "Theory of Matrix Structural Analysis," McGraw-Hill. Wilkinson, "The Algebraic Eigenvalue Problem," Oxford.

As a result of the vast improvements in large-scale, high-speed digital computers, numerical methods are being used at an exponentially increasing rate to analyze complex structural vibration problems. The numerical methods most widely used in structural mechanics are the finite element method and the finite difference method. In the finite element method, the actual structure is represented by a finite collection, or assemblage, of structural components whereas, in the finite difference method, spatial derivatives in the differential equations governing the motion of the structure are approximated by finite difference quotients. With either method, the continuous structure with an infinite number of degrees of freedom is, in effect, approximated by a discrete system with a finite number of degrees of freedom (unknowns). The linear, undamped free and forced vibration analysis of such a discrete system is discussed herein. The analysis is most conveniently carried out with the aid of matrix algebra, which is well suited for theoretical and computational purposes. The following discussion will be brief, and the reader is referred to the cited references for more comprehensive treatments and examples.

Equation of Motion

Consider a conservative discrete system undergoing small motion about a state of equilibrium (neutral or stable). Let $q_1(t), \ldots, q_n(t)$, where t is time, be the minimum number of independent coordinates (linear or rotational) that completely define the general dynamical configuration of the system and that are compatible with any geometrical constraints imposed on the system. Then the n coordinates $q_1, \ldots, q_n$ are called generalized coordinates, and the system is said to have n degrees of freedom. Let $\{Q(t)\}$ be the vector of generalized forces, and let $\{q(t)\}$ be the vector of generalized displacements, $\{q(t)\} = \{q_1(t), \ldots, q_n(t)\}$. Corresponding elements of the generalized force and displacement vectors are to be conjugate in the energy sense, which simply means that if one of the elements of $\{q\}$ is a rotation (for instance), then the corresponding element of $\{Q\}$ must be a moment, so that their product represents work or energy. Methods of computing $\{Q\}$ are discussed in the references (e.g., Meirovitch). The matrix equation governing the small undamped motion of the system is given by

$$[M]\{\ddot{q}\} + [K]\{q\} = \{Q\} \qquad (11.6.7)$$

In this equation a dot denotes single differentiation in time, and $[M]$ and $[K]$ are the mass and stiffness matrices, respectively. $[M]$ and $[K]$ are real and constant matrices, which are assumed to be symmetric. Such symmetry will arise whenever an energy approach (e.g., Lagrange's equations) is employed to derive the equation of motion. The mass matrix is assumed always to be positive definite (and hence nonsingular, or can be made to be positive definite; see Przemieniecki).

Free-Vibration Analysis

The following equation governing the free vibration motion of the conservative system is obtained from Eq. (11.6.2) with $\{Q\} = \{0\}$: $[M]\{\ddot{q}\} + [K]\{q\} = \{0\}$. To obtain a solution of this equation, we note that, for a conservative system, periodic motion may be possible. In particular, let us assume a harmonic solution for $\{q\}$ in the form $\{q\} = \{A\}\cos(\omega t - \alpha)$. That is, we inquire whether it is possible for all coordinates to vibrate harmonically with the same angular frequency ω and the same phase angle α. This means that all points of the system will reach their extreme positions at precisely the same instant of time and pass through the equilibrium position at the same time. Substitution of this trial solution into the free vibration equation yields the eigenvector equation $[K]\{A\} = \omega^2[M]\{A\}$. In this equation $\{A\}$ is called the eigenvector and ω^2 is the eigenvalue (the square of a natural frequency). This homogeneous equation always admits the trivial solution $\{A\} = \{0\}$.

Nontrivial solutions are possible provided that ω^2 takes on certain discrete values, which are obtained from the requirement that the determinant of the coefficient matrix of $\{A\}$ vanishes. This yields the eigen-equation (frequency equation) $\Delta(\omega^2) = |[K] - \omega^2[M]| = 0$. It is an nth-degree polynomial in ω^2, and its n roots are denoted by ω_r^2, $r = 1, \ldots, n$. The eigenvector $\{A\}_r$, $r = 1, \ldots, n$, corresponding to each eigenvalue is determined from $[K]\{A\}_r = \omega_r^2[M]\{A\}_r$. However, the homogeneous nature of this equation precludes the possibility of obtaining the explicit values of all the elements of $\{A\}_r$. It is possible to determine only their relative values or ratios in terms of one of the components. Thus, we see that for each eigenvalue ω_r^2 there is a corresponding eigenvector $\{A\}_r$, which has a unique shape (based on the relative values of its components) but which is determined only to within a scalar multiplicative factor that may be regarded as being the amplitude of the shape. This unique shape is called the mode shape, and it can be thought of as being the position of the system when it is in its extreme position, and hence momentarily stationary, as in a static problem. Each frequency along with its corresponding mode shape is said to define a so-called natural mode of vibration. Methods of determining these natural modes are presented in the references (e.g., Wilkinson). If the initial conditions are prescribed in just the right way, it is possible to excite just one of the natural modes. However, for arbitrary initial conditions all modes of vibration are excited. Thus, the general solution of the free-vibration problem is given by a linear combination of the natural modes, i.e.,

$$\{q\} = \sum_{r=1}^{n} C_r\{A\}_r \cos(\omega_r t - \alpha_r)$$

The $2n$ arbitrary constants C_r and α_r are determined through specification of the $2n$ initial conditions $\{q(0)\}$, and $\{\dot{q}(0)\}$. An explicit representation for $\{q\}$ in terms of $\{q(0)\}$ and $\{\dot{q}(0)\}$. is given later.

Some of the properties of the natural modes of vibration will now be listed. Unless stated otherwise, the following properties are all consequences of the assumptions that $[M]$ and $[K]$ are real and symmetric and that $[M]$ is positive definite. (1) The n eigenvalues and eigenvectors are real. (2) If $[K]$ is positive definite (corresponding to vibration about a stable state of equilibrium), all eigenvalues are positive. (3) If $[K]$ is positive semidefinite (corresponding to vibration about a neutral state of equilibrium), there is at least one zero eigenvalue. All zero eigenvalues correspond to rigid-body modes. (4) Eigenvectors $\{A\}_r$, $\{A\}_s$ corresponding to different eigenvalues ω_r^2, $\omega_s^2 (\omega_r^2 \neq \omega_s^2; r \neq s; r, s = 1, \ldots, n)$ are orthogonal by pairs with respect to the mass and stiffness matrices (weighting matrices), that is, $\{A\}_r^T[M]\{A\}_s = 0$ and $\{A\}_r^T[K]\{A\}_s = 0$ for $r \neq s; r, s = 1, \ldots, n$. Since the eigenvectors are orthogonal, and hence independent, they form a basis of the n-dimensional vector space. Thus, any vector in the space, such as the solution vector $\{q\}$, can be represented as a linear combination of the eigenvectors (base vectors). This theorem (expansion theorem) is of vital importance for the response problem to be considered next.

Response Analysis

The response of the system due to arbitrary deterministic excitations in the form of initial displacements $\{q(0)\}$, initial velocities (impulses)

$\{\dot{q}(0)\}$, or forcing functions $\{Q(t)\}$ may be obtained by means of the above expansion theorem. That is, the displacement solution of the response problem is expanded with respect to the eigenvectors of the corresponding free-vibration problem. The time-dependent coefficients in this expansion are the so-called normal coordinates of the system. The equations of motion when expressed in terms of the n normal coordinates are uncoupled, and each one of the equations is mathematically identical in form to the equation governing the motion of a simple one-degree-of-freedom spring-mass system. Hence solutions of these equations are readily obtained for the normal coordinates. The normal coordinates are then inserted into the eigenvector expansion to obtain the following explicit solution for displacement response $\{q(t)\}$ (Przemieniecki) for undamped constrained or unconstrained (free) structures:

$$
\begin{aligned}
\{q(t)\} = & \sum_{r=1}^{m} \frac{\{A\}_{ro}\{A\}_{ro}^T}{\{A\}_{ro}^T[M]\{A\}_{ro}} \Bigg([M](\{q(0)\} + t\{q(0)\}) \\
& + \int_{\tau2=0}^{\tau2=t}\int_{\tau1=0}^{\tau1=\tau2} \{Q(\tau_1)\}\, d\tau_1\, d\tau_2 \Bigg) \\
& + \sum_{r=m+1}^{n} \frac{\{A\}_{re}\{A\}_{re}^T}{\{A\}_{re}^T[M]\{A\}_{re}} \Bigg([M]\Big(\{q(0)\}\cos\omega_r t + \frac{1}{\omega_r}\{\dot{q}(0)\}\sin\omega_r t\Big) \\
& + \frac{1}{\omega_r}\int_{\tau=0}^{\tau=t}\{Q(\tau)\}\sin\omega_r(t-\tau)\, d\tau \Bigg)
\end{aligned}
\tag{11.6.8}
$$

From this equation, it is seen that the displacement response is expressed directly in terms of the arbitrary excitations and the free-vibration frequencies and eigenvectors. The eigenvectors have been separated into two types denoted by $\{A\}_{ro}$ and $\{A\}_{re}$. The "rigid-body" eigenvectors $\{A\}_{ro}$ correspond to the m rigid-body modes (if there are any) for which $\omega_r = 0$, $r = 1, \ldots, m$. The "elastic" eigenvectors $\{A\}_{re}$ correspond to the remaining $n - m$ eigenvectors, for which $\omega_r \neq 0$, $r = m + 1, \ldots, n$. The general solution given by Eq. (11.6.3) is very useful provided that the second integral (Duhamel's integral) can be evaluated. (The first integral is easier to evaluate.) Closed-form solutions of this integral for some of the simpler types of elements of $\{Q(t)\}$, such as step functions, ramps, etc., are available (see Przemieniecki). Solutions of Duhamel's integral for more complicated forcing functions can be obtained through superposition of solutions for the simpler cases or, of course, from direct numerical integration.

The displacement response obtained from Eq. (11.6.8) may be employed to obtain other response variables of interest, such as velocities, accelerations, stresses, strains, etc. Velocities and accelerations are obtained from successive time differentiations of Eq. (11.6.8). The method used to evaluate the stresses depends on the type of spatial discretization employed to approximate the structural continuum. For example, difference quotients are used in the finite difference approach. Note that the rigid-body components do not contribute to the stress or strain response and hence the first summation in Eq. (11.6.8) can be omitted for computation of these responses.

SPACE PROPULSION
by Henry O. Pohl
NASA

The propulsion system provides the maneuverability, speed, and range of a space vehicle. Modern propulsion systems have diverse forms, however, they all convert an energy source into a controlled high velocity stream of particles to produce force. Among the many possible energy sources, five are considered useful for space-vehicle propulsion:

1. Chemical reaction
2. Solar energy
3. Nuclear energy
4. Electrical energy
5. Stored energy (cold gas)

Accordingly, the various propulsion devices are categorized into chemical energy propulsion, nuclear energy propulsion, stored energy propulsion, solar energy propulsion and electrical energy propulsion.

Chemical propulsion can be further categorized into liquid propulsion or solid propulsion systems. Many combinations of fuels and oxidizers are available. Likewise, electrical energy propulsion can be subdivided into systems using resistive heating, arc heating, plasma, or ion propulsion. Nuclear energy and solar energy are external energy sources used to heat a gas (usually hydrogen).

All space propulsion systems, with the exception of ion propulsion, use nozzle expansion and acceleration of a heated gas as the mechanism for imparting momentum to a vehicle. The fundamental relationships of space propulsion systems are commonly expressed by a number of parameters as follows: The basic performance equation can be derived from the classical newtonian equation $F = MA$. It is defined as

$$ F = \left(\frac{\dot{W}}{g} \right) V_e + A_e (P_e - P_a) $$

where F = thrust, lbf; $\dot{W}$ = propellant weight flow rate; g = gravitational acceleration; V_e = exhaust velocity; A_e = nozzle exit area; P_e = pressure at nozzle exit; and P_a = ambient external pressure.

The specific impulse is one of the more important parameters used in rocket design. It is defined as $I_s = F/W$, where I_s = specific impulse in seconds, F = thrust, and W = propellant weight flow rate per second of operation. I_s increases in engines using gas expansion in a nozzle as the pressure and temperature increase and/or the molecular weight of the gas decreases. The specific impulse values for representative space propulsion systems are shown in Table 11.6.12.

The total impulse $I_t = \int F(t) \, dt$ is useful to determine candidate rocket firing duration and thrust combinations required to satisfy a given change in momentum, where F = thrust and t = time.

Mass fraction is the ratio of propellant to total weight of the vehicle and is defined as $\gamma = (W_i - W_f)/W_i$, where γ = mass fraction; W_i = total weight of the vehicle including the propellant, structure, engine, etc., at the start of operation; and W_f = total weight of the vehicle less the weight of propellant consumed.

In the simplified case of a vehicle operating free of atmospheric drag and gravitational attractions, the velocity increment obtained from a propulsion system is defined as $\Delta V = I_s g \ln(W_i/W_f)$, where ΔV = velocity increment obtained; W_i = total weight of vehicle including propellant, structure, engine, etc., at start of operation; and W_f = total weight of vehicle less weight of propellant consumed during operations. This equation brings together the mission parameters, vehicle parameters, and rocket-engine parameters.

Figure 11.6.49 graphically shows the effect specific impulse and mass fraction have on the ability of a propulsion system to change the velocity of a payload in space.

Fig. 11.6.49 Performance curves for a range of specific impulse and mass fractions.

Chemical Propulsion Only chemical energy has found wide acceptance for use in space propulsion systems. Modern materials and current technologies have made it possible to operate rockets at relatively high pressures. When these rockets are fitted with very large expansion-ratio nozzles (nozzle exits 40 to 200 times the area of the throat) the force produced per unit of propellant consumed is greatly increased when operating in the vacuum of space.

Solid-propellant rocket motors are used for moderate total impulse requirements, in spite of their low specific impulse, provided that the total impulse requirements can be precisely set before the rocket is built. The low specific impulse for these applications is largely offset by the high mass fraction (ratio of propellant weight to total propulsion system weight including propellant weight). They are simple to start, have good storage characteristics and multiple velocity requirements can be accommodated by staging, i.e., using a separate motor for each burn.

Liquid-propellant systems are generally used for moderate to high total impulse requirements, multiple burns, velocity-controlled thrust terminations, variable thrust requirements or for applications where the total impulse requirement is likely to change after the rocket is built or launched.

Bipropellant systems are the most commonly desired systems to satisfy moderate to high total impulse requirements. A fuel and oxidizer are injected into a combustion chamber where they burn, producing large volumes of high-temperature gases. Oxygen and hydrogen produce the highest specific impulse of any in-service chemical rocket (440 to 460 s). These systems also have good mass fractions (0.82 to 0.85). Oxygen and hydrogen, as stored in the rocket propellant tanks, are cryogenic, with boiling points of $-296.7°F$ and $-442°F$, respectively. Storing these propellants for long periods of time requires thermally insulated propellant containers and specially designed thermodynamic vents or refrigeration systems to control boil-off losses. The low density of hydrogen (4.4 lb/ft³) requires the use of bulky containers. Because of these characteristics, oxygen/hydrogen rocket systems can best satisfy moderate to high total-impulse requirements. Where designs are constrained by volume, very long durations of time between firings, many short burns, or precise impulse requirements, nitrogen tetroxide and monomethylhydrazine are often preferred, if the total impulse requirements are moderate.

Monopropellant systems generally use a single propellant that decomposes in the presence of a catalyst. These are usually low total-impulse

Table 11.6.12 Specific Impulse for Representative Space Propulsion Systems

Engine type	Working fluid	Specific impulse
Chemical (liquid)		
Monopropellant	Hydrogen peroxide-	110–140
	hydrazine	220–245
Bipropellant	O_2-H_2	440–480
	O_2-hydrocarbon	340–380
	N_2O_4-Monomethyl	300–340
	hydrazine	
Chemical (solid)	Fuel and oxidizer	260–300
Nuclear	H_2	600–1000
Solar heating	H_2	400–800
Arc jet	H_2	400–2000
Cold gas	N_2	50–60
Ion	Cesium	5000–25,000

systems used to provide attitude control of a spacecraft or vernier velocity corrections. For attitude control, small thrustors are arranged in clusters around the perimeter of the vehicle. These small thrustors may, in the course of a single mission, produce over 1,000,000 impulses, to keep the vehicle oriented or pointed in the desired direction.

Solar-heating propulsion would use a solar collector to heat the working fluid which is exhausted through a conventional nozzle. The concentrated solar energy heats the rocket propellant directly through the heat exchanger. The heated propellant is then exhausted through the nozzle to produce thrust. Many design problems make the practical development of the solar-heating rocket quite difficult. For example, the solar collector must be pointed toward the sun at all times and the available solar energy varies inversely with the square of the distance from the sun. Using hydrogen as a working fluid, this type system would have a specific impulse of 400 to 800 s and could satisfy very high total impulse requirements at low thrust levels.

Nuclear propulsion is similar to solar-heating propulsion except that the energy source is replaced by a nuclear reactor. A propellant is injected into the reactor heat exchanger, where the propellant is heated under pressure and expanded through the nozzle. Liquid hydrogen is the best choice for the nuclear rocket's working fluid because its low molecular weight produces the highest exhaust velocity for a given nozzle-inlet temperature. For example, for a given entrance nozzle pressure of 43 atm and temperatures of 1650, 3300, and 4950 K the following specific impulses are obtainable: 625, 890, and 1,216 s, respectively. The nuclear rocket engine is started by adjusting the reactor neutron-control drums to increase the neutron population. Propellant flow is initiated at a low reactor-power level and is increased in proportion to the increasing neutron population until the design steady-state reactor power output is obtained.

For shutdown of the engine, control drums are adjusted to poison the core and decrease the neutron population. Steady-state reactor thermal power, in megawatts, is determined by the relation $P1 = Kw (h_{out} - h_{in})$, where w and h are core flow rate and enthalpy, respectively, and K is the appropriate conversion factor to megawatts.

The reactor is a high-power-density, self-energizing heat exchanger which elevates the temperature of the hydrogen propellant to the limit of component materials.

SPACECRAFT LIFE SUPPORT AND THERMAL MANAGEMENT
by Walter W. Guy

NASA

References: *NASA SP-3006,* "Bioastronautics Data Book." Chapman, "Heat Transfer," Macmillan. Purser, Faget, and Smith, "Manned Spacecraft: Engineering Design and Operation," Fairchild.

Life Support—General Considerations

To sustain human life in a spacecraft two functions must be provided. The first is to replenish those substances which humans consume and the second is to control the conditions of the environment at levels consistent with human existence. Humans consume oxygen, water, and food in the metabolism process. As a by-product of this metabolism, they generate CO_2, respired and perspired water, urine, and feces (Table 11.6.13). The provisioning of man's necessities (and the elimination of his wastes) is a primary function of life support. However, an equally important function is the maintenance of the environment in which he lives.

There are three aspects of environmental control—ambient pressure, atmospheric composition, and thermal condition of the environs. In a typical spacecraft design the ambient pressure is maintained generally in the range between $\frac{1}{3}$ atm and sea-level pressure, with the oxygen partial pressure varying in accordance with Fig. 11.6.50 in order to provide the appropriate partial pressure of oxygen in the lungs for breathing.

Fig. 11.6.50   Atmospheric oxygen pressure requirements for humans.

The other constituents of concern in the atmosphere are the diluent, CO_2, water vapor, and trace amounts of noxious and toxic gases. The percent of the diluent in the atmosphere (or indeed whether a diluent is present at all) is selectable after proper consideration of the design parameters. For instance, short-duration spacecraft can take advantage of human adaptability and use atmospheric pressures and compositions significantly different from those at sea level, with the attendant advantage of system design simplicity. The Mercury, Gemini, and Apollo spacecrafts utilized 5-lb/in² pure oxygen as the basic cabin atmosphere. Skylab used an oxygen-nitrogen mixture although at a reduced pressure, while the shuttle is designed for an essentially sea-level equivalent atmosphere. When selecting the pressure and composition, it is important to remember that although a two-gas system adds complexity, it can eliminate the concerns of (1) flammability in an oxygen-rich environment, (2) oxygen toxicity of the crew, and (3) decreased atmospheric cooling potential for electronic equipment in a reduced-pressure cabin. In the selection of a diluent, nitrogen should be considered first, since its properties and effects on humans and equipment are well understood. However, other diluents, such as helium, which have characteristics attractive for a special application, are available.

The other atmospheric constituents—CO_2, water vapor, and trace gases—must be controlled within acceptable ranges for comfort and well-being. Although the partial pressure of CO_2 is less than half a millimetre of mercury in an ambient earth environment, humans are generally insensitive to CO_2 levels up to 1 percent of the atmosphere (or 7.6 mmHg) unless mission durations increase to more than a month. As spacecraft missions are further extended, the partial pressure of CO_2 should approach the earth ambient level to ensure physiological acceptance. Atmospheric water vapor (or humidity) should be maintained high enough to prevent drying of the skin, eyes, and mucous membranes, but low enough to facilitate evaporation of body perspiration, maximizing crew comfort. The generally accepted range for design is 40 to 70 percent relative humidity. In the closed environment of the spacecraft, cabin noxious and toxic gases are of particular concern. It should be noted that 8-h/day industrial exposure limits are not appropriate for the continuous exposure which results from a sealed spacecraft cabin. Therefore, materials which can offgas should be screened and limited in the crew compartment. The small quantities which

Table 11.6.13   Metabolic Balance for Humans

Consumed		Produced	
Oxygen	1.84	Carbon dioxide	2.20
Drinking water	5.18	Perspiration	2.00
Food (⅔ H_2O)	3.98	Respiration	2.00
		Urine	4.53
		Feces	0.27
	11.00		11.00

Note: Metabolic rate = 11,200 Btu/human-day and RQ = 0.87; RQ (respiratory quotient) = vol CO_2 produced/vol O_2 consumed. All values are in units of lb/human-day.

cannot be completely eliminated, as well as certain undesirable products of metabolism, must be actively removed by the life-support system to maintain an acceptable atmosphere for the crew.

The third of the environmental control aspects—atmospheric thermal condition—is subject to several design variables, all of which must be taken into account. Thermal comfort is a function of the temperature and relative humidity of the surrounding atmosphere, the velocity of any ventilation present, and the temperature and radiant properties of the cabin walls and equipment. The interrelationship of these parameters is shown in Fig. 11.6.51 for a reasonable range of flight clothing ensembles. Because of the effective insulations available to the vehicle designers, the problem of thermal control in the spacecraft cabin is generally one of heat removal. Utilizing increased ventilation rates to enhance local film coefficients is one means of improving thermal comfort within the cabin. However, the higher power penalties associated with gas circulation blowers, as compared to liquid pumps, should suggest using the spacecraft thermal control liquid circuit in reducing the temperature of the radiant environment as potentially a lower-power alternative. However, if the walls or equipment are chilled below the dew point, condensation will occur.

Fig. 11.6.51 Thermal comfort envelope for humans. (Note: metabolic rate 300–600 Btu/h; air velocity 15–45 ft/min; pressure 14.7 lb/in².)

As an additional benefit, maintaining an acceptable thermal environment through radiative means reduces the psychological annoyance of a drafty, noisy environment. Ventilation rates between 15 and 45 ft/min are generally considered comfortable, and exceeding this range should be avoided.

Life Support—Subsystems Selection

In designing the life-support system for a spacecraft, subsystems concepts must be selected to accomplish each of the required functions. This process of selection involves many design criteria. Size, weight, power consumption, reliability of operation, process efficiency, service and maintenance requirements, logistics dependency, and environmental sensitivity may each be very important or inconsequential depending on the spacecraft application. In general, for life support subsystems, the dominant selection criterion is mission duration. Various subsystem choices for accomplishing the life support functions are shown in Fig. 11.6.52.

To illustrate the relationship between mission duration and subsystem selection, the various concepts for removal of CO_2 from the atmosphere shown in Fig. 11.6.52b will be discussed. The simplest method of controlling the level of CO_2 in the spacecraft cabin is merely to purge a small amount of the cabin atmosphere overboard. This concept is simple and reliable and requires almost no subsystem hardware, although it does incur a fairly significant penalty for atmospheric makeup. For missions such as Alan Sheppard's first Mercury flight,

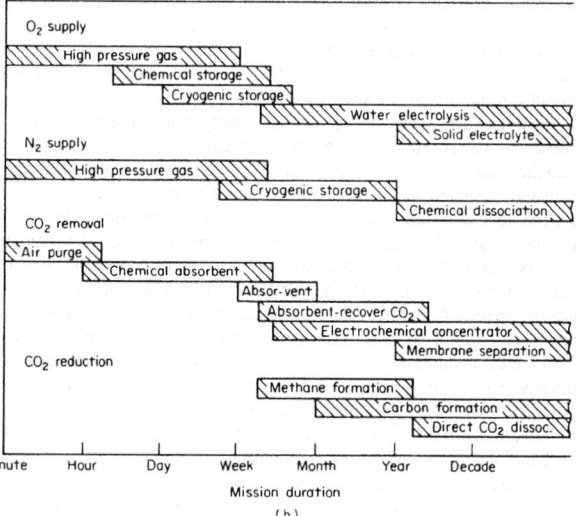

Fig. 11.6.52 Concepts to accomplish the life-support function. (a) Water recovery; (b) atmospheric revitalization.

this concept is probably optimum. However, even a minor increase in mission duration will result in a sizable weight penalty to the spacecraft. By adding a simple chemical absorption bed to the spacecraft ventilation loop (with a chemical such as lithium hydroxide as the absorbent), the expendable penalty for CO_2 removal can be reduced to the weight of the chemical canisters consumed in the absorption process. This concept was used on the Apollo spacecraft. The penalty for this type of concept also becomes prohibitive if mission durations are further extended. For missions such as Skylab, the expendable chemical absorbent can be replaced by a regenerable adsorbent. This change eliminates the lithium hydroxide weight penalty but incurs a small atmospheric ullage loss during the vacuum regeneration of the adsorbent. In addition, the regenerable system is more complex, requiring redundant beds (one for adsorption while the alternate bed desorbs), and mechanical-timing circuits and switching valves. This result is typical in that reducing expendable requirements is generally at the expense of system complexity.

The three CO_2 removal concepts discussed above all prevent the subsequent use of the collected CO_2 after removal from the atmosphere. Should the CO_2 be required for reclaiming the metabolic oxygen (such as will be the case for a space station design), the regenerable concepts can be adapted to accommodate retrieval of the CO_2 during desorption. However, this enhancement further complicates the subsystem design. There is an alternative CO_2 removal concept available which utilizes a fuel-cell-type reaction to concentrate the CO_2 in the atmosphere through electrochemical means. This process is mechanically simple (as compared to the adsorption-desorption system) but does require more electrical power. However, it has an integration advantage in that it provides the CO_2 collected from the atmosphere mixed with hydrogen, which makes its effluent a natural feedstock for either one of several CO_2 reduction subsystems (which will be required for oxygen recovery on a space station). Although Fig. 11.6.52 identifies the various life-support subsystem choices for candidate spacecraft designs of different mission durations, this information should be used only as a guide. The crew size as well as other design parameters can skew the range of applicability for each subsystem (e.g., a large crew can render expendable dependent subsystems inappropriate even for relatively short missions).

There is another major aspect of life support system design for spacecraft (in addition to subsystem selection) and that is process integration and mass balance. For example, a space-station-class life-support system design must accommodate wastewater with a wide range of contamination levels—from humidity condensate which includes only minor airborne contamination to various wash and rinse waters (including personal hygiene, shower, clothes washing, and dish washing) and urine. The degree of contamination generally influences the reclamation process selection; therefore, multiple contaminated sources tend to result in multiple subsystem processes. For example, highly contaminated water, such as urine, is most reliably reclaimed through a distillation process since evaporation and condensation can be more effectively utilized to separate dissolved and particulate material from the water, while less contaminated wash waters can be effectively reclaimed through osmotic and mechanical filtration techniques at significantly less power penalties in watts/lb H_2O recovered. Regardless of the techniques selected, water must be made available for drinking, food preparation, and cleaning the crew and the equipment they use, as well as for generating the oxygen they breathe. In fact, the space station life-support design process can be looked upon, in one respect, as a multifaceted water balance since the dissociation of CO_2 and generation of O_2 both involve water in the respective processes. Figure 11.6.53 is a typical space station schematic depicting the interrelationship of the subsystems and the resulting mass balance.

Thermal Management—General Considerations

Although there is an aspect of life support associated with thermal comfort, the area of spacecraft thermal management encompasses a much broader scope. In the context of this section, thermal management includes the acquisition and transport of waste heat from the various sources on the spacecraft, and the disbursing of this thermal energy at vehicle locations requiring it, as well as the final rejection of the net waste heat to space. Although there are thermal implications to the design of most spacecraft hardware, the Life Support and Thermal Management section only deals with the integration of the residual thermal requirements of the various vehicle subsystems to provide an energy-efficient spacecraft design (i.e., one which utilizes waste heat to the maximum and therefore minimizes requirements for the vehicle heat rejection system).

Thermal management systems are typically divided into two parts. The first, associated with the spacecraft cabin, has design requirements associated with maintaining the crew environment and equipment located in the cabin at acceptable temperatures, and in addition has special requirements imposed on its design because of the location of the system in the pressurized compartment and its proximity to humans. These additional requirements are usually dominant. Any cooling medium which is used in the crew compartment must have acceptable toxicological and flammability characteristics in order to not impose a hazard to the crew in the event of malfunction or leakage. Although there are thermal control fluids which have low toxicity potential, reasonable flammability characteristics and adequate thermal transport properties, none are better than water.

The second aspect of the thermal management system involves those spacecraft systems located outside the cabin. The dominant design criteria in thermal control fluid selection for these systems are thermal transport efficiency and low-temperature characteristics. The thermal transport efficiency is important since the size, weight, and power requirements of the pumping and distribution system depend on this parameter. However, in the environs of deep space the potential exists for very low temperatures. Therefore, for maximum system operating flexibility, the characteristics of the thermal transport fluid must be such that operation of the system can continue even in a cold environment; thus, low-temperature properties become a fluid selection constraint. Although the range of acceptable fluids is broader here, several fluids in the Freon family meet these requirements. For single-phase, pumped-fluid systems, the temperature levels of the heat sources are

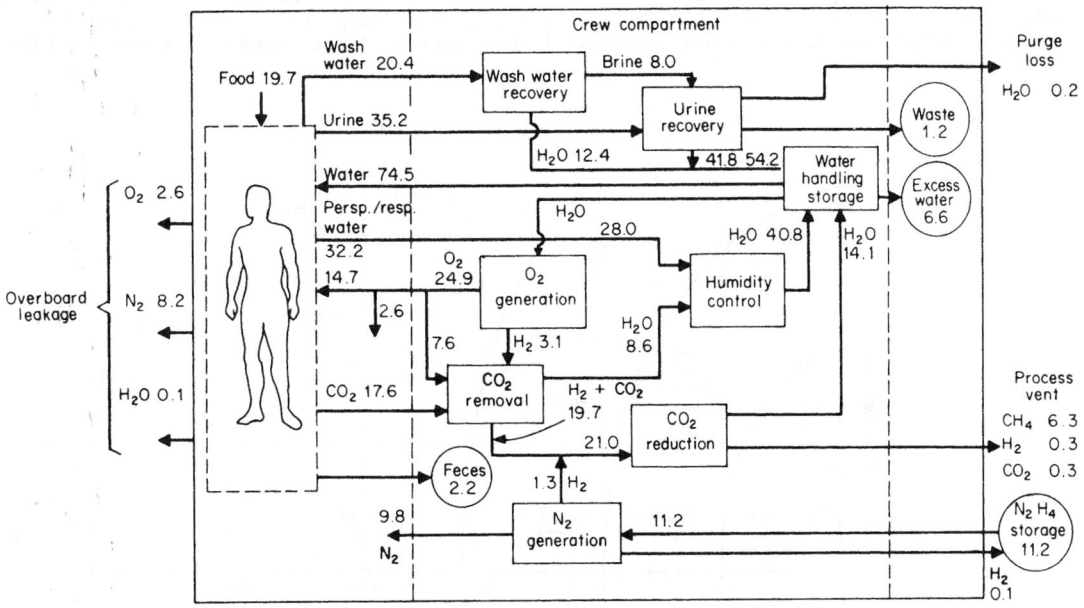

Fig. 11.6.53 Life-support systems mass balance. (All values are in pounds per day for an eight-person crew.)

important in determining the optimum sequence for locating equipment in the thermal circuit. The equipment with the lowest-temperature heat sink requirements are accommodated by placing them immediately after the heat rejection devices, while equipment which can accept a higher-temperature heat sink are located downstream. Also, those subsystems which require heating must be located at a point in the circuit which can provide the thermal energy at the required temperature level.

Typically spacecraft heat sources include cabin-atmosphere heat exchangers, the avionics system as well as other electronic black boxes, power generation and storage equipment, etc., while heat sinks include propulsion system components, various mechanical equipment requiring intermittent operation, dormant fluid systems which have freezing potential, etc. The ultimate elimination of the net thermal energy (i.e., sources minus sinks) is by rejection to space. This is accomplished through expendables or radiation.

Recent developments in thermal management technology have resulted in the consideration of a two-phase acquisition, transport, and rejection system. This concept uses the heat of evaporation and condensation of the transport fluid to acquire and reject thermal energy, reducing by several orders of magnitude the amount of fluid required to be pumped to the various heat sources and sinks within the spacecraft. This concept has the added advantage of operating nearly isothermally, permitting the idea of a thermal "bus" to be included in future spacecraft design, with indiscriminate sequencing of the heat sources and sinks within the vehicle.

Thermal Management Subsystem Selection

The subsystem selection process associated with the thermal management area depends largely on the magnitude of thermal energy being managed and the duration of the mission. Figure 11.6.54 graphically depicts the areas of applicability for various types of heat-rejection devices as a function of these parameters.

Small heat-rejection requirements can sometimes be met by direct heat sinking to the structure (or a fluid reservoir). Often, passive heat sinks need to be supplemented with fusible heat sinks (to take advantage of the latent heat of fusion) or evaporative heat sinks (to take advantage of the latent heat of evaporation). Of the three types of heat sinks, the evaporative heat sink is the only one which has been used as the primary vehicle heat-rejection device in the manned spaceflight program to date. Both the Mercury spacecraft and the lunar module utilized a water evaporative heat sink in this manner. However, the Apollo spacecraft and the shuttle have also used evaporative heat sinks

and Skylab used a fusible heat sink as supplemental heat rejection devices.

Since space is the ultimate heat sink for any nonexpendable heat rejection, devices which allow the thermal energy which has been collected in spacecraft coolant circuits to be radiated to space become the mainstay of the thermal management system. These "radiators" take many forms. The simplest design involves routing the cooling circuits to the external skin of the spacecraft so that the thermal energy can be radiated to deep space. Because of spacecraft configuration restrictions, only partial use can be made of the external skin of the spacecraft, and the amount of thermal energy which can be rejected by this technique is thereby limited. Although the Gemini and Apollo spacecraft successfully used integral skin radiators, the unique characteristics of the outer surface of the orbiter required a different technique be adopted for its radiator design. Accordingly, the radiators were designed to be deployed from inside of the payload bay doors. This not only provides good exposure to space, but also, in the case of the forward panels, allows two-sided exposure of the panels, thereby doubling the active radiating area.

Heat rejected Q_{rej}, according to the Stefan-Boltzmann law, is given by the following equation:

$$Q_{rej} = \varepsilon \gamma A T^4$$

where γ = constant, 0.174×10^{-8} Btu/(h · ft² · °R⁴); ε = emittance property of surface (termed emissivity), A = area available for radiation; and T = radiating temperature. From this equation it is obvious that a radiator with both sides exposed will reject twice as much heat to space as one of the same dimensions which is integral with the skin of the vehicle. A space radiator can absorb as well as reject thermal energy. The net heat rejected Q_{net} from a radiator is given by the following equation:

$$Q_{net} = Q_{rej} - \sum_{N}^{A=1} Q_{abs}$$

The energy absorbed by the radiator, Q_{abs}, can come from various sources (see Fig. 11.6.55).

The most common heat sources for an earth-orbiting spacecraft are direct solar impingement, reflected solar energy from the earth (i.e., albedo), direct radiation from the earth, radiation from other parts of the spacecraft, and radiation from other spacecraft in the immediate vicinity. These energy sources are absorbed on the radiator surface in direct proportion to the absorptance characteristic of the surface,

Fig. 11.6.54 Range of application of various heat-rejection methods.

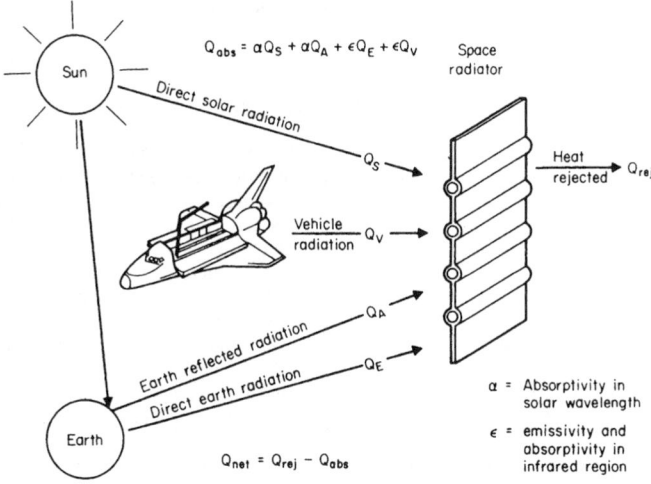

Fig. 11.6.55 Space radiator energy balance.

termed absorptivity. Emissivity and absorptivity are unique to each surface material but are variables directly linked to the temperature of the source of the radiation (i.e., wavelength of the radiation). Therefore, since absorptivity and emissivity are characteristics of the surface, the careful selection of a surface coating can significantly enhance the effectiveness of the radiator. The coating should have a high emissivity in the wavelength of the rejected energy from the radiator (for good radiation efficiency) and a low absorptivity in the wavelength of the impinging radiation (for minimal absorbed energy). Since generally the earth and the surrounding spacecraft are producing thermal energy of essentially the same wavelength as the radiator, it is necessary to restrict the view of the radiating surface to these objects since absorptivity and emissivity of a surface are equal for the same wavelength energy. Thus, radiators which can be located so that they are not obstructed from viewing deep space by elements of the spacecraft or other spacecraft in the vicinity will be much more effective in rejecting waste heat. The sun, however, emits thermal energy of a dramatically different wavelength and, therefore, coatings can be selected which absorb less than 10 percent of the solar energy while radiating nearly 90 percent of the energy at the emitting temperature of the radiator.

For applications which require large radiating surface areas as compared to the vehicle's size, integral skin radiators offer insufficient area and thus space radiators which use deployment schemes must be con-

sidered. However, as radiators become larger the mechanisms become very complicated. To circumvent the mechanical problems of deployment, a new radiator concept has been developed which allows in-orbit construction. This concept utilizes a single-tube heat pipe as the primary element (see Fig. 11.6.56).

There are two distinct advantages of heat pipe radiators. The first is that assembly of large surface areas in-orbit can be effected without breaking into the thermal control circuit. This is practical since the heat pipe radiator element is completely self-contained. The second advantage is that in-orbit damage from accidental collision or meteoroid puncture affects only that portion of the radiator actually damaged and does not drain the entire radiating surface of fluid as it would the radiators which have been used on spacecraft to date. Because of this advantage, future missions which require either of the three types of radiators (i.e., integral skin, deployable, or space constructible) will utilize the heat pipe concept to transport the thermal energy to the exposed radiating surface.

The design of space radiators can take many forms, but the basic concept is an exposed surface made of a conducting material with regularly spaced fluid tubes to distribute the heat over the surface (see Fig. 11.6.57).

The selection of the material and configuration of the radiating surface (i.e., honeycomb, box structure, thin sheet, etc.) determine its thermal conductance. This conductance, when coupled with tube spacing and the internal heat transfer characteristics within the tube and fluid system, determines the temperature gradient on the radiating surface between adjacent tubes. This so-called "fin effectiveness" represents a loss in radiating potential for any particular radiator design; therefore, these parameters should be selected carefully to optimize the radiator design within the constraints of each application. Figure 11.6.58 demonstrates the effect of the configuration parameters of the radiator on panel weight.

Fig. 11.6.56 Basic heat-pipe operation.

Fig. 11.6.57 Typical radiator tube/fin configurations.

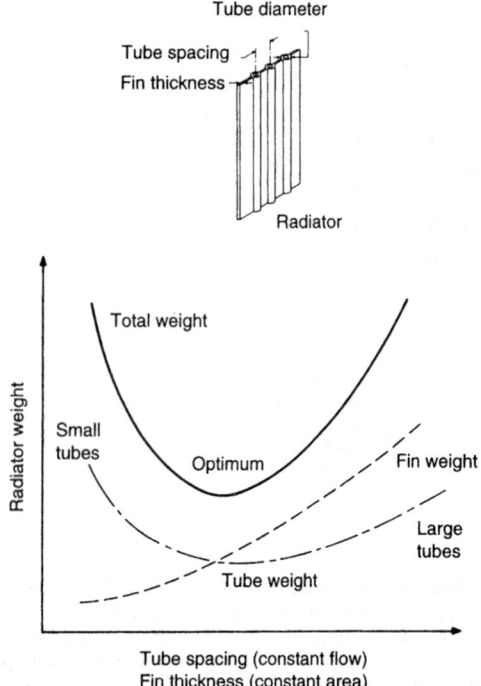

Fig. 11.6.58 Radiator configuration optimization.

DOCKING OF TWO FREE-FLYING SPACECRAFT
by Siamak Ghofranian and Matthew S. Schmidt

Rockwell Aerospace

REFERENCES: Ghofranian, Schmidt, Briscoe, and Shliesing, "Space Shuttle Docking to MIR, Mission-1"; Ghofranian, Schmidt, Briscoe, and Shliesing, "Simulation of Shuttle/MIR Docking"; Ghofranian, Schmidt, Lin, Khalessi, and Razi, "Probabilistic Analysis of Docking Mechanism Induced Loads for MIR/ Shuttle Mission"; *35th Structures, Structural Dynamics, and Materials Conf.* AIAA/ASME/ASCE/AHS/ASC. Haug, "Computer Aided Kinematics and Dynamics of Mechanical Systems," vol 1: "Basic Methods," Allyn and Bacon. Greenwood, "Principles of Dynamics," Prentice-Hall.

The information in this section covers the mechanics of docking two free-flying unconstrained vehicles in space. Different methods of mating are addressed with the emphasis applied to docking. The content does not encompass all facets of the procedure, but it does bring forth the physics of the operation which make the process an engineering challenge.

Types of Mating

The understanding of mating is a precursor to preliminary design. Different techniques of mating free-flying bodies are often discussed when planning the rendezvous of two vehicles in space. The mating of two structures can be performed in three ways: docking, berthing, or a combination of both. **Docking** relies on the transfer of vehicle momentum to provide the means to force the docking ports into alignment. If a significant amount of momentum is required for capture, jet firings can be used to introduce the impulse into the system as needed. **Berthing** relies on a secondary system, such as a robotic arm, to enforce the docking ports into alignment. The active vehicle grapples the other with a robot manipulator and guides the structural interfaces together. If manipulator usage is attempted, specific design factors need to be considered: arm configuration, gearbox stiffness, linkage flexibility, and controllability.

Mating of two structures in space is by every means considered a multibody dynamic event. Docking involves a significant level of momentum transfer, while berthing involves relatively little. The dynamics result from the deliberate impact of the two vehicles. As docking occurs the system responds with rigid-body motion and mechanical motion of the docking device. Depending on the relative states between the vehicles and the mode in which the maneuver is attempted, the response of the system and the physics vary considerably.

For discussion purposes the two docking vehicles can be considered as a target vehicle (passive) and a docking vehicle (active). The passive vehicle maintains attitude control as the active vehicle is navigated along an approach corridor. The active vehicle holds an attitude control mode with translational control authority. As the corridor narrows, sighting aids, range, and range-rate feedback on both vehicles are used to guide the docking ports into contact.

As the vehicles approach, a closing velocity exists. From a design standpoint the operation of a docking system involves various mechanism operational phases: mechanism deployment, interface contact and capture, attenuation of relative motion, mechanism retraction, and structural lockup. The **deployment phase** is where the mechanism is driven from its stowed position to its ready-to-dock position. Vehicle interface contact and capture are initiated when the docking ports of each vehicle are maneuvered into contact. Ideally, a positive closing velocity should be maintained from initial contact through capture. Once an opening velocity develops, the probability of a failed capture increases. As the vehicles interact the mechanism interfaces are indexed into alignment as the system responds to the impact loads. The compliance of the system occurs in two forms: mechanism movement and rigid body vehicle motion.

When the docking interfaces interact, the system responds in six degrees of freedom and the docking system is stroked as the mechanism absorbs the relative misalignments. Once the interfaces are fully aligned, capture latches lock the opposing interfaces together, and the vehicles are "soft-docked." After capture, residual energy in the system allows relative motion between the structures to continue. Depending on the location of the docking system relative to the vehicle center of masses, the motion may dissipate due to the inherent hysteresis in the docking system; otherwise, separate energy-dissipation devices may be necessary. After attenuation the vehicles are likely to be misaligned due to the new relative equilibrium position of the spacecraft. Separate procedures or devices can be used to bring the structural interfaces into alignment.

The **retraction process** is where the active mechanism is driven back to its stowed position. Once they are fully retracted, the structural interfaces are latched at hard interfaces, and the vehicles are "hard-docked."

The mass of each vehicle is an essential part of the docking system design. The momentum of the docking vehicle is the medium which provides the force required by the docking mechanism to comply with the relative misalignments between the vehicles at the docking interface. The operation can be complicated if a vehicle center of mass is offset from the longitudinal axis of the docking system. As a result, the vehicle effective mass at the docking interface is reduced. Furthermore, the center-of-mass offset induces relative rotation after contact which complicates the capture and attenuation process. Ideally two vehicles should dock such that the velocity vector and longitudinal axis of the docking system have minimum offset from the center of mass. The relative rotation of the vehicles after contact is known as *jackknifing.*

The docking mechanism center-of-mass offset and the vehicle size significantly influence the design of the docking interfaces, attenuation components, docking aids, piloting procedures, and capture-assist methods. The relative vehicle misalignments and rates at the moment the docking interfaces first come into contact (initial conditions) should be known in order to size the mechanism attenuation components, to establish piloting procedures and training criteria, and to study the mechanics of the system before, during, and after docking.

The initial conditions result from vehicle as-built tolerances, thermal and dynamic distortions, variation in piloting performance, and control system tolerances. The misalignments are composed of lateral misalignments in the plane of the docking interface and angular misalignments about three axes. The rates are composed of translational velocities along three axes and angular velocities about three axes.

The mass and motion of the system are the basic ingredients for providing the energy necessary to join the interfaces. Ideally the interfaces should capture before an opening velocity develops between the docking interfaces. Adequate momentum at the docking port is required to actuate the mechanism. The impulse introduced to the system during the impact must be less than that required to reduce the relative interface velocity to zero. If the relative interface velocity goes to zero before capture, the vehicles will separate. For low closing velocities, a small axial impulse due to initial contact can reduce the relative interface velocity to zero and cause a "no capture" condition.

A drive system along the longitudinal axis of the docking mechanism or reaction control jets increases the probability of capture. It extends the docking system toward the target vehicle once initial contact is made and thus improves capture performance for a docking vehicle with a low closing velocity. However, a drawback of a drive system is that the reaction has the tendency to push the vehicles apart. The action of jet thrusting has the opposite effect; it tends to push the vehicles together. If the momentum required to provide a low possibility of no capture becomes operationally prohibitive, thrusting can be used as a substitute and applied as necessary.

As energy introduced into the system increases, whether in the form of vehicle momentum or jets pulsing, a tradeoff exists between acceptable internal vehicle loads, allowable component stroke limits, vehicle jackknifing, and piloting capability.

Vehicle Mass and Inertia Effects

Concentrated spherical masses colliding in two dimensions can be used to describe the dynamic effects of an impact. The same class of examples are consistently used in beginning physics textbooks to describe the macroscopic concepts of impulse, momentum, and energy. The same theory should be applied to understand the requirements of mating two vehicles and perform conceptual and preliminary design studies. A single-degree-of-freedom mechanical equivalent exhibiting simple harmonic motion is a useful example for understanding the mechanics.

Consider two spacecraft in a two-dimensional plane on a planned course of docking without any misalignments and with some closing velocity. If only the docking-axis degree of freedom is considered, the effective mass of each body, M_1 and M_2, can be computed from

$$M_1 (\text{or } M_2) = M - (ML)^2/(I + ML^2)$$

where M is the vehicle mass, I is the vehicle mass moment of inertia about the center of mass, and L is the mass center offset from the docking axis. In this case the process of capture and attenuation can be thought of as a perfectly plastic impact of two bodies. The masses dock and stick together. After impact the two bodies continue their trajectory with some final velocity. The combined effective mass of the two bodies represented as a single degree of freedom can be computed from $M_{\text{eff}} = M_1 M_2/(M_1 + M_2)$. Since momentum is conserved during a perfectly plastic impact, kinetic energy loss during docking can be computed from $\frac{1}{2} M_{\text{eff}} v^2$, where v is the capture velocity. Since at capture we know the effective mass of the system and the capture velocity, the unknown parameter becomes the work required to bring the system to rest.

Design Parameters

The energy loss during docking is equal to the work done, $F \times d$, by the mechanism-attenuation components. If a hysteresis brake is used to bring the system to rest, then a trade can be performed on the force-distance relationship by using

$$\frac{1}{2} M_{\text{eff}} v^2 = Fd$$

where F represents the force necessary to bring the system to rest over distance d. Using the hysteresis brake as a starting point provides the average about which other devices would oscillate. For an underdamped spring/damper attenuation system, the relative motion and maximum force can be computed from

$$x = v/\omega_d \times e^{-\zeta \omega} \sin \omega_d t \qquad \text{and} \qquad F = Kx + C\, dx/dt$$

respectively. In this representation,

$$\omega_d = \omega(1 - \zeta^2)^{1/2} \qquad \text{and} \qquad \omega^2 = K/M_{\text{eff}}$$

where ζ represents the damping factor and K and C represent the attenuation component characteristics. As can be seen, minimal knowledge of the basic system characteristics can be used to design the type and arrangement of the attenuation components. During the design of the attenuation components the following components may prove useful: preloaded centering springs, staged springs, ON/OFF dampers, and variable-level hysteresis brakes. The material in this section is only a first look at understanding the design of a docking system. As more detail is developed, the effects of relative vehicle motion, interface contact dynamics, mechanism performance, and vehicle flexibility need to be simulated, which goes beyond the scope of this section. For a comprehensive background, the reader should refer to the cited references.

11.7 PIPELINE TRANSMISSION
by. G. David Bounds

REFERENCES: AGA publication "2001 Gas Facts." Huntington, "Natural Gas and Gasoline," McGraw-Hill. Leeston, Crichton, Jacobs, "The Dynamic Natural Gas Industry," University of Oklahoma Press. Lester, Hydraulics for Pipeliners, *Oildom*. Bell, "Petroleum Transportation Handbook," McGraw-Hill. Bureau of Mines, "Mineral Yearbook," Vol. II, Fuels. "The Transportation of Solids in Steel Pipelines," Colorado School of Mines Research Foundation, Inc. Echterhoff, "Pipeline Economics Revisited," PCGA, 1981. *U.S. Crude Oil, Natural Gas, and Natural Gas Liquids Reserves, 1999 Annual Report,* Department of Energy, Energy Information Administration, December, 2000, PLATTS, The McGraw-Hill Companies, Pipeline Project Tracker, WorldWide Web. Federal Energy Regulatory Commission Forms 2 & 2A, "Annual Report of Major/Minor Gas Companies," 1999. System Capacity & Mileage: Federal Energy Regulatory Commission, FERC Annual Capacity Report (18CFR 284.12). Various news sources and natural gas trade press periodicals on the World Wide Web.

NATURAL GAS

General According to AGA Estimates for 2003, approximately 73 percent of the U.S. natural gas reserves occur in five states (Texas, Wyoming, New Mexico, Colorado, and Oklahoma) and the Gulf of Mexico Federal Offshore. Within these areas alone, estimated production in 2003 exceeded 15 trillion ft³ (425 billion m³). Total exports, interstate and international, from these states in 2002 were estimated at approximately 9.1 trillion ft³ (258 billion m³), while imports were approximately 4.6 trillion ft³ (130 billion m³). Offshore production is becoming increasingly significant. According to an INGAA Foundation report referenced by PLATTS in May, 2002, the Gulf of Mexico is the strongest growth area for U.S. gas production through the year 2020.

Citing the growing interest in deepwater resources, current production, averaging 5 trillion ft³ (142 billion m³) per year, is predicted to increase to 7.2 trillion ft³ (204 billion m³) per year by 2020. In 2003, approximately 12 percent of the total proven U.S. natural gas reserves and 22 percent of natural gas production were located in the Gulf of Mexico Federal Offshore Area. U.S. natural gas reserves in the state of Wyoming are second only to those in Texas, due to the approximately 114 trillion ft³ (3.2 trillion m³) of recoverable gas resources in the Appalachian Basin, of which almost 20 trillion ft³ (566 billion m³) is from coal bed methane (CBM). Industry leaders are excited about this supply of gas but are concerned that the future of this resource will not be determined by how much is produced, but rather by how much pipeline capacity is available to carry it to market. Actual production from these CBM resources is only about 50 billion ft³ (1.4 billion m³) per year because of insufficient pipeline capacity, onsite power generation and water handling facilities.

Natural gas consumption in the United States increased by 17 percent in the past decade; from 19.0 trillion ft³ (538 billion m³) in 1991 to 22.3 trillion ft³ (631 billion m³) in 2000. This demand is expected to grow even more rapidly over the next 20 years. To meet this increasing demand, natural gas pipeline transmission companies have transformed the way they do business. In addition to the extensive growth in the U.S. natural gas pipeline network, the growth of natural gas market centers integrating underground storage facilities and the consolidation of corporations through acquisitions and mergers have been the key factors in this transformation.

In 2002, production in the United States declined by 750 billion ft³ (21.2 billion m³) compared to 2001. In the same year, consumption increased by 860 billion ft³ (24.4 billion m³). Hurricanes in the Gulf of Mexico played a role in the reduced production, which was partially offset by increased production of 11.4 billion ft³ (3.2 billion m³) in Colorado and Wyoming. The increased demand was primarily met, however, by increased withdrawals from storage. In 2002, natural gas storage capacity in the United States was approximately 8.2 trillion ft³ (232 billion m³). Over 40 percent of this capacity and 35 percent of all storage facilities in the nation were located in four sates: Michigan, Illinois, Pennsylvania, and Texas. Ranked by storage capacity, the common types of natural gas storage fields are: depleted fields (82 percent), acquifers (15 percent), and salt formations (3 percent). At the beginning of 2002, natural gas storage inventory was 2.9 trillion ft³ (82 billion m³), the largest volume since 1990. At the end of 2002, natural gas storage volume was 35 billion ft³ (1 billion m³) less than the previous 5-year average of 2.4 trillion ft³ (68 billion m³).

Implementation of FERC Order 636, issued in 1992, has had a significant impact on natural gas transmission pipelines. Order 636 mandated interstate natural gas pipeline companies to transform themselves from natural gas marketers to strictly transporters of natural gas. Pipeline companies that built and operated gathering systems to support contractual obligations are now restricted from including costs for these activities with the costs of buying, transporting, and selling gas, because of the mandate that gathering be provided as a separate service. Open-access marketing has replaced bundled sales services to meet the instantaneous needs of buyers at competitive prices. When open-access marketing was first tried in the early 1980s, competitive bidding resulted in curtailment of gas flow to approximately 18 trillion ft³ (510 billion m³) per year through a national pipeline grid that had moved 24 trillion ft³ (680 billion m³) per year in the 1970s. The swing in market demand and compliance with FERC Order 637 has magnified the need for pipeline companies to provide "parking" and "lending" type services. Under guidance from the Gas Industry Standards Board (GISB), FERC now requires all natural gas pipeline companies to have an Internet site providing information on its rates (tariffs), operational capabilities, and availability. Pipeline companies typically offer interruptible service supported by system linepack and storage facilities in both gathering and market areas to provide the flexibility required to operate efficiently in an open access market. Market centers and hubs have evolved to provide gas shippers the physical and administrative capabilities formerly provided as "bundled" services by the interstate pipelines. The two primary functions of these hubs are to provide interconnection with other pipelines and to provide short-term receipt/delivery balancing needs. Many of the hubs also developed Internet-based access to gas trading platforms and capacity release programs that have improved the efficiency of the entire gas transportation and marketing processes. In 2003, there were 37 operational market centers in the United States and Canada (see Table 11.7.1). The collapse of Enron in 2001 precipitated financial difficulties for many gas marketing companies and the shutdown of most major Internet-based energy trading platforms. However, the impact on gas market center activity was short-term. By the middle of 2003, most market centers reported business levels that matched or exceeded previous levels.

Increasing competition brought about as a consequence of open-access marketing has been accompanied by major shifts in pipeline ownership and changes in the way pipeline companies do business. In addition to a large number of mergers and acquisitions, many pipeline companies created affiliated marketing arms, or subsidiaries, and transferred merchant functions to these new divisions to comply with FERC Order 636. Many of these marketing entities engage in marketing other types of energy as well, and some are much larger now than the pipeline divisions from which they were originally developed. Corporate expansion strategies have helped pipeline companies to maintain market share and reduce overhead costs as they search to find more efficient ways to operate. A snapshot of the major gas pipeline companies in the United States, showing shifts in ownership as of the year 2001, is shown in Table 11.7.2.

The need to transport natural gas from the producing regions to consuming areas throughout the continent has generated a burgeoning network of transmission pipelines. In 1950, there were 314,000 miles (505,000 km) of gas pipelines in this country. By 2000, with natural gas continuing to supply 24 percent of the nation's energy requirements, this mileage had increased to 1,369,000 miles (2,204,000 km), an increase of 435 percent in 50 years. Approximately 250,000 miles (402,500 km) of this pipeline is devoted to transmission service. Field gathering pipelines amount to about 40,000 miles (64,400 km). The remaining 1,079,000 miles (1,737,000 km) are in distribution service. In 2002, an additional 3,571 miles (5,750 km) of pipelines was added to the pipeline grid in the United States, incorporating a record increase of 12.8 billion ft³ (362,000 m³) per day to the national pipeline capacity.

Modern gas pipelines in the United States consist of diameters ranging from 2 to 48 in (60 to 1,219 mm) and grades of steel with specified minimum yield strengths (SMYS) ranging from 35,000 to 80,000 lb/in² (2,460 to 5,625 kg/cm²). The type of longitudinal seams available are seamless (SMLS) in sizes ranging from 2 to 20 in (60 to 508 mm) diameter, electric resistance weld (ERW) in sizes ranging from 2 to 24 in (60 to 610 mm) diameter, and double submerged arc weld (DSAW) in sizes ranging from 16 to 48 in (406 to 1,219 mm) diameter. Higher yield strengths are usually more cost-effective in sizes larger than 10 in (273 mm) diameter.

Although the trend has been toward the use of larger pipe diameters and pipe steel of greater tensile strength, the economics of pipe production and coating and the ease of handling, including transportation as well as installation, has led to greater utilization of 36-in-diameter (914-mm) pipe with a specified minimum yield strength of 70,000 lb/in² (4,922 kg/cm²).

In most new construction, mainline valves are usually ball or gate valves and are of the full-opening variety to allow for the passage of cleaning pigs and inspection tools. Ball and gate valves are also used in addition to check and plug valves on piping other than mainline piping.

Pipe bends are usually made on site by a contractor; however, on short-distance pipelines, or in cases where a large degree of bend is required, shop-fabricated bends are produced by induction heating.

Repairs to a pipeline segment according to current regulations of the Department of Transportation (DOT) are to either replace the segment, or use alternative methods that have been shown by reliable testing and analysis to restore pipeline serviceability. The installation of steel repair sleeves over localized pipeline defects has been used extensively to effect permanent repairs. Repair sleeves may be installed by welding or bolting in place. A newer repair method being used by many pipeline

Table 11.7.1   Operational Natural Gas Market Centers in the United States and Canada, 2003

State/province	Market center	Administrator	Online service	Operation	Year started	Associated storage
		United States				
Colorado	Cheyenne Hub	Colorado Interstate Gas Co.	CIG-Express	Market hub	2000	Huntsman/Young/Litigo
Wyoming	Opal Hub	Williams Field Services	Gaskit	Production hub	1999	None
Kansas	Mid-Continent Center	Oneok Gas Transportation	Fax or phone only	Market center	1995	Brehm/Richfield
Illinois	ANR Joliet Hub	ANR Pipeline Co.	Gems	Market center	2003	System linepack only
Illinois	Chicago Hub	Enerchange Inc.	"Gas Exchange"	Market center	1993	All NICOR sites
New York	Iroquois Center	Iroquois Gas Trans. Co.	Iroquois Online	Market center	1996	System linepack
Pennsylvania	Dominion Hub	Dominion Transmission Inc.	EScript	Market center	1994	All Dominion sites
Pennsylvania	Ellisburg-Leidy Center	National Fuel Gas Supply Co.	SBSPRD	Market center	1993	NFGS sites/Stagecoach/Seneca Lake
Louisiana	Egan Hub	Egan Hub Partners LP	LINK System	Market hub	1995	Egan Storage
Louisiana	Henry Hub	Sabine Hub Services Inc.	Yes	Market hub	1988	Jefferson Island/Sorrento
Louisiana	Nautilus Hub	Shell Gas Trans. Co.	Caminus	Production hub	2000	None
Louisiana	Perryville Center	Centerpoint Energy Gas Trans. Co.	ServiceLynx	Market center	1994	Ruston/Ada/Childes/Bistineau
New Mexico	Blanco Hub	Transwestern Gas Pipeline	HotTap	Market center	1993	System linepack only
East Texas	Agua Dulce Hub	ConocoPhillips Inc.	Fax or phone only	Production hub	1990	None
East Texas	Carthage Hub	Duke Energy Field Services	AltraWeb	Market hub	1990	Indirect only
East Texas	Katy (DEFS) Hub	Duke Energy Field Services	Fax or phone only	Market hub	1995	None
East Texas	Katy Storage Center	Enstor Inc.	Yes	Market hub	1993	Katy/Stratton Ridge
East Texas	Moss Bluff Hub	Moss Bluff Hub Partners LP	LINK System	Market hub	1994	Moss Bluff
East Texas	Spindletop Storage Hub	Centana Intrastate Pipeline	Fax or phone only	Market hub	1998	Spindletop
West Texas	Waha (EPGT) Texas Hub	EI Paso Texas Pipeline LP	StarWeb	Market hub	1995	Boling Site
West Texas	Waha (DEFS) Hub	Duke Energy Field Services	Fax or phone only	Market hub	1995	None
West Texas	Waha (Encina) Hub	Sid Richardson Gas Co.	Fax or phone only	Production hub	1995	None
West Texas	Waha (Lone Star) Hub	TXU Lone Star Gas Co.	Caminus	Market hub	1995	Keystone
California	California Energy Hub	Southern California Gas Co.	ENVOY	Market center	1994	All SoCal fields
California	Golden Gate Center	Pacific Gas & Electric	PipeRanger	Market center	1996	All PG&E fields and linepack
Idaho	Kingsgate Center	PG&E Gas Trans.—NW	ETran system	Market hub services	1994	System linepack only
Oregon	Malin Center	PG&E Gas Trans.—NW	ETran system	Market hub services	1994	System linepack only
Oregon	Stanfield Center	Pacific Gas Transmission Co.	ETran system	Market hub services	1994	System linepack only
		Canada				
Alberta	AECO-C Hub	EnCana Energy Co.	AECO-Link	Market center	1990	Suffield/Dunvegan/Carbon
Alberta	Alberta Hub	Enstor–PPM Energy Inc.	Fax or phone only	Market hub	1997	Alberta Hub
Alberta	Alberta Market Centre	Atco Gas Services Ltd	Fax or phone only	Market hub	1998	Carbon Facility
Alberta	Crossfield Hub	Crossalta Gas Stor. & Srvcs.	Fax or phone only	Market hub	1995	East Crossfield
Alberta	Empress Center	Transcanada Gas Pipeline	NRGhighway	Market hub	1986	Linepack
Alberta	Intra-Alberta Center	Transcanada Gas Pipeline	NGX Trading Sys	Market center	1994	Indirect only
British Columbia	Sumas Center	Westcoast Pipeline Co.	WestFlo	Market hub	1994	Aitken Creek
Alberta Quebec	TransCanada Center	Transcanada Gas Pipeline	NRGhighway	Market hub	1998	Indirect only
Ontario	Dawn Market Center	Union Gas Ltd.	UnionOnline	Market center	1995	Dawn (20 fields)

operators is a polymer composite reinforcement warp (Clock Spring® or equivalent) which is wrapped around the repair area. For this type of repair, welding is not required.

The rapidly growing network of large-diameter pipelines, along with increasing pressure capabilities, has brought forth a considerable rise in prime mover and compressor requirements. Natural gas pipelines in the United States have experienced an increase from about 6 million installed hp (4.5 million kW) in 1960 to 15 million hp (11.2 million kW) by the end of 2003, an increase of 250 percent. More than 1,600 compressor stations are now installed on interstate gas pipelines at a spacing of between 50 to 75 miles (80 to 120 km). Natural gas–fueled engines, natural gas–fueled combustion gas turbines, and electric motors are employed as prime movers on gas pipelines. Approximately 60 percent of the existing 15 million hp is created by 5,700 reciprocating engines driving large-bore, slow-speed reciprocating compressors. The greatest part of the remaining 6 million hp (4.5 million kW) is developed by gas turbines and electric motors driving high-speed centrifugal compressors. Gas-fueled engines range in size from 2,500 to 13,500 hp (1,865 to 10,070 kW). Combustion gas turbines range in size from 1,000 to 35,000 hp (745 to 26,110 kW), and electric-motor drivers range in size from 2,500 to 20,000 hp (1,865 to 14,920 kW). Modified aircraft-type jet engines are also employed as prime movers, driving centrifugal compressors and range in size from 800 to 45,000 hp (600 to 33,570 kW).

Much progress has been made in automated operations of all types of pipeline facilities. Valves, measuring and regulating stations, and other facilities are operated remotely or automatically. Factory-packaged compressor station assemblies, most of which are highly automated, are now commonplace. Whole compressor stations with thousands of horsepower are operated unattended by coded dispatching systems handled by high-speed communications.

Automatic supervision and operation of pipeline systems are becoming almost commonplace in today's climate of microprocessor-based remote terminal units (RTUs). These ultrahigh-speed data-gathering systems can be used to gather operational data, do calculations, provide proportional integral derivative (PID) process control, and undertake other auxiliary functions. Data may be communicated to a central dispatch office for real-time control. A system of several hundred locations may be scanned and reported out in a few seconds.

Table 11.7.2   Major Interstate Natural Gas Pipeline Companies—Mergers as of 2001

Parent Pipeline name	Marketer affiliate Prior ownership	Merger year	System mileage	System capacity, MMCFD
CMS Energy Corp.	**CMS–MST**			
Panhandle Eastern Pipeline Co.	PanEnergy Corp.	2000	6,467	2,765
Sea Robin Pipeline Co.	SONAT Corp.	2000	470	1,241
Trunkline Gas Co.	PanEnergy Corp.	2000	4,134	1,884
Dominion Resources Corp.				
Dominion Gas Transmission Co.	Consolidated Natural Gas Co.	2000	9,950	6,257
Duke Energy Corp.	**Duke Energy Trading Co.**			
Algonquin Gas Transmission Co.	PanEnergy Corp.	1999	1,092	1,586
East Tennessee Natural Gas Co.	Tenneco Energy Corp.	2000	1,100	700
Maritimes & Northeast PL Co.	Built in 1999	n/a	304	440
Texas Eastern Transmission Corp.	PanEnergy Corp.	1999	12,100	5,939
El Paso Corp.	**El Paso Merchant Energy**			
ANR Pipeline Co.	Coastal Corp.	2000	9,553	5,846
Colorado Interstate Gas Co.	Coastal Corp.	2000	4,123	2,350
El Paso Natural Gas Co.	El Paso Energy Co.	n/a	10,009	5,344
Mojave Pipeline Co.	Built in 1993	n/a	362	550
Portland Gas Transmission Co.	Built in 1998	n/a	242	178
South Georgia Natural Gas Co.	SONAT Corp.	1999	909	129
Southern Natural Gas Co.	SONAT Corp.	1999	7,612	2,536
Tennessee Gas Pipeline Co.	Tenneco Energy Corp.	1999	9,270	5,587
Wyoming Interstate Gas Co.	Coastal Corp.	2000	425	1,175
Enron Corp	**Enron Online Trading**			
Florida Gas Transmission Co. (50%)	No Change	n/a	5,203	1,405
Northern Natural Gas Co.	No Change	n/a	15,637	3,800
Transwestern Gas Co.	No Change	n/a	2,532	2,700
Great Lakes Gas LP				
Great Lakes Gas Transmission Co.	No Change	n/a	2,101	2,483
Iroquois Pipeline LP				
Iroquois Gas Transmission System	Built in 1991	n/a	375	850
Kinder-Morgan Corp.	**KM Gas Services Division**			
Kinder-Morgan Interstate PL Co.	KN Energy Corp.	1999	6,081	1,075
Kinder-Morgan Texas PL Co.	MidCon Corp.	1998	2,101	na
Natural Gas Pipeline Co. of America	MidCon Corp.	1998	10,076	5,001
Trailblazer Pipeline Co.	MidCon Corp.	1998	436	605
Koch Corp.	**Koch Energy Trading**			
Gulf South Pipeline Co.	United Gas Corp.	n/a	7,252	3,476
Mobile Bay Pipeline Co.	Built in 1993	1998	26	600
Leviathan Gas Pipeline Partners				
High Island Offshore System	KN Energy Corp.	2000	247	1,800
UT Offshore System	KN Energy Corp.	2000	30	1,040
NiSource Corp.				
Columbia Gas Transmission Corp.	Columbia Energy Corp.	2000	11,215	7,276
Columbia Gulf Transmission Co.	Columbia Energy Corp.	2000	4,200	2,317
Crossroads Pipeline Co.	Built in 1995	n/a	205	250
Granite States Gas Transmission Co.	Northern Utilities Inc.	1999	na	na
Northern Border Partners				
Midwestern Gas Transmission Co.	Tenneco Energy Corp.	2001	350	785
Northern Border Pipeline Co.	No Change	n/a	1,214	2,355
Questar Corp.				
Overthrust Pipeline Co.	No Change	n/a	88	227
Questar Pipeline Co.	No Change	n/a	2,000	1,400
TransColorado Gas Transmission Co.	Built in 1998	n/a	295	300
Reliant Energy Corp.	**Reliant Energy Wholesale Group**			
Mississippi River Transmission Co.	No Change	n/a	1,976	1,670
Reliant Energy Gas Transmission Co.	No Change	n/a	6,228	2,797
Williams Companies, Inc.	**Williams Energy Services**			
Cove Point LNG LP	Columbia Energy Group	2001	87	585
Kern River Transmission Co.	Built in 1992	n/a	922	800
Northwest Pipeline Co.	No Change	n/a	3,932	2,900
Transcontinental Gas Pipeline Co.	Transco Energy Corp.	1997	10,562	7,000
Williams Gas Pipeline–Central	No Change	n/a	5,926	2,800
Williams Gas Pipeline–South Central	Transco Energy Corp.	1997	5,573	2,000
Xcel Corp.				
Viking Gas Transmission Co.	Northern States Power Co.	2000	662	516

NOTE: All interstate pipelines operating in the U.S. are not included. MMCFD = million cubic feet per day; n/a denotes not applicable and na means not available.

SOURCES: **Total gas deliveries**: Federal Energy Regulatory Commission Forms 2 & 2A, "Annual Report of Major/Minor Natural Gas Companies," 1999. **System capacity and mileage**: Federal Energy Regulatory Commission, FERC Annual Capacity Report (18CFR 284.12). **Ownership and affiliations**: Complied from various news sources and natural gas trade press periodicals.

Natural gas measurement stations usually consist of pressure control valves and gas measurement devices—either AGA dimensioned orifices, turbine meters, or ultrasonic meters. Flow and pressure control valves, as well as overpressure protection, may also be required. If a large temperature drop is anticipated across any of the pressure-reducing valves, then a radiant heater or indirect-fired heater may be required. A rough estimate of temperature drop is 6°F per 100 lb/in^2 (3.3°C per 690 kPa) of pressure drop. Electronic flow measurement has become commonplace, meeting the requirements of current regulations. Microprocessors calculate, remotely monitor, and control flow in real time using the latest calculation methods of the AGA and American Petroleum Institute (API) reports. The quality of the gas is also measured by one of several means. For small-volume stations, a continuous gas sampler accumulates a small volume for laboratory testing. Medium- and large-volume stations will use a gas chromatograph.

Flow   From the thermodynamic energy balance equations for the flow of compressible fluids (Sec. 4), the general flow formula is derived and expressed in *BuMines Monograph 6* (1935) as

$$Q = K \frac{T_0}{P_0} \left[\frac{(P_1^2 - P_2^2)d^5}{GT_f L_f} \right]^{1/2} \qquad (11.7.1)$$

where Q is the rate of flow, ft^3/day; T_0 is the temperature base, °R; P_0 is the pressure base, psia; P_1 is the initial gas pressure, psia; P_2 is the final gas pressure, psia; d is the internal pipe diameter, in; G is the specific gravity of the gas, which is dimensionless (for air, $G = 1.0$); T_f is the gas flowing temperature, °R; and L_f is the length of the pipeline, mi. The general flow equation for constant-volume (steady-state), isothermal, horizontal flow of natural gas is given by *U.S. BuMines Monograph 9* (1956) as

$$Q = 38.7744 \frac{T_b}{P_b} \left[\frac{(P_1^2 - P_2^2)D^5}{fZGTL} \right]^{0.5} \qquad (11.7.2)$$

where Q = rate of flow, ft^3/day; T_b = temperature base, °R; P_b = pressure base, psia; P_1 = inlet pressure, psia; P_2 = outlet pressure, psia; D = internal pipe diameter, in; f = friction factor, dimensionless; Z = average gas compressibility, dimensionless; G = gas gravity (for air, $G = 1.0$); T = average flowing temperature, °R; and L = length of pipe, mi. Monograph 9 also gives a good discussion of friction factors and elevation corrections.

Other flow equations may be derived from the general equation by substitution of the proper expression for the friction factor. The applies to the Panhandle A equation, which has the form

$$Q = 435.87 \left(\frac{T_b}{P_b} \right)^{1.0788} \left(\frac{P_1^2 - P_2^2}{G^{0.8539} TL} \right)^{0.5394} D^{2.6182} E \qquad (11.7.3)$$

where E = pipeline flow efficiency, a dimensionless decimal fraction (design values of 0.88 to 0.96 are common). In the Panhandle A equation, the compressibility is included in the pipeline efficiency, and the other symbols are as previously defined. This formula may be solved graphically as described by C. W. Marvin, *Oil and Gas Journal*, Sept. 20, 1954.

An AGA paper, Steady Flow in Gas Pipelines (1965), gives a good general review of constant-volume gas flow equations, both with and without elevation corrections. Equations and test data are included for both the partially turbulent (Reynolds number–dependent) flow regime and the fully turbulent flow regime using effective internal pipe roughness. In this reference, the drag factor is introduced to account for pressure losses for fittings such as bends and valves and is applied in the partially turbulent flow regime.

When the volume of gas flowing varies with time and position (transient or unsteady-state flow), the simultaneous momentum and mass balance equations are usually solved by use of numerical methods and the aid of computers. An AGA computer program, **Pipetran**, is available for this purpose, as are several related texts on transient flow. A paper entitled Gas Transportation System Modeling appears in the *1972 AGA Operating Proceedings*. Gas flowing in a pipeline undergoes a number of transient, or unsteady-state, flow conditions caused by such things as prime movers starting and stopping, valves opening and closing, and flows either into or out of the pipeline. DOS-based computer programs, such as **WinFlow** (Gregg Engineering) and others, allow the engineer to model simple as well as complex pipeline situations. These programs can be used to test proposed system expansions such as loops or additional compressor stations, to perform capacity studies, and to evaluate the effects of the transients mentioned above.

Compression   The theoretical horsepower requirements for a station compressing natural gas may be calculated by polytropic compression formulas (see Sec. 4). The change of state that takes place in almost all reciprocating compressors is close to polytropic. Heat of compression is taken away by the jacket cooling and by radiation, and a small amount of heat is added by piston-ring friction. Compression by centrifugal compressors is even closer to polytropic (see Sec. 14).

For general design, the horsepower requirements for a station compressing natural gas with reciprocating compressors can be calculated as follows:

$$\text{Reciprocating station horsepower} = \text{hp} \, Z_s \frac{T_s}{520} \qquad (11.7.4)$$

The value of hp can be taken from Fig. 11.7.1, typical manufacturer's curve; T_s is suction temperature, °R; and Z_s can be taken from Fig. 11.7.2.

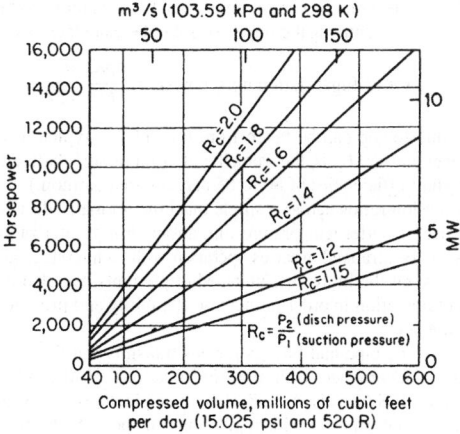

Fig. 11.7.1   Reciprocating-compressor horsepower (MW) graph. $R_c = P_2/P_1 =$ discharge pressure, lb/in^2 abs ÷ suction pressure, lb/in^2 abs.

Fig. 11.7.2   Supercompressibility factor Z_s for natural gas. (Natural gas, 1 percent N_2.)

Fig. 11.7.3 Centrifugal-compressor horsepower (MW) graph. $R_c = P_2/P_1 =$ discharge pressure, lb/in² abs ÷ suction pressure, lb/in² abs.

Similarly, the horsepower requirements for a station compressing natural gas with centrifugal compressors can be calculated as follows:

$$\text{Centrifugal station horsepower} = \frac{\text{hp}Z_s}{E_c}\frac{T_s}{520} \qquad (11.7.5)$$

The value for hp can be taken from Fig. 11.7.3, typical centrifugal horsepower curve; E_c is the centrifugal-compressor shaft efficiency. (Design shaft efficiencies of 80 to 85 percent are common.)

Compression ratios across a single machine normally do not exceed 3.0 because of outlet temperature limitations. For greater ratios, compressors can be arranged in series operation with an interstage gas cooler placed between the individual units. Multiple units can also be placed in parallel operation in order to increase flexibility and provide for routine maintenance.

Design   The pipe and fittings for gas transmission lines are manufactured in accordance with the specifications of the API, ANSI, ASTM, and MSS (see also Sec. 8). Interstate pipelines are installed and operated in conformance with regulations of the U.S. Department of Transportation (DOT). The ANSI B31.8 Code and the *Gas Measurement Committee Report 3* (Orifice) and *Gas Measurement Committee Report 9* (Ultrasonic) of the AGA provide supplemental design guidelines. The equation for pipeline design pressure, as dictated by the DOT in Part 192 of Title 49, Code of Federal Regulations, is

$$P = \frac{2St}{D}FET \qquad (11.7.6)$$

where P = design pressure, psig; S = specified minimum yield strength, lb/in²; t = nominal wall thickness, in; D = nominal outside diameter, in; F = design factor based on population density in the area through which the line passes, dimensionless; E = longitudinal joint factor, dimensionless; and T = temperature derating factor (applies above 250°F only), dimensionless.

The selection of the diameter, steel strength, and wall thickness of a line and the determination of the optimum station spacing and sizing are based on transportation economics. Figures 11.7.4 through 11.7.6 show schematically methods of analysis commonly used to relate design factors to the cost of transportation. A perspective of various elements of the cost of transportation for a long-distance, large-diameter, fully developed pipeline system is presented in Fig. 11.7.7.

The cost of installed pipelines varies greatly with the location, the type of terrain, the design pressure, the total length to be constructed, and many other factors. In general, figures from $32,000 to $90,000 per in (per 2.54 cm) of OD represent the range of approximate costs per mile (per 1.61 km) of installed pipeline. Offshore pipeline installed costs vary with such items as length, water depth, bottom conditions,

Fig. 11.7.4   Schematic relationship of cost of transportation versus design capacity for a given length of line with various pipeline diameters at equal design pressures.

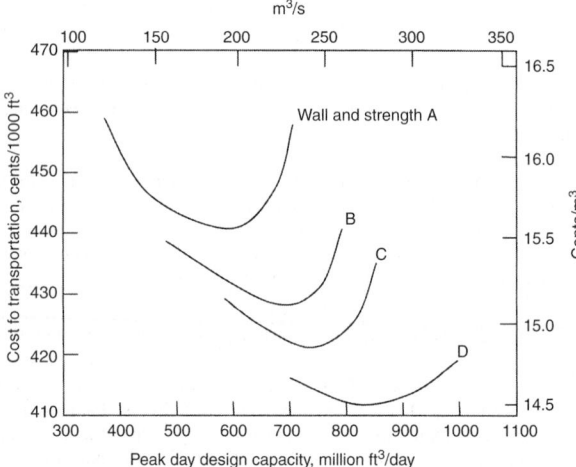

Fig. 11.7.5   Schematic relationship of cost of transportation versus pipeline steel strength and wall thickness.
A = 36-in dia., 0.385-in wall thickness, 52,000-lb/in² min. yield (800-lb/in² operating pressure).
B = 36-in dia., 0.385-in wall thickness, 60,000-lb/in² min. yield (924-lb/in² operating pressure).
C = 36-in dia., 0.404-in wall thickness, 65,000-lb/in² min. yield (1,050-lb/in² operating pressure).
D = 36-in dia., 0.438-in wall thickness, 70,000-lb/in² min. yield (1,220-lb/in² operating pressure).

pipeline crossings, depth of bury, and many other factors. These costs range from $53,000 to $76,000 per inch (per 2.54 cm) of OD per mile (per 1.61 km) for a typical offshore pipeline. Compressor-station installed costs vary widely with the amount and type of horsepower to be installed and with the location, the type of construction, and weather conditions. The installed costs of gas turbine stations range from $800 to $3,300 per installed hp ($1,070 to $4,425 per kW); gas engine stations, from $1,700 to $7,800 per hp ($2,280 to $10,455 per kW); and electric motor stations, from $1,300 to $5,000 per hp ($1,740 to $6,700 per kW).

Fig. 11.7.6 Schematic relationship of cost of transportation versus design capacity for a given pipeline diameter and length, with various spacings of compressor stations.

After a pipeline system is constructed and fully utilized, expansion of delivery capacity can be accomplished by adding pipeline loops (connecting segments of line parallel to the original line), by adding additional horsepower at compressor stations, or by a combination of these two methods.

CRUDE OIL AND OIL PRODUCTS

General Approximately 91 percent of the U.S. proven liquid hydrocarbon reserves are situated in the seven states of Texas, Alaska, California, New Mexico, Oklahoma, Wyoming, and Louisiana, or in the marine areas off their coasts. Offshore production is becoming increasingly significant. The refinement of multiphase pumping systems has enhanced crude oil recovery from formerly marginal offshore fields. In 2003, approximately 23 percent of the total proven crude oil reserves were located in federal waters offshore of California, Louisiana, and Texas. The major markets in which petroleum products are consumed are remote from the proven reserves. About 61 percent of all petroleum consumed in the United States is imported and must be moved from ports to the centers of consumption. Forty percent of all movements of petroleum and petroleum products is made by pipeline. The interstate pipelines alone transport over 2 billion barrels (318 million m³) of petroleum a year. There are about 177,000 mi (285,000 km) of operating interstate liquid-petroleum trunk lines in the United States.

Consumption of refined petroleum products in the United States has been divided into five Petroleum Administration for Defense Districts, or PADDs, based on historical, demographic, geological, and topographical characteristics, by the Energy Information Administration. The East Coast (PADD 1) is the most heavily populated and has the highest petroleum consumption. With virtually no local natural petroleum resources and limited refineries, the Northeast area must cover its deficit in refined product supply through pipeline deliveries from the Gulf Coast and imports of refined products by tanker. The Midwest (PADD II) has a better balance of supply and demand to meet the needs of its less heavily populated area. Logistics hubs as well as refining hubs in the region, such as the Chicago hub, depend on deliveries from pipelines to provide their services. The Gulf Coast (PADD III) is the nation's primary oil supply region. The design and sophistication of refineries in this region are unrivaled in the world. Production from this region alone is greater than all but two members of the Organization of Petroleum Exporting Countries (OPEC). Low regional demand enables it to provide for the needs of the other regions. The Rocky Mountain States (PADD IV) are thinly populated and more than 80 percent of the consumption within the area is low-sulfur diesel fuel for highway

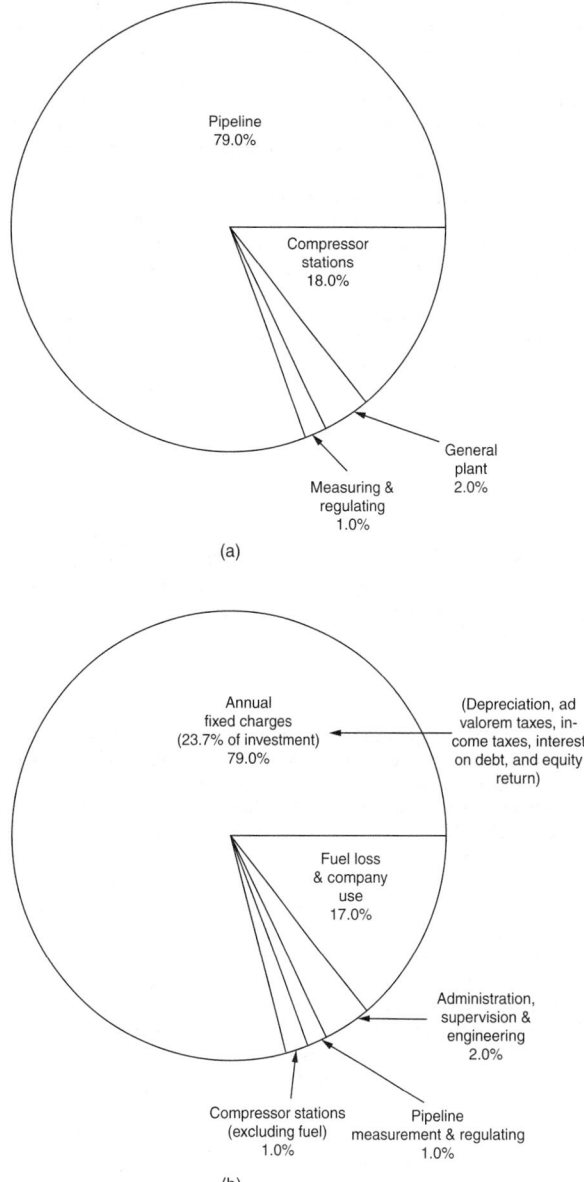

Fig. 11.7.7 Analysis of (*a*) investment and (*b*) cost of transportation.

consumption. On a per capita basis, the consumption of diesel fuel is approximately 60 percent higher than the national average, but its consumption per square mile is less than 30 percent of the national average. Pipelines serving this region distribute products from refineries in Denver, CO, Billings, MT, and Casper, WY. The West Coast (PADD V) oil market is separated from the rest of the country geographically as well as politically. For many years, the Rockies have been a challenge to the transport of oil in and out of the region. More recently, California's stringent oil product specifications have exacerbated its isolation from oil supplies. There is high demand in the region, due to the high population in California. Kinder Morgan is one of the key pipeline companies in the region, redistributing products from refineries in the northwest corner of Washington and from its own system in Arizona, which connects to supplies in PADD III.

Centrifugal pumps have been used extensively. Generally, these units are multistaged, installed in series or parallel, depending on the application. Many operators install smaller multiple units instead of larger single units to improve operating efficiency and flexibility.

Reciprocating pumps are installed in parallel; i.e., a common suction and a common discharge header are utilized by all pumps. There are instances where rotary, gear, and vane-type pumps have been employed to fulfill specific requirements.

Electrically driven centrifugal pumps are now widely employed for the liquid pipeline industry because they are environmentally friendly and because of their compactness, low initial costs, and ease of control. Centrifugal pumps driven by gas-turbine engines have also come into routine use.

Positive-displacement and turbine meters are in general use on products systems (see Sec. 16). Totalizers for multimeter installations and remote transmission of readings are an industry standard. Greater application of meters with crude-oil gathering facilities is being made as automatic-custody units are installed.

Flow   The **Darcy formula,**

$$h_f = \frac{fLV^2}{D2g} \tag{11.7.7}$$

where h_f = friction loss, ft; f = friction factor (empirical values), dimensionless; L = length of pipe, ft; D = ID, ft; V = velocity flow, ft/s; and g = 32.2 ft/s^2, is in general use by engineers making crude-oil pipeline calculations. The Darcy formula, when stated in a form utilizing conventional pipeline units, is

$$P = 34.87 \frac{fB^2S}{d^5} \tag{11.7.8}$$

where P = friction pressure drop, lb/(in^2 · mi); f = friction factor (empirical values), dimensionless; B = flow rate, bbl/h (42 gal/bbl); S = specific gravity of the oil, dimensionless; and d = ID, in.

The **Williams and Hazen formula,** which has wide acceptance for product pipeline calculations, can be stated in a form employing conventional pipeline terms as follows:

$$P = \frac{2340B^{1.852}S}{C^{1.852}d^{4.870}} \tag{11.7.9}$$

where P = friction pressure drop, lb/in^2 per 1,000 ft of pipe length; B = flow rate, bbl/h; S = specific gravity of the oil product; C = friction factor, dimensionless; and d = ID, in. The friction factor C includes the effect of viscosity and differs with each product (C for gasoline is 150; for No. 2 furnace oil, 130; for kerosene, 134).

The brake horsepower for pumping oil is

$$\mathrm{hp} = \frac{RSh}{3960E} \tag{11.7.10a}$$

or

$$\mathrm{hp} = \frac{BP}{2450E} \tag{11.7.10b}$$

where R = flow rate, gal/min; S = specific gravity of the oil; h = pump head, ft; E = pump efficiency, decimal friction; B = flow rate, bbl/h; and P = pump differential pressure, lb/in^2.

Design   The pipe and fittings for oil transmission lines are manufactured in accordance with specifications of the API and are fabricated, installed, and operated in the United States in conformance with Federal Regulations, DOT, Part 195 (see also Sec. 8). Many factors influence the working pressure of an oil line, and the ANSI B31.4 code for pressure piping is used as a guide in these matters. A study of the hydraulic gradient, land profile, and static-head conditions is made in conjunction with the selection of the mainline pipe (Sec. 3).

Oil pipelines typically carry multiple products or grades of product in the same pipeline. This is accomplished by introducing the different products sequentially at the origin of the pipeline so that a given "batch" of the product occupies a section of the pipeline during the transport.

A 25,000-barrel (4,000-m^3) product will occupy nearly 50 miles (80 km) of 10-in-diameter (273-mm) pipeline. Product batches are typically transported without any special means or devices to separate them. Turbulent flow minimizes interface mixing, whereas flows that start and stop will generate greater interface contamination. In most pipelines, interface contamination is relatively minor. It is typical for 150 barrels (24 m^3) of mixed material to be generated in a 10-in (273-mm) pipeline over a shipment of 100 miles (161 km). The mixed material is not lost or wasted, as it reflects the two materials from which it is generated. The interface mix between batches of different grades of the same material is generally blended back into the lower-grade material. The interface mix between batches of different products which cannot be blended is segregated and reprocessed in a full-scale refinery or special facility built for that purpose.

Terminals receiving products typically distinguish between batches by monitoring the specific gravity of the product. The pipeline operator "swings" batches from one pipeline or storage facility to another based on this signal of the interface location. This method of separating batches of low-sulfur diesel fuel may become inadequate, however, as ultralow-sulfur diesel (ULSD) fuel specifications are mandated. The ULSD may be contaminated by the sulfur in the adjacent product by the time the shift in specific gravity is discernable. This is one of the challenges facing products pipelines as product specifications become more critical.

Oil and refined petroleum products pipelines that are referred to as *trunk* lines transport high volumes of product over long distances. Pipelines that transport smaller volumes over shorter distances are called *delivering* pipelines. Both types of pipelines deliver batches of product based on contract deliveries. These delivery contracts also fall into two modes of operation: batched service and fungible service. Under *batched service,* a specific volume of refined petroleum products tracked throughout the transportation process. The product delivered is the same material designated for the contract at the origin of the shipment. Under *fungible service,* the carrier does not deliver the same material that is designated for the contract at the origin of the shipment, but instead delivers material that has the same product specifications. Operating and economic efficiency are the bases for which mode of operation is used. In general, the fungible mode is most efficient since it tends to minimize the generation of interface contamination and permits split-stream deliveries to multiple customers from the same batch. Trunk lines in fungible service provide greater flexibility than delivering pipelines operating in batch mode. The chances for operating lockouts increase as a trunk line reaches its furthest delivery capillaries. Such a lockout could occur if a terminal does not have the capacity to accept a scheduled shipment and there are no alternative terminals available to accept it. The pipeline would be stalled until the product can be delivered.

Determining the pipe size and station spacing to transport a given oil or oil product at a specified rate of flow and to provide the lowest cost of transportation is complex matter. Usually the basic approach is to prepare (1) a series of pipeline cost estimates covering a range of pipe diameters and wall thicknesses and (2) a series of station cost estimates for evaluation of the effect of station spacing. By applying capital charges to the system cost estimates and estimating the system operating costs, a series of transportation-cost curves similar to those in Fig. 11.7.8 can be drawn.

Estimates of $30,000 to $120,000 per in (per 2.54 cm) of OD represent the cost per mile (1.61 km) of installed trunk pipeline. The tabulation of pump-station investment costs for automated electric-motor, centrifugal-type installations presented in Table 11.7.3 can be used for preliminary evaluations.

Careful design and good operating practice must be followed to minimize contamination due to interface mixing in oil-product pipelines. When the throughput capacity of an oil line becomes fully utilized, additional capacity can be obtained by installing a parallel pipeline or a partial-loop line. A partial loop is a parallel line that runs for only part of the distance between stations. In the case of product lines, careful attention must be paid to the design of the facilities where the loop line is tied into the original pipeline to prevent commingling of products.

Fig. 11.7.8 Schematic relationship of pipeline capacity to cost of transportation.

Table 11.7.3 Pump-Station Investment Costs

Pump-station installed hp, total	Cost per installed hp (1 hp = 0.746 kW)
250–500	$800–1,400
500–750	$600–1,050
750–1,000	$500–850
1,000–2,000	$350–600
Above 2,000	$300

SOLIDS

General Solids may be transported by pipeline by a number of different methods. In *capsule pipelining*, dry solids are placed in a circular capsule, which is then pumped down the pipeline. This method can have economic and other advantages in many cases, especially where water is scarce, because the water is never contaminated and can be returned easily in a closed loop.

Another method being developed for the transport of coal is by *coal log pipelines* (CLPs). Pulverized coal is mixed with a binder (usually petroleum-based) to form a solid log, designed to resist water absorption, that can be transported down the pipeline intact. No dewatering is required with this method of coal transportation, and because the binder is combustible it enhances burning. Considerable research, funded by the Department of Energy (DOE), the National Science Foundation, the Missouri Department of Economic Development and the Electric Power Research Institute (EPRI) as well as numerous private companies, is being conducted at the Capsule Pipeline Research Center (CPRC), University of Missouri [both the Columbia Campus (UMC) and the Rolla Campus (UMR)]. A number of coal companies and electric utilities have proposed potential demonstration projects to refine this methodology so that it may be considered commercially viable. In addition to economic advantages, CLP offers environmental benefits that may improve the chances for state-level eminent domain legislation that restricts the development of coal slurry pipelines in the United States. These advantages include considerably less water usage (25 percent) than slurry pipelines; reduced water contamination compared to slurry pipelines; reduced air and noise pollution, due to fewer coal trucks and coal trains; reduced traffic congestion and accidents; less coal dust; and cleaner power plants. It is also predicted that commercial development of this methodology will lead to greater use of capsule pipelines for transportation of grain, agricultural products, solid wastes, and other solids.

The most prevalent method of transporting solids by pipeline is by *slurry*. A slurry is formed by mixing the solid in pulverized form with a true liquid, usually water, to a consistency that allows it to be pumped down the pipeline as a liquid. Nonvolatile gases, such as carbon dioxide, may also be used as the carrying fluid. In some instances, the carrying fluid may also be marketed after separation from the solid.

Slurry Transport Some of the solids being transported in slurry from considerable distances by pipeline are coal, coal refuse, gilsonite, phosphate rock, tin ore, nickel ore, copper ore, gold ore, kaolin, limestone, clay, borax, sand, and gravel. Solids pipelines differ from pipelines for oil, gas, and other true fluids in that the product to be transported must be designed and prepared for pipeline transportation. After much research in the field of slurry hydrotransportation, methods for predicting the energy requirements for pipeline systems have vastly improved and slurries are transported today at much higher concentrations than 20 years ago. In the early, 1980s, slurry designs ranging from 45 to 58 percent concentration, by weight, were common. Today, ultrafine coal-water mixtures with concentrations, by weight, ranging up to 75 percent, or more, are used. Dispersants and chemical additives, such as **Carbogel** and **Densecoal** are used to reduce drag and facilitate pumping. These additives significantly reduce the operating costs of slurry pipelines. Additional savings are realized by direct burning of these high-concentration coal-water mixtures through a nozzle, since dewatering is not required.

Current trends in the construction of solids pipelines and those proposed for construction indicate that future transportation of minerals and other commodities will bear more consideration. Some of the commercially successful slurry pipelines throughout the world are listed in Table 11.7.4.

Advances in slurry rheology studies and technology related to homogeneous and heterogeneous slurry transport have made the potential for slurry pipelining unlimited. Immediate advantages for slurry pipelines exist where the solids to be transported are remotely located; however, each potential slurry transport project possesses some other advantages. Materials must satisfy the following conditions to be transported by long-distance pipelines: (1) the largest particle size must be limited to that which can readily pass through commercially available pumps, pipes, and other equipment; (2) solids must mix and separate easily from the carrying fluids or be consumable in the slurry state; (3) material degradation must be negligible or beneficial to pipelining and utilization; (4) material must not react with the carrying fluid or become contaminated in the pipeline; (5) the solids-liquid mixture must not be excessively corrosive to pipe, pumps, and equipment; (6) solids must not be so abrasive as to cause excessive wear at carrying velocities.

Coal Slurry Uncleaned coal can be transported by pipeline, but a more stable and economic long-distance pipeline operation can be achieved with clean coal. The use of clean coal allows a variety of supply sources; produces a slurry with a lower, more uniform friction-head loss; reduces the pipe-wall wear, and increases the system capacity. Cleaning pipeline coal costs considerably less than cleaning coals for other forms of transportation, since drying can be eliminated after the cleaning operation.

Coal-slurry preparation normally consists of a wet grinding process for reducing clean coal to the proper particle size and size-range distribution. The final slurry-water concentration can be adjusted before it enters the pipeline with quality control of the slurry holding tanks. The cost of slurry preparation is only slightly affected by changes in concentration or size consist. The normal slurry-preparation-plant practice is to hold the concentration within 1 percent of design to provide a slurry with a uniform pressure drop. Continued research has made it possible to predict accurately the anticipated pressure drop in a pipeline by using a laboratory and computer correlation.

Optimum characteristics for a slurry are (1) maximum concentration, (2) low rate of settling, and (3) minimum friction-head loss. Variables that can be manipulated to control and optimize slurry design are particle shape, size distribution, concentration, and ash content. Established computer techniques are used to evaluate these variables.

Design Because of the many slurry variations, there is no single formula available by which the friction-head loss can be readily determined for all slurry applications. A general approach is described by Wasp, Thompson, and Snoek (The Era of Slurry Pipelines, *Chemtech*, Sept. 1971, pp. 552–562), and by Wasp, Kenny, and Gandhi in their book, "Solid-Liquid Flow Slurry Pipeline Transportation," giving calculation methods that have been developed for predicting slurry pressure drops. The described methods are quite accurate and can be used for preliminary design. Additional calculation procedures and methods for specifying slurry size consist (solid size–range distribution) and

Table 11.7.4   Commercial Slurry Pipelines

Location	Length,* mi	Diameter,† in	Capacity,‡ Millions of tons/year	Concentration, % by weight
Coal slurry				
Black Mesa, Arizona	273	18	4.8	45–50
Consolidation, Ohio	108	10	1.3	50
Belovo–Novosibirsk, Russia	155	20	3.0	60
Limestone slurry				
Rugby, England	57	10	1.7	50–60
Calaveras, California	17	7	1.5	70
Gladstone, Australia	18	8	1.8	62–64
Trinidad	6	8	0.6	60
Copper concentrate slurry				
Bougainville, New Guinea	17	6	1.0	56–70
KBI, Turkey	38	5	1.0	45
Pinto Valley, Arizona	11	4	0.4	55–65
China	7	5	0.6	60
Iron concentrate slurry				
Savage River, Tasmania	53	9	2.3	55–60
Samarco, Brazil	245	20	12.0	60–70
Bailadila, India	166	14/16	6.8	66
Kudremukh, India	44	18	7.5	60–70
La Perla–Hercules, Mexico	53/183	8/14	4.5	60–68
Phosphate concentrate slurry				
Valep, Brazil	74	9	2.0	60–65
Chevron Resources, U.S.A.	95	10	2.9	53–60
Goiasfertil, Brazil	9	6	0.9	62–65
Kaolin clay slurry				
Huber, U.S.A.	21	6	0.7	40
Cornwall, England	25	10	2.9	55

* 1 mi = 1.61 km.
† 1 in = 25.4 mm.
‡ 1 ton = 907 kg.

velocity have been developed by continuing research and may be found in the following books: Shook and Roco, "Slurry Flow Principles and Practice," Butterworth Heinemann, 1991, and Wilson, Addie, and Clift, "Slurry Transport Using Centrifugal Pumps," Elsevier, 1992. Slurry hydraulics should be verified with bench-scale tests and compared with commercial operating data reference points or pilot-plant operations before constructing a full-scale long-distance pipeline.

Slurry velocities are generally limited to a minimum critical velocity to prevent solids from settling to the bottom of the pipe and a maximum velocity to prevent excessive friction-head loss and pipe-wall wear. These velocity limitations and the throughput volume determine the diameter of the mainline pipe. The relationship of volume and velocity to the pipe diameter is expressed as $ID = 0.6392 \sqrt{(gal/min)/V}$, where the ID is in inches, gal/min is the volume rate of slurry flow (gallons per minute), and V is flow velocity, ft/s.

Pipe-wall thickness in inches is computed by the modified Barlow formula as $[(PD/2SF) + t_c]/t_m$, where P = maximum pressure, lb/in²; D = pipe OD, in; S = minimum yield strength of the pipe, lb/in²; F = safety factor; t_m = pipe mill tolerance, usually 0.875; and t_c = internal erosion/corrosion allowances, in, for the life of the system. The internal metal loss results from the erosion of corrosion products from the pipe wall and thus can be limited with chemical inhibitors and velocity control. This metal loss occurs in all pipeline steels and according to Cowper, Thompson, et al., Processing Steps: Keys to Successful Slurry-Pipeline Systems, *Chem. Eng.*, Feb. 7, 1972, pp. 58–67, must be evaluated on a case-by-case basis. A minimum pipe wall of 0.250 in (0.635 cm) is often used to provide mechanical strength for large lines, even where pressures and wear permit the use of a thinner wall. Coal pipelines employ a graduated wall thickness to meet the demands of the hydraulic gradient, which saves pipe costs.

Pump-station hydraulic horsepower is calculated as $(P_d - P_s)(Q)/1,716$, where $(P_d - P_s)$ = pressure across the station, lb/in²; Q = maximum flow rate through the station in gallons per minute, and 1,716 is a constant, gal/(ft in²). Stations usually employ positive-displacement

reciprocating pumps of the type used for oil-field mud pumps, as they attain high pressures with low slurry velocities. This allows fewer stations and lower overall costs than with centrifugal pumps. Stations are located where the friction-head loss and the effects of terrain have reduced the pressure head to 100 ft (30.48 m). Centrifugal pump usage is advantageous in low-pressure, high-velocity applications such as coarse coal transportation for short distances.

The annual tonnage throughput for any diameter line can be varied within the velocity limits or by intermittent operation near the low velocity limit. The optimum mainline size for high-load-factor systems is found by comparing the relative cost of transportation for a series of pipe sizes, as in Fig. 11.7.9.

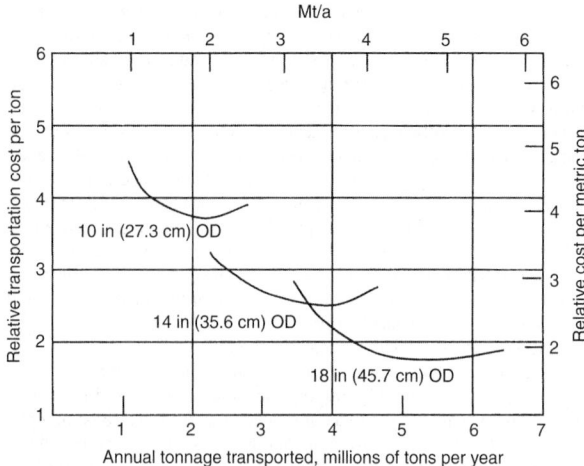

Fig. 11.7.9 System relative transportation cost for various mainline pipe sizes.

When a main-line size is selected, the cost of transportation for that line size at various rates of throughput is weighed against the projected system growth and compared with other line sizes to determine the size which gives the best overall system economics. Coal pipelines cannot be expanded by line looping because of the velocity limitations. Some expansion can be attained by oversizing the line and operating it intermittently near the minimum velocity in the earlier years of life.

11.8 CONTAINERIZATION
(Staff Contribution)

REFERENCE: Muller, "Intermodal Freight Transportation," 4th ed., Eno Foundation for Transportation. "The Container Handbuch," Web address: www.containerhandbuch.de.

Shipping of freight has always suffered from pilferage and other losses associated with the handling of cargo and its repeated exposure to weather conditions and to human access. In addition, the extremely labor-intensive loading and unloading practices of the past became more and more expensive. In the late 1920s and early 1930s the forerunners of present-day container service were offered. They varied from containers on trucks to services in which railroad cars were loaded onto rubber-tired trailers and delivered to the customer's door with no intermediate handling of the cargo. Starting in 1956, containerization as we know it today came into being.

The immense savings in shipping costs resulted in the rapid growth of containerization throughout the world in the 1960s and 1970s and in international standardization. The time savings obtainable through the use of containers is particularly noticeable in maritime service. It is estimated that a container-handling ship can achieve a 4 : 1 sea-to-port time ratio compared to 1 : 1 for a conventional break-bulk ship.

CONTAINER SPECIFICATIONS

There are many **standards** which apply to containers, among which are ISO, ANSI, the International Container Bureau (BIC), and the American Bureau of Shipping (ABS).

In order for a container to be used on a ship or a plane, a **CSC plate** is required. CSC stands for the International Convention for Safe Containers. CSC is also defined as standing for Container Safety Convention. A CSC plate is good for 5 years after construction. After that, the container must be reinspected at least every 2 years. Alternatively, a container may come under an Approved Continuous Examination Program (ACEP) under which registered owners examine, maintain, and repair their own containers.

Every conceivable aspect about containers, their construction, usage, labeling, biological-pest treatment, marking and numbering, is covered by detailed international standards. "The Container Handbuch," which is in both German and English and has Web address www.containerhandbuch.de, describes many of these.

Types Ninety percent of containers are dry, nonrefrigerated units with double doors on one end. Other types include refrigerated and/or insulated units, ventilated units, tank containers, and frames with folding ends or corner posts for carrying large items such as automobiles.

Condensation can be a problem inside a closed container. While a good unit can protect against outside elements, it cannot protect against condensation inside the container unless special steps are taken. Condensation typically arises from the moisture in the product being shipped; temperature changes during the trip; or moisture from dunnage, packing, or the container itself.

Construction materials usually consist of a steel, aluminum, or fiberglass outer cover on an internal steel frame. Containers are designed to be stacked from six to ten units high. Since containers are designed with oceangoing shipment in mind where the container, and those stacked on top of it, will be rising and falling with the movement of the ship, the additional acceleration forces have to be taken into account. This is usually done by rating the maximum stacking load using a g force of 1.8.

Figure 11.8.1 shows a typical 20-ft container. At each of the eight corners there are special fittings which facilitate the lifting, stacking, interconnecting, and tying down of units. These fittings have specially shaped slots and holes which enable standard crane and sling hooks and clevises to be used as well as special hooks and twist-locks.

Fig. 11.8.1 Typical 20-ft container.

Sizes The most widely used containers are 20 ft (6.1 m) or 40 ft (12.2 m) long by 8 ft (2.5 m) wide by 8 ft (2.5 m) high. The usual height of 102 in (2.6 m) includes an 8-in-deep (200-mm) underframe. This frame often has two, or sometimes four, transverse slots for forklift forks. Standard lengths are: 20, 35, 40, 45, and 48 ft (6.1, 10.7, 12.2, 13.7, and 14.6 m).

Typical specifications for a 20-ft (6.1-m) container are: maximum gross weight of 52,910 lb (24,000 kg), tare of 5,070 lb (2,300 kg), payload of 47,840 lb (21,700 kg), cubic capacity of 1,171 ft^3 (33.2 m^3), and allowable stacked weight for a g force of 1.8 of 423,280 lb (192,000 kg).

Typical specifications for a 40-ft (12.2-m) container are: maximum gross weight of 67,200 lb (30,480 kg), tare of 8,223 lb (3,730 kg), payload of 58,977 lb (26,750 kg), allowable stacked weight for a g force of 1.8 of 304,170 lb (182,880 kg), cubic capacity of 2,300 ft^3 (65 m^3). Thus a 40-ft (12.2-m) container has roughly the same volume as a 40-ft (12.2-m) highway semitrailer. A typical 40-ft (12.2-m) dry-cargo container with an aluminum outer skin weighs about 6,000 lb (2,720 kg).

Specifications for a typical 40-ft (12.2-m) **refrigerated** container are: maximum gross weight of 67,200 lb (30,480 kg), tare of 8,770 lb (3,980 kg), payload of 58,430 lb (26,500 kg), allowable stacked weight for a g force of 1.8 of 423,307 lb (192,000 kg), cubic capacity of 2,076 ft^3 (58.8 m^3). The refrigeration system specifications are: minimum inside temperature of 0°F (−18°C), maximum outside temperature of 100°F (38°C). The refrigeration system is powered through cables connected to the loading site, the ship, or to a generator mounted in the bed of the container chassis or trailer.

Some companies have a sufficiently large volume of shipping to have containers differing from these specifications. For example, the refrigerated containers used to ship Chiquita bananas are 43 ft (13.1 m) long. Their specifications are: maximum gross weight of 67,200 lb (30,480 kg), tare of 8,270 lb (3,750 kg), payload of 58,930 lb (26,730 kg), allowable stacked weight for a *g* force of 1.8 is 335,980 lb (152,400 kg), and a cubic capacity of 2,324 ft³ (65.8 m³). The height of the floor above the bottom of the underframe is 11 in (280 mm). The refrigeration rating is 16,721 Btu/h (4,214 kcal/h). Electric power supplied is three-phase at 460 V, 60 Hz (or 380 V, 50 Hz). The overall heat-transfer rate is 7,965 Btu/h (2,007 kcal/h).

ROAD WEIGHT LIMITS

The gross weight of a typical 40-ft (12.2-m) dry-cargo container is about 67,000 lb (30,000 kg). The special-purpose semitrailer for carrying containers on the highway is called a **container chassis** and weighs about 10,000 lb (4,500 kg). The towing tractor weighs about 16,000 lb (7,256 kg). Many U.S. highways do not allow such heavy loads. Typical allowable axle loads are 12,000 lb (5,442 kg) on the front wheels, 34,000 lb (15,420 kg) on the tractor rear wheels, and 34,000 lb (15,420 kg) on the chassis wheels. So if the container is to be forwarded by truck in the United States, the load must be limited to 54,000 lb (24,500 kg) and, on many secondary roads, even less. Similar constraints exist elsewhere in the world as well. As with any closed shipping space, depending on the density of the material being shipped, the weight or the volume may be the controlling factor.

CONTAINER FLEETS

As of 1988 about half of the world's containers were owned by container-leasing companies. Most of the remainder were owned by carriers. About 4 percent were owned by shippers. Container shipment figures and fleet sizes are usually given in TEUs (20-ft equivalent units). A **TEU** is the equivalent of a 20-ft (6.1-m) by 8-ft (2.5-m) by 8-ft (2.5-m) unit.

Containers are easy to make. As a result the competition among manufacturers is intense. The low-cost manufacturers are in Japan, Taiwan, the Peoples Republic of China, and South Korea. In 2004, prices for new standard 20-ft and 40-ft dry-freight containers were in the range of $1,800 to $2,200 and $2,500 to $5,200, respectively. For 20-ft and 40-ft refrigerated containers, prices were $5,000 to $9,000 and $10,000, respectively.

CONTAINER TERMINALS

A typical marine container terminal may have 50 acres (20 hectares) of yard space. It can load and unload 150 to 200 ships a year, and in so doing handle 60,000 to 80,000 containers. Under such circumstances the requirements for record keeping and fast, efficient handling are very severe. Specialized container-handling cranes and carriers have been developed to facilitate safe, fast, damage-free handling.

Handling Equipment By its design and purpose, the container is intended to go from shipper to receiver without any access to the contents. The container with its load may, however, travel by truck trailer from the shipper to the railroad, by rail to a marine terminal, by ship across the ocean, and again by rail and by truck trailer to the receiver. While the container is on the truck trailer, rail car, or ship, little or no handling needs to be done. When a container changes from one mode of transportation to another, at intermodal or transfer points, the container requires handling.

The general transfer tasks of **container-handling machinery** are: ship to/from rail, ship to/from truck, and rail to/from truck. Since a ship may have a capacity of up to 4,400 20-ft containers, extensive yard storage space has to be available. A 50-car double-stacked railroad train can handle up to 200 containers. Container-handling equipment varies from a simple crane with some slings or a large forklift truck at low-volume sites to special-purpose gantry and boom cranes and straddle carriers at high-volume sites.

Terminal designs come in two basic types: decked and wheeled systems. With a **wheeled** system, the containers are stored on their highway trailers. They are transferred to/from the loading crane by terminal tractors, which are also called *yardhorses*. In a **decked** system, the containers are removed from their highway trailers, rail cars, or ships and stacked in rows on the paved yard surface. When the containers are to be loaded, they are picked up at the stacked location, if possible by straddle carriers or mobile cranes, and loaded onto outgoing highway trailers, railcars, or ships. When the full area of the container storage yard is not accessible to the cranes, containers may be loaded onto yard trailers and transported from their stack location to the cranes. An equivalent procedure is also used for incoming containers where they must be transferred by yard trailer from the cranes to their yard storage location.

OTHER USES

As containers age and lose their CSC certification, their use as shipping containers becomes more restricted. Many businesses are finding that containers no longer acceptable for shipping service make excellent low-cost storage facilities and construction- and other-site offices. They are weather- and rodent-proof.

Section **12**

Building Construction and Equipment

BY

VINCENT M. ALTAMURO *President, VMA, Inc.*
WILLIAM L. GAMBLE *Professor Emeritus of Civil and Environmental Engineering, University of Illinois at Urbana-Champaign*
NORMAN GOLDBERG *Consulting Engineer, Economides and Goldberg*
ABRAHAM ABRAMOWITZ *Professor of Electrical Engineering, Emeritus, The City College of The City University of New York*
ROBIN O. CLEVELAND *Associate Professor of Aerospace and Mechanical Engineering, Boston University*
BENJAMIN C. DAVENNY *Vibration Consultant, Acentech, Inc.*
JONATHAN D. KEMP *Vibration Consultant, Acentech, Inc.*

12.1 INDUSTRIAL PLANTS
by Vincent M. Altamuro

REFERENCES: Hodson, "Maynard's Industrial Engineering Handbook," 4th ed., McGraw-Hill. Cedarleaf, "Plant Layout and Flow Improvement," McGraw-Hill. Immer, "Materials Handling," McGraw-Hill. Rosaler, "Standard Handbook of Plant Engineering," 2d ed., McGraw-Hill. Merritt and Ricketts, "Building Design and Construction Handbook," McGraw-Hill.

PURPOSES

Industrial plants serve many **functions.** They can:

1. Protect people, products, and equipment from the weather.
2. Preserve and conserve energy.
3. Condition the inside environment to be suitable for the processes, products, and people engaged therein.
4. Protect the outside environment from any fumes, dust, noise, or other contaminants their processes produce.
5. Provide physical security for their contents.
6. Block access, visual and/or physical, of those not authorized to see inside or enter them.
7. Provide the stable, strong, and smooth platform or surface required for operations.
8. Be the frameworks for the distribution networks of the services needed—electric power, lights, fuels, compressed air, gases, steam, air conditioning, fire protection, water, drainage piping, communications.
9. Support the cranes, hoists, racks, and other lifting and holding equipment attached to their superstructures.
10. Be integral parts of equipment (such as drying ovens) by having one or more walls serve as sides of them, and the like.
11. Be safe, pleasant, and efficient places to work for employees, impress visitors, and reflect positively on the company.
12. Fit the surroundings, blend in, be aesthetically attractive, make architectural statements.

This list of design criteria can be expanded to include functions other than those cited and which may be specific to a given proposed plant.

After a plant's expected present and future activities and contents are defined, it is designed to provide the desired functions, and then built to provide spaces for operating personnel and to house equipment and services. An existing plant considered for purchase or lease must be selected so that it will provide its user a competitive advantage. Generally, the decision to own or rent a plant depends on factors such as the expected life of the project, prudent use of capital, the possible need for early occupancy, the availability of a rentable building that matches the requirements, and the possibility of either a very rapid growth or complete failure of the enterprise. Some **commercial real estate** developers will erect a building to meet the specifications of the future operators of the industrial plant and then lease it to them under a long-term contract, often with an option to buy it at the end of the lease. A new structure to be designed and built must have its specifications clearly established so that it will serve its intended use and fit its surroundings.

The creation of an **industrial plant** involves the commitment of sizable resources for many years. The plant is therefore a valuable asset, both present and future. Its design and construction must be timely and within budget, and once undertaken, the project must, at most, be subject to minor modifications only, and preferably none. Once the structure is in place, it is very difficult or not economically feasible to move or change the plant. Even if not truly irreversible, design decisions can be corrected or changed only with great difficulty and added expense. There is often little difference in the cost of designing a plant and configuring its equipment and services layout one way versus another. The difference arises in the operating expenses and efficiencies, i.e., the wrong way versus the best way.

When designing a plant, one should estimate how long it is expected to last—both economically and physically. When making this determination, one accounts for the expected lives of the products manufactured, the processes and manufacturing methods used, the machines used, and the extant technology, as well as the duration of the market, the supply of labor, and so forth. Plant life and production life should coincide to the maximum extent possible. Major difficulties arise when a plant becomes obsolete and may warrant either abandonment or, at best, major modification for conversion. Often the manufacturing methods employed in production have so far outgrown the original methods that the conversion of an existing plant cannot be justified at all.

Obsolescence can be deferred by making the factory versatile, adaptable, and flexible. Possible changes in conditions can be anticipated and the ability to alter the plant can be included in the original design. An industrial plant is said to be *versatile* if it can do more than one thing, if its equipment can switch over and make more than one variation of the product, or if its output can be raised or lowered easily to match demand. It is said to be *adaptable* if it can make other related products with only programming and tooling changes to its equipment, without the need to alter their layout. It has *flexibility* to the degree that it is easy to move its services, machines, and other equipment to change the layout and flow paths, and thereby accommodate different products and changing product mixes. Such a building will have a minimum of permanent walls, barriers, and other fixed features. A plant can be made to be general-purpose or special-purpose, or more one than the other. A general-purpose plant is more generic and can be used for a wide variety of purposes. It is usually easier and less expensive to design, build, and convert to new use and is more salable when no longer needed. A special-purpose plant is designed for a specific task. It is more efficient and can make a lower-cost product, as long as the specifications stay within narrow limits. With modular machines, quick-change tooling, and programmable automation, such as robots, it is possible to gain the advantages of both general- and special-purpose construction in one plant. It can be built with the versatility, adaptability, and flexibility to be configured one way for one set of requirements, then reconfigured when conditions change. This enables the manufacturer to produce the wide variety of products that consumers want, and still make them with the lower per unit costs of a special-purpose plant. Both economy of scale and economy of scope are possible in the same facility.

Most industrial buildings are intended to be permanent structures—or at least to last many years. In cases where a permanent building is not essential, an industrial type of tent may be an alternative. **Tent technology** has advanced markedly. The advantages are that tents can be erected more quickly, more easily, and less expensively. They can be made with aluminum arch struts which support membrane envelopes made of flame-retardant vinyl, resulting in large poleless open spaces. The stronger designs can withstand winds of up to 80 mi/h (128.7 km/h). Naturally, they are not as strong as conventional structures, and security may be a problem; vandals can cut right though the vinyl skin. They find applications as churches, schools, prisons, and other entities as well as industrial buildings. In some cases, they are used to handle expansion or additional space needs until more permanent conventional buildings can be built. Models are now available with two floors, stairs, elevators, mezzanines, windows with screens, double doors with locks, wood flooring or carpeting, chandeliers, tiled toilets, lockers, HVAC, etc.

There are trends in the design of some products which are causing plants to be constructed differently, in addition to the demand for the benefits of versatility, adaptability, and flexibility cited above. Some products are getting smaller: computers, electronic circuits, TV cameras, and the like. For other products, greater manufacturing precision and freedom from processing contamination require the establishment of ultraclean facilities such as clean rooms. These and other trends are causing some plants to be built with the ratio of manufacturing space to administrative space different from that in the past. In some plants with large engineering, design, test, quality control, research and development, documentation, clerical, and other departments, the prior ratios have been reversed, so that manufacturing and warehousing spaces constitute a rather small percent of the total plant.

In addition to the basic building structure and its supporting services, an industrial plant comprises people, raw materials, piece parts, work in process, finished products, warehouse stock, machines and supporting equipment, systems and services, office furniture and files, computers, supplies, and a host of other things. While each is an individual entity, the plant must be designed so that all can work together in an **integrated and balanced system.** For a new facility, the best way to achieve this end is to design it in a progressive manner, going from the general to the specific, with each step based on prior decisions, until a set of detailed specifications is established. The sequence of design decisions is not linear, wherein one is finalized before the next one starts. Rather, it is more like a series of loops, with the output of one feeding back and possibly modifying prior decisions. For example, plant size could influence the site chosen, which then could influence the amount of air conditioning and other services needed, all of which, in turn, require space that could change the prior estimate of plant size. Almost all decisions have an impact on one or more interrelated plant specifications. The design of the product influences the degree of automation, which sets in motion a chain of decisions, starting with the number of employees required, down through the number of toilets, the size of the cafeteria, medical department, parking lot, personnel department, etc., and ultimately even placement of emergency exit doors.

PLANT DESIGN

An industrial plant must fulfill its intended functions efficiently and economically. Its design must consider and account for the basic operating conditions to be served. A detailed discussion along those lines follows.

Prerequisite Data, Decisions, and Documentation

The decision to build or buy an individual plant is made by the company's senior executives. It involves the commitment of large sums of money and is properly based on major considerations: need for additional production capacity; introduction of a new product; entering new markets; availability of new and/or better technology and machinery; building a new plant to replace an existing inefficient one or refurbishing and tailoring the existing plant for new production processes and cycles; relocation to a different area, especially if closer proximity to markets and availability of specialized labor are involved; making products in house which were previously bought for resale. Studies, analyses, and projections provide input data to help determine the proper site, size, shape, and specifications of a plant. Economic analyses, technological forecasts, and market surveys are used to justify the investment in a plant and to calculate the approximate amount of money required, break-even point, payback period, and profitability. Detailed product design and engineering documentation includes drawings, parts lists, test points, inspection standards, and specifications. Product variations, models, sizes, and options are defined. Sales forecast data addresses expected unit volumes, seasonality, peaks and valleys, growth projections, and other patterns. In addition to projections, constraints on the plant must also be understood at the outset. These may include location restrictions, budgetary limitations, timing deadlines, degree of mechanization, the type of building and its appearance, limits on its effluents, exhaust fumes, and noise, and other effects on the neighborhood and environment.

Activities and Contents

The determination of the activities and contents of a plant is facilitated by a series of analyses and management decisions. Make/buy decisions determine which items or component piece parts are to be made in the plant and which are to be purchased and stored until needed. Some purchased items are used as received, others need work (e.g., painting, plating, cutting to size). Parts purchased on a just-in-time basis will reduce the amount of in-process storage space required.

Processes are classified as **continuous** (e.g., refineries, distilleries, paper mills, chemical and plastic resin production); **repetitive** (e.g., automobiles, air conditioners, appliances, computers, telephones, toys); or **intermittent,** custom job shop, or to order (e.g., elevators, ships, airliners). Some plants include combinations of these processes when they make several types of products. In such mixed, or **hybrid,** situations, one product and processing method usually dominates, but there are cases where a plant set up for mass production work has a special-order shop to make small quantities of variations of the basic product, such as a vending machine that accepts only specific foreign coins. The manufacturing methods used in these processes may include casting, shearing, bending, drawing, forming, welding, machining, assembly, etc. (see Sec. 13). **Engineering documentation** used to facilitate manufacturing analysis includes operations analyses, flow process charts, precedence diagrams, "gozinto" charts, bills of materials, and exploded views of subassemblies and final assemblies. With computer-aided design (CAD) and other graphics, these documents can often be constructed, updated, and stored in a common database.

These documents help define the operations, their sequence and interrelationships, inspection points, storage points, and the points in the process where materials and parts join others to form subassemblies and the finished product. (See Hodson, "Maynard's Industrial Engineering Handbook.") Special characteristics of the operations may be noted on these documents, such as "Very noisy operation," "Piece parts can be stacked, but after assembly, they cannot," and the like. Product movement is determined. People/tools/things must move in an industrial plant. Workers and their tools go to the stationary work in process, as in shipbuilding; the worker and the work go to stationary tools, as in a general machine shop; the work in process goes to the workers and the tools, as in an assembly line. Other combinations of the relative movement of workers, tools, and work in progress are used. (See Cederleaf, "Plant Layout and Flow Improvement.")

An industrial plant should be designed to facilitate, not hamper, the **smooth movement of people, material, and information.** The efficiency of flow is also influenced by the layout of the equipment and other contents of the building. A multilevel building implies movement between the levels, which usually takes more time, energy, and expense than having the same activities on one level. Some operations are performed better in high structures, where raw material is elevated and then gravity-fed down through the lower levels as it evolves into a finished product. A plant built into the side of a hill receives material directly into an upper level without the need to elevate it. In most cases, a single-level plant affords the opportunity for the most efficient flow of materials and product. Even a single-level plant should have all its floor at the same elevation so that material moving from one area to another need not go up or down steps, ramps, or inclines. Features of the building such as toilets and utilities should be located where they will not interfere with the most efficient plant layout, and will not have to be moved in relayouts or expansions.

To the maximum extent possible, the **material flow** through an industrial plant should be smooth, straight, unidirectional, and coupled (the output of one machine should be the input to another, without a large cushion of inventory between them); require the least rehandling; be at constant speed over the shortest direct route, with the least energy and expense; and be always directed toward the shipping dock. Raw materials and piece parts should move continuously through the plant as they are converted progressively by a combination of people and equipment into finished products of the desired quantity and quality. Materials must flow with few interruptions, side excursions, or stops, so that they are in the plant for the shortest time possible and thereby facilitate expeditious product completion. Where possible, movements should be combined with operations, such as having a mobile robot or an automatic guided vehicle (AGV) work on (inspect, sort, mark, pack, etc.) the item while transporting it. An analysis of the plant's activities, including a list of its contents and an analysis of their relative movements, results in a rational determination of how they must flow through it. This must be done for materials (raw materials, purchased parts, work in process, finished goods, scrap, and supplies), people, data, and services.

Space and Size Calculations

Annual sales forecasts plotted by the month sometimes show peaks and valleys due to **seasonal variations** in demand. A plant with the capacity to produce the highest month's demand has much of its capacity idle or underutilized in the other months. The ideal plant would have a constant output every month, with long production runs of each item to minimize tool changes and setup times. This is not always practicable, for either very high inventory buildups or shortages could result. The compromise is the calculation of a production schedule that is more level than the sales forecast, builds as little inventory as possible without risking shortages, and allows for rejects, equipment breakdowns, vacations, bad weather, and other interruptions and inefficiencies. This will result in a plant with less production equipment and space but with more inventory space (and associated material handling equipment) to accumulate finished product to meet shipping demands in response to sales. Obviously, the warehouse and other storage areas must accommodate the maximum amount of inventory expected to be stored.

The proper size of a plant is determined by **calculating the total space required** (immediate and future) by all its contents: material at all stages of production, people (employees and visitors), equipment (production, materials handling, support, and services) and the ancilliary spaces required (working room, aisles, lobbies). At first, only the **general types of equipment are specified** (spot welders, overhead chain conveyors, forklift trucks, etc.). Later, **specific items are selected** and listed by manufacturer, model number, capacity, speed, size, weight, power, and other requirements. (See Sec. 10.) Finally, one model may be substituted for another to gain a particular advantage. A simplified typical equipment selection calculation follows:

1. Define the operation: Joining.
2. Decide the method: Welding.
3. Note personnel available: Semiskilled.
4. Determine general machine type: Spot welder.
5. Calculate required production output:

Spot Welds Required

	Product			
	A	B	C	Total
Number of spots	120	60	80	
Units per year	10,000	40,000	20,000	
Spots per year, millions	1.2	2.4	1.6	5.2
Spots per day (250 days/year)				20,800
Spots per hour (8 h/day)				2,600
Spots per minute (60 min/h)				44

6. Select a particular candidate machine, considering type of metal, thickness of metal, diameter of spot. From catalog, choose manufacturer (e.g., Hobart) and model (e.g., series 1500 rocker arm SW-V).

7. Calculate capacity of one machine in minutes per spot.

Operating time	0.03
Material handling time	0.06
Setup time allocated	0.01
Minutes per spot	0.10

8. Calculate the number of machines needed.

$$N = \frac{TP}{60HU}$$

where N = number of machines required; T = standard time to perform the operation, minutes; P = production needed per day, operations;

H = standard working hours per day; U = use factor—up-time of the machine, its utilization (percent of time it is producing), or its efficiency. Example:

$$N = \frac{0.10 \times 20,800}{60 \times 8 \times 0.80} = 5.4 \text{ machines}$$

9. Round off to next higher number: 5.4 becomes 6 machines.

10. Alternatively, return to catalog to see if there is a faster model. If so, 5.4 could become 5 of a different model machine.

11. Record specifications of selected machine on an equipment card or in a computer file. Appropriate notations should be made therein, such as:

Crane must go over this machine.
Must be near outside wall for venting.
Will require supplemental lighting.
Requires a special foundation.
Requires a drain for water.
Allow room on side for gear changes.
Allow room for largest-size sheet-metal parts.

12. Use data for all machines needed to help establish specifications for equippage of the plant:

Sum of space required (including space for workers, material, aisles, services) becomes size of that department or function.

Sum of utilities and supplies (electricity, water, steam, compressed air, oxygen, nitrogen, acetylene, coolants, lubricants, air conditioning) addresses requirements for them.

Sum of weights affects floor loading capacities and specifications. For some parameters, the extreme value is important in establishing the building specifications; i.e., the height of the tallest machine may set the required clear height inside the plant, locally or throughout. Equipment mounted on columns or roof beams will influence their size. Specific equipment may require special foundations or subfloor access pits.

Sums of the prices of the machines and the required tooling, the number of workers, their required skills, and their wages and benefits provide information regarding the investment and operating expenses for the plant.

The same procedure is used to determine the **material handling equipment**, its required space, and its influence on the plant's specifications. (See Immer, "Materials Handling.") Equipment may be capable of moving materials horizontally, vertically, or in both directions; some have a fixed path of travel, while the paths of others can be varied. The general types of equipment that move materials horizontally in a fixed path include conveyors, monorails, and carts pulled by trucks, dragged by chains in floor troughs, or that follow buried wires. Conveyors are overhead or floor-level type. Overhead conveyors clear the floors of some congestion but add to the loads carried by the building's columns and beams. Other types of conveyors are installed at floor level or at working height, and include belt conveyors and powered or unpowered roller conveyors. The general types of material handling equipment that travel in variable horizontal paths include hand trucks, powered trucks, pallet trucks, and automatic guided vehicles. Those that travel in fixed vertical paths include elevators, skip hoists, chutes, and lift tables. Those that travel in both horizontal and vertical fixed paths include traveling bridge cranes, gantry cranes, jib cranes, and pneumatic tubes. Those that travel in both horizontal and vertical variable paths include robots and forklift trucks. Forklift trucks may be powered by either battery, gasoline, or liquefied petroleum (LP) gas. Those powered by batteries will require the installation of charging facilities within the plant, and those with internal combustion engines will require safe fuel-storage space.

The installation of each of the above-listed material handling equipment types will influence the plant's specifications in some way. In computing loads on the building's structural members, all static and dynamic loads arising from material handling equipment must be factored into their design, as applicable.

With respect to **raw material,** its form, weight, size, temperature, ruggedness, and other qualities are considered, as are the expected distances to be moved, speed of transit, number of trips, and amount carried per trip. For **warehouses,** the important considerations are cubic space and density of use for various package sizes, weights, and stacking patterns.

The same basic approach is used for **support functions** and **plant services.** The space required for people can be estimated by listing them by functions, jobs, and categories (male versus female, plant versus office, department, location, shift, etc.), then allocating space to each. For example, office floor space allocation in ft^2 (m^2) per person may be: plant manager, from 150 to 300 (13.9 to 27.9) depending on size and type of plant; assistant plant manager, 125 to 250 (11.6 to 23.2); department heads, 100 to 200 (9.3 to 18.6); section heads, engineers, specialists, 100 to 150 (9.3 to 13.9); general personnel, 50 to 100 (4.6 to 9.3). When accounting for all employees in an industrial plant, the space per person can range from about 200 to 4,000 ft^2 (18.6 to 370 m^2), depending on the type of industry and the degree of automation employed. An estimate of the male/female ratio often must be made early in the planning stage, for many localities will issue a building permit only if adequate sanitary and rest facilities are included for each gender.

The space required for each department and function is compiled and added to show the approximate total plant size. See Fig. 12.1.1, which

also shows the percent of the total plant floor space allotted to functional subtotals. The size of the plant should not be deemed final without including the manner in which to allow for anticipated growth and/or plant expansion. Anticipated growth, if realistic, must enter into the decisions reached as to the initial size of the plant to be built. When dealing with the initial plant size vis-à-vis any anticipated future expansion, some questions usually posed are: Should it be large enough to accommodate expected growth, knowing that it will be too big initially? Should it meet only present needs, then be expanded as and if required? Should a midsize compromise be made? A plant too big for present needs will cost more to build and operate and may place the firm in an uncompetitive position. Expenses for taxes, insurance, heating, air conditioning, etc. for a partially vacant plant are almost the same as for one fully occupied. On the other hand, a plant built just to satisfy present requirements of a growing business can quickly become cramped and inefficient and may lead to rental of external storage and warehouse space, with all the added expense and loss of control entailed thereby. Any future expansion of the original plant will not only be disruptive to operations, but also will cost significantly more than building it bigger in the first place. The correct choice is made by analyzing all options and comparing expected costs and profits and the realistic probability of growth.

If a building is to be expanded or upgraded easily and economically after it is built, that capability must be included at the outset. This is done by deciding in advance the direction and degree of possible future expansion, knowing that if the business grows, not all sections of the plant may have to be expanded to the same degree. After the basic form of the building shape and envelope have been addressed, attention is then directed to other matters: location of the building on the property; materials of construction for internal walls that may have to be rearranged; location of spaces and equipment that will be difficult or impossible to move in the future, such as toilets, shipping/receiving docks, transformers, heavy machinery, and bridge cranes.

Superstructure members, pipes, monorails, ducts, and so on can be terminated at the proposed expansion side of the plant, with suitable end caps or terminations which may be removed easily and activated quickly when required. Installations of major utilities are prudently sized to handle future expansion. If the initial design strategy includes anticipation of vertical expansion via mezzanines, balconies, or additional stories, foundations, footings, superstructure, and other structural elements must be designed with that expansion in mind. To accommodate that requirement at some later date will prove to be prohibitively expensive and will disrupt ongoing operations to a degree unimaginable. By the same token, if future operations require a contraction of active space in the proposed plant, space would be available, at worst, for sublease to others. Regardless of whether possible expansion or contraction is contemplated, the initial plant design should consider both possibilities and seek to have either occur with minimum disruption of production.

Department or function	Area	
	ft^2	m^2
Main receiving	5,700	530
Incoming inspection	900	84
Main stockroom	9,600	892
Large-component storage	6,000	557
Coil steel storage	1,840	171
Steel coil line	2,360	219
Sheet metal fabrication	18,000	1,672
Welding	3,000	279
Sheet-steel storage	13,800	1,282
Copper tubing storage	600	56
Tubing fabrication	1,200	111
Cleaning and painting	9,600	892
Insulating	1,200	111
Electrical subassembly	1,200	111
Final assembly	33,600	3,121
Finished goods storage	40,200	3,735
Wood storage and fabrication	2,400	222
Shipping	9,600	892
Maintenance and tool room	600	56
Print shop	600	56
Plant offices	1,200	111
Engineering laboratory	12,000	1,115
Reproduction room	420	39
Canteen/eating area	3,600	334
Medical/first aid	330	31
Offices and lobby	21,600	2,007
Toilets	1,650	154
Total	202,800	18,840

Function	% of total plant	ft^2	m^2
All receiving, incoming inspection, storage, and shipping	44.2	89,440	8,309
All production	35.3	71,360	6,629
Engineering lab and repro	6.2	12,420	1,154
Support areas: Maintenance and tool, print, first aid, canteen, toilets, plant offices	4.0	7,980	741
General offices and lobby	10.3	21,600	2,007
Total	100.0	202,800	18,840

Fig. 12.1.1 Space required by department, subtotals by function, and for the entire facility for a plant designed and built to manufacture roof-mounted commercial air conditions. (*Source: VMA, Inc.*)

Plant Emplacement and Site Selection

The emplacement of a plant is defined by its geographical location, site, position, and orientation, in that order. Geographical location is determined by the country, region, area, state, county, city, and municipality in which the plant is situated. The site is that particular plot of land which can be identified by street address or block and lot numbers in that geographical location. The position and orientation speak to the exact placement and compass heading of the building on the site.

Those responsible for finding a place to build an industrial plant should be aware that places once polluted, unbuildable, and uninhabitable can sometimes be cleaned and cleared for occupancy after a time and treatments. The famous Love canal site in upstate New York had to be evacuated in 1978–1980 after more than 21,000 short tons of buried chemicals started oozing into basements and contaminating groundwater. Now, 20 years later, people are living safely in new homes there. Therefore, it might be productive to look at previously condemned places for the possible reclassification by the EPA as removed from the **Superfund** list and now decontaminated and buildable. Places cleaned up just enough that the remaining risk is compatible with the way the land is to be used are called **brown fields,** such that one may be designated as "clean," or not, if, say, a parking lot, an industrial plant, or homes are to be built on it.

Typical considerations entering each stage of the placement decision, from broad to specific, follow.

Location: Political and economic environment. Stability of currency and government. Market potential. Raw materials supply. Availability and cost of labor. Construction or rental costs. Environmental regulations. Climate. Applicable laws within the governing jurisdiction. Port, river, rail, highway, airport quality. Utilities and their relative costs. Taxes. Incentives offered. Local commercial services. Housing, schools, hospitals, shops, libraries, museums, restaurants, parks, houses of worship, and other community attractions. Crime rate, police and fire protection. General ambiance of the locality. Also, the trends of all the above.

Site: Availability. Price. Incentives offered. Building and zoning codes. Near highway entrance/exit, on railroad spur, waterway, and major road. Public transportation for employees. Availability of water, sewer, and other utilities. Topography, elevation, soil properties, subsurface conditions, drainage, flood risk, earthquake faults. Neighbors. Visibility.

Position: Placement for possible expansion with sides of building that may be extended facing an open area or parking lot and the sides not to be extended close to the property line. Proximity to railroad tracks, road, utilities or other fixed features. Positions of buildings on neighboring property.

Orientation: Turned toward or away from winter's wind and summer's sun, as desired, considering the frequently open large shipping and receiving doors. Needs for heating, air conditioning, natural light, color matching, and the like that would be affected by compass heading.

Proximity to competitors may be desirable if the locale has developed into a well-recognized center for that industry, where are found skilled people and a wide variety of suppliers and supporting services, (e.g., testing laboratories, local centers of higher learning with faculty available for consultation on an ad hoc basis). Without these conditions, and especially if the product is heavy and expensive to ship, it is desirable to locate close to market areas and distant from competitors.

The amount of land required depends on plant size, the number of employees and their need for parking space, the probability of storing some raw materials and finished goods outdoors, the need to turn large trucks around on the property, the possibility of future expansion of the plant, zoning, and the possible desire to create an open park-like ambiance. Total land area of from 4 to 8 times the plant size is usually adequate; the lower end of the range applies in built-up areas, and the higher end applies in suburban areas. In some locations the land can cost more than the plant.

Before placement of a plant becomes final, all conditions and extremes should be simulated, including winter storms, rainy seasons, floods, several consecutive days of 100°F heat, power outages, labor disruptions, and the like. It is unlikely that the perfect site will be found, thus sought-after attributes must be ranked in order of importance. As an incentive to get new industry, some governments will build roads and schools and reroute buses if necessary. Test borings are often taken at a candidate site before a commitment is made to buy or lease it to ensure that it is suitable. A plot plan (see Fig. 12.1.2) can be made with an approximation of the planned plant on the candidate site indicating land size, topography, drainage, position on the plot, anticipated expansion, compass orientation, location of power lines, railroad tracks, roads, rivers, open fields, neighbors, etc.

Plot Plan

Fig. 12.1.2   Plot plan of an industrial plant. *(Source: VMA, Inc.)*

Configuration

The configuration of a plant is determined ideally by the optimum layout of its contents. Compromises are usually made, however, often resulting in a conventionally shaped building. A building with several extended branches is very expensive. For a given floor area, a square building requires less total wall length than other quadrilateral shapes, but rectangular buildings are the most common compromise between cost and efficiency. For the lease or purchase of an existing building, the best approach is to design the best layout and then seek an existing building within the desired area in which that layout may be accommodated. For a new building, the layout is set down before the location is selected and then adjusted for the particular features of the site. A layout prepared after the site has been selected can be tailored to fit the site, all the while maintaining the desired features of that layout.

Before the configuration and layout become final, certain broad and tentative choices about the type of building should be made. These include general-purpose versus special-purpose, multilevel versus single level, use of mezzanines and balconies, with windows or windowless, etc. Setting aside considerations of material flow, a multilevel building enjoys the advantages, on a unit floor area basis, of using less land and costing less to build. The columns, however, will generally be spaced closer together, thereby presenting an impediment to desired flow paths and a reduction in options for overall layout. A single-level building, on the other hand, can more easily support heavier floor loads by virtue of its concrete slab on grade floor construction.

The matter of the shipping/receiving floor level vis-à-vis the truck bed level must be resolved satisfactorily, keeping in mind that loading and unloading ought to be effected with small hoists or forklifts for maximum efficiency. To that end the loading floor level can be built to be flush with the bed height of most of the anticipated truck traffic, or the truck parking apron can be inclined downward to align the two. If those cases when trucks with other bed heights have to be accommodated, small demountable ramps can be employed, the same being built sufficiently rigid to permit forklift traverse.

Balconies and mezzanines may be added to gain more storage space, to locate offices with a view of the production areas below, for raw materials or subassembly work that can be gravity-fed and drop-delivered to the operations below, and for the placement of service equipment (hot-water tanks, compressors, air conditioners, and the like). Basements may be included for the placement of heavy equipment, pumps, compressors, furnaces, the lower portion of very tall machines, supplies, and employee parking.

Windows in an industrial plant lower lighting bills, permit truer color matching in natural light, aid cooling and ventilation when opened, may provide a means of egress in case of fire, and may lower fire insurance rates. The advantages of a windowless plant are easy control of the intensity and direction of light (elimination of glare, contrasts, shadows, diffusion, and changing direction); (sometimes) less expensive construction; and less heat transfer, dust and dirt infiltration, maintenance expense to wash and repair glass, and worker distraction; better security; lower theft insurance rates. It is generally easier and more economical to design a windowless building. The absence of fenestration allows more flexibility in interior layouts by virtue of uninterrupted wall space. The absence of low windows, in particular, obviates the need for blinds or drapes to keep the interior private from the passing viewer. Elimination of an enticing target to vandals and terrorists is not to be dismissed lightly; often damage to machinery and equipment, as well as injury to personnel, results from flying missiles launched at and through windows.

The relative positions and detailed layouts of each department's machines, support equipment, services, and offices are based on analyses of their functions, contents, activities, operations, flow, relationships, and frequency of contacts. The layouts may be product- or process-oriented, or a combination of the two. In a product-oriented layout, machines are arranged in the sequence that the production process requires. This permits the product to advance in a direct path, such as on an assembly line, and with smooth material flow. In a process-oriented layout, the machines are grouped by type, such as a welding or drilling area. This requires that the products requiring those operations be brought to that area. It is used where products vary and the output of each is low, such as in a job shop, and allows production to continue even when a machine breaks down.

Another early decision involves the preferred placement of the receiving and shipping departments and the related basic pattern of material flow through the plant. If it is preferred to have the material enter at one end of the plant and exit at the other (Fig. 12.1.3a and b), then separate receiving and shipping facilities, equipment, and personnel will be required. Capital investment and operating economies ensue from combining these functions, with shared personnel, equipment, supervision, and space, but then the basic material flow will loop back to allow the finished product to exit from the same location where raw materials entered (Fig. 12.1.3c).

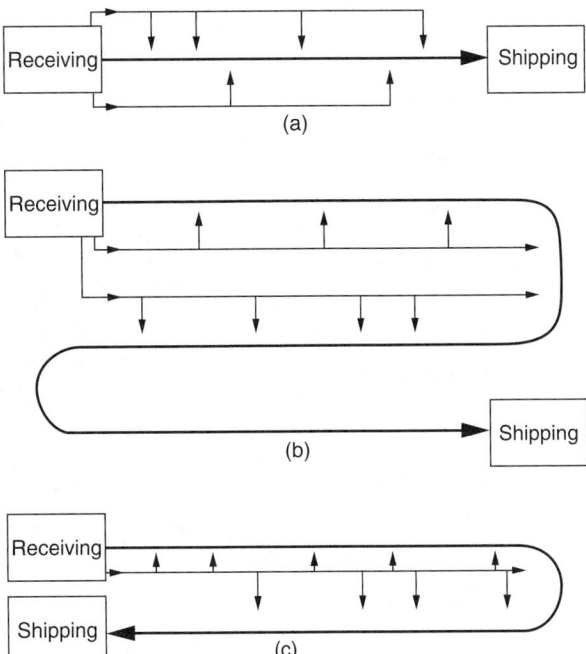

Fig. 12.1.3 Three different relative positions of a plant's receiving and shipping areas. (*Source: VMA, Inc.*)

The drawing showing the size, shape, and position of each department or area of an industrial plant is called a **block diagram**. It is developed by constructing a series of **analysis documents**. These include the frequency of relationship chart, proximity preference matrix, relative position block arrangement, sized block arrangement, initial block diagram, refined block diagram, and final block diagram (or simply block diagram). The **frequency of relationship chart** (also called a **from/to chart**) is constructed from an analysis of the engineering and manufacturing documentation and the activities of the plant. It shows the frequency and magnitude of movement, flow, and contacts between the entries. It may be weighted to include the importance of high-priority factors. The **proximity preference matrix** (also called a **relationship chart**) is constructed the same way. It shows (Fig. 12.1.4) which functions, departments, equipment, and people should be close to each other, how close, and why, and which should be away from which other and why. It is constructed with a diamond-shaped box at the intersection of every two entries. The notations made in each box show the importance of their need to be either close or separated by a number or code. The reasons can also be shown by entering a code in the lower half of the box, as shown in Fig. 12.1.4. The matrix may be made at the function

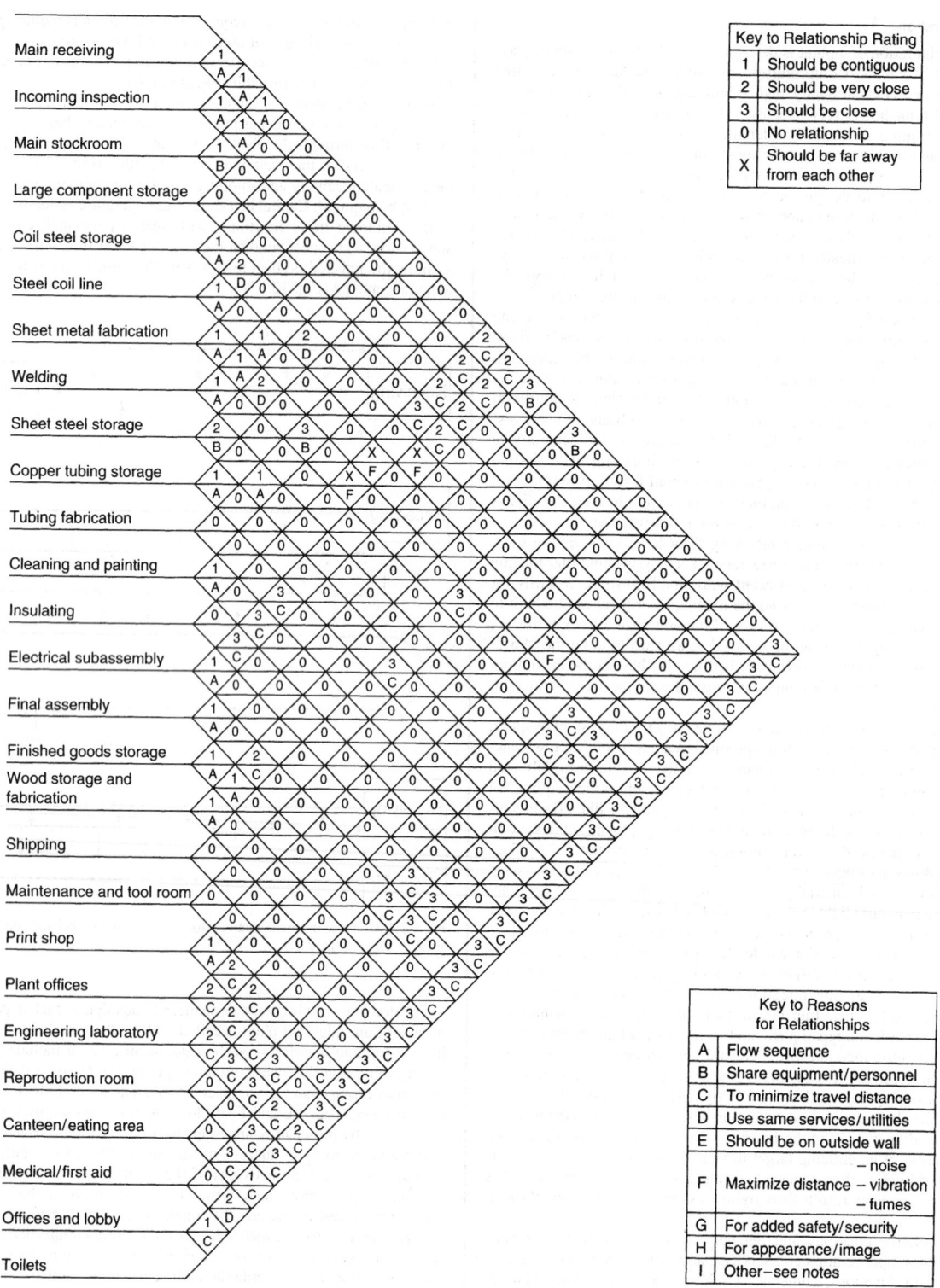

Main receiving
Incoming inspection
Main stockroom
Large component storage
Coil steel storage
Steel coil line
Sheet metal fabrication
Welding
Sheet steel storage
Copper tubing storage
Tubing fabrication
Cleaning and painting
Insulating
Electrical subassembly
Final assembly
Finished goods storage
Wood storage and fabrication
Shipping
Maintenance and tool room
Print shop
Plant offices
Engineering laboratory
Reproduction room
Canteen/eating area
Medical/first aid
Offices and lobby
Toilets

Key to Relationship Rating	
1	Should be contiguous
2	Should be very close
3	Should be close
0	No relationship
X	Should be far away from each other

Key to Reasons for Relationships	
A	Flow sequence
B	Share equipment/personnel
C	To minimize travel distance
D	Use same services/utilities
E	Should be on outside wall
F	Maximize distance — noise — vibration — fumes
G	For added safety/security
H	For appearance/image
I	Other—see notes

Fig. 12.1.4 A proximity preference matrix listing the departments, functions, and areas, and ranking how near or far each should be relative to the others, and why. *(Source: VMA, Inc.)*

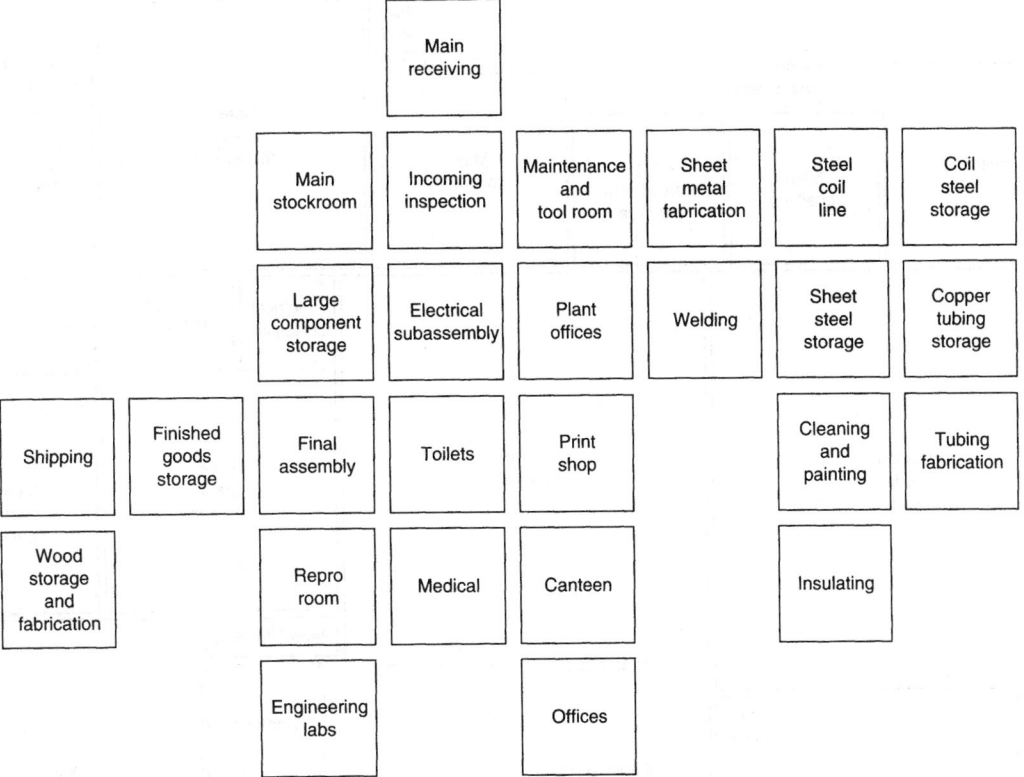

Fig. 12.1.5   A relative position block arrangement that attempts to satisfy and optimize the dictates of the prior proximity preference matrix. *(Source: VMA, Inc.)*

or department level and, later, at the individual machine or person level. The objective of the matrix and chart is to lay out the plant so that things are as close to (or as far from) other things to satisfy the criteria established, and ensure that those with the highest number of contacts are so located to minimize the time, distance, and energy required.

When this is done, a **relative position block arrangement** (Fig. 12.1.5) may be made; this shows the various arrangements possible by shifting around pieces representing the departments. The arrangement selected is the one that best satisfies the relationships (in decreasing order of importance) and degrees of contacts, as previously determined. The size of each model (or computer graphic representation) is the same because only the relative positions of the departments or areas is of interest at this stage of the design. Assigning a different color or background pattern to each adds to clarity and, when carried through to final and detailed drawings, helps visualize quickly the totals of separated areas; e.g., the total of inventory storage areas spread throughout the plant can be understood quickly if all are shaded the same color on the drawings. The **sized block arrangement diagram** (Fig. 12.1.6) is the relative position block arrangement with the size of each area scaled (but still square) to represent its square footage as determined by its expected contents and room for expansion (unless the expansion is to be handled by extending the building, in which case the department should be placed where expansion is proposed. The **initial block diagram** converts the square representation of each department into a shape that is more suitable for its contents and activities, but of the same square footage. The **refined block diagram** (Fig. 12.1.7) adjusts the initial shapes to effect a compromise between the advantages of having them be the best operational shape and those of fitting them in an economical rectangular building. L-, T-, U-, H-, and E-shaped buildings are often the result of such configuration compromises.

After the main aisles are drawn, as straight as possible, to serve as dividers between departments, the size and shape of each area can be fine-tuned, and a **final block diagram,** or simply **block diagram,** is drawn, keeping the same color code scheme used throughout. Detailed layouts are constructed for each department and function to fit within the spaces allocated in the block diagram. When assembled onto one drawing, the block diagrams become the detailed layout for the entire plant.

There are several aids to constructing both block diagrams and detailed layouts. These range from scaled templates to three-dimensional models of machines and equipment. They are available as plain blocks and as highly detailed plastic or cast-metal models. Computer-based optimization programs are also available. Some computer-aided design packages contain libraries of plant components and equipment that can be selected, positioned, rotated, and moved on the screen until a satisfactory layout is achieved. See Figs. 12.1.8 and 12.1.9 for two- and three-dimensional, respectively, computer-generated detailed layout drawings. Arrows are added to these diagrams to show flow paths. Copies of these drawings can be annotated with dimensions and machinery and furniture descriptors and given to vendors and contractors for them to submit bids to supply and install the items. They are also kept on file for use in future maintenance and/or revisions.

Features not expected to be moved in the future or that will become permanent parts of the building should be located first, e.g., stairways, doors, toilets, transformers, steam boilers, fuel tanks, pumps, compressors, piping, and permanent walls. Accurately dimensioned definitive drawings must be made to guide installers of equipment and services.

In addition to the layouts of the production areas, the support functions must also be planned. Support groups assist the production departments; they include research and development, testing laboratories, engineering, production planning and control, quality control,

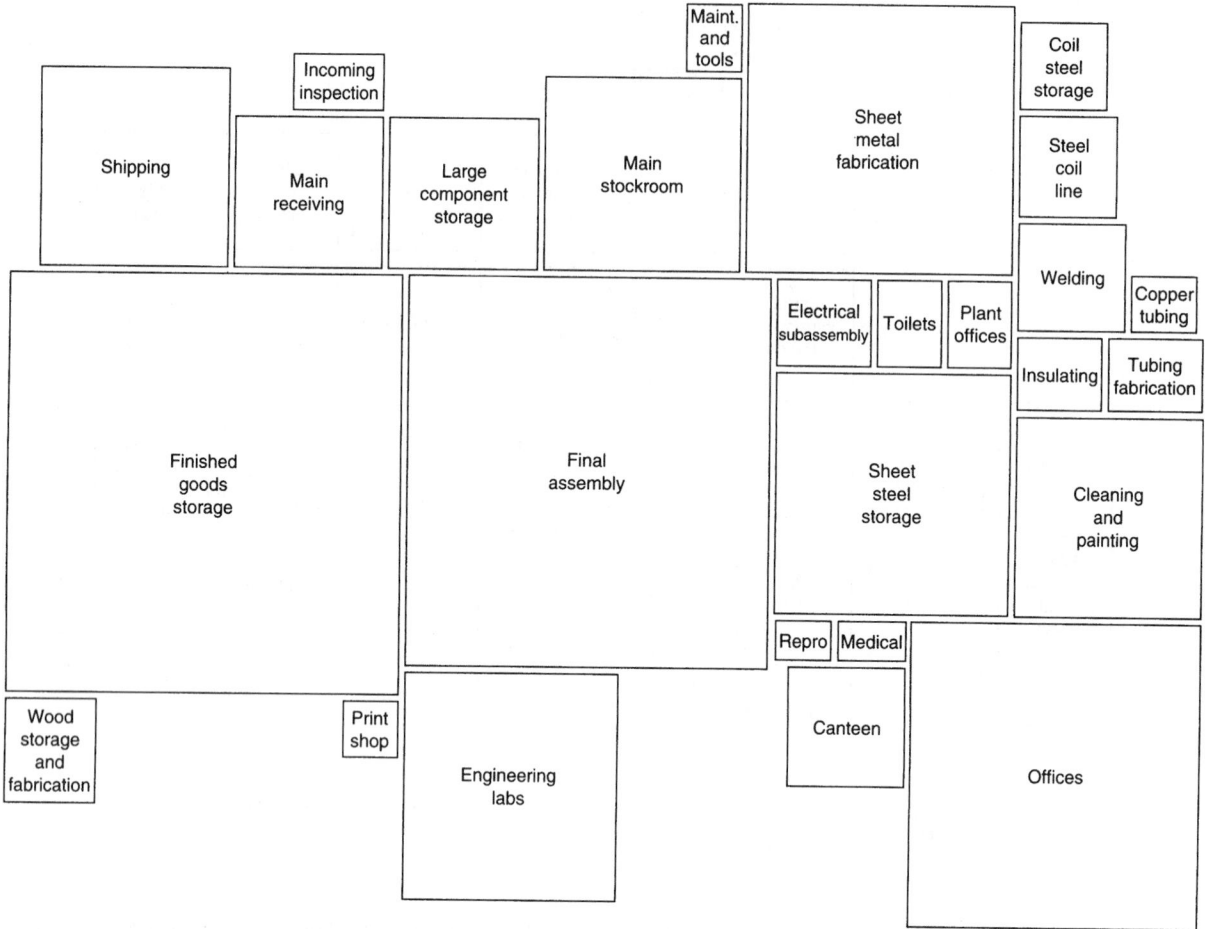

Fig. 12.1.6 A sized block arrangement that converts the prior relative position block arrangement into one that shows the required size of each department, function, and area, while maintaining the preferred relative position of each. *(Source: VMA, Inc.)*

machine shop, sales and advertising, technical literature, purchasing, data processing, accounting and finance, files storage, personnel, medical, administration and management, general office, supply storage, reception lobby, conference rooms, training rooms, library, mail room, copying and reproduction, cafeteria, vending machines, water fountains, washrooms, toilets, lockers, custodial and maintenance, security, and so on. Adequate space must be provided for the people working in these areas and their workstations must be designed with ergonomics, lighting, air quality, comfort, and safety in mind.

Many industrial plants have the supervisors' offices situated on the factory floor for better contact with and control of their people. Likewise, offices for quality control inspectors, industrial engineers, and manufacturing or process engineers often are located in the production areas. Such offices may be built with the building or may be purchased and installed as preengineered, prefabricated units. General and administrative offices usually are distant from the plant's machinery and equipment, whose noise and vibrations would otherwise affect the performance of the occupants therein and their ancillary equipment (computers, for example).

Office furniture may be arranged in military style (orderly straight rows), the open plan method (fewer and movable partitions), landscaped (free form with plants and curved dividers), individual offices (glass, wallboard, steel, or masonry walls), or any combination of these. In those arrangements wherein the partitions and dividers are classified

as furniture instead of parts of the building, they may be depreciated over a much shorter period of time than the permanent structure. Space is allocated on the basis of position in the organization chart, with the senior executives getting windows (if any) and the most senior getting a large corner office. Floor coverings, ranging from tile to carpeting, again depends on position, as do the amount and type of furniture.

The plant's reception lobby should measure approximately 160 ft^2 (14.9 m^2) if seating for four persons is required, and at least 200 ft^2 (18.6 m^2) for 10 visitors. Add 60 to 100 ft^2 (5.6 to 9.3 m^2) if a receptionist is to be seated there. Cloak rooms require 6 ft^2 (0.6 m^2) per 10 garments.

If a cafeteria is included, 20 ft^2 (1.9 m^2) per person of expected occupancy should be provided if it is equipped with tables and chairs, plus space required by any vending machines. Conference and meeting rooms should provide about 20 ft^2 (1.9 m^2) per person of expected attendance, and training rooms with theater-type seating should be 400 ft^2 (37.2 m^2) for groups of 20, 600 (55.8) for 40, and 1,000 (92.9) for 80. A plant library will range from 400 to 1,000 ft^2 (37.2 to 92.9 m^2), depending on its contents. Record storage requires 6 ft^2 (0.6 m^2) per file cabinet. Storage space must be provided for stationery and supplies. Slop sink and mop closets should be 12 to 15 ft^2 (1.1 to 1.4 m^2). Facilities should be placed as close as possible to those who will use them; those for universal use must be located conveniently. In very large plants, spaces and facilities for universal use must be supplied in

Fig. 12.1.7 A refined block diagram that converts the prior sized block arrangement onto one that makes the shape of each department, function, and area be such that, together, they fit into a building of a more conventional shape, while maintaining the size and relative position of each. (*Source: VMA, Inc.*)

multiples, and include toilets, clothes closets, lockers, time clocks, emergency exit doors, copy machines, vending machines, and drinking fountains. Aisle locations and widths are critical elements in the management of internal traffic, and are based on: use only for pedestrians or by material handling vehicles as well, in which case load widths must also be included as a design parameter; whether traffic is one-way or two-way, with loaded vehicles passing each other; traversing vehicular traffic only or with dropoff and/or pickup points along the route; and provision of turnaround space for vehicles (forklifts and the like) or restrictions on maneuvering within aisles. For efficient traffic flow, the configuration that will work best most often is one with one main aisle and a number of smaller feeder aisles, with straight, well-marked aisles intersecting at right angles. Aisles located at exterior walls will result in loss of storage space, and generally will be remote from the central area meant to be served. While necessary and functional, aisle space can occupy from 15 to 30 percent of the plant's total floor space. Planning for aisle space and the location of exits must defer to any applicable labor laws or local ordinances.

Services

An industrial plant's services are those utilities that power and supply the production and support functions. They include electric power and

backup emergency power; heating, ventilating, and air conditioning (HVAC); water for processes, drinking, toilets, and cleaning; smoke, fume, and fire detection and fire fighting; natural gas and/or fuel oil, process liquids and gases, compressed air, steam; battery chargers for electric forklift trucks, AGVs, and mobile robots; communication networks and links for telephone, facsimile, computers, and other database nodes; surveillance and security systems; and waste, scrap, and effluent disposal drains and piping. The design of these requires not only the specification of the equipment, but also the layout of the distribution networks, with drops to where needed and points of interface to the apparatus served.

Flexibility is increased and unsightly and dangerous wires are eliminated by installing buried electric raceways in the floor before concrete is poured. Wires channeled thus are connected to floor-mounted equipment through access caps; changes in machine layout or additions to the complement of machinery are facilitated via connections into the raceways. Definitive, current records of buried raceways document exact locations of raceways and any modifications made thereto over periods of time, and while they are archival in nature, they serve to prevent confusion and guesswork. Other raceways and wire conduits are dispersed through the plant to accommodate electric convenience outlets, communication equipment, and similar services.

N↑

Scale: 1/8" = 1'- 0"

Fig. 12.1.8   A two-dimensional computer-generated detailed layout of one area of an industrial plant, showing furniture, fixtures, and flow path. *(Source: Cederleaf, "Plant Layout and Flow Improvement," McGraw-Hill.)*

Electric power is usually transmitted over the utility's lines at between 22,000 and 115,000 V. The plant usually includes transformers to reduce voltage to 2,300–13,000 V. Most building codes require that these transformers be installed outside the building. The next level of voltage reduction, down to 120–480 V, is provided by transformers usually located inside the building and dispersed to strategic locations to provide balanced service with minimum-length runs of service connections. Alternatively, the plant may arrange for the local utility to supply electric power already stepped down to the levels needed; 240/120-V single phase, three-wire; 208/120-V three-phase, four-wire; 480/277-V three-phase, four wire. The last is more economical for motors and industrial lighting.

Natural light entering the plant through windows and skylights is erratic and difficult to control. All areas of the plant must be illuminated adequately for the activities conducted therein. Section 12.5 lists typical illumination levels. Density of light, illuminance, is designated in footcandles, fc (lumens per square foot) or lux, lx (lumens per square meter). One fc = 10.76 lx. In most cases, tasks requiring illuminance more than 100 to 150 fc (1,080 to 1,600 lx) also require directed supplemental illumination. No area should be illuminated at a level less than 20 percent of nor more than 5 times that of adjacent areas because eyes have trouble adjusting rapidly to drastic differences in illuminance. Proper illumination is also a function of the type and form of the lighting fixtures. Luminaires are selected to shed direct, indirect, or diffused light. Lamp types include incandescent, fluorescent, mercury vapor, metal halide (multivapor), and high-pressure sodium vapor. The number and type of lamps per luminaire, height, spacing, and percent

reflectance of the floor, walls, and ceiling are contributing factors. The plant should have a portion of its lights attached to an emergency power source that switches on when the regular power fails. (See Rosaler, "Standard Handbook of Plant Engineering.") A stationary diesel-, gas-, or gasoline-powered electric generator is usually installed to provide emergency electric power. The available fault current at all points in the electric distribution system should be determined so that protective devices can be installed to interrupt it. Selected circuits should have uninterruptible power supplies (UPSs), isolators, and regulators to protect against outages, voltage surges, sags, spikes, frequency drifts, and electrical noise. Exit signs and a clear path to the exits should be capable of being energized by emergency power, from either a standby generator or batteries.

Water, water distribution, and fixtures are essential for many plant processes, including paint booth "waterfalls"; for cooling machines, welders, and the like; for adding water as an ingredient in some products; in toilets, showers, cafeterias, and drinking fountains; for sprinkler systems and fire fighting; and for custodial work and landscaping maintenance. If large amounts of hot and/or cold water are required by the process, boilers and/or chillers are provided, along with associated piping and pumps for fuel and water distribution. The number of toilet fixtures as required by building codes is based on the number and gender of expected building occupants. Dispensers that chill and/or heat water are useful to prepare beverages and soups. Water consumption per person per 8-hr shift in personnel facilities ranges from 30 to 80 gal (114 to 303 L).

Fig. 12.1.9 A three-dimensional computer-generated detailed layout of the same area shown in Fig. 12.1.8. *(Source: Cederleaf, "Plant Layout & Flow Improvement," McGraw-Hill.)*

Fire protection In cases of large fires, fireproofing materials can prevent, lessen, or slow down the heating of a building's steel structural components, lest they lose strength, weaken, deform, and possibly collapse. Further studies of the collapse of the World Trade Center suggest that gaps in the sprayed-on insulation could act as gateways to actually direct heat onto the steel, rather than keep it away. Lightweight, fluffy fireproofing material is popular because it is relatively inexpensive and easy to spray onto the steel. But it is also easily damaged during the maintenance work on ducts, cabling, wiring, sprinkler piping, etc.; and it becomes friable over time such that the slightest touch causes it to fall off, creating the above-mentioned gaps.

Sprinklers are installed according to the recommendations of the National Fire Protection Association (NFPA). Automatic sprinklers can be the wet type, wherein water is always in the pipes up to each head, or the dry type, wherein the pipes are filled with air under pressure to restrain water until the fusible link in the head melts and releases the air pressure, allowing water to flow.

The dry type is used outdoors and in unheated areas where water could freeze. Sprinkler heads are either standard or deluge type. Standard heads have a fusible element in each head that is melted by the heat of a fire, thereby opening the head and releasing water. Deluge heads do not have fusible elements, but a deluge valve which is opened in response to a signal from any of several heat sensors situated in the protected area. In the standard type, only the heads whose elements are melted release water, whereas deluge heads act simultaneously. A deluge system is more likely to extinguish sparks at the periphery of a fire, but it also may ruin materials that are doused unnecessarily with water. A preaction system includes sensors and an alarm which gives plant personnel a chance to deal with the fire before sprinklers actuate and douse valuable merchandise. The alarm may also be wired to signal the local fire department. The sprinkler heads must be spaced in accordance with the prevailing code: generally one head per 200 ft² (18.6 m²) for low-hazard areas, 90 ft² (8.4 m²) for high-hazard areas, and about 120 ft² (11.2 m²) for general manufacturing; about 8 to 12 ft (2.44 to 3.66 m) between heads and height of the highest head not over 15 ft (4.57 m) are typical. (See also Sec. 18.3). Drains should be installed to carry away sprinkler water. The supply of water for all of the above listed needs must be adequate and reliable. For fire fighting, the greatest fire hazard and the size of the water supply required to deal with it must be estimated. The expected flow from all hoses and sprinklers, the static pressure, and the minimum flow available at a given residual pressure must be considered. Again, codes and insurance company requirements control. Sources of water include city mains, gravity tanks on towers, reservoirs on roofs or in decorative ponds that are part of the landscaping, wells, nearby lakes, and rivers.

Heat The **heat** required for personnel comfort depends on the plant's location, which in turn, influences the types and capacities of heaters selected. Generally, factory areas can be kept a little cooler, about 65°F than the 72°F recommended for office areas. Heating systems include warm air, hot water, steam, electric, and radiant. Warm air requires a furnace to heat the air and ducts and blowers to distribute it. Circulating hot water and steam heat also require boilers and furnaces and a network of distribution pipes and radiators. Electric space heaters require fixed wiring and local outlets. Unit radiant heats may be fueled by gas, steam, hot water, or electricity and are placed above doors and work areas, oriented to direct heat where it is required. Piping can be embedded in the concrete slab on grade, and in other floors, walls, and ceilings to provide radiant heat from circulating hot water.

Ventilation, the introduction of fresh outside air, the exhaust of stale inside air, or merely the movement of otherwise still inside air, can be effected naturally or mechanically. Natural ventilation requires windows, skylights, louvers, or other openings. Mechanical ventilation requires fans and blowers (and sometimes ducts) to draw air in, circulate it, and exhaust it. The number of cubic feet per minute (cubic metres per minute) of air per person required by people working in an office is about 10 (0.3), and between 25 and 50 (0.7 to 1.4) for those working on the factory floor. The total amount of air needed and the number of air changes per hour determine the number and sizes of fans, motors, and ducts. One change per hour is too little, with no discernible difference in air quality; 50 to 60 changes of air per hour are too many, and the resulting high-velocity drafts cause sensations of chilling and accompanying discomfort. Minimally, 5 changes of air per hour are required.

Whether a building is air conditioned or not, large high-mounted slow-speed **fans** can aid in expelling hot air in the summer when vents are opened. When vents are closed in the winter, they redirect risen warm air to the occupied spaces below.

Air conditioning adds to employee comfort, increases their productivity, and is essential for the manufacture of certain products. The calculation of the expected heat gains required to specify an **air-conditioning** installation is similar to the procedure for calculating heat loss in the design of a heating system. For air conditioning, the required cooling load is calculated in Btu per hour and converted to required tons of refrigeration by dividing by 12,000. (See Sec. 12.4.) The tonnage required would be one basis for the selection of the type of water-cooled or evaporative condensers, compressors, air handling units, motors, pumps, ducts, registers, circuitry, panels, and controls to be installed. Packaged air-conditioning units mounted on window sills, floors, ceilings, or roofs are often used. (See Merritt and Ricketts, "Building Design and Construction Handbook.")

The design and installation of a **compressed-air system** includes the summation of the volume and pressure of air required at all plant locations. The pipe diameters and total pressure drops between compressor and the most remote point of delivery must be determined. Pressure drop is a function of pipe and hose friction and the requirements of air-powered devices. Piping may be fixed, by and large, but the number and types of air-powered devices will change from time to time as production processes are changed or rearranged. Thus, there is some variability to be expected in the calculations to determine compressor discharge volume and pressure. Suitable allowances for unknown quantities are usually factored into the final design selection.

Internal and external communications systems must be installed. With autofacturing (see Sec. 17), the design, manufacture, production control, inventory control, storekeeping, warehousing, sales, and shipping records are all integrated and tied to the same information database. An internal local-area network (LAN), with computers, terminals, displays, and printers distributed throughout the plant and offices, will be required to plan, analyze, order, receive and accept material, record, feed back data, update files, and control and correct operating conditions. AGVs and mobile robots can be programmed to navigate off beacons installed throughout the plant.

Parking for employees and visitors must include spaces for handicapped persons. Parking lots are usually paved with 3 in (7.6 cm) of asphaltic cover over 6 in (15.2 cm) of gravel base. Lines are painted to designate spaces, which may be "straight in" (at a 90° angle to the curb), or at a smaller angle, typically 45° to 60°. The angle used influences the width of the aisles, depth (distance from the edge of the aisle to the curb) of each space, and the amount of curb length required for each car. For example, parking at a 45° angle requires an aisle width of about 13 to 15 ft (4.0 to 4.6 m), a space depth of about 20 to 21 ft (6.1 to 6.4 m) and uses about 13 ft (4.0 m) of curb per space (because of the overlaps); a 90° layout requires an aisle width of about 24 to 27 ft (7.3 to 8.2 m) and a space depth of about 19 to 20 ft (5.8 to 6.1 m) and uses about 9 to 10 ft (2.7 to 3.1 m) of curb per space. The width of each space (except those for the handicapped, which are wider) ranges from

9 to 10 ft (2.7 to 3.1 m) and their lengths from 19 to 20 ft (5.8 to 6.1 m). The number of parking spaces to be provided depends on the number of people expected and the extent to which public transportation is available and used. It is expected that there will be more than one employee per vehicle. Factors ranging from 1.2 to 2.5 persons per car can be used, depending on the firm's best estimate of the practices of its employees. For multiple-shift operations, it must be recognized that those working the second shift will arrive and need parking spaces before those on the first shift leave and vacate them. For those using public transportation, a shelter against the weather may be erected; often it is provided by the bus company.

In addition to parking space for automobiles, **space must be provided for trucks** that bring material to the plant and carry products away. Sizes of trucks, truck tractors, and their trailers vary widely, as do their turning radii. Many trucks will drive in frontward, maneuver to turn around, and back into the loading docks and platforms. Space must be provided for them to do this, even when there are other trucks present. The minimum size apron (measured by the distance from the outermost obstruction, whether it be a part of the building, the front of another truck already in the loading dock, or anything else that is in the way), required to maneuver a tractor-trailer into or out of a loading position, in one maneuver and with no driver error, varies with the length of the tractor-trailer expected and the width of the position to be provided. For example:

Tractor-trailer length	Position width	Minimum apron size
35 ft (10.7 m)	10 ft (3.1 m)	46 ft (14.0 m)
	12 ft (3.7 m)	43 ft (13.1 m)
	14 ft (4.3 m)	39 ft (11.9 m)
40 ft (12.2 m)	10 ft (3.1 m)	48 ft (14.7 m)
	12 ft (3.7 m)	44 ft (13.4 m)
	14 ft (4.3 m)	42 ft (12.8 m)
45 ft (13.7 m)	10 ft (3.1 m)	57 ft (17.4 m)
	12 ft (3.7 m)	49 ft (15.0 m)
	14 ft (4.3 m)	48 ft (14.6 m)

Truck heights range from about 8 ft (2.4 m) for a panel truck to around 13.5 ft (4.1 m) for double-axle semis and others. Doorways must be sized accordingly. Seals and pads may be added around the doors to fill the gap between the back of trucks and the building. Besides conventional rollup doors, plastic strips and air curtains may be used to provide a partial barrier between the plant's environment and the weather.

The **plant's security** measures should include structural protection against wind, rain, flood, lightning, earthquakes, and fire. Security against intruders, breakins, and vandalism can include fencing, gates, perimeter lights, surveillance cameras, and sensors. Security guard booths may be purchased as prefabricated units, complete with lights, heat, and air conditioners.

Emergency **egress** should provide for more than one route to safety, minimal distances to doors, doors that are locked from the outside but can be opened with a push from the inside (by panic bars), all to be accessible and usable by the handicapped. When designing a plant for ease of use by handicapped employees and visitors, all physical barriers must be eliminated and aids installed, such as ramps, braille signs, wheelchair-width doors and toilet booths, wheelchair-height sinks and drinking fountains, and the like. Applicable local building codes and federal regulations will guide and control the final designs.

Most of the above services and systems are built into the structure or influence its design and, therefore, must be determined before the specifications of the building become final.

Building Structure

The specifications for an industrial plant's **structure** depend on its contents, intended use, general location, and specific site. In the usual order in which they are constructed are foundation and footings, columns (and bay sizes), beams and roof trusses, exterior walls, floor slab, roof decking and covering, exterior doors and fenestration, interior partitions, walls, doorways, services (electric power, compressed air, etc.), interior decor, and other special features.

Firm foundations and footings are required to support the columns, peripheral and some interior walls, machinery, and other equipment. Especially heavy machines, or those subject to large dynamic loads (often repetitive), may require special design that incorporates measures to isolate vibrations. Depth, size, and foundation reinforcement depends on the subsurface conditions. Installation of piles may be dictated by unsuitable load-bearing soil. Almost all foundations consist of cast-in-place concrete; in rare instances, load-bearing concrete blocks are set onto a previously poured concrete base. Site topography and location of the building may require the construction of retaining walls; these may be cast concrete or assembled with precast concrete sections which may serve also to provide a decorative treatment. The subgrade on which the concrete floor slabs will be poured must be well-compacted and made level. Often, the floor slab is not cast until heavy equipment (used to erect superstructure) is removed from the site.

Steel superstructure members are designed to provide the clear spans delineated by the selected **bay sizes**. They also support intermediate floors (if any); roof-mounted equipment; material handling equipment such as monorails; and lighting fixtures, piping, ductwork, and other utilities. The design must account for all dead and live loads, usually in accordance with applicable local building codes and other accepted structural codes. (See Secs. 5, 6, and 12.)

The sizes and locations of **columns and beams** establish the bay sizes. The longer the span, the larger the bay size, but this reflects back into construction with heavier columns, deeper beams and trusses, and concomitant higher costs. Inasmuch as larger bay sizes permit greater freedom in plant layouts and ease material handling, the added expense of large bay sizes is often worth it. Accordingly, the trend in plant design is to incorporate large bays. While any bay size is feasible, a cost/benefit tradeoff may set reasonable limits. Typical bay sizes are 30×40 ft (9×12 m), 40×60 (12×18), and 40×80 (12×24). There are warehouses whose bay size is dictated by the requirement to accommodate stacks of standard pallets with minimum waste space. Aircraft manufacturing plants and commercial hangars enclose enormous column-free cavernous spaces; lengths of 300 to 400 ft (91 to 122 m) are the norm in those applications.

When columns must support heavy traveling bridge cranes, a second row of columns is incorporated into the main columns to provide support for rails and to effect a stiffer structural configuration.

Beams are set at the top of columns and establish the clear height of the building interior; 15, 20, and 25 ft (4.5, 6.1, and 7.6 m) are typical in manufacturing areas, and 30 to 40 ft and more (9.1 to 12.2 m) are employed in warehouse space. When a two-story office is part of a single-level plant (Fig. 12.1.7) the overall height of the building is often set to the second-story height to permit a simple flat roof. Factory clear heights should be at least the height of the highest machine, with a generous margin added; for large products, the clear height is often designed to be twice the height of the largest product. Anything suspended from the beams or girders reduces the clear height. Deeper open roof trusses permit some piping, wiring, and other services to be woven through them and light fixtures placed between them, thereby not reducing clear working space. Roof decking and roofing is affixed to roof trusses or beams.

Exterior walls can be load bearing (support one end of a beam) or nonbearing. If nonbearing, all beam loads are transferred to columns spaced at intervals along the building perimeter. Walls are constructed of masonry [brick or concrete block masonry units (CMUs), sometimes incorporating glass block]; metal panels, usually integrally stiffened; wood for special applications (storage of highway deicing salts); natural stone or stone veneer/precast-concrete panels; or stone veneer on masonry backup walls. Where conditions allow, poured concrete walls are cast on the flat (at grade), then tilted up and secured into place; this technique is attractive especially for warehouses with repetitive wall construction devoid of windows and other openings. Metal panels are usually galvanized steel, with or without paint, or painted aluminum.

Factory floors may be required to sustain live loads from less than 100 pounds per square foot (psf) (0.5 kPa) to over 2,000 psf (96 kPa). A general-purpose light-assembly floor ordinarily will have no less than 100 psf (0.5 kPa) floor load capacity. A special purpose plant floor may be built with floor load capacities which vary from place to place in accordance with the requirements for different load-carrying capacity. Certainly, in the extreme, to build in a maximum floor loading capacity of, say, 2,000 psf (96 kPa) throughout a plant when there is minimal likelihood for that requirement other than in discrete areas, would not be cost-effective. Concentrated live loads (most often from wheeled material handling vehicles) are accounted for in the design of the floor slabs as required. The final design of the plant floor will meld the above factors with other requirements so that the sum of all dead and live loads can be safely sustained in the several areas of the floor.

Virtually all slab on-grade concrete floors are cast in place; in rare cases, precast concrete sections are used. Wooden plank flooring is rarely used on new construction. Upper floors are most often constructed with precast concrete planks overlaid with a cement mortar topping coat, or employ ribbed decking. When upper-floor loading capacities require it, those floors can also be cast in place; in those instances, concrete is cast into ribbed metal decking which acts as wet form work and remains in place. End grain wood blocks may be set atop a concrete floor to mitigate against unavoidable spillage of liquids. (See Sec. 12.2.)

The load-carrying capacity of concrete slabs on grade is ultimately limited not only by the strength of the concrete itself, but also by the ability of the subgrade material to resist deformation. The proper design of the floor, especially at grade, will account for all the parameters and result in a floor which will safely sustain design floor loading. Automatic guided vehicle systems and mobile robots (see Sec. 10) require level and smooth floors. The floor surface may be roughened slightly with a float to reduce slipping hazards, or have a steel trowel finish to accommodate AGV and mobile robot navigation. If it is contemplated that products or equipment will be moved by raising and floating them on films of air, the floor must be crack-free. If a slight floor slope can be tolerated from an operational point of view, that will ease cleaning by allowing water to flow to drains. The slab's surface may be coated, treated, or tiled for protection, comfort, ease of maintenance, and/or aesthetics. Expansion joints around column bases will inhibit propagation of small cracks resulting from differential settlement between column footings and floor slabs.

At this stage, the building looks like the one shown in Fig. 12.1.10.

A flat roof is always slightly pitched to drain water toward scuppers, gutters, and downspouts. It is constructed with ribbed metal decking or precast concrete planks, topped with a vapor barrier, insulation, a topping surface, flashing, sealants or caulking, and roof drains. The covering of a **built-up roof (BUR)** comprises three to five layers of roofing felt (fiberglass, polyester fabric, or other organic base material), each layer mopped with hot bitumen (asphalt or coal tar pitch). The wearing surface of roofing may include a cap sheet with fine mineral aggregate, reflective coating, or stone aggregate. When extensive roof traffic is anticipated, pavers or wood walkways are installed. A recent successful roof surface is composed of a synthetic rubber sheet, seamless except at lapped, adhesively bonded joints. It has proved to be a superior product, providing long, carefree life.

The seal between roofing and parapet walls or curbed openings is effected with **flashing** bent and cemented in place with bituminous cement. The flashing material may be galvanized steel aluminum, copper, stainless steel, or a membrane material.

Flat roofs lend themselves to ponding water to help cool the plant. Roof ponds (and water sprays) can reduce interior temperatures by 10 to 15°F (6 to 9°C) in the summer. Roof ponds can serve as a backup source of water for fire fighting. If water is ponded on a roof, roof construction must be handled expressly for that purpose. There are structural implications which must be considered as well.

Doors and ramps should be made 2 ft (0.6 m) larger than the largest equipment or material expected and large enough to allow ingress of fire fighting apparatus. It is often convenient to place very large pieces of equipment such as molding machines and presses inside the building

Fig. 12.1.10   A view of a plant during construction. Shown are its columns, roof trusses, and exterior masonry walls. The reinforced-concrete floor slab is being poured over compacted soil. *(Source: VMA, Inc.)*

envelope before the walls are completed. After exterior walls and roof are in place, the building interior is sufficiently secure for assembling materials for the remaining construction and to receiving smaller equipment.

Interior walls and **partitions** (masonry, wood, wallboard, and metal) follow, after which office flooring and framing are installed (Fig. 12.1.11). **Steel framing** (called *steel for stick* by some) is sometimes an alternative to using wood studs. Walls framed with steel studs look the same as those made of wood when they are finished with any material externally, and with regular drywall inside. Metal is longer-lasting, stronger, and able to have larger spans, creating clearer spaces with fewer interior supports, which offers greater design and architectural flexibility and styling options. The studs are bolted and screwed together, versus the nails used with wood. The sizes and gage of the members can range from lighter and thinner galvanized steel to thicker, heavier "red iron" members. They come with prepunched access holes for ease of installing wire and cable and of adding cross supports to make their assemblage more rigid. **Metal framing** requires fewer wall studs, thus creating fewer uninsulated thermal bridges. Metal frames can also be more energy-efficient when dimensioned to provide deeper wall cavities to hold thicker insulation. They are impervious to insects, mold, rot, splits, shrinkage, and warpage. Many are made from recycled metal. The price of sheet metal studs may be higher or lower than for wood studs; but for a true cost comparison, the total cost (e.g., material, time, labor) of building with metal versus wood must be considered. The remainder of the construction involves completion of wiring, lighting, and other services; application of paint or wall coverings, floor coverings, and decor; and so on.

Other Considerations

Industrial plants must be built in conformance with all state and local codes. The International Code Council has issued—and periodically updates—an "International Building Code (affixed with the year of the latest version)." It applies to the design and construction of all buildings except one- and two-family houses, which are covered by other codes. It also has subcodes, such as for plumbing, electrical, HVAC, fire protection, and most recently energy conservation. Each individual state in the United States can and does select parts of or all the International Building Code to create its own pertinent code.

The **applicable standards, regulations, and procedures** of the 1971 Williams-Steiger Occupational Safety and Health Act (OSHA), as amended, must be followed in the construction and operation of the building and the design and use of the plant's equipment; likewise, the regulations of the Environmental Protection Agency (EPA), as amended, and the design standards of the Americans With Disabilities Act (ADA), as amended, must be followed.

Color can serve several purposes in an industrial plant. As an aid to safety, it can be used to identify contents of pipes, dangerous areas, aisles, moving parts of equipment, emergency switches, and fire fighting and first aid equipment. It also can be used to improve illumination, conserve energy, reduce employee fatigue and influence on morale positively.

OSHA, American National Standards Institute (ANSI) specification Z53.1, and specific safety regulations require the use of specific **identity colors** in certain applications. As a general guide, yellow or wide yellow-and-black bands are used to indicate the need for caution and to

Fig. 12.1.11 A view of framing for the first- and second-story offices of a plant, with a roof beam and the steelwork supporting the second floor visible. *(Source: VMA, Inc.)*

highlight physical hazards that persons might trip on, strike against, or fall into, such as edges of platforms, low beams, stairways, and trafficked aisles along which equipment moves. Orange designates dangerous parts of machines or energized equipment that may cut, crush, shock, or otherwise injure, such as the inside of gear boxes, gear covers, and exposed edges of gears, cutting devices, and power jaws. Red is the basic color for the identification of fire protection equipment and apparatus, danger and stop signs, fire alarm boxes, fire exit signs, and sprinkler piping. Green designates safety items, such as first aid kits, gas masks, and safety deluge showers, or their locations. Blue designates caution and is limited to warning against starting, use of, or movement of equipment under repair, and utilizes tags, flags, or painted barriers. Purple designates radiation hazards. Black, white, or a combination of both designates traffic control and housekeeping markings, such as aisleways, drinking fountains, and directional signs.

There are also color conventions for the identification of fluids conducted in pipes. Instead of being painted their entire length, pipes are painted with colored bands and labels at intervals along the lines, at valves, or where pipes pass through walls. ANSI Standard A 13.1-1975 specifies the following colors for pipe identification: red for fire protection; yellow for dangerous materials; blue for protective materials; green and white, black, gray, or aluminum for safe materials.

The **noises** created during the operation of a plant can be dealt with at their **sources, along their paths,** and at their **receivers.** At the sources, noise is minimized or eliminated by: changing the process; replacing (with quieter) equipment; modifying equipment by redesign or component changes; moving to another location; installing mufflers and shock mounts; using isolation pads; and shielding and enclosing equipment with material having a high sound transmission loss (STL) value. STL is a measure of the reduction in sound pressure as it passes through a material. Along the paths, attenuation of noise is enhanced by increasing the distance between sources and receivers; introducing discontinuities in the transmission path, such as barriers and baffles which interrupt direct transmission; installing acoustical barriers with an STL of at least 24 dB and with sufficient absorption to prevent reflection of the noise back to its source; and by placing sound-absorbent material on surfaces along its path, such as acoustical ceilings and floor and wall coverings. At the receivers, enclosures, workstation partitions of sound-absorbing material, earplugs, and "white-noise" generators can be used. (See Secs. 12.6 and 18.2.)

Provisions must be made for **waste disposal,** including: disposal of refuse and garbage, treatment of process fluids and solids, and waste and recyclables recovery, all in accordance with applicable EPA and other regulations.

Among many other things, the plant design process must address matters such as fuel storage, storm and surface drainage, snow removal facilities and procedures (if the local climate warrants), design for ease of maintenance of the building as well as the grounds and associated landscaping, the articulation of the architect's idiom, energy efficiency in the materials of construction and the equipment installed to service the plant, and diligence in compliance with environmental impact studies (if required). Buildings, like other designed and made products, can—to varying degrees—help or hurt the environment. Once pristine sites can become air-, soil-, and water-fouled by sewerage, animal waste, toxic materials dumping, lawn chemicals, etc. Governments, attempting to preserve the environment, tend to make it more and more difficult for prospective builders to get permits. A **green** industrial building is one whose design, construction, and operation have a positive—or at least a minimal negative—impact on the environment. The builders might have selected an "environmentally friendly" site; used salvaged and recycled materials in their construction and operations; minimized consumption of water, energy, and other resources; provided for good indoor air quality; treated their effluents; and done all other things possible to maintain a healthier ("greener") presence. Builders that do such things to a required degree may be awarded an **LEED** (Leadership in Energy and Environment) rating by the U.S. Green Building Council, There are comparable organizations in some other countries.

After the plant is designed, but before a commitment is made to build it, its operations should be simulated to bring forth and correct any errors or omissions. In addition to testing its specifications for normal conditions, extremes (severe storms, power outages, truckers' strikes, etc.) should also be evaluated. This simulation is similar to that done before the plant's site is made final, but comes after the structure and layout have been designed. A sensitivity analysis should be made to see the effect on operations of changes in inputs and operating conditions. Depending on these analyses and best guesses as to the future, adjustments may have to be made to ensure that the plant will operate effectively and economically under all reasonable conditions, and that it has no features that will harm people or do damage to the products or the environment.

CONTRACT PROCEDURES

Cost Estimates

Cost estimates fall into three general classes:

1. Preliminary estimates made usually from sketch drawings and brief outline specifications to determine the approximate total cost of a project.

2. Comparative estimates made usually during the progress of design to determine the relative cost of two or more alternative arrangements of equipment, type of building, type of floor framing, and the like.

3. Detail estimates made from final plans and specifications and based on a careful quantity survey of each component part of the work.

Primary quantity estimates are usually more accurate when comparison with comparable projects is lacking or not feasible. The procedure entails computations based on quantity takeoffs from preliminary plans and specifications, and can include the gross area of exterior walls, interior partitions, floors, and roof. Each is multiplied by known (or estimated) unit cost factors. Other items, such as number of electrical outlets and number of plumbing fixtures and sprinklers, are estimated, and their cost computed. Equipment costs can be based on preliminary quotations from manufacturers.

The following items should be included in preliminary estimates (or in other more detailed estimates which may follow as plans develop to the final stage): land cost; fees to real estate brokers, lawyers, architects, engineers, and contractors; interest during construction; building permits; taxes, including local sales or use taxes; demolition of existing structures including removal of old foundations; yard work including leveling, drainage, fencing, roads, walks, landscaping, yard lighting, and parking spaces; transportation facilities including railroad tracks, wharves, and docks; power supply and source; water supply and source; sewer and industrial waste disposal. A judicious **contingency factor** is included to provide for unforeseen conditions that may arise during the development of the project. It may amount to 5 to 15 percent (or more) depending on the character and accuracy of the estimate, whether the proposed design and construction will follow established procedures or incorporate new state-of-the-art features, and the exact purpose for which the estimate is made.

Especially for complex new or altered construction, estimates may be updated continually as a matter of course. This will serve to monitor the evolving cost of the project and the time constraints placed on construction.

Working Drawings, Specifications, and Contracts

The technical staff of a given industry has special knowledge of trade practices, process requirements, operating conditions, and other fundamentals affecting successful operation in that field and is best fitted to determine the basic factors of process design and plant expansion. Unless the company is extremely large, it is unlikely that they will have the specialized staff or that their own staff will have the time available to undertake the complete layout. The efficient transformation of these requirements into completed construction usually requires experience of a different nature. Therefore, it is generally advisable and economical to employ engineers or architects who specialize in this particular field.

Three general procedures are in common use. First, the employment on a percentage or fixed-fee basis of an engineering and construction organization skilled in the industrial field to prepare the necessary working drawings and specifications, purchase equipment and materials, and execute the work. This has become known as **design/build.** For the duration of the project, such an organization becomes a part of the owner's organization, working under the latter's direction and cooperating closely with the owner's technical staff in the development of the design and in the purchase and installation of equipment and materials. This procedure permits construction work to start as soon as basic arrangements and costs have been determined, but before the time required to complete all working drawings and specifications. This is often termed **fast tracking.** Such a program will result in the earliest possible completion consistent with economical construction.

A second procedure is to employ engineers or architects with wide experience in the industrial field to prepare working drawings and specifications and then to obtain **competitive lump-sum bids** and award separate contracts for each or a combination of several subdivisions of the work, such as foundations, structural steel, and brickwork. This method provides direct competition restricted to units of like character and permits intelligent consideration of the bids received. Provision must be made for proper coordination of these separate contracts by experienced and skilled field supervision. In recent years, a **construction manager** has been hired for this purpose. This procedure requires more time than that first described, since all work of a given class should be completely designed before bids for that subdivision are sought. Where construction conditions are uncertain or hazardous, the first method is likely to be more economical. A combination of the first method for uncertain conditions and the second method for the balance may prove most advantageous at times.

A third procedure is to employ engineers or architects to complete all plans and specifications and then **award a lump-sum contract** for the entire work, or one for all building work, and one or more supplementary major contracts to furnish and install equipment. This method is particularly useful where there are no serious complications or hazards affecting construction operations, but it requires considerably more time, since most of the working drawings and specifications must be complete before construction is started. It has the significant advantage, however, of fixing the total cost within narrow limits before the work starts, provided the contracts cover the complete scope of work intended and no major changes ensue.

CONSTRUCTION

The design, construction, occupancy, and operation of an industrial plant is a complex endeavor involving many people and organizations. **Project control tools and techniques** such as the critical path method (CPM), the program evaluation and review technique (PERT), bar charts, dated start/stop schedules, and daily "do lists" and "hot lists" of late items are all helpful in keeping construction on time and within budget.

Checkpoints and milestones, with appropriate feedback as the project progresses, are established to monitor progress. Bailout points are established in the event the project must be aborted sometime after the beginning of construction. On the other hand, contingency plans should be in hand and include planned reassignment of equipment, materials, and financial and personnel resources to keep the project on schedule. Figure 12.1.12 shows the scheduled and actual dates of completion of the phases noted for a small plant depicted in several of the previous figures.

Some portion of the organization (often the construction manager's office), is charged with the delicate task of coordinating contractors and the multiplicity of trades at the job site, and as part of its function, it will generate "punch lists" of deficient and/or incomplete work which must be completed to the owner's satisfaction before payment therefore is approved.

	Scheduled	Actual
Project planning	March	March
Basic layout	April	April
Structural design	April	April
Equipment specifications	April	April
Construction contract	May	May
Ground breaking	June	June
Detailed office layout	June	June
Detailed plant layout	July	July
Equipment purchases	August	August
Furniture purchases	September	September
Organization firmed	September	September
Steelwork completed	October	October
Masonry completed	October	October
Roof completed	November	November
Staffing and training	November	November
Systems designed	November	November
Forms designed	December	December
Offices completed	December	December
Equipment received	December	December
Building occupied	December	December

Fig. 12.1.12 The project schedule and completion dates for the single-level 202,800 ft² industrial plant used as an example in Figs. 12.1.1 to 12.1.7, 12.1.10, and 12.1.11. (*Source: VMA, Inc.*)

12.2 STRUCTURAL DESIGN OF BUILDINGS
Staff Contribution

REFERENCES: "Manual of Steel Construction—Allowable Stress Design," American Institute of Steel Construction. "National Design Specification for Wood Construction," American Forest and Paper Association. "Design Values for Wood Construction," American Forest and Paper Association. "Uniform Building Code" (as applicable), International Conference of Building Officials. SEI/ASCE 7-05, "Minimum Design Loads for Buildings and Other Structures," American Society of Civil Engineers. Blodgett, "Design of Welded Structures," J. F. Lincoln Arc Welding Foundation. "Masonry Designers Guide," The Masonry Society. "Building Code Requirements for Structural Concrete and Commentary," American Concrete Institute. "Wood Handbook," Forest Products. Laboratory, USDA. "International Building Code" (IBC), International Code Council. "Cold Formed Steel Framing Design Guide," AISI. *Note:* The editions of the references that currently apply and that are cited in the applicable codes must be used.

LOADS AND FORCES

Buildings and other structures should be designed and constructed to support safely all loads of both permanent and transient nature, without exceeding the allowable stresses for the specified materials of construction. Usually, there are governing local codes, either unique to the jurisdiction or incorporating accepted codes (International Building Code, Building Code Requirements for Structural Concrete, Uniform Building Code, etc.,—sometimes amended locally), that will guide the selection of appropriate loads. In general, the following will enter into the design considerations: **dead loads,** generally agreed to include the weight of all permanent construction; **live loads,** produced by the use or occupancy of the building; **external** (or **environmental**) **loads** resulting from wind, snow, rain, ice, and earthquakes. The individual types of loading may be used singly or in combination, depending upon the use and location of the building and its inherent design features. For instance, the building may be designed to impound water on the roof for the purpose of keeping the roof deck (and the building below) cooler. In that case the collected and retained water has a major impact on the design of the roof structure. The governing code will address the matter of loading per se, or the use of combined loads by reference to other definitive sources whose information has standing. Absent specific guidance in the local governing code, for areas not so regulated, the data cited in the paragraphs below have been culled from many years of successful application.

Live loads on floors are generally regulated by the building codes in cities or states. For areas not regulated, the following values will serve as a guide for live loads in lb/ft^2 (kPa): rooms for habitation, 40 (1.92); offices, 50 (2.39); halls with fixed seats, 60 (2.87); corridors, halls, and other spaces where a crowd may assemble, 100 (4.76); light manufacturing or storage, 125 (5.98); heavy manufacturing or storage, 250 (11.95); foundries, warehouses 200 to 300 (9.58 to 14.37). Floor decks and beams that support only a small floor area must also be designed for any local concentrations of load that may be placed upon them. Girders, columns, and members that support large floor areas, except in buildings such as warehouses where the full load may extend over the whole area, may often be designed for live loads progressively reduced as the supported area becomes greater. Where live loads, such as cranes and machinery, produce **impact** or **vibration,** static loads should be increased as follows: elevator machinery, 100 percent; reciprocating machinery, 50 percent; others, 25 percent.

Roof live loads should be taken as a minimum of 20 lb per horizontal ft^2 (957.6 kPa) for essentially flat roofs (rise less than 4 in/ft), varying linearly with increasing slope to 12 lb/ft^2 for steep slopes (rise greater than 12 inches per foot). Reductions may also be made on the basis of tributary area greater than 200 ft^2 (18.58 m^2) of the member under consideration, to a maximum of 40 percent for tributary areas greater

than 600 ft^2. The minimum roof live load shall be 12 lb/ft^2 after all reduction factors have been applied. Where **snow loads** occur, appropriate design values should be based on the local building codes.

Dead loads are due to the weight of the structure, partitions, finishes, and all permanent equipment not included in the live load. The weights of common building materials used in floors and roofs are given below (see also Sec. 6).

Material	Weight, lb/ft^2
Asphalt and felt, 4-ply	3
Corrugated asbestos board	5
Glass, corrugated wire	5–6
Glass, sheet, $\frac{1}{8}$ in thick	2
Lead, $\frac{1}{8}$ in thick	8
Plaster ceiling (suspended)	10
Acoustical tile	1–2
Sheet metal	1–2
Shingles, wood	3
Light-weight concrete over metal deck	30–45
Sheathing, 1 in wood	3
Skylight, $\frac{3}{16}$ to $\frac{1}{4}$ in, glass and frame	6–8
Slate, $\frac{3}{16}$ to $\frac{1}{2}$ in thick	8–20
Tar and gravel, 5-ply	6
Tar and slag, 5-ply	5
Roof tiles, plain, $\frac{5}{8}$ in thick	20

NOTE: $\text{lb/ft}^2 \times 0.04788 = \text{kPa}$.

Earthquake Effects Two distinct methods of designing structures for seismic loads are currently accepted. The more familiar method, employed by the Uniform Building Code (UBC), yields equivalent loading for use with the allowable stress design approach. The National Earthquake Hazard Reduction Program (NEHRP) has developed Recommended Provisions for the Development of Seismic Regulations for New Buildings, which is based on the ultimate strength design approach. The NEHRP approach is the basis for model codes, such as the IBC. SEI/ASCE 7-05 has an extended and detailed presentation applicable to determination of rational earthquake loads.

Wind Pressures on Structures Every building and component of buildings should be designed to resist wind effects, determined by taking into consideration the geographic location, exposure, and both the shape and height of the structure. For structures sensitive to dynamic effects, such as buildings with a high height-to-width ratio, structures sensitive to wind-excited oscillations, such as vortex shedding or icing, and very tall buildings, special consideration should be given to design for wind effects and procedures used should be in accordance with approved national standards. Wind loads should not be reduced for the shading effects of adjacent buildings.

Wind pressure on walls of buildings should be assumed to be a minimum of 15 lb/ft^2 (0.72 kPa) on surfaces less than 50 ft above the ground. For buildings in exposed locations and in locations with high wind velocity, pressures should be calculated based on the procedures outlined in SEI/ASCE 7-05, or as dictated by the governing code. Boundary-layer wind tunnels, which simulate the variation of wind speed with height and gusting, are used to better estimate the design wind loading.

The effect of the sudden application of **gust loads** has sometimes been blamed for peculiar failures due to wind. In most cases, these failures can be explained from the pressure distributions in a steady wind. If a relatively flexible structure such as a radio tower, chimney, or skyscraper with a natural period of 1 to 5 s is set into vibration, the stresses in the structure may be increased over those calculated from a static-load analysis. Provision should be made for this effect by increasing the static design wind. A rational analysis should be performed to calculate the magnitude of this increase, which will depend on the flexibility of the structure.

Even a steady wind may give rise to periodic forces which may build up into large vibrations and lead to failure of the structure when the frequency of the exciting force coincides with one of the natural frequencies of vibration of the structure. The periodic exciting force may be due to the separation of a system of **Kármán vortices** (Fig. 12.2.1) in the wake of the body. The exciting frequency n in cycles per second is related to d, the dimension of the body normal to the wind velocity V. Dangerous vibrations related to the "flutter" of airplane wings may arise on bridges and similar flat bodies. These self-induced vibrations may be caused (1) by a negative slope of the curve of lift against angle of attack (Den Hartog, op. cit.) or (2) by a dynamic instability which arises when a body having two or more degrees of freedom (such as bending and torsion) moves in such a manner as to extract energy out of the air stream. The first of these vibrations will occur at one of the natural frequencies of the structure. The second type of vibration will occur at a frequency intermediate between the natural frequencies of the structure.

Fig. 12.2.1   Von Kármán vortices.

The wind forces and the **pressure distribution** over a structure corresponding to a design wind V can be determined by **model testing** in a wind tunnel. Extrapolation from model to full scale is based on the fact that at every other point on the body, the pressure p is proportional to the stagnation pressure q (see Sec. 11) and thus the ratio p/q = constant for a fixed point on the body, as the scale of the model or the velocity of the wind is changed. Since the principal component of the wind force is due to the pressures, the force F acting on the surface S is

$$F \approx p_{avg}S = (p/q)_{avg}qS$$

and denoting $(p/q)_{avg}$ by a normal force coefficient or **shape factor** C_N

$$F = C_N\tfrac{1}{2}\rho V^2 S = C_N qS$$

The shape factors so obtained apply to full scale for structures with sharp edges whose principal resistance is due to the pressure forces. For bodies that do not have any sharp edges perpendicular to the flow, such as spheres or streamlined bodies, the factor C_N is not constant. It depends upon the Reynolds number (see Sec. 3). For such bodies, the law for variation of the shape factor C_N must be determined experimentally before safe predictions of full-scale forces can be made from model measurements.

Radio Towers and Other Framed Structures   For open frame towers, by using round structural members instead of flat and angular sections, a substantial reduction in wind force can be effected.

Load Combinations   Methods of combining types of loading vary with the governing local codes. Dead loads are usually considered to act all the time in combination with either full or reduced live, wind, earthquake, and temperature loads. The reductions in live loads are based on the improbability of fully loaded tributary areas, generally when the tributary area exceeds 150 ft². Most codes consider that wind and earthquake loads need not be taken to act simultaneously. There are two basic design methods in use: allowable stress or working stress design and limit states or ultimate strength design. Whereas allowable stress design compares the actual working stresses with an allowable value, the limit states approach determines the adequacy of each element by comparing the ultimate strength of the element with the factored design loads. For load combinations see SEI/ASCE 7-05, or as dictated by the governing code.

DESIGN OF STRUCTURAL MEMBERS

Members are usually proportioned so that stresses do not exceed allowable **working stresses** which are based on the strength of the material and, in the case of compressive stresses, on the stiffness of the element under compression. Internal forces and moments in **simple beams**, columns, and pin-connected truss bars are obtained by means of the equations of static equilibrium. **Continuous beams**, rigid frames, and other members characterized by practically rigid joints require for analysis additional equations derived from consideration of deflections and rotations.

Design may also be on the basis of the **ultimate strength** of members, the factor of safety being embodied in stipulated increases in the design loads. In steel-frame construction, the procedures of **plastic design** determine points where the material may be allowed to yield, forming *plastic hinges,* and the resulting redistribution of internal forces permits a more efficient use of the material.

Floors and Roofs

Except in reinforced-concrete flat-slab construction, floors and roofs generally consist of flat decks supported upon beams, girders, or trusses. The decks may usually be considered a series of beamlike strips spanning between beams and themselves designed as beams. The design of a beam consists chiefly in proportioning its cross section to resist the maximum bending and shear and providing adequate connections at its supports, without exceeding the unit stresses allowed in the materials (see Sec. 5) and limiting maximum live load deflection at midspan to $\frac{1}{360}$ of the span.

Up to spans of 20 to 30 ft (6.1 to 9.1 m), either wood or steel beams of uniform section are generally more economical than trusses, while for spans above 50 to 70 ft (15.2 to 21.3 m), trusses are usually more economical. Between these limits, the line of economy is not well-defined. Conditions that favor the use of trusses are as follows: (1) identical trusses are repeated many times, (2) the height of the building need not be increased for the greater depth of the truss, and (3) fire protection of wood or metal is not required.

Roof trusses often have their top chords sloped with the roof. Common trusses for steeply pitched roofs are shown in Figs. 12.2.2 to 12.2.6, the top chord panels equal in each truss. The members shown by heavy lines are in compression under ordinary loads, those in light lines in tension. The trusses of Figs. 12.2.6 to 12.2.9 are adapted for either steel or wood. In wooden trusses, the tensile web members may be steel rods with plates, nuts, and threaded ends. The truss of Fig. 12.2.6 is usually made of steel.

The forces in any member of these trusses under a vertical load uniformly distributed may be found by multiplying the coefficients in Tables 12.2.1 to 12.2.5 by the panel load P on the truss. For other slopes, types of trusses, or loads, see "General Procedure" below. Trusses for flat roofs are commonly of one of the types shown in Figs. 12.2.7 to 12.2.13 except that the top chords conform to the slope of the roof.

Floor trusses normally have parallel chords. Common types are shown in Figs. 12.2.7 to 12.2.13 in which heavy lines indicate members in compression, light lines in tension, and dash lines members with only nominal stress, under equal vertical panel loads. The panel lengths l in each truss are equal. The stress in each member is written next to the member in the figure, in terms of the panel load and the lengths of members. For a truss like one of the figures turned upside down, the stresses in the chords and diagonals remain the same in magnitude but reversed in sign. Forces in verticals must be computed (compare Figs. 12.2.7 and 12.2.8). For other loads and other types of trusses see "General Procedure" below.

Weights of Trusses   The approximate weight in pounds of a wooden roof truss may be taken as $W = LS(L/25 + L^2/6,000)$, where L is the span and S the spacing of trusses in feet. The approximate weight in pounds of a steel roof truss may be taken as $W = \frac{1}{5}LS(\sqrt{L} + \frac{1}{8}L)$.

Choice of Roof Trusses

WOODEN TRUSSES.   For pitched roofs with spans up to 20 ft (6.1 m) the simple king-post truss (Fig. 12.2.2) may be used. For spans up to 40 ft (12.2 m) the trusses of Fig. 12.2.2 and Fig. 12.2.4 are good. For spans up to 60 ft (18.3 m) the trusses of Figs. 12.2.3 and 12.2.5 are good. The number of panels rarely exceeds eight or the panel length, 10 ft (3m). For flat roofs, the Howe truss (Figs. 12.2.7, 12.2.10, and 12.2.12) is built of wood with steel rods for verticals; the depth is one-eighth to one-twelfth the span. Wooden trusses are usually spaced 10 to 15 ft (3.0 to 4.6 m) on centers. Wood is rarely used in roof trusses with span over 60 ft (18.3 m) long. Steel trusses for pitched roofs may well take the form shown in Figs. 12.2.3 to 12.2.6 for spans up to 100 ft (30.5 m). For flat roofs, the Warren truss (Figs. 12.2.9, 12.2.11, and 12.2.13) is usually constructed in steel; the depth of the trusses ranges from one-eighth to one-twelfth the span, with trusses spaced from 15 to 25 ft (4.6 to 7.6 m) on centers.

Stresses in Trusses

An ideal truss is a framework consisting of straight bars or members connected at their ends by frictionless ball-and-socket joints. The external forces are applied only at these ball-and-socket joints. Internal forces and

Fig. 12.2.2

Fig. 12.2.3

Fig. 12.2.4

Fig. 12.2.5

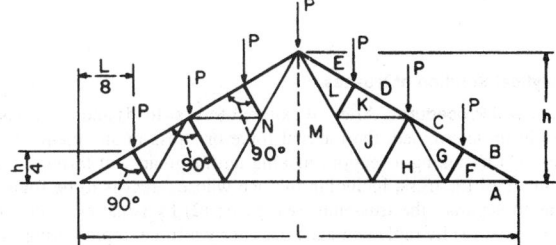

Fig. 12.2.6

Fig. 12.2.2 to 12.2.6 Types of steep roof trusses.

Fig. 12.2.7

Fig. 12.2.8

Fig. 12.2.9

Fig. 12.2.10

Fig. 12.2.11

Fig. 12.2.12

Fig. 12.2.13

Fig. 12.2.7 to 12.2.13 Types of floor and roof trusses.

Table 12.2.1   Coefficients for Truss Shown in Fig. 12.2.2

Pitch h/L	Coefficients of P for force in				
	AD	BD	CE	DE	EF
$\frac{1}{3}$	2.25	2.71	1.80	0.90	1.00
0.288*	2.60	3.00	2.00	1.00	1.00
$\frac{1}{4}$	3.00	3.35	2.24	1.12	1.00
$\frac{1}{5}$	3.75	4.03	2.69	1.35	1.00

*30° Slope.

Table 12.2.2   Coefficients for Truss Shown in Fig. 12.2.3

Pitch h/L	Coefficients of P for force in								
	AE	AG	BE	CF	DH	EF	FG	GH	HJ
$\frac{1}{3}$	3.75	3.00	4.50	2.7	3.60	0.83	0.72	1.25	2.00
0.288*	4.33	3.46	5.00	3.0	4.00	0.88	0.73	1.32	2.00
$\frac{1}{4}$	5.00	4.00	5.59	3.35	4.47	0.94	0.75	1.42	2.00
$\frac{1}{5}$	6.25	5.00	6.74	4.04	5.39	1.07	0.79	1.60	2.00

*30° slope.

Table 12.2.3   Coefficients for Truss Shown in Fig. 12.2.4

Pitch h/L	Coefficients of P for force in					
	AD	AF	BD	CE	DE	EF
$\frac{1}{3}$	2.25	1.50	2.70	2.15	0.83	0.75
0.288*	2.60	1.73	3.00	2.50	0.87	0.87
$\frac{1}{4}$	3.00	2.00	3.35	2.91	0.89	1.00
$\frac{1}{5}$	3.75	2.50	4.04	3.67	0.93	1.25

*30° slope.

Table 12.2.4   Coefficients for Truss Shown in Fig. 12.2.5

Pitch h/L	Coefficients of P for force in							
	AE	AH	BE	CF	DG	EF	FG	GH
$\frac{1}{3}$	3.75	2.25	4.51	3.91	4.51	0.83	1.42	2.50
0.288*	4.33	2.60	5.00	4.33	5.00	0.88	1.45	2.65
$\frac{1}{4}$	5.00	3.00	5.59	4.84	5.59	0.94	1.49	2.83
$\frac{1}{5}$	6.25	3.75	6.73	5.83	6.73	1.07	1.57	3.20

*30° slope.

Table 12.2.5   Coefficients for Truss Shown in Fig. 12.2.6

Pitch h/L	Coefficients of P for force in						FG KL	GH JK			
	AF	AH	AM	BF	CG	DK	EL	FG KL	GH JK	HJ	JM	LM
$\frac{1}{3}$	5.25	4.50	3.00	6.31	5.75	5.20	4.65	0.83	0.75	1.66	1.50	2.25
0.288*	6.06	5.20	3.46	7.00	6.50	6.00	5.50	0.87	0.87	1.73	1.73	2.60
$\frac{1}{4}$	7.00	6.00	4.00	7.83	7.38	6.93	6.48	0.89	1.00	1.79	2.00	3.00
$\frac{1}{5}$	8.75	7.50	5.00	9.42	9.05	8.68	8.31	0.93	1.25	1.86	2.50	3.75

*30° slope.

stresses in such straight bars are axial, either tension or compression, without bending. Since frictionless ball-and-socket joints are impossible, and the ends of bars are often bolted or welded, the ideal truss is never realized. For purposes of analysis, the primary stresses, which are always axial, are determined on the assumption that the truss under consideration conforms to the ideal. Secondary stresses are additional stresses, generally flexural or bending, brought about by all the factors that make the actual truss different from the ideal. In the following discussion, only primary forces and stresses will be considered.

Analytical Solution of Trusses

General Procedure   After all external forces (loads and reactions) have been determined, the internal force or stress in any member is found (1) by taking a section, making an imaginary cut through the members of the truss, including the one whose stress is to be found, so as to separate the truss into two parts; (2) by isolating either of these parts; (3) by replacing each bar cut by a force, representing the force in the bar; and (4) by applying the equations of statics to the part isolated.

The various ways in which sections can be taken and the equations used to determine the forces are illustrated by a solution of the truss in Fig. 12.2.14.

A section 1–1 (Fig. 12.2.14) may be taken around a joint, L_0. Isolate the forces inside the section (Fig. 12.2.15a). Assume the unknown forces to be tension. Since the forces are concurrent and coplanar, two independent equations of statics will establish equilibrium. These may be either $\Sigma x = 0$ and $\Sigma y = 0$ or $\Sigma M = 0$ taken about two axes perpendicular to the plane of the forces and passing through two points selected so that neither point is the intersection of the forces, or so that the line joining the two points is not coincident with either of the unknown forces.

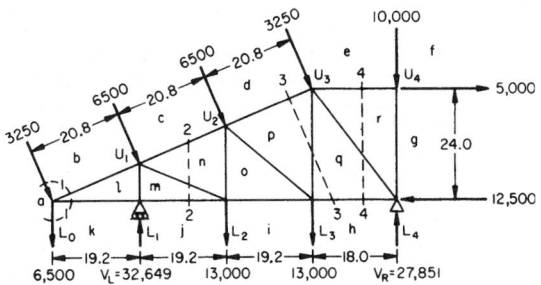

Fig. 12.2.14

Using the first set of equations and taking components of all forces along horizontal and vertical axes; e.g., the horizontal component of the 3,250-lb force is 1,250 and the vertical component is 3,000 lb.

$$\Sigma x = 0 = s_{lk} + s_{bl}(19.2/20.8) + 1,250$$
$$\Sigma y = 0 = s_{bl}(8/20.8) - 9,500$$

From these equations, $s_{lk} = -24,050$ and $s_{bl} = 24,700$.

The minus sign indicates that the force acts opposite to the assumed direction. If all the unknown forces are assumed to be in tension, then a plus sign in the result indicates that the force is tension and a minus sign indicates compression.

Hence, s_{lk} is 24,050 lb compression and s_{bl} is 24,700 lb tension.

Using the $\Sigma M = 0$ twice,

$$\Sigma M \text{ about } U_1 = 0$$
$$= -s_{lk} \times 8 - 1,250 \times 8 - 9,500 \times 19.2$$

from which $S_{lk} = 24,050$ lb compression.

$$\Sigma M \text{ about } L_1 = 0 = s_{bl}(8/20.8)19.2 - 9,500 \times 19.2$$

from which $s_{bl} = 24,700$ tension.

Instead of taking a section around a joint, a cut may be made vertically or inclined, cutting a number of bars such as Fig. 12.2.15b or Fig. 12.2.15c. If only three members are cut, and they are neither concurrent nor parallel, the forces can be found by taking moments of all forces on either side of the section about axes passing through the intersections of any two members.

Considering the part to the left of section 2–2, the forces in the three members (Fig. 12.2.15b) may be determined by taking moments about L_0, U_1, and L_2 of all forces acting on the part on either side of the section, e.g., on the left of the section because it has the fewer forces:

$$\Sigma M \text{ about } L_0 = s_{mn}(8/20.8)38.4 + 2,500$$
$$\times 8 - (32,649 - 6,000)19.2$$
$$\Sigma M \text{ about } U_1 = 0 = -s_{mj} \times 8 - 1,250 \times 8 - 9,500 \times 19.2$$
$$\Sigma M \text{ about } L_2 = 0 = s_{cn}(19.2/20.8)8 + 2,500$$
$$\times 8 - 9,500 \times 38.4 + 26,649 \times 19.2$$

from which $s_{mn} = 33,290$ tension, $s_{mj} = 24,050$ compression, and $s_{cn} = 11,298$ compression.

This method is sometimes called the "method of moments."

Considering the part to the right of section 3–3, the forces in the three members (Fig. 12.2.15c) may be determined by taking moments about L_0, U_3, L_3 of all the forces acting on part or either side of the section, e.g., on the right of the section because it has the fewer forces:

$$\Sigma M \text{ about } L_0 = 0 = s_{pq} \times 57.6 + 6,250 \times 24 + 3,000$$
$$\times 57.6 - (27,851 - 10,000) \times 75.6$$
$$\Sigma M \text{ about } L_3 = 0 = -s_{dp}(19.2/20.8)24 + 6,250 \times 24$$
$$-17,851 \times 18$$
$$\Sigma M \text{ about } U_3 = 0 = s_{qh} \times 24 + 12,500 \times 24 - 17,851 \times 18$$

from which $s_{pq} = 17,825$ tension, $s_{dp} = 7,733$ compression, and $s_{qh} = 888$ tension.

Considering the part to the right of section 4–4, the forces in the three members (Fig. 12.2.15d) may be found by taking moments about axes where two unknowns intersect, e.g., about U_3 and L_4. Since two unknown forces are parallel, their lines of action do not intersect. However, the equation $\Sigma y = 0$ will enable one to find the force in the member which is not parallel to the other two:

$$\Sigma M \text{ about } U_3 = 0 = s_{qh} \times 24 - (27,851 - 10,000)18 + 12,500 \times 24$$
$$\Sigma M \text{ about } L_4 = 0 = -s_{er} \times 24 + 5,000 \times 24$$
$$\Sigma y = 0 = s_{rq}(24/30) + 27,851 - 10,000$$

from which $s_{qh} = 888$ tension, $s_{er} = 5,000$ tension, and $s_{rq} = 22,314$ compression.

Columns and Walls

Vertical elements in building construction consist of columns, posts, or partitions that transmit concentrated loads; walls or partitions that transmit linearly distributed loads (and may, if so designed, transmit lateral loads from story to story); rigid frames that transmit lateral as well as vertical loads; and braced frames that, ideally, transmit only lateral loads from story to story.

Columns may be of timber, steel, or reinforced concrete and should be proportioned for the allowable stresses permitted for the material used and for the flexibility of the column. Care must be taken in the framing of beams and girders to avoid or to provide for the additional stresses due to connections which transfer loads to columns with large eccentricities. For instance, a beam framing to a flange of a steel column, with a seat or web connection, will produce an eccentric moment in the column equal to $Rd/2$, where R is the beam reaction and d the depth of the column section. Columns not adequately restrained against deflection of the top may be subject to considerable additional moment because of the resulting eccentricity of otherwise axial loads.

Rigid frames consist of columns and beams welded, bolted, or otherwise connected so as to produce continuity at the joints and permit the

All loads and stresses in pounds

Fig. 12.2.15   Resolution of forces for truss shown in Fig. 12.2.14.

entire frame to behave as a unit, capable of resisting both vertical and lateral loads. Advantages of rigid frames are the ease and simplicity of erection, increased headroom, and more open floor plans free of braced frames. Rigid frames composed of rolled sections are commonly used for spans up to 100 ft (30.5 m) in length. Built-up members have been utilized on spans to 250 ft (76.2 m). Welded fabrication offers particular advantages for frames utilizing variable-depth members and on parabolic-shape roofs. The distribution of moments in the statically indeterminate rigid frame is effected by the relationship of the column height to span and roof rise to column height, as well as the relative stiffness of the various members. The solution for the moments in the frame is obtained from the usual equations of statics plus one or more additional equations pertaining to the elastic deformations of the frame under load.

Bearing walls, which are intended to transmit gravity loads from story to story, are designed as vertical elements of unit width such that the combined effects of axial and/or bending stresses do not exceed their allowable values according to the formula $fa/Fa + fb/Fb \leq 1.0$, where f is the actual stress in the axial or bending mode, and F is the allowable stress in the appropriate mode.

Shear walls, either concrete or masonry, which are intended to transmit earthquake, wind, or other lateral loads from story to story parallel to the plane of the wall, are designed as cantilevered vertical shear beams. Loads are distributed among the various wall elements, created by door and window openings, on the basis of their relative rigidities. Unless special conditions indicate otherwise, the assumption is made, in calculating individual stiffnesses, that the wall is fixed against rotation, in the plane of the wall, at the bottom. The individual wall elements are then designed to resist both their share of the wall shear and the moments it induces as well as any vertical loads due to bearing wall action and/or overall wall overturning. Individual wall elements, then, will have special vertical reinforcing at each side (trim reinforcement) to resist bending moments. Diagonal trim reinforcements, 96 bar diameters in total length, where possible, are placed at corners of openings in concrete walls to inhibit cracking. Trim reinforcing should consist of a minimum of two ⅝-in-diameter (1.6-cm) bars (No. 5 bars).

Uplift and downward forces at opposite ends of the shear walls, due to the tendency of the wall to overturn in its own plane when acted upon by lateral forces, must be resisted by vertical reinforcement at those locations. Special boundary elements are required for shear walls in areas of high seismicity (UBC Zones 3 and 4). Only minimum tributary dead loads may be counted upon to help resist uplift while maximum dead plus live loads should be assumed acting simultaneously with downward overturning loads. **Load reversal,** due to wind or seismic loads, must be considered.

Shear walls should be positioned throughout the plan area of the building in such a manner as to distribute the forces uniformly among the individual walls with approximately one-half of the wall area oriented in each of the two major directions of the building. At least one major wall should be positioned, if possible, near each exterior face of the building. Shear walls should be continuous, where possible, from top to bottom of the building; avoid offsetting walls from floor to floor, and particularly avoid discontinuing walls from one level to the level below. Positioning shear walls near major floor openings that would preclude proper anchorage of floor (or roof) to wall should be avoided so that diaphragm forces in the plane of the floor may have an adequate force path to the wall.

Where **rigid diaphragms** (such as concrete slabs or metal deck with concrete fill) exist, shears are distributed to the individual walls on the basis of their relative rigidity taking into account the additional forces due to the larger of either accidental or actual horizontal torsion resulting from the eccentricity of the building seismic shear (located at the center of mass) or eccentricity of applied wind load with respect to the center of rigidity of the walls. Where **flexible diaphragms** (such as metal deck with nonstructural fill or plywood) exist, shears are distributed to walls on the basis of tributary area. When diaphragms are semirigid, based on type, construction, or spacing between supporting elements, shears could be distributed by both methods (relative rigidity

and tributary area) with individual walls designed to resist the larger of the two shears.

Braced frames, which are intended to transmit lateral loads from floor to floor, are designed as trusses that are loaded horizontally instead of vertically and are principally used in steel construction. Lateral loads are distributed to them on the same basis as to shear walls; therefore, the same general rules apply to their ideal placement. Bracing of individual bays may take the form of X-bracing, single diagonal bracing, or K-bracing. Where K-braces are used, the additional bracing forces associated with vertical floor loads to the braced beam must be considered. Alternative floor framing to avoid large beam loads to the braced beam should be considered where possible.

X-bracing may be designed so that either the tension diagonal takes all the lateral load or so that the tension and compression diagonals share the load. When the tension diagonal resists all the lateral load, the minimum slenderness ratio (kl/r) of the member should not exceed 300; the additional axial load in the column to which the top of the brace is connected, the connection load at each end of the brace and the horizontal footing load at the bottom of the brace, is double that obtained when braces share the load. In X-bracing systems where the tension and compression members share the lateral load, maximum slenderness ratios should be limited to 200 (the members may be considered to brace each other about axes that are normal to the plane of the bay), but connection and individual horizontal foundation loads are decreased. With either method of design, it is important to follow the path of the force from its origin through the structure until it is transferred to the ground. With any bracing system, it is wise to sketch to scale the major connection details prior to finalization of calculations; the desirable intersection of member axes is often more difficult to achieve than a single line drawing would indicate. Connection of braces to columns at intermediate points between floors should be avoided because plastic hinges in columns lead to structural instability.

Stud walls consist of wood or steel sheet metal studs with one or more lines of horizontal bridging. The allowable vertical load on the wall is a function of the maximum permissible load on each stud as a column and the spacing of the studs.

Corrugated or flat sheet steel or aluminum, are commonly employed for walls of industrial or mill buildings. These sheets are usually supported on steel girts framing horizontally between columns and supported from heavier eave struts by one or more lines of vertical steel sag rods.

Reinforced-concrete or masonry walls should be designed so that the allowable bending and/or axial stresses are not exceeded, but the minimum thicknesses of such walls should not be less than the following:

Material	Max ratio, unsupported height or length to thickness	Nominal min thickness, in*
Reinforced concrete	25	6
Plain concrete	22	7
Reinforced brick masonry	25	6
Grouted brick masonry	20	6
Plain solid masonry	20	8
Hollow-unit masonry	18	8
Stone masonry (ashlar)	14	16
Interior nonbearing concrete or masonry (reinforced)	48	2
Interior nonbearing concrete or masonry (unreinforced)	36	2

* × 25.4 = mm.

Foundations

Bearing Pressure of Soils The bearing pressure which may be allowed on soil may vary over a large range. For important structures, the nature of the underlying soil should be ascertained by borings or test pits. If the soil consists of medium or soft clay, a settlement analysis based on consolidation tests of undisturbed soil samples from the foundation strata is necessary. Structures founded upon mud, soft clay, silt, peat, or artificial filling will almost certainly settle, and no foundation for a permanent structure should rest on or above such material without adequate provision for the resulting settlement. Table 12.2.6 gives a general classification of soils and typical safe pressures which they may support.

These values approximate the pressures allowed by the building law in most cities. Actual allowable bearing pressures should be based on the quality and engineering characteristics of the material obtained from analysis of borings, standard penetration tests, and rock cores. Other laboratory tests, when the cost thereof is warranted by the magnitude of the project, may show higher values to be safe. The foundation for a building housing heavy vibrating machinery such as power hammers, heavy punches, and shears should receive some allowance for possible compression and rearrangement of soil due to the vibrations transmitted through it. The foundation for a tall chimney should be designed with a comparatively low pressure upon the soil, because of the disastrous results which might occur from local settlement.

Footings The purpose of footings is to spread the concentrated loads of building walls and columns over an area of soil so that the unit pressure will come within allowable limits. Footings are usually constructed of concrete, placed in open excavations with or without sheeting and bracing. In the past, stone, where available in quantity and the proper quality, has been used economically for residential building footings.

Table 12.2.6 Safe Bearing of Soils

Nature of soil	Safe bearing capacity	
	tons/ft^2	MPa
Solid ledge of hard rock, such as granite, trap, etc.	25–100	2.40–9.56
Sound shale and other medium rock, requiring blasting for removal	10–15	0.96–1.43
Hardpan, cemented sand, and gravel, difficult to remove by picking	8–10	0.76–0.96
Soft rock, disintegrated ledge; in natural ledge, difficult to remove by picking	5–10	0.48–0.96
Compact sand and gravel, requiring picking for removal	4–6	0.38–0.58
Hard clay, requiring picking for removal	4–5	0.38–0.48
Gravel, coarse sand, in natural thick beds	4–5	0.38–0.48
Loose, medium, and coarse sand; fine compact sand	1.5–4	0.15–0.38
Medium clay, stiff but capable of being spaded	2–4	0.20–0.38
Fine loose sand	1–2	0.10–0.20

Allowable bearing pressures for granular materials are typically determined as 10 percent of the standard penetration resistances (N values) obtained in the field from standard penetration tests (SPTs) and are in tons/ft^2.

Concrete footings may be either plain or reinforced. Plain concrete footings are generally limited to one- or two-story residential buildings. The center of pressure in the wall or column should always pass through the center of the footing. Where columns, because of fixity, impose a bending moment on the footing, the maximum soil pressures, due to the combined axial and moment components, shall be positive throughout the area of the footing (i.e., no net uplift) and shall be less than the allowable values. Footings in ground exposed to freezing should be carried below the possible penetration of frost.

Deep foundations are required when a suitable bearing soil is deep below the surface, typically more than 10 to 15 ft. (3.0 to 4.6 m). Sometimes deep foundations are necessary even if the suitable bearing materials are shallower, but groundwater conditions make excavation for footings difficult. The most common types of deep foundations include caissons or drilled piers and piles. Where the bearing soil is clay stiff enough to stand with undercutting, and the material immediately above it is peat or silt, the **open-caisson method** may be economical.

In this method, cylindrical steel casings 3 ft and more in diameter are sunk as excavation proceeds, the casings having successively smaller diameters. At the bottom of the shaft thus formed, the soil is undercut to obtain sufficient bearing area. The shaft and the enlargement at the base are then filled with concrete, the cylinders sometimes being withdrawn as the concrete is placed. The open-caisson method cannot be used where groundwater flows too freely into the excavation. Where large foundations under very heavy buildings must be carried to great depth to reach rock or hardpan, particularly where groundwater flows freely, the **pneumatic-caisson method** is used.

Drilled-in piers are formed by drilling with special power augers up to 5 ft (1.5 m) diameter or greater. The holes are drilled to the desired bearing level with or without metal casings, depending on the soil conditions. Belling of the bottom may also be performed mechanically from the surface. In poor soils, or where groundwater is present, the hole may be retained with bentonite clay slurry which is displaced as concrete is placed in the caisson by the **tremie method.**

Pile Foundations Piles for foundations may be of wood, concrete, steel, or combinations thereof. Wood piles are generally dressed and, if required, treated offsite; concrete piles may be prepared offsite or cast in place; steel piles are mill-rolled to section. They can be driven, jacked, jetted, screwed, bored, or excavated. Wood piles are best suited for loads in the range of 15 to 30 tons (133.4 to 177.9 kN) per pile and lengths of 20 to 45 ft (6.6 to 15 m). They are difficult to splice and may be driven untreated when located entirely below the permanent water table; otherwise piles treated with creosote should be used to prevent decay. Wood piles should be straight and not less than 6 in in diameter under the bark at the tip. Concrete piles are less destructible, and hence are adaptable to many conditions, including driving in dense gravels, and can be up to approximately 120 ft (40 m) long. Concrete piles are divided into two classes: those poured in place and those precast, cured, and driven. Cast-in-place piles are made by driving a mandrel into the ground and filling the resulting hole with concrete. In one well-known pile of this type (Raymond), a thin sheet-steel corrugated shell is fitted over a tapered mandrel before driving. This shell, which is left in the ground when the mandrel is removed, is filled with concrete. Prestressing of precast-concrete piles provides greater resistance to handling and driving stresses. With a concrete pile, 25 to 60 tons (222.4 to 533.8 kN) or more per pile are carried. Structural steel H columns and steel pipe with capacities of 40 to 300 tons (350 to 1,800 kN) per pile have been driven. Experience has shown that corrosion is seldom a practical problem in natural soils, but if otherwise, increasing the steel section to allow for corrosion is a common solution.

Methods of Driving Piles The drop hammer and the steam or pneumatic hammer are usually employed in driving piles. The steam or pneumatic hammer, with its comparatively light blows delivered in rapid succession, is of advantage in a plastic soil, the speed with which the blows are delivered preventing the readjustment of the soil. It is also of advantage in soft soils where the driving is easy, but a light hammer may fail to drive a heavy pile satisfactorily. The water jet is sometimes used in sandy soils. Water supplied under pressure at the point of the pile through a pipe or hose run alongside it erodes the soil, allowing the pile to settle into place. To have full capacity, jetted piles should be driven after jetting is terminated particularly if the pile is to resist uplift loads. Diesel hammers are sometimes used if the equipment is available. They do not require a steam boiler or an auxiliary engine for pneumatic power and thus generally render more economical service. When driving piles (or sheet piling) into relatively sandy soils, vibratory hammers have proved successful; the vibrations imparted to the hammer by an engine-driven unbalanced eccentric serve to set up dilatory action in the pile, making penetration faster and much simpler.

In applications where piles must be pulled, or extracted, the reverse operation is performed by the hammer; i.e., the pile is subjected to a tensile force for removal. Vibratory hammers offer exemplary performance in such applications.

Determination of Safe Loads for Piles Piles may obtain their supporting capacity from friction on the sides or from bearing at the point. In the latter case, the bearing capacity may be limited by the strength of the pile, considered as a column, to which, however, the surrounding soil affords some lateral support. In the former case, no precise determination of the bearing capacity can be made. Many formulas have been developed for determining the safe bearing capacity in terms of the weight of the hammer, the fall, and the penetration of the pile per blow, the most generally accepted of which is that known as the ***Engineering News* formula:** $R = 2wh/(s + 1.0)$ for drop hammers, $R = 2wh/(s + 0.1)$ for single-acting steam or pneumatic hammers, $R = 2E/(s + 0.1)$ for double-acting steam or pneumatic hammers, where $R =$ safe load, lb; $w =$ weight of hammer, lb; $h =$ fall of hammer, ft; $s =$ penetration of last blow, in; $E =$ rated energy, ft · lb per blow. This formula and similar ones are based on the determination of the energy in the falling hammer, and from this the pressure which it must exert on the top of the pile. The *Engineering News* formula is currently used typically for timber piles with capacities not exceeding 30 tons. It assumes a factor of safety of 6. It is a wise practice to drive index piles and determine their bearing capacity through pile load tests typically carried to twice the service load of the pile before proceeding with the final design of important structures to be supported on piles. The design then is based on the safe service load capacity so determined, and the piles are driven to the same penetration resistance to which the successfully tested index piles where driven by the same driving hammer.

Spacing of Piles Wood piles are preferably spaced not closer than 2½ ft (0.76 m), and concrete piles 3 ft (0.91 m) on centers. If driven closer than this, one pile is liable to force another up. Piles in a group must not cause excessive pressure in soil below their tips. The efficiency, or supporting value of friction piles when driven in groups, by the Converse-Labarre method, is

$$\frac{1 - d/s[(n - 1)m + (m - 1)n]}{90mn}$$

where $d =$ pile diameter, in (cm); $s =$ spacing center to center of piles, in (cm); $m =$ number of rows; $n =$ number of piles in a row.

Capping of Piles Piles are usually capped with concrete; wood piles sometimes with pressure-treated timber. Concrete is the most usual material and the most satisfactory for the reason that it gets a full bearing on all piles. The piles should be embedded 4 to 6 in (10 to 15 cm) in the concrete. In the event some pile(s) may be subjected to tension, a means of anchoring the pile top(s) into the concrete must be used.

Retaining Walls A wall used to sustain the pressure of earth behind it is called a retaining wall. Retaining walls which depend for their stability upon the weight of the masonry are classed as gravity walls. Such walls built on firm soil will usually be stable when they have the following proportions: top of fill level, back vertical, base, 0.4 height; top of fill level, back battered, base, 0.5 height; top of fill steeply inclined, back vertical, base, 0.5 height; top of fill steeply inclined, back battered, base, 0.6 height. An additional factor of safety is obtained by building the face on a batter. Care should be taken in the design of a wall that the allowable soil pressure is not exceeded and that drainage is provided for the back of the wall. The foundations of retaining walls should be placed below the level of frost penetration. Retaining walls of reinforced concrete are made thin, with a broad base, and the wall either cantilevered from the base or braced with buttresses or counterforts.

It is impossible to derive formulas for the earth pressure on the back of the wall which will take account of all the actual conditions. Assuming the earth to be a loose, homogeneous, granular mass, and the coefficient of friction to be independent of the pressure, Rankine deduced the following formula for a wall with vertical back:

$$P = \left(\frac{1}{2}wh^2 + vh\right)\cos d \left[\frac{\cos d - \sqrt{\cos^2 d - \cos^2 a}}{\cos d + \sqrt{\cos^2 d - \cos^2 a}}\right]$$

the center of pressure being at a height $\frac{1}{3}H(wh + 3v)/(wh + 2v)$ above the base, where $P =$ earth pressure per lin ft (m) of the wall, lb (kg); $h =$ height of the wall, ft (m); $w =$ weight of earth per ft³ (m³), lb (kg); $v =$ weight of superposed load per ft² (m²) of surface, lb (kg); $d =$ the angle with the horizontal of the earth surface behind the wall; and $a =$ angle of internal friction of the earth, deg. (For sands, $a = 30$ to $38°$.) The direction of the pressure is parallel to the earth's surface. The retaining

wall should have sufficient thickness at the base so that the resultant of the earth pressure P combined with the weight of the wall falls well within the base. If this resultant falls at the outside edge of the middle third, the maximum vertical pressure on the foundation (at the outer edge of the base) will be equal to $2W/T$ lb/ft² (Pa), where W is the total vertical pressure on the base of 1 ft (m) length of wall and T the thickness of the wall at the base, ft (m).

In the design of walls of buildings which must withstand earth pressure and low independent walls, where refinement is not necessary, the earth pressure is frequently assumed to be that of a fluid weighing 40 to 45 lb/ft³ (641 to 721 kg/m³).

MASONRY CONSTRUCTION

The term **masonry** applies to assemblages consisting of fired-clay, concrete, or stone units, mortar, grout, and steel reinforcement (if required). The choice of materials and their proportions is based on the required strength (compressive, flexural, and shear), fire rating, acoustics, durability, and aesthetics. The strength and durability of masonry depends on the size, shape, and quality of the unit, the type of mortar, and the workmanship. Resulting bond strength is affected by the initial rate of absorption, texture, and cleanliness of the masonry units.

Brick Common red bricks are made of clay burned in a kiln. Quality characteristics are hardness and density. Light-colored brick is apt to be soft and porous. Brick for masonry exposed to the weather or where strength is desired should have a crushing strength of not less than 2,500 lb/in² (17.3 MPa) and should absorb not over 20 percent of water by weight, after 5-h immersion in boiling water (see Sec. 6).

Mortar Mortar is the bonding agent that holds the individual units together as an assembly. All mortars contain cement, sand, lime, and water in varying proportions. Mortars are classified as portland-cement-lime mortar or masonry cement mortar. (See Sec. 6.9.)

Laying and Bonding Brick should be laid in a full bed of mortar and shoved laterally into place to secure solid bearing and a bed of even thickness and to fill the vertical joints. Clay brick should be thoroughly wet before laying, except in freezing weather. Concrete bricks/blocks should not be wetted before placement. Brick laid with long dimension parallel to the face of the work are called stretchers, perpendicular to the face, headers. Bats (half-brick) should not be used except where necessary to make corners or to form patterns on the face of the wall. Each continuous vertical section of a wall one masonry unit thick is called a **wythe**. Multiwythe walls, of brick or block faced with brick, are now commonly tied together with ties or anchors between each wythe of brick and block. Such ties are typically provided every other course. When the wythes are adequately bonded or tied; the wall may be considered as a composite wall for strength purposes. Walls may also be tied together longitudinally by overlapping stretchers in successive courses. Transverse bond is obtained by making every sixth course headers, the headers themselves overlapping in successive courses in the interior of thick walls. Variations in the arrangement of headers are often used in the face of walls for appearance. The area of cross section of full-length headers should not be less than one-twelfth the face of the wall, in bonding each pair of transverse courses of brick. Three examples of bond are shown in Fig. 12.2.16.

Arches over windows and doorways are laid in concentric rings of headers on edge, with radial joints. The radius of the arch should be 1 to 1¼ times the width of the opening.

Lateral Support Brick walls should be supported laterally by bonding to transverse walls or buttresses, or by anchoring to floors, at intervals not exceeding 20 times the thickness. Floors and anchors must be capable of transmitting wind pressure and earthquake forces, acting outward, to transverse walls or other adequate supports, and thus to the ground. The height of piers between lateral supports should not exceed 12 times their least dimension.

Concrete Masonry Units (CMUs) Hollow or solid concrete blocks are used in building walls, often faced with brick or stone on the exterior walls. Standard block sizes are based on a brick module, nominally 4 in high.

Fig. 12.2.16 Bonds used in bricklaying.

Allowable compressive stresses in masonry are related to the masonry unit strength and the type of mortar as follows:

Kind of masonry	Type M or S mortar		Type N mortar	
	lb/in²	MPa	lb/in²	MPa
Stone ashlar	360	2.48	320	2.20
Rubble	120	0.83	100	0.69
Clay brick (2500 lb/in²)	160	1.10	140	0.97
Hollow CMU (1000 lb/in²)	75	0.52	70	0.48
Solid CMU (2000 lb/in²)	160	1.10	140	0.97

Minimum thickness of load-bearing masonry walls shall be 6 in for up to one story and 8 in for more than one story.

Reinforced Masonry The design of masonry with vertical reinforcing bars placed and fully grouted in some of the cells provides increased capacity to resist flexure and axial loads. In addition, in areas where seismic design is required, due to the brittle nature of unreinforced masonry, all masonry must be reinforced to prevent a brittle failure mode. The design principles for reinforced masonry are similar to the design for reinforced concrete.

Reinforced Concrete See Sec. 12.3.

TIMBER CONSTRUCTION

Floors The framing of wooden floors may be divided into two general types: joist construction and solid, or mill, construction. The first consists of **joists** 2 to 6 in wide, of the necessary depth, and spaced about 12 to 16 in (30 to 40 cm) on centers. The wall ends should rest on and be anchored to walls and the interior ends carried by a line of girders on columns. These joists should be securely cross-bridged not over 8 ft (2.4 m) apart in each span to prevent twisting and to assist in distributing concentrated loads. Solid blocking should be provided at ends and at each point of support. The floor is formed of a thickness of rough boarding on which the finish flooring is laid. **Solid** or **mill-construction floors** are designed to do away with the small pockets which exist in joist construction and thus reduce the fire hazard. They are generally framed with beams spaced 8 to 12 ft (2.4 to 3.6 m) on centers and spanning 18 to 25 ft (5.5 to 7.6 m). The wall ends of beams rest on and are anchored to the wall, and the interior ends are carried on columns and tied together to form a continuous tie across the building. Ends of timbers in masonry walls should have metal bearing plates and ½ in space at sides and end for ventilation, to prevent rot. The ends should be beveled and the anchors placed low to avoid overturning the wall if the beams drop in a fire. In all cases, care should be taken to provide sufficient bearing at the points of support so that the allowable intensity of compression across the grain is not exceeded. In case it is desirable to omit columns, or the floor load requires a closer spacing of beams, girders are run lengthwise of the building over the columns to take the beams, the ends of which are hung in hangers or stirrup irons and tied together, over or through the girders. This is called **intermediate framing.** Steel beams are sometimes used in place of wooden beams in this type of construction, in which case a wooden strip is bolted to the top flange of the beam to take the nailing of the plank, or the plank is laid directly on top of the beam and secured by spikes driven from below and clinched over the flange. The floor is formed of 3 or 4 in (7.5 or 10 cm) plank grooved in each edge, put together with splines and securely spiked to beams. On top of the plank is laid flooring, with a layer of sheathing paper between. In case the floor loads require an excessive thickness of plank, or in localities where heavy plank is not easily obtainable, the floor is built up of 3 × 6 in (7.5 × 15 cm), or other sized pieces, placed on edge, and securely nailed together. (See Table 12.2.7.)

The roofs of buildings of joist and mill construction are framed in a manner similar to the floors of each type and should be securely anchored to the walls and columns. In case columns are not desired in the top story, steel beams or trusses of either steel or wood are used. For spans up to 35 ft (10.7 m), trussed beams can often be used to advantage.

Table 12.2.7 Properties of Plank and Solid Laminated Floors
(b = breadth = 12 in, f = fiber stress)

Nominal thickness or depth, in (1)	Actual thickness d, in (S4S) (2)	Area of section $A = bd$, in² (3)	Moment of inertia $I = bd^3/12$, in⁴ (4)	Section modulus $S = bd^2/6$, in³ (5)	Safe load, lb/ft² on 1-ft span*		Coef of deflection, uniform load‡	
					$f = 1,000$ lb/in²† (6)	$f = 1,600$ (7)	$E = 1,000,000$ (8)	$E = 1,760,000$ (9)
1	¾	9.00	0.422	1.13	753	1,205	53.4	93.98
1½	1¼	15.00	1.95	3.13	2,085	3,336	11.51	20.26
2	1½	18.00	3.38	4.50	3,000	4,800	6.66	11.72
2½	2	24.00	8.00	8.00	5,334	8,534	2.82	4.96
3	2½	30.00	15.60	12.50	8,334	13,334	1.441	2.54
4	3½	42.00	42.9	24.5	16,334	26,134	0.524	0.922
5	4½	54.00	91.1	40.5	27,000	43,200	0.247	0.1404
6	5½	66.00	166.4	60.5	40,300	64,500	0.1348	0.0765
8	7½	90.00	422	112.5	75,000	120,000	0.0533	0.0303
10	9½	114.00	857	180.5	120,400	192,500	0.0263	0.0149
12	11½	138.00	1,521	264.5	176,400	282,000	0.0148	0.0084

* Divide tabular value by square of span in feet.
† For other fiber stress f, multiply tabular value by $f/1,000$.
‡ For deflection in, multiply coefficient by load, lb/ft², and by fourth power of span in ft, and divide by 1,000,000. For other modulus of elasticity E, multiply coefficient of col. 8 by 1,000,000, and divide by E.
NOTE: 1 in = 2.54 cm; 1 lb = 4.45 N; 1 lb/in² = 6.89 kPa.

For **unit stresses in timber,** see Sec. 6.7. For unit stresses in wooden columns, see Table 12.2.9. Table 12.2.7 gives the properties of mill floors made of dressed plank, and of laminated floors made of planks of edge, laid close.

Timber Beams

Properties of Timber Beams Table 12.2.8 presents those properties of wooden timbers most useful in computing their strength and deflection as beams. (The "nominal size" of a timber is indicated by the breadth and depth of the section in inches. The "actual size" indicates the size of the dressed timber, according to National Lumber Manufacturers Assoc. The moment of inertia and section modulus are with the neutral axis perpendicular to the depth at the center. The **safe bending moment** in inch-pounds for a given beam is determined from the section modulus S by multiplying the tabular value by the allowable fiber stress. To select a beam to withstand safely a given bending moment, divide the

Table 12.2.8   Properties of Wooden Beams (Surfaced Size)

Nominal size, in (1)	Actual size $b \times d$, in, dressed (S4S) size (2)	Area of section bd, in^2 (3)	Weight at 40 lb/ft^3, lb/ft (4)	Moment of inertia $I = bd^3/12$, in^4 (5)	Section modulus $S = bd^2/6$, in^3 (6)	Max safe uniform load, lb, based on		Coef‡ of deflection, uniform load $E = 1,000,000$ (9)
						Bending on 1 ft span,* $f = 1,000$ lb/in^2 (7)	Shear at 100† lb/ in^2 (8)	
2 × 4	1½ × 3½	5.25	1.46	5.36	3.06	2,040	700	4.20
3 × 4	2½ × 3½	8.75	2.43	8.93	5.10	3,400	1,166	2.52
4 × 4	3½ × 3½	12.25	3.40	12.51	7.15	4,760	1,632	1.80
2 × 6	1½ × 5½	8.25	2.29	20.8	7.56	5,040	1,100	1.082
3 × 6	2½ × 5½	13.75	3.82	34.7	12.60	8,390	1,835	0.648
4 × 6	3½ × 5½	19.25	5.35	48.5	17.65	11,760	2,570	0.464
6 × 6	5½ × 5½	30.3	8.40	76.3	27.7	18,490	4,040	0.295
2 × 8	1½ × 7¼	10.87	3.02	47.6	13.14	8,760	1,445	0.473
3 × 8	2½ × 7¼	18.12	5.04	79.4	21.9	14,600	2,410	0.284
4 × 8	3½ × 7¼	25.4	7.05	111.1	30.7	20,500	3,380	0.202
6 × 8	5½ × 7½	41.3	11.4	193	51.6	34,400	5,500	0.1162
8 × 8	7½ × 7½	56.3	15.6	264	70.3	46,900	7,500	0.0852
2 × 10	1½ × 9¼	13.87	3.85	98.9	21.4	14,290	1,850	0.227
3 × 10	2½ × 9¼	23.1	6.42	164.9	35.7	23,700	3,080	0.1364
4 × 10	3½ × 9¼	32.4	8.93	231	49.9	33,300	4,310	0.0974
6 × 10	5½ × 9½	52.3	14.5	393	82.7	55,200	6,970	0.0573
8 × 10	7½ × 9½	71.3	19.8	536	113	75,200	9,500	0.0421
10 × 10	9½ × 9½	90.3	25.0	679	143	95,300	12,030	0.0332
2 × 12	1½ × 11¼	16.87	4.69	178	31.6	21,100	2,250	0.1264
3 × 12	2½ × 11¼	28.1	7.81	297	52.7	35,100	3,750	0.0757
4 × 12	3½ × 11¼	39.4	10.94	415	73.9	49,300	5,250	0.0543
6 × 12	5½ × 11½	63.3	17.5	697	121	80,800	8,430	0.0323
8 × 12	7½ × 11½	86.3	23.9	951	165	110,200	11,510	0.0237
10 × 12	9½ × 11½	109.3	30.3	1,204	209	139,600	14,570	0.01864
12 × 12	11½ × 11½	132.3	36.7	1,458	253	169,000	17,620	0.01543
4 × 14	3½ × 13¼	46.4	12.88	678	102.4	68,300	6,180	0.0332
6 × 14	5½ × 13½	74.3	20.6	1,128	167	111,400	9,900	0.01987
8 × 14	7½ × 13½	101.3	28.0	1,538	228	152,000	13,500	0.01462
10 × 14	9½ × 13½	128.3	35.6	1,948	289	192,400	17,120	0.01153
12 × 14	11½ × 13½	155.3	43.1	2,360	349	233,000	20,700	0.00953
14 × 14	13½ × 13½	182.3	50.6	2,770	410	273,000	24,300	0.00812
6 × 16	5½ × 15½	85.3	23.6	1,707	220	146,800	11,380	0.01315
8 × 16	7½ × 15½	116.3	32.0	2,330	300	200,000	15,530	0.00967
10 × 16	9½ × 15½	147.3	40.9	2,950	380	254,000	19,610	0.00762
12 × 16	11½ × 15½	178.3	49.5	3,570	460	307,800	23,800	0.00630
14 × 16	13½ × 15½	209	58.1	4,190	541	360,000	27,900	0.00539
16 × 16	15½ × 15½	240	66.7	4,810	621	414,000	32,000	0.00468
8 × 18	7½ × 17½	131.3	36.4	3,350	383	255,000	17,500	0.00672
10 × 18	9½ × 17½	166.3	46.1	4,240	485	323,000	22,200	0.00531
12 × 18	11½ × 17½	201	55.9	5,140	587	391,000	26,800	0.00438
14 × 18	13½ × 17½	236	65.6	6,030	689	459,000	31,500	0.00373
16 × 18	15½ × 17½	271	75.3	6,920	791	528,000	36,200	0.00325
18 × 18	17½ × 17½	306	85.0	7,820	893	595,000	40,800	0.00288
12 × 20	11½ × 19½	224	62.3	7,110	729	485,000	29,900	0.00316
20 × 20	19½ × 19½	380	106	12,050	1,236	824,000	50,700	0.00187
24 × 24	23½ × 23½	552	153	25,400	2,160	1,440,000	73,400	0.000888
26 × 26	25½ × 25½	650	180.6	35,200	2,760	1,840,000	86,700	0.000639
28 × 28	27½ × 27½	756	210	47,700	3,470	2,320,000	100,600	0.000472
30 × 30	29½ × 29½	870	242	63,100	4,280	2,850,000	116,000	0.000356

* For total safe uniform load, pounds, on beam of span L, feet, divide tabular value by L. For fiber stress f other than 1,000 lb/in^2 multiply by f and divide by 1,000.

† For shearing stress other than 100 lb/in^2, multiply by stress and divide by 100.

‡ For deflection, inches, multiply coefficient by total load, pounds, and by cube of span, feet, and divide by 1,000,000. For other modulus of elasticity E, multiply coefficient by 1,000,000 and divide by E.

NOTE: 1 in = 2.54 cm; 1 ft = 0.305 m; 1 lb = 4.45 N; 1 lb/in^2 = 6.89 kPa.

bending moment in inch-pounds by the allowable fiber stress, and choose a beam whose section modulus S is equal to or larger than the quotient thus obtained. For formulas for computing bending moments, see Sec. 5.2. Note that the allowable fiber stress must be modified by adjustment factors: C_D, load duration factor; C_M, wet-service factor; C_t, temperature factor; C_F, size factor; and other factors which apply to laminated, curved, round, and/or flat use. (See also Sec. 6.7, "Properties of Lumber Products.")

Maximum loads in Table 12.2.8, cols. 7 and 8, are for uniform loading. Use half the values of col. 7 for a single load concentrated at midspan; for other loadings compute the bending moment and use the section modulus, col. 6. The values of col. 8 apply to all symmetrical loadings. For unsymmetrical loading, compute the maximum shear, which must not exceed one-half the tabular value.

The **coefficients of deflection** listed in Table 12.2.8 can be used to deduce deflection as indicated in the footnotes to the table. Coefficients of deflection under concentrated loads applied at the middle of the span may be obtained by multiplying the values in the table by 1.6. The results are only approximate, as the modulus of elasticity varies with the moisture content of the wood.

The deflection due to live load of beams intended to carry plastered ceilings should not exceed $1/360$ of the span (span in inches).

A convenient rule may be derived by assuming that the modulus of elasticity is 1,000 times the allowable fiber stress, which applies to all woods with sufficient accuracy for the purpose. Beams loaded uniformly to capacity in bending will then deflect $1/360$ of the span when the depth in inches is 0.90 times the span in feet; and beams with central concentration, when the depth is 0.72 times the span in the same units. For such beams, the deflection in inches is, for uniform load, $0.03L^2/d$; for central concentration, $0.024L^2/d$, where L is the span, ft, and d the depth, in. Variation in type of loading affects this result comparatively little.

Timber Columns

Timber columns may be either square or round and should have metal bases, usually galvanized steel, to cut off moisture and prevent lateral displacement. For supporting beams, they should have caps which, at roofs, may be of steel, or wood designed for bearing across the grain. At intermediate floors, caps should be of steel, although in some cases hardwood bolsters may be used. Except when caps or beams are of steel, columns should run down and rest directly on the baseplate. Table 12.2.9 gives **working unit stresses for wood columns** recommended where the building laws do not prescribe lower stresses. Use actual, not nominal, dimension of timbers. The column capacity $P_{col} = F'_c A_{net}$, where F'_c is interpolated from Table 12.2.9, and A_{net} is the net cross-sectional area of the column. The values for F'_c in Table 12.2.9 are generally conservative and are based on the factors cited in the table footnotes. In the event $E < 1,000F_c$, the value of F'_c must be computed to include the value of C_P as follows:

$$F'_c = F_c C_D C_M C_t C_F C_P$$

(Note that actual design stress $f_c \le F'_c$.)

$$C_p = \text{column stability factor} = \frac{1 + F_{cE}/F^*_c}{2c}$$
$$- \sqrt{\left(\frac{1 + F_{cE}/F^*_c}{2c}\right)^2 - \frac{F_{cE}/F^*_c}{c}}$$

$$F^*_c = F_c C_D C_M C_t C_F$$

$F_{cE} = \dfrac{K_{cE}E}{(l_e/d)^2}$ = critical buckling design stress in compression parallel to the grain for a given wood species and geometrical configuration of column

where F_c = tabulated allowable design value for compression parallel to the grain for the species (Sec. 6.7); $K_{cE} = 0.3$ for visually and mechanically graded lumber and 0.418 for glulam members; l_e = effective length of column; d = least dimension of the column cross section; E = modulus of elasticity for the species (Sec. 6.7); c = constant for the type member: 0.8 for sawn lumber, 0.85 for timber piles, 0.9 for glulam members; for factors C_D, etc., sec Sec. 6.7.

Glued Laminated Timber   Structural glued laminated timber, commonly called glulam, refers to members which are fabricated by pressure gluing selected wood laminations of either $3/4$ or $1\frac{1}{2}$ in (19 or 38 mm) surfaced thickness. The grain of all the laminations is approximately parallel longitudinally, with exterior laminations being of generally higher-quality wood since bending stresses are greater at the outer fibers. Curved and tapered structural members are available with the recommended minimum radii of curvature being 9 ft 4 in (2.84 m) for $3/4$-in laminations and 27 ft 6 in (8.4 m) for $1\frac{1}{2}$-in lamination thickness. Laminations should be parallel to the tension face of members; sawn tapered cuts are permitted on the compression face.

Available net (surfaced) widths of members in inches are $2\frac{1}{4}$, $3\frac{1}{8}$, $5\frac{1}{8}$, $8\frac{3}{4}$, $10\frac{3}{4}$, $12\frac{1}{4}$, and $14\frac{1}{4}$; depths are determined by stress requirements. Economical spans (see "Timber Construction Manual," American Institute of Timber Construction) for roof framing range from 10 to 100 ft (3 to 30 m) for simple spans. Floor framing, which is designed for much heavier live loads, economically spans from 6 to 40 ft (1.8 to 12 m) for simple spans and from 25 to 40 ft (7.5 to 12 m) for continuous spans.

Glued laminated members are generally fabricated from either Douglas fir and larch, Douglas fir (coast region), southern pine, or California redwood, depending on availability. Allowable design stresses depend on whether the condition of use is to be wet (moisture content in service of 16 percent or more) or dry (as in most covered structures), the species and grade of wood to be used, the manner of loading, and the number of laminations as well as the usual factors for duration of loading. The cumulative reduction factors described above also apply to glulam beams. Additional factors including C_v, volume factor; C_{fu}, flat use factor; and C_c, curvature factor, also must be applied to glued laminated beams. Refer to the National Design Specification for Wood Construction for further information on glued laminated timber design. (See Sec. 6.7.)

Table 12.2.9   Values of F'_c, Working Stresses for Square or Rectangular Timber Columns, lb/in^2
(Compression parallel to grain.)*

F_c	l_e/d, in/in												
	10	15	20	25	30	35	40	45	50	55†	60†	70†	80†
1,000	680	615	508	389	293	225	176	141	116	96	81	60	46
1,300	884	799	660	505	381	292	229	184	150	125	106	78	60
1,600	1,088	984	812	622	469	359	282	226	185	154	130	96	74
1,900	1,292	1,168	964	739	557	427	335	268	220	183	154	114	88

* Values of F'_c in the table are based on $C_D = 0.9$ (long-duration, permanent, 50-year loading); $C_M = 0.8$ (moisture content $> 16\%$); $C_t = 1.0$ (operating temperature $< 100°F$); $C_F = 1.0$ (cross section up to 5 in wide $\times$ 12 in deep); $K_{cE} = 0.3$; $E \ge 1,000 F_c$. If $E < 1,000 F_c$, see text for procedure to compute F'_c; $c = 0.8$.
† Columns should be limited to $l/d = 50$, except for individual members in stud walls, which should be limited to $l/d = 80$.

A summary of allowable unit stresses may be found in Sec. 6.7 for glued laminated timber.

Connections

Bolted Joints   Compression may be transmitted by merely butting the timbers, with splice pieces bolted to the sides to keep alignment and resist incidental bending and shear. The same detail (Fig. 12.2.17)

Fig. 12.2.17   Bolted splices for timber framing.

serves in tension, but the entire stress must then be transmitted through the bolts and splice pieces. If of wood, these should have a thickness h equal to $\frac{1}{2}b$. In light, unimportant work, splice pieces may be spiked. Table 12.2.10 gives the allowable load in pounds for one bolt loaded at both ends (double shear) when h is at least equal to $\frac{1}{2}b$. When steel side plates are used for side members, the tabulated loads may be increased 25 percent for parallel-to-grain loading, but no increase should be made for perpendicular-to-grain loads. When a joint consists of two members (single shear), one-half the tabulated load for a piece twice the thickness of the thinner member applies. The safe load for bolts loaded at an angle θ with the grain of the wood is given by the formula $N = PQ/(P \sin^2 \theta + Q \cos^2 \theta)$, where N = allowable load per bolt in a direction at inclination θ with the direction of the grain, lb; P = allowable load per bolt in compression parallel to the grain, lb; and Q = allowable load per bolt in compression perpendicular to the grain, lb.

The size, arrangement, and **spacing of bolts** must be such that tension on the net section of the timber through the bolt holes and shear along the grain do not exceed allowable values. Bolts should be at least 7 diameters from the end of the timber for softwoods and 5 diameters for hardwoods and spaced at least 4 diameters on center parallel to the grain. Crossbolting, to prevent splitting the timber end, is sometimes desirable.

The efficiency of bolted timber connections may be greatly increased by the use of **ring connectors.** Split rings and shear plates are fitted into circular grooves, concentric with the bolt, in the contact surfaces, and transmit shear stresses across the joint. Grooves for split rings and shear plates are cut with a special tool, while toothed rings are usually seated by drawing together the timbers with high-strength bolts. Allowable loads for these various connectors are given in the "Design Values for Wood Construction," published by the National Lumber Manufacturers Assoc. Selected values are given in Table 12.2.11.

The **holding power of wire nails** is as follows ("Design Values for Wood Construction"): The resistance to withdrawal is proportional to the length of embedment, to the diameter of the nail (where the wood does not split), and to $G^{2.5}$, where G is the ovendry specific gravity of the wood (see Sec. 6 for G values of various species). The safe resistance to withdrawal of common wire nails driven into the side grain of seasoned wood is given by Table 12.2.12. Nails withdrawn from green wood have generally slightly higher resistance, but nails driven into green wood may lose much of their resistance when the wood seasons; the allowable withdrawal load should be one-fourth of that given in Table 12.2.12. Cement and other coatings on nails may add materially to their resistance in softwoods. Drilling lead holes slightly smaller than the nail adds somewhat to the resistance and reduces danger of splitting. The structural design should be such that nails are not loaded in withdrawal from end grain.

The safe **lateral resistance** of common wire nails driven in side grain to the specified penetrations is given in Table 12.2.12 and is proportional to $D^{1.5}$ where D is the diameter, in. These values are for seasoned wood and should be reduced 25 percent for woods which will remain wet or will be loaded before seasoning. For nails driven into end grain, values should be reduced one-third.

Common wire **spikes** are larger for their lengths than nails. Their resistance to withdrawal and lateral resistance are given by the same formulas as for nails, but greater precautions need to be taken to avoid splitting.

The **resistance of wood screws to withdrawal** from side grain of seasoned wood is given by the formula $P = 2,850G^2D$, where P = the allowable load on the screw, lb/in penetration of the threaded portion; G = specific gravity of ovendry wood; D = diameter of screw, in. Wood screws should not be designed to be loaded in withdrawal from end grain.

The **allowable safe lateral resistance** of wood screws embedded 7 diameters in the side grain of seasoned wood is given by the formula $P = KD^2$, where P is the lateral resistance per screw, lb; D is the diameter, in; and K is 4,800 for oak (red and white), 3,960 for Douglas fir (coast region) and southern pine, and 3,240 for cypress (southern) and Douglas fir (inland region).

Table 12.2.10   Allowable Load in Pounds on One Bolt Loaded at Both Ends (Double Shear)
(For additional values and for conditions other than normal, see "Design Values for Wood Construction.")

Length of bolt in main member, in	Diam of bolt, in	Douglas fir-larch		California redwood (open grain)		Oak, red and white		Western spruce, pine, fir, cedars	
		Parallel to grain	Perpendicular to grain	Parallel to grain	Perpendicular to grain	Parallel to grain	Perpendicular to grain	Parallel to grain	Perpendicular to grain
$1\frac{1}{2}$	$\frac{1}{2}$	1,050	470	780	310	1,410	730	760	290
	$\frac{3}{4}$	1,580	590	1,170	370	2,110	890	1,140	360
	1	2,100	680	1,560	440	2,810	1,020	1,520	420
$2\frac{1}{2}$	$\frac{1}{2}$	1,230	730	990	510	1,530	960	980	490
	$\frac{3}{4}$	2,400	980	1,950	620	2,890	1,480	1,900	600
	1	3,500	1,130	2,590	730	4,690	1,700	2,530	700
$3\frac{1}{2}$	$\frac{1}{2}$	1,230	730	990	580	1,530	960	980	560
	$\frac{3}{4}$	2,400	1,170	2,010	740	2,890	1,770	1,990	720
	1	4,090	1,350	3,110	870	4,820	2,040	3,040	840
$5\frac{1}{2}$	$\frac{3}{4}$	2,400	1,170	2,010	740	2,890	1,770	1,990	720
	1	3,180	1,260	2,690	810	3,780	1,920	2,660	790
	$1\frac{1}{4}$	4,090	1,350	3,110	870	4,820	2,040	3,040	840

NOTE: 1 in = 2.54 cm; 1 lb = 4.45 N.

Table 12.2.11 Allowable Load in Pounds for One-Connector Unit in Single Shear*
(For additional values and for conditions other than normal, see "Design Values for Wood Construction.")

Connector unit (diam)	Number of faces of piece with connectors on the same bolt	Net thickness of lumber, in	Min. edge distances, in	Group A Douglas fir-larch and southern pine (dense), oak, red and white ∥ to grain	⊥ to grain	Group B Douglas fir-larch, southern pine (med. grain) ∥ to grain	⊥ to grain	Group C California redwood (close grain), western hemlock, southern cypress ∥ to grain	⊥ to grain
2½-in split ring, ½-in bolt	1	1 min,	1¾	2,630	1,900	2,270	1,620	1,900	1,350
		1½ or more		3,160	2,280	2,730	1,940	2,290	1,620
	2	1½ min,		2,430	1,750	2,100	1,500	1,760	1,250
		2 or more		3,160	2,280	2,730	1,940	2,290	1,620
4-in split ring, ¾-in bolt	1	1 min,	2¾	4,090	2,840	3,510	2,440	2,920	2,040
		1½ or more		6,020	4,180	5,160	3,590	4,280	2,990
	2	1½ min,		4,110	2,980	3,520	2,450	2,940	2,040
		3 or more		6,140	4,270	5,260	3,660	4,380	3,050
2⅝-in shear plate, ¾-in bolt†	1	1½ min	1¾	3,110	2,170	2,670	1,860	2,220	1,550
	2	1½ min,		2,420	1,690	2,080	1,450	1,730	1,210
		2½ or more		3,330	2,320	2,860	1,990	2,380	1,650
4-in shear plate, ¾-in or ⅞-in bolt†	1	1½ min,	2¾	4,370	3,040	3,750	2,620	3,130	2,170
		1¾ or more		5,090	3,540	4,360	3,040	3,640	2,530
	2	1¾ min,		3,390	2,360	2,910	2,020	2,420	1,680
		2½		4,310	3,000	3,690	2,550	3,080	2,140
		3½ or more		5,030	3,500	4,320	3,000	3,600	2,510

* One connector unit consists of one split ring with its bolt in single shear or two shear plates back to back in the contact faces of a timber-to-timber joint with their bolt in single shear.

† Allowable loads for all loadings, except wind, should not exceed 2,900 lb for 2⅝-in shear plates; 4,400 and 6,000 lb for 4-in shear plates with ¾ and ⅞-in bolts, respectively; multiply values by 1.33 for wind loading.

NOTE: 1 m = 2.54 cm; 1 lb = 4.45 N.

Table 12.2.12 Allowable Loads in Pounds for Common Nails in Side Grain* of Seasoned Wood

Type of load	Specific gravity G	Size of nail d	6	8	10	12	16	20	30	40	50	60
		Length, in	2	2½	3	3¼	3½	4	4½	5	5½	6
		Diam, in	0.113	0.131	0.148	0.148	0.162	0.192	0.207	0.225	0.244	0.263
Withdrawal load per in penetration	0.31		9	10	12	12	13	15	16	18	20	21
	0.40		16	18	20	20	22	27	28	31	33	35
	0.44		20	23	26	26	29	34	37	40	43	46
	0.47		24	27	31	31	34	40	43	47	51	55
	0.51		29	34	38	38	42	49	53	58	63	68
	0.55		34	41	46	46	50	59	64	70	76	81
	0.67		57	66	75	75	82	97	105	114	124	133
Lateral load*†	0.60–0.75		78	97	116	116	132	171	191	218	249	276
	0.50–0.55		63	78	94	94	107	139	154	176	202	223
	0.42–0.50		51	64	77	77	88	113	126	144	165	182
	0.31–0.41		41	51	62	62	70	91	101	116	132	146

* The allowable lateral load for nails driven in end grain is two-thirds the values shown above.

† The minimum penetration for full lateral resistance for the four groups listed is 10, 11, 13, and 14 diam from higher to lower specific gravities, respectively. Reduce by interpolation for lesser penetration; minimum penetration is one-third the above.

NOTE: 1 in = 2.54 cm; 1 lb = 4.45 N.

The following rules should be observed: (1) the size of the lead hole in soft (hard) woods should be about 70 (90) percent of the core or root diameter of the screw; (2) lubricants such as soap may be used without great loss in holding power; (3) long, slender screws are preferable generally, but in hardwood too slender screws may reach the limit of their tensile strength; (4) in the screws themselves, holding power is favored by thin sharp threads, rough unpolished surface, full diameter under the head, and shallow slots.

The allowable **withdrawal load of lag screws** in side grain is given by the formula $p = 1,800D^{3/4}G^{3/2}$, allowable load per inch of penetration of threaded portion of lag screw into member receiving the point, lb;

D = shank diameter of lag screw, in; G = specific gravity of ovendry wood. Use of lag screws loaded in withdrawal from end grain should be avoided. The allowable load in such cases should not exceed 75 percent of that for side grain (see also Sec. 8).

The allowable **lateral resistance of lag screws** for parallel-to-grain loading with screws in side grain is proportional to D^2 and is dependent on species and type of side member. Selected values are given in Table 12.2.13 for one lag screw in single shear in a two-member joint.

Lead holes for lag screws (approximately 75 percent of shank diameter) should be prebored for the threaded portion. Lead holes for the shank should be of the same diameter and length as that of the

Table 12.2.13   Allowable Lateral Loads in Pounds on Lag Bolts or Lag Screws

Side member	Length of bolt, in	Diam of bolt at shank, in	Overdry specific gravity of species							
			0.60–0.75		0.51–0.55		0.42–0.50		0.31–0.41	
			∥	⊥	∥	⊥	∥	⊥	∥	⊥
1½-in wood	4	¼	200	190	170	170	130	120	100	100
	4	½	390	250	290	190	210	140	170	110
	6	⅜	480	370	420	320	360	280	290	220
	6	⅝	860	510	710	430	510	310	410	250
2½-in wood	6	½	620	410	470	310	340	220	270	180
	6	1	1,040	520	790	390	560	280	450	230
	8	¾	1,430	790	1,080	600	780	430	620	340
	8	1	1,800	900	1,360	680	970	490	780	390
½-in metal	3	¼	240	185	210	160	155	120	125	100
	3	½	550	285	415	215	295	155	240	125
	6	½	1,100	570	945	490	770	400	615	320
	6	¾	1,970	865	1,480	650	1,060	460	850	370
	10	⅞	3,420	1,420	2,960	1,230	2,340	970	1,890	785
	12	1	4,520	1,810	3,900	1,560	3,290	1,320	2,630	1,050
	16	1¼	7,120	2,850	6,150	2,460	5,500	2,200	4,520	1,810

NOTE: 1 in = 2.54 cm; 1 lb = 4.45 N.

unthreaded shank. Soap or other lubricant should be used to facilitate insertion and to prevent damage to the screw. Where steel-plate side pieces are used, the allowable loads given by the formula for parallel-to-grain loading may be increased by 25 percent.

The ultimate **withdrawal load per linear inch of penetration of a round drift bolt or pin** from side grain when driven into a prebored hole having a diameter ⅛ in less than that of the bolt diameter may be determined from the formula $p = 6,000G^2D$, where p = ultimate withdrawal load of penetration, lb/lin in; G = specific gravity of ovendry wood; D = diameter of drift bolt, in. A safety factor of about 5 is suggested for general use. The allowable load in lateral resistance for a drift bolt should ordinarily be taken as less than that for a common bolt.

STEEL CONSTRUCTION

(Note: In the design of steel structures, 1,000 lb is frequently designated as a kilopound or "kip," and a stress of 1 kip per square inch is designated as 1 ksi.)

Structural steel design was based only on the allowable stress design (ASD) approach until the introduction of the load and resistance factor design (LRFD) technique in the mid-1980s. The LRFD approach is an ultimate strength design approach, similar to that adopted by the American Concrete Institute for concrete design. Both ASD and LRFD are accepted in current codes. The ASD method is still the most commonly used for design.

Specifications   The following are in part condensed excerpts from the Specifications of the American Institute of Steel Construction.

Material   Ordinary steel for rolled shapes, plates, and bars is typically specified by ASTM A36, with a yield stress of 36,000 lb/in² (248.2 MPa).

However, advances in mill production methods have resulted in most steels satisfying the higher strength requirements of ASTM A572, Grade 50, leading to increased use of higher-strength steel at little or no cost premium. Other higher-strength steels used in structures are A440, A441, A588, and A242. Steel materials for pipe and tube, specified by ASTM A53 (welded-seam pipe) and A500 (cold-formed), have yield strengths of 33,000 to 50,000 lb/in² (227.5 to 344.7 MPa).

Ordinary unfinished machine bolts are specified by A307. Bolts used for structural steel connections are typically high-strength bolts specified by A325 or A490. Riveting is now rarely used, but may often be encountered in older structures. The most common rivets are A502, Grade 1.

Allowable Stresses in A36 Steel

	lb/in²	MPa
Tension F_t:		
On gross section	22,000	151.6
On net section, except at pinholes	29,000	200
On net section, at pinholes	16,000	110.2
Compression F_c: See Table 12.2.14		
Bending tension and compression on extreme fibers F_b:		
Basic stress, reduced in certain cases	22,000	151.6
Compact, adequately braced beams	24,000	165.3
Rectangular bearing plates	27,000	186.0
Shear F_v: Web of beams, gross section	14,500	99.9

Table 12.2.14   Allowable Stress, ksi, for Compression Members of A36 Steel

Main and secondary members, Kl/r not more than 120						Main members, Kl/r, 121–200			
$\dfrac{Kl}{r}$	F_a	$\dfrac{Kl}{r}$	F_a	$\dfrac{Kl}{r}$	F_a	$\dfrac{Kl}{r}$	F_a	$\dfrac{Kl}{r}$	F_a
1	21.56	41	19.11	81	15.24	121	10.14	161	5.76
5	21.39	45	18.78	85	14.79	125	9.55	165	5.49
10	21.16	50	18.35	90	14.20	130	8.84	170	5.17
15	20.89	55	17.90	95	13.60	135	8.19	175	4.88
20	20.60	60	17.43	100	12.98	140	7.62	180	4.61
25	20.28	65	16.94	105	12.33	145	7.10	185	4.36
30	19.94	70	16.43	110	11.67	150	6.64	190	4.14
35	19.58	75	15.90	115	10.99	155	6.22	195	3.93
40	19.19	80	15.36	120	10.28	160	5.83	200	3.73

NOTE: 1 ksi = 6.89 MPa.

Allowable Stresses* in Riveted and Bolted Connections

	lb/in^2	MPa
Bearing: A36 steel		
Pins in reamed, drilled, or bored holes	32,400	223.3
Bolts and rivets	69,000	475.7
Roller, lb/lin in (N/lin cm)	760 × diam	1,131 × diam
	(in)	(cm)
Shear: bearing-type connections†		
A502, grade 1 hot-driven rivets	17,500	120.6
A307 bolts	10,000	68.9
A325 bolts when threading is	30,000	206.8
excluded from shear planes		
(std. holes)		
A325 bolts when threading is not	21,000	144.7
excluded from shear planes		
(std. holes)		
Shear: friction-type connections†		
(with threads included or excluded		
from shear plane)		
A325 bolts in standard holes	17,500	120.6
A325 bolts in oversized or short	15,000	103.4
slotted holes		
A325 bolts in long slotted holes	12,000	82.6
Tension:		
A502, grade 1, hot-driven rivets	23,000	158.5
A307 bolts	20,000	137.9
A325 bolts	44,000	303.3
Bending in pins of A36 steel	27,000	186.1

* Allowable stresses are based on nominal body area of fasteners unless indicated.

† Rivets or bolts may not share loads with welds on bearing-type connections but may do so in friction-type connections.

Proportion of Parts

Most simple beams, columns, and truss members are proportioned to limit the actual stresses to the allowable stresses stipulated above. Other stability or serviceability criteria may control the design. **Deflection** may govern in such members as cantilevers and lightly loaded roof beams. **Buckling,** rather than strength, may govern the design of compression members. The slenderness ratio Kl/r, where Kl is the effective length of the member and r is its radius of gyration, should be limited to 200 in compression members, and L/r limited to 300 in tension members. Kl should not be less than the actual unbraced length l in columns of a frame which depends on its bending stiffness for lateral stiffness. **Width-thickness ratios** are specified for projecting elements under compression. Repeated fluctuations in stress leading to fatigue may be a controlling factor. Rules are given for **combined stresses** of tension, compression, bending, and shear.

Tension members should be proportioned for the gross and net section, deducting for bolt or rivet holes ⅛ in (0.3 cm) larger than the nominal diameter of the fastener.

Columns and other **compression members** subject to eccentric load or to axial load and bending are governed by special rules. A long-established rule is that $f_a/F_a + f_b/F_b$ should be equal to or less than unity, where f_a is the axial stress, f_b the bending stress, and F_a and F_b are the corresponding allowable stresses if axial or bending stress alone exist. This is still considered valid when f_a/F_a is less than 0.15. Joints shall be fully spliced, except that where reversal of stress is not expected and the joint is laterally supported, the ends of the members may be milled to plane parallel surfaces normal to the stresses and abutted with sufficient splicing to hold the connected members accurately in place. Column bases should be milled on top for the column bearing, except for rolled-steel bearing plates 4 in (10 cm) or less in thickness.

Beams and girders, of rolled section or built-up, should in general be sized such that the bending moment M divided by the section modulus S is less than the allowable bending stress F_b. For built-up sections a rule of thumb is, for A36 steel, flanges in compression should have a thickness of 1/16 the projecting half width, and webs should have a thickness of 1/320 the maximum clear distance between flanges. Web stiffeners

should be provided at points of high concentrated loads; additional web stiffeners are required in plate girders. Splices in the webs of plate girders should be made by plates on both sides of the web. When two or more rolled beams or channels are used side by side to form a beam, they should be connected at separators spaced no more than 5 ft (1.52 m); beams deeper than 12 in (30 cm) are to have at least two bolts to each separator.

The **lateral force on crane runways** due to the effect of moving crane trolleys may be assumed as 20 percent of the sum of the weights of the lifted loads and of the crane trolley (but exclusive of the other parts of the crane) applied at the top of the rail. When two or more rolled beams or channels are used side by side to form a beam, they way, and shall be considered as acting in either direction normal to the runway rail. The **longitudinal force** may be assumed as 10 percent of the maximum wheel reactions of the crane applied at the top of the rail.

Bolted or riveted connections carrying calculated stress, except lacing and sag bars, should be designed to support not less than 6,000 lb (27.0 kN). Rivets or high-strength bolts are preferred in all places, and both are implied in these paragraphs wherever "bolting" is mentioned; unfinished bolts, A307, may be used in the shop or in field connections of lightly loaded structures, or for secondary members, bracing, and the like.

Members in tension or compression, meeting at a joint, shall have their lines of center of gravity pass through a point, if practicable; if not, provision shall be made for the eccentricity. A group of bolts transmitting stress to a member shall have its center of gravity in the line of the stress, if practicable; if not, the group shall be designed for the resulting eccentricity. Where stress is transmitted from one member to another by bolts through a loose filler greater than ¼ in thickness, except in slip critical connections using high-strength bolts, the filler shall be extended beyond the connected member and the extension secured by enough bolts or sufficient welding to distribute the total stress in the member uniformly over the combined sections of the member and the filler.

Most bolted connections transmit shearing forces by developing the shearing or bearing values of the bolts, but bolts in certain connections, such as shelf angles and brackets, are required to transmit tension forces.

Bolts shall be proportioned by the nominal diameter. Rivets and A307 bolts whose grip exceeds 5 diam shall be allowed 1 percent less safe stress for each 1/16 in (0.16 cm) excess length. The minimum distance between centers of bolt holes shall be 2⅔ diam of the bolt; but preferably not less than 3 diam.

The minimum distance from the center of any bolt hole to a sheared edge shall be 2¼ in (5.7 cm) for 1¼-in (32-mm) bolts, 2 in (5.1 cm) for 1⅛-in (28-mm) bolts, 1¾ in (4.4 cm) for 1-in (25-mm) bolts, 1½ in (3.8 cm) for ⅞-in (22-mm) bolts, 1¼ in (3.2 cm) for ¾-in (19-mm) bolts, 1⅛ in (2.8 cm) for ⅝-in (16-mm) bolts, and ⅞ (2.22 cm) for ½-in (13-mm) bolts. The distance from any edge shall not exceed 12 times the thickness of the plate and shall not exceed 6 in (15 cm).

Design of Members

Properties of Standard Structural Shapes Detailed dimensions, geometric properties, and other data for standard rolled sections are listed in the AISC Manual of Steel Construction. This reference is usually the starting point when one is embarking on a design utilizing steel shapes. For sizes not standard, or for *unusual (generally much higher)* geometric properties required for special applications, an endless variety of beam and column shapes can be fabricated by welding the desired sections from plate stock.

A great variety of tees is produced by shearing or gas-cutting standard beams (S shapes) or wide-flange sections (W shapes) lengthwise at midheight of the web, making two T sections.

Welding The main advantage of assembling and connecting steel frames by welding is the reduction in the amount of metal used. The saving in metal is achieved by (1) elimination of bolt holes which reduce the net section of tension members, (2) simplification of details, and (3) elimination of splice plate and gusset plate material. (See also Sec. 13.3.)

Allowable stresses in welds depend on the type of weld, the manner of loading, and the relative strengths of the weld metal and the base metal. For **complete-penetration groove welds** in which the full edge thickness of the thinner part to be joined is beveled in preparation for welding, allowable stresses due to tension or compression normal to the effective area or parallel to the axis of the weld are the same as the base metal, and the allowable shear stress on the effective area is 0.3 times the nominal tensile strength of the weld metal (limited by 0.4 times yield stress in the base metal). The effective area for a complete-penetration groove weld is the width of the part joined times the thickness of the thinner part.

For **partial-penetration groove welds**, allowable stresses due to compression normal to the effective area and tension or compression parallel to the axis of the weld are the same as the base metal. For tension normal to the effective area, the allowable stress is 0.3 times the nominal tensile strength of the weld metal (limited by 0.6 times the yield stress of the base metal); for shear parallel to the axis of the weld, the allowable stress on the effective area is 0.3 times the nominal tensile strength of weld metal (limited by 0.4 times yield stress of the base metal). The effective thickness of partial-penetration groove welds depends on the welding process, welding position, and the included angle of the groove but may be safely taken as the depth of the chamfer less 1/8 in.

For **fillet welds**, allowable shear stresses on the effective area are taken as 0.3 times nominal tensile strength of the weld (limited by 0.4 times yield stress of the base metal), and for tension or compression parallel to the axis of the weld, allowable stresses are the same as the base metal. Fillet welds are not to be loaded in tension or compression normal to the effective area; they transfer loads between members in shear only. The effective area of a fillet weld is the overall length of the full-size weld times the shortest distance from the root to the face (normally leg size $\times \sin 45°$). Allowable shear in a fillet weld is taken as 930 lb/in of length (163 N/mm) for each 1/16 in (1.59 mm) of leg size. (See Sec. 13.) Typical details of welded connections are indicated in Fig. 12.2.18.

Fig. 12.2.18   Welded connections. (*a*) Moment connection; (*b*) bracing connection.

Safe Loads for Steel Beams

To determine the safe load uniformly distributed, as limited by bending, for a structural steel beam on a given span, apply the formula $W = 8F_b s/l$, where W is the total load, lb.; F_b is the allowable fiber stress (24,000 lb/in^2 or any other); S is the section modulus for the beam in question, and l is the span, in. (This formula may also be used with equivalent metric units.) The safe load concentrated at midspan is one-half this amount. For other safe loads, note that $F_b S$ is the safe resistance to bending in inch-pounds (or newton-metres) afforded by the beam. Compute the load, of whatever type or distribution, which will produce a maximum bending moment equal to safe moment of resistance (see Sec. 5 for bending-moment formulas).

To select a beam to support a given load, compute the maximum bending moment in inch-pounds, divide by the allowable fiber stress F_b, and refer to the table for a beam having a section modulus which is not smaller than the quotient.

Formulas for the safe loads and deflections of beams with various methods of support and of loading are given in Sec. 5.

Short beams should be investigated for crippling of the web. In the tables are given the safe end reactions for beams of A36 steel resting on a seat 3½ in (9 cm) long along the axis of the beam. Short beams should

also be investigated for shear, by dividing the maximum shear, in pounds, by the area of the web, excluding the flanges.

Single angles used as beams and loaded in the plane of axis *X-X* or *Y-Y* tend to deflect laterally as well as in the plane of the loads. Unless this is prevented, as by pairing the angles back to back and securing them together, the unit fiber stress due to bending may be as much as 40 percent above that computed by dividing the bending moment by S for the axis perpendicular to the plane of the loads. The relation $f = M/S$ does not hold for single angles and for Z bars, which are unsymmetrical about both axes.

Deflection of Beams   Table 12.2.15 gives **coefficients of deflection** for steel shapes under uniformly distributed loads, and is based on the formula; deflection in inches = $30fL^2/Ed$, the table giving the values of $30fL^2/E$. (f = fiber stress, lb/in^2; L = span, ft; d = depth of section, in; E = modulus of elasticity = 29,000,000 lb/in^2.)

To find the deflection in inches of a section symmetrical about the neutral axis, such as a beam, channel, etc., divide the coefficient in the table corresponding to given span and fiber stress by the depth of the section in inches.

To find the deflection in inches of a section which is not symmetrical about the neutral axis but which is symmetrical about an axis at right angles thereto, such as a tee or pair of angles, divide the coefficient corresponding to given span and fiber stress by twice the distance of extreme fiber from neutral axis obtained from table of elements of sections.

To find the deflection in inches of a section for any other fiber stress than that given, multiply this fiber stress by either of the coefficients in the table for the given span and divide by the fiber stress corresponding to the coefficients used.

I beams and channels loaded to a fiber stress of 24,000 lb/in^2 (165.5 MPa) will not deflect in excess of 1/360 of the span (allowed for plastered ceilings) if the depth in inches (cm) is not less than 0.74 (6.21) times the span in feet (m) for uniform loads and 0.60 (5.00) times the span for central concentration.

Beam Supports   Steel beams are supported at the ends generally (1) by means of web connections to girders and columns, (2) by resting on structural-steel seats, or (3) by resting on masonry. Limiting values of end reactions of the second type, for seats 3½ in (9 cm) long, are listed in the AISC Manual of Steel Construction. Standard AISC web connections of the first type are called **framed beam** connections and are designated by the number of rows of bolts. Examples of connections are given in Fig. 12.2.19. These connections may be specified as "Standard 3 row, 4 row, etc., connections." Connections must always be designated and detailed for the calculated design reaction.

The capacity of web connections is governed by the shearing of the fastener, or the bearing of the fastener on the web or on the material to which the beam is connected, or by the strength of the connecting angles. For example, the supporting values of standard framed beam connections, using 7/8 in fasteners in members of A36 material, are given in Table 12.2.16. For fasteners in webs thicker than 0.34 in use the values in the column headed Double Shear; for thinner webs, bearing limits the value, and the coefficients for web bearing are to be used. For ¾-in fasteners, multiply tabular bearing values by 6/7 and shear values by 36/49. Fasteners connecting the outstanding legs to the supporting metal are in double shear if two beams are framed opposite or in single shear if a beam is connected on one side only. If the supporting material of A36 steel is thinner than 0.34 in in double-shear connections or thinner than 0.17 in in single-shear connections, the capacity is limited by bearing. The value of any 7/8-in fastener in bearing on A36 material is 60,900t, where t is the thickness of the plate. The value of 7/8 in A502, grade 1 rivets or A325 HS bolts (slip-critical connections) is 9,020 lb in single shear and 20,400 lb in double shear. The corresponding values for A307 unfinished bolts are 6,000 lb and 12,000 lb, respectively.

Cast-iron columns were often used in the past instead of wood, to save space, in the lower stories of heavy buildings, but are obsolete. Occasionally they are encountered in repair and alteration work in older buildings. The ratio of length to least radius of gyration l/r should not exceed 70, and the average unit stress under axial compression should not exceed $9,000 - 40l/r$ lb/in^2.

Table 12.2.15   Coefficients of Deflection for Simply Supported Steel Beams under Uniformly Distributed Loads

Span, ft	Fiber stress of 24,000 lb/in²	Span, ft	Fiber stress of 24,000 lb/in²	Span, ft	Fiber stress of 24,000 lb/in²	Span, ft	Fiber stress of 24,000 lb/in²
1	0.026	14	4.87	27	18.1	39	37.7
2	0.098	15	5.59	28	19.5	40	39.8
3	0.223	16	6.36	29	20.9	41	41.8
4	0.398	17	7.18	30	22.4	42	43.9
5	0.621	18	8.04	31	23.9	43	45.8
6	0.892	19	8.97	32	25.4	44	48.0
7	1.23	20	9.93	33	27.0	45	50.4
8	1.59	21	10.9	34	28.7	46	52.6
9	2.01	22	12.1	35	30.5	47	54.7
10	2.48	23	13.1	36	32.2	48	57.1
11	3.00	24	14.3	37	34.1	49	59.5
12	3.58	25	15.6	38	35.8	50	62.2
13	4.20	26	16.8				

NOTE: For a load concentrated at midspan, use ⅘ of the coefficient given. 1 ft = 0.305 m; 1 lb/in² = 6.89 kPa.

Fig. 12.2.19   Framed beam connections.

Steel joists consisting of lightweight rolled sections, thin for their height, or open-web trussed members fabricated by welding or otherwise, are used with economy in buildings where spans are long and loads are light, and where a plaster ceiling affords sufficient fire protection. They are rarely used in industrial buildings, except to support roof loads.

Steel pipe is often used for columns under light loads. Table 12.2.17 gives the safe loads on standard size pipes (ASTM A501 or A53, grade B) used as columns. For extra-strong and double extra-strong

Table 12.2.16   Values of Standard Framed-Beam Connections
(⅞-in A325 HS bolts in standard holes,* A36 members)

AISC designation	Two angles thickness × length	Shear 1,000 lb	Bearing on beam web (t), 1,000 lb
10 rows	5/16 × 2′5½″	204	609t
9 rows	5/16 × 2′2½″	184	548t
8 rows	5/16 × 1′11½″	164	487t
7 rows	5/16 × 1′8½″	143	426t
6 rows	5/16 × 1′5½″	123	365t
5 rows	5/16 × 1′2½″	102	304t
4 rows	5/16 × 0′11½″	81.8	243t
3 rows	5/16 × 0′8½″	61.3	182t
2 rows	5/16 × 0′5½″	40.9	121t

* Values indicated are for slip-critical connections or bearing type where threads are not excluded from the shear plane. For bearing-type connections where threads are excluded from the shear plane, shear values may be increased by 1.47.

† If the web of the supporting beam is thinner than 0.17 in (0.42 in if beams frame on both sides), bearing must also be investigated.

NOTE: 1 in = 2.54 cm; 1 lb = 4.45 N.

pipe used as columns, the safe loads will increase approximately in the same proportion as the weight per foot. (See Sec. 8.7.)

Structural steel tubing (ASTM A500, grade B), in square or rectangular cross section with F_y = 46 ksi (317.1 MPa), is also used for columns, bracing members, and unbraced members subjected to large torsional loads. The closed box shape makes tube sections especially suited for resisting torsional loads.

Corrugated metal deck and siding is used for roofs and walls, respectively, to span between purlins for roof loads or between girts for wind loads. The decking is sized to resist the bending caused by these loads. Roof decking is often also used as a diaphragm to transfer wind or seismic loads to the lateral bracing system below. Load tables specifying safe loads for different spans are available from metal deck manufacturers.

Lightweight, cold-rolled, galvanized steel sheet metal structural sections are used when the design will allow. A popular use is as a substitute for wood framing (wall studs and light beams). Assembly is commonly through use of self-drilling, self-tapping screws with corrosion-resistant plating, although the thicker product is amenable to welding. The most common cross sections rolled are shown in Fig. 12.2.20. Wall stud sections are manufactured with punch-outs along their length to facilitate easy installation of electric wiring, piping, etc. The sections are fabricated per ASTM 653 specifications and come in minimum yield strengths of 33 and 50 ksi. Some of the larger sections are used as beams and columns when used as-rolled or when assembled into stiffer cross sections. (See Steel Stud Manufacturers Association literature.)

Fire Resistance   The resistance to fire of building materials has been tested extensively by various agencies. Table 12.2.18 gives the fire-resistance rating of a few of the common building materials and methods of construction as established by the Uniform Building Code from standard fire tests.

Table 12.2.17 Safe Axial Loads for Standard (Schedule 40) Steel Pipe Columns, kips
(Stress according to AISC specification for A501 pipe*)

Nominal pipe size, in	Outside diam, in	Wall thickness, in	6	7	8	9	10	11	12	14	16	18	20
						Effective length of column K1, ft							
3	3.500	0.216	38	36	34	31	28	25	22	16	12	10	
3½	4.000	0.226	48	46	44	41	38	35	32	25	19	15	12
4	4.500	0.237	59	57	54	52	49	46	43	36	29	23	19
5	5.563	0.258	83	81	78	76	73	71	68	61	55	47	39
6	6.625	0.280	110	108	106	103	101	98	95	89	82	75	67
8	8.625	0.322	171	168	166	163	161	158	155	149	142	135	127
10	10.750	0.365	246	243	241	238	235	232	229	223	216	209	201
12	12.750	0.375	303	301	299	296	293	291	288	282	275	268	261

* Yield stress is 36 ksi (248.2 MPa). Pipe ordered to ASTM A53, type E or S grade B, or to API standard 5L grade B will have a yield point of 35 ksi (241.3 MPa) and may be designed at stresses allowed for A501 pipe.

NOTE: 1 in = 2.54 cm; 1 ft = 0.305 m. For dimensions of standard pipe see Sec. 8. Safe loads above underscore lines are for values of Kl/r more than 120 but not over 200.

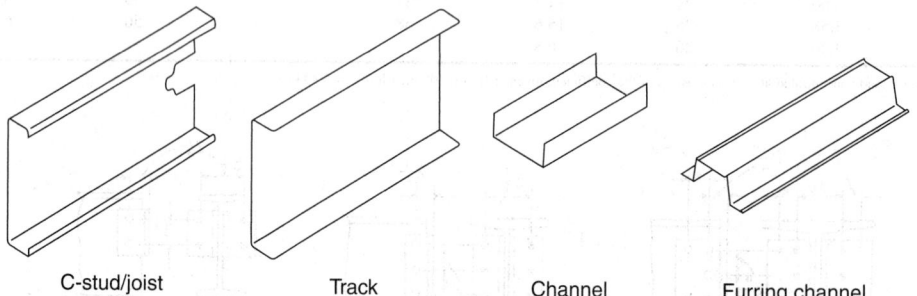

| C-stud/joist | Track | Channel | Furring channel |

Fig. 12.2.20 Light-gage sheet metal cross sections.

Table 12.2.18 Selected Fire-Resistance Ratings

Type	Details of construction	Rating	Type	Details of construction	Rating
Reinforced-concrete beams and girders	Grade A concrete, 1½ in clear to reinforcement	4 h	Wood joists	Wood floor; 1 in tongue-and-groove subfloor and 1 in finish floor with asbestos paper between. Ceiling of ⅝ in Underwriters' Laboratories listed wallboard	1 h
	Grade B concrete, 1½ in clear to reinforcement	3 h			
Steel beams, girders, and trusses	2½ in cover to steel	4 h	Brick walls	Solid walls, unplastered, with no combustible members framed in wall:	
	1 in gypsum-perlite plaster on metal lath, 1¼ in clear of steel	3 h		8 in nominal	4 h
	Ceiling of 1⅛ in gypsum-perlite plaster on metal lath with 2½ in min air space between lath and structural members	4 h		4 in nominal	1 h
Reinforced concrete columns	Grade A concrete 1½ in clear to reinforcement; 12-in columns or larger	4 h	Concrete masonry units	8 in Underwriters' Laboratories listed concrete blocks, laid as specified in Underwriters' Laboratories listing	4h
	Grade B concrete 2 in clear to reinforcement; 12-in columns or larger	4 h		4 in Underwriters' Laboratories listed concrete blocks; laid as specified in Underwriters' Laboratories listing	3 h
Steel columns, 8 × 8 in or larger	Concrete (siliceous gravel):		Steel-stud partitions	¾ in gypsum-perlite plaster both sides on metal lath	2 h
	2½ in clear to steel	4 h		⅝ in gypsum wallboard on 3⅝-in steel studs; attached with 6 d nails; joints taped and cemented	2 h
	2 in clear to steel	3 h			
	1 in clear to steel	2 h			
	1⅛ in gypsum-perlite plaster on metal lath spaced from flanges with 1¼-in steel furring channels	4 h	Wooden-stud partitions	Exterior walls: one side covered with ½ in gypsum sheathing and wood siding; other side faced with ½ in gypsum-perlite plaster on ⅜-in perforated gypsum lath	1 h
	⅞ in portland-cement plaster on metal lath over ¾-in channels	1 h		Interior Walls: 2 × 4 in studs with ⅝ in gypsum wallboard on each side	1 h
Reinforced concrete slabs	5 in concrete (expanded clay, shale, slate, or slag) 1 in clear to reinforcement	4 h	Plain or reinforced concrete walls	Solid, unplastered:	
				7 in thick	4 h
	6½ in concrete (all other aggregate) 1 in clear to reinforcement	4 h		6½ in thick	3 h
				5 in thick Grade B Grade A	2 h
Heavy-timber floors	3 in tongue-and-groove plank floor with 1 in finish flooring	1 h		3½ in thick	1 h

Grade A concrete is made with aggregates such as limestone, calcareous gravel, trap rock, slag, expanded clay, shale, or slate or any other aggregates possessing equivalent fire-resistance properties.

Grade B concrete is all concrete other than Grade A concrete and includes concrete made with aggregates conaining more than 40 percent quartz, cherts, or flint.

NOTE: 1 in = 2.54 cm.

SOURCE: Uniform Building Code.

12.3 REINFORCED CONCRETE DESIGN AND CONSTRUCTION
by William L. Gamble

REFERENCES: Breen et al., "Reinforced Concrete Fundamentals," Wiley. Nilson, "Design of Concrete Structures," McGraw-Hill. Lin and Burns, "Design of Prestressed Concrete Structures," Wiley. Park and Gamble, "Reinforced Concrete Slabs," Wiley. "Building Code Requirements for Structural Concrete," "Commentary on Building Code Requirements for Structural Concrete," "Formwork for Concrete, SP-4," and "Manual of Standard Practice for Detailing Reinforced Concrete Structures," American Concrete Institute. "Standard Specifications for Highway Bridges," American Association of State Highway and Transportation Officials (AASHTO). "Minimum Design Loads for Buildings and Other Structures (SEI/ASCE 7)," American Society of Civil Engineers. "Uniform Building Code (UBC)," International Conference of Building Officials. "PCI Design Handbook—Precast and Prestressed Concrete," Precast/Prestressed Concrete Institute.

The design, theory, and notation of this chapter are in general accord with the 2005 Building Code Requirements for Structural Concrete of the American Concrete Institute, although many detailed provisions have been omitted.

Standard Notation

Load Factors

D = dead load of structure or force caused by dead load
E = earthquake load or force
L = live load of structure or force caused by live load
W = wind load or force
U = required strength of structure to resist design ultimate loads
ϕ = understrength or capacity reduction factor

Beams and General Notation

a = depth of compression zone, using approximate method
A_b = area of individual reinforcing bar, in^2
A_{ps} = area of prestressing steel
A_s = area of tension reinforcement
A_s' = area of compression reinforcement
A_v = area of steel in one stirrup
b = width of compression face of beam
b_w = width of stem of T beam
c = $k_u d$ = depth to neutral axis at ultimate, from compression face
d = effective depth of beam, compression face to centroid of tension steel
d' = depth of compression steel, from compression face
d_b = diam of individual reinforcing bar, in
E_c = Young's modulus of concrete
E_s = Young's modulus of steel
f_c' = compressive strength of concrete from tests of 6 × 12 in cylinders, lb/in^2
$\sqrt{f_c'}$ = measure of shear and tensile strength of concrete, lb/in^2, i.e., if f_c' = 4,900 lb/in^2, then $\sqrt{f_c'}$ = 70 lb/in^2
f_{pu} = ultimate stress of prestressing steel, lb/in^2
f_y = yield stress of reinforcing steel, lb/in^2
h = overall height or thickness of member
k_u = ratio of ultimate neutral axis depth to effective depth
l_d = development length of reinforcing bar, in
s = spacing of shear reinforcement
M_n = nominal moment capacity of ideal section
M_u = ultimate moment of section = ϕM_n or required ultimate moment
v_c = shear stress in concrete
V_c = shear force resisted by concrete
V_s = shear force resisted by web reinforcement
V_u = shear strength of section or required ultimate shear

β_1 = factor relating neutral axis position to depth of equivalent approximate stress block (see Fig. 12.3.2)
ε_{su} = reinforcement strain at time of failure of member
ε_y = yield strain of reinforcement
ρ = tension reinforcement ratio = A_s/bd
ρ' = compression reinforcement ratio = A_s'/bd
ρ_{bal} = balanced reinforcement ratio

Columns

$A_c = A_g - A_s$ = net area of concrete in cross section
A_{core} = area within spiral
A_g = gross area of column
A_s = area of steel in column
e = eccentricity of axial load on column
$M = Pe$ = applied bending moment
P = axial thrust
P_0 = failure load of short column under concentric load
ρ_g = gross steel ratio = A_s/A_g
ρ_s = spiral steel ratio = volume of spiral steel/volume of core
ε_t = strain, extreme steel layer at nominal strength

Floor Systems

b_0 = effective shear perimeter around column in flat plate, flat slab, or footing
c_1 = width of supporting column or capital, in direction of span being considered
c_2 = width of supporting column, in transverse direction
I_b = moment of inertia of beam
K_b = flexural stiffness of beam, moment per unit rotation
K_c = flexural stiffness of columns at joint, moment per unit rotation
K_s = flexural stiffness of slab of width l_2, moment per unit rotation
l_1 = span, center to center of supports, in direction considered
l_2 = span, center to center of supports, in transverse direction
$l_n = l_1 - c_1$ = clear span in direction considered
M_0 = static moment
q = distributed design load, including load factors
α_1 = EI of beam in direction 1/EI of slab of width l_2

Footings

A_1 = loaded area
A_2 = area of same shape and concentric with A_1
f_b = ultimate bearing stress on concrete
β = ratio of long side/short side of footing

Walls

l_c = unbraced height of wall

Prestressed Concrete

A_{ps} = area of prestressed reinforcement
A_t = area of anchorage-zone reinforcement
f_{ci}' = compressive strength of concrete at time of prestressing
f_{ps} = stress in reinforcement at time of failure of member
f_{pu} = ultimate stress of prestressing reinforcement
$f_{se} = f_{si} - \Delta f_s$ = stress in reinforcement at service load
f_{si} = reinforcement stress at time of tensioning steel
Δf_s = loss of prestress from initial tensioning value
$\rho_p = A_{ps}/bd$

MATERIALS

Reinforced concrete is a combination of concrete and steel acting as a unit because of bond between the two materials. Concrete has a high compressive strength but a relatively low tensile strength. Beams of plain concrete fail by tension at very low stresses, but if properly reinforced by embedment of steel in their tensile regions, they may be loaded to utilize the much higher compressive strength of the concrete. Reinforced concrete structures are practically monolithic, are more rigid than steel structures of the same strength, and are inherently fire-resistant. Reinforcement in the concrete also controls cracking caused by temperature changes and shrinkage.

Prestressed concrete is a form of reinforced concrete in which initial stresses opposite those caused by the applied loads are induced by tensioning high-strength steel embedded in the concrete. Members may be *pretensioned,* in which the steel is tensioned and the concrete then cast around it, or *posttensioned,* in which the concrete is cast and cured, after which steel placed in ducts through the concrete is tensioned.

Concrete For reinforced concrete work, only high-quality portland cement concrete may be used, and the aggregates must be carefully selected. The proportions are governed by the required strength, durability, economy, and the quality of the aggregates. Concretes for building construction normally have compressive strengths of 3,000 to 5,000 lb/in² (20 to 35 MPa), except that concrete for columns may be considerably stronger, with 12,000 to 14,000 lb/in² common in some geographic areas. Most concrete for prestressed members will have 5,000 lb/in² or higher compressive strength. The higher-strength concretes require thorough quality control if the strengths are to be consistently obtained. Generally, lean, harsh mixes should be avoided because the bond with the steel will be poor, permeability will be high, and durability of the concrete may be poor. A consistency of concrete that will flow sluggishly but not so wet as to produce segregation of the materials when transported must be used for all reinforced concrete work in order to embed the steel and completely fill the molds or forms. The use of vibration is almost mandatory, and enables the use of stiffer, more economical, concretes than would otherwise be possible (see also Sec. 6.9).

Steel Reinforcing steel may be deformed or plain bars, welded-wire fabric, or high-strength wire and strand for prestressed concrete. Bars with deformations on their surfaces are designed to produce mechanical bond and greater adhesion between the concrete and steel, and are used almost universally in the United States. Welded-wire fabric is suitable in many cases for slabs and walls, and may result in cost savings through labor savings. Fabric is made with both smooth and deformed wires, and the deformed fabric may have some advantage in terms of better crack control. Deformed reinforcing bars having minimum yield stresses from 40,000 to 75,000 lb/in² (275 to 517 MPa) are currently manufactured. The higher-strength steels have the advantage of allowing higher working stresses, but their ductility may be less and it may be difficult to cold-bend them successfully, especially in the larger sizes. The chemical makeup of most reinforcing steels is such that it is not readily weldable without special techniques, including careful preheating and controlled cooling. Furthermore, the variation of the material from batch to batch is so great that separate procedures must be devised for each batch. Tack welding in assembling bar cages can be particularly troublesome because of the stress raisers introduced. A weldable steel is produced with the specification ASTM A-706, but it is not widely available. Prestressing steel is heat-treated high-carbon steel, and 7-wire strand will have a breaking stress of 250,000 or 270,000 lb/in² (1,720 or 1,860 MPa). The usual steel is Grade 270, low-relaxation strand. The strength of solid wire for prestressing is slightly less. ASTM specifications cover the various steels, and all steel used as reinforcement should comply with the appropriate specification.

Reinforcing Steel Sizes The sizes of reinforcing bars have been standardized, and the designation numbers are approximately the bar diameter, in 1/8-in units. Table 12.3.1 gives the nominal diameter, cross-sectional area, and perimeter for each bar size. Bars are now

marked in nominal millimeter diameters. Other marks on the bars identify the type of steel, the yield stress (grade), and the manufacturer. The areas of the four most common 7-wire prestressing strands are given below:

Diam, in	Area, in²	
	Grade 250	Grade 270
½	0.144	0.153
0.6	0.216	0.217

Table 12.3.1 Dimensions of Deformed Bars

No.	Diam in	mm	Area, in²	Perimeter, in
3	0.375	10	0.11	1.178
4	0.500	13	0.20	1.571
5	0.625	16	0.31	1.963
6	0.750	19	0.44	2.356
7	0.875	22	0.60	2.749
8	1.000	25	0.79	3.142
9	1.128	29	1.00	3.544
10	1.270	32	1.27	3.990
11	1.410	36	1.56	4.430
14	1.693	43	2.25	5.32
18	2.257	57	4.00	7.09

Plain and deformed wires are used as reinforcement, usually in the form of welded mats. Plain wires are made in sizes W0.5 through W45, where the number designates the cross-sectional area of the wire in hundredths of square inches. Deformed wires are made in sizes D1 through D45, and the number has the same meaning. Not all sizes are made by all manufacturers, and local suppliers should be consulted about availabilities of sizes. The formerly used wire gage numbers are no longer used for size specifications.

Moduli of Elasticity For concrete the modulus of elasticity E_c may be taken as $w^{1.5}33\sqrt{f'_c}$ in lb/in², where w is the weight of the concrete between 90 and 155 lb/ft³. Normal weight concrete may be assumed to weigh 145 lb/ft³. For steel the modulus of elasticity may be taken as 29,000,000 lb/in² (200 GPa), except for prestressing steel for which the modulus shall be determined by tests or supplied by the manufacturer, but is usually about 28,000,000 lb/in².

The **modular ratio** $n = E_s/E_c$ is of importance in designing reinforced concrete. It may be taken as the nearest whole number but not less than six. The value of n for lightweight concrete may be taken as the same for normal weight concrete of the same strength, except in calculations for deflections.

Protection of Reinforcement Reinforcement, for both regular and prestressed concrete, must be protected by the concrete so as to prevent corrosion. The amount of cover needed for various degrees of exposure is as follows:

Member and exposure	Cover, in
Concrete surface deposited against the ground	3
Concrete surface to come in contact with the ground after casting:	
Reinforcement larger than No. 5	2
Reinforcement smaller than No. 5	1½
Beams and columns not exposed to weather:	
Main steel, stirrups, and ties	1½
Joists, slabs, and walls not exposed to weather	¾
Column spirals and ties	1½

The clear cover and clear bar spacings should ordinarily exceed 4/3 times the maximum sized aggregate used in the concrete.

The amount of protection recommended is a minimum, and when corrosive environments or other severe exposure occurs, the cover should be increased. The concrete in the cover should be made as impermeable as possible. Fire-resistance requirements may also control the cover requirements.

LOADS

The dead and live loads are combined in determining the cross sections of the members. The dead load includes the weight of the structure, all finishing materials, and usually the installed equipment. The live load includes the contents and ordinarily refers to the movable items.

The **live load** (pounds per square foot) to be used for design depends upon the loadings that will occur in the particular structure as well as on the requirements of the local building code. The Boston Building Code illustrates good practice and is as follows for floor loads:

Heavy manufacturing, sidewalks, heavy storage, truck garages, 250; public garages, intermediate manufacturing, and hangars, 150; stores, heavy merchandise, light storage, 125; armories, assembly halls, gymnasiums, grandstands, public portions of hotels, theaters, and public buildings, corridors and fire escapes from public assembly buildings, light merchandising stores, stairs, first and basement floors of office buildings, theater stages, 100; upper floors of public buildings, office portions of public buildings, stairs, corridors and fire escapes except from public assembly buildings, theater and assembly halls with fixed seats, light manufacturing, locker rooms, stables, 75; church auditoriums, 60; office buildings above first floor including corridors, classrooms with fixed seats, 50; residence buildings and residence portions of hotels, apartment houses, clubs, hospitals, educational and religious institutions, 40. [Note: 100 lb/ft^2 = 4.79 kPa.]

Live loads affecting structural members supporting considerable tributary floor areas are sometimes reduced in recognition of the low probability that the entire area will be loaded to the full design load at the same time. Roofs are commonly designed to support live loads of 20 to 30 lb/ft^2, with many special provisions for snow loads in SEI/ASCE 7. Wind loads are commonly from 15 to 30 lb/ft^2 (higher in areas subject to hurricanes) of vertical projection. In the case of heavy moving loads, allowance should be made for impact by increasing the live load by 25 to 100 percent.

Dead loads include the weight of both structure and finishing materials, and the weights of some typical wall, floor, and ceiling materials are as follows:

Description	Weight, lb/ft^2
Granolithic finish, per in of thickness	12
$^7/_8$-in hardwood, $1^1/_8$-in plank intermediate floor, and tar base	16
3-in wood block in coal-tar pitch	10
Lightweight concrete fill, 2 in thick	14
Plaster on concrete, tile, or concrete block, two coats	5
Plaster on lath	10
6-in concrete block wall	25
Suspended ceiling	12

SEISMIC LOADINGS

Earthquakes induce forces in structures because of inertial forces which resist the ground motions. These forces have damaged or destroyed large numbers of structures throughout the world. The art and science of designing to resist seismic effects are still evolving, and may be thought of in several steps.

1. Determine the ground motion to be expected at the site of interest. This usually requires consulting a map in ASCE-7 or the prevailing local building code, such as the Uniform Building Code (UBC). Several steps may be required, including finding the expected acceleration of the bedrock, followed by an accounting for the soil types between the bedrock and the structure. There is an implicit or explicit assumption of a probability that this ground motion will not be exceeded in some particular time interval, such as 50 or 100 years.

2. Determine the effects of the ground accelerations on the structure which is being analyzed. This will require consideration of the natural period of vibration of the structure and of structural characteristics such as damping and ductility, and more elaborate analyses will be required for large and important structures than for smaller, less crucial cases.

Member forces, moment, shear, thrust, torsion, and deformations such as story drift and member end rotations will result from this analysis. The major building codes all have specific requirements and provisions for this analysis, and the UBC is the most widely used in seismically active regions. In most cases, an equivalent static horizontal load will be applied to the structure, with the force in the range of 10 percent or less to as much as 30 percent of the mass of the structure.

3. Design and detail members and connections for the imposed forces. Chapter 21 of the ACI Code has many requirements for the detailing of the reinforcement in the members and joints. When compared with designs for static loadings which include only dead, live, and wind loads, seismic designs will have much more lateral reinforcement in columns, and often much heavier shear reinforcement (especially near the joints), and will usually have extra reinforcement within the joint regions. Much of this added reinforcement is referred to as **confinement reinforcement**, and the intent is to utilize the fact that triaxially confined concrete may be able to sustain much greater strains before failure than unconfined concrete. Obviously, steps 2 and 3 are parts of an iterative process which converges to an acceptable structural design solution.

The current seismic design philosophy includes the assumption that it is not economically possible to design structures to resist extremely large earthquakes without damage. It is expected that structures will resist smaller earthquakes, such as might be expected several times during the life of a structure, with minimal damage, while a very large earthquake which might be expected to occur once during a structure's design life would cause significant damage but would leave the structure still standing. The assignment of an importance factor for buildings is one consequence of this design philosophy. Hospitals, school buildings, and fire stations, for example, would be designed to a higher seismic standard than would a three-story apartment building.

LOAD FACTORS FOR REINFORCED CONCRETE

The current ACI Building Code for Reinforced Concrete is based on a **strength design** concept, in which the strength of a member or cross section is the basis of design. This approach has been adopted because of the difficulty in assigning reasonable and consistent allowable stresses to the concrete and steel. Factors of safety are expressed in terms of *overload* and *understrength* factors. Overload factors reflect the uncertainty of the applied loads. The ACI Building Code and SEI/ASCE 7 have many loading combinations. The most common, omitting some of the lesser loads (snow, rain, roof live load), are as follows:
With no wind or earthquake loads:

$$U = 1.2D + 1.6L \tag{12.3.1}$$

With wind acting, use the larger of Eq. (12.3.2) or (12.3.3):

$$U = 1.2D + 0.8W \tag{12.3.2}$$

$$U = 1.2D + 1.6W + 1.0L \quad \text{or} \quad U = 0.9D + 1.6W \tag{12.3.3}$$

No section may be weaker than required by Eq. (12.3.1). Other load combinations include seismic (earthquake) forces, liquids, and earth pressure loadings. The understrength factors reflect the ductility of failure in the mode considered, and the consequences of a failure on the rest of the structure. These are expressed as ϕ factors with the following values:

Bending and tension	$\phi = 0.90$
Shear and torsion	$\phi = 0.75$
Spiral columns	$\phi = 0.70$
Tied columns and bearing	$\phi = 0.65$

These factors are used to reduce the computed ideal strengths of members, reflecting possible weaknesses related to materials, dimensions, and workmanship.

In addition to strength requirements, serviceability checks must be made to ensure freedom from excessive cracking, deflections, vibrations,

Fig. 12.3.1 Typical arrangement of reinforcement in beams.

etc., at working loads. Prestressed-concrete members must also satisfy a set of allowable stresses at this load level.

Deflections are usually controlled by specifying minimum member depths. The values of $l/16$ and $l/21$, for beams which are simply supported and continuous at both ends, respectively, are typical, unless an investigation is conducted to show that shallower members will result in acceptable deflections. Crack control is obtained by careful distribution of the steel through the tensile zone. In this regard, several small bars are superior to one large bar.

The approach to design is not a "limit design" or "plastic design" concept, as the ultimate forces are derived from elastic analyses of structures, using maximums for each critical section. The plastic collapse loads and mechanisms currently used in the "plastic design" of some steel structures in the United States are not considered.

Bridges may be designed using the same general concepts, but different overload and understrength factors are used, and the serviceability requirements include checks on the fatigue strength.

REINFORCED CONCRETE BEAMS

Concrete beams are reinforced to resist both flexural and shear forces. A typical reinforcement scheme for a simply supported beam is shown in Fig. 12.3.1.

As the load on a reinforced concrete beam is increased, vertical tension cracks appear in the maximum moment regions, and gradually grow in length, width, and number. By the time the working load is reached, some of the cracks extend to the neutral axis and the contribution of the tensile strength of the concrete to the flexural capacity of the beam has become very small. As the load is increased further, the reinforcement eventually yields. This load is nearly the maximum the member can sustain. Further attempts at loading produce large increases in deflection and crack widths with very small increases in load and a gradual reduction in the remaining concrete compression area. When a limiting concrete strain of about 0.003 is reached at the compression face of the beam, the concrete starts crushing and the capacity of the beam starts dropping. For static design purposes, the achievement of the 0.003 strain is usually regarded as the end of the useful life of the beam.

Within limits to be checked later, at the time of flexural failure of a beam the stress in the steel is equal to the yield stress, which greatly simplifies the analysis of the member. The strain and stress distributions in a beam are shown in Fig. 12.3.2, at failure.

It is assumed that plane sections remain plane, that adequate bond exists, that the stress-strain relationships for concrete and steel are known, and that tension in the concrete has a negligible influence.

The **flexural strength** of a cross section may be written as, including the understrength factor and rounding the terms in the parentheses to one significant figure:

$$\begin{aligned} M_u &= \phi A_s f_y d (1 - 0.4 k_u) \\ &= \phi A_s f_y d (1 - 0.6 \rho f_y / f_c') \end{aligned} \tag{12.3.4}$$

where $\rho = A_s/bd$ = reinforcement ratio. The reinforcement ratio will ordinarily be between 0.005 and 0.02.

Fig. 12.3.2 Stress and strain distribution in reinforced concrete beam at failure. (*a*) Section; (*b*) strains; (*c*) stresses; (*d*) approximate stresses.

A satisfactory approximate stress distribution is shown in Fig. 12.3.2d. Since $T = C = 0.85f_c'ba$, then $a = A_sf_y/0.85f_c'b$ and the flexural capacity is

$$M_u = \phi A_sf_y(d - a/2) \qquad (12.3.5)$$

It must be demonstrated for each case that the stress in the reinforcement has reached f_y, and the simplest approach is to show that $\varepsilon_{su} \geq \varepsilon_y$. From Fig. 12.3.2b, $\varepsilon_{su} = 0.003(1 - k_u)/k_u = 0.003(d - c)/c$. The ACI Code requires that $\varepsilon_{su} \geq 0.004$ to ensure that at least some yielding, with resultant large deflections, occurs before a member fails.

Beams may contain both compression and tension reinforcement, especially when the beam must be kept as small as possible or when the long-term deflections must be minimized. The forces at ultimate are shown in Fig. 12.3.3, and the moment may be calculated as

$$M_u = \phi[A_s'f_y(d - d') + (A_s - A_s')f_y(d - a/2)] \qquad (12.3.6)$$

where $a = (A_s - A_s')f_y/0.85f_c'b$. These equations assume that the compression steel yields. If it does not, the computed a value will be wrong, but the computed M_u value will be close to the correct value. Again, $\varepsilon_{su} \geq 0.004$ is required. The compression steel strain is $\varepsilon_{su}' = 0.003(k_ud - d')/k_ud$, and the generalized steel stress is $f_s = \varepsilon_sE_s \leq f_y$. A general solution requires adjusting k_ud so that $T = C_c + C_s$ and then summing moments of forces about some convenient axis.

Many reinforced concrete beams are flanged sections, or T beams, by virtue of having a slab cast monolithically with the beam, and such a cross section is shown in Fig. 12.3.4. It will be found that the neutral axis lies within the flange in most instances, and if $k_ud \leq t$, then the beam is treated as a rectangular beam of width b. This is checked using $k_ud = A_sf_y/0.85\beta_1f_c'b$, developed from equilibrium (Fig. 12.3.2). If $k_ud > t$, the flexural capacity is computed by

$$M_u = \phi[(A_s - A_{sw})f_y(d - a/2) + A_{sw}f_y(d - t/2)] \quad (12.3.7)$$

where $A_{sw} = (b - b_w)t0.85f_c'/f_y$; and $a = (A_s - A_{sw})f_y/0.85f_c'b_w$.

Continuous T beams are treated as rectangular beams of width b_w in the negative moment regions where the lower surface of the beam is in compression. Most continuous beams will have compression steel in the negative moment regions, as some bottom steel is always continued into the support regions.

Both T beams and beams with compression steel may be thought of in terms of dividing the beam into two components—one a rectangular beam containing part of the tension steel and the other a couple with the rest of the tension steel at the bottom and either the compression steel or the T-beam flanges at the top. Both components of the beam must satisfy horizontal equilibrium.

The reinforcement ratio should not be less than $\rho_{min} = 3\sqrt{f_c'}/f_y \geq 200/f_y$. This is to ensure that the ultimate moment is somewhat greater than the initial cracking moment. For T beams, $\rho = A_s/b_wd$ for purposes of the minimum steel requirement. This minimum steel ratio applies for concrete strengths up to about 4,000 lb/in². Stronger concretes require slightly larger steel quantities.

Fig. 12.3.3 Stress and strain distributions in a doubly reinforced concrete beam. (a) Section; (b) strains (c) approximate stresses.

Fig. 12.3.4 Strain and stress distribution in a T beam. (a) Section; (b) strains; (c) approximate stresses.

In the selection of a cross section, it is convenient to transform Eq. (12.3.4) by substituting pbd for A_s and rearranging to obtain

$$M_u/\phi bd^2 = \rho f_y(1 - 0.6\rho f_y/f'_c) \qquad (12.3.8)$$

With given materials f_y and f'_c selection of a ρ value enables calculation of the required bd^2, from which a cross section may be selected. For convenience, values of $M_u/\phi bd^2$, are plotted against ρf_y in Fig. 12.3.5, for several values f'_c.

The strengths of prestressed-concrete beams are treated much the same as reinforced concrete beams, and the differences will be noted later.

Fig. 12.3.5 Design factors for single reinforced concrete beams.

In addition to the tension stresses caused by bending forces, **shear forces** cause inclined tension stresses which may lead to inclined cracking such as is sketched in Fig. 12.3.1. Unless web reinforcement is present, the formation of such a crack usually leads directly to the complete collapse of the member at the load which caused the crack, or only slightly higher.

As a result of this undesirable behavior, shear reinforcement is required in all major beams, and the normal design procedure is to proportion the beam for the flexural requirements and then add shear steel, usually in the form of stirrups, to make the shear strength adequate.

The concrete can be assumed to resist a shear stress of $v_c = \phi 2\sqrt{f'_c}$, or a shear force of

$$V_c = \phi 2\sqrt{f'_c}\, bd \qquad (12.3.9)$$

(or $V_c = \phi 2\sqrt{f'_c}\, b_w d$ for T beams). The shear reinforcement must resist the force in excess of V_c, so that

$$V_u - V_c = V_s \qquad (12.3.10)$$

The area of shear reinforcement is then selected to satisfy

$$V_s = \phi A_v f_y d/s \qquad (12.3.11)$$

Shear reinforcement to satisfy this requirement is provided at every section of the beam, except that the region between the support and the section at the distance d from the support is supplied with the steel required at d from the support.

The stirrup spacing should not exceed $d/2$, and is reduced to a maximum of $d/4$ if $V_u > \phi 6\sqrt{f'_c}\, bd$. V_u must not exceed $\phi 10\sqrt{f'_c}\, bd$. The minimum area of shear reinforcement allowed is $A_v = 50 b_w s/f_y$. Closed stirrups of the form shown in Fig. 12.3.1 are recommended, and are essential in areas subjected to earthquake loadings.

Shear in prestressed-concrete members is handled in a similar manner, and the designer is referred to the ACI Code for details. However, it will be found that most prestressed members designed for buildings

and which do not support major concentrated loads will have adequate shear strength if the minimum steel given by the following expression is supplied:

$$A_v = \frac{A_{ps} f_{pu}}{80\ f_y}\frac{s}{d}\sqrt{\frac{d}{b_w}} \qquad \text{in}^2 \qquad (12.3.12)$$

The maximum stirrup spacing is 0.75 of the member depth, or 24 in, and for constant-depth members, d is measured at the section of maximum moment.

At least the minimum area of shear reinforcement must be supplied over the full length of most reinforced and prestressed members unless it can be shown, by a test acceptable to the building official, that the members can sustain the required ultimate loads without the steel.

Recent ACI Codes have undergone major revisions to reflect the effects of bar spacing and bar cover on development length, l_d, and on splice lengths. In some common cases the current development lengths will be much greater than in ACI Codes from 1983 and earlier. The current development lengths for straight bars are given in Code Sec. 12.2.2, and a more complex, less conservative, set of requirements are in Sec. 12.2.3. When the available anchorage lengths are less than l_d, hooks are often used. The conservative Sec. 12.2.2 requirements may be summarized as

$$l_d = K d_b \frac{f_y \psi_t \psi_e \lambda}{\sqrt{f'_c}} \qquad (12.3.13)$$

in which K is as follows:

	No. 6 and smaller bars and deformed wire	No. 7 and larger bars
1. Clear spacing of bars being developed or spliced $\geq d_b$ and clear cover $\geq d_b$ with minimum ties or stirrups or	$1/25$	$1/20$
2. Clear spacing $\geq 2d_b$ and clear cover $\geq d_b$	$1/25$	$1/20$
3. Other cases	$3/50$	$3/40$

The other terms are defined as

d_b = diameter of bar or wire, in
ψ_t = bar location factor
   = 1.3 for top bars (horizontal with more than 12 in of concrete cast below)
   = 1.0 for other bars
ψ_e = coating factor
   = 1.5 for epoxy-coated bars or wires with cover $\leq 3d_b$, or clear spacing $\leq 6d_b$
   = 1.2 for all other epoxy-coated bars or wires
   = 1.0 for uncoated reinforcement
λ = lightweight concrete factor
   = 1.3 when light-weight aggregate concrete is used
   = 1.0 when normal weight aggregate is used

The product $\psi_t \psi_e$ need not be taken greater than 1.7.

Bars No. 11 and smaller are commonly spliced by simply lapping them for a distance. These splices should be avoided in regions of high computed stress, and should be spread out so that not many bars are spliced near the same section. If the computed tensile stress is less than $0.5f_y$ at the design ultimate load and not more than half the bars are spliced at one section, the lap length is $1.0l_d$. For other cases the minimum lap length is $1.3l_d$. For lower stress levels, a lap of l_d is sufficient.

Development lengths in compression, important in columns and compression reinforcement, are somewhat shorter.

Requirements for reinforcement of beams subjected to torsional moments are contained in the ACI Code. The requirements are too complex for discussion here, but the basic reinforcement scheme consists of closed stirrups with longitudinal bars in each corner of the stirrup. The most efficient method of dealing with torsion in many instances will be to rearrange the structure so as to reduce or eliminate the torsional moments.

REINFORCED CONCRETE COLUMNS

Reinforced concrete compression members are proportioned taking into account the applied thrust, the bending moments, and the relationship between length and thickness for the member. The strength of a cross section or a short column can conveniently be shown with the aid of a *moment-thrust interaction diagram* such as is shown in Fig. 12.3.6a, which is drawn without considering the ϕ factor. The variation in ϕ with thrust is shown in Fig. 12.3.6b for the same column. The ϕ factor is a function of the tensile strain in the steel layer farthest from the neutral axis at nominal strength ε_t. The value of $\phi = 0.65$ (for a tied column, 0.7 for a spiral column) when $\varepsilon_t \le 0.002$ tension, and $\phi = 0.9$ when $\varepsilon_t \ge 0.005$. Linear interpolation is used for intermediate strains.

Fig. 12.3.6 Column capacity diagrams. (*a*) Typical moment-thrust interaction diagram; (*b*) variation of ϕ with P.

The load P_0 is the strength of a short column under a concentric load, and is expressed as

$$P_0 = 0.85A_c f'_c + A_s f_y \qquad (12.3.14)$$

The contribution of the concrete is slightly less than the cylinder strength because of differences in workmanship, curing, and position of casting. M_n is simply the strength in flexure, as was discussed for beams. The $M_b - P_b$ point is the "balance point," at which simultaneous crushing of the concrete and yielding of the reinforcement occur. Failure is initiated by crushing of the concrete at loads higher than P_b, and by yielding of the reinforcement in tension at lower loads. Only the reinforcement contributes to the tensile capacity of $T_n = A_s f_y$.

The balance-point moment and thrust are found using the strain and stress distributions shown in Fig. 12.3.7, for a symmetrical section with steel in two faces. From equilibrium,

$$P_b = C_s + C_c - T \qquad (12.3.15)$$

Summing moments about the centroidal axis gives

$$M_b = A_s f_y (d - d')/2 + 0.85\beta_1 f'_c k_u db(h/2 - \beta_1 k_u d/2) \qquad (12.3.16)$$

This assumes that the compression steel has yielded, and this will be true except for very small members or members with exceptionally large cover over the steel. The compression steel strain and stress can be evaluated as noted below Eq. (12.3.6).

The portion of the interaction diagram below P_b may be constructed in a point-by-point manner. For any value of $\varepsilon_s > \varepsilon_y$, or for any value of $k_u d < k_u d_{bal}$, the strain and stress distributions are defined. Once the forces are defined, the moments and thrusts are computed using the same formulas as for the balance point. A modification will have to be made when the compression-steel stress is less than f_y. Likewise, points above the balance point can be found by assuming several values of $\varepsilon_s < \varepsilon_y$ or $k_u d > k_u d_{bal}$; finding the strains, stresses, and forces; and then summing forces and moments of forces. The straight-line approximation is always conservative.

Care must be used in the selection of the axis about which moments are to be summed, especially if the section is not symmetrical or symmetrically reinforced, since the force system is not a pure couple. The most important thing is to remain consistent, and to be certain that the internal and external moments are summed about the same axis.

In practice, either the moment-thrust curve can be reduced by the appropriate ϕ factor, or the computed ultimate M and P increased by dividing by ϕ. If the required $M - P$ point lies on or slightly inside the interaction curve, the design is acceptable. The maximum permitted factored thrust is $0.80\phi P_0$, which limits the applied thrust capacity in cases where the computed moments are relatively small. This limitation is intended to provide resistance to accidental eccentricities. Column reinforcement consists of longitudinal bars and lateral ties or spiral bars. The ratio of longitudinal steel ρ_g should be between 0.01 and 0.08. Ties are usually No. 3 or No. 4 bars which are bent to enclose the longitudinal bars, and are spaced at not more than 16 longitudinal bar diam, 48 tie diam, or the least thickness of the column. Ties are arranged to bind each corner bar and alternate intermediate bars.

Fig. 12.3.7 Stress and strain distribution in a column at balanced moment and thrust. (*a*) Section; (*b*) strains; (*c*) stresses.

Several typical arrangements are shown in Fig. 12.3.8. Ties hold the longitudinal bars in place during construction, and may provide some shear resistance and improve the behavior of the column at loads near failure.

Spiral bars serve to confine the concrete core of the section as well as holding the longitudinal bars. The minimum amount of spiral steel, if advantage is to be taken of the higher ϕ factor for spiral columns, is

$$\rho_s = 0.45(A_g/A_{core} - 1)f_c'/f_y \qquad (12.3.17)$$

where f_y is the yield stress of the spiral, but not more than 60,000 lb/in². The clear spacing of the turns of the spiral must be between 1 and 3 in.

Fig. 12.3.8   Typical arrangement of steel in columns.

The strengths of compression members may be reduced below the cross-sectional strengths by length effects. Most columns in unbraced frames (frames in which the columns resist all horizontal forces) will have some strength reduction because of length effects, and most columns in braced frames will not, but this depends on the precise details of the length, width, restraint by other frame members, reinforcement, and amount of creep expected. A comprehensive method of taking column length into account is contained in the ACI Code.

Columns are also made by encasing structural steel sections in concrete, in which case the covering concrete must contain at least some steel in order to control cracks and maintain the integrity of the concrete in case of fire. Heavy steel pipes filled with concrete may also be used as columns. These are also used as piling, in which case the pipe is filled with concrete, usually after being driven to the final location.

REINFORCED CONCRETE FLOOR SYSTEMS

Several types of reinforced concrete floors are used, with the choice depending on a number of factors such as span, live load, deflection limits, cost, story-height limitations, local custom, the nature of the rest of the structural frame or system, and the probability of future alterations.

The floor systems may be divided, somewhat arbitrarily, into one-way and two-way systems. One-way systems include solid and hollow slabs and joists spanning between parallel supporting beams or walls. Floors in which panels are subdivided into a grid by subbeams spanning between main girders have usually been designed as one-way slabs when the grid length is several times the width.

Two-way systems include slabs supported on all four sides on beams or walls, traditionally called *two-way slabs*. Slabs supported only on columns located at the corners of the panels also carry loads by developing stresses in the two major directions, and are usually termed *flat plates* if the slab is supported directly on the columns and *flat slabs* if there are capitals on the columns to increase the effective support size.

A waffle slab is usually designed as a flat plate or flat slab, with pockets of concrete omitted from the lower surface, and the slab appears as a series of crossing joists.

One-way slabs and joists are designed as beams. A 12-in or other convenient width of slab is selected, analyzed as an isolated beam, and a depth and steel are picked. The main steel is perpendicular to the supports. Additional steel, parallel to the supports, is placed to control cracking and help distribute minor concentrated loads. This steel is usually one-fourth to one-third of the main steel, but not less than a gross steel ratio of 0.0018 for grade 60 and 0.002 for lower-grade steels. These minimums govern in both directions, and in two-way systems as well.

One-way joists, such as that shown in Fig. 12.3.9, are usually cast with reusable sheet metal or fiberglass forms owned or rented by the contractor. Standard form sizes range from about 20 to 36 in wide and 8 to 20 in deep. The web thicknesses are made to suit the shear and fireproofing requirements, and special tapered end sections may be available to increase the web widths in the regions of high shear near the supports. Joists are exempted from the requirement that web steel be supplied regardless of the concrete shear stress. The allowable shear stress for the concrete is 1.1 times that for beams. The top slabs usually range from 2.5 to 4.5 in thick, and are reinforced to span from rib to rib. The joists are essentially designed as isolated T beams, and may be supported on girders or walls. Joist systems are suitable for reasonably long spans and heavy loads, and have low dead weights for the effective depths attainable.

Fig. 12.3.9   Cross section of concrete floor joist.

Slabs spanning in two directions are all designed taking into account the shape of the panel and the relative stiffnesses of the supporting beams, if any. The choice of types is a matter of loadings and economics. For residential and light office loadings, the flat plate is frequently the choice, as the very simple formwork may lead to substantial economy and the story height is minimized. For heavier loads or longer spans, punching shear around the columns becomes a limiting factor, and the flat slab, with its column capitals, may be the most suitable.

In case of extremely heavy floor loads or very stringent limits on deflections, slabs with beams on all four sides of each panel will be most satisfactory. The formwork is more complex than for the other slabs, but there will be some compensating savings in the amounts of steel because of the greater depths of the beams. In addition, the two-way slab may be much more efficient if the building is to resist major lateral loads by frame action alone, because of the difficulties in transferring large moments between columns and flat plates or slabs. The design procedure is the same for slabs with and without beams. The basic steps, for each direction of span in each panel, are:

1. Compute static moment M_0.
2. Distribute M_0 to positive and negative moment sections.
3. Distribute section moments to column and middle strips and beams.

Most building slabs are designed for uniformly distributed loads, and the *static moment*, defined as the absolute sum of the midspan positive plus average negative moments, is

$$M_0 = ql_2(l_n)^2/8 \qquad (12.3.18)$$

The dimensions are illustrated in Fig. 12.3.10.

Relatively simple rules for the distribution of the static moment to various parts of the panel exist as long as the structure meets several simple limits:

1. Minimum of three spans in each direction.
2. Panel length no more than twice panel width.
3. Successive spans differ by not more than one-third the longer span.
4. Columns on a rectangular grid, or offset no more than 10 percent of the span.
5. Live load not more than two times the dead load.
6. If beams are used, they are used on all four sides of each panel and are approximately the same size, except that spandrel beams only are acceptable.

Fig. 12.3.10 Arrangement of typical slab panel.

The following is for slabs meeting these restrictions. Information on other cases is contained in the ACI Code.

The positive-negative moment distribution for interior spans is

$$+M = 0.35M_0 \qquad (12.3.19)$$
$$-M = 0.65M_0 \qquad (12.3.20)$$

For end spans, the stiffness of the exterior support must be taken into account. Table 12.3.2 gives the fractions of M_0 to be assigned to the three critical sections for five common cases.

Once the section moments have been determined, they are distributed to the column and middle strips and beams, taking into account the panel shape and the beam stiffness. The beam relative stiffness coefficient α_1 is calculated using an effective beam cross section as shown in Fig. 12.3.11, and the full width of the slab panel l_2. The locations of the column and middle strips are shown in Fig. 12.3.10.

The interior negative and positive moments are distributed to the column strips in the proportions shown in Fig. 12.3.12, with the remainder of the moment going to the middle strip. Linear interpolations are made

for intermediate beam stiffnesses, but in most instances where there are beams, they will be found to have

$$\alpha_1 l_2/l_1 \geq 1.0$$

At the exterior supports, the distribution of the negative moments is a complex function of the flexural and torsional stiffnesses of the beams and of the panel shape, but satisfactory designs can usually be achieved by assigning all the negative moment to the column strip and detailing the edge beam or edge strip of the slab for torsion, by using closed stirrups at relatively small spacings, and by placing bars parallel to the edge of the structure in each corner of the stirrups. At fully restrained edges, use the distribution for an interior negative moment section.

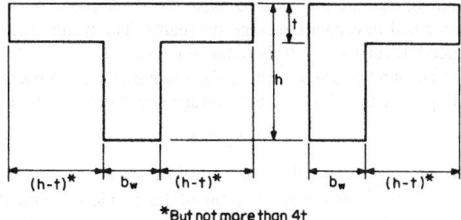

Fig. 12.3.11 Beam sections for calculation of I_b and α_1.

Fig. 12.3.12 Percentage of interior negative and positive moments assigned to a column strip.

Beam moments are found by dividing the column strip moment between beam and slab. If $\alpha_2 l_2/l_1 \geq 1.0$, the beam moment is 85 percent of the column strip moment. This moment is reduced linearly to zero as $\alpha_1 l_2/l_1$ approaches zero.

This design method assumes that all panels are loaded with the same uniformly distributed load at all times. This is obviously a gross simplification, and the requirement that the live load be no greater than twice the dead load limits the potential overstress caused by partial loadings.

Table 12.3.2   Moments in End Spans

	Exterior edge unrestrained (1)	Slab with beams between all supports (2)	Slab without beams between interior supports		Exterior edge fully restrained (5)
			Without edge beam (3)	With edge beam (4)	
Interior negative factored moment	0.75	0.70	0.70	0.70	0.65
Positive factored moment	0.63	0.57	0.52	0.50	0.35
Exterior negative factored moment	0	0.16	0.26	0.30	0.65

In addition, there is a requirement for column design moments, and unless a more complete analysis is made, the following moment, divided between the columns above and below the slab in proportion to their stiffnesses, must be provided for:

$$M = 0.07[(q_d + 0.5q_l)l_2 l_n^2 - q_d' l_2' (l_n')^2] \qquad (12.3.21)$$

The loads q_d and q_l are the distributed dead and live loads including the overload factors. The terms q_d', l_2', and l_n' are for the shorter of the two spans meeting at the column considered.

The shear strength of slab structures must always be checked, and shear stresses often govern the thickness of beamless slabs, especially flat plates.

If there are beams with $\alpha_1 l_2 / l_1 \geq 1.0$, all shear is assigned to the beams, and stirrups are provided to make the shear capacity adequate, as was described in earlier coverage on beams. The beam shear is linearly reduced to zero as $\alpha_1 l_2 / l_1$ is reduced to zero.

For the case of no beams, punching shear around the columns becomes a controlling factor. In this case take the average shear stress, calculated as

$$v_v = V_u / b_0 d \qquad (12.3.22)$$

must not exceed the smaller of $\phi 4\sqrt{f_c'}$, $\phi(2 + 4/\beta_c)\sqrt{f_c'}$, or $\phi(\alpha_s d/b_0 + 2)\sqrt{f_c'}$, where β_c = ratio of long side of critical shear perimeter to short side and α_s = 40 for interior columns, 30 for edge columns, and 20 for corner columns.

The critical shear perimeter b_0 is defined by a section located $d/2$ away from and extending all around the column, as shown in Fig. 12.3.13. It is very important that holes in the slab in the vicinity of the column be taken into account in reducing the value of b_0, and that no unauthorized holes, such as for piping, be made either during or after construction.

It is possible to increase the shear resistance by the use of properly designed shear reinforcement, but this is not often done and is not recommended as a standard practice.

Fig. 12.3.13 Two-way reinforced concrete footing.

Transfer of moments between columns and slabs sets up shear and torsional stresses which must also be considered in the analysis of the shear strength.

FOOTINGS

Footings (Fig. 12.3.13) may be classified as wall footings, isolated column footings, and combined column footings. The bending moments, shears, and bond stresses in such footings should be determined by the principles of statics on the basis of assumed or known soil-pressure distribution over the area of the footing. The bending moment on any projecting portion of a footing may be computed as the moment of the forces acting on the area to one side of a vertical plane through the critical section.

The critical section for bending in a concrete footing supporting a concrete column, pedestal, or wall should be taken at the face of the column, pedestal, or wall. For footings under metallic column bases or under masonry walls where bond with the footings is reduced to the friction value, the critical section is assumed midway between the middle and edge of the base or wall.

Shear stresses must be considered on two sections. The footing may act as a wide beam, for which the critical section is a vertical plane located d away from the critical section for moment, and the stresses must satisfy those for a beam. Punching shear will often govern, and the critical section lies at a distance $d/2$ from the face of the column or other critical section for moment, as shown in Fig. 12.3.13, and as was the case for flat plates and slabs. Footings supported on a small number of high-capacity piles present special shear problems since the conventional critical sections for shear may not be meaningful.

The critical section for bond should be taken at the same plane as for bending. Other vertical planes where abrupt changes of section occur should also be investigated for bond and shear stresses.

In sloped or stepped footings, sections other than the critical ones may require consideration. A square footing, reinforced in two directions, should have the reinforcement uniformly distributed across the entire width. Rectangular footings, reinforced in two directions, should have the reinforcement in the long direction uniformly distributed; in the short direction a portion, Eq. (12.3.23), should be uniformly distributed across a strip equal in width to the short side and centered on the structural element supported and the remainder distributed uniformly in the outer portions. The amount included in the center strip may be computed as follows:

Reinforcement in center strip

$$= \frac{2(\text{total reinforcement in short direction})}{\beta + 1} \qquad (12.3.23)$$

where β is the ratio of the long side to the short side.

Combined Footings Footings supporting two or more columns may be designed with sufficient accuracy by assuming uniform soil pressure and applying the laws of statics. The footing shape must be such that the center of gravity coincides with the center of gravity of the superimposed loads; otherwise unequal settlement may occur. The longitudinal and diagonal tension reinforcement should be designed by the ordinary rules of beam design. Lateral reinforcement should be designed as for isolated footings and should preferably be concentrated in bands under and near the columns proportionate in area to the column loads. The lateral reinforcement at each column should be uniformly distributed within a width centered on the column and should not be greater than the width of the column plus twice the effective depth of the footing.

Spread or raft foundation, consisting of a slab extending over the entire area under the columns or of a slab supported by beams, may be considered as loaded by a uniform upward reaction of the ground. The principle of design is exactly the same as that applied to a floor system, except that the load acts upward instead of downward.

Concrete piles of various types are widely used for foundations as they have larger carrying capacity and greater durability under many conditions of exposure than wooden piles. Precast piles are designed as columns with allowance for driving and handling stresses. Cast-in-place piles are constructed either by driving a steel shell and filling it with concrete or by filling the shell formed as it is withdrawn. Another method forms a bulb at the bottom by means of a ram which forces the concrete into the ground. The design load or capacity of cast-in-place concrete piles is largely empirical, being based on load-test data. The concrete for precast piles is usually over 5,000 lb/in^2 strength and for cast-in-place piles, over 3,000 lb/in^2 strength.

Dowels and Bearing Plates The stress in the longitudinal reinforcement of concrete columns should be transferred to the footing by means of dowels, equal in number and area to the column rods and of sufficient length to transfer the stress as in a lap splice in the column.

Bearing stresses in concrete, under design ultimate loads, should not exceed the following values:

Entire surface loaded: $f_b = 0.85\phi f_c'$ \qquad (12.3.24)

Part of surface loaded: $f_b = 0.85\phi f_c' \sqrt{A_2/A_1}$ \qquad (12.3.25)

where A_1 = loaded area; and A_2 = surface area of same shape and concentric with A_1. $\sqrt{A_2/A_1}$ should not exceed 2.0.

WALLS AND PARTITIONS

Reinforced concrete is well suited to the construction of walls, especially where they have to withstand heavy pressures, such as the retaining walls of a cellar or basement, walls for coal pockets, silos, reservoirs, or grain elevators. Such walls must be designed for flexural shear and bond stresses as well as stability against overturning, sliding, and soil pressure. Drainage should be provided for by weep holes or drains. Partitions may be built of solid concrete 4 to 6 in thick, reinforced to control temperature and shrinkage cracks. Reinforced concrete walls need to be anchored by reinforcement to adjacent structural members. All walls must be reinforced for temperature with steel placed horizontally and vertically.

The horizontal reinforcement shall not be less than 0.25 percent and the vertical reinforcement not less than 0.15 percent of the area of the wall when bars are used and three-fourths of these amounts when welded fabric is used. Adequate reinforcement must be provided around all openings for windows and doors.

Retaining walls of reinforced concrete are used to resist the pressures of earth, water, and other retained materials and are usually of T or L shape. The base must be so proportioned that there is sufficient resistance to sliding and overturning and that the safe bearing strength of the soil is not exceeded. The dimensions of the concrete section and the position and amount of steel reinforcement are determined by the moments and shears at critical vertical and horizontal sections at the junction of the wall and the base. Particular attention should be given to drainage to prevent excessive water pressure behind walls retaining earth or other materials. Walls retaining water, such as tanks, should have steel tensile stresses limited to 12,000 lb/in² unless special consideration is given to controlling cracks and should have ample reinforcement to provide for effects caused by shrinkage of the concrete and temperature change.

Bearing Walls   The allowable compressive force for a reinforced concrete wall that is restrained against lateral motion at both top and bottom and is restrained against rotation at one or both ends is given by Eq. (12.3.26). If neither end is restrained against rotation, the 40 is replaced with 32.

$$P_u = 0.55\phi f'_c A_g [1 - (l_c/40h)^2] \qquad (12.3.26)$$

For the case of concentrated loads, the effective width for computational purposes can be considered as the width of the bearing plus four times the wall thickness but not greater than the distance between loads. The wall thickness should be at least ¹⁄₂₅ of the unsupported height or width, whichever is smaller. For the upper 15 ft, bearing walls must be at least 6 in thick and increase at least 1 in in thickness for each successive 25 ft downward, except that walls of a two-story dwelling need to be only 6 in thick over the entire height, provided that the strength is adequate.

PRESTRESSED CONCRETE

Prestressed concrete members have initial internal stresses, set up by highly stressed steel tendons embedded in the concrete, which are generally opposite those caused by applied loads. Prestressed members are constructed in one of two ways: (1) *Pretensioned* members are factory precast products, made by tensioning steel tendons between abutments and then casting concrete directly around the steel. After the concrete has reached sufficient strength, the steel is cut and the force transferred to the concrete by bond. (2) *Posttensioned* members, either factory or site cast, contain steel in ducts cast in the concrete. After the concrete has cured, the steel is tensioned and mechanically anchored against the concrete. The ducts are preferably pumped full of grout after tensioning to provide bond and corrosion protection, or the tendons may be coated with corrosion inhibitors.

Very high strength steel is used for prestressing in order to overcome the losses of steel stress due to creep and shrinkage of concrete, and as a result of the strength, relatively small amounts of steel are required. The concrete for pretensioned members will usually be at least 5,000 lb/in² compressive strength, and at least 4,000 lb/in² for posttensioned members.

Because of the initial stress conditions, prestressed concrete members are generally crack-free at working loads and consequently are quite suitable for water-containing structures. Circular tanks and pipes are posttensioned by wrapping them with highly stressed wires, using specialized equipment.

Losses of prestress occur with time owing to creep and shrinkage of concrete and relaxation of steel stress. Pretensioned members also have an initial elastic shortening loss accompanying transfer of force to the concrete, and posttensioned members have losses due to friction between ducts and tendons and anchor set. These losses must be taken into account in the design of members.

Prestressed concrete members are checked for both strength and stresses at working loads. Because of the built-in stresses, the condition of dead load only may also govern. The flexural strength is computed using the same equations as for reinforced concrete which were developed earlier. The steel does not have a well-defined yield stress, and the steel stress at failure can be predicted from the following expressions:

Bonded tendons, low relaxation steel and no compression steel:

$$f_{ps} = f_{pu}\left(1 - \frac{0.28}{\beta_1}\rho_p\frac{f_{pu}}{f'_c}\right) \qquad (12.3.27)$$

Unbonded, $l/h \leq 35$:

$$f_{ps} = f_{se} + 10,000 + f'_c/100\rho_p \qquad (12.3.28)$$

but not more than f_{py} or $f_{se} + 60,000$.

The stresses are used directly in Eq. (12.3.4), (12.3.6), or (12.3.7), as appropriate, substituting f_{ps} for f_y and A_{ps} for A_s.

Allowable Stresses at Working Load

Steel:

Maximum jacking stress, but not to exceed recommendation by steel or anchorage manufacturer	$0.8 f_{pu}$
Immediately after transfer or posttensioning	$0.7 f_{pu}$

Concrete:

Temporary stresses immediately after prestressing	
Compression	$0.6 f'_{ci}$
Tension in areas without reinforcement	$3\sqrt{f'_{ci}}$
Design load stresses (after losses)	
Compression	$0.45 f'_c$
Tension in precompressed tensile zones	$7.5\sqrt{f'_c}$

The allowable tension is limited to $6\sqrt{f'_c}$ in two-way slabs. Larger stresses, to $12\sqrt{f'_c}$ or more, are permitted, but with increasing checks on deflections, corrosion protection, and fatigue as the stresses increase. Details are given in the ACI Code.

The final steel stress, at working loads, will usually be 30,000 to 45,000 lb/in² less than the initial stress for pretensioned members. Posttensioned members will have slightly lower losses. Losses, from initial tensioning values, for pretensioned members may be predicted satisfactorily using the following expressions from the AASHTO, Bridge Specifications, Sec. 9.16.2;

$$\Delta f_s = SH + ES + CR_c + CR_s = \text{ prestress loss} \qquad \text{lb/in}^2 \quad (12.3.29)$$

where SH = shrinkage loss = $17,000 - 150RH$; ES = elastic shortening loss = $(E_s/E_c)f_{cir}$; CR_c = creep loss = $12f_{cir} - 7f_{cds}$; CR_s = relaxation loss = $5,000 - 0.1ES - 0.05(SH + CR_c)$, for low relaxation strands; f_{cir} = concrete stress at level of center of gravity of steel (cgs) at section considered, due to initial prestressing force and dead load; f_{cds} = change in concrete stress at cgs due to superimposed composite or noncomposite dead load; and RH = relative humidity, percent. The loss calculations are carried out for each critical moment section. The average annual relative humidity of the service environment should be used in the SH calculation.

The shear reinforcement requirements for simple cases are covered in an earlier section. In addition to the shear steel, a few stirrups or ties should be placed transverse to the member axis as close to the ends as possible, to control potential splitting cracking. The area of steel, from the AASHTO Bridge Specification, should be $A_t = 0.04 f_{si} A_{ps}/20,000$. In posttensioned beams, end blocks will often have to be used to provide space for anchorage bearing plates.

The minimum clear spacing between strands in pretensioned members is three times the strand diameter near the ends of the beam, but many plants are set up to handle only 2-in spacings. Strands may be closer together in central positions of members, which will help in maximizing member effective depths and steel eccentricity.

Few precast, pretensioned members are solid rectangular sections, and double T beams and hollow floor slab units are used extensively in buildings. Hollow box beams and I-section beams are used extensively in bridges and in buildings with heavy design loads. Square piling with the prestressing strands arranged in a circular pattern is widely used. Because of the large number of possible sections, it is necessary to check availability of any particular section with local producers before designing any precast structure.

PRECAST CONCRETE

Precast slabs, beams, walls, and partitions as well as piles, retaining wall units, light standards, railroad crossings, and bridge slabs are being increasingly used because of the saving in time and labor cost. Such units vary in size from small slabs for use in floors of residences to large frames for industrial buildings. The small units, such as roof slabs, are cast in steel forms at central plants. Some of the larger units, such as bridge or highway slabs and wall units, are cast in wood forms at or near the place of use. A method by which wall or partition slabs are cast so that they are simply tilted into position has found wide use in housing and industrial construction. Another special adaptation is the method of casting complete floors on top of each other, then lifting into position vertically at the columns.

Particular attention must be given in the design of precast units to reduction in weight and to details to minimize the cost of erection and installation. Reduction in weight is obtained by the use of lightweight aggregates, high-strength concrete, and hollow units. Precast reinforced concrete units are seldom designed for concrete strengths of less than 4,000 lb/in². They are often combined with cast-in-place concrete so as to obtain the advantages of continuity. The combination of precast beams with cast-in-place slabs gives the advantage of T-beam action. Wall units are tied together by interlocking joints or by bolts. Care must be taken in shipping and handling to avoid damage to the precast units, and the design must take care of the stresses that come from such causes. All lifting devices built into the units should be designed for 100 percent impact. All units must be identified as to proper location and orientation in the structure.

Because precast units are made under conditions which allow good control of dimensions, certain restrictions can be relaxed that must be observed for cast-in-place concrete. Cover over the reinforcement for members not exposed to freezing need not be more than the nominal diameter of the steel but not less than ⅝ in. The maximum size of the coarse aggregate can be as large as one-third of the smaller dimension of the member. Precast wall panels are not limited to the minimum thickness requirements for cast-in-place walls.

To reduce the number of connections, precast units should generally be cast as large as can be properly handled. However, some joints will be needed to transfer moments, torsion, shear, and axial loads from one member to another. The integrity of the structure depends on the adequacy of the design of the various joints and connections. They may be made by use of bolts and pins or clips and keys, by welding the reinforcement or steel insert, or by a number of other methods limited only by the ingenuity of the designer. The connections should not be the weak links in the structure. Thought as to their location will avoid many problems. The PCI Handbook shows many proven connection details.

JOINTS

Contraction and expansion joints may be needed at intervals in a structure to help care for movement due to temperature changes and shrinkage. Joints at 20- to 30-ft intervals provide good crack control. A weakened plane, formed in the tension side of the member by a slot ¼ in wide and ½ in deep, will induce the formation of contraction cracks at selected points. Structures over 200 ft in length should have special consideration given to contraction provisions.

Construction joints are necessary in most structures because all sections cannot be cast continuously. They should be made at points of minimum shearing stress and reinforced across the joint with a steel area of not less than 0.5 percent of the area of the section cut. Provision must be made for the transfer of shear and other forces through the construction joint. Joints in columns should be made at the underside of the floor members, haunches, T beams, and column capitals.

The hardened concrete at a joint should be properly prepared for bonding with the new concrete by being cleaned, roughened, and wetted. On this surface, a coat of neat cement grout or other bonding agent should be applied just before depositing the new concrete.

FORMS

Forms are usually built of wood or metal but in special cases may be made of plastic or fiberglass reinforced plastic. Wooden forms may be the most economical unless the construction allows for the repeated use of the same forms. Plywood and compressed wood fiber sheets, specially treated to make them waterproof, are frequently used for form faces where good surfaces are required. Forms must be designed so that they can be easily erected, removed, and reerected. The usual order of removing forms is (1) column sides, (2) joists, (3) girder and beam sides, (4) slab bottoms, and (5) girder and beam bottoms. Column forms are held together by clamps made of wood or steel, the spacing of which is smallest at the bottom and increases with the decrease in pressure. Beam forms consist of the bottom and two sides held together by clamps or cleats and supported by posts. Slab forms consist of boards or other form material supported by joists spaced 2 or 3 ft apart or other means. The joists either rest on a horizontal joist bearer fastened to the clamps of the beam or girder form or are supported by stringers, or posts, or both.

Special consideration must be given to forms for prestressed concrete members. For pretensioned members the form must be constructed such that it will permit movement of the member during release of the prestressing force. For posttensioned members, the form should provide a minimum of resistance to shortening of the member. It is also necessary to consider the deflection of the members due to the stressing force.

Design in Formwork   The formwork is an appreciable portion of the cost of most concrete structures. Any efforts, however, to reduce the cost of the forms must not go beyond the point of safe design to prevent failures which would in themselves raise the cost of construction.

All forms must conform to the dimensions and shape of the members and must be sufficiently tight to prevent leakage of the mortar. They must be properly braced and tied together to maintain their position and shape during the construction procedure.

The formwork must support all the vertical and lateral loads that may be applied until these loads can be carried by the concrete structure. Loads on the form include the weight of the forms, reinforcing steel, fresh concrete, and various construction live loads. The construction live load varies with conditions but is often assumed to be 75 lb/ft² of floor area. The formwork should also be designed to resist lateral loads produced by wind and movement of construction equipment. Most frequently the steel and concrete will not be placed in a symmetrical pattern and frequently large impact loads will occur. Because of the many varied conditions, it is frequently impossible to determine with any great precision the loads which the form must carry. The designer must therefore make safe assumptions by which the forms can be designed such that failure will not result.

Lateral pressures in forms for walls and columns are influenced by a number of factors: weight of concrete, height of placing, vibration, temperature, size and shape of form, amount and distribution of reinforcing steel, and several other variables. Formulas have been suggested for computing safe lateral pressures to be used in form design. However, because of limited test data, they are not generally accepted by all engineers.

Form Liners   Absorbent form liners are occasionally applied to the surface of forms to extract the water from the surface of the concrete, eliminate air and water voids, and produce a concrete of uniform appearance with surfaces which are superior in durability and resistance to abrasion.

The vacuum process, whereby water is absorbed from the concrete through a special form liner made of two layers of screen or wire mesh covered by a layer of cloth, has a similar effect and if properly used reduces the water content of the concrete to a depth of several inches.

Neoprene and other types of rubber have been successfully used as liners in precasting work in which a number of units are made from one form. Rubber is particularly suited for patterned work.

Plastic form liners make it easy to obtain a textured surface or a glossy smooth surface. Generally speaking, plastic liners are easily cleaned and if not too thin are suitable for a number of reuses.

Removal of Forms   The time that forms should remain in place depends on the character of the members and weather conditions. The strength of concrete must be ascertained before removing the forms. Unless special precautions are taken, concrete should not be placed below 40°F. Fresh concrete should never be subjected to temperatures below freezing. As an approximate guide for the minimum time for form removal, the following rules, which assume moist curing at not less than 70°F for the first 24 h, may be observed.

WALLS IN MASS WORK.   In summer, 1 day; in cold weather, 3 days.

THIN WALLS.   In summer, 1 day; in cold weather, 5 days.

COLUMNS.   In summer, 1 day; in cold weather, 4 days, provided girders are shored to prevent appreciable weight reaching the columns.

SLABS UP TO 7-FT SPAN.   In summer, 4 days; in cold weather, 2 weeks.

BEAMS AND GIRDER SIDES.   In summer, 1 day; in cold weather, 5 days.

BEAMS AND GIRDERS AND LONG-SPAN SLABS. In summer, 7 days; in cold weather, 2 weeks.

CONDUITS.   2 or 3 days, provided there is not a heavy fill upon them.

ARCHES.   If a small size, 1 week; large arches with a heavy dead load, 3 weeks.

Forms for prestressed members may be removed when sufficient prestressing has been applied to enable them to carry their dead loads and the expected construction loads.

EVALUATION OF EXISTING CONCRETE STRUCTURES

This section is intended to give some guidance to the persons responsible for inspections of structures, whether these are done routinely, or before changing a loading or use, or during remodeling, or when something suspicious is found. Serious problems clearly need to be evaluated by a structural engineer experienced in such evaluations, testing, and renovation; such problems should not be evaluated by one who is primarily a structural designer.

The following are some signs of distress. Any cracks wider than about 0.02 in (0.5 mm) are potentially serious, particularly if there are more than a few. If the cracks are inclined like the shear crack shown in Fig. 12.3.1, they are an additional warning sign and should be investigated promptly. At interior supports and other locations where the beams are continuous with the columns, these cracks may also start at the top surface. Any cracking parallel to the member axis is potentially serious.

Rust stains and streaks from cracks indicate corrosion problems. Corrosion, especially from salt, disrupts the concrete surrounding the steel long before the bar areas are significantly reduced, because rust occupies much more volume than the steel it replaces. This rusting causes internal cracking, which can often be detected by tapping on the concrete surface with a hammer (1-lb size), which produces a distinctive "hollow" sound. "Stalactites" growing from the bottoms of members may be either salt or lime, but both indicate water-penetration problems and the need for waterproofing work.

Concrete exposed to freezing and thawing or to attack by some chemicals may crumble and disintegrate, leading to loss of section area and protective cover on the reinforcement. Large deflections and/or slopes in members may be indicative of impending distress or disaster, but sometimes members were not built very straight, and in such cases the warning sign is actually a *change* in the conditions.

Any reinforced concrete building designed before about 1956 has a potential weakness in the shear strength of the beams and girders because of a deficiency in the codes of that era, and major changes in loading and seemingly minor signs of distress should be investigated carefully. This problem may also exist in highway bridges designed before 1974. Prestressed members should not have this problem.

Repairs are made by injecting epoxy into cracks, by surface patching, by chipping away significant volumes of concrete and replacing it, and by other means. The repair materials range from normal concretes to highly modified concrete and polymer materials.

12.4   AIR CONDITIONING, HEATING, AND VENTILATING
By Norman Goldberg

REFERENCES: ASHRAE Handbooks "Fundamentals," "Systems and Equipment," and "Applications." Stamper and Koral, "Handbook of Air Conditioning, Heating and Ventilating," Industrial Press. Carrier, "Handbook of Air Conditioning Design," McGraw-Hill.

Air conditioning is the process of treating air to meet the requirements of a conditioned space by controlling its temperature, humidity, cleanliness, and distribution. This section presents standards, basic data, and physical laws for use in the design of air conditioning and related heating and ventilating systems.

COMFORT INDEXES

The human body generates heat and disipates that heat to the surrounding air by sensible flow and the evaporationg of moisture.

Effective Temperature   Effective temperature (ET) combines the effect of ambient temperature and humidity into a single index.

ASHRAE Comfort Chart   The American Society of Heating, Refrigeration, and Air Conditioning Engineers (ASHRAE) comfort envelope shown in Fig. 12.4.1 is based on clothing and activity. Comfort varies with skin temperature and skin wettedness.

Temperature-Humidity Index

The term **temperature-humidity index (THI)** is used to describe the combined effects of temperature and humidity on comfort experienced by people. Since individual reactions can vary considerably from person to person, this quantity should be considered as a guide rather than as an absolute. However, relatively few people will feel discomfort when the THI is 70 or below. By the time it reaches 75, about half the people will be uncomfortable. An index of 80 in a work area may cause a decrease in workers' efficiency. To compute THI, any one of the following formulas may be used:

$$\begin{aligned} \text{THI} &= 0.4(t_d + t_w) + 15 \\ &= 0.55t_d + 0.2t_{dp} + 17.5 \\ &= t_d - (0.55 - 0.55\,\text{RH})(t_d - 58) \end{aligned} \quad (12.4.1)$$

where t_d = dry-bulb temperature, °F; t_w = wet-bulb temperature, °F; t_{dp} = dew-point temperature, °F; and RH = relative humidity, expressed as a decimal. To convert to degrees Celsius, $t_c = \frac{5}{9}(t_f - 32)$.

Fig. 12.4.1 Standard effective temperatures and ASHRAE comfort zones. *(ASHRAE "Handbook of Fundamentals," 1993.)*

Wind-Chill Index

The **wind-chill index (WCI)** attempts to describe how much heat the body will lose under certain conditions of wind and temperature. It is determined empirically by an equation which is used to describe the rate of heat loss from a litre cylinder of water at 33°C (91.4°F) as a function of ambient temperature and wind velocity. The formulas for the wind-chill index are

$$\text{WCI} = (10.45 - V + 10\sqrt{V})(33 - t_a) \qquad \text{kcal/(m}^2 \cdot \text{h)} \quad (12.4.2)$$

where V = metres per second, t_a = °C.

$$\text{WCI} = (10.45 - 0.447V + 6.6854\sqrt{V})(91.4 - t_a) \quad (12.4.3)$$

where V = miles per hour, t_a = °F.

Instead of using the WCI to express the severity of a cold environment, meteorologists use an index derived from the WCI called the *equivalent wind-chill temperature.* This is the ambient temperature that would produce, in a calm wind (defined for this application as 4 mi/h), the same WCI as the actual combination of air temperature and wind velocity. Equivalent wind-chill temperature $t_{eq,wc}$ can be calculated by

$$t_{eq,wc} = -0.0818(\text{WCI}) + 91.4 \quad (12.4.4)$$

where $t_{eq,wc}$ is expressed as a temperature, °F. (See Table 12.4.1.)

INDOOR DESIGN CONDITIONS

Indoor design conditions are determined by the application, usually either the comfort of the people occupying the space or the control of conditions within that space to facilitate a process.

Comfort Conditions   Typical indoor design conditions for summer and winter comfort are shown in Table 12.4.2.

Conditions for Process Design   These vary widely, depending on the process. See the manufacturer's requirements and/or the ASHRAE Handbook "Applications."

OUTDOOR DESIGN CONDITIONS

Outdoor design conditions are based primarily on the geographical location but they are affected to some extent by the criticality of the application and by the economics of maintaining indoor conditions at specified design levels during unusual outdoor weather conditions.

Outdoor design conditions for comfort applications are usually not the most severe conditions that have been experienced in a locality, and recommended levels may be exceeded for an average of 5 percent of the hours during the cooling season.

Table 12.4.1   Equivalent Wind-Chill Temperature of Cold Environments*

Wind speed, mi/h	Actual thermometer reading, °F											
	50	40	30	20	10	0	−10	−20	−30	−40	−50	−60
	Equivalent chill temperature, °F											
0	50	40	30	20	10	0	−10	−20	−30	−40	−50	−60
5	48	37	27	16	6	−5	−15	−26	−36	−47	−57	−68
10	40	28	16	3	−9	−21	−34	−46	−58	−71	−83	−95
15	36	22	9	−5	−18	−32	−45	−59	−72	−86	−99	−113
20	32	18	4	−11	−25	−39	−53	−68	−82	−96	−110	−125
25	30	15	0	−15	−30	−44	−59	−74	−89	−104	−119	−134
30	28	13	−3	−18	−33	−48	−64	−79	−94	−110	−125	−140
35	27	11	−4	−20	−36	−51	−67	−83	−98	−114	−129	−145
40†	26	10	−6	−22	−38	−53	−69	−85	−101	−117	−133	−148

Little danger: In less than 5 h, with dry skin, Maximum danger from false sense of security. (WCI less than 1,400)	**Increasing danger:** Danger of freezing exposed flesh within one minute. (WCI between 1,400 and 2,000)	**Great danger:** Flesh may freeze within 30 s. (WCI greater than 2,000)

* Cooling power of environment expressed as an equivalent temperature under calm conditions.
† Winds greater than 43 mi/h have little added chilling effect.
SOURCE: U.S. Army Research Institute of Environmental Medicine.

Table 12.4.2   Recommended Inside Design Conditions

Type of application	Summer			Winter					
				With humidification			Without humidification		
	Dry bulb, °F (°C)	Rel. hum., %	Temp. swing, °F (°C)*	Dry bulb, °F (°C)	Rel. hum., %	Temp. swing, °F (°C)‡	Dry bulb, °F	Temp. swing, °F (°C)‡	
General comfort (apartment, house, hotel, office, hospital, school, etc.)	74–76 (23–24)	50–45	2 (1.1)	68–72 (20–22)	35–30	3 to 4 (1.6–2.2)	70–74	4 (2.2)	
Retail shops (short-term occupancy) (bank, barber or beauty shop, department store, supermarket, etc.)	74–76 (23–24)	50–45	2 (1.1)	68–72 (20–22)	35–30†	3 to 4 (1.6–2.2)	70–74	4 (2.2)	
Low-sensible-heat-factor applications (high latent load) (auditorium, church, bar, restaurant, kitchen, etc.)	74–76 (23–24)	55–50	2 (1.1)	68–72 (20–22)	40–35	2 to 3 (1.1–1.7)	70–74	4 (2.2)	
Factor comfort (assembly areas, machining rooms, etc.)	76–78 (24–26)	55–45	2 (1.1)	68–70 (18–21)	35–30	4 to 6 (2.2–3.3)	65–70	6 (3.3)	

* Temperature swing is above the thermostat setting at peak summer load conditions.
† Winter humidification in retail clothing stores is recommended to maintain the quality texture of goods.
‡ Temperature swing is below the thermostat setting at peak winter load conditions (no lights, people, or sun).
°C = 5⁄9(°F − 32)
Source: Adapted from Carrier Corp. "System Design Manual," with permission.

Frequently used winter outside dry-bulb design temperature are shown in Fig. 12.4.2. Summer outdoor design dry-bulb, wet-bulb, and dew-point temperatures for comfort applications are shown in Figs. 12.4.3 to 12.4.5, respectively.

Design of critical air-conditioning systems and selection of atmospheric evaporative cooling equipment are normally based on more severe outdoor conditions than for comfort applications. Figure 12.4.6 shows wet-bulb temperatures in the United States that are exceeded 1 percent of the time during the cooling season.

Ventilation and Infiltration

Ventilation   Ventilation rates must be adequate to control levels of indoor contaminants and provide makeup air for exhaust systems. It must be recognized that the introduction of outdoor air into a conditioned

Fig. 12.4.2   Isotherms of winter outdoor temperatures, °F. [*E. Stamper and L. Koral (eds.), "Handbook of Air Conditioning, Heating and Ventilating," 3d ed., 1979, Industrial Press.*]

Fig. 12.4.3   Summer dry-bulb temperature data. *(Marley Co.)*

space represents a significant load on the system since it affects both temperature and moisture levels.

Mechanical ventilation is controllable and is the desirable way to introduce outdoor air into a space. Infiltration is uncontrolled airflow through cracks and other openings and, where feasible, it should be limited. ASHRAE Standard 62-1989, "Ventilation for Indoor Air Quality," presents recommendations for outdoor air rates to maintain acceptable indoor air quality for a wide variety of applications. Requirements of local building codes may differ and may require more, or accept less, ventilation air. An extract from the ASHRAE Standard 62-1989, "Table of Air Requirements," is shown in Table 12.4.3.

Infiltration   Infiltration during the summer is due primarily to wind velocity creating a positive pressure on the windward side of a building. During the winter, when the density of indoor and outdoor air differ significantly, stack effect adds substantially to the wind effect. Completely offsetting infiltration by the introduction of sufficient outdoor air through the air-conditioning system is costly. Consequently, outdoor air infiltration should be curtailed by the use of vestibules and, if feasible, revolving doors at frequently used entrances, and tightly fitting windows and/or storm sash throughout. See Tables 12.4.4 and 12.4.5.

Flow of Water Vapor through Building Construction Components
Water vapor flow through building construction adds to the latent load, and although it is usually a minor load factor in comfort applications, it can affect the integrity of insulation and other building components.

Vapor transmission is determined by the differences in vapor pressure and the permeability of the construction, but can be reduced by the use of vapor barriers on the high-vapor-pressure side of the construction.

In applications where interior design dew points are either high or low, the vapor transmission should be calculated.

The permeance of some materials that will affect the transmission of water vapor is shown in Table 12.4.6.

Infiltration of Moisture

Infiltration of air, and particularly moisture, into a conditioned space is frequently a source of sizable heat gain or loss.

The moisture entering a building as water vapor may be expressed as

$$W_t = W_{trans} + W_{inf} + W_{vent} \qquad (12.4.5)$$

where W_t = total weight of vapor, grains (grams). Assuming that W_{trans} is negligible as a load factor,

$$W_1 = W_{inf} + W_{vent} \qquad (12.4.6)$$
$$W_{inf} = \text{air-infiltrated vapor} = W(M_o - M_i) \qquad \text{gr(g)}$$
$$= W(M_o - M_i)\,\Delta p \qquad \text{gr(g)} \qquad (12.4.7)$$
$$W_{vent} = \text{ventilation air vapor}$$
$$= W(M_o - M_i) \qquad \text{gr(g)} \qquad (12.4.8)$$

where Δp = vapor-pressure difference through flow path, inHg (N/m²); W = weight of air, lb (kg); M_o = moisture content of outside air, g/lb (g/kg); M_i = moisture content of inside air, g/lb (g/kg).

For a given locality, the moisture content of the outdoor air may be determined from Fig. 12.4.5 (summer dew-point temperature data) and the appropriate psychrometric chart.

Conductance Coefficients   Tables 12.4.7 to 12.4.9 list conductance heat-transfer coefficients U for various building components. Table 12.4.10 shows conductance heat transfer coefficients for reflective and

Design dry-bulb temperatures at the
5 percent level. Each line passes through
the localities at which the noted dry-bulb
temperature is equaled or exceeded 5 percent
of the time during the average 4-month summer season.

Fig. 12.4.4 Summer wet-bulb temperature data. *(Marley Co.)*

Dew-point temperatures at the 5 percent
level. Each line passes through the localities
at which the noted dew-point temperature is
equaled or exceeded 5 percent of the time
during the average 4-month summer season.

Fig. 12.4.5 Summer dew-point temperature data. *(Marley Co.)*

Design wet-bulb temperatures at the 1 percent level.
Each line passes through the localities at which the
noted wet-bulb temperature is equaled or exceeded
1 percent of the time during the average 4-month
summer season.

Fig. 12.4.6 High wet-bulb temperature data. *(Flour Products Co.)*

Table 12.4.3   Outdoor Air Requirements for Ventilation*

Application	Estimated maximum† occupancy, persons/1,000 ft^2 or 100 m^2	cfm/person	L/(s · person)	cfm/ft^2	L/(s · m^2)	Comments
			Commercial facilities			
Hotels, motels, resorts, dormitories						
Bedrooms				30	15	Per room.
Living rooms				30	15	Per room.
Baths				35	18	Per room. Installed capacity for intermittent use.
Lobbies	30	15	8			
Conference rooms	50	20	10			
Assembly rooms	120	15	8			
Dormitory sleeping areas	20	15	8			See also food and beverage services, merchandising, barber and beauty shops, garages.
Gambling casinos	120	30	15			Supplementary smoke-removal equipment may be required.
Offices						
Office space	7	20	10			Some office equipment may require local exhaust.
Reception areas	60	15	8			
Telecommunication centers and data entry areas	60	20	10			
Conference rooms	50	20	10			Supplementary smoke-removal equipment may be required.
Public spaces						
Corridors and utilites				0.05	0.25	
Public restrooms, cfm/wc or cfm/urinal		50	25			Normally supplied by transfer air.
Locker and dressing rooms				0.5	2.5	Local mechanical exhaust with no recirculation recommended.
Smoking lounge	70	60	30			
Food and beverage service						
Dining rooms	70	20	10			
Cafeteria, fast food	100	20	10			Supplementary smoke-removal equipment may be required.
Bars, cocktail lounges	100	30	15			
Kitchens (cooking)	20	15	8			Exhaust min. 1.5 cfm/ft^2
Retail stores, sales floors, and show room floors						
Basement and street	30			0.30	1.50	
Upper floors	20			0.20	1.00	
Storage rooms	15			0.15	0.75	
Dressing rooms				0.20	1.00	
Malls and arcades	20			0.20	1.00	
Shipping and receiving	10			0.15	0.75	
Warehouses	5			0.05	0.25	
Smoking lounge	70	60	30			Normally supplied by transfer air, local mechanical exhaust; exhaust with no recirculation recommended.
Sports and Amusement						
Spectator areas	150	15	8			When internal combustion engines are operated for maintenance of playing surfaces, increased ventilation rates may be required.
Game rooms	70	25	13			
Ice arenas (playing areas)				0.50	2.50	
Swimming pools (pool and deck area)				0.50	2.50	Higher values may be required for humidity control.
Playing floors (gymnasium)	30	20	10			
Ballrooms and discos	100	25	13			
Bowling alleys (seating areas)	70	25	13			
Theaters						
Ticket booths	60	20	10			Special ventilation will be needed to eliminate special stage effects (e.g., dry ice vapors, mists, etc.)
Lobbies	150	20	10			
Auditorium	150	15	8			
Stages, studios	70	15	8			
Miscellaneous						
Photo studios	10	15	8			
Darkrooms	10			0.50	2.50	
Pharmacy	20	15	8			
Bank vaults	5	15	8			
Duplicating, printing				0.50	2.50	Installed equipment must incorporate positive exhaust and control (as required) of undesirable contaminants (toxic or otherwise).

Table 12.4.3   Outdoor Air Requirements for Ventilation*   (*Continued*)

Application	Estimated maximum† occupancy, persons/1,000 ft² or 100 m²	Outdoor air requirements				Comments
		cfm/person	L/(s · person)	cfm/ft²	L/(s · m²)	
		Institutional facilities				
Education						
Classroom	50	15	8			
Laboratories	30	20	10			Special contaminant control systems
Training shop	30	20	10			may be required for processes or
Music rooms	50	15	8			functions including laboratory
Libraries	20	15	8			animal occupancy.
Locker rooms				0.50	2.50	
Corridors				0.10	0.50	
Auditoriums	150	15	8			
Smoking lounges	70	60	30			Normally supplied by transfer air. Local mechanical exhaust with no recirculation recommended.
Hospitals, nursing and convalescent homes						
Patient rooms	10	25	13			Special requirements or codes and
Medical procedure	20	15	8			pressure relationships may determine
Operating rooms	20	30	15			minimum ventilation rates and filter
Recovery and ICU	20	15	8			efficiency. Procedures generating contaminants may require higher rates.
Autopsy rooms				0.50	2.50	Air shall not be recirculated into other
Physical Therapy	20	15	8			spaces.
Correctional facilities						
Cells	20	20	10			
Dining halls	100	15	8			
Guard stations	40	15	8			

* This table prescribes supply rates of acceptable outdoor air required for acceptable indoor air quality. These values have been chosen to control CO_2 and other contaminants with an adequate margin of safety and to account for health variations among people, varied activity levels, and a moderate amount of smoking. Rationale for CO_2 control is presented in Appendix D of Standard 62-1989.
† Net occupable space.
SOURCE: Abstracted from ASHRAE Standard 62-1989, with permission.

Table 12.4.4   Infiltration through Doors—Winter*
15 mi/h (24 km/h) wind velocity†; doors on one or adjacent windward sides‡

	ft³/min per ft² area (m³/min per m²)§									
					Average use					
	Infrequent use		1 and 2 story buildings		Tall buildings, ft (m)					
	ft³/min per ft²	m³/min per m²	ft³/min per ft²	m³/min per m²	50	(15.25 m)	100	(30.5 m)	200	(61 m)
Revolving door	1.6	0.48	10.5	3.15	12.6	3.78	14.2	4.26	17.3	5.19
Glass door [³⁄₁₆ in (5 mm) crack]	9.0	2.70	30.0	9.00	36.0	10.80	40.5	12.15	49.5	14.85
Wood door 3 × 7 ft (0.9 × 2 m)	2.0	0.60	13.0	3.90	15.5	4.65	17.5	5.25	21.5	6.45
Garage and shipping-room door	4.0	1.2	9.0	2.70						

* All values are based on the wind blowing directly at the window or door. When the prevailing wind direction is oblique to the window or doors, multiply the values by 0.60 and use the total window and door areas on the windward side(s).
† Based on a wind velocity of 15 mi/h (24 km.p.a). For design wind velocities different from the base, multiply the table values by the ratio of velocities.
‡ Stack effect in tall buildings may also cause infiltration on the leeward side. To evaluate this, determine the equivalent velocity V_i and subtract the design velocity V. The equivalent velocity is

$$V_i = \sqrt{V^2 - 1.75a} \text{ (upper section)}$$
$$= \sqrt{V^2 - 1.75b} \text{ (lower section)}$$

where a and b are the distances above and below the midheight of the building, respectively, in feet.
   Multiple the table values by the ratio $(V_i - V)/15$ for one-half of the windows and doors on the leeward side of the building. (Use values under one- and two-story building for doors on leeward side of tall buildings.)
§ Doors on opposite sides increase values 25%.
SOURCE: Carrier Corporation "System Design Manual," Part 1, Load Estimating, 1970.

Table 12.4.5   Infiltration Through Windows and Doors—Crack Method—Summer-Winter*
Double-hung windows, unlocked on windward side.

	CFM per linear foot of crack											
	Wind velocity, mi/h											
	5		10		15		20		25		30	
Type of double-hung window	No W strip	W strip	No W strip	W strip	No W strip	W strip	No W strip	W strip	No W strip	W strip	No W strip	W strip
Wood sash												
Average window	0.12	0.07	0.35	0.22	0.65	0.40	0.98	0.60	1.33	0.82	1.73	1.05
Poorly fitted window	0.45	0.10	1.15	0.32	1.85	0.57	2.60	0.85	3.30	1.18	4.20	1.53
Poorly fitted—with storm sash	0.23	0.05	0.57	0.16	0.93	0.29	1.30	0.43	1.60	0.59	2.10	0.76
Metal sash	0.33	0.10	0.78	0.32	1.23	0.53	1.73	0.77	2.3	1.00	2.8	1.27

* Infiltration caused by stack effect must be calculated separately during the winter.
† No allowance has been made for usage.
Source: Carrier "Handbook of Air Conditioning Design," McGraw-Hill, 1965.

Table 12.4.6   Water Vapor Transmission through Various Materials

Description of material or construction	Permeance, Btu/(h · 100 ft²) (gr/lb diff) latent heat
Packaging materials	
Cellophane, moisture-proof	0.01–0.25
Glassine (1 ply waxed or 3 ply plain)	0.0015–0.006
Kraft paper soaked with paraffin wax, 4.5 lb per 100 ft²	1.4–3.1
Pliofilm	0.01–0.025
Paint films	
2 coats aluminum paint, estimated	0.05–0.2
2 coats asphalt paint, estimated	0.05–0.1
2 coats lead and oil paint, estimated	0.1–0.6
2 coats water emulsion, estimated	5.0–8.0
Papers	
Duplex or asphalt laminae (untreated)	
30-30-30, 3.1 lb per 100 ft²	0.15–0.27
30-60-30, 4.2 lb per 100 ft²	0.051–0.091
Kraft paper	
1 sheet	8.1
2 sheets	5.1
Aluminum foil on one side of sheet	0.016
Aluminum foil on both sides of sheet	0.012
Sheathing paper	
Asphalt impregnated and coated, 7 lb per 100 ft²	0.02–0.10
Slaters felt, 6 lb per 100 ft², 50% saturated with tar	1.4
Roofing felt, saturated and coated with asphalt	
25 lb/ft²	0.015
50 lb/ft²	0.011
Tin sheet with 4 holes $\frac{1}{16}$-in diameter	0.17
Crack 12 in long by $\frac{1}{32}$ in wide (approximate from above)	5.2

Painted surfaces: Two coats of a good vapor seal paint on a smooth surface give a fair vapor barrier. More surface treatment is required on a rough surface than on a smooth surface. Data indicates that either asphalt or aluminum paint is good for vapor seals.

Aluminum foil on paper: This material should also be applied over a smooth surface and joints lapped and sealed with asphalt. The vapor barrier should always be placed on the side of the wall having the higher vapor pressure if condemnation of moisture in wall is possible.

Application: The heat gain due to water vapor transmission through walls may be neglected for the normal air-conditioning or refrigeration job. This latent gain should be considered for air-conditioning jobs where there is a great vapor pressure difference between the room and the outside, particularly when the dew point inside must be low. Note that moisture gain due to infiltration usually is of much greater magnitude than moisture transmission through building structures.

Conversion factors: To convert above table values to:
grain/(hr) (sq ft) (inch mercury vapor pressure difference), multiply by 9.8. grain/(hr) (sq ft) (pounds per sq inch vapor pressure difference), multiply by 20.0.
To convert Btu latent heat to grains, multiply by 7000/1060 = 6.6.
Source: Carrier "Air Conditioning Design Manual," 1965.

Table 12.4.7   Overall Heat-Transfer Coefficient *U*

Outside air 15 mi/h wind, inside still air.

Example	Construction	Btu/(h · ft^2 · °F)	W/(m^2 · K)
Frame walls	Wood siding, insulation board, air space, gypsum board	0.19	1.079
Frame partition	Gypsum board, air space, gypsum board	0.34	1.931
Frame construction ceilings and floors	Linoleum or tile, felt, plywood, wood subfloor, air space, metal lath, plaster	0.23	1.306
Pitched roofs	Asphalt shingles, building paper, wood sheathing, air space, gypsum, lath, plaster	0.28	1.590
Masonry wall	Face brick 4 in, common brick 4 in	0.48	2.725
	Face brick 4 in, common brick 4 in, air space, gypsum lath, plaster	0.29	1.647
Masonry partition	Cement block (cinder aggregate), plaster on both sides	0.31	1.760
Flat masonry roof	Built-up roofing, roof insulation 1 in, concrete slab 4 in, air space, metal lath, plaster	0.18	1.022

Source: ASHRAE "Handbook of Fundamentals," 1981.

Table 12.4.8   Coefficient Heat Transmission *U* for Windows and Skylights

Air-to-air heat transfer, Btu/(h · ft^2 · °F) and W/(m^2 · K); outside air 0°F, 15 mi/h (24 km/h) wind, no solar radiation; inside still air.

	Vertical glass sheets				Horizontal glass sheets			
	Outdoor exposure		Indoor exposure		Outdoor exposure		Indoor exposure	
	Btu/ (h · ft^2 · °F)	W/(m^2 · K)	Btu/ (h · ft^2 · °F)	W/(m^2 · K)	Btu/ (h · ft^2 · °F)	W/(m^2 · K)	Btu/ (h · ft^2 · °F)	W/(m^2 · K)
Common window glass, single sheet	1.13	6.416	0.75	4.259	1.22	6.927	0.96	5.450
Common window glass, two sheets, 1-in air space (2.54 cm)	0.53	3.009	0.45	2.555	0.63	3.577	0.56	3.180

Source: ASHRAE "Handbook of Fundamentals," 1981.

Table 12.4.9   Coefficient of Heat Transmission *U* for Wood Doors

	Single		With glass storm door	
Construction	Btu/(h · ft^2 · °F)	(m^2 · K)	Btu/(h · ft^2 · °F)	W/(m^2 · K)
1-in-thick solid door ($^{25}/_{32}$ in) (2 cm)	0.64	3.634	0.37	2.101
2-in-thick solid door (1⅝ in) (4 cm)	0.43	2.442	0.28	1.590
Door containing wood or glass panels	0.85	4.826	0.39	2.214

Source: ASHRAE "Handbook of Fundamentals," 1981.

Table 12.4.10   Surface Conductances and Resistances for Air

		Surface emissivity, reflective											
Position of surface	Direction of heat flow	Nonreflective $\varepsilon = 0.90$				$\varepsilon = 0.20$				Reflective $\varepsilon = 0.05$			
		C	C'	R	R'	C	C'	R	R'	C	C'	R	R'
Still air:													
Horizontal	Upward	1.63	9.26	0.61	0.108	0.91	5.17	1.10	0.193	0.76	4.32	1.32	0.232
Vertical	Horizontal	1.46	8.29	0.68	0.121	0.74	4.20	1.35	0.238	0.59	3.35	1.70	0.299
Horizontal	Downward	1.08	6.76	0.92	0.148	0.37	2.10	2.70	0.476	0.22	1.25	4.55	0.800
Moving air (any position):													
15 mi/s (24 km/h) wind (for winter)	Any	6.00	34.07	0.17	0.029								
7½ mi/h (12 km/h) wind (for summer)	Any	4.00	22.71	0.25	0.044								

Notes: *C* = conductance, Btu/(h · ft^2 · °F temp. diff.)
  C' = conductance W/(m^2 · K).
  R = resistance = 1/*C*.
  R' = resistance = 1/*C'*.
Source: Adapted from ASHRAE "Handbook of Fundamentals," 1993.

Table 12.4.11   Determination of U Value Resulting from Addition of Insulation to Any Given Building Section

Given building section property* †				Added R‡ § (R')										
				R = 4 R' = 0.70		R = 6 R' = 1.05		R = 8 R' = 1.41		R = 12 R' = 2.13		R = 16 R' = 2.86		R = 20 R' = 3.57		R = 24 R' = 4.17	
U	U'	R	R'	U	U'	U	U'	U	U'	U	U'	U	U'	U	U'	U	U'
1.00	5.68	1.00	0.18	0.20	1.13	0.14	0.79	0.11	0.62	0.08	0.45	0.06	0.34	0.05	0.28	0.04	0.23
0.90	5.11	1.11	0.20	0.20	1.13	0.14	0.79	0.11	0.62	0.08	0.45	0.06	0.34	0.05	0.28	0.04	0.23
0.80	4.54	1.25	0.22	0.19	1.08	0.14	0.79	0.11	0.62	0.08	0.45	0.06	0.34	0.05	0.28	0.04	0.23
0.70	3.97	1.43	0.25	0.19	1.08	0.13	0.74	0.11	0.62	0.07	0.40	0.06	0.34	0.05	0.28	0.04	0.23
0.60	3.41	1.67	0.29	0.19	1.08	0.13	0.74	0.10	0.57	0.07	0.40	0.06	0.34	0.05	0.28	0.04	0.23
0.50	2.84	2.00	0.35	0.18	1.02	0.13	0.74	0.10	0.57	0.07	0.40	0.06	0.34	0.05	0.28	0.04	0.23
0.40	2.27	2.50	0.44	0.16	0.91	0.12	0.68	0.10	0.57	0.07	0.40	0.05	0.28	0.05	0.28	0.04	0.23
0.30	1.70	3.33	0.59	0.14	0.79	0.11	0.62	0.09	0.51	0.07	0.40	0.05	0.28	0.04	0.23	0.04	0.23
0.20	1.14	5.00	0.88	0.11	0.62	0.09	0.51	0.08	0.45	0.06	0.34	0.05	0.28	0.04	0.23	0.03	0.17
0.10	0.57	10.00	1.75	0.06	0.34	0.06	0.34	0.06	0.34	0.05	0.28	0.04	0.23	0.04	0.23	0.03	0.17
0.08	0.45	12.50	2.22	0.06	0.34	0.05	0.28	0.05	0.28	0.04	0.23	0.04	0.23	0.03	0.17	0.03	0.17

* For U and R values not shown in table, interpolate as necessary.
† Enter column 1 with U or R of the design building section.
‡ Under appropriate column heading for Added R, find U value of resulting design section.
§ If the insulation occupies a previously considered air space, an adjustment must be made in the given building section R value.
NOTE: U and R expressed in U.S. customary units. U' and R' in SI units.
SOURCE: Abstracted from ASHRAE "Handbook of Fundamentals," 1981, with permission.

nonreflective surfaces. Table 12.4.11 facilitates the computation of the enhancement of the U value by the addition of insulation. Table 12.4.12 presents an example of calculation of U for an assembly of various building components.

AIR CONDITIONING

Cooling Load

Cooling load Q_t, the total simultaneous cooling load, expressed in Btu/hour or (watts):

$$Q_t = Q_{ext} + Q_{int} + Q_{outside\ air} \qquad (12.4.9)$$

$$Q_{ext} = \text{external heat gains}$$
$$= Q_{transmission} + Q_{solar} \qquad (12.4.10)$$

For walls and roofs

$$Q_{tr-sol} = AU(Sa\ \Delta t) \qquad \text{Btu/h (W)} \qquad (12.4.11)$$

where Sa Δt = sol-air equivalent temperature differential, °F (°C), Tables 12.4.13 and 12.4.14.

Table 12.4.12   Example of Calculation of Coefficient of Transmission U for Composite Building Components
U and R expressed in U.S. Customary units, U' and R' in SI units

Example 1. Coefficients of transmission U of flat masonry roofs with built-up roofing

Construction (heat flow up)	Resistance	
	R	R'
1. Outside surface (15 mi/h wind 24 km/h)	0.17	0.03
2. Built-up roofing ⅜ in (9 mm)	0.33	0.06
3. Roof insulation (none)		
4. Concrete slab (lt. wt. agg.) (2 in) (50 mm)	2.22	0.39
5. Corrugated metal	0	0
6. Air space	0.85	0.15
7. Metal lath and ¾ in (19 mm) plaster (lt. wt. agg.)	0.47	0.08
8. Inside surface (still air)	0.61	0.11
Total resistance	4.65	0.82
$U = 1/R = 1/4.65 =$	0.22	
$U' = 1/R' = 1/0.82 =$		1.22

Example 2. Coefficients of transmission U of masonry walls

Construction (heat flow up)	Resistance	
	R	R'
1. Outside surface (15 mi/h wind) (24 km/h)	0.17	0.03
2. Face brick (4 in) (100 mm)	0.44	0.08
3. Cement mortar (½ in) (12.5 mm)	0.10	0.02
4. Concrete block (cinder agg.) (8 in) (200 mm)	1.72	0.30
5. Air space (reflective)	2.80	0.49
6. Gypsum wallboard, foil back (½ in) (12.5 mm)	0.45	0.08
7. Inside surface (still air)	0.68	0.12
Total resistance	6.36	1.12
$U = 1/R = 1/6.36 =$	0.36	
$U' = 1/R' = 1/1.12 =$		0.89

SOURCE: Abstracted from ASHRAE "Handbook of Fundamentals," 1981, with permission.

Table 12.4.13   Total Equivalent Temperature Differentials for Calculating Heat Gain through Sunlit Walls

North latitude wall facing	8 D	8 L	10 D	10 L	12 D	12 L	2 D	2 L	4 D	4 L	6 D	6 L	8 D	8 L	10 D	10 L	12 D	12 L	South latitude wall facing
	A.M.						P.M.												
	Exterior color of wall: D = dark, L = light																		
Group A																		
NE	27	16	31	18	26	17	24	17	24	18	23	17	20	15	17	13	15	11	SE
E	32	18	41	24	37	22	29	20	28	20	26	19	23	16	20	14	18	13	E
SE	25	15	36	21	38	23	33	21	28	20	26	18	22	16	19	14	18	12	NE
S	14	9	20	13	28	18	33	22	31	21	25	18	20	15	17	13	15	11	N
SW	17	11	20	13	24	16	34	22	42	27	41	26	28	19	20	14	18	12	NW
W	17	11	20	13	24	16	30	20	42	27	48	30	33	22	22	15	19	13	W
NW	14	9	17	11	21	14	23	17	31	21	38	25	28	19	18	13	16	11	SW
N	14	9	15	10	18	12	20	15	21	16	21	16	18	14	14	11	12	9	S
Group B																		
NE	12	7	27	14	31	17	30	19	31	21	30	22	27	20	21	17	16	13	SE
E	14	8	34	18	45	24	43	25	39	25	35	24	30	22	23	18	17	14	E
SE	9	5	25	13	39	21	44	26	41	26	37	25	31	23	24	18	17	14	NE
S	4	3	7	4	18	11	32	19	41	26	39	27	33	24	25	19	18	15	N
SW	5	3	7	4	11	7	23	15	41	26	54	34	51	33	38	25	26	19	NW
W	6	4	7	4	11	4	18	12	35	23	55	34	59	37	43	28	30	20	W
NW	5	3	6	4	11	7	17	12	26	18	41	27	47	31	36	24	25	18	SW
N	6	4	9	5	12	8	18	12	22	17	25	20	27	21	22	17	16	14	S
Group C																		
NE	9	6	19	10	26	15	28	17	29	18	29	20	28	20	24	19	20	16	SE
E	10	7	22	12	36	19	40	23	39	23	36	24	33	23	28	20	22	17	E
SE	8	6	16	9	29	16	38	21	39	24	37	24	34	23	28	21	23	17	NE
S	7	5	7	4	12	7	22	14	32	20	36	24	34	24	29	21	23	17	N
SW	9	6	8	5	10	6	16	10	28	18	42	26	48	30	42	28	35	22	NW
W	10	7	9	5	10	6	14	9	24	16	40	25	52	32	47	30	37	24	W
NW	8	6	8	5	9	6	13	9	19	14	30	20	40	27	38	26	30	21	SW
N	7	5	8	5	10	7	14	9	18	13	22	16	25	19	23	18	19	16	S
Group D																		
NE	8	5	19	10	28	15	29	17	30	19	30	21	28	21	24	19	19	16	SE
E	9	6	23	12	38	20	42	24	40	24	37	24	33	23	27	20	21	17	E
SE	7	5	16	9	30	16	40	22	41	25	38	25	34	24	28	21	22	17	NE
S	5	4	6	4	12	7	23	14	34	21	38	25	35	24	29	21	23	17	N
SW	8	5	7	4	9	6	16	10	30	19	44	28	51	32	43	28	33	22	NW
W	8	6	7	5	9	6	14	9	25	16	42	27	55	34	49	31	37	25	W
NW	7	5	7	4	9	6	13	9	20	14	31	21	42	28	40	27	31	21	SW
N	6	4	8	5	10	6	14	10	19	14	23	17	25	19	24	19	19	16	S
Group E																		
NE	10	6	23	12	30	16	30	18	30	20	30	21	28	21	23	18	18	14	SE
E	11	6	28	15	42	22	43	24	39	24	36	24	32	23	25	19	19	15	E
SE	8	5	20	11	35	19	42	24	41	25	38	25	33	23	26	20	20	16	NE
S	4	3	6	4	15	9	28	17	38	24	39	26	34	24	27	20	20	16	N
SW	6	4	7	4	10	6	19	12	35	22	49	31	52	33	41	27	30	21	NW
W	7	5	7	4	10	6	16	11	30	20	48	31	57	36	47	30	34	23	W
NW	6	4	6	4	10	6	15	10	23	16	36	24	45	30	38	26	28	20	SW
N	6	4	8	5	11	7	16	11	21	15	24	18	26	20	23	18	18	15	S
Group F																		
NE	9	7	14	9	21	12	25	15	27	17	29	19	28	20	26	19	23	17	SE
E	10	8	17	10	28	15	35	19	37	22	37	23	35	23	31	22	26	19	E
SE	10	7	13	8	22	12	31	17	36	21	37	23	35	23	32	22	27	19	NE
S	9	7	7	5	10	6	17	10	26	16	32	20	33	22	31	22	27	19	N
SW	12	9	10	6	9	6	13	8	22	14	33	21	42	27	42	27	37	25	NW
W	14	9	11	7	10	6	12	8	19	12	31	20	43	27	46	29	41	27	W
NW	12	8	9	6	9	6	11	8	16	11	24	16	33	22	36	24	33	23	SW
N	8	7	8	6	9	6	12	8	15	11	19	14	22	17	23	18	21	17	S

NOTE: To convert temperature differentials in °F to °C, multiply by ⅝.
SOURCE: Adapted from Carrier "System Design Manual," 1970.

Heat Transmission

Heat gain through exterior walls and roofs is highly variable and is caused by a combination of solar heat and the temperature difference between indoor and outdoor air. Total heat flow is affected by the type and weight of the construction, exposure, sun time, latitude, and design conditions. The equivalent temperature difference from Tables 12.4.13 and 12.4.14 combined with the appropriate heat-transfer coefficients U for the construction, Tables 12.4.15 and 12.4.16, permits estimating the total heat flow through walls and roofs.

Heat flow through interior construction is essentially constant and may be determined by using the actual temperatures on either side of the construction and the appropriate U values.

Solar Heat Gain

Heat gain through glass is a combination of the heat gains by conduction and solar radiation. Both factors are highly variable and one or both are affected by the type of glass, shading, day of the year, time of day, location and orientation angle to the sun, and outside and inside temperatures.

For glass:

$$Q_{glass} = Q_u + Q_{solar} \qquad (12.4.12)$$
$$= A_1 U(t_o - t_i) + A_2(MSHGF)(CLF)(SC) \qquad (12.4.13)$$

where A_1 = area of total glass, ft² (m²); A_2 = area of sunlit glass, ft² (m²); U = heat-transfer coefficient (Table 12.4.16); t_0 = outside design temperature, °F(°C) (Fig. 12.4.4); t_i = inside design temperature, °F(°C) (Table 12.4.2); MSHGF = maximum solar heat gain factor (Table 12.4.17); CLF = cooling load factor (Tables 12.4.18, 12.4.19, and 12.4.20); SC = shading coefficient (Tables 12.4.21, 12.4.22, 12.4.23, and 12.4.24).

Note that the data in Table 12.4.17 is for vertical and horizontal glass; the data in Tables 12.4.18 to 12.4.24 addresses the effect of heat storage within the space.

Internal Heat Gains

Internal heat gains are primarily from lighting, equipment, and occupants. They are independent of outdoor temperature and solar effect and include both sensible and latent heat.

$$\begin{aligned} Q_{int} &= \text{internal heat gains} \\ &= Q_{people} + Q_{lighting} + Q_{equipment} \end{aligned} \qquad (12.4.14)$$

In some applications such as general office space, internal heat gains are relatively constant; in others where equipment cycles and/or occupancy is not constant, these loads will vary widely.

Heat gains from people vary with the activity and the proportion of men and women. The proportion of sensible latent heat varies with the temperature of the space. Table 12.4.25 lists adjusted heat gains from people for different activities and typical applications.

Lighting heat gains are based on the total electrical wattage of the operating lights, including ballasts, converted to Btu/h. Note that under some circumstances the entire electrical input to the lights may not immediately become part of the cooling load. This depends on the type and arrangement of the lights, type of air supply and return, space furnishings, the thermal characteristics of the space, and length of time the lights will be on. Unless all of these can be forseen with a high degree of certainty, and lighting represents a significant portion of the cooling load, most designers will use the full electrical input of the operating lights as the lighting load.

Representative heat gains at various lighting levels for the most common lighting fixtures are shown in Fig. 12.4.7. Curves A, B, and C are drawn for fixtures with newer electronic ballasts and high-efficiency lamps. For a given footcandle level of illumination, they use about 40 percent of the energy of earlier fixtures with magnetic ballasts and standard 40-W lamps; see curves AA, BB, and CC. Curves D and E show heat gains from recessed incandescent fixtures.

Fig. 12.4.7 Heat gain from typical lighting fixtures. *(Courtesy: S. Ricketts.)*

Table 12.4.14   Total Equivalent Temperature Differentials for Calculating Heat Gain through Flat Roofs

Description of roof construction*· †	Wt., lb/ft²	kg/m²	U value Btu/(h · ft² · °F)	U value W/(m² · K)
2 in (50 mm) insulation + steel siding	7.8	38.06	0.125	0.710
2 in (50 mm) insulation + 1 in (25 mm) wood‡	8.5	41.48	0.122	0.693
2 in (50 mm) insulation + 2.5 in (62.5 mm) wood‡	13.1	63.93	0.117	0.664
2 in (50 mm) insulation + 4 in (100 mm) wood‡	17.8	88.86	0.113	0.642
2 in (50 mm) insulation + 2 in (50 mm) h.w. concrete	28.8	140.54	0.122	0.693
4 in (100 mm) l.w. concrete	17.8	86.86	0.213	1.209
2 in (50 mm) insulation + 4 in (100 mm) h.w. concrete	52.1	254.25	0.120	0.681
2 in (50 mm) insulation + 6 in (150 mm) h.w. concrete	75.4	367.95	0.117	0.664

Explanation: Total heat transmission from solar radiation and temperature difference between outdoor and room air, Btu/(h)(ft²)(W/m²) of roof area = equivalent temperature differential from above table × heat-transmission coefficient for summer, Btu/(h · ft² · °F) [W/(m² · K)].

Application. These values may be used for all normal air conditioning estimated; usually without correction (except as noted below in latitude 0 to 50° north or south when the load is calculated for the hottest weather.

Corrections. The values in the table were calculated for an inside temperature of 75°F (24°C) and an outdoor maximum temperature of 95°F (35°C) with an outdoor daily range of 21°F (12°C). The table remains approximately correct for other outdoor maximum (93 to 102°F) (34 to 39°C) and other outdoor daily ranges (16 to 34°F) (9 to 19°C) provided the outdoor daily average temperature remains approximately 85°F (30°C). If the room air temperature is different from 75°F (24°C) and/or the outdoor daily average temperature is different from 85°F (30°C), the following rules can be applied:

   1. For room air temperature less than 75°F (24°C) add the difference between 75°F (24°C) and room air temperature; if greater than 75°F (24°C), subtract the difference.

   2. For outdoor daily average temperature less than 85°F (30°C) subtract the difference between 85°F (30°C) and the daily average temperature, if greater than 85°F (30°C), add the difference.

Attics or other spaces between the roof and ceiling. If the ceiling is insulated and a fan is used for positive ventilation in the space between the ceiling and roof, the total temperature differential for calculating the room load may be decreased by 25 percent. If the attic space contains a return duct or other air plenum, care should be taken in determining the portion of the heat gain that reaches the ceiling.

Light color. Credit should not be taken for light-colored roofs except where the permanence of light color is established by experience as in rural areas or where there is little smoke.

For solar transmission in other months. The table values of temperature differentials that were calculated for July 21 will be approximately correct for a roof in the following months.

North latitude Latitude, deg	North latitude Months	South latitude Latitude, deg	South latitude Months
0	All months	0	All months
10	All months	10	All months
20	All months except Nov., Dec., Jan.	20	All months except May, June, July
30	Mar., Apr., May, June, July, Aug., Sept.	30	Sept., Oct., Nov., Dec., Jan., Feb., March
40	April, May, June, July, Aug.	40	Oct., Nov., Dec., Jan., Feb.
50	May, June, July	50	Nov., Dec., Jan.

* Includes outside surface resistance, ½ in (12.5 mm) membrane and ⅜ in (9 mm) felt on the top and inside surface resistance on the bottom.
† Dark roof (D) = 0.30; light roof (L) = 0.15.
‡ Nominal thickness of wood.

Heat gain from equipment located within the conditioned space is often a significant portion of the air-conditioning load. Lists of equipment that will be installed, manufacturers' heat release (or wattage) data, and anticipated use and load factors should be obtained from the user early in the design process. Tables 12.4.26, 12.4.27, and 12.4.28 show anticipated gains from a variety of office equipment, restaurant appliances and electric motors. For restaurant cooking equipment, a properly designed, mechanically exhausted hood will reduce the total load of the hooded equipment by about 50 percent.

Miscellaneous Heat Gains   Additional heat gains that add to the system load are the heat added at the supply and return fans and heat transferred from unconditioned areas to the distribution ductwork. An additional load on a chilled-water system is the heat equivalent of the chilled-water pump motor work.

Fan heat is the heat added by the fan motor, and manifests itself as the total power input to the fan motor when the motor is located within the airstream or conditioned space, or the net output of the fan motor when it is located out of the airstream or conditioned space. Since air can be considered to be a perfect gas, all the fan heat is added at the fan and the conditioned air temperature will increase as it flows through the fan. If the motor is in the airstream, its inefficiency will add to fan heat.

The contribution of fan heat to temperature rise Δt (°F) is computed as follows:

$$\Delta t = Q_{motor}/1.08 \qquad \text{ft}^3/\text{min} \qquad (12.4.15)$$

where Q_{motor} is obtained from data in Table 12.4.28.

						Sun time													
	A.M.						P.M.												
8		10		12		2		4		6		8		10		12		λ	δ
D	L	D	L	D	L	D	L	D	L	D	L	D	L	D	L	D	L		
Light construction roofs exposed to sun																			
24	8	61	29	88	46	96	53	81	46	48	30	10	8	2	2	−3	−3	0.99	1
8	0	41	18	72	36	90	48	88	49	65	38	30	19	9	7	1	0	0.93	2
1	−2	19	6	43	20	65	33	76	41	72	40	53	31	33	20	18	11	0.68	4
Medium construction roofs exposed to sun																			
6	1	13	4	28	12	45	22	58	30	63	34	56	31	43	25	32	18	0.48	5
2	−2	23	9	49	23	70	36	79	43	71	40	49	29	28	17	15	9	0.73	3
1	−3	28	11	59	28	82	43	88	48	74	42	44	27	19	12	6	4	0.87	3
Heavy construction roofs exposed to sun																			
7	2	15	6	30	13	46	23	58	30	61	33	54	30	41	23	31	17	0.45	5
15	7	17	7	25	11	36	17	46	23	51	27	50	27	43	24	36	20	0.30	6

Table 12.4.15   Descriptions of Wall Construction

		Weight		U value	
Group	Components	lb/ft^2	kg/m^2	Btu/(h · ft^2 · °F)	W/(m^2 · K)
A	1 in (2.5 cm) stucco + 4 in (10 cm) l.w. concrete block + air space	28.6	139.63	0.267	1.516
	1 in (2.5 cm) stucco + air space + 2 in (5 cm) insulation	16.3	79.58	0.106	0.601
B	1 in (2.5 cm) stucco + 4 in (10 cm) common brick	55.9	272.90	0.393	2.232
	1 in (2.5 cm) stucco + 4 in (10 cm) l.w. concrete	62.5	305.13	0.481	2.731
C	4 in (10 cm) face brick + 4 in (10 cm) l.w. concrete block + 1 in (2.5 cm) insulation	62.5	305.13	0.158	0.897
	1 in (2.5 cm) stucco + 4 in (10 cm) h.w. concrete + 2 in (5 cm) insulation	62.9	307.08	0.114	0.647
D	1 in (2.5 cm) stucco + 8 in (20 cm) l.w. concrete block + 1 in (2.5 cm) insulation	41.4	202.11	0.141	0.801
	1 in (2.5 cm) stucco + 2 in (5 cm) insulation + 4 in (10 cm) h.w. concrete block	36.6	178.68	0.111	0.630
E	4 in (10 cm) face brick + 3 in (10 cm) l.w. concrete block	62.2	303.66	0.333	1.891
	1 in (2.5 cm) stucco + 8 in (20 cm) h.w. concrete block	56.6	276.32	0.349	1.982
F	4 in (10 cm) face brick + 4 in (10 cm) common brick	89.5	436.94	0.360	2.044
	4 in (10 cm) face brick + 2 in (5 cm) insulation + 4 in (10 cm) l.w. concrete block	62.5	305.13	0.103	0.585

Table 12.4.16   Coefficients of Transmission U of Windows, Skylights, and Light Transmitting Partitions
These values are for heat transfer from air to air, Btu/(h · ft^2 · °F).

Part A—Vertical panels (exterior windows, sliding patio doors, and partitions)
—flat glass, glass block, and plastic sheet

Part B—Horizontal panels (skylights)—flat glass, glass block, and plastic domes

Description	Exterior[a]		Interior	Description	Exterior[a]		Interior[l]
	Winter	Summer			Winter[i]	Summer[j]	
Flat glass[b]				Flat glass[e]			
Single glass	1.10	1.04	0.73	Single glass	1.23	0.83	0.96
Insulating glass—double[c]				Insulating glass—doubles[c]			
0.1875-in airspace[d]	0.62	0.65	0.51	0.1875-in airspace[d]	0.70	0.57	0.62
0.25-in airspace[d]	0.58	0.61	0.49	0.25-in airspace[d]	0.65	0.54	0.59
0.5-in airspace[e]	0.49	0.56	0.46	0.5-in airspace[c]	0.59	0.49	0.56
0.5-in airspace, low-emittance coating[f]				0.5-in airspace, low-emittance coating[f]			
$e = 0.20$	0.32	0.38	0.32	$e = 0.20$	0.48	0.36	0.39
$e = 0.40$	0.38	0.45	0.38	$e = 0.40$	0.52	0.42	0.45
$e = 0.60$	0.43	0.51	0.42	$e = 0.60$	0.56	0.46	0.50
Insulating glass—triple[c]				Glass block[h]			
0.25-in airspaces[d]	0.39	0.44	0.38	11 × 11 × 3 in thick with cavity divider	0.53	0.35	0.44
0.5-in airspaces[g]	0.31	0.39	0.30	12 × 12 × 4 in thick with cavity divider	0.51	0.34	0.42
Storm windows							
1-in to 4-in airspace[d]	0.50	0.50	0.44				
Plastic sheet				Plastic domes[k]			
Single glazed				Single-walled	1.15	0.80	—
0.125 in thick	1.06	0.98	—	Double-walled	0.70	0.46	—
0.25 in thick	0.96	0.89	—				
0.5 in thick	0.81	0.76	—				
Insulating unit—double[c]							
0.25-in airspace[d]	0.55	0.56	—				
0.5-in airspace[c]	0.43	0.45	—				

Part C—adjustment factors for various window and sliding patio door types
(multiply U values in parts A and B by these factors)

Description	Single glass	Double or triple glass	storm windows
Windows			
All Glass[l]	1.00	1.00	1.00
Wood sash—80% glass	0.90	0.95	0.90
Wood sash—60% glass	0.80	0.85	0.80
Metal sash—80% glass	1.00	1.20[m]	1.20[m]
Sliding patio doors			
Wood frame	0.95	1.00	—
Metal frame	1.00	1.10[m]	—

(Glass block section in Part A:)

Description	Winter	Summer	Interior
Glass block[h]			
6 × 6 × 4 in thick	0.60	0.57	0.46
8 × 8 × 4 in thick	0.56	0.54	0.44
—with cavity divider	0.48	0.46	0.38
12 × 12 × 4 in thick	0.52	0.50	0.41
—with cavity divider	0.44	0.42	0.36
12 × 12 × 2 in thick	0.60	0.57	0.46

[a] See Part C for adjustment for various window and sliding patio door types.
[b] Emittance of uncooled glass surface = 0.84.
[c] Double and triple refer to the number of lights of glass.
[d] 0.125-in glass.
[e] 0.25-in glass.
[f] Coating on either glass surface facing airspace; all other glass surfaces uncoated.
[g] Window design: 0.25-in glass–0.125-in glass–0.25-in glass.
[h] Dimensions are nominal.
[i] For heat flow up.
[j] For heat flow down.
[k] Based on area of opening, not total surface area.
[l] Refers to windows with negligible opaque area.
[m] Values will be less than these when metal sash and frame incorporate thermal breaks. In some thermal break designs, U values will be equal to or less than those for the glass. Window manufacturers should be consulted for specific data.
SOURCE: Abstracted from ASHRAE "Handbook of Fundamentals," 1977, with permission.

Table 12.4.17   Maximum Solar Heat Gain Factor (MSHGF), Btu/(h · ft²), for Sunlit Glass

0°N Lat

	N	NNE/NNW	NE/NW	ENE/WNW	E/W	ESE/WSW	SE/SW	SSE/SSW	S	Hor.
Jan.	34	34	88	177	234	254	235	182	118	296
Feb.	36	39	132	205	245	247	210	141	67	306
Mar.	38	87	170	223	242	223	170	87	38	303
Apr.	71	134	193	224	221	184	118	38	37	284
May	113	164	203	218	201	154	80	37	37	265
June	129	173	206	212	191	140	66	37	37	255
July	115	164	201	213	195	149	77	38	38	260
Aug.	75	134	187	216	212	175	112	39	38	276
Sep.	40	84	163	213	231	213	163	84	40	293
Oct.	37	40	129	199	236	238	202	135	66	299
Nov.	35	35	88	175	230	250	230	179	117	293
Dec.	34	34	71	164	226	253	240	196	138	288

8°N Lat

	N	NNE/NNW	NE/NW	ENE/WNW	E/W	ESE/WSW	SE/SW	SSE/SSW	S	Hor.
Jan.	32	32	71	163	224	250	242	203	162	275
Feb.	34	34	114	193	239	248	219	165	110	294
Mar.	37	67	156	215	241	230	184	110	55	300
Apr.	44	117	184	221	225	195	134	53	39	289
May	74	146	198	220	209	167	97	39	38	277
June	90	155	200	217	200	141	82	39	39	269
July	77	145	195	215	204	162	93	40	39	272
Aug.	47	117	179	214	216	186	128	51	41	282
Sep.	38	66	149	205	230	219	176	107	56	290
Oct.	35	35	112	187	231	239	211	160	108	288
Nov.	33	33	71	161	220	245	233	200	160	273
Dec.	31	31	55	149	215	246	247	215	179	265

16°N Lat

	N	NNE/NNW	NE/NW	ENE/WNW	E/W	ESE/WSW	SE/SW	SSE/SSW	S	Hor.
Jan.	30	30	55	147	210	244	251	223	199	248
Feb.	33	33	96	180	231	247	233	188	154	275
Mar.	35	53	140	205	239	235	197	138	93	291
Apr.	39	99	172	215	227	204	150	77	45	289
May	52	132	189	218	215	179	115	45	41	282
June	66	142	194	217	207	167	99	41	41	277
July	55	132	187	214	210	174	111	44	42	277
Aug.	41	100	168	209	219	196	143	74	46	282
Sep.	36	50	134	196	227	224	191	134	93	282
Oct.	33	33	95	174	223	237	225	183	150	270
Nov.	30	30	55	145	206	241	247	220	196	246
Dec.	29	29	41	132	198	241	254	233	212	234

24°N Lat

	N	NNE/NNW	NE/NW	ENE/WNW	E/W	ESE/WSW	SE/SW	SSE/SSW	S	Hor.
Jan.	27	27	41	128	190	240	253	241	227	214
Feb.	30	30	80	165	220	244	243	213	192	249
Mar.	34	45	124	195	234	237	214	168	137	275
Apr.	37	88	159	209	228	212	169	107	75	283
May	43	117	178	214	218	190	132	67	46	282
June	55	127	184	214	212	179	117	55	43	279
July	45	116	176	210	213	185	129	65	46	278
Aug.	38	87	156	203	220	204	162	103	72	277
Sep.	35	42	119	185	222	225	206	163	134	266
Oct.	31	31	79	159	211	237	235	207	187	244
Nov.	27	27	42	126	187	236	249	237	224	213
Dec.	26	26	29	112	180	234	247	247	237	199

32°N Lat

	N	NNE/NNW	NE/NW	ENE/WNW	E/W	ESE/WSW	SE/SW	SSE/SSW	S	Hor.
Jan.	24	24	29	105	175	229	249	250	246	176
Feb.	27	27	65	149	205	242	248	232	221	217
Mar.	32	37	107	183	227	237	227	195	176	252
Apr.	36	80	146	200	227	219	187	141	115	271
May	38	111	170	208	220	199	155	99	74	277
June	44	122	176	208	214	189	139	83	60	276
July	40	111	167	204	215	194	150	96	72	273
Aug.	37	79	141	195	219	210	181	136	111	265
Sep.	33	35	103	173	215	227	218	189	171	244
Oct.	28	28	63	143	195	234	239	225	215	213
Nov.	24	24	29	103	173	225	245	246	243	175
Dec.	22	22	22	84	162	218	246	252	252	158

40°N Lat

	N (Shade)	NNE/NNW	NE/NW	ENE/WNW	E/W	ESE/WSW	SE/SW	SSE/SSW	S	Hor.
Jan.	20	20	20	74	154	205	241	252	254	133
Feb.	24	24	50	129	186	234	246	244	241	180
Mar.	29	29	93	169	218	238	236	216	206	223
Apr.	34	71	140	190	224	223	203	170	154	252
May	37	102	165	202	220	208	175	133	113	265
June	48	113	172	205	216	199	161	116	95	267
July	38	102	163	198	216	203	170	129	109	262
Aug.	35	71	135	185	216	214	196	165	149	247
Sep.	30	30	87	160	203	227	226	209	200	215
Oct.	25	25	49	123	180	225	238	236	234	177
Nov.	20	20	30	71	151	201	237	248	250	132
Dec.	18	18	18	60	135	188	232	249	253	113

48°N Lat

	N (Shade)	NNE/NNW	NE/NW	ENE/WNW	E/W	ESE/WSW	SE/SW	SSE/SSW	S	Hor.
Jan.	15	15	15	53	118	175	216	239	245	85
Feb.	20	20	36	103	168	216	242	249	250	138
Mar.	26	26	80	154	204	234	239	232	228	188
Apr.	31	61	132	180	219	225	215	194	186	226
May	35	97	158	200	218	214	192	163	150	247
June	46	110	165	204	215	206	180	148	134	252
July	37	96	156	196	214	209	187	158	146	244
Aug.	33	61	128	174	211	216	208	188	180	223
Sep.	27	27	72	144	191	223	228	223	230	182
Oct.	21	21	35	96	161	207	213	241	242	136
Nov.	15	15	15	52	115	172	212	234	240	85
Dec.	13	13	13	36	91	156	195	225	233	65

56°N Lat

	N (Shade)	NNE/NNW	NE/NW	ENE/WNW	E/W	ESE/WSW	SE/SW	SSE/SSW	S	Hor.
Jan.	10	10	10	21	74	126	169	194	205	40
Feb.	16	16	21	71	139	184	223	239	244	91
Mar.	22	22	65	136	185	224	238	241	241	149
Apr.	28	58	123	173	211	223	223	213	210	195
May	36	99	149	195	215	218	206	187	181	222
June	53	111	160	199	213	213	196	174	168	231
July	37	98	147	192	211	214	201	183	177	221
Aug.	30	56	119	165	203	216	215	206	203	193
Sep.	23	23	58	126	171	211	227	230	231	144
Oct.	16	16	20	68	132	176	213	229	234	91
Nov.	10	10	10	21	72	122	165	190	200	40
Dec.	7	7	7	7	47	92	135	159	171	23

60°N Lat

	N (Shade)	NNE/NNW	NE/NW	ENE/WNW	E/W	ESE/WSW	SE/SW	SSE/SSW	S	Hor.
Jan.	7	7	7	7	46	88	130	152	164	21
Feb.	13	13	13	58	118	168	204	225	231	68
Mar.	20	20	56	125	173	215	234	241	242	128
Apr.	27	59	118	168	206	222	225	220	218	178
May	43	98	149	192	212	220	211	198	194	208
June	58	110	162	197	213	215	202	186	181	217
July	44	97	147	189	208	215	206	193	190	207
Aug.	28	57	114	161	199	214	217	213	211	176
Sep.	21	21	50	145	160	202	222	229	231	123
Oct.	14	14	14	56	111	159	193	215	221	67
Nov.	7	7	7	7	45	86	127	148	160	22
Dec.	4	4	4	4	16	51	76	100	107	9

64°N Lat

	N (Shade)	NNE/NNW	NE/NW	ENE/WNW	E/W	ESE/WSW	SE/SW	SSE/SSW	S	Hor.
Jan.	3	3	3	3	15	45	67	89	96	8
Feb.	11	11	11	43	89	144	177	202	210	45
Mar.	18	18	47	113	159	203	226	236	239	105
Apr.	25	59	113	163	201	219	225	225	224	160
May	48	97	150	189	211	220	215	207	204	192
June	62	114	162	193	213	216	208	196	193	203
July	49	96	148	186	207	215	211	202	200	192
Aug.	27	58	109	157	193	211	217	217	217	159
Sep.	19	19	43	103	148	189	213	224	227	101
Oct.	11	11	11	40	83	135	167	191	199	46
Nov.	4	4	4	4	15	44	66	87	93	8
Dec.	0	0	0	0	1	5	11	14	15	1

SOURCE: Abstracted from ASHRAE "Handbook of Fundamentals," 1989, with permission.

Table 12.4.18 Cooling Load Factors (CLFs) for Glass without Interior Shading, in North Latitude Spaces Having Carpeted Floors

Dir.	Room mass	Solar time																						
		0100	0200	0300	0400	0500	0600	0700	0800	0900	1000	1100	1200	1300	1400	1500	1600	1700	1800	1900	2000	2100	2200	2300	2400
N	L	0.00	0.00	0.00	0.00	0.01	0.64	0.73	0.74	0.81	0.88	0.95	0.98	0.98	0.94	0.88	0.79	0.79	0.55	0.31	0.12	0.04	0.02	0.01	0.00
	M	0.03	0.02	0.02	0.02	0.02	0.64	0.69	0.69	0.77	0.84	0.91	0.94	0.95	0.91	0.86	0.79	0.79	0.56	0.32	0.16	0.10	0.07	0.05	0.04
	H	0.10	0.09	0.08	0.07	0.07	0.62	0.64	0.64	0.71	0.77	0.83	0.87	0.88	0.85	0.81	0.75	0.76	0.55	0.34	0.22	0.17	0.15	0.13	0.11
NE	L	0.00	0.00	0.00	0.00	0.01	0.51	0.83	0.88	0.72	0.47	0.33	0.27	0.24	0.23	0.20	0.18	0.14	0.09	0.03	0.01	0.00	0.00	0.00	0.00
	M	0.01	0.01	0.00	0.00	0.01	0.50	0.78	0.82	0.67	0.44	0.32	0.28	0.26	0.24	0.22	0.19	0.15	0.11	0.05	0.03	0.02	0.02	0.01	0.01
	H	0.03	0.03	0.03	0.02	0.03	0.47	0.71	0.72	0.59	0.40	0.30	0.27	0.26	0.25	0.23	0.20	0.17	0.13	0.08	0.06	0.05	0.05	0.04	0.04
E	L	0.00	0.00	0.00	0.00	0.00	0.42	0.76	0.91	0.90	0.75	0.51	0.30	0.22	0.18	0.16	0.13	0.11	0.07	0.02	0.01	0.00	0.00	0.00	0.00
	M	0.01	0.01	0.01	0.00	0.01	0.41	0.72	0.86	0.84	0.71	0.48	0.30	0.24	0.21	0.18	0.16	0.13	0.09	0.04	0.03	0.02	0.01	0.01	0.01
	H	0.03	0.03	0.03	0.02	0.02	0.39	0.66	0.76	0.74	0.63	0.43	0.29	0.24	0.22	0.20	0.18	0.15	0.12	0.08	0.06	0.05	0.05	0.04	0.04
SE	L	0.00	0.00	0.00	0.00	0.00	0.27	0.58	0.81	0.93	0.93	0.81	0.59	0.37	0.27	0.21	0.18	0.14	0.09	0.03	0.01	0.00	0.00	0.00	0.00
	M	0.01	0.01	0.01	0.00	0.01	0.26	0.55	0.77	0.88	0.87	0.76	0.56	0.37	0.29	0.24	0.20	0.16	0.11	0.05	0.04	0.03	0.02	0.02	0.01
	H	0.04	0.04	0.03	0.03	0.03	0.26	0.51	0.69	0.78	0.78	0.68	0.51	0.35	0.29	0.25	0.22	0.19	0.15	0.09	0.08	0.07	0.06	0.05	0.05
S	L	0.00	0.00	0.00	0.00	0.00	0.07	0.15	0.23	0.39	0.62	0.82	0.94	0.93	0.80	0.59	0.38	0.26	0.16	0.09	0.04	0.01	0.00	0.00	0.00
	M	0.01	0.01	0.01	0.01	0.01	0.07	0.14	0.22	0.38	0.59	0.78	0.88	0.88	0.76	0.57	0.38	0.28	0.18	0.13	0.06	0.04	0.03	0.02	0.02
	H	0.05	0.05	0.04	0.04	0.03	0.09	0.15	0.21	0.35	0.54	0.70	0.79	0.79	0.69	0.52	0.37	0.29	0.21	0.19	0.10	0.09	0.08	0.07	0.06
SW	L	0.00	0.00	0.01	0.01	0.01	0.05	0.09	0.13	0.16	0.19	0.23	0.38	0.60	0.82	0.89	0.94	0.81	0.54	0.19	0.07	0.04	0.03	0.00	0.00
	M	0.02	0.02	0.01	0.01	0.01	0.07	0.11	0.14	0.16	0.18	0.22	0.35	0.60	0.78	0.89	0.89	0.77	0.52	0.20	0.10	0.07	0.05	0.04	0.03
	H	0.07	0.06	0.05	0.05	0.04	0.03	0.07	0.10	0.13	0.15	0.21	0.39	0.55	0.71	0.80	0.79	0.69	0.48	0.25	0.14	0.11	0.10	0.08	0.07
W	L	0.00	0.02	0.01	0.01	0.01	0.03	0.07	0.10	0.13	0.15	0.16	0.18	0.31	0.55	0.78	0.92	0.93	0.73	0.24	0.10	0.04	0.01	0.01	0.00
	M	0.02	0.02	0.01	0.01	0.01	0.04	0.07	0.10	0.13	0.14	0.16	0.17	0.30	0.53	0.74	0.87	0.88	0.69	0.23	0.12	0.07	0.05	0.04	0.03
	H	0.06	0.06	0.05	0.05	0.04	0.06	0.09	0.11	0.13	0.15	0.16	0.17	0.28	0.49	0.67	0.78	0.79	0.62	0.28	0.14	0.11	0.09	0.08	0.07
NW	L	0.00	0.00	0.01	0.00	0.00	0.04	0.09	0.14	0.17	0.20	0.22	0.23	0.24	0.31	0.53	0.75	0.88	0.81	0.26	0.10	0.04	0.02	0.01	0.00
	M	0.02	0.02	0.01	0.00	0.01	0.05	0.10	0.13	0.17	0.19	0.21	0.22	0.23	0.30	0.52	0.68	0.79	0.77	0.23	0.12	0.07	0.05	0.04	0.03
	H	0.06	0.05	0.05	0.04	0.04	0.07	0.11	0.14	0.17	0.19	0.20	0.21	0.22	0.28	0.48	0.64	0.71	0.69	0.23	0.14	0.10	0.09	0.08	0.07
Hor.	L	0.00	0.00	0.00	0.00	0.00	0.08	0.25	0.45	0.64	0.80	0.91	0.97	0.97	0.91	0.80	0.63	0.44	0.23	0.08	0.03	0.01	0.00	0.00	0.00
	M	0.02	0.02	0.01	0.01	0.01	0.08	0.24	0.43	0.60	0.75	0.86	0.92	0.92	0.87	0.77	0.63	0.45	0.26	0.12	0.07	0.05	0.04	0.03	0.02
	H	0.07	0.06	0.05	0.05	0.04	0.11	0.25	0.41	0.56	0.68	0.77	0.83	0.83	0.80	0.71	0.59	0.44	0.28	0.17	0.13	0.11	0.10	0.09	0.08

Values for nominal 15 ft × 15 ft × 10 ft high space, with ceiling, and 50% or less glass in exposed surface at listed orientation.
L = lightweight construction.
M = mediumweight construction.
H = heavyweight construction.
SOURCE: Abstracted from ASHRAE "Handbook of Fundamentals," 1989, with permission.

Table 12.4.19 Cooling Load Factors (CLFs) for Glass without Interior Shading, in North Latitude Spaces Having Uncarpeted Floors

Dir.	Room mass	Solar time																							
		0100	0200	0300	0400	0500	0600	0700	0800	0900	1000	1100	1200	1300	1400	1500	1600	1700	1800	1900	2000	2100	2200	2300	2400
N	L	0.00	0.00	0.00	0.00	0.01	0.64	0.73	0.74	0.81	0.88	0.95	0.98	0.98	0.94	0.88	0.79	0.79	0.55	0.31	0.12	0.04	0.02	0.01	0.00
	M	0.12	0.09	0.07	0.06	0.05	0.33	0.45	0.53	0.61	0.69	0.76	0.82	0.85	0.86	0.85	0.81	0.80	0.70	0.60	0.43	0.32	0.24	0.19	0.15
	H	0.24	0.21	0.19	0.18	0.16	0.43	0.48	0.51	0.56	0.61	0.66	0.71	0.73	0.74	0.73	0.71	0.71	0.62	0.52	0.42	0.36	0.32	0.29	0.26
NE	L	0.00	0.00	0.00	0.00	0.01	0.51	0.83	0.88	0.72	0.47	0.33	0.27	0.24	0.23	0.20	0.18	0.14	0.09	0.03	0.01	0.00	0.00	0.00	0.00
	M	0.03	0.02	0.02	0.02	0.02	0.24	0.45	0.57	0.58	0.49	0.41	0.36	0.32	0.29	0.27	0.24	0.21	0.17	0.13	0.10	0.07	0.06	0.05	0.04
	H	0.08	0.07	0.07	0.06	0.06	0.27	0.43	0.49	0.45	0.37	0.32	0.29	0.28	0.27	0.26	0.24	0.22	0.19	0.16	0.14	0.12	0.11	0.10	0.09
E	L	0.00	0.00	0.00	0.00	0.00	0.42	0.76	0.91	0.90	0.75	0.51	0.30	0.22	0.18	0.16	0.13	0.11	0.07	0.02	0.01	0.00	0.00	0.00	0.00
	M	0.03	0.02	0.02	0.02	0.01	0.20	0.41	0.57	0.65	0.64	0.55	0.44	0.36	0.31	0.26	0.23	0.19	0.16	0.12	0.09	0.07	0.06	0.04	0.04
	H	0.08	0.08	0.07	0.06	0.06	0.24	0.40	0.50	0.53	0.50	0.41	0.33	0.30	0.28	0.26	0.24	0.22	0.19	0.16	0.14	0.13	0.11	0.10	0.09
SE	L	0.00	0.00	0.00	0.00	0.00	0.27	0.58	0.81	0.93	0.93	0.81	0.59	0.37	0.27	0.21	0.18	0.14	0.09	0.03	0.01	0.00	0.00	0.00	0.00
	M	0.04	0.03	0.02	0.02	0.02	0.13	0.31	0.48	0.62	0.69	0.69	0.61	0.50	0.41	0.35	0.30	0.25	0.20	0.15	0.12	0.09	0.07	0.06	0.05
	H	0.10	0.09	0.08	0.08	0.07	0.18	0.32	0.45	0.53	0.56	0.54	0.47	0.39	0.35	0.32	0.29	0.26	0.23	0.19	0.17	0.15	0.14	0.12	0.11
S	L	0.00	0.00	0.00	0.00	0.00	0.07	0.15	0.23	0.39	0.62	0.82	0.94	0.93	0.80	0.59	0.38	0.26	0.16	0.06	0.02	0.01	0.00	0.00	0.00
	M	0.05	0.04	0.04	0.03	0.02	0.05	0.09	0.14	0.24	0.38	0.53	0.65	0.72	0.71	0.63	0.52	0.42	0.33	0.24	0.18	0.14	0.11	0.09	0.07
	H	0.13	0.12	0.10	0.09	0.09	0.11	0.14	0.17	0.25	0.36	0.47	0.55	0.58	0.56	0.49	0.41	0.36	0.30	0.25	0.21	0.19	0.17	0.16	0.14
SW	L	0.00	0.00	0.00	0.00	0.00	0.04	0.09	0.13	0.16	0.19	0.23	0.39	0.62	0.82	0.94	0.94	0.81	0.54	0.19	0.07	0.03	0.01	0.00	0.00
	M	0.08	0.07	0.05	0.04	0.03	0.05	0.07	0.09	0.12	0.15	0.17	0.26	0.40	0.54	0.66	0.73	0.72	0.61	0.43	0.31	0.23	0.17	0.13	0.10
	H	0.15	0.14	0.12	0.11	0.10	0.11	0.12	0.14	0.15	0.17	0.18	0.26	0.37	0.48	0.56	0.59	0.57	0.47	0.33	0.27	0.23	0.21	0.19	0.17
W	L	0.00	0.00	0.00	0.00	0.00	0.03	0.07	0.10	0.13	0.15	0.16	0.18	0.31	0.55	0.78	0.92	0.93	0.73	0.25	0.10	0.04	0.01	0.01	0.00
	M	0.08	0.07	0.05	0.04	0.04	0.04	0.06	0.08	0.10	0.12	0.13	0.15	0.21	0.35	0.50	0.63	0.71	0.67	0.46	0.33	0.24	0.18	0.14	0.11
	H	0.14	0.13	0.12	0.11	0.10	0.10	0.11	0.12	0.13	0.14	0.15	0.16	0.21	0.33	0.45	0.54	0.58	0.52	0.33	0.26	0.22	0.19	0.18	0.16
NW	L	0.00	0.00	0.00	0.00	0.00	0.04	0.09	0.14	0.17	0.20	0.22	0.23	0.24	0.31	0.53	0.78	0.92	0.81	0.28	0.10	0.04	0.02	0.01	0.00
	M	0.08	0.06	0.05	0.04	0.03	0.05	0.07	0.10	0.13	0.15	0.17	0.19	0.20	0.24	0.36	0.51	0.64	0.66	0.46	0.32	0.23	0.17	0.13	0.10
	H	0.13	0.12	0.11	0.10	0.09	0.10	0.12	0.13	0.15	0.16	0.17	0.18	0.19	0.23	0.33	0.46	0.55	0.53	0.33	0.25	0.21	0.18	0.16	0.15
Hor.	L	0.00	0.00	0.00	0.00	0.00	0.08	0.25	0.45	0.64	0.80	0.91	0.97	0.97	0.91	0.80	0.64	0.44	0.23	0.08	0.03	0.01	0.00	0.00	0.00
	M	0.07	0.06	0.05	0.04	0.03	0.06	0.14	0.26	0.40	0.53	0.64	0.73	0.78	0.80	0.77	0.70	0.59	0.45	0.33	0.24	0.19	0.14	0.11	0.09
	H	0.16	0.15	0.13	0.12	0.11	0.13	0.20	0.29	0.39	0.48	0.56	0.61	0.65	0.65	0.63	0.57	0.49	0.40	0.32	0.28	0.25	0.22	0.20	0.18

Values for nominal 15 ft × 15 ft × 10 ft high space, with ceiling, and 50% or less glass in exposed surface at listed orientation.
L = lightweight construction.
M = mediumweight construction.
H = heavyweight construction.
SOURCE: Abstracted from ASHRAE "Handbook of Fundamentals," 1989, with permission.

Table 12.4.20 Cooling Load Factors (CLFs) for Glass with Interior Shading, North Latitudes (All Room Constructions)

Fenes-tration facing	Solar time																							
	0100	0200	0300	0400	0500	0600	0700	0800	0900	1000	1100	1200	1300	1400	1500	1600	1700	1800	1900	2000	2100	2200	2300	2400
N	0.08	0.07	0.06	0.06	0.07	0.73	0.66	0.65	0.73	0.80	0.86	0.89	0.89	0.86	0.82	0.75	0.78	0.91	0.24	0.18	0.15	0.13	0.11	0.10
NNE	0.03	0.03	0.02	0.02	0.03	0.64	0.77	0.62	0.42	0.37	0.37	0.37	0.36	0.35	0.32	0.28	0.23	0.17	0.08	0.07	0.06	0.05	0.04	0.04
NE	0.03	0.02	0.02	0.02	0.02	0.56	0.76	0.74	0.58	0.37	0.29	0.27	0.26	0.24	0.22	0.20	0.16	0.12	0.06	0.05	0.04	0.04	0.03	0.03
ENE	0.03	0.02	0.02	0.02	0.02	0.52	0.76	0.80	0.71	0.52	0.31	0.26	0.24	0.22	0.20	0.17	0.15	0.11	0.06	0.05	0.04	0.04	0.03	0.03
E	0.03	0.02	0.02	0.02	0.02	0.47	0.72	0.80	0.76	0.62	0.41	0.27	0.24	0.22	0.20	0.18	0.14	0.11	0.06	0.05	0.05	0.04	0.03	0.03
ESE	0.03	0.03	0.02	0.02	0.02	0.41	0.67	0.79	0.80	0.72	0.54	0.34	0.27	0.24	0.21	0.19	0.15	0.12	0.07	0.06	0.05	0.04	0.04	0.03
SE	0.03	0.03	0.02	0.02	0.02	0.30	0.57	0.74	0.81	0.79	0.68	0.49	0.33	0.28	0.25	0.22	0.18	0.13	0.08	0.07	0.06	0.05	0.04	0.04
SSE	0.04	0.03	0.03	0.03	0.02	0.12	0.31	0.54	0.72	0.81	0.81	0.71	0.54	0.38	0.32	0.27	0.22	0.16	0.09	0.08	0.07	0.06	0.05	0.04
S	0.04	0.04	0.03	0.03	0.03	0.09	0.16	0.23	0.38	0.58	0.75	0.83	0.80	0.68	0.50	0.35	0.27	0.19	0.11	0.09	0.08	0.07	0.06	0.05
SSW	0.05	0.04	0.04	0.03	0.03	0.09	0.14	0.18	0.22	0.27	0.43	0.63	0.78	0.84	0.80	0.66	0.46	0.25	0.13	0.11	0.09	0.08	0.07	0.06
SW	0.05	0.05	0.04	0.04	0.03	0.07	0.11	0.14	0.16	0.19	0.22	0.38	0.59	0.75	0.83	0.81	0.69	0.45	0.16	0.12	0.10	0.09	0.07	0.06
WSW	0.05	0.05	0.04	0.04	0.03	0.07	0.10	0.12	0.14	0.16	0.17	0.23	0.44	0.64	0.78	0.84	0.78	0.55	0.16	0.12	0.10	0.09	0.07	0.06
W	0.05	0.05	0.04	0.04	0.03	0.06	0.09	0.11	0.13	0.15	0.16	0.17	0.31	0.53	0.72	0.82	0.81	0.61	0.16	0.12	0.10	0.08	0.07	0.06
WNW	0.05	0.05	0.04	0.03	0.03	0.07	0.10	0.12	0.14	0.16	0.17	0.18	0.22	0.43	0.65	0.80	0.84	0.66	0.16	0.12	0.10	0.08	0.07	0.06
NW	0.05	0.04	0.04	0.03	0.03	0.07	0.11	0.14	0.17	0.19	0.20	0.21	0.22	0.30	0.52	0.73	0.82	0.69	0.16	0.12	0.10	0.08	0.07	0.06
NNW	0.05	0.05	0.04	0.03	0.03	0.11	0.17	0.22	0.26	0.30	0.32	0.33	0.34	0.34	0.39	0.61	0.82	0.76	0.17	0.12	0.10	0.08	0.07	0.06
Hor.	0.06	0.05	0.04	0.04	0.03	0.12	0.27	0.44	0.59	0.72	0.81	0.85	0.85	0.81	0.71	0.58	0.42	0.25	0.14	0.12	0.10	0.08	0.07	0.06

SOURCE: Abstracted from ASHRAE "Handbook or Fundamentals," 1989, with permission.

Table 12.4.21   Shading Coefficients for Single Glass and Insulating Glass*

			Shading coefficient	
	Single glass			
Type of glass	Nominal thickness,† in (mm)	Solar trans.†	$b_0 = 4.0$	$b_0 = 3.0$
Regular sheet	3/32, 1/8 (2, 3)	0.87	1.00	1.00
Regular plate/float	1/4 (6)	0.80	0.95	0.97
	3/8 (9)	0.75	0.91	0.93
	1/2 (12)	0.71	0.88	0.91
Gray sheet	1/8 (3)	0.59	0.78	0.80
	3/16 (5)	0.74	0.90	0.92
	7/32 (6)	0.45	0.66	0.70
	7/32 (6)	0.71	0.88	0.90
	1/4	0.67	0.86	0.88
Heat-absorbing	3/16 (5)	0.52	0.72	0.75
plate/float§	1/4 (6)	0.47	0.70	0.74
	3/8 (9)	0.33	0.56	0.61
	1/2 (12)	0.24	0.50	0.57

		Insulating glass*			
		Solar trans.†		Shading coefficient	
Type of glass	Nominal thickness,‡ in (mm)	Outer pane	Inner pane	$b_0 = 4.0$	$b_0 = 3.0$
Regular sheet out, regular sheet in	3/32, 1/8 (2, 3)	0.87	0.87	0.90	0.90
Regular plate/float out, regular plate/float in	1/4 (6)	0.80	0.80	0.83	0.83
Heat-absorbing plate/float out, regular plate/float in	1/4 (6)	0.46	0.80	0.56	0.58

* Refers to factory-fabricated units with 3/16 (5), 1/4 (6), or 1/2 (12) in (mm) airspace or to prime windows plus storm windows.
† Refer to manufacturer's literature for values.
‡ Thickness of each pane of glass, not thickness of assembled unit.
§ Refers to gray-, bronze-, and green-tinted heat-absorbing plate/float glass.
SOURCE: Abstracted from ASHRAE "Handbook of Fundamentals," 1993, with permission.

Table 12.4.22   Shading Coefficients for Single Glass with Indoor Shading by Venetian Blinds and Roller Shades

			Type of shading				
			Venetian blinds		Roller shade		
					Opaque		Translucent
Type of glass	Nominal thickness,* in (mm)	Solar trans.†	Medium	Light	Dark	White	Light
Regular sheet	3/32–1/4 (2–6)	0.87–0.80					
Regular plate/float	1/4–1/2 (6–12)	0.80–0.71					
Regular pattern	1/8–9/32 (3–7)	0.87–0.79	0.64	0.55	0.59	0.25	0.39
Heat-absorbing pattern	1/8 (3)						
Gray sheet	3/16, 7/32 (5, 6)	0.74, 0.71					
Heat-absorbing plate/float‡	3/16, 1/4 (5, 6)	0.46					
Heat-absorbing pattern	3/16, 1/4 (5, 6)		0.57	0.53	0.45	0.30	0.36
Gray sheet	1/8, 7/32 (3, 5)	0.59, 0.45					
Heat-absorbing plate/float or pattern		0.44–0.30	0.54	0.52	0.40	0.28	0.32
Heat-absorbing plate/float‡	3/8 (10)	0.34					
Heat-absorbing plate or pattern		0.29–0.15	0.42	0.40	0.36	0.28	0.31
		0.24					
Reflective coated glass							
SC§ = 0.30			0.25	0.23			
0.40			0.33	0.29			
0.50			0.42	0.38			
0.60			0.50	0.44			

* Refer to manufacturer's literature for values.
† For vertical blinds with opaque white and beige louvers in the tightly closed position. SC is 0.25 and 0.29 when used with glass of 0.71 to 0.80 transmittance.
‡ Refers to gray-, bronze-, and green-tinted heat-absorbing plate/float glass.
§ Shading coefficient for glass with no shading device.
SOURCE: Abstracted from ASHRAE "Handbook of Fundamentals," 1993, with permission.

Table 12.4.23   Shading Coefficients for Insulating Glass* with Indoor Shading by Venetian Blinds and Roller Shades

Type of glass	Nominal thickness, each light in (mm)	Solar trans†		Type of shading				
				Venetian blinds‡		Roller shade		
		Outer pane	Inner pane			Opaque		Translucent
				Medium	Light	Dark	White	Light
Regular sheet out, regular sheet in	³⁄₁₆, ⅛ (2, 3)	0.87	0.87	0.57	0.51	0.60	0.25	0.37
Regular plate/float out, regular plate/float in	¼ (6)	0.80	0.80					
Heat absorbing plate/float§ out, regular plate/float in	¼ (6)	0.46	0.80	0.39	0.36	0.40	0.22	0.30
Reflective coated glass								
SC¶ = 0.20				0.19	0.18			
0.30				0.27	0.26			
0.40				0.34	0.33			

* Refers to factors-fabricated units with ³⁄₁₆ (2), ¼ (3), or ½ (12), in (mm) airspace, or to prime windows plus storm windows.
† Refer to manufacturer's literature for exact values.
‡ For vertical blinds with opaque white or beige louvers, tightly closed, SC is approximately the same as for opaque white roller shades.
§ Refers to bronze- or green-tinted heat-absorbing plate/flat glass.
¶ Shading coefficient for glass with no shading device.
SOURCE: Abstracted from ASHRAE "Handbook of Fundamentals," 1993, with permission.

Table 12.4.24   Shading Coefficients for Double Glazing with Between-Glass Shading

Type of glass	Nominal thickness, each pane	Solar trans.*		Description of airspace	Type of shading		
		Outer pane	Inner pane		Venetian blinds		Louvered sun screen
					Light	Medium	
Regular sheet out, regular sheet in	³⁄₁₆, ⅛ (2, 3)	0.87	0.87	Shade in contact with glass or shade separated from glass by airspace	0.33	0.36	0.43
Regular plate out, regular plate in	¼ (6)	0.80	0.80	Shade in contact with glass void-filled with plastic			0.49
Heat-absorbing, plate/float† out, Regular plate in	¼ (6)	0.46	0.80	Shade in contact with glass or shade separated from glass by airspace	0.28	0.30	0.37
				Shade in contact with glass voids filled with plastic			0.41

* Refer to manufacturer's literature for exact values.
† Refers to gray-, bronze-, and green-tinted heat-absorbing plate/float glass.
SOURCE: Abstracted from ASHRAE "Handbook of Fundamentals," 1993, with permission.

Table 12.4.25   Rates of Heat Gain from Occupants of Conditioned Spaces*†

Degree of activity	Typical application	Total heat adult male, Btu/h	Total heat adjusted,‡ Btu/h	Sensible heat, Btu/h	Latent heat, Btu/h
Seated at theater	Theater—Matinee	390	330	225	105
Seated at theater	Theater—Evening	390	350	245	105
Seated, very light work	Offices, hotels, apartments	450	400	245	155
Moderately active office work	Offices, hotels, apartments	475	450	250	200
Standing, light work; walking	Department store, retail store	550	450	250	200
Walking; standing	Drug store, bank	550	500	250	250
Sedentary work	Restaurant§	490	550	275	275
Light bench work	Factory	800	750	275	475
Moderate dancing	Dance hall	900	850	305	545
Walking 3 mi/h; light machine work	Factory	1,000	1,000	375	625
Bowling¶	Bowling alley	1,500	1,450	580	870
Heavy work	Factory	1,500	1,450	580	870
Heavy machine work; lifting	Factory	1,600	1,600	635	965
Athletics	Gymnasium	2,000	1,800	710	1,090

* Tabulated values are based on 75°F room dry-bulb temperature. For 80°F room dry-bulb, the total heat remains the same, but the sensible heat values should be decreased by approximately 20%, and the latent heat values increased accordingly.
† All values are rounded to nearest 5 Btu/h.
‡ *Adjusted heat gain* is based on normal percentage of men, women, and children for the application listed, with the postulate that the gain from an adult female is 85% of that for an adult male, and that the gain from a child is 75% of that for an adult male.
§ Adjusted total heat gain for *Sedentary work, Restaurant,* includes 60 Btu/h for food per individual (30 Btu/h sensible and 30 Btu/h latent).
¶ For *Bowling,* figure one person per alley actually bowling, and all others as sitting (400 Btu/h) or standing and walking slowly (550 Btu/h).
SOURCE: Abstracted from ASHRAE "Handbook of Fundamentals," 1989, with permission.

Table 12.4.26 Rate of Heat Gain from Selected Office Equipment

Appliance	Size	Maximum input rating, Btu/h	Standby input rating, Btu/h	Recommended rate of heat gain, Btu/h
Cheek processing workstation	12 pockets	16,400	8,410	8,410
Computer devices				
Card puncher		2,730 to 6,140	2,200 to 4,800	2,200 to 4,800
Card reader		7,510	5,200	5,200
Communication /transmission		6,140 to 15,700	5,600 to 9,600	5,600 to 9,600
Disk drives/mass storage		3,410 to 34,100	3,412 to 22,420	3,412 to 22,420
Magnetic ink reader		3,280 to 16,000	2,600 to 14,400	2,600 to 14,400
Microcomputer	16 to 640 Kbyte*	340 to 2,050	300 to 1,800	300 to 1,800
Minicomputer		7,500 to 15,000	7,500 to 15,000	7,500 to 15,000
Optical reader		10,240 to 20,470	8,000 to 17,000	8,000 to 17,000
Plotters		256	128	214
Printers				
Letter quality	30 to 45 char/min	1,200	600	1,000
Line, high speed	5,000 or more lines/min	4,300 to 18,100	2,160 to 9,040	2,500 to 13,000
Line, low speed	300 to 600 lines/min	1,540	770	1,280
Tape drives		4,090 to 22,200	3,500 to 15,000	3,500 to 15,000
Terminal		310 to 680	270 to 600	270 to 600
Copiers/Duplicators				
Blue print		3,930 to 42,700	1,710 to 17,100	3,930 to 42,700
Copiers (large)	30 to 67* copies/min	5,800 to 22,500	3,070	5,800 to 22,500
Copiers (small)	6 to 30* copies/min	1,570 to 5,800	1,020 to 3,070	1,570 to 5,800
Feeder		100	—	100
Microfilm printer		1,540	—	1,540
Sorter/collator		200 to 2,050	—	200 to 2,050
Electronic equipment				
Cassette recorders/players		200	—	200
Receiver/tuner		340	—	340
Signal analyzer		90 to 2,220	—	90 to 2,220
Mail processing				
Folding machine		430	—	270
Inserting machine	3,600 to 6,800 pieces/h	2,050 to 1,300	—	1,330 to 7,340
Labeling machine	1,500 to 30,000 pieces/h	2,050 to 22,500	—	1,330 to 14,700
Postage meter		780	—	510
Word processors/typewriters				
Letter-quality printer	30 to 45 char/min	1,200	600	1,000
Phototypesetter		5,890	—	5,180
Typewriter		270	—	230
Word processor		340 to 2,050	—	300 to 1,800
Vending machines				
Cigarette		250	51 to 85	250
Cold food/beverage		3,920 to 6,550	—	1,960 to 3,280
Hot beverage		5,890	—	2,940
Snack		820 to 940	—	820 to 940
Miscellaneous				
Bar-code printer		1,500	—	1,260
Cash register		200	—	160
Coffee maker	10 cups	5,120	—	3,580 sensible 1,540 latent
Microfiche reader		290	—	290
Microfilm reader		1,770	—	1,770
Microfilm reader/printer		3,920	—	3,920
Microwave oven	1 ft³	2,050	—	1,360
Paper shredder		850 to 10,240	—	680 to 8,250
Water cooler	32 qt/h	2,390	—	5,970

* Input is not proportional to capacity.
SOURCE: Abstracted from ASHRAE "Handbook of Fundamentals," 1993, with permission.

Heat transfer to the distribution duct depends on the duct area exposed, the amount of insulation on the duct, and the temperature differential between the air in the duct and the surroundings.

Outside Air

Outside air loads depend on the amount of outside air intake and its condition. Table 12.4.3 shows recommended outdoor air ventilation rates for different occupancies. Tables 12.4.4 and 12.4.5 can be used to determine the quantity of outside air that will infiltrate because of wind effects. In all cases, sufficient outside air should be provided to balance

mechanical exhaust systems.

$$Q_{\text{outside air}} = q_{\text{sensible}} + q_{\text{latent}} \quad \text{Btu/h (W)}$$
$$= AU(t_o - t_i) \text{ ft}^3/\text{min}$$
$$+ 0.68(M_o - M_i) \text{ ft}^3/\text{min} \quad \text{Btu/h} \quad (12.4.16a)$$

where M_o = outside moisture content at design wet bulb, °F, gr/lb; and M_i = inside moisture content at design relative humidity, gr/lb. A comparable equation in SI units is

$$Q_{\text{outside air}} = 1.2(t_o' - t_i') \text{ m}^3/\text{s}$$
$$+ 42.6(M_o' - M_i') \text{ m}^3/\text{s} \quad \text{kW} \quad (12.4.16b)$$

Table 12.4.27   Heat Gain from Restaurant Appliances
Not hooded,* gas burning and steam heated.

Appliance	Type of control	Miscellaneous data	Manufacturer's max. rating, Btu/h	Maintaining rate, Btu/h	Recommended heat gain from average use		
					Sensible heat, Btu/h	Latent heat, Btu/h	Total heat, Btu/h
		Gas burning					
Coffee brewer ½ gal	Man.	Combination brewer and warmer	3,400		1,350	350	1,700
Warmer, ½ gal	Man.		500	500	400	100	500
Coffee brewer unit with tank		4 brewers and 4½-gal tank			7,200	1,800	9,000
Coffee urn, 3 gal	Auto.	Black finish	3,200	3,900	2,900	2,900	5,800
Coffee urn, 3 gal	Auto.	Nickel plated		3,400	2,500	2,500	5,000
Coffee urn, 5 gal	Auto.	Nickel plated		4,700	3,900	3,900	7,800
Food warmer, values per sq. ft top surface	Man.	Water bath type	2,000	900	850	450	1,300
Fry kettle, 15 lb fat	Auto.	Frying area 10 × 10	14,250	3,000	4,200	2,800	7,000
Fry kettle, 28 lb fat	Auto.	Frying area 11 × 16	24,000	4,500	7,200	4,800	12,000
Grill, Broil-O-Grill Top burner Bottom burner	Man.	Insulated 22,000 Btu/h 15,500 Btu/h	37,000		14,400	3,600	18,000
Stoves, short order—closed top; values per sq ft top surface	Man.	Ring-type burners 12,000 to 22,000 Btu each	14,000		4,200	4,200	8,400
Stoves, short order—closed top; values per sq ft top surface	Man.	Ring-type burners 10,000 to 12,000 Btu each	11,000		3,300	3,300	6,600
Toaster, continuous	Auto.	2 slices wide, 360 slices/h	12,000	10,000	7,700	3,300	11,000
		Steam heated					
Coffee urn, 3 gal	Auto.	Black finish			2,900	1,900	4,800
3 gal	Auto.	Nickel plated			2,400	1,600	4,000
5 gal	Auto.	Nickel plated			3,400	2,300	5,700
Coffee urn, 3 gal	Man.	Black finish			3,100	3,100	6,200
3 gal	Man.	Nickel plated			2,600	2,600	5,200
5 gal	Man.	Nickel plated			3,700	3,700	7,400
Food warmer, per sq ft top surface	Auto.				400	500	900
Food warmer, per sq ft top surface	Man.				450	1,150	1,500
Coffee brewer, ½ gal	Man.		2,240	306	900	220	1,120
Warmer, ½ gal	Man.		306	306	230	90	320
4 coffee brewing units with 4½-gal tank	Auto.	Water heater, 2,000 W Brewers, 2,960 W	16,900		4,800	1,200	6,000
Coffee urn, 3 gal	Man.	Black finish	11,900	3,000	2,600	1,700	4,300
3 gal	Auto.	Nickel plated	15,300	2,600	2,200	1,500	3,700
5 gal	Auto.	Nickel plated	17,000	3,600	3,400	2,300	5,700
Doughnut machine	Auto.	Exhaust system to outdoors, ½-hp motor	16,000		5,000		5,000
Egg boiler	Man.	Med. ht., 550 W Low ht., 275 W	3,740		1,200	800	2,000
Food warmer with plate warmer, per sq ft top surface	Auto.	Insulated, separate heating unit for each pot, plate warmer in base	1,350	500	350	350	700
Food warmer without plate warmer, per sq ft top surface	Auto.	Insulated, separate heating unit for each pot	1,020	400	200	350	550
Fry kettle, 11½ lb fat	Auto.		8,840	1,100	1,600	2,400	4,000
Fry kettle, 25 lb fat	Auto.	Frying area 12 × 14 in	23,800	2,000	3,800	5,700	9,500
Griddle, frying	Auto.	Frying top 18 × 14 in	8,000	2,800	3,100	1,700	4,800
Grille, meat	Auto.	Cooking area 10 × 12 in	10,200	1,900	3,900	2,100	6,000
Grille, sandwich	Auto.	Grill area 12 × 12 in	5,600	1,900	2,700	700	3,400
Roll warmer	Auto.	One drawer	1,500	400	1,100	100	1,200
Toaster, continuous	Auto.	2 slices wide, 360 slices/h	7,500	5,000	5,100	1,300	6,400
Toaster, continuous	Auto.	4 slices wide, 720 slices/h	10,200	6,000	6,100	2,600	8,700
Toaster, pop-up	Auto.	2 slices	4,150	1,000	2,450	450	2,900
Waffle iron	Auto.	One waffle 7-in diam.	2,480	600	1,100	750	1,850
Waffle iron for ice cream sandwich	Auto.	12 cakes, each 2½ × 3¾ in	7,500	1,500	3,100	2,100	5,200

* If properly designed positive exhaust hood is used, multiple recommended value by 0.50.
SOURCE: Carrier manual.

Table 12.4.28   Heat Gain from Typical Electric Motors

Motor nameplate or rated horsepower	Motor type	Nominal, r/min	Full load motor efficiency, %	Location of motor and driven equipment with respect to conditioned space or airstream		
				A	B	C
				Motor in, driven equipment in, Btu/h	Motor out, driven equipment in, Btu/h	Motor in, driven equipment out, Btu/h
0.05	Shaded pole	1,500	35	360	130	240
0.08	Shaded pole	1,500	35	580	200	380
0.125	Shaded pole	1,500	35	900	320	590
0.16	Shaded pole	1,500	35	1,160	400	760
0.25	Split phase	1,750	54	1,180	640	540
0.33	Split phase	1,750	56	1,500	840	660
0.50	Split phase	1,750	60	2,120	1,270	850
0.75	3-phase	1,750	72	2,650	1,900	740
1	3-phase	1,750	75	3,390	2,550	850
1.5	3-phase	1,750	77	4,960	3,820	1,140
2	3-phase	1,750	79	6,440	5,090	1,350
3	3-phase	1,750	81	9,430	7,640	1,790
5	3-phase	1,750	82	15,500	12,700	2,790
7.5	3-phase	1,750	84	22,700	19,100	3,640
10	3-phase	1,750	85	29,900	24,500	4,490
15	3-phase	1,750	86	44,400	38,200	6,210
20	3-phase	1,750	87	58,500	50,900	7,610
25	3-phase	1,750	88	72,300	63,600	8,680
30	3-phase	1,750	89	85,700	76,300	9,440
40	3-phase	1,750	89	114,000	102,000	12,600
50	3-phase	1,750	89	143,000	127,000	15,700
60	3-phase	1,750	89	172,000	153,000	18,900
75	3-phase	1,750	90	212,000	191,000	21,200
100	3-phase	1,750	90	283,000	255,000	28,300
125	3-phase	1,750	90	353,000	318,000	35,300
150	3-phase	1,750	91	420,000	382,000	37,800
200	3-phase	1,750	91	569,000	509,000	50,300
250	3-phase	1,750	91	699,000	636,000	62,900

SOURCE: Adapted from ASHRAE "Handbook of Fundamentals," 1993, with permission.

where M'_o = outside moisture content at design wet bulb, °C, g/kg; and M'_i = inside moisture content at design relative humidity, g/kg. t'_o, t'_i are expressed in °C.

Moisture Load

Moisture as a cooling load is a latent heat gain expressed in Btu per hour and is computed with the following equations:

$$Q_t = Q_{inf} + Q_{vent} \qquad (12.4.17)$$
$$Q_{inf} = 0.68 \times \text{ft}^3/\text{min} \times (M_o - M_i) \quad \text{Btu/h} \quad (12.4.18a)$$
$$Q_{vent} = 0.68 \times \text{ft}^3/\text{min} \times (M_o - M_i) \quad \text{Btu/h} \quad (12.4.19a)$$

where 0.68 = 60/13.5 × 1,076/7,000; 60 = min/h; 13.5 = specific volume of moist air at 70°F dB and 50 percent RH; 1,076 = average heat removal required to condense 1 lb of water vapor from the room air; and 7,000 = grains per pound.

Comparable equations in SI units are:

$$Q_{inf} = 42.6 \text{ m}^3/\text{s} (M'_o - M'_i) \quad \text{kW} \quad (12.4.18b)$$
$$Q_{vent} = 42.6 \text{ m}^3/\text{s} (M'_o - M'_i) \quad \text{kW} \quad (12.4.19b)$$

M'_o = moisture content of outside air, g/kg; and M'_i = moisture content of inside air, g/kg.

HEATING

Heating Load

$$Q_t = Q_{tr} + Q_{inf} + Q_{vent} \qquad (12.4.20a)$$

where Q_t = total heating load, Btu/h; Q_{tr} = transmission load = $AU(t_i - t_o)$, Btu/h; Q_{inf} = infiltration load = 1.08 (ft^3/min), $(t_i - t_o)$, Btu/h; Q_{vent} = ventilation load = 1.08 (ft^3/min)$(t_i - t_o)$, Btu/h; A = area through which heat flow occurs, ft^2; U = overall heat-transfer

coefficient = $1/R_t$ Btu/(h)(ft^2)(°F); t_o = outside-air design dry-bulb temperature, °F (Fig. 12.4.1); t_i = inside design dry-bulb temperature, °F (Table 12.4.1); R_t = thermal resistance, °F/(Btu · h · ft^2) = $R_1 + R_2 + R_3 + \cdots + R_n$; and R_1, R_2, . . . = resistances to heat flow to the individual components of a composite construction. See Fig. 12.4.2 and Table 12.4.2.

A comparable equation in SI units is

$$Q_1 = Q_{tr} + Q_{inf} + Q_{vent} \qquad (12.4.20b)$$

where Q_t = total heating load, kW; $Q_{tr} = AU(t_i - t_o)$, kW; $Q_{inf} = 1.2$ (m^3/s)$(t_i - t_o)$, kW; $Q_{vent} = 1.2$ (m^3/s)$(t_i - t_o)$, kW; A = area through which heat flow occurs, m^2; U = overall heat-transfer coefficient = $1/R_t$, W/(m^2 · K); t_o = outside-air design dry-bulb temperature, °C; t_i = inside design dry-bulb temperature, °C; m^3/s = cubic meters per second, air; and R_t = thermal resistance, K/(W · m^2).

PSYCHROMETRICS

The psychrometric chart can be used to simplify the analysis of air-conditioning processes. The processes can be shown simply and visualized clearly, and the chart lends itself to quick graphical computations. In addition, the state of any air–water vapor mixture can be fixed on the psychrometric chart with reasonable accuracy by means of any two psychrometric characteristics. Usually dry-bulb temperature and wet-bulb temperature are used because they are the simplest to measure.

Figures 12.4.8 and 12.4.9 are psychrometric charts for normal temperatures in inch-pound and SI units, respectively.

Psychrometric charts at different temperatures and barometric pressures are useful for problems falling outside the ranges shown in Figs. 12.4.8 and 12.4.9. A collection of different psychrometric charts, in both USCS and SI units, for low, normal, and high temperatures, at sea level and at

Fig. 12.4.8 Psychrometric chart for normal temperatures in USCS units. *(Carrier Corp.)*

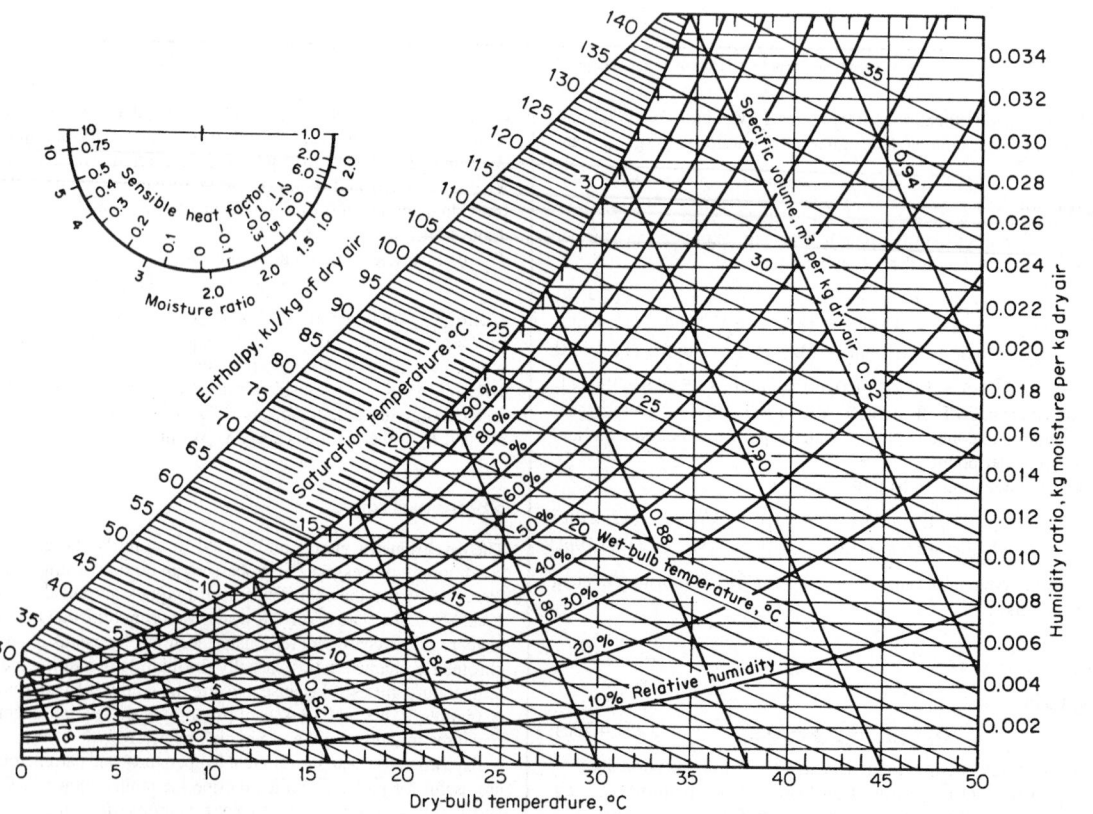

Fig. 12.4.9 Psychrometric chart for normal temperatures in SI units. *(Adapted from material published by Business News Publishing Co., 1975.)*

four elevations above sea level, are available from ASHRAE, the Carrier Corp. of Syracuse, NY, and from other equipment manufacturers.

Figure 12.4.10 shows a psychrometric chart simplified in skeleton form.

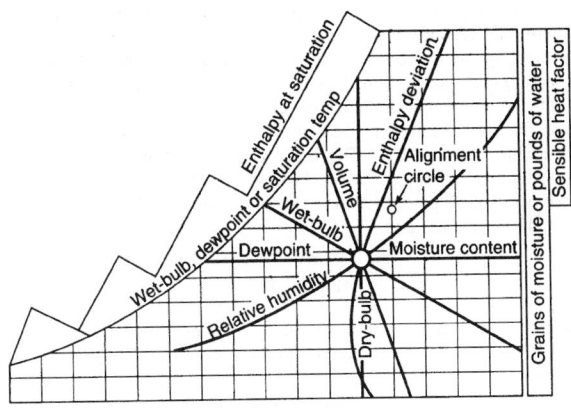

Dry-bulb temperature

Fig. 12.4.10   Skeleton psychrometric chart.

Terminology

Dry-bulb temperature: The temperature of air as registered by an ordinary thermometer.

Wet-bulb temperature: The temperature registered by a thermometer whose bulb is covered by a wetted wick and exposed to a current of rapidly moving air.

Dew-point temperature: The temperature at which condensation of moisture begins when the air is cooled.

Relative humidity: Ratio of the actual water vapor pressure of the air to the saturated water vapor pressure of the air at the same temperature.

Specific humidity or moisture content: The weight of water vapor in grains (or pounds) of moisture per pound of dry air.

Enthalpy: A thermal property indicating the quantity of heat in the air above an arbitrary datum, in Btu per pound of dry air. The datum for dry air is 0°F and, for the moisture content, 32°F water.

Enthalpy deviation: Enthalpy should be corrected by the enthalpy deviation due to the air not being in the saturated state. On normal air-conditioning estimates it is omitted.

Specific volume: The cubic feet of the mixture per pound of dry air.

Sensible heat factor: The ratio of sensible to total heat.

Alignment circle: Used in conjunction with the sensible heat factor to plot the various air-conditioning process lines.

Pounds of dry air: The basis for all psychrometric calculations. Remains constant during all psychrometric processes.

The dry-bulb, wet-bulb, and dew-point temperatures and the relative humidity are so related that, if two properties are known, all other properties shown may then be determined. When **air is saturated,** dry-bulb, wet-bulb, and dew-point temperatures are all equal.

Room sensible heat factor is the ratio of room sensible heat to the total of room sensible and room latent heat.

A line drawn between the room design condition at the slope of the sensible heat factor is the **condition line** and it will intersect the saturation line at the apparatus dew point. See Fig. 12.4.11. In order to maintain design conditions, supply air must be provided to the space at a temperature and moisture level that, when plotted, must fall on the condition line. If air is provided at conditions of apparatus dew point, the minimum quantity of conditioned air would be required to obtain design conditions.

Supply Air Temperature   The room cooling load is the sum of external and internal sensible and latent heat gains plus the difference in enthalpy between outside and room air for that portion of outside air

Fig. 12.4.11   Apparatus dew point and condition line.

which does not contact the cooling-coil surfaces. The percentage of air that passes through a cooling coil untreated is the numerical value of the *coil bypass factor;* e.g., a bypass factor of 20 percent represents a cooling-coil saturation efficiency of 80 percent.

The ratio of room sensible heat gains to total room sensible and latent heat gains is the **room sensible heat ratio (RSHR)**.

$$\text{RSHR} = \frac{Q_{rs}}{Q_{rs} + Q_{rl}} \qquad (12.4.21)$$

It represents the ratio of sensible cooling capacity to the total cooling capacity required of the supply air to satisfy room conditions. It is used to plot the slope of the *room-condition line* on a psychrometric chart for the determination of the *apparatus dew point* (ADP). See Fig. 12.4.11.

The actual supply air temperature and off coil wet-bulb temperature will depend on the bypass characteristic of the selected cooling coil (Fig. 12.4.11).

Supply Air Rate   The rate of supply air required is expressed by

$$Q_{sa} = Q_{rs}/1.08(t_r - t_s) \qquad \text{ft}^3/\text{min} \qquad (12.4.22a)$$

In SI units:

$$Q'_{sa} = Q'_{rs}/1.2(t_r - t_s) \qquad \text{m}^3/\text{s} \qquad (12.4.22b)$$

where Q_{sa} = supply air, ft³/min (m³/s); Q'_{rs} = room sensible heat, kW; t_r = room design temperature, °F (K); t_s = supply air temperature, °F (K).

The bypass factor is used to express the effectiveness of a cooling coil. It is based on the artificial premise that a portion of the air that enters a cooling coil is cooled completely to saturation, that is, to the apparatus dew point, and the remaining air is not cooled at all. The actual condition of the air that leaves the coil is then represented to be a mixture of the fully conditioned and completely unconditioned air. The ratio of unconditioned air to total air is the **bypass factor.**

The bypass factor is a function of the type and amount of coil surface and the time available for contact as the air passes over the coil. Table 12.4.29 gives approximate bypass factors for various finned coil surfaces and air velocities. A larger bypass factor requires the circulation of more conditioned air; where outside air is introduced, the bypassed air increases the room air load.

Table 12.4.29   Typical Bypass Factors for Unsprayed Coils

Depth of coils (rows)	5-in spacing	
	8 fins/in	14 fins/in
	Face velocity, ft/min	
	300–700	300–700
2	0.42–0.55	0.22–0.38
3	0.27–0.40	0.10–0.23
4	0.19–0.30	0.05–0.14
5	0.12–0.25	0.02–0.09
6	0.08–0.18	0.01–0.06
8	0.03–0.08	

The temperature difference between conditions ① and ② is due to heat gains from the supply fan and heat transfer to the supply duct.

Dry-bulb temperature

Fig. 12.4.12 Typical cooling process.

Typical Air-Conditioning Cooling Process Figure 12.4.12 depicts a typical cooling process adjusted for fan heating and duct losses.

DUCT DESIGN

Air Velocity Supply and return air ducts and apparatus are sized on the basis of air quantity and are kept within the limitations of allowable friction losses, velocity, and noise. Conveying air at low velocity will be quieter and require less pressure, less fan power, and less carefully constructed ductwork than conveying air at medium or high velocity. Lower velocities are preferred where space is available.

Pressure Loss in Duct Systems Pressure losses in duct systems are due to friction of the air in contact with the sides of the duct and dynamic losses caused by changes of duct shape or direction and by obstructions to flow.

$$\text{Friction } H_f = f \frac{L}{D}\left(\frac{V}{4,005}\right)^2 \qquad (12.4.23a)$$

where H_f = head loss due to friction, in H_2O: L = length of duct, ft; D = diameter of duct, ft; V = velocity of air, ft/min; and f = nondimensional friction coefficient. A comparable equation in SI units is

$$H_f = f \frac{L}{d}\frac{(V)^2}{2}\rho \qquad (12.4.23b)$$

where H_f = head loss due to friction, N/m^2; L = length of duct, m; D = diameter of duct, m; V = velocity of air, m/s; f = nondimensional friction coefficient; and ρ = density of air, kg/m^3.

$$\text{Dynamic losses } H_v = C\left(\frac{V}{4,005}\right)^2 \qquad (12.4.24a)$$

where H_v = velocity-head loss, in H_2O; C = experimentally determined constant; V = Q/A = air velocity, ft/min; Q = airflow rate, ft^3/min; and A = cross-sectional area of duct, ft^2. A comparable equation in SI units is

$$H'_v = C'\frac{(V')^2}{2} \qquad (12.4.24b)$$

where H'_v = velocity-head loss, N/m^2; and V' = air velocity, m/s. When the equation for H'_v is written in the form above, $C' = C$.

Duct Design Methods Two methods are frequently used: **equal friction** and **static regain**.

Equal Friction The equal friction method is applicable primarily to systems using low or moderate velocities, where the velocity head is not an important factor. A friction drop per 100 ft of length is chosen, and

the duct mains and branches are all sized on the basis of this friction drop. This will invariably result in higher velocities in the mains, where they can be tolerated, and low velocities in the branches, where they are desirable. Limitations are established for friction loss per 100 ft and for velocities in the mains and branches.

Static Regain The static regain method is used for both conventional and high velocity systems. It is especially applicable in the latter, where the velocity head may be appreciable. In the static regain method, the static pressure required to give proper airflow through the system outlets is determined, and this pressure is maintained by reducing the velocity at each branch or takeoff, so that the recovery in pressure due to reduction of velocity balances the friction loss in the preceding section of duct. This is possible because of the convertibility of static and velocity pressures. For practical applications it is usually assumed that 50 percent of the velocity pressure available will be converted to static pressure.

$$H_R = 0.5\left(\frac{V_1}{4,005}\right)^2 - \left(\frac{V_2}{4,005}\right)^2 \qquad (12.4.25a)$$

where H_R = head recovered, inH_2O; V_1 = system inlet velocity, ft/min; and V_2 = system outlet velocity, ft/min. A comparable equation in SI units is

$$H'_R = \frac{0.5(V'_1)^2}{2} - \frac{(V'_2)^2}{2} \qquad (12.4.25b)$$

where H'_R = head recovered, N/m^2; V'_1 = system inlet velocity, m/s; and V'_2 = system outlet velocity, m/s.

Figure 12.4.13 may be used to determine directly the friction loss per 100 ft for round ducts. Figure 12.4.14 provides a conversion from

Fig. 12.4.13 Friction loss for usual conditions in round ducts. (This chart applies to smooth, round, galvanized steel ducts, and is based on air at 70°F and 29.96 inHg absolute pressure. For air of different density, the friction may be assumed to vary directly with the density.)

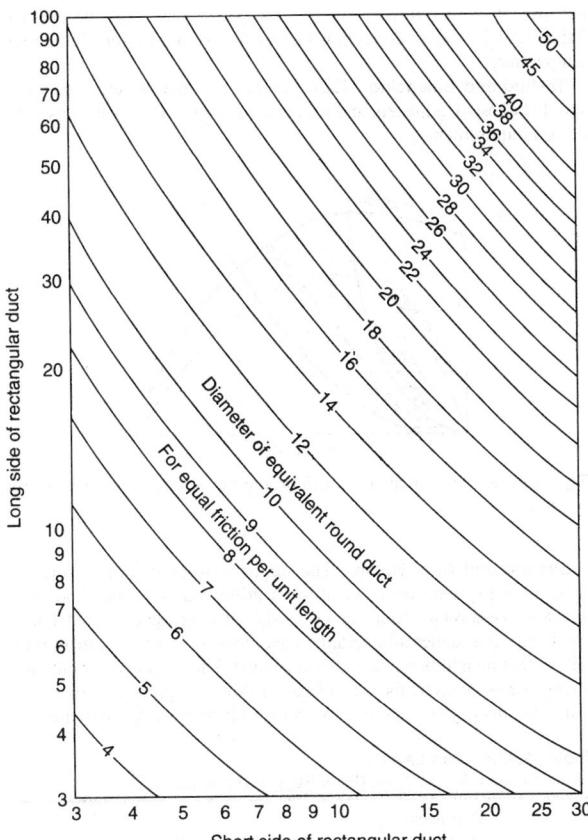

Fig. 12.4.14 Equivalent diameters of round ducts. Use with rectangular ducts, with capacity constant. *(Buffalo Forge.)*

round to rectangular ductwork with the same air capacity. Table 12.4.30 presents a range of friction loss per 100 ft and velocity limits for low- and high-velocity air systems. Typical velocities through some air system components are shown in Table 12.4.31.

Table 12.4.30   Duct System Pressure Losses and Design Velocities

	Pressure loss, in H$_2$O, gage, per 100 ft	Maximum velocity, ft/min
Low-velocity systems		
Noise critical	0.05–0.07	1,400
Usual design	0.08–0.1	1,800
Straight runs without takeoffs	0.08–0.12	2,400
High-velocity systems		
Usual design	0.4–0.6	2,600
Straight runs without takeoffs	0.6	3,400

Air Distribution

Air outlets are designed to control air motion, noise level, and temperature gradients caused by the introduction of air into a conditioned space and the removal of air from that space.

Supply Outlets Supply outlets (and diffusers) are usually located at near ceiling height with grilles or registers high on the sidewall. Supply outlets may be located below windows or in the floor as long as they satisfy the criteria for the space and are unobstructed. Factors which

Table 12.4.31   Typical Velocities through Air System Components

	Residences		Nonresidential buildings	
Designation	ft/min	m/min	ft/min	m/min
Outdoor-air intakes*	300	80	500	150
Filters*	250	75	300	90
Heating coils*	450	140	500	150
Cooling coils*	450	140	500	150
Air washers	500	150	500	150

* These velocities are for total face area, not the net free area; other velocities in the table are for net free area.
SOURCE: Adapted from ASHRAE "Handbook of Fundamentals," 1993, with permission.

usually affect the selection of supply outlets are noise, location of outlet, temperature of supply air, and area of diffusion. Manufacturers' performance data for supply outlets should be used as the basis for selection.

Return Outlets Selection of return outlets (and registers or grilles) is usually governed by face velocity to limit discomfort to occupants from drafts and/or noise. See Table 12.4.32.

Table 12.4.32   Recommended Return Face Velocities

	Velocity over gross area	
Intake location	ft/min	m/min
Above occupied zone	800 up	245 up
Within occupied zone, not near seats	600–800	185–245
Within occupied zone, near seats	400–600	120–185
Door or wall louvers	200–300	60–90
Undercutting of doors (through undercut area)	200–300	60–90

SOURCE: Adapted from ASHRAE, "Handbook of Fundamentals," 1993, with permission.

FANS

Supply and **return fans** must convey the conditioned air through the system efficiently and with a minimum of disturbing noise and vibrations.

Fans produce pressure by increasing airflow velocity. Velocity change is the result of tangential and radial velocity components in the case of centrifugal fans, and of axial and tangential velocity components in the case of axial-flow fans.

Centrifugal fans are most frequently used, but **axial-flow fans** have been applied successfully. Figures 12.4.15 to 12.4.17 show typical performance curves for forward-curved and backward-curved centrifugal

Fig. 12.4.15 Percentage performance curves for a forward-curved blade centrifugal fan.

Fig. 12.4.16 Percentage performance curves for a backward-curved blade centrifugal fan.

fans and axial-flow fans. Table 12.4.33 summarizes the relative characteristics of centrifugal fans. Figure 12.4.18 shows the zone of optimum performance for a backward-curved-blade centrifugal fan.

FAN LAWS

Fan laws in Table 12.4.34 relate the performance variables for any dynamically similar series of fans. (See also Sec. 14.5.) They can be

Fig. 12.4.17 Percentage performance curves for an axial fan.

used to predict the performance of any fan in a series when test data are available for a given fan in the series. Fan laws may also be used with a particular fan to determine the effect of speed change. Unless otherwise identified, fan performance data are based on dry air at standard conditions: 14.696 lb/in² abs and 70°F (0.075 lb/ft³). In applications where the fan may be required to handle air or gas at some other density,

Table 12.4.33 Relative Characteristics of Centrifugal Fans

Characteristic	Backward curved	Forward curved
First cost	High	Low
Efficiency	High	Low
Stability of operation	Good	Poor
Space required	Medium	Small
Tip speed	High	Low

because of temperature or altitude, fan performance will be affected. Table 12.4.35 shows correction factors to air volume for altitude and temperature.

    Testing and Balancing   Each fan and air duct system component should be tested, adjusted, and balanced to assure that it conforms to the design requirements.

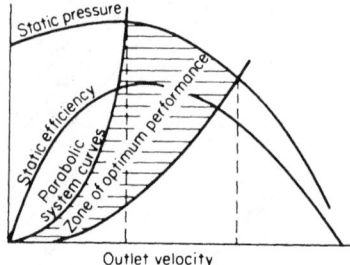

Fig. 12.4.18 Zone of optimum performance for backward-curved blade centrifugal fans.

    System and Duct Noise   The major sources of noise from air-conditioning systems are fans, diffusers, grilles, ducts, fittings, and vibration. **Sound control** for components consists of selecting devices that meet the design goal under all operating conditions and installing them properly so that no additional sound is generated. The sound power output of a fan is determined by the type of fan, airflow, and pressure. Sound control in the duct system requires proper duct layout, sizing, and provision

Table 12.4.34 Fan Laws*†

For all fan laws: $N_{t1} = N_{t2}$ and (Pt. of Rtg.)$_1$ = (Pt. of Rtg.)$_2$

No.	Dependent variables	Independent variables		
1a	$Q_1 = Q_2$	$\times \left(\dfrac{D_1}{D_2}\right)^3 \times \dfrac{N_1}{N_2}$		$\times 1$
1b	Press.$_1$ = Press.$_2$‡	$\times \left(\dfrac{D_1}{D_2}\right)^2 \times \left(\dfrac{N_1}{N_2}\right)^2$		$\times \dfrac{\rho_1}{\rho_2}$
1c	$H_1 = H_2$	$\times \left(\dfrac{D_1}{D_2}\right)^3 \times \left(\dfrac{N_1}{N_2}\right)^3$		$\times \dfrac{\rho_1}{\rho_2}$
2a	$Q_1 = Q_2$	$\times \left(\dfrac{D_1}{D_2}\right)^2 \times \left(\dfrac{\text{Press.}_1}{\text{Press.}_2}\right)^{1/2}$		$\times \left(\dfrac{\rho_2}{\rho_1}\right)^{1/2}$
2b	$N_1 = N_2$	$\times \left(\dfrac{D_2}{D_1}\right) \times \left(\dfrac{\text{Press.}_1}{\text{Press.}_2}\right)^{1/2}$		$\times \left(\dfrac{\rho_2}{\rho_1}\right)^{1/2}$
2c	$H_1 = H_2$	$\times \left(\dfrac{D_1}{D_2}\right)^2 \times \left(\dfrac{\text{Press.}_1}{\text{Press.}_2}\right)^{3/2}$		$\times \left(\dfrac{\rho_2}{\rho_1}\right)^{1/2}$
3a	$N_1 = N_2$	$\times \left(\dfrac{D_2}{D_1}\right)^3 \times \dfrac{Q_1}{Q_2}$		$\times 1$
3b	Press.$_1$ = Press.$_2$	$\times \left(\dfrac{D_2}{D_1}\right)^4 \times \left(\dfrac{Q_1}{Q_2}\right)^2$		$\times \dfrac{\rho_1}{\rho_2}$
3c	$H_1 = H_2$	$\times \left(\dfrac{D_2}{D_1}\right)^4 \times \left(\dfrac{Q_1}{Q_2}\right)^3$		$\times \dfrac{\rho_1}{\rho_2}$

* The subscript 1 denotes that the variable is for the fan under consideration.
† The subscript 2 denotes that the variable is for the tested fan.
‡ P_{total} or P_{static}.
NOTE: D = fan size, N = revolutions per minute, ρ = gas density, Q = volume flow rate, P = pressure, and H = horsepower.

Table 12.4.35   Air Volume Correction Factors for Altitude and Temperature

Altitude, ft (m) above sea level	0	1,000 (305)	2,000 (610)	3,000 (915)	4,000 (1,220)	5,000 (1,525)	6,000 (1,880)	7,000 (2,135)	8,000 (2,440)
Barometric pressure, in Hg (kN/m²)	29.92 (101.3)	28.86 (97.7)	27.82 (94.2)	26.81 (90.8)	25.84 (87.5)	24.89 (84.3)	23.98 (81.2)	23.09 (78.2)	22.2 (75.2)
Air temp, °F (°C)					Correction factors				
70 (21.1)	1.040	1.003	0.967	0.932	0.898	0.865	0.833	0.803	0.772
100 (38)	0.984	0.948	0.915	0.882	0.850	0.818	0.788	0.759	0.731
150 (66)	0.904	0.872	0.840	0.801	0.781	0.752	0.724	0.698	0.672
200 (93)	0.835	0.805	0.777	0.749	0.722	0.694	0.668	0.645	0.620
250 (121)	0.777	0.749	0.722	0.696	0.671	0.647	0.622	0.599	0.577
300 (149)	0.725	0.699	0.674	0.649	0.628	0.603	0.580	0.560	0.538
350 (177)	0.680	0.656	0.632	0.609	0.588	0.566	0.545	0.525	0.505
400 (204)	0.641	0.618	0.596	0.574	0.553	0.533	0.512	0.495	0.476
450 (232)	0.605	0.583	0.564	0.543	0.523	0.503	0.485	0.467	0.450
500 (260)	0.574	0.553	0.534	0.515	0.496	0.477	0.460	0.443	0.426
550 (288)	0.546	0.526	0.508	0.490	0.472	0.454	0.438	0.421	0.406
660 (316)	0.520	0.501	0.484	0.466	0.449	0.433	0.416	0.401	0.387
650 (343)	0.496	0.478	0.462	0.444	0.428	0.413	0.397	0.383	0.368
700 (371)	0.475	0.458	0.442	0.426	0.411	0.395	0.381	0.367	0.354

NOTE: Equivalent $= \dfrac{\text{ft}^3 \text{ min at actual conditions}}{\text{correction factor}}$.

SOURCE: "Bulletin 3576-B, Correction Factors for Temperature and Altitude," Buffalo Forge Co., Buffalo, NY.

for installing duct noise attenuators, if required. The noise generated in a system increases with both duct velocity and system pressure. The ASHRAE Handbook "Applications" and other related sources present methods for sound control in air-conditioning systems.

Group I
Panel-type filters of spun glass, open cell foams, expanded metal and screens, synthetics, textile denier woven and nonwoven, or animal hair.

Group II
Pleated panel-type filters of fine denier nonwoven synthetic and synthetic-natural fiber blends, or all natural fiber.

Group III
Extended surface supported and nonsupported filters of fine glass fibers, fine electret synthetic fibers, or wet-laid paper of cellulose glass or all-glass fibers.

Group IV
Extended-area pleated HEPA-type filters of wet-laid ultra-fine glass fiber paper. Biological grade air filters are generally 95% DOP efficiency; HEPA filters are 99.97 and 99.99%; and ULPA filters are 99.999%.

Notes:
1. Group numbers have no significance other than their use in this figure.
2. Correlation between the test methods shown are approximations for general guidance only.

Fig. 12.4.19a Comparative performance of viscous impingement and dry media filters. *(ASHRAE "Handbook of Fundamentals," 1993.)*

FILTRATION

Filters are provided in air-conditioning systems to assure that the conditioned air will satisfy space cleanliness criteria and to protect air treatment equipment from fouling due to dirt accumulation. Figure 12.4.19a describes types of dry media filters and shows their range of performance. In addition, it shows an approximate comparison of the filtration efficiencies obtained by the use of three different testing methods.

The filters designated as Group I are usually used in least critical applications and as a prefilter to extend the operating life of more costly, higher-efficiency filters. Filters designated as Groups II and III are used in good-quality comfort and process air-conditioning applications. The very high efficiency HEPA and ULPA filters are costly and result in high pressure losses. They are applied in clean rooms and for nuclear and toxic particulate filtration.

AIR TREATMENT CRITERIA

The purpose of an air-conditioning system is to control the indoor environment to provide comfortable conditions for occupants of a space or to facilitate a process to be carried on within the space.

To satisfy these criteria, the system should control temperature, humidity, air cleanliness, and air distribution in the space within established limits and accomplish this without excessive noise. The system should maintain indoor design conditions through the full range of outdoor temperature and humidity and indoor cooling or heating loads and accomplish this reliably and economically.

Performance criteria are often demanding for process systems, but many comfort systems operate for summer cooling and convenience only. They utilize filters that remove only coarse atmospheric dust and have no provision for adding moisture during the winter; during the summer, humidity is reduced, although not controlled, as part of the cooling process. These systems often serve a single room or a small area and may operate in conjunction with a separate building heating system.

Cooling Configurations Figure 12.4.19b illustrates some of the configurations of the cooling portion of air-conditioning systems. These can be provided in conjunction with a variety of air filtration, heating, humidification, air distribution, and automatic control components to make a complete system.

In most air-conditioning systems, peak summer loads are cooled by utilizing some form of compression or absorption cooling system, but there are systems that obtain cooling or precooling from other sources. Some examples are installations located where ample quantities of cold,

Direct Expansion

Chilled Water Coil

Sprayed Coil Dehumidifier

**Sprayed Coil Dehumidifier
with Winter Humidification**

Open Spray Air Washer

Fig. 12.4.19*b* Cooling configurations.

**Winter
Water Side Economizer**

good-quality well water are available and can be circulated through a cooling coil and areas whose summer wet-bulb temperatures are sufficiently low that substantial cooling can be achieved by evaporation of water into the supply airstream.

Direct Expansion Cooling   The most widely used method of air conditioning for cooling a single room or a small area utilizes mechanical compression and the direct expansion of the refrigerant in a cooling coil in the supply airstream. This usually involves a matched compressor and cooling coil in a factory-built, self-contained package. The high-pressure refrigerant leaves the condenser and expands into the cooling coil where it absorbs heat directly from the airstream. For small-capacity applications, the arrangement provides simplified, relatively quick installations and low first costs. Larger, more complex direct expansion systems utilize the same principles.

Chilled Water Cooling   Chilled water systems are the most widely used indirect systems. In these systems, the refrigerant cools a pumped closed water circuit which, in turn, is used to condition air from one or more air handlers.

The chilled water system permits serving widely separated, individually controlled air handlers from a central refrigeration system while limiting the extent of the refrigerant circulation.

Sprayed Coil Dehumidifiers   Sprayed coil systems utilize chilled water coils and a recirculating water spray. The system is similar to the chilled water system in that all the mechanical cooling is provided through the closed chilled water piping and coils. The spray circuit neither adds nor removes heat, but it improves controllability by maintaining the supply air near saturation. During the winter, in conjunction with a preheat coil, the spray system can be used to control relative humidity, and during all seasons it will remove some of the water-soluble contaminants from the airstream.

Open Spray Systems   Open spray systems utilizing cooled or heated spray water to cool, heat, and humidify supply air are now used primarily in industrial applications and in climates which are sufficiently dry that evaporative cooling alone can remove all or much of the sensible heat.

Nontoxic Water Treatment   For sprayed coil and open spray systems, nontoxic chemical water treatment is used to inhibit corrosion, control the growth of bacteria and slime, and limit the concentration of solids.

Winter Water-Side Economizers   An application that combines spray cooling with chilled water circulation is a water-side economizer system for air-conditioning applications where (1) cooling is required during cold outdoor weather, (2) the use of large quantities of outside air is not economical or otherwise unfeasible, and (3) continuous operation of the refrigeration system during cold weather is expensive.

In the example that is illustrated, with the refrigeration shut down, the building's central cooling tower is operated during the winter to provide water at temperatures approximating those of the chilled water system during normal cooling. A high-efficiency heat exchanger then permits the cooling tower water to cool the chilled water which continues to be circulated through the cooling coils.

The system reduces energy consumption and can substantially reduce the cost of winter cooling. Some of the factors that affect applicability of the system are the cold-weather capacity of the cooling tower; the hours of cold-weather cooling that are required compared with the cold-weather wet-bulb hours that can be anticipated; and how warm the chilled water can be and still satisfy the requirements of the winter cooling loads.

HEAT REJECTION APPARATUS

All the heat removed from a conditioned space and the heat equivalent of the work performed by the refrigeration equipment must be removed through the cooling system and dissipated via the refrigeration condenser. Table 12.4.36 lists approximate refrigeration equipment energy consumption values.

In most systems the heat is rejected to the atmosphere, but in some cases, water-cooled refrigeration condensers are used to transfer the rejected heat to surface water or to well water. Some examples of the most common systems used to transfer rejected heat to the atmosphere are shown in Fig. 12.4.20. The **air-cooled condenser and evaporative condenser**

are arranged with the refrigeration condenser in the outdoor airstream. Refrigerant is piped directly with no intervening heat transfer. The **air-cooled condenser** is the simplest arrangement, but the condenser must operate at temperatures sufficiently high so that condensing takes place well above the outdoor ambient dry-bulb temperature. The **evaporative condenser** adds a spray system to permit condensing above the outdoor ambient wet-bulb temperature.

Closed-Circuit Coolers   For larger systems and where extensive piping of refrigerant is either not desired or otherwise precluded, a water-cooled condenser is utilized. The rejected heat is transferred to water piped to an outdoor radiator—a closed-circuit dry cooler or evaporatively cooled device. The **closed-circuit dry cooler**, or **radiator**, is a simple indirect cooling arrangement, but can operate at temperatures higher than the air-cooled condenser because of the additional heat-transfer process involved.

Table 12.4.36   Refrigeration Energy Consumption per 12,000 Btu/h Cooling Capacity

Electric consumption, kW/ton cooling capacity		
	Type of heat rejection, kW/ton	
Item	Mechanical-draft cooling tower	Air-cooled condensers
Room-air conditioners		1.2–2.45*
Refrigeration compressors (motor-driven):		
Reciprocating (20 to 100 tons)	1.03–0.86	1.27–1.09
Centrifugal-hermetic	0.94–0.74	
Centrifugal-external drive (350 to 1,250 tons)	1.02–0.81	
Auxiliary equipment:		
Pump, cooling water	0.15	
Blower for air conditioner	0.10	0.10
Blower for rejection air	0.15	0.15

Steam-consumption rate lb/h/ton (kg/h/ton) cooling capacity		
	Steam rate	
Item	lb/h/ton	kg/h/ton
Centrifugal compressors (steam-driven):		
Condensing (27 in vacuum)	14.5	6.6
Noncondensing (14.7 psig)	30–35	13.6–15.9
Absorption chillers (100 to 1,200 ton) (low-pressure steam)	18–18.5	8.2–8.4
Absorption chillers (high-pressure steam)	11–11.5	5–5.2

Gas-consumption rate, Btu/ton (J/ton) cooling capacity		
	Gate rate	
Item	Btu/ton	kJ/ton
Refrigeration compressors (natural-gas-engine-driven)	8,500	8,970

* Depending on the EER (energy efficiency ratio) rating for the different manufacturers.

The dry cooler water circuit operating with antifreeze permits year-round operation.

The **closed-circuit evaporative cooler** has an indirect condenser water circuit, but uses a spray system and evaporative cooling to reduce the head pressure. In both the dry cooler and closed-circuit cooler there is no contact of either condenser water or antifreeze with the outdoor atmosphere; often this proves highly desirable.

Cooling Towers   In large systems and systems where maximum overall efficiency is important, a portion of the condenser water is evaporated into the atmosphere to cool the remaining condenser water. The water lost by evaporation must be made up, usually from a domestic water source. Figure 12.4.20 shows two cooling tower configurations. The **counterflow** arrangement requires a spray system and is usually taller than the crossflow type. **Crossflow** towers have simpler gravity distribution systems.

Air-Cooled Condenser

Closed Circuit
Dry Cooler

Evaporative Condenser

Closed Circuit
Evaporative Cooler

Counterflow
Cooling Tower

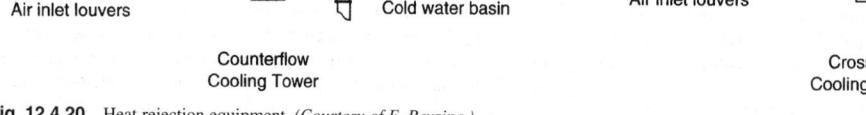

Crossflow
Cooling Tower

Fig. 12.4.20   Heat rejection equipment. *(Courtesy of E. Revzina.)*

All cooling towers and all evaporative spray systems that operate during subfreezing weather must be protected against freezing. A number of methods are available to add heat to the piping circuits and the basins or sumps.

Condenser Water System For electrically driven refrigeration compressors a temperature differential of 10°F (5.5°C) may be assumed, and for steam-driven equipment a temperature differential of 20°F (11°C) is usual. In the latter case the refrigeration and steam condensers are piped in series with a temperature rise of approximately 10°F (5.5°C) each.

Condenser water, gal/min

$$= \frac{\text{total heat rejected, Btu/h}}{500 \times \text{temperature differential, °F}} \quad (12.4.26a)$$

In SI units,

Condenser flow, m³/s

$$= \frac{\text{total heat rejected, kW}}{4,190 \times \text{temperature differential, °C}} \quad (12.4.26b)$$

Atmospheric Evaporative Cooling Equipment The lowest temperature to which water can be cooled in atmospheric cooling equipment is the wet-bulb temperature of the ambient air.

Water-Cooling Effectiveness in Percent

$$E = \frac{(\text{hot-water temp.} - \text{cold-water temp.}) \times 100}{\text{hot-water temp.} - \text{wet-bulb temp. entering}} \quad (12.4.27)$$

The cold-water temperature must be chosen to place the requirement within the effectiveness range of the equipment used.

Makeup Water for Atmospheric Cooling Equipment Makeup water is introduced to replace losses due to evaporation, drift, and blowdown. If all water were cooled by evaporation, the loss by evaporation for the usual 10°F (5.5°C) cooling range would be:

$$\text{Evaporation, \%} = \frac{Q \times 100}{8.3 \times \text{gal/min} \times hfg} \quad (12.4.28a)$$

where Q = total heat rejected. Btu/h; gal/min = total condenser water circulated; and hfg = evaporation heat of water, Btu/lb, at ambient design temperature. In SI units:

$$\text{Evaporation, \%} = \frac{Q' \times 100}{1,000 \times \text{m}^3/\text{s} \times hfg} \quad (12.4.28b)$$

where Q' = total heat rejected, kJ/min; m³/s = total condenser water circulated; and hfg = evaporation heat of water, kJ/kg, at ambient design temperature.

Drift losses depend on the tower design, but generally, from the cooling tower, they are limited to 0.2 percent of the circulated rate.

The makeup water replacing losses due to evaporation, drift, and blowdown introduces dissolved solids into the system.

To prevent excessive concentration, a portion of the circulating water is wasted. The quantity of blowdown depends on the original quantity of dissolved solids in the makeup water and the permissible concentration.

For most open systems, chemical water treatment is used to inhibit corrosion and minimize blowdown.

Water Distribution, Chilled-Water Systems' Temperature Differential Design temperature differentials for chilled-water systems are determined by the need to supply chilled water at an average temperature sufficiently low to provide adequate cooling and dehumidification and the desire to minimize the costs of piping, pumping energy, and refrigeration. These differentials normally range from 10°F (5.6°C) for average systems to as much as 15°F (8.3°C) or more for very large systems with widely separated loads.

Volume flow rate, gal/min

$$= \frac{(\text{total load} + \text{piping heat gains} + \text{pump heat}), \text{Btu/h}}{500 \times \text{temperature differential, °F}} \quad (12.4.29a)$$

A comparable equation in SI units is

Volume flow rate, m³/s

$$= \frac{(\text{total load} + \text{piping heat gains} + \text{pump heat}), \text{kW}}{4,190 \times \text{temperature differential, °C}} \quad (12.4.29b)$$

Water Distribution, Chilled-Water Systems Temperature Differential

	Temp rise	
Application	°C	°F
Close-coupled system on one floor	5–8	3–4.5
Two- or three-story building	8–11	4.5–6
Multistory building	12–20	6.5–11

Water Piping Design

There is a **friction loss** in any pipe through which water is flowing. This loss depends on the following factors:

1. Water velocity
2. Pipe diameter
3. Interior surface roughness
4. Pipe length

System pressure has no effect on the head loss of the equipment in the system. However, higher than normal system pressures may dictate the use of heavier pipe, fittings, and valves along with specially designed equipment.

To properly design a water piping system, the engineer must evaluate not only the pipe friction loss but the loss through valves, fittings, and other equipment. In addition to these friction losses, the use of diversity in reducing the water quantity and pipe size should be considered in designing the water piping system.

Most air-conditioning applications use either steel pipe or copper tubing in the piping system. For friction loss values in steel pipe or copper tubing, refer to Figs. 12.4.21 to 12.4.23.

Pipe size determination, especially for very long systems and/or unusual construction or operating cost considerations, is based on an economic analysis, but smaller systems with less critical economic factors are often based on average conditions. In a typical system pipe sizes are selected for a pressure loss of from 3 to 6 ft per 100 ft of pipe.

Pipe Length The total friction loss in a water piping system is the sum of losses in:

1. The total straight lengths of pipe
2. The equivalent lengths of pipe due to fittings, valves, and other elements in the piping system.

Figure 12.4.24 shows the resistance of some valves and fittings in terms of equivalent length of straight pipe.

Steam System Piping

Steam systems which operate below 15 psig are classified as low-pressure systems. Tables 12.4.37 and 12.4.38 may be used to determine the size of piping for low-pressure steam and condensate systems pitched in the direction of flow and for risers in which steam and condensate flow in opposite directions.

Table 12.4.39 lists total pressure drop limitations for long-length piping systems. In such systems, pipe sizes may have to be increased to keep total pressure loss in supply and return piping within the system limits. For systems in which pressure may vary, i.e., when steam is supplied by a boiler, the distribution pressure must be based on the minimum pressure that will be available.

Tables 12.4.40 and 12.4.41 are used to size piping for use in medium- and high-pressure systems, respectively, and assume piping is installed with proper pitch and adequate provisions to remove condensate.

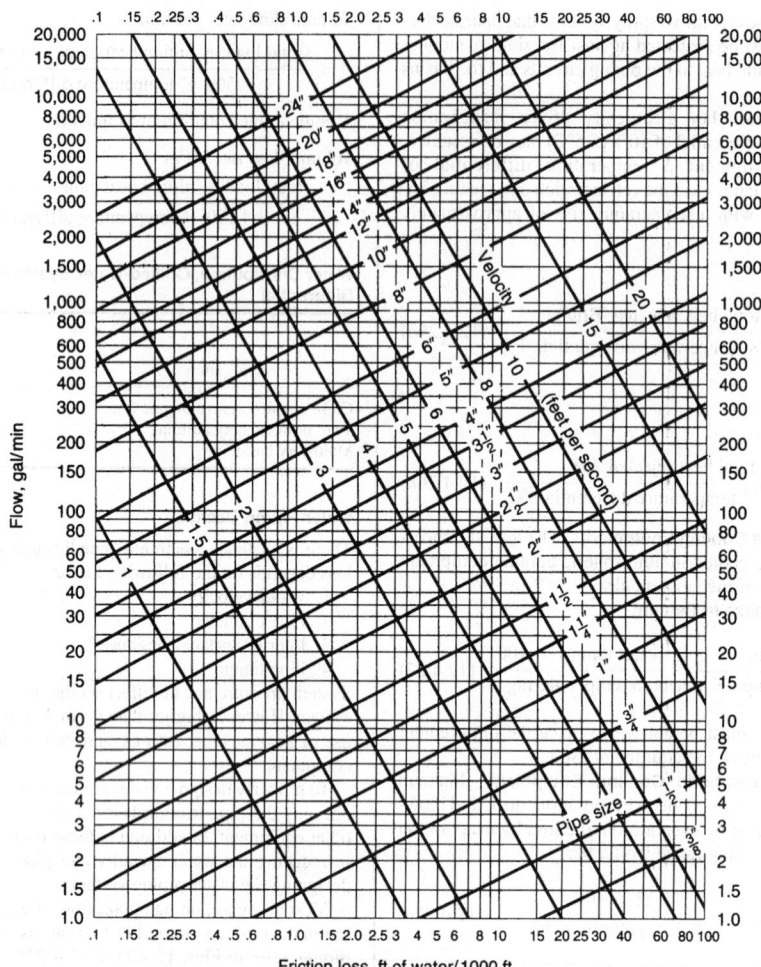

Fig. 12.4.21   Friction loss for open piping systems using schedule 40 pipe.

Table 12.4.37   Low-Pressure Steam System Pipe Capacities, lb/h, with Condensate Flowing with the Steam Flow

Nom. pipe size, in	¹⁄₁₆ psi (1 oz)		⅛ psi (2 oz)		¼ psi (4 oz)		½ psi (8 oz)		¾ psi (12 oz)		1 psi	
	Saturated pressure (psig)											
	3.5	12	3.5	12	3.5	12	3.5	12	3.5	12	3.5	12
¾	9	11	14	16	20	24	29	35	36	43	42	50
1	17	21	26	31	37	46	54	66	68	82	81	95
1¼	36	45	53	66	78	96	111	138	140	170	162	200
1½	56	70	84	100	120	147	174	210	218	260	246	304
2	108	134	162	194	234	285	336	410	420	510	480	590
2½	174	215	258	310	378	460	540	660	680	820	780	950
3	318	380	465	550	660	810	960	1,160	1,190	1,430	1,380	1,670
3½	462	550	670	800	990	1,218	1,410	1,700	1,740	2,100	2,000	2,420
4	726	800	950	1,160	1,410	1,690	1,980	2,400	2,450	3,000	2,880	3,460
5	1,700	1,430	1,680	2,100	2,440	3,000	3,570	4,250	4,380	5,250	5,100	6,100
6	1,920	2,300	2,820	3,350	3,960	4,850	5,700	7,000	7,200	8,600	8,400	10,000
8	3,900	4,800	5,570	7,000	8,100	10,000	11,400	14,300	14,500	17,700	16,500	20,500
10	7,200	8,800	10,200	12,600	15,000	18,200	21,000	26,000	26,200	32,000	30,000	37,000
12	11,400	13,700	16,500	19,500	23,400	28,400	33,000	40,000	41,000	49,500	48,000	57,500

NOTE: The weight flow rates at 3.5 psig can be used to cover saturated pressures from 1–6 psig and the weight flow rates at 12 psig can be used to cover saturated pressures from 8–16 psig with an error not exceeding 8%.

Fig. 12.4.22   Friction loss for closed piping systems using schedule 40 pipe.

Fig. 12.4.23   Friction loss for closed and open piping systems using copper tubing.

Example: The dotted line shows that the resistance of a 6-in. standard elbow is equivalent to approximately 16 ft of 6-in. standard pipe.

Note: For sudden enlargements or sudden contractions, use the smaller diameter, d, on the pipe size scale.

Fig. 12.4.24   Resistance of valves and fittings in terms of equivalent length of straight pipe. (*Crane Co.*)

Table 12.4.38   Return Main and Riser Capacities for Low-Pressure Steam Systems

Pipe size, in	Pressure drop per 100 ft														
	1/32 psi (1/2 oz)			1/24 psi (2/3 oz)			1/16 psi (1 oz)			1/8 psi (2 oz)			1/4 psi (4 oz)		
	Wet*	Dry	Vac	Wet*	Dry	Vac	Wet*	Dry	Vac	Wet*	Dry	Vac	Wet*	Dry	Vac
						Return mains									
3/4						42			100			142			200
1	125	62		145	71	143	175	80	175	250	103	249	350	115	350
1 1/4	213	130		248	149	244	300	168	300	425	217	426	600	241	600
1 1/2	338	206		393	236	388	475	265	475	675	340	674	950	378	950
2	700	470		810	535	815	1,000	575	1,000	1,400	740	1,420	2,000	825	2,000
2 1/2	1,180	760		1,580	868	1,360	1,680	950	1,680	2,350	1,230	2,380	3,350	1,360	3,350
3	1,880	1,460		2,130	1,560	2.180	2,680	1,750	2,680	3,750	2,250	3,800	5,350	2,500	5,350
3 1/2	2,750	1,970		3,300	2,200	3,250	4,000	2,500	4,000	5,500	3,230	5,680	8,000	3,580	8,000
4	3,880	2,930		4,580	3,350	4,500	5,500	3,750	5,500	7,750	4,830	7,810	11,000	5,380	11,000
5						7,880			9,680			13,700			19,400
6						12,600			15,500			22,000			31,000
						Return risers									
3/4		48			48	143		48	175		48	249		48	350
1		113			113	244		113	300		113	426		113	600
1 1/4		248			248	388		248	475		248	674		248	950
1 1/2		375			375	815		375	1,000		375	1,420		375	2,000
2		750			750	1,360		750	1,680		750	2,380		750	3,350
2 1/2						2,180			2,680			3,800			5,350
3						3,250			4,000			5,680			8,000
3 1/2						4,480			5,500			7,810			11,000
4						7,880			9,680			13,700			19,400
5						12,600			15,500			22,000			31,000

* Vacuum values may be used for wet return risers and mains.

Table 12.4.39   Total Pressure Drop for Two-Pipe Low-Pressure Steam Piping Systems

Initial steam pressure, psig	Total pressure drop in supply piping, psi	Total pressure drop in return piping, psi
2	1/2	1/2
5	1 1/4	1 1/4
10	2 1/2	2 1/2
15	3 3/4	3 3/4
20	5	5

Table 12.4.40   Medium-Pressure Steam System (30 psig) Pipe Capacities, lb/h

Pipe size, in	Pressure drop per 100 ft				
	1/8 psi (2 oz)	1/4 psi (4 oz)	1/2 psi (8 oz)	3/4 psi (12 oz)	1 psi (16 oz)
	Supply mains and risers	25–35 psig—max. error 8%			
3/4	15	22	31	38	45
1	31	46	63	77	89
1 1/4	69	100	141	172	199
1 1/2	107	154	219	267	309
2	217	313	444	543	627
2 1/2	358	516	730	924	1,033
3	651	940	1,330	1,628	1,880
3 1/2	979	1,414	2,000	2,447	2,825
4	1,386	2,000	2,830	3,464	4,000
5	2,560	3,642	5,225	6,402	7,390
6	4,210	6,030	8,590	10,240	12,140
8	8,750	12,640	17,860	21,865	25,250
10	16,250	23,450	33,200	40,625	46,900
12	25,640	36,930	52,320	64,050	74,000
	Return mains and risers	0–4 psig—max. return pressure			
3/4	115	170	245	308	365
1	230	340	490	615	730
1 1/4	485	710	1,025	1,285	1,530
1 1/2	790	1,155	1,670	2,100	2,500
2	1,575	2,355	3,400	4,300	5,050
2 1/2	2,650	3,900	5,600	7,100	8,400
3	4,850	7,100	10,250	12,850	15,300
3 1/2	7,200	10,550	15,250	19,150	22,750
4	10,200	15,000	21,600	27,000	32,250
5	19,000	27,750	40,250	55,500	60,000
4	31,000	45,500	65,500	83,000	98,000

Table 12.4.41   High-Pressure Steam System (150 psig) Pipe Capacities (lb/h)

Pipe size, in	⅛ psi (2 oz)	¼ psi (4 oz)	½ psi (8 oz)	¾ psi (12 oz)	1 psi (16 oz)	2 psi (32 oz)	5 psi
	Supply mains and risers			*130–180 psig—max, error 8%*			
¾	29	41	58	82	116	184	300
1	58	82	117	165	233	369	550
1¼	130	185	262	370	523	827	1,230
1½	203	287	407	575	813	1,230	1,730
2	412	583	825	1,167	1,650	2,000	3,410
2½	683	959	1,359	1,920	2,430	3,300	5,200
3	1,237	1,750	2,476	3,500	4,210	6,000	9,400
3½	1,855	2,626	3,715	5,250	6,020	8,500	13,100
4	2,625	3,718	5,260	7,430	8,400	12,300	19,200
5	4,858	6,875	9,725	13,750	15,000	21,200	33,100
6	7,960	11,275	15,950	22,550	25,200	36,500	56,500
8	16,590	23,475	33,200	46,950	50,000	70,200	120,000
10	30,820	43,430	61,700	77,250	90,000	130,000	210,000
12	48,600	68,750	97,250	123,000	155,000	200,000	320,000
	Return mains and risers			*1–20 psig—max, return pressure*			
¾	156	232	360	465	560	890	
1	313	462	690	910	1,120	1,780	
1¼	650	960	1,500	1,950	2,330	3,700	
1½	1,070	1,580	2,460	3,160	3,800	6,100	
2	2,160	3,300	4,950	6,400	7,700	12,300	
2½	3,600	5,350	8,200	10,700	12,800	20,400	
3	6,500	9,600	15,000	19,500	23,300	37,200	
3½	9,600	14,400	22,300	28,700	34,500	55,000	
4	13,700	20,500	31,600	40,500	49,200	78,500	
5	25,600	38,100	58,500	76,000	91,500	146,000	
6	42,000	62,500	96,000	125,000	150,000	238,000	

Pressure drop per 100 ft

12.5   ILLUMINATION
by Abraham Abramowitz, Revised by Staff

REFERENCES: Amick, "Fluorescent Lighting Manual," McGraw-Hill. "IESNA Lighting Handbook." Design publications of General Electric Co., North American Philips Co. (successor to Westinghouse Electric Co. Lamp Division), and GTE-Sylvania, and others.

Ordinarily, the term **illumination** encompasses the application of the science (and sometimes the art) of lighting. Electromagnetic radiation in the wavelength range of 380 to 760 nm elicits response in the normal human eye and thereby defines what we know as the visible spectrum. Illumination engineering deals with harnessing sources of lighting into the design of a system to distribute lighting, which will produce an environment enabling us to see.

BASIC UNITS

Candela, cd, is the unit of luminous intensity of a light source. One candela is defined as the luminous intensity in a given direction, of a source that emits monochromatic radiation of frequency 540×10^{12} Hz (approximately 555 nm) and of which the radiant intensity in that direction is $\frac{1}{683}$ W per steradian (W/sr) (Fig. 12.5.1).

Lumen, lm, is the unit of luminous flux ϕ. It is equal to the flux on a unit surface all points of which are one unit distant from a uniform point source of one candela. Such a point source emits 4π lumens.

Illuminance E is the density of luminous flux on a surface. If the foot is taken as the unit of length and the flux is uniformly distributed over the surface, the density in **lumens per square foot** is called **footcandles, fc**; in SI units **lumens per square metre, lux (lx)**, is used.

Luminance is the luminance intensity of any surface in a given direction per unit of projected area of the surface as viewed from that direction. The unit of luminance is cd/in^2; in SI units cd/m^2 is used.

Absorption, Reflection, and Transmission   Incident light flux interacts with a medium by **absorption**, whereby incident light flux is dissipated; by **reflection**, in which the incident light flux leaves a surface or medium from the incident side; by **transmission**, in which incident light flux leaves a surface or medium on a side other than the incident side.

Reflection may occur as from a mirror (specular reflection), or it may be reflected at angles different from that of the incident flux to incident plane (diffuse reflection), or it may be a combination of the two types of reflection.

In transmission, if the light ray is reduced only in intensity, the transmission is called regular; if the light ray emerges in all directions, transmission is called diffuse. Both modes may occur simultaneously.

The incident flux ϕ_i equals the flux absorbed ϕ_a, reflected ϕ_r, and transmitted ϕ_t. That is,

$$\phi_i = \phi_a + \phi_r + \phi_t$$

Dividing this equation by ϕ_i, we obtain

$$1 = \phi_a/\phi_i + \phi_r/\phi_i + \phi_t/\phi_i$$
or
$$1 = \alpha + \rho + \tau$$

α is the *absorptance*, ρ is the *reflectance*, and τ is the *transmittance*. In each case, the incident flux may be restricted to a single wavelength, a particular direction, and a given solid angle. These must be specified.

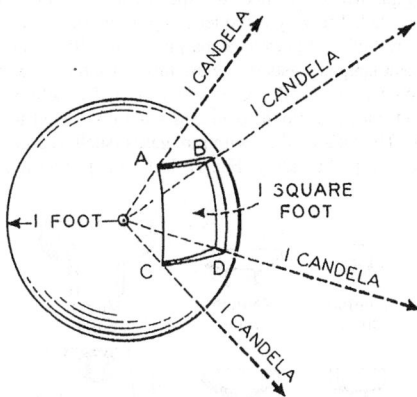

Fig. 12.5.1 Relationship between candela, lumen, and footcandle. [A uniform point source (luminous intensity or candlepower = 1 candela) is shown at the center of a sphere of 1-ft radius. It is assumed that the sphere is perfectly transparent (i.e., has 0 reflectance). The illumination at any point on the sphere is 1 ft · c (1 lumen per square foot). The solid angle subtended by the area $A\,B\,C\,D$ is 1 steradian. The flux density is, therefore, 1 lm/sr, which corresponds to a luminous intensity of 1 cd. The sphere has a total area of 12.57 (4π) ft², and there is a luminous flux of 1 lm falling on each square foot. Thus the source provides a total of 12.57 lm.]

The **wavelength** of electromagnetic radiation is measured in metres. For the frequencies involved in illumination, the wavelength is given in nanometres, nm, equal to 10^{-9} m.

VISION

Most engineering designs (bridges, structures, roads etc.) are based on strength and are not concerned with the way the human organism reacts. The **response of the eye** is central to illuminating engineering. The **lens** of the **eye** focuses an image on the **retina**. Here a photochemical process takes place which sends nerve impulses to the brain via the optic nerve. The amount of light entering the eye is controlled by the **pupil**. The normal eye automatically accommodates itself to focus on an object, while the pupil adjusts itself to allow for a high or low level of object luminance. The sensors in the eye are known as **rods** and **cones**. The cones are clustered in a small central part of the retina called the **fovea**. They transmit a sharp image to the brain and give color response. Outside the fovea the rods predominate. They give neither a sharp image nor a color response. When the luminance of the visual field is low, as at night, seeing is due to the rods only and is called **scotopic vision**. At higher levels, with the cones primarily involved, seeing is called **photopic vision**. There is an intermediate region called **mesopic vision**.

The response of the eye to colors of different wavelengths is given in Fig. 12.5.2. The **luminous efficacy** (lumen output per radiated watt) is 683 lm/W at the wavelength of maximum photopic response, 555 nm. For white light, radiation which has the characteristic of an equal energy spectrum with all the energy in the visual region, it is approximately 220 lm/W.

Spectral Lumen If the response curve of the eye for photopic vision, versus λ in nanometres, is expressed as $k(\lambda)$, and the spectral power function of the source in watts per nanometre is taken to be $Q_e(\lambda)$, then the luminous flux is given by the equation

$$\phi_{\text{lumens}} = 683 \int_{380}^{780} k(\lambda) Q_e(\lambda)\, d(\lambda) \qquad (12.5.1)$$

Fig. 12.5.2 Luminous efficiency versus wavelength for photoptic (day) vision.

LIGHT METERS

Early light meters compared the luminance of a diffuse highly reflecting surface with that obtained from a calibrated standard. The most common light meter in use today is similar to a photographic exposure meter. A photovoltaic cell is directly connected to a sensitive microammeter calibrated in footcandles (or dekalux). The best meters (called color-corrected) have a response similar to that of the eye in photopic vision. Special shapes are used on the cover to avoid total reflection of light from the glass surface of the cell. Such meters are said to be cosine law corrected. The microammeter is frequently replaced with an electronic amplifier using an analog or digital readout.

LIGHT SOURCES

The original and still major source of light is the sun. Next came fire, derived from candles, oil, and gas lamps. With the discovery of electricity came arc lamps, gas-discharge lamps, and hot-filament lamps. "Flame" or hot sources give a continuous spectrum. Gas-discharge devices such as sealed neon tubing and mercury-arc lamps give discrete, or line, spectra. The lines may be modified in various ways: by pressure broadening, use of phosphor coatings (to convert ultraviolet radiation into visible light), and using a mixture of gases. The continuous spectra of phosphors have colors which depend upon the mixture used. A light-emitting diode (LED) is a rectifying semiconductor device which can convert electric energy to electromagnetic radiation ranging from the near-ultraviolet to the near-infrared, i.e., from 400 to over 1,500 nm. They are extremely attractive because they are very energy-efficient, produce negligible heat, and contain no filament—and are thus highly resistant to breaking from shock, vibration, and temperature extremes. They can be inserted in standard sockets, including incandescent-type screw bases. For special applications, they can be configured as required. A major use is in display lighting or in indicators (traffic signals, panel backlighting, etc.), but they have begun to find favor as long-life energy-efficient substitutions for incandescent lighting. At this time (2005) the initial cost is quite high, but prices will probably decline as their use becomes more widespread.

Color Temperature and Luminance

In general, three quantities are required to specify the color of a light and its luminous level. However, an approximate designation is used by specifying the temperature of a hot (black-body) emitter whose color almost matches that of the light. The **color temperature** of daylight is about 6,000 K and that of tungsten lamps about 2,300 to 3,300 K.

Lamps

Electric lamps are the principal source of artificial light in common use. They convert electrical energy into light or radiant energy.

An **incandescent-filament lamp** contains a filament which is heated by the current passing through it. The filament is enclosed in a glass bulb which has a base suitable to connect the lamp to an electrical socket. To prevent oxidation of the filament at elevated temperature, the bulb is evacuated of air or filled with an inert gas. The bulb also serves to control the light from the incandescent filament, which is essentially a point source. High luminance of the source is typically reduced by acid etching to frost the inside surface of the bulb. Silica coating will also provide additional diffusion and can alter the color of the light emitted. Portions of the bulb's interior can be covered with reflecting material to give a predetermined direction to the emitted light. Chemical tinting of clear glass bulbs provides a variety of colors. Whenever the color that is normally produced by an incandescent filament is changed, the filtering process removes from the radiated light the energy of all wavelengths except those necessary to produce the desired color. This subtractive method of color alteration is less efficacious than the generation of light of varying colors by gaseous discharge.

Sizes and **shapes** of lamp bulbs are designated by a letter code followed by a numeral; the letter indicates the shape (Fig. 12.5.3), and the number indicates the diameter of the bulb in eighths of an inch. Thus a T-12 lamp has a tubular shape and is $1\frac{4}{8}$ or $1\frac{1}{2}$ in in diameter.

Fig. 12.5.3 Typical filament lamp shapes: S, straight; F, flame; G, globe; A, general service; T, tubular; PS, pear shape; PAR, parabolic; R, reflector.

Incandescent lamps are available with several **types of bases** (Fig. 12.5.4). Most general-service lamps have medium screw bases; larger or smaller screw bases are used depending on lamp wattage. Bipost and prefocus bases accurately position the filament, as in optical projection systems. Bipost lamps also serve where ruggedness and greater heat dissipation are required.

Fig. 12.5.4 Typical incandescent lamp bases.

Incandescent-lamp filaments are generally constructed of tungsten. Tungsten has a high melting point and a low vapor pressure, which permits high operating temperatures without evaporation: the higher the operating temperature, the higher the efficacy (lumens per watt) and the shorter the life. Filament evaporation throughout the life of the lamp causes blackening of the bulb and thinning of the filament with

consequent lower light output. Argon-nitrogen gas filling reduces the rate of evaporation.

Tungsten filaments are also placed in compact quartz tubes filled with a halogen atmosphere where the tungsten halide lighting source continuously returns evaporated tungsten particles to the filament. The inside walls do not blacken, and light output remains fairly constant throughout the life of the lamp.

For exposed illuminated outdoor lamp signs, **gas-filled incandescent lamps** are often used, especially when high-speed animation is to be depicted by the on-off sequencing of the lamps. The lamp envelope is filled with a gas which enhances the speed with which the filament is heated and cooled. The very fast quenching of the filament without the usual decay trail of light permits a sharply defined illusion of motion.

A **fluorescent lamp** consists of a glass tube coated on the inside with phosphor powders which fluoresce when excited by ultraviolet light; filament electrodes are mounted in end seals connected to base pins (Fig. 12.5.5). The tube is filled with rare gases (such as argon, neon, or krypton) and a drop of mercury (Fig. 12.5.6), and operates at a relatively

Fig. 12.5.5 Fluorescent lamp bases.

low pressure. In operation, electrons are emitted from the hot electrodes. These electrons are accelerated by the voltage across the tube until they collide with mercury atoms, causing them to be ionized and excited. When the mercury atom returns to its normal state, mercury spectral lines in both the visible and the ultraviolet region are generated. The low pressure enhances ultraviolet radiation. The ultraviolet radiation excites the phosphor coating to luminance. The resulting light output is not only much higher than that obtained from the mercury lines alone but also results in a continuous spectrum with colors dependent upon the phosphors used.

Fig. 12.5.6 Fluorescent lamp operation. *(Westinghouse.)*

As with all gas-discharge devices, these lamps have negative voltampere characteristics. Unless the voltage difference between the applied voltage and the lamp operating voltage is absorbed in some way, damaging currents will result. A reactor is used in series with the lamp. It may be capacitive or inductive. The supply voltage should be at least twice the lamp operating voltage. Where this is not the case, the supply voltage (up to 277 V is used) is stepped up by an autotransformer. The necessary reactance is frequently part of the transformer leakage inductance.

For starting purposes, voltages higher than twice lamp operating voltages may be used. For minimum line current a power-factor-correcting capacitor is used and assembled with the autotransformer. A capacitor is put across the lamp to reduce radio-frequency interference with nearby radio receivers. All these elements are placed inside a case filled with a potting compound. The assembly is called a **ballast.** The object of the compound is to reduce noise from the core lamination vibrations and to improve heat dissipation. Built-in thermal protectors, which deenergize the ballast when dangerous temperatures are reached, are now required in the United States. Ballast manufacturers rate their units by noise levels. Some ballasts use solid-state, high-frequency generators.

To start a fluorescent lamp, electron emission from the electrodes must be induced. Two methods are generally employed: (1) the filament electrodes are heated by passing current through them; (2) a high voltage, sufficient to start an electric discharge in the lamp, is applied across it. Once a discharge starts, mercury-ion bombardment keeps the filaments at a hot electron-emitting temperature. Lamps are designed for either type of operation. The first group is further divided into preheat lamps and rapid-start lamps. Some lamps can be used for both types of circuits. The lamp current is carried primarily by electrons emitted from the filaments. The end or weakening of electron emission is an important cause of the end of lamp life.

Preheat circuits contain starters which are really switches, closed when power is first applied, permitting current to flow and preheat the electrodes. After a predetermined period of time, the starter switch opens, throwing a potential across the lamp which starts the discharge. Lamps used for preheat circuits have bipin bases (Fig. 12.5.5).

Instant-start circuits have ballasts which apply sufficient voltage across the lamp to induce current flow without preheating the electrodes. **Slimline lamps** are the principal instant-start type. They have single-pin bases because no preheating is required, are available in sizes up to 8 ft in length, and can have varying lumen outputs depending upon the ballast current rating and wattage. Because of their high starting voltage, they generally employ spring-loaded push-pull lampholders which disconnect the ballast circuit unless the lamps are properly seated in position.

Instant-start lamps are sometimes available with bipin bases similar to those used in preheat lamps. In these instances the lead wires from the pins are connected together inside the lamp. These lamps, marked "instant start," are not interchangeable with rapid-start equipment. Cold-cathode lamps employ the instant-start principle, have cylindrical iron electrodes, and tend to be less efficacious at shorter lengths because of high wattage losses at the electrodes. They are limited to low current densities because the electrodes operate at temperatures below that necessary for thermionic emission. Cold-cathode lamps, whose operation is not affected by dimming or flashing, have long life and are generally used for custom-built shapes and patterns that require bending, such as for electric signs.

Rapid-start circuit ballasts have separate windings for the electrodes which are immediately and continuously heated when the circuit is energized. This rapid heating causes sufficient ionization in the lamp for a discharge to start from the voltage of the main ballast windings. Two-lamp rapid-start ballasts are of the series sequence type, in which the lamps start in sequence and, when fully lighted, operate in series.

A new type **screw-in fluorescent lamp** with built-in ballast can be used in a standard medium-screw socket. These lamps consume less power than incandescent lamps for the same luminance; accordingly, albeit their first cost is significantly higher, they are expected to prove to be more economical by virtue of their reduced power consumption and much longer life. A significant number of them have made way into household and commercial use.

Fluorescent lamps used in low-ambient-temperature applications, as in outdoor signs, are of the **high-output (HO)** type, and require special high-output ballasts to permit the lamps to maintain their luminance at lower operating temperatures.

High-intensity-discharge lamps consist of tubes in which electric arcs in a variety of materials are produced. Outer glass jackets provide thermal insulation in order to maintain the arc tube temperature. The temperature and amount of material is controlled so that the discharge operates in a vapor pressure of several atmospheres. This results in enhancing the radiation in the visible region.

Mercury-vapor lamps consist of mercury-argon-filled quartz tubes surrounded by a nitrogen-filled glass jacket. Clear lamps radiate the visible mercury lines (bluish green). Ultraviolet radiation is absorbed to some extent by the outer jackets. The color of the light and the lumen output is improved by coating the inside of the outer jackets with a phosphor. When excited by the ultraviolet radiation of the arc, the phosphors add light in the red part of the spectrum to the output. The resulting lamps are called white, color improved, or deluxe white. The lamps start by a discharge in argon between an electrode and a starting electrode (see Fig. 12.5.7). As the mercury vaporizes, the pressure builds up and the discharge transfers to a mercury discharge. This takes several minutes. After shutdown, the lamps cannot be restarted until the inner tube pressure drops so that an argon discharge can start.

Fig. 12.5.7 High-output lamps. *(GE.)*

Metal halide (multivapor) lamps use small quantities of sodium, thallium, scandium, dysprosium, and indium iodides in addition to the usual mercury-argon mix. Color is improved and output substantially increased over high-intensity-discharge lamps using mercury alone. While the construction is similar to mercury lamps, a bimetal switch is built into the lamp to short out the starting resistor after the lamps start. A vacuum jacket is used around the quartz discharge tube (see Fig. 12.5.7).

High-pressure sodium-vapor lamps use metallic sodium sealed in translucent aluminum oxide tubes. This material is used to withstand the corrosive effect of hot sodium vapor. For starting purposes a xenon fill gas and a sodium-mercury amalgam is used. Arc temperatures are maintained by an outer vacuum jacket. The lamp is started by generating a high-voltage pulse for about a microsecond (see Fig. 12.5.7).

High-pressure discharge lamps, like fluorescent lamps, require ballasts. These provide the necessary voltage, reactances, and power-factor-correcting capacitors.

Comparative lamp efficacies (lm/w) range from about 8 for tungsten incandescent to about 140 for high-pressure sodium.

Luminaires

Luminaires are generally categorized as **industrial, commercial,** or **residential.** Use within these categories usually determines the quality and ruggedness of materials of construction. Generally speaking, style, ornament, and in most cases low cost are prime considerations for residential fixtures. Industrial fixtures require low maintenance, low operating cost, efficiency, and durability. Commercial fixtures combine the elements of all of these and place heavy emphasis on visual comfort.

Luminaires are classified by the International Commission on Illumination (ICI) in accordance with the percentages of total luminaire output emitted above and below the horizontal (Fig. 12.5.8). Industrial fixtures usually are direct or semidirect.

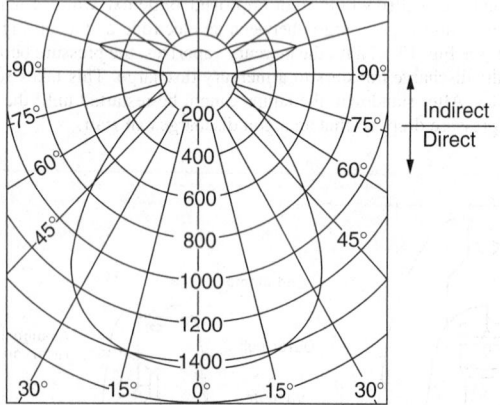

Fig. 12.5.8 Hypothetical light distribution curve for a luminaire providing both direct (below horizontal) and indirect (above horizontal) lighting. Manufacturers' literature provides such data in great detail for their lines of luminaires.

Luminaires control the source of light so that it can be better used for a given seeing task. Materials used in luminaires are designed to reflect, refract, diffuse, or obscure light.

Reflectors are commonly made of specular alzak aluminum, glass, baked-enamel steel, and porcelain-enamel steel.

Lenses with prismatic patterns refract light sources to disperse the rays or to direct them most effectively. Lenses are of glass, acrylic plastic (Plexiglas), or polystyrene. **Glass** is generally superior for incandescent fixtures because of its heat resistance. Fluorescent lamps with **acrylic** lenses have a life at least twice that of **polystyrene** because of their color stability. **Translucent** glass and plastic are used in bottom lenses, louvers, and side panels to diffuse light and to obscure light sources. Baked-enamel steel **louvers,** in egg-crate, concentric ring, or cellular configurations are widely used to shield light sources from normal viewing angles.

Fixture bodies, trims, and lens frames are commonly constructed of steel, electrogalvanized, and/or treated with a rust-inhibiting coating, and painted with several coats of baked enamel. Stamped, spun, cast, and die cast aluminum are also used.

Figure 12.5.9 shows two typical luminaires. Manufacturers' literature will provide complete details and light distribution curves for lighting design. The variety of luminaires available is large and quite diverse, offering the designer a large range of options.

Porcelain-enamel ventilated standard dome with incandescent lamp
(a)

4 lamp, 2' (610 mm) wide troffer with 45° white metal louver
(b)

Fig. 12.5.9 Two typical luminaires. (*a*) With incandescent lamp; (*b*) with fluorescent lamps and metal louver.

PRESCRIBING ILLUMINATION

The object of a **lighting design** is to provide sufficient illuminance for a given seeing task without introducing discomfort. Sufficient light is not difficult to obtain with modern light sources, but unless properly placed and controlled, uncomfortable, glaring light will result.

A given task has a size, luminance contrast, luminance, and color. The luminance of a perfectly diffusing reflecting surface is given by

$$L = E\rho \qquad (12.5.2)$$

where E is illuminance in footcandles, ρ its reflectance (ratio of reflected light flux to incident flux), and luminance L is in footlamberts. The contrast C between two adjacent areas is given by

$$C = (L_b - L_o)/L_b \qquad (12.5.3)$$

where L_b is the luminance of the larger or background area and L_o is the object or task luminance for a given illuminance E. Substituting Eq. (12.5.2) into Eq. (12.5.3),

$$C = \frac{E\rho_b - E\rho_o}{E\rho_b} = \frac{\rho_b - \rho_o}{\rho_b} \qquad (12.5.4)$$

Thus, contrast is basically a function of the task-to-background reflectance. Most surfaces are not perfectly diffuse and require the use of the luminance factor β (ratio of actual luminance for a given viewing angle to the illumination under actual conditions) instead of ρ. The contrast which can just be seen or detected (minimum perceptible contrast) is a function of background luminance for a given task.

Unfortunately most visual tasks are performed on mirrorlike (specular) or semimirrorlike surfaces. This results in reducing the work contrast. Where a highly luminous object such as an incandescent-lamp filament is reflected from a polished surface, "reflected glare" results. Reflections of a large luminous area by matte or semimatte material results in "veiling reflection." For work on a horizontal surface, the line of sight from the eye for most people is from 0 to 40° from the vertical, with a peak angle of about 25°. If light from the source is directly reflected into the eyes, contrast and resulting visibility are greatly reduced. When a highly luminous source is directly reflected into the eye, it is possible to completely obliterate a task. Veiling reflections are controlled by proper luminaire design and placement of fixtures. Some designs keep flux out of the 0 to 40° zone. As far as possible the work area should have a matte surface without shining details, and light should come from the side or behind the worker.

In addition to veiling reflectance, there is a reduction in contrast due to light directly entering the eye from the source. This is called the disability glare effect. It produces a light veil over the image of the task on the retina. It is not a serious problem in interior lighting, but it is important in roadway lighting and similar situations.

Visual-Comfort Criteria High luminances directly or reflected in the field of view can cause discomfort without necessarily interfering with seeing even though visual performance may be impaired. This discomfort glare can be caused by direct glare from sources which have too high a luminance, are inadequately shielded, or have too great an area. Lighting systems are rated by a **visual-comfort probability (VCP)**, expressed as a percentage of people who, if seated in the most undesirable location, will be expected to find it acceptable. (For a complete description of VCP, see the IESNA Handbook.) If the following conditions are met, direct glare will not be a problem in lighting installations:
  1. The VCP is 70 or more.
  2. The ratio of maximum-to-average luminaire luminance does not exceed 5:1 (preferably 3:1) at 45, 55, 65, 75, and 85° from the nadir crosswise and lengthwise.
  3. Maximum luminaire luminances crosswise and lengthwise do not exceed the following values:

Angle above nadir, deg	Maximum luminance cd/m²	fL
45	7,710	2,250
55	5,500	1,605
65	3,860	1,125
75	2,570	750
85	1,695	495

Design of Interior Lighting Systems

Lighting is as much an art as a science. While many studies have been made on what constitutes adequate lighting along with proper quality, the effect to be achieved depends upon the designer. In this section emphasis will be primarily on achieving adequate illumination.

The design approach is to consider the space to be lighted and the task to be performed. An illuminance is then selected. A suitable luminaire is picked, and calculations are made in order to determine the number and layout of the fixtures. The overall quality is then checked. If unsatisfactory, a new layout is made. An economic study is made to check costs. If these are too high, new layouts are studied until all design restraints are met.

Selection of illuminance Levels

An illuminance level guide for selected tasks is given in Table 12.5.1. The data are based on an assumption of average conditions for people, tasks, and visual performance requirements. (For other conditions see the "IESNA Lighting Handbook.")

Room, Furniture, and Equipment Finishes

The color and finish of rooms, furniture, and equipment are important in the overall lighting design. Best results are obtained when the lighting designer coordinates his or her work with the architect, interior decorator, or plant designer.

A color scheme should be selected to give light reflectance values as follows:

Area or unit	Percent reflectance
Ceilings	70–90
Floors	20–40*
Walls, draperies	40–60†
Bench top, desks, machine, and equipment	25–45

* In storage areas, keep reflectance of aisle floors as high as possible in order to reflect light onto the lower shelves. This should also be done where the underside of objects has to be seen.
† These values should be 30 to 40 where video display terminals (VDTs) are used to avoid veiling reflections in the VDT faces.

The color and finish of a space and equipment therein set the psychological feel of the space. For example, the trend is away from drab finishes on machinery and dark gray filing cabinets. Colors such as yellow, orange, red, and light gray seem to advance toward the eye. They tend to make large spaces feel smaller. Receding colors such as violet, blue, blue green, and dark grays make small spaces feel larger. Some colors are

Table 12.5.1 Illuminance Guide for Selected Tasks

	Footcandles (lm/ft)	Lux (lm/m²)
Commercial drafting		
Conventional	150*	1,600*
Libraries		
Reading good print, typed originals	30	320
Reading small print, handwriting, photocopies	75	800
Active stacks (vertical, illumination)	30	320
Offices		
Conference rooms—conferring	30	320
Conference rooms—typical visual tasks	75–100	800–1,080
Corridors, stairs, elevators	20	220
General tasks, varying difficulty	100	1,080
Lobbies, reception areas	30	320
Private	75	880
Rest rooms	30	320
Video display areas	75	800
School		
Classrooms, laboratories	75	800
Shops	100	1,080
Sight-saving rooms, hearing-impaired classes	150	1,600
Store		
Mass merchandizing, high activity	100	1,080
Self-service	200	2,200
Circulation, low activity	30	320
Feature displays, low, medium, high activity	150,* 300,* 500*	1,600,* 3,200,* 5,400*
Industrial		
Garages		
Repair	75	800
Active traffic areas	15	160
Loading platform	20	220
Machine shops and assembly areas		
Rough bench-machine work, simple assembly	50	540
Medium bench-machine work, moderately difficult assembly	100*	1,080*
Difficult machine work, assembly	150*	1,600*
Fine bench-machine work, assembly	300*	3,200*
Receiving and shipping	30	320
Warehouse storage rooms		
Active large items	15	160
Active small items, labels	30	320
Inactive	5	54
Outdoor areas		
Storage yards		
Active	20	220
Inactive	1	11
Parking areas		
Open, high activity	2	22
Open, medium activity	1	11
Covered parking, pedestrian areas	5	54
Covered night entrance	5	54
Covered day entrance	50	54

* Requires supplementary lighting. Care should be taken that the supplementary lighting does not introduce direct and reflected glare.
SOURCE: Adapted from General Electric Co. design data.

used for safety purposes. Various areas are painted to designate safe or hazardous locations in a fashion similar to piping identification discussed in Sec. 8. These colors have been carefully standardized in ANSI Z53.1.

Designating Color

In order to be able to obtain designed values of a lighting system, it is necessary to be able to specify the exact color wanted. Many methods have been devised for so doing. One method uses carefully controlled sets of colored chips, each one of which has a particular designation. The desired color is matched against these chips. The **Munsell system** uses *scales of hue* (the actual basic color such as red), *value* (a 10-step scale ranging from black through grays to white), and *chroma* (the amount of gray mixed in with the color). This system is used by many manufacturers to designate their colors. The **Ostwald system** describes color in terms of *color content, white content,* and *black content.* The **Inter-Society Color Council-National Bureau of Standards** (now NIST) (ISCC-NBS) **method** designates one-inch square samples with names. For color designation by the **ICI method,** a spectroradiometric curve of the source is determined together with a spectrophotometric curve of the reflecting or transmitting surface. By mathematical manipulation using spectral tristimulus values, chromaticity coordinates are obtained. See the IESNA Handbook for details. Chromaticity coordinates are extensively used for fluorescent lamp colors. These coordinates can be measured directly by photoelectric colorimeters. They are designed with filter photocell responses to be close to each of the ICI tristimulus values. Built-in logic circuitry results in direct reading of the chromaticity. Incandescent and vapor-type lamp colors are specified by color temperature.

LIGHTING DESIGN

Interior lighting is designed by the **lumen** method. This takes into account the interreflections of light inside a room. The average illuminance on the work plane equals the incident luminous flux ϕ divided by the area, or $E = \phi/A$. Lumens reaching the work plane is equal to lamp lumens multiplied by the **coefficient of utilization** CU. This factor is a function of room size, shape, and finish, mounting height of fixture, and type of luminaire used. The lumens ϕ_L initially available from the lamps may be reduced by ambient temperature, lower voltage, and the ballast used. As time goes by the room surfaces and luminaires become dirty, which further reduces the illuminance. In addition, lamp output falls, and some of them burn out. The total effect of all these factors is expressed by the *light-loss factor* LLF. The maintained illuminance E_m is the initial illumination times the LLF, or

$$E_m = (\phi_L \times CU \times LLF)/A \qquad (12.5.5)$$

The required maintained illuminance is selected from Table 12.5.1 or from the more extensive data in the IESNA Handbook. A fixture and lamp is selected, and Eq. (12.5.5) is solved for the necessary lamp flux ϕ_L. The number of luminaires N is found by dividing the total lamp lumens ϕ_L by the lumens per fixture ϕ_F. A trial layout is then made. A simple layout keeps spacing between units equal to twice the distance between fixtures and wall. Spacing is checked against the maximum allowable luminaire spacing from manufacturers' data to ensure uniform illumination. However, this criterion results in inadequate lighting near the walls. In order to light desks and benches along the walls, a spacing of 2½ ft from the luminaire center to the wall is used. The ends of fluorescent luminaire rows should be 6 to 12 in from the walls with a maximum distance of 2 ft.

Wall and ceiling cavity luminances can be obtained by using luminance coefficients (LC) for the fixtures (see manufacturers' literature). For interior areas, maximum luminance ratios should be 3:1 or 1:3 between tasks and immediate surround, and 10:1 or 1:10 between tasks and remote surfaces. To ensure eye comfort, the visual-comfort probability is investigated.

The **coefficient of utilization** is found by using the **zonal-cavity** method. In this method effects of the room proportion, luminaire suspension lengths, and work-plane height on the CU are found by dividing the room

into three cavities as shown in Fig. 12.5.10. For each cavity a cavity ratio is calculated:

$$\text{Cavity ratio} = \frac{5h(\text{room length} + \text{room width})}{(\text{room length}) \times (\text{room width})} \qquad (12.5.6)$$

where $h = h_{RC}$ for the room cavity ratio RCR; $= h_{CC}$ for the ceiling cavity ratio CCR; and $= h_{FC}$ for the floor cavity ratio FCR.

Fig. 12.5.10   The three cavities used in the zonal cavity method.

Supplemental tables for use of RCR, CCR, and FCR, together with a detailed description of the procedure to follow when applying the lumen method, are found in the IESNA Handbook.

While the lumen method is the accepted procedure for calculating interior lighting levels, it is often necessary to have a quick approximation of the quantity of lighting equipment needed to satisfy an illumination-level specification. There are several rules of thumb which serve that purpose.

Spacing Method   Table 12.5.2 indicates approximate *base*-maintained footcandle levels according to fixture spacing. For levels other than the base quantity, the level is changed inversely and proportionally to a change in spacing—i.e., doubling the spacing (in one direction) halves the level; halving the spacing doubles the level. Doubling the spacing in both directions reduces the level to one-fourth.

Lumens-per-Square-Foot Method   See footnote for Table 12.5.3.

Less accurate than the previous method, this one permits a fairly reasonable calculation by using the lamp lumens as given in the GE Lamp Catalog and substituting in the formula:

$$\text{Footcandle} = \frac{\text{total lamp lumens per fixture}}{2 \times \text{area per fixture}}$$

Or, for a given footcandle level, transposing the formula to determine area per fixture:

$$\text{Area per fixture} = \frac{\text{total lamp lumens per fixture}}{2 \times \text{footcandles}}$$

The 2 in the denominator assumes loss of half the lumens in fixture utilization, lamp depreciation and dirt accumulation. *(Source: General Electric.)*

Watts-per-Square-Foot Method   Table 12.5.3 shows another way to arrive at a quick approximation.

Point Method of Design

If uniformity of lighting is to be investigated, or if outdoor lighting is to be designed, the point method is used. Manufacturers furnish candlepower distribution curves for their fixtures. An average curve is given for symmetrical fixtures while curves in various planes are given for asymmetrical ones. The basic equation for calculating the illumination from such curves is

$$E_h = (I_\theta \cos \theta)/D^2$$
$$= I_\theta H/D^3 \qquad (12.5.7)$$

Table 12.5.2   Approximate Footcandle Levels According to Fixture Spacing*

Lighting system		Spacings†				
Lamp	Watts	10 × 10 ft	15 × 15 ft	20 × 20 ft	25 × 25 ft	30 × 30 ft
Lucalox	70	35	15	10	—	—
	100	55	25	15	10	—
	150	95	45	25	15	10
	250	180	80	45	30	20
	400	300	135	75	50	35
	1,000	—	—	210	135	95
Multivapor	175	85	35	20	15	10
	400	200	90	50	35	25
	1,000	—	300	165	105	75

Continuous rows of 2-lamp fixtures on spacing† of:					
Fluorescent (cool white)	6 ft	8 ft	10 ft	12 ft	15 ft
40W Rapid Start	120	90	70	60	50
75W Slimline	120	90	70	60	50
110W High Output	185	140	110	90	75
215W Power Groove	300	225	180	150	120

* See footnote for Table 12.5.3.
† Spacings assumed within maximums established by fixture manufacturer.
SOURCE: General Electric Co.

Table 12.5.3   Watts-per-Square-Foot Method*

A convenient, popular, quick approximation

Lamp	Per 100 fc
Lucalox	1.6 W/ft²
Multivapor	2.5 W/ft²
Fluorescent	3.0 W/ft²
Mercury-vapor	3.5 W/ft²
Incandescent (reflector lamp)	8.5 W/ft²

* Both Table 12.5.2 and lumen-per-square-foot method are based upon *large* industrial areas, where room width (*W*) is six times the fixture mounting height (MH). For medium-sized areas (where *W* = 3MH), reduce footcandles and increase wattage 15 percent. For small areas (where *W* = MH), reduce footcandles and increase wattage 50 percent. Lucalox fixtures and incandescent reflector lamps tend to be more efficient and have higher lumen-maintenance characteristics; thus, for these sources increase the footcandle value about 25 percent.
SOURCE: General Electric Co.

where E_h is the illumination on the horizontal plane, I_θ the candlepower in the given direction, and D the distance of the luminaire to the point P. See Fig. 12.5.11.

$$E_v = (I_\theta \sin \theta)D^2$$
$$= I_\theta R / D^3$$

Fig. 12.5.11   Footcandle calculation diagram.

For vertical surfaces

$$E_v = (I_\theta \sin \theta)D^2$$
$$= I_\theta R / D^3 \qquad (12.5.8)$$

Nomograms and graphical solutions are available for Eqs. (12.5.7) and (12.5.8).

THE ECONOMICS OF LIGHTING INSTALLATIONS

The cost of lighting is computed by summing the annual cost of energy; relamping; cost of labor for cleaning, relamping, and servicing; interest; and depreciation.

Different lighting systems can be evaluated by comparing the costs per million lumen hours per luminaire. This is given by Eq. (12.5.9),

$$U = \frac{10}{Q \times D} \left(\frac{P + h}{L} + W \times R + \frac{F + M}{H} \right) \qquad (12.5.9)$$

where U = unit cost of light in dollars per million lumen-hours; Q = mean lamp lumens; D = luminaire dirt depreciation (average between cleanings); P = price of lamp in cents; h = cost (in cents) to replace one lamp; L = average rated lamp life in thousands of hours; W = mean luminaire input watts (lamps plus ballast); R = energy cost in cents per kilowatthour; F = fixed or owning costs in cents per luminaire-year; M = cleaning costs in cents per luminaire-year; and H = annual hours of operation in thousands of hours.

In the above equations the area can be expressed in square metres. These equations are general and basic. With the advent of energy conservation, it has become important to use electric power effectively and efficiently. The IES, in its 1981 Handbook, established a watts per square foot (square metre) for various occupancies, called a *base unit power density* (UPD). These values are used to establish a power limit for a facility. For this purpose approximate values are used for CU, lamp plus ballast, lumens per watt, and LLF. For LLF, 0.7 is used.

Another way to compare installations is to compute the watts per square foot for each proposed installation. This is computed by either method:

$$\text{Watts/ft}^2 = \frac{\text{total lamp lumens}}{\text{area, ft}^2}$$
$$\times \frac{1}{\text{lumens/watt of lamp and ballast}} \qquad (12.5.10)$$

$$\text{Watts/ft}^2 = \frac{\text{designed illuminance}}{\text{CU} \times \text{LLF}}$$
$$\times \frac{1}{\text{lumens/watt of lamp and ballast}} \qquad (12.5.11)$$

Typical values are shown in Table 12.5.3.

DIMMING SYSTEMS

Dimming systems are used in theaters, auditoriums, ballrooms, etc. Originally, power-consuming rheostats were used. These have been replaced by continuously variable autotransformers, variable reactors, silicon controlled rectifiers (SCRs) and triacs. The development of controlled solid-state devices has resulted in small, reliable dimmers which can be readily programmed. Only incandescent and cold-cathode lamps can be dimmed easily. Fluorescent lamps require special ballasts which keep the electrodes hot at all times. Dimmers have been developed for high-intensity discharge (H.I.D.) lamps.

HEAT FROM LIGHTING

Lighting installations are a substantial source of heat, have long been a factor in the design of air-conditioning (cooling) systems, and are increasingly significant in the design of heating systems. The heating effect for 1 W is 3.413 Btu/h. Approximate wattage data for lighting systems at various lighting levels can be calculated by using watts per-square foot calculated from Eq. (12.5.10) or (12.5.11). Heat generated is delivered to surrounding areas in several ways, with energy distribution for fluorescent and incandescent lamps as illustrated in Fig. 12.5.12.

Fig. 12.5.12   Energy distribution of lamps.

With the prevalent high lighting intensities of modern buildings, it is essential to control the heat generated by a lighting system. Substantial portions of the energy which is not radiated into the room may be conducted away from the luminaire by an air stream or by water flowing through a coil attached to the luminaire. In the heating season, this heat energy is delivered to the perimeter of the building for effective space warming. In the cooling season, the heat is rejected to the exterior, thus reducing the load on the cooling system. Air-handling luminaires (Fig. 12.5.13) have been used for this purpose.

Fig. 12.5.13   Typical air handling system. *(Barber-Colman.)*

12.6 SOUND, NOISE, AND ULTRASONICS
by Robin O. Cleveland, Benjamin C. Davenny, and Jonathan D. Kemp

REFERENCES: Kinsler, et al., "Fundamentals of Acoustics," Wiley, 1982. Harris, "Noise Control in Buildings," INCE, 1997. Beranek and Ver, "Noise and Vibration Control Engineering," 1992. "2003 ASHRAE Handbook—HVAC Application," Chap. 47, Sound and Vibration Control, 2003.

SOUND

An **acoustic wave,** or a **sound wave,** is a mechanical disturbance of particles in a fluid from an equilibrium state which propagates as a wave through the medium with a speed of sound that depends on the properties of the fluid. In fluids the waves are longitudinal; i.e., the motion of the disturbed particles is in the direction of the wave propagation. Typical acoustic sources, such as audio speakers, vibrate backward and forward, generating sound waves consisting of compressional phases and rarefractional phases. The pressure, density, and particle velocity of the fluid are all affected by the passage of the wave. Similar waves exist in elastic solids, but in this case, two types of waves can propagate: **longitudinal waves** and **shear** (or **transverse) waves** in which the particle motion is transverse to the direction of propagation. Here, acoustic waves will refer to waves in fluids (liquids or gases) and elastic waves to those in elastic solids.

The **sound speed** c_0 is a material property, and for gases it is given by $c_0 = \sqrt{\gamma R T}$, where γ is the ratio of specific heats (1.4 for diatomic gases such as air), R is the gas constant [287 J/(kg · K) for air], and T is the absolute temperature. For air at 20°C (293 K) it is about 343 m/s

(1,125 ft/s). For liquids the expressions are more involved (see, e.g., Lovett, *Jour. Acoust. Soc. Am.*, **45**; 1969, pp. 1051–1053), and an approximate value for water at room temperature is 1,500 m/s (4,920 ft/s). (See Table 12.6.1.)

Acoustic waves vary in space and time, and the relationship between the temporal separation of points on an acoustic wave (Δt) and the spatial separation of the points (Δx) is the sound speed: $\Delta x = \Delta t c_0$. For a sinusoidal wave the spatial extent of one cycle of the wave is called the **wavelength** λ and the concomitant temporal extent the **period** T.

Table 12.6.1   Density and Sound Speed in Selected Materials

Material	Density		Velocity	
	SI unit, kg/m^3	USCS unit, lbf/ft^3	SI unit, m/s	USCS unit, ft/s
Steel	7,700	481	6,100	20,013
Aluminum	2,700	169	6,300	20,669
Concrete	2,600	162	3,100	10,171
Glass	2,300	144	5,600	18,373
Brick	2,000	125	3,650	11,975
Lucite	1,200	75	2,650	8,694
Water	1,000	62	1,500	4,921
Pine	450	28	3,500	11,483
Cork	240	15	500	1,640
Air	1.21	0.0755	343	1,125

The frequency of the sound is $f = 1/T = c_0/\lambda$ and is measured in hertz (Hz). In older texts the units may be stated as cycles per second (cps). Acoustic waveforms may be sinusoidal in shape, but most often are not. However, the use of Fourier analysis allows any acoustic wave to be represented by a combination of different frequencies. The **audible frequency range** is normally said to extend from 20 Hz to 20 kHz, although for frequencies above 10 kHz the sensitivity of the ear decreases dramatically. As an individual gets older, the upper frequency limit becomes progressively lower. Sound with frequencies less than 20 Hz is referred to as **infrasound,** and sound with frequencies above 20 kHz is referred to as **ultrasound.** (See Table 12.6.2.)

Table 12.6.2   Acoustic Spectrum

Lower range of elephant hearing	5 Hz
Infrasound/sound transition	20 Hz
Lowest note on a piano	27 Hz
E string on a bass guitar	41.3 Hz
Dominant frequency of thunder	100 Hz
Middle C	261.63 Hz
Upper limit of telephone bandwidth	3,400 Hz
Highest note on a piano	4,096 Hz
Sound/ultrasound transition	20 kHz
Upper range of bat sonar	100 kHz
Upper range of dolphin sonar	200 kHz

The density and sound speed of a material determine its **specific acoustic impedance** $Z_0 = \rho_0 c_0$. This term is often shortened to **acoustic impedance** or just **impedance.** The impedance of air is about 415 kg/(m$^2 \cdot$ s) [85 lbf/(ft$^2 \cdot$ s)] and of water is about 1.5×10^6 kg/(m$^2 \cdot$ s) [0.31×10^6 lbf/(ft$^2 \cdot$ s)]. The unit kg/(m$^2 \cdot$ s) is often referred to as MKS Rayls—although this is not an SI unit. For a plane wave moving in one direction (referred to as a progressive wave) the impedance relates the pressure p to the particle velocity u through $p = u Z_0$; i.e., acoustic impedance plays a role similar to electrical impedance in circuit theory. For waves that are not plane (such as focused waves) or not moving in one direction (such as standing waves), there is normally not a simple relation between the pressure and particle velocity.

A propagating acoustic wave carries energy, and this is most commonly described by the **intensity** I, the power per unit area. The intensity can be calculated by the following integral:

$$I = \frac{1}{T_{av}} \int_{t_0}^{t_0 + T_{av}} (p \cdot u)\, dt \qquad (12.6.1)$$

where the integration is done over a period for a periodic wave, over the duration of the pulse for a pulse, or over an appropriately long time for a noise signal. The units for I are watts per square metre (W/m^2). For a progressive wave the particle velocity is related to the acoustic pressure, and the intensity can be determined from measurements of the root mean square (rms) pressure and the impedance:

$$I = p_{rms}^2 / Z_0 \qquad (12.6.2)$$

When an acoustic wave encounters an interface, the wave is partially reflected and partially transmitted. In the case of **normal incidence,** the intensity transmission and reflection coefficients are given by

$$R_I = \frac{(Z_2 - Z_1)^2}{(Z_2 + Z_1)^2} \qquad (12.6.3)$$

$$T_I = \frac{4 Z_1 Z_2}{(Z_2 + Z_1)^2} \qquad (12.6.4)$$

that is, they depend predominantly on the difference in impedance. The coefficients sum to unity, and individually they must range between 0 and 1. In noise control applications, interfaces are often employed to reduce the transmission of sound, and it is common to use transmission loss TL, which is related to the intensity transmission coefficient by

$$\mathrm{TL} = -10 \log T_I \qquad (12.6.5)$$

A large transmission loss is normally desired.

For sound that is incident at oblique angles, or where there are multiple interfaces, the expressions for reflection and transmission coefficients become more complicated, and it often becomes more practical to measure the coefficients.

Attenuation of Sound   When sound propagates through a medium, most of the energy remains in the sound wave, but a small amount of it is absorbed by the medium. The amplitude of an acoustic wave will therefore decay or **attenuate** as it propagates through a medium. The attenuation can be due to geometric spreading effects, scattering by inhomogeneities in the medium, and direct absorption of the energy by the medium. For a plane wave the intensity will decay exponentially with distance

$$I = I_0 \exp(-2\alpha x) \qquad (12.6.6)$$

where α is the attenuation coefficient and x is the distance. The attenuation coefficient has units of nepers per metre (Np/m). Often when attenuation is reported to dB/m, it can be converted to Np/m by making use of the conversion 1 Np = 8.686 dB.

Absorption in air is dominated by the vibration of nitrogen and oxygen molecules. It varies with frequency (absorption increases with frequency) and is very sensitive to changes in temperature and humidity. Propagation distances typically need to be on the order of hundreds of metres for the attenuation to be significant. International Standard ISO 9613-1:1993 describes how to determine the attenuation of sound in air.

Absorption in pure water is very weak and normally negligible; however, it increases with the square of the frequency and can be important for frequencies above 10 MHz. In saltwater, absorption is much stronger and is dominated by chemical dissociation of salts (see Francois and Garrison, *Jour. Acoust. Soc. Am.* **72**; 1982, pp. 896–907 and 1879–1890).

In solids there are few quantitative formulas for determining attenuation, and most data are determined empirically. The attenuation in most solids typically follows a power law of the form $\alpha = \alpha_0 f^n$, where α_0 and n are material properties and n typically varies between 1 and 2. A comprehensive list of acoustic properties can be found at http://www.ondacorp.com/tecref_acoustictable.html.

Transducers

A **transducer** is a device that converts energy between two forms. In acoustics, transducers are used to measure sound (convert sound to an electric signal) and to generate sound (convert electric signals to acoustic waves).

Generating Sound   Sound is generated by an object or a surface vibrating in a fluid. For a sound source to be efficient, it is necessary that the dimensions of the vibrating surface of the source be much larger than the wavelength of sound. This is often not practical; e.g., at 300 Hz the wavelength in air is about 1 m (3 ft). Efficiency is also improved if the source is mounted in a wall (baffle) or a chamber so that there is no direct path from the front and back surfaces of the vibrating surface. The presence of a path between the front and back surfaces of the source allows some of the energy to "slosh" back and forth and not radiate as a sound wave.

Moving coil loudspeakers are perhaps the most common transducer for sound generation. The heart of the transduction is a cylindrical coil of wire (voice coil) and a magnet. When a current is passed through the coil, the interaction with the magnetic field results in a force on the voice coil. The voice coil is connected to a large diaphragm which is vibrated back and forth as the coil oscillates, and this results in the generation of an acoustic wave. The diaphragm is provided with some mechanical stiffness to produce a restoring force.

The attachment of a horn to the loudspeaker improves the impedance match between the speaker and the air since it is essentially an acoustic transformer. The dimensions and flare of the horn contribute to its matching ability. The operation of the speaker may be influenced by its enclosure, the baffle which separates the front from the back radiation, or its resonances.

Whistles and sirens may also be used to produce intense sound fields in gases and liquids. In a whistle a flow of air is passed over a sharp edge; turbulent vortices are produced which result in oscillation of the air. If a resonant chamber is placed behind the edge, then the vortices will excite the resonance and produce a tone. These oscillations can often result in high-amplitude signals. An example is flow over a car with an open sunroof, which will result in turbulence that can excite a resonance in the air in the car. The frequency of the resonance is typically less than 10 Hz (infrasound) and can be disturbing to passengers in the car. Sirens are devices in which a revolving disk or drum is perforated such that the holes alternately block and unblock a steam of air. Early sirens employed a series of holes on the outer edge of a disc with a jet of air impinging thereon. Modern sirens use a centrifugal fan inside a drum. The fan is rotated and forces air through the holes in the drum, to generate a sound.

In underwater applications, most sound sources employ piezoelectric materials. A piezoelectric material is one which generates a strain when a voltage is applied across the material. The change in thickness of the material launches an acoustic wave in the water. Typically the impedance of piezoelectric materials is much higher than that of water, and to ensure efficient sound transmission, quarter-wavelength impedance matching layers are used on the face of the transducer. Impulsive noise sources used underwater include explosions and implosions of hollow spheres.

Measuring Sound   The acoustic field is typically measured in air by using a microphone and in water by using a hydrophone. Most microphones are either condensor microphones or electret microphones. The active element of a condensor microphone consists of a capacitor where one plate is free to move with the particle motion of the acoustic wave, which results in a change in bias voltage on the capacitor. An electret uses a similar concept except that the dielectric is a permanently electrically charged material so that no biasing voltage need be employed. Electret microphones are becoming the dominant transducer for measuring audible sound even for quality measurement microphones. Micro-Electro-Mechanical-Systems (MEMS) based microphones, which typically employ micrometre-sized capacitors and may have an advantage in terms of mass production and direct integration with electronic circuits, have the potential to become competitive with electrets in the near future.

Other microphones use piezoelectric materials, where the strain or stress associated with the acoustic wave induces a voltage in the material. The majority of hydrophones employ piezoelectric transducers.

Sound Pressure Level   Most measurements report the acoustic disturbance in terms of pressure which has units of N/m^2 or Pa. The threshold of hearing in people with acute hearing is an rms pressure of approximately 20 μPa (2.9×10^{-9} psi) at around 4 kHz. The threshold of pain is an rms pressure of 63 Pa (9.3×10^{-3} psi). In comparison the atmospheric pressure at sea level is approximately 100,000 Pa (15 psi).

Although sound is measured as a pressure, it is normally reported in decibels as either a **sound pressure level (SPL)** or an **intensity level (IL)** which are given by

$$\text{SPL} = 20 \log 10(p_{\text{rms}}/p_{\text{ref}})$$
$$\text{IL} = 10 \log 10(I/I_{\text{ref}}) \qquad (12.6.7)$$

where in air $p_{\text{ref}} = 20\ \mu$Pa and $I_{\text{ref}} = 10^{-12}$ W/m^2 and are chosen so that the two levels are almost identical (less than 0.4-dB discrepancy). In terms of SPL, the threshold of hearing is 0 dB, and the threshold of pain is 130 dB. Table 12.6.3 gives typical SPLs for sound in the audible range.

Table 12.6.3   Examples of Representative SPLs for Various Environments

Rustling of leaves	15 dB
Whisper	30 dB
Normal conversation	60 dB
Average street noise	70 dB
Vacuum cleaner	80 dB
Heavy truck at 15 m (45 ft)	90 dB
Home stereo at full volume	100 dB
Rock concert (threshold of feeling)	120 dB
Jet engine at 25 m (75 ft)	140 dB

Note in underwater acoustics the **reference level** for an SPL is 1 μPa, and in older underwater acoustics literature a value of 0.1 Pa was used. It is good practice to report the reference pressure level when reporting SPL in dB to avoid confusion.

Perception of Sound

Although human hearing can extend from 20 Hz to 20 kHz, the auditory system is most sensitive to sounds in the range of 1 to 5 kHz. Due to the frequency dependence of human hearing, the SPL used for determining human response to sound is normally weighted so that sounds in the 1- to 5-kHz range are given stronger weighting. A number of algorithms have been developed that apply different weighting to sound in specific frequency bands that are either one octave or one-third-octave apart. An octave refers to a doubling of frequency; that is, 1,000 Hz is one octave above 500 Hz and two octaves above 250 Hz. An octave band centered at f_0 covers the range of frequencies from $f_0/\sqrt{2}$ to $f_0 \sqrt{2}$; e.g., for 1,000 Hz it covers the range of 707 to 1,414 Hz. The most common is the A weighting (dBA) which applies 0-dB weighting at 1,000 Hz, reduces signals at 63 Hz by 26.2 dB, and adds 1.0 dB at 4,000 Hz. The A weighting is appropriate for determining the perceived sound level for relatively quiet sounds, as it accounts for the strong frequency-dependent sensitivity of the ear. For more intense sounds the ear does not display such a strong frequency sensitivity, and in this case for C weighting (dBC) is more appropriate, which is effectively a flat bandpass filter with -3-dB frequencies at 31.5 Hz and 8 kHz. The C weighting is also often appropriate for impulsive sounds. Table 12.6.4 lists the octave band corrections for the A and C weightings.

Table 12.6.4   Correction, dB, to Be Applied to an SPL Measured in Octave Bands Centered at the Specified Frequencies.

Frequency, Hz	dBA	dBC
31.5	-39.4	-3.0
63	-26.2	-0.8
125	-16.1	-0.2
250	-8.6	0
500	-3.2	0
1,000	0	0
2,000	$+1.2$	-0.2
4,000	$+1.0$	-0.8
8,000	-1.1	-3.0

OSHA regulations (Standard 1910.95) require that when workers are exposed to an average weighted sound level of 85 dBA or more in an 8-h day, then monitoring and hearing conservation practices must be implemented. These practices include regular measurement of sound levels and regular auditory testing of employess. A SPL of 90 dBA is the maximum permissible for an 8-h workday, and the exposure time goes down by 50 percent for each 5-dB increase in noise levels; i.e., workers can only be exposed to an SPL of 95 dBA for 4 h, and a level of 100 dBA reduces exposure time to 2 h. Acoustic engineering can be employed to reduce noise levels and allow for longer exposure times.

NOISE

Noise measurements are usually made with a sound level meter, comprised of a microphone, an attenuator, an amplifier, a frequency-weighting network, and an indicator. See ANSI S1.4, "Specification for Sound Level Meters," and ANSI S1.42, "Design Response of Weighting Networks for Acoustical Measurements." A sound analyzer indicates sound pressure as a function of frequency by using a filter set designed in accordance with ANSI S1.11, "Specification for Octave-Band and Fractional-Octave-Band Analog and Digital Filters." Octave band filters are typically included with sound analyzers with additional one-third octave band filter sets available on some analyzers. Modern sound analyzers permit simultaneous real-time measurement across many frequency bands. In addition, narrowband analysis with fast Fourier transform (FFT) is available on some sound level meters. By using these instruments the components of the heterogeneous noise may be identified, and correlated with the noise sources. Once the noise source is located, techniques for noise control can be used.

Contact transducers (vibration pickups) may be used to locate sources of noise, such as in partitions and machine parts. They may be piezoelectric or magnetic and may be used to supplement information from the microphone and sound level meter.

Where the frequency distribution of the noise is significant, an analyzer may be used since the overall level meter tells nothing about frequency distribution. For most noise control applications, one-third octave band filters provide sufficient frequency resolution. However, narrowband FFT analyzers are helpful when finer frequency resolution is required.

Transmission Isolation

If the vibrations of noisy machinery cannot be suppressed at the source, their transmission to the listener should be impeded. The most effective vibration isolation method is the introduction of elastic discontinuities in the structure transmitting the noise. The discontinuities may be obtained by the use of rubber, neoprene, or springs in machinery mountings, or by the introduction of neoprene isolation sheets in expansion joints of concrete slabs. The isolation treatment should be applied as close to the source as possible, to eliminate sound radiation from the structures transmitting the vibrations. Where this is not possible, the listening space itself may be isolated. Thus quiet rooms, constructed especially for noisy measurements, are usually built as separate structures isolated from the main building.

Filtration

Some problems of noise transmission through air lend themselves to solution by methods of filtration. Typical examples are the transmission of sound in ventilating ducts and the noise production at engine exhaust pipes. In each of these cases, the steady flow of gas must not be impeded, but the alternating flow, representing sound transmission, must be effectively suppressed.

Noise in ventilating ducts is typically broadband and consists of fan noise (low and middle frequency) and turbulence noise (middle and high frequency) generated by airflow. Fans are capable of generating tones at a blade passage frequency determined by their rotational speed and number of blades. However, most fans used in HVAC are large centrifugal fans, which do not generate a tremendous amount of tonal noise. Some of this sound energy can be converted to heat by using dissipative glass or cotton fiber elements such as duct silencers or internal duct lining. Both approaches tend to give the most attenuation at middle frequencies around 1 kHz. Duct silencers, consisting of parallel baffles, are used to reduce low- and middle-frequency noise. For ventilating ducts, noise suppression may be obtained by lining the ducts with 1-in-thick glass fiber or cotton fiber sheets. Low-frequency propagation along a duct can be reduced by means of complaint duct walls that radiate noise away from the duct, or by discontinuities such as plenums and outlets. High-frequency noise can be reduced with elbows. Reactive (also known as no-fill, or no-media) silencers are commercially available and are commonly used where no fibrous fill is allowed. These silencers work on the same principle as the muffler on a car. They consist of a few pipe segments that interconnect a number of large chambers. The change in acoustic properties of the pipes (masslike) and the chambers (springlike) produce resonance effects which result in reflection of the noise back to the source.

For noise with strong tonal components, resonators can be fitted to the duct to filter specific tones. The filters are normally placed as close to the source as possible. However, a difficulty with using resonators in duct systems is that a separate resonator is needed for each tone, and tuning the resonator to the correct frequency can be difficult. Other problems include excess pressure drop and flow noise generation from duct discontinuities. Most noise control approaches incorporate broadband attenuation to reduce the entire noise spectrum below a target **noise criteria (NC)** level.

Ducts should be sized to limit airflow noise and duct drumming to acceptable levels in the spaces they serve. The volumetric flow rate is determined by the amount of air changes per hour required for the space of interest. Duct sizing and volumetric flow rate determine the average airflow velocity in the duct, which in turn affects the amount of turbulence noise that can be generated. Rules of thumb are typically used for setting appropriate airflow velocities. Main system ductwork that is usually located in building shafts away from noise-sensitive spaces is typically sized for medium-pressure standards with airflow velocities from 6 to 10 m/s [1,200 to 2,000 feet per minute (fpm)]. The main ductwork connects to branches that are sized for low-pressure standards. In general, velocity ventilation branches serving noise-sensitive rooms such as music performance spaces should be on the order of 0.5 to 1.5 m/s (100 to 300 fpm). For regular occupied spaces, such as offices, airflow velocities in duct branches should be on the order of 2 to 3 m/s (400 to 600 fpm). Spaces that have high airflow requirements such as laboratories permit higher noise levels, and airflow velocities in duct branches are on the order of 4.5 to 6 m/s (900 to 1,200 fpm).

Shielding

Noise can be reduced by the placement of sound-opaque partitions between the source and the receivers. In most buildings the panels are thin in comparison to a wavelength, and the transmission loss for normally incident sound through a thin panel is given by

$$TL_0 = 20 \log (\pi f \rho l / Z_1) \qquad (12.6.8)$$

where f is frequency, ρ is the mass density of the wall material, l is the thickness of the wall, and Z_1 is the impedance of air [415 Rayls or 85 lb/(ft^2·s)]. There is a 6 dB/octave increase in transmission loss as frequency increases. The transmission loss of sound can be increased by increasing the mass per unit area of the partition, either a denser material or a thicker panel. For sound that is not normally incident the TL is reduced, and for sound incident over a range of angles, typical transmission loss is approximately derated by 5 dB, that is, TL = $TL_0 - 5$ dB.

The sound transmission loss of single panels in the audible frequency range is often further reduced by coincidence effects associated with bending (or flexural) waves in the panel. This occurs when the frequency of obliquely incident sounds equals the bending wave frequency of the panel, which is determined by the thickness and stiffness of the panel. If the panel had no damping, then the sound transmission loss would be reduced to zero at this coincidence frequency; but the damping inherent in most panels reduces this degradation of transmission loss. Single panels can also have low-frequency resonances due to their mass and stiffness. However, this resonance is determined by the size of the panel, is mitigated by natural panel damping, and is usually not a concern in most noise control applications.

Typically mass-based insulation is too large and heavy to be practical. Thin panels separated by an airspace can achieve similar transmission loss to massive monolithic constructions. The sound transmission loss of double-panel constructions can be reduced by mass-air-mass resonance. Adding sound-absorptive material such as glass or mineral fiber batt insulation reduces this resonance effect and improves the general sound transmission loss of the partition. The deeper the airspace, the lower the mass-air-mass resonance frequency. Coincidence phenomena affecting each of the panels can also affect the total transmission loss of the assembly.

Double walls must be constructed carefully to avoid loss of vibration isolation through mechanical bridging between the opposite surfaces. Higher-density sound-absorbing fillers such as mineral fiber batt insulation can reduce sound insulation if they are in contact with both interior surfaces; but lower-density fill such as glass fiber batt insulation can touch both sides without degrading the isolation. A single bridging nail may alter significantly the insulating efficiency. For maximum effectiveness, one of the wall surfaces should be hung structurally free at all four edges, with the boundary cracks sealed against sound leakage with an elastometric sealant such as silicone or acrylic latex.

Given the factors that make the performance of partitions deviate from the mass law, it is advisable to use one of the many tabulations of measured sound transmission loss when selecting a construction. Sound transmission loss of partitions is measured by comparing the average sound level on each side of the partition under standardized conditions per ASTM E90, "Standard Test Method for Laboratory Measurement of Airborne Sound Transmission Loss of Building Partitions and Elements." The frequency-dependent transmission loss data from this test procedure can be reduced to a single-number **sound transmission class (STC)** by means of ASTM E413, "Classification for Rating Sound Insulation." The STC most heavily weights the transmission loss

data in the 500- to 2,000-Hz frequency band, and ratings for typical partitions are shown in Table 12.6.5. Additional data on transmission loss for a wide variety of building materials and structures are available in several publications (see References).

Transmission through composite walls can be approximated as an average intensity transmission coefficient

$$T_{av} = \frac{1}{S_{tot}} \sum_n T_n S_n \qquad (12.6.9)$$

where the sum is over all the different materials in the wall, T_n is the intensity transmission coefficient, and S_n is the surface area of the nth

Table 12.6.5 Transmission Loss Data for Various Building Constructions

Data are provided at 7 different octave bands. The final column gives the STC rating.

Material	Transmission loss, dB						STC rating
	125 Hz	250 Hz	500 Hz	1,000 Hz	2,000 Hz	4,000 Hz	
Walls							
Monolithic:							
3/8-in plywood (1 lb/ft²)	14	18	22	20	21	26	22
26-gage sheet metal (1.5 lb/ft²)	12	14	15	21	21	26	20
1/2-in gypsum board (2 lb/ft²)	15	20	25	31	33	27	28
2 layers 1/2-in gypsum board, laminated with joint compound (4 lb/ft²)	19	26	30	32	29	37	31
Interior:							
2 by 4 wood studs 16 in oc with 1/2-in gypsum board both sides (5 lb/ft²)	17	31	33	40	38	36	33
2 by 4 staggered wood studs 16 in oc each side with 1/2-in gypsum board both sides (8 lb/ft²)	23	28	39	46	54	44	39
6-in dense concrete block, 3 cells, painted (34 lb/ft²)	37	36	42	49	55	58	45
3 5/8-in steel channel studs 16 in oc with 1/20-in gypsum board both sides (5 lb/ft²)	26	36	43	51	48	43	43
Construction no. 19 with 3-in mineral-fiber insulation in cavity	28	45	54	55	47	54	48
Exterior:							
4 1/2-in face brick (50 lb/ft²)	32	34	40	47	55	61	45
Two wythes of 4 1/2-in face brick, 2-in airspace with metal ties (100 lb/ft²)	37	37	47	55	62	67	50
Two wythes of plastered 4 1/2-in brick, 2-in airspace with glass-fiber insulation in cavity	43	50	52	61	73	78	59
6-in solid concrete with 1/2-in plaster both sides (80 lb/ft²)	39	42	50	58	64	67	53
Floor-Ceilings							
2 by 10 word joints 16 in oc with 1/2-in on floor side and 5/8-in gypsum board nailed to joists on celing side (10 lb/ft²)	23	32	36	45	49	56	37
4-in reinforced concrete slab (54 lb/ft²)	38	42	45	56	57	66	44
6-in reinforced concrete slab (75 lb/ft²)	38	43	52	59	67	72	55
6-in reinforced concrete slab with 3/4-in T&G wood flooring on 1 1/2 by 2 wooden battens floated on 1-in glass fiber (83 lb/ft²)	38	44	52	55	60	65	55
Roofs							
3 by 8 wood beams 32 in oc with 2 by 6 T&G planks, asphalt felt build-up roofing, and gravel topping	29	33	37	44	55	63	43
As above with 2 by 4s 16 in oc between beams, 1/2-in gypsum board supported by metal channels on ceiling side with 4-in glass-fiber insulation in cavity	35	42	49	62	67	79	53
Doors							
1 3/4-in hollow-core wood door, no gaskets, 1/4-in air gap at sill (1.5 lb/ft²)	14	19	23	18	17	21	19
Construction no. 43 with gaskets and drop seal	19	22	25	19	20	29	21
1 3/4-in hollow-core 16 gage steel door, glass-fiber filled, with gaskets and drop seal (7 lb/ft²)	23	28	36	41	39	44	38
Glass							
1/8-in monolithic glass (1.4 lb/ft²)	18	21	26	31	33	22	26
1/4-in monolithic float glass (2.9 lb/ft²)	25	28	31	34	30	37	31
1/4-in laminated glass. 30-mil plastic interlayer (3.6 lb/ft²)	25	28	32	35	36	43	35
Double glass: 1/4-in laminated + 3/16-in monolithic glass with 2-in airspace (5.9 lb/ft²)	25	34	44	47	48	55	45

SOURCE: Adapted from data published in M. David Egan, "Architectural Acoustics," McGraw-Hill, 1988, pp. 204–206.

material. For air gaps $T_n = 1$ (perfect transmission), and so the presence of air gaps in walls should be removed as much as possible. For example, a 3-mm (⅛-in)-thick glass window offers 41 dB of transmission loss at 2,000 Hz. If the window is opened by 2.5 mm (0.1 in), the transmission loss is reduced to 26 dB.

Holes for piping penetrations should be slightly oversized to allow the penetration to be filled with fibrous insulation and to be sealed on both sides with an elastomeric sealant, to ensure that the transmission loss is not degraded by the presence of air gaps.

The benefit of shielding can be reduced by the presence of flanking paths, which are alternative ways for sound to travel from source to receiver other than the partition between them. Flanking transmission can occur when gaps are left around the perimeter of the partition. These gaps should be sealed with an elastometric sealant. Flanking transmission can also occur when lightweight partitions perpendicular to the separating partition are continuous across both rooms. The flanking noise issue can be addressed by cutting a joint in the flanking partition or by increasing the mass of the flanking partition so that its sound transmission loss approaches that of the separating partition.

Quieting

When a noise source is in a room, reflections of the noise from the wall (reverberation) can result in a higher sound level than would occur if the sound source were in free space. Placing sound-absorbing material on the interior walls will reduce the reverberation in the room and hence reduce noise levels. Sound absorption coefficients are the commonly used measurement of these panels. The sound absorption coefficient is the fraction of the total energy that reflects from a panel. The coefficient varies from 0 (for an open window where no energy returns to the room) to 1 (for a rigid wall where all the energy is reflected).

Absorbing panels typically employ highly porous materials which are very effective at converting sound vibrations to heat by friction and viscous resistance. The effectiveness of sound absorbents varies with frequency, usually being greater for high and intermediate than for low frequencies. Sound absorption coefficients are typically measured in accordance with ASTM C423, "Standard Test Method for Sound Absorption and Sound Absorption Coefficients by the Reverberant Room Method," and specific normal acoustic impedance is measured in accordance with ASTM C384, "Standard Test Method for Impedance and Absorption of Acoustical Materials by Impedance Tube Method," or ASTM E1050, "Standard Test Method for Impedance and Absorption of Acoustical Materials by Impedance Tube Method." The absorption coefficient will depend on the impedance of the panel and the attenuation of the sound within the panel material. However, it is not a property of the material alone, but depends on the mounting of the test sample per ASTM E795, "Standard Practices for Mounting Test Specimens During Sound Absorption Tests." For example, a sound-absorptive material is typically measured with higher sound absorption coefficients when backed by a deep airspace than when tested lying against a rigid surface.

The principal material employed is acoustic tile, which can be obtained as prefabricated units for use on walls and ceilings. The tile typically has an outer layer that has been mechanically perforated and conceals a porous absorbing material underneath. Note that painting or sealing of acoustic panels can have a dramatic deleterious effect on the performance of the panels, as the holes that allow the sound to permeate into the absorbing fill are blocked. This can be of particular concern in environments such as hospitals where sealing of pores and holes is desired for sterilization reasons. An alternative to tile is to hang acoustic blankets, e.g., hanging heavy drape curtains, on walls. Data on sound absorption coefficients for a wide variety of building materials are shown in Table 12.6.6.

Floors can also be considered in terms of quieting. A carpeted floor will provide improved absorption over concrete or linoleum. Carpeting will also have a positive effect on improving acoustic isolation of impact-generated noise to rooms below. Underlays made of felt, cork, or rubber can further improve the acoustic performance of the floor.

Finally, absorption can be enhanced by the presence of furniture in the room; e.g., upholstered chairs can absorb a nontrivial amount of sound.

Active Noise Control (ANC)

Recent advancements in computational technology have allowed many practical applications of active noise control. Human interaction with transportation and equipment noise thus far has been the focus of this technology. As of today, successful global ("whole room") active control of **tonal disturbances** has been accomplished in aircraft, submarines, and automobiles. Civilian examples of this technology are currently limited, but include the Saab 340B aircraft and Honda hybrid-powered vehicles. Military examples abound. For nontonal, steady-state noise, such as that heard in an aircraft cabin at cruising speed, active noise control has been accomplished only for "local" or small regions near the ears of passengers, and is typically limited to less than 1 kHz in frequency. Examples include the ubiquitous ANC headsets. No published ANC system to date is capable of readily controlling transient, unpredictable noise. Transient noise would include human conversation, music, and impulsive noises. All existing control algorithms require a feed-forward sensor capable of predicting noise at the receiver and/or a feedback sensor capable of monitoring a steady-state noise.

Active noise control for low-frequency noise in ducts is possible, but operating conditions of active duct silencers are generally limited to flow velocities of 7.5 m/s (1,500 fpm) or less (2003 ASHRAE Applications Handbook, p. 47.18). To the best of the author's knowledge, there are currently no large-scale commercially available, off-the-shelf lines of active duct silencers. Any active noise control approach to ductborne noise control is likely to be a custom-made endeavor, more expensive than commercially available passive silencers.

ULTRASONICS

REFERENCES: Ultrasonic Cleaning, in "ASM Handbook," vol. 5, "Surface Engineering," 1994, pp. 44–47. "Nondestructive Testing Handbook," 2d ed., vol. 7, "Ultrasonic Testing," Birks and Green (eds.), American Society for Nondestructive Testing, Columbia, Ohio, 1991. Johnson, The New Wave in Acoustic Testing, *Materials World, Jour. Inst. Materials* **7**, 1999, pp. 544–546.

Sound waves in the frequency range from 10 kHz to hundreds of megahertz have numerous industrial applications. The majority of the devices operate above 20 kHz and are therefore ultrasonic. One motivation in the choice of ultrasonic frequencies is to avoid impact on human hearing. High-intensity waves are used to mechanically alter materials, such as ultrasonic cleaners or welders. Low-intensity waves are used for nondestructive evaluation or testing. The list of applications is large, and just a small subset will be described here. The Ultrasonics Industry Association (UIA) provides a referral website which lists the majority of the industrial applications of ultrasound: http://ultrasonics.org/referral.html.

High-Intensity Applications

Ultrasonic cleaners employ a water bath that is agitated with sound waves with frequencies normally around 40 kHz (the majority of devices employ frequencies in the range of 20 to 80 kHz). The agitation is generated by bonding piezoelectric or magnetostrictive transducers to the outer surface of the tank. Ultrasonic cleaners work through acoustic cavitation. Acoustic cavitation occurs when the negative pressure of the tensile phase of the wave is greater than 1 atm (100 kPa or 15 psi), a pressure relatively easily obtained in water, at which level the fluid will tear apart, creating a cavity or bubble. The cavity will grow during the tensile phase of the sound wave, but during the compressive phase of the wave it will be driven to collapse. The collapse is extraordinarily violent with pressures inside the bubble reaching hundreds of atmospheres and temperatures reaching 5,000 K (9,000°R). These phenomena are localized to a small volume around the collapse but result in sonochemistry (e.g., free-radial production) and sonoluminescence (production of light). When the cavity collapses near a surface, the bubble collapses asymmetrically and a microjet of fluid pierces the cavity and strikes the surface at velocities in excess of 100 m/s (330 ft/s). These microjets act as intense microscopic scrubbers to remove particles from the surface.

Table 12.6.6   Sound Absorption Coefficients for Various Building Constructions
Data are provided at 7 different octave bands. The final column gives the NRC number.

Material	Sound absorption coefficient						NRC number
	125 Hz	250 Hz	500 Hz	1,000 Hz	2,000 Hz	4,000 Hz	
Walls							
Sound-Reflecting:							
Brick, unglazed	0.02	0.02	0.03	0.04	0.05	0.07	0.05
Brick, unglazed and painted	0.01	0.01	0.02	0.02	0.02	0.03	0
Glass, ordinary window	0.35	0.25	0.18	0.12	0.07	0.04	0.15
Gypsum board, ½ in thick (nailed to 2 × 4s, 16 in oc)	0.29	0.1	0.05	0.04	0.07	0.09	0.05
Gypsum board, 1 layer, ⅝ in thick (screwed to 1 × 3s, 16 in oc with airspaces filled with fibrous insultaion)	0.55	0.14	0.08	0.04	0.12	0.11	0.1
Plywood, ⅜-in paneling	0.28	0.22	0.17	0.09	0.1	0.11	0.15
Steel	0.05	0.1	0.1	0.1	0.07	0.02	0.1
Venetian blinds, metal	0.06	0.05	0.07	0.15	0.13	0.17	0.1
Wood, ¼-in paneling, with airspace behind	0.42	0.21	0.1	0.08	0.06	0.06	0.1
Sound-Absorbing:							
Lightweight drapery, 10 oz/yd², flat on wall	0.03	0.04	0.11	0.17	0.24	0.35	0.15
Heavyweight drapery, 18 oz/yd², draped to half area	0.14	0.35	0.55	0.72	0.7	0.65	0.6
Thick, fibrous material behind open facing	0.6	0.75	0.82	0.8	0.6	0.38	0.75
Carpet, heavy, on ⅝-in perforated mineral fiberboard with airspace behind	0.37	0.41	0.63	0.85	0.96	0.92	0.7
Wood, ½-in paneling perforated ³⁄₁₆-in-diameter holes, 11% open area, with 2½-in glass fiber in airspace behind	0.4	0.9	0.8	0.5	0.4	0.3	0.65
Floors							
Sound-Reflecting:							
Concrete or terrazzo	0.01	0.01	0.02	0.02	0.02	0.02	0
Linoleum, rubber, or asphalt tile on concrete	0.02	0.03	0.03	0.03	0.03	0.02	0.05
Wood	0.15	0.11	0.1	0.07	0.06	0.07	0.1
Sound-Absorbing:							
Carpet, heavy, on concrete	0.02	0.06	0.14	0.37	0.6	0.65	0.3
Carpet, heavy, on foam rubber	0.08	0.24	0.57	0.69	0.71	0.73	0.55
Ceilings							
Sound-Reflecting:							
Concrete	0.01	0.01	0.02	0.02	0.02	0.02	0
Gypsum board, ½ in thick	0.29	0.1	0.05	0.04	0.07	0.09	0.05
Plywood, ⅜ in thick	0.28	0.22	0.17	0.09	0.1	0.11	0.15
Sound-Absorbing:							
Acoustical board, ¾ in thick, in suspension system (mtg. E)	0.76	0.93	0.83	0.99	0.99	0.94	0.95
Thin, porous sound-absorbing material ¾ in thick (mtg. B)	0.1	0.6	0.8	0.82	0.78	0.6	0.75
Polyurethane foam, 1 in thick, open cell, reticulated	0.07	0.11	0.2	0.32	0.6	0.85	0.3
Seats and Audience							
Fabric, well-upholstered sets, with perforated seat pans, unoccupied	0.19	0.37	0.56	0.67	0.61	0.59	
Leather covered upholstered seats, unoccupied	0.44	0.54	0.6	0.62	0.58	0.5	
Chair, metal or wood seat, unoccupied	0.15	0.19	0.22	0.39	0.38	0.3	

SOURCE: Adapted from data published in M. David Egan, "Architectural Acoustics," McGraw-Hill, 1988, pp. 52–54.

Typically a detergent will be added to the water bath of an ultrasonic cleaner. The detergent allows the water to fully wet surfaces of the component that may be hydrophobic. For cavitation to be effective, it is important that the entire surface of the component be fully wetted. Sometimes solvents may be added to the water to chemically enhance the cleaning process. Most companies that produce ultrasonic cleaners also market proprietary ultrasonic cleaning solutions. Ultrasonic cleaners are particularly useful for cleaning objects with odd shapes and hidden structure. Types of objects that can be cleaned include metal parts (e.g., castings, valves, injectors), electronic components (particularly connector pins), printed-circuit boards, lenses, optical fibers, television tubes, and filters. This list is far from exhaustive.

For typical sized tanks (a few gallons) standing waves can be excited which can result in hot and cold spots in the tank. This is so because the speed of sound in water 1,500 m/s, and so the wavelength at 40 kHz is 150 mm (5.9 in)—on the order of the tank. Modern ultrasonic cleaners will normally employ a swept frequency signal which, even for variations of just a few kilohertz, will ensure even distribution of cavitation throughout the tank.

Ultrasonic welders are employed to bond materials together. They are most commonly used on plastics, but metal, textiles, films, and other materials can also be successfully welded. Ultrasonic welding is accomplished by placing the two pieces to be welded in direct mechanical contact. One piece is held stationary, and then an ultrasonic probe is brought into contact with the second piece. When the probe is excited, it vibrates the piece that it is in contact with, which produces frictional heat at the contact point between the two pieces. This, in turn, results in a temperature rise that softens or melts the plastic on both surfaces. The heated plastic interface melts, and when the probe is removed, the plastic cools and a homogeneous solid weld results.

The ultrasonic probe consists of a stack of piezoelectric discs which are operated at frequencies varying from 20 to 40 kHz, although research devices operating above 100 kHz have be reported. The stack is mounted to a tapered horn (or sonotrode), and as the sound energy propagates down the horn, the vibration amplitude is amplified because of the reduction in cross-sectional area of the horn. Vibrations at the tip are on the order of 100 μm. With ultrasonic welding it is possible to achieve spot welds, lap welds, butt welds, and steam welds.

Ultrasonic Atomization When ultrasound in air is incident on the surface of water, it will excite capillary waves on the surface of the water. When the amplitude of the capillary waves reaches the critical value, the waves become unstable and droplets of water are ejected into the air. The droplets are typically tens of micrometres in diameter and appear as a mist above the surface. Under the right conditions an acoustic fountain can be generated. This type of ultrasonic atomization is commonly employed in dehumidifiers. Nozzles have been created by using this principle to produce an atomized jet.

Low-Intensity Applications

Nondestructive Evaluation or Testing of Materials This application is perhaps the most widespread use of ultrasound in an industrial setting, and the most common manifestation is a pulse-echo configuration. An ultrasonic transducer is used to launch a short pulse into a sample, and whenever the pulse interacts with an interface, it is partially reflected and the echoes can be detected by the same transducer that generated the pulse. If the sound speed in the sample is known, then the distance to the interface can be determined from the time of flight. Common applications employ piezoelectric transducers with center frequencies that range from 1 to 30 MHz. The transducers are typically wideband, and so when excited with a short electrical pulse (with an amplitude on the order of 100 V), they generate an acoustic pulse that is about 1.5 cycles long. The spatial length of the pulse governs the spatial resolution of the system. At 1 MHz the pulse length is about 11 mm (0.45 in) in steel and 4 mm (0.16 in) in Lucite. At 30 MHz it is 0.4 mm (0.016 in) in steel and 0.13 mm (0.005 in) in Lucite. High frequencies therefore give better spatial resolution but typically are absorbed more strongly by the sample and so cannot "see" deeply into the material.

Frequencies of hundreds of megahertz have been employed for specialized high-resolution applications.

If the transducer is focused, it will generate a pencil beam of sound, and from the received echo signal (known as an A-line) it is possible to identify the position of interfaces to within the resolution of the system. The transducer can be scanned in a line to produce two-dimensional images (a B-scan) or in a two-dimensional raster pattern to produce a three-dimensional image (a C-scan). The position of interfaces can be used to size an object or to detect the presence of flaws or inclusions. Array transducers can be used to electronically scan the beam rather than relying on mechanical translation. The same principles described here for materials testing are used in diagnostic ultrasound imaging of the body.

Other modes of ultrasonic testing include through transmission and resonant ultrasound. In through transmission, two transducers are used in a pitch and catch configuration. The amplitude of the received signal will depend on the number of interfaces and flaws in the intervening material. Through transmission can also be used to ascertain flow velocities through the Doppler effect. In this case a short tone burst (approximately 10 cycles) is transmitted through the flow. If there is a net flow from the transmitter to the receiver, then the frequency of the tone will be shifted higher; if there is flow in the opposite direction, the frequency will be shifted down. Based on the frequency shift the flow velocity can be calculated. In **resonance ultrasound spectroscopy (RUS)** a transducer is used to excite a resonance in the component to be investigated, and the frequency and Q of specific resonances can be used to extract information about the structure and flaws in the components. A new and promising development for flaw detection in components is **nonlinear resonant ultrasound spectroscopy (NRUS)** and variations thereof. In this case, the ultrasonic excitation is of large enough amplitude that the material exhibits nonlinear response, e.g., the production of harmonic overtones of the driving signal. The nonlinear response is critically dependent on flaws in the material and can be used to evaluate damage in most construction materials.

Section **13**

Manufacturing Processes

BY

CHUCK FENNELL *Program Manager, Dalton Foundries*
RAJIV SHIVPURI *Professor of Industrial, Welding, and Systems Engineering,*
   Ohio State University
OMER W. BLODGETT *Senior Design Consultant, Lincoln Electric Co.*
DUANE K. MILLER *Manager, Engineering Services, Lincoln Electric Co.*
SEROPE KALPAKJIAN *Professor Emeritus of Mechanical and Materials Engineering,*
   Illinois Institute of Technology
ALI M. SADEGH *Professor of Mechanical Engineering, The City College of The City University*
   of New York
RICHARD W. PERKINS *Professor Emeritus of Mechanical, Aerospace, and Manufacturing*
   Engineering, Syracuse University
CHARLES OSBORN *Business Manager, Precision Cleaning Division, PTI Industries, Inc.*

13.1 FOUNDRY PRACTICE AND EQUIPMENT
by Chuck Fennell

REFERENCES: Publications of the American Foundrymen's Society: "Cast Metals Handbook," Alloy Cast Irons Handbook," "Copper-base Alloys Foundry Practice," Foundry Sand Handbook." "Steel Castings Handbook," Steel Founders' Society of America. Current publications of ASM International. Current publications of the suppliers of nonferrous metals relating to the casting of those metals; i.e., Aluminum Corp. of America, Reynolds Metal Co., Dow Chemical Co., INCO Alloys International, Inc., RMI Titanium Co., and Copper Development Assn. Publications of the International Lead and Zinc Research Organization (ILZRO).

BASIC STEPS IN MAKING SAND CASTINGS

The basic steps involved in making sand castings are:

1. **Patternmaking.** Patterns are required to make molds. The mold is made by packing molding sand around the pattern. The mold is usually made in two parts so that the pattern can be withdrawn. In horizontal molding, the top half is called the **cope,** and the bottom half is called the **drag.** In vertical molding, the leading half of the mold is called the **swing,** and the back half is called the **ram.** When the pattern is withdrawn from the molding material (sand or other), the imprint of the pattern provides the cavity when the mold parts are brought together. The mold cavity, together with any internal cores (see below) as required, is ultimately filled with molten metal to form the casting.

2. If the casting is to be hollow, additional patterns, referred to as **core boxes,** are needed to shape the sand forms, or **cores,** that are placed in the mold cavity to form the interior surfaces and sometimes the external surfaces as well of the casting. Thus the void between the mold and core eventually becomes the casting.

3. **Molding** is the operation necessary to prepare a mold for receiving the metal. It consists of ramming sand around the pattern placed in a support, or **flask,** removing the pattern, setting cores in place, and creating the gating/feeding system to direct the metal into the mold cavity created by the pattern, either by cutting it into the mold by hand or by including it on the pattern, which is most commonly used.

4. **Melting** and **pouring** are the processes of preparing molten metal of the proper composition and temperature and pouring this into the mold from transfer **ladles.**

5. **Cleaning** includes all the operations required to remove the **gates** and **risers** that constitute the gating/feeding system and to remove the adhering sand, scale, parting fins, and other foreign material that must be removed before the casting is ready for shipment or other processing. Inspection follows, to check for defects in the casting as well as to ensure that the casting has the dimensions specified on the drawing and/or specifications. Inspection for internal defects may be quite involved, depending on the quality specified for the casting (see Sec. 5.4). The inspected and accepted casting sometimes is used as is, but often it is subject to further processing which may include heat treatment, painting, rust preventive oils, other surface treatment (e.g., hot-dip galvanizing), and machining. Final operations may include electrodeposited plated metals for either cosmetic or operational requirements.

PATTERNS

Since patterns are the forms for the castings, the casting can be no better than the patterns from which it is made. Where close tolerances or smooth casting finishes are desired, it is particularly important that patterns be carefully designed, constructed, and finished.

Patterns serve a variety of functions, the more important being (1) to shape the mold cavity to produce castings, (2) to accommodate the characteristics of the metal cast, (3) to provide accurate dimensions, (4) to provide a means of getting liquid metal into the mold (gating system), and (5) to provide a means to support cores by using core prints outside of the casting.

Usual allowances built into the pattern to ensure dimensional accuracy include the following: (1) **Draft,** the taper on the vertical walls of the casting which is necessary to extract the pattern from the mold without disturbing the mold walls and is also required when making the core. (2) **Shrinkage allowance,** a correction to compensate for the solidification shrinkage of the metal and its contraction during cooling. These allowances vary with the type of metal and size of casting. Typical allowances for cast iron are $1/10$ to $5/32$ in/ft; for steel, $1/8$ to $1/4$ in/ft; and for aluminum, $1/16$ to $5/32$ in/ft. A designer should consult appropriate references (AFS, "Cast Metals Handbook"; ASM, "Casting Design Handbook"; "Design of Ferrous Castings") or the foundry. These allowances also include a size tolerance for the process so that the casting is dimensionally correct. (See also Secs. 6.1, 6.3, and 6.4.) Table 13.1.1 lists additional data for some commonly cast metals. (3) **Machine finish allowance** is necessary if machining operations are to be used so that stock is provided for machining. Tabulated data are available in the references cited for shrinkage allowances. (4) If a casting is prone to distortion, a pattern may be intentionally distorted to compensate. This is a **distortion allowance.**

Patterns vary in complexity, depending on the size and number of castings required. **Loose patterns** are single prototypes of the casting and are used only when a few castings are needed. They are usually constructed of wood, but metal, plaster, plastics, urethanes, or other suitable material may be used. With advancements in solids modeling utilizing computers, CAD/CAM systems, and laser technology, rapid prototyping is possible and lends itself to the manufacture of prototype patterns from a number of materials, including dense wax paper, or via stereolithographic processes wherein a laser-actuated polymerized plastic becomes the actual pattern or a prototype for a pattern or a series of patterns. The gating system for feeding the casting is cut into the sand by hand. Some loose patterns may be split into two parts to facilitate molding.

Gated patterns incorporate a gating system along with the pattern to eliminate hand cutting.

Match-plate patterns have the cope and drag portions of the pattern mounted on opposite sides of a wooden or metal plate, and are designed to speed up the molding process. Gating systems are also usually attached. These patterns are generally used with some type of molding machine and are recommended where a large number of castings are required.

Table 13.1.1 Average Linear Shrinkage of Castings

Bar iron, rolled	1:55	Cast iron	1:96	Steel, puddled	1:72
Bell metal	1:65	Gun metal	1:134	Steel, wrought	1:64
Bismuth	1:265	Iron, fine grained	1:72	Tin	1:128
Brass	1:65	Lead	1:92	Zinc, cast	1:624
Bronze	1:63	Steel castings	1:50	8 Cu + 1 Sn (by weight)	1:13

For fairly large castings or where an increase in production rate is desired, the patterns can be mounted on separate pattern plates, which are referred to as **cope- and drag-pattern** plates. They are utilized in horizontal or vertical machines. In horizontal molding machines, the pattern plates may be used on separate machines by different workers, and then combined into completed molds on the molding floor prior to pouring. In vertical machines, the pattern plates are used on the same machine, with the flaskless mold portions pushed out one behind the other. Vertical machines result in faster production rates and provide an economic edge in overall casting costs.

Special Patterns and Devices For extremely large castings, **skeleton** patterns may be employed. Large molds of a symmetric nature may be made for forming the sand mold by **sweeps,** which provide the contour of the casting through the movement of a template around an axis.

Follow boards are used to support irregularly shaped, loose patterns which require an irregular parting line between cope and drag. A **master pattern** is used as an original to make up a number of similar patterns that will be used directly in the foundry.

MOLDING PROCESSES AND MATERIALS

(See Table 13.1.2.)

Molding Processes

Green Sand Most castings are made in **green sand,** i.e., sand bonded with clay or bentonite and properly tempered with water to give it **green strength.** Miscellaneous additions may be used for special properties. This method is adaptable to high production of small- or medium-sized castings because the mold can be poured immediately after forming, and the sand can be reused and reprocessed after the casting has solidified.

Dry Sand Molds These molds are made with green sand but are baked prior to use. The surface is usually given a refractory wash before baking to prevent erosion and to produce a better surface finish. Somewhat the same effects are obtained if the mold is allowed to **air-dry** by leaving it open for a period of time before pouring, or it is **skin-dried** by using a torch, infrared lamps, or heating elements directed at the mold cavity surface.

Core molding makes use of assembled cores to construct the mold. The sand is prepared by mixing with oil, or cereal, forming in core boxes, and baking. This process is used where the intricacy of the casting requires it.

Carbon Dioxide Process Molds These molds are made in a manner similar to the green sand process but use sand bonded with sodium silicate. When the mold is finished, carbon dioxide gas is passed through the sand to produce a very hard mold with many of the advantages of dry sand and core molds but requiring no baking.

Floor and Pit Molding When large castings are to be produced, these may be cast either directly on the floor of the foundry or in pits in the floor which serve as the flask. **Loam molding** is a variation of floor molding in which molding material composed of 50 percent sand and 50 percent clay (approx) is troweled onto a brickwork surface and brought to dimension by use of patterns, sweeps, or templates.

Shell Molding Sand castings having close dimensional tolerances and smooth finish can be produced by a process using a synthetic resin binder. The sand and resin mixture is dumped onto a preheated metal pattern, which causes the resin in the mixture to set as a thin shell over the pattern. When the shell has reached the proper thickness, the excess sand is removed by rotating the pattern to dump out the sand. The remaining shell is then cured on the pattern and subsequently removed by stripping it off, using mold release pins which have been properly spaced and that are mechanically or hydraulically made to protrude through the pattern. Mating shell halves are bonded, suitably backed by loose sand or other material, and then ready for metal to be poured. Current practice using shell molds has produced castings in excess of 1,000 lb, but often the castings weigh much less.

Plaster Molds Plaster or plaster-bonded molds are used for casting certain aluminum or copper base alloys. Dimensional accuracy and excellent surface finish make this a useful process for making rubber tire molds, match plates, etc.

A variation of this method of molding is the **Antioch process,** using mixtures of 50 percent silica sand, 40 percent gypsum cement, 8 percent talc, and small amounts of sodium silicate, portland cement, and magnesium oxide. These dry ingredients are mixed with water and poured over the pattern. After the mixture is poured, the mold is steam-treated in an autoclave and then allowed to set in air before drying in an oven. When the mold has cooled it is ready for pouring. Tolerances of ± 0.005 in (± 0.13 mm) on small castings and ± 0.015 in (± 0.38 mm) on large castings are obtained by this process.

A problem presented by plaster molds lies in inadequate permeability in the mold material consistent with the desired smooth mold cavity surface. A closely related process, the **Shaw process,** provides a solution.

Table 13.1.2 Design and Cost Features of Basic Casting Methods

Design and cost features	Process					
	Sand casting	Shell-mold casting	Permanent-mold casting	Plaster-mold casting	Investment casting	Die casting
Choice of materials	Wide—ferrous and nonferrous	Wide—except for low-carbon steels	Restricted—brass, bronze, aluminum, some gray iron	Narrow—brass, bronze, aluminum	Wide—includes hard-to-forge and machine materials	Narrow—zinc, aluminum, brass, magnesium
Complexity	Considerable	Moderate	Moderate	Considerable	Greatest	Considerable
Size range	Great	Limited	Moderate	Moderate	Moderate	Moderate
Minimum section, in	$3/32$	$1/16$	0.100	0.010	0.010	0.025
Tolerances, in ft*	$1/16$–$1/8$	$1/32$–$3/32$	$1/32$–$7/64$	$1/32$–$5/64$	0.003–0.006	$1/32$–$1/16$
Surface smoothness, μin, rms	250–300	150–200	90–125	90–125	90–125	60–125
Design feature remarks	Basic casting method of industry	Considered to be good low-cost casting method	Production economics with substantial quantities	Little finishing required	Best for parts too complicated for other casting methods	Most economical where applicable
Tool and die costs	Low	Low to moderate	Medium	Medium	Low to moderate	High
Optimum lot size	Wide—range from few pieces to huge quantities	More required than sand castings	Best when requirements are in thousands	From one to several hundred	Wide—but best for small quantities	Substantial quantities required
Direct labor costs	High	Moderate	Moderate	High	Very high	Low to medium
Finishing costs	High	Low	Low to moderate	Low	Low	Low
Scrap costs	Moderate	Low	Low	Low	Low	Low

* Closer at extra cost.
SOURCE: Cook, "Engineered Castings," McGraw-Hill.

In this process, a refractory aggregate is mixed with a gelling agent and then poured over the pattern. Initial set of the mixture results in a rubbery consistency which allows it to be stripped from the pattern but which is sufficiently strong to return to the shape it had when on the pattern. The mold is then ignited to burn off the volatile content in the set gel and baked at very high heat. This last step results in a hard, rigid mold containing microscopic cracks. The permeability of the completed mold is enhanced by the presence of the so-called microcrazes, while the mold retains the high-quality definition of the mold surface.

Two facts are inherent in the nature of sand molds: First, there may be one or few castings required of a given piece, yet even then an expensive wood pattern is required. Second, the requirement of removal of the pattern from the mold may involve some very intricate pattern construction. These conditions may be alleviated entirely by the use of the **full mold process,** wherein a foamed polystyrene pattern is used. Indeed, the foamed pattern may be made complete with a gating and runner system, and it can incorporate the elimination of draft allowance. In actual practice, the pattern is left in place in the mold and is instantly vaporized when hot metal is poured. The hot metal which vaporized the foam fills the mold cavity to the shape occupied previously by the foam pattern. This process is ideal for casting runs of one or a few pieces, but it can be applied to production quantities by mass-producing the foam patterns. There is extra expense for the equipment to make the destructible foam patterns, but often the economics of the total casting process is quite favorable when compared with resorting to a reusable pattern. There are particular instances when the extreme complexity of a casting can make a hand-carved foam pattern financially attractive.

The **lost-wax, investment, or precision-casting, process** permits the accurate casting of highly alloyed steels and of nonferrous alloys which are impossible to forge and difficult to machine. The procedure consists of making an accurate metal die into which the wax or plastic patterns are cast. The patterns are assembled on a sprue and the assembly sprayed, brushed, or dipped in a slurry of a fine-grained, highly refractory aggregate, and a proprietary bonding agent composed chiefly of ethyl silicate. This mixture is then allowed to set. The pattern is coated repeatedly with coarser slurries until a shell of the aggregate is produced around the pattern. The molds are allowed to stand until the aggregate has set, after which they are heated in an oven in an inverted position so that the wax will run out. After the wax is removed, the molds are baked in a preheat furnace. The molds may then be supported with loose sand and poured in any conventional manner.

There have been attempts in the past to use frozen mercury as a pattern. While mercury is a viable pattern material and can be salvaged totally for reuse, the inherent hazards of handling raw mercury have mitigated against its continued use to make patterns for investment castings.

All dimensions can be held to a tolerance of ±0.005 in (±0.13 mm) with some critical dimensions held to 0.002 in (0.05 mm). Most castings produced by this process are relatively small.

Faithful reproduction and accurate tolerances can also be attained by the **Shaw process;** see above. It combines advantages of dimensional control of precision molds with the ease of production of conventional molding. The process makes use of wood or metal patterns and a refractory mold bonded with an ethyl silicate base material. Since the mold is rubbery when stripped from the pattern, some back draft is permissible.

In the **cement-sand process** portland cement is used as the sand binder. A typical mixture has 11 percent portland cement, 89 percent silica sand, and water 4½ to 7 percent of the total sand and cement. New sand is used for facing the mold and is backed with ground-up sand which has been rebonded. Cores are made of the same material. The molds and cores must airdry 24 to 72 h before pouring. The process can be used for either ferrous or nonferrous castings. This molding mixture practically eliminates the generation of gases, forms a hard surface which resists the erosive action of the metal, and produces castings with good surfaces and accurate dimensions. This process is seldom used, and then only for specific castings wherein the preparation of this type of mold outweighs many of its disadvantages.

Permanent-Mold Casting Methods

In the **permanent-mold casting method,** fluid metal is poured by hand into metal molds and around metal cores without external pressure. The

molds are mechanically clamped together. Of necessity, the complexity of the cores must be minimal, inasmuch as they must be withdrawn for reuse from the finished casting. Likewise, the shape of the molds must be relatively simple, free of reentrant sections and the like, or else the mold itself will have to be made in sections, with attendant complexity.

Metals suitable for this type of casting are lead, zinc, aluminum and magnesium alloys, certain bronzes, and cast iron.

For making iron castings of this type, a number of metal-mold units are usually mounted on a turntable. The individual operations, such as coating the mold, placing the cores, closing the mold, pouring, opening the mold, and ejection of the casting, are performed as each mold passes certain stations. The molds are preheated before the first casting is poured. The process produces castings having a dense, fine-grained structure, free from shrink holes or blowholes. The tool changes are relatively low, and better surface and closer tolerances are obtained than with the sand-cast method. It does not maintain tolerances as close or sections as thin as the die-casting or the plaster-casting methods.

Yellow brasses, which are high in zinc, should not be cast by the permanent-mold process because the zinc oxide fouls the molds or dies.

The **semipermanent mold casting method** differs from the permanent mold casting in that sand cores are used, in some places, instead of metal cores. The same metals may be cast by this method. This process is used where cored openings are so irregular in shape, or so undercut, that metal cores would be too costly or too difficult to handle. The structure of the metal cast around the sand cores is like that of a sand casting. The advantages of permanent mold casting in tolerances, density, appearance, etc., exist only in the section cast against the metal mold.

Graphite molds may be used as short-run permanent molds since they are easier to machine to shape and can be used for higher-melting point alloys, e.g., steel. The molds are softer, however, and more susceptible to erosive damage. Steel railroad wheels may be made in these molds and can be cast by filling the mold with **low-pressure casting** methods.

In the **slush casting** process, the cast metal is allowed partially to solidify next to the mold walls to produce a thin section, after which the excess liquid metal is poured out of the permanent mold.

In **centrifugal casting** the metal is under centrifugal force, developed by rotating the mold at high speed. This process, used in the manufacture of bronze, steel, and iron castings, has the advantage of producing sound castings with a minimum of risers. In **true centrifugal castings** the metal is poured directly into a mold which is rotated on its own axis. Obviously, the shapes cast by this method must have external and internal geometries which are surfaces of revolution. The external cast surface is defined by the internal surface of the water-cooled mold; the internal surface of the casting results from the effective core of air which exists while the mold is spun and until the metal solidifies sufficiently to retain its cast shape. Currently, all cast-iron pipe intended for service under pressure (e.g., water mains) is centrifugally cast. The process is extended to other metals falling under the rubric of **tubular goods.**

In **pressure casting,** for asymmetrical castings which cannot be spun around their own axes, the mold cavities are arranged around a common sprue located on the neutral axis of the mold. The molds used in the centrifugal-casting process may be metal cores or dry sand, depending on the type of casting and the metal cast.

Die casting machines consist of a basin holding molten metal, a metallic mold or die, and a metal transferring device which automatically withdraws molten metal from the basin and forces it under pressure into the die. Two forms of die casting machines are in general use. Lead, tin, and zinc alloys containing aluminum are handled in **piston machines.** Aluminum alloys and pure zinc, or zinc alloys free from aluminum, rapidly attack the iron in the piston and cylinder and require a different type of casting machine. The pressures in a piston machine range from a few hundred to thousands of lb/in^2.

The **gooseneck machine** has a cast-iron gooseneck which dips the molten metal out of the melting pot and transfers it to the die. The pressure is applied to the molten metal by compressed air after the gooseneck is brought in contact with the die. This machine, developed primarily for aluminum alloys, is sometimes used for zinc-aluminum alloys, especially for large castings, but, owing to the lower pressure, the casting is likely to be less dense than when made in the piston machine. It is seldom used for magnesium alloys.

In **cold chamber machines** the molten-metal reservoir is separated from the casting machine, and just enough metal for one casting is ladled by hand into a small chamber, from which it is forced into the die under high pressure. The pressures, quite high, ranging from the low thousands to in excess of 10,000 lb/in^2, are produced by a hydraulic system connected to the piston in the hot metal chamber. The alloy is kept so close to its melting temperature that it is in a slushlike condition. The process is applicable to aluminum alloys, magnesium alloys, zinc alloys, and even higher-melting-point alloys like brasses and bronzes, since the pouring well, cylinder, and piston are exposed to the high temperature for only a short time.

All metal mold external pressure castings have close tolerances, sharp outlines and contours, fine smooth surface, and high rate of production, with low labor cost. They have a hard skin and a soft core, resulting from the rapid chilling effect of the cold metal mold.

The dies usually consist of two blocks of steel, each containing a part of the cavity, which are locked together while the casting is being made and drawn apart when it is ready for ejection. One-half of the die (next to the ejector nozzle) is stationary; the other half moves on a carriage. The dies are preheated before using and are either air- or water-cooled to maintain the desired operating temperature. Die life varies with the alloy and dimensional tolerances required. Retractable and removable metal cores are used to form internal surfaces. Inserts can be cast into the piece by placing them on locating pins in the die.

A wide range of sizes and shapes can be made by these processes, including threaded pieces and gears. Holes can be accurately located. The process is best suited to large-quantity production.

A historic application of the process was for typesetting machines such as the linotype. Although now they are obsolescent and rarely found in service, for a long time the end products of typesetting machines were a prime example of a high-quality die-cast metal product.

MOLDING EQUIPMENT AND MECHANIZATION

Flasks may be filled with sand by hand shoveling, gravity feed from overhead hoppers, continuous belt feeding from a bin, sand slingers, and, for large molds, by an overhead crane equipped with a grab bucket.

Hand ramming is the simplest method of compacting sand. To increase the rate, pneumatic rammers are used. The method is slow, the sand is rammed in layers, and it is difficult to gain uniform density.

More uniform results and higher production rates are obtained by **squeezing machines.** Hand-operated squeezers were limited to small molds and are obsolete; **air-operated machines** permit an increase in the allowable size of molds as well as in the production rate. These machines are suitable for shallow molds. Squeezer molding machines produce greatest sand density at the top of the flask and softest near the parting line of pattern. Air-operated machines are also applied in vertical molding processes using flaskless molds. Horizontal impact molding sends shock waves through the sand to pack the grains tightly.

In **jolt molding machines** the pattern is placed on a platen attached to the top of an air cylinder. After the table is raised, a quick-release port opens, and the piston, platen, and mold drop free against the top of the cylinder or striking pads. The impact packs the sand. The densities produced by this machine are greatest next to the parting line of the pattern and softest near the top of the flask. This procedure can be used for any flask that can be rammed on a molding machine. As a separate unit, it is used primarily for medium and large work. Where plain jolt machines are used on large work, it is usual to ram the top of the flask manually with an air hammer.

Jolt squeeze machines use both the jolt and the squeeze procedures. The platen is mounted on two air cylinders: a small cylinder to jolt and a large one to squeeze the mold. They are widely used for small and medium work, and with match-plate or gated patterns. Pattern-stripping devices can be incorporated with jolt or squeezer machines to permit

mechanical removal of the pattern. Pattern removal can also be accomplished by using jolt-rockover-draw or jolt-squeeze-rollover-draw machines.

The **sand slinger** is the most widely applicable type of ramming machine. It consists of an impeller mounted on the end of a double-jointed arm which is fed with sand by belt conveyors mounted on the arm. The impeller rotating at high speed gives sufficient velocity to the sand to ram it in the mold by impact. The head may be directed to all parts of the flask manually on the larger machines and may be automatically controlled on smaller units used for the high-speed production of small molds.

Vibrators are used on all pattern-drawing machines to free the pattern from the grip of the sand before drawing. Their use reduces mold damage to a minimum when the pattern is removed, and has the additional advantage of producing castings of more uniform size than can be secured by hand rapping the pattern. Pattern damage is also kept to a minimum. Vibrators are usually air-operated, but some electrically operated types are in use.

Flasks generally consist of two parts: the upper section, called the **cope,** and the bottom section, the **drag.** When more than two parts are used, the intermediate sections are called **cheeks.** Flasks are classified as tight, snap, and slip. **Tight flasks** are those in which the flask remains until the metal is poured. Snap flasks are hinged on one corner and have a locking device on the diagonally opposite corner. In use, these flasks are removed as soon as the mold is closed. **Slip flasks** are of solid construction tapered from top to bottom on all four sides so that they can be removed as soon as the mold is closed. Snap or slip flasks permit the molder to make any number of molds with one flask. Before pouring snap- or slip-flask molds, a wood or metal pouring jacket is placed around the mold and a weight set on the top to keep the cope from lifting. The cope and drag sections on all flasks are maintained in proper alignment by flask pins and guides.

Tight flasks can be made in any size and are fabricated of wood, rolled steel, cast steel, cast iron, magnesium, or aluminum. Wood, aluminum, and magnesium are used only for small- and medium-sized flasks. Snap and slip flasks are made of wood, aluminum, or magnesium, and are generally used for molds not over 20 by 20 in (500 mm by 500 mm).

Mechanization of Sand Preparation

In addition to the various types of molding machines, the modern foundry makes use of a variety of equipment to handle the sand and castings.

Sand Preparation and Handling   Sand is prepared in **mullers,** which serve to mix the sand, bonding agent, and water. **Aerators** are used in conjunction to loosen the sand to make it more amenable to molding. **Sand cutters** that operate over a heap on the foundry floor may be used instead of mullers. Delivery of the sand to the molding floor may be by means of dump or scoop trucks or by belt conveyors. At the molding floor the molds may be placed on the floor or delivered by conveyors to a pouring station. After pouring, the castings are removed from the flasks and adhering sand at a **shakeout** station. This may be a mechanically operated jolting device that shakes the loose sand from flask and casting. The used sand, in turn, is returned to the storage bins by belt conveyor or other means. Small castings may be poured by using **stackmolding** methods. In this case, each flask has a drag cavity molded in its upper surface and a cope cavity in its lower surface. These are stacked one on the other to a suitable height and poured from a common sprue.

There is an almost infinite variety of equipment and methods available to the foundry, ranging from simple, work-saving devices to completely mechanized units, including completely automatic molding machines. Because of this wide selection available, the degree to which a foundry can be mechanized depends almost entirely on the economics of the operations, rather than the availability or lack of availability of a particular piece of equipment.

MOLDING SAND

Molding sand consists of silica grains held together by some bonding material, usually clay or bentonite.

Grain size greatly influences the surface finish of a casting. The proper grain size is determined by the size of the casting, the quality of surface required, and the surface tension of the molten metal. The grain size should be approximately uniform when maximum permeability is desired.

Naturally bonded sands are mixtures of silica and clay as taken from the pits. Modification may be necessary to produce a satisfactory mixture. This type of sand is used in gray iron, ductile iron, malleable iron, and nonferrous foundries (except magnesium).

Synthetically bonded sands are produced by combining clay-free silica sand with clay or bentonite. These sands can be compounded to suit foundry requirements. They are more uniform than naturally bonded sands but require more careful mixing and control. Steel foundries, gray iron and malleable iron foundries, and magnesium foundries use this type of sand.

Special additives may be used in addition to the basic sand, clay, and water. These include cereals, ground pitch, sea coal, gilsonite, fuel oil, wood flour, silica flour, iron oxide, pearlite, molasses, dextrin, and proprietary materials. These all serve the purpose of altering specific properties of the sand to give desired results.

The properties of the sand that are of major interest to the foundry worker are **permeability**, or the venting power, of the sand; **green compressive strength; green shear strength; deformation,** or the sand movement under a given load; **dry compressive strength;** and **hot strength,** i.e., strength at elevated temperatures. Several auxiliary tests are often made, including moisture content, clay content, and grain-size determination.

The foundry engineer or metallurgist who usually is entrusted with the control of the sand properties makes the adjustments required to keep it in good condition.

Facing sands, for giving better surface to the casting, are used for gray iron, malleable iron, steel, and magnesium castings. The iron sands usually contain **sea coal,** a finely ground coal which keeps the sand from adhering to the casting by generating a gas film when in contact with the hot metal. Steel facings contain silica flour or other very fine highly refractory material to form a dense surface which the metal cannot readily penetrate.

Mold washes are coatings applied to the mold or core surface to improve the finish of the casting. They are applied either wet or dry. The usual practice is to brush or spray the wet mold washes and to brush or rub on the dry ones. Graphite or silica flour mixed with clay and molasses water is frequently used. The washes are mixed usually with waterbase or alcohol-base solvent solutions that require oven drying time, during which not only does the wash set, but also the excess moisture is removed from the washed coating.

Core Sands and Core Binders

Green sand cores are made from standard molding-sand mixtures, sometimes strengthened by adding a binder, such as dextrin, which hardens the surface. Cores of this type are very fragile and are usually made with an arbor or wires on the inside to facilitate handling. Their collapsibility is useful to prevent hot tearing of the casting.

Dry sand cores are made from silica sand and a binder (usually oil) which hardens under the action of heat. The amount of oil used should be the minimum which will produce the necessary core strength.

Core binders are either organic, such as core oil, which are destroyed under heat, or inorganic, which are not destroyed.

Organic Binders   The main organic binder is **core oil.** Pure linseed oil is used extensively as one of the basic ingredients in blended-oil core binders. These consist primarily of linseed oil, resin, and a thinner, such as high-grade kerosene. They have good wetting properties, good workability, and better oxidation characteristics than straight linseed oil.

Corn flour produces good green strength and dry strength when used in conjunction with oil. Cores made with this binder are quick drying in the oven and burn out rapidly and completely in the mold.

Dextrin produces a hard surface and weak center because of the migration of dextrin and water to the surface. Used with oil, it produces a hard smooth surface but does not produce a green bond as good as that with corn flour.

Commercial **protein binders,** such as gelatin, casein, and glues, improve flowability of the sand, have high binding power, rapid drying, fair resistance to moisture, and low burning-out point, with only a small volume of gas evolved on burning. They are used where high collapsibility of the core is essential.

Other binders include paper-mill by-products, which absorb moisture readily, have high dry strength, low green strength, high gas ratio, and high binding power for clay materials.

Coal tar pitch and **petroleum pitch** flow with heat and freeze around the grains on cooling. These compounds have low moisture absorption rates and are used extensively for large iron cores. They can be used effectively with impure sands.

Wood and **gum rosin, plastic resins,** and rosin by-products are used to produce collapsibility in cores. They must be well ground. They tend to cake in hot weather, and large amounts are required to get desired strength.

Plastics of the **urea-** and **phenol-formaldehyde** groups and **furan resins** are being used for core binders. They have the advantage of low-temperature baking, collapse readily, and produce only small amounts of gas. These can be used in **dielectric baking ovens** or in the **shell molding, hot box, or air setting** processes for making cores.

Inorganic binders include fire clay, southern bentonite, western bentonite, and iron oxide.

Cores can also be made by mixing sand with sodium silicate. When this mixture is in the core box, it is infiltrated with CO_2, which causes the core to harden. This is called the **CO_2 process.**

Core-Making Methods

Cores are made by the methods employed for sand molds. In addition, **core blowers** and **extrusion machines** are used.

Core blowers force sand into the core box by compressed air at about 100 lb/in². They can be used for making all types of small- and medium-sized cores. The cores produced are very uniform, and high production rates are achieved.

Screw feed machines are used largely for plain cylindrical cores of uniform cross section. The core sand is extruded through a die onto a core plate. The use of these machines is limited to the production of stock cores, which are cut to the desired length after baking.

Core Ovens   Core oven walls are constructed of inner and outer layers of sheet metal separated by rock wool or Fiberglas insulation and with interlocked joints. Combustion chambers are refractory-lined, and the hot gases are circulated by fans. They are designed for operating at temperatures suitable for the constituents in the core body. Time at baking temperatures will, likewise, vary with the composition of the core.

Core driers are light skeleton cast iron or aluminum boxes, the internal shape of which conforms closely to the cope portion of the core. They are used to support, during baking, cores which cannot be placed on a flat plate.

Chaplets are metallic pieces inserted into the mold cavity which support the core. Long unsupported cores will be subject to flotation force as the molten metal fills the mold and may break if the resulting flexural stresses are excessive. Likewise, the liquid forces imposed on cores as metal flows through the mold cavity may cause cores to shift. The chaplets interposed within the mold cavity are placed to alleviate these conditions. They are generally made of the same material as that being cast; they melt and blend with the metal as cast, and they remain solid long enough for the liquid forces to equilibrate through the mold cavity.

CASTING ALLOYS

In general, the types of alloys that can be produced as wrought metals can also be prepared as castings. Certain alloys, however, cannot be forged or rolled and can only be used as cast.

Ferrous Alloys

Steel Castings   (See Sec. 6.3.) Steel castings may be classified as:

1. Low carbon (C < 0.20 percent). These are relatively soft and not readily heat-treatable.

2. Medium carbon (0.20 percent < C < 0.50 percent). These castings are somewhat harder and amenable to strengthening by heat treatment.

3. High carbon (C > 0.50 percent). These steels are used where maximum hardness and wear resistance are desired.

In addition to the classification based on carbon content, which determines the maximum hardness obtainable in steel, the castings can be also classified as **low alloy** content (≤ 8 percent) or **high alloy** content (> 8 percent).

Low-alloy steels behave essentially as plain carbon steels but have a higher **hardenability**, which is a measure of ability to be hardened by heat treatment. High-alloy steels are designed to produce some specific property, like corrosion resistance, heat resistance, wear resistance, or some other special property.

Malleable Iron Castings   The carbon content of malleable iron ranges from about 2.00 to 2.80 percent and may reach as high as 3.30 percent if the iron is melted in a cupola. Silicon ranging from 0.90 to 1.80 percent is an additional alloying element required to aid the annealing of the iron. As cast, this iron is hard and brittle and is rendered soft and malleable by a long heat-treating or annealing cycle. (See also Sec. 6.3.)

Gray Iron Castings   Gray iron is an alloy of iron, carbon, and silicon, containing a higher percentage of these last two elements than found in malleable iron. Much of the carbon is present in the elemental form as graphite. Other elements present include manganese, phosphorus, and sulfur. Because the properties are controlled by proper proportioning of the carbon and silicon and by the cooling rate of the casting, it is usually sold on the basis of specified properties rather than composition. The carbon content will usually range between 3.00 and 4.00 percent and the silicon will be between 1.00 and 3.00 percent, the higher values of carbon being used with the lower silicon values (usually), and vice versa. As evidence of the fact that gray iron should not be considered as a material having a single set of properties, the ASTM and AFS codify gray cast iron in several classes, with accompanying ranges of tensile strengths available. The high strengths are obtained by proper adjustment of the carbon and silicon contents or by alloying. (See also Sec. 6.3.)

An important variation of gray iron is **nodular iron,** or **ductile iron,** in which the graphite appears as nodules rather than as flakes. This iron is prepared by treating the metal in the ladle with additives that usually include magnesium in alloy form. Nodular iron can exceed 100,000 lb/in^2 (690 MN/m^2) as cast and is much more ductile than gray iron, measuring about 2 to 5 percent elongation at these higher strengths, and even higher percentages if the strength is lower. (See Sec. 6.3.)

Nonferrous Alloys

Aluminum-Base Castings   Aluminum is alloyed with copper, silicon, magnesium, zinc, nickel, and other elements to produce a wide variety of casting alloys having specific characteristics of foundry properties, mechanical properties, machinability, and/or corrosion resistance. Alloys are produced for use in sand casting, permanent mold casting, or die casting. Some alloys are heat-treatable using **solution** and **age-hardening** treatments. (See also Sec. 6.4.)

Copper-Base Alloys   The alloying elements used with copper include zinc (brasses), tin (bronzes), nickel (nickel bronze), aluminum (aluminum bronze), silicon (silicon bronze), and beryllium (beryllium bronze). The brasses and tin bronzes may contain lead for machinability. Various combinations of zinc and tin, or of tin or zinc with other elements, are also available. With the exception of some of the aluminum bronzes and beryllium bronze, most of the copper-base alloys cannot be hardened by heat treatment. (See also Sec. 6.4.)

Special Casting Alloys   Other metals cast in the foundry include **magnesium-base alloys** for light weight, **nickel-base alloys** for high-temperature applications, **titanium-base alloys** for strength-to-weight ratio, etc. The magnesium-base alloys require special precautions during melting and pouring to avoid burning. (See also Sec. 6.4.)

MELTING AND HEAT TREATING FURNACES

There are several types of **melting furnaces** used in conjunction with metal casting. Foundry furnaces used in melting practice for ferrous castings are predominantly **electric arc** (direct and indirect), **induction,** and **crucible** for small operations. For cast iron, **cupolas** are still employed, although in ever-decreasing quantities. The previous widespread use of open-hearth furnaces is now relegated to isolated foundries and is essentially obsolete. In general, ferrous foundries' melting practice has become based largely on electric-powered furnaces. **Duplexing** operations are still employed, usually in the form of cupola/induction furnace, or cupola/electric arc furnace.

In nonferrous foundries, electric arc, induction, and crucible furnaces predominate. There are some residual installations which use air furnaces, but they are obsolete and found only in some of the older, small foundries which cater to unique clients.

Vacuum melting and metal refining were fostered by the need for extremely pure metals for high-temperature, high-strength applications (e.g., gas-turbine blades). Vacuum melting is accomplished in a furnace located in an evacuated chamber; the source of heat is most often an electric arc and sometimes induction coils. Gases entrained in the melt are removed, the absence of air prevents oxidation of the base metals, and a high degree of metal purity is retained in the molten metal and in the casting ultimately made from that vacuum-melted metal. The mold is also enclosed in the same evacuated chamber.

The vacuum melting and casting process is very expensive because of the nature of the equipment required, and quantities of metal handled are relatively small. The economics of the overall process are justified by the design requirements for highest-quality castings for ultimately very demanding service.

Annealing and **heat-treating furnaces** used to process castings are the type usually found in industrial practice. (See Secs. 7.3. and 7.5.)

CLEANING AND INSPECTION

Tumbling barrels consist of a power-driven drum in which the castings are tumbled in contact with hard iron stars or balls. Their impact removes the sand and scale.

In **air-blast cleaning units,** compressed air forces silica sand or chilled iron shot into violent contact with the castings, which are tumbled in a barrel, rotated on a table, or passed between multiple orifices on a conveyor. Large rooms are sometimes utilized, with an operator directing the nozzle. These machines are equipped with hoppers and elevators to return the sand or shot to the magazine. Dust-collecting systems are required.

In **centrifugal-blast cleaning units,** a rotating impeller is used to impart the necessary velocity to the chilled iron shot or abrasive grit. The velocities are not so high as with air, but the volume of abrasive is much greater. The construction is otherwise similar to the air blast machine.

Water in large volume at pressures of 250 to 600 lb/in^2 is used to remove sand and cores from medium and large castings.

High-pressure water and sand cleaning (Hydroblast) employs high pressure water mixed with molding sand which has been washed off the casting. A sand classifier is incorporated in the sand reclamation system.

Pneumatic chipping hammers may be used to clean large castings where the sand is badly burned on and for deep pockets.

Removal of Gates and Risers and Finishing Castings   The following tabulation shows the most generally used methods for removing gates and risers (marked R) and for finishing (marked F).

Steel Oxyacetylene (R) hand hammer or sledge (R), grinders (F), chipping hammer, (F), and machining (F).

Cast iron Chipping hammer (R, F), hand hammer or sledge (R), abrasive cutoff (R), power saw (R), and grinders (F).

Malleable iron Hand hammer or sledge (R), grinders (F), shear (F), and machining (F).

Brass and bronze Chipping hammer (R, F), shear (R, F), hand hammer or sledge (R), abrasive cutoff (R), power saw (R), belt sanders (F), grinders (F), and machining (F).

Aluminum Chipping hammer (R), shear (R), hand hammer or sledge (R), power saw (R), grinders (F), and belt sander (F).

Magnesium Band saw (R), machining (F), and flexible-shaft machines with steel burr cutters (F).

Casting Inspection

(See Sec. 5.4.)

Castings are inspected for dimensional accuracy, hardness, surface finish, physical properties, internal soundness, and cracks. For **hardness** and for **physical properties,** see Sec. 6.

Internal soundness is checked by cutting or breaking up pilot castings or by nondestructive testing using X-ray, gamma ray, etc.

Destructive testing tells only the condition of the piece tested and does not ensure that other pieces not tested will be sound. It is the most commonly used procedure at the present time.

X-ray, gamma ray, and other methods have made possible the nondestructive checking of castings to determine internal soundness on all castings produced. Shrinks, cracks, tears, and gas holes can be determined and repairs made before the castings are shipped.

Magnetic powder tests (Magnaflux) are used to locate structural discontinuities in iron and steel except austenitic steels, but they are not applicable to most nonferrous metals or their alloys. The method is most useful for the location of surface discontinuities, but it may indicate subsurface defects if the magnetizing force is sufficient to produce a leakage field at the surface.

In this test a magnetic flux is induced in ferromagnetic material. Any abrupt discontinuity in its path results in a local flux leakage field. If finely divided particles of ferromagnetic material are brought into the vicinity, they offer a low reluctance path to the leakage field and take a position that outlines approximately its effective boundaries. The casting to be inspected is magnetized and its surface dusted with the magnetic powder. A low velocity air stream blows the excess powder off and leaves the defect outlined by the powder particles. The powder may be applied while the magnetizing current is flowing (**continuous method**) or after the current is off (**residual method**). It may be applied dry or suspended in a light petroleum distillate similar to kerosene. Expert interpretation of the tests is necessary.

CASTING DESIGN

Design for the best utilization of metal in the cast form requires a knowledge of metal solidification characteristics, foundry practices, and the metallurgy of the metal being used. Metals exhibit certain peculiarities in the formation of solid metal during freezing and also undergo shrinkage in the liquid state during the freezing process and after freezing, and the casting must be designed to take these factors into consideration. Knowledge concerning the freezing process will also be of assistance in determining the fluidity of the metal, its resistance to **hot tearing,** and its tendency to evolve dissolved gases. For economy in production, casting design should take into consideration those factors in molding and coring that will lead to the simplest procedures. Elimination of expensive cores, irregular parting lines, and deep drafts in the casting can often be accomplished with a slight modification of the original design. Combination of the foregoing factors with the selection of the right metal for the job is important in casting design. Consultation between the design engineer and personnel at the foundry will result in well-designed castings and cost-effective foundry procedures. Initial guidance may be had from the several references cited and from updated professional literature, which abounds in the technical journals. Trade literature, as represented by the publications issued by the various generic associations, will be useful in assessing potential problems with specific casting designs. Generally, time is well spent in these endeavors before an actual design concept is reduced to a set of dimensional drawings and/or specifications.

13.2 PLASTIC WORKING OF METALS
by Rajiv Shivpuri

(See also Secs. 5 and 6.)

REFERENCES: Crane,"Plastic Working of Metals and Power Press Operations," Wiley. Woodworth, "Punches, Dies and Tools for Manufacturing in Presses," Henley. Jones, "Die Design and Die Making Practice," Industrial Press. Stanley, "Punches and Dies," McGraw-Hill. DeGarmo, "Materials and Processes in Manufacturing," Macmillan. "Modern Plastics Encyclopedia and Engineers Handbook," Plastics Catalogue Corp., New York. "The Tool Engineers Handbook," McGraw-Hill. Bridgman, "Large Plastic Flow and Fracture," McGraw-Hill. "Cold Working of Metals," ASM. Pearson, "The Extrusion of Metals," Wiley.

STRUCTURE

Yieldable structural forces between the particles composing a material to be worked are the key to its behavior. Simple internal structures contain only a single element, as pure copper, silver, or iron. Relatively more difficult to work are the solid solutions in which one element tends to distribute uniformly in the structural pattern of another. Thus silver and gold form a continuous series of solid-solution alloys as their proportions vary. Next are alloys in which strongly bonded molecular groups dispersed through or along the grain boundaries of softer metals offer increasing resistance to working, as does iron carbide (Fe_3C) in solution in iron.

Bonding forces are supplied by electric fields characteristic of individual atoms. These forces in turn are subject to modification by temperature as energy is added, increasing electron activity.

The **particles** which constitute an atom are so small that most of its volume is empty space. For a similar energy state, there is some rough uniformity in the outside size of atoms. In general, therefore, the more complex elements have their larger number of particles more densely packed and so are heavier. For each element, the energy pattern of its electric charges in motion determines the field characteristics of that atom and which of the orderly arrangements it will seek to assume with relation to others like it in the orderly crystalline form.

Space lattice is the term used to describe the orderly arrangement of rows and layers of atoms in the crystalline form. This unstressed state is also described as balanced, unstrained, or **annealed.** The working or deforming of materials distorts the orderly arrangement, unbalancing the forces between atoms. Cubic patterns or space lattices characterize the more ductile or workable materials. Hexagonal and more complex patterns tend to be more brittle or more rigid. Flaws, irregularities, or distortions, with corresponding unbalanced strains among adjacent atoms, may occur in the pattern or along grain boundaries. **Slip-plane** movements in working to new shapes tend to slide the once orderly layers of atoms within the grain-boundary limitations of individual crystals. Such sliding movement tends to take place at 45° to the direction of the applied load because much higher stresses are required to pull atoms directly apart or to push them straight together.

Chemical combinations, in liquid or solid solutions, or molecular compounds depend upon relative field patterns of elements or upon actual displacement of one or more electrons from the outer orbit of a donor element to the outer orbit of a receptor element. Thus the molecules of hard iron carbide, Fe_3C, may be held in solid solution in soft pure iron (ferrite) in increasing proportions up to 0.83 percent of carbon in iron, which is described as **pearlite.** Zinc may occupy solid-solution positions in the copper space lattice up to about 45 percent, the range of the ductile red and yellow brasses.

Thermal Changes Adding heat (energy) increases electron activity and therefore also the mobility of the atom. Probability of brittle failure at low temperatures usually becomes less as temperature increases. Transition temperatures from one state to another differ for different elements. Thermal transitions therefore become more complex as such differing elements are combined in alloys and compounds. As temperatures rise, a **stress-relieving** range is reached at which the most severely strained atoms are able to ease themselves around into less strained positions. At somewhat higher temperatures, **annealing** or **recrystallization** of worked or distorted structure takes place. Old grain boundaries disappear and small new grains begin to grow, aligning nearby atoms into their orderly lattice pattern. The more severely the material has been worked, the lower is the temperature at which recrystallization begins. Grain growth is more rapid at higher temperatures. In working materials above their recrystallization range, as in forging, the relief of interatomic strains becomes more nearly spontaneous as the temperature is increased. **Creep** takes place when materials are under some stress above the recrystallization range, and the thermal mobility permits individual atoms to ease around to relieve that stress with an accompanying gradual change of shape. Thus a wax candle droops due to gravity on a hot day. Lead, which recrystallizes below room temperature, will creep when used for roofing or spouting. Steels in rockets and jet engines begin to creep around 1,300 to 1,500°F (704 to 815°C). Creep is more rapid as the temperature rises farther above the recrystallization range.

PLASTICITY

Plasticity is that property of materials which commends them to the mass-production techniques of pressure-forming desired shapes. It is understood more easily if several types of plasticity are considered.

Crystoplastic describes materials, notably metals, which can be worked in the stable crystalling state, below the recrystallization range. Metals which crystallize in the cubic patterns have a wider plastic range than those of hexagonal pattern. Alloying narrows the range and increases the resistance to working. Tensile or compressive testing of an annealed specimen can be used to show the plastic range which lies between the initial yield point and the point of ultimate tensile or compressive failure.

The **plastic range,** as of an annealed metal, is illustrated in Fig. 13.2.1. Changing values of **true stress** are determined by dividing the applied load at any instant by the cross-section area at that instant. As material is worked, a progressive increase in elastic limit and yield point registers the slip-plane movement or work hardening which has taken place and the consequent reduction in residual plasticity. This changing yield point or resistance, shown in Fig. 13.2.1, is divided roughly into three characteristic ranges. The contour of the lower range can be varied by nonuniformity of grain sizes or by small displacements resulting from prior direction of working. Random large, soft grains yield locally under slight displacement, with resulting **surface markings,** described as *orange peel, alligator skin,* or *stretcher strain markings.* These can be prevented by preparatory roller leveling, which gives protection in the case of steel for perhaps a day, or by a 3 to 5 percent temper pass of cold-rolling, which may stress relieve in perhaps 3 months, permitting recurrent trouble. The middle range covers most drawing and forming operations. Its upper limit is the point of normal tensile failure. The upper range requires that metal be worked

primarily in compression to inhibit the start of tensile fracture. Severe extrusion, spring-temper rolling, and music-wire drawing use this range.

Dispersion hardening of metal alloys by heat treatment (see Fig. 13.2.2) reduces the plastic range and increases the resistance to work hardening. Figure 13.2.2 also shows the common methods of plotting change of true stress against **percentage of reduction**—e.g., reduction of thickness in rolling or compressive working, of area in wire drawing, ironing, or tensile testing, or of diameter in cup drawing or reducing operations—and against **true strain,** which is the natural logarithm of change of area, for convenience in higher mathematics.

Fig. 13.2.2 High-range plasticity (dotted) of an SAE 4140 steel, showing the effect of dispersion hardening. Two plotting methods. *(ASME, 1958, Crane and Wagner, E. W. Bliss Co.)*

For metals, thermoplastic working is usually described as **hot working,** except for tin and lead, which recrystallize below room temperature. Hot-worked samples may be etched to show **flow lines,** which are usually made up of old-grain boundaries. Where these show, recrystallization has not yet taken place, and some work hardening is retained to improve physical properties. Zinc and magnesium, which are typical of the hexagonal-structure metals, take only small amounts of cold working but can be drawn or otherwise worked severely at rather moderate temperatures [Zn, 200 to 400°F (90 to 200°C); Mg, 500 to 700°F (260 to 400°C)]. Note that, although hexagonal-pattern metals are less easily worked than cubic-pattern metals, they are for that same reason structurally more rigid for a similar relative weight. Advantageous forging temperatures change with alloy composition: copper, 1,800 to 1,900°F (980 to 1,040°C); red brass, Cu 70, Zn 30, 1,600 to 1,700°F (870 to 930°C); yellow brass, Cu 60, Zn 40, 1,200 to 1,500°F (650 to 815°C). See Sec. 6 for general physical properties of metals.

Substantially pure iron shows an increasing elastic limit and decreasing plasticity with increasing amounts of work hardening by cold-rolling. The rate at which such work hardening takes place is greatly increased, and the remaining plasticity reduced, as alloying becomes more complex.

In **steels,** the mechanical working range is conventionally divided into cold, warm, and hot working. Figure 13.2.3 is a plot of flow stress, limit strain, scale factor, and dimensional error for different values of forging temperature and for two different strain rates. The **flow stress** is the resistance to deformation. As the temperature rises from room temperature to 2,072°F (1,100°C), the flow stress decreases first gradually and then rapidly to about 25 percent of its value [cold working 114 ksi (786 MPa) and hot working 28 ksi (193 MPa) at a strain of 0.5 and strain rate of 40 per second].

One measure of workability is the **strain limit.** As the temperature rises, the strain limit for the 70-in (in·s) strain rate (typical of mechanical

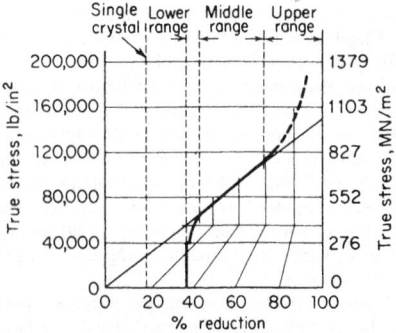

Fig. 13.2.1 Three ranges of crystoplastic work hardening of a low-carbon steel. *(ASME, 1954, W. S. Wagner, E. W. Bliss Co.)*

Fig. 13.2.3 Effect of forging temperature on forgeability and material properties. Material: AISI 1015 steel. ϕ = strain rate; ϕ^* = limiting strain; σ_f = flow stress; S_{tot} = dimensional error; Fe_L = scale loss. *(K. Lange, "Handbook of Metal Forming," McGraw-Hill, 1985.)*

press forging) decreases slightly up to 500°C (932°F), rises until 750°C (1,382°F), drops rapidly at 800°C (1,472°F) (often called *blue brittleness*), and beyond 850°C (1,562°F) increases rapidly to hot forging temperature of 1,100°C (2,012°F). Therefore, substantial advantages of low material resistance (low tool pressures and press loads) and excellent workability (large flow without material failure) can be realized in the hot-working range. Hot-working temperatures, however, also mean poor dimensional tolerance (total dimensional error), poor surface finish, and material loss due to scale buildup. Forging temperatures above 1,300°C (2,372°F) can lead to **hot shortness** manifested by melting at the grain boundaries.

MATERIAL RESPONSE IN METAL FORMING

The deformation conditions in metalworking processes span a range of deformation parameters, including strain and strain rates (Fig. 13.2.4) that are much higher than those encountered in conventional testing methods (Fig. 13.2.5). In machining, the strains are high and the strain rates can

Fig. 13.2.4 Range of deformation parameters for various metalworking processes. *(Source: P. F. Bariani, S. Bruschi, and T. Dal Negro, Enhancing Performances of SHPB for Determination of Flow Curves, Annals of the CIRP,* **50**, *no. 1, pp. 153–156.)*

reach 10⁵/s, while in explosive forming, strains are small at high strain rates providing extremely small response times. Forging and extrusion cover a wide range of strains and strain rates. Sheet forming carried out as small strains and strain rates differs from superplastic forming at extremely low strain rates but high strains. Consequently, different methods have been developed to test material response for different ranges of deformation parameters, i.e., strain and strain rate (Fig. 13.2.5).

PLASTIC WORKING TECHNIQUES

In the **metalworking** operations, as distinguished from metal cutting, material is forced to move into new shapes by plastic flow. **Hot-working** is carried on above the recovery temperature, and spontaneous recovery, or annealing, occurs about as fast as the properties of the material are altered by the deformation. This process is limited by the chilling of the material in the tools, scaling of the material, and the life of the tools at the required temperatures. **Cold-working** is carried on at room temperature and may be applied to most of the common metals. Since, in most cases, no recovery occurs at this temperature, the properties of the metal are altered in the direction of increasing strength and brittleness throughout the working process, and there is consequently a limit to which cold-working may be carried without danger of fracture.

A convenient way of representing the action of the common metals when cold-worked consists of plotting the actual stress in the material against the percentage reduction in thickness. Within the accuracy required for shop use, the relationship is linear, as in Fig. 13.2.6. The lower limit of stress shown is the yield point at the softest temper, or anneal, commercially available, and the upper limit is the limit of tensile action, or the stress at which fracture, rather than flow, occurs. This latter value does not correspond to the commercially quoted "tensile strength" of the metal, but rather to the "true tensile strength," which is the stress that exists at the reduced section of a tensile specimen at fracture and which is higher than the nominal value in inverse proportion to the reduction of area of the material.

As an example of the construction and use of the cold-working plots shown in Fig. 13.2.6, the action of a very-low-carbon deep-drawing steel has been shown in Fig. 13.2.7. Starting with the annealed material with a yield point of 35,000 lb/in² (240 MN/m²), the steel was drawn to successive reductions of thickness up to about 58 percent, and the

Fig. 13.2.5 Testing methods used to determine mechanical behavior of materials under various deformation regimes. (*Source: J. E. Field, W. G. Proud, S. M. Walley, and H. T. Goldrein, Review of Experimental Techniques for High Rate Deformation and Shock Studies, in "New Experimental Methods in Material Dynamics and Impact," AMAS, Warsaw, 2001.*)

Fig. 13.2.6 Plastic range chart of commonly worked metals.

corresponding stresses plotted as the heavy straight line. The entire graph was then extrapolated to 100 percent reduction, giving the **modulus of strain hardening** as indicated, and to zero stress so that all materials might be plotted on the same graph. Lines of equal reduction are slanting lines through the point marking the modulus of strain hardening at theoretical 100 percent reduction. Starting at any initial condition of previous cold work on the heavy line, a percentage reduction from this condition will be indicated by a horizontal traverse to the slanting reduction line of corresponding magnitude and the resulting increase in stress by the vertical traverse from this point to the heavy line.

The traverse shown involved three draws from the annealed condition of 30, 25, and 15 percent each, and resulting stresses of 53,000, 63,000, and 68,000 lb/in² (365, 434, and 469 MN/m²). After the initial 30 percent

reduction, the next 25 percent uses $(1.00 - 0.30) \times 0.25$, or 17.5 percent more of the cold-working range; the next 15 percent reduction uses $(1.00 - 0.30 - 0.175) \times 0.15$, or about 8 percent of the original range, totaling $30 + 17.5 + 8 = 55.5$ percent. This may be compared with the test value percent reduction in area for the particular material. The same result might have been obtained, die operation permitting, by a single reduction of 55 percent, as shown. Any appreciable reduction beyond this point would come dangerously close to the limit of plastic flow, and consequently an anneal is called for before any further work is done on the piece.

Figure 13.2.8 shows the approximate **true stress** vs. **true strain** plot of common plastic range values, for comparison with Fig. 13.2.6. In metal forming, a convenient way of representing the resistance of metal to

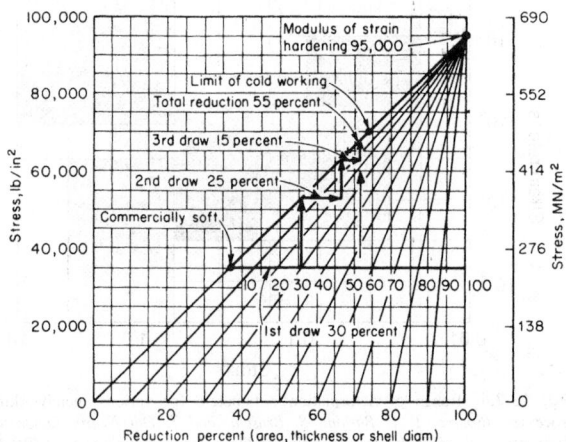

Fig. 13.2.7 Graphical solution of a metalworking problem.

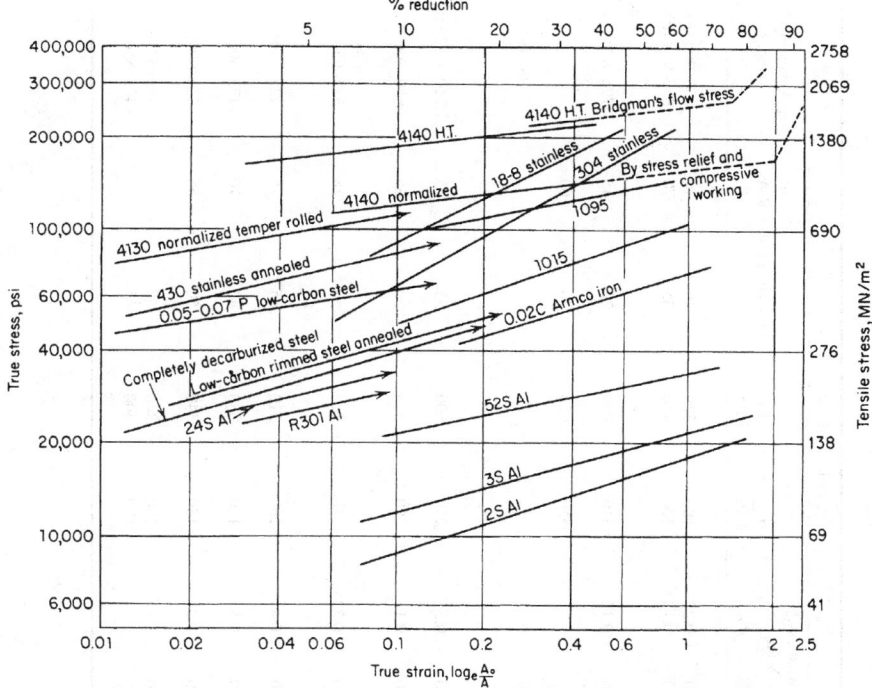

% reduction

Fig. 13.2.8   True stress vs. true strain curves for typical metals. *(Crane and Hauf, E. W. Bliss. Co.)*

deformation and flow is the **flow stress** σ, also known as the **logarithmic stress** or **true stress**. For most metals, flow stress is a function of the amount of deformation at cold-working temperatures (strain $\bar{\varepsilon}$) and the deformation rate at hot-working temperatures (strain rate $\dot{\bar{\varepsilon}}$) This relationship is often given as a power-law curve; $\sigma = K\bar{\varepsilon}^n$ for cold forming and $\sigma = C\dot{\bar{\varepsilon}}^m$ for hot forming. For commonly used materials, the values of the strength coefficients K and C and hardening coefficients n and m are given in Tables 13.2.1a and b.

A practical manufacturing method of judging relative plasticity is to compute the ratio of initial yield point to the ultimate tensile strength as developed in the tensile test. Thus a General Motors research memo listed steel with a 0.51 **yield/tensile ratio** [22,000 lb/in² (152 MN/m²) yp/ 43,000 lb/in² (296 MN/m²) ultimate tensile strength] as being suitable for really severe draws of exposed parts. When the ratio reaches about 0.75, the steel should be used only for flat parts or possibly those with a bend of not more than 90°. The higher ratios obviously represent a narrowing range of workability or residual plasticity.

Advanced High-Strength Sheet Steels   With greater emphasis being placed on weight reduction, many new grades of steel sheets for automotive bodies have been developed. Interstitial free (IF) steels were developed for applications requiring high ductility, BH bake hardening (BH) steels for dent resistance, dual phase (DP), transformation-induced plasticity (TRIP), complex phase (CP), and ferritic-bainitic (FB) steels for high-strength applications such as body panels and pillars (Fig. 13.2.9). There is a tradeoff between formability and strength in these steels. The steel industry is trying to develop steel grades that would improve both these properties simultaneously. For example, DP500, DP600, DP750, and TRIP800 grades have maximum strengths of 600, 650, 825, and 1,000 MPa, respectively. This is much higher than the 500 MPa expected from HSLA360, the most common sheet steel for automotive bodies.

ROLLING OPERATIONS

Rolling of sheets, coils, bars, and shapes is a primary process using plastic ranges both above and below recrystallization to prepare metals for further working or for fabrication. Metal squeezed in the bite area of

Fig. 13.2.9   New high-strength sheet steels for automotive bodies. *(Source: Ultra Light Steel Autobody Consortium, USA.)*

the rolls moves out lengthwise with very little spreading in width. This compressive working above the yield point of the metal may be aided in some cases by maintaining a substantial tensile strain in the direction of rolling.

A cast or forged billet or slab is preheated for the preliminary breakdown stage of rolling, although considerable progress has been made in continuous casting, in which the molten metal is poured continuously into a mold in which the metal is cooled progressively until it solidifies (albeit still at high temperature), whence it is drawn off as a quasi-continuous billet and fed directly into the first roll pass of the rolling mill. The increased speed of operation and production and the increased efficiency of energy consumption are obvious. Most new mills, especially minimills, have incorporated continuous casting as the normal method of operation. A reversing hot mill may achieve 5,000 percent elongation of an original billet in a series of manual or automatic passes. Alternatively, the billet may pass progressively through, say, 10 hot mills in rapid succession. Such a production setup requires precise control so that each mill stand will run enough faster than the previous one to

Table 13.2.1a Manufacturing Properties of Steels and Copper-Based Alloys*
(Annealed condition)

Designation and composition, %	Liquidus/ solidus, °C	Hot-working Usual temp., °C	Flow stress,† MPa at °C	Flow stress,† MPa C	Flow stress,† MPa m	Workability‡	Cold-working Flow stress,§ MPa K	Cold-working Flow stress,§ MPa n	$\sigma_{0.2}$, MPa	TS, MPa	Elongation, %	q R.A., %	Annealing temp.,¶ °C
Steels:													
1008 (0.08 C), sheet		<1,250	1,000	100	0.1	A	600	0.25	180	320	40	70	850–900 (F)
1015 (0.15 C), bar		<1,250	800	150	0.1	A	620	0.18	300	450	35	70	850–900 (F)
1045 (0.45 C)		<1,150	1,000	120	0.1	A	950	0.12	410	700	22	45	790–870 (F)
			1,200	50	0.17								
			800	180	0.07								
~8620 (0.2 C, 1 Mn 0.4 Ni, 0.5 Cr, 0.4 Mo)			1,000	120	0.13	A			350	620	30	60	
			1,000	120	0.1								
D2 tool-steel (1.5 C, 12 Cr, 1 Mo)		900–1,080	1,000	190	0.13	B	1,300	0.3					880 (F)
H13 tool steel (0.4 C, 5 Cr 1.5 Mo, 1 V)			1,000	80	0.26	B							
302 ss (18 Cr, 9 Ni) (austenitic)	1,420/1,400	930–1,200	1,000	170	0.1	B	1,300	0.3	250	600	55	65	1,010–1,120 (Q)
410 ss (13 Cr) (martensitic)	1,530/1,480	870–1,150	1,000	140	0.08	C	960	0.1	280	520	30	65	650–800
Copper-base alloys:													
Cu (99.94%)	1,083/1,065	750–950	600	130	0.06	A	450	0.33	70	220	50	78	375–650
				(48)	(0.17)								
			900	41	0.2								
Cartridge brass (30 Zn)	955/915	725–850	600	100	0.24	A	500	0.41	100	310	65	75	425–750
			800	48	0.15								
Muntz metal (40 Zn)	905/900	625–800	600	38	0.3	A	800	0.5	120	380	45	70	425–600
			800	20	0.24								
Leaded brass (1 Pb, 39 Zn)	900/855	625–800	600	58	0.14	A	800	0.33	130	340	50	55	425–600
			800	14	0.20								
Phosphor bronze (5 Sn)	1,050/950		700	160	0.35	C	720	0.46	150	340	57		480–675
Aluminum bronze (5 Al)	1,060/1,050	815–870				A			170	400	65		425–850

* Compiled from various sources; most flow stress data from T. Altan and F. W. Boulger, *Trans. ASME, Ser. B, J. Eng. Ind.* **95**, 1973, p. 1009.
† Hot-working flow stress is for a strain of $\varepsilon = 0.5$. To convert to 1,000 lbf/in², divide calculated stresses by 7.
‡ Relative ratings, with A the best, corresponding to absence of cracking in hot rolling and forging.
§ Cold-working flow stress is for moderate strain rates, around $\varepsilon = 1\ s^{-1}$. To convert to 1,000 lbf/in², divide stresses by 7.
¶ Furnace cooling is indicated by F, quenching by Q.
Source: Adapted from John A. Schey, *Introduction to Manufacturing Processes*, McGraw-Hill, New York, 1987.

Table 13.2.1b Manufacturing Properties of Various Nonferrous Alloys[a]
(Annealed condition, except 6061-T6)

Designation and composition, %	Liquidus/ solidus, °C	Usual temp., °C	Hot-working — Flow stress,[b] MPa: at °C	C	m	Workability[f]	Cold-working — Flow stress,[c] MPa: K	n	$\sigma_{0.2}$, MPa	TS,[d] MPa	Elongation,[d] %	R.A.,[q] %	Annealing temp.,[e] °C
Light metals:													
1100 Al (99%)	657/643	250–550	300, 500	60, 14	0.08, 0.22	A	140	0.25	35	90	35		340
Mn alloy (1 Mn)	649/648	290–540	400	35	0.13	A	380	0.15	100	130	14		370
~2017 Al (3.5 Cu, 0.5 Mg, 0.5 Mn)	635/510	260–480	400, 500	90, 36	0.12, 0.12	B			100	180	20		415 (F)
5052 Al (2.5 Mg)	650/590	260–510	480	35	0.13	A	210	0.13	90	190	25		340
6061-0 (1 Mg, 0.6 Si, 0.3 Cu)	652/582	300–550	400, 500	50, 37	0.16, 0.17	A	220	0.16	55	125	25	65	415 (F)
6061-T6	NA[g]	NA	NA	NA	NA	NA	450	0.03	275	310	8		
~7075 Al (6 Zn, 2 Mg, 1 Cu)	640/475	260–455	450	40	0.13	B	400	0.17	100	230	16	45	415
Low-melting metals:													
Sn (99.8%)	232	100–200	100	10	0.1	A				15	45	100	150
Pb (99.7%)	327	20–200	75	260	0.1	A				12	35	100	20–200
Zn (0.08% Pb)	417	120–275	225	40	0.1	A				130/170	65/50		100
High-temperature alloys:													
Ni (99.4 Ni + Co)	1,446/1,435	650–1,250	1,150	~140	0.2	A			140	440	45	65	650–760
Hastelloy X (47 Ni, 9 Mo, 22 Cr, 18 Fe, 1.5 Co, 0.6 W)	1,290	980–1,200				C			360	770	42		1,175
Ti (99%)	1,660	750–1,000	600, 900	200, 38	0.11, 0.25	C			480	620	20		590–730
Ti-6 Al-4 V	1,660/1,600	790–1,000	600, 900	550, 140	0.08, 0.4	A			900	950	12		700–825
Zirconium	1,852	600–1,000	900	50	0.25	C			210	340	35		500–800
Uranium (99.8%)	1,132	~700	700	110	0.1	A			190	380	4	10	

[a] Empty spaces indicate unavailability of data. Compiled from various sources; most flow stress data from T. Altan and F. W. Boulger, *Trans. ASME, Ser. B. J. Eng. Ind.* **95**, 1973, p. 1009.
[b] Hot-working flow stress is for a strain of $\varepsilon = 0.5$. To convert to 1,000 lb/in², divide calculated stresses by 7.
[c] Cold-working flow stress is for moderate strain rates, around $\varepsilon = 1$ s⁻¹. To convert to 1,000 lb/in², divide stresses by 7.
[d] Where two values are given, the first is longitudinal, the second transverse.
[e] Furnace cooling is indicated by F.
[f] Relative ratings, with A the best, corresponding to absence of cracking in hot rolling and forging.
[g] NA = Not applicable to the −T6 temper.
SOURCE: Adapted from John A. Schey, *Introduction to Manufacturing Processes*, McGraw-Hill, New York, 1987.

make up for the elongation of the metal that has taken place. Hot-rolled steel may be sold for many purposes with the black mill scale on it. Alternatively, it may be acid-pickled to remove the scale and treated with oil or lime for corrosion protection. To prevent scale from forming in hot-rolling, a nonoxidizing atmosphere may be maintained in the mill area, a highly special plant design.

Pack rolling of a number of sheets stacked together provides means of retaining enough heat to hot-roll thin sheets, as for high-silicon electric steels.

Cold-rolling is practical in production of thin coil stock with the more ductile metals. The number of passes or amount of reduction between anneals is determined by the rate of work hardening of the metal. Successive stands of cold-rolling help to retain heat generated in working. Tension provided by mill reels and between stands helps to increase the practical reduction per step. Bright annealing in a controlled atmosphere avoids surface pockmarks, which are difficult to get out. For high-finish stock, the rolls must be maintained with equal finish.

Cold Rolling of Threads and Gears   Threaded parts, mostly fasteners, are cold-rolled with special tooling to impart a typical helical thread geometry to the part. The thread profile is most often a standard 60° vee, although thread profiles (i.e., Acme) for power threads are possible and have been produced. In one form, the tooling consists of two tapered reciprocating dies with the desired thread profile cut thereon. A blank of diameter smaller than the OD of the screw is positioned between the dies when they are at maximum separation. Then, as the dies reciprocate and decrease the gap between them, the blank is gripped, rolled, and plastically deformed to the desired screw profile. The indenting dies displace metal upward to form the upper part of the thread profile. There is no waste metal, and the fastener suffers no tears at the root. The resulting cold work and plastic displacement of metal results in a superior product. Subsequent heat treatment of the fastener may follow, depending on the strength properties desired for a particular application.

Thread rolling is also accomplished by using a nest of three profiled rotating rollers. In that case, the blank is fed axially when the rollers are at maximum gap and then is plastically deformed as the roll gap closes during rotation.

Gear profiles also can be rolled. The action is similar to that described for rolling threads, except that the dies or rollers are profiled to impart the desired involute gear profile. The gear blank is caught between the rollers or dies, and conjugate action ensues between the blank and tooling as the blank progresses through the operating cycle. In some applications, the rolled gear is produced slightly oversize to permit a finer finish by subsequent hobbing. The advantage lies in the reduction of metal cut by the hob, thereby increasing the production rate as well as maintaining the beneficial cold-worked properties imparted to the metal by cold rolling.

Thread and gear rolling enables high production rates; most threaded fasteners in production are of this type.

Protective coating is best exemplified by high-speed tinplate mills in which coil stock passes continuously through the necessary series of cleaning, plating, and heating steps. Zinc and other metals are also applied by plating but not on the same scale. **Clad** sheets (high-strength aluminum alloys with pure aluminum surface for protection against electrolytic oxidation) are produced by rolling together; an aluminum alloy billet is hot-rolled together with plates of pure aluminum above and below it through a series of reducing passes, with precautions to ensure clean adhesion.

On the other hand, prevention of adhesion, as by a separating film, is essential in the final stages of **foil rolling,** where two coils may have to be rolled together. Such foil may then be **laminated** with suitable adhesive to paper backing materials for wrapping purposes. (See also Sec. 6.)

Shape-rolling of structural shapes and rails is usually a hot operation with roll-pass contours designed to distribute the displacement of metal in a series of steps dictated largely by experience. **Contour rolling** of relatively thin stock into tubular, channel, interlocking, or varied special cross sections is usually done cold in a series of roll stands for lengthwise bending and setting operations. There is also a wide range of simple bead-rolling, flange-rolling, and seam-rolling operations in relatively thin materials, especially in connection with the production of barrels, drums, and other containers.

Oscillating or **segmental rolling** probably developed first in the manually fed contour rolling of agricultural implements. In some cases, the suitably contoured pair of roll inserts or roll dies oscillates before the operator, to form hot or cold metal. In other cases, the rolls rotate constantly, toward the operator. The working contour takes only a portion of the circumference, so that a substantial clearance angle leaves a space between the rolls. This permits the operator to insert the blank to the tong grip between the rolls and against a fixed gage at the back. Then, as rotation continues, the roll dies grip and form the blank, moving it back to the operator. This process is sometimes automated; such units as **tube-reducing mills** oscillate an entire rolling-mill assembly and feed the work over a mandrel and into the contoured rolls, advancing it and possibly turning it between reciprocating strokes of the roll stands for cold reduction, improved concentricity, and, if desired, the tapering or forming of special sections.

Spinning operations (Fig. 13.2.10) apply a rolling-point pressure to relatively limited-lot production of cup, cone, and disk shapes, from floor lamps and TV tube housings to car wheels and large tank ends. Where substantial metal thickness is required, powerful machines and hydraulic servo controls may be used. Some of the large, heavy sections and difficult metals are spun hot.

Fig. 13.2.10   Spinning operations.

Rolling operations are distinguished by the relatively rapid and continuous application of working pressure along a limited line of contact. In determining the working area, consider the lineal dimension (width of coil), the bite (reduction in thickness), and the roll-face deflection, which tends to increase the contact area. Approximations of rolling-mill load and power requirements have been worked out in literature of the AISE and ASME.

SHEARING

The shearing group of operations includes such **power press operations** as blanking, piercing, perforating, shaving, broaching, trimming, slitting, and parting. Shearing operations traverse the entire plastic range of metals to the point of failure.

The maximum pressure P, in pounds, required in shearing operations is given by the equation $P = \pi DtS = Lts,$ where s is the resistance of the material to shearing, lb/in^2; t is the thickness of the material, in; L is the length of cut, in, which is the circumference of a round blank πD or the periphery of a rectangular or irregular blank. Approximate values of s are given in Table 13.2.2.

Shear (Fig. 13.2.11) is the advance of that portion of the shearing edge which first comes in contact with the material to be sheared over the last portion to establish contact, measured in the direction of motion. It should be a function of the thickness t. Shear reduces the maximum pressure because, instead of shearing the whole length of cut

Table 13.2.2 Approximate Resistance to Shearing in Dies

Material	Annealed state		Hard, cold-worked	
	Resistance to shearing, lb/in²*	Penetration to fracture, percent	Resistance to shearing, lb/in²*	Penetration to fracture, percent
Lead	3,500	50	Anneals at room temperature	
Tin	5,000	40	Anneals at room temperature	
Aluminum 2S, 3S	9,000–11,000	60	13,000–16,000	30
Aluminum 52S, 61S, 62S	12,000–18,000	...	24,000–30,000	
Aluminum 75S	22,000	...	46,000	
Zinc	14,000	50	19,000	25
Copper	22,000	55	28,000	30
Brass	33,000–35,000	50–55	52,000	25–30
Bronze 90–10	...	...	40,000	
Tobin bronze	36,000	25	42,000	
Steel 0.10C	35,000	50	43,000	38
Steel 0.20C	44,000	40	55,000	28
Steel 0.30C	52,000	33	67,000	22
Steel 0.40C	62,000	27	78,000	17
Steel 0.60C	80,000	20	102,000	9
Steel 0.80C	97,000	15	127,000	5
Steel 1.00C	115,000	10	150,000	2
Stainless steel	57,000	39		
Silicon steel	65,000	30		
Nickel	35,000	55		

*1,000 lb/in² = 6.895 MN/m².

NOTE: Available test data do not agree closely. The above table is subject to verification with closer control of metal analysis, rolling and annealing conditions, and die clearances. In dinking dies, steel-rule dies, hollow cutters, etc., cutting-edge resistance is substantially independent of thickness: cotton glove cloth (stack, 2 or 3 in thick), 240 lb/in; kraft paper (stack tested, 0.20 in thick), 385 lb/in, celluloid [1½₂ in thick, warmed in water to 120 to 150°F (49 to 66°C)], 300 lb/in.

at once, the shearing action takes place progressively, shearing at only a portion of the length at any instant. The maximum pressure for any case where the shear is equal to or greater than t is given by $P_{max} = P_{av} t/shear$, where P_{av} is the average value of the pressure on a punch, with shear $= t$, from the time it strikes the metal to the time it leaves.

Distortion results from shearing at an angle (Fig. 13.2.11) and accordingly, in blanking, where the blank should be flat, the punch should be flat, and the shear should be on the die. Conversely, in hole punching, where the scrap is punched out, the die should be flat and the shear should be on the punch. Where there are a number of punches, the effect of shear may be obtained by stepping the punches.

Crowding results during the plastic deformation period, before the fracture occurs, in any shearing operation. Accordingly, when small delicate punches are close to a large punch, they should be stepped shorter than the large punch by at least a third of the metal thickness.

Clearance between the punch and die is required for a clean cut and durability. An old rule of thumb places the clearance all around the punch at 8 to 10 percent of the metal thickness for soft metal and up to 12 percent for hard metal. Actually, hard metal requires less clearance for a clean fracture than soft, but it will stand more. In some cases, with delicate punches, clearance is as high as 25 percent. Where the hole diameter is important, the punch should be the desired diameter and the clearance should be added to the die diameter. Conversely, where the

blank size is important, the die and blank dimensions are the same and clearance is deducted from the punch dimensions.

The **work per stroke** may be approximated as the product of the maximum pressure and the metal thickness, although it is only about 20 to 80 percent of that product, depending upon the clearance and ductility of the metal. Reducing the clearance causes secondary fractures and increases the work done. With sufficient clearance for a clean fracture, the work is a little less than the product of the maximum pressure, the metal thickness, and the percentage reduction in thickness at which the fracture occurs. Approximate values for this are given in Table 13.2.2. The **power required** may be obtained from the work per stroke plus a 10 to 20 percent friction allowance.

Shaving   A sheared edge may be squared up roughly by shaving once, allowing for the shaving of mild sheet steel about 10 percent of the metal thickness. This allowance may be increased somewhat for thinner material and should be decreased for thicker and softer material. In making several cuts, the amount removed is reduced each time. For extremely fine finish a round-edged burnishing die or punch, say 0.001 or 0.0015 in tight, may be used. Aluminum parts may be blanked (as for impact extrusion) with a fine finish by putting a 30° bevel, approx one-third the metal thickness on the die opening, with a near metal-to-metal fit on the punch and die, and pushing the blank through the highly polished die.

Squaring shears for sheet or plate may have their blades arranged in either of the ways shown in Fig. 13.2.12. The square-edged blades in Fig. 13.2.12a may be reversed to give four cutting edges before they are reground. Single-edged blades, as shown in Fig. 13.2.12b, may have a clearance angle on the side where the blades pass, to reduce the working friction. They may also be ground at an angle or rake, on the face which comes in contact with the metal. This reduces the bending and consequent distortion at the edge. Either type of blade distorts also in the other direction owing to the angle of shear on the length of the blades (see Fig. 13.2.11).

Circular cutters for slitters and circle shears may also be square-edged (on most slitters) or knife-edged (on circle shears). According to one rule, their diameter should be not less than 70 times the metal thickness.

Knife-edge hollow cutters working against end-grain maple blocks represent an old practice in cutting leather, rubber, and cloth in multiple thicknesses. **Steel-rule dies,** made up of knife-edge hard-steel strip economically mounted against a steel plate in a wood matrix with rubber

(a)   (b)   (c)

α = shearing angle
$t = h$ = thickness of sheet
L = shearing length

Fig. 13.2.11 Shearing forces can be reduced by providing a rake or shear on (a) the blades in a guillotine, (b) the die in blanking, (c) the punch in piercing. (*J. Schey, "Introduction to Manufacturing Processes," McGraw-Hill, 1987.*)

Fig. 13.2.12 Squaring shears.

strippers and cutting against hard saw-steel plates, extend the practice to corrugated-carton production and even some limited-lot metal cutting.

Higher precision is often required in **finish shearing** operations on sheet material. For ease of subsequent operations and assembly, the cut edges should be clean (acceptable burr heights and good surface finish) and perpendicular to the sheet surface. The processes include precision or fine blanking, negative clearance blanking, counterblanking, and shaving, as shown in Fig. 13.2.13. By these methods either the plastic behavior of material is suppressed or the plastically deformed material is removed.

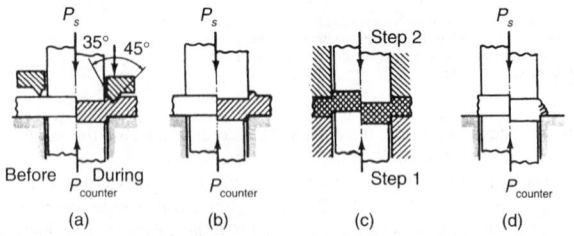

Fig. 13.2.13 Parts with finished edges can be produced by (*a*) precision blanking, (*b*) negative-clearance blanking, (*c*) counterblanking, (*d*) shaving a previously sheared part. (*J. Schey, "Introduction to Manufacturing Processes," McGraw-Hill, 1987.*)

BENDING

The bending group of operations is performed in **presses** (variety), **brakes** (metal furniture, cornices, roofing), **bulldozers** (heavy rolled sections), **multiple-roll forming machines** (molding, etc.), **draw benches** (door trim, molding, etc.), **forming rolls** (cylinders), and **roll straighteners** (strips, sheets, plates).

Spring back, due to the elasticity of the metal and amount of the bend, may be compensated for by overbending or largely prevented by striking the metal at the radius with a **coining** (i.e., squeezing, as in production of coins) pressure sufficient to set up compressive stresses to counterbalance surface tensile stresses. A very narrow bead may be used to localize the pinch where needed and minimize danger to the press on a large area. Under such conditions, good sharp bends in V dies have been obtained with two to four times the pressure required to shear the metal across the same section.

These are illustrated in Fig. 13.2.14, where P_b is the **bending load** on the press brake, W_b is the width of the die support, and $P_{counter}$ is the counterload. The bending load can be obtained from

$$P_b = wt^2(\text{UTS})/W_b$$

where t and w are the sheet thickness and width, respectively, and UTS is the ultimate tensile strength of the sheet material.

Bending Allowance The thickness of the metal over a small radius or a sharp corner is 10 or 15 percent less than before bending because the metal moves more easily in tension than in compression. For the same reason the neutral axis of the metal moves in toward the center of the corner radius. Therefore, in figuring the length of blank L to be allowed for the bend up to an inside radius r of two or three times the

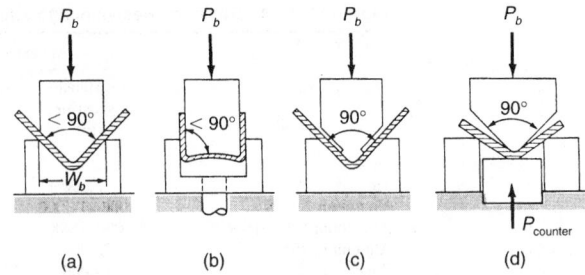

Fig. 13.2.14 Springback may be neutralized or eliminated by (*a*), (*b*) over-bending; (*c*) plastic deformation at the end of the stroke; (*d*) subjecting the bend zone to compression during bending. [*Part (d) after V. Kupka, T. Nakagawa, and H. Tyamoto, CIRP 22:73–74 (1973).*] (*Source: J. Schey, "Introduction to Manufacturing Processes," McGraw-Hill, 1987.*)

metal thickness t, the length may be figured closely as along a neutral line at $0.4t$ out from the inside radius. Thus, with reference to Fig. 13.2.15, for any angle a in deg and other dimensions in inches, $L = (r + 0.4t)2\pi a/360 = (r + 0.4t)a/57.3$.

The factor $0.4t$, which locates the neutral axis, is subject to some variation (say 0.35 to $0.45t$) according to radius, condition of metal, and angle. In figuring allowances for sharp bends, note that the metal builds up on the compression side of the corner. Therefore, in locating the neutral axis, consider an inside radius r of about $0.05t$ as a minimum.

Roll straighteners work on the principle of bending the metal beyond its elastic limit in one direction over rolls small enough in diameter, in proportion to the metal thickness, to give a permanent set, and then

Fig. 13.2.15 Bending allowance.

taking that bend out by repeatedly reversing it in direction and reducing it in amount. Metal is also straightened by gripping and stretching it beyond its elastic limit and by hammering; the results of the latter operation depend entirely upon the skill of the operator.

For approximating **bending loads,** the beam formula may be used but must be very materially increased because of the short spans. Thus, for a span of about 4 times the depth of section, the bending load is about 50 percent more than that indicated by the beam formula. It increases from this to nearly the shearing resistance of the section where some **ironing** (i.e., the thinning of the metal when clearance between punch and die is less than the metal thickness) occurs. Where hit-home dies do a little coining to "set" the bend, the pressure may range from two or three times the shearing resistance, and with striking beads and proper care, up to very much higher figures.

The work to roll-bend a sheet or plate t in thick with a volume of V in³, into curved shape of radius r in, is given as $W = CS(t/r)V/48$ ft·lb, in which S is the tensile strength and C is an experience factor between 1.4 and 2.

The **equipment for bending** consists of mechanical presses for short bends, press brakes (mechanical and hydraulic) for long bends, and roll formers for continuous production of profiles. The bends are achieved by bending between tools, wiping motion around a die corner, or bending between a set of rolls. These bending actions are illustrated in Fig. 13.2.16. Complex shapes are formed by repeated bending in simple tooling or by passing the sheet through a series of rolls which progressively bend it into the desired profile. Roll forming is economical for continuous forming for large volume production. Press brakes can be computer-controlled with synchronized feeding and bending as well as spring-back compression.

DRAWING

Drawing includes operations in which metal is pulled or drawn, in suitable containing tools, from flat sheets or blanks into cylindrical cups or rectangular or irregular shapes, deep or shallow. It also includes reducing

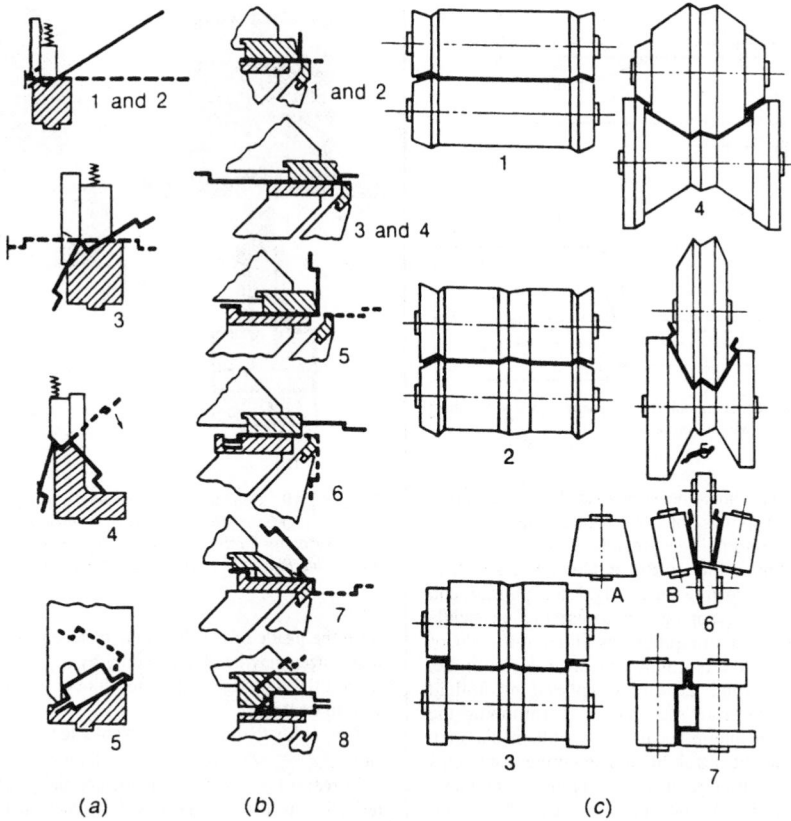

Fig. 13.2.16 Complex profiles formed by a sequence of operations on (*a*) press brakes, (*b*) wiping dies, (*c*) profile rolling. *(After Oehler; Biegen, Hanser Verlag, Munich, 1963.) (Source: J. Schey, "Introduction to Manufacturing Processes," McGraw-Hill, 1987.)*

operations on shells, tube, wire, etc., in which the metal being drawn is pulled through dies to reduce the diameter or size of the shape. All drawing and reducing operations, by an applied tensile stress in the material, set up circumferential compressive stresses which crowd the metal into the desired shape. The relation of the shape or diameter before drawing to the shape or diameter after drawing determines the magnitude of the stresses. Excessive draws or reductions cause thinning or tearing out near the bottom of a shell. Severe cold-drawing operations require very ductile material and, in consequence of the amount of plastic deformation, harden the metal rapidly and necessitate annealing to restore the ductility for further working.

The **pressure used in drawing** is limited to the load to shear the bottom of the shell out, except in cases where the side wall is ironed thinner, when wall friction makes somewhat higher loads possible. It is less than this limit for round shells which are shorter than the limiting height and also for rectangular shells. Drawing occurs only around the corner radii of rectangular shells, the straight sides being merely free bending.

A holding pressure is required in most initial drawing and some redrawing, to prevent the formation of wrinkles due to the circumferential compressive stresses. Where the blank is relatively thin compared with its diameter, the blank-holding pressure for round work is likely to vary up to about one-third of the drawing pressure. For material heavy enough to provide sufficient internal resistance to wrinkling, no pressure is required. Where a drawn shape is very shallow, the metal must be stretched beyond its elastic limit in order to hold its shape, making it necessary to use higher blank-holding loads, often in excess of the drawing pressure. To grip the edges sufficiently to do this, it is often advisable to use **draw beads** on the blank-holding surfaces if sufficient pressure is available to form these beads.

In sheet/deep-drawing practice, the punch force P can be approximated by

$$P = \pi D_p t_0 (\text{UTS})(D_0/D_p - 0.7)$$

where t_0 is the blank thickness and D_0 and D_p are the diameters of the blank and the punch.

The blank holder pressure for avoiding defects such as wrinkling of bottom/wall tear-out is kept at 0.7 to 1.0 percent of the sum of the yield and the UTS of the material. Punch/die clearances are chosen to be 7 to 14 percent greater than the sheet blank thickness t_0. The die corner radii are chosen to avoid fracture at the die corner from puckering or wrinkling. Recommended values of D_0/t_0 for deep-drawn cups are 6 to 15 for cups without flange and 12 to 30 for cups with flange. These values will be smaller for relatively thick sheets and larger for very thin blank thickness. For deeper-drawn cups, they may be redrawn or reverse-drawn, the latter process taking advantage of strain softening on reverse drawing. When the material has marked strain-hardening propensities, it may be necessary to subject it to an intermediate annealing process to restore some of its ductility and to allow progression of the draws to proceed.

Some shells, which are very thick or very shallow compared with their diameter, do not require a blank holder. Blank-holding pressure may be obtained through toggle, crank, or cam mechanisms built into the machine or by means of air cylinders, spring-pressure attachments, or rubber bumpers under the bolster plate. The length of car springs should be about 18 in/in (18 cm/cm) of draw to give a fairly uniform drawing pressure and long life. The use of car springs has been largely superseded by hydraulic and pneumatic cushions. Rubber bumpers may be figured on a basis of about 7.5 lb/in² (50 kN/m²) of cross-sectional area per 1 percent

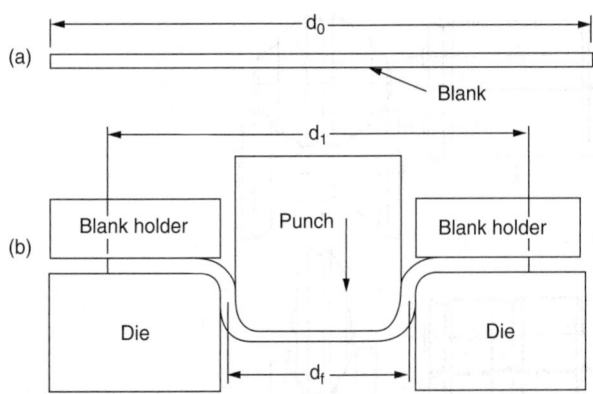

(a)

(b)

Fig. 13.2.17 Schematic of the conventional deep drawing process for sheet metal parts. (*a*) Initial blank, (*b*) drawing in process.

of compression. In practice they should never be loaded beyond 20 percent compression, and as with springs, the greater the length relative to the working stroke, the more uniform is the pressure.

Deep Drawing and Hydroforming of Sheet Metal Parts Sheet metal parts are conventionally deep drawn using rigid steel punches and dies (Fig. 13.2.17). An alternative approach is to use flexible media (Fig. 13.2.18) such as water, gas, or rubber as either the male or female die, and to perform the hydroforming process in a closed die. In hydromechanical deep drawing, the die is replaced by a fluid (Fig. 13.2.18) while in high-pressure sheet metal hydroforming the punch is replaced by the fluid medium. The use of flexible media often permits greater drawability and the possibility of combining many steps in one operation, such as permitting joining and trimming simultaneously with forming (Fig. 13.2.19). Complex parts can be made in a single step by using thin sheets, thus reducing the cost and weight of the shell structure.

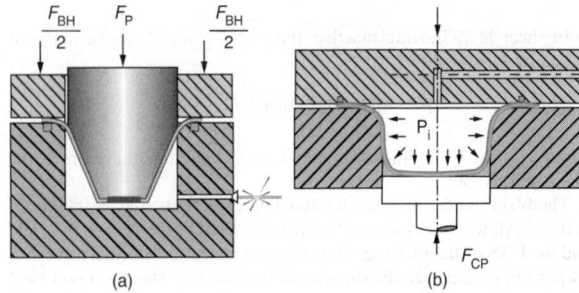

(a)   (b)

Fig. 13.2.18 Forming by flexible media. (*a*) Hydromechanical deep drawing (*Source: K. Seigert and M. Aust, Hydromechanical Deep-Drawing, Production Engineering, VII/2, Annals of the German Academic Society for Production Engineering, pp. 7–12.*) (*b*) High-pressure hydroforming. *M. Kliener, W. Homberg, and A. Brosius, Processes and Control of Sheet Metal Hydroforming. International Conference on Advanced Technology of Plasticity, Germany, 2, 1999, pp. 1243–1252.*)

Dimensions of Drawn Shells The smallest and deepest round shell that can be drawn from any given blank has a diameter d of 65 to 50 percent of the blank diameter D. The height of these shells is $h = 0.35d$ to $0.75d$, approximately. Higher shells have occasionally been drawn with ductile material and large punch and die radii. Greater thickness of material relative to the diameter also favors deeper drawing.

The area of the bottom and of the side walls added together may be considered as equal to the area of the blank for approximations. If the punch radius is appreciable, the area of a neutral surface about $0.4t$ out

(a)   (b)

Fig. 13.2.19 Combined hydroforming, joining, and trimming of sheet metal parts. (*a*) In process; (*b*) typical parts. (*Source: P. Hein and M. Geiger, Advanced Process Control Strategies for the Hydroforming of Sheet Metal Pairs, Int. Conf. Adv. Techn. Plasticity, Germany, 2, 1999, pp. 1267–1272.*)

from the inside of the shell may be taken for approximations. Accurate blank sizes may be obtained only by trial, as the metal tends to thicken toward the top edge and to get thinner toward the bottom of the shell wall in drawing.

Approximate diameters of blanks for shells are given by the expression $\sqrt{d^2 + 4dh}$, where d is the diameter and h the height of the shell.

In **redrawing** to smaller diameters and greater depths the amount of reduction is usually decreased in each step. Thus in double-action redrawing with a blank holder, the successive reductions may be 25, 20, 16, 13, 10 percent, etc. This progression is modified by the relative thickness and ductility of the metal. Single-action redrawing without a blank holder necessitates smaller steps and depends upon the shape of the dies and punches. The steps may be 19, 15, 12, 10 percent, etc. Smaller reductions per operation seem to make possible greater total reductions between annealings.

Rectangular shells may be drawn to a depth of 4 to 6 times their corner radius. It is sometimes desirable, where the sheet is relatively thin, to use draw beads at the corners of the shell or near reverse bends in irregular shapes to hold back the metal and assist in the prevention of wrinkles.

Work in drawing is approximately the product of the length of the draw, and the maximum punch pressure, as the load rises quickly to the peak, remains fairly constant, and drops off sharply at the end of the draw unless there is stamping or wall friction. To this, add the work of blank holding which, in the case of cam and toggle pressure, is the product of the blank-holding pressure and the spring of the press at the pressure (which is small). For single-action presses with spring, rubber, or air-drawing attachments it is the product of the average blank-holding pressure and the length of draw.

Rubber-die forming, especially of the softer metals and for limited-lot production, uses one relatively hard member of metal, plaster, or plastic with a hard powder filler to control contour. The mating member may be a rubber or neoprene mattress or a hydraulically inflatable bag, confined and at 3,000 to 7,000 lb/in^2 (20 to 48 MN/m^2). Babbitt, oil, and water have also been used directly as the mobile member. A large hydraulic press is used, often with a sliding table or tables, and even static containers with adequate pumping systems.

Warm Forming of Aluminum and Magnesium Sheets Aluminum and magnesium are used to decrease the weight of automotive and aerospace parts. Aluminum and magnesium exhibit increased ductility at elevated temperatures (Fig. 13.2.20). Magnesium does have many limitations, but its use for structural parts is growing.

(a)

(b)

Fig. 13.2.20 Drawing of magnesium sheets at elevated temperatures. (a) Flow stress of magnesium AZ61. [*Chabbi, et al., 2000, Hot and Cold Forming Behavior of Magnesium Alloys AZ31 and AZ61, in K. U. Kainer (ed.), "Magnesium Alloys and Their Applications," Wiley-VCH, D, pp. 621–627.*] (b) Drawn AZ31B cups at room temperature (left) and at 230°C (right). (*S. Novotni, Innenhochdruk-Umformen von Belchen aus Aluminium-und Magnesiumlegierungen bei erhohter Temperatur, Ph.D. thesis, Elrangen University, Germany.*)

Tailor Welded Blanks (TWB) in Forming With the demand for reducing both automotive structure weight and manufacturing costs many new processes are being employed:

1. Different parts are formed separately and then joined by laser. Part forming is independently optimized, followed by trimming and weld assembly. Forming is easier but welding along curved lines is more complex (Fig. 13.2.21a).

2. The blanks are welded, and then the panel is formed in one die. Welding is simpler, but forming is considerably more complex. Dimensional tolerances are better controlled (Fig. 13.2.21b).

3. Blanks of varying thickness are tailor-welded using laser techniques to create a single blank that subsequently is formed into the required geometry (Fig. 13.2.22). The forming process of TWBs is very complex as the blank areas with different thicknesses flow differently during the drawing operation.

4. Friction stir welding is used to create a single blank made of different materials or different sheet thickness. This composite blank is then drawn to final geometry (Fig. 13.2.23).

Hot drawing above the recrystallization range applies single- and double-action drawing principles. For light gages of plastics, paper, and hexagonal-lattice metals such as magnesium, dies and punches may be heated by gas or electricity. For thick steel plate and heat-treatable alloys, the mass of the blank may be sufficient to hold the heat required.

High-Pressure Hydroforming of Tubes Tubes formed to various cross sections and bent to various shapes are widely used in automotive frames. There are a number of variants of this process (Fig. 13.2.24), including forming under (1) tensile and compressive conditions,

Fig. 13.2.21 (a) Many parts are formed separately and joined to form side panels and pillars; (b) single blank is trimmed and then formed. (*Source: Ruch et al., Grobserienfertigung von Aluminumkarosserien, in K. Siegert, "Neuere Entwicklungen inder Blechumformung," MATINFO, Frankfurt, D, 2000*).

Fig. 13.2.22 Use of tailor-welded blanks in automotive side panels using different materials of unequal thickness. (*Source: Ultra Light Steel Autobody Consortium, USA.*)

Fig. 13.2.23 Use of friction stir welding to create a blank. (*Source: The Welding Institute.*)

(2) bending conditions, and (3) shear conditions. Each variant is intended to impart a particular deformation to the tube by a predetermined motion of the tools. Motions include axial compression due to tool motion and circumferential expansion due to internal pressure. In high-pressure forming, pressurized force can reach 35,000 tons. Lead times can be very high due to the slow pressurization and depressurization required for each forming cycle. These high pressures lead to metal

Fig. 13.2.24 Process and tooling variations in the hydroforming of tubes with high-pressure media. *(Source: D. Schmoeckel, C. Hielscher, and R. Huber, Metal Forming of Tubes and Sheets with Liquid and Other Flexible Media, CIRP Annals, 48, no. 2, 1999, pp. 497–513.)*

flow-related defects, such as buckling, wrinkling, and bursting, and a safe working window has to be determined for success (Fig. 13.2.25).

Incremental Forming The incremental forming process employs local forming of the workpiece and then rotating the workpiece or the tools to form the entire surface. Since the contact area between the tools and the workpiece is small, the force required is small and the friction conditions are favorable. Consequently, high deformation levels are achieved that are not possible by conventional means. These processes include flow forming, radial forming, rotary forming, orbital forming, spinning, shear forming, and incremental sheet forming (single-point or two-point contact). Incremental sheet forming uses sheet blanks held in a fixture while rotating tools incrementally stretch the blank to the required shape (Fig. 13.2.26). In the flow forming process for thin axisymmetric parts, a roller deforms the metal and changes the thickness of the formed piece (Fig. 13.2.27).

Lubricants for Presswork Many jobs may be done dry, but better results and longer life of dies are obtained by the use of a lubricant. Lard or sperm oil is used when punching iron, steel, or copper. Petroleum jelly is used for drawing aluminum. A soap solution is commonly used for drawing brass, copper, or steel. One manufacturer uses 90 percent mineral oil, 5 percent rosin, and 5 percent oleic acid for light work and

Buckling   Wrinkling   Bursting

Typical failures in internal high-pressure forming

Work Diagram

Fig. 13.2.25 Typical failures in tube hydroforming and the safe working window for the process. *(Source: D. Schmoeckel, C. Hielscher, and R. Huber, Metal Forming of Tubes and Sheets with Liquid and Other Flexible Media, CIRP Annals, 48, no. 2, 1999, pp. 497–513.)*

Fig. 13.2.26 Incremental forming of thin sheet parts. *(a)* Process principle; *(b)* increased strains possible. *(Source: J. Jesweit and E. Hagan, Rapid Prototyping of a Headlight with Sheet Metal. Trans. NAMRI, Society of Manufacturing Engineers, May 2002.)*

Fig. 13.2.27 Flow forming a rim for an automotive wheel.

an emulsion of a mineral oil, degras, and a pigment consisting of chalk, sulfur, or lithopone for heavy work. (See also Sec. 6.)

For heavy drawing operations and extrusion, steels may have a zinc phosphate coating bonded on, and a zinc or sodium stearate bonded to that, to withstand pressures over 300,000 lb/in² (2,070 MN/m²). An anodized coating for aluminum may be used as a host for the lubricant. It is reported that such a chemical treatment plus a lacquer or plastic coating and a lubricant is effective for severe ironing operations.

The problem is to prevent local pressure welding from starting as galling or pickup, with resulting scratching, by maintaining a fluid film separation between metal surfaces. At moderate pressures, almost any viscous liquid lubricant will do the job. Rust protection and easy removal of the lubricant are often major factors in the choice.

Lubricants for metalworking are often classified based on their interface friction coefficient, which depends on the workpiece material and

the lubricant being used. The **interface friction coefficient** μ is often defined as the ratio of the friction force to the normal force at the interface. Typical values of μ for different workpiece-lubricant pairs are included in Table 13.2.3.

Shock-wave forming For short runs, it is applied in several ways. **Explosive forming,** especially for large-area drawn or formed shapes, usually requires one metal contour-control die immersed in a large container of fluid, or even in a lake or pond. Explosives manufacturers have developed means of computing the charge and the distance that it should be suspended above the blank to be formed. The space back of the blank in the die has to be evacuated. A blank-holding ring to minimize wrinkle formation in the flange area is bolted very tightly to the die, with an O-ring seal to prevent leakage.

Electrohydraulic forming is similar to explosive forming except that the shock wave is imparted electrically from a large battery of capacitors. **Magnetic forming** uses the same source of power but does not require a fluid medium. A flexible pancake coil delivers the magnetic shock pulse.

Electromagnetic Forming (EMF) In this process the energy of a pulsed magnetic field is used with a contact-free tool to join metals with good electrical conductivity and magnetic permeability, such as aluminum. The rapid discharge of a high-voltage capacitor through a coil in the tool generates an intense magnetic field, inducing eddy currents in the workpiece. The magnetic forces acting between the tool and the workpiece cause movement of the workpiece, thus permanently changing its shape (Fig. 13.2.28). This process can be used with tubular workpieces to join two tubes or to change tube shape.

BULK FORMING

The squeezing group of operations are those in which the metal is worked in compression. Resultant tensile strains occur, however; in cases where the metal is thin compared with its area and there is an

Table 13.2.3   Typical Lubricants* and Friction Coefficients in Plastic Deformation

Workplace material	Working	Forging Lubricant	μ	Extrusion† lubricant	Wire drawing Lubricant	μ	Rolling Lubricant	μ	Sheet metalworking Lubricant	μ
Sn, Pb, Zn alloys		FO-MO	0.05	FO or soap	FO	0.05	FA-MO or MO-EM	0.05 0.1	FO-MO	0.05
Mg alloys	Hot or warm	GR and/or MoS₂	0.1–0.2	None			MO-FA-EM	0.2	GR in MO or dry soap	0.1–0.2
Al alloys	Hot	GR or MoS₂	0.1–0.2	None			MO-FA-EM	0.2		
	Cold	FA-MO or dry soap	0.1 0.1	Lanolin or soap on PH	FA-MO-EM, FA-MO	0.1 0.03	1–5% FA in MO (1–3)	0.03	FO, lanolin, or FA-MO-EM	0.05–0.1
Cu alloys	Hot	GR	0.1–0.2	None (or GR)			MO-EM	0.2		
	Cold	Dry-soap, wax, of tallow	0.1	Dry soap or wax or tallow	FO-soap-EM, MO	0.1	MO-EM	0.1	FO-soap-EM or FO-soap	0.05–0.1
Steels	Hot	GR	0.1–0.2	GL (100–300), GR			None or GR-EM	ST‡ 0.2	GR	0.2
	Cold	EP-MO or soap, on PH	0.1 0.05	Soap on PH	Dry soap or soap on PH	0.05 0.03	10% FO-EM	0.05	EP-MO, EM, soap, or polymer	0.05–0.1
Stainless steel, Ni and alloys	Hot	GR	0.1–0.2	GL (100–300)			None	ST‡	GR	0.2
	Cold	CL-MO or soap on PH	0.1 0.05	CL-MO or soap on PH	Soap on PH or CL-MO	0.03 0.05	FO-CL-EM or CL-MO	0.1 0.05	CL-MO, soap, or polymer	0.1
Ti alloys	Hot	GL or GR	0.2	GL (100–300)					GR, GL	0.2
	Cold	Soap or MO	0.1	Soap on PH	Polymer	0.1	MO	0.1	Soap or polymer	0.1

* Some more frequently used lubricants (hyphenation indicates that several components are used in the lubricant):
 CL = chlorinated paraffin
 EM = emulsion; listed lubricating ingredients are finely distributed in water
 EP = "extreme-pressure" compounds (containing S, Cl, and P)
 FA = fatty acids and alcohols, e.g., oleic acid, stearic acid, stearyl alcohol
 FO = fatty oils, e.g., palm oil and synthetic palm oil
 GL = glass (viscosity at working temperature in units of poise)
 GR = graphite; usually in a water-base carrier fluid
 MO = mineral oil (viscosity in parentheses, in units of centipoise at 40°C).
 PH = phosphate (or similar) surface conversion, providing keying of lubricant
† Friction coefficients are misleading for extrusion and therefore are not quoted here.
‡ The symbol ST indicates sticking friction.
SOURCE: John A. Schey, *Introduction to Manufacturing Processes*, McGraw-Hill, New York, 1987.

Joining inside   Joining outside   Joining outside
                 (one side)   (two sides)

Fig. 13.2.28 Electromagnetic forming and joining of tubes. *(Source: C. Beerwald, A. Brosius, and M. Kliener, Determination of Flow Stress at Very High Strain Rates by a Combination of Magnetic Forming and FEM Calculation, International Workshop on Friction and Flow Stress in Cutting and Forming, Paris, pp. 175–182.)*

appreciable movement of the metal, there results a pyramiding of pressure toward the center of the die which may prove serious. The metal is incompressible (beyond about 1 percent), and consequently, to reduce the thickness of any volume of metal in the center of the blank, its area must be increased, which involves spreading or stretching all the metal around it. The surrounding metal acts like shrunk bands and offers a resistance increasing toward the center, and often many times the compressive resistance of the material.

Squeezing operations and particularly the squeezing of steel are practically the severest of all press operations. They may be divided into four general classifications according to severity, although in every group there will be found examples of working to the limit of what the die steels will stand, which may be taken at about 100 tons/in². The severer operations, such as cold bottom extrusion and wall extrusion, are limited to the softer metals. Squeezing operations ordinarily require pressure through a very short distance, the pressure starting at the compressive yield strength of the material over the surface being squeezed and rising to a maximum at bottom stroke. This maximum is greatest when the metal is thin compared with its area or when the die is entirely closed as for coining. Care must be taken, on all squeezing operations, in the setting of presses and avoidance of double blanks or extra-heavy blanks as the presses must be stiff. In squeezing solidly across bottom center the mechanical advantage is such that a small difference in thickness or setup can make a very large difference in pressure exerted. For this reason high-speed self-contained hydraulic presses, with automatic pressure-control and size blocks, are now finding favor for some of this work.

Sizing, or the flattening or surfacing of parts of forgings or castings, is usually the least severe of the squeezing group. Tolerances are ordinarily closer than for the milling operations which are supplanted. When extremely close tolerances are required, say plus or minus one-thousandth of an inch (0.025 mm), arrange substantial size blocks to take half or two-thirds of the total load. These take up uniformly the bearing-oil films and any slight deflection of the bed and bolster and minimize the error in springback due to variation in thickness, hardness, and area of the rough forging or casting. The usual amount left for squeezing is ¹⁄₃₂ to ¹⁄₁₆ in (0.8 to 1.6 mm). Presses may be selected for this service on a basis of 60 to 80 tons/in² (830 to 1,100 MN/m²), although 100 tons/in² (1,380 MN/m²) is more often used in the automobile trade for reserve capacities. When figuring from experimental results obtained in testing machines, the recorded loads are usually doubled in selecting a press, in order to allow for the difference among the speed of the machines, the positive action, and a safety margin.

Swaging or **cold forging** involves squeezing of the blank to an appreciably different shape. Success in performing such operations on steel usually depends upon squeezing a relatively small area with freedom to flow without restraint. Dies for this work must usually be substantially

         (a)   (b)

Fig. 13.2.29 Comparison of (*a*) machined and (*b*) cold-forged gearbox endpiece. *(Source: D. Landgrebe, Precision Forming and Machining, "Recent Developments in Metalforming Technology," ERC/NSM, The Ohio State University, Columbus, 2002.)*

backed up with hardened steel plates. The edge of the blank after coining is usually ragged and must be trimmed for appearance.

Cold-Forging of Steel Parts More and more parts in machinery and automotive drive components are being cold-forged to net shape. Due to the work-hardening characteristics of the cold metal, the tool pressures can be several times the flow stress of the material (700 to 2,100 MPa for steels). Consequently, precise control of metal flow, good lubrication, and specially designed and shrink-fitted carbide dies are required. An example of a cold-forged transmission part is shown in Fig. 13.2.29. A machined part is also included for comparison purposes. Note that the hobbed part on the left requires longer run-out clearance that increases the part length. No such clearance is required for the cold-forged spline.

Cold-Forging of Aluminum Parts Cold-forged aluminum parts (Fig. 13.2.30) are increasingly replacing machined steel parts, primarily due to excellent formability and the lower weight of aluminum. A cold-forged aluminum column part is backward can-extruded, followed by successive ironing operations, reducing the wall thickness to 1.5 mm and enabling final forming of the bellow.

Fig. 13.2.30 Automotive steering column manufactured by cold-forging aluminum alloy. *(Source: N. Bay, Cold Forming of Aluminum—State of Art. Jour. Mater. Processing Technol., 71, 1997, pp. 76–90.)*

Hot forging is similar in certain respects to the above but permits much greater movement of metal. When dies are used, hot forging may be done in drop hammers, percussion presses, power presses, or forging machines. Steam or pneumatic hammers, helve hammers, or hydraulic presses, most often employ plain anvils.

Drop hammers are rated according to weight of ram. For carbon steel they may be selected on a basis of 50 to 55 lb of ram weight per square inch of projected area (3.5 to 3.9 kg/cm^2) of the forging, including as much of the flash as is squeezed. This allowance should be increased to 60 lb/in^2 (4.2 kg/cm^2) for 0.20 carbon steel, 70 lb/in^2 (4.9 kg/cm^2) for 0.30 carbon steel, and up to about 130 lb/in^2 (9.2 kg/cm^2) for tungsten steel.

In figuring the **forging pressure,** multiply the projected area of the forging, including the portion of the flash that is squeezed, by approximately one-third of the cold compressive strength of the material. Another method gives the forging pressure at three to four times the compressive strength of the material at forging temperatures times the projected area, for presses; or at ten times the compressive strength at forging temperatures times the projected area, for hammers. The pressure builds up to a rather high figure at bottom stroke owing to the cooling of the metal particularly in the flash and to the small amount of relief for excess metal which the flash allows.

In **heading operations,** hot or cold, the length of wire or rod that can be gathered into a head, without side restraint, in a single operation, is limited to three times the diameter. In coining and then cold heading large heads, wire of about 0.08 carbon must be used to avoid excessive strain-hardening.

Forging Dies   Drop-forge dies are usually of steel or steel castings. A good all-around grade of **steel** is a 0.60 percent carbon open-hearth. Dies of this steel will forge mild steel, copper, and tool steel satisfactorily if the number of forgings required is not too large. For a large number of tool-steel forgings, tool-steel dies of 0.80 to 0.90 percent carbon may be used and for extreme conditions, 3½ percent nickel steel.

Die blocks of alloy steels have special value for the production of drop forgings in large quantities. Widely used die materials and their recommended hardness are listed in Table 13.2.4. For a typical hot-working steel, the relationship between the hardness and the UTS is

HRC	UTS, ksi (MPa)
30	140 (960)
40	185 (1,250)
50	250 (1,700)
60	350 (2,400)

The pressure on dies can be kept as high as 80 percent of the above values. For higher tool pressures, carbide (tungsten carbide is the most common) die material is used. Carbides can withstand very high compressive pressures but have poor tensile properties. Consequently, carbide dies and inserts are always kept under compression, often by the use of shrink rings. These are some design guidelines for punches and dies:

1. *Long punch.* The punch pressure should be kept below the buckling stress σ_b

$$\sigma_b = \sigma_y \left[1 - \left(\frac{4\sigma_y}{\pi^2 E} \right) \left(\frac{L_p}{D_p} \right)^2 \right]$$

where σ_y is the yield stress of the punch material, E the elastic modulus 30,000 ksi (210 GPa) for steel and 50,000 ksi (350 GPa) for tungsten carbide, L_p is the punch length and D_p the punch diameter.

2. *Short punch.* The failure mode is plastic upset. Therefore, the punch pressure should be kept less than the yield stress of the punch material [180 ksi (1,200 MPa) for steel and 500 ksi (3,000 MPa) for tungsten carbide].

3. *Flat platen.* Flat platens fail by plastic yielding. A common formula for calculating acceptable maximum platen pressure p is

$$p = \sigma_y \text{ (diameter of platen/diameter of workpiece)}$$

Table 13.2.4   Typical Die Materials for Deformation Processes*

| Process | Die material† and HRC for working | | | |
	Al, Mg, and Cu alloys		Steels and Ni alloys	
Hot forging	6G	30–40	6G	35–45
	H12	48–50	H12	40–56
Hot extrusion	H12	46–50	H12	43–47
Cold extrusion:				
Die	W1, A2	56–58	A2, D2	58–60
	D2	58–60	WC	
Punch	A2, D2	58–60	A2, M2	64–65
Shape drawing	O1	60–62	M2	62–65
	WC		WC	
Cold rolling	O1	55–65	O1, M2	56–65
Blanking	Zn alloy		As for Al, and	
	W1	62–66	M2	60–66
	O1	57–62	WC	
	A2	57–62		
	D2	58–64		
Deep drawing	W1	60–62	As for Al, and	
	O1	57–62	M2	60–65
	A2	57–62	WC	
	D2	58–64		
Press forming	Epoxy/metal powder		As for Al	
	Zn alloy			
	Mild steel			
	Cast iron			
	O1, A2, D2			

* Compiled from "Metals Handbook," 9th ed., vol. 3, American Society for Metals, Metals Park, Ohio, 1980.
† Die materials mentioned first are for lighter duties, shorter runs. Tool steel compositions, percent (representative members of classes):
  6 G (prehardened die steel): 0.5 C, 0.8 Mn, 0.25 Si, 1 Cr, 0.45 Mo, 0.1 V
  H12 (hot-working die steel): 0.35 C, 5 Cr, 1.5 Mo, 1.5 W, 0.4 V
  W1 (water-hardening steel): 0.6–1.4 C
  01 (oil-hardening steel): 0.9 C, 1 Mn, 0.5 Cr
  A2 (air-hardening steel): 1 C, 5 Cr, 1 Mo
  D2 (cold-working steel): 1.5 C, 12 Cr, 1 Mo
  M2 (Mo high-speed steel): 0.85 C, 4 Cr, 5 Mo, 6.25 W, 2 V
  WC (tungsten carbide)
SOURCE: John A. Schey, *Introduction to Manufacturing Processes,* McGraw-Hill, New York, 1987.

4. *Die cavity.* The die pressures in a deep cavity (impression) are much higher than in a flat platen (die), resulting in die burst-out. A more conservative value of 150 ksi (1,000 MPa) is used for acceptable die pressure. If shrink rings are used, higher values of 250 ksi (1,700 MPa) for steels and 400 ksi (2,700 MPa) for tungsten carbide can be used for the inner insert.

For large massive dies or for intermittent service, chrome-nickel-molybdenum alloys are preferred. In closed die work, where the dies must dissipate considerable heat, the tungsten steels are preferred, with resulting increase in wear resistance but decrease in toughness.

For very large pieces with deep impressions, **cast-steel** dies are sometimes used. For large dies likely to spring in hardening, 0.85 **carbon steel** high in manganese is sometimes used unhardened.

Good **die-block proportions** for width and depth are as follows:

Width, in (cm)	8 (20)	10 (24)	12 (30)	14 (36)
Depth, in (cm)	6 (15)	7 (18)	7 (18)	7 or 8 (18 or 20)

For ordinary work, 1½ in (4 cm) of metal between impression and edge of block is sufficient.

Dimensions of **dovetailed die shanks:** for hammers up to 1,200 lb (550 kg) 4 in (10 cm) wide and 1⅛ in (3 cm) deep, with sides dovetailed at angles of 6° with the vertical; for hammers from 1,200 to 3,000 lb (550 to 1,360 kg) in size, 6 in (15 cm) wide and 1½ in (4 cm) deep, with 6° angles.

The **minimum draft for the impressions** is 7°, although for parts difficult to draw this may be increased up to 15°. It is not uncommon to have several drafts in the same impression.

Carbon steel and tool-steel **dies** are **hardened** by heating in a carbonizing box packed in charcoal and dipping face downward over a jet of brine. The jet is allowed to strike into the impression, thus freeing the face of steam and producing uniform hardness. After hardening, they are drawn in an oil bath to a temperature of 500 to 550°F (260 to 290°C).

The forging production per pair of dies is largely affected by the size and shape of the impression, the material forged, the material in the dies, the quality of heating of stock to be forged, and the care exercised in use. It may vary from a few hundred pieces to 50,000 or more.

Coining, Stamping, and Embossing   The metal is well confined in closed dies in which it is forced to flow to fill the shape. The U.S. Mint gives the following pressures; silver quarter, 100 tons/in² (1,380 MN/m²); nickel (0.25 Ni, 0.75 Cu), 90 tons/in² (1,240 MN/m²); copper cent, 40 tons/in² (550 MN/m²). In stamping designs, lettering, etc., in sheet metal the thickness is so little compared with the area that there is practically no relief for excess pressure. Where sharp designs are required, as in stamping panels, the dies should be arranged to strike on a narrow line [say ¹⁄₃₂ in (0.8 mm)] around the outline. If a sharp design is not obtained, it is often best to correct deflection in the machine by shimming or more substantial backing. Increasing the pressure only aggravates the condition and may break the press. General practice for light overall stamping is to allow 5 to 10 tons/in² (68 to 132 MN/m²) of area that is to be stamped, except in areas where the yield point must be exceeded.

Extrusion is the severest of the squeezing processes. The metal is forced to flow rapidly through an orifice, being otherwise confined and subject largely to the laws of hydraulics, with allowances for restraint of flow and for work hardening. Power-press **impact extrusion** began with tin and lead collapsible tubes. It has been extended to the backward and forward extrusion of aluminum, brass, and copper in pressure ranges of 30 to 60 or more tons/in² (413 to 825 MN/m²), and mild steel at pressures up to 165 tons/in² (2,275 MN/m²). Hot impact extrusion of steel, as in projectile piercing, ranges from about 25 to 50 tons/in² (345 to 690 MN/m²). **Forward extrusion** of long tubes, rods, and shapes usually performed hot in hydraulic presses has been extended from the softer metals to the extrusion of steels. Most work is done horizontally because of the lengths of the extrusions. Some vertical mechanical-press equipment is used in hot extrusion of steel tubing.

EQUIPMENT FOR WORKING METALS

The **mass production** industries use an extremely wide variety of machines to force materials to flow plastically into desired shapes (as compared with the more gradual methods of obtaining shapes by cutting away surplus material on machine tools). The application of working pressure may use hydraulic, pneumatic, mechanical, or electric means to apply pressing, hammering, or rolling forces. Mechanically and hydraulically actuated devices cover much the same range. In general, the mechanical equipment is faster, easier to maintain, and more efficient to operate by reason of energy-storing flywheels. The hydraulic equipment is more flexible and more easily adjusted to limited lots in pressure, positions, and strokes. Mechanical handling or feeding devices incorporated in or serving many of these more or less specialized machines further extend their productivity.

Power presses consist of a frame or substantial construction with devices for holding the dies or tools and a moving member or slide for actuating one portion of the dies. This slide usually receives its movement from a crankshaft furnished with a clutch for intermittent operation and a flywheel to supply the sudden power requirement. **Hydraulic presses** have no crankshaft, clutch, or flywheel but employ rams actuated by pumps.

The crankshaft is ordinarily the limiting factor in the pressure capacity of the machine and accordingly is often taken as the basis for tonnage ratings. There is no uniform basis for this rating, owing to variations in shaft proportions and materials and in the different relative severity of various press operations. The following valuation is tentative and is based on the shaft diameter in the main bearings. The bending strength is figured at a section through the center of the crankpin and the combined bending and torsional strength at the inside ends of the main bearings, taking the bending fulcrum at a distance out from these points equal to one-third the length of the main bearings. In the case of double-crank presses and twin-drive arrangements, the relative proportion of the torsional load must be varied to suit, but, except in the cases of long strokes, it is usually small. The working strength is based upon a stress in the extreme fibers of 28,000 lb/in² (193 MN/m²). The limit bearing capacities are taken approximately at 5,000 lb/in² (35 MN/m²) on the crankpins and 2,500 lb/in² (18 MN/m²) average over the main bearings for ordinary steel on cast-iron press bearings with proper grooving. On the knuckle-joint-type presses with hardened tool-steel bearing surfaces and flood lubrication, the bearings will take up to about 30,000 lb/in² (207 MN/m²). On eccentric-type shafts where the main bearings support right up to the oversize pin on each side, the limiting factor is the bearing load. The shaft is practically in shear, so that it has a considerable overload capacity (about $7d^2$ tons). In Table 13.2.5, uniform-diameter single crankshafts are those in which for manufacturing reasons the diameter is the same as at the crankpin and at the main bearings. Other crank shafts have an oversize crankpin to balance the bending load at the center with the combined bending and torsional load at the side. The strength of the shaft is figured at midstroke, and the stroke and tonnage capacity are given in terms of the diameter d at the main bearings. Where the working load comes on only near the bottom stroke, the shaft press capacity may be figured as if the stroke were shorter in proportion.

Table 13.2.5 gives the rated capacities of a series of power presses as a function of the shaft diameter.

The speed of operation of the press depends upon the energy requirement and the crankpin velocity. The latter determines the velocity of impact on the tools. In blanking, the blow varies directly with the contact speed and the thickness and hardness of the material. In drawing operations the variation depends upon contact speed, ductility of material, lubrication, etc.

The energy required per stroke is practically the product of the average load and the working distance, plus friction allowance, assumed at about 16 percent. On short-stroke operations, such as blanking, the working energy is supplied almost entirely by slowing down the flywheel; motor and belt pull serve merely to return the flywheel to speed during the large part of the cycle in which no work is done. In drawing operations, the working period is considerable, and in many cases the belt takes the largest part of the working load. In this case, add to the available flywheel energy, the work done by the belt. This amounts, for example, to the allowable belt loading (flat or Vee), multiplied by the ratio of the belt velocity to the crankpin velocity, multiplied by the length of the working stroke on the crank circle in feet. The maximum flywheel slowdown has been assumed to be up to 10 percent for continuous

Table 13.2.5   Power-Press Shaft Capacities

Type of press crankshaft	Max stroke, in*	Capacity tons†
Single crank, single drive, uniform diameter	d‡	$2.8d^2$
Single crank, single drive, oversize crankpin	d	$3.5d^2$
Single crank, single drive, oversize crankpin	$2d$	$2.2d^2$
Single crank, single drive, oversize crankpin	$3d$	$1.6d^2$
Single crank, twin drive, oversize crankpin	$2d$	$3.5d^2$
Single crank, twin drive, oversize crankpin	$3d$	$2.7d^2$
Double crank, single drive, oversize crankpin	$0.75d$	$5.5d^2$
Double crank, single drive, oversize crankpin	d	$4.4d^2$
Double crank, single drive, oversize crankpin	$2d$	$2.5d^2$
Double crank, single drive, oversize crankpin	$3d$	$1.7d^2$
Double crank, twin drive, oversize crankpin	$1.5d$	$5.5d^2$
Double crank, twin drive, oversize crankpin	$3d$	$3.2d^2$
Single eccentric, single or twin drive	$0.5d$	$4.3d^2$

* 1 in = 2.54 cm.
† 1 ton = 8.9 kN.
‡ d = shaft diameter in main bearing.
SOURCE: F. W. Bliss Co.

operation and up to 20 percent for intermittent operation. The following formula is based upon average press-flywheel proportions and a slowdown of 10 percent. The result may be doubled for 20 percent slowdown.

The flywheel capacity per stroke at 10 percent slowdown in inch-tons equals $WD^2N^2/5,260,000,000$, where W is the weight of the flywheel, lb, D is the diameter, in, and N is rotation speed, r/min. (See Sec. 8.2.)

The difference between nongeared and geared presses is only in speed of operation and the relatively greater flywheel capacity.

Press frames are designed for stiffness and usually have considerable excess strength. Good practice is to figure cast-iron sections for a stress of about 2,000 to 3,000 lb/in² (13.8 to 20.6 MN/m²). **C-frame presses** are subjected to an appreciable arc spring amount ordinarily of between 0.0005 and 0.002 in/ton (1.5 and 6 mm/MN), because the center of gravity of the frame section is a considerable distance back of the working centerline of the press. **Straight-sided presses** eliminate that portion of the spring or deflection which is on an arc. **Built-up frame presses** are held together with steel tie rods shrunk in under an initial tension in excess of the working load so that they minimize stretch in that portion of the press.

Power presses are built in a very wide variety of styles and sizes with shafts ranging from 1- to 21-in (2.5- to 53-cm) diam. Over a large part of this range they are built with C frames for convenience, straight-sided frames for heavier and thinner work, eccentric shafts for heavy forgings and stampings, double crankshafts for wide jobs, four-point presses for large panel work, underdrive presses in high-production plants where repairs to presses would interfere with flow of production, and knuckle-join presses for intensely high pressures at the very bottom of the stroke. All these are classified as single-action presses and are used for most of the operations previously discussed.

Double-action presses combine the functions of blank holding with drawing. In the smaller sizes, such presses have cams mounted on the cheeks of the crankshaft to actuate the outer or blank-holding slide. In larger machines, toggle mechanisms are provided to actuate the outer slide, with the advantage that the blank-holding load is taken on the frame instead of the crankshaft. Both of these types afford a considerable power saving over single-action presses equipped with drawing attachments, because the latter must add the blank-holding pressure to the working load for the full depth of the draw.

Types of presses include **foot presses,** in which the pendulum type has the lowest mechanical advantage and the longest stroke; the **lever type,** which has higher mechanical advantage and shorter strokes; **toggle** or **knuckle type,** which has the highest mechanical advantage and works through the shortest stroke with considerable advantage obtainable from the use of tie rods on fine stamping or embossing work; long-stroke rack and pinion-driven presses; triple-action drawing presses; cam-actuated presses; etc.

Screw presses consist of a conventional frame and a slide which is forced down by a steep pitch screw on the upper end of which is a flywheel or weight bar. Hand-operated machines are used for die testing and for small production stamping, embossing, forming, and other work requiring more power than foot presses. Power-driven screw presses are built with a friction drive for the flywheel and automatic control to limit the stroke. Such presses are built in comparatively large sizes and used to a considerable extent for press forging. They lack the accuracy and speed of power presses built for this work but have a safety factor which power presses have not, in that their action is not positive. In this they closely resemble a drop hammer, although their motion is slower. The energy available for work in these presses is $\frac{1}{2}I_f v^2 + \frac{1}{2}I_s v^2$, in which I_f is the moment of inertia of the flywheel, I_s is the moment of inertia of the spindle, and v is the angular velocity of both.

Self-contained fast-acting **hydraulic presses** are being increasingly used. Equipped with motor-driven variable-displacement oil hydraulic pumps, the speed and pressure of the operating ram or rams are under instant and automatic control; this is particularly advantageous for deep drawing operations. The punch can be brought into initial contact with the work without shock and moved with a uniform controlled velocity through the drawing portion of the cycle. The cold drawing of stainless steels and

aluminum alloys (in which the control of drawing speed is vital), as well as the hot drawing of magnesium, is best done on hydraulic presses.

The hydraulic press is used on the rubber pad, or *Guerin,* process of blanking or forming metals, in which a laminated-rubber pad replaces one half—usually the female half—of a die. In forming aluminum the practice has developed of using inexpensive dies of soft metal, vulcanized fiber, plastic, wood, or plaster; and cast dies in industries which, like the aircraft industry, require short runs on many different sizes of shapes and parts.

The older accumulator type of hydraulic-press construction is still used for hot extrusion and some forging work.

Pneumatic Hammers   A self-contained type of pneumatic forge hammer (the Bêché) has an air-operated ram with an air-compressing cylinder integral with the frame. The ram is raised by admitting compressed air beneath the ram piston; at the same time a partial vacuum is caused above it. The ram is forced down by a reversal of this action.

The **terminal velocity** (velocity at ram-workpiece contact) of the steam, pneumatic, or hydraulic assisted hammer can be calculated as follows:

$$v = \left[2hg\left(1 + \frac{Ap_m}{H}\right)\right]^{0.5}$$

where A is the area of the piston and p_m the mean pressure in the drive cylinder. The hammer energy is often calculated by dividing the energy required for plastic work by the mechanical efficiency of the hammer.

Hydraulic presses are energized by pressurized liquid, usually oil. They can deliver high tonnage but are slow. Due to large die chilling (heat loss to dies) present in hydraulic press forging, they are not usually used for hot forging, except in isothermal forging, where slow speed and large die-workpiece contact times are not a major limitation. They are specially suited to sheet metal forming operations, where slower ram speeds produce lower impact loads, speeds can be varied during the stroke, and multiple actions can be obtained for blank holder and die cushion operation.

Hammers and presses are often selected based on their characteristics such as energy, ram mass (t_{up}), force or tonnage, ram speed (stroking rate), stroke length, bed area, and mechanical efficiency. The characteristics for various hammers and presses are summarized in Table 13.2.6.

Rotary motion is used **for working sheet metal** in a variety of machines, including bending rolls (three rolls); rolling straighteners with five, seven, or more rolls; roll forming machines, in which a series of rolls in successive pairs are used to bend the strip material step by step to some desired shape; a series of two-spindle and multiple-spindle machines used for rolling beads, threads, knurls, flanges, and trimming or curling the edges of drawn shells of cylinders; seaming machines for double seaming, crimping, curling, and other operations in the production of tin cans, pieced tinware, etc.; and spinning machines for spinning, burnishing, trimming, curling, shape forming, and thickness reduction. Various production spinning operations and tool arrangements are shown in Fig. 13.2.10.

Plate-Straightening Machines   The **horsepower required** for plate-straightening machines operating on steel plate is shown in Table 13.2.7.

Power required for angle-iron-straightening machines: for 4-in angles, 12 hp; for 6-in, 18 hp; for 8-in, 25 hp.

Power or hydraulic presses are used to straighten **large rolled sections.** The presses make 20 to 30 strokes per min, and the amount of flexure is regulated by inserting wedges or pieces of flat iron. The beams are supported on rolls so they can be easily handled. The **power required for presses** of this kind is as follows:

Depth of girder, in*	4	6	8	10	12	16	20	24
Horsepower (approx)†	3	4	7	11	13	19	23	35

* 1 in = 2.54 cm.
† 1 hp = 0.746 kW.

Horizontal plate-bending machines consist of two stationary rolls and a third vertical adjustable upper roll which can be fitted obliquely for

Table 13.2.6   Characteristics of Hammers and Presses*

Equipment type	Energy,† kN · m	Ram mass, kg	Force,‡ kN	Speed, m/s	Strokes per min	Stroke, m	Bed area, m × m	Mechanical efficiency
Hammers								
Mechanical	0.5–40	30–5,000		4–5	350–35	0.1–1.6	0.1 × 0.1 to 0.4 × 0.6	0.2–0.5
Steam and air	20–600	75–17,000 (25,000)		3–8	300–20	0.5–1.2	0.3 × 0.4 to (1.2 × 1.8)	0.05–0.3
Counterblow	5–200 (1,250)			3–5	60–7		0.3 × 0.4 to (1.8 × 5)	0.2–0.7
Herf	15–750			8–20	<2			0.2–0.6
Presses								
Hydraulic, forging			100–80,000 (800,000)	<0.5	30–5	0.3–1 (3)	0.5 × 0.5 to (3.5 × 8)	0.1–0.6
Hydraulic, sheet metalworking			10–40,000	<0.5	130–20	0.1–1	0.2 × 0.2 to 2 × 6	0.5–0.7
Hydraulic, extrusion			1,000–50,000 (200,000)	<0.5	<2	0.8–5	0.06–0.6 diam. container	0.5–0.7
Mechanical, forging			10–80,000	<0.5	130–10	0.1–1	0.2 × 0.2 to 2 × 3	0.2–0.7
Horizontal upsetter			500–30,000 (1–9 in diam)	<1	90–15	0.05–0.4	0.2 × 0.2 to 0.8 × 1	0.2–0.7
Mechanical, sheet metalworking			10–20,000	<1	180–10	0.1–0.8	0.2 × 0.2 to 2 × 6	0.3–0.7
Screw			100–80,000	<1	35–6	0.2–0.8	0.2 × 0.3 to 0.8 × 1	0.2–0.7

* From a number of sources, chiefly A. Geleji, *Forge Equipment, Rolling Mills and Accessories*, Akademiai Kiado, Budapest, 1967.
† Multiply number in column by 100 to get m·kg, by 0.73 to get 10^3 lbf·ft.
‡ Divide number by – 10 to get tons. Numbers in parentheses indicate the largest sizes, available in only a few places in the world.
Source: John A. Schey, *Introduction to Manufacturing Processes*, McGraw-Hill, New York, 1987.

Table 13.2.7   Power Requirement for Plate-Straightening Machines (Steel)

Thickness of plate, in*	0.25	0.4	0.6	0.8	1.0	1.2	1.4	1.6
Width of plate, in*	48.0	52.0	60.0	72.0	88.0	102.0	120.0	140.0
Diameter of rolls, in*	5.0	8.0	10.0	12.0	13.0	14.0	15.0	16.0
Horsepower (approx)†	6.0	8.0	12.0	20.0	30.0	55.0	90.0	130.0

* 1 in = 2.54 cm.
† 1 hp = 0.746 kW.

Table 13.2.8   Power Requirement for Plate and Sheet Bending Machines (Steel)

Thickness of plate, in*	0.5	0.6	0.8	1	1.2
Horsepower for plate 120 in wide†	10.0	12.0	18.0	27.0	40.0
Horsepower for plate 240 in wide†	30.0	30.0	40.0	55.0	75.0

* 1 in = 2.54 cm.
† 1 hp = 0.746 kW.

taper bending and is held in bearings with spherical seats. The diameter of the rolls can be determined approximately from the equation $r^2 = bt$, in which r is the radius of the roll, b the width of the plate or sheet, and t its thickness, all in inches.

The **power requirements** of horizontal plate and sheet bending machines are shown in Table 13.2.8.

Numerically Adaptive Bending of Tubes and Profiles (Fig. 13.2.31) Extrusions, profiles, and tubes find greater use in car and truck body structures due to their high stiffness under bending loads. The shapes are bent or curved to conform to the geometric needs of design and assembly. Therefore, new fixtures and numerically controlled machines and tooling (Fig. 13.2.31) have been developed to stretch bend the structural shapes.

Vertical plate-bending machines have a hydraulically operated piston which moves an upper and a lower pair of rolls between inclined surfaces of the stationary upright and the crosshead. The bending is done piece by piece against a second stationary upright. Heavy ship plates are rigidly clamped down and bent by a roll operated by two hydraulic pistons. For angular bends or for the production of warped surfaces, the pistons can be operated independently or together. In vertical machines, angles and other rolled shapes are bent between suitably shaped rolls. Pipes are filled with sand to prevent flattening when being bent. For some work, pipes are bent hot between suitable forms operated by hydraulic pressure.

The **rotary swaging machines** for tapering, closing in, and reducing tubes, rods, and hollow articles is essentially a cage carrying a number of rollers and revolving at high speed; e.g., 14 rolls in a cage revolving at 600 r/min will strike 8,400 blows per min on the work.

Fig. 13.2.31 Schematic of numerically controlled adaptive stretch bending machine. (*Source: M. Nock and M. Geiger, Flexible Kinematic 3D-Bending of Tubes and Profiles, International Conference on the Advanced Technology of Plasticity, Japan, **1**, 2002, pp. 643–648.*)

A rapid succession of light blows is applied to a considerable variety of commercial **riveting** operations such as pneumatic riveting. Another method of riveting, described as spinning, involves rotating small rollers rapidly over the top of the rivet and at the same time applying pressure. Neither of these methods involves pressures as intense as those used in riveting by direct pressure, either hot or cold. Power presses and C-frame riveters, employing hydraulic pressure or air pressure of 80 to 100 lb/in² (550 to 690 kN/m²), are designed to apply 150,000 lb/in² (1,035 MN/m²) on the cross section of the body of the rivet for hot-working and 300,000 lb/in² (2,070 MN/m²) for cold-working. The pieces joined should be pressed together by a pressure 0.3 to 0.4 times that used in riveting.

13.3 WELDING AND CUTTING
by Omer W. Blodgett and Duane K. Miller

REFERENCES: From the American Welding Society (AWS): "Welding Handbook" (six volumes); "Structural Welding Code"; "Filler Metal Specifications," "Welding Terms and Definitions"; "Brazing Manual"; "Thermal Spraying"; "Welding of Chromium-Molybdenum Steels." From the Lincoln Arc Welding Foundation: Blodgett, "Design of Weldments"; Blodgett, "Design of Welded Structures." "Procedure Handbook of Welding," The Lincoln Electric Co. "Safety in Welding and Cutting," Penton Pub. Co. "Welding and Fabrication Data Book," AISC. "Steel Construction Manual—ASD," "Steel Construction Manual—LRFD," latest editions.

INTRODUCTION

Welded connections and assemblies represent a very large group of fabricated metal components, and only a portion of the aspects of their design and fabrication is treated here. The welding process itself is complex, involving heat and liquid-metal transfer, chemical reactions, and the gradual formation of the welded joint through liquid-metal deposition and subsequent cooling into the solid state, with attendant metallurgical transformations. Some of these items are treated in greater detail in the references and other extensive professional literature, as well as in Secs. 6.2, 6.3, and 13.1.

The material in this section will provide the engineer with an overview of the most important aspects of welded design. In order that the resulting welded fabrication be of adequate strength, stiffness, and utility, the designer will often collaborate with engineers who are experts in the broad area of design and fabrication of weldments.

ARC WELDING

Arc welding is one of several **fusion processes** for joining metal. By the generation of intense heat, the juncture of two metal pieces is melted and mixed—directly or, more often, with an intermediate molten **filler metal.** Upon cooling and solidification, the resulting welded joint metallurgically bonds the former separate pieces into a continuous structural assembly (a **weldment**). When the pieces are properly designed and fabricated, the strength properties are basically those of the individual pieces before welding.

In arc welding, the intense heat needed to melt metal is produced by an **electric arc.** The arc forms between the workpieces and an electrode that is either manually or mechanically moved along the joint; conversely, the work may be moved under a stationary electrode. The **electrode** generally is a specially prepared rod or wire that not only conducts electric current and sustains the arc, but also melts and supplies **filler metal** to the joint; this constitutes a **consumable electrode. Carbon** or **tungsten electrodes** may be used, in which case the electrode serves only to conduct electric current and to sustain the arc between tip and workpiece, and it is not consumed; with these electrodes, any filler metal required is supplied by rod or wire introduced into the region of the arc and melted there. Filler metal applied separately, rather than via a consumable electrode, does not carry electric current.

Most steel arc welding operations are performed with consumable electrodes.

Welding Process Fundamentals

Heat and Filler Metal   An ac or dc power source fitted with necessary controls is connected by a work cable to the workpiece and by a "hot" cable to an electrode holder of some type, which, in turn, is electrically connected to the welding electrode (Fig. 13.3.1). When the circuit is energized, the flow of electric current through the electrode heats the electrode by virtue of its electric resistance. When the electrode tip is touched to the workpiece and then withdrawn to leave a gap between the electrode and workpiece, the arc jumping the short gap presents a further path of high electric resistance, resulting in the generation of an extremely high temperature in the region of the sustained arc. The temperature reaches about 6,500°F, which is more than adequate to melt most metals. The heat of the arc melts both the base and the filler metal, the latter being supplied via a consumable electrode or separately. The puddle of molten metal produced is called a **weld pool,** which solidifies as the electrode and arc move along the joint being welded. The resulting weldment is metallurgically bonded as the liquid metal cools, fuses, solidifies, and cools. In addition to serving its main function of supplying heat, the arc is subject to adjustment and/or control to vary the proper transfer of molten metal to the weld pool, remove surface films in the weld region, and foster gas-slag reactions or other beneficial metallurgical changes.

Filler metal composition is generally different from that of the weld metal, which is composed of the solidified mix of both filler and base metals.

Shielding and Fluxing   High-temperature molten metal in the weld pool will react with oxygen and nitrogen in ambient air. These gases will remain dissolved in the liquid metal, but their solubility significantly decreases as the metal cools and solidifies. The decreased solubility causes the gases to come out of solution, and if they are trapped in the metal as it solidifies, **cavities,** termed **porosity,** are left behind. This is always undesirable, but it can be acceptable to a limited degree depending on the specification governing the welding.

Fig. 13.3.1   Typical welding circuit.

Smaller amounts of these gases, particularly nitrogen, may remain dissolved in the weld metal, resulting in reduction in the physical properties of otherwise excellent weld metal. **Notch toughness** is degraded by nitrogen inclusions. Accordingly, the molten metal must be shielded from harmful atmospheric gas contaminants. This is accomplished by gas shielding or slag shielding or both.

Gas shielding is provided either by an external supply of gas, such as carbon dioxide, or by gas generated when the electrode flux heats up. **Slag shielding** results when the flux ingredients are melted and leave behind a slag to cover the weld pool, to act as a barrier to contact between the weld pool and ambient air. At times, both types of shielding are utilized.

In addition to its primary purpose to protect the molten metal, the shielding gas will affect arc behavior. The shielding gas may be mixed with small amounts of other gases (as many as three others) to improve arc stability, puddle (weld pool) fluidity, and other welding operating characteristics.

In the case of shielded-metal arc welding (SMAW), the "stick" electrode is covered with an extruded coating of flux. The arc heat melts the flux and generates a gaseous shield to keep air away from the molten metal, and at the same time the flux ingredients react with deleterious substances, such as surface oxides on the base metal, and chemically combine with those contaminants, creating **a slag** which floats to the surface of the weld pool. That slag crusts over the newly solidified hot metal, minimizes contact between air and hot metal while the metal cools, and thereby inhibits the formation of surface oxides on the newly deposited weld metal, or weld bead. When the temperature of the weld bead decreases, the slag, which has a glassy consistency, is chipped off to reveal the bright surface of the newly deposited metal. Minimal surface oxidation will take place at lower temperatures, inasmuch as oxidation rates are greatly diminished as ambient conditions are approached.

Fluxing action also aids in wetting the interface between the base metal and the molten metal in the weld pool edge, thereby enhancing uniformity and appearance of the weld bead.

Process Selection Criteria

Economic factors generally dictate which welding process to use for a particular application. It is impossible to state which process will always deliver the most economical welds, because the variables involved are significant in both number and diversity. The variables include, but are not limited to, steel (or other base metal) type, joint type, section thickness, production quantity, joint access, position in which the welding is to be performed, equipment availability, availability of qualified and skilled welders, and whether the welding will be done in the field or in the shop.

Shielded Metal Arc Welding

The **SMAW** process (Fig. 13.3.2), commonly known as **stick welding,** or **manual welding,** is a popular and widespread welding process. It is versatile, relatively simple to do, and very flexible in being applied. To

Fig. 13.3.2   SMAW process.

those casually acquainted with welding, **arc welding** usually means **shielded-metal arc welding.** SMAW is used in the shop and in the field for fabrication, erection, maintenance, and repairs. Because of the relative inefficiency of the process, it is seldom used for fabrication of major structures. SMAW has earned a reputation for providing high-quality welds in a dependable fashion. It is, however, inherently slower and generally more costly than other methods of welding.

SMAW may utilize either direct current (dc) or alternating current (ac). Generally speaking, direct current is used for smaller electrodes, usually less than $3/16$-in diameter. Larger electrodes utilize alternating current to eliminate undesirable arc blow conditions.

Electrodes used with alternating current must be designed specifically to operate in this mode, in which current changes direction 120 times per second with 60-Hz power. All ac electrodes will operate acceptably on direct current. The opposite is not always true.

Flux Cored Arc Welding (FCAW)

In **FCAW,** the arc is maintained between a continuous tubular metal electrode and the weld pool. The tubular electrode is filled with flux and a combination of materials that may include metallic powder(s). FCAW may be done automatically or semiautomatically. FCAW has become the workhorse in fabrication shops practicing semiautomatic welding. Production welds that are short, change direction, are difficult to access, must be done out of position (e.g., vertical or overhead), or are part of a short production run generally will be made with semiautomatic FCAW.

When the application lends itself to automatic welding, most fabricators will select the submerged arc process (see material under "SAW"). Flux cored arc welding may be used in the automatic mode, but the intensity of arc rays from a high-current flux cored arc, as well as a significant volume of smoke, makes alternatives such as submerged arc more desirable.

Advantages of FCAW   FCAW offers two distinct advantages over SMAW. First, the electrode is continuous and eliminates the built-in starts and stops that are inevitable with SMAW using stick electrodes. An economic advantage accrues from the increased operating factor; in addition, the reduced number of arc starts and stops largely eliminates potential sources of weld discontinuities. Second, increased amperages can be used with FCAW. With SMAW, there is a practical limit to the amount of current that can be used. The covered electrodes are 9 to 18 in long, and if the current is too high, electric resistance heating within the unused length of electrode will become so great that the coating ingredients may overheat and "break down." With continuous flux cored electrodes, the tubular electrode is passed through a contact tip, where electric current is transferred to the electrode. The short distance from the contact tip to the end of the electrode, known as electrode extension or **"stickout,"** inhibits heat buildup due to electric resistance. This electrode extension distance is typically 1 in for flux cored electrodes, although it may be as much as 2 or 3 in in some circumstances.

Smaller-diameter flux cored electrodes are suitable for all-position welding. Larger electrodes, using higher electric currents, usually are restricted to use in the flat and horizontal positions. Although the equipment required for FCAW is more expensive and more complicated than that for SMAW, most fabricators find FCAW much more economical than SMAW.

FCAW Equipment and Procedures   Like all wire-fed welding processes, FCAW requires a power source, wire feeder, and gun and cable assembly (Fig. 13.3.3). The power supply is a dc source, although either electrode positive or electrode negative polarity may be used. The four primary variables used to determine welding procedures are voltage, wire feed speed, electrode extension, and travel speed. For a given wire feed speed and electrode extension, a specified amperage will be delivered to maintain stable welding conditions.

As wire feed speed is increased, amperage will be increased. On some equipment, the wire feed speed control is called the amperage control, which, despite its name, is just a rheostat that regulates the speed of the dc motor driving the electrode through the gun. The most accurate way, however, to establish welding procedures is to refer to the

Fig. 13.3.3   FCAW and GMAW equipment.

wire feed speed (WFS), since electrode extension, polarity, and electrode diameter will also affect amperage. For a fixed wire feed speed, a shorter electrical stick-out will result in higher amperages. If procedures are set based on the wire feed speed, the resulting amperage verifies that proper electrode extensions are being used. If amperage is used to set welding procedures, an inaccurate electrode extension may go undetected.

Self-Shielded and Gas-Shielded FCAW   Within the category of FCAW, there are two specific subsets: **self-shielded flux core arc welding (FCAW-S)** (Fig. 13.3.4) and **gas-shielded flux core arc welding (FCAW-G)** (Fig. 13.3.5). Self-shielded flux cored electrodes require no external

Fig. 13.3.4   Self-shielded FCAW.

Fig. 13.3.5   Gas-shielded FCAW.

shielding gas. The entire shielding system results from the flux ingredients contained in the tubular electrode. The gas-shielded variety of flux cored electrode utilizes, in addition to the flux core, an externally supplied shielding gas. Often, CO_2 is used, although other mixtures may be used.

Both these subsets of FCAW are capable of delivering weld deposits featuring consistency, high quality, and excellent mechanical properties. Self-shielded flux cored electrodes are ideal for field welding operations, for since no externally supplied shielding gas is required, the process may be used in high winds without adversely affecting the quality of the weld metal deposited. With any gas-shielded processes, wind shields must be erected to preclude wind interference with the gas shield. Many fabricators with large shops have found that self-shielded flux core welding offers advantages when the shop door can be left open or fans are used to improve ventilation.

Gas-shielded flux cored electrodes tend to be more versatile than self-shielded flux cored electrodes and, in general, provide better arc action. Operator acceptance is usually higher. The gas shield must be protected from winds and drafts, but this is not difficult for most shop fabrication. Weld appearance is very good, and quality is outstanding. Higher-strength gas-shielded FCAW electrodes are available, but current practice limits self-shielded FCAW deposits to a tensile strength of 80 ksi or less.

Submerged Arc Welding (SAW)

Submerged arc welding differs from other arc welding processes in that a blanket of fusible granular flux is used to shield the arc and molten metal (Fig. 13.3.6). The arc is struck between the workpiece and a bare-wire electrode, the tip of which is submerged in the flux. The arc is completely covered by the flux and it is not visible; thus the weld is made without the flash, spatter, and sparks that characterize the open-arc processes. The flux used develops very little smoke or visible fumes.

Fig. 13.3.6   SAW process.

Typically, the process is operated fully automatically, although semi-automatic operation is possible. The electrode is fed mechanically to the welding gun, head, or heads. In semiautomatic welding, the welder moves the gun, usually equipped with a flux-feeding device, along the joint.

Flux may be fed by gravity flow from a small hopper atop the torch and then through a nozzle concentric with the electrode, or through a nozzle tube connected to an air-pressurized flux tank. Flux may also be applied in advance of the welding operation or ahead of the arc from a hopper run along the joint. Many fully automatic installations are equipped with a vacuum system to capture unfused flux left after welding; the captured, unused flux is recycled for reuse.

During welding, arc heat melts some of the flux along with the tip of the electrode. The electrode tip and the welding zone are always shielded by molten flux and a cover layer of unfused flux. The electrode is kept a short distance above the workpiece. As the electrode progresses along the joint, the lighter molten flux rises above the molten metal to form slag. The weld metal, having a higher melting (freezing) point, solidifies while the slag above it is still molten. The slag then freezes over the newly solidified weld metal, continuing to protect the metal from contamination while it is very hot and reactive with atmospheric oxygen and nitrogen. Upon cooling and removal of any unmelted flux, the slag is removed from the weld.

Advantages of SAW High currents can be used in SAW, and extremely high heat input can be developed. Because the current is applied to the electrode a short distance above the arc, relatively high amperages can be used on small-diameter electrodes. The resulting extremely high current densities on relatively small-cross-section electrodes permit high rates of metal deposition.

The insulating flux blanket above the arc prevents rapid escape of heat and concentrates it in the welding zone. Not only are the electrode and base metal melted rapidly, but also fusion is deep into the base metal. Deep penetration allows the use of small welding grooves, thus minimizing the amount of filler metal to be deposited and permitting fast welding speeds. Fast welding, in turn, minimizes the total heat input to the assembly and thus tends to limit problems of heat distortion. Even relatively thick joints can be welded in one pass with SAW.

Versatility of SAW SAW can be applied in more ways than other arc welding processes. A single electrode may be used, as is done with other wire feed processes, but it is possible to use two or more electrodes in submerged arc welding. Two electrodes may be used in parallel, sometimes called **twin arc welding**, employing a single power source and one wire drive. In **multiple-electrode SAW**, up to five electrodes can be used thus, but most often, two or three arc sources are used with separate power supplies and wire drives. In this case, the lead electrode usually operates on direct current while the trailing electrodes operate on alternating current.

Gas Metal Arc Welding (GMAW)

Gas metal arc welding utilizes the same equipment as FCAW (Figs. 13.3.3 and 13.3.7); indeed, the two are similar. The major differences are: (1) GMAW uses a solid or metal cored electrode, and (2) GMAW leaves no residual slag.

GMAW may be referred to as **metal inert gas (MIG)**, solid wire and gas, **miniwire** or **microwire welding**. The shielding gas may be carbon dioxide or blends of argon with CO_2 or oxygen, or both. GMAW is usually applied in one of four ways: short arc transfer, globular transfer, spray arc transfer, and pulsed arc transfer.

Fig. 13.3.7 GMAW welding process.

Short arc transfer is ideal for welding thin-gage materials, but generally is unsuitable for welding on thick members. In this mode of transfer, a small electrode, usually of 0.035- to 0.045-in diameter, is fed at a moderate wire feed speed at relatively low voltages. The electrode contacts the workpiece, resulting in a short circuit. The arc is actually quenched at this point, and very high current will flow through the electrode, causing it to heat and melt. A small amount of filler metal is transferred to the welding done at this time.

The cycle will repeat itself when the electrode short-circuits to the work again; this occurs between 60 and 200 times per second, creating a characteristic buzz. This mode of transfer is ideal for sheet metal, but results in significant fusion problems if applied to thick sections, when **cold lap** or **cold casting** results from failure of the filler metal to fuse to the base metal. This is unacceptable since the welded connection will have virtually no strength. Caution must be exercised if the short arc transfer mode is applied to thick sections.

Spray arc transfer is characterized by high wire feed speeds at relatively high voltages. A fine spray of molten filler metal drops, all smaller in diameter than the electrode, is ejected from the electrode toward the work. Unlike with short arc transfer, the arc in spray transfer is maintained continuously. High-quality welds with particularly good appearance are obtained. The shielding gas used in spray arc transfer is composed of at least 80 percent argon, with the balance either carbon dioxide or oxygen. Typical mixtures would include 90-10 argon-CO_2, and 95-5 argon-oxygen. Relatively high arc voltages are used with spray arc transfer. Gas metal spray arc transfer welds have excellent appearance and evidence good fusion. However, due to the intensity of the arc, spray arc transfer is restricted to applications in the flat and horizontal positions.

Globular transfer is a mode of gas metal arc welding that results when high concentrations of carbon dioxide are used. Carbon dioxide is not an inert gas; rather, it is active. Therefore, GMAW that uses CO_2 may be referred to as **MAG**, for **metal active gas**. With high concentrations of CO_2 in the shielding gas, the arc no longer behaves in a spraylike fashion, but ejects large globs of metal from the end of the electrode. This mode of transfer, while resulting in deep penetration, generates relatively high levels of spatter, and weld appearance can be poor. Like the spray mode, it is restricted to the flat and horizontal positions. Globular transfer may be preferred over spray arc transfer because of the low cost of CO_2 shielding gas and the lower level of heat experienced by the operator.

Pulsed arc transfer is a newer development in GMAW. In this mode, a background current is applied continuously to the electrode. A pulsing peak current is applied at a rate proportional to the wire feed speed. With this mode of transfer, the power supply delivers a pulse of current which, ideally, ejects a single droplet of metal from the electrode. The power supply then returns to a lower background current to maintain the arc. This occurs between 100 and 400 times per second. One advantage of pulsed arc transfer is that it can be used out of position. For flat and horizontal work, it will not be as fast as spray arc transfer. However, when it is used out of position, it is free of the problems associated with gas metal arc short-circuiting mode. Weld appearance is good, and quality can be excellent. The disadvantages of pulsed arc transfer are that the equipment is slightly more complex and is more costly.

Metal cored electrodes comprise another newer development in GMAW. This process is similar to FCAW in that the electrode is tubular, but the core material does not contain slag-forming ingredients. Rather, a variety of metallic powders are contained in the core, resulting in exceptional alloy control. The resulting weld is slag-free, as are other forms of GMAW.

The use of metal cored electrodes offers many fabrication advantages. Compared to spray arc transfer, metal cored electrodes require less amperage to obtain the same deposition rates. They are better able to handle mill scale and other surface contaminants. When used out-of-position, they offer greater resistance to the cold lapping phenomenon so common with short arc transfer. Finally, metal cored electrodes permit the use of amperages higher than may be practical with solid electrodes, resulting in higher metal deposition rates.

The weld properties obtained from metal cored electrode deposits can be excellent, and their appearance is very good. Filler metal manufacturers are able to control the composition of the core ingredients, so that mechanical properties obtained from metal cored deposits can be more consistent than those obtained with solid electrodes.

Electroslag/Electrogas Welding (ESW/EGW)

Electroslag and **electrogas welding** (Figs. 13.3.8 and 13.3.9) are closely related processes that allow high deposition welding in the vertical plane. Properly applied, these processes offer tremendous savings over alternative, out-of-position methods and, in many cases, savings over flat-position welding. Although the two processes have similar applications and mechanical setup, there are fundamental differences in the arc characteristics.

Electroslag and electrogas are mechanically similar in that both utilize **copper dams,** or shoes, that are applied to either side of a square-edged butt joint. An electrode or multiple electrodes are fed into the joint. Usually, a starting sump is applied for the beginning of the weld. As the electrode is fed into the joint, a puddle is established that progresses vertically. The water-cooled copper dams chill the weld metal and prevent its escape from the joint. The weld is completed in one pass.

Fig. 13.3.8   ESW process.

Fig. 13.3.9   EGW process.

Gas Tungsten Arc Welding (GTAW)

The **gas tungsten arc welding** process (Fig. 13.3.10), colloquially called **TIG welding,** uses a nonconsumable tungsten electrode. An arc is established between the tungsten electrode and the workpiece, resulting in heating of the base metal. If required, a filler metal is used. The weld area is shielded with an inert gas, usually argon or helium. GTAW is ideally suited to weld nonferrous materials such as stainless steel and aluminum, and is very effective for joining thin sections.

Fig. 13.3.10   Gas tungsten arc welding (GTAW).

Highly skilled welders are required for GTAW, but the resulting weld quality can be excellent. The process is often used to weld exotic materials. Critical repair welds as well as root passes in pressure piping are typical applications.

Plasma Arc Welding (PAW)

Plasma arc welding is an arc welding process using a constricted arc to generate very high, localized heating. PAW may utilize either a transferred or a nontransferred arc. In the **transferred arc mode,** the arc occurs between the electrode and the workpiece, much as in GTAW, the primary difference being the constriction afforded by the secondary gases and torch design. With the **nontransferred arc** mode, arcing is contained within the torch between a tungsten electrode and a surrounding nozzle.

The constricted arc results in higher localized arc energies than are experienced with GTAW, resulting in faster welding speeds. Applications for PAW are similar to those for GTAW. The only significant disadvantage of PAW is the equipment cost, which is higher than that for GTAW.

Most PAW is done with the transferred arc mode, although this mode utilizes a nontransferred arc for the first step of operation. An arc and plasma are initially established between the electrode and the nozzle. When the torch is properly located, a switching system will redirect the arc toward the workpiece. Since the arc and plasma are already established, transferring the arc to the workpiece is easily accomplished and highly reliable. For this reason, PAW is often preferred for automated applications.

GAS WELDING AND BRAZING

The heat for **gas welding** is supplied by burning a mixture of oxygen and a suitable combustible gas. The gases are mixed in a torch which controls the welding flame.

Acetylene is almost universally used as the combustible gas because of its high flame temperature. This temperature, about 6,000°F (3,315°C), is so far above the melting point of all commercial metals that it provides a means for the rapid localized melting essential in welding. The oxyacetylene flame is also used in cutting ferrous metals.

A **neutral flame** is one in which the fuel gas and oxygen combine completely, leaving no excess of either fuel gas or oxygen. The neutral flame has an inside portion, consisting of a brilliant cone 1/16 to 3/4 in (1.6 to 19.1 mm) long, surrounded by a faintly luminous envelope flame. When fuel gas is in excess, the flame consists of three easily recognizable zones: a sharply defined inner cone, an intermediate cone of whitish color, and the bluish outer envelope. The length of the intermediate cone is a measure of the amount of excess fuel gas. This **flame is reducing,** or **carburizing.**

When oxygen is in excess in the mixture, the flame resembles the neutral flame, but the inner cone is shorter, is "necked in" on the sides, is not so sharply defined, and acquires a purplish tinge. A slightly **oxidizing flame** may be used in braze welding and bronze surfacing, and a more strongly oxidizing flame is sometimes used in gas-welding brass, bronze, and copper. A disadvantage of a strongly oxidizing flame is that it can oxidize the surface of the base metal and thereby prevent fusion of the filler metal to the base metal.

In **braze welding,** coalescence is produced by heating above 840°F (450°C) and by using a nonferrous filler metal having a melting point below that of the base metals. Braze welding with brass (bronze) rods is used extensively on cast iron, steel, copper, brass, etc. Since it operates at temperatures lower than base metal melting points, it is used where control of distortion is necessary or lower base metal temperatures during welding are desired. Braze-welded joints on mild steel, made with rods of classifications RCuZn-B and RBCuZn-D, will show transverse tensile values of 60,000 to 70,000 lb/in^2 (414 to 483 MPa). Joint designs for braze welding are similar to those used for gas and arc welding.

In braze welding it is necessary to remove rust, grease, scale, etc., and to use a suitable flux to dissolve oxides and clean the metal. Sometimes rods are used with a flux coating applied to the outside. Additional flux may or may not be required, notwithstanding the flux coating on the rods. The parts are heated to red heat [1,150 to 1,350°F (621 to 732°C)], and the rod is introduced into the heated zone. The rod melts first and "tins" the surfaces, following which additional filler metal is added.

Welding rods for oxyacetylene braze welding are usually of the copper-zinc (60 Cu-40 Zn) analysis. Additions of tin, manganese, iron, nickel, and silicon are made to improve the mechanical properties and usability of the rods.

Brazing is another one of the general groups of welding processes, consisting of the torch, furnace, induction, dip, and resistance brazing. Brazing may be used to join almost all metals and combinations of dissimilar metals, but some combinations of dissimilar metals are not compatible (e.g., aluminum or magnesium to other metals). In brazing, coalescence is produced by heating above 840°F (450°C) but below the melting point of the metals being joined. The nonferrous filler metal used has a melting point below that of the base metal, and the filler metal is distributed in the closely fitted lap or butt joints by capillary attraction. Clean joints are essential for satisfactory brazing. The use of a flux or controlled atmosphere to ensure surface cleanliness is necessary. Filler metal may be hand-held and fed into the joint (face feeding), or preplaced as rings, washers, shims, slugs, etc.

Brazing with the silver-alloy filler metals previously was known as **silver soldering and hard soldering. Braze welding** should not be confused with brazing. Braze welding is a method of welding employing a filler metal which melts below the welding points of the base metals joined, but the filler metal *is not* distributed in the joint by capillary attraction. (See also Sec. 6.)

Torch brazing uses acetylene, propane, or other fuel gas, burned with oxygen or air. The combination employed is governed by the brazing temperature range of the filler metal, which is usually above its liquidus. Flux with a melting point appropriate to the brazing temperature range and the filler metal is essential.

Furnace brazing employs the heat of a gas-fired, electric, or other type of furnace to raise the parts to brazing temperature. Fluxes may be used, although reducing or inert atmospheres are more common since they eliminate postbraze cleaning necessary with fluxes.

Induction brazing utilizes a high-frequency current to generate the necessary heat in the part by induction. Distortion in the brazed joint can be controlled by current frequency and other factors. Fluxes or gaseous atmospheres must be used in induction bearing.

Dip brazing involves the immersion of the parts in a molten bath. The bath may be either molten brazing filler metal or molten salts, which most often are brazing flux. The former is limited to small parts such as electrical connections; the latter is capable of handling assemblies weighing several hundred pounds. The particular merit of dip brazing is that the entire joint is completed all at one time.

Resistance brazing utilizes standard resistance-welding machines to supply the heat. Fluxes or atmospheres must be used, with flux predominating. Standard spot or projection welders may be used. Pressures are lower than those for conventional resistance welding.

RESISTANCE WELDING

In resistance welding, coalescence is produced by the heat obtained from the electric resistance of the workpiece to the flow of electric current in a circuit of which the workpiece is a part, and by the application of pressure.

The specific processes include **resistance spot welding, resistance seam welding, and projection welding.** Figure 13.3.11 shows diagrammatic outlines of the processes.

The resistance of the welding circuit should be a maximum at the interface of the parts to be joined, and the heat generated there must reach a value high enough to cause localized fusion under pressure.

Electrodes are of copper alloyed with such metals as molybdenum and tungsten, with high electrical conductivity, good thermal conductivity, and sufficient mechanical strength to withstand the high pressures to which they are subjected. The electrodes are water-cooled. The resistance at the surfaces of contact between the work and the electrodes must be kept low. This may be accomplished by using smooth, clean work surfaces and a high electrode pressure.

In **resistance spot welding** (Fig. 13.3.11), the parts are lapped and held in place under pressure. The size and shape of the electrodes control the size and shape of the welds, which are usually circular.

Designing for spot welding involves six elements: tip size, edge distance, contacting overlap, spot spacing, spot weld shear strength, and electrode clearance. For mild steel, the diameter of the tip face, in terms of sheet thickness t, may be taken as $0.1 + 2t$ for thin material, and as $\sqrt{t}$ for thicker material; all dimensions in inches. Edge distance should be sufficient to provide enough metal around the weld to retain it when in the molten condition. Contacting overlap is generally taken as the diameter of the weld nugget plus twice the minimum edge distance. Spot spacing must be sufficient to ensure that the welding current will not shunt through the previously made weld.

Resistance spot welding machines vary from small, manually operated units to large, elaborately instrumented units designed to produce high-quality welds, as on aircraft parts. Portable gun-type machines are available for use where the assemblies are too large to be transported to a fixed machine. Spot welds may be made singly or in multiples, the latter generally made on special purpose machines. Spacing of electrodes is important to avoid excessive shunting of welding current.

Fig. 13.3.11   (*a*) Resistance spot; (*b*) resistance seam; (*c*) projection welding.

The **resistance seam welding process** (Fig. 13.3.11) produces a series of spot welds made by circular or wheel type electrodes. The weld may be a series of closely spaced individual spot welds, overlapping spot welds, or a continuous weld nugget. The weld shape for individual welds is rectangular, continuous welds are about 80 percent of the width of the roll electrode face.

A **mash weld** is a seam weld in which the finished weld is only slightly thicker than the sheets, and the lap disappears. It is limited to thicknesses of about 16 gage and an overlap of 1½ times the sheet thickness. Operating the machine at reduced speed, with increased pressure and noninterrupted current, a strong quality weld may be secured that will be 10 to 25 percent thicker than the sheets. The process is applicable to mild steel but has limited use on stainless steel; it cannot be used on nonferrous metals. A modification of this technique employs a straight butt joint. This produces a slight depression at the weld, but the strength is satisfactory on some applications, e.g., for the production of some electric-welded pipe and tubing.

Cleanliness of sheets is of even more importance in seam welding than in spot welding. Best results are secured with cold-rolled steel, wiped clean of oil; the next best with pickled hot-rolled steel. Grinding or polishing is sometimes performed, but not sand- or shot-blasting.

In **projection welding** (Fig. 13.3.11), the heat for welding is derived from the localization of resistance at predetermined points by means of projections, embossments, or the intersections of elements of the assembly. The projections may be made by stamping or machining. The process is essentially the same as spot welding, and the projections seem to concentrate the current and pressure. Welds may be made singly or in multiple with somewhat less difficulty than is encountered in spot welding. When made in multiple, all welds may be made simultaneously. The advantages of projection welding are (1) the heat balance for difficult assemblies is readily secured, (2) the results are generally more uniform, (3) a closer spacing of welds is possible, and (4) electrode life is increased. Sometimes it is possible to projection-weld joints that could not be welded by other means.

OTHER WELDING PROCESSES

Electron Beam Welding (EBW)

In **electron beam welding**, coalescence of metals is achieved by heat generated by a concentrated beam of high-velocity electrons impinging on the surfaces to be joined. Electrons have a very small mass and carry a negative charge. An electron beam gun, consisting of an emitter, a bias electrode, and an anode, is used to create and accelerate the beam of electrons. Auxiliary components such as beam alignment, focus, and deflection coils may be used with the electron beam gun; the entire assembly is referred to as the electron beam gun column.

The advantages of the process arise from the extremely high energy density in the focused beam which produces deep, narrow welds at high speed, with minimum distortion and other deleterious heat effects. These welds show superior strength compared with those made utilizing other welding processes for a given material. Major applications are with metals and alloys highly reactive to gases in the atmosphere or those volatilized from the base metal being welded.

A disadvantage of the process lies in the necessity for providing precision parts and fixtures so that the beam can be precisely aligned with the joint to ensure complete fusion. Gapped joints are not normally welded because of fixture complexity and the extreme difficulty of manipulating filler metal into the tiny, rapidly moving, weld puddle under high vacuum. When no filler metal is employed, it is common to use the keyhole technique. Here, the electron beam makes a hole entirely through the base metal, which is subsequently filled with melted base metal as the beam leaves the area. Other disadvantages of the process arise from the cost, complexity, and skills required to operate and maintain the equipment, and the safety precautions necessary to protect operating personnel from the X-rays generated during the operation.

Laser Beam Welding and Cutting

By using a laser, energy from a primary source (electrical, chemical, thermal, optical, or nuclear) is converted to a beam of coherent electromagnetic radiation at ultraviolet, visible, or infrared frequency. Because high-energy laser beams are coherent, they can be highly concentrated to provide the high energy density needed for welding, cutting, and heat-treating metals.

As applied to welding, **pulsed** and continuously operating solid-state **lasers** and lasers that produce **continuous-wave (cw) energy** have been developed to the point that multikilowatt laser beam equipment based on CO_2 is capable of full-penetration, single-pass welding of steel to ¾-in thickness.

Lasers do not require a vacuum in which to operate, so that they offer many of the advantages of electron beam welding but at considerably lower equipment cost and higher production rates. Deep, narrow welds are produced at high speeds and low total heat input, thus duplicating the excellent weld properties and minimal heat effects obtained from electron beam welding in some applications. The application of lasers to metals—for cutting or welding—coupled with computerized control systems, allows their use for complex shapes and contours.

Solid-State Welding

Solid-state welding encompasses a group of processes in which the weld is effected by bringing clean metal surfaces into intimate contact under certain specific conditions. In **friction welding,** one part is rotated at high speed with respect to the other, under pressure. The parts are heated, but not to the melting point of the metal. Rotation is stopped at the critical moment of welding. Base metal properties across the joint show little change because the process is so rapid.

Friction stir welding (FSW) is a recently developed solid-state welding process that utilizes a cylindrical, shouldered tool that is mounted in a machine having the appearance of a vertical mill. The tool is rotated at a high speed and pressed into the joint (Fig. 13.3.12). Friction causes the material to heat and soften, but not melt. The plasticized material is moved from the leading edge of the tool to the trailing edge, leaving behind a solid-state bond. The low temperatures involved make FSW ideal for many aluminum alloys where arc welding processes result in softened regions adjacent to the weld.

Ultrasonic welding employs mechanical vibrations at ultrasonic frequencies plus pressure to effect the intimate contact between faying surfaces needed to produce a weld. (See also Sec. 12.) The welding tool is essentially a transducer that converts electric frequencies to ultra-high-frequency mechanical vibrations. By applying the tip of the tool, or anvil, to a small area in the external surface of two lapped parts, the vibrations and pressure are transmitted to the faying surfaces. Foils, thin-gage sheets, or fine wires can be spot- or seam-welded to each other or to heavier parts. Many plastics lend themselves to being joined by ultrasonic welding. Also see Secs. 6 and 12.

Explosion Welding

Explosion welding utilizes extremely high pressures to join metals, often with significantly different properties. For example, it may be used to clad a metal substrate, such as steel, with a protective layer of a dissimilar metal, such as aluminum. Since the materials do not melt, two metals with significantly different melting points can be successfully welded by explosion welding. The force and speed of the explosion are directed to cause a series of progressive shock waves that deform the faying surfaces at the moment of impact. A magnified section of the joint reveals a true weld with an interlocking waveshape and, usually, some alloying.

THERMAL CUTTING PROCESSES

Oxyfuel Cutting (OFC)   Oxyfuel cutting (Fig. 13.3.13) is used to cut steels and to prepare bevel and vee grooves. In this process, the metal is heated to its ignition temperature, or kindling point, by a series of preheat flames. After this temperature is attained, a high-velocity stream of pure oxygen is introduced, which causes oxidation or "burning" to occur. The force of the oxygen steam blows the oxides out of the joint, resulting in a clean cut. The oxidation process also generates additional thermal energy, which is radially conducted into the surrounding steel, increasing the temperature of the steel ahead of the cut. The next portion of the steel is raised to the kindling temperature, and the cut proceeds.

Carbon and low-alloy steels are easily cut by the oxyfuel process. Alloy steels can be cut, but with greater difficulty than mild steel. The level of difficulty is a function of the alloy content. When the alloy content reaches the levels found in stainless steels, oxyfuel cutting cannot be used unless the process is modified by injecting flux or iron-rich powders into the oxygen stream. Aluminum cannot be cut with the oxyfuel

Fig. 13.3.12   (1) Schematic of friction stir welding; (2) detail of probe.

process. Oxyfuel cutting is commonly regarded as the most economical way to cut steel plates greater than ½ in thick.

A variety of **fuel gases** may be used for oxyfuel cutting, with the choice largely dependent on local economics; they include natural gas, propane, acetylene, and a variety of proprietary gases offering unique advantages. Because of its role in the primary cutting stream, oxygen is always used as a second gas. In addition, some oxygen is mixed with the fuel gas in proportions designed to ensure proper combustion.

Plasma Arc Cutting (PAC) The plasma arc cutting process (Fig. 13.3.14) was developed initially to cut materials that do not permit the use of the oxyfuel process: stainless steel and aluminum. It was found, however, that plasma arc cutting offered economic advantages when applied to thinner sections of mild steel, especially those less than 1 in thick. Higher travel speed is possible with plasma arc cutting, and the volume of heated base material is reduced, minimizing metallurgical changes as well as reducing distortion.

PAC is a thermal and mechanical process. To utilize PAC, the material is heated until molten and expelled from the cut with a high-velocity stream of compressed gas. Unlike oxyfuel cutting, the process does not rely on oxidation. Because high amounts of energy are introduced through the arc, PAC is capable of extremely high-speed cutting. The thermal energy generated during the oxidation process with oxyfuel cutting is not present in plasma; hence, for thicker sections, PAC is not economically justified. The use of PAC to cut thick sections usually is restricted to materials that do not oxidize readily with oxyfuel.

Air Arc Gouging (AAG) The air carbon arc gouging system (Fig. 13.3.15) utilizes an electric arc to melt the base material; a high-velocity jet of compressed air subsequently blows the molten material away. The air carbon gouging torch looks much like a manual electrode holder, but

Fig. 13.3.14   Plasma arc cutting process.

it uses a carbon electrode instead of a metallic electrode. Current is conducted through the base material to heat it. A valve in the torch handle permits compressed air to flow through two air ports. As the air hits the molten material, a combination of oxidation and expulsion of metal takes place, leaving a smooth cavity behind. The air carbon arc gouging system is capable of removing metal at a much higher rate than can be deposited by most welding processes. It is a powerful tool used to remove metal at low cost.

Plasma Arc Gouging A newer development is the application of plasma arc equipment for gouging. The process is identical to plasma arc cutting, but the small-diameter orifice is replaced with a larger one,

Fig. 13.3.13   Oxyfuel cutting.

Fig. 13.3.15   Air arc gouging.

resulting in a broader arc. More metal is heated, and a larger, broader stream of hot, high-velocity plasma gas is directed toward the workpiece. When the torch is inclined to the work surface, the metal can be removed in a fashion similar to air carbon arc gouging. The applications of the process are similar to those of air carbon arc gouging.

DESIGN OF WELDED CONNECTIONS

A **welded connection** consists of two or more pieces of base metal joined by weld metal. Design engineers determine joint type and generally specify weld type and the required throat dimension. Fabricators select the specific joint details to be used.

Joint Types

When pieces of steel are brought together to form a **joint,** they will assume one of the five configurations presented in Fig. 13.3.16. Joint types are descriptions of the relative positions of the materials to be joined and do not imply a specific type of weld.

Weld Types

Welds fall into three categories: fillet welds, groove welds, and plug and slot welds (Fig. 13.3.17). Plug and slot welds are used for connections that transfer small loads.

Many engineers will see or have occasion to use standard welding symbols. A detailed discussion of their proper use is found in AWS documents. A few are shown in Fig. 13.3.18.

Fillet Welds   Fillet welds have a triangular cross section and are applied to the surface of the materials they join. By themselves, fillet welds do not fully fuse the cross-sectional areas of parts they join,

although it is still possible to develop full-strength connections with fillet welds. The size of a fillet weld is usually determined by measuring the leg, even though the weld is designed by specifying the required throat. For equal-legged, flat-faced fillet welds applied to plates that are oriented 90° apart, the throat dimension is found by multiplying the leg size by 0.707 (for example, sin 45°).

Groove Welds   Groove welds comprise two subcategories: complete joint penetration (CJP) groove welds and partial joint penetration (PJP) groove welds (Fig. 13.3.19). By definition, CJP groove welds have a throat dimension equal to the thickness of the material they join; a PJP groove weld is one with a throat dimension less than the thickness of the materials joined.

An **effective throat** is associated with a PJP groove weld. This term is used to differentiate between the depth of groove preparation and the probable depth of fusion that will be achieved. The **effective throat** on a PJP groove weld is abbreviated by E. The required **depth of groove preparation** is designated by a capital S. Since the designer may not know which welding process a fabricator will select, it is necessary only to specify the dimension for E. The fabricator then selects the welding process, determines the position of welding, and applies the appropriate S dimension, which will be shown on the shop drawings. In most cases, both the S and E dimensions will appear on the welding symbols of shop drawings, with the effective throat dimension shown in parentheses.

Sizing of Welds

Overwelding is one of the major factors of welding cost. Specifying the correct size of weld is the first step in obtaining low-cost welding. It is important, then, to have a simple method to figure the proper amount of weld to provide adequate strength for all types of connections.

Fig. 13.3.16   Joint types.

Fig. 13.3.17   Major weld types.

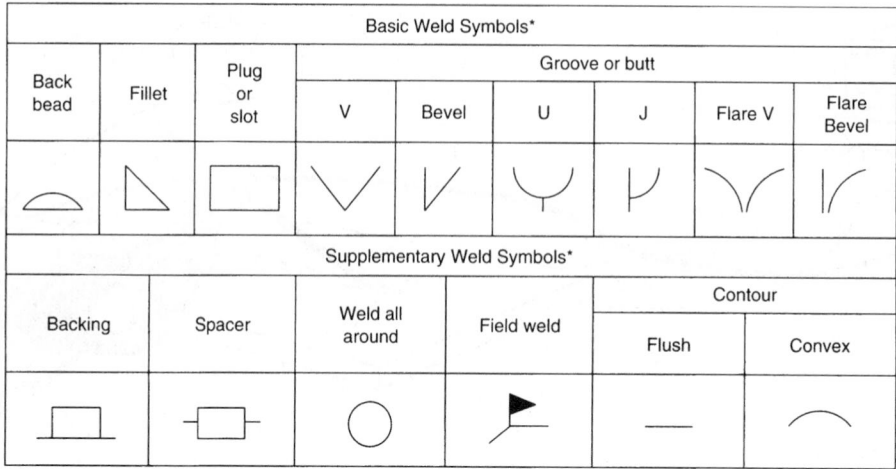

Basic Weld Symbols*								
Back bead	Fillet	Plug or slot	Groove or butt					
			V	Bevel	U	J	Flare V	Flare Bevel
⌒	◺	▭	∨	⊻	∪	⊍	∨	⎸⌵

Supplementary Weld Symbols*					
Backing	Spacer	Weld all around	Field weld	Contour	
				Flush	Convex
⊓	⊟	○	⚑	—	⌒

* See AWS publications for a complete listing.

Fig. 13.3.18 Some welding symbols commonly used. *(AWS.)*

Complete joint
penetration groove welds

Partial joint penetration
groove welds

Fig. 13.3.19 Types of groove welds.

In terms of their application, welds fall into two general types: primary and secondary. **Primary welds** are critical welds that directly transfer the full applied load at the point at which they are located. These welds must develop the full strength of the members they join. Complete joint penetration groove welds are often used for these connections. **Secondary welds** are those that merely hold the parts together to form a built-up member. The forces on these welds are relatively low, and fillet welds are generally utilized in these connections.

Filler Metal Strength **Filler metal strength** may be classified as matching, undermatching, or overmatching. **Matching filler metal** has the same, or slightly higher, minimum specified yield and tensile strength as the base metal. CJP groove welds in tension require the use of matching weld metal—otherwise, the strength of the welded connection will be lower than that of the base metal. **Undermatching filler metal** deposits welds of a strength lower than that of the base metal. Undermatching filler metal may be deposited in fillet welds and PJP groove welds as long as the designer specifies a throat size that will compensate for the reduction in weld metal strength. An **overmatching filler metal** deposits weld metal that is stronger than the base metal; this is undesirable unless, for practical reasons, lower-strength filler metal is unavailable for the application. When overmatching filler metal is used, if the weld is stressed to its maximum allowable level, the base metal can be overstressed, resulting in failure in the fusion zone. Designers must ensure that connection strength, including the fusion zone, meets the application requirements.

In welding high-strength steel, it is generally desirable to utilize undermatching filler metal for secondary welds. High-strength steel may require additional preheat and greater care in welding because there is an increased tendency to crack, especially if the joint is restrained. Undermatching filler metals such as E70 are the easiest to

use and are preferred, provided the weld is sized to impart sufficient strength to the joint.

Allowable Strength of Welds under Steady Loads A structure, or **weldment**, is as strong as its weakest point, and "allowable" weld strengths are specified by the American Welding Society (AWS), the American Institute of Steel Construction (AISC), and various other professional organizations to ensure that a weld will deliver the mechanical properties of the members being joined. Allowable weld strengths are designated for various types of welds for steady and fatigue loads.

CJP groove welds are considered **full-strength welds,** since they are capable of transferring the equivalent capacity of the members they join. In calculations, such welds are allowed the same stress as the plate, provided the proper strength level of weld metal is used (e.g., matching filler metal). In such CJP welds, the mechanical properties of the weld metal must at least match those of the base metal. If the plates joined are of different strengths, the weld metal strength must at least match the strength of the weaker plate.

Figure 13.3.20 illustrates representative applications of PJP groove welds widely used in the economical welding of very heavy plates. PJP groove welds in heavy material will usually result in savings in weld metal and welding time, while providing the required joint strength. The faster cooling and increased restraint, however, justify establishment of a minimum effective throat t_e (see Table 13.3.1).

Compression

Tension-parallel

℀ girder
in bending

Shear

Lifting
lug

Tension-transverse

Fig. 13.3.20 Applications of partial joint penetration (PJP) groove welds.

Table 13.3.1   Minimum Fillet Weld Size ω or Minimum Throat of PJP Groove Weld t_e

Material thickness of thicker part joined, in	ω or t_e, in
*To ¼ incl.	⅛
Over ¼ to ½	3⁄16
Over ½ to ¾	¼
†Over ¾ to 1½	5⁄16
Over 1½ to 2¼	⅜
Over 2¼ to 6	½
Over 6	⅝

Not to exceed the thickness of the thinner part.
* Minimum size for bridge application does not go below 3⁄16 in.
† For minimum fillet weld size, table does not go above 5⁄16-in fillet weld for over ¾-in material.

Other factors must be considered in determining the allowable stress on the throat of a PJP groove weld. Joint configuration is one. If a V, J, or U groove is specified, it is assumed that the welder can easily reach the bottom of the joint, and the effective weld throat t_e equals the depth of the groove. If a bevel groove with an included angle of 45° or less is specified and SMAW is used, ⅛ in is deducted from the depth of the prepared groove in defining the effective throat. This does not apply to the SAW process because of its deeper penetration capabilities. In the case of GMAW or FCAW, the ⅛-in reduction in throat only applies to bevel grooves with an included angle of 45° or less in the vertical or overhead position.

Weld metal subjected to compression in any direction or to tension parallel to the axis of the weld should have the same allowable strength as the base metal. Matching weld metal must be used for compression, but is not necessary for tension parallel loading.

The existence of tension forces transverse to the axis of the weld or shear in any direction requires the use of weld metal allowable strengths that are the same as those used for fillet welds. The selected weld metal may have mechanical properties higher or lower than those of the metal being joined. If the weld metal has lower strength, however, its allowable strength must be used to calculate the weld size or maximum allowable weld stress. For higher-strength weld metal, the weld allowable strength may not exceed the shear allowable strength of the base metal.

The AWS has established the **allowable shear value** for weld metal in a fillet or PJP bevel groove weld as

$$\tau = 0.30 \times \text{electrode min. spec. tensile strength} = 0.30 \times EXX$$

and has proved it valid from a series of fillet weld tests conducted by a special Task Committee of AISC and AWS.

Table 13.3.2 lists the allowable shear values for various weld metal strength levels and the more common fillet weld sizes. These values are for equal-leg fillet welds where the effective throat t_e equals $0.707 \times$ leg size ω. With the table, one can calculate the allowable unit force per lineal inch f for a weld size made with a particular electrode type. For example, the allowable unit force per lineal inch f for a ½-in fillet weld made with an E70 electrode is

$$f = 0.707\omega\tau = 0.707(\tfrac{1}{2}\,\text{in})(0.30)(70\,\text{ksi}) = 7.42\,\text{kips/lin in}$$

The minimum allowable sizes for fillet welds are given in Table 13.3.1. When materials of different thickness are joined, the minimum fillet weld size is governed by the thicker material; but this size need not exceed the thickness of the thinner material unless it is required by the calculated stress.

Connections under Simple Loads   For a **simple tensile, compressive, or shear load,** the imposed load is divided by weld length to obtain applied force, f, in pounds per lineal inch of weld. From this force, the proper leg size of the fillet weld or throat size of groove weld is found.

For **primary welds in butt joints,** groove welds must be made through the entire plate, in other words, **100 percent penetration.** Since a butt joint with a properly made CJP groove has a strength equal to or greater than that of the plate, there is no need to calculate the stress in the weld or to attempt to determine its size. It is necessary only to utilize matching filler metal.

With fillet welds, it is possible to have a weld that is either too large or too small; therefore, it is necessary to be able to determine the proper weld size.

Parallel fillet welds have forces applied parallel to their axis, and the throat is stressed only in shear. For an equal-legged fillet, the maximum shear stress occurs on the 45° throat.

Transverse fillet welds have forces applied transversely, or at right angles to their axis, and the throat is stressed by combined shear and normal (tensile or compressive) stresses. For an equal-legged fillet weld, the maximum shear stress occurs on the 67½° throat, and the maximum normal stress occurs on the 22½° throat.

Connections Subject to Horizontal Shear   A weld joining the flange of a beam to its web is stressed in horizontal shear (Fig. 13.3.21). A designer may be accustomed to specifying a certain size fillet weld for a given plate thickness (e.g., leg size about three-fourths of the plate thickness) in order that the weld develop full plate strength. This

Table 13.3.2   Allowable Loads for Various Size Fillet Welds

	60*	70*	80	90*	100	110*	120
	Strength level of weld metal (EXX)						

	60*	70*	80	90*	100	110*	120
Allowable shear stress on throat, ksi (1,000 lb/in²), of fillet weld or PJP weld							
$\tau =$	18.0	21.0	24.0	27.0	30.0	33.0	36.0
Allowable unit force on fillet weld, kips/lin in							
$f =$	12.73ω	14.85ω	16.97ω	19.09ω	21.21ω	23.33ω	25.45ω
Leg size ω, in	**Allowable unit force for various sizes of fillet welds, kips/lin in**						
1	12.73	14.85	16.97	19.09	21.21	23.33	25.45
⅞	11.14	12.99	14.85	16.70	18.57	20.41	22.27
¾	9.55	11.14	12.73	14.32	15.92	17.50	19.09
⅝	7.96	9.28	10.61	11.93	13.27	14.58	15.91
½	6.37	7.42	8.48	9.54	10.61	11.67	12.73
7⁄16	5.57	6.50	7.42	8.35	9.28	10.21	11.14
⅜	4.77	5.57	6.36	7.16	7.95	8.75	9.54
5⁄16	3.98	4.64	5.30	5.97	6.63	7.29	7.95
¼	3.18	3.71	4.24	4.77	5.30	5.83	6.36
3⁄16	2.39	2.78	3.18	3.58	3.98	4.38	4.77
⅛	1.59	1.86	2.12	2.39	2.65	2.92	3.18
1⁄16	0.795	0.930	1.06	1.19	1.33	1.46	1.59

*Fillet welds actually tested by the joint AISC-AWS Task Committee.

Fig. 13.3.21   Examples of welds stressed in horizontal shear.

particular joint between flange and web is an exception to this rule. In order to prevent web buckling, a lower allowable shear stress is usually used, which results in the requirement for a thicker web. The welds are in an area next to the flange where there is no buckling problem, and no reduction in allowable load is applied to them. From a design standpoint, these welds may be very small; their actual size sometimes is determined by the minimum size allowed by the thickness of the flange plate, in order to ensure the proper slow cooling rate of the weld on the heavier plate.

General Rules about Horizontal Shear   Aside from joining the flanges and web of a beam, or transmitting any unusually high force between the flange and web at right angles to the assembly (e.g., bearing supports, lifting lugs), the weld between flange and web serves to transmit the horizontal shear forces; the weld size is determined by the magnitude of the shear forces. In the analysis of a beam, a shear diagram is useful to depict the amount and location of welding required between the flange and web (Fig. 13.3.22).

Figure 13.3.22 shows that (1) members with applied transverse loads are subject to bending moments; (2) changes in bending moments cause horizontal shear forces; and (3) horizontal shear forces require welds to transmit them between the flange and web of the beam.

NOTE: (1) Shear forces occur only when the bending moment is changing. (2) It is quite possible for portions of a beam to have little or no

shear—i.e., the middle portions of the beams 1 and 2, within which the bending moment is constant. (3) When there is a difference in shear along the length of the beam, the shear forces are usually greatest at the ends of the beam (see beam 3), so that when web stiffeners are used, they are welded continuously when placed at the ends and welded intermittently when placed elsewhere along the length of the beam. (4) Fixing beam ends will alter the moment diagram to reduce the maximum moment; i.e., the bending moment is lower in the middle, but is now introduced at the ends. For the uniform loading configuration in beam 3, irrespective of the end conditions and their effect on bending moments and their location, the shear diagram will remain unchanged, and the amount of welding between flange and web will remain the same.

Application of Rules to Find Weld Size   Horizontal shear forces acting on the weld joining flange and web (Fig. 13.3.23) may be found from the following formula:

$$f = \frac{Vay}{In} \qquad \text{lb/lin in}$$

where f = force on weld, lb/lin in; V = total shear on section at a given position along beam, lb; a = area of flange held by weld, in^2; y = distance between center of gravity of flange area and neutral axis of whole section, in; I = moment of inertia of whole section, in^4; and n = number of welds joining flange to web.

Locate Welds at Point of Minimum Stress   In Fig. 13.3.24a, shear force is high because the weld lies on the neutral axis of the section, where the horizontal shear force is maximum. In Fig. 3.3.24b, the shear force is resisted by the channel webs, not the welds. In this last case, the shear formula above does not enter into consideration; for the configuration in Fig. 13.3.24b, full-penetration welds are not required.

Determine Length and Spacing of Intermittent Welds   If intermittent fillet welds are used, read the weld size as a decimal and divide this by the actual size used. Expressed as a percentage, this will give the

Fig. 13.3.23   Area of flange held by weld.

Fig. 13.3.22   Shear and moment diagrams.

1	2	3
Simply supported concentrated loads (lb)	Fixed ends, concentrated loads (lb)	Simply supported uniformly distributed load (lb/ft)

Beam support; loading diagrams (lb or lb/ft)

Shear diagrams (lb)

Moment diagrams (lb•ft)

Fig. 13.3.24   Design options for placement of welds. (a) Welds at neutral axis; (b) welds at outer fibers.

Table 13.3.3 Length and Spacing of Intermittent Welds

Continuous weld, %	Length of intermittent welds and distance between centers, in		
75	—	3–4	—
66	—	—	4–6
60	—	3–5	—
57	—	—	4–7
50	2–4	3–6	4–8
44	—	—	4–9
43	—	3–7	—
40	2–5	—	4–10
37	—	3–8	—
33	2–6	3–9	4–12
30	—	3–10	—
25	2–8	3–12	—
20	2–10	—	—
16	2–12	—	—

length of weld to be used per unit length. For convenience, Table 13.3.3 lists various intermittent weld lengths and distances between centers for given percentages of continuous welds; or

$$\% = \frac{\text{calculated leg size (continuous)}}{\text{actual leg size used (intermittent)}}$$

Connections Subject to Bending or Twisting The problem here is to determine the properties of the welded connection in order to check the stress in the weld without first knowing its leg size. One approach suggests assuming a certain weld leg size and then calculating the stress in the weld to see if it is over- or understressed. If the result is too far off, the assumed weld leg size is adjusted, and the calculations repeated. This iterative method has the following disadvantages:

1. A decision must be made as to throat section size to be used to determine the property of the weld. Usually some objection can be raised to any throat section chosen.

2. The resulting stresses must be combined, and for several types of loading, this can become rather complicated.

Proposed Method The following is a simple method used to determine the correct amount of welding required to provide adequate strength for either a bending or a torsion load. In this method, the weld is treated as a line, having no area but having a definite length and cross section. This method offers the following advantages:

1. It is not necessary to consider throat areas.

2. Properties of the weld are easily found from a table without knowledge of weld leg size.

3. Forces are considered per unit length of weld, rather than converted to stresses. This facilitates dealing with combined-stress problems.

4. Actual values of welds are given as force per unit length of weld instead of unit stress on throat of weld.

Visualize the welded connection as a line (or lines), following the same outline as the connection but having no cross-sectional area. In Fig. 13.3.25, the desired area of the welded connection A_w can be represented by just the length of the weld. The stress on the weld cannot be determined unless the weld size is assumed; but by following the proposed procedure which treats the weld as a line, the solution is more direct, is much simpler, and becomes basically one of determining the force on the weld(s).

Use Standard Formulas to Find Force on Weld Treat the **weld as a line.** By inserting this property of the welded connection into the standard design formula used for a particular type of load (Table 13.3.4*a*), the unit force on the weld is found in terms of pounds per lineal inch of weld.

Normally, use of these standard design formulas results in a **unit stress**, lb/in², but with the weld treated as a line, these formulas result in a **unit force** on the weld, in lb/lin in.

For problems involving bending or twisting loads, Table 13.3.4*c* is used. It contains the section modulus S_w and polar moment of inertia J_w of some 13 typical welded connections with the weld treated as a line.

Fig. 13.3.25 Treating the weld as a line.

For any given connection, two dimensions are needed: width b and depth d. Section modulus S_w is used for welds subjected to bending; polar moment of inertia J_w for welds subjected to twisting. Section modulus S_w in Table 13.3.4*c* is shown for symmetric and unsymmetric connections. For unsymmetric connections, S_w values listed differentiate between top and bottom, and the forces derived therefrom are specific to location, depending on the value of S_w used.

When one is applying more than one load to a welded connection, they are combined vectorially, but must occur at the same location on the welded joint.

Use Allowable Strength of Weld to Find Weld Size Weld size is obtained by dividing the resulting unit force on the weld by the allowable strength of the particular type of weld used, obtained from Table 13.3.5 (steady loads) or Table 13.3.6 (fatigue loads). For a joint which has only a transverse load applied to the weld (either fillet or butt weld), the allowable transverse load may be used from the applicable table. If part of the load is applied parallel (even if there are transverse loads in addition), the allowable parallel load must be used.

Applying the System to Any Welded Connection

1. Find the position on the welded connection where the combination of forces will be maximum. There may be more than one which must be considered.

2. Find the value of each of the forces on the welded connection at this point. Use Table 13.3.4*a* for the standard design formula to find the force on the weld. Use Table 13.3.4*c* to find the property of the weld treated as a line.

3. Combine (vectorially) all the forces on the weld at this point.

4. Determine the required weld size by dividing this value (step 3) by the allowable force in Table 13.3.5 or 13.3.6.

Sample Calculations Using This System The example in Fig. 13.3.26 illustrates the application of this procedure.

Summary

The application of the following guidelines will ensure effective welded connections:

1. Properly select weld type.

2. Use CJP groove welds only where loading criteria mandate.

3. Consider the cost of joint preparation vs. welding time when you select groove weld details.

4. Double-sided joints reduce the amount of weld metal required. Verify welder access to both sides and that the double-sided welds will not require overhead welding.

5. Use intermittent fillet welds where continuous welds are not required.

6. On corner joints, prepare the thinner member.

7. Strive to obtain good fit-up and do not overweld.

8. Orient welds and joints to facilitate flat and horizontal welding wherever possible.

9. Use the minimum amount of filler metal possible in a given joint.

10. Always ensure adequate access for the welder, welding apparatus, and inspector.

Allowable Fatigue Strength of Welds

The performance of a weld under **cyclic stress** is an important consideration, and applicable specifications have been developed following extensive research by the American Institute of Steel Construction (AISC). Although sound weld metal has about the same fatigue strength as unwelded metal, the change in section induced by the weld may lower the fatigue strength of the welded joint. In the case of a CJP groove weld, reinforcement, any undercut, incomplete penetration, or a crack will act as a notch; the notch, in turn, is a **stress raiser** which results in reduced fatigue strength. A fillet weld used in lap or tee joints provides an abrupt change in section; that geometry introduces a stress raiser and results in reduced fatigue strength.

The initial AISC research was directed toward bridge structure components; Table 13.3.6 illustrates a few such combinations. Similar details arise with other classes of fabricated metal products subjected to repeated loading, such as presses, transportation equipment, and material handling devices. The principles underlying fatigue performance are relatively independent of a particular application, and the data shown can be applied to the design of weldments other than for bridge construction.

Table 13.3.6 is abstracted from an extensive tabulation in the AISC "Manual of Steel Construction," 9th ed. The table also lists the variation of allowable range of stress vs. number of stress cycles for cyclic loading. A detailed discussion of the solution of fatigue-loaded welded joints is beyond the scope of this section. The reader is referred to the basic reference cited above and to the references at the head of this section in pursuing the procedures recommended to solve problems involving welded assemblies subjected to fatigue loads.

Figure 13.3.27 is a **modified Goodman diagram** for a CJP groove butt weld with weld reinforcement left on. The category is C, and the life is 500,000 to 2 million cycles (see Table 13.3.6). The vertical axis shows maximum stress $\sigma_{\max}$, and the horizontal axis shows minimum stress $\sigma_{\min}$, either positive or negative. A steady load is represented by the 45° line to the right, and a complete reversal by the 45° line to the left. The region to the right of the vertical line ($K = 0$) represents tensile loading. The fatigue formulas apply to welded butt joints in plates or other

Table 13.3.4   Treating a Weld as a Line

Type of loading		Standard design formula	Treating the weld as a line
		Stress lb/in²	Force, lb/in
Primary welds transmit entire load at this point			
	Tension or compression	$\sigma = \dfrac{P}{A}$	$f = \dfrac{P}{A_w}$
	Vertical shear	$\sigma = \dfrac{V}{A}$	$f = \dfrac{V}{A_w}$
	Bending	$\sigma = \dfrac{M}{S}$	$f = \dfrac{M}{S_w}$
	Twisting	$\sigma = \dfrac{TC}{J}$	$f = \dfrac{TC}{J_w}$
Secondary welds hold section together—low stress			
	Horizontal shear	$\tau = \dfrac{VA_y}{It}$	$f = \dfrac{VA_y}{In}$
	Torsion horizontal shear*	$\tau = \dfrac{T}{2At}$	$f = \dfrac{T}{2A}$

A = area contained within median line.
* Applies to closed tubular section only.

(*a*) Design formulas used to determine forces on a weld

b = width of connection, in
d = depth of connection, in
A = area of flange material held by welds in horizontal shear, in²
y = distance between center of gravity of flange material and N.A. of whole section, in
I = moment of inertia of whole section, in.⁴
C = distance of outer fiber, in
t = thickness of plate, in
J = polar moment of inertia of section, in.⁴
P = tensile or compressive load, lb
V = vertical shear load, lb

M = bending moment, in · lb
T = twisting moment, in · lb
A_w = length of weld, in
S_w = section modulus of weld, in²
J_w = polar moment of inertia of weld, in³
N_x = distance from x axis to face
N_y = distance from y axis to face
S = stress in standard design formula, lb/in²
f = force in standard design formula when weld is treated as a line, lb/in
n = number of welds

(*b*) Definition of terms

Table 13.3.4   Treating a Weld as a Line   (*Continued*)

Outline of welded joint b = width d = depth	Bending (about horizontal axis $x - x$)	Twisting
	$S_w = \dfrac{d^2}{6} \quad \text{in}^2$	$J_w = \dfrac{d^3}{12} \quad \text{in}^3$
	$S_w = \dfrac{d^2}{3}$	$J_w = \dfrac{d(3b^2 + d^2)}{6}$
	$S_w = bd$	$J_w = \dfrac{b^3 + 3bd^2}{6}$
$N_y = \dfrac{b^2}{2(b+d)}$ $N_x = \dfrac{d^2}{2(b+d)}$	$S_w = \dfrac{4bd + d^2}{6} = \dfrac{d^2(4b+d)}{6(2b+d)}$ top bottom	$J_w = \dfrac{(b+d)^4 - 6b^2d^2}{12(b+d)}$
$N_y = \dfrac{b^2}{2\,b+d}$	$S_w = bd + \dfrac{d^2}{6}$	$J_w = \dfrac{(2b+d)^3}{12} - \dfrac{b^2(b+d)^2}{2b+d}$
$N_x = \dfrac{d^2}{b+2d}$	$S_w = \dfrac{2bd + d^2}{3} = \dfrac{d^2(2b+d)}{3(b+d)}$ top bottom	$J_w = \dfrac{(b+2d)^3}{12} - \dfrac{d^2(b+d)^2}{b+2d}$
	$S_w = bd + \dfrac{d^2}{3}$	$J_w = \dfrac{(b+d)^3}{6}$
$N_y = \dfrac{d^2}{b+2d}$	$S_w = \dfrac{2bd + d^2}{3} = \dfrac{d^2(2b+d)}{3(b+d)}$ top bottom	$J_w = \dfrac{(b+2d)^3}{12} - \dfrac{d^2(b+d)^2}{b+2d}$
$N_y = \dfrac{d^2}{2(b+d)}$	$S_w = \dfrac{4bd + d^2}{3} = \dfrac{4bd^2 + d^3}{6b + 3d}$ top bottom	$J_w = \dfrac{d^3(4b+d)}{6(b+d)} + \dfrac{b^3}{6}$
	$S_w = bd + \dfrac{d^2}{3}$	$J_w = \dfrac{b^3 + 3bd^2 + d^3}{6}$
	$S_w = 2bd + \dfrac{d^2}{3}$	$J_w = \dfrac{2b^3 + 6bd^2 + d^3}{6}$
	$S_w = \dfrac{\pi d^2}{4}$	$J_w = \dfrac{\pi d^3}{4}$
	$I_w = \dfrac{\pi_d}{2}\left(D^2 + \dfrac{d^2}{2}\right)$ $S_w = \dfrac{I_w}{c}$ where $c = \dfrac{\sqrt{D^2 + d^2}}{2}$	

(*c*) Properties of welded connections

Table 13.3.5a   Allowable Stresses on Weld Metal

Type of weld stress	Permissible stress*	Required strength level†‡
Complete penetration groove welds		
Tension normal to effective throat	Same as base metal	Matching weld metal must be used. See table below.
Compression normal to effective throat	Same as base metal	Weld metal with a strength level equal to or one classification (10 ksi) less than matching weld metal may be used.
Tension or compression parallel to axis of weld	Same as base metal	Weld metal with a strength level equal to or less than matching weld metal may be used.
Shear on effective throat	$0.30 \times$ nominal tensile strength of weld metal (ksi) except stress on base metal shall not exceed $0.40 \times$ yield stress of base metal	
Partial penetration groove welds		
Compression normal to effective throat	Designed not to bear—$0.50 \times$ nominal tensile strength of weld metal (ksi) except stress on base metal shall not exceed $0.60 \times$ yield stress of base metal Designed to bear. Same as base metal	Weld metal with a strength level equal to or less than matching weld metal may be used.
Tension or compression parallel to axis of weld§	Same as base metal	
Shear parallel to axis of weld	$0.30 \times$ nominal tensile strength of weld metal (ksi) except stress on base metal shall not exceed $0.40 \times$ yield stress of base metal	
Tension normal to effective throat¶	$0.30 \times$ nominal tensile strength of weld metal (ksi) except stress on base metal shall not exceed $0.60 \times$ yield stress of base metal	
Fillet welds§		
Stress on effective throat, regardless of direction of application of load	$0.30 \times$ nominal tensile strength of weld metal (ksi) except stress on base metal shall not exceed $0.40 \times$ yield stress of base metal	Weld metal with a strength level equal to or less than matching weld metal may be used
Tension or compression parallel to axis of weld	Same as base metal	
Plug and slot welds		
Shear parallel to faying surfaces	$0.30 \times$ nominal tensile strength of weld metal (ksi) except stress on base metal shall not exceed $0.40 \times$ yield stress of base metal	Weld metal with a strength level equal to or less than matching weld metal may be used

* For matching weld metal, see AISC Table 1.17.2 or AWS Table 4.1.1 or table below.
† Weld metal, one strength level (10 ksi) stronger than matching weld metal may be used when using alloy weld metal on A242 or A588 steel to match corrosion resistance or coloring characteristics (Note 3 of Table 4.1.4 or AWS D1.1).
‡ Fillet welds and partial penetration groove welds joining the component elements of built-up members (ex. flange to web welds) may be designed without regard to the axial tensile or compressive stress applied to them.
§ Cannot be used in tension normal to their axis under fatigue loading (AWS 2.5). AWS Bridge prohibits their use on any butt joint (9.12.1.1), or any splice in a tension or compression member (9.17), or splice in beams or girders (9.21), however, are allowed on corner joints parallel to axial force of components of built-up members (9.12.1.2 (2). Cannot be used in girder splices (AISC 1.10.8).
¶ AWS D1.1 Section 9 Bridges–reduce above permissible stress allowables of weld by 10%.
SOURCE: Abstracted from AISC and AWS data, by permission. Footnotes refer to basic AWS documents as indicated.

Table 13.3.5b   Matching Filler and Base Metals*

Weld metal	60 or 70	70	80	100	110
Type of steel	A36; A53, Gr. B; A106, Gr. B; A131, Gr. A, B, C, CS, D, E; A139, Gr. B; A381, Gr. Y35; A500, Gr. A, B; A501; A516, Gr. 55, 60; A524, Gr. I, II; A529; A570, Gr. D.E; A573, Gr. 65; A709, Gr. 36; API 5L, Gr. B; API 5LX Gr. 42; ABS, Gr. A, B, D, CS, DS, E	A131, Gr. AH32, DH32, EH32, AH36, DH36, EH36; A242; A441; A516, Gr. 65; 70; A537, Class 17; A572, Gr. 42, 45, 50, 55; A588 (4 in and under); A595, Gr. A, B, C; A606; A607, Gr. 45, 50, 55; A618; A633, Gr. A, B, C, D (2½ in and under); A709, Gr. 50, 50W; API 2H; ABS Gr. AH32, DH32, EH32, AH36, DH36, EH36.	A572, Gr. 60, 65; A537, Class 2; A63, Gr. E	A514 [over 2½ in (63 mm)]; A709, Gr. 100, 100W [2½ to 4 in (63 to 102 mm)]	A514 [2½ in (63 mm) and under]; A517; A709, Gr. 100, 100W [2½ in (63 mm) and under]

* Abstracted from AISC and AWS data, by permission

joined members. The region to the left of this line represents cycles going into compressive loading.

The fatigue formulas are meant to reduce the allowable stress as cyclic loads are encountered. The resulting allowable fatigue stress should not exceed the usual steady load allowable stress. For this reason, these fatigue curves are cut off with horizontal lines representing the steady load allowable stress for the particular type of steel used. In Fig. 13.3.27, A36 steel with E60 or E70 weld metal is cut off at 22 ksi; A441 steel with E70 weld metal at 30 ksi; and A514 with E110 weld metal at either 54 or 60 ksi, depending on plate thickness.

Figure 13.3.28 is a modification of Fig. 13.3.27, where the horizontal axis represents the range K of the cyclic stress. Here, two additional

Table 13.3.6   AISC Fatigue Allowable Stresses for Cyclic Loading

Base metal–no attachments–rolled or clean surfaces

(A)

Base metal–built-up plates or shapes–connected by continuous complete penetration groove welds or fillet welds–without attachments. Note: don't use this as a fatigue allowable for the fillet weld to transfer a load. See (F) for that case.

(B)

Base metal and weld metal at full penetration groove welds–changes thickness or width not to exceed a slope of 1 in 2½ (22°). Ground flush and inspected by radiography or ultrasound (B).

For 514 steel (B')

Base metal and weld metal at full penetration groove welds–changes in thickness or width not to exceed a slope of 1 in 2½ (22°). Weld reinforcement not removed inspected by radiography or ultrasound.

(C)

Longitudinal loading
Base metal–full penetration groove weld
Weld termination ground smooth
Weld reinforcement not removed
Not necessarily equal thickness

$2 < a < 12b$ or 4 in   (D)
$a > 12b$ or 4 in when $b \pm 1$ in   (E)
$a > 12b$ or 4 in when $b > 1$ in   (E')

Weld metal of continuous or intermittent longitudinal or transverse fillet welds

(F)

| | Allowable Stress Range, σ_{sr} ksi | | | |
Category	20,000 to 100,000 ~	100,000 to 500,000 ~	500,000 to 2×10^6 ~	Over 2×10^6 ~
A	63	37	24	24
B	49	29	18	16
B'	39	23	15	12
C	35	21	13	10 (Note 1)
D	28	16	10	7
E	22	13	8	5
E'	16	9	6	3
F	15	12	9	8

NOTE 1: Flexural stress range of 12 ksi permitted at toe of stiffener welds on flanges.

Allowable fatigue stress:

$$\sigma_{max} = \frac{\sigma_{sr}}{1 - K} \quad \text{for normal stress } \sigma \qquad \tau_{max} = \frac{\tau_{sr}}{1 - K} \quad \text{for shear stress } \tau$$

but shall not exceed steady allowables

σ_{max} or τ_{max} = maximum allowable fatigue stress
σ_{sr} or τ_{sr} = allowable range of stress from table above

$$K = \frac{\sigma_{min}}{\sigma_{max}} = \frac{M_{min}}{M_{max}} = \frac{F_{min}}{F_{max}} = \frac{\tau_{min}}{\tau_{max}} = \frac{V_{min}}{V_{max}}$$

where S = shear, T = tension, R = reversal, M = stress in metal, and W = stress in weld.

SOURCE: Abstracted from AISC "Manual of Steel Construction," 9th ed., by permission.

strength levels of weld metal have been added—E80 and E90—along with equivalent strength levels of steel. Note that for a small range in stress of $K = 0.6$ to 1.0, higher-strength welds and steels show increased allowable fatigue stress. However, as the stress range increases—lower values of K—the increase is not as great, and below $K = + 0.35$ all combinations of weld and steel strengths exhibit the same allowable fatigue stress.

Figure 13.3.29 represents the same welded joint as in Fig. 13.3.28, but with a lower life of 20,000 to 100,000 cycles. Here, the higher-strength welds and steels have higher allowable fatigue stresses and over a wider range. A conclusion can be drawn that the wider the range of cycling, the less useful the application of a high-strength steel. When there is a complete stress reversal, there is not much advantage in using a high-strength steel.

Step 1

Properties of weld treated as a line
(Table 13.3.4c)

$$N_Y = \frac{b^2}{2b + d} = \frac{5^2}{2 \times 5 + 10} = 1.25 \text{ in}$$

Combined forces are maximum at point a

Twisting (vertical component)

$$f_v = \frac{T\,C_v}{J_w} = \frac{(180,000)(3.75)}{(385.9)} = 1750 \text{ lb/in}$$

Step 2 (Table 13.3.4a)

Vertical shear

$$f_s = \frac{P}{A_w} = \frac{18,000}{20} = 900 \text{ lb/in}$$

Twisting (horizontal component)

$$J_w = \frac{(2b + d)^3}{12} - \frac{b^2(b + d)^2}{(2b + d)} = \frac{(2 \times 5 + 10)^3}{12} - \frac{5^2(5 + 10)^2}{(2 \times 5 + 10)} = 385.9 \text{ in}^3$$

$$A_w = 20 \text{ in}$$

$$f_h = \frac{T\,C_h}{J_w} = \frac{(180,000)(5)}{(385.9)} = 2340 \text{ lb/in}$$

Step 3

$$f_R = \sqrt{f_v^2 + f_h^2} = \sqrt{.2340^2 + 2650^2} = 3540 \text{ lb/in}$$

$f_h = 2340$

$f_v = 1750$

$f_R = 3540$

$f_s = 900$

2650

Step 4 (Assume E60 electrodes; Table 13.3.2)

$$\omega = \frac{3540}{12,730} = 0.278 \text{ in or } \tfrac{5}{16} \text{ in leg fillet weld}$$

Fig. 13.3.26 Sample problem: Find the fillet weld size required for the connection shown.

Fig. 13.3.27 Modified Goodman diagram for butt weld. [Butt weld and plate, weld reinforcement left on. Category C; 500,000 to 2,000,000 cycles (see Table 13.3.6).]

Fig. 13.3.28 Fatigue allowable for groove weld. [Butt weld and plate, weld reinforcement left on. Category C; 500,000 to 2,000,000 cycles (see Table 13.3.6).]

Fig. 13.3.29 Fatigue allowable for groove weld. [Butt weld and plate, weld reinforcement left on. Category C; 20,000 to 100,000 cycles (see Table 13.3.6).]

BASE METALS FOR WELDING

Introduction

When one is considering welding them, the nature of the base metals must be understood and recognized, i.e., their chemical composition, mechanical properties, and metallurgical structure. Cognizance of the mechanical properties of the base metal will guide the designer to ensure that the weld metal deposited will have properties equal to those of the base metal; knowledge of the chemical composition of the base metal will affect the selection of the filler metal and/or electrode; finally, the metallurgical structure of the base metal as it comes to the welding operation (hot-worked, cold-worked, quenched, tempered, annealed, etc.) will affect the weldability of the metal and, if it is weldable, the degree to which the final properties are as dictated by design requirements.

Welding specifications may address these matters, and base metal suppliers can provide additional data as to the weldability of the metal.

In some cases, the identity of the base metal is absolutely not known. To proceed to weld such metal may prove disastrous. Identification may be aided by some general characteristics which may be self-evident: carbon steel (oxide coating) vs. stainless steel (unoxidized); brush-finished aluminum (lightweight) vs. brush-finished Monel metal (heavy); etc. Ultimately, it may become necessary to subject the unknown metal to chemical, mechanical, and other types of laboratory tests to ascertain its exact nature.

Steel

Low-Carbon Steels (Carbon up to 0.30 percent) Steels in this class are readily welded by most arc and gas processes. Preheating is unnecessary unless parts are very heavy or welding is performed below 32°F (0°C). Torch-heating steel in the vicinity of welding to 70°F (21°C) offsets low temperatures. Postheating is necessary only for important structures such as boilers, pressure vessels, and piping. GTAW is usable only on killed steels; rimmed steels produce porous, weak welds. Resistance welding is readily accomplished if carbon is below 0.20 percent; higher carbon requires heat-treatment to slow the cooling rate and avoid hardness. Brazing with BAg, BCu, and BCuZn filler metals is very successful.

Medium-Carbon Steels (Carbon from 0.30 to 0.45 percent) This class of steel may be welded by the arc, resistance, and gas processes. As the rapid cooling of the metal in the welded zone produces a harder structure, it is desirable to hold the carbon as near 0.30 percent as possible. These hard areas are proportionately more brittle and difficult to machine. The cooling rate may be diminished and hardness decreased by preheating the metal to be welded above 300°F (149°C) and preferably to 500°F (260°C). The degree of preheating depends on the thickness of the section. Subsequent heating of the welded zone to 1,100 to 1,200°F (593 to 649°C) will restore ductility and relieve thermal strains. Brazing may also be used, as noted for low-carbon steels above.

High-Carbon Steels (Carbon from 0.45 to 0.80 percent) These steels are rarely welded except in special cases. The tendency for the metal heated above the critical range to become brittle is more pronounced than with lower- or medium-carbon steels. Thorough preheating of metal, in and near the welded zone, to a minimum of 500°F is essential. Subsequent annealing at 1,350 to 1,450°F (732 to 788°C) is also desirable. Brazing is often used with these steels, and is combined with the heat treatment cycle.

Low-Alloy Steels The weldability of low-alloy steels is dependent upon the analysis and the hardenability, those exhibiting low hardenability being welded with relative ease, whereas those of high hardenability requiring preheating and postheating. Sections of ¼ in (6.4 mm) or less may be welded with mild-steel filler metal and may provide joint strength approximating base metal strength by virtue of alloy pickup in the weld metal and weld reinforcement. Higher-strength alloys require filler metals with mechanical properties matching the base metal. Special alloys with creep-resistant or corrosion-resistant properties must be welded with filler metals of the same chemical analysis. Low-hydrogen-type electrodes (either mild- or alloy-steel analyses) permit the welding of alloy steels, minimizing the occurrence of underhead cracking.

Stainless Steel

Stainless steel is an iron-base alloy containing upward of 11 percent chromium. A thin, dense surface film of chromium oxide which forms on stainless steel imparts superior corrosion resistance; its passivated nature inhibits scaling and prevents further oxidation, hence the appellation "stainless." (See Sec. 6.2.)

There are five types of stainless steels, and depending on the amount and kind of alloying additions present, they range from fully austenitic to fully ferritic. Most stainless steels have good weldability and may be welded by many processes, including arc welding, resistance welding, electron and laser beam welding, and brazing. With any of these, the joint surfaces and any filler metal must be clean.

The coefficient of thermal expansion for the austenitic stainless steels is 50 percent greater than that of carbon steel; this must be taken into account to minimize distortion. The low thermal and electrical conductivity of austenitic stainless steel is generally helpful. Low welding heat is required because the heat is conducted more slowly from the joint, but low thermal conductivity results in a steeper thermal gradient and increases distortion. In resistance welding, lower current is used because electric resistivity is higher.

Ferritic Stainless Steels Ferritic stainless steels contain 11.5 to 30 percent Cr, up to 0.20 percent C, and small amounts of ferrite stabilizers, such as Al, Nb, Ti, and Mo. They are ferritic at all temperatures, do not transform to austenite, and are not hardenable by heat treatment. This group includes types 405, 409, 430, 442, and 446. To weld ferritic stainless steels, filler metals should match or exceed the Cr level of the base metal.

Martensitic Stainless Steels Martensitic stainless steels contain 11.4 to 18 percent Cr, up to 1.2 percent C, and small amounts of Mn and Ni. They will transform to austenite on heating and, therefore, can be hardened by formation of martensite on cooling. This group includes types 403, 410, 414, 416, 420, 422, 431, and 440. Weld cracks may appear on cooled welds as a result of martensite formation. The Cr and C content of the filler metal should generally match these elements in the base metal. Preheating and interpass temperature in the 400 to 600°F range is recommended for welding most martensitic stainless steels. Steels with over 0.20 percent C often require a postweld heat treatment to avoid weld cracking.

Austenitic Stainless Steels Austenitic stainless steels contain 16 to 26 percent Cr, 10 to 24 percent Ni and Mn, up to 0.40 percent C, and small amounts of Mo, Ti, Nb, and Ta. The balance between Cr and Ni + Mn is normally adjusted to provide a microstructure of 90 to 100 percent austenite. These alloys have good strength and high toughness over a wide temperature range, and they resist oxidation to over 1,000°F. This group includes types 302, 304, 310, 316, 321, and 347. Filler metals for these alloys should generally match the base metal, but for most alloys should also provide a microstructure with some ferrite to avoid hot cracking. Two problems are associated with welding austenitic stainless steels: sensitization of the weld-heat-affected zone and hot cracking of weld metal.

Sensitization is caused by chromium carbide precipitation at the austenitic grain boundaries in the heat-affected zone when the base metal is heated to 800 to 1,600°F. Chromium carbide precipitates remove chromium from solution in the vicinity of the grain boundaries, and this condition leads to intergranular corrosion. The problem can be alleviated by using low-carbon stainless-steel base metal (types 302L, 316L, etc.) and low-carbon filler metal. Alternately, there are **stabilized stainless-steel** base metals and filler metals available which contain alloying elements that react preferentially with carbon, thereby not depleting the chromium content in solid solution and keeping it available for corrosion resistance. Type 321 contains titanium and type 347 contains niobium and tantalum, all of which are stronger carbide formers than chromium.

Hot cracking is caused by low-melting-point metallic compounds of sulfur and phosphorus which penetrate grain boundaries. When present in the weld metal or heat-affected zone, they will penetrate grain boundaries and cause cracks to appear as the weld cools and shrinkage stresses develop. Hot cracking can be prevented by adjusting the composition of the base metal and filler metal to obtain a microstructure with a small amount of ferrite in the austenite matrix. The ferrite provides ferrite-austenite boundaries which control the sulfur and phosphorus compounds and thereby prevent hot cracking.

Precipitation-Hardening Stainless Steels Precipitation-hardening (PH) stainless steels contain alloying elements such as aluminum which permit hardening by a solution and aging heat treatment. There are three categories of PH stainless steels: martensitic, semiaustenitic, and austenitic. Martensitic PH stainless steels are hardened by quenching from the austenitizing temperature (around 1,900°F) and then aging between 900 and 1,150°F. Semiaustenitic PH stainless steels do not transform to martensite when cooled from the austenitizing temperature because the martensite transformation temperature is below room temperature. Austenitic PH stainless steels remain austenitic after quenching from the solution temperature, even after substantial amounts of cold work.

If maximum strength is required of martensitic PH and semiaustenitic PH stainless steels, matching, or nearly matching, filler metal should be used, and before welding, the work pieces should be in the annealed or solution-annealed condition. After welding, a complete solution heat treatment plus an aging treatment is preferred. If postweld solution treatment is not feasible, the components should be solution-treated before welding and then aged after welding. Thick sections of highly restrained parts are sometimes welded in the overaged condition. These require a full heat treatment after welding to attain maximum strength properties.

Austenitic PH stainless steels are the most difficult to weld because of hot cracking. Welding is preferably done with the parts in solution-treated condition, under minimum restraint and with minimum heat input. Filler metals of the Ni-Cr-Fe type, or of conventional austenitic stainless steel, are preferred.

Duplex Stainless Steels Duplex stainless steels are the most recently developed type of stainless steel, and they have a microstructure of approximately equal amounts of ferrite and austenite. They have advantages over conventional austenitic and ferritic stainless steels in that they possess higher yield strength and greater stress corrosion cracking resistance. The duplex microstructure is attained in steels containing 21 to 25 percent Cr and 5 to 7 percent Ni by hot-working at 1,832 to 1,922°F, followed by water quenching. Weld metal of this composition will be mainly ferritic because the deposit will solidify as ferrite and will transform only partly to austenite without hot working or annealing. Since hot-working or annealing most weld deposits is not feasible, the metal composition filler is generally modified by adding Ni (to 8 to 10 percent); this results in increased amounts of austenite in the as-welded microstructure.

Cast Iron

Even though cast iron has a high carbon content and is a relatively brittle and rigid material, welding can be performed successfully if proper precautions are taken. Optimum conditions for welding include the following: (1) A weld groove large enough to permit manipulation of the electrode or the welding torch and rod. The groove must be clean and free of oil, grease, and any foreign material. (2) Adequate preheat, depending on the welding process used, the type of cast iron, and the size and shape of the casting. Preheat temperature must be maintained throughout the welding operation. (3) Welding heat input sufficient for a good weld but not enough to superheat the weld metal; i.e., welding temperature should be kept as low as practicable. (4) Slow cooling after welding. Gray iron may be enclosed in insulation, lime, or vermiculite. Other irons may require postheat treatment immediately after welding to restore mechanical properties. ESt and ENiFe identify electrodes of steel and of a nickel-iron alloy. Many different welding processes have been used to weld cast iron, the most common being manual shielded metal-arc welding, gas welding, and braze welding.

Aluminum and Aluminum Alloys

(See Sec. 6.4.)

The **properties** that distinguish the **aluminum alloys** from other metals determine which welding processes can be used and which particular procedures must be followed for best results. Among the welding processes that can be used, choice is further dictated by the requirements of the end product and by economic considerations.

Physical properties of aluminum alloys that most significantly affect all welding procedures include low-melting-point range, approx 900 to 1,215°F (482 to 657°C), high thermal conductivity (about two to four times that of mild steel), high rate of thermal expansion (about twice that of mild steel), and high electrical conductivity (about 3 to 5 times that of mild steel). Interpreted in terms of welding, this means that, when compared with mild steel, much higher welding speeds are demanded, greater care must be exercised to avoid distortion, and for arc and resistance welding, much higher current densities are required.

Aluminum alloys are not quench-hardenable. However, weld cracking may result from excessive shrinkage stresses due to the high rate of thermal contraction. To offset this tendency, welding procedures, where possible, require a fast weld cycle and a narrow-weld zone, e.g., a highly concentrated heat source with deep penetration, moving at a high rate of speed. Shrinkage stresses can also be reduced by using a filler metal of lower melting point than the base metal. The filler metal ER4043 is often used for this purpose.

Welding procedures also call for the removal of the thin, tough, transparent film of aluminum oxide that forms on and protects the surface of these alloys. The oxide has a melting point of about 3,700°F (2,038°C) and can therefore exist as a solid in the molten weld. Removal may be by chemical reduction or by mechanical means such as machining, filing, rubbing with steel wool, or brushing with a stainless-steel wire brush. Most aluminum is welded with GTAW or GMAW. GTAW usually uses alternating current, with argon as the shielding gas. The power supply must deliver high current with balanced wave characteristics, or deliver high-frequency current. With helium, weld penetration is deeper, and higher welding speeds are possible. Most welding, however, is done using argon because it allows for better control and permits the welder to see the weld pool more easily.

GMAW employs direct current, electrode positive in a shielding gas that may be argon, helium, or a mixture of the two. In this process, the welding arc is formed by the filler metal, which serves as the electrode. Since the filler metal is fed from a coil as it melts in the arc, some arc instability may arise. For this reason, the process does not have the same precision as the GTAW process for welding very thin gages. However, it is more economical for welding thicker sections because of its higher deposition rates.

Copper and Copper Alloys

In welding **commercially pure copper,** it is important to select the correct type. Electrolytic, or "tough-pitch," copper contains a small percentage of copper oxide, which at welding heat leads to oxide embrittlement. For welded assemblies it is recommended that deoxidized, or oxygen-free, copper be used and that welding rods, when needed, be of the same analysis. The preferred processes for welding copper are GTAW and GMAW; manual SMAW can also be used. It is also welded by oxy-acetylene method and braze-welded; brazing with brazing filler metals conforming to BAg, BCuP, and RBCuZn-A classifications is also employed. The high heat conductivity of copper requires special consideration in welding; generally higher welding heats are necessary together with concurrent supplementary heating. (See also Sec. 6.4.)

Copper alloys are extensively welded in industry. The specific procedures employed are dependent upon the analysis, and reference should be made to the AWS Welding Handbook. Filler metals for welding copper and its alloys are covered in AWS specifications.

SAFETY

Welding is safe when sufficient measures are taken to protect the welder from potential hazards. When these measures are overlooked or ignored, welders can be subject to electric shock; overexposure to radiation, fumes, and gases; and fires and explosion. Any of these can be fatal. Everyone associated with welding operations should be aware of the potential hazards and help ensure that safe practices are employed. Infractions must be reported to the appropriate responsible authority.

ANSI Z49.1:2005, "Safety in Welding, Cutting, and Allied Process," available as a free download from AWS (http://www.aws.org/technical/facts), should be consulted for information on welding safety. A printed copy is also available for purchase from Global Engineering Documents (www.global.ihs.com, telephone 1-800-854-7179). From the same website, a variety of AWS Safety & Health Fact Sheets also can be downloaded.

NOTE: Oxygen is incorrectly called **air** in some fabricating shops. Air from the atmosphere contains only 21 percent oxygen and obviously is different from the 100 percent pure oxygen used for cutting. The unintentional confusion of oxygen with air has resulted in fatal accidents. When compressed oxygen is inadvertently used to power air tools, e.g., an explosion can result. While most people recognize that fuel gases are dangerous, the case can be made that oxygen requires even more careful handling.

Information about welding safety is available from American Welding Society, P.O. Box 351040, Miami, FL 33135.

13.4 MACHINING PROCESSES AND MACHINE TOOLS
by Serope Kalpakjian

REFERENCES: Arnone, "High Performance Machining," Hanser, 1998. "ASM Handbook," Vol. 16: "Machining," ASM International, 1989. Brown, "Advanced Machining Technology Handbook," McGraw-Hill, 1998. El-Hofy, "Advanced Machining Processes," McGraw-Hill, 2005. Erdel, "High-Speed Machining," Society of Manufacturing Engineers, 2003. Kalpakjian and Schmid, "Manufacturing Engineering and Technology," 5th ed., Prentice-Hall, 2006. Krar, "Grinding Technology," 2d ed., Delmar, 1995. "Machining Data Handbook," 3d ed., 2 vols., Machinability Data Center, 1980. "Metal Cutting Handbook," 7th ed., Industrial Press, 1989. Meyers and Slattery, "Basic Machining Reference Handbook," Industrial Press, 1988. Salmon, "Modern Grinding Process Technology," McGraw-Hill, 1992. Shaw, "Metal Cutting Principles," 2d ed., Oxford, 2005. Shaw, "Principles of Abrasive Processing," Oxford, 1996. Sluhan (ed.), "Cutting and Grinding Fluids: Selection and Application," Society of Manufacturing Engineers, 1992. Stephenson and Agapiou, "Metal Cutting: Theory and Practice," Dekker 1996. Trent and Wright, "Metal Cutting," 4th ed., Butterworth Heinemann, 2000. Walsh, "Machining and Metalworking Handbook," McGraw-Hill, 1994. Webster et al., "Abrasive Processes," Dekker, 1999. Weck, "Handbook of Machine Tools," 4 vols., Wiley, 1984.

INTRODUCTION

Machining processes, which include cutting, grinding, and various non-mechanical chipless processes, are desirable or even necessary for the following basic reasons: (1) Closer dimensional tolerances, surface roughness, or surface-finish characteristics may be required than are available by casting, forming, powder metallurgy, and other shaping processes; and (2) part geometries may be too complex or too expensive to be manufactured by other processes. However, machining processes inevitably waste material in the form of chips, production rates may be low, and unless carried out properly, the processes can have detrimental effects on the surface properties and performance of parts.

Traditional machining processes consist of turning, boring, drilling, reaming, threading, milling, shaping, planing, and broaching, as well as abrasive processes such as grinding, ultrasonic machining, lapping, and honing. Advanced processes include electrical and chemical means of material removal, as well as the use of abrasive jets, water jets, laser beams, and electron beams. This section describes the principles of these operations, the processing parameters involved, and the characteristics of the machine tools employed.

BASIC MECHANICS OF METAL CUTTING

The basic mechanics of chip-type machining processes (Fig. 13.4.1) are shown, in simplest two-dimensional form, in Fig. 13.4.2. A tool with a certain **rake angle** α (positive as shown) and **relief angle** moves along the surface of the workpiece at a depth t_1. The material ahead of the tool is sheared continuously along the **shear plane**, which makes an angle of ϕ with the surface of the workpiece. This angle is called the **shear angle** and, together with the rake angle, determines the chip thickness t_2. The ratio of t_1 to t_2 is called the **cutting ratio** r. The relationship between the shear angle, the rake angle, and the cutting ratio is given by the equation $\tan \phi = r \cos \alpha /(1 - r \sin \alpha)$. It can readily be seen that the shear angle is important in that it controls the thickness of the chip. This, in turn, has great influence on cutting performance. The **shear strain** that the material undergoes is given by the equation $\gamma = \cot \phi + \tan (\phi - \alpha)$. Shear strains in metal cutting are usually less than 5.

Fig. 13.4.2   Basic mechanics of metal cutting process.

Investigations have shown that the shear plane may be neither a plane nor a narrow zone, as assumed in simple analysis. Various formulas have been developed which define the shear angle in terms of such factors as the rake angle and the friction angle β. (See Fig. 13.4.3.)

Because of the large shear strains that the chip undergoes, it becomes hard and brittle. In most cases, the chip curls away from the tool. Among possible factors contributing to chip curl are nonuniform normal stress distribution on the shear plane, strain hardening, and thermal effects.

Regardless of the type of machining operation, some basic types of chips or combinations of these are found in practice (Fig. 13.4.4).

Continuous chips are formed by continuous deformation of the workpiece material ahead of the tool, followed by smooth flow of the chip along the tool face. These chips ordinarily are obtained in cutting ductile materials at high speeds.

Fig. 13.4.1   Examples of chip-type machining operations.

Fig. 13.4.3   Force system in metal cutting process.

Fig. 13.4.4 Basic types of chips produced in metal cutting: (*a*) continuous chip with narrow, straight primary shear zone; (*b*) secondary shear zone at the tool-chip interface; (*c*) continuous chip with large primary shear zone; (*d*) continuous chip with built-up edge; (*e*) segmented or nonhomogeneous chip, (*f*) discontinuous chip. (*Source: After M. C. Shaw.*)

Discontinuous chips consist of segments which are produced by fracture of the metal ahead of the tool. The segments may be either loosely connected to each other or unconnected. Such chips are most often found in the machining of brittle materials or in cutting ductile materials at very low speeds or low or negative rake angles.

Inhomogeneous (serrated) chips consist of regions of large and small strain. Such chips are characteristic of metals with low thermal conductivity or metals whose yield strength decreases sharply with temperature. Chips from titanium alloys frequently are of this type.

Built-up edge chips consist of a mass of metal which adheres to the tool tip while the chip itself flows continuously along the rake face. This type of chip is often encountered in machining operations at low speeds and is associated with high adhesion between chip and tool and causes poor surface finish.

The **forces** acting on the cutting tool are shown in Fig. 13.4.3. The resultant force R has two components, F_c and F_t. The cutting force F_c in the direction of tool travel determines the amount of work done in cutting. The thrust force F_t does no work but, together with F_c, produces deflections of the tool. The resultant force also has two components on the shear plane: F_s is the force required to shear the metal along the shear plane, and F_n is the normal force on this plane. Two other force components also exist on the face of the tool: the friction force F and the normal force N.

Whereas the cutting force F_c is always in the direction shown in Fig. 13.4.3, the thrust force F_t may be in the opposite direction to that shown in the figure. This occurs when both the rake angle and the depth of cut are large, and friction is low.

From the geometry of Fig. 13.4.3, the following relationships can be derived: The **coefficient of friction** at the tool-chip interface is given by $\mu = (F_t + F_c \tan \alpha)/(F_c - F_t \tan \alpha)$. The **friction force** along the tool is $F = F_t \cos \alpha + F_c \sin \alpha$. The **shear stress** in the shear plane is $\tau = (F_c \sin \phi \cos \phi - F_t \sin^2 \phi)/A_0$, where A_0 is the cross-sectional area that is being cut from the workpiece.

The coefficient of friction on the tool face is a complex but important factor in cutting performance; it can be reduced by such means as the use of an effective cutting fluid, higher cutting speed, improved tool material and condition, or chemical additives in the workpiece material.

The net **power** consumed at the tool is $P = F_c V$. Since F_c is a function of tool geometry, workpiece material, and process variables, it is difficult reliably to calculate its value in a particular machining operation. Depending on workpiece material and the condition of the tool, **unit power** requirements in machining range between 0.2 hp·min/in³ (0.55 W·s/mm³) of metal removal for aluminum and magnesium alloys, to 3.5 for high-strength alloys. The power consumed is the product of unit power and rate of metal removal: $P = $ (unit power)(vol/min).

The power consumed in cutting is transformed mostly to **heat.** Most of the heat is carried away by the chip, and the remainder is divided between the tool and the workpiece. An increase in cutting speed or feed will increase the proportion of the heat transferred to the chip. It has been observed that, in turning, the average interface **temperature** between the tool and the chip increases with cutting speed and feed, while the influence of the depth of cut on temperature has been found to be limited. Interface temperatures to the range of 1,500 to 2,000°F (800 to 1,100°C) have been measured in metal cutting. Generally the use of a cutting fluid removes heat and thus avoids temperature buildup on the cutting edge.

In cutting metal at high speeds, the chips may become very hot and cause safety hazards because of long spirals which whirl around and become entangled with the tooling. In such cases, **chip breakers** are introduced on the tool geometry, which curl the chips and cause them to break into short sections. Chip breakers can be produced on the face of the cutting tool or insert, or are separate pieces clamped on top of the tool or insert.

A phenomenon of great significance in metal cutting is **tool wear.** Many factors determine the type and rate at which wear occurs on the tool. The major critical variables that affect wear are tool temperature, type and hardness of tool material, grade and condition of workpiece, abrasiveness of the microconstituents in the workpiece material, tool geometry, feed, speed, and cutting fluid. The type of wear pattern that develops depends on the relative role of these variables.

Tool wear can be classified as (1) flank wear (Fig. 13.4.5); (2) crater wear on the tool face; (3) localized wear, such as the rounding of the cutting edge; (4) chipping or thermal softening and plastic flow of the cutting edge; (5) concentrated wear resulting in a deep groove at the edge of a turning tool, known as **wear notch.**

In general, the wear on the flank or relief side of the tool is the most dependable guide for **tool life.** A wear land of 0.060 in (1.5 mm) on high-speed steel tools and 0.015 in (0.4 mm) for carbide tools is usually used as the endpoint. The cutting speed is the variable which has the greatest influence on tool life. The relationship between tool life and cutting speed is given by the Taylor equation $VT^n = C$, where V is the cutting speed; T is the actual cutting time to develop a certain wear land, min; C is a constant whose value depends on workpiece material and process variables, numerically equal to the cutting speed that gives a tool life of 1 min; and n is the exponent whose value depends on workpiece material and other process variables.

Fig. 13.4.5 Types of tool wear in cutting.

The recommended cutting speed for a high-speed steel tool is generally the one which produces a 60- to 120-min tool life. With carbide tools, a 30- to 60-min tool life may be satisfactory. Values of n typically range from 0.08 to 0.2 for high-speed steels, 0.1 to 0.15 for cast alloys, 0.2 to 0.5 for uncoated carbides, 0.4 to 0.6 for coated carbides, and 0.5 to 0.7 for ceramics.

When tool-life equations are used, caution should be exercised in extrapolation of the curves beyond the operating region for which they are derived. In a log-log plot, tool life curves may be linear over a short cutting-speed range but are rarely linear over a wide range of cutting speeds. In spite of the considerable data obtained to date, no simple formulas can be given for quantitative relationships between tool life and various process variables for a wide range of materials and conditions.

An important aspect of machining on computer-controlled equipment is **tool-condition monitoring** while the machine is in operation with little or no supervision by an operator. Most state-of-the-art machine controls are now equipped with tool-condition monitoring systems. Two common techniques involve the use of (1) transducers that are installed on the tool holder and continually monitor torque and forces and (2) acoustic emission through a piezoelectric transducer. In both methods the signals are analyzed and interpreted automatically for tool wear or chipping, and corrective actions are taken before any significant damage is done to the workpiece.

A term commonly used in machining and comprising most of the items discussed above is **machinability**. This is best defined in terms of (1) tool life, (2) power requirement, and (3) surface integrity. Thus, a good machinability rating would indicate a combination of long tool life, low power requirement, and a good surface. However, it is difficult to develop quantitative relationships between these variables. Tool life is considered as the important factor and, in production, is usually expressed as the number of pieces machined between tool changes. Various tables are available in the literature that show the machinability rating for different materials; however, these ratings are relative. To determine the proper machining conditions for a given material, refer to the machining recommendations given later in this section.

The major factors influencing **surface finish** in machining are (1) the profile of the cutting tool in contact with the workpiece, (2) fragments of built-up edge left on the workpiece during cutting, and (3) vibration and chatter. Improvement in surface finish may be obtained to various degrees by increasing the cutting speed and decreasing the feed and depth of cut. Changes in cutting fluid, tool geometry, and tool material are also important; the microstructure and chemical composition of the material have great influence on surface finish.

As a result of mechanical working and thermal effects, **residual stresses** are generally developed on the surfaces of metals that have been machined or ground. These stresses may cause warping of the workpiece as well as affect the resistance to fatigue and stress corrosion. To minimize residual stresses, sharp tools, medium feeds, and medium depths of cut are recommended.

Because of plastic deformation, thermal effects, and chemical reactions during machining processes, alterations of machined surfaces may take place which can seriously affect the **surface integrity** of a part. Typical detrimental effects may be lowering of the fatigue strength of the part, distortion, changes in stress-corrosion properties, burns, cracks, and residual stresses. Improvements in surface integrity may be obtained by post-processing techniques such as polishing, sanding, peening, finish machining, and fine grinding.

Vibration in machine tools, a very complex behavior, is often the cause of premature tool failure or short tool life, poor surface finish, damage to the workpiece, and even damage to the machine itself. Vibration may be **forced** or **self-excited**. The term **chatter** is commonly used to designate self-excited vibrations in machine tools. The excited amplitudes are usually very high and may cause damage to the machine. Although there is no complete solution to all types of vibration problems, certain measures may be taken. If the vibration is being forced, it may be possible to remove or isolate the forcing element from the machine. In cases where the forcing frequency is near a natural frequency, either the forcing frequency or the natural frequency may be raised or lowered. Damping will also greatly reduce the amplitude. Self-excited vibrations are generally controlled by increasing the stiffness and damping of the machine tool. (See also Secs. 3 and 5.)

Good machining practice requires a rigid setup. The machine tool must be capable of providing the **stiffness** required for the machining conditions used. If a rigid setup is not available, the depth of cut must be reduced. Excessive tool overhang should be avoided, and in milling, cutters should be mounted as close to the spindle as possible. The length of end mills and drills should be kept to a minimum. Tools with large nose radius or with a long, straight cutting edge increase the possibility of chatter.

CUTTING-TOOL MATERIALS

A wide variety of **cutting-tool materials** are available. The selection of a proper material depends on such factors as the cutting operation involved, the machine to be used, the workpiece material, production requirements, cost, and surface finish and accuracy desired. The major qualities required in a cutting tool are (1) hot hardness, (2) resistance to mechanical impact and thermal shock, (3) wear resistance, and (4) chemical stability and inertness to the workpiece material being machined. (See Table 13.4.1 and Figs. 13.4.6 and 13.4.7.)

Materials for cutting tools include high-speed steels, cast alloys, carbides, ceramics or oxides, cubic boron nitride, and diamond. Understanding the different types of **tool steels** (see Sec. 6.2) requires knowledge of the role of different alloying elements. These elements are added to (1) obtain greater hardness and wear resistance, (2) obtain greater impact toughness, (3) impart hot hardness to the steel such that its hardness is maintained at high cutting temperatures, and (4) decrease distortion and warpage during heat treating.

Table 13.4.1   Characteristics of Cutting-Tool Materials

	High-speed steels	Cast cobalt alloys	Carbides	Coated carbides	Ceramics	Polycrystalline cubic boron nitride	Diamond
Hot hardness	——————————→	increasing	——————————————————————————————→				
Toughness	←——————————	increasing	——————————————————————————————→				
Impact strength	←——————————	increasing	——————————————————————————————→				
Wear resistance	——————————→	increasing	——————————————————————————————→				
Chipping resistance	←——————————	increasing	——————————————————————————————→				
Cutting speed	——————————→	increasing	——————————————————————————————→				
Thermal shock resistance	←——————————	increasing	——————————————————————————————→				
Tool material cost	——————————→	increasing	——————————————————————————————→				

NOTE: These tool materials have a wide range of compositions and properties; thus overlapping characteristics exist in many categories of tool materials.
SOURCE: After R Komanduri.

Fig. 13.4.6 Hardness of tool materials as a function of temperature.

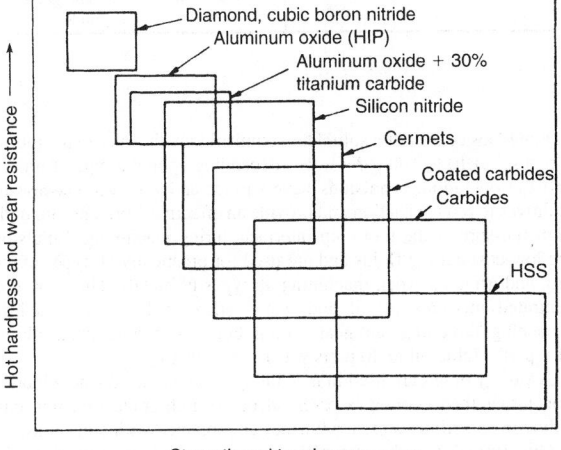

Fig. 13.4.7 Ranges of properties of various groups of tool materials.

Carbon forms a carbide with iron, making it respond to hardening and thus increasing the hardness, strength, and wear resistance. The carbon content of tool steels ranges from 0.6 to 1.4 percent. **Chromium** is added to increase wear resistance and toughness; the content ranges from 0.25 to 4.5 percent. **Cobalt** is commonly used in high-speed steels to increase hot hardness so that tools may be used at higher cutting speeds and still maintain hardness and sharp cutting edges; the content ranges from 5 to 12 percent. **Molybdenum** is a strong carbide-forming element and increases strength, wear resistance, and hot hardness. It is always used in conjunction with other alloying elements, and its content ranges to 10 percent. **Tungsten** promotes hot hardness and strength; content ranges from 1.25 to 20 percent. **Vanadium** increases hot hardness and abrasion resistance; in high-speed steels, it ranges from 1 to 5 percent.

High-speed steels are the most highly alloyed group among tool steels and maintain their hardness, strength, and cutting edge. With suitable procedures and equipment, they can be fully hardened with little danger of distortion or cracking. High-speed steel tools are widely used in operations using form tools, drilling, reaming, end-milling, broaching, tapping, and tooling for screw machines.

Cast alloys maintain high hardness at high temperatures and have good wear resistance. Cast-alloy tools, which are cast and ground into any desired shape, are composed of cobalt (38 to 53 percent), chromium (30 to 33 percent), and tungsten (10 to 20 percent). These alloys are recommended for deep roughing operations at relatively high speeds and feeds. Cutting fluids are not necessary and are usually used only to obtain a special surface finish.

Carbides have metal carbides as key ingredients and are manufactured by powder-metallurgy techniques. They have the following properties which make them very effective cutting-tool materials: (1) high hardness over a wide range of temperatures; (2) high elastic modulus, 2 to 3 times that of steel; (3) no plastic flow even at very high stresses; (4) low thermal expansion; and (5) high thermal conductivity. Carbides are used in the form of inserts or tips which are clamped or brazed to a steel shank. Because of the difference in coefficients of expansion, brazing should be done carefully. The mechanically fastened tool tips are called **inserts** (Fig. 13.4.8); they are available in different shapes, such as square, triangular, circular, and various special shapes.

There are three general groups of carbides in use: (1) tungsten carbide with cobalt as a binder, used in machining cast irons and nonferrous abrasive metals; (2) tungsten carbide with cobalt as a binder, plus a solid solution of WC-TiC-TaC-NbC, for use in machining steels;

(b)

Fig. 13.4.8 (*a*) Insert clamped to shank of a toolholder; (*b*) insert clamped with wing lockpins.

and (3) titanium carbide with nickel and molybdenum as a binder, for use where cutting temperatures are high because of high cutting speeds or the high strength of the workpiece material. Carbides are classified by ISO and ANSI, as shown in Table 13.4.2 which includes recommendations for a variety of workpiece materials and cutting conditions. (See also Sec. 6.4.)

Coated carbides consist of conventional carbide inserts that are coated with a thin layer of titanium nitride, titanium carbide, titanium carbonitride, ceramic, polycrystalline diamond, or diamondlike carbon. The coating provides additional wear resistance while maintaining the strength and toughness of the carbide tool. Coatings are also applied to high-speed steel tools, particularly drills and taps. The desirable properties of individual coatings can be combined and optimized by using **multiphase coatings.** Carbide tools are now available with, e.g., a layer of titanium carbide over the carbide substrate, followed by aluminum oxide and then titanium nitride. Various alternating layers of coatings are also used, each layer being on the order of 80 to 400 μin (2 to 10 μm) thick.

Stiffness is of great importance when using carbide tools. Light feeds, low speeds, and chatter are deleterious. No cutting fluid is needed, but if one is used for cooling, it should be applied in large quantities and continuously to prevent heating and quenching.

Ceramic, or **oxide,** inserts consist primarily of fine aluminum oxide grains which have been bonded together. Minor additions of other elements help to obtain optimum properties.

Other ceramics include silicon nitride, with various additives such as aluminum oxide, yttrium oxide, and titanium carbide. Silicon-nitride-based ceramics include **sialon** (from silicon, aluminum, oxygen, and nitrogen) which has toughness, hot hardness, and good thermal-shock resistance. More recent developments include **whisker-reinforced** cutting tools, with enhanced toughness, cutting-edge strength, and thermal-shock resistance. A common whisker material is silicon carbide. Ceramic tools have very high abrasion resistance, are harder than carbides, and have less tendency to weld to metals during cutting. However, they generally lack impact toughness, and premature tool failure can result by chipping or general breakage. Ceramic tools have been found to be

Table 13.4.2 Classification of Tungsten Carbides According to Machining Applications

ISO standard	ANSI classification no. (grade)	Materials to be machined	Machining operation	Type of carbide	Characteristics of	
					Cut	Carbide
K30–K40	C1	Cast iron, nonferrous metals, and nonmetallic materials requiring abrasion resistance	Roughing	Wear-resistant grades; generally straight WC-Co with varying grain sizes	Increasing cutting speed ↓	Increasing hardness and wear resistance ↑
K20	C2		General purpose			
K10	C3		Light finishing			
K01	C4		Precision finishing		Increasing feed rate	Increasing strength and binder content
P30–P50	C5	Steels and steel alloys requiring crater and deformation resistance	Roughing	Crater-resistant grades; various WC-Co compositions with TiC and/ or TaC alloys	Increasing cutting speed ↓	Increasing hardness and wear resistance ↑
P20	C6		General purpose			
P10	C7		Light finishing			
P01	C8		Precision finishing		Increasing feed rate	Increasing strength and binder content

NOTE: The ISO and ANSI comparisons are approximate.

effective for high-speed, uninterrupted turning operations. Tool and setup geometry is important. Tool failures can be reduced by the use of rigid tool mountings and rigid machine tools. Included in oxide cutting-tool materials are **cermets** (such as 70 percent aluminum oxide and 30 percent titanium carbide), combining the advantages of ceramics and metals.

Polycrystalline diamond is used where good surface finish and dimensional accuracy are desired, particularly on soft nonferrous materials that are difficult to machine. The general properties of diamonds are extreme hardness, low thermal expansion, high heat conductivity, and a very low coefficient of friction. The polycrystalline diamond is bonded to a carbide substrate. **Single-crystal diamond** is also used as a cutting tool to produce extremely fine surface finish on nonferrous alloys, such as copper-base mirrors.

Next to diamond, **cubic boron nitride (cBN)** is the hardest material presently available. Polycrystalline cBN is bonded to a carbide substrate and used as a cutting tool. The cBN layer provides very high wear resistance and edge strength. It is chemically inert to iron and nickel at elevated temperatures; thus it is particularly suitable for machining high-temperature alloys and various ferrous alloys. Both diamond and cBN are also used as abrasives in grinding operations.

CUTTING FLUIDS

Cutting fluids, frequently referred to as lubricants or coolants, comprise those liquids and gases which are applied to the cutting zone in order to facilitate the cutting operation. A cutting fluid is used (1) to keep the tool cool and prevent it from being heated to a temperature at which the hardness and resistance to abrasion are reduced; (2) to keep the workpiece cool, thus preventing it from being machined in a warped shape to inaccurate final dimensions; (3) through lubrication to reduce friction and power consumption, wear on the tool, and generation of heat; (4) to provide a good finish on the workpiece; (5) to aid in providing a satisfactory chip formation; (6) to wash away the chips (this is particularly desirable in deep-hole drilling, hacksawing, milling, and grinding); and (7) to prevent corrosion of the workpiece and machine tool.

Classification Cutting fluids may be classified as follows: (1) emulsions, (2) oils, and (3) solutions (semisynthetics and synthetics). Cutting fluids are also classified as light-, medium-, and heavy-duty; light-duty fluids are for general-purpose machining. Induced **air blast** may be used with internal and surface grinding and polishing operations. Its main purpose is to remove the small chips or dust, although some cooling is also obtained, especially in machining of plastics.

Emulsions consist of a soluble oil emulsified with water in the ratio of 1 part oil to 10 to 100 parts water, depending upon the type of product and the operation. Emulsions have surface-active or **extreme-pressure** additives to reduce friction and provide an effective lubricant film under high pressure at the tool-chip interface during machining. Emulsions are low-cost cutting fluids and are used for practically all types of cutting and grinding when machining all types of metals. The more concentrated mixtures of oil and water, such as 1:10, are used for broaching, threading, and gear cutting. For most operations, a solution of 1 part soluble oil to 20 parts water is satisfactory.

A variety of **oils** are used in machining. They are used where lubrication rather than cooling is essential or on high-grade finishing cuts, although sometimes superior finishes are obtained with emulsions. Oils generally used in machining are mineral oils with the following compositions: (1) straight mineral oil, (2) with fat, (3) with fat and sulfur, (4) with fat and chlorine, and (5) with fat, sulfur, and chlorine. The more severe the machining operation, the higher the composition of the oil. Broaching and tapping of refractory alloys and high-temperature alloys, for instance, require highly compounded oils. In order to avoid staining of the metal, aluminum and copper, for example, inhibited sulfur and chlorine are used.

Solutions are a family of cutting fluids that blend water and various chemical agents such as amines, nitrites, nitrates, phosphates, chlorine, and sulfur compounds. These agents are added for purposes of rust prevention, water softening, lubrication, and reduction of surface tension. Most of these chemical fluids are coolants but some are lubricants.

The **selection** of a cutting fluid for a particular operation requires consideration of several factors: cost, the workpiece material, the difficulty of the machining operation, the compatibility of the fluid with the workpiece material and the machine tool components, surface preparation, method of application and removal of the fluid, contamination of the cutting fluid with machine lubricants, and the treatment of the fluid after use. Also important are the **biological** and **ecological** aspects of the cutting fluid used. There may be potential health hazards to operating personnel from contact with or inhalation of mist or fumes from some fluids. Recycling and waste disposal are also important problems to be considered.

Methods of Application The most common method is **flood cooling** in quantities such as 3 to 5 gal/min (about 10 to 20 L/min) for single-point tools and up to 60 gal/min (230 L/min) per cutter for multiple-tooth cutters. Whenever possible, multiple nozzles should be used. In **mist cooling** a small jet equipment is used to disperse water-base fluids as very fine droplets in a carrier that is generally air at pressures 10 to

80 lb/in² (70 to 550 kPa). Mist cooling has a number of advantages, such as providing high-velocity fluids to the working areas, better visibility, and improving tool life in certain instances. The disadvantages are that venting is required and also the cooling capability is rather limited.

High-pressure refrigerated coolant systems are very effective in removing heat at high rates, particularly in computer-controlled machine tools. The fluid is directed generally at the relief angle of the cutting tools and at pressures as high as 5,000 lb/in² (35,000 kPa). Continuous filtering of the fluid is essential to eliminate any damage to workpiece surfaces due to the impact of any contaminants that may be present in the coolant system. More recent methods of application include delivering the coolant to the cutting zone through the tool and the machine spindle.

For economic as well as environmental reasons, an important trend is **near-dry** and **dry machining.** In near-dry machining, the cutting fluid typically consists of a fine mist of air containing a very small amount of cutting fluid (including vegetable oil) and is delivered through the machine spindle. Dry machining is carried out without any fluids but using appropriate cutting tools and processing parameters. Unlike other methods, however, dry machining cannot flush away the chips being produced; an effective means to do so is to use pressurized air.

MACHINE TOOLS

The general types of **machine tools** are lathes; turret lathes; screw, boring, drilling, reaming, threading, milling, and gear-cutting machines; planers and shapers; broaching, cutting-off, grinding, and polishing machines. Each of these is subdivided into many types and sizes. General items common to all machine tools are discussed first, and individual machining processes and equipment are treated later in this section.

Automation is the application of special equipment to control and perform manufacturing processes with little or no manual effort. It is applied to the manufacturing of all types of goods and processes, from the raw material to the finished product. Automation involves many activities, such as handling, processing, assembly, inspecting, and packaging. Its primary objective is to lower manufacturing cost through controlled production and quality, lower labor cost, reduced damage to work by handling, higher degree of safety for personnel, and economy of floor space. Automation may be partial, such as gaging in cylindrical grinding, or it may be total.

The conditions which play a role in decisions concerning automation are rising production costs, high percentage of rejects, lagging output, scarcity of skilled labor, hazardous working conditions, and work requiring repetitive operation. Factors which must be carefully studied before deciding on automation are high initial cost of equipment, maintenance problems, and type of product. (See also Sec. 16.)

Mass production with modern machine tools has been achieved through the development of self-contained **power-head** production units and the development of **transfer** mechanisms. Power-head units, consisting of a frame, electric driving motor, gearbox, tool spindles, etc., are available for many types of machining operations. Transfer mechanisms move the workpieces from station to station by various methods. Transfer-type machines can be arranged in several configurations, such as a straight line or a U pattern. Various types of machine tools for mass production can be built from components; this is known as the **building-block** principle. Such a system combines flexibility and adaptability with high productivity. (See **machining centers.**)

Numerical control (NC), which is a method of controlling the motions of machine components by numbers, was first applied to machine tools in the 1950s. Numerically controlled machine tools are classified according to the type of cutting operation. For instance, in drilling and boring machines, the positioning and the cutting take place sequentially (point to point), whereas in die-sinking machines, positioning and cutting take place simultaneously. The latter are often described as **continuous-path** machines, and since they require more exacting specifications, they give rise to more complex problems. Machines now perform over a very wide range of cutting conditions

without requiring adjustment to eliminate chatter, and to improve accuracy. Complex contours can be machined which would be almost impossible by any other method. A large variety of programming systems has been developed.

The control system in NC machines has been converted to computer control with various software. In **computer numerical control** (CNC), a microcomputer is a part of the control panel of the machine tool. The advantages of computer numerical control are ease of operation, simpler programming, greater accuracy, versatility, and lower maintenance costs.

Further developments in machine tools are **machining centers.** This is a machine equipped with as many as 200 tools and with an automatic tool changer (Fig. 13.4.9). It is designed to perform various operations on different surfaces of the workpiece, which is placed on a pallet capable of as much as five-axis movement (three linear and two rotational). Machining centers, which may be vertical or horizontal spindle, have flexibility and versatility that other machine tools do not have, and thus they have become the first choice in machine selection in modern manufacturing plants and shops. They have the capability of tool and part checking, tool-condition monitoring, in-process and postprocess gaging, and inspection of machined surfaces. **Universal machining centers** are the latest development, and they have both vertical and horizontal spindles. **Turning centers** are a further development of computer-controlled lathes and have great flexibility. Many centers are now constructed on a **modular** basis, so that various accessories and peripheral equipment can be installed and modified depending on the type of product to be machined.

Fig. 13.4.9 Schematic of a horizontal spindle machining center, equipped with an automatic tool changer. Tool magazines can store 200 different cutting tools.

An approach to optimize machining operations is **adaptive control.** While the material is being machined, the system senses operating conditions such as forces, tool-tip temperature, rate of tool wear, and surface finish, and converts these data into feed and speed control that enables the machine to cut under optimum conditions for maximum productivity. Combined with numerical controls and computers, adaptive controls are expected to result in increased efficiency of metalworking operations.

With the advent of sophisticated computers and various software, modern manufacturing has evolved into **computer-integrated manufacturing (CIM).** This system involves the coordinated participation of computers in all phases of manufacturing. **Computer-aided design** combined with **computer-aided manufacturing (CAD/CAM)** results in a much higher productivity, better accuracy and efficiency, and reduction in design effort and prototype development. CIM also involves the management of the factory, inventory, and labor, and it integrates all these activities, eventually leading to untended factories.

The highest level of sophistication is reached with a **flexible manufacturing system (FMS).** Such a system is made of **manufacturing cells** and an automatic materials-handling system interfaced with a central computer. The manufacturing cell is a system in which CNC machines are used to make a specific part or parts with similar shape. The workstations, i.e., several machine tools, are placed around an **industrial robot** which automatically loads, unloads, and transfers the parts. FMS has the capability to optimize each step of the total manufacturing operation, resulting in the highest possible level of efficiency and productivity.

The proper design of **machine-tool structures** requires analysis of such factors as form and materials of structures, stresses, weight, and manufacturing and performance considerations. The best approach to obtain the ultimate in machine-tool accuracy is to employ both improvements in structural stiffness and compensation of deflections by use of special controls. The C-frame structure has been used extensively in the past because it provides ready accessibility to the working area of the machine. With the advent of computer control, the box-type frame with its considerably improved static stiffness becomes practical since the need for manual access to the working area is greatly reduced. The use of a box-type structure with thin walls can provide low weight for a given stiffness. The light-weight-design principle offers high dynamic stiffness by providing a high natural frequency of the structure through combining high static stiffness with low weight rather than through the use of large mass. (Dynamic stiffness is the stiffness exhibited by the system when subjected to dynamic excitation where the elastic, the damping, and the inertia properties of the structure are involved; it is a frequency-dependent quantity.)

TURNING

Turning is a machining operation for all types of metallic and nonmetallic materials and is capable of producing circular parts with straight or various profiles. The cutting tools may be single-point or form tools. The most common machine tool used is a **lathe;** modern lathes are computer-controlled and can achieve high production rates with little labor. The basic operation is shown in Fig. 13.4.10, where the workpiece is held in a chuck and rotates at N r/min; a cutting tool moves along the length of the piece at a feed f (in/r or mm/r) and removes material at a radial depth d, reducing the diameter from D_0 to D_f.

Fig. 13.4.10   A turning operation on a round workpiece held in a three-jaw chuck.

Lathes generally are considered to be the oldest member of machine tools, having been first developed in the late eighteenth century. The most common lathe is called an engine lathe because it was one of the first machines driven by Watt's steam engine. The basic lathe has the following main parts: bed, headstock, tailstock, and carriage. The types of lathes available for a variety of applications may be listed as follows: engine lathes, bench lathes, horizontal turret lathes, vertical lathes, and automatics. A great variety of lathes and attachments are available within each category, also depending on the production rate required.

It is common practice to specify the size of an engine lathe by giving the **swing** (diameter) and the **distance between centers** when the tailstock

is flush with the end of the bed. The maximum swing over the ways is usually greater than the nominal swing. The **length of the bed** is given frequently to specify the overall length of the bed. A lathe size is indicated thus: 14 in (356 mm) (swing) by 30 in (762 mm) (between centers) by 6 ft (1,830 mm) (length of bed). Lathes are made for light-, medium-, or heavy-duty work.

All geared-head lathes, which are single-pulley (belt-driven or arranged for direct-motor drive through short, flat, or V belts, gears, or silent chain), increase the power of the drive and provide a means for obtaining 8, 12, 16, or 24 spindle speeds. The teeth may be of the spur, helical, or herringbone type and may be ground or lapped after hardening.

Variable speeds are obtained by driving with adjustable-speed dc shunt-wound motors with stepped field-resistance control or by electronics or motor-generator system to give speed variation in infinite steps. AC motors driving through infinitely variable speed transmissions of the mechanical or hydraulic type are also in general use.

Modern lathes, most of which are now computer-controlled (**turning centers),** are built with the speed capacity, stiffness, and strength capable of taking full advantage of new and stronger tool materials. The main drive-motor capacity of lathes ranges from fractional to more than 200 hp (150 kW). Speed preselectors, which give speed as a function of work diameter, are introduced, and variable-speed drives using dc motors with panel control are standard on many lathes. Lathes with contour facing, turning, and boring attachments are also available.

Tool Shapes for Turning

The standard **nomenclature** for single-point tools, such as those used on lathes, planers, and shapers, is shown in Fig. 13.4.11. Each tool consists of a shank and point. The point of a single-point tool may be formed by grinding on the end of the shank; it may be forged on the end of the shank and subsequently ground; a tip or insert may be clamped or brazed to the end of the shank (see Fig. 13.4.8). The **best tool shape** for each material and each operation depends on many factors. For specific information and recommendations, the various sources listed in the References should be consulted. See also Table 13.4.3.

Fig. 13.4.11   Standard nomenclature for single-point cutting tools.

Table 13.4.3 Recommend Tool Geometry for Turning, deg

	High-speed steel and cast-alloy tools					Carbide tools (inserts)				
Material	Back rake	Side rake	End relief	Side relief	Side and end cutting edge	Back rake	Side rake	End relief	Side relief	Side and end cutting edge
Aluminum alloys	20	15	12	10	5	0	5	5	5	15
Magnesium alloys	20	15	12	10	5	0	5	5	5	15
Copper alloys	5	10	8	8	5	0	5	5	5	15
Steels	10	12	5	5	15	−5	−5	5	5	15
Stainless steels, ferritic	5	8	5	5	15	0	5	5	5	15
Stainless steels, austenitic	0	10	5	5	15	0	5	5	5	15
Stainless steels, martensitic	0	10	5	5	15	−5	−5	5	5	15
High-temperature alloys	0	10	5	5	15	5	0	5	5	45
Refractory alloys	0	20	5	5	5	0	0	5	5	15
Titanium alloys	0	5	5	5	15	−5	−5	5	5	5
Cast irons	5	10	5	5	15	−5	−5	5	5	15
Thermoplastics	0	0	20−30	15−20	10	0	0	20−30	15−20	10
Thermosetting plastics	0	0	20−30	15−20	10	0	15	5	5	15

SOURCE: "Matchining Data Handbook," published by the Machinability Data Center, Metcut Research Associates, Inc.

Positive **rake angles** improve the cutting operation with regard to forces and deflection; however, a high positive rake angle may result in early failure of the cutting edge. Positive rake angles are generally used in lower-strength materials. For higher-strength materials, negative rake angles may be used. **Back rake** usually controls the direction of chip flow and is of less importance than the **side rake**. The purpose of **relief angles** is to avoid interference and rubbing between the workpiece and tool flank surfaces. In general, they should be small for high-strength materials and larger for softer materials. Excessive relief angles may weaken the tool. The **side cutting-edge angle** influences the length of chip contact and the true feed. This angle is often limited by the workpiece geometry, e.g., the shoulder contour. Large angles are apt to cause tool chatter. Small **end cutting-edge angles** may create excessive force normal to the workpiece, and large angles may weaken the tool point. The purpose of the **nose radius** is to give a smooth surface finish and to obtain longer tool life by increasing the strength of the cutting edge. The nose radius should be tangent to the cutting-edge angles. A large nose radius gives a stronger tool and may be used for roughing cuts; however, large radii may lead to tool chatter. A small nose radius reduces forces and is therefore preferred on thin or slender workpieces.

Turning Recommendations Recommendations for tool materials, depth of cut, feed, and cutting speed for turning a variety of materials are given in Table 13.4.4. The cutting speeds for high-speed steels for turning, which are generally M2 and M3, are about one-half those for uncoated carbides. A general **troubleshooting guide** for turning operations is given in Table 13.4.5. The range of applicable cutting speeds and feeds for a variety of tool materials is shown in Fig. 13.4.12.

High-Speed Machining To increase productivity and reduce machining costs, there is a continuing trend to increase cutting speeds, especially in turning, milling, boring, and drilling. High-speed machining is a general term used to describe this trend, where speeds typically range as follows: High speed: up 6,000 ft/min (1,800 m/min): very high speed: up to 60,000 ft/min (18,000 m/min); and ultrahigh-speed, higher than this range. Because of the high speeds involved, important considerations in these operations include inertia effects, spindle design, bearings, and power; stiffness and accuracy of the machine tools; selection of appropriate cutting tools; and chip removal systems.

Hard Turning and Machining As workpiece hardness increases, its machinability decreases and there may be difficulties with traditional machining operations regarding surface finish, surface integrity, and tool life. With advances in cutting tools and the availability of rigid and powerful machine tools and work-holding devices, however, it is now possible to machine hard materials, including heat-treated steels, with high dimensional accuracy. Hard machining can compete well with grinding processes and has been shown to be economical for parts such as shafts, gears, pinions, and various automotive components.

Ultraprecision Machining To respond to increasing demands for special parts with surface finish and dimensional accuracies on the order of a nanometre (10^{-9} m), several important developments have been taking place in advanced machining. A common example of ultraprecision machining is **diamond turning**, typically using a single-crystal diamond cutting tool and rigid machine tools. Applications for such parts and components are in the computer, electronic, nuclear, and defense industries.

Turret Lathes

Turret lathes are used for the production of parts in moderate quantities and produce interchangeable parts at low production cost. Turret lathes may be chucking, screw machine, or universal. The universal machine may be set up to machine bar stock as a screw machine or have the work held in a chuck. These machines may be semiautomatic, i.e., so arranged that after a piece is chucked and the machine started, it will complete the machining cycle automatically and come to a stop. They may be horizontal or vertical and single- or multiple-spindle; many of these lathes are now computer-controlled and have a variety of features.

The basic principle of the turret lathe is that, with standard tools, setups can be made quickly so that combined, multiple, and successive cuts can be made on a part. By **combined** cuts, tools on the cross slide operate simultaneously with those on the turret, e.g., facing from the cross slide and boring from the turret. **Multiple** cuts permit two or more tools to operate from either or both the cross slide or turret. By **successive** cuts, one tool may follow another to rough or finish a surface; e.g., a hole may be drilled, bored, and reamed at one chucking. In the tool-slide machine only roughing cuts, such as turn and face, can be made in one machine.

Ram-type turret lathes have the turret mounted on a ram which slides in a separate base. The base is clamped at a position along the bed to suit a long or short workpiece. A cross slide can be used so that combined cuts can be taken from the turret and the cross slide at the same time. Turret and cross slide can be equipped with manual or power feed. The short stroke of the turret slide limits this machine to comparatively short light work, in both small and quantity-lot production.

Saddle-type turret lathes have the turret mounted on a saddle which slides directly on the bed. Hence, the length of stroke is limited only by the length of bed. A separate square-turret carriage with longitudinal and transverse movement can be mounted between the head and the hex-turret saddle so that combined cuts from both stations at one time are possible. The saddle type of turret lathe generally has a large hollow vertically faced turret for accurate alignment of the tools.

Screw Machines

When turret lathes are set up for bar stock, they are often called **screw machines.** Turret lathes that are adaptable only to bar-stock work are

Table 13.4.4   General Recommendations for Turning Operations

Workpiece material	Cutting tool	General-purpose starting conditions			Range for roughing and finishing		
		Depth of cut, mm (in)	Feed, mm/r (in/r)	Cutting speed, m/min (ft/min)	Depth of cut, mm (in)	Feed, mm/r (in/r)	Cutting speed, m/min (ft/min)
Low-C and free-machining steels	Uncoated carbide	1.5–6.3 (0.06–0.25)	0.35 (0.014)	90 (300)	0.5–7.6 (0.02–0.30)	0.15–1.1 (0.006–0.045)	60–135 (200–450)
	Ceramic-coated carbide	1.5–6.3 (0.06–0.25)	0.35 (0.014)	245–275 (800–900)	0.5–7.6 (0.02–0.30)	0.15–1.1 (0.006–0.045)	120–425 (400–1,400)
	Triple-coated carbide	1.5–6.3 (0.06–0.25)	0.35 (0.014)	185–200 (600–650)	0.5–7.6 (0.02–0.30)	0.15–1.1 (006–045)	90–245 (300–800)
	TiN-coated carbide	1.5–6.3 (0.06–0.25)	0.35 (0.014)	105–150 (350–500)	0.5–7.6 (0.02–0.30)	0.15–1.1 (0.006–0.045)	60–230 (200–750)
	Al_2O_3 ceramic	1.5–6.3 (0.06–0.25)	0.25 (0.010)	395–440 (1,300–1,450)	0.5–7.6 (0.02–0.30)	0.15–1.1 (0.006–0.045)	365–550 (1,200–1,800)
	Cermet	1.5–6.3 (0.06–0.25)	0.30 (0.012)	215–290 (700–950)	0.5–7.6 (0.02–0.30)	0.15–1.1 (0.006–0.045)	105–455 (350–1,500)
Medium- and high-C steels	Uncoated carbide	1.2–4.0 (0.05–0.20)	0.30 (0.012)	75 (250)	2.5–7.6 (0.10–0.30)	0.15–0.75 (0.006–0.03)	45–120 (150–400)
	Ceramic-coated carbide	1.2–4.0 (0.05–0.20)	0.30 (0.012)	185–230 (600–750)	2.5–7.6 (0.10–0.30)	0.15–0.75 (0.006–0.03)	120–410 (400–1,350)
	Triple-coated carbide	1.2–4.0 (0.050–0.20)	0.30 (0.012)	120–150 (400–500)	2.5–7.6 (0.10–0.30)	0.15–0.75 (0.006–0.03)	75–215 (250–700)
	TiN-coated carbide	1.2–4.0 (0.05–0.20)	0.30 (0.012)	90–200 (300–650)	2.5–7.6 (0.10–0.30)	0.15–0.75 (0.006–0.03)	45–215 (150–700)
	Al_2O_3 ceramic	1.2–4.0 (0.05–0.20)	0.25 (0.010)	335 (1,100)	2.5–7.6 (0.10–0.30)	0.15–0.75 (0.006–0.03)	245–455 (800–1,500)
	Cermet	1.2–4.0 (0.05–0.20)	0.25 (0.010)	170–245 (550–800)	2.5–7.6 (0.10–0.30)	0.15–0.75 (0.006–0.03)	105–305 (350–1,000)
Cast iron, gray	Uncoated carbide	1.25–6.3 (0.05–0.25)	0.32 (0.013)	90 (300)	0.4–12.7 (0.015–0.5)	0.1–0.75 (0.004–0.03)	75–185 (250–600)
	Ceramic-coated carbide	1.25–6.3 (0.05–0.25)	0.32 (0.013)	200 (650)	0.4–12.7 (0.015–0.5)	0.1–0.75 (0.004–0.03)	120–365 (400–1,200)
	TiN-coated carbide	1.25–6.3 (0.05–0.25)	0.32 (0.013)	90–135 (300–450)	0.4–12.7 (0.015–0.5)	0.1–0.75 (0.004–0.03)	60–215 (200–700)
	Al_2O_3 ceramic	1.25–6.3 (0.05–0.25)	0.25 (0.010)	455–490 (1,500–1,600)	0.4–12.7 (0.015–0.5)	0.1–0.75 (0.004–0.03)	365–855 (1,200–2,800)
	SiN ceramic	1.25–6.3 (0.05–0.25)	0.32 (0.013)	730 (2,400)	0.4–12.7 (0.015–0.5)	0.1–0.75 (0.004–0.03)	200–990 (650–3,250)
Stainless steel, austenitic	Triple-coated carbide	1.5–4.4 (0.06–0.175)	0.35 (0.014)	150 (500)	0.5–12.7 (0.02–0.5)	0.08–0.75 (0.003–0.03)	75–230 (250–750)
	TiN-coated carbide	1.5–4.4 (0.06–0.175)	0.35 (0.014)	85–160 (275–525)	0.5–12.7 (0.02–0.5)	0.08–0.75 (0.003–0.03)	55–200 (175–650)
	Cermet	1.5–4.4 (0.06–0.175)	0.30 (0.012)	185–215 (600–700)	0.5–12.7 (0.02–0.5)	0.08–0.75 (0.003–0.03)	105–290 (350–950)
High-temperature alloys, nickel base	Uncoated carbide	2.5 (0.10)	0.15 (0.006)	25–45 (75–150)	0.25–6.3 (0.01–0.25)	0.1–0.3 (0.004–0.012)	15–30 (50–100)
	Ceramic-coated carbide	2.5 (0.10)	0.15 (0.006)	45 (150)	0.25–6.3 (0.01–0.25)	0.1–0.3 (0.004–0.012)	20–60 (65–200)
	TiN-coated carbide	2.5 (0.10)	0.15 (0.006)	30–55 (95–175)	0.25–6.3 (0.01–0.25)	0.1–0.3 (0.004–0.012)	20–85 (60–275)
	Al_2O_3 ceramic	2.5 (0.10)	0.15 (0.006)	260 (850)	0.25–6.3 (0.01–0.25)	0.1–0.3 (0.004–0.012)	185–395 (600–1,300)
	SiN ceramic	2.5 (0.10)	0.15 (0.006)	215 (700)	0.25–6.3 (0.01–0.25)	0.1–0.3 (0.004–0.012)	90–215 (300–700)
	Polycrystalline cBN	2.5 (0.10)	0.15 (0.006)	150 (500)	0.25–6.3 (0.01–0.25)	0.1–0.3 (0.004–0.012)	120–185 (400–600)
Titanium alloys	Uncoated carbide	1.0–3.8 (0.04–0.15)	0.15 (0.006)	35–60 (120–200)	0.25–6.3 (0.01–0.25)	0.1–0.4 (0.004–0.015)	10–75 (30–250)
	TiN-coated carbide	1.0–3.8 (0.04–0.15)	0.15 (0.006)	30–60 (100–200)	0.25–6.3 (0.01–0.25)	0.1–0.4 (0.004–0.015)	30–60 (30–325)
Aluminum alloys free-machining	Uncoated carbide	1.5–5.0 (0.06–0.20)	0.45 (0.018)	490 (1,600)	0.25–8.8 (0.01–0.35)	0.08–0.62 (0.003–0.025)	200–670 (650–2,000)
	TiN-coated carbide	1.5–5.0 (0.06–0.20)	0.45 (0.018)	550 (1,800)	0.25–8.8 (0.01–0.35)	0.08–0.62 (0.003–0.025)	60–915 (200–3,000)
	Cermet	1.5–5.0 (0.06–0.20)	0.45 (0.018)	490 (1,600)	0.25–8.8 (0.01–0.35)	0.08–0.62 (0.003–0.025)	215–795 (700–2,600)
	Polycrystalline diamond	1.5–5.0 (0.06–0.20)	0.45 (0.018)	760 (2,500)	0.25–8.8 (0.01–0.35)	0.08–0.62 (0.003–0.025)	305–3,050 (1,000–10,000)
High-silicon copper alloys	Polycrystalline diamond	1.5–5.0 (0.06–0.20)	0.45 (0.018)	530 (1,700)	0.25–8.8 (0.01–0.35)	0.08–0.62 (0.003–0.025)	365–915 (1,200–3,000)
	Uncoated carbide	1.5–5.0 (0.06–0.20)	0.25 (0.010)	260 (850)	0.4–7.5 (0.015–0.3)	0.15–0.75 (0.006–0.03)	105–535 (350–1,750)

Table 13.4.4   General Recommendations for Turning Operations   (*Continued*)

Workpiece material	Cutting tool	General-purpose starting conditions			Range for roughing and finishing		
		Depth of cut, mm (in)	Feed, mm/r (in/r)	Cutting speed, m/min (ft/min)	Depth of cut, mm (in)	Feed, mm/r (in/r)	Cutting speed, m/min (ft/min)
High-silicon copper alloys (*cont.*)	Ceramic-coated carbide	1.5–5.0 (0.06–0.20)	0.25 (0.010)	365 (1,200)	0.4–7.5 (0.015–0.3)	0.15–0.75 (0.006–0.03)	215–670 (700–2,200)
	Triple-coated carbide	1.5–5.0 (0.06–0.20)	0.25 (0.010)	215 (700)	0.4–7.5 (0.015–0.3)	0.15–0.75 (0.006–0.03)	90–305 (300–1,000)
	TiN-coated carbide	1.5–5.0 (0.06–0.20)	0.25 (0.010)	90–275 (300–900)	0.4–7.5 (0.015–0.3)	0.15–0.75 (0.006–0.03)	45–455 (150–1,500)
	Cermet	1.5–5.0 (0.06–0.20)	0.25 (0.010)	245–425 (800–1,400)	0.4–7.5 (0.015–0.3)	0.15–0.75 (0.006–0.03)	200–610 (650–2,000)
	Polycrystalline diamond	1.5–5.0 (0.06–0.20)	0.25 (0.010)	520 (1,700)	0.4–7.5 (0.015–0.3)	0.15–0.75 (0.006–0.03)	275–915 (900–3,000)
Tungsten alloys	Uncoated carbide	2.5 (0.10)	0.2 (0.008)	75 (250)	0.25–5.0 (0.01–0.2)	0.12–0.45 (0.005–0.018)	55–120 (175–400)
	TiN-coated carbide	2.5 (0.10)	0.2 (0.008)	85 (275)	0.25–5.0 (0.01–0.2)	0.12–0.45 (0.005–0.018)	60–150 (200–500)
Thermoplastics and thermosets	TiN-coated carbide	1.2 (0.05)	0.12 (0.005)	170 (550)	0.12–5.0 (0.005–0.20)	0.08–0.35 (0.003–0.015)	90–230 (300–750)
	Polycrystalline diamond	1.2 (0.05)	0.12 (0.005)	395 (1,300)	0.12–5.0 (0.005–0.20)	0.08–1.35 (0.003–0.015)	150–730 (500–2,400)
Composites, graphite-reinforced	TiN-coated carbide	1.9 (0.075)	0.2 (0.008)	200 (650)	0.12–6.3 (0.005–0.25)	0.12–1.5 (0.005–0.06)	105–290 (350–950)
	Polycrystalline diamond	1.9 (0.075)	0.2 (0.008)	760 (2,500)	0.12–6.3 (0.005–0.25)	0.12–1.5 (0.005–0.06)	550–1,310 (1,800–4,300)

NOTE: Cutting speeds for high-speed-steel tools are about one-half those for uncoated carbides.
SOURCE: Based on data from Kennametal Inc.

Table 13.4.5   General Troubleshooting Guide for Turning Operations

Problem	Probable causes
Tool breakage	Tool material lacks toughness; improper tool angles; machine tool lacks stiffness; worn bearings and machine components; cutting parameters too high
Excessive tool wear	Cutting parameters too high; improper tool material; ineffective cutting fluid; improper tool angles
Rough surface finish	Built-up edge on tool; feed too high; tool too sharp, chipped, or worn; vibration and chatter
Dimensional variability	Lack of stiffness of machine tool and work-holding devices; excessive temperature rise; tool wear
Tool chatter	Lack of stiffness of machine tool and work-holding devices; excessive tool overhang; machining parameters not set properly

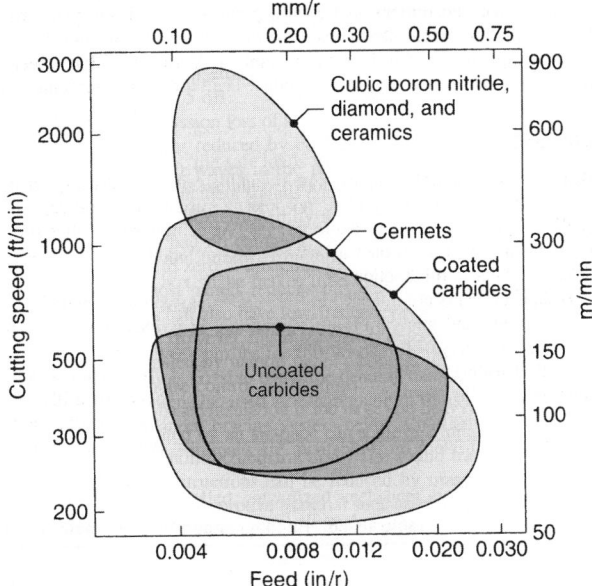

Fig. 13.4.12   Range of applicable cutting speeds and feeds for various tool materials.

constructed for light work. As with turret lathes, they have spring collets for holding the bars during machining and friction fingers or rolls to feed the bar stock forward. Some bar-feeding devices are operated by hand and others semiautomatically.

Automatic screw machines may be classified as single-spindle or multiple-spindle. Single-spindle machines rotate the bar stock from which the part is to be made. The tools are carried on a turret and on cross slides or on a circular drum and on cross slides. Multiple-spindle machines have four, five, six, or eight spindles, each carrying a bar of the material from which the piece is to be made. Capacities range from ⅛ to 6 in (3 to 150 mm) diam of bar stock.

Feeds of forming tools vary with the width of the cut. The wider the forming tool and the smaller the diameter of stock, the smaller the feed. On multiple-spindle machines, where many tools are working simultaneously, the feeds should be such as to reduce the actual cutting time to a minimum. Often only one or two tools in a set are working up to capacity, as far as actual speed and feed are concerned.

BORING

Boring is a machining process for producing internal straight cylindrical surfaces or profiles, with process characteristics and tooling similar to those for turning operations.

Boring machines are of two general types, horizontal and vertical, and are frequently referred to as horizontal boring machines and vertical boring and turning mills. A classification of boring machines comprises horizontal boring, drilling, and milling machines; vertical boring and turning mills; vertical multispindle cylinder boring mills; vertical cylinder boring mills; vertical turret boring mills (vertical turret lathes); carwheel boring mills; diamond or precision boring machines (vertical and horizontal); and jig borers.

The **horizontal type** is made for both precision work and general manufacturing. It is particularly adapted for work not conveniently revolved, for milling, slotting, drilling, tapping, boring, and reaming long holes, and for making interchangeable parts that must be produced without jigs and fixtures. The machine is universal and has a wide range of speeds and feeds, for a face-mill operation may be followed by one with a small-diameter drill or end mill.

Vertical boring mills are adapted to a wide range of faceplate work that can be revolved. The advantage lies in the ease of fastening a workpiece to the horizontal table, which resembles a four-jaw independent chuck with extra radial T slots, and in the lessened effect of centrifugal forces arising from unsymmetrically balanced workpieces.

A **jig-boring machine** has a single-spindle sliding head mounted over a table adjustable longitudinally and transversely by lead screws which roughly locate the work under the spindle. Precision setting of the table may be obtained with end measuring rods, or it may depend only on the accuracy of the lead screw. These machines, made in various sizes, are used for accurately finishing holes and surfaces in definite relation to one another. They may use drills, rose or fluted reamers, or single-point boring tools. The latter are held in an adjustable **boring head** by which the tool can be moved eccentrically to change the diameter of the hole.

Precision-boring machines may have one or more spindles operating at high speeds for the purpose of boring to accurate dimensions such surfaces as wrist-pin holes in pistons and connecting-rod bushings.

Boring Recommendations Boring recommendations for tool materials, depth of cut, feed, and cutting speed are generally the same as those for turning operations (see Table 13.4.4). However, tool deflections, chatter, and dimensional accuracy can be significant problems because the boring bar has to reach the full length to be machined and space within the workpiece may be limited. Boring bars have been designed to dampen vibrations and reduce chatter during machining.

DRILLING

Drilling is a commonly employed hole-making process that uses a **drill** as a cutting tool for producing round holes of various sizes and depths. Drilled holes may be subjected to additional operations for better surface finish and dimensional accuracy, such as reaming and honing, described later in this section.

Drilling machines are intended for drilling holes, tapping, counterboring, reaming, and general boring operations. They may be classified into a large variety of types.

Vertical drilling machines are usually designated by a dimension which roughly indicates the diameter of the largest circle that can be drilled at its center under the machine. This dimensioning, however, does not hold for all makes of machines. The sizes begin with about 6 and

Fig. 13.4.13 Straight shank twist drill.

continue to 50 in. Heavy-duty drill presses of the vertical type, with allgeared speed and feed drive, are constructed with a box-type column instead of the older cylindrical column.

The size of a **radial drill** is designated by the length of the arm. This represents the radius of a piece which can be drilled in the center.

Twist drills (Fig. 13.4.13) are the most common tools used in drilling and are made in many sizes and lengths. For years they have been grouped according to numbered sizes, 1 to 80, inclusive, corresponding approximately to Stub's steel wire gage; some by lettered sizes A to Z, inclusive; some by fractional inches from $\frac{1}{64}$ up, and the group of millimetre sizes.

Straight-shank twist drills of fractional size and various lengths range from $\frac{1}{64}$ in diam to $1\frac{1}{2}$ in by $\frac{1}{64}$ in increments; to $1\frac{1}{2}$ in by $\frac{1}{32}$ in; and to 2 in by $\frac{1}{16}$ in. **Taper-shank drills** range from $\frac{1}{8}$ in diam to $1\frac{3}{4}$ in by $\frac{1}{64}$ increments; to $2\frac{1}{4}$ in by $\frac{1}{32}$ in; and to $3\frac{1}{2}$ in by $\frac{1}{16}$ in. Larger drills are made by various drill manufacturers. Drills are also available in metric dimensions.

Tolerances have been set on the various features of all drills so that the products of different manufacturers will be interchangeable in the user's plants.

Twist drills are decreased in diameter from point to shank (back taper) to prevent binding. If the web is increased gradually in thickness from point to shank to increase the strength, it is customary to reduce the helix angle as it approaches the shank. The shape of the groove is important, the one that gives a straight cutting edge and allows a full curl to the chip being the best. The **helix angles** of the flutes vary from 10 to 45°. The standard **point angle** is 118°. There are a number of **drill grinders** on the market designed to give the proper angles. The point may be ground either in the **standard** or the **crankshaft** geometry. The drill geometry for high-speed steel twist drills for a variety of workpiece materials is given in Table 13.4.6.

Among the common **types of drills** (Fig. 13.4.14) are the combined drill and countersink or **center drill** used to center shafts before squaring and turning: the **step drill**, with two or more diameters; the **spade drill** which has a removable tip or bit clamped in a holder on the drill shank, used for large and deep holes; the **trepanning tool** used to cut a core from a piece of metal instead of reducing all the metal

Table 13.4.6 Recommended Drill Geometry for High-Speed Steel Twist Drillls

Material	Point angle, deg	Lip relief angle, deg	Chisel edge angle, deg	Helix angle, deg	Point grind
Aluminum alloys	90–118	12–15	125–135	24–48	Standard
Magnesium alloys	70–118	12–15	120–135	30–45	Standard
Copper alloys	118	12–15	125–135	10–30	Standard
Steels	118	10–15	125–135	24–32	Standard
High-strength steels	118–135	7–10	125–135	24–32	Crankshaft
Stainless steels, low-strength	118	10–12	125–135	24–32	Standard
Stainless steels, high-strength	118–135	7–10	120–130	24–32	Crankshaft
High-temperature alloys	118–135	9–12	125–135	15–30	Crankshaft
Refractory alloys	118	7–10	125–135	24–32	Standard
Titanium alloys	118–135	7–10	125–135	15–32	Crankshaft
Cast irons	118	8–12	125–135	24–32	Standard
Plastics	60–90	7	120–135	29	Standard

SOURCE: "Machining Data Handbook," published by the Machinability Data Center, Metcut Research Associates Inc.

Fig. 13.4.14 (*a*) Various types of drills and drilling and reaming operations; (*b*) spade drill; (*c*) trepanning tool with four cutting-tool inserts.

removed to chips; the **gun drill,** run at a high speed under a light feed, and used to drill small long holes; the **core drill** used to bore out cored holes; the **oil-hole drill,** having holes or tubes in its body through which oil is forced to the cutting lips; **three-** and **four-fluted drills,** used to enlarge holes after a leader hole has been cored, punched, or drilled with a two-fluted drill; **twist drills** made from flat high-speed steel or drop-forged to desired shape and then twisted. Drills are also made of solid carbide or of high-speed steel with an insert of carbide to form the chisel edge and both cutting edges. They are used primarily for drilling abrasive or very hard materials.

Drilling Recommendations   The most common tool material for drills is high-speed steel M1, M7, and M10. General recommendations for speeds and feeds in drilling a variety of materials are given in Table 13.4.7. Hole depth is also a factor in selecting drilling parameters. A general **troubleshooting guide** for drilling is given in Table 13.4.8.

Table 13.4.8   General Troubleshooting Guide for Drilling Operations

Problem	Probable causes
Drill breakage	Dull drill; drill seizing in hole because of chips clogging flutes; feed too high; lip relief angle too small
Excessive drill wear	Cutting speed too high; ineffective cutting fluid; rake angle too high; drill burned and strength lost when sharpened
Tapered hole	Drill misaligned or bent; lips not equal; web not central
Oversize hole	Same as above; machine spindle loose; chisel edge not central; side pressure on workpiece
Poor hole surface finish	Dull drill; ineffective cutting fluid; welding of workpiece material on drill margin; improperly ground drill; improper alignment

REAMING

A **reamer** is a multiple-cutting edge tool used to enlarge or finish holes, and to provide accurate dimensions as well as good finish. Reamers are of two types: (1) rose and (2) fluted.

The **rose reamer** is a heavy-bodied tool with end cutting edges. It is used to remove considerable metal and to true up a hole preparatory to flute reaming. It is similar to the three- and four-fluted drills. Wide cylindrical lands are provided back of the flute edges.

Fluted reamers cut principally on the periphery and remove only 0.004 to 0.008 in (0.1 to 0.2 mm) on the bore. Very narrow cylindrical margins are provided back of the flute edges, 0.012 to 0.015 in (0.3 to 0.4 mm) wide for machine-finish reaming and 0.004 to 0.006 in (0.1 to 0.15 mm) for hand reaming, to provide free cutting of the edges due to the slight body taper and also to pilot the reamer in the hole. The hole to be flute- or finish-reamed should be true. A rake of 5° is recommended for most reaming operations. A reamer may be straight or helically fluted. The latter provides much smoother cutting and gives a better finish.

Expansion reamers permit a slight expansion by a wedge so that the reamer may be resharpened to its normal size or for job shop use; they provide slight variations in size. **Adjustable reamers** have means of adjusting inserted blades so that a definite size can be maintained through numerous grindings and fully worn blades can be replaced with new ones. **Shell reamers** constitute the cutting portion of the tool which fits interchangeably on arbors to make many sizes available or to make replacement of worn-out shells less costly. Reamers float in their holding fixtures to ensure alignment, or they should be piloted in guide bushings above and below the work. They may also be held rigidly, such as in the tailstock of a lathe.

The **speed** of high-speed steel reamers should be two-thirds to three-quarters and **feeds** usually are two or three times that of the corresponding drill size. The most common tool materials for reamers are M1, M2, and M7 high-speed steels and C2 carbide.

Table 13.4.7   General Recommendations for Drilling

| Workpiece material | Surface speed | | Feed, mm/r (in/r) | | r/min | |
| | m/min | ft/min | Drill diameter | | | |
			1.5 mm (0.060 in)	12.5 mm (0.5 in)	1.5 mm	12.5 mm
Aluminum alloys	30–120	100–400	0.025 (0.001)	0.30 (0.012)	6,400–25,000	800–3,000
Magnesium alloys	45–120	150–400	0.025 (0.001)	0.30 (0.012)	9,600–25,000	1,100–3,000
Copper alloys	15–60	50–200	0.025 (0.001)	0.25 (0.010)	3,200–12,000	400–1,500
Steels	20–30	60–100	0.025 (0.001)	0.30 (0.012)	4,300–6,400	500–800
Stainless steels	10–20	40–60	0.025 (0.001)	0.18 (0.007)	2,100–4,300	250–500
Titanium alloys	6–20	20–60	0.010 (0.0004)	0.15 (0.006)	1,300–4,300	150–500
Cast irons	20–60	60–200	0.025 (0.001)	0.30 (0.012)	4,300–12,000	500–1,500
Thermoplastics	30–60	100–200	0.025 (0.001)	0.13 (0.005)	6,400–12,000	800–1,500
Thermosets	20–60	60–200	0.025 (0.001)	0.10 (0.004)	4,300–12,000	500–1,500

NOTE: As hole depth increases, speeds and feeds should be reduced. Selection of speeds and feeds also depends on the specific surface finish required.

THREADING

Threads may be formed on the outside or inside of a cylinder or cone (1) with single-point threading tools (see Fig. 13.4.1), (2) with threading chasers, (3) with taps, (4) with dies, (5) by thread milling, (6) by thread rolling, and (7) by grinding. There are numerous types of taps, such as hand, machine screw, pipe, and combined pipe tap and drill. Small taps usually have no radial relief. They may be made in two, three, or four flutes. Large taps may have still more flutes.

The **feed** of a tap depends upon the lead of the screw thread. The **cutting speed** depends upon numerous factors: Hard tough materials, great length of hole, taper taps, and full-depth thread reduce the speed; long chamfer, fine pitches, and a cutting fluid applied in quantity increase the speed. Taps are cut or formed by grinding. The ground-thread taps may operate at much higher speeds than the cut taps. Speeds may range from 3 ft/min (1 m/min) for high-strength steels to 150 ft/min (45 m/min) for aluminum and magnesium alloys. Common high-speed steel tool materials for taps are M1, M7, and M10.

Threading dies, used to produce external threads, may be solid, adjustable, spring-adjustable, or self-opening die heads. Replacement chasers are used in die heads and may be of the fixed or self-opening type. These chasers may be of the radial type, hobbed or milled; of the tangenital type; or of the circular type. Emulsions and oils are satisfactory for most threading operations.

For thread rolling, see Sec. 13.2.

MILLING

Milling is one of the most versatile machining processes and is capable of producing a variety of shapes involving flat surfaces, slots, and contours (Fig. 13.4.15). **Milling machines** use cutters with multiple teeth in contrast with the single-point tools of the lathe and planer.

Milling-machine classification is based on design, operation, or purpose. **Knee-and-column** type milling machines have the table and saddle supported on the vertically adjustable knee gibbed to the face of the column. The table is fed longitudinally on the saddle, and the latter transversely on the knee to give three feeding motions.

Knee-type machines are made with horizontal or vertical spindles. The **horizontal** universal machines have a swiveling table for cutting helices. The plain machines are used for jobbing or production work, the universal for toolroom work. **Vertical** milling machines with fixed or sliding heads are otherwise similar to the horizontal type. They are used for face or end milling and are frequently provided with a rotary table for making cylindrical surfaces.

The **fixed-bed** machines have a spindle mounted in a head dovetailed to and sliding on the face of the column. The table rests directly on the bed. They are simple and rigidly built and are used primarily for high-production work. These machines are usually provided with work-holding fixtures and may be constructed as plain or multiple-spindle machines, simple or duplex.

Planer-type millers are used only on the heaviest work. They are used to machine a number of surfaces on a particular part or group of parts arranged in series in fixtures on the table.

Milling Cutters

Milling cutters are made in a wide variety of shapes and sizes. The nomenclature of tooth parts and angles is standardized as in Fig. 13.4.16. Milling cutters may be classified in various ways, such as purpose or use of the cutters (Woodruff keyseat cutters, T-slot cutters, gear cutters, etc.); construction characteristics (solid cutters, carbide-tipped cutters, etc.); method of mounting (arbor type, shank type, etc.); and relief of teeth. The latter has two categories: profile cutters which produce flat, curved, or irregular surfaces, with the cutter teeth sharpened on the land; and formed cutters which are sharpened on the face to retain true cross-sectional form of the cutter.

(a)

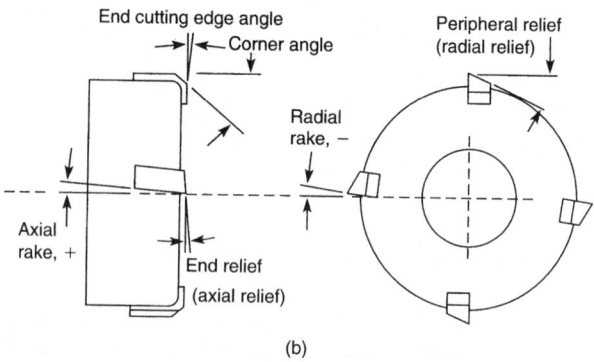

(b)

Fig. 13.4.16 (a) Plain milling cutter teeth; (b) face milling cutter.

Two kinds of milling are generally considered to represent all forms of milling processes: **peripheral (slab)** and **face** milling. In the peripheral-milling process the axis of the cutter is parallel to the surface milled, whereas in face milling, the cutter axis is generally at a right angle to the surface. The peripheral-milling process is also divided into two types: **conventional (up) milling** and **climb (down) milling.** Each has its advantages, and the choice depends on a number of factors such as the type and condition of the equipment, tool life, surface finish, and machining parameters.

Milling Recommendations Recommendations for tool materials, feed per tooth, and cutting speed for milling a variety of materials are

Fig. 13.4.15 Basic types of milling cutters and operations. (a) Peripheral milling; (b) face milling; (c) end milling.

Table 13.4.9   General Recommendations for Milling Operations

Workpiece material	Cutting tool	General-purpose starting conditions		Range of conditions	
		Feed, mm/tooth (in/tooth)	Speed, m/min (ft/min)	Feed, mm/tooth (in/tooth)	Speed, m/min (ft/min)
Low-C and free-machining steels	Uncoated carbide, coated carbide, cermets	0.13–0.20 (0.005–0.008)	120–180 (400–600)	0.085–0.38 (0.003–0.015)	90–425 (300–1,400)
Alloy steels Soft	Uncoated, coated, cermets	0.10–0.18 (0.004–0.007)	90–170 (300–550)	0.08–0.30 (0.003–0.012)	60–370 (200–1,200)
Hard	Cermets, PcBN	0.10–0.15 (0.004–0.006)	180–210 (600–700)	0.08–0.25 (0.003–0.010)	75–460 (250–1,500)
Cast iron, gray Soft	Uncoated, coated, cermets, SiN	0.10–0.20 (0.004–0.008)	120–760 (400–2,500)	0.08–0.38 (0.003–0.015)	90–1,370 (300–4,500)
Hard	Cermets, SiN, PcBN	0.10–0.20 (0.004–0.008)	120–210 (400–700)	0.08–0.38 (0.003–0.015)	90–460 (300–1,500)
Stainless steel, austenitic	Uncoated, coated, cermets	0.13–0.18 (0.005–0.007)	120–370 (400–1,200)	0.08–0.38 (0.003–0.015)	90–500 (300–1,800)
High-temperature alloys, nickel base	Uncoated, coated, cermets, SiN, PcBN	0.10–0.18 (0.004–0.007)	30–370 (100–1,200)	0.08–0.38 (0.003–0.015)	30–550 (90–1,800)
Titanium alloys	Uncoated, coated, cermets	0.13–0.15 (0.005–0.006)	50–60 (175–200)	0.08–0.38 (0.003–0.015)	40–140 (125–450)
Aluminum alloys Free-machining	Uncoated, coated, PCD	0.13–0.23 (0.005–0.009)	610–900 (2,000–3,000)	0.08–0.46 (0.003–0.018)	300–3,000 (1,000–10,000)
High-silicon	PCD	0.13 (0.005)	610 (2,000)	0.08–0.38 (0.003–0.015)	370–910 (1,200–3,000)
Copper alloys	Uncoated, coated, PCD	0.13–0.23 (0.005–0.009)	300–760 (1,000–2,500)	0.08–0.46 (0.003–0.018)	90–1,070 (300–3,500)
Thermoplastics and thermosets	Uncoated, coated, PCD	0.13–0.23 (0.005–0.009)	270–460 (900–1,500)	0.08–0.46 (0.003–0.018)	90–1,370 (300–4,500)

NOTE: Depths of cut, d, usually are in the range of 1–8 mm (0.04–0.3 in). PcBN: polycrystalline cubic boron nitride; PCD: polycrystalline diamond.
SOURCE: Based on data from Kennametal Inc.

Table 13.4.10   General Troubleshooting Guide for Milling Operations

Problem	Probable causes
Tool breakage	Tool material lacks toughness; improper tool angles; cutting parameters too high
Tool wear excessive	Cutting parameters too high; improper tool material; improper tool angles; improper cutting fluid
Rough surface finish	Feed too high; spindle speed too low; too few teeth on cutter; tool chipped or worn; built-up edge; vibration and chatter
Tolerances too broad	Lack of spindle stiffness; excessive temperature rise; dull tool; chips clogging cutter
Workpiece surface burnished	Dull tool; depth of cut too low; radial relief angle too small
Back striking	Dull cutting tools; cutter spindle tilt; negative tool angles
Chatter marks	Insufficient stiffness of system; external vibrations; feed, depth, and width of cut too large
Burr formation	Dull cutting edges or too much honing; incorrect angle of entry or exit; feed and depth of cut too high; incorrect insert geometry
Breakout	Lead angle too low; incorrect cutting edge geometry; incorrect angle of entry or exit; feed and depth of cut too high

given in Table 13.4.9. A general **troubleshooting guide** for milling operations is given in Table 13.4.10.

GEAR MANUFACTURING

(See also Sec. 8.)

Gear Cutting   Most gear-cutting processes can be classified as either **forming or generating.** In a forming process, the shape of the tool is reproduced on the workpiece; in a generating process, the shape produced on the workpiece depends on both the shape of the tool and the relative motion between the tool and the workpiece during the cutting operation. In general, a generating process is more accurate than a forming process.

In the **form cutting** of gears, the tool has the shape of the space between the teeth. For this reason, form cutting will produce precise tooth profiles only when the cutter is accurately made and the tooth space is of constant width, such as on spur and helical gears. A form cutter may cut or finish one of or all the spaces in one pass. Single-space cutters may be disk-type or end-mill-type milling cutters. In all single-space operations, the gear blank must be retracted and indexed, i.e., rotated one tooth space, between each pass.

Single-space form milling with disk-type cutters is particularly suitable for gears with large teeth, because, as far as metal removal is concerned, the cutting action of a milling cutter is more efficient than that of the tools used for generating. Form milling of spur gears is done on machines that retract and index the gear blank automatically.

For the same tooth size (pitch), the shape (profile) of the teeth on an involute gear depends on the number of teeth on the gear. Most gears have active profiles that are wholly, partially, or approximately involute,

Table 13.4.11

No. of cutter	1	2	3	4	5	6	7	8
No. of teeth	135–∞	55–134	35–54	27–34	21–26	17–20	14–16	12 and 13

For more accurate gears, 15 cutters are available

No. of cutter	1	1½	2	2½	3	3½	4	4½
No. of teeth	135–∞	80–134	55–79	42–54	35–41	30–34	26–29	23–25
No. of cutter	5	5½		6	6½	7	7½	8
No. of teeth	21 and 22	19 and 20		17 and 18	15 and 16	14	13	12

and, consequently, accurate form cutting would require a different cutter for each number of teeth. In most cases, satisfactory results can be obtained by using the eight cutters for each pitch that are commercially available. Each cutter is designed to cut a range of tooth numbers; the no. 1 cutter, for example, cuts from 135 teeth to a rack, and the no. 8 cuts 12 and 13 teeth. (See Table 13.4.11.)

In a **gear generating machine,** the generating tool can be considered as one of the gears in a conjugate pair and the gear blank as the other gear. The correct relative motion between the tool arbor and the blank arbor is obtained by means of a train of indexing gears within the machine.

One of the most valuable properties of the involute as a gear-tooth profile is that if a cutter is made in the form of an involute gear of a given pitch and any number of teeth, it can generate all gears of all tooth numbers of the same pitch and they will all be conjugate to one another. The generating tool may be a pinion-shaped cutter, a rack-shaped (straight) cutter, or a hob, which is essentially a series of racks wrapped around a cylinder in a helical, screwlike form.

On a **gear shaper,** the generating tool is a pinion-shaped cutter that rotates slowly at the proper speed as if in mesh with the blank; the cutting action is produced by a reciprocation of the cutter parallel to the work axis. These machines can cut spur and helical gears, both internal and external; they can also cut continuous-tooth helical (herringbone) gears and are particularly suitable for cluster gears, or gears that are close to a shoulder.

On a **rack shaper** the generating tool is a segment of a rack that moves perpendicular to the axis of the blank while the blank rotates about a fixed axis at the speed corresponding to conjugate action between the rack and the blank; the cutting action is produced by a reciprocation of the cutter parallel to the axis of the blank. Since it is impracticable to have more than 6 to 12 teeth on a rack cutter, the cutter must be disengaged from the blank at suitable intervals and returned to the starting point, the blank meanwhile remaining fixed. These machines can cut both spur and helical external gears.

A **gear-cutting hob** (Fig. 13.4.17) is basically a worm, or screw, made into a generating tool by cutting a series of longitudinal slots or "gashes" to form teeth; to form cutting edges, the teeth are "backed off," or relieved, in a lathe equipped with a backing-off attachment. A hob may have one, two, or three threads; on involute hobs with a single thread, the generating portion of the hob-tooth profile usually has straight sides (like an involute rack tooth) in a section taken at right angles to the thread.

In addition to the conjugate rotary motions of the hob and workpiece, the hob must be fed parallel to the workpiece axis for a distance greater than the face width of the gear. The feed, per revolution of the workpiece, is produced by the feed gears, and its magnitude depends on the material, pitch, and finish desired; the feed gears are independent of the indexing gears. The hobbing process is continuous until all the teeth are cut.

The same machines and the same hobs that are used for cutting spur gears can be used for helical gears; it is only necessary to tip the hob axis so that the hob and gear pitch helices are tangent to one another and to correlate the indexing and feed gears so that the blank and the hob are advanced or retarded with respect to each other by the amount required to produce the helical teeth. Some hobbing machines have a differential gear mechanism that permits the indexing gears to be selected as for spur gears and the feed gearing to be chosen independently.

The threads of worms are usually cut with a disk-type milling cutter on a thread-milling machine and finished, after hardening, by grinding.

Fig. 13.4.17 A gear-cutting hob.

Worm gears are usually cut with a hob on the machines used for hobbing spur and helical gears. Except for the gashes, the relief on the teeth, and an allowance for grinding, the hob is a counterpart of the worm. The hob and workpiece axes are inclined to one another at the shaft angle of the worm and gear set, usually 90°. The hob may be fed in to full depth in a radial (to the blank) direction or parallel to the hob axis.

Although it is possible to approximate the true shape of the teeth on a straight bevel gear by taking two or three cuts with a form cutter on a milling machine, this method, because of the taper of the teeth, is obviously unsuited for the rapid production of accurate teeth. Most straight bevel gears are roughed out in one cut with a form cutter on machines that index automatically and then finished to the proper shape on a generator.

The generating method used for straight bevel gears is analogous to the rack-generating method used for spur gears. Instead of using a rack with several complete teeth, however, the cutter has only one straight cutting edge that moves, during generation, in the plane of the tooth of a basic crown gear conjugate to the gear being generated. A crown gear is the rack among bevel gears; its pitch surface is a plane, and its teeth have straight sides.

The generating cutter moves back and forth across the face of the bevel gear like the tool on a shaper; the "generating roll" is obtained by rotating the gear slowly relative to the tool. In practice two tools are used, one for each side of a tooth; after each tooth has been generated, the gear must be retracted and indexed to the next tooth.

The machines used for cutting spiral bevel gears operate on essentially the same principle as those used for straight bevel gears; only the cutter is different. The spiral cutter is basically a disk that has a number of straight-sided cutting blades protruding from its periphery on one side to form the rim of a cup. The machines have means for indexing, retracting, and producing a generating roll; by disconnecting the roll gears, spiral bevel gears can be form cut.

Gear Shaving For improving the surface finish and profile accuracy of cut spur and helical gears (internal and external), gear shaving, a free-cutting gear finishing operation that removes small amounts of metal from the working surfaces of the teeth, is employed. The teeth on the shaving cutter, which may be in the form of a pinion (spur or helical) or a rack, have a series of sharp-edged rectangular grooves running from tip to root. The intersection of the grooves with the tooth profiles creates cutting edges; when the cutter and the workpiece, in tight mesh, are caused to move relative to one another along the teeth, the cutting edges remove metal from the teeth of the work gear. Usually the cutter drives the workpiece, which is free to rotate and is traversed past the cutter parallel to the workpiece axis. Shaving requires less time than grinding, but ordinarily it cannot be used on gears harder than approximately 400 HB (42 HRC).

Gear Grinding Machines for the grinding of spur and helical gears utilize either a forming or a generating process. For form grinding, a disk-type grinding wheel is dressed to the proper shape by a diamond held on a special dressing attachment; for each number of teeth a special index plate, with V-type notches on its periphery, is required. When grinding helical gears, means for producing a helical motion of the blank must be provided.

For grinding-generating, the grinding wheel may be a disk-type, double-conical wheel with an axial section equivalent to the basic rack of the gear system. A master gear, similar to the gear being ground, is attached to the workpiece arbor and meshes with a master rack; the generating roll is created by rolling the master gear in the stationary rack.

Spiral bevel and hypoid gears can be ground on the machines on which they are generated. The grinding wheel has the shape of a flaring cup with a double-conical rim having a cross section equivalent to the surface that is the envelope of the rotary cutter blades.

Gear Rolling The cold-rolling process is used for the finishing of spur and helical gears for automatic transmissions and power tools; in some cases it has replaced gear shaving. It differs from cutting in that the metal is not removed in the form of chips but is displaced under heavy pressure. (See Sec. 13.2.)

There are two main types of cold-rolling machines, namely, those employing dies in the form of racks or gears that operate in a parallel-axis relationship with the blank and those employing worm-type dies that operate on axes at approximately 90° to the workpiece axis. The dies, under pressure, create the tooth profiles by the plastic deformation of the blank.

When racks are used, the process resembles thread rolling; with gear-type dies the blank can turn freely on a shaft between two dies, one mounted on a fixed head and the other on a movable head. The dies have the same number of teeth and are connected by gears to run in the same direction at the same speed. In operation, the movable die presses the blank into contact with the fixed die, and a conjugate profile is generated on the blank. On some of these machines the blank can be fed axially, and gears can be rolled in bar form to any convenient length.

On machines employing worm-type dies, the two dies are diametrically opposed on the blank and rotate in opposite directions. The speeds of the blank and the dies are synchronized by change gears, like the blank and the hob on a hobbing machine; the blank is fed axially between the dies.

PLANING AND SHAPING

Planers are used to rough and finish large flat surfaces, although arcs and special forms can be made with proper tools and attachments. Surfaces to be finished by scraping, such as ways and long dovetails and, particularly, parts of machine tools, are, with few exceptions, planed. With fixtures to arrange parts in parallel and series, quantities of small parts can be produced economically on planers. **Shapers** are

used for miscellaneous planing, surfacing, notching, key seating, and production of flat surfaces on small parts. The tool is held in a holder supported on a clapper on the end of a ram which is reciprocated hydraulically or by crank and rocker arm, in a straight line.

BROACHING

Broaching is a production process whereby a cutter, called a **broach,** is used to finish internal or external surfaces such as holes of circular, square, or irregular section, keyways, the teeth of internal gears, multiple spline holes, and flat surfaces. Broaching round holes gives greater accuracy and better finish than reaming, but since the broach may be guided only by the workpiece it is cutting, the hole may not be accurate with respect to previously machined surfaces. Where such accuracy is required, it is better practice to broach first and then turn other surfaces with the workpiece mounted on a mandrel. The broach is usually long and is provided with many teeth so graded in size that each takes a small chip when the tool is pulled or pushed through the previously prepared leader hole or past the surface.

The main features of the broach are the pitch, degree of taper or increase in height of each successive tooth, relief, tooth depth, and rake.

The **pitch** of the teeth, i.e., the distance from one tooth to the next, depends upon tooth strength, length of cut, shape and size of chips, etc. The pitch should be as coarse as possible to provide ample chip clearance, but at least two teeth should be in contact with the workpiece at all times. The formula $p = 0.35 \sqrt{l}$ may be used, where p is pitch of the roughing teeth and l the length of hole or surface, in. An average pitch for small broaches is ⅛ to ¼ in (3.175 to 6.35 mm) and for large ones ½ to 1 in (12.7 to 25.4 mm). Where the hole or other surface to be broached is short, the teeth are often cut on an angle or helix, so as to give more continuous cutting action by having at least two teeth cutting simultaneously.

The degree of **taper,** or increase in size per tooth, depends largely on the hardness or toughness of the material to be broached and the finish desired. The degree of taper or feed for broaching cast iron is approximately double that for steel. Usually the first few teeth coming in contact with the workpiece are undersize but of uniform taper to take the greatest feeds per tooth, but as the finished size is approached, the teeth take smaller and smaller feeds with several teeth at the finishing end of nearly zero taper. In some cases, for soft metals and even cast iron, the large end is left plain or with rounded lands a trifle larger than the last cutting tooth so as to burnish the surface. For medium-sized broaches, the taper per tooth is 0.001 to 0.003 in (0.025 to 0.076 mm). Large broaches remove 0.005 to 0.010 in (0.127 to 0.254 mm) per tooth or even more. The teeth are given a **front rake** angle of 5 to 15° to give a curl to the chip, provide a cleaner cut surface, and reduce the power consumption. The **land** back of the cutting edge, which may be ¹⁄₆₄ to ¹⁄₁₆ in (0.4 to 1.6 mm) wide, usually is provided with a land relief varying from 1 to 3° with a clearance of 30 to 45°.

The heavier the feed per tooth or the longer the surface being broached, the greater must be the **chip clearance** or space between successive teeth for the chips to accumulate. The root should be a smooth curve.

Broaches are generally made of M2 or M7 high-speed steel; carbide is also used for the teeth of large broaches. Broaches of complicated shape are likely to warp during the heat-treating process. For this reason, in hardening, they are often heated in a vertical cylindrical furnace and quenched by being hung in an air blast furnished from small holes along the side of pipes placed vertically about the broach.

Push broaches are usually shorter than pull broaches, being 6 to 14 in (150 to 350 mm) long, depending on their diameter and the amount of metal to be removed. In many cases, for accuracy, four to six broaches of the push type constitute a set used in sequence to finish the surface being broached. Push broaches usually have a large cross-sectional area so as to be sufficiently rigid. With **pull broaches,** pulling tends to straighten the hole, whereas pushing permits the broaches to follow any irregularity of the leader hole. Pull broaches are attached to the crosshead of the broaching machine by means of a key slot and key, by a threaded connection, or by a head that fits into an automatic broach puller. The threaded connection is used where the broach is not removed from the drawing head while the workpiece is placed over the cutter, as in cutting a keyway. In enlarging holes, however, the small end of the pull broach must first be extended through the reamed, drilled, or cored hole and then fixed in the drawing head before being pulled through the workpiece.

Broaching Machines   Push broaching is done on machines of the press type with a sort of fixture for holding the workpiece and broach or on presses operated by power. They are usually vertical and may be driven hydraulically or by screw, rack, or crank. The pull type of broach may be either vertical or horizontal. The ram may be driven hydraulically or by screw, rack, or crank. Both are made in the duplex- and multiple-head type.

Processing Parameters for Broaching   Cutting speeds for broaching may range from 5 ft/min (1.5 m/min) for high-strength materials to as high as 30 to 50 ft/min (9 to 15 m/min) for aluminum and magnesium alloys. The most common tool materials are M2 and M7 high-speed steels and carbides. An emulsion is often used for broaching for general work, but oils may also be used.

CUTTING OFF

Cutting off involves parting or slotting bars, tubes, plate, or sheet by various means. The machines come in various types such as a lathe (using a single-point cutting tool), hacksaws, band saws, circular saws, friction saws, and thin abrasive wheels. Cutting off may also be carried out by shearing and cropping, as well as using flames and laser beams.

In **power hacksaws,** the frame in which the blade is strained is reciprocated above the workpiece which is held in a vise on the bed. The cutting feed is effected by weighting the frame, with 12 to 50 lb (55 to 225 N) of force from small to large machines; adding weights or spring tension giving up to 180 lb (800 N); providing a positive screw feed or a friction screw feed; and by a hydraulic feed mechanism giving forces up to 300 lb (1.34 kN) between the blade and workpiece. With highspeed steel blades, cutting speeds range from about 30 strokes per minute for high-strength materials to 180 strokes per minute for carbon steels.

Hacksaw **blades** for hand frames are made 8, 10, and 12 in long, $\frac{7}{16}$ to $\frac{9}{16}$ in wide, and 0.025 in thick. Number of teeth per inch for cutting soft steel or cast iron, 14; tool steel and angle iron, 18; brass, copper, and heavy tubing, 24; sheet metal and thin tube, 32. Blades for power hacksaws are made of alloy steels and of high-speed steels. Each length is made in two or more widths. The coarsest teeth should be used on large workpieces and with heavy feeds.

Band saws, vertical, horizontal, and universal, are used for cutting off. The kerf or width of cut is small with a consequently small loss in expensive material. The teeth of band saws, like those of hacksaws, are set with the regular alternate type, one bent to the right and the next to the left; or with the alternate and center set, in which one tooth is bent to the right, the second to the left, and the third straight in the center. With high-speed steel saw blades, band speeds range from about 30 ft/min (10 m/min) for high-temperature alloys to about 400 (120) for carbon steels. For aluminum and magnesium alloys the speed ranges up to about 1,300 ft/min (400 m/min) with high-carbon blades. The band speed should be decreased as the workpiece thickness increases.

Circular saws are made in a wide variety of styles and sizes. Circular saws may have teeth of several shapes, as radial face teeth for small fine-tooth saws; radial face teeth with a land for fine-tooth saws; alternate bevel-edged teeth to break up the chips, with every other tooth beveled 45° on each side with the next tooth plain; alternate side-beveled teeth; and one tooth beveled on the right, the next on the left, and the third beveled on both sides, each leaving slightly overlapping flats on the periphery of the teeth. The most common high-speed steels for circular saws are M2 and M7; saws also may have welded carbide teeth for better performance.

Abrasive cutoff wheels are made of thin resinoid or rubber bonded abrasives. The wheels operate at surface speeds of 12,000 to 16,000 ft/min (3,600 to 4,800 m/min). These wheels are used for cutting off tubes, shapes, and hardened high-speed steel.

Friction sawing machines are used largely for cutting off structural shapes. Peripheral speeds of about 20,000 ft/min (6,000 m/min) are used. The wheels may be plain on the periphery, V-notched, or with milled square notches.

ABRASIVE PROCESSES

(See also Sec. 6.)

Abrasive processes consist of a variety of operations in which the tool is made of an abrasive material, the most common examples of which are grinding (using wheels, known as **bonded abrasives**), honing, and lapping. An abrasive is a small, nonmetallic hard particle having sharp cutting edges and an irregular shape. Abrasive processes, which can be performed on a wide variety of metallic and nonmetallic materials, remove material in the form of tiny chips and produce surface finishes and dimensional accuracies that are generally not obtainable through other machining or manufacturing processes.

Grinding wheels have characteristics influenced by (1) type of abrasive; (2) grain size; (3) grade; (4) structure; and (5) type of bond. (See Figs. 13.4.18 and 13.4.19.)

Selection of Abrasive   Although a number of natural abrasives are available, such as emery, corundum, quartz, garnet, and diamond, the most commonly used abrasives in grinding wheels are **aluminum oxide** and **silicon carbide,** the former being more commonly used than the latter. Aluminum oxide is softer than silicon carbide, and because of its friability and low attritious wear it is suitable for most applications. Silicon carbide is used for grinding aluminum, magnesium, titanium, copper, tungsten, and rubber. It is also used for grinding very hard and brittle materials such as carbides, ceramics, and stones. **Diamond** and **cubic boron nitride** grains are used to grind very hard materials and are known as **superabrasives.**

Selection of **grain size** depends on the rate of material removal desired and the surface finish. Coarse grains are used for fast removal of stock; fine grain for low removal rates and for fine finish. Coarse grains are also used for ductile materials and a finer grain for hard and brittle materials.

The **grade** of a grinding wheel is a measure of the strength of its bond. The force that acts on the grain in grinding depends on process variables (such as speeds, depth of cut, etc.) and the strength of the work material. Thus a greater force on the grain will increase the possibility of dislodging the grain; if the bond is too strong, the grain will tend to get dull, and if it is too weak then wheel wear will be high. If glazing occurs, the wheel is **acting hard;** reducing the wheel speed or increasing the work speed or the depth of cut causes the wheel to **act softer.** If the wheel breaks down too rapidly, reversing this procedure will make the wheel act harder. Harder wheels are generally recommended for soft work materials, and vice versa.

A variety of **bond** types are used in grinding wheels; these are generally categorized as **organic** or **inorganic.** Organic bonds are materials such as resin, rubber, shellac, and other similar bonding agents. Inorganic materials are glass, clay, porcelain, sodium silicate, magnesium oxychloride, and metal. The most common bond type is the vitrified bond which is composed of clay, glass, porcelain, or related ceramic materials. This type of bond is brittle and produces wheels that are rigid, porous, and resistant to oil and water. The most flexible bond is rubber which is used in making very thin, flexible wheels. Wheels subjected to bending strains should be made with organic bonds. In selecting a bonding agent, attention should be paid to its sensitivity to temperature, stresses, and grinding fluids, particularly over a period of time. The term **reinforced** as applied to grinding wheels indicates a class of organic wheels which contain one or more layers of strengthening fabric or filament, such as fiberglass. This term does not cover wheels with reinforcing elements such as steel rings, steel cup backs, or wire or tape winding. Fiberglass and filament reinforcing increases the ability of wheels to withstand operational forces when cracked.

The **structure** of a wheel is important in two aspects: It supplies a clearance for the chip, and it determines the number of cutting points on the wheel.

Fig. 13.4.18 Standard marking system chart for aluminum oxide and silicon carbide bonded abrasives.

In addition to wheel characteristics, grinding wheels come in a very large variety of shapes and dimensions. They are classified as **types,** such as type 1: straight wheels, type 4: taper side wheels, type 12: dish wheels, etc.

The **grinding ratio** is defined as the ratio of the volume of material removed to the volume of wheel wear. The grinding ratio depends on parameters such as the type of wheel, workpiece speed, wheel speed, cross-feed, down-feed, and the grinding fluid used. Values ranging from a low of 2 to over 200 have been observed in practice. A high grinding ratio, however, may not necessarily result in the best surface integrity of the part.

Wheel Speeds Depending on the type of wheel and the type and strength of bond, wheel speeds for standard applications range between 4,500 and 16,000 surface ft/min (1,400 and 4,800 m/min). The lowest speeds are for low-strength, inorganic bonds whereas the highest speeds are for high-strength organic bonds. The majority of surface grinding operations are carried out at speeds from 5,500 to 6,500 ft/min (1,750 to 2,000 m/min). The trend is toward high-efficiency grinding where

wheel speeds from 12,000 to 18,000 ft/min (3,600 to 5,500 m/min) are employed.

It has been found that, by increasing the wheel speed, the rate of material removal can be increased, thus making the process more economical. This, of course, requires special grinding wheels to withstand the high stresses. Design changes or improvements involve items such as a composite wheel with a vitrified bond on the outside and a resinoid bond toward the center of the wheel; elimination of the central hole of the wheel by providing small bolt holes; and clamping of wheel segments instead of using a one-piece wheel. Grinding machines for such high-speed applications have requirements such as rigidity, high work and wheel speeds, high power, and special provisions for safety.

Workpiece speeds depend on the size and type of workpiece material and on whether it is rigid enough to hold its shape. In surface grinding, table speeds generally range from 50 to 100 ft/min (15 to 30 m/min); for cylindrical grinding, work speeds from 70 to 100 ft/min (20 to 30 m/min), and for internal grinding they generally range from 75 to 200 ft/min (20 to 60 m/min).

Fig. 13.4.19 Standard marking system chart for diamond tool and cubic boron nitride (cBN) bonded abrasives.

Cross-feed depends on the width of the wheel. In roughing, the workpiece should travel past the wheel ¾ to ⅞ of the width of the wheel for each revolution of the work. As the workpiece travels past the wheel with a helical motion, the preceding rule allows a slight overlap. In finishing, a finer feed is used, generally ⅒ to ¼ of the width of the wheel for each revolution of the workpiece.

Depth of Cut In the roughing operation, the depth of cut should be all the wheel will stand. This varies with the hardness of the material and the diameter of the workpiece. In the finishing operation, the depth of cut is always small: 0.0005 to 0.001 in (0.013 to 0.025 mm). Good results as regards finish are obtained by letting the wheel run over the workpiece several times without cross-feeding.

Grinding Allowances From 0.005 to 0.040 in (0.13 to 1 mm) is generally removed from the diameter in rough grinding in a cylindrical machine. For finishing, 0.002 to 0.010 in (0.05 to 0.25 mm) is common. Workpieces can be finished by grinding to a tolerance of 0.0002 in (0.005 mm) and a surface roughness of $50+$ μin R_q (1.2 μm).

In situations where grinding leaves unfavorable surface **residual stresses,** the technique of **gentle** or **low-stress** grinding may be employed. This generally consists of removing a layer of about 0.010 in (0.25 mm) at depths of cut of 0.0002 to 0.0005 in (0.005 to 0.013 mm) with wheel speeds that are lower than the conventional 5,500 to 6,500 ft/min.

Truing and Dressing In **truing,** a diamond supported in the end of a soft steel rod held rigidly in the machine is passed over the face of the wheel two or three times to remove just enough material to give the wheel its true geometric shape. **Dressing** is a more severe operation of removing the dull or loaded surface of the wheel. Abrasive sticks or wheels or steel star wheels are pressed against and moved over the wheel face.

Safety If not stored, handled, and used properly, a grinding wheel can be a very dangerous tool. Because of its mass and high rotational speed, a grinding wheel has considerable energy and, if it fractures, it can cause serious injury and even death to the operator or to personnel nearby. A safety code B7.1 entitled "The Use, Care, and Protection of Abrasive Wheels" is available from ANSI; other safety literature is available from the Grinding Wheel Institute and from the National Safety Council.

The salient features of safety in the use of grinding wheels may be listed as follows: Wheels should be stored and handled carefully; a wheel that has been dropped should not be used. Before it is mounted on the machine, a wheel should be visually inspected for possible cracks; a simple "ring" test may be employed whereby the wheel is tapped gently with a light nonmetallic implement and if it sounds cracked, it should not be used. The wheel should be mounted properly with the required blotters and flanges, and the mounting nut tightened not excessively. The label on the wheel should be read carefully for maximum operating speed and other instructions. An appropriate guard should always be used with the machine, whether portable or stationary.

Newly mounted wheels should be allowed to run idle at the operating speed for at least 1 min before grinding. The operator should always wear safety glasses and should not stand directly in front of a grinding wheel when a grinder is started. If a grinding fluid is used, it should be turned off first before stopping the wheel to avoid creating an out-of-balance condition. Because for each type of operation and workpiece material there usually is a specific type of wheel recommended, the operator must make sure that the appropriate wheel has been selected.

Grinders

Grinding machines may be classified as to purpose and type as follows: for **rough removal of stock,** the swinging-frame, portable, flexible shaft, two-wheel stand, and disk; **cutting off** or parting, the cutting-off machine; **surface finishing,** band polisher, two-wheel combination, two-wheel polishing machine, two-wheel buffing machine, and semiautomatic polishing and buffing machine; **precision grinding,** tool post, cylindrical (plain and universal), crankshaft, centerless, internal, and surface (reciprocating table with horizontal or vertical wheel spindle, and rotary table with horizontal or vertical wheel spindle); **special form** grinders, gear or worm, ball-bearing balls, cams, and threads; and **tool and cutter** grinders for single-point tools, drills, and milling cutters, reamers, taps, dies, knives, etc.

Grinding equipment of all types (many with computer controls) has been improved during the past few years so as to be more rigid, provide more power to the grinding wheel, and provide automatic cycling, loading, clamping, wheel dressing, and automatic feedback.

Centerless grinders are used to good advantage where large numbers of relatively small pieces must be ground and where the ground surface has no exact relation to any other surface except as a whole; the work is carried on a support between two abrasive wheels, one a normal grinding wheel, the second a rubber-bonded wheel, rotating at about ¹⁄₂₀th the grinding speed, and is tilted 3 to 8° to cause the work to rotate and feed past the grinding wheel (see also Sec. 6).

The **cylindrical grinder** is a companion machine to the engine lathe; shafts, cylinders, rods, studs, and a wide variety of other cylindrical parts are first roughed out on the lathe, then finished accurately to size by the cylindrical grinder. The work is carried on centers, rotated slowly, and traversed past the face of a grinding wheel.

Universal grinders are cylindrical machines arranged with a swiveling table so that both straight and taper internal and external work can be ground. **Drill grinders** are provided with rests so mounted that by a simple swinging motion, correct cutting angles are produced automatically on the lips of drills; a cupped wheel is usually employed. **Internal grinders** are used for finishing the holes in bushings, rolls, sleeves, cutters, and the like; spindle speeds from 15,000 to 30,000 r/min are common.

Horizontal surface grinders range from small capacity, used mainly in tool making or small production work, to large sizes used for production work.

Vertical surface grinders are used for producing flat surfaces on production work. **Vertical** and **horizontal disk grinders** are used for surfacing. Grinding machines are used for **cutting off** steel, especially tubes, structural shapes, and hard metals. A thin resinoid or rubber-bonded wheel is used, with aluminum oxide abrasive for all types of steel, aluminum, brass, bronze, nickel, Monel, and Stellite; silicon carbide for cast iron, copper, carbon, glass, stone, plastics, and other nonmetallic materials; and diamond for cemented carbides and ceramics.

Belt grinders use a coated abrasive belt running between pulleys. Belt grinding is generally considered to be a roughing process, but finer finishes may be obtained by using finer grain size. Belt speeds generally range from 2,000 to 10,000 ft/min (600 to 3,000 m/min) with grain sizes ranging between 24 and 320, depending on the workpiece material and the surface finish desired. The process has the advantage of high-speed material removal and is applied to flat as well as irregular surfaces.

Although grinding is generally regarded as a finishing operation, it is possible to increase the rate of stock removal whereby the process becomes, in certain instances, competitive with milling. This type of grinding operation is usually called **creep-feed grinding.** It uses equipment such as reciprocating table or vertical-spindle rotary table surface grinders with capacities up to 300 hp (220 kW). The normal stock removal may range up to ¼ in (6.4 mm) with wheel speeds between 3,400 and 5,000 surface ft/min (1,000 and 1,500 m/min).

Finishing Operations

Polishing is an operation by which scratches or tool marks or, in some instances, rough surfaces left after forging, rolling, or similar operations are removed. It is not a precision operation. The nature of the polishing process has been debated for a long time. Two mechanisms appear to play a role: One is fine-scale abrasion, and the other is softening of surface layers. In addition to removal of material by the abrasive particles, the high temperatures generated because of friction soften the asperities of the surface of the workpiece, resulting in a smeared surface layer. Furthermore, chemical reactions may also take place in polishing whereby surface irregularities are removed by chemical attack.

Polishing is usually done in stages. The first stage is rough polishing, using abrasive grain sizes of about 36 to 80, followed by a second stage, using an abrasive size range of 80 to 120, a third stage of size 150 and finer, etc., with a final stage of buffing. For the first two steps the polishing wheels are used dry. For finishing, the wheels are first worn down a little and then coated with tallow, oil, beeswax, or similar substances. This step is partly polishing and partly buffing, as additional abrasive is often added in cake form with the grease. The cutting action is freer, and the life of the wheel is prolonged by making the wheel surface flexible. Buffing wheels are also used for the finishing step when tallow, etc., containing coarse or fine abrasive grains is periodically rubbed against the wheel.

Polishing wheels consisting of wooden disks faced with leather, turned to fit the form of the piece to be polished, are used for flat surfaces or on work where it is necessary to maintain square edges. A large variety of other types of wheels are in common use. Compress wheels are used extensively and are strong, durable, and easily kept in balance. They consist of a steel center the rim of which holds a laminated surface of leather, canvas, linen, felt, rubber, etc., of various degrees of pliability. Wheels of solid leather disks of walrus hide, buffalo hide, sheepskin, or bull's neck hide, or of soft materials such as felt, canvas, and muslin, built up of disks either loose, stitched, or glued, depending on the resiliency or pliability required, are used extensively for polishing as well as buffing. **Belts** of cloth or leather are often charged with abrasive for polishing flat or other workpieces. Wire brushes may be used with no abrasive for a final operation to give a satin finish to nonferrous metals.

For most polishing operations speeds range from 5,000 to 7,500 surface ft/min (1,500 to 2,250 m/min). The higher range is for high-strength steels and stainless steels. Excessively high speeds may cause burning of the workpiece and glazing.

Mirrorlike finishes may be obtained by **electropolishing,** a process that is the reverse of electroplating; it is particularly useful for polishing irregularly shaped workpieces which otherwise would be difficult to polish uniformly. A more recent specialized process is **magnetic-field polishing,** in which fine abrasive polishing particles are suspended in a magnetic fluid. This process is effective for polishing ceramic balls, such as ball bearings. The **chemical-mechanical polishing** process combines the actions of abrasive particles, suspended in a water-based solution, with a chemistry selected to cause controlled corrosion. It produces exceptionally flat surfaces with very fine surface finish, particularly important in polishing silicon wafers for the semiconductor industry.

Buffing is a form of finish polishing in which the surface finish is improved; very little material is removed. The powdered abrasives are applied to the surface of the wheel by pressing a mixture of abrasive and tallow or wax against the face for a few seconds. The abrasive is replenished periodically. The wheels are made of a soft pliable material, such as soft leather, felt, linen, or muslin, and rotated at high speed.

A variety of buffing compounds are available: aluminum oxide, chromium oxide, soft silica, rouge (iron oxide), pumice, lime compounds, emery, and crocus. In cutting down nonferrous metals, Tripoli is used; and for steels and stainless steels, aluminum oxide is the common abrasive. For coloring, soft silica, rouge, and chromium oxide are the more common compounds used. Buffing speeds range from 6,000 to 10,000 surface ft/min (1,800 to 3,000 m/min); the higher speeds are for steels, although the speed may be as high at 12,000 surface ft/min (3,600 m/min) for coloring brass and copper.

Lapping is a process of producing extremely smooth and accurate surfaces by rubbing the surface which is to be lapped against a mating form which is called a **lap.** The lap may either be charged with a fine abrasive and moistened with oil or grease, or the fine abrasive may be introduced with the oil. If a part is to be lapped to a final accurate dimension, a mating form of a softer material such as soft close-grained cast iron, copper, brass, or lead is made up. Aluminum oxide, silicon carbide, and diamond grits are used for lapping. Lapping requires considerable time. No more than 0.0002 to 0.0005 in (0.005 to 0.013 mm) should be left for removal by this method. Surface plates, rings, and plugs are common forms of laps. For most applications grit sizes range between 100 and 800, depending on the finish desired. For most efficient lapping, speeds generally range from 300 to 800 surface ft/min (150 to 240 m/min) with pressures of 1 to 3 lb/in^2 (7 to 21 kPa) for soft materials and up to 10 lb/in^2 (70 kPa) for harder materials.

Honing is an operation similar to lapping. Instead of a metal lap charged with fine abrasive grains, a honing stone made of fine abrasives is used. Small stones of various cross-sectional shapes and lengths are manufactured for honing the edges of cutting tools. Automobile cylinders

are honed for fine finish and accurate dimensions. This honing usually follows a light-finish reaming operation or a precision-boring operation using diamonds or carbide tools. The tool consists of several honing stones adjustable at a given radius or forced outward by springs or a wedge forced mechanically or hydraulically and is given a reciprocating (25 to 40 per min) and a rotating motion (about 300 r/min) in the cylinder which is flooded with kerosine.

Hones operate at speeds of 50 to 200 surface ft/min (15 to 60 m/min) and use universal joints to allow the tool to center itself in the workpiece. The automatic pressure-cycle control of hone expansion, in which the pressure is reduced in steps as the final finish is reached, removes metal 10 times as fast as with the spring-expanded hone. Rotational and reciprocating movements are provided to give an uneven ratio and thus prevent an abrasive grain from ever traversing its own path twice.

Superfinishing is a honing process. Formed honing stones bear against the workpiece previously finished to 0.0005 in (0.013 mm) or at the most to 0.001 in (0.025 mm) by a very light pressure which gradually increases to several pounds per square inch (1 lb/in^2 = 0.0069 MPa) of stone area in proportion to the development of the increased area of contact between the workpiece and stone. The workpiece or tool rotates and where possible is reciprocated slowly over the surface which may be finished in a matter of 20 s to a surface quality of 1 to 3 μin (0.025 to 0.075 μm). Superfinishing is applied to many types of workpieces such as crankshaft pins and bearings, cylinder bores, pistons, valve stems, cams, and other metallic moving parts.

Deburring involves removing burrs (thin ridges, generally triangular, resulting from operations such as punching and blanking of sheet metals, and from machining and drilling) along the edges of a workpiece. Several deburring operations are available, the most common ones being manual filing, wire brushing, using abrasives (emery paper, belts, abrasive blasting), and vibratory and barrel finishing. Deburring operations can also be carried out using programmable robots.

MACHINING AND GRINDING OF PLASTICS

The low strength of thermoplastics permits high cutting speeds and feeds, but the low heat conductivity and greater resilience require increased reliefs and less rake in order to avoid undersize cutting. Hard and sharp tools should be used. Plastics are usually abrasive and cause the tools to wear or become dull rapidly. Dull tools generate heat and cause the tools to cut to shallow depths. The depth of cut should be small. When high production justifies the cost, diamond turning and boring tools are used. Diamond tools maintain sharp cutting edges and produce an excellent machined surface. They are particularly advantageous when a more abrasive plastic such as reinforced plastic is machined.

A cutting fluid, such as a small blast of air or a stream of water, improves the **turning** and cutting of plastics as it prevents the heating of the tool and causes the chips to remain brittle and to break rather than become sticky and gummy. A zero or slightly negative back rake and a relief angle of 8 to 12° should be used. For thermoplastics cutting speeds generally range from 250 to 400 ft/min (75 to 120 m/min) and for thermosetting plastics from 400 to 1,000 ft/min (120 to 300 m/min). Recommended tool materials are M2 and T5 high-speed steels and C2 carbide.

In **milling** plastics, speeds of 400 to 1,000 ft/min (120 to 300 m/min) should be used, with angles similar to those on a single-point tool. From 0 to 10° negative rake may be used. Good results have been obtained by hobbing plastic gears with carbide-tipped hobs. Recommended tool materials are M2 and M7 high-speed steels and C2 carbide.

In **drilling,** speeds range from 150 to 400 ft/min (45 to 120 m/min), and the recommended drill geometry is given in Table 13.4.6. Tool materials are M1, M7, and M10 high-speed steel. Usually the drill cuts undersize; drills 0.002 to 0.003 in (0.05 to 0.075 mm) oversize should be used.

In **sawing** plastics, either precision or buttress tooth forms may be used, with a pitch ranging from 3 to 14 teeth/in (1.2 to 5.5 teeth/cm), the thicker the material the lower the number of teeth per unit length of saw. Cutting speeds for thermoplastics range from 1,000 to 4,000 ft/min (300 to 1,200 m/min) and for thermosetting plastics from about 3,000 to 5,500 ft/min (900 to 1,700 m/min), with the higher speeds for thinner stock. High-carbon-steel blades are recommended. An air blast is helpful in preventing the chips from sticking to the saw. Abrasive saws operating at 3,500 to 6,000 ft/min (1,000 to 1,800 m/min) are also used for cutting off bars and forms.

Plastics are tapped and threaded with standard tools. Ground M1, M7, or M10 high-speed steel **taps** with large polished flutes are recommended. **Tapping speeds** are usually from 25 to 50 ft/min (8 to 15 m/min); water serves as a good cutting fluid as it keeps the material brittle and prevents sticking in the flutes. **Thread cutting** is generally accomplished with tools similar to those used on brass.

Reaming is best accomplished in production by using tools of the expansion or adjustable type with relatively low speeds but high feeds. Less material should be removed in reaming plastics than in reaming other materials.

Polishing and **buffing** are done on many types of plastics. Polishing is done with special compounds containing wax or a fine abrasive. Buffing wheels for plastics should have loose stitching. Vinyl plastics can be buffed and polished with fabric wheels of standard types, using light pressures.

Thermoplastics and thermosets can be **ground** with relative ease, usually by using silicon carbide wheels. As in machining, temperature rise should be minimized.

MACHINING AND GRINDING OF CERAMICS

The technology of machining and grinding of ceramics, as well as composite materials, has advanced rapidly, resulting in good surface characteristics and product integrity. Ceramics can be machined with carbide, high-speed steel, or diamond tools, although care should be exercised because of the brittle nature of ceramics and the resulting possible surface damage. **Machinable ceramics** have been developed which minimize machining problems. Grinding of ceramics is usually done with diamond wheels.

ADVANCED MACHINING PROCESSES

In addition to the mechanical methods of material removal described above, there are a number of other important processes which may be preferred over conventional methods. Among the important factors to be considered are the hardness of the workpiece material, the shape of the part, its sensitivity to thermal damage, residual stresses, tolerances, and economics. Some of these processes produce a heat-affected layer on the surface; improvements in surface integrity may be obtained by postprocessing techniques such as polishing or peening. Almost all machines are now computer-controlled.

Electric-discharge machining (EDM) is based on the principle of erosion of metals by spark discharges. Figure 13.4.20 gives a schematic diagram of this process. The spark is a transient electric discharge through the space between two charged electrodes, which are the tool and the workpiece. The discharge occurs when the potential difference between the tool and the workpiece is large enough to cause a breakdown in the medium (which is called the **dielectric fluid** and is usually a hydrocarbon) and to procure an electrically conductive spark channel. The breakdown

potential is usually established by connecting the two electrodes to the terminals of a capacitor charged from a power source. The spacing between the tool and workpiece is critical; therefore, the feed is controlled by servomechanisms. The dielectric fluid has the additional functions of providing a cooling medium and carrying away particles produced by the electric discharge. The discharge is repeated rapidly, and each time a minute amount of workpiece material is removed.

The rate of metal removal depends mostly on the average current in the discharge circuit; it is also a function of the electrode characteristics, the electrical parameters, and the nature of the dielectric fluid. In practice, this rate is normally varied by changing the number of discharges per second or the energy per discharge. Rates of metal removal may range from 0.01 to 25 in^3/h (0.17 to 410 cm^3/h), depending on surface finish and tolerance requirements. In general, higher rates produce rougher surfaces. Surface finishes may range from 1,000 μin R$_q$ (25 μm) in roughing cuts to less than 25 μin (0.6 μm) in finishing cuts.

The response of materials to this process depends mostly on their thermal properties. Thermal capacity and conductivity, latent heats of melting and vaporization are important. Hardness and strength do not necessarily have significant effect on metal removal rates. The process is applicable to all materials which are sufficiently good conductors of electricity. The tool has great influence on permissible removal rates. It is usually made of graphite, copper-tungsten, or copper alloys. Tools have been made by casting, extruding, machining, powder metallurgy, and other techniques and are made in any desired shape. Tool wear is an important consideration, and in order to control tolerances and minimize cost, the ratio of tool material removed to workpiece material removed should be low. This ratio varies with different tool and workpiece material combinations and with operating conditions. Therefore, a particular tool material may not be best for all workpieces. Tolerances as low as 0.0001 to 0.0005 in (0.0025 to 0.0127 mm) can be held with slow metal removal rates. In machining some steels, tool wear can be minimized by reversing the polarity and using copper tools. This is known as "no-wear EDM."

The electric-discharge machining process has numerous applications, such as machining cavities and dies, cutting small-diameter holes, blanking parts from sheets, cutting off rods of materials with poor machinability, and flat or form grinding. It is also applied to sharpening tools, cutters, and broaches. The process can be used to generate almost any geometry if a suitable tool can be fabricated and brought into close proximity to the workpiece.

Thick plates may be cut with **wire EDM** (Fig. 13.4.21). A slowly moving wire travels a prescribed path along the workpiece and cuts the metal with the sparks acting like saw teeth. The wire, usually about 0.01 in (0.25 mm) in diameter, is made of brass, copper, or tungsten and is generally used only once. The process is also used in making tools and dies from hard materials, provided that they are electrically conducting.

Electric-discharge grinding (EDG) is similar to the electric-discharge machining process with the exception that the electrode is in the form of

Fig. 13.4.20 Schematic diagram of the electric-discharge machining process.

Fig. 13.4.21 Schematic diagram of the wire EDM process.

Fig. 13.4.22 Schematic diagram of the electrochemical machining process.

a grinding wheel. Removal rates are up to 1.5 in³/h (25 cm³/h) with practical tolerances on the order of 0.001 in (0.025 mm). A graphite or brass electrode wheel is operated around 100 to 600 surface ft/min (30 to 180 m/min) to minimize splashing of the dielectric fluid. Typical applications of this process are in grinding of carbide tools and dies, thin slots in hard materials, and production grinding of intricate forms.

The **electrochemical machining (ECM)** process (Fig. 13.4.22) uses electrolytes which dissolve the reaction products formed on the workpiece by electrochemical action; it is similar to a reverse electroplating process. The electrolyte is pumped at high velocities through the tool. A gap of 0.005 to 0.020 in (0.13 to 0.5 mm) is maintained. A dc power supply maintains very high current densities between the tool and the workpiece. In most applications, a current density of 1,000 to 5,000 A is required per in² of active cutting area. The rate of metal removal is proportional to the amount of current passing between the tool and the workpiece. Removal rates up to 1 in³/min (16 cm³/min) can be obtained with a 10,000-A power supply. The penetration rate is proportional to the current density for a given workpiece material.

The process leaves a burr-free surface. It is also a cold machining process and does no thermal damage to the surface of the workpiece. Electrodes are normally made of brass or copper; stainless steel, titanium, sintered copper-tungsten, aluminum, and graphite have also been used. The electrolyte is usually a sodium chloride solution up to 2.5 lb/gal (300 g/L); other solutions and proprietary mixtures are also available. The amount of overcut, defined as the difference between hole diameter and tool diameter, depends upon cutting conditions. For production applications, the average overcut is around 0.015 in (0.4 mm). The rate of penetration is up to 0.750 in/min (20 mm/min).

Very good surface finishes may be obtained with this process. However, sharp square corners or sharp corners and flat bottoms cannot be machined to high accuracies. The process is applied mainly to round or odd-shaped holes with straight parallel sides. It is also applied to cases where conventional methods produce burrs which are costly to remove. The process is particularly economical for materials with a hardness above 400 HB.

The **electrochemical grinding (ECG)** process (Fig. 13.4.23) is a combination of electrochemical machining and abrasive cutting where most of the metal removal results from the electrolytic action. The process consists of a rotating cathode, a neutral electrolyte, and abrasive particles in contact with the workpiece. The equipment is similar to a

Fig. 13.4.23 Schematic diagram of the electrochemical grinding process.

conventional grinding machine except for the electrical accessories. The cathode usually consists of a metal-bonded diamond or aluminum oxide wheel. An important function of the abrasive grains is to maintain a space for the electrolyte between the wheel and workpiece.

Surface finish, precision, and metal-removal rate are influenced by the composition of the electrolyte. Aqueous solutions of sodium silicate, borax, sodium nitrate, and sodium nitrite are commonly used as electrolytes. The process is primarily used for tool and cutter sharpening and for machining of high-strength materials.

A combination of the electric-discharge and electrochemical methods of material removal is known as **electrochemical discharge grinding (ECDG)**. The electrode is a pure graphite rotating wheel which electrochemically grinds the workpiece. The intermittent spark discharges remove oxide films that form as a result of electrolytic action. The equipment is similar to that for electrochemical grinding. Typical applications include machining of fragile parts and resharpening or form grinding of carbides and tools such as milling cutters.

In **chemical machining (CM)** material is removed by chemical or electrochemical dissolution of preferentially exposed surfaces of the workpiece. Selective attack on different areas is controlled by masking or by partial immersion. There are two processes involved: chemical **milling** and chemical **blanking**. Milling applications produce shallow cavities for overall weight reduction, and are also used to make tapered sheets, plates, or extrusions. Masking with paint or tapes is common. Masking materials may be elastomers (such as butyl rubber, neoprene, and styrene-butadiene) or plastics (such as polyvinyl chloride, polystyrene, and polyethylene). Typical blanking applications are decorative panels, printed-circuit etching, and thin stampings. Etchants are solutions of sodium hydroxide for aluminum, and solutions of hydrochloric and nitric acids for steel.

Ultrasonic machining (USM) is a process in which a tool is given a high-frequency, low-amplitude oscillation, which, in turn, transmits a high velocity to fine abrasive particles that are present between the tool and the workpiece. Minute particles of the workpiece are chipped away on each stroke. Aluminum oxide, boron carbide, or silicone carbide grains are used in a water slurry (usually 50 percent by volume), which also carries away the debris. Grain size ranges from 200 to 1,000 (see Sec. 6 and Figs. 13.4.18 and 13.4.19).

The equipment consists of an electronic oscillator, a transducer, a connecting cone or toolholder, and the tool. The oscillatory motion is obtained most conveniently by magnetostriction, at approximately 20,000 Hz and a stroke of 0.002 to 0.005 in (0.05 to 0.13 mm). The tool material is normally cold-rolled steel or stainless steel and is brazed, soldered, or fastened mechanically to the transducer through a toolholder. The tool is ordinarily 0.003 to 0.004 in (0.075 to 0.1 mm) smaller than the cavity it produces. Tolerances of 0.0005 in (0.013 mm) or better can be obtained with fine abrasives. For best results, roughing cuts should be followed with one or more finishing operations with finer grits. The ultrasonic machining process is used in drilling holes, engraving, cavity sinking, slicing, broaching, etc. It is best suited to materials which are hard and brittle, such as ceramics, carbides, borides, ferrites, glass, precious stones, and hardened steels.

In **water jet machining (WJM)**, water is ejected from a nozzle at pressures as high as 200,000 lb/in² (1,400 MPa) and acts as a saw. The process is suitable for cutting and deburring of a variety of materials such as polymers, paper, and brick in thicknesses ranging from 0.03 to 1 in (0.8 to 25 mm) or more. The cut can be started at any location, wetting is minimal, and no deformation of the rest of the piece takes place. Abrasives can be added to the water stream to increase material removal rate, and this is known as **abrasive water jet machining (AWJM)**.

In **abrasive-jet machining (AJM)**, material is removed by fine abrasive particles (aluminum oxide or silicon carbide) carried in a high-velocity stream of air, nitrogen, or carbon dioxide. The gas pressure ranges up to 120 lb/in² (800 kPa), providing a nozzle velocity of up to 1,000 ft/s (300 m/s). Nozzles are made of tungsten carbide or sapphire. Typical applications are in drilling, sawing, slotting, and deburring of hard, brittle materials such as glass.

In **laser-beam machining (LBM)**, material is removed by converting electric energy into a narrow beam of light and focusing it on the workpiece. The high energy density of the beam is capable of melting and vaporizing all materials, and consequently, there is a thin heat-affected zone. The most commonly used laser types are CO_2 (pulsed or continuous-wave) and Nd:YAG. Typical applications include cutting a variety of metallic and nonmetallic materials, drilling (as small as 0.0002 in or 0.005 mm in diameter), and marking. The efficiency of cutting increases with decreasing thermal conductivity and reflectivity of the material. Because of the inherent flexibility of the process,

programmable and computer-controlled laser cutting is now becoming important, particularly in cutting profiles and multiple holes of various shapes and sizes on large sheets. Cutting speeds may range up to 25 ft/ min (7.5 m/min).

The **electron-beam machining (EBM)** process removes material by focusing high-velocity electrons on the workpiece. Unlike lasers, this process is carried out in a vacuum chamber and is used for drilling small holes, scribing, and cutting slots in all materials, including ceramics.

13.5   SURFACE TEXTURE DESIGNATION, PRODUCTION, AND QUALITY CONTROL
by Ali M. Sadegh

REFERENCES: American National Standards Institute, "Surface Texture," ANSI/ASME B 46.1-1985, and "Surface Texture Symbols," ANSI Y 14.36-1978. Broadston, "Control of Surface Quality," Surface Checking Gage Co., Hollywood, CA. ASME, "Metals Engineering Design Handbook," McGraw-Hill SME, "Tool and Manufacturing Engineers Handbook," McGraw-Hill.

Rapid changes in the complexity and precision requirements of mechanical products since 1945 have created a need for improved methods of determining, designating, producing, and controlling the surface texture of manufactured parts. Although standards are aimed at standardizing methods for measuring by using stylus probes and electronic transducers for surface quality control, other descriptive specifications are sometimes required, i.e., interferometric light bands, peak-to-valley by optical sectioning, light reflectance by commercial glossmeters, etc. Other parameters are used by highly industrialized foreign countries to solve their surface specification problems. These include the high-spot counter and bearing area meter of England (Talysurf); the total peak-to-valley, or R_1, of Germany (Perthen); and the R or average amplitude of surface deviations of France. In the United States, peak counting is used in the sheet-steel industry, instrumentation is available (Bendix), and a standard for specification, SAE J-911, exists.

Surface texture control should be considered for many reasons, among them being the following:

1. Advancements in the technology of metal-cutting tools and machinery have made the production of higher-quality surfaces possible.

2. Products are now being designed that depend upon proper quality control of critical surfaces for their successful operation as well as for long, troublefree performance in service.

3. Remote manufacture and the necessity for controlling costs have made it preferable that finish requirements for all the critical surfaces of a part be specified on the drawing.

4. The design engineer, who best understands the overall function of a part and all its surfaces, should be able to determine the requirement for surface texture control where applicable and to use a satisfactory standardized method for providing this information on the drawing for use by manufacturing departments.

5. Manufacturing personnel should know what processes are able to produce surfaces within specifications and should be able to verify that the production techniques in use are under control.

6. Quality control personnel should be able to check conformance to surface texture specifications if product quality is to be maintained and product performance and reputation ensured.

DESIGN CRITERIA

Surfaces produced by various processes exhibit distinct differences in texture. These differences make it possible for honed, lapped, polished, turned, milled, or ground surfaces to be easily identified. As a result of its unique character, the surface texture produced by any given process can be readily compared with other surfaces produced by the same process

through the simple means of comparing the average size of its irregularities, using applicable standards and modern measurement methods. It is then possible to predict and control its performance with considerable certainty by limiting the range of the average size of its characteristic surface irregularities. Surface texture standards make this control possible.

Variations in the texture of a critical surface of a part influence its ability to resist wear and fatigue; to assist or destroy effective lubrication; to increase or decrease its friction and/or abrasive action on other parts, and to resist corrosion, as well as affect many other properties that may be critical under certain conditions.

Clay has shown that the load-carrying capacity of nitrided shafts of varying degrees of roughness, all running at 1,500 r/min in diamond-turned lead-bronze bushings finished to 20 μin (0.50 μm), varies as shown in Fig. 13.5.1. The effects of roughness values on the friction between a flat slider on a well-lubricated rotating disk are shown in Fig. 13.5.2.

Surface texture control should be a normal design consideration under the following conditions:

1. For those parts whose roughness must be held within closely controlled limits for optimum performance. In such cases, even the process may have to be specified. Automobile engine cylinder walls, which should be finished to about 13 μin (0.32 μm) and have a circumferential (ground) or an angular (honed) lay, are an example. If too rough, excessive wear occurs, if too smooth, piston rings will not seat properly, lubrication is poor, and surfaces will seize or gall.

2. Some parts, such as antifriction bearings, cannot be made too smooth for their function. In these cases, the designer must optimize the tradeoff between the added costs of production and various benefits derived from added performance, such as higher reliability and market value.

Fig. 13.5.1 Load-carrying capacity of journal bearings related to the surface roughness of a shaft. (*Clay, ASM Metal Progress, Aug. 15, 1955.*)

Fig. 13.5.2 Effect of surface texture on friction with hydrodynamic lubrication using a flat slider on a rotating disk. Z = oil viscosity, cP; N = rubbing speed, ft/min; P = load, lb/in^2.

3. There are some parts where surfaces must be made as smooth as possible for optimum performance regardless of cost, such as gages, gage blocks, lenses, and carbon pressure seals.

4. In some cases, the nature of the most satisfactory finishing process may dictate the surface requirements to attain production efficiency, uniformity, and control even though the individual performance of the part itself may not be dependent on the quality of the controlled surface. Hardened steel bushings, e.g., which must be ground to close tolerance for press fit into housings, could have outside surfaces well beyond the roughness range specified and still perform their function satisfactorily.

5. For parts which the shop, with unjustified pride, has traditionally finished to greater perfection than is necessary, the use of proper surface texture designations will encourage rougher surfaces on exterior and other surfaces that do not need to be finely finished. Significant cost reductions will accrue thereby.

It is the designer's responsibility to decide which surfaces of a given part are critical to its design function and which are not. This decision should be based upon a full knowledge of the part's function as well as of the performance of various surface textures that might be specified. From both a design and an economic standpoint, it may be just as

unsound to specify too smooth a surface as to make it too rough—or to control it at all if not necessary. Wherever normal shop practice will produce acceptable surfaces, as in drilling, tapping, and threading, or in keyways, slots, and other purely functional surfaces, unnecessary surface texture control will add costs which should be avoided.

Whereas each specialized field of endeavor has its own traditional criteria for determining which surface finishes are optimum for adequate performance, Table 13.5.1 provides some common examples for design review, and Table 13.5.6 provides data on the surface texture ranges that can be obtained from normal production processes.

DESIGNATION STANDARDS, SYMBOLS, AND CONVENTIONS

The precise definition and measurement of surface texture irregularities of machined surfaces are almost impossible because the irregularities are very complex in shape and character and, being so small, do not lend themselves to direct measurement. Although both their shape and length may affect their properties, control of their average height and direction usually provides sufficient control of their performance. The standards do not specify the surface texture suitable for any particular application, nor the means by which it may be produced or measured. Neither are the standards concerned with other surface qualities such as appearance, luster, color, hardness, microstructure, or corrosion and wear resistance, any of which may be a governing design consideration.

The standards provide definitions of the terms used in delineating critical surface-texture qualities and a series of symbols and conventions suitable for their designation and control. In the discussion which follows, the reference standards used are "Surface Texture" (ANSI/ASME B46.1-1985) and "Surface Texture Symbols" (ANSI Y 14.36-1978).

The basic ANSI symbol for designating surface texture is the checkmark with horizontal extension shown in Fig. 13.5.3. The symbol with the triangle at the base indicates a requirement for a machining allowance, in preference to the old f symbol. Another, with the small circle in the base, prohibits machining; hence surfaces must be produced without the removal of material by processes such as cast, forged, hot- or cold-finished, die-cast, sintered- or injection-molded, to name a few. The surface-texture requirement may be shown at A; the machining allowance at B; the process may be indicated above the line at C;

Table 13.5.1 Typical Surface Texture Design Requirements

(250 μin) 6.3	Clearance surfaces Rough machine parts	(16 μin) 0.40 Motor shafts Gear teeth (heavy loads) Spline shafts
(125 μin) 3.2	Mating surfaces (static) Chased and cut threads Clutch-disk faces Surfaces for soft gaskets	O-ring grooves (static) Antifriction bearing bores and faces Camshaft lobes Compressor-blade airfoils Journals for elastomer lip seals
(63 μin) 1.60	Piston-pin bores Brake drums Cylinder block, top Gear locating faces Gear shafts and bores Ratchet and pawl teeth Milled threads Rolling surfaces Gearbox faces Piston crowns Turbine-blade dovetails	(13 μin) 0.32 Engine cylinder bores Piston outside diameters Crankshaft bearings (8 μin) 0.20 Jet-engine stator blades Valve-tappet cam faces Hydraulic-cylinder bores Lapped antifriction bearings (4 μin) 0.10 Ball-bearing races Piston pins Hydraulic piston rods Carbon-seal mating surfaces
(32 μin) 0.80	Broached holes Bronze journal bearings Gear teeth Slideways and gibs Press-fit parts Piston-rod bushings Antifriction bearing seats Sealing surfaces for hydraulic tube fittings	(2 μin) 0.050 Shop-gage faces Comparator anvils (1 μin) 0.025 Bearing balls Gages and mirrors Micrometre anvils

Fig. 13.5.3 Application and use of surface texture symbols.

the roughness width cutoff (sampling length) at D, and the lay at E. The ANSI symbol provides places for the insertion of numbers to specify a wide variety of texture characteristics, as shown in Table 13.5.2.

Control of **roughness,** the finely spaced surface texture irregularities resulting from the manufacturing process or the cutting action of tools or abrasive grains, is the most important function accomplished through the use of these standards, because roughness, in general, has a greater effect on performance than any other surface quality. The roughness-height index value is a number which equals the arithmetic average deviation of the minute surface irregularities from a hypothetical perfect surface, expressed in either millionths of an inch (microinches, μin, 0.000001 in) or in micrometres, μm, if drawing dimensions are in metric, SI units. For control purposes, roughness-height values are taken from Table 13.5.3, with those in boldface type given preference.

The term *roughness cutoff,* a characteristic of tracer-point measuring instruments, is used to limit the length of trace within which the asperities of the surface must lie for consideration as roughness. Asperity spacings greater than roughness cutoff are then considered as waviness.

Waviness refers to the secondary irregularities upon which roughness is superimposed, which are of significantly longer wavelength and are usually caused by machine or work deflections, tool or workpiece vibration, heat treatment, or warping. Waviness can be measured by a dial indicator or a profile recording instrument from which roughness has been filtered out. It is rated as maximum peak-to-valley distance and is indicated by the preferred values of Table 13.5.4. For fine waviness control, techniques involving contact-area determination in percent (90, 75, 50 percent preferred) may be required. Waviness control by interferometric methods is also common, where notes, such as "Flat

within XX helium light bands," may be used. Dimensions may be determined from the precision length table (see Sec. 1).

Lay refers to the direction of the predominant visible surface roughness pattern. It can be controlled by use of the approved symbols given in Table 13.5.5, which indicate desired lay direction with respect to the boundary line of the surface upon which the symbol is placed.

Flaws are imperfections in a surface that occur only at infrequent intervals. They are usually caused by nonuniformity of the material, or they result from damage to the surface subsequent to processing, such as scratches, dents, pits, and cracks. Flaws should not be considered in surface texture measurements, as the standards do not consider or classify them. Acceptance or rejection of parts having flaws is strictly a matter of judgment based upon whether the flaw will compromise the intended function of the part.

To call attention to the fact that surface texture values are specified on any given drawing, a note and typical symbol may be used as follows:

$\sqrt{}$ Surface texture per ANSI B46.1

Values for nondesignated surfaces can be limited by the note

$\overset{xx}{\sqrt{}}$ All machined surfaces except as noted

MEASUREMENT

Two general methods exist to measure surface texture: **profile methods** and **area methods.** Profile methods measure the contour of the surface in a plane usually perpendicular to the surface. Area methods measure an area of a surface and produce results that depend on area-averaged properties.

Table 13.5.2   Application of Surface Texture Values to Surface Symbols

(63) $\underset{\sqrt{}}{1.6}$	Roughness average rating is placed at the left of the long leg. The specification of only one rating shall indicate the maximum value and any lesser value shall be acceptable. Specify in micrometres (microinches).
(63) 1.6 (32) $\underset{\sqrt{}}{0.8}$	The specification of maximum value and minimum value roughness average ratings indicates permissible range of value rating. Specify in micrometres (microinches).
(32) $0.8\underset{\sqrt{}}{\overline{0.05}}$	Maximum waviness height rating is placed above the horizontal extension. Any lesser rating shall be acceptable. Specify in millimetres (inches).
(32) $0.8\underset{\sqrt{}}{\overline{0.05 - 100}}$	Maximum waviness spacing rating is placed above the horizontal extension and to the right of the waviness height rating. Any lesser rating shall be acceptable. Specify in millimetres (inches).
(63) $\underset{3.5\,\triangledown}{1.6}$	Machining is required to produce the surface. The basic amount of stock provided for machining is specified at the left of the short leg of the symbol. Specify in millimetres (inches).
(63) $\underset{\text{Q}}{1.6}$	Removal of material by machining is prohibited.
(32) $0.8\underset{\perp}{/}$	Lay designation is indicated by the lay symbol placed at the right of the long leg.
(32) $0.8\underset{\sqrt{2.5}}{}$ (0.100)	Roughness sampling length or cutoff rating is placed below the horizontal extension. When no value is shown, 0.80 mm is assumed. Specify in millimetres (inches).
(32) $0.8\underset{\sqrt{\perp 0.5}}{}$	Where required, maximum roughness spacing shall be placed at the right of the lay symbol. Any lesser rating shall be acceptable. Specify in millimetres (inches).

Table 13.5.3   Preferred Series Roughness Average Values R_a, μm and μin

μm	μin	μm	μin	μm	μin	μm	μin	μm	μin
0.012	0.5	0.125	5	0.50	20	2.00	80	8.0	320
0.025	1	0.15	6	0.63	25	2.50	100	10.0	400
0.050	**2**	**0.20**	**8**	**0.80**	**32**	**3.20**	**125**	**12.5**	**500**
0.075	3	0.25	10	1.00	40	4.0	160	15.0	600
0.10	**4**	0.32	13	1.25	50	5.0	200	20.0	800
		0.40	**16**	**1.60**	**63**	**6.3**	**250**	**25.0**	**1,000**

Table 13.5.4   Preferred Series Maximum Waviness Height Values

mm	in	mm	in	mm	in
0.0005	0.00002	0.008	0.0003	0.12	0.005
0.0008	0.00003	0.012	0.0005	0.20	0.008
0.0012	0.00005	0.020	0.0008	0.25	0.010
0.0020	0.00008	0.025	0.001	0.38	0.015
0.0025	0.0001	0.05	0.002	0.50	0.020
0.005	0.0002	0.08	0.003	0.80	0.030

Another categorization is by **contact methods** and **noncontact methods.** Contact methods include stylus methods (tracer-point analysis) and capacitance methods. Noncontact methods include light section microscopy, optical reflection measurements, and interferometry.

Replicas of typical standard machined surfaces provide less accurate but often adequate reference and control of rougher surfaces with R_a over 16 μin.

The United States and 25 other countries have adopted the **roughness average R_a** as the standard measure of surface roughness. (See ANSI/ASME B46.1-1985.)

PRODUCTION

Various production processes can produce surfaces within the ranges shown in Table 13.5.6. For production efficiency, it is best that critical areas requiring surface texture control be clearly designated on drawings so that proper machining and adequate protection from damage during processing will be ensured.

SURFACE QUALITY VERSUS TOLERANCES

It should be remembered that surface quality and tolerances are distinctly different attributes that are controlled for completely separate purposes. **Tolerances** are established to limit the range of the size of a part at the time of manufacture, as measured with gages, micrometres,

Table 13.5.5   Lay Symbols

Lay symbol	Interpretation	Example showing direction of tool marks
▬	Lay approximately parallel to the line representing the surface to which the symbol is applied	
⊥	Lay approximately perpendicular to the line representing the surface to which the symbol is applied	
X	Lay angular in both directions to line representing the surface to which symbol is applied	

Lay symbol	Interpretation	Example showing direction of tool marks
M	Lay multidirectional	
C	Lay approximately circular relative to the center of the surface to which the symbol is applied	
R	Lay approximately radial relative to the center of the surface to which the symbol is applied	
P	Lay particulate, nondirectional, or protuberant	

Table 13.5.6 Surface-Roughness Ranges of Production Processes

Process	Roughness height rating, μm (μin) R_a
	50 (2000) 25 (1000) 12.5 (500) 6.3 (250) 3.2 (125) 1.8 (63) 0.80 (32) 0.40 (16) 0.20 (8) 0.10 (4) 0.05 (2) 0.025 (1) 0.012 (0.5)
Flame cutting	
Snagging	
Sawing	
Planing, shaping	
Drilling	
Chemical milling	
Elect. discharge mach.	
Milling	
Broaching	
Reaming	
Electron beam	
Laser	
Electro-chemical	
Boring, turning	
Barrel finishing	
Electrolytic grinding	
Roller burnishing	
Grinding	
Honing	
Electro-polish	
Polishing	
Lapping	
Superfinishing	
Sand casting	
Hot rolling	
Forging	
Perm. mold casting	
Investment casting	
Extruding	
Cold rolling, drawing	
Die casting	

Smoother texture higher cost →   ← *Rougher texture lower cost*

The ranges shown above are typical of the processes listed.
Higher or lower values may be obtained under special conditions.

■ Average application
▨ Less frequent application

or other traditional measuring devices having anvils that make contact with the part. **Surface quality** controls, on the other hand, serve to limit the minute surface irregularities or asperities that are formed by the manufacturing process. These lie under the gage anvils during measurement and **do not use up tolerances.**

QUALITY CONTROL (SIX SIGMA)

Quality control is a system that outlines the policies and procedures necessary to improve and control the various processes in manufacturing that will ultimately lead to improved business performance.

Six Sigma is a quality management program to achieve "six sigma" levels of quality. It was pioneered by Motorola in the mid-1980s and has spread to many other manufacturing companies. In statistics,

sigma refers to the standard deviation of a set of data. Therefore, **six sigma** refers to six standard deviations. Likewise, **three sigma** refers to three standard deviations. In probability and statistics, the standard deviation is the most commonly used measure of statistical dispersion; i.e., it measures the degree to which values in a data set are spread. The standard deviation is defined as the square root of the variance, i.e., the root mean square (rms) deviation from the average. It is defined in this way to give us a measure of dispersion.

Assuming that defects occur according to a standard normal distribution, this corresponds to approximately 2 quality failures per million parts manufactured. In practical application of the six sigma methodology, however, the rate is taken to be 3.4 per million.

Initially, many believed that such high process reliability was impossible, and **three sigma** (67,000 defects per million opportunities, or DPMO) was considered acceptable. However, market leaders have measurably reached six sigma in numerous processes.

13.6 WOODCUTTING TOOLS AND MACHINES
by Richard W. Perkins

REFERENCES: Davis, Machining and Related Characteristics of United States Hardwoods, *USDA Tech. Bull.* 1267. Harris, "A Handbook of Woodcutting," Her Majesty's Stationery Office, London. Koch, "Wood Machining Processes," Ronald Press. Kollmann, Wood Machining, in Kollmann and Côté, "Principles of Wood Science and Technology," chap. 9, Springer-Verlag.

SAWING

Sawing machines are classified according to basic machine design, i.e., band saw, gang saw, chain saw, circular saw. Saws are designated as **ripsaws** if they are designed to cut along the grain or **crosscut** saws if they are designed to cut across the grain. A **combination** saw is designed to cut reasonably well along the grain, across the grain, or along a direction at an angle to the grain (**miter**). Sawing machines are often further classified according to the specific operation for which they are used, e.g., **headsaw** (the primary log-breakdown saw in a sawmill), **resaw** (saw for ripping cants into boards), **edger** (saw for edging boards in a sawmill), **variety saw** (general-purpose saw for use in furniture plants), **scroll saw** (general-purpose narrow-band saw for use in furniture plants).

The thickness of the saw blade is designated in terms of the Birmingham wire gage (BWG) (see Sec. 8.2). Large-diameter [40 to 60 in (1.02 to 1.52 mm)] circular-saw blades are tapered so that they are thicker at the center than at the rim. Typical headsaw blades range in thickness from 5 to 6 BWG [0.203 to 0.220 in (5.16 to 5.59 mm)] for use in heavy-duty applications to 8 to 9 BWG [0.148 to 0.165 in (3.76 to 4.19 mm)] for lighter operations. Small-diameter [6 to 30 in (152 to 762 mm)] circular saws are generally flat-ground and range from 10 to 18 BWG [0.049 to 0.134 in (1.24 to 3.40 mm)] in thickness. Band-saw and gang-saw blades are flat-ground and are generally thinner than circular-saw blades designed for similar applications. For example, typical wide-band-saw blades for sawmill use range from 11 to 16 BWG [0.065 to 0.120 in (1.65 to 3.05 mm)] in thickness. The thickness of a band-saw blade is determined by the cutting load and the diameter of the band wheel. Gang-saw blades are generally somewhat thicker than band-saw blades for similar operations. Narrow-band-saw blades for use on scroll band saws range in thickness from 20 to 25 BWG [0.020 to 0.035 in (0.51 to 0.89 mm)] and range in width from ⅛ to about 1¾ in (3.17 to 44.5 mm) depending upon the curvature of cuts to be made.

The considerable amount of heat generated at the cutting edge results in compressive stresses in the rim of the saw blade of sufficient magnitude to cause mechanical instability of the saw blades. Circular-saw blades and wide-band-saw blades are commonly prestressed (or **tensioned**) to reduce the possibility of buckling. Small circular-saw blades for use on power-feed rip-saws and crosscut saws are frequently provided with **expansion slots** for the same purpose.

The **shape of the cutting portion of the sawtooth** is determined by specifying the hook, face bevel, top bevel, and clearance angles. The optimum tooth shape depends primarily upon cutting direction, moisture content, and density of the workpiece material. Sawteeth are, in general, designed in such a way that the portion of the cutting edge which is required to cut across the fiber direction is provided with the maximum effective rake angle consistent with tool strength and wear considerations. Ripsaws are designed with a hook angle between some 46° for inserted-tooth circular headsaws used to cut green material and 10° for solid-tooth saws cutting dense material at low moisture content. Ripsaws generally have zero face bevel and top bevel angle; however, spring-set ripsaws sometimes are provided with a moderate top bevel angle (5 to 15°). The hook angle for crosscut saws ranges from positive 10° to negative 30°. These saws are generally designed with both top and face bevel angles of 5 to 15°; however, in some cases top and face bevel angles as high as 45° are employed. A compromise design is used for combination saws which embodies the features of both ripsaws and crosscut saws in order to provide a tool which can cut reasonably well in all directions. The clearance angle should be maintained at the smallest possible value in order to provide for maximum tooth strength. For ripsawing applications, the clearance angle should be about 12 to 15°. The minimum satisfactory clearance angle is determined by the nature of the work material, not from kinematical considerations of the motion of the tool through the work. In some cases of cutoff, combination, and narrow-band-saw designs where the tooth pitch is relatively small, much larger clearance angles are used in order to provide the necessary gullet volume.

A certain amount of clearance between the saw blade and the generated surface (**side clearance** or **set**) is necessary to prevent frictional heating of the saw blade. In the case of solid-tooth circular saws and band or gang saws, the side clearance is generally provided either by deflecting alternate teeth (**spring-setting**) or by spreading the cutting edge (**swage-setting**). The amount of side clearance depends upon density, moisture content, and size of the saw blade. In most cases, satisfactory results are obtained if the side clearance S is determined from the formula S [in (mm)] $= A/2[f(g - 5) - f(g)]$, where g = gage number (BWG) of the saw blade, $f(n)$ = dimension in inches (mm) corresponding to the gage number n, and A has values from Table 13.6.1. Certain specialty circular saws such as planer, smooth-trimmer, and miter saws are hollow-ground to provide side clearance. Inserted-tooth saws, carbide-tipped saws, and chain-saw teeth are designed so that sufficient side clearance is provided for the life of the tool; consequently, the setting of such saws is unnecessary.

Table 13.6.1   Values of *A* for Computing Side Clearance

	Workpiece material			
	Specific gravity less than 0.45		Specific gravity greater than 0.55	
Saw type	Air dry	Green	Air dry	Green
Circular rip and combination	0.90	1.00	0.85	0.95
Glue-joint ripsaw	0.80	—	0.60	—
Circular crosscut	0.95	1.05	0.90	1.00
Wide-band saw	0.55	0.65	0.30	0.40
Narrow-band saw	0.65	—	0.55	

The **tooth speed** for sawing operations ranges from 3,000 to 17,000 ft/min (15 to 86 m/s) approx. Large tooth speeds are in general desirable in order to permit maximum work rates. The upper limit of permissible tooth speed depends in most cases on machine design considerations and not on considerations of wear or surface quality as in the case of metal cutting. Exceptionally high tooth speeds may result in charring of the work material, which is machined at slow feed rates.

In many sawing applications, **surface quality** is not of prime importance since the sawed surfaces are subsequently machined, e.g., by planing, shaping, sanding; therefore, it is desirable to operate the saw at the largest feed per tooth consistent with gullet overloading. Large values of feed per tooth result in lower amounts of work required per unit volume of material cut and in lower amounts of wear per unit tool travel. Large-diameter circular saws, wide-band saws, and gang saws for ripping green material are generally designed so that the feed per tooth should be about 0.08 to 0.12 in (2.03 to 3.05 mm). Small-diameter circular saws are designed so that the feed per tooth ranges from 0.03 in (0.76 mm) for dense hardwoods to 0.05 in (1.27 mm) for low-density softwoods. Narrow-band saws are generally operated at somewhat smaller values of feed per tooth, e.g., 0.005 to 0.04 in (0.13 to 1.02 mm).

Smaller values of feed per tooth are necessary for applications where surface quality is of prime importance, e.g., glue-joint ripsawing and variety-saw operations. The degree of gullet loading is measured by the **gullet-feed index (GFI),** which is computed as the feed per tooth times the depth of face divided by the gullet area. The maximum GFI depends primarily upon species, moisture content, and cutting direction. It is generally conceded that the maximum GFI for ripsawing lies between 0.3 for high-density, low-moisture-content material and 0.4 for low-density, high-moisture-content material. For specific information, see Telford, *For. Prod. Res. Soc. Proc.,* 1949.

Saws vary considerably in design of the **gullet shape.** The primary design considerations are gullet area and tooth strength; however, special design shapes are often required for certain classes of workpiece material, e.g., for ripping frozen wood.

Materials Saw blades and the sawteeth of solid-tooth saws are generally made of a nickel tool steel. The bits for inserted-tooth saws were historically plain carbon tool steel; however, high-speed steel bits or bits with a cast-alloy inlay (e.g., Stellite) are sometimes used in applications where metal or gravel will not be encountered. Small-diameter circular saws of virtually all designs are made with cemented-carbide tips. This type is almost imperative in applications where highly abrasive material is cut, namely, in plywood and particleboard operations.

Sawing Power References: Endersby, The Performance of Circular Plate Ripsaws, *For. Prod. Res. Bull.* 27, Her Majesty's Stationery Office, London, 1953. Johnston, Experimental Cut-off Saw, *For. Prod. Jour.,* June 1962. Oehrli, Research in Cross-cutting with Power Saw Chain Teeth, *For. Prod. Jour.,* Jan. 1960. Telford, Energy Requirements for Insert-point Circular Headsaws, *Proc. For. Prod. Res. Soc.,* 1949.

An approximate relation for computing the power P, ft · lb/min (W), required to saw is

$$P = kvb(A + Bt_a)/p$$

where k is the kerf, in (m); v is the tooth speed, ft/min (m/s); p is the tooth pitch, in (m); A and B are constants for a given sawing operation, lb/in (N/m) and lb/in^2 (N/m^2), respectively; and t_a is the average chip thickness, in (m). The average chip thickness is computed from the relation $t_a = \gamma f_t \times d/b$, where f_t is the feed per tooth; d is the depth of face; b is the length of the tool path through the workpiece; and γ has the value 1 except for saws with spring-set or offset teeth, in which case γ has the value 2. The constants A and B depend primarily upon cutting direction (ripsawing, crosscutting), moisture content below the fiber-saturation point and specific gravity of the workpiece material, and tooth shape. The values of A and B (see Table 13.6.2) depend to some degree upon the depth of face, saw diameter, gullet shape, gullet-feed

index, saw speed, and whether the tool motion is linear or rotary; however, the effect of these variables can generally be neglected for purposes of approximation.

Computers are now utilized in sawmills where raw logs are first processed into rough-cut lumber. With suitable software, the mill operator can input key dimensions of the log and receive the cutting pattern which provides a mix of cross sections of lumber so as to maximize the yield from the log. The saving in waste is sizable, and this technique is especially attractive in view of the decreasing availability of large-caliper old stand timber, together with the cost of same.

PLANING AND MOLDING

Machinery Planing and molding machines employ a rotating cutter-head to generate a smooth, defect-free surface by cutting in a direction approximately along the grain. A **surfacer** (or **planer**) is designed to machine boards or panels to uniform thickness. A **facer** (or **facing planer**) is designed to generate a flat (plane) surface on the wide faces of boards. The **edge jointer** is intended to perform the same task on the edges of boards in preparation for edge-gluing into panels. A **planer-matcher** is a heavy-duty machine designed to plane rough boards to uniform width and thickness in one operation. This machine is commonly used for dressing dimension lumber and producing millwork. The **molder** is a high-production machine for use in furniture plants to generate parts of uniform cross-sectional shape.

Recommended Operating Conditions It is of prime importance to adjust the operating conditions and knife geometry so that the machining defects are reduced to a satisfactory level. The most commonly encountered defects are torn (chipped) grain, fuzzy grain, raised and loosened grain, and chip marks. **Torn grain** is caused by the wood splitting ahead of the cutting edge and below the generated surface. It is generally associated with large cutting angle, large chip thickness, low moisture content, and low workpiece material density. The **fuzzy-grain** defect is characterized by small groups of wood fibers which stand up above the generated surface. This defect is caused by incomplete severing of the wood by the cutting edge and is generally associated with small cutting angles, dull knives, low-density species, high moisture content, and (often) the presence of abnormal wood known as reaction wood. The **raised-grain** defect is characterized by an uneven surface where one portion of the annual ring is raised above the remaining part. **Loosened grain** is similar to raised grain; however, loosened grain is characterized by a separation of the early wood from the late wood which is readily discernible to the naked eye. The raised- and loosened-grain defects are attributed to the crushing of springwood cells as the

Table 13.6.2   Constants for Sawing-Power Estimation

Species	Material Specific gravity	Material Moisture content, %	Tool Angles[e]	Tool Sawing situation	Constants A, lb/in	Constants A, N/m	Constants B, lb/in^2 × 10^{-3}	Constants B, N/m^2 × 10^{-6}
Beech, European[a]	0.72	12	20, 0, 12	SS, R	27.8	4,869	5.760	39.71
Birch, yellow[b]	0.55	FSP	−30, 10, 10	SS,CC	19.7	3,450	4.100	28.27
Elm, wych[a]	0.67	12	20, 0, 12	SS, R	23.2	4,063	4.840	33.37
Maple, sugar[c]	0.63	FSP	41, 0, 0	IT,R	85.6	15,991	2.995	20.65
Maple, sugar[c,f]	0.63	FSP	41, 0, 0	IT,R	48.0	8,406	4.400	30.34
Pine, northern white[c]	0.34	FSP	41, 0, 0	IT,R	27.1	4,746	1.675	11.55
Pine, northern white[c,f]	0.34	FSP	41, 0, 0	IT,R	28.2	4,939	2.085	14.38
Pine, northern white[b]	0.34	FSP	−30, 10, 10	SS,CC	0.0	0	3.300	22.75
Pine, ponderosa[d]	0.38–0.40	15–40	28, 25, 0	OFT, R	29.3	5,131	1.700	11.72
Pine, ponderosa[d]	0.38–0.40	15–40	28, 25, 0	OFT, CC	0.0	0	2.120	14.62
Poplar (*P. serotina*)[a]	0.48	12	20, 0, 12	SS, R	18.6	3,257	3.290	22.68
Redwood, California[a]	0.37	12	20, 0, 12	SS,R	15.0	2,627	2.260	15.58
Spruce, white[b]	0.32	FSP	−30, 10, 10	SS, CC	0.0	0	4.680	32.27

[a] Endersby.
[b] Johnston.
[c] Hoyle, unpublished report, N.Y. State College of Forestry, Syracuse, NY, 1958.
[d] Oehrli.
[e] The numbers represent hook angle, face bevel angle, and top bevel in degrees.
[f] Cutting performed on frozen material.
NOTE: FSP = moisture content greater than the fiber saturation point; CC = crosscut; IT = insert-tooth; OFT = offset-tooth; R = rip; SS = spring-set.

knife passes over the surface. (Edge-grain material may exhibit a defect similar to the raised-grain defect if machining is performed at a markedly different moisture content from that encountered at some later time.) Raised and loosened grains are associated with dull knives, excessive jointing of knives [the jointing land should not exceed ⅟32 in (0.79 mm)], and high moisture content of workpiece material. **Chip marks** are caused by chips which are forced by the knife into the generated surface as the knife enters the workpiece material. Chip marks are associated with inadequate exhaust, low moisture content, and species (e.g., birch, Douglas fir, and maple have a marked propensity toward the chip-mark defect).

Depth of cut is an important variable with respect to surface quality, particularly in the case of species which are quite prone to the torn-grain defect (e.g., hard maple, Douglas fir, southern yellow pine). In most cases, the depth of cut should be less than ⅟16 in (1.59 mm). The number of **marks per inch (marks per metre)** (reciprocal of the feed per cutter) is an important variable in all cases; however, it is most important in those cases for which the torn-grain defect is highly probable. The marks per inch (marks per metre) should be between 8 and 12 (315 and 472) for rough planing operations and from 12 to 16 (472 to 630) for finishing cuts. Slightly higher values may be necessary for refractory (brittle) species or for situations where knots or curly grain are present. It is seldom necessary to exceed a value of 20 marks per inch (787 marks per metre). The **clearance angle** should in all cases exceed a value of 10°. When it is desired to hone or joint the knives between sharpenings, a value of about 20° should be used. The optimum **cutting angle** lies between 20 and 30° for most planing situations; however, in the case of interlocked or wavy grain, low moisture content, or species with a marked tendency toward the torn-grain defect, it may be necessary to reduce the cutting angle to 10 or 15°.

BORING

Machinery The typical general-purpose wood-boring machine has a single vertical spindle and is a hand-feed machine. Production machines are often of the vertical, multiple-spindle, adjustable-gang type or the horizontal type with two adjustable, independently driven spindles. The former type is commonly employed in furniture plants for boring holes in the faces of parts, and the latter type is commonly used for boring dowel holes in the edges and ends of parts.

Tool Design A wide variety of tool designs is available for specialized boring tasks; however, the most commonly used tools are the taperhead drill, the spur machine drill, and the machine bit. The **taperhead drill** is a twist drill with a point angle of 60 to 90°, lip clearance angle of 15 to 20°, chisel-edge angle of 125 to 135°, and helix angle of 20 to 40°. Taper-head drills are used for drilling screw holes and for boring dowel holes along the grain. The **spur machine** drill is equivalent to a twist drill having a point angle of 180° with the addition of a pyramidal point (instead of a web) and spurs at the circumference. These drills are designed with a helix angle of 20 to 40° and a clearance angle of 15 to 20°. The **machine bit** has a specially formed head which determines the configuration of the spurs. It also has a point. Machine bits are designed with a helix angle of 40 to 60°, cutting angle of 20 to 40°, and clearance angle of 15 to 20°. Machine bits are designed with spurs contiguous to the cutting edges (**double-spur machine bit**), with spurs removed from the vicinity of the cutting edges (**extension-lip machine bit**), and with the outlining portion of the spurs removed (**flat-cut machine bit**).

The purpose of the spurs is to aid in severing wood fibers across their axes, thereby increasing hole-wall smoothness when boring across the grain. Therefore, drills or bits with spurs (double-spur machine drill and bit) are intended for boring across the grain, whereas drills or bits without spurs (taper-head drill, flat-cut machine bit) are intended for boring along the grain or at an angle to the grain.

Taper-head and spur machine drills can be sharpened until they become too short for further use; however, machine bits and other bit styles which have specially formed heads can only be sharpened a limited number of times before the spur and cutting-face configuration is significantly altered. Since most wood-boring tools are sharpened by

filing the clearance face, it is important to ensure that sufficient clearance is maintained. The clearance angle should be at least 5° greater than the angle whose tangent (function) is the feed per revolution divided by the circumference of the drill point.

Recommended Operating Conditions The most common defects are tearing of fibers from the end-grain portions of the hole surface and charring of hole surfaces. Rough hole surfaces are most often encountered in low-density and ring-porous species. This defect can generally be reduced to a satisfactory level by controlling the chip thickness. Charring is commonly a problem in high-density species. It can be avoided by maintaining the peripheral speed of the tool below a level which depends upon density and moisture content and by maintaining the chip thickness at a satisfactory level. Large chip thickness may result in excessive tool temperature and therefore rapid tool wear; however, large chip thickness is seldom a cause of hole charring. The following recommendations pertain to the use of spur-type drills or bits for boring material at about 6 percent moisture content across the grain. For species having a specific gravity less than 0.45, the chip thickness should be between 0.015 and 0.030 in (0.38 and 0.76 mm), and the peripheral speed of the tool should not exceed 900 ft/min (4.57 m/s). For material of specific gravity between 0.45 and 0.65, satisfactory results can be obtained with values of chip thickness between 0.015 and 0.045 in (0.38 and 1.14 mm) and with peripheral speeds less than 700 ft/min (3.56 m/s). For material of specific gravity greater than 0.65, the chip thickness should lie between 0.015 and 0.030 in (0.38 and 0.76 mm) and the peripheral speed should not exceed 500 ft/min (2.54 m/s). Somewhat higher values of chip thickness and peripheral speed can be employed when the moisture content of the material is higher.

SANDING

(See also Secs. 6.7 and 6.8.)

Machinery Machines for production sanding of parts having flat surfaces are multiple-drum sanders, automatic-stroke sanders, and wide-belt sanders. **Multiple-drum** sanders are of the endless-bed or rollfeed type and have from two to six drums. The drum at the infeed end is fitted with a relatively coarse abrasive (40 to 100 grit), takes a relatively heavy cut [0.010 to 0.015 in (0.25 to 0.38 mm)], and operates at a relatively slow surface speed [3,000 to 3,500 ft/min (15.24 to 17.78 m/s)]. The drum at the outfeed end has a relatively fine abrasive paper (60 to 150 grit), takes a relatively light cut [about 0.005 in (0.13 mm)], and operates at a somewhat higher surface speed [4,000 to 5,000 ft/min (20.3 to 25.4 m/s)]. **Automatic-stroke** sanders employ a narrow abrasive belt and a reciprocating shoe which forces the abrasive belt against the work material. This machine is commonly employed in furniture plants for the final white-sanding operation prior to finish coating. The automatic-stroke sander has a relatively low rate of material removal (about one-tenth to one-third of the rate for the final drum of a multiple-drum sander) and is operated at a belt speed of 3,000 to 7,500 ft/min (15.2 to 38.1 m/s). **Wide-belt** sanders are commonly used in board plants (plywood, particle board, hardboard). They have the advantage of higher production rates and somewhat greater accuracy than multiple-drum sanders [e.g., feed rates up to 100 ft/min (0.51 m/s) as opposed to about 35 ft/min (0.18 m/s)]. Wide-belt sanders operate at surface speeds of approximately 5,000 ft/min (25.4 m/s) and are capable of operating at depths of cut of 0.006 to 0.020 in (0.15 to 0.51 mm) depending upon workpiece material density.

Abrasive Tools The abrasive tool consists of a **backing** to carry the **abrasive** and an **adhesive** coat to fix the abrasive to the backing. Backings are constructed of paper, cloth, or vulcanized fiber or consist of a cloth-paper combination. The adhesive coating (see also Sec. 6) is made up of two coatings; the first coat (**make coat**) acts to join the abrasive material to the backing, and the second coat (**size coat**) acts to provide the necessary support for the abrasive particles. Coating materials are generally animal glues, urea resins, or phenolic resins. The choice of material for the make and size coats depends upon the required flexibility of the tool and the work rate required of the tool. Abrasive materials (see also Sec. 6) for woodworking applications are garnet, aluminum oxide, and silicon carbide. **Garnet** is the most commonly used abrasive mineral

Table 13.6.3   Recommendations for Common Whitewood Sanding Operations

		Backing		Adhesive		Mineral			Grit size		
		Material	Weight	Make coat	Size coat	1st drum	2d drum	3d drum	1st drum	2d drum	3d drum
Multiple drum	Softwood	Paper	E	Glue	Resin	G	G	G or S	50	80	100
	Hardwood	Paper	E	Glue	Resin	A	A	A	60	100	120
	Particleboard	Paper	E	Glue	Resin	G	G	G or A	40	60	80
		Fiber	0.020	Resin	Resin	S	S	S	40	60	80
	Hardboard	Paper	E	Glue	Resin	S	S	S	60	80	120
		Fiber	0.020	Resin	Resin	S	S	S	60	80	120
Wide-belt	Softwood	Paper	E	Glue	Resin	G	or	S	80–220*		
		Cloth	X	Resin	Resin	G	or	S	80–220*		
	Hardwood	Paper	E	Glue	Resin		A		80–220*		
		Cloth	X	Resin	Resin		A		80–220*		
	Burnishing	Paper	E	Glue	Glue		A		280–400		
	Particleboard	Cloth	X	Resin	Resin		S		24–150*		
	Hardboard	Cloth	X	Resin	Resin		S		100–150		
Stroke sanding	Softwood	Paper	E	Glue	Resin		G		80; 120†		
	Hardwood	Paper	E	Glue	Resin		A		100; 150–180†		
		Cloth	X	Glue	Resin		A		100; 150–180†		
	Particleboard	Paper	E	Glue	Resin	A	or	S	80; 120†		
		Cloth	X	Resin	Resin	A	or	S	80; 120†		
	Hardboard	Paper	E	Glue	Resin	A	or	S	100; 150†		
		Cloth	X	Resin	Resin	A	or	S	100; 150†		
Edge sanding	Softwood	Cloth	X	Glue	Resin		G		60; 100†		
	Hardwood	Cloth	X	Glue	Resin		A		60–150*		
		Cloth	X	Resin	Resin		A		60–150*		
Mold sanding	Softwood	Cloth	J	Glue	Glue		G		80–120*		
		Cloth	J	Glue	Resin		G		80–120*		
	Hardwood	Cloth	J	Glue	Glue	G	or	A	80–120*		
		Cloth	J	Glue	Resin	G	or	A	80–120*		

* May be single- or multiple-grit operation.
† First number for cutting-down operations, second number for finishing operations.
NOTE: G = garnet; A = aluminum oxide; S = silicon oxide.
SOURCE: Graham, *Furniture Production,* July and Aug., 1961; and Martin, *Wood Working Digest,* Sept. 1961.

because of its low cost and acceptable working qualities for low-work-rate situations. It is generally used for sheet goods, for sanding softwoods with all types of machines, and for sanding where the belt is loaded up (as opposed to worn out). **Aluminum oxide** abrasive is used extensively for sanding hardwoods, particleboard, and hardboard. **Silicon carbide** abrasive is used for sanding and polishing between coating operations and for machine sanding of particleboard and hardboard. Silicon carbide is also frequently used for the sanding of softwoods where the removal of raised fibers is a problem. The size of the abrasive particles is specified by the mesh number (the approximate number of openings per inch in the screen through which the particles will pass). (See Commercial Std. CS217-59, "Grading of Abrasive Grain on Coated Abrasive Products," U.S. Government Printing Office.) Mesh numbers range from about 600 to 12. Size may also be designated by an older system of symbols which range from 10/0 (mesh no. 400) through 0 (mesh no. 80) to 4½ (mesh no. 12). Some general recommendations for common white wood sanding operations are presented in Table 13.6.3.

13.7   PRECISION CLEANING
by Charles Osborn

IMPORTANCE OF CLEANLINESS

Long ignored, and still somewhat underestimated, the importance of part **cleanliness** is rapidly coming to the fore in today's high-tech products, especially as we head further into nanotechnologies. Historically many manufacturers have dismissed part cleaning as an insignificant part of the process. Many learn all too late that their highly engineered, closely toleranced device is rendered inoperable by a tiny particle, often so small that it can't be seen with the naked eye. Suddenly they are faced with a steep learning curve, for myriad equipment, chemistry, staffing, and environmental issues await them. This material will help you become familiar with some of the procedures, equipment, and terms that you will encounter on blueprint requirements.

Parts cleaning covers a wide spectrum, ranging from "gross" cleaning of heavy industrial components to "critical" cleaning for the space program components. Many factors must be considered in deciding which processes best fit your particular situation. Applications run the gamut from outer-space/aerospace (oxygen delivery systems to both rocket motors and breathing apparatus for astronauts) to inner space (pharmaceuticals and hip replacements in the medical field). Increasingly the automotive industry is requiring verifiable cleanliness levels for fuel injector components and other metering devices used in antilock brake components.

SELECTING A CLEANLINESS LEVEL

Ultimately, fit, form, and function determine the type and amounts of **contaminates** that can cause improper operation of moving parts or can restrict the flow of fluids. If the limits governing your parts have been established for you, it remains only to institute processes to ensure compliance. If, however, *you* are asked to establish cleanliness limits for a given part, the task is more challenging. You begin by gathering an array of parts, some that "work" well in the application and others that don't. Perform cleanliness testing on each to establish the current level of cleanliness for each case. With this information you now have benchmarks from which you can establish minimum levels of acceptance. This can be done by referring to an industry standard such as International Standard ISO 14952 Part 2, "Surface Cleanliness of Fluid Systems—Cleanliness Levels."

There are two main contaminate types, **particles** and **nonvolatile residues.** There are two types of particles: **hard** and **soft**.

Hard particles are defined as (but not limited to) metal chips, scale from heat treatment, grinding debris, e.g., aluminum oxide, rust, and weld slag. Hard particles, by their very nature, will be detrimental to precision assemblies. When probed with a dental pick, they will maintain their shape and mass and not crumble or smear into separate smaller parts. If small orifices are present, the absolute maximum acceptable particle size is $\cong 70$ percent of the diameter of the smallest downstream office.

Soft particles are fragile and break up when gently probed with a dental pick. They may also exhibit a pasty or gummy consistency. They are generally not considered as solids and may not be of great concern in your application (you will need to evaluate this for your application). Natural fibers such as hair or bits of shop rags are called **linters** and are a subclass of soft particle. Linters are organic fibers whose length exceeds the width by a 10:1 ratio. Linters would be of concern if you have small metering orifices that could become clogged or obstructed by an accumulation of linters without regard to size.

The second contaminate classification **nonvolatile residue,** may consist of both particles and hydrocarbons but is usually thought of in terms of residual hydrocarbon films. If either is present in oxygen service applications, fire and/or explosion could result. Highly sensitive systems may ignite if there is even a slight residue left from inadequate rinsing of aqueous cleaning solution. A reference covering nonvolatile residue is published by the SAE: "ARP 1176, Oxygen System and Component Cleaning and Packaging."®

There are three levels of cleanliness: **gross precision,** and **critical. Gross cleaning** is performed primarily to remove scale from heat treating, other rust, and metal chips. Cleanliness is generally verified by a visual examination and may be aided with a 2 to 10 × loupe under normal lighting conditions. Borescopes are often used to gain access to cored passageways and cavities.

Precision cleaning is performed in a clean-room environment of ≥ class 100,000 but ≤ 1,000 to remove minute contaminants. Verification involves use of equipment to both quantify and qualify the cleanliness level.

Critical cleaning is performed in clean rooms rated at class 100 or better, and only on the most sensitive components.

ENVIRONMENT

Selection of the environment that best suits your application is not always obvious. A common statement found in many industry specifications is the following: **The environment you clean and test in must be compatible with the requirements of the parts.** Put another way, the environment must be clean enough to allow you to achieve the cleanliness level that you need on your parts.

Gross cleaning is typically performed in a **clean area,** which is defined as space physically separated from other manufacturing operations, e.g., machining, finishing, and welding, with means to prevent entry of airborne particles and fumes. A portable laminar flow bench or hood is acceptable in most cases. The following are basic **clean-area protocols:**

Sticky mats at every entrance
No food, drink, or smoking
No oils or greases
No cardboard or wood
No wool sweaters
No sandpaper or abrasives
Clean-room gloves required when handling cleaned parts

Precision cleaning is performed in a **clean-room** environment of ≥ class 100,000 but ≤ 1,000. A reference that outlines the exact parameters of a clean room and how to test for them is FED-STD-209, "Airborne Particulate Cleanliness Classes in Clean Rooms and Clean Zones."

Class name	Class limits				
	0.1 μm/ft³	0.2 μm/ft³	0.3 μm/ft³	0.5 μm/ft³	5 μm/ft³
1	35	7.5	3.0	1.0	—
10	350	75	30	10	—
100	—	750	300	100	—
1,000	—	—	—	1,000	7.0
10,000	—	—	—	10,000	70
100,000	—	—	—	100,000	1,750

The limits shown designate specific concentrations of particles per unit volume of airborne particles with sizes equal to and larger than the particle sizes shown.

The following are basic **clean-room protocols:**

All those that apply to a clean room, plus the following:
No pencils are to be used—pens only.
No drawings or paper is used unless necessary and only in sealed or laminated plastic sheets.
Everyone entering shall wear clean room garments of monofilament polyethylene such as Tyvek® or GoreTex®. Basic coverage would be full-body suits, booties, and head covering.
Gloves that touch anything other than cleaned parts must be changed.

SELECTION OF A CLEANING METHOD

Many factors must be considered in selecting an appropriate cleaning method. Influencing factors include part geometry and sizes, presence of blind holes and cored passageways, all of which may be compounded by the presence of delicate components such as microcircuitry and highly polished surfaces. Component material is important especially if non-corrosion-resistive or highly reactive materials are involved.

As important as the component features are, the nature of the contaminates will have a direct bearing on how they are best removed. There are two type of contaminates, **polar** and **nonpolar.** Example of **polar soil** are inks, metals, and their oxides, and they are best removed in water-based solutions (see Table 13.7.1) **Nonpolar soils** such as mineral oil,

Table 13.7.1   Aqueous Cleaner Effectiveness

Type of cleaner	pH	Contaminates
Acidic	0–2	Metal oxides, scales
Mild acidic	2–5.5	Inorganic salts, soluble metal
Neutral	5.5–8.5	Light oils, small particles
Mild alkaline	8.5–11	Oils, particles, films
Alkaline	11–12.5	Oils, fats, proteins
Corrosive alkaline	12.5–14	Heavy grease

greases, and wax are more easily removed in organic solvents. There are many cleaning methods, complicated methods may not be necessary to achieve high levels of cleanliness. Surprising results can be achieved with simple equipment if it is set up in the proper environment with controls in place. Regardless of the method chosen, there are some basic elements that remain constant in the production of clean parts. Environment, already discussed, is paramount in achieving cleanliness. Filtration of all fluids at the point of use is essential during rinsing and drying. Deionized (DI) water of at least 50,000 Ω ranging up to 18,000,000 Ω must be used for rinsing to ensure that parts dry spot-free.

Drying can be the biggest problem in the process, especially when aqueous-based solution are used. Commons methods of drying include compressed air (provided there are adequate filtering systems), vacuum chambers, convection ovens fitted with filtration devices, and centrifugal units utilizing spinning baskets. Care of parts after cleaning is just as important as the cleaning process itself; don't place clean parts in dirty containers. Packaging components in 4- to 6-mil-thick polyethylene bags sealed closed with an impulse heat sealer are industry standard. Most bags purchased from catalog vendors are not clean; insist on certified bags to at least class 100 to ensure cleanliness is maintained.

Components sensitive to static electricity will need to be packaged in antistatic material, and metallized or Saran coated films may be necessary as a vapor barrier in very humid conditions.

Parts made form non-corrosion-resistant metals may require additions to the bagging material to ensure parts don't oxidize during shipping and storage. Using vacuum to evacuate the air or back-filling with inert gas is effective. Insertion of clean-room-quality desiccant packets is also acceptable, as well as dipping or spraying with preservative oils, provided that the oil delivery systems are properly cleaned and verified. If nonvolatile residue (NVR) is a concern, ultralow out-gassing material will be needed to ensure no residues are redeposited onto the parts.

Manual washing is the oldest and most basic cleaning method and can be as simple as scrubbing a part with a brush, using solvents or aqueous solutions, and drying with compressed shop air.

Advantages Equipment is readily available and inexpensive; there is minimal operator training required.

Disadvantages It may not be adequate on components with inaccessible passageways; direct exposure of operators to the chemicals is possible. Brush bristle size may be larger than the pores holding the contaminates.

Mechanical blasting consists of a cabinet outfitted with doors on one or both sides for moving the work in and out, with a viewing panel placed just above a port. Rubber gloves are used to manipulate the workpiece and protect the operator's hands and arms. A hose and nozzle system delivers the **media** propelled by compressed air on demand which the operator activates with foot pedal. The physical impact of the media upon the workpiece imparts a cleaning action along the substrate. There are numerous types of media ranging from aluminum oxide with a Moh's hardness of 9 to sodium bicarbonate with Moh's hardness of 2.5.

Advantages It is excellent at removing heavy oxides and scale; very versatile; removes a wide variety of soils; and is able to change and impact desirable surface finishes.

Disadvantages Inherently loud and messy, it is undesirable to be near clean areas. It can result in warped and damaged parts if exposed too long or if wrong media are used. Additional cleaning methods are required to remove the residual media and achieve high levels of cleanliness.

Vapor Degreasing The process uses a holding tank with a controlled heat source that is used to boil the solvent. The vapor from the heated solvent rises, surrounding the part, penetrating the pores, and washes away the debris. A water-cooled chill ring at the top of the unit condenses the vapor back to liquid. It settles into a collection tank, is filtered, and flows back to the boiling tank.

Advantages It is a one-step process; parts are cleaned and dried in the same operation.

Disadvantages Even with good ventilation there is always a possibility that some vapors may escape, exposing workers. Traditional chlorinated solvents—trichloroethylene, perchloroethylene, and carbon tetrachloride—are heavily regulated by the EPA. N-propyl bromide is a more user-friendly substitute.

Ultrasonics Ultrasonics couple a tank with an electronic transducer that convert electric energy to intense high-frequency vibrations. These vibrations form bubbles in the tank solution that expand and then collapse against the surface of the part, dislodging contaminants. This method is very effective in penetrating blind holes and cavities. These units can be operated with aqueous or semiaqueous solutions, and other solvents are best suited to cleaning hard substrates. Often they are coupled with sonic rinse tanks and drying chambers. There units run at from 20 to 110 kHz.

Advantages They are very effective in removing small particulates down to ≅ 25 μm, especially in inaccessible cavities and blind holes. Megasonic units (700 kHz to 1.2 MHz) can remove particles down to the 0.15-μm range.

Disadvantages Cavitation can occur on softer substrates and erode the parent material, usually in the form of a "starburst" pattern. When solvents are used, the entire system will need to be "explosion-proof," which dramatically increases the cost of equipment.

CO_2 Cleaning High-purity CO_2 is contained in special siphon cylinders and metered through a specially designed nozzle, producing a jet of fine "dry ice" and effectively becoming a mechanical blast system with a medium that disappears as it sublimes from a solid to a gas. When first emitted from the nozzle, the crystals of dry ice impact the surface of the workpiece, imparting mechanical cleaning actions. At the same time, as the crystals sublime from a solid to a gas, a solvency action takes place that dissolves light oils.

Advantages Capital equipment is low-cost and easy to install. There are very few negative health issues and so very little regulation. There are virtually no waste disposal issues. It is effective in removing solid particles down to 0.1 μm. The system employs no water, so there are no drying issues.

Disadvantages It is only effective on very light oils and particles.

TEST METHODS AND ANALYSIS

There are many test methods, but in general they can be categorized as **gross, precision,** and **critical.**

Examples of **gross** testing are as follows:

Visibly clean parts are nominally free of oil, grease, moisture, slag, rust, and other foreign material when viewed with a naked eye by a person with normal visual acuity, corrected or uncorrected, in normal room light.

Water break testing is a basic method of quickly and inexpensively determining the presence of oils and films. DI water is sprayed onto the subject part; if it "beads up" (as on a freshly waxed car surface), the work is considered to have residual oils or films still present and will require further cleaning. If the water forms large "puddles" or "sheets" for 10 to 15 s, it is considered clean

Black light or **ultraviolet light** from 3,200 to 3,900 Å will cause most oils and solid particles to fluoresce. Simply place the workpiece in a dark room and expose it to the light; contaminants will glow, ranging in color from bright white in the case of lint to yellow-green for hydrocarbons.

Conductivity is an inexpensive and quick method of determining the presence of contaminates, especially nonvolatile residue. The procedure is to submerge the study piece in a clean container of 18-MΩ water at 100 to 120°F for 5 min. The conductivity of the soak water is measured with a meter, and if the reading is less than 5 μmho, it is considered free of hydrocarbons.

In the **precision** class, the most widely used technique is the **extraction** method; commonly referred to as **Millipore** or **patch testing.**

This method can detect contaminates down to 0.01 mg when weight tests are performed and 5 μm using microscopic techniques. The part to be tested is sprayed with filtered solvent from a pressurized container, and the resulting effluent is collected and strained through a preweighed filter, using vacuum. The filter is dried and then weighed or scanned with a microscope to qualify the nature and quantify the amount of the debris. Variations in collection techniques are allowed to accommodate inaccessible areas, such as tubing and vessels, by filing them with solvent and manually agitating, then draining the solution into the filter funnel assembly.

International Standard ISO 14952-3, "Analytical Procedures for the Determination of Nonvolatile Residues and Particulate," is a excellent reference regarding these techniques.

A short list of equipment needed to collect and analyze a sample follows:

50X to 100X microscope with measuring device
Analytical balance accurate to 0.1 mg
$<$5-μm membrane filters (with grids if counting particles)
Vacuum flask, filter holder base, grounded clamp, and funnel
Petri dishes
Pressurized solvent dispenser
Vacuum pump capable of 26 inHg
Convection oven
Forceps with flat blades
Collection container

If the application is approaching the **critical** classification, exotic verification methods may be necessary: optical particle counting, scanning electron microscope, total organic carbon analysis, and contact angle measurement. Details are available in the literature.

INTERPRETATION AND USE OF DATA

There is a "cost of cleanlines," and it almost becomes exponential as the requirements tighten. Time spent identifying the contamination and preventing it at the source is definitely worthwhile. Identification of contaminates requires a good microscope. Establishing and monitoring cleanliness levels of the cleaning processes can ensure repeatable results and serve as a baseline for continuous improvements and maximization of the cleaning process.

Even after precision cleaning, a part may still not be absolutely clean. Nothing is perfectly clean, and some contaminates are nearly impossible to remove. Small amounts of contamination may not harm your device; the challenge is to determine an acceptable level and to repeatably achieve it.

REGULATORY CONSIDERATIONS

Many of the chemicals used for cleaning and testing are highly regulated substances, and they are far too numerous to even begin to list here. Before you go too far in setting up a process, be sure to check with local officials to ensure operation within the codes, and learn who has oversight: local, state, or federal authorities. Setting up and registering for a "waste stream" permit can also take a considerable amount of time and money; it's best to have the appropriate authorities involved at the beginning so that the proposed operations and processes may proceed smoothly and without delay.

Fans, Pumps, and Compressors

BY

TERRY L. HENSHAW   *Consulting Engineer, Magnolia, TX*
HEIMIR FANNAR   *Chief Design Engineer, Ariel Corp.*
MARSBED HABLANIAN   *Retired Manager of Engineering and R&D,*
   Varian Vacuum Technologies
ROBERT JORGENSEN   *Engineering Consultant*

14.1 DISPLACEMENT PUMPS

by Terry L. Henshaw

REFERENCES: Buse, "Positive-Displacement Pumps," Marks' Handbook, 9th ed., McGraw-Hill. Henshaw, "Reciprocating Pumps," Van Nostrand Reinhold. "Hydraulic Institute Standards," 14 ed., Hydraulic Institute. Chesney, Water Injection—Pump Development, paper 68-Pet-11, ASME. Lees, Designing for High Pressures, paper 93-DE-9, ASME. "Positive Displacement Pumps—Reciprocating," API Standard 674, American Petroleum Institute.

NOTE: Much of the text and some of the illustrations are from the author's book: Henshaw, "Reciprocating Pumps," Van Nostrand Reinhold, 1987.

GENERAL

A **displacement pump** (also called **positive-displacement,** or just **p-d**) is a pump which imparts energy to the pumpage (the material pumped) by trapping a fixed volume at suction (inlet) conditions, compressing it to discharge pressure, then pushing it into the discharge (outlet) line. A displacement pump does not rely on velocity to achieve pumping action, as does a centrifugal pump or ejector.

Displacement pumps fall into two major classes: reciprocating and rotary, as illustrated in Fig. 14.1.1.

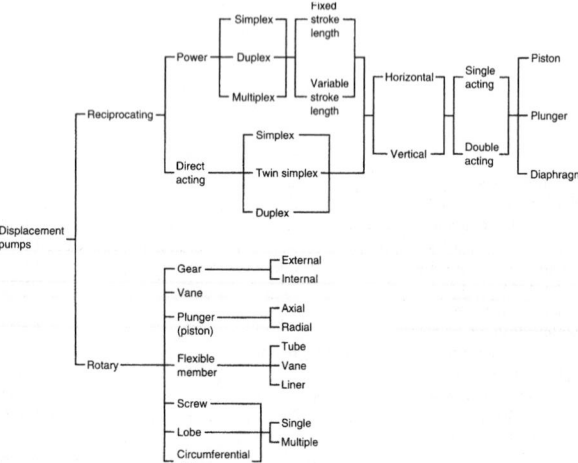

Fig. 14.1.1 Classification diagram of displacement pumps.

Uses and Applications

Displacement pumps serve primarily in applications of low capacity and high pressure, those mostly beyond the capabilities of centrifugal pumps. Some of these services could be performed by centrifugals, but not without an increase in power requirements and/or maintenance. Because displacement pumps achieve high pressures with low pumpage velocities, they are well-suited for abrasive-slurry and high-viscosity services. A reciprocating pump must have special fittings to be suitable for most slurries.

Net Positive Suction Head

Net positive suction head (NPSH), also called **net positive inlet pressure (NPIP)** and **net inlet pressure (NIP),** is the difference between suction pressure and vapor pressure, at the pump suction nozzle, when the pump is running. In a reciprocating pump, NPSH is required to push the suction valve from its seat and to overcome the friction losses and acceleration head within the pump liquid end. In a rotary pump, NPSH

is required to push the pumpage into the cavities created by the pumping elements.

If sufficient NPSH is not provided by the system, the pumpage will begin to flash (boil) as it flows into the pump. The vapor will cause a deterioration of pump performance. As the vapor flows into regions of higher pressure in the pump, it condenses back to a liquid. This collapse of the vapor bubbles occurs with significant impact, impinging upon metal surfaces with enough energy to break out small pieces of the metal. This is called **cavitation** damage. The shock created by the bubble collapse may be severe enough to crack a fluid cylinder or break a crankshaft.

All pumps require the system to provide some NPSH. The NPSH provided by the system is called NPSHA (the A is for available). The NPSH required by the pump is called NPSHR. For displacement pumps, NPSHR is normally expressed in pressure units (lb/in^2 or kPa).

Since water usually contains dissolved air, the vapor pressure of the solution is higher than for deaerated water, but this is often overlooked when NPSH calculations are performed. The Hydraulic Institute recommends an NPSH margin of 3 lb/in^2 (20 kPa) for power pumps in systems where the pumpage has been exposed to a gas other than the liquid's own vapor. A liquid (such as propane) at its bubble point in the suction vessel requires no such margin.

RECIPROCATING PUMPS

A **reciprocating pump** is a displacement pump which reciprocates the pumping element (piston, plunger, or diaphragm). The capacity of a reciprocating pump is proportional to its speed, and is relatively independent of discharge pressure.

A **power pump** is one that reciprocates the pumping element with a crankshaft or camshaft (see Figs. 14.1.2 and 14.1.3). It requires a driver which has a rotating shaft, such as a motor, engine, or turbine. A constant-speed power pump will deliver essentially the same capacity at any pressure within the capability of the driver and the strength of the pump.

A **direct-acting** pump is a reciprocating pump driven by a fluid which has a differential pressure (see Fig. 14.1.4). The motive fluid pushes on

Fig. 14.1.2 Horizontal power pump.

a piston (or diaphragm) which pushes the pumping element through a rod (or directly on the pumpage).

Reciprocating pumps are classified by the following features:

Drive end (power or direct-acting)
Orientation of centerline of the pumping element (horizontal or vertical)
Number of discharge strokes per cycle of each drive rod (single-acting or double-acting)
Pumping element (piston, plunger, or diaphragm)
Number of drive rods (simplex, duplex, triplex, quintuplex, etc.)

Fig. 14.1.3   Vertical power pump. *(Ingersoll-Rand Company.)*

Uses and Applications for Reciprocating Pumps

The justification for selecting a reciprocating pump instead of a centrifugal or rotary is cost, including costs of power and maintenance.

Reciprocating pumps are best suited for high-pressure/low-capacity services. Such services include high-pressure water-jet cleaning and cutting; glycol injection and charge, as well as amine and lean oil charge, in gas processing; ammonia and carbamate charging for fertilizer production; nuclear-reactor charging and standby control; oil-field

saltwater injection and disposal; oil well blowout preventers; pipelines (slurries and crude oil, and injection of ammonia and light hydrocarbons); steel and aluminum mill hydraulic systems; knockout drums in process plants; hydrostatic testing; process slurries; metering; food and chemical homogenizing; well-drilling mud; and car washes.

Fig. 14.1.4   Horizontal direct-acting gas-driven pump.

Applications that practically mandate reciprocating units are abrasive and/or viscous slurries above about 500 lb/in² (3500 kPa). Examples of these services range from powdered coal to peanut butter.

High-Pressure Applications

Hand-powered pumps, for pressures to 40,000 lb/in² (300 MPa), are used for small jacks, deadweight testers, and hydrostatic testing of small components. Small, air-driven direct-acting pumps are used for low-flow, high-pressure systems such as hydrostatic testing. The liquid pressure is approximately the air pressure multiplied ("intensified") by the ratio of the air piston area to the liquid plunger area. They are very simple and will stall and hold a fixed pressure without using power; but when pumping, they consume relatively large amounts of air. Available up to 10 hp (7 kW), the most popular sizes are up to 2 hp (1.5 kW). Pressures range to 100,000 lb/in² (700 MPa).

Hydraulically driven **intensifiers** operate on the same principle, but are more efficient, and because of the higher pressures commonly available from hydraulic systems, can produce high pressures with lower intensification ratios. Intensifiers are available to 300,000 lb/in² (2100 MPa) for laboratory-scale hydrostatic applications. Intensifier systems for water jetting are available from 30,000 to 60,000 lb/in² (210 MPa to 410 MPa) up to 200 hp (150 kW). They are powered by electric motors or diesel engines. Figure 14.1.5 shows the pumping end of an intensifier.

Fig. 14.1.5   Pumping end of a direct-acting, liquid-driven, "intensifier" pump. The check valves are in the tubing.

Power pumps (usually triplex or quintuplex) are available from about 5 to 150 hp (5 to 100 kW) with pressures to 40,000 lb/in^2 (280 MPa), and up to 2,000 hp (1,500 kW) at 20,000 lb/in^2 (140 MPa). Power pumps are efficient and mechanically simple. Figure 14.1.6 shows a power pump liquid end for a 36,000 lb/in^2 (250 MPa) water-jetting pump.

Fig. 14.1.6 Power pump liquid end for 36,000 lb/in^2 (250 MPa) water jetting. *(NLB Corporation.)*

Slurry Applications

The standard reciprocating pump is not designed to handle slurries. Modifications to standard designs, and in some cases special designs, are required to achieve satisfactory operation and component life.

To achieve satisfactory packing and plunger life, abrasive slurry must be prevented from entering the packing. Methods include a wiper ring between the pumpage and packing, a long throat bushing, injection of a clean liquid into the throat area, insertion of a diaphragm or floating piston between the plunger and the pumpage, and complete removal of the valve assembly from the stuffing box area. This last arrangement requires a liquid column between the valve assembly and the stuffing box, which increases the clearance volume and acceleration head within the pump.

Special pump valves are usually required for slurries. Depending on the nature of the solids, they can be ball, bell, bevel seat with elastomer insert, wing-guided with reduced seating area, or disk with special seating surfaces. Special construction can prevent the slurry from contacting the packing, but the valves cannot avoid contact with the slurry.

Problems and How to Avoid Them

Reciprocating pumps have some disadvantages, the most common being pulsating flow. Because of the pulsation, special consideration must be given to system design. Guidelines are provided later in this section.

In most applications, the initial and maintenance costs for a reciprocating pump will be greater than for a centrifugal or rotary pump. The packing in a typical power pump lasts about 2,500 hours, less than a mechanical seal on a rotating shaft.

Most problems with reciprocating pumps can be minimized by selecting pumps to operate at conservative speeds, by carefully designing the pumping system, by careful operating procedures, and by maintenance practices which preserve the alignment of the plunger (or rod) with the stuffing box.

Liquid-End Components

All reciprocating pumps contain one or more pumping elements (pistons, plungers, or diaphragms) that reciprocate into and out of the pumping chambers. (In reciprocating pump terminology, a **piston** is a cylindrical disk which mounts on a rod. A **plunger** is just a smooth rod.)

Each pumping chamber contains at least one suction and one discharge valve.

The liquid cylinder is the major pressure-retaining part of the liquid end, and forms the major portion of the pumping chamber.

A piston pump is normally equipped with a replaceable liner (sleeve) that absorbs the wear from the piston rings. Because a plunger contacts only stuffing-box components, plunger pumps do not require liners.

Sealing between the pumping chamber and atmosphere is accomplished by a **stuffing box** (Fig. 14.1.7). The stuffing box contains rings of packing that conform to and seal against the stuffing box bore and the rod (or plunger).

Fig. 14.1.7 Reciprocating pump stuffing box (nonlubricated) showing pressure gradient during discharge stroke.

The valves in a reciprocating pump are simply check valves which are opened by the liquid differential pressure. Most valves are springloaded. They have a variety of shapes, including spheres, hemispheres, and disks. Disks may be center-guided, outside-diameter-guided, or wing-guided. Figures 14.1.8 through 14.1.13 illustrate some of the valves used in reciprocating pumps.

Fig. 14.1.8 Reciprocating pump disk valve. *(Ingersoll-Rand Company.)*

Fig. 14.1.9 Reciprocating pump disk valve. *(Durabla Manufacturing Company.)*

Fig. 14.1.10 Reciprocating pump elastomeric-insert valve. *(TRW-Mission Manufacturing Co.)*

Fig. 14.1.11 Reciprocating pump double-ported disk valve.

Fig. 14.1.12 Reciprocating pump conical-faced (bevel-seat) wing-guided valve.

Fig. 14.1.13 Reciprocating pump ball valve. *(Ingersoll-Rand Company.)*

The Pumping Cycle

As the pumping element reverses from the discharge stroke to the suction stroke, liquid within the pumping chambers expands, and the pressure drops. Since most liquids are relatively incompressible, little movement of the element is required to reduce the pressure. When the

pressure drops sufficiently below suction pressure, the differential pressure (suction pressure minus chamber pressure) pushes the suction valve open. This occurs when the element is moving slowly, so that the valve opens slowly and smoothly as the velocity of the element increases. Liquid then flows through the valve and follows the element on its suction stroke. As the element decelerates near the end of the stroke, the suction valve returns to its seat. When the element stops, the suction valve closes.

The pumping element then reverses and starts its discharge stroke. The liquid trapped in the pumping chamber is compressed until chamber pressure exceeds discharge pressure by an amount sufficient to push the discharge valve from its seat. The action of the discharge valve, during the discharge stroke, is then the same as that of the suction valve during the suction stroke.

Power Pumps

The drive end of a power pump is called a **power end** (see Fig. 14.1.2). Its function is to convert rotary motion from a driver into reciprocating motion for the liquid end. The main component of the power end is the **power frame,** which supports all other power end parts and, usually, the liquid end. The second major component of the power end is the **crankshaft** (sometimes a camshaft). The crankshaft transmits power from the driver to the connecting rods.

The **connecting rod** is driven by a throw (journal) of the crankshaft on one end, and drives a crosshead on the other. The crankshaft moves in pure rotating motion, the crosshead in pure reciprocating motion. The connecting rod is the link between the two. **Main bearings** support the shaft in the power frame.

The **crosshead** transmits the force from the connecting rod into the pumping element. It is fastened directly to a plunger or piston rod, or to a rod called an **extension, stub,** or **pony rod.** The other end of this rod is fastened to the pumping element.

Power pump overall efficiencies normally range from 85 to 94 percent, higher than any other type of pump.

Selection Graphs Different manufacturers produce these graphs in different configurations. All graphs consist of straight lines since displacement is directly proportional to speed. Figure 14.1.14 shows one such graph for a triplex power pump. Displacement is plotted against crankshaft speed for the different plunger diameters available in the pump. The pressures listed are those at which each plunger size would operate to load the power end to its rated value (4,500 lbf or 20,000 N).

Fig. 14.1.14 Power pump selection graph. *(Ingersoll-Rand Company.)*

NPSH Curves Figure 14.1.15 provides insight into power pump NPSHR characteristics, valve action, and the effect of valve springs on pump performance. These curves are for a 3-in (76-mm) stroke, horizontal, triplex plunger power pump with 2¼-in-diameter (57-mm) plungers, and with suction valves that operate vertically. The valves are wing-guided, and have a diameter approximately equal to the plunger diameter.

Because the axis of the suction valve is vertical, the valve can operate without a spring if the pump speed is kept low. Curve A represents the NPSH requirements with no springs on the suction valves. NPSHR at 100 r/min is only 0.7 lb/in² (5 kPa) (1.6 ft or 0.5 m of water), less than for most centrifugals.

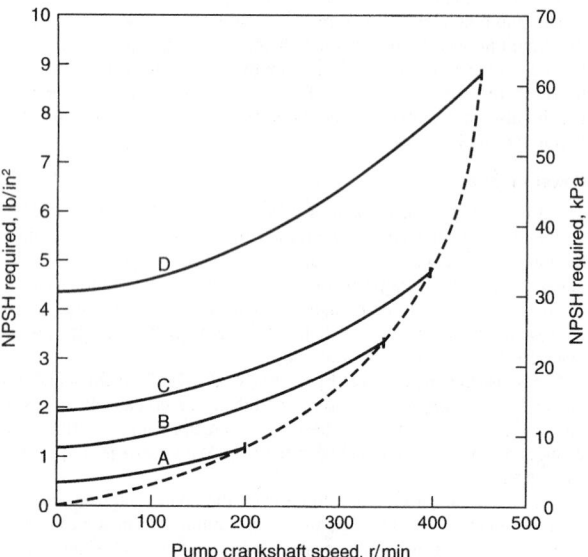

A = no suction valve springs (std. disch.)
B = weak suction valve springs (std. disch.)
C = standard suction valve springs (std. disch.)
D = heavy valve springs — suction and discharge

Fig. 14.1.15   NPSH curves for a 3-in- (75-mm-) stroke triplex power pump.

The speed of the pump in this configuration is limited by the ability of the suction valve to keep up with the plunger. Since there is no spring to push the valve back onto its seat, gravity is the only force tending to close the valve against the entering fluid. If the pump is run too fast, the valve will still be off its seat when the plunger reverses to the discharge stroke. The liquid will then momentarily flow backward through the seat, and the valve will be slammed onto the seat, sending a shock wave into the suction manifold and piping. At that moment, the plunger is moving at a finite velocity, but the discharge valve is still closed. The pressure in the pumping chamber will quickly exceed discharge pressure, and the discharge valve will be driven from its seat. A shock wave will be transmitted from the pumping chamber through the discharge manifold, and into the discharge line. The vertical line at the end of Curve A indicates the maximum speed for proper suction valve operation.

Curve B is for weak springs added to the suction valves. Because both spring force and valve weight must now be overcome to open the suction valve, NPSHR has increased about 100 percent over Curve A. These springs get the suction valves onto their seats quicker, so that operation is smooth at higher speeds.

If speeds beyond the end of Curve B are desired, stronger suction valve springs are required. Stronger springs allow operation to 400 r/min. NPSH requirements are about 3 times those of Curve A, ranging from about 2 to 5 lb/in² (14 to 35 kPa).

Curves A through C represent a pump equipped with the same (standard) discharge valve spring. Only the suction spring has been changed.

If operation is required at speeds exceeding the limits of Curve C, extra-strong springs are required on both suction and discharge valves. As shown by Curve D, NPSHR is about double that required for the standard springs, ranging from about 4.5 to 9 lb/in² (31 to 62 kPa).

The dashed line passing through the ends of the NPSH curves illustrates the dramatic increase of NPSHR as pump speed is increased. When it is necessary to change suction springs, NPSHR increases approximately as the **square** of speed.

To minimize the problem of dissolved air, the NPSH tests that produced these curves were performed with water at, or near, its boiling point in the suction vessel.

Reciprocating pumps, under the correct conditions, can operate with a suction pressure below atmospheric. Such a situation, though, can lead to air being drawn through the stuffing box packing and into the pumping chamber on the suction stroke. This air will cause as many problems as air entrained in the pumpage. Capacity will drop, the pump may operate noisily, the system may vibrate, and damage may occur to pump and system components.

This inward air leakage can be reduced by an external sealing liquid, such as lubricating oil, being directed onto the plunger surface or into the packing.

Test Criteria for NPSH   Power pump NPSH tests are performed by holding the pump speed and discharge pressure constant, and varying the NPSH available (NPSHA) in the system. Capacity remains constant for all NPSHA values above a certain point. Below this NPSHA value, capacity beings to fall. NPSHR is defined as the NPSH available when the capacity has dropped 3 percent.

Acceleration Head   Because the velocities in the suction and discharge piping are not constant, the pumpage must accelerate a number of times for each revolution of the crankshaft. Since the liquid has mass and, therefore, inertia, energy is required to produce the acceleration. This energy is returned to the system upon deceleration. Sufficient pressure, however, must be provided to accelerate the liquid on the suction side of the pump to prevent cavitation in the suction pipe and/or pumping chambers. The drop in pressure, below the average, on the suction side, caused by this acceleration, is called **acceleration head.**

An approximation of acceleration head for power pumps is given by the equation $h_a = LvNC/gK$, where h_a = acceleration head, ft (m) of liquid being pumped; L = actual length of suction line (not equivalent length), ft (m); v = average liquid velocity in suction line, ft/s (m/s); N = speed of pump crankshaft, r/min; C = constant, depending on type of pump; g = gravitational constant, ft/s² (m/s²); and K = fluid compressibility correction factor. Values for the constants C and K are:

Pump type	C
Simplex, single-acting	0.400
Duplex, single-acting	0.200
Triplex	0.066
Quintuplex	0.040
Septuplex	0.028
Nonuplex	0.022

Fluid	K
Slightly compressible liquids such as deaerated water	1.4
Most liquids	1.5
Highly compressible liquids (such as ethane)	2.5

Note that increasing the pump speed, without changing the suction line, increases h_a as the square of speed, because both v and N increase proportionately with speed.

Because this equation does not adequately compensate for pumpage and system elasticity, it is recommended only for short, rigid suction lines.

To reduce acceleration head in a suction line, a bottle or suction stabilizer may be installed in the pipe, adjacent to the pump.

Figure 14.1.16 shows the theoretical flow rates for various types of power pumps during one revolution of the crankshaft. Pumps with an odd number of cranks have the same flow variations for both single-acting and double-acting liquid ends (except the simplex). Flow curves for septuplex and nonuplex pumps are similar to the quintuplex curve, with

A = Simplex - double acting ... 1
Max. rate of flow above mean   59.96%
Min. rate of flow below mean   100.0 %
B = Duplex - double acting - cranks 90° apart ... 2
Max. rate of flow above mean   26.72%
Min. rate of flow below mean   21.56%
C = Triplex - single acting - cranks 120° apart ... 3
Max. rate of flow above mean   6.64%
Min. rate of flow below mean   18.42%
D = Quintuplex - single acting - cranks 72° apart ... 5
Max. rate of flow above mean   1.88%
Min. rate of flow below mean   5.63%
Septuplex - single acting - cranks 51 $\frac{3}{7}$° apart ... 7
Max. rate of flow above mean   1.2 %
Min. rate of flow below mean   2.8%
Nonuplex - single acting - cranks 40° apart ... 9
Max. rate of flow above mean   0.7%
Min. rate of flow below mean   1.5%

Fig. 14.1.16 Flow-rate variations in reciprocating power pumps.

more but smaller variations, as indicated in the tabulation. The values tabulated will vary slightly with variations in the ratio of connecting rod length to stroke length. These curves indicate flow-rate variations, not **pressure pulsations.** Pressure pulsations are a product of the acceleration created by the pump and the mass of the pumpage in the system.

Actual flow rates may vary considerably from these curves, particularly with high-speed pumps, where valve closing and opening may lag the crankshaft rotation as much as 15°, and at high pressures, where liquid compression may cause a further lag of 5° to 10° (or more for a large clearance volume).

Power Pump Speeds   Probably the most controversial factor in the selection of power pumps is the maximum allowable speed. Some manufacturers, for example, offer 3-in- (76-mm-) stroke triplex pumps for operation at 500 r/min. One builder of portable water-jetting units routinely operates this size pump at 600 r/min. Most users of this size pump, in continuous-duty applications, prefer to operate it no faster than about 400 r/min.

Top speeds quoted by one manufacturer for the 3-in-stroke triplex increased over a period of years from 150 to 520 r/min. Although improvements in power end and liquid end construction made the pump capable of operating at higher speeds, albeit with shorter packing life, a major obstacle to higher speeds is system design. Initially, some thought that a smaller pump, running faster to achieve the same capacity, would produce less pulsation in the suction and discharge pipes. Unfortunately, the opposite is true.

The velocity variation in the piping for a triplex pump is 25 percent, regardless of the pump size or speed (see Fig. 14.1.16). For the same capacity, a smaller pump, running faster, will produce the same maximum and minimum flow rates and more pulses per second. Since the acceleration head is proportional to frequency, doubling the pump speed doubles the acceleration head, thereby reducing the NPSH available from the system. Also, doubling the speed requires a much stiffer valve spring, thereby significantly increasing the NPSHR of the pump. If the NPSHA drops below the NPSHR, cavitation and pounding will occur.

The best solution for excessive acceleration is an effective dampener in the suction line; but such devices often are less effective at higher frequencies, due to the inertia of their moving parts and inertia of the liquid inside them, which must oscillate for them to function.

Through field observations, the author developed a set of **maximum recommended speeds** for plunger-type power pumps in continuous-duty services. These speeds, as adopted by API 674, are shown in Fig. 14.1.17, along with the Hydraulic Institute basic speeds. Intermittent and cyclic operation is often satisfactory above these speeds. Lower speeds may be dictated by factors such as pump construction, system design, low NPSHA, entrained gas, high temperature, entrained solids, high viscosity, and requirements for a low sound level.

Fig. 14.1.17 Maximum recommended speeds for plunger-type power pumps in continuous-duty services.

The **minimum speed** of a power pump is determined by its ability to provide sufficient lubrication to all bearing surfaces in the power end. Some units can run satisfactorily at 20 r/min. Others must be maintained at 100 r/min or more.

Plunger load (also called **frame load** and **rod load**) is the force transmitted to the power end by one plunger. For a single-acting pump, the discharge plunger load is calculated by multiplying discharge pressure by the cross-sectional area of one plunger. Suction plunger load is the suction pressure multiplied by the plunger area.

A power pump is rated by the maximum discharge plunger load that the power end is capable of absorbing when the suction pressure is zero. Some units are rated for continuous operation, some for intermittent operation, and some are rated both ways.

Torque Characteristics   For fixed suction and discharge pressures, a power pump requires an **average** input torque that is independent of speed (except for increases at very low and very high speeds). The torque actually has a variation that mirrors the discharge flow-velocity curve. A power pump will require the same torque at one-half or one-quarter rated speed and, therefore, will require one-half or one-quarter rated power, respectively. Figure 14.1.18 illustrates the variation of average torque with speed for a typical triplex power pump.

The curve for full-load torque in Fig. 14.1.18 is for starting against full discharge pressure. Breakaway torque is about 150 percent of average full-load running torque. As speed is increased, and proper lubrication is established in the power end and packing, torque drops to the full-load, full-speed value, and is thereafter constant up to full speed.

For a start-up that is easier on equipment, pump discharge is piped back to the suction vessel, making the discharge pressure near suction pressure. The no-load curve in Fig. 14.1.18 illustrates the resulting torque requirements imposed by the pump on the drive train. Breakaway torque is approximately 25 percent of full-load torque. This will vary, depending on the type of packing and bearings in the pump and the length of time the pump has been idle. As speed increases, the torque drops to less than 10 percent of the full-load torque.

Packing   The component normally requiring the most maintenance on reciprocating pumps is packing. Although the life of packing in a power pump is typically about 2,500 h (3 months), some installations, with special stuffing box arrangements, have experienced lives of more than 18,000 h (2 years), at discharge pressures to 4,000 lb/in^2 (28 MPa).

Fig. 14.1.18 Torque-speed curves for a typical triplex power pump.

Short packing life can result from any of the following conditions: (1) misalignment of plunger (or rod) with stuffing box; (2) worn plunger, rod, stuffing-box bushings, or stuffing-box bore; (3) improper packing for the application; (4) insufficient or excessive lubrication; (5) packing gland too tight or too loose; (6) excessive speed or pressure; (7) high or low temperature of pumpage; (8) excessive friction (too much packing); (9) packing running dry (pumping chamber gas-bound); (10) shock conditions caused by entrained gas or cavitation, broken or weak valve springs, or system problems; (11) solids from the pumpage, environment, or lubricant; (12) improper packing installation or break-in; and (13) icing caused by volatile liquids that refrigerate and form ice crystals on leakage to atmosphere, or by pumpage at temperatures below 32°F (0°C). As is evident from this list, short packing life can indicate problems elsewhere in the pump or system.

To achieve a low leakage rate, the clearance between the plunger (or rod) and packing must be essentially zero. This requires that the sealing rings be relatively soft and pliant. Because the packing is pliant, it tends to flow into the stuffing box clearances, especially between the plunger and follower bushing. If this bushing does not provide an effective barrier, the packing will extrude, and leakage will increase.

A set of square or V-type packing rings will experience a pressure gradient, during operation, as indicated in Fig. 14.1.7. The last ring of packing, adjacent to the gland-follower bushing, will experience the largest axial loading, resulting in greater deformation, tighter sealing and, therefore, the largest pressure drop. The gap between the plunger and the follower must be small enough to prevent packing extrusion. Most packing failures originate at this critical sealing point.

Because this last ring of packing is the most critical, does the most sealing, and generates the most friction, it requires more lubrication than the others. In a nonlubricated arrangement (Fig. 14.1.7), this ring must rely on the plunger to drag some of the pumpage back to it in order to provide cooling and lubrication. Therefore, to maximize packing life, the overall stack height of the packing should not exceed the stroke length of the pump.

Because the last ring of packing requires more lubrication than do the others, lubrication of the packing from the atmospheric side is more effective than injecting oil into a lantern ring located in the center of the packing. Care must be exercised to get the lubricant onto the plunger surface and close enough to the last ring, so that the stroke of the plunger will carry the lubricant under the ring.

During the first few hours of pump operation with new packing, each stuffing box should be monitored for temperature. It is normal for some boxes to run warmer than others—as much as 50°F (30°C) above the pumping temperature. Only if this exceeds the maximum temperature rating of the packing are steps required to reduce box temperature.

The best lubricant for packing in most services has been found to be steam cylinder oil.

The concepts that a higher discharge pressure requires more rings of packing and that a larger number of rings lasts longer are true for long-stroke, low-speed machines, but have been disproven in a number of moderate- to high-speed power-pump applications. Unless they are profusely lubricated, the larger number of rings create additional frictional heat and wipe lubricant from the plunger surface, thus depriving some rings of lubrication. On numerous saltwater injection pumps operating at pressures above 4,000 lb/in^2 (28 MPa), Chesney reported that packing life was only two weeks with twelve rings of packing in each stuffing box. With three rings in each box, packing life was approximately 6 months.

Plungers Next to packing, the plunger is the component of a power pump that typically requires most frequent replacement. The high speed of the plunger and the friction load of the packing tend to wear the plunger surface. For longer life, plungers are sometimes hardened, although it is more common to apply a hard coating. Such coatings are of chrome, various ceramics, nickel-based alloys, or cobalt-based alloys. Desired characteristics of the coatings include hardness, smoothness, high bond strength, low porosity, corrosion resistance, and low cost. No one coating optimizes all characteristics.

Ceramic coatings are harder than the metals, but are brittle, porous, and sometimes lower in bond strength. Porosity contributes to shorter packing life. Mixing of hard particles, such as tungsten carbide, into the less-hard nickel or cobalt alloys has resulted in longer plunger life at the expense of shorter packing life.

Stuffing Boxes Various stuffing box designs (various lubrication and bleed-off arrangements to minimize leakage and extend packing life) are shown in Fig. 14.1.19.

The most significant advance in packing arrangements in recent years has been spring loading. Spring loading is applied most to V-ring (chevron) packing but also works well with square packing. The spring must be located on the pressure side of the packing. Springs of various types can be used, including single-coil, multiple coil, wave-washer, belleville, and thick-rubber-washer. The force provided by the spring is small compared to the force imposed on the packing by discharge pressure. The major functions of the spring are to provide a small preload to help set the packing, and to hold all bushings and packing in place during operation.

Spring loading of packing has the following advantages:

Requires no adjustment of the gland. The gland is tightened until it bottoms, then is locked. This removes one of the biggest variables in packing life: personnel skill.

Alloys expansion. If the packing expands during the initial break-in, the spring allows for the expansion.

Adjusts for wear. As the packing wears, adjustment automatically occurs from inside the box. The problem of transmitting gland movement through the top packing ring is eliminated.

Provides a cavity. The spring cavity provides an annular space for the injection of a clean liquid for slurry applications.

A Standard Lubricated Stuffing Box

Lubricant injected under pressure

Movement of lubricant

Last ring of packing does most of sealing-bites hardest into plunger.

1. Good design.
2. Most of lubricant migrates into pumpage.
3. Packing may be square, chevron, or nonadjustable.

B Alternate Lubricated Stuffing Box

Lubricant fed to plunger on atmospheric side of packing

Movement of lubricant

Last ring of packing does most of sealing-bites hardest into plunger.

1. Good design—performs better than standard arrangement in most cases.
2. Puts lubricant under last ring—where it's needed most.
3. Allows use of low-pressure and drip-type lubricators.
4. Very little lubricant migrates into pumpage.
5. Packing may be square, chevron, or nonadjustable.

C Spring-Loaded Packing

Lubricant fed to plunger on atmospheric side of packing

Movement of lubricant

Last ring of packing does most of sealing-bites hardest into plunger.

1. Good design—long life—minimal leakage.
2. Puts lubricant under last ring—where it's needed most.
3. Allows use of low-pressure and drip-type lubricators.
4. No adjustment required (self adjusting).

D Standard Box Used to Bleed Pumpage to Low-Pressure Point—Bad Application

Total ΔP occurs across these 3 rings

Bleed-off to low pressure point

All 3 rings are fully loaded mechanically—bite hard into plunger.

1. High friction causes excess heat.
2. Short life of packing and plungers.
3. Improper use of standard box—poor application.

E Modified Gland Follower to Allow Bleed-Off to Low Pressure Point

Bleed-off

Unloaded mechanically. Only load is low-pressure.

Only one ring highly loaded.

1. Less friction and lower temp. than "D".
2. Longer life of packing and plungers.
3. Big improvement over "D", but secondary packing not adjustable to compensate for wear.

F Two-Gland Stuffing Box—Lubricated

Bleed-off to low-pressure point (suction).

Lubricant injection

Large gland adjusts primary packing

Movement of lubricant

Primary (hi-press.) packing

This ring does most of sealing.

Secondary (low-press.) packing

Small gland adjusts secondary packing

1. The old standard for high-pressure, critical services.
2. Provides independent adjustment of primary and secondary packing. (Proper adjustment requires skilled mechanic.)
3. Full-size secondary packing.
4. Positive packing lubrication.
5. Negligible leakage to atmosphere.
6. Long packing and plunger life.
7. Excellent for volatile fluids.

G Double Tandem Spring-Loaded Packing

Bleed-off to low-pressure point (suction).

Lubricant injection

Movement of lubricant

1. The best design available for most high-pressure, critical services.
2. Combines best features of two-gland box and spring-loaded packing—negligible leakage, long life, and the packing is self-adjusting.

Fig. 14.1.19 Various stuffing box designs for reciprocating pumps.

Eliminates the need for a gland, if allowed by pump design. The stuffing box assembly (if a separate component) can be disassembled and reassembled on a workbench.

Disadvantages of spring-loading packing are associated with the cavity created by the spring. Since this cavity communicates directly with the pumping chamber, the additional clearance volume will cause a reduction in volumetric efficiency if the pumpage is sufficiently compressible. This cavity also provides a place for vapors to accumulate. If the pump design does not provide for venting this space, a further reduction in volumetric efficiency may occur.

Spring-loaded packing is the reciprocating pump's equivalent to the mechanical seal for rotating shafts. Leakage is low, life is extended, and adjustments are eliminated. Packing sets can be stacked in tandem (they must be independently supported) for a stepped pressure reduction, or to capture leakage from the primary packing that should not escape to the environment.

Controlled-Volume Pumps   A **controlled-volume** pump (also called a **metering, proportioning,** or **chemical-injection** pump) is a power pump with an adjustable stroke length (or adjustable *effective* stroke length). It is used to provide an accurate, and adjustable, capacity. The capacity can be changed manually, as shown in Fig. 14.1.20, or with an electric, pneumatic, or hydraulic controller.

Fig. 14.1.20   Controlled-volume pump with mechanically actuated diaphragm and manual-adjustment stroke length. (*Reprinted from James P. Poynton, "Metering Pumps: Selection and Application," Marcel Dekker, Inc., Copyright 1983. Used by permission.*)

Such pumps can develop pressures to 30,000 lb/in^2 (200 MPa) with capacity controlled with ± 1 percent. The piston or plunger may come in direct contact with the pumpage, or may be separated from the pumpage by a diaphragm. The diaphragm may be flat, tubular, or conical. It can be actuated mechanically, as shown in Fig. 14.1.20, or hydraulically, as shown in Fig. 14.1.21. Mechanically actuated diaphragms are generally limited to capacities of 25 gal/h (100 L/h) and pressures of 250 lb/in^2 (1,700 kPa).

Direct-Acting Pumps

The direct-acting pump (see Fig. 14.1.4) has some of the same advantages as the power pump, plus others. These units are also well suited for high-pressure, low-flow applications. Discharge pressures normally range from 100 to 5,000 lb/in^2 (0.7 to 35 MPa), but may exceed 10,000 lb/in^2 (70 MPa). Capacity is proportional to speed from stall to maximum speed, affected little by discharge pressure. Speed is controlled by controlling the motive fluid. The unit is normally self-priming, particularly the low-clearance-volume type.

Direct-acting pumps are negligibly affected by hostile environments, such as corrosive fumes, because of the absence of a bearing housing, crankcase, or oil reservoir (except for units requiring lubricators). Some direct-acting pumps, inundated by floodwater, have continued to operate without adverse effects. The direct-acting pump is quiet, simple to maintain, and its low speed often results in a long life.

Fig. 14.1.21   Controlled-volume pump with hydraulically actuated double diaphragms. (*Reprinted from James P. Poynton, "Metering Pumps: Selection and Application," Marcel Dekker, Inc., Copyright 1983. Used by permission.*)

Since the direct-acting pump was originally designed to be driven by steam (initially, most pumps were), it was known as a **steam pump**—not for the pumped fluid, but for the motive fluid. Other fluids are now also used to drive direct-acting pumps. Fuel gas, which would otherwise be throttled through a pressure regulator for plant use, is often piped through a direct-acting pump to provide "free" pumping. Compressed air is frequently used to drive small pumps for services such as hydrostatic testing and chemical metering. Hydraulic oil is used to drive **intensifiers** for high-pressure water-jetting.

Direct-acting air-driven diaphragm pumps handle a variety of liquids at pressures up to the pressure of the driving air. This pump, as shown in Fig. 14.1.22, has no stuffing box or packing, and therefore has no leakage, unless a diaphragm fails. Failure of a diaphragm may result in pumpage in the exhaust air and/or air in the pumpage.

Fig. 14.1.22   Direct-acting, air-driven diaphragm pump. (*Used by permission of Warren Rupp, Inc., a unit of IDEX Corp., Mansfield, OH.*)

Direct-acting pumps operate at speeds that normally range from 0 to 50 cycles/min, depending on the stroke length. The ability of these units to operate at speeds down to zero (i.e., stall conditions) is desirable for some applications.

The direct-acting pump has a low thermal efficiency when driven by a gas (such as steam or air). The mechanical efficiency (output force divided by input force) is high; but because the unit has no device to store energy (such as a flywheel), the motive gas must remain at full inlet pressure in the drive cylinder through the entire stroke. At the end of the stroke, the gas expands to exhaust pressure, but no work is performed during this expansion. Hence, the thermal energy of the gas is lost to friction. Steam pumps consume about 100 lb/h of steam for each hydraulic horsepower (hhp) (60 kg/h · kW) developed on the liquid end. When natural gas or air is the motive fluid, consumption is about 3,500 std. ft³/h · hhp (130 m³/h · kW).

The **drive end** (or steam end, or gas end) of a direct-acting pump converts the differential pressure of the motive fluid to reciprocating motion for the liquid end. It is similar in construction to the liquid end, containing a piston (or diaphragm) and valving. The major difference is that the valve is mechanically actuated by a control system which senses the location of the drive piston, causing the valve to reverse the flow of the motive fluid when the drive piston reaches the end of its stroke.

The main component of the drive end is the drive cylinder. This cylinder forms the major portion of the pressure boundary, and supports the other drive end parts.

Mechanical Efficiency—Direct-Acting Pumps For a direct-acting pump, mechanical efficiency is the ratio of the force transmitted to the liquid by the piston (or plunger or diaphragm) to the force transmitted to the drive piston (or diaphragm) by the motive fluid. For double-acting pumps, the differential pressures are used at both ends of the pump, and since fluid friction losses in valves and ports must be charged to the pump, the pressures are as measured at the inlet and outlet ports. In equation form:

$$\eta_m = \frac{A_L \, \Delta p_L}{A_{dr} \, \Delta p_{dr}} \qquad (14.1.1)$$

where η_m = mechanical efficiency of pump, A_L = area of liquid piston or plunger, Δp_L = differential pressure across liquid end, A_{dr} = area of drive piston, and Δp_{dr} = differential pressure across drive end.

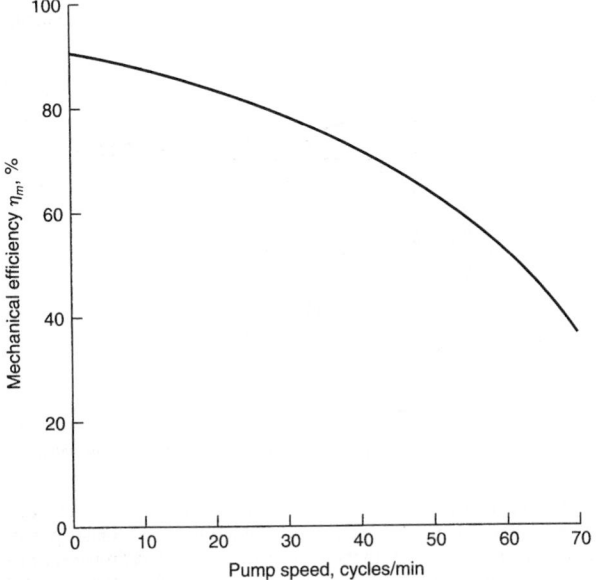

Fig. 14.1.23 Mechanical efficiency of a direct-acting pump.

The area of the piston rod is usually small relative to the piston area, and is often ignored. However, it must be considered when the rod area becomes a significant portion of the piston area.

The mechanical efficiency of a typical direct-acting pump is shown in Fig. 14.1.23. Note that η_m rises as speed is reduced. It is this characteristic that allows the unit to be controlled by throttling the motive fluid. The reduction in the available differential driving pressure forces the pump to operate more efficiently, i.e., at a lower speed.

ROTARY PUMPS

Rotary pumps are displacement pumps which have rotating pumping elements, such as gears, lobes, screws, vanes, or rollers. They do not contain inlet and outlet check valves, as do reciprocating pumps. They are built for capacities from a fraction of a gallon per minute (cubic metre per hour) (for domestic oil burners) to about 15,000 gal/min (3,000 m³/h) (for marine cargo service). Though used for pressures up to 20,000 lb/in² (140 MPa), their particular field is for pressures of 25 to 500 lb/in² (170 to 3,500 kPa).

Most rotary pumps rely on close running clearances to prevent the pumpage from leaking from the discharge side back to the suction side of the pump. Because of the close clearances, the pumpage must be clean. Most rotary pumps are noted for their ability to handle viscous liquids, and many actually *require* a viscous liquid to achieve peak performance. Viscosity affects the mechanical efficiency, volumetric efficiency, and NPSHR of a rotary pump.

The **rotor** is the pumping element of the rotary pump, and is usually the feature by which the pump is classified.

Gears The teeth of the gear trap and displace the pumpage. If the teeth are on the outside of the gear, as shown in Fig. 14.1.24, it is called an **external gear**. As the teeth disengage, suction pressure pushes liquid

Fig. 14.1.24 External-gear rotary pump. *(Reprinted from "Hydraulic Institute Standards for Centrifugal, Rotary and Reciprocating Pumps," 14th ed., copyright 1983, with the permission of Hydraulic Institute.)*

into the cavities between the teeth. The teeth then carry the liquid around to the discharge side of the pump. As the teeth engage, the fluid is pushed into the discharge line. In the most common type of gear pump, one of the pumping gears drives the other pumping gear. This direct contact between the teeth requires the pumpage to be a good lubricant. In some units, though, the two pumping gears are driven by gears (called **timing gears**) mounted on the two shafts external to the pumpage. With this arrangement, the two pumping gears do not contact each other, and the unit is more suitable for liquids with low lubricity. This arrangement does, however, require a larger, more complex pump.

If the teeth are on the inside of a ring, as shown in Fig. 14.1.25, it is called an **internal gear**. In this unit, the larger internal-tooth gear

Fig. 14.1.25 Internal-gear rotary pump.

drives the smaller external-tooth gear. Both gears transfer liquid from the suction to the discharge.

Both external- and internal-gear rotary pumps are used in lubrication systems of engines, compressors, and larger pumps.

Lobe pumps are similar in construction and pumping action to external gear pumps, but one lobe does not drive the other. They do not even touch each other, and therefore must have their shafts independently driven by external timing gears. The lobes are often made of elastomers, and operate at low speeds. These units are used to transfer delicate items such as cherries and other foods—and even live fish. Figure 14.1.26 shows a rotary pump with three-lobe rotors.

Fig. 14.1.26 Three-lobe rotary pump. *(Reprinted from "Hydraulic Institute Standards for Centrifugal Rotary and Reciprocating Pumps," 14th ed., copyright 1983, with the permission of Hydraulic Institute.)*

Screw pumps are constructed with one, two, or three screws. The single-screw pump is more commonly called a progressing-cavity pump, and is discussed separately below. The two-screw pump, as illustrated by Fig. 14.1.27, is used to handle liquids with viscosities to greater than 1,000,000 SSU (200,000 mm²/s), with capacities to about 15,000 gal/min (3,000 m³/h), pressures to 3,500 lb/in² (25 MPa), and temperatures to 600°F (300°C). It is particularly suited for high-viscosity liquids.

Fig. 14.1.27 Two-screw rotary pump.

The three-screw rotary pump, as illustrated in Fig. 14.1.28, is available, in the smaller sizes, for operation up to 8,000 r/min for use in lubrication systems on turbines, compressors, and centrifugal pumps. Larger units are found in such services as crude oil pipelines.

Fig. 14.1.28 Three-screw rotary pump.

A **progressing-cavity** pump is illustrated in Fig. 14.1.29. The rotor is made of polished steel, and in the shape of a single-lead screw. The stator,

Fig. 14.1.29 Progressing-cavity (single-screw) rotary pump. *(Hydraulic Institute.)*

usually made of elastomer, is in the shape of a double-lead screw. As the rotor turns, its axis prescribes a circle. This characteristic requires the rotor to be driven by an internal universal joint (or equivalent). The pumpage moves through the pump in a spiral. The rubber stator makes this unit well-suited for abrasive services.

Figure 14.1.30 illustrates a **sliding vane** rotary pump. The single rotor contains multiple vanes which slide in radial slots. The rotor and casing are eccentric. The vanes maintain contact with the casing by centrifugal force and pressure. These units are typically available for capacities to 1,000 gal/min (200 m³/h), viscosities to 500,000 SSU (100,000 mm²/s), and pressures to 350 lb/in² (2.5 MPa). Some sliding-vane pumps are suitable for low-lubricity liquids such as light hydrocarbons.

Fig. 14.1.30 Sliding-vane rotary pump. *(Reprinted from "Hydraulic Institute Standards for Centrifugal, Rotary and Reciprocating Pumps," 14th ed., copyright 1983, with the permission of Hydraulic Institute.)*

Flexible Member Some rotary pumps are built with flexible vanes, liners, and tubes. The flexible tube pump, also called a **peristaltic** pump, is shown in Fig. 14.1.31.

Fig. 14.1.31 Flexible-tube (peristaltic) rotary pump. *(Hydraulic Institute.)*

A **radial-plunger** rotary pump is shown in Fig. 14.1.32. An **axial-plunger** rotary pump (also called an **axial-piston** or **swash-plate** pump) is shown in Fig. 14.1.33. In these pumps, the inner body rotates, causing each plunger to alternately accept pumpage from the inlet and deliver it to the discharge port. In the radial pump, the length of stroke

Fig. 14.1.32 Radial-plunger rotary pump.

Fig. 14.1.33 Axial-plunger (swash-plate) rotary pump.

of each plunger is established by the eccentricity of the rotor. In the axial pump, the stroke length is established by the angle of the plate or shaft. On some pumps, the eccentricity (or angle) is adjustable, providing variable displacement. Rotary plunger pumps are used in hydraulic systems to provide power to hydraulic motors and cylinders.

Circumferential-Piston Pump   Although sometimes considered a lobe pump, this unit (shown in Fig. 14.1.34) differs from the lobe unit in that there is no close clearance between the two rotors. Close clearance does exist, though, between each rotor and adjacent stationary parts.

Fig. 14.1.34 Circumferential-piston rotary pump. *(Hydraulic Institute.)*

The **casing** (also called a **body** or **housing**) is the main pressure-retaining component of a rotary pump. It supports the rotor, contains the suction and discharge nozzles, and usually forms a portion of the boundaries of the pumping chambers.

The **bearings** of a rotary pump may be internal or external. If internal, they are lubricated by the pumpage. If the pumpage is not suitable for bearing lubrication, the bearings must be external. External bearings require two pumpage seals on each shaft. Some rotary pumps contain four seals.

Figure 14.1.35 shows typical **rotary pump performance curves.** Capacity, mechanical efficiency, and power are plotted against pressure.

Fig. 14.1.35 Rotary pump performance curves for a herringbone gear pump at 600 r/min with a 400-SSU (90-mm²/s) liquid.

The difference between displacement and capacity is called **slip.** Power input is the sum of power output and power losses (due to friction and slip).

As defined by the Hydraulic Institute, the NPSH required by a rotary pump is that value of NPSHA when cavitation noise, a sharp drop in capacity, or a 5 percent reduction in capacity occurs, whichever occurs first.

Increased Running Clearances   An increase in the clearances between the rotating parts, and/or between the rotating and stationary parts, of a rotary pump will cause the slip to increase. An increase in slip causes the volumetric efficiency to drop, the mechanical efficiency to drop, and the NPSHR to rise. If the unit is running at a constant speed, the lower volumetric efficiency (lower capacity) will result in a lower discharge pressure if the system head varies with capacity.

VOLUMETRIC EFFICIENCY

Volumetric efficiency η_v of a displacement pump is the ratio of capacity Q to displacement D: $\eta_v = Q/D$. Capacity is the volume flow rate in the suction pipe. Displacement is the volume swept by the pumping elements per unit time. For a single-acting triplex plunger pump, displacement may be calculated as follows: $D = 3(\pi/4)d^2LN$, where D = displacement; d = diameter of plunger; L = stroke length; and N = crankshaft speed, r/min.

To predict the capacity of a reciprocating pump, in the selection process, it is necessary to predict the volumetric efficiency. Volumetric efficiency may be calculated with the following equation:

$$\eta_v = 1 - \frac{S}{D'} - \frac{C}{D'}\left(\frac{v_s}{v_d} - 1\right) \qquad (14.1.2)$$

where η_v = volumetric efficiency; S = slip through both suction and discharge valves, typically about 3 percent of D'; D' = displacement of one pumping element during one stroke; C = clearance volume (dead space) in the pumping chamber (when the pumping element is at the end of the discharge stroke); v_s = specific volume of pumpage at suction conditions; and v_d = specific volume of the pumpage at discharge conditions.

Although volumetric efficiency has an effect on mechanical efficiency, in reciprocating pumps the two do not necessarily move in unison. It is possible to have a high η_m with a low η_v (due to liquid compressibility), or a low η_m with a high η_v (such as would occur in a high-suction-pressure, low-differential-pressure application).

The same equation can be applied to rotary pumps, but usually $v_s = v_d$, so that Eq. (14.1.2) reduces to $\eta_v = 1 - S/D'$. For a rotary pump, S = leakage of pumpage from the discharge side of the pump to the suction side. Because rotary pumps have no check valves, they must rely on close clearances to restrict slip. Slip is a function of the differential pressure, the viscosity of the pumpage, and the clearance between the parts. Slip in a rotary pump is usually a higher fraction of displacement than in a reciprocating pump.

POWER OUTPUT AND INPUT

The **power output** (also called **hydraulic power, hydraulic horsepower,** and **water horsepower**) of any pump is the usable power it imparts to the pumpage. It is the product of capacity and differential pressure.

In U.S. units, $P_o = Q \times \Delta p/1{,}714 = Q(p_d - p_s)/1{,}714$, where P_o = power output, hp; Q = pump capacity, gal/min; Δp = pump differential pressure, lb/in²; p_d = pump discharge pressure, lb/in²; and p_s = pump suction pressure, lb/in².

In metric units, $P_o = Q \times \Delta p/3{,}600 = Q(p_d - p_s)/3{,}600$, where P_o = power output, kW; Q = pump capacity, m³/h; Δp = pump differential pressure kPa; p_d = pump discharge pressure, kPa; and p_s = pump suction pressure, kPa.

The **power input** (sometimes called **brake horsepower**) is the power required to drive power and rotary pumps. Normally, the power input includes the losses in an integral gear unit, but not the losses in a separate gear unit or variable-speed drive. V-belt losses may be included. If the mechanical efficiency is known, power input may be calculated by dividing the power output by the mechanical efficiency.

MECHANICAL EFFICIENCY—POWER AND ROTARY PUMPS

Mechanical efficiency (also called **pump efficiency** and **overall efficiency**) of a power-driven pump is the ratio of output power P_o to input power P_i: $\eta_m = P_o/P_i$.

As shown in Fig. 14.1.36, mechanical efficiency drops as the frame load is reduced, because the power output falls faster than friction losses, and becomes a smaller part of the power input. Mechanical efficiency is zero when the differential pressure is zero.

Fig. 14.1.36 Mechanical (overall) efficiency of a triplex power pump.

PULSATION DAMPENERS

A **pulsation dampener** is a device which reduces the pressure pulsations created, in a suction or discharge pipe, by the interaction of a pump and a system. The dampener performs this function by reducing the velocity variations of the pumpage in the pipe.

Most reciprocating pumps, and many rotary pumps, create a variation in the velocity of the pumpage in the piping. Unless the suction and discharge vessels are very close to the pump, these velocity variations will be converted by the system into pressure variations. To reduce these pulses, a small vessel can be installed in the suction and/or discharge line. Such a vessel mounted on the discharge line is called a **pulsation dampener.** If the suction pressure is high enough, this same type of device can be used in the suction line, but a special dampener, called a **suction stabilizer** will perform the dual function of reducing pulses and separating free gas from the pumpage.

For optimum effectiveness, each dampener must be installed close to the pipe and adjacent to the pump. There must be minimal restriction between the dampener and pipe. For a low suction pressure, a simple air-filled standpipe is often adequate. For the discharge side, and for high suction pressures, the dampener often contains a diaphragm or bladder that is precharged with nitrogen to about 70 percent of system pressure.

SYSTEM DESIGN

A properly designed system is necessary for a satisfactory installation. Improper design will result in a system that vibrates and is noisy. Pulsations may be severe enough to damage pump components and instrumentation.

Field experience and information from the Hydraulic Institute standards are condensed below into a list of general design guidelines for the (1) suction vessel, (2) suction piping, and (3) discharge piping. The asterisks in the following lists show features that *do not* apply to rotary pumps that produce a uniform flow rate (constant velocity).

The suction vessel should:

Be large enough to provide sufficient retention time to allow free gas to rise to the liquid surface.

Have the feed and return lines enter below the minimum liquid level.

Include a vortex breaker over the outlet (pump suction) line and/or provide sufficient submergence to preclude vortex formation.

Contain a weir plate to force gas bubbles toward the surface. The top of the weir must be sufficiently below the minimum tank level to avoid disturbance.

The suction piping should:

Be as short and direct as possible.

Be one or two pipe sizes larger than the pump suction connection.

Have an average liquid velocity less than the values in Fig. 14.1.37.*

Contain a minimum number of turns, which should be long-radius elbows or laterals.*

Fig. 14.1.37 Suction line velocities for constant acceleration head, for reciprocating power pumps.

Be designed to preclude the collection of vapor in the piping. (There should be no high points, unless vented. The reducer at the pump should be the eccentric type, installed with the straight side up.)

Be designed so that NPSHA, allowing for acceleration head, exceeds NPSHR.

Include a suitable suction stabilizer/gas separator located in the suction pipe adjacent to the pump liquid end if the acceleration head is excessive or the pumpage contains free gas.

Contain a full-opening block valve so that flow to the pump is not restricted.

Not include a strainer or filter unless regular maintenance is assured. (The starved condition resulting from a plugged strainer can cause more damage to a pump than solids.)

The discharge piping should:

Be one or two pipe sizes larger than the pump discharge connection.

Have an average velocity less than 3 times the maximum suction line velocity.

Contain a minimum number of turns, using long-radius elbows and laterals where practical.

Include a suitable pulsation dampener (or provisions for adding one) adjacent to the pump liquid end.[*]

Contain a relief valve, sized to pass full pump capacity at a pressure that does not exceed 110 percent of cracking pressure (the opening pressure of the relief valve). The discharge from the relief valve should be piped back to the suction vessel so that gases liberated through the valve are not fed back into the pump.

Contain a bypass line and valve so that the pump may be started against negligible discharge pressure.

Contain a check valve to prevent the imposition of system pressure on the pump during start-up.

The features of a good system are illustrated in Fig. 14.1.38.

Remedies for Low NPSHA/High NPSHR

In designing the suction system for a displacement pump, if NPSHA is found to be less than NPSHR, a remedy may be found in the following list:

Increase the diameter of the suction line.

Reduce the length of the suction line by providing a more direct route, or by moving the pump closer to the suction vessel.

Install a suction bottle or stabilizer adjacent to the pump liquid end. A fabricated bottle has often been successfully used at pressures below about 50 lb/in^2 (300 kPa), but maintenance of a liquid level is required.

A section of rubber hose in the suction line, adjacent to the pump, will often reduce the acceleration head.

Elevate the suction vessel or the level of liquid in the suction vessel.

Reduce temperature of pumpage. (If the pumpage is at bubble point in the vessel, it must be cooled after it leaves the vessel.)

Reduce speed of the pump, or select a larger pump running slower.

Fig. 14.1.38 A good system for a reciprocating pump.

At a lower speed, it may be possible to operate a reciprocating pump with light suction valve springs, or none at all.)

If the above steps are insufficient, impractical, or impossible, a booster pump must be provided. A booster for a displacement pump is normally a centrifugal, although direct-acting reciprocating and rotary pumps are sometimes used. The NPSHR of the booster must be less than that available from the system. The head of the booster should exceed the main pump NPSHR plus suction-line losses by about 20 percent. The booster should be installed near the suction vessel, and a pulsation bottle or stabilizer should be installed in the suction line, adjacent to the main pump, to protect the booster from pulsating flow.

14.2 CENTRIFUGAL PUMPS
by Terry L. Henshaw

REFERENCES: Stepanoff, "Centrifugal and Axial Flow Pumps," Wiley. Wislicenus, "Fluid Mechanics of Turbomachinery," McGraw-Hill, and Marks' Handbook, 6th ed., McGraw-Hill. Pfleiderer, "Die Kreiselpumpen," Springer. Spannhake, "Centrifugal Pumps, Turbines and Propellers," Technology Press. Hicks, "Pump Selection and Application" and "Pump Operation and Maintenance," McGraw-Hill. Standards of the Hydraulic Institute. Karassik, "Centrifugal Pump Clinic," Marcel Dekker.

NOMENCLATURE AND MECHANICAL DESIGN

The term "centrifugal pumps" as used here includes radial-flow, mixed-flow, and axial-flow (propeller) pumps. Centrifugal pumps are sometimes called **impeller pumps.**

A centrifugal pump consists primarily of a vane-carrying wheel, called the **impeller,** and a **casing** (Fig. 14.2.1). The impeller imparts velocity to the **pumpage,** with most of that velocity being converted to pressure inside the impeller and casing. The casing guides the pumpage to and from the impeller.

If a pump contains a single impeller, it is called a **single-stage** pump (Figs. 14.2.1 and 14.2.2). When two or more impellers are arranged in series (for higher pressure), it is called a **multistage** pump (Figs. 14.2.9 and 14.2.12).

Fig. 14.2.1 Single-stage single-suction overhung pump with volute casing.

Size designations of centrifugal pumps usually contain the nominal sizes of the discharge and suction nozzles, and sometimes the approximate maximum diameter of the impeller—or the number of stages. For example, in U.S. units, a $2 \times 3 \times 8$ pump would have a 2-in discharge

Fig. 14.2.2 Horizontal single-stage double-suction between-bearings pump with a volute casing. (Numbers refer to parts listed in Table 14.2.1.)

nozzle, a 3-in suction nozzle, and an 8-in (approximately) maximum-diameter impeller.

If the impeller is mounted near the middle of the shaft, with a bearing near each end of the shaft, the pump is said to be of the **between-bearings** design (Fig. 14.2.2). If the impeller is mounted at the end of the shaft, with both bearings between the impeller and the coupling, the pump is called **overhung** or **cantilevered** (Fig. 14.2.1).

If the pump shaft is horizontal, it is a **horizontal pump.** If the shaft is vertical, it's a **vertical pump.**

Table 14.2.1 is a list of names used to describe centrifugal pump parts that are discussed below.

Table 14.2.1   Recommended Names of Centrifugal-Pump Parts
(These parts are called out in Figs. 14.2.2, 14.2.4, and 14.2.7)

Item no.	Name of part	Item no.	Name of part
1	Casing	33	Bearing housing (outboard)
1A	Casing (lower half)	35	Bearing cover (inboard)
1B	Casing (upper half)	36	Propeller key
2	Impeller	37	Bearing cover (outboard)
4	Propeller	39	Bearing bushing
6	Pump shaft	40	Deflector
7	Casing ring	42	Coupling (driver half)
8	Impeller ring	44	Coupling (pump half)
9	Suction cover	46	Coupling key
11	Stuffing-box cover	48	Coupling bushing
13	Packing	50	Coupling lock nut
14	Shaft sleeve	52	Coupling pin
15	Discharge bowl	59	Handhole cover
16	Bearing (inboard)	68	Shaft collar
17	Gland	72	Thrust collar
18	Bearing (outboard)	78	Bearing spacer
19	Frame	85	Shaft-enclosing tube
20	Shaft-sleeve nut	89	Seal
22	Bearing lock nut	91	Suction bowl
24	Impeller nut	101	Column pipe
25	Suction-head ring	103	Connector bearing
27	Stuffing-box-cover ring	123	Bearing end cover
29	Lantern ring	125	Grease (oil) cup
31	Bearing housing (inboard)	127	Seal piping (tubing)
32	Impeller key		

Impellers

An impeller is either **single-suction** (Fig. 14.2.1) or **double-suction** (Fig. 14.2.2). A single-suction impeller can have a **radial-flow** (Fig. 14.2.1), a **mixed-flow** (Fig. 14.2.3), or an **axial-flow** (Fig. 14.2.4) profile. A **closed impeller** has full **shrouds** (sidewalls) enclosing the vanes (Figs. 14.2.1 and 14.2.2), an **open impeller** has no shrouds, and a **semi-open impeller** has partial shrouds (Figs. 14.2.8 and 14.2.11). A semi-open impeller is often called an open impeller.

Fig. 14.2.3   End-suction pump with a mixed-flow impeller and removable suction and stuffing-box heads.

Fig. 14.2.4   Vertical wet-pit diffuser pump bowl assembly with an axial-flow impeller (Numbers refer to parts listed in Table 14.2.1.)

Some impellers are designed for specific applications. Impellers for sewage have blunt edges and large passages to minimize clogging from rags and other fibrous matter. Paper stock impellers are open and may have screw conveyor vanes that extend into the suction nozzle. Impellers for abrasive slurries have thick vanes for extended life.

Casings

The casing usually is either of the **volute** (Fig. 14.2.1) or **diffuser** (guide-vane) (Fig. 14.2.5) design, although some casings are circular. Diffuser pumps were once commonly called turbine pumps, but this term has recently been more selectively applied to vertical diffuser pumps, now called **vertical turbine pumps** (Fig. 14.2.9).

Fig. 14.2.5   Single-stage single-suction pump with a multivane diffuser in a volute casing.

If the casing separates along a plane that is parallel to the axis of the shaft, the casing is said to be **axially split** (Fig. 14.2.2). If it parts in a plane that is at a right angle to the axis, it is called **radially split** (Figs. 14.2.6 and 14.2.7).

Fig. 14.2.6 Single-stage end-suction pump with a radially split casing.

Fig. 14.2.7 Vertical end-suction pump with a double-volute radially split casing. (Numbers refer to parts listed in Table 14.2.1.)

Multistage pumps also are made with both axially split casings (Figs. 14.2.12 and 14.2.13) and radially split casings (Fig. 14.2.14). The axially split units are limited to approximately 2,000 lb/in² (15 MPa). Higher-pressure pumps consist of a radially split outer casing (**barrel**) that contains a radially split (Fig. 14.2.14) or axially split inner casing assembly.

The location of the suction (inlet) nozzle determines whether the pump is classified as **end suction, side suction, bottom suction,** or **top suction.**

End-suction pumps usually have radially split casings. A cover (head) is bolted to the casing on the suction side and/or the back side. A **suction cover (casing suction head)** contains the suction nozzle. A back cover contains the seal or packing, and is called a **stuffing-box cover, casing cover,** or **stuffing-box head.** Both designs are illustrated in Figs. 14.2.3 and 14.2.7.

The discharge nozzle of end-suction single-stage horizontal pumps is usually in a top-vertical position (Fig. 14.2.6), although with some units the casing can be rotated so the discharge nozzle is top horizontal, bottom horizontal, or bottom vertical. Most double-suction, axially split pumps (Fig. 14.2.2) have a side discharge nozzle and either a side or bottom suction nozzle. The bottom-suction units are rarely made in sizes below a 10-in (25-mm) discharge nozzle.

If the casing is supported by a foot directly below the casing, it is said to be **foot mounted** (Fig. 14.2.2). If the support is under the bearing housing, it is called **frame mounted** (Fig. 14.2.8). If the casing is supported at the centerline of the shaft, it is a **centerline-mounted** pump.

Fig. 14.2.8 End-suction frame-mounted pump with a semi-open impeller and a removable stuffing-box head.

Centerline support is provided for high-temperature services, so that coupling alignment is retained as the casing expands at the higher temperature.

With most pumps the liquid flows to the unit through piping, but some vertical pumps are submerged in their suction supply. Vertical pumps are therefore called either **dry-pit** or **wet-pit** type (Fig. 14.2.9). The pumpage from a wet-pit pump may discharge above or below its support surface, creating designations of **above-grade discharge** (Fig. 14.2.9) and **below-grade discharge.**

Wearing Rings

Wearing rings (wear rings) (Figs. 14.2.2 and 14.2.7) are used with closed impellers to restrict the leakage of the pumpage from the discharge side to the suction side of the impeller. These rings may be integral with the casing and/or impeller, or may be separate parts. The stationary ring is called a **casing ring** or **cover ring.** The rotating ring, on the impeller, is called the **impeller ring.** Normally only small pumps have integral casing rings. Integral impeller rings are common on small pumps and on high-speed, high-energy pumps. High-energy, and some lower-energy multistage, pumps have experienced loosening and breakage of separate impeller rings, resulting in pump failures. Therefore, these units are often made with integral impeller rings, with hard surfaces to extend the life.

Although available in a variety of configurations, the most common ring configurations are the flat type (Fig. 14.2.6) and the L type (Fig. 14.2.3).

The simple semi-open-impeller pump of Fig. 14.2.8 minimizes leakage from the discharge side to the suction side with a small axial clearance between the impeller and casing. **Pump-out vanes** on the back of the impeller reduce axial thrust and pressure on the seal.

Fig. 14.2.9 Vertical turbine multistage pump with closed impellers and oil-lubricated enclosed line shaft.

Shafts

The shaft in a centrifugal pump supports and drives the impeller and other rotating parts, and is, in turn, supported by the bearings. Because rubbing contact can lead to accelerated wear, or seizure, of wearing rings, the shaft may be larger than needed to transmit the torque, so as to reduce radial deflection of rotating parts. The diameter may be large enough to prevent contact between wearing rings and between the shaft and internal bushings.

The "dry" first **critical speed** of a rotor is related to its static deflection in the horizontal position. A shaft for a multistage pump, designed for a static deflection of, for instance, 0.006 in. (0.15 mm), will have a dry first critical speed of about 2,400 r/min. If operated within 20 percent of 2,400 r/min, excessive vibration and damage can occur. Such a pump, operating at 3,600 r/min, or faster, will be running above this dry critical speed. Smooth wearing rings and internal bushings can increase the actual ("wet") critical speed to a value above running speed. Subsequent wear of these surfaces (increased radial clearances), or grooves in the surfaces, will cause the wet critical speed to drop. If the wet critical speed is within 20 percent of running speed, excessive vibration and damage can result.

Shaft Sleeves

Pump shafts are often protected from erosion, corrosion, and wear at the sealing device, at leakage joints, and at liquid passages by **shaft sleeves** (Fig. 14.2.2). The most common application is at the stuffing box. Shaft sleeves serving other purposes are given specific names that indicate their purposes. For example, a sleeve between two impellers, which, in conjunction with a bushing, limits leakage between the impellers, is called an **interstage** (or distance) **sleeve.**

Bearings

(See also Sec. 8.)

Centrifugal pumps are provided with **rolling-contact** (ball and roller) **bearings** (Figs. 14.2.2, 14.2.6, and 14.2.7), and **fluid-film** (hydrodynamic) sliding **bearings** (Fig. 14.2.14). Their functions are to position the rotor and to absorb residual radial and axial forces created by the rotor.

The typical horizontal pump contains ball bearings (Fig. 14.2.2), lubricated by either grease or oil. If the rating of the ball bearings is exceeded (or by user preference), the pump is provided with fluid-film bearings. Some pumps are available with a combination of sleeve radial- and ball axial-thrust bearings, or sleeve radial and tilting-pad axial-thrust bearings (Fig. 14.2.14).

A **tilting-pad** (axial) **thrust bearing** requires a pressure-lubrication system, with its attendant components of pumps, filters, coolers, gages, and switches, which significantly increases the complexity and cost of the installation.

Some pumps contain internal fluid-film bearings that are lubricated by the pumpage. Examples include vertical turbine pumps (Fig. 14.2.9) and sealless pumps. These bearings can be damaged by solids or gas in the pumpage.

For a between-bearings pump, the bearing nearer the coupling is called the **inboard bearing.** The one farther from the coupling is called the **outboard bearing.** For an overhung pump, the bearing nearer the impeller is the inboard bearing; the one farther from the impeller is the outboard bearing.

Couplings

(See also Sec. 8.)

Some centrifugal pumps are constructed with the impeller mounted directly on the end of the driver shaft (Fig. 14.2.10). These units are called **close-coupled,** and require no coupling.

If the pump shaft and driver shaft are separate pieces, the two shafts must be connected by a mechanical coupling. If the pump and driver each contains its own bearing system, the coupling must be flexible. **Flexible couplings** are available in designs such as gear or grid, which require lubrication, and disk or diaphragm, which require no lubrication.

Fig. 14.2.10   Close-coupled (motor-mounted) pump.

If the pump shaft relies on the driver shaft for support, the coupling must be a **rigid coupling.**

Sealing Devices

(See also Sec. 8.)

Packing is just fancy square rope, available in a variety of fibers, or foil, cut and formed into rings that fit snugly into the bore of the **stuffing box** and onto the surface of the shaft or sleeve (Figs. 14.2.2 and 14.2.6). Packing is held in the stuffing box by a **gland** (Figs. 14.2.2 and 14.2.6), which, in turn is retained and adjusted by gland bolts or studs and nuts.

Packed stuffing boxes have the primary function of minimizing leakage at the point where the shaft passes through the casing, although packing also provides radial support to the shaft, reducing vibration and radial deflection thereby reducing bending of the shaft (which reduces the probability of a fatigue failure of the shaft).

Packing on a rotating shaft (or sleeve) must be allowed to leak, or it will overheat and fail. Leakage must be a tiny stream, or at least a fast drip. If the pressure at the "throat" of the stuffing box (normally suction pressure) is below atmospheric pressure, the packing must prevent air from leaking into the pump. The rings of packing must therefore be separated by a **lantern ring (seal cage)** (Figs. 14.2.2 and 14.2.6), and clean, cool liquid must be provided to the lantern ring at a positive pressure. This "sealing liquid" flows along the shaft (or sleeve) both into the pump and to the atmosphere. If the pumpage is clean, cool water, the sealing liquid can be fed from the discharge of the pump into the lantern ring; either through external piping (Fig. 14.2.2), or through an internal passage (Fig. 14.2.6).

If the pumpage is hot, a slurry, toxic, flammable, or otherwise it should not be allowed to flow into the packing or escape to the atmosphere. A clean, cool sealing liquid must be provided to a lantern ring, at a pressure higher than throat pressure.

Mechanical seals are used in centrifugal pumps when the leakage rate from packing is unacceptable, the requirement for manual adjustment is undesirable, or a longer-life sealing device is desired. A mechanical seal provides no radial support for the pump rotor, so should be used only when the shaft and bearings provide adequate rigidity and support.

All mechanical seals leak. The leakage may be so small as to be visibly undetectable. In fact, a seal *must* leak, or it will overheat and fail prematurely. The sealing faces rely on a thin film of liquid between the faces for cooling and lubrication. Some of this liquid leaks past the faces.

For decades, a mechanical seal was required to fit inside the bore of a stuffing box that was designed for packing. This required the seal cross section to be small, and provided an inadequate environment for the seal, both contributing to shorter seal life. Newer pump designs provide larger cavities for seals, allowing more robust cross sections and a better environment, resulting in longer seal life.

A centrifugal pump seal chamber normally contains only one mechanical seal, but safety and environmental concerns have lead to seal chambers containing two seals. These dual arrangements can be provided in the **double seal** arrangement, in which the seals are mounted back to back (Fig. 14.2.11) or face to face, or in the **tandem seal**

Fig. 14.2.11   Single-stage end-suction pump with semi-open impeller and double mechanical seal.

arrangement, in which both seals face the same direction. The back-to-back double seal requires a liquid between the seals at a pressure above the throat pressure. The barrier liquid between tandem seals is normally at atmospheric pressure.

Power plant boiler feed pumps are often equipped with a long labyrinth-type throttle bushing for a seal. Cool condensate is injected near the center of the bushing, flowing both into the pump and to the atmosphere. The atmospheric leakage is collected and returned to the system. Such a seal has high leakage, but a longer life than packing or a mechanical seal.

Sealless Pumps

Sealless centrifugal pumps can be provided for services where no leakage to the atmosphere can be tolerated. These pumps are available in both **canned-motor** and **magnet-drive** designs. In the canned-motor design, the motor rotor is inside the pump, surrounded by the pumpage, separated from the motor stator by a thin "can." The motor rotor is connected directly to the pump impeller. In the magnet-drive design, there are two sets of magnets, separated by a thin can. The inner set is connected to the pump impeller. The outer set is coupled to the pump driver. As the outer set rotates, lines of flux pass through the can and rotate the inner set. Both designs utilize fluid-film bearings that are lubricated by the pumpage.

HYDRAULIC THRUST

Radial thrust on an impeller is created by a nonuniform pressure around the periphery of the impeller. In a **single-volute** casing (Fig. 14.2.1), when the pump is operating at or near its capacity correspondent to peak efficiency (**best efficiency point, bep**), both velocity and pressure are uniform around the impeller, resulting in zero hydraulic radial thrust. At capacities both above and below bep, a nonuniform pressure distribution creates a radial force (thrust), which will increase shaft deflection and bearing loads. To reduce radial thrust, casings may be designed with a **double volute** (Fig. 14.2.7) or a **diffuser** (Fig. 14.2.5).

Axial thrust is created by pressures acting axially on impeller shrouds and vanes. A double-suction impeller creates very little axial thrust. A single-suction impeller can create a significant axial thrust. This thrust is created by near-discharge pressure acting on a back shroud that is larger than the front shroud, and by vanes that twist and extend into the impeller eye. The thrust can be balanced by a back wearing ring and **balance holes** (Figs. 14.2.6 and 14.2.10) drilled through the back shroud, allowing the pumpage to flow from the back balance chamber into the impeller eye. For suction pressures near atmospheric, the back wearing ring diameter is normally equal to or larger than the front (eye) ring. For higher suction pressures on overhung pumps, the back ring may be reduced in diameter to compensate for the thrust created by the higher suction pressure pushing on the end of the shaft.

Most multistage pumps are built with single-suction impellers. To balance the axial thrust of these impellers, two arrangements are used:

(1) The impellers all face in the same direction and are mounted in the ascending order of the stages. The axial thrust is balanced by a hydraulic balancing device (Figs. 14.2.12 and 14.2.14). (2) An even number of single-suction impellers is used, one half of these facing in a direction opposite to the second half (Fig. 14.2.13). This mounting of single-suction impellers back to back is frequently called **opposed impellers.**

Fig. 14.2.12   Multistage pump with axially split casing, single-suction impellers facing one direction, and hydraulic axial balancing device.

Fig. 14.2.13   Four-stage pump with opposed impellers.

Fig. 14.2.14   Double-casing multistage pump with radially split inner and outer casing.

Fig. 14.2.15   Balancing drum.

Hydraulic balancing devices may take the form of (1) a **balancing drum**, (2) a **balancing disk**, or (3) a combination of these two. The **balancing drum** is illustrated in Fig. 14.2.15. The balancing chamber at the back of the last-stage impeller is separated from the pump interior by a drum mounted on the shaft. The drum is separated by a small radial clearance from the stationary portion of the balancing device, called the **balancing drum head,** which is fixed to the pump casing. The balancing chamber is connected either to the pump suction or to the vessel from which the pump takes its suction. The forces acting on the balancing drum are (1) toward the discharge end—the discharge pressure multiplied by the front balancing area (area *B*) of the drum; (2) toward the suction end—the back pressure in the balancing chamber multiplied by the back balancing area (area *C*) of the drum. The first force is greater than the second, thereby counterbalancing the axial thrust exerted upon the single-suction impellers. The drum diameter can be selected to balance the axial thrust completely or to balance 90 to 95 percent of this thrust, depending on whether a slight thrust load in a specific direction on the thrust bearing is desirable.

The operation of the simple **balancing disk** is illustrated in Fig. 14.2.16. The rotating disk is separated from the balancing-disk head by a small axial clearance. The leakage through this clearance flows into the balancing chamber and from there either to the pump suction or to the

Fig. 14.2.16   Simple balancing disk.

suction vessel. The back of the balancing disk is subject to the balancing-chamber back pressure, whereas the disk face experiences a range of pressures. These vary from discharge pressure at its smallest diameter to back pressure at its periphery. The inner and outer disk diameters are chosen so that the difference between the total force acting on the disk face and that acting on its back will balance the impeller axial thrust. If the axial thrust of the impellers should exceed the thrust acting on the disk during operation, the latter is moved toward the disk head, reducing the axial clearance. The amount of leakage through this clearance is reduced so that the friction losses in the leakage return line are also reduced, lowering the back pressure in the balancing chamber. This automatically increases the pressure difference acting on the disk and moves it away from the disk head, increasing the clearance. Now the pressure builds up in the balancing chamber, and the disk is again moved toward the disk head until an equilibrium is reached. To ensure proper balancing-disk operation, the change in back pressure must be of an appreciable magnitude. This is accomplished by introducing a restricting orifice in the leakage return line.

The **combination disk and drum** (Fig. 14.2.14) incorporates portions rotating within radial clearances of stationary portions and a disk face rotating within an axial clearance of another portion of the stationary part. The radial clearance remains constant regardless of any axial displacement of the rotor within the casing. Such displacement, however, changes the axial clearance within the balancing device. These changes cause changes in the leakage, which in turn change the pressure drop across the radial clearances and thus increase or decrease the average value of the pressure acting on the disk face. These changes in the intermediate pressure on the disk face act to move the balancing device in whichever direction is required to restore equilibrium and axial balance.

HYDRAULIC PERFORMANCE

The performance of a centrifugal pump is generally described in terms of the following **characteristics:** (1) rate of flow, or **capacity** *Q*, expressed in units of volume per unit of time, most frequently ft³/s, gal/min, or m³/h (1 ft³/s = 449 gal/min; 1 m³/h = 4.403 gal/min); (2) increase of energy content in the fluid pumped, or **head** *H*, expressed in units of energy per unit of mass, usually ft · lb/lb or, more simply, ft, or m; (3) input **power** *P*, expressed in units of work per unit of time, bhp or kW; (4) **efficiency** η, the ratio of useful work performed to power input, (5) **rotative speed** *N*, in r/min.

Since the parameters indicated are all mutually interdependent, it is customary to represent the performance of a centrifugal pump by means of **characteristic curves** similar to that shown in Fig. 14.2.17. While it is possible, within certain limits, for the pump designer to regulate the shape of these curves to suit the needs of a particular application, this is essentially a function of design capacity, head, and speed and hence for

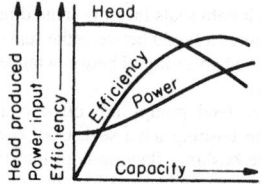

Fig. 14.2.17   Typical characteristic curves at constant speed.

a standard production unit is determined by test and is not subject to modification. For any given capacity on such a characteristic curve, the relationship between performance characteristics is expressed by the equation $\eta = \gamma QH/3{,}960P$, where η is efficiency expressed as a decimal, γ is the specific gravity of the fluid pumped, *Q* is in gal/min, *H* in ft, and *P* in bhp. In metric units, $\eta = \gamma QH/367P$, where *Q* is in m³/h and *H* in metres.

Inasmuch as the actual performance of centrifugal pumps is determined by testing, it is highly desirable to be able to use the results of past tests as a basis for predicting the performance of future designs. To this end, an interesting and widely used characteristic number known as **specific speed,** $N_s = N\sqrt{Q}/H^{3/4}$, has been developed. In this expression, values of *N*, *Q*, and *H* are all for the point of best efficiency. While not dimensionless, N_s can be made dimensionless by the appropriate selection of units. Specific speed is of interest to both the pump designer and the pump user essentially in two ways: (1) All geometrically similar pumps, regardless of their size, will have identical specific speeds (but all pumps of the same specific speed will not necessarily be geometrically similar). (2) Within reasonable limits, pump geometry and performance can be predicted as a function of N_s and *Q*. For *N* in r/min, *Q* in gal/min, and *H* in ft, the practical range of N_s is approximately 500 to 15,000. In metric units *Q* is in m³/s and *H* is in metres, so that this range becomes approximately 10 to 300 ($N_{sm} = N_s/51.66$) The shape of the profiles for impellers varies over the range of specific speeds, as shown in Fig. 14.2.18. The commercially attainable efficiencies also vary with

Fig. 14.2.18   Approximate impeller shapes versus specific speed.

the specific speed (Fig. 14.2.19). It must be remembered that these efficiencies are attainable at the capacity at best efficiency. The actual capacity may differ from this best-efficiency flow. In addition, the general shape of the characteristic curves will vary widely from one end of this range to the other, as illustrated by Figs. 14.2.24 and 14.2.25.

Fig. 14.2.19 Efficiencies of single-stage end-suction and double-suction centrifugal pumps (solid lines) and wet pit pumps (dashed lines).

Hydraulic Design

The basic purpose of a centrifugal pump in any fluid-handling system is to add energy to the fluid, and since it is a dynamic machine, the pump depends entirely on changes in velocity to provide the energy. While the measurable evidence of energy addition is in most cases largely in the form of static pressure, this is partially the result of velocity reductions and constraints occurring in the diffuser or casing and to this extent represents a conversion from the velocity energy produced by the impeller. Thus, any discussion of centrifugal-pump theory generally becomes a discussion of velocities occurring at various points within the pump.

The true velocity relationships existing within a pump are extremely complex; but for practical purposes, a one-dimensional analysis serves to illustrate the basic concepts and, indeed, has served as the basis of design for virtually all centrifugal pumps ever built.

For radial and mixed-flow impellers ($500 \leq N_s \leq 7,500$ or $10 \leq N_{sm} \leq 150$) the velocities at the inlet and outlet of the impeller are shown by the vector diagrams in Fig. 14.2.20. The head produced by such an impeller is represented by

$$H = \eta_H (U_2 V_{u2} - U_1 V_{u1})/g \qquad (14.2.1)$$

where H = head, ft (m); U = circumferential velocity of impeller at radius being considered, ft/s (m/s), V_u = average value of the circumferential component of absolute fluid velocity, ft/s (m/s), and $g = 32.174$ ft/s^2

Fig. 14.2.20 Velocity diagram of a radial-flow impeller.

(9.80 m/s^2). Subscript 1 refers to the impeller inlet section, and subscript 2 to the impeller discharge section. The coefficient η_H is the hydraulic efficiency of the rotating-vane system and, for the range of specific speeds indicated above, will generally fall between 0.85 and 0.95. This hydraulic efficiency is considerably higher than pump efficiency η since it does not include mechanical losses due to bearing or packing friction, volumetric losses due to internal wearing-ring clearances, impeller-disk friction, or fluid-friction losses due to velocity conversion or boundary-layer considerations ahead of or following the impeller. For pumps in this specific-speed range, the hydraulic losses $1 - \eta_H$ will generally be between one-quarter and three-quarters of total pump losses $1 - \eta$.

For pumps arranged with an axial inlet to the impeller (such as that shown in Fig. 14.2.7), it is generally assumed that the entering flow will have no rotational component, and V_{u1} is therefore zero. Equation (14.2.1) can thus be reduced to

$$H = \eta_H U_2 V_{u2}/g \qquad (14.2.2)$$

In practice, Eq. (14.2.2) will provide a close approximation to total head for any pump up to a specific speed of 2,000 ($N_{sm} \approx 40$) since the term $U_1 V_{u1}$ in Eq. (14.2.1) is very small compared with $U_2 V_{u2}$.

It should also be noted in Fig. 14.2.20 that the relative velocity v_2 does not coincide in direction with the vane angle at the impeller discharge. The angular difference between v_2 and the direction of the vane is due to the irrotational nature of the flow between the vanes. The effect of this difference can be taken into account as the vector difference $V_{u2}{}^* - V_{u2} = Na/229$, where a is the shortest distance in inches taken in the radial plane between the discharge tip of any vane and the upper surface of the following vane. ($V_{u2}{}^* - V_{u2} = Na/19,100$, where a is in mm and velocities are in m/s.)

The foregoing relationships provide the basis for determining impeller diameters and vane angles required to produce a given total head requirement at a specified rotative speed. It is equally necessary, of course, to provide in the design for handling a specified flow volume, and this is readily accomplished by providing the necessary cross-sectional area A between vanes to pass the required flow at velocities previously determined. A useful relationship for this purpose, in light of the units commonly employed in pump design, is $Q = AV/0.321$, where Q is in gal/min, V in ft/s, and A in in^2 ($Q = AV/278$, where Q is in m^3/h, V in m/s, and A in mm^2).

Upon leaving the impeller, the liquid pumped enters either (1) a system of diffusing vanes surrounded by an outer casing or (2) directly into a casing designed to contain the fluid and control its velocities. Where diffusing vanes are used, they are designed on the basis of velocity relationships very similar to those employed in impeller design but with the objective being to slow the fluid down to convert velocity energy to pressure energy and, further, to reduce frictional losses in the discharge system following the diffuser. Where the impeller discharges directly into the casing, this component of the machine is most frequently designed in the form of a volute to provide constant velocity all around the impeller periphery up to the point of entry into the discharge nozzle. From this point, commonly called the **casing throat,** to the discharge flange or to the inlet of a succeeding stage, the velocity is gradually reduced. Special circumstances related to pump design or application often result in modifications to the constant-velocity design of such a casing, and variations may be found covering the entire range from constant velocity to constant area. In addition, many casings are now designed with one or more spiral vanes placed in such a way as to approximate a condition of geometric similarity in relation to the impeller, which is advantageous when a pump is operated at capacities other than those for which it is designed. A casing of this nature represents an effort by the designer to obtain an optimum balance between the desirable geometric similarity of the diffuser discharge and the manufacturing simplicity and generally high efficiency of the volute-type casing. The most common form of such a casing is the double-volute type discussed under Hydraulic Thrust.

For **axial-flow impellers** ($7,500 \leq N_s \leq 15,000$ or $150 \leq N_{sm} \leq 300$) velocity relationships can be approximated in a manner similar to that used for lower-specific-speed pumps, but refinement of these approximations is approached in a somewhat different manner, largely because of the considerable body of knowledge available in the form of airfoil data which can be applied. Velocity diagrams for pumps of this type are shown in Fig. 14.2.21.

Fig. 14.2.21   Velocity diagram of an axial-flow vane system.

Considering a cylindrical stream tube intersecting the vanes of an axial-flow impeller, we can rewrite Eq. (14.2.1) in the form

$$H = \eta_H U \Delta V_u / g \qquad (14.2.3)$$

where ΔV_u represents the increase in the tangential component of the absolute velocity as the fluid passes through the impeller. For pumps in this specific-speed range, η_H will generally fall between 0.80 and 0.90 and $1 - \eta_H$ will generally be between one-half and three-quarters of $1 - \eta$.

As in the case of radial-flow impellers, the relative velocity v_2 at the impeller exit does not coincide with the vane angle. In this case, the necessary correction can be applied by means of the expression

$$\Delta V_u = 2 \Delta V_u^* / [(t/l)(2/\pi K)(1/\sin \beta) + 1] \qquad (14.2.4)$$

where t is vane spacing, l is vane length, β is the discharge vane angle, and K is the coefficient determined from Fig. 14.2.22, which provides

Fig. 14.2.22   Lattice-effect coefficient. *(Weinig.)*

for the fact that the impeller blades are, in effect, arranged in a continuous lattice.

In the design of axial-flow pumps, it is generally assumed that the head developed by the blade elements within all the cylindrical stream tubes between the impeller hub and its outer diameter will be the same. For this condition to be achieved, it will be evident from Eq. (14.2.3) that since U will vary directly with the radius, ΔV_u must vary inversely with the radius. Thus vane camber (or curvature) will be greater near the hub than at the periphery. It is further generally assumed that the axial velocity is constant throughout the impeller, and to satisfy this condition, the blade angles will be greater at the hub than at the periphery, giving rise to the twist of the vane.

By designing the vane element at the hub, the periphery, and any reasonable number of radial stations between, it is possible to define the vane over its entire surface to produce the total head desired. The design capacity can be simply established from the constant axial velocity and the annular area between the hub and the periphery.

Axial-flow impellers will almost invariably discharge into a vaned diffuser designed primarily to convert into pressure the tangential component of the absolute velocity leaving the impeller, thus producing a uniform, nonrotating velocity profile at the pump discharge or at the entrance to any succeeding stage.

Pump Application

In applying any centrifugal pump to a fluid-handling system, the engineer generally has two variables in the pump which may be used to effect a match between it and the system, namely, speed of operation and impeller diameter. (In axial-flow pumps, impeller diameter cannot conveniently be changed, but similar results can be accomplished within a limited range by reducing vane length.) To take advantage of these variables, it is necessary to understand the similarity relationships which govern pump behavior.

The effect of change in speed can be most readily explained by referring back to Eq. (14.2.1). For a fixed impeller diameter, it will immediately be evident that any increase in rotative speed will result in a directly proportional increase in the peripheral velocities of the impeller at both inlet and outlet, U_1 and U_2, respectively. Furthermore, since the directions of both the relative and absolute velocities v and V, respectively, are controlled by the vane angles, it follows that both the inlet and outlet velocity diagrams remain geometrically similar, in other words, as pump speed changes, all fluid velocities change in direct proportion. Thus the effect of a speed change on pump capacity can be represented for two speeds, N_1 and N_2 by

$$Q_1 / Q_2 = N_1 / N_2$$

Referring again to Eq. (14.2.1), it will be evident that the change in head will be proportional to the square of the change in speed, resulting in

$$H_1 / H_2 = (N_1 / N_2)^2$$

Also, since P is proportional to QH, then

$$P_1 / P_2 = Q_1 H_1 / Q_2 H_2 = (N_1 / N_2)^3$$

The effect of a reduction in impeller diameter is for practical purposes, identical to that of a reduction in speed, and the above equations can all be rewritten in the same form, substituting D_1 and D_2 for N_1 and N_2. In this case, however, there is somewhat less latitude for change, the practical limit for cut-down being approximately 25 to 30 percent in the case of low-specific-speed impellers and decreasing from this value as specific speed increases.

From the foregoing it should be noted that running a pump at speeds far below its normal rated speed would be uneconomical since the pump would obviously be overdesigned for the service. Conversely, the same pump could not be run at speeds far above its rating since it would soon exceed its power capability. Thus the normal approach to pump application calls for selection of a unit which will meet the system requirements at or near the pump's maximum rated speed and preferably as close to maximum impeller diameter as possible.

A further aspect of the similarity laws, of interest primarily to pump designers, is the matter of geometrically similar pumps, or as they are commonly called, **factors of each other.** Assuming two such pumps with a size ratio f, then **at the same rotative speed** $H_1/H_2 = f^2$, which is the same as for a change in speed of the same ratio. In the case of capacity, however, we have, in addition to the linear increase due to velocity change, a further increase due to the change in the cross-sectional area of all fluid channels which is proportional to the square of the size factor, resulting in the relationship $Q_1Q_2 = f^3$. The ratio of power requirements therefore becomes $P_1/P_2 = f^5$.

To simplify selection of pumps, it is customary to plot, for any given speed, performance curves to show pump characteristics over the available range of impeller diameters rather than at the single diameter which would be implicit in a curve of the type shown in Fig. 14.2.17. A typical rating curve of this nature is shown in Fig. 14.2.23. Such rating

Fig. 14.2.23   Rating curve of 10-in double-suction single-stage pump.

curves will vary widely in their nature with changes in N_s, and an understanding of these variations is essential to the selection of properly sized pump drivers and, in many cases, proper design of discharge piping and/or the determination of limiting ranges of pump operation. Some idea of the diversity of characteristics which is available can be obtained by a comparison between Figs. 14.2.24 and 14.2.25, both of which are plotted entirely in percentages of design values occurring at the point of best efficiency.

Fig. 14.2.24   Typical characteristics for $N_s = 1,550$ ($N_{sm} = 30$).

In the case of the lower-specific-speed pump shown in Fig. 14.2.24, a driving motor selected for the rated condition would be more than adequate at lower capacities, and discharge piping would be subjected to only modest overpressures. On the other hand, if the pump shown in

Fig. 14.2.25   Typical characteristics for $N_s = 10,000$ ($N_{sm} = 194$).

Fig. 14.2.25 is to be operated at reduced capacities, then both motor size and discharge piping must be designed for the minimum flows to be encountered.

NPSH

In order for a pump to deliver its rated output, it is obviously necessary that it be supplied with fluid at its inlet at the same rate. It is further necessary that the absolute pressure (including velocity head, $V^2/2g$) of the fluid at the inlet exceed the vapor pressure by an amount sufficient to overcome (1) any entrance or frictional losses between the point of entry into the pump and the impeller, and (2) the shock losses occurring at the impeller inlet. This gives rise to the definition of **net positive suction head (NPSH)** which is the absolute pressure at the pump inlet expressed in feet (metres) of liquid, plus velocity head, minus the vapor pressure of the fluid at pumping temperature, and corrected to the elevation of the pump centerline in the case of horizontal pumps or to the entrance to the first-stage impeller for vertical pumps.

NPSH required is determined by the pump manufacturer and is a function of both pump speed and pump capacity. **NPSH available** represents the energy level of the fluid over the vapor pressure at the pump inlet and is determined entirely by the system preceding the pump. Unless NPSH available at least equals NPSH required at any condition of operation, some of the fluid will vaporize in the pump inlet and bubbles of vapor will be carried into the impeller. These bubbles will collapse violently at some point downstream of the pump inlet (usually at some point within the impeller) and produce very sharp, crackling noises, frequently accompanied by physical damage of adjacent metal surfaces. This phenomenon is known as **cavitation** and is generally highly undesirable.

A term similar to that used for pump specific speed has been developed for pump NPSHR characteristics and has been identified as **suction specific speed,** $S = N\sqrt{Q}/H_{sv}^{3/4}$, where H_{sv} is NPSH. For double-suction impellers, $S = N\sqrt{Q/2}/H_{sv}^{3/4}$. The value S, like N_s, is expressed simply as a number and, for general-purpose pumps, does not generally exceed 10,000, where N is in r/min, Q is in gal/min, and H_{sv} is in ft. When Q is in m^3/s and H_{sv} is in metres, the equivalent value S_m is 194.

Within the limits of NPSH capabilities, it is desirable to select a pump of the highest specific speed since this will produce the highest pump speed and consequently use the smallest pump. As a convenience in accomplishing this, the Hydraulic Institute has published two charts indicating the recommended NPSH as a function of pump capacity and operating speed and, conversely, the maximum recommended operating speed for different capacities if the NPSH value is known or first selected. These charts are based on a recommended value of suction specific speed of 8,500 ($S_m = 165$). The chart for double-suction pumps is reproduced in Fig. 14.2.26.

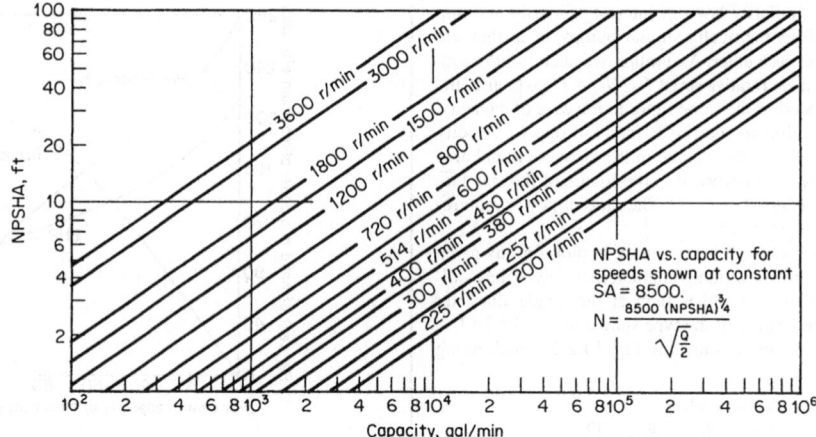

Fig. 14.2.26 Recommended maximum operating speeds for double-suction pumps. *(Hydraulic Institute Standards.)*

Recirculation A better understanding has been recently reached of the flow pattern in a pump impeller when the pump is operated at capacities below that at its best efficiency. All pumps exhibit the phenomenon of recirculation which is a flow reversal at the inlet or at the discharge tips of the impeller. The capacities at which flow recirculation occurs at the suction or discharge are not necessarily coincidental. It is now possible to predict the flow patterns that must exist to produce this flow reversal. Depending on the size and speed of the pump, the effects of recirculation can be very damaging not only to the operation but also to the life of the impeller and casing. The symptoms associated with recirculation and a method for calculating the capacity at which it takes place at both the suction and discharge of the impeller are given in Fraser, "Recirculation in Centrifugal Pumps," ASME Winter Annual Meeting, Nov. 16, 1981, Washington, D.C.

System Head Curve

Just as the head characteristic of a pump is represented by a curve indicating reducing head developed with increasing capacity, the head requirements of the system served by the pump can be depicted by a curve showing increased head requirements for increased flow, i.e., a **system-head curve** (Fig. 14.2.27). The capacity at which a pump will operate in this system is that at which the pump head curve intersects the system-head curve.

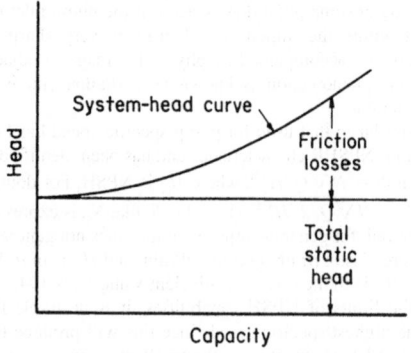

Fig. 14.2.27 System-head curve.

Every system-head curve consists of (1) a total static head (which may in some cases be equal to zero) plus (2) friction losses. The total static head $H_{stat} = H_d - H_s + P_d - P_s$, where H_d and H_s are the elevations at the system's termination and origin, respectively, and P_d and P_s are the gage pressures at the same points.

The **friction losses** are the sum of the line losses and fitting losses.

In complex pumping systems, such as municipal water-supply operations, both the total static-head and the friction-loss components may be variable, e.g., the former by changes in reservoir levels and the latter by the particular combination of lines being served at any given moment. This results in upper and lower limits of system-head requirements, and the intersection of these limiting curves with the pump head curve will define the capacity range over which the pump will be required to operate.

Since the capacity at which a pump will operate corresponds to the intersection of the system-head curve with the pump head-curve, any changes in pump capacity can only be obtained by varying one or other of these two curves (Fig. 14.2.28). Thus the capacity of a centrifugal pump operating in a system can be regulated by (1) changing the pump speed or (2) throttling in the discharge piping. The former method is preferable whenever the driver permits it, because throttling always involves an appreciable waste of power.

Fig. 14.2.28 Pump operation in a system.

Pump capacity should never be permitted to be reduced to zero because the fluid within the pump would have to absorb the entire power input and therefore would heat up rapidly with injurious effects on the pump. A bypass line should be provided from the pump discharge line to permit a predetermined amount of liquid to flow through the pump if the discharge valve is closed entirely. This bypass may be operated manually or automatically.

Centrifugal pumps may sometimes be operated in **parallel** or in **series.** To construct the head-capacity curve of two pumps in parallel, it is merely necessary to add the capacities of the individual pumps for various total heads. The head-capacity curve of two pumps in series is constructed by adding the individual pump heads for various capacities.

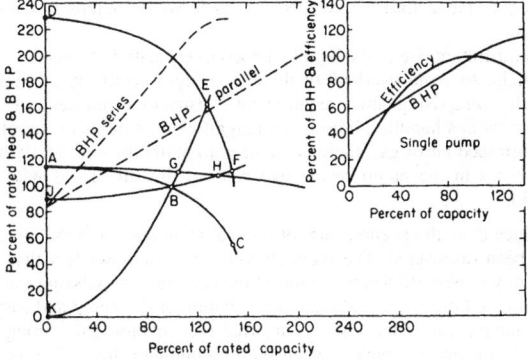

Fig. 14.2.29 Series and parallel operation of centrifugal pumps.

Figure 14.2.29 shows series and parallel operation of two centrifugal pumps with both flat and steep system-head curves.

Performance on Other Than Water

Selection of drivers, as well as pumps, is also influenced by **properties of the fluid** handled. The required pump power varies directly with the specific gravity of the liquid and is influenced in a more complex manner by viscosity. The effect of viscosity is illustrated in Fig. 14.2.30. For the example shown in the chart, the pump in question has a best-efficiency-point (bep) of 750 gal/min (170 m³/h) at 100-ft (30-m) head. When handling a fluid having a viscosity of 1,000 SSU (220 cS), its capacity is reduced to 95 percent of that of water for all capacities. Its head is reduced to 96, 94, 92, and 89 percent of its head on cold water at 60, 80, 100, and 120 percent of the bep capacity, and its efficiency is reduced to 63.5 percent of its efficiency on water for all capacities. Thus, if the pump had a bep efficiency on cold water of 70 percent, it would now deliver 712.5 gal/min (161.8 m³/h) at 92-ft (28.0-m) head with an efficiency of 44.5 percent at the bep.

MATERIALS OF CONSTRUCTION

Centrifugal pumps are available in many common metals and alloys, as well as porcelain, glass, and synthetics. A listing of materials recommended for various services can be found in the Standards of the Hydraulic Institute. Table 14.2.2 has some of the materials commonly used for pump parts.

INSTALLATION, OPERATION, MAINTENANCE

Proper installation, operation, and maintenance of centrifugal pumps will vary widely over the complete range of services to which the pumps may be applied, and satisfactory results in these areas can be fully achieved only by following the manufacturer's instructions for the size

Fig. 14.2.30 Performance-correction chart for effect of viscosity. *(Hydraulic Institute.)*

and type of unit involved. There are, however, certain general considerations that should be observed and will seldom need to be modified under any circumstances.

In general, the location selected for installation should be as close to the source of the fluid as possible, consistent with the requirement that adequate space be made available to provide accessibility for operation, inspection, and maintenance.

Shaft Alignment

Alignment of the driver shaft with the pump shaft is crucial for component longevity. Misaligned shafts will result in vibration and shorter life of bearings and seals. For a horizontal **separately coupled pump** (one in

Table 14.2.2 Materials for Various Fittings
(Materials for bearing housings, bearings, and other parts are not usually affected by the liquid handled)

Part	Standard fitting	All-iron fitting	All-bronze fitting
Casing	Cast iron	Cast iron	Bronze
Suction head	Cast iron	Cast iron	Bronze
Impeller	Bronze	Cast iron	Bronze
Impeller ring	Bronze	Cast iron or steel	Bronze
Casing ring	Bronze	Cast iron	Bronze
Diffuser	Cast iron or bronze	Cast iron	Bronze
Stage piece	Cast iron or bronze	Cast iron	Bronze
Shaft, with sleeve	Steel	Steel	Steel, bronze, or Monel
Without sleeve	Stainless steel or steel	Stainless steel or steel	Bronze or Monel
Shaft sleeve	Bronze	Steel or stainless steel	Bronze
Gland	Bronze	Cast iron	Bronze

which the pump and driver bolt independently to a baseplate), the coupling can be precisely aligned when the unit is idle, but become misaligned when the pump runs. Running misalignment can be caused by temperature changes in the driver, pump, and connected piping; by hydraulic forces produced by the piping; and by the torque between the driver and the pump. Piping should be supported so as to minimize piping loads on the pump, and any expansion joint should be installed so as not to transmit any force to the pump.

Supporting the Pump

To retain coupling alignment, a separately coupled unit must be supported by a rigid **baseplate** (and pedestals) and by a **foundation** of concrete (or an equivalent support).

A single-stage in-line pump is typically designed to be supported solely by the piping, like a valve. The driver mounts directly on the pump, and moves with the pump as the pump moves with the piping, thereby retaining shaft alignment. Such a unit does not require a baseplate or foundation, although the piping support must be adequate for the additional weight of the pump and driver.

Priming

A centrifugal pump is primed when the casing, all impellers, and the seal chambers are filled with the liquid to be pumped. Failure to evacuate all air, or other gases, from the pump can result in wear or seizure of wear rings and bushings, and accelerated wear of mechanical seals. If the suction supply is above atmospheric pressure, priming is accomplished by venting the pump at all high points in the casing and seal chambers. If the suction supply is below atmospheric pressure, priming must be accomplished by providing a vacuum at all the high points; by providing a foot valve, below the suction liquid level, and filling the pump and suction line with liquid; or by providing a priming chamber in the suction line.

Starting The normal starting sequence is as follows: (1) open valves in all auxiliary sealing, cooling, flushing, and bypass lines; (2) open suction valve; (3) close discharge valve for low-specific-speed pumps where no check valve is installed after the pump, or open discharge valve for high-specific-speed pumps or wherever a discharge check valve has been provided; (4) prime or vent as necessary; (5) energize the driver; and (6) open discharge valve if it was previously closed in step 3.

Following start-up and until stable operation has been adequately established, it is desirable to monitor bearing temperature, stuffing-box leakage, and other outward symptoms of the unit's behavior. Securing of the pump is accomplished by a reversal of the start-up sequence, encompassing steps 6, 5, 3, and 1, in that order.

Minimum Flow

Until the early 1970s there were only four factors to consider when setting an acceptable minimum flow for centrifugal pumps:

Higher radial thrust developed by single-volute pumps at reduced flows.

Temperature rise in the liquid pumped as capacity is reduced.

Desire to avoid overload of drivers on high-specific-speed pumps, which have a rising power requirement as capacity is reduced.

For pumps handling liquids with significant amounts of dissolved or entrained air or gas, there is the need to maintain sufficiently high velocities in the pump casing to wash out this air or gas with the liquid.

Since then, the phenomenon of internal recirculation in the impeller has been discovered. The recirculation occurs at flows less than the capacity at best efficiency. Problems include pressure pulsations at the suction and discharge and rapid deterioration of the pump and casing. The unfavorable effects of this internal recirculation have brought a fifth factor into the picture when setting minimum flows. This is discussed more fully above under "Recirculation."

Each of the effects mentioned above may dictate a different minimum operating capacity. Obviously, the final decision must be based on the greatest of the individual minimums. The internal recirculation usually sets the recommended minimum. Assuming that the pump is properly furnished with the necessary instrumentation, such as flowmeters, pressure gages with sufficient sensitivity to show pulsations, and vibration- and noise-monitoring equipment, an experienced test engineer should be able to pinpoint the onset of internal recirculation.

If the flow demand of the service in which the pump is installed is expected to fall below the minimum permissible flow, some means must be provided to accommodate the difference between the minimum permissible flow and that required by the service. This is accomplished by installing a bypass in the discharge line from the pump, located on the pump side of the check and gate valves and leading to some lower pressure point in the installation where excess heat absorbed through operation at low flows may be dissipated. A valve is located in this bypass line. For small installations, the bypass valve is operated as fully open or fully closed. For large installations, a modulating valve is usually used to bypass the difference between the required flow and the minimum permissible.

Unnecessary Maintenance

A generally accepted cardinal rule is that as long as operation continues normal, the unit should be left alone. Thus, except in special circumstances, periodic overhauls are not recommended. The amount and degree of maintenance likely to be required are influenced primarily by the nature of the service to which the pump is applied, and maintenance practices must therefore be determined largely by the user as a result of personal experience.

14.3 COMPRESSORS

by Heimir Fannar

REFERENCES: Chlumsky, "Reciprocating and Rotary Compressors," E&FN Spons Ltd., Loomis, "Compressed Air and Gas Data," 3rd ed., Ingersoll-Rand. Rollins, "Compressed Air and Gas Handbook," 5th ed., Compressed Air and Gas Institute. Sheppard, "Principles of Turbomachinery," Macmillan. Stepanoff, "Turboblowers," Wiley. Koppers Research and Engineering Staff, "Engineers Handbook of Piston Rings, Seal Rings, Mechanical Shaft Seals," 8th ed., Koppers Company, Inc. Lundberg and Glanvall, "A Comparison of SRM and Globoid Type Screw Compressors at Full Load," Proceedings of the 1978 Purdue Compressor Technology Conference, Purdue University, W. Lafayette, IN. Von Nimitz, "Pulsation and Vibration Control Requirements in the Design of Reciprocating Compressor and Pump Installations," Proceedings of the 1982 Purdue Compressor Technology Conference, Purdue University, W. Lafayette, IN. Mechanics, Simulation and Design of Compressor Valves, Gas Passages and Pulsation Mufflers, short course, July 12–14, 1992, Purdue Univeristy, West Lafayette, IN.

Notation

a = speed of sound
A = piston area
A_v = valve flow area, cell area
c_p = constant pressure specific heat
c_v = constant volume specific heat
C = clearance volume (decimal)
d = depth of ring section, diameter
D = diameter
e = eccentricity
E = modulus of elasticity (lb/in^2)
f = valve resistance, in velocity heads
g = ring free gap less end clearance (in)

g_c = gravitational constant
ghp = gas horsepower
h = enthalpy
H_p = polytropic head (ft · lb/lbm or kJ/kg)
k = ratio of specific heats
K = area constant (decimal)
L = length, leakage coefficient, connecting rod length
L_{eq} = equivalent length of pipe
m = mass
MW = molecular weight
n = polytropic coefficient
N_m = number of lobes on main rotor
n_s = number of stages
n_v = number of vanes
N = rotational speed
N_s = specific speed
P = pressure
Q = heat flow, volume flow at inlet conditions
r = compression ratio (P_2/P_1)
r_s = pressure ratio per stage
r_t = pressure ratio across entire compressor
r_δ = ratio of cylinder pressure at bottom dead center to inlet pressure
R = gas constant [for air 53.34 ft · lbf/(lbm · R)], crank radius
s = entropy, stress
sg = specific gravity relative to air
T = absolute temperature, thrust
T_v = vane thickness
U = piston velocity, peripheral velocity
V = volume
V_i = built-in volume ratio
$\overline{V}$ = velocity
W = work, mass flow
Z = compressibility factor

Greek

α = ratio of piston area over valve flow area
β = blade angle, degrees
λ = wave length
Δ = pressure loss, differential pressure (lb/in^2)
η_c = adiabatic efficiency
η_p = polytropic efficiency
η_v = volumetric efficiency
μ = pressure coefficient (gH_p/U^2)
ϕ = flow coefficient (Q/ND^3)
ω = angular speed (radians/second)

Subscripts

1 = inlet conditions
2 = discharge conditions
A = axial
R = radial

Compressed-Air and Gas Usage

Compressed-air is used for machine and tool operation, drilling, painting, soot blowing, pneumatic conveying, food processing, instrument operations, and in situ operations (e.g., underground combustion). Pressures range from 25 psig (172 kPa) to 60,000 psig (413,790 kPa). The largest usage is at 90 to 110 psig which is the normal plant air-pressure range.

Gas compressors are used for refrigeration, air conditioning, heating, pipeline conveying, natural gas gathering, catalytic cracking, polymerization, and in other chemical processes.

Standard Units and Conditions

In the ISO system, the standard unit of pressure for compressors is the kilopascal (kPa). In some countries this is the only unit which can appear on the compressor pressure gages by law. In Europe, the European Committee of Manufacturers of Compressors, Vacuum Pumps and Pneumatic Tools (PNEUROP) and, in the United States, the Compressed Air and Gas Institute (CAGI) prefer the bar as the standard unit of pressure. PNEUROP and CAGI have selected as standard conditions 1 bar (14.5 lb/in^2) (100 kPa), 20°C (68°F), and 0 percent relative humidity. The unit of flow in the ISO system is m^3/s. Other units still in common usage are m^3/h, m^3/min, and L/s. In the United States the most commonly used units are ft^3/min (cfm) and ft^3/h (cfh). Power is normally expressed in kilowatts (PNEUROP) and horsepower (CAGI).

Thermodynamics of Compression

Most compressors are analyzed using the ideal-gas law and assuming constant specific heat. Real-gas deviations are handled by applying a compressibility factor (also called *super compressibility*). The ideal-gas law will give satisfactory results for nonhydrocarbon gases for pressures to approximately 1,000 psig (6,900 kPa) at normal temperatures. Most hydrocarbon gases and refrigerants deviate strongly from the ideal-gas laws even at moderate pressures. For these cases, thermodynamics property tables, Mollier Charts, or compressibility charts should be used. Unfortunately these are not available for many gases of industrial importance. In this case a generalized compressibility chart is used. Compressibility charts for various gases can be found in Loomis, "Compressed Air and Gas Data," 3d ed., Ingersoll-Rand, and Rollins, "Compressed Air and Gas Handbook," 5th ed., Compressed Air and Gas Institute.

Adiabatic Analysis

The equation of state for an ideal gas is:

$$PV = mRT \qquad (14.3.1)$$

The first law of thermodynamics for a steady-flow process is (per unit mass):

$$Q = (h_2 - h_1) + \frac{\overline{V}_2^2 - \overline{V}_1^2}{2g_c} + W \qquad (14.3.2)$$

Neglecting the kinetic energy of the gas and assuming constant specific heat, for a reversible adiabatic process, Eq. (14.3.2) becomes:

$$W = c_p(T_1 - T_2') \qquad (14.3.3)$$

Substituting $(P_2/P_1)^{(k-1)/k}$ for T_2'/T_1 and $k/(k-1)$ for c_p/R, Eq. (14.3.3) becomes:

$$W = P_1V_1\frac{k}{1-k}\left[\left(\frac{P_2}{P_1}\right)^{(k-1)/k} - 1\right] \qquad (14.3.4)$$

The same results can be obtained by evaluating $W = \int V\,dp$ on an idealized P-V diagram (Fig. 14.3.1).

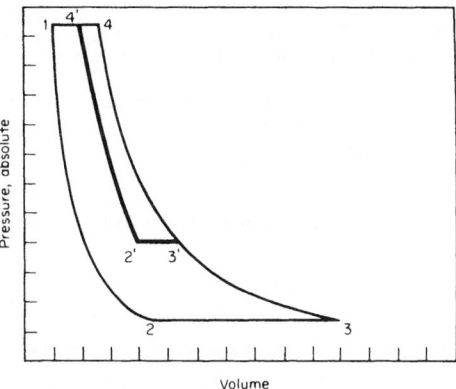

Fig. 14.3.1 Idealized indicator diagram for a positive-displacement compressor.

The isentropic efficiency (ratio of isentropic work to actual work) is given by:

$$\eta_c = T_1 \frac{(P_2/P_1)^{(k-1)/k} - 1}{T_2 - T_1} \qquad (14.3.5)$$

Polytropic Process

As an alternative to the adiabatic analysis, a polytropic process can be defined as

$$PV^n = C \qquad (14.3.6)$$

All the adiabatic equations apply if $(n - 1)/n$ is substituted for $(k - 1)/k$ where

$$\frac{n-1}{n} = \frac{k-1}{k} \frac{1}{\eta_p}$$

If the inlet and discharge temperatures and pressures are known, n can be calculated from:

$$n = \frac{1}{1 - \ln (T_2/T_1)/\ln (P_2/P_1)} \qquad (14.3.7)$$

where n is directly related to the heat transferred during the process by

$$n = \frac{C_p(T_2 - T_1) - Q}{C_v(T_2 - T_1) - Q} \qquad (14.3.8)$$

The reversible work done along the polytropic path is called the **polytropic work.**

The polytropic efficiency η_p (ratio of polytropic work to actual work) is given by:

$$\eta_p = \frac{\frac{k-1}{k} \ln (P_2/P_1)}{\ln (T_2/T_1)} \qquad (14.3.9)$$

The polytropic efficiency and the isentropic efficiency can be related by:

$$\eta_c = \frac{r^{(k-1)/k} - 1}{r^{(k-1)/(k\eta_p)} - 1} \qquad (14.3.10)$$

Equation (14.3.10) is strictly valid only for ideal gases with constant specific heats. However, because the real-gas errors occur in both the numerator and denominator, the equation gives reasonably accurate results for real gases which deviate moderately from ideal gases.

Positive-displacement compressors are almost universally analyzed using the adiabatic model. Dynamic compressors are analyzed using both the adiabatic and polytropic models with the trend toward the polytropic model. The advantage of the polytropic model is that the discharge temperature is immediately available and η_p is essentially constant for different gases. The polytropic efficiency is independent of the thermodynamic state of the gas. Also, the total polytropic head is the sum of the polytropic heads for the stages. This is not the case for isentropic heads. The advantage of the isentropic model is that the isentropic work can be read immediately from thermodynamic tables or charts. Comparing adiabatic efficiencies at different pressure ratios is not a valid procedure.

Real-Gas Effects

Deviations from the ideal-gas law are accounted for by introducing a compressibility factor

$$Z = \frac{PV}{RT} \qquad (14.3.11)$$

The isentropic work of compression for a real gas becomes:

$$W = P_1 V_1 \frac{k}{1-k} \left[\left(\frac{P_2}{P_1}\right)^{(k-1)/k} - 1 \right] \frac{Z_1 + Z_2}{2Z_1} \qquad (14.3.12)$$

where Z_1 and Z_2 are the compressibilities at conditions 1 and 2. Compressibility will also affect the volume flow of the compressor because of the reexpansion of the clearance volume gas.

Multistaging and Intercooling

Temperature rise and mechanical stresses limit the maximum pressure differential across a single stage of any compressor type. The pressure rise across a dynamic compressor stage is further limited by the available polytropic head which the stage can develop. The volumetric efficiency of a reciprocating compressor decreases with increasing pressure ratio placing a practical limit on the maximum ratio per stage.

Multistaging is used to overcome these limitations and to save power. With perfect intercooling and no pressure losses between stages, theoretically the minimum power is obtained when

$$r_s = \sqrt[n_s]{r_t} \qquad (14.3.13)$$

where n_s = the number of stages; r_s = the pressure ratio per stage; and r_t = the ratio across the compressor.

For every 10°F of imperfect intercooling, the horsepower increases approximately 1 percent. Perfect intercooling is achieved when the temperature of the gas leaving the intercooler equals the temperature of the gas at the compressor inlet. The maximum theoretical work that can be saved by perfect intercooling is:

$$W = \left(\frac{k}{1-k} P_1 V_1 \right) \left[\left(\frac{P_2}{P_1}\right)^{(k-1)/k} \right.$$
$$\left. - n_s \left(\frac{P_2}{P_1}\right)^{(k-1)/n_s k} + n_s - 1 \right] \qquad (14.3.14)$$

On the ideal indicator diagram (Fig. 14.3.1), this is area 3'4 4'2'.

In practice, perfect intercooling is seldom obtained. Normal practice is to cool the interstage gas to within 15°F to 20°F of the inlet gas. For water-cooled machines, the coldest water available should be used for intercooling.

Positive-Displacement Compressors Versus Dynamic Compressors

Positive-displacement compressors collect a fixed volume of gas within a chamber and compress it by reducing the chamber volume. The ideal work for such a process is given by Eq. (14.3.4). The work could also be obtained from the area enclosed by an ideal indicator diagram (Fig. 14.3.1). The actual horsepower, taking into account real-gas deviations, is

$$\text{hp} = \frac{144 \, P_1 V_1 k}{33,000(1 - k)} \left[\left(\frac{P_2}{P_1}\right)^{(k-1)/k} - 1 \right]$$
$$\times \frac{1}{\eta_c} \left(\frac{Z_1 + Z_2}{2Z_1} \right) \qquad (14.3.15)$$

with pressures in lbf/in² and volumes in ft³/min. With the volume rate V in m³/s and the pressures in pascals (dropping the 144/33,000), the power will be in watts.

For a given volume flow, the ideal power is affected by inlet pressure, the k value of the gas, and the compression ratio. The ideal power is not affected by the inlet temperature of the gas or its molecular weight.

The actual power is increased due to losses through the intake and discharge valves or ports as can be seen on the indicator diagram Fig. 14.3.2.

Additional losses are caused by turbulence within the compression chamber, preheating of the inlet gas, and leakage from the compression chamber. The isentropic compression efficiency (ratio of isentropic work to actual compression work) can be determined from the indicator diagram or calculated from Eq. (14.3.5) provided that there is no liquid injection. In addition to the thermodynamic losses, the mechanical losses must be added to the power requirement.

Fig. 14.3.2 Typical indicator diagram showing intake and discharge losses.

Volumetric efficiency is defined for positive-displacement compressors as

$$\eta_v = \frac{\text{actual delivery at intake conditions}}{\text{compressor displacement}}$$

For multistaging compressors, only the first-stage displacement is used.

Positive-displacement compressors are essentially constant-volume, variable-pressure machines as shown in Fig. 14.3.3. For fixed inlet and discharge pressures and varying speed, positive-displacement compressors are essentially constant-torque machines.

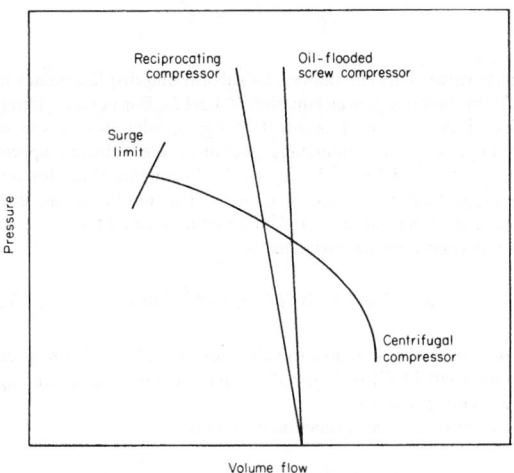

Fig. 14.3.3 Pressure capacity curves for different compressor types.

Dynamic compressors operate by transferring momentum to the gas via a high-speed rotor. The process is steady flow with the work given by Eq. (14.3.4). Head is defined as the energy per unit mass of the fluid. Head is imparted to the fluid through the change in magnitude and radius of the velocity components of the fluid as it passes through the rotor. The head developed by the rotor is

$$H = (U_1 V_{U1} - U_2 V_{U2})/g_c \qquad (14.3.16a)$$

where U_1 and U_2 are the linear velocity of the rotor at radii 1 and 2. V_{U1} and V_{U2} are the fluid tangential velocity at radii 1 and 2.

The polytropic head required for a given compression ratio is

$$H_p = \frac{Z_{ave} R T_1 [r^{(n-1)/n} - 1]}{(n-1)/n} \qquad (14.3.16b)$$

The gas horsepower on a CFM basis is

$$\text{ghp} = Q_1 P_1 [r^{(n-1)/n} - 1] \frac{n}{229\eta_p(n-1)} \left(\frac{Z_1 + Z_2}{2Z_1}\right) \qquad (14.3.17)$$

and on a mass flow basis:

$$\text{ghp} = \frac{WH_p}{33,000\eta_p} \qquad (14.3.18)$$

With H_p in J/kg and W in kg/s, the power in kilowatts is

$$\text{kW} = H_p \left(\frac{W}{\eta_p}\right) \qquad (14.3.19)$$

Mechanical losses must be added to the gas power to obtain the shaft power. Volumetric efficiency is not defined for dynamic compressors.

For a fixed impeller design and a fixed speed, the energy that is transferred to a unit mass of fluid as it passes through the impeller (i.e., the head) is constant. Pressure rise and power vary directly with inlet gas density independent of the cause of the density change. An increase in k will cause the pressure rise to decrease but will not affect the power.

A centrifugal compressor is essentially a constant-pressure, variable-capacity machine (see Fig. 14.3.3). An axial compressor is a constant-capacity, variable-pressure machine over a significant discharge pressure range. Dynamic compressors at least qualitatively follow the fan laws.

$$\frac{Q_1}{Q_2} = \frac{N_1}{N_2} \qquad \frac{H_{p1}}{H_{p2}} = \left(\frac{N_1}{N_2}\right)^2 \qquad \frac{\text{ghp}_1}{\text{ghp}_2} = \left(\frac{N_1}{N_2}\right)^3$$

The geometry of a dynamic compressor is dictated by the required flow, polytropic head, and speed. The specific speed is a dimensionless parameter used to select the proper geometry.

$$N_s = \frac{NQ^{(1/2)}}{H_p^{(3/4)}} \qquad (14.3.20)$$

For specific speeds from 400 to 900, radial impellers are used. Between 800 and 1,400, mixed-flow impellers are preferred. For specific speeds above 1,400, axial impellers are normally used.

Surging

All dynamic compressors have a minimum flow point called the **surge limit** below which the operation of the machine is unstable. The surge limit is a function of the compressor type, design pressure ratio, gas properties, inlet temperature, blade angle, and speed. Operation at or below the surge limit must be avoided. The existence of the surge limit makes it essential to know the demand curve for the particular installation involved.

Reciprocating Compressors

Reciprocating compressors as shown in Figs. 14.3.4 and 14.3.5 range in size from fractional cfm to 15,000 cfm (25,485 m³/h) with discharge pressures as high as 60,000 psig (413,790 kPa). The majority of

Fig. 14.3.4 Cross section of gas-compressor cylinder, showing water jackets, clearance pocket with pneumatic control, and suction-valve lift devices.

Fig. 14.3.5   Single-stage double-acting reciprocating compressor.

applications fall in the pressure range of 10 to 300 psig (690 to 2,069 kPa) and capacities less than 2,500 cfm (4,250 m³/h). Single-acting compressors (which compress gas on one side of the piston only) have their widest application below 50 hp (37 kW). Larger compressors are usually double-acting (both sides of the piston are used to compress gas). Most plant air systems operate at 100 to 150 psig. Oil-flooded screw compressors dominate this market. Above 200 psig, reciprocating compressors dominate.

Single-stage compressors of 100 hp (75 kW) operating at 125 psig were once common. Modern practice is to use multistaging with intercooling for pressures above 80 psig and sizes as low as 10 hp (7.5 kW). The intercooling saves power, and the reduced discharge temperatures improve safety and prolong compressor life.

The capacity of a reciprocating compressor is determined by the first-stage displacement multiplied by the volumetric efficiency.

$$\eta_v = (1 + C)(r_\delta)^{(1/k)} - \frac{C(r)^{(1/k)}}{(Z_2/Z_1)} - L \qquad (14.3.21)$$

where $r_\delta = P_s/P_1$; $r = P_2/P_1$; and $L =$ a leakage coefficient $= 0.05$ for lubricated compressors and 0.10 for nonlubricated. $P_s =$ the pressure that exists in the cylinder when the piston is at bottom dead center. C is the clearance volume as a decimal fraction and includes valve passage volumes up to the sealing elements (point P_2V_c in Fig. 14.3.2). Clearance volume for normal service is 5 to 15 percent. High-pressure machines have as small a clearance volume as practical whereas pipeline compressors may have 100 percent clearance. The indicator volumetric efficiency from Fig. 14.3.2 is V_s/V_d. In general, this will be greater than the value obtained by Eq. (14.3.21).

The compression efficiency varies between 0.85 and 0.95. The mechanical efficiency falls in the range 0.88 to 0.95. The overall efficiency is the product of the compression and mechanical efficiencies.

The discharge temperature can be calculated from Eq. (14.3.22):

$$T_2 = T_1\left[1 + \frac{(P_2/P_1)^{(k-1)/k} - 1}{\eta_c}\right] \qquad (14.3.22)$$

If the inlet pressure is varied while holding the discharge pressure constant, the horsepower of the compressor will pass through a

maximum value. This can be seen by substituting displacement times η_v for V_1 in the horsepower equation (14.3.12). For correct sizing of the driver, it is necessary to know the range of inlet pressures as well as discharge pressure. Interstage pressures are selected approximately by Eq. (14.3.13) with empirical adjustments made for interstage piping, intercooler, and valve losses. The loss of volume due to condensation in the intercoolers should be accounted for.

Valve losses can be approximated by:

$$\Delta = 1.26 \times 10^{-6}\alpha f U_{\text{ave}}^2 \, \text{sg}\left(\frac{P}{Z}\right) \text{lbs/in}^2 \qquad (14.3.23)$$

where $\alpha =$ ratio of piston area to valve flow area; $f =$ valve resistance in velocity heads ($\cong 4$); sg $=$ specific gravity relative to air; and $U_{\text{ave}} =$ average piston speed, ft/s.

The piston velocity as a function of time is:

$$U = R\omega\left(\sin \omega t + \frac{R}{2L}\sin 2\,\omega t\right) \qquad (14.3.24)$$

where $R =$ crank radius; $L =$ connecting rod length; and $\omega =$ crank speed (rad/s).

If there were no sealing elements in the valves, the velocity through the valves would be:

$$U_v = \left(\frac{A}{A_v}\right)U$$

This pulsating velocity is amplified by the valve sealing elements so that pulsating flow is inherent in reciprocating compressors. The full wavelength for a double-acting cylinder is $\lambda = 60\,a/2N$, where a is the speed of sound. For a single-acting cylinder, it is two times this value. Pipes with total equivalent lengths equal to multiples of $\lambda/4$ should be avoided.

For a double-acting cylinder, the major problems occur at $\lambda/4$ and $3\lambda/4$ although resonance can occur at the other multiples of $\lambda/4$. Equivalent length L_{eq} is given by

$$L_{\text{eq}} = \frac{\lambda}{360}\tan^{-1}\left(\frac{2\pi V}{S\lambda}\right) \qquad (14.3.25)$$

where V = equivalent cylinder and passage volume; and S = cross-sectional flow area of the pipe.

If it is not feasible to avoid the designated lengths, a volume bottle or an orifice can be installed in the piping system. If an orifice is used, it should be $1/2$ the pipe diameter and located at a node (one node occurs at the open end of the pipe). A volume bottle should be located as close to the cylinder as possible. Von Nimitz ("Pulsation and Vibration Control Requirements in the Design of Reciprocating Compressor and Pump Installations," Proceedings of the 1982 Purdue Compressor Technology Conference, Purdue University) recommends the following equations for sizing:

$$V_s = 4\text{PD}\left(\frac{kT_s}{\text{MW}}\right)^{1/4} \qquad V_d = \frac{V_s}{R^{1/k}} \qquad (14.3.26)$$

PD is the total double-acting displacement volume of all cylinders to be manifolded to the surge bottle (ft^3). Subscript s refers to suction, d to discharge. R is the compression ratio. Temperatures are in degrees Rankine.

Compressor Valves

The valve lift area is defined as the product of the sealing element lift and the sum of the valve seat peripheries. The flow coefficient can be determined by measuring the pressure drop through the valve in a wind tunnel. Once the flow coefficient is determined, the "equivalent," or effective, flow area can be derived. The flow coefficient equals the square of the lift area divided by the square of the equivalent area. Poppet valves (Fig. 14.3.6) are ideal for low-speed compressors used in low compression ratios, such as pipeline applications. Concentric ring valves (Fig. 14.3.7) are used in process, natural gas, and high-pressure applications, and can better tolerate certain types of contamination than some of the other valve designs. Plate valves (Fig. 14.3.8) are used in a wide variety of applications including gas gathering, compressed natural gas, pipeline, gas storage, process, and fuel booster applications.

Valve lift ranges from approximately 0.039 in (1.0 mm) to 0.157 in (4.0 mm) in plate valve and concentric ring valves, and approximately 0.098 in (2.5 mm) to 0.25 in (6.3 mm) in poppet valves.

Springs are used to close the sealing element against the valve seat in the last few degrees of crank angle, before the piston momentarily stops

Fig. 14.3.6 Poppet valve, illustrating seat, sealing elements and springs, and guard. *(Hoerbiger Corporation.)*

Fig. 14.3.7 Concentric ring valve, illustrating concentric rings and seat. *(Hoerbiger Corporation.)*

Fig. 14.3.8 Plate valve, illustrating guard, cushion plate and wafer spring, sealing element, and seat. *(Hoerbiger Corporation.)*

and reverses direction. This prevents backflow from slamming the valve closed, which can damage or destroy the sealing element. A single sealing element design, such as a plate valve, is the simplest to control.

Most compressor valve designs now use sealing elements made of nylon, peek, or other nonmetallic materials. Some designs, however, still use steel sealing elements. These are less common.

Adequate valve lift and spring load may be calculated on the basis of pressure, specific gravity of the gas, speed of the compressor, and mean velocity of the gas through the valve.

Piston Rings

Reciprocating compressors employ compression rings, oil rings, and rider rings. Compression rings are always present, but whether oil rings or rider rings are used will depend on the type of compressor and its service.

The standard compression ring is the one-piece square-cut or angle-cut ring although two-piece and segmental rings are sometimes used. The one-piece ring usually has a rectangular cross section, but occasionally the cylinder contact face is crowned to prevent edge loading and to improve lubrication. Cast-iron rings are still widely used but are being displaced in many applications by filled TFE rings. Carbon-filled TFE rings are preferred for process compressors. For air compressors, carbon, glass, and bronze fillers are used. TFE rings can be used from -450 to $500°F$ (-268 to $260°C$). Bronze-filled TFE is good for higher-temperature operations because of its higher thermal conductivity. Carbon-filled TFE is preferred for oil-free service. Many compressors that handle gases with corrosive components (e.g., H_2S) use piston rings made of cloth-reinforced thermoset laminates with molybdenum disulfide added to improve the lubricity. Polyamide, polyimide, and poly(amide-imide) rings work well with marginal or no lubrication.

A one-piece ring must be stretched over the piston to install it in the piston groove. The maximum stress occurs $180°$ from the joint and is given by Koppers Research and Engineering Staff ("Engineers Handbook of Piston Rings, Seal Rings, Mechanical Shaft Seals," 8th ed., Koppers Company, Inc.) as:

$$S_o = \frac{0.424E(8d - g)/d}{(D/d - 1)^2} \qquad \text{lb/in}^2 \qquad (14.3.27)$$

When closed to the cylinder diameter, the ring is stressed to supply the initial seating pressure. The maximum stress ($180°$ from the joint) is given by the Koppers "Engineers Handbook" as:

$$S_c = \frac{0.482E(g/d)}{(D/d - 1)^2} \qquad \text{lb/in}^2 \qquad (14.3.28)$$

Oil rings are used in lubricated compressors to scrape oil from the cylinder wall and to meter a small quantity of oil to the compression rings. The oil rings may be above or below the wrist pin. Operating conditions for oil rings are less severe than for compression rings.

Nonlubricant compressors use rider rings to prevent piston cylinder contact. To prevent pressure buildup in front of the rider rings, longitudinal grooves are cut into the rings. Usually two or more rider rings are required.

Below 300 lb/in² differential, two compression rings and one or two oil rings are normally used. From 300 to 600 lb/in² differential, three to four compression rings are used. From 600 to 1,500 lb/in² differential, four to five rings are used. Above 1,500 lb/in², six or more rings may be required.

Piston-Rod Packing

Double-acting compressors are driven through a crosshead (Fig. 14.3.5) and require a rod packing to seal the piston rod connecting the piston to the crosshead. The operating conditions for rod packings are more severe than those for piston rings. In a double-acting compressor, the piston rings seal against the pressure differential across the stage whereas the rod packing seals the differential between discharge pressure and atmospheric pressure.

Rod packings use the same materials as piston rings. Metallic packing can tolerate rod wear of 0.15 percent of the rod diameter, with less tolerance above 2,000 lb/in² (13,790 kPa). The rod should be hardened to Rockwell 40 C and ground to a 10-rms or less finish.

Nonlubricated Cylinders

NL-1 class does not permit any direct lubrication and tolerates the minute contamination introduced along the piston rod. This can be eliminated by an additional spacer and rod wiper, qualifying the cylinder as a class NL-2. The pistons for this service use various Teflon rings and riders. The life of carbon rings is reduced by low humidities. The cylinder should have a ground 10- to 16-rms finish and a minimum hardness of Rockwell 20 C.

Cylinder Lubrication

Lubricants serve to (1) prevent wear by providing a supporting film between the rubbing surfaces, (2) seal close clearances, (3) protect against corrosion, and (4) transmit heat of friction and minute wear particles away from points of contact. Industrial cylinders and packings require force-feed lubricators. Viscosity is the best index of suitability: general service at pressures below 500 lb/in² (3,448 kPa) requires 400 SSU (86 centistokes) at 100°F (38°C) (SAE 40); 700 SSU (151 centistokes) for 2,000 lb/in² (13,793 kPa); and 1,000 SSU (216 centistokes) (SAE 50) for 8,000 lb/in² (55,000 kPa) operations. These oils have very high viscosity at 40°F (4.5°C), which will result in poor lubrication. Good winterizing practice provides continuous circulation of warm jacket water and an immersed electric heater in the oil reservoir. Crankcase oils should include a foam inhibitor, sludge dispersant, and rust inhibitor. Phosphorous extreme-pressure agents are added to high-pressure lubricants to avoid wear and scuff damage. Castor-bean and rapeseed oils are additives resistant to solvent action of condensing hydrocarbons.

Cellulube, Houghto-safe, Pydraul AC, Anderol, and Fluorolubes are used to resist the hazard of exothermic reaction in air compression. The crankcase lubricant consumption for 10 plants over a score of years averaged 20,000 bhp/(h)(gal). Table 14.3.1 shows lubricant requirements for general-process and natural-gas operations.

Table 14.3.1   Lubrication Required for Compressor Cylinders

Range of cylinder diam, in	Film thickness, μm	10³ hp/(h) per gal	Pints per cylinder per day per 100 hp of load	Oil drops/min
24–36	0.40	13	1.5	13
15–23	0.45	19	1.0	9
10–14	0.55	24	0.8	7
7–9	0.60	32	0.6	5
4–7	1.20	24	0.8	7
3–5	1.40	27	0.7	6

NOTE: Based on 12,500 drops per pint of SAE 40 oil at 75°F with vacuum-type lubricator. Reduced drop count to roughly one-half for pressure-type lubricator. Example: 500 bhp, 20-in cylinder, requires 5 pint/day and feed of 45 drops/min. Each installation will vary on lubrication requirements. Manufacturers' recommendations on oil type and quantity should be followed.

General Lubricant Issues

The choice of lubricant is important, in particular in those compressor designs (screw, scroll, rolling piston, rotary vane, etc.) where the lubricant comes into direct contact with the working fluid. Some gases, such as heavy hydrocarbons and chlorinated refrigerants (HCFCs) are highly soluble in oil, and can significantly affect operational viscosity. They can also affect compressor capacity, as the gas comes out of solution in the lubricant during cyclic pressure fluctuations. Synthetic lubricants such as polyalkylene glycol (PAG), polyalfaolefins (PAO), and polyolesters (POE) are commonly used. Care must be taken to select oil such that its viscosity remains within limits after full saturation of gas in the oil.

Because of the adverse environmental effects of chlorinated refrigerants (HCFCs), chlorine-free hydrofluorocarbon (HCF) refrigerants (R410A etc.) have been made in order to reduce global warming and ozone depletion. The removal of the chlorine causes the refrigerant/lubricant mix to lose some lubricity, and consequently an increase in component wear can occur.

The elastomer material in sealing elements (O rings, lip seals, etc.) is commonly affected by both the operating gas and the lubricant. The results include shrinkage, hardening, or chemical breakdown, and can cause failure to seal if compatibility is not thoroughly checked.

Chemical compounds (additives) are added to lubricants in order to enhance their properties. These include anticorrosion agents, EP (extreme pressure) agents, oxidation inhibitors, stabilizers, and antiwear additives. Care should be taken to ensure that these chemicals do not attack materials such as bearings in the compressor.

The choices of lubricant, operating temperatures, viscosity range, and replacement intervals are usually clearly specified by the compressor manufacturer.

Compressor Accessories

Most state laws and safe practice require a relief valve ahead of the first stop valve in every positive-displacement compressor. It is set to release at 1.25 times normal discharge or at the maximum working pressure of the cylinder, whichever is lower. The relief-valve piping system sometimes includes a manual vent valve and/or a bypass valve to the suction to facilitate start-up and shutdown operations. Quick line-sizing equations are (1) line connection, $d/1.75$; (2) bypass, $d/4.5$; (3) vent, $d/6.3$; (4) relief-valve port, $d/9$.

Volume flow is controlled by variable-speed drivers; steam engines can operate at 20 percent of rated speed and gas engines at 60 percent, and electric-motor speed can be varied by means of eddy-current and hydraulic couplings and special wound rotors with rheostats, which are both costly and inefficient. Unloading can be applied by means of valve-lift and clearance pockets, as shown in Figs. 14.3.4 and 14.3.9. Utility air plants are throttled with suction unloaders, as shown in Fig. 14.3.10. Suction throttling and bypass controls are used in process operations.

Fig. 14.3.10   Air suction unloader. Piston E moves to closed position as line A pressure reaches design point.

Cylinder Cooling

Single-acting compressors, especially smaller ones, are usually air cooled by fins cast into the cylinder. Cooling air is drawn across the fins by a fan. Double-acting compressors usually have cylinders that are water jacketed.

Fig. 14.3.9   Cards from a compressor with combination clearance and suction-valve bypass control.

The principal purpose of a jacket-water system is to normalize cylinder-casting strains. It has some effect on compressor efficiency due to reduced preheating of the air. This effect is reduced as the compressor speed is increased. The heat rejection to the jacket water is approximated by:

$$H_j = 4(t_{ag} - t_{aw}) + 100 \text{ Btu/(bhp} \cdot \text{h)} \qquad (14.3.29)$$

where t_{ag} and t_{aw} represent the average gas and water temperature, respectively, usually 155 and 140°F, respectively, and a rejection of 160 Btu/(bhp · h). Where the temperature rise is between 2 and 5°F, the jacket-water requirement is $160 \text{ bhp}/(500 \times 3.2 \times t_w)$, or 0.1 gal/(min · bhp). The use of cold water in jacket-water systems causes condensation on the cylinder walls, washing of lubricant, and excessive ring wear.

Rotary-Vane Compressors

Figure 14.3.11 is a section through a typical rotary-vane compressor. The displacement can be calculated using the following formula.

$$V_{th} = 2eL(\pi D - T_v N_v)N \qquad (14.3.30)$$

Fig. 14.3.11   Section through a typical vane compressor with N_v vanes.

Rotary-vane compressors have a built-in volume ratio V_i. This means that they will compress the gas within the compression chamber to $P_1(V_i)^n$ before opening the compression chamber to the discharge port. P_1 is the inlet pressure, and V_i is the ratio of the volume of the compression chamber at the inlet cutoff point to the volume at the point just prior to the opening of the discharge port. If the operating-pressure ratio does not match the built-in ratio, overcompression or undercompression will take place as indicated in Fig. 14.3.12. For this case the ideal work should be calculated as:

$$W_{\text{ideal}} = \frac{P_1 V_1}{1 - n}[V_i^{n-1} - n] - P_2\frac{V_1}{V_i} \qquad (14.3.31)$$

Fig. 14.3.12   Idealized *PV* diagram for a compressor with a built-in volume ratio (also referred to as a *built-in pressure ratio*).

The ratio of this work over the ideal work required for a perfectly matched compressor can be read from Fig. 14.3.16. For example, the theoretical work for an air compressor with $V_i = 2$ operating at $P_2 = 0.8P_i$ theoretically will require 2.5 percent more work than it would if it were designed so that the built-in ratio matched the operating ratio.

However, normal practice is to base efficiencies on a power calculated as though the compressor built-in ratio matches the operating ratio.

Rotary-vane compressors may be dry, lubricated, or oil-flooded. Prior to 1960 two-stage, oil-flooded vane compressors up to 900 cfm (1,530 m³/h) compressing to 150 psig (10.3 bar) (1,034 kPa) were used in portable compressors. They are now usually limited to less then 100 cfm (170 m³/h) air compressors and to small refrigeration compressors and boosters.

Vane materials may be phenolic (micarta, polyamid-imides, polyimides, Ryton) or metal (aluminum-silicon alloy).

The major limitations of vane compressors are imposed by vane-tip friction, vane-bending stress, and vane-length limitations. Maximum vane tip speeds are approximately 65 ft/s (20 m/s). Normally 6 to 8 vanes are used although as many as 20 vanes have been used. The trade-off is between reduced pressure differential per cell and increased friction. The eccentricity ratio is normally $e = 0.07D$ for low pressure (25 psig) (172 kPa) and $e = 0.05D$ for higher pressures (50 psig) (345 kPa). For service over 50 psig, the compressors are usually multistaged.

Assuming an injected oil temperature of 140°F (60°C) and no intercooling, a two-stage vane air compressor operating at 100 psig will have an adiabatic efficiency (including mechanical losses) of 60 to 72 percent. The discharge temperature will be approximately 90 to 95°C or 200°F. Oil-flooded vane compressors normally use synthetic lubricants. The high shearing rates and vane tip temperatures rapidly oxidize petroleum based lubricants.

Rolling-Piston Compressors

Superficially, the rolling-piston compressor (Fig. 14.3.13) resembles the sliding-vane compressor. However, its kinematics and performance are quite different from those of the sliding-vane compressor. The rolling-piston compressor consists of an eccentric, a rolling piston, a vane, a cylinder, and a discharge valve. The motion of the rolling piston with respect to the eccentric and cylinder is controlled by the various friction forces action on the piston. Compression and suction occur

Fig. 14.3.13   Section through a typical rolling piston compressor.

simultaneously in different chambers. A cycle covers two revolutions of the eccentric. Rolling-piston compressors are usually used in refrigeration systems, typically 0.25 hp (180 W) to 0.75 hp (550 W). Typical values for volumetric and power efficiencies are 0.90 and 0.84 to 0.86, respectively. The eccentricity is normally about $0.09 \times D$. The piston diameter is approximately $0.81 \times D$ and the length-to-diameter ratio is approximately 0.5. The compressor displacement can be calculated from Eq. (14.3.30) with $N_v = 1$.

Rotary, Twin-Screw, Oil-Flooded Compressors

Figure 14.3.14 is a cross section of an oil-flooded, rotary-screw compressor. The principle of operation can be determined from Fig. 14.3.15. The displacement of a twin-screw compressor is

$$V_{th} = KD^3 \frac{L}{D} C_{th} N \qquad (14.3.32)$$

where $K = N_m(A_{mg} + A_{fg})D^2$; D = the main rotor diameter; L = the rotor length; N_m = the number of lobes on the main rotor; A_{mg} = the area of one groove on the main rotor; and A_{fg} = the area of one groove on the secondary rotor. C_{th} is an overlap constant (between 0.95 and 1.0 for most designs) which corrects for the overlapping of the filling and compressing processes.

Fig. 14.3.14 Cross section of a typical single-stage, oil-flooded, twin-screw compressor.

Fig. 14.3.15 Operating principle of a twin-screw compressor.

C_{th} depends on the wrap angle, the profile, and the number of lobes. Lobe combinations of (4 and 5), (4 and 6), (5 and 6), and (5 and 7) are used. The most common design uses the (4 and 6) combination. Typical values for C_{th} and K are given in Table 14.3.2. A complete cycle (filling

Table 14.3.2 Typical Constants for Different Lobe Combinations (300° Male Wrap)

N_m/N_f	3/4	4/5	4/6	5/6	5/7
C_{th}	0.954	0.973	0.969	0.984	0.980
K	0.530	0.439	0.507	0.454	0.468

and compressing) takes place over approximately 750° of main rotor rotation. The number of compression cycles per revolution of the main rotor equals the number of lobes on the male rotor. The discharge process for adjacent grooves overlaps by approximately 40°, giving an inherently smooth discharge process.

Rotary-screw compressors have a built-in volume ratio V_i (typically 4.4 for single-stage air compressors and 1.9 to 3.5 for refrigeration compressors). For an inlet pressure of P_1, the compressor will have a built-in pressure of $P_1 V_i^n$. The effect on the P-V diagram and on the theoretical power can be found from Fig. 14.3.12 and Fig. 14.3.16. In practice, the increased discharge port losses associated with higher built-in volume ratios (smaller port area) skew the results in Fig. 14.3.16. In practice, it is more efficient, at least in air compressors, to have the built-in ratio slightly lower than the theoretical ratio. Single-stage, twin-screw air compressors rarely have built-in pressure ratios greater than 9, and most are designed with ratios of 7 to 8. An air compressor designed for 100 psig can operate at 175 psig with a loss in adiabatic efficiency of approximately 5 and 4 percent loss of volumetric efficiency.

Fig. 14.3.16 The effect of the built-in volume ratio on theoretical work of compression for a compressor operating off the design point.

Adiabatic efficiency (including mechanical losses in the air end) vary from 70 percent for small sizes (80 mm) to 88 percent for large (300 mm) slow-running compressors. An average value for a midsize compressor is 0.78 to 0.75 based on an injection temperature of inlet temperature plus 50°F. Accuracy in manufacturing the rotors is critical to achieving good efficiency. The heat absorbed by the cooling oil can be estimated by:

$$Q_h = \text{power} \left[1 - \frac{\eta_c(T_2 - T_1)}{T_1(r^{(k-1)/k} - 1)} \right] \qquad (14.3.33)$$

where power = the shaft power to the compressor. T_2 is normally 70 to 90°F over T_1.

For most applications, tip speeds range from 15 to over 60 m/s, with rotor sizes from 40 to 600 mm, and L/D ratio from 1 to 2.2. Oil-flooded screw compressors dominate the portable-compressor market and the low-pressure plant air market, while reciprocating compressors dominate the higher-pressure applications. In industrial refrigeration the screw compressor finds its widest application in the 300 to 1,000 kW range.

In heat-pump applications, the screw compressor's zero-clearance volume is an advantage over reciprocating compressors. Reciprocating compressors lose volumetric efficiency as the outside temperature drops (the compression ratio increases) requiring supplemental heat to compensate for the lost capacity. Refrigeration screw compressors are now available with a secondary suction port known as an **economizer connection.** This allows for two-temperature-level operation. At full-load operation, the system COP is improved significantly. As the compressor is unloaded, the improvement is lost.

Capacity can be controlled by on-line/off-line operation, inlet throttling, slide-valve or turn-valve operation, or speed regulation. Stationary air-compressors are normally controlled by on-line/off-line operation or inlet throttling with unload below 50 percent capacity. The air-oil sump is usually blown down to atmospheric pressure at unload.

Portable compressors are normally controlled by a combination of inlet throttling and speed regulation; the air-oil sumps are not blown down.

Refrigeration compressors are controlled by inlet throttling or more often by a slide valve or turn valve. A compressor with a slide valve is shown in Fig. 14.3.17. The effect on power consumption for each type of capacity control can be found from Fig. 14.3.18.

Fig. 14.3.17 Principle of slide-valve operation. As the valve is moved toward the discharge end of the compressor, the point of inlet cutoff is delayed.

Fig. 14.3.18 Comparison of different capacity control methods for oil-flooded screw compressors operating at constant speed.

The discharge temperature for an air compressor must be kept high enough to prevent condensation in the lubricant during part-load operation. This is usually accomplished by a thermal valve set to maintain the minimum injection temperature at 140 to 165°F (60 to 76°C). Standard oil separators limit oil carryover to 5 ppm or less. Optional treatment can reduce the levels further.

The capacity loss of an oil-flooded, twin-screw compressor with altitude is quite small compared to a normal reciprocating compressor. The approximate capacity correction for altitude can be read from Fig. 14.3.19.

Fig. 14.3.19 Altitude correction for capacity, single-stage, oil-flooded screw compressor.

Rotary Single-Screw Compressors

The operating principle of the single-screw compressor can be determined from Fig. 14.3.20. Lundberg and Glanvall give the displacement as:

$$\beta = \arcsin \frac{B - D/2}{R}$$

$$\Omega = \pi - 2\beta$$

$$C = 2R \sin \frac{\Omega}{2}$$

$$\gamma = \frac{\alpha \Omega}{\xi}$$

$$t = R - \frac{B - D/2}{\sin (\beta + \gamma)}$$

$$u = B - \frac{D}{2} + \frac{t}{2} \sin (\beta + \gamma)$$

$$V = \sum_{\alpha = \alpha 1}^{\xi} AU \, \Delta\alpha$$

The rotary single-screw compressor is an oil-flooded compressor with a built-in volume ratio. The compressions for each star wheel are almost in phase so the gross effect as far as the discharge piping is that there are six discharges per revolution of the main rotor.

The single-screw compressor is used in the refrigeration industry where the low bearing load on the main rotor is an advantage. It is also being used as an air compressor in the smaller sizes.

Adiabatic efficiency of a single-screw compressor will be 2 to 5 percent lower than that of a comparable twin-screw compressor. The effect of off design operation on theoretical power can be found from Fig. 14.3.16. Capacity control is identical to the twin-screw compressor.

Dry Rotary Twin-Screw Compressors

A dry rotary, twin-screw compressor has timing gears to separate the rotors which may or may not be coated. The typical compressor is water-jacketed. Seals prevent gas leakage along the shafts and seal the oil to the bearings and timing gears. Sleeve bearings are most common, but some units have antifriction bearings. The displacement is the same as for the oil-flooded screw. Tip speeds vary from 80 to 140 m/s. Sizes range from approximately 400 to 20,000 cfm, and in special applications as high as 60,000 ft³/min. As a result of practical limitations, the discharge temperature is limited to 400°F, which determines the maximum permitted pressure ratio across each stage. Two-stage compressors are almost always intercooled. With intercooling to 15°F they approach

Fig. 14.3.20   Principle of operation and geometry of a single-screw compressor. *(Lundberg and Glanvall, "A Comparison of SRM and Globoid Type Screw Compressors at Full Load," Proceedings of the 1978 Purdue Compressor Technology Conference, Purdue University, W. Lafayette, IN.)*

adiabatic efficiency (based on zero intercooling) including mechanical-loss ranges from 72 to 80 percent. While oil-flooded screw compressors can be controlled to very low capacities by inlet throttling, dry screw compressors cannot. The resulting pressure ratio causes excessive temperature increase across the compressor.

Orbiting Scroll Compressors

This is a positive-displacement compressor, which is common in air conditioning and heat pump applications, primarily in the 5 to 50 hp (3 to 37 kW) range.

The compression is achieved by two identical scrolls (typically involutes of circles phased 180° apart), whereby one of the scrolls is stationary, and the other orbits without rotating. The filling, compression, and discharging process usually takes 900 to 1100° of shaft rotation, depending on the number of wraps. The orbiting scroll is driven by the crankshaft and prevented from rotating by an Oldham ring. A typical scroll is shown in Fig. 14.3.21.

Fig. 14.3.21   View of a typical scroll element.

The movement of the orbiting scroll causes pockets between flanks of the scrolls to get progressively smaller, until the volume reaches its internal limit (internal volume ratio, see "Screw Compressors"), finishing the process by discharging the gas. Different stages of the compression process are illustrated in Fig. 14.3.22.

Fig. 14.3.22 Scroll compression stages (moving scroll orbits clockwise).

The flow rate of a scroll compressor given by the equation

$$Q = 2A_{fv}H_vN\eta_v$$

where A_{fv} is the area of one of a full-volume cell (Fig. 14.3.22) and H_v is the height of the scroll vanes.

In the noncompliant version, both scrolls are held rigidly, with a small clearance between the flanks and tips. This design is susceptible to damage from particle throughput and hydraulic locking. The compliant version overcomes these issues by "floating" the scrolls into contact with each other. Axial contact is achieved by back pressure on the stationary scroll, while contact on the flanks is achieved by centrifugal force. This allows the scrolls to yield to the excessive forces without causing damage. There is, however, a potential for increased scroll wear, in particular if used with some of the latest chlorine-free refrigerants, which reduce the lubricity of the oil.

The internal leakage across the tip of the scroll vanes and along the vane flanks affects the compressor performance, more so in the noncompliant case. The leakage across the vane tips is significantly greater and damaging to performance than that along the flanks. For this reason, a groove is usually machined into the tip of the scroll vanes, and a sealing strip inserted.

Dynamic Compressors

A typical centrifugal compressor is shown in Fig. 14.3.23. The optimum range of operation can be found from Fig. 14.3.24. An estimate of the polytropic efficiency can be obtained from Fig. 14.3.25 or calculated from

$$\eta_p = 0.014 \ln(Q) + 0.600 \qquad (14.3.34)$$

where Q is in cubic feet per minute.

Fig. 14.3.23 Single-stage, cantilever-design centrifugal compressor.

Fig. 14.3.24 Range of speed versus inlet capacity for near-optimum dynamic compressor efficiency. *(Loomis, "Compressed Air and Gas Data," 3rd ed., Ingersoll-Rand, pp. 4–28.)*

Fig. 14.3.25 Approximate polytropic efficiency versus compressor inlet capacity. *(Rollins, "Compressed Air and Gas Handbook," 4th ed., Compressed Air and Gas Institute, pp. 3–71.)*

For normal industrial compressors, $H_p = 10,000$ ft · lb/lb (30 kJ/kg) per stage although much higher values can be obtained.

The required speed can be estimated from:

$$N = \frac{(g_cH_p/\mu)^{0.5}}{\pi D} \qquad (14.3.35)$$

The pressure coefficient ($\mu = g_cH_p/U^2$) varies from 0.4 to 0.7 (in the high-efficiency range). In the absence of specific data, use 0.55 for estimating.

Fig. 14.3.26 Basic characteristics of centrifugal compressor, showing effects of backward-leaning impeller blades. *(After Balje, ASME 51-F-12.)*

The diameter is determined by the required flow and can be estimated from the flow coefficient $\varphi = Q/ND^3 \propto V/U$, where V is the axial inlet velocity. The normal range of φ is 0.15 to 0.4 with the optimum at 0.3.

The slope of the μ-φ curve (and the $H_p \cdot Q$ curve) is strongly dependent on the backward lean of the impeller blades as shown in Fig. 14.3.26. $\beta_2 = 90°$ is a radial blade. Normal closed industrial impeller blades will have an angle of 65 to 55°. It should be noted that an increase in density of the gas will flatten the slope.

A typical axial compressor is shown in Fig. 14.3.27. The range of application can be found from Fig. 14.3.18. Axial compressors are more efficient than centrifugals (as much as 5 percent) but have a much narrower range of operation. The surge limit will vary from 85 to 95 percent of design flow. For a given stage, the polytropic head will be about one-half that developed by a centrifugal stage, i.e., $\mu = 0.3$. Axial compressors are used only for air or clean gas. They find their major

application in refineries, butadiene plants, and ethylene oxide plants. Jet-engine compressors are axial-flow compressors.

Thrust Pressures

A double-inlet and an open impeller have no axial thrust, whereas a semienclosed impeller, with a frontal shroud, can impose substantial thrust, evaluated by:

$$T_A = Y(A_1 - A_2)\Delta\,(\text{lb/in}^2)$$
$$+ 0.8(A_2 - A_3)\Delta\,(\text{lb/in}^2) - P_1 A_3$$

where A_1 = full impeller area; A_2 = impeller area less the area of the inlet-eye seal; A_3 = shaft area through the seal, all in in²; and Δ (lb/in²) = pressure differential per stage. Y is the percentage of Δ (lb/in²) acting in back of a single-stage disk: (1) Without a frontal shroud and a plain back, $Y = 0.35$; (2) with 0.060-in back ribs, $Y = 0.28$; (3) with $1/4$-in equalizer holes at half radius and about 2 in apart, $Y = 0.22$; and (4) with deep scallops, $Y = 0.18$. For fully shrouded impellers, as used in multiple stage, $Y = 1.00$. These thrusts are balanced by opposing impeller inlet ends or, more generally, by a balancing drum wherein an equal and opposite thrust is created, equal to the product of the compressor Δ (lb/in²) and the area of the drum. Stepanoff ("Turboblowers," Wiley) shows that radial impellers create a radial thrust, caused by the uneven volute-pressure distribution, which is:

$$T_R = 0.36\,\Delta\,(\text{lb/in}^2)\,db$$

where d = the impeller diameter; and b = the width, including the shrouds, both in inches.

Applicable Standards

Many national and international standards are available for guidance regarding construction, testing, packaging, and reliability. These include API, ANSI, ASME, CAGI, SAE, and international standards. Examples of coverage include construction (API 616, API 618, and API 619), testing (ASME PTC 9), and safety and packaging (API 670, ASME 19.3).

Compressor manufacturers commonly make a list of exceptions to stipulations in the various standards, or offer these as options at an additional cost. Therefore, it is important—if guidelines are to be drawn from specific standards—that the customer and manufacturer agree on compliance or exceptions to all relevant sections of the applicable standards, including construction, testing, test interpretation, and packaging features.

Fig. 14.3.27 Multistage axial compressor.

14.4 HIGH-VACUUM PUMPS

By Marsbed Hablanian

REFERENCES: Van Atta, "Vacuum Science and Engineering," McGraw-Hill. O'Hanlon, "A User's Guide to Vacuum Technology," Wiley. Weissler and Carlson (eds.), "Vacuum Physics and Technology," Academic Press. Roth, "Vacuum Technology," North-Holland. Hoffman et al., "Handbook of Vacuum Science and Technology," Academic Press. Lafferty (ed.), "Foundations of Vacuum Science and Technology," Wiley. Hablanian, "High-Vacuum Technology," Marcel Dekker. Weston, "Ultrahigh Vacuum Practice," Butterworths. Berman, "Vacuum Engineering Calculations, Formulas, and Solved Exercises," Academic Press. Redhead et al., "The Physical Basis of Ultrahigh Vacuum," Chapman & Hall. Welch, "Capture Pumping Technology," Pergamon. Wutz et al., "Theory and Practice of Vacuum Technique," Vieweg & Sons/Wiley.

INTRODUCTION

High-vacuum pumps can be broadly divided into two categories: transfer pumps and capture pumps. In regard to transfer pumps, i.e., pumps that transfer the gas from a reservoir into the atmosphere, it can be said that the world *pump* is a historical misnomer. Such pumps could easily be called *rarefied gas compressors*. After all, to turn a device from a compressor into a vacuum pump, it is only necessary to place the reservoir at the inlet instead of at the discharge. However, because high-vacuum systems deal with small pressure differences (usually 1 atmosphere), enormous compression ratios, and very small mass flows, specialized designs are necessary. Because of very low mass flows involved at the very low pressures (or densities), it is possible to employ devices that do not exhaust the pumped gas but keep it out of the gas phase inside the pump.

Vacuum can be produced by a few basic methods: by mechanical displacement using pistons, vanes, lobes, and so on; by fluid ejectors and turbomachinery devices; by chemical reactions leaving solid residue; by ionizing the gas and driving the ions into a solid target with high electric fields; and by freezing the gas at cryogenic temperatures. In addition, high-vacuum engineering deals with gas flows ranging from the "normal" continuum conditions to the region of molecular flow. In the latter range, the gas molecules collide mainly with the walls of the vacuum vessel or conduits rather than with each other. This produces high compression ratios across short sections of conductors (pipes, elbows, etc.), but at extremely small pressure differences. In molecular flow conditions, there is no genuine pressure gradient inside the pipe, and molecules can flow as easily upstream as downstream. In such conditions it is often more useful to think in terms of densities rather than pressures.

Before proceeding to the description of the most common devices for producing high vacuum, it may be useful to note that high vacuum engineers, as a result of an independent historical development, use concepts and terminology differing from those associated with compressors. For example, *throughput*, a product of volumetric flow and pressure, is used instead of *mass flow*, because the gas is assumed to be always at room temperature. Instead of *volumetric flow*, vacuum engineers use *pumping speed*, which is associated with the rate of pressure reduction in a reservoir during evacuation. Also, when displaying the performance of the pump, they plot pumping speed versus inlet pressure (which is the dependent variable), rather than vice versa as is the custom in other mechanical engineering fields. Instead of speaking about resistance of lines, pipes, conduits, etc., they use the reciprocal concept of conductance.

The usual pascal unit for pressure can be used for designating the level of high vacuum, but millibar and torr are also used. Currently, in United States, the use of torr (equivalent of old mm Hg) is more common, while in Europe, mbar (1/1,000 of an atmosphere) is more frequent. One torr is equal to 133.3 Pa or 1.333 mbar. The pumping speed is measured in litres per second (L/s) and conductance in torr · L/s · torr (torr · L/s of throughput per torr of pressure difference), which, abbreviated, is also L/s (1 L/s = 2.12 ft³/s).

Pumping speed S which primarily involves the size of the pump, is defined by the following equation:

$$t = V/S(\ln P_1/P_2) \qquad \text{or} \qquad t = V/S(2.3 \log P_1/P_2)$$

where t is the time period of evacuation, V is the volume of the vacuum chamber, P_1 the initial pressure in the chamber, and P_2 the reduced pressure (see Fig. 14.4.1). This relationship works well in typical systems when leaks and gas evolution are neglected. In the presence of a leak from atmosphere into the chamber, the equation should be modified into:

$$t = V/S[\ln(P_1 - P_f)/(P_2 - P_f)]$$

Fig. 14.4.1 Schematic view of a vacuum chamber with three pressure gages and a pump.

where P_f is the final pressure due to the leak. In practice, vacuum pumps are never used at zero flow condition, because there always remains some residual gas evolution from internal surfaces exposed to high vacuum. With that in mind, the "ultimate" pressure achieved in a vacuum system after a long period of pumping can be written as follows:

$$P = \sum(Q_i/S_i)_{\text{ext}} + \sum(Q_i/S_i)_{\text{int}} + \sum(P_{2i}/k_i)$$

The first term is the summation of contributions of gas evolutions from external sources, where Q_i and S_i are throughputs and pumping speeds for various gases. The second term is the summation of corresponding gas evolutions in the inlet areas of the pump itself, P_{2i} represents corresponding discharge pressures, k_i is the limiting compression ratio for each gas, and the subscript i means "for each gas."

PUMP SELECTION

There are a few common ways vacuum systems are used: rapid, repetitive evacuation to a given pressure for a process of short duration; long term processing of some material in a vacuum environment in the presence of substantial continuous or slowly diminishing gas evolution; or evacuation and complete isolation of a vacuum chamber or a container from the outside world. Each of these requires a different pumping system in regard to the type and size of pumps used. Figure 14.4.2 is a graph that can be used for the approximate evaluation of time required to reach a certain pressure depending on the volume of the vacuum chamber and the pumping speed of the pump. Because of the great range of variation of pressure (or density) during an evacuation of a container to high vacuum levels, it is usually necessary to pump initially with coarse vacuum pumps in order to remove the bulk of the gas. Then, when the amount of mass flow (throughput) is reduced to a tolerable level, pumps designed for pumping at high-vacuum conditions are engaged. This is reflected in Fig. 14.4.2 with the label "High Vacuum Valve Opens."

At the initial conditions in continuum flow, transfer-type pumps tend to have the same pumping speed for all gases, but when the molecular flow is approached, the rate of pumping begins to be different for various gases. This is especially true of capture-type pumps. Many technical articles and textbooks often omit this word of caution. In addition to the

Fig. 14.4.2 Evacuation process, from atmosphere to the high-vacuum region, shown on a log-log graph. Letters A, B, and C correspond to sand-blasted mild steel, 100-grit-finish stainless steel, and prebaked stainless steel chamber walls. The point called "high vacuum chamber opens" indicates the start of high-vacuum engagement. (*Reprinted from Hablanian, "High-Vacuum Technology," p. 133, Marcel Decker.*)

differences in pump performance, conductance values of conduits are also different for different gases. In common vacuum systems, engineering experience can be used for design with reasonable accuracy, but in unusual installations, laborious estimations must be made to evaluate the expected combined performance of pumps and systems.

The most significant performance aspects of pumps are: pumping speed (related to volumetric flow), maximum steady-state throughput

Fig. 14.4.3 General basic limitations of pump (or compressor) performance.

Fig. 14.4.4 Traditional way of presenting performance of a high-vacuum/rough-vacuum pump combination. Q_{max} is the highest permissible steady-state gas flow for the high-vacuum pump.

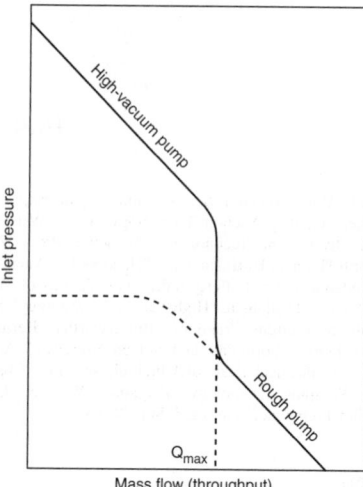

Fig. 14.4.5 Typical performance of a high vacuum pump (see Fig. 14.4.4), where the achieved vacuum level is plotted versus mass flow (inlet pressure increasing downward).

capacity (related to mass flow and power), and the ultimate pressure capability (related to limiting compression ratios and system gas evolution rates, called *outgassing*). This is indicated in Fig 14.4.3. A typical performance curve for a transfer-type high-vacuum pumping system is shown in Fig. 14.4.4 in the traditional mode, and the same data are shown in Fig. 14.4.5, where the degree of achieved vacuum is shown on the vertical axis versus the imposed mass flow (throughput).

COARSE-VACUUM PUMPS

For the last 100 years, the common pump for producing high vacuum, beginning at atmosphere, has been the oil-sealed rotary-vane, positive-displacement-type pump shown in Fig. 14.4.6. To achieve high compression ratios, as high as 100,000 to 1 in a single stage, oil is introduced continuously into the pumping space and exhausted with the pumped gas. Oil is used for several functions: for lubrication, for flooding the exhaust space under the exit valve (so-called dead space), to seal the tight spaces between the rotor and stator, to serve as a heat transfer medium for cooling the rotor, to flush particular matter out of the pump, and sometimes to actuate auxiliary valves. A two-stage pump, such as

Fig. 14.4.6 (*a*) Two-stage-pump flow diagram. (*b*) Cross section of a two-stage, oil-sealed, rotary-vane rough-vacuum pump. (*Courtesy of Alcatel Vacuum Technology.*)

the one shown in Fig. 14.4.6, when attached to a small reservoir, can produce inlet pressures below 1×10^{-3} torr while exhausting to atmosphere. However, it should be noted that the pumping action itself does not cease at that point for gases that are not present in high quantities in atmosphere. For example, helium from the vacuum chamber is pumped by such pumps at partial pressures of at least 10^{-10} torr.

Pumps of that type are not made in very large sizes (they become expensive, noisy, and power consuming). To achieve lower pressures, such pumps are often used in series with pumps of high pumping speed such as Roots blowers, screw pumps, and other cam-type pumps. In many applications, the presence of oil films at the inlet side of the pump is objectionable, because at molecular flow conditions, oil molecules can travel into the vacuum system and produce contamination. To prevent this, "oil-free" vacuum pumps have been developed, which do not include deliberate introduction of oil into the vacuum space. This is usually achieved by using several stages in series. Typically, it takes three or four stages in series to produce performance similar to that of a single-stage oil-sealed pump. For smaller "dry" pumps, multistaged scroll and diaphragm pumps are also used.

TURBOMOLECULAR PUMPS

There are two types of pumps in this category: so-called drag pumps and turbomolecular pumps. **Drag pumps** function by adding momentum to "stationary" gas molecules during a collision with the fast-moving solid surface of the rotor, thereby moving the molecules toward the discharge side of the pump. Three early versions of drag pumps are shown in Fig. 14.4.7. There is no basic difference between the three types; they differ only in the general geometric arrangement. All three concepts are utilized in modern pumps in a variety of combinations. **Turbomolecular pumps** are essentially axial-flow compressors designed for pumping rarefied gases, typically intended for molecular flow conditions at the inlet to the pump. The basic structure in shown in Fig. 14.4.8, and a typical

Fig. 14.4.9 Cross section of a typical axial-flow pump (with 13 bladed rotors).

pump construction in shown in Fig. 14.4.9. The rotors of such pumps are usually made of strong aluminum alloys to reduce the rotational moment of inertia. The peripheral velocities of the rotor are usually in the 200 to 300 m/s range. Usually, the rotor is supported by ceramic ball bearings lubricated by special low-vapor-pressure grease. To avoid even a suspicion of possible contamination, magnetic bearings are often used, with auxiliary dry bearings engaged only during starting and stopping periods.

For the drag pump, the pumping speed of a stationary channel placed in close proximity to the moving surface of the rotor can be estimated by comparing the forward flow, due to the velocity of the rotor, to the backflow, resulting from the developing pressure difference. Neglecting leakages, the maximum compression ratio K obtained a channel of certain size and length is derived to be $K_{max} = e^{S_o/C}$, where S_o is pumping speed and C is conductance. In molecular flow, this indicates the dependency of compression ratios on the conductance, and thus on the molecular weight of the gas. The obtained compression ratios are much smaller for lighter gases. In practice, such drag channels work best for small flows, a few litres per second. In viscous flow conditions, the pressure differences obtained in small pumps are in the range of a few torr per circular loop (which may be called a "stage").

The mechanism of pumping of the turbomolecular pumps is not completely different from that of axial-flow compressors, but in molecular flow, because of the reduced amount of backflow, the compression ratios obtained in high vacuum are much higher than those customary in compressor technology (as high as 10:1 per rotor-stator pair). The volumetric displacement of the bladed rotor can be estimated from the velocity in the pumping direction, times the projected pumping area, times a capture coefficient (ratio of pumped molecules to reflected ones). For pumping speed estimates of typical turbomolecular pumps, with first-row blade angles at 45°, the capture coefficient is near 0.6 for air or nitrogen.

Because of the high compression ratios obtained in high-vacuum pumps, the volumetric flow at the discharge side of the pump is rather low. Also, the flow tends to get into the viscous range. Therefore, it makes sense to combine the characteristics of drag pumps and turbine-type pumps into a hybrid design. An example of such a pump is shown in Fig. 14.4.10. Figure 14.4.11 shows the comparative performances of a conventional turbomolecular pump for helium (left side) and a hybrid pump (right side). A significant improvement is obtained without an increase of pump size or motor power.

Fig. 14.4.7 Three early versions of molecular drag high-vacuum pumps.

Fig. 14.4.8 Partial cross section of an axial-flow pump. Two rotors and one stator are shown.

Fig. 14.4.10 A comparison of an axial flow pump and a hybrid pump. On the left side of the vertical axis there are 11 bladed rotors (interspersed with 10 bladed stators). On the right side is the hybrid pump with 7 bladed rotors and 4 solid discs rotating in 4 drag channels connected in series. (Varian V-60.)

Fig. 14.4.11 Compression ratios for two pump types (Fig. 14.4.10), for helium, depending on the pressure of helium at the exit. Upper curve, hybrid pump (Varian V-250); bottom curve, all-axial pump (V-200.)

Turbomolecular pumps are usually manufactured in sizes from 50 to 2,000 L/s, but they can be made larger.

VAPOR JET PUMPS (DIFFUSION PUMPS)

Vapor jet pumps (often called *diffusion pumps*) have been the most commonly used pumps for production of high vacuum throughout the major part of the twentieth century and into the twenty-first century. The principle of operation of vapor jet pumps is illustrated in Fig. 14.4.12. The pumping action is due to the momentum transfer from a high-velocity

Fig. 14.4.12 Pumping effect of a fluid jet. Solid lines are for the motive fluid, dashed lines for pumped gas; P_0, P_1, are initial pressures and P_f is final pressure.

fluid jet to ambient gas molecules. The device is basically similar to steam ejectors but is highly specialized to function at molecular flow conditions at its inlet port. Figure 14.4.13 is a schematic drawing of a modern pump that has three downward-directed nozzles and one horizontal one, called an *ejector,* all connected in series. The motive fluid used is usually an oil-like substance having a molecular weight in the range of 400 to 500 (for example: tetramethyl-tetraphenyl-trisiloxane). Such fluids must have a very low vapor pressure at room temperature but be capable of producing a few torr of boiler pressure without undergoing excessive thermal decomposition at typical temperatures of 200 to 240°C). They should be also nontoxic and noncorrosive and have a viscosity range at water-cooling temperature that permits draining back into the boiler by gravity.

Fig. 14.4.13 Schematic cross section of a four-stage vapor jet pump (Varian).

The practical pressure difference that can be sustained by high-vacuum jet pumps is 0.5 to 1 torr, so they must be operated only when attached to a backing pump that provides necessary primary vacuum. Also, in order to prevent a possibility of rapid deterioration of and even combustion of the fluid, vapor jet pumps cannot be exposed to atmospheric air while at operating temperature. Therefore, they must be protected from an accidental air release by appropriate valves. An example of a system arrangement is shown in Fig. 14.4.14. In operation, the fluid is vaporized in the boiler by electric heaters, the vapor issues from the nozzles at velocities of a few hundred metres per second, and then condenses at the water-cooled walls and returns to the boiler.

Fig. 14.4.14 Typical arrangement of valves for a vapor-jet pump system.

A typical performance curve of a vapor-jet pump is shown in Fig. 14.4.15. Such performance curves are usually measured by using total-pressure gages, with air as the pumped gas. The decay of the pumping speed at the left of the graph represents the onset of compression ratio limits. The region called "overload" in the graph should never

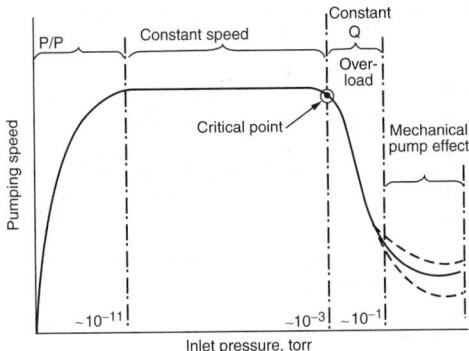

Fig. 14.4.15 Pumping speed versus inlet pressure in four different performance regions.

be used in any continuous operation. Here, the reduction of pumping speed is caused by a complete collapse of function in one of the jets (usually one near the inlet). The overloaded vapor stream may not reach the wall and may produce a serious amount of backstreaming into the vacuum chamber. At the inlet of the pump, a small number of vapor molecules can leave the condensed liquid and migrate into the chamber. To prevent this possibility of contamination, water-cooled baffles and traps cooled to cryogenic temperature are used, as shown in Fig. 14.4.14.

Vapor-jet pumps are made in sizes ranging from 50 to 50,000 L/s, corresponding to inlet diameters from 2 to 36 in. The larger pumps may use five or six stages instead of the four shown in Fig. 14.4.13.

There also exist so-called diffusion-ejector pumps, which utilize motive fluids of lower molecular weight and have higher boiler pressures. These pumps extend the full pumping speed region of operation toward higher inlet pressures. Such pumps are made in sizes from 4,000 to 12,000 L/s.

CRYOGENIC PUMPS

One way of creating high vacuum is to lower the temperature of the gas until it condenses into a liquid or a solid. In industrial practice, instead of cooling the entire chamber, it is sufficient to use a cryocooled pump. The vapor pressure curves in Fig. 14.4.16 demonstrate that it is not practical to cool gases of low molecular weight by condensation alone. Therefore, some internal surfaces of cryopumps are usually cooled by gaseous helium refrigeration to temperatures of 10 to 20 K combined with surfaces covered with absorbent layers. A schematic drawing of a typical cryopump is shown in Fig. 14.4.17. Inside the pump is a two-stage cryogenerator consisting of two connected reciprocating pistons filled with heat exchange materials. The pump is connected to a special helium compressor, which is used to circulate helium through the expander assembly and to recompress it. Industrial cryopumps utilize specially modified refrigeration

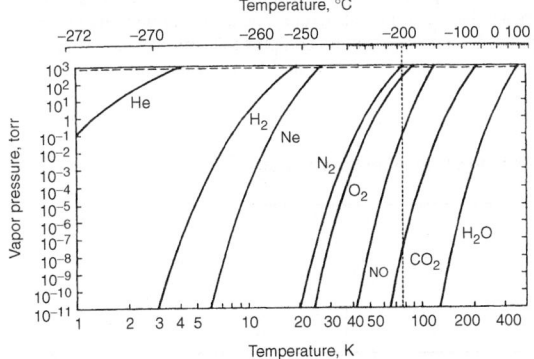

Fig. 14.4.16 Vapor pressure curves of some common gases.

Fig. 14.4.17 Schematic drawing of a typical cryopump.

compressors. Effective filtering devices must be used to prevent lubricating oil from reaching areas cold enough to freeze the lubricant.

The pumping surfaces of a cryopump are arranged in three distinct varieties. The first is a metal surface kept at 50 to 80 K. It may be shaped as a cylinder with the closed end attached to the first stage of refrigeration and the open end connected to a baffle structure. This part of the pump can condense the water vapor, which is one of the main gases left in a vacuum system after an initial evacuation by coarse vacuum pumps. The second section is a metal surface attached to the second stage of the refrigerator, which is kept at 10 to 20 K. This section is used for pumping such gases as nitrogen, oxygen, and argon. The third section is at the same temperature as the second but includes a layer of charcoal (or activated alumina) bonded to the inner surfaces of the second-stage cryoarrays. Charcoal may have internal surface area approaching 1,000 m^2/g as well as small enough pores to hold molecules of light gases.

The pumping speed of a condensation and sorption pump depends on the arrival rate of the gas, capture probability, and surface area. Figure 14.4.18 shows an example of the performance of an industrial cryopump having an inlet diameter of 20 cm. From the kinetic theory of gases, the maximum pumping speed can be calculated to be for hydrogen 44.6 L/s for each square centimetre of surface; the corresponding value for water vapor is 14.9 and for nitrogen 11.9, provided that access to the surface is not impeded by conductance limitations. In practice, full pumping speed may be realized for water vapor because of the open exposure of the cooled baffle at the inlet, but the speed for nitrogen is typically 3 L/s · cm^2 and for hydrogen 5.6 L/s · cm^2. Because the gas is not removed form the pump, the pumping speed will decline in time. For example, for the pump of Fig. 14.4.18, the decline of pumping speed for hydrogen begins after about 7,000 torr-L of gas are pumped. The values given for pumping speed are measured under ideal conditions. In use, the actual speed may depend on amounts and types of gases previously pumped. Cryopumps do not have a continuously functioning exhaust port, therefore they become saturated after a certain period of operation.

At high-vacuum inlet pressures, they can function continuously for several months, but eventually regeneration becomes necessary. The need for regeneration arises when the pump cannot maintain expected

Fig. 14.4.18   Pumping speed of a cryopump for different gases.

Fig. 14.4.20   Pumping speeds of gettering pumps of various sizes.

pressure or evacuation time. Regeneration requires warming of the pump and may take several hours. It involves several steps of gradual warming and purging so as not to drive certain unwanted gases into the inlet space, and so as to properly manage the disposal of dangerous gases.

GETTER AND ION PUMPS

When gases are rarefied to high-vacuum and ultrahigh-vacuum levels, a surface can hold larger quantities of the gases, as compared to the amount of the gas in space. Properly prepared surfaces (with and without additional absorbing layers) can then be used as a pumping medium. Many chemically active materials can be used to enhance the quantity of gas held on the surface, a process called *gettering*. Often the material used is titanium. To produce a pumping surface, all that is needed is to form a fresh (unoxidized) layer of material on a sufficiently large surface. An example of a simple arrangement of a gettering pump is shown in Fig. 14.4.19. As is the case with cryopumps, the performance

of a gettering pump cannot be predicted exactly because of its dependency on the mixture of gases, the presence of chemically inert gases, the temperature, and the previous history of the surface. Approximate pumping speeds are shown in Table 14.4.1, where it can be seen that the inert gases are not pumped. In Fig. 14.4.20 actual pumping speeds are shown for air for pumps of various sizes.

There also exist so-called nonevaporated getter pumps. In such pumps the active material is not evaporated, but remains in a specially prepared, very porous intermetallic compound. The pumping action involves an adsorption event followed by diffusion of the pumped gas into the material. The material is in the form of thick films deposited on a ribbon of nickel-iron or stainless steel. A common composition for the pumping compound is 84% zirconium and 16% aluminum and may include other transition metals. In operation, the gettering action is activated by heating, which produces structural transformation in the material, making it porous. Then, it may have real physical area 100 times greater than the projected area. Figure 14.4.21 is an example of a pumping cartridge placed in a cylinder that can be attached to a vacuum chamber at the top flange.

To provide pumping action for inert gases, ion-getter or sputter-ion pumps were developed. A basic cell of such a pump is shown in Fig. 14.4.22. Electrons are emitted from the essentially cold cathode by the action of the electric field of 5 to 7 kV applied between the electrodes. Long trajectories of electrons are produced by the action of the magnet, to improve chances of collision and subsequent ionization. Positive ions bombard the anode at high energies, producing the phenomenon called sputtering, which is an ejection of atoms from the titanium cathode surface. The deposited titanium film reacts with active gases; inert gases are pumped when highly accelerated ions collide with and penetrate the surface structure. The details of pumping mechanisms of sputter-ion pumps are more complex than outlined here, but it is clear that various gases are pumped in different ways, at different pumping

Fig. 14.4.19   Typical arrangement of titanium gettering pump.

Table 14.4.1   Typical* Pumping Speeds per Square Centimetre of Titanium Sublimation Surface for Various Gases (L/s)

Surface temperature (°C)	H_2	N_2	O_2	CO	CO_2	H_2O	CH_4	A	He
20	20	30	60	60	50	20	0	0	0
−195	65	65	70	70	60	90	0	0	0

*Actual speeds vary with the condition of the deposits; usable speed is reduced by the conductance between the pumping area and the process area in the vacuum chamber.

Fig. 14.4.21 The view of a nonevaporated getter pump. H and 1—electric heater, W—water cooling (for large pumps), C and 3—the getter ribbon; 2—inlet flange.

Fig. 14.4.22 Basic cell of an ion-getter pump.

speeds and different saturation effects. Some pumping speed values are shown in Table 14.4.2. It is also clear that eventually cathodes become exhausted and must be replaced.

For practical designs of sputter-ion pumps the cells are between 15 and 25 mm, which is suitable for magnetic fields of 1 to 1.5 kG. The pumping speed of a cell is between 0.3 and 2 L/s. Multiple cell structures are used for larger pumps. Pumps are commonly manufactured in speeds up to 200 L/s. They are delivered prepumped and sealed to keep internal surfaces clean. To achieve ultrahigh vacuum, the entire vacuum system, including the pumps, is backed for several hours at temperatures up to 400°C, while being pumped by some clean auxiliary pumps.

Table 14.4.2   Relative Pumping Speeds of Ion-Getter Pumps for Various Gases*

Gas	Diode pump (3.5 kV)	Triode pump (5.2 kV)	Diode pump with slotted cathode (7.3 kV)	Diode pump with Mo and Ti cathodes
H_2	2.5–3.0	2.0	2.7	
CH_4	2.7	0.9–1.05		
NH_3	1.7			
H_2O	1.3		1.0	
Air	1.0	1.0	1.0	1.0
N_2	0.98	0.95	1.0	
CO	0.86		1.0	
CO_2	0.82	1.0	1.0	
O_2	0.55	0.57	0.57	
He	0.11	0.3	0.1	0.25
Ar	0.01	0.3	0.06	0.2

*Taken from manufacturers' catalogs. Air speed taken as unity.

14.5 FANS
by Robert Jorgensen

REFERENCES: Jorgensen, "Fan Engineering," 9th ed., Howden Buffalo. "Standards Handbook," AMCA Publication 99-86 (R 1998). "Fans and Systems," AMCA Publication 201:2002. "Field Performance Measurement of Fan Systems," AMCA Publication 203-90. "Balance Quality and Vibration Levels for Fans," AMCA Standard 204-96. "Laboratory Methods of Testing Fans for Aerodynamic Rating," American National Standard ANSI/AMCA 210-1999 and ANSI/ASHRAE 51-1999. "Reverberant Room Method for Sound Testing of Fans," AMCA Standard 300-96. "Laboratory Methods of Sound Testing of Fans Using Sound Intensity," Proposed American National Standard, BSR/AMCA Standard 320. "Laboratory Method of Testing to Determine the Sound Power in a Duct," ANSI/ASHRAE Standard 68-1997 and ANSI/AMCA 330-97. "Industrial Process/Power Generation Fans: Specification Guidelines," AMCA Publication 801-01. "Industrial Process/Power Generation Fans: Establishing Performance Using Laboratory Models," AMCA Publication 802-02. "Industrial Process/Power Generation Fans: Site Performance Test Standard," AMCA Publication 803:2002. "ASME Performance Test Codes, Fans," ANSI/ASME PTC 11-1984.

Symbols

A = area, ft² (m²)
b = barometric pressure, in Hg (Pa)
D = diameter, ft (m)
D_s = specific diameter, ft (m)
H = fan power input, hp (W)
H_0 = fan power output, hp (W)
h = head, ft (m)
K_p = compressibility factor, dimensionless
L_p = sound pressure level, dB
L_w = sound power level, dB
L_{ws} = specific sound power level, dB
log = logarithm to base 10
M = molecular weight, dimensionless
N = speed of rotation, r/min
N_s = specific speed, r/min
n = number or polytropic exponent
p_s = fan static pressure, in wg (Pa)
p_t = fan total pressure, in wg (Pa)
p_v = fan velocity pressure, in wg (Pa)
p_{sx} = static pressure at plane x, in wg (Pa)
P_{tx} = total pressure at plane x, in wg (Pa)
P_{ux} = velocity pressure at plane x, in wg (Pa)
Q = fan capacity, ft³/min (m³/s)
Q_x = volumetric flow rate at plane x, ft³/min (m³/s)
T = absolute temperature, °R (K)
W = power, W
x = factor used to determine K_p, dimensionless
z = factor used to determine K_p, dimensionless
γ = isentropic exponent, dimensionless
η_s = fan static efficiency, per unit
η_t = fan total efficiency, per unit
ρ = fan air density, lbm/ft³ (kg/m³)
ρ_x = air density at plane x, lbm/ft³ (kg/m³)

Subscripts

1 = fan inlet plane
2 = fan outlet plane
a = absolute
b = basic known conditions
c = calculated condition
r = reading
x = plane 1, 2, 3, or other

FAN TYPES AND NOMENCLATURE

Any device which produces a current of air may be called a fan. This discussion will be limited to fans which have a rotating impeller to produce the flow and a stationary casing to guide the flow into and out of the impeller (see Fig. 14.5.1). The form of the casing or the impeller may vary widely.

Fig. 14.5.1 Elements of fans and preferred terminology. (*a*) Centrifugal fans; (*b*) axial fans. (*Based on AMCA Publ. 201.*)

Fan Classification

One of the characteristics by which fans are classified is the nature of the flow through the blade passages of the impeller. Axial flow, radial flow, mixed flow, and cross flow are all possible in fan impellers. Certain fan names result from these classifications. Other fan names derive from other characteristics.

Propeller fans, tube-axial fans, and vane-axial fans all utilize **axial-flow impellers,** but their casings differ. Propeller fans may be mounted in a ring or panel. Tube-axial fans and vane-axial fans both use tubular casings, but for vane-axial fans they are equipped with stationary guide vanes. A great deal of the energy transferred to the air in axial-flow machines is in kinetic form. Some of this kinetic energy can be transformed into pressure energy by straightening the swirl, e.g., with vanes, or by reducing the exit velocity, e.g., with a diffuser. Propeller fans effect very little transformation and hence have very low pressure-producing capability. Vane-axial fans can be equipped for maximum transformation as well as high transfer of energy and hence have high pressure-producing potential depending on tip speed and blade angles. High hub ratios promote high energy transfer.

Centrifugal fans and tubular centrifugal fans both utilize **radial-flow impellers.** Centrifugal fans usually employ a volute or scroll-type casing, the flow entering the casing axially and leaving tangentially. Tubular centrifugals use tubular casings so that the flow both entering and leaving the casing is axial. A considerable portion of the energy transferred to the air in a radial-flow machine is due to centrifugal action; hence the name centrifugal fan. Since centrifugal action varies with blade depth, the pressure-producing capability of radial-flow fans will vary with this factor as well as tip speed and blade angles.

Mixed-flow impellers can be used in either axial or scroll-type casings. They are characterized as mixed flow because both axial and radial flow take place in the blading. Mixed-flow impellers used in axial-flow casings have a hub similar to a pure axial-flow impeller, but the inlet portion of the blading extends down over the face of the hub, thereby giving some radial guidance. Mixed-flow impellers used in scroll-cased fans have blades which give most of the axial guidance in the inlet portion and most of the radial guidance in the discharge portion.

In a **cross-flow impeller,** the air passes through the blading twice, entering more or less tangentially through the tip, passing across the impeller and out the other side. The casings are designed to provide this transverse flow. Cross-flow fans are also known as tangential fans or transverse-flow fans. Pressure-producing potential is low and depends on the formation of a vortex as the air leaves the impeller.

Fan Details

Propeller fans and other **axial fans** may use blades shaped to airfoil sections or blades of uniform thickness. Blading may be fixed, adjustable at standstill, or variable in operation. Propeller fans have very small hubs. Hub-to-tip diameter ratios ranging from 0.4 to 0.7 are common in vane-axial fans. The larger the hub, the more important it is to have an inner cylinder approximately the hub size located downstream of the impeller. The guide vanes of a vane-axial fan are located in the annular space between the tubular casing and the inner cylinder. Diffusers are generally used between the fan and the discharge ductwork.

Centrifugal fans use various types of blading. Forward-curved blades are shallow and curved so that both the tip and the heel point in the direction of rotation. Radial and radial-tip blades both are radial at the tip, but the latter are curved at the heel to point in the direction of rotation. Backward-curved and backward-inclined blades point in the direction opposite rotation at the tip and in the direction of rotation at the heel. All the above blades are of uniform thickness and are designed for radial flow. Airfoil blades have backward-curved chord lines so that the leading edge of the airfoil is at the heel pointing forward and the trailing edge at the tip pointing backward with respect to rotation. Impellers for all blade shapes are usually shrouded and may have single or double inlets. Blade widths are related to the inlet-to-tip-diameter ratio. Tip angles may vary widely, but heel angles should be set to minimize entrance losses. Scroll casings may be fitted with a streamlined inlet bell, an inlet cone, or simply a collar.

Tubular centrifugals may be designed for backward-curve, air foil, or mixed-flow impellers. An inlet bell and discharge guide vanes are required for good performance.

Cross-flow fans utilize impellers with blading similar to that of a forward-curved centrifugal, but the end shrouds have no inlet holes. Blade-length-to-tip-diameter ratios are limited only by structural considerations.

Power roof ventilators may use either axial- or radial-flow impellers. The casings will include either a propeller fan mounting ring or a tubular casing to guide the flow for an axial-flow impeller. If a radial-flow impeller is used, an inlet bell is required, but the scroll case may be replaced by the ventilator hood.

Plenum fans and **plug fans** are fans without traditional casings. Plenum fans are used to move air through a system. They utilize centrifugal impellers and their drives are inside the plenum. Plug fans are used to circulate air or gas within a chamber or system. They may have either centrifugal or axial impellers and their drives are outside the chamber.

All fan types may have direct or indirect drive arrangements. Various standards including arrangement numbers have been adopted by AMCA. Arr. 1 signifies an overhung impeller on a fan shaft with two bearings on a base; Arr. 3 signifies an impeller on a fan shaft between bearings; Arr. 4 is for overhung impeller on a motor shaft; Arr. 7 is Arr. 3 with a motor base; Arr. 8 is Arr. 1 with an extended base for motor; Arr. 9 is Arr. 1 with motor mounted on side of unit; Arr. 10 is Arr. 1 with motor mounted inside of base. Rotation is specified as cw or ccw when viewed from the drive side. All fans can be equipped with variable-speed drives, variable inlet vanes, or dampers, but controllable-pitch axials do not need speed or vane control.

FAN PERFORMANCE AND TESTING

The conventional terms used to describe fan performance in the United States are defined below.

Fan Air Density   Air density is the mass per unit volume of the air. The density of a perfect gas is a function of its molecular weight, temperature, and pressure as indicated by

$$\rho_x = \frac{M}{386.7} \frac{529.7}{T_x} \frac{b_x}{29.92}$$

where the subscript x indicates the plane of the measurements. For dry air this reduces to

$$\rho_x = 1.325 b_x / T_x$$

This expression will usually be accurate enough even when moist air is involved. Fan air density is the density of the air corresponding to the total pressure and total temperature at the fan inlet

$$\rho = \rho_1$$

Fan Capacity   Volume flow rate is usually determined from pressure measurements, e.g., a velocity-pressure traverse taken with a Pitot static tube or a pressure drop across a flowmeter. The average velocity pressure for a Pitot traverse is

$$p_{vx} = (\sum \sqrt{p_{vxr}} / n)^2$$

where subscript x indicates the plane of the measurements, subscript r indicates a reading at one station, n is the number of stations, and $\sum$ is the summation sign. The corresponding capacity is:

$$\dot{Q}_x = 1{,}097 A_x \sqrt{p_{vx}/\rho_x}$$

For a flowmeter

$$\dot{Q}_x = 1{,}097 C A_x Y \sqrt{\Delta p / \rho_x / F}$$

where C = coefficient of discharge of meter, Y = expansion factor for gas, Δp = measured pressure drop, and F = velocity-of-approach factor for the meter installation. Fan capacity is the volumetric flow rate at fan air density

$$\dot{Q} = \dot{Q}_x \rho_x / \rho$$

Fan Total Pressure   Fan total pressure is the difference between the total pressure at the fan outlet and the total pressure at the fan inlet

$$p_t = p_{t2} - p_{t1}$$

When the fan draws directly from the atmosphere.

$$p_{t1} = 0$$

When the fan discharges directly to the atmosphere,

$$p_{t2} = p_{v2}$$

If either side of the fan is connected to ductwork, etc., and the measuring plane is remote, the measured values should be corrected for the approximate pressure drop

$$p_{t2} = p_{tx} + \Delta p_{2-x} \qquad p_{t1} = p_{tx} + \Delta p_{x-1}$$

Fan Velocity Pressure   Fan velocity pressure is the pressure corresponding to the average velocity at the fan outlet

$$p_v = \left(\frac{\dot{Q}_2/A_2}{1,097}\right)^2 \rho_2$$

Fan Static Pressure   Fan static pressure is the difference between the fan total pressure and the fan velocity pressure. Therefore, fan static pressure is the difference between the static pressure at the fan outlet and the total pressure at the fan inlet

$$p_s = p_t - p_v = p_{s2} - p_{t1}$$

Fan Speed   Fan speed is the rotative speed of the impeller.

Compressibility Factor   The compressibility factor is the ratio of the fan total pressure p_t' that would be developed with an incompressible fluid to the fan total pressure p_t that is developed with a compressible fluid, all other conditions being equal:

$$K_p = \frac{p_t'}{p_t} = \frac{[n/(n-1)][(p_{t2a}/p_{t1a})^{(n-1)/n} - 1]}{(p_{t2a}/p_{t1a}) - 1}$$

Compressibility factor can be determined from test measurements using

$$x = p_t/p_{t1a}$$
$$z = [(\gamma - 1)/\gamma]\,6356H/\dot{Q}p_{t1a}$$
$$K_p = [z \log (1 + x)]/[x \log (1 + z)]$$

Fan Power Output   Fan power output is the product of fan capacity and fan total pressure and compressibility factor

$$H_o = \dot{Q}p_t K_p /6,356$$

Fan Power Input   Fan power input is the power required to drive the fan and any elements in the drive train which are considered a part of the fan. Power input can be calculated from appropriate measurements for a dynamometer, torque meter, or calibrated motor.

Fan Total Efficiency   Fan total efficiency is the ratio of the fan power output to the fan power input

$$\eta_t = \dot{Q}p_t K_p /6,356H$$

Fan Static Efficiency   Fan static efficiency is the fan total efficiency multiplied by the ratio of fan static pressure to fan total pressure

$$\eta_s = \eta_t p_s /p_t$$

Fan Sound Power Level   Fan sound power level is 10 times the logarithm (base 10) of the ratio of the actual sound power in watts to 10^{-12} watts,

$$L_w = 10 \log (W/10^{-12})$$

The total sound power level of a fan is usually assumed to be 3 dB higher than either the inlet or outlet component. The casing component varies with construction but will usually range from 15 to 30 dB less than the total.

The frequency content of fan noise is usually expressed in octave bands, but may be given in narrow bands or even as an overall value. The sound power level can be calculated from sound pressure measurements or sound intensity measurements. The reverberant room method utilizes a known sound source to calibrate the room and a microphone to measure the sound pressures. The sound intensity method utilizes two microphones in a sound-intensity probe to measure the sound intensities over an enveloping surface. The in-duct method utilizes anechoic terminations to facilitate sound pressure measurements without end reflections.

Head   The difference between head and pressure is important in fan engineering. Both are measures of the energy in the air. Head is energy per unit weight and can be expressed in ft · lb/lb, which is often abbreviated to ft (of fluid flowing). Pressure is energy per unit volume and can be expressed in ft · lb/ft³, which simplifies to lb/ft² or force per unit area. The use of the inch water gage (in wg) is a convenience in fan engineering reflecting the usual methods of measurement. It sounds like a head measurement but is actually a pressure measurement corresponding

to 5.192 lb/ft², the pressure exerted by a column of water 1 inch high. Pressures can be converted into heads and vice versa:

$$p = \rho h/5.192 \qquad h = 5.192 p/\rho$$

For instance, for air at 0.075 lbm/ft³ density, 1 in wg corresponds to 69.4 ft of air. That is, a column of air at that density would exert a pressure of 5.192 lb/ft². Lighter air would exert less pressure for a given head. For a given pressure, the head would be higher with lighter air. Although it is not widely used in the United States, head is commonly used in a number of European countries.

Fans, like other turbomachines, can be considered constant-head-constant-capacity (volumetric) machines. This means that a fan will develop the same head at a given capacity regardless of the fluid handled, all other conditions being equal. Of course, this also means that a fan will develop a pressure proportional to the density at a given capacity, all other conditions being equal.

All the preceding equations are based on the U.S. Customary Units listed under Symbols. If **SI units** are to be used, certain numerical coefficients will have to be modified. For instance, substitute $\sqrt{2}$ for 1,097 in the flow equations when using Pa for pressure, kg/m³ for density, and m³/s for capacity. Similarly, substitute 1.0 for 6,356 in the power equations when using Pa for pressure, m³/s for capacity, and W for power.

Common metric practice in fan engineering leads to other numerical coefficients. When mm wg is used for pressure, m³/s for capacity, kW for power, and kg/m³ for density, substitute 4.424 for 1,097 in the flow equations and 102.2 for 6,356 in the power equations.

The preceding discussion utilized what may be termed the **volume-flow-rate-pressure** approach to expressing fan performance. An alternative, the **mass-flow-rate-specific-energy** approach is equally valid but not generally used in the United States. Refer to ASME PTC-11 for additional details.

FAN AND SYSTEM PERFORMANCE CHARACTERISTICS

The performance characteristics of a fan are best described by a graph. The conventional method of graphing fan performance is to plot a series of curves with capacity as abscissa and all other variables as ordinates. System characteristics can be plotted in a similar manner.

System Characteristics

Most systems served by a fan have characteristics which can be described by a parabola passing through the origin; i.e., the energy required to produce flow through the system (which can be expressed as pressure or head) varies approximately as the square of the flow. In some cases, the system characteristics will not pass through the origin because the energy required to produce flow through an element of the system may be controlled at a particular value, e.g., with venturi scrubbers. In some cases, the system characteristic will not be parabolic because the flow through an element of the system is laminar rather than turbulent, e.g., in some types of filters. Whatever the case, the system designer should establish the characteristics by determining the energy requirements at various flow rates. The energy requirement (pressure drops or head losses) for each element can be determined by reference to handbook or manufacturer's literature or by test.

The true measure of the energy requirement for a system element is the total pressure drop or the total head loss. Only if the entrance velocity for the element equals the exit velocity will the change in static pressure equal the total pressure drop. There are some advantages in using static pressure change, but the system designer is usually well advised to use total pressure drops to avoid errors in fan selection. The sum of the total pressure losses for elements on the inlet side of the fan will equal $- p_{t1}$. This should include energy losses at the entrance to the system but not the energy to accelerate the air to the velocity at the fan inlet, which is chargeable to the fan. The sum of the total pressure losses for elements on the discharge side of the fan will equal p_{t2}. This should include the kinetic energy of the stream issuing from the system. The total system requirement will be the arithmetic sum of all the appropriate losses or the algebraic difference $p_{t2} - p_{t1}$, which is also p_t.

A system characteristic curve based on total pressure is plotted in Fig. 14.5.2. A characteristic based on static pressure is also shown. The latter recognizes the definition of fan static pressure so that the only difference is the velocity pressure corresponding to the fan outlet velocity. This system could operate at any capacity provided a fan delivered the exact pressure to match the energy requirements shown on the system curve for that capacity. The advantages of plotting the system characteristics on the fan graph will become evident in the following discussions.

Fig. 14.5.2 Fan and system characteristics.

Fan Characteristics

The constant-speed performance characteristics of a fan are illustrated in Fig. 14.5.2. These characteristics are for a particular size and type of fan operating at a particular speed and handling air of a particular density. The fan can operate at any capacity from zero to the maximum shown, but when applied on a particular system the fan will operate only at the intersection of the system characteristics with the appropriate fan pressure characteristic. For the case illustrated, the fan will operate at $Q = 27,300$ ft³/min and $p_t = 3.4$ in wg or $p_s = 3$ in wg, requiring $H = 18.3$ hp at the speed and density for which the curves were drawn. The static efficiency at this point of operation is 73 percent, and the total efficiency is 80 percent. If the system characteristic had been lower, it would have intersected the fan characteristic at a higher capacity and the fan would have delivered more air. Contrariwise, if the system characteristic had been higher, the capacity of the fan would have been less. Capacity reduction can be accomplished, in fact, by creating additional resistance, as with an outlet damper.

Figure 14.5.3 illustrates the characteristics of a fan with damper control, variable inlet vane control, and variable speed control. These particular characteristics are for a backward-curved centrifugal fan, but the general principles apply to all fans. Outlet dampers do not affect the flow to the fan and therefore can alter fan performance only by adding resistance to the system and producing a new intersection. Point 1 is for wide-open dampers. Points 2 and 3 are for progressively closed dampers. Note that some power reduction accompanies the capacity reduction for this particular fan. Variable-inlet vanes produce inlet whirl, which reduces pressure-producing capability. Point 4 is for wide-open inlet vanes and corresponds to point 1. Points 5 and 6 are for progressively closed vanes. Note that the power reduction at reduced capacity is better for vanes than for dampers. Variable speed is the most efficient means of capacity control. Point 7 is for full speed and points 8 and 9 for progressively reduced speed. Note the improved power savings over other methods. Variable speed also has advantages in terms of lower noise and reduced erosion potential but is generally at a disadvantage regarding first cost.

Figure 14.5.4 illustrates the characteristics of a fan with variable pitch control. These characteristics are for an axial-flow fan, and comparisons to the centrifugal-fan ratings in Fig. 14.5.3 can be made. Point 10 corresponds to points 1, 4, or 7. Reduced ratings are shown at points 11 and 12 obtained by reducing the pitch. Power savings can be almost as good as with speed control. Notice that in all methods of capacity control, operation is at the intersection of a system characteristic with a fan characteristic. With variable vanes, variable speed, and variable pitch, the fan characteristic is modified. With damper control, the effect could be considered a change in fan characteristics, but it generally makes more sense to consider it a change in system characteristics.

Fan-System Matching

It has already been observed in the discussions of both system and fan characteristics that the point of operation for **one fan** on a particular system will be at the intersection of their characteristics. Stated another way, the energy required by the system must be provided by the fan exactly. If the fan delivers too much or too little energy, the capacity will be more or less than desired. The effects of utilizing dampers, variable speed, variable-inlet vanes, and variable pitch were illustrated, but in all cases the fan was preselected. The more general case is the one involving the selection or design of a fan to do a particular job. The

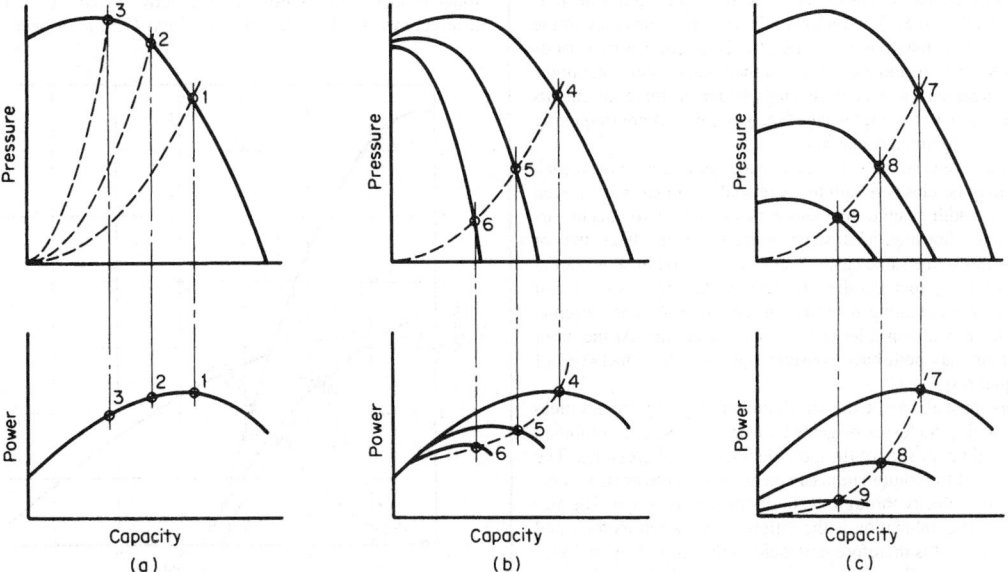

Fig. 14.5.3 Fan characteristics with (*a*) damper control, (*b*) variable-inlet vane control, (*c*) variable speed control.

Fig. 14.5.4 Characteristics of an axial-flow fan with variable-pitch control.

design of a fan from an aerodynamicist's point of view is beyond the scope of this discussion. Fortunately, most fan problems can be solved by selecting a fan from the many standard lines available commercially. Again, the crux of the matter is to match fan capability with system requirements.

Most fan manufacturers have several standard lines of fans. Each line may consist of various sizes all resembling each other. If the blade angles and proportions are the same, the line is said to be **homologous.** There is only one fan size in a line of homologous fans that will operate at the maximum efficiency point on a given system. If the fan is too big, operation will be at some point to the left of peak efficiency on a standard characteristic plot. If the fan is too small, operation will be to the right of peak. In either case it will be off peak. A slightly undersized fan is often preferred for reasons of cost and stability.

Selecting and rating a fan from a catalog is a matter of fan-system matching. When it is recognized that many of the catalog sizes may be able to provide the capacity and pressure, selection becomes a matter of trying various sizes and comparing speed, power, cost, etc. The choice is a matter of evaluation.

There are various reasons for using **more than one fan** on a system. Supply and exhaust fans are used in ventilation to avoid excessive pressure build-up in the space being served. Forced and induced-draft fans are used to maintain a specified draft over the fire. Two fans may fit the available space better than one larger fan. Capacity control by various fan combinations may be more economical than other control methods. Multistage arrangements may be necessary when pressure requirements exceed the capabilities of a single-stage fan. Standby fans are frequently required to ensure continuous operation.

When two fans are used, they may be located quite remote from each other or they may be close enough to share shaft and bearings or even casings. Double-width, double-inlet fans are essentially two fans in parallel in a common housing. Multistage blowers are, in effect, two or more fans in series in the same casing. Fans may also be in series but at opposite ends of the system. Parallel-arrangement fans may have almost any amount of their operating resistance in common. At one extreme, the fans may have common inlet and discharge plenums. At the other extreme, the fans may both have considerable individual ductwork of equal or unequal resistance.

Fans in **series** must all handle the same amount of gas by weight measurements, assuming no losses or gains between stages. The combined total pressure will be the sum of the individual fans' total pressures. The velocity pressure of the combination can be defined as the pressure corresponding to the velocity through the outlet of the last stage. The static pressure for the combination is the difference between its total and velocity pressures and is therefore not equal to the sum of the individual fan static pressures. The volumetric capacities will differ whenever the inlet densities vary from stage to stage. Compression in one stage

will reduce the volume entering the next if there is no reexpansion between the two. As with any fan, the pressure capabilities are also influenced by density.

The combined total pressure-capacity characteristic for two fans in series can be drawn by using the volumetric capacities of the first stage for the abscissa and the sum of the appropriate total pressures for the ordinate. Because of compressibility, the volumetric capacities of the second stage will not equal the volumetric capacities of the first stage. The individual total pressures must be chosen accordingly before they are combined. If the gas can be considered incompressible, the pressures for the two stages may be read at the same capacity. In the area near free delivery, it may be necessary to estimate the negative pressure characteristics of one of the fans in order to combine values at the appropriate capacity.

Fans in **parallel** must all develop sufficient pressure to overcome the losses in any individual ductwork, etc., as well as the losses in the common portions of the system. When such fans have no individual ductwork but discharge into a common plenum, their individual velocity pressures are lost and the fans should be selected to produce the same fan static pressures. If fan velocity pressures are equal, the fan total pressures will be equal in such cases. When the fans do have individual ducts but they are of equal resistance and joined together at equal velocities, the fans should be selected for the same fan total pressures. If fan velocity pressures are equal, fan static pressures will be equal in this case. If the two streams join together at unequal velocities, there will be a transfer of momentum from the higher-velocity stream to the lower-velocity stream. The fans serving the lower-velocity branch can be selected for a correspondingly lower total pressure. The other fan must be selected for a correspondingly higher total pressure than if velocities were equal.

The combined pressure-capacity curves for two fans in parallel can be plotted by using the appropriate pressures for ordinates and the sum of the corresponding capacities for abscissa. Such curves are meaningful only when a combined-system curve can be drawn. In the area near shutoff, it may be necessary to estimate the negative capacity characteristics of one of the fans in order to combine values at the appropriate pressure.

Figure 14.5.5 illustrates the combined characteristics of two fans with slightly different individual characteristics (*A-A* and *B-B*). The combined characteristics are shown for the two fans in series (*C-C*) and in parallel (*D-D*). Only total-pressure curves are shown. This is always correct for series arrangements but may introduce slight errors for parallel arrangements. An incompressible gas has been assumed. The questionable areas near shutoff or free delivery have been omitted. Two different system characteristics (*E-E* and *F-F*) have been drawn on the

Fig. 14.5.5 Combined characteristics of fans in series and parallel.

chart. With the two fans in series, operation will be at point *EC* or *FC* if the fan is on system *E* or *F*, respectively. Parallel arrangement will lead to operation at *ED* or *FD*. Single-fan operation would be at the point indicated by the intersection of the appropriate fan and system curves provided the effect of an inoperative second fan is negligible. Some sort of bypass is required around an inoperative fan in series whereas an inoperative fan in parallel need only be dampered shut. Parallel operation yields a higher capacity than series operation on system *F*, but the reverse is true on system *E*. For the type of fan and system characteristics drawn there is only one possible point of operation for any arrangement.

Fan Laws

The fan laws are based on the experimentally demonstrable fact that any two members of a homologous series of fans which have performance curves which are homologous. At the same point of rating, i.e., at similarly situated points of operation on their characteristic curves, efficiencies are equal and other variables are interrelated according to the fan laws. If size and speed are considered independent variables and if compressibility effects are ignored, the fan laws can be written as follows:

$$\eta_{tc} = \eta_{tb}$$
$$\dot{Q}_c = \dot{Q}_b (D_c/D_b)^3 (N_c/N_b)$$
$$p_{tc} = p_{tb} (D_c/D_b)^2 (N_c/N_b)^2 (\rho_c/\rho_b)$$
$$H_c = H_b (D_c/D_b)^5 (N_c/N_b)^3 (\rho_c/\rho_b)$$
$$L_{wc} = L_{wb} + 70 \log (D_c/D_b) + 50 \log (N_c/N_b)$$
$$+ 20 \log (\rho_c/\rho_b)$$

The above laws are useful, but they are dangerous if misapplied. The calculated fan must have the same point of rating as the known fan. When in doubt, it is best to reselect the fan rather than attempt to use the fan laws.

The fan designer utilizes the fan laws in various ways. Some of the more useful relationships in addition to those above derive from considering fan capacity and fan total pressure as the independent variables. This leads to specific diameter, specific speed, and specific sound power level:

$$D_s = D(p_t/\rho)^{1/4}/\dot{Q}^{1/2}$$
$$N_s = N\dot{Q}^{1/2}/(p_t/\rho)^{3/4}$$
$$L_{ws} = L_w - 10 \log \dot{Q} - 20 \log p_t$$

D_s, N_s, and L_{ws} are the diameter, speed, and sound power level of a homologous fan which will deliver 1 ft³/min at 1 in wg at the same point of rating as Q and p_t for D and N. D_s and N_s can be used to advantage in fan selection. They can also be used by a designer to determine how well a line will fit in with other lines. This is illustrated in Fig. 14.5.6. Each segment represents a particular fan line. Note the trends for each kind of fan. Incidentally, some fan engineers utilize different formulas for specific diameter and specific speed

$$D_{se} = Dp_{te}^{1/4}/\dot{Q}^{1/2} \qquad N_{se} = N\dot{Q}^{1/2}/p_{te}^{3/4}$$

where p_{te} is the equivalent total pressure based on standard air. This makes $D_{se} = D_s \times 1.911$ and $N_{se} = N_s \times 6.978$.

Specific sound power level is useful in predicting noise levels as well as in comparing fan designs. There appears to be a lower limit of L_{ws} in the vicinity of 45 dB for the more efficient types ranging to 70 dB or more for cruder designs. Actual sound power levels can be figured from

$$L_w = L_{ws} + 10 \log \dot{Q} + 20 \log p_t$$

Another useful parameter which derives from the fan laws is orifice ratio

$$R_o = \dot{Q}/D^2(p_t/\rho)^{1/2}$$

This ratio can be plotted on a characteristic curve for a known fan. If the ratio is determined for a calculated homologous fan, the point of rating can be established by inspection. Other ratios can be used in the same manner including p_v/p_t, $p_t/\dot{Q}^2$, D_s, and N_s.

Fig. 14.5.6 Specific diameter and efficiency versus specific speed for single-inlet fan types.

Stability Considerations

The flow through a system and its fan will normally be steady. If the fluctuations occasioned by a temporary disturbance are quickly damped out, the fan system may be described as having a **stable** operating characteristic. If the unsteady flow continues after the disturbance is removed, the operating characteristic is **unstable.**

To ensure stable operation the slopes of the pressure-capacity curves for the fan and system should be of opposite sign. Almost all systems have a positive slope; i.e., the pressure requirement or resistance increases with capacity. Therefore, for stable operation the fan curve should have a negative slope. Such is the case at or above the design capacity.

When the slopes of the fan and system characteristics are of opposite sign, any system disturbance tending to produce a temporary decrease in flow is nullified by the increase in fan pressure. When the slopes are of the same sign, any tendency to decrease flow is strengthened by the resulting decrease in fan pressure. When fan and system curves coincide over a range of capacities, the operating characteristics are extremely unstable. Even if the curves exactly coincide at only one point, the flow may vary over a considerable range.

There may or may not be any obvious indication of unstable operation. The pressure and power fluctuations that accompany unsteady flow may be so small and rapid that they cannot be detected by any but the most sensitive instruments. Less rapid fluctuations may be detected on the ordinary instruments used in fan testing. The changes in noise which occur with each change in flow rate are easily detected by ear as individual beats if the beat frequency is below about 10 Hz. In any event, the overall noise level will be higher with unsteady flow than with steady flow.

The conditions which accompany unsteady flow are variously described as pulsations, hunting, surging, or pumping. Since these conditions occur only when the operating point is to the left of maximum pressure on the fan curve, this peak is frequently referred to as the **surge point** or **pumping limit.**

Pulsation can be prevented by rating the fan to the right of the surge point. Fans are usually selected on this basis, but it is sometimes necessary to control the volume delivered to the value below that at the surge point. This may lead to pulsation, particularly if the fan pressure exceeds 10 in wg.

If the required capacity is less than that at the pumping limit, pulsation can be prevented in various ways, all of which in effect provide a negatively sloping fan curve at the actual operating point. To accomplish this effect the required pressure must be less than the fan capabilities at the required capacity. One method is to bleed sufficient air for actual operation to be beyond the pumping limit. Other possible methods are the use of pitch, speed, or vane control for volume reduction. In any of these cases, the point of operation on the new fan curve must be to the right of the new surge point. Although in the section on Capacity Control dampers were considered a part of the system, they may also be considered a part of the fan if located in the right position. Accordingly, pulsations may be eliminated in a supply system if the damper is on the inlet of the blower. Similarly, dampering at the outlet of the exhauster may control pulsation in exhaust systems.

Another condition frequently referred to as instability is associated with flow separation in the blade passages of an impeller and is evidenced by slight discontinuities in the performance curve. There may be a small range of capacities at which two distinctly different pressures may be developed depending on which of the two flow patterns exists. Such a condition usually occurs at capacities just to the left of peak efficiency. Still another condition of unsteady flow may develop at extremely low capacities. This is known as **blowback** or **puffing** because air puffs in and out of a portion of the inlet. Operation in the blowback range should be avoided, particularly with high-energy fans.

A different type of unsteady flow may occur when two or more fans are used in parallel. If the individual fan characteristics exhibit a dip in pressure between shutoff and design, the combined characteristic will contain points where the point of operation of the individual fans may be widely separated even for identical fans. If the system characteristic intersects the combined-fan characteristic at such a point, the individual fans may suddenly exchange loads. That is, the fan operating at high capacity may become the one operating at low capacity and vice versa. This can produce undesirable shocks on motors and ducts. Careful matching of fan to system is required for either forward-curved centrifugals or most axials for this reason.

Fan Applications

The selection of a particular size and type of fan for a particular application involves considerations of aerodynamic, economic, and functional suitability. Many of the factors involved in aerodynamic suitability have been discussed above. Determination of economic suitability requires an evaluation of first cost and operating costs. The functional suitability of various types of fans with respect to certain applications is discussed below.

Heating, ventilating, and **air-conditioning systems** may require supply and exhaust or return air fans. Historically, high-efficiency centrifugal fans, using either backward-curved or airfoil blades have been used for supply on duct systems. In low-pressure applications these types can be used without sound treatment. In high-pressure applications, sound treatment is almost always required. Axials have long been used for shipboard ventilation because they generally can be made smaller than centrifugals. Both adjustable axials and tubular centrifugals have proved popular on duct systems for exhaust service in building ventilation. Centrifugals with variable inlet vanes and axials with pitch control are being used for supply in variable-air-volume systems. Propeller fans and power roof ventilators are used for either supply or exhaust systems when there is little or no ductwork. Heating, ventilating, and air-conditioning applications are considered clean-air service. Various classes of construction are available in standard lines for different pressure ranges. Both direct and indirect drive are used, the latter being most common.

Industrial exhaust systems generally require fans that are less susceptible to the unbalance that may result from dirty-gas applications than the clean-air fans used for heating, ventilating, and air conditioning. Simple, rugged, industrial exhausters are favored for applications up to 200 hp. They have a few radial blades and relatively low efficiency. Most

are V-belt-driven. Extra-heavy construction may be required where significant material passes through the fan.

Process air requirements can be met with either centrifugal or axial fans; the latter may be used in single or double stage. The higher pressure ratings are usually provided by a single-stage centrifugal fan with radial blades, known as a pressure blower. These units are generally direct connected to the driver. They not only compete with centrifugal compressors but resemble them.

Large industrial process and **pollution-control systems** involving more than about 200 hp are generally satisfied with a somewhat more sophisticated fan than an industrial exhauster or pressure blower. Centrifugal fans with radial-tip blades are frequently used on the more severe service. For the less severe requirements, backward-curved or airfoil blades may be used. Rugged fixed-pitch axials have also been used. Industrial fans are usually equipped with inlet boxes and independently mounted bearings and are usually direct-driven. Journal bearings are usually preferred. Inlet-box damper control can be used to approximate the power saving available from variable-inlet vane control. Variable-speed hydraulic couplings may be economically justified in some cases. Special methods or special construction may be required to provide protection against corrosion or erosion. These fans tend to take on the name of the application such as sintering fan, scrubber exhaust fan, etc.

Mechanical draft systems may utilize any of the fan types described in connection with the above applications. Ventilating fans, industrial exhausters, and pressure blowers have been used for forced draft, induced draft, and primary air service on small steam-generating units. The large generating units are generally equipped with the most efficient fans available consistent with the erosion-corrosion potential of the gas being handled.

Both centrifugals and axials are used for forced and induced draft. Axials have predominated in Europe, and there is a growing trend throughout the rest of the world toward axials. Centrifugals have been used almost exclusively in the United States until very recently.

Forced-draft centrifugals invariably have airfoil blade impellers. Induced-draft centrifugals may have airfoil-blade impellers, but for scrubber exhaust, radial-tip blades are more common. This is due, in part, to the high pressures required for scrubber operation and in part to the erosion-corrosion potential downstream of a scrubber. Gas recirculating fans are usually of the radial-tip design. Forced-draft control may be by variable speed but is more likely to be variable vanes on centrifugal fans. Variable inlet vanes or inlet-box dampers may be used to control induced-draft fans. All large centrifugals are direct-connected and have independent pedestal-mounted bearings. Journal bearings are almost always used.

Forced-draft axials are likely to be of the variable-pitch full-airfoil-section design. Hydraulic systems are almost always used for pitch control, but pneumatic and mechanical systems have been tried. Bearings are usually of the antifriction type. Fixed-pitch axials are usually used for induced-draft duty. Control is by variable-inlet vanes. Either journal or antifriction bearings may be used.

Other Systems

Fans are incorporated in many different kinds of machines. Electronic equipment may require cooling fans to prevent hot spots. Driers use fans to circulate air to carry heat to, and moisture away from, the product. Air-support structures require fans to inflate them and maintain the supporting pressure. Ground-effect machines use fans to provide the lift pressure. Air conditioners and other heat exchangers incorporate fans. Aerodynamic, economic, and functional considerations will dictate the type and size of fan to be used.

Tunnel ventilation can be achieved by using either axial or centrifugal fans. Transverse ventilation utilizes supply and exhaust fans connected to the tunnel by ductwork. These fans are rated in the manner described above. Another method utilizes specially tested and rated axial fans called **jet fans.** The fans, which are placed in the tunnel, increase the momentum of some of the air flowing with the traffic. This air induces additional flow.

Section **15**

Electrical and Electronics Engineering

BY

C. JAMES ERICKSON *Retired Principal Consultant, E. I. du Pont de Nemours & Co., Inc.*
NICHOLAS R. RAFFERTY *Retired Technical Associate, E. I. du Pont de Nemours & Co., Inc.*
BYRON M. JONES *Consulting Engineer, Assistant Professor of Electrical Engineering,*
  University of Wisconsin—Platteville
TIMOTHY M. COCKERILL *Senior Project Manager, University of Illinois*

15.1 ELECTRICAL ENGINEERING
by C. James Erickson
revised by Nicholas R. Rafferty

REFERENCES: Knowlton, "Standard Handbook for Electrical Engineers," McGraw-Hill. Pender and Del Mar, "Electrical Engineers' Handbook," Wiley. Dawes, "Course in Electrical Engineering," Vols. I and II, McGraw-Hill. Gray, "Principles and Practice of Electrical Engineering," McGraw-Hill. Laws, "Electrical Measurements," McGraw-Hill. Karapetoff-Dennison, "Experimental Electrical Engineering and Manual for Electrical Testing," Wiley. Langsdorf, "Principles of Direct-current Machines," McGraw-Hill. Hehre and Harness, "Electric Circuits and Machinery," Vols. I and II, Wiley. Timbie-Higbie, "Alternating Current Electricity and Its Application to Industry," Wiley. Lawrence, "Principles of Alternating-current Machinery," McGraw-Hill. Puchstein and Lloyd, "Alternating-current Machinery," Wiley. Lovell, "Generating Stations," McGraw-Hill. Underhill, "Coils and Magnet Wire" and "Magnets," McGraw-Hill. Abbott, "National Electrical Code Handbook," McGraw-Hill. Dyke, "Automobile and Gasoline Engine Encyclopedia," The Goodheart-Wilcox Co., Inc. Fink and Beaty, "Standard Handbook for Electrical Engineers," McGraw-Hill.

ELECTRICAL AND MAGNETIC UNITS

System of Units The International System of Units (**SI**) is being adopted universally. The SI system has its roots in the metre, kilogram, second (mks) system of units. Since a centimetre, gram, second (**cgs**) system has been widely used, and may still be used in some instances, Tables 15.1.1 and 15.1.2 are provided for conversion between the two systems. Basic SI units have been adopted by the General Conference on Weights and Measures (CPGM) as base quantities, that is, quantities that are not derived from other quantities. The base units are metre, kilogram (mass), second, ampere, kelvin (thermodynamic temperature), and candela (luminous intensity). Other SI units are derived from these basic units.

Electrical Units

(See Table 15.1.1.)

Current (*I*, *i*) The SI unit of current is the **ampere**, which is equal to one-tenth the absolute unit of current (**abampere**). The abampere of current is defined as follows: if 0.01 metre (1 centimetre) of a circuit is bent into an arc of 0.01 metre (1 centimetre) radius, the current is 1 abampere if the magnetic field intensity at the center is 0.01257 ampere per metre (1 oersted), provided the remainder of the circuit produces no magnetic effect at the center of the arc. One **international ampere** (9.99835 amperes) (dc) will deposit 0.001118 gram per second of silver from a standard silver solution.

Quantity (*Q*) The **coulomb** is the quantity of electricity transported in one second by a current of one ampere.

Potential Difference or Electromotive Force (*V*, *E*, emf) The **volt** is the difference of electric potential between two points of a conductor carrying a constant current of one ampere, when the power dissipated between these points is equal to one watt.

Resistance (*R*, *r*) The **ohm** is the electrical resistance between two points of a conductor when a constant difference of potential of one volt, applied between these two points, produces in this conductor a current of one ampere, this conductor not being the source of any electromotive force.

Resistivity (*ρ*) The resistivity of a material is the dc resistance between the opposite parallel faces of a portion of the material having unit length and unit cross section.

Conductance (*G*, *g*) The **siemens** is the electrical conductance of a conductor in which a current of one ampere is produced by an electric potential difference of one volt. One siemens is the reciprocal of one ohm.

Conductivity (*γ*) The conductivity of a material is the dc conductance between the opposite parallel faces of a portion of the material having unit length and unit cross section.

Capacitance (*C*) is that property of a system of conductors and dielectrics which permits the storage of electricity when potential difference exists between the conductors. Its value is expressed as a ratio of a quantity of electricity to a potential difference. A capacitance value is always positive. The **farad** is the capacitance of a capacitor between the plates of which there appears a difference of potential of one volt when it is charged by a quantity of electricity equal to one coulomb.

Permittivity or dielectric constant (ε_0) is the electrostatic energy stored per unit volume of a vacuum for unit potential gradient. The permittivity of a vacuum or free space is 8.85×10^{-12} farads per metre.

Relative permittivity or dielectric constant (ε_r) is the ratio of electrostatic energy stored per unit volume of a dielectric for a unit potential gradient to the permittivity (ε_0) of a vacuum. The relative permittivity is a number.

Self-inductance (*L*) is the property of an electric circuit which determines, for a given rate of change of current in the circuit, the emf induced in the same circuit. Thus $e_1 = -L di_1/dt$, where e_1 and i_1 are in the same circuit and *L* is the coefficient of self-inductance.

The **henry** is the inductance of a closed circuit in which an electromotive force of one volt is produced when the electric current varies uniformly at a rate of one ampere per second.

Mutual inductance (*M*) is the common property of two associated electric circuits which determines, for a given rate of change of current in one of the circuits, the emf induced in the other. Thus $e_1 = -M di_2/dt$ and $e_2 = -M di_1/dt$, where e_1 and i_1 are in circuit 1; e_2 and i_2 are in circuit 2; and *M* is the mutual inductance.

The **henry** is the mutual inductance of two separate circuits in which an electromotive force of one volt is produced in one circuit when the electric current in the other circuit varies uniformly at a rate of one ampere per second.

If *M* is the mutual inductance of two circuits and *k* is the coefficient of coupling, i.e., the proportion of flux produced by one circuit which links the other, then $M = k(L_1 L_2)^{1/2}$, where L_1 and L_2 are the respective **self-inductances** of the two circuits.

Energy (*J*) in a system is measured by the amount of work which a system is capable of doing. The **joule** is the work done when the point of application of a force of one newton is displaced a distance of one metre in the direction of the force.

Power (*W*) is the time rate of transferring or transforming energy. The **watt** is the power required to do work at the rate of one joule per second.

Active power (*W*) at the points of entry of a single-phase, two-wire circuit or of a polyphase circuit is the time average of the values of the instantaneous power at the points of entry, the average being taken over a complete cycle of the alternating current. The value of active power is given in **watts** when the rms currents are in amperes and the rms potential differences are in volts. For sinusoidal emf and current, $W = EI \cos \theta$, where *E* and *I* are the rms values of volts and currents, and *θ* is the phase difference of *E* and *I*.

Reactive power (*Q*) at the points of entry of a single-phase, two-wire circuit, or for the special case of a sinusoidal current and sinusoidal potential difference of the same frequency, is equal to the product obtained by multiplying the rms value of the current by the rms value of the potential difference and by the sine of the angular phase difference by which the current leads or lags the potential difference. $Q = EI \sin \theta$. The unit of *Q* is the **var** (volt-ampere-reactive). One kilovar = 10^3 var.

Apparent power (*VA*) at the points of entry of a single-phase, two-wire circuit is equal to the product of the rms current in one conductor multiplied by the rms potential difference between the two points of entry. Apparent power = *VA*.

Table 15.1.1 Electrical Units

Quantity	Symbol	Equation	SI unit	SI unit symbol	CGS unit	Ratio of magnitude of SI to cgs unit
Current	I, i	$I = E/R$; $I = E/Z$; $I = Q/t$	Ampere	A	Abampere	10^{-1}
Quantity	Q, q	$Q = it$; $Q = CE$	Coulomb	C	Abcoulomb	10^{-1}
Electromotive force	E, e	$E = IR$; $E = W/Q$	Volt	V	Abvolt	10^{8}
Resistance	R, r	$R = E/I$; $R = \rho l/A$	Ohm	Ω	Abohm	10^{9}
Resistivity	ρ	$\rho = RA/l$	Ohm-metre	$\Omega \cdot m$	Abohm-cm	10^{11}
Conductance	G, g	$G = \gamma A/l$; $G = A/\rho l$	Siemens	S	Abmho	10^{-9}
Conductivity	γ	$\gamma = 1/\rho$; $\gamma = l/RA$	Siemens/metre	S/m	Abmho/cm	10^{-11}
Capacitance	C	$C = Q/E$	Farad	F	Abfarad*	10^{-9}
Permittivity	ε		Farads/metre	F/m	Stat farad*/cm	8.85×10^{-12}
Relative permittivity	ε_r	$\varepsilon_r = \varepsilon/\varepsilon_0$	Numerical		Numerical	1
Self-inductance	L	$L = -N(d\phi/dt)$	Henry	H	Abhenry	10^{9}
Mutual inductance	M	$M = K(L_1 L_2)^{1/2}$	Henry	H	Abhenry	10^{9}
Energy	J	$J = eit$	Joule	J	Erg	10^{7}
	kWh	kWh = kW/3,600; 3.6 MJ	Kilowatthour	kWh		36×10^{12}
Active power	W	$W = J/t$; $W = EI \cos \theta$	Watt	W	Abwatt	10^{7}
Reactive power	jQ	$Q = EI \sin \theta$	Var	var	Abvar	10^{7}
Apparent power	VA	$VA = EI$	Volt-ampere	VA		
Power factor	pf	pf $= W/VA$; pf $= W/(W + jQ)$				1
Reactance, inductive	X_L	$X_L = 2\pi fL$	Ohm	Ω	Abohm	10^{9}
Reactance, capacitive	X_C	$X_C = 1/(2\pi fC)$	Ohm	Ω	Abohm	10^{9}
Impedance	Z	$Z = E/I$; $Z = R + j(X_L - X_C)$	Ohm	Ω	Abohm	10^{9}
Conductance	G	$G = R/Z^2$	Siemens	S	Abmho	10^{-9}
Susceptance	B	$B = X/Z^2$	Siemens	S	Abmho	10^{-9}
Admittance	Y	$Y = I/E$; $Y = G + jB$	Siemens	S	Abmho	10^{-9}
Frequency	f	$f = 1/T$	Hertz	Hz	Cps, Hz	1
Period	T	$T = 1/f$	Second	s	Second	1
Time constant	T	L/R; RC	Second	s	Second	1
Angular velocity	ω	$\omega = 2\pi f$	Radians/second	rad/s	Radians/second	1

* 1 Abfarad (EMU units) $= 9 \times 10^{-20}$ stat farads (ESU units).

Power factor (pf) is the ratio of power to apparent power. pf $= W/VA = \cos \theta$, where θ is the phase difference between E and I, both assumed to be sinusoidal.

The **reactance** (X) of a portion of a circuit for a sinusoidal current and potential difference of the same frequency is the product of the sine of the angular phase difference between the current and potential difference times the ratio of the rms potential difference to the rms current, there being no source of power in the portion of the circuit under consideration. $X = (E/I) \sin \theta = 2\pi fL$ ohms, where f is the frequency, and L the inductance in henries; or $X = 1/2\pi fC$ ohms, where C is the capacitance in farads.

The **impedance** (Z) of a portion of an electric circuit to a completely specified periodic current and potential difference is the ratio of the rms value of the potential difference between the terminals to the rms value of the current, there being no source of power in the portion under consideration. $Z = E/I$ ohms.

Admittance (Y) is the reciprocal of impedance. $Y = I/E$ siemens. Conductance is the real part of admittance.

The **susceptance** (B) of a portion of a circuit for a sinusoidal current and potential difference of the same frequency is the product of the sine of the angular phase difference between the current and the potential difference times the ratio of the rms current to the rms potential difference, there being no source of power in the portion of the circuit under consideration. $B = (I/E) \sin \theta$. Susceptance is the imaginary part of admittance.

Magnetic Units

(See Table 15.1.2.)

Magnetic flux (Φ, ϕ) is the magnetic flow that exists in any magnetic circuit.

The **weber** is the magnetic flux which, linking a circuit of one turn, produces in it an electromotive force of one volt as it is reduced to zero at a uniform rate in one second.

Magnetic flux density (β) is the ratio of the flux in any cross section to the area of that cross section, the cross section being taken normal to the direction of flux.

Table 15.1.2 Magnetic Units

Quantity	Symbol	Equation*	SI unit	SI unit symbol	CGS unit	Ratio of magnitude of SI to cgs unit
Magnetic flux	Φ, ϕ	$\phi = F/R$	Weber	wb	Maxwell	10^{8}
Magnetic flux density	β	$\beta = \phi/A$	Tesla	T	Gauss	10^{4}
Pole strength	Q_m	$Q_m = F/\beta$; $Q_m = Fl/NI\mu_0\mu_r$	Ampere-turns-metre	A·m		
			Unit pole		Unit pole	0.7958×10^{7}
Magnetomotive force	$\mathscr{F}$	$\mathscr{F} = NI$	Ampere-turns	A	Gilbert	1.257
Magnetic field intensity	H	$H = \mathscr{F}/l$	Ampere-turns per metre	A/m	Oersted	0.01257
Permeability air	μ_0	$\mu_0 = \beta/H$	Henry per metre	H/m	Gilbert per oersted	1.257×10^{-6}
Relative permeability	μ_r	$\mu_r = \mu/\mu_0$	Numeric		Numeric	1
Reluctivity	γ	$\gamma = 1/\mu_r$	Numeric		Numeric	1
Permeance	P	$P = \mu_0\mu_r A/l$	Henry	H		7.96×10^{7}
Reluctance	R	$R = l/\mu_0\mu_r A$	1/Henry	1/H		1.257×10^{-8}

* l = length in metres; A = area in square metres; F = force in newtons; N = number of turns.

The **tesla** is the magnetic flux density given by a magnetic flux of one weber per square metre.

Unit magnetic pole, when concentrated at a point and placed one metre apart in a vacuum from a second unit magnetic pole, will repel or attract the second unit pole with a force of one **newton.**

The **weber** is the magnetic flux produced by a unit pole.

Magnetomotive force ($\mathscr{F}$, mmf) produces magnetic flux and corresponds to electromotive force in an electric circuit.

The **ampere** (turn) is the unit of mmf.

Magnetic field intensity (H) at a point is the vector quantity which is measured by a mechanical force which is exerted on a unit pole placed at the point in a vacuum.

An **ampere per metre** is the unit of field intensity.

Permeability (μ) is the ratio of unit magnetic flux density to unit magnetic field intensity in air (B/H). The permeability of air is 1.257×10^{-6} **henry per metre.**

Relative permeability (μ_r) is the ratio of the magnetic flux in any element of a medium to the flux that would exist if that element were replaced with air, the magnetomotive force (mmf) acting on the element remaining unchanged ($\mu_r = \mu/\mu_0$).

The relative permeability is a number.

Permeance (P) of a portion of a magnetic circuit bounded by two equipotential surfaces, and by a third surface at every point of which there is a tangent having the direction of the magnetic induction, is the ratio of the flux through any cross section to the magnetic potential difference between the surfaces when taken within the portion under consideration. The equation for the permeance of the medium as defined above is $P = \mu_0\mu_r A/l$. Permeance is the **reciprocal of reluctance.**

Reluctivity (γ) of a medium is the reciprocal of its permeability.

Reluctance (R) is the **reciprocal of permeance.** It is the resistance to magnetic flow. In a homogeneous medium of uniform cross section, reluctance is equal to the length divided by the product of the area and permeability, the length and area being expressed in metre units. $R = l/A\mu_0\mu_r$, where $\mu_0 = 1.257 \times 10^{-6}$.

CONDUCTORS AND RESISTANCE

Resistivity, or **specific resistance,** is the resistance of a sample of the material having both a length and cross section of unity. The two most common resistivity samples are the centimetre cube and the cir mil·ft. If l is the length of a conductor of uniform cross section a, then its resistance is

$$R = \rho l/a \qquad (15.1.1)$$

where ρ is the resistivity. With a cir mil·ft ρ is the resistance of a cir mil·ft and a is the cross section, cir mils. Since $v = la$ is the volume of a conductor,

$$R = \rho l^2/v = \rho v/a^2 \qquad (15.1.2)$$

A **circular mil** is a unit of area equal to that of a circle whose diameter is 1 mil (0.001 in). It is the unit of area which is used almost entirely in this country for wires and cables. To obtain the cir mils of a solid cylindrical conductor, square its diameter expressed in mils. For example, the diameter of 000 AWG solid copper wire is 410 mils and its cross section is $(410)^2 = 168,100$ cir mils. The diameter in mils of a solid cylindrical conductor is the square root of its cross section expressed in cir mils.

A **cir mil·ft** is a conductor having a length of 1 ft and a uniform cross section of 1 cir mil. In terms of the copper standard the resistance of a cir mil·ft of copper at 20°C is 10.371 Ω. As a first approximation 10 Ω may frequently be used.

At 60°C a cir mil·in of copper has a resistance of 1.0 Ω. This is a very convenient unit of resistivity for magnet coils since the resistance is merely the length of copper in inches divided by its cross section in cir mils.

Temperature Coefficient of Resistance The resistance of the pure metals increases with temperature. The resistance at any temperature t°C is

$$R = R_0(1 + \alpha t) \qquad (15.1.3)$$

where R_0 is the resistance at 20°C and α is the **temperature coefficient of resistance.** For copper, $\alpha = 0.00393$.

With any initial temperature t_1, the resistance at temperature t°C is

$$R = R_1[1 + \alpha_1(t - t_1)] \qquad (15.1.4)$$

where R_1 is the resistance at temperature t_1°C and α_1 is the temperature coefficient of resistance at temperature t_1 [see Eq. (15.1.5)].

For any initial temperature t_1 the value of α_1 is

$$\alpha_1 = 1/(234.5 + t_1) \qquad (15.1.5)$$

Inferred Absolute Zero Between 100 and 0°C the resistance of copper decreases at a rate which is practically uniform and which if continued would give a resistance of zero at -234.5°C (an easy number to remember). If the resistance at t_1°C is R_1 and the resistance at t_2°C is R_2, then

$$R_2/R_1 = (234.5 + t_2)/(234.5 + t_1) \qquad (15.1.6)$$

EXAMPLE. The resistance of a copper coil at 25°C is 4.26 Ω. Determine its resistance at 45°C. Using Eq. (15.1.4) and $\alpha_1 = 1/(234.5 + 25) = 0.00385$, $R = 4.26[1 + 0.00385(45 - 25)] = 4.59$ Ω. Using Eq. (15.1.6) $R = 4.26(234.5 + 45)/(234.5 + 25) = 4.26 \times 1.077 = 4.59$ Ω.

The inferred absolute zero for aluminum is -228°C.

Materials The materials generally used for the transmission and distribution of electrical energy are copper, aluminum, and sometimes iron and steel. For resistors and heaters, iron, steel, commercial alloys, and carbon are most used.

Copper is the most widely used electrical conductor. It has high conductivity, relatively low cost, good resistance to oxidation, is readily soldered, and has good mechanical characteristics such as tensile strength, toughness, and ductility. Its tensile strength together with its low linear temperature coefficient of expansion are desirable characteristics in its use for overhead transmission lines. The international copper standard for 100 percent conductivity annealed copper is a density of 8.89 g/cm³ (0.321 lb/in³) and resistivity is given in Table 15.1.3. ASTM specifications for minimum conductivities of copper wire are as follows:

Conductor diam, in	Soft or annealed	Medium hard drawn	Hard drawn
0.040–0.324	98.16%	96.60%	96.16%
0.325–0.460	98.16%	97.66%	97.16%

Aluminum is used to considerable extent for high-voltage transmission lines, because its weight is one-half that of copper for the same conductance. Moreover, the greater diameter reduces corona loss. As it has 1.4 times the linear temperature coefficient of expansion, changes in sag with temperature are greater. Because of its lower melting point, spans may fail more readily with arc-overs. In aluminum cable steel-reinforced (ACSR), the center strand is a steel cable, which gives added tensile strength. Aluminum is used occasionally for bus bars because of its large heat-dissipating surface for a given conductance. The greater cross section for a given conductance requires a greater volume of insulation for a given voltage. When the ratio of the cost of aluminum to the cost of copper becomes economically favorable, aluminum is often used for insulated wires and cables. The international aluminum standard for 62 percent conductivity aluminum is a density of 2.70 g/cm³ (0.0976 lb/in³) and resistivity as given in Table 15.1.3.

Steel, either **galvanized** or **copper-covered** ("copperweld"), is used for high-voltage transmission spans where tensile strength is more important than high conductance. Steel is also used for third rails.

Copper alloys and **bronzes** are of increasing importance as electrical conductors. They have lower electrical conductivity but greater tensile strength and are resistant to corrosion. **Hitenso, Calsum bronzes, Signal bronze, Phono-electric,** and **Everdur** are bronzes containing phosphorus, silicon, manganese, or zinc. Their conductivities vary from 20 to 85

Table 15.1.3   Properties of Metals and Alloys
(See Table 15.1.27 for properties of resistor alloys)

Metals	Resistivity, 20°C		Temperature coefficient of resistance at 20°C
	$\mu\Omega \cdot cm$	$\Omega \cdot cir\ mil/ft$	
Aluminum	2.828	17.01	0.00403
Antimony	42.1	251.0	0.0036
Bismuth	111.0	668.0	0.004
Brass	6.21	37.0	0.0015
Carbon: amorphous	3,800–4,100	—	(−)
Retort (graphite)	720–812*	—	(−)
Copper (drawn)	1.724	10.37	0.00393
Gold	2.44	14.7	0.0034
Iron: electrolytic	10.1	59.9	0.0064
Cast	75.2–98.8	448–588	
Wire	97.8	588	
Lead	22.0	132	0.00387
Molybdenum	5.78	34.8	
Monel metal	43.5	262	0.0019
Mercury	96.8	576	0.00089
Nickel	8.54	50.8	0.0041
Platinum	10.72	63.8	0.003
Platinum silver, 2Ag + 1Pt	24.6†	148.0	0.00031
Silver	1.628	9.8	0.0038
Steel: soft	15.9	95.8	0.0016
Glass hard	45.7	275	
Silicon (4 percent)	51.18	308	
Transformer	11.09	66.7	
Trolley wire	12.7	76.4	
Tin	11.63	70	0.0042
Tungsten	5.51	33.2	0.005
Zinc	5.97	35.58	0.0037

NOTE: Max working temperature: Cu, 260°C; Ni, 600°C; Pt, 1,500°C.
* Furnace electrodes, 3,000°C.
† 0°C.

percent of 100 percent conductivity copper, and they have tensile strengths up to 130,000 lb/in², about twice that of hard-drawn copper. Such alloys were frequently used for trolley wires. Copper alloys having lower conductivity are usually classified as resistor materials.

In Table 15.1.3 are given the electrical properties of some of the pure metals and alloys.

American Wire Gage (AWG)   The AWG (formerly Brown & Sharpe gage) is based on a constant ratio between diameters of successive gage numbers (Table 15.1.4). The ratio of any diameter to the next smaller is 1.123, and the corresponding ratio of cross sections is $(1.123)^2 = 1.261$, or

1¼ approximately. $(1.123)^6$ is 2.0050, so that diameters differing by 6 gage numbers have a ratio of approximately 2; cross sections differing by 3 gage numbers also have a ratio of approximately 2. The ratio of cross sections differing by 2 numbers is $(1.261)^2 = 1.590$, or 1.6 approximately. The ratio of cross sections differing by 10 numbers is approximately 10. The gage ordinarily extends from no. 40 to 0000 (4/0). Wires larger than 0000 must be stranded, and their cross section is given in cir mils.

The diameter of no. 10 wire is 102.0 mils. As an approximation this may be considered as being 100 mils; the cross section is 10,000 cir mils; the resistance is 1 Ω per 1,000 ft; and the weight of 1,000 ft is

Table 15.1.4   Working Table, Standard Annealed Copper Wire, Solid
[American wire gage (B & S)]

Gage no.	Diameter at 20°C		Cross section at 20°C			Ohm per 1,000 ft		Ohm/mi at	Weight
	mils	mm	mil²	cir mils	mm²	25°C	65°C	25°C	per 1,000 ft, lb
0000	460	11.68	166,200	211,600	107.2	0.05	0.0577	0.264	641
000	410	10.40	131,800	167,800	85.01	0.063	0.0727	0.333	508
00	365	9.27	104,500	133,100	67.43	0.0795	0.917	0.42	403
0	325	8.25	82,291	105,600	53.49	0.1	0.116	0.528	319
1	289	7.35	65,730	83,690	42.41	0.126	0.146	0.665	253
2	258	6.54	52,120	66,360	33.62	0.159	0.184	0.839	201
4	204	5.19	32,780	41,740	21.15	0.253	0.292	1.335	126
6	162	4.12	20,610	26,240	13.3	0.403	0.465	2.13	79.5
8	129	3.26	12,970	16,510	8.367	0.641	0.739	3.38	50
10	102	2.59	8,155	10,380	5.261	1.02	1.18	5.38	31.4
12	80	2.05	5,130	6,530	3.31	1.62	1.87	8.55	19.8
14	64	1.63	3,230	4,110	2.08	2.58	2.97	13.62	12.4
16	51	1.29	2,030	2,580	1.31	4.09	4.73	21.6	7.82
18	40	1.02	1,280	1,620	0.823	6.51	7.51	34.4	4.92
20	32	0.81	804	1,020	0.519	10.4	11.9	54.9	3.09
22	25	0.64	503	640	0.324	16.5	19	87.1	1.94
24	20	0.51	317	404	0.205	26.2	30.2	138.3	1.22
26	16	0.40	199	253	0.128	41.6	48	220	0.769
28	13	0.32	125	159	0.0804	66.2	76.4	350	0.484
30	10	0.25	78.5	100	0.0507	105	121	554	0.304

Table 15.1.5 Bare Concentric Lay Cables of Standard Annealed Copper

| AWG no. | cir mils | Ω per 1,000 ft | | Weight per 1,000 ft, lb | Standard concentric standing | | |
		25°C (=77°F)	65°C (=149°F)		No. of wires	Diam of wires, mils	Outside diam, mils
	2,000,000	0.00539	0.00622	6,180	127	125.5	1,631
	1,700,000	0.00634	0.00732	5,250	127	115.7	1,504
	1,500,000	0.00719	0.00830	4,630	91	128.4	1,412
	1,200,000	0.00899	0.0104	3,710	91	114.8	1,263
	1,000,000	0.0108	0.0124	3,090	61	128.0	1,152
	900,000	0.0120	0.0138	2,780	61	121.5	1,093
	850,000	0.0127	0.0146	2,620	61	118.0	1,062
	750,000	0.0144	0.0166	2,320	61	110.9	998
	650,000	0.0166	0.0192	2,010	61	103.2	929
	600,000	0.0180	0.0207	1,850	61	99.2	893
	550,000	0.0196	0.0226	1,700	61	95.0	855
	500,000	0.0216	0.0249	1,540	37	116.2	814
	450,000	0.0240	0.0277	1,390	37	110.3	772
	400,000	0.0270	0.0311	1,240	37	104.0	728
	350,000	0.0308	0.0356	1,080	37	97.3	681
	300,000	0.0360	0.0415	926	37	90.0	630
	250,000	0.0431	0.0498	772	37	82.2	575
0000	212,000	0.0509	0.0587	653	19	105.5	528
000	168,000	0.0642	0.0741	518	19	94.0	470
00	133,000	0.0811	0.0936	411	19	83.7	418
0	106,000	0.102	0.117	326	19	74.5	373
1	83,700	0.129	0.149	258	19	66.4	332
2	66,400	0.162	0.187	205	7	97.4	292
3	52,600	0.205	0.237	163	7	86.7	260
4	41,700	0.259	0.299	129	7	77.2	232

NOTE: See Table 15.1.21 for the carrying capacity of wires.
SOURCE: From *NBS Cir.* 31.

31.4(10π) lb. Also the weight of 1,000 ft of no. 2 is 200 lb. These facts give many short cuts in estimating resistances and weights of various gage numbers.

Lay Cables In order to obtain sufficient flexibility, wires larger than 0000 are stranded, and they are designated by their circular mils (Table 15.1.5). Smaller wires may be stranded also since sizes as small as no. 4 when insulated are usually too stiff for easy handling. Lay cables are made up geometrically as shown in Fig. 15.1.1. Six strands will just fit around the single central conductor; the number of strands in each succeeding layer increases by 6. The number of strands that can thus be laid up are 1, 7, 19, 37, 61, 91, 127, etc. In order to obtain sufficient flexibility with large cables, the strands themselves frequently consist of stranded cable.

Fig. 15.1.1 Makeup of a 19-strand cable.

The **resistance of cables** is readily computed from Eq. (15.1.1), using the cir mil ft as the unit of resistivity.

EXAMPLE. Determine the resistance of 3,500 ft of 800,000 cir mil cable at 20°C. *Answer:* ρ (of a cir mil · ft) = 10.37. R = 10.37 × 3,500/800,000 = 0.0454 Ω.

$\rho = 10$ Ω/cir mil · ft is often sufficiently accurate for practical purposes.

ELECTRICAL CIRCUITS

Figure 15.1.2 shows standard symbols for electrical circuit diagrams.

Ohm's law states that, with a steady current, the current in a circuit is **directly** proportional to the **total** emf acting in the circuit and is **inversely** proportional to the total resistance of the circuit. The law may be expressed by the following three equations:

$$I = E/R \qquad (15.1.7)$$
$$E = IR \qquad (15.1.8)$$
$$R = E/I \qquad (15.1.9)$$

where E is the emf, V; R the resistance, Ω; and I the current, A.

Series Circuits The combined resistance of a number of series-connected resistors is the sum of their separate resistances. When batteries or other sources of emf are connected in series, the total emf of the combination is the sum of the separate emfs. The open-circuit emf of a battery is the total generated emf and can be measured at the battery terminals only when no current is being delivered by the battery. The internal resistance is the resistance of the battery alone. The current in a circuit connected in series with a source of emf is $I = E/(R + r)$, where E is the open-circuit emf, R the external resistance, and r the internal resistance of the source of emf.

Parallel Circuits The combined conductance of a number of parallel-connected resistors is the sum of their separate conductances.

$$G = G_1 + G_2 + G_3 + \cdots \qquad (15.1.10)$$

$$\frac{1}{R} = \frac{1}{R_1} + \frac{1}{R_2} + \frac{1}{R_3} + \cdots \qquad (15.1.11)$$

The equivalent resistance for two parallel resistors having resistances R_1, R_2 is

$$R = R_1 R_2/(R_1 + R_2) \qquad (15.1.12)$$

The equivalent resistance for three parallel resistors having resistances R_1, R_2, R_3 is

$$R = \frac{R_1 R_2 R_3}{R_1 R_2 + R_2 R_3 + R_3 R_1} \qquad (15.1.13)$$

and for four parallel resistors having resistances R_1, R_2, R_3, R_4

$$R = \frac{R_1 R_2 R_3 R_4}{R_1 R_2 R_3 + R_2 R_3 R_4 + R_3 R_4 R_1 + R_4 R_1 R_2} \qquad (15.1.14)$$

Meters and Instruments

A letter or a letter combination from the following list shall be placed within the circle to indicate the function of the meter or instrument unless some other identification is provided in the circle and explained in the diagram.

A	Ammeter.	F	Frequency meter	PF	Power Factor meter	V	Voltmeter
AH	Amp-hr meter	G	Galvanometer	REC	Recording meter	VA	Volt-ammeter
CRO	Cath. Ray Oscill.	μA or UA	Microammeter	SY	Synchroscope	VAR	Varmeter
DM	Demand meter	MA	Milliammeter	+°	Temperature meter	W	Wattmeter
DB	Decibel meter	OHM	Ohmmeter	VARH	Varhour meter	WH	Watthour meter

Fig. 15.1.2 Diagrammatic symbols for electrical machinery and apparatus. (*American Standard, "Graphic Symbols for Electrical and Electronic Diagrams," ANS/IEEE, 315, 1975.*)

To obtain the resistance of combined series and parallel resistors, the equivalent resistance of each parallel portion is obtained separately and then these equivalent resistances are added to the series resistances according to the principles stated above.

Kirchhoff's laws (derived from Ohm's law) make it possible to solve many circuit networks that would otherwise be difficult to solve. The first law states that: *In any branching network of wires the algebraic sum of the currents in all the wires that meet at a point is zero.* The second law states that: *The sum of all the electromotive forces acting around a complete circuit is equal to the sum of the resistances of its separate parts multiplied each by the strength of the current in it, or the total change of potential around any closed circuit is zero.*

In applying Kirchhoff's laws the following rules should be observed. Currents going toward a junction should be preceded by a plus sign. Currents going away from a junction should be preceded by a minus sign. A rise in potential should be preceded by a plus sign. (This occurs in going through a source of emf from the negative to the positive terminal, and in going through resistance in opposition to the direction of current.) A drop in potential should be preceded by a minus sign. (This occurs in going through a source of emf from the positive to the negative terminal and in going through resistance in conjunction with the current.)

The application of Kirchhoff's laws is illustrated by the following example.

EXAMPLE.   Determine the three currents I_1, I_2, and I_3 in the circuit network (Fig. 15.1.3). The arrows show the assumed directions of the three currents.

Applying Kirchhoff's second law to circuit *abcdea*,

$$+4 + 0.2I_1 + 0.5I_1 - 3I_2 + 2 - 0.1I_2 + I_1 = 0$$

or

$$+6 + 1.7I_1 - 3.1I_2 = 0 \qquad (I)$$

and for *edcfge*,

$$-2 + 0.1I_2 + 3I_2 + I_3 + 3 + 0.3I_3 = 0$$

or

$$+1 + 3.1I_2 + 1.3I_3 = 0 \qquad (II)$$

Applying Kirchhoff's first law to junction *c*,

$$-I_1 - I_2 + I_3 = 0 \qquad (III)$$

Solving (I), (II), and (III) simultaneously gives $I_1 = -2.56$, $I_2 = +0.53$, and $I_3 = -2.03$. The minus signs before I_1 and I_3 show that the actual directions of these two currents are opposite the assumed directions.

Fig. 15.1.3   Electric network and Kirchhoff's laws.

Electrical Power   With direct currents the electrical power is given by the product of the volts and amperes. That is,

$$P = EI \quad \text{W} \qquad (15.1.15)$$

Also, by substituting for E and I Eqs. (8) and (7),

$$P = I^2R \quad \text{W} \qquad (15.1.16)$$
$$P = E^2/R \quad \text{W} \qquad (15.1.17)$$

The watt is too small a unit for many purposes. Hence, the **kilowatt** (kW) is used. 746 watts = 1 hp = 0.746 kW; 1 kW = 1.340 hp. The **kilowatthour** (**kWh**) is the common engineering unit of electrical energy.

Joule's Law   When an electric current flows through resistance, the number of heat units developed is proportional to the square of the current,

directly proportional to the resistance, and directly proportional to the time that the current flows. $h = i^2rt$, where h = number of joules; i = current, A; r = resistance, Ω; and t = time, s. h (in Btu) = $0.0009478i^2rt$.

MAGNETISM

Magnetic Circuit   The magnetic circuit is analogous to the electric circuit in that the flux Φ is proportional to the magnetomotive force $\mathscr{F}$ and inversely proportional to the reluctance $\mathscr{R}$ or magnetic resistance. Thus

$$\Phi = \mathscr{F}/\mathscr{R} \qquad (15.1.18)$$

Compare with Eq. (15.1.7). Φ is in webers, where the weber is the SI unit of flux, $\mathscr{F}$ in ampere-turns, and $\mathscr{R}$ in SI reluctance units. In the cgs system, ϕ is in maxwells, $\mathscr{F}$ is in gilberts, and $\mathscr{R}$ is in cgs reluctance units.

$$\mathscr{R} = l/\mu_r\mu_v A \qquad (15.1.19)$$

where μ_r is **relative permeability** (commonly called permeability, μ), a property of the magnetic material, and μ_v is the permeability of evacuated space = $4\pi \times 10^{-7}$, and A is in square metres. In the cgs system $\mu_v = 1$

$$\mathscr{R} = \frac{l}{\mu_r(4\pi \times 10^{-7})A} = \frac{l}{\mu_r(1.257 \times 10^{-6})A} \qquad (15.1.20)$$

l is in metres and A in square metres.

The unit of **flux density** in the SI system is the tesla, which is equal to the number of webers per square metre taken perpendicular to their direction. One ampere-turn between opposite faces of a metre cube of a magnetic medium produces μ_r tesla. For air, $\mu_r = 4\pi \times 10^{-7}$. In the cgs system the unit of flux density is gauss = 10^4T (see Table 15.1.2).

Magnetic-circuit calculations cannot be made with the same degree of accuracy as electric-circuit calculations because of several factors. The cross-sectional dimensions of the magnetic circuit are large relative to its length; magnetic paths are irregular, and their geometry can only be approximated as with the air gap of electric machines, which usually have slots on one or both sides of the gap.

Magnetic flux cannot be confined to definite magnetic paths, but a considerable proportion usually takes paths external to the circuit giving magnetic leakage (see Fig. 15.1.7). The relative permeability of iron varies over wide ranges with the flux density and with the previous magnetic condition (see Fig. 15.1.5). These variations of relative permeability cannot be expressed by any simple equation. Although the foregoing factors prevent the obtaining of extremely high accuracy in magnetic calculations, yet, with experience, it is possible to design magnetic circuits with a precision that is satisfactory for all practical purposes.

The **magnetomotive force** $\mathscr{F}$ in Eq. (15.1.18) is expressed in ampere-turns = NI, where N is the number of turns linked with the circuit and I is the current, A. The unit of reluctance is the reluctance of a 1-m cube of air. The total reluctance is proportional to the length and inversely proportional to the cross-sectional area of the magnetic circuit, which is analogous to electrical resistance. Hence the reluctance of any given path of uniform cross section A is $l/A\mu$, where l = length of path, cm; A = its cross section, cm²; and μ = permeability. Reluctances in series are added to obtain their combined reluctance. Ohm's law of the magnetic circuit becomes

$$\Phi = \frac{NI}{l_1/A_1\mu_1 + l_2/A_2\mu_2 + l_3/A_3\mu_3 \cdots} \text{Mx} \qquad (15.1.21)$$

where l_1, A_1, μ_1, etc., are the lengths, cross sections, and relative permeabilities of each series part of the circuit.

Fig. 15.1.4   Magnetic circuit.

EXAMPLE. In Fig. 15.1.4 is shown a magnetic circuit of cast steel with a 0.4-cm air gap. The cross section of the core is 4 cm square. There are 425 turns wound on the core and the current is 10 A. The relative permeability of the steel at the operating flux density is 1,100. Assume that the path of the flux is as shown, the average path at the corners being quarter circles. Neglect fringing at the air gap and any leakage. Determine the flux and the flux density.

Using the SI system, the length of the iron is 0.522 m, the length of the air gap is 0.004 m, and the cross section of the iron and air gap is 0.0016 m^2.

$$\Phi = \frac{425 \times 10}{\dfrac{0.522}{1{,}100 \times 4\pi \times 10^{-7} \times 0.0016} + \dfrac{0.004}{4\pi \times 10^{-7} \times 0.0016}}$$
$$= 0.00191 \text{ Wb}$$

Using the cgs system, the length of the magnetic path in the iron = $12 + 8 + 8 + 5.8 + 5.8 + 4\pi = 52.2$ cm. From Eq. (15.1.21),

$$\Phi = \frac{0.4\pi \times 425 \times 10}{[52.2/(16 \times 1{,}100)] + (0.4/16)} = 191{,}000 \text{ Mx}$$
$$B = \frac{191{,}000}{16} = 11{,}940 \text{ G}$$

Magnetization and Permeability Curves The magnetic permeability of air is a constant and is taken as unity. The relative permeability of iron and other magnetic substances varies with the flux density. In Fig. 15.1.5 is shown a magnetization curve for cast steel in which the flux density B in tesla is plotted as a function of the field intensity, amperes

Fig. 15.1.5 Magnetization and relative-permeability curves for cast steel.

per metre, H. Also the relative permeability $\mu_r = B/H$ is plotted as a function of the flux density B. Note the wide range over which the relative permeability varies. No satisfactory equation has been found to express the relation between magnetizing force and flux density and between relative permeability and flux density. If an attempt is made to solve Eq. (15.1.21) for flux, the factors μ_1, μ_2, etc., are unknown since they are functions of the flux density, which is being determined. The simplest method is one of trial and error, i.e., a value of flux, and the corresponding permeability, is first assumed, the equation solved for the flux, and if the computed flux differs widely from the assumed flux, a second approximation is made, etc. In nearly all magnetic designs either the flux or flux density is the independent variable, and it is required to find the necessary ampere-turns to produce them. Let the flux $\Phi = BA$ where B is the flux density, G. Then

$$\Phi = BA = 0.4\pi NI(l/A\mu_r)$$
and
$$NI = Bl/\mu_0\mu_r = 0.796Bl/\mu_r \times 10^6 \qquad (15.1.22)$$

Equation (15.1.22) shows that the necessary ampere-turns are proportional to the **flux density** and the length of path and are inversely proportional to the relative permeability.

With air and nonmagnetic substances μ_r [Eq. (15.1.22)] becomes unity, and

$$NI = 0.796Bl \times 10^6 \qquad (15.1.23)$$

in metre units. With inch units

$$NI = 0.313B'l' \qquad (15.1.24)$$

where B' is the flux density, Mx/in^2; and l' the length of the magnetic path, in.

EXAMPLE. The average flux density in the air gap of a generator is 40,000 Mx/in^2, and the effective length of the gap is 0.2 in. How many ampere-turns per pole are necessary for the gap?

$$NI = 0.313 \times 40{,}000 \times 0.2 = 2{,}500$$

Since the relation of μ_r to flux density B in Eq. (15.1.22) is not simple, the relation of ampere-turns per unit length of magnetic circuit to flux density is ordinarily shown graphically. Typical curves of this character are shown in Fig. 15.1.6, inch units being used although scales of tesla, and ampere turns per metre are also given. To determine the number of ampere-turns necessary to produce a given total flux in a magnetic circuit composed of several parts in series having various lengths, cross sections, and relative permeabilities, determine the flux density if the cross section is fixed, or otherwise choose a cross section to give a suitable flux density. From the magnetization curve obtain the ampere-turns necessary to drive this **flux density** through a unit length of the portion of the circuit considered and multiply by the length. Add together the ampere-turns required for each series part of the magnetic circuit to obtain the total ampere-turns necessary to give the assumed flux.

Fig. 15.1.6 Typical magnetization curves.

It is desirable to operate magnetic circuits at as high flux densities as is practicable in order to reduce the amount of iron and copper. The air gaps of dynamos are operated at average densities of 40,000 to 50,000 Mx/in^2. Higher densities increase the exciting ampere-turns and tooth losses. At 45,000 Mx/in^2 the flux density in the teeth may be as high as 120,000 to 130,000 Mx/in^2. The flux densities in transformer cores are limited as a rule by the permissible losses. At 60 Hz and with silicon steel the maximum density is 60,000 to 70,000 Mx/in^2, at 25 Hz the density may run as high as 75,000 to 90,000 Mx/in^2. With laminated cores, the net iron is approximately 0.9 the gross cross section.

Magnetic Leakage It is impossible to confine all magnetic flux to any desired path since there is no known insulator of magnetic flux. Figure 15.1.7 shows the magnetic circuit of a modern four-pole dynamo.

Fig. 15.1.7 Magnetic circuit of a four-pole dynamo with leakage flux.

A considerable proportion of the useful magnetic flux leaks between the pole shoes and cores, rather than across the air gap. The ratio of the maximum flux, which exists in the field cores, to the useful flux, i.e., the flux that crosses the air gap, is the **coefficient of leakage**. This coefficient must always be greater than unity and in carefully designed dynamos may be as low as 1.15. It is frequently as high as 1.30. Although the geometry of the leakage-flux paths is not simple, the leakage flux may be determined by approximations with a fair degree of accuracy.

Magnetic Hysteresis   The magnetization curves shown in Figs. 15.1.5 and 15.1.6 are called **normal curves.** They are taken with the magnetizing force continuously increased from zero. If at any point the magnetizing force be decreased, a greater value of flux density for any given magnetizing force will result. The effect of carrying iron through a complete cycle of magnetization, both positive and negative, is shown in Fig. 15.1.8.

Fig. 15.1.8   Hysteresis loop for dynamo steel.

The curve *OKB*, taken with increasing values of magnetizing force per centimetre *H*, is the **normal induction** curve. If after the magnetizing force has reached the value *OA*, it is decreased, the magnetic flux density *B* will decrease in accordance with curve *BCD*, between *A* and *O* the values being much greater than those given by the normal curve, i.e., the flux density lags the magnetizing force. At zero magnetizing force, the flux density is *OC*, call the **remanence.** A negative magnetizing force *OD*, called the **coercive force,** is required to bring the flux density to zero. If the magnetizing force is increased negatively to *OA′*, the flux density will be given by the curve *DE*. If the magnetizing force is increased positively from *A′* to *A*, the flux density will be given by the curve *EFGB*, which is similar to the curve *BCDE*. *OF* is the negative remanence and *OG* again is the coercive force. The complete curve is called a **hysteresis loop.** When the normal curve reaches the point *K*, if the magnetizing force is then decreased, another hysteresis loop, a portion of which is shown at *KL*, will be obtained. It is seen that the flux density lags the magnetizing force throughout.

The energy dissipated per cycle is proportional to the area of the loop and is equal to $(1/4\pi)\int H\,dB$ ergs/(Hz)(cm³). For moderately high densities the energy loss per cycle varies according to the **Steinmetz law**

$$W = 10\eta B_m^{1.6} \qquad \text{W}\cdot\text{s/m}^3 \qquad (15.1.25)$$

where B_m is the maximum value of the flux density, T (Fig. 15.1.8). Table 15.1.6 gives values of the Steinmetz coefficient η for common magnetic steels.

Table 15.1.6   Steinmetz Coefficients

Hard tungsten steel	0.058	Ordinary sheet iron	0.004
Hard cast steel	0.025	Pure iron	0.003
Forged steel	0.020	Annealed iron sheet	0.002
Cast iron	0.013	Best annealed sheet	0.001
Electrolytic iron	0.009	Silicon steel sheet	0.00046
Soft machine steel	0.009	Permalloy	0.0001
Annealed cast steel	0.008		

A permanent increase in the hysteresis constant occurs if the temperature of operation remains for some time above 80°C. This phenomenon is known as **aging** and may be much reduced by proper annealing of the iron. Silicon steels containing about 3 percent silicon have a lower hysteresis loss, somewhat larger eddy-current loss, and are practically nonaging.

Eddy-current losses, also known as Foucault-current losses, occur in iron subjected to cyclic magnetization. Eddy-current losses are reduced by laminating the iron, which subdivides the emf and increases greatly the length of path of the parasitic currents. Eddy currents also have a screening effect, which tends to prevent the flux penetrating the iron. Hence laminating also allows the full cross section of the iron to be utilized unless the frequency is too high.

Eddy-current loss in sheets is given by

$$P_e = (\pi t f B_m)^2/6\rho 10^{16} \qquad \text{W/cm}^3 \qquad (15.1.26)$$

where t = thickness, cm; f = frequency, Hz; B_m = the maximum flux density, G; ρ = the resistivity, $\Omega \cdot$ cm.

Relations of Direction of Magnetic Flux to Current Direction   The direction of the magnetizing force of a current is at right angles to its direction of flow. Magnetic lines about a cylindrical conductor carrying current exist in circular planes concentric with and normal to the conductor. This is illustrated in Fig. 15.1.9a. The ⊕ sign, corresponding to the feathered end of the arrow, indicates a direction of current away from the observer; a ⊙ sign, corresponding to the tip of an arrow, indicates a direction of current toward the observer.

(a)   (b)

Fig. 15.1.9   Currents in (a) opposite directions, (b) in the same direction.

Corkscrew Rule   The direction of the current and that of the resulting magnetic field are related to each other as the forward travel of a corkscrew and the direction in which it is rotated.

Hand Rule   Grasp the conductor in the right hand with the thumb pointing in the direction of the current. The fingers will then point in the direction of the lines of flux.

The applications of these rules are illustrated in Fig. 15.1.9. If the currents in parallel conductors are in opposite directions (Fig. 15.1.9a), the conductors tend to move apart; if the currents in parallel conductors are in the same direction (Fig. 15.1.9b), the conductors tend to come together. The magnetic lines act like stretched rubber bands and, in attempting to contract, tend to pull the two conductors together.

The relation of the direction of current in a solenoid helix to the direction of flux is shown in Fig. 15.1.10. Figure 15.1.11 shows the

Fig. 15.1.10   Direction of current and poles in a solenoid.

(a)   (b)

Fig. 15.1.11   Effect of a current on a uniform magnetic field.

effect on a uniform field of placing a conductor carrying current in that field and normal to it. In (*a*) the direction of the current is toward the observer. By applying the corkscrew rule it is seen that the current weakens the field immediately above it and strengthens the field immediately below it. The reverse is true in (*b*), where the direction of the current is away from the observer.

Figure 15.1.11 is illustrative of the force developed on a conductor carrying current in a magnetic field. In (*a*) the conductor will tend to move upward owing to the stretching of the magnetic lines beneath it. Similarly, the conductor in (*b*) will tend to move downward. This principle is the basis of motor action. (See also "Magnets.")

BATTERIES

In an **electric cell,** or **battery,** chemical energy is converted into electrical energy. The word *battery* may be used for a single cell or for an assembly of cells connected in series or parallel. A battery utilizes the potential difference which exists between different elements. When two different elements are immersed in electrolyte an emf exists tending to send current within the cell from the negative pole, which is the more highly electropositive, to the positive pole. The **poles,** or **electrodes** of a battery form the junction with the external circuit.

If the external circuit is closed, current flows from the battery at the **positive electrode,** or **anode,** and enters the battery at the **negative electrode,** or **cathode.**

In a **primary battery** the chemically reacting parts **require renewal;** in a **secondary battery,** the electrochemical processes **are reversible** to a high degree and the chemically reacting parts are restored after partial or complete discharge by reversing the direction of current through the battery. See Table 15.1.7 for a summary of battery types and applications.

Electromotive force of a battery is the total potential difference existing between the electrodes on open circuit. When current flows, the potential difference across the terminal drops because of the resistance drop within the cell and because of **polarization.**

Polarization   When current flows in a battery, hydrogen is deposited on the cathode. This produces two effects, both of which reduce the terminal voltage of the battery. The hydrogen in contact with the cathode constitutes a hydrogen battery which opposes the emf of the battery; the hydrogen bubbles reduce the contact area of the electrolyte with the cathode, thus increasing the battery resistance. The most satisfactory method of reducing polarization is to have present at the cathode some compound that supplies negative ions to combine with the positive hydrogen ions at the plate. In the Leclanché cell, manganese

peroxide in contact with the carbon cathode serves as a depolarizer, its oxygen ion combining with the hydrogen ion to form water.

If E is the emf of the cell, E_p the emf of polarization, r the internal resistance, V the terminal voltage, when current I flows, then

$$V = (E - E_p) - Ir \qquad (15.1.27)$$

Primary Batteries

Dry Cells   A dry cell is one in which the electrolyte exists in the form of a jelly, is absorbed in a porous medium, or is otherwise restrained from flowing from its intended position, such a cell being completely portable and the electrolyte nonspillable. The Leclanché cell consists of a cylindrical zinc container which serves as the negative electrode and is lined with specially prepared paper, or some similar absorbent material, to prevent the mixture of carbon and manganese dioxide, which is tamped tightly around the positive carbon electrode, from coming in contact with the zinc. The absorbent lining and the mixture are moistened with a solution of zinc chloride and sal ammoniac. In smaller cells (Fig. 15.1.12) the manganese-carbon mixture is often

Fig. 15.1.12   Cross section of a standard round zinc-carbon cell. *(From "Standard Handbook for Electrical Engineers," Fink and Carrol, McGraw-Hill, NY, copyright 2000.)*

Table 15.1.7   Battery Types and Applications

Battery type	Cell type	Nominal cell voltage	Capacity, wH/kg	Applications
Primary				
Leclanché (zinc-carbon)	Dry	1.5	22–44	Flashlights, emergency lights, radios
Zinc-mercury (Ruben)	Dry	1.34	90–110	Medical, marine, space, laboratory, and emergency devices
Zinc-alkaline-manganese dioxide	Dry	1.5	66	Models, cameras, shavers, lights
Silver or cuprous chloride-magnesium	Wet		55–120	Disposable devices: torpedoes, rescue beacons, meteorological balloons
Secondary				
Lead-acid	Wet	2		Automotive, industrial trucks, railway, station service
Lead-calcium	Wet	2		Standby
Edison (nickel-iron)	Wet	1.2		Industrial trucks; boat and train lights
Nickel-cadmium	Wet	1.2	28	Engine starting, emergency lighting, station service
Silver oxide-cadmium	Wet	1.4	45–65	Space
Silver-zinc	Wet	1.55	90–155	Models, photographic equipment, missiles

molded into a cylinder around the carbon electrode, the whole is then set into the zinc cup, and the space between the molded mixture and the zinc is filled with electrolyte made into a paste in such a manner that it can be solidified by either standing or heating. The top of the cell is closed with a sealing compound, and the cell is placed in a cardboard container. The emf of a dry cell when new is 1.4 to 1.6 V.

In **block assembly** the dry cells, especially in the smaller sizes, are assembled in series and sealed in blocks of insulating compound with only two terminals and, sometimes, intermediate taps brought out. This type of battery is used for radio B and C batteries. Another construction is to build the battery up of layers in somewhat the manner of the old voltaic pile. Each cell consists of a layer of zinc, a layer of treated paper, and a flat cake of the manganese-carbon mixture. The cells are separated by layers of a special material which conducts electricity but which is impervious to electrolyte. A sufficient number of such cells are built up to give the required voltage and the whole battery is sealed into the carton.

Leclanché cells (carbon-zinc dry cells), in use for over 100 years, are still the most widely used of all dry-cell batteries because of their low cost, reliable performance, and ready availability. They are available in sizes ranging from small penlight batteries to large assemblies of cells in series or parallel connections for special high-voltage or high-current applications.

The **efficiency** of a standard-size dry battery depends on the rate at which it is discharged. Up to a certain rate the lower the discharge rate, the greater the efficiency. Above this rate the efficiency decreases (see *Natl. Bur. Stand. Circ.* 79, p. 39).

When used efficiently, a 6-in dry cell will give over 30 A · h of service. As ordinarily used, however, the dry cell give no more than 8 to 10 A · h of service and at times even less. The 1¼- by 2¼-in flashlight battery is usually employed with a lamp taking 0.25 to 0.35 A. Under these conditions 3 A · h or thereabouts may be expected if the battery is used for not more than an hour or so a day. The so-called "heavy-duty" radio battery will give about 8 to 10 A · h when efficiently used.

For the best results 6-in dry cells should not be used for current drains of over 0.5 A except for very short periods of time. Flashlight batteries should not be used for higher than the preceding current drain, and heavy-duty radio batteries will give best results if the current drain is kept below 25 mA.

Dry cells should be stored in a cool, dry place. Extreme heat during storage will shorten their life. The cell will not be injured by being frozen but will be as good as new after being brought back to normal temperature. In extreme cold weather dry cells may not give more than half of their normal service. At a temperature of about −30°F they freeze solid and give neither voltage nor current.

The amperage of a dry cell by definition is the current that it will give when it is short-circuited (at about 70°F) through an ammeter which with its leads has a resistance of 0.01 Ω.

The **Ruben cell** (Ruben, Balanced Alkaline Dry Cells, *Trans. Electrochem. Soc.,* **92,** 1947) was developed jointly by the Ruben Laboratories and P. R. Mallory & Company during World War II for the operation of radar equipment and other electronic devices which require a high ratio of ampere-hour capacity to the volume of the cell at higher current densities than were considered practicable for the Leclanché type. The anode is of amalgamated zinc, and the cathode is a mercuric oxide depolarizing material intimately mixed with graphite in order to reduce its electrical resistivity. The electrolyte is a solution of potassium hydroxide (KOH) containing potassium zincate. The cell is made in three forms as shown in Fig. 15.1.13, the wound-anode type (*a*), the button type (*b*), and the cylindrical type (*c*).

The no-load emf of the cell is 1.34 V and remains essentially constant irrespective of time and temperature. Advantages of the cell are long shelf life, which enables them to be stored indefinitely; long service life, about four times that of the Leclanché dry cell of equivalent volume; small weight; a flat voltage characteristic which is advantageous for electronic uses in which the characteristics of tubes vary widely with voltage; adaptability to operating at high temperatures without deterioration; high resistance to shock.

Fig. 15.1.13 Ruben cells. (*a*) Wound-anode; (*b*) button; (*c*) cylindrical. (*From "Standard Handbook for Electrical Engineers," Fink and Carrol, McGraw-Hill, NY, copyright 1968.*)

The **zinc-alkaline-manganese dioxide cell** is a cell especially useful in applications that require a dry cell with relatively heavy or continuous drain. The anode is of amalgamated zinc, and the cathode is a manganese dioxide depolarizing material mixed with graphite for conductivity. The electrolyte is a solution of highly alkaline potassium hydroxide immobilized in cellulosic-type separators. These cells are available in standard-size cylindrical construction and wafer (flat) construction for cassette and tape recorder applications.

Wet Cells The **silver** or **cuprous chloride-magnesium cell** is a one-shot battery with a life of days after the electrolyte is added. A wet cell may be stored for years in a dry state. The cathode is either compacted copper chloride and graphite or sheet silver chloride, while the cathode is a thin magnesium sheet. The electrolyte is a solution of sodium chloride. The silver chloride cells are more expensive and are available in more and larger ratings.

The **Weston cell** is a primary cell used as a standard of emf. It consists of a glass H tube in the bottom of one leg of which is mercury which forms the cathode; in the bottom of the other leg is cadmium amalgam forming the anode. The electrolytes consist of mercurous sulfate and cadmium sulfate. There are two forms of the Weston cell: the saturated or normal cell and the unsaturated cell. In the normal cell the electrolyte is saturated. This is the official standard since it is more permanent than the unsaturated type and can be reproduced with far greater accuracy. When carefully made, the emfs of cells agree within a few parts per million. There is, however, a small temperature coefficient. Although the unsaturated cell is not so reliable as the normal cell and must be standardized, it has a negligible temperature coefficient and is more convenient for general use. The manufacturers recommend that the temperature be not less than 4°C and not more than 40°C and the current should not exceed 0.0001 A. The emf is between 1.0185 and 1.0190 V. Since no appreciable current can be taken from the cell, a null method must be used to utilize its emf.

Storage (Secondary) Batteries

In a **storage battery** the electrolytic action must be **reversible** to a high degree. There are three types of storage batteries; the lead-lead-acid type, the nickel-iron-alkaline type (**Edison** battery), and the nickel-cadmium-alkali type (**Nicad**). In addition, there are various specialized types of cells for scientific and military purposes, and there is continuous development work in the search for higher capacities.

In the manufacture of the **lead-lead-acid cells** there are three general types of plates, or electrodes. In the **Planté type** the active material is electrically formed of pure lead by repeated reversals of the charging current. In the **Faure,** or **pasted-plate,** type, the positive and negative plates are formed by applying a paste, largely of lead oxides (PbO_2, Pb_3O_4), to lead-antimony or lead calcium supporting grids. A current is passed through the plates while they are immersed in weak sulfuric acid, the positive plates being connected as anodes and the negative ones as cathodes. The paste on the positive plates is converted into lead peroxide while that on the negative plate is reduced to spongy lead. The **tubular plate** (iron-clad) type has lead-alloy rods surrounded by perforated dielectric tubes with powdered-lead oxides packed between the rod and tube for the positive plate.

In order to obtain high capacity per unit weight it is necessary to expose a large plate area to the action of the acid. This is done in the Planté plate by "ploughing" with sharp steel disks, and by using corrugated helical inserts as active positive material (Manchester plate). In the pasted plate a large area of the material is necessarily exposed to the action of the acid.

The chemical reactions in a lead cell may be expressed by the following equation, based on the double sulfation theory:

$$PbO_2 \ + \ Pb \ + \ 2H_2SO_4 \ = \ 2PbSO_4 \ + \ 2H_2O$$

| positive plate | negative plate | sulfuric acid | positive and negative plates | water |

Between the extremes of complete charge and discharge, complex combinations of lead and sulfate are formed. After complete discharge a hard insoluble sulfate forms slowly on the plates, and this is reducible only by slow charging. This sulfation is objectionable and should be avoided.

Specific Gravity   Water is formed with discharge and sulfuric acid is formed on charge, consequently the specific gravity must decrease on discharge and increase on charge. The variation of the specific gravity for a stationary battery is shown in Fig. 15.1.14. With starting and

Fig. 15.1.14   Variations of specific gravity in a stationary battery.

vehicle batteries it is necessary to operate the electrolyte from between 1.280 to 1.300 when fully charged to as low as 1.100 when completely discharged. The condition of charge of a battery can be determined by its specific gravity.

Battery electrolyte may be made from concentrated sulfuric acid (oil of vitriol, sp gr 1.84) by **pouring the acid into the water** in the following proportions:

Parts Water to 1 Part Acid

Specific gravity	1.200	1.210	1.240	1.280
Volume	4.3	4.0	3.4	2.75
Weight	2.4	2.2	1.9	1.5

Freezing Temperature of Sulfuric Acid and Water Mixtures

Specific gravity	1.180	1.200	1.240	1.280
Freezing temp, °F	− 6	− 16	− 51	− 90

Voltage   The emf of a lead cell when fully charged and idle is 2.05 to 2.10 V. Discharge lowers the voltage in proportion to the current. When charging at constant current and normal rate, the terminal voltage gradually increases from 2.14 to 2.3 V, then increases rapidly to between 2.5 and 2.6 V (Fig. 15.1.15). This latter interval is known as the **gassing period.** When this period is reached, the charging rate should be reduced in order to avoid waste of power and unnecessary erosion of the plates.

Fig. 15.1.15   Voltage curves on charge and discharge for a lead cell.

Practically all batteries have a **normal rating** based on the 8-h rate of discharge. Thus a 320 A · h battery would have a normal rate of 40 A. The ampere-hour capacity of batteries falls off rapidly with increase in discharge rate.

Effect of Discharge Rate on Battery Capacity

Discharge rate, h	8	5	3	1	⅓	¹⁄₁₀
Percentage of rated capacity,						
Planté type	100	88	75	55.8	37	19.5
Pasted type	100	93	83	63	41	25.5

The following rule may be observed in **charging a lead battery.** The charging rate in amperes should be less than the number of ampere-hours out of the battery. For example, if 200 A · h are out of a battery,

Fig. 15.1.16   Connections for charging a storage battery from (*a*) 110-V dc mains, (*b*) copper oxide rectifier.

a charging rate of 200 A may be used until the ampere-hours out of the battery are reduced appreciably.

There are two common methods of charging: the **constant-current method** and the **constant-potential method.** Figure 15.1.16*a* shows a common method of charging with constant current, provided a low-voltage dc power supply is available. The resistor connected in series may be adjusted to give the required current. Several batteries may be connected in series. Figure 15.1.16*b* shows a more common method, using a copper oxide or silicon rectifier, since ac power supply is more common than dc. The rectifier disks, mounted in a stack, are bridge-connected, the directions of rectification being indicated. The polarity of the two wires can readily be determined by means of a dc voltmeter.

The constant-potential method is to be preferred since the rate automatically tapers off as the cell approaches the charged condition. Without resistance the terminal voltage should be 2.3 V per cell, but it is preferable to use 2.4 to 2.5 V per cell with low resistance in series.

When a battery is being charged, its terminal voltage

$$V = E + Ir \qquad (15.1.28)$$

Compare with Eq. (15.1.27).

When a battery is fully charged, any rate will produce gassing, but the rate may be reduced to such a low value that gassing is practically harmless. This is called the **finishing rate.**

Portable batteries for automobile starting and lighting, airplanes, industrial trucks, electric locomotives, train lighting, and power boats employ the pasted-type plates because of their high discharge rates for a given weight and size. The separators are either of treated grooved wood, perforated hard rubber, glass-wool mats, perforated rubber and grooved wood, or ribbed microporous rubber. In low-priced short-lived batteries for automobiles, grooved wood alone is used; in the better types, the wood is reinforced with perforated hard rubber. Containers for the low-priced short-lived automobile-type starting batteries are of asphaltic compound; for other portable types they are usually of hard rubber.

The **Exide iron-clad** battery is a portable type designed for propelling electric vehicles. The positive plate consists of a lead-antimony frame supporting perforated hard-rubber tubes. An irregular lead-antimony core runs down the center of each tube, and the lead peroxide paste is packed into these tubes so that shedding of active material from the positive plate cannot occur. Pasted negative plates are used. The separators are flat microporous rubber.

Stationary Batteries   The tanks of stationary batteries are made of hard rubber or plastics. When the battery is used for regulating or cycling duty, the positive plates may be of the Planté type because of their long life. However, in most modern installations thick pasted plates are used. Because of the tight fit of the plate assembly within the container and the resulting pressure of the separator against the plate surfaces, shedding of active material is reduced to a minimum and long life is obtained. Pasted negative plates are used in almost all batteries.

A lead storage battery **removed from service** for less than 9 months should be charged once a month if possible; if not, it should be given a

heavy overcharge before discontinuing service. If removed for a longer period, siphon off acid (which may be used again) and fill with fresh water. Allow to stand 15 h and siphon off water. Remove and throw away the wood separators. The battery will now stand indefinitely. To put in service again, install new separators, fill with acid (sp gr 1.210) and charge at normal rate 35 h or until gravity has ceased to rise over a period of 5 h. Charge at a low rate a few hours longer.

The **ampere-hour efficiency** of lead batteries is 85 to 90 percent. The **watthour efficiency** obtained from full charge to discharge at the normal rate and at rated amp-hour is 75 to 80 percent. Batteries which do regulating duty only may have a much higher watthour efficiency.

The **Edison storage cell** when fully charged has a positive plate of nickel pencils filled with a higher nickel oxide and a negative plate of flat nickel-plated-steel stampings containing metallic iron in finely divided form. The active material for the positive plate is nickel hydrate and for the negative plate, iron oxide. The electrolyte is a 21 percent solution of potassium hydrate with lithium hydroxides. The initial emf is about 1.4 V and the average emf about 1.1 V throughout discharge. In Fig. 15.1.17 are shown typical voltage characteristics on charge and discharge for an Edison cell. On account of the higher internal resistance of the cell the battery is not so efficient from the energy standpoint as the lead cell. The jar is welded nickel-plated steel. The battery is compact and extremely light and strong and for these reasons is particularly adapted for propelling electric vehicles and for boat- and train-lighting systems. The battery is rugged, and since there is no opportunity for the growth of active material on the plates or flaking of active material, the battery has long life.

Fig. 15.1.17   Voltage during charge and discharge of an Edison cell.

Nickel-Cadmium-Alkali (Nicad) Battery   The positive active material is nickelic (black) hydroxide mixed with graphite to give it high conductivity. The negative active material is cadmium oxide. Both materials are used in powdered form and are contained within flat perforated steel pockets. These pockets are locked into steel plates, the positive and negative being alike in construction. All steel parts are nickel-plated. A complete plate group consists of a number of positive and negative plates assembled on bolts and terminal posts common to plates of the same polarity. The separators are thin strips of polystyrene, and all other battery insulation is also polystyrene. The entire plate assembly is contained within a welded-steel tank. The electrolyte is potassium hydroxide (KOH), specific gravity 1.210 at 72°F (22°C); it does not enter into any chemical reactions with the electrode materials, and its specific gravity remains constant during charge and discharge, neglecting any slight change due to the small amount of gassing. On charge, the voltage is 1.4 to 1.5 V until near the end when it rises to 1.8 V. On discharge, the voltage is nearly constant at 1.2 V.

Nicad batteries are strong mechanically and are not damaged by overcharge; they hold their charge over long periods of idleness, the active material cannot flake off, the internal resistance is low, there is no corrosion, and the battery has an indefinitely long life. It is a general-purpose battery.

In the **Sonotone** nickel-cadmium battery the positive plates are nickel oxide when the battery is charged, and the negative plates are metallic cadmium. On discharge the positive plates are reduced to a state of lower oxidation, and the negative plates regain oxygen. The electrolyte is a 30 percent solution of potassium hydroxide, the specific gravity of which is 1.29 at room temperature. The case is a transparent plastic. The terminal voltage at the normal discharge rate is 1.2 V per cell.

Rechargeable batteries, exemplified by Gould Nicad cells (Alkaline Battery Division, Gould National Batteries, Inc.), are hermetically sealed nickel-cadmium cells that contain no free alkaline electrolyte. Since there is no spillage or leakage, they can operate in any position, have long life, and require no maintenance or servicing, and their weight is small for their output. They are thus well adapted to power many types of cordless appliances such as tools, hedge shears, cameras, dictating equipment, electric razors, radios, and television sets. The electrodes consist of a plaque of microporous sintered nickel having an extremely high surface area. The electrochemical reactions differ from those of the conventional vented-type alkaline battery, a type which at the end of a charge liberates both oxygen and hydrogen gases as well as electrolytic fumes that must be vented through a valve in the top of the cell. In the sealed nickel-cadmium cell, the negative electrode (at the time that the cell is sealed) never becomes fully charged, and the evolution of hydrogen is completely suppressed. On charging, when the positive electrode has reached its full capacity, the oxygen which has evolved is channeled through the porous separator to the negative electrode and oxidizes the finely divided cadmium of the microporous plate to cadmium hydroxide, which at the same time is reduced to metallic cadmium. The cells are constructed in three different forms: the button type, the cylindrical type, and the prismatic type. Their ratings range from 20 mA · h to 23 A · h. Their average discharge voltage is 1.22 V, and they require 14 h of charge at the normal rate (one-tenth A · h rating), which for a 3.5 A · h cell is 0.35 A.

Precautions in the **care of storage batteries:** An ammeter should not be connected directly across the terminals to test the condition of a cell; a battery should not be left to stand in a discharged condition; a flame should not be brought in the vicinity of a battery that is being charged; the battery should not be allowed to become heated when charging; water should never be added to the concentrated acid—always acid to the water; acid should never be equalized except when the battery is in a charged condition; a battery should never be exposed to the influence of external heat; voltmeter tests should be made when the current is flowing; batteries should always be kept clean. To replace acid lost through slopping, use a solution of 2 parts concentrated sulfuric acid in 5 parts water by weight, unless a hydrometer is at hand to enable the solution to be made up according to the specifications of the makers of the cell.

DIELECTRIC CIRCUITS

Dynamic and Static Electricity   Electricity in motion such as an electric current is dynamic electricity; electricity at rest is static electricity. The two are identical physically. Since static electricity is frequently produced at high voltage and small quantity, the two are frequently considered as being two different types of electricity.

Capacitors

Capacitors (formerly condensers)   Two conducting bodies, or electrodes, separated by a dielectric constitute a capacitor. If a positive charge is placed on one electrode of a capacitor, an equal negative charge is induced on the other. The medium between the capacitor

plates is called a **dielectric.** The dielectric properties of a medium relate to its ability to conduct **dielectric lines.** This is in distinction to its **insulating** properties which relate to its property to conduct **electric current.** For example, air is an excellent insulator but ruptures dielectrically at low voltage. It is not a good dielectric so far as breakdown strength is concerned.

With capacitors

$$Q = CE \qquad (15.1.29)$$
$$C = Q/E \qquad (15.1.30)$$
$$E = Q/C \qquad (15.1.31)$$

where Q = quantity, C; C = capacitance, F; and E = voltage. The unit of capacitance in the practical system is the **farad.** The farad is too large a unit for practical purposes, so that either the **microfarad** (μF) or the **picofarad** (pF) are used. However, in voltage, current, and energy relations the capacitance must be expressed in farads.

The energy stored in a capacitor is

$$W = \tfrac{1}{2}QE = \tfrac{1}{2}CE^2 = \tfrac{1}{2}Q^2/C \qquad \text{J} \qquad (15.1.32)$$

Capacitance of Capacitors   The capacitance of a **parallel-electrode** capacitor (Fig. 15.1.18) is

$$C = \varepsilon_r A/(4\pi d \times 9 \times 10^3) \qquad \mu\text{F} \qquad (15.1.33)$$

where ε_r = relative capacitivity; A = area of one electrode, m²; and d = distance between electrodes, m.

Fig. 15.1.18   Parallel-electrode capacitor.

Fig. 15.1.19   Coaxial-cylinder capacitor.

The capacitance of **coaxial cylindrical** capacitors (Fig. 15.1.19) is

$$C = 0.2171\varepsilon_r l/[9 \times 10^5 \log{(R_2/R_1)}] \qquad \mu\text{F} \qquad (15.1.34)$$

where ε_r is the relative capacitivity and l the length, m. Also

$$C = 0.03882\varepsilon_r/\log(R_2/R_1) \qquad \mu\text{F/mi} \qquad (15.1.35)$$

Equation (15.1.35) is useful in that it is applicable to cables.

The capacitance of two **parallel cylindrical conductors** D m between centers and having radii of r m is

$$C = 0.01941/\log(D/r) \qquad \mu\text{F/mi} \qquad (15.1.36)$$

In practice, the capacitance to neutral or to an infinite conducting plane midway between the conductors and perpendicular to their plane is usually used. The capacitance to neutral is

$$C = 0.03882/\log(D/r) \qquad \mu\text{F/mi} \qquad (15.1.37)$$

Equations (15.1.36) and (15.1.37) are used for calculating the capacitance of overhead transmission lines. When computing charging current, use voltage between lines in (15.1.36) and to neutral in (15.1.37).

Capacitances in Parallel   The equivalent capacitance of capacitances in parallel (Fig. 15.1.20) is

$$C = C_1 + C_2 + C_3 \qquad (15.1.38)$$

Capacitances in parallel are all across the same voltage. If the voltage is E, then the total quantity $Q = CE$ and $Q_1 = C_1 E$, etc.

Fig. 15.1.20   Capacitances in parallel.

Capacitances in Series   The equivalent capacitance C of capacitances in series (Fig. 15.1.21) is found as follows:

$$1/C = 1/C_1 + 1/C_2 + 1/C_3 \qquad (15.1.39)$$

Fig. 15.1.21   Capacitances in series.

If the capacitances are not leaky, the charge Q is the same on each. $Q = CE$, $E_1 = Q/C_1$, $E_2 = Q/C_2$, etc.

Insulators and Dielectrics   Insulating materials are applied to electric circuits to prevent the leakage of current. Insulating materials used with high voltage must not only have a high resistance to leakage current, but must also be able to resist dielectric puncture; i.e., in addition to being a good insulator, the material must be a good dielectric. Insulation resistance is usually expressed in $M\Omega$ and the resistivity given in $M\Omega \cdot cm$. The dielectric strength is usually given in terms of voltage gradient, common units being V/mil, V/mm, and kV/cm. Insulation resistance decreases very rapidly with increase in temperature. Absorbed moisture reduces the insulation resistance, and moisture and humidity have a large effect on surface leakage. In Table 15.1.8 are given the insulating and dielectric properties of several common insulating materials (see also Sec. 6). Dielectric heating of materials is described in Sec. 7.

TRANSIENTS

Induced EMF   If a flux ϕ webers linking N turns of conductor changes, an emf

$$e = -N(d\phi/dt)   \text{V}   (15.1.40)$$

is induced.

Self-inductance   Let a flux ϕ link N turns. The linkages of the circuit are $N\phi$ weber-turns. If the permeability of the circuit is assumed constant, the number of these linkages per ampere is the **self-inductance** or **inductance** of the circuit. The unit of inductance is the **henry.** The inductance is

$$L = N\phi/(i)   \text{H}   (15.1.41)$$

If the permeability changes with the current

$$L = N(d\phi/di)   \text{H}   (15.1.42)$$

The energy stored in the magnetic field

$$W = \frac{1}{2}Li^2   \text{J}   (15.1.43)$$

EMF of Self-induction   If Eq. (15.1.41) is written $Li = N\phi$ and differentiated with respect to the time t, $L(di/dt) = N(d\phi/dt)$ and from Eq. (15.1.40)

$$e = -L(di/dt)   \text{V}   (15.1.44)$$

e is the emf of self-induction. If a rate of change of current of 1 A/s induces an emf of 1 V, the inductance is then 1 H.

Current in Inductive Circuit   If a circuit containing resistance R and inductance L in series is connected across a steady voltage E, the voltage E must supply the iR drop in the circuit and at the same time overcome the emf of self-induction. That is $E = Ri + L \, di/dt$. A solution of this differential equation gives

$$i = (E/R)(1 - e^{-Rt/L})   \text{A}   (15.1.45)$$

where e is the base of the natural system of logarithms.

Fig. 15.1.22   Rise of current in an inductive circuit.

Figure 15.1.22 shows this equation plotted when $E = 10$ V, $R = 20\ \Omega$, $L = 0.6$ H. It is to be noted that inductance causes the current to rise slowly to its Ohm's law value, $I_0 = E/R = {}^{10}\!/_{20} = 0.5$ A. When $t = L/R$, the current has reached 63.2 percent of its Ohm's law value. L/R is the **time constant** of the circuit. In the foregoing circuit, the time constant $L/R = 0.6/20 = 0.03$ s. The initial rate of rise of current is $\tan \alpha = E/L$. If current continued at this rate, it would reach $\alpha = E/R$ in L/R s $[(E/L) \times (L/R) = E/R]$.

Table 15.1.8   Electrical Properties of Insulating Materials

Material	Volume resistivity, $M\Omega \cdot cm$	Dielectric constant, 60 Hz	Dielectric strength	
			V/mil	V/mm
Asbestos board (ebonized)	10^7		55	2×10^3
Bakelite	5–30×10^{11}	4.5–5.5	450–1,400	$(17$–$55) \times 10^3$
Epoxy	10^{14}	3.5–5	300–400	$(12$–$16) \times 10^3$
Fluorocarbons:				
Fluorinated ethylene propylene	10^{18}	2.1	500	20×10^3
Polytetrafluoroethylene	10^{18}	2.1	400	16×10^3
Glass	17×10^9	5.4–9.9	760–3,800	$(3$–$15) \times 10^4$
Magnesium oxide		2.2	300–700	$(12$–$27) \times 10^3$
Mica	10^{14}–10^{17}	4.5–7.5	1,000–4,000	$(4$–$16) \times 10^4$
Nylon	10^{14}–10^{17}	4–7.6	300–400	$(12$–$16) \times 10^3$
Neoprene		7.5	600	23.5×10^3
Oils:				
Mineral	21×10^6	2–4.7	300–400	$(12$–$16) \times 10^3$
Paraffin	10^{15}	2.41	410–550	$(16$–$22) \times 10^3$
Paper		1.7–2.6	110–230	$(4$–$9) \times 10^3$
Paper, treated		2.5–4	500–750	$(20$–$30) \times 10^3$
Phenolic (glass filled)	10^{12}–10^{13}	5–9	140–400	$(5.5$–$16) \times 10^3$
Polyethylene	10^{15}–10^{18}	2.3	450–1,000	$(17$–$40) \times 10^3$
Polyimide	10^{16}–10^{17}	3.5	400	16×10^3
Polyvinyl chloride (flexible)	10^{11}–10^{15}	5–9	300–1,000	$(12$–$40) \times 10^3$
Porcelain	3×10^8	5.7–6.8	240–300	$(9.5$–$12) \times 10^3$
Rubber	10^{14}–10^{16}	2–3.5	500–700	$(20$–$27) \times 10^3$
Rubber (butyl)	10^{18}	2.1		

If a circuit containing inductance and resistance in series is short-circuited when the current is I_0, the equation of current becomes

$$i = I_0 e^{-Rt/L} \quad \text{A} \quad (15.1.46)$$

Figure 15.1.23 shows this equation plotted when $I_0 = 0.5$ A, $R = 20$ Ω, $L = 0.6$ H. It is seen that inductance opposes the decay of current. Inductance always opposes change of current.

Fig. 15.1.23 Decay of current in an inductive circuit.

Mutual Inductance If two circuits having inductances L_1 and L_2 henrys are so related to each other geometrically that any portion of the flux produced by the current in one circuit links the other circuit, the two circuits possess **mutual inductance**. It follows that a change of current in one circuit causes an emf to be induced in the other. Let ε_2 be induced in circuit 2 by a change di_1/dt in circuit 1. Then

$$e_2 = -M\,di_1/dt \quad \text{V} \quad (15.1.47)$$

M is the mutual inductance of the two circuits.

$$M = k\sqrt{L_1 L_2} \quad (15.1.48)$$

where k is the **coefficient of coupling** of the two circuits, or the proportion of the flux in one circuit which links the other. Also a change of current di_2/dt in circuit 2 induces an emf e_1 in circuit 1, $e_1 = -M\,di_2/dt$.

The stored energy is

$$W = \tfrac{1}{2}L_1 I_1^2 + \tfrac{1}{2}L_2 I_2^2 + MI_1 I_2 \quad \text{J} \quad (15.1.49)$$

where I_1 and I_2 are the currents in circuits 1 and 2.

Current in Capacitive Circuit If capacitance C farads and resistance R ohms are connected in series across the steady voltage E, the current is

$$i = (E/R)e^{-t/CR} \quad \text{A} \quad (15.1.50)$$

If a capacitor charged to voltage E is discharged through resistance R, the current is

$$i = -(E/R)e^{-t/CR} \quad \text{A} \quad (15.1.51)$$

Except for sign, these two equations are identical and are of the same form as Eq. (15.1.46).

Fig. 15.1.24 Transient current to a capacitor.

In Fig. 15.1.24 is shown the transient current to a capacitor in series with a resistor when $E = 200$ V, $C = 4.0$ μF, $R = 2$ kΩ. When $t = CR$, the current has reached $1/e = 0.368$ its initial value. CR is the **time constant** of the circuit. The initial rate of decrease of current is tan $\alpha = -E/CR^2$. If the current continued at this rate it would reach zero when the time is CR s. If, in its fully charged condition, the capacitor of Fig. 15.1.24 is discharged through the resistor R, the curve will be the negative of that shown in Fig. 15.1.24.

Resistance, Inductance, and Capacitance in Series If a circuit having resistance, inductance, and capacitance in series is connected across a source of steady voltage, a transient condition results. If $R > \sqrt{4L/C}$, the circuit is nonoscillatory or overdamped.

The current is

$$i = \frac{EC}{\sqrt{R^2C^2 - 4LC}}\left(e^{(-\alpha+\beta)t} - e^{(-\alpha-\beta)t}\right) \quad \text{A} \quad (15.1.52)$$

where $\alpha = R/2L$ and $\beta = (\sqrt{R^2C^2 - 4LC})/2LC$.

In Fig. 15.1.25 is shown the curve corresponding to Eq. (15.1.52). When $R = \sqrt{4L/C}$, the system is **critically damped** and the transient dies out rapidly without oscillation. The current is

$$i = (E/L)te^{-Rt/2L} \quad \text{A} \quad (15.1.53)$$

Figure 15.1.25 shows also the curve corresponding to Eq. (15.1.53).

Fig. 15.1.25 Transient current in nonoscillatory circuits.

If $R < \sqrt{4L/C}$ the transient is oscillatory, being a logarithmically damped sine wave. The current is

$$i = \frac{2EC}{\sqrt{4LC - R^2C^2}}\,e^{-Rt/2L}\sin\frac{\sqrt{4LC - R^2C^2}}{2LC}t \quad \text{A} \quad (15.1.54)$$

The transient oscillates at a frequency very nearly equal to $1/(2\pi\sqrt{LC})$ Hz. This is the **natural frequency** of the circuit.

In Fig. 15.1.26 is shown the curve corresponding to Eq. (15.1.54). If the capacitor, after being charged to E V, is discharged into the foregoing series circuits, the currents are given by Eqs. (15.1.52) to (15.1.54) multiplied by -1. Equations (15.1.52) to (15.1.54) are the same types obtained with dynamic mechanical systems with friction, mass, and elasticity.

Fig. 15.1.26 Transient current in an oscillatory circuit.

ALTERNATING CURRENTS

Sine Waves In the following discussion of alternating currents, sine waves of voltage and current will be assumed. That is, $e = E_m \sin \omega t$ and $i = I_m \sin(\omega t - \theta)$, where E_m and I_m are maximum values of voltage and current; ω, the angular velocity, in rad/s, is equal to $2\pi f$, where f is the frequency; θ is the angle of phase difference.

Cycle; Frequency When any given armature coil has passed a pair of poles, the emf or current has gone through 360 electrical degrees, or 1 **cycle**. An **alternation** is one-half cycle. The **frequency** of a synchronous machine in cycles per second (hertz) is

$$f = NP/120 \quad \text{Hz} \quad (15.1.55)$$

where N is the speed in r/min and P the number of poles. In the United States and Canada the frequency of 60 Hz is almost universal for general lighting and power. For the ac power supply to dc transit systems, and for railroad electrification, a frequency of 25 Hz is used in many installations. In most of Europe and Latin America the frequency of 50 Hz is in general use. In aircraft the frequency of 400 Hz has become standard.

Static inverters make it possible to obtain high and variable frequencies to drive motors at greater than the 3,600 r/min limitation on 60-Hz circuits, and to vary speeds. The textile industry has small motors operating at 12,000 r/min (200 Hz), and larger motors have been run at 6,000 r/min (100 Hz). Many large mainframe computers have been powered at 400 Hz.

The **root-mean-square (rms)**, or **effective, value of a current wave** produces the same heating in a given resistance as a direct current of the same ampere value. Since the heating effect of a current is proportional to i^2r, the rms value is obtained by squaring the ordinates, finding their average value, and extracting the square root, i.e., the rms value is

$$I = \sqrt{1/T \int_0^T i^2\, dt} \qquad A \qquad (15.1.56)$$

where T is the time of a cycle. The rms value I of a sine wave equals $(1/\sqrt{2})I_m = 0.707I_m$.

Average Value of a Wave The average value of a sine wave over a complete cycle is zero. For a half cycle the average is $(2/\pi)I_m$, or $0.637\,I_m$, where I_m is the maximum value of the sine wave. The average value is of importance only occasionally. A dc measuring instrument gives the average value of a pulsating wave. The average value is of use (1) when the effects of the current are proportional to the number of coulombs, as in electrolytic work and (2) when converting alternating to direct current.

Form Factor The form factor of a wave is the ratio of rms value to average value. For a sine wave this is $\pi/(2\sqrt{2}) = 1.11$. This factor is important in that it enters equations for induced emf.

Inductive reactance, $2\pi fL$ or ωL, opposes an alternating current in inductance L. It is expressed in Ω. Reactance is usually denoted by the symbol X. Inductive reactance is denoted by X_L.

The **current** in an inductive reactance X_L when connected across the voltage E is

$$I = E/X_L = E/(2\pi fL) \qquad A \qquad (15.1.57)$$

This current lags the voltage by 90 electrical degrees. Inductance absorbs no energy. The energy stored in the magnetic field during each half cycle is returned to the source during the same half cycle.

Capacitive reactance is $1/(2\pi fC) = 1/\omega C$ and is denoted by X_C, where C is in F. If C is given in μF, $X_C = 10^6\,2\pi fC$. The current in a capacitive reactance X_C when connected across voltage E is

$$I = E/X_C = 2\pi fCE \qquad A \qquad (15.1.58)$$

This current leads the voltage by 90 electrical degrees. Pure capacitance absorbs no energy. The energy stored in the dielectric field during each half cycle is returned to the source during the same half cycle.

Impedance opposes the flow of alternating current and is expressed in Ω. It is denoted by Z. With resistance and inductance in series

$$Z = \sqrt{R^2 + X_L^2} = \sqrt{R^2 + (2\pi fL)^2} \qquad \Omega \qquad (15.1.59)$$

With resistance and capacitance in series

$$Z = \sqrt{R^2 + X_C^2} = \sqrt{R^2 + [1/(2\pi fC)]^2} \qquad \Omega \qquad (15.1.60)$$

With resistance, inductance, and capacitance in series

$$Z = \sqrt{R^2 + (X_L - X_C)^2}$$
$$= \sqrt{R^2 + [2\pi fL - 1/(2\pi fC)]^2} \qquad \Omega \qquad (15.1.61)$$

The current is

$$I = E/\sqrt{R^2 + [2\pi fL - 1/(2\pi fC)]^2} \qquad A \qquad (15.1.62)$$

Phasor or Vector Representation Sine waves of voltage and current can be represented by phasors, these phasors being proportional in magnitude to the waves that they represent. The angle between two phasors is also equal to the time angle existing between the two waves that they represent.

Phasors may be combined as forces are combined in mechanics. Both graphical methods and the methods of complex algebra are used.

Impedances and also admittances may be similarly combined, either graphically or symbolically. The usual method is to resolve series impedances into their component resistances and reactances, then combine all resistances and all reactances, from which the resultant impedance is obtained. Thus $Z_1 + Z_2 = \sqrt{(r_1 + r_2)^2 + (x_1 + x_2)^2}$, where r_1 and x_1 are the components of Z_1, etc.

Phase Difference With resistance only in the circuit, the current and the voltage are in phase with each other; with inductance only in the circuit, the current lags the voltage by 90 electrical degrees; with capacitance only in the circuit, the current leads the voltage by 90 electrical degrees.

With resistance and inductance in series, the voltage leads the current by angle θ where $\tan\theta = X_L/R$. With resistance and capacitance in series, the voltage lags the current by angle θ where $\tan\theta = -X_C/R$.

With resistance, inductance, and capacitance in series, the voltage may lag, lead, or be in phase with the current.

$$\tan\theta = (X_L - X_C)/R = (2\pi fL - 1/2\pi fC)/R \qquad (15.1.63)$$

If $X_L > X_C$ the voltage leads; if $X_L < X_C$ the voltage lags; if $X_L = X_C$ the current and voltage are in phase and the circuit is in resonance.

Power Factor In ac circuits the power $P = I^2R$ where I is the current and R the effective resistance (see below). Also the power

$$P = EI\cos\theta \qquad W \qquad (15.1.64)$$

where θ is the phase angle between E and I. Cos θ is the **power factor** (pf) of the circuit. It can never exceed unity and is usually less than unity.

$$\cos\theta = P/EI \qquad (15.1.65)$$

P is often called the true power. The product EI is the volt-amp (V · A) and is often called the apparent power.

Active or **energy** current is the projection of the total current on the voltage phasor. $I_e = I\cos\theta$. Power $= EI_e$.

Reactive, quadrature, or **wattless current** $I_q = I\sin\theta$ and is the component of the current that contributes no power but increases the I^2R losses of the system. In power systems it should ordinarily be made low.

The **vars** (volt-amp-reactive) are equal to the product of the voltage and reactive current. **Vars** $= EI_q$. **Kilovars** $= EI_q/1,000$.

Effective Resistance When alternating current flows in a circuit, the losses are ordinarily greater than are given by the losses in the ohmic resistance alone. For example, alternating current tends to flow near the surface of conductors (skin effect). If iron is associated with the circuit, eddy-current and hysteresis losses result. These power losses may be accounted for by increasing the ohmic resistance to a value R, where R is the **effective resistance**, $R = P/I^2$. Since the iron losses vary as $I^{1.8}$ to I^2, little error results from this assumption.

Fig. 15.1.27 Resistor, inductor, and capacitor in series.

SOLUTION OF SERIES-CIRCUIT PROBLEM. Let a resistor R of 10 Ω, an inductor L of 0.06 H, and a capacitor C of 60 μF be connected in series across 120-V 60-Hz mains (Fig. 15.1.27). Determine (1) the impedance, (2) the current, (3) the voltage across the resistance, the inductance, the capacitance, (4) the power factor, (5) the power, and (6) the angle of phase difference.

(1) $\omega = 2\pi 60 = 377$. $X_L = 0.06 \times 377 = 22.6\ \Omega$; $X_C = 1/(377 \times 0.000060) = 44.2\ \Omega$; $Z = \sqrt{(10)^2 + (22.6 - 44.2)^2} = 23.8\ \Omega$; (2) $I = 120/23.8 = 5.04$ A; (3) $E_R = IR = 5.04 \times 10 = 50.4$ V; $E_L = IX_L = 5.04 \times 22.6 = 114.0$ V; $E_C = IX_C = 5.04 \times 44.2 = 223$ V; (4) $\tan\theta = (X_L - X_C)/R = -21.6/10 = -2.16$, $\theta = -65.2°$, $\cos\theta = $ pf $= 0.420$; (5) $P = 120 \times 5.04 \times 0.420 = 254$ W; $P = I^2R = (5.04)^2 \times 10 = 254$ W (check); (6) from (4) $\theta = -65.2°$. Voltage lags. The phasor diagram to scale of this circuit is shown in Fig. 15.1.28. Since the current is common for all elements of the circuit, its phasor is laid horizontally along the axis of reference.

Fig. 15.1.28   Phasor diagram for a series circuit.

Resonance   If the voltage E and the resistance R [Eq. (15.1.62)] are fixed, the maximum value of current occurs when $2\pi fL - 1/2\pi fC = 0$. The circuit so far as its terminals are concerned behaves like a nonductive resistor. The current $I = E/R$, the power $P = EI$, and the power factor is unity.

The voltage across the inductor and the voltage across the capacitor are opposite and equal and may be many times greater than the circuit voltage. The frequency

$$f = 1/(2\pi\sqrt{LC}) \quad \text{Hz} \qquad (15.1.66)$$

is the **natural frequency** of the circuit and is the frequency at which it will oscillate if the circuit is not acted upon by some external frequency. This is the principle of radio sending and receiving circuits. Resonant conditions of this type should be avoided in power circuits, as the piling up of voltage may endanger apparatus and insulation.

EXAMPLE.   For what value of the inductance in the circuit (Fig. 15.1.27) will the circuit be in resonance, and what is the voltage across the inductor and capacitor under these conditions?

From Eq. (15.1.66) $L = 1/(2\pi f)^2 C = 0.1173$ H. $I = E/R = 120/10 = 12$ A. $L\omega I = I/C\omega = 0.1173 \times 377 \times 12 = 530$ V. This voltage is over four times the line voltage.

Parallel Circuits   Parallel circuits are used for nearly all power distribution. With several series circuits in parallel it is merely necessary to find the current in each and add all the current phasors vectorially to find the total current. Parallel circuits may be solved analytically.

A series circuit has resistance r_1 and **inductive reactance** x_1. **The conductance** is

$$g_1 = r_1/(r_1^2 + x_1^2) = r_1/Z_1^2 \quad \text{S} \qquad (15.1.67)$$

and the **susceptance** is

$$b_1 = x_1/(r_1^2 + x_1^2) = x_1/Z_1^2 \quad \text{S} \qquad (15.1.68)$$

Conductance is not the reciprocal of resistance unless the reactance is zero; susceptance is not the reciprocal of reactance unless the resistance is zero. With inductive reactance the susceptance is **negative;** with capacitive reactance the susceptance is **positive.**

If a second circuit has resistance r_2 and **capacitive reactance** x_2 in series, $g_2 = r_2/(r_2^2 + x_2^2) = r_2/Z_2^2; b_2 = x_2/(r_2^2 + x_2^2) = x_2/Z_2^2$. The total conductance $G = g_1 + g_2$; the total susceptance $B = -b_1 + b_2$. The **admittance** is

$$Y = \sqrt{G^2 + B^2} = 1/Z \quad \text{S} \qquad (15.1.69)$$

The energy current is EG; the reactive current is EB; the power is

$$P = E^2G \qquad \text{W} \qquad (15.1.70)$$
$$\text{vars} = E^2B \qquad \text{W} \qquad (15.1.71)$$

The power factor is

$$\text{pf} = G/Y \qquad (15.1.72)$$

Also the following relations hold:

$$r = g/(g^2 + b^2) = g/Y^2 \qquad \Omega \qquad (15.1.73)$$
$$x = b/(g^2 + b^2) = b/Y^2 \qquad \Omega \qquad (15.1.74)$$

SOLUTION OF A PARALLEL-CIRCUIT PROBLEM.   In the parallel circuit of Fig. 15.1.29 it is desired to find the joint impedance, the total current, the power in each branch, the total power, and the power factor, when $E = 100, f = 60, R_1 = 2\,\Omega, R_2 = 4\,\Omega, L_1 = 0.00795$ H, $X_1 = 2\pi fL_1 = 3\,\Omega, C_2 = 1,326\,\mu\text{F}, X_2 = 1/2\pi fC_2 = 2\,\Omega, Z_1 = \sqrt{2^2 + 3^2} = 3.6\,\Omega$, and $Y_1 = 1/3.6 = 0.278$ S.

Solution: $g_1 = R_1/(R_1^2 + X_1^2) = 2/13 = 0.154$; $b_1 = -3/13 = -0.231$; $Z_2 = \sqrt{16 + 4} = 4.47$; $Y_2 = 1/4.47 = 0.224$; $g_2 = R_2/(R_2^2 + X_2^2) = 4/(16 + 4) = 0.2$ S; $b_2 = 2/20 = 0.1$ S; $G = g_1 = g_2 = 0.154 + 0.2 = 0.354$ S; $B = b_1 + b_2 = -0.231 + 0.1 = -0.131$ S; $Y = \sqrt{G^2 + B^2} = \sqrt{0.354^2 + (-0.131)^2} = 0.377$ S, and joint impedance $Z = 1/0.377 = 2.65\,\Omega$. Phase angle $\theta = \tan^{-1}(-0.131/0.354) = -20.3°$. $I = EY = 100 \times 0.377 = 37.7$ A; $P_1 = E^2g_1 = 100^2 \times 0.154 = 1,540$ W; $P_2 = E^2g_2 = 100^2 \times 0.2 = 2,000$ W; total power $= E^2G = 100^2 \times 0.354 = 3,540$ W. Power factor $= \cos\theta = 3,540/(100 \times 37.7) = 93.8$ percent.

With parallel circuits, unity power factor is obtained when the algebraic sum of the quadrature currents is zero. That is, $b_1 + b_2 + b_3 \cdots = 0$.

Three-Phase Circuits   Ac generators are usually wound with three armature circuits which are spaced 120 electrical degrees apart on the armature. Hence these coils generate emfs 120 electrical degrees apart. The coils are connected either in Y (star) or in Δ (mesh) as shown in Fig. 15.1.30. Whether Y- or Δ-connected, with a balanced load, the three coil emfs E_c and the three coil currents I_c are equal. In the Y connection the line and coil currents are equal, but the line emfs E_{AB}, E_{BC}, E_{CA} are $\sqrt{3}$ times in magnitude the coil emfs E_{OA},

Fig. 15.1.30   Three-phase connections. (_a_) Y connection; (_b_) Δ connection.

E_{OB}, E_{OC}, since each is the phasor difference of two coil emfs. In the delta connection the line and coil emfs are equal, but I, the line current, is $\sqrt{3}I_c$, the coil current, i.e., it is the phasor difference of the currents in the two coils connected to the line. The power of a coil is $E_cI_c \cos\theta$, so that the total power is $3E_cI_c \cos\theta$. If θ is the angle between coil current and coil voltage, the angle between line current and line voltage will be $30° \pm \theta$. In terms of line current and emf, the power is $\sqrt{3}EI \cos\theta$. A fourth or neutral conductor connected to O is frequently used with the Y connection. The neutral point O is frequently grounded in transmission and distribution circuits. The coil

Fig. 15.1.29   Parallel circuit and phasor diagrams.

emfs are assumed to be sine waves. Under these conditions they balance, so that in the delta connection the sum of the two coil emfs at each instant is balanced by the third coil emf. Even though the third, ninth, fifteenth . . . harmonics, $3(2n + 1)f$, where $n = 0$ or an integer, exist in the coil emfs, they cannot appear between the three external line conductors of the three-phase Y-connected circuit. In the delta circuit, the same harmonics $3(2n + 1)f$ cause local currents to circulate around the mesh. This may cause a very appreciable heating. In a three-phase system the power

$$P = \sqrt{3}\,EI\cos\theta \qquad W \qquad (15.1.75)$$

the power factor is

$$P/\sqrt{3}\,EI \qquad (15.1.76)$$

and the kV · A

$$\sqrt{3}\,EI/1{,}000 \qquad (15.1.77)$$

where E and I are line voltages and currents.

Two-Phase Circuits Two-phase generators have two windings spaced 90 electrical degrees apart on the armature. These windings generate emfs differing in time phase by 90°. The two windings may be independent and power transmitted to the receiver though the two single-phase circuits are entirely insulated from each other. The two circuits may be combined into a two-phase three-wire circuit such as is shown in Fig. 15.1.31, where OA and OB are the generator circuits

Fig. 15.1.31 Two-phase, three-wire circuit.

(or transformer secondaries) and $A'O'$ and $B'O'$ are the load circuits. The wire OO' is the common wire and under balanced conditions carries a current $\sqrt{2}$ times the current wires AA' and BB'. For example, if I_c is the coil current, $\sqrt{2}\,I_c$ will be the value of the current in the common conductor OO'. If E_c is the voltage across OA or OB, $\sqrt{2}\,E_c$ will be the voltage across AB. The power of a two-phase circuit is twice the power in either coil if the load is balanced. Normally, the voltages OA and OB are equal, and the current is the same in both coils. Owing to nonsymmetry and the high degree of unbalancing of this system even under balanced loads, it is not used at the present time for transmission and is little used for distribution.

Four-Phase Circuit A four-phase or quarter-phase circuit is shown in Fig. 15.1.32. The windings AC and BD may be independent or

Fig. 15.1.32 Four-phase or quarter-phase circuit.

connected at O. The voltages AC and BD are 90 electrical degrees apart as in two-phase circuits. If a neutral wire O-O' is added, three different voltages can be obtained. Let E_1 = voltage between O-A, O-B, O-C, O-D. Voltages between A-B, B-C, C-D, D-A = $\sqrt{2}\,E_1$. Voltages between A-C, B-D = $2E_1$. Because of this multiplicity of voltages and the fact that polyphase power apparatus and lamps may be connected at the same time, this system is still used to some extent in distribution.

Advantages of Polyphase Power The advantages of polyphase power over single-phase power are as follows: The output of synchronous generators and most other rotating machinery is from 60 to 90 percent greater when operated polyphase than when operated single phase; pulsating fluxes and corresponding iron losses which occur in many common types of machinery when operated single phase are negligible when operated polyphase; with balanced polyphase loads polyphase power is constant whereas with single phase the power fluctuates over wide limits during the cycle. Because of its minimum number of wires and the fact that it is not easily unbalanced, the three-phase system has for the most part superseded other polyphase systems.

ELECTRICAL INSTRUMENTS AND MEASUREMENTS

Electrical measuring devices that merely indicate, such as ammeters and voltmeters, are called **instruments;** devices that totalize with time such as watthour meters and ampere-hour meters are called **meters.** (See also Sec. 16.) Most types of electrical instruments are available with digital read out.

DC Instruments Direct current and voltage are both measured with an indicating instrument based on the principle of the D'Arsonval galvanometer. A coil with steel pivots and turning in jewel bearings is mounted in a magnetic field produced by permanent magnets. The motion is restrained by two small flat coiled springs, which also serve to conduct the current to the coil. The deflections of the coil are read with a light aluminum pointer attached to the coil and moving over a graduated scale. The same instrument may be used for either current or voltage, but the method of connecting in circuit is different in the two cases. Usually, however, the coil of an instrument to be used as an ammeter is wound with fewer turns of coarser wire than an instrument to be used as a voltmeter and so has lower resistance. The instrument itself is frequently called a **millivoltmeter.** It cannot be used alone to measure voltage of any magnitude since its resistance is so low that it would be burned out if connected across the line. Hence a resistance r' in series with the coil is necessary as indicated in Fig. 15.1.33a in which r_c is the resistance of the coil. From 0.2 to 750 V this resistance is usually within the instrument. For higher voltages an external resistance R, called an extension coil or multiplier (Fig. 15.1.33b), is necessary. Let e be the reading of the instrument, in volts (Fig. 15.1.33b), r the internal resistance of the instrument, including r' and r_c in Eq. (15.1.33a), R the resistance of the multiplier. Then the total voltage is

$$E = e(R + r)/r \qquad (15.1.78)$$

It is clear that by using suitable values of R a voltmeter can be made to have several scales.

Fig. 15.1.33 Voltmeters. (a) Internal resistance; (b) with multiplier.

Instruments themselves can only carry currents of the magnitudes of 0.01 to 0.06 A. To measure larger values of current the instrument is provided with a shunt R (Fig. 15.1.34). The current divides inversely as the resistances r and R of the instrument and the shunt. A low resistance r' within the instrument is connected in series with the coil. This permits some adjustment to the deflection so that the instrument can be adapted to its shunt. Usually most of the current flows through the shunt, and the current in the instrument is negligible in comparison. Up to 50 and 75 A the shunt can be incorporated within the instrument. For larger currents it is usually necessary to have the shunt external to

Fig. 15.1.34   Millivoltmeter with shunt.

the instrument and connect the instrument to the potential terminals of the shunt by means of leads. Any given instrument may have any number of ranges by providing it with a sufficient number of shunts. The range of the usual instrument of this type is approximately 50 mV. Although the same instrument may be used for voltmeters or ammeters, the moving coils of voltmeters are usually wound with more turns of finer wire. They take approximately 0.01 A so that their resistance is approximately 100 Ω/V. Instruments used as ammeters alone operate with 0.01 to 0.06 A.

Permanent-magnet moving-coil instruments may be used to measure unidirectional pulsating currents or voltages and in such cases will indicate the average value of the periodically varying current or voltage.

AC Instruments   Instruments generally used for alternating currents may be divided into five types: **electrodynamometer, iron-vane, thermocouple, rectifier,** and **electronic.** Instruments of the **electrodynamometer** type, the most precise, operate on the principle of one coil carrying current, turning in the magnetic field produced by a second coil carrying current taken from the same circuit. If these circuits or coils are connected in series, the torque exerted on the moving system for a given relative position of the coil system is proportional to the square of the current and is not dependent on the direction of the current. Consequently, the instrument will have a compressed scale at the lower end and will usually have only the upper two-thirds of the scale range useful for accurate measurement. Instruments of this type ordinarily require 0.04 to 0.08 A or more in the moving-coil circuit for full-scale deflection. They read the rms value of the alternating or pulsating current. The wattmeter operates on the electrodynamometer principle. The fixed coil, however, is energized by the current of the circuit, and the moving coil is connected across the potential in series with high resistance. Unless shielded magnetically the foregoing instruments will not, in general, indicate so accurately on direct as on alternating current because of the effects of external stray magnetic fields. Also reversed readings should be taken. **Iron vane** instruments consist of a fixed coil which actuates magnetically a light movable iron vane mounted on a spindle; they are rugged, inexpensive, and may be had in ranges of 30 to 750 V and 0.05 to 100 A. They measure rms values and tend to have compressed scales as in the case of electrodynamometer instruments.

The compressed part of the scale may, however, be extended by changing the shape of the vanes. Such instruments operate with direct current and are accurate to within 1 percent or so. AC instruments of the **induction type** (Westinghouse Electric Corp.) must be used on ac circuits of the frequency for which they have been designed. They are rugged and relatively inexpensive and are used principally for switchboards where a long-scale range and a strong deflecting torque are of particular advantage. **Thermocouple instruments** operate on the **Seebeck effect.** The current to be measured is conducted through a heater wire, and a thermojunction is either in thermal contact with the heater or is very close to it. The emf developed in the thermojunction is measured by a permanent-magnet dc type of instrument. By controlling the shape of the air gap, a nearly uniform scale is obtained. This type of instrument is well adapted to the measurement of high-frequency currents or voltages, and since it operates on the heating effect of current, it is convenient as a transfer instrument between direct current and alternating current.

In the **rectifier-type instrument** the ac voltage or current is rectified, usually by means of a small copper oxide or a selenium-type rectifier, connected in a bridge circuit to give full-wave rectification (Fig. 15.1.35). The rectified current is measured with a dc permanent-magnet-type instrument M. The instrument measures the **average** value of the half waves that have been rectified, and with the sine waves, the average

Fig. 15.1.35   Rectifier-type instrument.

value is 0.9 the rms value. The scale is calibrated to indicate rms values. With nonsinusoidal waves the ratio of average to rms may vary considerably from 0.9 so that the instrument may be in error up to $\pm$ 5 percent from this cause. This type of instrument is widely used in the measurement of high-frequency voltages and currents. **Electronic voltmeters** operate on the principle of the amplification which can be obtained with a transistor. Since the emf to be measured is applied to the base, the instruments take practically no current and hence are adapted to measure potential differences which would change radically were any appreciable current taken by the measuring device. This type of instrument can measure voltages from a few tenths of a volt to several hundred volts, and with a potential divider, up to thousands of volts. They are also adapted to frequencies up to 100 MHz.

Particular care must be used in selecting instruments for measuring the nonsinusoidal waves of rectifier and controlled rectifier circuits. The **electrodynamic, iron vane,** and **thermocouple** instruments will read rms values. The **rectifier** instrument will read average values, while the **electronic** instrument may read either rms or average value, depending on the type.

Power Measurement in Single-Phase Circuits   Wattmeters are not rated primarily in W, but in A and V. For example, with a low power factor the current and voltage coils may be overloaded and yet the needle be well on the scale. The current coil may be carrying several times its rated current, and yet the instrument reads zero because the potential circuit is not closed, etc. Hence it is desirable to use both an ammeter and a voltmeter in conjunction with a wattmeter when measuring power (Fig. 15.1.36a). The instruments themselves consume appreciable power, and correction is often necessary unless these losses are negligible compared with the power being measured. For example, in Fig. 15.1.36a, the wattmeter measures the $I^2 R$ loss in its own current coil and in the ammeter (1 to 2 W each), as well as the loss in the voltmeter ($= E^2/R$ where R is the resistance of the voltmeter). The losses in the ammeter and voltmeter may be eliminated by short-circuiting the

Fig. 15.1.36   Connections of instruments to single-phase load.

ammeter and disconnecting the voltmeter when reading the wattmeter. If the wattmeter is connected as shown in Fig. 15.1.36b, it measures the power taken by its own potential coil (E^2/R_p) which at 110 V is 5 to 7 W. (R_p is the resistance of the potential circuit.) Frequently correction must be made for this power.

Power Measurement in Polyphase Circuits; Three-Wattmeter Method   Let ao, bo, and co be any Y-connected three-phase load (Fig. 15.1.37). Three wattmeters with their current coils in each line and their potential circuits connected to neutral measure the total power, since the power in each load is measured by one of the wattmeters. The connection oo' may, however, be broken, and the total power is still the sum of the three readings; i.e., the power $P = P_1 + P_2 + P_3$. This method is applicable to any system of n wires. The current coil of one

Fig. 15.1.37   Three-wattmeter method.

wattmeter is connected in each of the n wires. The potential circuit of each wattmeter is connected between its own phase wire and a junction in common with all the other potential circuits. The wattmeters must be connected symmetrically, and the readings of any that read negative must be given the negative sign.

In the general case any system of n wires requires at least $n - 1$ wattmeters to measure the power correctly. The $n - 1$ wattmeters are connected in series with $n - 1$ wires. The potential circuit of each is connected between its own phase wire and the wire in which no wattmeter is connected (Fig. 15.1.38).

Fig. 15.1.38   Power measurement in an n-wire system.

The thermal watt converter is also used to measure power. This instrument produces a dc voltage proportional to three-phase ac power.

Three-Phase Systems   The three-wattmeter method (Fig. 15.1.37) is applicable to any three-phase system. It is commonly used with the three-phase four-wire system. If the loads are balanced, $P_1 = P_2 = P_3$ and the power $P = 3P_1$.

The **two-wattmeter** method is most commonly used with three-phase three-wire systems (Fig. 15.1.39). The current coils may be connected in *any* two wires, the potential circuits being connected to the third. It will be recognized that this is adapting the method of Fig. 15.1.38 to three wires. With balanced loads the readings of the wattmeters are $P_1 = Ei \cos(30° + \theta)$, $P_2 = Ei \cos(30° - \theta)$, and $P = P_2 \pm P_1$. θ is the angle of phase difference between coil voltage and current. Since

$$P_1/P_2 = \cos(30° + \theta)/\cos(30° - \theta) \qquad (15.1.79)$$

Table 15.1.9   Ratio P_1/P_2 and Power Factor

P_1/P_2	Power factor	P_1/P_2	Power factor	P_1/P_2	Power factor	P_1/P_2	Power factor
+1.0	1.000	+0.4	0.804	−0.1	0.427	−0.6	0.142
+0.9	0.996	+0.3	0.732	−0.2	0.360	−0.7	0.102
+0.8	0.982	+0.2	0.656	−0.3	0.296	−0.8	0.064
+0.7	0.956	+0.1	0.576	−0.4	0.240	−0.9	0.030
+0.6	0.918	0.0	0.50	−0.5	0.188	−1.0	0.000
+0.5	0.866						

Fig. 15.1.39   Two-wattmeter method.

the power factor is a function of P_1/P_2. Table 15.1.9 gives values of power factor for different ratios of P_1/P_2.

$$P = P_2 + P_1 \text{ when } \theta < 60°.$$

When $\theta = 60°$, pf $= \cos 60° = 0.5$, $P_1 = \cos(30° + 60°) = 0$, $P = P_2$. When $\theta > 60°$, pf < 0.5, $P = P_2 - P_1$. Also,

$$\tan\theta = \sqrt{3}(P_2 - P_1)/(P_2 + P_1) \qquad (15.1.80)$$

In a **polyphase wattmeter** the two single-phase wattmeter elements are combined to act on a single spindle. Hence the adding and subtracting of the individual readings are done automatically. The total power is indicated on one scale. This type of instrument is almost always used on switchboards. The connections of a portable type are shown in Fig. 15.1.40.

Fig. 15.1.40   Connections for a polyphase wattmeter in a three-phase circuit.

In the foregoing instrument connections, Y-connected loads are shown. These methods are equally applicable to delta-connected loads. The two-wattmeter method (Fig. 15.1.39) is obviously adapted to the two-phase three-wire system (Fig. 15.1.31).

Measurement of Energy

Watthour meters record the energy taken by a circuit over some interval of time. Correct registration occurs if the angular velocity of the rotating element at every instant is proportional to the power. The method of accomplishing this with dc meters is illustrated in Fig. 15.1.41. The meter is in reality a small motor. The field coils FF are in series with the line. The armature A is connected across the line, usually in series with a resistor R. The movable field coil F' is in series with the armature A and serves to compensate for friction. C is a small commutator, either of copper or of silver, and the two small brushes are usually of silver. An aluminum disk, rotating between the poles of permanent magnets M, acts as a magnetic brake the retarding torque of which is proportional to the angular velocity of the disk. A small worm and the gears G actuate the recording dials.

Fig. 15.1.41 DC watthour meter.

The following relation, or an equivalent, holds with most types of meter. With each revolution of the disk, K Wh are recorded, where K is the **meter constant** found usually on the disk. It follows that the average watts P over any period of time t s is

$$P = 3,600KN/t \qquad (15.1.81)$$

where N is the revolutions of the disk during that period. Hence, the meter may be calibrated by connecting standardized instruments to measure the average power taken by the load and by counting the revolutions N for t s. Near full load, if the meter registers fast, the magnets M should be moved outward radially; if it registers slow, the magnets should be moved inward. If the meter registers fast at light (5 to 10 percent) load, the starting coil F' should be moved further away from the armature; if it registers slow, F' should be moved nearer the armature. A meter should not register more than 1.5 percent fast or slow, and with calibrated standards it can be made to register to within 1 percent of correct.

The **induction watthour meter** is used with alternating current. Although the dc meter registers correctly with alternating current, it is more expensive than the induction type, the commutator and brushes may cause trouble, and at low power factors compensation is necessary. In the induction watthour meter the driving torque is developed in the aluminum disk by the joint action of the alternating magnetic flux produced by the potential circuit and by the load current. The driving torque and the retarding torque are both developed in the same aluminum disk, hence no commutator and brushes are necessary. The rotating element is very light, and hence the friction torque is small. Equation (15.1.81) applies to this type of meter. When calibrating, the average power W for t s is determined with a calibrated wattmeter. The friction compensation is made at light loads by changing the position of a small hollow stamping with respect to the potential lug. The meter should also be adjusted at low power factor (0.5 is customary). If the meter is slow with lagging current, resistance should be cut out of the compensating circuit; if slow with leading current, resistance should be inserted.

Power-Factor Measurement The usual method of determining power factor is by the use of voltmeter, ammeter, and wattmeter. The wattmeter gives the watts of the circuit, and the product of the voltmeter reading and the ammeter reading gives the volt-amperes. The power factor is the ratio of the two [see Eqs. (15.1.65) and (15.1.76)]. Also single-phase and three-phase power-factor indicators, which can be connected directly in circuit, are on the market.

Instrument Transformers

With voltages higher than 600 V, and even at 600 V, it becomes dangerous and inaccurate to connect instruments and meters directly into power lines. It is also difficult to make potential instruments for voltages in excess of 600 V and ammeters in excess of 60-A ratings. To insulate such instruments from high voltage and at the same time to permit the use of low-range instruments, instrument transformers are used. **Potential transformers** are identical with power transformers except that their volt-ampere rating is low, being 40 to 500 W. Their primaries are wound for line voltage and their secondaries for 110 V. **Current transformers** are designed to go in series with the line, and the rated secondary current is 5 A. The secondary of a current transformer **should always be closed** when current is flowing; it should never be allowed to become open circuited under these conditions. When open-circuited the voltage across the secondary becomes so high as to be dangerous and

the flux becomes so large in magnitude that the transformer overheats. Semiconductors that break down at safe voltages and short current-transformer secondaries are available to ensure that the secondary is closed. The secondaries of both potential and current transformers should be well grounded at one point (Figs. 15.1.42 and 15.1.43). Instrument transformers introduce slight errors because of small variations in their ratio with load. Also there is slight phase displacement in both current and potential transformers. The readings of the instruments must be multiplied by the instrument transformer ratios. The scales of switchboard instruments are usually calibrated to take these ratios into account.

Fig. 15.1.42 Single-phase connections of instruments with transformers.

Figure 15.1.42 shows the use of instrument transformers to measure the voltage, current, power, and kilowatthours of a single-phase load. Figure 15.1.43 shows the connections that would be used to measure the voltage, current, and power of a 26,400-V 600-A three-phase load.

Fig. 15.1.43 Three-phase connections of instruments and instrument transformers.

Measurement of High Voltages Potential transformers such as those shown in Figs. 15.1.42 and 15.1.43 may be used even for very high voltages, but for voltages above 132 kV they become so large and expensive that they are used only sparingly. A convenient method used with testing transformers is the employment of a voltmeter coil, which consists of a coil of a few turns interwoven in the high-voltage winding and insulated from it. The voltage ratio is the ratio of the turns in the high-voltage winding to those in the voltmeter coil. A capacitance voltage divider consists of two or more capacitors connected in series across the high voltage to be measured. A high-impedance voltmeter, such as an electronic one, is connected across the capacitor at the grounded end. The high voltage $V = V_m C_m / C$ V, where V_m is the voltmeter reading, C the capacitance (in μF) of the entire divider, and C_m the equivalent capacitance (in μF) of the capacitor at the grounded end. A bushing potential device consists of a high-voltage-transformer bushing having a capacitance tap brought out from one of the metallic electrodes within the bushing which is near ground potential. This device is obviously a capacitance voltage divider. For testing, **sphere gaps** are used for the very high voltages. Calibration data for sphere gaps are given in the ANSI/IEEE Std. 4-1978 Standard Techniques for High Voltage Testing. Even when it is not being used for the measurement of voltage, it is frequently advisable to connect a sphere gap in parallel with the specimen being tested so as to prevent overvoltages. The gap is set to a slightly higher voltage than that which is desired.

Measurement of Resistance

Voltmeter-Ammeter Method A common method of measuring resistance, known as the voltmeter-ammeter or fall-in-potential method, makes use of an ammeter and a voltmeter. In Fig. 15.1.44, the resistance to be measured is R. The current in the resistor R is I A, which is measured by the ammeter A in series. The drop in potential across the resistor R is measured by the voltmeter V. The current shunted by the voltmeter is so small that it may generally be neglected. A correction

Fig. 15.1.44 Voltmeter-ammeter method for resistance measurement.

may be applied if necessary, for the resistance of the voltmeter is generally given with the instrument. The potential difference divided by the current gives the resistance included between the voltmeter leads. As a check, determinations are generally made with several values of current, which may be varied by means of the controlling resistor r. If the resistance to be measured is that of the armature of a dc machine and the voltmeter leads are placed on the brush holders, the resistance determined will include that of the brush contacts. To measure the resistance of the armature alone, the voltmeter leads should be placed directly on the commutator segments on which the brushes rest but not under the brushes.

Insulation Resistance Insulation resistance is so high that it is usually given in megohms (10^6 ohms, MΩ) rather than in ohms. Insulation resistance tests are important, for although they may not be conclusive they frequently reveal flaws in insulation, poor insulating material, presence of moisture, etc. Such tests are applied to the insulation of electrical machinery from the windings to the frame, to underground cables, to insulators, capacitors, etc.

For moderately low resistances, 1 to 10 MΩ, the voltmeter method given in Fig. 15.1.45, which shows insulation measurement to the frame of the field winding of a generator, may be used. To measure the current when a voltage E is impressed across the resistor R, a high-reading voltmeter V is connected in series with R. The current under this condition with the switch connecting S and A is $E/(R + r)$, where r is the resistance of the voltmeter. A high-resistance voltmeter is necessary, since the method is in reality a comparison of the unknown insulation resistance R with the known resistance r of the voltmeter. Hence, the

Fig. 15.1.45 Voltmeter method for insulation resistance measurement.

resistance of the voltmeter must be comparable with the unknown resistance, or the deflection of the instrument will be so small that the results will be inaccurate. To determine the impressed voltage E, the same voltmeter is used. The switch S connects S and B for this purpose. With these two readings, the unknown resistance is

$$R = r(E - e)/e \qquad (15.1.82)$$

where e is the deflection of the voltmeter when in series with the resistance to be measured as when S is at A. If a special voltmeter, having a resistance of 100 kΩ per 150 V, is available, a resistance of the order of 2 to 3 MΩ may be measured very accurately.

When the insulation resistance is too high to be measured with a voltmeter, a sensitive galvanometer may be used. The connections for measuring the insulation resistance of a cable are shown in Fig. 15.1.46. The battery should have an emf of at least 100 V. Radio B batteries are convenient for this purpose. The method involves comparing the unknown resistance with a standard 0.1 MΩ. To calibrate the galvanometer the cable is short-circuited (dotted line) and the switch S is thrown to position (a). Let the galvanometer deflection be D_1 and the reading of the

Fig. 15.1.46 Measurement of insulation resistance with a galvanometer.

Ayrton shunt S_1. The short circuit is then removed. The 0.1 MΩ is left in circuit since it is usually negligible in comparison with the unknown resistance X. Let the reading of the galvanometer now be D_2 and the reading of the shunt S_2. Then

$$X = 0.1S_2D_1/S_1D_2 \qquad \text{M}\Omega \qquad (15.1.83)$$

When the switch S is thrown to position (b), the cable is short-circuited through the 0.1 MΩ and becomes discharged.

The **Megger** insulation tester is an instrument that indicates insulation resistance directly on a scale. It consists of a small hand or motor-driven generator which generates 500 V, 1,000 V, 2,500 V, or 5,000 V. A clutch slips when the voltage exceeds the rated value. The current through the unknown resistance flows through a moving element consisting of two coils fastened rigidly together, but which move in different portions of the magnetic field. A pointer attached to the spindle of the moving element indicates the insulation resistance directly. These instruments have a range up to 10,000 MΩ and are very convenient where portability and convenience are desirable.

The insulation resistance of electrical machinery may be of doubtful significance as far as dielectric strength is concerned. It varies widely with temperature, humidity, and cleanliness of the parts. When the insulation resistance falls below the prescribed value, it can (in most cases of good design) be brought to the required standard by cleaning and drying the machine. Hence it may be useful in determining whether or not the insulation is in proper condition for a dielectric test. IEEE Std. 62-1978 specifies minimum values of insulation resistance in MΩ = (rated voltage)/(rating in kW + 1,000). If the operating voltage is higher than the rated voltage, the operating voltage should be used. The rule specifies that a dc voltage of 500 be used in testing. If not, the voltage should be specified.

Wheatstone Bridge Resistors from a fraction of an Ω to 100 kΩ and more may be measured with a high degree of precision with the Wheatstone bridge (Fig. 15.1.47). The bridge consists of four resistors

Fig. 15.1.47 Wheatstone bridge.

ABCX connected as shown. *X* is the unknown resistance; *A* and *B* are ratio arms, the resistance units of which are in even decimal Ω as 1, 10, 100, etc. *C* is the rheostat arm. A battery or low-voltage source of direct current is connected across *ab*. A galvanometer *G* of moderate sensitivity is connected across *cd*. The values of *A* and *B* are so chosen that three or four significant figures in the value of *C* are obtained. As a first approximation it is well to make *A* and *B* equal. When the bridge is in balance,

$$X/C = A/B \qquad (15.1.84)$$

The positions of the battery and galvanometer are interchangeable. There are many modifications of the bridge which adapt it to measurements of very low resistances and also to ac measurements.

Kelvin Double Bridge   The simple Wheatstone bridge is not adapted to measuring very low resistances since the contact resistances of the test specimen become comparable with the specimen resistance. This error is avoided in the Kelvin double bridge, the diagram of which is shown in Fig. 15.1.48. The specimen *X*, which may be a short length of copper wire or bus bar, is connected in series with an adjustable calibrated resistor *R* whose resistance is comparable with that of the specimen. The arms *A* and *B* of the bridge are ratio arms usually with decimal values of 1, 10, 100 Ω. One terminal of the galvanometer is connected to *X* and *R* by means of two resistors *a* and *b*. If these resistors are set so that *a/b* = *A/B*, the contact resistance *r* between *X* and *R* is eliminated in the measurement. The contact resistances at *c* and *d* have no effect since at balance the galvanometer current is zero. The contact resistances at *f* and *e* need only be negligible compared with the resistances of arms *A* and *B* both of which are reasonably high. By means of the variable resistor *Rh* the value of current, as indicated by ammeter *A*, may be adjusted to give the necessary sensitivity. When the bridge is in balance,

$$X/R = A/B \qquad (15.1.85)$$

Fig. 15.1.48 Kelvin double bridge.

Potentiometer   The principle of the potentiometer is shown in Fig. 15.1.49. *ab* is a slide wire, and *bc* consists of a number of equal individual resistors between contacts. A battery *Ba* the emf of which is approximately 2 V supplies current to this wire through the adjustable rheostat *R*. A slider *m* makes contact with *ab*, and a contactor *m'* connects with the contacts in *bc*. A galvanometer *G* is in series with the wire connecting to *m*. By means of the double-throw double-pole switch *Sw,* either the standard cell or the unknown emf (*EMF*) may

Fig. 15.1.49 Potentiometer principle.

be connected to *mm'* through the galvanometer *G*. The potentiometer is standardized by throwing *Sw* to the standard-cell side, setting *mm'* so that their positions on *ab* and *bc* correspond to the emf of the standard cell. The rheostat *R* is then adjusted until *G* reads zero. (In commercial potentiometers a dial which may be set directly to the emf of the standard cell is usually provided.) The unknown emf is measured by throwing *Sw* to *EMF* and adjusting *m* and *m'* until *G* reads zero. The advantage of this method of measuring emf is that when the potentiometer is in balance no current is taken from either the standard cell or the source of emf. Potentiometers seldom exceed 1.6 V in range. To measure voltage in excess of this, a **volt box** which acts as a multiplier is used. To measure current, the voltage drop across a standard resistor of suitable value is measured with the potentiometer. For example, with 50 A a 0.01-Ω standard resistance gives a voltage drop of 0.5 V which is well within the range of the potentiometer.

Potentiometers of low range are used extensively with thermocouple pyrometers. Figure 15.1.49 merely illustrates the principle of the potentiometer. There are many modifications, conveniences, etc., not shown in Fig. 15.1.49.

DC GENERATORS

All electrical machines are comprised of a magnetic circuit of iron (or steel) and an electric circuit of copper. In a generator the armature conductors are rotated so that they cut the magnetic flux coming from and entering the field poles. In the dc generator (except the unipolar type) the emf induced in the individual conductors is alternating, but this is rectified by the commutator and brushes, so that the current to the external circuit is unidirectional.

The induced emf in a generator (or motor)

$$E = \phi ZNP/60P' \qquad \text{V} \qquad (15.1.86)$$

where ϕ = flux in webers entering the armature from one north pole; Z = total number of conductors on the armature; N = speed, r/min; P = number of poles; and P' = number of parallel paths through the armature.

Since with a given generator, Z, P, P' are fixed, the induced emf

$$E = K\phi N \qquad \text{V} \qquad (15.1.87)$$

where K is a constant. When the armature delivers current, the terminal volts are

$$V = E - I_a R_a \qquad (15.1.88)$$

where I_a is the armature current and R_a the armature resistance including the brush and contact resistance, which vary somewhat.

There are three standard types of dc generators: the **shunt generator,** the **series generator,** and the **compound generator.**

Shunt Generator   The field of the shunt generator in series with its rheostat is connected directly across the armature as shown in Fig. 15.1.50. This machine maintains approximately constant terminal voltage over its working range of load. An external characteristic of the

Fig. 15.1.50 Shunt generator.

generator is shown in Fig. 15.1.51. As load is applied the terminal voltage drops owing to the armature-resistance drop [Eq. (15.1.88)] and armature reaction which decreases the flux. The drop in terminal voltage reduces the field current which in turn reduces the flux, hence the induced emf, etc. At some point B, usually well above rated current, the foregoing reactions become cumulative and the generator starts to break down. The current reaches a maximum value and then decreases to nearly zero at short circuit. With large machines, point B is well above rated current, the operating range being between O and A. The voltage may be maintained constant by means of the field rheostat. Automatic regulators which operate through field resistance are frequently used to maintain constant voltage.

Shunt generators are used in systems which are all tied together where their stability when in parallel is an advantage. If a generator fails to build up, (1) the load may be connected; (2) the field resistance may be too high; (3) the field circuit may be open; (4) the residual magnetism may be insufficient; (5) the field connection may be reversed.

Series Generator In the series generator (Fig. 15.1.52) the entire load current flows through the field winding, which consists of relatively few turns of wire of sufficient size to carry the entire load current without undue heating. The field excitation, and hence the terminal voltage, depends on the magnitude of the load current. The generator

Fig. 15.1.51 Shunt generator characteristic.

Fig. 15.1.52 Series generator.

supplies an essentially constant current and for years was used to supply series arc lamps for street lighting requiring direct current. Except for some special applications, the series generator is now obsolete.

Compound-Wound Generators By the addition of a series winding to a shunt generator the terminal voltage may be automatically maintained very nearly constant, or, by properly proportioning the series turns, the terminal voltage may be made to increase with load to compensate for loss of voltage in the line, so that approximately constant voltage is maintained at the load. If the shunt field is connected outside the series field (Fig. 15.1.53), the machine is **long shunt;** if the shunt field is connected inside the series field, i.e., directly to the armature terminals, it is **short shunt.** So far as the operating characteristic is concerned, it makes little difference which way a machine is connected. Table 15.1.10 gives performance characteristics.

Fig. 15.1.53 Compound-wound dc generator.

Table 15.1.10 Approximate Test Performance of Compound-Wound DC Generators with Commutating Poles

kW	Speed, r/min	Volts	Amperes	Efficiencies, percent		
				¼ load	½ load	Full load
5	1,750	125	40	77.0	80.5	82.0
10	1,750	125	80	80.0	83.0	85.0
25	1,750	125	200	84.0	86.5	88.0
50	1,750	125	400	83.0	86.0	88.0
100	1,750	125	800	87.0	88.5	90.0
200	1,750	125	1,600	88.0	90.5	91.0
400	1,750	250	1,600	91.7	91.9	91.7
1,000	1,750	250	4,000	92.1	92.6	92.1

SOURCE: Westinghouse Electric Corp.

Compound-wound generators are chiefly used for small isolated plants and for generators supplying a purely motor load subject to rapid fluctuations such as in railway work. When first putting a compound generator in service, the shunt field must be so connected that the machine builds up. The series field is then connected so that it aids the shunt field. Figure 15.1.54 gives the characteristics of an overcompounded 200-kW 600-V compound-wound generator.

Amplidynes The amplidyne is a dc generator in which a small amount of power supplied to a control field controls the generator output, the response being nearly proportional to the control field input. The amplidyne is a dc amplifier which can supply large amounts of power. The amplifier operates on the principle of armature reaction. In Fig. 15.1.55, NN and SS are the conventional north and south poles of a dc generator with central cavities. BB are the usual brushes placed at right angles to the pole axes of NN and SS. A control winding CC of small rating, as low as 100 W, is wound on the field poles. In Fig. 15.1.55, for simplicity, the control winding is shown as being wound on one pole

Fig. 15.1.54 Characteristics of a 200-kW compound-wound dc generator.

only. The brushes BB are short-circuited, so that a small excitation mmf in the control field produces a large short-circuit current along the brush axis BB. This large short-circuit current produces a large armature-reaction flux AA along brush axis BB. The armature rotating in this field produces a large voltage along the brush axis B'B'. The load or working current is taken from brushes B'B' as shown. In Fig. 15.1.55 the working current only is shown by the crosses and dots in the circles. The short-circuit current would be shown by crosses in the conductors to the left of brushes BB and by dots in the conductors to the right of brushes BB.

A small current in the control winding produces a high output voltage and current as a result of the large short-circuit current in brushes BB.

In order that the brushes B'B' shall not be short-circuiting conductors which are cutting the flux of poles NN and SS, cavities are cut in these poles. Also the load current from brushes B'B' produces an armature reaction mmf in opposition to flux A'A' produced by the control field CC. Were this mmf not compensated, the flux A'A' and the output

Fig. 15.1.55   Amplidyne.

of the machine would no longer be determined entirely by the control field. Hence there is a compensating field FF' in series with the armature, which neutralizes the armature-reaction mmf which the load current produces. For simplicity the compensating field is shown on one field pole only.

The amplidyne is capable of controlling and regulating speed, voltage, current, and power with accurate and rapid response. The amplification is from 10,000 to 250,000 times in machines rated from 1 to 50 kW. Amplidynes are frequently used in connection with **selsyns** and are employed for gun and turret control and for accurate controls in many industrial power applications.

Parallel Operation of Shunt Generators   It is desirable to operate generators in parallel so that the station capacity can be adapted to the load. Shunt generators, because of their drooping characteristics (Fig. 15.1.51), are inherently stable when in parallel. To connect shunt generators in parallel it is necessary that the switches be so connected that like poles are connected to the same bus bars when the switches are closed. Assume one generator to be in operation; to connect another generator in parallel with it, the incoming generator is first brought up to speed and its terminal voltage adjusted to a value slightly greater than the bus-bar voltage. This generator may then be connected in parallel with the other without difficulty. The proper division of load between them is adjusted by means of the field rheostats and is maintained automatically if the machines have similar voltage-regulation characteristics.

Parallel Operation of Compound Generators   As a rule, compound generators have either flat or rising voltage characteristics. Therefore, when connected in parallel, they are inherently unstable. Stability may, however, be obtained by using an equalizer connection, Fig. 15.1.56, which connects the terminals of the generator at the junctions of the series fields. This connection is of low resistance so that any increase of current divides proportionally between the series fields of the two machines. The equalizer switch (E.S.) should be closed first and opened last, if possible. In practice, the equalizer switch is often one blade of a

Fig. 15.1.56   Connections for compound-wound generators operating in parallel.

three-pole switch, the other two being the bus switch S, as in Fig. 15.1.56. When compound generators are used on a three-wire system, two series fields—one at each armature terminal—and two equalizers are necessary. It is possible to operate any number of compound generators in parallel provided their characteristics are not too different and the equalizer connection is used.

DC MOTORS

Motors operate on the principle that a conductor carrying current in a magnetic field tends to move at right angles to that field (see Fig. 15.1.11). The ordinary dc generator will operate entirely satisfactorily as a motor and will have the same rating. The conductors of the motor rotate in a magnetic field and therefore must generate an emf just as does the generator. The induced emf

$$E = K\phi N \qquad (15.1.89)$$

where K = constant, ϕ = flux entering the armature from one north pole, and N = r/min [see Eq. (15.1.87)]. This emf is in opposition to the terminal voltage and tends to oppose current entering the armature. Its value is

$$E = V - I_a R_a \qquad (15.1.90)$$

where V = terminal voltage, I_a = armature current, and R_a = armature resistance [compare with Eq. (15.1.88)]. From Eq. (15.1.89) it is seen that the speed

$$N = K_s E/\phi \qquad (15.1.91)$$

when $K_s = 1/K$. This is the fundamental speed equation for a motor. By substituting in Eq. (15.1.90)

$$N = K_s(V - 1_a R_a)/\phi \qquad (15.1.92)$$

which is the general equation for the speed of a motor.

The internal or electromagnetic torque developed by an armature is proportional to the flux and to the armature current; i.e.,

$$T_t = K_t \phi I_a \qquad (15.1.93)$$

when K_t is a constant. The torque at the pulley is slightly less than the internal torque by the torque necessary to overcome the rotational losses, such as friction, windage, eddy-current and hysteresis losses in the armature iron and in the pole faces.

The total mechanical power developed internally

$$P_m = EI_a \quad W \qquad (= EI_a/746 \quad hp) \qquad (15.1.94)$$

The internal torque thus becomes

$$T = EI_a 33,000/(2\pi \times 746N) = 7.04 EI_a/N \qquad (15.1.95)$$

Let VI be the motor input. The output is $VI\eta$ where η is the efficiency. The horsepower is

$$P_H = VI\eta/746 \qquad (15.1.96)$$

and the torque is

$$T = 33,000 P_H/2\pi N = 5,260 P_H/N \quad lb \cdot ft \qquad (15.1.97)$$

where N is r/min.

Shunt Motor   In the shunt motor (Fig. 15.1.57) the flux is substantially constant and $I_a R_a$ is 2 to 6 percent of V. Hence from Eq. (15.1.92), the speed varies only slightly with load (Fig. 15.1.58), so that the motor is adapted to work requiring constant speed. The speed regulation of constant-speed motors is defined by ANSI/IEEE Std. 100-1992 Standard Dictionary of Electrical and Electronic terms as follows:

> The speed regulation of a constant-speed direct-current motor is the change in speed when the load is reduced gradually from the rated value to zero with constant applied voltage and field rheostat setting expressed as a percent of speed at rated load.

In Fig. 15.1.61 the speed regulation under each condition is $100(ac - bc)/bc$ (Fig. 15.1.58a). Also from Eq. (15.1.93) it is seen that the torque

Fig. 15.1.57 Connections for shunt dc motors and starters. (a) Three-point box; (b) four-point box.

is practically proportional to the armature current (see Fig. 15.1.58b). The motor is able to develop full-load torque and more on starting, but the ordinary starter is not designed to carry the current necessary for starting under load. If a motor is to be started under load, the starter

Fig. 15.1.58 Speed and torque characteristics of dc motors. (1) Shunt motor (2) cumulative compound motor; (3) differential compound motor; (4) series motor.

should be provided with resistors adapted to carry the required current without overheating. A controller is also adapted for starting duty under load.

Commutating poles have so improved commutation in dc machines that it is possible to use a much shorter air gap than formerly. Since, with the shorter air gap, fewer field ampere-turns are required, the armature becomes magnetically strong with respect to the field. Hence, a sudden overload might weaken the field through armature reaction, thus causing an increase in speed; the effect may become cumulative and the motor run away. To prevent this, modern shunt motors are usually provided with a **stabilizing winding**, consisting of a few turns of the field in series with the armature and aiding the shunt field. The resulting increase of field ampere-turns with load will more than compensate for any weakening of the field through armature reaction. The series turns are so few that they have no appreciable compounding effect. The shunt motor is used to drive constant-speed line shafting, for machine tools, etc. Since its speed may be efficiently varied, it is very useful when **adjustable speeds** are necessary, such as individual drive for machine tools.

Shunt-Motor Starters At standstill the counter emf of the motor is zero and the armature resistance is very low. Hence, except in motors of very small size, series resistance in the armature circuit is necessary on starting. The field must, however, be connected across the line so that it may obtain full excitation.

Figure 15.1.57 shows the two common types of starting boxes used for starting shunt motors. The armature resistance remains in circuit only during starting. In the **three-point box** (Fig. 15.1.57a) the starting lever is held, against the force of a spring, in the running position, by an electromagnet in series with the field circuit, so that, if the field circuit is interrupted or the line voltage becomes too low, the lever is released

and the armature circuit is opened automatically. In the **four-point starting box** the electromagnet is connected directly across the line, as shown in Fig. 15.1.57b. In this type the arm is released instantly upon failure of the line voltage. In the three-point type some time elapses before the field current drops enough to effect the release. Some starting rheostats are provided with an overload device so that the circuit is automatically interrupted if too large a current is taken by the armature. The four-point box is used where a wide speed range is obtained by means of the field rheostat. The electromagnet is not then affected by changes in field current.

In large motors and in many small motors, automatic starters are widely used. The advantages of the automatic starter are that the current is held between certain maximum and minimum values so that the circuit does not become opened by too rapid starting as may occur with manual operation; the acceleration is smooth and nearly uniform. Since workers can stop and start a motor merely by the pushing of a button, there results considerable saving by the shutting down of the motor when it is not needed. Automatic controls are essential to elevator motors so that smooth rapid acceleration with frequent starting and stopping may be obtained. Also automatic starting is very necessary with multiple-unit operation of electric-railway cars and with rolling-mill motors which are continually subjected to rapid acceleration, stopping, and reversing.

Series Motor In the series motor the armature and field are in series. Hence, if saturation is neglected, the flux is proportional to the current and the torque [Eq. (15.1.93)] varies as the current squared. Therefore any increase in current will produce a much greater proportionate increase in torque (see Fig. 15.1.58b). This makes the motor particularly well adapted to traction work, cranes, hoists, fork-lift trucks, and other types of work which require large starting torques. A study of Eq. (15.1.92) shows that with increase in current the numerator changes only slightly, whereas the change in the denominator is nearly proportional to the change in current. Hence the speed of the series motor is practically inversely proportional to the current. With overloads the speed drops to very low values (see Fig. 15.1.58a). With decrease in load the speed approaches infinity, theoretically. Hence the series motor should always be connected to its load by a direct drive, such as gears, so that it cannot reach unsafe speeds (see Speed Control of Motors). A series-motor starting box with no-voltage release is shown in Fig. 15.1.59.

Differential Compound Motors The cumulative compound winding of a generator becomes a differential compound winding when the machine is used as a motor. Its speed may be made more nearly constant

Fig. 15.1.59 Series motor starter, no-voltage release.

Table 15.1.11   Test Performance of Compound-Wound DC Motors

Power, hp	Speed, r/min	115 V		230 V		550 V	
		Current, A	Full-load efficiency, %	Current, A	Full-load efficiency, %	Current, A	Full-load efficiency, %
1	1,750	8.4	78	4.3	79	1.86	73.0
2	1,750	16.0	80	8.0	81	3.21	82.0
5	1,750	40.0	82	20.0	83	8.40	81.0
10	1,750	75.0	85.6	37.5	85	15.4	86.5
25	1,750	182.0	87.3	91.7	87.5	38.1	88.5
50	850	—	—	180.0	89	73.1	90.0
100	850	—	—	350.0	90.5	149.0	91.0
200	1,750	—	—	700.0	91	295.0	92.0

SOURCE: Westinghouse Electric Corp.

than that of a shunt motor, or, if desired, it may be adjusted to increase with increasing load.

The speed as a function of armature current is shown in Fig. 15.1.58*a* and the torque as a function of armature current is shown in Fig. 15.1.58*b*.

Since the speed of the shunt motor is sufficiently constant for most purposes and the differential motor tends toward instability, particularly in starting and on overloads, the differential motor is little used.

Cumulative compound motors develop a more rapid increase in torque with load than shunt motors (Fig. 15.1.58*b*); on the other hand, they have much poorer speed regulation (Fig. 15.1.58*a*). Hence they are used where larger starting torque than that developed by the shunt motor is necessary, as in some industrial drives. They are particularly useful where large and intermittent increases of torque occur as in drives for shears, punches, rolling mills, etc. In addition to the sudden increase in torque which the motor develops with sudden applications of load, the fact that it slows down rapidly and hence causes the rotating parts to give up some of their kinetic energy is another important advantage in that it reduces the peaks on the power plant. Performance data for compound motors are given in Table 15.1.11.

Commutation   The brushes on the commutator of either a motor or generator should be set in such a position that the induced emf in the armature coils undergoing commutation and hence short-circuited by the brushes, is zero. In practice, this condition can at best be only approximately realized. Frequently conditions are such that it is far from being realized. At no load, the brushes should be set in a position corresponding to the geometrical neutral of the machine, for under these conditions the induced emf in the coils short-circuited by the brushes is zero. As load is applied, two factors cause sparking under the brushes. The mmf of the armature, or **armature reaction**, distorts the flux; when the current in the coils undergoing commutation reverses, an emf of self-induction $-L\,di/dt$ tends to prolong the current flow which produces sparking. In a generator, armature reaction distorts the flux in the direction of rotation and the brushes should be advanced. In order to neutralize the emf of self-induction the brushes should be set a little ahead of the neutral plane so that the emf induced in the short-circuited coils by the cutting of the flux at the fringe of the next pole is opposite to this emf of self-induction. In a motor the brushes are correspondingly moved backward in the direction opposite rotation.

Theoretically, the brushes should be shifted with every change in load. However, practically all dc generators and motors now have **commutating poles** (or **interpoles**) and with these the brushes can remain in the no-load neutral plane, and good commutation can be obtained over the entire range of load. Commutating poles are small poles between the main poles (Fig. 15.1.60) and are excited by a winding in series with the armature. Their function is to neutralize the flux distortion in the **neutral plane** caused by armature reaction and also to supply a flux that will cause an emf to be induced in the conductors undergoing commutation, opposite and equal to the emf of self-induction. Since armature reaction and the emf of self-induction are both proportional to the armature current, saturation being neglected, they are neutralized theoretically at every load. Commutating poles have made possible dc

Fig. 15.1.60   Commutating poles in motor.

generators and motors of very much higher voltage, greater speeds, and larger kW ratings than would otherwise be possible.

Occasionally, the commutating poles may be connected incorrectly. In a motor, passing from an N main pole in the direction of rotation of the armature, an N commutating pole should be encountered as shown in Fig. 15.1.60. In a generator under these conditions an *S* commutating pole should be encountered. The test can easily be made with a compass. If poor commutation is caused by too strong interpoles, the winding may be shunted. If the poles are too weak and the shunting cannot be reduced, they may be strengthened by inserting sheet-iron shims between the pole and the yoke thus reducing the air gap.

Although the emfs induced in the coils undergoing commutation are relatively small, the resistance of the coils themselves is low so that unless further resistance is introduced, the short-circuit currents would be large. Hence, with the exception of certain low-voltage generators, carbon brushes that have relatively large contact resistance are almost always used. Moreover, the graphite in the brushes has a lubricating action, and the usual carbon brush does not score the commutator.

Speed Control of Motors

Shunt Motors   In Eq. (15.1.91) the speed of a shunt motor $N = K_s E/\phi$, where K_s is a constant involving the design of the motor such as conductors on armature surface and number of poles. Obviously, in order to change the speed of a shunt motor, without changing its construction, two factors may be varied, the counter emf E and the flux ϕ.

Armature-Resistance Control   The counter emf $E = V - I_a R_a$, where V is the terminal voltage, assumed constant. R_a must be small so that the armature heating can be maintained within permissible limits. Under these conditions the speed change with load is small. By inserting an external resistor, however, into the armature circuit the counter emf E may be made to decrease rapidly with increase in load; that is, $E = V - I_a(R_a + R)$ [see Eq. (15.1.90)] where R is the resistance of the external resistor. The resistor R must be inserted in the **armature** circuit only. The advantages of this method are its simplicity, the full torque of the motor is developed at any speed, and the method introduces no commutating difficulties. Its disadvantages are the increased speed regulation with change of load (Fig. 15.1.61), the low efficiency, particularly at the lower speeds, and the fact that provision must be made to dissipate the comparatively large power losses in the series resistor. Figure 15.1.61 shows typical speed-load curves without and with series

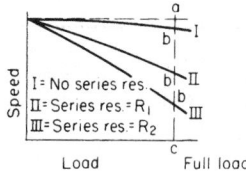

Fig. 15.1.61 Speed-load characteristics with armature resistance control.

resistors in the armature circuit. The armature efficiency is nearly equal to the ratio of the operating speed to the no-load speed. Hence at 25 percent speed the armature efficiency is practically 25 percent. Frequently the controlling and starting resistors are one, and the device is called a **controller.** Starting rheostats themselves are not designed to carry the armature current continuously and must not be used as controllers. The armature-resistance method of speed control is frequently used to regulate the speed of ventilating fans where the power demand diminishes rapidly with decrease in speed.

Control by Changing Impressed Voltage   From Eq. (15.1.92) it is evident that the speed of a motor may be changed if V is changed by connecting the armature across different voltages. Speed control by this method is accomplished by having mains (usually four), which are maintained at different voltages, available at the motor.

The shunt field of the motor is generally permanently connected to one pair of mains, and the armature circuit is provided with a controller by means of which the operator can readily connect the armature to any pair of mains. Such a system gives a series of distinct and widely separated speeds and generally necessitates the use of field-resistance control, in combination, to obtain intermediate speeds. This method, known as the **multivoltage method,** has the disadvantage that the system is expensive, for it requires several generating machines, a somewhat complicated switchboard, and a number of service wires. The system is used somewhat in machine shops and is extensively used for dc elevator starting and speed control.

In the **Ward Leonard system,** the variable voltage is obtained from a separately excited generator whose armature terminals are connected directed to the armature terminals of the working motor. The generator is driven at essentially constant speed by a dc shunt motor if the power supply is direct current, or by an induction motor or a synchronous motor if the power supply is alternating current. The field circuit of the generator and that of the motor are connected across a constant-voltage dc supply. The terminal voltage of the generator, and hence the voltage applied to the armature of the motor, is varied by changing the generator-field current with a field rheostat. The rheostat has a wide range of resistance so that the speed of the motor may be varied smoothly from 0 to 100 percent. Since three machines are involved the system is costly, somewhat complicated, and has low power efficiency. However, because the system is flexible and the speed can be smoothly varied over wide ranges, it has been used in many applications, such as elevators, mine hoists, large printing presses, paper machines, and electric locomotives.

The **Ward-Leonard system** has been largely replaced by a **static converter system** where a silicon-controlled-rectifier bridge is used to convert three-phase alternating current, or single-phase on smaller drives, to dc voltage that can be smoothly varied by **phase-angle firing** of the rectifiers from full voltage to zero. This system is smaller, lighter, and less expensive than the motor-generator system. Care must be taken to assure that the dc motor will accept the **harmonics** present in the dc output without overheating.

Control by Changing Field Flux   Equation (15.1.91) shows that the speed of a motor is inversely proportional to the flux ϕ. The flux can be changed either by varying the shunt-field current or by varying the reluctance of the magnetic circuit. The variation of the **shunt-field current** is the simplest and most efficient of all the methods of speed control.

With the ordinary motor, speed variation of 1.5 to 1.0 is obtainable with this method. If attempt is made to obtain greater ratios, severe

sparking at the brushes results, owing to the field distortion caused by the armature mmf becoming large in comparison with the weakened field of the motor. Speed ratios of 5 : 1 and higher are, however, obtainable with motors which have commutating poles. Since the field current is a small proportion of the total current (1 to 3 percent), the rheostat losses in the field circuit are always small. This method is efficient. Also for any given speed adjustment the speed regulation is excellent, which is another advantage. Because of its simplicity, efficiency, and excellent speed regulation, the control of speed by means of the field current is by far the most common method. Output power remains constant when the field is weakened, so output torque varies inversely with motor speed.

Speed Control of Series Motors   The series motor is fundamentally a variable-speed motor, the speed varying widely from light load to full load and more (see Fig. 15.1.58a). From Eq. (15.1.92) the speed for any value of ϕ, or current, can be changed by varying the impressed voltage. Hence the speed can be controlled by inserting resistance in series with the motor. This method, which is practically the same as the armature-resistance control method for shunt motors, has the same objections of low efficiency and poor regulation with fluctuating loads. It is extensively used in controlling the speed of hoist and crane motors.

The **series-parallel** system of series-motor speed control is almost universally used in electric traction. At least two motors are necessary. The two motors are first connected in series with each other and with the starting resistor. The starting resistor is gradually cut out and, since each motor then operates at half line voltage, the speed of each is approximately half speed. Both motors take the same current, and each can develop full torque. This condition of operation is efficient since there is no external resistance in circuit. When the controller is moved to the next position, the motors are connected in parallel with each other and each in series with starting resistors. Full speed of the motors is obtained by gradually cutting out these resistors. Connecting the two motors in series on starting reduces the current to one-half the value that would be required for a given torque were both motors connected in parallel on starting. The power taken from the trolley is halved, and an intermediate running speed is efficiently obtained.

In the **multiple-unit** method of speed control which is used for electric railway trains, the starting contactors, reverser, etc., for each car are located under that car. The relays operating these control devices are actuated by energy taken from the train line consisting usually of seven wires. The train line runs the entire length of the train, the connections between the individual cars being made through the couplers. The train line is energized by the action of the motorman operating any one of the small master controllers which are located in each car. Hence corresponding relays, contactors, etc., in every car all operate simultaneously. High accelerations may be reached with this system because of the large tractive effort exerted by the wheels on every car.

SYNCHRONOUS GENERATORS

The synchronous generator is the only type of ac generator now in general use at power stations.

Construction   In the usual synchronous generator the armature or stator is the stationary member. This construction has many advantages. It is possible to make the slots any reasonable depth, since the tooth necks increase in cross section with increase in depth of slot; this is not true of the rotor. The large slot section which is thus obtainable gives ample space for copper and insulation. The conductors from the armature to the bus bars can be insulated throughout their entire lengths, since no rotating or sliding contacts are necessary. The insulation in a stationary member does not deteriorate as rapidly as that on a rotating member, for it is not subjected to centrifugal force or to any considerable vibration.

The **rotating member** is ordinarily the field. There are two general types of field construction: the **salient-pole type** and the **cylindrical,** or **nonsalient-pole, type.** The salient-pole type is used almost entirely for slow and moderate-speed generators since this construction is the least expensive and permits ample space for the field ampere-turns.

Table 15.1.12   Performance Data for Synchronous Generators

80% pf, 3 phase, 60 Hz, 240 to 2,400 V, horizontal-coupled or belted-type engine							
		Speed,	Excitation,	Efficiencies, %			Approx. net weight,
kVA	Poles	r/min	kW	$\frac{1}{2}$ load	$\frac{3}{4}$ load	Full load	lb
25	4	1,800	0.8	81.5	85.7	87.6	900
93.8	8	900	2	87	89.5	90.9	2,700
250	12	600	5	90	91.3	92.2	6,000
500	18	400	8	91.7	92.6	93.2	10,000
1,000	24	300	14.5	92.6	93.4	93.9	16,100
3,125	48	150	40	93.4	94.2	94.6	52,000

Industrial-size turbine generators, direct-connected type, 80% pf, 3 phase, 60 Hz, air-cooled										
		Speed,	Excitation		Efficiency, %			Volume of air,		Approx. weight, including
kVA	Poles	r/min	kW	V	$\frac{1}{2}$ load	$\frac{3}{4}$ load	Full load	ft^3/min	Voltage	exciter, lb
1,875	2	3,600	18	125	95.3	96.1	96.3	3,500	480–6,900	21,900
2,500	2	3,600	22	125	95.3	96.1	96.3	5,000	2,400–6,900	22,600
3,125	2	3,600	24	125	95.3	96.3	96.5	5,500	2,400–6,900	25,100
3,750	2	3,600	24	125–250	95.3	96.3	96.6	6,500	2,400–6,900	27,900
5,000	2	3,600	29	125–250	95.3	96.3	96.6	11,000	2,400–6,900	40,100
6,250	2	3,600	38	125–250	95.3	96.3	96.7	12,000	2,400–13,800	43,300
7,500	2	3,600	42	125–250	95.5	96.5	96.9	15,000	2,400–13,800	45,000
9,375	2	3,600	47	125–250	95.5	96.5	96.9	16,500	2,400–13,800	61,200

Central-station-size turbine generators, direct-connected type, 85% pf, 3 phase, 60 Hz, 11,500 to 14,400 V										
		Speed,	Excitation		Efficiency, %			Volume of air,		Approx. weight, including
kVA	Poles	r/min	kW	V	$\frac{1}{2}$ load	$\frac{3}{4}$ load	Full load	ft^3/min	Ventilation	exciter, lb
13,529	2	3,600	70	250	96.3	97.1	97.3	22,000	Air-cooled	116,700
17,647	2	3,600	100	250	97.7	97.9	97.9	22,000	H$_2$-cooled	115,700
23,529	2	3,600	115	250	98.0	98.2	98.2	25,000	H$_2$-cooled	143,600
35,294	2	3,600	145	250	98.1	98.3	98.3	34,000	H$_2$-cooled	194,800
47,058	2	3,600	155	250	98.3	98.5	98.5	42,000	H$_2$-cooled	237,200
70,588	2	3,600	200	250	98.4	98.7	98.7	50,000	H$_2$-cooled	302,500

SOURCE: Westinghouse Electric Corp.

It is not practicable to employ salient poles in high-speed turboalternators because of the excessive windage and the difficulty of obtaining sufficient mechanical strength. The **cylindrical type** consists of a cylindrical steel forging with radial slots in which the field copper, usually in strip form, is placed. The fields are ordinarily excited at low voltage, 125 and 250 V, the current being conducted to the rotating member by means of slip rings and brushes. An ac generator armature to supply **field voltage** can be mounted on the generator shaft and supply dc to the motor field through a static rectifier bridge, also mounted on the shaft, eliminating all slip rings and brushes. The field power is ordinarily only 1.5 percent and less of the rated power of the machine (see Table 15.1.12).

Classes of Synchronous Generators   Synchronous generators may be divided into three general classes: (1) the slow-speed engine-driven type; (2) the moderate-speed waterwheel-driven type; and (3) the high-speed turbine-driven type. In (1) a hollow box frame is used as the stator support, and the field consists of a spider to which a larger number of salient poles are attached, usually bolted. The speed seldom exceeds 75 to 90 r/min, although it may run as high as 150 r/min. Waterwheel generators also have salient poles which are usually dovetailed to a cylindrical spider consisting of steel plates riveted together. Their speeds range from 80 to 900 r/min and sometimes higher, although the 9,000-kVA Keokuk synchronous generators rotate at only 58 r/min, operating at a very low head. The speed rating of direct-connected waterwheel generators decreases with decrease in head. It is desirable to operate synchronous generators at the highest permissible speed since the weight and costs diminish with increase in speed.

Waterwheel-driven generators must be able to run at double speed, as a precaution against accident, should the governor fail to shut the gate sufficiently rapidly in case the circuit breakers open or should the governing mechanism become inoperative.

Turbine-driven generators operate at speeds of 720 to 3,600 r/min. Direct-connected exciters, belt-driven exciters from the generator shaft, and separately driven exciters are used. In large stations separately driven (usually motor) exciters may supply the excitation energy to excitation bus bars. Steam-driven exciters and storage batteries are frequently held in reserve. With slow-speed synchronous generators, the belt-driven exciter is frequently used because it can be driven at higher speed, thus reducing the cost.

Synchronous-Generator Design   At the present time single-phase generators are seldom built. For single-phase service two phases of a standard three-phase Y-connected generator are used. A single-phase load or unbalanced three-phase load produces flux pulsations in the magnetic circuits of synchronous generators, which increase the iron losses and introduce harmonics into the emf wave. Two-phase windings consist of two similar single-phase windings displaced 90 electrical space degrees on the armature and ordinarily occupying all the slots on the armature. The most common type of winding is the three-phase lap-wound two-layer type of winding. In three-phase windings three windings are spaced 120 electrical space degrees apart, the individual phase belts being spaced 60° apart. Usually, all the slots on the armature are occupied. Standard voltages are 550, 1,100, 2,200, 6,600, 13,200, and 20,000 V. It is much more difficult to insulate for 20,000 V than it is for the lower voltages. However, if the power is to be transmitted at this voltage, its use would be justified by the saving of transformers. In machines of moderate and larger ratings it is common to generate at 6,600 and 13,200 V if transformers must be used. The higher voltage is preferable, particularly for the higher ratings, because it reduces the cross section of the connecting leads and bus bars.

The **standard frequency** in the United States for lighting and power systems is 60 Hz; the few former 50-Hz systems have practically all been converted to 60 Hz. The frequency of 25 Hz is commonly used in street-railway and subway systems to supply power to the synchronous converters and other ac-dc conversion apparatus; it is also commonly used in railroad electrification, particularly for single-phase series-motor locomotives (see Sec. 11). At 25 Hz incandescent lamps have noticeable flicker. In European (and most other) countries 50 Hz is standard. The frequency of a synchronous machine

$$f = P \times r/\min/120 \qquad \text{Hz} \qquad (15.1.98)$$

where P is the number of poles. Synchronous generators are rated in kVA rather than in kW, since heating, which determines the rating, is dependent only on the current and is independent of power factor. If the kilowatt rating is specified, the power factor should also be specified.

Induced EMF   The induced emf per phase in synchronous generator is

$$E = 2.22k_b k_p \Phi f Z \qquad \text{V/phase} \qquad (15.1.99)$$

where k_b = breadth factor or belt factor (usually 0.9 to 1.0), which depends on the number of slots per pole per phase, 0.958 for three-phase, four slots per pole per phase; k_p = pitch factor = 1.0 for full pitch, 0.966 for $\frac{5}{6}$ pitch; Φ = total flux Wb, entering armature from one north pole and is assumed to be sinusoidally distributed along the air gap; f = frequency; and Z = number of series conductors per phase.

Synchronous generators usually are Y-connected. The advantages are that for a given line voltage the voltage per phase is $1/\sqrt{3}$ that of the delta-connected winding; third-harmonic currents and their multiples cannot circulate in the winding as with a delta-connected winding; third-harmonic emfs and their multiples cannot exist in the line emfs; a neutral point is available for grounding.

Regulation   The terminal voltage of synchronous generator at constant frequency and field excitation depends not only on the current load but on the power factor as well. This is illustrated in Fig. 15.1.62, which shows the voltage-current characteristics of a synchronous generator with lagging current, leading current and in-phase current (pf = 1.00). With leading current the voltage may actually rise with increase in load;

Fig. 15.1.62   Synchronous generator characteristics.

the rate of voltage decrease with load becomes greater as the lag of the current increases. The regulation of a synchronous generator is defined by the ANSI/IEEE Std. 100-1992 Standard Dictionary of Electrical and Electronic Terms as follows:

The voltage regulation of a synchronous generator is the rise in voltage with constant field current, when, with the synchronous generator operated at rated voltage and rated speed, the specified load at the specified power factor is reduced to zero, expressed as a percent of rated voltage.

For example, in Fig. 15.1.62 the regulation under each condition is

$$100(ac - bc)/bc \qquad (15.1.100)$$

With leading current the regulation may be negative.

Three factors affect the regulation of synchronous generators; the **effective armature resistance,** the **armature leakage reactance,** and the **armature reaction.** With alternating current the armature loss is greater than the value obtained by multiplying the square of the armature current

by the ohmic resistance. This is due to hysteresis and eddy-current losses in the iron adjacent to the conductor and to the alternating flux-producing losses in the conductors themselves. Also the current is not distributed uniformly over conductors in the slot, but the current density tends to be greatest in the top of the slot. These factors all have the effect of increasing the resistance. The ratio of effective to ohmic resistance varies from 1.2 to 1.5. The **armature leakage reactance** is due to the flux produced by the armature current linking the conductors in the slots and also the end connections.

The armature mmf reacts on the field to change the value of the flux. With a single-phase generator and with an unbalanced load on a polyphase generator, the armature mmf is pulsating and causes iron losses in the field structure. With polyphase machines under a constant balanced load, the armature mmf is practically constant in magnitude and fixed in its relation to the field poles. Its direction with relation to the field-pole axis is determined by the power factor of the load.

A component of current in phase with the no-load induced emf, or the excitation emf, merely distorts the field by strengthening the trailing pole tip and weakening the leading pole tip. A component of current lagging the excitation emf by 90° weakens the field without distortion. A component of current leading the excitation emf by 90° strengthens the field without distortion. Ordinarily, both cross magnetization and one of the other components are acting simultaneously.

The foregoing effects are called **armature reaction.** Frequently the effects of armature reactance and armature reaction can be combined into a single quantity.

It is difficult to determine the regulation of synchronous generator by actual loading, even when in service, owing to the difficulty of obtaining, controlling, and absorbing the large balanced loads. Hence methods of **predetermining regulation without actually loading** the machine are used.

Synchronous Impedance Method   Both armature reactance and armature reaction have the same effect on the terminal voltage. In the synchronous impedance method the generator is considered as having no armature reaction, but the armature reactance is increased a sufficient amount to account for the effect of armature reaction. The phasor diagram for a current I lagging the terminal voltage V by an angle θ is shown in Fig. 15.1.63. In a polyphase generator the phasor diagram is applicable to one phase, a balanced load almost always being assumed.

Fig. 15.1.63   Phasor diagram for synchronous impedance method.

The power factor of the load is $\cos \theta$; IR is the effective armature resistance drop and is parallel to I; IX_s is the synchronous reactance drop and is at right angles to I and leading it by 90°. IX_s includes both the reactance drop and the drop in voltage due to armature reaction. That part of IX_s which replaces armature reaction is in reality a fictitious quantity. The synchronous impedance drop is given by IZ_s. The no-load or open-circuit (excitation) voltage

$$E = \sqrt{(V\cos \theta + IR)^2 + (V\sin \theta \pm IX_s)^2} \qquad \text{V} \qquad (15.1.101)$$

All quantities are per phase. The negative sign is used with leading current.

$$\text{Regulation} = 100(E - V)/V \qquad (15.1.102)$$

With leading current E may be less than V and a negative regulation results.

The synchronous impedance is determined from an open-circuit and a short-circuit test, made with a weak field. The voltage E' on open circuit is divided by the current I' on short circuit for the same value of field current.

$$Z_s = E'/I' \qquad X_s = \sqrt{Z_s^2 - R^2} \qquad \Omega \qquad (15.1.103)$$

R is so small compared with X_s that for all practical purposes $X_s = Z_s$. R may be determined by measuring the ohmic resistance per phase and multiplying by 1.4 to 1.5 to obtain the effective resistance. This value of R and the value of X_s obtained from Eq. (15.1.103) may then be substituted in Eq. (15.1.101) to obtain E at the specified load and power factor.

Since the synchronous reactance is determined at low saturation of the iron and used at high saturation, the method gives regulations that are too large; hence it is called the **pessimistic method.**

MMF Method In the mmf method the generator is considered as having no armature reactance but the armature reaction is increased by an amount sufficient to include the effect of reactance. That part of armature reaction which replaces the effect of armature reactance is in reality a fictitious quantity. To obtain the data necessary for computing the regulation, the generator is short-circuited and the field adjusted to give rated current in the armature. The corresponding value of field current I_a is read. The field is then adjusted to give voltage E' equal to rated terminal voltage + IR drop ($= V + IR$, as phasors, Fig. 15.1.64) on open circuit and the field current I' read.

I_a is 180° from the current phasor I, and I' leads E' by 90° (Fig. 15.1.64). The angle between I' and I_a is $90 - \theta + \phi$, but since ϕ is small, it can usually be neglected. The phasor sum of I_a and I' is I_o. The open-circuit

Fig. 15.1.64 Phasor diagram for the mmf method.

voltage E corresponding to I_o is the no-load voltage and can be found on the saturation curve. The regulation is then found from Eq. (15.1.102). This method gives a value of regulation less than the actual value and hence is called the **optimistic method.** The actual regulation lies somewhere between the values obtained by the two methods but is more nearly equal to the value obtained by the mmf method.

ANSI Method The ANSI method (American Standard 50, Rotating Electrical Machinery) which has become the accepted standard for the predetermination of synchronous generator operation, eliminates in large measure the errors due to saturation which are inherent in the synchronous impedance and mmf methods. In Fig. 15.1.65a is shown the saturation curve OAF of the generator. The axis OP is not only the field-current axis but also the axis of the current phasor I as well. V the terminal voltage is drawn θ deg from I or OP, where θ is the power factor angle. The effective resistance drop IR and the **leakage reactance** drop IX are drawn parallel and perpendicular to the current phasor. E_a, the phasor sum of V, IR, and IX, is the internal **induced** emf. Arcs are swung with O as the center and V and E_a as radii to intercept the axis of ordinates at B and C. OK, tangent to the straight portion of the saturation curve, is the **air-gap line.** If there is no saturation, I_v is the field current necessary to produce V, and CK is the field current necessary to produce E_a. The field current I_s is the increase in field current necessary to take into account the saturation corresponding to E_a.

The corresponding phasor diagram to a larger scale is shown in Fig. 15.1.65b. I'_f, the field current necessary to produce rated current at short circuit, corresponding to I_a (Fig. 15.1.64), is drawn horizontally. The field current I_v is drawn at an angle θ to the right of a perpendicular erected at the right-hand end of I'_f. I_r is the resultant of I'_f and I_v. I_s is added to I_r giving I_f the resultant field current. The no-load emf E is found on the saturation curve, Fig. 15.5.65a, corresponding to $I_f = OD$.

Excitation is commonly supplied by a small dc generator driven from the generator shaft. On account of commutation, except in the smaller sizes, the dc generator cannot be driven at 3,600 r/min, the usual speed for turbine generators, and belt or gear drives are necessary. The use of

(a)

(b)

Fig. 15.1.65 ANSI method of synchronous generator regulation.

the silicon rectifier has made possible simpler means of excitation as well as voltage regulation. In one system the exciter consists of a small rotating-armature synchronous generator (which can run at high speed) mounted directly on the main generator shaft. The three-phase armature current is rectified by six silicon rectifiers and is conducted directly to the main generator field without any sliding contacts. The main generator field current is controlled by the current to the stationary field of the exciter generator. In another system there is no rotating exciter, the generator excitation being supplied directly from the generator terminals, the 13,800 V, three-phase, being stepped down to 115 V, three-phase, by small transformers and rectified by silicon rectifiers. Voltage regulation is obtained by saturable reactors actuated by potential transformers connected across the generator terminals.

Most regulators such as the following operate through the field of the exciter. In the **Tirrill regulator** the field resistance of the exciter is short-circuited temporarily by contacts when the bus-bar voltage drops. Actually, the contacts are vibrating continuously, the time that they are closed depending on the value of the bus-bar voltage. The General Electric Co. manufactures a direct-acting regulator in which the regulating rheostat is part of the regulator itself. The rheostat consists of stacks of graphite plates, each plate being pivoted at the center. Tilting the plates changes the path of the current through the rheostat and thus changes the resistance. The plates are tilted by a sensitive torque armature which is actuated by variations of voltage from the normal value (for regulators employing silicon rectifiers).

Parallel Operation of Synchronous Generators The kilowatt division of load between synchronous generators in parallel is determined entirely by the speed-load characteristics of their prime movers and not by the characteristics of the generators themselves. No appreciable adjustment of kilowatt load between synchronous generators in parallel can be made by means of their field rheostats, as with dc generators. Consider Fig. 15.1.66, which gives the speed-load characteristics in terms of frequency, of two synchronous generators, no. 1 and no. 2, these characteristics being the speed-load characteristics of their prime movers. These speed-load characteristics are drooping, which is necessary for stable parallel operation. The total load on the two machines is

Fig. 15.1.66   Speed-load characteristics of synchronous generators in parallel.

$P_1 + P_2$ kW. Both machines must be operating at the same frequency F_1. Hence generator 1 must be delivering P_1 kW, and generator 2 must be delivering P_2 kW (the small generator losses being neglected). If, under the foregoing conditions, the field of either machine is strengthened, it cannot deliver a greater kilowatt load, for its prime mover can deliver more power only by dropping its speed. This is impossible, for both generators must operate always at the same frequency f_1. For any fixed total power load, the division of kilowatt load between synchronous generators can be changed only by modifying in some manner the speed-load characteristics of their prime movers, such, for example, as changing the tension in the governor spring. Synchronous generators in parallel are of themselves in stable equilibrium. If the driving torque of one machine is increased, the resulting electrical reactions between the machines cause a circulating current to flow between machines. This current puts more electrical load on the machine whose driving torque is increased and tends to produce motor action in the other machines. In an extreme case, the driving torque of one prime mover may be removed entirely, and its generator will operate as a synchronous motor, driving the prime mover mechanically.

Variations in driving torques cause currents to circulate between synchronous generators, transferring power which tends to keep the generators in synchronism. If the power transfer takes the form of recurring pulsations, it is called **hunting,** which may be reduced by building heavy copper grids called **amortisseurs,** or **damper windings,** into the pole faces. Turbine- and waterwheel-driven synchronous generators are much better adapted to parallel operation than are synchronous generators which are driven by reciprocating engines, because of their uniformity of torque.

Increasing the field current of synchronous generators in parallel with others causes it to deliver a greater lagging component of current. Since the character of the load determines the total current delivered by the system, the lagging components of current delivered by the other generators must decrease and may even become leading components. Likewise if the field of one generator is weakened, it delivers a greater leading component of current and the other machines deliver components of current which are more lagging. These leading and lagging currents do not affect appreciably the division of **kilowatt** load between the synchronous generators. They do, however, cause unnecessary heating in their armatures. The fields of all synchronous generators should be so adjusted that the heating due to the quadrature components of currents is a minimum. With two generators having equal armature resistances, this occurs when both deliver equal quadrature currents.

Armature reactance in the armature of machines in parallel is desirable. If not too great, it stabilizes their operation by producing the synchronizing action. Synchronous generators with too little reactance are sensitive, and if connected in parallel with slight phase displacement or inequality of voltage, considerable disturbance results. Armature reactance also reduces the current on short circuit, particularly during the first few cycles when the short-circuit current is a maximum. Frequently, external power-limiting reactances are connected in series to protect the generators and equipment from injury that would result from the tremendous short-circuit currents. For these reasons, poor regulation in large synchronous generators is frequently considered to be an advantage rather than a disadvantage.

Ground Resistors   Most power systems operate with a grounded neutral. When the station generators deliver current directly to the system (without intervening transformers), it is customary to ground the neutral (of the Y-connected windings) or one generator in a station; this is usually done through a grounding resistor of from 2 to 6 Ω. If the neutral of more than one generator is grounded, third-harmonic (and multiples thereof) currents can circulate between the generators. The ground resistor reduces the short-circuit currents when faults to ground occur, and hence reduces the violence of the short circuit as well as the duty of the circuit breakers. Grounding reactors are sometimes used but have limited application owing to the danger of high voltages resulting from resonant conditions.

INDUCTION GENERATORS

The **induction generator** is an induction motor driven above synchronous speed. The rotor conductors cut the rotating field in a direction to convert shaft mechanical power to electrical power. Load increases as speed increases, so the generator is self-regulating and can be used without governor control. On short circuits the induction generator will deliver current to the fault for only a few cycles because, unlike the synchronous generator, it is not self-exciting. Since it is not self-excited, an induction generator must always be used in parallel with an electrical system where there are some synchronous machines, or capacitor banks, to deliver lagging current (VARs) to the induction generator for excitation.

Induction generators have found favor in industrial **cogeneration** applications and as **wind-driven** generators where they provide a small part of the total load.

CELLS

Fuel cells convert chemical energy of fuel and oxygen directly to electrical energy. **Solar cells** convert solar radiation to electrical energy. At present, these conversion methods are not economically competitive with historic generating techniques, but they have found applications in isolated areas, such as microwave relaying stations, satellites, and residential lighting, where power requirements are small and costs for transmitting electrical power from more conventional sources is prohibitive. As solar technology continues to improve, many other applications will be evaluated.

TRANSFORMERS

Transformer Theory   The transformer is a device that transfers energy from one electric circuit to another without change of frequency and usually, but not always, with a change in voltage. The energy is transferred through the medium of a magnetic field: it is supplied to the transformer through a primary winding and is delivered by means of a secondary winding. Both windings link the same magnetic circuit. With no load on the secondary, a small current, called the exciting current, flows in the primary and produces the alternating flux. This flux links both primary and secondary windings and induces the same volts per turn in each. With a sine wave the emf is

$$E = 4.44\Phi_m nf \quad V \qquad (15.1.104)$$

where Φ_m = maximum instantaneous flux in webers, n = turns on either winding, and f = frequency. Equation (15.1.104) may also be written

$$E = 4.44B_m Anf \quad V \qquad (15.1.105)$$

B_m = maximum instantaneous flux density in iron and A = net cross section of iron. If B_m is in T, A is in m²; if B_m is in Mx/in², A is in in².
In English units, Eq. (15.1.104) becomes

$$E = 4.44\Phi_m nf 10^{-8} \quad V \qquad (15.1.104a)$$

where Φ_M is in maxwells; Eq. (15.1.105) becomes

$$E = 4.44B_m Anf 10^{-8} \quad V \qquad (15.1.105a)$$

where B_m is in Mx/in² and A is in in².
B_m is practically fixed. In large transformers with silicon steel it varies between 60,000 and 75,000 Mx/in² at 60 Hz and between 75,000 and 90,000 Mx/in² at 25 Hz. It is desirable to operate the iron at as high

density as possible in order to minimize the weight of iron and copper. On the other hand, with too high densities the eddy-current and hysteresis losses become too great, and with low frequency the exciting current may become excessive. It follows from Eq. (15.1.104) that

$$E_1/E_2 = n_1/n_2 \qquad (15.1.106)$$

where E_1 and E_2 are the primary and secondary emfs and n_1 and n_2 are the primary and secondary turns. Since the impedance drops in ordinary transformers are small, the terminal voltages of primary and secondary are also practically proportional to their number of turns. As the change in secondary terminal voltage in the ordinary constant-potential transformer over its range of operation is small (1.5 to 3 percent), the flux must remain substantially constant. Therefore, the added ampere-turns produced by any secondary load must be balanced by opposite and equal primary ampere-turns. Since the exciting current is small compared with the load current (1.5 to 5 percent) and the two are usually out of phase, the exciting current may ordinarily be neglected. Hence,

$$n_1 I_1 = n_2 I_2 \qquad (15.1.107)$$
$$I_1/I_2 = n_2/n_1 \qquad (15.1.108)$$

where I_1 and I_2 are the primary and secondary currents.

When load is applied to the secondary of a transformer, the secondary ampere-turns reduce the flux slightly. This reduces the counter emf of the primary, permitting more current to enter and thus supply the increased power demanded by the secondary.

Both primary and secondary windings must necessarily have resistance. All the flux produced by the primary does not link the secondary; the counter ampere-turns of the secondary produce some flux which does not link the primary. These **leakage fluxes** produce reactance in each winding. The combined effect of the resistance and reactance produces an impedance drop in each winding when current flows. These impedance drops produce a slight drop in the secondary terminal voltage with load.

Transformer Testing Transformer regulation and losses are so small that it is far more accurate to compute the regulation and efficiency than to determine them by actual measurement. The necessary measurements and computations are comparatively simple, and little power is involved in making the tests. In the **open-circuit** test, the power input to either winding is measured at its rated voltage. Usually it is more convenient to make this test on the low-voltage winding, particularly if it is rated at 110, 220, or 550 V. The open-circuit power practically all goes to supply the core losses, consisting of eddy-current and hysteresis losses. Let this value of power be P_0. The eddy-current loss varies as the square of the voltage and frequency; the hysteresis loss varies as the 1.6 power of the voltage, and directly as the frequency. In the **short-circuit** test one winding is short-circuited, and the current in the other is adjusted to near its rated value. The voltage V_c, the current I_1, and the power input P_c are measured. When one winding of a transformer is short-circuited, the voltage across the other is only 3 to 4 percent of rated value when rated current flows. Since a voltage range of from 110 to 250 V is best adapted to measuring instruments, that winding whose rated voltage, multiplied by 0.03 or 0.04, is closest to this voltage range should be used for making the short-circuit test, the other winding being short-circuited. Practically all the power on short circuit goes to supply the copper loss of primary and secondary. If the measurements are made on the primary,

$$R_{01} = P_c/I_1^2 \qquad (15.1.109)$$
$$Z_{01} = V_c/I_1 \qquad (15.1.110)$$
$$X_{01} = \sqrt{Z_{01}^2 - R_{01}^2} \qquad (15.1.111)$$

where R_{01}, Z_{01}, and X_{01} are the equivalent resistance, impedance, and reactance referred to the primary. Also $R_{02} = R_{01}(n_2/n_1)^2$; $Z_{02} = Z_{01}(n_2/n_1)^2$; $X_{02} = X_{01}(n_2/n_1)^2$, these quantities being the equivalent resistance, impedance, and reactance referred to the secondary. If the dc resistances R_1 and R_2 of the primary and secondary are measured,

$$R_{01} = R_1 + (n_1/n_2)^2 R_2 \qquad (15.1.112)$$
$$R_{02} = R_2 + (n_2/n_1)^2 R_1 \qquad (15.1.113)$$

The ac or effective resistances, determined from Eq. (15.1.109), are usually 10 to 15 percent greater than these values.

Regulation The regulation may be computed from the foregoing data as follows:

$$V_1' = \sqrt{(V_1 \cos \theta + I_1 R_{01})^2 + (V_1 \sin \theta \pm I_1 X_{01})^2} \qquad (15.1.114)$$
$$\text{Regulation} = 100(V_1' - V_1)/V_1 \qquad (15.1.115)$$

V_1 = rated primary terminal voltage; $\cos \theta$ = load power factor; I_1 = rated primary current; R_{01} = equivalent resistance referred to primary [from Eq. (15.1.109)]; X_{01} = equivalent reactance referred to primary. The + sign is used with lagging current and the − sign with leading current. Equations (15.1.114) and (15.1.115) are equally applicable to the secondary if the subscripts are changed.

Efficiency The only two losses in a constant-potential transformer are the core loss in W, P_0, which is practically independent of load, and P_c the copper loss in W, which varies as the load current squared. The efficiency for any current I_1 is

$$\eta = V_1 I_1 \cos \theta / (V_1 I_1 \cos \theta + P_0 + I_1^2 R_{01}) \qquad (15.1.116)$$

Equation (15.1.116) applies equally well to the secondary if the subscripts are changed. The maximum efficiency occurs when the core and copper losses are equal.

All-Day Efficiency Since transformers must usually be on the line 24 h/day, part of which time the load may be very light, the all-day efficiency is important. This is equal to the total energy or watthour output divided by the total energy or watthour input for the 24 h. That is,

$$\eta = \frac{(V_1 I_1 \cos \theta_1)t_1 + \cdots}{(V_1 I_1 \cos \theta_1)t_1 + \cdots + (I_1^2 R_{01})t_1 + \cdots + 24P_0} \qquad (15.1.117)$$

where t_1 = time in hours that load $V_1 I_1 \cos \theta_1$ is being delivered, etc.

Polyphase Transformer Connections Three-phase transformer banks may be connected Δ-Δ, Δ-Y, Y-Y, and Y-Δ. The Δ-Δ connection is very common, particularly at the lower voltages, and has the important advantage that the bank will operate V-connected if one transformer is disabled. The Δ-Y connection is advantageous for stepping up to high voltages since the secondary of the transformers need be wound only for 58 percent $(1/\sqrt{3})$ of the line voltage; it is also necessary when a four-wire three-phase system is obtained from a three-wire three-phase system since "a floating neutral" on the secondary cannot occur. This connection has found increased application in step-down distribution at 600 V and lower because of the relative ease of applying sensitive fault protection. The Y-Y system may be used for stepping up voltage. It should not be used for obtaining a three-phase four-wire system from a three-phase three-wire system, because of the "floating neutral" on the secondary and the resulting high degree of unbalance of the secondary voltages. The Y-Δ system may be used to step down high voltages, the reverse of the Δ-Y connection. Y-connected windings also preclude third harmonic, and their multiples, circulating currents in the transformer windings. In the Δ-Y and Y-Δ systems the ratio of line voltage is obviously not that of the individual transformers. Because of different phase displacement between primaries and secondaries, a Δ-Δ bank cannot be connected in parallel (on both sides) with a Δ-Y bank, etc., even if they both have the correct voltage ratios between lines.

Three-phase transformers combine the magnetic circuits of three single-phase transformers so that they have parts in common. A material saving in cost, in weight, and in space results, the greatest saving occurring in the core and oil. The advantages of three-phase transformers are often outweighed by their lack of flexibility. The failure of a single phase shuts down the entire transformer. With three single units, one unit may be readily replaced with a single spare. The primaries of single-phase transformers may be connected in Y or Δ at will and the secondaries properly phased. The primaries, as well as the secondaries of three-phase transformers, must be phased.

For the transformation of moderate amounts of power from three-phase to three-phase, two transformers employing either the **open delta** or **T connection** (Fig. 15.1.67) may be used. With each connection the ratio of line voltages is the same as the transformer ratios. In the figure,

Fig. 15.1.67 Transformer connections for transforming moderate amounts of three-phase power. (*a*) Open delta connection; (*b*) T connection.

ratios 10 : 1 are shown. In the T connection the primary and the secondary of the main transformer must be provided with a center tap to which one end of the teaser transformer is connected. The ratings of these systems are only 58 percent of the rating of the system using three similar transformers, one for each phase. Owing to dissymmetry, the terminal voltages become somewhat unbalanced even with a balanced load.

To transform from two- to three-phase or the reverse, the **T connection** of Fig. 15.1.68 is used. To make the secondary voltages symmetrical a tap (called a **Scott tap**) is brought out at 86.6 percent ($\sqrt{3}/2$) of the primary winding of the teaser transformer as shown in Fig. 15.1.68. With balanced no-load voltages the voltages become slightly unbalanced even under a symmetrical load, owing to unequal phase differences in the individual coils. The three-phase neutral O is one-third the winding of the teaser transformer from the junction. In Fig. 15.1.68*a* the transformation is from three-phase to a two-phase three-wire system. In Fig. 15.1.68*b* the transformation is from three-phase to a four-phase, five-wire system. The voltages are given on the basis of 100-V primaries with 1 : 1 transformer ratios.

An **autotransformer,** also called **compensator,** consists essentially of a single winding linking a magnetic circuit. Part of the energy is transformed, and the remainder flows through conductively. Suitable taps are provided so that, if the primary voltage is applied to two of the taps, a voltage may be taken from any other two taps. The ratio of voltages is equal practically to the ratio of the turns between their taps. An autotransformer should be installed only when the ratio of transformation is not large. The ratio of power transformed to total power is $1 - n$, where n is the ratio of low-voltage to high-voltage emf. This gives the saving over the ordinary transformer and is greatest when the ratio is not far from unity. Figure 15.1.69*a* shows 100 kW being changed from 3,300 to 2,300 V; 30.3 kW only are being actually transformed, and the remainder of the power flows through conductively. Figure 15.1.69*b* shows how an ordinary 10 : 1, 10-kW lighting transformer may be connected to boost 110 kW 10 percent in voltage. In Fig. 15.1.69*b*, however, the 230-V secondary must be insulated for 2,300 V to the core and ground. The voltage may likewise be reduced by reversing the 230-V coil. An autotransformer should never be used when it is desired to keep dangerous primary potentials from the secondary. It is used for starting induction motors (Fig. 15.1.71) and for a number of similar purposes.

Fig. 15.1.69 Autotransformer.

Data on Transformers Single-phase 55° self-cooled oil-insulated transformers for 2,300-V primaries, 230/115-V secondaries, and in sizes from 5 to 200 kVA for 60(25) Hz have efficiencies from one-half to full load of about 98 (97 to 98.7) percent and regulation of 1.5 (1.1 to 2.1) percent with pf = 1, and 3.5 (2.7 to 4.1) percent with pf = 0.8. Power transformers with 13,200-V primaries and 2,300-V secondaries in sizes from 667 to 5,000 kVA and for both 60 and 25 Hz have efficiencies from one-half to full load of about 99.0 percent and regulation of about 1.0 (4.2) percent with pf = 1 (0.8).

AC MOTORS

Polyphase Induction Motor The polyphase induction motor is the most common type of motor used. It may be classified as squirrel cage or wound rotor and may be single-speed or multispeed type. Squirrel-cage motors are further classified on the basis of torque, speed, and current characteristics as Designs A, B, C, and D and for locked-rotor current in kVA/hp by Codes A to V. It consists ordinarily of a stator which is wound in the same manner as the synchronous-generator stator. If two-phase current is supplied to a two-phase winding or three-phase current to a three-phase winding, a rotating magnetic field is produced in the air gap. The number of poles which this field has is the same as the number of poles that a synchronous generator employing the same stator winding would have. The speed of the rotating field, or the **synchronous speed,**

$$N = 120f/P \qquad \text{r/min} \tag{15.1.118}$$

where f = frequency and P = number of poles.

Fig. 15.1.68 Connections for transforming from three-phase to two- and four-phase power.

There are two general types of rotors. The **squirrel-cage type** consists of heavy copper or aluminum bars short-circuited by end rings, or the bars and end rings may be an integral aluminum casting. The **wound rotor** has a polyphase winding of the same number of poles as the stator, and the terminals are brought out to slip rings so that external resistance may be introduced. The rotor conductors must be cut by the rotating field, hence the rotor cannot run at synchronous speed but must slip. The per unit **slip** is

$$s = (N - N_2)/N \qquad (15.1.119)$$

where N_2 = the rotor speed, r/min. The rotor frequency

$$f_2 = sf \qquad (15.1.120)$$

The torque is proportional to the air-gap flux and the components of rotor current in space phase with it. The rotor currents tend to lag the emfs producing them, because of the rotor-leakage reactance. From Eq. (15.1.120) the rotor frequency and hence the rotor reactance ($X_2 = 2\pi f_2 L_2$) are low when the motor is running near synchronous speed, so that there is a large component of rotor current in space phase with the flux. With large values of slip the increased rotor frequency increases the rotor reactance and hence the lag of the rotor currents behind their emfs, and therefore considerable space-phase difference between these currents and the flux develops. Consequently even with large values of current the torque may be small. The torque of the induction motor increases with slip until it reaches a maximum value called the **breakdown torque,** after which the torque decreases (see Fig. 15.1.72). The breakdown torque varies as the square of the voltage, inversely as the stator impedance and rotor reactance, and is independent of the rotor resistance.

The squirrel-cage motor develops moderate torque on starting ($s = 1.0$) even though the current may be three to seven times rated current. For any value of slip the torque of the induction motor varies as the square of the voltage. The torque of the squirrel-cage motor which, on starting, is only moderate may be reduced in the larger motors because of starting voltage drop from inrush and possible necessity of applying reduced-voltage starting.

Polyphase squirrel-cage motors are used mostly for constant-speed work; however, recent developments in variable-frequency pulse-width modulation (PWM) ac drives have seen these motors used in variable-speed applications where the ratio of maximum to minimum speed does not exceed 4 : 1. They are used widely on account of their rugged construction and the absence of moving electrical contacts, which makes them suitable for operation when exposed to flammable dust or gas. General-purpose squirrel-cage motors have starting torques of 100 to 250 percent of full load torque at rated voltage. The highest torques occur at the higher rated speeds. The locked rotor currents vary between four and seven times full-load current. In the **double-squirrel-cage** type of motor there is a high-resistance winding in the top of the rotor slots and a low-resistance winding in the bottom of the slots. The low-resistance winding is made to have a high leakage reactance, either by separating the windings with a magnetic bridge, Fig. 15.1.70a, or by making the slot very narrow in the area between the two windings, Fig. 15.1.70b. On starting, because of the high reactance of the low-resistance winding, most of the rotor current will flow in the high-resistance winding, giving the motor a large starting torque. As the rotor approaches the low value of slip at which it normally operates, the rotor frequency and hence the rotor reactance become low and most of the rotor current now flows in the low-resistance winding. Cage bars can be shaped so that one winding gives comparable results. See Fig. 15.1.70c. The rotor operates with a low value of slip. The high starting torque of

the high-resistance motor and the excellent constant-speed operating characteristics of the low-resistance squirrel-cage rotor are combined in one motor.

The **single shaped** cage bar is more economical, and almost any shape for required characteristics can be extruded from aluminum. Single bars also eliminate the problems of differential expansion with double cage bars.

Nameplates of polyphase integral-hp squirrel-cage induction motors carry a *code* letter and a *design* letter. These provide information about motor characteristics, the former on locked rotor or starting inrush current (see Table 15.1.26) and the latter on torque characteristics. National Electrical Manufacturers Association (NEMA) standards publication No. MG1-2006 defines four design letters: A, B, C, and D. In all cases the motors are designed for full voltage starting. Locked rotor current and torque, pull-up torque and breakdown torque are tabulated according to horsepower and speed. Designs A, B, and C have full load slips less than 5 percent and design D more than 5 percent. Design B motors are the most widely used, having starting-torque and line-starting current characteristics suitable for most power systems. The nature of the various designs can be understood by reference to the full voltage values for a 100-hp, 1,800-r/min motor which follow:

	Design			
	A	B	C	D
Locked rotor torque	125*	125*	200*	275*
Pull-up torque	100†	100†	140†	—
Breakdown torque	200†	200†	190†	—
Locked rotor current	—	600*	600*	600*
Full load slip (%)	5§	5§	5§	5‡

NOTE: All quantities, except slip, are percent of full load value.
* Upper limit. [See NEC® Table 430.7(B).]
† Not less than.
‡ Greater than.
§ Less than.

Starting   Polyphase induction motor controllers are commonly available for full-voltage (across-the-line) or reduced-voltage starting for single-speed or multispeed applications and may include reversing control. It is desirable to start induction motors by direct connection across the line, since reduced voltage starters are expensive and almost always reduce the starting torque. The capacity of the distribution system dictates when reduced voltage starting must be used to limit voltage dips on the system. On stiff industrial systems 25,000-hp motors have been successfully started **across-the-line.**

In Fig. 15.1.71a is shown an **"across-the-line" starter** which may be operated from different push-button stations. The START push button closes the solenoid circuit between phases C and A through three bimetallic strips in series. This energizes solenoid S, which attracts armature D, which in turn closes the starting switch and the auxiliary blade G. This blade keeps the solenoid circuit closed when the START push button is released. Pressing the STOP push button opens the solenoid circuit, permitting the starting switch to open. A prolonged heavy overload raises the temperature of the heaters by an amount that will cause at least one of the overload sensing elements to open the solenoid circuit, releasing the starting switch.

A common method of applying reduced-voltage start is to use a **compensator** or **autotransformer** or **autostarter** (Fig. 15.1.71b). When the switch is in the starting position, the three windings AB of the three-phase autotransformer are connected in Y across the line and the motor terminals are connected to the taps which supply reduced voltage. When the switch is in the running position, the starter is entirely disconnected from the line. In modern practice, motors are protected by thermal or magnetic overload relays (Fig. 15.1.71) which operate to trip the circuit breaker. Since a time element is involved in the operation of such relays, they do not respond to large starting currents, because of their short duration. To limit the current to as low a value as possible, the lowest taps that will give the motor sufficient voltage to supply the

Fig. 15.1.70   Types of slots for squirrel-cage windings.

(a) Across-the-line-starter

(b) Autostarter

Fig. 15.1.71 Starters for squirrel-cage induction motor. (*a*) Across-the-line starter; (*b*) autostarter.

required starting torque should be used. As the torque of an induction motor varies as the square of the voltage, the compensator produces a very low starting torque.

Resistors in series with the stator may also be used to start squirrel-cage motors. They are inserted in each phase and are gradually cut out as the motor comes up to speed. The resistors are generally made of wire-type resistor units or of graphite disks enclosed within heat-resisting porcelain-lined iron tubes. The disadvantage of resistors is that if the motor is started slowly the resistor becomes very hot and may burn out. Resistor starters are less expensive than autotransformers. Their application is to motors that start with light loads at infrequent intervals.

A **phase-controlled,** silicon-controlled rectifier (SCR) may be used to limit the motor-starting current to any value that will provide sufficient starting torque by reducing voltage to the motors. The SCR can also be used to start and stop the motors. A positive opening device such as a contactor or circuit breaker should be used in series with the SCR to stop the motor in case of a failed SCR, which will normally fail shorted and apply full voltage to the motor. The SCR starter has been combined with a microprocessor to provide gradual motor terminal voltage increases during an adjustable acceleration period. This is commonly called *soft start*. Other features include the ability to limit motor starting current and save energy when the motor is lightly loaded.

By **introducing resistance into the rotor circuit** through slip rings, the rotor currents may be brought nearly into phase with the air-gap flux and, at the same time, any value of torque up to maximum torque obtained. As the rotor develops speed, resistance may be cut out until there is no external resistance in the rotor circuit. The speed may also be controlled by inserting resistance in the rotor circuit. However, like the armature-resistance method of speed control with shunt motors, this method is also inefficient and gives poor speed regulation. Figure 15.1.72 shows graphically the effect on the torque of applying reduced voltage (curves *b, c*) and of inserting resistance in the rotor circuit (curve *d*). As shown by curves *b* and *c*, the torque for any given slip is proportional to the square of the line voltage. The effect of introducing resistance into the rotor circuit is shown by curve *d*. The point of maximum torque is shifted toward higher values of slip. The maximum torque at starting (slip = 1.0) occurs when the

Fig. 15.1.72 Speed-torque curve for 10-hp, 60-Hz, 1,140-r/min induction motor.

rotor resistance is equal to the rotor reactance at standstill. The wound-rotor motor is used where large starting torque is necessary as in railway work, hoists, and cranes. It has better starting characteristics than the squirrel-cage motor, but, because of the necessarily higher resistance of the rotor, it has greater slip even with the rotor resistance all cut out. Obviously, the wound rotor, controller, and external resistance make it more expensive than the squirrel-cage type. See Table 15.1.13 for performance data.

One **disadvantage of induction motors** is that they take lagging current, and the power factor at half load and less is low. The speed- and torque-load characteristics of induction motors are almost identical with those of the shunt motor. The speed decreases slightly to full load, the slip being from 10 percent in small motors to 2 percent in very large motors. The torque is almost proportional to the load nearly up to the breakdown torque. The power factor is 0.8 to 0.9 at full load. The direction of rotation of any three-phase motor may be reversed by interchanging any two stator wires.

Speed Control of Induction Motors   The induction motor inherently is a constant-speed motor. From Eqs. (15.1.118) and (15.1.119) the rotor speed is

$$N_2 = 120f(1 - s)/P \qquad (15.1.121)$$

The speed can be changed only by changing the frequency, poles, or slip. By employing two distinct windings or by reconnecting a single winding by switching it is possible to change the number of poles. Complications prevent more than two speeds being readily obtained in this manner. Another objection to changing the number of poles is the fact that the design is a compromise, and sacrifices of desirable characteristics usually are necessary at both speeds. Three-phase, two-speed squirrel-cage motors of either consequent pole or separate winding types are commonly used for applications such as machine tools, fans, blowers, and refrigeration compressors in designs for constant horse-power and for constant or variable torque.

The change of slip by introducing resistance into the rotor circuit has been discussed under the wound-rotor motor.

Table 15.1.13   Performance Data for Drip-Proof Motors

			Power factor, %			Efficiency, %*		
hp	Weight, lb	Current, A	½ load	¾ load	Full load	½ load	¾ load	Full load
	3 phase, 230 V, 60 Hz, 1,750 r/min, squirrel cage

hp	Weight, lb	Current, A	Power factor, % ½ load	Power factor, % ¾ load	Power factor, % Full load	Efficiency, %* ½ load	Efficiency, %* ¾ load	Efficiency, %* Full load
	3 phase, 230 V, 60 Hz, 1,750 r/min, squirrel cage							
1	40	3.8	45.3	64.8	66.2	63.8	71.4	75.5
2	45	6.0	54.3	67.6	76.5	75.2	79.9	81.5
5	65	14.2	61.3	73.7	80.7	77.0	80.9	81.5
10	110	26.0	66.3	77.7	83.1	83.5	86.0	86.5
20	190	53.6	69.5	78.4	80.9	85.0	87.0	86.5
40	475	101.2	69.8	79.4	83.6	86.8	88.3	88.5
100	830	230.0	83.7	88.2	89.0	92.2	92.4	91.7
200	1,270	446.0	82.5	87.8	89.2	93.5	94.2	94.1
	3 phase, 230 V, 60 Hz, 1,750 r/min, wound rotor							
5	155	15.6	50.7	64.1	74.0	77.6	81.0	81.4
10	220	27.0	66.8	77.5	82.5	84.3	85.4	84.8
25	495	60.0	79.2	86.6	89.4	88.1	88.6	87.8
50	650	122.0	72.4	82.3	86.8	86.7	87.9	87.7
100	945	234.0	78.4	86.2	89.4	90.0	90.4	89.8
200	2,000	446.0	79.8	87.6	90.8	90.5	92.1	92.5
	3 phase, 2,300 V, 60 Hz, 1,750 r/min, squirrel cage							
300	2,300	70.4	76.2	83.4	86.0	90.8	92.5	92.8
700	3,380	155.0	85.5	89.4	90.0	91.7	93.4	93.7
1,000	4,345	221.0	86.2	89.5	90.0	92.1	93.8	94.1
	3 phase, 2,300 V, 60 Hz, 1,750 r/min, wound rotor							
300	3,900	68	84.4	86.2	89.9	90.8	92.5	92.8
700	5,750	154	82.7	87.7	90.9	91.7	93.4	93.7
1,000	8,450	218	82.9	87.9	91.1	92.1	93.8	94.1

* High-efficiency motors are available at premium cost.
SOURCE: Westinghouse Electric Corp.

Solid-state ac drives, also referred to as **adjustable-speed drives (ASDs),** using **pulse-width modulation (PWM)** frequency converters have become widely used for speed control of ac induction motors and are available for powers up to 10,000 horsepower.

AC drives using **line-commutated rectifiers (LCI)** are used for speed control of synchronous motors and, to a lesser degree, induction motors. PWM drives are widely used in industrial applications for pumps, blowers, fans, and compressors. Benefits in energy savings, reduced maintenance, and reduced mechanical and thermal stresses can be realized. Important issues to be considered in applying PWM drives include, but are not limited to: turndown ratio, harmonic effects on electrical networks, impact on the mechanical system, condition of existing motors retrofitted to PWM control, use of insulating bearings to minimize circulating currents in motors, distance from drive to motor, and increased frame size when motors are operated at less than rated speed. AC drive suppliers may have specific installation requirements or recommendations on motor feeder cable type and grounding practices. In many applications the motor/drive combination is engineered and supplied as a system.

Polyphase voltages should be evenly **balanced** to prevent phase current **unbalance.** If voltages are not balanced, the motor must be derated in accordance with National Electrical Manufacturers Association (NEMA), Publication No. MG-1-14.34.

Approximately 5 percent voltage unbalance would cause about 25 percent increase in temperature rise at full load. The input current unbalance at full load would probably be 6 to 10 times the input voltage unbalance.

Single phasing, one phase open, is the ultimate unbalance and will cause overheating and burnout if the motor is not disconnected from the line.

Single-Phase Induction Motor   Single-phase induction motors are usually made in fractional horsepower ratings, but they are listed by

NEMA in integral ratings up to 10 hp. They have relatively high rotor resistances and can operate in the single-phase mode without overheating. Single-phase induction motors are not self-starting.

However, the single-phase motor runs in the direction in which it is started. There are several methods of starting single-phase induction motors. Short-circuited turns, or **shading coils,** may be placed around the pole tips which retard the time phase of the flux in the pole tip, and thus a weak torque in the direction of rotation is produced. A high-resistance starting winding, displaced 90 electrical degrees from the main winding, produces poles between the main poles and so provides a rotating field which is weak but is sufficient to start the motor. This is called the **split-phase** method. In order to minimize overheating this winding is ordinarily cut out by a centrifugal device when the armature reaches speed. In the larger motors a repulsion-motor start is used. The rotor is wound like an ordinary dc armature with a commutator, but with short-circuited brushes pressing on it axially rather than radially. The motor starts as a repulsion motor, developing high torque. When it nears its synchronous speed, a centrifugal device pushes the brushes away from the commutator, and at the same time causes the segments to be short-circuited, thus converting the motor into a single-phase induction motor.

Capacitor Motors   Instead of splitting the phase by means of a high-resistance winding, it has become almost universal practice to connect a capacitor in series with the auxiliary winding (which is displaced 90 electrical degrees from the main winding). With capacitance, it is possible to make the flux produced by the auxiliary winding lead that produced by the main field winding by 90° so that a true two-phase rotating field results and good starting torque develops. However, the 90° phase relation between the two fields is obtainable at only one value of speed (as at starting), and the phase relation changes as the motor comes up to speed. Frequently the auxiliary winding is disconnected either by a centrifugal switch or a relay as the motor approaches full speed, in which case the motor is called a **capacitor-start** motor. With proper design the auxiliary winding may be left in circuit permanently (frequently with additional capacitance introduced). This improves both the power factor and torque characteristics. Such a motor is called **permanent-split capacitor motor.**

Phase Converter   If a polyphase induction motor is operating single-phase, polyphase emfs are generated in its stator by the combination of stator and rotor fluxes. Such a machine can be utilized, therefore, for converting single-phase power into polyphase power and, when so used, is called a **phase converter.** Unless corrective means are utilized, the polyphase emfs at the machine terminals are somewhat unbalanced. The power input, being single-phase and at a power factor less than unity, not only fluctuates but is negative for two periods during each cycle. The power output being polyphase is steady, or nearly so. The cyclic differences between the power output and the power input are accounted for in the kinetic energy stored in the rotating mass of the armature. The armature accelerates and decelerates, but only slightly, in accordance with the difference between output and input. The phase converter is used principally on railway locomotives, since a single trolley wire can be used to deliver single-phase power to the locomotive, and the converter can deliver three-phase power to the three-phase wound-rotor driving motors. PWM-type adjustable-speed drives are also available for small motor applications that can provide variable-speed control of a three-phase motor from a one- or two-phase voltage source.

AC Commutator Motors   Inherently simple ac motors are not adapted to high starting torques and variable speed. There are a number of types of commutating motor that have been developed to meet the requirement of high starting torque and adjustable speed, particularly with single phase. These usually have been accompanied by compensating windings, centrifugal switches, etc., in order to overcome low power factors and commutation difficulties. With proper compensation, commutator motors may be designed to operate at a power factor of nearly unity or even to take leading current.

One of the simplest of the **single-phase commutator motors** is the ac series railway motor such as is used on the erstwhile New York, New Haven, and Hartford Railroad. It is based on the principle that the torque of the dc series motor is in the same direction irrespective of the polarity of its line terminals. This type of motor must be used on low frequency, not over 25 Hz, and is much heavier and more costly than an equivalent dc motor. The torque and speed curves are almost identical with those of the dc series motor. Unlike most ac apparatus the power factor is highest at light load and decreased with increasing load. Such motors operate with direct current even better than with alternating current. For example, the New Haven locomotives also operate from the 600-V dc third-rail system (two motors in series) from the New York City line (238th St.) into Grand Central Station. (See also Sec. 11.)

On account of difficulties inherent in ac operation such as commutation and high reactance drops in the windings, it is economical to construct and operate such motors only in sizes adaptable to locomotives, the ratings being of the order of 300 to 400 hp. **Universal motors** are small simple series motors, usually of fractional horsepower, and will operate on either direct or alternating current, even at 60 Hz. They are used for vacuum cleaners, electric drills, and small utility purposes.

Synchronous Motor   Just as dc shunt generators operate as motors, a synchronous generator, connected across a suitable ac power supply, will operate as a motor and deliver mechanical power. Each conductor on the stator must be passed by a pole of alternate polarity every half cycle so that at constant frequency the rpm of the motor is constant and is equal to

$$N = 120 f/P \quad \text{r/min} \quad (15.1.122)$$

and the speed is independent of the load.

There are two types of synchronous motors in general use: the **slip-ring** type and the **brushless** type. The motor field current is transmitted to the motor by brushes and slip rings on the slip-ring type. On the brushless type it is generated by a shaft-mounted exciter and rectified and controlled by shaft-mounted static devices. Eliminating the slip rings is advantageous in dirty or hazardous areas.

The synchronous motor has the desirable characteristic that its power factor can be varied over a wide range merely by changing the field excitation. With a weak field the motor takes a lagging current. If the load is kept constant and the excitation increased, the current decreases (Fig. 15.1.73) and the phase difference between voltage and current becomes less until the current is in phase with the voltage and the power factor is unity. The current is then at its minimum value such as I_0, and the corresponding field current is called the **normal excitation.** Further increase in field current causes the armature current to lead and the power factor to decrease. Thus **underexcitation** causes the current to **lag;** **overexcitation** causes the current to **lead.** The effect of varying the field current at constant values of load is shown by the V curves (Fig. 15.1.73). Unity power factor occurs at the minimum value of armature current, corresponding to normal excitation. The power factor for any point such as P is I_0/I_1, leading current. Because of its adjustable power factor, the motor is frequently run light merely to improve power factor or to control the voltage at some part of a power system. When so used the motor is called a **synchronous condenser.** The motor may, however, deliver mechanical power and at the same time take either leading or lagging current.

Fig. 15.1.73   V curves of a synchronous motor.

Synchronous motors are used to drive **centrifugal** and **axial compressors,** usually through speed increasers, **pumps, fans,** and other high-horsepower applications where **constant speed, efficiency,** and **power factor** correction are important. **Low-speed** synchronous motors, under 600 r/min, sometimes called **engine** type, are used in driving **reciprocating compressors** and in **ball mills** and in other slow-speed applications. Their low length-to-diameter ratio, because of the need for many poles, gives them a high moment of inertia which is helpful in smoothing the **pulsating torques** of these loads.

If the motor **field current** is separately supported by a **battery** or **constant voltage transformer,** the synchronous motor will maintain speed on a lower voltage dip than will an induction motor because torque is proportional to voltage rather than voltage squared. However, if a synchronous motor drops **out of step,** it will normally not have the ability to reaccelerate the load, unless the driven equipment is automatically unloaded.

The synchronous motor is usually not used in smaller sizes since both the motors and its controls are more expensive than induction motors, and the ability of a small motor to supply VARs to correct power factor is limited.

If situated near an inductive load the motor may be overexcited, and its leading current will neutralize entirely or in part the lagging quadrature current of the load. This reduces the I^2R loss in the transmission

Table 15.1.14   Performance Data for Coupled Synchronous Motors

Power, hp	Poles	Speed, r/min	Current, A	Excitation, kW	Efficiencies, %			Weight, lb
					½ load	¾ load	Full load	
Unity power factor, 3 phase, 60 Hz, 2,300 V								
500	4	1,800	100	3	94.5	95.2	95.3	5,000
2,000	4	1,800	385	9	96.5	97.1	97.2	15,000
5,000	4	1,800	960	13	96.5	97.3	97.5	27,000
10,000	6	1,200	1,912	40	97.5	97.9	98.0	45,000
500	18	400	99.3	5	92.9	93.9	94.3	7,150
1,000	24	300	197	8.4	93.7	94.6	95.0	15,650
4,000	48	150	781	25	94.9	95.6	95.6	54,000
80% power factor, 3 phase, 60 Hz, 2,300 V								
500	4	1,800	127	4.5	93.3	94.0	94.1	6,500
2,000	4	1,800	486	13	95.5	96.1	96.2	24,000
5,000	4	1,800	1,212	21	95.5	96.3	96.5	37,000
10,000	6	1,200	2,405	50	96.8	97.3	97.4	70,000
500	18	400	125	7.2	92.4	93.4	93.6	9,500
1,000	24	300	248	11.6	93.3	94.2	94.4	17,500
4,000	48	150	982	40	94.6	95.3	95.5	11,500

SOURCE: Westinghouse Electric Corp.

lines and also increases the kilowatt ratings of the system apparatus. The **synchronous condenser and motor** can also be used to control voltage and to stabilize power lines. If the condenser or motor is overexcited, its leading current flowing through the line reactance causes a rise in voltage at the motor; if it is underexcited, the lagging current flowing through the line reactance causes a drop in voltage at the motor. Thus within limits it becomes possible to control the voltage at the end of a transmission line by regulating the fields of synchronous condensers or motors. Long 220-kV lines and the 287-kV Hoover Dam–Los Angeles line require several thousand kVa in synchronous condensers floating at their load ends merely for voltage control. If the load becomes small, the voltage would rise to very high values if the synchronous condensers were not underexcited, thus maintaining nearly constant voltage. See Table 15.1.14 for characteristics.

A **salient-pole synchronous motor** may be started as an induction motor. In **laminated-pole** machines conducting bars of copper, copper alloy, or aluminum, **damper** or **amortisseur windings** are inserted in the pole face and short-circuited at the ends, exactly as a squirrel-cage winding in the induction motor is connected. The bars can be designed only for starting purposes since they carry no current at synchronous speed and have no effect on efficiency. In solid-pole motors a block of steel is bolted to the pole and performs the current-carrying function of the damper winding in the laminated-pole motor. At times the pole faces are interconnected to minimize starting-pulsating torques. When the synchronous motor reaches 95 to 98 percent speed as an induction motor, the motor field is applied by a **timer** or **slip frequency** control circuit, and the motor pulls into step at 100 percent speed. While accelerating, the motor field is connected to resistances to minimize induced voltages and currents.

Two-pole motors are built as **turbine** type or **round-rotor motors** for mechanical strength and do not have the thermal capacity or space for starting windings, so they must be started by supplementary means.

One such supplementary starting means is the use of a variable-frequency source, either a **variable-speed generator** or more commonly a static **converter-inverter.** The motor is brought up to speed in synchronism with a slowly increasing frequency. One common application is the starting of the large motor-generators used in **pump-storage** utility systems.

Variable frequency may be used to start salient-pole machines also. Requirements for a start without high torques and pulsations or high voltage drops on small electrical systems may dictate the use of something other than full voltage starting.

The **synchronous reluctance motor** is similar to an induction machine with salient poles machined in the rotors. Under light loads the motor

will synchronize on reluctance torque and lock in step with the rotating field at synchronous speed. These motors are used in small sizes with variable-frequency inverters for speed control in the paper and textile industry.

The **synchronous-induction motor** is fundamentally a wound-rotor slip-ring induction motor with an air gap greater than normal, and the rotor slots are larger and fewer. On starting, resistance is inserted in the rotor circuit to produce high torque, and this is cut out as the speed increases. As synchronism is approached, the rotor windings are connected to a dc power source and the motor operates synchronously.

Timing or **clock motors** operate synchronously from ac power systems. Figure 15.1.74a illustrates the Warren Telechron motor which operates on the hysteresis principle. The stator consists of a laminated element with an exciting coil, and each pole piece is divided, a short-circuited shading turn being placed on each of the half poles so formed. The rotor consists of two or more hard-steel disks of the shape shown, mounted on a small shaft. The shaded poles produce a 3,600 r/min rotating magnetic field (at 60 Hz), and because of hysteresis loss, the disk follows the field just as the rotor of an induction motor does. When the rotor approaches the synchronous speed of 3,600 r/min, the rotating magnetic field takes a path along the two rotor bars and locks the rotor in with it. The rotor and the necessary train of reducing gears rotate in oil sealed in a small metal can. Figure 15.1.74b shows a subsynchronous motor. Six squirrel-cage bars are inserted in six slots of a solid cylindrical iron rotor, and the spaces between the slots form six salient poles. The motor, because of the squirrel cage, starts as an induction motor, attempting to attain

Fig. 15.1.74 Synchronous motors for timing. (a) Warren Telechron motor; (b) Holtz induction-reluctance subsynchronous motor.

Table 15.1.15   Relationships for AC-DC Conversion Static Devices

Device description	Voltages, %		Currents, %		Ripple, %
	E_{ac}	E_{dc}	I_{ac}	I_{dc}	
1ϕ, half wave	100	45	100	100	121
1ϕ, full wave	100	90	100	90	48
3ϕ, full wave	100	135	100	123	4.2

the speed of the rotating field, or 3,600 r/min (at 60 Hz). However, when the rotor reaches 1,200 r/min, one-third synchronous speed, the salient poles of the rotor lock in with the poles of the stator and hold the rotor at 1,200 r/min.

AC-DC CONVERSION

Static Rectifiers   Silicon devices, having replaced gas tubes, are the primary means of ac to dc or dc to ac conversion in modern installations. They are advantageous when compared to synchronous converters or motor generators because of efficiency, cost, size, weight, and reliability. Various bridge configurations for single-phase and three-phase applications are shown in Fig. 15.1.75a. Table 15.1.15 shows the relative outputs of rectifier circuits. The use of two three-phase bridges fed from an ac source consisting of a three-winding transformer with both a Δ and Y secondary winding so that output voltages are 30° out of phase will reduce dc ripple to approximately 1 percent.

The use of **silicon-controlled rectifiers** (SCRs) to replace rectifiers in the various bridge configurations allows the output voltages to be varied from rated output voltage to zero. The output voltage wave will not be a sine wave but a series of square waves, which may not be suitable for some applications. Dc to ac conversions are shown in Fig. 15.1.75b.

A new technology of ac-to-dc conversion commonly called *switch mode power supplies* (SMPSs) is found in almost all microprocessor-based modern electronic equipment for dc supplies between 3 and 15 V. The supply voltage (120 V ac) is rectified by a single-phase full-wave bridge circuit. See Fig. 15.1.75. The output is stored in a capacitor. A switcher will then switch the dc voltage from the capacitor on and off at a high frequency, usually between 10 and 100 kHz. These high-frequency pulses are stepped down in voltage by a transformer and

rectified by diodes. The diodes' output is filtered to the dc supply required. These SMPSs are small in size, have higher efficiency, and are lower in cost. They do create harmonic and power quality problems that have to be addressed.

SYNCHRONOUS CONVERTERS

The synchronous converter is essentially a dc generator with slip rings connected by taps to equidistant points in the armature winding. Alternating current may also be taken from and delivered to the armature. The machine may be single-phase, in which case there are two slip rings and two slip-ring taps per pair of poles; it may be three-phase, in which case there are three slip rings and three slip-ring taps per pair of poles, etc. Converters are usually used to convert alternating to direct current, in which case they are said to be operating **direct;** they may equally well convert direct to alternating current, in which case they are said to be operating **inverted.** A converter will operate satisfactorily as a dc motor, a synchronous motor, a dc generator, a synchronous generator, or it may deliver direct and alternating current simultaneously, when it is called a **double-current generator.**

The rating of a converter increases very rapidly with increase in the number of phases owing, in part, to better utilization of the armature copper and also because of more uniform distribution of armature heating.

Because of the materially increased rating, converters are nearly all operated six-phase. The rating decreases rapidly with decrease in power factor, and hence the converter should operate near unity power factor. The diametrical ac voltage is the ac voltage between two slip-ring taps 180 electrical degrees apart. With a two-pole closed winding, i.e., a winding that closes on itself when the winding is completed, the diametrical ac voltage is the voltage between any two slip-ring taps diametrically opposite each other.

With a sine-voltage wave, the dc voltage is the peak of the diametrical ac voltage wave. The voltage relations for sine waves are as follows: dc volts, 141; single phase, diametrical, 100; three-phase, 87; four-phase, diametrical, 100; four phase, adjacent taps, 71; six phase, diametrical, 100; six phase, adjacent taps, 50. These relations are obtained from the sides of polygons inscribed in a circle having a diameter of 100 V, as shown in Fig. 15.1.76.

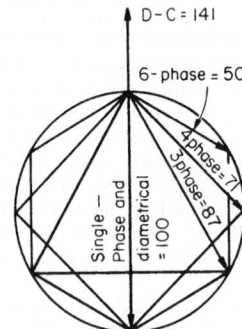

Fig. 15.1.76   EMF relations in a converter.

Selsyns   The word **selsyn** is an abbreviation of self-synchronizing and is applied to devices which are connected electrically, and in which an angular displacement of the rotating member of one device produces an equal angular displacement in the rotating member of the second device. There are several types of selsyns and they may be dc or ac, single-phase or polyphase. A simple and common type is shown in Fig. 15.1.77. The two stators S_1, S_2 are phase-wound stators, identical electrically with synchronous-generator or induction-motor stators. For simplicity Gramme-ring windings are shown in Fig. 15.1.77. The two stators are connected three-phase and in parallel. There are

Fig. 15.1.75   AC-DC conversion with static devices.

Fig. 15.1.77   Selsyn system.

also two bobbin-type rotors R_1, R_2, with single-phase windings, each connected to a single-phase supply such as 115 V, 60 Hz. When R_1 and R_2 are in the same angular positions, the emfs induced in the two stators by the ac flux of the rotors are equal and opposite, there are no interchange currents between stators, and the system is in equilibrium. However, if the angular displacement of R_1, for example, is changed, the magnitudes of the emfs induced in the stator winding of S_1 are correspondingly changed. The emfs of the two stators then become unbalanced, currents flow from S_1 to S_2, producing torque on R_2. When R_2 attains the same angular position as R_1, the emfs in the two rotors again become equal and opposite, and the system is again in equilibrium.

If there is torque load on either rotor, a resultant current is necessary to sustain the torque, so that there must be an angular displacement between rotors. However, by the use of an auxiliary selsyn a current may be fed into the system which is proportional to the angular difference of the two rotors. This current will continue until the error is corrected. This is called **feedback.** There may be a master selsyn, controlling several secondary units.

Selsyns are used for position indicators, e.g., in bridge–engine-room signal systems. They are also widely used for fire control so that from any desired position all the turrets and guns on battleships can be turned and elevated simultaneously through any desired angle with a high degree of accuracy. The selsyn itself rarely has sufficient power to perform these operations, but it actuates control through power multipliers such as amplidynes.

RATING OF ELECTRICAL MOTORS

Application of Electrical Motors

In addition to motor Design and Code classifications, motors are assigned other classifications to meet specific industry standards and practices such as the **temperature rating** of the motor at which the materials in the machine can be operated for long periods of time without deterioration.

Standard ANSI/IEEE 100-200 defines seven different **classes** (Class 90, 105, 130, 155, 180, 220, and Class over 220) for **insulating materials** based on a not-to-exceed temperature rating in degrees Celsius. For example, Class 155 insulation includes materials or combinations of materials such as silicone elastomer, mica, glass fiber, and asbestos, with bonding substances suitable for continuous operation at 155°C. Temperature class information would normally be used by wire, motor, or electrical apparatus designers in selecting insulating materials.

A system of motor **insulation classes** is used to establish maximum temperature rise in the motor winding (determined by winding resistance) based on a 40°C ambient, 3,300 ft (1006 m) elevation. In practice the motor manufacturer specifies the ambient temperature and insulation class. The temperature rise shall not exceed the limit for the insulation

system when the motor is loaded to its rating or to its service factor load. Temperature rise by insulation class for single and polyphase motors is as follows:

Motor type	Class of motor insulation system			
	A	B	F	H
Integral horsepower				
All motors with 1.15 service factor or higher	70°C	90°C	115°C	—
Totally enclosed fan-cooled motors (TEFC)	60°C	80°C	105°C	125°C
Totally enclosed nonventilated motors (TENV)	65°C	85°C	110°C	135°C
Motors with encapsulated windings, 1.0 service factor	65°C	85°C	110°C	—
All other motors	60°C	80°C	105°C	125°C
Fractional horsepower				
All motors with 1.15 service factor or higher	70°C	90°C	115°C	—
Totally enclosed motors (TEFC and TENV)	65°C	85°C	110°C	135°C
Any motor in frame size smaller than 42 frame	65°C	85°C	110°C	135°C
All other motors	60°C	80°C	105°C	125°C

NOTE: Manufacturers may offer insulation systems that exceed the above classes to provide additional measures of protection—for example, corona-resistant magnet wire for use in PWM ac drive applications.
SOURCE: Table 5-20, *Standard Handbook for Electrical Engineers,* McGraw-Hill.

General-purpose fractional horsepower and integral-horsepower motors are assigned a **service factor,** which allows the motor to be operated above its rated horsepower, when operated at its rated voltage and frequency, without damaging its insulation system. The normal service factors range from 1.40 to 1.25 for fractional to 1-hp motors, 1.15 for motors greater than 1 to 200 hp, and 1.0 for motors from 250 to 500 hp.

The starting kVA of a squirrel-cage motor is indicated by a **locked-rotor indicating code letter** (A to V) included on the motor nameplate (see Table 15.1.16). For additional information refer to NEMA Standard MG-1 or NEC®-2005 Table 430.7(B) for values and application. The locked-rotor current is used for selecting the size of the motor disconnecting means and motor controller (starter).

Required information on the **motor nameplate** includes manufacturer, rated volts, full-load amperes, rated frequency, number of phases if an ac motor, rated full-load speed for each speed, rated temperature rise for the insulation class (A, B, F, or H) at rated ambient temperature, time

Table 15.1.16   Maximum Rating or Setting of Motor Branch-Circuit Protective Devices and Starting-Inrush Code Letters

	Percent of full-load current					
Type of motor	Non-time delay fuse	Dual-element (time delay) fuse	Instantaneous breaker	Inverse time breaker	Code letter	kVA/hp with locked rotor*
Single-phase, all types, no code letter	300	175	700	250	A	0–3.14
AC motors:					B	3.15–3.54
Polyphase other than wound-rotor	300	175	800	250	C	3.55–3.99
Polyphase Design E or energy-efficient Design B	300	175	1100	250	D	4.0–4.49
High reactance squirrel-cage:					E	4.5–4.99
Synchronous	300	175	800	250	F	5.0–5.59
Wound rotor	150	150	800	150	G	5.6–6.29
DC motors:					H	6.3–7.09
Constant voltage	150	150	250	150	J	7.1–7.99
					K	8.0–8.99
					L	9.0–9.99
					M	10.0–11.19
					N	11.2–12.49
					P	12.5–13.99
					R	14.0–15.99
					S	16.0–17.99
					T	18.0–19.99
					U	20.0–22.39
					V	22.4 and up

* This is a general conversion. Refer to NEC® 2005, Tables 430-251(A) and (B) for conversion tables of single and polyphase motor locked-rotor currents for selection of disconnecting means and controllers as determined by motor horsepower and voltage rating.
SOURCE: NEC 1993, Tables 430-152 and 430-7(b). Reprinted with permission from NFPA 70-1993, National Electrical Code®, Copyright © 1992, National Fire Protection Association, Quincy, Massachusetts 02269. This reprinted material is not the complete and official position of the NFPA on the referenced subject, which is represented only by the standard in its entirety.

rating (5, 15, 30, 60 minutes, or continuous), rated horsepower for each speed, locked-rotor amperes of code letter (A to V), design letter (B, C, D, or E). Other required information includes identification of secondary volts for wound-rotor motors, field voltage and current for dc-excited synchronous motors, winding type if a dc motor, and whether motor is thermally or impedance protected. Motors are required to be marked with any certifying agency approvals (listing) for hazardous (classified) area use. Large or special motors may also include nominal efficiency, power factor, or other information such as marine or inverter duty, as specified by the manufacturer.

The **motor branch-circuit protection device** provides both short-circuit and ground-fault protection. Once motor hp is known, the approximate motor full load current can be determined from Table 15.1.17 and the maximum setting of motor branch-circuit protection device (fuses, circuit breakers) can be selected using Table 15.1.16. The motor branch-circuit protective device must be capable of carrying the motor locked-rotor current. Where the selected value of the branch-circuit protection device does not correspond to the size of a standard and fuse or nonadjustable circuit breaker, the next standard size (ampere rating) may be used.

Running overload protection is required for each motor unless the motor is impedance protected or provided with an integral thermal protector. A separate overload device is required that is responsive to motor current and selected on the basis of motor nameplate full-load current to trip at no more than 125 percent of the motor nameplate full-load rating for motors with a service factor of 1.15 or greater and motors with a marked temperature rise of 40°C. All other motors are set at 115 percent. Overload devices include electronic current-sensing relays, thermal and eutectic electromechanical relays, and, for motors larger than 1,500 hp, embedded temperature detectors. Thermal and electronic overload relays are supplied with either a fixed or adjustable class of trip curves responsive to motor current. The standard overload trip class is Class 20 and is suitable for most applications of Design B motors. Trip Class 10 is used for quick-trip applications and Class 30 for slow-trip applications.

Motor circuit conductors for a single motor in a continuous duty application are required to have an ampacity of not less than 125 percent of the motor full-load current as determined by Table 15.1.17. Conductors

supplying several motors and other loads are sized at 125 percent of the largest load plus the full-load current of other motors, plus ampacity of other loads.

Efficiency of Electrical Motors

Methods of determining efficiency are by **direct measurement** or by **segregated losses**. Methods are outlined in Standard Test Procedure for Polyphase Induction Motors and Generators, ANSI/IEEE Std. 112-1996; Test Procedure for Single-Phase Induction Motors, ANSI/IEEE Std. 114-2001; and Test Procedures for Synchronous Machines, IEEE Std. 115-2002.

Direct measurements can be made by using calibrated **motors, generators,** or **dynamometers** for input to generators and output from motors, and precision electrical motors for input to motors and output from generators.

$$\text{Efficiencies} = \frac{\text{output}}{\text{input}} \qquad (15.1.123)$$

The segregated **losses** in motors are classified as follows: (1) Stator I^2R (shunt and series field I^2R for dc); (2) rotor I^2R (armature I^2R for dc); (3) core loss; (4) stray-load loss; (5) friction and windage loss; (6) brush-contact loss (wound rotor and dc); (7) brush-friction loss (wound rotor and dc); (8) exciter loss (synchronous and dc); and (9) ventilating loss (dc). Losses are calculated separately and totaled. Measure the electrical output of the generator; then

$$\text{Efficiency} = \frac{\text{output}}{\text{output} + \text{losses}} \qquad (15.1.124)$$

Measure the electrical input of the motors; then

$$\text{Efficiency} = \frac{\text{input} - \text{losses}}{\text{input}} \qquad (15.1.125)$$

When testing dc motors, compensation should be made for the **harmonics** associated with rectified ac used to provide the variable dc voltage to the motors. Instrumentation should be chosen to accurately reflect the rms value of currents.

Temperature rise under full-load conditions may be measured by tests as outlined in the IEEE Standards referred to above. Methods of loading

Table 15.1.17   Approximate Full-Load Currents of Motors,* A
(See NEC® 2005, Art. 430, Tables 430-248, 430-249, 430-250 for more complete information.)

hp	Three-phase ac motors squirrel-cage and wound-rotor induction types				Synchronous type, unity power factor				Single-phase ac motors		DC motors†	
	230 V	460 V	575 V	2,300 V	230 V	460 V	575 V	2,300 V	115 V‡	230 V‡	120 V	240 V
½	2.2	1.1	0.9	—	—	—	—	—	9.8	4.9	5.4	2.7
¾	3.2	1.6	1.3	—	—	—	—	—	13.8	6.9	7.6	3.8
1	4.2	2.1	1.7	—	—	—	—	—	16	8	9.5	4.7
1½	6.0	3.0	2.4	—	—	—	—	—	20	10	13.2	6.6
2	6.8	3.4	2.7	—	—	—	—	—	24	12	17	8.5
3	9.6	4.8	3.9	—	—	—	—	—	34	17	25	12.2
5	15.2	7.6	6.1	—	—	—	—	—	56	28	40	20
7½	22	11	9	—	—	—	—	—	80	40	58	29
10	28	14	11	—	—	—	—	—	100	50	76	38
15	42	21	17	—	—	—	—	—	—	—	—	55
20	54	27	22	—	—	—	—	—	—	—	—	72
25	68	34	27	—	53	26	21	—	—	—	—	89
30	80	40	32	—	63	32	26	—	—	—	—	106
40	104	52	41	—	83	41	33	—	—	—	—	140
50	130	65	52	—	104	52	42	—	—	—	—	173
60	154	77	62	16	123	61	49	12	—	—	—	206
75	192	96	77	20	155	78	62	15	—	—	—	255
100	248	124	99	26	202	101	81	20	—	—	—	341
125	312	156	125	31	253	126	101	25	—	—	—	425
150	360	180	144	37	302	151	121	30	—	—	—	506
200	480	240	192	49	400	201	161	40	—	—	—	675

* The values of current are for motors running at speeds customary for belted motors and motors having normal torque characteristics. Use name-plate data for low-speed, high-torque, or multispeed motors. For synchronous motors of 0.8 pf multiply the above amperes by 1.25, at 0.9 pf by 1.1. The motor voltages listed are rated voltages. Respective nominal system voltages would be 220 to 240, 440 to 480, and 550 to 600 V. For full-load currents of 208-V motors, multiply the 230-V amperes by 1.10; for 200-V motors, multiply by 1.15.

† Ampere values are for motors running at base speed.

‡ Rated voltage. Nominal system voltages are 120 and 240.

are: (1) Load motor with **dynamometer** or **generator** of similar capacity and run until temperatures stabilize; (2) load generator with **motor-generator** set or **plant** load and run until temperature stabilizes; (3) alternately apply **dual frequencies** to motor until it reaches rated temperature; (4) synchronous motor may be operated as synchronous **condenser** at no load with **zero power factor** at rated current, voltage, and frequency until temperatures stabilize.

Industrial Applications of Motors

Alternating or Direct Current   The induction motor, particularly the squirrel-cage type, is preferable to the dc motor for constant-speed work, for the initial cost is less and the absence of a commutator reduces maintenance. Also there is less fire hazard in many industries, such as sawmills, flour mills, textile mills, and powder mills. The use of the induction motor in such places as cement mills is advantageous since with dc motors the grit makes the maintenance of commutators difficult.

For variable-speed work like cranes, hoists, elevators, and for adjustable speeds, the dc motor characteristics are superior to induction-motor characteristics. Even then, it may be desirable to use induction motors since their less desirable characteristics are more than balanced by their simplicity and the fact that ac power is available, and to obtain dc power conversion apparatus is usually necessary. Where both lights and motors are to be supplied from the same ac system, the 208/120-V four-wire three-phase system is now in common use. This gives 208 V three-phase for the motors, and 120 V to neutral for the lights.

Full-load speed, temperature rise, efficiency, and power factor as well as breakdown torque and starting torque have long been parameters of concern in the application and purchase of motors. Another qualification is service factor.

Special enclosures, fittings, seals, ventilation systems, electromagnetic design, etc., are required when the motor is to be operated under unusual service conditions, such as exposure to (1) combustible, explosive, abrasive, or conducting dusts, (2) lint or very dirty conditions where the accumulation of dirt might impede the ventilation, (3) chemical fumes or flammable or explosive gases, (4) nuclear radiation, (5) steam, salt laden air, or oil vapor, (6) damp or very dry locations, radiant heat, vermin infestation, or atmosphere conducive to the growth of fungus, (7) abnormal shock, vibration, or external mechanical loading, (8) abnormal axial thrust or side forces on the motor shaft, (9) excessive departure from rated voltage, (10) deviation factors of the line voltage exceeding 10 percent, (11) line voltage unbalance exceeding 1 percent, (12) situations where low noise levels are required, (13) speeds higher than the highest rated speed, (14) operation in a poorly ventilated room, in a pit, or in an inclined attitude, (15) torsional impact loads, repeated abnormal overloads, reversing or electric braking, (16) operation at standstill with any winding continuously energized, and (17) operation with extremely low structureborne and airborne noise. For dc machines, a further unusual service condition occurs when the average load is less than 50 percent over a 24-h period or the continuous load is less than 50 percent over a 4-h period.

The standard direction of rotation for all nonreversing dc motors, ac single-phase motors, synchronous motors, and universal motors is counterclockwise when facing the end of the machine opposite the drive end. For dc and ac generators, the rotation is clockwise.

Further information may be found in Publication No. MG-1 of the National Electrical Manufacturers Association.

In applying motors the temperature rating of the branch-circuit wiring conductors must be sufficient to meet the operating environment inside the motor terminal box to prevent possible failure of the insulation inside the motor terminal box.

ELECTRIC DRIVES

Cranes and Hoists   The dc series motor is best adapted to cranes and hoists. When the load is heavy the motor slows down automatically and develops increased torque thus reducing the peaks on the electrical system. With light loads, the speed increases rapidly, thus giving a lively crane. The series motor is also well adapted to moving the bridge itself and also the trolley along the bridge. Where alternating current only is available and it is not economical to convert it, the slip-ring type of induction motor, with external-resistance speed control, is the best type of ac motor. Squirrel-cage motors with high resistance end rings to give high starting torque (design D) are used (design D motors; also see Ilgner system).

Constant-Torque Applications Piston pumps, mills, extruders, and **agitators** may require constant torque over their complete speed range. They may require high starting torque design C or D squirrel-cage motors to bring them up to speed. Where speed is to be varied while running, a variable armature voltage **dc motor** or a **variable-frequency** squirrel-cage induction-motor drive system may be used.

Centrifugal Pumps Low WK^2 and low starting torques make design B general-purpose squirrel-cage motors the preference for this application. When **variable flow** is required, the use of a **variable-frequency** power supply to vary motor speed will be energy efficient when compared to changing flow by **control-valve** closure to increase head.

Centrifugal Fans High WK^2 may require high starting torque design C or D squirrel-cage motors to bring the fan up to speed in a reasonable period of time. When **variable flow** is required, the use of a **variable-frequency** power supply or a multispeed motor to vary fan speed will be energy efficient when compared to closing **louvers.** For large fans, **synchronous-motor** drives may be considered for high efficiency and improved power factor.

Axial or Centrifugal Compressors For smaller compressors, say, up to 100 hp, the squirrel-cage induction motor is the drive of choice. When the WK^2 is high, a design C or D high-torque motor may be required. For larger compressors, the synchronous motor is more efficient and improves power factor. Where **variable flow** is required, the **variable-frequency** power supply to vary motor speed is more efficient than controlling by **valve** and in some applications may eliminate a **gearbox** by allowing the motor to run at compressor operating speed.

Pulsating-Torque Applications Reciprocating compressors, rock crushers, and **hammer mills** experience widely varying torque pulsations during each revolution. They usually have a **flywheel** to store energy, so a high-torque, high-slip design D motor will accelerate the high WK^2 rapidly and allow energy recovery from the flywheel when high torque is demanded. On larger drives a slow-speed, engine-type synchronous motor can be directly connected. The motor itself supplies significant WK^2 to smooth out the torque and current pulsations of the system.

SWITCHBOARDS

Switchboards may, in general, be divided into four classes: direct-control panel type; remote mechanical-control panel type; direct-control truck type; electrically operated. With **direct-control panel-type boards** the switches, rheostats, bus bars, meters, and other apparatus are mounted on or near the board and the switches and rheostats are operated directly, or by operating handles if they are mounted in back of the board. The voltages, for both direct current and alternating current, are usually limited to 600 V and less but may operate up to 2,500 V ac if oil circuit breakers are used. Such panels are not recommended for capacities greater than 3,000 kVA. **Remote mechanical-control panel-type boards** are ac switchboards with the bus bars and connections removed from the panels and mounted separately away from the load. The oil circuit breakers are operated by levers and rods. This type of board is designed for heavier duty than the direct-control type and is used up to 25,000 kVA. **Direct-control truck-type switchboards** for 15,000 V or less consist of equipment enclosed in steel compartments completely assembled by the manufacturers. The high-voltage parts are enclosed, and the equipment is interlocked to prevent mistakes in operation. This equipment is designed for low- and medium-capacity plants and auxiliary power in large generating stations. **Electrically operated switchboards** employ solenoid or motor-operated circuit breakers, rheostats, etc., controlled by small switches mounted on the panels. This makes it possible to locate the high-voltage and other equipment independently of the location of switchboard.

In all large stations the switching equipment and buses are always mounted entirely either in separate buildings or in outdoor enclosures. Such equipment is termed **bus structures** and is electrically operated from the main control board.

Switchboards should be erected at least 3 to 4 ft from the wall. Switchboard frames and structures should be grounded. The only exceptions are effectively insulated frames of single-polarity dc switchboards. For low-potential work, the conductors on the rear of the switchboard are usually made up of flat copper strip, known as **bus-bar** copper. The size required is based upon a current density of about 1,000 A/in². Figure 15.1.78 gives the approximate continuous dc carrying capacity of copper bus bars for different arrangements and spacings for 35°C temperature rise.

Switchboards must be individually adapted for each specific electrical system. Space permits the showing of the diagrams of only three boards each for a typical electrical system (Fig. 15.1.79). Aluminum bus bars are also frequently used.

Equipment of Standard Panels Following are enumerated the various parts required in the equipment of standard panels for varying services:

Generator or synchronous-converter panel, dc two-wire system: 1 circuit breaker; 1 ammeter; 1 handwheel for rheostat; 1 voltmeter; 1 main switch (three-pole single throw or double throw) or 2 single-pole switches.

Generator or synchronous-converter panel, dc three-wire system: 2 circuit breakers; 2 ammeters; 2 handwheels for field rheostats; 2 field switches; 2 potential receptacles for use with voltmeter; 3 switches; 1 four-point starting switch.

Generator or synchronous-motor panel, three-phase three-wire system: 3 ammeters; 1 three-phase wattmeter; 1 voltmeter; 1 field ammeter; 1 double-pole field switch; 1 handwheel for field rheostats; 1 synchronizing receptacle (four-point); 1 potential receptacle (eight-point); 1 field

Fig. 15.1.78 Current-carrying capacity of copper bus bars.

Fig. 15.1.79 Switchboard wiring diagrams for generators. (*a*) 125-V or 250-V dc generator; (*b*) three-phase, synchronous generator and exciter for a small or isolated plant; (*c*) three-wire dc generator for a small or isolated plant. A, ammeter; AS, three-way ammeter switch; CB, circuit breaker, CT, current transformer; DR, ground detector receptacle; L, ground detector lamp; OC, overload coil; OCB, oil circuit breaker; PP, potential ring; PR, potential receptacle; PT, potential transformer; Rheo, rheostat; RS, resistor; S, switch; Sh, shunt; V, voltmeter; WHM, watthour meter.

rheostat; 1 triple-pole oil switch; 1 power-factor indicator; 1 synchronizer; 2 series transformers; 1 governor control switch.

Synchronous-converter panel, three-phase: 1 ammeter; 1 power-factor indicator; 1 synchronizing receptacle; 1 triple-pole oil circuit breaker; 2 current transformers; 1 potential transformer; 1 watthour meter (polyphase); 1 governor control switch.

Induction motor panel, three-phase: 1 ammeter; series transformers; 1 oil switch.

Feeder panel, dc, two-wire and three-wire: 1 single-pole circuit breaker; 1 ammeter; 2 single-pole main switches; potential receptacles (1 four-point for two-wire panel; 1 four-point and 1 eight-point for three-wire panel).

Feeder panel, three-wire, three-phase and single-phase: 3 ammeters; 1 automatic oil switch (three-pole for three-phase, two-pole for single-phase); 2 series transformers; 1 shunt transformer; 1 wattmeter; 1 voltmeter; 1 watthour meter; 1 handwheel for control of potential regulator.

Exciter panel (for 1 or 2 exciters): 1 ammeter (2 for 2 exciters); 1 field rheostat (2 for 2 exciters); 1 four-point receptacle (2 for 2 exciters); 1 equalizing rheostat for regulator.

Switches The current-carrying parts of switches are usually designed for a current density of 1,000 A/in². At contact surfaces, the current density should be kept down to about 50 A/in².

Circuit Breakers Switches equipped with a tripping device constitutes an elementary load interrupter switch. The difference between a load interrupter switch and a circuit breaker lies in the interrupting capacity. A circuit breaker must open the circuit successfully under short circuit conditions when the current through the contacts may be several orders of magnitude greater than the rated current. As the circuit is being opened, the device must withstand the accompanying mechanical forces and the heat of the ensuing arc until the current is permanently reduced to zero.

The opening of a metallic circuit while carrying electric current causes an electric arc to form between the parting contacts. If the action takes place in air, the air is ionized (a plasma is formed) by the passage of current. When ionized, air becomes an electric conductor. The space between the parting contacts thus has relatively low voltage drop and the region close to the surface of the contacts has relatively high voltage drop. The thermal input to the contact surfaces (VI) is therefore relatively large and can be highly destructive. A major aim in circuit breaker design is to quench the arc rapidly enough to keep the contacts in a reusable state. This is done in several ways: (1) lengthening the arc mechanically, (2) lengthening the arc magnetically by driving the current-carrying plasma sideways with a magnetic field, (3) placing barriers in the arc path to cool the plasma and increase its length, (4) displacing and cooling the plasma by means of a jet of compressed air or inert gas, and (5) separating the contacts in a vacuum chamber.

By a combination of shunt and series coils the circuit breaker can be made to trip when the energy reverses. Circuit breakers may trip unnecessarily when the difficulty has been immediately cleared by a local breaker or fuse. In order that service shall not be thus interrupted unnecessarily, **automatically reclosing breakers** are used. After tripping, an automatic mechanism operates to reclose the breaker. If the short circuit still exists, the breaker cannot reclose. The breaker attempts to reclose two or three times and then if the short circuit still exists it remains permanently locked out.

Metal-clad switch gears are highly developed pieces of equipment that combine buses, circuit breakers, disconnecting devices, controlling devices, current and potential transformers, instruments, meters, and interlocking devices, all assembled at the factory as a single unit in a compact steel enclosing structure. Such equipment may comprise truck-type circuit breakers, assembled as a unit, each housed in a separate steel compartment and mounted on a small truck to facilitate removal for inspection

Fig. 15.1.80   Three equivalent symmetrical transmission, or distribution, systems. (*a*) Single-phase; (*b*) three-phase; (*c*) four-phase.

and servicing. The equipment is interlocked to prevent mistakes in operation and in the removal of the unit; the removal of the unit breaks all electrical connections by suitable disconnecting switches in the rear of the compartment, and all metal parts are grounded. This design provides compactness, simplicity, ease of inspection, and safety to the operator.

High-voltage circuit breakers can be oil type, in which the contacts open under **oil, air-blast** type, in which the arc is extinguished by a powerful blast of air directed through an orifice across the arc and into an arc chute, H_2S type, or vacuum contact type. The tripping of high-voltage circuit breakers is initiated by an abnormal current acting through the secondary of a current transformer on an inverse-time relay in which the time of closing the relay contacts is an inverse time function of the current; i.e., the greater the current the shorter the time of closing. The breaker is tripped by a dc tripping coil, the dc circuit being closed by the relay contacts. Modern circuit breakers should open the circuit within 3 cycles from the time of the closing of the relay contacts.

Vacuum circuit breakers have received wide acceptance in all fields in recent years, both for indoor work and for outdoor applications. Indoor breakers are available up to 40 kV and interrupting capacities up to 2.5 GVA. Outdoor breakers are available in ratings up to that of the EHV (extra-high voltage 765-kV three-pole breaker capable of interrupting 55 GVA, or 40,000-A symmetrical current. Its operating rating is 3,000 A, 765 kV. The arc is extinguished in a vacuum. Switching stations, gas-insulated and operating at 550 kV, are also in use.

POWER TRANSMISSION

Power for long-distance transmission is usually generated at 6,600, 13,200, and 18,000 V and is stepped up to the transmission voltage by Δ-Y-connected transformers. The transmission voltage is roughly 1,000 V/mi. Preferred or standard transmission voltages are 22, 33, 44, 66, 110, 132, 154, 220, 287, 330, 500, and 765 kV. High-voltage lines across country are located on private rights of way. When they reach urban areas, the power must be carried underground to the substations which must be located near the load centers in the thickly settled districts. In many cases it is possible to go directly to underground cables since these are now practicable up to 345 kV between three-phase line conductors (200 kV to ground). High-voltage cables are expensive in both first cost and maintenance, and it may be more economical to step down the voltage before transmitting the power by underground cables. Within a city, alternating current may be distributed from a substation at 13,200, 6,600, or 2,300 V, being stepped down to 600, 480, and 240 V, three-phase for power and 240 to 120 V single-phase three-wire for lights, by transformers at the consumers' premises. **Direct current** at 1,200 or 600 V for railways, 230 to 115 V for lighting and power, is supplied by motor-generator sets, synchronous converters, and rectifiers. **Constant current** for series street-lighting systems is obtained through constant-current transformers.

Transmission Systems

Power is almost always transmitted three-phase. The following fundamental relations apply to any transmission system. The weight of conductor required to transmit power by any given system with a given percentage power loss varies directly with the power, directly as the square of the distance, and inversely as the square of the voltage. The cross-sectional area of the conductors with a given percentage power loss varies directly with the power, directly with the distance, and inversely as the square of the voltage.

For two systems of the same length transmitting the same power at different voltages and with the same power loss for both systems, the cross-sectional area and weight of the conductors will vary inversely as the square of the voltages. The foregoing relations between the cross section or weight of the conductor and transmission distance and voltage hold for all systems, whether dc, single-phase, three-phase, or four-phase. With the power, distance, and power loss fixed, all symmetrical systems having equal voltages to neutral require equal weights of conductor. Thus, the three symmetrical systems shown in Fig. 15.1.80 all deliver the same power, have the same power loss and equal voltages to neutral, and the transmission distances are all assumed to be equal. They all require the same weight of conductor since the weights are inversely proportional to all resistances. (No actual neutral conductor is used.) The respective power losses are (1) $2I^2R$ W; (2) $3(2I/3)^2(3R/2) = 2I^2R$ W; (3) $4(I/2)^2(2R) = 2I^2R$ W, which are all equal.

Size of Transmission Conductor   Kelvin's law states, "The most economical area of conductor is that for which the annual cost of energy wasted is equal to the interest on that portion of the capital outlay which can be considered proportional to the weight of copper used." In Fig. 15.1.81 are shown the annual interest cost, the annual cost of I^2R loss, and the total cost as functions of circular mils cross section for both typical overhead conductors and three-conductor cables. Note that the total-cost curves have very flat minimums, and usually other

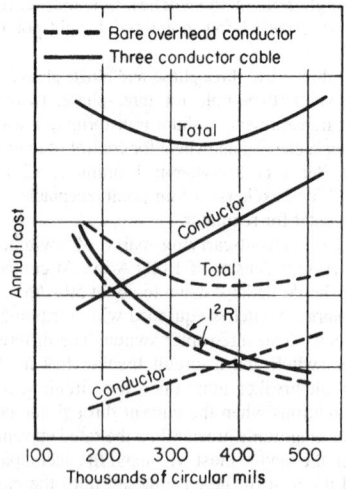

Fig. 15.1.81   Most economical sizes of overhead and underground conductors.

Fig. 15.1.82 Three-phase power system.

factors such as the character of the load and the voltage regulation, are taken into consideration.

In addition to resistance, overhead power lines have inductive reactance to alternating currents. The inductive reactance

$$X = 2\pi f\{80 + 741.1 \log [(D - r)/r]\}10^{-6}$$
$$\Omega/\text{conductor mile} \quad (15.1.126)$$

where f = frequency, D = distance between centers of conductors (in), and r their radius (in). Table 15.1.18 gives the inductive reactance per mile at 60 Hz and the resistance of stranded and solid copper conductor. (See Table 15.1.22.)

Any **symmetrical system** having n conductors can be divided into n equal single-phase systems, each consisting of one wire and a return circuit of zero impedance and each having as its voltage the system voltage to neutral.

Figure 15.1.82 shows a symmetrical three-phase system, with one phase detached. The load or received voltage between line conductors is E_R' so that the receiver voltage to *neutral* is $E_R = E_R'/\sqrt{3}$ V. The current is IA, the load power factor is $\cos \theta$, and the line resistance and reactance are R and X Ω per wire, and the sending-end voltage is E_S. The phasor diagram is shown in Fig. 15.1.83 (compare with Fig. 15.1.63). Its solution is

$$E_S = \sqrt{(E_R \cos \theta + IR)^2 + (E_R \sin \theta + IX)^2} \quad (15.1.127)$$

[see Eq. (15.1.101)].

Fig. 15.1.83 Phasor diagram for a power line.

Figure 15.1.84 (Mershon diagram) shows the right-hand portion of Fig. 15.1.83 plotted to large scale, the arc 00 corresponding to the arc ab (Fig. 15.1.83). The abscissa 0 (Fig. 15.1.84) corresponds to point b (Fig. 15.1.83) and is the load voltage E_R taken as 100 percent. The concentric circular arcs 0–40 are given in percentage of E_R. To find the sending-end voltage E_S for any power factor $\cos \theta$, compute first the resistance drop IR and the reactance drop IX in percentage of E_R. Then follow the ordinate corresponding to the load power factor to the inner arc 00 (a, Fig. 15.1.83). Lay off the percentage IR drop horizontally to the right, and the percentage IX drop vertically upward. The arc at which the IX drop terminates (c, Fig. 15.1.83) when added to 100 percent gives the sending-end voltage E_S in percent of the load voltage E_R.

EXAMPLE. Let it be desired to transmit 20,000 kW three-phase 80 percent power factor lagging current, a distance of 60 mi. The voltage at the receiving end is 66,000 V, 60 Hz and the line loss must not exceed 10 percent of the power delivered. The conductor spacing must be 7 ft (84 in). Determine the sending-end voltage and the actual efficiency. $I = 20,000,000/(66,000 \times 0.80 \times \sqrt{3}) = 218.8$ A. $3 \times 218.8^2 \times R' = 0.10 \times 20,000,000$. $R' = 13.9 \Omega = 0.232 \Omega/\text{mi}$. By referring to Table 13.1.18, 250,000 cir mils copper having a resistance of 0.2278

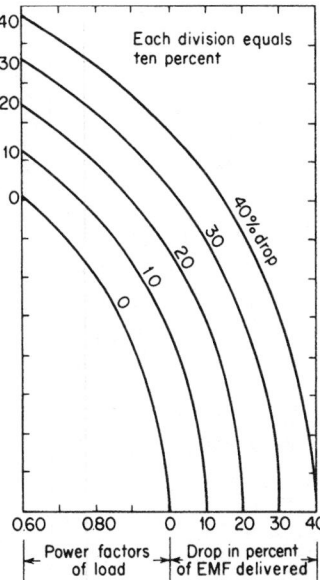

Fig. 15.1.84 Mershon diagram for determining voltage drop in power lines.

Ω/mi may be used. The total resistance $R = 60 \times 0.2278 = 13.67 \Omega$. The reactance $X = 60 \times 0.723 = 43.38 \Omega$. The volts to neutral at the load, $E_R = 66,000/\sqrt{3} = 38,100$ V. $\cos \theta = 0.80$; $\sin \theta = 0.60$. Using Eq. (15.1.127), $E_S = \{[(38,100 \times 0.80) + (218.8 \times 13.67)]^2 + [(38,100 \times 0.60) + (218.8 \times 43.38)]^2\}^{1/2} = 46,500$ V to neutral or $\sqrt{3} \times 46,500 = 80,500$ between lines at the sending end. The line loss is $3(218.8)^2 \times 13.67 = 1,963$ kW. The efficiency $\eta = 20,000/(20,000 + 1,963) = 0.911$, or 91.1 percent. This same line is solved by means of the Mershon diagram as follows. Let $E_R = 38,100$ V = 100 percent. $IR = 218.8 \times 13.67 = 2,991$ V = 7.85 percent. $IX = 218.8 \times 43.38 = 9,490$ V = 24.9 percent. Follow the 0.80 power-factor ordinate (Fig. 15.1.84) to its intersection with the arc 00; from this point go 7.85 percent horizontally to the right and then 24.9 percent vertically. (These percentages are measured on the horizontal scale.) This last distance terminates on the 22.5 percent arc. The sending-end voltage to neutral is then $1.225 \times 38,100 = 46,500$ V, so that the sending-end voltage between line conductors is $E_s' = 46,500\sqrt{3} = 80,530$ V.

In Table 15.1.18 the spacing is the distance between the centers of the two conductors of a single-phase system or the distance between the centers of each pair of conductors of a three-phase system if they are equally spaced. If they are not equally spaced, the geometric mean distance (*GMD*) is used, where $GMD = \sqrt[3]{D_1 D_2 D_3}$ (Fig. 15.1.85a). With the flat horizontal spacing shown in Fig. 15.1.85b, $GMD = \sqrt[3]{2D^3} = 1.26D$.

Fig. 15.1.85 Unequal spacing of three-phase conductors. (*a*) GMD = $(D_1 D_2 D_3)^{1/3}$; (*b*) flat horizontal spacing; GMD = 1.26D.

In addition to copper, aluminum cable steel-reinforced (ACSR), Table 15.1.19, is used for transmission conductor. For the same resistance it is lighter than copper, and with high voltages the larger diameter reduces corona loss.

Until 1966, 345 kV was the highest operating voltage in the United States. The first 500 kV system put into operation (1966) was a 350-mi

Table 15.1.18 Resistance and Inductive Reactance per Single Conductor

| | | | | 60 Hz | | | | | | | | | | | | |
| | | | | Spacing, ft | | | | | | | | | | | | |
Size, cir mils or AWG	No. of strands	OD, in	Resistance, Ω/mi	1	2	3	4	5	6	7	8	10	12	15	20	30
Hard-drawn copper, stranded																
500,000	37	0.814	0.1130	0.443	0.527	0.576	0.611	0.638	0.660	0.679	0.695	0.722	0.745	0.772	0.807	0.856
400,000	19	0.725	0.1426	0.458	0.542	0.591	0.626	0.653	0.675	0.694	0.710	0.737	0.760	0.787	0.822	0.871
300,000	19	0.628	0.1900	0.476	0.560	0.609	0.644	0.671	0.693	0.712	0.728	0.755	0.778	0.805	0.840	0.889
250,000	19	0.574	0.2278	0.487	0.571	0.620	0.655	0.682	0.704	0.723	0.739	0.766	0.789	0.816	0.851	0.900
0000	19	0.528	0.2690	0.497	0.581	0.630	0.665	0.692	0.714	0.733	0.749	0.776	0.799	0.826	0.861	0.917
000	7	0.464	0.339	0.518	0.602	0.651	0.686	0.713	0.735	0.754	0.770	0.797	0.820	0.847	0.882	0.931
00	7	0.414	0.428	0.532	0.616	0.665	0.700	0.727	0.749	0.768	0.784	0.811	0.834	0.861	0.896	0.945
0	7	0.368	0.538	0.546	0.630	0.679	0.714	0.741	0.763	0.782	0.798	0.825	0.848	0.875	0.910	0.959
Hard-drawn copper, solid																
0000	—	0.4600	0.264	0.510	0.594	0.643	0.678	0.705	0.727	0.746	0.762	0.789	0.812	0.839	0.874	0.923
000	—	0.4096	0.333	0.524	0.608	0.657	0.692	0.719	0.741	0.760	0.776	0.803	0.826	0.853	0.888	0.937
00	—	0.3648	0.420	0.538	0.622	0.671	0.706	0.733	0.755	0.774	0.790	0.817	0.840	0.867	0.902	0.951
0	—	0.3249	0.528	0.552	0.636	0.685	0.720	0.747	0.769	0.788	0.804	0.831	0.854	0.881	0.916	0.965
1	—	0.2893	0.665	0.566	0.650	0.699	0.734	0.761	0.783	0.802	0.818	0.845	0.868	0.895	0.930	0.979

Table 15.1.19   Properties of Aluminum Cable Steel-Reinforced (ACSR)

Cir mils or AWG		No. of wires			Cross section, in^2		Total	Ω/mi of single conductor at 25°C				
									200 A		600 A	
Aluminum	Copper equivalent	Aluminum	Steel	OD, in	Aluminum	Total	lb/mi	0 A dc	25 Hz	60 Hz	25 Hz	60 Hz
1,590,000	1,000,000	54	19	1.545	1.249	1.4071	10,777	0.0587	0.0589	0.0594	0.0592	0.0607
1,431,000	900,000	54	19	1.465	1.124	1.2664	9,699	0.0652	0.0654	0.0659	0.0657	0.0671
1,272,000	800,000	54	19	1.382	0.9990	1.1256	8,621	0.0734	0.0736	0.0742	0.0738	0.0752
1,192,500	750,000	54	19	1.338	0.9366	1.0553	8,082	0.0783	0.0785	0.0791	0.0787	0.0801
1,113,000	700,000	54	19	1.293	0.8741	0.9850	7,544	0.0839	0.0841	0.0848	0.0843	0.0857
1,033,500	650,000	54	7	1.246	0.8117	0.9170	7,019	0.0903	0.0906	0.0913	0.0908	0.0922
954,000	600,000	54	7	1.196	0.7493	0.8464	6,479	0.0979	0.0980	0.0985	0.0983	0.0997
874,500	550,000	54	7	1.146	0.6868	0.7759	5,940	0.107	0.107	0.108	0.107	0.109
795,000	500,000	26	7	1.108	0.6244	0.7261	5,770	0.117	0.117	0.117	0.117	0.117
715,500	450,000	54	7	1.036	0.5620	0.6348	4,859	0.131	0.131	0.133	0.131	0.133
636,000	400,000	54	7	0.977	0.4995	0.5642	4,319	0.147	0.147	0.149	0.147	0.149
556,500	350,000	26	7	0.927	0.4371	0.5083	4,039	0.168	0.168	0.168	0.168	0.168
477,000	300,000	26	7	0.858	0.3746	0.4357	3,462	0.196	0.196	0.196	0.196	0.196
397,500	250,000	26	7	0.783	0.3122	0.3630	2,885	0.235	0.235	0.235	0.235	0.235
336,400	0000	26	7	0.721	0.2642	0.3073	2,442	0.278	0.278	0.278	0.278	0.278
266,800	000	26	7	0.642	0.2095	0.2367	1,936	0.350	0.350	0.350	0.350	0.350
0000	00	6	1	0.563	0.1662	0.1939	1,542	0.441	0.443	0.446	0.447	0.464
000	0	6	1	0.502	0.1318	0.1537	1,223	0.556	0.557	0.561	0.562	0.579
00	1	6	1	0.447	0.1045	0.1219	970	0.702	0.703	0.707	0.706	0.718
0	2	6	1	0.398	0.0829	0.0967	769	0.885	0.885	0.889	0.887	0.893

SOURCE: Aluminum Co. of America.

transmission loop of the Virginia Electric and Power Company; the longest transmission distance was 170 mi. The towers, about 94 ft high, are of corrosion-resistant steel, and the conductors are 61-strand cables of aluminum alloy, rather than the usual aluminum cable with a steel core (ACSR). The conductor diameter is 1.65 in with two "bundled" conductors per phase and 18-in spacing. The standard span is 1,600 ft, the conductor spacing is flat with 30-ft spacing between phase-conductor centers, and the minimum clearance to ground is 34 to 39 ft. To maintain a minimum clearance of 11 ft to the towers and 30 ft spacing between phases, vee insulator strings, each consisting of twenty-four 10-in disks, are used with each phase. The highest EHV system in North American is the 765-kV system in the midwest region of the United States. A dc transmission line on the west coast of the United States operating at ± 450 kV is transmitting power in bulk more than 800 miles.

High-voltage dc transmission has a greater potential for savings and a greater ability to transmit large blocks of power longer distances than has three-phase transmission. For the same crest voltage there is a saving of 50 percent in the weight of the conductor. Because of the power stability limit due to inductive and capacitive effects (inherent with ac transmission), the ability to transmit large blocks of power long distances has not kept pace with power developments, even at the present highest ac transmission voltage of 765 kV. With direct current there is no such power stability limit.

Where cables are necessary, as under water, the capacitive charging current may, with alternating current become so large that it absorbs a large proportion, if not all, of the cable-carrying capability. For example, at 132 kV, three-phase (76 kV to ground), with a 500 MCM cable, at 36 mi, the charging current at 60 Hz is equal to the entire cable capability so that no capability remains for the load current. With direct current there is no charging current, only the negligible leakage current, and there are no ac dielectric losses. Furthermore, the dc voltage at which a given cable can operate is twice the ac voltage.

The high dc transmission voltage is obtained by converting the ac power voltage to direct current by means of mercury-arc rectifiers; at the receiving end of the line the dc voltage is inverted back to a power-frequency voltage by means of mercury-arc inverters. The maximum dc transmission voltage in use in the United States, as of 1994, is 500 kV.

Alternating to alternating current is nonsynchronous transmission of electric power. It can be overhead or under the surface. In the early 1970s the problem of bulk electric-power transmission over

high-voltage transmission lines above the surface developed the insistent discussion of land use and environmental cost.

While the maximum ac transmission voltage in use in the United States (1994) is 765 kV, ultrahigh voltage (UHV) is under consideration. Transmission voltages of 1,200 to 2,550 kV are being tested presently. The right-of-way requirements for power transmission are significantly reduced at higher voltages. For example, in one study the transmission of 7,500 MVA at 345 kV ac was found to require 14 circuits on a corridor 725 ft (221.5 m) wide, whereas a single 1,200-kV ac circuit of 7,500-MVA capacity would need a corridor 310 ft (91.5 m) wide.

Continuing research and development efforts may result in the development of more economic high-voltage underground transmission links. Sufficient bulk power transmission capability would permit **power wheeling,** i.e., the use of generating capacity to the east and west to serve a given locality as the earth revolves and the area of peak demand glides across the countryside.

Corona is a reddish-blue electrical discharge which occurs when the voltage-gradient in air exceeds 30 kV peak, 21.1 kV rms, at 76 cm pressure. This electrical discharge is caused by ionization of the air and becomes more or less concentrated at irregularities on the conductor surface and on the outer strands of stranded conductors. Corona is accompanied by a hissing sound; it produces ozone and, in the presence of moisture, nitrous acid. On high-voltage lines corona produces a substantial power loss, corrosion of the conductors, and radio and television interference. The fair-weather loss increases as the square of the voltage above a critical value e_0 and is greatly increased by fog, smoke, rainstorms, sleet, and snow (see Fig. 15.1.86). To reduce corona, the diameter of high-voltage conductors is increased to values much greater than would be required for the necessary conductance cross section. This is accomplished by the use of hollow, segmented conductors and by the use of aluminum cable, steel-reinforced (ACSR), which often has inner layers of jute to increase the diameter. In extra-high-voltage lines (400 kV and greater), corona is reduced by the use of bundled conductors in which each phase consists of two or three conductors spaced about 16 in (0.41 m) from one another.

Underground Power Cables

Insulations for power cables include heat-resisting, low-water-absorptive synthetic rubber compounds, varnished cloth, impregnated paper, cross-linked polyethylene thermosetting compounds, and thermoplastics

Fig. 15.1.86 Corona loss with snowstorm.

such as polyvinyl chloride (PVC) and polyethylene (PE) compounds (see Sec. 6).

Properly chosen **rubber-insulated cables** may be used in wet locations with a nonmetallic jacket for protective covering instead of a metallic sheath. Commonly used jackets are flame-resisting, such as neoprene and PVC. Such cables are relatively light in weight, easy to train in ducts and manholes, and easily spliced. When distribution voltages exceed 2,000 V phase to phase, an ozone-resisting type of compound is required. Such rubber insulation may be used in cables carrying up to 28,000 V between lines in three-phase grounded systems. The insulation wall will be thicker than with varnished cloth, polyethylene, ethylene propylene rubber, or paper.

Varnished-cloth cables are made by applying varnish-treated closely woven cloth in the form of tapes, helically, to the metallic conductor. Simultaneously a viscous compound is applied between layers which fills in any voids at laps in the taping and imparts flexibility when the cable is bent by permitting movement of one tape upon another. This type of insulation has higher dielectric loss than impregnated paper but is suitable for the transmission of power up to 28,000 V between phases over short distances. Such insulated cables may be used in dry locations with flame-resisting fibrous braid, reinforced neoprene tape, or PVC jacket and are often further protected with an interlocked metallic tape armor; but in wet locations these cables should be protected by a continuous metallic sheath such as lead or aluminum. Since varnish-cloth-insulated cable has high ozone resistance, heat resistance, and impulse strength, it is well adapted for station or powerhouse wiring or for any service where the temperature is high or where there are sudden increases in voltage for short periods. Since the varnish is not affected by mineral oils, such cables make excellent leads for transformers and oil switches.

PVC is readily available in several fast, bright colors and is often chosen for color-coded multiconductor control cables. It has inherent flame and oil resistance, and as single conductor wire and cable with the proper wall thickness for a particular application, it usually does not need any outside protective covering. On account of its high dielectric constant and high power factor, its use is limited to low voltages, i.e., under 1,000 V, except for series lighting circuits.

Polyethylene, because of its excellent electrical characteristics, first found use when it was adapted especially for high-frequency cables used in radio and radar circuits; for certain telephone, communication, and signal cables; and for submarine cables. Submarine telephone cables with built-in repeaters laid first in the Atlantic Ocean and then in the Pacific are insulated with polyethylene. Because of polyethylene's thermal characteristics, the standard maximum conductor operating temperature is 75°C. It is commonly used for power cables (including large use for underground residential distribution), with transmissions up to 15,000 V. Successful installations have been in service at 46 kV and some at 69 kV. The upper limit has not been reached, inasmuch as work is in progress on higher-voltage polyethylene power cables as a result of advancements in the art of compounding.

Cross-linked polyethylene is another insulation which is gaining in favor in the process field. For power cable insulations, the cross-linking process is most commonly obtained chemically. It converts polyethylene from a thermoplastic into a thermosetting material; the result is a compound with a unique combination of properties, including resistance to heat and oxidation, thus permitting an increase in maximum conductor operating temperature to 90°C. The service record with this compound has been good at voltages which have been gradually increased to 35 kV. Another thermosetting material, ethylene propylene rubber (EPR), has found wide acceptance in the 5- to 35-kV range.

Impregnated-paper insulation is used for very-high-voltage cables whose range has been extended to 345 kV. To eliminate the detrimental effects of moisture and to maintain proper impregnation of the paper, such cables must have a continuous metallic sheath such as lead or aluminum or be enclosed within a steel pipe; the operation of the cable depends absolutely on the integrity of that enclosure. In three-conductor belted-type cables the individual insulated conductors are surrounded by a belt or wall of impregnated paper over which the lead sheath is applied. When all three conductors are within one sheath, their inductive effects practically neutralize one another and eddy-current loss in the sheath is negligible. In the type-H cable, each of the individual conductors is surrounded with a perforated metallic covering, either aluminum foil backed with a paper tape or thin perforated metal tapes wound over the paper. All three conductors are then enclosed within the metal sheath. The metallic coverings being grounded electrically, each conductor acts as a single-conductor cable. This construction eliminates "tangential" stresses within the insulation and reduces pockets or voids. When paper tapes are wound on the conductor, impregnated with an oil or a petrolatum compound, and covered with a lead sheath, they are called **solid type.**

Three-conductor cables are now operating at 33,000 V, and single-conductor cables at 66,000 V between phases (38,000 V to ground). In New York and Chicago, special hollow-conductor oil-filled single-conductor cables are operating successfully at 132,000 V (76,000 V to ground). In France, cables are operating at 345 kV between conductors.

Other methods of installing underground cables are to draw them into steel pipes, usually without the sheaths, and to fill the pipes with oil under pressure (**oilstatic**) or **nitrogen** under 200 lb pressure. The ordinary medium-high-voltage underground cables are usually drawn into duct lines. With a straight run and ample clearance the length of cable between manholes may reach 600 to 1,000 ft. Ordinarily, the distance is more nearly 400 to 500 ft. With bends of small radius the distance must be further reduced.

Cable ratings are based on the permissible operating temperatures of the insulation and environmental installation conditions. See Table 15.1.20 for ampacities.

POWER DISTRIBUTION

Distribution Systems   The choice of the system of power distribution is determined by the type of power that is available and by the nature of the load. To transmit a given power over a given distance with a given power loss (I^2R), the weight of conductor varies inversely as the square of the voltage. Incandescent lamps will not operate economically at voltages much higher than 120 V; the most suitable voltages for dc motors are 230, 500, and 550 V; for ac motors, standard voltages are 230, 460, and 575 V, three-phase. When power for lighting is to be distributed in a district where the consumers are relatively far apart, alternating current is used, being distributed at high voltage (2,400, 4,160, 4,800, 6,900, and 13,800 V) and transformed at the consumer's premises, or by transformers on poles or located in manholes or vaults under the street or sidewalks, to 240/120 V three-wire for lighting and domestic customers, and to 208, 240, 480, and 600 volts, three-phase, for power.

The first central station power systems were built with dc generation and distribution. The economical transmission distance was short. Densely populated, downtown areas of cities were therefore the first sections to be served. Growth of electric service in the United States

Table 15.1.20 Ampacities of Insulated High-Voltage Cables in Underground Raceways
[Based on type MV-90 conductor temperature of 90°C, ambient earth temperature of 20°C, 100 percent load factor, thermal resistance (RHO) of 90, and three circuits in group]

Conductor size, AWG or MCM	Copper				Aluminum			
	3-1/C conductors per raceway		1-3/C cable per raceway		3-1/C conductors per raceway		1-3/C cable per raceway	
	2,001–5,000-V ampacity	5,001–35,000-V ampacity	2,001–5,000-V ampacity	5,001–35,000-V ampacity	2,001–5,000-V ampacity	5,001–35,000-V ampacity	2,001–5,000-V ampacity	5,001–35,000-V ampacity
8	56		59		44		46	
6	73	77	78	88	57	60	61	69
4	95	99	100	115	74	77	80	89
2	125	130	135	150	96	100	105	115
1	140	145	155	170	110	110	120	135
1/0	160	165	175	195	125	125	140	150
2/0	185	185	200	220	145	145	160	170
3/0	210	210	230	250	160	165	180	195
4/0	235	240	265	285	185	185	205	220
250	260	260	290	310	205	200	230	245
350	315	310	355	375	245	245	280	295
500	375	370	430	450	295	290	340	355
750	460	440	530	545	370	355	425	440
1,000	525	495	600	615	425	405	495	510

NOTE: This is a general table. For other temperatures and installation conditions, see NFPA 70, 2005, National Electrical Code®, Tables 310.77, through 310.80 and associated notes.

was phenomenal in the last two decades of the nineteenth century. After 1895 when ac generation was selected for the development of power from Niagara Falls, the expansion of dc distribution diminished. The economics overwhelmingly favored the new ac system.

Direct current service is still available in small pockets in some cities. In those cases, ac power is generated, transmitted, and distributed. The conversion to dc takes place in rectifiers installed in manholes near the load or in the building to be served. Some dc customers resist the change to ac service because of their need for motor speed control. Elevator and printing press drives and some cloth-cutting knives are examples of such needs. (See Low-Voltage AC Network.)

Series Circuits These **constant-current** circuits were widely used for **street lighting.** The voltage was automatically adjusted to match the number of lamps in series and maintain a constant current. With the advent of **HID lamps** and individual **photocells** on street lighting fixtures which require parallel circuitry, this system fell into disuse and is rarely seen.

Parallel Circuits **Power** is usually distributed at **constant potential,** and all the devices or receivers in the circuit are connected in parallel, giving a constant-potential system, Fig. 15.1.87*a*. If conductors of constant cross section are used and all the loads, L_1, L_2, etc., are operating, there will be a greater voltage *IR* drop per unit length of wire in the portion of the circuit *AB* and *CD* than in the other portions; also the voltage will not be the same for the different lamps but will decrease along the mains with distance from the generating end.

Loop Circuits A more nearly equal voltage for each load is obtained in the loop system, Fig. 15.1.87*b*. The electrical distance from one generator terminal to the other through any receiver is the same as

that through any other receiver, and the voltage at the receivers may be maintained more nearly equal, but at the expense of additional conductor material.

Series-Parallel Circuit For incandescent lamps the power must be at low voltage (115 V) and the voltage variations must be small. If the transmission distance is considerable or the loads are large, a large or perhaps prohibitive investment in conductor material would be necessary. In some special cases, lamps may be operated in groups of two in series as shown in Fig. 15.1.88. The transmitting voltage is thus doubled, and, for a given number of lamps, the current is halved, the permissible voltage drop (*IR*) in conductors doubled, the conductor resistance quadrupled, the weight of conductor material thus being reduced to 25 percent of that necessary for simple parallel operation.

Fig. 15.1.88 Series-parallel system.

Three-Wire System In the series-parallel system the loads must be used in pairs and both units of the pair must have the same power rating. To overcome these objections and at the same time to obtain the economy in conductor material of operating at higher voltage, the three-wire system is used. It consists merely of adding a third wire or neutral to the system of Fig. 15.1.88 as shown in Fig. 15.1.89.

Fig. 15.1.89 Three-wire system.

If the neutral wire is of the same cross section as the two outer wires, this system requires only 37.5 percent of the copper required by an equivalent two-wire system. Since the neutral ordinarily carries less current than the outers, it is usually smaller and the ratio of copper to that of the two-wire system is even less than 37.5 percent (see Table 15.1.21).

When the loads on each half of the system are equal, there will be no current in the middle or neutral wire, and the condition is the same as that shown in Fig. 15.1.88. When the loads on the two sides are unequal,

Fig. 15.1.87 (*a*) Parallel circuit; (*b*) loop circuit.

Table 15.1.21   Resistance and 60-Hz Reactance for Wires with Small Spacings, Ω, at 20°C
(See also Table 15.1.18.)

AWG and size of wire, cir mils	Resistance in 1,000 ft of line (2,000 ft of wire), copper	Reactance in 1,000 ft of line (2,000 ft of wire) at 60 Hz for the distance given in inches between centers of conductors										
		½	1	2	3	4	5	6	9	12	18	24
14– 4,107	5.06	0.138	0.178	0.218	0.220	0.233	0.244	0.252	0.271	0.284	0.302	
12– 6,530	3.18	0.127	0.159	0.190	0.210	0.223	0.233	0.241	0.260	0.273	0.292	
10– 10,380	2.00	0.116	0.148	0.180	0.199	0.212	0.223	0.231	0.249	0.262	0.281	
8– 16,510	1.26	0.106	0.138	0.169	0.188	0.201	0.212	0.220	0.238	0.252	0.270	0.284
6– 26,250	0.790	0.095	0.127	0.158	0.178	0.190	0.201	0.209	0.228	0.241	0.260	0.272
4– 41,740	0.498	0.085	0.117	0.149	0.167	0.180	0.190	0.199	0.217	0.230	0.249	0.262
2– 66,370	0.312	0.074	0.106	0.138	0.156	0.169	0.180	0.188	0.206	0.220	0.238	0.252
1– 83,690	0.248	0.068	0.101	0.132	0.151	0.164	0.174	0.183	0.201	0.214	0.233	0.246
0– 105,500	0.196	0.063	0.095	0.127	0.145	0.159	0.169	0.177	0.196	0.209	0.228	0.241
00– 133,100	0.156	0.057	0.090	0.121	0.140	0.153	0.164	0.172	0.190	0.204	0.222	0.236
000– 167,800	0.122	0.052	0.085	0.116	0.135	0.148	0.158	0.167	0.185	0.199	0.217	0.230
0000– 211,600	0.098	0.046	0.079	0.111	0.130	0.143	0.153	0.161	0.180	0.193	0.212	0.225
250,000	0.085	—	0.075	0.106	0.125	0.139	0.148	0.157	0.175	0.189	0.207	0.220
300,000	0.075	—	0.071	0.103	0.120	0.134	0.144	0.153	0.171	0.185	0.203	0.217
350,000	0.061	—	0.067	0.099	0.118	0.128	0.141	0.149	0.168	0.182	0.200	0.213
400,000	0.052	—	0.064	0.096	0.114	0.127	0.138	0.146	0.165	0.178	0.197	0.209
500,000	0.042	—	—	0.090	0.109	0.122	0.133	0.141	0.160	0.172	0.192	0.202
600,000	0.035	—	—	0.087	0.106	0.118	0.128	0.137	0.155	0.169	0.187	0.200
700,000	0.030	—	—	0.083	0.102	0.114	0.125	0.133	0.152	0.165	0.184	0.197
800,000	0.026	—	—	0.080	0.099	0.112	0.122	0.130	0.148	0.162	0.181	0.194
900,000	0.024	—	—	0.077	0.096	0.109	0.119	0.127	0.146	0.159	0.178	0.191
1,000,000	0.022	—	—	0.075	0.094	0.106	0.117	0.125	0.144	0.158	0.176	0.188

NOTE: For other frequencies the reactance will be in direct proportion to the frequency.

there will be a current in the neutral wire equal to the difference of the currents in the outside wires.

AC Three-Wire Distribution   Practically all energy for lighting and small motor work is distributed at 1,150, 2,300, or 4,160 V ac to transformers which step down the voltage to 240 and 120 V for three-wire domestic and lighting systems as well as 208, 240, 480, and 600 V, three-phase, for power. For the three-wire systems the transformers are so designed that the secondary or low-voltage winding will deliver power at 240 V, and the middle or neutral wire is obtained by connecting to the center or midpoint of this winding (see Figs. 15.1.90 and 15.1.91).

Fig. 15.1.90   Three-wire generator.

Disturbances of utility power have been present since the inception of the electric utility industry; however, in the past equipment was much more forgiving and less affected. Present electronic data processing equipment is microprocessor-based and consequently requires a higher quality of ac power. Although the quality of utility power in the United States is very good, sensitive electronic equipment still needs to be protected against the adverse and damaging effects of transients, surges, and other power-system aberrations.

Fig. 15.1.91   Three-wire 230/115-V ac system.

This has led to the development of technologies which condition power for use by this sensitive equipment. Power-enhancing devices (surge suppressors, isolation transformers, and line voltage regulators) modify the incoming waveform to mitigate some aberrations. Power-synthesizing devices (motor-generator sets, magnetic synthesizers, and uninterruptible power supplies) use the incoming power as a source of energy to generate a new, completely isolated waveform. Each type has advantages and limitations.

Grounding   The neutral wire of the secondary circuit of the transformer should be grounded on the pole (or in the manhole) and at the service switch in the building supplied. If, as a result of a lightning stroke or a fault in the transformer insulation, the transformer primary circuit becomes grounded at *a* (Fig. 15.1.91) and the transformer insulation between primary and secondary windings is broken down at *b* and if there were no permanent ground connection in the secondary neutral wire, the potential of wire 1 would be raised 2,300 V above ground potential. This constitutes a very serious hazard to life for persons coming in contact with the 120 V system. The National Electrical Code® requires the use of a ground wire not smaller than 8 AWG copper. With the neutral grounded (Fig. 15.1.91), voltages to ground on the secondary system cannot exceed 120 V. (See National Electrical Code® 2005, Art. 250.)

Feeders and Mains   Where power is supplied to a large district, improved voltage regulation is obtained by having **centers of distribution.** Power is supplied from the station bus at high voltage to the centers of distribution by large cables known as **feeders.** Power is distributed from the distribution centers to the consumers through the mains and transformed to a usable voltage at the user's site. As there are no loads connected to the feeders between the generating station and centers of distribution, the voltage at the latter points may be maintained constant. **Pilot wires** from the centers of distribution often run back to the station, allowing the operator or automatic controls to maintain constant voltage at the centers of distribution. This system provides a means of maintaining very close **voltage regulation** at the consumer's premises.

A common and economical method of supplying business and thickly settled districts with high load densities is to employ a 208/120-V, three-phase, four-wire **low-voltage ac network.** The network operates with 208 V between outer wires giving 120 V to neutral (Fig. 15.1.92). Motors are connected across the three outer wires operating at 208 V, three-phase. Lamp loads are connected between outer wires and the grounded neutral. The network is supplied directly from 13,800-V feeders by 13,800/208-V three-phase transformer units, usually located in manholes, vaults, or outdoor enclosures. This system thus eliminates the necessity for transformation in the substation. A large number of such units feed the network, so that the secondaries are all in parallel. Each transformer is provided with an overload reverse-energy circuit breaker (network protector), so that a feeder and its transformer are isolated if trouble develops in either. This system is flexible since units can be easily added or removed in accordance with the rapid changes in local loads that occur particularly in downtown business districts.

Fig. 15.1.92 208/120-V secondary network (single unit) showing voltages.

Voltage Drops In ac distribution systems the voltage drop from transformer to consumer in lighting mains should not exceed 2 percent in first-class systems, so that the lamps along the mains can all operate at nearly the same voltage and the annoying flicker of lamps may not occur with the switching of appliances. This may require a much larger conductor than the most economical size. In transmission lines and in feeders where there are no intermediate loads and where means of regulating the voltage are provided, the drop is not limited to the low values that are necessary with mains and the matter of economy may be given consideration.

WIRING CALCULATIONS

These calculations can be used for dc, and for ac if the reactance can be neglected. The determination of the proper size of conductor is influenced by a number of factors. Except for short distances, the minimum size of conductor shown in Table 15.1.23, which is based on the maximum permissible current for each type of insulation, cannot be used; the size of conductor must be larger so that the voltage drop IR shall not be too great. With branch circuits supplying an incandescent-lamp load, this drop should not be more than a small percentage of the voltage between wires. The National Electric Code® 2005, Article 215.2, FPN 2 recommends that conductors for feeders, i.e., from the service equipment to the final branch circuit overcurrent device, be sized to prevent (1) a voltage drop of more than 3 percent at the farthest outlet of power, heating, and lighting loads or combinations thereof, and (2) a maximum voltage drop on combined feeders and branch circuits to the farthest outlet of more than 5 percent.

The resistance of 1 cir mil · ft of commercial copper may be taken as 10.8 Ω. The resistance of a copper conductor may be expressed as $R = 10.8l/A$, where l = length, ft and A = area, cir mils. If the length is expressed in terms of the transmission distance d (since the two wires are usually run parallel), the voltage drop IR to the end of the circuit is

$$e = 21.6Id/A \qquad (15.1.128)$$

and the size of conductor in circular mils necessary to give the permissible voltage drop e is

$$A = 21.6Id/e \qquad (15.1.129)$$

If e is expressed as a percentage x of the voltage E between conductors, then

$$A = 2,160Id/xE \qquad (15.1.130)$$

EXAMPLE.   Determine the size of conductor to supply power to a 10-hp, 220-V dc motor 500 ft from the switchboard with 5 V drop. Assume a motor efficiency of 86 percent. The motor will then require a current of $(10 \times 746)/(0.86 \times 220) = 39.4$ A. From Eq. (15.1.129), $A = 21.6 \times 39.4 \times 500/5 = 85,100$ cir mils. The next largest wire is no. 0 AWG.

The calculation of the size conductor for three-wire circuits is made in practically the same manner. With a balanced circuit there is no current in the neutral wire, and the current in each outside wire will be equal to one-half the sum of the currents taken by all the receiving devices connected between neutral and outside wires plus the sum of the currents taken by the receivers connected between the outside wires. Using this total current and neglecting the neutral wire, make calculations for the size of the outside wires by means of Eq. (15.1.129). The neutral wire should have the same cross section as the outside wires in interior wiring.

EXAMPLE.   Determine the size wire which should be used for the three-wire main of Fig. 15.1.93. Allowable drop is 3 V and the distance to the load center 40 ft; circuit loaded with two groups of receivers each taking 60 A connected between the neutral and the outside wires, and one group of receivers taking 20 A connected across the outside wires.

Fig. 15.1.93 Three-wire 230/115-V main.

Solution: load $= (60 + 60)/2 + 20 = 80$ A. Substituting in Eq. (15.1.129), cir mils $= 21.6Id/e = 21.6 \times 80 \times 40/3 = 23,030$ cir mils.

From Tables 15.1.21 and 15.1.23, no. 6 wire, which has a cross section of 26,250 cir mils, is the next size larger. This size of wire would satisfy the voltage-drop requirements, but No. 6 wire w/type TW insulation has a safe carrying capacity of 55 A at 60°C. The current in the circuit is 80 A. Therefore, No. 3 wire w/type TW insulation which has a carrying capacity of 85 A should be used. The neutral wire should be the same size as the outside wires.

See also examples in the National Electrical Code® 2005.

Wiring calculations for ac circuits require some consideration of power factor, reactance, and skin effect. Skin effect becomes pronounced only when very large conductors are used for alternating current. For interior wiring, conductors larger than 700,000 cir mils should not be used, and many prefer not to use conductors larger than 300,000 cir mils. Should the required copper cross section exceed these values, a number of conductors may be operated in parallel.

For voltages under 5,000 the effect of line capacitance may be neglected. With ordinary single-phase interior wiring, where the effect of the line reactance may be neglected and where the power factor of the load (incandescent lamps) is nearly 100 percent, the calculations are made the same as for dc circuits. Three-wire ac circuits of ordinary length with incandescent lamp loads are also determined in the same manner. When the load is other than incandescent lamps, it is necessary to know the power factor of the load in order to make calculations. When the exact power factor cannot be accurately determined, the following approximate values may be used: incandescent lamps, 0.95 to 1.00; lamps and motors, 0.75 to 0.85; motors 0.5 to 0.80. Equation (15.1.131) gives the value of current in a single-phase circuit. See also Table 15.1.17.

$$I = (P \times 1,000)(E \times pf) \qquad (15.1.131)$$

where I = current, A; P = kW; E = load voltage; and pf = power factor of the load. The size of conductor is then determined by substituting this value of I in Eq. (15.1.129) or (15.1.130).

For three-phase three-wire ac circuits the current per wire

$$I = 1,000P/\sqrt{3}Epf = 580P/Epf \qquad (15.1.132)$$

Computations are usually made of voltage drop **per wire** (see Fig. 15.1.94). Hence, if reactance can be neglected, the conductor cross section in cir mils is one-half that given by Eq. (15.1.129). That is,

$$A = 10.8Id/e \qquad \text{cir mils} \qquad (15.1.133)$$

where e in Eq. (15.1.133) is the voltage drop per wire. The voltage drop between any two wires is $\sqrt{3}e$. The percent voltage drop should be in terms of the voltage to **neutral**. That is, percent drop = $[e/(E/\sqrt{3})]100 = [\sqrt{3}e/E]100$ (see Fig. 15.1.82).

Fig. 15.1.94 Three-phase lamp and induction motor load.

EXAMPLE. In Fig. 15.1.94, load 10 kW; voltage of circuit 230; power factor 0.85; distance 360 ft; allowable voltage drop per wire 4 V. Substituting in Eq. (15.1.132) $I = (580 \times 10)/(230 \times 0.85) = 29.7$ A. Substituting in Eq. (15.1.133), $A = 10.8 \times 29.7 \times 360/4 = 28,900$ cir mils.

The next larger commercially available standard-size wire (see Table 15.1.21) is 41,700 cir mils corresponding to AWG no. 4. From Table 15.1.23 the allowable ampacity is 70 A with type TW insulation, and is therefore ample in section for 29.7 A. Three no. 4 wires would be used for this circuit.

From Table 15.1.21 the resistance of 1,000 ft of no. 4 copper wire is 0.249 Ω. Hence, the voltage drop per conductor, $e = 29.7 \times (360/1,000)0.249 = 2.66$ V. Percent voltage drop = $\sqrt{3} \times 2.66/230 = 2.00$ percent.

Where all the wires of a circuit, two wires for a single-phase circuit, four wires for a four-phase circuit (see Fig. 15.1.32 and Fig. 15.1.80c), and three wires for a three-phase circuit, are carried in the same conduit or where the wires are separated less than 1 in between centers, the effect of line (inductive) reactance may ordinarily be neglected. Where circuit conductors are large and widely separated from one another and the circuits are long, the inductive reactance may increase the voltage drop by a considerable amount over that due to resistance alone. Such problems are treated using IR and IX phasors. Line reactance decreases somewhat as the size of wire increases and decreases as the distance between wires decreases.

Fig. 15.1.95 Single-phase induction motor load on branch circuit.

EXAMPLE. Determine the size of wire necessary for the branch to the 50-hp, 60-Hz, 250-V single-phase induction motor of Fig. 15.1.95. The name-plate rating of the motor is 195 A, and its full-load power factor is 0.85. The wires are run open and separated 4 in; length of circuit, 600 ft. Assume the line drop must not exceed 7 percent, or $0.07 \times 250 = 17.5$ V. The point made by this example is emphasized by the assumption of an outsize motor.

Solution. To ascertain approximately the size of conductor, substitute in Eq. (15.1.129) giving cir mils, $X = 21.6 \times 195 \times 600/17.5 = 144,400$. Referring to Table 15.1.21, the next larger size wire is no. 000 or 167,800 cir mils. This size would be ample if there were no line reactance. In order to allow for reactance drop, a larger conductor is selected and the corresponding voltage drop determined. Inasmuch as this is a motor branch, the code rules require that the carrying capacity be sufficient for a 25 percent overload. Therefore the conductor should be capable of carrying $195 \times 1.25 = 244$ A. From Table 15.1.23, 250,000-cir mil conductor Type THHN insulated cable would be required to carry 255 A at 75°C. Resistance drop (see Table 15.1.21), $IR = 195 \times 0.061 \times 0.6 = 7.14$ V. $7.14/250 = 2.86$ percent. From Table 15.1.21, $X = 0.128 \times 0.6 = 0.0768$ Ω. $IX = 195 \times 0.768 = 14.98$ V. $14.98/250 = 5.99$ percent. Using the Mershon diagram (Fig. 15.1.84), follow the ordinate corresponding to power factor, 0.85, until it intersects the smallest circle. From this point, lay off horizontally the percentage resistance drop, 2.86. From this last point, lay off vertically the percentage reactance drop 5.99. This last point lies about on the 6.0 percent circle, showing that with

195 amp the difference between the sending-end and receiving-end voltages is $0.06 \times 250 = 15.0$ V, which is within the specified limits.

Also Eq. (15.1.127) may be used. $\cos \theta = 0.85$; $\sin \theta = 0.527$.

$$E_s = \{[(250 \times 0.85) + 7.14]^2 + [(250 \times 0.527) + 14.98]^2\}^{1/2} = 264.3 \text{ V}$$
$$264.3/250 = 105.7 \text{ percent}$$

In the calculation of three-phase three-wire circuits where line reactance must be considered, the method found above under Power Transmission may be used. The system is considered as being three single-phase systems having a ground return the resistance and inductance of which are zero, and the voltages are equal to the line voltages divided by $\sqrt{3}$. When the three conductors are spaced unequally, the value of GMD given in Fig. 15.1.85 should be used in Tables 15.1.18 and 15.1.21. (When the value of resistance or reactance per 1,000 ft of **conductor** is desired, the values in Table 15.1.21 should be divided by 2.)

The National Electrical Code® specifies that the size of conductors for branch circuits should be such that the voltage drop should not exceed 3 percent to the farthest outlet for power, heating, lighting, or combination thereof, requiring further that the total voltage drop for feeders and branch circuits should not exceed 5 percent overall (NEC, Article 210.19, FPN No. 4). For examples of calculations for interior wiring, see National Electrical Code® (Chap. 9).

INTERIOR WIRING

Interior wiring requirements are based, for the most part, on the **National Electrical Code®** (NEC®), which has been adopted by the National Fire Protection Association, American National Standards Institute (ANSI), and the Occupational Safety and Health Act (OSHA).

The Occupational Safety and Health Act of 1970 (OSHA) made the National Electrical Code® a national standard. Conformance with the NEC became a requirement in most commercial, industrial, agricultural, etc., establishments in the United States. Some localities may not accept the latest edition of NEC® standards. In those cases, local rules must be followed.

NEC® authority starts at the point where the connections are made to the conductor of the **service drop** (overhead) or **lateral** (underground) from the electricity supply system. The **service equipment** must have a rating not less than the load to be carried (computed according to NEC® methods). Service equipment is defined as the necessary equipment, such as circuit breakers or fused switches and accompanying accessories. This equipment must be located near the point of entrance of supply conductors to a building or other structure or an otherwise defined area. Service equipment is intended to be the main control and means of cutoff of the supply.

Service-entrance conductors connect the electricity supply to the service equipment. Service-entrance conductors running along the exterior or entering a building or other structure may be installed (1) as separate conductors, (2) in approved cables, (3) as cable bus, or (4) enclosed in rigid conduit. Also, for voltages less than 600 V, the conductors may be installed in electrical metallic tubing; IMC, ENT, and MC cable wireways; auxiliary gutters; or busways. Service-entrance cables which are subject to physical damage from awnings, swinging signs, coal chutes, etc., shall be of the protected type or be protected by conduit, electrical metallic tubing, etc. Service heads must be raintight. Thermoplastic thermoset insulation is required in overhead services. A grounded conductor may be bare. If exposed to the weather or embedded in masonry, raceways must be raintight and arranged to drain. Underground service raceway or duct entering from an underground distribution system must be sealed with a suitable compound (spare ducts, also).

NEC® rules permit **multiple service** to a building for various reasons, such as: (1) fire pumps, (2) emergency light and power, (3) multiple occupancy, (4) when the calculated load is greater than 2,000 A, (5) when the building extends over a large area, and (6) where different voltages, frequencies, number of phases, or classes of use are required.

Ordinary service drops (overhead) and lateral (underground) must be large enough to carry the load but not smaller than no. 8 copper or no. 6 aluminum. As an exception, for installations to supply only limited loads of a single branch circuit, such as small polyphase power, etc., service drops must not be smaller than no. 12 hard-drawn copper or equivalent, and service laterals must be not smaller than no. 12 AWG copper or no. 10 AWG aluminum.

The phrase **large enough to carry the load** requires elaboration. The various conductors of public-utility electric-supply systems are sized according to the calculations and decisions of the personnel of the specific public utility supplying the service drop or lateral. At the load end of the drop or lateral, the NEC® rules apply, and from that point on into the consumer's premises, NEC® rules are the governing authority. There is a discontinuity at this point in the calculation of combined load demand for electricity and allowable current (ampacity) of conductors, cables, etc. This discontinuity in calculations results from the fact that the utility company operates locally, whereas the NEC® is a set of national standards and therefore cannot readily allow for regional differences in electrical coincident demand, ambient temperature, etc. The NEC's aim is the assurance of an electrically "safe" human environment. This will be fostered by following the NEC® rules.

Service-entrance cables are conductor assemblies which bear the type codes SE (for overhead services) and USE (for underground services). Under specified conditions, these cables may also be used for interior feeder and branch-circuit wiring.

The service-entrance equipment must have the **capability of safely interrupting** the current resulting from a short circuit at its terminals. **Available short-circuit current** is the term given to the maximum current that the power system can deliver through a given circuit to any negligible-impedance short circuit applied at a given point. (This value can be in terms of symmetrical or asymmetrical, momentary or clearing current, as specified.)

In most instances, the available short-circuit current is limited by the impedance of the last transformer in the supply system and connecting wiring. **Large power users,** however, must become aware of changes in the electricity supply system which, because of growth of system capacity or any other reason, would increase the short-circuit current available to their service-entrance equipment. If this current is too great, explosive failure can result in a hazard to personnel and equipment.

♦ Fuses, non-adjustable or adjustable circuit breakers
♦ The switches in the above diagrams are often a part of the circuit breaker or fuse box.

Fig. 15.1.96   Motor and wiring protection.

Kilowatthour and sometimes demand-metering equipment are connected to the service-entrance conductors. Proceeding toward the utilization equipment, the power-supply system fans out into feeders and branch circuits (see Fig. 15.1.96). Each of the feeders, i.e., a run of untapped conductor or cable, is connected to the supply through a switch and fuses or a circuit breaker. At a point, usually near that portion of the electrical loads which are to be supplied, a **panelboard** or perhaps a **load-center assembly** of switching and/or control equipment is installed. From this panel box or load-center assembly, circuits originate; i.e., circuits are installed to extend into the area being served to connect electrical machinery or devices or make available electric receptacles connected to the source of electric power.

Each feeder and each branch circuit will have its own **over-current protection** and **disconnect** means in the form of a fuse and switch combination or a circuit breaker.

Wiring Methods

There is a provision in the NEC® rules for the following types of feeder and branch circuit wiring:

1. Open Wiring on Insulators (NEC® 2005, Art. 398). This wiring method uses approved cleats, knobs, tubes, and flexible tubing for the protection and support of insulated conductors run on or in buildings and not concealed by the building structure. It is permitted only in industrial or agricultural establishments.

2. Concealed Knob-and-Tube Work (NEC® 2005, Art. 394). Concealed knob-and-tube work may be used in the hollow spaces of walls and ceilings. It may be used only for the extension of existing installations.

3. Flat Conductor Cable, Type FCC (NEC® 2005, Art. 324). Type FCC cable may be installed under carpet squares. It may not be used outdoors or in wet locations, in corrosive or hazardous areas, or in residential, school, or hospital buildings.

4. Mineral-Insulated Metal-Sheathed Cable, Type MI (NEC® 2005, Art. 332). Type MI cable contains one or more electrical conductors insulated with a highly compressed refractory mineral insulation and enclosed in a liquid- and gastight continuous copper of steel sheath. Appropriate approved fittings must be used with it. It may be used for services, feeders, and branch circuits either exposed or concealed and dry or wet. It may be used in Class I, II, or III hazardous locations. It may be used for under-plaster extensions and embedded in plaster finish or brick or other masonry. It may be used where exposed to weather or continuous moisture, for underground runs and embedded in masonry, concrete or fill, in buildings in the course of construction or where exposed to oil, gasoline, or other conditions. If the environment would cause destruction of the sheath, it must be protected by suitable materials.

5. Power and Control Tray Cable (NEC® 2005, Art. 336). Type TC cable is a factory assembly of two or more insulated conductors with or without associated bare or covered-grounding conductors under a nonmetallic jacket, approved for installation in cable trays, in raceways, or where supported by messenger wire.

6. Metal-Clad Cable, Type MC and AC Series (NEC® 2005, Art. 320 and 330). These are metal-clad cables, i.e., an assembly of insulated conductors in a flexible metal enclosure. Type **AC cables** are manufactured in sizes 14 AWG through 1 AWG copper and 12 AWG through 1 AWG aluminum conductors. Type **MC cables** are manufactured in sizes 14 AWG and larger copper and 12 AWG and larger aluminum conductors. Optical fibers are permitted in multiconductor-type MC cables. Type AC cables utilize the armor as the path for equipment grounding. Armor of type MC cable is not recognized as the sole grounding path and includes a separate equipment-grounding conductor. Metal-clad cables may generally be installed where not subject to physical damage, for feeders and branch circuits in exposed or concealed work, with qualifications for wet locations, direct burial in concrete, etc. The use of Type AC cable is prohibited (1) in motion-picture studios, (2) in theaters and assembly halls, (3) in hazardous locations, (4) where exposed to corrosive fumes or vapors, (5) on cranes or hoists except where flexible connections to motors, etc., are required, (6) in storage-battery rooms, (7) in hoistways or on elevators except (*i*) between risers and limit switches, interlocks, operating buttons, and similar devices in hoistways and in escalators and moving walkways and (*ii*) short runs on elevator cars, where free from oil, and if securely fastened in place, or (8) in commercial garages in hazardous locations.

7. Nonmetallic-Sheathed Cable, Types NM, NMC, and NMS (NEC® 2005, Art. 334). These are assemblies of two or more insulated conductors (nos. 14 through 2 for copper, nos. 12 through 2 for aluminum) having an outer sheath of moisture-resistant, flame-retardant, nonmetallic material. In addition to the insulated conductors, the cable may have an approved size of insulated or bare conductor for equipment grounding purposes only. The outer covering of NMC cable is flame retardant and corrosion resistant. The use of this type of cable, commonly called Romex is permitted in one- or two-family dwellings, multifamily dwellings, and other structures of type III, IV, and V construction. These cables are **not** permitted as open runs in dropped or suspended ceilings, as service entrance cable, in commercial garages having hazardous (classified) locations, in theaters, motion picture studios, storage battery rooms, hoistways, escalators, or embedded in concrete.

8. Service Entrance Cable, Types SE and USE (NEC® 2005, Art. 338). These cables, containing one or more individually insulated conductors, are primarily used for electric services. Type SE has a flame-retardant, moisture-resistant covering and is not required to have built-in protection against mechanical abuse. Type USE is recognized for use underground. It has a moisture-resistant covering, but not necessarily a flame-retardant one. Like the SE cable, USE cable is not required to have inherent protection against mechanical abuse. Where all circuit conductors are fully insulated, type USE and SE cables can be used for branch circuits or feeders. The equipment grounding conductor is the only conductor permitted to be bare within service entrance cables used for branch circuits.

9. Underground Feeder and Branch-Circuit Cable, Type UF (NEC® 2005, Art. 340). This cable is made in sizes 14 through 4/0, and the insulated conductors are Types TW, RHW, and others approved for the purpose. As in the NM cable, the UF type may contain an approved size of uninsulated or bare conductor for grounding purposes only. The outer jacket of this cable shall be flame-retardant, moisture-resistant, fungus-resistant, corrosion-resistant, and suitable for direct burial in the ground.

10. Other Installation Practices. The NEC® details rules for nonmetallic circuit extensions and underplaster extensions. It also provides detailed rules for installation of electrical wiring in (*a*) rigid metal conduit (which may be used for all atmospheric conditions and locations with due regard to corrosion protection and choice of fittings), (*b*) rigid nonmetallic conduit (which is essentially corrosion-proof), in electrical metallic tubing (which is lighter-weight than rigid metal conduit), (*c*) flexible metal conduit, (*d*) liquidtight flexible metal conduit, (*e*) surface raceways, (*f*) underfloor raceways, (*g*) multioutlet assemblies, (*h*) cellular metal floor raceways, (*i*) structural raceways, (*j*) cellular concrete floor raceways, (*k*) wireways (sheet-metal troughs with hinged or removable covers), (*l*) flat, Type FC (NEC® 2005, Art. 322), cable assemblies installed in a surface metal raceway (Type FC cable contains three or four no. 10 special stranded copper wires), (*m*) busways, and (*n*) cable-bus. Busways and cable-bus installations are permitted for exposed work only.

In all installation work, only approved outlets, switch and junction boxes, fittings, terminal strips, and dead-end caps shall be used, and they are to be used in an approved fashion (see NEC® 2005, Art. 314).

Conductors

Table 15.1.22 relates American wire gage (AWG) AWG wire sizes to metric units. Table 15.1.23 lists the allowable ampacities of copper and aluminum conductors. Table 15.1.24 lists various conductor insulation systems approved by the 2005 NEC® for conductors used in interior wiring. Adjustment factors for number of conductors in a raceway or cable are listed in Table 15.1.25. For installations not covered by the tables presented here, review the NEC®. Estimated full-load currents of motors can be taken from Table 15.1.17.

Table 15.1.22 Wire Table for Standard Annealed Copper at 20°C in SI Units

AWG size	No. wires	Diameter, mm	kg/km	m/Ω	Area, mm²
14	1	1.628	18.50	120.7	2.08
12	1	2.053	29.42	191.9	3.31
10	1	2.588	46.77	305.1	5.261
8	1	3.264	74.37	485.2	8.367
6	7	4.115	118.2	771.5	13.30
4	7	5.189	188.0	1227	21.15
2	7	6.544	299.0	1951	33.62
1	19	7.348	377.0	2460	42.41
0	19	8.252	477.6	3102	53.49
00	19	9.266	599.5	3911	67.43
000	19	10.40	755.7	4932	85.01
0000	19	11.68	935.2	6219	107.2

Protection of conductors Unless excepted by the NEC®, other than fixture cords, flexible cables, and fixture wires, conductors shall be protected against overcurrent according to their rated ampacities (see Table 15.1.23). A fuse or overcurrent trip element of circuit breakers is connected in series with each ungrounded conductor and is normally located at the point where the conductor receives its supply.

Standard ratings for fuses and circuit-breakers (inverse time) are 15, 20, 25, 30, 40, 45, 50, 60, 70, 80, 90, 100, 125, 150, 175, 200, 225, 250, 300, 250, 400, 450, 500, 600, 700, 800, 1,000, 1,200, 1,600, 2,000, 2,500, 3,000, 4,000, 5,000, and 6,000 amperes. Additional standard sizes for fuses are 1, 3, 6, 10, and 601. Listed nonstandard sizes of fuses and circuit-breakers are permitted.

Allowable **conductor ampacities** (Table 15.1.23) are subject to limitations and adjustments that may limit or reduce the initial conductor ampacity. One guiding limitation is that the conductor shall be selected so as not to exceed the lowest temperature rating of any connected termination, conductor, or device. Unless the equipment (termination) is listed for a higher temperature, the following applies:

• Termination provisions for circuits rated 100 A or less, or marked for 14 AWG through 1 AWG conductors, shall be conductors rated for 60°C. Conductors with higher temperature ratings can be used as long as the ampacity is based on the 60°C ampacity.

• Termination provisions for circuits rated over 100 A, or marked for conductors larger than 1 AWG, shall be rated for conductors rated for 75°C. Conductors with higher temperature ratings can be used as long as the ampacity is based on the 75°C ampacity.

• The above temperature limitations are most commonly applied at circuit breakers at motor starter terminals. Listing temperatures are normally identified on each device.

• For additional information see NEC® Article 110.14(C)(1).

Unless the equipment is listed and marked otherwise, conductor ampacities apply after any required derating for ambient temperature or number of conductors in a raceway. Conductor ampacity tables should be carefully checked, since the basis can include ambient temperatures of 20°C, 30°C, or 40°C depending on the application. Refer to Table 15.1.23 for examples of temperature correction factors. For complete ampacity information refer to NEC® Article 310.

Where the number of **current-carrying conductors** in a raceway or cable exceeds three the allowable ampacity of each conductor in Table 15.1.23 must be reduced. Table 15.1.25 establishes derating factors to be applied to current-carrying conductors after any required derating for ambient temperature. The factors in Table 15.1.25 are based on conductors with no **load diversity,** meaning all conductors are fully loaded. For conductors with **load diversity,** adjustment factors for more than three current-carrying conductors in a raceway or cable are found in NEC® 2005, Annex B, Table B.310.11. For information on ampacity derating of cables in cable trays, refer to NEC® 2005, Article 392.

After the final ampacity is derived, the basic rule for selecting overcurrent protection is that all conductors, other than flexible cords, shall be protected against overcurrent in accordance with their ampacities. For devices rated 800 A or less, the next higher standard overcurrent device rating (above the ampacity of the conductors being protected) is permitted to be used. Refer to NEC® Article 240.4(A) for additional information and exceptions.

The **number and size of conductors in raceways** must not be more than will permit dissipation of heat and ready installation or removal of conductors without damage. Lookup tables for conduit fill by conductor insulation are often used. Because of the large variety of available insulation types, raceway systems, and measurement units (U.S. versus metric), the NEC® approaches fill by several methods. If all of the conductors are the same size and insulation type, conduit fill can be determined by lookup tables based on the type of raceway; refer to NEC® 2005, Annex-C, Tables C1 through C12A. If the conductors are mixed sizes, then the process has two steps:

1. Determine the total area occupied by the conductors (in² or mm²), based on the wire insulation type and size, using NEC®, Chap. 9, Table 5 or 5A.

Table 15.1.23 Allowable Ampacities of Insulated Conductors Rated 0 through 2,000 Volts, 60°C through 90°C (140°F through 194°F), Not More Than Three Current-Carrying Conductors in Raceway, Cable, or Earth (Directly Buried), Based on Ambient Temperature of 30°C (86°F)

Size AWG or kcmil	Temperature rating of conductor						Size AWG or kcmil
	60°C (140°F)	75°C (167°F)	90°C (194°F)	60°C (140°F)	75°C (167°F)	90°C (194°F)	
	Types TW, UF	Types RHW, THHW, THW, THWN, XHHW, USE, ZW	Types TBS, SA, SIS, FEP, FEPB, MI, RHH, RHW-2, THHN, THHW, THW-2, THWN-2, USE-2, XHH, XHHW, XHHW-2, ZW-2	Types TW, UF	Types RHW, THHW, THW, THWN, XHHW, USE	Types TBS, SA, SIS, THHN, THHW, THW-2, THWN-2, RHH, RHW-2, USE-2, XHH, XHHW, XHHW-2, ZW-2	
	Copper			Aluminum or copper-clad aluminum			
18	—	—	14	—	—	—	—
16	—	—	18	—	—	—	—
14*	20	20	25	—	—	—	—
12*	25	25	30	20	20	25	12*
10*	30	35	40	25	30	35	10*
8	40	50	55	30	40	45	8
6	55	65	75	40	50	60	6
4	70	85	95	55	65	75	4
3	85	100	110	65	75	85	3
2	95	115	130	75	90	100	2
1	110	130	150	85	100	115	1
1/0	125	150	170	100	120	135	1/0
2/0	145	175	195	115	135	150	2/0
3/0	165	200	225	130	155	175	3/0
4/0	195	230	260	150	180	205	4/0
250	215	255	290	170	205	230	250
300	240	285	320	190	230	255	300
350	260	310	350	210	250	280	350
400	280	335	380	225	270	305	400
500	320	380	430	260	310	350	500
600	355	420	475	285	340	385	600
700	385	460	520	310	375	420	700
750	400	475	535	320	385	435	750
800	410	490	555	330	395	450	800
900	435	520	585	355	425	480	900
1,000	455	545	615	375	445	500	1,000
1,250	495	590	665	405	485	545	1,250
1,500	520	625	705	435	520	585	1,500
1,750	545	650	735	455	545	615	1,750
2,000	560	665	750	470	560	630	2,000

Correction factors

Ambient temp., °C	For ambient temperatures other than 30°C (86°F), multiply the allowable ampacities shown above by the appropriate factor shown below.						Ambient temp.,°F
21–25	1.08	1.05	1.04	1.08	1.05	1.04	70–77
26–30	1.00	1.00	1.00	1.00	1.00	1.00	78–86
31–35	0.91	0.94	0.96	0.91	0.94	0.96	87–95
36–40	0.82	0.88	0.91	0.82	0.88	0.91	96–104
41–45	0.71	0.82	0.87	0.71	0.82	0.87	105–113
46–50	0.58	0.75	0.82	0.58	0.75	0.82	114–122
51–55	0.41	0.67	0.76	0.41	0.67	0.76	123–131
56–60	—	0.58	0.71	—	0.58	0.71	132–140
61–70	—	0.33	0.58	—	0.33	0.58	141–158
71–80	—	—	0.41	—	—	0.41	159–176

* Overcurrent protection shall not exceed 15 A for 14 AWG, 20 A for 12 AWG, 30 A for 10 AWG copper, or 15 A for 12 AWG and 25 A for 10 AWG aluminum or copper-clad aluminum after any correction factors for ambient temperature and numbers of conductors in a raceway have been applied.

NOTE: This table is but one example of many provided for various installation methods. Refer to NEC Art. 310 for additional installations that include conductors in air, underground, and others.

SOURCE: NEC Table 310-16. Reprinted with permission from NFPA 70-2002, National Electric Code®, Copyright © 2001, National Fire Protection Association, Quincy, MA 02269. This printed material is not the complete and official position of the NFPA on the referenced subject, which is represented only by the standard in its entirety.

Table 15.1.24   Conductor Type and Application

Type letter	Insulation, trade name (see NEC®, Art. 310 for complete information)	Environment	Outer covering[a]
Max operating temperature = 60°C (140°F)			
TW	Moisture-resistant thermoplastic	Dry and wet	None
TF	Thermoplastic-covered, solid or 7-strand	[c,d]	None
TFF	Thermoplastic-covered, flexible stranding	[c,d]	None
MTW	Moisture-, heat-, and oil-resistant thermoplastic machine-tool wiring (NFPA Stand. 79, NEC 1975, Art. 670)	Wet	None or nylon
UF	Moisture-resistant, underground feeder	Dry and wet	None
Max operating temperature = 75°C (167°F)			
RHW	Moisture-resistant thermoset	Dry and wet	1,2
THW	Moisture- and heat-resistant thermoplastic	Dry and wet	None
THWN	Moisture- and heat-resistant thermoplastic	Dry and wet	Nylon
XHHW	Moisture-resistant thermoset	Wet	None
RFH-1 and 2	Heat-resistant rubber-covered solid or 7-strand	[b–d]	None
FFH-1 and 2	Heat-resistant rubber-covered flexible stranding	[b–d]	None
UF	Moisture-resistant and heat-resistant	Dry and wet	None
USE	Heat- and moisture-resistant	Dry and wet	4
ZW	Modified ethylene tetrafluoroethylene	Wet	None
Max operating temperature = 90°C (194°F)			
MI	Mineral-insulated (metal-sheathed)	Dry and wet	Copper or alloy steel
Max operating temperature = 90°C (194°F)			
RHH	Heat-resistant thermoset	Dry and damp	1,2
THHN	Heat-resistant thermoplastic	Dry and damp	Nylon
THW	Moisture- and heat-resistant thermoplastic	[f]	None
XHHW	Moisture-resistant thermoset	Dry	None
FEP	Fluorinated ethylene propylene	Dry	None
FEPB	Fluorinated ethylene propylene	Dry	3
TFN	Heat-resistant thermoplastic covered, solid or 7-strand	[c,d]	Nylon
TFFN	Heat-resistant thermoplastic flexible stranding		Nylon
MTW	Moisture-, heat-, and oil-resistant thermoplastic machine-tool wiring (NFPA Stand. 79, NEC 1975, Art. 670)	Dry	None or nylon
PFA	Perfluoralkoxy	Dry and damp	None
SA	Silicone	Dry	Glass
Max operating temperature = 150°C (302°F)			
Z, ZW	Modified ethylene tetrafluoroethylene	Dry	None
Max operating temperature = 200°C (392°F)			
FEP, FEPB	Fluorinated ethylene propylene	Dry	None
	Special applications		3
PFA	Perfluoralkoxy	Dry	None
PF, PGF	Fluorinated ethylene propylene	[c,d]	None or glass braid
PFA	Perfluoralkoxy	Dry	None
SF-2	Silicone rubber, solid or 7-strand	[c,d]	Nonmetallic
Max operating temperature = 250°C (482°F)			
MI	Mineral-insulated (metal-sheathed), for special applications	Dry and wet	Copper or alloy steel
TFE	Extruded polytetrafluoroethylene, only for leads within apparatus or within raceways connected to apparatus, or as open wiring (nickel or nickel-coated copper only)	Dry	None
PFAH	Perfluoroalkoxy (special application) (nickel or nickel-coated copper)	Dry	None
PTF	Extruded polytetrafluoroethylene, solid or 7-strand (nickel or nickel-coated copper only)	[c,d]	None

[a] 1: Moisture-resistant, flame-retardant nonmetallic; 2: outer covering not required when thermoset insulation has been specifically approved for the purpose; 3: no. 14-8 glass braid, no. 6-2 asbestos braid; 4: moisture-resistant nonmetallic.

[b] Limited to 300 V.

[c] No. 18 and no. 16 conductor for remote controls, low-energy power, low-voltage power, and signal circuits; NEC® 2005 Sec. 725.27, Sec. 760.27.

[d] Fixture wire no. 18-16.

[e] For over 2,000 V, the insulation shall be ozone-resistant.

[f] Special applications within electric discharge lighting equipment. Limited to 1,000 V open-circuit volts or less.

Source: NEC® 2005 Tables 310.13 and 402.3. Reprinted with permission from NFPA 70-1993, National Electrical Code®, Copyright © 1992, National Fire Protection Association, Quincy, Massachusetts 02269. This reprinted material is not the complete and official position of the NFPA on the referenced subject, which is represented only by the standard in its entirety.

Note: Not all types of approved wires are listed.

Table 15.1.25 Adjustment Factors for More Than Three Current-Carrying Conductors in a Raceway or Cable

Number of current-carrying conductors	Percent of values in Table 15.1.23 as adjusted for ambient temperature if necessary
4–6	80
7–9	70
10–20	50
21–30	45
31–40	40
41 and above	35

SOURCE: NEC® Table 310.15 (A)(2)(a). Reprinted with permission from NFPA 70-2005. National Electric Code®, Copyright © 2001, National Fire Protection Association, Quincy, MA. This reprinted material is not the complete and official position of the NFPA on the referenced subject, which is represented only by the standard in its entirety.

2. Then, using the appropriate table for the raceway system from NEC® 2005, Table 4, find a raceway equal to or greater than the maximum allowable fill for the number of conductors.

Generally, insulated conductors in premises wiring systems and circuits are required to have a **grounded conductor** (commonly referred to as the *neutral*) identified by a continuous **white or gray finish.** Where the wiring method does not include a white colored conductor, a conductor (other than green) can be identified by three continuous white stripes along its length, or for conductor sizes larger than 6 AWG, white tape or paint can be applied at terminations. Refer to NEC® 2005, Art. 200.6, for additional information.

The insulation color of **ungrounded circuit conductors** is not restricted other than that it be different from a grounded (white) or grounding conductor (green). A common identification practice for circuit conductors in 208Y120-V, three-phase, three-wire systems is black, red, and blue, and for three-phase, four-wire systems is **black, red, blue, and white.** For 480Y277-V, three-phase, three-wire systems, brown, orange, and yellow (BOY) is widely used, and for three-phase, four-wire systems, **brown, orange, yellow, and white** is common practice. Where more than one nominal voltage system exists in a building (or premises), unique identification either by conductor color or tape is required to be applied at each termination.

Switching Arrangements   Small quick-break switches must be set in or on an approved box or fitting and may be of the push, tumbler, or rotary type. The following types of switches are used to control lighting circuits: (1) single-pole, (2) double-pole, (3) three-point or three-way, (4) four-way, in combination with three-way switches to control lights from three or more stations.

In all metallic protecting systems, such as conduit, armored cable, or metal raceways, joints and splices in conductors must be made only in junction boxes or other proper fittings; therefore, these fittings can be located only in accessible places and never concealed in partitions. Splices or joints in the wire must never be in the conduit piping, raceway, or metallic tubing itself, for the splices may become a source of trouble as a result of corrosion or grounding if water should enter the conduit.

All conductors of an ac system must be placed in the same metallic casing so that their resultant magnetic field is nearly zero. If this is not done, eddy currents are set up causing heating and excessive loss. With single conductors in a casing, or multiple conductors of the same phase, an excessive reactance drop may result.

Services   A **service disconnecting** means is required to disconnect all ungrounded conductors in buildings or other structures from the service entrance conductors. The disconnecting means must be installed at a readily accessible location either outside a building or structure or inside, nearest the point of entrance of the service conductors. The length of **service entrance conductors** should be kept to a minimum inside buildings. Some local jurisdictions have ordinances that limit the distance of service entrance conductors or the location of service equipment inside a building. The service drop or entrance conductors must be insulated or covered, except a grounded conductor is permitted to be bare when part of a service cable, underground lateral, and certain other installations. (Refer to NEC® 2005, Art. 230, for additional information on electrical service requirement.)

The minimum size of service entrance conductors (before any adjustment or correction factors) shall not be less than the sum of the non-continuous load plus 125 percent of the continuous load. The service disconnecting means shall have a rating of not less than the load to be carried. Calculating the **service size** requires accounting for (1) all fixed loads (motors, appliances, heating devices, etc.), adjusted for any applicable demand factors (ratio of maximum demand to connected load); (2) general lighting load (lights and general-use receptacles) based on type of building (dwelling, hospital, hotel, apartments, warehouse, etc.), adjusted for applicable demand factor; and (3) additional capacity for the future, if applicable. (Refer to NEC® 2005, Art. 220, for additional information and NEC® 2005, Annex D, for examples.)

The computed load of a feeder shall not be less than the sum of all the loads on branch circuits after any applicable demand factors have been applied. In no case shall the service rating be less than 15 A for loads of a single branch circuit, 30 A for loads consisting of two 2-wire circuits, 100 A for a one-family dwelling. For all other installations the service disconnecting means shall be rated for a minimum of 60 A.

Unless specifically excepted, **alternating-current circuits and systems** operating from 50 to 1,000 V supplying premises wiring and premises wiring systems are required to be **grounded.** Also, two-wire **direct-current systems** operating from 50 to 300 V and all three-wire direct-current systems supplying premises wiring are required to be **grounded.** Grounding of electrical systems to earth is done in a manner that will limit the voltage caused by lightning, surges, or unintentional contact with higher voltage sources and acts to stabilize the voltage to earth during normal operation.

System grounding is normally accomplished by connecting a **grounding electrode conductor** from a neutral terminal in the service equipment to a **grounding electrode.** The types of grounding electrodes permitted include driven ground rods, water pipe, metal frames of buildings, ground rings, and others. Refer to NEC® 2005, Art. 250, Part III, for additional information on application and conductor sizing.

Electrical equipment, wiring, and other electrically conductive materials likely to become energized must be installed in a manner that will provide an **effective ground path** capable of carrying the maximum ground-fault current likely in any part of the system where a ground fault may occur to the electrical supply. All non-current-carrying conductive materials enclosing conductors (raceways, wireways, shields, etc.), enclosing equipment, or forming part of electrical equipment (motor frames, etc) are required to be **bonded** (connected together) in a manner that establishes an effective path for ground-fault current.

Equipment grounding conductors are used to provide the grounding of non-current-carrying metal parts likely to become energized for fixed, stationary, or portable equipment served by fixed wiring or cord-and-plug connected. Equipment grounding conductors can be a bare or insulated (green or green/yellow stripe) wire where properly sized, rigid metal or intermediate conduit, copper sheath of MI cables, and cable trays. Under certain conditions electrical metallic tubing, type MC cable, wireways/auxilliary gutters, flexible metal conduit, and liquid-tight flexible metal conduit can be used as an equipment-grounding conductor. See NEC® 2005, Art. 250.118, for additional information.

Ground-Fault Protection   There are several types of ground-fault protection used for protection of either personnel or equipment.

Ground-fault circuit interrupters (GFCIs) have been required by the NEC® since 1973 to provide personnel protection in certain areas of dwellings, manufactured housing, hotels, swimming pools, health-care facilities, recreational vehicles, and other locations to prevent electrocution. The GFCI monitors the current in the two conductors of a circuit (120 or 240 V, single phase) and automatically trips the circuit breaker on a differential current in excess of 4 to 6 mA (nominally 5 mA). During the time to trip a fault a person can be exposed to a shock sensation that can cause secondary accidents such as falls.

Ground-fault equipment protection (GFEP) is required for **electric heating circuits** as defined by NEC® 2005, Arts. 426 and 427. The requirement can be met by use of a ground-fault equipment protection circuit interrupter (GFEPCI) with a nominal trip value of 30 mA or a heating controller or ground-fault relay with adjustable trip settings.

Ground-fault protection of equipment has been required since 1971 in **solidly grounded wye systems** of more than 150 V to ground but not exceeding 600 V phase-to-phase at each **service disconnect rated 1,000 A** or more. Adjustable current-sensing relays are commonly used for equipment protections; they monitor either all phase conductors or the main bonding jumper. When the stray current exceeds the setting of the ground-fault relay, a shunt-trip breaker is operated to open all ungrounded circuit conductors.

Arc-fault interrupter protection is designed to detect and interrupt arcing faults. **Arc-fault circuit interrupter (AFCI)** circuit breakers are required in 15- and 20-A, 125-V circuits that supply outlets (receptacles, lighting, etc.) in dwelling-unit bedrooms to provide protection from arcing faults occurring in extension cords or cord-connected lights or appliances.

RESISTOR MATERIALS

For use in rheostats, electric furnaces, ovens, heaters, and many electrical appliances, a resistor material with high melting point and high resistivity which does not disintegrate or corrode at high temperatures is necessary. These requirements are met by the nickel-chromium and nickel-chromium-iron alloys. For electrical instruments and measuring apparatus, the resistor material should have high resistivity, low temperature coefficient, and, for many uses, low thermoelectric power against copper. The properties of resistor materials are given in Table 15.1.26. Most of these materials are available in ribbon as well as in wire form. Cast-iron and steel wire are efficient and economical resistor materials for many uses, such as power-absorbing rheostats and motor starters and controllers. (See also Table 15.1.3.)

Advance has a low temperature coefficient and is useful in many types of measuring instrument and precision equipment. Because of its high thermoelectric power to copper, it is valuable for thermoelements and pyrometers. It is noncorrosive and is used to a large extent in industrial and radio rheostats. **Hytemco** is a nickel-iron alloy characterized by a high temperature coefficient and is used advantageously where self-regulation is required as in immersion heaters and heater pads. **Magno** is a manganese-nickel alloy used in the manufacture of incandescent lamps and radio tubes. **Manganin** is a copper-manganese-nickel alloy which, because of its very low temperature coefficient and its low thermal emf with respect to copper, is very valuable for high-precision electrical measuring apparatus. It is used for the resistance units in bridges, for shunts, multipliers, and similar measuring devices. **Nichrome V** is a nickel-chromium alloy free from iron, is noncorrosive, nonmagnetic, withstands high temperatures, and has high

resistivity. It is recommended as material for heating elements in electric furnaces, hot-water heaters, ranges, radiant heaters, and high-grade electrical appliances. **Kanthal** is used for heating applications where higher operating temperatures are required than for Nichrome. Mechanically, it is less workable than Nichrome. **Platinum** is used in specialized heating applications where very high temperatures are required. **Tungsten** may be used in very high temperature ovens with an inert atmosphere. **Pure nickel** is used to satisfy the high requirements in the fabrication of radio tubes, such as the elimination of all gases and impurities in the metal parts. It also has other uses such as in incandescent lamps, for combustion boats, laboratory accessories, and resistance thermometers.

Carbon withstands high temperatures and has high resistance; its temperature coefficient is negative; it will safely carry about 125 A/in^2. Amorphous carbon has a resistivity of 3,800 and 4,100 $\mu\Omega \cdot$ cm, retort carbon about 720 $\mu\Omega \cdot$ cm, and graphite about 812 $\mu\Omega \cdot$ cm. The properties of any particular kind of carbon depend on the temperature at which it was fired. Carbon for rheostats may best be used in the form of compression rheostats. **Silicon carbide** is used to manufacture heating rods that will safely operate at 1,650°C (3,000°F) surface temperature. It has a negative coefficient of resistance up to 650°C, after which it is positive. It must be mechanically protected because of inherent brittleness.

MAGNETS

A **permanent magnet** is one that retains a considerable amount of magnetism indefinitely. Permanent magnets are used in electrical instruments, telephone receivers, loudspeakers, magnetos, tachometers, magnetic chucks, motors, and for many purposes where a constant magnetic field or a constant source of magnetism is desired. The magnetic material should have high retentivity, a high remanence, and a high coercive force (see Fig. 15.1.8). These properties are usually found with hardened steel and its alloys and also in ceramic permanent magnet materials.

Since permanent magnets must operate on the molecular mmf imparted to them when magnetized, they must necessarily operate on the portion *CDO* of the hysteresis loop (see Fig. 15.1.8). The area *CDO* is proportional to the **stored energy** within the magnet and is a criterion of its usefulness as a permanent-magnetic material. In the left half of Fig. 15.1.97 are given the *B-H* characteristics of several permanent-magnetic materials; these include 5 to 6 percent tungsten steel (curve 1); 3½ percent chrome magnet steel (curve 2); cobalt magnet steel, containing 16 to 36 percent cobalt and 5 to 9 percent chromium and in some alloys tungsten (curve 3); and the carbon-free aluminum-nickel-cobalt-steel alloys called Alnico. There are many grades of Alnico; the

Table 15.1.26   Properties of Metals, Alloys, and Resistor Materials

Material	Composition	Sp gr	Microhms-cm at 20°C	Ohms cir-mil-ft at 20°C	Temp coef of resistance per deg C	Temp range, °C	Max safe working temp, °C	Approx melting point, °C
				Resistivity				
Advance	Cu 0.55; Ni 0.45	8.9	48.4	294	± 0.00002	20–100	500	1210
Comet	Ni 0.30; Cr 0.05; Fe 0.65	8.15	95	570	0.00088	20–500	600	1480
Bronze, commercial	Cu; Zn	8.7	4.2	25	0.0020	0–100	—	1040
Hytemco-Balco	Ni 0.50; Fe 0.50	8.46	20	120	0.0045	20–100	600	1425
Kanthal A	Al 0.055; Cr 0.22; Co 0.055; Fe 0.72	7.1	145	870	0.00002	0–500	1330	1510
Magno	Ni 0.955; Mn 0.045	8.75	20	120	0.0036	20–100	400	1435
Manganin	Cu 0.84; Mn 0.12; Ni 0.04	8.19	48.2	290	± 0.000015	15–35	100	1020
Monel metal	Ni 0.67; Cu 0.28	8.9	42.6	256	0.0001	0–100	425	1350
Nichrome	Ni 0.60; Fe 0.25; Cr 0.15	8.247	112	675	0.00017	20–100	930	1350
Nichrome V	Ni 0.80; Cr 0.20	8.412	108	650	0.00013	20–100	1100	1400
Nickel, pure	Ni 0.99	8.9	10	60	0.0050	0–100	400	1450
Platinum	Pt	21.45	10.616	63.80	0.003	0–100	1200	1773
Silver	Ag	10.5	1.622	9.755	0.00361	0–100	650	960
Tungsten	W	19.3	5.523	33.22	0.0045	—	*	3410

* Tungsten subject to rapid oxidation in air above 150°C.

characteristics of three of them are shown by curves 4, 5, 6. Their composition is as follows:

Curve	Alnico no.	Composition, percent					
		Al	Ni	Co	Cu	Ti	Fe
4	5	8	14	24	3	. . .	51
5	6	8	14	24	3	1.25	49.75
6	12	12	6	18	35	. . .	8.0	33

All the Alnicos can be made by the sand or the precision-casting (lost-wax) process, but the most satisfactory method is by the sintering process.

If the alloys are held in a magnetic field during heat-treatment, a **magnet grain** is established and the magnetic properties in the direction of the field are greatly increased. The alloys are hard, can be formed only by casting or sintering, and cannot be machined except by grinding.

The curves in the right half of Fig. 15.1.97 are "external energy" curves and give the product of B and H. The optimum point of operation is at the point of maximum energy as is indicated at A_1 on curve 5.

Considering curve 5, if the magnetic circuit remained closed, the magnet would operate at point B. To utilize the flux, an air gap must be introduced. The air gap acts as a demagnetizing force, $H_1 (= B_1 A_1)$, and the magnet operates at point A_1 on the HB curve. The line OA_1 is called the *air-gap line* and its slope is given by $\tan \theta_1 = B_1/H_1$ where $H_1 = B_g l_g/l_m$ and $B_1/B_g = A_g/A_m$, where B_g = flux density in gap, T; l_g = length of gap, m; l_m = length of magnet, m; A_g, A_m = areas of the gap and magnet, m^2.

If the air gap is lengthened, the magnet will operate at A_2 corresponding to a lesser flux density B_3 and the new air-gap line is OA_2. If the gap is now closed to its original value, the magnet will not return to operation at point A_1 but will operate at some point C on the line OA_1. If the air gap is varied between the two foregoing values, the magnet will operate along the **minor hysteresis loop** A_2C. Return to point A_1 can be accomplished only by remagnetizing and coming back down the curve from B to A_1.

Alnico magnets corresponding to curves 4 and 5 are best adapted to operation with short air gaps, since the introduction of a long air gap will demagnetize the magnet materially. On the other hand, a magnet with a long air gap will operate most satisfactorily on curve 6 on account of the high coercive force H_2. With change in the length of the air gap, the operation will be essentially along that curve and the magnet will lose little of its original magnetization.

There are several other grades of Alnico with characteristics between curves 4, 5, and 6. Ceramic PM materials have a very large coercive force.

The steels for permanent magnets are cut in strips, heated to a red-hot temperature, and forged into shape, usually in a "bulldozer." If they are to be machined, they are cooled in mica dust to prevent air hardening. They are then ground, tumbled, and tempered. Alnico and ceramic types are cast and then finish-ground.

Permanent magnets are magnetized either by placing them over a bus bar carrying a large direct current, by placing them across the poles of a powerful electromagnet, or by an ampere-turn pulse.

Unless permanent magnets are subjected to artificial aging, they gradually weaken until after a long period they become stabilized usually at from 85 to 90 percent of their initial strength. With magnets for electrical instruments, where a constant field strength is imperative, artificial aging is accomplished by mechanical vibration or by immersion in oil at 250°F for a period of a few hours.

In an **electromagnet** the magnetic field is produced by an electric current. The core is usually made of soft iron or mild steel because, the permeability being higher, a stronger magnetic field may be obtained. Also since the retentivity is low, there is little trouble due to the sticking of armatures when the circuit is opened. Electromagnets may have the form of simple solenoids, iron-clad solenoids, plunger electromagnets, electromagnets with external armatures, and lifting magnets, which are circular in form with a flat holding surface.

A **solenoid** is a winding of insulated conductor and is wound helically; the direction of winding may be either right or left. A **portative electromagnet** is one designed only for holding material brought in contact with it. A **tractive electromagnet** is one designed to exert a force on the load through some distance and thus do work. The **range** of an electromagnet is the distance through which the plunger will perform work when the winding is energized. For long range of operation, the plunger type of tractive magnet is best suited, for the length of core is governed practically by the range of action desired, and the area of the core is

Fig. 15.1.97 Characteristics of permanent magnet materials.

Table 15.1.27   Maximum Pull per Square Inch of Core for Solenoids with Open Magnetic Circuit

Length of coil l, in	Length of plunger, in	Core area A, in^2	Total ampere-turns $I \times n$	Max pull P, psi	$1,000 \times C$	Length of coil l, in	Length of plunger, in	Core area A, in^2	Total ampere-turns $I \times n$	Max pull P, psi	$1,000 \times C$
6	Long	1.0	15,900	22.4	9.0	12	Long	1.0	11,200	8.75	9.4
9	Long	1.0	11,330	11.5	9.1	12	Long	1.0	20,500	16.75	9.8
9	Long	1.0	14,200	14.6	9.2	18	36	1.0	18,200	9.8	9.7
10	10	2.76	40,000	40.2	10.0	18	36	1.0	41,000	22.5	9.8
10	10	2.76	60,000	61.6	10.3	18	18	1.0	18,200	9.8	9.7
10	10	2.76	80,000	80.8	10.1	18	18	1.0	41,000	22.5	9.8

SOURCE: From data by Underhill, *Elec. World*, 45, 1906, pp. 796, 881.

determined by the pull. **Solenoid and plunger** is a solenoid provided with a movable iron rod or bar called a plunger. When the coil is energized, the iron rod becomes magnetized and the mutual action of the field in the solenoid on the poles created on the plunger causes the plunger to move within the solenoid. This force becomes zero only when the magnetic centers of the plunger and solenoid coincide. If the load is attached to the plunger, work will be done until the force to be overcome is equal to the force that the solenoid exerts on the plunger. When the iron of the plunger is not saturated, the strength of magnetic field in the solenoid and the induced poles are both proportional to the exciting current, so that the pull varies as the **current squared**. When the plunger becomes highly saturated, the pull varies almost **directly** with the current.

The **maximum uniform pull** occurs when the end of the plunger is at the center of the solenoid and is equal to

$$F = CAnI/l \quad \text{lb} \quad (15.1.134)$$

where A = cross-sectional area of plunger, in^2; n = number of turns; I = current, A; l = length of the solenoid, in; and C = pull, (lb/in^2)/(A-turn/in). C depends on the proportions of the coil, the degree of saturation, the length, and the physical and chemical purity of the plunger. Table 15.1.27 gives values of C for several different solenoids.

Curve 1, Fig. 15.1.98, shows the characteristic pull of an open-magnetic circuit solenoid, 12 in long, having 10,000 A-turns or 833 A-turns/in.

Fig. 15.1.98   Pull on solenoid with plunger. (1) Coil and plunger; (2) coil and plunger with stop; (3) iron-clad coil and plunger; (4) and (5) same as (3) with different lengths of stop.

When a **strong pull** is desired at the end of the stroke, a stop may be used as shown in Fig. 15.1.99. Curve 2, Fig. 15.1.98, shows the pull obtained by adding a stop to the plunger. It will be noted that, except when the end of the plunger is near the stop, the stop adds little to the solenoid pull. The pull is made up of two components: one due to the attraction between plunger and winding, the other to the attraction between plunger and stop. The equation for the pull is

$$P = AIn[(In/l_a^2 C_1^2) + (C/l)] \quad (15.1.135)$$

where A = area of the core, in^2; n = number of turns; l_a = length of gap between core and stop; and C, C_1 = constants. At the beginning of the stroke the second member of the equation is predominant, and at the end

Fig. 15.1.99   Solenoid with stop.

of the stroke the first member represents practically the entire pull. Approximate values of C and C_1 are $C_1 = 2,660$ (for l greater than $10d$), $C = 0.0096$, where d is the diameter of the plunger, in. In SI units

$$P = 1.7512AnI\left(\frac{2.54nI}{l_a^2 C_1^2} + \frac{C}{l}\right) \quad \text{N} \quad (15.1.135a)$$

where A is in cm^2, l and l_a in cm, and the pull P in N. All other quantities are unchanged.

The **range of uniform** pull can be extended by the use of conical ends of stop and plunger, as shown in Fig. 15.1.100. A stronger magnet mechanically can be obtained by using an iron-clad solenoid, Fig. 15.1.101, in which an iron return path is provided for the flux. Except for low flux densities and short air gaps the dimensions of the iron

Fig. 15.1.100   Conical plunger and stop.

return path are of no practical importance, and the fact that an iron return path is used does not affect the pull curve except at short air gaps. This is illustrated in Fig. 15.1.98 where curves 3, 4, and 5 are typical pull curves for this same solenoid when it is made iron-clad, each curve corresponding to a different position of the stop.

Fig. 15.1.101   Iron-clad solenoid.

Mechanical jar at the end of the stroke may be prevented by leaving the end of the solenoid open. The plunger then comes to equilibrium when its middle is at the middle of the winding, thus providing a magnetic cushion effect. Electromagnets with external armatures are best adapted for **short-range** work, and the best type is the horseshoe magnet. The pull for short-range magnets is expressed by the equation

$$F = B^2A/72,134,000 \quad \text{lb} \quad (15.1.136)$$

where B = flux density, Mx/in^2, and A = area of the core, in^2.

In SI units

$$F = 397,840B^2A \quad \text{N} \quad (15.1.136a)$$

where the flux density B is in Wb/m^2; A = area, m^2; and F = force, N.

A greater holding power is obtained if the surfaces of the armature and core are not machined to an absolutely smooth contact surface. If the surface is slightly irregular, the area of contact A is reduced but the flux density B is increased approximately in proportion (if the iron is being operated below saturation) and the pull is increased since it varies as the square of the density B. Nonmagnetic stops should be used if it is desired that the armature may be released readily when the current is interrupted.

Lifting magnets are of the portative type in that their function is merely to hold the load. The actual lifting is performed by the hoisting apparatus. The magnet is almost toroidal in shape. The coil shield is of manganese steel which is very hard and thus resists wear and is practically nonmagnetic. The holding power is given by Eq. (15.1.136), where $A =$ area of holding surface, in². It is difficult to calculate accurately the holding force of a lifting magnet for it depends on the magnetic characteristics of the load, the area of contact, and the manner in which the load is applied.

Rapid action in a magnet can be obtained by reducing the time constant of the winding and by subdividing the metal parts to reduce induced currents which have a demagnetizing effect when the circuit is closed. The movement of the plunger through the winding causes the winding and its bobbin to be cut by a magnetic field; if the bobbin is of metal and not slotted longitudinally, it is a short-circuited turn linked by a changing magnetic field and hence currents are induced in it. These currents oppose the flux and hence reduce the pull during the transient period. They also cause some heating. Where it is found impossible to reduce the time constant sufficiently, an electromagnet designed for a voltage much lower than normal is often used. A resistor is connected in series which is short-circuited during the stroke of the plunger. At the completion of the stroke the plunger automatically opens the short circuit, reducing the current to a value which will not overheat the magnet under continuous operation. The extremely short time of overload produces very rapid action but does not injure the winding. The solenoids on many automatic motor-starting panels are designed in this manner.

When **slow action** is desired, it can be obtained by using solid cores and yoke and by using a heavy metallic spool or bobbin for the winding. A separate winding short-circuited on itself is also used to some extent.

Sparking at switch terminals may be reduced or eliminated by neutralizing the inductance of the winding. This is accomplished by winding a separate short-circuited coil with its wires parallel to those of the active winding. (This method can be used with dc magnets only.) This is not economical, since one-half the winding space is wasted. By connecting a capacitor across the switch terminals, the energy of the inductive discharge on opening the circuit may be absorbed. For the purpose of neutralizing the inductive discharge and causing a quick release, a small **reverse current** may be sent through the coil winding automatically on opening the circuit. Sleeves of tin, aluminum, or copper foil placed over the various layers of the winding absorb energy when the circuit is broken and reduce the energy dissipated at the switch terminals. This scheme can be used for dc magnets only. **Sticking** of the parts of the magnetic circuit due to residual magnetism may be prevented by the use of nonmagnetic stops. In the case of lifting magnets subjected to rough usage and hard blows (as in a steel works), these stops usually consist of plates of manganese steel, which are extremely hard and nonmagnetic.

AC Tractive Magnets   Because of the iron losses due to eddy currents, the magnetic circuits of ac electromagnets should be composed of laminated iron or steel. The magnetic circuit of large magnets is usually built up of thin sheets of sheet metal held together by means of suitable clamps. Small cores of circular cross section usually consist of a bundle of soft iron wires. Since the iron losses increase with the flux density, it is not advisable to operate at as high a density as with direct current. The current instead of being limited by the resistance of the winding is now determined almost entirely by the inductive reactance as the resistance is small. With the removal of the load the current rises to high values. The pull of ac magnets is nearly constant irrespective of the length of air gap.

In a **single-phase magnet** the pull varies from zero to a maximum and back to zero twice every cycle, which may cause considerable chattering of the armature against the stop. This may be prevented by the use of a spring or, in the case of a solenoid coil, by allowing the plunger to seek its position of equilibrium in the coil. **Chattering** may also be prevented by the use of a short-circuited winding or shading coil around

one tip of the pole piece or by the use of polyphase. In a **two-phase magnet** the pull is constant and equal to the maximum instantaneous pull produced by one phase so long as the voltage is a sine function. In a **three-phase magnet** under the same conditions the pull is constant and equal to 1.5 times the maximum instantaneous pull of one phase. Should the load become greater than the minimum instantaneous pull, there will be chattering as in a single-phase magnet.

Heating of Magnets   The lifting capacity of an electromagnet is limited by the permissible current-carrying capacity of the winding, which, in turn, is dependent on the amount of heat energy that the winding can dissipate per unit time without exceeding a given temperature rise. **Enamels, synthetic varnishes,** and **thermoplastic wire insulations** are available to allow a wide variety of temperature rises.

Design of Exciting Coil   Let $n =$ number of turns, $l =$ mean length of turn, in ($l = 2\pi r$, where r is the mean radius, in), $A =$ cross section of wire, cir mils. The resistance of 1 cir mil · ft of copper is practically 12 Ω at 60°C, or 1 Ω/cir mil · in. Hence the resistance, $R = nl/A \ \Omega$; the current, $I = EA/nl$; the ampere-turns, $nI = EA/l$; the power to be dissipated, $P = E^2A/nl$ W. From the foregoing equations the cross section of wire and the number of turns can be calculated.

Fig. 15.1.102   Winding space factor.

Space Factor of Winding   Space factor of a coil is the ratio of the space occupied by the conductor to the total volume of the coil or winding. Only in the theoretical case of uninsulated square or rectangular conductor may the space factor be 100 percent. For wire of circular section with insulation of negligible thickness, wound as shown in Fig. 15.1.102a, the space factor will be 78.5 percent. When the turns of wire are "bedded," as shown in Fig. 15.1.102b (the case in most windings, particularly with smaller wires), there is a theoretical gain of about 7 percent in space factor. Experiments have shown that in most cases this gain is about neutralized in practice by the flattening out of the insulation of the wire due to the tension used in winding. When wound in a haphazard manner, the space factors of magnet wires vary according to size, substantially as follows:

	Double covered				Single covered		
Size, AWG	0	5	10	15	20	25	30	35
Space factor, %	60	53.8	45.5	35.1	32.2	32	25.7	16

Magnet wire is usually a soft annealed copper wire of high conductivity. It can be obtained in square, rectangular, and circular conductors. Round or cylindrical wire is almost entirely used in smaller sizes; ribbons are frequently used in the larger sizes. Common round wire sizes for copper magnet wire include No. 42 AWG (0.0025-in bare wire diameter) to 8 AWG (0.1285 in). Common round wire sizes for aluminum magnet wire include No. 26 AWG to No. 4 AWG. For use in very small devices, ultrafine round magnet wire is available in sizes as small as No. 60 AWG copper and No. 52 aluminum.

Magnet wire insulation has evolved to become high in electrical, physical, and thermal performance. Polymers are most commonly used for film-insulated magnet wire and are based on polyvinyl acetates, polyesters, polyamideimides, polyimides, polyamides, and polyurethanes. It is common to use different layers of polymers to obtain the best performance for a specific application. Film-insulated magnet wires are available in temperature classes from 105 to 200°C, each with specific advantages such as solderability, thermal stability, transformer resistance, chemical resistance, windability, and toughness.

Magnet wire is also produced with fabric layers served over bare or conventional film-insulated wire. Fabric materials include fiberglass, polyester (Dacron), aromatic polyamide (Nomex) paper, and aromatic polyamide film (Kapton). Fibrous insulated magnet wires are available in temperature classes from 105 to 220°C, each with specific advantages such as high dielectric strength, positive separation of conductors, high thermal and chemical stability, physical protection, and high-temperature performance.

AUTOMOBILE SYSTEMS

The evolution of automobile electrical systems toward microprocessors has transformed all aspects of automotive engineering, operation, and repair. Engine operation, emission controls, drive-train control, operator interfaces to vehicle operation and drive controls, entertainment systems, and communication and navigational systems, are commonplace and will continue to evolve. At the heart of understanding the operation of the car/truck for-cycle combustion engine is a basic electrical system that has supported engine operation for generations. The following information describes electrical-based automotive ignition, starting, lighting, and accessory systems commonly used prior to microprocessor-based systems.

Automobile Ignition Systems

The ignition system in an automobile produces the spark which ignites the combustible mixture in the engine cylinders. This is accomplished by a high-voltage, or high-tension, spark between metal points in a spark plug. (A spark plug is an insulated bushing screwed into the cylinder head.) Spark plugs usually have porcelain insulation, but for some special uses, such as in airplane engines, mica may be used. There are two general sources for the energy necessary for ignition; one is the electrical system of the car which is maintained by the generator and the battery (**battery ignition**), and the other is a **magneto.** Battery ignition systems have traditionally operated electromechanically, using a spark coil, a high-voltage distributor, and low-voltage breaker points. Electronic ignition systems, working from the battery, became standard on U.S. cars in 1975. These vary in complexity from the use of a single transistor to reduce the current through the points to pointless systems triggered by magnetic pulses or interrupted light beams. Capacitor discharge into a pulse transformer is used in some systems to obtain the high voltage needed to fire the spark plugs.

Battery ignition is most widely used since it is simple, reliable, and low in cost, and the electrical system is a part of the car equipment. The high voltage for the spark is obtained from an ignition coil which consists of a primary coil of relatively few turns and a secondary coil of a large number of turns, both coils being wound on a common magnetic core consisting of either thin strips of iron or small iron wires. In a 6-V system the resistance of the primary coil is from 0.9 to 2 Ω and the inductance is from 5 to 10 mH. The number of secondary turns varies from 9,000 to 25,000, and the ratio of primary to secondary turns varies from 1 : 40 to 1 : 100.

The coil operates on the following principle. It stores energy in a magnetic field relatively slowly and then releases it suddenly. The power developed ($p = dw/dt$) is thus relatively large (w = stored energy). The high emf e_2 which is required for the spark is induced by the sudden change in the flux ϕ in the core of the coil when the primary current is suddenly interrupted, $e_2 = -n_2(d\phi/dt)$, where n_2 is the number of secondary turns. For satisfactory ignition, peak voltages from 10 to 20 kV are desirable. Figure 15.1.103 shows the relation between the volts required to produce a spark and pressure with compressed air.

Fig. 15.1.103   Pressure-voltage curve for spark plug.

A battery ignition system for a four-cylinder engine is shown diagrammatically in Fig. 15.1.104. The primary circuit supplied by the battery consists of the primary coil P and a set of contacts, or "points" operated by a four-lobe cam, in series. In order to reduce arcing and burning of the contacts and to produce a sharp break in the current, a capacitor C is connected across the contacts. The contacts, which are of pure tungsten, are operated by a four-lobe cam which is driven at one-half engine speed. A strong spring tends to keep the contacts closed.

Fig. 15.1.104   Battery-ignition system.

In the Delco-Remy distributor (Fig. 15.1.105) two breaker arms are connected in parallel; one coil and one capacitor are used. One set of contacts is open when the other is just breaking but closes a few degrees after the break occurs. This closes the primary of the ignition coil immediately after the break and increases the time that the primary of the ignition coil is closed and permits the flux in the iron to reach its full value. The interrupter shown in Fig. 15.1.105 is designed for an eight-cylinder engine.

Fig. 15.1.105   Delco-Remy dual-point eight-cylinder interrupter.

Electronic ignition systems have no problem operating at the required speed.

The spark should advance with increase in engine speed so as to allow for the time lag in the explosion. To take care of this automatic timers are equipped with centrifugally operated weights which advance the breaker cam with respect to the engine drive as the speed increases.

Automobile Lighting and Starting Systems

Automobile lighting and starting systems initially operated at 6 V, but at present nearly all cars operate at 12 V because larger engines, particularly V-8s, are now common and require more starting power. With 12 V, for the same power, the starting current is halved, and the effect of resistance in the leads, connections, and brushes is materially reduced. In some systems the positive side of the system is grounded, but more often the negative side is grounded.

A further development is the application of an **ac generator,** or alternator, combined with a rectifier as the generating unit rather than the usual dc generator. One advantage is the elimination of the commutator, made up of segments, which requires some maintenance due to the sparking and wear of the carbon brushes. With the alteraator the dc field rotates, the brushes operating on smooth slip rings require almost no maintenance. Also, the system is greatly simplified by the fact that rectifiers are "one-way" devices, and the battery **cannot** deliver current back to the generator when its voltage drops below that of the battery. Thus, no cutout relay, such as is required with dc generators, is necessary. This ac development is the result of the development of reliable, low-cost germanium and silicon semiconductor rectifiers.

Figure 15.1.106 shows a schematic diagram of the Ford system (adapted initially to trucks). The generator stator is wound three-phase

Fig. 15.1.106   Schematic diagram of Ford lighting and starting system.

Y-connected, and the field is bipolar supplied with direct current through slip rings and brushes. The rectifier diodes are connected full-wave bridge circuit to supply the battery through the ammeter.

Regulator   The function of the regulator is to control the generator current so that its value is adapted to the battery voltage which is related to the condition of charge of the battery (see Fig. 15.1.15). Thus, when the battery voltage drops (indicating a lowered condition of charge), the current should be increased, and, conversely, when the battery voltage increases (indicating a high condition of charge), the current should be decreased.

Neglecting for the moment the starting procedure, when the ignition switch is thrown to the normal "on" position at c, the coil actuating the field relay is connected to the battery + terminal and causes the relay contacts a to close. This energizes the regulator circuits, and, if the two upper voltage regulator contacts b are closed as shown, the rotating field of the alternator is connected directly to the battery + terminal, and the field current is then at its maximum value and produces a high generator voltage and large output current. At the same time the voltage regulator coil in series with the 0.3- and 14-Ω resistors is connected between the battery + terminal and ground. If the voltage of the battery rises owing to its higher condition of charge, the current to the voltage regulator coil increases, causing it to open the two upper contacts at b. Current from the battery now flows through the 0.3-Ω resistor and divides, some going through the 10-Ω resistor and dividing between the field and the 50-Ω resistor, and the remainder going to the voltage regulator coil. The current to the rotating field is thus reduced, causing the alternator output to be reduced. Because of the 0.3 Ω now in circuit, the current to the coil of the voltage regulator is reduced to such a value that it holds the center contact at b in the mid, or open, position. If the battery voltage rises to an even higher value, the regulator coil becomes strong enough to close the lower

contacts at b; this short-circuits the field, reducing its current almost to zero, and thus reducing the alternator output to zero. On the other hand when the battery voltage drops, the foregoing sequence is reversed, and the contacts at b operate to increase the current to the alternator field.

As was mentioned earlier, the battery cannot supply current to the alternator because of the "one-way" characteristic of the rectifier. Thus, when the alternator voltage drops below that of the battery and even when the alternator stops running, its current automatically becomes zero. The alternator has a normal rectifier open-circuit voltage of about 14 V and a rating of 20 A.

Starting   In most cases, for starting, the ignition key is turned far to the right and held there until the motor starts. Then, when the key is released, the ignition switch contacts assume a normal operating position. Thus, in Fig. 15.1.106, when the ignition-switch contact is in the starting position S, the starter relay coil becomes connected by a lead to the battery + terminal and thus becomes energized, closing the relay contacts. The starter motor is then connected to the battery to crank the engine. At the time that the contact closes it makes contact with a small metal brush e which connects the battery + terminal to the primary terminal of the ignition coil through the protective resistor R. After the motor starts, the ignition switch contacts spring to the normal operating position C, and the starter relay switch opens, thereby breaking contact with the small brush e. However, when contact C is closed, the ignition coil primary terminal is now connected through leads to the battery + terminal. The interrupter, the ignition coil, and the distributor now operate in the manner described earlier (see Fig. 15.1.104); the system shown in Fig. 15.1.106 is that for a six-cylinder engine.

The connection of **accessories** to the electric system is illustrated in Fig. 15.1.106 for the horn, head and other lights, and temperature and fuel gages.

15.2 ELECTRONICS

by Byron M. Jones
revised by Timothy M. Cockerill

REFERENCES: "Reference Data for Radio Engineers," Howard Sams & Co. "Transistor Manual," General Electric Co. "SCR Manual," General Electric Co. "The Semiconductor Data Book," Motorola Inc. Fink, "Television Engineering," McGraw-Hill. "Industrial Electronics Reference Book," Wiley. Mano, "Digital and Logic Design," Prentice-Hall. McNamara, "Technical Aspects of Data Communication," Digital Press. Fletcher, "An Engineering Approach to Digital Design," Prentice-Hall. "The TTL Data Book," Texas Instruments, Inc. "1988 MOS Products Catalog," American Microsystems, Inc. "1988 Linear Data Book," National Semiconductor Corp. "CMOS Standard Cell Data Book," Texas Instruments, Inc. "Power MOSFET Transistor Data," Motorola, Inc. Franco, "Design with Operational Amplifiers and Analog Integrated Circuits," McGraw-Hill. Soclof, "Applications of Analog Integrated Circuits," Prentice-Hall. Gibson, "Computer Systems Concepts and Design," Prentice-Hall. Sedra and Smith, "Microelectronic Circuits," HRW Saunders. Millman and Grabel, "Microelectronics," McGraw-Hill. Ghausi, "Electronic Devices and Circuits: Discrete and Integrated," HRW Saunders. Savant, Roden, and Carpenter, "Electronic Design Circuits and Systems," Benjamin Cummings. Stearns and Hush, "Digital Signal Analysis," Prentice-Hall. Kassakian, Schlecht, and Verghese, "Principles of Power Electronics," Addison Wesley. Yariv, "Optical Electronics," HRW Saunders.

The subject of electronics can be approached from the standpoint of either the design of devices or the use of devices. The practicing mechanical engineer has little interest in designing devices, so the approach in this article will be to describe devices in terms of their external characteristics.

COMPONENTS

Resistors, capacitors, reactors, and **transformers** are described earlier in this section, along with basic circuit theory. These explanations are equally applicable to electronic circuits and hence are not repeated here. A description of additional components peculiar to electronic circuits follows.

A **rectifier,** or **diode,** is an electronic device which offers unequal resistance to forward and reverse current flow. Figure 15.2.1 shows the schematic symbol for a diode. The arrow beside the diode shows the

Anode   Cathode

Fig. 15.2.1 Diode schematic symbol.

direction of current flow. Current flow is taken to be the flow of positive charges, i.e., the arrow is counter to electron flow. Figure 15.2.2 shows typical forward and reverse volt-ampere characteristics. Notice that the scales for voltage and current are not the same for the first and third quadrants. This has been done so that both the forward and reverse characteristics can be shown on a single plot even though they differ by several orders of magnitude.

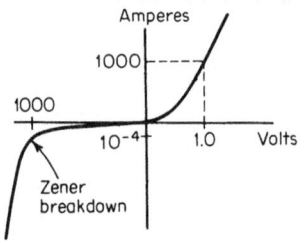

Fig. 15.2.2 Diode forward-reverse characteristic.

Diodes are rated for forward current capacity and reverse voltage breakdown. They are manufactured with maximum current capabilities ranging from 0.05 A to more than 1,000 A. Reverse voltage breakdown varies from 50 V to more than 2,500 V. At rated forward current, the forward voltage drop varies between 0.7 and 1.5 V for silicon diodes. Although other materials are used for special-purpose devices, by far the most common semiconductor material is silicon. With a forward current of 1,000 A and a forward voltage drop of 1 V, there would be a power loss in the diode of 1,000 W (more than 1 hp). The basic diode package shown in Fig. 15.2.3 can dissipate about 20 W. To maintain an acceptable temperature in the diode, it is necessary to mount the diode on a **heat sink.** The manufacturer's recommendation should be followed

Fig. 15.2.3 Physical diode package.

very carefully to ensure good heat transfer and at the same time avoid fracturing the silicon chip inside the diode package.

The selection of **fuses** or **circuit breakers** for the protection of **rectifiers** and rectifier circuitry requires more care than for other electronic devices. Diode failures as a result of circuit faults occur in a fraction of a millisecond. Special semiconductor fuses have been developed specifically for semiconductor circuits. Proper protective circuits must be provided for the protection of not only semiconductors but also the rest of the circuit and nearby personnel. Diodes and diode fuses have a short-circuit rating in amperes-squared-seconds (I^2t). As long as the I^2t rating of the diode exceeds the I^2t rating of its protective fuse, the diode and its associated circuitry will be protected. Circuit breakers may be used to protect diode circuits, but additional line impedance must be provided to limit the current while the circuit breaker clears. Circuit breakers do not interrupt the current when their contacts open. The fault is not cleared until the line voltage reverses at the end of the cycle of the applied voltage. This means that the **clearing time for a circuit breaker** is about $\frac{1}{2}$ cycle of the ac input voltage. Diodes have a 1-cycle overcurrent rating which indicates the fault current the diode can carry for circuit breaker protection schemes. Line inductance is normally provided to limit fault currents for breaker protection. Often this inductance is in the form of leakage reactance in the transformer which supplies power to the diode circuit.

A **thyristor,** often called a **silicon-controlled rectifier (SCR),** is a rectifier which blocks current in both the forward and reverse directions. Conduction of current in the forward direction will occur when the anode is positive with respect to the cathode and when the gate is pulsed positive with respect to the cathode. Once the thyristor has begun to conduct, the gate pulse can return to 0 V or even go negative and the thyristor will continue to pass current. To stop the cathode-to-anode current, it is necessary to reverse the cathode-to-anode voltage. The thyristor will again be able to block both forward and reverse voltages until current flow is initiated by a gate pulse. The schematic symbol for an SCR is shown in Fig. 15.2.4. The physical packaging of thyristors is

Fig. 15.2.4 Thyristor schematic symbol.

Fig. 15.2.5 Transistor schematic symbol.

similar to that of rectifiers with similar ratings except, of course, that the thyristor must have an additional gate connection.

The gate pulse required to fire an SCR is quite small compared with the anode voltage and current. Power gains in the range of 10^6 to 10^9 are easily obtained. In addition, the power loss in the thyristor is very low, compared with the power it controls, so that it is a very efficient power-controlling device. Efficiency in a thyristor power supply is usually 97 to 99 percent. When the thyristor blocks either forward or reverse current, the high voltage drop across the thyristor accompanies low current. When the thyristor is conducting forward current after having been fired by its gate pulse, the high anode current occurs with a forward voltage drop of about 1.5 V. Since high voltage and high current never occur simultaneously, the power dissipation in both the on and off states is low.

The thyristor is rated primarily on the basis of its forward-current capacity and its voltage-blocking capability. Devices are manufactured to have equal forward and reverse voltage-blocking capability. Like diodes, thyristors have I^2t ratings and 1-cycle surge current ratings to allow design of protective circuits. In addition to these ratings, which the SCR shares in common with diodes, the SCR has many additional specifications. Because the thyristor is limited in part by its average current and in part by its rms current, forward-current capacity is a function of the duty cycle to which the device is subjected. Since the thyristor cannot regain its blocking ability until its anode voltage is reversed and remains reversed for a short time, this time must be specified. The time to regain blocking ability after the anode voltage has been reversed is called the **turn-off time.** Specifications are also given for minimum and maximum gate drive. If forward blocking voltage is reapplied too quickly, the SCR may fire with no applied gate voltage pulse. The maximum safe value of rate of reapplied voltage is called the *dv/dt* rating of the SCR. When the gate pulse is applied, current begins to flow in the area immediately adjacent to the gate junction. Rather quickly, the current spreads across the entire cathode-junction area. In some circuits associated with the thyristor an extremely fast rate of rise of current may occur. In this event localized heating of the cathode may occur with a resulting immediate failure or in less extreme cases a slow degradation of the thyristor. The maximum rate of change of current for a thyristor is given by its *di/dt* rating. Design for *di/dt* and *dv/dt* limits is not normally a problem at power-line frequencies of 50 and 60 Hz. These ratings become a design factor at frequencies of 500 Hz and greater. Table 15.2.1 lists typical thyristor characteristics.

A **triac** is a bilateral SCR. It blocks current in either direction until it receives a gate pulse. It can be used to control in ac circuits. Triacs are widely used for light dimmers and for the control of small universal ac motors. The triac must regain its blocking ability as the line voltage crosses through zero. This fact limits the use of triacs to 60 Hz and below.

A **transistor** is a semiconductor amplifier. The schematic symbol for a transistor is shown in Fig. 15.2.5. There are two types of transistors, *p-n-p* and *n-p-n*. Notice that the polarities of voltage applied to these devices are opposite. In many sizes matched *p-n-p* and *n-p-n* devices are available. The most common transistors have a collector dissipation rating of 150 to 600 mW. Collector to base breakdown voltage is 20 to 50 V. The amplification or gain of a transistor occurs because of two facts: (1) A small change in current in the base circuit causes a large change in current in the collector and emitter leads. This current amplification is designated *hfe* on most transistor specification sheets. (2) A small change in base-to-emitter voltage can cause a large change in either the collector-to-base voltage or the collector-to-emitter voltage. Table 15.2.2 shows basic ratings for some typical transistors. There is a great profusion of transistor types so that the choice of type depends upon availability and cost as well as operating characteristics.

The gain of a transistor is independent of frequency over a wide range. At high frequency, the gain falls off. This cutoff frequency may be as low as 20 kHz for audio transistors or as high as 1 GHz for radio-frequency (rf) transistors.

The schematic symbols for the field effect transistor (FET) is shown in Fig. 15.2.6. The flow of current from source to drain is controlled by an electric field established in the device by the voltage applied between the gate and the drain. The effect of this field is to change the resistance of the transistor by altering its internal current path. The FET has an extremely high gate resistance (10^{12} Ω), and as a consequence, it is used for applications requiring high input impedance. Some FETs have been designed for high-frequency characteristics. The two basic constructions used for FETs are **bipolar junctions** and **metal oxide semiconductors**. The schematic symbols for each of these are shown in Fig. 15.2.6*a* and 15.2.6*b*. These are called JFETs and MOSFETs to distinguish between them. JFETs and MOSFETs are used as stand-alone devices and are also widely used in integrated circuits. (See below, this section.)

A MOSFET with higher current capacity is called a **power MOSFET.** Table 15.2.3 shows some typical characteristics for power MOSFETs. Power MOSFETs are somewhat more limited in maximum power than are thyristors. Circuits with multiple power MOSFETs are limited to about 20 kW. Thyristors are limited to about 1500 kW. As a point of interest, there are electric power applications in the hundreds of megawatts which incorporate massively series and parallel thyristors. An advantage of power MOSFETs is that they can be turned on and turned off by means of the gate-source voltage. Thus low-power electric

Table 15.2.1 **Typical Thyristor Characteristics**

| Voltage | Current, A | | I^2t, A^2 · s | 1-cycle surge, A | di/dt, A/μs | dv/dt, V/μs | Turn-off time, μs |
	rms	avg					
400	35	20	165	180	100	200	10
1,200	35	20	75	150	100	200	10
400	110	70	4,000	1,000	100	200	40
1,200	110	70	4,000	1,000	100	200	40
400	235	160	32,000	3,500	100	200	80
1,200	235	160	32,000	3,500	75	200	80
400	470	300	120,000	5,500	50	100	150
1,200	470	300	120,000	5,500	50	100	150

Table 15.2.2   Typical Transistor Characteristics

JEDEC number	Type	Collector-emitter volts at breakdown, BV_{CE}	Collector dissipation, $P_c(25°C)$	Collector current, I_c	Current gain, h_{fe}
2N3904	*n-p-n*	40	310 mW	200 mA	200
2N3906	*p-n-p*	40	310 mW	200 mA	200
2N3055	*n-p-n*	100	115 W	15 A	20
2N6275	*n-p-n*	120	250 W	50 A	30
2N5458	JFET	40	200 mW	9 mA	*
2N5486	JFET	25	200 mW		†

* JFET, current gain is not applicable.
† High-frequency JFET—up to 400 MHz.

control can turn the device on and off. Thyristors can be turned on only by their gate voltage. To turn a thyristor off it is necessary for its high-powered anode-to-cathode voltage to be reversed. This factor adds complication to many thyristor circuits. Another power device which is

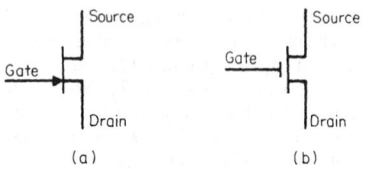

Fig. 15.2.6   Field-effect transistor. (*a*) Bipolar junction type (JFET); (*b*) metal-oxide-semiconductor type (MOSFET).

Table 15.2.3   Typical Power MOSFET Characteristics

Drain voltage	Device	Drain amperes	Power dissipation, W	Case type
600	MTH6N60	6.0	150	TO-218
400	MTH8N40	8.0	150	TO-218
100	MTH25N10	25	150	TO-218
50	MTH35N05	35	150	TO-218
600	MTP1N60	1.0	75	TO-220
400	MTP2N40	2.0	75	TO-220
100	MTP10N10	10	75	TO-220
200	MTE120N20	120	500	346-01
100	MTE150N10	150	500	346-01
50	MTE200N05	200	500	346-01

similar to the power MOSFET is the **insulated gate-bipolar transistor (IGBT)**. This device is a **Darlington** combination of a MOSFET and a bipolar transistor (see Fig. 15.2.13). A low-power MOSFET first transistor drives the base of a second high-power bipolar transistor. These two transistors are integrated in a single case. The IGBT is applied in high-power devices which use high-frequency switching. IGBTs are available in ratings similar to thyristors (see Table 15.2.1), so they are power devices. The advantage of the IGBT is that it can be switched on and off by means of its gate. The advantage of an IGBT over a power MOSFET is that it can be made with higher power ratings.

The **unijunction** is a special-purpose semiconductor device. It is a pulse generator and widely used to fire thyristors and triacs as well as in timing circuits and waveshaping circuits. The schematic symbol for a unijunction is shown in Fig. 15.2.7. The device is essentially a silicon resistor. This resistor is connected to base 1 and base 2. The emitter is fastened to this resistor about halfway between bases 1 and 2. If a positive voltage is applied to base 2, and if the emitter and base 1 are at zero, the

Fig. 15.2.7   Unijunction.

emitter junction is back-biased and no current flows in the emitter. If the emitter voltage is made increasingly positive, the emitter junction will become forward-biased. When this occurs, the resistance between base 1 and base 2 and between base 2 and the emitter suddenly switches to a very low value. This is a regenerative action, so that very fast and very energetic pulses can be generated with this device.

Before the advent of semiconductors, electronic rectifiers and amplifiers were **vacuum tubes** or **gas-filled tubes**. Some use of these devices still remains. If an electrode is heated in a vacuum, it gives up surface electrons. If an electric field is established between this heated electrode and another electrode so that the electrons are attracted to the other electrode, a current will flow through the vacuum. Electrons flow from the heated cathode to the cold anode. If the polarity is reversed, since there are no free electrons around the anode, no current will flow. This, then, is a vacuum-tube rectifier. If a third electrode, called a **control grid**, is placed between the cathode and the anode, the flow of electrons from the cathode to the anode can be controlled. This is a basic vacuum-tube amplifier. Additional grids have been placed between the cathode and anode to further enhance certain characteristics of the vacuum tube. In addition, multiple anodes and cathodes have been enclosed in a single tube for special applications such as radio signal converters.

If an inert gas, such as neon or argon, is introduced into the vacuum, conduction can be initiated from a cold electrode. The breakdown voltage is relatively stable for given gas and gas pressure and is in the range of 50 to 200 V. The **nixie** display tube is such a device. This tube contains 10 cathodes shaped in the form of the numerals from 0 to 9. If one of these cathodes is made negative with respect to the anode in the tube, the gas in the tube glows around that cathode. In this way each of the 10 numerals can be made to glow when the appropriate electrode is energized.

An **ignitron** is a vapor-filled tube. It has a pool of liquid mercury in the bottom of the tube. Air is exhausted from the enclosure, leaving only mercury vapor, which comes from the pool at the bottom. If no current is flowing, this tube will block voltage whether the anode is plus or minus with respect to the mercury-pool cathode. A small rod called an **ignitor** can form a cathode spot on the pool of mercury when it is withdrawn from the pool. The ignitor is pulled out of the pool by an electromagnet. Once the cathode spot has been formed, electrons will continue to flow from the mercury-pool cathode to the anode until the anode-to-cathode voltage is reversed. The operation of an ignitron is very similar to that of a thyristor. The anode and cathode of each device perform similar functions. The ignitor and gate also perform similar functions. The thyristor is capable of operating at much higher frequencies than the ignitron and is much more efficient since the thyristor has 15 V forward drop and the ignitron has 15 V forward drop. The ignitron has an advantage over the thyristor in that it can carry extremely high overload currents without damage. For this reason ignitrons are often used as electronic "crowbars" which discharge electrical energy when a fault occurs in a circuit.

DISCRETE-COMPONENT CIRCUITS

Several common rectifier circuits are shown in Fig. 15.2.8. The waveforms shown in this figure assume no line reactance. The presence of line reactance will make a slight difference in the waveshapes and the conversion factors shown in Fig. 15.2.8. These waveshapes are equally applicable for loads which are pure resistive or resistive and inductive.

Type	Circuit	Output voltage waveform	E_{dc} (avg)	Ripple fundamental frequency	% ripple	Peak inverse voltage
Half-wave 1φ		e_{load} / E_M / e_{ac}	0.318 E_M 0.45 E_{ac}	F	121	3.14 E_{dc}
Full-wave 1φ		e_{load} / E_M / e_{ac}	0.636 E_M 0.9 E_{ac}	2F	48	3.14 E_{dc}
Bridge 1φ		e_{load} / E_M / e_{ac}	0.636 E_M 0.9 E_{ac}	2F	48	1.57 E_{dc}
Half-wave 3φ		e_{load} / E_M / e_{ac}	0.827 E_M 1.17 E_{ac}	3F	18	2.09 E_{dc}

E_M = maximum value of e_{ac}
E_{ac} = effective value of e_{ac}
E_{dc} = average value of d-c load voltage
F = line frequency
% ripple = 100 x rms ripple / E_{dc}

Fig. 15.2.8 Comparison of rectifier circuits.

In a resistive load the current flowing in the load has the same waveshape as the voltage applied to it. For inductive loads, the current waveshape will be smoother than the voltage applied. If the inductance is high enough, the ripple in the current may be indeterminantly small. An approximation of the ripple current can be calculated as follows:

$$I = \frac{E_{dc}PCT}{200\pi fNL} \qquad (15.2.1)$$

where I = rms ripple current, E_{dc} = dc load voltage, PCT = percent ripple from Fig. 15.2.8, f = line frequency, N = number of cycles of ripple frequency per cycle of line frequency, L = equivalent series inductance in load. Equation (15.2.1) will always give a value of ripple higher than that calculated by more exact means, but this value is normally satisfactory for power-supply design.

Capacitance in the load leads to increased regulation. At light loads, the capacitor will tend to charge up to the peak value of the line voltage and remain there. This means that for either the single full-wave circuit or the single-phase bridge the dc output voltage would be 1.414 times the rms input voltage. As the size of the loading resistor is reduced, or as the size of the parallel load capacitor is reduced, the load voltage will more nearly follow the rectified line voltage and so the dc voltage will approach 0.9 times the rms input voltage for very heavy loads or for very small filter capacitors. One can see then that dc voltage may vary between 1.414 and 0.9 times line voltage due only to waveform changes when **capacitor filtering** is used.

Four different **thyristor rectifier circuits** are shown in Fig. 15.2.9. These circuits are equally suitable for resistive or inductive loads. It will be noted that the half-wave circuit for the thyristor has a rectifier across the load, as in Fig. 15.2.8. This diode is called a **freewheeling diode** because it freewheels and carries inductive load current when the thyristor is not conducting. Without this diode, it would not be possible to build up current in an inductive load. The gate-control circuitry is not shown in Fig. 15.2.9 in order to make the power circuit easier to see. Notice the location of the thyristors and rectifiers in the single-phase full-wave circuit. Constructed this way, the two diodes in series perform the function of a free-wheeling diode. The circuit can be built with a thyristor and rectifier interchanged. This would work for resistive loads but not for inductive loads. For the full three-phase bridge, a freewheeling diode is not required since the carryover from the firing of one SCR to the next does not carry through a large portion of the negative half cycle and therefore current can be built up in an inductive load.

Capacitance must be used with care **in thyristor circuits.** A capacitor directly across any of the circuits in Fig. 15.2.9 will immediately destroy the thyristors. When an SCR is fired directly into a capacitor with no series resistance, the resulting di/dt in the thyristor causes extreme local heating in the device and a resultant failure. A sufficiently high series resistor prevents failure. An inductance in series with a

(a) 1φ Half wave

(b) 1φ Full-wave bridge

(c) 3φ Half bridge

(d) 3φ Full bridge

Fig. 15.2.9   Basic thyristor circuits.

capacitor must also be used with caution. The series inductance may cause the capacitor to "ring up." Under this condition, the voltage across the capacitor can approach twice peak line voltage or 2.828 times rms line voltage.

The advantage of the thyristor circuits shown in Fig. 15.2.9 over the rectifier circuits is, of course, that the thyristor circuits provide variable output voltage. The output of the thyristor circuits depends upon the magnitude of the incoming line voltage and the phase angle at which the thyristors are fired. The control characteristic for the thyristor power supply is determined by the waveshape of the output voltage and also by the phase-shifting scheme used in the firing-control means for the thyristor. Practical and economic power supplies usually have control characteristics with some degree of nonlinearity. A representative characteristic is shown in Fig. 15.2.10. This control characteristic is usually given for nominal line voltage with the tacit understanding that variations in line voltage will cause approximately proportional changes in output voltage.

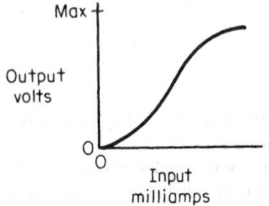

Fig. 15.2.10   Thyristor control characteristic.

Transistor amplifiers can take many different forms. A complete discussion is beyond the scope of this handbook. The circuits described here illustrate basic principles. A basic **single-stage amplifier** is shown in Fig. 15.2.11. The transistor can be cut off by making the input terminal sufficiently negative. It can be saturated by making the input terminal sufficiently positive. In the linear range, the base of an *n-p-n* transistor will be 0.5 to 0.7 V positive with respect to the emitter. The collector voltage will vary from about 0.2 V to V_c (20 V, typically). Note that there is a sign inversion of voltage between the base and the collector; i.e., when the base is made more positive, the collector becomes less positive. The resistors in this circuit serve the following functions. Resistor *R*1 limits the input current to the base of the transistor so that it is not harmed when the input signal overdrives. Resistors *R*2 and *R*3 establish the transistor's operating point with no input signal. Resistors *R*4 and *R*5 determine the voltage gain of the amplifier. Resistor *R*4 also serves to stabilize the zero-signal operating point, as established by resistors *R*2

and *R*3. Usual practice is to design single-stage gains of 10 to 20. Much higher gains are possible to achieve, but low gain levels permit the use of less expensive transistors and increase circuit reliability.

Figure 15.2.12 illustrates a basic **two-stage transistor amplifier** using complementary *n-p-n* and *p-n-p* transistors. Note that the first stage is

Fig. 15.2.11   Single-stage amplifier.

Fig. 15.2.12   Two-stage amplifier.

identical to that shown in Fig. 15.2.11. This *n-p-n* stage drives the following *p-n-p* stage. Additional alternate *n-p-n* and *p-n-p* stages can be added until any desired overall amplifier gain is achieved.

Figure 15.2.13 shows the **Darlington** connection of transistors. The amplifier is used to obtain maximum current gain from two transistors. Assuming a base-to-collector current gain of 50 times for each transistor, this circuit will give an input-to-output current gain of 2,500. This high level of gain is not very stable if the ambient temperature changes, but in many cases this drift is tolerable.

Fig. 15.2.13   Darlington connection.

Figure 15.2.14 shows a circuit developed specifically to minimize temperature drift and drift due to power supply voltage changes. The **differential amplifier** minimizes drift because of the balanced nature of the circuit. Whatever changes in one transistor tend to increase the output are compensated by reverse trends in the second transistor. The input signal does not affect both transistors in compensatory ways, of course,

Fig. 15.2.14   Differential amplifier.

and so it is amplified. One way to look at a differential amplifier is that twice as many transistors are used for each stage of amplification to achieve compensation. For very low drift requirements, matched transistors are available. For the ultimate in differential amplifier performance, two matched transistors are encapsulated in a single unit. **Operational amplifiers** made with discrete components frequently use differential amplifiers to minimize drift and offset. The operational amplifier is a low-drift, high-gain amplifier designed for a wide range of control and instrumentation uses.

Oscillators are circuits which provide a frequency output with no signal input. A portion of the collector signal is fed back to the base of the transistor. This feedback is amplified by the transistor and so maintains a sustained oscillation. The frequency of the oscillation is determined by parallel inductance and capacitance. The oscillatory circuit consisting of an inductance and a capacitance in parallel is called an *LC* **tank circuit.**

This frequency is approximately equal to

$$f = 1/2\pi \sqrt{CL} \qquad (15.2.2)$$

where f = frequency, Hz; C = capacitance, F; L = inductance, H. A 1-MHz oscillator might typically be designed with a 20-μH inductance in parallel with a 0.05-μF capacitor. The exact frequency will vary from the calculated value because of loading effects and stray inductance and capacitance. The **Colpitts** oscillator shown in Fig. 15.2.15 differs from the **Hartley** oscillator shown in Fig. 15.2.16 only in the way energy is fed back to the emitter. The Colpitts oscillator has a capacitive voltage

Fig. 15.2.15 Colpitts oscillator.   **Fig. 15.2.16** Hartley oscillator.

divider in the resonant tank. The Hartley oscillator has an inductive voltage divider in the tank. The **crystal oscillator** shown in Fig. 15.2.17 has much greater frequency stability than the circuits in Figs. 15.2.15 and 15.2.16. Frequency stability of 1 part in 10^7 is easily achieved with a crystal-controlled oscillator. If the oscillator is temperature-controlled by mounting it in a small temperature-controlled oven, the frequency

Fig. 15.2.17 Crystal-controlled oscillator.

stability can be increased to 1 part in 10^9. The resonant *LC* tank in the collector circuit is tuned to approximately the crystal frequency. The crystal offers a low impedance at its resonant frequency. This pulls the collector-tank operating frequency to the crystal resonant frequency.

As the desired operating frequency becomes 500 MHz and greater, **resonant cavities** are used as tank circuits instead of discrete capacitors and inductors. A rough guide to the relationship between frequency and resonant-cavity size is the wavelength of the frequency

$$\lambda = 300 \times 10^6/f \qquad (15.2.3)$$

where λ = wavelength, m; 300×10^6 = speed of light, m/s; f = frequency, Hz. The resonant cavities will be smaller than indicated by Eq. (15.2.3) because in general the cavity is either one-half or one-fourth wavelength and also, in general, the electromagnetic wave velocity is less in a cavity than in free space.

The operating principles of these devices are beyond the scope of this article. There are many different kinds of **microwave tubes** including **klystrons, magnetrons,** and **traveling-wave** tubes. All these tubes employ moving electrons to excite a resonant cavity. These devices serve as either oscillators or amplifiers at microwave frequencies.

Lasers operate at a frequency of light of approximately 600 THz. This corresponds to a wavelength of 0.5 μm, or, the more usual measure of visual-light wavelength, 5×10^3 Å. Most laser oscillators are basically a variation of the Fabry-Perot etalon, or interferometer.

Most light is disorganized insofar as the axis of vibration and the frequency of vibration are concerned. When radiation along different axes is attenuated, as with a polarizing screen, the light is said to be *polarized*. White light contains all visible frequencies. When white light is filtered, the remaining light is colored, or frequency-limited. A single color of light still contains a broad range of frequencies. Polarized, colored light is still so disorganized that it is difficult to focus the light energy into a narrow beam. Laser light is inherently a single-axis, single-frequency light.

Light-emitting diodes (LEDs) and semiconductor **laser diodes** (LDs) are specialty diodes that emit light and have the same current-voltage characteristic shown in Fig. 15.2.2. These diodes are made from various alloy compositions of compound semiconductor materials of gallium arsenide and indium phosphide that are capable of highly efficient light emission, in contrast to the silicon diodes described previously in this section. The devices are fabricated on 2-, 3-, 4-, and 6-in wafers via a layered deposition technique such as **molecular beam epitaxy** (MBE) or **metalorganic vapor phase epitaxy** (MOVPE) followed by standard semiconductor fabrication processes. Wavelength selection is controlled by engineering the alloy composition and thickness of the diode materials in the deposition process. Typical diode thickness is on the order of a couple of micrometres, with critical wavelength-control layers, known as quantum wells, as thin as a few nanometres. Diodes can be engineered to cover a broad wavelength range from ultraviolet (UV) through infrared (IR). LEDs and lasers have very similar materials, but differ in their fabrication process. For lasers, the fabrication process creates a Fabry-Perot cavity, or gain region, with two highly reflective mirrors, which are required for operation, as described in Section 19, Optics. LEDs have a simpler, yet similar, design that incorporates one or two lower-reflectivity mirrors that enhance the light output, but are not sufficiently reflective to create laser emission. Individual dice, on the order of 500 μm square, are cut from the wafers and then packed in a variety of methods to suit the particular application.

LEDs and lasers have become prevalent in our everyday lives. Early use of LED technology was for indicator lights on industrial electronic equipment. LEDs are now found nearly everywhere, including kitchen appliances, toys, home computers, remote control units, alarm clocks, and solar-powered yard lights. More recently, LEDs have been implemented in vehicle brake lights, traffic signals, and large outdoor video displays, and are starting to break into the overhead lighting market to replace fluorescent lighting. Advantages of LEDs over incandescent or fluorescent technologies are many:

1. Higher electrooptical conversion efficiency results in lower operating cost per lumen.

2. Long life, typically on the order of 10 years, and longer time between replacement, so costs for personnel and equipment for replacement are lower.

3. For transportation applications, improved safety because, with the high number of LEDs used in a stoplight or brake light, about 50 percent can burn out before replacement is required, whereas a single incandescent bulb failure makes the light inoperative.

The semiconductor laser has also become part of our everyday lives, and is more of an enabling technology, as opposed to the LED, which replaced incandescent technology in many of its applications. The laser diode has enabled novel products and services such as:

1. Fiber-optic communications
2. Read/write optical disk technology (CD/DVD)
3. Portable handheld bar code scanners
4. Laser pointers
5. Laser printers

Additionally, arrays of laser diodes are used for replacing older solid-state lasers in high-power applications such as industrial cutting, machining, and marking, with advantages of much smaller footprint and lower power consumption.

A radio wave consists of two parts, a **carrier** and an information signal. The carrier is a steady high frequency. The information signal may be a voice signal, a video signal, or telemetry information. The carrier wave can be modulated by varying its amplitude or by varying its frequency. **Modulators** are circuits which impress the information signal onto the carrier. A **demodulator** is a circuit in the receiving apparatus which separates the information signal from the carrier. A simple amplitude modulator is shown in Fig. 15.2.18. The transistor is base-driven with the carrier input and emitter driven with the information signal. The modulated carrier wave appears at the collector of the transistor.

Fig. 15.2.18 AM modulator.

An FM modulator is shown in Fig. 15.2.19. The carrier must be changed in frequency in response to the information signal input. This is accomplished by using a saturable ferrite core in the inductance of a Colpitts oscillator which is tuned to the carrier frequency. As the collector current in transistor $T1$ varies with the information signal, the saturation level in the ferrite core changes, which in turn varies the inductance of the winding in the tank circuit and alters the operating frequency of the oscillator.

Fig. 15.2.19 FM modulator.

The **demodulator** for an AM signal is shown in Fig. 15.2.20. The diode rectifies the carrier plus information signal so that the filtered voltage appearing across the capacitor is the information signal. Resistor $R2$ blocks the carrier signal so that the output contains only the information signal. An FM **demodulator** is shown in Fig. 15.2.21. In this circuit, the carrier plus information signal has a constant amplitude. The information is in the form of varying frequency in the carrier wave. If inductor $L1$ and capacitor $C1$ are tuned to near the carrier frequency but

Fig. 15.2.20 AM demodulator.

Fig. 15.2.21 FM discriminator.

not exactly at resonance, the current through resistor $R1$ will vary as the carrier frequency shifts up and down. This will create an AM signal across resistor $R1$. The diode, resistors $R2$ and $R3$, and capacitor $C2$ demodulate this signal as in the circuit in Fig. 15.2.20.

The waveform of the basic **electronic timing circuit** is shown in Fig. 15.2.22 along with a basic timing circuit. Switch $S1$ is closed from time $t1$ until time $t3$. During this time, the transistor shorts the capacitor and holds the capacitor at 0.2 V. When switch $S1$ is opened at time $t3$, the transistor ceases to conduct and the capacitor charges exponentially due to the current flow through resistor $R1$. Delay time can be measured to any point along this exponential charge. If the time is measured until time $t6$, the timing may vary due to small shifts in supply voltage or slight changes in the voltage-level detecting circuit. If time is

Fig. 15.2.22 Basic timing circuit.

measured until time $t4$, the voltage level will be easy to detect, but the obtainable time delay from time $t3$ to time $t4$ may not be large enough compared with the reset time $t1$ to $t2$. Considerations like these usually dictate detecting at time $t4$. If this time is at a voltage level which is 63 percent of V_c, the time from $t3$ to $t4$ is one time constant of $R1$ and C. This time can be calculated by

$$t = RC \qquad (15.2.4)$$

where t = time, s; R = resistance, Ω; C = capacitance, F. A timing circuit with a 0.1-s delay can be constructed using a 0.1-μF capacitor and a 1.0-MΩ resistor.

An improved timing circuit is shown in Fig. 15.2.23. In this circuit, the unijunction is used as a level detector, a pulse generator, and a reset means for the capacitor. The transistor is used as a constant current source for charging the timing capacitor. The current through the transistor is determined by resistors $R1$, $R2$, and $R3$. This current is adjustable by means of $R1$. When the charge on the capacitor reaches approximately 50 percent of V_c, the unijunction fires, discharging the capacitor and generating a pulse at the output. The discharged capacitor is then recharged by the transistor, and the cycle continues to repeat. The pulse rate of this circuit can be varied from one pulse per minute to many thousands of pulses per second.

Fig. 15.2.23 Improved timing circuit.

INTEGRATED CIRCUITS

Table 15.2.4 lists some of the more common physical packages for discrete component and integrated semiconductor devices. Although discrete components are still used for electronic design, **integrated circuits (ICs)** are becoming predominant in almost all types of electronic equipment. Dimensions of common dual in-line pin (DIP) integrated-circuit

Table 15.2.4   Semiconductor Physical Packaging

Signal devices	
Plastic	TO92
Metal can	TO5, TO18, TO39
Power devices	
Tab mount	TO127, TO218, TO220
Diamond case	TO3, TO66
Stud mount	
Flat base	
Flat pak (Hockey puck)	
Integrated circuits	
Dip (dual in-line pins)	(See Fig. 15.2.24)
Flat pack	
Chip carrier (50-mil centers)	

devices are shown in Fig. 15.2.24. An IC costs far less than circuits made with discrete components. Integrated circuits can be classified in several different ways. One way to classify them is by complexity. **Small-scale integration (SSI), medium-scale integration (MSI), large-scale integration (LSI),** and **very large scale integration (VLSI)** refer to this kind of classification. The cost and availability of a particular IC are more dependent upon the size of the market for that device than on the level of its internal complexity. For this reason, the classification by circuit complexity is not as meaningful today as it once was. The literature still refers to these classifications, however. For the purpose of this text, ICs will be separated into two broad classes: linear ICs and digital ICs.

The trend in IC development has been toward greatly increased complexity at significantly reduced cost. Present-day ICs are manufactured with internal spacings as low as 0.5 μm. The limitation of the contents of a single device is more often controlled by external connections than by internal space. For this reason, more and more complex combinations of circuits are being interconnected within a single device. There is also a tendency to accomplish functions digitally that were formerly done by analog means. Although these digital circuits are much more complex than their analog counterparts, the cost and reliability of ICs make the resulting digital circuit the preferred design. One can expect these trends will continue based on current technology. One can also

Number of pins	W	P	L	Approximate height
64	0.8	0.1	3.3	0.24
40	0.6	0.1	2.0	0.2
28	0.6	0.1	1.4	0.2
24	0.6	0.1	1.3	0.2
22	0.4	0.1	1.1	0.2
20	0.3	0.1	1.0	0.2
18	0.3	0.1	0.9	0.2
16	0.3	0.1	0.87	0.2
14	0.3	0.1	0.78	0.2
8	0.3	0.1	0.4	0.2

Fig. 15.2.24 Approximate physical dimensions of dual in-line pin (DIP) integrated circuits. All dimensions are in inches. Dual in-line packages are made in three different constructions—molded plastic, cerdip, and ceramic.

anticipate further declines in price versus performance. It has been demonstrated again and again that digital IC designs are much more stable and reliable than analog designs.

For years, the complexity of large-scale integrated circuits doubled each year. This meant that, over a 10-year period, the complexity of a single device increased by 2^{10} times, or over 1,000 times, and, over the period from 1960 to 1980 grew from one transistor on a chip to one million transistors on a chip. More recently the complexity increase has fallen off to only 1.7 times, or 1.7^{10}, or over 200 times in a 10-year period. Manufacturers also have developed application-specific integrated circuits (ASICs). These devices allow a circuit designer to design integrated circuits almost as easily as printed board circuits.

LINEAR INTEGRATED CIRCUITS

The basic building block for many linear ICs is the **operational amplifier.** Table 15.2.5 lists the basic characteristics for a few representative IC operational amplifiers. In most instances, an adequate design for an operational amplifier circuit can be made assuming an "ideal" operational amplifier. For an ideal operational amplifier, one assumes that it has infinite gain and no voltage drop across its input terminals. In most designs, feedback is used to limit the gain of each operational amplifier. As long as the resulting closed-loop gain is much less than the open-loop gain of the operational amplifier, this assumption yields results that are within acceptable engineering accuracy. Operational amplifiers use a balanced input circuit which minimizes input voltage offset. Furthermore, specially designed operational amplifiers are available which have extremely low input offset voltage. The input voltage must be kept low because of temperature drift considerations. For these reasons, the assumption of zero input voltage, sometimes called a "virtual ground," is justified. Figure 15.2.25 shows three operational amplifier circuits and the equations which describe their behavior. In this figure S

Table 15.2.5   Operational Amplifiers

Type	Purpose	Input bias current, nA	Input res., Ω	Supply voltage, V	Voltage gain	Unity-gain bandwidth, MHz
LM741	General purpose	500	2×10^6	+20	25,000	1.0
LM224	Quad gen. purpose	150	2×10^6	3 to 32	50,000	1.0
LM255	FET input	0.1	10^{12}	+22	50,000	2.5
LM444A	Quad FET input	0.005	10^{12}	+22	50,000	1.0

Fig. 15.2.25 Operational amplifier circuits.

is the **Laplace transform** variable. In these equations the input and output voltages are functions of S. The equations are written in the frequency domain. A simple transformation to steady variable-frequency behavior can be obtained by simply substituting cos ωt for the variable S wherever it appears in the equation. By varying ω, the frequency in radians per second, one can obtain the steady-state frequency response of the circuit.

The operational amplifier circuits shown in Figs. 15.2.25, 15.2.26, 15.2.27, and 15.2.28, show only the signal wires. There are additional connections to a dc power supply, and in some instances, to stabilizing circuits and guard circuits. These connections have been omitted for conceptual clarity.

Fig. 15.2.26 Difference amplifier.

One of the most useful analog integrated circuits is the **difference amplifier.** This is a balanced input amplifier that is a fundamental component in instrumentation and control applications. The difference amplifier is shown in Fig. 15.2.26. This amplifier maximizes the voltage gain for $V_{\text{output}}/V_{\text{dif}}$, the **difference gain,** and minimizes the voltage gain for $V_{\text{output}}/V_{\text{CM}}$, the **common-mode gain.** There are two resistors in the circuit shown as $R1$ and two resistors shown as $R2$. The resistance values of the two $R1$ resistors are equal, and similarly the resistance values of the two $R2$ resistors are equal. The difference gain is given by

$$\frac{V_{\text{output}}}{V_{\text{dif}}} = -\frac{R2}{R1} \quad (15.2.5)$$

Fig. 15.2.27 Instrumentation amplifier.

The ratio of the common-mode gain to the difference gain is called the **common-mode rejection ratio (CMRR).** In a well-designed difference amplifier, the CMRR is −80 dB. Stated another way, the common-mode gain is 10,000 times smaller than the difference gain. The difference amplifier is the most common input to instrumentation amplifiers and analog-to-digital converter circuits. Bridge-connected transducers have a common-mode voltage that is 100 times the difference voltage, or more. Such transducers require a difference amplifier. The common-mode gain is highly dependent on the matching of the $R1$ resistors and the $R2$ resistors. A 1 percent difference in these resistors causes the CMRR to be degraded to −30 or −40 dB.

An **instrumentation amplifier** is a high-grade difference amplifier. Although there are many implementations of the instrumentation amplifier, the three op-amp circuit is quite common, and will illustrate this device. Figure 15.2.27 shows the three op-amp instrumentation amplifier. This is a two-stage amplifier. The first stage is composed of amplifiers $X1$ and $X2$ and resistors $R1$, $R2$, and $R2$. The second stage is composed of amplifier $X3$ and resistors $R3$, $R3$, $R4$, and $R4$. The CMRR is usually much lower for the instrumentation amplifier, typically −120 dB. The differential mode gain is much higher for the instrumentation amplifier. For stability and frequency response considerations, the difference gain of a normal difference amplifier is usually 10 or less. For the instrumentation amplifier, the difference gain is typically 1,000. In the circuit shown in Fig. 15.2.27, the difference gain of the first stage is given by

$$\frac{V_{\text{internal}}}{V_{\text{dif}}} = \frac{2 \times R2 + R1}{R1} \quad (15.2.6)$$

The difference gain of the second is given by

$$\frac{V_{\text{output}}}{V_{\text{internal}}} = -\frac{R4}{R3} \quad (15.2.7)$$

The overall gain for the amplifier is the product of the individual stage gains. In the typical case the gain of the first stage is 100, and for the second stage is 10, giving an overall gain of 1,000. Integrated circuit instrumentation amplifiers are available which include all of the circuitry in Fig. 15.2.27. The resistors are matched so that the circuit designer does not have to contend with component matching.

A word about cost is in order. In 1994, high-grade operational amplifiers cost $0.50 for four op-amps in a single IC chip, or about $0.125 for each amplifier. The instrumentation amplifier is somewhat more expensive, but can be obtained for less than $5.00.

Filtering of electronic signals is often required. A filter passes some frequencies and suppresses others. Filters may be classed as low pass, high pass, band pass, or band stop. Figure 15.2.28 shows a low-pass active filter circuit which is a modification of the Sallen-Key circuit. In this circuit, the two resistors labeled R are matched and the two capacitors labeled C are matched. The frequency response of this circuit is given by its transfer function:

$$\frac{V_{\text{output}}}{V_{\text{input}}} = \frac{1}{S^2 + \dfrac{\omega_N}{Q}S + \omega_N^2} \tag{15.2.8}$$

Fig. 15.2.28   Modified Sallen-Key filter circuit.

In this equation, ω_N is the resonant frequency of this circuit, in radians per second, and Q is a quality factor that indicates the amount of amplitude increase at the resonant frequency. For the circuit shown in Fig. 15.2.28, ω_N is given by

$$\omega_N = \frac{1}{RC} \tag{15.2.9}$$

and Q is given by

$$Q = \frac{1}{2 - (\text{RB/RA})} \tag{15.2.10}$$

The filter shown in Fig. 15.2.28 is a two-pole low-pass filter. Higher-order filters—four pole, six, etc.—can be constructed by cascading sections of two-pole filters. Thus, a six-pole low-pass filter would consist of three circuits of the configuration shown in the figure. The components R, R, C, C, RA, and RB would vary in each filter section.

Continuing with the low-pass filter, there are three common variations in filter response: Butterworth response, Chebychev response, and elliptical response. The Butterworth filter has no ripple in either the pass band or the stop band. The Chebychev filter has equal ripple variations in the pass band, but is flat in the stop band, and the elliptical filter has equal ripple variations in both the pass band and the stop band. There is a transition band of frequencies between the pass band and the stop band. The Butterworth filter has the greatest transition band. The Chebychev response has a sharper cutoff frequency characteristic than the Butterworth response, and the elliptical response has the sharpest transition from pass band to stop band. The circuits shown in Figs. 15.2.27 and 15.2.28 can be used to realize Butterworth and Chebychev filters. The elliptical filter requires a more complex circuit.

The high-pass filter can be formulated by substituting $1/S$ for S in Eq. (15.2.8). The circuit configuration for a high-pass filter is shown in Fig. 15.2.29. Notice that only the position of the Rs and Cs have changed. In general, the values of Rs, Cs, RA, and RB will change for the high-frequency filter, so it would be inappropriate to design a low-pass filter and simply reverse the positions of the Rs and Cs.

Fig. 15.2.29   High-pass filter section.

Band-pass and band-stop filters can be made by combinations of low-pass and high-pass filter sections. Figure 15.2.30a shows the diagram configuration of filters to realize a band-pass filter. The low-pass filter would be designed for the high-frequency transition, and the high-pass filter would be designed for the low-frequency transition. Figure 15.2.30b shows the block diagram configuration for a band-stop filter. The band-stop filter and the low-pass filter would be designed for the low-frequency transition, and the high-pass filter would be designed for the high-frequency transition. Band-pass and band-stop filters may also be realized by means of special high-Q circuits. A more complete discussion of filter technology can be found in the references at the beginning of this section.

Table 15.2.6 lists some typical linear ICs, most of which contain operational amplifiers with additional circuitry. The **voltage comparator** is an operational amplifier that compares two input voltages, $V1$ and $V2$. Its output voltage is positive when $V1$ is greater than $V2$, and negative when $V2$ is greater than $V1$. The **sample-and-hold** circuit samples an analog input voltage at prescribed intervals, which are determined by an input clock pulse. Between clock pulses the circuit holds the sample voltage level. This circuit is useful in converting from an analog voltage to a digital number whose value is proportional to the analog voltage. An **analog-to-digital (A/D) converter** is a signal-converting device which changes several analog signals into digital signals. It consists of an input difference amplifier, an analog time multiplexer, a sample-and-hold amplifier, a digital decoder, and the necessary logic to interface with a digital computer. The A/D converter is programmable, in that a computer can set up the device for the requisite number of input channels and whether these input channels are differential inputs or single-ended inputs. The A/D converter typically takes eight differential analog input signals and converts them to 10-, 12-, or 16-bit digital signals. Typical A/D converters convert signals at a rate of 200,000 samples per second. Other types of A/D converters, such as flash converters, can convert over 10,000,000 samples per second.

A companion circuit to the A/D converter is the **digital-to-analog (D/A) converter,** which is used to convert from digital signals to analog signals. This is also a programmable device, but in general the D/A converter is less complex than is the A/D converter. **Voltage-regulators** and **voltage references** are electronic circuits which create precision dc voltage sources. The voltage reference is more precise than the voltage regulator. The **voltage-controlled oscillator** is a circuit which converts from a dc signal to a proportional ac frequency. The output of a voltage-controlled oscillator is usually a rectangular ac wave rather than a sine wave. The **NE555 timer/oscillator** is a general-purpose timer/oscillator which has been integrated into a single IC chip. It can function as a **monostable multivibrator,** a **free-running multivibrator,** or as a **synchronized multivibrator.** It can also be used as a **linear ramp generator,** or for time delay or sequential timing applications.

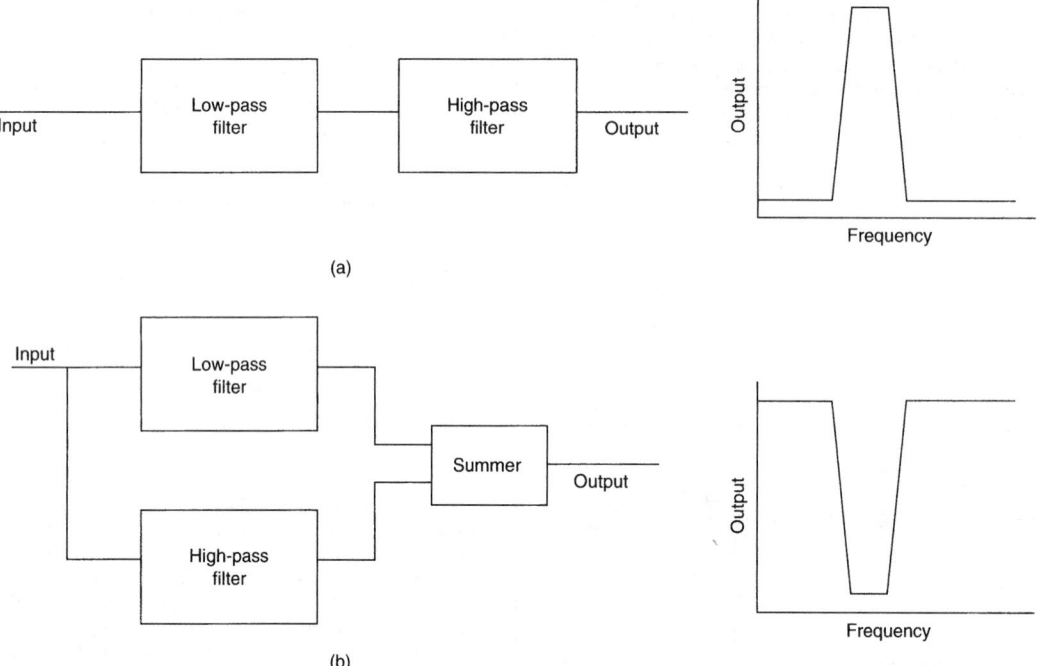

Fig. 15.2.30 (*a*) Band-pass and (*b*) band-stop filters.

Table 15.2.6   Linear Integrated-Circuit Devices

Operational amplifier	Voltage comparator
Sample and hold	Digital-to-analog converter
Analog-to-digital converter	Voltage reference
Voltage regulator	NE555 Timer/oscillator
Voltage-controlled oscillator	

Table 15.2.7 lists linear ICs that are used in audio, radio, and television circuits. The degree of complexity that can be incorporated in a single device is illustrated by the fact that a complete AM-FM radio circuit is available in a single IC device. The **phase-locked loop** is a device that is widely utilized for accurate frequency control. This device produces an output frequency that is set by a digital input. It is a highly accurate and stable circuit. This circuit is often used to demodulate FM radio waves.

Table 15.2.8 lists linear IC circuits that are used in telecommunications. These circuits include digital circuits within them and/or are used with digital devices. Whether these should be classed as linear ICs or digital

Table 15.2.7   Audio, Radio, and Television Integrated-Circuit Devices

Audio amplifier	Tone-volume-balance circuit
Dolby filter circuit	Phase-locked loop (PLL)
Intermediate frequency circuit	AM-FM radio
TV chroma demodulator	Digital tuner
Video-IF amplifier-detector	

Table 15.2.8   Telecommunication Integrated-Circuit Devices

Radio-control transmitter-encoder
Radio-control receiver-decoder
Pulse-code modulator–coder-decoder (PCM CODEC)
Single-chip programmable signal processor
Touch-tone generators
Modulator-demodulator (modem)

ICs may be questioned. Several manufacturers include them in their linear device listings and not with their digital devices, and for this reason, they are listed here as linear devices. The radio-control **transmitter-encoder** and **receiver-decoder** provide a means of sending up to four control signals on a single radio-control frequency link. Each of the four channels can be either an on-off channel or a **pulse-width-modulated (PWM)** proportional channel. The **pulse-code modulator–coder-decoder (PWM CODEC)** is typical of a series of IC devices that have been designed to facilitate the design of digital-switched telephone circuits.

Integrated circuits are normally divided into two classes: linear or analog ICs and digital ICs. There are a few hybrid circuits which have both analog and digital characteristics. Some representative hybrid integrated circuits are listed in Table 15.2.9, and shown in Fig. 15.2.31. In the **superdiode,** there are two modes of operation. When the input

Table 15.2.9   Hybrid Circuits

Superdiode	Limiters
Schmitt trigger circuits	Dead-band circuits
Switched-capacitor filters	Logarithmic amplifiers

signal is positive, the output of the op-amp is negative, causing diode $D2$ to conduct and diode $D1$ to block current flow. When the input is negative, diode $D2$ blocks and diode $D1$ conducts. The output of the circuit is zero when the input is negative, and the output is proportional to the input when the input is positive. A normal diode has 0.5 to 1.0 V of forward voltage drop. This circuit is called a super diode because its switching point is at zero volts. There is a voltage inversion from input to output, but this can be reversed by means of an additional op-amp inverter following the super diode. The action of the circuit can be reversed by reversing diodes $D1$ and $D2$. The **limiter** circuit, shown in Fig. 15.2.31*b*, provides linear operation at low output voltages, either positive or negative, and neither diode is conducting. When the input voltage goes sufficiently negative, diode $D1$ begins to conduct, limiting the negative output of the op-amp. Similarly when the input goes sufficiently positive, diode $D2$ begins to conduct, limiting the positive

Fig. 15.2.31 Hybrid circuits. (*a*) Superdiode circuit; (*b*) limiter circuit; (*c*) Schmitt trigger circuit.

output of the op-amp. The **Schmitt trigger** circuit, shown in Fig. 15.2.31*c*, has positive feedback to the op-amp, through resistors $R1$ and $R2$. If the amplifier output is initially negative, the output will not change until the input becomes as negative as the noninverting terminal of the op-amp. When the input becomes slightly more negative, the op-amp output will suddenly switch from negative to positive due to the positive feedback. The op-amp will remain in this condition until the input becomes sufficiently positive, causing the op-amp to switch to a negative output.

The **dead-band** circuit is similar to the limiter circuit except that for input voltages around zero, there is no output voltage. Above a threshold input voltage, the output voltage is proportional to the input voltage. The logarithmic amplifier relies upon the fact that the voltage drop across a diode causes logarithmically varying current to flow through

the diode. This relationship is remarkably true for current changes of 10^6 to 1. The logarithmic amplifiers can have their outputs added together to effect a multiplication of the input signals.

DIGITAL INTEGRATED CIRCUITS

The basic circuit building block for digital ICs is the gate circuit. A gate is a switching amplifier that is designed to be either on or off. (By contrast, an operational amplifier is a proportional amplifier.) For 5-V logic levels, the gate switches to a 0 whenever its input falls below 0.8 V and to a 1 whenever its input exceeds 2.8 V. This arrangement ensures immunity to spurious noise impulses in both the 0 and the 1 state.

Several representative **transistor-transistor-logic (TTL) gates** are listed in Table 15.2.10. Gates can be combined to form logic devices of two fundamental kinds: combinational and sequential. In **combinational logic,** the output of a device changes whenever its input conditions change. The basic gate exemplifies this behavior.

Table 15.2.10   Digital Integrated-Circuit Devices

Type 54/74*	No. circuits per device	No. inputs per device	Function
00	4	2	NAND gate
02	4	2	NOR gate
04	6	1	Inverter
06	6	1	Buffer
08	4	2	AND gate
10	3	3	NAND gate
11	3	3	AND gate
13	2	4	Schmitt trigger
14	6	1	Schmitt trigger
20	2	4	NAND gate
21	2	4	AND gate
30	1	8	NAND gate
74	2		D flip-flop
76	2		JK flip-flop
77	4		Latch
86	4	2	EXCLUSIVE OR gate
174	6		D flip-flop
373	8		Latch
374	8		D flip-flop

NOTE: Example of device numbers are 74LS04, 54L04, 5477, and 74H10. The letters after the series number denote the speed and loading of the device.
* 54 series devices are rated for temperatures from − 55 to 125°C. 74 series devices are rated for temperatures from 0 to 70°C.

A number of gates can be interconnected to form a **flip-flop** circuit. This is a bistable circuit that stays in a particular state, a 0 or a 1 state, until its "clock" input goes to a 1. At this time its output will stay in its present state or change to a new state depending upon its input just prior to the clock pulse. Its output will retain this information until the next time the clock goes to a 1. The flip-flop has memory, because it retains its output from one clock pulse to another. By connecting several flip-flops together, several sequential states can be defined permitting the design of a **sequential logic** circuit.

Table 15.2.11 shows three common flip-flops. The **truth table,** sometimes called a **state table,** shows the specification for the behavior of each circuit. The present output state of the flip-flop is designated $Q(t)$. The next output state is designated $Q(t + 1)$. In addition to the truth table, the Boolean algebra equations in Table 15.2.11 are another way to describe the behavior of the circuits. The **JK flip-flop** is the most versatile of these three flip-flops because of its separate J and K inputs. The **T flip-flop** is called a *toggle*. When its T input is a 1, its output toggles, from 0 to 1 or from 1 to 0, at each clock pulse. The **D flip-flop** is called a *data cell*. The output of the D flip-flop assumes the state of its input at each clock pulse and holds this data until the next clock pulse. The JK flip-flop can be made to function as a T flip-flop by applying the T input to both the J and K input terminals. The JK flip-flop can be made to function as a D flip-flop by applying the data signal to the J input and applying the inverted data signal to the K input. Some common IC flip-flops are listed in Table 15.2.10.

Table 15.2.11   Flip-Flop Sequential Devices

Name		Graphic symbol	Algebraic function	Truth table			
JK flip-flop	Clock	S / J Q / < / K $\overline{Q}$ / R	$Q(t + 1) = JQ'(t) + K'Q(t)$	J K Q(t) | Q(t + 1) 0 X 0 | 0 1 X 0 | 1 X 0 1 | 1 X 1 1 | 0			
T flip-flop	Clock	T Q / < / $\overline{Q}$	$Q(t + 1) = TQ'(t) + T'Q(t)$	T Q(t) | Q(t + 1) 0 0 | 0 0 1 | 1 1 0 | 1 1 1 | 0			
D flip-flop	Clock	D Q / < / $\overline{Q}$	$Q(t + 1) = D$	D Q(t) | Q(t + 1) 0 0 | 0 0 1 | 0 1 0 | 1 1 1 | 1			

Various types of gates are shown in Table 15.2.12. Combinational logic defined by means of these various gates is used to define the input to flip-flops, which serve as memory devices. At each clock pulse, these flip-flops change state in accordance with their respective inputs. These new states are retained in the flip-flop and also applied to the gates. The output of the gates change (with only a small delay due to propagation time), and at the next clock pulse the flip-flops will change to the next state as directed by the gates.

The gates shown in Table 15.2.12 have only two inputs. As was seen in Table 15.2.10, gates may have as many as eight inputs. In the case of an AND **gate**, all its inputs must be 1 in order for its output to be a 1. For an OR gate, if any of its inputs become a 1, then its output will become a 1. The NAND and NOR gates function in a similar way.

Boolean algebra is the branch of mathematics used to analyze logic circuits. Boolean algebra has two operators: $\cdot$, which indicates an AND operation, and $+$, which indicates an OR operation: The $=$ has the same meaning in Boolean algebra as in ordinary algebra. The symbol for "X not" is X' (or sometimes $\overline{X}$). The identity element for the AND operation is 0; the identity element for the OR operation is 1. The rules for Boolean algebra can be derived from set theory applied to a system in which only two numbers exist, i.e., zero and one. These rules are summarized in Huntington's postulates and DeMorgan's theorem and are listed in Table 15.2.13.

Table 15.2.12   Combinational Gate Logic

Name	Graphic symbol	Algebraic function	Truth table
AND	x / y — Q	$Q = xy$	x y | Q 0 0 | 0 0 1 | 0 1 0 | 0 1 1 | 1
OR	x / y — Q	$Q = x + y$	x y | Q 0 0 | 0 0 1 | 1 1 0 | 1 1 1 | 1
Inverter	x — Q	$Q = x'$	x | Q 0 | 1 1 | 0
Buffer	x — Q	$Q = x$	x | Q 0 | 0 1 | 1
NAND	x / y — Q	$Q = (xy)'$	x y | Q 0 0 | 1 0 1 | 1 1 0 | 1 1 1 | 0
NOR	x / y — Q	$Q = (x + y)'$	x y | Q 0 0 | 1 0 1 | 0 1 0 | 0 1 1 | 0
EXCLUSIVE-OR	x / y — Q	$Q = xy' + x'y$ $= x + y$	x y | Q 0 0 | 0 0 1 | 1 1 0 | 1 1 1 | 0

Table 15.2.13   Rules for Boolean Algebra

$X + 0 = X$	$X \cdot 1 = X$
$X + 1 = 1$	$X \cdot X' = 0$
$X + X' = 1$	$X \cdot X = X$
$(X')' = X$	$X \cdot 0 = 0$
$X + Y = Y + X$	$X \cdot Y = Y \cdot X$
$X + (Y + Z) = (X + Y) + Z$	$X \cdot (YZ) = (X \cdot Y)Z$
$X + X \cdot Y = X$	$X \cdot (X + Y) = X$

DeMorgan's theorem:

$(X + Y)' = X' \cdot Y'$	$(X \cdot Y)' = X' + Y'$

To facilitate the analysis of digital circuits and to aid in the application of the rules given in Table 15.2.13, **Karnaugh maps** are used. Typical two-variable and four-variable Karnaugh maps are shown in Fig. 15.2.32 along with the algebraic expressions represented by each map.

Fig. 15.2.32   Karnaugh maps of typical logic functions.

Table 15.2.14   Large-Scale Digital Integrated Circuits

Adder	Accumulator
Arithmetic logic unit	Parity generator-checker
Shift register	Encoder
Multiplexer	Decoder
Demultiplexer	Counters
Display controller-driver	

Looking at Fig. 15.2.32b and realizing each of the ones individually would lead to the Boolean expression:

$$OUT = W'X'YZ + WXYZ + WX'YZ + WXYZ' \quad (15.2.11)$$

Realizing this expression directly would require four four-input AND gates, and one four-input OR gate. The reduced expression

$$OUT = WXY + X'YZ \quad (15.2.12)$$

requires only two three-input AND gates and one two-input OR gate, a simpler and less expensive logic circuit.

Some of the more common large-scale integrated circuits are listed in Table 15.2.14. The adder can add two binary numbers together and output the result. For a single bit adder which adds a bit from word A, a bit from word B, and a carry bit, C, the Boolean expressions for the two outputs of the adder are

$$SUM = AB'C' + A'BC' + A'B'C + ABC \quad (15.2.13)$$
$$CO = AB + BC + AC \quad (15.2.14)$$

Adders are available, not for a single bit, but rather for 4 bits, 8 bits, or 16 bits. There is only one carry bit for the whole device. The devices in Table 15.2.14 will be described in words, but the specifications for each of these devices has a truth table as a part of the product description. An arithmetic logic unit (ALU) or accumulator can perform several simple logical or arithmetic operations, such as add, complement, shift left, and shift right. In addition, the set of instructions which command these operations will also be decoded by the ALU. As one might expect, the Boolean expression for a 16-bit or 32-bit ALU can be formidable. A parity generator/checker is a circuit that can make a cyclic redundancy code which can be used to check the accurate transmission of data. Shift registers perform only the shifting functions of ALUs. Counters are made in many configurations—up-down counters, binary-coded-decimal counters, Johnson counters, etc. A decoder can best be explained by a simple example. A three- to eight-line decoder will be described. The three input lines can have eight binary variations, 000, 001, 010, 011, 100, 101, 110, and 111. For each of these input conditions one, and only one, of the output lines will be set. The input signal is said to have been decoded. The encoder performs the related operation of encoding from eight to three lines. Depending on which one input line is set, an appropriate output bit pattern is set. Encoders and decoders are available in 2 to 4, 3 to 8, and 4 to 16 line configurations. The multiplexer/demultiplexer performs a similar function. The multiplexer has a data input terminal and control input terminal and several data output terminals. It switches a single train of information, ones and zeros, on an input line, and routes this information to one of many output lines depending on the bit pattern placed on a control input. The demultiplexer performs the reverse operation. Multiplexers/demultiplexers are also configured in 2 to 4, 3 to 8, and 4 to 16 line devices.

There are a wide variety of display controllers and drivers. These are devices which interface between digital circuits and displays which humans can use. Light-emitting diodes, liquid crystals, and fluorescent displays have different drive requirements. The digital device may perform in binary, but the display requires decimal. In many cases, the display driver is separate from the display itself, but more recently the display driver has been incorporated in the display.

The complexity of integrated circuits is growing continually. The simple gates and flip-flops discussed so far are still used to some extent. Logically, these gates and flip-flops can be integrated into larger single-chip circuits called *large-scale integrated circuits* (**LSI**). Combinations of these circuits are grouped together on single chips called *very large scale integrated circuits* (**VLSI**). VLSI chips contain 500,000 to more than 2,000,000 gates. VLSI chips have enabled the explosive growth of personal computers, by allowing ever-increasing functionality at lower cost. These circuits are custom-designed VLSI circuits which are developed for large-volume applications. Bridging the gap between LSI and custom VLSI circuits are a series of linear and digital user-designed circuits.

The first of these user-designed circuits is called programmable logic arrays (**PLAs**) or programmable array logic (**PAL**). The PLA grew out of the programmable read-only memory (**PROM**). A programmable read-only memory is a computer memory which can be programmed one time, and then can be read many times. The PROM consists of a series of gates, initially all ones, each of which can be electrically altered to be a zero. A PROM might be logically organized to be 1 bit wide by 64K bits long, or 4 bits by 16K bits, 8 bits by 8K bits. The example just given is a series of 64K-bit PROMs. PROMs are available up to 4M bits on a single chip. The PLA is also field programmable, but has a more limited programming capability than the PROM. The limited capability makes it less costly to program. The PLA is adaptable to replacing Boolean functions in both combinational logic and sequential logic circuits. The PLA can replace 10 to 40 gate and flip-flop chips, resulting in a simpler, more reliable, less costly design.

There are two other approaches to application-specific integrated circuits (**ASICs**). These are field-programmable gate arrays (**FPGAs**) and standard cells. The FPGA is sometimes referred to as a "sea of gates." The chip is designed with a series of identical gates, but without any interconnections between them. The user specifies the interconnections, and thus makes the chip specific to the application. The manufacture of the chip cannot be completed until the interconnections have been specified, however. There was initially some concern that the circuit designer would be losing design control, but this has largely disappeared now that alternative sources of supply are available. The customizing production drawings and specifications can be pulled from one manufacturer, and sent to another. The standard cell device is similar to an FPGA, except that some interconnections are specified to connect gates to form subcircuits. The design by the user consists of selecting the individual cells, AND gates, OR gates, flip-flops, etc. and specifying the interconnections of these standard cells. The standard cells may be as simple as an AND gate or as complex as a 16K-bit memory or a microprocessor. The standard cell also allows the user to specify light-current cells or heavy-current cells. The standard cell gives the user much more flexibility in the design. Standard cells have also been extended to include linear circuits, op-amps, difference amplifiers, limiters, etc. Some standard cell chips are suitable for hybrid digital and linear circuits on a single chip.

Two problems arise as more complex circuitry is incorporated in ASICs, and also in microprocessors (which will be discussed in the next section). These devices require more input and output data pins than are available in dual in-line packages (see Fig. 15.2.24), and they generate more heat on the chip. One solution to the pin problem is the pin array package, shown in Fig. 15.2.33. These packages still maintain 0.1-in spacing between pins, but the pin count has been increased to 132 and 208 in the examples shown. The package size has increased from dual in-line packages, and the force necessary to install and remove the package is not appreciably greater than for dual in-line packages. There are additional microcircuit packages at various stages of development and standardization that have up to 500 pins and 0.05-in pin spacing.

Higher-speed and higher-density digital electronics lead to higher heat dissipation in devices. One solution is to provide heat sinks and fans to remove heat. A more recent development is the **heat pipe**. The heat pipe consists of boiling fluid/vapor refrigerant in a closed chamber. The semiconductor boils the fluid to change it to the vapor state. A pipe conducts the vapor to a vapor-to-air heat exchanger, where the vapor is condensed. The heat pipe also provides a return path for the fluid to return to the semiconductor. An attractive feature of the heat pipe is that there is no requirement for a mechanical compressor.

A further problem for complex electronic circuits is proof of design and proof of manufacture testing. To adequately test a device, it is

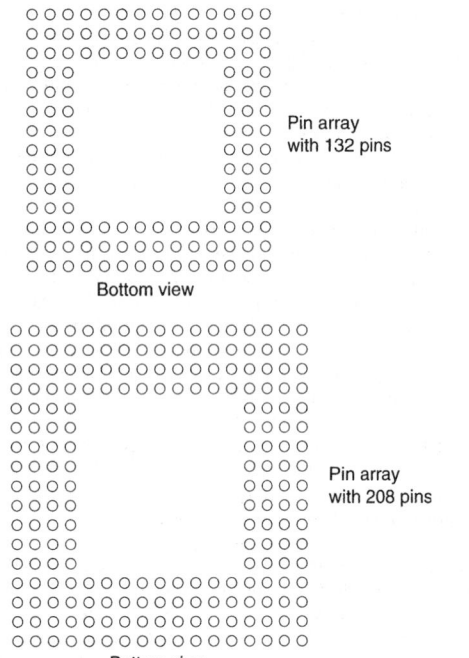

Fig. 15.2.33   Typical pin grid array packages. (All pins are on a 0.1 in grid.)

necessary to have access to additional test points in the circuit, beyond those needed to interface in actual operation. With a limited number of output pins and highly complex internal logic, the design of the circuit must include a fairly long test vector that will excite the various circuit functions. In addition, a fairly long results vector will be generated by the circuit. One approach to test design is called **boundary scan** design, described in specification IEEE 1149.1 of the Institute of Electrical and Electronics Engineers. This specification outlines a four- or five-wire test circuit, and the internal circuitry, which allows a test vector to be entered serially onto the chip, and allows a results vector to be serially unloaded from the chip. Internal circuitry is required to accept each bit of the input test vector and to generate each bit of the results vector. The serial transfer is a tradeoff between the time required to enter and retrieve the test vectors and the cost of additional input/output pins for the device. Serial transfer would be unacceptable for normal operations, but is all right for test purposes.

COMPUTER INTEGRATED CIRCUITS

One of the devices which has become feasible as a result of VLSI is the **microprocessor.** A complete computer can be built in a single IC device. In most cases, however, several devices are employed to build a complete computer system. In most computers, the cost of the microprocessor is negligible compared with the total system cost. Not long ago, the **central processing unit (CPU)** was the most expensive part of a computer. The microprocessor provides the total CPU function at a fraction of the earlier cost.

The power of a microprocessor is a function of its clock speed and the size and number of its registers. Clock rates vary from 1 to 200 MHz. Common register sizes are 8, 16, 24, 32, and 64 bits. Most personal computers currently use 32-bit registers. Commonly microprocessors have 12 to 16 registers. Other factors affect processor power and speed. Most microprocessors have a data bus, an address bus, and a set of control and power supply terminals. The number of each of these buses may vary. A small inexpensive microprocessor might have 8 data lines, 16 address lines, and 8 control/power supply lines. A large sophisticated microprocessor might have 64 data lines, 24 address lines, and 20

control/power supply lines. Some microprocessors have 32-bit internal registers and only 16 data bit terminals. The 16 data terminals are time-multiplexed to allow the complete 32-bit word to be entered onto the microprocessor chip in two read cycles. Once in the microprocessor, the 32-bit registers communicate with each other over a data bus which is a full 32 bits wide. This scheme reduces the number of terminals on the microprocessor, and reduces its cost, with a slight timing penalty.

Clock speed is limited by clock skew and parasitic capacitance. There is a finite time required for electric pulses to move along wires and for gates to change state. Typically electric pulses move along wires at about half the speed of light. Gates change state in a few nanoseconds. Parasitic capacitance is the stray capacitance that exists between any circuit and ground. When the gate changes state, this capacitance must be charged. The charging current increases linearly with switching rate and voltage. Because of the small circuit components and short wire lengths, the parasitic capacitance is much smaller for circuits on a chip compared with circuits on printed boards. Modern microprocessors use an on-chip clock rate that is 3 times as fast as the off-chip clock.

The **microcontroller** is a device similar to the microprocessor, in that it is also a CPU. Microcontrollers are configured for controlling applications. In general, they have greater input and output capability. It is very easy to control a single output bit for turning on an electric relay. With a microprocessor, 8 bits are switched together, but with a microcontroller individual bits can be switched. Some microcontrollers have 256 internal registers, rather than the 15 to 20 of a microprocessor. Microcontrollers are often stand-alone single-chip devices, and are a less expensive system than a microcomputer. Typical applications for microcontrollers are microwave oven controllers, washer and dryer controls, and automobile engine controls.

Memory is an important part of a computer system. Memory can be divided into two broad classes: volatile and nonvolatile. Volatile memory forgets what it contains when the power is removed from it. Nonvolatile memory retains its contents when power is removed. Random-access memory (**RAM**) is volatile memory that is sometimes called *main memory.* This is the memory that the CPU uses for most of its operations. RAM is a semiconductor memory. RAM may be dynamic memory or static. In a static RAM the memory elements are flip-flops, and retain their memory for as long as power is applied. Dynamic RAMs are also semiconductor memories, but hold the information as charges on capacitors. This charge will leak off even though the power is still applied to the memory. To retain the memory in a dynamic RAM, it is necessary to continually refresh the memory. This is typically done every 10 ms. Dynamic RAMs are much cheaper and have less loss than static RAMs. Large banks of memory are made with dynamic RAMs. Small memories, or memories with special requirements, are made with static RAMs. Typical dynamic RAMs are 256K, 1M, and 4M bytes. Typical static RAM may be as small as 1K bytes and can be as large as 1M bytes.

Nonvolatile memory can take several forms: ROMs, disk storage, compact disks, and tapes. A read-only memory (ROM) is a set of flip-flops which are programmed to take a particular state. This programming may be installed at the time of manufacture. Some devices are programmable by the user (**PROMs**). A PROM is typically all ones as shipped. The individual memory cells may be programmed to a zero or left a one. Programming memory cell is done by applying a high voltage to a programming pin when that cell's address is set. Normally the ROM or PROM is written once and read many times. An **EPROM** is an erasable, programmable read-only memory. The EPROM is programmed in the same way as a PROM. To erase the memory, the EPROM is placed under a high-intensity ultraviolet light. When the EPROM is erased, all of the memory is lost, so the entire program must be reentered.

Disks and tapes are magnetic storage devices. When power is removed from these devices, the memory is retained due to magnetic retentivity. Magnetic memory can only be read (or written) by moving the magnetic medium under a read/write head. **Compact disks** are a form of read-only memory. In the compact disk, the information is stored optically rather than magnetically.

The magnetic tape is a removable nonvolatile memory device. Hard disks are usually not removable. Floppy disks are removable, but for large quantities of data, the tape medium is cheaper and more convenient. Tape is either in a cartridge or on a reel. Newer tape cartridges have large volumes, 50M bytes to 1,000M bytes. The trend is that cartridge tape is replacing reel tape as a storage means. Magnetic tape can store computer data for over a year, and with a periodic refreshing can store data indefinitely.

In addition to the CPU and storage devices, other integrated circuits have been designed for computer systems. These are listed in Table 15.2.15. The **tristate buffer** is a switching amplifier which can be a zero, a one, or high impedance. When a computer bus line can be driven by several devices, all of the devices are high impedance except the one which is controlling the bus line at that time. Typically eight tristate buffers are contained in one IC package. The **tristate transceiver** is similar to the tristate buffer except that it allows the bus signal to be

Table 15.2.15   Digital Computer Integrated-Circuit Devices

Tristate buffer	Tristate transceiver
Parallel interface adapter	Programmable peripheral interface
Universal asynchronous transmitter-receiver	Floppy/hard disk controller
	Digital-to-analog converter
Analog-to-digital converter	Direct memory access controller
Programmable timer	
Programmable interrupt controller	

received as well as written. The **parallel interface adapter** is a device that permits a parallel output device, such as a printer, to be connected to a computer bus. There are several variations of the parallel interface adapter. Some control two, three, four, or more parallel devices. Some have been designed to work with Intel microprocessors, others to work with Motorola microprocessors. The **programmable peripheral interface** is a parallel interface adapter designed for use with Intel computers. The **universal asynchronous receiver-transmitter (UART)** is a device which will allow a serial output device to be connected to a computer bus. The UART has a parallel connection to the computer bus, but provides a serial output to the peripheral device. The UART is programmable, so that it can be set for 7-bit or 8-bit transmission. It can be set for even or odd parity checking, and the transmitting/receiving speed can be set. The **floppy/hard disk controller** is a programmable device which can handle the interfacing requirements for floppy and/or hard disks. Analog-to-digital and digital-to-analog converters allow the digital computer to interface with analog signals. The **programmable timer** is a timing circuit which can be programmed by the computer either to generate an interrupt periodically, or to generate an interrupt after a programmed time. An interrupt is a signal on one of the control lines of the computer bus. The **direct memory access controller (DMA)** is a CPU type of device which can control the computer bus. The direct memory access controller generates a DMA request signal on a computer control line. The main CPU will complete the portion of the instruction that it is executing, and generate a DMA grant signal. The DMA controller then transfers data into or out of main memory or disk memory. Since the CPU is not involved in the memory transfer, it can proceed with the instruction it is executing, as long as that instruction does not require the computer bus that the DMA controller is using. DMA controllers can off-load the CPU, allowing the overall system to have more power. The **programmable interrupt controller** is a general-purpose semiconductor which can be used in various types of computer printed boards to service interrupts.

COMPUTER APPLICATIONS

The computer is universally used, and a complete discussion of computer usage is beyond the scope of this section. Computer techniques are used in electronic design so extensively that some description of these electronic applications of computers are necessary in order to understand modern electronic design.

Computers can be used as control devices. The computer might be a controlling device to simply turn a pump on and off. On the other hand, a variable displacement pump might incorporate a microcontroller in its feedback control loop in place of pilot hydraulic circuits. Entire processes may be controlled by computers. For example, modern paper machines use computers throughout for control purposes. Computer terminals at

various locations show the operation conditions throughout the machine. Entire chemical-producing plants are under computer control.

Some automotive applications include control of air, fuel mixture, and spark timing in an automobile engine. This is typical of an application in which there are several input variables that must be measured, and several controlled outputs. The input variables that must be measured are air temperature, manifold temperature, crankcase temperature, engine speed, fuel flow, etc. The outputs which are controlled by the computer are ignition timing and fuel/air mixture. Automobile brakes, transmissions, and suspensions are automotive subsystems that incorporate microcontrollers for improved automobile performance. Microcomputers may also be used to reduce the complexity of automobile wiring. In such a system, battery power is distributed to all parts of the car along a single wire. Intelligence for the control of lights and accessories flows along a single control wire. Relays are provided at each lamp, switch, etc. to interface between the control and battery wire. Messages along the control wire are time-multiplexed so that all devices share the same control wire.

Testing, data acquisition, and analysis is another use of computers. Modern testing systems make extensive use of computers. Either digital or analog signals can be incorporated into a test system. Analog-to-digital converters convert analog signals into digital signals which the computer can use. Within the computer, a model of the process being tested can be constructed, and monitored by the test data being accumulated. Within the computer, signals can be generated which are not being actually measured. For example, the velocity of a mass may be measured, but the acceleration or the position of the mass may be calculated within the computer. Highly reliable systems can be developed using electronic control and computers, if redundancy is incorporated into the measuring and computer systems. In the aerospace industry, these systems are called **fly-by-wire** control systems. The space shuttle and supersonic fighters are examples of craft using fly-by-wire. In such systems, failure of a computer, a controller, or a sensor must not cause a failure of the craft.

Data acquisition systems give the user much greater insight into the data being taken. Limitations and anomalies of the measured data become much more apparent than with analog data. The need for signal conditioning becomes imperative in a digital data acquisition system. Signal processing is an integral part of any digital data system. Analog signals can be filtered as a part of signal conditioning, but after the signal has been converted to digital data, digital filtering may also be used. Digital filters can be broken into two classes: **finite impulse response (FIR)** and **infinite impulse response (IIR)** filters. As can be seen in Fig 15.2.34, the FIR filter does not have any feedback from a later part of the filter to a previous part. The IIR filter incorporates feedback from a later part of the filter to an earlier part. The FIR filter normally requires a greater length to accomplish the filtering objective than does the IIR filter. Because the IIR filter incorporates feedback, it is possible for it to be unstable. The FIR filter cannot be unstable. Both FIR and IIR filters employ a clock to generate a fixed delay from point to point in the filter. The clock delays are shown as ΔT in the figure. Each delay is a clock period. FIR filters are often 45 to 75 delays long. IIR filters are usually 8 to 10 delays long. The effective delay for an IIR filter is longer than the actual number of delays, because of the feedback in these filters. Although the infinite response suggests that the filter never completes its response, in a practical sense the IIR filter response is finite. The design approach for FIR filters and IIR filters is quite different. For an FIR filter, the filter coefficients A_0 through A_n are designed to optimally fit a desired filter response curve. The response of this optimal filter often has excessive sidebands, that is, high output in the nonpass frequency range. The sidebands can be reduced by applying a window function, which modifies the filter coefficients. The effect of the window function is to broaden the transition region between the passband and the stopband, and reduce the magnitude of the sidebands. The response equation, or transfer function, for the FIR filter with even symmetry is

$$H(\omega) = \sum_{i=0}^{i=n} A_i \cos\ (\omega i) \qquad (15.2.15)$$

In this equation, the frequency ω is a continuous variable in radians per second. Although the filter is a discrete device, its frequency response

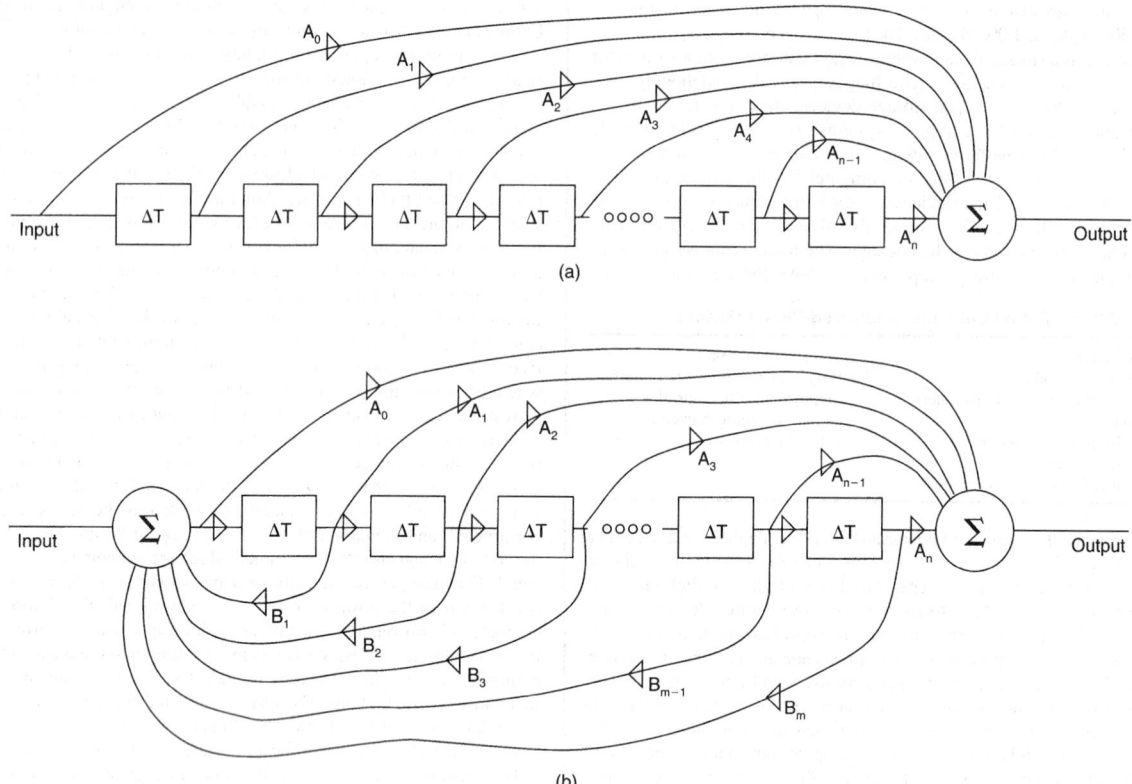

Fig. 15.2.34   (*a*) Finite impulse response (FIR) and (*b*) infinite impulse response (IIR) digital filters.

can be viewed as continuous. The digital signal is discrete, and when a signal is converted from digital to analog, this conversion adds its own modification of the signal.

The design approach for an IIR filter is quite different. For an IIR filter, the first step is to design a low-pass analog filter for a particular filter response, such as a Butterworth, Chebychev, or Cauer response. The analog filter can then be converted to a low-pass digital filter using the bilinear transform. Finally, the low-pass digital filter can be converted to a high-pass, band-pass, or band-stop filter. The resulting filter has a transfer function given by

$$H(\omega) = \frac{\displaystyle\sum_{i=0}^{i=n} A_i \cos{(\omega i)}}{1 - \displaystyle\sum_{i=1}^{i=m} B_i \cos{(\omega i)}} \qquad (15.2.16)$$

A more complete discussion of digital filter technology can be found in the references at the beginning of this section. Further help in digital filter design can be obtained by using a computer program, such as MATLAB.

Digital filter calculations consist of a series of mathematical operations which follow a consistent pattern. Each calculation is performed by retrieving a number, multiplying it by a coefficient, adding it to another number, and storing the result. This is the operation at each of the summation points in the FIR and the IIR. A microprocessor called a **digital signal processor (DSP)** has been developed for this purpose. DSPs complete a retrieve, multiply, add, and store in a single clock pulse (30 ns), several times faster than a general-purpose microprocessor. DSP microprocessors are widely used in test equipment, telephone equipment, television, and military equipment. Because of the wide usage, DSP microprocessors are quite inexpensive. DSPs were developed in order to increase the bandwidth of digital signal processing. It should be further noted that digital filters can employ parallel processing, in

which several DSP chips can operate in parallel to further increase bandwidth.

Digital signal processing is not limited to analog and digital filtering. Digital data may also be improved by using smoothing algorithms, or by removing inconsistent data points, such as outliers. Insight into data content can also be obtained by statistical analysis. Digital signal processing is also used to improve the quality of voice signals, and to enhance video signals.

DIGITAL COMMUNICATIONS

The bandwidth required to transmit information depends on the digital pulse rate. A fundamental relationship exists for the bandwidth required for the transmission of digital or analog information. For analog signals, the bandwidth depends on the highest frequency component in the signal. For digital circuits, the bandwidth depends on the maximum pulse rate. The relationship between the two is known as the Nyquist criterion. The minimum pulse rate pr necessary to transmit an analog signal digitally must be twice the highest frequency f:

$$pr = 2f \qquad (15.2.17)$$

This is a bilateral relationship. The frequency bandwidth BW of a transmission system must be equal to at least half the pulse rate:

$$BW = pr/2 \qquad (15.2.18)$$

These conditions are minimum requirements. A transmission system that has greater bandwidth can support slower pulse rates, and similarly, a high-pulse-rate system can be used to pass a low-frequency signal. A higher pulse rate than the Nyquist criterion will approximate an analog signal waveform more accurately, but is not necessary to pass the information. Voice-grade lines have a bandwidth of 4,000 Hz, and will pass 8,000 bits per second (bps). Twisted-pair cable, properly terminated, can pass about 1 MHz or 2 Mbps. Coaxial cable can pass 100 to 200 MHz.

To obtain greater bandwidths, light is being used to transmit information, rather than electricity. Inexpensive fiber-optic devices can be used for distances up to several hundred feet. For longer distances, phase-locked lasers and single-mode fiber-optic cable is being employed. Single-mode fiber is quite small and somewhat hard to terminate. Tooling to facilitate connecting single-mode cable is becoming more readily available.

Electric instrumentation has several communication standards which facilitate the connection of measuring systems. A low-speed instrument bus, IGIB, has been developed and described in a standard, IEEE 488. For higher communication rates, a CAMAC crate has been standardized. These standards give physical characteristics, such as connectors and plugs, mounting dimensions, printed board sizes, etc. These standards also define communications protocol, bus timing, and connector pin assignment.

In most computers, the interconnection of computer printed boards is accomplished using a "mother board." The mother board may contain the CPU, semiconductor memory, and several input/output devices. While these devices are somewhat arbitrary, it is essential that the mother board contain a computer bus structure. The computer bus is defined by a standard which defines the type connector, the connector circuit configuration, and the computer communication protocol. There are several agencies which define computer bus standards in the United States and around the world. Some of the more common IEEE bus standards are listed in Table 15.2.16.

Table 15.2.16   IEEE Computer Bus Standards

Identifier	Common name	Description
IEEE 488	GPIB	General-purpose instrument bus
IEEE 796	Multibus	General-purpose computer bus
IEEE 961	STD bus	Standard computer bus
IEEE 1014	VMEBUS	European computer bus
IEEE Proposal	VXIBUS	Instrumentation adaptation of VMEBUS
IEEE Proposal	MXIBUS	Extension of VXIBUS

POWER ELECTRONICS

Power electronics can be divided into three major groups: motor drives, power supplies, and power amplifiers. Motor drives can be divided into dc motor drives and ac motor drives. Power supplies include power supplies for electronic circuits, battery back-up systems, electric power applications, and induction/dielectric heating supplies. Power amplifiers cover other power conditioning topics.

The power for dc motor armatures can be derived from thyristor circuits like those shown in Fig. 15.2.9. Single-phase bridge circuits are used for 5-hp drives and smaller. Three-phase bridge circuits are used for drives larger than 5 hp. A single set of six thyristors can supply power for about 300 hp. Above 300 hp, multiple sets of thyristors must be used in parallel. Mill drives have been built with more than 10,000 hp provided by thyristors.

The control of dc motors whether powered by thyristors or by dc generators is accomplished electronically. Control of individual drives can be accomplished by tachometer feedback or by armature voltage feedback. The speed-regulation accuracy for armature feedback is 5 percent; for tachometer feedback speed-regulation accuracy is from 0.1 to 1.0 percent. When two drives must be coordinated with each other, as in a continuous-web processing machine, they can be regulated to control torque, speed, position, draw, or a combination of these parameters. Torque controls can be achieved using dc motor armature current for a feedback signal. Speed-control signals are derived as for single motors. Position or draw control can be accomplished by using selsyn ties or dancer rolls. A **dancer roll** is a weight- or spring-loaded roll that rides on the web. It is free to move up and down, and as it does, a signal is taken from its position to serve as a feedback for the drive regulator before the dancer or after it.

Coordination of the motions of two or more drives requires tracking of the drives in both steady-state and transient conditions. Linearity of the control and feedback signals determine steady-state tracking. Provision must be made for both low-speed and high-speed matching signals. Transient matching requires that signals not only be the right magnitude but also arrive at the right time. An example will serve to illustrate this point. Suppose it is desired to have two drives with tachometer feedback which have a continuous web between them. One way to accomplish this would be to designate one drive as a master and the other as a slave. The tachometer on the master drive would serve as its own feedback signal and as the reference or command signal for the slave drive. The slave drive would have its own feedback from its own tachometer and so its regulator would try to minimize the difference between the two tachometer signals. On a transient basis the master drive will always start before the slave. An alternate and more common arrangement is to provide a common reference for both drives and let each drive receive its command signals at the same instant.

Digital computers are being used on-line in mills and continuous processing industries. DC motors can be controlled by either analog or digital regulators. With the greatly reduced cost of integrated circuits, digital regulators are being increasingly used.

For digital speed control, feedback is taken from a digital tachometer, which generates a pulse frequency proportional to motor speed. An adjustable frequency reference is compared with the feedback, and the difference is applied to an up/down counter. The output of the counter is converted to a dc control signal for the motor controller. In this system the dc drive functions as a form of phase-locked loop, and in fact, a phase-locked loop can be substituted for the up/down counter in digital speed control. For digital position control, a shaft encoder is used for feedback. In a shaft encoder, a single turn of the motor shaft is encoded in a binary code. For example, an eight-track shaft encoder can define 2^8, or 256, parts of a revolution. The output of the shaft encoder is compared with a digitally derived reference code. The difference between the feedback and the reference is converted into a dc signal which controls the motor, and holds the desired position. Sometimes the position encoder is mounted on the driven machine. The accuracy of sensing the machine position is improved. Accurate milling machines have feedback from machine position, rather than from the drive motor. Computers can be used to derive a speed reference or a position reference for several motors, and thus, an entire process line can be controlled. Modern high-speed paper making machines are one example of coordinated computer control of several motors. Graphic terminals, which display the entire paper making machine, are located at various positions allowing several operators to coordinate the operation of the machine.

DC motors can also be supplied from a fixed dc voltage such as a battery or a fixed dc power supply. Inherently, fixed voltage on the field and on the armature of a dc motor implies constant-speed operation. The armature voltage can be controlled by a proportional device such as a transistor. Proportional devices are limited to low-power applications. At low speeds, the dc motor requires low voltage and full current to produce full torque. The losses in the control device are proportional to the product of voltage and current. The full battery voltage is applied to the control device and the motor in series. Thus, the control device has a high voltage drop across it at low motor armature voltages and low motor speeds. To reduce controller losses, a **chopper** may be used. The chopper is thyristor, power MOSFET, or IGBT circuit which can rapidly switch on and off. When the chopper is on, the loss in the control device is low because the voltage drop is small. When the chopper is off, the loss in the control device is small because the current through the device is very small. Large amounts of power can be controlled with low losses by means of a switching controller. For large dc motors, the switching rate of the chopper is made high enough that the inherent inductance of the motor armature maintains a nearly constant motor current.

The speed range of a variable-speed dc motor is limited by its armature resistance, which produces IR drop. For a very small motor, the IR drop can be as much as 20 percent of rated armature voltage. The resistance-to-inductance ratio of small motors is much greater than for large motors. To extend the speed range of small motors, a low-frequency chopper can be used which will cause the armature current to be

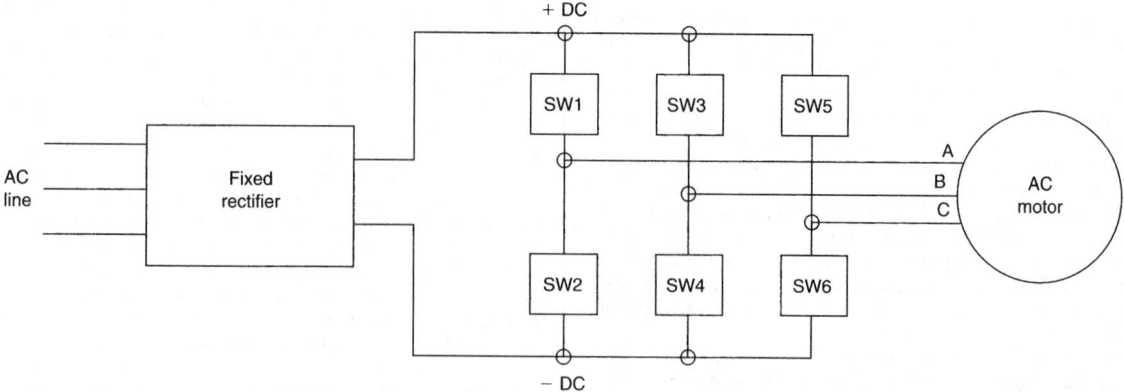

Fig. 15.2.35 Pulse width modulation inverter.

discontinuous. Discontinuous conduction reduces the effective *IR* drop for the motor allowing its speed range to be extended.

DC motors have been widely used for variable-speed applications because of their excellent characteristics. AC motors have been used primarily for constant-speed applications. The control schemes described above are equally applicable to ac motors (except of course for armature voltage and armature current feedback). If power circuitry is properly handled, the control of an ac motor is just as flexible and versatile as that of a dc motor.

AC motors can be supplied either from **phase-controlled circuits** or from **inverter circuits**. Phase control uses a simpler electronic circuit than the inverter, but it results in high loss in the ac motor. The phase control switches at line frequency, and so the frequency applied to the motor is constant. This means that the synchronous speed of the induction motor is constant. The difference between the operating speed of the motor and synchronous speed produces losses in the rotor of the motor proportional to the torque demanded of the motor. For this reason, phase-controlled ac motor drives are limited to applications with slight speed change requirements or to loads that require much lower torque at low speed, such as pumps or fans. Inverters are used for ac motor drives which require constant torque operation over a wide speed range. Several different schemes have been used for ac motor inverters. Some of these are **pulse width modulation (PWM), six-step inverters,** and **current link inverters.** Most modern drives are PWM inverters. In a PWM inverter, the ac line voltage is converted into a fixed dc voltage, and then this dc voltage is inverted into a variable frequency and a variable voltage. Figure 15.2.35 shows a block diagram of PWM inverter. Since the motor is an ac motor, both positive and negative voltages must be applied to each motor phase. The motor shown in Fig. 15.2.35 has three phases. Consider the operating conditions at only one of the phases of the motor. To apply a positive voltage to phase A-B of the motor, switches SW1 and SW4 are turned on. To apply a negative voltage to phase A-B, SW2 and SW3 are turned on. To apply zero volts to phase A-B, either SW2 and SW4 or SW1 and SW3 are turned on. A similar operation takes place for phase B-C and phase C-A of the motor. The dc voltage in the PWM inverter is constant. To generate an approximate sine wave of voltage, the inverter must produce a low voltage during part of the cycle and high voltage during another part of the cycle, and it must do this on both the positive half of the sine wave as well as the negative half. On the positive half-cycle, switching occurs alternately between switches SW1 and SW4 and switches SW1 and SW3. To control the speed of the ac motor and keep its losses low, one must vary both the frequency of the motor voltage and its magnitude. The motor requires an approximately constant volts per hertz rate for full torque and low losses. Although this control scheme seems very complicated, PWM inverter drives have been developed which are far more dependable than dc drives. The commutator and brushes required for a dc motor limit its reliability. The motor frequency for an ac motor

inverter drive varies from nearly zero to a maximum of 120 Hz typically. To reverse the rotation of the ac motor, the phase sequence applied to the motor is reversed. The actual switching frequency of the inverter is approximately 10,000 Hz, allowing the inverter switch on and off several times during a cycle applied to the motor. The pulse width is varied or modulated, to change the output voltage, hence the name pulse-width-modulated inverter. The inductance of the motor keeps the current variations in the motor from being too great.

Switching power supplies use technology similar to ac motor inverters. In Fig. 15.2.35, SW1 and SW2 are called an *inverter leg.* The inverter leg is fundamental to switching power supplies. The dc chopper described earlier could be made with a single inverter leg. Regulated dc power supplies are also made using an inverter leg. A higher intermediate voltage is produced in an ac-to-dc rectifier. This voltage is then chopped to make a lower output voltage. Feedback is taken from the load point of the power supply to control the pulse width of the inverter leg, or chopper. Precise computer power supplies are often switching power supplies of this type. They maintain precise dc output voltage under changing load conditions and changing supply voltage. Because the power supplies switch, the loss in these power supplies is much lower than for proportionally controlled power supplies.

The switching behavior of the inverter leg is actually more complex than simply turning on one switch and turning off another. Each switching action involves a resonant transient. At each switching point an *R-L-C* circuit rings from one voltage to another. At very high switching rates, a second transient starts before the first transient is over. This leads to a power supply with a parade of transients. Such power supplies are called *resonant converters.* These power supplies may be used for dc-to-dc, dc-to-ac, ac-to-dc, and ac-to-ac conversions. By using the ring-up inherent in resonant circuits, voltage levels can be increased as well as decreased. Induction heating power supplies are typical of these kinds of circuits. Induction heating is produced by the magnetic losses generated in nonmagnetic and magnetic metals. If the frequency and voltage are increased sufficiently, dielectric heating may also be produced in high-frequency power supplies. High-frequency, high-voltage power supplies are also used to produce a corona discharge to treat the surface of plastic materials. Plastics which are normally hydrophobic can be made hydrophilic by corona discharge treatment.

COMMUNICATIONS

The Federal Communications Commission (FCC) regulates the use of radio-frequency transmission in the United States. This regulation is necessary to prevent interfering transmissions of radio signals. Some of the frequency allocations are given in Table 15.2.17. The frequency bands are also classified as shown in Table 15.2.18. Very low frequencies are used for long-distance communications across the surface of the earth. Higher frequencies are limited to line-of-sight transmission. Because of bandwidth considerations, high frequencies are used for high-density

Table 15.2.17 Partial Table of Frequency Allocations
(For a complete listing of frequency allocations, see "Reference Data for Radio Engineers," published by Howard Sams & Co.)

Frequency, MHz	Utilization
0.535–1.605	Commercial broadcast band
27.255	Citizen's personal radio
54–72	Television channels 2–4
76–88	Television channels 5–6
88–108	Frequency-modulation broadcasting
174–216	Television channels 7–13
460–470	Citizens' personal radio
470–890	Television channels 14–83

Table 15.2.18 Frequency Bands

Designation	Frequency	Wavelength
VLF, very low frequency	3–30 kHz	100–10 km
LF, low frequency	30–300 kHz	10–1 km
MF, medium frequency	300–3,000 kHz	1,000–100 m
HF, high frequency	3–30 MHz	100–10 m
VHF, very high frequency	30–300 MHz	10–1 m
UHF, ultra-high frequency	300–3,000 MHz	100–10 cm
SHF, super-high frequency	3,000–30,000 MHz	10–1 cm
EHF, extremely high frequency	30,000–300,000 MHz	10–1 mm

NOTE: Wavelength in metres = $300/f$, where f is in megahertz.

communication links. Orbiting **satellites** allow the use of high-frequency transmission for long-distance high-density communications.

A **radio transmitter** is shown in Fig. 15.2.36. It consists of four basic parts: an rf oscillator tuned to the carrier frequency, an information-input device (microphone), a modulator to impress the input signal on the carrier, and an antenna to radiate the modulated carrier wave.

Fig. 15.2.36 Radio transmitter.

A radio wave can be modulated in several ways. It can be amplitude-, frequency-, or phase-modulated. These are examples of analog signal modulation. A radio wave may also be digitally modulated. Examples of digital modulation include on-off, binary phase-shift keying (BPSK), frequency-shift keying (FSK), quadrature phase-shift keying (QPSK), and quadrature amplitude modulation. Regardless of which method of modulation is used, the frequency content in the signal requires a bandwidth of frequencies to support the transmission. If a 10-kHz signal is to be broadcast, the transmitted radio wave will consist of a center frequency, called the *carrier frequency,* and an upper and lower sideband, for a frequency bandwidth of 2 times 10 kHz, or 20 kHz. It is possible to suppress the upper or lower sideband leading to single-sideband transmission, but this is a special case. Regardless of the modulation scheme, the bandwidth required for signal transmission is the same for a given signal bandwidth. Higher frequencies or higher digital pulse rates require greater bandwidth for transmission.

A **radio receiver** is shown in Fig. 15.2.37. This is called a **superheterodyne** receiver because it utilizes a frequency-mixing scheme. The tuned radio-frequency amplifier is tuned to receive the desired radio signal. The local oscillator is adjusted by the same tuning control to a lower frequency. The mixer produces an output frequency which is the difference between the incoming radio-signal frequency and the local-oscillator frequency. Since this difference frequency is constant for all tuning positions, the intermediate-frequency amplifier always operates

TRF = tuned radio-frequency amplifier
IF = intermediate-frequency amplifier
AF = audio-frequency amplifier

Fig. 15.2.37 Superheterodyne radio receiver.

with a constant frequency. This allows optimum design of the intermediate-frequency (IF) amplifiers since they are constant-frequency amplifiers. The IF frequency signal is modulated in just the same way as the radio signal. The demodulator separates this audio signal, which is then amplified so that the loudspeaker can be driven.

The term **radar** is **derived** from the first letters of the words "*radio detection and ranging*." It is essentially an echo system in which the location of an object is determined by sending out short pulses of radio waves and observing and measuring the time required for their reflections or echoes to return to the sending point. The time interval is a measure of the distance of the object from the transmitter. The velocity of radio waves is the same as the velocity of light, or 984 ft/μs, so that each microsecond interval corresponds to a distance of 492 ft. The direction of an object can be determined by the position of the directional transmitting and receiving antenna. Radio waves penetrate darkness, fog, and clouds, and hence are able to detect objects that otherwise would remain concealed. Radar can be used for the automatic "tracking" of objects such as airplanes.

A block diagram of a radar system is shown in Fig. 15.2.38. The transmitting system consists of an rf oscillator which is controlled by a modulator, or pulser, so that it sends to the antenna intermittent trains of rf waves of relatively high power but of very short duration, corresponding to the pulses received by the modulator. The energy of the

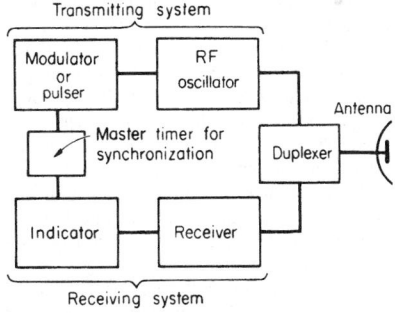

Fig. 15.2.38 Block diagram of radar system.

oscillator is transmitted through the duplexer and to the antenna through either coaxial cable or waveguides. The **receiver** is an ordinary heterodyne-type radio receiver which has high sensitivity in the band width corresponding to the frequency of the oscillator. For low frequencies the local oscillator is an ordinary oscillator for frequencies of 2,000 MHz; and higher a reflex **klystron** (hf cavity oscillator) is used. A common intermediate frequency is 30 MHz but 15 and 60 MHz are also frequently used.

In most radar systems the same antenna is used for receiving as for transmitting. This requires the use of a **duplexer** which cuts off the receiver during the intervals when the oscillator is sending out pulses and disconnects the transmitter during the periods between these pulses when the echo is being received.

The antenna is highly directional. By noting its angular position, the direction of the object may be determined. In the PPI (plan position indicator), the angle of the sweep of the cathode-ray beam on the screen of the oscilloscope is made to correspond to the azimuth angle of the antenna.

The receiver output is delivered to the indicator which consists of a cathode-ray tube or oscilloscope. The pulses which are received, corresponding to echoes from the target, must be synchronized with the sending pulses in order that the distance to the target may be determined. This is accomplished by synchronization of the sweep circuit of the oscilloscope with the pulses by the master timer.

Displays Conversion of the received radar signals to usable display is accomplished by a cathode-ray oscilloscope. The simplest type, called the **A presentation,** is shown in Fig. 15.2.39*a*. When the pulser operates, a sawtoothed wave produces a linear sweep voltage (Fig. 15.2.39*b*), is so rapid that it is not detectable by the sweep plates of the cathode-ray tube; at the same time a transmitter pulse is impressed on the deflection plates and the return echoes appear as AM pulses, or "pips," on the screen, as shown in Fig. 15.2.39*a*. The distance on the screen between the transmitter pulse and the pip caused by the echo is proportional to the distance to the target, and the screen can be calibrated in distance such as miles.

Fig. 15.2.39 Type A presentation.

[The return of the spot to its initial starting position, produced by the sweep interval *cd* (Fig. 15.2.39*b*), is so rapid that it is not detectable by the eye.] The direction of the target may be determined by the angular position of the antenna, which can be transmitted to the operator by means of a selsyn. Different objects, such as airplanes, ships, islands, and land approaches, have characteristic pips, and operators become skilled in their interpretation. A bird in flight can be recognized on the screen. Also a portion of the scale such as *ab* can be segregated and amplified for close study of the characteristics of the pips.

Plan Position Indicator (PPI) In the PPI (Fig. 15.2.40) the direction of a radial sweep of the electron beam is synchronized with the azimuth sweep of the antenna. The sweep of the beam is rotated continuously in synchronism with the antenna, and the received signals intensity-modulate the electron beam as it sweeps from the center of the oscilloscope screen radially outward. In this way the direction and range position of an object can be determined from the pattern on the screen of the oscilloscope, as shown in Fig. 15.2.40.

There are two methods by which the angular direction of the cathode spot is made to correspond with the angular position of the antenna. In one method, used on board ship, two magnetic deflecting coils are rotated around the neck of the tube in synchronism with the antenna, by means of a selsyn. In the other method, used on aircraft, two fixed magnetic deflecting coils at right angles to each other and placed at the neck of the tube are supplied with current from a small two-phase synchronous generator whose rotor is driven by the antenna. Thus a rotating field, similar to that produced by the stator of an induction motor, is produced by the magnetic deflecting coils. These two rotating fields, although produced by different means, are equivalent and cause the cathode beam to sweep radially in synchronism with the antenna. Circular coordinates spaced radially corresponding to distance are obtained by impressing on the control electrode short positive pulses synchronized with the transmitted pulse but delayed by time values corresponding to the desired distances. These coordinates appear as circles on the screen. Since the time of rotation of the antenna is relatively slow, it is necessary that a persistent screen be used in order that the operator may view the entire pattern. In Fig. 15.2.32 is shown a line drawing of a PPI presentation of Cape Cod, Mass., on a radar screen, taken from an airplane.

Fig. 15.2.40 Plan position indicator (PPI) of southeastern Massachusetts.

The applications of radar to war purposes are well known, such as detecting enemy ships and planes, aiming guns at them, and locating cities, rivers, mountains, and other landmarks in bombing operations. In peacetime, radar is used to navigate ships in darkness and poor visibility by locating navigational aids such as buoys and lighthouses, as well as protruding ledges, islands, and other landmarks. It can be similarly used in air navigation, as well as to operate altimeters for determining the height of the plane above ground. It is also used for aerial mapping.

Sonar is closely related to radar. The difference is that sound is used as the transmitting signal. Since the velocity of sound in water is so much slower than radio waves in free space, the timing for sonar signals is much slower than for radar signals. A 1-s delay of a sonar return echo implies that the target is 2,500 ft away. A 1-s delay of a radar return echo would imply that the target is 193,000 miles away. Although A displays and PPI displays are used with sonar signals, much reliance is placed on hearing. Transmitting a sonar pulse is an excellent way of disclosing your presence to an enemy vessel. Today more reliance is placed on passive detection of enemy vessels, by **signature analysis**. Signature analysis is signal analysis using fast Fourier transforms and filters.

Television is accomplished by systematically scanning a scene or the image of a scene to be reproduced and transmitting at each instant a current or a voltage which is proportional to the light intensity of the elementary area of the scene which at the instant is being scanned. The varying voltage or current is amplified, modulated on a carrier wave, and then transmitted as a radio wave. At the receiver the radio wave enters the antenna, is amplified, and demodulated to give a voltage or a current wave similar to the original wave. This voltage or current wave is then used to control the intensity of a cathode-ray beam which is focused on a fluorescent screen in a cathode-ray reproducing tube. The cathode-ray beam is caused to move over the screen in the same pattern as the scanning beam at the transmitter and in synchronism with it. Thus each small area of the receiver screen is illuminated instantaneously with light intensity corresponding to that of a similarly placed area in the original scene. This process is conducted so rapidly that owing to persistence of vision of the eye, the reproduction of each instantaneous scene appears to be a complete picture and the effect with successive scenes is similar to that produced by the projection of successive frames of a motion picture.

Scanning and Blanking In the United States the ratio of width to height of a standard television picture is 4 : 3, and the picture is composed of 525 lines repeated 30 times a second, this last factor being one-half 60, the prevalent electric power frequency in the United States. The scanning sequence along the individual lines is from left to right and the

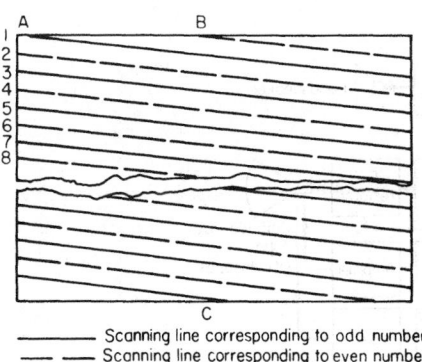

Fig. 15.2.41 Pattern of interlaced scanning.

——————— Scanning line corresponding to odd numbers
— — — — Scanning line corresponding to even numbers

sequence of the lines is from top to bottom. Also, interlacing is employed, the general method of which is shown in Fig. 15.2.41. The cathode-ray spot starts at 1 in the upper left-hand corner and is swept rapidly from left to right either by a sawtooth emf wave applied to the sweep plates or by the sawtooth current wave applied to the sweep coils of the tube. When the spot arrives at the right-hand side of the picture, the sawtooth wave of either emf or current in the sweep circuit acts to return the cathode-ray spot rapidly to point 3 at the left-hand side of the picture. However, during this period the cathode-ray is blanked, or entirely eliminated, by the application of a negative potential to the control grid of the tube. At the end of the return period, the blanking effect ceases and the spot appears at point 3, from which it again is swept across the picture and this process is repeated for 262.5 lines until the spot reaches a midpoint C at the bottom of the picture. It is then carried vertically and rapidly to B, the midpoint of the top of the picture, the beam also being blanked during this period. This process of scanning is then repeated, a second set of lines corresponding to the even numbers 2, 4, 6 being established between the lines designated by the odd numbers. These lines are shown dashed in Fig. 15.2.41. This method or pattern of scanning is called **interlacing**. The two sets of lines taken together produce a frame of 525 lines, which are repeated 30 times each second. However, owing to interlacing, the flicker frequency is 60 Hz which is not noticeable, 50 Hz having been determined as the threshold of flicker noticeable to the average eye. In Fig. 15.2.41, for the sake of clarity, the distances between horizontal lines are greatly exaggerated and no attempt is made to maintain proportions.

Frequency Band   In order to obtain the necessary resolution of pictures, television frequencies must be high. In the United States, VHF frequencies from 54 to 88 MHz (omitting 72 to 76 MHz) and 174 to 216 MHz are assigned for television broadcasting. A UHF band of frequencies for commercial television use is also allocated and consists of the frequencies from 470 to 890 MHz (see also Tables 15.2.17 and 15.2.18).

In order to obtain the 525 lines repeated 30 times per second, a bandwidth of 6 MHz is necessary. The video, or picture, signal with the superimposed scanning and blanking pulses is amplitude-modulated, amplified, and transmitted. The carrier frequency associated with the sound transmitter is 4.5 MHz higher than the video carrier frequency and is frequency-modulated with a maximum frequency deviation of 25 kHz.

In scanning motion-picture films a complication arises because standard film rate is 24 frames per second, while the television rate is 30 frames per second. This difficulty is overcome by scanning the first of two successive film frames twice and the second frame three times at the 60-Hz rate, making the total time for the two frames $1/12$ ($2/60 + 3/60$) or $1/24$ s average per frame.

Kinescope   The kinescope (Fig. 15.2.42) is the terminal tube in which the televised picture is reproduced. It is relatively simple, being not unlike the cathode-ray oscilloscope tube. It has an electric gun operating at 8,000 to 20,000 V which produces an electron beam focused on a

fluorescent surface within the front wall of the tube. The picture is viewed at the front wall. The horizontal and vertical deflections of the beam are normally controlled by deflection coils, as shown in Fig. 15.2.42.

Fig. 15.2.42 Kinescope for television receiver.

Television Receivers   A block diagram for a television receiver is given in Fig. 15.2.43. It is in reality a superheterodyne receiver with tuned rf amplification, the separating of the sound and video or picture channels taking place at the intermediate frequency in the mixer. The sound channel is then conventional, a discriminator being used to demodulate the FM wave (Fig. 15.2.21). The object of the dc restorer is to make the picture reproduction always positive, and it consists of applying a dc voltage at least equal in magnitude to the maximum values of the negative loops of the ac waves. The synchronizing pulses for both the vertical and the horizontal deflections are delivered by the dc restorer to an amplifier and the two pulses are then divided into the V and H components. The integrating and differentiating circuits are necessary to separate horizontal and vertical synchronizing signals.

As stated earlier, at any instant the magnitude of the current from the pickup tube varies in accordance with the light intensity of the part of the scene being scanned at that instant. This current is amplified and, together with the sound and synchronizing current, is broadcast and received by the circuit shown in Fig. 15.2.43. The video current is detected by rectification, is amplified, and is then made to control the intensity of the kinescope electron beam. Tubes produce a scanning pattern, identical with that in the pickup tube, and these tubes are triggered by the synchronizing pulses which are transmitted in the broadcast wave. Hence, the original televised scene is reproduced on the fluorescent screen of the kinescope.

Color-television transmission is similar to black-and-white television, and the two signals must be compatible with each other. The kinescope for color TV has three electron guns, one for each primary color. The fluorescent screen has a matrix of three different colors of phosphor and a mask with many small holes in it. The intensity signals for each color are phase-shifted from each other so that the proper phosphors are excited by each electron stream at each mask point over the entire screen. A black-and-white signal does not have the same synchronizing signal as a color signal. The color receiver has circuits which recognize this state and switch it to black and white reception.

TELEPHONE COMMUNICATIONS

Telephone communications is an important aspect of electronics because it provides a source of low-cost electronic devices which can be adapted for electronic circuit design. One of the big problems in telephone system design is switching. Each long-distance phone call requires that several circuits be joined together, and thousands of phone calls take place simultaneously. Older telephone exchange buildings had rack upon rack of telephone relays. These are being replaced by semiconductor switching. Older trunk lines were coaxial cable or microwave radio links. These are being replaced by fiber-optic cables. Fiber-optic cables have a tremendous bandwidth. A telephone voice circuit has a bandwidth of 4 kHz. Thousands of these circuits can be routed through one fiber-optic cable. Separating these voice circuits into communication channels by frequency separation and filtering is not feasible. Communication on fiber-optic cables is done digitally, and in packets. Each packet is a few hundred bytes long, and includes a header with an address and other control information, the message, and a trailer with error-correcting codes. The address is a session address to identify where the packet is from and

Fig. 15.2.43 Block diagram for television receiver. TRF: tuned radio frequency; IF: intermediate frequency.

where it is going. Users time-share the fiber-optic cable, and all of this signal processing must be done at a high enough rate to appear to be instantaneous to the subscriber.

Any form of signal modulation can be considered to be data encryption. The information is coded or encrypted onto the transmitting medium. At the transmitting end, a clear signal with no noise is sent, but at the receiving end of the medium, the signal contains the signal plus unwanted noise. Since the coding method is known, and since the noise is known to be random, separation of signal and noise is possible. This process of separating signal from noise will be repeated several times for long-distance communications. Each time, the noise is discarded and only the signal is sent on. Such a scheme of data encryption, or modulation, is also warranted in data telemetry, in order to allow noise to be removed from the desired signal.

Long-distance communications takes several different forms. Historically, cables with hundreds of twisted-pair circuits were used for long-distance communication. These are used for local subscriber service today. The next step was a cable made up of many wideband microwave coaxial cables. The coaxial cables were replaced by microwave radio links. Then, fiber-optic cables replaced microwave links. Wireless communication, using laser-generated signals, is the latest form of high volume data transmission. In a telephone system, all of these systems must coexist and work together.

Two technologies are replacing traditional telephone communications: **cellular communication** and **voice over Internet protocol** (VOIP).

The first **cell phones** used analog voice transmission technology, which consumed a lot of power. The cost and lack of versatility limited wide acceptance of the technology. Early cell phones were powered by the electrical system in a car, and the first batteries were bigger than the phone and thus were not an integrated handset but a collection of individual parts (handset, keypad, battery, and antenna all separate) with antennas mounted on the exterior of the vehicle. Evolution of several technologies allowed widespread use. Digital encoding technologies such as code division multiple access (CDMA), time division multiple access (TDMA), and global system for mobile communications (GSM) reduced the power consumption dramatically, as did the use of gallium arsenide compound semiconductor materials for the power amplifier that sends the signal to the cell tower. Improvements in nickel metal hydride (NiMH) and lithium ion technologies enabled smaller and more powerful batteries that could be integrated directly into the handset. These developments enabled ubiquitous mobile cellular communications. You could take your phone anywhere and not be tethered to a vehicle or large power source. Additionally, expansion of cellular communications is lower in cost and faster compared to the traditional copper or optical lines, which has enabled communication in remote and underprivileged areas of the world.

VOIP, also referred to as Internet telephony, is another replacement technology for traditional telephone communications. This technology is piggybacked onto a high-speed Internet connection, such as a digital subscriber line (DSL) or high-speed cable modem. This convergence of data and voice enables a single connection to the home that allows multiple concurrent users. By bundling voice with the Internet, companies are able to offer very attractive flat-fee rates for nationwide calling, email, and Internet access.

WIRELESS TECHNOLOGY

In addition to cellular phones, many other applications are taking advantage of wireless technology. Two standards have gained broad acceptance: **Bluetooth (IEEE 802.15)** and **WiFi (IEEE 802.11)**. Bluetooth is finding its niche in **personal area networking (PAN)**, whereas WiFi is making its mark in **local area networking (LAN)**. Bluetooth enables wireless interconnection of devices, such as hands-free headsets and cell phones, while WiFi replaces wired computer networks such as Ethernet. Another emerging standard is **WiMax/EV-DO (IEEE 802.16)**, an enabling technology for wireless broadband access, which would allow computers Internet access analogous to the current cell phone system.

WiFi is becoming ubiquitous, with deployment branching out from universities and businesses, now into homes, hotels, airports, coffee shops, and restaurants. Laptop computers rival their desktop counterparts for performance, and prices have dropped considerably, such that the demand for mobile Internet access is driving the rapid acceptance of WiFi. The IEEE 802.11 standard has three "flavors"—a, b, and g—with respective peak speeds of 11 Mbps, 22 Mbps, and 108 Mbps and ranges of 330 ft indoors and 1,300 ft line-of-sight. While these speeds are slower than those of wired Ethernet, they are fast enough, and the desire for mobility outweighs the need for speed. Peripherals such as wireless cameras and integrated video phones are just a couple of examples of the limitless additional capabilities that can be attached to WiFi networks. As demand increases, WiMax, which has a 30-mi range and 1.5-Mbps speed, is seeing initial deployment in some small cities, and speeds will increase as the technology is accepted and evolves.

While WiFi and WiMax enable multiple computers and peripherals to communicate on a network, Bluetooth, on the other hand, is a device-to-device direct interconnect technology which enables such applications as hands-free cell phone use in vehicles. It is a close-range wireless connection technology that enables a myriad of personal devices to intercommunicate, e.g., cell phones and laptop computers, laptop computers and digital cameras, or personal digital assistants and cell phones.

Other wireless applications, while not as cutting edge, are still benefitting from technology advancements that provide one- or two-chip designs for low-cost consumer-level products. Two-way radios with ranges up to 3 mi that do not require an amateur radio license are in wide use today, and wireless intercom systems for the home can be installed without the need to pull wires through walls. Commuters in high-traffic areas are benefitting from wireless RF identification (**RFID**), as tollways implement systems whereby commuters have an RFID device in their vehicle that allows them to proceed through a special **EZPass** lane at a tollbooth without stopping. The RFID is linked to an account, and the toll is debited each time the vehicle passes through the system. Additional RFID applications are to implant a chip containing critical medical information into a chronically ill person's arm. This can be extended to all people, which of course has extreme privacy concerns, which are being debated. Other more mundane uses of RFID are retail product tagging and automated inventory tracking.

DISPLAY TECHNOLOGY

Display technology has evolved beyond the curved monitor cathode-ray tube (CRT), with emphasis on **flat-screen technology,** wide viewing angles, large screens, and outdoor capability.

While CRT flat screens are widely available at reasonable prices, the tube still means a distinct depth of several inches. Other technologies, such as LCD, LED and plasma, are enabling displays to converge toward being truly flat, almost two-dimensional, by shrinking the depth of the display to about an inch. Each of these technologies has its strengths and weaknesses, all are relatively immature display technologies, and as a result all are currently very expensive. Plasma is the technology leader for brightness and large screen size; however, leading CRT manufacturers are dropping production of smaller (less than 20-in) CRTs in favor of LCD displays, which will help drive down price and increase production, leading to new technology developments that will enable wider availability of LCD flat-screen displays. For outdoor displays, LED technology is the best solution for its outstanding brightness capability, simple scalability to larger screens, and robustness in all weather conditions.

GLOBAL POSITIONING INFORMATION

Navigational aids take several different forms. **Shoran** (short-range navigation) and **loran** (long-range navigation) are methods by which ships or planes can locate their position. A loran station consists of a master transmitter and slave transmitter located about 200 mi apart. A pulse is broadcast from the master transmitter and, when received by the slave transmitter, causes it to also broadcast a pulse. The loran receiver, located on a ship or airplane, receives both the master and the slave pulses. The receiver allows the user to measure the delay between receiving the master and slave pulses. The delay defines a line across the surface of the earth. By picking three loran stations, determining the delay of each of the transmitter pairs, and using loran charts, three lines can be defined, allowing the user to determine the triangle which defines the receiver's position. **Very high frequency omnirange (VOR)** is used for short-range aircraft navigation, often in conjunction with **direction measuring equipment (DME).** VOR stations are located approximately 100 mi apart. The receiver in the aircraft receives a signal from the VOR transmitter with two pieces of information, which allows the pilot to determine the bearing from the VOR station. The first piece of information is a 30-Hz amplitude-modulated signal which is broadcast from a rotating directional antenna. The antenna rotates 30 times per second, and broadcasts in the 108- to 112-MHz range. The second piece of information is a frequency-modulated signal which varies between 9,480 and 10,440 Hz. This FM signal changes frequency at 30 Hz, and forms a reference for the VOR system. A receiver on the aircraft compares the 30-Hz amplitude-modulated signal and the 30-Hz frequency-modulated reference, and translates the difference in phase between the FM signal and the AM signal to give the airplane its bearing. The VOR also broadcasts station identification information in the 112- to 118-MHz range.

Most VOR stations can also broadcast a **distance measuring equipment (DME).** The aircraft has a DME transmitter which broadcasts a pulse which activates a ground mounted transponder. The aircraft sends an FM signal in the 1,025- to 1,150-MHz range. The DME ground station responds with a delayed signal in the 962- to 1,024-MHz or 1,151- to 1,213-MHz range. Distance is determined by the time delay required for the propagation of the signals. The depressed, straight line distance from the aircraft is measured by the DME equipment. To relieve the pilot of the job of correcting for altitude in the DME signal and to allow the pilot to fly any heading, a computer system called RNAV was introduced. RNAV uses the VOR and DME information, plus wind velocity and airspeed, to calculate the required aircraft heading and the estimated time of arrival at the destination.

Most major airports have an **instrument low approach system (ILS).** This produces a set of signals in the 108- to 112-MHz region. There a right-hand antenna lobe, relative to the aircraft, which is modulated at 150 Hz, and a left-hand lobe modulated at 90 Hz. In addition, in the 329- to 335-MHz range there is a vertical set of antenna lobes, with the upper lobe modulated at 90 Hz and lower lobe at 150 Hz. The display on the aircraft is an azimuth needle and a glide path needle. The pilot simply flies the aircraft to center the needles to determine a correct landing approach. There are also ILS marker beacons to tell the pilot how close the receiver is to the airport. Other aircraft navigation systems are based on ground-operated radar systems. Still others are based on aircraft-mounted Doppler radar. Doppler radar is a velocity measuring system.

Satellite positioning systems are the newest development for long-range and short-range navigation. The **Global Positioning System (GPS)** is a system of 24 satellites that broadcast accurate time marks from a known position in space, which enable accurate positioning determination by triangulation. GPS was initially developed for military applications, for which it is still in use. By knowing which satellite has broadcast the time mark, and measuring its delay, a spherical surface in space centered on the satellite is determined. By determining the delay from four satellites, four spherical surfaces in space can be determined, with the intersection providing positioning accuracy to within a meter. Technologically advanced GPS electronics can determine position within a centimetre. Technology advancements have lowered the cost of GPS receivers, enabling widespread consumer use. GPS receivers have been miniaturized to just a few integrated circuits and are now used in cars, boats, planes, construction equipment, movie-making gear, farm machinery, laptop computers, and most recently cell phones. The lower cost has enabled an increase in the complexity and integration of GPS systems, by including significant memory and simple microprocessors for a variety of applications including integrated traffic maps and real-time driving instructions. Farmers combine laboratory-analyzed soil samples that are mapped into a GPS system with a fertilizer delivery system for efficient, economic, and ecologically sound soil treatments. By introducing GPS into cell phones, the 911 emergency system can be utilized by anyone, anytime, anywhere. Fishermen and hunters use GPS receivers with integrated memory to archive their favorite locations so they can return time and time again. Vehicle tracking and control is another interesting application, not only for commercial vehicle fleets, but for private citizens as well. Service providers offer the ability to track a vehicle's location and speed, and make the information available via the Internet, with parents of teens being the largest target market. Typical prices for these integrated GPS products start at around $300, making them widely available to consumers.

Instruments and Controls

BY

WALTER H. BOYES, JR. *Editor-in-Chief Publisher, Control Magazine*
GREGORY V. MURPHY *Professor, Department of Electrical and Computer Engineering, College of Engineering, University of Alabama*
W. DAVID TETER, *P.E., Retired Professor, Department of Civil Engineering, College of Engineering, University of Delaware*

16.1 INSTRUMENTS

by Walter H. Boyes, Jr.

REFERENCES: ASME publications: "Instruments and Apparatus Supplement to Performance Test Codes (PTC 19.1–19.20)"; "Fluid Meters, pt. II, Application." ASTM, "Manual on the Use of Thermocouples in Temperature Measurement," STP 470B. ISA publications: "Standards and Recommended Practices for Instrumentation and Controls," 11 ed. Spitzer, "Flow Measurement." Preston-Thomas, The International Temperature Scale of 1990 (ITS-90), *Metrologia*, **27**, 3–10 (1990), Springer-Verlag. NIST Monograph 175, "Temperature-Electromotive Force Reference Functions and Tables for the Letter-Designated Thermocouple Types Based on the ITS-90," Government Printing Office, April 1993. Schooley, (ed.), "Temperature, Its Measurement and Control in Science and Industry," Vol. 6, Pts. 1 and 2, American Institute of Physics. Time and frequency services offered by the National Institute of Standards and Technology (NIST) include the Automated Computer Time Service (ACTS) (www.boulder.nist.gov/timefreg/service/acts.htm) and the significantly more popular and useful Internet Time Service's (www.boulder.nist.gov/timefreg/service/its.htm). Liptak, "Instrument Engineer's Handbook," 3rd and 4th ed., Chilton/ISA. Boyes, "Instrumentation Reference Book," 3rd ed., Butterworth-Heineman. Berge, "Fieldbushes for Process Control," ISA. Verhappen, "Foundation Fieldbus, A Pocket Guide," ISA. Beckwith, et al., "Mechanical Measurements," Addison-Wesley. Considine, "Encyclopedia of Instrumentation and Control," Krieger reprint. Considine, "Handbook of Applied Instrumentation," McGraw-Hill, Krieger reprint. Dally, et al., "Instrumentation for Engineering Measurements," Wiley. Doebelin, "Measurement Systems, Application and Design," McGraw-Hill. Erikson and Graber, Harris et al., "Shock and Vibration Control Handbook," McGraw-Hill. Holman, "Experimental Methods for Engineers," McGraw-Hill. Jones (ed.), "Instrument Science and Technology, Vol. 1, Measurement of Pressure, Level, Flow and Temperature," Heyden. Lion, "Instrumentation in Scientific Research, Electrical Input Transducers," McGraw-Hill. Sheingold, (ed.), "Transducer Interfacing Handbook," Analog Devices, Inc. Norwood, MA. Snell, "Nuclear Instruments and Their Uses," Wiley. Spink, "Principles and Practice of Flow Meter Engineering," Foxboro Co. Stout, "Basic Electrical Measurements," Prentice-Hall. *Periodicals: InTech*, monthly, ISA. *Sensors*, monthly, Alvanstar. *Test & Measurement World*, Cahners. *Control*, monthly, Putman Media. *Control Engineering*, monthly, Reed.

INTRODUCTION TO MEASUREMENT

An **instrument,** as referred to in the following discussion, is a device for determining the value or magnitude of a quantity or variable. The variables of interest are those which help describe or define an object, system, or process. Thus, in a manufacturing operation, product quality is related to measurements of its various dimensions and physical properties such as hardness and surface finish. In an industrial process, measurement and control of temperature, pressure, flow rates, etc., determine quality and efficiency of production.

Measurements may be direct, e.g., using a micrometer to measure a dimension, or indirect, e.g., determining moisture in steam by measuring the temperature in a throttling calorimeter.

Because of physical limitations of the measuring device and the system under study, practical measurements always have some error. The **accuracy** of an instrument is the closeness with which its reading approaches the true value of the variable being measured. Accuracy is commonly expressed as a percentage of measurement span, measurement value, or full-scale value. Span is the difference between the full-scale and the zero-scale value. **Uncertainty,** the sum of the errors at work to make the measured value different from the true value, is the accuracy of measurement standards. Uncertainty is expressed in parts per million (ppm) of a measurement value. **Precision** refers to the reproducibility of the measurements, i.e., with a fixed value of the variable, how much successive readings differ from one another. **Sensitivity** is the ratio of output signal or response of the instrument to a change in input or measured variable. **Resolution** relates to the smallest change in measured value to which the instrument will respond.

Error may be classified as systematic or random. Systematic errors are those due to assignable causes. These may be static or dynamic.

Static errors are caused by limitations of the measuring device or the physical laws governing its behavior. Dynamic errors are caused by the instrument not responding fast enough to follow the changes in measured variable. Random errors are those due to causes which cannot be directly established because of random variations in the system.

Standards for measurement are established by the National Institute of Standards and Technology. Secondary standards are prepared by very precise comparison with these primary standards and, in turn, form the basis for calibrating instruments in use. A well-known example is the use of precision gage blocks for the calibration of measuring instruments and machine tools.

There are three essential parts to an instrument: the **sensing element,** the **transmitting means,** and the **output** or **indicating element.** The sensing element responds directly to the measured quantity, producing a related motion, pressure, or electrical signal. This is transmitted by linkage, tubing, wiring, etc., to a device for display, recording, and/or control. Displays include motion of a pointer or pen on a calibrated scale, chart, oscilloscope screen, or direct numerical indication. Recording forms include writing on a chart and storage on magnetic tape or disk. The instrument may be actuated by mechanical, hydraulic, pneumatic, electrical, optical, or other energy medium. Often a combination of several energy modes is employed to obtain the accuracy, sensitivity, or form of output desired.

The transmission of measurements to distant indicators and controls is industrially accomplished by using the standardized electrical current signal of 4 to 20 mA; 4 mA represents the zero-scale value and 20 mA the full-scale value of the measurement range. A pressure of 3 to 15 lb/in^2 is commonly used for pneumatic transmission of signals.

COUNTING EVENTS

Event counters are used to measure the number of items passing on a conveyor line, the number of operations of a machine, etc. Coupled with time measurements, they yield measures of average rate or frequency. They find important application, therefore, in inventory control, production analysis, and in the sequencing control of automatic machines.

Choice of the proper counting device depends on the kind of events being counted, the necessary counting speed, and the disposition of the measurement; i.e., whether it is to be indicated remotely, used to actuate a machine, etc. Errors in the total count may be introduced by events being too close together or by too much nonuniformity in the items being counted.

The **mechanical counter** is shown in Fig. 16.1.1. Motion of the event being counted deflects the arm, which through an appropriate linkage

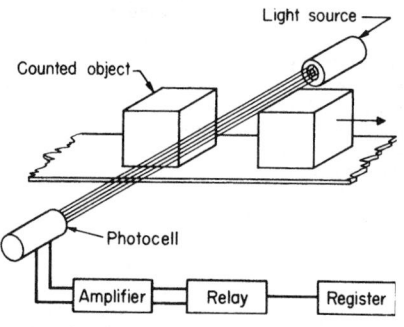

Fig. 16.1.1 Mechanical counter.

advances the count register one unit. Alternatively, motion of the actuating arm may close an electrical switch which energizes a relay coil to advance the count register one step.

Where there is a desire to avoid contact or close proximity with the object being counted, the photoelectric cell or diode, in conjunction with a lamp, a light-emitting diode (LED), or a laser light source, is employed in the transmitted or reflected light mode (Fig. 16.1.2). A signal to a counter is generated whenever the received light level is altered by the passing objects. Objects may be very small and very high counting speeds may be achieved with electronic counters.

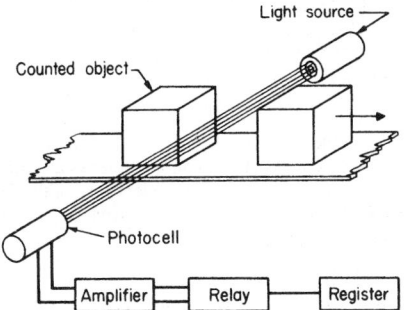

Fig. 16.1.2 Photoelectric counter.

Sensing methods based on electrical capacitance, magnetic, and eddy-current effects are extremely sensitive and fast acting, and are suitable for objects in close proximity to the sensor. The capacitive probe senses dielectrics other than air, such as glass and plastic parts. The magnetic pickup, by induction, responds to the motion of iron and nickel. The eddy-current sensor, by energy absorption, detects nonmagnetic conductors. All are suitable for counting machine operations.

The count is displayed by either a mechanical register as in Fig. 16.1.1, a dial-type register (as on the household watthour meter), or an electronic pulse counter with either number indicators or digital printing output. **Electronic counters** can operate accurately at rates exceeding 1 million counts per second.

TIME AND FREQUENCY MEASUREMENT

Measurement of time is basic to time and motion studies, time program controls, and the measurements of velocity, frequency, and flow rate. (See also Sec. 1.)

Mechanical clocks, chronometers, and stopwatches measure time in terms of the natural oscillation period of a system such as a pendulum, or hairspring balance-wheel combination. The minimum resolution is one-half period. Since this period is somewhat affected by temperature, precise timepieces employ a compensating element to maintain timing accuracies over long periods. Stopwatches may be obtained to read to better than 0.1 s. The major limitation, however, is in the response time of the user.

Electric timers are simple, inexpensive, and readily adaptable to remote-control operations. The majority of these are ac synchronous motors geared in the proper ratio to the indicator. These depend for their accuracy on the frequency of the line voltage. Consequently, care must be exercised in using such devices for precise short-time measurements.

Electronic timers are started and stopped by electrical pulses and hence are not limited by the observer's reaction time. They may be made extremely accurate and capable of measuring to less than 1 μs. These measure time by counting the number of cycles in a high-frequency signal generated internally by means of a quartz crystal. Stopwatch versions read at 0.01 s. Commercial instruments offer one or more functions: counting, measurement of frequency, period, and time intervals. Microprocessor-equipped versions increase versatility.

There are a variety of timing devices designed to indicate or control to a fixed time. These include timers based on the charging time of a condenser (e.g., type 555 integrated circuit), and the flow of oil or other fluid through a restriction. Most of these timing devices are now obsolete in favor of solid-state timing devices, but some mechanical or electromechanical timing devices are still in use.

Timing devices can be **calibrated** by comparison with a standard instrument, or by reference to the National Institute of Standards and Technology (NIST) timed radio signals (stations WWV, WWVB, and WWVH), telephone modem synchronization system (ACTS), or Internet Time Service (ITS). The most up-to-date information on time standards can be found on the Internet at www.boulder.nist.gov/timefreq/index.html. Frequency devices can also be calibrated by referring to the information on that page. The GOES Time Service, using geostationary satellites, was ended on January 1, 2005. See the References section at the beginning of this chapter for information on ACTS and ITS services.

Fast-moving, repetitive motions may be timed and studied by the use of stroboscopes which generate brilliant, very brief flashes of light at an adjustable rate.

The frequency of the observed motion is measured by adjusting the stroboscopic frequency until the system appears to stand still. The frequency of the motion is then equal to the stroboscope frequency or an integer multiple of it.

Many other means exist for **measuring vibrational or rotational frequencies.** These include timing a fixed number of rotations or oscillations of the moving member. Contact sensing can be done by an attached switch, or noncontact sensing can be done by magnetic or optical means. The pulses can be counted by an electronic counter or displayed on an oscilloscope or recorder and compared with a known frequency. Also used are reeds which vibrate when the measured oscillation excites their natural frequencies, flyball devices which respond directly to angular velocity, and generator-type tachometers which generate a voltage proportional to the speed.

MASS AND WEIGHT MEASUREMENT

Mass is the measure of the quantity of matter. The fundamental unit is the kilogram. The U.S. customary unit is the pound; 1 lb = 0.4536 kg (see Sec. 1.2, "Measuring Units"). Weight is a measure of the force of gravity acting on a mass (see "Units of Force and Mass" in Sec. 4).

A general equation relating weight W and mass M is $W/g = M/g_c$, where g is the local acceleration of gravity, and $g_c = 32.174$ lbm · ft/(lbf) (s²) [(1 kg · m/(N) (s²)] is a property of the unit system. Then $W = Mg/g_c$. The specific weight w and the mass density p are related by $w = pg/g_c$. Masses are conveniently compared by comparing their weights, and masses are often loosely referred to as weights. Indeed, almost all practical measures of mass are based upon weight.

Weighing devices fall into two major categories: balances and force-deflection systems. The device may be batch or continuous weighing, automatic or manual. Accuracies are expected to be of the order of 0.1 to better than 0.0001 percent, depending on the type and application of the scale. Calibration is normally performed by use of standard weights (masses) with calibrations traceable to the National Institute of Standards and Technology.

The **equal arm balance** compares the weight of an object with a set of standard weights. The laboratory balance shown in Fig. 16.1.3 is used

Fig. 16.1.3 Laboratory balance.

for extreme precision and sensitivity. A chain poise provides fine adjustment of the final balance weight. The magnetic damper causes the balance to come to equilibrium quickly.

Large weighing scales operate on the same principle; however, the arms are unequal to allow multiplication between the tare and the measured weights. In this group are platform, track, hopper, and tank scales. Here balance is achieved by adjusting the position of one or more balance weights along a beam directly calibrated in weight units. In dial-indicating-type scales, balance is achieved automatically through the deflection of calibrated pendulum weights from the vertical. The deflection is greatly magnified by the pointer-actuating mechanism, providing a direct-reading weight indication on the dial.

Since the deflection of a spring (within its design range) is directly proportional to the applied force, a calibrated spring serves as a simple and inexpensive weighing device. Applications include the **spring scale** and **torsion balance.** These are subject to hysteresis and temperature errors and are not used for precise work.

Other force-sensing elements are adaptable to weight measurement. Strain-gage load cells eliminate pivot maintenance and moving parts and provide an electrical output which can be used for direct recording and control purposes. Pneumatic pressure cells are also used with similar advantages.

In production processes, **continuous and automatic operating scales** are employed. In one type, the balancing weight is positioned by a reversible electric motor. Deflection of the beam makes an electrical contact which drives the motor in the proper direction to restore balance. The final balance position is translated by means of a potentiometer or digital encoding disk into a signal which is used for recording or control purposes.

The **batch-type scale** (Fig. 16.1.4) is adaptable to continuous flow streams of either liquids or solid particles. Material flows from the feed hopper through an adjustable gate into the scale hopper. When the

Fig. 16.1.4 Automatic batch-weighing scale.

weight in the scale hopper reaches that of the tare, the trip mechanism operates, closing the gate and opening the door. As soon as the scale hopper is empty, the weight of the tare forces the door closed again, resets the trip, and opens the gate to repeat the cycle. The agitator rotates while the gate is open, to prevent the solids from packing. Also, a "dribble" (partial closing of the gate just before the mechanism trips) is employed to minimize the error from the falling column of material at the instant balance is achieved. Since each dump of the scale represents a fixed weight, a counter yields the total weight of material passing through the scale.

In **continuous weighers,** a section of conveyor belt is balanced on a weigh beam (Fig. 16.1.5). The belt is driven at a constant speed; hence, if the total weight is held constant, the weight rate of material fed through the scale is fixed. Unbalance of the weight beam causes the rate of material flow onto the belt to be changed in the direction of restoring balance. This is accomplished by a mechanical adjustment of the feed gate or by varying the speed of a belt or screw feeder drive.

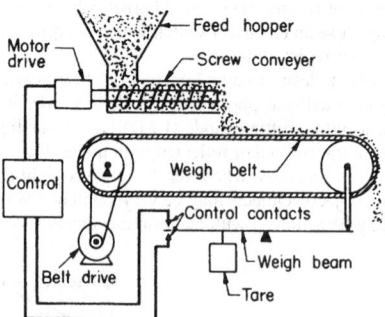

Fig. 16.1.5 Continuous-weighing scale.

If the density of the material is constant, **volume measurements** may be used to determine the mass. Thus, calibrated tanks are frequently used for liquids and vane and screw-type feeders for solids. Though often simpler to apply, these are not generally capable of as high accuracies as are common in weighing.

MEASUREMENT OF LINEAR AND ANGULAR DISPLACEMENT

Displacement-measuring devices are employed to measure dimension, distances between points, and some derived quantities such as velocity, area, etc. These devices fall into two major categories: those based on comparison with a known or reference length and those based on some fixed physical relationship.

The **measurement of angles** is closely related to displacement measurements, and indeed, one is often converted into the other in the process of measurement. The common unit is the degree, which represents $1/360$ of an entire rotation. The radian is used in mathematics and is related to the degree by π rad = 180°; 1 rad = 57.3°. The grad is an angle unit = $1/400$ rotation.

Figure 16.1.6 illustrates some methods of **rotary to linear conversion.** Figure 16.1.6a is a simple link and lever, Fig. 16.1.6b is a flexible link and sector, and Fig. 16.1.6c is a rack-and-pinion mechanism. These can be used to convert in either direction according to the relationship $D = RA/57.3$, where R = mean radius of the rotating element, in; D = displacement, in; and A = rotation, deg. (This equation holds for the link and lever of Fig. 16.1.6a only if the angle change from the perpendicular is small.)

Comparative devices are generally of the indicating type and include ruled or graduated devices such as the machinist's scale, folding rule, tape measure, digital caliper (Fig. 16.1.7), digital micrometer (Fig. 16.1.8), etc. These vary widely in their accuracy, resolution, and measuring span, according to their intended application. The manual readings depend for their accuracy on the skill and care of the operator.

The digital caliper and digital micrometer provide increased sensitivity and precision of reading. The stem of the digital caliper carries an embedded encoded distance scale. That scale is read by the slider. The distance so found shows on the digital display. The device is battery-operated and capable of displaying in inches or millimetres. Typical resolution is 0.0005 in or 0.01 mm.

The **digital micrometre,** employing rotation and translation to stretch the effective encoded scale length, provides resolution to 0.0001 in or 0.003 mm.

Dial gages are also used to magnify motion. A rack and pinion (Fig. 16.1.6c) converts linear into rotary motion, and a pointer moves over a calibrated scale.

Various modifications of the above-mentioned devices are available for making special kinds of measurements; e.g., **depth gages** for measuring the depth of a hole or cavity, **inside and outside calipers** (Fig. 16.1.7) for measuring the internal and external dimensions, respectively, of an object, **protractors** for angular measurement, etc.

Fig. 16.1.6 Linear-rotary conversion mechanisms.

Fig. 16.1.7 Digital caliper.

Fig. 16.1.8 Digital micrometer.

For line production and inspection work, **go no-go gages** provide a rapid and accurate means of dimension measurement and control. Since the measured values are fixed, the dependence on the operator's skill is considerably reduced. Such gages can be very complex in form to embrace a multidimensional object. They can also take the more general forms of the **feeler, wire, or thread-gage** sets. Of particular importance are **precision gage blocks,** which are used as standards for calibrating other measuring devices.

Displacement can be measured electrically through its effect on the resistance, inductance, reluctance, or capacitance of an appropriate sensing element.

The **potentiometer** is comparatively inexpensive, accurate, and flexible in application. It consists of a fixed linear resistance over which slides a rotating contact keyed to the input shaft (Fig. 16.1.9). The resistance or voltage (assuming constant voltage across terminals 1 and 3) measured across terminals 1 and 2 is directly proportional to the angle A. For straight-line motion, a mechanism of the type shown in Fig. 16.1.6 converts to rotary motion (or a rectilinear-type potentiometer can be

Fig. 16.1.9 Potentiometer.

used directly). (See also Sec. 15.) Versions with multiturns, straight-line motion, and special nonlinear resistance vs. motion are available.

The **synchro,** the **linear variable differential transformer (LVDT),** and the **E transformer** are devices in which the input motion changes the inductive coupling between primary and secondary coils. These avoid the limitations of wear, friction, and resolution of the potentiometer, but they require an ac supply and usually an electronic amplifier for the output. (See also Sec. 15.)

The **synchro** is a rotating device which is used to transmit rotary motions to a remote location for indication or control action. It is particularly useful where the rotation is continuous or covers a wide range. They are used in pairs, one transmitter and one receiver. For measurement of difference in angular position, the **control-transmitter** and **control-transformer** synchros generate an electrical error signal useful in control systems. A synchro differential added to the pair serves the same purpose as a gear differential.

The **linear variable differential transformer (LVDT)** consists of a primary and two secondary coils wound around a common core (Fig. 16.1.10). An armature (iron) is free to move vertically along the axis of the coils. An ac voltage is applied to the primary. A voltage is

Fig. 16.1.10 Linear variable differential transformer (LVDT).

induced in each secondary coil proportional to the relative length of armature linking it with the primary. The secondaries are connected to oppose each other so that when the armature is centered, the output voltage is zero. When the armature is displaced off center by an amount D, the output will be proportional to D (and phased to show whether D is above or below the center). These devices are very linear near the centered position, require negligible actuating force, and have spans ranging from 0.1 to several inches (0.25 cm to several centimetres).

The **E transformer** is very similar to the above except that the coils are wound around a laminated iron core in the shape of an E (with the primary and secondaries occupying the center and outside legs respectively). The magnetic path is completed through an armature whose motion, either rotary or translational, varies the induced voltage in the secondaries, as in the device of Fig. 16.1.10. This, too, is sensitive to extremely small motions.

A method that is readily applied, if a strain-gage analyzer is handy, is to measure the deflection of a cantilever spring with strain gages bonded to its surface (see Strain Gages, Sec. 5).

The **change of capacitance** with the displacement of the capacitor plates is extremely sensitive and suitable to very small displacements or large rotation. Often, one plate is fixed within the instrument; the other is formed or rotated by the object being measured. The capacitance can be measured by an impedance bridge, by determining the resonant frequency of a tuned circuit or using a relaxation oscillator.

Many optical instruments are available for obtaining precise measurements. The **transit and level** are used in surveying for measuring angles and vertical distances (see Sec. 16.3). A telescope with fine cross hairs permits accurate sighting. The angle scales are generally equipped with verniers. The **measuring microscope** permits measurement of very small displacements and dimensions. The microscope table is equipped with micrometer screws for sensitive adjustment. In addition, templates of scales, angles, etc., are available to permit measurement by comparison. The **optical comparator** projects a magnified shadow image of an object on a screen where it can readily be compared with a reference template.

Light can be used as a standard for the measurement of distance, straightness, and related properties. The wavelength of light in a medium is the velocity of light in vacuum divided by the index of refraction n of the medium. For dry air $n - 1$ is closely proportional to air density and is about 0.000277 at 1 atm and 15°C for 550-nm green light. Since the wavelength changes about $+ 1$ ppm/°C, and about -0.36 ppm/mmHg, density gradients bend light slightly. A temperature gradient of 1°C/m (0.5°F/ft) will cause a deviation from a tangent line of about 0.05 mm (0.002 in) at 10 m (33 ft).

Optical equipment to establish and test alignment, plumb lines, squareness, and flatness includes **jig transits, alignment telescopes, collimators,** optical squares, mirrors, targets, and scales.

Interference principles can be used for distance measurements. An optical flat placed in close contact with a polished surface and illuminated perpendicular to the surface with a monochromatic light will show interference bands which are contours of constant separation distance between the surfaces. Adjacent bands correspond to separation differences of one-half wavelength. For 550-nm wavelength this is 275 nm (10.8 μin). This test is useful in examining surfaces for flatness and in length comparisons with gage blocks.

Laser beams can be used over great distances. Surveying instruments are available for measurements up to 40 mi (60 km). Accuracy is stated to be about 5 mm (0.02 ft) + 1 ppm. These instruments take several measurements which are processed automatically to display the distance directly. Momentary interruptions of the light beam can be tolerated.

A laser system for machine tools, measurement tables, and the like is available in modular form (Agilent Corp.). It can serve up to eight axes by using beam splitters with a combined range of 200 ft (60 m). Normal resolution of length is about one-fourth wavelength, with a digital display least count of 10 μin (0.1 μm). Angle-measurement display resolves 0.1 s of arc. Accuracy with proper environmental compensation is stated to be better than 1 ppm + 1 count in length measurement. Velocities up to 720 in/min (0.3 m/s) can be followed. Accessories are

available for measuring straightness, parallelism, squareness, and flatness and for automatic temperature compensation. Various output options include displays and automatic computation and plots. The system can be used directly in measurement and control or to calibrate lead screws and other conventional measuring devices.

Pneumatic gaging previously found an important place in line inspection and quality control and is still seen. The device (Fig. 16.1.11) consists of a nozzle fixed in position relative to a stop or jig. Air at constant supply pressure passes through a restriction and discharges through the

Fig. 16.1.11   Pneumatic gage.

nozzle. The nozzle back pressure P depends on the gap G between the measured surface and the nozzle opening. If the measured dimension D increases, then G decreases, restricting the discharge of air, increasing P. Conversely, when D decreases, P decreases. Thus, the pressure gage indicates deviation of the dimension from some normal value. With proper design, this pressure is directly proportional to the deviation, limited, however, to a few thousandths of an inch span. The device is extremely sensitive [better than 0.0001 in (0.003 mm)], rugged, and, with periodic calibration against a standard, quite accurate. The gage is adaptable to automatic line operation where the pressure signal is recorded or used to actuate "reject" or "accept" controls. Further, any number of nozzles can be used in a jig to check a multiplicity of dimensions. In another form of this device, the flow of air is measured with a rotameter in place of the back pressure. The linear-variable differential transformer (LVDT) is also applicable.

The advent of automatically controlled machine tools has brought about the need for very accurate displacement measurement over a wide range. Most commonly applied for this purpose is the calibrated **lead screw** which measures linear displacement in terms of its angular rotation. **Digital** systems greatly extend the resolution and accuracy limitations of the lead screw. In these, a uniformly spaced optical or inductive grid is displaced relative to a sensing element. The number of grid lines counted is a direct measure of the displacement (see discussion of lasers above).

Measurement of strip thickness or coating thickness is achieved by **X-ray** or **beta-radiation**-type gages (Fig. 16.1.12). A constant radiation source (X-ray tube or radioisotope) provides an incident intensity I_0; the radiation intensity I after passing through the absorbing material is measured by an appropriate device (scintillation counter, Geiger-Müller tube, etc.). The thickness t is determined by the equation $I = I_0 e^{-kt}$, where k is a constant dependent on the material and the measuring device. The major advantage here is that measurements are continuous

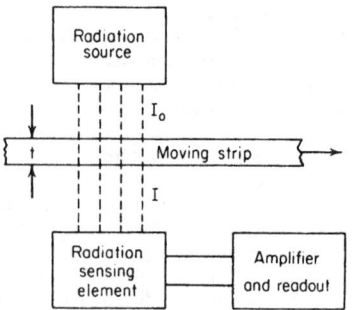

Fig. 16.1.12   Radiation-type thickness gage.

and nondestructive and require no contact. The method is extended to measure liquid level and density.

MEASUREMENT OF AREA

Area measurements are made for the purpose of determining surface area of an object or area inside a closed curve relating to some desired physical quantity. Dimensions are expressed as a length squared; e.g., in^2 or m^2. The areas of simple forms are readily obtained by formula. The area of a complex form can be determined by subdividing into simple forms of known area. In addition, various numerical methods are available (see **Simpson's rule**, Sec. 2) for estimating the area under irregular curves.

Area measuring devices include various mechanical, electrical, or electronic **flow integrators** (used with flowmeters) and the **polar planimeter.** The latter consists of two arms pivoted to each other. A tracer at the end of one arm is guided around the boundary curve of the area, causing rotation of a recorder wheel proportional to the area enclosed.

MEASUREMENT OF FLUID VOLUME

For a liquid of known density, volume is a quick and simple means of measuring the amount (or mass) of liquid present. Conversely, measuring the weight and volume of a given quantity of material permits calculation of its density. Volume has the dimensions of length cubed; e.g., cubic metres, cubic feet. The volume of simple forms can be obtained by formula.

A volumetric device is any container which has a known and fixed calibration of volume contained vs. the level of liquid. The device may be calibrated at only one point (**pipette, volumetric flask**) or may be graduated over its entire volume (**burette, graduated cylinder, volumetric tank**). In the case of the tank, a sight glass may be calibrated directly in liquid volume.

Volumetric measure of continuous flow streams is obtained with the **displacement meter.** This is available in various forms: the nutating disk, reciprocating piston, rotating vane, etc. The **nutating-disk meter** (Fig. 16.1.13) is relatively inexpensive and hence is widely used (water meters, etc.). Liquid entering the meter causes the disk to nutate or "roll" as the liquid makes its way around the chamber to the outlet. A

Fig. 16.1.13 Nutating-disk meter.

pin on the disk causes a counter to rotate, thereby counting the total number of rolls of the disk. Meter accuracy is limited by leakage past the disk and friction. The **piston meter** is like a piston pump operated backward. It is used for more precise measure (available to 0.1 percent accuracy).

Volumetric gas measurement is commonly made with a **bellows meter.** Two bellows are alternately filled and exhausted with the gas. Motion of the bellows actuates a register to indicate the total flow. Various liquid-sealed displacement meters are also available for this purpose. Recent innovations in gas flow measurement have made possible the use of Coriolis mass flowmeters, transit time ultrasonic flowmeters, and thermal mass flowmeters for volumetric gas measurements. It is likely that one or more of these methods will replace the bellows meter for many applications within the life of this edition.

For precise volume measurements, corrections for temperature must be made (because of expansion of both the material being measured and the volumetric device). In the case of gases, the pressure also must be noted.

FORCE AND TORQUE MEASUREMENT

Force may be measured by the deflection of an elastic element, by balancing against a known force, by the acceleration produced in an object of known mass, or by its effects on the electrical or other properties of a stress-sensitive material. The common unit of force is the pound (newton). **Torque** is the product of a force and the perpendicular distance to the axis of rotation. Thus, torque tends to produce rotational motion and is expressed in units of pound feet (newton metres). Torque can be measured by the angular deflection of an elastic element or, where the moment arm is known, by any of the force measuring methods.

Since weight is the force of gravity acting on a mass, any of the weight-measuring devices already discussed can be used to measure force. Common methods employ the deflection of springs or cantilever beams.

The strain gage is an element whose electrical resistance changes with applied strain (see Sec. 5). Combined with an element of known force-strain, motion-strain, or other input-strain relationship it is a transducer for the corresponding input. The relation of gage-resistance change to input variable can be found by analysis and calibration. Measure of the resistance change can be translated into a measure of the force applied. The gage may be bonded or unbonded. In the bonded case, the gage is cemented to the surface of an elastic member and measures the strain of the member. Since the gage is very sensitive to temperature, the readings must be compensated. For this purpose, four gages are connected in a Wheatstone-bridge circuit such that the temperature effect cancels itself. A four-element unbonded gage is shown in Fig. 16.1.14. Note that as the applied force increases, the tension on two of the elements increases while that on the other two decreases. Gages subject to strain change of the same sign are put in opposite arms of the bridge. The zero adjustment permits balancing the bridge for zero output at any desired input. The e_1 and e_2 terminal pairs may be used interchangeably for the input excitation and the signal output.

Fig. 16.1.14 Unbonded strain-gage board.

The **piezoelectric** effect is useful in measuring rapidly varying forces because of its high-frequency response and negligible displacement characteristics. Quartz rochelle salt, and barium titanate are common piezoelectric materials. They have the property of varying an output charge in direct proportion to the stress applied. This produces a voltage inversely proportional to the circuit capacitance. Charge leakage produces drifting at a rate depending on the circuit time constant. The voltage must be measured with a device having a very high input resistance. Accuracy is limited because of temperature dependence and some hysteresis effect.

Forces may also be measured with any of the pressure devices described in the next section by balancing against a fluid pressure acting on a fixed area.

PRESSURE AND VACUUM MEASUREMENT

Pressure is defined as the force per unit area exerted by a fluid. Pressure devices normally measure with respect to atmospheric pressure (mean value = 14.7 lb/in^2), $p_a = p_g + 14.7$, where p_a = total or **absolute** pressure and p_g = **gage** pressure, both lb/in^2. Conventionally, gage pressure and vacuum refer to pressures above and below atmospheric, respectively.

Common units are lb/in², in Hg, ftH₂O, kg/cm², bars, and mmHg. The mean SI atmosphere is 1.013 bar.

Pressure devices are based on (1) measure of an equivalent height of liquid column; (2) measure of the force exerted on a fixed area; (3) measure of some change in electrical or physical characteristics of the fluid.

The manometer measures pressure according to the relationship $p = wh = \rho gh/g_c$, where h = height of liquid of density ρ and specific weight w (assumed constants) supported by a pressure p. Thus, pressures are often expressed directly in terms of the equivalent height (head) of manometer liquid, e.g., inH₂O or inHg. Usual manometer fluids are water or mercury, although other fluids are available for special ranges.

The **U-tube manometer** (Fig. 16.1.15a) expresses the pressure difference $p_1 - p_2$ as the difference in levels h. If p_2 is exposed to the atmosphere, the manometer reads the gage pressure of p_1. If the p_2 tube is evacuated and sealed ($p_2 = 0$), the absolute value of p_1 is indicated. A common modification is the **well-type manometer** (Fig. 16.1.15b). The scale is specially calibrated to take into account changes of level inside the well so that only a single tube reading is required. In particular, Fig. 16.1.15b illustrates the form usually applied to measurement of atmospheric pressure **(mercury barometer).**

Fig. 16.1.15   Manometers. (a) U tube; (b) well type.

The sensitivity of readings can be increased by inclining the manometer tubes to the vertical **(inclined manometer)**, by use of low-specific-gravity manometer fluids, or by application of optical-magnification or level-sensing devices. Accuracy is influenced by surface-tension effects (reading of the meniscus) and changes in fluid density (due to temperature changes and impurities).

By definition, pressure times the area acted upon equals the force exerted. The pressure may act on a diaphragm, bellows, or other element of fixed area. The force is then measured with any force-measuring device, e.g., spring deflection, strain gage, or weight balance. Very commonly, the unknown pressure is balanced against an air or hydraulic pressure, which in turn is measured with a gage. By use of unequal-area diaphragms, the pressure can thus be amplified or attenuated as required. Further, it permits isolating the process fluid which may be corrosive, viscous, etc.

The **Bourdon-tube gage** (Fig. 16.1.16) is the most commonly used pressure device. It consists of a flattened tube of spring bronze or steel bent into a circle. Pressure inside the tube tends to straighten it. Since one end of the tube is fixed to the pressure inlet, the other end moves proportionally to the pressure difference existing between the inside and outside of the tube. The motion rotates the pointer through a pinion-and-sector mechanism. For amplification of the motion, the tube may be bent through several turns to form spiral or helical elements as are used in pressure recorders.

Fig. 16.1.16   Bourdon-tube gage.

In the **diaphragm gage,** the pressure acts on a diaphragm in opposition to a spring or other elastic member. The deflection of the diaphragm is therefore proportional to the pressure. Since the force increases with the area of the diaphragm, very small pressures can be measured by the use of large diaphragms. The diaphragm may be metallic (brass, stainless steel) for strength and corrosion resistance, or nonmetallic (leather, neoprene, silicon, rubber) for high sensitivity and large deflection. With a stiff diaphragm, the total motion must be very small to maintain linearity.

The **bellows gage** (Fig. 16.1.17) is somewhat similar to the diaphragm gage, with the advantage, however, of providing a much wider range of motion. The force acting on the bottom of the bellows is balanced by the deflection of the spring. This motion is transmitted to the output arm, which then actuates a pointer or recorder pen.

Fig. 16.1.17   Bellows gage.

The motion (or force) of the pressure element can be converted into an electrical signal by use of a differential transformer or strain-gage element or into an air-pressure signal through the action of a nozzle and pilot. The signal is then used for transmission, recording, or control.

The **dead-weight tester** is used as a standard for calibrating gages. Known hydraulic or gas pressures are generated by means of weights loaded on a calibrated piston. The useful range is from 5 to 5,000 lb/in² (0.3 to 350 bar). For low pressures, the water or mercury manometer serves as a reference.

For many applications (fluid flow, liquid level), it is important to measure the **difference between two pressures.** This can be done directly with the manometer. Other pressure devices are available as differential devices where (1) the case is made pressure-tight so that the second pressure can be applied external to the pressure element; (2)

two identical pressure elements are mounted so that their outputs oppose each other.

Similar devices to those discussed are used to measure **vacuum**, the only difference being a shift in range or at most a relocation of the zeroing spring. When the vacuum is high (absolute pressure near zero) variations in atmospheric pressure become an important source of error. It is here that absolute-pressure devices are employed.

Any of the differential-pressure elements can be converted to an **absolute-pressure device** by sealing one pressure side to a perfect vacuum. A common instrument for the range 0 to 30 inHg employs two bellows of equal area set back to back. One bellows is completely evacuated and sealed; the other is connected to the measured pressure. The output is a bellows displacement, as in Fig. 16.1.17.

There are many instruments for high-vacuum work (0.001 to 10,000 μm range). These kinds of devices are based on the characteristic properties of gases at low pressures. The **McLeod gage** amplifies the pressure to be measured by compressing the gas a known amount and then measuring its pressure with a mercury manometer. The ratio of initial to final pressure is equal to the ratio of final to initial volume (for common gases). This gage serves as a standard for low pressures.

The **Pirani gage** (Fig. 16.1.18) is based on the change of heat conductivity of a gas with pressure and the change of electrical resistance of a wire with temperature. The wire is electrically heated with a constant current. Its temperature changes with pressure, producing a voltage across the bridge network. The compensating cell corrects for room-temperature changes.

Fig. 16.1.18   Pirani gage.

The **thermocouple gage** is similar to the Pirani gage, except that a thermocouple is used to measure the temperature difference between the resistance elements in the measuring and compensating cells, respectively.

The **ionization gage** measures the ion current generated by bombardment of the molecules of the gas by the electron stream in a triode-type tube. This gage is limited to pressures below 1 μm. It is, however, extremely sensitive.

LIQUID-LEVEL MEASUREMENT

The past 30 years have seen significant advancement in the measurement of liquid level. Level instruments are used for determining or controlling the height of liquid in a vessel, or finding the location of the interface between two liquids of different specific gravity. Liquid level measurement can be made both continuously, or as a point level measurement (or switch).

In process measurement, typical level devices include differential pressure and hydrostatic pressure transmitters of various types; capacitance and radio-frequency (RF) admittance device; a wide variety of floats and other mechanical level measurement systems; magnetostrictive gages; ultrasonic level sensors; pulse and continuous wave radar level sensors; guided wave radar sensors; laser level sensors; and nuclear level gages. Typically, the applications for these devices fit along a continuum from very easy to measure (water) to very hard to measure (molten plastic inside a jacketed reactor).

Each of these level measurement techniques has advantages and limitations, dependent on the application. For example, a mechanical gage is not advisable in an application where the fluid is sticky or viscous.

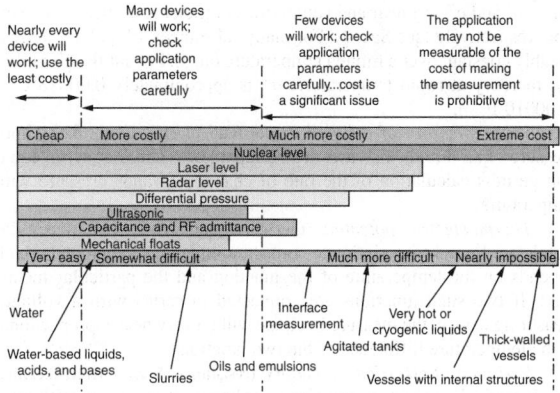

Fig. 16.1.19   Level measurement continuum.

Ultrasonic level meters are extremely sensitive to temperature changes in the vapor space and to density changes of the vapor, and should normally only be used at atmospheric pressure. Capacitance or RF admittance–type gages may be used in a wide variety of applications, although long measurement spans make them expensive. Radar level gages of either type can be used in pressurized vessels, since the changing dielectric of the vapor space is usually too small a change to affect the measurement. Nuclear level gages are the "last resort" in part because of their expense, but also because of the licensing and training required to operate them and the disposal and storage requirements of the nuclear source.

Differential pressure measurement is still used as a common means of measuring level in vessels, although the use of this technique is declining. As indicated in the discussion on manometers, the pressure difference is the height times the specific weight of the liquid.

Ultrasonic and pulse radar level devices operate by measuring the transit time of a signal reflected off the liquid surface and converting that signal into a level measurement. Continuous-wave and guided-wave radar instruments measure the signal return from the largest peak of the reflected signals in the vessel. Ultrasonic level devices are also commonly used as open-channel flowmeters when combined with a physical primary device such as a weir or flume.

Capacitive or RF admittance level devices measure the change in dielectric that occurs at the level surface or at the interface between two liquids. Like ultrasonics, these can be supplied as either point or continuous level transmitters, and like ultrasonics, they have applications in bulk solids and powders as well as liquid level.

Nuclear level gages are also available in point or continuous level designs. These instruments work by the attenuation of a beam of gamma energy by the surface of the liquid (which is much more dense than the vapor phase in the vessel). The point at which the measured radiation drops to near zero is the level.

TEMPERATURE MEASUREMENT

The common temperature scales (Fahrenheit and Celsius) are based on the freezing and boiling points of water (see Sec. 4 for discussion of temperature standards, units, and conversion equations).

Temperature is measured in a number of different ways. Some of the more useful are as follows.

1. *Thermal expansion of a gas* (**gas thermometer**). At constant volume, the pressure p of an (ideal) gas is directly proportional to its absolute temperature T. Thus, $p = (p_0/T_0)T$, where p_0 is the pressure at some known temperature T_0.

2. *Thermal expansion of a liquid or solid* (**mercury thermometer, bimetallic element**). Substances tend to expand with temperature. Thus, a change in temperature $t_2 - t_1$ causes a change in length $l_2 - l_1$ or a change in volume $V_2 - V_1$, according to the expressions

$$l_2 - l_1 = a'(t_2 - t_1)l_1 \quad \text{or} \quad V_2 - V_1 = a'''(t_2 - t_1)V_1$$

where a' and a''' = linear and volumetric coefficients of thermal expansion, respectively (see Sec. 4). For many substances, a' and a''' are reasonably constant over a limited temperature range. For solids, $a''' = 3a'$. For mercury at room temperature, a''' is approximately $0.00018°C^{-1}$ ($0.00010°F^{-1}$).

3. *Vapor pressure of a liquid* (**vapor-bulb thermometer**). The vapor pressure of all liquids increases with temperature. The Clapeyron equation permits calculation of the rate of change of vapor pressure with temperature.

4. *Thermoelectric potential* (**thermocouple**). When two dissimilar metals are brought into intimate contact, a voltage is developed which depends on the temperature of the junction and the particular metals used. If two such junctions are connected in series with a voltage-measuring device, the measured voltage will be very nearly proportional to the temperature difference of the two junctions.

5. *Variation of electrical resistance* (**resistance thermometer, thermistor**). Electrical conductors experience a change in resistance with temperature which can be measured with a Wheatstone- or Mueller-bridge circuit, or a digital ohmmeter. The platinum resistance thermometer (PRT) can be very stable and is used as the temperature scale interpolation standard from -160 to $660°C$. Commercial resistance temperature detectors (RTD) using copper, nickel, and platinum conductors are in use and are characterized by a polynomial resistance-temperature relationship, such as

$$t = A + B \times R_t + C \times R_t^2 + D \times R_t^3 + E \times R_t^4$$

where R_t = resistance at prevailing temperature t in $°C$. A, B, C, D, and E are range- and material-dependent coefficients listed in Table 16.1.1. R_0, also shown in the table, is the base resistance at $0°C$ used in the identification of the sensor.

The thermistor has a large, negative temperature coefficient of resistance, typically -3 to -6 percent/$°C$, decreasing as temperature increases. The temperature-resistance relation is approximated (to perhaps $0.01°$ in range 0 to 100°C) by:

$$R_t = \exp\left(A_0 + A_1/t + \frac{A_2}{t^2} + \frac{A_3}{t^3}\right)$$

and

$$\frac{1}{t} = a_0 + a_1 \ln R_t + a_2 (\ln R_t)^2 + a_3 (\ln R_t)^3$$

with the constants chosen to fit four calibration points. Often a simpler form is given:

$$R = R_0 \exp\left\{\beta\left[\left(\frac{1}{t}\right) - \left(\frac{1}{t_0}\right)\right]\right\}$$

Typically β varies in the range of 3,000 to 5,000 K. The reference temperature t_0 is usually 298 K($= 25°C$, 77°F), and R_0 is the resistance at that temperature. The error may be as small as $0.3°C$ in the range of $0°$ to $50°C$. Thermistors are available in many forms and sizes for use from -196 to $+450°C$ with various tolerances on interchangeability and matching. (See "Catalog of Thermistors," Thermometrics, Inc.) The AD590 and AD592 integrated circuit (Analog Devices, Inc.) passes a current of $1\ \mu A/K$ very nearly proportional to absolute temperature. All these sensors are subject to self-heating error.

6. *Change in radiation* (**radiation and optical pyrometers**). A body radiates energy proportional to the fourth power of its absolute temperature. This principle is particularly adaptable to the measurement of very high temperatures where either the total quantity of radiation or its intensity within a narrow wavelength band may be measured. In the former type (radiation pyrometer), the radiation is focused on a heat-sensitive element, e.g., a thermocouple, and its rise in temperature is measured. In the latter type (optical pyrometer) the intensity of the radiation is compared optically with a heated filament. Either the filament brightness is varied by a control calibrated in temperature or a fixed brightness filament is compared with the source viewed through a calibrated optical wedge.

The infrared thermometer accepts radiation from an object seen in a definite field of view, filters it to select a portion of the infrared spectrum, and focuses it on a sensor such as a blackened thermistor flake, which warms and changes resistance. Electronic amplification and signal processing produce a digital display of temperature. Correct calibration requires consideration of source emissivity, reflection, and transmission from other radiation sources, atmospheric absorption between the source object and the sensor, and compensation for temperature variation at the sensor's immediate surroundings.

Electrical nonconductors generally have fairly high (about 0.95) emissivities, while good conductors (especially smooth, reflective metal surfaces), do not; special calibration or surface conditioning is then needed. Very wide band (0.7 to 20 μm) instruments gather relatively large amounts of energy but include atmospheric absorption bands which reduce the energy received from a distance. The band 8 to 14 μm is substantially free from atmospheric absorption and is popular for general use with source objects in the range 32 to 1,000°F (0 to 540°C). Other bands and two-color instruments are used in some cases. See Bonkowski, Infrared Thermometry, *Measurements and Control*, Feb. 1984, pp. 152–162.

Fiber-optic probes extend the use of radiation methods to hard-to-reach places.

Important relationships used in the design of these instruments are the Wien and Stefan-Boltzmann laws (in modified form):

$$\lambda_m = k_1/T \qquad q = k_2 \varepsilon A (T_2^4 - T_1^4)$$

where λ_m = wavelength of maximum intensity, μm (nm); q = radiant energy flux, Btu/h (W); A = radiation surface, ft^2 (m^2); ε = mean emissivity of the surfaces; T_2, T_1 = absolute temperatures of radiating and receiving surfaces, respectively, °R (K); $k_1 = 5,215\ \mu m$, °R (2,898 $\mu m \cdot$ K); $k_2 = 0.173 \times 10^{-8}$ Btu/(h $\cdot ft^2 \cdot$ °R^4) [5.73 $\times 10^{-8}$ W/($m^2 \cdot$ K^4)]. The emissivity depends on the material and form of the surfaces involved (see Sec. 4). Radiation sensors with scanning capability can produce maps, photographs, and television displays showing temperature-distribution patterns. They can operate with resolutions to under 1°C and at temperatures below room temperature.

7. *Change in physical or chemical state* (**Seger cones, Tempilsticks**). The temperatures at which substances melt or initiate chemical reaction are often known and reproducible characteristics. Commercial products are available which cover the temperature range from about 120 to 3,600°F (50 to 2,000°C) in intervals ranging from 3 to 70°F (2 to 40°C).

Table 16.1.1   Polynomial Coefficients for Resistance Temperature Detectors

Material of conductor	ID R_0, Ω	Useful range, °C	Polynomial coefficients					Typical accuracy,* °C
			A, °C	B, °C/Ω	C, °C/Ω^2	D, °C/Ω^3	E, °C/Ω^4	
Copper 10Ω @ 25°C	9.042	-70 to 0	-225.64	23.30735	$+0.246864$	-0.00715		1.5
		0 to 150	-234.69	25.95508				1.5
Nickel	120	-80 to 320	-199.47	1.955336	-0.00266	1.88E $-$ 6		1
Platinum DIN/IEC $\alpha = 0.00385/°C$	100	-200 to 0	-241.86	2.213927	0.002867	-9.8E $-$ 6	1.64E $-$ 8	1
		0 to 850	-236.06	2.215142	0.001455			0.5

* For higher accuracy consult the table or equation furnished by the manufacturer of the specific RTD being used. Temperatures per ITS-90, resistances per SI-90.

The temperature-sensing element may be used as a solid which softens and changes shape at the critical temperature, or it may be applied as a paint, crayon, or stick-on label which changes color or surface appearance. For most the change is permanent; for some it is reversible. Liquid crystals are available in sheet and liquid form: these change reversibly through a range of colors over a relatively narrow temperature range. They are suitable for showing surface-temperature patterns in the range 20 to 50°C (68 to 122°F).

An often used temperature device is the **mercury-in-glass thermometer.** As the temperature increases, the mercury in the bulb expands and rises through a fine capillary in the graduated thermometer stem. Useful range extends from $-$ 30 to 900°F ($-$ 35 to 500°C). In many applications of the mercury thermometer, the stem is not exposed to the measured temperature; hence a correction is required (except where the thermometer has been calibrated for **partial immersion**). Recommended formula for the correction K to be added to the thermometer reading is $K = 0.00009D(t_1 - t_2)$, where D = number of degrees of exposed mercury filament, °F; t_1 = thermometer reading, °F; t_2 = the temperature at about middle of the exposed portion of stem, °F. For Celsius thermometers the constant 0.00009 becomes 0.00016.

For **industrial applications** the thermometer or other sensor is encased in a metal or ceramic protective well and case (Fig. 16.1.20). A threaded union fitting is provided so that the thermometer can be installed in a line or vessel under pressure. Ideally the sensor should have the same

Fig. 16.1.20   Industrial thermometer.

temperature as the fluid into which the well is inserted. However, heat conduction to or from the pipe or vessel wall and radiation heat transfer may also influence the sensor temperature (see ASME PTC 19.3-1974 Temperature Measurement, on well design). An approximation of the conduction error effect is

$$T_{\text{sensor}} - T_{\text{fluid}} = (T_{\text{wall}} - T_{\text{fluid}})E$$

For a sensor inserted to a distance $L - x$ from the tip of a well of insertion length L, $E = \cosh[m(L - x)]/\cosh mL$, where $m = (h/kt)^{0.5}$; x and L are in ft (m); h = fluid-to-well conductance, Btu/(h) (ft²)(°F) [J/(h) (m²)(°C)]; k = thermal conductivity of the well-wall material. Btu/(h)(ft)(°F) [J/(h)(m)(°C)]; and t = well-wall thickness, ft (m). Good thermal contact between the sensor and the well wall is assumed. For $(L - x)/L = 0.25$:

mL	1	2	3	4	5	6	7
E	0.67	0.30	0.13	0.057	0.025	0.012	0.005

Radiation effects can be reduced by a polished, low-emissivity surface on the well and by radiation shields around the well. Concern with mercury contamination has made the bimetal thermometer the most commonly used expansion-based temperature measuring device. Differential thermal expansion of a solid is employed in the simple **bimetal** (used in thermostats) and the bimetallic helix (Fig. 16.1.21). The bimetallic

Fig. 16.1.21   Bimetallic temperature gage.

element is made by welding together two strips of metal having different coefficients of expansion. A change in temperature then causes the element to bend or twist an amount proportional to the temperature. A common bimetallic pair consists of invar (iron-nickel alloy) and brass.

For control or alarm indications at fixed temperatures, thermometers may be equipped with electrical contacts such that when the temperature matches the contact point, an external relay circuit is energized.

A popular industrial-type instrument employs the deflection of a **pressure-spring** to indicate (or record) the temperature (Fig. 16.1.22). The sensing element is a metal bulb containing some specific gas or liquid. The bulb connects with the pressure spring (in the form of a spiral or helix) through a capillary tube which is usually enclosed in a protective sheath or armor. Increasing temperature causes the fluid in the bulb to expand in volume or increase in pressure. This forces the pressure spring to unwind and move the pen or pointer an appropriate distance upscale.

Fig. 16.1.22   Pressure-spring element.

The **bulb fluid** may be mercury (mercury system), nitrogen under pressure (gas system), or a volatile liquid (vapor-pressure system). Mercury and gas systems have linear scales; however, they must be compensated to avoid ambient temperature errors. The capillary may range up to 200 ft in length with, however, considerable reduction in speed of response. The use of mercury is being phased out due to environmental and health concerns over the use of the metal.

The **pneumatic transmitter** (Fig. 16.1.23) was at one time the standard for long-distance transmission of temperature. Although its place has been taken by various electronic means, it is still commonly used in many parts of the world, especially in refining and other hydrocarbon processing applications. The bulb is filled with gas under pressure, which acts on the diaphragm. An increase in bulb temperature increases the upward force acting on the main beam, tending to rotate it clockwise. This causes the baffle or flapper to move closer to the nozzle, increasing the nozzle back pressure. This acts on the pilot, producing an increase in output pressure, which increases the force exerted by the feedback bellows. The system returns to equilibrium when the increase in bellows pressure exactly balances the effect of the increased diaphragm pressure. Since the lever ratios are fixed, this results in a direct proportionality between bulb temperature and output air pressure. For precision, compensating elements are built into the instrument to correct for the effects of changes in barometric pressure and ambient temperature.

Fig. 16.1.23 Pneumatic temperature transmitter.

Electrical systems based on the thermocouple or resistance thermometer are particularly applicable where many different temperatures are to be measured, where transmission distances are large, or where high sensitivity and rapid response are required. The thermocouple is used with high temperatures; the resistance thermometer for low temperatures and high accuracy requirements.

The **choice of thermocouple** depends on the temperature range, desired accuracy, and the nature of the atmosphere to which it is to be exposed. The temperature-voltage relationships for the more common of these are

given by the curves of Fig. 16.1.24. Table 16.1.2 gives the recommended temperature limits, for each kind of couple. Table 16.1.3 gives polynomials for converting thermocouple millivolts to temperature. The thermocouple voltage is measured by a digital or deflection millivoltmeter or null-balance type of potentiometer. Completion of the thermocouple circuit through the instrument immediately introduces one or more additional junctions. Common practice is to connect the thermocouple (hot junction) to the instrument with special lead wire (which may be of the

Fig. 16.1.24 Thermocouple voltage-temperature characteristics [reference junction at 32°F (0°C)].

Table 16.1.2   Limits of Error on Standard Wires without Selection*†

ANSI symbol‡	Materials and polarities Positive	Negative	°F: − 150 °C: − 101	−75 −59	32 0	200 93	530 277	600 316	700 371	1,000 538	1,400 760	2,300 1,260	2,700 1,482
T	Cu	Constantan§		2%	1.5°F (0.8°C)		¾%						
E	Ni-Cr	Constantan			3°F (1.7°C)					½%			
J	Fe	Constantan			4°F (2.2°C)			¾%					
K	Ni-Cr	Ni-Al			4°F (2.2°C)				¾%				
N	Ni-Cr-Si	Ni-SiMn			4°F (2.2°C)				¾%				
R	Pt-13% Rh	Pt			3°F (1.5°C)						¼%		
S	Pt-10% Rh	Pt			3°F (1.5°C)						¼%		

* Protect copper from oxidation above 600°F; iron above 900°F. Protect Ni-A1 from reducing atmospheres. Protect platinum from nonreducing atmospheres. Type B (Pt-30% Rh versus Pt-6%) is used up to 3,200°F(1,700°C). Its standard error is ½ percent above 1,470°F (800°C).

† Closer tolerances are obtainable by selection and calibration. Consult makers' catalogs. Tungsten-rhenium alloys are in use up to 5,000°F (2,760°C). For cryogenic thermocouples see Sparks et al., Reference Tables for Low-Temperature Thermocouples. *Natl. Bur. Stand. Monogr.* 124.

‡ Individual wires are designated by the ANSI symbol followed by P or N; thus iron is JP.

§ Constantan is 55% Cu, 45% Ni. The nickel-chromium and nickel-aluminum alloys are available as Chromel and Alumel, trademarks of Hoskins Mfg. Co.

Table 16.1.3   Polynomial Coefficients for Converting Thermocouple emf to Temperature*

Range	Type E	Type J	Type K	Type N	Type S	Type T
mV	0 to 76.373	0 to 42.919	0 to 20.644	0 to 47.513	1.874 to 11.95	0 to 20.872
°C	0 to 1,000°	0 to 760°	0 to 500°	0 to 1,300°	250 to 1,200°	0 to 400°
°F	32 to 1,832°	32 to 1,400°	32 to 932°	32 to 2,372°	482 to 2,192°	32 to 752°
α_0	0	0	0	0	1.291507177E + 01	0
α_1	1.7057035E + 01	1.978425E + 01	2.508355E + 01	3.8783277E + 01	1.466298863E + 02	2.592800E + 01
α_2	−2.3301759E − 01	−2.001204E − 01	7.860106E − 02	−1.1612344E + 00	−1.534713402E + 01	−7.602961E − 01
α_3	6.5435585E − 03	1.036969E − 02	−2.503131E − 01	6.9525655E − 02	3.145945973E + 00	4.637791E − 02
α_4	−7.3562749E − 05	−2.549687E − 04	8.315270E − 02	−3.0090077E − 03	−4.163257839E − 01	−2.165394E − 03
α_5	−1.7896001E − 06	3.585153E − 06	−1.228034E − 02	8.8311584E − 05	3.187963771E − 02	6.048144E − 05
α_6	8.4036165E − 08	−5.344285E − 08	9.804036E − 04	−1.6213839E − 06	−1.291637500E − 03	−7.293422E − 07
α_7	−1.3735879E − 09	5.099890E − 10	−4.413030E − 05	1.6693362E − 08	2.183475087E − 05	
α_8	1.0629823E − 11		1.057734E − 06	−7.3117540E − 11	−1.447379511E − 07	
α_9	−3.2447087E − 14		−1.052755E − 08		8.211272125E − 09	
Maximum deviation	± 0.02°C	± 0.04°C	−0.05 to + 0.04°C	± 0.06°C	± 0.01°C	± 0.03°C

*$T(°C) = \sum_{i=0}^{n} a_i \times (mV)^i$.

All temperatures are ITS-1990 and all voltages are SI-1990 values. Maximum deviation is that from the ITS-1990 tables; thermocouple wire error is additional. Computed temperature deviates greatly outside of given ranges. Consult source for thermocouple types B and R and for other millivolt ranges.

SOURCE: NIST Monograph 175, April 1993.

same materials as the thermocouple itself). This assures that the **cold junction** will be inside the instrument case, where compensation can be effectively applied. Cold junction compensation is typically achieved by measuring the temperature of the thermocouple wire to copper wire junctions or terminals with a resistive or semiconductor thermometer and correcting the measured terminal voltage by a derived equivalent millivolt cold junction value. Figure 16.1.25 shows a digital temperature indicator with correction for different ANSI types of thermocouple voltage to temperature nonlinearities being stored in and applied to the analog-to-digital converter (A/D) by a read-only memory (ROM) chip.

Fig. 16.1.25 Temperature measurement with thermocouple and digital millivoltmeter.

The **resistance thermometer** employs the same circuitry as described above, with the resistance element (RTD) being placed external to the instrument and the cold junction being omitted (Fig. 16.1.26). Three types of RTD connections are in use: two wire, three wire, and four wire. The two-wire connection makes the measurement sensitive to lead wire temperature changes. The three-wire connection, preferred in industrial applications, eliminates the lead wire effect provided the leads are of the same gage and length, and subject to the same environment. The four-wire arrangement makes no demands on the lead wires and is preferred for scientific measurements.

Fig. 16.1.26 Three-wire resistance thermometer with self-balancing potentiometer recorder.

The resistance bulb consists of a copper or platinum wire coil sealed in a protective metal tube. The **thermistor** has a very large temperature coefficient of resistance and may be substituted in low-accuracy, low-cost applications.

By use of a **selector switch,** any number of temperatures may be measured with the same instrument. The switch connects in order each thermocouple (or resistance bulb) to the potentiometer (or bridge circuit) or digital voltmeter. When balance is achieved, the recorder prints the temperature value, then the switch advances on to the next position.

Optical pyrometers are applied to high-temperature measurement in the range 1,000 to 5,000°F (540 to 2,760°C). One type is shown in Fig. 16.1.27. The surface whose temperature is to be measured (target) is focused by the lens onto the filament of a calibrated tungsten lamp.

Fig. 16.1.27 Optical pyrometer.

The light intensity of the filament is kept constant by maintaining a constant current flow. The intensity of the target image is adjusted by positioning the optical wedge until the image intensity appears exactly equal to that of the filament. A scale attached to the wedge is calibrated directly in temperature. The red filter is employed so that the comparison is made at a specific wavelength (color) of light to make the calibration more reproducible. In another type of optical pyrometer, comparison is made by adjusting the current through the filament of the standard lamp. Here, an ammeter in series is calibrated to read temperature directly. **Automatic operation** may be had by comparing filament with image intensities with a pair of photoelectric cells arranged in a bridge network. A difference in intensity produces a voltage, which is amplified to drive the slide wire or optical wedge in the direction to restore zero difference.

The **radiation pyrometer** is normally applied to temperature measurements above 1,000°F. Basically, there is no upper limit; however, the lower limit is determined by the sensitivity and cold-junction compensation of the instrument. It has been used down to almost room temperature. A common type of radiation receiver is shown in Fig. 16.1.28. A lens focuses the radiation onto a thermal sensing element. The temperature rise of this element depends on the total radiation received and the

Fig. 16.1.28 Radiation pyrometer.

conduction of heat away from the element. The radiation relates to the temperature of the target; the conduction depends on the temperature of the pyrometer housing. In normal applications the latter factor is not very great; however, for improved accuracy a compensating coil is added to the circuit. The sensing element may be a thermopile, vacuum thermocouple, or bolometer. The **thermopile** consists of a number of thermocouples connected in series, arranged so that all the hot junctions lie in the field of the incoming radiation; all of the cold junctions are in thermal contact with the pyrometer housing so that they remain at ambient temperature. The **vacuum thermocouple** is a single thermocouple whose hot junction is enclosed in an evacuated glass envelope. The **bolometer** consists of a very thin strip of blackened nickel or platinum foil which responds to temperature in the same manner as the resistance thermometer. The **thermal sensing element** is connected to a potentiometer or bridge network of the same type as described for the self-balance thermocouple and resistance-thermometer instruments. Because of the nature of the radiation law, the scale is nonlinear.

Accuracy of the optical- and radiation-type pyrometers depends on:

1. *Emissivity of the surface being sighted on.* For closed furnace applications, blackbody conditions can be assumed (emissivity = 1). For other applications corrections for the actual emissivity of the surface must be made (correction tables are available for each pyrometer model).

Multiple color or wavelength sensing is used to reduce sensitivity to hot object emissivity. For measuring hot fluids, a target tube immersed in the fluid provides a target of known emissivity.

2. *Radiation absorption between target and instrument.* Smoke, gases, and glass lenses absorb some of the radiation and reduce the incoming signal. Use of an enclosed (or purged) target tube or direct calibration will correct this.

3. *Focusing of the target on the sensing element.*

MEASUREMENT OF FLUID FLOW RATE

(See also Secs. 3 and 4.)

Flow is expressed in volumetric or mass units per unit time. Thus gases are generally measured in ft³/min (m³/min) or ft³/h (m³/h), steam in lb/h (kg/h), and liquid in gal/min (L/min) or gal/h (L/h). Conversion between volumetric flow Q and mass flow m is given by $m = K\rho Q$, where ρ = density of the fluid and K is a constant depending on the units of m, Q, and ρ. Flow rate can be measured directly by attaching a rate device to a volumetric meter of the types previously described, e.g., a tachometer connected to the rotating shaft of the nutating-disk meter (Fig. 16.1.13).

There are many kinds of flow instruments that serve in different applications. Depending on the application, the engineer may choose from a dozen different technologies. Issues, such as accuracy, repeatability, and cost, as well as suitability for the service, make selecting a flowmeter a demanding exercise. The basic flowmeter technologies include:

Positive displacement
Differential pressure
Variable area and rotameter
Turbine
Electromagnetic
Ultrasonic
Coriolis mass flowmeter
Vortex shedding and other fluidic
Thermal dispersion
Open-channel primary elements

Because of their use in domestic water systems worldwide, positive displacement meters (nutating disk, rotary piston, multijet) are the most widely used method of measurement of flow rate and total. They are, as the name implies, a means of measuring volumetric flow rate directly, one small volume at a time. Positive displacement meters are limited in the materials of construction, pressure, amount of air entrained, and the amount and size of solids they can pass.

They must be compensated for temperature and viscosity changes. Typically, they are used for totalization of clean fluids from water to petroleum byproducts. Differential pressure flowmeters operate by the conversion of fluid velocity to pressure (head). Thus, if the fluid is forced to change its velocity from V_1 to V_2, its pressure will change from p_1 to p_2 according to the equation (neglecting friction, expansion, and turbulence effects):

$$V_2^2 - V_1^2 = \frac{2g_c}{\rho}(p_1 - p_2) \qquad (16.1.1)$$

where g = acceleration due to gravity, ρ = fluid density, and g_c = 32.184 lbm · ft/(lbf)(s²) [1.0 kg · m/(N)(s²)]. **Caution:** If the flow pulsates, the average value of $p_1 - p_2$ will be greater than that for steady flow of the same average flow.

See "ASME Pipeline Flowmeters," and "Pitot Tubes" in Sec. 3 for coverage of venturi tubes, flow nozzles, compressible flow, orifice meters, ASME orifices, and Pitot tubes.

The tabulation orifice coefficients apply only for **straight pipe** upstream and downstream from the orifice. In most cases, satisfactory results are obtained if there are no fittings closer than 25 pipe diameters upstream and 5 diameters downstream from the orifice. The upstream limitation can be reduced a bit by employing **straightening vanes.** Reciprocating pumps in the line may introduce serious errors and require special efforts for their correction.

A wide variety of differential pressure meters is available for measuring the orifice (or other primary element) pressure drop.

Figure 16.1.29 shows an orifice plate and differential-pressure flow transmitter. An electronic device is shown, but in relatively old plants, and in some very special cases in the refinery industry, pneumatic devices still are present. The differential across the orifice plate produces a pressure differential on either side of the diaphragm, thus causing an electrical signal to be transmitted. Modem differential-pressure flow transmitters generally include a microprocessor, which permits temperature compensation, pressure compensation, and the conversion of the differential pressure (the square root of flow) to a linear flow rate. The absolute accuracy of a differential pressure flow device is fixed by the accuracy of the primary device.

Fig. 16.1.29 Orifice plate and differential-pressure transmitter.

The meters described thus far are termed **variable-head** because the pressure drop varies with the flow, orifice ratio being fixed. In contrast, the **variable-area** meter maintains a constant pressure differential but varies the orifice area with flow.

The **rotameter** (Fig. 16.1.30) consists of a float positioned inside a tapered tube by action of the fluid flowing up through the tube. The flow restriction is now the annular area between the float and the tube (area increases as the float rises). The pressure differential is fixed, determined by the weight of the float and the buoyant forces. To satisfy

Fig. 16.1.30 Rotameter.

the volumetric flow equation then, the annular area (hence the float level) must increase with flow rate. Thus the rotameter may be calibrated for direct flow reading by etching an appropriate scale on the surface of the glass tube. The calibration depends on the float dimensions, tube taper, and fluid properties. The equation for volumetric flow is

$$Q = C_R(A_T - A_F)\left[\frac{2gV_F}{\rho A_F}(\rho_F - \rho)\right]^{1/2}$$

where A_T = cross-sectional area of tube (at float position), A_F = effective float area, V_F = float volume, ρ_F = float density, ρ = fluid density, and C_R = rotameter coefficient (usually between 0.6 and 0.8). The coefficient varies with the fluid viscosity; however, special float designs are available which are relatively insensitive to viscosity effects. Also, fluid density compensation can be obtained.

The **rotameter reading may be transmitted** for recording and control purposes by affixing to the float a stem which connects to an armature or permanent magnet. The armature forms part of an inductance bridge whose signal is amplified electronically to drive a pen-positioning motor. For pneumatic transmission, the magnet provides magnet coupling to a pneumatic motion transmitter external to the rotameter tube. This generates an air pressure proportional to the height of the float.

The **area meter** is similar to the rotameter in operation. Flow area is varied by motion of a piston in a straight cylinder with openings cut into the wall. The piston position is transmitted as above by an armature and inductance bridge circuit.

Turbine, propeller, and paddlewheel flowmeters operate by converting velocity into rotational energy by means of a flighted rotor. These flowmeters are primarily used to measure water, and water-based liquids, and come in a very wide variety of designs, styles, and materials of construction. Some of these devices are mechanical, using gears and cams to convert the rotation of the rotor to increment an electromechanical counter. Others use solid-state pickups such as magnetic induction coils and Hall effect sensors to sense the speed of rotation of the rotor.

Fig. 16.1.31   Propeller-type flowmeter.

These flowmeters are limited by the velocity of the fluid (they can be overdriven, with disastrous results), by the size and composition of solids they can pass, and by temperature and viscosity changes, and they are subject to significant wear in operation. Some turbine meters can be calibrated to be very highly accurate and are often used to meter expensive additives and blends.

One of the most widely used flowmeter types is the electromagnetic flowmeter, often called a **magmeter**. The device operates by means of Faraday's law, converting velocity of a conductive fluid to a proportional current signal by passing the fluid at right angles to an electromagnetic field. Magmeters are inherently the most accurate of the velocimetric flow technologies, because the current signal is directly proportional to the average velocity of the conductive fluid. Magmeters are therefore restricted to conductive fluids (generally above 5 microsiemens). Magmeters are supplied in a very wide range of sizes (generally from 1 to 2,000 mm), and the larger meters are among the most economical devices. Magmeters also have the advantage that they can be made out of a wide variety of materials, including ones suitable for ultrapure service and highly corrosive or abrasive service as well. Their final large advantage over other flowmeter types is that they are obstructionless, having no moving parts in the flow stream, and therefore do not increase the pressure drop to the system over the pressure drop of an equivalent length of pipe.

Four different types of flowmeter using ultrasonic technology are in common use. These are Doppler flowmeters, transit time flowmeters, sonar flowmeters, and ultrasonic level devices used in conjunction with open-channel primary flow devices such as weirs and flumes.

Doppler flowmeters operate by means of the Doppler effect. A signal of known frequency (typically 660 kHz) is beamed into a pipe. The return echo is frequency-shifted proportionally to the velocity of the fluid in the pipe. Because homogenous fluids generally do not reflect the signal well, Doppler flowmeters are restricted to fluids with relatively large quantities of entrained particulate (usually a minimum of 100 mg/L of plus-100-μm solids). They are commonly used in mining, dredging, and wastewater sludge applications.

Transit time flowmeters operate by beaming an ultrasonic signal into the pipe in the direction of flow, and a duplicate signal downstream. The difference in time for the transmitted signal to reach the receiver between the upstream and downstream signals is proportional to the velocity of the fluid. The difference in transit time is $2vl/(c^2 + v^2)$, where l is the path length. For $v \ll c$, the factor $(c^2 + v^2)$ is nearly constant. Like magmeters, transit time flowmeters have no pressure drop. Transit time flowmeters can be provided in similar sizes to magmeters, and are used in all types of fluids, except slurries. Transit time flowmeters can be used in gas flow measurement. Increasing the number of paths of signal transmission can increase the accuracy of the flowmeter, and transit time flowmeters are increasingly used for transmission and custody transfer of petroleum products and gas.

Sonar flowmeters are acoustic, rather than strictly ultrasonic, and operate by "listening" to the sonic properties of the fluid and determining velocity from the sonic profile.

Ultrasonic open-channel flowmeters are actually ultrasonic level transmitters used at the measurement point of a flume or weir, and supplied with electronic conversion from measured height to flow rate. Because of their noncontacting nature, ultrasonic open-channel flowmeters have become the standard technology for this measurement, replacing floats and other mechanical devices. The limiting accuracy of these devices is not the accuracy of the sensor, but rather the accuracy of the primary device.

Coriolis mass flowmeters are now common in many process industries. Generally available from fractional inch to 4 in (100 mm), at least one manufacturer produces up to 12-in sizes. Using the principle of measurement of the Coriolis force, these devices natively measure mass flow, and can be used to compute volumetric flow rate and total, and fluid density. Recent advances in Coriolis mass flowmeter design have permitted the device to also be used for measurement of gases.

Another extremely common measurement devices is the **fluidic flowmeter**, of which the best known examples are the vortex-shedding, the vortex-precession (or swirlmeter), and the Coanda effect flowmeter. The **vortex-shedding flowmeter** operates by sensing in a variety of ways the vortices shed by a bluff body in the flow stream. The frequency of the shed vortices is proportional to the fluid velocity. The **swirlmeter**, an ancestor of the vortex-shedder, uses the combination of a stationary turbine rotor to impart a twist and a flow restriction to produce a standing wave, the frequency of which is proportional to velocity. The **Coanda effect flowmeter** uses the Coanda effect to produce a hydraulic switch. The frequency of the switching operation is proportional to the velocity of the fluid. Vortex-shedding flowmeters and swirlmeters are commonly used in chemical and steam applications, while Coanda effect flowmeters are normally found in district heating and gas applications in relatively small sizes. Vortex-shedding flowmeters are limited in turndown at low velocities, and should not be used in applications with large-diameter solids such as wastewater or mining.

Thermal dispersion flowmeters operate by measuring the temperature differential caused by the passage of a fluid between two sensors. One sensor is heated to a constant temperature, and imparts heat to the fluid. The other sensor measures the temperature of the fluid. The temperature differential produces a power demand on the heater circuit that is proportional to the mass flow of the fluid. Thermal dispersion flowmeters are commonly used in gas flow measurement, but can also be used in some liquid flow measurements, generally as switches. These devices can be extremely accurate, and at least one manufacturer produces ultralow-flow thermal dispersion flowmeters that measure in the millilitre per hour range.

The most common open-channel primary flow elements in current usage are the **v-notch** and rectangular **weirs**, and the **Parshall** and

Palmer-Bowlus flumes. All of these devices produce a flow restriction that raises the fluid behind the restriction to a predictable height. The change in height H produces a measurement of flow rate by the formula $Q = K/n$, where K is a constant related to flow rate units and n is a constant related to the design of the specific primary element. The constant n is often written as a fraction, as in $^3/_2$. Other forms of primary elements for open-channel flow exist, and sometimes the **Manning formula** and its derivatives are used to measure the flow in open conduits without a primary device. Manning's formula is: $v = K/n \, R^{2/3} \, S^{1/2}$ (English units) [SI units], where v = fluid velocity (ft/s) or [m/s], K = 1.486 (English), K =1.00 [SI], R = hydraulic radius (ft) or [m] = A/P_w, A = cross-sectional area of flow (ft^2) or [m^2], P_w = wetted perimeter (ft) or [m], S = channel slope, and n = Manning's roughness coefficient. Then $Q = vA$.

Most flowmeters measure rate of flow. To measure the total quantity of fluid flowing during a specific time interval, it is necessary to integrate the flow rate over that interval. The integration, prior to the advent of microprocessor-based flow transmitters, was often done manually, or with the use of specially shaped cams. Almost all flow transmitters now integrate mathematically, and display total flow on an electronic or electromechanical register.

POWER MEASUREMENT

Power is defined as the rate of doing work. Common units are the horsepower and the kilowatt: 1 hp = 33,000 ft · lb/min = 0.746 kW. The power input to a rotating machine in hp (W) = $2\pi nT/k$, where n = r/min of the shaft where the torque T is measured in lbf · ft (N · m), and k = 33,000 ft · lbf/hp · min [60 N · m/(W · min)]. The same equation applies to the power output of an engine or motor, where n and T refer to the output shaft. Mechanical power-measuring devices (**dynamometers**) are of two types: (1) those absorbing the power and dissipating it as heat and (2) those transmitting the measured power. As indicated by the above equation, two measurements are involved: shaft speed and torque. The speed is measured directly by means of a tachometer. Torque is usually measured by balancing against weights applied to a fixed lever arm; however, other force measuring methods are also used. In the **transmission dynamometer,** the torque is measured by means of strain-gage elements bonded to the transmission shaft.

There are several kinds of **absorption dynamometers.** The **Prony brake** applies a friction load to the output shaft by means of wood blocks, flexible band, or other friction surface. The **fan brake** absorbs power by "fan" action of rotating plates on surrounding air. The **water brake** acts as an inefficient centrifugal pump to convert mechanical energy into heat. The pump casing is mounted on antifriction bearings so that the developed turning moment can be measured. In the **magnetic-drag** or **eddy-current brake**, rotation of a metal disk in a magnetic field induces eddy currents in the disk which dissipate as heat. The field assembly is mounted in bearings in order to measure the torque.

One type of **Prony brake** is illustrated in Fig. 16.1.32. The torque developed is given by $L(W - W_0)$, where L is the length of the brake arm, ft; W and W_0 are the scale loads with the brake operating and with the brake free, respectively. The brake horsepower then equals $2\pi nL(W - W_0)/33,000$, where n is shaft speed, r/min.

Fig. 16.1.32   Prony brake.

In addition to eddy-current brakes, **electric dynamometers** include calibrated generators and motors and cradle-mounted generators and motors. In calibrated machines, the efficiency is determined over a range of operating conditions and plotted. Mechanical power measurement can then be made by measuring the electrical power input (or output) to the machine. In the electric-cradle dynamometer, the motor or generator stator is mounted in trannion bearing so that the torque can be measured by suitable scales.

The **engine indicator** is a device for plotting cylinder pressure as a function of piston (or volume) displacement. The resulting p-v diagram (Fig. 16.1.33) provides both a measure of the work done in a reciprocating engine, pump, or compressor and a means for analyzing its performance (see Secs. 4, 9, and 14). If A_d is the area inside the closed curve drawn by the indicator, then the indicated horsepower for the cylinder under test = KnA_pA_d where K is a proportionality factor determined by the scale factors of the indicator diagram, n = engine speed, r/min, A_p = piston area.

Fig. 16.1.33   Indicator diagram.

Completely mechanical indicators can be used only for low-speed machines. They have largely been superseded by electrical transducers using strain gages, variable capacitance, piezoresistive, and piezoelectric principles which are suitable for **high-speed** as well as low-speed pressure changes (the piezoelectric principle has low-speed limitations). The usual diagram is produced on an oscilloscope display as pressure vs. time, with a marker to indicate some reference event such as spark timing or top dead center. Special transducers can be coupled to a crank or cam shaft to give an electrical signal representing piston motion so that a p-v diagram can be shown on an oscilloscope.

ELECTRICAL MEASUREMENTS
(See also Sec. 15.)

Electrical measurements serve two purposes: (1) to measure the electrical quantities themselves, e.g., line voltage, power consumption, and (2) to measure other physical quantities which have been converted into electrical variables, e.g., temperature measurement in terms of thermocouple voltage.

In general, there is a sharp distinction between ac and dc devices used in measurements. Consequently, it is often desirable to transform an ac signal to an equivalent dc value, and vice versa. An ac signal is converted to dc (rectified) by use of **selenium rectifiers, silicon or germanium diodes, or electron-tube diodes**. Full-wave rectification is accomplished by the **diode bridge,** shown in Fig. 16.1.34. The rectified signal may be passed through one or more low-pass filter stages to smooth the waveform to its average value. Similarly, there are many ways of modulating a dc signal (converting it to alternating current). The most common method used in instrument applications is a solid-state oscillator.

Fig. 16.1.34   Full-wave rectifier.

The **galvanometer** (Fig. 16.1.35), recently supplanted by the direct-reading **digital voltmeter** (DVM), is basic to dc measurement. The input signal is applied across a coil mounted in jeweled bearings or on a taut-band suspension so that it is free to rotate between the poles of a

Fig. 16.1.35 D'Arsonval galvanometer.

permanent magnet. Current in the coil produces a magnetic moment which tends to rotate the coil. The rotation is limited, however, by the restraining torque of the hairsprings. The resulting deflection of the coil θ is proportional to the current I:

$$\theta = \frac{NBWL}{K} I$$

where N = number of turns in coil; W, L = coil width and length, respectively; B = magnetic field intensity; K = spring constant of the hairsprings. Galvanometer deflection is indicated by a balanced pointer attached to the coil. In very sensitive elements, the pointer is replaced by a mirror reflecting a spot of light onto a ground-glass scale; the bearings and hairspring are replaced by a torsion-wire suspension.

The galvanometer can be converted into a **dc voltmeter, ammeter, or ohmmeter** by application of Ohm's law, $IR = E$, where I = current, A; E = electrical potential, V; and R = resistance, Ω.

For a **voltmeter**, a fixed resistance R is placed in series with the galvanometer (Fig. 16.1.36a). The current i through the galvanometer is proportional to the applied voltage E: $i = E/(r + R)$, where r = coil resistance. Different voltage ranges are obtained by changing the series resistance.

An **ammeter** is produced by placing the resistance in parallel with the galvanometer or DVM (Fig. 16.1.36b). The current then divides between the galvanometer coil or DVM and the resistor in inverse ratio to their resistance values (r and R, respectively); thus, $i = IR/(r + R)$, where i = current through coil and I = total current to be measured. Different current ranges are obtained by using different shunt resistances.

Fig. 16.1.36 (a) Voltmeter; (b) ammeter.

The common ohmmeter consists of a battery, a galvanometer with a shunt rheostat, and resistance in series to total Ri Ω. The shunt is adjusted to give a full-scale (0 Ω) reading with the test terminals shorted. When an unknown resistance R is connected, the deflection is $Ri/(Ri + R)$ fraction of full scale. The scale is calibrated to read R directly. A half-scale deflection indicates $R = Ri$. Alternatively, the galvanometer is connected to read voltage drop across the unknown R while

a known current flows through it. This principle is used for low-value resistances and in digital ohmmeters.

Digital instruments are available for all these applications and often offer higher resolution and accuracy with less circuit loading. Fluctuating readings are difficult to follow, however.

Alternating current and voltage must be measured by special means. A dc instrument with a rectifier input is commonly used in applications requiring high input impedance and wide frequency range. For precise measurement at power-line frequencies, the electrodynamic instrument is used. This is similar to the galvanometer except that the permanent magnet is replaced by an electromagnet. The movable coil and field coils are connected in series; hence they respond simultaneously to the same current and voltage alternations. The pointer deflection is proportional to the square of the input signal. The moving-iron-type instruments consist of a soft-iron vane or armature which moves in response to current flowing through a stationary coil. The pointer is attached to the iron to indicate the deflection on a calibrated nonlinear scale. For measuring at very high frequencies, the **thermocouple voltmeter or ammeter** is used. This is based on the heating effect of the current passing through a fixed resistance R. Heat is liberated at the rate of E^2/R or $I^2 R$ W.

DC **electrical power** is the product of the current through the load and the voltage across the load. Thus it can be simply measured using a voltmeter and ammeter. AC power is directly indicated by the wattmeter, which is similar to the electrodynamic instrument described above. Here the field coils are connected in series with the load, and the movable coil is connected across the load (to measure its voltage). The deflection of the movable coil is then proportional to the effective load power.

Precise voltage measurement (direct current) can be made by balancing the unknown voltage against a measured fraction of a known reference voltage with a **potentiometer** (Fig. 16.1.9). Balance is indicated by means of a sensitive current detector placed in series with the unknown voltage. The potentiometer is calibrated for angular position vs. fractional voltage output. Accuracies to 0.05 percent are attainable, dependent on the linearity of the potentiometer and the accuracy of the reference source. The reference standard may be a Weston standard cell or a regulated voltage supply (based on diode characteristics). The balance detector may be a galvanometer or electronic amplifier.

Precision resistance and general impedance measurements are made with bridge circuits (Fig. 16.1.37) which are adjusted until no signal is detected by the null detector (bridge is balanced). Then $Z_1Z_3 = Z_2Z_4$. The basic Wheatstone bridge is used for resistance measurement where all the impedances (Z's) are resistances (R's). If R_1 is to be measured, $R_1 = R_2R_4/R_3$, when balanced. A sensitive galvanometer for the null detector and dc voltage excitation is usual. All R_2, R_3, R_4 must be calibrated, and some adjustable. For general impedance measurement, ac voltage excitation of suitable frequency is used. The null detector may be a sensitive ac meter, oscilloscope, or, for audio frequencies, simple earphones. The basic balance equation is still valid, but it now requires also that the sum of the phase angles of Z_1 and Z_3 equal the sum of the phase angles of Z_2 and Z_4. As an example, if Z_1 is a capacitor, the bridge can be balanced if Z_2 is a known capacitor while Z_3 and Z_4 are resistances. The phase-angle condition is met, and $Z_1Z_3 = Z_2Z_4$ becomes $(1/2\pi fC_1)R_3 = (1/2\pi fC_2)R_4$ and $C_1 = C_2R_3/R_4$. Variations on the basic

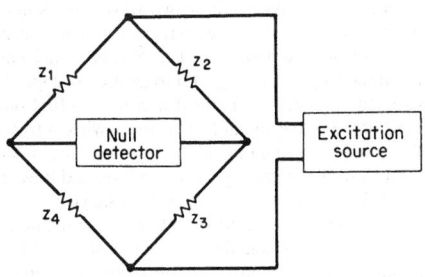

Fig. 16.1.37 Impedance bridge.

principle include the Kelvin bridge for measurement of low resistance, and the Mueller bridge for platinum resistance thermometers.

Voltage measurement requires a meter of substantially higher impedance than the impedance of the source being measured. The vacuum-tube cathode follower and the field-effect transistor are suitable for high-impedance inputs. The following circuitry may be a simple amplifier to drive a pointer-type meter, or may use a digital technique to produce a digital output and display. Digital counting circuits are capable of great precision and are widely adapted to measurements of time, frequency, voltage, and resistance. Transducers are available to convert temperature, pressure, flow, length and other variables into signals suitable for these instruments.

The charge amplifier is an example of an operational amplifier application (Fig. 16.1.38). It is used for outputs of piezoelectric transducers in which the output is a charge proportional to input force or other input converted to a force. Several capacitors switchable across the feedback path provide a range of full-scale values. The output is a voltage.

Fig. 16.1.38   Charge amplifier application.

The **cathode-ray oscilloscope** (Fig. 16.1.39) is an extremely useful and versatile device characterized by high input impedance and wide frequency range. An electron beam is focused on the phosphor-coated face of the cathode-ray tube, producing a visible spot of light at the point of impingement. The beam is deflected by applying voltages to vertical

Fig. 16.1.39   Cathode-ray tube.

and horizontal deflector plates. Thus, the relationship between two varying voltages can be observed by applying them to the vertical and horizontal plates. The horizontal axis is commonly used for a linear time base generated by an internal sawtooth-wave generator. Virtually any desired sweep speed is obtainable as a calibrated sweep. Sweeps which change value part way across the screen are available to provide localized time magnification. As an alternative to the time base, any arbitrary voltage can be applied to drive the horizontal axis. The vertical axis is usually used to display a dependent variable voltage. **Dual-beam** and **dual-trace** instruments show two waveforms simultaneously. Special long-persistence and storage screens can hold transient waveforms for from seconds to hours. Greater versatility and unique capabilities are afforded by use of **digital-storage** oscilloscope. Each input signal is sampled, digitized, and stored in a first-in–first-out memory. Since a record of the recent signal is in memory when a trigger pulse is received, the timing of the end of storing new data into memory determines how much of the stored signal was before, and how much after, the trigger. Unlike storage screens, the stored signal can be amplified and shifted on the screen for detailed analysis, accompanied by numerical display of voltage and time for any point. Care must be taken that enough samples are taken in any waveform; otherwise aliasing results in a false view of the waveform.

The stored data can be processed mathematically in the oscilloscope or transferred to a computer for further study. Accessories for microcomputers allow them to function as digital oscilloscopes and other specialized tasks.

VELOCITY AND ACCELERATION MEASUREMENT

Velocity or speed is the time rate of change of displacement. Consequently, if the displacement measuring device provides an output signal which is a continuous (and smooth) function of time, the velocity can be measured by **differentiating** this signal either graphically or by use of a differentiating circuit. The accuracy may be very limited by noise (high-frequency fluctuations), however. More commonly, the output of an accelerometer is integrated to yield the velocity of the moving member. **Average speed** over a time interval can be determined by measuring the time required for the moving body to pass two fixed points a known distance apart. Here photoelectric or other rapid sensing devices may be used to trigger the start and stop of the timer. **Rotational speed** may be similarly measured by counting the number of rotations in a fixed time interval.

The **tachometer** provides a direct measure of angular velocity. One form is essentially a small permanent-magnet-type generator coupled to the rotating element; the voltage induced in the armature coil is directly proportional to the speed. The principle is also extended to rectilinear motions (restricted to small displacements) by using a straight coil moving in a fixed magnetic field.

Angular velocity can also be measured by magnetic drag-cup and **centrifugal-force** devices (flyball governor). The force may be balanced against a spring with the resulting deflection calibrated in terms of the shaft speed. Alternatively, the force may be balanced against the air pressure generated by a pneumatic nozzle-baffle assembly (similar to Fig. 16.1.23).

Vibration velocity pickups may use a coil which moves relative to a magnet. The voltage generated in the coil has the same frequency as the vibration and, for sine motion, a magnitude proportional to the product of vibration frequency and amplitude. Vibration **acceleration pickups** commonly use strain-gage, piezoresistive, or piezoelectric elements to sense a force $F = Ma/g_c$. The maximum usable frequency of an accelerometer is about one-fifth of the pickup's natural frequency (see Sec. 3.4). The minimum usable frequency depends on the type of pickup and the associated circuitry. The output of an accelerometer can be integrated to obtain a velocity signal; a velocity signal can be integrated to obtain a displacement signal. The **operational amplifier** is a versatile element which can be connected as an integrator for this use.

Holography is being applied to the study of surface vibration patterns.

MEASUREMENT OF PHYSICAL AND CHEMICAL PROPERTIES

Physical and chemical measurements are important in the control of product quality and composition. In the case of manufactured items, such properties as color, hardness, surface, roughness, etc., are of interest. Color is measured by means of a **colorimeter,** which provides comparison with color standards, or by means of a **spectrophotometer,** which analyzes the color spectrum. The **Brinell and Rockwell testers** measure surface hardness in terms of the depth of penetration of a hardened steel ball or special stylus. Testing machines with **strain-gage** elements provide measurement of the strength and elastic properties of materials. **Profilometers** are used to measure surface characteristics. In one type, the surface contour is magnified optically and the image projected onto a screen or viewer; in another, a stylus is employed to translate the surface irregularities into an electrical signal which may be recorded in the form of a highly magnified profile of the surface or presented as an averaged roughness-factor reading.

For liquids, attributes such as density, viscosity, melting point, boiling point, transparency, etc., are important. Density measurements have already been discussed. **Viscosity** is measured with a **viscosimeter,** of which there are three main types: flow through an orifice or capillary

(Saybolt), viscous drag on a cylinder rotating in the fluid (MacMichael), damping of a vibrating reed (Ultrasonic) (see Secs. 3 and 4). **Plasticity** and consistency are related properties which are determined with special apparatus for heating or cooling the material and observing the temperature-time curve. The **photometer, reflectometer, and turbidimeter** are devices for measuring transparency or turbidity of nonopaque liquids and solids.

A variety of properties can be measured for determining chemical composition. **Electrical properties** include pH, conductivity, dielectric constant, oxidation potential, etc. **Physical properties** include density, refractive index, thermal conductivity, vapor pressure, melting and boiling points, etc. Of increasing industrial application are **spectroscopic measurements: infrared absorption spectra, ultraviolet and visible emission spectra, mass spectrometry,** and **gas chromatography.** These are specific to particular types of compounds and molecular configurations and hence are very powerful in the analysis of complex mixtures. As examples, infrared analyzers are in use to measure low-concentration contaminants in engine oils resulting from wear and in hydraulic oils to detect deterioration. **X-ray diffraction** has many applications in the analysis of crystalline solids, metals, and solid solutions.

Of special importance in the realm of composition measurements is the determination of **moisture content.** A common laboratory procedure measures the loss of weight of the oven-dried sample. More rapid methods employ electrical conductance or capacitance measurements, based on the relatively high conductivity and dielectric constant values for ordinary water.

Water vapor in air (humidity) is measured in terms of its physical properties or effects on materials (see also Secs. 4 and 12). (1) The **psychrometer** is based on the cooling effect of water evaporating into the airstream. It consists of two thermal elements exposed to a steady airflow; one is dry, the other is kept moist. See Sec. 4 for psychrometric charts. (2) The **dew-point recorder** measures the temperature at which water just starts to condense out of the air. (3) The **hygrometer** measures the change in length of such humidity-sensitive elements as hair and wood. (4) **Electric sensing elements** employ a wire-wound coil impregnated with a hygroscopic salt (one that maintains an equilibrium between its moisture content and the air humidity) such that the resistance of the coil is related to the humidity.

The **throttling calorimeter** (Fig. 16.1.40) is most commonly used for determining the moisture in steam. A sampling nozzle is located preferably in a vertical section of steampipe far removed from any fittings. Steam enters the calorimeter through a throttling orifice and into a well-insulated expansion chamber. The steam quality x (fraction dry steam) is determined from the equation $x = (h_c - h_f) / h_{fg}$, where h_c is the enthalpy of superheated steam at the temperature and pressure measured in the calorimeter; h_f and h_{fg} are, respectively, the liquid enthalpy

Fig. 16.1.40   Throttling calorimeter.

and the heat of vaporization corresponding to line pressure. The chamber is conveniently exhausted to atmospheric pressure; then only line pressure and temperature of the throttled steam need be measured. The range of the throttling calorimeter is limited to small percentages of moisture; a **separating calorimeter** may be employed for larger moisture contents.

The **Orsat** apparatus is generally used for chemical analysis of flue gases. It consists of a graduated tube or burette designed to receive and measure volumes of gas (at constant temperature). The gas is analyzed for CO_2, O_2, CO, and N_2 by bubbling through appropriate absorbing reagents and measuring the resulting change in volume. The reagents normally employed are KOH solution for CO_2, pyrogallic acid and KOH mixture for O_2, and cuprous chloride (Cu_2Cl_2) for CO. The final remaining unabsorbed gas is assumed to be N_2. The most common errors in the Orsat analysis are due to **leakage** and poor **sampling.** The former can be checked by simple test; the latter factor can only be minimized by careful sampling procedure. Recommended procedure is the taking of several simultaneous samples from different points in the cross-sectional area of the flue-gas stream, analyzing these separately, and averaging the results.

There are many instruments for **measuring CO_2 (and other gases) automatically.** In one type, the CO_2 is absorbed in KOH, and the change in volume determined automatically. The more common type, however, is based on the difference in **thermal conductivity** of CO_2 compared with air. Two thermal conductivity cells are set into opposing arms of a Wheatstone-bridge circuit. Air is sealed into one cell (reference), and the CO_2-containing gas is passed through the other. The cell contains an electrically heated resistance element; the temperature of the element (and therefore its resistance) depends on the thermal conductivity of the gaseous atmosphere. As a result, the unbalance of the bridge provides a measure of the CO_2 content of the gas sample.

The same principle can be employed for analyzing other constituents of gas mixtures where there is a significant thermal-conductivity difference. A modification of this principle is also used for determining CO or other combustible gases by mixing the gas sample with air or oxygen. The combustible gas then burns on the heated wire of the test cell, producing a temperature rise which is measured as above.

Many other physical properties are employed in the determination of specific components of gaseous mixtures. An interesting example is the **oxygen analyzer,** based on the unique paramagnetic properties of oxygen.

NUCLEAR RADIATION INSTRUMENTS
(See also Sec. 9.)

Nuclear radiation instrumentation is increasing in importance with two main areas of application: (1) measurement and control of radiation variables in nuclear reactor-based processes, such as nuclear power plants and (2) measurement of other physical variables based on radioactive excitation and tracer techniques. The instruments respond in general to electromagnetic radiation in the gamma and perhaps X-ray regions and to beta particles (electrons), neutrons, and alpha particles (helium nuclei).

Gas Ionization Tubes   The **ion chamber, proportional counter,** and **Geiger counter** are common instruments for radiation detection and measurement. These are different applications of the gas-ionization tube distinguished primarily by the amount of applied voltage.

A simple and very common form of the instrument consists of a gas-filled cylinder with a fine wire along the axis forming the anode and the cylinder wall itself (at ground potential) forming the cathode, as shown in Fig. 16.1.41. When a radiation particle enters the tube, its collision with gas molecules causes an ionization consisting of electrons (negatively charged) and positive ions. The electrons move very rapidly toward the positively charged wire; the heavier positive ions move relatively slowly toward the cathode. The above activity is detected by the resulting current flow in the external circuitry.

When the voltage applied across the tube is relatively low, the number of electrons collected at the anode is essentially equal to that produced by the incident radiation. In this voltage range, the device is

Fig. 16.1.41   Gas-ionization tube.

called an **ion chamber.** The device may be used to count the number of radiation particles when the frequency is low; when the frequency is high, an external integrating circuit yields an output current proportional to the radiation intensity. Since the amplification factor of the ion chamber is low, high-gain electronic amplification of the current signal is necessary.

If the applied voltage is increased, a point is reached where the radiation-produced ions have enough energy to collide with other gas molecules and produce more ions which also enter into collisions so that an "avalanche" of electrons is collected at the anode. Thus, there is a very considerable amplification of the output signal. In this region, the device is called a **proportional counter** and is characterized by the voltage or current pulse being proportional to the energy content of the incident radiation signal.

With still further increase in the applied voltage, a point of saturation is reached wherein the output pulses have a constant amplitude independent of the incident radiation level. The resulting **Geiger counter** is capable of producing output pulses up to 10 V in amplitude, thus greatly reducing the requirements on the external circuitry and instrumentation. This advantage is offset somewhat by a lower maximum counting rate and more limited ability to differentiate among the various types of radiation as compared with the proportional counter.

The **scintillation counter** is based on the excitation of a phosphor by incident radiation to produce light radiation which is in turn detected by a photomultiplier tube to yield an output voltage. The signal output is greatly amplified and nearly proportional to the energy of the initial radiation. The device may be applied to a wide range of radiations, it has a very fast response, and, by choice of phosphor material, it offers a large degree of flexibility in applications.

Applications to the Measurement of Physical Variables   The ready availability of radioactive isotopes of long half-life, such as cobalt 60, make possible a variety of industrial and laboratory measuring techniques based on radiation instruments of the type described above. Most applications are based on (1) radiation absorption, (2) tracer identification, and (3) other properties. These techniques often have the advantages of isolation of the measuring device from the system, access to a variable not observable by conventional means, or measurement without destruction or modification of the system.

In the utilization of **absorption** properties, a radioactive source is separated from the radiation-measuring device by that part of the system to be measured. The measured radiation intensity will depend on the fraction of radiation absorbed, which in turn will depend on the distance traveled through the absorbing medium and the density and nature of the material. Thus, the instrument can be adapted to measuring thickness (see Fig. 16.1.12), coating weight, density, liquid or solids level, or concentration (of certain components).

Tracer techniques are effectively used in measuring flow rates or velocities, residence time distributions, and flow patterns. In flow measurement, a sharp pulse of radioactive material may be injected into the flow stream; with two detectors placed downstream from the injection point and a known distance apart, the velocity of the pulse is readily measured. Alternatively, if a known constant flow rate of tracer is injected into the flow stream, a measure of the radiation downstream is easily converted into a measure of the desired flow rate. Other applications

of tracer techniques involve the use of tagged molecules embedded in the process to provide measures of wear, chemical reactions, etc.

Other applications of radiation phenomena include level measurements based on a floating radioactive source, level measurements based on the back-scattering effect of the medium, pressure measurements in the high-vacuum region based on the amount of ionization caused by alpha rays, location of interface in pipeline transmission applications, and certain chemical analysis applications.

INDICATING, RECORDING, AND LOGGING

Rapid change has overtaken the ways in which process information is presented and stored. The control panel of the 1970s, with banks of analog pointer indicators and circular and strip chart recorders, has given way to digital indicators, dataloggers, and recorder simulations that are entirely software based. Some industrial sectors continue the use of paper chart recorders, so typical strip and circular chart recorders are shown here.

Fig. 16.1.42   Circular-chart recorder.

Fig. 16.1.43   Strip-chart recorder.

These include the nuclear power, food, and pharmaceuticals industry sectors. However, even these last holdouts are now moving to more modem means of indicating and recording information about measured variables.

With the advent of liquid crystal displays (LCDs) that can display graphical elements, even field instruments now have digital displays, some of which mimic analog indicators with "dials" or "bargraphs" or other even more sophisticated displays.

Replacing recorders in most applications now are digital dataloggers with cathode-ray tube (CRT) or LCD displays that simulate the visual aspect of an analog trend recorder. These are designed with the same form factor as older mechanical recorders so that they can be inserted into precut holes in existing control panels.

Most HMI packages for control systems contain a software system called a "data historian," which provides for logging and display of large amounts of process measurement data. Data historian software is often used to provide data to advanced process control software packages and to enterprise software systems for quality control and supply chain integration.

Field dataloggers have become small and are operated by very low power internal batteries, and can even be inserted into, for example, a single carton of food for long-term monitoring. These dataloggers can be accessed by connection to a laptop, another type of host computer, or remotely by means of wireless connections such as Bluetooth, Zigbee, and the various IEEE 802.11 standards.

INFORMATION TRANSMISSION

In the process plant, information is transmitted from field devices and control elements to control stations and the process control computer system by means of either analog or digital signals.

With the exception of some refinery and petrochemical applications, where pneumatic instrumentation, using a 3 to 15 lb/in^2 variable pressure range, is still used, the vast majority of analog information transmission is by means of either variable voltage, current, or frequency.

ANSI/ISA50 standardized the analog transmission protocol to be 4 to 20 mA dc (1 to 5 V dc) into 1,000 ohms resistive as equal to 0 to 100 percent scale. Another transmission protocol in less common use is 0 to 20 mA dc, 10 to 50 V dc, among many others. Two-wire instruments receive their operational power as well as transmit their signal on the "current output loop." Four-wire instruments require separate power supplies for operation.

ANSI/ISA50 chose a "live zero" at 4 mA dc so that a reading of 0 mA would indicate a diagnostic failure condition. Numerous manufacturers provide field converters to take voltages and frequencies and convert them to current loop outputs, and vice versa.

While many field sensors continue to have analog outputs (voltage, current, frequency, or pressure), many field instruments are now supplied with digital communications systems built into the device. The most ubiquitous in process automation is the **HART protocol** (www.hartcomm.org), which is a digital communications system carried on existing 4- to 20-mA dc current loop outputs. The HART protocol permits remote readout, calibration, diagnostics, and reprogramming of many field instruments and control valves.

Other bus systems are in use, some relatively old, such as IEEE 488 GPIB (or BPIB), developed as a general-purpose interface bus. Other buses such as Modbus are often found in machine control applications in discrete automation. Another digital communications bus, BACNet, has found wide acceptance in building automation systems worldwide.

Growing reliance is being placed, at least in new plant construction, on two other, more fully featured communications methods, known by the common name of **fieldbuses**. **Profibus** (www.profibus.com) and **Foundation Fieldbus** (www.fieldbus.org) provide complete bidirectional communications and control between field instruments and final control elements and the process control system. The design of fieldbuses is beginning to converge with the design of enterprise LAN and WAN (local- and wide-area networks) using TCP/EP protocols over Ethernet configured networks. Recently, IEEE released a standard for power over Ethernet, but it is not yet workable for field instruments.

Despite intellectual property challenges, the use of embedded web servers in field instruments and controllers is proliferating. This permits the use of standard Internet-based communication protocols by even a pressure transmitter or flowmeter. The control system then is a virtual control system architected as a series of World Wide Web pages, rather than a physical hardware set.

16.2 AUTOMATIC CONTROLS

by Gregory V. Murphy

REFERENCES: Thaler, "Elements of Servomechanism Theory," McGraw-Hill. Shinskey, "Process Control Systems: Application, Design and Tuning," McGraw-Hill. Zoss, "Applied Instrumentation in the Process Industries," Vol. IV, Gulf Publishing Co. Truxal, "Control Engineers Handbook," McGraw-Hill. A. W. Langill, "Automatic Control Systems Engineering," Prentice-Hall. Kuo, "Automatic Control Systems," Prentice-Hall. Phillips and Nagle, "Digital Control System Analysis and Design," Prentice-Hall. Lewis, "Applied Optimal Control and Estimation: Digital Design and Implementation," Prentice-Hall. Cochin and Plass, "Analysis and Design of Dynamic Systems," HarperCollins. Astrom and Hagglund, "Automatic Tuning of PID Controllers," Instrument Society of America. Maciejowski, "Multivariable Feedback Design," Addison-Wesley. Doyle and Stein, "Robustness with Observers," IEEE Trans. Automatic Control, vol. AC-24, pp. 607–611, August 1979. Murphy and Bailey, LQG/LTR Control System Design for a Low-Pressure Feedwater Heater Train with Time Delay, Proc. IECON, 1990. Murphy and Bailey, Evaluation of Time Delay Requirements for Closed-Loop Stability Using Classical and Modern Methods, IEEE Southeastern Symp. on System Theory, 1989. Murphy and Bailey, "LQG/LTR Robust Control System Design for a Low-Pressure Feedwater Heater Train," Proc. IEEE Southeastcon, 1990. Kazerooni and Narayanan, "Loop Shaping Design Related to LQG/LTR for SISO Minimum Phase Plants," IEEE American Control Conf., Vol. 1, 1987. Murphy and Bailey, "Robust Control Technique for Nuclear Power Plants," ORNL-10916, March 1989. Seborg, Edgar, and Mellichamp, "Process Dynamics and Control," Prentice-Hall. Buckley, "Techniques of Process Control," Wiley. Birdwell, Crockett, Bodenheimer, and Chang, The CASCADE Final Report: Vol. II, "CASCADE Tools and Knowledge Base," University of Tennessee. Wang and Birdwell, A Nonlinear PID-Type Controller Utilizing Fuzzy Logic, Proc. Joint IEEE/IFAC Symp. on Controller-Aided Control System Design, 1994. Upadhyaya and Eryurek, Application of Neural Networks for Sensor Validation and Plant Monitoring, Nuclear Technology, 97, no. 2, Feb. 1992. Vasadevan et al., Stabilization and Destabilization Slugging Behavior in a Laboratory Fluidized Bed, International Conf. on Fluidized Bed Combustion, 1995. Doyle and Stein, "Robustness with Observers." IEEE Trans. Automatic Control, AC-24, 1979. Upadhaya et al. "Development and Testing of an Integrated Validation System for Nuclear Power Plants," Report prepared for the U.S. Dept. of Energy. Vols. 1–3, DOE/NE/37959-34, 35, 36, Sept. 1989.

INTRODUCTION

The purpose of an **automatic control** on a system is to produce a desired output when inputs to the system are changed. Inputs are in the form of commands, which the output is expected to follow, and disturbances, which the automatic control is expected to minimize. The usual form of an automatic control is a **closed-loop feedback control** which Ahrendt defines as "an operation which, in the presence of a disturbing influence, tends to reduce the difference between the actual state of a system and an arbitrarily varied desired state and which does so on the basis of this difference." The general theories and definitions of automatic control have been developed to aid the designer to meet primarily three basic specifications for the performance of the control system, namely, stability, accuracy, and speed of response.

The **terminology** of automatic control is being constantly updated by the ASME, IEEE, and ISA. Redundant terms, such as *rate, preact,* and *derivative,* for the same controller action are being standardized. Common terminology is still evolving. The introduction of the digital computer as a control device has necessitated the introduction of a whole new subset of terminology. The following terms and definitions have been selected to serve as a reference to a complex area of technology whose breadth crosses several professional disciplines.

Adaptive control system: A control system within which automatic means are used to change the system parameters in a way intended to improve the performance of the system.

Amplification: The ratio of output to input, in a device intended to increase this ratio. A gain greater than 1.

Attenuation: A decrease in signal magnitude between two points, or a gain of less than 1.

Automatic-control system: A system in which deliberate guidance or manipulation is used to achieve a prescribed value of a variable and which operates without human intervention.

Automatic controller: A device, or combination of devices, which measures the value of a variable, quantity, or condition and operates to correct or limit deviation of this measured value from a selected command (set-point) reference.

Bode diagram: A plot of log-gain and phase-angle values on a log-frequency base, for an element, loop, or output transfer function.

Capacitance: A property expressible by the ratio of the time integral of the flow rate of a quantity (heat, electric charge) to or from a storage, divided by the related potential charge.

Command: An input variable established by means external to, and independent of, the automatic-control system, which sets the ideal value of the controlled variable. See *set point.*

Control action: Of a control element or controlling system, the nature of the change of the output affected by the input.

Control action, derivative: That component of control action for which the output is proportional to the rate of change of input.

Control action, floating: A control system in which the rate of change of the manipulated variable is a continuous function of the actuating signal.

Control action, integral (reset): Control action in which the output is proportional to the time integral of the input.

Control action, proportional: Control action in which there is a continuous linear relationship between the output and the input.

Control system, sampling: Control using intermittently observed values of signals such as the feedback signal or the actuating signal.

Damping: The progressive reduction or suppression of the oscillation of a system.

Deviation: Any departure from a desired or expected value or pattern. Steady-state deviation is known as *offset.*

Disturbance: An undesired variable applied to a system which tends to affect adversely the value of the controlled variable.

Error: The difference between the indicated value and the accepted standard value.

Gain: For a linear system or element, the ratio of the change in output to the causal change in input.

Load: The material, force, torque, energy, or power applied to or removed from a system or element.

Nyquist diagram: A polar plot of the loop transfer function.

Nichols diagram: A plot of magnitude and phase contours using ordinates of logarithmic loop gain and abscissas of loop phase angle.

Offset: The steady-state deviation when the command is fixed.

Peak time: The time for the system output to reach its first maximum in responding to a disturbance.

Proportional band: The reciprocal of gain expressed as a percentage.

Resistance: An opposition to flow that results in dissipation of energy and limitation of flow.

Response time: The time for the output of an element or system to change from an initial value to a specified percentage of the steady-state.

Rise time: The time for the output of a system to increase from a small specified percentage of the steady-state increment to a large specified percentage of the increment.

Self-regulation: The property of a process or a machine to settle out at equilibrium at a disturbance, without the intervention of a controller.

Set point: A fixed or constant command given to the controller designating the desired value of the controlled variable.

Settling time: The time required, after a disturbance, for the output to enter and remain within a specified narrow band centered on the steady-state value.

Time constant: The time required for the response of a first-order system to reach 63.2 percent of the total change when disturbed by a step function.

Transfer function: A mathematical statement of the influence which a system or element has on a signal or action compared at input and output terminals.

BASIC AUTOMATIC-CONTROL SYSTEM

A **closed-loop control system** consists of a process, a measurement of the controlled variable, and a controller which compares the actual measurement with the desired value and uses the difference between them to automatically adjust one of the inputs to the process. The **physical system** to be controlled can be electrical, thermal, hydraulic, pneumatic, gaseous, mechanical, or described by any other physical or chemical process. Generally, the control system will be designed to meet one of two **objectives**. A **servomechanism** is designed to follow changes in set point as closely as possible. Many electrical and mechanical control systems are servomechanisms. A **regulator** is designed to keep output constant despite changes in load or disturbances. Regulatory controls are widely used on chemical processes. Both objectives are shown in Fig. 16.2.1.

The **control components** can be actuated pneumatically, hydraulically, electronically, or digitally. Only in very few applications does actuation affect controllability. **Actuation** is chosen on the basis of economics.

The **purpose** of the control system must be defined. A large capacity or inertia will make the system sluggish for servo operation but will help to minimize the error for regulator operation.

PROCESS AS PART OF THE SYSTEM

Figure 16.2.1 shows the **process** to be part of the control system either as load on the servo or process to be controlled. Thus the process must be designed as part of the system. The process is **modeled** in terms of its static and dynamic gains in order that it be incorporated into the system diagram. Modeling uses Ohm's and Kirchhoff's laws for electrical systems, Newton's laws for mechanical systems, mass balances for fluid-flow systems, and energy balances for thermal systems.

Fig. 16.2.1 Feedback control loop showing operation as servomechanism or regulator.

Consider the **electrical system** in Fig. 16.2.2

$$i = \frac{E_1 - E_2}{R}$$
$$i = C\frac{dE_2}{dt}$$

(16.2.1)

Combining

$$RC\frac{dE_2}{dt} + E_2 = E_1$$

(16.2.2)

where $\left(R\,\dfrac{\text{volt-second}}{\text{coulomb}}\right)\left(C\,\dfrac{\text{coulomb}}{\text{volt}}\right) = \tau\,\text{s} = \text{time constant}$

Fig. 16.2.2 Electrical system where current flows upon closing switch.

Consider the **mass balance** of the vessels shown in Figs. 16.2.3 and 16.2.4:

$$\text{Accumulation} = \text{input} - \text{output}$$
$$\frac{d(\rho V)}{dt} = w_1 - w_2 \quad \text{lb/min}$$

(16.2.3)

Fig. 16.2.3 Liquid level process.

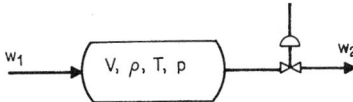

Fig. 16.2.4 Gas pressure process.

For the liquid level process (Fig. 16.2.3):

$$V = AL$$

(16.2.4)

Linearizing

$$w_2 = \frac{w_2}{x}\bigg|_{ss} x$$

Substituting

$$\rho A\frac{dL}{dt} = w_1 - \frac{w_2}{x}\bigg|_{ss} x$$

(16.2.5)

where ρA = analogous capacitance.
    For the **gas pressure process** (Fig. 16.2.4):

$$V\frac{d\rho}{dt} = w_1 - w_2$$

(16.2.6)

From thermodynamics

$$\left(\frac{\rho}{p}\right)^n = \text{constant}$$

(16.2.7)

Linearizing

$$w_2 = \frac{w_2}{x}\bigg|_{ss} x + \frac{w_2}{2(p - p_2)}\bigg|_{ss}\, p - p_2$$

(16.2.8)

Substituting

$$RC\frac{dp}{dt} + p = \frac{2(p - p_2)}{x}\bigg|_{ss} x + p_2$$

(16.2.9)

where $\quad C = \dfrac{V}{nRT} \qquad R = \dfrac{2(p - p_2)}{w_2} \qquad RC = \tau\,\text{min}$

and the terms are defined: V = volume, ft^3 (m^3); A = cross-sectional area, ft^2 (m^2); L = level, ft (m); ρ = density, lb/ft^3 (g/ml); x = valve stem position (normalized 0 to 1); T = temperature, °F (°C); p = pressure, lb/in^2 (kPa); and w = mass flow, lb/min (kg/min).
    The **thermal process** of Fig. 16.2.5 is modeled by a heat balance (Shinskey, "Process Control Systems," McGraw-Hill):

$$Mc\frac{dT}{dt} = wc(T_0 - T) - UA(T - T_w)$$

(16.2.10)

$$\frac{Mc}{wc + UA}\frac{dT}{dt} + T = \frac{wc}{wc + UA}T_0 + \frac{UA}{wc + UA}T_w$$

(16.2.11)

where $\quad C = Mc \qquad R = \dfrac{1}{wc + UA} \qquad RC = \tau\,\text{min}$

$$RC\frac{dT}{dt} + T = \frac{wc}{wc + UA}T_0 + \frac{UA}{wc + UA}T_w$$

(16.2.12)

The terms are defined: M = weight of process fluid in vessel, lb (kg); c = specific heat, Btu/lb · °F (J/kg · °C); U = overall heat-transfer coefficient, Btu/ft^2 · min · °F (W/m^2 · °C); and A = heat-transfer area, ft^2 (m^2).

Fig. 16.2.5 Thermal process with heat from vessel being removed by cold water in jacket.

Newton's laws can be applied to the manometer shown in Fig. 16.2.6.

$$\text{Inertia force} = \text{restoring force} - \text{flow resistance}$$

$$\frac{Al\rho}{g_c}\frac{d^2h}{dt^2} = A\left(p - 2h\rho\frac{g}{g_c}\right) - RA\frac{dh}{dt}$$

(16.2.13)

Fig. 16.2.6 Filled manometer measuring pressure P.

Flow resistance for laminar flow is given by the Hagen-Poiseuille equation:

$$R = \frac{\text{driving force}}{\text{rate of transfer}} = \frac{321\mu}{d^2 g_c} \qquad (16.2.14)$$

Substituting

$$\frac{1}{2g}\frac{d^2h}{dt^2} + \frac{161\mu}{\rho d g^2}\frac{dh}{dt} + h = h_i \qquad (16.2.15)$$

In standard form

$$\tau_c^2 \frac{d^2h}{dt^2} + 2\tau_c \zeta \frac{dh}{dt} + h = h_i \qquad (16.2.16)$$

where τ_c = characteristic response time, min, = $1/[(60)(2\pi)\omega_n]$ where ω_n = natural frequency, Hz; and ζ = damping coefficient (ratio), dimensionless.

The variables τ_c and ζ are very valuable design aids since they define **system response** and **stability** in terms of system parameters.

TRANSIENT ANALYSIS OF A CONTROL SYSTEM

The stability, accuracy, and speed of response of a control system are determined by analyzing the **steady-state** and the **transient** performance. It is desirable to achieve the steady state in the shortest possible time, while maintaining the output within specified limits. Steady-state performance is evaluated in terms of the accuracy with which the output is controlled for a specified input. The transient performance, i.e., the behavior of the output variable as the system changes from one steady-state condition to another, is evaluated in terms of such quantities as maximum overshoot, rise time, and response time (Fig. 16.2.7).

Fig. 16.2.7 System response to a unit step-function command.

Transient-Producing Disturbances A number of factors affect the **quality of control,** among them **disturbances** caused by set-point changes and process-load changes. Both set point and process load may be defined in terms of the setting of the final control element to maintain the controlled variable at the set point. Thus both cause the final control element to reposition. For a purely mechanical system the disturbance may take the form of a vibration, a displacement, a velocity, or an acceleration. A process-load change can be caused by variations in the rate of energy supply or variations in the rate at which work flows through the process. Reference to Fig. 16.2.5 and Eq. (16.2.12) shows disturbances to be variations in inlet process fluid temperature and cooling-water temperature. Further linearization would show variations in process flow and the overall heat-transfer coefficient to also be disturbances.

The Basic Closed-Loop Control To illustrate some characteristics of a basic closed-loop control, consider a mechanical, rotational system composed of a prime mover or motor, a total system inertia J, and a viscous friction f. To control the system's output variable θ_o, a command signal θ_i must be supplied, the output variable measured and compared to the input, and the resulting signal difference used to control the flow of energy to the load. The basic control system is represented schematically in Fig. 16.2.8.

Fig. 16.2.8 A basic closed-loop control system.

The differential equation of this basic system is readily obtained from the idealized equations

$$\text{Load torque } T_L = J\frac{d^2\theta_o}{dt^2} + f\frac{d\theta_o}{dt} \qquad (16.2.17)$$

$$\text{Developed torque } T_D = K\varepsilon \qquad (16.2.18)$$

$$\text{Error } \varepsilon = \theta_i - \theta_o \qquad (16.2.19)$$

The above equations combine to yield the system differential equation:

$$J\frac{d^2\theta_o}{dt^2} + f\frac{d\theta_o}{dt} + K\theta_o = K\theta_i \qquad (16.2.20)$$

Step-Input Response of a Viscous-Damped Control If the control system described in Fig. 16.2.8 by Eq. (16.2.20) is subjected to a step change in the input variable θ_i, a solution $\theta_o = \theta_o(t)$ can be obtained as follows. (1) Let the ratio $\sqrt{K/J}$ be designated by the symbol ω_n and be called the **natural frequency.** (2) Let the quantity $2\sqrt{JK}$ be designated by the symbol f_c and be called the **friction coefficient** required for critical damping. (3) Let f/f_c be designated by the symbol ζ and be called the **damping ratio.** Equation (16.2.20) can then be written as

$$\frac{d^2\theta_o}{dt^2} + 2\zeta\omega_n\frac{d\theta_o}{dt} + \omega_n^2\theta_o = \omega_n^2\theta_i \qquad (16.2.21)$$

For $\theta_i = 1$:

$$\theta_o = 0 \quad \text{and} \quad \frac{d\theta_o}{dt} = 0 \text{ at } t = 0$$

The complete solution of Eq. (16.2.21) is

$$\theta_o = 1 - \frac{\exp(-\zeta\omega_n t)}{\sqrt{1 - \zeta^2}}$$

$$\sin\left(\sqrt{1 - \zeta^2}\omega_n t + \tan^{-1}\frac{\sqrt{1 - \zeta^2}}{\zeta}\right) \qquad (16.2.22)$$

Equation (16.2.22) is plotted in dimensionless form for various values of damping ratio in Fig. 16.2.9. The curves for $\zeta = 0.1$, 2, and 1 illustrate the underdamped, overdamped, and critically damped case, where any further decrease in system damping would result in overshoot.

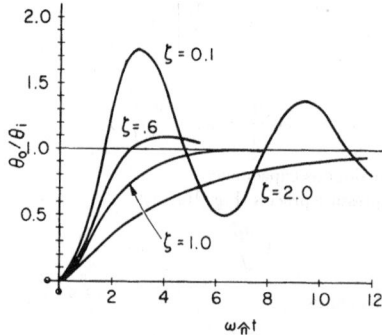

Fig. 16.2.9 Transient response of a second-order viscous-damped control to unit-step input displacement.

Damping is a property of the system which opposes a change in the output variable.

The immediately apparent features of an observed transient performance are (1) the existence and magnitude of the maximum overshoot, (2) the frequency of the transient oscillation, and (3) the response time.

Maximum Overshoot   When an automatic-control system is underdamped, the output variable overshoots its desired steady-state condition and a transient oscillation occurs. The first overshoot is the greatest, and it is the effect of its amplitude which must concern the control designer. The primary considerations for limiting this maximum overshoot are (1) to avoid damage to the process or machine due to excessive excursions of the controlled variable beyond that specified by the command signal, and (2) to avoid the excessive settling time associated with highly underdamped systems. Obviously, exact quantitative limits cannot generally be specified for the magnitude of this overshoot. However, experience indicates that satisfactory performance can generally be obtained if the overshoot is limited to 30 percent or less.

Transient Frequency   An undamped system oscillates about the final steady-state condition with a frequency of oscillation which should be as high as possible in order to minimize the response time. The designer must, however, avoid resonance conditions where the frequency of the transient oscillation is near the natural frequency of the system or its component parts.

Rise Time T_n, Peak Overshoot P, Peak Time T_p   These quantities are related to ζ and ω_n in Figs. 16.2.10 and 16.2.11. Some useful formulas are listed below:

$$\omega_n T_n \approx 1.02 + 0.48\zeta + 1.15\zeta^2 + 0.76\zeta^3 \qquad 0 \leq \zeta \leq 1$$

$$\omega_n T_s = \left.\begin{array}{ll} 17.6 - 19.2\zeta & 0.2 \leq \zeta \leq 0.75 \\ -3.8 + 9.4\zeta & 0.75 \leq \zeta \leq 1 \end{array}\right\} \; 2\% \text{ tolerance band}$$

$$P = \exp\left(\frac{-\pi\zeta}{\sqrt{1 - \zeta^2}}\right) \qquad T_p = \frac{\pi}{\omega_n \sqrt{1 - \zeta^2}}$$

Although these quantities are defined for a second-order system, they may be useful in the early design states of higher-order systems if the response of the higher-order system is dominated by roots of the characteristic equation near the imaginary axis.

Derivative and Integral Compensation   (Thaler) Four common compensation methods for improving the steady-state performance of a proportional-error control without damaging its transient response are shown in Fig. 16.2.12. They are (1) error derivative compensation, (2) input derivative compensation, (3) output derivative compensation, (4) error integral compensation.

Fig. 16.2.10   Rise time T_r as a function of ζ and ω_n.

Fig. 16.2.11   Peak overshoot P and peak time T_p as functions of ζ and ω_n.

Error Derivative Compensation   The torque equilibrium equation is

$$J\frac{d^2\theta_o}{dt^2} + f\frac{d\theta_o}{dt} = K_2\varepsilon + K_1\frac{d\varepsilon}{dt} \qquad (16.2.23)$$

Writing Eq. (16.2.23) in terms of the input and output variables yields

$$J\frac{d^2\theta_o}{dt^2} + (f + K_1)\frac{d\theta_o}{dt} + K_2\theta_o = K_2\theta_i + K_1\frac{d\theta_i}{dt} \qquad (16.2.24)$$

Fig. 16.2.12   Derivative and integral compensation of a basic closed-loop system. (*a*) Error derivative compensation; (*b*) input derivative compensation; (*c*) output derivative compensation; (*d*) error integral compensation. (*Thaler.*)

By adjusting K_1 and reducing f so that the quantity $f + K_1$ is equal to f in the uncompensated system, the system performance is affected as follows: (1) ε resulting from a constant-first-derivative input is reduced because of the reduction in viscous friction; (2) the transient performance of the uncompensated system is preserved unchanged.

Derivative Input Compensation   The torque equilibrium equation is

$$J \frac{d^2\theta_o}{dt^2} + f \frac{d\theta_o}{dt} = K_2\varepsilon + K_1 \frac{d\theta_i}{dt} \qquad (16.2.25)$$

Writing Eq. (16.2.25) in terms of the input and output variables yields

$$J \frac{d^2\theta_o}{dt^2} + f \frac{d\theta_o}{dt} + K_2\theta_o = \theta_i + K_1 \frac{d\theta_i}{dt} \qquad (16.2.26)$$

Examination of Eq. (16.2.26) yields the following information about the compensated system's performance: (1) since the characteristic equation is unchanged from that of the uncompensated system, the transient performance is unaltered; (2) the steady-state solution to Eq. (16.2.26) is

$$\theta_o = \theta_i - \frac{f}{K_2}\left(1 - \frac{K_1}{K_2}\right)\frac{d\theta_i}{dt} \qquad (16.2.27)$$

Therefore the input derivative signal can reduce the steady-state error by adjusting K_1 to equal K_2.

Derivative Output Compensation   The torque equilibrium equation is

$$J \frac{d^2\theta_o}{dt^2} + f \frac{d\theta_o}{dt} = K_2\varepsilon \pm K_1 \frac{d\theta_o}{dt} \qquad (16.2.28)$$

Writing Eq. (16.2.28) in terms of the input and output variables yields

$$J \frac{d^2\theta_o}{dt^2} + (f \pm K_1)\frac{d\theta_o}{dt} + K_2\theta_o = K_2\theta_i \qquad (16.2.29)$$

Examination of Eq. (16.2.29) yields the following information about the compensated system's performance. (1) Output derivative feedback produces the same system effect as the viscous friction does. This compensation therefore damps the transient performance. (2) Under conditions where $\theta_i = ct$, the steady-state error is increased.

Error Integral Compensation   Error integral compensation is used where it is necessary to eliminate steady-state errors resulting from input signals with constant first derivatives or under conditions of externally applied loads. The torque equilibrium equation is

$$J \frac{d^2\theta_o}{dt^2} + f \frac{d\theta_o}{dt} \pm \text{external load torque} = K_2\varepsilon + K_1 \int_0^t \varepsilon\, dt \quad (16.2.30)$$

Writing Eq. (16.2.30) in terms of the input variable and the error yields

$$\text{External load torque } + J \frac{d^2\theta_i}{dt^2} + f \frac{d\theta_i}{dt}$$

$$= J \frac{d^2\varepsilon}{dt^2} + f \frac{d\varepsilon}{dt} + K_2\varepsilon + K_1 \int_0^t \varepsilon\, dt \quad (16.2.31)$$

At steady state for $t \gg 0$ and with a step change in the input derivative

$$\frac{d^2\theta_i}{dt^2} = \frac{d\varepsilon}{dt} = \frac{d^2\varepsilon}{dt^2} = 0 \qquad \frac{d\theta_i}{dt} = \text{const} \quad (16.2.32)$$

Equation (16.2.31) assumes the form

$$\pm \text{ External load torque } + f \frac{d\theta_i}{dt} = K_2\varepsilon + K_1 \int_0^t \varepsilon\, dt \quad (16.2.33)$$

Since the sum of the load torque and the term $f(d\theta_i/dt)$ is finite, $\varepsilon = 0$ for

$$\int_0^\infty \varepsilon\, dt \to \infty \qquad (16.2.34)$$

If the integrating coefficient is a small number, additional torque produced by the integration action is developed very slowly, and, although the steady-state error is eventually eliminated, the transient performance is essentially unchanged. However, if K_1 is a large value, a large torque is produced in a short period of time, increasing the effect T/J ratio, and thereby decreasing the damping. The general effects of error integral compensation within its useful range are (1) steady-state error is eliminated; and (2) transient response is adversely effected, resulting in increased overshoot and the attendant increase in response time.

TIME CONSTANTS

The **time constant** τ, the **characteristic response time** (τ_c), and the **damping coefficient** are combined with operational calculus to design a control system without solving the classical differential equations. Note that the electrical, Eq. (16.2.2), mass, Eq. (16.2.9), and energy, Eq. (16.2.12) processes are all described by analogous first-order differential equations of the form

$$\tau \frac{dc}{dt} + c = Ka \qquad (16.2.35)$$

Solving with initial conditions equal to zero, the **time-domain response** to a step disturbance is

$$c = Ka(1 - e^{t/\tau}) \qquad (16.2.36)$$

Plotting is shown in Fig. 16.2.13.

Figure 16.2.13 shows the **time constant** to be defined by the point at 63.2 percent of the response, while response is essentially complete at three time constants (95 percent).

Fig. 16.2.13   Response of first-order system showing time constant relationship.

Operational calculus provides a systematic and simple method for handling linear differential equations with constant coefficients. The **Laplace operator** is used in control system analysis because of the straightforward transformation among the domains of interest:

Time domain	Complex domain	Frequency domain
$\dfrac{d}{dt}$	s	$j\omega$

Extensive tables of transforms are available for converting between the time and complex domains and back again. Transformation from the **complex** or **frequency domains** to **time** is difficult, and it is seldom attempted. Time-domain solutions are obtained only from computer simulations, which can solve the finite difference equations fast enough to serve as a design tool. Control system analysis and design proceed with a number of analytical and graphical techniques which do not require a time-domain solution.

The **root locus method** is a graphical technique used in the complex domain which provides substantial insight into the system. It has the weaknesses of handling dead time poorly and of the graph's being very tedious to plot. It is used only when computerized plotting is available.

BLOCK DIAGRAMS

The **physical diagram** of the system is converted to a **block diagram** in order that the different components of the system (all the way from a steam boiler to a thermocouple) can be placed on a common mathematical footing for analysis as a system. The block diagram shows the functional relationship among the parts of the system by depicting the action of the variables in the system. The circle represents an algebraic function of addition. Each system component is represented by a **block** which has one input and one output. The block represents a dynamic function such that the output is a function of time and also of the input. The dynamic function is called a **transfer function**—the ratio of the Laplace transform of the output variable to the input variable with all initial conditions equal to zero. The input and output variables are considered as signals, and the blocks are connected by arrows to show the flow of information in the system.

Rewriting Eq. (16.2.12) for the thermal system,

$$RC\frac{dT}{dt} + T = K_1 T_0 + K_2 T_w \qquad (16.2.37)$$

Transforming

$$(\tau s + 1)T = K_1 T_0 + K_2 T_w \qquad (16.2.38)$$

for which the block diagram is shown in Fig. 16.2.14. Two conditions are specified: (1) the components must be described by linear differential equations (or nonlinear equations linearized by suitable approximations), and (2) each block is unilateral. What occurs in one component may not affect the components preceding.

Fig. 16.2.14   Block diagram of thermal process.

Block-Diagram Algebra

The block diagram of a single-loop feedback-control system subjected to a command input $R(s)$ and a disturbance $U(s)$ is shown in Fig. 16.2.15.

Fig. 16.2.15   Single-loop feedback control system.

When $U(s) = 0$ and the input is a reference change, the system may be reduced as follows:

$$E(s) = -\theta_o(s)H(s)$$
$$\theta_o(s) = E(s)[G_1(s)G_2(s)] \qquad (16.2.39)$$

Therefore

$$\frac{\theta_o(s)}{\theta_i(s)} = \frac{G_1(s)G(s)}{1 + G_1(s)G_2(s)H(s)}$$

When $\theta_i(s) = 0$ and the input is a disturbance, the system may be reduced as follows:

$$E(s) = -\theta_o(s)H(s)$$
$$[E(s)\,G_1(s) + U(s)]G_2(s) = \theta_o(s)$$
$$\frac{\theta_o(s)}{U(s)} = \frac{G_2(s)}{1 + G_1(s)G_2(s)H(s)} \qquad (16.2.40)$$

Closed-loop transfer function Eq. (16.2.40) can be determined by observation from

$$\frac{\theta_o(s)}{U(s)} = \frac{\text{feedforward functions}}{1 + \text{complete-loop functions}}$$

The denominator of Eq. (16.2.40) is the **characteristic equation** which determines **stability.** Most systems are designed on the basis of the characteristic equation since it sets both response time τ_c and damping ζ. The equation is undefined (unstable) if $G_1(s)G_2(s)H(s)$ equals -1. But -1 is a vector of magnitude 1 and a phase of $-180°$. This fact is used to determine stable parameter adjustments in the graphical techniques to be discussed later.

The input disturbance $[U(s)]$ can be any time function in actual operation. The **step input** is widely used for analysis and testing since it is easily implemented; it results in a simple Laplace transform; it is the most severe type of disturbance; and the response to a step change shows the maximum error that could occur.

In many complex control systems, especially in the nonmechanical process-control field, auxiliary feedback paths are provided in order to adjust the system's performance. Figure 16.2.16a illustrates such a condition. In analyzing such a system it is usually best to combine secondary loops into the main control loop to form an equivalent series block and transfer function. The system of Fig. 16.2.16a might be reduced in the following sequence.

(a)

(b)

(c)

Fig. 16.2.16   Reduction of a closed-loop control system with multiple secondary loops.

1. Replace $K_3G_3(s)$ and $K_4H_1(s)$ with a single equivalent element

$$\frac{\theta_o}{\theta_2} = \frac{K_3G_3(s)}{1 + K_4H_1(s)K_3G_3(s)} = K_6G_6 \qquad (16.2.41)$$

The result of this first reduction is shown in Fig. 16.2.16b:

2. Figure 16.2.16b can be treated in a similar fashion and a single block used to replace K_2G_2, K_6G_6, and K_5H_2

$$\frac{\theta_o}{\theta_1} = \frac{K_2G_2K_6G_6(s)}{1 + K_5H_2K_2G_2K_6G_6(s)} = K_7G_7 \qquad (16.2.42)$$

The result of this second reduction is shown in Fig. 16.2.16c. The resulting open-loop transfer function is

$$\theta_o/\varepsilon = K_1G_1K_7G_7(s) \qquad (16.2.43)$$

The closed-loop or frequency response function is

$$\frac{\theta_o}{\theta_i} = \frac{K_1G_1K_7G_7(s)}{1 + K_1G_1K_7G_7(s)} \qquad (16.2.44)$$

Equation (16.2.44) can, of course, be expanded to include the terms of the system's secondary loops.

SIGNAL-FLOW REPRESENTATION

An alternate graphical representation of the mathematical relationships is the signal-flow graph. For complicated systems it allows a more compact representation and more rapid reduction techniques than the block diagram.

In Fig. 16.2.17, the nodes represent the variables θ_i, ε, θ_1, ..., θ_o, and the branches the relationships between the nodes, of the system shown in Fig. 16.2.16. For example,

$$\theta_1(s) = \varepsilon K_1G_1(s) - K_5H_2(s)\theta_o(s) \qquad (16.2.45a)$$

and

$$\varepsilon = \theta_i - \theta_o \qquad (16.2.45b)$$

Fig. 16.2.17 Signal flow graph of the closed-loop control system shown in Fig. 16.2.16.

Signal-flow terminology follows:

Source: node having only outgoing branches, for example, θ_i
Sink: node having only incoming branches, θ_o
Path: series of branches with the same sense of direction, for example, *abcd, cdf*
Forward path: path originating at a source and ending at a sink, with no node encountered more than once, for example, *abcd*
Path gain: product of the coefficients along a path, for example, $1 [K_1G_2(s)][K_2G_2(s)][K_3G_3(s)]$
Feedback loop: path starting at a node and ending at the same node, for example, *bcdg*
Loop gain: product of coefficients along a feedback loop, for example, $[K_1G_2(s)][K_2G_2(s)][K_3G_3(s)] (-1)$

The overall gain of the system can be calculated from

$$G = \frac{\Sigma_i G_i \Delta_i}{\Delta} \qquad (16.2.46)$$

where G_i = gain of ith forward path and

$$\Delta = 1 - \Sigma L_1 + \Sigma L_2 - \Sigma L_3 + \cdots + (-1)^k \Sigma L_k$$

where ΣL_1 = sum of the gains of each forward loop; ΣL_2 = sum of products of loop gains for nontouching loops (no node is common), taken two at a time; ΣL_3 = sum of products of loop gains for nontouching loops taken three at a time; Δ_i = value of Δ for signal flow graph resulting when ith path is removed.

From Fig. 16.2.17 there is only one forward path, *abcd*.

$$\therefore G_1 = K_1G_1(s)K_2G_2(s)K_3G_3(s)$$

Closed loops are *de, cdf,* and *bcdg*, with gains $-K_1H_1(s)K_3G_3(s)$, $-K_2G_2(s)K_3G_3(s)K_5H_2(s)$, and $-K_1G_1(s)K_2G_2(s)K_3G_3(s)$.

There are no nontouching closed loops;

$$\therefore \Delta = 1 + K_1H_1(s)K_3G_3(s) + K_2G_2(s)K_3G_3(s)K_5H_2(s)$$
$$+ K_1G_1(s)K_2G_2(s)K_3G_3(s)$$

There are no loops remaining if the forward path *abcd* is removed;

$$\therefore \Delta_1 = 1$$

Thus

$$\theta_o/\theta_1 = K_1G_1(s)K_2G_2(s)K_3G_3/[1 + K_1H_1(s)K_3G_3(s)$$
$$+ K_2G_2(s)K_3G_3(s)K_5H_2(s)$$
$$+ K_1G_1(s)K_2G_2(s)K_3G_3(s)] \qquad (16.2.47)$$

which is identical with Eq. (16.2.44).

CONTROLLER MECHANISMS

The **controller** modifies the error signal in a desired manner to produce an output pressure which is used to actuate the valve motor. The several controller modes used singly or in combination are (1) the proportional mode in which $P_{out}(t) = K_c E(t)$, (2) the integral mode, in which $P_{out}(t) = 1/T_1 \int E(t) \, dt$, and (3) the rate mode, in which $P_{out}(t) = T_2 \, dE(t)/dt$. In these expressions $P_{out}(t)$ = controller output pressure, $E(t)$ = input error signal, K_c = proportional gain, $1/T_1$ = reset rate, and T_2 = rate time.

A **pneumatic controller** consists of an **error-detecting mechanism, control modes** made up of proportional (P), integral (I), and derivative (D) actions in almost any combination, and a **pneumatic amplifier** to provide output capacity. The error-detecting mechanism is a differential link, one end of which is positioned by the signal proportional to the controlled variable, and the other end of which is positioned to correspond to the command set point. The proportional action is provided by a flapper nozzle (Fig. 16.2.18), where the flapper is positioned by the error signal. A motion of 0.0015 in by the flapper is sufficient for nearly full output range. Nozzle back pressure is inversely proportional to the distance between nozzle opening and flapper.

Fig. 16.2.18 Gain reduction of pneumatic amplifier by means of feedback bellows. *(Raven, "Automatic Control Engineering," McGraw-Hill.)*

The controller employs a power-amplifying pilot for providing a larger quantity of air than could be provided through the small restriction shown in Fig. 16.2.18. The nozzle back pressure, instead of operating the final control element directly, is transmitted to a bellows chamber where it positions the pilot valve.

The combination of flapper-nozzle amplifier and power relay shown in Fig. 16.2.18 has a very high gain since small flapper displacements can result in the output traversing the full range of output pressure. Negative feedback is employed to reduce the gain. Controller output is connected to a feedback bellows which operates to reposition the flapper. With the feedback bellows, a movement of the flapper toward the nozzle increases back pressure, causing output pressure to decrease and the feedback bellows to move the flapper away from the nozzle. Thus the mechanism is stabilized. Figure 16.2.19 shows controller gain being varied by adjusting the point at which the feedback bellows bears on the flapper.

The **rate mode**, called "anticipatory," can take large corrective action when errors are small but have a high rate of change. The mode resists not only departures from the set point but also returns and so provides a stabilizing action. Since the rate mode cannot control to a set point, it is not used alone. When used with the proportional mode, its stabilizing influence (90° phase lead; see Table 16.2.3) may allow an increase in gain K_c and a consequent decrease in steady-state error.

Fig. 16.2.19 Proportional controller with negative feedback. *(Reproduced by permission. Copyright © John Wiley & Sons, Inc. Publishers, 1958. From D. P. Eckman, "Automatic Process Control.")*

Proportional-derivative (PD) control is obtained by adding an adjustable feedback restriction as shown in Fig. 16.2.20. This results in delayed negative feedback. The restriction delays and reduces the feedback, and, since the feedback is negative, the output pressure is momentarily higher and leads, instead of lags, the error signal.

Fig. 16.2.20 Proportional-derivative controller. *(Reproduced by permission. Copyright © John Wiley & Sons, Inc. Publishers, 1958. From D. P. Eckman, "Automatic Process Control.")*

Since the proportional mode requires an error signal to change output pressure, set point and load changes in a proportionally controlled system are accompanied by a steady-state error inversely proportional to the gain. For systems which, because of stability considerations, cannot tolerate high gains, the integral mode added to the proportional will eliminate the steady-state error since the output from this mode is continually varying so long as an error exists. The addition of the integral mode to a proportional controller has an adverse effect on the relative stability of the control because of the 90° phase lag introduced.

Proportional-integral (PI) control is obtained by adding a positive feedback bellows and an adjustable restriction (Fig. 16.2.21). The addition of the positive feedback bellows cancels the gain reduction brought about by the negative feedback bellows at a rate determined by the adjustable restriction.

Electronic controllers which are analogous to the pneumatic controller have been developed. They have the advantages of elimination of time lags; compatability with computers; being less expensive to install (although more expensive to purchase); being more energy efficient; and being immune to low temperatures (water in pneumatic lines freezes).

Fig. 16.2.21 Proportional-integral controller. *(Reproduced by permission. Copyright © John Wiley & Sons, Inc. Publishers, 1958. From D. P. Eckman, "Automatic Process Control.")*

The heart of the electronic controller is the **high-gain operational amplifier** with passive elements at the input and in the feedback. Figure 16.2.22 shows the classical control elements, though circuits in commercially available controllers are far more sophisticated. The current drawn by the dc amplifier is negligible, and the amplifier gain is around 1,000, so that the junction point is essentially at ground potential. The reset amplifier has delayed negative feedback, similar to the pneumatic circuit in Fig. 16.2.21. The derivative amplifier has advanced negative feedback, similar to Fig. 16.2.20. Gains are adjusted by changing the ratios of resistors or capacitors, while adjustable time constants are made up of a variable resistor and a capacitor *(RC)*.

Regardless of actuation, the commonly applied **controller modes** are as follows:

Designation	Transform	Symbol
Proportional	K (gain)	P
Reset	$1 + \dfrac{1}{\tau_i s}$	I
Derivative	$\dfrac{\tau_d s + 1}{(t_d/\alpha)s + 1}$	D
Floating	$\dfrac{1}{\tau_F s}$	F

The modes are combined as needed into P, PI, PID, PD, and F. Choice depends on the application and can be made from the Bode diagram. The block diagram for a PID controller is shown in Fig. 16.2.23.

The **derivative mode** can be placed on the measured variable signal (as shown), on the error, or on the controller output. The arrangement of Fig. 16.2.23 is preferred for regulation since the derivative does not affect set-point changes, and the derivative acts to reduce overshoot on start-up (Zoss, "Applied Instrumentation in the Process Industries," Vol. IV, Gulf Publishing Co.). Regardless of a sequence, the block diagram of a PID controller is commonly drawn as shown in Fig. 16.2.24.

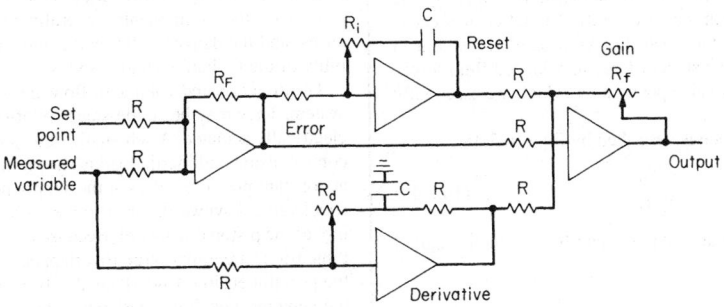

Fig. 16.2.22 Simplified circuit for a typical electronic controller.

Fig. 16.2.23 Block diagram of PID controller.

Fig. 16.2.24 Combined controller block diagram.

FINAL CONTROL ELEMENTS

The **final control element** is a mechanism which alters the value of the **manipulated variable** in response to an output signal from the automatic-control device. It will typically receive a signal from the controller and manipulate a flow of material or energy to the process. The final control element can be a control valve, an electrical motor, a servovalve, or a damper. Servovalves are discussed in the next section, "Hydraulic-Control Systems."

The final control element often consists of two parts: first, an **actuator** which translates the controller signal into a command for the manipulating device, and, second, a **mechanism** to adjust the manipulated variable. Figure 16.2.25 shows a control valve made up of an actuator and an adjustable orifice.

Fig. 16.2.25 Control valve and pneumatic actuator. *(Reproduced by permission. Copyright © John Wiley & Sons, Inc. Publishers, 1958. From D. P. Eckman, "Automatic Process Control.")*

The most commonly used **actuator** is a diaphragm motor in which the output pressure from the controller is counteracted not only by the spring but also by fluid forces in the valve body. The latter may cause serious deviation from linear static behavior with deleterious control effects. High friction at the valve stem or large unbalanced fluid forces at the plug can be overcome by valve positioners, which are essentially proportional controllers.

A **control valve** in liquid service is described by Eq. (16.2.48):

$$q = C_v \sqrt{\frac{\Delta p}{G}} \qquad (16.2.48)$$

And C_v is described by its relationship to stem lift x, which is called the **valve-flow characteristic.**

The linear characteristic is

$$C_v = C_v|_{max} x \qquad (16.2.49)$$

and the equal-percentage characteristic is

$$C_v = C_v|_{max}(r)^{x-1} \qquad (16.2.50)$$

where q = flow, gpm (m³/min); $C_v|_{max}$ = valve sizing coefficient; Δp = pressure drop across valve, lb/in² (kPa); G = specific gravity (density, g/ml); r = valve rangeability; and x = stem position, normalized to 0 to 1.

The flow characteristic relationship is not necessarily the lift-flow characteristic of the valve when installed since the valve is but one component in a piping system in which pressure drops vary with the flow rate. The change in characteristic as less percentage of total system drop is taken across the control valve is shown in Fig. 16.2.26.

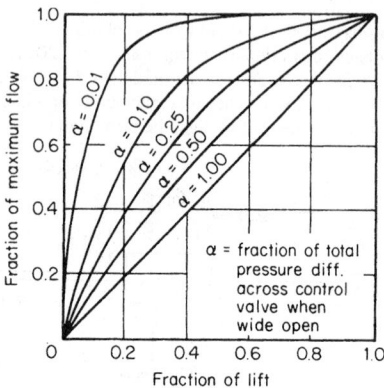

Fig. 16.2.26 Control valve linear flow characteristic. *(Reproduced by permission. Copyright © John Wiley & Sons, Inc. Publishers, 1958. From D. P. Eckman, "Automatic Process Control.")*

Valve characteristics are generally selected according to their ability to compensate for nonlinearities in the system. Nonlinearities typically represent a change in gain which in turn changes the character of the control response for a given controller setting when load- or set-point changes occur or when there is a variable overall pressure drop across the system. Equal-percentage valves, for example, tend to control over widely varying operating conditions, since the change in flow is always proportional to flow rate. The selection of the proper valve therefore depends on study of the particular system.

HYDRAULIC-CONTROL SYSTEMS

Hydraulic systems are used for rapid-response servomechanisms at high power levels. Operating system pressures are from 50 to 100 lb/in² for slower-acting systems and up to 5,000 lb/in² where lightweight and fast responses are required. Compared with **electrical systems** the major advantages are a rapid response in the large horsepower ranges and the capability of operating at high power-density levels since the fluid can transmit dissipated energy from the point of generation. Compared with **pneumatic systems,** hydraulic systems are faster because the fluid is essentially incompressible. Major disadvantages are vulnerability to dirt, since the components generally require close machining tolerances, and the danger of fire and explosion resulting from the flammability of the hydraulic fluids used.

The direction and volume of flow are controlled by **servo valves** in the system. They may be single-stage (pilot-operated) and mechanically or electrically actuated. A schematic of a spool-type four-way single-stage control piston and inertia load is shown in Fig. 16.2.27. Hydraulic fluid at constant pressure enters at the supply port. With displacement of the spool valve downward, for example, inflow to the top side of the piston moves the piston downward. Because of machining tolerances the spool dimensions are either large (overlapped) or smaller (underlapped) than the port dimension. Underlapped valves permit leakage to the piston in the centered position; overlapped valves result in a dead zone, where motion x results in no flow until a port is opened.

Fig. 16.2.27 Four-way valve-piston circuit. *(Truxal, "Control Engineer's Handbook," McGraw-Hill.)*

The transfer function of this circuit is given as (Truxal, "Control Engineers' Handbook")

$$\frac{y}{x} = \frac{C_1 \dfrac{1}{1 + \alpha(C_1/C_2)}}{s\left[\dfrac{VM}{2BA^2}\dfrac{1}{1 + \alpha(C_1/C_2)}s^2 + \dfrac{C_1 m/C_2 + V\alpha/2BA^2}{1 + \alpha(C_1/C_2)}s + 1\right]}$$

where C_1 = servo velocity gradient, in/(s)(in), C_2 = servo force gradient, lb/in, α = viscous friction of load and piston, lb/(in)(s), B = bulk modulus of fluid, lb/in^2 M = mass of load and piston, lb/(in)(s^2), A = piston area, in^2, and V = effective entrained fluid volume, in^3 (one-half of total entrained volume between valve and piston).

The velocity of the output is proportional to the input resulting in a **velocity-control** servo. To convert this system to a **position-control** servo, mechanical, hydraulic, or electrical feedback may be employed. A valve-piston position servo with **mechanical feedback** is shown in Fig. 16.2.28. Any difference between the input D and the piston position y causes a motion x, which causes the piston to move in a direction opposite to D, that is, in a direction to reduce x. The lever ratio establishes the relationship between y and D.

Fig. 16.2.28 Valve-piston position servomechanical feedback. *(Truxal, "Control Engineer's Handbook," McGraw-Hill.)*

Most commercially available servo valves are two stages, permitting electrohydraulic action. The pilot stage can be operated by a low-power, short-travel electrical device, with a concomitant increase in flexibility. A typical pilot-operated servo valve is shown in Fig. 16.2.29. In this case the pilot is a double-flapper valve rather than a spool valve. (In general, small, accurate low-leakage spool valves are costly.) Upward movement of the flapper by the actuating motor results in increased pressure to the right end of the power spool. Hydraulic feedback occurs because of the increased flow across restrictor a. The power spool moves to the left until the unbalanced pressure is matched by the spring resistance. The disadvantage of this valve is the continual leakage flow through the flapper nozzle, but the torque motor has a low-power requirement and is inexpensive.

The major disadvantages of spool-type valves are (1) high cost, because of high-tolerance requirements between the valve lands, (2) high static friction and inertia, and (3) susceptibility to dirt. The **flapper** valve is less expensive to manufacture than a spool valve of equivalent characteristics and is not so susceptible to damage by dirt particles.

Fig. 16.2.29 Two-stage electrohydraulic servo valve. The first-stage is a four-way flapper valve with a calibrated pressure output driving a second-stage, spring-loaded, four-way spool valve. *(Moog Servocontrols, Inc.)*

In Fig. 16.2.30, P_1, the supply pressure, is constant. Input motion of the flapper toward the nozzle increases P_2 and drives the piston toward the right. The steady-state characteristic P_2 vs. x is shown in Fig. 16.2.31.

Fig. 16.2.30 Flapper valve. *(Raven.)*

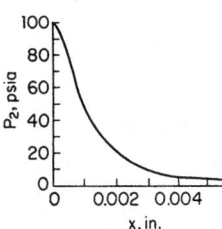

Fig. 16.2.31 Equilibrium curve of P_2 versus x for a flapper valve. *(Raven, "Automatic Control Engineering," McGraw-Hill.)*

STEADY-STATE PERFORMANCE

The steady-state error of a control system can be determined by using the final value theorem (see Laplace Transforms, Sec. 2) in which

$$\theta_o(t)\Big|_{t\to\infty} = s\theta_o(s)\Big|_{s\to0} \quad \text{provided } \theta_o(t) \text{ is stable}$$

The classification of control systems according to the form of the open-loop transfer function facilitates the determination of the steady-state errors when the system is subjected to various inputs.

The open-loop transfer function $\theta_o(s)/\varepsilon(s) = KG(s)$ may be written

$$
\begin{aligned}
KG(s) &= \frac{\theta_o(s)}{\varepsilon(s)} \\
&= \frac{K(1 + \tau_a s)(1 + \tau_a s)(1 + \tau_o s) \cdots}{s^N(1 + \tau_1 s)(1 + \tau_2 s)(1 + \tau_3 s) \cdots}
\end{aligned}
\tag{16.2.51a}
$$

The **system type** is given according to the value of N as

$$
\begin{aligned}
&\text{Type 0 system} \quad N = 0 \\
&\text{Type 1 system} \quad N = 1 \\
&\text{Type 2 system} \quad N = 2 \\
&\text{Type 3 system} \quad N = 3
\end{aligned} \qquad (16.2.51b)
$$

Error coefficients, based on a system with unity feedback $[H(s) = 1]$, are defined as

$$\text{Positional error constant} = K_o = \lim_{s \to 0} KG(s) \qquad (16.2.52a)$$

$$\text{Velocity error constant} = K_v = \lim_{s \to 0} sKG(s) \qquad (16.2.52b)$$

$$\text{Acceleration error constant} = K_a = \lim_{s \to 0} s^2 KG(s) \qquad (16.2.52c)$$

A summary of error coefficients for systems of different types is given in Table 16.2.1.

Table 16.2.1   Summary of Error Coefficients

N	K_o	K_v	K_a
0	const	0	0
1	∞	const	0
2	∞	∞	const

A summary of the errors for types 0, 1, and 2 systems, when subjected to various inputs is given in Table 16.2.2.

Table 16.2.2   Summary of Errors

Input error	$\theta_i(t) = A$ ε_0/A	$\theta_i(t) = vt$ ε_v/v	$\theta_i(t) = at^2$ ε_a/a
$N = 0$	$1/(1 + K_0)$	∞	∞
1	0	$1/K_v$	∞
2	0	0	$1/K_a$

The higher the system type, the better is the output able to follow the higher degrees of input. Higher-type systems, however, are more difficult to stabilize, and a compromise must be made between the steady-state error and the settling time of the response.

CLOSED-LOOP BLOCK DIAGRAM

The **transfer functions** of the process, controller, and final control element can be combined with the measuring transducer into a **block diagram** of the complete control loop. Consider the thermal system of Fig. 16.2.5 and the heat balance on the vessel jacket:

$$M_j c \frac{dT_w}{dt} = w_w c (T_{w1} - T_w) \qquad (16.2.53)$$

Linearizing and rearranging:

$$\frac{M_j}{w_w} T_w s + T_w = \left[\frac{T_{w1} - T_w}{w_w} \right] \left[\frac{w_w}{x} \right] x \qquad (16.2.54)$$

where M_j = weight of cooling water in jacket, lb (kg); c = specific heat of water, Btu/lb · °F (J/kg · °C); T = temperature, °F; x = valve stem position (normalized 0 to 1); and w_w = water flow, lb/min (kg/min).

For simplicity, let the jacket time constant M_j/w_w be so small as to be negligible (not typically the case). The block diagram of Fig. 16.2.32 then represents process diagram Fig. 16.2.5. The **gain** K in each case is an incremental change in output divided by an incremental change in input. For example, if the temperature transmitter has a temperature span of 100°F and an output of 4 to 20 mA, the gain would be

$$K_T = \frac{20 - 4}{100} = \frac{16}{100} = 0.16 \text{ mA/°F}$$

The **open-loop gain** is defined as the product of all the gains in the control loop:

$$K_o = K_c K_v K_T \left(\frac{T_{w1} - T_w}{x} \right) \left(\frac{UA}{wc + UA} \right)$$

Combining the **closed-loop system** of Fig. 16.2.32 results in the closed-loop equation

$$T = \frac{\left(\dfrac{wc}{wc + UA} \right) \left(\dfrac{1}{\tau s + 1} \right) T_0}{1 + \dfrac{K_o(\tau_i s + 1)(\tau_d s + 1)}{\tau_i s (\tau s + 1)\left(\dfrac{\tau_d}{\alpha} s + 1 \right)(\tau_T s + 1)}} \qquad (16.2.55)$$

The time-dependent modes on the controller have been designed to compensate for the dynamics of the rest of the loop. Let

$$\tau_i = \tau \quad \text{and} \quad \tau_d = \tau_T$$

After cancellation, Eq. (16.2.55) becomes

$$T = \frac{\left(\dfrac{wc}{wc + UA} \right) T_0}{(\tau s + 1)\left[1 + \dfrac{K_c}{\tau_i s \left(\dfrac{\tau_d}{\alpha} s + 1 \right)} \right]} \qquad (16.2.56)$$

Equation (16.2.56) can be multiplied out to become

$$T = \frac{\left(\dfrac{wc}{wc + UA} \right)(\tau_i s)\left(\dfrac{\tau_d}{\alpha} s + 1 \right) T_0}{(\tau s + 1)\left(\dfrac{\tau_i \tau_d}{\alpha} s^2 + \tau_i s + K_c \right)} \qquad (16.2.57)$$

Fig. 16.2.32   Block diagram of a thermal system.

Table 16.2.3   Process Characteristics versus Mode of Control

| Number of process capacities | Process reaction rate | Process time lags | | Load changes | | Suitable mode of control |
		Resistance-capacitance (RC)	Dead time (transportation)	Size	Speed	
Single	Slow	Moderate to large	Small	Any	Any	Two-position. Two-position with differential gap
				Moderate	Slow	Multiposition. Proportional input
Single (self-regulating)	Fast	Small	Small	Any	Slow	Floating modes: Single-speed Multispeed
					Moderate	Proportional-speed floating
Multiple	Slow to moderate	Moderate	Small	Small	Moderate	Proportional position
Multiple	Moderate	Any	Small	Small	Any	Proportional plus rate
Multiple	Any	Any	Small to moderate	Large	Slow to moderate	Proportional plus reset
Multiple	Any	Any	Small	Large	Fast	Proportional plus reset plus rate
Any	Faster than that of the control system	Small or nearly zero	Small to moderate	Any	Any	Wideband proportional plus fast reset

SOURCE: Considine, "Process Instruments and Controls Handbook," McGraw-Hill.

Equation (16.2.57) can be written in the standard form of Eqs. (16.2.16) and (16.2.21):

$$T = \frac{\left(\dfrac{wc}{wc + UA}\right)(\tau_i s)\left(\dfrac{\tau_d}{\alpha} s + 1\right) T_0}{K_c(\tau s + 1)(\tau_c^2 s^2 + 2\tau_c \zeta s + 1)} \qquad (16.2.58)$$

Controller gain K_c can then be chosen to give the best response τ_c with stability ζ. More rigorous techniques for determining controller parameter values are shown under Bode diagrams and Routh's criterion. Table 16.2.3 provides preliminary guidance for the selection of control modes.

FREQUENCY RESPONSE

Although it is the time response of the control system that is of major importance, study of the effect on transient response of changes in system parameters, either in the process or controller, is more conveniently made from a **frequency-response** analysis of the system. The **frequency response** of a system is the steady-state output of the system to input sinusoids of varying frequency. The output for a linear system can be completely described in terms of the amplitude ratio of the output sinusoid to the input sinusoid. The amplitude ratio or gain, and phase, are functions of the frequency of the input sinusoid.

The use of **sinusoidal methods** to analyze and test dynamic systems has gained widespread popularity because system response is obtained easily from the response of the individual elements, no matter how many elements are included. By contrast, transient analysis is quite tedious, with only three dynamic components in the system, and is too difficult to be worthwhile for four or more components.

Frequency-response analysis provides quantitative information concerning the system on maximum gain K_o for stability, time response τ_c, ω_n, and the controller parameter adjustments τ_i, τ_d. In most cases, this information is adequate for assurance of stability, for comparing alternatives, or for judging the merits of a proposed control system. System parameters are generally obtained from open-loop frequency response, which is the response with the loop broken between the controller and the final control element (Fig. 16.2.32).

Frequency response also lends itself to system identification by testing for systems not readily amenable to mathematical analysis by subjecting the system to input sinusoids of varying frequency.

The shift from the **complex domain** to the **frequency domain** (linear systems, zero initial conditions) is accomplished by simply substituting $j\omega$ for s. Consider Eq. (16.2.38).

$$(\tau s + 1)T = K_1 T_0 \qquad (16.2.59)$$

Substituting

$$\frac{T}{T_0} = \frac{K_1}{\tau j\omega + 1}$$

Multiply by the complex conjugate

$$\frac{T}{T_0} = \frac{K_1}{\tau j\omega + 1}\frac{\tau j\omega - 1}{\tau j\omega - 1}$$

Collecting terms

$$\frac{T}{T_0} = K_1\left[\frac{1}{\tau^2\omega^2 + 1} - j\frac{\tau\omega}{\tau^2\omega^2 + 1}\right] \qquad (16.2.60)$$

The first term is the real part of the solution, and the second term, the imaginary part. These terms define a vector, shown in Fig. 16.2.33,

Fig. 16.2.33   Polar plot showing vector.

which has magnitude M and phase angle θ as determined either graphically or by the Pythagorean theorem.

$$M = \left[\frac{1}{\tau^2\omega^2 + 1}\right]^{1/2} \qquad (16.2.61)$$

$$\theta = \tan^{-1}\left(\frac{-\zeta\omega}{1}\right) \qquad (16.2.62)$$

GRAPHICAL DISPLAY OF FREQUENCY RESPONSE

Sinusoidal response is **plotted** in three different ways: (1) the rectangular plot with the log of the amplitude versus the log of the frequency, called a **Bode plot**; (2) a phase-margin plot with magnitude shown versus a function of phase with frequency as a parameter, called a **Nichols plot;** and (3) a polar plot with magnitude and phase shown in vector form with frequency as a parameter, called a **Nyquist plot.** The Nichols plot is actually a plot of phase margin $(180° - \theta)$ versus frequency and is used to define system performance. The polar plot enables the absolute stability of the system to be determined without the need for obtaining the roots of the characteristic equation. The logarithmic (Bode) plot has the advantage of ease in plotting, especially in design, since the individual effects of cascaded elements can be gaged by superposition. All the graphical procedures use the "minus 1" point [discussed with Eq. (16.2.38)] as the criterion for stability (magnitude = 1, phase = $-180°$).

NYQUIST PLOT

The **Nyquist diagram** is a graphical method of determining stability. The diagram is essentially a mapping into the $G(s)$ plane of a contour enclosing the major portion of the right half of the s plane. Figure 16.2.34 shows the Nyquist plot of a typical system.

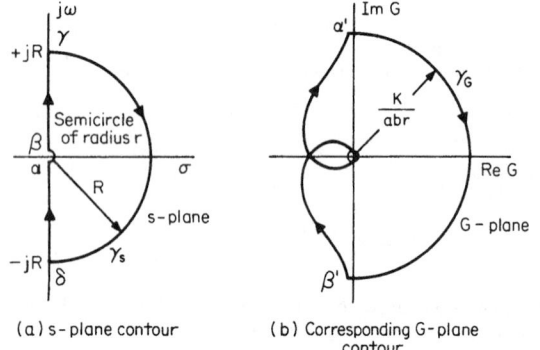

(a) s-plane contour   (b) Corresponding G-plane contour

Fig. 16.2.34 Nyquist diagram for $G(s) = 1/[s(s + a)(s + b)]$. *(Truxal, "Control System Synthesis," McGraw-Hill.)*

The frequency response of a closed-loop system can also be derived from the **polar plot** of the direct transfer function $KG(s)$. Systems with dynamic elements in the feedback can be transformed into direct systems by block-diagram manipulation. In Fig. 16.2.35, the amplitude ratio θ_o/θ_i at any frequency ω is the ratio of the lengths of the vectors $O\beta$ and $\alpha\beta$. The angle formed by the vectors $\alpha\beta$ and $O\beta$ is the phase angle of the frequency response $\angle\theta_o/\theta_i(j\omega)$. The numerator of the transfer function is a constant representing the closed-loop gain. Changing the gain proportionately changes the length of the vector $O\beta$. Figure 16.2.36 shows another typical Nyquist plot.

The Nyquist diagram is widely used in the design of electrical systems, but plots are very tedious to make and are generated via computer programs.

Fig. 16.2.35 Typical $(\theta_o/\varepsilon)/(j\omega)$ plot.

Fig. 16.2.36 Typical Nyquist diagram showing loop transfer function $|\theta_o/\theta_i|$. *(ASME, "Terminology for Automatic Control," Pub. ASA C85.1-1963.)*

BODE DIAGRAM

The most common method of presenting the **response data at various frequencies** is to use the log-log plot for amplitude ratios, accompanied by a semilog plot for phase angles. These rectangular plots are called **Bode diagrams,** after H. W. Bode, who did basic work on the theory of feedback amplifiers. A major advantage of this method is that the numerical computation of points on the curve is simplified by the fact that log magnitude versus log frequency can be approximated to engineering accuracy with straight-line approximations. In the **sinusoidal testing** of a system, the system dynamics will be faster than a very low frequency test signal, and the output amplitude has time to recover fully to the input amplitude. Thus the magnitude ratio is 1. Conversely, at high frequency, the system does not have time to reach equilibrium, and output amplitude will decrease. Magnitude ratio goes to a very small value under this condition. Thus the system will show a decrease in magnitude ratio as frequency increases. By similar reasoning, the difference between input and output phase angles becomes progressively more negative as frequency increases.

Consider the equations for magnitude, Eq. (16.2.61), and phase, Eq. (16.2.62). If $\tau_w \ll 1$,

$$M = \left(\frac{1}{1}\right)^{1/2} = 1 \qquad \theta = \tan^{-1}(0) = 0$$

and if $\tau_w \gg 1$,

$$M = \left(\frac{1}{\tau^2\omega^2}\right)^{1/2} = \frac{1}{\tau\omega} \approx 0 \qquad \theta = \tan^{-1}\left(\frac{-\tau_w}{1}\right) \approx -90°$$

The two **straight-line asymptotes** of magnitude, if extended, would intersect at $\tau\omega = 1$. Thus a relationship between frequency and time constant is established. The point at which $\tau\omega = 1$ is called the *corner frequency,* or the *break frequency.* The asymptote past the break frequency is easily plotted from a point at the break frequency $\omega = 1/\tau$ and a point at 10 times the frequency and 0.1 times the magnitude ratio.

The frequency response of systems containing different dynamic elements may be calculated by suitably employing the magnitude and phase characteristics of each element separately. Combined magnitudes are calculated

$$|M_1 M_2| = |M_1| \times |M_2|$$

Or, on the log-log Bode diagram,

$$\log |M_1 M_2| = \log |M_1| + \log |M_2|$$

The logarithmic scale permits **multiplication** by adding vertical distances. For phase

$$\underline{/\theta_1\theta_2} = \angle\theta_1 + \angle\theta_2$$

Thus **phase addition** is also made by adding vertical distances. Figure 16.2.37 shows the combination of the straight-line asymptotes for the following transfer function:

$$G(s) = \frac{1}{\tau s(\tau s + 1)} \qquad (16.2.63)$$

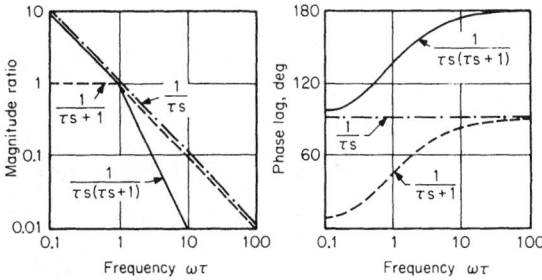

Fig. 16.2.37 Combined transfer functions. *(Reproduced by permission. Copyright © John Wiley & Sons, Inc. Publishers, 1958. From D. P. Eckman, "Automatic Process Control.")*

Table 16.2.4 lists additional frequency-response relationships for system elements. The noninteracting controller, item 8, has the advantage of no interaction among tuning parameters but has the disadvantages of being more difficult to plot on the Bode diagram and of not offering the direct compensation of the controller shown in Fig. 16.2.24.

Figure 16.2.38 shows a typical combined Bode diagram. The ordinate has a scale in decibels in addition to the magnitude ratio scale. Decibels (dB) are sometimes used to provide a linear scale. They are defined

$$dB = -20 \log M$$

The discontinuity at the break frequency due to the meeting of the straight-line asymptotes will not be found in actual test data. Instead, the corner will be rounded, as shown dotted in Fig. 16.2.39. The error due to the approximation is some 3 dB at the break frequency, as shown in the figure.

Quadratic terms (Table 16.2.4) also lend themselves to asymptotic approximation, but the form of the magnitude and phase plots around the break frequency depends on the damping coefficient ζ. The slope of the magnitude ratio attenuation past the break point is twice that of the first-order system.

Fig. 16.2.38 Typical Bode diagram showing loop-transfer function B/E. *(ASME, "Terminology for Automatic Control," Pub. ASA C85.1-1963.)*

Table 16.2.4 Frequency-Response Equations for Some Common Control-System Elements

Description	Transfer function $G(s)$	Frequency response $G(j\omega)$	Magnitude ratio	Phase angle
1. Dead time	$\varepsilon^{-T_L s}$	$\varepsilon^{-j\omega T_L}$	1	$-\omega T_L$ radians
2. First-order lag	$\dfrac{1}{Ts+1}$	$\dfrac{1}{j\omega T+1}$	$\dfrac{1}{\sqrt{\omega^2 T^2+1}}$	$-\tan^{-1}(\omega T)$
3. Second-order lag	$\dfrac{1}{(Ts+1)(aTs+1)}$	$\dfrac{1}{-a\omega^2 T^2+j(1+a)\omega T+1}$	$\dfrac{1}{\sqrt{(1-a\omega^2 T^2)^2+(1+a)^2\omega^2 T^2}}$	$-\tan^{-1}\left[\dfrac{(1+a)\,\omega T}{1-aT^2\omega^2}\right]$
4. Quadratic (underdamped)	$\dfrac{1}{\left(\dfrac{s}{\omega_n}\right)^2+\dfrac{2\zeta}{\omega_n}s+1}$	$\dfrac{1}{-\left(\dfrac{\omega}{\omega_n}\right)^2+j2\zeta\dfrac{\omega}{\omega_n}+1}$	$\dfrac{1}{\sqrt{\left(1-\dfrac{\omega^2}{\omega_n^2}\right)^2+4\zeta^2\left(\dfrac{\omega}{\omega_n}\right)^2}}$	$-\tan^{-1}\left[\dfrac{2\zeta\dfrac{\omega}{\omega_n}}{1-\left(\dfrac{\omega}{\omega_n}\right)^2}\right]$
5. Ideal proportional controller	K	K	K	0
6. Ideal proportional-plus-reset controller $T_i=\dfrac{1}{r}$ r = reset rate	$K\left(1+\dfrac{1}{T_i s}\right)$ or $K\dfrac{T_i s+1}{T_i s}$	$K\left(1+\dfrac{1}{j\omega T_i}\right)$ or $K\dfrac{j\omega T_i+1}{j\omega T_i}$	$K\sqrt{1+\left(\dfrac{1}{\omega T_i}\right)^2}$	$-\tan^{-1}\left(\dfrac{1}{\omega T_i}\right)$
7. Ideal proportional-plus-rate controller	$K(1+T_d s)$	$K(1+j\omega T_d)$	$K\sqrt{1+\omega^2 T_d^2}$	$\tan^{-1}(\omega T_d)$
8. Ideal proportional-plus-reset-plus-rate controller	$K\left(1+T_d s+\dfrac{1}{T_i s}\right)$	$K\left(1+j\omega T_d+\dfrac{1}{j\omega T_i}\right)$ or $K\dfrac{j\omega T_i-\omega^2 T_d T_i+1}{j\omega T_i}$	$K\sqrt{(\omega T_i)^2+(1-\omega^2 T_d T_i)^2}$	$\tan^{-1}\left(\omega T_d-\dfrac{1}{\omega T_i}\right)$

SOURCE: Considine, "Process Instruments and Controls Handbook," McGraw-Hill.

Fig. 16.2.39   Bode plot of term $(j\omega\tau_a + 1)$.

CONTROLLERS ON THE BODE PLOT

Controller modes, gain, reset, and derivative are plotted individually (in the form of Fig. 16.2.24) on the Bode diagram and summed as required. The functions for reset (integral) and derivative modes are moved horizontally on the diagram to determine proper parameter adjustment. The combined magnitude ratio curve for the whole system is moved vertically to determine open-loop gain.

After all the dynamic elements of the system are plotted and combined, reset and derivative are set in accordance with the following **criteria:**

Reset: The reset phase curve is positioned horizontally so that it contributes 10° of phase lag where the system phase lag is 170°.

Derivative: The derivative phase curve is positioned horizontally so that it contributes 60° of phase lead where the system phase lag is 180°.

The magnitude ratio curves are positioned with the phase-angle curves, and parameter values for reset and derivative can be read at the break points.

STABILITY AND PERFORMANCE OF AN AUTOMATIC CONTROL

An automatic-control system is stable if the amplitude of transient oscillations decreases with time and the system reaches a steady state. The stability of a system can be evaluated by examining the roots of the differential equation describing the system. The presence of positive real roots or complex roots with positive real parts dictates an unstable system. Any stability test utilizing the open-loop transfer function or its plot must utilize this fact as the basis of the test.

The Nyquist Stability Criterion   The $KG(j\omega)$ locus for a typical single-loop automatic-control system plotted for all positive and negative frequencies is shown in Fig. 16.2.40. The locus for negative values of ω is the mirror image of the positive ω locus in the real axis. To complete the diagram, a semicircle (or full circle if the locus approaches $-\infty$ on the real axis) of infinite radius is assumed to connect in a positive sense, the $+$ locus at $\omega \to 0$ with the negative locus at $\omega \to -0$. If this locus is traced in a positive sense from $\omega \to \infty$ to $\omega \to 0$, around the circle at ∞, and then along the negative-frequency locus the following may be concluded: (1) if the locus **does not** enclose the $-1 + j0$ point, the system is stable; (2) if the locus **does** enclose the $-1 + j0$ point, the system is unstable. The Nyquist criterion can also be applied to the log magnitude of $KG(j\omega)$ and phase-vs.-log ω diagrams. In this method of display, the criterion for stability reduces to the requirement that the log magnitude of $KG(j\omega)$ must cross the 0-dB axis at a frequency less than the frequency at which the phase curve crosses the $-180°$ line. Two stability conditions are illustrated in Fig. 16.2.41. The Nyquist criterion not only provides a simple test for the stability of an automatic-control system but also indicates the degree of stability of the system by indicating the degree to which $KG(j\omega)$ locus avoids the $-1 + j0$ point.

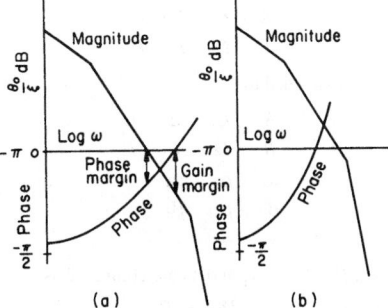

Fig. 16.2.41   Nyquist stability criterion in terms of log magnitude $KG(j\omega)$ diagrams. (*a*) Stable; (*b*) unstable. (*Porter, "An Introduction to Servomechanism," Wiley.*)

The concepts of **phase margin** and **gain margin** are employed to give this quantitative indication of the degree of stability of an automatic-control system. **Phase margin** is defined as the additional negative phase shift necessary to make the phase angle of the transfer function $-180°$ at the frequency where the magnitude of the $KG(j\omega)$ vector is unity. Physically, phase margin can be interpreted as the amount by which the unity KG vector has to be shifted to make a stable system unstable.

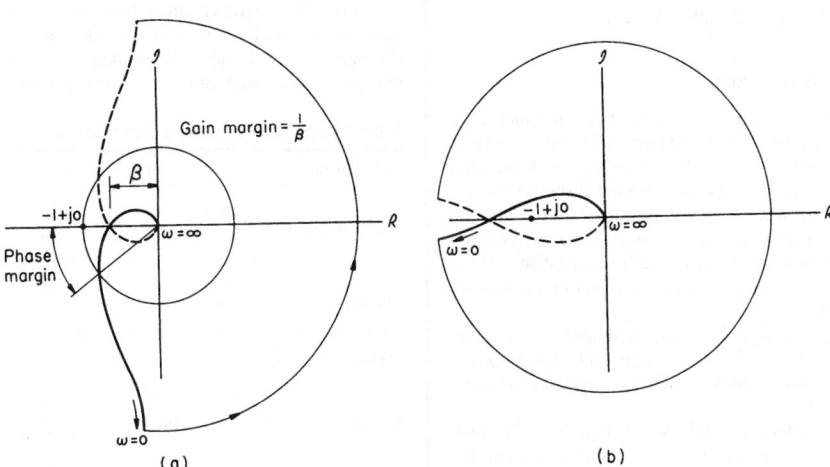

Fig. 16.2.40   Typical $KG(j\omega)$ loci illustrating application of Nyquist's stability criterion. (*a*) Stable; (*b*) unstable.

In a similar manner, **gain margin** is defined as the reciprocal of the magnitude of the KG vector at $-180°$. Physically, gain margin is the number by which the gain must be multiplied to put the system to the limit of stability. Thaler suggests satisfactory results can be obtained in most control applications if the phase margin is between 40 and 60° while the gain margin is between 3 and 10 (10 to 20 dB). These values will ensure a small transient overshoot with a single cycle in the transient. The margin concepts are qualitatively illustrated in Figs. 16.2.40 and 16.2.41.

Routh's Stability Criterion The frequency-response equation of a closed-loop automatic control is

$$\frac{\theta_o}{\theta_i} = \frac{KG(j\omega)}{1 + KG(j\omega)} \qquad (16.2.64)$$

The characteristic equation obtained therefrom has the algebraic form

$$A(j\omega)^n + B(j\omega)^{n-1} + C(j\omega)^{n-2} + \cdots = 0 \quad (16.2.65)$$

The purpose of Routh's method is to determine the existence of roots of this equation which are positive or which are complex with positive real parts and thus identify the resulting instability. To apply the criterion the coefficients are written alternately in two rows as

$$\begin{array}{cccc} A & C & E & G \\ B & D & F & H \end{array}$$

This array is then expanded to

$$\begin{array}{cccc} A & C & E & G \\ B & D & F & H \\ \alpha_1 & \alpha_2 & \alpha_3 \\ \beta_1 & \beta_2 & \beta_3 \\ \gamma_1 & \gamma_2 \end{array}$$

where α_1, α_2, α_3, β_1, β_2, β_3, γ_1 and γ_2 are computed as

$$\alpha_1 = \frac{BC - AD}{B} \qquad \beta_1 = \frac{D\alpha_1 - B\alpha_2}{\alpha_1}$$

$$\alpha_2 = \frac{BE - AF}{B} \qquad \beta_2 = \frac{F\alpha_1 - B\alpha_3}{\alpha_1} \qquad \gamma_1 = \frac{\beta_1\alpha_2 - \alpha_1\beta_2}{\beta_1}$$

$$\alpha_3 = \frac{BG - AH}{B} \qquad \beta_3 = \frac{H\alpha_1 - B_o}{\alpha_1} \qquad \gamma_2 = \frac{\beta_1\alpha_3 - \alpha_1\beta_3}{\beta_1}$$

When the array has been computed, the left-hand column (A, B, α_1, β_1, γ_1) is examined. If the signs of all the numbers in the left-hand column are the same, there are no positive real roots. If there are changes in sign, the number of positive real roots is equal to the number of changes in sign. It should be recognized that this is a test for instability; the absence of sign changes does not guarantee stability.

SAMPLED-DATA CONTROL SYSTEMS

Definition Sampled-data control systems are those in which continuous information is transformed at one or more points of the control system into a series of pulses. This transformation may be performed intentionally, e.g., the flow of information over long distances to preserve the accuracy of the data during the transmission, or it may be inherent in the generation of the information flow, e.g., radiating energy from a radar antenna which is in the form of a train of pulses, or the signals developed by a digital computer during a direct digital control of machine-tool operation.

Methods of analysis analogous to those for continuous-data systems have been developed for the sampled-data systems. Discussed herein are (1) sampling, (2) the z transformation, (3) the z-transfer function, and (4) stability of sampled data systems.

Sampling The ideal sampler is a simple switch (Fig. 16.2.42) which is closed only instantaneously and opens and closes at a constant frequency. The switch, which may or may not be a physical component in a sampled-data feedback system, indicates a sampled signal. Such a sampled-data feedback system is shown in Fig. 16.2.43. The error signal in continuous form is $\epsilon(t)$, and the sampled error signal is $\epsilon^*(t)$. Figure 16.2.42 shows the relationship between these signals in a graphical form.

Fig. 16.2.42 Ideal sampler, showing continuous input and sampled output.

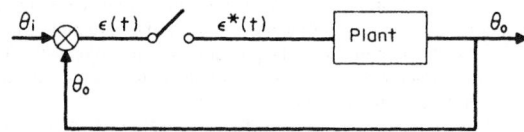

Fig. 16.2.43 Sampled data feedback system.

z Transformation In the analysis of continuous-data systems, it has been shown that the Laplace transformation can be used to reduce ordinary differential equations to algebraic equations. For sampled-data systems, an operational calculus, the z transform, can be used to simplify the analysis of such systems.

Consider the sampler as an impulse modulator; i.e., the sampling modulates an infinite train of unit impulses with the continuous-data variable. Then the Laplace transformation can be shown to be

$$F^*(s) = \sum_{n=0}^{\infty} f(nT)e^{-nTs} \qquad (16.2.66)$$

(See A. W. Langill, "Automatic Control Systems Engineering.")

In terms of the z transform, this becomes

$$F^*(s) = F(z) = \sum_{n=0}^{\infty} f(nT)z^{-n} \qquad (16.2.67)$$

where $z^{-n} = e^{-nTs}$.

Tables in Sec. 2 list the Laplace transforms for a number of continuous time functions. Table 16.2.5 lists the z transforms for some of these functions.

It should be noted that unlike the Laplace transforms, the z-transform method imposes the restriction that the sampled-data-system response can be determined only at the sampling instant. The same z transform may apply to different time functions which may have the same value at

Table 16.2.5 Laplace and z Transforms

Time function	$f(t)$	$F(s) = \pounds[f(t)]$	$F^*(z) = F(z)$
Unit ramp function	t	$\dfrac{1}{s^2}$	$\dfrac{Tz}{(z-1)^2}$
Unit acceleration function	$\dfrac{t^2}{2}$	$\dfrac{1}{s^3}$	$\tfrac{1}{2}T^2\dfrac{z(z+1)}{(z-1)^3}$
Exponential function	$e^{-\alpha t}$	$\dfrac{1}{s+\alpha}$	$\dfrac{z}{z-e^{-\alpha T}}$
Sinusoidal function	$\sin \beta t$	$\dfrac{\beta}{s^2+\beta^2}$	$\dfrac{z\sin\beta T}{z^2-2z\cos\beta T+1}$
Cosinusoidal function	$\cos \beta t$	$\dfrac{s}{s^2+\beta^2}$	$\dfrac{z(z-\cos\beta T)}{z^2-2z\cos\beta T+1}$

Table 16.2.6   Typical Block Diagrams of Sampled-Data Control Systems and Their Transforms

System	Laplace transform of the output, $C(s)$	z-transform of the output, $C^*(z)$
1. [block diagram]	$\dfrac{G_1(s)}{1 + G_1 G_2^*(z)} R^*(z)$	$\dfrac{G_1^*(z)}{1 + G_1 G_2^*(z)} R^*(z)$
2. [block diagram]	$\dfrac{G_1(s)}{1 + G_1(s)G_2(s)} R^*(z)$	$\left[\dfrac{G_1(s)}{1 + G_1(s)G_2(s)}\right]^* R^*(z)$
3. [block diagram]		$\dfrac{G_1^*(z)}{1 + G_1^*(z)G_2^*(z)} R^*(z)$
4. [block diagram]	$G_1(s)\left[R(s) - G_2(s)\dfrac{RG_1^*(z)}{1 + G_1 G_2^*(z)}\right]$	$\dfrac{RG_1^*(s)}{1 + G_1 G_2^*(s)}$
5. [block diagram]	$G_1(s)\left[R^*(z) - \dfrac{G_1^*(z)G_2^*(z)R^*(z)}{1 + G_1^*(z)G_2^*(z)}\right]$	$\dfrac{G_1^*(z)}{1 + G_1^*(z)G_2^*(z)} R^*(z)$
6. [block diagram]	$G_1(s)\left[R^*(z) - G_2(s)\dfrac{R^*(z)G_1^*(z)}{1 + G_1 G_2^*(z)}\right]$	$\dfrac{G_1^*(z)}{1 + G_1 G_2^*(z)} R^*(z)$
7. [block diagram]	$\dfrac{G_2(s)}{1 + G_1 G_2 G_3^*(z)} RG_1^*(z)$	$\dfrac{G_2^*(z)}{1 + G_1 G_2 G_3^*(z)} RG_1^*(z)$
8. [block diagram]	$\dfrac{G_2(s)}{1 + G_2^*(z) + G_1 G_2^*(z)} RG_1^*(z)$	$\dfrac{G_2^*(z)}{1 + G_2^*(z) + G_1 G_2^*(z)} RG_1^*(z)$

SOURCE: John G. Truxal (ed.), "Control Engineers' Handbook," McGraw-Hill.

the instant of sampling. The z-transform function is not defined, therefore, in a continuous sense, and the inverse z transform is not unique.

z-Transfer Function The ratio of the sampled output function of a discrete network to the sampled input function is the z-transfer function. A discrete network is one which has both a sampled input and output. A table of block diagrams of a number of sampled-data control systems with their associated transfer functions is presented in Table 16.2.6.

Stability of Sampled-Data Systems The stability of sampled-data systems can be demonstrated utilizing frequency-response methods which have been discussed in this section. See "Stability and Performance of an Automatic Control." The Nyquist stability criterion again applies with the same conclusions relative to the -1 point, and the methods of generalizing the open- and closed-loop frequency response plots remain the same.

MODERN CONTROL TECHNIQUES

Many engineers and researchers have been actively pursuing the application and development of advanced control strategies. In recent years this effort has included extensive work in the area of robust control theory. One of the primary attractions of this theory is that it generalizes the **single input-single system output (SISO)** concept of gain and phase margins, and its effect on system sensitivity to multivariable control system.

In this section the design technique for a **linear quadratic gaussian with loop transfer recovery (LQG/LTR)** robust controller design method will be summarized. The concepts required to understand this method will be reviewed. A linearized model of a simple process has been chosen to illustrate this technique. The simple process has three state variables, one input, and one output. Three control system design methods are compared: LQG, a LQG/LTR, and a proportional plus integral controller (PI).

Previous Work and Results

Most applications of modern control techniques to process control have appeared in the form of **optimal control strategies** which were in the form of full-state feedback control with and without state estimation [linear quadratic regulator (LQR)]. The regulator problem, coupled with a state estimation problem, was shown not to have desirable robustness properties. Doyle and Stein (Doyle and Stein, 1979) proposed the multivariable robust design philosophy known as LQG/LTR to eliminate the shortcomings of the LQR and LQG design techniques. The LQG/LTR design technique has been applied to control submersible vehicles, engine speed, helicopters, ship steering, and large, flexible space structures.

The LQG/LTR method requires the plant to be **minimum phase,** which cannot be guaranteed in a plant process. Typically **nonminimum phase** conditions in a plant process are due to time delay. The LQG/LTR design procedure offers no guarantees when a nonminimum-phase plant is used. However recent work develops a method to obtain a robust controller design for plants with time delay (Murphy and Bailey, IEEE 1989 and IECON 1990).

The LQG/LTR synthesis method for minimum phase plants achieves desired loop shapes and maximum robustness properties in the design of feedback control systems. The LQG/LTR design procedure uses **singular value analysis** and design procedures to obtain the desired performance and stability robustness. The use of singular values yields a frequency-domain design and analysis method generally expected by control engineers to determine system robustness. The LQG/LTR synthesis method is applicable to both **multiple-input/multiple-output (MIMO)** and SISO control systems. The performance limitations of this design method are analogous to those of a SISO system.

Definition of Robustness

A system is considered to be robust and have **good robustness properties** if it has a large stability margin, good disturbance attenuation, and/or low sensitivity to parameter variations. The term **stability margin** refers to the gain margin and the phase margin, which are quantitative measures of stability. A proven method of obtaining good robustness properties is the use of a feedback control system, which can be designed to allow for variations in the system dynamics. Some causes of **variation** in system dynamics are

Modeling and data errors in the nominal plant and system.

Changes in environmental conditions, manufacturing tolerances, wear due to aging, and noncritical material failures.

Errors due to calibration, installation, and adjustments.

Feedback control systems with good feedback properties have been synthesized for SISO systems. Classical frequency domain techniques such as Nyquist, root-locus, Bode, and Nichols plots have been used to obtain the feedback control system for the SISO system. These design techniques have allowed the synthesis of feedback control systems yielding **insensitivity** to bounded parameter variations and a large stability margin. The success of the feedback control system for the SISO system has led to the direct extension of the classical frequency domain technique to the design of a multivariable feedback control system. This extension to the multivariable design problem examines an individual feedback loop as the phase and gain margins are varied, while the nominal phase and gain values in the remaining feedback paths are held constant. This technique, however, fails to consider the results of simultaneous variation of gain and phase in all paths, which is a real-world possibility and needs to be considered. A method of obtaining a feedback control system with good robustness properties that takes into consideration simultaneous gain and phase variation is the LQG/LTR.

The LQG/LTR technique can be applied to MIMO systems or SISO systems. The LQG/LTR technique not only has the good robustness properties of the classical frequency domain techniques but also is capable of minimizing the effects of unmodeled high-frequency dynamics, neglected nonlinearities, and a reduced-order model. **Computer control system design tools,** for instance, developed by Integrated Systems Incorporated (ISI) and The MathWorks can be used in synthesizing the LQG/LTR technique and other controller design techniques. The use of computer-aided design tools eliminates the numerical programming burden of programming the complex LQG/LTR algorithm.

In what follows, the needed concepts are defined and the LQG/LTR procedure for development of a **model-based compensator** (MBC) is described. A unity-feedback MBC is selected for the controller structure, because of its similarity to the SISO unity-feedback control system well known to classical control designers. The MBC closed-loop

control system has proven to offer great practical considerations in the design of automatic control systems.

Robustness Concepts of the LQG and LQG/LTR Control Systems

The LQG/LTR design procedure is based on the system configuration of the LQG controller shown in Fig. 16.2.44. The LQG controller consists of a Kalman filter state estimator and a linear quadratic regulator. The Kalman filter state estimator has good robustness properties for plant perturbations at the plant output. The linear quadratic regulator (LQR) has good robustness properties for perturbations at the plant input. Even though its components separately have good robustness properties, the LQG controller is found to have no guaranteed robustness properties at either the input (point 2) or the output (point 3) of the plant.

- LQG guaranteed no robustness properties at the input or output of the plant.
- Point 1 has the good robustness properties of the full-state feedback system.
- Point 4 has the good robustness properties of the Kalman filter.
- Point 2 has no guaranteed robustness properties.
- Point 3 has no guaranteed robustness properties.
- The LQG/LTR design method permits recovery of the robustness properties of point 1 at point 2 or the robustness properties of point 4 at point 3.

Fig. 16.2.44 Summary of the robustness properties of the LQG block diagram.

The LQG/LTR design procedure allows us to recover robustness properties at either the input or the output of the plant. If robustness is desired at the input to the plant, first a nominal robust LQR design is made to satisfy the design constraints. Next, an LTR step is made to design a Kalman filter gain that recovers the robustness at the input to the plant of the LQG controller that is approximately that of the nominal LQR design. This implies from Fig. 16.2.44 that the robustness properties at points 1 and 2 are approximately the same.

If robustness is desired at the **output of the plant,** first a nominal robust Kalman filter design is made to satisfy the performance constraints. Next, an LTR step is made to design an LQR gain that recovers the robustness at the output of the plant that is approximately that of the nominal Kalman filter design. This implies from Fig. 16.2.44 that the robustness properties at points 3 and 4 are approximately the same.

The block diagram of the unity-feedback MBC is shown in Fig. 16.2.45. This control system structure allows tracking and regulation of a reference input at the output of the plant. The filter and regulator gains used in the controller $R(s)$ are obtained appropriately, depending on whether robustness is desired at the input or the output of the plant. Note that the robustness properties of the **unity-feedback MBC** (UFMBC) are the same as those of the LQG system, since the UFMBC is just an alternative structure of the LQG system.

Fig. 16.2.45 Unity feedback model-based compensator (UFMBC).

MATHEMATICS AND CONTROL BACKGROUND

The use of **matrix algebra** is very important in the study of MIMO control system design methods. Some MIMO control system design methods use matrix algebra to decouple the system so that the plant matrix transfer function consists only of individual **decoupled transfer functions,** each represented conventionally as the ratio of two polynomials. This simplification then allows the use of conventional methods such as the Bode plot, Nyquist plot, and Nichols chart as a means of designing a controller for each loop of the MIMO plant. The LQG/LTR control system design method, however, does not require **decoupling** of the plant but rather preserves the coupling between the individual loops of the MIMO plant. Each step of the LQG/LTR design technique therefore requires the use of some matrix algebra, vector and matrix norms, and singular values during the control system design procedure. For instance, matrix algebra is used to derive a suitable representation of the closed-loop system for stability and robustness analysis. Matrix norms and singular values are used to examine the matrix magnitude of the MIMO transfer function. The **matrix magnitude** or **singular values** of a system allows preservation of the coupling between the various loops of the MIMO control system.

The concepts of matrix norm, singular values, controllability, observability, stabilizability, and detectability are very important in applying MIMO design procedures. It is therefore important that these concepts be well understood.

The purpose of this section is to provide **background material** in mathematics and control systems that has proved useful in deriving the procedures and analysis required to implement the control system design procedure.

Matrix Norm and Singular Values

The matrix norm of $||\mathbf{A}||_2$ is defined for the $m \times n$ matrix $\mathbf{A}$. The matrix $\mathbf{A}$ defines a linear transformation from the vector space V, called the domain to a vector space W, which is called the range. Thus the linear transformation

$$\mathbf{A}(x) = \mathbf{A}x \qquad (16.2.68)$$

transforms a vector x in $V \in C^n$ into a vector $\mathbf{A}(x)$ in $W \in C^m$. The matrix norm of $||\mathbf{A}||_2$ is defined as

$$||\mathbf{A}||_2 = \max_{x \neq 0} \frac{||\mathbf{A}||_2}{||x||_2} = l_2 \text{ norm} \qquad (16.2.69)$$

Since the l_2 **norm** is a useful tool in control system analysis, further insight into its definition is given. The l_2 norm of Eq. (16.2.69) can be written as

$$||\mathbf{A}||_2 = \max_i [\lambda_i(\mathbf{A}^H\mathbf{A})]^{1/2} \qquad i = 1, 2, \ldots, n \qquad (16.2.70)$$

$$||\mathbf{A}||_2 = \max_i [\lambda_i(\mathbf{A}\mathbf{A}^H)]^{1/2} \qquad i = 1, 2, \ldots, n \qquad (16.2.71)$$

The matrices $\mathbf{A}^H\mathbf{A}$ and $\mathbf{A}\mathbf{A}^H$ are Hermitian and positive semidefinite. The eigenvalues of these matrices are real and nonnegative. The definition of the **singular value** of a complex matrix $\mathbf{A} \in C^{n \times n}$ is

$$\sigma_i(\mathbf{A}) = [\lambda_i(\mathbf{A}^H\mathbf{A})]^{1/2}$$
$$= [\lambda_i(\mathbf{A}\mathbf{A}^H)]^{1/2} \geq 0 \qquad i = 1, 2, \ldots, n \qquad (16.2.72)$$

If $\mathbf{A} \in C^{m \times n}$, which indicates that $\mathbf{A}$ is a nonsquare matrix, then

$$\sigma_i(\mathbf{A}) = [\lambda_i(\mathbf{A}^H\mathbf{A})]^{1/2} = [\lambda_i(\mathbf{A}\mathbf{A}^H)]^{1/2} \qquad (16.2.73)$$

for $1 \leq i \leq k$, where k is the number of singular values $= \min(m, n)$. The maximum and minimum singular values are defined, respectively, as

$$\overline{\sigma}(\mathbf{A}) = ||\mathbf{A}||_2 \qquad (16.2.74)$$

and

$$\underline{\sigma}(\mathbf{A}) = \max_{x \neq 0} \frac{||\mathbf{A}x||}{||x||_2} = \frac{1}{||\mathbf{A}^{-1}||_2} \qquad (16.2.75)$$

provided $\mathbf{A}$ has an inverse.

Controllability, Observability, Stabilizability, and Detectability

The LQG/LTR control system design method has a certain **requirement** of the plant for which the controller is being designed: the plant must be at least stabilizable and detectable. Therefore it is important that the conditions of controllability, observability, and the less restricted conditions of stabilizability and detectability be understood. A description of these concepts follows.

Given the state equation for the linear time invariant (LTI) system described as $\dot{x}(t) = \mathbf{A}x(t) + \mathbf{B}u(t)$ and $y(t) = \mathbf{C}x(t)$. The matrices $\mathbf{A}$, $\mathbf{B}$, and $\mathbf{C}$ are constant. Note that $x \in R^n$, $n \in R^p$, and $y \in R^p$ are, respectively, the state, input, and output vectors. Variables n and p are the dimensions of the states and the input and output vectors. The input and output vectors are of the same dimension.

A system is said to be controllable if it is possible to transfer the system from any initial state x_0 in state space to any other state x_F using the input in a finite period of time. The system is said to be uncontrollable otherwise. The condition of controllability places **no constraint** on the input in the system.

The algebraic test for controllability of the LTI system state equation uses the controllability matrix

$$\mathbf{Q} = [\mathbf{B} \ \mathbf{AB} \cdots \mathbf{A}^{n-1}\mathbf{B}] \qquad (16.2.76)$$

where n is the **number of states.** The LTI system is controllable if the rank of $\mathbf{Q}$ is equal to n. If the rank of $\mathbf{Q}$ is less than n, the system is not controllable. An uncontrollable system indicates that some modes or poles are not affected by the control input. Uncontrollable systems with **unstable modes** imply that a controller design is not possible to ensure system stability. If the uncontrollable modes of the uncontrollable system are stable, then the system is stabilizable.

An LTI system is considered **observable** at initial time t_0 if there exists a finite time t_1 greater than t_0 such that for any initial state x_0 in state space using knowledge of the input u and output y over the time interval $t_0 < t > t_1$, the state x_0 can be determined. The system is said to be unobservable otherwise.

The algebraic test for observability of the LTI system uses the observability matrix

$$\mathbf{N} = [\mathbf{C}^T \ \mathbf{A}^T\mathbf{C}^T \cdots (\mathbf{A}^T)^{n-1}\mathbf{C}^T] \qquad (16.2.77)$$

The LTI system is observable if the rank of $\mathbf{N}$ is equal to n, where n is the number of states. If the rank of $\mathbf{N}$ is less than n, the system is unobservable. An unobservable system indicates that not all system modes contribute to the output of the system. If the **unobservable modes** (states) are stable, then the system is considered to be detectable.

EVALUATING MULTIVARIABLE PERFORMANCE AND STABILITY ROBUSTNESS OF A CONTROL SYSTEM USING SINGULAR VALUES

Performance Robustness

The requirements for the control system design are formulated by placing certain restrictions on the singular values of the **return ratio, return difference,** and **inverse return difference** of the control system. The block diagram in Fig. 16.2.46 shows the controller $K(s)$ and plant $G(s)$ with input $d_i(s)$ and output $d_o(s)$ disturbances and sensor noise $n(s)$.

Using Fig. 16.2.46, a **frequency-domain transfer function** representation for the plant output is first obtained. These transfer function representations, used in conjunction with concepts of singular values, allow the formulation of requirements in the frequency domain to obtain a control system with good performance and stability robustness.

In this design, performance robustness is of concern at the output of the plant. Therefore it is required that the plant output transfer function be derived in order to analyze and assist in obtaining a control system design to meet certain performance robustness requirements. Using Fig. 16.2.46,

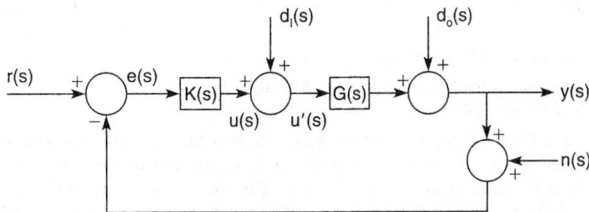

Fig. 16.2.46   Model-based unity feedback MIMO control system.

the plant output transfer function is derived and performance robustness requirements established. From Fig. 16.2.46 it can be shown that

$$y(s) = d_o(s) + G(s)u'(s) \qquad (16.2.78)$$
$$e(s) = r(s) - y(s) - n(s) \qquad (16.2.79)$$
and
$$u'(s) = d_o(s) + u(s) = d_I(s) + K(s)e(s). \qquad (16.2.80)$$

where $G(s) = C[sI - A]^{-1}B$ is the plant transfer function and **A**, **B**, and **C** are system matrices. The transfer functions $y(s)$, $d_o(s)$, $e(s)$, $r(s)$, $n(s)$, $u(s)$, $d_I(s)$, and $K(s)$ are, respectively, the output, output disturbance, error, reference, noise, input, input disturbance, and controller transfer functions.

Using Eqs. (16.2.79) and (16.2.80) it can be shown that Eq. (16.2.78) becomes

$$
\begin{aligned}
y(s) &= [G(s)K(s)][r(s) - y(s) - n(s)] \\
&\quad + d_o(s) + G(s)d_I(s) \\
&= G(s)K(s)r(s) - G(s)K(s)y(s) - G(s)K(s)n(s) \\
&\quad + d_o(s) + G(s)d_I(s) \qquad (16.2.81)
\end{aligned}
$$

Therefore,

$$
\begin{aligned}
y(s) &= [I + G(s)K(s)]^{-1}[G(s)K(s)r(s) - G(s)K(s)n(s) \\
&\qquad + d_o(s) + G(s)d_I(s)] \\
&= [I + G(s)K(s)]^{-1}\{G(s)K(s)[r(s) - n(s)]\} \\
&\quad + [I + G(s)K(s)]^{-1}[d_o(s) + G(s)d_I(s)] \\
&= y_r(s) + y_{d_I}(s) + y_{d_o}(s) + y_n(s) \qquad (16.2.82)
\end{aligned}
$$

where $y_r(s)$, $y_{d_I}(s)$, $y_{d_o}(s)$, and $y_n(s)$ are, respectively, the contributions to the plant output due to the reference input, input disturbance, output disturbance, and noise input to the system. The requirements for good command following, disturbance rejection, and insensitivity to sensor noise are defined by using singular values to characterize the magnitude of MIMO transfer functions. These requirements are also **applicable to SISO systems** (Kazerooni and Narayan; Murphy and Bailey, April 1989).

Command Following   The conditions required to obtain a MIMO control system design with good command following is described. Good command following of the reference input $r(s)$ by the plant output $y(s)$ implies that

$$y(s) \approx r(s) \qquad \forall s \in s_R \qquad (16.2.83)$$

where $s = j\omega$ and $s_R = j\omega_R$ and ω_R is the frequency range of the input $r(s)$. Letting $d_o(s) = d_I(s) = n(s) = 0$ in Eq. (16.2.82) results in

$$y_r(s) = [I + G(j\omega)K(j\omega)]^{-1}G(j\omega)K(j\omega)r(j\omega) \qquad \forall \omega \in \omega_R$$

$$= \{I + [G(j\omega)K(j\omega)]^{-1}\}^{-1}r(j\omega) \qquad \forall \omega \in \omega_R \qquad (16.2.84)$$

where $G(j\omega)K(j\omega)$ is defined as the return ratio. Therefore, to obtain good command following, the system **inverse return difference** $I + [G(j\omega)K(j\omega)]^{-1}$ must have the property such that

$$I + [G(j\omega)K(j\omega)]^{-1} \approx I \qquad \forall \omega \in \omega_R \qquad (16.2.85)$$

which implies that

$$\underline{\sigma}[G(j\omega)K(j\omega)] \gg 1 \qquad \forall \omega \in \omega_R \qquad (16.2.86)$$

Thus the magnitude of the minimum singular value of the **return ratio** $\underline{\sigma}[G(j\omega)K(j\omega)]$ must be large over the frequency range of the input to obtain good command following of the control system.

Disturbance Rejection   To minimize the effect of the output disturbance, $y_{d_o}(s)$, on the output signal $y(s)$, Eq. (16.2.82) is used with $n(s) = r(s) = d_I(s) = 0$, which results in

$$y_{d_o}(s) = [I + G(s)K(s)]^{-1}d_o(s) \qquad (16.2.87)$$

Rejection of this output disturbance at the plant output requires that the magnitude of the **return difference I** $+ G(s)K(s)$ to meet the following condition

$$\underline{\sigma}[I + G(j\omega)K(j\omega)] \gg 1 \qquad \forall \omega \in \omega_{d_o} \qquad (16.2.88)$$

which implies

$$\underline{\sigma}[G(j\omega)K(j\omega)] \gg 1 \qquad \forall \omega \in \omega_{d_o} \qquad (16.2.89)$$

where y_{d_o} is the frequency range of the output disturbance $d_o(s)$.

With $n(s) = r(s) = d_o(s) = 0$ in Eq. (16.2.82), the effect of input disturbance on the plant output is examined. This effect is seen to be

$$y_{d_I}(s) = [I + G(j\omega)K(j\omega)]^{-1}[G(j\omega)d_I(j\omega)] \qquad \forall \omega \in \omega_{d_I} \qquad (16.2.90)$$

where ω_{d_I} is the frequency range of the input disturbance $d_I(s)$. To minimize the effect of $y_{d_I}(s)$ on the plant output requires that the magnitude of the return difference be large (in general, the larger the better). This requirement implies that

$$\underline{\sigma}[G(j\omega)K(j\omega)] \gg 1 \qquad \forall \omega \in \omega_{d_I} \qquad (16.2.91)$$

to reject the input disturbance. It should be noted that this equation gives only general conditions for rejection. The control engineer must determine how much greater than 1 the return ratio magnitude must be.

Noise Rejection   The requirements to minimize the effects of sensor noise on the plant output must be determined. First letting $n(s) = d_I(s) = d_o(s) = 0$ in Eq. (16.2.82) results in

$$y_n(s) = [I + G(s)K(s)]^{-1}[G(s)K(s)]n(s) \qquad \forall s \in s_n$$

$$= \{I + [G(s)K(s)]^{-1}\}^{-1}n(s) \qquad \forall s \in s_n \qquad (16.2.92)$$

where $s_n = j\omega_n$ and ω_n is the frequency range of the sensor noise. Thus to minimize the effects of sensor noise on the plant output it is desired to have the magnitude of $y_n(s)$ small over the frequency range of the noise. Thus

$$\overline{\sigma}[G(j\omega)K(j\omega)] \gg 1 \qquad \forall \omega \in \omega_n \qquad (16.2.93)$$

is required to minimize the effect of noise on the plant output.

Any overlapping of any of the frequency ranges ω_R, ω_{d_I}, ω_{d_o}, ω_n will require system design specification tradeoffs to be made. This will definitely occur when the lower frequency ranges of ω_R and/or ω_{d_I} and ω_{d_o} overlap the higher frequency range ω_n.

Stability Robustness

All linear control system design methods are based on a linear model of the plant. Because this design model only approximates the actual plant, an error exists between the design model and the actual plant. This error can be evaluated as **absolute** or **relative.** The method of evaluation of this error (model uncertainty) depends on the nature of the differences between the actual design model and the actual plant. The evaluation of an absolute error defines an additive model uncertainty, whereas the evaluation of a relative error defines a multiplicative model uncertainty. The **additive** and multiplicative model uncertainties can be further classified as either structured or unstructured model uncertainties (Murphy and Bailey, ORNL 1989).

A **structured model uncertainty** is usually a low-frequency phenomenon characterized by variations in the parameters of the linear time-invariant plant design model. These parameter variations are caused by changes in the plant operating conditions, wear due to aging, and environmental conditions. A structured uncertainty also indicates that the gain and phase information concerning the uncertainty is known. Additive model uncertainties are generally considered structured uncertainties.

An **unstructured uncertainty** is typically a high-frequency phenomenon in which the only information known about the uncertainty is its magnitude. Unstructured uncertainties usually have the characteristic of becoming significant at high frequencies. Multiplicative uncertainties are characterized as unstructured. Unstructured uncertainties are usually caused by the truncation of high-frequency plant dynamics or modes due to the linearization process, and the neglect of plant dynamics such as actuators and sensors. Some plant dynamics that characterize unstructured uncertainty are flexible-body dynamics, electrical and mechanical resonances, and transport delays. Most linear control-system design methods neglect these high-frequency plant dynamics. The result is that if a high-frequency bandwidth controller is implemented, the high-frequency modes could be excited, resulting in an unstable control system.

Model uncertainty clearly is seen to impose limitations on the achievable performance of a feedback control system design. Hence this section will focus on the limitations imposed by uncertainty, not on the difficult problem of exact representation of the system uncertainty.

Multiplicative Uncertainty   Multiplicative uncertainty (perturbation) can be represented at the input or output of the plant, and is called respectively the input multiplicative uncertainty or the **output multiplicative uncertainty.** The model uncertainty due to actuator and sensor dynamics is modeled, respectively, as an input or output multiplicative perturbation. A reduced-order model of an already **linearized model** results in a multiplicative model uncertainty. **Time delay** at the input or output of a plant yields, respectively, input and output multiplicative uncertainties in the plant. Even though other multiplicative model uncertainties exist among various types of control system design problems, the model uncertainties mentioned above are a common consideration in most control problems.

Output Multiplicative Uncertainty

The output multiplicative uncertainty $\Delta G(s)$ is defined as

$$\Delta G(s) = [G'(s) - G(s)]G^{-1}(s) \qquad (16.2.94)$$

where
$$G'(s) = L(s)G(s) \qquad (16.2.95)$$

is the perturbed transfer function, $L(s)$ represents dynamics not included in the nominal plant $G(s)$. Thus resulting in

$$\Delta G(s) = [L(s) - \mathbf{I}] \qquad (16.2.96)$$

A block diagram of a control system with output multiplicative perturbation is shown in Fig. 16.2.47. The conditions of stability for a control system with output multiplicative uncertainty require that

$$\underline{\sigma}\left\{\mathbf{I} + [G(j\omega)K(j\omega)]^{-1}\right\} > \overline{\sigma}\left[\Delta G(j\omega)\right] \qquad \forall\,\omega > 0 \quad (16.2.97)$$

Fig. 16.2.47   Control system with output multiplicative perturbation.

Additive Uncertainty   The block diagram of a control system with a plant additive model uncertainty is shown in Fig. 16.2.48. The representation of the additive model uncertainty using an absolute error criterion is defined as

$$\Delta G(s) = G'(s) - G(s) \qquad (16.2.98)$$

where in this case the perturbed transfer function $G'(s)$ is

$$G'(s) = [\mathbf{C} + \Delta \mathbf{C}]\{s\mathbf{I} - [\mathbf{A} + \Delta \mathbf{A}]\}^{-1}[\mathbf{B} + \Delta \mathbf{B}] \qquad (16.2.99)$$

and the nominal plant transfer function $G(s)$ is

$$G(s) = \mathbf{C}[s\mathbf{I} - \mathbf{A}]^{-1}\mathbf{B} \qquad (16.2.100)$$

and where $\mathbf{A}$, $\mathbf{B}$, and $\mathbf{C}$ are the nominal system matrices and $\Delta \mathbf{A}$, $\Delta \mathbf{B}$, and $\Delta \mathbf{C}$ are the perturbation matrices that indicate the respective deviations from the nominal system matrices. The condition for stability of the additively perturbed control system requires that

$$\underline{\sigma}\left[G(j\omega)K(j\omega)\right] > \overline{\sigma}\left[\Delta G(j\omega)K(j\omega)\right] \qquad \forall\,\omega > 0 \quad (16.2.101)$$

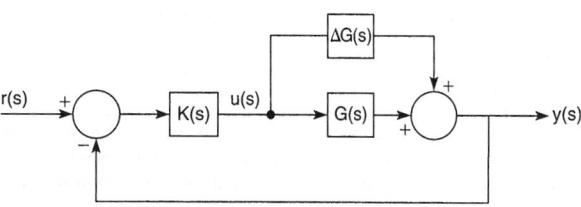

Fig. 16.2.48   Control system with additive perturbation.

Table 16.2.7 contains a summary of the general design requirements presented in this section to obtain a control system with good performance and stability robustness.

REVIEW OF OPTIMAL CONTROL THEORY

Linear Quadratic Regulator

Consider the linear time-invariant state-space system, first-order vector differential equation model

$$\dot{\mathbf{x}}(t) = \mathbf{A}x(t) + \mathbf{B}u(t) \qquad (16.2.102)$$
$$y(t) = \mathbf{C}x(t) \qquad (16.2.103)$$

where $\mathbf{A}$, $\mathbf{B}$, and $\mathbf{C}$ are constant system matrices, and $x(t)$, $u(t)$, and $y(t)$ are, respectively, the system state, input, and output vectors. The system is stabilizable and controllable. The goal of this procedure is to minimize the finite performance index

$$
\begin{aligned}
J &= \int_0^\infty [x^{\mathrm{T}}(t)\mathbf{Q}x(t) + u^{\mathrm{T}}(t)\mathbf{R}u(t)]\,dt \\
&= \int_0^\infty [x^{\mathrm{T}}\mathbf{C}^{\mathrm{T}}\mathbf{C}x + u^{\mathrm{T}}\mathbf{R}u]\,dt \\
&= \int_0^\infty [y^{\mathrm{T}}y + u^{\mathrm{T}}\mathbf{R}u]\,dt \qquad (16.2.104)
\end{aligned}
$$

where $\mathbf{Q}$ is a symmetric positive semidefinite matrix and $\mathbf{R}$ is positive definite.

Table 16.2.7   General System Design Specifications

System requirements		Range
Good command-following	$\underline{\sigma}[G(j\omega)K(j\omega)] \gg 0$ dB	$\forall\,\omega \in \omega_R$
Good disturbance rejection	$\underline{\sigma}[G(j\omega)K(j\omega)] \gg 0$ dB	$\forall\,\omega \in \omega_d$
Good immunity to noise	$\overline{\sigma}[G(j\omega)K(j\omega)] \ll 0$ dB	$\forall\,\omega \in \omega_n$
Good system response to high-frequency modeling error*	$\underline{\sigma}\{I + [G(j\omega)K(j\omega)]^{-1}\} > \|\Delta G(j\omega)\|$	$\forall\,\omega > 0$ rad/s
Good insensitivity to parameter variations at low frequencies*	$\underline{\sigma}[G(j\omega)K(j\omega)] > \overline{\sigma}[\Delta G(j\omega)K(j\omega)]$	

* The $\Delta G(j\omega)$ indicated corresponds to the applicable multiplicative or additive model uncertainty.

The **performance index** J, referred to as the quadratic performance index, implies that a control $u(t)$ is sought to facilitate the minimization of J. The weighting matrices $\mathbf{Q}$ and $\mathbf{R}$ are selected to reflect the importance of particular states and control inputs. In general the weighting of the diagonal elements of $\mathbf{Q}$ can be determined by the importance of the state, as observed by the $\mathbf{C}$ matrix of the output equation. The effect of $\mathbf{Q}$ is the ability to control the **transient response** of the LQR. The effect of $\mathbf{R}$ is the ability to **control** the **energy** resulting from u.

Therefore a linear control law that minimizes J is defined as

$$u(t) = -\mathbf{K}x(t) \qquad (16.2.105)$$

where
$$\mathbf{K} = \mathbf{R}^{-1}\mathbf{B}^{\mathrm{T}}\mathbf{S} \qquad (16.2.106)$$

where, in turn, $\mathbf{S}$ is a constant symmetric positive semidefinite matrix that satisfies the algebraic Riccati equation

$$\mathbf{Q} - \mathbf{SBR}^{-1}\mathbf{B}^{\mathrm{T}}\mathbf{S} + \mathbf{A}^{\mathrm{T}}\mathbf{S} + \mathbf{SA} = 0 \qquad (16.2.107)$$

The closed-loop regulator is given by

$$\dot{\mathbf{x}}(t) = [\mathbf{A} - \mathbf{BK}]x(t) = [\mathbf{A} - \mathbf{BR}^{-1}\mathbf{B}^{\mathrm{T}}\mathbf{S}]x(t) \qquad (16.2.108)$$

and is asymptotically stable provided the state equation (16.2.102) and the output equation (16.2.103) are detectable.

Kalman Filter

What follows is a method of reconstructing an estimate of the states using only the output measurements of the system. The method of reconstruction must be applicable when **process noise** and **measurement noise,** respectively, corrupt the plant state equations and the output measurement equations.

Let us consider the stochastic linear system state space vector differential equation model

$$\dot{\mathbf{x}}(t) = \mathbf{A}x(t) + \mathbf{B}u(t) + \boldsymbol{\Gamma}d(t) \qquad (16.2.109)$$
$$y(t) = \mathbf{C}x(t) + n(t) \qquad (16.2.110)$$

where $\mathbf{A}$, $\mathbf{B}$, $\mathbf{C}$, and $\boldsymbol{\Gamma}$ are constant system matrices. Vectors $s(t)$, $u(t)$, and $y(t)$ are, respectively, the state, input, and output vectors. Vector $d(t)$ is a process noise random vector and vector $n(t)$ is a measurement noise random vector. Assuming that $d(t)$ and $n(t)$ are zero-mean, uncorrelated, gaussian white noises,

$$E[d(t)E[n(t)] = 0 \qquad \text{for all } t \qquad (16.2.111)$$
$$E[d(t)d^{\mathrm{T}}(\tau)] = \mathbf{D}_o\delta(t - \tau) \qquad (16.2.112)$$
$$E[n(t)n^{\mathrm{T}}(\tau)] = \mathbf{N}\delta(t - \tau) \qquad \text{for all } t < \tau \qquad (16.2.113)$$
$$E[d(t)n^{\mathrm{T}}(\tau)] = 0 \qquad (16.2.114)$$

where τ is the difference between two points in time, $E[\ ]$ is mean value, δ is an impulse function, and $\mathbf{N}$ and $\mathbf{D}_o$ are constant symmetric, positive definite, and positive semidefinite matrices, respectively. Note that $\mathbf{N}$ and $\mathbf{D}_o$ are constant because it is assumed that $d(t)$ and $n(t)$ are wide-sense stationary.

The **state estimate** $\hat{\mathbf{x}}$ **of the actual measured state** x is obtained from the noisy measurement y. Therefore we can define a state error vector

$$e(t) = x(t) - \hat{\mathbf{x}}(t) \qquad (16.2.115)$$

where we desire to minimize the mean square error

$$\bar{\mathbf{e}} = E[e^{\mathrm{T}}(t)e(t)] \qquad (16.2.116)$$

The estimator can thus be shown to take the form

$$\dot{\hat{\mathbf{x}}} = \mathbf{A}x(t) + \mathbf{B}u(t) + \mathbf{F}[y(t) - \mathbf{C}x(t)] \qquad (16.2.117)$$

where $\mathbf{F}$ is the filter gain which minimizes Eq. (16.2.116); $\mathbf{F}$ is defined as

$$\mathbf{F} = \Sigma\mathbf{C}^{\mathrm{T}}\mathbf{N}^{-1} \qquad (16.2.118)$$

where Σ is a constant-variance matrix of the error $e(t)$. It is assumed here that $e(t)$ is wide-sense stationary. The matrix Σ is obtained by solving the algebraic variance Riccati equation

$$\mathbf{D} - \Sigma\mathbf{C}^{\mathrm{T}}\mathbf{N}^{-1}\mathbf{C}\Sigma + \mathbf{A}\Sigma + \Sigma\mathbf{A}^{\mathrm{T}} = 0 \qquad (16.2.119)$$

where
$$\mathbf{D} = \boldsymbol{\Gamma}\mathbf{D}_o\boldsymbol{\Gamma} \qquad (16.2.120)$$

A sufficient condition to obtain a unique and positive definite Σ from Eq. (16.2.117) requires that $[\mathbf{A}, \mathbf{C}]$ be completely observable. However, if the pair $[\mathbf{A}, \mathbf{C}]$ is required to be detectable, then Σ can be positive semidefinite.

The **reconstruction error** $e(t)$ satisfies the differential equation

$$\dot{e}(t) = [\mathbf{A} - \mathbf{FC}]e(t) + [\boldsymbol{\Gamma} - \mathbf{F}]\begin{bmatrix} d(t) \\ n(t) \end{bmatrix} \qquad (16.2.121)$$

such that

$$e(t) \to 0 \quad \text{as} \quad t \to \infty \quad \forall\, t \geq t_0$$

if and only if the **observer** is asymptotically stable. The poles of the observer (filter) are found by using the closed-loop dynamics matrix $[\mathbf{A} - \mathbf{FC}]$. To obtain asymptotic stability of the filter, it is required that the pair $[\mathbf{A}, \boldsymbol{\Gamma}]$ be completely controllable. The necessary and sufficient condition of **stabilizability** of the pair $[\mathbf{A}, \boldsymbol{\Gamma}]$ will ensure **stability** also.

PROCEDURE FOR LQG/LTR COMPENSATOR DESIGN

An LQG/LTR compensator will be designed for a **minimum phase plant (no right half plane zeroes).** The control system so designed will be guaranteed to be stable and will be robust at the plant output.

First, it is assumed that if **zero steady-state error** is desired in the control system design, the plant inputs will be augmented to obtain the desired **integral action.** In some cases the dynamics of the plant may have poles that are nearly zero, which establishes a **near integral action** in the plant. In such a case the addition of integrals at the plant input causes a much greater than **20-dB/decade roll-off,** thus yielding unreasonable plant dynamics characterized by the return ratio singular-value plots. Therefore caution is required in deciding how or whether to augment the plant dynamics. In the following description of the compensator design it is assumed that the nominal plant transfer function $G(s)$ has been augmented as required by the control engineers' desires.

Step 1   The first step requires selecting the appropriate transfer function $[G_{\mathrm{FOL}}(s)]$ that satisfies the desired **performance and stability robustness** of the final control system. Therefore, the appropriate scalar $\mu > 0$ and a constant matrix $\mathbf{L}$ such that the singular values of

$$G_{\mathrm{FOL}}(s) = (1/\sqrt{\mu})[\mathbf{C}(s\mathbf{I} - \mathbf{A})^{-1}\mathbf{L}] \qquad (16.2.122)$$

approximately meet the desired control system requirements. The system matrices $\mathbf{C}$ and $\mathbf{A}$ of the minimum phase component are assumed to have been augmented already to obtain the necessary integral action. The selection of $\mathbf{L}$ is made so that at low frequencies ($s = j\omega \to 0$), high frequencies ($s = j\omega \to j\omega$), or intermediate frequencies ($s = j\omega \to j\omega_I$) the singular values of $G_{\mathrm{FOL}}(s)$ are approximately the same. After $\mathbf{L}$ is selected, then μ is adjusted to obtain the desired crossover frequency $\sigma[G_{\mathrm{FOL}}(j\omega)]$; therefore μ functions as a gain parameter of the transfer function. It should be noted that it is desired to balance the singular values of $G_{\mathrm{FOL}}(s)$ in the region of the crossover frequency in order to obtain similar responses for each loop of the system. Also, typically it is desired to have the return ratio singular values roll off at a rate no greater than -20 dB/decade in the region of gain crossover.

Step 2   In this step the Kalman filter transfer function is defined as

$$G_{\mathrm{KF}}(s) = \mathbf{C}(s\mathbf{I} - \mathbf{A})^{-1}\mathbf{F} \qquad (16.2.123)$$

where $\mathbf{F}$ is the Kalman filter gain and $\mathbf{C}$ and $\mathbf{A}$ are the system matrices of the linear time-invariant state-space system.

To compute the **filter gain F,** first the following Kalman filter **algebraic Riccati equation (ARE)** is solved

$$\mathbf{LL}^{\mathrm{T}} - (1/\sqrt{\mu})\Sigma\mathbf{C}^{\mathrm{T}}\mathbf{C} + \mathbf{A}\Sigma + \Sigma\mathbf{A}^{\mathrm{T}} = 0 \qquad (16.2.124)$$

where $\mathbf{L}$ and μ are as obtained in step 1 and Σ (the covariance matrix) is the solution to the equation. The filter gain is thus computed to be

$$\mathbf{F} = (1/\sqrt{\mu})\Sigma\mathbf{C}^{\mathrm{T}} \qquad (16.2.125)$$

Fig. 16.2.49 Block diagram of unity-feedback LQG/LTR control system.

Thus as seen above, **L** and μ are tunable parameters used to produce the necessary filter gain **F** to meet the desired system performance and stability robustness constraints. This step completes the LQG/LTR design requirement of designing the target Kalman filter transfer function matrix with the desired properties which will be recovered at the plant output during the loop transfer recovery step. The general block diagram of the LQG/LTR control system is shown in Fig. 16.2.49.

Step 3 The transfer function of the LQG/LTR controller is defined as

$$K(s) = \mathbf{K}[s\mathbf{I} - (\mathbf{A} - \mathbf{BK} - \mathbf{FC})]\mathbf{F} \qquad (16.2.126)$$

where **A**, **B**, and **C** are the system matrices of Eqs. (16.2.102) and (16.2.103). The filter gain **F** is as stated in Eq. (16.2.125). The **regulator gain K** will be computed in this step to achieve loop transfer recovery. The selection of **K** will yield a return ratio at the plant output such that

$$G(s)K(s) \to G_{KF}(s) \qquad (16.2.127)$$

where $G(s)$ is the plant transfer function, and $K(s)$ is the controller transfer function.

The first step in selecting **K** requires solving the LQR problem, which has the following ARE:

$$\mathbf{Q}(q) - \mathbf{SBR}_0^{-1}\mathbf{B}^T\mathbf{S} + \mathbf{SA} + \mathbf{A}^T\mathbf{S} = 0 \qquad (16.2.128)$$

where $\mathbf{R}_0 = \mathbf{I}$ is the control weighting matrix and $\mathbf{Q}(q) = \mathbf{Q}_0 + q^2\mathbf{CC}^T$ is the modified state weighting matrix. The scalar q is a free design parameter. The resulting regulator gain **K** is computed as

$$\mathbf{K} = \mathbf{B}^T\mathbf{S} \qquad (16.2.129)$$

As $q \to 0$ Eq. (16.2.128) becomes the ARE to the optimal regulator problem. As $q \to \infty$ the LQG/LTR technique guarantees that, since the design model $G(s) = (\mathbf{C}(s)\mathbf{I} - \mathbf{A})^{-1}\mathbf{B}$ has no nonminimum phase zeroes, then pointwise in s

$$\lim_{q \to \infty} G(s)K(s) \to G_{KF}(s) \qquad (16.2.130)$$

$$\lim_{q \to \infty} \{\mathbf{C}(s\mathbf{I} - \mathbf{A})^{-1}\mathbf{BK}[s\mathbf{I} - (\mathbf{A} - \mathbf{BK} - \mathbf{FC})]^{-1}\mathbf{F}\} \to$$
$$\mathbf{C}(s\mathbf{I} - \mathbf{A})^{-1}\mathbf{F} \qquad (16.2.131)$$

This completes the LQG/LTR design procedure. Now that the loop transfer recovery step is complete, the singular-value plots of the return ratio, return difference, and inverse return difference must be examined to verify that the desirable loop shape has been obtained.

EXAMPLE CONTROLLER DESIGN FOR A DEAERATOR

In this section, three linear control methods are used to obtain a level-control system design for a deaerator. The **three linear control system design methods** that will be used are proportional plus integral, linear quadratic gaussian, and linear quadratic gaussian with loop transfer recovery. In this investigation, the dual of the procedure developed for the LQG/LTR design at the plant input will be used to obtain a **robust control system design** at the **plant output** (Murphy and Bailey, ORNL 1989).

The **Bode gain** and **phase plots** of the resulting control system design will be presented for each controller. Also, the singular-value plots of the **return ratio, return difference,** and **inverse return difference** for opening the loop at the plant output (point 3 in Fig. 16.2.50) will be presented.

Fig. 16.2.50 Block diagram of LQG system.

The deaerator was chosen because of its **simple mathematical structure;** one input, one output, and three state variables. Thus the model is easy to follow but complex enough to illustrate the design technique.

The Linearized Model

The mathematical model of the deaerator is **nonlinear,** with a process flow diagram as shown in Fig. 16.2.51. This nonlinear plant is **linearized** about a nominal operating condition, and the resulting linearized plant

Fig. 16.2.51 General process flow diagram of the deaerator.

model will be used to obtain the controller design. The linear **deaerator model** is described by the state space linear time-invariant (LTI) differential equation

$$\dot{x}(t) = \mathbf{A}x(t) + \mathbf{B}u(t) \qquad (16.2.132)$$

$$y(t) = \mathbf{C}x(t) \qquad (16.2.133)$$

where **A**, **B**, and **C** are the system matrices given, respectively, as

$$\mathbf{A} = \begin{bmatrix} -53.802 & 1.7093 & 9.92677 \\ 0.0001761 & -0.0009245 & -0.0053004 \\ -0.001124 & -0.0243636 & -0.139635 \end{bmatrix}$$

$$\mathbf{B} = \begin{bmatrix} -9526.25 \\ 0.265148 \\ -1.72463 \end{bmatrix}$$

$$\mathbf{C} = [0.0 \quad 3.2833 \quad 0.06373]$$

The system states are defined as

$$x = \begin{bmatrix} \delta P \\ \delta \rho \\ \delta H \end{bmatrix}$$

where δP is operating pressure between the pump and the extraction line, $\delta \rho$ is fluid density, and δH is internal energy of the tank level. Plant output y = change in deaerator tank level, and the plant input u = change in control valve.

LQG Controller Design

The design considerations for the tracking LQG control system will first be discussed. The block diagram of the LQG control system is shown in Fig. 16.2.50. It is apparent that output tracking of a reference input will not occur with the present LQG controller configuration. An alternative compensator structure, shown in Fig. 16.2.52, is therefore used to obtain a tracking LQG controller. The controller $K(s)$ uses the same filter gain **F** and regulator gain **K** computed by the LQG design procedure. The detailed block diagram is the same structure that will be used for the LQG/LTR controller shown in Fig. 16.2.49.

Fig. 16.2.52   Alternative compensator structure.

Using a **control system design package** MATRIX$_X$, the regulator gain **K** and filter gain **F** are computed as

$$\mathbf{K} = 10^5 \times [0.0233 \quad 3.145 \quad -0.1024]$$

and

$$\mathbf{F} = \begin{bmatrix} 0.1962 \\ 9.9998 \\ -0.0083 \end{bmatrix}$$

given **user-defined weighting matrices.** The frequency response of the open-loop transfer function of the closed-loop compensated system

$$\frac{y(s)}{r(s)} = \frac{\text{tank level}}{\text{reference input}}$$

where $y(s)$ is the **tank level** and $r(s)$ is the **reference input,** as shown in Fig. 16.2.53.

LQG/LTR Controller Design

This LQG/LTR control system has the structure shown in Fig. 16.2.52. This LQG/LTR control system design is a two-step process. First, a Kalman filter (KBF) is designed to obtain **good command following** and **disturbance rejection** over a specified low-frequency range. Also, the KBF design is made to meet the required robustness criteria with system

Fig. 16.2.53   Open-loop frequency response of the PI, LQG, LQG/LTR control system. (*a*) Frequency response of system gain; (*b*) frequency response of system phase.

uncertainty of $\Delta G(s)$ at the system's output. To obtain good command following and disturbance rejection requires that

$$\underline{\sigma}(G_4(s)) \geq 20 \text{ dB} \qquad \forall \, \omega < 0.1 \text{ rad/s} \qquad (16.2.134)$$

where $G_4(s)$ is the loop transfer function at the output of the KBF (point 4 of Fig. 16.2.50). Assuming the **high-frequency uncertainty** $\Delta G(s)$ for this system becomes significant at 5 rad/s, then for robustness

$$\bar{\sigma}\{\mathbf{I} + [G_4(s)]^{-1}\} \geq \|\Delta G(j\omega)\| = 5 \text{ dB}$$

$$\forall \, \omega \geq 5 \text{ rad/s} \qquad (16.2.135)$$

In this application, the frequency at which system uncertainty becomes significant, 5 rad/s, and the magnitude of the uncertainty, 5 dB at 5 rad/s, are assumed values. This assumption is required because of the lack of information on the high-frequency modeling errors of the process being studied.

The **second step** is to **recover the good robustness properties** of the loop transfer function $G_4(s)$ of point 4 at the output node y (point 3). This will be accomplished by applying the loop transfer recovery (LTR) step at the output node (point 3). The LTR step requires the computation of a regulator gain (**K**) to obtain the robustness properties of the loop transfer function $G_4(s)$ at the output node y.

The tool used to obtain the LQG/LTR design for this system, considering the low- and high-frequency bound requirements is CASCADE, a computer-aided system and control analysis and design environment that synthesizes the LQG/LTR design procedure (Birdwell et al.). Note that more recent computer-aided control system design tools synthesize this design procedure.

The resulting filter gain **F** and regulator **K** gain for this system are, respectively,

$$\mathbf{F} = \begin{bmatrix} -0.0107 \\ 0.472 \\ -0.02299 \end{bmatrix}$$

and

$$\mathbf{K} = [0.660 \quad 27703.397 \quad 589.4521]$$

The resulting open-loop frequency response of the compensated system is shown in Fig. 16.2.53.

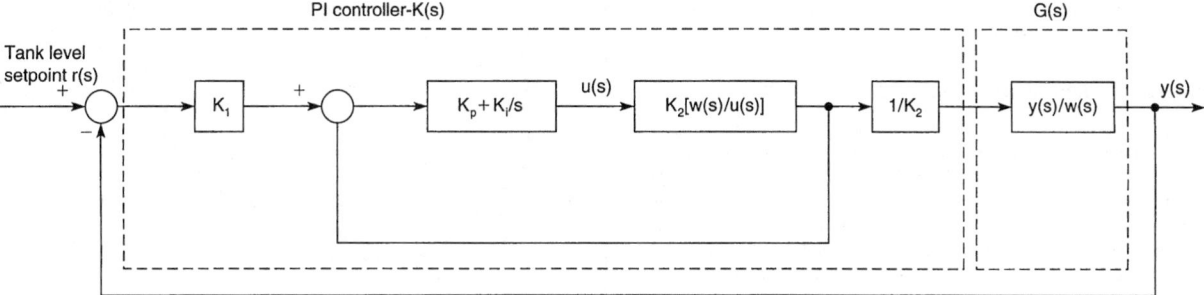

Fig. 16.2.54 Block diagram of a PI control system.

PI Controller Design

The PI controller for this system design takes on a structure similar to the three-element controller prevalent in process control. Therefore the PI control system structure takes the form shown in Fig. 16.2.54. The classical design details of the deaerator are shown in Murphy and Bailey, ORNL 1989.

The transfer functions used in this control structure are defined as

$$\frac{w(s)}{u(s)} = \frac{\text{condensate flow}}{\text{controller output}} \quad (16.2.136)$$

$$= 4.5227 \times 10^8 \times \frac{s + 0.142}{(s + 0.1405)(s + 53.8015)}$$

and

$$\frac{y(s)}{w(s)} = \frac{\text{tank level}}{\text{condensate}} \quad (16.2.137)$$

$$= 1.6804 \times 10^{-9} \times \frac{(s + 0.1986)(s + 47.458)}{s(s + 0.142)}$$

The constants K_1 and K_2 are used for unit conversion and defined, respectively, as 1.0 and 10^{-6}. The terms K_p and K_i are defined, respectively, as the proportional gain and the integral gain.

Using the practical experience of a prior three-element controller design, the proportional and integral gains are selected to be $K_p = 0.1$ and $K_i = 0.05$. The resulting open-loop frequency response plot of the compensated system $y(s)/r(s)$ is shown in Fig. 16.2.54.

Precompensator Design

The transient responses of the PI, LQG, and LQG/LTR control systems are shown in Fig. 16.2.55. The transient response of the LQG/LTR controller is seen to have a faster time to peak than the LQG and PI transient responses. Considering the practical physical limitations of the plant, it appears unlikely that the required rise time dictated by the transient response of the LQG/LTR control system is possible. The obvious solution would be to redesign the LQG/LTR control system to obtain a slower, more practical response. In the case of the LQG/LTR control system, the possibility of a redesign is eliminated, because the control system was designed to meet certain command-tracking, disturbance-rejection, and stability-robustness requirements.

Therefore, to eliminate this difficulty, a second-order precompensator will be placed in the forward path of the reference input signal. This precompensator will shape the output transient response of the system, since the control system design requires the output to follow the reference input. The precompensator in this investigation will be required to have a time to peak of 30 s with a maximum overshoot of about 3 percent. The requirements are selected to emulate somewhat the transient response of the PI controller.

The transfer function of the second-order precompensator used is as follows:

$$\frac{r(s)}{r_i(s)} = \frac{\omega_n^2}{s^2 + 2\xi\omega_n s + \omega_n^2} \quad (16.2.138)$$

Fig. 16.2.55 PI, LQG, LQG/LTR, LQG/LTR (with precompensator) control system closed-loop transient responses.

where ξ is the damping ratio, ω_n is the natural frequency, $r(s)$ is the output of the precompensator, and $r_i(s)$ is the actual reference input. The precompensator Eq. (16.2.138) can be represented in state equation form as follows:

$$\begin{bmatrix} x_1 \\ x_2 \end{bmatrix} = \begin{bmatrix} 0 & 1 \\ -\omega_n^2 & -\zeta\omega_n \end{bmatrix} \begin{bmatrix} x_1 \\ x_2 \end{bmatrix} + \begin{bmatrix} 0 \\ 1 \end{bmatrix} r_i(t) \quad (16.2.139)$$

and

$$r(t) = [\omega_n^2 \ 0] \begin{bmatrix} x_1 \\ x_2 \end{bmatrix} \quad (16.2.140)$$

where $r_i(t)$ is the time-domain representation of the reference input signal and $r(t)$ is the time-domain signal at the output of the precompensator.

Considering the 3 percent overshoot requirement, then ξ must be 0.7. Using the expression

$$t_{\text{max}} = \frac{\pi}{\omega_n \sqrt{1 - (0.7)^2}} \quad (16.2.141)$$

where t_{max} must be 30 s, gives

$$\omega_n = \frac{\pi}{30\sqrt{1 - (0.7)^2}} \quad (16.2.142)$$

The complete state equation of the precompensator is

$$\begin{bmatrix} x_1 \\ x_2 \end{bmatrix} = \begin{bmatrix} 0 & 1 \\ -0.0215 & -0.2052 \end{bmatrix} \begin{bmatrix} x_1 \\ x_2 \end{bmatrix} + \begin{bmatrix} 0 \\ 1 \end{bmatrix} r_i(t) \quad (16.2.143)$$

and

$$r(t) = [0.0215 \ 0] \begin{bmatrix} x_1 \\ x_2 \end{bmatrix} \quad (16.2.144)$$

The resulting transient response of the LQG/LTR control system using the precompensator is shown in Fig. 16.2.55.

ANALYSIS OF SINGULAR-VALUE PLOTS

Systems with large stability margins, good disturbance rejection/command following, low sensitivity to plant parameter variations, and stability in the presence of model uncertainties are described as being robust and having good robustness properties. The singular-value plots for the PI, LQG, and LQG/LTR control systems are examined to evaluate the robustness properties. The singular-value plots of the return ratio, return difference, and the inverse return difference are shown in Figs. 16.2.56, 16.2.57, and 16.2.58.

Fig. 16.2.56   Singular-value plots of return ratio.

Fig. 16.2.57   Singular-value plots of return difference.

Fig. 16.2.58   Singular-value plots of inverse return difference.

Deaerator Study Summary and Conclusions

Summary of Control System Analysis   The design criteria used for analysis in this system is defined as shown in Table 16.2.8. The performance and robustness results summarized from the analysis using the singular-value plots are shown in Table 16.2.9, from which it appears that the LQG/LTR control system has the widest low-frequency range of disturbance rejection and insensitivity to parameter variations. The PI control system has the worst disturbance rejection property and the most sensitivity to parameter variations.

All of the control systems are capable of maintaining a stable robust system in the presence of high-frequency modeling errors, but the PI control system is best. Though all the control systems also have good immunity to noise at frequencies significantly greater than 5 rad/s, the PI control system has the best immunity.

Unlike the other systems, the LQG/LTR control system meets all the design criteria. Its design was obtained in a systematic manner, in contrast to the trial-and-error method used for the LQG control system. The LQG controller design was obtained by shaping the output transient response using the system state weighting matrix, then examining the system's singular-value plots. The PI control system used is simply a

Table 16.2.8   System Design Specifications

System requirements	Range
Good command-following/ disturbance rejection	$\underline{\sigma}[L(j\omega)] > 20$ dB $\forall \omega \leq 0.1\,\mathrm{rad/s}$
Good system response to high- frequency modeling error	$\overline{\sigma}(I + [L(j\omega)]^{-1}) \geq \|\Delta L\| = 5$ dB $\forall \omega \leq 5\,\mathrm{rad/s}$
Good insensitivity to parameter variations at low frequencies	$\underline{\sigma}[I + L(j\omega)] \geq 26$ dB for low frequencies
Good immunity to noise, $\omega \geq 5\,\mathrm{rad/s}$	$\overline{\sigma}(L(j\omega)] \ll 0$ dB for high frequencies

Table 16.2.9   Performance and Robustness Results

System properties	PI control system	LQG control system	LQG/LTR control system
Command following/disturbance	$\underline{\sigma}[L(j\omega)] \geq 20$ dB $\forall \omega \leq 0.01\,\mathrm{rad/s}$	$\underline{\sigma}[L(j\omega)] \geq 20$ dB $\forall \omega \leq 0.017\,\mathrm{rad/s}$	$\underline{\sigma}[L(j\omega)] \geq 20$ dB $\forall \omega \leq 0.1\,\mathrm{rad/s}$
System response to high-frequency modeling error ($\omega \geq 5$ rad/s)	$\overline{\sigma}(I + [L(j\omega)]^{-1})$ ≥ 43.0 dB	$\overline{\sigma}(I + [L(j\omega)]^{-1}) \geq 30$ dB	$\underline{\sigma}(I + [L(j\omega)]^{-1}) \geq 16.7$ dB
Insensitivity to parameter variations at low frequencies	$\underline{\sigma}[I + L(j\omega)] \geq 26$ dB $\forall \omega \leq 0.0055\,\mathrm{rad/s}$	$\underline{\sigma}[I + L(j\omega)] \geq 26$ dB $\forall \omega \leq 0.008\,\mathrm{rad/s}$	$\underline{\sigma}[I + L(j\omega)] \geq 26$ dB $\forall \omega \leq 0.0055\,\mathrm{rad/s}$
Immunity to noise ($\omega \geq 5$ rad/s)	$\overline{\sigma}[L(j\omega)] \leq -42.5$ dB	$\overline{\sigma}[L(j\omega)] \leq -30$ dB	$\overline{\sigma}[L(j\omega)] \leq -16$ dB

three-element control system strategy that has previously been used in the simulation study of the deaerator flow control system.

Conclusions It should be evident that performance characteristics such as a suitable transient response do not imply that the system will have good robustness properties. Obtaining suitable performance characteristics and a stable robust system are two separate goals. Classical unity-feedback design methods have been used for transient response shaping, with little consideration for robustness properties. The main advantage of a closed-loop feedback system is that good performance and stability robustness properties are obtainable. As has been shown, transient response shaping can be obtained easily using a precompensator. Therefore it appears that the goal of the control system designer is first to design a stable robust system and then use prefiltering to obtain the desired transient response.

Optimal control methods as demonstrated by the LQG control system do not guarantee good robustness properties when applied systematically to meet minimization requirements of a performance index. Methods such as **pole placement** emphasize transient response characteristics without regard to robustness, which could be detrimental to the system integrity in the presence of model uncertainties.

Classical and Modern Representation of Process Models

A dynamic unsteady state model will be developed for a heated stirred-tank process shown in Fig. 16.2.59. This process is common to the chemical and other industries. This process will be used to show that the classical frequency domain models (Buckley, "Techniques of Process Control," Wiley) presently and previously used in classical control design can also easily be represented in a state differential equation form common for use with the modern control techniques.

Fig. 16.2.59 Process diagram of heated stirred-tank.

The first step in developing the model is to obtain the system dynamic energy balance as follows

$$C\frac{d[V\rho(T_{out} - T_{tank})]}{dt} = w_{in}C(T_{in} - T_{tank})$$
$$- w_{out}C(T_{out} - T_{tank}) + Q \quad (16.2.145)$$

where T_{in}, T_{out}, and T_{tank} are, respectively, the liquid temperatures in the inlet and outlet lines and the tank for enthalpy calculations (assumed constant). The terms w_{in}, w_{out}, C, V, ρ, and Q are, respectively, the inlet and outlet mass flow rate, the specific heat of the liquid, the tank liquid volume, density, and the heat input to the tank from the electrical heating unit. Equation (16.2.145) defines the rate of energy accumulation for the tank contents.

The dynamic unsteady-state mass balance for the tank contents will now be defined as

$$\frac{d(V\rho)}{dt} = w_{in} - w_{out} \quad (16.2.146)$$

Equation (16.2.146) gives the rate of mass accumulation of liquid in the tank (Seborg, Edgar, Mellichamp, "Process Dynamics and Control," Prentice-Hall). The simplifying assumptions for this model are that ρ and C are constants for Eqs. (16.2.145) and (16.2.146). Thus it follows that Eq. (16.2.146) becomes

$$\frac{\rho\, dV}{dt} = w_{in} - w_{out} \quad (16.2.147)$$

and the differentiation of the left-hand side of Eq. (16.2.145) yields

$$C\frac{d[V\rho(T_{out} - T_{tank})]}{dt} = \rho C(T_{out} - T_{tank})\frac{dV}{dt}$$
$$+ V\rho C\frac{d(T_{out} - T_{tank})}{dt} \quad (16.2.148)$$

Substituting the mass balance relation of Eq. (16.2.147) into Eq. (16.2.148) yields

$$C\frac{d[V\rho(T_{out} - T_{tank})]}{dt} = V\rho C\frac{dT_{out}}{dt}$$
$$+ C(T_{out} - T_{tank})(w_{in} - w_{out}) \quad (16.2.149)$$

where $dT_{tank}/dt = 0$. Now substituting the right-hand side of Eq. (16.2.149) into the left-hand side of Eq. (16.2.145) yields with simplification

$$V\rho C\frac{dT_{out}}{dt} = C(T_{in} - T_{out})w_{in} + Q \quad (16.2.150)$$

The resulting equations of the heated stirred-tank process model are

$$\frac{dV}{dt} = (w_{in} - w_{out})/\rho \quad (16.2.151)$$

$$V\rho C\frac{dT_{out}}{dt} = C(T_{in} - T_{out})w_{in} + Q \quad (16.2.152)$$

where Eq. (16.2.152) is nonlinear.

This set of equations gives the flexibility to define various operating conditions for the heated stirred-tank process. Next we will formulate the classical and modern representation for the process model that can be used for control system design.

Linearization of the Nonlinear Model

Typical classical and more of the popular modern control techniques of control system design use linear process models. Equation (16.2.152) above is nonlinear. We must make some assumptions to obtain a linear equation model. The assumptions are: the process has a constant liquid volume and the inlet and outlet flow rates are equal and constant ($w = w_{in} = w_{out}$). The initial steady-state conditions of the heated stirred-tank process at time $t = 0$ are assumed as follows: $T_{outi} = T_{out}(0)$, $T_{ini} = T_{in}(0)$, and $Q_{int} = Q(0)$.

Considering that the dynamic components of the system variables are more important for the control system designer, the initial steady-state component will be subtracted from the process variables. Applying this normalization to Eq. (16.2.152) yields the following perturbation model equation

$$V\rho C\frac{d(T_{out} - T_{outi})}{dt} = C[(T_{in} - T_{ini}) - (T_{out} - T_{outi})]w_{in}$$
$$+ (Q - Q_{int}) \quad (16.2.153)$$

Let us define $\delta T_{in} = T_{in} - T_{ini}$, $\delta T_{out} = T_{out} - T_{outi}$, and $\delta Q = Q - Q_{int}$. Therefore Eq. (16.2.153) becomes

$$\tau\frac{d(\delta T_{out})}{dt} = \delta T_{in} - \delta T_{out} + \delta Q/Cw \quad (16.2.154)$$

where $\tau = V\rho/w$ and Eq. (16.2.154) is a linear perturbed differential equation describing the dynamics of the heated stirred-tank process.

Classical Transfer Function Model

Equation (16.2.154) can be used to obtain the frequency domain process transfer function. Let's assume that at time $t = 0$, $\delta T_{out} = 0$, and the inlet temperature is constant $[\delta T_{in}(t) = T_{in}(t) - T_{ini}(0) = 0$, for $t \geq 0]$. Thus Eq. (16.2.154) becomes

$$\tau \frac{d(\delta T_{out})}{dt} = -\delta T_{out} + \frac{\delta Q}{Cw} \qquad (16.2.155)$$

and, taking the Laplace transform and simplifying, the heated stirred-tank transfer function is found to be

$$\frac{\delta T_{out}(s)}{dQ(s)} = \frac{1}{Cw(\tau s + 1)} \qquad (16.2.156)$$

Assuming δQ, the input heat, makes a step change from $\delta Q = Q - Q_{int}$, where $Q = Q_{int}$, to $\delta Q = Q - Q_{int}$, where $Q = \Delta Q + Q_{int}$ at $t = 0$, then for $t \geq 0$, $\delta Q = \Delta Q$. For these conditions the heated stirred-tank transfer function now becomes

$$\delta T_{out}(s) = \frac{\Delta Q}{Cw(\tau s + 1)(s)} \qquad (16.2.157)$$

A general block diagram of the perturbed transfer function process model using Eq. (16.2.156) with closed-loop feedback to control the output temperature is shown in Fig. 16.2.60.

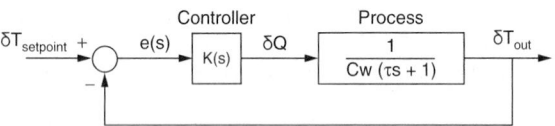

Fig. 16.2.60 Block diagram for heated stirred-tank closed-loop temperature feedback control using process transfer function.

Modern State Differential Equation Process Model

The state differential equation typically used in modern control for the process model takes the general form of

$$\dot{x}(t) = \mathbf{A}x(t) + \mathbf{B}u(t) \qquad (16.2.158)$$

$$y(t) = \mathbf{C}x(t) \qquad (16.2.159)$$

The perturbed state differential equation is easily obtained for the heated stirred-tank process by using Eq. (16.2.155), resulting in

$$\dot{\delta T}_{out} = [-1/\tau]\delta T_{out} + [1/\tau Cw]\delta Q \qquad (16.2.160)$$

where $\mathbf{A} = -1/\tau$, $\mathbf{B} = 1/\tau Cw$, $\dot{x} = \dot{\delta T}_{out}$, $x = \delta T_{out}$, $u = \delta Q$, $C = 1$, and $y = \delta T_{out}$. A block diagram of the process model using Eq. (16.2.160) with closed-loop feedback to control the output temperature is shown in Fig. 16.2.61.

Fig. 16.2.61 Block diagram for heated stirred-tank closed-loop temperature feedback control using state differential equations.

Conclusion

This heated stirred-tank example provides a detailed look at the derivation of the process transfer function and the process differential equation that would be used, respectively, for the frequency domain classical and modern control system design methods. It is clear that obtaining the linearized process model is a crucial step in this process. It should be noted that obtaining the transfer function model is an easy step from the state differential equation used in modern control system design. Finally, it should be observed that the time constant obtained during this example is directly related to the design properties of the process. Therefore it is clear that the best control engineer cannot overcome any inherent characteristic of the process that limits, for instance, the desire for a fast process response. This makes a clear argument for the need to include the control engineer in the initial design efforts for a new process or the modification of an existing process to ensure the desired system performance is obtained in the final process.

TECHNOLOGY REVIEW

Fuzzy Control

Fuzzy controllers are a popular method for construction of simple control systems from **intuitive knowledge of a process.** They generally take the form of a set of IF . . . **THEN** . . . rules, where the conditional parts of the rules have the structure "Variable i is <fuzzy value 1> AND Variable j is <fuzzy value 2> . . . . " Here the <fuzzy value> refers to a **membership function,** which defines a range of values in which the variable may lie and a number between zero and one for each element of that range specifying the possibility (where certainty is one) that the variable takes that value. Most fuzzy control applications use an **error signal** and the rate of change of an error signal as the two variables to be tested in the conditionals.

The actions of the rules, specified after THEN, take a similar form. Because only a finite collection of membership functions is used, they can be named by mnemonics which convey meaning to the designer. For **example,** a fuzzy controller's rule base might contain the following rules:

> **IF temperature-error is high and temperature-error-rate is slow**
> **THEN fuel-rate is negative.**
> **IF temperature-error is zero and temperature-error-rate is slow**
> **THEN fuel-rate is zero.**

Fuzzy controllers are an attractive control design approach because of the ease with which fuzzy rules can be interpreted and can be made to follow a designer's **intuitive knowledge** regarding the control structure. They are inherently nonlinear, so it is easy to incorporate effects which mimic variable gains by adjusting either the rules or the definitions of the membership functions. Fuzzy controller technology is attractive, in part because of its intuitive appeal and its **nonlinear nature,** but also because of the frequent claim that fuzzy rules can be utilized to approximate arbitrary continuous functional relationships. While this is true, in fact it would require substantial complexity to construct a fuzzy rule base to mimic a desired function sufficiently well for most engineers to label it an approximation. Herein also lie the potential liabilities of fuzzy controller technology: there is very little theoretical guidance on how rules and membership functions should be specified. Fuzzy controller design is an **intuitive process;** if intuition fails the designer, one is left with very little beyond trial and error.

One remedy for this situation is the use of machine, or **automated learning technology** to **infer** fuzzy controller **rules** from either collected data or simulation of a process model. This process is inference, rather than deductive reasoning, in that proper operation of the controller is **learning by example,** so if the examples do not adequately describe the process, problems can occur. This inference process, however, is quite similar to the methods of system identification, which also infer model structure from examples, and which is an accepted method of model construction. Machine learning technology enables utilization of more complex controller structures as well, which introduce additional degrees of freedom in the design process but also enhance the capabilities of the controller. One **example** of this approach is the **Fuzzy PID controller** structure (Wang and Birdwell), which utilizes a fuzzy rule base to generate gains for a PID controller. Other similar approaches have been reported in the literature. An advantage of this approach is the ability to establish theoretical results on closed-loop stability, which are otherwise lacking for fuzzy controllers.

Signal Validation Technology

Continuous monitoring of instrument channels in a process industry facility serves many purposes during plant operation. In order to achieve the desired operating condition, the system states must be measured accurately. This may be accomplished by implementing a **reliable signal validation procedure** during both normal and transient operations. Such a system would help reduce challenges on control systems, **minimize plant downtime,** and help plan maintenance tasks. Examples of measurements include pressure, temperature, flow, liquid level, electrical parameters, machinery vibration, and many others. The performance of control, safety, and plant monitoring systems depends on the accuracy of signals being used in these systems. Signal validation is defined as the detection, isolation, and characterization of faulty signals. This technique must be applicable to systems with redundant or single sensor configuration.

Various methods have been developed for signal validation in aerospace, power, chemical, metals, and other process industries. Most of the early development was in the aerospace industry. The signal validation techniques vary in complexity depending on the level of information to be extracted. Both **model-based** and **direct data-based techniques** are now available. The following is a list of **techniques** often implemented for signal validation:

Consistency checking of redundant sensors
Sequential probability ratio testing for incipient fault detection
Process empirical modeling (static and dynamic) for state estimation
Computational neural networks for state estimation
Kalman filtering technique for state estimation in both linear and nonlinear systems
Time-series modeling techniques for sensor response time estimation and frequency bandwidth monitoring

PC-based signal validation systems (Upadhyaya, 1989) are now available and are being implemented on-line in various industries. The computer software system consists of one or more of the above signal processing modules, with a decision maker that provides sensor status information to the operator.

An example of **signal validation** (Upadhyaya and Eryurek) in a power plant is the monitoring of water level in a steam generator. The objective is to estimate the water level using other related measurements and compare this with the measured level. An empirical model, a neural network model, or the Kalman filtering technique may be used. For example, the inputs to an empirical model consist of steam generator main feedwater flow rate, steam generator pressure, and inlet and outlet temperatures of the primary water through the steam generator. Such a model would be developed during normal plant operation, and is generally referred to as the *training phase*. The signal estimation models should be updated according to the plant operational status. The use of multiple signal validation modules provides a high degree of confidence in the results.

Chaos

Recent advances in dynamical process analysis have revealed that nonlinear interactions in simple deterministic processes can often result in highly complex, aperiodic, sometimes apparently "random" behavior.

Processes which exhibit such behavior are said to be exhibit *deterministic chaos*. The term *chaos* is perhaps unfortunate for describing this behavior. It evokes images of purely erratic behavior, but this phenomenon actually involves highly structured patterns. It is now recognized that deterministic chaos dominates many engineering systems of practical interest, and that, in many cases, it may be possible to exploit previously unrecognized deterministic structure for improved understanding and control.

Deterministic chaos and nonlinear dynamics both apply to phenomena that arise in process equipment, motors, machinery, and even control systems as a result of nonlinear components. Since nonlinearities are almost always present to some extent, nonlinear dynamics and chaos are the rule rather than the exception, although they may sometimes occur to such a small degree that they can be safely ignored. In many cases, however, nonlinearities are sufficiently large to cause significant changes in process operation. The effects are typically seen as unstable operation and/or apparent noise that is difficult to diagnose and control with conventional linear methods. These instabilities and noise can cause reduced operating range, poor quality, excessive downtime and maintenance, and, in worst cases, catastrophic process failure.

The apparent erratic behavior in chaotic processes is caused by the extreme sensitivity to initial conditions in the system, introduced by nonlinear components. Though the system is deterministic, this sensitivity and the inherent uncertainty in the initial conditions make long-term predictions impossible. In addition, small external disturbances can cause the process to behave in unexpected ways if this sensitivity has not been considered. Conversely, if one can learn how the process reacts to small perturbations, this inherent sensitivity can be taken advantage of. Once the system dynamics are known, small changes in system parameters can be used to drive the system to a desired state and to keep it there. Thus, great effects can be achieved through minimal input. Various control algorithms have been developed that use this principle to achieve their goals. A recent example from engineering is the control of a slugging fluidized bed by Vasadevan et al.

The following example illustrates the use of chaos analysis and control which involves a laboratory fluidized-bed experiment. Several recent studies have recognized that some modes of fluidization in fluidized beds can be classified as deterministic chaos. These modes involve large-scale cyclic motions where the mass of particles move in a pistonlike manner up and down in the bed. This mode is usually referred to as *slugging* and is usually considered undesirable because of its inferior transfer properties. Slugging is notoriously difficult to eradicate by conventional control techniques. Vasadevan et al. have shown that the inherent sensitivity to small perturbations can be exploited to alleviate the slugging problem. They did this by adding an extra small nozzle in the bed wall at the bottom of the bed. The small nozzle injected small pulses of process gas into the bed, thus disrupting the "normal" gas flow. The timing of the pulses was shown to be crucial to achieve the desired effect. Different injection timing caused the slugging behavior to be either enhanced or destroyed.

16.3 SURVEYING
by W. David Teter

REFERENCES: Moffitt, Bouchard, "Surveying," HarperCollins. Wolf and Brinker, "Elementary Surveying," Harper & Rowe. Kavanaugh and Bird, "Surveying Principles and Applications," Prentice-Hall. Kissam, "Surveying for Civil Engineers," McGraw-Hill. Anderson and Mikhail, "Introduction to Surveying," McGraw-Hill. McCormac, "Surveying Fundamentals," Prentice-Hall.

INTRODUCTION

Surveying is often defined as the art and science of measurement for location or establishment of position above, on, or beneath the earth's surface. The principles of surveying practice have remained consistent from their earliest inception, but in recent years the equipment and technology have changed rapidly. The emergence of the total station device, electronic distance measurements (EDM), Global Positioning System (GPS), and geographic information systems (GIS) is of significance in comparing modern surveying to past practice.

The information required by a surveyor remains basically the same and consists of the measurement of direction (angle) and distance, both horizontally and vertically. This requirement holds regardless of the type of survey such as land boundary description, topographic, construction, route, or hydrographic.

HORIZONTAL DISTANCE

Most surveying and engineering measurements of distance are horizontal or vertical. Land measurement referenced to a map or plat is reduced to horizontal distance regardless of the manner in which the field measurements are made.

The methods and devices used by the surveyor to measure distance include pacing, odometers, tachometry (stadia), steel tape, and electronic distance measurement. These methods produce expected precision ranging from ±2 percent for pacing to +0.0003 percent for EDM. The surveyor must be aware of the necessary precision for the given application.

Tapes

Until the advent of EDM technology, the primary distance-measuring instrument was the steel tape, usually 100 ft or 30 m (or multiples) in length, with 1 ft or 1 dm on the end graduated to read with high precision (0.01 ft or 1 mm). Cloth tapes can be used when low precision is acceptable.

Corrections Measurements with a steel tape are subject to variations that normally must be corrected for when high precision is required. Tapes are manufactured to a standard length at usually 68°F (20°C), for a standard pull of between 10 and 20 lb (4.5 and 9 kg), and supported over the entire length. The specifications for the tape are provided by the manufacturer. If any of the conditions vary during field measurement, corrections for temperature, pull, and sag will have to be made according to the manufacturer's instructions.

Tape Use Most surveying measurements are made with reference to the horizontal, and if the terrain is level, involves little more than laying down the tape in a sequential manner to establish the total distance. Care should be taken to measure in a straight line and to apply constant tension. When measuring on a slope, either of two methods may be employed. In the first method, the tape is held horizontal by raising the low end and employing a plumb bob for location over the point. In cases of steep terrain a technique known as *breaking tape* is used, where only a portion of the total tape length is used. With the second method, the slope distance is measured and converted by trigonometry to the horizontal distance. In Fig. 16.3.1 the horizontal distance $D = S \cos(\alpha)$ where S is the slope distance and α is the slope angle which must be estimated or determined.

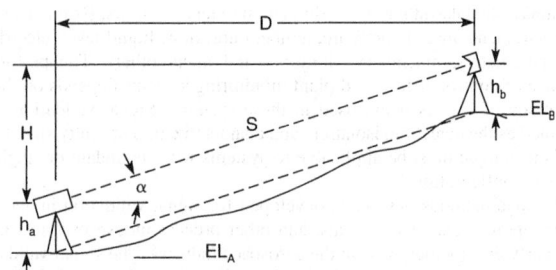

Fig. 16.3.1 Slope-reduction measurements used with EDM devices.

Electronic Distance Measurement

Modern surveying practice employs EDM technology for precise and rapid distance measurement. A representation of an EDM device is shown in Fig 16.3.2. The principle of EDM is based on the comparison of the modulated wavelength of an electromagnetic energy source (beam) to the time required for the beam to travel to and return from a

Fig. 16.3.2 Electronic distance meter (EDM). *(PENTAX Corp.)*

point at an unknown distance. The EDM devices may be of two types: (1) electrooptical devices employing light transmission within or just beyond the visible region or (2) devices transmitting in the microwave spectrum. Electrooptical devices require an active transmitter and a passive reflector, while microwave devices require an active transmitter and an identical unit for reception and retransmission at the endpoint of the measured line. Ideally the EDM beam would propagate at the velocity of light; however, the actual velocity of propagation is affected by the index of refraction of the atmosphere. As a result, the atmospheric conditions of temperature, pressure, and humidity must be monitored in order to apply corrections to the measurements. Advances in electrooptical technology have made microwave devices, which are highly sensitive to atmospheric relative humidity, nearly obsolete.

Atmospheric Corrections EDM devices built prior to about 1982 require a manual computation for atmospheric correction; however, the more modern instruments allow for keystroke entry of values for temperature and pressure which are processed and applied automatically as a correction for error to the output reading of distance. The relationships between temperature, pressure, and error are shown in Fig. 16.3.3. Microwave devices require an additional correction for relative humidity. A psychrometer wet-dry bulb temperature difference of 3°F (2°C) causes an error of about 0.001 percent in distance measurement.

Instrument Corrections Electrooptical devices may also be subject to error attributed to *reflector constant*. This results from the physical center of the reflector not being coincident with the effective or optical center. The error can range to 30 or 40 mm and is unique, but is usually known, for each reflector. The older EDM devices required this value to

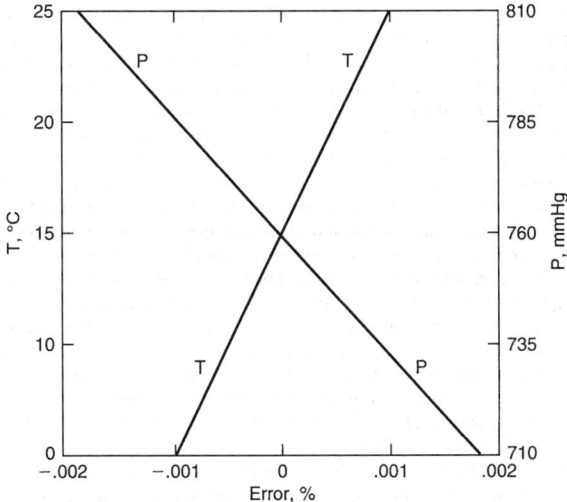

Fig. 16.3.3 Relationship of error to atmospheric pressure and temperature.

Fig. 16.3.4 Determination of elevation of an inaccessible point.

be subtracted from each reading, but in more modern devices the reflector constant can be preset into the instrument and the compensation occurs automatically. Finally, microwave EDM devices are subject to error from ground reflection which is known as *ground swing*. Care must be taken to minimize this effect by deploying the devices as high as possible and averaging multiple measurements.

EDM Field Practice Practitioners have developed many variations in employing EDM devices. Generally the field procedures are governed by the fact that the distance readout from the device is a slope measurement which must be reduced to the horizontal equivalent. Newer EDM devices may allow for automatic reduction by keystroke input of the vertical angle, if known, as would be the case when the EDM device is employed with a theodolite or if the device is a total station (see later in the section). Otherwise the slope reduction can be accomplished by using elevation differences. With reference to Fig. 16.3.1, it is seen that $H = (\text{EL}_A + h_a) - (\text{EL}_B + h_b)$ and that the horizontal distance $D = (S^2 - H^2)^{1/2}$. EDM devices are sometimes mounted on a conventional theodolite. In such cases the vertical angle as measured by the theodolite must be adjusted to be equivalent to the angle seen by the EDM device. It is common to sight the theodolite at a point below the reflector equal to the vertical distance between the optical centers of the theodolite and the EDM device.

VERTICAL DISTANCE

The acquisition of data for vertical distance is also called **leveling.** Methods for the determination of vertical distance include direct measurement, tachometry or stadia leveling, trigonometric leveling, and differential leveling. Direct measurement is obvious, and stadia leveling is discussed later.

Trigonometric Leveling This method requires the measurement of the vertical angle and slope distance between two points and is illustrated in Fig. 16.3.1. The vertical distance $H = S \sin(\alpha)$. A precise value for the vertical angle and slope distance will not be obtained if the EDM device is mounted on the theodolite or the instrument heights h_a and h_b are not equal. See "EDM Field Practice," above. Before the emergence of EDM devices, the slope distance was simply taped and the vertical angle measured with a theodolite or transit.

If the elevation of an inaccessible point is required, the EDM reflector cannot be placed there and a procedure employing two setups of a transit or theodolite may be used, as shown in Fig. 16.3.4. Angle $Z_3 = Z_1 - Z_2$, and the application of the law of sines shows that distance $A/\sin Z_2 = C/\sin Z_3$. C is the measured distance between the two instrument setups. The vertical distance $H = A \sin Z_1$. The height of the stack above

the instrument point is equal to $H + h_1$, where h_1 is the height of the instrument.

Differential Leveling This method requires the use of an instrument called a **level,** a device that provides a horizontal line of sight to a rod graduated in feet or metres. Figure 16.3.5 shows a vintage level, and Fig. 16.3.6 shows a modern "automatic" level. The difference is that with the older device, four leveling screws are used to center the spirit-level bubble in two directions so that the instrument is aligned with the horizontal. The modern device adjusts its own optics for precise levelness once the three leveling screws have been used to center the rough-leveling/circular level bubble within the scribed circle. The instrument legs should be planted on a firm footing. The eyepiece is checked and adjusted to the user's eye for clear focus on the crosshairs, and the objective-lens focus knob is used to obtain a clear image of the graduated rod.

Fig. 16.3.5 Y level.

Fig. 16.3.6 Automatic (self-adjusting) level.

To determine the difference in elevation between two points, set the level nearly midway between the points, hold a rod on one, look through the level and see where the line of sight, as defined by the eye and horizontal cross wire, cuts the rod, called **rod reading.** Move the rod to the second point and read. The difference of the readings is the difference in level of the two points. If it is impossible to see both points from a single setting of the level, one or more intermediate points, called turning points, are used. The readings taken on points of known or assumed elevation are called **plus sights,** those taken on points whose elevations are to be determined are called **minus sights.** The elevation of a point plus the rod readings on it gives the elevation of the line of sight; the elevation of the line of sight minus the rod reading on a point of unknown elevation gives that elevation. In Fig. 16.3.7, I_1 and I_2 are intermediate points between A and B. The setups are numbered. Assuming A to be of known elevation, the reading on A is a + sight; the reading on I_1 from 1 is a – sight; the reading on I_1 from 2 is a + sight and on I_2 is a – sight. The algebraic sum of the plus and minus sights is the difference of elevation between A and B. Target rods can usually be

Fig. 16.3.7   Determining elevation difference with intermediate points.

read by vernier to thousandths of a foot. In grading work the nearest tenth of a foot is good; in lining shafting the finest possible reading is none too good. It is desirable that the sum of the distances to the plus sights approximately equal the sum of the distances to the minus sights to ensure compensation of errors of adjustment. On a side hill this can be accomplished by zigzagging. When the direction of pointing is changed, the position of the level bubble should be checked. In the case of a vintage instrument, the bubble should be adjusted back to a centered position along the spirit level. An automatic level requires the bubble to be repositioned within the centering circle.

To Make a Profile of a Line   A benchmark is a point of reasonably permanent character whose elevation above some surface—such as sea level—is known or assumed and used as a reference point for elevation. The level is set up either on or a little off the line some distance—not more than about 300 ft (90 m)—from the starting point or a convenient benchmark (BM), as at K in Fig. 16.3.8. A reading is taken on the BM and added to the known or assumed elevation to get the height of the instrument, called HI. Readings are then taken at regular intervals (or **stations**) along the line and at such irregular points as may be necessary to show change of slope, as at B and C between the regular points. The regular points are marked by stakes previously set "on line" at distances of 100 ft (30 m), 50 ft (15 m), or other distance suitable to the character of

Fig. 16.3.8   Making a profile.

the ground and purpose of the work. When the work has proceeded as far as possible—not more than about 300 ft (90 m) from the instrument for good work—a **turning point** (TP) is taken at a regular point or other convenient place, the instrument moved ahead and the operation continued. The first reading on the BM and the first reading on a TP after a new setup are plus sights (+S); readings to points along the line and the first reading on a TP to be established are minus sights (−S).

The **notes** are taken in the form shown in Fig. 16.3.9. The elevation of a given point, both sights taken on it, and the HI determined from it all appear on a line with its station (Sta) designation. In plotting the **profile,** the vertical scale is usually exaggerated from 10 to 20 times.

Inspection and Adjustment of Levels   Leveling instruments are subject to maladjustment through use and should be checked from time to time. Vintage instruments require much more attention than modern devices. Generally, the devices should be checked for verticality of the crosshair, proper orientation of the leveling bubble, and the optical line of sight. The casual user would normally not attempt to physically adjust an instrument, but should be familiar with methods for inspection for maladjustment as follows:

1. *Verticality of crosshair.* Set up the level and check for coincidence of the vertical crosshair on a suspended plumb line or vertical corner of a building. The crosshair ring must be rotated if this condition is not satisfied. This test applies to all types of instruments.

2. *Alignment of the bubble tube.* This check applies to older instruments having a spirit level. In the case of a Y level, the instrument is carefully leveled and the telescope removed from the Ys and turned end for end. In the case of a dumpy level, the instrument is carefully leveled and then rotated 180°. In each case, if the bubble does not remain centered in the new position, the bubble tube requires adjustment.

3. *Alignment of the circular level.* This check applies to tilting levels and modern automatic levels equipped with a circular level consisting of a bubble to be centered within a scribed circle. The instrument is carefully adjusted to center the bubble within the leveling circle and then rotated through 180°. If the bubble does not remain centered, adjustment is required.

4. *Line of sight coincident with the optical axis.* This check (commonly called the *two-peg* test) applies to all instruments regardless of construction. The check is performed by setting the instrument midway between two graduated rods and taking readings on both. The instrument is then moved to within 6 ft of one of the rods, and again readings are taken on both. The difference in readings should agree with the first set. If they do not, an adjustment to raise or lower the crosshair is necessary.

ANGULAR MEASUREMENT

The instruments used for measurement of angle include the transit (Fig. 16.3.10), the optical theodolite, and the electronic total station (Fig. 16.3.11). Angular measurements in both horizontal and vertical planes are obtained with the quantitative value derived in one of three

	Left-hand page			Right-hand page
This space for a heading, telling what the work is, who does it, and the date on which it is done.				
Sta.	+ S	H.I.	− S	Elev.
B.M.	6.42	506.42	....	500.0
$A = 0$.........	....	....	10.4	496.0
1.........	....	....	8.2	98.2
2.........	....	....	6.1	500.3
+30.........	....	....	5.5	0.9
3.........	....	....	6.1	0.3
4.........	....	....	7.9	498.5
$B = + 40$.....	....	....	8.4	98.0
5.........	....	....	7.5	98.9
6.........	....	....	5.1	501.3
7.........	....	....	3.2	3.2
+ 10 T.P......	4.27	509.13	1.56	504.86
8.........	....	....	2.2	506.9

This page is for remarks describing B.M.s and T.P.s or other important particulars.

Fig. 16.3.9   Form for surveyor's notes.

Fig. 16.3.10   Vintage transit.

Top handle locking knob

Top handle

Collimator

Objective lens

Display panel

Optical plummet

Circular vial

Leveling screw
Bottom plate

Tribrach locking lever

Focusing knob
Eyepiece lens

On-board battery

Telescope tangent screw

Telescope clamp screw

Plate vial

Plate vial
Lower tangent screw
Lower clamp screw

Data-out terminal

External battery connector

Fig. 16.3.11   Electronic total station. *(PENTAX Corp.)*

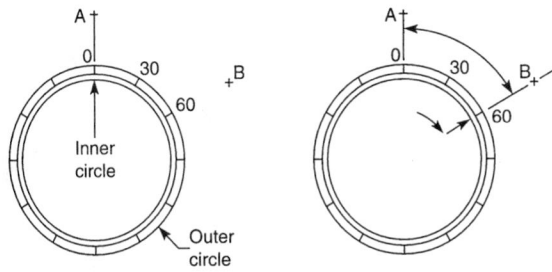

Fig. 16.3.12   Use of graduated circles to measure the angles between points.

ways depending on the era in which the instrument was manufactured. In principle, the effective manner in which these devices operate is shown in Fig. 16.3.12. The graduated outer circle is aligned at zero with the inner-circle arrow, and an initial point *A* is sighted on. Then, with the outer circle held fixed by means of clamps on the instrument, the inner circle (with pointer and moving with the telescope) is rotated to a position for sighting on point *B*. The relative motion between the two circles describes the direction angle which is read to precision with the aid of the scale's vernier. In the actual case, only the transit operates in the mechanical manner described. The operation of an optical theodolite occurs internally with the user viewing internal scales that move relative to each other. Many optical theodolites have scale microscopes or scale micrometres for precise reading. The electronic total station (discussed later) is also dependent on internal optics, but in addition it gives its angle values in the form of a digital display.

Examination of these devices reveals the presence of an *upper clamp* (controlling the movement of the inner circle described before) and a *lower clamp* (controlling the movement of the outer circle). Both clamps have associated with them a *tangent* screw for very fine adjustment. Setting up the instrument involves leveling the device and, unlike the procedure with a level, locating the instrument exactly over a particular point (the apex of the angle). This is accomplished with a suspended plumb (in the case of older transits) or an optical (line-of-sight) plummet in modern instruments. Prior to any measurement the clamps are manipulated so as to zero the initial angular reading prior to turning the angle to be read. In the case of the digital output total station, the initial angular reading is zeroed by pushing the zero button on the display. A total station, in effect, has only one motion clamp.

Angle Specification   Figure 16.3.13 shows several ways in which direction angles are expressed. The angle describing the direction of a line may be expressed as a **bearing,** which is the angle measured east or west from the north or south line and not exceeding 90°, e.g.: N45°E, S78°E, N89°W. Another system for specifying direction is by **azimuth** angle. Azimuths are measured clockwise from north (usually) up to 360°. Still other systems exist where direction is expressed by angle to the right (or left), deflection angles, or interior angles.

To Produce a Straight Line   Set up the instrument over one end of the line; with the lower motion clamp and tangent screw bring the telescopic line of sight to the other end of the line marked by a flag, a pencil, a pin, or other object; transit the telescope, i.e., plunge it by revolving on its horizontal axis, and set a point (drive a stake and "center" it with a tack or otherwise) a desired distance ahead in line with the telescopic line of sight; loosen the lower motion clamp and turn the instrument in azimuth until the line of sight can be again pointed to the other end of the line; again transit and set a point beside the first point set. If the instrument is in adjustment, the two points will coincide; if not, the point marking the projection of the line lies midway between the two established points.

To Measure a Horizontal Angle   Set up the instrument over the apex of the angle; with the lower motion bring the line of sight to a distant point in one side of the angle; unclamp the upper motion and bring the line of sight to a distant point in the second side of the angle, clamp and set exactly with the tangent screw; read the angle as displayed using scale micrometre or vernier if required.

To Measure a Vertical Angle   Set up the instrument over a point marking the apex of the angle *A* (see Fig. 16.3.14) by the lower motion and the motion of the telescope on its horizontal axis, bring the intersection of the vertical and horizontal wires of the telescope in line with a point as much above the point defining the lower side of the angle as

a. Bearing angles

b. Azimuth angles

c. Deflection angles

d. Angles to the right (left)

e. Interior angles

Fig. 16.3.13   Diagram showing various methods of specifying direction angles.

the telescope is above the apex; read the vertical angle, turn the telescope to a point which is the height of the instrument above the point marking the upper side of the angle and read the vertical angle. How to combine the readings to find the angle will be obvious.

Fig. 16.3.14   Measuring a vertical angle.

To Run a Traverse   A traverse is a broken line marking the line of a road, bank of a stream, fence, ridge, or valley, or it may be the boundary of a piece of land. The bearing or azimuth and length of each portion of the line are determined, and this constitutes "running the traverse."

To Establish Bearing   The bearing of a line may be specified relative to a north-south line (meridian) that is true, magnetic, or assumed, and whose angle is less than 90°. Whatever the reference meridian, the instrument is set over one end of the line, and the horizontal angle readout is set to zero with the instrument pointing in the direction of the north-south meridian. This is shown in Fig. 16.3.12 when point A is in the direction of the meridian (north). Using appropriate clamps, the angle is turned by pointing the telescope to the other end of the desired line. In Fig. 16.3.12 the bearing angle would be N60°E. Note that modern electron total stations generally do not have a magnetic needle and, as a result, the bearings or azimuths are typically measured relative to an arbitrary (assumed) meridian.

To establish azimuth the same procedure as described for the determination of bearing may be used. An azimuth may have values up to 360°. When the preceding line of a traverse is used to orient the instrument, the azimuth of this line is known as the **back azimuth** and its value should be set into the instrument as it is pointed along the preceding line of the traverse. The telescope is turned clockwise to the next line of the traverse to establish the **forward azimuth.**

In traverse work, azimuths and bearings are not usually measured except for occasional checks. Instead, deflection angles or angles to the right from one course to the next are measured. The initial line (course) is taken as an assumed meridian, and the bearings or azimuths of the other courses with respect to the initial course are calculated. The calculations are derived from the measured deflection angles or angles to the right. If the magnetic or true meridian for the initial course is determined, then the derived bearings or azimuths can likewise be adjusted.

In Fig. 16.3.15, the bearing of a is N40°E, of b is N88° 30′E, of c is S49°20′E = 180° − (40° + 48°30′ + 42°10′), of d is S36°40′W = 86° − 49°20′, or 40° + 48°30′ + 42°10′ + 86° − 180°, of e is N81°20′W = 180° − (36°40′ + 62°). No azimuths are shown in Fig. 16.3.15, but note that the azimuth of a is 40° and the back azimuth of a is 220° (40° + 180°). The azimuths of b and c are 88°30′ and 130°40′ respectively.

Fig. 16.3.15   Bearings derived from measured angles.

Inspection and Adjustment of the Instrument   The transit, optical theodolite, and total station devices are subject to maladjustment through use and should be checked from time to time. Regardless of the type of instrument, the basic relationships that should exist are (1) the plate level-bubble tubes must be horizontal, (2) the line of sight must be perpendicular to the horizontal axis, (3) the line of sight must move in the vertical plane, (4) the bubble (if there is one) of the telescope tube must be centered when the scope is horizontal, and (5) the reading of vertical angles must be zero when the instrument is level and the telescope is level.

All modern devices are subject to the following checks: (1) adjustment of the plate levels, (2) verticality of the crosshair, (3) line of sight perpendicular to the horizontal axis of the telescope, (4) the line of sight must move in a vertical plane, (5) the telescope bubble must be centered (two-peg test), (6) the vertical circle must be indexed, (7) the circular level must be centered, (8) the optical plummet must be vertical, (9) the reflector constant should be verified, (10) the EDM beam axis and the line of sight must be closely coincident, and (11) the vertical and horizontal zero points should be checked.

To Measure Distances with the Stadia In the transit and optical theodolite telescope are two extra horizontal wires so spaced (when fixed by the maker) that they are 1/100 of the focal length of the objective lens apart. When looking through the telescope at a rod held in a vertical position, 100 times the rod length S intercepted between the two extra horizontal wires plus an instrumental constant C is the distance D from the center of the instrument to the rod if the line of sight is horizontal, or $D = 100S + C$ (see Fig. 16.3.16). If the line of sight is inclined by a vertical angle A, as in Fig. 16.3.17, then if S is the space intercepted on the rod and C is the instrumental constant, the distance is given by the formula $D = 100S \cos^2 A + C \cos A$. For angles less than 5 or 6°, the distance is given with sufficient exactness by $D = 100S$. Although theory would indicate that distances can thus be determined to within 0.2 ft, in practice it is not well to rely on a precision greater than the nearest foot for distances of 500 ft or less.

Fig. 16.3.16 Horizontal stadia measurement.

Fig. 16.3.17 Sloping stadia measurement.

Only the oldest transits, known as *external-focusing devices,* have an instrument constant. They can be recognized by noting that the objective lens will physically move as the focus knob is turned. Such devices have not been manufactured for at least 50 years, and therefore it is most probable that, for stadia applications, the value of $C = 0$ should be used in the equations. Likewise, it is certain that the stadia interval factor K in the following equation will be exactly 100. Stadia surveying falls into the category of low-precision work, but in certain cases where low precision is acceptable, the speed and efficiency of stadia methods is advantageous.

Stadia Leveling This method is related to the previously described trigonometric leveling. The elevation of the instrument is determined by sighting on a point of known elevation with the center crosshair positioned at the height of the instrument (HI) at the setup position. Record the elevation angle A from the setup point to the distant point of known elevation, read the rod intercept S, and apply the following equation to determine the difference in elevation H between the known point and the instrument:

$$H = KS \cos A \sin A + C \sin A$$

Since K is sure to be equal to 100 and that the stadia constant C for most existing instruments will be equal to zero, the value of H is easily determined and the elevation of the instrument HI is known. With known HI, sights can now be performed on other points and the above equation applied to determine the elevation differences between the instrument and the selected points. Note also that the horizontal distance D can be determined by a similar equation:

$$D = \frac{1}{2}KS \sin 2A + C \cos A$$

Stadia Traverse A stadia traverse of low precision can be quickly performed by recording the stadia data for rod intercept S and vertical angle A plus the horizontal direction angle (or deflection angle, angle to the right, or azimuth) for each occupied point defining a traverse. Application of the equation for distance D eliminates the need for laborious taping of a distance. Note that modern surveying with EDM devices makes stadia surveying obsolete.

Topography Low-precision location of topographic details is also quickly performed by stadia methods. From a known instrument setup position, the direction angle to the landmark features, along with the necessary values of rod intercept S and vertical angle A, allows computation of the vertical and horizontal location of the landmark position from the foregoing equations for H and D. If sights are made in a regular pattern as described in the next paragraph, a contour map can be developed.

Contour Maps A contour map is one on which the configuration of the surface is shown by lines of equal elevation called **contour lines.** In Fig. 16.3.18, contour lines varying by 10 ft in elevation are shown. H, H are hill peaks, R, R ravines, S, S saddles or low places in the ridge $HSHSH$. The horizontal distance between adjacent contours shows the distance for a fall or rise of the contour interval—10 ft in the figure. A profile of any line as AB can be made from the contour map as shown in the lower part of the figure. Conversely, a contour map may be made from a series of profiles, properly chosen. Thus, a profile line run along the ridge $HSHSH$ and radiating profile lines from the peaks down the hills and from the saddles down the ravines would give data for projecting points of equal elevation which could be connected for contour lines. This is the best method for making contour maps of very limited areas, such as city squares, or very small parks. If the ground is not too much broken, the small tract is divided into squares and elevations are taken at each square, corner, and between two corners on some lines if necessary to get correct profiles.

Fig. 16.3.18 Contour map.

SPECIAL PROBLEMS IN SURVEYING AND MENSURATION

Volume of Earth in Foundation and Area Grading The volume of earth removed from a foundation pit or in grading an area can be computed in several ways, of which two follow.

1. The area (Fig. 16.3.19) is divided into squares or rectangles, elevations are taken at each corner before and after grading, and the volumes are computed as a series of prisms. If A is the area (ft²) of one of the squares or rectangles—all being equal—and b_1, b_2, b_3, b_4 are corner heights (ft) equal to the differences of elevation before and after grading, the subscripts referring to the number of prisms of which b is a corner, then the volume in cubic yards is

$$Q = A(\Sigma h_1 + 2\Sigma h_2 + 3\Sigma h_3 + 4\Sigma h_4)/(4 \times 27)$$

In Fig. 16.3.19 the h's at $A_0, D_0, D_3, C_5,$ and A_5, would be h_1's; those at $B_0, C_0, D_1, D_2, C_4, B_5, A_4, A_3, A_2,$ and A_1 would be h_2's; that at C_3 an h_3; and the rest h_4's. The rectangles or squares should be of such size that their tops and bottoms are practically planes.

Fig. 16.3.19 Estimating volume of earth by squares.

2. A large-scale profile of each line one way across the area is carefully made, as the $A, B, C,$ and D lines of Fig. 16.3.19, the final grade line is drawn on it, and the areas in excavation and embankment are separately measured with a planimeter or by estimation from the drawing. The excavation area of profile A is averaged with that of profile B, and the result multiplied by the distance AB and divided by 27 to reduce to cubic yards. Similarly, the material between B and C is found.

To Pass an Obstacle   Four cases are shown in Fig. 16.3.20. If the obstacle is large, as a building, (1) turn right angles at *B, C, D,* and *E,* making *BC = DE* when *CD = BE.* All distances should be long enough to ensure sufficiently accurate sighting. (2) At *B* turn the angle *K* and measure *BC* to a convenient point. At *C* turn left = 360° − 2*K;* measure *CD = BC.* At *D* turn *K* for line *DE. BD = 2BC* cos (180° − *K*). (3) At *B* lay off a right angle and measure *BC.* At *C* measure any angle to clear object and measure *CD = BC/*cos *C.* At *D* lay off *K* = 90° + *C* for the line *DE. BC = BC* × tan *C.* If the obstacle is small, as a tree. (4) at *A,* some distance back, turn the small angle *a* necessary to pass the obstacle and measure *AB.* At *B* turn the angle 2*a* and measure *BC = AB.* At *C* turn the small angle *a* for the line *AC,* and transit, or turn the large angle *K* = 180° − *a.* If *a* is but a few minutes of arc, *AC = AB + BC* with sufficient exactness. If only a tape is available, the right-angle method (1) above given may be used, or an equilateral triangle, *ABC* (Fig. 16.3.21) may be laid out, *AC* produced a convenient distance to *F,* the similar triangle *DEF* laid out, *FE* produced to *H* making *FH = AF,* and the similar triangle *GHI* then laid out for the line *GH. AH = AF.*

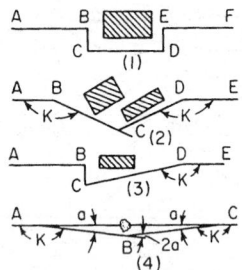

Fig. 16.3.20   Surveying past an obstacle.

To Measure the Distance across a Stream   To measure *AB,* Fig. 16.3.22, *B* being any established point, tree, stake, or building corner: (1) Set the instrument over *A;* turn a right angle from *AB* and measure any distance *AC;* set over *C* and measure the angle *ACB. AB = AC* tan *ACB.* (2) Set over *A,* turn any convenient angle *BAC'* and measure *AC';* set over *C'* and measure *AC'B.* Angle *ABC' =* 180° − *AC'B − BAC".* *BA = AC'* × sin *AC' B/*sin *ABC'.* (3) Set up on *A* and produce *BA* any measured distance to *D;* establish a convenient point *C* about opposite *A* and measure *BAC* and *CAD;* set over *D* and measure *ADC;* set over *C,* and measure *DCA* and *ACB;* solve *ACD* for *AC,* and *ABC* for *AB.* For best results the acute angles of either method should lie between 30 and 60°.

Fig. 16.3.21   Surveying past an obstacle by using an equilateral triangle.

Fig. 16.3.22   Measuring across a stream.

To Measure a Visible but Inaccessible Distance   (as *AB* in Fig. 16.3.23) Measure *CD.* Set the instrument at *C* and measure angles *ACB* and *BCD;* set at *D* and measure angles *CDA* and *ADB. CAD* = 180° − (*ACB + BCD + CDA*). *AD = CD* × sin *ACD/*sin *CAD. CBD* = 180° − (*BCD + CDA + ADB*). *BD = CD* × sin *BCD/*sin *CBD.* In the triangle *ABD,* ¹/₂(*B + A*) = 90° − ¹/₂ *D,* where *A, B,* and *D* are the angles of the triangle; tan ¹/₂(*B − A*) = cot ¹/₂ *D*(*AD − BD*) / (*AD + BD*); *AB = BD* sin *D/*sin *A = AD* sin *D/*sin *B.*

Random Line   On many surveys it is necessary to run a random line from point *A* to a nonvisible point *B* which is a known distance away. On the basis of compass bearings, a line such as *AB* is run. The distances *AB* and *BC* are measured, and the angle *BAC* is found from its calculated tangent (see Fig. 16.3.24).

Fig. 16.3.23   Measuring an inaccessible distance.

Fig. 16.3.24   Running a random line.

To Stake Out a Simple Horizontal Curve   A simple horizontal curve is composed of a single arc. Usually the curve must be laid out so that it joins two straight lines called tangents, which are marked on the ground by PT (points on tangent). These tangents are run to intersection, thus locating the PI (point of intersection). The plus of the PI and the angle *I* are measured (see Fig. 16.3.25). With these values and any given value of *R* (the radius desired for the curve), the data required for staking out the curve can be computed.

$$R = \frac{5,729.58}{D} \qquad L = 100 \frac{\Delta}{D} \qquad T = R \tan \frac{\Delta}{2}$$

where *R* = radius, *T* = tan distance, *L* = curve length, *C* = long chord. In a sample computation, assume that Δ = 8°24' and a 2° curve is required. *R* = 5,729.58/2 = 2,864.79 ft (873 m); *L* = (100) (8.40/2) = 420.0 ft (128 m); *T* = 2,864.79 × 0.07344 = 210.39 ft (64 m). The degree of the curve is always twice as great as the deflection angle for a chord of 100 ft (30 m).

Setting Stakes for Trenching   A common way to give line and grade for trenching (see Fig. 16.3.26) is to set stakes *K* ft from the center line, driving them so that the near face is the measuring point and the top is some whole inch or tenth of a foot above the bottom grade or grade of the center or top of the pipe to be laid. The top of the pipe barrel is perhaps the better line of reference. If preferred, two stakes can be driven on opposite sides and a board nailed across, on which the centerline is marked and the depth to pipeline given. When only one stake is used, a graduated pole sliding on one end of a level board at right angles is convenient for workmen and inspectors. On long grades, the grade stakes are set by "shooting in." Two grade stakes are set, one at each end of the grade, the instrument is set over one, its height above grade determined, and a rod reading calculated for the distance stake such as to make the line of sight parallel to the grade line; the line of sight is then set at this rod reading; when the rod is taken to any intermediate stake, the height of instrument above grade less the rod reading will be the height of the top of the stake above grade. If the ground is uniform, the stakes may all be set at the same height above grade by driving them so as to give the same rod readings throughout.

Fig. 16.3.25   Staking out a horizontal curve.

Fig. 16.3.26   Setting stakes for trenching.

To Reference a Point   The point *P* (Fig. 16.3.27), which must be disturbed during construction operations and will be again required as a line point in a railway, pipeline, or other survey, is referenced as follows: (1) Set the instrument over it and set four points, *A, B,* and *C, D* on two intersecting lines. When *P* is again required, the transit is set over *B* and, with foresight on *A,* two temporary points close together near *P* but on opposite sides of the line *DC* are set; the instrument is then set on *D* and, with foresight on *C,* a point is set in the lines *DC* and *BA* by setting it in *DC* under a string stretched between the two temporary points on *BA.* (2) Points *A* and *E* and *C* and *F* may be established instead of *A, B, C, D.* (3) If the ground is fairly level and is not to be much disturbed, only points *A* and *C* need be located, and these by simple tape measurement from *P.* They should be less than a tape length from

P. When *P* is wanted, arcs struck from *A* and *C* with the measured distance for radii will give *P* at their intersection.

Foundations The corners and lines of a foundation are preserved by setting stakes outside the area to be disturbed, as in Fig. 16.3.28. Cords stretched around nails in the stakes marking the reference points will give the referenced corners at their intersections and the main lines of the building. These corners can be plumbed down to the level desired if the height of the stakes above grade is given. It is well to nail boards across the stakes at *AB,* putting nails in the top edge of the board to mark the points *A* and *B,* and if the ground permits, to put all the boards at the same level.

Fig. 16.3.27 Referencing a point which will be disturbed.

Fig. 16.3.28 Reference points for foundation corners.

To Test the Alignment and Level of a Shaft Having placed the shaft hangers as closely in line as possible by the use of a chalk line, the shaft is finally adjusted for line by hanging plumb lines over one side of the shaft at each hanger and bringing these lines into a line found by stretching a cord or wire or by setting a theodolite at one end and adjusting at each hanger till its plumb line is in the line of sight. The position of the line will be known either on the floor or on the ceiling rafters or beams to which the hangers are attached. If the latter, the instrument may be centered over a point found by plumbing down, and sighted to a plumb line at the farther end.

To level the shaft, an ordinary carpenter's level may be used near each hanger, or, better, a pole with an improvised sliding target may be hung over the shaft at each hanger by a hook in one end. The target is brought to the line of sight of a leveling instrument set preferably about under the middle of the shaft, by adjusting the hanger.

When the hangers are attached to inclined roof rafters, the two extreme hangers can be put in a line at right angles to the vertical planes of the rafters by the use of a square and cord. The other hangers will then be put as nearly as possible without instrumental test in the same line. The shaft being hung, the two extreme hangers, which have been attached to the rafters about midway between their limits of adjustment, are brought to line and level by trial, using an instrument with a well-adjusted telescope bubble, a plumb line, and inverted level rod or target pole. Each intermediate hanger is then tested and may be adjusted by trial.

To Determine the Verticality of a Stack If the stack is not in use and its top is accessible, a board can be fitted across the top, the center of the opening found, and a plumb line suspended to the bottom, where its deviation from the center will show any leaning. If the stack is in use or its top not accessible and its sides are battered, the following procedure may be followed. Referring to Fig. 16.3.29, set up an instrument at any point *T* and measure the horizontal angles between vertical planes tangent, respectively, to both sides of the top and the base and also the angle *a* to a second point T_1. On a line through *T* approximately at right angles to the chimney diameter, set the transit at T_1 and perform the same operations as at T_1, measuring also *K* and the angle *b*. On the drawing board, lay off *K* to as large a scale as convenient, and from the plotted *T* and T_1 lay off the several angles shown in the figure. By trial, draw circumferences tangent to the two quadrilaterals formed by the intersecting tangents of the base and top, respectively. The line joining the centers of these circumferences will be the deviation from the vertical in direction and amount. If the base is square, *T* and T_1 should be established opposite the middle points of two adjacent sides, as in Fig. 16.3.30.

To Determine Land Area and Boundaries Two procedures may be followed to gather data to establish and define land area and boundaries. Generally the data is gathered by the traverse method or the radial/radiation-survey method.

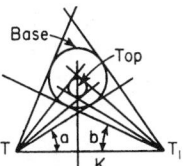

Fig. 16.3.29 Determining the verticality of a stack.

Fig. 16.3.30 Determining the verticality of a stack.

The **traverse** method entails the occupation of each end point on the boundary lines for a land area. The length of each boundary is measured (in more recent times with EDM) and the direction is measured by deflection angle or angle to the right. Reference sources explain how direction angles and boundary lengths are converted to **latitudes** and **departures** (north-south and east-west change along a boundary). The latitudes and departures of each course/boundary allow for a simple computation of double meridian distance (DMD), which in turn is directly related by computation to land area. In Fig. 16.3.31 each of the points 1 through 4 would be occupied by the instrument for direction and distance measurement.

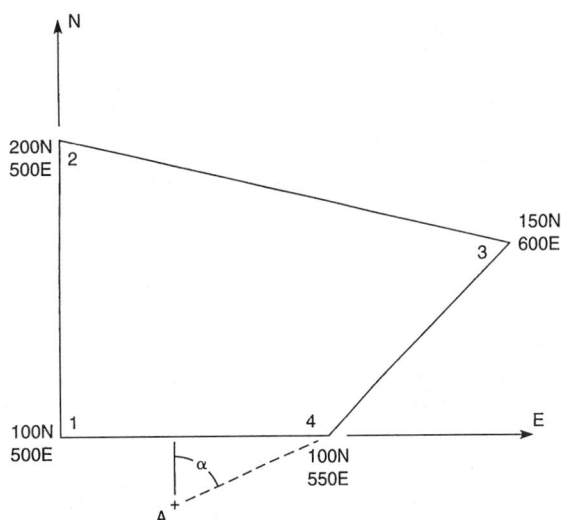

Fig. 16.3.31 Land area survey points.

The **radial** method requires, ideally, only a single instrument setup. In Fig. 16.3.31 the setup point might be at *A* or any boundary corner with known or assumed coordinates and with line of sight available to all other corners. This method relies on gathering data to permit computation of boundary corners relative to the instrument coordinates. The data required is the usual direction angle from the instrument to the point (e.g., angle α in Fig. 16.3.31) and the distance to the point. The references detail the computation of the coordinates of each corner and the computation of land area by the **coordinate method.** The coordinate method used with the radial survey technique lends itself to the use of modern EDM and total station equipment. In very low precision work (approximation) the older methods might be used.

Stream flow estimates may be obtained by using leveling methods to determine depth along a stream cross section and then measuring current velocity at each depth location with a current meter. As shown in Fig. 16.3.32, the total stream flow *Q* will be the sum of the products of area *A* and velocity *V* for each (say) 10-ft-wide division of the cross section.

To define limits of cut and fill once the route alignment for the roadway has been established (see Fig. 16.3.15), the surveyor uses leveling methods to define transverse ground profiles. From this information and the required elevation of the roadbed, the distance outward from the roadway center is determined, thus defining the limits of cut and fill. The surveyor then implants slope stakes to guide the earthmoving equipment operators. See Fig. 16.3.33.

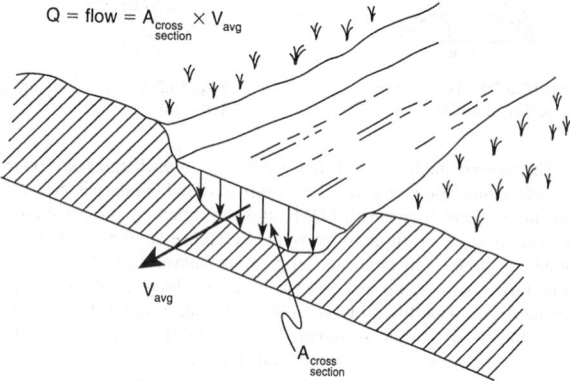

Fig. 16.3.32 Determination of stream flow.

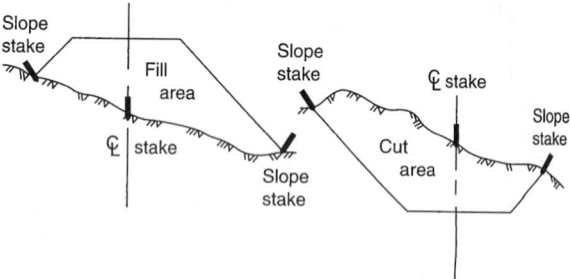

Fig. 16.3.33 Stakes showing limits of roadway cut and fill.

Datum of Reference The establishment of horizontal position, elevation, and direction is meaningless unless a known reference or datum is specified. It is fundamentally the question of "where is the point?," "which way is north?," and "what is the basis for specified latitude, longitude, and direction?" Plane surveying is based on the concept that all distances and directions are as projected upon a horizontal plane which at a single point is tangent to the earth spheroid. This practice serves nicely for local surveys but as the survey expands, the distances and angles measured on the earth spheroid will depart more from the corresponding data as projected upon the horizontal plane. The deviation can become unacceptable as the limits of plane surveying are reached and the accuracy of geodetic surveying is required. However, the methods of geodetic surveying (dealing with spherical angles and coordinates) are complex and require difficult computations, expense, and expertise.

Mapping of coordinates and direction to establish geodetic points has been carried out by the U.S. Coast and Geodetic Survey (USCGS, now NGS and NOAA), originating many years ago. The mapping of geodetic position is represented on the Lambert conformal projection or the transverse Mercator projection. The Lambert projection is appropriate in most states and areas having their long dimension in the east-west direction, while the Mercator projection is best used in places having their longer extents in the north-south direction. The Lambert projection utilizes a developed conic surface and the Mercator projection uses a developed cylinder. The methods to produce the flat/developed cone or cylinder are found in the references.

Simplification has been accomplished through evolution of the **State Plane Coordinate System** (SPCS), which is a rectangular coordinate definition. The USCGS has developed the SPCS for every state, providing a common reference for all points lying in the state plane. On the SPCS, grid points are located by x-y values and all longitudinal lines are parallel to a central meridian (unlike converging true geodetic meridians). Thus the traditional thinking of azimuth direction as (a) geographic north-south, (b) magnetic north-south, and (c) assumed north-south is now expanded to include grid north-south. Low-order surveys can be conducted throughout the state plane with no adjustment in distance. For higher-order surveys, however, a *grid factor* (1.000 or very close to 1.000) is used. The *grid factor* is the product of a *scale factor* and an *elevation factor*. These values are available in tables for each state.

The adoption of SPCS within a state or region is gaining support within the profession. Lawmakers and the courts have been slow to incorporate the principles of SPCS. Among the advantages of the system are the following:

1. When the SPCS (x and y) coordinates of a point are lost they may be regained from relationship to nearby known SPCS points within a degree of accuracy of its original specification.

2. Lengthy statewide surveys (right-of-way for highways, railroads, utilities) may be conducted without loop closure and with near geodetic accuracy.

3. Maps referenced to SPCS will always "fit" when joined regardless of the agency producing the map.

4. There is a common *grid north* direction definition (see Resurvey and Retracement below).

Resurvey and Retracement On occasion it may become necessary to verify and reestablish boundaries described in a deed description dating from many years ago. At least two problems arise in the effort to find the original monuments, corners, and ties referenced in the original survey. Equipment used in the older surveys does not approach the precision of that used in recent times and the basic measurements of angle and distance must be examined and interpreted carefully.

Angles are almost always based upon a magnetic reference and usually specified to about 15′ to 30′, consistent with the precision of the compass box of the era. Subsequent measured angles are limited to the refinement of the graduated circle of the transit often specified to no better than 30′. The magnetic reference is subject to adjustment to current declination which is arguably an inexact process.

Additionally, distance measurement may be subject to the diligence of the original crew of chain men and the surveyor. The possibility also exists that the distances were measured on the slope and not converted to horizontal equivalent. It is known that often the early chain men were paid "by the foot" or "by the chain," which is a practice that does not lend itself to due diligence.

A resurvey/retracement requires the consideration of several quantities. Research is required to determine all the relevant information from deeds, legal records, historic documents, etc. The research goal is to identify references that include monuments (natural or artificial), identification of corner markers, established lines (fence lines, stone walls, trails, streams), acreage, and referenced angles and distances.

After the thorough research, the subsequent visit to the field is likely to reveal that features and data required for definition of the survey may be missing or conflicting. In the case of conflict there is an accepted order of preference to be applied. Courts tend to follow precedence, as listed, in the absence of intent that is contrary. Priorities are as follows:

1. Natural monuments or landmarks. These are objects of specific reference or set by the early surveyor and found along the boundary.

2. Secondary or artificial monuments which are references documented by other maps or deeds and established by other previous surveyors.

3. Boundaries or lines of record to adjoining tracts documented by maps, plats, or deeds.

4. Bearing or distance references (ties) to on-site or off-site objects.

5. Distances are deemed to be more reliable than angle bearing references.

6. Angle measurement related to direction of a line.

7. Quantity or area may be cited in the old survey but is inferior to all the above cited items.

With the continued development of the State Plane Coordinate System, many surveying groups argue for the inclusion of coordinates in the hierarchy of calls listed above.

GLOBAL POSITIONING SYSTEM

The GPS concept is based on a system of 24 satellites in six orbital planes with four satellites per orbit. Two elements of data support the underlying principle of GPS surveying. First, radio-frequency signals are transmitted from the satellites and received by GPS receiver units at or near the earth's surface. The signal transmission time is combined with the velocity of propagation (approximately the speed of light) to result in a pseudodistance $d = Vt$, where V is the velocity of propagation and t is the time interval from source to receiver. A second requirement is the precise location of the satellites as published in satellite ephemerides.

The satellite data (ephemeride and time) is transmitted on two frequencies referred to as L1 (1,575.42 MHz) and L2 (1,227.60 MHz). The transmissions are band-modulated with codes that can be interpreted by the GPS receiver-signal processor. The L1 frequency modulation results in two signals: the precise positioning service (PPS) P code and the C/A or standard positioning service (SPS) code. The L2 frequency contains the P code only. The most recent GPS receivers use dual-frequency receiving and process the C/A code from L1 and the P code from L2.

Single-Receiver Positioning   A single point position can be established using one GPS receiver where C/A or SPS code is processed from several satellites (at least four for both horizontal and vertical position). The reliability depends on the uncertainty of the satellite position, which is +15 m in the ephemeride data transmitted. This handicap can be circumvented through the application of Loran C technology. The surveyor may also obtain high precision through the use of two receivers and differential positioning. The U.S. government reserves the right to degrade ephemeride data for reasons of national security.

Differential Positioning   Multiple GPS receivers may be used to attain submeter accuracy. This is done by placing one receiver at a known location and multiple receivers at locations to be determined. Virtually all error inherent in the transmissions is eliminated or cancels out, since the difference in coordinates from the known position is being determined. By averaging data taken over time and postprocessing, the precision of this technique can approach millimetre precision.

Field Practice   A procedure, sometimes called *leapfrogging,* can be used with three GPS receivers (R1, R2, and R3) operating simultaneously (see Fig. 16.3.34). Receiver R1 is placed at a known control point *P,* while R2 and R3 are placed at *L* and *M.* The three units are operated for a minimum of 15 min to obtain time and position data from at least four satellites. The R1 at *P* and R2 at *L* are leapfrogged to *N* and *O,*

respectively. This procedure is repeated until all points have been occupied. It is also desirable to include a partial leapfrog to obtain data for the *OP* line if a closed traverse is wanted.

A second procedure requires two receivers to be initially located at two known control points. The units are operated for at least 15 min in order to establish data defining the relative location of the two points. Then one of the GPS receivers is moved to subsequent positions until all points have been occupied. As in the previous procedure, signals from a minimum of four satellites must be acquired.

GPS Computing   The data for time and position of a satellite acquired by a GPS station is subsequently postprocessed on a computer. The object is to convert the time and satellite-position data to the earth-surface position of the GPS receiver. At least four satellites are needed for a complete (three-dimensional) definition of coordinate location and elevation of a control point. The computer program typically requires input of the ephemerides-coordinate location of the (at least) four satellites, the propagation velocity of the radio transmission from the satellite, and a satellite-clock offset determined by calibration or by receipt of correction data from the satellite.

General Survey Computing   The microcomputer is now an integral part of the processing of survey data. Software vendors have developed many programs that perform virtually every computation required by the professional surveyor. Some programs will also produce the graphic output required for almost every survey: site drawings, plat drawings, profile and transverse sections, and topographic maps.

The Total Station   The construction of the total station device is such that it performs electronically all the functions of a transit, optical theodolite, level, and tape. It has built-in EDM capability. Thus the total data-gathering requirements are available in a single instrument: horizontal and vertical angles plus the distance from the instrument's optical center to the EDM reflector. A built-in microprocessor converts the slope distance to the desired horizontal distance, and also to vertical distance for leveling requirements. If the height of the instrument's optical center and the height of the reflector are keyed in, the on-board computer can display that actual elevation difference between ground points with correction for earth curvature and atmospheric refraction. A typical total station is shown in Fig. 16.3.11.

Another feature found in some total stations is data-collection capability. This permits data as read by the instrument (angle, distance) to be electronically output to a data collector for transfer to a computer for later processing. A total station operating in conjunction with a remote processing unit allows a survey to be conducted by a single person. The variety of features found on total station devices is enormous.

The total station changes some of the traditional procedures in surveying practice. An example is the laying out of a route curve as shown in Fig. 16.3.24. The references describe the newer methods.

Fig. 16.3.34   Determining positions by leapfrogging.

Section **17**

Industrial Engineering

BY

ERWIN M. SANIGA   *Dana Johnson Professor of Information Technology and Professor of Operations Management, University of Delaware.*
SCOTT JONES   *Professor, Department of Accounting & MIS, Alfred Lerner College of Business and Economics, University of Delaware.*
ASHLEY C. COCKERILL   *Vice President and Event Coordinator, nanoTech Business, Inc.*
VINCENT M. ALTAMURO   *President, VMA, Inc., Toms River, NJ.*
ANDREW M. DONALDSON   *Project Director, Parsons E&C, Reading, PA.*
ROBERT F. GAMBON   *Power Plant Design and Development Consultant.*
EZRA S. KRENDEL   *Emeritus Professor of Operations Research and Statistics, Wharton School, University of Pennsylvania.*

17.1 OPERATIONS MANAGEMENT

by Erwin M. Saniga

REFERENCES: Meredith and Shafer, "Operations management for MBAs," 2d ed., John Wiley & Sons. Chase, Aquilano, and Jacobs, "Operations Management for Competitive Advantage," 9th ed., McGraw-Hill Irwin. Stevenson, "Production/ Operations Management," 5th ed., Irwin. Montgomery, "Introduction to Statistical Quality Control," 3d ed., John Wiley and Sons. Hoerl, "Six Sigma and the Future of the Quality Profession," *Quality Progress*, June 1998, pp. 35–42. Mentch, "Manufacturing Process Quality Optimization Studies," *Journal of Quality Technology*, vol. 12, no. 3, July 1980, pp. 119–129. Dilworth, "Operations Management: Providing Value in Goods and Services," 3d ed., Dryden Press. Box, Jenkins, and Reinsel, "Time Series Analysis," 3d ed., Prentice Hall. Deming, "Out of the Crisis," MIT Center for Advanced Engineering Study.

Operations management is concerned with the production of the goods and services of an organization. In particular, operations management problems focus on determining the way an organization will produce high-quality goods or services in an efficient manner. We define efficiency as producing goods and services in a timely and cost-efficient manner. In the past, the field was limited to the study of manufacturing processes, but in the last several decades the emphasis has included the study of service processes, which now number more organizations than manufacturing in the United States. Many of the same principles of effective operations management that have been found to work in manufacturing organizations will optimize the efficiency and quality of service organizations.

Operations management has generated increased interest among organizations today because of the competitive advantage that accrues to the organization that has a high-quality product or service and manages its efficiency of production in an optimal way. The terminology that is common in business today is due to an emphasis on operations management strategies. Some examples include six sigma, lean manufacturing, supply chain management, and just-in-time inventory methods.

FORECASTING

There are three general types of forecasting models used in organizations today. These are time series models, regression models, and subjective models.

Time series models are models that are based upon the assumptions that the past behavior of the variable to be forecasted, say sales, is the best indicator of the future behavior of sales. Some of the common models that have been used in many firms include moving-average methods and exponential smoothing methods. More complex models due to Box and Jenkins are used when simpler models do not satisfy.

Moving-average methods are based upon simply averaging the past n observations of a stream of data to obtain a forecast for the next period. For example, a three-period moving average of sales would be obtained by averaging the past three periods of sales. This would be the forecast for the next period (or future periods) sales. The formula for a moving average forecast of sales is simply

$$S_{t+1} = \sum_{i=t-N+1} X_i / N$$

where S_t is the forecast in period t, X_i is sales in period i, and N is the number of periods in the moving average.

An obvious weakness of moving-average forecasts is that equal weight is placed on all observations. It seems more plausible to weigh the more recent observations higher in determining a forecast. **Exponential smoothing** is a common method that allows more recent observations to have more weight in the averaging scheme. The model for exponential smoothing is:

$$S_{t+1} = \alpha X_t + (1 - \alpha)S_t$$

where α is the smoothing coefficient.

Note that the model is dependent upon the choice of α. If α is near 1, there is very little smoothing taking place and the forecast follows the rapid changes in the data. Note that if α equals 1, the forecast for the next period is simply the actual observation for the previous period. If α is near zero, the forecast approaches the average of the past data. In practice, common values of α are between 0.1 and 0.3.

Trend and seasonalities are not accounted for in the exponential smoothing model presented above. One can account for trend in a model by using a model called **double exponential** smoothing, which is a simple extension of the model presented above. **Seasonalities** are common in many industries. Consider the demand for the sale of turkeys. One would expect there would be peaks around the Thanksgiving and Christmas holidays in the United States. The usual exponential smoothing or double exponential smoothing models would yield forecasts that lag the actual demand. One simple and effective method for handling these seasonalities is to first **deseasonalize** the data by calculating the seasonal component for the data. An example of this would be to calculate the percentage of sales occurring in each month by averaging the last 3 years of monthly data (assuming there is a monthly seasonality) and then dividing the data by this monthly index. One can then develop a forecast on the deseasonalized data and, when the forecast is obtained, multiply the forecast by the seasonal index for the month in which a forecast is desired.

More complicated methods for forecasting time series are due to George Box and Gwilym Jenkins and are based upon a set of models called ARIMA, or mixed autoregressive and moving-average models. These are more general and powerful models of the class mentioned above and have been shown to be extremely effective in forecasting in practice. Consult their book for a detailed explanation of this class of models.

A second class of forecasting models is based upon using explanatory variables in a **regression** model to forecast sales. The regression model may be represented by the function F, where

$$Y = F(X_1, X_2, \ldots, X_k) + \varepsilon$$

The variables labeled X_j, $j = 1, 2, \ldots, k$ are the explanatory variables, and the variable Y is called the dependent variable. The term ε represents the error. For example, we may hypothesize that sales (Y) can be explained by price (X_1) and average delivery time (X_2). We can build the model by entering data from the past where we have recorded average delivery time in a month, the price in a month, and sales in a month.

Human judgment is the basis for a third class of models. These are commonly used in practice and, at times, yield accurate forecasts. Perhaps the most common types of judgmental forecasts are those based upon management or sales force opinions obtained from discussions with customers. Surveys of customers are also used to build these types of forecasts.

INVENTORY MANAGEMENT

Inventory decisions have a major impact on the profitability of an organization. Organizations that carry large inventories incur large holding costs, which impact the bottom line. On the other hand, maintaining low inventories may prevent an organization from meeting the variability in demand. Additionally, low inventories may not allow flexibility in scheduling, may not allow the organization to take account of economic order quantities, or may not allow for variation that occurs in vendor-supplied inputs. Thus, the inventory decision is very important in organizations. Essentially, the inventory decision can be broken down into two major components. These are (*a*) determining when items should be ordered and (*b*) determining how much to order.

The standard criterion on which to base inventory decision is cost. Four costs are generally considered. These are holding or carrying costs, ordering costs, setup costs, and shortage costs. **Holding costs** include as a major component the cost of the capital tied up in the inventory and additional costs include the cost of storage, insurance, deterioration, obsolescence, etc. **Ordering costs** refer to the costs incurred when a purchase order is placed or when a production order is given. **Setup costs** include the clerical costs in changing production. **Shortage costs** are difficult to measure since they include a component for lost sales, lost customers, or penalties for late delivery.

When analyzing inventory decisions, analysts place inventory in two categories: independent demand inventory and dependent demand inventory. Independent demand items are items whose demand does not depend on the demand for another item. Dependent demand items are ones for which the demand depends on the demand for another item. For example, the demand for bicycles in an organization may be an independent demand item. Wheels for the bicycle, on the other hand, are dependent demand items, since their demand depends on the demand for bicycles. That is, if the demand estimate for bicycles is 500 units, the demand for wheels is 1,000 units.

The most common inventory model is labeled the **economic order quantity (EOQ)** model. It is used when one is interested in determining the optimal order quantity where the criterion is cost minimization and where the inventory demand item is independent. The assumptions of the EOQ model are that demand is satisfied on time, price is fixed (there are no price breaks), stock is depleted linearly, and there are no stock-outs. Under these assumptions one can show that the EOQ is

$$EOQ = (2 R c_p/c_h)^{1/2}$$

where c_p is the ordering costs, R is the yearly demand, and c_h is the holding costs. In addition, it can be shown that the reorder point is when the inventory reaches a level of P, where

$$P = L \times R/52$$

where L is the lead time in weeks. Other similar models are developed for the situations where price breaks are available or where backorders are allowable.

An equivalent model to the EOQ is available for the situation where the analyst wishes to determine the **economic production quantity** or lot size **(EPQ)**. Here, one can find the EPQ as

$$EPQ = \{2c_p R/[c_h(1 - r/p)]\}^{1/2}$$

where r = daily usage and p = production rate per day.

When demand is not constant, organizations must provide a level of safety stock if they wish to maintain a certain service level. Safety stock, then, is the excess inventory carried over expected demand. In the standard EOQ model, which is also called the *fixed order quantity* model, the EOQ is calculated in the same way if the demand is constant or varies. But if the demand varies, the **reorder point** differs. If the demand varies, the reorder point must be calculated by using some knowledge of the distribution of demand. For example, if demand is normally distributed with standard deviation, the reorder point becomes

$$P = dL + z$$

where d is the average daily demand, L is the lead time in days, and z is the number of standard deviations associated with a particular service level probability.

Another way to address this problem is to use a fixed time period model. Here, order quantities vary from time period to time period but the reorder point remains the same. An advantage of this model is that one does not have to constantly account for the inventory; all one must keep track of is the time period. A disadvantage of this model is that it requires the maintenance of a higher level of inventory. Most textbooks present the models and discuss the implications of the fixed time period model.

There are other simpler systems that enjoy popularity in applications because they do not rely on any assumptions such as a normal distribution of demand or constant usage. Some common systems include the ABC system of classifying inventory and one- and two-bin systems.

The **ABC system** for inventory management essentially classifies inventory into one of three levels: A, B, and C. The classification is based upon the dollar value of inventory, which is determined by multiplying annual usage times the cost. Then, the A items are defined as the ones with the top 75 to 80 percent of dollar value. B items are the items with the next 15 percent of value and C items are those with the next 5 to 10 percent of value. Once the items are classified, management uses this information to determine where to apply the greatest control.

The **one-bin system** is very simple to manage but has a disadvantage in its cost that may be far from minimal compared to more complex systems. To use the one-bin system the analyst simply replenishes inventory up to a maximum level on a regular basis, say weekly for A items. The **two-bin system** is one in which there are two bins of the same inventory. Inventory is used from the first bin until it is used up; subsequently an order is placed for replenishment and inventory is used from the second bin. The second bin should contain enough inventory to cover needs during the order lead time. The amount in the second bin thus should equal P, where P is the order point defined above.

The methods discussed above constitute some of the more valuable methods of inventory management available to managers but, at times, various industries employ other methods that are more usable because of the difficulty in managing a large number of items. For example, shoes are distributed according to batches organized by sizes. Department stores use a method called "open to buy," which refers to the budgeted value of inventory minus the amount that was currently spent for various departments. That is, instead of managing individual inventory items, the manager focuses on managing the cost of inventory for a department as a whole.

As mentioned earlier, the demand for inventory can be independent or dependent. When demand for an inventory item is dependent upon the demand for another item, such as the case for bicycle wheels examined before, the methods of managing inventory differ from the case where demand is independent. A common technique for managing dependent demand inventories is the method of **materials requirements planning (MRP)**, which is an accounting information system technique used for keeping track of inventory needs for these items. MRP has been popular for about 30 years, and computer programs and systems to perform the cumbersome calculations associated with MRP are readily available.

MRP is based upon knowledge of the bill of materials for an item, the master production schedule, and a record of the inventories currently in stock. The bill of materials is simply a list of the components of a particular product. For our bicycle example, the bill of materials would contain two wheels, two tires, a set of handlebars, a set of rear brakes, front brakes, and various other items. The master production schedule provides information on when to produce these bicycles and what quantity to produce. The MRP then uses these inputs along with lead times to provide the production manager with planned order releases along with various other kinds of information, depending upon the needs of the manager. In our simple example, suppose our forecasted demand is such that we need to produce 3,000 bicycles in March and in June; this would be the master production schedule. Suppose we have 3,000 wheels in inventory and the lead time for wheels is 1 month. The MRP would calculate that the organization needs 6,000 wheels in March and in June, but since the organization has 3,000 on hand it would show the necessary order releases are for 3,000 wheels on February 1 and for 6,000 wheels on May 1.

Of course, in practice MRP is much more complicated than the simple example presented here, for a number of reasons. For example, some firms produce many products and some of these have common parts, or dependent items. Order releases must account for these factors. Also, firms might want to take advantage of price breaks for larger lots or by ordering less often to minimize ordering costs. In MRP, this problem is called **lot sizing**. A variety of techniques for lot sizing are available and range from the simplest, called *lot-for-lot ordering*, to complex ones such as dynamic programming techniques—for example, the Wagner Whitin model. The lot-for-lot model, for example, indicates that orders are to be placed when they are needed for each period; here, holding costs are minimized.

Related to MRP is an approach to broaden the scope of MRP to the firm as a whole. That is, all of the factors going into production are considered together on a larger scale than just the production system. This includes the overall planning of the business. Marketing and finance functions, for example, might take part in the determination of the master production schedule, forecasting, the financing requirements of the process, etc. This approach is called **MRP II** and has been popular for about 20 years.

An approach to managing production systems that includes the management of its inventory is called a **just-in-time** approach. This approach was initially used at Toyota in Japan and is very much related to the teachings of W. Edwards Deming in that it requires that the production system be in a state of statistical control, that suppliers be kept to a minimum, that scrap and rework are to be avoided as a policy rather than as a cost consideration, that preventive maintenance be performed, that continuous improvement is in place, and that workers are team based. Because the system is in control and there is a minimum of scrap or rework, an innovative method of managing the inventory is possible. This method is to manage inventory, whether purchased from a vendor or work-in-process inventory, so that large stocks of inventory are not necessary to be held, thereby minimizing holding costs. Other common elements of just-in-time systems are that setups are quick, and therefore not costly, and that lot sizes are small. Additionally, just-in-time systems move work in a pull versus push way. Push systems work with less regard for the forecast or the demands of higher upstream stations. Pull systems, on the other hand, work in a just-in-time method such that the upstream demand is pulled from downstream. As a simple example, a firm might produce according to expected demand only rather than by efficient lot sizes.

A drawback encountered in just-in-time systems is that, since the supplier is penalized for late delivery, it often prepares and stores finished product which is then shipped to arrive at the customer just in time. The carrying cost for this storage, which the customer thinks has been avoided, is passed on to the customer in the form of an increased price.

One way in which to manage the information in a pull system is with a **kanban sheet,** which is simply a record of what work is taking place. For example, a kanban sheet might be attached to a lot of parts. When upstream demand requires that this lot is needed for production, the kanban sheet is placed in an area where the need for replenishment is indicated.

JIT systems are conceptually very attractive and have been used to advantage by a number of companies, especially those in Japan.

SCHEDULING

Scheduling the accomplishment of tasks within the operations of a manufacturing or service environment is dependent upon a number of factors. First, there are differences between services and manufacturing in that the concept of an inventory of service has little meaning. Second, within manufacturing there are different types of production systems used for different types of products. More specifically these can be broken down into systems in which the same product is made many times (flow shop), a system in which several products are made by the same manufacturer on the same production system, a system in which similar products are made to order (job shop), and finally a system in which a single product is made (project). Some examples of these are respectively automobile production, a private press, a machine shop producing milled products, and construction of a skyscraper.

In a service environment, operations scheduling is a matter of providing the service when it is demanded by the customer. Analytical approaches involve developing forecasts of when customers arrive and the probability distributions of these arrivals and using these to build service capacity so that desired service levels are achieved. These service levels are usually expressed in terms of the probability or percentage of the time customers are serviced within a time limit. For example, the credit card company MBNA answers more than 90 percent of its phone calls from customers within two rings.

Some organizations such as doctors' offices schedule by appointment. The demand is then known and the service supply can be provided to meet that customer demand. Airlines use reservations to accomplish much of the same objective. Of course the determination of feasible schedules involves forecasting as well.

Complex service operations such as hospitals have compound scheduling problems encompassing the issues mentioned above in addition to the consideration that multiple service providers of various types are needed. For example, scheduling surgery involves scheduling surgeons, nurses, anesthesiologists, operating rooms, etc.

Scheduling for the situation when a product is produced in a series of workstations, or a flowshop, is called assembly line balancing. The product spends a period of time in each workstation; this time is referred to as the *cycle time*. The objective of assembly line balancing is to allocate all of the tasks needed to complete a product into workstations. Criteria for evaluation of assembly line balances include the cycle time required, the number of workstations required, and the amount of slack in the assembly line, which is the sum of the idle time in the workstations.

To perform assembly line balancing for producing a particular product, one needs to establish the precedence relationships between the tasks and determine the task times for each of the tasks. In the situation where cycle time is specified, one can calculate the theoretical minimum number of workstations; it is the next highest integer obtained from the ratio S/C, where S is the sum of the task times and C is the cycle time. In assembly line balancing, one then assigns tasks to workstations such that precedence relationships are not violated and, in addition, the sum of the task times for each station is less than the cycle time.

When one assigns tasks to workstations, one finds that there usually will be alternative assignments. Researchers have investigated the performance of a number of heuristics to use in these situations. Some of the more common are to assign tasks to workstations such that the longest tasks are assigned first without, of course, violating precedence relationships, and to assign tasks according to positional weights, where positional weight is the sum of the task times and the times of all of the tasks that follow.

A graph showing the precedence relationships for an 8-task assembly line is shown in Fig. 17.1.1. The times for the various tasks are:

Task	1	2	3	4	5	6	7	8	Total
Time	2	4	3	4	5	4	6	4	32

Thus, if the cycle time is 7, the theoretical minimum number of workstations is $32/7 = 4.55 = 5$.

If we allocate tasks to stations such that the longer tasks are assigned first, we have the following balance:

Station	1	2	3	4	5	6
Tasks	1,2	5	3,6	4	7	8
Time	6	5	7	4	6	4
Slack	1	2	0	3	1	2

In practice the problem is usually more complex because task times are stochastic variables and actual results may differ from the expected. Additionally, some tasks may require special training, which may change the way in which tasks are assigned to workstations. Many other

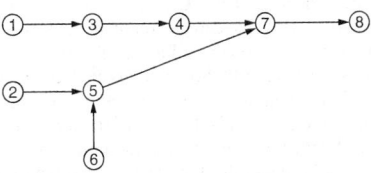

Fig. 17.1.1 Network showing precedence relationships for an 8-task assembly line. Code numbers within nodes signify events. Connecting lines with directional arrows indicate operations that are dependent on prerequisite operations.

situations occur that make the procedures described above a starting point in the assembly line balancing decision.

When a firm produces a number of products using similar equipment and locations, the scheduling problem becomes more complex because the production manager needs to make decisions on the size of the lot produced and the sequencing of the lots. In some cases the economic lot size can be determined by using methods discussed in the section on inventory management; methods are also available for multiple products. Sequencing becomes an issue of satisfying the needs of the customers, whether they be internal or external, and minimizing the cost of holding inventory.

A common tool used for evaluating various types of schedules is called the **Gantt chart.** This is a simple chart that provides a visual record of the order of processing tasks over time (Fig. 17.1.2). Here, M_1 and M_2 represent two machines in a job shop. The numbers within the blocks show the processing order, or schedule, to complete three jobs, each of which requires processing time on each of the two machines. Note that all the jobs can be done first on machine 1 or on machine 2 but not on both simultaneously.

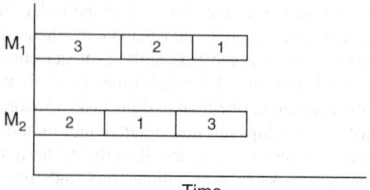

Fig. 17.1.2 Sample Gantt chart.

In job shop scheduling the object is to determine the sequence in which machines will process jobs. The evaluation criteria are varied. Some common criteria include **makespan,** which is the total time to process all jobs; **mean flowtime,** which is the average time a job is in the shop; and **average lateness** per job, which is the average time a job takes longer than the promised time.

If arrivals are static and there is one machine to process jobs, an optimal schedule in terms of minimizing mean flow time is to schedule jobs by processing time, with the shortest jobs being processed first. Note that makespan for this situation is the same for all processing orders.

When there are two machines and n jobs to process on these machines and the order is the same for each job, a method called Johnson's rule can be used to minimize makespan. A variant, Jackson's rule, can be used if jobs can be performed on machine 1 only, machine 2 only, first machine 1 then machine 2, or first machine 2 then machine 1. This rule minimizes the makespan of the sequence.

With three machines and n jobs, Johnson's rule can be used by employing a trick in the solution procedure. If there are m machines and n jobs, Campbell's heuristic can be used to minimize makespan. This is not an optimal procedure since there is not an efficient method available to find an exact solution.

If, on the other hand, jobs arrive intermittently throughout a time period, which is the usual case in a job shop, one must rely on schedules based on priority dispatch rules. In these cases, jobs with the smallest priority are scheduled first. One common rule is the shortest processing time rule. This rule can be dominant in terms of minimizing lateness, makespan, and flowtime. First come, first serve is another common priority dispatch rule. Some others include: random priority, least work remaining, total work fewest operations remaining, due date, slack, and critical ratio, which is the ratio of the time remaining before delivery divided by the remaining processing time. One can evaluate the effectiveness of all of these rules for a particular situation by constructing Gantt charts of each of the resulting schedules and comparing these on the basis of makespan, mean flow time, and average lateness. Implementation involves choosing that one that best fits the needs of the

organization. Subsequently, one should monitor the results of the solution over time to see, if, in fact, it is providing the best solution in terms of optimizing the solution criteria.

Scheduling projects is somewhat different from scheduling for assembly lines for a single or multiple product or for job shops. Projects are different in that the firm is producing a single unit (or small number) of the product. Some examples include implementing an MRP or MRP II system in an organization, building a skyscraper, or doing a consulting project for an organization. For many projects the Gantt chart is a useful tool in that it lists tasks to be accomplished, the time it should take to accomplish them, and the sequence and procedure of the tasks.

Researchers have made improvements to this method with the use of PERT and CPM charts to help them schedule and better manage projects. **PERT** is an acronym for program evaluation and review technique and **CPM** is an acronym for critical path method. These methods were initially different in that PERT used probabilistic times for tasks and CPM used deterministic times, but after the more than 50 years in which these methods have been applied there is little to distinguish the two methods.

PERT and CPM rely on a network diagram, which is a graph which contains a number of nodes and arrows. Usually the arrows designate the tasks to be performed and the nodes represent the beginning and end of these tasks. See Fig. 17.1.1. The value of the network diagram is that it clearly displays all of the tasks that must be completed to complete the project, the precedence relationships between these tasks, and, in some graphs, the time to complete each task. Of special interest among the various paths from the beginning to the ending node is the critical path. The **critical path** is the path that is the longest in terms of the time it takes to go from the beginning to the ending node. It is of interest because any delay in accomplishing tasks on the critical path will result in a delay of the completion of the project. To shorten the length of the project, tasks on the critical path will have to be shortened.

One advantage of PERT and CPM are that the graphical display clearly indicates tasks and precedences. Once the critical path has been determined managers can closely monitor tasks on the critical path and, further, monitor tasks not on the critical path that have **slack,** which is the difference between the time it takes to complete a task and the time allowable to complete the task. Another advantage of the use of these techniques is that they help the manager to organize all information and data required to complete the project.

AGGREGATE PLANNING

Aggregate planning is the task of determining how much to produce of a product, what workforce level to use, and what inventory to carry for a period of time. One use of the aggregate plan is as an input into the master production schedule. It is dependent, of course, on the forecast for the product for each of the periods for which the plan is to be made and the amount on inventory on hand along with the costs of the various components.

Factors that can be used to match production to demand include the ability to hire employees and lay off employees; the ability to vary the number of shifts, the length of shifts and the amount of overtime; the ability to subcontract; and the ability to vary the amount of inventory held or backordered.

Aggregate plans are usually evaluated on the basis of objective factors such as cost and meeting demand on time, as well as subjective factors such as lost sales potential. Costs to be considered in most approaches include the basic cost of producing the product (fixed and variable costs); costs associated with hiring, layoffs, and overtime; costs of inventory (holding and backorder costs); and the cost of subcontracting.

There are a number of approaches used to develop aggregate plans in industry, and these range from simple trial-and-error evaluations of plans on a spreadsheet to sophisticated mathematical programming or artificial intelligence models to develop plans that are optimal in terms of particular criteria.

The most simple planning method, trial and error using a spreadsheet, consists of trying various plans and seeing what they yield in terms of meeting production schedules and, in addition, what costs are incurred in the various categories mentioned above. One of the important reasons for using this type of planning method is that a number of subjective considerations can be incorporated into the solution. These include constraints such as, say, maintaining a stable workforce, eliminating shortages or backorders, eliminating subcontracting, etc. Some of these may be hard to model when one is using sophisticated modeling methods but are relatively easy to incorporate into a trial and error method.

Sophisticated methods of aggregate planning include methods such as mathematical programming, of which linear programming is a subset (covered later in this section), and expert systems approaches to the problem. Mathematical programming methods involve deciding upon what criterion to optimize and what constraints are to be placed upon the system. The objective function may be one where the costs of producing the product (which include costs of hiring, layoffs, subcontracting, inventory holding and shortage costs, etc.) are minimized. Constraints are then written that ensure that, say, the amount of inventory available for a particular period is such that demand is satisfied in that period. Note that the inventory can be obtained from inventory on hand, from production in a variety of ways, and from subcontracting. Linear constraints such as these become part of the mathematical model. The mathematical program built in the way above then is optimized using a computer program and the outputs of this program are the optimal plan and the costs of this plan.

One of the advantages of the mathematical programming approach is that cost is minimized. A disadvantage is that some constraints are not easily modeled in some cases. For example, subcontractors may offer price breaks for certain quantities of products or inventory produced. In practice, most firms use the trial-and-error approach.

QUALITY MANAGEMENT

In the past, much of the emphasis on quality management was placed on inspection. An organization produced a product and inspected samples of that product, usually at the end of production. Technology consisted of implementing inspection sampling schemes.

In the early 1980s, many organizations began to modify their methods to adhere to the teachings of proponents such as W. Edwards Deming. Deming emphasized strategic changes in the management of quality. He argued that quality was not a cost to be minimized but rather a constraint that had to be met in order for a firm to remain in business. Some of his more revolutionary contributions were that rework should be eliminated, that high quality is associated with high efficiency, and that an organization should always try to get better. Deming labeled this **continuous improvement.** Deming had a list of 14 points that he called the basis of transformation of American industry. These are (Deming, 1993, p. 23):

1. Create constancy of purpose toward improvement of product and service, with the aim to become competitive and to stay in business, and to provide jobs.

2. Adopt the new philosophy. We are in a new economic age. Western management must awaken to the challenge, must learn its responsibilities, and take on leadership for change.

3. Cease dependence on inspection to achieve quality. Eliminate the need for inspection on a mass basis by building quality into the product in the first place.

4. End the practice of awarding business on the basis of price tag. Instead, minimize total cost. Move toward a single supplier for any one item, on a long-term relationship of loyalty and trust.

5. Improve constantly and forever the system of production and service, to improve quality and productivity, and thus constantly decrease costs.

6. Institute training on the job.

7. Institute leadership. The aim of supervision should be to help people and machines and gadgets to do a better job. Supervision of management is in need of overhaul, as is supervision of production workers.

8. Drive out fear, so that everyone may work effectively for the company.

9. Break down barriers between departments. People in research, design, sales, and production must work as a team, to foresee problems of production and in use that may be encountered with the product or service.

10. Eliminate slogans, exhortations, and targets for the workforce asking for zero defects and new levels of productivity. Such exhortations only create adversarial relationships, as the bulk of the causes of low quality and low productivity belong to the system and thus lie beyond the power of the workforce.

11. a. Eliminate work standards (quotas) on the factory floor. Substitute leadership.

   b. Eliminate management by objective. Eliminate management by numbers and numerical goals. Substitute leadership.

12. a. Remove barriers that rob the hourly worker of the right to pride of workmanship. The responsibility of supervisors must be changed from sheer numbers to quality.

   b. Remove barriers that rob people in management and in engineering of their right to pride of workmanship. This means, *inter alia*, abolishment of the annual or merit rating and of management by objective.

13. Institute a vigorous program of education and self-improvement.

14. Put everybody in the company to work to accomplish the transformation. The transformation is everybody's job.

One of the essential requirements of the Deming idea is that if a problem in the production of a product occurs, it should be fixed before any more of the product is produced. This concept, of course, can be applied to any organization, as all organizations produce either a product or a service. In service organizations, if the service that is provided is poor, one risks losing the customer forever. Moreover, some marketing researchers claim that a dissatisfied customer tells several other customers about the bad experience and this may magnify the cost of a poor service encounter.

STATISTICAL PROCESS CONTROL

The primary set of tools in managing the production of a good or service so that quality is optimized is called **statistical process control** or **SPC.** SPC has as its purpose the identification of the quality of a process, the identification of the factors that influence the quality of a process, the elimination of factors that cause excess variability in the process, the monitoring of a process so that problems are quickly detected and eliminated, and the continuous improvement of a process. Most of the tools in SPC are graphical; these have the added advantage of ease of use, understanding, and communication. Perhaps the most important of these tools is the **control chart,** which is a graph of a quality measurement over time. This quality measurement has been calculated from a sample taken from the process because sampling is more efficient than 100 percent inspection in most cases, when one considers cost. The control chart is used to determine the current performance of a process, to determine the capability of a process, and to monitor a process.

The concept behind the control chart is that it is a tool for separating the variation of the process quality into two parts. Deming calls these two parts **special causes of variability** and **common causes of variability.** Some other common usage is assignable causes (special) and chance or natural (common) variability. Special-cause variability is due to problems with machines, operators, or raw materials and is usually large compared to common-cause variability, which is the variability in the process due to all of the other factors of production. When a process has had all of the special causes of variability removed, it is said to be stable or in a state of statistical control. Further reduction in the remaining common-cause variability is called process improvement but should not be attempted until the process is stable.

Other common tools for SPC include graphs such as histograms, scatterplots, stem and leaf diagrams, Pareto charts, and boxplots. Most textbooks in statistical quality control give in-depth coverage of these topics. Section 17.3, Engineering Statistics and Quality Control, covers some of the tools mentioned here as well as others in more detail.

There are several steps in SPC. The first step is to define quality. In the past product quality was measured by specifications determined by engineers. A more contemporary view of the measurement of quality is to use customer input into the design of the product or service and subsequently to use these specifications of the dimensions of quality as targets.

The second step in SPC is to measure the current quality of a process; this step is called process performance evaluation. One accurate method of measuring quality is to take random samples from the process over time. In SPC, though, one is interested in eliminating special causes of variability after one does a process performance evaluation so the most common form of sampling is by rational subgroups. These are samples chosen so that any differences due to machines, operators, or raw materials will be apparent in the data. While not as accurate in terms of providing an estimate of quality as a random sample, rational subgrouping provides a very accurate measure of quality and, more important, allows an analyst to have some insight into what special causes of variability are present. For example, a process performance evaluation might show that quality depends upon the supplier of raw material, as in a case where the product quality obtained from shifts in which a particular supplier's inputs are used is much worse than the quality when other suppliers' raw materials are used. The value of rational subgrouping is that these types of differences are apparent if the rational subgroups are cleverly chosen.

The process of removing special causes of variability is called *process capability analysis*. In the example above, the analyst might work with the particular supplier to find and eliminate the causes of poor-quality input.

After a process has undergone a process capability analysis and special causes of variability are removed, the process should be monitored. Monitoring involves taking samples from the process at particular times and calculating a statistic that measures quality. Some common statistics are the sample mean, the sample range, the sample proportion defective, and the sample number of defects per unit. These statistics are plotted on a control chart; for these statistics the control charts are respectively the X bar chart, the R chart, the p chart and the c chart. A control chart is used to visually determine if a special cause of poor quality has occurred; if it has, ideally the process is shut down and the special cause removed.

The last step in the SPC is process improvement, which ideally happens continuously. Thus, it has the name *continuous improvement*. In this step, the analyst tries to continuously reduce the common-cause variability through observational studies with rational subgrouping or experimental designs.

In practice there have been many general attempts to push the idea of quality in the workplace. Some advocates such as Deming, Juran, and Crosby came up with platforms for convincing managers that their methods would ensure successful quality efforts within an organization. Other general schemes to manage quality have appeared over the past several decades and include total quality management, ISO certification, the Malcom Baldridge national quality award, and reengineering. Perhaps the most successful is the current platform called six sigma which originated at Motorola under the leadership of Bob Galvin and at GE under Jack Welch.

Six sigma is a method that fully uses the methods discussed above in SPC. It gets its name from the fact that six sigma means there are not more than 3.4 defects per million, an admirable goal. Some of the major ideas behind six sigma include teaching everyone involved the standard statistical tools of SPC such as those discussed above, teaching a structured process such as SPC as a general problem-solving tool, making decisions analytically by using data rather than intuition or experience, and striving for a continuous need to reduce variation in a process. One of the advantages of six sigma is that its use is suggested in all areas of an organization, not just manufacturing.

Six sigma has generated admirable results because it stands on a strong theoretical foundation and is plausible. It is also not beyond the scope of anyone in its complexity. And, empirically, it has had a major financial impact on many firms.

LINEAR PROGRAMMING

At the heart of management's responsibility is the best or optimum use of limited resources including money, personnel, materials, facilities, and time. Linear programming, a mathematical technique, permits determination of the best use which can be made of available resources. It provides a systematic and efficient procedure which can be used as a guide in decision making.

As an example, imagine the simple problem of a small machine shop that manufactures two models, standard and deluxe. Each standard model requires 2 h of grinding and 4 h of polishing. Each deluxe model requires 5 h of grinding and 2 h of polishing. The manufacturer has three grinders and two polishers; therefore, in a 40-h week there are 120 h of grinding capacity and 80 h of polishing capacity. There is a profit of $3 on each standard model and $4 on each deluxe model and a ready market for both models. The management must decide on: (1) the allocation of the available production capacity to standard and deluxe models and (2) the number of units of each model in order to maximize profit.

To solve this linear programming problem, the symbol X is assigned to the number of standard models and Y to the number of deluxe models. The profit from making X standard models and Y deluxe models is $3X + 4Y$ dollars. The term *profit* refers to the **profit contribution,** also referred to as **contribution margin** or **marginal income.** The profit contribution per unit is the selling price per unit less the unit variable cost. Total contribution is the per-unit contribution multiplied by the number of units.

The restrictions on machine capacity are expressed in this manner: To manufacture one standard unit requires 2 h of grinding time, so that making X standard models uses $2X$ h. Similarly, the production of Y deluxe models uses $5Y$ h of grinding time. With 120 h of grinding time available, the grinding capacity is written as follows: $2X + 5Y \leq 120$ h of grinding capacity per week. The limitation on polishing capacity is expressed as follows: $4X + 2Y \leq 80$ h per week. In summary, the basic information is:

	Grinding time	Polishing time	Profit contribution
Standard model	2 h	4 h	$3
Deluxe model	5 h	2 h	4
Plant capacity	120 h	80 h	

Two basic linear programming techniques, the graphic method and the simplex method, are described and illustrated using the above capacity-allocation-profit-contribution maximization data.

Graphic Method

| Operations | Hours available | Hours required per model | | Maximum number of models | |
		Standard	Deluxe	Standard	Deluxe
Grinding	120	2	5	$\frac{120}{2} = 60$	$\frac{120}{5} = 24$
Polishing	80	4	2	$\frac{80}{4} = 20$	$\frac{80}{2} = 40$

The lowest number in each of the two columns at the extreme right measures the impact of the hours limitations. The company can produce 20 standard models with a profit contribution of $60 (20 × $3) or 24 deluxe models at a profit contribution of $96 (24 × $4). Is there a better solution?

To determine production levels in order to maximize the profit contribution of $3X + $4Y when:

$$2X + 5Y \leq 120 \text{ h} \quad \text{grinding constraint}$$
$$4X + 2Y \leq 80 \text{ h} \quad \text{polishing constrain}$$

a graph (Fig. 17.1.3) is drawn with the constraints shown. The two-dimensional graphic technique is limited to problems having only two variables—in this example, standard and deluxe models. However, more than two constraints can be considered, although this case uses only two, grinding and polishing.

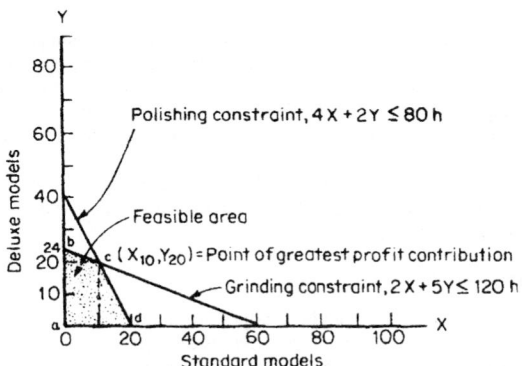

Fig. 17.1.3 Graph depicting feasible solution.

The **constraints** define the solution space when they are sketched on the graph. The solution space, representing the area of feasible solutions, is bounded by the corner points a, b, c, and d on the graph. Any combination of standard and deluxe units that falls within the solution space is a feasible solution. However, the *best* feasible solution, according to mathematical laws, is in this case found at one of the corner points. Consequently, all corner-point variables must be tried to find the combination which maximizes the profit contribution: $3X + $4Y.

Trying values at each of the corner points:

$$a = (X = 0, Y = 0); \$3(0) + \$4(0) = \$ \ 0 \text{ profit}$$
$$b = (X = 0, Y = 24); \$3(0) + \$4(24) = \$ 96 \text{ profit}$$
$$c = (X = 10, Y = 20); \$3(10) + \$4(20) = \$110 \text{ profit}$$
$$d = (X = 20, Y = 0); \$3(20) + \$4(0) = \$ 60 \text{ profit}$$

Therefore, in order to maximize profit the plant should schedule 10 standard models and 20 deluxe models.

Simplex Method   The simplex method is considered one of the basic techniques from which many linear programming techniques are directly or indirectly derived. The method uses an iterative, stepwise process which approaches an optimum solution in order to reach an objective function of maximization (for profit) or minimization (for cost). The pertinent data are recorded in a tabular form known as the **simplex tableau.** The components of the tableau are as follows (see Table 17.1.1):

The **objective row** of the matrix consists of the coefficients of the objective function, which is the profit contribution per unit of each of the products.

The **variable row** has the names of the variables of the problem including slack variables. **Slack variables** S_1 and S_2 are introduced in order to transform the set of inequalities into a set of equations. The use of slack variables involves simply the addition of an arbitrary variable

to one side of the inequality, transforming it into an equality. This arbitrary variable is called **slack variable,** since it takes up the slack in the inequality. The simplex method requires the use of equations, in contrast to the inequalities used by the graphic method.

The **problem rows** contain the coefficients of the equations which represent constraints upon the satisfaction of the objective function. Each constraint equation adds an additional problem row.

The **objective column** receives different entries at each iteration, representing the profit per unit of the variables. In this first tableau (the only one illustrated due to space limitations) zeros are listed because they are the coefficients of the slack variables of the objective function. This column indicates that at the very beginning every S_n has a net worth of zero profit.

The **variable column** receives different notations at each iteration by replacement. These notations are the variables used to find the profit contribution of the particular iteration. In this first matrix a situation of no (zero) production is considered. For this reason, zeros are marked in the objective column and the slacks are recorded in the variable column. As the iterations proceed, by replacements, appropriate values and notations will be entered in these two columns, objective and variable.

The **quantity column** shows the constant values of the constraint equations.

Based on the data used in the graphic method and with a knowledge of the basic components of the simplex tableau, the first matrix can now be set up.

Letting X and Y be respectively the number of items of the standard model and the deluxe model that are to be manufactured, the system of inequalities or the set of constraint equations is

$$2X + 5Y \leq 120$$
$$4X + 2Y \leq 80$$

in which both X and Y must be positive values or zero ($X \geq 0$; $Y \geq 0$) for this problem.

The objective function is $3X + 4Y = P$; these two steps were the same for the graphic method.

The set of inequalities used by the graphic method must next be transformed into a set of equations by the use of slack variables. The inequalities rewritten as equalities are

$$2X + 5Y + S_1 = 120$$
$$4X + 2Y + S_2 = 80$$

and the objective function becomes

$$3X + 4Y + 0S_1 + 0S_2 = P \quad \text{to be maximized}$$

The first tableau with the first solution would then appear as shown in Table 17.1.1.

The tableau carries also the first solution which is shown in the **index row.** The index row carries values computed by the following steps:

1. Multiply the values of the quantity column and those columns to the right of the quantity column by the corresponding value, by rows, of the objective column.

2. Add the results of the products by column of the matrix.

3. Subtract the values in the objective row from the results in step 2. For this operation the objective row is assumed to have a zero value in the quantity column. By convention the profit contribution entered in the cell lying in the quantity column and in the index row is zero, a condition valid only for the first tableau; in the subsequent matrices it will be a positive value.

Index row:

Steps 1 and 2:	Step 3:
120(0) + 80(0) = 0	0 − 0 = 0
2(0) + 4(0) = 0	0 − 3 = −3
5(0) + 2(0) = 0	0 − 4 = −4
1(0) + 0(0) = 0	0 − 0 = 0
0(0) + 1(0) = 0	0 − 0 = 0

Table 17.1.1   First Simplex Tableau and First Solution

		0	3	4	0	0	Objective row
	Mix	Quantity	X	Y	S_1	S_2	Variable row
0	S_1	120	2	5	1	0	Problem rows
0	S_2	80	4	2	0	1	
		0	−3	−4	0	0	Index row
Objective column	Variable column	Quantity column					

In this first tableau the slack variables were introduced into the product mix, variable column, to find *a feasible* solution to the problem. It can be proven mathematically that beginning with slack variables assures a feasible solution. One possible solution might have S_1 take a value of 120 and S_2 a value of 80. This approach satisfies the constraint equation but is undesirable since the resulting profit is zero.

It is a rule of the simplex method that the optimum solution has not been reached if the index row carries any negative values at the completion of an iteration in a maximization problem. Consequently, this first tableau does not carry the optimum solution since negative values appear in its index row. A second tableau or matrix must now be prepared, step by step, according to the rules of the simplex method.

Duality of Linear Programming Problems and the Problem of Shadow Prices   Every linear programming problem has associated with it another linear programming problem called its **dual.** This duality relationship states that for every maximization (or minimization) problem in linear programming, there is a unique, similar problem of minimization (or maximization) involving the same data which describe the original problem. The possibility of solving any linear programming problem by starting from two different points of view offers considerable advantage. The two paired problems are defined as the dual problems because both are formed by the same set of data, although differently arranged in their mathematical presentation. Either can be considered to be the **primal;** consequently the other becomes its dual.

Shadow prices are the values assigned to one unit of capacity and represent economic values per unit of scarce resources involved in the restrictions of a linear programming problem. To maximize or minimize the total value of the total output it is necessary to assign a quantity of unit values to each input. These quantities, as cost coefficients in the dual, take the name of "shadow prices," "accounting prices," "fictitious prices," or "imputed prices" (values). They indicate the amount by which total profits would be increased if the producing division could increase its productive capacity by a unit. The shadow prices, expressed by monetary units (dollars) per unit of each element, represent the least cost of any extra unit of the element under consideration, in other words, a kind of marginal cost. The real use of shadow prices (or values) is for management's evaluation of the manufacturing process.

QUEUING THEORY

Queuing theory or **waiting-line** theory problems involve the matching of servers, who provide, to randomly arriving customers, services which take random amounts of time. Typical questions addressed by queuing theory studies are: how long does the average customer wait before being waited on and how many servers are needed to assure

that only a given fraction of customers waits longer than a given amount of time.

In the typical problem applicable to queuing theory solution, people (or customers or parts) arrive at a server (or machine) and wait in line (in a queue) until service is rendered. There may be one or more servers. On completion of the service, the person leaves the system. The rate at which people arrive to be serviced is often considered to be a random variable with a Poisson distribution having a parameter λ. The average rate at which services can be provided is also generally a Poisson distribution with a parameter μ. The symbol k is often used to indicate the number of servers.

MONTE CARLO

Monte Carlo simulation can be a helpful method in gaining insight to problems where the system under study is too complex to describe or the model which has been developed to represent the system does not lend itself to an analytical solution by other mathematical techniques. Briefly, the method involves building a mathematical model of the system to be studied which calculates results based on the input variables, or **parameters.** In general the variables are of two kinds: decision parameters, which represent variables which the analyst can choose, and stochastic, or random, variables, which may take on a range of values and which the analyst cannot control.

The random variables are selected from specially prepared probability tables which give the probability that the variable will have a particular value. All the random variables must be independent. That means that the probability distribution of each variable is **independent** of the values chosen for the others. If there is any correlation between the random variables, that correlation will have to be built into the system model.

For example, in a model of a business situation where market share is to be calculated, a decision-type variable representing selling price can be selected by the analyst. A variable for the price of the competitive product can be randomly selected. Another random variable for the rate of change of market share can also be randomly selected. The purpose of the model is to use these variables to calculate a market share suitable to those market conditions. The algebra of the model will take the effects of all the variables into account. Since the rate-of-change variable can take many values which cannot be accurately predicted, as can the competitive price variable, many runs will be made with different randomly selected values for the random variables. Consequently a range of probable answers will be obtained. This is usually in the form of a **histogram.** A histogram is a graph, or table, showing values of the output and the probability that those values will occur. The results, when translated into words, are expressed in the typical Monte Carlo form of: if such a price is chosen, the following probability distribution of market shares is to be expected.

17.2 COST ACCOUNTING
by Scott Jones

REFERENCES: Horngren and Foster, "Cost Accounting—A Managerial Emphasis," Prentice-Hall. Anthony and Govindarajan, "Management Control Systems," Irwin. Cooper and Kaplan, "The Design of Cost Management Systems—Text, Cases, and Readings," Prentice-Hall.

ROLE AND PURPOSE OF COST ACCOUNTING

Cost measurements and reporting procedures are integral components of management information systems, providing financial measurements of economic value and reports useful to many and varied objectives. In a functional organization, where lines of authority are drawn between engineering, production, marketing, finance, and so on, cost accounting information is primarily used by managers for control in guiding departmental units toward the attainment of specific organizational goals. In team-oriented organizations, authority is distributed to multifunctional teams empowered as a group to make decisions. The focus of cost accounting in this organizational structure is not so much on control, but on supporting decisions through the collection of relevant information for decision making. Cost accounting also provides insight for the attainment of competitive advantage by providing an information set and analytical framework useful for the analysis of product or process design, or service delivery alternatives.

The basic purpose of cost measurements and reporting procedures can be organized into a few fundamental areas. These are: (1) identifying and measuring the economic value to be placed on goods and services for reporting periodic results to external information users (creditors, stockholders, and regulators); (2) providing control frameworks for the implementation of specific organizational objectives; (3) supporting operational and strategic decision making aimed at achieving and sustaining competitive advantage.

MEASURING AND REPORTING COSTS TO STOCKHOLDERS

Accounting in general and cost accounting in particular are most visible to the general public in the role of **external reporting.** In this role, cost accounting is geared toward measuring and reporting periodic results, typically annual, to users outside of the company such as stockholders, creditors, and regulators. The annual report presents to these users management forecasts for the coming period and results of past years operations as reflected in **general-purpose financial statements.** (See also Sec. 17.1, "Operations Management.") Performance is usually captured through the presentation of three reports: the income statement, the balance sheet, and the statement of cash flows (see Fig. 17.2.1). These statements are audited by independent certified public accountants who attest that the results reported present a fair picture of the financial position of the company and that prescribed rules have been followed in preparing the reports. The accounting principles and procedures that guide the preparation of those reports are governed by **generally accepted accounting principles** (GAAP).

The underlying principles that guide GAAP financial statements encourage comparability among companies and across time. Therefore, these rules are usually too constraining for reports destined for internal use. External reports such as the balance sheet and income statement are prepared on the **accrual basis.** Under this basis, revenues are recognized (reported to stockholders) when earned, likewise costs are expensed against (matched with) revenue when incurred. This is to be contrasted with the **cash basis,** which recognizes revenues and expenses when cash is collected or paid. For example, if a drill press is purchased for use in the factory, a cash outflow occurs when the item is paid for. The cash basis would recognize the purchase price of the drill press as an expense when the payment is made. On the other hand, the accrual basis of accounting would capitalize the price paid, and this amount would be the cost (or basis) of the asset called *drill press.* In accrual accounting, an asset is something having future economic benefit, and therefore the cost of this asset must be distributed among the periods of time when it is used to generate revenue. The cost of the drill press would be expensed periodically by deducting a small amount of that cost from revenue as the drill press is used over its economic life, which may be several years. This periodic charge is called *depreciation.* To capture the effects that revenue-generating activities have on cash, GAAP financial statements also include the statement of cash flows. The statement of cash flows is not prepared on an accrual basis; rather, it reflects the amount of cash flowing into a company during a period, as well as the cash outflow. The first section of that statement, "Cash flows from operating activities," is essentially an income statement prepared on the cash basis.

Another application of cost accounting measurements for external users involves the preparation of reports such as income tax returns for governmental agencies. Federal, state, and local tax authorities prescribe specific accounting procedures to be applied in determining taxable income. These rules are conceptually similar to general-purpose financial reporting but differ mainly in technical aspects of the computations, which are modified to support whatever public finance goals may exist for a particular period. Whereas GAAP financial statements allow for the analysis of credit and investment opportunities, Internal Revenue Service regulations are designed to raise revenue, stimulate the economy, or both. Regulations may be primarily aimed at reducing the federal deficit and hence assign rather long "useful lives" to depreciable assets; at other times in history useful lives were shortened to stimulate investment and economic growth. For GAAP, management usually estimates the useful life of an asset for purposes of depreciation. For IRS purposes, the amount of depreciation is based on the **class life** of the asset, and the depreciation system elected by the taxpayer. Class life is based on whether the asset is specialized or general purpose, and in what industry the asset is employed.

CLASSIFICATIONS OF COSTS

The purposes of cost accounting require classifications of costs so that they are recognized (1) by the nature of the item (a natural classification), (2) in their relation to the product, (3) with respect to the accounting period to which they apply, (4) in their tendency to vary with volume or activity, (5) in their relation to departments, (6) for control and analysis, and (7) for planning and decision making.

Direct material and direct labor may be listed among the items which have a **variable** nature. Factory overhead, however, must be carefully examined with regard to items of a variable and a fixed nature. It is impossible to budget and control factory-overhead items successfully without regard to their tendency to be fixed or variable; the division is a necessary prerequisite to successful budgeting and intelligent cost planning and analysis.

In general, **variable expenses** show the following characteristics: (1) variability of total amount in direct proportion to volume, (2) comparatively constant cost per unit or product in the face of changing volume, (3) easy and reasonably accurate assignments to operating departments, and (4) incurrence controllable by the responsible department head.

The characteristics of **fixed expenses** are (1) fixed amount within a relative output range, (2) decrease of fixed cost per unit with increased output, (3) assignment to departments often made by managerial

Balance Sheet (Illustrative)
Black Carbon, Inc.
December 31, 20-
Assets

Current Assets
Cash		$5,050,000	
Accounts Receivable (net)		6,990,000	
Inventories:			
Raw Materials and Supplies	$1,000,000		
Work in Process	1,800,000		
Finished Goods	2,900,000	5,700,000	
Investments		1,000,000	
Deferred Charges		340,000	
Total Current Assets			$19,080,000

Property, Plant and Equipment:
Land		$4,000,000	
Buildings and Equipment	$75,500,000		
Less: Allowance for Depreciation	47,300,000	28,200,000	
Total Fixed Assets			32,200,000
Total Assets			$51,280,000

Liabilities

Current Liabilities:
Accounts Payable and Accruals		$3,580,000	
Provision for Income Taxes:			
Federal	$2,250,000		
State	65,000	2,315,000	
Total Current Liabilities			$5,895,000
Long-term Debt			5,300,000
Total Liabilities			$11,195,000

Stockholders Equity:
Common Stock—no par value			
Authorized—2,000,000 shares			
Outstanding—1,190,000 shares		$11,900,000	
Earnings retained in the business		28,185,000	
Total Stockholders' Equity			$40,085,000
Total Liabilities and Stockholders' Equity			$51,280,000

Income Statement (Illustrative)
Black Carbon Inc.
for the year 20-

Net Sales		$50,087,000
Cost of products:		
Material, Labor, and Overhead (excluding depreciation)	$32,150,000	
Depreciation	5,420,000	37,570,000
Gross Profit		$12,517,000
Less: Selling and Administrative Expenses		3,220,000
Profit from Operations		9,297,000
Other Deductions		305,000
		8,992,000
Other Income		219,000
Income before Federal and State Income Taxes		9,211,000
Less: Provision for Federal and State Income Taxes		4,055,000
Total Net Income		5,156,000
Dividends paid to shareholders		2,200,000
Income retained in the business		$ 2,956,000

Statement of Cash Flows (Illustrative)
Black Carbon Inc.
December 31, 20-

Cash flows from operating activities		
Cash received from customers	$49,525,000	
Cash paid to suppliers and employees	(40,890,000)	
Cash paid for interest	(1,050,000)	
Cash paid for income taxes	(2,860,000)	
Net cash provided by operating activities		$ 4,725,000
Cash flows from investing activities		
Purchase of equipment	(1,970,000)	
Net cash from investing activities		(1,970,000)
Cash flows from financing activities		
Principal payment on loans	(2,780,000)	
Dividend payments	(2,200,000)	
Net cash used by financing activities		(4,980,000)
Net increase (decrease) in cash		(2,225,000)
Cash at beginning of year		7,275,000
Cash at end of year		$5,050,000

Fig. 17.2.1 Examples of the balance sheet, income statement, and statement of cash flows based on the published annual report.

decisions or cost-allocation methods, and (4) control for incurrence resting with top management rather than departmental supervisors. Whether an expense is classified as fixed or variable may well be the result of managerial decisions.

Some **factory overhead** items are semivariable in nature; i.e., they vary with production but not in direct proportion to the volume. For practical purposes, it is desirable to resolve each semivariable expense item into its variable and fixed components.

A factory is generally organized along departmental lines for production purposes. This factory departmentalization is the basis for the important classification and subsequent accumulation of costs by departments to achieve (1) cost control and (2) accurate costing. The departments of a company generally fall into two categories: (1) producing, or productive, departments, and (2) nonproducing, or service, departments. A producing department is one in which manual and machine operations are performed directly upon any part of the product manufactured. A service department is one that is not directly engaged in production but renders a particular type of service for the benefit of other departments. The expense incurred in the operation of service departments represents a part of the total factory overhead that must be absorbed in the cost of the product.

For **product costing,** the factory may be divided into departments, and departments may also be subdivided into cost centers. As a product passes through a cost center or department, it is charged with a share of the indirect expenses on the basis of a departmental factory-overhead rate. For cost-control purposes, budgets are established for departments and cost centers. Actual expenses are compared with budget allowances in order to determine the efficiency of a department and to measure the manager's success in controlling expenses.

Factory overhead, which is charged to a product or a job on the basis of a predetermined overhead rate, is considered indirect with regard to the product or the job to which the expense is charged. Service-department expenses are prorated to other service departments and/or to the producing departments. The proration is accomplished by using some rational basis such as area occupied or number of workers. The prorated costs are termed **indirect departmental charges.** When all service-department expenses have been prorated to the producing departments, each producing department's total factory overhead will consist of its own direct departmental expense and the indirect (or prorated, or apportioned) charges. This total cost is charged to the product or the job on the basis of the predetermined factory-overhead rate.

A company's cost system provides the data required for establishing **standard costs** and for the preparation and operation of a **budget.**

The **budget** program enlists all members of management in the task of creating a workable and acceptable plan of action, welds the plan into a homogeneous unit, communicates to the managerial levels differences between planned activity and actual performance, and points out unfavorable conditions which need corrective action. The budget not only will help promote coordination of people, clarification of policy, and crystallization of plans, but with successful use will create greater internal harmony and unanimity of purpose among managers and workers.

The established **standard-cost** values for material, labor, and factory overhead form the foundation for the budget. Since standard costs are an invaluable aid in the process of setting prices, it is essential to set these standard costs at realistic levels. The measurement of deviations from established standards or norms is accomplished through the use of variance accounts.

Costs as a basis for planning are estimated costs which may be incurred if any one of several alternative courses of action is adopted. Different types of costs involve varying kinds of consideration in managerial planning and decision making.

METHODS OF ACCUMULATING COSTS IN RECORDS OF ACCOUNT

The balance sheet lists the components of inventory as raw materials, work in process, and finished goods. These accounts reflect the cost of unsold production at various stages of completion. The costs in work in process and finished goods are accumulated or tabulated in the record of accounts according to one of two methods:

1. The Job-Order Cost Method   When orders are placed in the factory for specific jobs or lots of product, which can be identified through all manufacturing processes, a job cost system is appropriate. This method has certain characteristics. A manufacturing order often corresponds to a customer's order, though sometimes a manufacturing order may be for stock. The customer's order may be obtained on the basis of a bid price computed from an estimated cost for the job. The goods in each order are kept physically separate from those of other jobs. The costs of a manufacturing order are entered on a job cost sheet which shows the total cost of the job upon completion of the order. This cost is compared with the estimated cost and with the price which the customer agreed to pay.

2. The Process Cost Method   When production proceeds in a continuous flow, when units of product are not separately identifiable, and when there are no specific jobs or lots of product, a process cost system is appropriate, for it has certain characteristics: work is ordered through the plant for a specific time period until the raw materials on hand have all been processed or until a specified quantity has been produced; goods are sold from the stock of finished goods on hand since a customer's order is not separately processed in the factory; the cost-of-production sheet is a record of the costs incurred in operating the process—or a series of processes—for a period of time. It shows the quantity produced in pounds, tons, gallons, or other units, and the cost per unit is obtained by dividing the total costs of the period by the total units produced. Performance is indicated by comparing the quantity produced and the cost per unit of the current period with similar figures of other periods or with standard cost figures.

ELEMENTS OF COSTS

The main items of costs shown on the income statement are factory costs which include direct materials, direct labor and factory overhead; and selling and administrative expenses. A breakdown of costs is shown in Figure 17.2.2.

Materials   The cost of materials purchased is recorded from purchase invoices. When the materials are used in the factory, an assumption must be made as to **cost flow,** that is, whether to charge them to operations at average prices, at costs based on the first-in, first-out method of costing, or at costs based on the last-in, first-out method of costing. Each method will lead to a different cost figure, depending on how prices change. Each situation must be studied individually to determine which practice will give a maximum of accuracy in cost figures with a minimum of accounting and clerical effort. Once the choice has been made, records must be set up to charge materials to operations based on requisitions. Indirect material is necessary to the completion of the product, but its consumption with regard to the final product is either so small or so complex that it would be futile to treat it as a direct-material item.

Labor   Labor also consists of two categories: direct and indirect. Direct labor, also called **productive labor,** is expended immediately on the materials comprising the finished product. Indirect labor, in contrast to direct labor, cannot be traced specifically to the construction

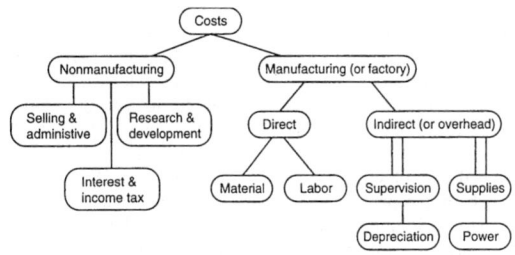

Figure 17.2.2   Summary diagram of cost relationships.

or composition of the finished product. The term includes the labor of supervisors, shop clerks, general helpers, cleaners, and those employees engaged in maintenance work.

Factory Overhead Indirect materials or factory supplies and indirect labor constitute an important segment of factory overhead. In addition, costs of fuel, power, small tools, depreciation, taxes on real estate, patent amortization, rent, inspection, supervision, social security taxes, health and accident insurance, workers' compensation insurance, and many others fall into this large category. These expenses must be collected and allocated to all jobs or units produced. Many expenses are definitely applicable to a specific department and are easily assigned thereto. Other expenses relate to the entire plant and must be prorated to departments on some suitable basis. For instance, heat might be prorated to departments on the basis of volume of space occupied. The expenses of the service departments are prorated to the producing departments on some basis such as service rendered in the case of a maintenance department or per dollar of payroll processed in the case of a cost department.

The charging of factory overhead to jobs or products is accomplished by means of an overhead or burden rate. This rate is essentially a ratio computed to show the relationship of the total burden of a department to some other easily measurable total figure for the department. For example, the total burden cost of a department may be divided by its direct-labor cost to give a percentage-of-direct-labor rate. This percentage applied to the direct-labor cost of a job or a product gives the amount of overhead chargeable thereto. Other common types of burden rates are the labor-hour rate (departmental expenses ÷ total direct-labor hours) and the machine-hour rate (departmental expenses ÷ total machine hours available). Labor rates are most commonly used. When, however, machines perform the greater amount of the work, machine-hour rates give better results. It must be clearly understood that these rates are computed in advance of production, generally at the beginning of the year. They are used throughout the fiscal period unless seasonal fluctuations or unusual changes in expense amounts necessitate the creation of a new rate. The determination of the overhead rate is closely tied up with overhead budgets.

Departmental Classification As mentioned above, the establishment of departmental lines is important not only for costing purposes but also for budgetary control purposes. Departmental lines are set up in order to (1) segregate basically different processes of production, (2) secure the smoothest possible flow of production, and (3) establish lines of responsibility for control over production and costs. When the costing methods are designed to fit in with the departmentalization of factory and office, costs can be accumulated within a department with production being on either the job-order or process cost method.

ACTIVITY-BASED COSTING

The method of assigning overhead to products based on labor hours or machine hours is referred to as **volume-based** overhead absorption because the amount assigned will vary strictly with the volume of either labor or machine time consumed. In applications where production costs may not be strictly driven by volume of labor hours, **activity-based** or **transaction** costing is appropriate. This situation typically occurs when there are many options or alternatives available to the customer. Typically, these products are produced in low volumes and have a high degree of complexity. An example would be the option of a premium radio in an automobile. Though the actual purchase cost of that radio would not be overlooked in pricing the automobile, the indirect costs would be overlooked in a volume-based system because the indirect costs associated with the premium radio, such as holding an additional item of inventory, documenting and producing separate receiving orders, added clerical and assembly coordination effort, increased engineering complexity, and so on would be grouped under the general category of costs. On the other hand, in an activity-based system, indirect costs would be assigned to a unique cost pool, such as shown in Fig. 17.2.3 and 17.2.4, which compare the procedures of volume and activity-based costing. The striking difference is that the overhead cost pool used in volume systems is not present. In activity-based systems, many more cost pools are used and are closely related to some **causal aspect** of the

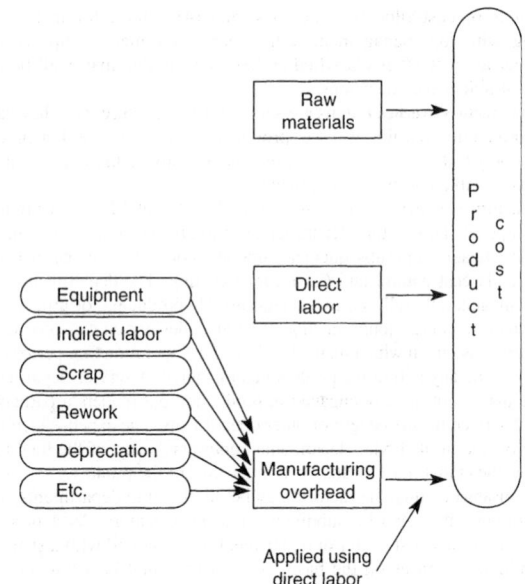

Fig. 17.2.3 Volume-based costing.

process such as machine setup, receiving orders, or material movement. The costs are assigned to products based on the relative amounts of each cost driver consumed by that product. Therefore, low-volume options such as the premium radio receive a larger share of indirect costs relative to the actual volume of labor used to insert the component. The amount of each cost pool attributed to a product can then be spread over the number of units produced and a more accurate assignment of costs obtained.

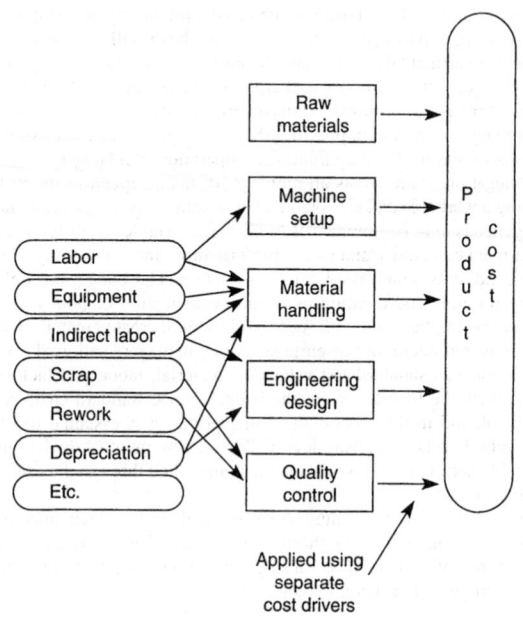

Fig. 17.2.4 Activity-based costing.

MANAGEMENT AND THE CONTROL FUNCTION

To be successful, management must integrate its own knowledge, skills, and practices with the know-how and experience of those who are entrusted with the task of carrying out company objectives. Management, together with its employees and workers, can achieve its objectives through performance of the three managerial functions: (1) planning and setting objectives, (2) organizing, and (3) controlling.

Planning is a basic function of the management process. Without planning there is no need to organize or control. However, planning must precede doing, and the budget is the most important planning tool of an enterprise.

Organizing is essentially the establishment of the framework within which the required activities are to be performed, together with a list of who should perform them. Creation of an organization requires the establishment of organizational or functional units generally known as departments, divisions, sections, floors, branches, etc.

Controlling is the process or procedure by which management ensures operative performance which corresponds with plans. The control process is pictured diagrammatically in Fig. 17.2.5. Recognition of accounting as an important tool in the controlling phase is evidenced through the role of performance reports in pointing out areas and jobs or tasks which require corrective action. These reports should make possible "management by exception."

The effectiveness of the control of costs depends upon proper communication through control and action reports from the accountant to the various levels of operating management. An organization chart is essential to the development of a cost system and cost reports which parallel the responsibilities of individuals for implementing management plans. The coordinated development of a company's organization with the cost and budgetary system will lead to "responsibility accounting." Responsibility accounting plays a key role in determining the type of cost system used.

Fig. 17.2.5 The control process.

TYPES OF COST SYSTEMS

The construction of a cost system requires a thorough understanding of (1) the organizational structure of the company, (2) the manufacturing procedure, and (3) the type of information which management requires of the cost system.

1. The organization chart gives a graphic picture of the ranking authority of superintendents, department heads, and managers who are responsible for (a) providing the detailed information needed by the accounting division in order to install a successful system; (b) incurring expenditures in personnel, materials, and other cost elements, which the cost accountant must segregate and report to those in charge. The cost system with its operating accounts must correspond to organizational divisions of authority so that the individual supervisor, department head, or executive can be held "accountable" for the costs incurred in the department.

2. The manufacturing procedure and shop methods lead to a consideration of the type of pay (piece rate, incentive, day rate, etc.); the method of collecting hours worked; the control of inventories; the problem of costing tools, dies, jigs, and machinery; and many other problems connected with the factory.

3. The organizational setup on the one hand and the manufacturing procedure on the other form the background for the design of a cost system that is based on (a) recognition of the various cost elements, (b) departmentalization of factory and office, and (c) the chart of accounts.

Any cost system should be perfected so that it will (1) aid in the control and management of the company; (2) measure the efficiency of personnel, materials, and machines; (3) help in eliminating waste; (4) provide comparison within individual industries; (5) provide a means of valuing inventories; and (6) aid in establishing selling prices. In an organization departmentalized or segmented along product lines, it is often arbitrary to allocate certain indirect costs especially when common facilities or personnel are shared. This is because there is no objective basis to compute a division of common costs. Control methods for evaluating performance often rely on the **segment margin** statement.

Example of a Segment Margin Statement

	Product A	Product B	Total
Sales	$9,000	$11,000	$20,000
Direct material & labor	(4,000)	(5,000)	(9,000)
Contribution margin	5,000	6,000	11,000
Product specific overhead	(1,000)	(2,500)	(3,500)
Segment margin	4,000	3,500	7,500
Common costs			(2,000)
Operating income			$5,500

The cost system's value is greatly enhanced when it is interlocked with a **budgetary control system.** When budget figures are based upon standard costs, the greatest benefit will be derived from such a combination.

Basically, two types of cost systems exist: (1) the **actual** (or **historical**) and (2) the **standard** (or **predetermined**). The actual cost system accumulates and summarizes costs as they occur and determines a final product cost after all manufacturing operations have been completed. The job is charged with actual quantities and costs of materials used and labor expended; the overhead or burden is allocated on the basis of some predetermined overhead rate. This predetermined overhead rate shows that even the so-called **actual system** does not entirely live up to its name. Under a standard cost system all costs are predetermined in advance of production. Both the actual (historical) and the standard cost system may be used in connection with either (1) the job-order cost method or (2) the process cost method.

BUDGETS AND STANDARD COSTS

A budget provides management with the information necessary to attain the following major objectives of budgetary control: (1) an organized procedure for planning; (2) a means for coordinating the activities of the various divisions of a business; (3) a basis for cost control. The planning phase provides the means for formalizing and coordinating the plans of the many individuals whose decisions influence the conduct of a business. Sales, production, and expense budgets must be established. Their establishment leads necessarily to the second phase of coordination. Production must be planned in relation to expected sales, materials and labor must be acquired or hired in line with expected production requirements, facilities must be expanded only as foreseeable future needs justify, and finances must be planned in relation to volume of sales and production. The third phase of cost control is predicated on the idea that actual costs will be compared with budgeted costs, thus relating what actually happened with what should have happened. To accomplish this purpose, a good measure of what costs should be under any given set of conditions must be provided. The most important condition affecting costs is volume or rate of activity. By predetermining, through the use of the flexible budget, the expenses allowed for any given rate of activity and comparing it with the actual expense, a better measurement of the performance of an individual department is achieved and the control of costs is more readily accomplished.

In the construction of overhead budgets the volume or activity of the entire organization as well as of the individual department is of considerable importance in their relationship to existing capacity. Capacity must be looked upon as that fixed amount of plant, machinery, and personnel to which management has committed itself and with which it expects to conduct the business. Volume or activity is the variable factor in business related to capacity by the fact that volume attempts to make the best use of the existing capacity. To find a profitable solution to this relationship is one of the most difficult problems faced by business management and the accountant who tries to help with appropriate cost data. Volume, particularly of a department, is often expressed in terms of direct-labor hours. With different rates of capacity, a different cost per hour of labor will be computed. This relationship can be demonstrated in the following manner:

Percentage of productive capacity	60%	80%	100%
Direct-labor hours	600	800	1,000
Factory overhead			
Fixed overhead	$1,200	$1,200	$1,200
Variable overhead	600	800	1,000
Total	$1,800	$2,000	$2,200
Overhead rate per direct-labor hour	$3.00	$2.50	$2.20

The existence of fixed overhead causes a higher rate at lower capacity utilization. It is desirable to select that overhead rate which permits a full recovery of production costs by the end of the business cycle. The above tabulation reveals another important axiom with respect to fixed and variable overhead. Fixed overhead remains constant in total but varies in respect to cost per unit or hour. Variable overhead varies in total but remains fixed in relationship to the unit or hour.

Standard Costs   The budget, as a statement of expected costs, acts as a guidepost which keeps business on a charted course. Standards, however, do not tell what the costs are expected to be but rather what they will be if certain performances are attained. In a well-managed business, costs never exceed the budget. They should constantly approach predetermined standards. The uses of standard costs are of prime importance for (1) controlling and reducing costs, (2) promoting and measuring efficiencies, (3) simplifying the costing procedures, (4) evaluating inventories, (5) calculating and setting selling prices. The success of a standard cost system depends upon the reliability and accuracy of the standards. To be effective, standards should be established for a definite period of time so that control can be exercised and variances from standards computed. Standards are set for materials, labor, and factory overhead. When actual costs differ from standard costs with respect to material and labor, two causes can generally be detected. (1) The price may be higher or lower or the rate paid a worker may be different; the difference is called a material price or a labor rate variance. (2) The quantity of the material used may be more or less than the standard quantity or the hours used by the worker may be more or less. The difference is called material-quantity variance or labor-efficiency variance, respectively. For factory overhead, the computation is somewhat more elaborate. Actual expenses are compared not only with standard expenses but also with budget figures. Various methods are in vogue, resulting in different kinds of overhead variances. Most accountants compute a controllable and a volume variance. The controllable variance deals chiefly with variable expenses and measures the efficiency of the manager's ability to hold costs within the budget allowance. The volume variance portrays fixed overhead with respect to the use or nonuse of existing capacity. It measures the success of management in its ability to fill capacity with sales or production volume. These two variances can be analyzed further into an expenditure and efficiency variance for the controllable variances and into an effectiveness and capacity variance for the volume variance. Such detailed analyses might bring forth additional information which would help management in making decisions. Of absolute importance for any cost system is the fact that the information must reach management promptly, with regularity, and in a report that is analytical, permitting quick comparison with targets and goals. Only in this manner can management, which includes all echelons from the foreman to the president, exercise control over costs and therewith over profits.

TRANSFER PRICING

A **transfer price** refers to the selling price of a good or service when both the buyer and seller are within the same organization. For example, one division of a company may produce a component, such as an engine, and transfer this component to an assembly division. For purposes of control, these organizational units may be treated as **profit centers** (responsible for earning a specified profit or return on investment). Accordingly, the transfer price is a revenue for the seller and a cost to the buyer. Because organizational control is at issue whenever interdivisional transfers are made, companies must often specify a policy to dictate the basis for determining a transfer price. Transfer prices should be based on market prices when available. Most taxing authorities require intercompany transfers to be made at market price as well. To solve situations of suboptimal resource usage (e.g., idle capacity) it is often possible to construct transfer prices based on manufacturing cost plus some allowance for profit. If the producing division is a **cost center** (responsible for controlling costs to achieve a certain budgeted level), in order to promote efficiency transfers are usually made on the basis of **standard cost.**

SUPPORTING DECISION MAKING

The analytical phase of cost accounting has become more important and influential in the last few years. Management must make many decisions, some of a short-range, others of a long-range nature. To base judgment upon good, reliable data and analyses is a major task for controllers and their staffs. Cost analysis comprises such matters as analysis of distribution costs, gross-profit analysis, break-even analysis, profit-volume analysis, differential-cost analysis, direct costing, capital-expenditure analysis, return on capital employed, and price analysis. A detailed discussion of each phase mentioned lies beyond the scope of this section, but a short description is appropriate.

Distribution-cost analysis deals with allocation of selling expenses to territories, customers, channels of distribution, products, and sales representatives. Once so allocated it might be possible to determine the most profitable and the least profitable commodity, product, territory, or customer. Segment margin statements are useful for this analysis. Standards have been introduced recently in these analyses. The Robinson-Patman Act, an amendment to Section 2 of the Clayton Act, gave additional impetus to the analytical phase of distribution costs. This act prohibits pricing the same product at different amounts when the amounts do not reflect actual cost differences (such as distribution or warranty).

Gross-profit analysis attempts to determine the causes for an increase or decrease in the gross profit. Any change in the gross profit is due to one or a combination of the following: (1) changes in the selling price of the products; (2) changes in the volume sold; (3) changes in the types of products sold, called the sales-mix; (4) changes in the cost elements. Cost elements are analyzed through budgetary control methods. Sales figures must be scrutinized to unearth the changes from the contemplated course and therewith from the final profit.

Break-even analysis, generally presented in the form of a break-even chart, constitutes one of the briefest and most easily understood devices for data presentation for policy-making decisions. The name "break-even" implies that point at which the company neither makes a profit nor suffers a loss from the operations of the business. A break-even chart can be defined as a portrayal in graphic form of the relation of production and sales to profit or, more briefly, a graphic variable income statement. The computation of the break-even point can be made by the following formula.

$$\text{Break-even sales volume} = \frac{\text{total fixed expenses}}{1 - \dfrac{\text{total variable expenses}}{\text{total sales volume}}}$$

EXAMPLE. Assume fixed expenses, $13,800,000; variable expenses, $27,000,000; total sales volume, $50,000,000. Computation: Break-even sales volume = $13,800,000/[1 − ($27,000,000/$50,000,000)] = $30,000,000.

Results can be obtained in chart form (Fig. 17.2.6).

Fig. 17.2.6 Illustrative break-even chart.

Cost-volume-profit analysis deals with the effect that a change of volume, cost, price, and product-mix will have on profits. Managements of many enterprises attempt to stimulate the public to purchase their products by conducting intensive promotion campaigns in radio, press, mail, and television. The customer, however, makes the final decision. What management wants to know is which product or model will yield the most profitable margin; which is the least profitable; what effect a reduction in sales price will have on final profit; what effect a shift in volume or product-mix will have on product costs and profits; what the new break-even point will be under such changing conditions; what the effect of expected increases in wages or other operating costs on profit will be; what the effect will be on costs, profit, and sales volume should there be an expansion of the plant. Cost-volume-profit analysis can also be presented graphically in a so-called *volume-profit-analysis graph.* Using the same data as in the break-even chart, a volume-profit analysis graph takes the form shown in Fig. 17.2.7.

Fig. 17.2.7 Illustrative cost-volume-profit analysis graph.

Differential-cost analysis treats differences, as the title suggests. These differences, also called alternative courses, arise when management wants to know whether or not to take business at a special price, to risk a decline in price of total sales, to sacrifice volume for price, to shut down part of the plant, or to enlarge plant capacity. While accountants generally use the term **"differential,"** economists speak of **"marginal"** and engineers of **"incremental"** costs in connection with such a study. As in any of the previously discussed analyses, the classification of costs into their fixed and variable components is absolutely essential. However, while in break-even analysis the emphasis rests upon the fixed expenses, differential-cost

studies stress the variable costs. The differential-cost statement presents only the differences in the following manner:

	Present business	Additional business	Total
Sales	$100,000	$10,000	$110,000
Variable costs	60,000	6,000	66,000
Marginal income	40,000	4,000	44,000
Fixed expenses	30,000	none	30,000
Profit	$ 10,000	$ 4,000	$ 14,000

This statement shows that additional business is charged with the variable expenses only because present business is absorbing all fixed expenses.

Direct costing is a costing method which charges the products with only those costs that vary directly with volume. Variable or direct costs such as direct materials, direct labor, and variable manufacturing expenses are examples of costs chargeable to the product. Costs that are a function of time rather than of production are excluded from the cost of the product. The only costs assignable to inventories are variable costs, and because they should vary in proportion to increases or decreases in production, the unit cost assigned to inventories should be uniform.

CAPITAL-EXPENDITURE DECISIONS

The preparation of a capital-expenditure budget must be preceded by an analytical and decision-making process by management. This area of managerial decisions not only is important to the success of the company but also is crucial in case of errors. Financial requirements, present and anticipated costs, profits, tax considerations, and legal, personnel, and market problems must be studied and reviewed before making the final decision.

Five evaluation techniques are generally accepted as representative tools for decision making: (1) **payback-** or **payout-period method;** (2) **average-return-on-investment method;** (3) **present-value method;** (4) **discounted-cash-flow method;** and (5) **economic value added (EVA).** None of these methods serves every purpose or every firm. The methods should, however, aid management in exercising judgment and making decisions. Of significance in the evaluation of a capital expenditure is the time value of money which is employed in the present value and the discounted-cash-flow methods. The **present value** means that a dollar received a year hence is not the equivalent of a dollar received today, because the use of money has a value. For this reason, the estimated results of an investment proposal can be stated as a cash equivalent at the present time, i.e., its present value. Present-value tables have been devised to facilitate application of present-value theory.

In the present-value method the discount rate is known or at least predetermined. In the **discounted-cash-flow** (DCF) method the rate is to be calculated and is defined as the rate of discount at which the sum of positive present values equals the sum of negative present values. The DCF method permits management to amortize corporate profits by selecting proposals with the highest rates of return as long as the rates are higher than the company's own **cost of capital** plus management's allowance for risk and uncertainty. Cost of capital represents the expected return for a given level of risk that investors demand for investing their money in a given firm or venture. However, when related to capital-expenditure planning, the cost of capital refers to a specific cost of capital from a particular financing effort to provide funds for a specific project or numerous projects. Such use of the concept connotes the marginal cost of capital point of view and implies linkage of the financing and investment decisions. It is, therefore, not surprising that the cost of capital differs, depending upon the sources. A company could obtain funds from (1) bonds, (2) preferred and common stock, (3) use of retained earnings, and (4) loans from banks. If a company obtains funds by some combination of these sources to achieve or maintain a particular capital structure, then the cost of capital (money) is the weighted average cost of each money source.

Return-on-Capital Concept   This aids management in making decisions with respect to proposed capital expenditures. This concept can also be used for (1) measuring operating performance, (2) profit planning and decision making, and (3) product pricing. The return on capital may be expressed as the product of two factors: the percentage of profit to sales and the rate of capital turnover. In the form of an equation, the method appears as

$$\frac{\text{Profit}}{\text{Sales}} = \text{profit margin}$$
$$\times \rightarrow = \quad \begin{array}{l} \text{return on} \\ \text{capital} \\ \text{investment} \end{array}$$
$$\frac{\text{Sales}}{\text{Capital}} = \frac{\text{Investment}}{\text{turnover}}$$

Whether for top executive, plant or product manager, plant engineer, sales representative, or accountant, the concept of return on capital employed tends to mesh the interest of the entire organization. An understanding and appreciation of the return-on-capital concept by all employees help in building an organization interested in achieving fair profits and an adequate rate of return.

COST MANAGEMENT

Often, programs of **continuous improvement** require that costs be computed according to the activity-based method. That method facilitates

identifying and setting priorities for the elimination of non-value-adding activities. **Non-value-added** activities decrease cycle time efficiency, where cycle time efficiency is the sum of all value-added activity times divided by total cycle time. The engineer may redesign the product using common parts or through process redesign so as to eliminate those activities or cost pools that add to product cost without adding to value. Some examples of these activities are material movement, run setup, and queue time.

Efforts to manage product costs by eliminating non-value-adding activities are frequently the result of a need to attain a specific target cost. Traditionally selling price was determined by adding a required markup to total cost. Global competition has forced producers to accept a market price determined by competitive forces:

Target cost = market selling price − required return on investment

Accordingly, the company that stays in business is the one that can accept this price and still earn a return on investment. **Target costing** is a concerted effort to design, produce, and deliver the product at a cost that will assure long-term survival.

Engineers may also focus on process throughput as a means of cost management. The **theory of constraints** proposes that profit is maximized when process throughput is maximized. The underlying assumption is that costs are fixed in the short run and cost per unit is lowest when fixed costs are distributed over the largest possible volume. This goal is accomplished by reducing cycle time through the elimination of process bottlenecks.

17.3   ENGINEERING STATISTICS AND QUALITY CONTROL
by Ashley C. Cockerill

REFERENCES: Brownlee, "Statistical Theory and Methodology in Science and Engineering," Wiley. Conover, Two *k*-sample slippage tests, *Journal of the American Statistical Association,* **63:** 614–626. Conover, "Practical Nonparametric Statistics," Wiley. Duncan, "Quality Control and Industrial Statistics," Richard D. Irwin. Gibbons, "Nonparametric Statistical Inference," McGraw-Hill. Olmstead, A corner test for association, *Annals of Mathematical Statistics,* **18:** 495–513. Owen, "Handbook of Statistical Tables," Addison-Wesley. Pearson, "'Biometrika Tables for Statisticians," vol. 1, 2d ed., Cambridge University Press. Tukey, "A quick compact two sample test to Duckworth's specifications," *Technometrics,* **1:** 31–48. Wilks, "Mathematical Statistics," Wiley.

**ENGINEERING STATISTICS
AND QUALITY CONTROL**

Statistical models and statistical methods play an important role in modern engineering. Phenomena such as turbulence, vibration, and the strength of fiber bundles have statistical models for some of their underlying theories. Engineers now have available to them batteries of computer programs to assist in the analysis of masses of complex data. Many textbooks are needed to cover fully all these models and methods; many are areas of specialization in themselves. On the other hand, every engineer has a need for easy-to-use, self-contained statistical methods to assist in the analysis of data and the planning of tests and experiments. The sections to follow give methods that can be used in many everyday situations, yet they require a minimum of background to use and need little, if any, calculation to obtain good results.

STATISTICS AND VARIABILITY

One of the primary problems in the interpretation of engineering and scientific data is coping with **variability.** Variability is inherent in every physical process; no two individuals in any population are the same. For example, it makes no real sense to speak of the tensile strength of a synthetic fiber manufactured under certain conditions; some of the fibers

will be stronger than others. Many factors, including variations of raw materials, operation of equipment, and test conditions themselves, may account for differences. Some factors may be carefully controlled, but variability will be observed in any data taken from the process. Even tightly designed and controlled laboratory experiments will exhibit variability.

Variability or variation is one of the basic concepts of statistics. Statistical methods are aimed at giving objective, quantitative, and reproducible ways of assessing the effects of variability. In particular, they aim to provide measures of the uncertainty in conclusions drawn from observational data that are inherently variable.

A second important concept is that of a **random sample.** To make valid inferences or conclusions from a set of observational data, the data should be able to be considered a random sample. What does this mean? In an operational sense it means that everything we are interested in seeing should have an equal chance of being represented in the observations we obtain. Some examples of what not to do may help. If machine setup is an important contributor to differences, then all observations should not be taken from one setup. If instrumental variation can be important, then measurements on the same item should not be taken successively—time to "forget" the last reading should pass. A random sample of n items in a warehouse is not the first n that you can find. It is the n that is selected by a procedure guaranteed to give each item of interest an equal chance of selection. One should be guided by generalizations of the fact that the apples on top of a basket may not be representative of all apples in the basket.

**CHARACTERIZING OBSERVATIONAL DATA:
THE AVERAGE AND STANDARD DEVIATION**

The two statistics most commonly used to characterize observational data are the **average** and the **standard deviation.** Denote by $x_1, x_2, \ldots, x_n$

the n individual observations in a random sample from some process. Then the average and standard deviation are defined as follows:
Average:

$$\bar{x} = \sum_{i=1}^{n} x_i/n$$

Standard deviation:

$$s = \left[\sum_{i=1}^{n} (x_i - \bar{x})^2/(n-1) \right]^{1/2}$$

Clearly, the average gives one number around which the n observations tend to cluster. The standard deviation gives a measure of how the n observations vary or spread about this average. The square of the standard deviation is called the **variance**. If we consider a unit mass at each point x_i, then the variance is equivalent to a moment of inertia about an axis through $\bar{x}$. It is readily seen that for a fixed value of $\bar{x}$, greater spreads from the average will produce larger values of the standard deviation s. The average and the standard deviation can be used jointly to summarize where the observations are concentrated. Tchebysheff's theorem states: A fraction of at least $1 - (1/k^2)$ of the observations lie within k standard deviations of the average. The theorem guarantees lower bounds on the percentage of observations within k (also known as z in some textbooks) standard deviations of the average.

Interval	Lower bound on % of measurements
$\bar{x} - 2s$ to $\bar{x} + 2s$	75%
$\bar{x} - 3s$ to $\bar{x} + 3s$	89%
$\bar{x} - 4s$ to $\bar{x} + 4s$	94%
$\bar{x} - 5s$ to $\bar{x} + 5s$	96%
$\bar{x} - 6s$ to $\bar{x} + 6s$	97%

Since the average and the standard deviation are computed from a sample, they are themselves subject to fluctuation. However, if μ, is the long-term average of the process and σ is the long-term standard deviation, then:

Average $(\bar{x}) = \mu$, process average
Average $(s) = \sigma$, process standard deviation

Furthermore, the intervals $\mu \pm k\sigma$ contain the same percentage of all values, as do the intervals $\bar{x} \pm ks$ for the sample; that is, at least 89 percent of all the long-term values will be contained in the interval $\mu - 3\sigma$ to $\mu + 3\sigma$, etc.

Range Estimate of the Standard Deviation

For $n \leq 20$ it is more convenient to compute the range r to estimate the standard deviation σ. The range is $x_{(n)} - x_{(1)}$, where $x_{(n)}$ is the largest value in a random sample of size n and $x_{(1)}$ is the smallest value. For example, if $n = 10$ and the observations are 310, 309, 312, 316, 314, 303, 306, 308, 302, 305, the range is $r = 316 - 302 = 14$. An estimate of the standard deviation σ is obtained by multiplying r by the factor f_n in Table 17.3.1. The average value of $r \cdot f_n$ is σ. Thus, in the example above, an estimate of σ and a value that can be used for s is $0.3249r = 0.3249(14) = 4.5486$.

Table 17.3.1   Average of Range $f_n = s$

Sample size	f_n	Sample size	f_n
2	0.8862	11	0.3152
3	0.5908	12	0.3069
4	0.4857	13	0.2998
5	0.4299	14	0.2935
6	0.3946	15	0.2880
7	0.3698	16	0.2831
8	0.3512	17	0.2787
9	0.3367	18	0.2747
10	0.3249	19	0.2711
		20	0.2677

PROCESS VARIABILITY—HOW MUCH DATA?

Since the output of all processes is variable, one can make reasonable decisions about the output only if one can obtain a measure of how much variability or spread one can expect to see under normal conditions. Variability cannot be measured accurately with a small amount of data. Methods for assessing how much data are needed are given for two general situations.

Specified Tolerances

A convenient statement about the variability or spread of a process can be based on the smallest and largest values in a random sample of the output. There are no practical limitations on its use. Suppose that we have a random sample of n values from our process. Denote the values by $X_1, X_2, \ldots, X_n$. After obtaining the values we find the smallest, $X_{(1)}$, and the largest, $X_{(n)}$. Now we want to assess what percent of all future values that this process might generate will be covered by $X_{(1)}$ and $X_{(n)}$. In statistics, $X_{(1)}$ and $X_{(n)}$ are called **tolerance limits**. If the process generates bolts and X is the diameter, then the engineering concept of tolerance and the statistical concept of tolerance are seen to be quite similar.

Let p be the percentage of all the process values that on a long-term basis will be between $X_{(1)}$ and $X_{(n)}$. Let P be a lower bound for this percentage p. Now consider the probability statement: Probability $(p \geq P) = C$. The quantity C we call **confidence**. Since it is a probability its value is between 0 and 1. As C approaches 1 our confidence in the percentage P increases. The interpretation of P and C can be explained in terms of Table 17.3.2.

Suppose that we take a random sample of size $n = 269$ values from our process output; the smallest value is 10 and the largest is 54. In Table 17.3.2 we see that 269 is the entry for $P = 99$ and $C = 0.75$. This tells us that at least $P = 99$ percent of all future values that this process will generate will be between 10 and 54, the smallest and largest values seen in the sample of 269. The confidence $C = 0.75 = {}^3/_4$ tells us that the chances are 3 out of 4 that our statement is correct. As we increase the sample size n, we increase the chances that our statement is correct. For example, if our sample size had been $n = 473$, then $C = 0.95$ and the chances are 95 in 100 that we are correct in making the statement that at least 99 percent of all process values will be between the 10 and 54 seen in the sample. Similarly, if the sample size had been 740, then the chances of being correct increase to 995 in 1000. If sample size n is decreased sufficiently, the confidence $C = 0.50 = {}^1/_2$ indicates that the chances are one in two of being right, *and* one in two of being *wrong*. Therefore, it is important to select n so as to keep C as large as possible. The cost of acquiring the data will determine the upper limit for n.

Further information on tolerance limits can be found in Wilks (1962) and Duncan (1986).

Wear-Out and Life Tests

A special case of coverage occurs if our interest is in a wear-out phenomenon or a life test. For example, suppose we put a number of incandescent light bulbs on test; our interest is in the length of time to failure. Clearly we do not want to wait until all specimens fail to draw a conclusion; it

Table 17.3.2

P, %	Confidence, C			
	0.995	0.99	0.95	0.75
99.9	7427	6636	4742	2692
99.5	1483	1325	947	538
99	740	661	473	269
98	368	330	235	134
97	245	219	156	89
96	183	163	117	67
95	146	130	93	53
90	71	64	46	26
80	34	31	22	13
75	27	24	18	10

NOTE: Sample size r required to have a confidence C that at least P percent of all future values will be included between the smallest and largest values in a random sample.

Table 17.3.3

	Confidence, C			
Q, %	0.995	0.99	0.95	0.75
0.1	5296	4603	2995	1386
1	528	459	299	138
2	263	228	149	69
3	174	152	99	46
4	130	113	74	34
5	104	90	59	28
10	51	44	29	14
15	33	29	19	9
20	24	21	14	7
25	19	17	11	5

NOTE: Sample size r required to have a confidence C that fewer than Q percent of future units will fail in a time shorter than the shortest life in the sample. For a more extensive table of values, see Owen (1962).

Table 17.3.4

i	X	Y	i	X	Y
1	10.45	4.1	18	9.65	3.8
2	13.81	2.7	19	7.44	5.4
3	12.22	1.6	20	10.70	7.6
4	9.05	4.3	21	13.38	6.0
5	17.86	2.6	22	13.00	10.4
6	14.54	0.1	23	13.90	10.7
7	19.99	3.7	24	11.94	9.4
8	8.73	3.5	25	14.11	10.7
9	4.66	5.3	26	0.93	12.9
10	13.88	3.9	27	3.18	12.5
11	5.10	4.4	28	13.13	6.5
12	3.98	4.1	29	13.45	11.7
13	8.12	6.3	30	12.70	9.6
14	12.26	6.6	31	15.95	8.5
15	10.30	6.5	32	7.30	16.6
16	5.40	11.9	33	7.78	8.8
17	10.39	5.8			

NOTE: Data are on paper samples. X is proportional to reciprocal of light transmission. Y is proportional to tensile strength.

might take an inordinate length of time for the last one to fail. From a practical point of view we would probably be interested in those that fail first anyway. If the sample size is properly chosen, there will be important information as soon as we obtain the first failure.

In a random sample of size n let the failure times be $T_1 \geq T_2 \geq \ldots \geq T_n$. The value T_1 is thus the smallest value in a random sample of size n. Now let q be the percentage of future units that can be expected to fail in a time less than T_1, the smallest value in the sample. As before we can make a probability statement about q. Let Q be an upper bound to q. Then we can compute: Probability $(q \leq Q) = C$.

For example, suppose that we put a random sample of 299 items on test and the first one fails in time $T_1 = 151$ h. From Table 17.3.3 we see that 299 is an entry for $Q = 1$ and $C = 0.95$. Thus we can conclude that not more than 1 percent of future units should fail in a time less than 151 h. The confidence in the statement is 95 chances in 100 of being correct. Again referring to Table 17.3.3, we see that if $T_1 = 151$ were based on a sample of 528, then the confidence would be increased to 995 chances in 1000. Most importantly, Table 17.3.3 tells us that we need to test a very large sample if we want to have high confidence that only a small percentage of future units will fail in a time less than the smallest observed. The theory behind Table 17.3.3 can be found in Wilks (1962). For a more extensive table of values see Owen (1962). If $Q' = Q/100$, use

$$r = [\log (1 - C)]/[\log (1 - Q')]$$

CORRELATION AND ASSOCIATION

One of the most common problems in the analysis of engineering data is to determine if a correlation or an association exists between two variables X and Y, where the data occur in pairs (X_i, Y_i). The "corner test of association" developed by Olmstead and Tukey (1947) is a quick and simple test to assist in making this determination.

Corner Test

Conditions for Use Each pair (X_i, Y_i) should have been obtained independently; there are no other practical assumptions for its use. Of course, the user should consider the physical process generating the data when interpreting any correlation or association that is determined to exist.

Procedure

1. Make a scatter plot on graph paper of the data pairs (X_i, Y_i), with the usual convention that X is the horizontal scale and Y the vertical.

2. Determine the median X_m of the X_i values. Determine the median Y_m of the Y_i values.

The median splits the data into two parts so that there is an equal number of values above and below the median. Let N denote the total number of points. If N is odd, then N can be written as $2k + 1$ and the median is the $(k + 1)$st value as the values are ordered from the smallest to the

largest. If N is even, then N can be written as $2k$. Then the median is taken to be midway between the kth and $(k + 1)$st values.

3. Draw a vertical line through X_m.

4. Draw a horizontal line through Y_m.

5. The lines in (3) and (4) divide the graph into four quadrants. Label the upper right and lower left as plus. Label the upper left and lower right as minus.

6. Begin at the right side of the plot. Count the values, in order of decreasing X, until forced to cross the horizontal median Y_m. Give the count a plus sign if the values are in a plus quadrant, a minus sign if they are in a minus quadrant.

7. Repeat the procedure in (6), moving down from above until you have to cross the vertical median, moving from left to right until you have to cross the horizontal median, and moving up from below until you have to cross the vertical median.

8. Compute the algebraic sum of the four counts obtained in (6) and (7). Denote the sum by T.

Test If $|T| \geq 11$, then there is evidence of correlation between X and Y; $|T|$ is the value of T ignoring the sign.

EXAMPLE. Table 17.3.4 gives 33 pairs of values obtained from samples of a paper product. The X coordinate is proportional to the reciprocal of light transmitted by the sample. The Y coordinate is proportional to tensile strength.

1. The 33 pairs of values are plotted in Fig. 17.3.1.

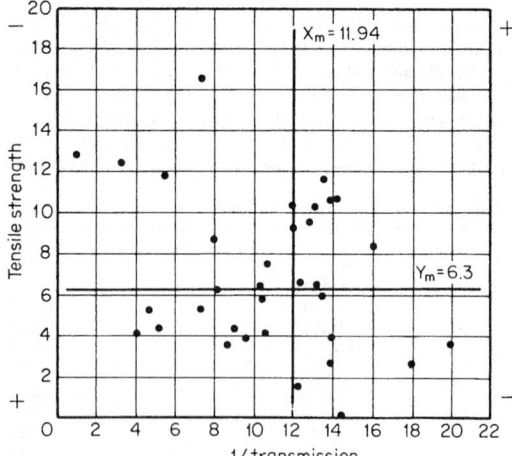

Fig. 17.3.1 Plot of example data used in conjunction with the Corner Test.

2. The median of X values is $X_m = 11.94$. The median of Y values is $Y_m = 6.3$.

3 to 5. The medians are shown in Fig. 17.3.1, and the quadrants are labeled.

6 to 7. The counts are as follows:

Right to left:	-2	(points at 19.99 and 17.86 on X)
Top to bottom:	-4	(points at 16.6, 12.9, 12.5, 11.7 on Y)
Left to right:	-2	(points at 0.93 and 3.18 on X)
Bottom to top:	-4	(points at 0.1, 1.6, 2.6, 2.7 on Y)

8. The algebraic sum of the counts is -12. Hence $T = -12$. And since $|T| \geq 11$, one can conclude that there is evidence of correlation or association between the variables X and Y.

COMPARISON OF METHODS OR PROCESSES

A common problem in engineering investigations is that of using experimental or observational data to assess the performance of two processes, two treatment methods, or the like. Often one process or treatment is a standard or the one in current use. The other is an alternative that is a candidate to replace the standard. Sometimes it is cheaper, and one hopes to see no performance difference. Sometimes it is supposed to offer superior performance, and one hopes to see a measurable difference in the variable of interest. In either case we know that the variable of interest will have a distribution of values; and if the two processes are to be measurably different the distribution of values should not overlap too much. For an objective assessment we need to have some way to calibrate the overlap. A quick and easy-to-use test for this purpose is the **outside count test** developed by Tukey (1959).

Two Methods—Outside Count Test

Conditions for Use   Given two groups of measurements taken under conditions 1 and 2 (treatments, methods, etc.), we identify the direction of difference by insisting that the two groups have minimum overlap. Use 1 to denote the group with the smaller number of measurements and let N_1 be the number of measurements for that group. Let N_2 be the number of measurements for the other group. The number of observations for each group should be about the same.

The conditions to be satisfied are:

$$4 \leq N_1 \leq N_2 \leq 30$$
$$N_2 \leq (4N_1/3) + 3$$

Procedure

1. Count the number of values in the one group *exceeding* all values in the other.

2. Count the number of values in the other group *falling below* all those in the other.

3. Sum the two counts in (1) and (2). (It is required that neither count be zero. One group must have the largest value and the other the smallest.)

Test   If the sum of the two counts in (3) is 7 or larger, there is sufficient evidence to conclude that the two methods are measurably different.

EXAMPLE.   The following data represent the results of a trial of two methods for increasing the wear resistance of a grinding wheel. The data are proportional to wear:

Method 1: 13.06**, 9.52, 9.98, 8.83, 12.78, 9.00, 11.56, 8.10*, 12.21.

Method 2: 8.44, 9.64, 9.94, 7.30, 8.74, 6.30*, 10.78**, 7.24, 9.30, 6.66.

The smallest value for each method is marked with an asterisk; the largest value is marked with two asterisks.

Count 1: The largest value is 13.06 for method 1. The values 13.06, 12.78, 12.21, and 11.56 for method 1 exceed the largest value for method 2, viz., 10.78. Hence the count is 4.

Count 2: The smallest value for method 1 is 8.10. For method 2 the values 7.30, 6.30, 7.24, and 6.66 are less than 8.10. Hence the count is 4.

Count 3: The total count is $4 + 4 = 8 > 7$.

The data support the conclusion that method 2 produces measurably less wear than method 1.

Several Methods

The problem outlined in the preceding section can be generalized so that one can make a comparison of several processes, treatments,

Table 17.3.5

	Supplier			
Sample no.	1	2	3	4
1	45.37	30.05	41.30	46.21
2	21.68	36.04	31.09	36.01
3	43.91	18.04	24.31	46.28
4	47.76*	32.91	15.64	21.80
5	23.81	41.67	54.85*	28.57
6	19.90	37.40	32.96	48.45
7	44.68	46.67*	45.48	33.49
8	11.81	27.93	45.14	53.07*
9	35.42	45.20	45.49	35.65
10	39.85	29.54	52.82	14.95

NOTE: The data are proportional to the time to failure of a standard cutting tool. Asterisks denote largest value in each group.

methods, or the like. Again, if there are differences among the methods, the values that we see should not overlap too much. We give you two easy-to-use tests. The first is for the situation where there is an equal amount of data for each method. For the second, the amount of data may differ. Each method will be demonstrated using the data in Table 17.3.5.

Several Methods—Overlap Test

Conditions for Use   Independent data should be obtained for each of the k methods. *The number of values n should be the same* for each method.

Procedure

1. For each of the k methods, determine the *largest* value. Label it with an asterisk.

2. Scan the largest values. Label the group with the *largest largest* value as BIG. Label the group with the *smallest* largest value as SMALL, and its largest value as S.

3. In the group labeled BIG *count* the number of values that are larger than S, the largest value in SMALL. Denote this count by C.

4. Enter Table 17.3.6 for n values of k groups. If C exceeds the tabled value, then the data support a conclusion that the methods are different. Otherwise, they do not.

EXAMPLE.   1. In Table 17.3.5 the largest value for each of the four groups is marked with an asterisk.

2. Group 3 is BIG. Group 2 is SMALL; the largest value in Group 2 is $S = 46.67$.

3. The number of values in Group 3 larger than 46.67 is 2 (52.82, 54.85).

4. Enter Table 17.3.6 with $n = 10$ and $k = 4$. The entry is 5 which is greater than 2. Hence, the data do not support a conclusion that the time to failure for the cutting tools of the four suppliers is measurably different.

Several Methods—Rank Test

Conditions for Use   Independent data should be obtained for each of the methods. The amount of data for each method may be different.

Procedure

1. Let n_i be the number of values in Group i.

2. Let $N = \sum_1^k n_i$ be the total number of values.

3. Rank each value from 1 to N beginning with the smallest. (If there are ties among t values, divide the successive ranks equally among them.)

4. Compute r_i, the sum of the ranks for the ith group. [*Note:* For a check $\sum_1^k r_i = N(N + 1)/2$.]

Table 17.3.6   95% Point for *k*-Sample Problems

				k				
n	3	4	5	6	7	8	9	10
5	4	4	4	4	4	4	4	4
6	4	4	4	5	5	5	5	5
7	4	5	5	5	5	5	5	5
8	4	5	5	5	5	5	5	5
9	5	5	5	5	5	5	6	6
10	5	5	5	5	6	6	6	6
12	5	5	5	6	6	6	6	6
14	5	5	6	6	6	6	6	6
16	5	5	6	6	6	6	6	6
18	5	6	6	6	6	6	6	7
20	5	6	6	6	6	6	7	7
25	5	6	6	6	6	7	7	7
30	5	6	6	6	7	7	7	7
40	5	6	6	7	7	7	7	7
>40	6	6	7	7	7	7	8	8

NOTE: k is the number of groups; n is the number of values per group. For other n, k, and percentage points see Conover (1968).

Test

1. Compute

$$T = [12/N(N + 1)]\left[\sum_1^k (r_i^2/n_i)\right] - 3(N + 1)$$

2. Go to Table 17.3.7; find the entry under $k - 1$.

If T exceeds the entry, then the data support the conclusion that the groups are different. Otherwise, they do not.

Table 17.3.7   Chi-Square Distribution

k	w	k	w
1	3.841	16	26.30
2	5.991	17	27.59
3	7.815	18	28.87
4	9.488	19	30.14
5	11.07	20	31.41
6	12.59	22	33.92
7	14.07	24	36.42
8	15.51	26	38.89
9	16.92	28	41.34
10	18.31	30	43.77
11	19.68	40	55.76
12	21.03	50	67.50
13	22.36	60	79.08
14	23.68	70	90.53
15	25.00	80	101.9

NOTE: Entries are $P(W > w) = p = 0.05$. For other values of k and p, see Pearson and Hartley (1962).

EXAMPLE. We again use the data shown in Table 17.3.5. In Table 17.3.8 the numerical values representing times to failure have been replaced by their ranks. To facilitate such ranking it is convenient to order the values in each group from smallest to largest. Then all values are ranked from smallest to largest. In Table 17.3.8 the values have been reordered this way. The ranks are in parentheses.

1. The number of values in each group is 10. Hence $n_i = 10$ for each value of i.
2. The total number of values $N = 40$.
3 and 4. The sum of the ranks r_i is shown for each group. [Note that $\Sigma r_i = 820 = (40)(41)/2$.]

$$
\begin{aligned}
T &= [12 /N(N + 1)][\Sigma(r_i^2/10)] - 3(N + 1) \\
&= [12/(40)(41)][170182/10] - 3(41) \\
&= 124.523 - 123. = 1.523
\end{aligned}
$$

Now go to Table 17.3.7 and obtain the entry under $k = 4 - 1 = 3$. The entry is 7.815, which is larger than 1.523. Hence, the data do not support a conclusion that the time to failure for the cutting tools for the four suppliers is measurably different.

Table 17.3.8

	Supplier*		
1	2	3	4
11.81 (1)	18.04 (4)	15.64 (3)	14.95 (2)
19.90 (5)	27.93 (10)	24.31 (9)	21.80 (7)
21.68 (6)	29.54 (12)	21.09 (14)	28.57 (11)
23.81 (8)	30.05 (13)	32.96 (16)	33.49 (17)
35.42 (18)	32.91 (15)	41.30 (24)	35.65 (19)
39.85 (23)	36.04 (21)	45.14 (28)	36.01 (20)
43.91 (26)	37.40 (22)	45.48 (31)	46.21 (33)
44.68 (27)	41.67 (25)	45.49 (32)	46.28 (34)
45.37 (30)	45.20 (29)	52.82 (38)	48.45 (37)
47.76 (36)	46.67 (35)	54.85 (40)	53.07 (39)
180	186	235	219 r_i
32400	34596	55225	47961 r_i^2

* These are the data of Table 17.3.5 with the values for each supplier listed from smallest to largest. The values in parentheses are the ranks of the time to failure values from smallest to largest.

GO/NO-GO DATA

Quite often the data that we encounter will be **attribute** or **go/no-go data**; that is, we will not have quantitative measurements but only a characterization as to whether an item does or does not have some attribute. For example, if a manufactured part has a specification that it should not be shorter than 2 in, we might construct a template; and if a part is to meet the specification, it should not fit into the template. After inspecting a series of units with the template our data would consist of a tabulation of "gos" and "no-gos"—those that did not meet the specification and those that did.

If the items that are checked for an attribute are obtained by random sampling, the resulting data will follow what is known as the **binomial distribution.** Its standard form is as follows:

p is the long-term fraction of failures
$q = 1 - p$ is the long-term fraction of successes
n is the size of the random sample.

Then the probability that our sample gives x failures and $n - x$ successes is

$$f(x;n,p) = \binom{n}{x}p^x q^{n-x}; x = 0,1,2, \ldots , n$$

where

$$\binom{n}{x} = n!/x!(n - x)!$$

From $f(x; n, p)$, for a given n and p, we can calculate the probability of x failures in a sample size n. Similarly, by summing terms for different values of x, we can calculate the probability of having more than w failures or fewer than r failures, etc. Here we are not going to try to be so precise; rather we are going to try to show the general picture of the relationship between x, p, and n by the use of examples and the graph in Fig. 17.3.2.

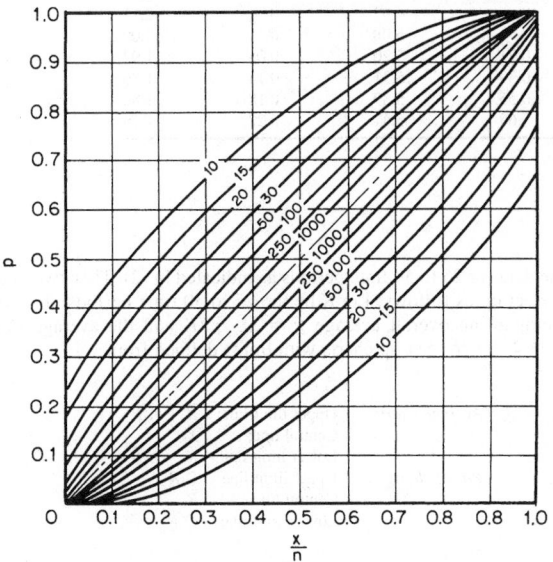

Fig. 17.3.2 Binomial distribution, 95 percent confidence bands. *(Reproduced with the permission of the Biometrika Trust from C. J. Clopper and E. S. Pearson, "The Use of Confidence or Fiducial Limits Illustrated in the Case of the Binomial,"* Biometrika, 26 (1934), p. 410.)

Estimating the Failure Rate

In a manufacturing process a general index of quality is the fraction of items which fail to pass a certain test. Suppose that we take a random sample of size $n = 100$ from the process and observe $x_0 = 10$ failures. Clearly we have met the conditions for a binomial distribution and an estimate of p, the long-term failure rate is $\hat{p} = x_0/n = 10/100 = 0.1$. However, we would also like to know the accuracy of the estimate. In other words, if we operate the process for a long time under these conditions and obtain a large sample, what might be the value of p? One simple way to assess the estimate of $\hat{p}$ is to find values p_1 and p_2 ($p_1 < p_2$) such that the following probabilities are satisfied for a fixed value of α:

$$\Pr[x \geq x_0 | p_1] = \alpha/2$$
$$\Pr[x \leq x_0 | p_2] = \alpha/2$$

These values are the solutions for p of the two equations.

$$\sum_{x=x_0}^{n} \binom{n}{x} p_1^x (1 - p_1)^{n-x} = \alpha/2$$

$$\sum_{x=0}^{x_0} \binom{n}{x} p_2^x (1 - p_2)^{n-x} = \alpha/2$$

General solutions for these equations for $\alpha = 0.05$ are shown in Fig. 17.3.2. If we go to Fig. 17.3.2 with $x/n = 0.1$ and read where the lines for $n = 100$ intersect the ordinate or p scale, we see that $p_1 = 0.07$ and $p_2 = 0.18$. We can then state that we have reasonable confidence (the probability is 0.95) that the long-term failure rate for the process is between 0.07 and 0.18.

Estimating the Sample Size

It should be evident that Fig. 17.3.2 can also be used to determine how large a sample is needed to estimate a proportion or a percentage with a specific accuracy or tolerance. Suppose that the proportion of interest is assumed to be around $p = 0.20$. Now enter Fig. 17.3.2 with $x/n = 0.20$. From the figure we see that if we take a sample of size 50, our estimate will have a range of about ± 0.10 (actually -0.10, $+0.13$). On the other hand, if the sample size is 250, the estimate will have a range of about ± 0.06.

Often one wants to compare the performance of two processes. As above, suppose that the rate p for our process is 0.20. We have a modification that we want to test; however, to be economical the modification has to bring the rate p down to 0.15 or less. If the modification is going to be assessed on the p's for the standard and the modification, then we do not want the uncertainty in their estimates to overlap and the uncertainty should be less than half of 0.05 where $0.05 = (0.20 - 0.15)$. Figure 17.3.2 shows that we would have to use a sample size of at least 1000. This demonstrates that attribute sampling is effective only when the items and their characterization are not expensive. Otherwise, it is best to go to measurements where smaller sample sizes can be used to assess differences.

A more detailed exposition of the binomial distribution and its uses can be found in Brownlee (1970).

Statistical Software Packages

There are numerous statistical software packages available that may be used to determine sample sizes, design experiments, fit curves to data, test for goodness of fit, and perform basic statistical calculations. The packages include both standardized and customized menus, features for importing and exporting data to and from external files, and well-presented analysis summaries and reports. For example, a widely used statistical package called JMP was developed and is periodically upgraded by the SAS Institute, Inc.

CONTROL CHARTS

When an industrial process is under control it is in a state of "statistical equilibrium." By equilibrium we mean that we can characterize its output by a fixed average μ and a fixed standard deviation σ. The variation in output is what one would expect to see from that μ and σ, as bounded by the values given in Tchebysheff's theorem, let us say. However, if control is lost, we tend to get a greater spread in values. In effect, the average μ, or the standard deviation σ is changing because of some cause. The causes of lack of control are manifold—it can be a change in raw materials, tool wear, instrumentaton failure, operator error, etc. The important thing is that one wants to be able to detect when this lack of control occurs and take the appropriate steps to make corrections.

One of the most frequently used statistical tools for analyzing the state of an industrial process is the control chart. The two most commonly used charts are those for the **average** and the **range.** The control chart procedure consists of these steps:

1. Choose a characteristic X which will be used to describe the product coming from the process.

2. At time t_i, take a small number of observations n on the process. The number n should be small enough so that it is reasonable to assume that conditions will not change during the course of obtaining the observstions.

3. For each set of n observations, compute the average $\bar{x}_i$ and the range r_i as defined in the section "Characterizing Observational Data." There are two different control situations of interest.

4a. *Control standards given.* Suppose that from past operation of the process or from the need to meet certain specifications, a goal average μ^* and a goal standard deviation σ^* are specified. Then we set up two charts as follows:

Average chart:	Upper limit line:	$\mu^* + A\sigma^*$
	Central line:	μ^*
	Lower limit line:	$\mu^* - A\sigma^*$
Range chart:	Upper limit line:	$D_2\sigma^*$
	Central line:	$d\sigma^*$
	Lower limit line:	$D_1\sigma^*$

Table 17.3.9   Factors for Control-Chart Limits

Sample size n	For averages		For ranges				
	A	A_2	d	D_1	D_2	D_3	D_4
2	2.12	1.88	1.128	0	3.69	0	3.27
3	1.73	1.02	1.693	0	4.36	0	2.57
4	1.50	0.73	2.059	0	4.70	0	2.28
5	1.34	0.58	2.326	0	4.92	0	2.11
6	1.22	0.48	2.534	0	5.08	0	2.00
7	1.13	0.42	2.704	0.21	5.20	0.08	1.92
8	1.06	0.37	2.847	0.39	5.31	0.14	1.86
9	1.00	0.34	2.970	0.55	5.39	0.18	1.82
10	0.95	0.31	3.078	0.69	5.47	0.22	1.78

The values of A, d, D_1, and D_2 depend upon n and can be found in Table 17.3.9.

Plot the values of $\bar{x}_i$ and r_i obtained in (3) on the two charts as shown in Fig. 17.3.3. Whenever a value falls outside the limit lines, there is an indication of lack of control, and one is justified in seeking the causes for a change.

4b. *Control, no standards given.* Often one has no prior information about the process μ and σ, and one wants to determine if the process behaves as though it is in statistical equilibrium, and if not, take actions to get it there. In this case one has to determine the central lines for the charts from process data. To do this one first accumulates the data for 25 to 50 time periods as indicated in (2). Then two charts are set up as follows: Let K be the 25 to 50 time periods observed. Compute an overall average $\bar{X} = \Sigma_{i=1}^k \bar{x}_i/K$ and the average range $\bar{R} = \Sigma_{i=1}^k r_i/K$. Set up charts with limits defined from:

Average chart:	Upper limit line:	$\bar{X} + A_2\bar{R}$
	Central line:	$\bar{X}$
	Lower limit line:	$\bar{X} - A_2\bar{R}$
Range chart:	Upper limit line:	$D_4\bar{R}$
	Central line:	$\bar{R}$
	Lower limit line:	$D_3\bar{R}$

The values of A_2, D_3, and D_4 depend upon n and can be found in Table 17.3.9. The individual $\bar{x}_i$ and r_i are plotted on the charts, and again a value outside the limits is an indication of lack of control and is justification for seeking the cause for a change.

Process Capability Indices

If the process is in statistical control, an estimate for the process standard deviation can be obtained by using

$$\hat{\sigma} = \bar{R}/d$$

In turn, $\hat{\sigma}$ is used to calculate the **process capability indices** C_p and C_{pk}. These two indices compare the actual spread of the data with the desired range (usually as specified by a customer). C_p is used when the actual process average is equal to the goal average. C_{pk} is used when the actual process average is not equal to the goal average. The desired range is called the **specification tolerance** (ST) and is equal to the upper specification limit minus the lower specification limit, namely, $2A\sigma^*$.

C_p is given by the following equation:

$$C_p = \text{ST}/6\hat{\sigma} = 2A\sigma^*/6\hat{\sigma}$$

If $C_p \geq 1$, the process is considered to be capable, which means that most or all of the data stayed within the desired range. If $C_p < 1$, the process is considered to be not capable and requires adjustment. Ideally, one should control the process variability so that $C_p \geq 2$.

The other index, C_{pk}, is given by

$$C_{pk} = C_p(1 - k)$$

where

$$k = 2(\mu^* - \bar{X})/\text{ST} = (\mu^* - \bar{X})/A\sigma^*$$

and

$$\bar{X} = \Sigma\bar{x}_i/K$$

The $(1 - k)$ factor modifies C_p so as to allow the actual average $\bar{X}$ to be different from the goal average μ^*. Ideally, one should control the process so that $C_{pk} \geq 1.5$. In process capability analysis, the indices should be calculated on a frequent basis, but the trends should be examined only monthly or quarterly in order to be meaningful.

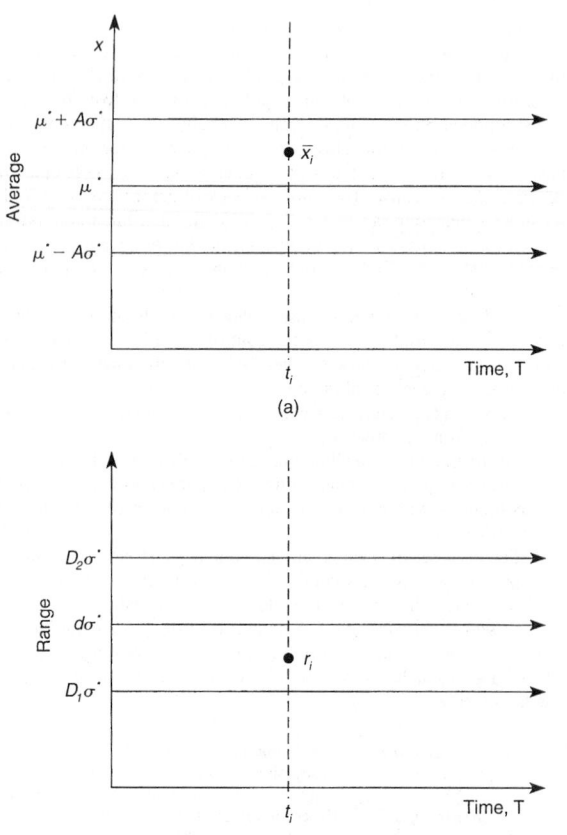

Fig. 17.3.3 Control chart. (*a*) Average chart; (*b*) range chart.

Charts for Go/No-Go Data

The control chart concept can also be used for attribute or go/no-go data. The procedures are, in general, the same as outlined for averages and ranges. Briefly, they are as follows:

1. Select a sample of size n from the process; for best results n should be in the range of 50 to 100.

2. Let x_i denote the number of defective units in the sample of size n at time t_i; then $\hat{p}_i = x_i/n$ is an estimate of the process fraction defective.

3. Set control limits for a standard fraction defective $p*$ at

$$p* \pm 3[p*(1 - p*)/n]^{1/2}$$

If no standard is given, then take $K = 25$ to 50 samples of size n to get a good estimate of the fraction-defective p. Define $\bar{p} = \Sigma_{i=1}^{K}\hat{p}_i/K$. In this case set control limits at

$$\bar{p} \pm 3[\bar{p}(1 - \bar{p})/n]^{1/2}$$

4. Interpret a $\hat{p}_i$ outside the limits as an indication of a change worthy of investigation.

Further information on control charts can be found in Duncan (1986).

17.4   METHODS ENGINEERING
by Vincent M. Altamuro

REFERENCES: ASME Standard Industrial Engineering Terminology. Barnes, "Motion and Time Study," Wiley. Krick, "Methods Engineering," Wiley. Maynard, Stegemerten, and Schwab, "Methods-Time-Measurement," McGraw-Hill.

SCOPE OF METHODS ENGINEERING

Methods engineering is concerned with the selection, development, and documentation of the methods by which work is to be done. It includes the analysis of input and output conditions, assisting in the choice of the processes to be used, operations and work flow analyses, workplace design, assisting in tool and equipment selection and specifications, ergonomic and human factors considerations, workplace layout, motion analysis and standardization, and the establishment of work time standards. A primary concern of methods engineering is the integration of humans and equipment in the work processes and facilities.

PROCESS ANALYSIS

Process analysis is that step in the conversion of raw materials to a finished product at which decisions are made regarding what methods, machines, tools, inspections and routings are best. In many cases, the product's specifications can be altered slightly, without diminishing its function or quality level, so as to allow processing by a preferred method. For this reason, it is desirable to have the product's designer and the process engineer work together before specifications are finalized.

WORKPLACE DESIGN

Material usually flows through a facility, stopping briefly at stations where additional work is done on it to bring it closer to a finished product. These workstations, or workplaces, must be designed to permit performance of the required operations, to contain all the tooling and equipment needed to fit the capabilities and limitations of the people working at them, to be safe and to interface smoothly with neighboring workplaces.

Human engineering and ergonomic factors must be considered so that all work, tools, and machine activation devices are not only within the comfortable reach of the operator but are designed for safe and efficient operation. A workplace chart (Fig. 17.4.1) which analyzes the required actions of both hands is an aid in workplace design.

Fig. 17.4.1   Workplace layout chart.

METHODS DESIGN

Methods design is the analysis of the various ways a task can be done so as to establish the one best way. It includes *motion analysis*—the study of the actions the operator can use and the advantages and/or disadvantages of each variation—and *standardization of procedure*—the selection and recording of the selected and authorized work methods.

While "time and motion study" is the more commonly used term, it is more correct to use "motion and time study," as the motion study to establish the standard procedure must be done prior to the establishment of a standard time to perform that work.

According to ASME Standard Industrial Engineering Terminology, motion study is defined as

. . . the analysis of the manual and the eye movements occurring in an operation or work cycle for the purpose of eliminating wasted movements and establishing a better sequence and coordination of movements.

In the same publication, time study is defined as

. . . the procedure by which the actual elapsed time for performing an operation or subdivisions or elements thereof is determined by the use of a suitable timing device and recorded. The procedure usually but not always includes the adjustment of the actual time as the result of performance rating to derive the time which should be required to perform the task by a workman working at a standard pace and following a standard method under standard conditions.

Attempts have been made to separate the two functions and to assign each to a specialist. Although motion study deals with method and time study deals with time, the two are nearly inseparable in practical application work. The method determines the time required, and the time determines which of two or more methods is the best. It has, therefore, been found best to have both functions handled by the same individual.

ELEMENTS OF MOTION AND TIME STUDY

Figure 17.4.2 presents graphically the steps which should be taken to make a good motion and time study and shows their relation to each other and the order in which they must be performed.

METHOD DEVELOPMENT

The first steps are concerned with the development of the method. Starting with the drawing of the product, the operations which must be performed are determined and tools and equipment are specified. In large companies, this is usually done by a specialist called a process engineer. In smaller companies, processing is commonly done by the time study specialist.

Next, the detailed method by which each operation should be performed is developed. The procedures used for this are known as operation analysis and motion study.

OPERATION ANALYSIS

Operation analysis is the procedure employed to study all major factors which affect a given operation. It is used for the purpose of uncovering possibilities of improving the method. The study is made by reviewing the operation with an open mind and asking either of oneself or others questions which are likely to lead to methods-improving ideas. If this is done systematically, so that the possibility of overlooking factors which should be considered is minimized, worthwhile improvements are almost certain to result.

The 10 major factors explored during operation analysis, together with typical questions which should be asked about each factor, are as follows:

1. Purpose of operation
   a. Is the result accomplished by the operation necessary?
   b. Can the purpose of the operation be accomplished better in any other way?

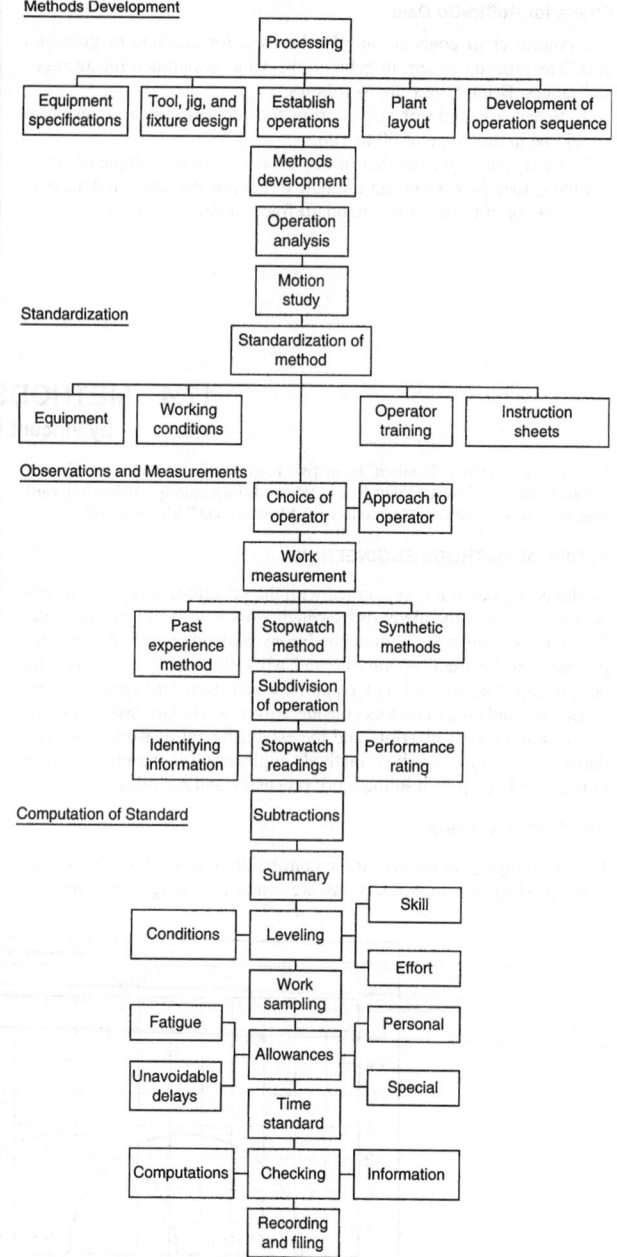

Fig. 17.4.2   Graphic analysis of the elements of motion and time study.

2. Design of part
   a. Can motions be eliminated by design changes which will not affect the functioning and other desirable characteristics of the product?
   b. Is the design satisfactory for automated assembly?
3. Complete survey of all operations performed on part
   a. Can the operation being analyzed be eliminated by changing the procedure or the sequence of operations?
   b. Can it be combined with another operation?
4. Inspection requirements
   a. Are tolerance, allowance, finish, and other requirements necessary?

b. Will changing the requirements of a previous operation make this operation easier to perform?

5. Material
   a. Is the material furnished in a suitable condition for use?
   b. Is material utilized to best advantage during processing?
6. Material handling
   a. Where should incoming and outgoing material be located with respect to the work station?
   b. Can a progressive assembly line be set up?
7. Workplace layout, setup, and tool equipment
   a. Does the workplace layout conform to the principles of motion economy?
   b. Can the work be held in the machine by other means to better advantage?
8. Common possibilities for job improvement
   a. Can "drop delivery" be used?
   b. Can foot-operated mechanisms be used to free the hand for other work?
9. Working conditions
   a. Has safety received due consideration?
   b. Are new workers properly introduced to their surroundings, and are sufficient instructions given them?
10. Method
   a. Is the repetitiveness of the job sufficient to justify more detailed motion study?
   b. Should full automation be considered?

When the method has been developed, conditions are standardized, and the operators are trained to follow the approved method.

At this time, not before, the job is ready for time study. Suitable operators are selected, the purposes of the study are carefully explained to them, and the time study observations are made. During the study, time study specialists rate the performance being given by operators either by judging the skill and effort they are exhibiting or by assessing the speed with which motions are made as compared with what they consider to be a normal working pace. The final step is to compute the standard.

PRINCIPLES OF MOTION STUDY

Operation analysis is a primary analysis which eliminates inefficiencies. Motion study is a secondary analysis which refines the method still further. Motion study may and often does suggest further improvements in the factors considered during operation analyses, such as tools, material handling, design, and workplace layouts. In addition, it studies the human factors as well as the mechanical and sets up operations in conformance with the limitations, both physical and psychological, of those who must perform them.

The technique of motion study rests on the concept originally advanced by Frank B. and Lillian M. Gilbreth that all work is performed by using a relatively few basic operations in varying combinations and sequences. These Gilbreth Basic Elements have also been called "therbligs" and "basic divisions of accomplishment."

The basic elements together with their symbols (for definitions see ASME Industrial Engineering Terminology), grouped in accordance with their effect on accomplishment, are as follows:

| Group 1 | |
Accomplishes	
Reach	R
Move	M
Grasp	G
Position	P
Disengage	D
Release	RL
Examine	E
Do	DO

| Group 2 | |
Retards accomplishment	
Change direction	CD
Preposition	PP
Search	S
Select	SE
Plan	PL
Balancing delay	BD

| Group 3 | |
Does not accomplish	
Hold	H
Avoidable delay	AD
Unavoidable delay	UD
Rest to overcome fatigue	F

Group 1 is the useful group of basic elements or the ones that accomplish work. They do not necessarily accomplish it in the most effective way, however, and a study of these elements will often uncover possibilities for improvement.

Group 2 contains the basic elements that tend to retard accomplishment when present. In most cases, they do this by slowing down the group 1 basic elements. They should be eliminated wherever possible.

Group 3 is the nonaccomplishment group. The greatest improvements in method usually come from the elimination of the group 3 basic elements from the cycle. This is done by rearranging the motion sequence, by providing mechanical holding fixtures, and by improving the workplace layout.

An operation may be analyzed into its basic elements either by observation or by making a micromotion study of a motion picture of the operation.

Methods improvement may be made on any operation by eliminating insofar as possible the group 2 and group 3 basic elements and by arranging the workplace so that the group 1 basic elements are performed in the shortest reasonable time. In doing this, certain laws of motion economy are followed. The following, derived from the laws originally stated by the Gilbreths, are the most important.

1. When both hands begin and complete their motions simultaneously and are not idle during rest periods, maximum performance is approached.

2. When motions of the arms are made simultaneously in opposite directions over symmetrical paths, rhythm and automaticity develop most naturally.

3. The motion sequence which employs the fewest basic elements is the best for performing a given task.

4. When motions are confined to the lowest practical classification, maximum performance and minimum fatigue are approached. Motion classifications are: Class 1, finger motions; Class 2, finger and wrist motions; Class 3, finger, wrist, and forearm motions; Class 4, finger, wrist, forearm, and upper-arm motions; Class 5, finger, wrist, forearm, upper-arm, and body motions.

STANDARDIZING THE JOB

When an acceptable method has been devised, equipment, materials, and conditions must be standardized so that the method can always be followed. Information and records describing the standard method must be carefully made and preserved, for experience has shown that, unless this is done, minor variations creep in which may in time cause a major problem. In the case of repetitive work, a job is not standardized until each piece is delivered to operators in the same condition, and it is possible for them to perform their work on each piece by completing a set cycle of motions, doing a definite amount of work with the same equipment under uniform working conditions.

The operator or operators must then be taught to follow the approved method. Operator training is always important if reasonable production is to be obtained, but it is an absolute necessity where methods have been devised by motion study. It is quite apparent that the operators cannot be expected to discover for themselves the method which the time-study specialist developed as the result of hours of concentrated study. They must, therefore, be carefully trained if they are to be expected to reach standard production. In addition, an accurate time study cannot be made until the operator is following the approved method with reasonable proficiency.

WORK MEASUREMENT

Work measurement is the calculation of the amount of time it should take to do a standardized job. It utilizes the concept of a standard time. The standard time to perform a task is the agreed-upon and reproducible calculated time that a hypothetical typical person working at a normal rate of speed should take to do the job using the specified method with the proper tools and materials. It is the normal time determined to be required to complete one prescribed cycle of an operation, including noncyclic tasks, allowances, and unavoidable delays.

Time Study Methods There are several bases upon which time standards may be calculated. They include:

1. *Application of past experience.* The time required to do the operation in the past, either recorded or remembered, may be used as the present standard or as a basis for estimating a standard for a similar operation or the same operation being done under changed conditions.

2. *Direct observation and measurement.* The operation may be observed and its time recorded as it is actually performed and adjustments may be made to allow for the estimated pace rate of the operator and for special allowances. A stop watch or other recording instrument may be used or work sampling may be employed, which makes statistical inferences based upon random observations.

3. *Synthetic techniques.* A time standard for an actual or proposed operation may be constructed from the sum of the times to perform its several components. The times of the components are extracted from standard charts, tables, graphs, and formulas in manuals or in computer databases and totaled to arrive at the overall time for the entire operation. Standard data or predetermined motion times may be used. Methods-time measurement (MTM), basic-motion times (BMT), work factor (WF), and others are some of the systems available.

TIME STUDY OBSERVATIONS

The time study specialist can study any operator he or she wishes so long as that operator is using the accepted method. By applying what is known as the leveling procedure, the time study specialist should arrive at the same final time standard regardless of whether studying the fastest or the slowest worker.

The manner in which the operator is approached at the beginning of the study is important. This is particularly true if the operator is not accustomed to being studied. Time study specialists should be courteous and unassuming and should show a recognition of and a respect for the problems of the operator. They should be frank in their dealings with the operator and should be willing at any time to explain what they are doing and how they do it.

The first step in making time study observations is to subdivide the operation into a number of smaller operations which will be studied and timed separately. These subdivisions are known as elements, or elemental operations. Each element is exactly described in a few well-chosen words, which are recorded on the top of the time study form. Figure 17.4.3 shows how this is done. The beginning and ending points

Fig. 17.4.3 Face of a time study form.

of these elements must be clearly recognizable so that the chances of overlapping watch readings will be minimized.

The timing is done with the aid of a stopwatch, or less frequently, with a special type of "time study machine." There are several types of stopwatches as well as several methods of recording watch readings in common use. The study illustrated by Fig. 17.4.3 was made using a decimal-hour stopwatch that reads directly in ten-thousandths of an hour. The readings were recorded using what is known as the continuous method of recording. In this method, the watch runs continuously from the beginning of the study to the end. Thus every moment of time is accounted for, something that may be important if the correctness of the study is ever questioned. The watch is read at the end of each elemental operation, and the reading is recorded in the "R" column under the proper element description. The elapsed time for each element is later secured by subtracting successive readings. This observation procedure gives results as accurate as any other and more accurate than some.

Occasionally variations from the regular sequence of elemental operations occur. The time study specialist must be prepared to handle such situations when they happen. These variations may be divided into four general classes as follows: (1) elements performed out of order, (2) elements missed by the time study specialist, (3) elements omitted by the operator, (4) foreign elements.

The time study illustrated by Fig. 17.4.3 contains examples of each of these kinds of irregularity. Elements 12 and 1 on lines 12 and 13 were performed out of order. On line 3, the time study specialist missed obtaining the watch readings for elements 9 and 10. On line 6, element 12 was omitted by the operator. Foreign elements A, B, C, and D occurred during regular elements 2, 5, 1, and 7, respectively. A study of these examples will show how the time study specialist handles

variations from the regular sequence of elements which occur during the making of a time study.

A time study to be of value for future use must tell the whole story of a job in such a way that it will be understood by anyone familiar with the time study procedure. This will not be possible unless all identifying and other pertinent information is recorded at the time the study is made. Records should be made to show complete identification of the operator; the part or assembly; the machines, tools, and equipment used; the operation; the department in which the operation was performed; and the conditions existing at the time the study was made. Sketches are generally a desirable part of this description. Figure 17.4.4 shows the information which would be recorded on the reverse side of the time study form illustrated by Fig. 17.4.3.

PERFORMANCE RATING

The objective of a time study is to determine the time which a worker giving average performance will require to do the job under average or normal conditions. It is important to understand that when the time study specialist speaks of average performance, he or she is not referring to the mathematical average of all human beings, or even the average of all persons engaged in a given occupation. Average performance is established by definition and not statistically. It represents the time study specialist's conception of the normal, steady, but unhurried performance which may reasonably be expected from anyone qualified for the work. If sufficient inducement is offered by incentives or otherwise, this performance may be considerably surpassed.

If all operators available for study worked at the average performance level, the task of establishing a standard would be easy. It would

Fig. 17.4.4 Back of a time study form.

Skill			Conditions			Effort		
+0.15	A1	Superskill	+0.06	A	Ideal	+0.13	A1	Excessive
+0.13	A2		+0.04	B	Excellent	+0.12	A2	
+0.11	B1	Excellent	+0.02	C	Good	+0.10	B1	Excellent
+0.08	B2		0.00	D	Average	+0.08	B2	
+0.06	C1	Good	−0.03	E	Fair	+0.05	C1	Good
+0.03	C2		−0.07	F	Poor	+0.02	C2	
0.00	D	Average				0.00	D	Average
−0.05	E1	Fair				−0.04	E1	Fair
−0.10	E2					−0.08	E2	
−0.16	F1	Poor				−0.12	F1	Poor
−0.22	F2					−0.17	F2	

Fig. 17.4.5 Leveling factors for performance rating.

be necessary merely to average the elapsed elemental times determined from time study and add an allowance for fatigue and personal and unavoidable delays. It is seldom, however, that a performance is observed which is rated throughout as average. Therefore, to establish a standard which represents the time which would be taken had an average performance been observed, it is necessary to use some method of adjusting the recorded elemental times when other than average performance is timed.

One of the well-known methods of doing this is the **leveling procedure.** When properly applied it gives excellent results. It must be correctly understood, of course, and the time study specialist who uses it must be thoroughly trained to apply it correctly.

The procedure recognizes that when the correct method is being followed, skill, effort, and working conditions will affect the level at which the operator works. These factors are judged during the making of the time study. Skill is defined as *proficiency at following a given method.* This is not subject to variation at will by the operator but develops with practice over a period of time. Effort is defined as *the will to work.* It is controllable by the operator within the limits imposed by skill. Conditions are those conditions which affect the operator and not those which affect the method.

Definitions have been established for different degrees of skill, effort, and conditions. Numerical factors have been established by extensive research for each degree of skill, effort, and conditions. These are shown by Fig. 17.4.5. The algebraic sum of these numerical values added to 1.0 gives the leveling factor by which all actual elemental times are multiplied to bring them to the average or normal level. The leveling factor represents in effect the amount in percent which actual performance times are above and below the average performance level.

ALLOWANCES FOR FATIGUE AND PERSONAL AND UNAVOIDABLE DELAYS

The leveled elemental time values are net elapsed times adjusted to the average performance level. They do not provide for delays and other legitimate allowances. Something, therefore, must be added to take care of such things as fatigue, and special conditions of the work.

Fatigue allowances vary according to the nature of the work. Flat percentages are determined for each general class of work, such as bench work, machine-tool operation, hard physical labor, and so on. Personal allowances are the same for most classes of work. Unavoidable delay allowances vary with the nature of the work and the conditions under which it is performed. Peculiar conditions surrounding specific jobs sometimes require additional special allowances.

It is apparent, therefore, that the proper allowance factor to use can only be determined by a study of the class of work to which it is to be applied. Allowances are determined either by a series of all-day time studies or by a statistical method known as work sampling, or both. When an allowance factor has once been established, it is then applied to all time studies made on that class of work thereafter.

DEVELOPING THE TIME STANDARD

When time study observations have been completed, a series of calculations are made to develop the time standard. Elapsed times are determined by subtracting successive watch readings. Each subtraction is recorded between the two watch readings that determine its value. Elapsed time is noted in ink to ensure a permanent record and to distinguish it from the watch readings which are usually recorded in pencil. A study of Fig. 17.4.3 will show how subtractions are entered on the time study form and later summarized.

The several elapsed times for each element are next carefully compared and examined for abnormal values. If any are found, they are circled so that they can be distinguished and excluded from the summary.

The remaining elapsed times for each element are added and are averaged by dividing by the number of elapsed time readings. The results are average elapsed times which represent the time taken by the operator during that particular study. These times must be adjusted by multiplying them by a leveling factor to bring them to the average performance level. This factor is determined by the rating of skill, effort, and conditions made during the period of observation.

Each average elapsed time is multiplied by the leveling factor, except when the element is not controlled by the operator. An element that is outside the control of the operator, such as element 7 in Fig. 17.4.3 which is a cut with power feed, should not be leveled, because it is unaffected by the ability of the operator. As long as the proper feed and speed are used, the time for performing this element will be the same whether the worker is an expert or a learner.

If workers were able to work continuously, the leveled time would be the correct value to allow for doing the operation studied, but constant application to the job is neither possible nor desirable. In the course of a day, there are certain to be occasional interruptions and delays, for which due allowance must be made in establishing the final standard. Therefore, each elemental time is increased by an allowance which covers time that will be consumed by personal and unavoidable delays, fatigue, and any special factors that may affect the job.

The numbers and descriptions of the elemental operations together with their allowed time are transcribed on the back of the time study form as shown by Fig. 17.4.4. The number of times an elemental operations occurs on each piece or cycle of the operation is taken into account, and the total time allowed for each element is determined and recorded. The final standard for the operation is the sum of the amounts recorded in the "time-allowed" column. When all computations have been checked and all supporting records have been properly identified and filed, the task of developing the time standard is complete.

TIME FORMULAS AND STANDARD DATA

On repetitive work, time study is a satisfactory tool of work measurement. A single time study may be sufficient to establish a standard which will cover the work of one or more operators for a long period of time.

As quantities become smaller, however, the cost of establishing standards by individual time study increases until at length it becomes prohibitive. In the extreme case, where products are manufactured in quantities of one, it would require at least one time study specialist for each operator if standards were established by detailed time study, and the standards would not be available until after the jobs had been completed.

In order to simplify the task of setting standards on a given class of work and in order to improve the consistency of the standards, standard data are frequently used by time study specialists. A compilation of standard data in its simplest form is merely a list of all the different elements that have occurred during all the time studies made on a given class of work, with representative time values for each element. Every element that differs even slightly from any other element has its own time value.

When a job comes into the shop on which no standard has previously been established, time study specialists analyze the job either mentally or by direct observation and determine the elements required to perform it. They then select time values from the standard data for each element. Their sum gives the standard for the job.

This method, although a decided improvement from a time, cost, and consistency standpoint over individual time study, is capable of further refinement and improvement. On a given class of work, certain elements will be performed—for example, "pick up part"—on every piece produced, while others—such as "secure in steady rest"—will be performed only when a piece has certain characteristics. In some cases, the performance of a certain element will always require the performance of another element, e.g., "start machine" will always require the subsequent performance of the element "stop machine." Then again, the time for performing certain elements—for example, "engage feed"—will be the same regardless of the characteristics of the part being worked upon, while the time for performing certain other elements—like "lay part aside"—will be affected by the size and shape of the part.

Thus it is possible to make certain combinations and groupings which will simplify the task of applying standard data. Time study specialists construct various charts, and tables which they still call standard data, or, in the ultimate refinement, develop time formulas. A time formula is a convenient arrangement of standard data which simplifies their accurate application. Much of the analysis which is necessary when applying standard data is done once and for all at the time the formula is derived. The job characteristics which make the performance of certain elements or groups of elements necessary are determined, and the formula is expressed in terms of these characteristics.

Figure 17.4.3 illustrates a detailed time study made to establish a standard on a simple milling machine operation. The same standard can be derived much more quickly from the following time formula:

$$\text{Curve } A + \text{Table } T = \text{each piece time}$$

Curve A combines the times for the variable elements "pick up part from table," "place in vise," and "lay aside part in totepan" with the times for the constant elements "tighten vise," "start machine," "run table forward 4 in," "engage feed," "stop machine," "release vise," and "brush vise." Table T combines the times for "mill slot" and "return table." The standard time for milling a slot in a brass clamp of any size is computed by determining the variable characteristics of the job from the drawing—in this case, the volume of the clamp and the perimeter of the cut—and adding together the time read from curve A and the time read from Table T.

The amount of time which the use of time formulas will save the time study specialist is readily apparent. It takes a certain amount of time and no little know-how to develop a time formula, but once it is available, the job of establishing accurate standards becomes a simple, fairly routine task. The time required to make and work up a time study will be from 1 to 100 or more hours, depending upon the length of the operation cycle studied. The time required to establish a standard from a time formula will, in the majority of cases, range from 1 to 15 min, depending upon the complexity of the formula and the amount of time required to determine the characteristics of the job. Where all the necessary information may be obtained from the drawing of the part, the standard may generally be computed in less than 5 min.

USES OF TIME STANDARDS

Some of the more common uses of time standards are in connection with
1. Wage incentive plans
2. Plant layout
3. Plant capacity studies
4. Production planning and control
5. Standard costs
6. Budgetary control
7. Cost reduction activities
8. Product design
9. Tool design
10. Top-management controls
11. Equipment selection
12. Bidding for new business
13. Machine loading
14. Effective labor utilization
15. Material-handling studies

Time standards can be established not only for direct labor operations but also for indirect work, such as maintenance and repair, inspection, office and clerical operations, engineering, and management. They can also be set for machinery and equipment, including robots.

17.5 COST OF ELECTRIC POWER
by Andrew M. Donaldson and Robert F. Gambon

REFERENCES: Department of Energy (DOE), "Electric Plant Cost and Power Production Expenses," 1991. Electric Power Research Institute, "Technical Assessment Guide," 1993. Oak Ridge National Laboratory, "CONCEPT-V, A Computer Code for Conceptual Cost Estimates of Steam Electric Power Plants." "Handy-Whitman Index of Public Utility Construction Costs," Whitman, Requardt and Associates. Grant and Ireson, "Principles of Engineering Economy," Ronald Press.

In simplest terms, the cost of electric power is the cost of the initial energy source plus the cost of converting that energy into electricity (generation), plus the cost of delivering that electricity to the consumer (transmission and distribution).

That initial energy source is the potential and kinetic energy of hydroelectric power; the chemical energy of fossil, biomass, and waste power; or the atomic energy of nuclear power. In addition, there are other energy sources for the conversion to electricity. These other sources include energy derived from solar radiation, wind, and ocean waves, but these sources are limited, inefficient, and currently not cost-effective without subsidy. These alternative generation approaches are not addressed herein.

The cost of the initial energy and its conversion to electricity constitute generation and is the major factor in electric power cost and is the primary concern of this section. The largest contributor to electricity cost of fossil fuel plants is delivered fuel followed by plant capital cost amortization, and operations personnel and maintenance expenses. The higher the fuel cost and the lower the plant efficiency, the greater the effect of fuel cost on the final electric price. For natural gas, which is generally the most expensive fossil fuel, a combination of lower plant capital cost and/or higher plant efficiency is necessary for cost competitiveness. Electricity production with natural gas in a 33 percent efficient Rankine cycle steam boiler and turbine makes little sense today because the same fuel can be used in a large-frame combustion turbine combined-cycle plant and approaches 60 percent efficiency. The traditionally lower delivered cost of coal allows it to be used economically in the Rankine cycle. However, for many recent years, low gas prices and restrictive environmental regulations combined with the high efficiency and lower construction cost of combustion turbine combined-cycle plants have almost stopped coal plant construction. Only larger electricity producers conscious of the need for fuel diversity in their generation base have even considered coal in the last 5 to 10 years.

Environmental regulations have severely restricted coal use, but improving emission control equipment and increasing natural gas prices are likely to result in a resurgence of coal plants in the early twenty-first century.

For hydroelectric, geothermal, nuclear, and waste fuel plants, the highest cost contributor to electricity price is paying for the capital investment that is the power plant, followed by operations and maintenance.

Of course, economics have been overshadowed by public opinion and politics when it comes to nuclear and large hydroelectric plants, which have not been constructed in the United States in 20 years.

Generation costs consist of production expenses (fuel plus operating and maintenance expenses) and fixed charges on investment (cost of investment capital, depreciation or amortization, taxes, and insurance).

Interconnected power supply systems must price power to include transmission costs, distribution costs, and commercial expenses. Power price savings are realized by scheduling the installation and operation of a range of plant types to optimize overall generation costs. Generally, efficient plants burning lowest-cost fuels or operating on river water flows that are available the year round are assigned to continuous base-load operation at or near full capacity. Plants bearing low unit-kilowatt investment costs and lower efficiencies typically are installed to provide peaking capacity. Operation of these units during short daily periods of peak load limits their energy output and thus minimizes characteristic cost penalties associated with their poorer station efficiency or their requirement for higher-priced fuels. These higher unit cost per kilowatt but higher energy density **peaking plants** supply peak energy needs but are idle most of the time. This approach is preferable to building high capital cost but more efficient central stations that would sit idle most of the time until load growth catches up to capacity.

Power generating stations also participate in regional **transmission grid** systems supplementing capacity and reserve needs of neighboring systems, and by the daily and seasonal exchange of off-peak, low-incremental-cost energy. Historically, there have been significant regional differences in the price of power. These differences have reflected such factors as availability of hydroelectric energy, the cost of fossil fuels, labor costs, ownership type and investment composition, local tax structure, and the opportunities pooling and coordination provide to exploit the economies of scale. This exchange of capacity and reserves to serve neighboring grid and spread the benefits of economies of scale, combined with regional power cost differences, led to the advent of state-by-state deregulation of the power industry, and proliferation of "nonutility" generators.

The early industrial areas of this country in the northern Midwest and New England areas were increasingly served by aging and decreasingly efficient power plants, often burning oil or requiring long-distance coal deliveries. These fuels have shown volatility on the world market and escalating delivery costs. The flight of heavy power-demanding industry from these areas allowed residential and light industrial growth to be picked up by the aging plants without new, more efficient capacity being added. The population density increased but the same old plants were kept in operation, often past their original life expectancy.

Since the early 1980s regulators and well-intentioned environmentalists have opposed nearly every new power plant. Existing plants, with their higher cost of replacement, were under no replacement pressure, since installing new, efficient, cost-effective power plants was delayed because of this environment. Regulators that refused requests from utilities for price increases, which would have allowed eventual new construction, further exacerbated the situation, preserving inefficiency.

Regional retail power price differences like 12 cents per kilowatt in areas of New York versus 6 cents per kilowatt in Georgia contributed to industry and job flight away from areas with high-priced electricity. As a result, business continues to look for lower electric rates, while politicians search for opportunities to keep jobs and voters in the area. Out of this dilemma, the idea of deregulation, in search of lower-cost electric supplies, was born. Opening some markets produced temporary downward pressure on prices even in regulated areas but also produced chaos

and threatened the reliability of the national electric system that powers the U.S. economy.

The deregulation promises of lower prices were often temporary, as were the companies offering the lower prices. The California deregulation situation, with tremendous electricity price swings and state-assisted bankruptcy of one of California's largest utilities, and the Enron scandal exposed in 2001, have severely reduced the number of nonutility generators and the likelihood of deregulation. However, some states, such as Pennsylvania, have had some success in the deregulation of electricity. The electrical supply, distribution, reliability, and price chaos of the early twentieth century electrical system, which led to utility regulation, was just repeated for entry into the twenty-first century.

This section, while aware of the short-term volatility resulting from regulation, deregulation, and politics will concentrate on the long-term aspects affecting the cost of power.

Cost or price is a basic characteristic of power supply. Along with relative abundance, reliability, and high quality of service, low prices have encouraged the widespread use of electric energy that has come to be associated with our national way of life. Industrial use of electric energy is particularly heavy in the electroprocess and metallurgical industries. The price of electricity has a significant effect on the end-product cost in these industries. In many manufacturing and process industries, quality of service in terms of voltage regulation, frequency control, and reliability is a major concern. Uninterrupted supply is of crucial importance in many process industries where a power failure causes material waste or damage to equipment, in addition to loss of production revenue. Because of the flexibility and convenience of electric power, future increases in industrial, residential, and commercial consumption are anticipated despite deregulation, fuel price increases, environmental considerations, and energy conservation efforts.

Various types of power-generating units are owned and operated by industry and private sector investors to provide power (and thermal energy) to industrial facilities and often sell excess power to the local utility at the avoided cost of incremental power for the utility. Inside-the-fence cogenerators at paper mills and other industrial facilities that are electrical and thermal consumers but provide only a fraction of their electric generation to the grid are not considered in this section. However, the calculations and information in this section can assist in deciding the advisability of purchasing versus generating electric power. Larger cogenerating facilities built under the "qualifying facility" (QF) sections of the **Public Utilities Regulatory Policies Act (PURPA)** that have electricity production for wholesale to the grid or local utility as their main objective are included in this section. With the PURPA regulations no longer restricting the development of **nonutility generators (NUGs),** a significant portion of new generation in the 1990s was provided by NUGs. Of course, many "nonutility generators" had regulated utility parents.

CONSTRUCTED PLANT COSTS

A **central station** serving a transmission system is designed to meet not only the existing and prospective loads of the system in which it is to function but also the pooling and integration obligations to adjacent systems. Service requirements which establish station size, type, location, and design characteristics ultimately affect cost of delivered power. Selection of plant type and the overall philosophy followed in design must accommodate a combination of objectives which may include high operating efficiency, minimum investment, high reliability and availability, maximum reserve capability margins, rapid load change capability, quick-start capability, or service adaptability as spinning reserve.

For a given plant, the design must account for siting factors such as environmental impacts, subsoil conditions, local meteorology and air quality, quality and quantity of available water supply, access for construction, transmission intertie, fuel delivery and storage, and maintainability. In addition, plant siting and design will be significantly affected by legal restrictions on effluents which may have adverse impacts on the environment. In the case of **nuclear plants,** siting must also consider the proximity of population centers and the size of exclusion areas. Reactor plant design must bear the investments required to control radioactive releases and to provide safeguard systems which protect against accidents. No new nuclear power plants are contemplated to begin the licensing process in the United States until 2010 or beyond. **Fossil-fueled power plants** may be sited adjacent to fuel supplies or in proximity to load centers, thereby increasing transmission costs on the one hand or fuel delivery charges on the other. Depending primarily on climate, plants may be enclosed, semienclosed, or of the outdoor type. Spare auxiliary components can be installed to improve reliability. Increased investments in sophisticated heat cycles and controls, for improved equipment performance, can achieve higher plant efficiencies. **Combustion-turbine, simple and combined-cycle plant** outputs are restricted by both increased elevation and high ambient air temperatures. Hydroelectric sites are frequently very distant from load centers and thus require added costs for extensive transmission facilities. Also, **hydroelectric facilities** may provide for flood control, navigation, or recreation as by-products of power production. In such instances, total cost should be properly allocated to the various product elements of the multipurpose project.

An **industrial power plant** provided to meet the requirements of an isolated load entails design considerations and exhibits cost characteristics which differ from those of a central station power plant assigned an integrated role within a connected generating system. Industrial power plants often produce both thermal and electric power. Industrial facilities must often accommodate both base- and peak-load requirements. They may be designed to provide for on-site reserve capacity or spinning reserve capacity. Frequency control and voltage regulation must be viewed as a special problem because of the limited capability of a single plant to meet load changes.

Table 17.5.1 provides typical installed cost data for central station generating plants. The figures represent costs of facilities in place, excluding interest during construction. The cost of land, waste-disposal facilities, and fuel is not included. Costs apply to plants completed in 2003. Interest during construction can be estimated by multiplying the simple interest rate per year by the construction period in years and dividing by 2 to reflect the carrying costs on the average commitment of capital toward equipment and labor during construction. Escalation effects for plants to be completed beyond this date may be extrapolated in accordance with anticipated cost trends for labor, material, and equipment. Historical cost trends, by region, as experienced in the power industry, which can be helpful in forecasting future costs, may be determined by use of the figures in Table 17.5.2. Escalation can significantly affect plant costs on future projects, especially in view of the 10- to 12-year engineering and construction periods historically experienced for nuclear facilities and the corresponding 5 to 6 years required for fossil-fueled power plants. In addition to rising equipment, construction labor, and material costs, major factors influencing the upward trend in plant costs include increased investment in environmental control systems, an emphasis on improved quality assurance and plant reliability, and a concern for safety, particularly in the nuclear field.

Table 17.5.1 Typical Investments Costs* (2003 Price Level)

Plant description			Total investment cost in $/net kW
Type	Net capacity, MW	Fuel	
Combustion turbine			
Simple cycle	200	Gas	250–400
Combined cycle	500	Gas	400–600
Conventional steam plant			
Fossil	500	Coal	1,200–1,500
	500	Oil	1,100–1,300
	500	Gas	900–1,000

* Capital investments exclude costs for the following: initial fuel supply, cost of decommissioning for nuclear plants, main transformers, switchyard, transmission facilities, waste disposal, land and land rights, and interest during construction.

Conventional Steam-Electric Plants Conventional fossil plant investment costs given in Table 17.5.1 are for 500-MW nominal units which are deemed to be representative of future central station fossil units. Costs will vary from those in the table due to equipment arrangement, pollution control systems, foundations, and cooling-water-system designs dictated by plant site conditions. The variety of cycle arrangements and steam conditions selected also affects plant capital cost. Plant designers try to economically balance investment and operating costs for each plant. Present-day parameters, in the face of current economics, call for a drum boiler, regenerative reheat cycles at initial steam pressures of 2,400 psig with superheat and reheat temperatures of 1,000°F (539°C). Similar temperature levels are employed for 3,500-psig initial steam pressure supercritical-reheat and double-reheat cycles which require higher investment outlays in exchange for efficiency or heat rate improvements of between 5 and 10 percent. Increased investment costs are also caused by more generous boiler-furnace sizing, and larger fuel and ash-handling, precipitator, and scrubber facilities required for burning poor quality coals. Investment increments are also required to provide partial enclosure of the turbine building and furnace structure and to fully enclose the boiler by providing extended housing to weatherprotect duct work and breeching.

Nuclear Plants The construction of nuclear power plant capacity in the United States has been suspended for over 20 years. A new generation of smaller light-water reactor nuclear power plants is currently in the conceptual design stage. Costs associated with these units, including siting, licensing, and fuel cycle, remain unclear, as they are precommercial.

In general, light-water reactor nuclear plants require considerably higher investments than do fossil-fueled plants, reflecting the need for leakproof reactor pressure containment structures, radiation shielding, and a host of reactor plant safety-related devices and redundant equipment. Also, light-water reactor plants operate at lower initial steam conditions than do fossil-fueled plants and, because of their poorer turbine

Table 17.5.2 U.S. Cost Trends of Electric Plant Construction by Region (1973 Index is 100)

	North Atlantic	South Atlantic	North Central	South Central	Plateau	Pacific
	Total: Steam generating plants*					
1993	329	308	318	304	317	333
1994	341	321	331	319	331	347
1995	354	332	345	329	340	357
1996	361	339	352	333	346	366
1997	373	350	362	347	357	376
1998	380	357	368	351	365	382
1999	385	361	374	355	369	388
2000	396	372	386	365	379	397
2001	420	391	404	391	394	419
2002	431	397	417	395	400	424
2003	445	412	438	412	423	445

*As of January 1 of each year.
SOURCE: "Handy-Whitman Index of Public Utility Construction Costs," compiled and published by Whitman, Requardt & Associates, LLP, 801 Caroline St., Baltimore, MD 21231.

cycle efficiency, require larger steam flows and increased equipment sizes at added investment.

Combustion Turbine/Combined-Cycle Plants Simple and combined-cycle plants are offered by a number of vendors in the 100- to 1000-MW size range. These plants consist of multiple installations of combustion gas turbines arranged to exhaust to waste-heat steam generators which may be equipped for supplementary firing of fuel. Steam produced is supplied to a conventional steam turbine cycle. Advantages of combustion turbine–based plants are lower unit investment costs, efficient thermal performance, increased flexibility (which allows independent operation of the gas-turbine portion of the plant), shorter installation schedules, reduced cooling-water requirements, and the reduction in sulfur oxide and particulate emissions characteristic of gaseous fuels. The disadvantages of combustion turbines are the high costs of fuel and combustion turbine maintenance.

Hydroelectric Plants Hydroelectric generation offers unique advantages. Fuel, a heavy contributor to thermal plant operating costs, is eliminated. Also, hydro facilities last longer than do other plant types; thus they carry lower depreciation rates. They have lower maintenance and operating expenses, eliminate air and thermal discharges, and because of their relatively simple design, exhibit attractive availability and forced-outage rates. Quick-start capability and rapid response to load change ideally suit hydro turbines to spinning reserve and frequency-control assignments.

The constructed cost of a hydroelectric station is strongly site dependent. Overall costs fluctuate significantly with variations in dam costs, intake and discharge system requirements, pondage required to firm up capacity, and with the cost of relocating facilities within the areas inundated by the impoundments. For a given investment in structures, available head and flow quantity may vary considerably, resulting in a wide range of outputs and unit investment costs. Installed plant costs are competitive with other facilities. Cost prediction for future hydroelectric construction is difficult, particularly in view of the decreasing availability of economical sites and restrictions imposed by concern for the ecological and social consequences of disrupting the natural flow patterns of rivers and streams. As with nuclear, no new large hydroelectric plants have been constructed in over 20 years. A number of dams and associated power plants have been demolished or otherwise taken out of service to save fish or restore waterways to some past condition. Any new capacity in this area has come from upgrading existing hydroturbines for greater output and efficiency.

Pumped-Storage Plants Pumped-storage plants involve a special application of hydroelectric generation, allowing the use of off-peak energy supplied at incremental charges by low-operating-cost thermal stations to elevate and store water for the daily generation of energy during peak-load hours. Pumped hydro projects must justify the inefficiencies of storage pumping and hydroelectric reconversion of off-peak thermal plant energy by investment cost savings over competing peaking plants. Installation of a pumped hydro station calls for a suitable high head site which minimizes required water storage and upper and lower reservoir areas and an available makeup source to supply the evaporative losses of the closed hydraulic loop. Despite the added complications of installing both pumping and generating units, or of utilizing reversible motor-generator pump-turbines, costs for pumped hydro stations generally fall below those for conventional hydroelectric stations. Differences in gross head, impoundment, and siting make plant cost comparisons difficult.

Geothermal Plants Geothermal generation utilizes the earth's heat by extracting it from steam or hot water found within the earth's crust. Prevalent in geological formations underlying the western United States and the Gulf of Mexico, geothermal energy is predominantly unexploited, but it is receiving increased attention in view of escalating demands on limited worldwide fossil-fuel supplies. Because natural geothermal heat supplants fuel, the atmospheric release of combustion products is eliminated. Nevertheless, noxious gases and chemical residues, usually contained in geothermal steam and hot water, must be treated when geothemral resources are tapped. There is a current lack of significant cost data covering geothermal plants. Because boiler and

associated fuel-handling facilities are eliminated, investment in these generating plants is considerably less than the cost of comparable fossil-fueled units. However, overall investment chargeable to geothermal facilities includes significant exploration and drilling costs which are site dependent and cannot be accurately predicted without extensive geophysical investigation.

Environmental Considerations Environmental protection has become a dominant factor in the siting and design of new power generating stations. Both stack emissions to the atmosphere and thermal discharges to natural water courses must be significantly reduced in order to meet increasingly stringent environmental criteria. In many cases older plants are being required to reduce emission levels to achieve legislated ambient air-quality standards and to control thermal discharges by the use of closed cooling systems to prevent aquatic thermal pollution.

Control of **air pollution** in fossil-fueled power plants includes the reduction of particulates, sulfur oxides, and nitrogen oxides in flue gas emissions. Particulate collection can be achieved by electrostatic precipitators, baghouse filters, or as part of stack gas scrubbing. Stack gas scrubbing is required for all new coal-fired power plants, regardless of sulfur content of the fuel. Fossil-fueled power plants are believed to be a major contributor to acid rain.

Scrubbers reduce sulfur oxide emissions by contacting flue gas with a sorbate composed of metal (usually sodium, magnesium, or calcium) hydroxides in solution which act as bases to produce sulfate and sulfite precipitates when they contact sulfur oxides in the flue gas. Wet scrubbers contact flue gas with a sorbate in solution. Dry scrubbers evaporate sorbate solution into the gas stream.

Fossil-fueled power plants are required to burn low-sulfur fuels. Restrictions on the use of oil in new power plants and its high cost have virtually eliminated new central station steam power plants designed to utilize oil.

Control of nitrogen oxides NO_x is attained primarily through modifications to flame propagation and the combustion processes.

NO_x control for combustion turbines is typically staged combustion. New burner designs and the use of water or steam injection into the combustion zone have effectively reduced emissions to acceptable rates. For steam boilers, changes to the burners and combustion zone were deemed insufficient and introduction of ammonia or urea spray has been incorporated with or without a catalyst [selective catalytic reduction (SCR) and selective noncatalytic reduction (SNCR), respectively]. The two systems, SCR and SNCR, operate at different temperatures. The SCR system includes ammonia injection across the flue gas in an area of the gas path at approximately 700°F. The flue gas then passes through a vanadium pentoxide (V_2O_5) catalyst. This combination removes NO_x with a removal efficiency of approximately 90 percent. SNCR operates at approximately 1600 to 2100°F but does not require a catalyst. Its removal efficiency is approximately 70 percent.

In order to avoid plant discharges of waste heat to the aquatic environment, evaporative-type closed-cooling cycles are employed in lieu of once-through cooling system designs. Closed-loop cooling systems in current use employ evaporative **cooling towers** or **cooling ponds.**

More advanced closed-cooling-loop designs use dry and wet-dry cooling towers. These represent alternates to conventional evaporative systems where makeup water is in short supply or where visible vapor plumes or ice formed by vapor discharge present hazards. The penalties for dry-tower cooling are significantly higher than those for conventional evaporative designs. Large, more costly water-to-air heat-transfer surfaces are required, and characteristically higher condensing temperatures result in higher turbine backpressure, restricting plant capability and reducing efficiency.

Evaporative cooling systems (wet towers) have a significant advantage over hybrid (wet and dry sections) and all-dry cooling systems in initial installed and operating costs. A dry cooling tower system can be $200 to $400 per kW more expensive. For a large facility the operating penalty can be $1,000 per kW or more because of the additional fan power requirements. Hybrid systems, which are a combination of an evaporative section and a dry heat exchanger section, are often used in areas where water usage is a concern or fogging issues are present. The wet/dry hybrid arrangement utilizes the required amount of wet and dry surface to meet the site constraints. In the Northeast, many smaller facilities (under 100 MW) have used the wet/dry arrangement to minimize fogging or icing of nearby highways. In the western half of the United States, wet/dry towers have been used to conserve the precious and sometimes scarce water resources. Another consideration for the addition of any dry surface beyond the typical evaporative cooling system is the effect on efficiency. Use of a dry cooling system can easily increase the condensing temperature for the Rankine cycle by 20 to 30°F. The efficiency and output penalty of the increase in heat rejection temperature can approach 5 to 10 percent, which can reduce overall efficiency by several percent or more.

FIXED CHARGES

Costs that are established by the amount of capital investment in plant and which are fixed regardless of production level are termed **fixed charges.** Annual fixed charges are ordinarily expressed as a percentage of investment and include interest or the cost of money, funds applied to amortize investment or to allow for replacement of depreciated plant, and charges covering property taxes and insurance. Additionally, fixed charges may include an interim replacement allowance to cover the replacement cost of plant equipment not expected to last the full life of the plant.

For investor-owned utilities, the **cost of money** employed for plant expansion depends upon financial market conditions in general and upon the attitude of investors with regard to a particular utility enterprise or specific project. Funds for investor-owned utility expansion are derived from both the risk capital (equity) and debt capital (bond) markets. Prevailing return rates for investments in utility plant facilities are influenced by the rate-setting practices of public utility regulatory agencies and by supply-and-demand factors in the investment market.

Public utility facilities owned by state and municipal government organizations are generally financed by long-term revenue bonds, most of which qualify for tax-free-income status. Interest rates currently fall between 4 and 6 percent and reflect the tax relief on interest income enjoyed by the bondholders.

Generating facilities are often financed by industrial concerns whose primary business is the production of a manufactured product. In these instances, the annual cost of money invested in power facilities will be established by considering alternate investment of the required funds in manufacturing plant. Inasmuch as returns on equity capital invested in manufacturing industries are usually on the order of 10 to 20 percent, the rate of return for industrial power plants will tend to be set at these higher levels.

Nonutility Generators (NUGs) generally form a project-specific limited liability corporations, or LLC, and borrow through nonrecourse financing at or above market rates. They put up a minimum of capital. In the early days of NUGs, the project would get equity contributions from the equipment vendors and contractors to cover the equity required by the banks as "down payment." The plant and its revenue stream would collateralize the balance. Later revenue from earlier operating plants and real equity was required to satisfy the loan. As we enter the twenty-first century, project financing has seen numerous project defaults and the decline of many independent NUGs and utility-spawned NUGs. Equity

Table 17.5.3 Representative Useful Life of Alternate Utility Facilities

Facility	Representative useful service life, years
Steam-electric generating plant	30
Hydroelectric plant	50
Combustion turbine-combined cycle	30
Nuclear plant	30
Transmission and distribution plant	40

requirements for new projects have increased dramatically, as well as the expected interest rates. Lenders expect multiple layers of guarantee from all participants. The short-lived electricity trading market has shrunk and the expectation of huge short-term profits that could justify and secure merchant generating has disappeared. New NUG generation will require guaranteed fuel costs, a reasonable expectation of electric sales through renewable sales contracts, and/or an electric price tied profitably to fuel and operating costs.

Corporate federal taxes are levied on equity capital income. Thus, corporate earnings on equity investment must exceed the return paid the investor by an amount sufficient to cover the tax increment.

The amortization of debt capital or the provision for **depreciation** over the physical life of utility plant facilities may be effected by several methods. Straight-line depreciation requiring uniform charges in each year over a predetermined period of useful service is commonly applied because of its simplicity. The percentage method of depreciation assumes a constant percentage decrease in the value of capital investment from its value the previous year, thereby resulting in annual depreciation charges which progressively diminish. The sinking-fund method of economic analysis assumes equal annual payments which, when invested at a given interest rate, will accumulate the capital value of facilities less their salvage value, over a predetermined useful service life. Table 17.5.3 illustrates the representative useful life for alternate utility facilities.

Use may be made of interest tables which show, for any rate of interest and any number of years, the equal annual payment (sinking-fund) rate which will amortize an investment and additionally will yield an annual return on investment equal to the interest rate.

$$\text{Equal annual payment} = \frac{i(1 + i)^n}{(1 + i)^n - 1}$$

where n = number of years of life, and i = interest rate or rate of return.

Property taxes and property insurance premiums are normally established as a function of plant investment and thus are properly included as fixed charges. Property taxes vary with the location of installed facilities and with the rates levied by the various governmental authorities having jurisdiction. In general, public power authorities will be free of taxes, although public enterprises often render payments to government in lieu of taxes. Annual property tax rates for private enterprise will amount to perhaps 2 to 4 percent of investment, while property insurance may account for annual costs of between 0.3 and 0.5 percent.

A representative makeup of fixed charges on investment in a conventional steam-electric station having a useful service life of 30 years is shown in various classes of ownership in Table 17.5.4.

Table 17.5.4 Typical of Fixed-Charge Rate for Conventional Fossil-Fueled Steam-Electric Plant with 30-Year Economic Life

	Investor-owned utility, %	Government-owned utility, %	Industrial-commercial ownership, %
Rate of return or interest	6.4	5.0	8.0
Amortization or depreciation	1.2	1.5	0.9
Federal income tax	1.5	0.0	4.3
Local taxes (or payment in lieu of taxes)	2.0	2.0	2.0
Insurance	0.3	0.3	0.3
Total	11.4	8.8	15.5

Table 17.5.5   Operating Expense for Fuel for Representative Heat Rates and Fuel Prices

	Nominal size, MW	Typical heat rate, Btu/kWh	Fuel price, ¢/MBtu	Fuel cost, mills/kWh
Conventional coal fired	500	9,800	200	19.6
Advanced light-water reactor	600	10,700	60	6.4
Combustion turbine/combined cycle	500	7,000	400	28
Combustion turbine/simple cycle	200	8,000	400	32

NOTE: Operating fuel cost, mills/kWh = (Btu/kWh) × (¢/Btu) × 10^{-5}.

OPERATING EXPENSES

Fossil Fuels   Currently, fossil fuels contribute approximately three-fourths of the primary energy consumed by the United States in the production of electric energy. Generating station demands for fossil fuels are continuing to increase as the electric utility and industrial power markets grow. Price comparisons of fossil fuel are generally made on the basis of delivered cost per million Btu. This cost includes mine-mouth or well-head price, plus the cost of delivery by pipeline or carrier. Price comparisons must recognize that solid and, to a lesser extent, liquid fuels require plant investments and operating expenditures for fuel receipt, storage, handling and processing facilities, and for ash collection and removal. High transportation cost contributions on a Btu basis will be incurred by high-moisture and ash-content coals with low heating values. It is, therefore, advantageous to fire lignite and subbituminous coals at mine-mouth generating plants.

Coal represents our most abundant indigenous energy resource, with enough economically recoverable supplies at current use rates to last well into the twenty-first century. About half of the recoverable coal reserves have a sulfur content above 1 percent and are considered high-sulfur coal. Low-sulfur coals are found chiefly in the low-load areas of the mountainous West, and delivered cost at the major markets east of the Mississippi include high transportation charges. Increasingly, coal production is bearing the cost of more rigid enforcement of stringent mine safety regulations, coal washing environmental compliance, and the charges associated with strip-mine land restoration. Delivered price depends upon transportation economies as may be affected by barging, unit train haulage, or pumping in slurry pipelines. Delivered price levels for coal fuel vary with plant location. Plants conveniently located with respect to eastern coal reserves report delivered-coal prices in the general range of $1.25 to $2.50 per million Btu, depending upon sulfur and ash content. Low-sulfur western subbituminous coals have delivered prices of between $1.00 to $2.00 per million Btu.

The domestic supply of **petroleum** is now outstripped by nationwide demand. The United States has become dependent on overseas sources to meet growing energy demands. The consequences of this trend are a continued unfavorable balance of payments and dependence on foreign oil supplies from politically unstable areas in the Middle East and North Africa. Residual and distillate oil prices have risen because of short supply and pressure by the major oil producers on worldwide market price levels. Blends of low-sulfur oils delivered during 2004 to generating plants along the eastern seaboard range from $3.50 per million Btu to as high as $4.50 per million Btu for the 0.3 percent sulfur fuel required for firing in some metropolitan areas. During this same period distillate oil commanded a nationwide price ranging from approximately $5.00 to $7.00 per million Btu.

Consumption of **natural gas** as a power fuel is increasing because it is clean burning, convenient to handle, and generally requires smaller and cheaper furnaces. Historically, its limited supply made it best-suited for consumption by residential and commercial users and meeting industrial process needs. Present-day power plant use of natural gas is on the rise. The proliferation of larger, more efficient combustion turbines like the 150+ MW "F series" machines designed for combined-cycle application has increased the consumption of natural gas for power generation. During the 1990s almost all new generation was natural gas-fired combustion turbine combined-cycle.

Fuel prices for 2004 ranged from $4.00 to $6.00 per million Btu. Price levels of natural gas, severely regulated in the past by government controls imposed at the well, have been falling sharply in response to current supply-and-demand factors and gas deregulation.

Nuclear Fuel   Reactor plant fuel costs present a special case. Actually, the initial core loading which will support operation of a nuclear plant over its early years of life requires a single purchase prior to commissioning of the plant. For comparison with fossil-fuel prices, therefore, nuclear-fuel cycle costs, including first-core investment, periodic charges for reload fuel, and spent-fuel shipment and processing costs, are ordinarily extrapolated at assigned load factors over the life of the plant and are converted to an economic equivalent expressed in dollars per million Btu of released fission heat. Nuclear-fuel costs are not only influenced by ore prices and by fuel fabrication and processing costs, they are also sensitive to investment and uranium-enrichment costs. Future costs are subject to inflationary pressures and cost-saving technology changes such as extended burn-up cycles. Charges up to $1 per million Btu are representative of the levelized fuel prices for nuclear plants.

Operating Costs of Fuel   Fuel price contributions to energy-generation costs will reflect start-up and will depend upon plant efficiencies which, at low loads, show considerable departure from the best-point performance achieved at or near full unit loadings. These factors significantly affect the operating expenses of load-following utility system units as well as plants assigned fluctuating demands in manufacturing or industrial service. As an example, calculated performance for a nominal 500-MW, 2,400-psig coal-fired regenerative reheat steam unit shows a best-point heat rate of 9,800 Btu/net kWh at rated output. Load reduction yields heat rates of 10,500 and 12,400 Btu/kWh at loadings of 250 MW and 125 MW, respectively. Typical operating-expense ranges for given fuel prices and estimated full-load heat rates may be determined directly where continuous operation at or near unit rating is assumed (Table 17.5.5).

Operating Labor and Maintenance   In addition to fuel costs, operating expenses include labor costs for plant operation and maintenance, plus charges for operating supplies and maintenance materials, general administrative expenses, and other costs incidental to normal plant operation. Operating labor and maintenance costs vary considerably with unit size, operating regimen, plant-design conditions, type of facility, and the local labor market. Representative figures appear in Table 17.5.6.

Table 17.5.6   Representative Operating and Maintenance Costs (2004 Price Level)

Type of plant	Nominal size, MW	Fuel	Operating and maintenance costs, mills/kWh
Conventional fossil	500	Coal	5–9
Advanced lightwater reactor	600	Nuclear	5–12
Combustion turbine/ combined cycle	500	Gas	6–11
Combustion turbine/ simple cycle	200	Gas	7–12
Conventional hydro	300	—	6–9

NOTE: Unit kilowatt-hour costs include labor, maintenance materials, operating supplies, and incidental expenses. Costs shown assume base-load operation.

Environmental Controls   Systems and equipment required for air and water pollution abatement generally carry increased fuel and maintenance labor and materials costs. Reductions in plant output resulting from the higher condensing pressures associated with cooling-tower operation or the added auxiliary power for stack gas clean-up systems lower plant efficiency and increase fuel consumption.

OVERALL GENERATION COSTS

The total cost of power generation may now be estimated by reference to the preceding material assuming type of ownership, capital structure, plant type, fuel, and loading regimen. Table 17.5.7 comprises an illustrative tabulation of the factors determining the overall generation cost of an investor-owned generating facility.

It should be noted that capacity factor, or the ratio of average-actual to peak-capable load carried by a given generating facility, will have a significant effect on generation expenses. In addition to the effects of part-load operation on fuel costs as previously discussed, capacity factor will determine the plant generation which will support fixed charges. High-capacity-factor operation will spread fixed charges over a large number of kilowatt-hours of output, thereby reducing unit generation costs. During its initial life, a thermal plant is usually operated at high-capacity factor. As inevitable obsolescence brings newer and more efficient equipment into service, a unit's baseload position on the utility system load duration curve is relinquished, and capacity factor tends to drop. This decline in capacity factor must be acknowledged in estimating output and generation costs over the life of a given facility.

Most modern electric utilities incorporate computerized systems designed to economically dispatch power generated at each production plant feeding the load. Individual generating-unit loads are assigned in a manner that can be demonstrated to result in minimum overall cost; i.e., at each system power level, load is shared between units so that all operate at the same incremental production cost. Telemetered data reflecting system load and generation is transmitted to central dispatch computers by multiplexing via power line carrier, microwave, or telephone lines. The communication schemes include channels for transmitting load adjustment commands developed by the computer to on-line generating units. Loading instructions account for the unit production efficiencies and transmission losses associated with each dispatch assignment.

TRANSMISSION COSTS

Because of a growing scarcity of urban sites, increasing emphasis on environmental protection, and public attitudes, it has become more and more difficult to site major power stations near centers of load. As a consequence, added cost of transmission along with attendant resistive power losses add significantly to the overall cost of service. Because of the distances involved, these costs are generally greatest for nuclear facilities, mine-mouth stations, and hydroelectric plants where remoteness or remote resources strongly govern siting. Transmission plant investment also reflects a trend toward interconnection of neighboring utilities. Designed to improve service reliability by the pooling of reserves and to effect savings by capacity and energy interchanges, such

Table 17.5.7   Total Cost of Generating Power
(Plant type: 500-MW conventional fossil; plant net heat rate: 9,800 Btu/kWh)

	Generating costs, mills/kWh
Coal fuel @ $2.00/mBtu	19.6
Operating and maintenance	7.0
Fixed charges @ 11.4% per annum	24.0
Total operating costs	50.6

NOTES: Fixed charges are based on assumed initial plant investment of 1,580/kW and 7,500 h/year of operation. Values shown are typical and could vary significantly for individual plants.

interties must carry substantial ratings so that emergency power transfers can be accommodated without exceeding system stability limits. Costs for major overhead transmission ties (1,000 MVA and up) are estimated to range between $400,000 and $900,000 per circuit mile depending on terrain and voltage level. Investments are also sensitive to factors of climate and proximity to urban areas that can increase right-of-way costs significantly. Where underground transmission is elected, installed investment costs can be as much as 10 times the cost of overhead lines.

Transmission and distribution systems have seen a large-scale change in ownership in deregulated areas of the country as regional grids were sold, divested, or handed to new owner/operators. These grid systems that had been maintained by high-cash-flow integrated utilities were now the responsibility of lower-cash-flow entities receiving only a small percentage of the retail price of electricity. Extensive blackouts, like the August 2003 blackout that darkened Ohio to New York City and parts of Canada, can result from insufficient transmission line investment, publicly restricted transmission line additions, and capacity increases, as was demonstrated in this situation.

The choice of **transmission voltage** level and whether ac or dc is used for bulk power transfer depends on the amount of power transmitted, the transmission distance involved, and at each voltage level, the cost of line and substation equipment. Voltages generally employed are 230, 345, 500, and 765 kV ac and 5000 + kV dc. Where long distances on the order of 400 mi (650 km) or more are encountered, dc transmission becomes economically attractive. For shorter lines, however, savings in fewer conductors and lighter transmission towers are nullified by the high cost of dc-ac conversion equipment at both line terminals.

POWER PRICES

The price of power delivered must account for the production costs at each generating station in addition to transmission and distribution costs. As previously noted, these costs depend upon labor rates, fuel prices, and material charges. They reflect investment levels in generating plant and the fixed-charge rates established by funding patterns, type of ownership, and expected equipment service life. Overall system production costs are affected by the investment requirements of specific mixes of generating equipment types, and by the manner in which load is shared by units, i.e., how production is allocated between highly efficient base-load stations and the less-efficient peaking equipment which normally runs for only a few hours each day. Also, important cost reductions are achieved in hydro systems by controlling natural and stored water flows to allow optimized sizing and scheduling of hydroelectric output, thereby reducing needs for thermal peaking capacity and decreasing the generation requirements of high fuel cost fossil plants.

Power prices cover the investment charges, maintenance costs, and capacity and energy losses chargeable to the transmission and distribution plant. They also include the administrative costs incurred to maintain corporate enterprise and the commercial expense of metering and billing. The advertising and public relations cost of competition or the cost of preparing for competition in a deregulated market must be considered.

Generally, power prices must provide a return to cover the average cost of power production throughout a given system. For smaller utility systems and commercial and industrial companies, each plant must provide an adequate return based solely on the individual plant's costs rather than an average as discussed above. Rate schedules and supply contracts, however, are drawn to reflect the reduced cost of off-peak energy produced by available generating units during periods of low system load. Additionally, prices for high load factor service often recognize the cost reductions effected by spreading fixed charges over increased units of energy output. Large blocks of capacity and energy supplied for industrial use are often priced by establishing an annual charge for capacity which equals the fixed charges on investment in committed generation and transmission plant, plus charges for energy representing the sum of the variable kilowatt-hour production costs for fuel, maintenance, and operation.

Historically, utilities have applied rate schedules which promote consumption by applying progressively lower rates to blocks of increased energy usage. Rationale for such pricing is the savings that load growth can

Table 17.5.8   Average Cost, in Cents per Kilowatt-Hour, for Consumers by Sector, Census Division, and State, 2002

Census division, State	Residential	Commercial	Industrial	Other	All Sectors
New England	**11.18**	**9.91**	**8.52**	**10.50**	**10.16**
Connecticut	10.96	9.35	7.68	10.36	9.73
Maine	11.98	10.47	11.24	32.82	11.36
Massachusetts	10.97	10.14	8.77	9.79	10.18
New Hampshire	11.77	10.09	8.83	12.07	10.49
Rhode Island	10.21	8.84	8.04	8.09	9.19
Vermont	12.78	11.10	7.90	19.26	10.87
Middle Atlantic	**11.32**	**10.14**	**6.03**	**9.45**	**9.59**
New Jersey	10.38	8.87	7.83	14.04	9.31
New York	13.58	12.46	5.16	9.05	11.29
Pennsylvania	9.71	8.03	6.06	11.10	8.01
East North Central	**8.06**	**7.20**	**4.58**	**6.09**	**6.50**
Illinois	8.39	7.49	5.01	5.56	6.97
Indiana	6.91	5.98	3.95	9.75	5.34
Michigan	8.28	7.36	4.95	10.43	6.92
Ohio	8.29	7.68	4.68	5.70	6.66
Wisconsin	8.18	6.54	4.43	8.08	6.28
West North Central	**7.37**	**6.01**	**4.22**	**5.72**	**5.97**
Iowa	8.35	6.56	4.06	4.92	6.01
Kansas	7.67	6.28	4.53	9.30	6.31
Minnesota	7.49	5.88	4.19	7.36	5.84
Missouri	7.06	5.88	4.42	6.20	6.09
Nebraska	6.73	5.62	3.89	6.37	5.55
North Dakota	6.39	5.85	3.98	3.68	5.45
South Dakota	7.40	6.24	4.54	3.63	6.26
South Atlantic	**7.90**	**6.43**	**4.24**	**6.42**	**6.56**
Delaware	8.70	6.98	5.11	10.62	7.05
District of Columbia	7.82	7.38	4.95	6.60	7.37
Florida	8.16	6.64	5.23	7.43	7.31
Georgia	7.63	6.46	3.95	8.31	6.24
Maryland	7.71	6.09	3.88	10.18	6.21
North Carolina	8.19	6.51	4.70	6.70	6.74
South Carolina	7.72	6.48	3.85	6.44	5.83
Virginia	7.79	5.87	4.13	5.15	6.23
West Virginia	6.23	5.41	3.81	10.01	5.11
East South Central	**6.57**	**6.33**	**3.71**	**6.32**	**5.39**
Alabama	7.12	6.63	3.82	7.46	5.71
Kentucky	5.65	5.30	3.09	4.61	4.26
Mississippi	7.28	6.83	4.40	8.76	6.24
Tennessee	6.41	6.45	4.15	8.92	5.72
West South Central	**7.70**	**6.69**	**4.48**	**6.31**	**6.33**
Arkansas	7.25	5.68	4.01	6.52	5.61
Louisiana	7.10	6.64	4.42	7.05	5.99
Oklahoma	6.73	5.75	3.81	5.06	5.59
Texas	8.05	6.95	4.66	6.55	6.62
Mountain	**7.87**	**6.64**	**4.86**	**5.57**	**6.52**
Arizona	8.27	7.28	5.20	4.56	7.21
Colorado	7.37	5.67	4.52	6.64	6.00
Idaho	6.59	5.71	4.34	5.18	5.58
Montana	7.23	6.53	3.70	7.14	5.75
Nevada	9.43	9.06	7.25	6.54	8.42
New Mexico	8.50	7.22	4.48	6.23	6.73
Utah	6.79	5.60	3.84	4.69	5.39
Wyoming	6.97	5.71	3.55	5.93	4.68
Pacific Contiguous	**10.46**	**11.38**	**8.26**	**6.29**	**10.28**
California	12.90	13.22	10.83	6.68	12.50
Oregon	7.12	6.59	4.72	9.44	6.32
Washington	6.29	6.11	4.56	4.94	5.80
Pacific Noncontiguous	**14.20**	**12.46**	**10.26**	**14.63**	**12.35**
Alaska	12.05	10.13	7.65	14.04	10.46
Hawaii	15.63	14.11	11.02	16.85	13.39
U.S. Total	**8.46**	**7.86**	**4.88**	**6.73**	**7.21**

SOURCE: DOE website www.eia.doe.gov/cneaf/electricity/esr/table1abcd.xls#A238.

realize through economies of scale, as well as the improved utilization of existing utility plant. However, regulatory pressure, reflecting a policy of minimizing the cost to the consumer, impact on the environment, and the critical need to conserve high-cost imported fuel, has favored a marginal cost-pricing system more nearly reflecting the actual cost of production and transmission of a particular user's supply of power. Rate setting under this conservationist approach calls for flat, rather than reduced, rates as usage increases and for high unit energy charges during peak-load periods.

Cogeneration facilities designed for the **dual-purpose production of power and process steam** permit investment savings, principally in steam-generation plants, which result in combined production charges falling below the total cost of separate, single-purpose production of power and of steam. Such savings can permit proportionate decreases in the prices ordinarily charged for separate single-purpose production of each of the products, or they may be assigned in total to reduce the price of one or the other by-product. This latter option is often exercised in the case of dual-purpose water product plants arranged for seawater flash evaporation using power-turbine extraction as a process steam source. Where severe shortages of fresh water exist, social considerations favor the total assignment of dual-purpose savings to the water product. Thus, minimal prices for desalted product water are achieved, while dual-purpose power is marketed at prices competing with single-purpose power generation costs. Similarly, assignment of the total savings of dual-purpose power and process steam production to the power product may justify on-site industrial plant power generation in preference to outside purchases of higher-priced utility system power supplies.

Since the 1980s, some facilities qualifying under the PURPA regulations have provided very low-priced steam to a host industrial plant as a means to satisfy the PURPA regulations and sell electricity to the local utility, whether the utility wanted the facility or not. Changes to the pricing structures payable by the local utilities to the "qualifying facility" and the eventual end of the PURPA statute eliminated the "free steam" approach from cogeneration.

Where hydro facilities supply power in combination with irrigation, flood control, navigation, or recreational benefits, power costs are largely sensitive to the allocation of investment charges against each of the multipurpose project functions. Should generation be treated as a by-product, power prices can be reduced drastically to reflect equipment operating expenses and the limited fixed charges covering investment in only the generating plant itself.

Cheap fuel, advances in design, the economies of scale, and the economic application of alternate generating unit types produced downward trends in the price of electric service in the 1960s. In the 1970s these trends were reversed by inflationary effects on plant costs, high interest rates, and the increased fuel prices. A dwindling number of favorable plant sites, licensing delays, and the added costs of environmental impact controls combined to cause further upward pressure on power prices in the 1980s. However, falling interest rates and fuel prices began to slow the rate of increase in the cost of electric power. These stabilizing influences have continued into the 1990s, tempered by stricter environmental regulation costs. Increases in electric rates are expected into the foreseeable future.

The twenty-first century has started with good and bad deregulation cases, corporate financial problems (such as Enron), and increased terrorist attacks on the U.S. infrastructure. All of these factors, coupled with a significant rise in the price of natural gas in the first several years of the new millennium, have resulted in continual escalation of the price of electricity. Table 17.5.8 shows the electricity prices by sector for each region and state in the United States for the year 2002.

17.6   HUMAN FACTORS AND ERGONOMICS
by Ezra S. Krendel

REFERENCES: "Aviation Safety and Pilot Control," NRC, National Academy Press, Washington, DC, 1997. Allen, McRuer, et al., "Computer-Aided Procedures for Analyzing Man-Machine System Dynamic Interactions," Vol. I, "Methodology and Application Examples," Vol. II, "Simplified Pilot-Modeling for Divided Attention Operations," Vol III, "Users Guide," WADC TR-89-3070, June 1989. Badler, Phillips, and Webber, "Simulating Humans: Computer Graphics, Animation and Control," Oxford University Press, 1993. Boff, Kaufman, and Thomas (eds.), "Handbook of Perception and Human Performance," Vols. I and II, John Wiley & Sons, 1986. Card, Moran, and Newell, "The Psychology of Human-Computer Interaction," Lawrence Erlbaum Associates, 1983. Kleinman, Baron, and Levison, "An Optimal Control Model of Human Response, Part I: Theory and Validation," *Automatica,* **6,** 1970. Konz and Johnson, "Work Design: Occupational Ergonomics," 6th ed., Holcomb Hathaway, 2004. Krendel and Wodinsky, "Search in an Unstructured Visual Field," *Jour. Optical Soc.,* **50,** 1960. McRuer, Pilot-Induced Oscillations and Human Dynamic Behavior, *NASA CR-4683* July 1995. McRuer, Clement, Thompson, and Magdaleno, "Minimum Flying Qualities," Vol. II, "Pilot Modeling for Flying Qualities Applications," Systems Technology, Inc. Hawthorne, CA, 1990. McRuer, "Human Dynamics in Man-Machine Systems," *Automatica,* **16,** 1980. McRuer and Krendel, "Mathematical Models of Human Pilot Behavior," AGARDograph No. 188, 1974. Salvendy (ed.), "Handbook of Human Factors," 2d ed., John Wiley & Sons, 1997. Sundin, Örtengren, and Sjöberg, "Proactive Human Factors Engineering Analysis in Space Station Design Using the Computer Manikin Jack," SAE Technical Paper 2000-01-2166. Thompson, "Program CC Version 5.0 for Windows," Systems Technology Inc., Hawthorne, CA, 2001.

SCOPE

The mission of human factors engineering/ergonomics (HFE/E) is to improve the performance, reliability, efficiency, and the risk management of systems in which humans work in concert with machines, man/machine/systems (MMS). This discipline developed many of its techniques during and after World War II. The life-or-death stakes and the advances in military technology made even minor improvements in the performance and reliability of manned military systems highly desirable and major improvements essential. The design of MMS interact with experimental psychology, physiology, and physical anthropology and with aeronautical, electrical, industrial, mechanical, systems, computer, and cognitive engineering.

MMS cover a wide range of enterprises, from piloting a jet aircraft to microsurgery. Achieving the goals of HFE/E in MMS requires a knowledge of humans as sensors, manipulators, responders, information processors, and decision makers. The expected performance of the "man" in MMS is affected by: training and motivation, variability of behavior among humans, personal stress, work load, age, fatigue, substance abuse, and by environmental conditions: vibration, noise, temperature, gravity, and distractions from the task. Some of the references for this section are sources for details and data on the impact of these many human-related variables. Other references provide engineering theory and practice for the design of MMS. What follows is an introduction to HFE/E findings that are likely to be helpful to mechanical engineers.

VISION

The visual pathway in psychomotor tasks, written or iconic messages, and searching for targets is primary in most MMS. Other senses, in declining order of importance to MMS design, are: hearing, which is superior to vision for alerting or warning signals; kinesthesis and vestibular senses, which facilitate positional awareness; and touch and pain.

High-resolution vision requires that the image of the object seen be projected by the eye's lens upon an approximately 2° central sector of the retina, known as the *fovea*, composed entirely of photoreceptors called *cones.* Cones generate color vision and respond to light intensity from the limit of brightness tolerance to that of candlelight reflected from a newspaper. As it scans its field of view, the eye continually moves in pulses of approximately 50 ms, pausing to fixate its fovea on objects in this field. Each fixation pause is about 250 to 300 ms, but can be as long as 1 minute when the viewer is uncertain about the image on his fovea.

The density of cones on the retina falls steeply beyond its area of concentration on the fovea and reaches a flat minimum at an eccentricity of about 20°. Beyond the fovea a second set of retinal photoreceptors, rods, appears among the cones. Rods provide vision from low brightness levels to the darkest conditions for which vision is still possible. Rods reach their maximum sensitivity to low light levels after about 30 min in the dark. In a twilight level of brightness comparable to snow under a full moon, rods and cones operate together. The number of rods on the retina increases in density until reaching a maximum at an eccentricity of about 20°. Although much greater than that of cones, rod density gradually declines as it approaches the limiting eccentricity of 80°. This peripheral vision supports orientation and motion detection.

Visual perception is the brain's interpretation of visual images. The perception of the size and distance of an object begins when the eyes' lenses, controlled by the ciliary muscles, accommodate so as to focus upon the object. Estimates of an object's apparent distance are influenced by: this muscle action, the visual cues leading to the object, and the anticipated size of the object. These cues may be inadequate or they may be inconsistent with one another, and as a consequence a perception of distance that differs from reality may emerge. The vision of an automobile driver may be momentarily focused on the dashboard displays at a distance of 0.5 m. Refocusing on an object appearing unexpectedly in the road ahead may take as long as 0.4 s. The object may be unfamiliar and helpful visual cues obscured by darkness or fog. The probability of an accident is increased by this briefly held misperception of object size or distance.

Some of the characteristics of vision can support a two-parameter model for the probability of success over time in searching with the naked eye for a target alone in a visual field. This simplified model must be elaborated when the proposed application differs greatly from the experimental conditions under which it was validated. This is a common condition when most HFE/E models or data are used.

Assuming the search is random and p_{sg} is the probability of finding the target during a single glimpse or one fixation, then the probability of finding the target after k glimpses is:

$$P_k = 1 - (1 - p_{sg})^k \tag{17.6.1}$$

Each glimpse plus movement time is T seconds, therefore the elapsed search time is $t = kT$; the probability of finding the target by this time is:

$$P(t) = 1 - e^{-mt} \qquad t_{mean} = 1/m \qquad m \approx p_{sg}/T \tag{17.6.2}$$

This model was confirmed for $0.006 \le p_{sg} \le 0.80$, different size sectors of the sky, different target sizes, and different target contrasts. In unpracticed random searches for a command in a menu on a monitor, the single-glimpse probability of detection, p_{sg}, was 0.08. After practice, p_{sg} increased to 0.60. Were the search systematic with no overlap of fixation areas the probability of detection, P_{sys}, for search for a limited time t would be:

$$P_{sys}(t) = p_{sg} t/T, \tag{17.6.3}$$

where $P_{sys} \le 1$.

This model for the probability of finding targets becomes more elaborate when the target must be found from among camouflaged decoys as well as from among decoys differing from it by contrast, shape, area, color, and velocity as well as by area of search, numbers and spatial density of decoys and of targets, and the use of auditory aids.

PSYCHOMOTOR PERFORMANCE

A program of skilled psychomotor behavior can be constructed in several ways: by incrementally aggregating a sequence of discrete reaction times and movements, by continuous closed-loop control known as manual control or tracking, by complex open-loop activities, and by various combinations.

Manual reaction time (RT), like the eyes' fixation time, increases with uncertainty. This increase in RT can be quantified when a subject must select the correct manual response out of n possible responses to a corresponding visual signal out of n probable signals. For situations in which the independent or the sequential probability, p_i, for the occurrence of each of these n signals can be estimated, uncertainty can be expressed as information theory entropy, H:

$$H = \sum_{i}^{n} p_i \log_2 (1/p_i + 1) \tag{17.6.4}$$

For the range $0 \le H \le 3$, there is an empirical equation, known as the Hick-Hyman law for RT, for these manual responses:

$$RT = a + bH \tag{17.6.5}$$

The constants $a \approx 200$ ms and $b \approx 150$ ms are influenced by practice and by the compatibility between the geometry of the physical presentation of the stimulus and that of the mechanism for responding.

For rapid discrete or repetitive actions, usually by the hands, movement time (MT) is proportional to the difficulty of the task as defined by Fitts' law. Equation (17.6.6) defines the index of difficulty, I_D, in terms of target width W and movement amplitude A. MTs from a wide range of rapid movements, as in operating a key pad or sorting items into bins, can be described. In applications, d is about 100 ms and c, which depends on the movement geometry, is approximately 200 ms. Estimates of MT enable comparisons to be made among different operating procedures.

$$I_D = \log_2 (2A/W) \tag{17.6.6}$$
$$MT = c + d\, I_D \tag{17.6.7}$$

To be useful, measurements of RT and MT must come from a stable plateau of skilled performance that can be determined by units of error or of time to complete the given task. For many psychomotor tasks, skilled learning follows a power function

$$T_n = T_1 n^{-\beta} \tag{17.6.8}$$

where T_1 is the time to perform the task on the first trial. T_n the time to perform the task on the nth trial, and n is the number of trials. In the development of psychomotor skill, $0.2 \le \beta \le 0.6$.

SKILL AND ERRORS

The highest skill level in tracking is attained after extensive training when an operator fully familiar with the dynamics of the machine, the manipulator under her control, and the appropriate responses to the input signals can extract what coherence is present in these signals, develop pathways that reorganize the perceptual system, adapt her behavior to create a repertory of special responses, and select from this repertory the appropriate response for the best MMS performance. The procedure resulting in this performance is known as successive organization of perception (SOP). SOP results in a virtual display that provides the operator with the information that would have been present in the actual display had it been augmented. In closed-loop control situations, the precognitive mode response can be anticipatory and compensate for inherent human control lags. In an open-loop mode, the response may be programmed as in executing sequential actions in preparing to stop a car on approaching a red light or otherwise responding to behavioral cues.

As psychomotor skills develop with training, errors diminish, and a plateau of stable, skilled performance is reached. Transient errors occur and are shed as training progresses. Posttraining errors are departures from the expected performance of a motivated, skilled operator. These errors may place the MMS in danger. Their number increases under the following pressures on human performance: divided attention, multitasking, and physiologically impairment such as fatigue, hypoxia, and alcohol or drug use. Errors can be divided into two classes: (1) mistakes, errors in the thought processes behind an action or decision, and (2) slips, errors in the detection or interpretation of sensory signals and unintended responses. The thinking processes which lead to mistakes have been reconstructed; for example, inappropriate expectations can result in misinterpreting a situation acting accordingly and making a malign error. The basic sources for such reconstructions are accident reports, anecdotes, interviews with participants, and introspection on the part of the analyst. If the mistake is made under time pressure, it is difficult for the operator to reexamine the situation, sort out his misconceptions, and examine his options in a deliberate manner. Mistakes can be decreased by training methods, personnel selection, and operating procedures that emphasize the examination of options and impose redundancy for critical actions to lessen the chance that operators will persevere in erroneous behavior.

By applying HFE/E findings in the selection and positioning of information displays, controls, monitors, and keyboards, and by accommodating the physical dimensions of the operator, the designer can make slips such as misreading a display or reaching for the wrong switch less likely. In the unlikely event that an error does occur, emergency corrective procedures and warnings must be part of the MMS design. Warnings can address different sensory modalities with audio signals, speech, annunciators, lights, and vibration, either separately or in various combinations determined by the task being performed and the environment of the person or persons to be warned.

The probabilities of slips occurring can be estimated from the existing databases in HFE/E and from manned and unmanned simulator studies. These estimates, together with the failure probabilities for the inanimate components of the MMS, can be incorporated into flowcharts for the MMS to determine those locations where further risk management procedures are necessary.

MANUAL CONTROL

The body of knowledge in manual control developed from the USAF's mid-twentieth-century interest in mathematically modeling the dynamic response of pilots in order to provide aeronautical engineers with data for designing aircraft of high performance, stability, and positive evaluations from their pilots. Classical control theory provided the structure for this model. The core was a quasi-linear transfer function or describing function for the pilot's dynamics represented as $Y_p(j\omega), Y_p(s)$, or Y_p, depending on the projected analysis. The quasi-linear human controller's response to an input comprises two parts: describing function

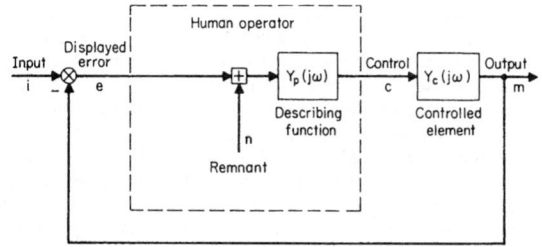

Fig. 17.6.1 Simplified block diagram for manual control.

components that correspond to the response of the equivalent linear elements driven by that input, and a "remnant" component that represents the difference between the response of the actual system and a system composed of equivalent linear elements. Test pilots operating fixed- and moving-base flight simulators as well as actual aircraft provided the empirical data for $Y_p(j\omega)$. Aircraft dynamics, the controlled element, $Y_c(j\omega)$, confine the dynamic response of motivated pilots to a limited range so that intersubject variability is suppressed. Consequently, adequate measurements for $Y_p(j\omega)$ can be obtained from a small number of skilled, motivated test subjects.

The compensatory mode, Fig. 17.6.1, the most common and extensively studied mode of closed loop control, is present in vehicle control and many other configurations. In this mode, the human tracks a moving visual signal so as to keep its position as close to a reference marker as she can. This signal is the MMS's error $e(t)$, which is the difference between $i(t)$, the input to the MMS, and $m(t)$, the MMS output. Detecting cyclical and coherent structures in $e(t)$ and from these developing estimates of $m(t)$ and $i(t)$ are the first step in the SOP progression of evolving virtual displays. Maintaining system stability and minimizing the closed-loop system's error are basic design goals for closed-loop systems. If the open-loop gain $|Y_p(j\omega)Y_c(j\omega)|$ in the compensatory mode is too high, instability can occur in the closed-loop task. An expert pilot may unwittingly regress from a higher level in the SOP procedure to the compensatory mode. In this mode the pilot may inadvertently generate excessive open-loop gain and destabilize an otherwise stable system. Potentially destructive pilot-induced oscillations of the aircraft about its flight path are the result.

Compensatory pursuit and precognitive modes of the SOP procedures are presented in Figure 17.6.2. In the pursuit mode the operator

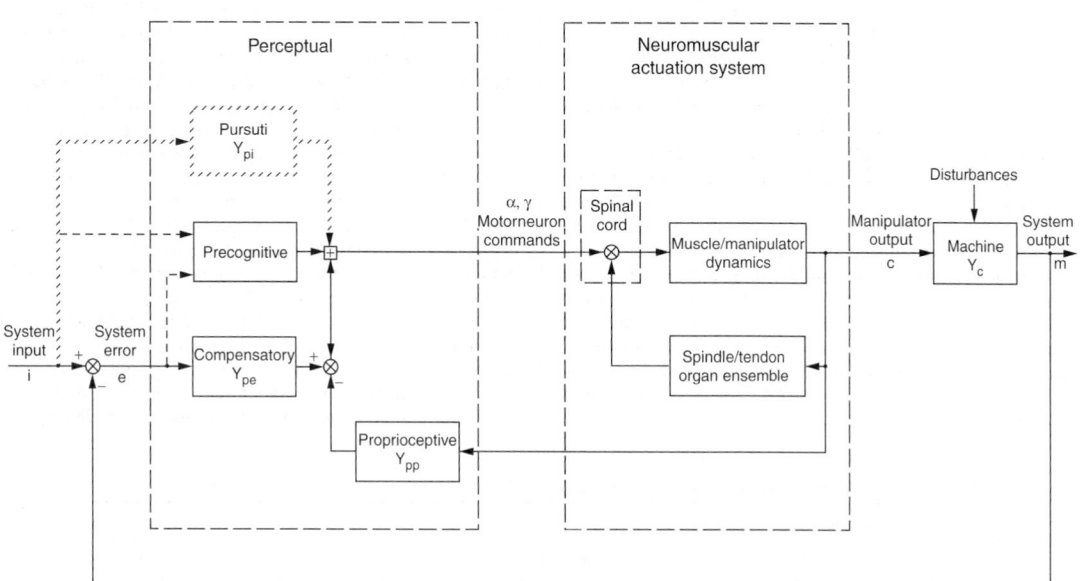

Fig. 17.6.2 Compensatory, pursuit, and precognitive pathways.

attempts to follow $i(t)$ with a cursor driven by the MMS dynamics. The perceptual pathways composing SOP for the human controller can be separated into sensory mechanisms that generate internal signals and to a central processor that integrates them, extracts coherence, and determines the human operator's dynamic responses. The describing functions and remnant depend explicitly on: the task variables, MMS inputs and disturbances, controlled element dynamics, and the operator-centered variables. The effects of these variables have been integrated into a set of rules for applying and adjusting human-operator describing functions for compensatory closed-loop control. The following conditions must apply before using these rules:

1. The forcing functions that act as inputs to the MMS are unpredictable, have bandwidth ω_i below 6.2 rad/s, and continuous waveforms.

2. The controlled element is a low-order system or can be so approximated and has no highly resonant modes over the input bandwidth.

3. The display and controls are reasonably well scaled and smooth so as to minimize the effects of thresholds, detents, and friction.

4. The other task demands are sufficiently light to permit the operator to devote the majority of his or her attention to minimizing the displayed error.

McRUER'S RULE

Under the foregoing conditions, which are commonly met in most operational control tasks, theory, experiments, and practice have shown that the operator compensates for the dynamics of the controlled element so that the open-loop characteristics of the MMS act like an integrator in series with an effective time delay in the frequency range most critical to system stability and performance. The frequency within this range where the open-loop gain of the MMS is unity is the crossover frequency ω_c. By compensating in this way, the operator creates a desirable, stable control system in accordance with good engineering design. This behavior by the human controller is known as **McRuer's rule,** and it is expressed by the crossover model:

$$Y_p(j\omega)Y_c(j\omega) \approx \frac{\omega_c e^{-j\omega\tau}}{j\omega}. \qquad (17.6.9)$$

Many of the basic qualities of feedback systems relate to ω_c, which is a close measure of the bandwidth of the closed-loop system. In the range of input bandwidth frequencies from very low up to and approaching ω_c, the output $m(t)$ of the closed loop system follows the system input $i(t)$, and the system error $e(t)$, is reduced. At increasingly higher input bandwidths beyond ω_c these properties are lost as $e(t)$ increases and $i(t)$ and $m(t)$ are no longer similar. For desirable closed-loop performance, ω_c must exceed the largest frequency in the input or in disturbances to the system that have appreciable power.

The effective time delay τ is the sum of perceptual and neuromuscular lags as affected by the demands of the tracking task.

As the controlled-element dynamics demand more lead from the operator to conform to the crossover model, the concomitant increased anticipation required of the operator imposes a cost. This is reflected in increases in τ, and affects the operator's evaluation of the MMS. Table 17.6.1 lists crossover model parameters.

As an illustration consider automobile heading control. At low to moderate speeds, this can be approximated by the controlled element $Y_c(j\omega) = K_c/j\omega$. Since the heading rate is proportional to the position of the steering wheel, the driver need only generate proportional control, i. e., zero lead. The effective time delay will then be 0.18 s and the maximum effective closed-loop system bandwidth will be the maximum crossover frequency or 4.7/rad/s.

Attitude control of a spacecraft with damper off can be approximated in the region of human control by $Y_c(j\omega) = K_c/(j\omega)^2$. Maintaining stability demands more of the operator than did automobile heading control, for the operator must generate a low-frequency lead. The cost incurred by control is an increase in the effective time delay τ to 0.32 s. A significant consequence is the reduction of the maximum system crossover frequency, and thus the maximum effective closed loop system bandwidth, to 3.3 rad/s.

For a single-loop control system of maximum crossover frequency ω_c, the attainable relative mean square error coherent with an input bandwidth, ω_i and variance, σ_i can be estimated:

$$\overline{e^2}_i/\sigma_i^2 \approx \frac{(\omega_i/\omega_c)^2}{3} \qquad (17.6.10)$$

Thus for similar MMS input characteristics, the estimated attainable mean square error for attitude control of a spacecraft is twice the estimated attainable mean square error for the less demanding heading control of an automobile.

MODELING THE HUMAN IN THE MMS

The crossover model is a first approximation in the process of elaborating models for human operators of increasingly more complex control systems in which there may be competing demands for the operator's attention. The process has been described in a reductionist fashion by analyzing the input-output behavior of the human into subsystem components. The sensory mechanism subsystems in Fig. 17.6.2 include optics, the retina, oculomotor muscles, eye dynamics, and the dynamics of linear acceleration and angular velocities acting on the vestibular and kinesthetic sensors. The central elements act to integrate and to equalize the outputs of these subsystems so as to generate commands to the neuromuscular actuation system, which is divisible into dynamic subsystems. These components are adjusted according to **McRuer's rule** so that the crossover model obtains and then is integrated to describe the operator's behavior in a structural-isomorphic model. In contrast, the algorithmic or optimal control model (OCM) proceeds in a holistic fashion and mimics the human's total response by the application of optimal control computational methods to those human properties subject to adaptation and hence optimization. The OCM proceeds by minimizing a quadratic performance index with an optimum linear predictor operating on estimated delayed state vectors to emulate the human's control actions. The computational procedure results in very high-order transfer functions for the human. PC-compatible programs, for example, Program CC Version 5, have been written for both the classical model and the OCM. Since **McRuer's rule** is the best understood, most widely applicable empirically tested description of human dynamics, the crossover model serves as the standard for comparing the classical model with the OCM model. This has been done for $Y_c = K_c/s$ and the comparison is good for frequencies less than 15 rad/s.

Manual control models present one perspective of the virtual human. Other perspectives are reach, strength, movements, and eye point of regard. Digital models exist which allow 3-D virtual humans to be placed in 3-D CAD-generated virtual environments. In the design of automobile, aircraft, spacecraft, or submarine interiors as well as factory work stations and large construction machinery, engineers are able to examine a human's activity in virtual environments and to sort through design options. One such digital human model, Jack/Jill, manufactured by EDS, has 69 segments, 68 interconnected joints, a 17-segment spine, 16-segment hands, coupled shoulder/clavicle joints, and 135 degrees of freedom to mimic the motions and positions of a wide range of humans.

Table 17.6.1   Crossover Model Parameters, ωi = 2.5 rad/s

Controlled-element dynamics	K_c	$K_c/(j\omega)$	$K_c/(j\omega)^2$
Operator lead unit to establish control	−1 (integral)	0 (proportional)	+1 (low frequency lead)
Effective time delay τ, seconds	0.14	0.18	0.32
Maximum crossover frequency ω_c, rad/s	5.7	4.7	3.3

Joint limits as derived from NASA studies and sizes and shapes based on SAE measurements and ANSUR 88 data round out the engineer's ability to select digital manikins for specific problems and representative of 5th, 50th, or 95th percentile human actors. The virtual human operator can thus be used to simulate a population of potential users and determine the diversity permitted by a particular design.

Risk management of slips could be addressed by Jack/Jill simulating a control room operator confronted with an arrangement of manual controls such as knobs or levers. By selecting 5th and 95th percentile operators and adding statistical variability to their head position, eye point of regard, and reach, the probabilities for slip errors occurring for configurations of manual controls under different time pressures could be estimated.

17.7 AUTOMATIC MANUFACTURING
by Vincent M. Altamuro

The production of many products involves both **fabrication** and **assembly** activities. Fabrication is the making of the component piece parts that are later assembled into the final product. Fabrication methods include casting, molding, forging, forming, stamping, and machining. Assembly may be manual or automatic. Automatic manufacturing can range from **semiautomatic** to **fully automatic,** depending on the number of human operations required. Automatic manufacturing installations may also be called **fixed** and **flexible** according to how easily they can be altered to make product variations. These several available modes allow the creation of a mode that draws on all of them—the **autofacturing** mode, that combination of the manual and the specific types of automatic operations which best achieves the quality, operational, and economic objectives sought.

The selection of assembly mode must be consistent with the design of the product, the personnel skills, the equipment available, and other factors. Manual assembly is suitable for low volume output and relatively short production runs with high product-to-product variations. Fixed automatic assembly is suitable for high-speed, high-volume production of uniform products. Also called *hard, dedicated,* or *conventional* automation, it can produce low-unit-cost items of uniformly high quality once its machinery is perfected and until a product variation is wanted. If a product's assembly cost is the sum of its *setup cost* (the cost of getting everything ready to make it) and its *run cost* (the cost of making it), then manual assembly has a relatively low setup cost but a high run cost, while fixed automatic assembly has the reverse. Flexible automatic assembly is suitable for midrange-volume production with variations in product specifications. Also called *soft* or *programmable* automation, it uses robots and other computer-based equipment that are more easily movable and adjustable so as to reduce setup costs and still enjoy low run costs. It fits between fully manual and fully fixed automatic assembly, as shown in Fig. 17.7.1.

Autofacturing is a production system that is comprised primarily of automated equipment which is configured as several integrated subsystems, using one common database and computer controls to make, test, and transport specifically designed products at high and uniform quality levels meeting flexible specifications with a minimum of human effort. There are many levels of autofacturing from individual cells, or *islands,* all the way up to a complete and integrated system. Most situations are somewhere in between, but progressing toward a total system.

Autofacturing is a subset of the overall field of automation. It refers to automated manufacturing activities as distinguished from the many other things called automation, such as data processing, office automation, electronic funds transfers, automated banking, central station telephone operations, airline ticket reservations, and the like.

An autofacturing system is composed of several integrated subsystems. Also, along with other systems such as marketing and distribution systems, general accounting and cost accounting systems, management information systems (MISs), and others, it is a subsystem of the corporation's overall operating system. All subsystems should be interconnected, complementary, in balance, and integrated into one total system. Some of the prerequisites and subsystems of an autofacturing system are listed and described below.

DESIGN FOR ASSEMBLY

After management decisions have been made regarding policy, practices, and long-range objectives, decisions must be made about the best combination of a multitude of characteristics of the product. These include the goals to be achieved by it: its salability, functionality, safety, targeted life, repairability, recyclability, ease of use, size, shape, color, and many other considerations, some reinforcing others and some in conflict with others. Designing the product for ease and economy of fabrication, assembly, test, handling, shipping, and installation are some important considerations. Then, the basis of all of the above determinations, decisions can be made regarding the best levels and mix of autofacturing. That includes deciding which portions of the product should be made by people, hard automation, or programmable devices, since the size, shape, and features of its several parts may be different and better suited for one mode over the others.

Human assemblers, for example, have the advantage of three-dimensional vision, color sensitivity, two hands, eye-hand coordination, the facility to jiggle parts together, the ability to back off when things do not go together as they should and look for the reason rather than force them, the ability to pick out patterns which are varied or complex, etc. However, they cannot work in dangerous environments, apply large or exact forces, perform with precision consistently cycle after cycle, or handle very large, very small, or very fragile items, etc.

Hard automation requires the input of a steady, voluminous stream of very uniform parts, all positioned and oriented precisely so that the high-speed machines can operate without jamming or stopping. Machines and instrumentation are better than humans in monitoring and responding rapidly and consistently to many stimuli simultaneously— some beyond the range of human capabilities. They also can be made to do several things at once for long periods of time with little variance or deterioration of performance.

Soft, or programmable, automation permits variations in inputs, as it can sense and adjust for deviations and continue to operate. It is usually faster than manual work but slower than dedicated machines, and its output can permit variations from the standard product so as to achieve marketing and inventory advantages that often more than offset its added cost per unit of output.

In autofacturing, each assembly operation is analyzed to determine whether it is best done manually, with fixed, or with flexible automation. An autofacturing system, then, results in a mixed mode of operation, each with that portion of the product designed to be assembled in the most efficient way available. In it, the actions of conventional machines, automated equipment, human operators, and robots are all interrelated and in balance.

There are over 100 guidelines on how to configure a product and its component piece parts so that they can be assembled as easily and economically as possible. The most important of these rules are as follows:

1. *Minimize and simplify.* Reduce the number of piece parts as much as possible. Determine the essential functions of each part. Transfer functions to other parts. Combine parts. Eliminate as many parts as economically feasible. Simplify before automating.

2. *Modularize.* Design the product such that parts are grouped into modules or sections of the end product. Create modules, like building blocks, so that they can be selected and joined in various ways to make different products. This will aid in assembly, test, repair, and replacement of more manageable subassemblies. It will also make it possible to follow rule 3 below.

3. *Create families of products.* Products should be designed so as to have as much commonality with other products as possible in their modules and piece parts. Products related in a series of escalating capabilities or features can all be built with the same platform (base, power supply, etc.), case, panels, circuit boards, and other common and interchangeable parts. Product distinctions, such as extra features, should be the last components assembled so that the work in process is common for as long as possible. Parts that make a product distinctive should be grouped into the same module. Even seemingly different products can be designed to have the same power supply, switches, gears, and other internal components and modules.

4. *Design parts to have as many different uses as possible,* depending on how they are installed and which portion of them is used. Determine whether the extra cost of the added complexity is more than offset by the benefits.

5. *Use the best overall method to make each part.* That is, consider not only its fabrication costs, but the relative costs of handling, subsequent operations, assembly, inspection, scrap, etc.

6. *Design in layers.* Where possible, design the product so that it is made up of layers of components, such that making it requires the adding of one layer on top of another, built up from the base to the cover or outer case. See Fig. 17.7.2.

7. *Assemble in short, straight vertical strokes.* If rule 6, above, is observed, then this should be possible. Avoid the need for lateral, curved, or complex motion paths in assembling components. Minimize the number of directions of assembly. Minimize the amount of lifting, rotating, and other handling of the components, subassemblies, or product. See Fig. 17.7.2.

8. *Present parts appropriately.* Bring parts to the machines, robots, and human operators in the position and orientation best suited to use without the need for rehandling or repositioning. Hold them in trays, on reels, in connected strips, etc. rather than disoriented in a tote box or bin. Parts stamped from strips or molded should be left on their webs as long as possible, for they are orderly. Parts to be fed to assembly stations by vibratory bowl feeders should be designed so as to feed through easily and not jam, snag, tangle, nest, shingle, bridge, or interlock.

9. *Design for symmetry and orientation.* To the degree possible, design parts to be symmetrical so that no matter what position they are in, it is proper for assembly. If perfect symmetry is not possible, strive

Fig. 17.7.2 Selected design for assembly guidelines.

to have the parts go together in as many ways as possible. Where that is not possible, accentuate the asymmetry and give the parts an orientation feature, so that a person or robot can easily sense it and orient the part correctly. See Fig. 17.7.2.

10. *Make errors impossible.* Design the product and its parts so that the components physically cannot be assembled wrong. If a part has internal or hard-to-sense features, add a feature that can be sensed, even if it has no operational function. Do not have the dimensions of a part be too similar if the part's proper orientation and assembly depend on discerning a small difference.

11. *Give each part a positive location.* Through the use of shoulders, recesses, guides, tabs, and the like, create an exact place where each component is to go. See Fig. 17.7.2.

12. *Minimize fasteners.* Eliminate or reduce to the lowest essential number the product's discrete fasteners. Nuts, bolts, screws, and the like are difficult and expensive to handle and assemble. They should be used only where future disassembly, adjustments, or other needs make them necessary. Very short screws, bolts, and rivets are more difficult to position than are their longer counterparts, as the weight of their heads often makes them fall out of the holes. It is more difficult to place the blade of a driver in screws and bolts with round heads and slotted tops than in those whose tops are flat with closed patterns for the driver's matching blade tips.

13. *Snap in.* Where possible have the parts snap together instead of being held with discrete fasteners. Plastic tabs can be added to a part while it is being molded at negligible extra cost. The operations of placing the part in position for assembly and of assembling it become a single action when it is snapped into place. See Fig. 17.7.2.

14. *Reduce variations.* Where fasteners must be used, have as few variations as possible. Seven or eight different screw sizes can often be reduced to two or three. The added cost of overdesign where a slightly larger size than needed is used is more than made up in purchasing, handling, inventory, record keeping, and tooling savings.

15. *Chamfer.* Assembly can be speeded up and less expensive tooling can be used if holes are chamfered and pins are slightly pointed and rounded. See Fig. 17.7.3.

16. *Avoid problem parts.* Flimsy, amorphous, and long flexible parts such as wire, fabric, insulation, and the like are difficult to handle and should be avoided or provided for in the process. Some coil springs tangle because of their helix and wire diameter or if they are open at their ends.

17. *Help the robot.* Give different parts a common feature so that the same gripper can be used to handle them all. Changing grippers takes

Peg not chamfered
Hole not chamfered

Not Good

Peg not chamfered
Hole chamfered

Better

Peg chamfered
Hole not chamfered

Better

Peg chamfered
Hole chamfered

Best

Fig. 17.7.3   The use of chamfers to accommodate slight misalignments.

time. Extra or complex grippers add to expense. Allow space around components so that the gripper can get in to assemble the part. Design the grippers to fit the contours of the parts or to wrap around them rather than merely hold them with pressure and friction. This will minimize slippage and movement.

AUTOFACTURING SUBSYSTEMS

Material Holding, Feeding, and Metering   In the progression from raw materials and purchased parts to the finished product, the first autofacturing system subsystems are those whose functions are to hold, feed, and meter the correct amount of input to the operating stations at the correct time. In designating the holding system, the material must be described in detail. Then the required holding status must be defined. Is it to be held in bulk form or as discrete items? Is each item unique or the same as all the others? Must each item be segregated from the others or may they all be commingled at this stage? Then, the best-suited holding devices must be specified, e.g., bins, hoppers, reservoirs, reels, spools, bobbins, trays, racks. Subsequent to holding, the material or parts are to be fed into the next stage of the process. Again, a series of determinations must be made. Are parts to be fed individually or en masse, in metered amounts or at any rate, separated or connected, oriented or unaligned, selectively or randomly, pretested or untested, interruptible or nonstop? Then the most suitable feeding and metering devices must be specified, e.g., dispensers, pumps, applicators, vibratory feeders, escapements. See also Sec. 10.1.

Operations   The operating stations of the system include those machines, tools, and devices that convert the raw materials to parts or finished products, or bring parts together and assemble them into finished products. It is at these stations that the casting, molding, bending,

machining, soldering, welding, riveting, bolting, snapping together, painting, and the like are done. The operators of the machines may be humans or robots or the machines may be designed to operate automatically or under the direction of a computer.

Transferring   When more than one operation is required, means of transferring the work in process from one machine or station to the next must be provided. In many companies, the time that material is on a machine and being processed is much less than the time it is being unfixtured from the machine, moved to the next machine and fixtured in it, with several delays, storages, and handlings in between. Work can also be transferred while on a machine, but to other tooling or test stations of it. Some such machines are the in-line (or straight line) pallet (or platen) and plain (or nonpallet) transfer machines and the dial (or turret) rotary table machines. These may move continuously or intermittently (indexing). They may be of the trunnion, center column, or shuttle-table type and may be purchased as standard or special custom designs. Robots and other mechanisms are often used to transfer work from station to station within a machine and between machines. See Secs. 10.5 and 10.6. It is important that the inputs and outputs of the individual stations and machines be in balance, in order to avoid idle equipment and work-in-process buildups.

Sensing   There will be many points within an autofacturing system at which the sensing of conditions will be required. Sensing is a function which usually must be done prior to measurement, inspection/test, counting, sorting, computing, actuating/moving, feedback/control, and correction. Sensors can be used to measure position, presence or proximity, characteristics, or conditions of the parts, product, machines, tooling/fixture, equipment, people, and environment. Contact-type sensors include mechanical, electromechanical, electrical, chemical, and air jet. Noncontact-type sensors include permanent magnet, electromagnet, capacitance, inductance, reluctance (magnet and variable), sonic/ultrasonic, photoelectric/optoelectronic, laser, radio frequency, eddy current, thermal and relative expansion, Hall effect, air (pressure and fluidic), and inertial/accelerometer.

In deciding whether to use a sensor, where, and what type, the questions to ask are

1. What *should* happen?
2. What *could* happen?
3. What measurable phenomena, conditions, data, features, properties, changes, etc. will be present to distinguish item 2 from item 1?
4. When is the earliest that item 3 can be detected?
5. Where? How?
6. What is the best sensor or combination of sensors to do this?
7. What should be done with the information sensed?

Measurement, Inspection, and Test   In autofacturing, it is essential that the product be inspected while on the machines and while it is being made. The objective is to make only good products. To do this, critical parameters are sensed and measured in real time. If they begin to drift beyond preset statistical limits, error signals are fed back and adjustments in the machine are made automatically to bring the variables back in control before any bad products are made. It is important, therefore, that the product be designed so that there is access to inspection and test points during processing. It also means that the machines should include sensors and inspection stations amid the processing stations.

Equipment Monitoring and Performance Maintenance   In addition to the product, all of the machines and equipment must be monitored constantly to assure continued peak performance and to avoid breakdowns. Again, a multitude of sensors, clocks, and counters is a must for the measurement, feedback, correction, and control of the status of the system. For tooling, where wear is certain, automatic tool changers that pull the used tool out and replace it with a new one can be used. An operator or maintenance person then removes the used tool from the device and loads it with another new one so the tool changer is ready again. In some cases, duplicate modular workstations are kept on hand so that a malfunctioning one can be replaced rapidly.

Sorting, Counting, and Marking   After the product is made, its characteristics can be sensed and measured and it can be sorted according to whatever decision rules are set. Deflection devices—such as conveyor belt gates, air jets, and the like—can be used to divert each class of product to its own receptacle. They can then be counted (with the data going to the computer) and marked or labeled, as desired.

Wrapping and Packaging The wrapping and packaging of the product is just as much an integrated part of the overall system as is assembly, and should be just as automated. Even the packing cartons should be designed to facilitate automated operations. Instead of conventional flaps that tend to close and block the robot's path in trying to put products in, boxes should have straight sides and a lid so that both the products and the lid can be handled in straight, vertical motions.

Automated Material Handling and Automated Warehousing An aim of autofacturing is to have the raw material and purchased parts flow from the receiving department to the shipping department as smoothly and with as few stops, temporary storages, and handlings as possible as they are gradually converted to finished products. To do this, automated material handling and warehousing equipment must be installed whose capacities are in balance with the outputs of the production machines. The equipment available ranges from forklift trucks and automatic guided vehicles (AGV) to robots and automatic storage and retrieval systems (AS/RS). See Sec. 10.7 and the other sections cited above for descriptions of such equipment.

Order Picking, Packing, and Shipping The automation of the factory does not end with the production and storage of the products. As orders are received, the correct number of the correct items must be picked, packed, and shipped from inventory. A perfect autofacturing system would be one wherein the production and demand for items is so balanced that products are shipped directly from the final production and inspection station. In all cases, the receipt of orders should be entered into the same computer system that schedules production so that the match between what is being sold and what is being made is as close as possible.

Operator/Machine Interfaces Even the most automatic machines occasionally require the attention of a human attendant or maintenance person. The fields of study that examine the human/machine interfaces and the degrees of effort and responses of people at work are called *human engineering, human factors engineering, biomechanics, ergonomics,* and *cybernetics.* While each concentrates on a particular aspect of the subject, they are all concerned with matching the needs and capabilities of humans to the designs of the machines (and total environment) with which they interface. Humans control machines via buttons, knobs, handles, switches, keyboards, voice recognition devices, and the like, and these must all be designed, colored, located, etc. to minimize human effort and errors and to maximize effectiveness. Machines output information to humans via lights, sounds, dials, gages, video screens, and the like. These, too, must be designed to maximize response and minimize errors. At all times, the human/machine interface must be designed with safety in mind. See Sec. 17.6.

Communication, Computation, and Control The system's robots and other equipment have a multitude of microprocessors embedded in them. That, and the fact that they also have many sophisticated sensors and heuristic programs justifies calling them "smart" machines. They, in turn, are controlled by more powerful microcomputers that report to minicomputers, thence to a powerful central host computer. The host computer does not control each detail of the system; it merely integrates and balances a distributed and hierarchical communication, computation, and control capability that uses the same database and is linked by a local-area network (LAN).

Auxiliary Systems There are several ancillary systems that can be employed to enhance automated manufacturing and the autofacturing system. Some of these are group technology (GT), concurrent engineering or simultaneous engineering, and computer-aided design and computer-aided manufacturing (CAD/CAM) in the product design and prototyping stages; just-in-time (JIT) delivery, material requirements planning (MRP), and manufacturing resources planning (MRP II) in the production planning and inventory control stages; computer-aided testing (CAT) and total quality control (TQC) in the production, inspection, and testing stages; computer-integrated manufacturing (CIM) in the communication, computation, and control stages; and management information systems (MISs) in the reporting and analysis stages.

MICRON AND NANO AUTOFACTURING

As some products get smaller for operational, energy, space, and economic reasons, they can reach the point of size reduction to where human workers cannot see them, let alone manipulate them for fabrication, assembly, and test. In such cases, microelectromechanical systems (MEMS) and nanotechnology methods and processes, rather than human methods, must be used, and an autofacturing mode is essential. These very small, very thin, very elemental parts, components, sensors, activators, devices, and products are becoming prevalent in computers, electronic circuits, communications, medicine, warfare, automotive, appliances, robots, toys, games, and other products. Micron technology deals with things in the size of a millionth of a metre (0.00004 inches). Nanotechnology deals with things in the size of a billionth of a metre (0.00000004 inches).

MEMS processes are similar to those used to fabricate integrated circuit "chips" or microprocessors, such as photolithography, thin film deposition, and etching. But where only one side of a wafer is typically processed to create integrated circuits, MEMS fabrication typically processes both sides of a wafer and sometimes a stack of them to create 3-D items. The MEMS processes are still in development, but they may be divided into four technologies: (1) bulk, a subtractive process in which material is selectively removed until the desired configuration is achieved; (2) surface, an additive process in which layers of various materials are added to a substrate; (3) hybrid, a combination process in which bulk techniques are used on surface processed substrates; and (4) high-aspect ratio, which uses deep ultraviolet lithography and x-ray techniques to create micron-sized items out of polymers, metals, and ceramic materials. Some MEMS products in existence combine sensing capabilities with information processing to detect, interpret, activate, and control systems. They are being used in automobile air bags, robots, aircraft guidance systems, missiles, and elsewhere.

In nanotechnology, the goal is to produce items much much smaller than those micron sized, striving for the range of 10 to 100 nanometres. The dot at the end of this sentence can hold approximately 350 million nanometre-sized objects. The measurement of objects in the nanometre range is not new; what is new is nanomanipulation—the creating of objects in that range. To create new, better, and useful materials and devices of this size requires mastering the science and art of observing, moving, manipulating, rearranging, and positioning individual atoms and molecules of matter. To obtain the item sought, it is built atom by atom, molecule by molecule, and causing them to arrange themselves in the structure desired. To accomplish this not only is a full autofacturing mode essential, but special apparatus such as scanning probe microscopes (SPM), scanning tunnel microscopes (STM), atomic-force microscopes (AFM), in addition to photolithography and soft lithography having smaller feature sizes, and others.

There are two approaches being pursued: nanofabrication, also called nanoengineering, which is the building of the item with man-made tools; and self-assembly, which is causing the atoms and molecules to move together and adhere in the desired way and shape, based upon their size, shape composition, and physical properties—much like nature causes soil, water, air and the energy of the sun to combine to form trees and vegetation. Some areas where nano items may find application are in pharmaceuticals, chemicals, energy, agriculture, food, bar codes, communication, electronics, avionics, transportation, motors, military, materials, and a wide range of everyday products. One problem with creating such small objects is that atoms do not always act according the laws of classical physics; at the nano level quantum mechanics phenomena may occur.

The Regulatory Environment

BY

PASCAL M. RAPIER *Scientist III, Retired, Lawrence Berkeley Laboratory.*
ANDREW C. KLEIN *Professor of Nuclear Engineering, Oregon State University; Director of Training, Education and Research Partnership, Idaho National Laboratories.*
RAYMOND E. FARRELL, ESQ. *Partner; Carter, DeLuca, Farrell and Schmidt, LLP, Melville, N.Y.*
BERNADETTE M. BENNETT, ESQ. *Associate; Carter, DeLuca, Farrell and Schmidt, LLP, Melville, N.Y.*

18.1 ENVIRONMENTAL CONTROL
by Pascal M. Rapier and Andrew C. Klein

REFERENCES: "McGraw-Hill Encyclopedia of Environmental Science." Beckmann, "Health Hazards of Not Going Nuclear," Golem Press, Boulder, CO. Edinger et al., "Heat Exchange and Transport in the Environment," P.B. 240757 NTIS, USDA, Nov. 1974. Corbitt, "Standard Handbook of Environmental Engineering." McGraw-Hill. Current updated revisions of pertinent local and federal regulations.

PRINCIPLE OF ENVIRONMENTAL CONTROL

Nature has provided two almost inexhaustible sumps for maintaining a steady state environment on earth. The first of these is the 2.7 K background temperature of absolute space, which nature uses for heat rejection to close its heat balances. The second is the oceans, which serve to close the material balances of its cyclic processes by accepting the combined runoffs of the continents. The greatest engineering progress comes when people control their environmental activities so as to take maximum advantage at minimum cost of these sumps and of nature's cyclic processes. This is the **basic principle** upon which the science of **environmental control** is founded. With it, any environmental problem can be modeled on a conveniently small scale and then solved precisely by LeChatelier's controlled equilibria principle for any large-scale operation. When the principle is applied to obtain optimum yield of a desired end product or result, it will help predict the necessary input compositions, pressure, temperature, and concentration. This same principle is applied in other diverse fields of study such as genetics.

NATIONAL POLICY

The National Environmental Policy Act of 1969, Public Law 91-190 (NEPA), concerns air, water, and land quality and conservation of natural resources. Section 2 of NEPA declares a national policy which will encourage productive and enjoyable harmony between people and their environment and which will enrich their understanding of ecological systems and natural resources.

Section 101(b) (2) provides for the "widest range of beneficial uses of the environment without degradation, risk to health or safety, or other undesirable consequences."

Section 101(b) (5) provides for achieving "a balance between population and resource use which will permit high standards of living" (e.g., cost/benefit balancing).

Section 101(b) (6) provides for enhancing the quality of renewable resources and the maximum attainable recycling of depletable resources (e.g., developing breeder reactors to extend nuclear fuel reserves).

Section 102(c) requires that all agencies of the federal government shall file environmental impact statements. These detail in a published document for each case: (*a*) its adverse environmental effects, (*b*) its alternatives, (*c*) its enhancement of long-term environmental productivity, and (*c*) its irreversible and irretrievable commitments of natural resources.

Legislation most responsive to the NEPA objectives so far is the Magnetic Fusion Energy Act of 1980. Aiming for a working magnetic fusion system by the year 2000, this act set as a goal a controlled nuclear fusion economy by the middle to late twenty-first century. (See Considine, "Energy Technology Handbook," McGraw-Hill.)

AN OVERVIEW OF ECOLOGY
(See also Sec. 9.1.)

Ecology is the study of the **biosphere,** a hypothetical system comprising the surface of the earth and all the subsystems necessary to maintain a steady-state life-support system. The **thermal cycle** is open, isenthalpic, and irreversible. Solar energy is utilized, becomes degraded, and is finally rejected into outer space by means of long-wave radiation through the optical window of the earth's atmosphere. Major quasi-cyclic processes in the biosphere, necessary for life support, are the **hydrological cycle,** the **carbon cycle,** the **nitrogen cycle,** the **potassium cycle,** the **phosphorous cycle,** and the **sulfur cycle.** These are almost closed cycles, which are occasionally interrupted and renewed by techtonic geological upheavals.

In these cycles, vegetation and animal wastes decay, thereby furnishing nutrients to land, air, and water. New green plants grow, animals eat them, and the recycling continues. Ecologically, the bacteria which decompose all these wastes are called *destroyers,* the green plants which employ photosynthesis are called *producers,* and animals which eat the plants are called *consumers.* A hypothetically bounded small portion of the biosphere is called an **ecosystem.** All ecosystems are **open,** as their essential cycles extend into the biosphere.

An ecologically idealized community would be one in which the ecosystem, including the surrounding agricultural area, would operate steady-state, almost closed cycle. Tucson, Arizona, approaches this state, its crops being irrigated by tertiary sewage treatment so that water and all the plant nutrients, are recycled. Minerals entering the ecosystem through the Santa Cruz River and imported fuels are almost counterbalanced by exports of crops from the area and by irrigation runoff into the normally dry Gila River bed. These activities conform to nature's naturally occurring cycles, and maintain the fertility of the soil.

In larger, heavily overpopulated communities, however, people's activities are counter to nature's, and so humanity suffers. The earth is denuded of local vegetation, and crops are imported. All the nitrogen, potassium, sulfur, and phosphorus contained in the agricultural commodities and fuels, instead of being recycled to the farmlands, are burned or handled by the waste disposal systems of the communities and dumped into rivers and harbors, thereby reaching the oceans and being lost to the continents, despite the recycling and soil fixation methods intended by nature to prevent such losses. Severe pollution problems result, mainly from nonpoint sources and storm run-offs which are not amenable to control, no matter how much money is spent. But the major damage done is depletion of wetlands, forests, and farmlands and buildup of atmospheric CO_2 from fossil-fuel use. It is irreversible and irretrievable and will continue until our society finally faces the fact that the changes required to help stem these events must be made on a local as well as planetwide level.

Environmental control seeks to subdue and to utilize nature's ecological cycles in order to serve people's needs, thereby conserving natural energy and mineral resources, and to replenish desirable local flora and fauna populations by agriculture and cultivation to provide adequate food, clothing, and shelter. Environmental control seeks to extend depletable fuel supplies with clean, abundant forms of gravitational, solar, and nuclear energy (fission and fusion). Also, replenishable substitutes for other depletable resources are sought, as well as recovery and recycling means for scarce and irretrievable substances. Environmental control seeks to conserve land and water quality by diversion into adequately controlled drainage canals of concentrated runoffs, e.g., from irrigation, municipalities, industries and mining, so that these nonrecyclable, unsalvageable mixtures are ultimately discharged into the oceans, where they blend inconsequentially with the vastly larger amounts of runoff from the continents, occurring naturally as part of the hydrological cycle. The public standard of living is highest when all these things are done voluntarily by a responsible, cost-conscious citizenry.

The following subsections describe typical, cost-effective, and productive practices among the free nations. These should be considered to the extent permitted by local regulations.

COST/BENEFIT BALANCING AND ISO 9000 REGISTRATION

REFERENCES: Summary of UWAG-EEI Comments on EPA's Proposed Regulations under 304, 306, and 316(a) of Public Law 92-500, the Federal Water Pollution Control Act Amendments of 1972, June 1974. Effluent Limitations Guidelines for Steam Electric Power Generating Point Source Category 40 CFR, Chapter I, Subchapter N, Part 423. Craig, "The No Nonsense Guide to Achieving ISO 9000 Registration," ASME.

Cost/Benefit Calculations

In order to assess the feasibility [under Section 304b (1B) of the FWPCA Amendments of 1972] of any environmental control technology or augment, both the public cost and benefit of its operation must be considered. For demonstrating feasibility, this dimensionless cost/benefit ratio has the limits

$$0 \leq C/B \leq 1$$

The upper limit of unity means that $1 spent in the incremental public operational costs of the augment will yield $1 of return in public benefits. If an augment is required for which the cost/benefit ratio is greater than unity, it is unfeasible. (It is more feasible to accept the pollution.) If an augment is required for which the cost/benefit ratio is negative, the public benefit is negative and constitutes a public detriment. The more that is spent on such an augment, the worse the environmental degradation becomes.

NOTE: Financial items cited in the examples below are for illustrative purposes only, and will vary with location and time.

EXAMPLE. Once-through vs. recirculating cooling systems: A 1,000-MW(e) steam/electric generating plant is operating with once-through cooling on a river. The heat rate is 10,300 Btu/kWh (0.33 thermal efficiency), with about one-half of the heat, or 5,150 Btu/kWh, being discharged into the river. The public benefit of this thermal enrichment is a stable and abundant fish population, which is estimated to offset all public detriments except occasional fish kills caused by emergency shut-downs. These average each year not more than 8-h duration and $45,000 public detriment.

Proper environmental management to provide emergency heating of the water during unexpected outages will avoid the fish kills with an estimated 90 percent confidence factor. A cooling tower retrofit will reduce the thermal discharge by 90 percent but will substitute some undesirable atmospheric, noise, and stream pollution conditions. Neglect these problems, but otherwise compare the feasibility of an open-recirculation cooling tower retrofit with that of proper environmental management.

First alternative: Environmental management of existing plant. The annual public detriment for the fish kills is $0.045/kW(e) of installed capacity. For avoiding these by emergency heating of the river water, let the estimated operating cost to the consumer be $1 per million Btu. The annual environmental management cost for an 8-h outage per year becomes:

$$8 \text{ h} \times \$.00515/\text{kWh} = \$0.0412/\text{kW(e)}$$

The cost/benefit ratio becomes $0.412/$.045 or 0.916, so that 91.6¢ spent will yield $1 return in public benefits. This is a wise investment. For the existing plant the possibility of a fish kill is reduced to one-tenth its original value, and the remaining annual public detriment becomes $0.0045 per kW(e) of installed capacity.

Second alternative: Retrofit of an open recirculating cooling tower. The viability of a cooling tower retrofit must be assessed on the basis of this reduced public detriment for the existing plant. The estimated annual operating cost to the public for a cooling tower augment, based on public operational overhead, a 1970 overall plant investment, and a 3 percent incremental capacity loss, due to the retrofit, is $20/kW(e) of installed capacity. Dividing this by the $0.0045 expected public benefit yield a cost/benefit ratio of 4,444. This means that $4,444 must be spent (and the cost passed on to consumer) for every dollar of public benefit. As this is clearly not in the public interest, it is unfeasible. It is cheaper to replenish the fish supply. Moreover, the 3 percent capacity loss of power generation due to the augment results in a 3 percent wastage of fuel and a 3 percent increase in the total thermal discharge to the environment. Energy, in contrast to the fish population, is a nonreplenishable natural resource. It must be conserved, and it must be considered when assessing the public detriment. Valued at 3 percent of the heat rate, $1 per million Btu and 0.9 load factor, this waste of natural resources will cost $2.435 per kW(e) annually.

Loss of cooling water by evaporation in the cooling tower is also a public detriment that can be conservatively valued at 25¢ per 1,000 gal. At 5,150 Btu/kWh

thermal discharge rate and 0.9 evaporative load factor, this depletion of a natural resource will cost:

$$0.9 \times 5{,}150 \times \$0.25 \times 24 \times 365 \div 8.33 \times 10^6 = \$1.219/\text{kW(e)}$$

Subtracting these two detriments, amounting to $3.654, from the annual public benefit of $0.0045 for conserving the fish population yields a negative public benefit due to the augment of −$3.65. The overall cost/benefit of this installation therefore becomes

$$\$20/(-\$3.65) = -5.48$$

The negative sign indicates that money spent to improve the environment by this means simply makes matters worse. The expected public benefit is not possible, and therefore the technology is not economically achievable.

Achieving and Practicing ISO 9000 Registration

Both capital and operating costs of a new or reorganized operation or venture can be minimized through the mechanism of ISO 9000 registration. In this way parts, machinery, construction, management, and operating and servicing techniques are standardized locally, and ultimately internationally. The favorable economics that ensue are of immeasurable consequence in a competitive marketplace and contribute to the optimum utilization of labor, facilities, and raw materials. A large portion of the benefits which accrue are related to personnel involved in the business entity.

Direct Water Cooling

For large-scale power generation, natural surface cooling in open water bodies is most cost-effective. The intake and outfall should be so oriented as to minimize possibility of heat recirculation and short-circuiting, which can usually be done by locating the outfall sufficiently down current of the intake.

Surface discharge of the condenser water as a hot, buoyant plume requires the least area of water surface for heat dissipation, yields the largest value for the overall surface heat transfer coefficient K, and hence minimizes both the cost and environmental impact of the facility.

The surface heat-transfer coefficient K combines the heat loss contributions due to longwave radiation, evaporation, and convection into a function which can be reliably estimated by two independent equations (J. E. Edinger, 1974):

$$K = H_s (T_s - T_{dp})^{-1} \tag{18.1.1}$$

$$K = \frac{T_s - 183.16}{20} + (0.47 + B)f(W) \tag{18.1.2}$$

where K = overall surface heat-transfer coefficient, W/(m² · K); H_s = gross 24-h average sunshine factor, W/m²; T_s = water surface temperature, ambient, K; T_{dp} = dew point temperature, K; W = wind speed, m/s; $f(W)$ = wind speed function, $9.2 + 0.46W^2$ (Brady, 1969), W/(m² · mm Hg); B = slope of vapor pressure curve at $T_m = \frac{1}{2}(T_s - T_{dp})$, mm Hg/K.

The derivative of the 1967 ASME steam table saturation line B is given accurately over the temperature range 0 to 50°C by the Rapier log-hyperbolic curve fit

$$B = \frac{5{,}301.734}{T_m^2} \exp\left(20.94673 - \frac{5{,}301.734}{T_m}\right)$$

The overall heat transfer coefficient K can be most accurately obtained from Eq. (18.1.1), since all the variables can be directly measured; then the wind speed function $f(W)$ can be determined for the particular site by substitution of K into Eq. (18.1.2). On the other hand, if T_s is not known, both T_s and K can be approximated by iteration between the two equations, after adopting a reasonable wind speed function, such as the Brady parabola, given above.

Once the value of K has been determined and the temperature rise through the condensers has been decided upon, the initial temperature difference or excess above ambient lake surface temperature at the discharge point, and the final temperature difference or excess above ambient at the plant intake, can be readily calculated, since the effective area A_e in square metres from the discharge point up to the intake determines the log-mean temperature difference, as given by the equation

$$\text{LMTD, °C} = \frac{\text{thermal discharge rate, W}}{KA_e}$$

At locations where limited quantities of water are available, some heat recirculation occurs and, particularly during summer months, can cause temperatures in excess of 32°C at the inlet of the plant. When considering such a location, tradeoffs should be made to determine whether it is more economical to accept a higher temperature at the inlet or to install auxiliary cooling means such as towers or spray ponds.

EXAMPLE. During July, a proposed inland lake site has a gross sunshine factor of 387.3, a dew point of 13.9°C, and a wind speed of 4.07 m/s. Iteration yields 26.1°C for the water surface temperature and 31.8 W/(m² · K) for the heat-transfer coefficient. If thermally loading the lake will cause a temperature excess of more than 8° at the intake, tradeoffs should be considered here.

CONTROL OF THERMAL DISCHARGES
(See also Sec. 9.5.)

REFERENCES: Water Resources and Waste Heat, *Power Eng.,* pp. 26–33, June 1973.

The total solar flux reaching the earth is 182 trillion kW; 32.4 trillion kW is utilized in the hydrological cycle, whereby freshwater distilled from the oceans is utilized on the continents by the plant and animal kingdoms and finally drains back into the ocean to complete the cycle. The entire thermal demand estimated for the United States in 1980 was 5 billion kW, very small compared with the solar flux in the earth's biosphere. The temperature of the earth's biosphere is stabilized by reradiating the solar flux as long wave radiation back into absolute space through the "optical window" (7- to 13-μm wavelength) of the earth's atmosphere. For example, if once-through cooling for 25,000 MW(e) power generation were added to Lake Michigan, the average surface temperature rise would be limited to ¼°F by longwave radiation from the surface of the lake.

The Joint Committee on Atomic Energy, 93d Congress (1973), anticipated that by 1980 the total heat dissipated into the environment by the generation and use of energy in the United States will be equivalent to 43.2 million bbl of oil per day. The combined heat loss caused by electric and industrial power generation will be 11.5 million bbl per day, or 26.6 percent, which includes all the point sources. Distributed and diffuse sources, such as automotive, residential, commercial, and industrial heating and cooling systems, which constitute the balance, account for 73.4 percent of the heat dissipation, and create "heat islands" around every major metropolis. About half the heat loss from power generation is presently discharged into bodies of water via nonconsumptive once-through cooling systems and does not contribute to the "heat island." The use of consumptive evaporative cooling for this would add to the atmospheric discharges in the "heat island" and raise the ambient temperatures. By 1978, increased urbanization in the Los Angeles Basin will raise the ambient temperature an estimated 5°F. The surface temperatures of water bodies in the Basin would be raised over 5°F, which exceeds guideline limitations for thermal discharges. By the year 2000, if evaporative cooling towers are required for all the incremental power generating capacity, the New York-Philadelphia "heat island" will suffer a mean ambient temperature increase of about 15°F, and the cooling towers will consume the flow of the entire Colorado River Basin, 31,000 ft³/s, or 3,000 times their 1970 consumption of water. Not included is the "greenhouse" effect of excess CO_2 in the atmosphere from the burning of fossil fuels. An estimated trillion tons of CO_2 will have accumulated, and trapped solar radiation will have raised the earth's mean temperature 3.6°F (2 K). (See Plass, Carbon Dioxide and Climate, *Sci. Amer.,* 41–47, July 1959, pp. 41–47. Also pp. 173–175, Sept. 1983.)

For the year 2000, the percentages of total heat discharged to the environment from various human activities have been estimated as shown in Table 18.1.1.

Table 18.1.1   Heat Discharged into Environment; Year 2000 (Estimate)

Thermal discharge source	Coolant	Total heat, %
Residential and commercial	Atraospheric and water	31.75
Industrial	Atmospheric and water	27.00
Transportation (all forms)	Atmospheric	22.20
Electric power generation	Atmospheric and water	14.30
Electric power consumption	Atmosphere	4.75
Total heat into environment		100.00

The data and projections cited above are dated, and certainly the current situation is different from all projections, but only in degree. The data in Table 18.1.1 are useful in that they address a global problem. Increased use of nuclear fuel to power central-station electric generating plants reduces the heat discharged into the atmosphere from exhaust stacks, but there is almost the same amount of rejected heat dissipated to condenser cooling water or to the atmosphere in cooling towers. Likewise, there is much controversy among the polity in the matter of the greenhouse effect, ozone holes, mandated use of electric-driven vehicles, and so on. The solutions offered to ameliorate those problems are not universally acceptable, but recognition of the existence of problems in this regard is not usually contested. In certain quarters the solution prated is for so-called renewable sources of energy which are "benign" (i.e., solar energy), but the technology is not refined yet to make the extraction of solar energy competitive. Much intellectual effort and well-intentioned persuasion must be brought to bear on this and associated type problems as the population grows and the consumption of fossil fuels continues apace.

The largest thermal discharge, 31.75 percent, is primarily attributable to residential and commercial heating and ventilating systems, with air conditioning units making the most undesirable contributions to the environmental "heat island" effect. Cogenerative district heating, better thermal insulation, and reduction of electrical loads (such as by replacing electrical heating units with steam or gas-fired units and atmospheric air conditioners with once-through, water-cooled units) could substantially reduce the undesirable side effects.

The second largest thermal discharge is from industry. Because of keen marketplace competition, thermal processes have been selected to provide the most favorable cost/benefit ratio for the consumer. Regulatory control here, except to prevent malpractices, can do more public harm than good. Industrial/nuclear power complexes, where practicable, will produce the maximum public benefit.

Transportation is the third largest contributor to the "heat island" effect and the one with the worst side effects. Thermal discharge to city streets could be reduced to one-third its present value and petroleum stocks conserved if all-electric transportation could be substituted economically for the internal combustion engine. This would be particularly advantageous in the summertime.

The smallest contributor to the "heat island" effect is electric power generation, which accounts for 7.15 percent of the total atmospheric discharge, with another 7.15 percent thermal discharge to water. Irrigation canals, navigation canals, and artificial waterway complexes designed to include treated municipal effluents could, if implemented in the future, provide sufficient water for a significant number of thermal power generating units, save energy, and reduce the atmospheric discharge near cities by as much as 7.15 percent.

In short, the environmental impact of heat discharged by human activities can be minimized and fuel conserved by cascading processes to get the highest practicable thermodynamic efficiencies and the lowest practicable heat-sink temperatures. Greater efforts must be brought to bear in the matter of curtailment of CO_2 emissions from all sources, inasmuch as they are recognized to have a deleterious effect in polluting the global environment as well as bringing about higher ambient temperatures. Reforestation and/or the maintenance of green spaces and the increased utilization of nonfossil fuels are looked upon with favor in these regards.

Use of Cooling Towers
(See Sec. 9.5.)

REFERENCES: Nomographs for Thermal Pollution Control Systems, EPA 660/2-73-004, Sept. 1973.

In locales where water supplies are inadequate to provide nonconsumptive once-through cooling, evaporative open-recirculating cooling towers become necessary. These consume water at about 2 percent of the recirculation rate, but reduce circulation into rivers or streams by over 95 percent even when employing low concentrating cycles. Nomographs are available from EPA which provide cost comparisons of the cooling systems sampled by Table 18.1.2. Selection of the most feasible cooling system depends on specific site conditions. Economics dictate once-through cooling where possible, wet cooling towers or ponds where water supplies are limited, and dry cooling towers where water is unavailable. All of these will contribute significantly to the "heat island" effect unless specifically designed to provide buoyant plumes that will puncture inversion layers.

Table 18.1.2   Cooling System Augments

Cooling systems	Relative capital cost	Power consumption
Natural draft wet towers	3.2	Low
Mechanical draft wet towers	1	High
Spray ponds	2.3	High
Cooling ponds	1.4	Low
Natural/mechanical draft dry towers	7.3/6.7	Medium/high

Wet cooling towers substitute water pollution for thermal pollution by concentrating blowdown, require chemical treatment, and release heavy-metal corrosion products into the blowdown stream. Their operation requires an optimized water chemistry which balances the acid treatment and cycles of concentration to provide nonaggressive water with a slightly positive Langlier index, and to retard the buildup of scale. The Langlier (saturation) index is the difference between the actual measured pH and the calculated pH_s at saturation with $CaCO_3$. A zero Langlier index balances pH, alkalinity, and calcium hardness, and establishes the threshold for alkaline scale formation. (For further information on this and other water treatments, see Betz, "Handbook of Industrial Water Conditioning," Betz Laboratories.) For environmental reasons, pretreatments based upon zinc, chromate, and phosphate inhibitors should generally be avoided, while posttreatments to remove pollutants and corrosion products from the blowdown may have to be employed. With all constraints considered, the most economical design results from a series of tradeoffs. Dry cooling towers are justified only where water shortages preclude the use of evaporative coolers, as their cost, inefficiency, and environmental impact from abnormal heating of the air are otherwise excessive.

Cooling towers may be required by environmental regulations, and "zero discharge" of blowdown may also be required, with high concentrating cycles and "perpetual storage" of the resultant salty wastes. The salty plume will cause additional difficulties. Blowdown volume can be reduced by concentrating to 10 percent solids using vapor compression. The power requirement for this is 90 kWh/kgal evaporated. Further concentration requires open kettles. (For further information, see Kolflat, Cooling Tower Practices, *Power Eng.,* pp. 32–39, Jan. 1974, and Boles and Levin, Treating Cooling Tower Blowdown, *Power,* pp. 68–76, Aug. 1974.)

WASTEWATER CONTROL

REFERENCES: The Clean Water Act of 1972, as amended through 1981, PL 97-117. Sittig, "Resource Recovery and Recycling Handbook of Industrial Wastes," Noyes Data Corp., Park Ridge, NJ. Lund, "Industrial Pollution Control Handbook," McGraw-Hill. "Pollution Control Technology," Research and Education Association, New York. Current updated revisions of pertinent local and federal regulations.

Industrial Categorization by the 1972 FWPCA Amendments (Clean Water Act)

The Federal Water Pollution Control Act (FWPCA) Amendments of 1972, Public Law 92-500, preempt all existing federal regulations by authorizing guidelines for "effluent discharges," rather than by controlling "receiving water quality."

Title III, Section 306(b)(1)(A) categorizes sources. "Source" means any building, structure, facility, or installation from which there is or may be the discharge of pollutants for which effluent limitations guidelines and revisions will be developed. These are published periodically in the Federal Register, under the 40 CFR part numbers shown by Table 18.1.3. The limitations were to be achieved by July 1, 1977, for "best practicable control technology currently available" (BPCTCA) and by July 1, 1983, for "best available technology economically achievable" (BATEA).

Both BPCTCA and BATEA guidelines must meet feasibility and achievability criteria under cost/benefit testing and will be modified periodically to reflect current needs and technology.

Pretreatment guidelines have been established by EPA under Section 307(b) to control industries with effluent discharges into municipal systems. Industrial wastes not so regulated become part of the municipal wastewater flow and thus a municipal responsibility.

Permit and Application for Industrial Discharges

The Permit Program for Industrial Discharges under the provisions of the "Refuse Act of 1899" was previously administered by the Army Corps of Engineers. The Clean Water Act of 1972 incorporated this program and transferred control from the Army Corps of Engineers to the EPA.

A Permit Program Application plus an environmental report [see NEPA-102(c) for criteria] is filed by the discharger with the state, which then refers it to the Regional EPA permit section for review. The state draws a permit that includes limitations, compliance schedules, and state monitoring requirements, making the permit forms available to the public 30 days prior to issuing the permit. Hearings on the permit may cause much delay. After this procedure is completed, the permit may be issued for a maximum of 5 years.

Environmental Considerations for Construction Projects

Introduction   The environmental considerations which follow should be taken into account by industry when preparing an environmental report for construction projects, and before an effluent discharge

Table 18.1.3   Industries Categorized under 40 CFR, Chapter I, Subchapter N

Industrial groups	Part nos.	Stds. of performance
1. Food and grain processing	405–409	In general, BPCTCA 1977 guidelines limit pH* between 6 and 9, BOD,* TSS,* and COD,* to the ppm range, toxic substances, e.g., Ba, Pb, Hg, Ag, As, Cr^{+6}, F, to nontoxic concentration. BATEA 1983 guidelines limit these further on nonprocess streams, and seek "zero discharge" of process streams.
2. Textiles	410	
3. Feed lots; meat products	412; 432	
4. Chemical; petroleum metallurgy	410–424	
5. Power generation	423	
6. Leather, glass, and rubber products	425–427	
7. Wood products, paper, and pulp	428–431	
8. Oil, gas, coal, minerals and biochemicals	434–440	
9. Tars, asphalt, paint inks, and gums	443–458	
10. Pesticides, chems. and explosives	455–457	
11. Photographic, hospts.	459–460	

* BOD (biochemical oxygen demand); TSS (total suspended solids); COD (chemical oxygen demand); pH (log of the reciprocal hydrogen ion concentration).

permit can be granted. Cost/benefit factors for applicable environmental control technologies and regulations limiting effluent discharges should be obtained and included in the record of public hearings and should lead to rational decision-making based on a total short- and long-term project cost to the public consumer in relation to the worth of anticipated public benefits. (*Note:* Cost/benefit ratios for meeting effluent discharge regulations can be obtained from the comments submitted to EPA by the categories of industries listed by Table 18.1.3. Public costs generally range from 100 to 1,000 times greater than social or environmental benefits. The engineer will also find these reports a valuable source of design information for environmental projects.)

It should be recognized that the considerations identified should be individually tailored to respond to the unique guidelines for effluent discharge encountered on a particular project and in a particular state.

General Construction Projects

1. What site selection criteria will be used? Will they include cost/benefit balancing of all environmental worth factors such as the conservation of land and fuel reserves, air and water quality, noise, impacts on residents of the area, fish, wildlife, and vegetation? Will alternative sites and alternative orientations of the plant on the selected site be considered?

2. What disposition will be made of solid and liquid residues (ashes and chemical and industrial wastes)? Does the disposal method include adequate cassetting or neutralization to minimize the danger of soil or water pollution? What steps are planned to contain and reclaim ash dumps and chemical and oil spills to avoid pollution of surface and groundwater by runoff? What provision has been made for the disposal of waste-concentrated brines? If waste disposal into water bodies is planned, what will be the dollar value of its detriment to aquatic life? To what degree will tidal action and currents dilute plant effluents? What disposition will be made of potentially harmful and toxic corrosion products occurring in the plant effluent? Can they be injected into underground aquifers? What provision will be made for controlling the release of waste material into water bodies? If additional units are constructed, what will be the total load of waste materials? What downstream environmental effects can be anticipated with respect to humans, crops, forests, and wildlife? With what cost-effectiveness can any harmful effects be minimized by dilution and toxic materials be discharged below toxic concentrations to merge into the normal background of compositions?

3. What public benefits and detriments will thermal effluents have on the receiving waters? What temperature increase can be anticipated? Is there sufficient water motion in the receiving bodies to dissipate heat effectively? Has the use of cooling towers been explored? What is the probable cost of their public detriment, incurred by the cooling towers' consumptive use of water and energy, by producing fog and undesirable weather conditions through the dissipation of waste heat, and by producing blowdown which may require treatment?

4. What impact will the impoundment for a water supply have in terms of the destruction of agricultural, forest lands, and habitats for fish and wildlife? What measures are planned to mitigate the loss of natural habitats for fish and wildlife? To what degree will archaeological and scenic values be affected? How will the reservoir and downstream flow affect water quality parameters, i.e., temperature, dissolved oxygen, nutrients, nitrogen concentration, hydrogen sulfide, and color?

5. How vulnerable is the facility to surface subsidence, earthquakes, hunicanes, war, insurrection, and other catastrophes? What is the probable annual cost (e.g., liability insurance cost) of the environmental degradation which could be expected from such catastrophes? What preventive and remedial safeguards for downstream inhabitants will be incorporated in plant design and construction? What provisions are made for training plant operators in environmental protection?

6. Have the environmental consequences of transportation, pipeline, and power transmission been considered in site selection? What steps are planned to avoid soil erosion and the silting of streams as pipelines, transmission facilities, and access roads are constructed?

7. What provision has been made for recycling and for industrial by-product development associated with the facility? (See also Sec. 6.10.) What impact will that activity have on the environment?

Process Selection, Recycling, and Ultimate Disposal of Residual Wastes
(See also Sec. 6.10.)

In the design of an industrial installation, segregation of process and nonprocess streams, together with good housekeeping practices and diversion of storm waters will generally result in minimum volumes of water requiring special treatments before being discharged. Wastewater treatment processes involved are: chemical and physical treatments for removal of undesirable waste products; suspended and dissolved solids from process waste streams; removal of corrosion products and carbonaceous wastes (biological treatment) from nonprocess streams; oil-water separations; suspended-solids removal from storm runoffs; good housekeeping and protected stockpiling to prevent contamination of runoffs; and blending of treated effluents and impoundment to meet local end-of-pipe regulations (see Table 18.1.4).

Recycling of consumer and commercial wastes back to their sources (e.g., crankcase oil, glycol, fat, photographic and plating wastes, paper, rags, rubber, metals, and irrigational and fertilizer values of municipal wastes), following the good housekeeping practices of industry, is a very cost-effective method for reducing municipal stream pollution. However, within most industries recycling of salvageable material is almost always practiced to the point of optimum return, so that proposed benefits from further recycling in an effort to obtain "zero discharge" are usually minimal. Recycling, or "closed-cycle" water reuse schemes consume water and produce blow-down streams of concentrated pollutants in contrast to "open-cycle" schemes, whereby solids have traditionally been discharged at normal and below toxic levels, within the assimilative capacity of the receiving bodies of water. The latter, and not the former, conserves water and enhances the naturally occurring ecological cycles—provided, of course, that abnormal toxic substances are not discharged, but are removed by a posttreatment for ultimate disposal.

Descriptions, flow sheets, and costs for a variety of industrial water use and recovery schemes are found in "Water Reuse, Industry's Opportunity" (1973), published by the AIChE.

Table 18.1.4   Removal Process Guide for Selected Contaminants

Treatment processes	pH, acidic, alkaline	Suspended and precipitated solids	Bacteria, odor, organic, color	Nutrients (N, P)	Herbicides, pesticides, CN⁻	TDS	Oils, solvents	Heavy metals Cr^{6+}
Chemical*	C	B	A	A	A	A, B	B	A, B
Clarification† and settling		G, D, I	D, E, H	H, D			D	F, H, D	D, I
Selected‡ biological			J, K	J, K	K		J	
Desalting§			O, M	L, N		L	O, M	L, N
Sludge¶ dewatering		P	P				P	P

* A, selected chemical treatment; B, coagulation; C, neutralization.

† D, selected clarification; E, aeration, oxidation; F, flotation, decantation; G, filtration; H, lagooning; I, sedimentation.

‡ J, secondary (e.g., aerobic, anaerobic); K, tertiary (e.g., carbon adsorption; extended aeration); anaerobic dentrification.

§ L, selected desalting and brine blowdown; M, membranes; N, ion exchange; O, distillation.

¶ P, dewatering (e.g., gravity, centrifugal, vacuum filtration; drying).

Water reuse by consumptive processes always results in a stream of concentrated residual wastes for which an ultimate disposal site must be found. Thus, for brine from saline-water conversion processes, there are four major avenues: (1) ocean discharge; (2) underground discharge into saline aquifers; (3) dry lake beds, liner protected; and (4) solidification and use for landfill. Reference to current regulatory and trade publications should be made to determine which of these is permitted for a given situation.

ATMOSPHERIC POLLUTION

REFERENCES: Laws: Current Legislation, *Chem. Eng.,* June 21, 1971. Victor Sussman, New Priorities in Air Pollution Control, *Jour. Air Poll. Control Assoc.,* **21,** no. 4, April 1971, p. 201. Corbitt, "Standard Handbook of Environmental Engineering," McGraw-Hill. Current updated revisions of pertinent local and federal regulations.

Abstract of Federal Air Pollution Control Legislation

Clean Air Amendments, 1965 Authorized federal control of new motor vehicle emissions. Extended federal control to cover international pollution originating in the United States.

Clean Air Act of November 21, 1967 Under Public Law 90-148 the states are held responsible for establishing local air quality standards consistent with federal criteria.

Clean Air Act Amendments of 1977 Under Public Law 95-95, the EPA can require offending sources to be shut down immediately. Unsatisfactory state enforcement programs can be taken over and run by the EPA. New plants and new additions to old ones must have the best available pollution control technology, and meet emission standards, including zero levels for hazardous substances. A 90 percent reduction from baseline in automotive emissions is required.

The deadline set forth in the act requiring cities to comply with federal standards by 1987 was not possible to meet universally, although some cities did meet the criteria for SO_2 and ozone.

Clean Air Act of 1990 This was the first new air pollution legislation since the 1977 amendments. Briefly, its major requirements are to reduce SO_2 emission by 50 percent and to set emission standards for a number of chemicals which had not been addressed in previous legislation, as amended. The gradual phaseout of use of CFCs is set forth, with complete elimination set for 1996. Airborne pollution respects no national boundaries; accordingly, there has been much cooperation among sovereign nations to solve, among others, the serious problem of acid rain resulting from some forms of pollution.

Some Existing Federal Standards The EPA publishes in the Federal Register national ambient air quality standards covering six pollutants (see Table 18.1.5).

Effect of Energy Crisis on Legislation The energy crisis of the 1970s led Congress in 1974 to consider legislation to postpone the NEPA and to modify the Clean Air Act in order to provide relief from the shortage of petroleum and natural gas. Currently, there are abundant supplies of all types of fossil fuels available in the global market, and at competitive prices. There was no actual postponement of then-extant

Table 18.1.5 Ambient Air Quality Standards

Pollutant	Primary	Secondary, if not same as primary
Particulates, $\mu g/m^3$		
Annual geometric mean	75	60
Max. 24-h conc.*	260	150
Sulfur dioxide, $\mu g/m^3$		
Annual arith. aver.	80 (0.03 ppm)	
Max. 24-h conc.*	365 (0.14 ppm)	
Max. 3-h conc.*		1,300 (0.5 ppm)
Carbon monoxide, mg/m^3		
Max. 8-h conc.*	10 (9 ppm)	
Max. 1-h conc.*	40 (35 ppm)	
Ozone, μ/m^3		
1-h max.*	160 (0.08 ppm)	
Lead particulates, $\mu g/m^3$		
Calendar qtr., aver.	160 (0.24 ppm)	
Nitrogen dioxide, $\mu g/m^3$		
Annual arith. aver.	100 (0.05 ppm)	

* Not to be exceeded more than once a year.
NOTE: All values subject to periodic revision.

legislation, but concerted efforts to abate CO_2 pollution resulting from a fossil-fuel-based economy continue. (See discussion above under "Control of Thermal Discharges.") Electric power generation increases worldwide. Most of it is represented by fossil fuel plants as dictated by local resources available and the state of those local economies. Few unexploited very large water power sites exist, and of those, only a handful are economically viable in the current market. Nuclear-powered plants are being emplaced in many parts of the world, although no new plants have been ordered in the United States for almost 20 years. In some quarters it appears inevitable that the conflicting requirements of pollution abatement, reduction of CO_2 in the atmosphere, and reluctance to exploit nuclear power will ultimately result in the reemergence of nuclear power. The timing is not clear, and it will be a function, among other things, of the successful and timely resolution of the problems attendant disposal of high-level radioactive wastes, depletion of fossil fuel reserves, and possible development of advanced-type nuclear reactors which may prove economically and environmentally viable sometime in the twenty-first century.

It would be prudent to conserve irreplaceable natural resources by utilizing them in the most productive energy, materials, and ecological cycles. Often, alternative methods are at cross purposes with those objectives and result in inadvertent squandering of those resources.

SOURCES AND CONTROL OF AIR POLLUTION

REFERENCES: Magil, Holden, and Ackley, "Air Pollution Handbook," McGraw-Hill. Gibbs, "Clouds and Smoke," Churchill. White, "Industrial Electrostatic Precipitation," Addison-Wesley. Morgensen, "Fan Engineering," Buffalo Forge. Dallavalle, "Micrometrics," Pitman. Stern, "Air Pollution," Academic Press. "Air Pollution Engineering Manual," J. A. Danielson, EPA-Research Triangle Park, NC, May 1973. Current updated revisions of pertinent local and federal regulations.

Table 18.1.6 Principal Air Pollutants

		Particulate matter			
Gases	Vapors	Dusts	Fumes	Smokes	Mists
Acid gases	Alcohols	Alumina	Metallic halogens	Ash	Acid
Carbon monoxide	Esters	Calcium fluoride	Metallic oxides	Cinders	Chromic
Hydrogen chloride	Hydrocarbons	Cement	Silicon tetrafluoride	Organic compounds	Phosphoric
Hydrogen fluoride	Ketones	Coal		Soot	Sulfuric
Hydrogen sulphide	Mercaptans	Grain		Tobacco	Organic chemicals
Nitrogen oxides		Limestone			Oils
Sulfur dioxide		Metal			Salt spray
Alkaline gases		Ore			
Ammonia		Rock			
		Wood			

Table 18.1.7   Pounds of Contaminants Discharged Daily from Domestic Sources in a Metropolitan Area of 100,000 Persons

| | Pounds per day per 100,000 persons using each category of heating and refuse disposal | | | | | |
| | Fuel—domestic heating | | | Domestic incineration | | |
Principal contaminants	Coal	Oil	Gas	Backyard burning	Household* incinerator	Apartment incinerator
SO_x	42,000	17,000	0.4	180	1	12
NO_x	8,000	6,000	6	90	1,150	30
H_2S	1,000	500	0.1			24
NH_3	2,000	800	0.3	345		24
HCl	2,000	500	0.3			
Aldehydes	2,000	800	1	600	8,400	72
Organics	20,000	4,000	1	42,000	12,000	1,800
Organic acids	30,000	12,000	1	225	1,900	4,800
Solids	200,000	800	0.1	3,400	16,500	4,000
Total of above categories, lb	307,000	42,400	10	46,800	40,000	10,700
Total of above categories, kg	139,100	19,210	4.5	21,200	18,120	4,850

* Add 17 percent for cigarette smoking (1984 estimate).
NOTES: Each column shows estimates of the pollutants released to the atmosphere if the entire population used the noted fuel or method of incineration.
SO_x—oxides of sulfur—SO_2 and SO_3.
NO_x—oxides of nitrogen.
Aldehydes measured as formaldehyde.
Organic acids measured as acetic acid.
Total includes only above categories and does not imply total contamination in the ambient atmosphere.
SOURCES: Eliassen, Domestic and Municipal Sources of Air Pollution, *Proc. National Conference on Air Pollution*, 1958.

Domestic and Municipal Sources of Air Pollution

(See Tables 18.1.7 and 18.1.8.)

All **pollutants** are usually classified as gases, vapors, and **particulate matter** (see Table 18.1.6 and Fig. 18.1.1). Particulate matter includes: **Dusts**—a loose term applied to solid particles predominantly larger than colloidal and capable of temporary gas suspension. Dusts do not tend to flocculate except under electrostatic forces; they do not diffuse but settle under the influence of gravity. Dusts result from such operations as crushing, grinding, drilling, screening, and blasting. **Fumes**—the solid particles generated by condensation from the gaseous state, generally after volatilization from melted substances, and often accompanied by a chemical reaction such as oxidation. For example, zinc vapor will evaporate from the surface of heated liquid metal and will then condense upon contact with ambient air to form small, fluffy particles of zinc oxide. Fumes are usually less than 1 μm, although they may coalesce or flocculate to form larger particles. **Smokes**—gas-borne particles (usually less than 0.5 μm) resulting from incomplete combustion of materials such as wood, tobacco, coal, and oil. **Mists** and **plumes**—loose terms applied to dispersions of liquid particles (0.1 to 2.5 μm), the dispersion being of low concentration and the particles **large**.

Low-Emission Automotive Propulsion Systems

(See Secs. 9.6 and 11.1.)

The automobile is a major source of air pollution and a significant contributor to photochemical smog (caused by the products of photo-reactions in the atmosphere—organic peroxides, peracids, hydroxy peracids, peroxyacyl nitrate, and other compounds). For every 1,000 gal of gasoline consumed by an automobile engine, there are discharged the following air pollutants, in pounds: carbon monoxide, 3,000; hydrocarbons, 200 to 400; nitrogen oxides, 50 to 150; aldehydes, 5; sulphur compounds, 5 to 10; organic acids, 2; ammonia, 2; solids, 0.3.

Conventional Gasoline Engines A major thrust aimed at the reduction of air pollution from automotive exhausts resulted in enactment of legislation to increase the fuel economy obtained by passenger cars. Commonly known as the **corporate average fuel economy (CAFE)** program, the legislation sought to increase average fuel economy from 18 mi/gal (1978) to 27.5 mi/gal (1985). Current average fuel economy for passenger cars is slightly above 28 mi/gal (1995).

In the quest to reduce pollutants entrained in automotive exhaust gases, it was also recognized that the alarmingly high levels of lead in the local atmosphere, especially in large urban areas and their surrounds, had to be eliminated. This led to a dual approach: introduction of nonleaded gasoline (with the concommitant phaseout of leaded fuel) and incorporation of catalytic converters as original equipment by passenger car manufacturers. The two matters went hand in hand, for it was known that a practical catalytic converter would be rendered ineffective if it were contaminated by lead (or phosphorus). At this time, all passenger cars (with a few exceptions, e.g., some off-road vehicles on farms) are equipped with catalytic converters. The converter functions to reduce the levels of emission of carbon monoxide (CO), hydrocarbons (HC), and oxides of nitrogen (NO_x). Chemical reactions take place when the exhaust gases pass through the converter, resulting in conversion of the offending

Table 18.1.8   Sources of Air and Typical Loss Rates

Class	Aerosols	Gases and vapors	Typical loss rates
Combustion processes	Dust, fume	NO_2, SO_2, CO, organics, acids	0.05–1.5% by weight of fuel
Stationary engines	Fume	NO_2, CO, acids, organics	4–7% by weight of fuel (hydrocarbons)
Petroleum operations	Dust, mist	SO_2, H_2S, NH_3, CO, hydrocarbons, mercaptans	0.25–1.5% by weight of material processed
Chemical processes	Dust, mist, fume, spray	Process-dependent (SO_2, CO, NH_3, acids, organics, solvents, odors, sulfides)	0.5–2% by weight of material processed
Pyro- and electro-metallurgical processes	Dust, fume	SO_2, CO, fluorides, organics	0.5–2% by weight of material processed
Mineral processing	Dust, fume	Process-dependent (SO_2, CO, fluorides, organics)	1–3% by weight of material processed
Food and feed operations	Dust, mist	Odorous materials	0.25–1% by weight of material processed

SOURCE: Rose et al., Prevention and Control of Air Pollution by Process Changes and Equipment, W.H.O. Rept. Air Pollution, *Monograph Series* 46-307-343.

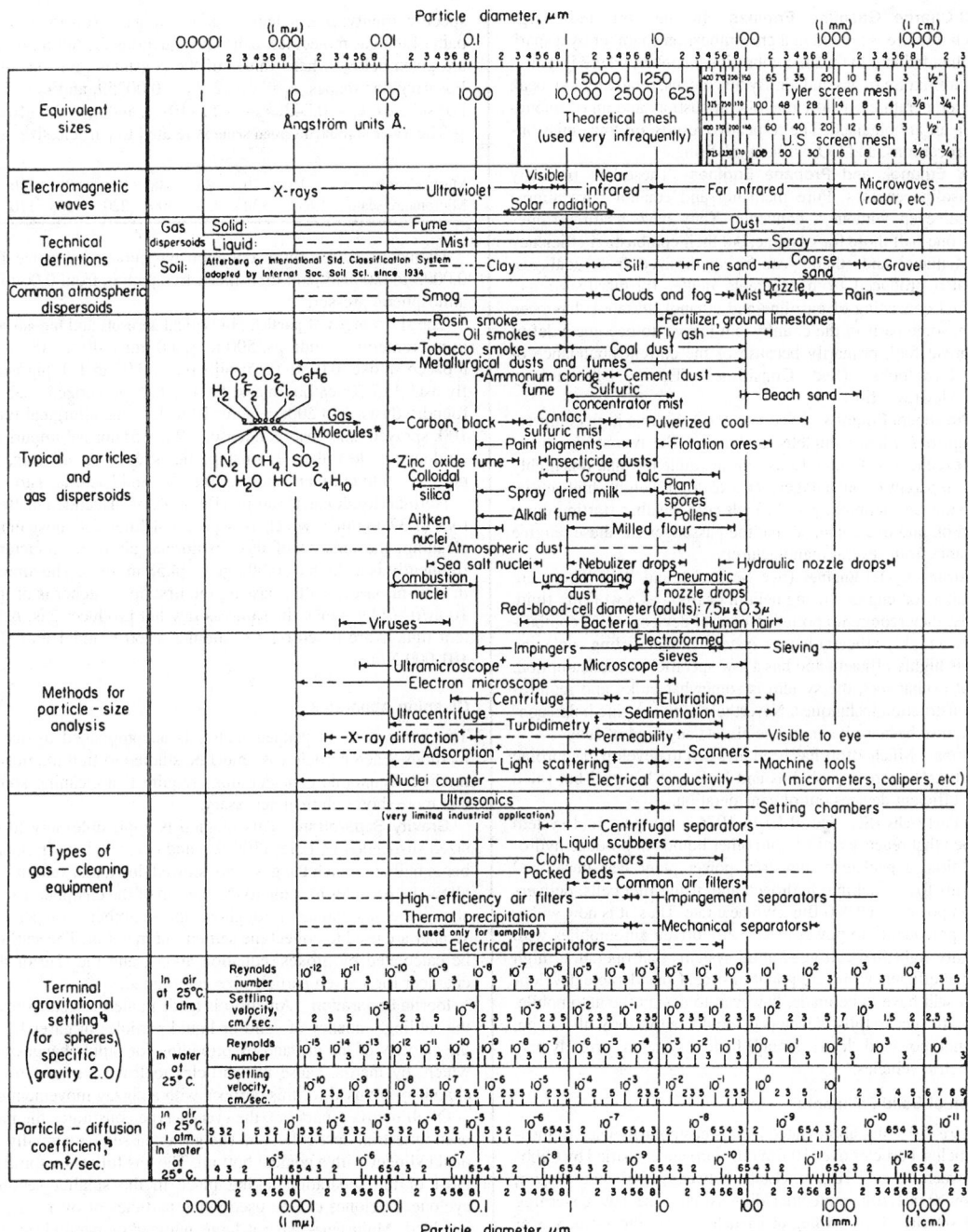

Fig. 18.1.1 Sizes and characteristics of airborne particles. (*From C. E. Lapple, Stanford Res. Inst. Jour., vol. 5, p. 94, Third Quarter 1961. Reproduced by permission.*)

pollutants to CO_2, water vapor, and nitrogen. The catalytic converter will function properly only if free of contaminants (lead and phosphorus, cited above) and if maintained properly to prevent operation at extremely high temperatures. Continued improvements in design and construction have proved effective in reducing pollutants ascribable to automotive exhaust gases, and have done so at no appreciable loss in fuel economy.

Fuel Injection Gasoline Engines   A typical fuel injection engine of about 1,700-cm³ displacement and burning 91 octane fuel, when compared with a conventional two-carburetor engine of the same displacement and using the same gasoline, will demonstrate an improved gasoline economy and substantially lower emissions because of improved combustion.

Stratified-Charge Gasoline Engines In the stratified-charge engine, a rich mixture is ignited in a precombustion chamber by a spark plug and is then fed into the main combustion chamber, where it ignites a much leaner mixture of gas and air. By this means, relatively low peak combustion temperatures, a slow rate of combustion, and an overabundance of free air are obtained. The respective results are low emissions of NO_x, hydrocarbons, and CO, without misfiring.

Methanol, Ethanol, and Propane Engines These are basically modified gasoline engines. Pure methanol and ethanol burn almost pollution-free, even without a converter. They have heating values slightly over one-half those for gasoline, but their combustion efficiency is higher. Methanol and ethanol can be synthesized from coal gas; ethanol is also produced from vegetable matter (primarily grains). Combined fuel consisting of gasoline and ethanol (gasahol) has been introduced in some parts of the country, but has yet to become a definitive widespread fuel, primarily because of the current abundance of petroleum feedstocks. (See Considine, "Energy Technology Handbook," McGraw-Hill.)

Two-Stroke Diesel Engines (See Sec. 9.6.) With its high compression ratio (up to 21) and complete combustion, the two-stroke diesel engine is an excellent performer. Emissions are quite low and it is highly efficient. Its apparent disadvantages, such as high weight, high initial cost, noise, slow acceleration, and difficult cold-weather starting, have been worked on and overcome, so that the passenger car diesel engine is now available from several manufacturers.

Philips (Stirling Cycle) Engines (See Secs. 4.1 and 9.1.) These are external combustion engines, using helium or freon as a working fluid. Consequently, they require no pollution control devices, since combustion occurs in a highly efficient burner system. In the Stirling cycle, the engine, too, is highly efficient and has a low specific fuel consumption. With its heat exchangers, the system is somewhat bulky and requires expensive construction techniques. Nevertheless, the engine holds considerable promise because it can utilize almost any kind of fuel.

Gas Turbines Much effort has been invested to develop gas turbine driven vehicles. Numerous problems and high cost have all but ruled these highly efficient devices out of consideration.

Hydrogen Fuel Cells (See Sec. 9.1.) In 1974 a compact fuel system was developed that reacts gasoline with other liquids to produce hydrogen. In addition, a packaged industrial power plant was developed which re-forms fossil fuel into hydrogen and, using fuel cells, delivers 26 MW of net power at a 9,000 Btu/kWh heat rate. Thus, it is now within the range of practicality to power electrically driven automobiles with compact, lightweight fuel cell generating systems that operate at high overall thermal efficiency, and are pollution-free. Considerable developmental effort will have to be made, however, to design the automobile. This system, if successfully developed, could outperform the huge, heavy, and underpowered electric storage battery systems currently proposed for electric vehicles.

Classification of Solid Pollutants

The **micrometre** (10^{-6} m), abbreviated μm, is customarily used to measure fine particles. Particles over 10 μm (10^{-5} m) are classified by Gibbs as **dusts;** between 0.1 and 10 μm, **clouds;** between 0.001 and 0.1 μm, **smokes;** below 0.001 μm, molecular dimensions. The law governing the **terminal velocity of the settling of particles** under the influence of gravity varies with the size of the particle. In the turbulent region (above 2,000 μm)

$$V_t = Ks^{1/2}p^{1/2}D^{1/2} = k_1\sqrt{SD}$$

In the intermediate region (between 1,000 and 100 μm), the law is $V_n = K's^{2/3}\rho^{-1/3}\eta^{-1/3}D = k_2 s^{2/3}D$. In the streamline region (2 to 50 μm), Stokes' law holds: $V_s = K''sD^2\eta^{-1} = k_3 sD^2$. From 0.1 to 1.0 μm, Cunningham's correction must be applied to Stokes' law:

$$V_c = V_s(1 + 1.72\lambda/D) = V_s(1 + 0.172/D)$$

Below 0.1 μm, the movement due to molecular shock (brownian motion) exceeds that due to velocity. In these equations, V is the terminal velocity, ft/min; D, particle diameter, μm; p, gas pressure, atm; s,

specific gravity; ρ, gas density, g/cm^3; η, gas viscosity, P; λ, mean free path of the gas molecules, μm. The equations for falling in standard air are given below, where values of the constants give velocities in m/s. For irregular shapes, $k_1 = 0.142$, $k_2 = 0.00259$, and $k_3 = 1.98\,(10)^{-5}$. For spheres, $k_1 = 0.243$, $k_2 = 4.11\,(10)^{-3}$, and $k_3 = 3.0\,(10)^{-5}$.

The **Tyler standard screen scale** is related to particle size as below:

Meshes per in	10	20	35	48	64	100	150	200	325
Micrometre scale	1,650	830	420	300	220	150	110	74	44

Diameters of the more common gas molecules range from about 0.0003 to 0.00045 μm; their mean free path is from 0.06 to 0.2 μm at atmospheric pressure.

The size ranges of **particles** in typical **aerosols** and **industrial dusts** are, in micrometres: raindrops, 500 to 5,000; mist, 40 to 500; fog, 1 to 40; tobacco smoke, 0.01 to 0.15; oil smoke, 0.03 to 1.0; pigments, 1 to 7; fly ash, 3 to 70; carbon black, 0.04 to 0.2; pulverized coal, 10 to 400; foundry dusts, 1 to 200; cement, 10 to 150; metallurgical fumes, 0.1 to 100; sprayed zinc dust (condensed), 2 to 15; normal impurities in quiet outdoor air, less than 1; dust particles causing silicosis, below 10; pollens, 20 to 60; plant spores, 10 to 30; and bacteria, 1 to 15.

Normal **city air** carries around 0.0006 g of suspended matter per cubic foot (1.37 mg/m^3), which is a practical limit for most industrial gas cleaning; the amount of dust in normal air in manufacturing plants frequently is as high as 0.002 gr/ft^3 (4.58 mg/m^3). The amount of dust in **blast-furnace gas** after passing the first dust catcher is of the order of 10 gr/ft^3 (22.9 g/m^3), the same as raw hot producer gas. All dust content figures are based on air volumes at 60°F and 1 atm (15.6°C and 101,000 N/m^2).

Cleaning Apparatus

The removal of suspended matter is accomplished by the following methods. Each particle case must be studied to find the most desirable method. Economic considerations require that cleaning should not be any more thorough than necessary.

Gravity Separation This method is applicable only to the larger-sized suspended particles (100 μm and over). A long horizontal chamber is built, in which the gases are slowed down to a velocity which will allow the particles to settle to the bottom of the chamber. Even gas flow should be maintained throughout the chamber. Hoppers or a drag scraper are used to collect the settled-out material. The settling rate can be calculated by Stokes' equation above, and the size of the settling chamber for a given particle size determined.

Inertia Separation An effort is made in inertia separators to magnify the settling tendency of solid or liquid particles in gases by increasing the velocity of the gas and by providing for rapid changes in direction which, by inertia, cause the particles to leave the gas stream. Baffle chambers are used for this purpose with a zigzag movement of the gas.

Of all inertia separators, the **cyclone** type constitutes the most widely used industrial dust collector. The gas is passed tangentially into a vertical cylinder with a conical bottom. The gas follows a spiral path, with most of the separation taking place in the smaller sections of the cyclone. Cyclones can be used when particles of over 5 μm diam are involved. **Multiclones** have a large number of parallel, small cyclone units. The gas pressure drop is about 4 times the entrance velocity head (1 to 4 in w.g.). The dust content of the cleaned gas is seldom less than 1 gr/ft^3 (2.29 g/m^3).

In the **Rotoclone** apparatus, the impeller is a concave disk with curved blades along which the dust slips and is discharged over the edge of the disk into a ring-shaped dust chamber. The main air stream leaves in a wide scroll on the side of this chamber. The apparatus is usually preceded by a cyclone for removal of coarse dust. By spraying water in the entering air, the precipitation efficiency of dust removal can be materially raised. Particles of minimum size 8 μm are claimed to be removed by the dry apparatus, 1 μm by the wet apparatus. The power for dust separation cannot readily be separated from that for compressing the gases. The power requirement is 3 bhp per 1,000 ft^3/min for an outlet pressure of 10 in w.g.; in the water-sprayed machine, the required power is at least 25 percent more.

Static spray scrubbers are usually of the tower type, the gas passing upward countercurrently to the descending liquid. Sets of sprays are placed in the top zone, with various materials used in layers to channel and mix the gas and water. Hurdles, cylindrical tiles, and random-packed ceramic tiles or metal spirals are common packing materials. With a gas flow-rate of about 350 ft^3/min per ft^2 of cross-sectional area and a water rate of 25 gal/Mft^3 of gas, a cleanliness of 0.1 to 0.3 gr/ft^3 can be obtained with blast furnace, coke oven, and producer gas. The pressure drop through a tower scrubber is in the range of 4 to 10 in w.g. The scrubber process is used as the primary cleaning and cooling stages before the cleaning of gases.

Dynamic Spray Scrubbers   Contact between water droplets and suspended matter in gas is improved by mechanical agitation of gas and water.

An example of dynamic spray scrubbers is the **Pease-Anthony venturi.** In this system, the gas is forced through a venturi throat in which the gas is mixed with high-pressure water sprays. A tank after the venturi is needed to cool and eliminate the moisture. Cleanliness in the range of 0.1 to 0.3 gr/ft^3 has been reported.

Filtration is comparatively little used for cleaning fuel gases; it is in extensive use in cleaning air and waste gases. Materials used for gas filtration are ordinarily thickly woven cotton or wool cloth for temperatures up to 250°F; for higher temperatures, metal cloth or woven glass cloth is advisable. The gases filtered should be well above their dew point, as condensation on the filter cloth plugs the pores. If necessary, saturated gas should be reheated. The cloth is frequently made into "bags," tubes of 6 to 12 in diameter and up to 40 ft long, suspended from a steel framework (baghouse). The gas inlet is at the lower end through a header to which the bags are connected in parallel; the outlet is through a housing surrounding all the bags. At frequent intervals, the operation of all or part of each unit is interrupted while the bags are pulsed, shaken, or reverse-air-cleaned to dislodge the accumulated dust, which drops back into the gas-inlet header and is removed by a conveyor screw. The dust content can be reduced to 0.01 gr/ft^3 or less at reasonable cost. The apparatus also is used for the recovery of valuable solids from gases.

Electrical Precipitation   This principle, although involving highly technical physioelectrical phenomena, is generally familiar to all. If suspended particles are placed in a high-voltage electric field, they receive an electric charge and move toward one or the other of the electrodes between which the electric field is established. If one electrode is a pipe or flat plate and the other electrode is a wire axially suspended in the center of the pipe or between two plates, the field is strongest around the wire and weakest near the surface of the pipe or plate. The charged dust particles move from the strong part of the field toward the weak part of the field.

Two types of precipitators are manufactured. **The vertical-round type** is usually built in a cylindrical tank having a header sheet near the inside top. In this header sheet are nested pipes which act as the collecting electrodes. The high-voltage ionizing electrodes are twisted square steel rods, about ¼ in in size, supported from above the header sheet and hung axially in the collecting electrode pipes. Water is introduced above the header sheet and flows over weir rings at the tops of the pipes to form a water film on the inside of the pipes. **The horizontal-plate type** has vertical parallel plates as collecting electrodes. The discharge electrodes are a series of wires suspended equidistantly between the plates, and can be operated as either a "wet" or "dry" unit. When used as a wet precipitator, weir boxes are built on the top of each plate to produce a wetting of both sides of the plate, washing the collected particles down the plate. As a dry precipitator, a rapping device operated on a time cycle shakes the plates, causing the collected particles to fall into a hopper under the plates. Dry collection requires injection of a conductive gas such as SO_3 to work efficiently. (See Marchello, "Control of Air Pollution Sources," Marcel Dekker, p. 182.)

The electrical apparatus provided with a precipitator to produce a dc potential of 30,000 to 90,000 V can be mechanical, vacuum tube, or solid state. Power packs are self-contained cabinet-type units, built in sizes from 2½ to 50 kVA. Rectifiers are mounted in a substation building close to the precipitator or can be mounted in weatherproof enclosures on the roof of the precipitator. Controls can be at a distance because the wiring is of low potential.

The efficiency of dust removal by precipitators is expressed as log $(1 - E) = t \log K$, where E = the efficiency as fraction removal; t = the time the gas is in the electric field, s; and K = an apparatus constant varying from 0.05 to 0.80 ("Cottrell Electrical Precipitators," Western Precipitator Corp.; see also Schmidt and Anderson, *Elec. Eng.*, 1938). The above function implies that small variations of t cause considerable variations of $(1 - E)$, the fraction of dust left in the gas. This indicates a sensitivity of the precipitators to overload. (See Table 18.1.12 for common applications).

Precipitators working on coke oven gas for tar removal, under favorable temperature and moisture conditions, show outlet figures of 0.005 gr of dry tar per ft^3 of gas with inlet of 0.25 gr/ft^3. The outlet tar content increases rapidly upon overloading the apparatus.

Precipitators for blast furnace gas reduce the dust to about 0.01 gr/ft^3. Electric precipitators are used for removal of suspended matter from gases and air from boiler gases, metallurgical fumes, cement and lime dusts, etc. Precipitators can be built for dust in practically all particle sizes, as well as mist. They cannot collect material in the gaseous or vapor state. Precipitators are not applicable where explosive gas or dust is involved in the presence of air or oxygen in explosive proportions because of the hazard of ignition by an electric spark.

Blast-Furnace Gas Cleaning   Modern equipment for this service represents a field in which gas cleaning has been developed to a high degree of efficiency. The method described here is used on most modern blast-furnace gas-cleaning plants.

The **dust catcher** is a large cylindrical steel tank, brick-lined, about 30 to 40 ft in diameter. The dust catcher is fundamentally a settling chamber. Dust is dropped out of the gas by reducing the velocity of the flow. Gas enters at 400°F approx and with 10 to 20 gr of dust per cubic foot. A good dust catcher will clean the gas to 3 gr of dust per cubic foot.

The **primary washer** is a cylindrical steel shell 16 to 21 ft in diameter, with a conical bottom section. Four to five banks of ceramic, round, or cross-partition tile are arranged in the tower. At the top, sprays cover the top bank of tile with an even distribution of water at the rate of 7 to 9 gal/($ft^2 \cdot$ min). Size of the tower is determined with a gas rate of 350 ft^3/min per ft^2 of cross-sectional area. The gas enters the washer at an angle of 45° downward, passes up through the tile and out a main at the top of the washer. The water passes down through the tile banks and collects the dust removed from the gas. The water collects in the bottom section and overflows through a pipe arranged with a gas seal. The water then is sent to a thickener for removal of the iron-ore dust. The gas enters the washer at 400°F approx and is cooled to 100°F approx. The dust content of the gas is reduced to about 0.15 to 0.30 gr/ft^3, which is satisfactory for entrance into a precipitator or Theisen disintegrator.

Electrostatic precipitators reduce the dust content of the gas to the cleanliness required by its final use. The following are common requirements for blast furnace gas in grains per cubic foot at standard conditions (1 atm, 70°F): hot blast stoves, 0.01; boilers, 0.01; soaking pits, 0.10; coke ovens, 0.005.

Effect of Sulfur Gases and Particulates on Vegetation

REFERENCES: Kamprath, E. J., Possible Benefits from Sulfur in the Atmosphere. *Combustion,* **44,** no. 4, 1972, pp. 16, 17; Grennard and Ross, Progress Report on Sulfur Dioxide, *Combustion,* **45,** no. 7, 1974, pp. 4–9; Wagner, TVA's Proposed Clean Air Strategy, *Public Utilities Fortnightly,* pp. 34–37, June 6, 1974. Current updated revisions of pertinent local and federal regulations.

Damage from Sulfur Fumigation   Vegetation is damaged by abnormal concentrations (above 0.3 ppm) of sulfur gases at ground level. Users of sulfur-bearing fuels may become involved with damage claims where ground-level concentrations are of consequent increasing importance. The contaminants causing damage to vegetation are sulfur trioxide (SO_3), sulfates, sulfur dioxide (SO_2), and particulate materials.

Particulate materials on leaf surfaces have no harmful effect other than to exclude sunlight, and to that extent, retard growth.

Sulfur trioxide is hygroscopic and occurs at ground level as sulfuric acid mist. Plant damage has been observed in the field. Sulfur trioxide is deleterious to plants, animals, and humans.

Impure sulfur dioxide (SO_2) has the greatest effect on plants. Concentrations greater than 0.3 ppm will damage the leaves of sensitive vegetation if maintained for over 5 h. Animals and humans are more resistant than plants. Concentrations above 0.5 to 2.0 ppm are recognizable by smell and taste. The concentration and the duration of the fumigation determine the degree of injury. (See Table 18.1.9.) Environmental conditions of temperature, humidity, soil moisture, soil fertility, nutrient supply, light intensity, age of plants, and moist-leaf conditions have a marked influence on the effect of sulfur dioxide on vegetation.

Table 18.1.9 Suspectibility of Vegetation to Injury by Sulfur Dioxide (from Field Observation)

Maximum susceptibility	Moderate susceptibility	Resistant
Weeds		
Plantain	Bracken fern	Goldenrod
Ragweed	Wild carrot	Small-leaved milkweed
Smartweed	Wild grape	Knotweed
Dewberry	Sweet clover	Milkweed
Greenbriar		
Dandelion		
Galinsoga		
Pigweed (redroot)		
Wildflowers		
Greenbriar	Blackberry	Goldenrod
Galinsoga	Witch hazel	Small-leaved milkweed
Dandelion	Huckleberry	Elderberry
	Blueberry	Mountain laurel
	Bracken fern	Yarrow
	Viburnum	Knotweed
	Wild carrot	Joe-Pye weed
	Sweet clover	Milkweed
Farm crops		
Alfalfa	Alsike clover	Potato
Oats	Corn	
Buckwheat	Cabbage	
Barley		
Garden vegetables		
Beet	Huckleberry	Onion
Endive	Blueberry	Cucumber
Bean	Tomato	Corn
Peas	Parsley	Potato
Brussels sprouts	Cauliflower	
Landscape materials		
Hawthorn	Huckleberry	Hazelnut
Sunflower	Blueberry	Mountain laurel
Cosmos	Nasturtium	
Sweet William	Dahlia	
	Gladiolus	
	Willow	
Trees		
White pine	Larch	Red pine
Hemlock	Scotch pine	Blue spruce
Hawthorn	Norway spruce	Black locust
Black birch	Sumac	Black oak
Yellow birch	Ash	Red oak
	Wild cherry	White oak
	Domestic apple	Sugar maple
	Willow	Swamp red maple
		Norway maple

Benefits from Sulfur in the Atmosphere   In normal atmospheric concentrations below 0.3 ppm, as occurs in metropolitan areas 95 percent of the time, sulfur dioxide is an essential plant nutrient. Vegetation utilizes it for growth by absorption from the atmosphere through leaves, and from rain-saturated earth through the roots. Fossil fuels, by virtue of their origin as former vegetation, contain from 0.4 to over 4 percent sulfur (Table 18.1.10). To maintain the natural ecological sulfur cycle for sustaining plant growth, this organic sulfur should be recycled back into the soil. As sulfur in stack gases is brought down by rainfall largely over a radius of about 15 mi from steam-electric plants in rural areas, averaging about 5 lb/(acre · yr) (0.55 g/m²), anywhere from 20 to 50 percent of the annual sulfur requirement for crops can be furnished by stack emissions, while the balance must be supplied by fertilizers. Agriculture requires annually 10 to 40 lb/acre (1.11 to 4.44 g/m²) of sulfur, which, in the absence of stack emissions, must be furnished by direct sulfur application for optimum yields of crops.

The sulfur dioxide problem, therefore, is mainly one of preventing excessive ground-level concentrations, whereby a normal plant nutrient can become a menace. Two methods of control are possible when burning fossil fuels: sulfur oxide removal by wet scrubbing, or dispersion by tall stacks with subsequent precipitation by normal rainfall.

Sulfur Oxide Removal Processes   Stack-gas scrubbing processes remove sulfur from the flue gas but also increase the buildup of atmospheric CO_2. In 1973 the seven most promising stack-gas scrubbing processes selected for further study by the Sulfur Oxide Control Technology Assessment Panel were

1. Catalytic oxidation of SO_2 to SO_3, to recover sulfuric acid as a saleable by-product
2. Wet lime/limestone scrubbing, which produces calcium sulfite sludge for disposal
3. Alkali scrubbing without regeneration, which produces 18 percent sodium sulfite brine for disposal
4. Alkali scrubbing with lime regeneration, which produces calcium sulfite sludge for disposal
5. Alkali scrubbing with thermal regeneration, which recovers SO_2 gas by-product, and a sodium sulfite/sulfate brine for disposal
6. Magnesium oxide scrabbing, with off-site sulfuric acid manufacture and regeneration of magnesium oxide by others
7. Alkali scrubbing, plus electrodialytic regeneration to recover SO_2 gas and a purge stream of impure brine

A comparison of these systems is given by Table 18.1.11. The sulfur dioxide emissions limitations (SDEL) case, with its low-cost and negligible environmental impact, is technically justifiable with a cost/benefit ratio of 0.19. The other designs are bulky, costly, noisy, and experimental. They consume energy, water, fuel, and chemicals, and require direct sulfur application to surrounding agricultural areas, plus waste sludge disposal. These adverse environmental impacts usually result in negative total public benefits, unjustifiable except under unusually favorable circumstances.

By far the most difficult and costly environmental hazard produced by sulfur removal is the volume of high chemical-oxygen-demand (COD) liquid and solid wastes produced. This contaminated thixotropic sludge is environmentally unacceptable for landfill or for any other conceivable use. Reduction to elemental sulfur for permanent underground storage by an inverse Frasch process would reduce the volume of wastes, but the cost is exorbitant. Moreover, even if sulfur is removed, the buildup of acid gases (mainly CO_2) in the global atmosphere will continue unabated, along with increasing acid rains, as the CO_2 content of the air increases. Substitution of nuclear fuels for fossil fuels will correct the situation.

Sulfur Dioxide Emission Limitation (SDEL) Programs

To avoid the water pollution difficulties and the wet sludge disposal difficulties introduced by wet scrabbing, Tennessee Valley Authority and others have developed various SDEL programs.

SDEL methods employ combinations of five or more sensing, dispersing, and limitation approaches simultaneously.

Table 18.1.10   Emission of Sulfur-Bearing Gases and Solids from a 1,000,000-kW (1,000-MW) Fossil Fuel Fired Power Station

Fuel:	Mid-western bit coal	Eastern bit coal	Central Ill. bit coal	High-sulfur fuel oil	Low-sulfur fuel oil
Firing method:	Pulv.-coal round burners	Pulv.-coal round burners	Cyclone burners	Round burners	Round burners
Sulfur content of fuel, %	4.14	1.15	4.65	2.41	0.40
Ash content of fuel, %	18.16	7.59	15.00	0.10	0.10
Sulfur-bearing gases leaving boiler:					
Sulfur dioxide, lb/h	69,000	15,100	97,500	23,200	3,620
Sulfur trioxide, lb/h	3,450	755	4,875	1,160	180
Total sulfur-bearing gases, lb/h	72,450	15,855	102,375	24,360	3,800
Solids in flue gas leaving boiler:					
Sulfur compounds, lb/h	4,600	920	920	2,520	420
Carbon, lb/h	8,280	2,760	1,190	4,200	4,200
Fly ash, lb/h	118,680	38,640	20,250	420	420
Total solids, lb/h	131,560	42,320	22,360	7,140	5,040
Collection efficiency,* %	99+‡	99+‡	99+‡	†	†
Solids in flue gas leaving stack, lb/h	360	360	15	†	†

* Based on clear stack.
† See text. Firebox temperatures are kept below 1,900°F to control NO_2 emissions.
‡ Total efficiency whether combination mechanical and electrostatic units or electrostatic alone, with SO_3 injection.
NOTE: 1 lb/h = $7.55(10)^{-3}$ kg/s.

Table 18.1.11   Comparison of SO₂ Control Systems

Process	Requirements	Throwaway or recovery	Additional cost of power generation, %	SO₂ control efficiency, %	Additional plant capacity investment, %
Sulfur dioxide emission limitation	Fuel-switching and tall stacks		1–2.35	99+	1–2.8
Low-sulfur fuel (coal or oil)			28–85	Varies	—
Dry fluidized-bed combustors	Limestone (200% of stoichiometric)	Throwaway CaSO₃/CaSO₄	25–38	50–90	25–90
Wet lime/limestone scrubbing	Lime (100–120% stoich.); limestone (120–150% stoich.)	Throwaway CaSO₃/CaSO₄	19–37	60–85	17–30
MgO scrubbing	MgO carbon and fuel (for regeneration and drying)	Recover conc. H₂SO₄	25–50	90	20–35
Catalytic oxidation (add-on)	V₂O₅ catalyst heat	Recover dilute H₂SO₄	25–45	85–90	25–40
Na₂SO₃ scrubbing	Sodium makeup heat	Recover conc. H₂SO₄ or sulfur	25–50	90	25–40
Double alkali	Sodium makeup plus lime (100–130% stoich.)	Throwaway CaSO₃/CaSO₄	17–35	90	15–30

BASIS: 80 percent plant load factor. Yearly fixed charges are 18 percent of capital investment. Cost ranges are for 200- to 1,000-MW plant sizes. Costs of sludge disposal ponds not included.
SOURCES: Burns and Roe data, TVA's Clean Air Strategy. *Public Utilities Fortnightly*, 34: June 6, 1974.

1. Very tall stacks (1,000 ft) for dispersing SO_2 gases high in the atmosphere, thereby minimizing unacceptable ground-level concentrations and occurrences. (See the Holland equation for stack height, below.)

2. Atmospheric (AFBC) and pressurized (PFBC) fluidized bed combustor units utilizing pulverized coal with limestone or dolomite added to the bed to absorb sulfur dioxide. The bed is cooled internally by steam coils. Ninety percent sulfur removal along with low NO_x is possible. Stack gases are cleaned mechanically and electrostatically. (See Energy Technology Conference VIII, Government Institutes, Rockville, MD, pp. 893–919.)

3. Computers, which quickly correlate meteorological and plant operational data, so as to anticipate and prevent unacceptable ground-level concentrations.

4. Sensors and monitors, which sense and record automatically the local ambient SO_2 concentrations, and pH's, thereby ensuring the effectiveness of the control operations. (Limestone application is made directly to lakes in the area to raise pH.)

5. Emission limitations, accomplished by fuel switching between high- and low-sulfur fuels during normal power plant operations, and by local power generation curtailment during emergency situations, when even switching to low-sulfur fuels might not provide acceptable ground-level concentrations.

6. Daily local plant weather observations and meteorological measurements by means of light aircraft and balloons. These approaches allow high sulfur fuels to be burned most of the time, thereby conserving scarce low sulfur fuels for domestic and other ground level uses not employing SDEL methods.

As opposed to wet scrubbers for SO_2 control, SDEL methods are estimated to require less than one-tenth the total investment cost and one-thirteenth the annual operating cost, exclusive of wet sludge disposal costs.

Atmospheric Dispersion of Pollutants from Stacks

Stacks can provide effective atmospheric dispersion of gaseous and particulate pollutants with acceptable ground-level concentrations. Theoretical and empirical formulas are available to estimate the dispersion of air borne pollutants continuously emitted from stacks. Effective stack height of a plume is given by $H = h_s + h_r$. (See notation below.)

Notation

Symbol	Definition	System of consistent units
b	Atmospheric-dispersal parameter	Dimensionless
d	Stack-exit diameter	m
H	Effective stack height	m
h_r	Rise of plume	m
h_s	Stack height	m
p	Atmospheric-dispersal parameter	Dimensionless
Q	Emission rate of pollutant	m³/s
Q_h	Heat emission rate	cal/s
q	Atmospheric-dispersal parameter	Dimensionless
u	Mean wind speed	m/s
v	Stack-exit velocity	m/s
X_g	Ground-level concentration of pollutant at a distance x from base of stack	ppm by vol
X_{mg}	Maximum ground-level concentration of pollutant at a distance x_m from base of stack	ppm by vol
x	Downwind distance from emission source in direction of mean wind	m
x_m	Distance from base of stack to maximum ground-level concentration	m
σ_A	Basic diffusion parameter related to azimuth plume angle in radians	m^{1-q}
σ_E	Basic diffusion parameter related to elevation plume angle in radians	m^{1-p}

The Holland formula, recommended for calculating h_r, is $h_r = (1.5vd + 4.09 \times 10^{-5}Q_h)/u$. A ratio of stack height h_s to building height of 2½ to 1 or more is commonly used to avoid entrapment of the plume in the vortex of adjacent buildings and the associated high values of groundlevel concentration X_g. Stack-exit velocity v should be 14 to 28 m/s (45 to 90 ft/s) to minimize plume entrapment in stack vortices. The lower the ratio of stack inside to outside diameter at its top, the higher the necessary v.

The **Cramer equation,** recommended for calculating X_g, is $X_g = 10^6 Q/(\pi u x^b \sigma_A \sigma_E)\exp[H^2/(2\sigma_E^2 x^{2p})]$. The equation is for sampling intervals of approximately 20 min, with X_g varying approximately inversely as the one-fifth power of the interval.

Simplified approximate formulas for X_{mg} and x_m are

$$X_{mg} = 2Q10^6\sigma_E/(\pi e^{uH^{1.9}}\sigma_A) \qquad \text{and} \qquad x_m = H/(1.35\sigma_E)$$

Atmospheric dispersal characteristics and hence the values of X_g, X_{mg}, and x_m are influenced by thermal stratification and atmospheric turbulence. A superadiabatic (adiabatic) (subadiabatic) lapse rate, i.e., a

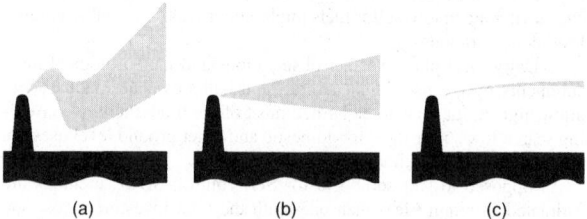

Fig. 18.1.2 Height and plume behavior for (*a*) unstable, (*b*) neutral, and (*c*) stable conditions.

drop in temperature with altitude greater than (equal to) (less than) 5.4°F per 1,000 ft (9.84 K/km) results in an unstable (neutral) (stable) atmosphere with much (moderate) (very little) vertical mixing. (See Fig. 18.1.2.)

The recommended parameters for use in the Cramer equations under the above conditions of thermal stratification are:

Thermal stratification	σ_A, rad	σ_E, rad	b	p	q
Unstable	0.39	0.13	2	1.1	0.9
Neutral	0.21	0.07	1.85	1.0	0.85
Stable	0.12	0.04	1.6	0.8	0.8

The above formulas deal only with steady-state conditions. There are transient meteorological conditions which can result in significant ground-level concentrations of pollutants and which cannot readily be analyzed by computations. These conditions include the breakup of a temperature inversion and the slow, steady buildup of pollutants during stagnations accompanied by deep ground fogs in a valley. For large power plants, high stacks (400 ft and more) (122 m), high stack-exit velocities [45 to 90 ft/s (13.7 to 27.4 m/s)], and adequate exit temperatures [250 to 300°F (394 to 422 K)] are usually effective in avoiding such fumigations.

Stack Emissions

Emission of sulfur-bearing gases and solids from a 1,000-MW fossilfuel fired power station is shown in Table 18.1.10. (See also Tables 18.1.12 and 18.1.13 for more data.) Reducing the remaining solids in stack emissions from bituminous coal fired (excluding stoker firing) power boilers to 0.02 gr/standard ft³ (45.8 mg/m³) and less results in a clear stack discharge. This requires collection efficiencies of 99+ percent, attainable with commercial equipment. These high collection-efficiency requirements demonstrate the need for optical instruments which will readily determine stack-emission quality. The **Ringelman chart** is frequently used to evaluate stack emission, but it is a crude and inaccurate method. A good indicator is a stack emission invisible to the naked human eye. With oil firing, present practice is to control stack emissions by furnace design, combustion control, use of additives, and multiple cyclone-type collectors located in the low- or high-temperature zones of the flue-gas system. For future plants of 1,000 MW and larger, use of high-temperature electrostatic precipitators located ahead of the air preheater may be considered.

Air Pollution Control in Various Industries
(See Tables 18.1.12 and 18.1.13.)

Industrial sources of air pollution and typical loss rates are given in Table 18.1.8.

Nonferrous metal smelting (e.g., copper, zinc, lead) and thermal operations (e.g., blast furnace, sintering, converters) are sources of air pollution and of potentially recoverable metallic materials. A large variety of collection equipment has been tried. Experience has proved that precipitators, cloth filters, and gas washers are best.

Paper industry sources of air pollution are (1) recovery furnace gases and (2) lime kiln exhaust gases. Noxious gases require scrubbing. **Recovery-furnace gases** in the sulfate and soda processes of manufacture emit solids [2 to 6 gr/standard ft³ (4.6 to 13.7 g/m³)] which are mainly sodium salts. The small particle size is particularly bothersome because of the large, obscuring plume in the stack tail. Chemically, the solids are destructive to painted and finished surfaces (e.g., automobiles, structures). Valuable constituents are recovered for return to the cycle. Electrostatic precipitators with 95 to 99 percent efficiency are favored for collection and return of solids. Corrosion-resistant materials (glazed tile, special cements) must be used to reduce maintenance. **Lime kilns** are used to calcine calcium carbonate and recover it as CaO for use in the process. The consequent air pollutant is recovered by gas washers and mechanical collectors.

Cement industry sources of air pollutants are (1) clinker kilns and (2) bagging and other mechanical operations. The **kilns** are the major source of air pollution, with large, unattractive exhaust plumes depositing cement-making solids on the landscape and on painted surfaces. Damage is reduced by the use of electrostatic precipitators with the wet process and cyclones plus cloth filters with the dry process. **Bagging and other mechanical losses** are reduced by use of cloth filters following cyclone separators.

Table 18.1.12 Common Applications of Industrial Precipitators

Industry	Application	Gas-flow range, ft³/min*	Temp. range, °F	Percent weight of dust (below 10)	Usual efficiencies, %	Dust conc. range in ppm† by wt in air	Temp. range, K
Electric power	Fly ash from pulverized-coalfired boilers	50,000–750,000	270–600	25–75	95–99 +	760–9,500	405–589
Portland cement	Dust from kilns	50,000–1,000,000	300–750	35–75	85–99 +	950–29,000	422–672
	Dust from dryers	30,000–100,000	125–350	10–60	95–99	1,905–29,000	325–450
	Mill ventilation	2,000–10,000	50–125	35–75	95–99	9,500–98,000	283–325
Steel	Cleaning blast-furnace gas for fuel	20,000–100,000	90–110	100	95–99	38–950	306–317
	Collecting tars from coke-oven gases	50,000–200,000	80–120	100	95–99	190–1,905	300–322
	Collecting fume from open-hearth and electric furnaces	30,000–400,000	300–700	95	90–99	95–5,700	422–644
Nonferrous metals	Fume from kilns, roasters, sintering machines, aluminum potlines, etc.	5,000–1,000,000	150–1100	10–100	90–98	5,700–95,000	339–867
Pulp and paper	Soda-fume recovery in kraft pulp mills	50,000–200,000	275–350	99	90–95	950–7,600	408–450
Chemical	Acid mist	2,500–20,000	100–200	100	95–99	38–1,905	311–367
	Cleaning hydrogen, CO_2, SO_2, etc.	5,000–20,000	70–200	100	90–99	19–1,905	295–367
	Separate dust from vaporized phosphorus	2,500–7,500	500–600	30–85	99 +	19–1,905	533–589
Petroleum	Powdered catalyst recovery	50,000–150,000	350–550	50–75	90–99.9	190–4,800	450–561
Rock products	Roofing, magnesite, dolomite, etc.	5,000–200,000	100–700	30–45	90–98	950–4,800	311–644
Gas	Tar from gas	2,000–50,000	50–150	100	90–98	19–3,800	283–339
Carbon black	Collecting and agglomerating carbon black	20,000–150,000	300–700	100	10–35	57–9,500	422–644
Gypsum	Dust from kettles, conveyors, etc.	5,000–20,000	250–350	95	90–98	2,850–9,500	394–450

* ft³/min = 4.72(10)⁻⁴ m³/s.
† 1 ppm = 0.525 gr/1,000 ft³ = 2 mg/m³, in air.
SOURCE: Holden and Ackley, "Air Pollution Handbook," McGraw-Hill.

Carbon-black production utilizes electrostatic precipitators, cloth filters, and scrubbers to collect the black from the gas stream and to control air pollution.

Iron and steel industry sources of air pollution are cupolas, oxygen-furnaces, electric-furnaces, and sintering operations. The dispersoids in these operations are mostly oxides of iron in particle sizes below 10 μm and with loadings ranging to 15 gr/standard ft³ (34.3 g/m³). Dust emittance is acceptably controlled generally by use of precipitators and scrubbers for open-hearth and basic oxygen processes, use of cloth filters and scrubbers for electric furnaces, and use of precipitators, scrubbers, and filters for sintering.

Petroleum-refining sources of air pollution are mainly gaseous in nature. Mercaptans are removed by scrubbing. Solid dispersoids from the fluid-cracking processes are removed by electrostatic precipitators and/or cyclone-type collectors.

Miscellaneous Dusts Four industrial dusts have been classified as causing severe respiratory ailments: asbestos, causing asbestosis; crystalline silica, causing silicosis; cotton and cottonseed dust, a powerful allergen causing bronchial asthma and byssonosis; and lead, causing plumbism.

In mining operations and foundries, cyclonic collectors, followed by baghouses, have been found most effective. Likewise, in textiles and in cottonseed-oil pressing, baghouses following the lint-collecting cyclones can remove almost 100 percent of the hazardous pollutant.

Large central incineration plants can produce odors, noxious gases, and particulates (see Sec. 7.4). Odors and noxious gases can be minimized by carefully controlled combustion, using modern traveling-grate continuous systems. Furnace temperatures between 1,600 and 1,900°F (1,150 to 1,300 K) are recommended. Particulates can be minimized by well-managed covered conveying of dust-bearing refuse, incinerator residues and fly ash, and the use of electrostatic precipitators and wet scrubbers to comply with increasingly stringent air quality standards. (See Table 18.1.13 .) Gaseous emissions of SO_x, CO, and NO_x do not cause problems with continuous incineration, as their concentrations are small, but HCl and HF from plastics are hazardous. Stack height should be sufficient to minimize ground effects of any residual pollution. Municipal refuse can also be fired into waste heat utility boilers, a fuel conservation measure which will be increasingly important in electric power cogeneration.

RADIOACTIVE WASTE MANAGEMENT

REFERENCES: Cochran and Tsoulfanidis, "The Nuclear Fuel Cycle: Analysis and Management," American Nuclear Society. "Radiological Health Handbook." Nuclear Waste Policy Act of 1982 and its 1987 amendments. Low Level Radioactive Waste Policy Act of 1980 and its 1985 amendments, 10 CFR 50 Appendix 1, and 10 CFR 20. Current updated revisions of pertinent local and federal regulations.

General Considerations

Under existing legislation and authorization the Nuclear Regulatory Commission (NRC) has the primary responsibility for developing and regulating waste management methods and practices, sites of operation, and enforcement of all applicable standards, including those developed adjunctively with EPA. Radioactive wastes are generally divided into five general categories by the Nuclear Waste Policy Act: (1) **high-level waste (HLW)**; (2) **transuranic (TRU) waste**; (3) **low-level waste (LLW)**; (4) **uranium mill tailings**; and (5) **naturally occurring and accelerator-produced radioactive material (NARM)**. HLW is "the highly radioactive material resulting from the reprocessing of spent nuclear fuel, including liquid waste produced directly in reprocessing, and any solid material derived from such liquid waste that contains fission products in sufficient concentrations, and other highly radioactive material that the NRC, consistent with existing law, determines by rule requires permanent isolation." TRU wastes are wastes that contain alpha-particle-emitting radioisotopes with atomic numbers greater than 92 and half-lives longer than 5 years. LLW is defined as material which is "not HLW, spent nuclear fuel, TRU or by-product material, and material which the NRC, consistent

Table 18.1.13   Characteristics of Air- and Gas-Cleaning Devices

Name of device — General class	Name of device — Specific type	Device most suitable for	Removable contaminants	Optimum size particle μm	Limits of gas temperature, °F	Opt. conc., ppm by wt	Lim. Gas temp., K
Odor adsorbers	Shallow bed		Malodors, gases	(Molecular)	0–100	<1.9	256–311
Air washers	Spray chamber			>20	40–700		
	Wet cell			>5	40–700	<9.5	278–644
Electro. Precip., low–voltage	Two–stage, plate	Atmospheric air cleaning	Lints, dusts, pollens, tobacco smoke	<1	0–250		
	Two–stage, filter			<1	0–180	<1.9	256–394
Air filters, viscous–coated	Throwaway			>5	0–180	<3.81	256–356
	Washable			>5	0–250	<3.81	256–394
Air filters, dry–fiber	5–10 μm			>3	0–180	<1.9	256–356
	2–5 μm			>0.5	0–180	<1.9	256–356
Absolute fibers	Paper		Special†	<1	0–1,800	<1.9	256–1,256
Industrial filters	Cloth bag			>0.3	0–180‡	>190	256–356
	Cloth envelope			>0.3	0–180‡	>190	256–356
Electro, precip., high–voltage	Single–stage, plate			<2	0–700	>190	256–644
	Single–stage, pipe			<2	0–700	>190	256–644
Dry inertial collectors	Settling chamber		Dusts, fumes, smokes, mists	>50	0–700	>9,520	256–644
	Baffled chamber			>50	0–700	>9,520	256–644
	Skimming chamber			>20	0–700	>1,905	256–644
	Cyclone			>10	0–700	>1,905	256–644
	Multiple–cyclone			>5	0–700	>1,905	256–644
	Impingement			>10	0–700	>1,905	256–644
	Dynamic	Stack gas cleaning		>10	0–700	>1,905	256–644
Scrubbers§	Cyclone			>10	40–700	>1,905	256–644
	Impingement			>5	40–700	>1,905	256–644
	Dynamic			>10	40–700	>1,905	256–644
	Fog			<2	40–700	>190	256–644
	Pebble bed			>5	40–700	>190	256–644
	Multidynamic			<1	40–700	>190	256–644
	Venturi			<2	40–700	>190	256–644
	Submerged nozzle			>2	40–700	>190	256–644
	Jet			<5	40–700	>190	256–644
Incinerators Afterburners	Direct		Gases, vapors, malodors	Any	2,000	Combustible	1,367
	Catalytic			(Molecular)	1,000	any	811
Gas absorbers	Spray tower			(Molecular)	40–100	>1.9	278–311
	Packed column			(Molecular)	40–100	>1.9	278–311
	Fiber cell			(Molecular)	40–100	>1.9	278–311
Gas adsorbers	Deep bed			(Molecular)	0–100	>1.9	256–311

* Based on std. air @ 0.075 lb/ft³ (1.2 kg/m³).
† Bacteria, radioactive, or highly toxic fumes.
‡ 500°F for glass (553 K), 450°F for Teflon (505 K), 275°F for dacron (408 K), and 240°F for orion (389 K).
§ Reheating of scrubbed stack is necessary to avoid plumes.
Source: Jorgensen, "Fan Engineering," Buffalo Forge Co. Used by permission.

with existing law, classifies as LLW." LLW is generated by the nuclear power industry, hospitals, and other medical facilities, universities, and research laboratories. Examples of LLW include contaminated clothing, tools, syringes, cotton wipes, paper, rags, and other trash. Uranium mill tailings also are treated as radioactive wastes and are not transported from their generation point. NARM is not regulated by the NRC, but is regulated by the individual states. Usually, most NARM qualifies as LLW and most often is treated as such.

HLW and TRU wastes are covered under the **Nuclear Waste Policy Act (NWPA),** which spells out the step-by-step procedures for all tasks related to their disposal. The NWPA details the process to be followed in the selection of repositories for the disposal of HLW, TRU waste, and spent nuclear fuel; the development of interim storage and monitored retrievable storage for HLW and spent fuel; and the responsibilities and obligations of the Department of Energy and the nuclear utilities.

LLW is governed by the **Low Level Radioactive Waste Policy Act (LLRWPA).** The LLRWPA dictates that every state is responsible for the disposal of all LLW generated within its borders, and assumes that the LLW can be stored safely on a regional basis. To carry out the disposal of LLW, many states have entered into "compacts" with other states to site and operate regional disposal facilities. Other states have chosen to develop their own disposal sites. The LLW generated by nuclear power plants usually consists of three types: gases, liquids, and solids.

Gases   From the buildings which contain the fuel-handling system, auxiliary equipment, and waste systems, ventilating air, filtered if necessary, is released through a vent which is monitored for radioactive gases, iodine, and airborne particulates. Automatic controls will shut down the ventilating system, thereby stopping any release of airborne radioactivity as the monitor set points dictate.

Ventilating air for the containment vessel, which holds the reactor, primary pumps, and steam generators, is drawn when it is necessary for inspections or for other reasons to purge the vessel. **Purge air** is filtered through high-efficiency particulate air (HEPA) and through charcoal filters before release. In case monitor set points are exceeded, purging is automatically curtailed.

The third source of gases is the nitrogen cover gas of the volume control system in the PWR, and the steam jet ejector system in the BWR. Krypton, xenon, and iodine are produced within the reactor and mix with the cover gas. To release them, the principle employed is to "delay and decay" by compressing them into decay tanks of about 45 days' storage capacity in the PWR case, and into decay pipes of 20- to 60-min storage capacity in the BWR case. All releases are carefully monitored to be within established limits.

Liquids   Liquid radwaste streams from boiling water reactors may be classified into two major systems. In one of these, relatively high-radioactivity, low-conductivity water is reclaimed by filtration and deionization, and is stored for recycling into plant makeup systems. In the other, relatively low-radioactivity, high-conductivity water from floor and laboratory drains, laundry drains, makeup demineralizer regeneration, and equipment decontamination is treated by various means, such as filtration, waste concentration, and deionization, is monitored and recycled for in-plant use. Pressurized water reactors can be characterized by three major liquid streams. The first of these consists of borated, relatively high-radioactivity demineralized water from the primary cooling loop. During the feed-and-bleed operating mode, borated water is removed and replaced with freshwater, which slowly reduces the amount of chemical shim (boric acid) present. A boric acid evaporator makes it possible to recover boric acid and to recycle a large portion of the distillate. The second major stream consists of floor and equipment drains from the containment vessel and the auxiliary building and from the radiochemistry laboratory, which go to a large **waste holdup tank.** These are subsequently transferred to an evaporator, and the distillate therefrom is deionized, passed through a radiation monitor, and then recycled. With both types of reactor, the heavy ends, crud, and dirt concentrate in the evaporator feed tank. This waste is pumped to the radwaste **drumming station,** where it is solidified with Portland cement or chemical grout. The third of the major streams is soapy water from laundry and showers. Because of foaming difficulties, this stream is not always processed through an evaporator, but instead is filtered or put through a reverse osmosis unit, monitored for radioactivity, and if satisfactory, released.

Solids   Solid wastes for BWRs and PWRs are handled similarly. Evaporator and filter sludges are pumped to the radwaste drumming station and solidified with Portland cement or chemical grout in 55-gal (or larger) drums. Spent ion-exchange resins that are not regenerated are sent to an accumulation tank for decay and are then flushed into a shipping cask and dewatered for removal to a burial ground. They may also be solidified. Balable wastes are drummed but, if highly radioactive, are set in concrete. Large, heavy, junk equipment items can sometimes be stored on-site more economically than cutting them up and packaging them for off-site shipment.

Reference to current federal regulations and trade publications must be made to keep abreast of the latest developments.

SOLID AND HAZARDOUS WASTE MANAGEMENT

REFERENCES: "Pollution Control Technology, Solid Waste Disposal," Research & Educational Assn., New York. Sittig, "Handbook of Toxic and Hazardous Chemicals"; also Sittig, "Resource Recovery and Recycling Handbook of Industrial Wastes," Noyes Data Corp., Park Ridge, NJ. Exner, "Detoxication of Hazardous Waste," Ann Arbor Science Publishers. The Resource Conservation and Recovery Act of 1976 (RCRA) (as amended in 1980), PL 96–482 and CERCLA, PL 96-510 (SW-171). Metry, "The Handbook of Hazardous Waste Management," Technomic Publishing, Westport, CT. Peirce and Vesilind, "Hazardous Waste Management," Ann Arbor Science Publishers, Ann Arbor, MI. Current EPA regulations. Current updated revisions of pertinent local and federal regulations.

Solid waste includes solids discarded permanently, or temporarily, and materials suspended in air or water. Also forming part of this group are wet solids with insufficient liquid content to be free-flowing. Total solid wastes (excluding sewage) produced currently in the United States amount to about 250 million tons/year, or about 1 ton/(person · year), and include all industrial, domestic, commercial, and agricultural wastes. Between 4 and 5 percent of that total constitutes **hazardous**

wastes subject to special handling and/or disposal. It is projected that the total solid waste stream will increase to 300 to 350 million tons by early in the twenty-first century. The increasing accumulation of dry waste which results from the high standard of living of modern society will deplete existing economically exploited natural resources and further aggravate the already costly waste disposal problem. Product and energy recovery, therefore, constitutes the preferable direction to follow rather than elaborate means of disposal. There is, however, a growth factor for solid residuals which is caused by environmental control legislation: unpredictable amounts of waste generated by air and water pollution control. Environmental regulations in effect simply transfer the pollution from those media to the solid state, and vice versa.

Federal Environmental Legislation   See RCRA and CERCLA following.

Solid Waste Disposal

Solid waste handling involves two main steps: collection and disposal. **Solid waste conditioning** involves recovering those materials of value already present while conditioning the rest for conversion, recycling, or ultimate disposal. **Solid waste segregation** prior to collection into refuse, combustible wastes, noncombustible wastes, and hazardous waste is always a beneficial preconditioning step.

The principal solid disposal methods used are **incineration, landfill, and ocean disposal.** Many mining and processing operations leave a process residue on the premises since disposal cost elsewhere is prohibitive. Examples are red mud from refining bauxite and sulfite sludge from SO_2 stack gas scrubbing processes.

Landfilling requires spreading and compacting the waste and covering it with soil. This system is applicable both to biodegradable and nondegradable types of wastes. Leaching contamination of ground water must be prevented by installing liners. Careful attention must be given to possible subsidence and gas formation in landfilling. Topsoil disposal is limited to biodegradable organic materials.

Ocean dumping takes advantage of the ocean bottom trench configuration and the dispersing effects of ocean currents. Until recently, ocean dumping was the ultimate form of ocean disposal, but the practice has effectively ceased. The cost/benefit of ocean disposal was very favorable compared to alternate methods, but is now highly regulated and, with rare exceptions, the practice is proscribed and illegal. Ecological considerations effectively rule out this form of waste disposal for the foreseeable future.

Incineration is a volume reduction process rather than solid waste disposal, resulting in a residue which still must ultimately be disposed of. Heating values of solid wastes range from 5,000 to 15,000 Btu/lb for ordinary trash and industrial wastes. The growth of incineration is affected by higher fuel costs, construction costs, and added expense for air pollution and hazardous waste control.

Another means of solid waste reduction is **composting,** not widely used because of the limited market for its products, but ultimately may become more widely spread to maintain a rural-urban ecological balance.

Product Recovery

The average solid waste (per 100 lb) produced daily in the United States is roughly 6 lb industrial (from process), 5 lb municipal, and 89 lb mine tailings, smelter slags, dredging spoil, and agricultural wastes. Materials salvaged from solid waste recycled back to their sources not only reduce disposal costs but also conserve farmland values and natural resources. Recovery of metal, paper, glass, plastics, and hazardous substances is an outstanding example of materials recycling.

Materials separation from solid waste may be performed by physical or chemical means. The most common types of equipment used are material separators (varies with application: magnetic, thermal, flotation, size reduction—shredders, crushers, grinders, and pulverizers), materials flow (conveyors, lifters), mixers, and blenders. (See Sec. 10.)

Basic chemical conversion processes for industrial waste are hydrogenation, oxidation, acetylation, cross-linking cellulosic materials. For more detail, see Engdahl, "Solid Waste Processing—A State of the Art Report on Unit Operations and Processes," Public Health Service Publication No. 1856, Supt. of Documents, Washington, DC.

Table 18.1.14 Cost/Benefit Evaluation of Recycled Sewage Sludge as Fertilizer (Denver)*

	Total cost delivery and application, $/dry ton	Current worth of nutrients, $/dry ton	Disposal cost savings, $/dry ton	Total benefit (worth and savings), $/dry ton	Cost/ benefit
Air-dried (trucked)	$ 4.00	$ 8.37	$11.34	$19.71	0.20
Liquid (5% solids) (trucked)	14.00	12.20	11.34	23.54	0.595
Liquid (5% solids) (pipelined)	9.00	12.20	11.34	23.54	0.382

* 1975 data.

Biochemical processing of wastes from the food and chemical industries is common practice. An example is recovery of B-complex vitamins and animal feed from solid brewery wastes.

Pyrolysis of organic wastes can economically utilize them as supplementary fuels depending upon waste types and waste energy conversion processes. New efforts are being made in that direction to process municipal refuse as well as industrial and hazardous wastes.

Dry sludge production from municipal sewage has been achieved with the development of activated sludge processes using high purity oxygen. These processes reduce the volume of activated sludge by two-thirds and improve its dewatering characteristics.

Table 18.1.14 itemizes the 1975 cost factors at Denver from which cost benefit ratios have been developed. (See Cost/Benefit Balancing.) The cost benefit for air-dried sludge is particularly favorable, inasmuch as for every 20¢ invested, the public consumer receives $1.00 in benefits (1990 data).

RCRA AND CERCLA

The Resource Conservation and Recovery Act of 1976 (RCRA), as amended in 1980, regulates current and future waste disposal practices, including both municipal and hazardous waste, while the **Comprehensive Environmental Response Compensation and Liability Act (CERCLA),** or "Superfund," cleans up old, hazardous waste sites. Also, see 39 CFR 158, parts 240 and 241. "Thermal Processing and Land Disposal of Solid Waste."

Hazardous Waste Management RCRA defines *hazardous waste* as a solid waste, or combination of solid wastes, which because of its quantity, concentration, or physical, chemical, or infectious characteristics may

cause, or significantly contribute to an increase in mortality or an increase in serious irreversible, or incapacitating reversible, illness; or pose a substantial present or potential hazard to human health or the environment when improperly treated, stored, transported, or disposed of, or otherwise managed.

The act requires EPA to promulgate regulations establishing criteria and standards for all aspects of hazardous waste management. The regulations developed by the EPA designate a large number of substances as hazardous and also provide specific characteristics to determine whether a particular waste is hazardous. The characteristics include ignitability, corrosivity, reactivity, and EP toxicity.

The EPA also promulgated regulations establishing standards for hazardous waste treatment, storage, and disposal facilities. Provisions include groundwater protection, development and implementation of contingency plans, closure and postclosure activities, and specific facility performance standards. The standards cover facilities such as tanks, surface impoundments, land treatment, landfills, incineration, underground injection, and physical, chemical, and biological treatment. Other regulations cover "cradle-to-grave" control of the collection, source separation, storage, transportation, processing, treatment, recovery, and disposal of such wastes. Generators of these wastes bear the ultimate legal responsibility and are required to keep specific records for the disposition of their hazardous wastes by means of a manifest system.

18.2 OCCUPATIONAL SAFETY AND HEALTH
by Pascal M. Rapier and Andrew C. Klein

REFERENCES: "General Industry Guide for Applying Safety and Health Standards," 29 CFR 1910 OSHA, U.S. Dept. of Labor, 1973, and Amendments. 29 CRF 1910.1–1910.309, *Fed. Reg.,* **37:**202, Oct. 18, 1972, and Amendments. Current updated revisions of pertinent local and federal regulations.

WORKER'S COMPENSATION LAWS

Worker's compensation laws in this country were introduced in 1911 and have now been enacted in all the states. The principle involved is that the worker injured or disabled in industry should be enabled, through proper medical treatment, to return to wage-earning capacity as promptly as possible and, while incapacitated, should receive compensation in lieu of wages, regardless of fault. The expense of medical treatment and compensation should properly be borne by industry and become a part of the cost of its products. The laws generally provide that workers injured in industry shall be furnished the necessary medical treatment, and, in addition, compensation based on a percentage of

their average weekly wages, payable periodically. Dependents of employees killed in industry are likewise compensated.

Many states have supplemented their worker's compensation laws by providing comparable benefits in cases of incapacity or death due to occupational disease. Some states make these provisions applicable only in case of specified occupational diseases; others make them of general application to cases of any disease directly attributable to the employment. At the present time, all states and the District of Columbia provide for compensation benefits in occupational-disease cases either by enlarging the scope of the worker's compensation law, by separate legislative enactment, or by judicial construction.

The enactment of worker's compensation laws has been followed in many jurisdictions by more stringent provisions relating to factory inspections for the prevention of accidents in industry and of occupational disease.

The enactment of worker's compensation and occupational-disease laws has increased materially the cost of insurance to industry. The increased cost and the certainty with which it is applied have put a premium on accident-prevention work. This cost can be materially reduced by the installation of safety devices. Experience has shown that approximately 80 percent of all industrial accidents are preventable.

SAFETY

REFERENCES: "General Industry Guide for Applying Safety and Health Standards," 29 CFR 1910 OSHA, U.S. Dept. of Labor, 1973. 29 CRF 1910.1–1910.309, *Fed. Reg.*, 37:202, Oct. 18, 1972. "Accident Prevention Manual for Industrial Operations." National Safety Council. "Manual of Accident Prevention in Construction," Associated General Contractors of America, Inc. American Conference of Governmental Industrial Hygienists, "Industrial Ventilation—A Manual of Recommended Practice." Blake, "Industrial Safety." Prentice-Hall. DeReamer, "Modern Safety Practices," Wiley. Heinrich, "Industrial Accident Prevention," McGraw-Hill. Patty (ed.), "Industrial Hygiene and Toxicology," Interscience. Simonds and Grimaldi, "Safety Management," Irwin. Stubbs (ed.), "Handbook of Heavy Construction," McGraw-Hill. Labor laws applicable in the jurisdiction where the working enterprise is located.

The person(s) responsible for overseeing the safety of all operating personnel must be cognizant of the latest laws and regulations pertaining to worker safety, which are changed and/or updated from time to time.

The logical time to install safety devices is when new machines are being built, while general construction work is being done, or when alterations or repairs are being made; results can be accomplished with a minimum of expense and delay at the time plans and specifications are being prepared.

Checking for Safety   To ensure that the question of safety will not be overlooked, it is well to have all plans, specifications, and drawings checked for safety, making special provision for this in each set of specifications and in the title plate of each drawing.

Buildings
(See 29 CFR 1910, Subparts D, E, and F, as amended.)

Plant Arrangement   A fundamental factor in effective accident prevention is the provision of ample ground space in the plant site to reduce crowding of buildings, congestion of plant traffic, unusual fire hazards, unsafe yard conditions, etc. One-story buildings have definite advantages in regard to fire hazards, building collapse, and natural lighting.

Where a plant is located near a main line of a railroad, consideration for safe access by employees and vehicular traffic should be given serious attention. Safe passageways from one building to another are also important. Blind corners, doorways opening onto yard railway tracks, etc., should be avoided as much as possible, and, where such conditions must necessarily exist, safety railings, gates, signs, or gongs should be installed to warn pedestrians of danger.

Because of their inherent **hazards of fire and explosion,** the storage, handling, and utilization of *flammable liquids* should be carefully controlled. The National Fire Codes, promulgated by the NFPA, are considered authoritative in establishing minimum safeguards necessary to control these hazards. It is necessary in any event to conform to the requirements of any applicable state and local code.

Buildings in which **dusty operations** are carried on should be designed to present a minimum area of projections, ledges, and resting places for dust accumulations.

Floors, Stairways, Aisles, etc.   Probably the most important factor to be considered in connection with floors, stairways, etc., concerns **slipperiness.** Floors and stairs should be free from projecting nails, bolt-heads, etc., as noiseless as possible, wear well, and be strong enough to carry safely any static or moving load. The weight of modern industrial machinery and material handling equipment should be carefully considered in checking floor-load calculations. Floors and stairways should be kept clear of unnecessary obstructions over which workers may trip.

Spilling of oil, water, acid, etc., should be prevented to eliminate slipping hazards. Splash guards, drip collectors, etc., can be designed in many instances to reduce slippage. Excessive spillage of dusty materials onto the floor is sometimes taken care of by installing floor gratings beneath which pits or conveyor systems are located to collect the falling material.

Ample aisle space is very important, especially in foundries where workers carry ladles of molten metal and where there is considerable shop traffic involving power-driven trucks, etc. Aisles should be clearly marked off to assist in keeping them clear. One-way traffic is often advantageous.

Stairways should be provided with **handrails** on both sides, and an intermediate middle handrail should be installed on stairways over 88 in in width. Nonslip treads on stairs are desirable. Stairways should be adequately lighted (see Sec. 12.5.)

Exits and Fire Escapes (See Life Safety Code NFPA.)   As far as practicable, all doors should open outward or with the natural direction of egress; they must not block passageways from other floors or parts of the building.

For factories, not less than two means of exit should be provided on every floor, including basements, of all buildings or sections; these exits should be separated in such a manner that they are not likely to be cut off by a single local fire. The location of stairways around or adjacent to passenger elevators is undesirable unless there is separation by fire walls.

Outside fire escapes are inferior to stairways as means of egress. Where they are used, they should be located on blank walls or arranged so that persons on them will be protected from flames issuing from windows or openings underneath by use of wired glass in standard metal frames, fire doors, etc. Outside fire escapes, to provide maximum protection for persons using them during fire, should be enclosed in non-combustible towers which will protect against weather, smoke, or fire, with access through or over some intermediate balcony or structure to the building proper.

Lighting   (See Sec. 12.5.) Adequate lighting has a definite bearing on the prevention of accidents. Workrooms should be well lighted to reduce eye strain and the possibility of permanent eye impairment and also to remove any danger of employees falling over obstructions or being caught in machinery in darkened areas.

Ventilation   (See Sec. 12.4.) A lack of adequate ventilation in a workroom tends to bring on fatigue and reduces the alertness of workers, thus making them more susceptible to accidents. Where injurious dusts or noxious vapors are encountered, it is necessary to provide for their removal by the installation of adequate local exhaust systems (see Sec. 12.4 and ACGIH "Industrial Ventilation" and ASA Standard, "Fundamentals Governing the Design and Operation of Local Exhaust Systems").

Identification of Piping   (See Sec. 8.7.) It is desirable that a plan of identifying the contents of various pipelines be adopted so that in case of an emergency it will be possible to determine quickly the service of all pipelines involved.

Mechanical Guarding in Machine Design
(29 CFR 1910, Subpart 0, as Amended.)

The most logical time at which to consider the safeguarding of a machine is during its design. In this stage, features of safe operation can be incorporated so that there will be a minimum of specific guarding required on the finished machine. The following points are fundamental considerations concerning accident prevention which should be taken into account in machine design.

Care should be taken in arranging clearances of moving parts to avoid shearing or crushing points in which hands or other parts of operator's body might be caught or injured.

Arrangements should be made so that adjustments, inspections, and hand lubrications can be done safely.

Machines should be so designed that operators are not required to stand in an uncomfortable position, reach over moving parts, or exert themselves in awkward positions.

Machines should be designed so that there will be little danger of the operator tripping over parts of the frame or striking against projecting parts during normal operation movements.

Careful attention should be given to strength of all parts whose failure might result in injury to operator.

All guards, covers, or enclosures should be designed strong enough to prevent the possibility of their giving way and permitting an accident in case the operator should fall or be thrown against them.

Point-of-Operation Guarding   The point of operation on a machine is taken to be that zone where the work of the machine is actually performed and where the operator, by manipulating the material being processed, is exposed to a hazard from moving parts of the machine. Guards for point-of-operation protection are placed on the machines as additional equipment. The first requirement for a successful guard of this type is that it shall be convenient and not interfere with the operator's movement or affect the output of the machine. The following statements describe several basic principles which may be utilized in point-of-operation guarding.

Where possible, the **danger point** should be completely **covered** by a barrier or enclosure before the hazardous operation of the machine begins. This may be accomplished for example on a treadle-controlled machine by having the treadle operate the guard which, when in proper location, will then activate the machine.

Operator's hands may be kept out of dangerous positions by installing **starting devices** so located that, to operate the machines, both hands of the operator must be out of the danger zone.

Feeding devices may be used so that material to be processed is placed in a feeding mechanism at a point where there is no exposure to moving parts. Special holders or feeding tongs can also be used to place work in hazardous positions.

Electronic controls, which operate by the interruption of a light beam or other energy source to protect the danger zone, may be used to start and stop machines.

Electrical interlocks on guarding devices may be utilized in operating circuits so that unless a guard is in proper position the circuit is open and no current will flow until the machine is safely protected by the guard being brought into proper position.

Automation has minimized the hazards associated with the manual handling of stock and has eliminated the need for repetitive exposure at the point of operation, but the urgency of making repairs or adjustments introduces the need for special precautions covering maintenance work, such as locking out the power source.

Electric Equipment
(See 29 CFR 1910, Subpart S, as amended.)

In considering electrical equipment for industrial establishments, it should be borne in mind that the **hazard** to human life **increases with increase in voltage.** Thus, economy in transmission wiring and copper parts, which is achieved through the use of high voltage motors, may be offset by increased danger of accident. For small motors, lights, and general service inside industrial plants, installations of 110 or 220 V are recommended.

All **switches, fuse boxes, terminals, starting rheostats, motors,** etc., located within 8 ft of a floor or working platform, should be enclosed or guarded in such a manner as will prevent accidental contact with live parts, irrespective of voltage. **Switches** should be arranged so that they can be locked in the open position, to guard against a switch being activated accidentally while workers are at work on the lines or equipment that the switch controls. **Equipment for operation at 550 V and higher** should be isolated from other operating equipment, in separate rooms or enclosures, with provision to lock these enclosures. **All metallic cases, frames, and supports** of such equipment should be permanently grounded; foundation bolts should not be depended upon for this purpose, but substantial ground conductors should be used. It is preferable to have these ground wires accessible for inspection. **All low voltage secondary circuits** of 300 V or less (such as light, motor, and meter circuits) should be permanently grounded whenever a neutral point is available for the purpose. Secondary circuits of 250 V or less should be permanently grounded even though a neutral point is not available. Whenever grounding has to be omitted, the frame or casing of the apparatus should be permanently grounded, and all live parts of the secondary circuit should be shielded to prevent accidental contact therewith. It is also desirable to have floors or platforms adjacent to switchboards built of nonconducting material, or suitably insulated. In the absence of such construction, rubber mats may be used.

Portable electric lamps, tools, and machines are subjected to severe operating conditions. Many electric shock fatalities have occurred on electric portables, even with 110-V lighting circuits. Cable used to service portable equipment should be high-grade insulated cord designed for the service, not ordinary twisted lamp cord. Its mechanical attachment to both the portable and the attachment plug or source of power should be designed to prevent sharp bending or chafing that would break down the insulation. All portable lamps should have nonmetallic sockets (such as rubber, plastic, or porcelain). The heavier portable tools and machines cannot practically have their metallic parts insulated from contact by the operator, and protection from shock due to accidental charge thereon must be provided by means of a special grounding wire. Such protective grounding should be in the form of an additional wire in the cable that feeds the portable device. One end should be permanently connected to the machine frame and the other connected to ground through an additional pole in both the attachment plug and the receptacle to which the device is to be attached.

Occupational-Disease Prevention
(See 29 CFR 1910, Subpart G, as amended.)

In any industrial operation where a toxic material is being processed in such a manner that those persons engaged in or working near the operation are exposed to appreciable quantities of dusts, fumes, vapor, or gas, it is important that adequate control measures be adopted. The following statements cover the major considerations involved in the application of effective control to industrial occupational disease.

Contaminants (See 29 CFR 1910, 93.)   The physical and chemical characteristics of a contaminant should be known. In the case of dusts or fumes, the chemical nature, particle size, solubility, etc., should be determined. For gases or vapors, the composition, vapor pressure, flash point, etc., are important factors. In all atmospheric contamination, the quantity of material in the worker's breathing zone must be known before the degree of hazard can be evaluated. The chemical characteristics are important in the selection of materials to be used in the construction of any control equipment where corrosion, etc., might be factors.

A careful investigation should be made to determine accurately the sources from which the contaminant is being produced or from which it is being dispersed. The most common types of dust-producing operations are crashing, screening, grinding, polishing, etc. Dispersion of dust is encountered in practically all dry handling operations of fine materials. Vapors and fumes are produced by chemical processes and reactions and are most commonly found in connection with the use of solvents.

A number of testing instruments and procedures have been developed for determining quantitatively the concentrations or amounts of various toxic material in the atmosphere of a work room (see ACGIH, "Air Sampling Instruments" and reports of U.S. Public Health Service and U.S. Bureau of Mines; see also AIHA Hygienic Guide Series and ACGIH, "Annual Threshold Limits"). In general, the seriousness of any exposure involving a health hazard is directly proportional to the dosage (concentration and length of time of the exposure). Engineering control should be directed to the reduction of these two factors.

Local Ventilation at the Source of Contaminant   Removal by means of exhaust hoods, enclosures, etc., so located as to prevent the escape into occupied areas of any appreciable amount of contaminant at its source is the most effective method of control. For details of hood design, piping, collectors, fan characteristics, and for general principles of exhaust design system, see Sec. 12.4. and ANSI Standard, "Fundamentals Governing the Design and Operation of Local Exhaust Systems"; AFS, "Engineering Manual for Control of In-plant Environment in Foundries."

Natural ventilation has a limited application in industrial occupational disease control. It is important with a natural ventilation system to maintain close supervision of adjustment as required by changes in temperature, wind directions, etc. Regular tests of air content should be made to check on the degree of dilution being obtained. This method of control is not recommended for exposures where severe hazards exist owing to extreme toxicity or high concentrations.

Isolation   Under some circumstances, the isolation of a hazardous operation, physically, or in point of time, is required. For example, a hazardous operation may be carried on in a separated room in which all contamination can be confined or it may be carried on outside regular working hours when no one except the persons engaged in the operation will be present. By isolating an operation, the number of persons exposed to any accompanying hazard may be reduced to a minimum.

Process Revision   The possibility of substituting a less toxic material should be borne in mind; e.g., high-boiling-point distillates for benzol in rubber cements, thinners, etc.; dolomite lime for quartz in foundry parting compounds. It may be possible to reduce the concentrations of objectionable contaminants and the time of exposure by changing handling operations, by substituting mechanical feeds or conveyors for manual operations, or by installing automatic machinery or controls so as to dispense with the presence of an attendant. An automatic temperature or pressure control device, in connection with a chemical process, may reduce materially the time of an operator's exposure as well as prevent the production of excess or undesirable fumes or gases. An enclosed conveyor system handling dry material and automatically weighing it may be utilized to eliminate a severe exposure that would result from hand shoveling.

Physical Hazards   In addition to exposures to toxic gases, vapors, dusts, fumes, etc., there are several physical conditions such as abnormal pressures, temperatures, and humidities as well as radiation (including ultraviolet; infrared; X-ray; α, β, and γ rays from radioactive substances; and radiation from handling radioactive isotopes), all of which may, in case of excessive exposure, prove detrimental to the health of exposed persons. Evaluation depends on physical methods of measurement. Protective measures include shielding from radiation and control of exposure time rather than ventilation, as in the case of atmospheric contaminations.

Ambient Noise Control   (See also Sec. 12.6.)   Outer property noise propagation is regulated by EPA, under provisions of the Noise Control Act of 1972; Section 6 applies to new-product noise emission standards; Section 17 is concerned with noise emission regulations for railroads and Section 18 for motor carriers. EPA is directed to publish noise emission guidelines in the *Federal Register.* The regulation for aircraft-airports are promulgated by the Federal Aviation Administration (FAA) consistent with Section 611 of the Federal Aviation Act of 1958 and guidelines proposed by EPA.

Occupational Noise Exposure   Mandatory protection of employees against noise is required by 29 CFR 1910.95 as follows:

(a) Protection against the effects of noise exposure shall be provided when the sound levels exceed those shown in Table 18.2.1 when measured on the A scale of a standard sound-level meter at slow response. When noise levels are determined by octave band analysis, the equivalent A-weighted sound level may be determined from Fig. 18.2.1. Octave band sound-pressure levels may be converted to the equivalent A-weighted sound level by plotting them on this graph and noting the A-weighted sound level corresponding to the point of highest penetration into the sound-level contours. This equivalent A-weighted sound level, which may differ from the actual A-weighted sound level of the noise, is used to determine exposure limits from Table 18.2.1.

Table 18.2.1   Permissible Noise Exposure*

Duration, h/day	Sound level, slow response, dBA
8	90
6	92
4	95
3	97
2	100
1½	102
1	105
½	110
¼ or less	115

* 29 CFR 1910.95.

Fig. 18.2.1   Equivalent sound-level contours.

(b) (1) When employees are subjected to sound levels exceeding those listed in Table 18.2.1, feasible administrative or engineering controls shall be utilized. If such controls fail to reduce sound levels within the levels of Table 18.2.1, personal protective equipment shall be provided and used to reduce sound levels within the levels of the table. (2) If the variations in noise level involve maxima at intervals of 1 s or less, it is to be considered continuous. (3) In all cases where the sound levels exceed the values shown herein, a continuing, effective hearing conservation program shall be administered.

When the daily noise exposure is composed of two or more periods of noise exposure of different levels, their combined effect should be considered, rather than the individual effect of each. If the sum of the fractions $C_1/T_1 + C_2/T_2 + \cdots + C_n/T_n$ exceeds unity, then the mixed exposure should be considered to exceed the limit value. C_n indicates the total time of exposure at a specified noise level, and T_n indicates the total time of exposure permitted at that level. Exposure to impulsive or impact noise should not exceed 140-dB peak sound-pressure level.

Personal Respiratory Protection   (See 29 CFR 1910, 134, as amended.)   Where it is not practicable to control air contamination in the breathing zone of workers by adequate exhaust ventilation, etc., and it is necessary for personnel to be exposed to harmful amounts of dust, smoke, fumes, or gases or to work in an atmosphere with a deficiency of oxygen, personal respiratory equipment must be provided. Such equipment is not, in general, suited for prolonged daily use because of inherent discomfort and inconvenience to the wearer. For emergency or temporary situations or until effective control of contamination can be developed and applied, personal respiratory protection should be given very careful consideration where harmful exposures are encountered (see "The Respirator Manual," prepared jointly by the American Industrial Hygiene Association and the American Conference of Governmental Industrial Hygienists). The U.S. Bureau of Mines (Schedules 19 and 21) has developed certain standards for approval of respiratory protection devices. The equipment may be (1) **supplied-air respirators** (hose type), devices that supply clean, respirable air to the wearer through a hose line extending from a source outside the contaminated zone; (2) **supplied-oxygen respirators** (self-contained type of oxygenbreathing apparatus), devices that supply oxygen to the wearer from a source of supply that he or she carries as part of the respirator; or (3) **filter-type respirators**, which may be chemical filter respirators that remove the harmful constituents from air passing through them by chemical reactions, absorption, and adsorption, including ordinary gas masks and chemical respirators; or which may be mechanical-filter respirators that remove the harmful constituents from air by mechanical filtration, including dust, mist, or fume respirators.

Safety Codes and Published Material: OSHA The American National Standards Institute (ANSI) publishes a series of ANSI Standards. State and local ordinances must be complied with as required.

The National Safety Council publishes a series of Industrial Data Sheets and Safe Practice Pamphlets covering the safeguarding of industrial operations, in addition to literature, posters, films, and other safety educational material. The annual publication "Accident Facts" reports current trends in accident experience on an overall basis.

United States government agencies, such as the Bureau of Mines, the Bureau of Labor Statistics, the National Bureau of Standards, and the Division of Labor Standards publish statistical information concerning industrial-accident occurrence and accident prevention. The National Fire Protection Association (NFPA) promulgates fire codes and publishes information of all types relating to fire protection.

The Williams Steiger Act of 1970 created the **Federal Occupational Safety and Health Administration** (OSHA), which has now codified and published blanket standards for national enforcement in all public safety and health areas. Updated revisions will continue to appear in the Federal Register under Title 29 CFR 1910 and 1518, and these standards make all preexisting standards obsolete.

Radiation Health Physics (Nuclear Regulatory Commission, Code of Federal Regulations 10 CFR 20; International Commission of Radiological Protection, Pub. 30, "Radiological Health Handbook") **Radiation,** which accompanies all nuclear reactions, can be advantageous or disadvantageous, depending generally on the manner in which contact is made and on the amount of energy involved. In its application to humans, the common terminology includes **radiation absorbed dose (rad)** and **dose equivalent,** measured in SI units as **gray (Gy)** and **sieverts (Sv),** respectively. In an older, though still common, usage the respective units are designated **rad** and **rem.** Absorbed dose, measured in grays (or rads), is a measure of the energy absorbed from radiation per unit mass of irradiated material. Absorbed dose is applicable to all types of ionizing radiation and irradiating material. The amount of biological damage to human tissue, however, depends on the type of radiation and its energy. Thus, absorbed dose measurements must be modified in order to relate the amount of energy absorbed to the biological effect. Dose equivalent, as measured in sieverts (or rems), accounts for these effects through the use of **radiation weighting factors** (formerly known as **quality factors)** Q. For radiation protection purposes, the **dose equivalent** is defined as the product of the absorbed dose and the radiation weighting factor. The currently accepted weighting factors are 1, for photons and electrons of all energies; 5, for protons with energies greater than 2 meV; 20, for alpha particles, fission fragments, and heavy nuclei; from 5 to 20, for neutrons, depending on the neutron energy.

Exposure to radiation may be divided into two general types: (1) external exposure, which is that resulting from radiation sources external to the body; and (2) internal exposure, which is that resulting from radionuclides within the body. Radiation-protection standards are established for both types of exposure. Limits for external exposure are given in units of sieverts per unit of time. Basic limits currently are based on a control point of 1 year. Local operational rules may specify control limits for shorter exposure times (day, week, month), such limits being based on the basic limits. Radiation-protection standards for controlling internal exposures are based on the same basic limits as those for external exposures. Controls, however, are affected by the use of **annual levels of intake (ALIs)** and **derived air concentration (DAC)** values established for all radionuclides. Many factors (physical, chemical, biological, and physiological) are involved in determining the ALI and DAC of a radionuclide. As such, the ALIs and DACs of the radionuclides have a very wide range of values. All **ALIs** and **DACs** are normally expressed in terms of **bequerels** and **bequerels per cubic metre,** respectively. Generally, ALI and DAC values are based on a 40-h workweek for persons exposed to radiation.

Monitoring is the periodic or continuous determination of the presence and extent of ionizing radiation and radioactive contamination. It provides information for radiation protection and normally is assigned to a health physics group, which establishes schedules, executes or supervises the operation, and evaluates their results. It is divided into these special types: personnel, area, source, surface, air, and water.

Regulations governing exposure to radiation are promulgated in the form of Nuclear Regulation Commission regulations found in 10 CFR 20. The current **occupational dose limit** is 0.050 Sv (or 5.0 rem) total effective dose equivalent per year. The total effective dose equivalent includes both external and internal doses. Additional radiation protection limits exist for radiation exposures to individual organs of the body, the lens of eye, and the extremities. Reference to the current full statement of 10 CFR 20 is mandatory for a complete understanding of the regulations.

18.3 FIRE PROTECTION
by Pascal M. Rapier and Andrew C. Klein

REFERENCES: "Fire Protection Handbook," "National Fire Codes," "Fire Inspection Manual," National Fire Protection Assn. "Handbook of Industrial Loss Prevention," McGraw-Hill. "Uniform Building Code 1976," International Conference of Building Officials, Building Offic. and Code Admin. International (BOCA). Fire Insurance Assn. (FIA), Factory Mutual (FM), Code of Federal Regulations: 29 CFR 1518, July 1976, and 29 CFR 1910, July 1976, Jan. 28, 1977, as amended. Current updated revisions of pertinent local and federal regulations.

IMPORTANCE OF FIRE PROTECTION
(29 CFR 1910, Subpart L)

The profitable use or availability for use of facilities is the aim of all business, whether industrial, mercantile, professional, scientific, or educational. Destruction of or damage to the facilities cripples the attainment of that purpose.

Monetary compensation for destroyed property and for resulting lost profits is obtained by the purchase of insurance. Reputation, goodwill, and other intangible factors may be irreparably damaged. Loss of life is not recoverable. Loss by fire is largely an avoidable, nonproductive tax resulting from carelessness, ignorance, apathy, or incompetence. The annual fire loss in the United States (refer to NFPA) has increased steadily since 1964 at about 8 to 12 percent per year, reflecting the population and economic growth and inflation. In recent years, the number of losses has exceeded 2 million. Building fires have made up in excess of 80 percent of the **amount** of loss but only about 40 percent of the **number.**

Fire-Loss Prevention Neither insurance organizations nor legally established regulatory bodies assume the responsibility of industry for conserving its own resources and operating its facilities safely, although they may detect deficiencies or a lack of compliance with regulations and may serve in a consulting capacity. Fire loss prevention is an indispensable element in industry and business. It exists only with top management direction and the support of labor.

The designation **fire protection** usually encompasses the entire field of **prevention of loss by fire,** including both the causes for the occurrence of fires and methods for minimizing their consequences. Included along with fire are other destructive agencies such as explosion, lightning, electric current, wind, earthquake, nuclear excursions, and radioactive contamination. OSHA and building codes mandate minimum standards of protection to prevent injury and loss of life. Higher standards to protect investment are a worthwhile management option.

Fire brigades (29 CFR 1910.164) are essential to the development and maintenance of an effective fire protection program at every job site. By prompt, immediate response and notification of the local fire department, every effort should be made to bring the fire quickly under control during the early minutes of an outbreak. The immediate availability of the correct fire protection and suppression equipment is essential.

Fire protection engineering involves the application of sound engineering principles to the reduction of loss by fire and related hazards. The Society of Fire Protection Engineers has established a well-defined scope for fire-protection engineering practice, both in building design and in safe operating practices.

Limitation of Loss   With large businesses it is difficult to limit single destructive occurrences to an amount that will not be catastrophic to the owner or to others having a financial interest. Principles and methods for preventing large losses by fire are well known to fire protection engineers.

Sources of Fire Protection Information   Much specific, detailed, and technical information is given in the standards, codes, and rules of the NFPA, Factory Mutual Engineering Division, Factory Insurance Assoc., American Insurance Association, ANSI, Underwriters Laboratories Inc., and may also be found in the procedures of many insurance companies and inspection bureaus.

The NFPA "Handbook of Fire Protection" and the Factory Mutual "Handbook of Industrial Loss Prevention" are especially comprehensive references. The most complete and readily available source of standards is the "National Fire Codes," prepared and published by the NFPA, currently in ten volumes as follows: 1—Flammable Liquids; 2—Gases; 3—Combustible Solids, Dusts, and Explosives; 4—Building Construction and Facilities; 5—Electrical; 6—Sprinklers, Fire Pumps and Water Tanks; 7—Alarms and Special Extinguishing Systems; 8—Portable and Manual Fire-control Equipment; 9—Occupancy Standards and Process Hazards; 10—Transportation. Many of the standards have been adopted by insurance inspection and advisory groups, such as the American Insurance Association, and by Federal, state, and municipal regulatory bodies and have been incorporated into building codes and other regulations. The Occupational Safety and Health Administration (OSHA) U.S. Dept. of Labor, is promulgating blanket national standards for fire protection, which appear in the Federal Register under Title 29, Code of Federal Regulations for the working place, and Part 1518 during construction. Where applicable, these supersede and make obsolete all existing standards.

CONSTRUCTION

Building construction and the protection of buildings and their contents against fire are frequently governed by local building codes and ordinances and by insurance standards. When new construction or changes are planned, the property owner should have the advice of a competent fire protection engineer and should consult local authorities and insurance carriers to avoid delay and the possibility of expensive changes later.

Some construction features which have increasing importance in fire-loss prevention are (1) the trend toward large fire areas which present high values to possible loss and make manual fire fighting difficult; (2) blank wall, air-conditioned, and artificially lighted buildings which interfere with ready access; and (3) mechanization of materials handling, resulting in conveyor systems that interconnect areas, and in larger storage areas with high-piled stock which make fires difficult to extinguish. These cause increased cost of fire protection and emphasize the need for automatic fire control methods, sources of dependable water supplies, and special extinguishing systems. A measure that should be adopted where practical is the use of outdoor, totally enclosed process equipment, with shelters provided only for control rooms, laboratories, and some maintenance functions.

Types of Construction   **Fire-resistive** refers to types that withstand considerable fire without serious damage, such as reinforced concrete or protected steel. **Noncombustible** refers to any construction that contains no elements of burnable material built which may be structurally

damaged by fire, such as unprotected metal. **Combustible** means structures entirely of combustible materials or having combustible elements of such character and distribution that a fire can spread and contribute fuel so that severe damage results. Combustible types of construction are frequently subdivided into (1) "heavy timber," also called "plank on timber," "mill" or "slow burning" that has masonry walls with floors and roof of plank on heavy timbers; (2) "ordinary construction" that has masonry walls with floors and roof made of boards on joists and is sometimes called "quick burning"; (3) "wood frame" that has all of its elements of wood, except that the exterior may be surfaced with a noncombustible sheathing.

Many types of construction have composite elements that may include combustible materials. Insulation, acoustic materials, and surface treatments may aid the start and contribute to the rapid spread of fire. A roof of interlocking metal sheets, with asphalt on its upper surface as part of a vapor barrier or as an adhesive for insulation or weather surfacing, can, if initially heated by a local interior exposing fire, furnish gaseous fuel through the joints to produce a spreading damaging fire. Owing to the difficulty and cost of protection, enclosed spaces at roofs, ceilings, in walls, and below floors should be avoided. Important steel structural members and steel supports for heavy equipment that may be exposed to severe fire should have heat-insulating protection. Economic tradeoffs, based on the probability of exposure, should be made.

Fire Performance of Building Materials   The kind and amount of fire protection needed are governed by the fire performance of materials used in construction and by the combustibility of building contents. Completely noncombustible construction avoids the need for sprinkler protection, provided that contents are also certain to be noncombustible. Knowledge of the fire performance of materials, based upon tests made by qualified laboratories, is essential before they are adopted (see Sec. 6.8). Fire performance is in two categories, fire resistance and fire hazard.

Fire resistance measures the susceptibility of materials to damage by exposure to fire and is usually measured as the time period of exposure, without significant damage, to a standard fire-exposure as specified by standard fire tests, such as "ASTM E-119 Fire Tests of Building Construction and Materials," and expressed as a **fire-resistance classification** in hours as ½ h, 2 h, 6 h, etc.

Fire-hazard indicators are also in two generally recognized categories used by the Underwriters Laboratories. The **listing** of specific appliances and materials indicates that the products comply with the Laboratories' test requirements, with particular reference to fire-preventive and fire protective capabilities when the manufacturer's instructions and limitations of use have been followed. The **fire-hazard classification** of materials and products is related mainly to burning characteristics. It has a numerical basis determined by (1) flame spread, (2) fuel contributed, (3) smoke-developed. A fire hazard classification of 0 indicates a material with a fire performance equivalent to that of cement board composition. A value of 100 corresponds to the behavior of red oak lumber. Building codes and other regulations generally prescribe limiting values for fire hazard classifications.

Protection against Exposure Fires   Protection of buildings or other structures against fires in nearby property must sometimes be provided. A practical barrier against a conflagration is the presence of fire-resistive buildings along the exposed side. Usually the most severe exposure is localized, and protection against it may be provided by a blank brick or concrete wall, by wired-glass metal-frame windows, or by open sprinklers alone or in combination. The relative value of various safeguards against exposure fires has been estimated to be roughly blank brick or concrete walls, 100; tin-clad shutters, 60; wired-glass windows, metal frames, glass block (4 in minimum thickness), 40; wired-glass windows, wooden frames, 20; plain glass windows, 5; additional value of open sprinklers with any of the foregoing, except blank fire walls, 30; open windows, 0.

Horizontal Cutoffs   Provide effective fire walls between important buildings, and subdivide large building areas in order to limit the probable maximum damage from a single fire.

Vertical Cutoffs   Enclose stairs, elevator wells, vertical conveyors, chutes, and other floor openings with adequate fire-resistive walls, having fire doors or their equivalent at openings to prevent the rapid spread of fire and heat upward from floor to floor.

Isolation of Hazards   Cut off hazardous occupancies by fire-resistive partitions or fire walls or, if the degree of hazard warrants, isolate them in separate buildings. Make provision for adequate ventilation and for explosion vents where needed.

Floor Leakage   Provide drained, watertight floors over large values susceptible to water damage.

Storehouses and Vaults   Provide suitable storehouses for large quantities of combustible raw stock, tools, and patterns, and for valuable finished goods. Provide reliable vaults for the storage of business records and valuable drawings.

SAFEGUARDS DURING BUILDING CONSTRUCTION
(20 CFR 1518, Subpart F.)

A building under construction is more vulnerable to fire than the same building completed and in use with normal fire protection. As construction progresses, concentrations of readily combustible materials appear at new locations from day to day. So do many potential ignition sources—temporary heaters, welders' torches, roofers' tar kettles, and workers' discarded matches or cigarettes. If fire starts, wall and floor openings create draft and help flames to spread rapidly. Many serious construction fires have taken place. Most can be prevented. Guard against cutting and welding hazards by proper supervision of operations. Plan in advance for effective notification of, and coordinated firefighting operations with, any available public fire department. (See 29 CFR 1518.24.)

Start installation of the fire-protective equipment—including water supplies, yard mains, and hydrants—as soon as construction of the building starts. Make layouts for sprinklers as soon as building drawings are completed, and let contracts without delay. It is sometimes possible to arrange for limited temporary protection by providing connections for hose from the piping system which furnishes water for construction uses. Have hose and nozzles available as soon as the yard piping, hydrants, and water supplies are ready. Distribute ample hand extinguishing equipment throughout the premises, including contractor's temporary buildings.

Temporary Heating   Arrange any needed heat safely. Use coke as fuel for salamanders, rather than wood trash or rubbish. Oil-fired heaters, if well-designed and properly located, are safer. Unit heaters are still better and are recommended if an adequate steam supply is available. Locate salamanders or oil heaters away from woodwork or tarpaulins. Keep the floor clean and free from combustible material. Flame-proofed, water-repellent tarpaulins are available, and their use is advised. The tarpaulin salamander wooden-form combination has caused many serious fires, particularly in reinforced concrete construction work during freezing weather. Scaffolding constructed of metal, or of fire-retardant-treated wood, largely avoids a serious fire hazard. Locate construction sheds a safe distance from a building under erection, never inside.

Temporary Storage   Keep combustible storage out of buildings under construction, as far as possible and at least 10 ft (3.05 m) from structures when stored outside. Under no conditions erect canvas tents or wooden shelters inside. A safe location and good fire protection are important for valuable equipment or machinery, delivered before the building is ready for its installation.

AUTOMATIC SPRINKLERS
(29 CFR 1910.195.)

Advantages of Automatic Sprinklers   Automatic sprinklers are the most dependable and effective means of fire protection. The advantages are (1) automatic sprinklers go into operation soon after a fire starts, and before it has gained dangerous proportions; (2) the automatic sprinklers that open are over, and in the vicinity of, the fire; (3) fires at all locations, including out-of-the-way places, are controlled as effectively as

fires easily seen and reached; (4) sprinklers operate where fire fighters could not enter or would be driven out by heat or smoke; (5) sprinklers are always ready; (6) only sprinklers that are needed to control a fire operate, so that water is used to better advantage than from hose streams.

Where Sprinklers Are Needed   Automatic sprinklers are needed for complete protection where there is an appreciable amount of combustible material in either building construction or in building contents, e.g., (1) throughout buildings having floors or roofs of sufficient combustibility to contribute to a spreading fire, whether or not the contents are combustible; (2) in wholly noncombustible buildings where contents are combustible, including the storage or use of flammable liquids; (3) in concealed spaces, such as attics and under low roofs, if they contain any combustibles, including heat insulation and exposed electric wiring; (4) in vacant spaces beneath a combustible first floor of a building, unless the floor is completely tight, the space sealed against entry, and no possible sources of ignition are present; (5) in dryers, large ducts, closets, and small offices, unless there is no fuel for a fire in either the construction or the contents; (6) under wide storage shelves or work tables over 4 ft in width having any combustible stock, and under canopies over platforms where combustible materials may be present.

Sprinklers over Electric Equipment   If the construction is fire-resistive or adequately fireproofed, sprinklers can generally be omitted in electric generator rooms and over switchboards. If voltages exceed 600, it is generally better to replace combustible roofs or ceilings with non-combustible construction or to protect all exposed woodwork with metal lath and cement plaster, rather than to install sprinklers. Metal shields or hoods may be used to protect generators, switchboards, or other important electrical equipment from possible water damage.

Automatic-Sprinkler Equipment   The responsibility for the design of automatic sprinkler systems should be given only to experienced and responsible parties. Approval of preliminary layouts and working plans are usually required by municipal or insurance inspection bureaus before installation is started. Sprinkler installation is a well-established trade by itself. Standards for the Installation of Sprinkler Systems are given in complete detail in a publication under that name by the National Fire Protection Assoc.

Dry Pipe Systems   The ordinary wet pipe sprinkler system cannot be used where subject to temperatures below freezing. A dry pipe system must be substituted. In the dry pipe system the sprinkler piping contains air under pressure instead of water. When a sprinkler is opened by fire, the air pressure falls and water is admitted automatically by the operation of a dry pipe valve.

Special Sprinkler Systems   The deluge system is a special type of sprinkler equipment frequently used in hazardous occupancies, such as airplane hangers and storage areas for flammable liquids or materials, where a flash fire could spread before the regular automatic sprinklers became operative, and where the prompt discharge of a large amount of water over a considerable area is needed. The deluge system has sprinkler heads with the fusible element removed. Water is controlled by a quick-opening valve (deluge valve) operated by heat sensitive elements distributed over the area protected. Piping and heads are arranged as in a standard sprinkler system, with larger pipe sizes.

Precision systems are actuated by heat-sensitive devices, smoke detectors, or ionization detectors, which trip a controlling valve and admit water to the otherwise dry sprinkler piping before the regular closed automatic sprinkler heads operate. The system has the advantage of giving an alarm before sprinklers operate. Accidental opening of automatic sprinklers or mechanical breakage of piping does not result in the discharge of water.

Deluge and other systems actuated by heat-sensitive or other special devices introduce additional mechanical equipment and lose the simplicity of the standard automatic sprinkler system. Some additional skilled maintenance is required. Various arrangements for mechanically supervising the condition of such equipment can be provided.

Nonfreeze sprinkler systems are small and are sometimes used in relatively unimportant locations where it is impracticable to provide heat.

They consist of special piping connections to wet-pipe systems, with a nonfreezing solution in the exposed piping. Antifreeze solutions (glycols) are acceptable for this purpose. They should be limited to systems with less than 20 heads. It is better to provide heat, to connect to existing dry pipe systems, or to provide small dry pipe valves. Nonfreeze systems are preferable to shutting off the sprinklers and draining the piping during the winter. No portion of an automatic sprinkler system connected to a public water supply should be filled with a nonfreeze liquid without determining that the arrangement will comply with water department health regulations.

Outside sprinklers are used for protection against fire from outside sources. Windows in outside walls present the most frequent problem. Outside sprinklers may be operated either manually or automatically. The effectiveness of these systems depends on the water supply valve opening in time. They can sometimes be adapted to automatic control by a thermostatically operated deluge valve.

Types of Automatic Sprinkler Heads   Approved automatic-sprinkler heads are well standardized as to the rate of water discharge, pattern of discharge, and temperature operating characteristics. The essential elements of an automatic sprinkler are a nozzle, a closing device that is released at a definite temperature, and a deflector that produces the desired distribution pattern. Temperature ratings appropriate for the location must be selected. The ratings for normal atmospheric temperatures, not substantially above 100°F, are 135 to 165°F. Ratings for other ceiling temperature ranges are as follows:

Max ceiling temp		Sprinkler rating	
°F	K	°F	K
101–150	311–338	175–212	352–373
151–225	339–380	250–286	394–414
226–300	381–442	325–360	436–455

Heads of higher rating are obtainable for ambient temperatures up to 500°F. Corrosion-resistant heads, usually wax or plastic coated, are obtainable for use in corrosive atmospheres.

Sprinkler Flow Alarms   The flow of water from a sprinkler system may be made to sound an alarm which gives warning of both a fire and accidental leakage from the system. This alarm may be either a hydraulic gong or an electric bell, or both. If the property has no constant attendance, or if added security is desired, central-station supervisory service is available in many industrial centers. This service transmits water flow alarms to public fire departments and can perform numerous other supervisory functions. Preaction alarms give advance notice of hazard.

Location and Spacing of Automatic Sprinklers   As the purpose of an automatic sprinkler system is to protect both building and contents, the water distribution pattern and the quantity of water applied must be held within close limits. Sprinklers customarily are installed so that the discharge from one head is allowed for each 60 to 130 ft² of floor area. The distance between adjacent heads should not exceed 15 ft. Spacing averages 80 to 120 ft² per head with 8 to 12 ft between heads.

Recent developments in sprinkler system design for the more severe hazards produced by high piled combustible goods or flammable liquids are leading to standards including a **discharge density** in terms of gallons per minute per square foot of floor area. In this consideration building contents, rather than building construction, are the determining factor. Densities called for commonly have been in a range of 0.15 to 0.3 gal/min per ft² of floor area. High-piled storage of combustible materials, such as rolled paper, imposes a severe hazard which may call for a water density of 0.6 gal/min per ft², or more. NFPA standards (1973) call for intermediate levels of sprinkler heads in high rack storage areas. Open and deluge sprinklers require hydraulically designed systems based on friction loss instead of standard pipe schedules (see NFPA No. 13, No. 13a, and No. 231c).

The number of heads, to be supplied through pipes of standard sizes, with ordinary hazards and reasonably uniform water discharge, is as follows:

Size of pipe, in	Max. no. of sprinklers	Size of pipe, in	Max. no. of sprinklers
1	2	3½	65
1¼	3	4	100
1½	5	5	160
2	10	6	275
2½	20	8	400
3	40		

Not more than eight sprinklers should be on one branch line on either side of a cross main. For deluge systems, with which all heads are open and must be supplied simultaneously, and for especially hazardous occupancies where a large proportion of the automatic heads are expected to operate, larger pipe sizes are required as follows:

Size of pipe, in	Max. no. of sprinklers	Size of pipe, in	Max. no. of sprinklers
1	1	3	27
1¼	2	3½	40
1½	5	4	55
2	8	5	90
2½	15	6	150

WATER SUPPLIES FOR FIRE PROTECTION

(See Sec. 3.3 for hydraulics, Sec. 8.7 for piping,
Sec. 9.6 for engines, Secs. 14.1 and 14.2 for pumps.)

Insurance companies or their inspection and engineering bureaus should be consulted concerning water supply requirements for fire service. In addition to providing water for automatic sprinklers, the supply should be adequate for hydrants and hose streams. Common water supplies are public water systems, gravity tanks or private reservoirs, and fire pumps taking suction from aboveground suction tanks, rivers, or ponds. Two or more independent supplies are usually needed for larger properties. Strong, dependable public water systems alone are adequate only for small plants, of good construction, having safe occupancies, and not dangerously exposed.

The "primary" water supply automatically maintains pressure on the property fire system at all times. The "secondary" supply supplements it as needed. The **primary supply** should not be less than 500 gal/min available at a pressure that will be effective for automatic sprinklers at the highest plant elevation. A **secondary water supply** serves several purposes. The most important usually is to provide more water, frequently at higher pressure than is available from the constant primary supply. A secondary supply is also of value in maintaining protection at any time the primary supply may be interrupted due to public supply deficiencies or to the necessity of taking tanks or reservoirs out of service for repairs and maintenance.

High water demands for special hazard protection, especially where large areas must have deluge system protection, can seldom be met by connections to public water systems. Fire pumps and suction reservoirs can give the added volume of water at desirably high pressure for the expected duration of the total demand, which is made up of the anticipated sprinkler demand and an allowance for the number of hose streams likely to be used. Large airplane hangars, large building areas containing flammable liquids, or very extensive properties that could have a general fire, approaching conflagration proportions, can have an estimated total demand of 5,000 gal/min or even more.

Gravity Water Tanks   The smallest gravity tank on a tower considered advisable for fire protection is 25,000 gal, which may be suitable for the protection of small properties. Water from tanks of such limited size should be reserved for automatic sprinklers only. Tanks of 50,000- to 100,000-gal capacity may provide either the primary or secondary supplies for large properties of moderate hazard. Tanks combining fire service and domestic or industrial supplies in a single structure are allowable, provided that the amount needed for the fire supply cannot be withdrawn into the industrial use system.

Gravity tanks should be at such elevation that the bottom of the tank will be at least 35 ft above the highest sprinkler to be supplied. The bottom of large tanks, intended to provide water for hose streams as well as sprinklers, should be at least 75 ft, and preferably 100 ft, above ground level.

Pressure Tanks Hydropneumatic steel tanks can be used for primary water supplies for demands of short duration, such as needed to bring automatic fire pumps into full operation. Under some conditions they can be used to facilitate the automatic control of the automatic fire pumps by providing a rapid drop in the control pressure. High cost and limited capacity deter the general use of pressure tanks.

Private Fire Pumps Well-located fire pumps with ample suction supplies and capable of maintaining high pressure over a long period provide a very satisfactory secondary supply. Centrifugal fire pumps of approved design are in common use. Practically all new installations use centrifugal pumps driven by electric motors, gasoline or diesel engines, or steam turbines. The electric motor driven centrifugal pump is most desirable because of simplicity of control and operation. Gasoline engine drives are used to a limited extent where other reliable sources of power are not available, or as auxiliary units in connection with other pumps.

Adequate suction and reliable power supply are essential for fire pumps. The suction supply should be sufficient to operate the pump at rated capacity for at least 1½ h at the smaller plants, and an inexhaustible suction supply is desirable for good protection of the largest industrial properties.

An automatically controlled, electrically driven centrifugal fire pump is sometimes used as a booster pump when the public water is of good volume but too low in pressure for direct use in sprinkler systems.

Fire pumps are used in capacities of 500, 750, 1,000, 1,500, and occasionally 2,000 or 2,500 gal/min. The 750 and 1,000 gal/min sizes are most common.

Fire Department Connections For properties where the public water supply is at comparatively low pressure but adequate in quantity, fire department connections to the plant fire-service piping can serve as a valuable auxiliary for pumpers to deliver water into the fire system at high pressures. In city properties, particularly, fire department connections are usually a part of the sprinkler system.

UNDERGROUND WATER PIPING FOR FIRE SERVICE

Complete specifications or standards covering all of the main features of underground piping for fire service are given in NFPA National Fire Codes, vol. 6, "Sprinklers, Fire Pumps, and Water Tanks," and the Factory Mutual "Handbook of Industrial Loss Prevention." The "Handbook of Cast-Iron Pipe," published by the Cast-Iron Research Association, contains much useful information.

Underground Mains Buried underground piping should be located so that hydrants and control valves are at a safe distance from possible falling building walls. A complete loop system around buildings or groups of buildings, with multiple supplies preferably connected at opposite sides and with prudently located division valves, affords the best hydraulic characteristics for heavy flows and for freedom from impairments. Pipe trenches should allow careful laying and uniform support for the pipe. Foreign materials should be kept out of the pipe. Pipe should be anchored at bends, dead ends, and branch connections. The system should be hydraulically tested for 2 h at a minimum pressure of 200 lb/in² or at least 50 lb/in² in excess of the maximum static pressure if it is to be more than 150 lb/in². Leakage at joints must be less than specified amounts. Underground piping should be thoroughly flushed before it is connected to indoor piping. Pipe should be buried well below the deepest frost penetration as shown by weather charts. In the coldest areas a cover of at least 5 ft is necessary. In the southern states 2½ ft may be adequate, except that under roadways a greater cover may be needed for protection against traffic loads. Clearance to prevent breakage by settlement is needed at foundation walls.

Soil Corrosion Rapid external corrosion of cast iron pipe may be expected if the mains pass under coal piles, through cinder fill, or where pickling liquors, acids, alkalies, or salts penetrate the soil. Cast iron pipe covered with heavy coating of asphalt, or other pipe suitably covered should be used to offset soil corrosion in such locations. Backfill should be of clear sand or gravel. (See Secs. 6.8 and 6.9.)

Type of Pipe Pipe for underground use is usually cast iron or steel. Ordinarily the working water pressure rating does not need to exceed 150 lb/in². Representative specifications for pipe suitable for fire service are: "Specifications for Cast-Iron Pit-Cast Pipe" (ANSI A21.2 or AWWA C102), "Specifications for Cast-Iron Pipe Centrifugally Cast in Sand-Lined Molds" (ANSI 21.8 or AWWA C108); asbestos-cement pipe should conform to "Tentative Standard Specifications for Asbestos-Cement Water Pipe" (AWWA C400) or "Commodity Federal Specification SS-P-55/A."

OUTSIDE HYDRANTS AND HOSE

Location of Hydrants Where space permits, outdoor hydrants and hose provide a supplemental to automatic sprinklers and afford means for fighting fires in combustible yard storage, in railroad cars and vehicles, and small combustible unsprinkled sheds. Recommended space between hydrants varies from 150 to 300 ft, depending upon the type of buildings and the character of outdoor combustibles. Hose for outdoor use, 2½ in in size, is made of woven cotton or modern synthetic fiber and rubber-lined.

STANDPIPES AND INSIDE HOSE
(29 CFR 1910.158.)

Standpipe systems furnish the best means of obtaining effective fire streams in the upper stories of buildings. They are designed for small hose streams used by the occupants and for large hose streams to be used by public or plant fire departments. Outlets may be designed to supply both. In buildings of unusual height, standpipes are sometimes supplied by a series of fire pumps and tanks at different elevations.

Small hose is of particular value in areas with hazardous occupancies, such as opener, picker, spinning, and woodworking rooms. Hose should be 1½-in cotton, rubber-lined, or unlined linen, with ⅜- or ½-in nozzles. Spray-type nozzles are usually best for the hazards mentioned. The water supply should preferably be taken from a connection independent of the sprinkler piping so that hose streams will be available when the sprinklers are shut off after a fire or during sprinkler repairs or changes. Small hose, for fire service only, in locations of ordinary hazards, may be connected to wet pipe sprinkler systems, but in no case should hose connections be made to sprinkler lines smaller than 2½-in diam. (See Sec. 8.7.)

SPECIAL FORMS OF FIRE PROTECTION

Special types of protection are adapted to the control of unusual hazards, such as flammable liquids. Equipment should be secured from makers specializing in the form of protection required, but such use does not supplant general building protection by automatic sprinklers.

Water Spray A dense strong spray of water from suitably designed nozzles is effective for controlling fires in flammable liquids of moderate hazard, for unusually flammable solid materials, and for surface fires of ordinary combustible materials. Such a system may be most appropriate for the protection of transformers and other oil-filled electric equipment and for systems handling fuel or lubricating oil under pressure. The entire zone to be protected must be within reach of the strong spray. Water pressures of 50 lb/in² prevail with properly designed equipment. (See Sec. 3.3.)

Foam Foams for fire control provide a blanketing action to exclude air and in some cases to give useful insulating effect. Foams are commonly designated as "chemical foam" and "air foam," depending upon their service or method of production. For flammable-liquid fires, foam is available for manual or automatic applications with a selection of characteristics appropriate for a wide variety of conditions. The amount of foam varies from ½ ft³ to several cubic feet per square foot of surface protected. Rate of application and means of distribution largely determine its effectiveness. A sprinkler system, designated as "Foam-water," provides an initial discharge of very fluid foam backed up by a sprinkler discharge of water. Special **alcohol type** foam is needed for application to alcohols, alcohol-type liquids, and organic solvents, all of which seriously break down the commonly used foams. **High expansion** foams are available. They are easily produced in a **high expansion foam generator** by blowing air through a screen wet with a continuous spray of water having a bubble producing additive. The foam is very light and fluid and can be applied to fill completely and quickly a room or other enclosed area of considerable size. One gallon of high-expansion foam-producing liquid can form as much as 1,000 gal of foam. Ordinary foam producing liquids yield only about 10 gal of foam per gal of liquid.

Carbon Dioxide   (29 CFR 1910.161) For (1) flammable liquids; (2) electric equipment, such as large enclosed electric generators; and (3) hazards where a space-filling effect by an inert atmosphere is needed, or where a nonwetting extinguishing agent is desired. Carbon dioxide for fire extinguishing is available in small cylinders for manual use, in banks of large cylinders, or in refrigerated storage tanks for piped extinguishing systems.

Dry chemical (29 CFR 1910.160) extinguishers and extinguishing systems are mainly used for flammable liquids and electrical fires. They are effective also on surface fires in combustible fibers. Multipurpose dry chemical extinguishers have special ingredients which make them suitable for fires in ordinary combustibles.

PORTABLE HAND EXTINGUISHERS
(29 CFR 1910.157)

Hand-operated extinguishers and small hose are effective when employees are on hand to attack fires immediately after discovery. They are frequently used to put out the final vestiges of fires brought under control by automatic sprinklers. Grouped by hazards, the hand extinguishers are selected from four classifications, as shown by Table 18.3.1.

Table 18.3.1   Portable Extinguishers by Hazard Classification

Extinguisher type	Class A general purpose	Class B flammable liquids	Class C electrical fires	Class D combustible metal
Foam	x			
Loaded stream	x	x		
Dry chemical		x	x	
CO₂ (plastic horn)		x	x	
Vaporizing liquid		x	x	
Dry powder				x

Source: "NFPA Handbook," vol. 17.

18.4   PATENTS, TRADEMARKS, AND COPYRIGHTS
by Raymond E. Farrell and Bernadette M. Bennett

PATENTS

REFERENCES: Chisum, "Patents," Matthew Bender, Patent Act of 1952, 35 U.S.C. Rules of Practice of U.S. Patent and Trademark Office, 37 CFR §1, et seq. U.S. Patent and Trademark Office website (www.uspto.gov).

United States
(All statutory references are to the Patent Act of 1952, 35 U.S.C., et seq.)

Types of Patents Granted in the United States   United States patents are classified as utility, design, and plant patents with the first category comprising the vast majority of patents granted in the United States. A design patent can be obtained for any "new, original and ornamental design for an article of manufacture" (35 U.S.C. §171). The emphasis in design patents is on the ornamental, rather than functional, aspects of the article of manufacture. A plant patent may be obtained for "whoever invents or discovers and asexually reproduces any distinct and new variety of plant" (35 U.S.C. §161), and gives that person the right to exclude others from asexually reproducing the plant or selling or using the plant so produced (35 U.S.C. §163).

The Patent Grant   In return for a full disclosure of an invention to the public in the patent document, the United States will grant to a patentee, his or her heirs or assigns, the **right to exclude** others from making, using, offering to sell, or selling the patented invention, including products made by a patented process, within the United States; or importing into the United States any patented invention, including products made by a patented process, during the term of the patent (35 U.S.C. §154). Patent rights may be assigned or licensed to other persons or entities. Multiple licenses can be granted under a single patent, giving each licensee nonexclusive rights under the patent. Geographic and use limitations are also possible in a license as a means of maximizing the patent owner's benefit from her or his patent. The right to exclude conferred by a patent may be exercised by the patent owner in the form of an infringement action in the federal court system which has exclusive jurisdiction over lawsuits relating to patents. Remedies that may be awarded to the patent owner in the event of a successful infringement action include injunction against further infringement, damages which may be increased up to 3 times the amount found in the case of willful infringement, and in exceptional cases attorneys' fees (35 U.S.C. §§281–289). A provisional right of a reasonable royalty is also available for infringement of the claims of a published patent application for the period beginning with the publication date of the application and ending with the patent issue date. This remedy is available if the published claims are substantially identical to the claims which ultimately issued in the patent (35 U.S.C. §154).

The exclusionary right granted by a patent *does not* grant to the patentee any express right to make, use, offer to sell, or sell the patented invention. The patentee's ability to perform any of these acts is the same as that of any one else; i.e., the performance of such acts is subject to the claims of any other enforceable U.S. patents which may cover any of the acts. Thus, it is possible for a patentee to be infringing another patent which is broader in scope. This could occur, e.g., in the situation where patent A was previously issued to another inventor and is a broader patent than patent B which was issued because it is a patentable improvement over the prior art including the disclosure of patent A. Therefore, prior to making, using, offering to sell, or selling the invention, it is highly recommended that a patent owner conduct a separate evaluation of his or her ability to commercially practice, the invention without infringing the rights of other patents.

Patent Term   The term for design patents is 14 years from the issue date of the patent (35 U.S.C. §173). The term for utility and plant patents depends upon its filing date and is subject to the payment of maintenance fees and any express disclaimers which may have been made for a portion of the patent term. The term for patents based on applications filed before June 8, 1995, is the greater of the 20-year term or 17 years from the patent issue date. Patents based on applications filed on or after June 8, 1995, have a term which begins on the date the patent issues and ends 20 years from the filing date of the application or from the filing date of the earliest application that the patent claims the benefit of. The 20-year term does not begin from the filing date of a prior filed U.S. provisional application or a national application of a foreign country either of which the patent may claim priority to.

Separate patent term adjustments or extensions may be available under certain circumstances for utility and plant patents. For applications filed between June 8, 1995, and May 28, 2000, patent term extensions are available for a maximum of 5 years for delays in the granting of a patent due to interference proceedings (35 U.S.C. §135) or secrecy orders (35 U.S.C. §181) in the case of *original* utility and plant patents, i.e., those which do not claim the benefit of the filing date of an earlier application, or successful appeals of any utility or plant patent to the U.S. Patent and Trademark Office (PTO) or the courts. The term for *original* utility and plant patents based on applications filed on or after May 29, 2000, may be adjusted for certain patent term guarantees afforded by the patent statute at 35 U.S.C. §154. In particular, the term of a utility or plant patent may be adjusted for certain delays on the part of the PTO during prosecution of the application; the failure of the PTO to issue a patent within 3 years from the actual filing date in the United States; delays to an original patent application due to interference proceedings (35 U.S.C. §135), secrecy orders (35 U.S.C. §181), and successful appeals to the PTO or courts. For a limited number of patents with very special types of claims delineated in the patent statute, the term of the patent may be extended or restored (35 U.S.C. §§155–156). After a patent expires, which may occur at the end of its term including any adjustments and extensions or upon the failure to pay required maintenance fees, the subject matter of the patent enters the public domain.

Determining What Is Patentable The basic statutory requirements for patentability are utility of the invention (35 U.S.C. §101), novelty (35 U.S.C. §102), and nonobviousness (35 U.S.C. §103). To meet the utility requirement, an inventor need only disclose in his or her patent an invention which is new, useful, and nonobvious in the field of applied technology as opposed to theoretical or abstract ideas.

The novelty and nonobviousness requirements of the patent laws are assessed in light of the prior art which generally comprises all written materials antedating the inventor's application for a patent and/or his or her actual date of invention (see section on interferences below for more detail on what constitutes date of invention). Also included in prior art are prior public uses and sales by others of the same or similar processes or products as that of the inventor. A prior public disclosure or offer to sell by the inventor made more than 1 year before the inventor's patent application also constitutes prior art which can prevent the grant of a patent [35 U.S.C. §102(b)]. Thus, it is crucial that patent applications be promptly filed (within a year) after any effort at commercial exploitation of an invention by or on behalf of the inventor. If, however, the inventor wishes to secure foreign patent protection, it is imperative that a U.S. or an international application under the Patent Cooperation Treaty be filed before the inventor makes any such public disclosure or offers to sell the invention. This is so because foreign countries require absolute novelty and do not provide any grace period in which to file a patent application. In assessing an invention against the novelty requirement of 35 U.S.C. §102, the courts have held that if the invention is different—without regard to how different—from each item of prior art, it meets that requirement. In other words, if no single piece of prior art discloses the invention, it is considered novel under 35 U.S.C. §102.

Since few inventions—at least not those for which patent protection is sought—are identical to any single item of prior art, the patentability of most patents must be assessed under 35 U.S.C. §103, i.e., whether "the differences between the subject matter sought to be patented and the prior art are such that the subject matter as a whole would have been obvious at the time the invention was made to a person having ordinary skill in the art to which said subject matter pertains." Such assessment requires a factual determination at the time of the invention of: the prior art, differences between the prior art and the invention, the level of skill in the art to which the invention pertains, and a further assessment whether those differences would have been obvious in light of the level of skill of the practitioners in the art. To assist in this determination, the Supreme Court has indicated that courts should look to important background facts such as long-felt need for the invention, failure of others to achieve the invention, and ultimately, any commercial success of the invention [*Graham v. John Deere Company of Kansas City*, 383 U.S. 1, 86 S.Ct. 684 (1966)].

Documentation of an Invention Recordation of developments is essential in ongoing research. A notebook is commonly used to record, explain, and date research work. The notebook entries should be signed by the researcher and witnessed by one or more persons who have direct knowledge of, and understand, what was being done during the making of the invention.

During patent proceedings, proof of the date of invention may become critical. The date of invention may become important to antedate a prior art reference or to determine priority of invention as between two inventors in a PTO interference proceeding. In both instances, the priority of invention usually must be established by contemporaneous, witnessed documentation of the work underlying the invention as contained in the researcher's notebook. Hence, a notebook is invaluable to prove the invention date.

In light of globalization, inventive activity is not limited to the borders of the United States. The patent statute at 35 U.S.C. §104 provides that evidence of inventive activity in the territory of a World Trade Organization (WTO) member country, Canada, or Mexico be treated the same as inventive activity in the United States. Thus, companies or individuals engaging in research in foreign countries may rely on that activity to prove a date of invention.

Preparing the Patent Application A recommended first step in preparing an application is to evaluate the prior art and its relation to the invention for which patent protection is to be sought, utilizing the obviousness guidelines of the Supreme Court noted above. The most comprehensive source of prior art patents is located at the PTO in Alexandria, Va., where millions of patents and publications are searchable in the public search room by using computer search tools which facilitate searching the PTO's patent databases including published applications, issued patents, as well as certain foreign applications and patents. Other proprietary search services exist whereby patent searching can be conducted via the Internet without having to physically go to the PTO. Compilations of U.S. patents can also be found in specific libraries designated as Patent and Trademark Depository Libraries (PTDLs) located across the country. Copies of U.S. patents can be viewed and downloaded via the Internet at www.uspto.gov. Search results should be evaluated by a competent patent attorney or agent who can advise whether the invention is patentable over the prior art uncovered and who can also focus on the differences between the invention and prior art when preparing a patent application covering he invention.

If the invention, after careful evaluation, is considered to be patentable, the next step is identification of the proper inventors. In the United States, patent applications are filed in the name of the individual(s) who contributed to all or part of the invention rather than the legal owner of the invention or application, e.g., the inventor's employer. Where two or more persons contributed to all or part of the invention (a joint invention), the respective efforts of these persons must have been made in collaboration with each other and the invention must be the product of their aggregate efforts.

The patent application comprises four essential parts: the specification or detailed description of the invention; claims, particularly pointing out the features which the inventor considers to be his or her invention; a drawing illustrating the invention, where applicable; and an oath or declaration of the inventor(s) stating a belief that he or she believes himself or herself to be the original and first inventor of the invention claimed in the application (35 U.S.C. §§111–115).

In addition to the standard patent application, applicants may elect to file a provisional application. A provisional application only requires the filing of a specification complying with 35 U.S.C. §112, paragraph 1, and drawings where necessary for the understanding of the invention.

The application must be made in the name of the inventors, include the appropriate filing fee and a suitable cover sheet. A provisional application does not require any claims, nor does it require an oath or declaration. A provisional application is not examined by the PTO, cannot claim priority of an earlier application, and cannot issue as a patent. In addition, provisional applications will automatically become abandoned 12 months after filing and can become abandoned prior to the expiration of that 12-month-period for failure to respond to a PTO requirement or to pay a required fee. A provisional application may be revived, but its pendency cannot extend beyond 12 months from its filing date. The provisional application is a form of domestic priority system which provides a quick and inexpensive method to establish an early effective filing date which establishes a constructive reduction to practice for any invention described in the application. However, this priority is only effective provided a suitable nonprovisional application is filed prior to the expiration of the provisional application.

Proceedings in the U.S. Patent and Trademark Office After the application is filed with the appropriate fee, an examiner in the PTO reviews the application to ensure compliance with all statutory requirements as to the form of the application and to independently assess the patentability of the claims in the application in light of the prior art. If the examiner concludes the invention claimed in the application is not patentable over prior art, all such claims will be rejected. The applicant then has the opportunity to rebut the rejection by arguing the novelty of the invention over the prior art and/or amending the claimed invention. If the examiner maintains the rejection of the claims, an administrative appeal to the Patent Office Board of Appeals and Interferences is available to challenge such rejection, and a further appeal to the federal courts is available from an adverse decision of the Board of Appeals (35 U.S.C. §§134, 141, 145). If the claims of the application are found to be patentably distinct from the prior art, a patent will be granted upon payment of an issue fee.

To maintain a patent, the owner must also make periodic maintenance payments to the PTO before the fourth, eighth, and twelfth anniversaries of the patent grant. These maintenance payments increase as the patent ages. Since all the above-mentioned fees change annually, the fees should be determined in advance to avoid the consequences of fee underpayment.

Reissue and Reexamination of Patent If through inadvertence, accident, or mistake, the original patent was defective or claimed too much or too little, the error may be repaired by a reissue proceeding provided the subject matter needed to correct the mistake appeared in the original patent (35 U.S.C. §251). The term of a reissued patent expires the day the original patent would have expired.

Also, if the patentability of a patent is put in question, either the patent owner or a third party may ask the PTO to reexamine and reevaluate the patent vis-à-vis the prior art. The PTO then proceeds to examine the patent much the same as it would with an original application.

Foreign Countries

International Convention for the Protection of Industrial Property The important provision of the convention is that any person who has duly applied for a patent in one of the contracting states shall have a right of priority for 12 months in making application in the other states.

Patent Cooperation Treaty (PCT) and European Patent Convention (EPC) To simplify the process of obtaining foreign patents, it is now possible to file a single application, e.g., in the United States, and have that application mature into a patent in each country worldwide which is a member of the Patent Cooperation Treaty. Similarly, if the focus of intended patent protection is in Europe, a single application in the European Patent Office located in Berlin, Munich, or the Hague can be issued as a patent in most of the countries on the European continent under the European Patent Convention.

Use of either the PCT or EPC provisions will generally be cost-effective versus separate filings in individual countries, since, for example, translation costs are usually minimized or eliminated. However, as in all such complex matters, knowledgeable patent professional counsel should be consulted.

TRADEMARKS

REFERENCES: Callman, "Unfair Competition, Trademarks and Monopolies," Callaghan and Co., Chicago. McCarthy, "Trademarks and Unfair Competition," Lawyers Cooperative, Bancroft-Whitney. Gilson, "Trademark Protection and Practice," Matthew Bender.

United States

[Section references (§) refer to Lanham Act of 1946, and 15 U.S.C., both section numbers being given.]

What Is a Trademark? The term *trademark* includes any word, name, symbol, or device or any combination thereof adopted and used by manufacturers or merchants to identify their goods and distinguish them from those manufactured or sold by others (§45, §1127). A *service mark* is defined as a mark used in the sale or advertising of services to identify the services of one person and distinguish them from the services of others (§45, §1127).

How Rights in a Mark Are Obtained Unlike a patent, but somewhat like a copyright, ownership of a mark is obtained by use. If use occurs only within a state, a mark may be registered in that state. Forms for registering marks under state law are obtainable from the Secretary of State of each state. If use occurs only in interstate or foreign commerce, a mark may be registered in the U.S. Patent and Trademark Office (PTO) by filing an application in the form prescribed (37 CFR Pt. IV). After federal registration the symbol ® may be used with the mark. Prior to federal registration, the symbols **TM** for a trademark or **SM** for a service mark may be used to give notice of a claim of common law or state law rights.

A person who has a bona fide, good-faith intention to use a trademark in commerce, but is not actually using the mark, may apply for federal registration by filing an Intent To Use application in the PTO. Thus, a person who is not actually using a mark may still obtain federal registration provided that person commences using the mark in interstate commerce within a specified period after allowance of the Intent To Use application.

Tests for Registrability and Infringement A mark may be registrable on the Principal Register of the PTO only if, when applied to the goods or used in connection with the services, it is not likely to cause confusion or mistake or to deceive. The same test is used to determine whether a mark infringes on an earlier mark. Marks that are descriptive, misdescriptive, geographical, or primarily merely a surname may not be registrable on the Principal Register, if at all, until after 5 years of substantially exclusive and continuous use in interstate commerce (§2[f], §1052[f]). Marks not registrable on the Principal Register may be registrable on the Supplemental Register if they are capable of becoming distinctive as to the applicant's goods or services (§23, §1091).

Term of Registration Federal registrations remain in force for 10 years unless canceled. However, unless an affidavit is filed during the sixth year after registration showing that the mark is still in use, the Commissioner of Patents and Trademarks will cancel the registration of the mark (§8, §1058). In addition, federal registrations only remain in force for marks which are actually in use. Nonuse of a registered mark can result in finding that the mark has been abandoned by its owner. The law currently provides that nonuse of a mark for 2 years or longer is prima facie evidence that the mark has been abandoned. Within the sixth year after registration on the Principal Register, a registered mark may become more secure from legal challenge, i.e., "incontestable" upon the filing of the required affidavit (§15[3], §1065[b]). During the last 6 months of the registration term, an application for renewal may be filed (§9, §1059). Unlike with patents, an unlimited number of renewals are available for trademark registrations, with the proviso that the mark is in use. The terms of state registrations vary.

Preliminary Search Before adopting a mark, it is advisable to have a search made in the PTO to determine whether the mark under consideration would conflict with any pending application, registrations, or others' use of the mark in connection with any similar goods or services.

Cost of Registering a Mark Government fees and time for payment are fixed by law (§31, §1113) but attorneys' fees vary.

Effect of Federal Registration Federal registration of a mark affords nationwide protection, and once the certificate has been issued, no person can acquire any additional rights superior to those obtained by the federal registrant. Federal registration of a mark establishes federal jurisdiction in an infringement action, can be the basis for treble damages, and is admissible as evidence of trademark rights. Registration on the Principal Register constitutes constructive notice, constitutes prima facie or conclusive evidence of the exclusive right to use the mark in interstate commerce, may become incontestable, and may be recorded with the U.S. Customs Department to bar importation of goods bearing an infringing trademark.

Assignment of a Mark   A mark can be assigned only in conjunction with the goodwill of the business or that portion thereof with which the mark is associated. Assignments of registered marks or of applications for registration should be recorded in the United States Patent and Trademark Office within 3 months after the date of assignment (§10, §1060).

Foreign Countries

The Madrid Protocol   Similar to the PCT and EPC processes to obtain foreign coverage of a patent, the Madrid Protocol enables an applicant to file a single trademark application and potentially gain registration worldwide. At present, the Madrid Protocol has 77 countries that are members (also referred to as Contracting Parties). The multistage process begins with the filing of a single international trademark application, e.g., with the PTO, with the possibility that the application will mature into a registered trademark in the member countries selected by the applicant.

If trademark registration is desired in multiple countries, use of the Madrid Protocol will generally be cost-effective compared to separate filings in each individual country. Due to the complexity of international trademark filing, foreign protection of a mark should be undertaken after consultation with an attorney.

COPYRIGHTS

REFERENCES: Nimmer, "Nimmer on Copyright," Matthew Bender. Copyright Act of 1976, 17 U.S.C. §101 et seq. Copyright Office Regulations, 37 CFR, Chapter 2.

United States

[Section references (§) refer to Copyright Act of 1976, 17 U.S.C.]

Subject Matter of Copyright   Neither publication nor registration at the Copyright Office is required to secure a copyright. The copyright is automatically established when the work is created. Copyright protection subsists in works of authorship in literary works; musical works including any accompanying words; dramatic works including any accompanying music; pantomimes and choreographic works; pictorial, graphic, and sculptural works; motion pictures and other audiovisual works; sound recordings, and architectural works but not in any idea, procedure, process, system, method of operation, concept, principle, or discovery [§102]. However, copyright protection may be obtained for computer programs, including both the source code and object code. Original designs of useful articles are also subject matter for copyright (§ 1301). Additionally, the law provides explicit protection for vessel hulls. An owner of an original vessel hull design is granted certain exclusive rights provides that the application for registration of the design is made within 2 years of the design being made public. Only designs for actual vessel hulls that are publicly exhibited, publicly distributed, or offered for sale or sold to the public on or after October 28, 1998, are eligible for protection. Issuance of a design patent for the same vessel design terminates protection of the design under Chapter 13 of the copyright law.

Registration   Copyright registration is effected by submitting an application to the Register of Copyrights along with the fee (currently $30). Deposit requirements exist for applicants to submit copies of the works. These requirements vary for different types of works to be registered. Forms for the registering the various types of works as well as informative circulars are available from the Library of Congress via the Internet at www.copyright.gov. Although a copyright may be registered any time during its term, in order to preserve the maximum available statutory remedies for infringement, the copyright should be registered within 3 months of first publication of the work, or in the case of unpublished works it should be registered as soon as the work is in its fixed form (§412).

Form of Copyright Notice   Use of copyright notice is no longer required in the United States. However, the notice should still be used because it informs the public that the work is protected by copyright, identifies the copyright owner, and shows the year of first publication. Should there be an infringement of the work, if a proper notice of copyright is used on the published copy or copies and the infringer had access to the copy, then a defense based on innocent infringement in mitigation of actual or statutory damages will be given no weight except as provided in §504(c)(2) of the copyright law.

If a notice is placed on the copies, it shall consist of three elements: (1) the symbol © (c in a circle), or the word *Copyright*, or the abbreviation *Copr.*; (2) the year of first publication of the work; and (3) the name of the copyright owner (§401). For phonorecords, instead of the © symbol, the notice shall include the symbol Ⓟ (P in a circle) as well as the other two elements (§402).

Who May Obtain Copyright   Only the author or those deriving their rights through the author can rightfully claim copyright protection. Where a copyrighted work is prepared by an employee within the scope of his or her employment (a "work made for hire"), the employer and not the employee is considered to be the author. The name of the author (creator of the work) is normally used in the copyright notice even though an assignee may file the application for registration.

Duration of Copyright   Copyright in a work created on or after January 1, 1978, subsists from its creation and, with certain exceptions, endures for a term consisting of the life of the author and 70 years after the author's death (§302). Where the work was not a work for hire and is a joint work prepared by two or more authors, the term lasts for 70 years after the last surviving author's death. For works made for hire, and anonymous and pseudonymous works where the author's identity is not revealed in Copyright Office records, the duration of copyright will be the shorter of 95 years from publication or 120 years from creation. Copyright in a work created before January 1, 1978, but not theretofore in the public domain or copyrighted, subsists from January 1, 1978, and endures for the term provided by §302. In no case, however, shall the copyright term in such a work expire before December 31, 2002; and, if the work is published on or before December 31, 2002, the term of copyright shall not expire before December 31, 2047 (§303). Any copyright, the first term of which is subsisting on January 1, 1978, shall endure for 28 years from the date it was originally secured, with certain exceptions (§304). All terms of copyright provided by §302–304 run to the end of the calendar year in which they would otherwise expire (§305).

Infringement of Copyright   A violation of any of the exclusive rights (§1) by a person not licensed could lead to liability (§101) for infringement. Actions for infringement of copyright are brought in the U.S. district courts.

Foreign Countries

The Berne Convention   Under this convention a U.S. citizen may obtain a copyright in most countries of the world simply by publishing within the United States.

Refrigeration, Cryogenics, and Optics

BY

A. J. RYDZEWSKI *Consultant, Du Pont Engineering, E. I. du Pont de Nemours and Company.*
F. J. EDESKUTY *Los Alamos National Laboratory, retired.*
PATRICK C. MOCK *Principal Electron Optical Scientist, Science and Engineering Associates, Inc.*

19.1 MECHANICAL REFRIGERATION
by A. J. Rydzewski

REFERENCES: Publications of the ASRE and ASHRAE; "Air-conditioning Refrigerating Data Books"; *Refrigerating Engineering;* ARI Equipment Standards; ASHRAE Handbooks. Merkel and Bosnjakovic, "Diagrammen und Tabellen zur Berechnung der Absorptionskältemaschinen," Springer, Berlin. Jordan and Priester, "Refrigeration and Air Conditioning," Prentice-Hall.

REFRIGERATION MACHINES AND PROCESSES

(For general theory of refrigeration, see Sec. 4; for air-conditioning uses, see Sec. 12.4; for cryogenics, see Sec. 19.2.)

Refrigeration is a term used to describe thermal systems which maintain a process space or material at a temperature less than available from ambient conditions. Heat is transferred from the materials to be cooled into a lower-temperature substance referred to as the **refrigerant.** The refrigerant is cooled by physical processes which pump heat into the ambient environment. These apply several physical phenomena such as phase changes, sensible heating, thermoelectric effects, work extraction by mechanical devices, and endothermic chemical reactions. Stored energy in the form of ice, sublimating solid carbon dioxide, and cold liquids are included with processes described by the word **refrigeration.**

Many refrigeration systems use a continuous cycle. This requires that the refrigerant be recovered for recycling after it has absorbed heat from the product or space to be cooled. The most commonly used system has a recirculating fluid that boils at a low temperature to remove heat. The vapor is compressed to a pressure such that the condensing temperature is elevated. The vapor is cooled and condensed by a heat exchanger which rejects the heat into the environment. The condensed liquid is passed through a restricting valve or orifice. The liquid changes into a mixture of liquid and vapor that is colder than the refrigerated space. This expansion at constant enthalpy is known as the *Joule-Thompson effect.* An expansive turbine can remove heat from a high-pressure vapor by extracting energy in the form of mechanical work. When the exhaust conditions are proper, the material will be a mixture of gas and condensate. This fluid can be used as a refrigerant or stored in liquid form as product. The remaining vapor can be recycled as needed to continue the process.

The equipment used in refrigeration processes is referred to as follows: high-pressure vapor heat exchangers are called *condensers,* low-pressure heat exchangers are called *evaporators,* throttling devices are called *expansion valves* or *capillaries,* and pressure-raising devices are called *compressors.*

With continuous recirculation of the cycle fluid, the low-temperature heat pickup must be pumped, via energy input, to a higher level so as to discharge it to atmospheric temperature. In this sense the system acts as an **energy-transport** device or **heat pump** or, from another viewpoint, as a temperature transformer. This philosophy of energy transport may equally well be applied to more remote systems, such as thermoelectric. It is, of course, a manifestation of the principles of the Second Law, and the effectiveness of the system's operation may be evaluated numerically.

Efficiency ratings of refrigeration equipment are expressed several ways. Some of the more common terms include **coefficient of performance** (COP) and **energy efficiency ratio** (EER). **COP** is defined as refrigerant effect divided by net work input, where the refrigerant effect is the absolute value of the heat transferred from the lower temperature source, and the net work input is the absolute value of heat transferred to the higher temperature sink minus the refrigerant effect. COP can alternatively be defined as the ratio calculated by dividing the total heating capacity in Btu per hour provided by the refrigeration system, including circulating fan heat, but excluding supplementary resistance, by the total electric input in watts × 3.412. This definition applies primarily to heat pumps. **EER** is a ratio calculated by dividing the cooling capacity in Btu per hour by the power input in watts at any set of rating conditions, expressed in Btu/W · h.

Because load on equipment used for comfort applications varies, additional criteria to rate efficiency at part load have been introduced. ARI Standard 550/590-1998 has **integrated part load value** (IPLV) and **nonstandard part load value** (NPLV). The IPLV uses a standard rating point with specific values for evaporator and condenser temperatures, water flows, and fouling factors at specific load percentages. The NPLV allows selection of these values to be more representative of the actual operating conditions. Other terms used primarily for residential and commercial air-conditioning units and heat pumps include **seasonal energy efficiency ratio** (SEER) and **heating season performance factor** (HSPF). These also adjust ratings for the variations encountered over a normal heating or cooling season.

The various association and society standards have specific conditions at which equipment is rated. In efficiency comparisons, familiarity with the specific rating method is important. The factors used in the calculation of manufacturer's published data may not be representative of the specific application. Part load conditions must be considered in any evaluation when the unit will not normally be fully loaded. If the load is constant and is known, comparisons should be made at that point. Other important factors include whether the stated efficiency is just the compressor efficiency, or if it includes motor, seal, bearing, and drive losses. It is also recommended that rated operating conditions be adjusted to actual conditions when known. For example, the actual condensing temperatures may be significantly higher or lower than rated conditions depending on the geographic region.

Units of Refrigeration In the United States, refrigeration capacity is defined in terms of the standard ton. A **standard ton of refrigeration** corresponds to a heat absorption at a rate of 288,000 Btu/day or 200 Btu/min (3.5168 kW). The heat absorption per day is approximately the heat of fusion of 1 ton (907.19 kg) of ice at 32°F (0°C). The **standard rating** of a refrigerating machine, using a condensable vapor, is the number of standard tons of refrigeration it can produce under the following conditions: (1) liquid only enters the expansion valve and vapor only enters the compressor or the absorber of an absorption system; (2) the liquid entering the expansion valve is subcooled 9°F (5°C) and the vapor entering the compressor or absorber is superheated 9°F (5°C), these temperatures to be measured within 10 ft (3.05 m) of the compressor cylinder or absorber; (3) the pressure at the compressor or absorber inlet corresponds to a saturation temperature of 5°F (−15°C); and (4) the pressure at the compressor or absorber outlet corresponds to a saturation temperature of 86°F (30°C).

The **British unit of refrigeration** corresponds to a heat-absorption rate of 237.6 Btu/min (4.175 kW) with inlet and outlet pressures corresponding to saturation temperatures of 23°F (−5°C) and 59°F (15°C), respectively. In locations outside the United States it is common practice to use kilowatts to specify refrigeration capacity. One kilowatt of refrigeration capacity is roughly equivalent to 0.2844 tons. Occasionally a unit of refrigeration capacity called the **frigorie** is used. A frigorie is approximately equivalent to 50 Btu/min (0.8786 kW), or one-quarter of a standard ton of refrigeration.

PROPERTIES OF REFRIGERANTS

(For detailed properties, see Sec. 4.)

Refrigerants are the transport fluids which convey the heat energy from the low-temperature level to the high-temperature level, where it can, in terms of heat transfer, give up its heat. In the broad sense, gases involved in liquefaction processes or in gas-compression cycles go through low-temperature phases and hence may be termed "refrigerants," in a way similar to the more conventional vapor-compression fluids.

Refrigerants are **designated by number.** The identifying number may be preceded by the letter R, the word "Refrigerant," or the manufacturer's trademark or trade name. The trade marks or trade names shall not be used to identify refrigerants on equipment or in specifications. ANSI/ ASHRAE Standard 34-2001, "Designation and Safety Classification of Refrigerants," and addenda cover the designation methodology and safety classifications for toxicity and flammability.

The **safety classification** is designated by a two-character alphanumeric system. The first character indicates toxicity, and is either an A or B, with lower toxicity indicated by an A. The second character indicates flammability, and is a 1, 2, or 3, with the lowest flammability indicated by a 1.

There are two terms widely used to describe the relative effect refrigerants and other chemicals have on the environment. The **ozone depletion potential** (ODP) is the ozone-destroying effect of a substance compared to a similar mass of Refrigerant 11. The **global warming potential** (GWP) is the relative measure of the warming caused by a substance versus a similar mass of carbon dioxide. Substances that increase global warming are referred to as "greenhouse gases."

Much of the damage to the atmospheric ozone layer is believed to be the result of decomposition of chlorofluorocarbon (CFC) and hydrochlorofluorocarbon (HCFC) chemicals. An international agreement known as the Montreal Protocol took effect in 1989. This agreement and subsequent amendments to it resulted in scheduled phaseouts of CFC and HCFC refrigerants. The schedule is dependent on the specific refrigerant and the country, with CFCs and developed nations having the most aggressive phaseouts.

International legislation has primarily addressed ozone depletion; however, global warming is an important and an emerging issue. There have been international meetings initiated by the United Nations Environmental Programme (UNEP) resulting in the Kyoto Protocol. Framework to reduce global warming has been developed, and legislation is being proposed. As with ODP, both CFC and HCFC refrigerants are affected. In addition, the replacement hydrofluorocarbon (HFC) refrigerants are under increased scrutiny, especially in Europe. Global warming is a significantly more complex issue since there are many sources of greenhouse gas emissions. A common debate is whether HFC equipment has a lower overall global warming impact due to its higher efficiency and reduced power consumption than refrigerants with lower GWP but higher power consumption. A listing of the ODP, GWP, and 8-hour toxicity limits (AEL or TLV) for various refrigerants is given in Table 19.1.1.

As a result of the legislation requiring phaseout of CFC and HCFC refrigerants, new refrigerants, primarily HFCs, have been commercialized

Table 19.1.1   Environmental Properties of Refrigerants

Refrigerant	ODP	GWP	Toxicity, ppm (v/v)	AEL,* TLV†	Flash point, lower/upper, vol. %
R-11	1	4600	1000	TLV	None
R-12	1	10600	1000	TLV	None
R-22	0.055	1700	1000	TLV	None
R-114	1	9800	1000	TLV	None
R-123	0.02	95	50	AEL	None
R-124	0.22	599	1000	AEL	None
R-125	0	2800	1000	AEL	None
R-134a	0	1300	1000	AEL	None
R-401A	0.037	1120	1000	AEL	None
R-401B	0.04	1220	1000	AEL	None
R-402A	0.021	2330	1000	AEL	None
R-402B	0.033	2080	1000	AEL	None
R-404A	0	3260	1000	AEL	None
R-407C	0	1530	1000	AEL	None
R-410A	0	1730	1000	AEL	None
R-502	0.334	4520	1000	TLV	None
R-508B	0	10350	1000	AEL	None

*AEL is the recommended time-weighted average concentration of an airborne chemical to which nearly all workers may be exposed during an 8-h and/or 12-h day, 40-h week without adverse effect. Determined by E. I. du Pont de Nemours and Company for compounds that do not have a TLV.

†TLV, established for industrial chemicals by the American Conference of Governmental Industrial Hygienists, is the recommended time-weighted average concentration of an airborne chemical to which nearly all workers may be exposed during an 8-h day, 40-h week without adverse effect.

SOURCE: Data from E. I. du Pont de Nemours and Company.

to replace them. There has also been a renewed interest in nonhalocarbon refrigerants such as ammonia and carbon dioxide. A listing of **replacement refrigerants** for existing CFC and HCFC compounds is given in Table 19.1.2.

Table 19.1.2   Replacement Refrigerant Compounds

Current refrigerant	Replacement refrigerant	Type	Application*
R-11 (CFC)	R-123 (HCFC)†	HCFC	Chillers [+25°F (−3.9°C)]
R-12 (CFC)	R-134a†	HFC	Medium temp [−5°F (−20.6°C)]
	R-401B	HCFC	
R-13 (CFC)	R-508B	PFC	Low temp [−80°F (−62.2°C)]
R-22 (HCFC)†	R-407C	HFC	Retrofit and new equipment
	R-410A	HFC	New Equipment
R-500 (CFC)	R-401B	HCFC	
R-503 (CFC)	R-508B	PFC	

*Lower optimum use temperature.
†See Figs. 19.1.1 to 19.1.3 for the thermodynamic properties of R-22, R-123, and R-134a.
SOURCE: Data from E. I. du Pont de Nemours and Company, Bulletin 300151B.

The term "replacement refrigerant" does not mean the replacement can simply be substituted for the existing refrigerant in the existing equipment. In many cases "replacement" refers to the application, such as temperature range, capacity range, or compressor type. Where existing equipment can be converted, the new refrigerant may not be miscible with the petroleum-based lubricant used with the CFC refrigerant. Synthetic oils such as polyol ester and polyalkylene glycol must be used with the new refrigerants. Seals, bushings, gaskets, and motor insulation (hermetic machines) may also need replacement. Centrifugal machines may require compressor wheel or speed changes to meet the head requirements of the replacement refrigerant. Controls may need to be replaced or reset. The resulting efficiency and capacity will likely be different from that of the original machine.

The owner must decide whether field conversion is justified. The following should be considered: the cost of conversion versus the cost of a new machine, the anticipated capacity and efficiency of the converted equipment versus that of a new machine, and the condition and remaining life of the converted equipment.

The refrigerant charge in a system must be handled in an environmentally acceptable manner to accomplish repairs requiring refrigerant removal. Venting of refrigerant is prohibited. Sections 608 and 609 of the Clean Air Act mandate the recovery of CFC, HCFC, HFC, and PFC refrigerants. The rules on recovery, recycling, repair of leaks, reporting, record keeping, and technician certification are strict, with the imposition of severe penalties for violation. For further information concerning regulations related to refrigerants, visit the EPA's website at http://www.epa.gov.

The **normal boiling point** is an important attribute of refrigerants for vapor compression systems. The desirable characteristic is a positive gage pressure throughout the refrigerating system. This will eliminate contamination of the refrigerant and resulting heat-transfer inefficiencies due to air inleakage. Table 19.1.3 shows some common fluids in order of boiling point; included are halocarbon compounds that offer reduced ozone depletion potential, hydrocarbons, sulfur compounds, and inorganic compounds.

The desirable thermal properties of a refrigerant are (1) convenient evaporation and condensation pressures, (2) high critical and low freezing temperatures, (3) high latent heat of evaporation and high vapor specific heat, (4) low viscosity and high film heat conductivity. Desirable practical properties include (1) low cost, (2) chemical and physical inertness under operating conditions, (3) noncorrosiveness toward ordinary construction materials, and (4) low explosive hazard both alone and mixed with air. The refrigerant should be nonpoisonous and nonirritating and should not cause deterioration in the lubricant used. Leakage should be detectable by simple tests, easily performed.

The **specific volume of refrigerant** to be handled determines the size of positive displacement compressors, but with centrifugal compressors a large volume is not objectionable and may be an advantage for small units.

A comparison of various refrigerants based on *ideal performance* is given in Table 19.1.4. The results are for the **standard temperature range**

Fig. 19.1.1 Thermodynamic properties of R-22. (*Reprinted with permission of E. I. du Pont de Nemours and Company.*)

19-4

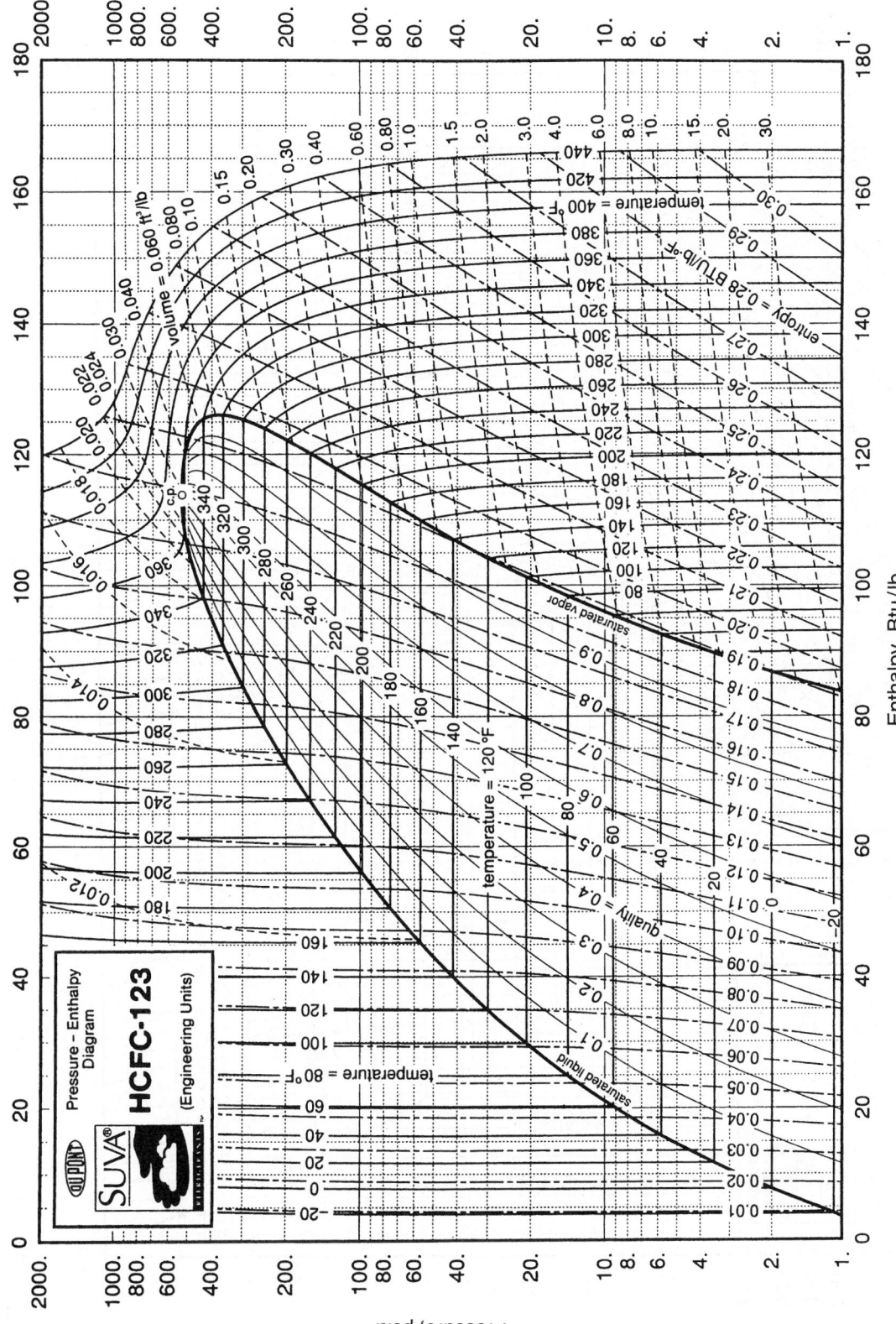

Fig. 19.1.2 Thermodynamic properties of HCFC-123. *(Reprinted with permission of E. I. du Pont de Nemours and Company.)*

Fig. 19.1.3 Thermodynamic properties of HCFC-134a. (*Reprinted with permission of E. I. du Pont de Nemours and Company.*)

Table 19.1.3 Selected Refrigerants and Gases in Low-Temperature Applications

Refrigerant	Refrigerant no.	Symbol	Molecular weight	Boiling point, °F	Critical point Temperature, °F	Critical point Pressure, lb/in² abs
Helium	704	He	4.0	−452	−450.0	33
Hydrogen	702	H_2	2.0	−423	−400.0	188
Nitrogen	728	N_2	28.0	−320	−232.0	492
Air	729	—	29.0	−312	−220.3	547
Oxygen	732	O_2	32.0	−297	−182.0	730
Methane	50	CH_4	16.0	−258.9	−115.8	673
Carbon tetrafluoride	14	CF_4	88.0	−198.4	−49.9	542
Ethylene	1,150	C_2H_4	28.0	−155.0	48.8	732
Ethane	170	C_2H_6	30.0	−127.5	90.1	708
Trifluoromethane	23	CHF_3	70.01	−115.66	78.6	701.4
Carbon dioxide	744	CO_2	44.0	−109.3	87.8	1,071
Propane	290	C_3H_8	44.1	−44.2	202	661
Monodichlorodifluoromethane	22	$CHClF_2$	86.5	−41.36	204.8	716
Ammonia	717	NH_3	17.0	−28.0	271.4	1,657
1,1,1,2-tetrafluoroethane	134a	CH_2FCF_3	102	−15	213.9	588.3
Methylchloride	40	CH_3Cl	50.5	−10.8	289.4	969
Isobutane	601	C_4H_{10}	58.1	10.3	272.7	537
2-chloro-1,1,1,2-tetrafluoroethane	124	$CHClFCF_3$	136.5	12.2	252	524.5
Sulfur dioxide	764	SO_2	64.1	14.0	314.8	1,142
Butane	600	C_4H_{10}	58.1	31.3	306	550
Dichloromonofluoromethane	21	$CHCl_2F$	102.9	48.1	353.3	750
2,2-dichloro-1,1,1-trifluoroethane	123	$CHCl_2CF_3$	152.9	82	362.8	533.2
Ethyl-ether	610	$C_4H_{10}O$	74.1	94.3	522.1	381
Methylene chloride	30	CH_2Cl_2	84.9	103.6	480	670
Water	718	H_2O	18.0	212	706.1	3,226

Table 19.1.4 Ideal Performance of Refrigerants for Various Temperature Ranges

Refrigerant and (number)	Operating temperature range, °F	Suction pressure, lb/in² abs	Head pressure, lb/in² abs	Ratio of head to suction pressure	With dry and saturated suction vapor, per ton — Weight of vapor, lb/min	Piston displacement, ft³/min	Theoretical hp	Temperature at end of compression, °F	Type of compressor
Air (729)	5–86	14.7	73.5	5.0	7.02	82.3	2.82	277.0	Recip. and exp. Cyl
Water (718)	32–86	0.0885	0.6152	6.95	0.1957	647.0	0.618	332	Centrif. or ejector
	32–100	0.0885	0.9492	10.73	0.1985	656.2	0.819	420	
	40–100	0.1217	0.9492	7.80	0.1978	483.4	0.687	366	
Carbon dioxide (CO_2) (744)	5–86	332.0	1,043.0	3.14	3.61	0.960	1.827	160.3	Recip.
Ammonia (NH_3)(717)	5–86	34.27	169.2	4.94	0.421	3.44	0.99	209.8	Recip.
	20–100	48.21	211.9	4.40	0.421	2.49	0.94	212.8	
	40–100	73.32	211.9	2.89	0.427	1.70	0.65	176.0	
SUVA 123 ($CHCl_2CF_3$) (HCFC-123)	5–86	2.303	15.9	6.9	3.26	45.6	0.95*	90.5	Recip.
	20–100	3.48	20.8	5.97	3.33	31.7	0.91	101.	Centrif.
	40–100	5.79	20.8	3.59	3.18	18.8	0.6*	100.	Centrif.
SUVA 134a (CH_2FCF_3) (HFC-134a)	5–86	23.77	111.8	4.7	3.14	6.1	1.04	98.	Centrif.
	20–100	33.13	139.0	4.19	3.28	4.62	1.0*	109.	
	40–100	49.77	139.0	2.79	3.14	3.0	0.68	106.	
SUVA HP 81 blend [R-402 (38/2/60)]	5–86	54.37	200.5	3.68	3.56	3.37	1.07	99.	Screw or rotary or recip.
	20–100	72.12	243.9	3.38	3.78	2.71	1.04	121.5	
	40–100	102.26	243.9	2.385	3.62	1.83	0.68*	114.	
R 22 ($CHClF_2$) (22)	5–86	43.02	174.5	4.06	2.926	3.65	1.03	131.7	Recip.
	20–100	57.98	212.6	3.67	3.023	2.83	0.99	152.7	
	40–100	83.72	212.6	2.54	2.936	1.93	0.68	131.0	
Methylene chloride (CH_2Cl_2)(30)	5–86	1.28	10.07	8.56	1.485	74.0	0.96*	205.1*	Centrif.
	20–100	1.92	13.25	6.90	1.520	47.72	0.91*	157.1*	
	40–100	3.38	13.25	3.92	1.493	27.76	0.63*	167.7*	
Methyl chloride (CH_3Cl) (40)	5–86	21.15	94.70	4.48	1.331	5.95	0.96	178.1	Recip.
	20–100	29.16	116.7	4.00	1.363	4.51	0.90	184.4	
	40–100	43.25	116.7	2.69	1.342	3.07	0.62	157.0	
Sulfur dioxide (SO_2) (764)	5–86	11.81	66.45	5.63	1.415	9.08	0.97	191.4	Recip.
	20–100	17.18	84.52	4.92	1.453	6.52	0.92	193.4	
	40–100	27.10	84.52	3.12	1.444	4.17	0.63	162.6	
Propane (C_3H_8) (290)	5–86	42.1	155.3	3.69	1.653	4.10	1.35*	92.9*	Recip.
	20–100	55.5	187.0	3.37	1.730	3.29	1.32*	103.9*	
	40–100	78.0	187.0	2.40	1.646	2.26	0.90*	101.5*	
Ethane (C_2H_6) (170)	5–86	236.0	675.9	2.87	3.41	1.82	2.180	105	Recip.
Ethyl chloride (C_2H_5Cl) (160)	5–86	4.65	27.10	5.83	1.405	24.0	0.95*	106.3*	Rotary
	20–100	6.80	34.79	5.12	1.425	17.19	0.92*	116.1*	
	40–100	10.79	34.79	3.22	1.375	10.73	0.63*	109.9*	

* Values may be slightly in error.

of 5 to 86°F (−15 to 30°C) and also for certain other ranges. For these calculations, zero piston clearance and expansion through a throttle valve were assumed. The use of an expansion cylinder instead of a valve would have yielded work (available for supplying some of the work of compression), but this is always a negligible quantity, except when the compression pressure approaches the critical pressure, as for carbon dioxide. Theoretical *coefficients of performance* may be obtained by dividing 4.72 by the theoretical horsepower given in the table.

Ammonia (R-717) is used extensively in commercial, industrial, and moderately low-temperature installations. It is toxic, and recent laws regulate the operation, maintenance, and design of ammonia systems that contain 10,000 lb of refrigerant or more. (See Document 29 CFR 1910.119 of the Federal Register for further information.) Ammonia is provided commercially in several purity grades. It is recommended that refrigerant-grade anhydrous ammonia of 99.95 percent purity be used. Moisture contamination from impure ammonia will accumulate in evaporator coils and inhibit performance. Ammonia is corrosive to copper and copper-bearing alloys. It is not miscible to any large extent with lubricating oils. Exposure to it in a confined space at concentrations below 400 ppm is rarely fatal. Ammonia is immediately fatal at 0.5 to 0.6 percent by volume. The presence of ammonia can be detected by burning a sulfur candle in the vicinity of a leak. A white cloud will form if ammonia vapors are present. Hydrochloric acid solutions and wetted litmus papers will also react with ammonia to give visible evidence of the presence of vapors. See Tables 19.1.5 and 19.1.6 for thermodynamic properties of R-717.

Table 19.1.5   Properties of Superheated Ammonia

(v = specific volume in ft^3/lb; h = enthalpy in Btu/lb; s = entropy; h_f and s_f are measured from −40°F.)

Pressure, lb/in^2 abs	Temperature of saturated vapor, °F		Temperature of superheated vapor, °F								
			−30	−20	−10	0	10	20	30	40	50
10	−41.34	v	26.58	27.26	27.92	28.58	29.24	29.90	30.55	31.20	31.85
		h	603.2	608.5	613.7	618.9	624.0	629.1	634.2	639.3	644.4
		s	1.4420	1.4542	1.4659	1.4773	1.4884	1.4992	1.5097	1.5200	1.5301
20	−16.64	v			13.74	14.09	14.44	14.78	15.11	15.45	15.78
		h			610.0	615.5	621.0	626.4	631.7	637.0	642.3
		s			1.3784	1.3907	1.4025	1.4138	1.4248	1.4356	1.4460
30	−0.57	v				9.250	9.492	9.731	9.966	10.20	10.43
		h				611.9	617.8	623.5	629.1	634.6	640.1
		s				1.3371	1.3497	1.3618	1.3733	1.3845	1.3953
40	11.66	v						7.203	7.387	7.568	7.746
		h						620.4	626.3	632.1	637.8
		s						1.3231	1.3353	1.3470	1.3583
50	21.67	v							5.838	5.988	6.135
		h							623.4	629.5	635.4
		s							1.3046	1.3169	1.3286

Pressure, lb/in^2 abs	Temperature of saturated vapor, °F		100	120	140	160	180	200	240	280	320
80	44.40	v	4.190	4.371	4.548	4.722	4.893	5.063	5.398	5.730	
		h	658.7	670.4	681.8	693.2	704.4	715.6	738.1	760.7	
		s	1.3199	1.3404	1.3598	1.3784	1.3963	1.4136	1.4467	1.4781	
100	56.05	v	3.304	3.454	3.600	3.743	3.883	4.021	4.294	4.562	
		h	655.2	667.3	679.2	690.8	702.3	713.7	736.5	759.4	
		s	1.2891	1.3104	1.3305	1.3495	1.3678	1.3854	1.4190	1.4507	
120	66.02	v	2.712	2.842	2.967	3.089	3.190	3.326	3.557	3.783	
		h	651.6	664.2	676.5	688.5	700.2	711.8	734.9	758.0	
		s	1.2628	1.2850	1.3058	1.3254	1.3441	1.3620	1.3960	1.4281	
140	74.79	v	2.288	2.404	2.515	2.622	2.727	2.830	3.030	3.227	3.420
		h	647.8	661.1	673.7	686.0	698.0	709.9	733.3	756.7	780.0
		s	1.2396	1.2628	1.2843	1.3045	1.3236	1.3418	1.3763	1.4088	1.4395
160	82.64	v	1.969	2.075	2.175	2.272	2.365	2.457	2.635	2.809	2.980
		h	643.9	657.8	670.9	683.5	695.8	707.9	731.7	755.3	778.9
		s	1.2186	1.2429	1.2652	1.2859	1.3054	1.3240	1.3591	1.3919	1.4229
180	89.78	v	1.720	1.818	1.910	1.999	2.084	2.167	2.328	2.484	2.637
		h	639.9	654.4	668.0	681.0	693.6	705.9	730.1	753.9	777.7
		s	1.1992	1.2247	1.2477	1.2691	1.2891	1.3081	1.3436	1.3768	1.4081
200	96.34	v	1.520	1.612	1.698	1.780	1.859	1.935	2.082	2.225	2.364
		h	635.6	650.9	665.0	678.4	691.3	703.9	728.4	752.5	776.5
		s	1.1809	1.2077	1.2317	1.2537	1.2742	1.2935	1.3296	1.3631	1.3947
220	102.42	v		1.443	1.525	1.601	1.675	1.745	1.881	2.012	2.140
		h		647.3	662.0	675.8	689.1	701.9	726.8	751.1	775.3
		s		1.1917	1.2167	1.2394	1.2604	1.2801	1.3168	1.3507	1.3825
240	108.09	v		1.302	1.380	1.452	1.521	1.587	1.714	1.835	1.954
		h		643.5	658.8	673.1	686.7	699.8	725.1	749.8	774.1
		s		1.1764	1.2025	1.2259	1.2475	1.2677	1.3049	1.3392	1.3712
260	113.42	v		1.182	1.257	1.326	1.391	1.453	1.572	1.686	1.796
		h		639.5	655.6	670.4	684.4	697.7	723.4	748.4	772.9
		s		1.1617	1.1889	1.2132	1.2354	1.2560	1.2938	1.3258	1.3608

SOURCE: Condensed from NBS Circ. 142, 1923.

Table 19.1.6 Properties of Saturated Ammonia
(h_f and s_f are measured from $-40°F$.)

Temp., °F, t	Pressure, lb/in² abs, p	Specific volume, ft³/lb		Enthalpy, Btu			Entropy		
		Sat. liquid, v_f	Sat. vapor, v_g	Sat. liquid, h_f	Vapor-ization, h_{fg}	Sat. vapor, h_g	Sat. liquid, s_f	Vapor-ization, s_{fg}	Sat. vapor, s_g
−40	10.41	0.02322	24.86	0.0	597.6	597.6	0.0000	1.4242	1.4242
−38	11.04	0.02326	23.53	2.1	596.2	598.3	0.0051	1.4142	1.4193
−36	11.71	0.02331	22.27	4.3	594.8	599.1	0.0101	1.4043	1.4144
−34	12.41	0.02335	21.10	6.4	593.5	599.9	0.0151	1.3945	1.4096
−32	13.14	0.02340	20.00	8.5	592.1	600.6	0.0201	1.3847	1.4048
−30	13.90	0.02345	18.97	10.7	590.7	601.4	0.0250	1.3751	1.4001
−28	14.71	0.02349	18.00	12.8	589.3	602.1	0.0300	1.3655	1.3955
−26	15.55	0.02354	17.09	14.9	587.9	602.8	0.0350	1.3559	1.3909
−24	16.42	0.02359	16.24	17.1	586.5	603.6	0.0399	1.3464	1.3863
−22	17.34	0.02364	15.43	19.2	585.1	604.3	0.0448	1.3370	1.3818
−20	18.30	0.02369	14.68	21.4	583.6	605.0	0.0497	1.3277	1.3774
−18	19.30	0.02374	13.97	23.5	582.2	605.7	0.0545	1.3184	1.3729
−16	20.34	0.02378	13.29	25.6	580.8	606.4	0.0594	1.3092	1.3686
−14	21.43	0.02383	12.66	27.8	579.3	607.1	0.0642	1.3001	1.3643
−12	22.56	0.02388	12.06	30.0	577.8	607.8	0.0690	1.2910	1.3600
−10	23.74	0.02393	11.50	32.1	576.4	608.5	0.0738	1.2820	1.3558
−8	24.97	0.02399	10.97	34.3	574.9	609.2	0.0786	1.2730	1.3516
−6	26.26	0.02404	10.47	36.4	573.4	609.8	0.0833	1.2641	1.3474
−4	27.59	0.02409	9.991	38.6	571.9	610.5	0.0880	1.2553	1.3433
−2	28.98	0.02414	9.541	40.7	570.4	611.1	0.0928	1.2465	1.3393
0	30.42	0.02419	9.116	42.9	568.9	611.8	0.0975	1.2377	1.3352
2	31.92	0.02424	8.714	45.1	567.3	612.4	0.1022	1.2290	1.3312
4	33.47	0.02430	8.333	47.2	565.8	613.0	0.1069	1.2204	1.3273
6	35.09	0.02435	7.971	49.4	564.2	613.6	0.1115	1.2119	1.3234
8	36.77	0.02440	7.629	51.6	562.7	614.3	0.1162	1.2033	1.3195
10	38.51	0.02446	7.304	53.8	561.1	614.9	0.1208	1.1949	1.3157
12	40.31	0.02451	6.996	56.0	559.5	615.5	0.1254	1.1864	1.3118
14	42.18	0.02457	6.703	58.2	557.9	616.1	0.1300	1.1781	1.3081
16	44.12	0.02462	6.425	60.3	556.3	616.6	0.1346	1.1697	1.3043
18	46.13	0.02468	6.161	62.5	554.7	617.2	0.1392	1.1614	1.3006
20	48.21	0.02474	5.910	64.7	553.1	617.8	0.1437	1.1532	1.2969
22	50.36	0.02479	5.671	66.9	551.4	618.3	0.1483	1.1450	1.2933
24	52.59	0.02485	5.443	69.1	549.8	618.9	0.1528	1.1369	1.2897
26	54.90	0.02491	5.227	71.3	548.1	619.4	0.1573	1.1288	1.2861
28	57.28	0.02497	5.021	73.5	546.4	619.9	0.1618	1.1207	1.2825
30	59.74	0.02503	4.825	75.7	544.8	620.5	0.1663	1.1127	1.2790
32	62.29	0.02508	4.637	77.9	543.1	621.0	0.1708	1.1047	1.2755
34	64.91	0.02514	4.459	80.1	541.4	621.5	0.1753	1.0968	1.2721
36	67.63	0.02521	4.289	82.3	539.7	622.0	0.1797	1.0889	1.2686
38	70.43	0.02527	4.126	84.6	537.9	622.5	0.1841	1.0811	1.2652
40	73.32	0.02533	3.971	86.8	536.2	623.0	0.1885	1.0733	1.2618
42	76.31	0.02539	3.823	89.0	534.4	623.4	0.1930	1.0655	1.2585
44	79.38	0.02545	3.682	91.2	532.7	623.9	0.1974	1.0578	1.2552
46	82.55	0.02551	3.547	93.5	530.9	624.4	0.2018	1.0501	1.2519
48	85.82	0.02558	3.418	95.7	529.1	624.8	0.2062	1.0424	1.2486
50	89.19	0.02564	3.294	97.9	527.3	625.2	0.2105	1.0348	1.2453
52	92.66	0.02571	3.176	100.2	525.5	625.7	0.2149	1.0272	1.2421
54	96.23	0.02577	3.063	102.4	523.7	626.1	0.2192	1.0197	1.2389
56	99.91	0.02584	2.954	104.7	521.8	626.5	0.2236	1.0121	1.2357
58	103.7	0.02590	2.851	106.9	520.0	626.9	0.2279	1.0046	1.2325
60	107.6	0.02597	2.751	109.2	518.1	627.3	0.2322	0.9972	1.2294
62	111.6	0.02604	2.656	111.5	516.2	627.7	0.2365	0.9897	1.2262
64	115.7	0.02611	2.565	113.7	514.3	628.0	0.2408	0.9823	1.2231
66	120.0	0.02618	2.477	116.0	512.4	628.4	0.2451	0.9750	1.2201
68	124.3	0.02625	2.393	118.3	510.5	628.8	0.2494	0.9676	1.2170
70	128.8	0.02632	2.312	120.5	508.6	629.1	0.2537	0.9603	1.2140
72	133.4	0.02639	2.235	122.8	506.6	629.4	0.2579	0.9531	1.2110
74	138.1	0.02646	2.161	125.1	504.7	629.8	0.2622	0.9458	1.2080
76	143.0	0.02653	2.089	127.4	502.7	630.1	0.2664	0.9386	1.2050
78	147.9	0.02661	2.021	129.7	500.7	630.4	0.2706	0.9314	1.2020

(Continued)

Table 19.1.6   Properties of Saturated Ammonia (*Continued*)
(h_f and s_f are measured from $-40°F$.)

Temp., °F, t	Pressure, lb/in² abs, p	Specific volume, ft³/lb		Enthalpy, Btu			Entropy		
		Sat. liquid, v_f	Sat. vapor, v_g	Sat. liquid, h_f	Vapor-ization, h_{fg}	Sat. vapor, h_g	Sat. liquid, s_f	Vapor-ization, s_{fg}	Sat. vapor, s_g
80	153.0	0.02668	1.955	132.0	498.7	630.7	0.2749	0.9242	1.1991
82	158.3	0.02675	1.892	134.3	496.7	631.0	0.2791	0.9171	1.1962
84	163.7	0.02684	1.831	136.6	494.7	631.3	0.2833	0.9100	1.1933
86	169.2	0.02691	1.772	138.9	492.6	631.5	0.2875	0.9029	1.1904
88	174.8	0.02699	1.716	141.2	490.6	631.8	0.2917	0.8958	1.1875
90	180.6	0.02707	1.661	143.5	488.5	632.0	0.2958	0.8888	1.1846
92	186.6	0.02715	1.609	145.8	486.4	632.2	0.3000	0.8818	1.1818
94	192.7	0.02723	1.559	148.2	484.3	632.5	0.3041	0.8748	1.1789
96	198.9	0.02731	1.510	150.5	482.1	632.6	0.3083	0.8678	1.1761
98	205.3	0.02739	1.464	152.9	480.0	632.9	0.3125	0.8608	1.1733
100	211.9	0.02747	1.419	155.2	477.8	633.0	0.3166	0.8539	1.1705
105	228.9	0.02769	1.313	161.1	472.3	633.4	0.3269	0.8366	1.1635
110	247.0	0.02790	1.217	167.0	466.7	633.7	0.3372	0.8194	1.1566
115	266.2	0.02813	1.128	173.0	460.9	633.9	0.3474	0.8023	1.1497
120	286.4	0.02836	1.047	179.0	455.0	634.0	0.3576	0.7851	1.1427

Carbon dioxide (R-744) was used for a long time as a "safety refrigerant." It is not toxic, and exposure to it is not dangerous unless concentrations are high. Like many other refrigerants, it is heavier than air, and caution must be exercised in confined spaces, as it may displace oxygen and result in asphyxiation. With cooling water at 70°F (21°C), the condensing pressure of CO_2 is high, and since its critical temperature is 87.8°F (31°C), condensation will not occur at water temperatures above this. Specific power consumption is high on CO_2 machines.

Sulfur dioxide (R-764) has rapidly gone out of use. It is extremely corrosive unless absolutely anhydrous. It is extremely irritating in small concentrations and toxic in high concentrations.

Methyl chloride (R-40) (CH_3Cl), an anesthetic in amounts of 5 to 10 percent by volume, has been used in air-cooled units of moderate or small sizes. It is miscible with mineral oils; small amounts of moisture in a methyl chloride system will cause trouble by freezing in expansion valves.

For industrial work, butane (C_4H_{10}), propane (C_3H_8), and, at low temperatures, ethane (C_2H_6) are used. **Dieline** (dichloroethylene, $C_2H_2Cl_2$) and **Carrene** (dichloromethane, CH_2Cl_2) have been used to some extent, usually in centrifugal compressors.

In order to create refrigerants with thermodynamic properties similar to those being phased out, some refrigerants are blends of two or more other refrigerants. Blends have 400 and 500 series designations. For example, R-407C is a blend of three HFC refrigerants, R-32, R-125, and R-134a. It is designed to be used in new equipment and as a service replacement for existing R-22 machines. The thermodynamic properties are shown in Fig. 19.1.4.

Use caution when using blended refrigerants in centrifugal compressors. The compressor impeller may act to separate the components of the blend by molecular weight. For safety purposes, blends are classified on the basis of worst-case fractionation.

Building codes often require monitoring of refrigerant levels in equipment spaces. In addition to a monitor, an alarm system and exhaust facilities should be installed to prevent exposure above the acceptable exposure limit (AEL). ANSI/ASHRAE 15-2004, "Safety Standard for Refrigeration Systems," and addenda have recommendations covering equipment room safety.

Leak Detection   Detection and repair of refrigerant leaks is an essential part of responsible refrigeration equipment operation. Worker safety, environmental impact, and refrigerant cost are the primary reasons for a detection and repair program. Section 608 of the Clean Air Act as amended July 24, 2003, has rules governing the keeping of service records documenting the date and type of service and the quantity

of refrigerant added. This currently applies to equipment that contains 50 or more pounds of refrigerant.

Area monitors are useful for determining that a leak has occurred. Handheld detectors are useful for determining a specific leak location. For many years, refrigerant leaks were detected with a halide torch. Copper gauze was kept hot by the essentially colorless flame of burning methyl alcohol. An intense green color was imparted to the flame by the presence of even minute amounts of chlorine. The alternative refrigerants have no chlorine or a lower amount of chlorine. Some refrigerants are flammable. New types of detection equipment are required.

Nonselective, halogen-specific, and compound-specific detectors are available. Each type has limitations, and selection must be based on the specific application. For example, the nonspecific-type detector is poorly suited for area monitoring because of its high detection limit and potential for false alarms, but is a reliable, low-cost choice for pinpointing leaks. The halogen-specific detectors can be applied for either leak pinpointing or area monitoring. They are best applied where only one refrigerant is in use. The compound-specific detectors are relatively expensive, but are very selective. They are best applied in environments where several refrigerants or other chemical compounds may be present.

The pressure-temperature relations for the saturated vapors of many of the commercially important refrigerants are given in Fig. 19.1.5. If the temperature at which the refrigeration is desired and the temperature at which heat can be discarded (condenser temperature) are known, the chart is convenient for determining for any chosen refrigerant the pressures which must be maintained. For instance, if the use of methyl chloride is contemplated in an air-cooled cabinet food freezer located in a room where the air temperature may rise to 90°F (32°C), the compressor discharge must be at least 100 psia (690 kPa absolute), and if the cabinet cooling coil is to be held at $-30°F$ ($-34°C$), compressor suction will be 9 psia (62 kPa absolute).

The large volumes required for HCFC-123 and dieline can be handled satisfactorily by centrifugal compressors. When the evaporating pressure is below atmospheric—as in the case of HCFC-123, dieline, ammonia in food freezers, sulfur dioxide, water vapor, and butane—air leaks are likely to occur. The air tends to accumulate in the condenser and the heat-transfer performance is degraded.

Refrigerating systems which operate with a positive pressure in the equipment will leak refrigerant into the atmosphere. Many HFC- and HCFC-type refrigerants are practically odorless. Dangerous concentrations can build up in unventilated enclosed rooms such as freezers should a significant leak occur. Leak detection devices are available commercially to provide warnings of equipment leakage before harmful

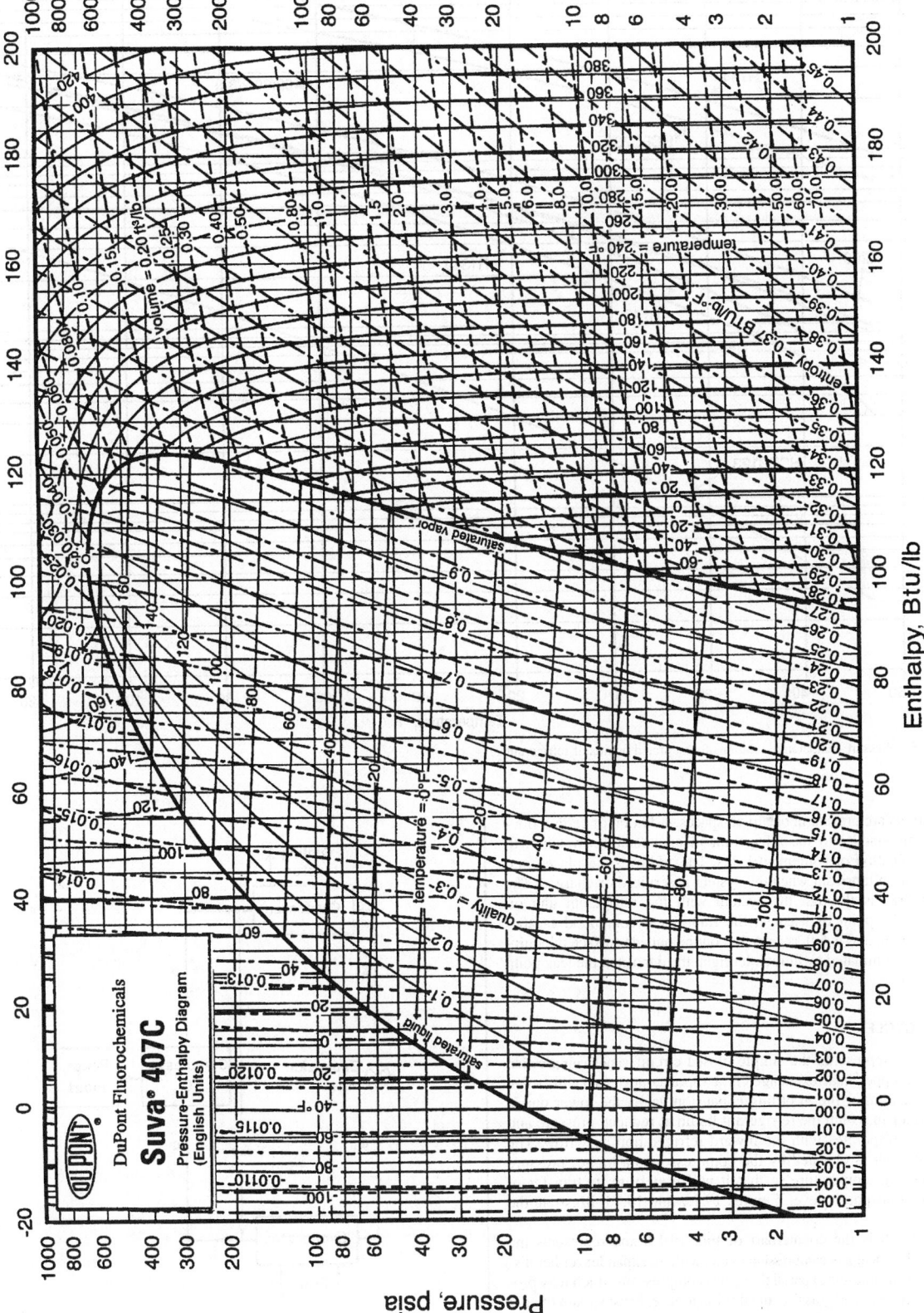

Fig. 19.1.4 Thermodynamic properties of SUVA® 407C (R-407C). (*Reprinted with permission of E. I. du Pont de Nemours and Company.*)

Fig. 19.1.5   Pressure-temperature relations for saturated liquids of refrigerants.

concentrations are present in occupied spaces. Addition of certain fluorescent compounds into the refrigerant in an operating system will allow identification of equipment leaks. The additives are placed into the refrigeration lubricant when the system is serviced or charged. The compound travels through the system with the refrigerant and oil mixture. At the point of leakage, the fluorescent compound escapes from the system, and is left as a deposit on the outside surface of equipment. When illuminated by ultraviolet light, the deposit at the leak point becomes visible.

OVERALL CYCLES

Figure 19.1.6 represents the simple closed-circuit **vapor-compression** system. The upper heat exchanger is a vapor condenser (with some superheat), and after the throttling expansion valve, the lower one is the evaporator in which the refrigerant liquid at reduced pressure and temperature evaporates with the inward refrigeration heat flow. The refrigerant vapor is elevated by the compressor to a higher pressure and condensing temperature so that it will liquefy in its transfer of heat to the atmospheric level. Figures 19.1.7 and 19.1.8 illustrate the cycle.

Figure 19.1.9 is the counterpart of Fig. 19.1.6 and represents the closed sensible-heat **gas-compression** cycle, with an expander (either displacement or turbine-type) parallel to a gas compression which may provide a compressor-work "assist" in addition to its exhaust stream of cold gas. This is the refrigeration version of the Brayton cycle (see Sec. 4),

Fig. 19.1.6   Simple closed-circuit vapor-compression system.

the upper exchanger serving to cool the hot compressed gas, and the lower exchanger handling the sensible-heat refrigeration effect. In place of the expander, an expansion valve could be used; this is particularly effective on very low-temperature open-cycle operations, such as with gas liquefaction processes incorporating secondary "cold-regenerating" heat exchangers.

An extended temperature-entropy diagram (Fig. 19.1.10) illustrates paths for gas-compression sensible-heat cycles as follows: (1) 1-2-3-4 isentropic expander; (2) 1-2-3-4' polytropic expander; (3) 1-2-5-6 throttling (isenthalpic); and (4) 1-2-7-8 throttling (low temperature).

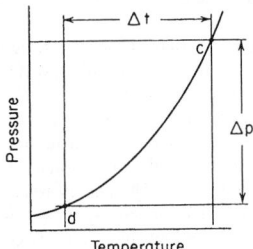

Fig. 19.1.7 Pressure-temperature curve illustrating overall differences.

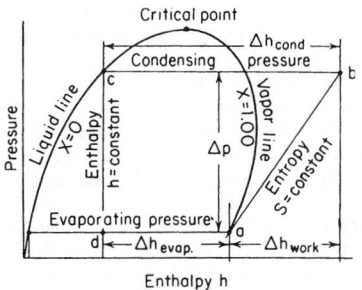

Fig. 19.1.8 Pressure-enthalpy diagram for simple ideal refrigeration cycle.

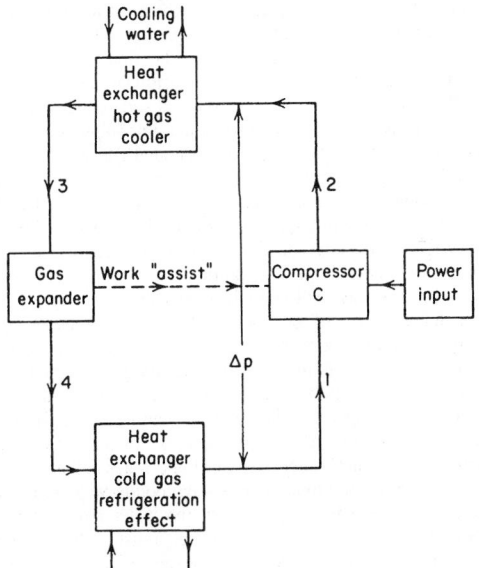

Fig. 19.1.9 Closed sensible-heat gas-compression cycle.

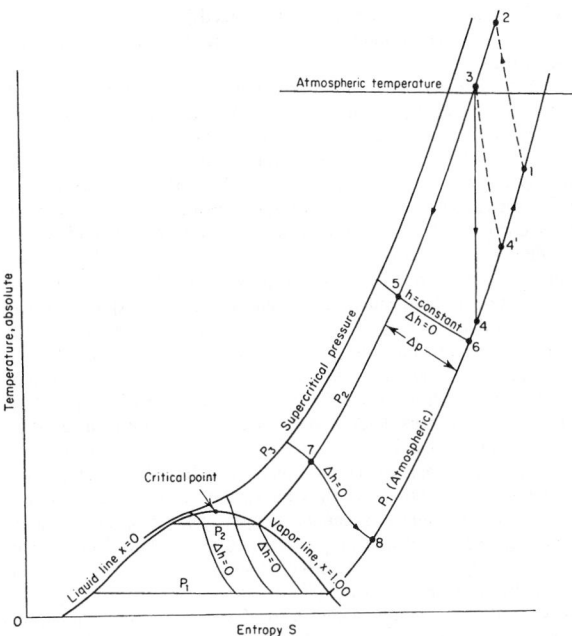

Fig. 19.1.10 Extended temperature-entropy diagram for air illustrating sensible-heat air-compression processes.

COMPONENTS OF COMPRESSION SYSTEMS

A refrigeration system is an energy-transport complex of assorted components. It may be a conventional system in which the same fluid, the refrigerant, recirculates continuously in a closed circuit, or it may entail a partially open system with discharge of some of the processed fluid either as a liquefied gas or as a solidified gas, with replacement makeup, e.g., by air liquefaction and solid carbon dioxide production. Components include vapor and gas compressors; liquid pumps; heat-transfer equipment (gas coolers, intercoolers, aftercoolers, exchangers, economizers); vapor condensers and the counterpart evaporators; liquid coolers and receivers; expanders; control valves and pressure-drop throttling devices (capillaries, refrigerant-mixture separating chambers, stream-mixing chambers); and connecting piping and insulation.

Codes and **standards** have been published by various associations and societies. These cover design, selection, designation, testing, sound ratings, safety, efficiency, and other areas relating to mechanical refrigeration. Publication of a comprehensive list is impractical because of frequent revisions. Several of the organizations include Air Conditioning Contractors of America (ACCA), American National Standards Institute (ANSI), Air-Conditioning and Refrigeration Institute (ARI), American Society of Heating, Refrigerating and Air-Conditioning Engineers, Inc. (ASHRAE), The American Society of Mechanical Engineers (ASME), Canadian Standards Association (CSA), International Institute of Ammonia Refrigeration (IIAR), and Underwriters Laboratories Inc. (UL). A list of current codes and standards can be obtained by contacting the particular organization.

Several of the more general refrigeration equipment standards include: ANSI/ASHRAE 15 "Safety Standard for Refrigeration Systems"; ANSI/ASHRAE 34 "Designation and Safety Classification of Refrigerants"; and CAN/CSA-C22.2 "General Standard on Refrigeration Equipment.

There are numerous additional standards relating to specific types of refrigeration equipment (air conditioners, heat pumps, chillers) or specific components (compressors, water- or air-cooled condensers, lubricating oil). Because refrigeration equipment requires auxiliaries (electric motors, steam turbines, piping, insulation) and is part of a larger system such as food processing or heating, ventilating, and air conditioning

(HVAC), other standards not relating directly to refrigeration may apply. Any installation must also comply with building and fire code requirements.

Compression may be single stage (Fig. 19.1.6) or multiple stage. Displacement compressors include reciprocating piston, screw, rotary vane, scroll, and trochoidal compressors. The volumetric capacity of the compressors is always less than the swept volume of pistons or the volume displaced in cyclic rotation by the other forms of displacement devices such as screws, nutating scrolls, or rotating vanes. The true measure of compressor capacity is the volumetric efficiency. It is influenced by cylinder, screw mesh, or cavity clearance volume; pressure ratio; nature of the gas; pressure drops and passage volume in the valves or vanes; suction heating; and the equivalent compression exponent n. The numerical value of n expresses the effect of heat transfer to or from the gas during compression of the gas. Normally it lies between $n = 1.0$ and $n = k$ (adiabatic), such as for the ideal (isentropic). Figure 19.1.11 indicates the comparative volumetric efficiency of a pair of reciprocating and screw compressors. **Reciprocating compressors** are disadvantaged by high discharge temperatures and excessive clearance volume losses associated with compression ratios greater than 12 to 1 in a single compression stage. Compound compression staging is a practical solution if reciprocating machinery is applied across cycle compression ratios greater than 10 to 1. **Screw** compressors with oil injection systems can operate reliably as a single stage with compression ratios greater than 20 to 1. **Rotary vane** compressors are fixed-volume-ratio machines and can be used across compression ratios of 18 to 1, vane material and lubricating oil permitting; however, in the larger-size machines, a 7 to 1 compression ratio is commonly used in order to maintain efficiency. **Scroll** compressors have been used in commercial practice for systems that have capacities less than 10 tons (35 kW). **Trochoidal** compressors are often called "Wankel" compressors because, in one form, the rotor is similar to the Wankel internal combustion engine rotor.

Fig. 19.1.11 Screw and reciprocating compressor volumetric and adiabatic efficiency versus pressure ratio. *(Courtesy of J. W. Pillis, "Development of a Variable Volume Ratio Screw Compressor," Frick Company, 1983.)*

Reciprocating-piston compressors may be horizontal or vertical. The pistons can be designed to be either single acting or double acting with compression occurring on either stroke of the piston. Commercial designs have evolved into multiple cylinder machines arranged in a V block similar to an automotive engine. Average piston speed may be as great as 800 ft/min (4 m/s). Design characteristics of the suction and discharge valves installed on the cylinder determine the energy efficiency, and maintenance frequency. Valves most frequently used are the free-floating reed, clamped reed, and ring valve. Gas velocity through the valves is limited to 12,000 ft/min (61 m/s) with R-717 and 9,000 ft/min (46 m/s) with HCFC-22. For additional information on reciprocating compressors in refrigeration service, refer to the ASHRAE Handbook, ASHRAE Handbook, 2004 Systems and Equipment, pp. 34.4 to 34.9.

Screw compressors are available in several designs, both single screw and twin screw, with oil-free and oil-injected designs in both types. Twin-screw oil-injected compressors are slightly more energy efficient at moderate compression ratios. Many of the commercially available screw compressors are fixed-volume-ratio machines. (See also Sec. 14.) If the compression ratio varies, the operational efficiency of the compressor also varies significantly. The energy reductions that are available by operating with cooler condensing pressures during spring, fall, and winter may not be fully achievable. Capacity control can be accomplished with a movable slide valve that varies the internal volume, valved suction ports, variable-speed drives, and hot-gas bypass. The slide valve is the most common method. Figure 19.1.11 illustrates the adiabatic efficiency of a twin-screw compressor with a movable slide stop control and a reciprocating compressor. Flash intercoolers (Fig. 19.1.13), also called *economizers*, can be provided to improve energy efficiency on the larger-capacity machines.

Air-cooled compressors are used where discharge temperatures are low; water cooling is used where discharge temperatures are high, as with ammonia, and on larger industrial units. **Oil separators** return to the compressor the lubricating oil carried over by the refrigerant vapors. **Rotary compressors** of the vane, eccentric, gear, and screw types are in use.

Kinetic compressors include high-speed centrifugal and axial flow machines, usually multistaged, and jet-entrainment devices. Centrifugal machines are especially adapted to high-volume flow (>500 ft³/min) (0.24 m³/s). (See also Sec. 14.)

The entire machine-compression operation may be replaced by a secondary absorber-pump-generator system, in which the complex is known as an absorption refrigeration system.

Dual (or **multiple effect**) **compression** may be used when refrigeration at two temperatures is desired. The compressor takes vapor from a lower temperature expansion coil during the first part of its intake stroke, and from a higher temperature expansion coil at or near the end of the stroke. The mixture is then compressed and condensed.

In **wet compression,** cooling is obtained during compression by spraying liquid refrigerant into the compressor cylinder. The desuperheating of the compressed vapors results in better heat transfer in condensers and more nearly isothermal compression; the disadvantages are reduced compressor capacity and the problem of control of the amount of injection.

Table 19.1.7 shows some ideal performance values for a single-state ammonia system under different conditions.

Condensers are usually shell-and-tube type, with the refrigerant passing outside the tubes. Older industrial installations still use **double-pipe condensers**. In double-pipe condensers, gas flows between the two pipes while water passes through the inner. The outside pipe diameter may be 2 in (5 cm); the inner, 1¼ in (3.2 cm). In some instances, the pipes are exposed to the atmosphere either with or without water drip. In an **evaporative condenser** the refrigerant vapor is condensed as it passes through tubes over which water is sprayed; the water is then evaporated by air flowing over the wet tubes. In this way, cooling-water requirements are reduced to from 5 to 15 percent of the water requirements of a nonevaporative condenser with no water reuse. The evaporative condenser combines in a single unit a refrigerant condenser and the atmospheric cooling tower or spray pond which is required if the

Table 19.1.7 Theoretical Horsepower and Theoretical Volume of Dry Ammonia Gas Pumped per Minute to Produce 1 Ton of Refrigeration

| Suction pressure and temperature | | Condenser pressure, lb/in² gage (temperature, °F) | | | | | | | | | |
| | | 103 (65) | | 127 (75) | | 153 (85) | | 182 (95) | | 215 (105) | |
lb/in² gage	°F	hp	ft³/min	hp	ft³/min	hp	ft³/min	hp	ft³/min	hp	ft³/min
4	−20	1.058	5.84	1.205	5.96	1.361	6.09	1.525	6.23	1.691	6.43
6	−15	0.997	5.35	1.145	5.46	1.300	5.58	1.461	5.70	1.546	5.83
9	−10	0.903	4.66	1.045	4.76	1.193	4.86	1.347	4.97	1.435	5.08
13	−5	0.818	4.09	0.954	4.17	1.094	4.25	1.244	4.35	1.321	4.44
16	0	0.735	3.59	0.865	3.66	1.002	3.74	1.147	3.83	1.219	3.91
20	5	0.666	3.20	0.795	3.27	0.928	3.34	1.066	3.41	1.138	3.49
24	10	0.592	2.87	0.726	2.93	0.854	2.99	0.991	3.06	1.060	3.12
28	15	0.541	2.59	0.664	2.65	0.792	2.71	0.922	2.76	0.994	2.82
33	20	0.474	2.31	0.592	2.36	0.715	2.41	0.842	2.46	0.903	2.51
39	25	0.410	2.06	0.523	2.10	0.599	2.15	0.767	2.20	0.829	2.24
45	30	0.351	1.85	0.461	1.89	0.576	1.93	0.694	1.97	0.759	2.01
51	35	0.300	1.70	0.410	1.74	0.521	1.77	0.640	1.81	0.701	1.85

cooling water is to be reused. Air-cooled finned-tube condensers with forced ventilation are widely used on small units, and shell-and-coil or double-tube water-cooled condensers on medium units.

Evaporators may have plain or finned surfaces. Fin density is available from 1 to 14 fins per inch. Fin density of 2 to 4 fins per inch is specified for air-cooling service below 28°F (− 2.2°C), because air can flow through the coil while thin layers of frost are forming. Four types of defrosting methods are used: hot gas from the compressor, electrical air heaters, hot water, and room air circulation. Flooded evaporators are operated practically full of refrigerant. The level is controlled by a float valve. Wet-expansion evaporators are operated with a level approaching flooded operation. Dry-expansion (once-through) units operate with an indefinite amount of liquid in the evaporator. Usually a thermal bulb measures the outlet pipe temperature and adjusts a thermostatic expansion valve to maintain 15°F (8.3°C) superheat in the outlet vapor. Flooded or wet operation gives high heat-transfer rates. Pumped recirculation of refrigerant is used to promote heat transfer in finned coil evaporators, and also on some shell and tube evaporators (liquid chillers) by completely wetting the surfaces with a rapidly moving layer of refrigerant. The most commonly used liquid feed ratio is 3 to 1. However, a feed ratio as high as 6 to 1 may be required to accommodate distribution imbalances in the piping system. Combined liquid and vapor **return piping** should be sloped at 0.25 in/ft (2.08 percent) to prevent formation of waves in the liquid-phase layer that can react with the flowing vapors to cause liquid hammer that, in turn, can severely damage piping components. **Receivers** and **recirculating pumps** must be carefully matched to the operating system to prevent pump cavitation, liquid droplet carryover, and gross liquid overflow due to surges of liquid that may occur as a result of equipment operations such as defrosting processes. Specifically, the tank drop leg liquid velocity should be limited to 100 ft/min (0.51 m/s) and the vapor velocity over the liquid level in the tank should also be limited to 100 ft/min (0.51 m/s).

For a low-cost unit operating in limited space under conditions of motion, dry operation is preferred.

Controls are required on the liquid level of the refrigerant and on the temperature of the refrigerated space. The liquid control regulates the flow of refrigerant into the evaporator and also serves as the pressure barrier between the high operating pressure of the condenser and the lower operating pressure of the evaporator. It may take the form of a **capillary tube** between the condenser (high side) and the evaporator (low side), in which case it is nonadjustable. Plugging due to dirt is a common difficulty. Capillary-tube liquid control is largely confined to relatively small units assembled and charged at the factory and particularly for hermetically sealed systems. The **constant-pressure expansion valve**, maintaining a constant evaporator pressure, and the **thermal expansion valve**, maintaining a constant superheat leaving the evaporator, are standard

liquid controls for most commercial applications. A **low-side float** liquid control, used with a flooded evaporator operating at evaporator (low) pressure, consists of a float-operated valve to admit liquid refrigerant to the evaporator in accordance with demand so that a constant liquid level is held in it. A **high-side float** liquid control is often used with a single flooded evaporator; the float operating the valve between the evaporator and the condenser is in a float chamber containing liquid refrigerant at the condenser (high-side) pressure. As the liquid level in the float chamber falls, the valve closes, thus preventing hot gaseous refrigerant from passing from the high to the low side as is possible when using a capillary or a low-side float.

Several types of noncondensable-gas removal devices have been developed to remove tramp air from refrigerating systems. The exhaust of the purge system contains a portion of the refrigerant. This refrigerant loss is detrimental to the environment, and the replacement of lost refrigerant is costly. Three types of refrigerant purge systems have been developed. One uses a scavenging condenser, cooled by system refrigerant, followed by a carbon absorption device to remove the last vestiges of refrigerant from the purge vapors. Other types, such as oil-absorption systems and reverse-osmosis systems, are developed but not offered commercially. The efficiency of purge systems is specified by the weight ratio of refrigerant lost to the air that is exhausted.

Vapor-Compression Circuits

Compound Low-Temperature Systems Multistaging is advantageous as it reduces operating temperatures by intercooling and power requirements if the fluid superheats considerably on adiabatic compression, as with ammonia; with F-12 refrigerant, propane, and butane, the superheat is negligible (see Figs. 19.1.12 to 19.1,15).

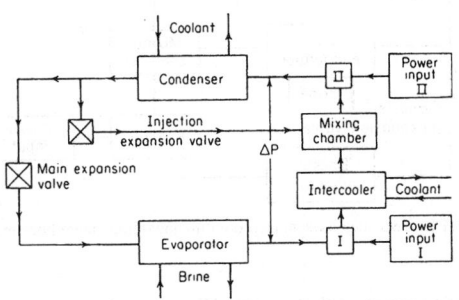

Fig. 19.1.12 Two-stage system with conventional intercooler plus refrigerant injection.

Fig. 19.1.13   Two-stage system with flash intercooler.

Gas Intercooler Plus Liquid Injection   Since gas temperature in the intercooler is limited by cooling water temperature, further favorable reduction of gas temperature, prior to entrance to the second stage, may be accomplished by direct injection, via an expansion valve, of refrigerant liquid (Fig. 19.1.12).

Flash Intercooler   With a flash intercooler separator located midway between successive expansion valves, the gas leaving the low-stage (I) compressor passes through the flash intercooler, gets cooled by direct contact with the liquid, and, augmented by the flash vapor and at reduced temperature, enters the second stage (II) (Fig. 19.1.13).

For positive-displacement compressors, selection of flash intercooler temperature is determined by the following relation:

$$P_{\text{intermediate}} = (P_{\text{I suction}} \times P_{\text{II discharge}})^{1/2}$$

where P = pressure in absolute units, lb/in² abs (Pa). This equation is valid for centrifugal machines with modification. The common practice is to limit single-impeller-stage adiabatic head to approximately 10,000 ft (3,000m) for low-molecular-weight gas and 6,000 ft (1,800 m) for the heavier HCFC and HFC refrigerants. Economical rotor design may require an interstage pressure that does not match the ideal flash intercooler pressure. Where choice is required, select a pressure slightly greater than the ideal pressure given by the above equation.

Low-Temperature Booster Plus Single Stage   For the addition of a second lower-temperature evaporator to an existing single-stage system, a booster for the low-temperature vapor may be employed, its output feeding into the single-stage suction. The booster is a centrifugal or rotary compressor, suitable for handling the large vapor volumes encountered at low temperatures (and pressures); the reciprocating compressor is more advantageous at high pressures (Fig. 19.1.14).

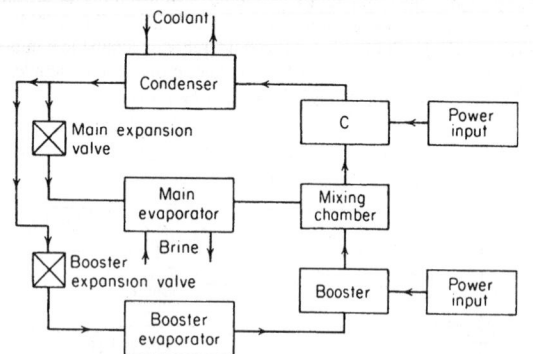

Fig. 19.1.14   Two-stage system incorporating low-temperature booster circuit.

Cascade Systems with Segregated Circuits   Here two separate circuits are employed with the low-stage vapor condensed in the second-stage evaporator. Two different refrigerants may be appropriately chosen, a lower-boiling-point refrigerant serving for the first stage. Thus, for low temperatures ($< -50°C$ or $-50°F$), CO_2 may be employed in the lower stage with NH_3 in the second stage condensing the CO_2. Or the same refrigerant may be used in both stages, limiting the migration of the lubricant to the low-temperature realm (Fig. 19.1.15).

Fig. 19.1.15   Two-stage system in cascade form.

Water-vapor refrigeration is necessarily limited to near-atmospheric temperature levels; it involves large suction volumes at high vacuum and utilizes a nontoxic refrigerant; this type of refrigeration was frequently used in the early days of space air conditioning. It utilizes (1) centrifugal compressors or (2) steam-jet compressors, entraining low-density water vapor from the evaporator. The centrifugal water-vapor compressor, usually multistage, must operate at high speeds (7,000 to 10,000 r/min). Operation at high vacuum in the evaporator requires an air pump at the condenser to eliminate leakage effects.

Steam-jet refrigeration systems are used where cooling to temperatures above 0°C is desired. Applications include industrial air conditioning and cooling of city gas to condense out tar and other objectionable impurities, of gas absorbers to increase efficiency of absorption, of reaction units where heat removal and temperature control are important during chemical transformations, and of wort and mash in the brewing and other fermentation industries particularly in the summer months. Jet refrigeration has been used for cooling passenger trains and is coming into popularity for marine installations, e.g., cooling of banana boats and large passenger vessels. The system is a compression-type refrigerator: it uses water as a refrigerant, a part of which is evaporated to produce cooling of the remainder, steam-jet ejectors to compress the water vapor resulting from evaporation, and a condenser, either the surface type, where refrigerant water is to be reused, or the barometric type, where refrigerant water can be discarded. The advantages of the system are: low installation cost; the absence of moving parts except for small liquid pumps; and safety, since no noxious or toxic refrigerants are used.

For considerable differences in temperatures between condensing-water temperature and refrigerating temperatures (10°C or more), cooling of the refrigerant water by evaporation in two stages operating at different pressures (and temperatures) will usually give better operating economy, but at somewhat higher installation cost.

Figure 19.1.16 shows the variation of refrigerating capacity with variation of the chilled-water temperature, capacity increasing as chilled-water temperature rises.

Household compression machines are designed for continuous automatic operation and for conservation of the charges of refrigerant and oil. These units are almost universally motor-driven air-cooled compressors, the principal exception being the Electrolux-Servel absorption unit. The compressor may be hermetically sealed, with the compressor and motor enclosed in the same casing, or may use a shaft seal, embodying some form of sylphon bellows, with an outside coil spring at the place where the crankshaft passes out of the casing.

Fig. 19.1.16 Variation of refrigerating capacity with the jet-chilled water temperature in steam refrigeration.

The compressor may be reciprocating, but rotary types are being increasingly used either with a floating piston ring driven by an eccentric or with sliding blades. Lubrication is by internal forced-feed circulation in most cases. The condenser may be of the radiator, coil, or plate type, cooled by natural draft, by a fan on the main motor, or by a separate motor. Refrigerant feed control to the evaporator may be by (1) float feed (flooded system), (2) pressure-actuated diaphragm valve, or most commonly (3) fixed orifice. The fixed orifice may be either a plate orifice or a capillary tube. The evaporator is usually made of aluminum. Controls are generally pressure-operated thermostatic switches. The power consumption of an average home refrigerator (18 ft³) (0.5 m²) is 133 kWh per month. A typical home deep freezer consumes approximately 152 kWh per month.

ABSORPTION SYSTEMS

Absorption refrigeration machines are essentially vapor-compression plants (Fig. 19.1.6) in which the mechanical compressor has been replaced by a thermally activated arrangement (Fig. 19.1.17). The basic element are the **absorber, pump, heat exchanger, throttle valve,** and **generator.**

In an absorption cycle, a secondary fluid, the **absorbent,** is used to absorb the primary fluid, the **refrigerant,** after the refrigerant has left the evaporator. The vaporized refrigerant is converted back to liquid phase in the absorber. Heat released in the absorption process must be rejected to cooling water. The solution of absorbent and refrigerant is then pumped to the generator. A heat exchanger may be used for heat recovery and a corresponding increase in efficiency. In the generator, heat is added, and the more volatile refrigerant is separated from the absorbent through distillation. The refrigerant continues to the condenser, expansion valve, and evaporator, while the absorbent returns to the absorber.

Although there are many refrigerant-absorbent combinations, the most common systems are **ammonia-water** and **water-lithium bromide.** A common variation of the absorption system is the **Electrolux-Servel** process (Platen-Munters Patent). It has had wide application in gas-fired household refrigeration machines and air-conditioning systems. The use of hydrogen in a hermetically sealed system eliminates the use of pumps and other moving parts.

Absorption machines represent only a small percentage of refrigeration systems in operation. This is because they are generally bulky and difficult to operate efficiently. There has been renewed interest in absorption machines with the advent of high energy costs. This is because an absorption machine is an excellent way to utilize waste heat. Examples of potential applications are varied. One example may be on

an exothermic process, or where extensive use of steam turbine with limited turndown results in vented steam especially during warm weather when air conditioning is required. A site where peak electrical demand is set by air-conditioning load and ratcheted demand charges are high may justify absorption refrigeration if adequate steam-generation facilities are already present.

Fig. 19.1.17 Elemental absorption system circuit.

A typical absorption machine requires about 20 lb (9 kg) of steam for 1 ton (3.5 kW) of refrigeration effect at 45°F (7°C).

A more detailed explanation of absorption systems is found in ASHRAE Handbook, 2005 Fundamentals, pp. 1.13 to 1.20.

AIR MACHINES

Compressed-air machines in which air is compressed, cooled, and then adiabatically expanded in an engine while doing work are obsolete for usual industrial refrigeration because of bulk and inefficiency. Their coefficient of performance is less than unity as compared with 4 or 5 for vapor machines. Operation may be closed cycle, in which the air is reused, or open cycle, in which the air is discarded after expansion. Compression ratios of about 3 seem best. In the **dense-air** closed system, the pressure at the compressor intake averages 75 lb/in² gage (560 kPa) and up to 250 lb/in² gage (1,700 kPa) at the outlet. The use of superatmospheric pressure throughout the cycle decreases the size, and the reuse of dried air eliminates the ice formation on engine valves which occurs with open-system operation. Air cycles are used for liquefaction of the so-called permanent gases and to an increasing extent for comfort cooling in high-speed planes where the small lightweight equipment and the safety of the cooling medium are important. High-speed (90,000 to 100,000 r/min) gas turbines are used to extract work from 5 to 100 lb/in² gage compressed air.

THERMOELECTRIC COOLING

Thermoelectric cooling utilizes the **Peltier effect** whereby a temperature differential will occur between two junctions of a closed loop of dissimilar metallic conductors when an electric current is imposed. It is variously called *Peltier effect cooling, thermoelectric refrigeration,* and *electronic cooling.* It is the inverse of the thermocouple operating principle whereby a temperature differential across the junctions of two dissimilar metals produces an electric current.

The capacity of a thermoelectric couple is small. System capacity can be increased by increasing the number of elements used. The temperature produced can be lowered by cascading the systems. Capacities approaching 25 tons (87.9 kW) cooling effect have been achieved, and temperatures as low as $-266°F$ ($-166°C$) have been attained.

Aside from limited practical capacities, the major disadvantage of a thermoelectric cooling system is its poor coefficient of performance. The absence of moving parts, silent operation, ability to function in zero gravity, and lack of pressurized vessels, has given thermoelectric cooling wide space-program application. Other applications include use on submarines, electronic equipment cooling, and small units such as drinking water coolers or units for use in recreational vehicles.

The elements of a thermoelectric cooler are quite simple and are equivalent to those of a thermoelectric generator (see Sec. 9). For development of various output capacities, the number of basic elements is proportionally increased; they all act in parallel, and their outputs add together. **Semiconductor** materials, such as bismuth-telluride-selenide and bismuth-antimony-telluride alloys, are employed.

With an input to the circuit of low-voltage direct current and with heat continuously abstracted at one junction at room temperature, the other junction will become cold. Essentially each of the two junctions becomes an "activity cell," in which at the lower refrigeration temperature level, heat is converted into electrical effect, and at the higher atmospheric temperature, the electrical effect is converted into heat. This establishes the basis of "heat-pump" operation: the lower-temperature heat requirement must be supplied, essentially as in the evaporator of a vapor-compression system, by heat withdrawal via heat transfer, from all connected surroundings. The energy-transport circuit thus parallels the machine-activated circuit of Fig. 19.1.6. The amount of heat for a single-component circuit depends on the circulating electric current, on the properties and dimensions of the two conductors (*p* and *n* types), and on the resultant temperature difference between the high side room temperature and the low side input heat that is divided into two groups, useless and useful. The **useless heat** is the heat conduction between the high and low temperature through the thermoelements plus the current-generated resistance-loss heat within the elements that flows to the colder junction. The **useful heat** is the refrigeration effect in a regular evaporator. All the heat must in turn be discharged via the Peltier effect at the higher room temperature, as in the condenser of a vapor-compression system. Ultimate performance of the circuit depends on the voltage-generating nature of the two dissimilar conductors and on their electrical resistance, thermal conductivity, and physical dimensions. The different properties are mathematically combined into a **figure of merit, Z.**

METHODS OF APPLYING REFRIGERATION

In **direct expansion systems** the evaporator is placed in the space which is to be cooled; in **indirect systems** a secondary fluid (**brine**) is cooled by contact with the evaporator surface, and the cooled brine goes to the space which is to be refrigerated. Brine systems require 40 to 60 percent more surface than do direct expansion; they have an equalizing effect due to the large heat capacity of cold brine, they are safer (particularly if the refrigerating effect must be carried considerable distance or widely distributed), and they permit closer temperature regulation than is possible with direct expansion. If two temperatures are to be held, the lower may be direct expansion, the higher by brine cooling. Development of better controls and newer piping methods has made direct expansion more attractive than previously.

Brines used for industrial refrigeration are usually aqueous solutions of calcium chloride, ethylene glycol, propylene glycol, or undiluted methylene chloride, and silicone-based alkylated fluids. Calcium chloride should not contain over 0.2 percent magnesia, calculated as magnesium chloride. Calcium chloride brines are recommended down to $-45°F$ ($-43°C$). Brines should be chemically neutral: acidic brines attack ferrous materials, alkaline brines attack zinc, ammonia in brine (resulting from leaks in the ammonia system) is especially harmful to most nonferrous metals. Corrosion by brine is increased by the presence of oxygen, air, or carbon dioxide and by galvanic action between dissimilar metals. Corrosion inhibitors are widely used, a satisfactory one being about 1,600 g of sodium chromate or dichromate per 1 m^3 of calcium chloride brine or 3,200 g per 1 m^3 of sodium chloride brine.

The properties of sodium and calcium chloride brines are given in Tables 19.1.8 to 19.1.10.

The density of brine is measured by a **salinometer** (or **salometer**), which is a simple hydrometer the indications on which are 4 times greater than on the corresponding Baumé scale (see Sec. 1).

It is undesirable to use a **strength of solution** of salt greater than is necessitated by its freezing temperature, as the specific heat (Table 19.1.9) decreases as the concentration of the brine increases, and consequently the stronger the brine, the less heat a given amount of it is able to convey between certain definite temperatures and the more power is required to pump the brine. Moreover, brine which is too strong may cause clogging of pipes, etc., by depositing salt. On the other hand, if the solution is too weak it may not be able to withstand the temperature existing in the expansion coil, so that a layer of thin ice will form around the latter and interfere with the absorption of heat from the brine. The surface of the expansion coils in the brine tank should be inspected from time to time to see if any ice has formed on them. In larger plants, it is customary to use a solution with a freezing point not less than 10°F (5.5°C) below the lowest temperature which will be obtained in the operation of the plant. In smaller isolated plants and where careful supervision is not ensured, it is customary to make the solution as strong as possible without being unstable, usually 1.240 to 1.250 sp gr.

Ethylene glycol brine is used in various strengths from 15 to 50 percent by weight for refrigeration temperatures down to $-1°F$ ($-18°C$). It is toxic, and therefore its usage is not recommended in food or beverage processing equipment where a leak can contaminate the product. A corrosion inhibitor is recommended to protect carbon steel equipment and piping. Specific heat ranges from 0.975 to 0.76 Btu/ lb · °F (4,080 to 3,180 J/kg · K) in typical systems. Specific gravity varies from 1.08 to 1.05.

Propylene glycol brine in 15 to 35 percent weight strength is considered nontoxic and is often used in brewing and other food and beverage applications. The brine is more viscous (10 centipoise at $-6.7°C$ for 30 percent weight solution) than ethylene glycol. Specific heat varies from 0.9 to 0.975 Btu/lb · °F. Specific gravity ranges from 1.05 to 1.01. Propylene glycol has often replaced ethylene glycol due to the latter's toxicity.

Methylene chloride brine is often used in systems requiring -20 to $-125°F$ (-30 to $-87°C$) low temperatures. It has low flammability. Hydrolysis and water contamination must be prevented to avoid equipment corrosion. Specific heat varies from 0.269 to 0.273 Btu/lb · °F (1,126 to 1,142 J/kg · K). Specific gravity ranges from 1.4 to 1.5.

Proprietary chemical blends, such as Dowtherm J and Syltherm, are used where the fluid properties are advantageous. Dowtherm J, a mixture of isomers of an alkylated aromatic compound, is usable between the temperatures of 358°F (181°C) and $-120°F$ ($-84°C$). The usual range of service is between 0°F ($-18°C$) and $-70°F$ ($-21°C$). Physical properties vary from 0.412 to 0.390 Btu/lb · °F (0.412 to 0.390 cal/g · °C) for specific heat; 0.894 to 0.926 for specific gravity (ref. 68°F water); 0.079 to 0.084 Btu · ft/h · ft² · °F (0.136 to 0.145 W/m · K) for thermal conductivity; and 1.72 to 4.61 cP in the range of usage.

Sodium chloride brine usage is being reduced due to its corrosivity and its relatively high freezing point.

Table 19.1.11 gives a comparison of the performance of the four major brines in a typical shell and tube chiller. This chart does not indicate effects of specific heat, ingredient cost, pumping power requirements, or system investment cost in the selection of brines.

Brine coolers may be of three types: shell-and-tube, shell-and-coil, and double pipe. The shell-and-tube type is the most widely used, the brine flowing through the tubes which are surrounded by the evaporating refrigerant. Tubes may be arranged for multipass operation. The effective heat-transfer surface varies from 8 to 15 ft² per ton (0.21 to 0.4 m²/kW), varying with temperature and brine velocity. A submerged coil in an open brine tank is used for ice making by the can process.

Table 19.1.8 Properties of Calcium Chloride Solutions
(For variation of specific gravity, with temperature, see Table 19.1.9.)

Parts of CaCl$_2$ by weight in 100 parts of the solution	Specific gravity at 60°F	Deg Baumé	Weight per gal, lb	Weight per cu ft, lb	Freezing point, °F	Specific heat at					
						−4°F	14°F	32°F	50°F	68°F	86°F
6	1.050	7.0	8.76	65.52	28.0	—	—	—	—	—	—
8	1.069	9.33	8.926	66.70	24.2	—	—	0.882	0.887	0.892	0.897
10	1.087	11.57	9.076	67.83	21.4	—	—	0.853	0.858	0.863	0.868
12	1.105	13.78	9.227	68.95	18.2	—	—	0.825	0.831	0.836	0.842
14	1.124	15.96	9.377	70.08	14.4	—	—	0.799	0.805	0.811	0.817
16	1.143	18.12	9.536	71.26	9.9	—	0.768	0.775	0.781	0.787	0.792
18	1.162	20.24	9.703	72.51	4.7	—	0.745	0.752	0.759	0.764	0.769
20	1.182	22.32	9.853	73.63	−1.0	—	0.723	0.731	0.738	0.744	0.749
22	1.202	24.38	10.04	75.0	−7.3	0.695	0.704	0.711	0.718	0.724	0.729
24	1.223	26.41	10.21	76.32	−14.1	0.678	0.686	0.693	0.700	0.706	0.712
26	1.244	28.41	10.38	77.56	−22.0	0.663	0.670	0.677	0.683	0.690	0.696
28	1.265	30.39	10.56	78.94	−32.0	0.649	0.656	0.662	0.669	0.675	0.682
30	1.287	32.34	10.75	80.35	−46.0	0.638	0.643	0.648	0.655	0.661	0.668

Table 19.1.9 Specific Gravities of Brines
(To change to lb/ft³ multiply by 62.43; to change to lb/gal multiply by 8.35.)

Parts by weight of salt in 100 parts of brine	Sodium chloride				Calcium chloride				Magnesium chloride			
	Temperature, °F											
	14	32	50	68	14	32	50	68	14	32	50	68
6	—	1.046	1.044	1.041	—	1.053	1.051	1.049	—	1.053	1.051	1.049
8	—	1.061	1.059	1.056	—	1.071	1.069	1.066	—	1.070	1.069	1.067
10	—	1.077	1.074	1.071	—	1.090	1.087	1.084	—	1.089	1.087	1.084
12	—	1.093	1.090	1.086	—	1.108	1.106	1.103	1.108	1.107	1.105	1.102
14	—	1.108	1.105	1.101	—	1.127	1.124	1.121	1.127	1.126	1.123	1.121
16	1.128	1.124	1.121	1.116	1.150	1.147	1.144	1.140	1.147	1.145	1.142	1.139
18	1.144	1.140	1.136	1.132	1.170	1.167	1.163	1.159	1.166	1.164	1.161	1.158
20	1.161	1.157	1.152	1.148	1.190	1.187	1.183	1.179	1.186	1.183	1.181	1.178
22	1.178	1.173	1.169	1.164	1.211	1.208	1.203	1.199	1.206	1.203	1.201	1.197
24	1.195	1.190	1.185	1.180	1.233	1.229	1.224	1.219	1.226	1.224	1.221	1.218

Table 19.1.10 Weight of Commercial Calcium Chloride in Brine

Sp gr	1.10	1.12	1.14	1.16	1.18	1.20	1.22	1.24	1.26	1.28	1.30	1.32
Wt per gal, lb	1.41	1.70	2.00	2.30	2.59	2.90	3.20	3.50	3.83	4.13	4.46	4.78
Wt per cu ft, lb	10.55	12.72	14.96	17.20	19.37	21.69	23.94	26.18	28.62	30.89	33.36	35.75

NOTE: Specific gravity is at 60°F for both brine and water. The weights are of 73 to 75 percent solid calcium chloride per gal of brine at 60°F. For flake 77 to 80 percent calcium chloride multiply the weights given by 0.94.

Table 19.1.11 Comparative Performance of Brines[†]

Brine	Heat-transfer coefficient, U_o*	Fouling factor		Refrigerant no.	Tube material
		Inner	Outer		
Calcium chloride, 24%	53	0.001	0.002	R717	Carbon steel
Propylene glycol, 40%	26.3	0.0005	0.0	HCFC 22	Copper
Ethylene glycol, 40%	44.7	0.0005	0.0	HCFC 22	Copper
Methylene chloride	111.9	0.0	0.0	HCFC 22	Copper

* Values presented in Btu/h · ft² · °F; to convert to W/m² · K, multiply by 5.678.
† Values calculated using ¾-in diameter tubes; integrally finned; 19 fins per inch; 7 ft/s velocity; brine is inside the tube; brine bulk temperature is 5°F.

Table 19.1.12 Capacity of Multipass Shelf-and-Tube Brine Coolers, Flooded
(Tons refrigeration)

Diameter of shell, in	Length of shell, ft	Velocity of brine through cooler, ft/min							
		200				400			
		Total brine, gal/min	Mean temp diff, brine and ammonia, °F			Total brine, gal/min	Mean temp diff, brine and ammonia, °F		
			7½	12½	17½		5	10	15
26	6	140	7.74	12.9	18.1	290	7.15	14.3	21.2
26	9	140	11.5	19.3	26.0	290	10.7	21.4	32.0
26	12	140	15.5	25.8	36.2	290	14.3	28.6	42.9
34	9	230	18.5	30.7	43.1	440	17.1	34.1	51.1
34	12	230	24.6	41.0	57.4	440	22.6	45.3	68
34	18	230	37.0	62.0	86.2	440	34.1	68.2	101
42	12	510	45.0	74.8	105	970	41.5	83	123
42	18	510	67.4	112	157	970	62.2	124	187

The **double-pipe cooler** is usually of 2-in (50-mm) inner or brine-flow pipe and 3-in (75-mm) outer pipe. The commercial rating is 15 to 20 ft (5 to 6 m) length of coil per ton of refrigeration.

The **shell-and-tube cooler** is used with closed heads and is erected both vertically and horizontally; brine flows through the tubes and ammonia is in the shell. It is made in sizes from 1 to 350 tons with ratings of 8 to 15 ft² effective surface per ton, varying with the temperature and brine velocities; tubes 1 to 2½ in (25 to 63 mm) arranged multipass. This type of cooler has largely displaced all other types in recent installations (Table 19.1.12).

REFRIGERANT PIPING

(See also Sec. 8.)

It is important to the proper operation of a refrigeration system that the piping or mains interconnecting the compressors, condensers, evaporators, and receivers be properly sized. This piping must be considered in three categories, viz., liquid lines, suction lines, and discharge lines. Essentially pipeline sizing is governed by pressure drop, first cost, operating cost, noise, and oil entrainment. Excessive pressure drops penalize compressor efficiencies and may affect control-valve operation adversely. Liquid-line velocities for most refrigerants are in the order of 60 to 400 ft/min (3.3 to 22 m/s), suction lines from 700 to 4,600 ft/min (38 to 250 m/s), and discharge lines from 1,000 to 5,000 ft/min (55 to 275 m/s). Pressure drops vary approximately as the square of the velocity (or tonnage) and directly as the length of the piping. For liquid lines, when evaporator is located above condenser, the following pressure drops in pounds per square inch per foot (kilopascals per metre) of static lift should be allowed: ammonia, 0.26 (5.9); R 12, 0.57 (12.9); R 22, 0.51 (11.5); R 11, 0.64 (14.5). See also Secs. 3 and 4.)

Ammonia Mains For the average installation, Table 19.1.13 shows the maximum tons of refrigeration normally allowed for various sizes of standard pipe, based on 100 ft (30 m) equivalent length of piping (measured length plus allowance for valves and fittings). The gas pressure drops per 100 ft (30 m) equivalent length upon which Table 19.1.13 is based are as follows: suction lines at 5 lb/in² (gage) (34.5 kPa) = 0.25 lb/in² (1.7 kPa); suction lines at 20 lb/in² (gage) (138 kPa) = 0.50 lb/in² (3.4 kPa); suction lines at 45 lb/in² (gage) (310 kPa) = 1 lb/in² (6.9 kPa); and discharge lines = 1 lb/in² (6.9 kPa).

Standard-weight (schedule 40) steel pipe is used for ammonia mains, except for liquid lines 1½ (38 mm) and smaller where extra-strong (schedule 80) pipe is used. Joints may be either screwed, flanged, or welded, but welding is preferred.

R 12 Mains Maximum tons refrigerant for various pressure drops per 100 ft equivalent length of R 12 mains are given in Table 19.1.14. For average conditions, allowable pressure drops in R 12 suction lines are 0.5 to 1.0 lb/in² (3.4 to 6.9 kPa) below 0°F (−18°C) evaporator temperature; 1.0 to 1.5 lb/in² (6.9 to 10.3 kPa) from 0 to 25°F (−18 to −4°C); and 2.0 to 2.5 lb/in² (13.8 to 17.2 kPa) from 25 to 50°F (−4 to 10°C). Discharge-line pressure drops upon which Table 19.1.14 is based are approximately 1 lb/in² (6.89 kPa) per 100 ft (30 m) equivalent length. Pressure drops of 1 to 4 lb/in² (6.9 to 27.6 kPa) are permissible. Liquid-line pressure drops of 3 to 5 lb/in² (20.7 to 34.5 kPa) per 100 ft (30 m) equivalent length are normal, with 10 lb/in² (69 kPa) usually considered maximum. To facilitate oil return in vertical suction lines with the evaporator located below the compressor, the velocity in the vertical section should be 1,000 ft/min (38 m/s) or more. Suction-line capacities in Table 19.1.14 may be increased 9 percent for 50°F (10°C) evaporator temperature and will decrease approximately 7 percent for each 10°F (5.6°C) below 40°F (4.4°C).

R 22 piping will, in general, be smaller for the same tonnage than R 12 piping. Suction lines will handle about one-third more tonnage for the same pressure drop. Liquid and discharge lines can be the same as for the equivalent tonnage of R 12 or slightly smaller.

Piping may be standard-weight steel pipe, but copper tubing is used in most cases. Medium-weight type L copper tubing is normally used for land installations, and the heavier type K copper tubing is used for marine work. Joints in the copper tubing may be flared compression fittings for tubing ¾ in (18 mm) OD and smaller. However, hard solder (silver-base alloy melting above 1000°F) (540°C) is preferred for most tubing connections. Copper tubing is almost always used for liquid lines. Steel pipe may be used for the larger suction and discharge lines but should be sandblasted on the inside and carefully cleaned before installation.

Piping flexibility is an important consideration, especially in low-temperature systems [0°F(−18°C) or colder] and where large-diameter or long runs of piping are utilized. The coefficient of thermal expansion is about $9 \times 10^{-6}/°F$ ($16 \times 10^{-6}/°C$) for carbon steel pipe. This means a 20-ft (6-m) length will shrink about 0.15 in (3.8 mm) when cooled from 70 to 0°F (21 to −18°C). If the piping configuration were such to exert a moment on a compressor flange, premature equipment failure could result. Long runs of cold brine lines should be analyzed for proper anchoring, guiding, and support to prevent system failure.

Flexibility is provided by a combination of changes in direction and use of properly positioned anchors, spring hangers, and guides. Refer to Sec. 5 for additional information on piping flexibility.

Table 19.1.13   Maximum Tons Refrigeration for Ammonia Mains
(100 ft equivalent pipe length)

Pipe size, in	Suction line — Suction pressure, lb/in² gage — 5(−17.2°F)	20(5.5°F)	45(30°F)	Discharge line	Liquid line — Condenser to receiver	Receiver to system
⅜	—	—	—	—	2.5	12.0
½	0.6	1.1	2.0	3.1	6.0	20.0
¾	1.2	2.2	4.1	6.0	14.0	75.0
1	2.2	4.0	7.5	11.4	24.0	137
1¼	4.4	8.0	15.0	22.4	50.0	245
1½	6.4	11.8	21.6	30.9	77.0	400
2	12.1	22.2	42.0	62.0	140	850
2½	19.1	35.5	65.0	97.5	220	1,475
3	31.5	59.0	108	160	375	2,400
3½	46.6	87.5	156	238	540	3,500
4	64.0	118	240	330	740	
5	117	208	385	560	1,320	
6	175	306	600	905	2,030	
8	362	650	1,200	1,810	4,200	
10	640	1,180	2,160	3,200		
12	940	1,850				

SOURCE: From ASRE, Air-Conditioning Refrigerating Data Book, 1955–1956 ed., based on ARI Equipment Standards; see ASHRAE Handbook, 2006 Refrigeration, Chapter 3, for more complete tables.

Table 19.1.14   Maximum Tons Refrigeration for R 12 Lines

Line sizes, in	Suction-line (based on 105°F condensing temp) pressure drop, lb/in² per 100 ft equivalent length at 40°F saturation						Discharge-line condensing temp.		Liquid-line (based on 105°F condensing temp; 25°F evaporating temp) pressure drop, lb/in² per 100 ft equivalent length (type L tubing)			
	½	1	2	3	4	5	115°F	90°F	3	5	10	20
⅜ OD	—	—	—	—	—	—	—	—	0.88	1.14	1.80	2.58
½ OD	0.14	0.20	0.28	0.35	0.41	0.45	—	—	2.89	3.64	5.56	8.50
⅜ IPS	0.17	0.24	0.34	0.42	0.49	0.54						
⅝ OD	0.25	0.35	0.51	0.62	0.73	0.81	1.43	1.15	4.86	6.81	10.2	15.8
½ IPS	0.35	0.45	0.65	0.79	0.93	1.03	1.87	1.50	4.86	6.81	10.2	15.8
⅞ OD	0.55	0.76	1.10	1.34	1.58	1.75	2.97	2.38	10.5	14.1	21.8	33.0
¾ IPS	0.68	0.94	1.35	1.65	1.92	2.12	3.26	2.62	9.73	12.6	18.5	27.0
1⅛ OD	1.26	1.80	2.57	3.17	3.76	4.15	5.05	4.05	21.4	28.2	41.3	60.8
1 IPS	1.43	2.01	2.89	3.54	4.17	4.60	5.29	4.25	21.4	28.2	41.3	60.8
1⅜ OD	2.21	3.12	4.45	5.50	6.38	7.05	7.72	6.19	36.9	48.1	70.5	101
1¼ OD	2.70	3.82	5.37	6.72	7.68	8.48	9.16	7.35	36.9	48.1	70.5	101
1⅝ OD	3.40	4.78	6.79	8.42	9.77	10.8	10.92	8.75	62.0	80.2	114	160
1½ IPS	4.05	5.75	8.10	10.12	11.6	12.8	12.5	10.0	62.0	80.2	114	160
2⅛ OD	6.12	8.60	12.1	15.1	17.4	19.2	19.2	15.3				
2 IPS	7.66	10.9	15.3	19.2	22.2	24.5	20.6	16.5	124	161	231	328
2⅝ OD	12.0	17.1	24.0	30.1	34.6	38.2	32.2	25.9				
2½ IPS	12.0	17.1	24.0	30.1	34.6	38.2	32.2	25.9	230	297	426	607
3⅛ OD	19.1	27.2	38.2	47.8	55.0	60.7	51.5	39.8				
3 IPS	20.9	29.4	42.3	51.8	60.0	66.2	54.5	43.8	364	469	676	972
3⅝ OD	27.8	39.7	55.7	69.8	80.3	88.7	72.0	57.6				
3½ IPS	30.2	43.2	61.0	76.1	87.0	96.0	78.8	62.3	539	704	1,005	1,430
4⅛ OD	38.6	55.2	78.0	97.3	111	123	95.8	77.1				
4 IPS	40.7	58.6	83.0	103	118	130	101.6	81.6	753	972	1,385	1,945
5 IPS	71.3	100	141	176	203	224	171.5	137.8				
6 IPS	126	183	257	322	366	403	266	214				
8 IPS	211	297	422	523	602	664	461	370				
10 IPS	352	503	712	887	1,024	1,130	725	582				
12 IPS	550	780	1,106	1,373	1,582	1,748	1,041	836				

SOURCE: From ASRE, Air-Conditioning Refrigerating Data Book, 1955–1956 ed., based on ARI Equipment Standards.

Mains of Other Refrigerants   For sizing piping for other refrigerants, reference may be made to the ASHRAE Handbook, 2006, Refrigeration, Chapters 2 and 3.

COLD STORAGE

The first consideration in cold-storage room design is the temperature to be maintained in the refrigerated space. The temperature can generally be divided into three ranges: above 32°F (0°C), 32°F (0°C) and below, and 0°F (−18°C) and below. Methods of construction and degree of insulation are dependent on the maintained temperature. Other more specialized design considerations are humidity control, ventilation requirements, controlled atmosphere (control of O_2, N_2, or CO_2), fire protection, and material-handling facilities (see Table 19.1.15).

Cold-storage rooms can either be site fabricated or prefabricated. The increasing availability of preinsulated panels with low U factors has led to increasing use of prefabricated rooms. Many of these are available as complete packages with the refrigeration system included.

Vapor Barrier   Although good insulation is important, a well-designed and installed vapor barrier is the key to maintaining satisfactory insulation performance and life. A poor vapor barrier allows moisture to condense in the insulation, greatly reducing its insulating value. A well-designed system is of little value if poorly installed. Poor workmanship, in fact, accounts for most vapor barrier inadequacies.

Common vapor barrier materials include metal foils, polyethylene, asphalted felts and kraft paper, hot-melt asphalt, and special vapor barrier paints. Vapor barriers must be installed on the warm side of the wall.

Table 19.1.15   Cold-Storage Usage and Transmission Factors

	Room volume, ft³									
	20	50	100	500	1,000	2,000	5,000	10,000	50,000	100,000
Usage load*										
Average	4.68	2.28	1.61	1.21	1.10	0.835	0.403	0.240	0.178	0.173
Heavy	5.51	3.55	2.52	1.87	1.67	1.29	0.625	0.408	0.305	0.295
Storage temp, °F	40 and above		25–40		15–25		0–15		−25–0	
Insulation thickness, in	3		4		5		6		8	
Theoretical U factor†	0.086		0.066		0.055		0.046		0.035	
Practical U factor‡	0.111		0.083		0.066		0.055		0.046	
Piping U factor§	2.5		2.2		2.0		1.8		1.6	

* Btu per cu ft per 24 h per °F temp diff between outside and inside. Usage load includes 10°F or less of product cooling, and infiltration losses, etc., but does not include any freezing of product or fan motor loads when unit coolers are used. Values are for normal conditions and should not be used for unusual product loads.
† Theoretical U factor is based on corkboard ($k = 0.30$) with plaster on both sides, Btu per ft² per h per °F temp diff.
‡ Practical U factor allows for inefficiency of joints and structural supports in insulation.
§ Piping U factor is Btu/ft² outside pipe surface per h per °F temp diff between cooling medium and air (gravity circulation). Allowance has been made for frosting at the lower temperatures.

Table 19.1.16   Product Storage Data

| Product | Quick-freeze temp, °F | Storage temp., °F | | Humidity, %R.H. | Specific heat | | Latent heat | Freezing point | Respiration, Btu per lb per day |
		Long	Short		Above freezing	Below freezing			
Apples	−15	30–32	38–42	85–88	0.86	0.45	121.0	28.4	0.75
Asparagus	−30	32	40	85–90	0.95	0.44	134.0	29.8	
Bacon, fresh	—	0–5	36–40	80	0.55	0.31	30.0	25.0	
Bananas	—	56–72	56–72	85–95	0.80	0.42	108.0	28.0	4.18
Beans, green	—	32–34	40–45	85–90	0.92	0.47	128	29.7	3.3
Beans, dried	—	36–40	50–60	70	0.30	0.237	18		
Beef, fresh, fat	−15	30–32	38–42	84	0.60	0.35	79		
Beef, fresh, lean	−15	30–32	38–42	85	0.77	0.40	100		
Beets, topped	—	32–35	45–50	95–98	0.86	0.47	129	31.1	2.0
Blackberries	−15	31–32	42–45	80–85	0.89	0.46	125	28.9	
Broccoli	—	32–35	40–45	90–95	0.92	0.47	130	29.2	
Butter	+15	—	40–45	—	0.64	0.34	15	15.0	
Cabbage	−30	32	45	90–95	0.93	0.47	130	31.2	
Carrots, topped	−30	32	40–45	95–98	0.87	0.45	120	29.6	1.73
Cauliflower	—	32	40–45	85–90	0.93	0.47	132	30.1	
Celery	−30	31–32	45–50	90–95	0.92	0.48	135	29.7	2.27
Cheese	+15	32–38	39–45	—	0.70				
Cherries	—	31–32	40	80–85	0.87	0.45	120	26.0	6.6
Chocolate coatings	—	45–50	—	—	0.3				
Corn, green	—	31–32	45	85–90	0.80	0.43	108	29.0	4.1
Cranberries	—	36–40	40–45	85–90	0.90	0.46	124	27.3	
Cream	—	34	40–45	—	0.88	0.37	84		
Cucumbers	—	45–50	45–50	80–85	0.97	0.47	137	30.5	
Dates, cured	—	28	55–60	50–60	0.83	0.44	104		
Eggs, fresh	−10	30–31	38–45	—	0.76	0.40	98	31.0	
Eggplants	—	45–50	46–50	85–90	0.94	0.47	132	30.4	
Flowers	—	35–40		85–90					
Fish, fresh, iced	−15	25	25–30	—	0.82	0.41	105	30.0	
Fish, dried	—	30–40		60–70	0.56	0.34	65		
Furs	—	32–34	40–42	40–60					
Furs, to shock	—	15	15						
Grapefruit	—	32	32	85–90	0.91	0.46	126	28.4	0.5
Grapes	—	30–32	35–40	80–85	0.86	0.44	116	27.0	0.5
Ham, fresh	—	28	36–40	80	0.68	0.38	87		
Honey	—	31–32	45–50	—	0.35	0.26	26		
Ice cream	−20	—	0–10	—	0.5–0.8	0.45	96		
Lard	—	32–34	40–45	80	0.52	0.31	90		
Lemons	—	55–58		80–85	0.92	0.46	127	28.1	0.4
Lettuce	—	32	45	90–95	0.96	0.48	136	31.2	8.0
Liver, fresh	—	32–34	36–38	83	0.72	0.42	94		
Lobster, boiled	—	25	36–40	—	0.81	0.42	105		
Maple syrup	—	31–32	45	—	0.24	0.215	7.0		
Meat, brined	—	31–32	40–45	—	0.75	0.36	75.0		
Melons	—	34–40	40–45	75–85	0.92	0.35	115	28.5	1.0
Milk	—	34–36	40–45	—	0.92	0.46	124	31.0	
Mushrooms	—	32–35	55–60	80–85	0.93	0.47	130	30.2	
Mutton	—	32–34	34–42	82	0.81	0.39	96	29.0	
Nut meats	—	32–50	35–40	65–75	0.30	0.24	14	20.0	
Oleomargarine	—	34–36	—	—	0.65	0.34	35	15.0	
Onions	—	32	50–60	70–75	0.91	0.46	120	30.1	1.0
Oranges	—	32–34	50	85–90	0.90	0.46	124	27.9	0.7
Oysters	—	—	32–35	—	0.85	0.45	120.0		
Parsnips	−30	32–34	34–40	90–95	0.82	0.45	120.0	28.9	
Peaches, fresh	—	31–32	50	85–90	0.90	0.46	126	29.4	1.0
Pears, fresh	—	29–31	40	85–90	0.86	0.45	118	28.0	6.6
Peas, green	—	32	40–45	85–90	0.80	0.42	108	30.0	
Peas, dried	—	35–40	50–60	—	0.28	0.22	14		
Peppers	—	32	40–45	85–90	0.94	0.47	—	30.1	2.35
Pineapples, ripe	—	40–45	50	85–90	0.88	0.45	122	29.9	
Plums	—	31–32	40–45	80–85	0.88	0.45	123	28.0	

Table 19.1.16 Product Storage Data (Continued)

Product	Quick-freeze temp, °F	Storage temp., °F Long	Storage temp., °F Short	Humidity, % R.H.	Specific heat Above freezing	Specific heat Below freezing	Latent heat	Freezing point	Respiration, Btu per lb per day
Pork, fresh	—	30	36–40	85	0.60	0.38	66	28.0	
Potatoes, white	−30	36–50	45–60	85–90	0.77	0.44	105	28.9	0.85
Poultry, dressed	−10	28–30	29–32	—	0.80	0.41	99	27	
Pumpkins	—	50–55	55–60	70–75	0.92	0.47	130	30.2	
Quinces	—	31–32	40–45	80–85	0.90	—	—	28.1	
Raspberries	—	31–32	40–45	80–85	0.89	0.46	125	30.0	3.3
Sardines, canned	—	—	35–40	—	0.76	0.410	101		
Sausage, fresh	—	31–36	36–40	80	0.89				
Sauerkraut	—	33–36	36–38	85	0.91	0.47	128	26	
Squash	—	50–55	55–60	70–75	0.92	0.47	130	29.3	
Spinach	—	32	45–50	85	0.94	0.48	132	30.8	
Strawberries	−15	31–32	42–45	80–85	0.92	0.48	129	30.0	3.3
Tomatoes, ripe	—	40–50	55–70	85–90	0.95	0.48	135	30.4	0.5
Turnips	—	32	40–45	95–98	0.93	0.40	137	30.5	1.0
Veal	−15	28–30	36–40	—	0.71	0.39	91	29	

Insulation   In recent years newer insulating materials have largely supplanted the use of plaster and corkboard in cold-storage room construction. Insulation has generally been used in three forms for this application: board form, panelized, and poured, sprayed, or foamed in place.

Board form insulation includes corkboard, rigid polystyrene, polyisocyanurate, and polyurethane, foamed, and fibrous glass.

Panelized insulation has gained acceptance, particularly in prefabricated installations. The outer metal skin can serve as a vapor barrier, and the panels are factory insulated with polystyrene, polyisocyanurate, polyurethane, or fibrous glass. U values as low as 0.035 have been attained in 4-in (100-mm) thick panels.

Poured, sprayed, or foamed in place insulation has its greatest applications on existing installations. It is primarily a urethane material. The application requires special equipment, and precautionary measures must be taken.

Cold Storage Temperatures   A great deal of research and experience are required to obtain authoritative information on optimum temperatures and humidities for various products in cold storage. Table 19.1.16 is representative of good practice. The safe storage period depends upon the product and the storage temperature, and operational techniques vary greatly. Modern cold-storage warehouses of the larger concerns are cooled by brine which is furnished at two different temperatures only. The higher temperature for the mild-temperature warehouses is 10 to 12°F (−12 to −11°C), and the low-temperature brine for the freezers is −10 to −12°F (−23 to −24°C). All temperatures above that of the brine are obtained by regulating the amount of brine circulated in any particular set of coils. In the low-temperature warehouses, the piping is arranged for two classes of service: (1) **sharp freezers,** where the goods which are to be frozen are kept in a blast of air whose temperature is −30°F (−34°C) while their temperature is brought down quickly to the holding temperature (say in from 6 to 10 h) after which they are stored in (2) **holding rooms** where the desired temperature is maintained.

The system of cooling which is now being installed in the highest type of warehouses consists of a coil room containing the necessary brine-type or wetted, direct-expansion-type coils, through which the air from the different rooms is circulated by a pressure blower. The inlet and outlet of each room are so arranged that the cooled circulating air will cover the entire room in its transit; this is usually accomplished by having the cold-air inlet in the center of the room and two cold air outlets—one at each end of the room. The piping ratio for the coil rooms in this system, assuming a high-grade insulation with 2 to 4 Btu transmission per ft² per 24 h per °F temp diff, should be 1 ft² of external pipe surface to 15 ft³ of space to be cooled [with brine at −10 to −12°F (−23 to −24°C)] for warehouses carrying temperatures of zero and below, and

1 ft² of pipe surface to 24 ft³ of space to be cooled [with brine at 10 to 12°F (−12 to −11°C)] in warehouses carrying mild temperatures of from 30 to 40° F (−1 to 4°C). Another system is used in which the blower and coils are supplemented by coils in the rooms. With this arrangement, it is possible to reduce temperatures quickly and hold them with the coils.

The average coil transmission in cold-storage rooms without forced circulation of the air is about 2 Btu per ft² of outside metal surface per h per °F temp diff with horizontal piping and 2.5 Btu with vertical piping. When forced air circulation is used, the transmission rate will increase to 20 Btu or more. In **brine circulation,** the brine, at same compressor back pressure, has a higher temperature than the ammonia, and consequently 1½ times as much pipe or more is used in brine circulation as is direct expansion for a given back pressure.

Piping of Rooms   The **size of pipe** usually employed for piping rooms varies from 1 to 2 in (25 to 50 mm) with either brine circulation or direct expansion.

The extra cost of liberal piping allowance will often be offset by the consequent improvement in the efficiency of operation of the compressor. An expansion valve should be provided for every 500-ft (150 m) length of 1-in (25 mm) pipe, every 650 ft (200 m) of 1¼-in (30 mm) pipe, and every 1,000 ft (300 m) of 2-in (50 mm) pipe when direct expansion is used.

Values of **overall coefficients of heat transfer** in Btu/h · ft² · °F for refrigerating practice are given in Tables 19.1.15 and 19.1.17. (See Sec. 4.)

Brine circulation is generally preferred to direct expansion so as to avoid danger from escaping ammonia or other refrigerant in case the pipes should leak. An advantage of the brine system is that there is always a considerable mass of refrigerated brine which can be drawn on in case the machinery should have to be stopped for any reason. In small plants, the general machinery may be stopped at night and only the brine pump be kept going to distribute the surplus refrigeration which has been accumulated in the brine during the day. Brine piping must consist of two lines, a flow and a return, usually of the same size. Brine storage is seldom used in large plants because of its bulk, its first cost, and the practical inability to store much refrigeration.

The **brine coils** in each cooled space are in parallel across the supply and return pipes. It has been common practice to allow 100 to 120 running ft (30 to 36 m) of **pipe** per circuit for low temperatures, and 400 to 440 (120 to 135 m) for high temperatures. The **tons refrigeration produced by brine** at various temperature differences and rates of pumping may be calculated approximately as follows: tons refrigeration = gal/min × °F range/28.

Air Conditioning of Low-Temperature Storage Rooms   Bacterial growth and chemical decomposition are retarded by lowering the

Table 19.1.17   Overall Coefficients of Heat Transfer
(Btu/ft^2 · h · °F)

Can ice-making piping		Brine coolers	
Old-style feed, non-flooded	12–15	Shell and tube	45–100
Flooded	20–40	Double pipe	150–300
High-velocity raceway trunk coils	80–110	Cooling coils	
Ammonia condensers		Boiling refrigerant to air in unit coolers	4–8
Submerged (obsolete)	30–40	Water to air in unit coolers	5–9
Atmospheric, gas entering at top	60–65	Brine to unagitated air	2.2–2.8
Atmospheric, drip or bleeder	125–200	Direct expansion to unagitated air*	1.6–2.5
Flooded	125–150	Water cooler, shell and coil	15–25
Shell and tube	150–300	Water cooler, shell and tube	50–150
Double pipe	150–250	Water cooler, shell and finned tube (Freon)	30–150
Baudelot coolers, counterflow,		Liquid-ammonia cooler, shell and coil	
atmospheric type		accumulator	45
Milk coolers	75	Air dehydrator	
Cream coolers	60	Shell and coil (brine in coil) { 1st coil	5.0
Oil coolers	10	{ 2nd coil	3.0
Water coolers		Double pipe	6–7
Direct expansion	60–150	Superheat remover, shell and tube	15–25
Flooded	60–200		

* U factor increases to 3.3 at 15°F temp diff with ammonia recirculation systems.
Note: Forced circulation of the air increases the coefficient to 1½ to 2½ times the values for still air. One inch of frost decreases the value 25 percent.

temperature of perishable goods. Too low a temperature will freeze the goods and may result in spoilage. A holding temperature above the freezing point of the article is best for storage conditions. The maintenance of a proper relative **humidity** in a storage space holding unwrapped goods is as important as the maintenance of the proper temperature. High humidity favors the growth of mold and bacteria. Low humidities rob the product of moisture resulting in losses in value through impaired appearance and lost weight. **Air motion** is of importance in maintaining uniform conditions throughout the storage space. Too high air velocities cause excessive drying, and stagnant air through high humidities will cause mold. An optimum air motion falling between stagnation on one side and excessive drying on the other must be selected.

Low-temperature conditioning for storage purposes is preferably accomplished by the use of **cold-diffuser methods.** The cold diffuser connected to or located in the storage space consists of a fan and cooling means. The cooling means may be either a brine or cold-water spray or a surface-type cooler using brine, or a volatile refrigerant such as ammonia, Freon, etc. Unitary equipment, performing the functions of the cold diffuser, is commercially available. For the requirements for various perishable products, see ASHRAE Handbooks, 1993 Fundamentals and 1995 Applications, 1994 Refrigeration, and Table 19.1.16.

Quick freezing is freezing in 2 h or less; as employed for foods the temperatures may be as low as −20 to −50°F (−29 to −46°C). The methods of freezing include direct immersion in cold brine or a brine spray with the commodity held in a metal container. Another procedure is the use of a freezing tunnel approximately 50 ft (15 m) long, equipped with a stainless-steel conveyor belt. The commodities to be frozen pass through this tunnel in 15 to 30 min. The speed of the belt is variable, and the heat transfer is accomplished by high-velocity air circulation at a number of points in the tunnel, the circulation being transverse to the belt. The circulating air is cooled by brine sprays or wetted, direct-expansion coils which maintain the required temperature and the high humidity necessary to prevent shrinkage. In another process, packages of the material to be quick-frozen are clamped between steel platens containing evaporating refrigerant. Successful results are obtained with temperatures as high as 0°F (−18°C). Storage may be at −5 to −10°F (−20 to −23°C) and transportation at 10°F (−12°C) or lower.

Cold-storage lockers for individual family use are used in large numbers, particularly in suburban and rural areas. Lockers are rented usually on a yearly basis. A locker plant may have several hundred rental steel lockers, 6-ft^3 (0.17-m^3) capacity each, placed in rooms at about 0°F (−18°C); a room for cutting and wrapping the meat; and a sharp freezer room held at about −20°F (−29°C). Two compressors are frequently specified or a single compressor using automatic control valves on the separate rooms. Ammonia and R 12, operating under direct expansion, are commonly used.

ICE MAKING

Ice refrigeration is economical for some applications. The use of natural ice is fast disappearing. Manufactured ice is made by several methods. In the can system, galvanized cans containing up to 400 lb (180 kg) of water are immersed in brine for 25 to 40 h for freezing. Air dissolved in the water separates as bubbles causing opacity unless the water is agitated during the early stages of cooling, and any dissolved impurities come out of solution to form a central off-color core in the ice block unless the core is removed and replaced with fresh water just before freezing. The time and investment required for making ice in cans are often uneconomical. In the **Flakice** method (Crosby Field, *Trans. ASME,* May 1951, pp. 347–357) refrigerant is sprayed against the inner wall of a flexible, slowly rotating cylinder which is partially immersed in a tank of water. As the cylinder rotates, a thin layer of ice forms on the outside and then is broken off as further rotation causes the cylinder surface to flex under the action of an internal cam. For a patent history of small ice machines of this type, refer to Crosby Field, *Refrig. Eng.,* **58,** Dec., 1950, p. 1163. In the **PakIce** process (Taylor, *Refrig. Eng.,* **22,** 1931, p. 307) liquid ammonia is evaporated in an annular space formed by two concentric cylinders, the inner being corrugated. Water sprayed on the inner corrugated surface quickly freezes in a thin layer about .25 mm thick and is continually removed by mechanical scrapers. In an **extrusion method** (Watt, *Trans. ASME,* 1949) a slightly tapered circular or rectangular vertical cylinder with large end up is open at the top, but closed by a ram at the bottom. This cylinder has a jacket in which refrigerant is evaporated directly against the cylinder walls. Water enters the cylinder and freezes to a tapered block at the start of the operation. As soon as this occurs, the ram lifts (about 6 mm), shearing the ice block from the cold cylinder walls. Water admitted to this space quickly freezes, the ram again operates and the block of ice is again lifted, and shearing occurs again between the freshly formed ice and the cylinder walls, but not at the fresh ice and old ice interface. In this way, a continuous cylinder of ice is formed by the series of nested ice shells. The installation and upkeep costs are small and so is the space required per ton of the ice produced; a pilot-plant 1-ton (907 kg) unit requires 3 ft^2 (0.3 m^2) of floor space.

Batch ice makers can manufacture up to 75 tons (68 tonnes) of ice per day. These machines use constant-diameter ice-forming tubes to produce ice in shell and toroidal forms. The ice forms on a bank of vertical tubes in either of two configurations: inside the tubes for pellets or outside of the tubes for shell ice. A water distribution system is provided at the top. An ice conveyor and mechanical size-reducing device is mounted below the tubes. The refrigerant is circulated through the annulus of concentric tubes when ice pellets are made, and inside of single tubes when shell ice is made. The refrigerant supply can be designed for flooded operation; however, equipment capacity is reduced. The cycle starts as water passes through the distributors and flows over the surface of the tubes as a film. The refrigerant freezes the water to form a thickening layer of ice as time passes. Unfrozen water drains out of the tube into a sump. Then the water is recirculated, by a pump, up to the distribution system at the top of the tube bank. A purge of water is taken from the sump to remove accumulated salts and other impurities. When harvest is initiated, the refrigerant system is drained, and hot gas from the condenser is introduced into the refrigerant side of the tube and the pressure is raised. This causes the ice to partially melt and release from the tube wall. The cylinder or shell of ice falls by gravity into the conveying and size-reducing machinery. The size of an ice pellet can vary from 0.75 in (19 mm) to 1.375 in (35 mm) outside diameter. Ice clarity can be controlled by flushing the ice with water before and during harvest. The machine is controlled by digital logic microprocessors. Energy consumption when using 65°F (18°C) feedwater varies from 1.94 to 2.33 tons (6.8 to 8.2 kW) refrigeration per ton (909 kg) of ice when using R-717 versus HCFC-22 in a recirculation system. Standard model ice makers can be operated with the refrigerant system either in the flooded condition or with circulating pumped refrigerant in the tubes. Similar units lose 18 percent capacity when operating in a flooded state compared to operating with pumped recirculating refrigerant.

Making clear ice using continuous methods is often accomplished by washing the newly frozen ice surface with cold water. This method can increase water requirements by up to 50 percent and energy usage by 15 percent.

The refrigeration tonnage required for making 1 ton (909 kg) of ice varies from 1.84 with 55°F (12.8°C) inlet water temperature to about 2.0 if the water to be frozen enters as high as 75°F (23.9°C).

SKATING RINKS

Rinks for ice skating, hockey, or curling vary in size from about 400 ft² (37 m²) area to 16,000 ft² (1,486 m²) or more. Construction of the ice floor is important. For a low-cost, efficient unit, brine pipes are laid directly in sand, which is kept wet during freezing to increase heat transfer. As freezing progresses, a layer of ice is built up by spraying water on the surface. The expense of maintaining such a floor in a satisfactory level condition may be considerable, and corrosion of the pipes, from the outside in, in contact with the wet sand may be excessive, particularly during shutdown periods. Moreover, it is not possible to use such a floor for other purposes. When a floor for diversified activities is needed, it is usual to provide a lower layer of insulation, primarily to prevent sweating on ceilings under the floor, surmounted by a concrete slab in which the steel pipes carrying the brine are imbedded and on which the ice layer is formed. This construction protects the pipe from external corrosion, affords fast freezing, and makes possible use of the floor in a few hours for purposes where an ice surface is not desired. Floor pipes can be mild steel or polyethylene and are usually 1 or 1¼ in (25 or 32 mm) spaced 3 to 6 in (75 to 150 mm) on centers; care must be exercised to ensure a uniform distribution of brine. Brine flow may be 10 gal/min per ton of refrigeration, corresponding to about 1 gal/min per 10 to 20 ft² of piping surface. In a few installations, cold brine is sprayed directly against the under side of a steel floor on which the ice layer is frozen. Close control of ice surface temperature is necessary for good skating, usually −3°C ± 0.5. Refrigeration capacity should be from 0.4 to 0.85 ton per 100 ft² (150 to 320 W per m²) of floor surface, although the load may be much higher for open-air rinks or other unusual conditions.

REFRIGERATION IN THE CHEMICAL INDUSTRIES

The use of refrigeration in petroleum and other industries is widespread. In petroleum processing, refrigeration is used (1) to control vapor pressure of highly volatile constituents (methane, ethane, propane, and butane),

such control being necessary during distillation, during processing, and for recovery of gasoline fractions from natural gas; (2) to shift the solubility relationship so that undesired constituents such as asphalt and wax in lubricating oils may be removed by precipitation; (3) to produce selective chemical reaction such as occurs when sulphuric acid is used to remove gum-forming constituents from light fuels or when an alkylate fuel fraction is formed by combining a low molecular weight unsaturated with a similar saturated hydrocarbon. Alkylate is a valuable and important constituent of aviation fuels.

For (1), temperature ranging from −35°F (−38°C) or lower to as high as 60°F (15°C) are required, the refrigerant being ammonia, propane, or butane. For the higher temperatures spray-cooled water or jet refrigerating units (see above) are sometimes satisfactory. For (2), it is possible to use propane as a combined solvent and refrigerant, evaporation of the propane causing the required cooling, temperature level [approximately −40°F (−40°C)] being controlled by the pressure held on the equipment. The propane vapor is recompressed, condensed, and reused as in any compression cycle. As much as ⅓ ton of refrigeration per barrel of lubricating oil dewaxed is needed. For (3), temperatures of up to 60°F (−16 to 15°C) are general, and ammonia is a satisfactory and widely used refrigerant.

The chemical industry uses refrigeration to control condensation polymerization processes in the manufacture of synthetic rubbers and plastics. Syntheses of refrigerants and other chemicals require systems to maintain temperatures which range from 5°C down to −85°C in process equipment. Large centralized refrigeration systems which use ethylene glycol and methylene chloride brine are common. Several modern chemical processes use the eventual product as a refrigerant (for example, the manufacture of ammonia). Cascade systems with segregated circuits are used to produce temperatures below −50°C.

Recently imposed regulations limit volatile organic vapor emissions to the atmosphere from storage tank vents and other process systems. The recovery efficiency required of cleanup systems has led to development of refrigerated liquid absorption processes to scrub vaporous effluents from organic chemical storage tanks. Recently installed systems have required liquid absorbent temperatures of −40°F (−40°C). The capacity of these systems varies from 2 tons (7 kW) to 500 tons (1760 kW). The absorbent liquid is often accumulated from the effluent stream over a period of time.

DEEP REFRIGERATION

Dry ice (solid CO_2) is useful as a refrigerating medium in special cases. The main objection to the general use of dry ice is its cost, but the ease in its handling, its low temperature [−110°F (−79°C) at atmospheric pressure], its noncorrosive and nontoxic properties, its high latent heat, and the absence of liquid drip make it desirable. The heat absorbed per pound of solid CO_2 during sublimation at atmospheric pressure and −110°F (−70°C) is 245 Btu approx. The specific heat of the gas at constant pressure is about 0.2.

Carbon dioxide is obtained commercially either by fermentation or by burning. The gas is compressed, usually in three stages, and cooled to atmospheric temperature, thereby forming a liquid at about 1,000 psia (690 kPa); the liquid is then throttled to below 5.3 atm pressure. During throttling part of the CO_2 solidifies and is compressed to form a dense ice. The fraction that returns to the vapor phase is recompressed.

Production of Solid Carbon Dioxide Figure 19.1.18 shows a simple cycle and Fig. 19.1.19 the elemental pressure-enthalpy diagram for the production of solid carbon dioxide. From the high pressure at atmospheric temperature the liquid is throttled (isenthalpic) to atmospheric pressure, the resultant point being within the sublimation solid-vapor zone below the triple-point realm. In this resultant throttling expansion, part of the CO_2 solidifies; it is then removed in the separator from the residual vapor and is compacted to form a dense CO_2 ice. The fraction representing the vapor phase is augmented by fresh makeup gas to continue the production cycle.

Beyond the manufacture of dry ice, operation at deep refrigerations, such as air liquefaction [at −312°F (−191°C) approx], lead into

Fig. 19.1.18 Elemental dry-ice production circuit.

Fig. 19.1.19 Pressure-enthalpy diagram illustrating dry-ice production circuit.

Fig. 19.1.20 Elemental Linde system circuit.

Fig. 19.1.21 Elemental Claude system circuit.

cryogenics (see Sec. 19). Illustrative systems are (1) Linde (throttling) and (2) Claude (expander). In the elemental **Linde system** (Fig. 19.1.20), by virtue of the "cold-effect" regenerator, the process air is reduced in temperature by the return fraction of the air from the separator. The throttling process thus ends in the "wet zone" at atmospheric pressure with generation of low-temperature liquid air that flows from the separator. In the **Claude system** (Fig. 19.1.21), use is made of an expander (whose work output reduces the net work requirement of the compressor) and of an expansion valve with the Joule-Thomson effect.

19.2 CRYOGENICS
by F. J. Edeskuty

REFERENCES: Timmerhaus (ed.), "Advances in Cryogenic Engineering" (annual), Plenum Press. Vance and Duke, "Applied Cryogenic Engineering," Wiley. Scott, "Cryogenic Engineering," Met Chem Research. Vance (ed.), "Cryogenic Technology," Wiley. Mendelssohn, "Cryophysics," Interscience. McClintock, "Cryogenics," Reinhold. Scott, Denton, and Nicholls (eds.), "Technology and Uses of Liquid Hydrogen," Pergamon. Hoare, Jackson, and Kurti, "Experimental Cryophysics," Butterworth. White, "Experimental Techniques in Low-temperature Physics," Oxford. Johnson (ed.), "Compendium of Properties of Materials at Low Temperature," WADD Technical Report 60–56, U.S. Dept. of Commerce. Durham, McClintock, and Reed, "Cryogenic Materials Data Handbook," PB 171–809, U.S. Dept. of Commerce. Cook (ed.), "Argon, Helium, and the Rare Gases," Interscience. Perry, Chilton, and Kirkpatrick, "Chemical Engineers' Handbook," McGraw-Hill. Barron, "Cryogenic Systems," McGraw-Hill. Bailey, "Advanced Cryogenics," Plenum Press. Haselden, "Cryogenic Fundamentals," Academic. Williamson and Edeskuty (eds.), "Liquid Cryogens," CRC. Timmerhaus and Flynn, "Cryogenic Process Engineering," Plenum Press. Van Sciver, "Helium Cryogenics," Plenum Press. Edeskuty and Stewart, "Safety in the Handling of Cryogenic Fluids," Plenum Press.

Notation

Some symbols in this section differ from the notation of engineering practice but are retained to facilitate use of the prevalent scientific sources of information.

A = area, ft^2 (m^2)
b.c.c. = body centered cubic lattice structure
C = specific heat, Btu/lbm · °R(J/kg · K)
C_p, C_v = specific heats at constant pressure and volume respectively, Btu/lbm · °R(J/kg · K)
f.c.c. = face centered lattic structure
H = enthalpy, Btu/lbm (J/kg)
H = magnetic field, $G(T)$
h.c.p. = hexagonal close-packed lattice structure
κ = thermal conductivity, Btu/h · ft · °R (W/m · K)
σ = electrical conductivity (reciprocal of resistivity) or pair separation for zero potential in Lenard-Jones equation
s = Boltzmann's constant
L = length, ft (m)
LH$_2$ = liquid hydrogen (H$_2$)
LHe = liquid helium (He)
LN$_2$ = liquid nitrogen (N$_2$)
LNG = liquefied natural gas
LO$_2$ = liquid oxygen (O$_2$)
n = number of moles
P = pressure, psia or torr (vacuum) (Pa)

R_e = electrical resistance, Ω
T = temperature, °R (K)
V = volume, ft³ (m³)
α = thermal linear-expansion coefficient, °R^{-1} (K^{-1}), or accommodation coefficient
η = viscosity, lbm/ft · h (N · s/m²)
ρ = density, lbm/ft³ (kg/m³)
ΔT = temp difference, °R(K)
   * = Denotes property divided by quantum parameter (Table 19.2.5) which is not the critical point value.

Cryogenics is the study, production, and utilization of low temperatures. Cryogenic temperatures have been defined so ambiguously that "upper limits" to the cryogenic range from 216 to 396° R (120 K to 220 K) may be found in the literature. In this section the cryogenic range for a property of a given substance is considered to embrace the scale between absolute zero and the temperature above which the property has the expected or normal behavior. Cryogenics thus embraces the unusual and unexpected variations which appear at low temperatures and make extrapolations of properties from ambient to low temperatures unreliable.

Progressively lower temperatures become increasingly difficult to attain in practice. A cryogenic refrigerator operating on the Carnot cycle (ideal) will require the least amount of input work to effect refrigeration (remove heat) at a lower temperature and reject it at a higher temperature (usually ambient temperature). As the working temperature of a refrigerator is lowered, the work required to transfer a given amount of heat increases as demonstrated by the Carnot limitation, to wit, $W = Q[(T_1 - T_2)/T_2]$, where W is the work required to extract the heat Q at a low temperature T_2 and reject it at a higher temperature T_1 (see Sec. 4). The actual work is always greater than this because of inefficiencies of mechanical equipment, thermal losses associated with finite temperature differences in heat exchangers, and heat leaks from the surroundings to the cold equipment. Cryogenic refrigerators built to date do not approach this limit very closely; typically less than 40 percent of Carnot efficiency for the largest refrigerators and progressively less as the size of the refrigerator becomes smaller.

REFRIGERATION METHODS

(See also Sec. 4 and Sec. 19.1, "Mechanical Refrigeration.")

Cryogenic refrigerators (cryocoolers) may be classified by (1) the functions they perform (e.g., the delivery of liquid cryogens, the separation of mixtures of gases, and the maintenance of spaces at cryogenic temperatures), (2) their refrigerating capacities and (3) the temperatures they reach. Large industrial-sized plants (a) deliver LNG (~120 K), LO₂ (~83 K), LN₂ (~77 K), LH₂ (~20 K), and LHe (~4 K), (b) separate gaseous mixtures, e.g. the constituents of the atmosphere, H₂ from petroleum refinery gases, H₂ and CO from coke oven and coal-water gas reactors, and He from natural gas, and (c) provide refrigeration to maintain spaces at low temperatures. For the latter, i.e., (c), a unit has been manufactured and installed that delivers 13 kW of refrigeration at less than 3.8 K.

An important area of refrigerator development that is being commercially exploited now is laboratory-sized cryogenic refrigerators for laboratory research and development at LHe temperatures (<5 K). These refrigerators are used for many different purposes, refrigerating for example: high-field superconducting electromagnets, LHe bubble chambers for high-energy (nuclear) particle research, experimental superconducting electric generators and motors and superconducting magnets for levitation of railroad trains. Another area of commercial exploitation that is important for the progress of cryogenic physics research is the development of refrigerators for the continuous production of refrigeration at temperatures below 1 K. These include L³He evaporation refrigerators and L³He-L⁴He dilution refrigerators that reach temperatures of 0.4 and 0.003 K, respectively.

Of various methods of refrigeration, the most commonly used to produce temperatures as low as 1 K are (1) the evaporation of a volatile liquid (referred to as the **cascade** method when applied in several successive stages using progressively lower-boiling liquids), (2) Joule-Thomson (isenthalpic) expansion of a compressed gas, and (3) an

adiabatic (isentropic) expansion of a compressed gas in an engine (reciprocating or turbine) or from a bomb or cylinder through a throttling valve (Simon expansion). Using L³He, method (1) is capable of reaching temperatures as low as ~0.4 K. Numerous refrigeration cycles have been devised utilizing various combinations of the above three methods. The final stage of refrigeration in plants that deliver liquid cryogens is generally a Joule-Thomson (isenthalpic) expansion. **Expansion engines** (reciprocating and turbine) are commonly used for producing the refrigeration needed to reach the final stage. Turbine expanders are preferred for the large refrigerators because their efficiencies (70 to 85 percent) are higher than for reciprocating expanders. The efficiencies of the heat exchangers for counterflowing "cold" and "warm" gases are an important factor in the overall refrigerator efficiency. In industrial plants heat exchanger efficiencies reach 98 percent. Their design represents a compromise between the attainment of a high heat-transfer coefficient, on the one hand, and a low resistance to the flow on the other.

Figure 19.2.1 typifies a commonly employed modified Brayton cycle which uses an expansion engine for generating the refrigeration. The theoretical figure of merit (the ratio of the work W_c done by the compressor to the heat Q, transferred to the refrigerant at the low temperature) is for this cycle

$$\frac{W_c}{Q} = \frac{RT_1 \ln (P_1/P_5)}{M(H_4 - H_3)}$$

where M is the molecular weight of the gas whose enthalpy per unit mass is H. The other symbols are defined by reference in Fig. 19.2.1.

Fig. 19.2.1 (a) Schematic of a modified (isothermal compression) Brayton cycle, and (b) the process path. W_c is the energy (work) to drive the compressor C. Q is the heat absorbed by the refrigerator from the refrigerated area R.A., where the working fluid passes from state 3 to state 4. The heat-exchanger (H.E.) processes are isobaric. The dashed line 2–3′ is the ideal isentropic expansion; 2–3 represents the actual expansion in the expander E.

The **Stirling refrigerating engine** and modifications of it are used in a number of commercial makes of laboratory and miniature-sized refrigerators. The Stirling-cycle refrigerator is well suited to (1) the liquefaction of air on a laboratory scale (~7 L/h), (2) the recondensation of evaporated liquid cryogens, and (3) the refrigeration of closed spaces where the refrigeration load is not very large. These laboratory scale refrigerators supply ~1 kW of refrigeration at ~80 K and ~2 kW at 160 K. They are used with liquid air fractionating columns for the production of LN₂ (~7 L/h) and LO₂ (~5 L/h).

The Stirling-cycle refrigerator consists of a piston for compressing isothermally the working fluid (usually He), and a displacer which can operate in the same cylinder with the piston. The displacer and piston are connected to the same electrically driven shaft but displaced in phase by 90°. The displacer pushes compressed gas isochorically from the warm region where it was compressed, through the regenerator into the cold region where the compressed gas is expanded isothermally doing work on the piston and producing refrigeration. The regenerator consists of a porous mass (packed metal wool of high heat capacity) in which a steep temperature gradient is established between the warm region of compression and the cold region of expansion. The displacer returns the gas after expansion to the region for compression through the regenerator in which it is warmed to the temperature of the isothermal compression. The refrigerant (He) is recycled. The regenerators in the larger refrigerators are usually placed around the outside of the engine cylinders, but in small capacity, lower temperature refrigerators they are put inside the displacers.

In Stirling refrigerators that reach temperatures of 17 K and produce 1 W of refrigeration at 25 K, the compressed He is expanded in two or three stages at temperatures intermediate between room temperature and the lowest temperature reached. There is only a single piston for compression. The expansion in stages allows part of the heat that leaks into the coldest region of a single stage engine to be absorbed and removed at a higher (intermediate) temperature where the efficiency for transferring heat is greater.

The **Vuilleumier refrigerating engine** is a modification of the Stirling refrigerator. It *resembles* two single Stirling engines placed back to back. One of the engines operates as a heat engine and the other as a refrigerator. The lower temperature at which the heat engine discharges its waste heat, is also the top temperature at which the refrigerator engine discharges its waste heat. Hence, the Vuilleumier refrigerator operates at elevated, intermediate and low temperature levels which may be 800, 300 (ambient), and 90 K, respectively. Each engine has a cylinder, a displacer, and a regenerator, and the two engines are connected to a common crankshaft but displaced by 90°. Very little external power is needed to operate the two displacers which are operated in sinusoidal motion, displaced 90° in phase. The working fluid (He) is recycled.

The **Gifford-McMahon refrigerator** is another modification of the Stirling-cycle refrigerator. It may be single stage for refrigeration at higher temperatures (~80 K), or multistage for lower temperatures. Each stage has a cylinder, a displacer and a regenerator. The displacer pushes gas (usually He) under a very small head of pressure from the top of the cylinder, where the temperature is ambient, through the regenerator, into the bottom of the cylinder which is the cold region of the refrigerator. Compressed He is supplied from an external source—the refrigerator has no piston for compressing gas. Only a small source of external energy is needed for driving the displacer which has very little work to do in overcoming the forces of friction. The valves that control the admission of compressed He to the top of the cylinder, and the expansion of the cold He from the bottom of the cylinder are externally operated by a drive mechanism. The cycling rate is < 2 Hz. The expansion of the cold He is of the adiabatic-Simon-bomb type. Expanded gas is discharged from the refrigerator.

Some applications of cryogenic refrigerators require very high reliability without maintenance. The fewer the working parts, the more likely is the needed reliability to be achieved. The **pulse-tube** and **thermoacoustic refrigerators** are advantageous in this respect in that they have only one moving part, and that part operates at ambient temperature. Pressure oscillations are introduced into a tube with a closed end and fitted with the proper arrangement of heat exchangers. For the optimum frequency of the pulsation, a temperature gradient is established between the heat exchangers, and this gradient can pump heat from a low temperature to a higher one. A number of different designs have resulted, including single-stage and multistage units, and operating frequencies have ranged from about 1 Hz up to acoustic frequencies. This type of refrigerator is capable of reaching low temperatures down to 28 K with a single-stage unit, 26 K with a two-stage unit, and 11.5 K with a three-stage unit, all with ambient temperature heat sinks. Present research is also focused on the use of an acoustically driven pulse-tube refrigerator that would have no moving parts and no close tolerances of the components, thus making it inexpensive and easy to produce.

Magnetic refrigeration, originally proposed in 1925, depends upon the ability of paramagnetic materials at low temperature (or ferromagnetic materials near their Curie temperatures) to warm up (or expel heat at constant temperature) with the application of a magnetic field. Conversely, these materials will cool (or absorb heat at constant temperature) upon removal or reduction of the magnetic field. With the use of a paramagnetic material magnetized at a higher temperature of about

1.5 K, temperatures of about 0.001 K have been reached by a single demagnetization. Magnetic refrigeration is applicable over a wide range of temperatures, and recent research work is leading to continuous application of this process by mechanisms that promise a considerable increase in efficiency over that of more conventional refrigerators.

³He-⁴He dilution refrigerators for reaching temperatures between 0.01 and 0.3 K and generating continuously as much as 750 ergs/s of refrigeration are commercially available. Refrigeration results from the solution of L³He in L⁴He, the heat of solution being negative. Figure 19.2.2 is a schematic diagram of a dilution refrigerator. ³He vapor from the "still" at ~1 K (the upper operating temperature of the refrigerator) is collected and returned by a pump to the "mixing" chamber where solution takes place. The ³He arrives at the mixing chamber as L³He near the mixing chamber temperature (the lower operating temperature of the refrigerator) after flowing through a heat exchanger counter to the flow

Fig. 19.2.2 Schematic diagram of the mixing chamber and still of a ³He-⁴He dilution refrigerator.

of cold ³He-⁴He solution on its way to the "still." In the electrically heated still, ³He is evaporated from the solution. The evaporated ³He is withdrawn by the collecting pump (diffusion) that returns the ³He to the mixing chamber. At 1 K the vapor pressure of L⁴He is negligible. L⁴He in the still, freed of ³He, returns to the mixing chamber by super-fluid flow, building up in the mixing chamber an osmotic pressure that drives ³He atoms towards the still against the forces of viscous flow.

In the mixing chamber there is a two-phase separation of ³He and ⁴He. A layer of nearly pure L³He of lower density rides on a heavier L⁴He-rich layer containing ~ 6 percent ³He. At temperatures lower than 0.87 K liquid solutions of ³He and ⁴He separate into two liquid phases, one ³He-rich and the other ⁴He-rich, in thermodynamic equilibrium. At 0 K, the ³He-rich phase is 100 percent ³He, whereas the ⁴He-rich phase (in equilibrium with L³He phase) contains ~ 6 percent ³He.

In the mixing chamber, L³He enters the upper ³He-rich phase; solution of ³He in ⁴He takes place at the interphase boundary. The lower (denser) ⁴He-rich phase connects with the still.

Starting at 50 mK (a temperature reached in a ³He-⁴He dilution refrigerator), temperatures in the 2- to 3-mK range are attainable with a **Pomeranchuk refrigerator.** Refrigeration is generated by compression of a mixture of liquid and solid phases of ³He at pressures in excess of 28.9 atm, the minimum P at which these phases can coexist in equilibrium. The entropy of solid ³He exceeds the entropy of L³He, which is contrary to the normal behavior for other substances. This occurs in ³He because of its nuclear magnetic properties. The ³He nucleus is magnetic and at temperatures in the mK range and above, in solid ³He, the nuclear moments are randomly oriented whereas in L^{3}He, at temperatures lower than ~ 0.3 K, the nuclear moments are paired or partially paired, antiferromagnetically. An adiabatic increase of P converts liquid to solid with a reduction in T, whereas an isothermal increase of P results in an absorption of heat. ^{3}He, condensed, is confined in a container with flexible metal walls in the "cold" region of a ^{3}He-^{4}He dilution refrigerator. The ^{3}He container is surrounded with L^{4}He which serves as the pressure transmitting fluid.

GAS LIQUEFACTION

A conventional method of gas liquefaction utilizes a gas compressor, a countercurrent heat exchanger, and a throttling valve through which the gas expands isenthalpically (Joule-Thomson) and cools if it has already been cooled below its inversion temperature. After expansion, the cold gas returns through the heat exchanger, in which it exchanges heat with the countercurrent compressed gas flowing to the throttling valve. It leaves the heat exchanger near the temperature of the entering compressed gas. The temperature decreases at the throttling valve until the condensing temperature is reached, and then liquefaction occurs at a rate determined by the rate of refrigeration.

The rate of liquefaction x in pounds per hour is $x = \{[H$ (expanded *gas* out) $- H$ (compressed *gas* in)$] w - q\}/[H$ (compressed *gas* in) $- H$ (*liquid* following valve)], where the H's are enthalpies in Btu per pound for the final heat exchanger, w is the flow rate in the same units as x, and q is the heat leak in Btu per hour from outside into the heat exchanger. For an ideal heat exchanger the expanded gas leaves at the temperature of the entering compressed gas. In practice, there is a small difference in temperature which represents a small loss of refrigeration. For ideal gases, H is independent of P for a given T, and hence no liquefaction of an ideal gas results. The H's of air, O$_2$, N$_2$, A, CO$_2$, the hydrocarbons, and the normal refrigerants decrease with increasing pressure at ambient T, except at very high P's, and these gases are liquefiable by this kind of isenthalpic (Joule-Thomson) expansion. The H's of H$_2$ and He at ambient T, however, increase with increasing P, even at low P's, and hence an auxiliary mechanism is required for the liquefaction of these gases. In one method, the stream of compressed gas is split in the liquefier, and a part goes to an engine in which it is cooled by an isentropic expansion with the performance of work. This part of the flow, thus cooled, is sent to a heat exchanger, where it flows counter to the other fraction of compressed gas, cooling it to a temperature below the inversion temperature (see Sec. 4). This flow of compressed gas, thus precooled, is run through another and final heat exchanger with a throttling valve at its lower end, with the result that the quantity x is liquefied.

Another method of precooling H$_2$ and He below their inversion temperatures makes use of a boiling liquid cryogen through which the compressed gas flows in a heat exchanger. LN$_2$ is used for precooling H$_2$, and LH$_2$ for He.

APPLICATIONS OF CRYOGENICS

Gases, such as O$_2$, N$_2$, natural gas, H$_2$, and He, are liquefied for transportation because of the gain in density at low pressure. Shipments have been made of LH$_2$ by trailer trucks in quantities of 13×10^3 gal (50 m^3) for thousands of miles and by railway tank cars holding as much as 28×10^3 gal (100 m^3). LHe is regularly shipped in quantities from 25 gal (0.1 m^3) to 10,000 gal (40 m^3). Liquid cryogens are used as liquid refrigerants. The aerospace, steel, and bottled-gas industries are the principal users of liquefied cryogens. Important applications as coolants can be found in vacuum technology, electronics, biology, medicine, and metal forming.

In recent years, large superconducting magnets have been undergoing development for a variety of applications including fusion reactors, energy storage, electric-power transmission, high-energy physics experiments, and magnetically levitated trains. Magnets of Nb-Ti are available with fields up to 80 kG (8 T) with large magnetic field volumes (1 m^3). Higher field magnets [up to 1500 kG (15 T)] of Nb$_3$Sn are available with smaller magnetic field volumes (10^{-4} m^3). The Los

Alamos National Laboratory installed a 30-MJ superconducting magnetic energy-storage system in the Bonneville Power Administration's Tacoma, Washington, substation to provide stabilization of utility long-line power-transmission systems. The response characteristics of the unit have been demonstrated on the utility's power grid, and the unit has been run successfully for extended periods. Two 100-m long superconducting ac transmission cables have been successfully tested at Brookhaven National Laboratory at 4,100 A and the equivalent of 138 kV three-phase. At the other end of the size spectrum, Josephson junction devices are being considered for applications such as radiation detectors, highly precise and accurate electrical measurements, geophysical instrumentation, medical instrumentation, and computer memory cells.

Cryogenics is also becoming increasingly important in the storage and transport of energy. Large quantities of natural gas are routinely transported as a liquid (LNG). Peak shaving in municipal gas systems is accomplished by storage of LNG. If, due to depletion of fossil fuels, hydrogen becomes an important synthetic portable fuel, cryogenic hydrogen will play an important role as it already does in space applications including the shuttle. Another potential application is aircraft fuel. This may be extended to surface transportation applications.

PROPERTIES OF SOLIDS AT LOW TEMPERATURES

(See also Sec. 4.)

Specific heats of solids in general decrease with decreasing T, becoming zero at 0°R (Fig. 19.2.3). The downward approach to $C_p = 0$ at 0° R is interrupted for some substances (principally compounds) by "bumps" on the curve (excess C_p that rises to a maximum and then decreases). Paramagnetic salts undergoing transitions to either a ferromagnetic or an antiferromagnetic state are examples. This excess specific heat of a paramagnetic salt is connected with the effectiveness of the salt for reaching low temperatures by the method of adiabatic demagnetization. There are also transitions in solids from a more orderly to a less orderly arrangement of atoms and molecules in the lattice that give rise to excess C_p. The transition in solid ortho- and normal H$_2$ below 20°R (11 K) is an example. In Fig. 19.2.3, C_p's are plotted for various materials.

Heat is transferred in dielectrics by lattice vibrations, or waves. In good electrical conductors, heat is transferred principally by the conduction electrons, and thermal and electrical conductivities are related by the Wiedemann-Franz law; $\kappa/\sigma T = $ const. This means that the ratio

Fig. 19.2.3   Specific heats of solids at low temperatures.

of the thermal and electrical conductivities at a given T is approximately the same for the good conductors. In the poor electrical conductors, alloys for example, both lattice waves and conduction electrons play important parts in the transfer of heat. The κ's of "pure" dielectrics and "pure" metals rise with increasing T from $\kappa = 0$ at 0°R (proportional to T for metals and to T^3 for dielectrics), reach a maximum, normally between 10 and 100°R (5 and 55 K), and then decrease to a value approximately independent of T (Fig. 19.2.5 and ice in Fig. 19.2.4). The κ's of alloys (Fig. 19.2.4) are an order of magnitude smaller than for "pure" metals and *do not* exhibit the maxima characteristic of the "pure" metals at low T. Lattice disorder introduced by alloying, even in small amounts, and working a metal, even a pure metal, reduces κ. Annealing, in general, raises κ.

Fig. 19.2.4 Thermal conductivity of solids at low temperatures, part 1.

At 0°R, an absolutely pure and perfect single crystal of metal would have zero electrical resistance, $R_e = 0$. **Electrical resistance** arises from the scattering of the conduction electrons as they move through the lattice of metal ions under the influence of an externally applied electric field. Scattering arises for two reasons: (1) the amplitudes of the thermal vibrations of the lattice which increase with T at a rate proportional to $\sqrt{C}$ of the metal, and (2) the imperfections in the regularity of the lattice

Fig. 19.2.5 Thermal conductivity of solids at low temperatures, part 2.

as caused by impurity atoms (solid-solution alloys included), lattice vacancies, dislocations, and grain boundaries. Resistances for "pure" metals increase with T and are roughly proportional to $C \cdot T$, except at very small T's where they are proportional to T^5 for absolutely pure crystals. The resistance due to impurities and imperfections is very

roughly independent of T (Matthiessen's rule). For very pure metals, R_e is approximately constant below 20°R (11 K). Magnetic impurities can give rise to a minimum in resistivity [usually below 36°R (20 K)] called the Kondo effect. Resistivity in $\Omega \cdot$ cm (Fig. 19.2.6) is the R_e of a 1-cm cube.

Fig. 19.2.6 Electric resistivity of solids at low temperatures.

Some metals, including elements (none from column 1 of the Mendeleeff table or the ferromagnetic elements), intermetallic compounds, metal alloys, and a number of oxides exhibit the phenomenon of **superconductivity,** characterized by a zero dc electrical resistance from 0°R up to a transition temperature T_c, at which normal resistance (value extrapolated from higher T) appears. In pure, single crystals the transition range ΔT, from $R_e = 0$ to the normal value, may be as small as a few millidegrees. Until the discovery of the so-called high-temperature superconductors (HTSCs), the higher T_c in a practical superconductor was that of the compound Nb_3Ge at 41.9°R (23.3 K). The search for higher T_c materials continues with T_c's that have already been reported as high as 239°R (133 K). Closed circuits consisting entirely of superconductors can support a persistent, resistanceless current without an external source of voltage. A superconducting circuit therefore maintains constant the value of the total magnetic flux that was enclosed by the circuit at the time it entered the superconducting state. Hence, superconducting materials influence the magnetic fields in their environment. For this reason, their use in the construction of equipment and apparatus must sometimes be avoided when expected temperatures could be below their T_c. Lead brasses and some solders, in particular Pb-Sn alloys, become superconductors.

Superconductivity, in materials designated Type 1 superconductor, is characterized by (1) perfect electric conductivity ($R_e = 0$) and (2) the internal magnetic induction $B = 0$ when H (external) is not zero (the Meissner effect). Persistent currents at the surface of the specimen shield the interior from H (external). H parallel to the surface of a specimen is continuous at the surface and falls off exponentially below the surface. The penetration depth of H for perfect (chemically and mechanically) specimens is of the order of 5×10^{-8} m. The penetration depth is larger for alloys, specimens with lattice imperfections, and all superconducting specimens near their superconducting transition temperatures T_c.

The normal-to-superconducting transition temperature T_c is also a function of H (external) as well as of the current being carried by the superconductor. For a Type I superconductor, the effect of H (external) is expressed by $T_c = T_{c0} [1 - (H_c/H_0)]^{1/2}$ where T_{c0} is the value of T_c at essentially zero H, and H_0 is the value of H_c at 0 K. For every $H < H_c$, a further decrease in T_c occurs when the superconductor is carrying a current, the maximum current density for maintaining the superconducting state being denoted J_c. Thus the superconducting state exists when $T < T_c$, $H < H_c$, and $J < J_c$, and the normal, or resistive, state exists when either $T > T_c$, $H > H_c$, or $J > J_c$. Because T_c decreases with increases in H and J, the practical application of a superconductor requires that it be maintained at a temperature considerably below its T_c.

Tables 19.2.1 and 19.2.2 are only representative. The number of superconductors including compounds and alloys runs into the hundreds. Of the chemical elements, 38 are known to become superconductors. Some that are not superconductors at normal pressures have high-pressure allotropic modifications that are superconductors, e.g., Bi, Si, and Ge at pressures of ~25, 130, and 120×10^8 Pa, respectively. Many of the superconducting alloys have nonsuperconducting constituents, and there are compounds all of whose constituent elements in their pure state are nonsuperconductors (see Table 19.2.2). It is interesting that good conductors Cu, Ag, and Au do not become superconductors to the lowest temperatures at which they have been tested (~ 0.01 K).

Table 19.2.1   Transition Temperature T_0 and Critical Magnetic Fields H_0 of Some Type I Superconductors

Superconductor	T_0, K	H_0, A/m*
Nb	9.25	1.57×10^5
Pb	7.23	6.4×10^4
V	5.31	8.8×10^4
Hg	4.154	3.3×10^4
Sn	3.722	2.4×10^4
In	3.405	2.2×10^4
Al	1.175	8.4×10^3
Mo	0.916	7.2×10^3
Zr	0.53	3.7×10^3
Ti	0.39	8.0×10^3
W	0.0154	9.2×10
Na Bi	2.2	The elements, in pure state at normal
Au_2Bi	1.7	pressure, are not superconductors.
Cu S	1.6	

* To convert to oersteds, divide by 79.57.

Table 19.2.2   Transition Temperatures T_0 and Upper Critical Fields H_{c2} of Some High-Field, Type II Superconductors

Superconductor	T_0, K	H_{c2}, A/m*	T, K, for H_{c2}
$Al_{0.75}Ge_{0.25}Nb_3$	18.5	3.4×10^7	4.2
Nb_3Sn	18.0	1.9×10^7	4.2
$N_{0.93}Nb$	15.9	1.3×10^7	0
GaV_3	14.8	1.9×10^7	0
NbZr	10.8	7.4×10^6	0
$Nb_{0.2}Ti_{0.8}$	7.5	6.4×10^6	4.2
$Ti_{0.6}V_{0.4}$	7.0	8.8×10^6	2

* To convert to oersteds, divide by 79.57.

Type I superconductor is the designation given to those superconductors that exhibit a complete Meissner effect for $0 < H < H_c$; i.e., B = 0 until H_c is reached and penetration of the specimen by B becomes complete. The superconducting elemental metals in the pure and mechanically perfect state are Type I superconductors. Superconducting alloys (intermetallic compounds, solid solutions, and mixed phases) and work-hardened (unannealed) metals are Type II superconductors. For Type II superconductors, the Meissner effect is complete only for $0 < H < H_{c1}$ where H_{c1} is smaller than H_c. For $H_{c1} < H < H_{c2}$, the magnetic field penetrates the body of the specimen in the form of current vortices, called *fluxons*, that generate a fixed quantized amount of magnetic flux. Each fluxon is a cylindrical region of circulating current centered on a resistive core of material in which the superconductivity is

suppressed. Hence the region defined by $H_{c1} < H < H_{c2}$ is called the *mixed state* of a Type II superconductor. The number of fluxons increases with H from zero at H_{c1} to a density at which complete overlap of the cores occurs along with the restoration of the normal state at H_{c2}. Thus H_{c2} is an upper limit to the superconducting state that increases as T decreases.

The alloys commonly used for high-field superconducting magnets (Nb_3Sn, NbTi, and NbZr, see Table 19.2.2) are Type II superconductors. H_c's for Type I superconductors are small, whereas magnet alloy wires have H_{c2} values that are 100 to 1,000 times higher.

The fluxons in the mixed state of a Type II superconductor transporting a current are acted on by a Lorentz force that is proportional to $i \times B$ and acts in a direction perpendicular to **B**, and to **i** (the current intensity). Unless the fluxons are pinned to crystal lattice sites they are propelled across the superconductor under the influence of the Lorentz force. This movement involves the performance of work by the current and results in (1) production of heat within the superconductor and (2) the appearance of electrical resistance to the flow of current and, eventually, the destruction of the superconducting state. Fluxons are pinned by lattice imperfections (chemical and mechanical) and their displacement is resisted until the Lorentz force exceeds a breakaway value, freeing the fluxons to move. Lattice imperfections are introduced in the wires for high-field superconducting magnets to pin fluxons as well as to increase H_{c2}. Currents avoid the normal core of a pinned fluxon and thus retain their resistanceless character.

A current exceeding the critical value destroys the superconducting state. This critical current is determined by the critical Lorentz force (the product of current and magnetic field) at which pinning forces are exceeded. Hence, the critical value of an applied field depends on the current transported by the superconductor and vice versa. In general, the larger the applied field, the smaller is the critical current and vice versa. For superconducting magnets, it is essential that the values of the limiting magnetic field and the limiting current be simultaneously large.

For the Types I and II superconductors discussed above, practical application requires operation in the temperature region of 12 K or lower. Thus the lower limit (ideal refrigeration) for the work to heat removal ratio is seen to be 24 for operation between ambient temperature and 12 K. Raising the operating temperature of a superconducting system to 20, 40, or 77 K (the temperature of liquid nitrogen boiling at normal atmospheric pressure) reduces this ratio to 14, 6.5, and 2.9 respectively. Although the actual refrigerator may take 2.5 or more times this ratio, the saving in input energy that can be effected by raising the operating temperature is obvious. The equipment for producing refrigeration at these higher temperatures is also less complicated and less expensive than that needed for producing refrigeration at the lower temperatures required for the operation of the low-temperature superconductors.

In 1986 a true breakthrough was achieved in producing superconductors able to enter the superconducting state at much higher temperatures. The high-temperature superconductors (HTSCs) are oxide compounds consisting of several elements. They are more like ceramics and have a complicated crystal structure consisting of layers of Cu-O separated by other oxides. All of them exhibit highly anisotropic superconducting properties and their performance is greatly influenced by the oxygen content of the compound. If the crystals are not properly oriented so that there is too great an angle between the crystal axes of adjacent crystals (e.g., greater than about 5°), intergranular supercurrent flow is supressed and the HTSCs will have sections of superconductor connected by rather poorly conducting junctions. This has been referred to as "having weak links." Therefore, it is desirable to prepare the HTSC material in such a way that it is biaxially textured, in order to minimize weak links. Also, the ability of these superconductors to conduct useful quantities of current can be severely limited by an imposed magnetic field at temperatures near the high end of their superconducting range. The magnetic field enters the superconductor in fluxons, or vortices. When these vortices move through the superconductor, electrical resistance is generated and the superconductor behaves much like a normal metal. Therefore it is necessary to "pin" the vortices so they cannot move. As with low-temperature superconductors, vortices are pinned by lattice imperfections, either chemical or mechanical.

At temperatures below 30 K, the extraordinarily high H_{c2} of the HTCSs allows them to carry high current densities at much higher magnetic fields than is possible with the conventional, low-temperature superconductors. At higher temperatures a new phenomenon has been observed, the "melting" of the vortex lattice that limits the ability of crystal imperfections to pin vortices at magnetic fields far below H_{c2}. The HTSCs are fabricated as thin films and multifilamentary wires. The necessity for optimum chemistry of the constituents, proper crystal orientation, and the fact that the ceramic-like nature of the superconductor makes it brittle complicate both the manufacture of the wires in useful lengths and the subsequent fabrication of equipment such as coils for motors and generators. There are two classes of compounds that have received the most attention in these research efforts. One class is referred to as BSCCO because it consists of the oxides of bismuth, strontium, calcium, and copper in varying ratios. The other class is referred to as YBCO because it consists of oxides of yttrium, barium, and copper.

A great deal of research is continuing in the attempt to improve the commercial usefulness of the HTSCs. This research has made enormous strides in finding production methods for practical superconductors and ways to produce them more economically. There are already fabrication methods that enable the production of kilometre lengths of superconductors capable of carrying very high currents (about 10^5 A/cm²) in self-field at 77 K. These efforts should make possible the large-scale utilization of HTSCs. Table 19.2.3 lists several of the HTSCs, along with the pertinent properties.

Table 19.2.3   Properties and Status of some HTSCs

HTSC*	T_c, K	Comments
$La_{2-x}Ba_xCuO_4$	35	First HTSC discovered
$YBa_2Cu_3O_7$ (YBCO or 1–2–3)	95	First HTSC with T_c above normal boiling point of liquid nitrogen, $H_{c2} > 30$ T at 77 K
$HgBa_2Ca_2Cu_3O_{8+\delta}$	133	Highest T_c to date
$Bi_2Sr_2Ca_2Cu_3O_8$ (2223 BSCCO)	110	$H_{c2} = 30$ T at 77 K, $J_c > 8 \times 10^4$ A/cm² at 77 K and $H_{ext} = 0$; industrial pilot-plant-scale production
$Tl_2Ba_2CaCu_2O_8$ (2212 TBCCO)	~115	$J_c = 7 \times 10^6$ A/cm² at 75 K and $H_{ext} = 0$

* These HTSCs are representative of classes of superconductors.

The discovery in 2001 of superconductivity in the relatively common intermetallic compound, magnesium diboride (MgB_2), with a transition temperature of about 40 K, opens new possibilities to the application of low-temperature superconductors (LTSCs). MgB_2 is easier than the HTSCs to produce, handle, and fabricate into wires, and MgB_2 reduces the energy requirement for refrigeration by about an order of magnitude compared to that of the previously known LTSCs. MgB_2 can be used for the manufacture of superconducting magnets, and its use should make more feasible the utilization of superconducting electric power transmission lines.

The **coefficient of linear expansion** α is $(1/L)(dL/dT)$. Thermal expansion of asymmetric crystals differs along different crystal axes. For an isotropic solid, the volume or cubical-expansion coefficient $(1/V)(dV/dT) = 3\alpha$. Expansion coefficients do not vary appreciably in the temperature region near ambient temperatures, but they all approach zero at 0°R, and the approach to zero is tangential to the T axis ($d\alpha/dT = 0$ at 0°R.) Some expansion coefficients are negative at low temperatures, e.g., stainless steel, some Invars, and fused quartz. The expansivities of some crystals are negative in some directions even though their volume coefficients are positive. Cold-working may produce differences in expansivity in different directions. Annealing restores isotropy. Figure 19.2.7 shows the relative change in length, the integral of α from T to 528°R (293 K).

Because the cold interiors of cryogenic equipment must at times be warmed to ambient temperature, provisions have to be made for those changes in dimensions of the interior that results from large changes in T. Even for equipment of ordinary size, these changes can be too large for accommodation within the elastic limits of the materials of construction or of the structure. In such situations, flexibility has to be designed into

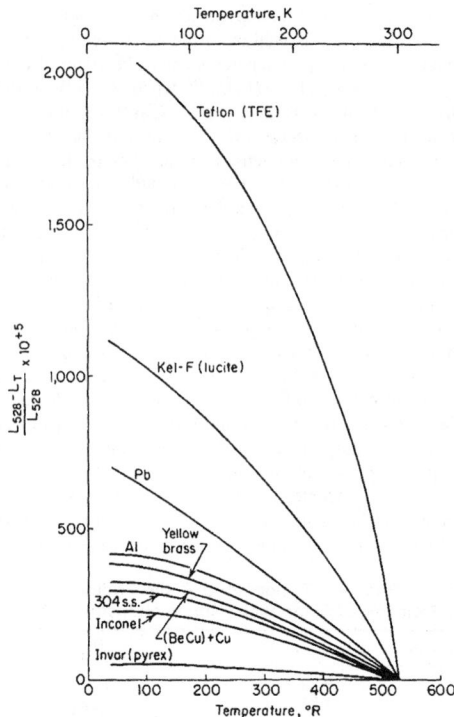

Fig. 19.2.7   Relative change in length from T to 528°R (293 K), equal to $\int_t^{528°R} \alpha \, dt$. Bracketed materials are within ~5 percent of the curve shown.

the structure. For example, bellows and U bends are commonly employed to obtain this flexibility in insulated transfer lines for liquid cryogens.

The **mechanical properties** important for the design and construction of cryogenic equipment are the same as for other temperatures (see also Sec. 5). Frequently, the choice of materials for the construction of cryogenic equipment will depend upon other considerations besides mechanical strength, e.g., lightness (density or weight), thermal conductivity (heat transfer along structural support members), and thermal expansivity (change of dimensions when cycling between ambient and low temperatures). Frequently, mechanical properties at low temperature are significantly different from properties at ambient temperature. This makes room-temperature data unreliable for engineering use at low temperatures.

It is not possible to make generalizations that would not have numerous exceptions about the temperature variations of the mechanical properties. In this discussion, Figs. 19.2.8 to 19.2.10 are only illustrative guides. There is no substitute for test data on a truly representative specimen when designing for the limit of effectiveness of a cryogenic material or structure. Just as the mechanical properties at ambient temperatures are dependent upon the impurities (metallic and nonmetallic), their chemical nature and concentration, the thermal history of the specimen, the amount and kind of working (microstructure and dislocations of the lattice), and the rate of loading and type of stress (uni-, bi-, and triaxial), so also are the changes, quantitative and qualitative, in the mechanical properties when changing the temperature from ambient to low.

Metals, nonmetallic solids (glass), plastics, and elastomers are discussed in order. The metals are classed by their lattice (crystal) symmetry.

The **f.c.c. metals** and their alloys are most commonly used for the construction of cryogenic equipment. Al, Cu, Ni, their alloys, and the austenitic stainless steels of the 18–8 type (300 series) are f.c.c. They do not exhibit an impact (or a notched-tensile) ductile-to-brittle transition at low temperatures. As a general rule, which has some exceptions, the mechanical properties of these metals improve as T is reduced: (1) Young's modulus at 40°R (22 K) is 5 to 20 percent larger than at 530°R (294 K), (2) the yield strength at 40°R (22 K) is considerably greater than the strength at 530°R (294 K). (Cu is an exception; see Fig. 19.2.9), and (3) the fatigue properties at low T are improved. There is a large

Fig. 19.2.8 Ultimate tensile strength of solids at low temperatures.

Fig. 19.2.9 Yield strength of solids at low temperatures.

difference between the yield strength and the ultimate tensile strength of these metals and alloys, especially when they have been annealed. Pb (f.c.c.) and In (face centered tetragonal) are used for low-T deformable gaskets because of their creep properties. Low-T creep data are meager, but the rate of creep decreases with T.

Beta brass (f.c.c.) is ductile down to 7°R, though it is like all alloys of Cu in being less ductile than Cu itself. Thin brass is sometimes porous, and this limits its usefulness for high-vacuum enclosures at low temperatures. Free-machining brass normally contains Pb, which even in low concentrations can render the brass superconducting at LHe temperatures. In the superconducting state, it may then affect the magnetic field in its vicinity (see electrical properties, above).

The b.c.c. **metals and alloys** are normally classed as undesirable. These include Fe, the martensitic steels (low carbon and the 400 series of stainless steels), Mo, and Nb. If not brittle at room temperature, they have a ductile-to-brittle transition at low T. Working can induce the austenite-to-martensite transition in some steels. AISI 301 austenitic stainless is an example. It remains moderately "tough" at very low T and is a valuable material for construction, but cold-drawing, in excess of 70 percent strain, induces a partial transformation of structure and reduces the elongation and notched-strength ratio to nearly zero. This same type of reduction in toughness is observed in the "heat-treatable" stainless steels. Improved tensile properties can be obtained by

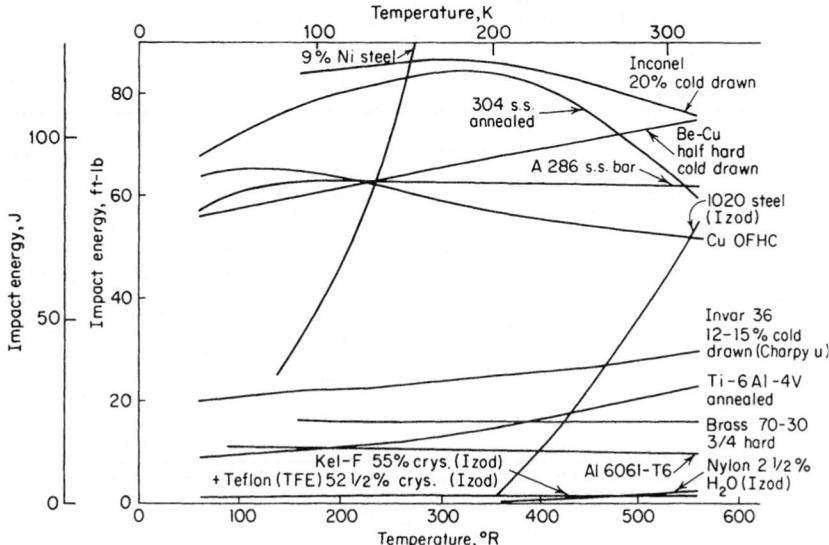

Fig. 19.2.10 Impact energy for solids at low temperatures (Charpy V unless noted). For Kel-F, nylon, and Teflon, the impact energy units are foot-pound per inch of notch width.

cold-working and by heat-treating. These alloys usually have decreased toughness at low temperatures. Alloys of V, Nb and Ta, although not f.c.c., behave well as regards brittleness at low temperatures. These alloys have the advantage of being suitable for use at high T. Carbon steels, though brittle, have found special uses at low temperatures, e.g., in the construction of the expansion engines of the Collins He liquefier and cryostat.

The **h.c.p. metals** exhibit properties intermediate between those of the f.c.c. and b.c.c. metals; e.g., Zn undergoes a brittle-to-ductile transition, whereas Zr and pure Ti do not. Ti and some Ti alloys, having a h.c.p. structure, remain moderately ductile at low T and are excellent for many applications. They have high strength-to-weight and strength-to-thermal-conductivity ratios. The low-T properties of Ti and its alloys are extremely sensitive to even small amounts of O, N, H, and C.

Brittle materials, when in very thin sheets, can have a high degree of flexibility, and this makes them useful for some cryogenic applications. **Flexibility** cannot be used as a criterion for ductility.

Ordinarily, normal welding practices are observed in the construction of cryogenic equipment. These practices, however, are modified in accordance with available knowledge of the performance of the metals at low T.

Nonmetal materials for construction are in many cases brittle, or they are susceptible to brittle fracture. The strength of **glass,** measured at a constant rate of loading, increases on going to low T. Failure occurs at a lower stress when the glass surface contains cracks and abrasions. The strength of glass can be improved by tempering the surface, i.e., by putting the surface under compression.

The **plastics** increase in strength as T is decreased, but this is accompanied by a rapid decrease in elongation in a tensile test and a decrease in impact resistance. Teflon and the glass-reinforced plastics (e.g., glass-reinforced epoxy resin) retain appreciable impact resistance as T is lowered. Teflon, which is polytetrafluoroethylene, can be deformed plastically at T's as low as 7°R (3.9 K). The amount is considerably less than at room temperature, but it is enough to make Teflon very useful for some cryogenic applications. The glass-reinforced epoxies, besides having appreciable impact resistance at low T's, also have high strength-to-weight and strength-to-thermal-conductivity ratios.

All the **elastomers** become brittle at low T. However, elastomers like natural rubber, nitrile rubber, Viton A, and plastics such as Mylar and nylon that become brittle at low T can be used for static seal gaskets when *highly compressed* at room temperature, prior to cooling.

PROPERTIES OF CRYOGENIC FLUIDS (CRYOGENS)

Table 19.2.4 indicates the **temperature ranges** accessible with liquid cryogen baths. There are two inaccessible ranges: (1) between helium and hydrogen [9.36 to 24.9°R (5.2 to 13.8 K)] and (2) between neon and oxygen [80 to 97.9°R (44.4 to 54.4 K)]. The limiting temperatures are set by the critical and triple points. Pumping on a cryogen bath to lower the pressure results in lower temperatures ultimately reaching the triple point; except for helium, further pumping leads to solidification and in practice poor heat-transfer. Raising the bath pressure results in higher boiling temperatures with the limit set by the critical temperature, above which liquid and vapor phases cannot coexist. Cryogen baths are most frequently operated near atmospheric pressure.

The algebraic sign of the Joule-Thomson coefficient determines whether a gas cools or warms upon free expansion. Figure 19.2.11, based on the law of corresponding states, allows rough calculations of inversion temperatures and pressures. Free (Joule-Thomson) expansion inside the "cooling" region results in cooling, i.e., refrigeration; outside this region, heating results. Upper inversion temperatures at 14.7 psia (0.101 MPa) are given in Table 19.2.4.

Fig. 19.2.11   Reduced inversion temperature versus reduced pressure.

Volumetric latent heats (Table 19.2.4) decrease with the lower-boiling cryogens and emphasize the importance of the insulation in storing and handling LH_2 and LHe.

Two forms of **hydrogen** and **deuterium** exist: **ortho** and **para.** The ortho and para molecules differ in the relative orientation of the nuclear spins of the two atoms composing the diatomic molecule. In the ortho form of H_2 the nuclear spins of the two atoms in the molecule are parallel (in the same direction), whereas in the para form the spins are antiparallel. The relative orientations for ortho and para D_2 differ from those for H_2. The thermodynamic equilibrium composition of ortho and para varieties is temperature dependent as shown in Fig. 19.2.12.

In liquid hydrogen, the uncatalyzed ortho-para reaction is second order (proportional to the square of the o-H_2 concentration) with a rate

Table 19.2.4   Common Cryogen Properties

Cryogen	Boiling point,* °R	Triple-point Temp., °R	Triple-point Pressure, lb/in² abs	Critical temp., °R	Critical pressure, lb/in² abs	Upper inversion temp.,* °R	Heat of vaporization* Btu/lbm	Heat of vaporization* Btu/ft³	Liquid density,† lbm/ft³	Vapor density,† lbm/ft³	Gas density,‡ lbm/ft³
He³	5.7			6	17	72	3.6	13.4	3.72	1.50	0.0084
He⁴	7.6			9.36	33.2	92	8.8	68.6	7.80	1.04	0.0111
H₂ (equilib.)	36.5	24.9	1.02	59.4	187.7	368	192	853	4.42	0.0837	0.00561
D₂ (normal)	42.6	33.7	2.49	69.0	240		131	1,343	10.25	0.14	0.0112
Ne	48.7	44.2	6.26	80.0	395.3	486	37.1	2,790	75.35	0.58	0.056
N₂	139.2	113.7	1.86	227.27	492.3	1,115	85.9	4,294	50.4	0.28	0.078
Air	(141.8)						88.22	4,813	54.56	0.28	0.0807
A	157.1	150.8	9.97	271.3	709.8	1,300	70.2	6,135	87.4	0.36	0.111
F₂	151	(96.4)§		259.3	808		74.1	6,965	94	0.50	0.106
O₂	162.3	97.9	0.022	278.59	736.3	1,605	91.5	6,538	71.24	0.28	0.089
CH₄	201.1	163.2	1.69	343.27	673		219.2	5,804	26.5	0.115	0.0448

NOTE: To convert °R to K, multiply °R by 0.556; to convert lb/in² to MPa, multiply by 0.00689; to convert Btu/lbm to J/g, multiply by 2.326; to convert Btu/ft³ to J/m³, multiply by 3.73 × 10⁴; to convert lbm/ft³ to kg/m³, multiply by 16.01.
* 14.7 lb/in² abs.
† At normal boiling point.
‡ At 14.7 lb/in² abs and 492°R.
§ Melting point, 14.7 lb/in² abs.

constant of 0.0114/h. Thus, in the course of uncatalyzed liquefaction, normal LH$_2$ (75 percent ortho) is produced. Since the heat of conversion of o-H$_2$ to p-H$_2$ at the normal boiling point is 302 Btu/lbm (702 J/g) (greater than the heat of vaporization; see Table 19.2.4), long-term storage of normal LH$_2$ is impractical. Therefore, in the commercial production of LH$_2$ a catalyst is used to produce LH$_2$ with more than 95 percent para which for practical purposes is equivalent to the equilibrium composition (99.8 percent para).

Fig. 19.2.12 Thermodynamic equilibrium composition of ortho and para varieties of H$_2$ and D$_2$; percent ortho = 100 − percent para.

For pressures of no more than 150 lb/in^2 (1.03 MPa) and temperatures at least twice the critical temperature, the ideal gas law ($PV = nRT$) enables one to calculate PVT data with sufficient accuracy for many engineering purposes. For cases in which the ideal gas law is not adequate or in which experimental data are not available, the procedures outlined in Sec. 4 may be used. The theoretical prediction of PVT data for liquids is considerably more difficult than for gases. Consequently no universal equation analogous to the perfect gas law exists for extensive ranges of P and T.

Reduced quantum mechanical correlations of saturated liquid densities, viscosities, and thermal conductivities for several cryogens are shown in Figs. 19.2.13 to 19.2.15. Dashed lines represent areas void of experimental data.

Fig. 19.2.13 Reduced density versus reduced temperature along the saturation curve for cryogenic liquids. Values of T^* are obtained by dividing the desired temperature (°R) by the value of T/T^* given in Table 19.2.5. Reduced density ρ^* must be multiplied by ρ/ρ^* given in Table 19.2.5 to obtain density in lbm/ft^3. Multiply lbm/ft^3 by 16.01 to obtain kg/m^3 and °R by 0.556 to obtain K.

Fig. 19.2.14 Reduced viscosity versus reduced temperature along the saturation curve for cryogenic liquids. Values of T^* are obtained by dividing the desired temperature (°R) by the value of T/T^* given in Table 19.2.5. Reduced viscosity η^* must be multiplied by η/η^* given in Table 19.2.5 to obtain viscosity in lbm/ft · h. Multiply lbm/ft · h by 0.000413 to obtain Pa · s.

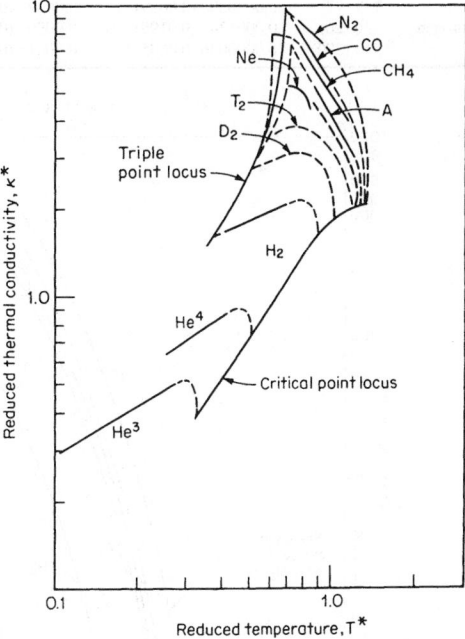

Fig. 19.2.15 Reduced thermal conductivity versus reduced temperature along the saturation curve for cryogenic liquids. Values of T^* are obtained by dividing the desired temperature (°R) by the value of T/T^* given in Table 19.2.5. Reuced thermal conductivity κ^* must be multiplied by κ/κ^* given in Table 19.2.5 to obtain thermal conductivity in Btu/h · ft · °R. Multiply by 1.73 to obtain W/m · K.

Utilization of Figs. 19.2.13 to 19.2.15 requires knowledge of the constants ε and σ for the Lennard-Jones function for the intermolecular potential energy $\varphi(r)$ of a pair of molecules separated by a distance r; so that $\varphi(r) = 4\varepsilon[(\sigma/r)^{12} - (\sigma/r)^6]$. Here ε is the depth of the minimum in the potential energy of a pair of molecules, and σ is the separation of the pair at which $\varphi = 0$. To facilitate use of Figs. 19.2.13 to 19.2.15, all required constants and conversion factors have been combined and are listed in Table 19.2.5.

Table 19.2.5 Multiplication Factors for Use with Figs. 19.2.13 to 19.2.15

Substance	T/T^*, °R	ρ/ρ^*, lbm/ft³	η/η^*, lb/ft-hr	κ/κ^*, Btu/(h · ft · °R)
Helium 3	18.4	18.71	0.0311	0.0204
Helium 4	18.4	24.84	0.0359	0.0178
Para-hydrogen	66.1	8.06	0.0360	0.0363
Deuterium	63.4	16.22	0.0500	0.0257
Tritium	62.1	24.37	0.0607	0.0210
Neon	64.1	100.66	0.1300	0.0128
Nitrogen	171.1	57.39	0.1382	0.0098
Argon	215.6	104.84	0.2183	0.0109
Krypton	297.4	177.1		0.0080
Xenon	397.8	197.4		

Thermal-expansion coefficient $(1/V)(\partial V/\partial T)_P$ and **compressibility** $(1/V)(\partial V/\partial P)_T$ data for the cryogenic liquids are obtained as derivatives of PVT data. Liquids hydrogen and helium have large thermal-expansion and compressibility coefficients (see Table 19.2.6).

Table 19.2.6 Compressibility and Thermal-Expansion Coefficients. Coefficients for Liquid Nitrogen, Hydrogen, and Helium at 14.7 lb/in² abs (0.101 MPa)

Cryogen	T, °R (K)	Compressibility, $(1/V)(\partial V/\partial P)_T$, 1/(lb/in² abs) (1/MPa)		Thermal expansion, $(1/V)(\partial V/\partial T)_p$, 1/°R (1/K)	
Helium 4	7 (4)	0.0028	(0.4)	0.085	(0.153)
Nitrogen	139 (77)	0.000020	(0.0029)	0.0032	(0.0058)
Para-hydrogen	36 (20)	0.000134	(0.0195)	0.0090	(0.0162)
Water	540 (300)	0.0000036	(0.000526)	0.00015	(0.00027)

Fig. 19.2.16 Viscosity of selected gases at low temperatures and 14.7 lb/in² abs (0.1013 MPa).

Fig. 19.2.17 Thermal conductivity of gases at low temperatures and 14.7 lb/in² abs (0.1013 MPa).

Fig. 19.2.18 Vapor pressures of the common cryogens.

For gases, kinetic theory predicts an increase of **viscosity** with rising temperature as shown in Fig. 19.2.16 ($\eta \sim \sqrt{T}$). Normally, the **thermal conductivities** of liquids decrease with increasing temperature, whereas the thermal conductivities of gases increase. Fig. 19.2.17 presents thermal conductivity for several gases as a function of temperature at 14.7 psia (0.1013 MPa).

Figure 19.2.18 is a graph of **vapor pressures** of the common cryogens.

Figure 19.2.19 presents **specific heats** at constant pressure for several of the cryogenic liquids, while Fig. 19.2.20 presents the specific heats for liquid hydrogen and helium along the saturation curve. Specific heats of gases are given in Fig. 19.2.21.

Fig. 19.2.19 Specific heat at constant pressure (C_p) for liquid cryogens at saturation pressure. Asterisk means C_p evaluated along an isobar at 14.7 lb/in² abs (0.1013 MPa) instead of saturation pressure.

Fig. 19.2.20 Specific heats at saturation C_s along the saturation curve versus temperature for ⁴He and para hydrogen.

Fig. 19.2.21 Specific heat C_p of cryogen gases at 14.7 lb/in² abs (0.1013 MPa) versus temperature.

INSTRUMENTATION

(See also Secs. 15 and 16.)

The usual instrumentation problems of engineering practice are further complicated by low-temperature problems which require special calibration procedures.

Pressures are measured with normally used apparatus, e.g., bourdon gages, transducers, and manometers. If the measuring device is located outside the cryogenic environment and insulated from it by a thermal barrier, no special problems are introduced. Occasionally response time requirements necessitate the placing of the transducer in the cryogen space. Then the problems of temperature compensation and calibration must be considered.

Level measurements currently utilize several methods. The differential-pressure method can in principle be used with all cryogens. With cryogens of low densities and low heats of vaporization (e.g., H_2 and He) pressure oscillation occurs in the liquid leg. In the case of LH_2 these can be eliminated by the proper design of the liquid leg and the introduction of He gas which does not condense.

Direct weighing is also used to determine "levels," although difficulties are encountered owing to the large tare-to-cryogen weight ratios (as large as 10 to 1 for a full container in the case of H_2) and extraneous loadings introduced by permanently connected piping, frost, and wind. Weighing systems have nevertheless been successfully used on 50,000-gal (190 m³) LH_2 dewars with an accuracy in level changes of 250 gal (1 m³).

With the **capacitance gage,** the sensing element consists of a concentric tube capacitor which can be calibrated to give liquid-level measurements accurate to a few tenths of a percent. An inexpensive, accurate **point-level sensor** is easily fabricated from a carbon resistor, typically ¹⁄₁₀ W, 1,500 Ω. Its use depends upon the fact that its resistance is a strong function of temperature (Fig. 19.2.22). A small current (50 to 80 mA) is passed through the resistor to heat it. Since heat from the resistor is dissipated more readily in the liquid than in the gas, the temperature of the resistor

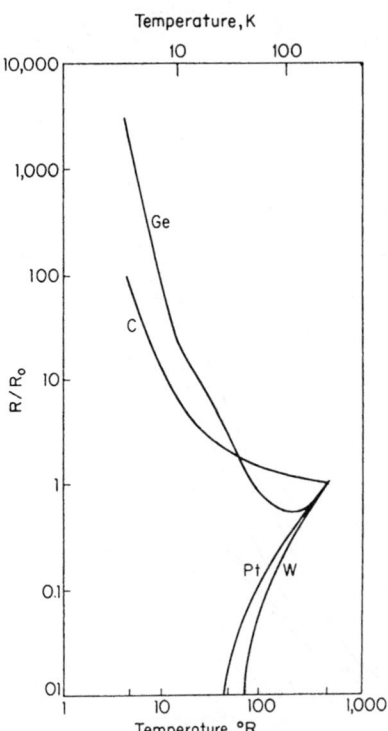

Fig. 19.2.22 Ratio of resistance R to the resistance R_0 at 492°R (273 K) versus temperature for several resistance thermometers. (Values for germanium and carbon are representative only.)

undergoes a step increase when the resistor environment changes from liquid to gas, whereas the resistance decreases stepwise.

Flow measurements utilize orifices, venturis, and turbine-type meters. Various devices for direct mass-flow measurement exist, but none are widely used. "Quality" meters have been built to determine the percentage liquid in two-phase flow. The determination of quality is complicated by nonequilibrium of temperature in the gas phase.

Temperature measurements are usually made with thermocouples, resistance thermometers, and vapor-pressure thermometers.

The **vapor-pressure thermometer** depends upon a vapor-pressure-temperature relationship (see Fig. 19.2.18). In general, a small cavity is filled with a gas which has a condensation temperature in the vicinity of the temperature to be measured. If sufficient gas is present, the pressure will be that of the liquid vapor pressure at the coldest part of the measuring system. Maximum speed of response is obtained if the total quantity of gas is minimized to permit only a thin film of liquid to form. In the case of H_2, a catalyst (e.g., iron hydroxide) to promote ortho-para conversion should be included in the bulb for an accurate measurement since the vapor pressure of H_2 is dependent upon its ortho-para composition (see above).

The pressure at the surface of a liquid cryogen in a dewar may, under conditions of thermal equilibrium, be used to measure the temperature of the cryogen and of apparatus immersed in it. Corrections for the pressure of the hydrostatic head of liquid may be required. If using this method, it should be realized that large vertical temperature gradients can exist in an unstirred liquid cryogen in a vessel with good insulation.

Thermocouples are favored for the measurement of temperatures because of their low cost, ease of application, and rapid response. The thermoelectric powers (temperature sensitivities) of the thermocouples commonly used at higher T's decrease with decreasing temperature, and spurious emfs generated in wires of non-uniform composition are troublesome. The proper use of a reference junction at a known fixed temperature close to the temperature being measured is often advantageous for accuracy. Variability in composition of thermocouple alloys makes individual calibrations necessary for accurate results. Figure 19.2.23

gives typical thermocouple emf-versus-temperature curves for some common thermocouples. A thermocouple consisting of gold with 0.07-at % iron versus chromel has been found useful over the entire temperature range from 2 to 300 K. Its sensitivity increases from 11 μV/K at 4 K to 17 μV/K at 77 K to 22 μV/K at 300 K.

Radio carbon resistors (*Allen Bradley* for 2 to 100 K and *Speer Carbon* for 0.01 to 4.2 K) are much used as temperature sensors because of their low cost and high sensitivity. However, the permanence and stability of their calibrations are not very good, and germanium (doped) resistance thermometers are preferred (0.01 to 100 K) for their good reproducibility of calibration, although they are considerably more expensive.

The carbon and germanium resistors (thermometers) in magnetic fields H exhibit a magnetoresistance (Hall) effect, proportional to H^2, which complicates the determination of temperatures. Recent developments include: (1) a carbon-glass electrical resistor which has a stable calibration; and (2) the electrical capacitance thermometer (dielectric: $SrTiO_3$, a perovskite in solution in a solid SiO_2 or Al_2O_3 glass) with very small (approaching zero) magnetic effect and a stable calibration. The capacitance thermometer is preferred when magnetic fields are present.

The Rh-0.5-at % Fe, wire-coil thermometers, usable in the range of 0.1 to 300 K, have a high sensitivity especially below 10 K.

INSULATION

(See also Sec. 6.)

The degree of thermal isolation required at low T's is normally greater than at elevated T's because it is more costly to remove heat leaking into a low-temperature system than to replace heat lost at elevated T's, as demonstrated by the Carnot cycle. Other differences between insulating for low and elevated temperatures are:

1. Condensation of moisture in low-temperature insulation is possible. When this occurs, the conductance of the insulation is significantly increased. This problem is avoided by a vapor barrier on the outside of the insulation.

2. Condensation of the atmosphere in the insulation and surface washing by the condensed liquid are possible when the insulating surfaces are below 150°R (83 K). This can result in a large added transfer of heat. It is avoided by using an outer cover or surface impermeable to air, and evacuating the insulating space or replacing the atmosphere in it with a noncondensable gas.

At low temperatures, as at other temperatures, the fundamental modes of heat transfer are conduction and radiation, and it is against these that insulation is used. Convection of heat is practically eliminated by the insulation.

Low-T insulation categories are (1) high vacuum, with or without multiple radiation shields, (2) powders, (3) rigid foams, and (4) low-conductivity solids, such as balsa wood and corkboard.

1. (i) **Heat is transferred across an evacuated space** by radiation and by conduction through the residual gas. The conductivity of a gas is almost independent of its density (pressure) as the pressure is reduced in an evacuated space until the free paths of the molecules are increased to an appreciable fraction of the separation of the containing walls.

The molecular *mean* free path at 1 atm (0.1013 MPa) and 492°R (273 K) in air is 4×10^{-6} in (10^{-7} m); in H_2, 6×10^{-6} in (1.5×10^{-7} m); and in He, 10×10^{-6} in (2.5×10^{-7} m). At pressures lower than about 10^{-3} torr (0.133 Pa), the rate of heat transfer Q by residual gas between parallel surfaces at T_1 and T_2 less than 1 in (0.0254 m) apart is approximately $(Q/A) \approx$ (const) $\alpha P(T_2 - T_1)$, where P is pressure *measured* in torr with a gage at *ambient T* and α is an *overall* accommodation coefficient of gas molecules in collision with the walls. Thus, $\alpha = \alpha_1 \alpha_2 / [\alpha_2 + \alpha_1 (1 - \alpha_2)]$, where α_1 and α_2 are accommodation coefficients of gas molecules at the two walls (see Table 19.2.7). Table 19.2.8 gives the values of the *(const)* in the equation for Q/A.

The **rate of radiative heat transfer** between parallel surfaces at T_1 and $T_2 > T_1$, where emissivities are e_1 and e_2, respectively, is $Q = e_1 e_2 s / [e_2 + (1 - e_2)e_1][T_2^4 - T_1^4]$, where s is the Stefan-Boltzmann constant whose value is 1.712×10^{-9} Btu/ft^2 · h · °R^4

Fig. 19.2.23 EMF versus temperature for several thermocouples.

Table 19.2.7 Approximate Values of Accomodation Coefficients, α

Temp, °R (K)	He	H$_2$	Air
540 (300)	0.3	0.3	0.8–0.9
140 (78)	0.6	0.5	1
35 (19)	0.6	1	1

Table 19.2.8 Constant in the Gas-Conduction Equation

$Q/A \approx$ (const) $\alpha P(T_2 - T_1)$

Gas	T_2 and T_1, °R (K)	Const, Btu/h · ft^2 · torr · °R (J/N · s · K)
N$_2$	<700 (389)	28 (1.19)
O$_2$	<540 (300)	26 (1.11)
H$_2$	540 and 140 (300 and 78)	93 (3.96)
H$_2$	140 and 36 (78 and 20)	70 (2.98)
He	Any	49 (2.09)

NOTE: T_1—inner wall (cold). T_2—outer wall (hot).

$(5.67 \times 10^{-8}$ W/m^{2}K$^4)$ (see also Sec. 4). Table 19.2.9 contains emissivities for 540°R (300 K) thermal radiation; these for most engineering purposes may be considered equivalent to minimal values of the e's for very carefully prepared and cleaned surfaces. Organic coatings have high emissivities approaching unity (> 0.9). Mechanically polished surfaces have higher e's than the minimal values. Handling a surface transfers an absorbing (high e) coating to it.

Table 19.2.9 Minimal Values of Emissivity of Metal Surfaces at Various Temperatures for 540°R (300 K) Thermal Radiation

Surfaces	Surface temp, °R (K)		
	7 (3.9)	140 (78)	540 (300)
Copper	0.005	0.008	0.018
Silver	0.0044	0.008	0.02
Aluminum	0.011	0.018	0.03
Chromium		0.08	0.08
Nickel		0.022	0.04
Brass	0.018		0.035
Stainless steel 18-8		0.048	0.08
50 Pb-50 Sn solder		0.032	
Glass, paints, carbon			>0.9
Nickel plate on copper		0.033	

(ii) **Refrigerated Radiation Shield** An LN$_2$ (140°R) (78 K) cooled surface interposed in the insulating vacuum space of an LH$_2$ or an LHe container reduces the heat transferred by radiation to the LH$_2$ or LHe by a factor of about 250 when compared with that transferred by a surface of the same emissivity at room temperature without the interposition of the refrigerated shield. Nearly all the heat radiated by the surface at room temperature is absorbed by the LN$_2$, and this results in evaporation of the relatively inexpensive LN$_2$.

(iii) **A floating radiation shield** is an opaque layer (e.g., a metal sheet), having surfaces of low emissivity suspended in the vacuum space with a minimum of thermal contacts with surfaces that are warmer or cooler. If the emissivity of the surfaces of the floating shield is the same as the emissivity of the vacuum-space walls, the floating shield decreases the *radiative* heat transfer by a factor of 2. For m floating shields arranged in series between the inner cold and outer warm surfaces, the rate of *radiative* heat transfer becomes $\dot{Q}_m$ (m floating shields) = $[1/(m+1)]\dot{Q}_0$ where $\dot{Q}_0$ = heat transfer for $m = 0$.

(iv) **Multilayer (Super) Insulation** The principle of multiple floating radiation shields has been extended to the use of many thin metal layers (up to 75 layers per inch or 3,000 layers per metre) separated by thin thermal insulation. Best results have been obtained with Al foil separated by glass-fiber paper. Other materials have been used successfully, e.g.,

nylon nets as separators and aluminized Mylar with no spacer material. The *apparent* thermal *conductivity* of Al foil multilayer insulation spaced with glass fiber paper is variable depending on the number of layers of foil, the thickness of the paper, and the thickness and compacting of the insulating layer. For $\Delta T = (540 - 36)$°R or $(300 - 20)$ K, apparent mean conductivities range from 2.0 to 4.0×10^{-5} Btu/h · ft · °R (3.5 to 7×10^{-5} J/s · m · K). For $\Delta T = [540 - 140]$°R $(300 - 78)$ K], the values are about one-third larger. Using nylon nets, glass fabric or glass fibers for separating the Al foil increases the conductance from three to seven times. The above conductivities are for the direction normal to the foil surfaces. Lateral conductivities are many thousands of times larger. An advantage of aluminized Mylar is its reduced lateral conductivity. Another advantage is its relatively low density (one-third to one-half of other multilayer insulations). If the residual gas pressure in multilayer insulation is less than 10^{-4} torr, the conductance is practically independent of the residual gas pressure. At 10^{-3} torr (0.133 Pa), the conductance is increased by the order of 50 percent.

2. (i) **Powder insulation** consists of finely divided materials. Conductances of evacuated powders may vary by as much as 100 percent with variations of particle size and *apparent* density (packing) of the powder. The most commonly used powders are perlite, expanded SiO$_2$ (aerogel), calcium-silicate, diatomaceous earth, and carbon black. The conductance of powder insulation is practically independent of gas pressure down to 10 torr (1.333 kPa), at which pressure it begins to decrease rapidly with decreasing pressure as the free path of the gas molecules becomes comparable with the space between powder particles. The most rapid decrease of conductance occurs between 1 and 10^{-1} torr (133 and 13.3 Pa). At 10^{-3} torr (0.133 Pa) the conductance reaches practically the lower limit. Even at 10^{-2} torr (1.33 Pa) the conductance is quite low. For this reason, good mechanical pumps are adequate for the evacuation of powder insulation (fine mesh filters are needed to protect the pump from being damaged by the powder). When evacuated to 10^{-4} torr (1.33×10^{-2} Pa) or less, the apparent mean thermal conductivity for perlite [540°R (300 K) to 140°R (78 K)] ranges from 0.6 to 1.2 mW/m · K. For nonevacuated perlite filled with He or H$_2$ an apparent mean thermal conductivity of 115 mW/m · K can be used. With N$_2$ a typical value is 30 mW/m · K.

The principal function of the powder is to impede the transfer of heat by radiation. Powders are used when radiation would constitute an important heat leak. If an insulating vacuum is bounded by highly reflecting walls, such as clean Cu, Ag, or Al, adding powder may result in an increase of heat transfer because of the paths of solid conduction through the powder.

(ii) **Al powder or flakes added to a powder insulation** reduces the radiative transfer through the insulation. Powdered Al is more effective than flaked Al. The conductance of highly evacuated perlite insulation [$\Delta T = 540 - 140$°R (300 − 78 K)] may be reduced 40 percent by the addition of 25 percent of Al powder. A similar addition of Al powder to Santocel and Cab-O-Sil (diatomaceous earth) reduces the conductance by 70 percent. Metal powder additions are most effective with those powders that are partially transparent to infrared radiation of wave lengths longer than 3 μm.

3. **Rigid-Foam insulation** The foams most used in cryogenic applications are the more or less closed-cell foams. Table 19.2.10 gives thermal conductivities of selected samples of insulating foams. The conductivity of a foam is dependent on the conduction through the intracellular gas, on the transfer by thermal radiation, and on solid conduction. Temperatures below the condensing T of the encased gas condense it and improve the insulation. Gases encased when polymer foams are blown gradually diffuse out of the cells and are replaced by ambient gases. The conductivities of many Freon and CO$_2$ blown foams increase in time as much as 30 percent because of the diffusion of air into the foam. Conductivities may increase by a factor of 3 or 4 when cells are permeated with H$_2$ or He. Glass and silica foams appear to be the only ones having fully closed cells. The **structural rigidity** of foams may be used to eliminate the need for mechanical supports. It is possible to use rigid foams without inner liners and outer casings. Polymer

**Table 19.2.10   Apparent Mean Thermal Conductivities
of Selected Foams, Balsa Wood, and Corkboard**

Material	Density lbm/ft³ (kg/m³)	Thermal conductivity,* Btu/h · ft · °R(W/K · m)
Polystyrene foam†	2.4 (38)	0.019 (0.033)
	2.9 (46)	0.015 (0.026)
Epoxy resin foam†	5.0 (80)	0.019 (0.033)
Polyurethane foam†	5.0–8.7 (80–139)	0.019 (0.033)
Rubber foam†	5.0 (80)	0.021 (0.036)
Silica foam†	10.0 (160)	0.032 (0.055)
Glass foam†	8.7 (139)	0.020 (0.035)
Balsa wood, across		
grain	7.3 (117)	0.027 (0.047)
	8.8 (141)	0.032 (0.055)
	20.0 (320)	0.048 (0.083)
Corkboard, no added		
binder	5.4 (86)	0.021 (0.036)
	7.0 (112)	0.0225 (0.039)
	10.6 (170)	0.025 (0.043)
	14.0 (224)	0.028 (0.048)
Corkboard with asphalt		
binder	14.5 (232)	0.027 (0.047)

* Test space pressure = 14.7 lb/in² abs (0.101 MPa).
† ΔT = 540 to 140°R (300 to 78 K).

foams have relatively high **thermal expansions,** even greater than the same polymer without voids. This makes cracking of the insulation a problem when it is bonded to metal. Sliding expansion joints have been used to overcome this. Mylar film bags (0.001 in or 0.025 mm thick) have been used to hold cryogenic liquids inside rigid-foam insulation and plastic films outside to keep the ambient atmosphere from the insulation.

4. **Balsa wood and corkboard** have been used for insulating cryogens that are not very cold, e.g., liquid methane (201°R or 112 K) and higher-boiling cryogens. The conductivities of these insulators (Table 19.2.10) are considerably higher than for the other types of low-T insulation discussed above. Their use, therefore, is restricted to special applications, as where their ability to support internal structures mechanically is important.

Conduction of heat along structural supports passing through the insulation merits special attention for an adequate realization of the benefits of superior insulation. Ideal materials of construction would have high mechanical strength and low thermal conductivity. In Table 19.2.11, the yield strengths, in tension, of construction materials are compared with their thermal conductivities. This kind of comparison favors the nonmetallic materials in Table 19.2.11. When large

quantities of cryogens are handled in the field, metals are commonly used for structural members because glass and plastics (Teflon excepted) are brittle at low T's.

Minimizing the transfer of heat through the supporting structure offers opportunities for ingenuity and inventiveness in design. The transfer is minimized by lengthening the supporting members. A stack of thin metal disks or washers, utilizing the contact resistance between adjacent disks, will increase thermal resistance without increase in length of support. A stack of 0.0008-in (0.02-mm) stainless steel disks under a compressive stress of 1,000 lb/in² (6.895 MPa), in vacuum, has the same thermal resistance as a solid rod 50 times the length of the stack.

Insulation Selection   Because of their smaller heats of vaporization and greater cost of production, the lowest-T cryogens (LHe and LH$_2$) ordinarily merit more effective insulation than the higher-boiling cryogens like LO$_2$ and LN$_2$, and these in turn are ordinarily better insulated than other cryogens boiling at still more elevated T's. The choice of insulation ordinarily involves the conditions of use of the equipment, as well as considerations of the tolerable loss of refrigeration and the cost of the insulation and its installation. If equipment is to have intermittent, rather than long, uninterrupted operating periods, the insulation heat capacity and time lag in reaching steady state are important. An insulation with large heat capacity (e.g., powders) will evaporate large quantities of refrigerant and take a long time to reach steady state. Powders are compacted by mechanical shocks and vibrations. Continued mechanical loading will compact both multilayer insulation and powders. Weight and bulkiness of the insulation can be important. The heat transported by the mechanical supports and vents connecting the cold interior with the warm exterior becomes the minimum refrigeration loss, attainable only with perfect insulation of the rest of the system. A large loss of refrigeration through supports and vents ordinarily reduces the attractiveness of costly insulation having the lowest conductance.

SAFETY

Precautions are needed for the safe handling and storage of cryogens because (1) air can contaminate the cryogen, creating a potential explosive, either directly as when air mixes with H$_2$ or CH$_4$ or indirectly by transforming an inert cryogen like LN$_2$ into an oxidant (a potential hazard for combustible insulation), and (2) moisture and air can freeze on cold surfaces, clogging vents and preventing normal operation of valves.

Vent systems should be sized for sufficient exhaust velocities to prevent back-diffusion of air into the cryogen space. Storage at slightly

Table 19.2.11   Comparison of Materials for Support Members

Material	E_y,*¶ kpsi (MPa)	κ,* Btu/h · ft · °R(W/K · m)	E_y/κ kpsi · h · ft · °R/Btu (MN · K/W · m)
Aluminum 2024	55 (379)	47 (81)	1.17 (4.68)
Aluminum 7075	70 (483)	50 (87)	1.4 (5.55)
Copper, ann	12 (83)	274 (474)	0.044 (0.18)
Hastelloy† B	65 (448)	5.4 (9.3)	12 (48)
Hastelloy† C	48 (331)	5.9 (10.2)	8.1 (32)
K Monel‡	100 (690)	9.9 (17.1)	10.1 (40)
Stainless steel 304, ann	35 (241)	5.9 (10.2)	5.9 (24)
Stainless steel (drawn 210,000 lb/in²)	150 (1034)	5.2 (9.0)	29 (115)
Titanium, pure	85 (586)	21 (36.3)	4.0 (16)
Titanium alloy (4A1–4Mn)	145 (1000)	3.5 (6.1)	41.4 (164)
Dacron§	20 (138)	0.088 (0.15)¶	227 (920)
Mylar§	10 (69)	0.088 (0.15)¶	113 (460)
Nylon§	20 (138)	0.18 (0.31)	111 (445)
Teflon§	2 (14)	0.14 (0.24)	14.3 (58)

* E_y is the yield stress, and κ is the average thermal conductivity between 36 and 540°R (20 and 300 K).
† Haynes Stellite Co.
‡ International Nickel Co.
§ E.I. du Pont de Nemours & Company
¶ Room-temperature value.

elevated pressure is preferable to prevent in-leakage of contaminants. Condensation can also be a problem on poorly insulated lines carrying low-boiling cryogens; frost is not a hazard but the oxygen-enriched condensed air is hazardous in the presence of combustibles. The insulation of liquid vent lines is a necessary precaution. Where this is not possible, appropriately placed guttering can prevent the contact of any condensed air with combustible objects.

Any space, either containing a cryogen or being refrigerated by a cryogen, should be protected with proper **safety-relief mechanisms,** i.e., relief valves or rupture disks. Such space includes less obvious ones such as the insulating vacuum space surrounding a cryogenic fluid or a section of line which can trap liquid between two valves. A heat leak to trapped liquid confined without a vent for escape of vapor may develop pressures sufficient to burst the containing vessel. Gas pressures necessary to maintain liquid density at ambient temperature are for helium, 18,000 lb/in² (124 MPa); for hydrogen, 28,000 lb/in² (193 MPa); and for nitrogen, 43,000 lb/in² (296 MPa).

The **selection of structural** materials calls for consideration of the effects of low temperatures on the properties of those materials. A number of otherwise suitable materials become brittle at low temperature (Fig. 19.2.10). Materials must perform satisfactorily over the complete range from cryogenic to room temperature, with hydrogen embrittlement particularly significant. While hydrogen embrittlement is not believed to be a problem at liquid-hydrogen temperatures, it has been shown to occur at room temperature. The 300-series stainless steels, aluminum alloys and BeCu, among other alloys, are more resistant to hydrogen embrittlement.

In calculating allowable stresses, room temperature yield and tensile strengths should be used because (1) pressure testing of containers at ambient temperature is frequently necessary, and (2) in the case of large vessels, large unknown temperature gradients can exist above the liquid level both in the ullage space and in the walls of the vessel. The 300-series stainless steels are normally austenitic (f.c.c.). However, in some 300-series stainless steels austenite is partially transformed by cold-working and possibly by temperature cycling to martensite. As martensite is brittle at low T's, the ductility of these steels is reduced by this kind of treatment. For this reason, it is recommended as an added safety factor that room temperature strength of the 300 series stainless steels be used for the design of structural members.

Careful **stress analysis** is essential and must include stresses due to (1) thermal contraction of equipment, and (2) radial, axial, and circumferential temperature gradients caused by non-uniform cooling rates. The latter are frequently encountered in the cool-down of cryogenic transfer lines.

Cryogenic equipment should be **purged** before being placed in service to remove unwanted condensables and gases that could form explosive mixtures. Commonly used methods are (1) evacuating and back-filling with purge gas and (2) flowing of purge gas through the system.

The cryogen can be a direct **hazard to personnel.** Cryogen "burns" can result from direct contact with either the cryogen or uninsulated equipment containing the cryogen. The large evolution of gases associated with cryogenic spills can result in asphyxiation. Asphyxiation is a hazard in entering a warmed cryogenic vessel. Further safety details are found in the references.

19.3 OPTICS
by Patrick C. Mock

REFERENCES: "AIP Handbook," McGraw-Hill. Born and Wolf, "Principles of Optics," Pergamon Press. "Photonics Handbook," Laurin Publishing Co. Feynman, "QED: The Strange Theory of Light and Matter," Princeton University Press. Smith, "Modern Optical Engineering," McGraw-Hill. Miller and Friedman, "Photonics Rules of Thumb," McGraw-Hill. "Diffraction Grating Handbook," Richardson Grating Laboratory.

FUNDAMENTAL THEORIES OF LIGHT

Optics is the science of light, and light is **photons.** The fundamental theory of photons is quantum electrodynamics, and Richard Feynman explains it with great clarity in his book for the layman, "QED: The Strange Theory of Light and Matter."

Photons are the particles that mediate the electromagnetic force, and they have three fundamental properties: energy, polarization, and constant velocity in a vacuum. The **speed of light** in a vacuum, c, is a fundamental physical constant with the defined value of 299,792,458 m/s (often approximated as 3×10^8 m/s). Every photon has an energy, which determines its frequency and wavelength. These are related by $E = hf$ and $\lambda = c/f$, where E is the photon energy, h is Plank's constant, f is the frequency, and λ is the **wavelength.** Polarization is fundamentally related to the quantum spin of the photon. The classical interpretation is that a photon is an electromagnetic wave packet with perpendicular electric and magnetic fields that are also perpendicular to the direction of propagation. The polarization is the direction of the electric field component.

Light is photons with wavelengths shorter than the resolution of the human eye and long enough that their behavior is primarily wavelike (roughly 0.1 to 300 μm). It is a small portion of the electromagnetic spectrum (Fig. 19.3.1). **Visible light** is the small band with wavelengths of 400 to 760 nm. Longer wavelength light is infrared radiation, and shorter wavelength light is ultraviolet radiation. The human eye is most

sensitive near 560 nm. The interactions that light can have with matter are reflection, refraction, interference, diffraction, polarization, absorption, and scattering; however, the last three are beyond the scope of this section.

GEOMETRIC OPTICS: REFLECTION AND REFRACTION

Reflection and refraction govern geometric optics. The speed of light in an optical material, c', is always smaller than the speed of light in a vacuum. The **index of refraction,** $n = c/c'$, is an important optical property of any material. The refractive index for most optical materials depends on the wavelength of the light, and it is typically between 1 and 4. This wavelength dependence is called **dispersion.** A beautiful example of the effects of refraction and dispersion is a rainbow.

The reflection and refraction of a light ray from a flat interface between two media is shown in Fig. 19.3.2. The plane of incidence is defined by the incident light ray and the surface normal at the point of incidence. The **law of reflection** states that the reflected ray lies in the plane of incidence, and that the angle of reflection is equal to the angle of incidence, $I = I''$, as measured from the surface normal. The **law of refraction** (also known as **Snell's law**) states that refracted ray lies in the plane of incidence, and the angle of refraction is related to the angle of incidence by $n' \sin I' = n \sin I$. Total internal reflection occurs when $n > n'$, and $(n/n') \sin I > 1$. The incident ray is reflected with no loss and the transmitted ray vanishes. The critical angle, $I_c = \arcsin(n'/n)$, is the minimum angle at which this occurs.

The refraction of a light ray at a spherical interface between two media with radius of curvature r is shown in Fig. 19.3.3. The optical axis passes through the center of curvature. The image distance l' is related to the object distance l by $(n'/l' - n/l) = (n' - n)/R$, when the

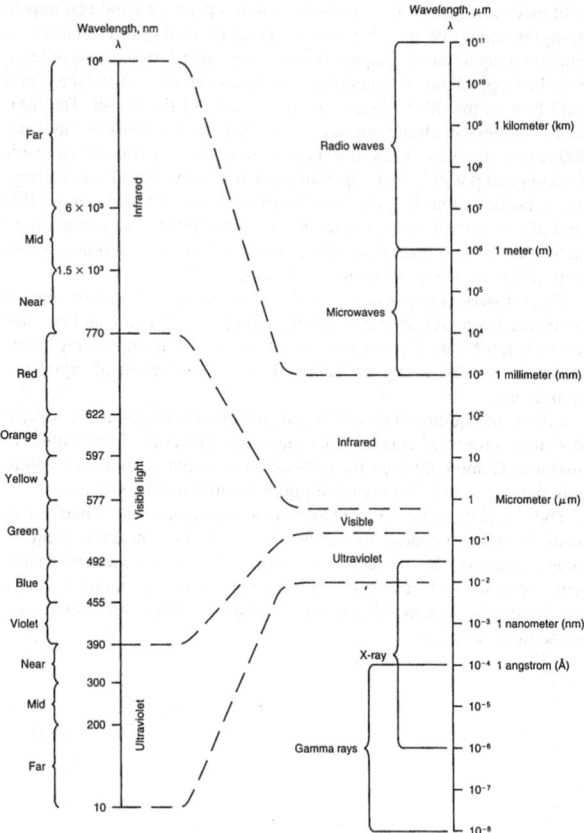

Fig. 19.3.1   Electromagnetic spectrum.

rays are close to the optical axis (the paraxial approximation). The **magnification** m is given by $m = h'/h = nl'/n'l$. If the image is inverted, the magnification is negative. A typical optical assembly contains several lenses, and each lens is two surfaces. The optical properties of the assembly can be found by starting with the object and determining the image for the first surface. The object for all subsequent surfaces is the image of the previous surface.

Fig. 19.3.2   Reflection and refraction at a plane surface.

Fig. 19.3.3   Refraction at a spherical surface; all parameters are shown in their positive sense.

A thin lens has a thickness much less than its diameter. The image distance l_i for a thin lens is related to the object distance l_O by $(1/l_O + 1/l_i) = (1/r_1 - 1/r_2)(n' - n)/n$, where r_1 and r_2 are the radii of curvature of the surfaces of the lens, and n and n' are the indices of refraction of the medium and the lens, respectively. It is derived by applying the refraction equation from above to the two surfaces of the thin lens. The **focal length** f of a lens is the image distance when the object is at infinity. This leads to the lens maker's equation $1/f = (1/r_1 - 1/r_2)(n' - n)/n$ and the gaussian lens equation $1/f = 1/l_O + 1/l_i$. The **optical invariant** (also known as the Lagrange invariant) is an invariant quantity at every surface of an optical system. In its simplest form, the product of the image size and ray angle is a constant. It implies that the field of view and useful aperture are inversely proportional, i.e., a large telescope has a very small field of view, and wide-angle camera has a small aperture.

The relative illumination of the image from the object is proportional to the area of the aperture and inversely proportional to the size of the image. This relationship is characterized by two optical parameters, the f **number** and the **numerical aperture** NA. The f number is the ratio of the focal length to the diameter of the clear aperture of a lens system, $f/\# = f/D$, and it typically is used to characterize systems with a large image or object distance. Since the image size is proportional to the square of the focal length, the square of the f number is inversely proportional to the relative intensity. For example, a telescope with a 1-m aperture and a 10-m focal length has an f number of 10, customarily written as $f/10$. Numerical aperture is defined as $NA = n \sin \phi$, where ϕ is maximum ray angle that passes through the optical system, and it is typically used to characterize systems with small image and object distances. When this angle is small, $f/\# \approx 1/(2\,NA)$.

THE WAVE NATURE OF LIGHT: INTERFERENCE AND DIFFRACTION

Interference is the interaction of waves. The interference is constructive when the waves are in phase and destructive when they are not. Diffraction is the interaction of a wave with an aperture, which often results in the bending of the wave at the edge of the aperture. Both phenomena depend on the wavelength of the wave. A common example of optical interference is the colors observed in a thin film of oil on water. Light reflected from the oil/water boundary interferes with light reflected from the air/oil boundary. The apparent color depends on the thickness of the oil layer and the viewing angle. As these change, different wavelengths interfere constructively and destructively along the path to the eye.

When the light source is monochromatic, the constructive and destructive interference leads to formation of light and dark regions known as **fringes.** If the fringe pattern is caused by the interference of two beams following different paths from the same source, the fringe pattern can be analyzed to determine path length differences with an accuracy smaller than the wavelength of the light (see "Photonics Handbook"). For example, the flatness of an unknown surface can be compared with the flatness of a reference surface. Instruments that make these measurements are known as **interferometers.** The most common types of interferometers are the Fizeau, Michelson, Twyman-Green, and Fabry-Perot.

Diffraction limits the resolution of any optical imaging system. The limit depends on the size of the aperture and the wavelength. The angular resolution of a circular aperture is given by $\phi = 2.44\lambda/D$, where D is the diameter of the aperture. For an aperture with an arbitrary shape, the far-field diffraction limit is approximately $\phi \cong \lambda P/\pi^2 A$, where P and A are the perimeter and area of the aperture, respectively.

A **diffraction grating** is a set of finely spaced narrow apertures with a spacing comparable to the wavelength of the light to be measured. The grating can be imposed on a reflective or transparent surface. The light diffracted by each aperture in the grating interferes with light from nearby apertures. If the surface in Figure 19.3.2 is a reflective diffraction grating, constructive interference occurs when $m\lambda = d(\sin I + \sin I'')$, where d is the spacing and m is an integer which indicates the diffraction order. Physically realizable diffraction orders exist for $|m\lambda/d| < 2$.

SPECIAL TOPICS: LASERS, FIBERS, AND OPTICAL DESIGN

Lasers have become a ubiquitous optical tool. The term *laser* is an acronym for *l*ight *a*mplification by the *s*timulated *e*mission of *r*adiation. Stimulated emission is the underlying physical process that all lasers exploit. The result is an intense, monochromatic optical source that is nearly parallel and spatially coherent. Lasers are used in a broad array of applications, including CD and DVD players, fiber-optic and free-space telecommunications, photolithography, machining, photomedicine, interferometry, and surveying.

An **optical fiber** exploits total internal reflection to transmit light with very little attenuation. A simple bare glass fiber suspended in air will transmit light, but surface contamination and surface defects cause high losses. An optical fiber achieves extremely low attenuation by surrounding a high index core with a low index cladding, which protects the surface of total internal reflection. Single mode optical fibers, used in telecommunications, achieve losses as low as 0.2 dB/km. Large optical fibers are effective light pipes that can illuminate hard-to-reach areas. Coherent bundles of optical fibers are used to transmit images from hard-to-reach areas.

Optical design is the science of precisely manipulating light. Simple optical problems can be solved with the formulas in this section or the references. However, these ideal formulas ignore distortions, aberrations, and tolerances that affect all real optical systems. Sophisticated optical computer-aided design tools use precise ray tracing to accurately predict the performance of an optical design. They include libraries of design templates that can be adapted to meet the requirements of common optical systems and powerful optimizers to maximize the performance of the optical system. The "Photonics Handbook" has more information on optical CAD tools.

Section **20**

Emerging Technologies and Miscellany

BY

LUC G. FRÉCHETTE *Professor, Mechanical Engineering Department, Université de Sherbrooke, Canada*
ALEXANDER COUZIS *Professor, Chemical Engineering, The City College of the City University of New York*
JACKIE JIE LI *Professor, Mechanical Engineering, The City College of the City University of New York*
FARID M. AMIROUCHE *Professor, Department of Mechanical and Industrial Engineering, University of Illinois at Chicago*
DOYLE KNIGHT *Professor, Department of Mechanical and Aerospace Engineering, Rutgers University*
YIANNIS ANDREOPOULOS *Professor, Department of Mechanical Engineering, The City College of the City University of New York*
ALI M. SADEGH *Professor, Department of Mechanical Engineering, The City College of the City University of New York*
SRIRANGAM KUMARESAN *Biomechanics Institute, Santa Barbara, Calif.*
ANTHONY SANCES, JR. *Biomechanics Institute, Santa Barbara, Calif.*
PAUL CAVALLARO *Department of Navy, NUWC, Newport, Rhode Island*
ANDREW GOLDENBERG *Professor, Mechanical and Industrial Engineering, University of Toronto, Canada; President and CEO of Engineering Services Inc. (ESI), Toronto*

20.1 AN INTRODUCTION TO MICROELECTROMECHANICAL SYSTEMS (MEMS)

by Luc G. Fréchette

INTRODUCTION

Trend toward Small Scales

The twentieth century was marked by mankind's accomplishments at ever increasing scales, from civil works such as bridges that span rivers to sophisticated machinery such as airplanes and spacecraft that span our planet and beyond. Although this quest for immensity will remain vibrant, the twenty-first century will likely be marked by achievements at smaller scales. This trend is best demonstrated by the information revolution experienced by our society over the past decades, which was a result of the remarkable development of integrated circuits (ICs), itself enabled by progress in semiconductor manufacturing. The microfabrication approach developed for ICs is based on parallel processing of large arrays of devices on a planar substrate, using photolithography and thin-film processing in a clean room environment. The semiconductor industry leverages this approach to make devices with impressive levels of performance, reliability, and complexity. Today, microprocessors integrating over 100 million transistors per chip can be batch-fabricated in large volumes and at low cost.

Such capabilities are expanding beyond microelectronics to allow the microfabrication of sensors, actuators, and microsystems that implement functions from other physical domains, such as mechanical, thermal, fluidic, magnetic, electrical, optical, chemical, or biological domains. The field of **microelectromechanical systems,** or MEMS, focuses on the miniaturization of such a diverse set of systems through the use of microfabrication technology. It is also referred to as **microsystems technology,** or MST, in Europe and **micromachines** in Asia.

Microsystems and Their Applications

The scale of a device or machine is set by the task it must accomplish as well as the fabrication technology available to implement it. For example, information storage devices, such as hard disk drives and solid-state memory, increase in storage density as technology enables smaller bit size, since bits of information do not have a physical scale associated with them. Similarly, devices pertaining of other physical domains could also be miniaturized to the extent that technology allows, if commercially beneficial. A prime example is **sensors,** i.e., devices that measure a physical quantity, often by converting it to an electrical signal. Most measurement applications would benefit from local, compact, nonintrusive, and inexpensive sensors, suggesting that batch microfabrication may be appropriate. Pressure sensors (Fig. 20.1.1a) and accelerometers (Fig. 20.1.7a) are common examples. **Actuators** that act on the environment by converting an electrical signal into a mechanical displacement or another physical domain can also potentially be miniaturized, as long as the power transmitted to the environment is small and the action is highly localized. Examples include inkjet printer heads, micromirror arrays for image projection (Fig. 20.1.1b), and microvalves for flow control (Fig. 20.1.7c). Microfabrication and MEMS technology can also be enabling to implement **microsystems** that would otherwise be impossible or impractical, such as microfluidic labs-on-a-chip that allow interaction with molecules or cells at their length scales (0.1–10 μm).

The annual market for MEMS is estimated at \$US 8.2 billion worldwide for 2007,[1] in the following areas: microfluidics (27 percent), inertial sensors (22 percent), optical MEMS (22 percent), pressure sensors (11 percent), RF MEMS (3 percent), and other sensors (10 percent) and actuators (5 percent). These devices are used in the automotive industry, in consumer electronics and computers, for telecommunications, as well as in instrumentation for biochemistry and medicine. The field has experienced an average growth of 16 percent annually since 2000, as a result of novel applications of MEMS and insertion of this disruptive technology into existing markets.

MICROFABRICATION TECHNOLOGY AND MATERIALS

The key enabling technology responsible for the rapid development of microsystems over the past decade is **microfabrication.** Most of today's computer chips are manufactured by the CMOS (complementary

Metal contact pads   Piezoresistors   Diaphragm   Electronics

Pressure

(a)

Array of 1000x1000 micromirrors

Redirected light

Mirror

(b)

Fig. 20.1.1 Commercial examples of microelectromechanical systems (MEMS). (a) Automotive piezoresistive pressure sensor and (b) micromirror array for display applications. The pressure sensor (a) measures the deflection of a thin silicon diaphragm due to an applied pressure by monitoring a strain-sensitive resistor fabricated on the diaphragm through on-chip circuitry. (*Photo: Bosch.*) The micromirror array (b) is an electrostatic actuator that redirects reflected light over 1 million independent pixels to digitally project an image on a screen. (*Photo: Texas Instruments.*)

metal-oxide semiconductor) process, which is a sequence of steps that repeatedly deposit and pattern thin films to form an integrated circuit with transistors. Silicon, typically used in microelectronics, has become the dominant material for MEMS because of its material properties (semiconductor, mechanical, thermal) and the extensive processing experience with the silicon material set pioneered by the semiconductor industry. This section describes the underlying microfabrication technologies and materials borrowed from this industry that are the basis for silicon MEMS processing. To gain further insight on microfabrication science and technology, beyond what will be presented here, the reader is referred to classic texts on the subject.[2,3]

Silicon and Its Properties

Most microfabrication processes create devices on the surface or through the thickness of thin, planar, circular substrates, or **wafers**. Single-crystal silicon wafers are the most common starting material, although glass, quartz (single-crystal and fused), and compound semiconductor wafers (GaAs, InP) are also of interest. As provided by the semiconductor industry, single-crystal silicon wafers offer unique properties, such as: (1) stable, defect-free bulk material of high chemical purity; (2) precise substrates with low wafer bow (>25 μm), minimal taper (>15 μm), and mirror polished surfaces (one or both sides); (3) semiconductor properties and the possibility of integrating MEMS with electronics.

Wafers are characterized by the orientation of the **crystal's cubic structure** with respect to the wafer surface. As illustrated in Fig. 20.1.2, principal planes and directions in single-crystal silicon can be identified by **Miller indices**.[2] Specific planes are denoted by (h, k, l), where the indices are components of the direction vector [**h, k, l**] that is normal to the plane. Since the lattice of silicon is cubic, hence symmetric, the three principal directions are identical and interchangeable. Therefore, we commonly refer to families of directions, written <100> to identify the [100], [010], and [001] directions, and equivalently to families of planes, written {100}, for the planes normal to these directions. Wafers are classified by the direction vector that is normal to the wafer surface,

so, for example, <100> wafers will have the cubic lattice parallel to the wafer surface. As will be discussed later, the crystalline structure of silicon has been leveraged to create useful structures such as the pressure sensor membrane in Fig. 20.1.1a by anisotropic wet etching.

Doping, by introducing impurities in the silicon lattice, controls the substrate's electrical conductivity (which can range from 0.001 to >10,000 $\Omega \cdot$ cm) and also defines the wafer type (Group III atoms such as boron produce **p-type** material, Group V atoms such as phosphorus produce **n-type** material). Mechanically, single-crystal silicon is a strong and light material (ρ = 2.33 kg/m^3, E = 130 to 187 GPa, ν = 0.06 to 0.28, and G = 57 to 79 GPa depending on orientation[2]), but exhibits a brittle behavior with a fracture strength that is process and size dependent (typically 1.2 GPa, but can vary from 0.5 to 7 GPa).[4] Traditional machining of silicon is therefore not common, since it introduces surface cracks that can significantly reduce the fracture strength; etching techniques are therefore used instead. Thermally, silicon is a good conductor (k = 150 W/m $\cdot$ K) and exhibits less thermal expansion and higher specific heat than most metals (CTE = 2.6 $\times$ 10^{-6}/K, c_p = 0.713 kJ/kg $\cdot$ K at 300 K), which tends to reduce thermal stresses. It can sustain relatively high temperatures (brittle-to-ductile transition above 550°C and melting at 1421°C), although the electronic properties of silicon are best conserved below 150°C. These electronic properties are interesting for electromechanical transduction in sensors and will be discussed later.

Thin-Film Processing

A broad range of thin films can be deposited or grown on the surface of a substrate to form mechanical or electrical elements, such as membranes, mechanisms, or circuits. Materials include silicon or silicon-based compounds, metals, and polymers. The most common thin films are summarized in Table 20.1.1 with their characteristics and typical microfabrication approaches. Conductive layers can be formed of metals or doped semiconductors, such as polycrystalline silicon (referred to as **polysilicon**) diffused with boron or phosphorous dopants. Electrical insulation is typically created with silicon dioxide (SiO$_2$) or silicon nitride (Si$_3$N$_4$) films. Materials can be deposited in thin films by various techniques, such as physical vapor deposition, chemical vapor

(a) Diamond unit cell of silicon

Corner Si atoms
Face centered Si atoms
Interior Si atoms

Tetrahedral bonding of silicon atoms

(b) Crystal planes

(100) plane   (110) plane   (111) plane

Fig. 20.1.2   Crystal lattice, and Miller indices for silicon. The unit cell (*a*) has the structure of diamond, with each silicon atom covalently bonded to its four adjacent neighbors. The major planes, directions, and corresponding Miller indices are illustrated in (*b*).

Table 20.1.1   Common Substrate and Thin-Film Materials Used for MEMS

Material	Typical application	Fabrication process and characteristics	Material characteristics
Silicon (Si) (substrate)	Most common substrate; compliant or rigid structures; electrical and electronic components	Single-crystal substrate; 300–700 μm thick; many micromachining techniques	Defect-free, high-strength, brittle, thermally conductive semiconductor
Quartz (substrate)	Bulk structural materials for active components	Crystalline or fused silica substrate; 400–700 μm thick; limited micromachining	Piezoelectric properties; visually transparent
Borosilicate glass (Pyrex) (substrate)	Insulating substrate; bulk structures for microfluidics	Amorphous glass substrate; 400–700 μm-thick wafers; anodic bonding to Si; limited micromachining	Thermal expansion matched with Si; visually transparent
Polycrystalline silicon (polySi)	Thin-film structures; piezoresistors; conductors	Polycrystalline; LPCVD; 0.2–2 μm-thick films	Process-dependent properties; semiconductor
Silicon dioxide (SiO$_2$)	Sacrificial thin film; insulator (thermal and electric)	Amorphous; thermal growth or PECVD; 0.1–10 μm-thick films	Typically high compressive strength; thermal insulator: dielectric
Silicon nitride (Si$_3$N$_4$)	Membrane; insulator (electric)	LPCVD or PECVD; 0.1–1 μm-thick films	High tensile strength; dielectric
Metals	Electrical conductors; structural layers	Evaporation or sputtering (Al, Au, Cr, Ti, Pt, W); 20 nm to 2 μm-thick films	Ductile; thermal and electrical conductors
	High-aspect-ratio structures	Electrodeposited (Ni); 1–1000 μm-thick films	
PDMS	Microfluidics; molded microstructures	Flexible polymer; cast; 10–1000 μm-thick films	Low stiffness, dielectric, biocompatible
SU-8	Mold for electroplating; thick structures	Photosensitive epoxy; spin cast; 5–500 μm-thick films	Moderate stiffness, dielectric

deposition, electrochemical deposition, or casting. Since thin-film materials are synthesized as they are deposited or grown, their properties are strongly related to the specific processing conditions. Material properties can be tailored within a certain range, which is useful to match design requirements but requires stringent process control for repeatability. Film thicknesses depend on the process and can vary from a few nanometres (nm) up to tens of micrometres (μm) [1 m = 10^6 μm = 10^9 nm = 10^{10} Å (angstroms)]. As a reference, the human hair is approximately 70 μm in diameter.

In **physical vapor deposition**, a solid piece of the material to be deposited is either vaporized or sputtered in line of sight with the substrate, such that atoms of the material will deposit uniformly on the substrate to form a thin film. Pure metals are easily evaporated, while sputtering can also deposit a wide range of metallic and nonmetallic compounds.

Chemical vapor deposition (CVD) involves a chemical reaction to synthesize a material, such as polycrystalline silicon, as it is deposited over the substrate. This process is often carried out in batches in high-temperature furnaces with a controlled flow of reactants and at low or atmospheric pressure (LPCVD or APCVD, respectively). In a similar environment, the silicon substrate can also be directly oxidized by reacting with a flow of oxygen to form SiO$_2$ on the surface (the addition of H$_2$ enhances the oxidation rate, and is referred to as **wet oxidation**, as opposed to **dry oxidation** when only oxygen is used). The chemical reactions for CVD can also be plasma enhanced (PECVD), which allows deposition at lower temperatures, good control of the film properties, and high deposition rates, but requires expensive serial processing.

An alternative approach for metal films, especially thick films (from micrometres to millimetres), is electrochemical deposition, or **electroplating.** The substrate acts as the cathode, where electrons combine with metallic ions in the liquid electrolyte, to cover exposed

areas of the wafer with the metal (commonly Ni, but also Cu, Au, NiFe, or NiW).

The substrate can also be covered with solvent-based solutions by directly pouring the solution on the wafer and using gravity or centrifugal forces (**spin casting**) to form a uniform thin film. Spin-on-glass and photosensitive polymers are commonly applied by spin casting and then heated to evaporate the solvents and solidify.

Photolithography

The photolithographic process is used to transfer a pattern onto a planar substrate, using light. As illustrated in Fig. 20.1.3, a thin photosensitive polymer film, or **photoresist**, is first spin-cased on the surface of the wafer. The pattern is then exposed over the film by shining UV light through a photolithographic mask. The mask consists of a glass plate coated with a patterned chrome layer to locally expose the photoresist: the mask blocks the light where chrome is present and allows light to shine through where chrome has been removed. The coated wafer is then developed in a liquid solution, physically dissolving the photoresist in the exposed areas, for *positive* photoresist (for *negative* resist, dark or unexposed areas would be dissolved instead).

This process allows replication of complex, micrometre-scale features over the entire wafer in a single or a few repeated exposures. This parallel fabrication approach is central to the low-cost, high-volume advantage of microfabrication. The feature size is limited to fractions of the wavelength of light used for exposure, but alternative **nanolithography** techniques have been developed to reach tens of nanometres, such as electron beam and nanoimprint lithography.

Etching and Other Patterning Approaches

The pattern created in the photoresist layer is typically transferred to the underlying thin film or into the substrate by etching. The wafer is

Fig. 20.1.3 Photolithography process: (*a*) spin-coat a layer of photoresist over the wafer and prebake the film; (*b*) expose certain areas to UV light; (*c*) develop photoresist, leaving a physical mask in the nonexposed areas.

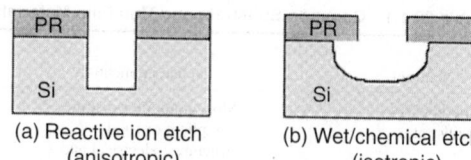

Fig. 20.1.4 Cross section profiles of etched structures: (*a*) reactive-ion-etched features with vertical side walls and (*b*) chemical-etch-dominated process (plasma or wet etching) that results in mask under cut and rounded features.

immersed in an etchant, which reacts and removes the exposed material, while the photoresist acts as a physical mask that protects the unexposed areas of the wafer. In **wet etching,** a chemical in liquid or vapor phase comes in direct contact with the wafer, preferentially reacting with the exposed material. In **dry etching,** a gas is ionized to create a plasma of free radicals and ions that chemically react with the exposed material or physically ablate the surface through ion bombardment, by driving the ions toward the wafer with an electrical field. This process is referred to as reactive ion etching (RIE) and is most commonly used in microelectronics to form fine patterns in thin films. Given the directionality created by the electrical field applied between the plasma and substrate, anisotropic features with relatively vertical side walls are possible (Fig. 20.1.4*a*). Chemically dominated etching, such as wet etching, is, however, characterized by isotropic removal of material that effectively undercuts the mask and widens the exposed features (Fig. 20.1.4*b*). In addition to etch profiles, etch rate and selectivity are important parameters that characterize an etch process. The etch rate is simply defined as the thickness of material removed per unit time, while selectivity is the ratio of etch rate of the material intentionally removed to that of a secondary material such as the photoresist mask or the underlying substrate.

Reactive ion etching exhibits poor selectivity to photoresist (near 1:1) while chemically dominated etching can show selectivity >100:1. Table 20.1.2 summarizes some of the most common combinations of materials and etchants, along with their properties. After etching, the photoresist is commonly removed in an oxygen plasma (ashing) or in a solution (acid or solvent). The wafer is then ready for subsequent processing.

Lift-off is an alternative patterning approach consists of depositing and patterning the photoresist layer first, then covering the wafer surface with a thin film, commonly a metal. The photoresist is then dissolved to effectively lift-off the metal that covers it while leaving the metal that was deposited in the exposed areas of the mask.

PROCESSING FOR MEMS

The basic microelectronics processing techniques introduced above have been extended to allow the fabrication of movable structures for sensors, actuators, and multiphysical microsystems. The most common approaches are presented next.

Surface Micromachining

One approach consists of depositing and patterning a structural material over a sacrificial thin film. The sacrificial material is then selectively etched to release the structures formed in the top layer. This approach creates thin planar structures that can move with respect to the substrate, normal or parallel to it. Figure 20.1.5 illustrates a surface micromachining process in which alternating layers of polysilicon (to form the structure) and silicon dioxide (as the sacrificial material) are deposited and patterned. Etched openings in the sacrificial layers allow mechanical and electrical connections between the polysilicon layers and attachment to the substrate. A hydrofluoric solution can be used to remove the silicon dioxide and release the structures, since it does not etch silicon significantly. This surface micromachining approach is the basis for successful commercial devices, such as automotive accelerometers,[5] and foundry services, such as polyMUMPs[6] and SUMMIT.[7] Other material combinations can be used, such as aluminum and photoresist[8] or silicon carbide and polysilicon/silicon dioxide.[9]

An alternative approach consists of forming the electromechanical structures into the thin films of a **CMOS wafer,**[10] leveraging the substantial

Table 20.1.2   Common combinations of Materials and Etching Approaches in Micromachining; Many Other Choices Are Available[2]

Material to pattern	Common patterning approaches	Characteristics	Masking materials and selectivity
Si, polySi	RIE (Cl-based)	Slow, directional	PR (5:1)
	DRIE (SF_6)	Fast, vertical profile	PR (70:1); SiO_2 (150:1)
	Wet (KOH, TMAH)	Anisotropic (54.74°)	SiN (>100:1)
SiO_2, glass, quartz	RIE (CF_4)	Directional; isotropic	PR (10:1)
	HF (diluted, BOE)	(quartz: anisotropic)	PR (>20:1)
Si_3N_4	RIE (CF_4)	Directional	PR (10:1); SiO_2 (50:1)
	Phosphoric acid	Isotropic, wet	SiO_2 (10:1); Si (30:1)
Metals	Lift-off (for Au)	Wet, thin film (<1 μm)	—
	RIE (Cl, F, Ar)	Directional, slow	PR (1:1)
	Etchant solutions	Isotropic, wet	PR (>100:1)

Fig. 20.1.5 Basic principles of surface micromachining: (a) thin silicon nitride (Si$_3$N$_4$) and silicon dioxide (SiO$_2$) films are deposited, and openings are etched in the SiO$_2$, using lithography and etching; (b) a polysilicon film is deposited (LPCVD) and (c) patterned; (d) the polysilicon structures are released by wet etching of the sacrificial SiO$_2$. Steps (a)–(c) can be repeated to create multilayer structures such as the gear train mechanism (Photo: *Sandia National Lab.*) shown in (e). Rotary structures are formed with a hub covering part of the rotor (f), separated by SiO$_2$ before release.

CMOS manufacturing resources and allowing natural integration with electronics. The structures can also extend into the silicon substrate by using bulk micromachining techniques.

Bulk Micromachining

Alternatively, the structures can be formed directly into the substrate, which is referred to as bulk micromachining. Three types of processing methods are available, described by the etchant state: wet, vapor, or plasma (the last two are referred to as dry etching). Although wet etching typically creates isotropic features (rounded etching in all directions), unique formulations can leverage the microstructure of single-crystal silicon to form anisotropic features. Hydroxide-based etchants (such as KOH and TMAH) etch much slower in one direction that the others, exposing the {111} planes of silicon. Depending on the crystallographic orientation of the wafer, {111} planes are either at 54.74° to the wafer surface (<100> silicon wafers) or perpendicular to it (<110> silicon wafers). Furthermore, the etch rate can also depend on dopant type and concentration, such that areas of the wafer that were previously diffused with a dopant can remain after wet etching. These characteristics allow the fabrication of a wide range of geometries, such as pyramidal pits, channels, cantilevers, or membranes as illustrated in Fig. 20.1.6a, but with geometric constraints to the crystal planes.

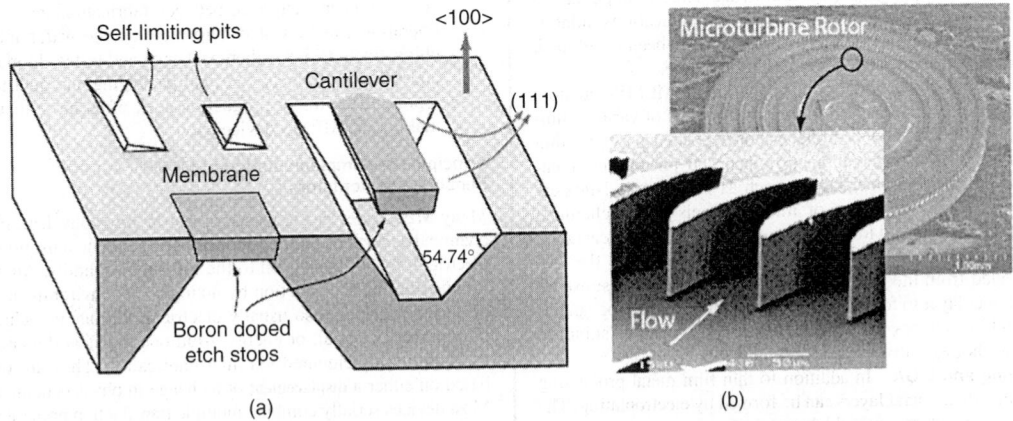

Fig. 20.1.6 Bulk micromachining of Si: (a) Anisotropic wet etching of Si. Typical anisotropic features that leverage the crystalline structure of Si. The {111} planes and boron-doped regions etch at significantly lower rates than the rest during wet etching with KOH, EDP, or TMAH; (b) deep reactive ion etch (DRIE) structures, showing that unconstrained 2-D shapes are possible, but only with vertical sidewalls.

Commercial pressure sensors are commonly fabricated as single-crystal silicon membranes by anisotropic wet etching, as illustrated in Fig. 20.1.1*a*.

Alternatively, plasma processing can allow **deep reactive ion etching** (DRIE) into silicon substrates of practically any lithographically defined geometry, as illustrated in Fig. 20.1.6*b*. The DRIE process combines isotropic plasma etching with sidewall passivation in order to generate deep trenches with vertical side walls,[11] reaching aspect ratios above 30:1 and fast enough etch rates (>3 μm/min) for through-wafer etching. Compared to wet etching, DRIE allows greater design freedom but requires expensive equipment and serial processing.

Wafer Bonding

In contrast to thin structures created by surface micromachining, thick multilayer devices can be formed by joining patterned or blank substrates together in a process referred to as wafer bonding. Three main bonding techniques are available: anodic bonding (glass to silicon), direct fusion bonding (silicon to silicon), and intermediate layer bonding. The planar substrates are first cleaned and brought into contact in a controlled ambient.

In **anodic bonding,** a potential (500 to 1,000 V) is applied across a contacted pair of sodium-bearing glass and silicon wafers, elevated to a moderate temperature (300 to 450°C). Migration of the sodium ions generates strong attraction forces and a permanent chemical bond is formed between the substrates. This approach is commonly used to create optically accessible flow channels in microfluidic devices and to thermally and mechanically isolate sensors from their package and environment. **Direct fusion bonding** relies on molecular diffusion at the interface of two substrates brought in intimate contact. Silicon wafers can be fusion bonded by aligning, pressing, and heating (800 to 1000°C) them. The bond strength can reach that of the crystalline substrate, but requires clean and polished surfaces (roughness < 1 nm) and precludes materials that would not withstand the high temperature process, such as most metals. Alternatively, an **intermediate layer** can be used to join two substrates, such as eutectic bonds (Au layer), polymers (resists, polyimide), solders, low-melting-temperature glass (frits) or thermocompression bonds. Some intermediate-layer bonds require only low temperatures, but may not allow long-term vacuum cavities due to outgassing (for glasses, polymers, and solders) and yield lower bond strengths.

Wafer bonding is commonly combined with bulk micromachining or thin-film processes to encapsulate sensors in vacuum or controlled environments (wafer-level packaging), form closed cavities for absolute pressure sensors, or create complex microfluidic devices (such as microturbines[12]).

Nonsilicon Materials and Processing

In order to meet the requirements for a growing range of applications such as biomedicine, chemistry, and wireless communications, nontraditional microfabrication materials and processes have been developed. Polymers and metal electroplating are highlighted here.

Polymers   The need for low-rigidity materials in MEMS actuators and polymers for **bioMEMS** has led to the development of various thin- and thick-film polymers. For example, vapor-deposited parylene thin films can serve as flexible membranes, joints, and pneumatic actuators.[13] Casting of thick polymer layers, such as polydimethylsiloxane (PDMS), allows rapid fabrication of microchannels for biochemical MEMS. In a process referred to as **soft lithography,**[14] PDMS is cast over a micromachined substrate that acts as a mold. After curing, the cast polymer is pealed from the mold and pressed against a glass substrate or another polymer layer to form microfluidic geometries. Microchannels and other thick structures can also be directly formed in standard photoresists or photosensitive epoxy, such as SU-8.[15]

Electroplating and LIGA   In addition to thin film metal processing described earlier, thick metal layers can be formed by electroplating. The most common approach to create thick metal microstructures is **selective electroplating,** in which a seed layer is covered by a patterned thick photoresist or SU-8 mask, such that the metal (e.g., Ni) is plated only in the exposed areas of the mask. The LIGA process (German acronym for

lithography, electroplating, and molding) implements this approach by using X-ray lithography to expose very thick photoresist (several millimetres) with vertical sidewalls.[16] This yields very high aspect ratio metal parts (up to 70:1) for MEMS and other micromachines.

ENGINEERING SCIENCE AND PRACTICE FOR MEMS DESIGN

Many of the engineering topics covered throughout this handbook can be applied for the design of microsystems. Engineering practice, however, differs as a result of miniaturization and convergence with the semiconductor world. Here we discuss the design approach for MEMS, the effects of scaling on engineering science, and the most common transduction and operating principles, along with examples of MEMS devices. The reader is referred to the literature[17] for an extended discussion of MEMS design.

MEMS Design Process

Depending on the device complexity and scope of the application, development of MEMS may entail:

1. *Definition of market needs and product specifications.* An assessment must be made to determine if MEMS technology is appropriate for the application, by answering the following questions:
   - Does a MEMS and microfabrication approach bring unique capabilities and performance levels due to: fundamental scaling, unique properties of micromachined materials, potential integration with microelectronics, or large array integration?
   - Is there an economic benefit at the component or system level for using MEMS and microfabrication, such as low unit cost due to batch fabrication, on-chip functionality, or miniaturization?
2. *Choice of operating principle and fabrication process, system-level design, and detailed design.* The conceptual device is defined by using previously demonstrated technology and creative thinking. The principles of operation and the microfabrication approach are intimately related since the implementation is constrained by available fabrication technologies. Analytical and numerical modeling of increasing complexity along with detailed process flow definition completes the device design. Traditional or specialized CAD software can facilitate this effort.[18] Packaging considerations are critical at the design phase to ensure appropriate mechanical, thermal, fluidic, biological, chemical, electrical, magnetic, and/or optical interfaces between the device and the environment.
3. *Process development, fabrication, packaging, and testing.* Lithographic masks are made with the designed patterns and used for device microfabrication. Each fabrication process is developed or improved to achieve the desired microstructures. Iterations with the design phase may be required, ranging from slight mask modifications to adopting alternative process flows. The process accuracy is characterized between fabrication steps. Basic device operation can be tested before dicing the wafer into individual chips. Each device is then packaged and tested. In many commercial MEMS products, the final packaging and qualification steps represent the major production cost, often more than the cost of making the MEMS device itself.

Principles and Implementation of MEMS Sensors and Actuators

Many MEMS devices are sensors that monitor physical, biological, or chemical aspects of their environment to generate a useful signal, often electrical, that is correlated to the measured quantity. An actuator will provide the inverse function by acting on its environment on the basis of an input signal, also usually electrical. Various transduction principles convert a signal, or energy, from one physical domain to another, and can be implemented via microfabrication. They are of two types, based on either a **displacement** or a change in physicochemical **properties.** Most devices serially combine multiple transduction principles on chip to achieve high levels of functionality and sensitivity. The most common approaches are presented here with device examples. The reader is referred to the literature[19] for a comprehensive discussion of transduction mechanisms and their MEMS implementations.

Mechanical Elements: Flexures, Mechanisms, and Other Microstructures

Most MEMS consist of micromechanical components and mechanisms that provide a displacement, in combination with other transduction principles. Some of the most common mechanical elements are depicted in Fig. 20.1.7, and can be classified as flexible structures, mechanisms, or static structures. These include: beams that act as cantilever beams or springs; plates that form membranes, bridges, mirrors, or masses; hub-and-pin joints for rotary gears and motors (see Fig. 20.1.5e–f); hinges for large deflections; and microchannels and other internal flow components to guide fluid flow (see Fig. 20.1.6b). Continuum elasticity modeling is generally applicable for such components, although the geometries can achieve extreme aspect ratios (up 1000:1) and the material properties can be process-dependent and anisotropic, as discussed previously.

These components can be used in static behavior, or for a desired dynamic response. Micromachined spring-mass systems can be excited at resonance and used as gyroscopes through the Coriolis effect or as chemical sensors by monitoring changes in stiffness or mass. Dynamic behavior, such as resonance and time response, are also affected by fluid film damping.[17] Such sensors are typically packaged under vacuum to reduce this effect, and therefore achieve a large quality factor Q, which is defined as the amplitude at resonance normalized by the displacement at low frequency.

Capacitive Sensing and Electrostatic Actuation

Displacements can be measured or induced with electrostatic charges. Capacitive sensing consists of measuring the change in capacitance between two electrodes (i.e., conductive structures), as for the commercial accelerometer illustrated in Fig. 20.1.7a and b. Neglecting edge effects, capacitance between parallel plates is simply expressed as $C = \varepsilon A/g$, where a change in capacitance C can result from a change in: overlapping plate area A, due to parallel motion of the plates (electrodes); gap g, due to normal motion of the plates; or even a change in permittivity ε, by insertion of a different material in the gap. When the sensor is combined with on-chip circuits for capacitance measurement, extreme sensitivity can be achieved since parasitic capacitance is reduced, especially when a differential change in capacitance is measured.

Electrostatic actuation is accomplished by applying an electrostatic field between structures attached to flexures or mechanisms that allow relative motion. For simple parallel plates, attraction forces are experienced normal to the plates and parallel to them (when they are offset), as illustrated in Fig. 20.1.8a. Since electrostatic forces are proportional to V^2 and vary inversely to the spacing, d or d^2, high voltages and small gaps are desirable. The maximum force is limited by breakdown across the air gap, which occurs above 300 V for micrometre-scale gaps at atmospheric pressure. Comb drive actuators provide in-plane motion (Fig. 20.1.8b), while micromirror devices for display applications use electrostatic actuation to tilt a mirror out of plane and redirect light that reflects on an array of individually actuated micromirrors (see Ref. 8 and Fig. 20.1.1b).

Piezoresistive Sensing

Stain gages are commonly used in mechanical testing to measure deformations and the related stress in structures. In MEMS, the same principle is used to measure inertial or pressure forces, by transmitting the force to a silicon structure and measuring the induced strain. Crystalline silicon offers the benefit of a large gage factor (GF > 100)

Fig. 20.1.7 Microstructures commonly used in MEMS. (*a*) Capacitive accelerometer. (*Photo: Analog Devices.*) The surface micromachined accelerometer consists of a mass supported on folded beams that act as springs. (*b*) Components and operation. Displacement is measured from the change in capacitance between the stationary and movable electrodes. (*c*) Thermopneumatic valve. Flexible membranes can be used in valves to control the flow through microchannels, as well as in pressure sensors.

Fig. 20.1.8 (*a*) Electrostatic attraction forces between electrodes with components normal and parallel to the structures. For example, the parallel force is used in (*b*) comb drive actuators to move an electrode structure held by a spring. This configuration is similar to capacitive accelerometers but it is used for actuation instead of sensing. (*Photo: Sandia National Laboratory.*)

compared to metals (GF ≈ 2), where the gage factor is defined as the relative change in resistance per unit strain, $GF = \frac{\Delta R/R}{\Delta L/L} = \Delta R/\varepsilon R$. Typical implementations for pressure sensors consist of resistances that are diffused on the surface of a bulk micromachined silicon membrane (Fig. 20.1.1*a*). The piezoresistors are connected in a Wheatstone bridge configuration to provide a voltage measurement proportional to the applied pressure difference across the membrane. Piezoresistive accelerometers can operate on a similar principle, by locating a resistor on a silicon beam that supports a proof mass etched from the substrate. Commercial devices can include on-chip temperature compensation, stress isolation with the package, and mechanical stops to prevent excessive stresses.[20]

Piezoelectric Sensing and Actuation

Certain crystalline materials, such as quartz, lead titanate zirconate (PZT), and zinc oxide (ZnO), exhibit an electrical polarization under mechanical strain due to a relative displacement of negative and positive charge centers in the material's microstructure, a phenomenon called **piezoelectricity**.[2] This behavior can be leveraged for force sensing, but also for actuation since an applied electric field will in turn induce mechanical deformation. Traditionally, piezoelectric materials can provide large forces or pressures (>10 MPa) and small displacements (<0.1 μm) at high voltages (>100 V), but piezoelectric thin films applied on flexible beams and membranes in MEMS devices allow larger deflections (up to 10 μm), although at proportionally lower forces. Some commercial ink-jet print heads integrate piezoelectric actuators to generate microdroplets at high frequencies.

Thermal Transduction and Applications

Microsystems can also integrate temperature measurement capabilities, use heat for actuation, or act on their thermal environment. Traditional approaches to measure temperature, such as thermocouples and thermistors, can be implemented via thin-film deposition and patterning to create a solid-state temperature sensor for integration within MEMS devices. For example, resistors can be formed by depositing a thin film of metal or doped polysilicon, characterized by its sheet resistivity $R_{sheet} = \rho/t$, where ρ is the material's electrical resistivity and t is the film thickness. Using photolithography and etching (or through lift-off), the resistor can be patterned into a line of total length l and width w, for a resistance of $R = R_{sheet}\, l/w$ (Fig. 20.1.9*a*). Since the material resistivity is a function of temperature, monitoring electrical resistance provides a measurement of temperature, according to $\Delta R/R_0 = \alpha \Delta T$, where α is the material's **temperature coefficient of resistance** (TCR). Thermocouple junctions have also been microfabricated by depositing and patterning multiple thin metal films. Heat flux can be measured by monitoring the temperature of two overlaid resistors or thermocouple

Fig. 20.1.9 Thermal sensing and actuation using thin films. A patterned thin-film resistor (*a*) can act as a temperature sensor by monitoring the change of resistance with temperature, or as a heat source through ohmic heating. A thermal bimorph cantilever (*b*) will curve when heated, so it can act as a thermal sensor or an actuator when electrically heated.

junctions separated by a thermally and electrically insulating film, such as a SiO_2, Si_3N_4, or polyimide.

A resistor can also be used as a heat source, converting electrical energy into heat at a rate of $\dot{Q} = P_{elec} = Vi = Ri^2$. This allows the integration of electrically controlled heaters in MEMS and microfluidic devices. Heat can be a driving mechanism for actuation by raising the local temperature of microstructures or trapped gases to induce thermal expansion, which then results in a membrane or beam deformation. Micropumps, microvalves, and switches can be based on this operating principle, as illustrated in Fig. 20.1.7*c*. Here, an enclosed fluid tends to expand because of resistive heating, increasing the pressure in the cavity and pushing the membrane upward to close the valve. Alternatively, out-of-plane deflections of thin-film cantilevers can be achieved by layering materials with different coefficients of thermal expansion (CTE), referred to as bimorph actuators, as illustrated in Fig. 20.1.9*b*.

To minimize the actuation power and provide accurate temperature measurements, thermal isolation can be achieved through the use of high-aspect-ratio structures and low-thermal-conductivity materials, such as SiO_2 or Si_3N_4 films and glass substrates. Microsystems for cooling are less advanced, but potential approaches include thermoelectrics and microchannel heat exchangers, with single or two-phase flow.

Optical Transduction and Applications

Semiconductor technology was also leveraged to create optoelectronic components such as photodiodes and light-emitting diodes. Silicon-based

MEMS can therefore integrate not only microelectronics but also optical elements for a broader range of transduction principles. When combined with mechanical and/or fluidic elements, optoelectronics can measure the displacement of a shutter that partly covers photodiodes, or detect biological species on-chip with an emitter-detector pair along a microfluidic channel. Light manipulation can also be accomplished by reflecting incident light onto discrete, controllable, movable micromirrors. Commercial devices include actuated micromirror arrays that redirect light onto a display[8] (Fig. 20.1.1b) and variable diffraction gratings to separate wavelengths of the light for analytical or telecommunication applications.[21]

Other Transduction Principles and Applications

The range of transduction principles extends beyond the most common ones presented above. It includes principles from material science, such as shape memory alloys; from physics, such as electron tunneling; from chemistry, such as electrochemical spectroscopy; and from biology. Novel approaches are generally spurred by specific needs for sensing, actuation, or miniaturization requirements, but are often limited by the microfabrication technologies and capabilities. The reader is referred to scientific journals for the latest advancements in MEMS science and technology (see references).

PHYSICS AND SCALING

To assess the potential benefits of MEMS technology for a given application or compare various operating principles, it is useful to consider the effects of miniaturization through scaling laws. Two main aspects are to be considered: the length-scale dependence of various physical phenomena and the effect of scale on physical properties and behavior.

Length-Scale Dependence

Most physical phenomena can be classified as **volumetric,** such as gravity or heat capacity, or **surface-dependent,** such as viscous forces and heat transfer. When a physical system is miniaturized, volume scales as $[l^3]$, whereas surface scales as $[l^2]$, where l is a characteristic dimension of the system. The ratio of volume to surface therefore scales as $[l^3/l^2] = [l]$, such that surface phenomena tend to prevail at small scales; this is referred to as the **cube-square law.** This first-order analysis is applied for various physical domains in Table 20.1.3, showing, for example, that electrostatic forces, viscous forces, capillary forces, and heat transfer become prevalent upon miniaturization. In particular at macroscale, electrostatic transduction is not common because of its low energy density compared to magnetic devices. As they are miniaturized, however, electrostatic forces scale as $[l^2]$ or remain constant, assuming constant field strength or constant voltage, respectively. In comparison, magnetic forces are more affected by scaling, and their three-dimensional windings are challenging to microfabricate.

True scaling may not follow these trends exactly, since the capabilities of micromachining allow aspect ratios and material properties that are significantly different from those possible with traditional machining, as discussed next.

Physical Properties and Behavior

Not only can commonly neglected forces and phenomena become important at small scales, but physical behavior can shift dramatically. In traditional engineering, the smallest device features are typically much greater than the atomic or molecular scales of the device's constituents, such that the material behavior can be represented by continuum mechanics. As device scales are reduced, this representation may not hold and discrete molecular interactions become important. In microfluidics, for example, gas flows become rarefied when the mean

Table 20.1.3   Scaling of Forces and Fluxes

Field, forces, fluxes	Basic relation	Scaling proportional to	Note
Fluid mechanics			
Inertial forces	$\frac{1}{2}\rho A U^2$	$[l^2]$	Reynolds number:
Viscous forces	$\tau A = \mu \dfrac{\partial U}{\partial y} A$	$[l^1]$	$\mathrm{Re} = \dfrac{\text{inertial}}{\text{viscous}} = \dfrac{\rho U L}{\mu} \propto [l^1]$
Heat transfer			
Conduction	$q_{\text{cond}} = -k \dfrac{\partial T}{\partial y} A$	$[l^1]$	Biot number:
Convection	$q_{\text{conv}} = h A \Delta Te \quad (h \propto l^{0.5})$	$[l^{2.5}]$	$Bi = \dfrac{\text{convection}}{\text{conduction}} \propto [l^{1.5}]$
Energy transport	$\dot{m}\, c_p T = \rho A U\, c_p T$	$[l^2]$	Stanton number:
			$St = \dfrac{\text{convection}}{\text{transport}} \propto [l^{0.5}]$
Mechanics		$[l^1]$	
Strain energy	$\frac{1}{2}\sigma\varepsilon\, AL = \frac{1}{2}\dfrac{\sigma}{E} AL$	$[l^3]$	
Beam bending	$F = kx \quad \left(k = \dfrac{Ewt^3}{L^3}\right)$	$[l^2]$	
Natural frequency	$\omega = \sqrt{k/m}$	$[l^{-1}]$	
Electromagnetism			
Electrostatic forces	$F = \dfrac{1}{2}\dfrac{\varepsilon V^2}{d^2} A$	$[l^0]$ $[l^2]$	For constant voltage V For constant field (V/d)
Magnetic forces	$F = \dfrac{1}{2}\dfrac{B^2}{\mu_0} A$	$[l^2]$	
Piezoelectric forces	$F = \dfrac{1}{2}\dfrac{\varepsilon V}{t} A$	$[l^1]$	
Physicochemistry		$[l^1]$	
Van der Walls		$[l^{1/4}]$	
Surface tension	$F = \gamma 2\pi r \cos(\theta_c)$	$[l^1]$	Bond number:
Gravity	$F = mg$	$[l^3]$	$Bo = \dfrac{\text{surface}}{\text{gravity}} \propto [l^{-1}]$

free path between molecules, λ, is comparable to channel widths d, as represented by the Knudsen number, Kn $= \lambda/d$. For Kn < 0.01, flow can be considered a continuum with zero relative velocity along surfaces. In the range $0.01 < $ Kn < 0.1, slip flow may occur at the surfaces, which can be accounted for in continuum modeling through appropriate boundary conditions.[22] For air at atmospheric conditions, this range corresponds to 0.5- to 5-μm gaps, which are relatively common in MEMS. For Kn > 0.1, however, the flow is best represented by statistical molecular interactions. Similar considerations lead to quantum effects for the heat transfer through thin films (less than 0.1 μm) or through nanostructured materials, such as superlattices. Modeling of heat conduction as phonon transport along or through these structures becomes necessary.[23]

Even in the continuum domain, physical phenomena can differ with scale. In microfluidics, flows are typically laminar because of low Reynolds numbers, Re $= \rho U d/\mu < 1000$–2000, where ρ is fluid density, U is a characteristic velocity, and μ is the fluid viscosity. Such flows are well represented by laminar flow theory, such as Poisseuille flow, Couette flow, and more generally, Stokes flow and the Reynolds equation for lubrication theory.[24] The absence of turbulence increases the challenges of mixing in biochemical microsystems, but can also be leveraged to form layered sheets of fluids with well-defined diffusion-based concentration profiles. For liquids, such as water and electrolytes, electrical surface phenomena can become noticeable in microchannels. A charged double layer can form at the surfaces, such that an electrical field applied along the channel will tend to drive the flow along the wall. Because of the small channel diameter, the core flow is entrained by viscous shear to create a plug-flow, uniform velocity profile. This approach requires high electrical fields, which are achievable in microsystems with reasonably high voltages (up to 1000 V) over short channels lengths (millimetres). This approach is commonly used to drive fluids in micro total analysis and lab-on-a-chip devices, and is referred to as **electro-osmosis**.

In electrostatics, the maximum field strength (V/d) at which breakdown occurs is a function of scale. At small gaps ($d < 5$ μm, atmospheric pressure), the voltage limit stops decreasing to reach a minimum near 300 V; it even increases at submicrometre gaps. This results from the lack of molecules in the air gap to create the avalanche of collisions that lead to arcing. This benefits electrostatic actuation by increasing its maximum energy density and force.

In micromechanics, the crystalline microstructure of silicon components formed by bulk micromachining introduces the need for anisotropic continuum mechanics to accurately predict the stress and strain.[25] Surface micromachined devices or other MEMS devices formed of thin films must be designed with consideration for their residual in-plane stress and stress gradient across the film thickness, which are highly dependent on the processing conditions. Material characterization and process control become critical aspects for the performance and reliability of such devices. In addition, the brittle nature of silicon-based materials excludes plastic deformation and their fracture strength is defined by maximum flaw size in the part. Fortunately, the probability of having a significant flaw in a brittle part increases with its volume and surface area, hence strength and reliability are favored in small scale structures.

BEYOND SENSORS AND ACTUATORS: INTEGRATED MICROSYSTEMS

Although most commercial MEMS can be classified as sensors or actuators, there is an increasing trend toward microsystems that integrate multiple components to perform specific applications. For example:

• Photonic and microoptoelectromechanical systems, or MOEMS, that direct, filter, and characterize light for optical telecommunication applications
• Radio frequency (RF) MEMS that generate and transmit high-frequency signals for wireless communication applications
• BioMEMS and lab-on-a-chip devices that perform high-throughput biological assays and revolutionize medical instrumentation

• Micro total analysis systems (μTAS) for synthesis and characterization of chemical and biochemical products
• Power MEMS to provide miniature power sources, cooling, and propulsion for consumer electronics, people, and small robots.

CONCLUSIONS

MEMS technology can be considered a toolbox to create miniature sensors, actuators, and integrated microsystems. The main tools are the materials and processes used in microfabrication, the engineering science with emphasis on small-scale phenomena, as well as the sensing and actuating principles. The parallel nature of microfabrication processes such as photolithography, thin-film deposition, and etching allows batch fabrication of devices with high accuracy and in large quantities. In-plane device complexity (in terms of number of components in a system) comes for free, allowing large arrays or complex networks to be implemented with micrometre-scale resolution. The devices can naturally integrate electrical and mechanical elements, as well as optical, fluidic, thermal, chemical, or biological functionality on chip.

The microfabrication infrastructure, equipment, process development, and control are, however, expensive, and are best amortized over large production quantities. To this cost must be added the costs of packaging and testing, which can exceed the cost of the chip itself, and the extensive engineering effort to achieve reproducibility and reliability, as required in commercial products. To be commercially successful, MEMS devices must typically offer increased functionality or performance at a lower cost than competing technologies, and in a high-volume market. Lower-volume applications can also be successful when the use of MEMS adds high value to the final product and the market share is protected by intellectual property or unique know-how. Whether commercially successful or not, an ever-increasing range of MEMS devices and technologies is being explored. This will contribute to our MEMS toolbox and enable novel solutions to our greatest challenges, ranging from nano to macro scales and spanning the physical, chemical, and biological worlds.

REFERENCES

1. *Got MEMS? 2003 Industry Overview and Forecast*, In-Stat/MDR, July 2003.
2. M. Madou, *Fundamentals of Microfabrication*, CRC Press, New York, 1997.
3. S. Wolf and R. N. Tauber, *Silicon Processing for the VLSI Era—Vol. 1: Process Technology*, 2nd ed., Lattice Press, Sunset Beach, CA, 2000.
4. O. M. Jadaan, N. N. Nemeth, J. Bagdahn, and W. N. Sharp Jr., "Probabilistic Weibull behavior and mechanical properties of MEMS brittle materials," *J. Materials Sci.*, **38**:4087–4113 (2003).
5. *ADXL* capacitive accelerometer, Analog Devices Inc. Norwood, MA.
6. *polyMUMPs* multiuser polysilicon surface micromachined process, MEMSCap, Bernin, France.
7. SUMMIT V polysilicon surface micromachined process, MEMX Albuquerque, New Mexico.
8. L. Hornbeck, "Digital Light Processing and MEMS: Timely Convergence for a Bright Future," *Proc. SPIE*, **2639**:2, Micromachining and Microfabrication Process Technology (1995) (www.dlp.com).
9. A. A. Yasseen, C. A. Zorman, and M. Mehregany, "Surface Micromachining of Polycrystalline SiC Films," *IEEE/ASME J. Microelectromechanical Systems* **8**(3):237–242 (1999).
10. G. Fedder et al., "Laminated High-Aspect-Ratio Microstructures in a Conventional CMOS Process," *Sensors & Actuators A*, **57**(2):103–110 (1996).
11. A. A. Ayón et al., "Characterization of a time multiplexed inductively coupled plasma etcher," *J. Electrochem. Soc.* **146**(1) (1999).
12. L. G. Fréchette et al., "High-Speed Microfabricated Silicon Turbomachinery and Fluid Film Bearings," *J. Microelectromechanical Systems*, **14**(1): 141–152 (2005).
13. Y. Lu and C. J. Kim, "Micro-finger Articulation by Pneumatic Parylene Balloons," *Proc. 12th Int'l Conf. on Solid State Sensors, Actuators and Microsystems*, Boston, June 8–12, 2003.
14. Y. Xia and G. M. Whitesides, "Soft Lithography," *Annu. Rev. Mater. Sci.*, **28**: 153–184 (1998).
15. H. Lorenz, M. Despont, P. Vettiger, P. Renaud, "Fabrication of photoplastic high-aspect ratio microparts and micromolds using SU-8 UV resist," *Microsystem Technologies*, **4**:143–146 (1998).

16. H. Guckel, "High-Aspect-Ratio Micromachining Via Deep X-Ray Lithography," *Proc. IEEE*, **86**(8):1586–1593 (1998).
17. S. D. Senturia, *Microsystem Design*, Kluwer Academic Publishers, Norwell, 2001.
18. S. D. Senturia, "CAD Challenges for Microsensors, Microactuators, and Microsystems," *Proc. IEEE*, **86**(8):1611–1626 (1998).
19. G. T. A. Kovacs, *Micromachined Transducers Sourcebook*, WCB McGraw-Hill, New York, 1998.
20. *Kulite Pressure Transducer Handbook*, Kulite Semiconductor Products, Inc., Leonia, NJ.
21. O. Solgaard, F. S. A. Sandejas, and D. M. Bloom, "Deformable grating optical modulator," *Optics Letters*, **17**:688–690 (1992).
22. G. E. Karniadakis and A. Beskok, *Micro Flows: Fundamentals and Simulation*, Springer, New York, 2002.
23. D. G. Cahill, W. K. Ford, K. E. Goodson, G. D. Mahan, A. Majumdar, H. J. Maris, R. Merlin, S. R. Phillpot, "Nanoscale Thermal Transport," *J. Applied Physics*, **93**:793–818 (2003).
24. V. N. Constantinescu, *Laminar Viscous Flow*, Springer, New York, 1995.
25. W. M. Lai, D. Rubin, and E. Krempl, *Introduction to Continuum Mechanics*, Butterworth–Heinemann, New York, 1996.

Scientific Journals

Journal of Microelectromechanical Systems, ASME/IEEE.
Sensors and Actuators, Elsevier.
Lab on a chip, RSC Publishing.
Journal of Micromechanics and Microengineering, Institute of Physics.
Sensors Journal, IEEE.

20.2   INTRODUCTION TO NANOTECHNOLOGY
by Alexander Couzis

Nanotechnology focuses on the fabrication of devices with atomic- or molecular-scale precision. Molecular nanotechnology is defined as the three-dimensional positional control of molecular structure to create materials and devices to molecular precision. Devices with minimum feature sizes less than 100 nanometres (nm) are considered to be products of nanotechnology. A nanometre (nm) is one billionth of a metre (10^{-9} m), that is, about 1/80,000 of the diameter of a human hair, or 10 times the diameter of a hydrogen atom. The nanometre is the unit of length that is generally most appropriate for describing the size of single molecules. The nanoscale marks the interface between the classical and quantum-mechanical worlds. Nanoscience is an interdisciplinary field that encompasses physics, biology, engineering, chemistry, and computer science, among others, and the prefix nano appears with increasing frequency in scientific journals and the news. Thus, as we increase our ability to fabricate computer chips with smaller features and improve our ability to cure disease at the molecular level, nanotechnology is at the ready. Scientists and engineers believe that the fabrication of nanomachines, nanoelectronics, and other nanodevices will help solve numerous problems faced by mankind today related to energy, health, and materials development. Nanotechnology, currently in its early stages, has already provided us with the ability to organize matter on the atomic scale, and there are already numerous products available as a direct result of our rapidly increasing ability to fabricate and characterize feature sizes less than 100 nm. Transparent nanoparticulate titania-based ultraviolet-absorbing coatings,[1–3] particle-based methods for delivery of highly potent anticancer drugs,[4] mirrors that don't fog, biomimetic* coatings with a water contact angle near 180°,[5,6] and gene and protein chips[7,8] are some of the first manifestations of nanotechnology. Breakthroughs in computer science and medicine will be where the real potential of nanotechnology will first be achieved; some breakthroughs are not that for off.

A BRIEF HISTORY OF NANOTECHNOLOGY

The amount of space available to us for information storage (or other uses) is enormous. Richard P. Feynman,[9] in a 1959 lecture titled "There's Plenty of Room at the Bottom," stated that there is nothing besides our clumsy size that keeps us from using this space. At the time of that lecture, it was not possible to manipulate single atoms or molecules because they were far too small for the existing tools, yet Feynman correctly predicted that the time would come when atomically precise manipulation of matter would become possible.

Feynman also described such atomic scale fabrication as a **bottom-up** approach, as opposed to the **top-down** approach that we are accustomed to. The current top-down method for manufacturing involves the construction of parts through methods such as cutting, carving, and molding. Using these methods, we have been able to fabricate a remarkable variety of machinery and electronics devices. However, the sizes at which we can make these devices is severely limited by our ability to cut, carve, and mold. The current set of top-down tools that allow us to achieve nanoscale-sized devices include chemical vapor deposition, reactive ion etching, optical or deep ultraviolet (UV) or X-ray lithography, electron-beam (e-beam) lithography, ion implantation, etc. All these tools are well developed and have been used for a number of years in the microelectronics industry for the fabrication of central processing units (CPUs), memory, and photonic devices.[10]

Bottom-up manufacturing, on the other hand, would provide components made of single molecules, which are held together by covalent forces that are far stronger than the forces that hold together macroscale components. Furthermore, the amount of information that could be stored in devices built from the bottom up would be enormous. The key obstacle in these approaches is the degree of our ability to exercise positional control over molecules and atoms. At the macroscopic scale, the idea that we can hold parts in our hands and assemble them by properly positioning them with respect to each other goes back to prehistory. At the molecular scale, the idea of holding and positioning molecules is new. One method that provides positional control is referred to as directed self-assembly. In a self-assembly-based process, the intermolecular interactions and the experimental conditions are manipulated precisely so that the parts themselves can spontaneously assemble into the desired structure. This is a well-established and powerful method of synthesizing complex molecular structures. A basic principle in self-assembly is selective affinity: if two molecular parts have complementary shapes and charge patterns—one part has a hollow where the other has a bump, and one part has a positive charge where the other has a negative charge—then they will tend to stick together in one singular way. By stirring these parts, aided by Brownian motion especially if the parts are in solution, the parts will eventually, purely by chance, be brought together in just the right way and combine into a bigger part. This bigger part can combine in the same way with other parts, letting us gradually build a complex whole from molecular pieces by stirring them together. Even though self-assembly is a path to the formation of nanostructures, by itself it would not be able to make the very wide range of products promised by nanotechnology. During self-assembly the molecular components move around and collide into each other in all kinds of ways, and if they aggregate in an undesirable manner, unwanted random structures will result. Many molecular components exhibit this problem, so self-assembly won't work for them.

*Systems that mimic biological equivalents.

We can avoid these issues if we can hold and position the parts; this leads to the concept of directed self-assembly. When two sticky parts do come into contact with each other, they'll do so in the right orientation because they are held in the right orientation. In short, positional control at the molecular scale should let us make things that would otherwise be difficult or impossible. Thus, in a competing approach, we develop positioning tools. One such tool is the scanning probe microscope (SPM).[11-13] In SPMs, a very sharp tip is brought down to the surface of the sample being scanned. By monitoring the interaction of the tip with the surface, we can tell that the tip is approaching the surface and so can "feel" the outlines of the surface in front of us. Many different types of physical interactions with the surface are used to detect its presence (see earlier references). Some scanning probe microscopes literally push on the surface and note how hard the surface pushes back. Others connect the surface and probe to a voltage source, and measure the current flow when the probe gets close to the surface. A host of other probe-surface interactions can be measured, and are used to make different types of SPMs. In all of them, the basic idea is the same: when the sharp tip of the probe approaches the surface, a signal is generated, a signal that allows us to map out the surface being probed. The SPM can not only map a surface, but also in many cases the probe-surface interaction can change the surface. This has already been used experimentally to spell out molecular words, and opportunities to modify the surface in a controlled way are being investigated both experimentally and theoretically.[14-16]

Since that initial preview of nanotechnology, several methods have been developed which prove that Feynman was correct. The most notable methods are scanning probe microscopy[1,14,17-22] and the corresponding advancements in supramolecular chemistry.[3,23,24] Scanning probe microscopy gives us the ability to position single atoms and/or molecules in the desired place exactly as Feynman had predicted. Feynman was critical of traditional chemistry's tedious procedures and the random nature of its reactions; recent advances have shown improved potential for use in nanotechnology.

SEMICONDUCTOR FABRICATION[25]

Computer chips (and the silicon-based transistors within them) are rapidly shrinking according to a predictable formula (by a factor of 4 every 3 years—Moore's law). According to the Semiconductor Industry Association's extrapolation of formulas such as this one (SIA roadmap), it is expected that the sizes of circuits within chips will reach the size of only a few atoms in about 20 years.

Since almost all of our modern computers are built from silicon semiconductor transistors patterned and carved by light (photolithography), the shrinking of circuits predicted by the SIA may not be economical in the future, despite the enormous investment made to constantly shrink and improve semiconductor electronics. Smaller circuits require less energy, operate more quickly and, of course, take up less space. Thus, Moore's law has held since computers first became commercially available. However, simple shrinking of components cannot continue much longer.

As a transistor such as the metal-oxide-semiconductor field-effect transistor (MOSFET; one of the primary components used in integrated circuits) is made smaller, both its properties and manufacturing cost change with scale. Currently, ultraviolet light is used to create the silicon circuits with a lateral resolution around 200 nm (the wavelength of ultraviolet light). As the circuits shrink below 100 nm, new fabrication methods must be created, but at increasing cost. Furthermore, once the circuit size reaches only a few nanometres, quantum effects such as tunneling must be reckoned with; this drastically affects the computer's ability to function normally. New methods for computer chip fabrication are constantly being sought by microchip manufactures.

ALTERNATIVE ARCHITECTURES FOR NANOCOMPUTING[26]

In addition to single electron transistors, two promising alternatives to traditional computers are molecular computing and quantum computing. These two methods are intimately related, yet deal with information on two different levels. Progress made in these areas during recent years has shown both to be feasible replacements for semiconductor chips.

Quantum computing seeks to write, process, and read information on the quantum level. It is at the nanoscale that quantum mechanical effects (such as wave-particle duality) become apparent. Numerous scientists are seeking ways to store information within the quantum mechanical realm. This is not a simple task. Since the laws of quantum mechanics involve unintuitive principles such as superposition and entanglement, a quantum computer would be able to violate some rules that limit our current computers. For instance, taking advantage of superposition allows a quantum bit of information, termed a qubit, to be used in several computations at the same time. Taking advantage of entanglement would allow information to be processed over long distances without the conventional requirement for wires.

Molecular computation is another method, complementary to quantum computing, that seeks to write, process, and read information within single molecules. One molecule that has proved most promising for molecular computation is deoxyribonucleic acid (DNA). DNA is a long polymer made of four different nucleotides that can be represented by the letters A, T, C, and G. The order or sequence of these nucleotides within DNA provides the information for making protein, the main components of the molecular-scale machinery used by living organisms to carry out life-sustaining functions.

Mathematicians have devised numerous ways to use DNA and the various proteins that come with it to carry out numerical computations that are notoriously difficult for silicon computers, namely NP-complete problems. The advantage that molecular computing using DNA has over conventional computing is that it is massively parallel. This means that each DNA molecule can function as a single processor, which greatly improves the speed of computation for complex problems.

MEDICAL APPLICATIONS

Molecular-scale manipulation of matter is receiving abundant attention in medicine. Since all living organisms are composed of molecules, molecular biology has become the primary focus of biotechnology. Countless diseases have been cured by our ability to synthesize small molecules, commonly know as "drugs," that interact with the protein molecules that make up the molecular machinery that keeps us alive. Our understanding of how proteins interact with DNA, phospholipids, and other biological molecules allows such progress.

Living systems exist because of the vast amount of highly ordered molecular machinery from which they are built. The central principle of molecular biology states that information required to build a living cell or organism is stored in DNA (which was described above for its use in molecular computation). This information is transferred from the DNA to the proteins by transcription and translation. These processes are all executed by various biomolecular components, mostly protein and nucleic acids.

Molecular biology and the study of these interactions have led to the discovery of numerous pharmaceuticals that have been enormously effective to combat disease. Understanding of molecular mechanisms including substrate recognition, energy expenditure, electron transport, membrane activity, and much more have greatly improved our medical technology.

This is related to nanotechnology for several reasons. First of all, it illustrates the capabilities of molecular-scale machinery. The goal of nanotechnology is molecular and atomic precision; thus nanotechnology has much to learn from nature. Copying, borrowing, and learning from nature is one of the primary techniques used by nanotechnology and has been termed biomimetics.[15,27-35] Secondly, our ability to design synthetic, semisynthetic, and natural molecular machinery gives us an enormous potential for curing disease and preserving life. See Refs. 36 and 37. The human body is composed of molecules, hence the availability of molecular nanotechnology will permit dramatic progress in human medical services. More than just an extension of "molecular medicine," nanomedicine will employ molecular machine systems to address medical problems, and will use molecular knowledge to maintain and improve human health at the molecular scale. Nanomedicine will have extraordinary and far-reaching implications for the medical profession, for the definition of disease, for the diagnosis and treatment of medical conditions including aging, and ultimately for the improvement and extension of natural human biological structure and function.

MOLECULAR SIMULATION[38,39]

A branch of computer science that is allowing rapid progress to be made in nanotechnology is the computer simulation of molecular-scale events. Molecular simulation is able to provide and predict data about molecular systems that would normally require enormous effort to obtain physically. By organizing virtual atoms in a molecular simulation environment, one can effectively model nanoscale systems.

Current limitations of molecular simulation techniques are the molecular simulation algorithm and computation time for complex systems. Force-field algorithms are currently quite efficient and are often used today. However, such models neglect electronic properties of the system. In order to calculate electron density, quantum mechanical models are required. However, as the number of atoms and electrons is increased, the computational complexity of the model quickly reaches the limits of our most modern supercomputers. Thus, as the computational abilities of our computers are improved (often with help from nanoscience), increasingly complex systems will be within the reach of molecular simulation.

CARBON NANOTUBES

Carbon nanotubes are one of nanotechnology's most mentioned building blocks. With one hundred times the tensile strength of steel, thermal conductivity better than all but the purest diamond, and electrical conductivity similar to copper, but with the ability to carry much higher currents, they seem to be a wonder material.

Carbon nanotubes come in a variety of flavors: long, short, single-walled, multiwalled, open, closed, with different types of spiral structure, etc. Each type has specific production costs and applications. Some have been produced in large quantities for years while others are only now being produced commercially with decent purity and in quantities greater than a few grams.

Carbon nanotubes were "discovered" in 1991 by Sumio Iijima of NEC and are effectively long, thin cylinders of graphite. Graphite is made up of layers of carbon atoms arranged in a hexagonal lattice, like chicken wire (see Fig. 20.2.1). Though the chicken wire structure itself is very strong, the layers themselves are not chemically bonded to each other but held together by weak Van der Waals forces.

Taking one of these sheets of chicken wire and rolling it up into a cylinder and joining the loose wire ends will result in a carbon nanotube. In addition to their remarkable tensile strength (usually quoted as

Fig. 20.2.1   Layer structure of graphite.

100 times that of steel at one-sixth the weight), carbon nanotubes have shown other surprising properties. They can conduct heat as efficiently as most diamond (only diamond grown by deposition from a vapor is better), conduct electricity as efficiently as copper, yet can also be semiconducting. They can produce streams of electrons very efficiently (field emission), which can be used to create light in displays for televisions or computers, or even in domestic lighting, and they can enhance the fluorescence of materials they are close to. Their electrical properties can be made to change in the presence of certain substances or as a result of mechanical stress. Nanotubes within nanotubes can act like miniature springs, and they can even be stuffed with other materials. Nanotubes and their variants hold promise for storing fuels such as hydrogen or methanol for use in fuel cells, and they make good supports for catalysts.

One of the major classifications of carbon nanotubes is into single-walled varieties (SWNTs), which have a single cylindrical wall, and multiwalled varieties (MWNTs), which have cylinders within cylinders (see Fig. 20.2.2).

The lengths of both types vary greatly, depending on the way they are made, and are generally microscopic rather than nanoscopic, i.e., greater than 100 nanometres. The aspect ratio (length divided by diameter) is typically greater than 100 and can be up to 10,000. Recent developments continue apace.

Fig. 20.2.2   Computational image of single- and multiwalled nanotubes.

Fig. 20.2.3   Representation of a multiwalled carbon nanotube.

Single-Walled Carbon Nanotubes (SWNTs)

These are the stars of the nanotube world, and are much harder to make than the multiwalled variety. The oft-quoted amazing properties generally refer to SWNTs. As previously described, they are basically tubes of graphite and are normally capped at the ends (see Fig. 20.2.2), although the caps can be removed. The caps are made by mixing in some pentagons with the hexagons and are the reason that nanotubes are considered close cousins to buckminsterfullerene, a roughly spherical molecule made of 60 carbon atoms that looks like a soccer ball and is named after the architect Buckminster Fuller. (The word *fullerene* is used to refer to the variety of such molecular cages, some with more carbon atoms than buckminsterfullerene, and some with fewer.)

The theoretical minimum diameter of a carbon nanotube is around 0.4 nm, which is about as wide as two silicon atoms side by side; nanotubes of this size have been made. Average diameters tend to be around 1.2 nm, depending on the process used to create them. SWNTs are more pliable than their multiwalled counterparts and can be twisted, flattened, and bent into small circles or around sharp bends without breaking.

Discussions of the electrical behavior of carbon nanotubes usually relate to experiments on the single-walled variety. They can be conducting, like metal (such nanotubes are often referred to as metallic nanotubes), or semiconducting. The latter property has given rise to dreams of using nanotubes to make extremely dense electronic circuitry, and the year 2005 saw major advances in creating basic electronic structures from nanotubes in the lab, from transistors up to simple logic elements. The gap between these experiments and commercial nanotube electronics is extremely wide.

Multiwalled Carbon Nanotubes (MWNTs)

Multiwalled carbon nanotubes are basically concentric SWNTs—cylindrical graphitic tubes. In these more complex structures, the different SWNTs that form the MWNT may have quite different structures (length and chirality). MWNTs are typically 100 times longer than they are wide and have outer diameters mostly in the tens of nanometres.

Although it is easier to produce significant quantities of MWNTs than SWNTs, their structures are less well understood than single-wall nanotubes because of their greater complexity and variety. Multitudes of exotic shapes and arrangements, often with imaginative names such as bamboo trunks, sea urchins, necklaces, and coils, have also been observed under different processing conditions. The variety of forms may be interesting but also has a negative side—MWNTs always (so far) have more defects than SWNTs, and these diminish their desirable properties.

Many of the nanotube applications now being considered or put into practice involve multiwalled nanotubes, because they are easier to produce in large quantities at a reasonable price and have been available in decent amounts for much longer than SWNTs. In fact, one of the current major manufacturers of MWNTs, Hyperion Catalysis, does not even sell the nanotubes directly, but only premixed with polymers for composite applications. The tubes typically have 8 to 15 walls and are around 10 nm wide and 10 μm long.

Focus on Structure

A few key studies have explored the structure of carbon nanotubes by using high-resolution microscopy techniques. These experiments have confirmed that nanotubes are cylindrical structures based on the hexagonal lattice of carbon atoms that forms crystalline graphite. Three types of nanotubes are possible, called armchair, zigzag, and chiral nanotubes, depending on how the two-dimensional graphene sheet is "rolled up."

The different types are most easily explained in terms of the unit cell of a carbon nanotube—in other words, the smallest group of atoms that defines its structure (see Fig. 20.2.4a). The so-called chiral vector of the nanotube, C_h, is defined by $C_h = n\bar{a}_1 + m\bar{a}_2$, where $\bar{a}_1$ and $\bar{a}_2$ are unit vectors in the two-dimensional hexagonal lattice and n and m are integers. Another important parameter is the chiral angle, which is the angle between C_h and $\bar{a}_1$.

When the graphene sheet is rolled up to form the cylindrical part of the nanotube, the ends of the chiral vector meet each other. The chiral vector thus forms the circumference of the nanotube's circular cross section, and different values of n and m lead to different nanotube structures (see Fig. 20.2.5). Armchair nanotubes are formed when $n = m$ and the chiral angle is 30°. Zigzag nanotubes are formed when either n or m are zero and the chiral angle θ, as shown in Fig. 20.2.4a, is 0°. All other nanotubes, with chiral angles intermediate between 0° and 30°, are known as chiral nanotubes.

The properties of nanotubes are determined by their diameter and chiral angle, both of which depend on n and m. The diameter d_t is simply the length of the chiral vector divided by ¼, and we find that $d_t = (\sqrt{3}/\pi)a_{c-c}(m^2 + mn + n^2)^{1/2}$, where a_{c-c} is the distance between neighboring carbon atoms in the flat sheet. In turn, the chiral angle is given by $\tan^{-1}\left[\sqrt{3}n/(2m + n)\right]$.

Measurements of the nanotube diameter and the chiral angle have been made with scanning tunneling microscopy and transmission electron microscopy. However, it remains a major challenge to determine d_t and θ at the same time as measuring a physical property such as resistivity. This is partly because the nanotubes are so small, and partly because the carbon atoms are in constant thermal motion. Also, the nanotubes can be damaged by the electron beam in the microscope.

How to Make Nanotubes

When a Rice University group found a relatively efficient way to produce bundles of ordered SWNTs in 1996, they opened new opportunities for quantitative experimental studies on carbon nanotubes. These ordered nanotubes are prepared by the laser vaporization of a carbon target in a furnace at 1,200°C. A cobalt-nickel catalyst helps the growth of the nanotubes, presumably because it prevents the ends from being "capped" during synthesis, and about 70 to 90 percent of the carbon target can be converted to SWNTs. By using two laser pulses 50 ns apart, growth conditions can be maintained over a larger volume and for a longer time. This scheme provides more uniform vaporization and better

(a)

(b)

Fig. 20.2.4 Basics of structure.

control of the growth conditions. Flowing argon gas sweeps the nanotubes from the furnace to a water-cooled copper collector just outside of the furnace.

At the University of Montpellier (France) Catherine Journet, Patrick Bernier, and colleagues later developed a carbon-arc method to grow similar arrays of SWNT. In this case, ordered nanotubes were also produced from ionized carbon plasma, and joule heating from the discharge generated the plasma. Several other groups are now making bundles of SWNTs, using variants of these two methods. However, the Rice group has had the largest impact on the field, largely because it was the first to develop an efficient synthesis method and has formed a lot of international collaborations to measure the properties of SWNTs.

In a scanning electron microscope, the nanotube material produced by either of these methods looks like a mat of carbon ropes. The ropes are between 10 and 20 nm across and up to 100 μm long. When examined in a transmission electron microscope, each rope is found to consist of a bundle of SWNTs aligned along a single direction. X-ray diffraction, which views many ropes at once, also shows that the diameters of the SWNTs have a narrow distribution with a strong peak.

For the synthesis conditions used by the Rice and Montpellier groups, the diameter distribution peaked at 1.38 ± 0.02 nm, very close

to the diameter of an ideal (10, 10) nanotube. X-ray diffraction measurements by John Fischer and coworkers at the University of Pennsylvania showed that bundles of SWNTs form a two-dimensional triangular lattice. The lattice constant is 1.7 nm and the tubes are separated by 0.315 nm at closest approach, which agrees with prior theoretical modeling by Jean-Christophe Charlier and others at the University of Louvain-la-Neuve (Belgium).

Chemical vapor deposition (CVD) can be used to make either SWNTs or MWNTs, and produces them by heating a precursor gas and flowing it over a metallic or oxide surface with a prepared catalyst (typically a nickel, iron, molybdenum, or cobalt catalyst).

In several gas-phase processes, such as a high-pressure carbon monoxide process, both SWNTs and MWNTs can be produced by using a reaction that takes place on a catalyst flowing in a stream, rather than bonded to a surface.

While MWNTs do not need a catalyst for growth, SWNTs can be grown only with a catalyst. However, the detailed mechanisms responsible for growth are not yet well understood. Experiments show that the width and peak of the diameter distribution depend on the composition of the catalyst, the growth temperature, and various other growth conditions. Great efforts are now being made to produce narrower

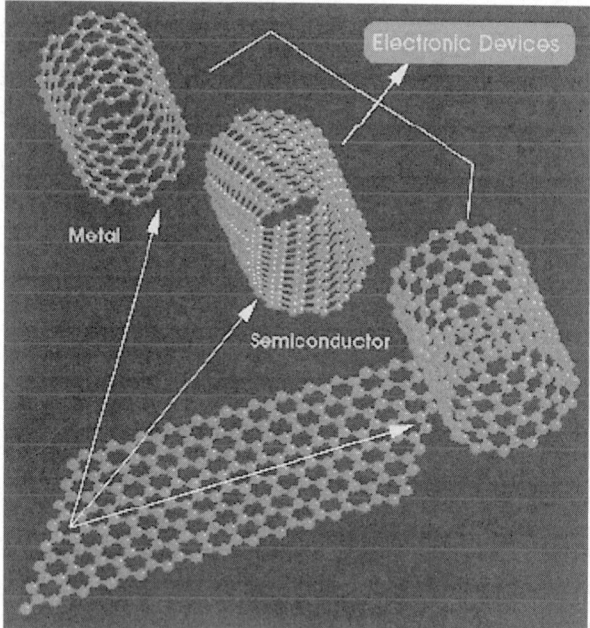

Fig. 20.2.5 Schematic representation of rolling graphite to create a carbon nanotube.

diameter distributions with different mean diameters, and to gain better control of the growth process. From an applications point of view, the emphasis will be on methods that produce high yields of nanotubes at low cost, and some sort of continuous process will probably be needed to grow carbon nanotubes on a commercial scale.

Elastic Nanotube Properties   Determining the elastic properties of SWNTs has been one of the most disputed areas of nanotube study in recent years. On the whole, SWNTs are stiffer than steel and are resistant to damage from physical forces. Pressing on the tip of the nanotube will cause it to bend without damage. When the force is removed, the tip of the nanotube will recover to its original state. Quantizing these effects, however, is rather difficult and an exact numerical value cannot be agreed on.

The Young's modulus (elastic modulus) of SWNTs lies close to 1 TPa. The maximum tensile strength is close to 30 GPa.

The results of various studies over the years have shown a large variation in the value reported. In 1996, researchers at NEC in Princeton and the University of Illinois measured the average modulus to be 1.8 TPa. This was measured by first allowing a tube to stand freely and then taking a microscopic image of its tip. The modulus is calculated from the amount of blur seen in the photograph at different temperatures. In 1997, G. Gao, T. Cagin, and W. Goddard III reported three variations on the Young's modulus to five decimal places that were dependent on the chiral vector. They concluded that a (10, 10) armchair tube had a modulus of 640.30 GPa, a (17, 0) zigzag tube had a modulus of 648.43 GPa, and a (12, 6) tube had a value of 673.94 GPa. These values were calculated from the second derivatives of potential.

Further studies were conducted. In 1998, Treacy reported an elastic modulus of 1.25 TPa using the same basic method used 2 years earlier. This compared well with the modulus of MWNTs (1.28 TPa), found by Wong in 1997. Using an atomic force microscope (AFM), they pushed the unanchored end of a free-standing nanotube out of its equilibrium position and recorded the force that the nanotube exerted back onto the tip. In 1999, E. Hernández and Angel Rubio showed, using tight-binding calculations, the Young's modulus was dependent on the size and chirality of the SWNT, ranging from 1.22 TPa for the (10, 0) and (6, 6) tubes to 1.26 TPa for the large (20, 0) SWNTs. However, using first-principle calculations, they calculated a value of 1.09 TPa for a generic tube.

The previous evidence would lead one to assume that the diameter and shape of the nanotube was the determining factor for its elastic modulus. However, when working with different MWNTs, Forró noted that the modulus measurements of MWNTs in 1999 (using AFM) did not strongly depend on the diameter, as had been recently suggested. Instead, they argued that the modulus of MWNTs correlates to the amount of disorder in the nanotube walls. However, the evidence showed that the value for SWNTs does in fact depend on diameter; an individual tube had a modulus of about 1 TPa while bundles (or ropes) of 15 to 20 nm in diameter had a modulus of about 100 GPa.

It has been suggested that the controversy into the value of the modulus is due to the interpretation of the thickness of the walls of the nanotube. If the tube were considered to be a solid cylinder, then it would have a lower Young's modulus. If the tube is considered to be hollow, the modulus gets higher, and the thinner the walls of the nanotube are treated, the higher the modulus will become.

Nanotube Applications

The materials markets are already seeing applications for composites based on multi-walled carbon nanotubes and nanofibers. In many ways this is an old market; carbon fibers have been around commercially for a couple of decades. The benefits of the new materials are the same as those of carbon fibers, but better, given that the main properties considered are strength and conductivity. Carbon fibers are quite large, typically about a tenth of a millimetre in diameter, and blacken the material to which they are added. MWNTs can offer the same improvements in strength to a polymer composite without the blackening and often with a smaller amount of added material (a filler). The greater aspect ratio (length compared to diameter) of the newer materials can make plastics conducting with a smaller filler load, one significant application being electrostatic painting of composites in products such as car parts. Additionally, the surface of the composite is smoother, which benefits more refined structures such as platens for computer disk drives.

When thinking about structural applications such as these, it should be remembered that, in general, as the fibers get smaller, the number of defects decreases, in a progression toward the perfection of the single-walled nanotube. The inverse progression is seen in terms of ease of manufacturing—the more perfect, and thus more structurally valuable, the material, the harder it is to produce in quantity at a good price. This relationship is not written in stone—there is no reason that near-perfect SWNTs should not be producible cheaply and in large quantities. When this happens, and it might not be too far off, the improvements seen in the strength-to-weight ratios of composite materials could soar, impacting a wide variety of industries from sports equipment to furniture, from the construction industry to kitchenware, and from automobiles to airplanes and spacecraft (the aerospace industry is probably the one that stands to reap the greatest rewards). In fact, a carbon nanotube composite has recently been reported that is 6 times stronger than conventional carbon fiber composites.

Just because the perfect nanotube is 100 times stronger than steel at a sixth of the weight doesn't mean one is going to be able to achieve those properties in a bulk material containing them. The chicken wire arrangement that makes up the layers in graphite does not adhere well to other materials, which is why graphite is used in lubricants and pencils. The same holds true of nanotubes—they are quite insular in nature, preferring not to interact with other materials. To capitalize on their strength in a composite, they need to latch on to the surrounding polymer, which they are not inclined to do (blending a filler in a polymer is difficult even without these issues—it took a decade to perfect this for the new nanoclay polymers now hitting the markets).

One way of making a nanotube interact with something else, such as a surrounding polymer, is to modify it chemically. This is called **functionalization** and is being explored not just for composite applications but also for a variety of others, such as biosensors. For structural applications, the problem is that functionalization can reduce the valuable qualities of the nanotubes that one is trying to capitalize on.

Of course, in theory there is no need to mix the nanotubes with another material. To make superstrong cables, for example, the best solution would be to use bundles of sufficiently long nanotubes with no other material added. For this reason, one of the dreams of nanotube production is to be able to spin them, like thread, to indefinite lengths. Such a technology would have applications from textiles (the U.S. military is in fact investigating the use of nanotubes for bulletproof vests) to the "space elevator." The space elevator concept, which sounds like something straight out of science fiction, involves anchoring one end of a huge cable to the earth and another to an object in space. The taut cable so produced could then support an elevator that would take passengers and cargo into orbit for a fraction of the cost of the rockets used today. Sounds too far out? In fact, it has been established by NASA as feasible in principle, given a material as strong as SWNTs.

The materials market is a big one, and there are others, which we'll come to, but smaller ones exist too. Nanotubes are already being shipped on the tip of atomic-force microscope probes to enhance atomic-resolution imaging. Nanotube-based chemicals and biosensors should be on the market soon (they face stiff competition from other areas of nanotechnology). The thermal conductivity of nanotubes shows promise in applications from cooling integrated circuits to aerospace materials. Electron beam lithography, which is a method of producing nanoscale patterns in materials, may become considerably cheaper thanks to the field-emission properties of carbon nanotubes. Recent developments show promise of the first significant change in X-ray technology in a century. Entering more speculative territory, nanotubes may one day be used as nanoscale needles that can inject substances into, or sample substances from, individual cells, or they could be used as appendages for miniature machines (the tubes in multiwalled nanotubes slide over each other like graphite, but have preferred locations that they tend to spring back to).

Big markets, apart from materials, in which nanotubes may make an impact, include flat panel displays (near-term commercialization is promised here), lighting, fuel cells, and electronics. This last is one of the most talked-about areas but one of the farthest from commercialization, with one exception, this being the promise of huge computer memories (more than 1,000 times greater in capacity than what you probably have in your machine now) that could, in theory, put a lot of the $40 billion magnetic disk industry out of business. Companies like to make grand claims, however, and in this area there is not just the technological hurdle to face but the even more daunting economic one, a challenge made harder by a host of competing technologies.

Despite an inevitable element of hype, the versatility of nanotubes does suggest that they might one day rank as one of the most important materials ever discovered. In years to come they could find their way into myriad materials and devices around us and quite probably make some of the leaders in this game quite rich.

REFERENCES

1. M. Bartz et al., "Stamping of monomeric SAMs as a route to structured crystallization templates: Patterned titania films," *Chemistry—A European Journal*, **6**(22):4149–4153 (2000).
2. B. T. Holland et al., "Synthesis of highly ordered, three-dimensional, macroporous structures of amorphous or crystalline inorganic oxides, phosphates, and hybrid composites," *Chemistry of Materials*, **11**(3):795–805 (1999).
3. J. D. Brennan, R. S. Brown, and U. J. Krull, "Transduction of analytical signals by supramolecular assemblies of amphiphiles containing heterogeneously distributed fluorophores," *Analytical Sciences*, **14**(1):141–149 (1998).
4. A. Miwa et al., "Development of novel chitosan derivatives as micellar carriers of taxol," *Pharmaceutical Research*, **15**(12):1844–1850 (1998).
5. T. Onda et al., "Super-Water-Repellent Fractal Surfaces," *Langmuir*, **12**(9):2125–2127 (1996).
6. E. A. Vogler, "Structure and reactivity of water at biomaterial surfaces," *Advances in Colloid and Interface Science*, **74**(1–3):69–117 (1998).
7. M. M. Ngundi et al., "Array Biosensor for Detection of Ochratoxin A in Cereals and Beverages," *Analytical Chemistry*, **77**(1):148–154 (2005).
8. F. X. Zhou, J. Bonin, and P. F. Predki, "Development of functional protein microarrays for drug discovery: Progress and challenges," *Combinatorial Chemistry & High Throughput Screening*, **7**(6):539–546 (2004).
9. R. P. Feynman, "There's Plenty of Room at the Bottom," in *American Physical Society*. December 29, 1959, California Institure of Technology.
10. S. G. Wolf and R. N. Tauber, *Silicon Processing for the VLSI Era*, Vol. 1. Lattice Press, New York, 1987.
11. E. Occhiello, G. Mara, and F. Garbassi, "STM and AFM microscopies: new techniques for the study of polymer surfaces," *Polymer News*, **14**(7):198–200 (1989).
12. C. A. J. Putman et al., "Tapping Mode Atomic Force Microscopy in Liquids," *Applied Physics Letters*, **64**:2454–2456 (1994).
13. J. Schneider et al., "Atomic force microscope image contrast mechanisms on supported lipid bilayers," *Biophysical Journal*, **79**(2):1107–1118 (2000).
14. N. A. Amro, S. Xu, and G. Y. Liu, "Patterning surfaces using tip-directed displacement and self-assembly," *Langmuir*, **16**(7):3006–3009 (2000).
15. H. Colfen and L. M. Qi, "A systematic examination of the morphogenesis of calcium carbonate in the presence of a double-hydrophilic block copolymer," *Chemistry—A European Journal*, **7**(1):106–116 (2001).
16. G. Liu, S. Xu, and Y. Qian, "Nanofabrication of Self Assembled Monolayers Using Scanning Probe Lithography," *Acc. Chem. Res.*, **33**:457–466 (2000).
17. A. R. Burns et al., "Molecular Level Friction As Revealed with a Novel Scanning Probe," *Langmuir*, **15**: 2922–2930 (1999).
18. P. Cacciafesta et al., "Human plasma fibrinogen adsorption on ultraflat titanium oxide surfaces studied with atomic force microscopy," *Langmuir*, **16**(21):8167–8175 (2000).
19. M. Callaway, D. J. Tildesley, and N. Quirke, "Recording of Surface Phases during Atomic Force Microscopy: A Simulation Study," *Langmuir*, **10**: 3350–3356 (1994).
20. J. J. Davis, H. A. O. Hill, and A. M. Bond, "The application of electrochemical scanning probe microscopy to the interpretation of metalloprotein voltammetry," *Coordination Chemistry Reviews*, **200**:411–442 (2000).
21. J. A. DeRose et al., Electrochemical deposition of nucleic acid polymers for scanning probe microscopy *J. Vac. Sci. Technol. B*, **9**(2, Pt. 2):1166–1170.
22. A. C. Dumaual, L. J. Jenski, and W. Stillwell, "Liquid crystalline/gel state phase separation in docosahexaenoic acid-containing bilayers and monolayers," *Biochimica et Biophysica Acta—Biomembranes*, **1463**(2): 395–406 (2000).
23. P. L. Anelli et al., "Towards synthetic supramolecular polymers," *Polym. Prepr. (Am. Chem. Soc., Div. Polym. Chem.)*, **32**(1):405–406 (1991).
24. S. I. Stupp et al., "Supramolecular Materials: Self-Organized Nanostructures," *Science*, **276**(5311): 384–389 (1997).
25. R. C. Jaeger, *Introduction to Microelectronic Fabrication*, 2nd ed, Prentice Hall, Upper Saddle River, N.J., 2002.
26. S. K. Shukla and R. I. Bahar, *Nano, Quantum, and Molecular Computing: Implications to High Level Design and Validation*, Kluwer Academic Publishers, Boston, 2004.
27. P. Calvert and P. Rieke, "Biomimetic Mineralization in and on Polymers," *Chemistry of Materials*, **8**(8):1715–1727 (1996).
28. N. Costa and P. M. Maquis, "Biomimetic processing of calcium phosphate coating," *Medical Engineering & Physics*, **20**(8):602–606 (1998).
29. N. U. Dan, "Synthesis of hierarchical materials," *Trends in Biotechnology*, **18**(9):370–374 (2000).
30. A. J. Golumbfskie, V. S. Pande, and A. K. Chakraborty, "Simulation of Biomimetic Recognition between Polymers and Surfaces," *Proceedings of the National Academy of Sciences of the United States of America*, **96**(21): 11707–11712 (1999).
31. A. H. Heuer et al., "Innovative materials processing strategies: A biomimetic approach," *Science*, **255**:1098–1105 (1992).
32. T. Hirai, S. Hariguchi, and I. Komasawa, "Biomimetic Synthesis of Calcium Carbonate Particles in a Pseudovesicular Double Emulsion," *Langmuir*, **13**: 6650–6653 (1997).
33. G. F. Xu et al., "Biomimetic synthesis of macroscopic-scale calcium carbonate thin films. Evidence for a multistep assembly process," *Journal of the American Chemical Society*, **120**(46):11977–11985 (1998).
34. M. Yada et al., "Biomimetic Surface Patterns of Layered Aluminum Oxide Mesophases Templated by Mixed Surfactant Assemblies," *Langmuir*, **13**(20): 5252–5257 (1997).
35. J. A. Zasadzinski, E. Kisak, and C. U. Evans, "Complex vesicle-based structures," *Current Opinion in Colloid & Interface Science*, **6**(1):85–90 (2001).
36. R. A. Freitas, *Nanomedicine*, Vol. 1, Landes Bioscience, Austin, TX, 1999.
37. J. S. Hall, *Nanofuture: What's Next for Nanotechnology*, Prometheus Books, Amherst, NY, 2005.
38. S. Bandyopadhyay, et al., "Surfactant aggregation at a hydrophobic surface," *Journal of Physical Chemistry B*, **102**(33):6318–6322 (1998).
39. M. R. Zachariah and M. J. Carrier, "Molecular dynamics computation of gas-phase nanoparticle sintering: A comparison with phenomenological models," *Journal of Aerosol Science*, **30**(9):1139–1151 (1999).

20.3 FERROELECTRICS/PIEZOELECTRICS AND SHAPE MEMORY ALLOYS

by Jacqueline J. Li

FERROELECTRICS/PIEZOELECTRICS

Ferroelectrics are materials exhibiting a spontaneous dielectric polarization and **hysteresis effects** in the relation between dielectric displacement and electric field. This ferroelectric behavior is observed only below a transition temperature T_c—the so-called Curie temperature. Above the Curie temperature, the materials are no longer ferroelectric and show normal dielectric behavior.

Ferroelectric phenomena were first observed in 1920 by Valasek in Rochelle salt [304-59-6], $KNaC_4H_4O_6 \cdot 4H_2O$. In his description, Valasek called attention to the analogy with ferromagnetic phenomena. *Ferro* is a prefix associated with iron-containing materials. But the hysteresis nature in the relation between dielectric displacement and electric field is remarkably similar to the induction-magnetic field relation for ferromagnetic materials. Ferromagnetic materials generally contain iron. Ferroelectrics are named after the similar hysteresis loop and rarely contain iron as a significant constituent.

Table 20.3.1 (Table I-1 in "Ferroelectric Crystals," Franco Jona and G. Shirane, Dover, 1993) provides a partial list of ferroelectric crystals arranged approximately in chronological order of their discovery.

Most ferroelectric materials are ceramics; some polymers also exhibit ferroelectric behaviors. At present, the number of known pure compounds with ferroelectric properties is about 200; they are classified into about 40 families. Examples of typical ferroelectrics are: potassium dihydrogen phosphate, KH_2PO_4 (commonly abbreviated as KDP), and a number of isomorphous phosphates and arsenates; barium titanate, $BaTiO_3$, and other isomorphous double oxides; Rochelle salt (sodium potassium tartrate tetrahydrate, $NaKC_4H_4O_6 \cdot 4H_2O$); and some other isomorphous crystals. The most widely used ferroelectrics now are lead zirconate titanate solid solutions, $Pb(Zr\text{-}Ti)O_3$ (commonly known as PZT). Before we introduce the general properties and structures of ferroelectric materials, we focus on barium titanate, which is one of the simplest ferroelectric materials.

Barium Titanate

In order to better understand the ferroelectric properties, let's concentrate on barium titanate ($BaTiO_3$), which is the most extensively investigated ferroelectric material and whose structure is far simpler than that of any other ferroelectrics. Its crystal structure is of perovskite type.

Table 20.3.1 Partial List of Ferroelectric Crystals

Name and chemical formula	Curie temperature, °C	Spontaneous polarization, 10^{-6} C/cm^2	Year in which reported
Rochelle salt $NaKC_4H_4O_6 \cdot 4H_2O$	+ 23	0.25	1921
Lithium ammonium tartrate $Li(NH_4)C_4H_4O_6 \cdot H_2O$	−170	0.20	1951
Potassium dihydrogen phosphate KH_2PO_4	−150	4.0	1935
Potassium dideuterium phosphate KD_2PO_4	− 60	5.5	1942
Potassium dihydrogen arsenate KH_2AsO_4	−177	5.0	1938
Barium titanate $BaTiO_3$	+120	26.0	1945
Lead titanate $PbTiO_3$	+490	>50	1950
Potassium niobate $KNbO_3$	+415	30	1951
Potassium tantalate $KTaO_3$	−260	?	1951
Cadmium (pyro) niobate $Cd_2Nb_2O_7$	− 85	~10	1952
Lead (meta) niobate $PbNb_2O_6$	+570	?	1953
Guanidinium aluminium sulfate hexahydrate $C(NH_2)_3Al(SO_4)_2 \cdot 6H_2O$	None	0.35	1955
Methylammonium aluminum alum $CH_3 NH_3Al(SO_4)_2 \cdot 12H_2O$	− 96	1.0	1956
Ammonium sulfate $(NH_4)_2SO_4$	− 50	0.25	1956
Triglycine sulfate $(NH_2CH_2COOH)_2 \cdot H_2SO_4$	+ 49	2.8	1956
Colemanite $CaB_3O_4(OH)_3 \cdot H_2O$	− 7	0.65	1956
Dicalcium strontium propionate $Ca_2Sr(CH_3CH_2COO)_6$	+ 8	0.3	1957
Lithium acid selenite $LiH_3(SeO_3)_2$	None	10.0	1959
Sodium nitrite $NaNO_2$	+160	7.0	1958

Fig. 20.3.1   (*a*) Cubic structure of a BaTiO$_3$ unit cell above the Curie point of 120°C. (*b*) Tetragonal structure below 120°C. (*c*) 90° reorientation under a compressive stress or a perpendicular electric field.

This structure is common to a large family of compounds with the general formula ABO$_3$, whose representative in nature is the mineral CaTiO$_3$, called perovskite. Figure 20.3.1*a* shows a unit cell of the cubic perovskite-type lattice of BaTiO$_3$. The Curie temperature of BaTiO$_3$ is at 120°C. Above 120°C, BaTiO$_3$ remains in the centrosymmetric cubic structure and nonpolar phase (nonpiezoelectric, or paraelectric). Upon cooling just below 120°C, BaTiO$_3$ undergoes a phase transformation to a tetragonal modification as shown in Fig. 20.3.1*b*. In this process, the positively charged titanium ion moves from its original central position to a position aligned along the polar axis (or *c* axis), and together with the negatively charged oxygen ions, forms a dipole moment. The total dipole moments in the material with a unit volume define the spontaneous polarization *P*. Between 5°C and 120°C, the crystal has a stable tetragonal structure and the lattice parameters are *a* = 0.3992 nm and *c* = 0.4032 nm. Such a poled crystal can undergo a 180° domain switch under a reversed electric field, or a 90° switch under a compressive stress or under an electric field perpendicular to its original polar direction, as shown in Fig. 20.3.1*c*.

Hysteresis Loop

The ferroelectric material can have zero overall **polarization** under zero applied electric field, due to a random orientation of microscopic-scale domains, regions in which the *c* axes of adjacent unit cells have a common direction. If we first apply a small electric field, we will have only linear relationship between polarization *P* and the electric field *E* because the field is not large enough to switch any of the domains and the crystal will behave like a normal paraelectric. Figure 20.3.2 shows schematic plot of *P* versus *E*. When the applied electric field increases, the dipole orientations are favored to the direction parallel to the applied field direction. In this case, domains with such orientations "grow" at the expense of other, less favorably oriented, ones. When the applied electric field increases more, more domains will switch over in the favored applied field direction and the polarization will increase rapidly, until it reaches a state in which all the domains are aligned in the applied electric field direction. This is a state of saturation and the crystal consists now of a single domain.

If now we decrease the applied field to zero, the polarization will generally not return to zero and some of the domains will remain aligned in the applied field direction. The polarization at this point is called remanant polarization P_r. The extrapolation of the linear portion of the top curve back to the polarization axis represents the value of the spontaneous polarization P_s.

In order to observe the overall polarization of a ferroelectric material, an electric field in the opposite direction needs to be applied. The value of the electric field required to reduce the polarization *P* to zero is called the *coercive field* E_c. Further increase of the field in the opposite direction will, of course, cause complete alignment of the dipoles in this direction, and the cycle can be completed by reversing the field direction once again.

From Fig 20.3.2, the hysteresis loop of the polarization *P* versus the applied field *E* is obtained, which is the most important characteristic of a ferroelectric material. The essential feature of a ferroelectric is not that it has a spontaneous polarization, but rather that this spontaneous polarization can be reversed by means of an electric field.

Electromechanical Coupling Behavior

Because of the nature of polarization and the reorientation or domain switch of a ferroelectric under a specific electric and/or mechanical loading, a coupling behavior between electric and mechanical can be observed. When a ferroelectric material is subjected to an electric field, it can be deformed up to 0.2 percent of strain. Figure 20.3.3 shows the relation between the axial strain and the electric field of a PLZT, which gives a typical butterfly-shaped strain versus electric field curve. On the other hand, under the applied compressive stress loading, a ferroelectric produces a measurable voltage change.

Piezoelectricity

Although ferroelectricity is an intriguing phenomenon, it does not have the practical importance that ferromagnetic materials display (in areas such as magnetic storage of information). The most common applications of ferroelectrics stem from a closely related phenomenon, piezoelectricity. The prefix *piezo* comes from the Greek word for pressure. Piezoelectricity is the ability of certain crystalline materials to develop an electric charge proportional to a mechanical stress. The piezoelectric phenomenon was first discovered by J. Curie and P. Curie in 1880. The direct piezoelectric effect is that electric polarization is produced by mechanical stress. Closely related to it is the converse effect, in which

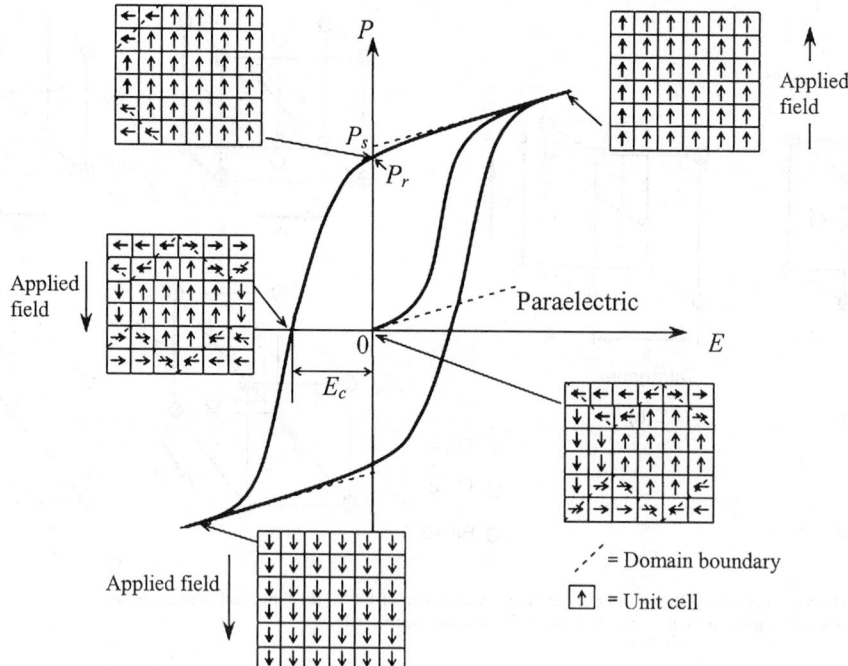

Fig. 20.3.2 Schematic diagram of ferroelectric hysteresis loop.

Fig. 20.3.3 The butterfly-shaped strain versus electric field relations of a PLZT.

a crystal becomes strained when an electric field is applied. We can write the two effects that occur in piezoelectrics as

Direct piezoelectric effect: $\qquad E = g\sigma \qquad$ (20.3.1)

Converse piezoelectric effect: $\qquad e = dE \qquad$ (20.3.2)

where E is the electric field (V/m), σ is the applied stress (Pa), e is the strain, and g and d are **piezoelectric constants**. The g constant is related to the d constant by the permittivity:

$$g = d/\varepsilon' = d/K\varepsilon_0$$

where ε' is permittivity and K is the relative dielectric constant, which equals to the ratio between the charge stored on an electroded slab of material brought to a given voltage and the charge stored on a set of identical electrodes separated by vacuum ($K = \varepsilon'/\varepsilon_0$). Some piezoelectric materials and their typical values for d are listed in Table 20.3.2.

All ferroelectric materials exhibit piezoelectric behavior in limited elastic range. But not all pyroelectrics that possess a spontaneous polarization are ferroelectric in the sense of possessing a reorientable electric moment. While ferroelectric materials exhibit nonlinear coupling behavior between electrical and mechanical systems and hysteresis loops as discussed above, piezoelectrics have only a linear electrical-mechanical coupling effect. The principal reason that ferroelectrics were discovered so much later was because the formation of domains of differently oriented polarization within virgin single crystals led to a lack of any net polarization and very small pyroelectric and piezoelectric response.

Piezoelectric Polymers

So far we have seen plenty of examples of ceramics showing piezoelectricity. It is known that some natural organic substances, e.g., rubber, wood, and hair, are piezoelectric. Bone also shows a piezoelectricity of the same order as wood. The symmetry of rubber and wood is the same as that for piezoelectric ceramics. In 1969, Kawai discovered that the piezoelectricity in polyvinylidene fluoride (PVDF or PVF$_2$) is several times larger than that of quartz. A thin PVDF film was extended and poled at 100°C under a high field, such as 300 kV/cm. Some improvements were made in the poling procedure and a high d constant, about 30 pC/N, was obtained.

PVDF is a polymer that has crystallinity of 40 to 50 percent. The PVDF crystal is **dimorphic,** with two phases (α and β phases).The β phase is polar and piezoelectric. To improve the piezoelectricity, some copolymers have been developed. The copolymer VDF/TrFE consists mostly of the piezoelectric β phase of PVDF with high crystallinity, about 90 percent. The magnitude of polarization in TrEF is only half of that in VDF, but the excellent crystallinity improves the piezoelectric coupling in VDF polymers.

Because of the flexibility of the piezoelectric polymer, it usually is supplied in the form of a thin film. This feature makes it suitable for applications in headphones and speakers. Furthermore, the thin-film transducer is applicable to a high-frequency ultrasonic generator. The performance of these materials greatly depends on the thermal and poling conditions. Typical characteristics of piezoelectric polymers are given in Table 20.3.3 (Table 9.4 in Ikeda's book).

Table 20.3.2   The Piezoelectric Constant d for Selected Materials

Material	Piezoelectric constant $d \times 10^{-12}$, C/Pa $\cdot$ m^2 = m/V
Piezoelectric crystals	
Quartz (SiO$_2$)	2.3
Rochelle salt (NaKC$_4$H$_4$O$_6 \cdot$ 4H$_2$O)	11
EDT (C$_6$H$_{12}$N$_2$O$_6$)	10–20
DKT (N$_2$C$_4$H$_4$O$_6 \cdot$ H$_2$O)	10–20
Amonium dihydrogen phosphate—ADP (NH$_4$H$_2$PO$_4$)	23
BaTiO$_3$	85.6
Lithium niobate (LiNbO$_3$)	6
Lithium tantalite (LiTaO$_3$)	8
Piezoelectric semiconductors	
BeO	0.24
ZnO	12.4
CdS	10.3
CdSe	7.8
AIN	5
Piezoelectric ceramics	
PZTs (PbZr O$_3$ and PbTiO$_3$)	250–365
PLZT	1188
BaTiO$_3$	191
PbNb$_2$O$_6$	80

Application of Piezoelectrics

Because of the electric and mechanical coupling behavior in piezoelectric materials, they can be used as transducers between electrical and mechanical energy. They also are active materials (or "smart" materials), which have the ability to perform both sensing and responding functions. This ability traditionally is associated only with living systems, as opposed to inanimate matter. The distinguishing feature of smart materials is, then, their ability to imitate this fundamental aspect of life. In practice, piezoelectric transducers are limited to devices involving only very small mechanical displacements and small amounts of electric charge per cycle. They are also the ideal candidates as transducers, actuators, or sensors in microelectromechanical systems (MEMS). A list of representative applications of piezoelectric materials is given below.

1. **Phonograph cartridges.** Phonograph pickups have used more piezoelectric ceramic elements than all other applications combined. In the period from 1935 to 1950, the phonograph pickup field was dominated by the first ferroelectric single crystal, Rochelle salt. A schematic sketch of phonograph pickup elements in the conventional configuration of a cemented sandwich of flat plates of ceramic on a brass vane is shown in Fig. 20.3.4. The forces available from the phonograph record groove are only 1 to 5 g. Typical output voltage for the ceramic is from 0.1 to 0.5 V. Modified lead titanate zirconates (PZTs) have largely replaced barium titanate, as they give approximately twice as much output.

2. **Ultrasonic transducers.** A common application of piezoelectric materials is in ultrasonic transducers. Figure 20.3.5 shows a typical cross section of an ultrasonic transducer. In this cutaway view, the piezoelectric crystal or element is encased in a convenient housing. The constraint of the backing material causes the reverse piezoelectric effect to generate a pressure when the faceplate is pressed against a solid material to be inspected. When the transducer is operated in this way (as an ultrasonic transmitter), an ac electrical signal produces an ultrasonic

Table 20.3.3   Characteristics of Piezoelectric Polymers

	Polymer PVDF	Copolymer P(VDF–TrFE)*	Copolymer P(VDF-TeFE)†	Copolymer P(VDCN–VAc)‡	Composite PZT/PVDF
ρ [10^3 kg/m^3]	1.78	1.88	1.90	1.20	5.5
$\varepsilon_{33}/\varepsilon_0$	6.2	6.0	5.5	6.0	30
$\tan \delta$	0.25	0.15			
d_{31} [pC/N]	28	12.5	8	6	
e_{33} [C/m^2]	−0.16	−0.23		−0.14	0.18
g_{33} [10^{-3} V $\cdot$ m/N]	−320	−380			26
k_t	0.20	0.30	0.21	0.22	0.07
$c_{33}{}^D$ [10^9 N/m^2]	9.1	11.3			26.6
v_{31} [km/s]	2.26	2.40			2.2
Q_m	10	20			
Z_0 [10^6 kg/m$^2 \cdot$ s]	4.0	4.5	4.4	3.2	12.1
T_C [°C]	170	130	141		
P_t [C/m^2]	0.055	0.082	0.071	0.035	
E_c [Kv/mm]	45	38	39		
p [μC/m$^2 \cdot$ K]§	35	50	40	30	

*VDF$_{0.75}$ TrFE$_{0.25}$.
† VDF$_{0.8}$ TeFE$_{0.2}$.
‡ VDCN$_{0.5}$ VAc$_{0.5}$.
§ Pyroelectric coefficient.
VDF: vinylidene fluoride; TrFE: trifluoroethylene; TeFE: tetrafluoroethylene; VDCN: vinylidene cyanide; VAc: vinylacetate.

Fig. 20.3.4 Ceramic phonograph pickup, series- and parallel-connected. (*From Jaffe, Cook, and Jaffe, Fig. 12.1.*)

Fig. 20.3.5 A cross section view of an ultrasonic transducer. (*From Metals Handbook, 8th ed. Vol. 11, American Society for Metals, Metals Park, Ohio,* 1976.)

signal of the same frequency (usually in megahertz). When the transducer is operated as an ultrasonic receiver, the direct piezoelectric effect is employed. In that case, the high-frequency elastic wave striking the faceplate generates a measurable voltage oscillation of the same frequency.

3. **Instrument transducers.** PZT ceramics have become the dominant material for piezoelectric accelerometers. In addition to their use in generating ultrasonic waves of frequencies in the order of 1 to 5 MHz to detect flaws and other inhomogeneities in solids or in joints between solid bodies, miniature ceramic transducers have been inserted into blood vessels to record periodic pressure changes connected with the heart beat. Special-purpose transducers such as blast gages and noise-sensing elements to detect boiling of cooling water in nuclear reactors have been served well by piezoelectrics. Much ingenuity has been expended on other schemes that utilize piezoelectrics for measurement of forces that can be related to input variables of aerospace, medical, and industrial instrumentation.

4. **Air transducers.** Air transducers have been widely used in microphones, especially in the hearing-aid field. But with the advent of the transistor, the market became dominated by electrodynamic microphones of low impedance. Progress in techniques for manufacturing thin ceramic bending elements has resulted in recouping some of the microphone market for piezoelectricity. Piezoelectrics also work well as tuned ultrasonic microphones in remote control of television sets.

5. **Actuators.** One example of piezoelectric materials as smart materials is the piezoelectric actuator used in an impact dot-matrix printer. A schematic illustration of such a ceramic actuator is shown in Fig. 20.3.6. In this case, of course, the signal is coming in the form of a computer text or image to be printed. Each character formed in a typical printer is a 24 × 24 dot matrix. A printing ribbon is impacted by the multiwire array. The printing element is composed of a multilayer piezoelectric device involving 100 thin ceramic sheets. Each sheet is 100 μm thick. The hinged lever provides a displacement magnification of 30 times, resulting in a net motion at the ink ribbon of 0.5 mm with an energy transfer efficiency of >50 percent. The advantages of the piezoelectric design application compared with conventional electromagnetic types are order-of-magnitude higher printing speed, order-of-magnitude lower energy consumption, and reduced printing noise. The reason for the last advantage is that the low level of heat generation by the piezoelectric allows greater sound shielding.

There are many other applications of piezoelectrics such as delay line transducers, wave filters, high-voltage sources, and sensors. Good sources of information on piezoelectric devices and their applications

(a)   (b)

Fig. 20.3.6 A schematic illustration of a piezoelectric actuator in an impact dot-matrix printer. (*a*) Overall structure of the printer head. (*b*) Close-up of the multilayer piezoelectric printer-head element. (*From K. Uchino, MRS Bulletin,* **18**; *42, 1993.*)

can be found in Katz's book (1959) which emphasizes circuit applications, Mason's book (1958), and some other materials science handbooks.

Glossary

Converse piezoelectric effect: A crystal becomes strained when an electric field is applied.

Curie temperature: The temperature at which such a transition from the polar into the nonpolar state occurs.

Direct piezoelectric effect: Electric polarization is produced by mechanical stress.

Domains: Regions with uniform alignment of the electric dipoles or uniform polarization.

Ferroelectric crystal: Pyroelectric crystal with reversible polarization.

Hysteresis loop: The loop traced out by the polarization as the electric field is cycled. A similar loop occurs in magnetic materials.

Paraelectric: A symmetric unit cell material; only a small polarization is possible, as the applied electric field causes a small induced dipole.

Piezoelectricity: A linear interaction between electrical and mechanical systems.

Polarization: Alignment of dipoles so that a charge can be permanently stored.

Pyroelectric: A crystal having a spontaneous polarization.

Pyroelectricity: An interaction between electrical and thermal systems.

Spontaneous polarization: Measured in terms of dipole moment per unit volume, or with reference to the charges induced on the surfaces perpendicular to the polarization, in terms of charge per unit area.

SHAPE MEMORY ALLOYS

Shape memory alloys (SMAs) are metals that, after being strained, at a certain temperature revert back to their original shape. A change in their crystal structure above their transformation temperature causes them to return to their original shape. They exhibit two very unique properties: pseudo-elasticity and the shape memory effect.

The shape memory effect was first discovered by the Swedish physicist Arne Olander in 1932, who used an alloy of gold and cadmium, but not until the 1960s were any serious research advances made in the field of shape memory alloys. In early 1960s, researchers at the U.S. Naval Ordnance Laboratory discovered the shape memory effect in nickel titanium alloy and named it Nitinol (an acronym for nickel titanium naval ordnance laboratory). Still, not until the early 1970s did several commercially available products appear that implemented shape memory alloys. The most effective and widely used SM alloys include NiTi (nickel-titanium), CuZnAl, and CuAlNi.

Most shape memory alloys also exhibit **pseudo-elastic behavior,** which is a generic name for both rubberlike behavior and superelasticity.

Both the shape memory effect and pseudo-elasticity are mainly due to the solid-state phase transformation of the material. There are two stable phases in shape memory alloys—martensite at low temperature and austenite at high temperature.

Martensite (after the German scientist Martens) was originally used to designate the hard microconstituent formed in quenched steels. Now it is used in a wider context to apply to products of all **martensitic transformations.** Such transformations take place in crystalline solids by coordinated displacements of atoms or molecules over distances smaller than interatomic distances in the parent phase. For this reason the transformations are described as being **diffusionless.** More descriptions of martensite can be found in other sections.

Austenite was the name originally given to the face-centered-cubic (FCC) crystal structure of iron. It is now used in a wider context for other metals, alloys with cubic structures.

In general, a shape memory alloy starts martensitic transformation at a temperature M_s during cooling; the transformation is complete when a lower temperature M_f is reached, where the material is said to be in the martensitic state. Above a certain temperature A_s, the material starts to transform back to austenite, and the transformation is completed at a

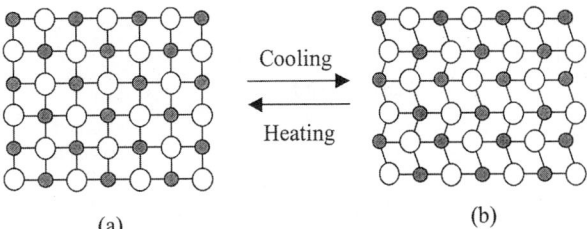

Fig. 20.3.7 Schematic sketch of NiTi microstructure. (*a*) Austenite; (*b*) martensite (twined).

higher temperature, A_f. A schematic illustration of two phases of nitinol, which is a typical shape memory alloy that contains a nearly equal mixture of nickel (55 wt.%) and titanium, is shown in Fig. 20.3.7. If no forces applied during heating or cooling, the microstructure of the alloy will change without macroscopically noticeable shape change.

In the martensitic condition, an alloy can be easily deformed into a new shape by nonconventional atomistic mechanisms. The alloy will remain in the new deformed shape as long as the temperature is kept constant. Heating will cause the material to transform back to the austenitic phase and restore its original shape. This restoring effect is called **shape memory effect** (Fig. 20.3.8).

Fig. 20.3.8 Schematic microstructure diagram of the shape memory effect.

In most cases, the memory effect is one-way, which means that, upon cooling, a shape memory alloy does not undergo any shape change, even though the structure changes back to martensite. When the martensite is strained up to several percent, however, that strain is retained until the material is heated again, at which time shape recovery occurs. Upon recooling, the material does not spontaneously change shape, but must be deliberately strained if shape recovery is again desired.

It is possible in some of the shape memory alloys to cause two-way shape memory. That is, shape change occurs upon both heating and cooling. The amount of this shape change is always significantly less than obtained with the one-way shape memory effect, and very little stress can be exerted by the alloy as it tries to assume its low-temperature shape. The heating shape change can still exert very high forces, as with the one-way memory.

Shape memory alloys also exhibit pseudo-elastic behavior. Unlike the shape memory effect, pseudo-elasticity occurs without a change of temperature. In the austenite condition, a SMA undergoes a phase transformation from austenite to deformed martensite (or stress-induced martensite) when a stress loading is applied. Once the stress is released, the material returns to the austenite state, which means its shape is restored. This is called **superelasticity.** A typical stress-strain curve is shown

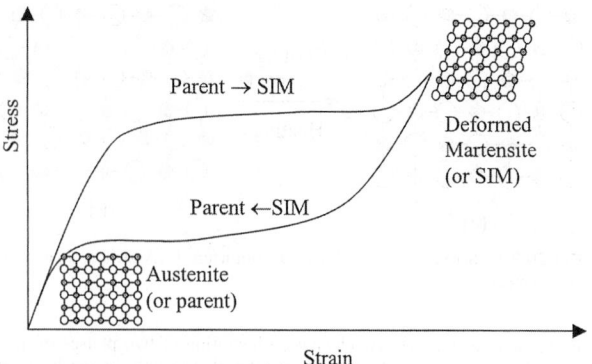

Fig. 20.3.9 Schematic stress-strain curve of a NiTi alloy loaded above M_s temperature and then unloaded. Stress-induced martensite (SIM) is formed during loading, which disappears upon unloading.

in Fig. 20.3.9, while the load-temperature diagram in Fig. 20.3.10 shows the transformation process during a loading and unloading cycle.

Rubberlike behavior was first observed in 1932 by Olander, who studied an Au-Cd alloy. This phenomenon has also been found in several other SMA alloy systems, such as Cu-Al-Ni, Cu-Au-Zn, and Cu-Zn.

Considering the Au-Cd alloy, it is interesting to note that the rubberlike behavior is found only after the martensite is aged at room

Fig. 20.3.10 Load diagram of the superelastic effect above A_f temperature.

temperature for at least a few hours. Freshly transformed specimens exhibit the shape memory effect, and bent rods will not spring back to their original shape without aging.

Rubberlike behavior also constitutes a mechanical type of shape memory, as does the process of superelasticity, and, in fact, these two phenomena cannot be distinguished on the basis of stress-strain curves alone. But rubberlike behavior is characteristic of a fully martensitic structure, whereas supperelastic behavior is associated with formation of martensite from the parent phase under stress. These two types of behavior collectively fall in the category of **pseudo-elasticity.**

The most commercially available SMAs are the NiTi alloys and the copper-base alloys. Properties of the two systems (see Table 20.3.4) are quite different. The NiTi alloys have greater shape memory strain (up to 10 percent versus 4 to 5 percent for the copper-base alloys), tend to be much more thermally stable, have excellent corrosion resistance, and have much higher ductility. On the other hand, the copper-base alloys are much less expensive, can be melted and extruded in air with ease, and have a wider range of potential transformation temperatures. The two alloy systems thus have advantages and disadvantages that must be considered in a particular application.

Applications of Shape Memory Alloys

Some examples of applications using shape memory effect are listed below.

• **Coffeepots:** A nickel-titanium spring in coffeepots marketed in Japan (Fig. 20.3.11) is trained to open a valve and release hot water at the proper temperature to brew a perfect pot of coffee.

• **Linear motor for remote control:** Locks, valves, screens, and other devices can be remotely controlled by a **shape memory linear motor.** A self-limiting electric heating element (PTC) is fixed onto a **shape memory bending element.** Applied heat causes shape recovery of the bending element that

Fig. 20.3.11 Coffeepot.

is transformed into linear motion of the actuator. As the electric current drops, the temperature falls and the actuator returns to its starting position. Energy consumption is low, and heating elements to operate at any desired voltage are easily available. The main advantage of this application is the direct linear movement, without the need to convert a rotary movement into a linear one. Another advantage is its noiseless operation.

• **Hubble space telescope:** In April 1990 the Hubble space telescope was launched into space aboard the shuttle *Discovery*. The telescope

Table 20.3.4 Some Selected Properties of Shape Memory Alloys

Alloy		Composition, wt.%	Transformation temperature A_f, °C	Modulus of elasticity, GPa	Recovery strain, %
Nitinol*	SE508 tubing	Ni: 55.8 Ti: balance O (max): 0.05 C (max): 0.02	<15	41–75	10
	SE508 wiring	Same as tubing	5–18	41–75	10
	SM495 wire	Ni: 54.5 Ti: balance O (max): 0.05 C (max): 0.02	60	28–41	10
Ni-Ti		Ni: 49–51 (at.%)	−200 to 110	28–83	8.5
Cu-Al-Ni		Al: 14–14.5 Ni: 3–4.5	<200	80–85	4
Cu-Zn-Al		Zn: 38.5–41.5 Al: A few wt.%	<120	70–72	4

* From Nitinol.com.

was put into orbit around the earth for observing planets and stars and with the aim to obtain pictures 50 times sharper compared to observations from the earth. Several systems of the Hubble space telescope were driven by solar energy. The solar panels that were folded during the launching were unfolded and positioned after the Hubble space telescope was brought into position. The mechanism, designed by Dornier, is reliable and space- and energy-efficient. It used a specially bent Ni-Ti shape memory element to keep the solar panels in the locked position during launch. At a temperature of 115°F, the shape memory element remembered its original straight form and unlocked the panels.

Fig. 20.3.12   Stent.

• **Vascular stents:** A vascular stent is a device that is implanted in a blood vessel (vein or artery) to provide structural reinforcement of the vessel wall (Fig. 20.3.12). The introduction of SMA to stent manufacture allows greater recovery strains for use in wider vessels, or stronger recovery force in narrower vessels.

• **Antiscald devices:** Basically these are an extension that fits between a shower head and water pipe, containing a Ti-based SMA element that expands if water becomes too hot, choking off the flow. A very cheap, easy, and effective alternative to more expensive technologies.

• **Connecting rings:** Rings and cylinders made of SMA that is austenite at ambient temperature can provide significant constraining force around the inner circumference, ensuring a strong join that can be released by cooling the coupling. For instance, NiTi rings on electrical connectors (Fig. 20.3.13) provide a secure joint at ambient temperatures. Yet, they can easily be released by chilling.

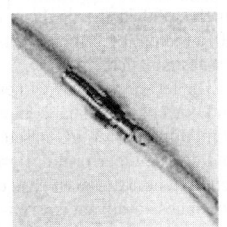

Fig. 20.3.13   Electric connector.

Some examples of applications in which pseudo-elasticity is used are:

• **Eyeglass frames:** Nitinol is extensively used in the manufacture of eyeglass frames. It is used in the superelastic state or in the combined shape memory and superelastic states. The frames hold prescription lenses in the proper position so that the wearer gets the maximum benefit from them.

• **Endodontic wire—high-fatigue superelastic wire:** Nickel-titanium alloy has entered the world of endodontics. The flexibility of nickel titanium makes it an ideal material for use in the manufacture of endodontic instruments. Endodontic files using NiTi alloys, both hand and rotary types, have the potential to greatly improve a clinician's ability to negotiate curved root canal systems. Current research is under progress to better define the limitations and strengths of NiTi, and to determine the file designs and techniques to take best advantage of this unique material.

• **Orthodontic arches—low-deformation-resistance superelastic wire:** Superelastic shape memory wires have found wide use as orthodontic wires. The difference between these shape memory wires and a normal wire is a large elastic deformation combined with a low plateau stress obtainable in the shape memory wire. The advantages for the patient are two-fold: fewer visits to the orthodontist, because of the larger elastic stroke, and more comfort, because of lower stress levels.

Glossary

A_s: The temperature at which austenite starts to form.

A_f: The temperature at which a material is completely transformed into austenite.

M_s: The temperature at which the martensitic transformation starts upon cooling.

M_f: The temperature at which the martensitic transformation is completed, where the material is said to be in the martensitic state.

Austenite: The name originally given to the FCC crystal structure of iron. It is a stronger phase with a cubic structure in shape memory alloys at high temperature.

Martensite: The name originally used to designate the hard microconstituent formed in quenched steels. Now it is a phase that forms as the result of a diffusionless solid-state transformation. It is a more deformable phase in shape memory alloys at lower temperature.

Pseudo-elasticity: A generic name for both rubberlike flexibility and superelasticity.

Rubberlike behavior: Some SME alloys show rubberlike flexibility. When bars are bent, they spontaneously unbend upon release of the stress. In order to obtain this behavior, the martensite, after its initial formation, usually must be aged for a period of time; unaged martensites show typical SME behavior. The rubberlike behavior, unlike superelasticity, is a characteristic of the martensite phase—not the parent phase.

Shape memory effect (SME): The ability of certain materials to develop microstructures that, after being deformed, can return to the material to its initial shape when heated. Upon cooling, the specimen does not return to its deformed shape.

Superelasticity (SE): The result of stress-induced phase transformation from austenite directly into deformed martensite under external loading. When the deforming stress is released, the material will transform back to its stable austenitic state and shape. The martensite disappears (reverses) when the stress is released, giving rise to a superelastic stress-strain loop with some stress hysteresis.

REFERENCES
Ferroelectrics/Piezoelectrics
Cady (1964). "Piezoelectricity," Dover.
Ferroelectrics in "Ullmann's Encyclopedia of Industrial Chemistry," Wiley, 2003.
"Ferroelectrics" Journal.
IEEE Standard on Piezoelectricity, 1987.
Ikeda, (1996). "Fundamentals of Piezoelectricity," Oxford.
Jaffe, Cook, and Jaffe (1971). "Piezoelectric Ceramics," Academic Press.
Jona and Shirane, (1993). "Ferroelectric Crystals," Dover.
Katz (1959). "Solid State Magnetic and Dielectric Devices," John Wiley, New York.
Lines and Glass (1997). "Principles and Applications of Ferroelectrics and Related Materials," Oxford.
Mason (1958). "Physical Acoustics and the Properties of Solids," Van Nostrand, Princeton, NJ.
Nye, (1979). "Physical Properties of Crystals," Oxford.
Shackelford (2000). "Introduction to Materials Science for Engineers," Sec. 15.4, Prentice Hall.
Valasek, *Phys. Rev.* **15** (1920): 537–538; **17**(1921):475–481.
Shape Memory Alloys
Bever, "Encyclopedia of Materials Science and Engineering," Vol. 4, Pergamon, MIT Press, 1986.
Chang and Read, *Trans. AIME*, **191**(47), 1951.
http://www.nitinol.com, http://www.amtbe.com
Wayman, "Shape Memory and Related Phenomena," *Progress in Materials Science*, Vol. 36, 203–224, 1992.

20.4 INTRODUCTION TO THE FINITE-ELEMENT METHOD

by Farid Amirouche

INTRODUCTION

The finite-element method (FEM) is essentially a technique used to obtain a solution to complex problems by discretizing a given physical or mathematical problem into smaller fundamental parts called *elements*. Then an analysis of the element is conducted using the required principles or law of mechanics. Finally, the solution to the problem as a whole is obtained through an assembly procedure of the individual solutions of the elements. These finite-element techniques have been used in many fields of engineering and science as numerical tools for solution of problems, using digital computers. For example, FEM can be used in computational fluid dynamics (CFD) where the most fundamental consideration is how one treats a continuous fluid in a discretized fashion on a computer. One method is to discretize the spatial domain into small cells to form a volume mesh or grid, and then apply a suitable algorithm to solve the equations of motion. In addition, such a mesh can be either irregular (for instance, consisting of triangles in 2-D, or pyramidal solids in 3-D) or regular; the distinguishing characteristic of the former is that each cell must be stored separately in memory. Lastly, if the problem is highly dynamic and occupies a wide range of scales, the grid itself can be dynamically modified in time, as in adaptive mesh refinement methods.

In numerical analysis, the **finite-element method** (FEM) is used for solving partial differential equations (PDE) approximately. Solutions are approximated by either eliminating the differential equation completely (steady-state problems) or rendering the PDE into an equivalent ordinary differential equation, which is then solved by using standard techniques such as finite differences. The use of the finite-element method in engineering for the analysis of physical systems is commonly known as finite-element analysis. Finite-element methods have also been developed to approximately solve integral equations such as the heat transport equation.

Finite-element analysis (FEA) or the finite-element method (FEM) is a numerical technique for solution of **boundary-value problems**. It was first developed for use in problems related to trusses and structural analysis in general. The system is represented by a similar model consisting of multiple linked discrete elements. From equations of equilibrium, as will be seen in this section, in conjunction with the appropriate boundary conditions applicable to each element and the system as a whole, a set of simultaneous equations is constructed. The system of equations is then solved for the unknown values by using the techniques of linear algebra. Although it is an approximate method, the accuracy of the FEA

Fig. 20.4.2 Car crash simulation results.

method can be improved by refining the mesh in the model by using more elements and nodes.

A common use of FEA is for the determination of stresses and displacements in mechanical objects and systems. However, it is also routinely used in the analysis of many other types of problems. Figures 20.4.1 and 20.4.2 show the finite-element analysis used for studying car **crash simulation**.

The last decade has witnessed an unprecedented increase in the speed and power of personal computers. Because of this revolution in personal computing power, PC-based FEM programs are becoming very popular and can now perform designs in minutes that not long ago would have required hours or even days of processing time. As computer technology continues to advance, even more sophisticated programs will become available to the designer/engineer at an affordable cost.

Fig. 20.4.3 FEM of total knee joint.

Fig. 20.4.1 Car crash simulation.

Three-dimensional FEM can handle greater detail and more complex characterizations of construction materials. Three-dimensional FEM can incorporate nonlinear and nonelastic material models not available in linear elastic analysis (LEA). In finite-element modeling, the structure is divided into a large number of smaller parts, or "elements." Elements may be of different sizes, and may be assigned different properties, depending on their location within the structure. By breaking down the structure into elements, the structural problem is transformed into a finite set of equations that can be solved on a computer. Because a structure can be broken down into finite elements, or "discretized," in any number of ways, it is essential that any design standard incorporating finite elements include rules for systematic three-dimensional meshing.

Finite-element modeling is also a tool used extensively in biomechanics for estimating the mechanical behavior of complex adaptive materials such as trabecular bone. Figure 20.4.3 shows the finite-element model of a knee with a tibial insert. The intrinsic complexities captured by FEM include the volume fraction, connectivity, and anisotropy of bone. In the analysis of total knee arthroplasty (TKA), balancing of the soft tissue can be helped by developing FEM models that can simulate the behavior of the knee during flexion extension and contact pressure in particular at the condyles. The objective of such a study is to optimize the design of the stabilizing post and to understand the mechanism of rollback.

BASIC CONCEPTS IN THE FINITE-ELEMENT METHOD

There are three phases in either using an FE code or simply developing the procedures for solving problems by FEM. Each phase is described below:

1. **Preprocessing phase**
   a. Create and discretize the solution domain into finite elements.
   b. Assume a shape function to represent the physical behavior of an element; that is, an approximate continuous function is assumed to represent the solution of an element.
   c. Develop relationships between input and output equations for an element.
   d. Assemble the elements to present the entire problem. Construct the **global stiffness matrix.**
   e. Apply boundary conditions, initial conditions, and loading conditions.
2. **Solution phase.** Solve a set of linear or nonlinear algebraic equations simultaneously to obtain nodal results, such as displacement values at different nodes or temperature values at different nodes.
3. **Postprocessor phase.** Obtain other important information such as stresses, energy, and any information needed to understand the problem solution.

In the **preprocessing** phase the FEM takes the object to be analyzed and performs a discretization of the body into a limited number of elements. There is usually a selection of the elements used that is best suitable for the problem solution. On the basis of the equation for a single element, the entire problem is formulated by an assembly of these element equations. Before a solution is adapted the boundary conditions, initial conditions, and loading conditions are introduced. At this stage the **solution phase** is adapted in solving a set of linear or nonlinear equations simultaneously. The solution yields values such as displacements or temperatures at the nodes. The final phase of the FEM is the **postprocessor,** where one chooses the type of output to be displayed. The latter might require additional steps in analysis to extract the information needed.

Consider an element of a continuum as shown in Fig. 20.4.4; let the element be represented by a small cube. The nodal points are defined as the end points of the edges of the cube, which are eight in this case.

All the elements are connected through the nodal points. Any deformation of the body caused by external loads or temperatures induces certain displacements at the nodes. In general, the displacements are the

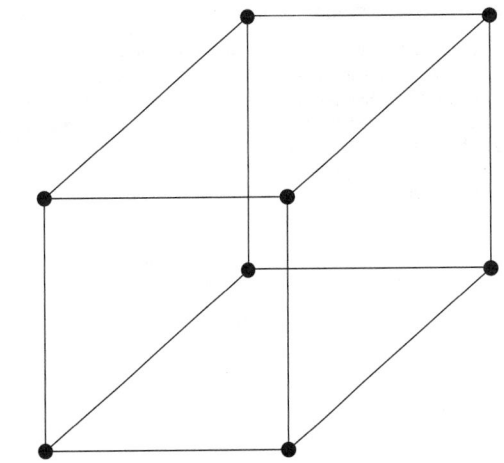

Fig. 20.4.4 Cubic continuum element.

unknowns and are related to the external loads or temperatures through a mathematical relationship of the form

$$f^e = k^e u^e + f^e_{\text{add}} \qquad (20.4.1)$$

where k^e (e being the element) is the local element stiffness, f^e is the external forces applied at each element, u^e represents the nodal displacements for the element, and f^e_{add} represents the additional forces. These additional forces could be a result of surface traction or of the material being initially under stress. For the moment, we will assume them to be zero. Hence, Equation (20.4.1) reduces to

$$f^e = k^e u^e \qquad (20.4.2)$$

It is important at times to express this equation by using the relationship between stresses and strains:

$$\sigma^e = s^e \varepsilon^e \qquad (20.4.3)$$

where σ^e is the element stress matrix and s^e is the constitutive relationship connecting the strain ε^e and *stress* σ^e. From Eq. (20.4.2), we can see that if f^e and k^e are known, u^e can be evaluated. Similarly, if u^e and k^e are known, we can compute f^e.

The finite-element method's most basic function is to automatically generate the **local stiffness matrix** knowing the element type and the properties of the material of the element. Several types of elements that are available in the finite-element library are given in Fig. 20.4.5. The well-known general-purpose FEM packages provide an element library that helps the user in the solution of particular problems. These codes and others can select any of these elements with the proper number of nodes. Thus, it is important at this stage to understand the basic mathematics involved in finite-element formulations.

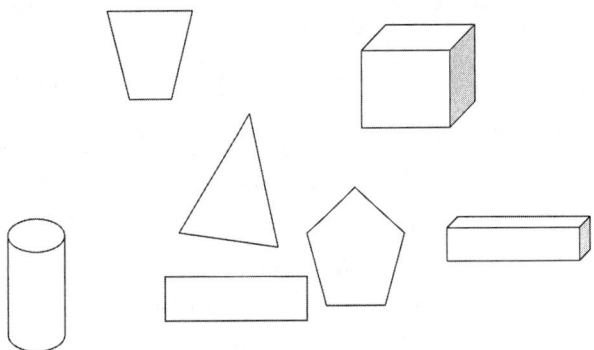

Fig. 20.4.5 Types of elements in the finite-element library.

POTENTIAL ENERGY FORMULATION

The finite-element method is based on the minimization of the total potential energy formulation. When a closed-form solution is not possible, approximation methods such as finite element are the most commonly used in solid mechanics. To illustrate how the finite-element method is formulated. We will consider an elastic body such as the one shown in Fig. 20.4.6, where the body is subjected to a loading force that causes it to deform.

Fig. 20.4.6

From Hooke's law we write

$$P = \frac{AE}{L}\delta \qquad (20.4.4)$$

Let δ be the displacement in the y direction, A be the cross-section area, E be Young's modulus, L be the length, and k be the stiffness. Hence $F = ky$, where $F = P$.

The work performed by the compressive force P is stored in the material in a form of strain energy. The finite-element method discretizes bodies into elements, hence if we can compute the strain energy associated with each element, we can then evaluate the total strain energy associated with all external forces applied to the body

Consider the element shown in Fig. 20.4.6. Let the element deformation be denoted as dy. The strain energy is given by the work of F over the deformed length y. The stored energy is

$$dE = \int_0^{y_1} F\,dy = \int_0^{y_1} ky_1\,dy_1 = \frac{1}{2}ky_1^2 \qquad (20.4.5)$$

The normal stress and strain are directly related to the stored energy if we make the following substitutions

$$\frac{F}{A} = \sigma_y \Rightarrow F = \sigma_y\,dx\,dz \qquad (20.4.6)$$

where σ_y is the stress in the y direction and ε is the strain. Hence

$$d\Lambda = \frac{1}{2}\sigma\varepsilon dV \qquad (20.4.7)$$

when the volume of the element is

$$dV = dx\,dy\,dz \qquad (20.4.8)$$

We can then write the strain energy for an element as

$$N^e = \int dE = \int \frac{\sigma\varepsilon}{2}dV \qquad (20.4.9)$$

$$\sigma_{(e)} = E\varepsilon \qquad (20.4.10)$$

$$\Lambda_s = \int_v \frac{E\varepsilon^2}{2}dV \qquad (20.4.11)$$

The total energy, which consists of the strain energy due to the deformation of the body and the work performed by the external forces, can be expressed as function of the potential energy

$$\Pi = \sum_{e=1}^n \Lambda^{(e)} - \sum_{i=1}^P F_i u_i \qquad (20.4.12)$$

where U denotes displacement along the force F_1 or potential energy Π.

It follows that a stable system requires that the potential energy be minimum at equilibrium

$$\frac{\partial \Pi}{\partial u_i} = \frac{\partial}{\partial u}\left[\sum \Lambda^{(e)} - \sum F_i u_i\right] = 0 \qquad (20.4.13)$$

We know that

$$\varepsilon = (u_{i+1} - u_i)/l,$$

Then

$$\frac{\partial \Lambda^e}{\partial u_i} = \frac{A_{avg}E}{l}(u_i - u_{i+1}) \qquad (20.4.14)$$

$$\frac{\partial \Lambda^e}{\partial u_{i+1}} = \frac{A_{avg}E}{l}(u_{i+1} - u_i) \qquad (20.4.15)$$

The above equations can be expressed in a matrix form as

$$\left\{\begin{array}{c} \dfrac{\partial \Lambda^{(e)}}{\partial u_i} \\ \dfrac{\partial \Lambda^{(e)}}{\partial u_{i+1}} \end{array}\right\} = k_{eq}\begin{bmatrix} 1 & -1 \\ -1 & 1 \end{bmatrix}\left\{\begin{array}{c} u_i \\ u_{i+1} \end{array}\right\} \qquad (20.4.16)$$

where $k_{eq} = (A_{avg}E)/l$.

Similarly we get

$$\frac{\partial}{\partial u_i}(F_i u_i) = F_i \qquad (20.4.17)$$

$$\frac{\partial}{\partial u_{i+1}}(F_{i+1}u_{i+1}) = F_{i+1} \qquad (20.4.18)$$

Combining Eqs. (20.4.16), (20.4.17), and (20.4.18), we obtain the elemental force/displacement relation

$$F^{(e)} = k^{(e)}u^{(e)} \qquad (20.4.19)$$

where $k^{(e)}$ is the element stiffness matrix, $u^{(e)}$ is the element displacement array associated with nodes i and $i + 1$, and $F^{(e)}$ denotes the forces acting at the nodes.

In summary, we see that the minimization of the concept of potential energy leads to an equation similar to the discrete method adopted early in this section. Other popular methods in finite-element analysis are presented for the reader to show how these methods lead to results similar to those of the minimization of potential energy approach.

CLOSED-FORM SOLUTION

It would be ideal if closed-form solutions of most of the engineering problems existed. This of course is not feasible, and the science of engineering is to obtain a solution that is as close as possible to the exact solution, relying on approximation techniques such as FEM and other numerical techniques. For the sake of validating our approximation resulting from discretizing the problem at hand, let's solve the above problem by means of direct integration, hence getting its closed-form solution.

From the equilibrium conditions, and using Newton's first law for a beam with varying cross-sectional area (see Fig. 20.4.7), we write

$$P - \sigma_{avg}(A(y)) = 0 \qquad (20.4.20)$$

where the stress is related to the strain as

$$\sigma = E\varepsilon \tag{20.4.21}$$

Then we express Eq. (20.4.20) further as

$$P - E\varepsilon A(y) = 0 \tag{20.4.22}$$

Recall that the strain is defined by

$$\varepsilon = \frac{du}{dy} \tag{20.4.23}$$

which makes Eq. (20.4.22) take the following form:

$$P - EA(y)\frac{du}{dy} = 0 \tag{20.4.24}$$

resulting in

$$du = \frac{P\,dy}{EA(y)} \tag{20.4.25}$$

The total deformation is obtained by integrating over the total length:

$$\int_0^u du = \int_0^L \frac{P\,dy}{EA(y)} \tag{20.4.26}$$

Note that the cross-sectional area for Fig. 20.4.7 is

$$A(y) = \left(\frac{r_2 - r_1}{L}\right)y + r_1 \tag{20.4.27}$$

Fig. 20.4.7 Free-body diagram showing constraints and loads.

Then

$$u(y) = \int_0^L \frac{P\,dy}{E\left[\left(\dfrac{r_2 - r_1}{2}\right)y + r_1\right]} \tag{20.4.28}$$

or

$$u(y) = \frac{P}{E}\int_0^L \frac{dy}{ay + r_1} \tag{20.4.29}$$

where

$$a = \frac{r_2 - r_1}{L} = \text{constant} \tag{20.4.30}$$

Let $z = ay + r_1 \Rightarrow dz = a\,dy$. Then

$$u(y) = \frac{P}{Ea}\int_{r_2}^{aL+r_1} \frac{dz}{z} = \frac{P}{Ea}[\ln(z)] \tag{20.4.31}$$

$$= \frac{P}{Ea}[\ln(ay + r_1)]^2 \tag{20.4.32}$$

And the final solution is found to be

$$u(y) = \frac{P}{Ea}[\ln(aL + r_1) - \ln(r_1)] \tag{20.4.33}$$

The above solution can be used to find displacement at various points along the length.

WEIGHTED RESIDUAL METHOD (WRM)

The WRM assumes an approximate solution to the governing differential equations. The solution criterion is one where the boundary conditions and initial conditions of the problem are satisfied. It is evident that the approximate solution leads to some marginal errors. If we require that the errors vanish over a given interval or at some given point, then we will force the approximate solution to converge to an accurate solution.

Consider the differential equation discussed in previous example, where

$$\begin{cases} A(y)E\dfrac{du}{dy} - P = 0 \\ \text{with} \quad u(0) = 0 \end{cases} \tag{20.4.34}$$

Let us choose a displacement field u to approximate the solution, which is

$$u(y) = c_0 + c_1 y + c_2 y^2 \tag{20.4.35}$$

where c_0, c_1, and c_2 are unknown coefficients.

If we substitute $u(y)$ and $A(y)$ into the differential equation, we get

$$\left[\left(\frac{r_2 - r_1}{L}\right)y + r_1\right]E(c_1 + 2c_2 y) - P = R \tag{20.4.36}$$

The equation above has two constants, c_1 and c_2. If we require that R vanishes at two points, we will get two equations, which can be used to solve for c_1 and c_2:

$$R\big|_{y=\frac{L}{2}} = 0$$

$$R\big|_{y=L} = 0$$

Solving for c_0, c_1, c_2 we get

$$u(y) = c_1 + c_2 y^2$$

The solution can be compared to the exact solution derived previously.

GALERKIN METHOD

The **Galerkin method** requires the integer of the error function over some selected interval to be forced to zero and the error be orthogonal to some weighting functions ϕ, according to the integral

$$\int \phi_i R\,dy = 0 \tag{20.4.37}$$

where $i = 1, 2, \ldots, n$

The ϕ are selected as part of the approximate solution. This is done simply by assigning the ϕ function to the terms to multiply the coefficients.

Since we assume $u(y) = c_1 y + c_2 y^2$, then $\phi_1 = y$ and $\phi_2 = y^2$. Substituting these two into equation (20.4.37), we get

$$\int_0^L y\left[(10 + y)(c_1 + 2c_2 y) - \frac{180}{E}\right]dy = 0 \tag{20.4.38}$$

$$\int_0^L y^2\left[(10 + y)(c_1 + 2c_2 y) - \frac{180}{E}\right]dy = 0 \tag{20.4.39}$$

Integrating the above two equations leads to two linear equations in terms of c_1 and c_2. Solving the above two equations, we get $c_1 = 0.2314 \times 10^{-4}$ and $c_2 = -0.50 \times 10^{-6}$. The solution is then approximated by $u(y) = c_1 + 2c_2 y^2$.

Note that the WRM and the Galerkin methods are very popular methods used in finite-element analysis to generate the approximate solution to the problem. One can see that these methods require numerics to be able to carry the integration further. If one leaves the loads as a variable P, then the solution is one that relates the deformation to the load. What follows is the introduction of FEM to trusses, where we can make use of the element displacement/force relation and extend it to the entire truss.

INTRODUCTION TO TRUSS ANALYSIS

Trusses are used in many engineering applications, including bridges, buildings, and towers and support structures. In a truss, we are faced with a structure where the displacements, translations, or compressions of any truss member vary linearly with the applied forces. That is, any increment in displacement is proportional to the force causing it to deform. All deformations are assumed small, so that the resulting displacements do not significantly affect the geometry of the structure and hence do not alter the forces in the members. In this case Hooke's law is preserved and the theory of elasticity is used to search for solutions of the truss. Most often, the truss design requires that its member be tested for tension, compression, stress, and strain relations. The applied loads are then tested against the possible yield stress to determine their evaluated limits and the overall stability of the truss.

For large and complicated truss structures, manual computation is often impractical to analyze the static equilibrium conditions, therefore a digital computer has to be used to accommodate the required computations. Over the past two decades, finite-element method codes have been developed for a large number of engineering applications, as trusses have always been a favorite topic for demonstrating FEM's usefulness. Consider the truss shown in Fig. 20.4.8, a 2-D planar truss with vertical loads.

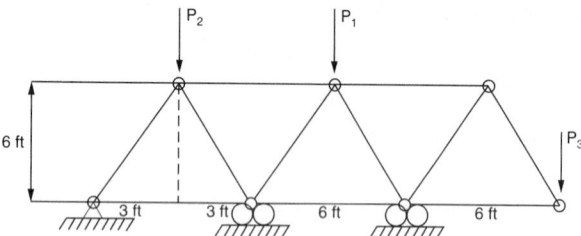

Fig. 20.4.8   A planar truss subject to vertical loads (P_1, P_2, P_3 represent arbitrary loads).

Finite-Element Formulation for a Truss

Trusses are typical structures in which the finite-element method can be best illustrated. We know that FEM relies on (1) discretizing the finite elements of the system; (2) developing the mathematical relationships between the forces and displacements, stresses and strains, etc., for a given element; and (3) formulating the general problem through an assembly procedure of all the elements to solve the given problem.

Consider an element of an arbitrary truss, as shown in Fig. 20.4.9. It is subjected to either tension or compression, as is the case for all the truss elements. Let us label the element's ends 1 and 2, and, consequently, call the corresponding forces f_1 and f_2.

We also know from static analysis that, for the element to be in a state of equilibrium, the following equation must be the true one:

$$f_1 + f_2 = 0 \qquad (20.4.40)$$

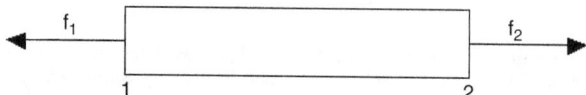

Fig. 20.4.9   A typical truss element.

This simply says that the forces at both ends are equal and opposite in direction. This is true for both tension and compression of the element. Using Hooke's law, we can write the force displacement relation as

$$u = f_{1,2} \frac{L}{AE} \qquad (20.4.41)$$

or

$$f_{1,2} = ku \qquad (20.4.42)$$

where $k = AE/L$ is the stiffness constant when the element is under tension or compression. Note that $f_{1,2}$ denotes either f_1 or f_2. The relative displacement between the nodal points of this truss element can be written as

$$u = u_2 - u_1 \qquad (20.4.43)$$

where u_1 and u_2 are the displacements at ends 1 and 2, respectively. We also refer to ends 1 and 2 as nodes 1 and 2 of the truss element.

Using Eq. (20.4.43), we can write

$$f_1 = ku_1 - ku_2 \qquad (20.4.44)$$

and

$$f_2 = -ku_1 + ku_2 \qquad (20.4.45)$$

Using matrix notation, we write Eq. (20.4.44) and (20.4.45) in combined form as

$$\begin{bmatrix} f_1 \\ f_2 \end{bmatrix} = \begin{bmatrix} k & -k \\ -k & k \end{bmatrix} \begin{bmatrix} u_1 \\ u_2 \end{bmatrix} \qquad (20.4.46)$$

or simply

$$\{f\} = [k]\{u\} \qquad (20.4.47)$$

where

$$\{f\} = \begin{Bmatrix} f_1 \\ f_2 \end{Bmatrix} = \text{the nodal force vector for the element}$$

$$\{u\} = \begin{Bmatrix} u_1 \\ u_2 \end{Bmatrix} = \text{the nodal displacement vector}$$

$$[k] = k\begin{bmatrix} 1 & -1 \\ -1 & 1 \end{bmatrix} = \text{the element stiffness matrix}$$

Thus, Eq. (20.4.46) shows that the nodal displacements and nodal forces are related by the element (local) **stiffness matrix**. As the orientation and loading of various elements in the structure (a truss in this case) vary from each other, we need to develop the local stiffness that will apply to any element orientation.

In the sequel, a more general local stiffness (2-D) will be developed. A uniform system for symbol notation will be adopted that is crucial for easy reading and understanding and subsequent computer implementation.

Let us consider an orientation of a truss element, as shown in Fig. 20.4.10. For $\theta = 0$, we develop a relationship between the forces

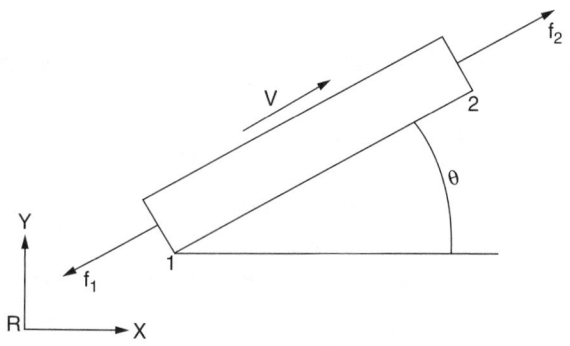

Fig. 20.4.10   A truss element making an angle θ with the x axis.

and the displacements given by Eq. (20.4.46), where v is a unit vector along the line of action of the forces.

Because we have to express all forces (tension or compression) in the global axes (x, y) given by the fixed inertial frame R, we define the element force components as follows:

$$f_{1x} = f_1 \cos \theta \quad f_{1y} = f_1 \sin \theta$$
$$f_{2x} = f_2 \cos \theta \quad f_{2y} = f_2 \sin \theta \quad (20.4.48)$$

Given that the relative displacement u is along the unit vector v, then

$$u = (u_2 - u_1)v \quad (20.4.49)$$

where

$$v = (\cos \theta)i + (\sin \theta)j \quad (20.4.50)$$

and

$$u_1 = u_{1,x}i + u_{1,y}j \quad u_2 = u_{2,x}i + u_{2,y}j \quad (20.4.51)$$

Then by substituting Eqs. (20.4.51) and (20.4.52) into Eq. (20.4.49) and making use of Eqs. (20.4.42), (20.4.48), and (20.4.49), we obtain an expression for each nodal force in terms of the local displacement, the orientation θ and the element stiffness K.

$$f_{1,x} = (k \cos^2 \theta)u_{1,x} + (k \sin \theta \cos \theta)u_{1,y} - (k \cos^2 \theta)u_{2x} - (k \sin \theta \cos \theta)u_{2y} \quad (20.4.52)$$

$$f_{1,y} = (k \sin \theta \cos \theta)u_{1,x} + (k \sin^2 \theta)u_{1,y} - (k \sin \theta \cos \theta)u_{2x} - (k \sin^2 \theta)u_{2y} \quad (20.4.53)$$

$$f_{2,x} = (-k \cos^2 \theta)u_{1,x} - (k \sin \theta \cos \theta)u_{1,y} + (k \cos^2 \theta)u_{2x} + (k \sin \theta \cos \theta)u_{2y} \quad (20.4.54)$$

$$f_{2,y} = (k \sin \theta \cos \theta)u_{2x} - (k \sin^2 \theta)u_{2,y} + (k \sin \theta \cos \theta)u_{2x} + (k \sin^2 \theta)u_{2y} \quad (20.4.55)$$

Writing Eqs. (20.4.52) through (20.4.55) in matrix form yields

$$\begin{bmatrix} f_{1,x} \\ f_{1,y} \\ f_{2,x} \\ f_{2,y} \end{bmatrix} = k \begin{bmatrix} c^2 & sc & -c^2 & -sc \\ sc & s^2 & -sc & -s^2 \\ -c^2 & -sc & c^2 & sc \\ -sc & -s^2 & sc & s^2 \end{bmatrix} \begin{bmatrix} u_{1x} \\ u_{1y} \\ u_{2x} \\ u_{2y} \end{bmatrix} \quad (20.4.56)$$

where s and c are abbreviations for $\sin \theta$ and $\cos \theta$, respectively, and k is the stiffness constant. We can write Eq. (20.4.56) in more compact form as

$$[f] = [k][u] \quad (20.4.57)$$

where the local forces, local displacement, and local stiffness are defined as follows:

$$[f] = \begin{bmatrix} f_{1,x} \\ f_{1,y} \\ f_{2,x} \\ f_{2,y} \end{bmatrix} \quad [u] = \begin{bmatrix} u_{1x} \\ u_{1y} \\ u_{2x} \\ u_{2y} \end{bmatrix} \quad (20.4.58)$$

$$[k] = \frac{AE}{L} \begin{bmatrix} c^2 & sc & -c^2 & -sc \\ sc & s^2 & -sc & -s^2 \\ -c^2 & -sc & c^2 & sc \\ -sc & -s^2 & sc & s^2 \end{bmatrix} \quad (20.4.59)$$

For $\theta = 0$, the truss element reduces to the one shown in Fig. 20.4.9. The local stiffness matrix is simply

$$[k] = \frac{AE}{L} \begin{bmatrix} 1 & 0 & -1 & 0 \\ 0 & 0 & 0 & 0 \\ -1 & 0 & 1 & 0 \\ 0 & 0 & 0 & 0 \end{bmatrix} \quad (20.4.60)$$

Which checks with Eq. (20.4.46). Note how the zero rows and columns are simply used to expand the local stiffness matrix given by Eq. (20.4.60) to account for the zero forces and displacements along the y axis.

Global Stiffness Matrix

The **global stiffness matrix** relates the global forces (external forces) and the global displacements (displacements associated with each joint). We developed the local stiffness matrix for an arbitrary element of the truss—what remains is to assemble all the local stiffness matrices associated with the truss elements. Later it will be apparent in our analysis that the efficiency of the computer program is greatly dependent on the technique developed to formulate the global stiffness matrix.

The classical approach for a truss is to take a free-body diagram of each joint and, using the equilibrium equations, group all the equations together and isolate the global stiffness matrix. This approach is tedious and requires a large computational effort from the computer. The method that is illustrated in what follows to obtain the global stiffness matrix is one that Huston and Passerelo developed. It shows how the building of the global stiffness matrix can be done by a simple strategy in which connectivity tables are used to identify the truss elements and their joints. The method is as follows.

Step 1. Consider an arbitrary truss. First, label the **truss elements** and joints in an arbitrary fashion, as shown in Fig. 20.4.11. There are four joints (1, 2, . . . , 4) and five elements ([1], [2], . . . , [5]).

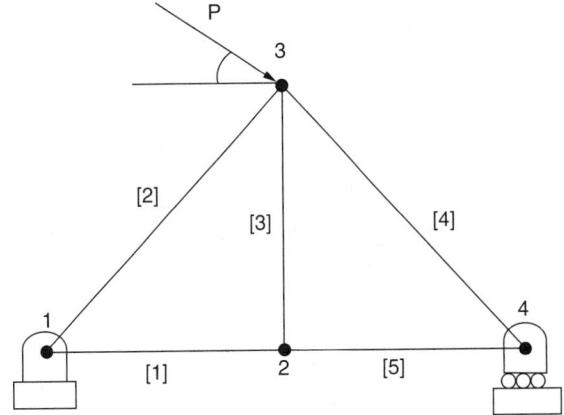

Fig. 20.4.11 A two-dimensional truss.

Step 2. We proceed to develop three tables that basically store geometrical information about the truss. Table 20.4.1 has the truss-joint/matrix-column matching, where pairs starting from 1 develop the column numbers, 2. Table 20.4.2 identifies the connection joints to all the elements of the truss. Using Tables 20.4.1 and 20.4.2, we construct Table 20.4.3, which will be used to assemble the global matrix.

Table 20.4.1   Truss-Joint/Matrix-Column Matrix

Joint	Column	Numbers
1	1	2
2	3	4
3	5	6
4	7	8

Table 20.4.2   Elements versus Joint Numbers

Truss element e	N_1	N_2
1	1	2
2	1	3
3	2	3
4	3	4
5	2	4

Table 20.4.3   Truss Elements versus k_{ij}

i, j from k_{ij}^{e}	Truss elements e				
	1	2	3	4	5
1	1	1	3	5	3
2	2	2	4	6	4
3	3	5	5	7	7
4	4	6	6	8	8

To see how Table 20.4.3 is constructed, consider truss element 1. From Table 20.4.2, we read the connecting joints to element 1, which are 1 and 2; with these joint numbers, we use Table 20.4.1 to extract the corresponding column numbers, which are 1, 2 and 3, 4. Therefore, in Table 20.4.3 number 1 (referring to the element number), we insert the values 1, 2, 3, and 4, as shown. We repeat the procedure for the remaining elements 2 to 7. Table 20.4.3 plays an important role in identifying the location where the local stiffness matrix terms will be inserted in the global matrix.

For the whole truss system, we write the equation relating the global forces and displacements:

$$\{F\} = [K]\,\{U\} \qquad (20.4.61)$$

where

$$\{F\} = \begin{bmatrix} F_{1x} \\ F_{1y} \\ F_{2x} \\ F_{2y} \\ F_{3x} \\ F_{3y} \\ F_{4x} \\ F_{4y} \end{bmatrix} \quad \text{and} \quad \{U\} = \begin{bmatrix} U_{1x} \\ U_{1y} \\ U_{2x} \\ U_{2y} \\ U_{3x} \\ U_{3y} \\ U_{4x} \\ U_{4y} \end{bmatrix} \qquad (20.4.62)$$

and

$$[K] = \begin{bmatrix} K_{1,1} & K_{1,2} & K_{1,3} & \cdots & K_{1,8} \\ K_{2,1} & K_{2,2} & K_{2,3} & \cdots & K_{2,8} \\ \vdots & \vdots & \vdots & \ddots & \vdots \\ K_{8,1} & K_{8,2} & K_{8,3} & \cdots & K_{8,8} \end{bmatrix} \qquad (20.4.63)$$

Note that the size of the global stiffness matrix $[K]$ is 10 by 10. This is equal to the number of nodal points (joints in this case) in the truss multiplied by the degrees of freedom permitted at each node (2 for the planar truss shown). Thus, the $[K]$ in Eq. (20.4.27) is the assembled global stiffness matrix obtained from the assembly of individual element stiffness matrices, that is,

$$f_i^e = k_{ij}^e u_j \qquad (e = 1, \ldots, \text{number of elements}; i, j = 1, 4)$$
$$(20.4.64)$$

where k_{ij}^e is the local element stiffness matrix and is given by Eq. (20.4.56).

For $\theta = 0$, k_{ij}^1 is given by Eq. (20.4.57). For $\theta = 90$, k_{ij}^3 is

$$k_{ij}^3 = \frac{AE}{L} \begin{bmatrix} 0 & 0 & 0 & 0 \\ 0 & 1 & 0 & -1 \\ 0 & 0 & 0 & 0 \\ 0 & -1 & 0 & 1 \end{bmatrix} \qquad (20.4.65)$$

In order to construct the global stiffness matrix, first evaluate all the local element stiffness matrices. Then Table 20.4.3 is used to transfer the terms from the local stiffness matrices to the locations in the global stiffness matrix.

For example, consider element 1 from Table 20.4.3. The element column identifies the location in the global stiffness matrix of its local stiffness. The variation by pair of the column numbers identifies the i, j entries of the local stiffness matrix in the global stiffness matrix, K_{ij}. For example, the column numbers for element 1 basically correspond to

the local stiffness matrix for element 1 when transferred to the location in the assembled global stiffness matrix looks like

$$K^1 = \begin{bmatrix} k_{11}^1 & k_{12}^1 & k_{13}^1 & k_{14}^1 & 0 & 0 & 0 & 0 \\ k_{21}^1 & k_{22}^1 & k_{23}^1 & k_{24}^1 & 0 & 0 & 0 & 0 \\ k_{31}^1 & k_{32}^1 & k_{33}^1 & k_{34}^1 & 0 & 0 & 0 & 0 \\ k_{41}^1 & k_{42}^1 & k_{43}^1 & k_{44}^1 & 0 & 0 & 0 & 0 \\ 0 & 0 & 0 & 0 & 0 & 0 & 0 & 0 \\ 0 & 0 & 0 & 0 & 0 & 0 & 0 & 0 \\ 0 & 0 & 0 & 0 & 0 & 0 & 0 & 0 \\ 0 & 0 & 0 & 0 & 0 & 0 & 0 & 0 \end{bmatrix} \qquad (20.4.66)$$

For element 2, we read the row and column entries, which are 1, 2, 5, and 6, from Table 20.4.3. Hence, its contribution to the global stiffness is given by

$$K^2 = \begin{bmatrix} k_{11}^2 & k_{12}^2 & 0 & 0 & k_{13}^2 & k_{14}^2 & 0 & 0 \\ k_{21}^2 & k_{22}^2 & 0 & 0 & k_{23}^2 & k_{24}^2 & 0 & 0 \\ 0 & 0 & 0 & 0 & 0 & 0 & 0 & 0 \\ 0 & 0 & 0 & 0 & 0 & 0 & 0 & 0 \\ k_{31}^2 & k_{32}^2 & 0 & 0 & k_{33}^2 & k_{34}^2 & 0 & 0 \\ k_{41}^2 & k_{42}^2 & 0 & 0 & k_{43}^2 & k_{44}^2 & 0 & 0 \\ 0 & 0 & 0 & 0 & 0 & 0 & 0 & 0 \\ 0 & 0 & 0 & 0 & 0 & 0 & 0 & 0 \end{bmatrix} \qquad (20.4.67)$$

Once all the elements are completed and their local stiffness matrices are inserted into the global stiffness matrix $[K]$, the general global matrix of the system is then obtained by adding all the entries in different locations if they are more than one.

$$K_{ij} = K_{ij}^1 + K_{ij}^2 + K_{ij}^3 + \cdots K_{ij}^5 \qquad (20.4.68)$$

The above outlined procedure for developing the global stiffness matrix could be automated by using computer coding.

Solution of the Truss Problem

The global force and displacement vectors are related through the global stiffness matrix:

$$[F] = [K]\,[U] \qquad (20.4.69)$$

After the development of $[K]$, the global stiffness matrix, the identification of the boundary forces and displacements is important. These are to be substituted into the $[F]$ and $[U]$ vectors.

For the truss shown in Fig. 20.4.12, the reaction forces at joint 1 in the x and y directions are the reaction forces R_{1x} and R_{1y}, respectively. The reaction at joint 2 is only in the y direction as it is the roller joint, that is, R_{2y}. Also, force P acts as an external force in the x direction at joint 3.

Thus,
$$F_{1x} = R_{1x} \qquad F_{1y} = R_{1y} \qquad F_{4y} = R_{4y}$$
$$F_{3x} = P_x \qquad F_{3y} = P_y \qquad (20.4.70)$$

Except for these, all other forces are equal to zero.

Substituting these values into force vector $[F]$, we get Eq (20.4.71):

$$[F] = \begin{bmatrix} R_{1x} \\ R_{1y} \\ 0 \\ 0 \\ P_x \\ P_y \\ 0 \\ R_{4y} \end{bmatrix} \qquad (20.4.71)$$

Similarly, the displacement boundary condition can be identified, where for joints 1 and 2 we have Eq. (20.4.72)

$$U_{1x} = 0$$
$$U_{1y} = 0 \qquad (20.4.72)$$
$$U_{4y} = 0$$

All others are nonzero. Substituting these zero displacements into the displacement vector $[U]$, we get Eq. (20.4.73)

$$[U] = \begin{bmatrix} 0 \\ 0 \\ U_{2x} \\ U_{2y} \\ U_{3x} \\ U_{3y} \\ U_{4x} \\ 0 \end{bmatrix} \qquad (20.4.73)$$

Substituting $[F]$ and $[U]$ from Eqs. (20.4.71) and (20.4.73) into Eq. (20.4.69), we obtain the general equations governing the truss force/displacement equilibrium conditions.

$$\begin{bmatrix} R_{1x} \\ R_{1y} \\ 0 \\ 0 \\ P_x \\ P_y \\ 0 \\ R_{4y} \end{bmatrix} = \begin{bmatrix} K_{1,1} & K_{1,2} & \cdots & K_{1,8} \\ K_{2,1} & K_{2,2} & \cdots & K_{2,8} \\ \vdots & \vdots & \cdots & \vdots \\ \vdots & \vdots & \cdots & \vdots \\ \vdots & \vdots & \cdots & \vdots \\ \vdots & \vdots & \cdots & \vdots \\ \vdots & \vdots & \cdots & \vdots \\ K_{8,1} & K_{8,1} & \cdots & K_{8,8} \end{bmatrix} \begin{bmatrix} 0 \\ 0 \\ U_{2x} \\ U_{2y} \\ U_{3x} \\ U_{3y} \\ U_{4x} \\ 0 \end{bmatrix} \qquad (20.4.74)$$

Note how in Eq. (20.4.39) the unknowns are in the global force array as well as in the joint displacement vector. A typical strategy to solve such a problem in which the unknowns are on both sides of the equation is to solve for the U's first by partitioning the matrices such that the force vector is completely in terms of the known forces. Eliminating the reaction forces does the partitioning. The resulting equation is Eq. (20.4.75).

$$\begin{bmatrix} 0 \\ 0 \\ P_x \\ P_y \\ 0 \end{bmatrix} = [K'] \begin{bmatrix} U_{2x} \\ U_{2y} \\ U_{3x} \\ U_{3y} \\ U_{4x} \end{bmatrix} \qquad (20.4.75)$$

where $[K']$ is the new stiffness matrix resulting from the global stiffness after eliminating the rows and columns corresponding to the zero displacements. Equation (20.4.75) constitutes a set of seven equations and seven unknowns, which can be solved using the gaussian elimination method, Cramer's rule, or simply the inverse of $[K']$ as Eq. (20.4.76)

$$\begin{bmatrix} U_{2x} \\ U_{2y} \\ U_{3x} \\ U_{3y} \\ U_{4x} \end{bmatrix} = [K']^{-1} \begin{bmatrix} 0 \\ 0 \\ P_x \\ P_y \\ 0 \end{bmatrix} \qquad (20.4.76)$$

Once the equations are solved for the displacements, reactions R_{1x}, R_{1y}, and R_{2y} can be evaluated by premultiplying the corresponding terms of $[K]$ and $[U]$ in Eq. (20.4.74). The solutions are found to be

$$R_{1x} = -P_x \qquad R_{4y} = \frac{P_y}{2} - \frac{b}{2a} P_x \qquad R_{1y} = \frac{P_y}{2} - \frac{b}{2a} P_x$$
$$(20.4.77)$$

The answers obtained from the FEM analysis as described before can be checked by simply taking the free-body diagram for the truss as shown in Fig. 20.4.12. Writing the equilibrium equations

$$\sum F_x = 0 \qquad \sum F_y = 0 \qquad \sum M = 0 \qquad (20.4.78)$$

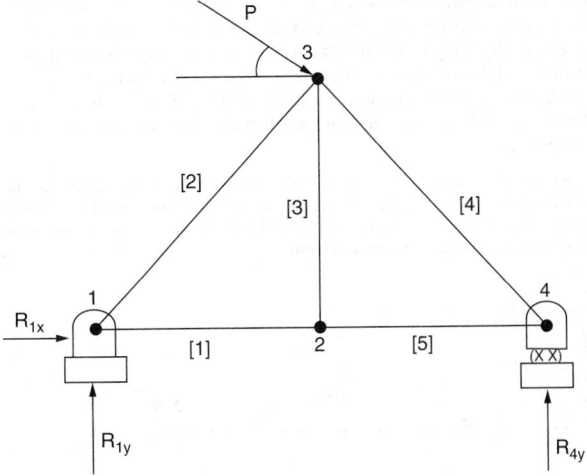

Fig. 20.4.12   A simple free-body diagram of 2-D truss subject to loading.

we get

$$\sum F_x = R_{1x} + P_x = 0$$
$$\sum F_y = R_{1y} + R_{4y} + P_y = 0 \qquad (20.4.79)$$
$$\sum M_1 = R_{4y}2a - P_xb - P_ya = 0$$

which checks with the FEM solution.

Evaluation of the Local Forces

The internal forces are those that are either compressing or tensioning the truss elements. To find the components of the forces acting at each end, we use the previously computed global displacements and local stiffness matrix.

From Eq. (20.4.56), we have

$$\begin{bmatrix} f^e_{1x} \\ f^e_{1y} \\ f^e_{2x} \\ f^e_{2y} \end{bmatrix} = \frac{AE}{l} \begin{bmatrix} c^2_e & s_e c_e & -c^2_e & -s_e c_e \\ s_e c_e & s^2_e & -s_e c_e & -s^2_e \\ -c^2_e & -s_e c_e & c^2_e & s_e c_e \\ -s_e c_e & -s^2_e & s_e c_e & s^2_e \end{bmatrix} \begin{bmatrix} u^e_{1x} \\ u^e_{1y} \\ u^e_{2x} \\ u^e_{2y} \end{bmatrix} \qquad (20.4.80)$$

The stiffness matrix in this equation (for the particular element at hand) is known. The nodal displacements are also known from the global displacement vector. The relationship between the global and local displacement is explained below.

Let the local displacement be written as u^e_{ij}, where e is the element number of the truss, i denotes either end 1 or end 2 of the element, and j assigns the direction x or y to the end displacements. Assuming that all elements of the truss undergo the same displacements at each joint, we then write the following.

For joint 1: $U_{1x} = u^1_{1x} = u^2_{1x} = u^4_{1x} \qquad U_{1y} = u^1_{1y} = u^2_{1y} = u^4_{1y}$
$$(20.4.81)$$

For joint 2: $U_{2x} = u^1_{2x} = u^5_{1x} = u^7_{1x} \qquad U_{2y} = u^1_{2y} = u^5_{1y} = u^7_{1y}$
$$(20.4.82)$$

For joint 3: $U_{3x} = u^2_{2x} = u^3_{1x} \qquad U_{3y} = u^2_{2y} = u^3_{1y} \qquad (20.4.83)$

For joint 4: $U_{4x} = u^3_{2x} = u^4_{2x} = u^6_{1x} = u^5_{2x}$
$$U_{4y} = u^3_{2y} = u^4_{2y} = u^6_{1y} = u^5_{2x} \qquad (20.4.84)$$

For joint 5: $U_{5x} = u^7_{2x} = u^6_{2x} \qquad U_{5y} = u^7_{2y} = u^6_{2y} \qquad (20.4.85)$

The nodal displacements for any particular element can be found from the relationships between the global displacements and the local displacement by using Eqs. (20.4.81) to (20.4.85). Subsequent substitution of these values for a particular element in Eq. (20.4.80) and multiplication by the corresponding stiffness matrix terms yields the nodal forces. The signs of these forces indicate whether the member is in tension or compression.

EXAMPLE: ANALYSIS OF A THREE-ELEMENT TRUSS   Use the finite-element method on the truss in Fig. 20.4.13. (1) Find the global stiffness matrix. (2) Solve for the reaction forces. (3) Solve for the member forces and determine whether a truss element is in tension or compression.

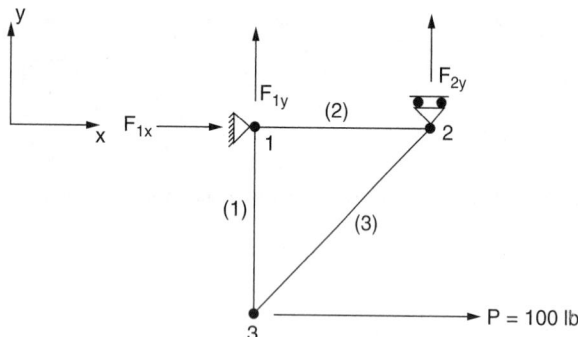

Fig. 20.4.13 A three-element truss subject to loading.

Solution: The first step in our analysis is to label the truss for the joint numbers and link numbers as shown in the figure. The second step is to compute the local stiffness matrices for each member, using Eq. (20.4.59) which gives

$$[k_{ij}^1] = \frac{A_1 E}{l}\begin{bmatrix} 0 & 0 & 0 & 0 \\ 0 & 1 & 0 & -1 \\ 0 & 0 & 0 & 0 \\ 0 & -1 & 0 & 1 \end{bmatrix} \qquad [k_{ij}^2] = \frac{A_2 E}{l\sqrt{3}}\begin{bmatrix} 1 & 0 & -1 & 0 \\ 0 & 0 & 0 & 0 \\ -1 & 0 & 1 & 0 \\ 0 & 0 & 0 & 0 \end{bmatrix}$$

$$[k_{ij}^3] = \frac{A_3 E}{8l}\begin{bmatrix} 3 & \sqrt{3} & -3 & -\sqrt{3} \\ \sqrt{3} & 1 & -\sqrt{3} & -1 \\ -3 & -\sqrt{3} & 3 & \sqrt{3} \\ -\sqrt{3} & -1 & \sqrt{3} & 1 \end{bmatrix}$$

where A_1, A_2, A_3 and L_1, L_2, L_3 are the areas of cross section and lengths of the members of the truss, respectively, and E is Young's modulus of elasticity. $A_1 = A_2 = A_3 = A$ and $L_1 = l$, such that $L_2 = l\sqrt{3}$ and $L_3 = 2l$.

Now let us construct Tables 20.4.4 to 20.4.6 (see earlier Tables 20.4.1 to 20.4.3), which help in arriving at the global stiffness matrix $[K]$. Using the local

Table 20.4.4

Joint number	Column number	
1	1	2
2	3	4
3	5	6

Table 20.4.5

Elements	Joint number	
1	1	3
2	1	2
3	2	3

Table 20.4.6

K_{ij}	1	2	3
1	1	1	3
2	2	2	4
3	5	3	5
4	6	4	6

stiffness matrices and transferring the entries with the help of Table 20.4.6 we can arrive at the global stiffness matrix:

$$\frac{AE}{L}\begin{bmatrix} \frac{1}{\sqrt{3}} & 0 & -\frac{1}{\sqrt{3}} & 0 & 0 & 0 \\ 0 & 1 & 0 & 0 & 0 & -1 \\ -\frac{1}{\sqrt{3}} & 0 & \frac{8+3\sqrt{3}}{8\sqrt{3}} & \frac{\sqrt{3}}{8} & -\frac{3}{8} & -\frac{\sqrt{3}}{8} \\ 0 & 0 & \frac{\sqrt{3}}{8} & \frac{1}{8} & -\frac{\sqrt{3}}{8} & -\frac{1}{8} \\ 0 & 0 & -\frac{3}{8} & -\frac{\sqrt{3}}{8} & \frac{3}{8} & \frac{\sqrt{3}}{8} \\ 0 & -1 & -\frac{\sqrt{3}}{8} & -\frac{1}{8} & \frac{\sqrt{3}}{8} & \frac{9}{8} \end{bmatrix}\begin{Bmatrix} U_{1x} \\ U_{1y} \\ U_{2x} \\ U_{2y} \\ U_{3x} \\ U_{3y} \end{Bmatrix} = \begin{Bmatrix} F_{1x} \\ F_{1y} \\ 0 \\ F_{2y} \\ P \\ 0 \end{Bmatrix}$$

Zeros in the force vector indicate that the forces in the x and y directions at joints 2 and 3 are zeroes because of the roller and free joint, respectively.

Applying the displacement boundary conditions, $U_{1x} = 0$, $U_{1y} = 0$, $U_{2x} = 0$, and eliminating the corresponding rows and columns, we get

$$\begin{bmatrix} 0.952 & -0.375 & -0.2165 \\ -0.375 & 0.375 & 0.2165 \\ -0.2165 & 0.2165 & 1.125 \end{bmatrix}\begin{Bmatrix} U_{2x} \\ U_{3x} \\ U_{3y} \end{Bmatrix} = \begin{Bmatrix} 0 \\ 100 \\ 0 \end{Bmatrix}$$

Solving for the unknowns, we obtain

$$U_{2x} = 173.31 \frac{l}{AE} \qquad U_{3x} = 473.13\frac{l}{AE} \qquad U_{3y} = -57.57\frac{l}{AE}$$

The reaction forces can be computed using Eq. (20.4.43) as

$$F_{1x} = \left(\frac{1}{\sqrt{3}}U_{1x} - \frac{1}{\sqrt{3}}U_{2x}\right)\frac{AE}{l} = -100.0 \text{ lb}$$

$$F_{1y} = (1.U_{1y} - 1.U_{3y})\frac{AE}{l} = 57.57 \text{ lb}$$

$$F_{2y} = \left(-\frac{\sqrt{3}}{8}U_{2x} + \frac{\sqrt{3}}{8}U_{3y} - \frac{1}{8}U_{3y}\right) = -57.57 \text{ lb}$$

These results can be verified by using the free-body diagram of the truss (Fig. 20.4.12)

$$\sum F_x = 0 \qquad \sum F_y = 0 \qquad \sum M = 0$$

$\sum F_x = 0$ gives $F_{1x} + 100.0 = 0$, that is, $F_{1x} = -100.0$ lb

$\sum F_y = 0$ gives $F_{1y} + F_{2y} = 0$, that is, $F_{1y} = -F_{2y}$

$\sum M = 0$ gives $pl + F_{2y}\sqrt{3}l = 0.0$, that is, $F_{2y} = \begin{cases} -57.57 \\ -F_{1y} \end{cases}$

Therefore, members 1 and 3 are in tension, whereas member 2 is in compression. The member forces are obtained from the local element force-displacement relationship:

$$\begin{Bmatrix} f_{1x}^e \\ f_{1y}^e \\ f_{2x}^e \\ f_{2y}^e \end{Bmatrix} = \frac{AE}{l}\begin{bmatrix} c^2 & sc & -c^2 & -sc \\ sc & s^2 & -sc & -s^2 \\ -c^2 & -sc & c^2 & sc \\ -sc & -s^2 & sc & s^2 \end{bmatrix}\begin{Bmatrix} u_{1x}^e \\ u_{1y}^e \\ u_{2x}^e \\ u_{2y}^e \end{Bmatrix}$$

The global and local displacement at each joint are related as follows:

$$U_{1x} = u_{1x}^0 = 0 \qquad U_{1y} = u_{1y}^1 = 0 \qquad U_{3x} = u_{2x}^1 \qquad U_{3y} = u_{2y}^1$$

Using the local stiffness already computed, we obtain

$$\begin{Bmatrix} f_{1x}^1 \\ f_{1y}^1 \\ f_{2x}^1 \\ f_{2y}^1 \end{Bmatrix} = \frac{AE}{l}\begin{bmatrix} 0 & 0 & 0 & 0 \\ 0 & 1 & 0 & -1 \\ 0 & 0 & 0 & 0 \\ 0 & -1 & 0 & 1 \end{bmatrix}\begin{Bmatrix} 0 \\ 0 \\ U_{3x} \\ U_{3y} \end{Bmatrix}$$

$$\Rightarrow f_{1y}^1 = -\left(\frac{AE}{l}\right)U_{3y} = -\left(\frac{AE}{l}\right)57.57\frac{l}{AE} = 57.57 \text{ lb}$$

$$f_{2y}^1 = \left(\frac{AE}{l}\right)U_{3y} = -57.57 \text{ lb}$$

The forces acting on element 1 clearly show that it is in tension as predicted. Similarly, we can obtain the magnitude of the forces and directions for elements 2 and 3.

$$F_{1x} = F_{2x} = 0$$

Stress Analysis

In the analysis of a truss, the main objective is to decide whether the truss elements are designed to sustain the load they support. For that we need to evaluate the stress or average stress in each element. The latter is an indication as to whether the tension or compression can be sustained.

Let the element stress be given by

$$\sigma^{(e)} = E\varepsilon^{(e)}$$
$$= E^{(e)} \frac{u_2^{(e)} - u_1^{(e)}}{l} \qquad (20.4.86)$$

Now we can compute the stress for each element of the truss in the previous example. For each element we can write

$$\sigma^{(1)} = E^{(1)} \frac{u_2^{(1)} - u_1^{(1)}}{l} = E^{(1)} \frac{U_{3y} - U_{1y}}{l}$$

$$\sigma^{(2)} = E^{(2)} \frac{u_2^{(2)} - u_1^{(2)}}{l} = E^{(2)} \frac{U_{2x} - U_{1x}}{\sqrt{3}l}$$

$$\sigma^{(3)} = E^{(3)} \frac{u_2^{(3)} - u_1^{(3)}}{l} = E^{(3)} \frac{U_{3x} - U_{2x}}{2l}$$

and the results are

$$\sigma^{(1)} = E\frac{U_{3y}}{l} = \frac{E}{l}\left(-57.57\frac{l}{AE}\right)$$

$$\sigma^{(2)} = E\frac{U_{2x}}{l\sqrt{3}} = \frac{E}{l\sqrt{3}}\left(173.31\frac{l}{AE}\right)$$

$$\sigma^{(3)} = E\frac{U_{2x} - U_{3x}}{2l} = -\frac{E}{2l}\left(299.82\frac{l}{AE}\right)$$

HEAT CONDUCTION ANALYSIS AND FEM

In most instances, the important problems of engineering involving an exchange of energy by the flow of heat are those in which there is a transfer of internal energy between two systems. In general, the internal energy transfer is called *heat transfer*. When such exchanges of internal energy or heat take place, the first law of thermodynamics requires that the heat given up by one body must equal that taken up by the other. The second law of thermodynamics demands that the transfer of heat take place from the hotter system to the colder system. It is customary to categorize the various heat transfer processes into three basic types or modes, although it is certainly a rare instance when one encounters a problem of practical importance that does not involve at least two, and sometimes all three, of these modes occurring simultaneously. The three modes are conduction, convection, and radiation. Heat conduction will be the focus of this section. Heat conduction is the term applied to the mechanism of internal energy exchange from one body to another, or from one part of a body to another part, by the exchange of kinetic energy.

When the relationship between force and displacement can be approximated by a linear function, the problem reduces to a one-dimensional analysis. This is seen in the study of trusses where the element deformation is expressed as a linear function in terms of the element force. A number of engineering problems, and heat conduction in particular, can be solved by using one-dimensional element and linear functions to approximate the solution. Here, we will extend the one-dimensional solution to heat conduction problems, and define the concept of shape functions for one and two dimensions in the finite-element method.

One-Dimensional Elements

Now we apply the finite-element method to the solution of heat flow in some simple one-dimensional steady-state heat conduction systems. Several physical shapes fall into the one-dimensional analysis, such as spherical and cylindrical systems, in which the temperature of the body is a function only of radial distance. Consider the straight bar of Fig. 20.4.14, where the heat flows across the end surfaces. Heat is also assumed to be generated internally by a heat source at a rate f per unit volume. The temperature varies only along the axial direction x, and we want to formulate a finite-element technique that would yield the temperature $T = T(x)$ along the position x in the steady-state condition.

Fig. 20.4.14 A typical bar with temperatures T_0 and T_f at each end.

In steady-state conditions, the net rate of heat flow into any differential element is zero. We know that for heat conduction analysis, the Fourier heat conduction equation is

$$q = -\kappa\frac{dT}{dx} \qquad (20.4.87)$$

This equation states that the heat flux q in direction x is proportional to the gradient of temperature in direction x. The conductivity constant is defined by κ.

From the differential element in Fig. 20.4.14, we can write the heat flux balance:

$$qA + fA\,dx - [qA + d(Aq)] = 0 \qquad (20.4.88)$$

With the differentiation of q, the heat flux equation becomes

$$qA + fA\,dx - \left[qA + \frac{dq}{dx}dx\,A\right] = 0 \qquad (20.4.89)$$

This reduces to a first-order differential equation of the form

$$\frac{dq}{dx} = f \qquad (20.4.90)$$

where A = cross-sectional area; f = heat source/unit volume; q = heat flux; T = temperature.

Substituting Eq. (20.4.87) into Eq. (20.4.90), we get the governing differential equation for the temperature:

$$\kappa\frac{d^2T}{dx^2} = -f \qquad (20.4.91)$$

The boundary conditions for the physical problem described in Fig. 20.4.14 are $T = T_0$ at $x = 0$ and $T = T_f$ at $x = L$.

Integrating (20.4.4) we get an explicit solution for the temperature at any point along the bar:

$$T(x) = \frac{fL}{2\kappa}\left(x - \frac{x^2}{L}\right) + \left(\frac{T_f - T_0}{L}\right)x + T_0 \qquad (20.4.92)$$

For one-dimensional problem the temperature at any point x can be found by using Eq. (20.4.91). Finite-element analysis is tailored for engineering problems where the closed-form solution can't be evaluated and numerical approximation is required. There are a number of scenarios in which the governing heat equation is nonlinear and the closed form is unattainable. What follow are the initial steps in applying finite-element methods to heat conduction problems. First we develop the solution to one-dimensional heat conduction problems and compare that to the solution given by Eq. (20.4.91).

Finite-Element Formulation

We must use either the principle of virtual work or energy to derive the necessary governing equations in the finite-element method. The method as shown in the previous two sections leads to the formulation of the element stiffness and stiffness matrix. We first develop the following energy equation as

$$I = \int \left[kA \frac{d^2T}{dx^2} + fA \right] dx \qquad (20.4.93)$$

which yields Eq. (20.4.91) for $dI = 0$ after the standard manipulation of calculus of variations. Equation (20.4.93) could be expressed further in two parts, I_1 and I_2, as

$$I = \underbrace{\int \frac{d}{dx}\left[Ak \frac{dT}{dx} \right] T\, dx}_{I_1} + \underbrace{\int fAT\, dx}_{I_2} \qquad (20.4.94)$$

Integrating the function I_1 by parts, we get

$$I_1 = TAk \frac{dT}{dx} \Big|_0^L - \int_0^L \frac{dT}{dx} Ak \frac{dT}{dx} dx \qquad (20.4.95)$$

The first term defines the boundary conditions' contributions. If we assume that the boundary conditions are such that

$$T_{(x=0)} = T_L \qquad \text{and} \qquad q|_{x=L} = h(T_L - T_\infty)$$

then the functional I becomes

$$I = -\int_0^L \frac{dT}{dx} Ak \frac{dT}{dx} dx - \int fAT\, dx + \frac{1}{2} h (T_L - T_\infty)^2 \qquad (20.4.96)$$

We have an expression for I ready to use if the temperature $T = T(x)$ has an explicit form in x that can be substituted into the equation so that we can carry on the integration.

Next, consider the function $I^{(e)}$ for an element rather than for the total system:

$$I^{(e)} = -\int_{x_1}^{x_2} \frac{dT}{dx} A\kappa \frac{dT}{dx} dx + TA\kappa \frac{dT}{dx}\Big|_{x_1}^{x_2} + \int_{x_1}^{x_2} fAT\, dx \qquad (20.4.97)$$

$$I^{(e)} = I_1^{(e)} + I_2^{(e)} + I_3^{(e)} \qquad (20.4.98)$$

where $I_1^{(e)}$, $I_2^{(e)}$, and $I_3^{(e)}$ correspond to the three terms of element $I^{(e)}$, respectively. To develop all the $I_1(e)$ terms we need to find an expression for the temperature T. Assume a linear interpolation for the temperature between x_1 and x_2, as the distance between these two points is assumed small. A representation of the temperature is shown in Fig. 20.4.15, where the temperature varies linearly as

$$T = ax + b \qquad (20.4.99)$$

Fig. 20.4.15   Linear interpolation of the temperature.

At each node, the temperature is assumed to be T_1 and T_2, respectively. We can write the temperature equation for each node as

$$T_1 = ax_1 + b \qquad \text{and} \qquad T_2 = ax_2 + b \qquad (20.4.100)$$

from which we can solve for a and b:

$$a = \frac{T_2 - T_1}{L_e} \qquad \text{and} \qquad b = T_1 - \frac{T_2 - T_1}{L_e} x_1$$

where L_e denotes the length of the element ($x_2 - x_1$). Substituting the values of a and b into Eq. (20.4.100), we write an expression for T by introducing shape functions:

$$T = T_1 N_1 + T_2 N_2 \qquad (20.4.101)$$

where

$$N_1 = \frac{x_2 - x}{L_e} \qquad \text{and} \qquad N_2 = \frac{x - x_1}{L_e} \qquad (20.4.102)$$

The latter are known as shape functions. These functions are linear in x and represent the characteristic of the function assumed in representing the temperature between x_1 and x_2.

In matrix form, the temperature from Eq. (20.4.101) can be expressed as

$$T = [T_1 \quad T_2] \begin{bmatrix} N_1 \\ N_2 \end{bmatrix} \qquad (20.4.103)$$

We note in Eq. (20.4.97) that the time derivative of T is also required, hence the derivative of T as given by Eq. (20.4.103) takes on the following form:

$$\frac{dT}{dx} = [T_1 \quad T_2] \begin{bmatrix} \dfrac{dN_1}{dx} \\ \dfrac{dN_2}{dx} \end{bmatrix} \qquad (20.4.104)$$

$$\frac{dN_1}{dx} = -\frac{1}{L_e} \qquad \text{and} \qquad \frac{dN_2}{dx} = \frac{1}{L_e} \qquad (20.4.105)$$

The functional $I^{(e)}$ then becomes

$$I^{(e)} = \int_{x_1}^{x_2} [T_1 \quad T_2] \begin{bmatrix} dN_1/dx \\ dN_2/dx \end{bmatrix} A\kappa \begin{bmatrix} \dfrac{dN_1}{dx} & \dfrac{dN_2}{dx} \end{bmatrix} \begin{bmatrix} T_1 \\ T_2 \end{bmatrix} dx$$
$$- \int_{x_1}^{x_2} fA[T_1 \quad T_2] \begin{bmatrix} N_1 \\ N_2 \end{bmatrix} dx + \frac{1}{2} h(T_L - T_\infty) \qquad (20.4.106)$$

Here the boundary conditions at both ends are defined by the last term in the above equation.

Let the first term be I_1^e and defined by

$$I_1^e = A\kappa \int_{x_1}^{x_2} [T_1 \quad T_2] \begin{bmatrix} \left(\dfrac{dN_1}{dx}\right)^2 & \dfrac{dN_1}{dx}\dfrac{dN_2}{dx} \\ \dfrac{dN_2}{dx}\dfrac{dN_1}{dx} & \left(\dfrac{dN_2}{dx}\right)^2 \end{bmatrix} \begin{bmatrix} T_1 \\ T_2 \end{bmatrix} dx \qquad (20.4.107)$$

Substituting the Eq. (20.4.105) derivatives into Eq. (20.4.107) and integrating yields

$$I_1^e = A\kappa [T_1 \quad T_2] \begin{bmatrix} \dfrac{1}{L_e} & -\dfrac{1}{L_e} \\ -\dfrac{1}{L_e} & \dfrac{1}{L_e} \end{bmatrix} \begin{bmatrix} T_1 \\ T_2 \end{bmatrix} \qquad (20.4.108)$$

Similarly, let $I_2^{(e)}$ denote the term defined in Eq. (20.4.108). Evaluating this term we obtain the term that involves the contribution of the heat source f:

$$I_2^{(e)} = \int_{x_1}^{x_2} fA[T_1 \quad T_2] \begin{bmatrix} N_1 \\ N_2 \end{bmatrix} dx = \frac{fAL_c}{2} [T_1 \quad T_2] \begin{bmatrix} 1 \\ 1 \end{bmatrix} \qquad (204.109)$$

Next, writing the steady-state condition for an element we get

$$\left\{\frac{\partial I^{(e)}}{\partial T_e}\right\} = \{0\} \quad \text{which yields} \quad k_c = \frac{kA}{L_e}\begin{bmatrix} 1 & -1 \\ -1 & 1 \end{bmatrix}$$

(20.4.110)

and the element loading vector from the second term $I_2^{(e)}$,

$$f_Q = \frac{fAL_e}{2}\begin{bmatrix} 1 \\ 1 \end{bmatrix}$$

(20.4.111)

Combining the last equations, we obtain the first step in the finite-element formulation where

$$[k_c]\{T_{X\text{out}}\} = f_2$$

(20.4.112)

Note how $[k_c]$ is analogous to a local stiffness matrix (we refer to it as the conductivity matrix), how $\{T\}$ is analogous to nodal displacements, and how $\{f_e\}$ is analogous to nodal forces. With this in mind, we can use a table analogous to Table 20.4.3 in the analysis of trusses, to develop the connectivity relations between elements. The latter will assist us in the formulation of the global stiffness matrix and the heat source array. The global problem can be stated as

$$[K]\{T\} = \{F\}$$

(20.4.113)

where $[K]$ is the global conductivity matrix (equivalent to the global stiffness) assembled from the element conductivity matrix k_c, $\{T\}$ the nodal temperatures, and $\{F\}$ the heat source contribution.

Boundary Condition Contribution

The term in the function I in Eq. (20.4.93) dealing with the convection can be written further as $\frac{1}{2}T_L hT_L - (hT_\infty)T_L + \frac{1}{2}hT_\infty^2$, where we see the last term drop out from the variational $\partial I/\partial T$.

We see that the hT_L term will be added to the K matrix at the (L, L) location and hT_∞ will be added to the F vector at the Lth location. The way the K and F will be formulated is shown below:

$$\begin{bmatrix} K_{11} & \cdots & \cdots & K_{1L} \\ K_{21} & \cdots & \cdots & K_{2L} \\ \vdots & & & \vdots \\ K_{L1} & & & K_{LL} + hT_L \end{bmatrix}\begin{Bmatrix} T_1 \\ T_2 \\ \vdots \\ T_L \end{Bmatrix} = \begin{Bmatrix} F_1 \\ F_2 \\ \vdots \\ F_L + hT_\infty \end{Bmatrix}$$

(20.4.114)

Handling of Additional Constraints

The handling of specified temperature boundary conditions such as $T_L = T_0$ can be accompanied by either the elimination or penalty approach. The procedure for elimination is demonstrated below.

Elimination Approach   This technique consists of eliminating rows and columns of the corresponding temperature and then modifying the force vector to include the boundary force displacement relation as described in the finite-element solution of trusses. In general, we write the global problem as

$$KU = F$$

(20.4.115)

Consider the constraint where the displacement is defined by $U_1 = c_1$. The global displacement vector is array of order $1 \times n$. $U = [U_1 \quad U_2 \quad U_3 \quad \cdots \quad U_n]^T$, and similarly the global force vector is $F = [F_1 \quad F_2 \quad F_3 \quad \cdots \quad F_n]^T$.

We first start by defining the potential energy π as function of elastic energy and the work associated with F:

$$\pi = \frac{1}{2}U^T KU - U^t F$$

(20.4.116)

This is one way of obtaining the balance equation of force/displacement as stated previously. The energy explicit matrix form is further shown

to be expressed as

$$\pi = \frac{1}{2}\begin{pmatrix} U_1K_{11}U_1 + U_1K_{12}U_2 + \cdots + U_2K_{1N}U_N \\ U_2K_{21}U_1 + U_2K_{22}U_2 + \cdots + U_2K_{2N}U_N \\ \vdots \\ U_NK_{N1}U_1 + U_NK_{N2}U_2 + \cdots + U_NK_{NN}U_N \end{pmatrix} \\ - (U_1F_1 + U_2F_2 + \cdots + U_NF_N)$$

(20.4.117)

Let us substitute the boundary condition $U_1 = C_1$. Then we get

$$\pi = \frac{1}{2}\begin{pmatrix} C_1K_{11}C_1 + C_1K_{12}U_2 + \cdots + C_1K_{1N}U_N \\ U_2K_{21}C_1 + U_2K_{22}U_2 + \cdots + U_2K_{2N}U_N \\ \vdots \\ U_NK_{N1}C_1 + U_NK_{N2}U_2 + \cdots + U_NK_{NN}U_N \end{pmatrix} \\ - (C_1F_1 + U_2F_2 + \cdots + U_NF_N)$$

(20.4.118)

To solve the problem at hand we need to minimize π, hence $\partial\pi/\partial U_i = 0$, where $i = 1, 2, 3, \ldots, N$, which yields

$$\begin{bmatrix} K_{22} & K_{23} & \cdots & K_{2N} \\ K_{32} & K_{33} & \cdots & K_{3N} \\ \vdots & & & \\ K_{N2} & K_{N3} & & K_{NN} \end{bmatrix}\begin{Bmatrix} U_2 \\ U_3 \\ \vdots \\ U_N \end{Bmatrix} = \begin{Bmatrix} F_2 - K_{21}C_1 \\ F_3 - K_{31}C_1 \\ \vdots \\ F_N - K_{N1}C_1 \end{Bmatrix}$$

(20.4.119)

The above matrix equation represents a set of $(N - 1) \times (N - 1)$ equations with the original K matrix being reduced by eliminating the first row and first column as well as the first element of the vector F. We see that the boundary conditions are subtracted from the force vector array by premultiplying it with the elements of the first row of K. This technique will be used in the solutions of heat transfer problems.

Penalty Approach   An alternative to the elimination approach is the **penalty approach.** In handling constraints, this might be easier to implement and works well for multiple constraints. The methods are designed to handle the boundary conditions once the global problem has been formulated. Once more, let the boundary conditions be given by the displacement at node 1 such that $U_1 = C_1$.

The total potential energy is then defined by adding an extra term to account for the additional boundary condition or simply to account for the additional energy contribution from the boundary conditions:

$$\pi = \frac{1}{2}u^T ku - u^T F + \frac{1}{2}Q(u_1 - c_1)^2$$

(20.4.120)

So, the energy term $\frac{1}{2}Q(U_1 - C_1)^2$ is significant only if the value of Q is large enough to emphasize the contribution of $(U_1 - C_1)$.

Minimization of π results in

$$\begin{bmatrix} K_{11} + Q & K_{12} & \cdots & K_{1N} \\ K_{21} & K_{22} & \cdots & K_{2N} \\ \vdots & \vdots & \vdots & \vdots \\ K_{N1} & K_{N2} & & K_{NN} \end{bmatrix}\begin{Bmatrix} U_1 \\ U_2 \\ \vdots \\ U_N \end{Bmatrix} = \begin{Bmatrix} F_1 + QC_1 \\ F_2 \\ \vdots \\ F_N \end{Bmatrix}$$

(20.4.121)

On examining the above equation, it is apparent that Q is added to K_{11} in the global K matrix and Q_c is added to the element force vector F_1. We can view Q as a stiffness value whose numerical values can be defined or selected by noting that

$$(K_{11} + Q)u_1 + K_{12}U_2 + K_{13}U_3 + \cdots + K_{1N}U_N = F_1 + QC_1$$

(20.4.122)

If we divide by Q we obtain

$$\left(\frac{K_{11}}{Q} + 1\right)U_1 + \frac{K_{12}}{Q}U_2 + \cdots + \frac{K_{1N}}{Q}U_N = \frac{F_1}{Q} + C_1$$

(20.4.123)

Observe how, if Q is chosen to be a large volume, the equation reduces to $U_1 = C_1$, which is the desired boundary condition. We also see further that Q is large in comparison to $K_{11}, K_{12}, \ldots, K_{1N}$, hence we need to select Q large enough to satisfy the condition of the equation above. A suggested value in previous work is K_{ij}, $Q = \max |K_{1j}| \times 10^4$.

EXAMPLE Determine the temperature distribution in the composite wall used to isolate the outside (Fig. 20.4.16). Convection heat transfer on the inner surface of the wall with $T_\infty = 500°C$ is given by $h = 25$ W/m^{2}C. The conductivity constants for each wall are $K_1 = 20$ W/m $\cdot$ °C, $K_2 = 30$ W/m $\cdot$ °C and $K_3 = 40$ W/m $\cdot$ °C respectively. Let the cross-sectional area $A = 1$ m^2 and $L_1 = 0.4$ m, $L_2 = 0.3$ m, $L_3 = 0.1$ m.

This example is used to demonstrate not only how to build the conductivity stiffness matrices and the loading vector F but how to implement the technique that describes how the boundary conditions are employed.

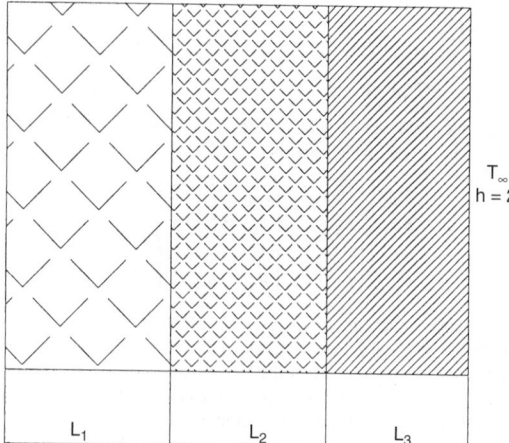

$T_\infty = 500°C$
$h = 25$ W/m°C

L_1 L_2 L_3

Fig. 20.4.16

Solution: Let the temperature at each wall be denoted by T_i and let the width of the wall represent the length of each element.

We need to compute the local conductivity stiffness for each element. Since the conductivity constant is given per unit length, we write

$$K^{(1)} = \frac{\kappa_1}{L_1}\begin{bmatrix} 1 & -1 \\ -1 & 1 \end{bmatrix} \qquad K^{(2)} = \frac{\kappa_2}{L_2}\begin{bmatrix} 1 & -1 \\ -1 & 1 \end{bmatrix} \qquad K^{(3)} = \frac{\kappa_3}{L_3}\begin{bmatrix} 1 & -1 \\ -1 & 1 \end{bmatrix}$$

(20.4.124)

The global K is found by assembling $K^{(1)}$, $K^{(2)}$, and $K^{(3)}$, which results in

$$K = 50\begin{bmatrix} 1 & -1 & 0 & 0 \\ -1 & 3 & -2 & 0 \\ 0 & -2 & 10 & -8 \\ 0 & 0 & -8 & 8 \end{bmatrix}$$

Since convection occurs at node 1, we add $h = 25$ to the (1, 1) location in K, which results in

$$K = 50\begin{bmatrix} 1.5 & -1 & 0 & 0 \\ -1 & 3 & -2 & 0 \\ 0 & -2 & 10 & -8 \\ 0 & 0 & -8 & 8 \end{bmatrix}$$

We have no heat source in the problem, so the F vector consists only of hT_∞: $\{F\} = [25 \times 500,0,0,0]$.

Applying the boundary condition $T_4 = 10°C$ can be handled by the penalty approach. Let us choose a value for Q from the previously proposed procedure, where $Q = \max |K_{ij}| \times 10^4 = 50 \times 10 \times 10^4$. As stated, in the penalty function we add the Q value to the K matrix in the (4×4) location, and QT_4 to the (4×1) location of the F vector, resulting in

$$K = 50\begin{bmatrix} 1.5 & -1 & 0 & 0 \\ -1 & 3 & -2 & 0 \\ 0 & -2 & 10 & -8 \\ 0 & 0 & -8 & 100008 \end{bmatrix}\begin{Bmatrix} T_1 \\ T_2 \\ T_3 \\ T_4 \end{Bmatrix} = \begin{Bmatrix} 25 \times 500 \\ 0 \\ 0 \\ 5 \times 10^7 \end{Bmatrix}$$

The solution of which is found to be

$$T = [229.6559 \quad 94.4839 \quad 26.8979 \quad 10.0014]°C$$

Finite-Difference Approach

Finite difference is discussed briefly in the following example for the purpose of validating the one-dimensional solution we have derived.

EXAMPLE A special design for a building wall consists of three studs containing the materials siding, sheathing, and insulation batting. The inside room temperature is maintained at 85°F and the outside air temperature is measured at 15°F. The area of the wall exposed to air is 180 ft^2. Determine the temperature distribution through the wall.

Table 20.4.7   Characteristics of the Wall

Items	Resistance, h $\cdot$ ft^2 $\cdot$ F/Btu	U factor, Btu/h $\cdot$ ft^2 $\cdot$ °F
Outside film resistance	0.17	5.88
Siding	0.81	1.23
Sheathing	1.32	0.76
Insulation	11.0	0.091
Inside film resistance	0.68	1.47

The steady-state condition of this system can be explained through Fourier's law: $q_x = -kA \, \partial T/\partial X$. We can express the gradient of temperature by $(T_{i+1} - T_i)/l$, and the heat transfer rate becomes $q = [kA(T_{i+1} - T_i)]/l$, or $q = UA(T_{i+1} - T_i)$, where U is defined by k/l.

The heat transfer between the surface and fluid is due to convection. Newton's law of cooling governs the heat transfer rate between the fluid and the surface: $q = hA(T_s - T_f)$, where h is the convection coefficient.

The heat loss through the wall due to conduction must equal the heat loss to the surrounding cold air by convection. That is, $-kA \, \partial T/\partial X = hA(T_s - T_f)$.

Expanding the equation on the temperature distribution at the edge of each wall leads to the following equations:

$$U_2A(T_3 - T_2) = U_1A(T_2 - T_1) \qquad U_3A(T_4 - T_3) = U_2A(T_3 - T_2)$$
$$U_4A(T_5 - T_4) = U_3A(T_4 - T_3) \qquad U_5A(T_6 - T_5) = U_4A(T_5 - T_4)$$

Expressing the above in a matrix form, we get

$$A\begin{bmatrix} u_1 + u_2 & -u_2 & 0 & 0 \\ -u_2 & u_2 + u_3 & -u_3 & 0 \\ 0 & -u_3 & u_3 + u_4 & -u_4 \\ 0 & 0 & -u_4 & u_4 + u_5 \end{bmatrix}\begin{bmatrix} T_2 \\ T_3 \\ T_4 \\ T_5 \end{bmatrix} = \begin{bmatrix} u_1 AT_1 \\ 0 \\ 0 \\ u_5 AT_6 \end{bmatrix}$$

For the numerical values given, we define the four equations in four unknown as

$$180\begin{bmatrix} 7.11 & -1.23 & 0 & 0 \\ -1.23 & 1.99 & -0.76 & 0 \\ 0 & -0.76 & 0.851 & -0.091 \\ 0 & 0 & -0.091 & 1.561 \end{bmatrix}\begin{bmatrix} T_2 \\ T_3 \\ T_4 \\ T_5 \end{bmatrix} = \begin{bmatrix} 15876 \\ 0 \\ 0 \\ 22491 \end{bmatrix}$$

The solution is found to be

$$[T_2 \quad T_3 \quad T_4 \quad T_5] = [15.8523 \quad 19.9266 \quad 26.5205 \quad 81.5909]°C$$

FORMULATION OF GLOBAL STIFFNESS MATRIX FOR N ELEMENTS

Let us consider a body discretized into N one-dimensional elements, as shown in Fig. 20.4.17. Let the boundary conditions be such that

$$T_1 = T_{N+1} = 0 \qquad (20.4.124)$$

The connectivity of the rod elements, which is determined by the same procedure outlined in Table 20.4.3 for trusses is used to define the

Fig. 20.4.17 Discretization of a heat conduction rod into N elements.

global conductivity matrix, which is of order $(N + 1) \times (N + 1)$. The ascending order of the elements helps the global K take a special form with a predictable bandwidth.

$$N\begin{bmatrix} 1 & -1 & 0 & & & 0 \\ -1 & 1+1 & -1 & & & \\ 0 & -1 & 1+1 & -1 & \cdots & 0 \\ \vdots & \vdots & \vdots & \vdots & \ddots & \vdots \\ & & & -1 & 1+1 & -1 \\ 0 & & & & -1 & 1 \end{bmatrix} \begin{bmatrix} T_1 \\ T_2 \\ T_3 \\ \vdots \\ T_{N+1} \end{bmatrix}$$

$$= \begin{bmatrix} \dfrac{1}{2N} \\ \dfrac{1}{2N} + \dfrac{1}{2N} \\ \dfrac{1}{2N} + \dfrac{1}{2N} \\ \vdots \\ \dfrac{1}{2N} \end{bmatrix} \qquad (20.4.125)$$

where f is assumed 1 for simplicity. Applying the boundary conditions reduces the problem to

$$\begin{bmatrix} 2 & -1 & 0 & & & 0 \\ -1 & 2 & -1 & & & \\ 0 & -1 & 2 & -1 & \cdots & 0 \\ \vdots & \vdots & \vdots & \vdots & \ddots & \vdots \\ & & & -1 & 2 & -1 \\ 0 & & & & -1 & 2 \end{bmatrix} \begin{bmatrix} T_2 \\ T_3 \\ \vdots \\ T_{N-1} \\ T_N \end{bmatrix} = \frac{1}{N^2} \begin{bmatrix} 1 \\ 1 \\ 1 \\ \vdots \\ 1 \\ 1 \end{bmatrix}$$

$$(20.4.126)$$

EXAMPLE   For the one-dimensional heat transfer problem given by $d^2T/dx^2 = -10$, where $A = 1$ ($0 \le x \le 1$ with $T(0) = 0$), find the temperature at $x = 0.2$, 0.4, 0.6, 0.8 and 1.0 m (Fig. 20.4.18).

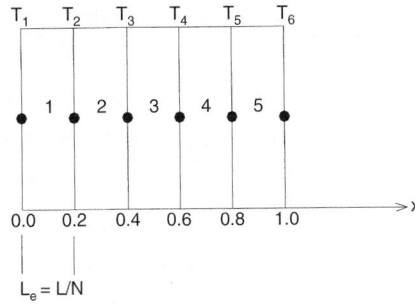

Fig. 20.4.18

Solution: Each element has an element conductivity matrix K_e of the form:

$$K_e = \frac{kA}{L_e} \begin{bmatrix} 1 & -1 \\ -1 & 1 \end{bmatrix}$$

Substituting $A = 1$ m^2 and $L_e = L/N = \frac{1}{5} = 0.2$ m, and assuming the conductivity constant to be $k = 1$, we evaluate the element conductivity matrix:

$$K_e = 5 \begin{bmatrix} 1 & -1 \\ -1 & 1 \end{bmatrix}$$

Using the connectivity table, we expand the local matrices K_s^e to an order of 5×5, and the global matrix $[K]$ is obtained by summation:

$$[K] = 5 \begin{bmatrix} 1 & -1 & 0 & 0 & 0 & 0 \\ -1 & 2 & -1 & 0 & 0 & 0 \\ 0 & -1 & 2 & -1 & 0 & 0 \\ 0 & 0 & -1 & 2 & -1 & 0 \\ 0 & 0 & 0 & -1 & 2 & -1 \\ 0 & 0 & 0 & 0 & -1 & 1 \end{bmatrix} \qquad T = \begin{bmatrix} 0 \\ T_2 \\ T_3 \\ T_4 \\ T_5 \\ 0 \end{bmatrix}$$

With application of the boundary conditions, the global temperature vector becomes $\{T\}$, shown above.

The forcing vector for an element is shown to be

$$F_e = \frac{fAL_e}{2} \begin{bmatrix} 1 \\ 1 \end{bmatrix}$$

where f is the heat generation per unit volume and is obtained from the relation $\kappa\, d^2T/dx^2 = -f$. Substituting $\kappa = 1$ and $d^2T/dx^2 = -10$ yields $f = 10$. Substituting into F^e those values, we get

$$F_e = \begin{bmatrix} 1 \\ 1 \end{bmatrix}$$

Assembling the global forcing vector, using the connectivity table, yields F, and using the relation $[K]\,\{T\} = \{F\}$, we write

$$5\begin{bmatrix} 1 & -1 & 0 & 0 & 0 & 0 \\ -1 & 2 & -1 & 0 & 0 & 0 \\ 0 & -1 & 2 & -1 & 0 & 0 \\ 0 & 0 & -1 & 2 & -1 & 0 \\ 0 & 0 & 0 & -1 & 2 & -1 \\ 0 & 0 & 0 & 0 & -1 & 1 \end{bmatrix} \begin{bmatrix} T_1 \\ T_2 \\ T_3 \\ T_4 \\ T_5 \\ T_6 \end{bmatrix} = \begin{bmatrix} 1 \\ 2 \\ 2 \\ 2 \\ 2 \\ 1 \end{bmatrix} = F = \begin{bmatrix} 1 \\ 1+1 \\ 1+1 \\ 1+1 \\ 1+1 \\ 1 \end{bmatrix}$$

By deleting the first and last rows together with their corresponding columns, and modifying the force vector, we obtain

$$\begin{bmatrix} 2 & -1 & 0 & 0 \\ -1 & 2 & -1 & 0 \\ 0 & -1 & 2 & -1 \\ 0 & 0 & -1 & 2 \end{bmatrix} \begin{bmatrix} T_2 \\ T_3 \\ T_4 \\ T_5 \end{bmatrix} = \begin{bmatrix} 0.4 \\ 0.4 \\ 0.4 \\ 0.4 \end{bmatrix}$$

Note that $T_2 = T_5$ and $T_3 = T_4$. From symmetry, we can solve the equation very easily. The solutions are as follows:

$$\begin{bmatrix} T_1 \\ T_2 \\ T_3 \\ T_4 \\ T_5 \\ T_6 \end{bmatrix} = \begin{bmatrix} 0 \\ 0.8 \\ 1.2 \\ 1.2 \\ 0.8 \\ 0 \end{bmatrix} °C$$

2-D HEAT CONDUCTION ANALYSIS

In a fashion similar to the one-dimensional analysis, the finite-element method can be used to analyze 2-D and 3-D heat conduction problems. Let us examine the 2-D case and bear in mind that, first, we need to develop the element temperature relationship and then expand it to the global problem.

The heat conduction problem is formulated by a variational boundary value problem as

$$\delta I = 0 \qquad (20.4.127)$$

where

$$I = \frac{1}{2} \int_\Omega [k(\nabla T)^2 + 2fT]\, d\Omega \qquad (20.4.128)$$

and where $k =$ thermal conductivity, which we assume is constant; $f =$ heat source; $\nabla T =$ temperature gradient; $(\nabla T)^2 = \nabla T \cdot \nabla T$ (a dot product); $\Omega =$ domain of interest. Note that the heat convection

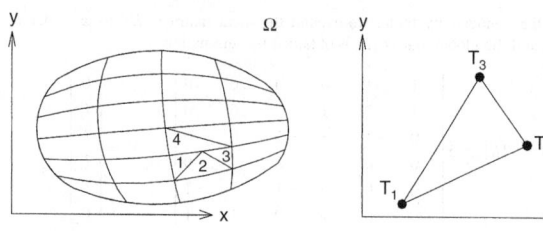

Fig. 20.4.19

part is neglected. If domain Ω is divided into N elements, as shown in Fig. 20.4.19, then

$$I = \sum_{e=1}^{N} I^e \qquad (20.4.129)$$

where

$$I^e = \frac{1}{2} \int_e [k(\nabla T^e)^2 + 2f^e T^e]\, d\Omega \qquad (20.4.130)$$

Let us consider the triangular element shown in Fig. 20.4.19. The local representation of the temperature can be expressed as

$$T(x,y) = T_1 N_1 + T_2 N_2 + T_3 N_3 \qquad (20.4.131)$$

where $N_i(x, y)$ ($i = 1, 2, 3$) are the shape functions given by

$$N_i^e = a_i^e + b_i^e x + c_i^e y \qquad (20.4.132)$$

The shape functions must satisfy the following conditions:
1. $N_i^e (x, y)$ are linear in both x and y.
2. $N_i^e (x, y)$ have the value 1 at node i and zero at other nodes.
3. $N_i^e (x, y)$ are zero at all points in Ω, except those of $N_i^e (x, y)$ can be written as

$$N_i(x, y) = [a_i \quad b_i \quad c_i] \begin{bmatrix} 1 \\ x \\ y \end{bmatrix} \qquad (20.4.133)$$

and for the three nodes of the triangular element, we can write

$$\begin{bmatrix} N_1 \\ N_2 \\ N_3 \end{bmatrix} = \begin{bmatrix} a_1 & b_1 & c_1 \\ a_2 & b_2 & c_2 \\ a_3 & b_3 & c_3 \end{bmatrix} \begin{bmatrix} 1 \\ x \\ y \end{bmatrix} \qquad (20.4.134)$$

For node 1, following condition 1, Eq. (20.4.134) yields

$$\begin{aligned} N_1 = 1 &= a_1 + b_1 x_1 + c_1 y_1 \\ N_2 = 0 &= a_1 + b_1 x_2 + c_1 y_2 \\ N_3 = 0 &= a_1 + b_1 x_3 + c_1 y_3 \end{aligned} \qquad (20.4.135)$$

Which can be written in matrix form as

$$\begin{bmatrix} 1 \\ 0 \\ 0 \end{bmatrix} = [A] \begin{bmatrix} a_1 \\ b_1 \\ c_1 \end{bmatrix} \qquad (20.4.136)$$

where

$$[A] = \begin{bmatrix} 1 & x_1 & y_1 \\ 1 & x_2 & y_2 \\ 1 & x_3 & y_3 \end{bmatrix} \qquad (20.4.137)$$

Solving for coefficients a, b, and c, we get

$$\begin{bmatrix} a_1 \\ b_1 \\ c_1 \end{bmatrix} = \begin{bmatrix} 1 & x_1 & y_1 \\ 1 & x_2 & y_2 \\ 1 & x_3 & y_3 \end{bmatrix}^{-1} \begin{bmatrix} 1 \\ 0 \\ 0 \end{bmatrix} = [A]^{-1} \begin{bmatrix} 1 \\ 0 \\ 0 \end{bmatrix} \qquad (20.4.138)$$

Similarly, for the interpolation functions N_2 and N_3, we get

$$\begin{bmatrix} a_2 \\ b_2 \\ c_2 \end{bmatrix} = [A]^{-1} \begin{bmatrix} 0 \\ 1 \\ 0 \end{bmatrix} \quad \text{and} \quad \begin{bmatrix} a_3 \\ b_3 \\ c_3 \end{bmatrix} = [A]^{-1} \begin{bmatrix} 0 \\ 0 \\ 1 \end{bmatrix} \qquad (20.4.139)$$

The inverse of matrix A is

$$[A]^{-1} = \frac{1}{2a} \begin{bmatrix} (x_2 y_3 - x_3 y_2) & (y_2 - y_3) & (x_3 - x_2) \\ (x_3 y_1 - x_1 y_3) & (y_3 - y_1) & (x_1 - x_3) \\ (x_1 y_2 - x_2 y_1) & (y_1 - y_2) & (x_2 - x_1) \end{bmatrix} \qquad (20.4.140)$$

where a is the area of the triangle. By combining Eqs. (20.4.138) and (20.4.139), we obtain the inverse of A as

$$[A]^{-1} = \begin{bmatrix} a_1 & b_1 & c_1 \\ a_2 & b_2 & c_2 \\ a_3 & b_3 & c_3 \end{bmatrix} \qquad (20.4.141)$$

Then the triangle element functions can be written in a more general form:

$$\{N^e\} = \begin{bmatrix} N_1 \\ N_2 \\ N_3 \end{bmatrix} = [A]^{-1} \begin{bmatrix} 1 \\ x \\ y \end{bmatrix} \qquad (20.4.142)$$

where

$$N_1 = \frac{1}{2a}[(x_2 y_3 - x_3 y_2) + x(y_2 - y_3) + y(x_3 - x_2)]$$

$$N_2 = \frac{1}{2a}[(x_3 y_1 - x_1 y_3) + x(y_3 - y_1) + y(x_1 - x_3)] \qquad (20.4.143)$$

$$N_3 = \frac{1}{2a}[(x_1 y_2 - x_2 y_1) + x(y_1 - y_2) + y(x_2 - x_1)]$$

Now that we have defined the shape function, we can proceed in the evaluation of the conductivity matrix of individual elements.

Element Conductivity Matrix

From Eq. (20.4.130), we write the variational equation in terms of elements. This defines the element equation as

$$I^e = \frac{1}{2} \int [k^e (\nabla T)^2 + f^e T^e]\, d\Omega \qquad (20.4.144)$$

The temperature at the nodes of the triangle element is expressed following the triangular element assumption developed in a previous section where

$$T(x, y) = T_1 N_1 + T_2 N_2 + T_3 N_3 \qquad (20.4.145)$$

From Eq. (20.4.143), we define the partial derivatives with respect to x and y as

$$\frac{\partial N_i}{\partial x} = b_i \quad \text{and} \quad \frac{\partial N_i}{\partial y} = c_i \qquad (20.4.146)$$

Hence, we can write the gradient of the temperature as follows

$$\nabla T = \begin{bmatrix} \dfrac{\partial T}{\partial x} \\ \dfrac{\partial T}{\partial y} \end{bmatrix} = \begin{bmatrix} \dfrac{\partial N_1}{\partial x} & \dfrac{\partial N_2}{\partial x} & \dfrac{\partial N_3}{\partial x} \\ \dfrac{\partial N_1}{\partial y} & \dfrac{\partial N_2}{\partial y} & \dfrac{\partial N_3}{\partial y} \end{bmatrix} \begin{bmatrix} T_1 \\ T_2 \\ T_3 \end{bmatrix} \qquad (20.4.147)$$

which, expressed in compact form, yields $\Delta T = BT$, where

$$B = \begin{bmatrix} b_1 & b_2 & b_3 \\ c_1 & c_2 & c_3 \end{bmatrix} \quad \text{and} \quad T = \begin{bmatrix} T_1 \\ T_2 \\ T_3 \end{bmatrix} \qquad (20.4.148)$$

Note how the coefficients of B are obtained from the partial derivatives of the shape functions given by Eq. (20.4.149).

Now we can define the $(\nabla T)^2$ needed in Eq. (20.4.150).

$$(\nabla T)^2 = \begin{bmatrix} \dfrac{\partial T}{\partial x} & \dfrac{\partial T}{\partial y} \end{bmatrix} \begin{bmatrix} \dfrac{\partial T}{\partial x} \\ \dfrac{\partial T}{\partial y} \end{bmatrix} \quad (20.4.149)$$

For a given element, this equation can be expressed as

$$[\nabla T^e]^2 = \{T^e\}^T[B^e]^T[B^e]\{T^e\} \quad (20.4.150)$$

This yields

$$I_1^e = \frac{1}{2}k\int_e \{T^e\}^T[B^e]^T[B^e]\{T^e\}\,d\Omega \quad (20.4.151)$$

Because $[B]$ and $[T]$ are constant matrices, Eq. (20.4.151) reduces to

$$I_1^e = \frac{1}{2}k\{T^e\}^T[B^e]^T[B^e]\{T^e\}\int_e d\Omega \quad (20.4.152)$$

or simply

$$I_1^e = \frac{1}{2}\{T^e\}^T[k^e]\{T^e\} \quad (20.4.153)$$

where $[k^e]$ denotes the element conductivity matrix:

$$[K^e] = \kappa a[B^e]^T[B^e] \quad (20.4.154)$$

This takes the final form

$$[K^e] = \kappa a \begin{bmatrix} (b_1^2 + c_1^2) & (b_1 b_2 + c_1 c_2) & (b_1 b_3 + c_1 c_3) \\ (b_2 b_1 + c_2 c_1) & (b_2^2 + c_2^2) & (b_2 b_3 + c_2 c_3) \\ (b_3 b_1 + c_3 c_1) & (b_3 b_2 + c_3 c_2) & (b_3^2 + c_3^2) \end{bmatrix}$$

$$(20.4.155)$$

and a is the area of the triangular element.

Element-Forcing Function

To complete the integration of Eq. (20.4.144), we need to evaluate the second term, I_2^e.

$$I_2^e = \int f^e T^e d\Omega \quad (20.4.156)$$

As we have done with temperature, the heat source f can be expressed in a similar fashion:

$$f = f_1 N_1 + f_2 N_2 + f_3 N_3 = [f_1 \quad f_2 \quad f_3] \begin{bmatrix} N_1 \\ N_2 \\ N_3 \end{bmatrix} \quad (20.4.157)$$

For an arbitrary element, this equation can be written in compact matrix form:

$$f^e = \{f^e\}\{N^e\} \quad (20.4.158)$$

Recall that

$$T = T_1 N_1 + T_2 N_2 + T_3 N_3 = [N_1 \quad N_2 \quad N_3] \begin{bmatrix} T_1 \\ T_2 \\ T_3 \end{bmatrix} \quad (20.4.159)$$

or

$$T^e = \{N^e\}^T\{T^e\} \quad (20.4.160)$$

Therefore, I_2^e after substitution becomes

$$I_2^e = \int \{f^e\}\{N^e\}\{N^e\}^T\{T^e\}\,d\Omega = \{g^e\}\{T^e\} \quad (20.4.161)$$

where

$$\{g^e\}^T = \{f^e\}^T \int \{N^e\}\{N^e\}^T d\Omega \quad (20.4.162)$$

The integrand $\{N^e\}\{N^e\}^T$ yields

$$\{N^e\}\{N^e\}^T = [A]^{-1} \begin{bmatrix} 1 \\ x \\ y \end{bmatrix} [1 \quad x \quad y] [A^T]^{-1}$$

$$= [A]^{-1} \begin{bmatrix} 1 & x & y \\ x & x^2 & xy \\ y & xy & y^2 \end{bmatrix} [A^T]^{-1} \quad (20.4.163)$$

Note that $[A]$, given by Eq. (20.4.137), is a constant matrix, therefore the integral of $\{g\}$ depends only on the matrix with variables x and y. Integrating each element of the matrix is rather tedious and time-consuming. An alternative is to use a method developed by Eisenberg and Malvern.

From this method, we have the following statement of the integral:

$$\int_e N_1^m N_2^n N_3^p\,d\Omega = \frac{m!n!p!}{(m+n+p+2)!}2a \quad (20.4.164)$$

$$\{g^e\} = \left[\int_e \{N^e\}\{N^e\}^T d\Omega\right]\{f^e\} \quad (20.4.165)$$

Hence,

$$\{g^e\} = \left\{\int \begin{bmatrix} N_1^2 & N_1 N_2 & N_1 N_3 \\ N_2 N_1 & N_2^2 & N_2 N_3 \\ N_3 N_1 & N_3 N_2 & N_3^2 \end{bmatrix} d\Omega\right\}\{f^e\} \quad (20.4.166)$$

which yields

$$\{g^e\} = \frac{a}{12}\begin{bmatrix} 2 & 1 & 1 \\ 1 & 2 & 1 \\ 1 & 1 & 2 \end{bmatrix}\{f^e\} \quad (20.4.167)$$

The element integral of the variational formulation is broken into two parts:

$$I^e = I_1^e + I_2^e \quad (20.4.168)$$

Which simplifies to

$$I^e = \frac{1}{2}\{T^e\}^T[k^e]\{T^e\} + \{g^e\}^T\{T^e\} \quad (20.4.169)$$

The "global integral" over the domain Ω of the entire body becomes

$$I = \sum_{e=1}^N I^e = \sum \left\{\frac{1}{2}\{T^e\}^T[k^e]\{T^e\} + \{g^e\}^T\{T^e\}\right\} \quad (20.4.170)$$

or

$$I = \frac{1}{2}\{T\}^T\left[\sum_{3=1}^N [k^e]\right]\{T\} - \left[\sum_{e=1}^N \{F^e\}^T\right]\{T\} \quad (20.4.171)$$

where $\{F^e\}^T = -\{g^e\}^T$ and $\{T\} = [T_1 T_2 \ldots T_n]^T$. Hence,

$$I = \frac{1}{2}\{T\}^T[k]\{T\} - \{F\}^T\{T\} \quad (20.4.172)$$

where the global conductivity matrix is defined by

$$[k] = \sum_{e=1}^N [k^e] \quad (20.4.173)$$

and the global function (equivalent to the global force in the analysis of a truss) is

$$\{F\} = \sum_{e=1}^N \{F^e\} \quad (20.4.174)$$

The variation $\delta I = 0$ is equivalent to

$$\frac{\partial I}{\partial T_i} = 0 \quad (i = 1, \ldots, n) \quad (20.4.175)$$

Applying Eq. (20.4.175) to Eq. (20.4.171) gives the global equation governing the temperature distribution and the heat source:

$$[k]\{T\} = \{F\} \qquad (20.4.176)$$

This equation is similar to our FEM application to the truss and the one-dimensional heat flow problems.

Analysis of 2-D heat conduction problems can be done by using the FEM procedures developed herein. One proceeds by identifying the element shape functions and then evaluate the corresponding local conductivity (stiffness) matrices. The global $[K]$ is then assembled, using Eq. (20.4.173), and the elements' forcing functions are computed, using Eq. (20.4.167). Finally the global array $\{F\}$ is assembled according to Eq. (20.4.174).

BIBLIOGRAPHY

Aboudi, J. "Mechanics of Composite Materials," Elsevier, Amsterdam, 1991, 328 pp.
Adeli, H., and Soegiarso, R. "High Performance Computing in Structural Engineering," CRC Press, Boca Raton, FL, 1998, 272 pp.
Akin, J. E. "Finite Element Analysis for Undergraduates," Academic Press, London, 1986, 319 pp; "Finite Elements for Analysis and Design," Academic Press, London, 1994.
Argyris, J. H., and Mlejnek, H. P. "Dynamics of Structures," North-Holland, Amsterdam, 1991.
Atanackovic, T., and Guran, A. "Theory of Elasticity for Scientists and Engineers," Birkhauser Verlag, Basel, 2000, 386 pp.
Atluri, S. N., et al., eds. "Hybrid and Mixed Finite Element Methods," J. Wiley & Sons, Chichester, 1983, 582 pp.
Aziz, A. K., ed. "The Mathematical Foundations of the Finite Element Method with Applications to Partial Differential Equations," Academic Press, New York, 1972, 797 pp.
Bathe, K. J. "Finite Element Procedures," Prentice-Hall, Englewood Cliffs, NJ, 1995, 1037 pp.
Bathe, K. J., Oden, J. T., and Wunderlich, W., eds. "Formulations, and Computational Algorithms in Finite Element Analysis," MIT Press, Cambridge, 1977, 1091 pp.
Becker, E. B., Carey, G. F., and Oden, J. T. "Finite Elements—An Introduction," Prentice-Hall, Englewood Cliffs, NJ, 1981, 255 pp.
Belytschko, T., Liu, W. K., and Moran, B. "Nonlinear Finite Elements for Continua and Structures," J. Wiley & Sons, New York, 2000, 600 pp.
Bergan, P. G., Bathe, K. J., and Wunderlich, W., eds. "Finite Element Methods for Nonlinear Problems," Springer Verlag, Berlin, 1986, 817 pp.
Bianchi, N. "Electrical Machine Analysis Using Finite Elements," CRC Press, Boca Raton, FL, 2005, 275 pp.
Bickford, W. "Advanced Mechanics of Materials," Addison Wesley, Reading, MA, 1998, 600 pp.
Boyle, J. T., Brown, D. K., Mair, B., Mehta, P., and Wood, J., eds. "Finite Element Analysis: Education and Training," Chapman & Hall, London, 1991, 244 pp.
Brebbia, C. A., ed. "Finite Element Systems—A Handbook," Springer Verlag, New York, 1981, 1983, 1985, 767 pp.
Carey, G. F. and Oden, J. T. "Finite Elements—Computational Aspects," Vol. III, Prentice-Hall, Englewood Cliffs, NJ, 1984, 350 pp.

Cheung, Y. K., and Hutton, S. G., eds. "Finite Element Methods in Engineering," AIAA, New York, 1977.
Cheung, Y. K., Lo, S. H., and Leung, Y. T. "Finite Element Implementation," Blackwell Science, Oxford, 1995, 384 pp.
Cook, R., and Young, W. "Advanced Mechanics of Materials," Prentice-Hall, Upper Saddle River, NJ, 1999, 496 pp.
Dym, C. L. "Structural Modeling and Analysis," Cambridge University Press, Cambridge, 1997, 281 pp.
Glowinski, R. et al., eds. "Energy Methods in Finite Element Analysis," J. Wiley & Sons, New York, 1979, 361 pp.
Hsu, T. R. "The Finite Element Method in Thermodynamics," Allen & Unwin, Boston, 1986, 391 pp.
Hughes, T. J. R., Onate, E., and Zienkiewicz, O. C., eds. Recent Developments in Finite Element Analysis, a Book Dedicated to Robert L. Taylor."
Kardestuncer, H., and Norrie, D. H., eds. "Finite Element Handbook," McGraw-Hill, New York, 1987, 1424 pp.
Kythe, P. K. K., and Wei, D. "An Introduction to Linear and Nonlinear Finite Element Analysis," Birkhauser Verlag, Basel, 2003, 456 pp.
Logan, D., "First Course in the Finite Element Method Using Algor," PWS Publishing, Boston, 1997, 908 pp.
Oden, J. T., and Reddy, J. N. "An Introduction to the Mathematical Theory of Finite Elements," J. Wiley & Sons, Chichester, 1976, 429 pp.
Ogden, R. W. "Non-Linear Elastic Deformations," Dover Publications, Mineola, NY, 1997 (2nd ed.), 544 pp.
Papadrakakis, M., and Topping, B. H. V., eds. "Advances in Finite Element Techniques," Saxe-Coburg Publ., Edinburgh, 1994, 235 pp.
Pozrikidis, C. "Introduction to Finite and Spectral Element Methods Using MATLAB," CRC Press, Boca Raton, FL, 2005, 653 pp.
Reddy, J. N. "An Introduction to Nonlinear Finite Element Analysis," Oxford University Press, Oxford, 2004, 482 pp.
Reddy, J. N., ed. "Mechanics of Composite Materials: Selected Works of Nicholas J. Pagano," Kluwer Academic Publ., Dordrecht, 1994, 460 pp.
Robinson, J., ed. "FEM Today and in the Future," Robinson and Assoc., Wimborne, Dorset, U.K., 1993.
Shames, I. H., and Dym, C. L. "Energy and Finite Element Methods in Structural Mechanics," SI Unit Edition Taylor & Francis, PA, 1995, 750 pp.
Stasa, F. L. "Applied Finite Element Analysis for Engineers," CBS Publ. Japan, Ltd., New York, 1985, 658 pp., distributed by Oxford University Press.
Steele, J. M. "Applied Finite Element Modeling—Practical Problem Solving for Engineers," Marcel Dekker Inc., New York, 1989, 361 pp.
Topping, B. H. V., ed. "Advances in Finite Element Procedures and Techniques," Saxe-Coburg Publ., Edinburgh, 1998, 298 pp.
White, R. E. "An Introduction to the Finite Element Method with Applications to Nonlinear Problems," J. Wiley & Sons, New York, 1985, 354 pp.
Wiberg, N. E., ed. "Advances in Finite Element Technology," CIMNE, Barcelona, 1995, 305 pp.
Yamada, Y., and Gallagher, R. H., eds. "Theory and Practice in Finite Element Structural Analysis," University of Tokyo Press, Tokyo, 1973, 733 pp.
Zahavi, E. "The Finite Element Method in Machine Design," Prentice-Hall, Englewood Cliffs, NJ, 1992, 320 pp.
Zienkiewicz, O. C., and Morgan, K. "Finite Elements and Approximation," J. Wiley & Sons, New York, 1983, 328 pp.
Zienkiewicz, O. C., and Taylor, R. L. "Finite Element Method—Basic Formulation and Linear Problems," Vol. 1, McGraw-Hill, New York, 1989, 648 pp.

20.5   COMPUTER-AIDED DESIGN, COMPUTER-AIDED ENGINEERING, AND VARIATIONAL DESIGN

by Farid Amirouche

INTRODUCTION

Computer-aided design (CAD) has gained tremendous popularity in the past decade, and it is becoming an absolutely necessary tool for any engineering task. Its application is broadening, and its success is due mostly to the mass production of powerful microprocessors and their low cost.

Computer-aided design is the blending of human and machine, working together to achieve the optimum design and manufacture of a product. The computer's graphics capability and computing power allow designers to fashion and test their ideas interactively in real time without having to create real prototypes as in conventional approaches to design.

CAD has revolutionized engineering over the past two decades, and engineering design has become the engine to a successful company. The road between the product in its initial design stages and the mature design ready to go into production is lined with a series of prototypes needed to evaluate the ongoing design. **Virtual reality** offers a novel approach in product development, especially in the early stages of a development process. Virtual models are CAD prototypes that can be displayed on a computer screen for the purpose of getting a view into what reality might be and how to prepare for the product use.

PARAMETRIC AND VARIATIONAL DESIGN

Designing and drafting are the preliminary steps carried out in the manufacture of a component. Traditional mechanical CAD systems are effective tools in improving the efficiency of the drafting process, but they cannot handle the task of design and drafting at the same time. This forces the engineer to think simultaneously about all the design requirements and engineering relationships for each design approach.

Several new CAD systems claim to meet the needs of engineers engaged in the preliminary design of mechanical products. These systems should enhance the level of productivity of engineers in creating or modifying designs. Two approaches that are gaining popularity for providing flexibility in the design process are parametric design and variational design.

Parametric Design Systems

In a parametric design, the engineer selects a set of geometric constraints that can be applied for creating the geometry of the component. The geometric elements include lines, arcs, circles, and splines. A set of engineering equations can also be used to define the dimensions of the component.

From a given set of geometric constraints, the basic dimensions of the component can be defined. The set of geometric constraints would be regarded as a set of instructions that when executed would create the desired geometry. Hence any change to the known entities would yield a different picture altogether. So the objective is to create a geometric relationship between entities to minimize time and effort spent on design.

Parametric design systems help users manipulate the length, angle, and pitch of a particular component. These systems also have the ability to record engineering relationships within a design: they enable users to pinpoint the complex relationships that exist among parts of an assembly. A simple example would be the meshing of two gears. Thus the use of parametric design clarifies users' understanding of the engineering aspects of a component. Moreover, the parametric system can produce designs that are more meaningful than those generated by traditional CAD systems. Designers, analysts, and production engineers can work with the same solid model, extracting and adding information according to need.

A drawback of this system is that it is constrained to the set of geometric or engineering relationships provided by the engineer for creating the design of a component. The limited set of geometric constraints can make it difficult to change a parametric design model once the initial conditions have been set. These systems are best suited for design tasks that do not involve many variations in the design approach. Software equipped with such an option is therefore an excellent tool for design. Such software gives the designer flexibility to avoid redrawing objects and minimizes the time involved in arriving at a final product design.

Variational Design Systems

This technique can simultaneously solve a set of geometric constraints and engineering equations with important relationships among the elements of design. Essentially, the design is governed by a set of engineering equations relating its geometry and functions.

Unlike the parametric approach, a variational system is able to determine the positions of geometric elements and constraints. In addition, it is structured to handle the coupling between parameters in the geometric constraints and the engineering equations. The variational design concept helps the engineer evaluate the design of a component in depth instead of considering only the geometric aspects to satisfy the design relationships. This can be illustrated by the following example.

In design, one knows the performances and characteristics wanted for a developing product but does not know the geometric dimensions,

material characteristic, etc. A good design starts with an appropriate set of parameters. The process of finding a set of parameters is done through optimization. Variational design has the ability to incorporate optimization into the design environment. The engineer can specify both equality and inequality design constraints. The objective function is a set of both dependent and independent variables that are optimized (minimized or maximized) relative to one or more independent variables.

Variational Design Applications

In their work Hofer and Pottmann[12] presented a solution for variational motion design of a rigid body in the presence of rigid obstacles of arbitrary shape that fully employs the available degrees of freedom in the design process. The main idea is to reduce the problem to a curve design task on a suitable barrier surface embedded in an appropriate kinematic image space (see Fig. 20.5.1).

Fig. 20.5.1   Motion of a rigid body passing through a narrow channel.

This is an investigation of a 3-D shape reconstruction from measurement data in the presence of constraints. The constraints may fix the surface type or set geometric relations between parts of an object's surface, such as orthogonality, parallelity, and others. The authors of this work (see Fig. 20.5.2) have proposed to use a combination of surface fitting and registration within the geometric optimization framework of **squared distance minimization** (SDM). In this way, one obtains a **quasi-Newtonian** like optimization algorithm, which in each iteration simultaneously registers the data set with a rigid motion to the fitting surface and adapts the shape of the fitting surface. Figure 20.5.2 highlights the applicability of the method to constrained 3-D shape fitting for reverse engineering of CAD models and to high-accuracy fitting with kinematic surfaces, which include surfaces of revolution (reconstructed from fragments of archeological pottery) and spiral surfaces, which are fitted to 3-D measurement data of shells.

Innovative Process in Product Development

The focus and ultimate goal of an engineer is to be able to answer the following questions: Does my design (geometry) and my choices of material allow my product to achieve the performance required by the technical specifications? What is the optimum achievable performance? Will my design be sensitive to manufacturing tolerances? Answering the first question requires us to solve a set of equations relating output

Fig. 20.5.2   Approximation of a broken archeological finding by using inverse engineering.

and input conditions. The known quantities are the various targeted performances of a **product** (for example, static or modal behavior), and the unknowns of that problem are of various kinds: geometric dimensions, material characteristics, shell thickness, number of stiffeners, and density of weld spots, for example. A good design is a design where parameters have the appropriate values. Usually, these good values are not unique. The conventional way to solve the inverse problem is to use optimization techniques. The solution of an optimization problem is one set of parameter values that satisfies mathematical criteria. Even if simple criteria such as maximum stress or first Eigen frequency are mathematical criteria, these criteria rarely take into account manufacturing process constraints or the invaluable know-how of the engineer. Furthermore, finding the best point does not give any information on the global quality of the solution and its stability. Most of the optimization algorithms used in common practice are local algorithms that only find the closest point to the initial design.

ENGINEERING ANALYSIS AND CAD

Analysis is of two types: analytical and experimental. Both are performed to assess product performance.

Analytical Methods

In the analytical approach, different types of analysis can be carried out by using various software packages available for computer-aided design. These include finite-element analysis, kinematics analysis and synthesis, and static and dynamic analysis.

Finite-Element Analysis   Finite-element analysis (FEA) is a computer-assisted technique for determining stresses and deflections in a structure. The method divides a structure into small elements with known shapes for which a mathematical solution can be found; then the global problem is solved by using an assembly procedure (see Sec. 20.4).

The first step in FEA is the creation of a model that breaks the structure into simple standardized shapes and plots it on a common coordinate grid system. The coordinate points are called *nodes* and serve as the locations in the model where output data is provided. Nodal stiffness properties are then calculated by the finite-element program for each element and arranged into matrices within the computer. These parameters are then processed with applied loads and boundary conditions or calculations of displacement, natural frequencies, strains, and engineering functions useful in interpreting the results. An application of FEA is biomechanics, as shown in Fig. 20.5.3.

In most cases, finite-element models are developed for prototype designs for which experimental data exist so that FEA results can be cross-checked and design modifications can be made. Modifications are also tested for actual prototypes.

Kinematics and Synthesis   Mechanisms are devices used to transfer motion and/or force from a given source to an output device. The efficiency with which this transfer is accomplished depends on the mechanism design. General-purpose programs like Linkages, IMP, ADAMS, Unigraphics, and MCADA are available today to assist engineers in the kinematic analysis and synthesis of those systems. Kinematics starts with the mechanism design. It calculates large displacements, velocities, and accelerations of mechanisms without regard to the forces acting on them. The mechanism assumes zero degrees of freedom. This means that each degree of freedom (coordinate) is constrained to a particular type of motion. The advantages of combining the study of mechanisms with CAD are that it enables the design's output to be optimized and it eliminates interferences between links.

Synthesis analysis is concerned with arriving at the best characteristics of a mechanism through a set of computer iterations. The designer provides the parameters of the mechanism and the program develops the alternatives.

The first step before doing a synthesis analysis is deciding various parameters of the mechanism, such as number of links, joints, joint types, and connectivity of links by specifying which links are fixed. The basic linkage types are defined as kinematic chains within the program. The synthesis analysis is then carried out, and the mechanism structures

Fig. 20.5.3   Application of FEA in biomechanics. Figure illustrates finite-element model of right half of human pelvic bone.

satisfying the design objectives are determined. These programs are generally used to design four-bar linkages, although they can be used to design six-bar linkages.

Some kinematic synthesis programs use graphics to make the software user-friendly. The user specifies points and motion vectors via interactive graphics that can be easily moved about the screen. After a number of points have been specified, the program displays all possible locations of additional points to complete the mechanism. Some software also allows the user to animate the mechanism on the screen to check for criteria such as clearances of moving links.

Static Analysis and Dynamic Analysis   Static analysis combines motion analysis with mass properties and force input data to determine positions and joint reaction forces of mechanisms at rest. Static analysis can also be done on mechanisms at various points in their range of movement if zero velocity is assumed. Static models can have multiple degrees of freedom. Motion and force are uncoupled in this type of analysis. In static analysis, we are mostly concerned with the reaction forces on the mechanical systems and their joints' interconnection forces. Knowing these latter forces is useful in stress analysis to determine engineering criteria such as performance, reliability, and fatigue.

Dynamic analysis uses mass properties and forces to calculate positions, velocities, accelerations, and joint and constraint reaction forces of all model parts when motion is coupled to forces in the system. Analysis is done in discrete steps within a specified time interval. Each degree of freedom in a dynamic model is associated with an independent coordinate for which the analyst has to specify both the initial position and velocity. The computer models for dynamic analysis include geometric data and mass properties of the structure as well as applied forces. The model is generally created through part, joint, marker, force, and generator statements that are input by the user.

Part statements define the geometry, mass, and moment of inertia of each rigid part of the structure. Joint statements describe contacts between moving parts that hold the assembly together. Joints can be specified as providing translational and rotational movement, including that of revolute, screw, spherical, universal, cylindrical, and translational joints. Marker statements provide a point or coordinate system fixed on each part, orienting it to other parts and together defining the overall configuration of the system. Internal reaction forces are selected from a library of standard force elements such as dampers and linear springs. User-written routines can also be used to define additional parameters. Basic geometric data from CAD programs can be sent through direct interfaces. Because complete geometric description is not required, mechanical analysis can be used early in the design cycle, before the final configuration is known. Engineers can design the system knowing only mass and inertia characteristics.

The dynamic analysis output is typically available in tabular form. Graphics can be produced showing output such as force-versus-time or force-versus-displacement characteristics. In some cases, animation is reproduced from the combination of dynamic analysis data and geometry representation of the object. Software capable of performing such a task includes ADAMS, DADS, DISCOS, DYAMUS, DARS, and DYNACOMBS.

Experimental Testing   The second aspect of analysis deals with experimental methods in which testing is conducted on prototypes and models either to extract material properties or to validate performance characteristics of a particular design.

Testing in the conventional design process is done to qualify the design after manufacturing. The goal in computer-aided design is to use testing throughout the development cycle and to make better use of the test data through more advanced analysis techniques and to integrate these with other disciplines involved in design.

Initial testing can be done on a prototype or on different components to understand model response to certain loading conditions. Experimental data are extremely useful in the analysis of models where analytical solutions are not reliable. The data can also be used to refine finite-element models for large-scale analysis. The integration of experimental testing with analytical tools results in more effective engineering analysis. Computer-aided design provides the ultimate tool for combining such methods with the graphics capabilities to allow the designer to arrive at realistic and effective designs in a minimum span of time.

Additional experimental analyses include fatigue testing to estimate part life from strain gage measurements and modal testing to extract natural frequency and mode shapes, response functions, and a number of engineering functions in heat transfer, fluid dynamics, and thermodynamics.

COMPUTER-AIDED ENGINEERING (CAE)

According to J. R. Lemon, computer-aided engineering (CAE) is a product design and development philosophy, which combines three key engineering elements in an integrated environment for all engineering functions within each stage of the product development cycle. These elements are applications (performance, structural integrity, reliability, costs, etc.), facilities (hardware/software), and technology (data and information management). Thus, the CAE approach results in the restructuring and streamlining of the entire engineering process itself instead of the automation of specific steps in the current process such as has been the case in computer-aided design (CAD) and computer-aided manufacturing (CAM) activities.

The CAE approach places emphasis in terms of time and money on the initial conceptual design phases. This revised allocation of resources results in extensive analysis and testing to guide product design iterations in the computer as opposed to making physical prototypes. This is a major departure from current methods in most mechanical industries, where prototype manufacture and tests are the fundamental means of evaluating product performance and quality. However, the need for prototypes is not eliminated, but the reduction in the number of prototypes

needed before full production can go ahead results in significant time and cost savings.

The CAE approach to mechanical product development emphasizes system analytical modeling and analysis techniques at the earliest phase of design, that is, conceptual design. The process starts with an integrated set of total system simulations of the entire product. Each alternative product concept is mathematically modeled as an entire system. At this point, overall product designers have the flexibility of defining significantly varying concepts in order to minimize weight, reduce energy consumption and maximize performance for acceptable product concepts.

Detailed component and subassembly design specifications are derived from system models by exercising computer simulations under severe environmental conditions and external product loadings. In addition, internal loads, duty cycles, and constraints acting on components and subassemblies at interfaces and connection points are derived from system models. System modeling and early development of product alternatives within the computer are essential differences compared with today's developments that rely on physical prototypes and build-and-test methods.

The new CAE approach attempts to integrate and automate various engineering functions in the entire product development process, which includes the following: (1) design, (2) analysis, (3) testing, (4) drafting, (5) documentation, (6) project management, (7) data management, (8) process planning (9) tool design, (10) numerical control, and (11) quality assurance. Improved product quality, increased market share, and greater profitability depend upon efficient and effective CAE integration and implementation. This is why CAE implementation has become a strategic issue in mechanical-related industries worldwide.

CAE IMPLEMENTATION

Implementation of an overall CAE system and its associated technologies must be planned carefully. Implementation should proceed step by step because it is not possible to switch off today's design, development, and manufacturing processes and switch on the CAE approach at the same time. As a first step toward CAE implementation, existing problems should be solved to gain knowledge and familiarity with the CAE system's capabilities and behavior.

A design audit is not simply a learning exercise; it is an important step to extend a basic product to a family of new products or to fix problems that may have developed. The next phase involves new product design and development. The CAE tools should be applied to a new and unique product design for which new concepts have the potential of providing strategic product quality and market share advantages. An optimum hardware system along with a similar integrated and distributed database management system should be configured for in-house usage and expected long-range expansion requirements. Software that performs efficiently should also be brought in house if it is economically justified.

Along with these tasks, outside CAE consultants are recommended to help guide and develop an overall CAE implementation plan. Educational programs and seminars in all aspects of CAE implementation are essential for all levels within an organization. Once CAE tools are implemented and the in-house organization structured to suit the particular CAE implementation philosophy, the outside consultant's role can be reduced to solving difficult problems.

VIRTUAL REALITY AND ITS APPLICATIONS TO PRODUCT DESIGN AND DEVELOPMENT

Virtual reality (VR) is a human-computer interface that aims at creating a realistic in-depth 3-D environment where manipulation and interaction of objects provides the user with an insight of the environment at the micro and macro scales through graphics and hepatics, providing the user a unique experience of a simulated reality.

Virtual reality has been used to visualize engineering and scientific concepts and offers new ways of training, enhancing the human

Fig. 20.5.4   Virtual reality simulation is used to develop an automated bus system and analyze the benefits of the automation process, such as decreased travel times and proper adherence to schedule.

Fig. 20.5.5   Shared virtual environments.

computer interface and constructing graphically supported real-time simulation systems. An example of VR simulation is shown in Fig. 20.5.4.

Virtual reality technology is defined by three major qualities: **immersion, presence,** and **interactivity:** The user perceives the presence of the environment as real and at any time can travel through space, exploring and navigating through the environment without fear of collision. Feedback can be provided on making sensors and controls to provide the necessary information to the human to measure his or her interaction with the environment.

Computer simulation has for years been the backbone for all input/output displays by means of graphics. With the advent of CAD software, realistic models are created and simulation represents the execution of mathematical algorithms based on the models developed under some conditions described in the problem at hand. Simulation, unlike virtual reality, does not allow the user to interact but to observe an executed dynamic situation.

VR Application to Engineering

The typical VR display is a wall-projection display that uses either a large computer screen or a rear projection display. These systems can provide a good sense of depth but the feeling of immersion is very low. They are relatively inexpensive and nonintrusive. A greater impression of immersion may be achieved through **head-mounted** or **helmet-mounted displays (HMDs),** which consist of a pair of liquid-crystal display or cathode-ray tube display devices, mounted on a helmet worn by the user. In some HMDs there is only one display that is connected to a system of lenses and mirrors that conveys the image into the retinas of the eyes.

The **Cave Automatic Virtual Environment** (CAVE) was developed at the University of Illinois at Chicago, provides the illusion of **immersion** by projecting **stereo images** on the walls and floor of a room-sized cube. Several persons wearing lightweight stereo glasses can enter and walk freely inside the CAVE. A head tracking system continuously adjusts the stereo projection to the current position of the leading viewer.

In the example illustrated in Fig. 20.5.5, three networked users at different locations (anywhere in the world) meet in the same virtual world by using a BOOM device (a video display counterbalanced on a boom

support), a CAVE system, and a head-mounted display, respectively. All users see the same virtual environment from their respective points of view. Each user is presented as a virtual human (avatar) to the other participants. The users can see each other, communicate with each other, and interact with the virtual world as a team.

As the technologies of virtual reality evolve, the applications of VR become literally unlimited. It is assumed that VR will reshape the interface between people and information technology by offering new ways for the communication of information, the visualization of processes, and the creative expression of ideas.

Note that a virtual environment can represent any three-dimensional world that is either real or abstract. This includes real systems such as buildings, landscapes, underwater shipwrecks, spacecraft, archaeological excavation sites, human anatomy, sculptures, crime scene reconstructions, and solar systems. Of special interest is the visual and sensual representation of abstract systems like magnetic fields, turbulent-flow structures, molecular models, mathematical systems, auditorium acoustics, stock market behavior, population densities, information flows, and any other conceivable system including artistic and creative work of abstract nature. These virtual worlds can be animated, interactive and shared, and can expose behavior and functionality.

Useful applications of VR include training in a variety of areas (military, medical, equipment operation, etc.), education, design evaluation (virtual prototyping), architectural walk-through, human factors and ergonomic studies, simulation of assembly sequences and maintenance tasks, assistance for the handicapped, study and treatment of phobias (e.g., fear of height), entertainment, and much more. An application of immersive virtual reality is shown in Fig. 20.5.6.

Applying Virtual Environments to Manufacturing

Even though virtual environment technologies are still difficult and expensive to use, people are doing real work. The following case studies represent examples of recent state of the art work that exemplifies the application of virtual environments in one or several aspects of manufacturing. Manufacturing, in this case, is taken to encompass issues relating to maintenance and training as well as the actual creation of parts and the assembly of systems. These examples are actual systems, not simply speculative fantasies.

Boeing   The Research and Technology organization of Boeing Computer Services is actively pursuing two projects using VR technology. Boeing uses a concept known as augmented reality.

Augmented reality is a term that refers to the ability to see through a computer-generated display. The generated images are superimposed

Fig. 20.5.6 Viewing a concept car in immersive virtual reality using the CAVE.

on top of reality. This is accomplished by projecting a computer image onto a half-silvered mirror that the user looks through. This technique provides an effective and intuitive way of **"annotating"** reality. The Boeing team is using a headset configured for augmented reality, which they call a HUD set (heads-up, see-through, head-mounted display).

The assembly of aircraft is a highly complex task that is difficult to automate. Many of the skills required demand dexterity not easily accomplished by robots. In addition, airplanes consist of many small-lot-size parts, and reprogramming robots for these quantities is an expensive prospect. Someone once said that a Boeing 747 is not really an airplane, but 5 million parts flying in close formation.

Caterpillar Caterpillar uses VR in its design and evaluation of heavy vehicles (Fig. 20.5.7). This technology allows the company to dramatically shorten the amount of time it takes to analyze a new design concept and incorporate it into the production process. It also represents a sizable cost savings because it eliminates the need to build prototype machines or make last-minute design changes. It takes 6 to 9 months to build full-scale models and make design changes, using conventional design methods. However, with the virtual reality approach, designs

usually can be evaluated in less than 1 month. Such methods are currently being employed, and a number of other companies are fully invested in VR research and development. Today customers can "field-test" new products by putting on a special helmet and operating the vehicles in a VR environment.

PROTOTYPING ANALYSIS

A pioneering application of virtual reality was the NASA wind tunnel. A 3-D workstation generates flow-field visualizations. The user interacts with the simulation via a BOOM display and a data glove. (A BOOM is a video display counterbalanced on a boom support which the user easily moves with the head.) The data glove effectively acts as a source of "smoke," allowing the user to observe local flow lines. The computation of unsteady flow visualization occurs in real time and the user can view the dynamic scene from many points of view and scales. Since fluid dynamic phenomena occur over a large range of scales (several orders of magnitude in space), navigation through the virtual environment is more difficult than, say, architectural walk-through. Thus, in addition to "standard" viewing capabilities sensitive to head position and head orientation, our virtual environment lets the user rapidly change his or her scale, or the scale of the environment. Relative to the environment, a user can shrink to become completely surrounded by some small vortex or grow so that the entire flow field fits in one hand.

The role of a CAD workstation is to provide the means of graphically displaying the huge amount of data for visual inspection and quick review of the analysis. Finite-element programs have two types of data: the input required to run the program and the generated output that describes the behavior of the system under the assumed conditions. To further enhance communication between the user and the machine in which a particular finite-element program resides, many vendors of CAD software are developing pre- and postprocessors that allow the user to graphically visualize their input and output (see Fig. 20.5.8). Therefore, CAD workstations equipped with such software assist the user to interact with finite-element programs, thus minimizing the time required to learn how to use the FEM program. Most important, the graphics capability of CAD workstations provides the user the means of visual display of huge amounts of output, making the interpretation faster and more convenient.

Fig. 20.5.7 Virtual backhoe project illustrates an "operator" driving the virtual equipment at the National Center for Supercomputing Applications. VR lab.

Fig. 20.5.8 NASA virtual wind tunnel product design.

THE USE OF VIRTUAL REALITY IN INTERACTIVE FINITE-ELEMENT ANALYSIS

Combining virtual reality (VR) with finite-element analysis (FEA) and executing interactive FEA provides an interface to implement comprehensive and intuitive analysis. A study of related research on and application of VR for interactive FEA has been completed. An overview of available VR systems and graphics APIs, and their suitability for interactive FEA/CAD, is presented here. FEA is quite developed in calculation with the improvement of computer techniques. There are two important procedures for FEA: preprocessing and the postprocessing. The preprocessor is used for setting up the model; the postprocessor is used for viewing the results. Three-dimensional visualization is already supported by currently available software, but 2-D is commonly used.

Engineers still need to imagine what is happening in a real 3-D world. Using VR in interactive FEA gives the preprocessor and the postprocessor a good method to describe the model and results in a 3-D world, which leads to a more intuitive and accurate understanding of design and construction. VR offers a real-time, full-scale interaction technique in the design period. For the iterative design procedure, real-time interactive FEA offers an excellent way to adjust the design. Apart from using VR in interactive FEA in the building and construction industry, it can also be used in the education of engineering students, to give students an intuitive understanding of the FEA technique.

BIBLIOGRAPHY

Bejczy, A. K., "Virtual reality in manufacturing," in *Re-engineering for Sustainable Industrial Production, Proceedings of the OE/IFIP/IEEE Int. Conf. on Integrated and Sustainable Industrial Production,* Lisbon, May 1997, pp. 48–60.
Broll, W., "Augmenting the Common Working Environment by Virtual Objects," *ERCIM News,* no. 44. January 2001, pp. 42–43.
Carrie, A. S., Adhami, E., Stephens, A., and Murdoch, I. C., and "Introducing a flexible manufacturing system," *Int. J. of Prod. Research,* vol. 22, no. 6, 1984, pp. 907–916.
Chuter, C., Jernigan, S., and Barber K., "A Virtual Environment Simulator For Reactive Manufacturing Schedules," *Proc. of Symposium on Virtual Reality in Manufacturing, Research and Education,* University of Illinois at Chicago, October 7–8, 1996.
Erdélyi, F., Tóth, T., and Rayegani, F., "Real-time Control of Manufacturing Systems on the Base of Multi-Layered Model," *Proceedings of the Second Mexican-Hungarian Workshop on Factory Automation and Material Sciences,* Mexico 1999, pp. 37–50.
Figueiredo, M., Böhm, K., and Teixeira, J., "Advanced Interaction Techniques in Virtual Environments," in *Computers & Graphics Journal,* vol. 17. no. 6., Pergamon Press, 1993, pp. 655–661.
Haidegger, G., and Kopácsi, S., "Telepresence with Remote Vision in Robotic Applications," *Proceedings of the 5th International Workshop on Robotics in Alpe-Adria-Danube Region* (RAAD 96), vol. 1, pp. 557–562.
Hofer, M., and Pottmann, H., "Designing energy-minimizing rigid body motions in the presence of obstacles," Technical Report 142, Geometry Preprint Series, Vienna University of Technology, June 2005.
http://www.iau.dtu.dk/~tdl/task52ab.html.
Liu, Y., Pottmann, H., and Wang, W., "Constrained 3D shape reconstruction using a combination of surface fitting and registration," Technical Report 144, Geometry Preprint Series, Vienna University of Technology, August 2005.
Mezgár, I., and Kovács, G. L., "Co-ordination of SME production through a co-operative network," *Journal of Intelligent Manufacturing,* vol. 9, 1998, pp. 167–172.
Ránky, P., "Components and Planning of Flexible Manufacturing Systems," *Introduction to CIM Systems,* Industrial Information Centre, Budapest, 1985 (in Hungarian).
Taylor Enterprise Dynamics, User Manual, F&H Simulation B.V., 2000.

20.6 INTRODUCTION TO COMPUTATIONAL FLUID DYNAMICS
by Doyle Knight

INTRODUCTION

Computational fluid dynamics (CFD) is one of the three branches of modern fluid dynamics, the other two being analytical and experimental. CFD emerged in the 1950s with the advent of digital computers, although a number of scientific papers appeared earlier in the twentieth century which constitute a not insignificant part of its theoretical foundation. Richardson's 1910 paper (Richardson, 1910) is generally cited as the progenitor of modern CFD because of its presentation of numerical methods for classical equations, some of which have found application in fluid dynamics (e.g., Laplace's equation), and Courant et al. (1928) established the concept of the domain of dependence of a numerical solution and its relationship to the stability of the numerical scheme. Roache (1972, 1988) provides additional historical detail. The appearance of major textbooks [e.g., Richtmyer and Morton (1957), Roche (1972), Kovenya and Yanenko (1981), and Peyret and Taylor (1983)] firmly established the role of CFD in modern fluid dynamics research and engineering.

Computational fluid dynamics is the science of solving the partial differential equations of fluid motion by discrete approximation. Although a formal development of CFD methodology is beyond the scope of this section, it is worthwhile to identify the principal theoretical concepts that form the foundation of CFD algorithms. *Accuracy* is the formal order of approximation of an individual discrete approximation to a partial derivative. Consider, for example, a uniformly spaced set of grid points $x_i = (i - 1)\Delta x$ for $i = 1, \ldots, n$ where Δx is a constant, and a function $f(x)$ defined at each point x_i. The discrete approximation to the derivative $\partial f/\partial x$, denoted as $\delta f/\delta x$, given by

$$\frac{\delta f}{\delta x} = \frac{f_{i+1} - f_{i-1}}{2\Delta x} \qquad (20.6.1)$$

can be shown by Taylor's series expansion to approximate the derivative of $f(x)$ as

$$\frac{f_{i+1} - f_{i-1}}{2\Delta x} = \frac{\partial f}{\partial x} + \mathbb{O}(\Delta x)^2 \qquad (20.6.2)$$

where $\mathbb{O}(\Delta x)^2$ indicates that the remaining terms are proportional to $(\Delta x)^2$ and therefore vanish as $\Delta x \to 0$. The discrete approximation $\delta f/\delta x$ is said to be second-order accurate (since the error is proportional to the second power of the grid spacing Δx). *Consistency* is the equivalence of the discrete approximation of the partial derivative to the partial derivative itself in the limit of $\Delta x \to 0$, irrespective of its order of accuracy. *Convergence* is the equivalence of the *solution* of the partial differential equation(s) (not merely the equivalence of an individual term) to the exact solution in the limit of vanishing grid spacing. There are examples of accurate but nonconvergent numerical algorithms for the partial differential equations of fluid dynamics (Roache, 1972). *Stability* is the property that the solution must remain bounded (Lax and Richtmyer, 1956).

The remainder of the section is organized into the following topics: (1) the governing equations of incompressible homogeneous fluid mechanics, (2) the concept of potential flow and the conditions for its existence, (3) a detailed development of the vortex panel method for solution of potential flow, (4) a numerical algorithm for solution of the boundary layer equations, and (5) an algorithm for the Navier-Stokes equations. Items (3) to (5) represent, therefore, the sequence of increasing complexity of fluid motions, and also represent a chronology of major advances in CFD. A list of references concludes the section.

GOVERNING EQUATIONS

The governing equations of an incompressible, laminar, constant-property, Newtonian viscous fluid in differential, dimensional form, using the Einstein summation convention,[*] are (Wilcox, 1997):

$$\frac{\partial \tilde{u}_i}{\partial \tilde{x}_i} = 0 \qquad (20.6.3)$$

$$\tilde{\rho}\frac{\partial \tilde{u}_i}{\partial \tilde{t}} + \tilde{\rho}\tilde{u}_j\frac{\partial \tilde{u}_i}{\partial \tilde{x}_j} = -\frac{\partial \tilde{p}}{\partial \tilde{x}_i} + \frac{\partial \tilde{\tau}_{ij}}{\partial \tilde{x}_j} \qquad (20.6.4)$$

where $\tilde{u}_i$ are the dimensional Cartesian velocity components in the $\tilde{x}_i$ directions, $\tilde{\rho}$ is the density, $\tilde{p}$ is the static pressure, and $\tilde{\tau}_{ij}$ is the viscous stress tensor,

$$\tilde{\tau}_{ij} = \mu\left(\frac{\partial \tilde{u}_i}{\partial \tilde{x}_j} + \frac{\partial \tilde{u}_j}{\partial \tilde{x}_i}\right) \qquad (20.6.5)$$

where μ is the molecular dynamic viscosity.

The equation may be made nondimensionalized by using a reference velocity U_∞ and length L,

$$\frac{\partial u_i}{\partial x_i} = 0 \qquad (20.6.6)$$

$$\frac{\partial u_i}{\partial t} + u_j\frac{\partial u_i}{\partial x_j} = -\frac{\partial p}{\partial x_i} + \frac{\partial \tau_{ij}}{\partial x_j} \qquad (20.6.7)$$

where

$$u_i = \frac{\tilde{u}_i}{U_\infty}$$

$$x_i = \frac{\tilde{x}_i}{L}$$

$$t = \frac{\tilde{t}U_\infty}{L}$$

$$p = \frac{\tilde{p}}{\rho U_\infty^2}$$

$$\tau_{ij} = \frac{\tilde{\tau}_{ij}}{\rho U_\infty^2} \qquad (20.6.8)$$

In particular, the viscous stress tensor becomes

$$\tau_{ij} = \frac{1}{\text{Re}}\left(\frac{\partial u_i}{\partial x_j} + \frac{\partial u_j}{\partial x_i}\right) \qquad (20.6.9)$$

where $\text{Re} = \rho U_\infty L/\mu$ is the *Reynolds number*.

VORTICITY AND KELVIN'S THEOREM

The cartesian components of the vorticity ω_i are defined as[†]

$$\omega_i = \epsilon_{ijk}\frac{\partial u_k}{\partial x_j} \qquad (20.6.10)$$

---

[*]The Einstein summation convention uses repeated indices to indicate summation over all values of the index. This, for example, Eq. (20.6.3) is

$$\frac{\partial \tilde{u}_1}{\partial \tilde{x}_1} + \frac{\partial \tilde{u}_2}{\partial \tilde{x}_2} + \frac{\partial \tilde{u}_3}{\partial \tilde{x}_3} = 0$$

† The indices i, j, k are cyclic if they are in increasing order. The possible combinations are $(i, j, k) = (1, 2, 3), (2, 3, 1)$, and $(3, 1, 2)$. The indices are anticyclic if they are in decreasing order. The possible combinations are $(i, j, k) = (3, 2, 1), (2, 1, 3)$, and $(1, 3, 2)$.

where ε_{ijk} is the cyclic tensor defined as

$$\varepsilon_{ijk} = \begin{cases} 1 & \text{if } i, j, k \text{ are cyclic} \\ -1 & \text{if } i, j, k \text{ are anticyclic} \\ 0 & \text{otherwise} \end{cases} \quad (20.6.11)$$

The vorticity is twice the effective local angular velocity of the fluid (Batchelor, 1970).

A flow field where $\omega_i = 0$ everywhere is denoted *irrotational flow* or *potential flow,* and the velocity u_i may be represented as the gradient of a scalar potential ϕ (Batchelor, 1970),

$$u_i = \frac{\partial \phi}{\partial x_i} \quad (20.6.12)$$

Kelvin's theorem (Wilcox, 1997) states that an inviscid, incompressible fluid with zero vorticity at some instant of time remains irrotational for all time.

IRROTATIONAL FLOW

At high Reynolds number Re, the viscous stresses [Eq. (20.6.9)] are negligible outside of a narrow region (denoted the *boundary layer*) adjacent to solid surfaces. Figure 20.6.1 shows an airfoil in cross section (without the boundary layer). Provided that the angle of attack of the airfoil is sufficiently small, the boundary layer at high Re remains affixed to the airfoil and the vorticity generated therein is convected downstream into a narrow wake. Thus, the flow field domain is almost entirely inviscid and the boundary layer can be ignored in determining the pressure distribution on the airfoil. Assuming the flow upstream is uniform (and hence has zero vorticity), the flow field outside the boundary layer is rotational.

Fig. 20.6.1 Airfoil.

For irrotational flow, the governing Eqs. (20.6.3) and (20.6.4) become (with ~ omitted for clarity),

$$\frac{\partial u_i}{\partial x_i} = 0 \quad (20.6.13)$$

$$\frac{\partial u_i}{\partial t} + u_j \frac{\partial u_i}{\partial x_j} = -\frac{1}{\rho} \frac{\partial p}{\partial x_i} \quad (20.6.14)$$

Using Eq. (20.6.12), we find the conservation of mass becomes

$$\frac{\partial^2 \phi}{\partial x_i^2} = 0 \quad (20.6.15)$$

which is known as *Laplace's equation*. The boundary conditions are

$$\frac{\partial \phi}{\partial x_i} = U_i \quad \text{as } |x_i| \to \infty$$
$$\frac{\partial \phi}{\partial n} = 0 \quad \text{on the surface} \quad (20.6.16)$$

The first boundary condition states that the fluid velocity approaches the free-stream velocity with cartesian components U_i as the distance from the body becomes large. The second boundary condition states that the velocity normal to the surface is zero (assuming the airfoil is at rest in the frame of reference). In addition, the Zhukovsky (Joukowski) hypothesis must be applied to achieve a unique solution (Milne-Thompson, 1958),

Joukowski's hypothesis is that the circulation[*] in the case of a properly designed aerofoil, in its working range of incidence, always adjusts itself so that the airspeed at the trailing edge is finite.

[*] See "Elementary Solutions," below.

Using the identities

$$u_j \frac{\partial u_i}{\partial x_j} = \frac{\partial}{\partial x_i} \left(\frac{1}{2} u_j u_j \right) - \epsilon_{ijk} u_j \omega_k \quad (20.6.17)$$

and

$$\frac{\partial u_i}{\partial t} = \frac{\partial}{\partial t} \left(\frac{\partial \phi}{\partial x_i} \right) = \frac{\partial}{\partial x_i} \left(\frac{\partial \phi}{\partial t} \right) \quad (20.6.18)$$

and noting that $\omega_i = 0$ for an irrotational flow, we can rewrite the conservation of momentum [Eq. (20.6.7)] as

$$\frac{\partial}{\partial x_i} \left(\frac{\partial \phi}{\partial t} + \frac{1}{2} u_j u_j + \frac{p}{\rho} \right) = 0 \quad (20.6.19)$$

and thus

$$\frac{\partial \phi}{\partial t} + \frac{1}{2} u_j u_j + \frac{p}{\rho} = g(t) \quad (20.6.20)$$

where $g(t)$ is a function of time. This is *Bernoulli's equation* and provides an expression for the pressure p in terms of the velocity potential and its gradient.

Elementary Solutions

Laplace's equation (and its inhomogenous counterpart, Poisson's equation) have been studied extensively because of their broad application in physics and engineering (Kellogg, 1953). For simplicity, we consider Laplace's equation in two-dimensional cartesian or cylindrical coordinates where $(x_1, x_2) = (x, y)$, $r = \sqrt{x_1^2 + x_2^2}$, and $\theta = \tan^{-1}(x_2/x_1)$. The cartesian and cylindrical polar forms of Laplace's equation are

$$\frac{\partial^2 \phi}{\partial x^2} + \frac{\partial^2 \phi}{\partial y^2} = 0 \quad \text{cartesian} \quad (20.6.21)$$

$$\frac{1}{r} \frac{\partial}{\partial r} \left(r \frac{\partial \phi}{\partial r} \right) + \frac{\partial^2 \phi}{\partial \theta^2} = 0 \quad \text{cylindrical} \quad (20.6.22)$$

There are three fundamental solutions of Eq. (20.6.15) that are commonly used in computational methods. The first is *uniform flow,* defined by

$$\phi = U_i x_i \quad \text{uniform flow} \quad (20.6.23)$$

where U_i is the component of the velocity along the x_i axis and $U = \sqrt{U_i U_i}$. The second is a two-dimensional *source/sink* defined by

$$\phi = \frac{q}{2\pi} \log r \quad \text{source/sink} \quad (20.6.24)$$

It may be directly verified that the volume flux of fluid crossing a circular boundary enclosing a source at $r = 0$ is

$$\int_0^{2\pi} u_i n_i r \, d\theta = \int_0^{2\pi} \frac{q}{2\pi r} r \, d\theta = q \quad (20.6.25)$$

where n_i are the cartesian components of the outward normals. The third is a *line vortex* defined by

$$\phi = \frac{\Gamma \theta}{2\pi} \quad \text{line vortex} \quad (20.6.26)$$

It may be directly verified that the line integral of the velocity about a contour enclosing the line vortex is

$$\int_0^{2\pi} u_i s_i r \, d\theta = \int_0^{2\pi} \frac{\Gamma}{2\pi r} r \, d\theta = \Gamma \quad (20.6.27)$$

where s_i are the components of the unit tangent vector (taken in the counterclockwise direction). The quantity Γ is denoted the *circulation*.

Panel Methods

The elementary solutions may be combined in a discrete manner to generate the solution for irrotational flow past an arbitrary number of closed boundaries (Hess and Smith, 1986; Hess, 1975). For simplicity,

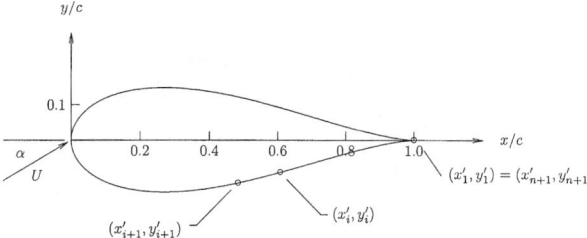

Fig. 20.6.2   Airfoil.

we consider a single body represented by the airfoil in Fig. 20.6.2. The airfoil surface is discretized into n segments with endpoints[*] (x_i', y_i'), $i = 1, \ldots, n + 1$. Note that $(x_1', y_1') = (x_{n+1}', y_{n+1}')$. The slope of the ith segment relative to the x axis is

$$\delta_i = \tan^{-1}\left[\frac{(y_{i+1}' - y_i')}{(x_{i+1}' - x_i')}\right] \qquad -\frac{\pi}{2} \le \delta_i \le \frac{\pi}{2} \qquad (20.6.28)$$

In particular, it is noted that δ_n is not necessarily equal to δ_1 (i.e., the trailing edge of the airfoil may have a finite angle).

Nonlifting Airfoil

Consider an infinite number of sources/sinks on the ith segment (Fig. 20.6.3) with strength per unit length q_i and length ds_i. From Eq. (20.6.24), the velocity potential is

$$d\phi_i = \frac{q_i\, ds_i}{2\pi} \log r_i \qquad (20.6.29)$$

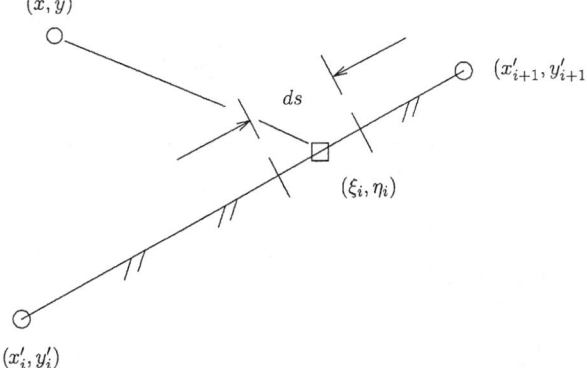

Fig. 20.6.3   Sources/sinks on segment.

where the summation on i is not performed and r is measured from the center of the infinitesimal segment (ξ_i, η_i) to the field point

$$r_i = \sqrt{(x - \xi_i)^2 + (y - \eta_i)^2} \qquad (20.6.30)$$

The velocity potential for the ith source/sink segment is therefore

$$\phi = \frac{q_i}{2\pi} \int \log r_i\, ds_i \qquad \text{(no sum on } i) \qquad (20.6.31)$$

[*]Note that the subscript i indicates the ith point, and should not be confused with the earlier cartesian notation.

Since Laplace's equation is linear, a linear superposition of elementary solutions is also a solution, and therefore the velocity potential is

$$\phi(x, y) = Ux \cos \alpha + Uy \sin \alpha + \sum_{j=1}^{j=n} \frac{q_j}{2\pi} \int \log r_j\, ds_j \qquad (20.6.32)$$

where U is the amplitude of the free-stream velocity and α is the angle of the free-stream velocity with respect to the x axis.

Define n control points (x_i, y_i) for $i = 1, \ldots, n$, where

$$x_i = \frac{1}{2}(x_{i+1}' + x_i')$$
$$y_i = \frac{1}{2}(y_{i+1}' + y_i') \qquad (20.6.33)$$

and apply the boundary condition that the velocity normal to the boundary at each control point is zero. The resultant equations are (Kellogg, 1953):

$$\frac{q_i}{2} + \sum_{j=1}^{j=n} \frac{q_j}{2\pi} \int \frac{\partial}{\partial n_i} \log r_j\, ds_j = -\frac{\partial}{\partial n_i}(Ux \cos \alpha + Uy \sin \alpha)$$
$$\text{for } i = 1, \ldots, n \qquad (20.6.34)$$

where n_i is the outward (i.e., into the fluid) normal direction at the ith control point and the derivative is evaluated at (x_i, y_i). Define the unit normal (n_1, n_2) and tangential vectors (t_1, t_2) at each segment i for $i = 1, \ldots, n$ by (no sum on i)

$$n_{1_i} = -(y_{i+1}' - y_i')\Delta s_i^{-1}$$
$$n_{2_i} = +(x_{i+1}' - x_i')\Delta s_i^{-1}$$
$$t_{1_i} = +(x_{i+1}' - x_i')\Delta s_i^{-1}$$
$$t_{2_i} = +(y_{i+1}' - y_i')\Delta s_i^{-1}$$
$$(20.6.35)$$

where Δs_i is the length of the ith segment

$$\Delta s_i = \sqrt{(x_{i+1}' - x_i')^2 + (y_{i+1}' - y_i')^2} \qquad (20.6.36)$$

Equation (20.6.34) can be rewritten as (Kellogg, 1953):

$$\frac{Q_i}{2} + \sum_{j=1, j\ne i}^{j=n} \frac{Q_j}{2\pi}(n_{1_i}G_{ij} + n_{2_i}H_{ij}) = -(n_{1_i}\cos \alpha + n_{2_i}\sin \alpha)$$
$$\text{for } i = 1, \ldots, n \qquad (20.6.37)$$

where

$$Q_i = \frac{q_i}{U} \qquad (20.6.38)$$

Equation (20.6.37) represents n linear equations for the n unknowns q_i. The $n \times n$ influence matrices G_{ij} and $H_{i,j}$ are defined as follows. Define the following (no sum on i or j):

$$a_{ij} = (x_i - x_j)t_{1_j} + (y_i - y_j)t_{2_j}$$
$$b_{ij} = \sqrt{(x_i - x_j)^2 + (y_i - y_j)^2 - a_{ij}^2}$$
$$c_{ij} = x_i - x_j - a_{ij}t_{1_j}$$
$$d_{ij} = \frac{1}{2}\Delta s_j - a_{ij} \qquad (20.6.39)$$
$$e_{ij} = \frac{1}{2}\Delta s_j + a_{ij}$$
$$f_{ij} = (x_i - x_j)^2 + (y_i - y_j)^2 + \Delta s_j a_{ij} + \frac{1}{4}\Delta s_j^2$$
$$g_{ij} = y_i - y_j - a_{ij}t_{2_j}$$

The influence matrices are

$$G_{ij} = \frac{c_{ij}}{b_{ij}}\left[\tan^{-1}\frac{d_{ij}}{b_{ij}} + \tan^{-1}\frac{e_{ij}}{b_{ij}}\right] - \frac{1}{2}t_{1_j}\log\frac{f_{i,j+1}}{f_{ij}}$$

$$H_{ij} = \frac{g_{ij}}{b_{ij}}\left[\tan^{-1}\frac{d_{ij}}{b_{ij}} + \tan^{-1}\frac{e_{ij}}{b_{ij}}\right] - \frac{1}{2}t_{2_j}\log\frac{f_{i,j+1}}{f_{ij}} \qquad (20.6.40)$$

The solution of Eq. (20.6.37) yields the source/sink values q_i. The velocity at the midpoint of the ith segment on the airfoil is tangential to the airfoil and given by

$$V_i = U(t_{1_i}\cos\alpha + t_{2_i}\sin\alpha) + \sum_{j=1, j\neq i}^{j=n} \frac{q_i}{2\pi}(t_{1_i}G_{ij} + t_{2_i}H_{ij}) \quad (20.6.41)$$

Evaluating Eq. (20.6.20) in the free stream and assuming steady flow, we obtain

$$\frac{1}{2}U^2 + \frac{p_\infty}{\rho} = \frac{1}{2}(u^2 + v^2) + \frac{p}{\rho} \quad (20.6.42)$$

Thus, the pressure coefficient c_p, defined by

$$c_p = \frac{p - p_\infty}{\frac{1}{2}\rho U^2} \quad (20.6.43)$$

becomes

$$c_p = 1 - \left(\frac{u^2 + v^2}{U^2}\right)^2 \quad (20.6.44)$$

and on the airfoil surface at the midpoint of the ith segment,

$$c_{p_i} = 1 - \left(\frac{V_i}{U}\right)^2 \quad (20.6.45)$$

The above method is applied to the flow past a cylinder as illustrated in Fig. 20.6.4. The computed pressure distribution using $n = 50$ segments is shown in Fig. 20.6.5 together with the exact solution (Anderson, 2001)

$$c_p(\theta) = 1 - 4\sin^2\theta \quad (20.6.46)$$

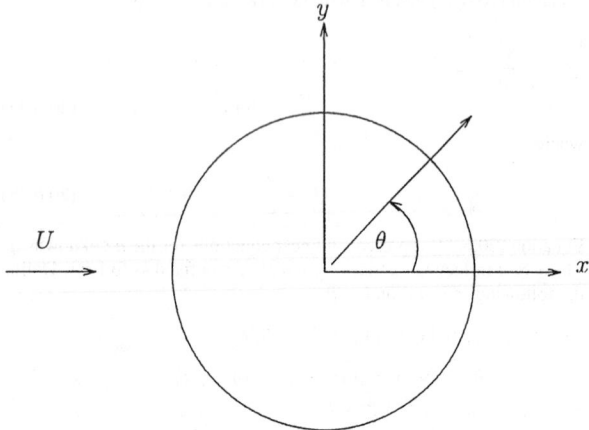

Fig. 20.6.4   Flow past a cylinder.

Lifting Airfoil

For a two-dimensional airfoil in steady flow, the Kutta-Zhukovsky theorem (Milne-Thompson, 1958) states

$$l = -\rho U\Gamma \quad (20.6.47)$$

where l is the lift per unit span, U is the free-stream velocity, and Γ is the circulation defined by Eq. (20.6.27). The lift is the force on the airfoil in a direction perpendicular to the free-stream velocity. It is therefore evident that the solution of Laplace's equation for a lifting airfoil must include distributed vortices.

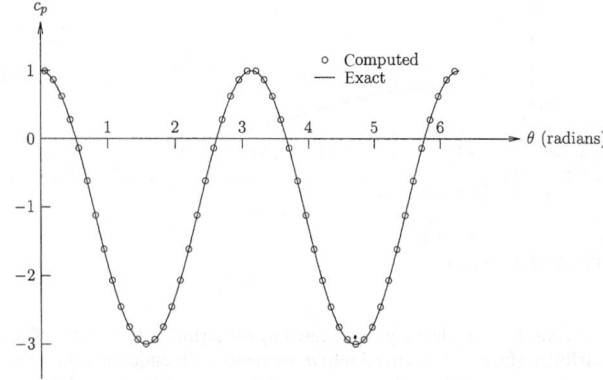

Fig. 20.6.5   Pressure coefficient on cylinder.

Consider an infinite number of line vortices on the ith segment (Fig. 20.6.6). For simplicity, the origin of the coordinate system is taken to be at the center of the segment. Each vortex has a circulation $\gamma d\xi$. The cartesian velocity components at point (x, y) due to the vortex at $(\xi, 0)$ are

$$\begin{aligned} u &= -\frac{\gamma\, d\xi}{2\pi r}\sin\theta \\ v &= +\frac{\gamma\, d\xi}{2\pi r}\cos\theta \end{aligned} \quad (20.6.48)$$

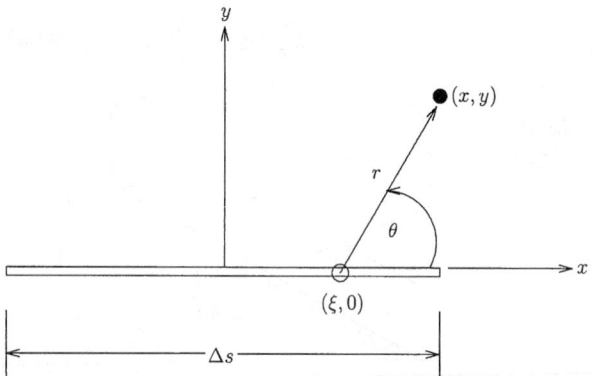

Fig. 20.6.6   Vortex on segment.

where

$$\begin{aligned} r &= \sqrt{(x - \xi)^2 + y^2} \\ \theta &= \tan^{-1}\left(\frac{y}{x - \xi}\right) \end{aligned} \quad (20.6.49)$$

Thus, the cartesian velocity components at point (x, y) due to the distribution of vortices on the segment is

$$u = -\frac{\gamma}{2\pi}\int_{-\frac{1}{2}\Delta s}^{+\frac{1}{2}\Delta s} \frac{y\, d\xi}{[(x - \xi)^2 + y^2]}$$

$$v = +\frac{\gamma}{2\pi}\int_{-\frac{1}{2}\Delta s}^{+\frac{1}{2}\Delta s} \frac{(x - \xi)\, d\xi}{(x - \xi)^2 + y^2} \quad (20.6.50)$$

These equations may be integrated to obtain

$$
u = -\frac{\gamma}{2\pi}\,\mathrm{sign}(y)\left\{\tan^{-1}\left(\frac{\frac{1}{2}\Delta s - x}{|y|}\right) + \tan^{-1}\left(\frac{\frac{1}{2}\Delta s + x}{|y|}\right)\right\}
$$

$$
v = -\frac{\gamma}{4\pi}\,\log\left[\frac{x^2 + y^2 - x\Delta s + (\frac{1}{2}\Delta s)^2}{x^2 + y^2 + x\Delta s + (\frac{1}{2}\Delta s)^2}\right] \tag{20.6.51}
$$

where sign (y) is the sign of y. Evaluating Eq. (20.6.51) at the midpoint of the segment yields

$$
\lim_{x=0,\,y\downarrow0} u(x,y) = -\frac{\gamma}{2}
$$
$$
\lim_{x=0,\,y\downarrow0} v(x,y) = 0 \tag{20.6.52}
$$

Therefore, the vectoral velocity at the midpoint of the ith segment due to the jth vortex segment is

$$
\vec{v}_{ij_v} = -\frac{\gamma_j}{2\pi}\frac{h_{ij}}{b_{ij}}\left[\tan^{-1}\frac{d_{ij}}{b_{ij}} + \tan^{-1}\frac{e_{ij}}{b_{ij}}\right]\vec{t}_j - \frac{\gamma_j}{4\pi}\log\frac{f_{i,j+1}}{f_{ij}}\,\vec{n}_j \tag{20.6.53}
$$

where

$$
h_{ij} = (x_i - x_j')n_{1_j} + (y_i - y_j')n_{2_j} \tag{20.6.54}
$$

Applying the boundary condition that the velocity normal to the boundary at each control point is zero yields

$$
\frac{Q_i}{2} + \sum_{j=1,\,j\neq i}^{j=n}\frac{Q_j}{2\pi}(n_{1_i}G_{ij} + n_{2_i}H_{ij}) + \frac{1}{U}\sum_{j=1,\,j\neq i}^{j=n}\vec{v}_{ij_v}\cdot\vec{n}_i
$$
$$
= -(n_{1_i}\cos\alpha + n_{2_i}\sin\alpha) \qquad \text{for } i = 1,\ldots,n \tag{20.6.55}
$$

The additional boundary condition is the Zhukovsky hypothesis described above.

It is evident that Eq. (20.6.55) provides n equations. Together with the Zhukovsky hypothesis, there are $n + 1$ equations for the $2n$ unknowns q_i and γ_i. It is evident, therefore, that a functional form must be assumed for γ_j involving only a single unknown variable (Hess and Smith, 1966; Hess, 1975). At a trailing edge, the vortex strength must vanish, otherwise the tangential velocity will be infinite. Thus, we assume the form

$$
\gamma_i = U\Lambda S_i \qquad \text{for } i = 1,\ldots,n \tag{20.6.56}
$$

where

$$
S_i = s_i(1 - s_i) \tag{20.6.57}
$$

with $s_1 = 0$ and

$$
s_i = \left[\sum_{j=2}^{j=i}\sqrt{(x_j - x_{j-1})^2 + (y_j - y_{j-1})^2}\right]
$$
$$
\times\left[\sum_{j=2}^{j=n}\sqrt{(x_j - x_{j-1})^2 + (y_j - y_{j-1})^2}\right]^{-1} \tag{20.6.58}
$$

Thus, $S_1 = 0$ and $S_n = 0$.

Define the influence matrices

$$
A_{ij} = S_j\left\{-\frac{h_{ij}t_{1_j}}{b_{ij}}\left[\tan^{-1}\frac{d_{ij}}{b_{ij}} + \tan^{-1}\frac{e_{ij}}{b_{ij}}\right] - \frac{1}{2}n_{1_j}\log\frac{f_{i,j+1}}{f_{ij}}\right\}
$$

$$
B_{ij} = S_j\left\{-\frac{h_{ij}t_{2_j}}{b_{ij}}\left[\tan^{-1}\frac{d_{ij}}{b_{ij}} + \tan^{-1}\frac{e_{ij}}{b_{ij}}\right] - \frac{1}{2}n_{2_j}\log\frac{f_{i,j+1}}{f_{ij}}\right\} \tag{20.6.59}
$$

Equation (20.6.55) becomes

$$
\frac{Q_i}{2} + \sum_{j=1,\,j\neq i}^{j=n}\frac{Q_i}{2\pi}(n_{1_i}G_{ij} + n_{2_i}H_{ij}) + \frac{\Lambda}{2\pi}\sum_{j=1,\,j\neq i}^{j=n}(n_{1_i}A_{ij} + n_{2_i}B_{ij})
$$
$$
= -(n_{1_i}\cos\alpha + n_{2_i}\sin\alpha) \tag{20.6.60}
$$

The tangential velocity at the ith control point becomes

$$
V_i = U(t_{1_i}\cos\alpha + t_{2_i}\sin\alpha) + \sum_{j=1,\,j\neq i}^{j=n}\frac{q_j}{2\pi}(t_{1_i}G_{ij} + t_{2_i}H_{ij})
$$
$$
+ U\Lambda\left[-\frac{1}{2}S_i + \frac{1}{2\pi}\sum_{j=1,\,j\neq i}^{j=n}(t_{1_i}A_{ij} + t_{2_i}B_{ij})\right] \tag{20.6.61}
$$

The Zhukovsky hypothesis indicates that the velocity at the training edge is finite (Fig. 20.6.7). For a sharp trailing edge, this implies that the velocity vector must be aligned with the bisector of the trailing edge. In order for this to occur, the pressure at control points $i = 1$ and $i = n$ must be the same, and therefore, by Bernoulli's equation, $V_1^2 = V_n^2$. Since the velocity is tangent to the surface at all control points, and since the tangential vectors $\vec{t}_1$ and $\vec{t}_n$ point toward and away from the trailing edge, respectively,

$$
V_1 = -V_n \tag{20.6.62}
$$

Thus,

$$
\sum_{j=2}^{j=n}\frac{Q_i}{2\pi}(t_{1_1}G_{1j} + t_{2_1}H_{1j}) + \sum_{j=1}^{j=n-1}\frac{Q_i}{2\pi}(t_{1_n}G_{nj} + t_{2_n}H_{nj})
$$
$$
+ \Lambda\left[-\frac{1}{2}(S_1 + S_n) + \frac{1}{2\pi}\sum_{j=2}^{j=n}(t_{1_1}A_{1j} + t_{2_1}B_{1j}) + \frac{1}{2\pi}\sum_{j=1}^{j=n}(t_{1_n}A_{nj} + t_{2_n}B_{nj})\right]
$$
$$
= -(t_{1_1} + t_{1_n})\cos\alpha - (t_{2_1} + t_{2_n})\sin\alpha \tag{20.6.63}
$$

Equations (20.6.55) and (20.6.63) are solved for the source/sink strengths q_i and vortex sheet parameter Λ.

The lift on the airfoil is obtained from the Kutta-Zhukovsky theorem [Eq, (20.6.47)]. The circulation is defined by Eq. (20.6.27), which yields

$$
l = -\rho U^2\Lambda\sum_{i=1}^{i=n}S_i\Delta s_i \tag{20.6.64}
$$

and the corresponding lift coefficient

$$
c_l = \frac{l}{\frac{1}{2}\rho U^2 c} = -\frac{2\Lambda}{c}\sum_{i=1}^{i=n}S_i\Delta s_i \tag{20.6.65}
$$

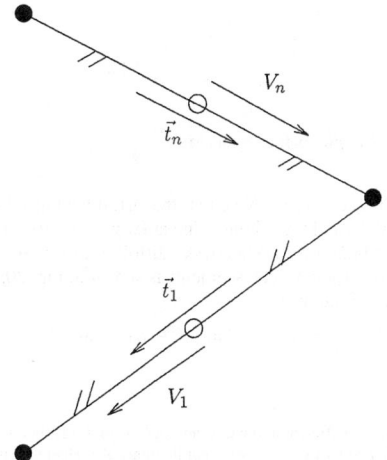

Fig. 20.6.7   Velocities at trailing edge.

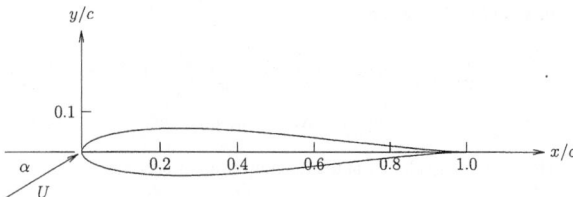

Fig. 20.6.8   Zhukovsky airfoil.

where c is the airfoil chord. The drag on the airfoil (i.e., the component of force acting in the direction of the freestream velocity) is zero.

The above method is applied to the flow past a Zhukovsky airfoil (Houghton and Carpenter, 1993) as illustrated in Fig. 20.6.8. The airfoil coordinates are defined as

$$x = \frac{1}{4\kappa_1}\left[\xi\left(1 + \frac{1}{\xi^2 + \eta^2}\right) + 2\kappa_2\right]$$
$$y = \frac{1}{4\kappa_1}\left[\eta\left(1 - \frac{1}{\xi^2 + \eta^2}\right)\right]$$

(20.6.66)

where

$$\xi = -e + (1 + e)\cos\theta'$$
$$\eta = (1 + e)\sin\theta'$$
$$\kappa_1 = (1 + 2e + e^2)(1 + 2e)^{-1}$$
$$\kappa_2 = (1 + 2e + 2e^2)(1 + 2e)^{-1}$$

(20.6.67)

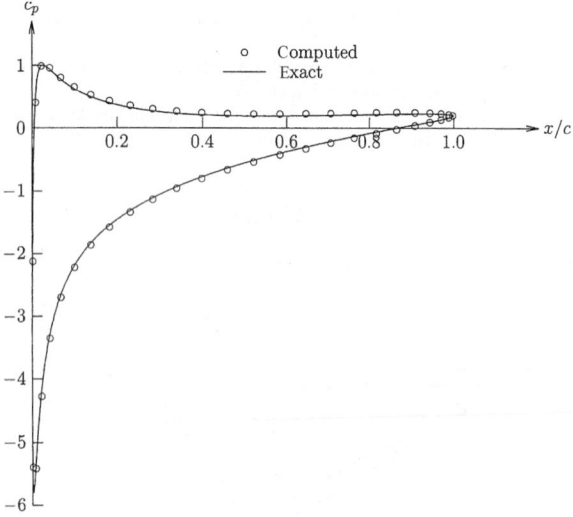

Fig. 20.6.9   Pressure coefficient on airfoil.

where e is the *eccentricity*. Note that the airfoil leading edge is located at $(x, y) = (0, 0)$ and the trailing edge at $(x, y) = (1, 0)$. The computed pressure distribution for a Zhukovsky airfoil with $e = 0.1$ an angle of attack $\alpha = 10°$ using $n = 50$ segments is shown in Fig. 20.6.9 together with the exact solution[*]

$$c_p = 1 - 4\kappa_3[\sin(\theta' - \alpha) + \sin\alpha]^2$$

(20.6.68)

[*]The pressure coefficient at the trailing edge $(x, y) = (1, 0)$, corresponding to $\theta' = 0$, is obtained by two successive applications of L'Hospital rule to yield

$$c_p\big|_{\theta'=0} = 1 - (1 + e)^{-2}\cos^2\alpha$$

where

$$\kappa_3 = \left[1 - 2\frac{1 - 2\sin^2\theta}{\kappa_4} + \frac{1}{\kappa_4^2}\right]^{-1}$$
$$\kappa_4 = [-e + (1 + e)\cos\theta']^2 + [(1 + e)\sin\theta']^2$$
$$\sin^2\theta = \frac{(1 + e)^2\sin^2\theta'}{\kappa_4}$$

(20.6.69)

BOUNDARY LAYER FLOW

At high Reynolds number Re and in the absence of strong pressure gradients,[†] the viscous stresses generate a thin *boundary layer* adjacent to solid surfaces wherein the flow velocity decreases to zero at the surface. A detailed history of the development of boundary layer theory is presented by Tani (1977).

Consider a body-oriented coordinate system (Fig. 20.6.10) where the x and y coordinates are locally tangential and normal to the surface, respectively. For laminar, steady, two-dimensional flow, the boundary layer equations are obtained from Eqs. (20.6.3) and (20.6.4) under the assumption that the normal y derivatives are large compared to the streamwise x derivatives.[‡] The steady boundary layer equations are (with ~ omitted for clarity):

$$\frac{\partial u}{\partial x} + \frac{\partial v}{\partial y} = 0$$

(20.6.70)

$$u\frac{\partial u}{\partial x} + v\frac{\partial u}{\partial y} = -\frac{1}{\rho}\frac{dp}{dx} + v\frac{\partial^2 u}{\partial y^2}$$

(20.6.71)

where $v = \mu/\rho$ is the kinematic viscosity. The boundary conditions are

$$\begin{aligned} u &= 0 && \text{on } y = 0 \\ v &= 0 && \text{on } y = 0 \\ u &= U_e(x) && \text{as } y \to \infty \end{aligned}$$

(20.6.72)

where $U_e(x)$ is the velocity outside of the boundary layer at the streamwise location x, and dp/dx is the streamwise pressure gradient. Both $U_e(x)$ and $dp(x)/dx$ are obtained from the solution of the irrotational flow (see "Irrotational Flow," above).

Mangler-Levy-Lees Transformation

The boundary layer equations may be rewritten by using the Mangler-Levy-Lee transformation (Cebeci and Cousteix, 1999):

$$d\xi = \rho\mu U_e(x)\, dx$$

(20.6.73)

$$d\eta = \frac{\rho U_e(x)}{\sqrt{2\xi}}\, dy$$

(20.6.74)

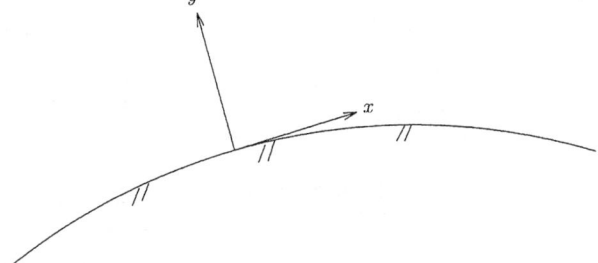

Fig. 20.6.10   Coordinate system.

[†] A sufficiently strong pressure gradient will cause the boundary layer to *separate* from the body surface. Under such circumstances, the boundary layer equations are no longer valid and the full Navier-Stokes equations (see "Navier-Stokes Flow," below) must be solved.

[‡]In addition, it is assumed that the local boundary layer thickness $\delta(x)$ is small compared to the local radius of curvature R of the boundary.

which may be integrated to obtain

$$\xi = \rho\mu \int_0^x U_e(x')\,dx' \qquad (20.6.75)$$

$$n = \frac{\rho U_e(x)y}{\sqrt{2\xi}} \qquad (20.6.76)$$

By introducing the stream function ψ according to

$$\rho u = \frac{\partial\psi}{\partial y}$$
$$\rho v = -\frac{\partial\psi}{\partial x} \qquad (20.6.77)$$

the conservation of mass, [Eq. (20.6.70)] is exactly satisfied. Introducing the dimensionless function $f(\xi, \eta)$ according to

$$\psi(\xi, \eta) = \sqrt{2\xi}\, f(\xi, \eta) \qquad (20.6.78)$$

we see that the velocity components become

$$u = U_e \frac{\partial f}{\partial n} \qquad (20.6.79)$$

$$v = -\left[\frac{f}{\sqrt{2\xi}} + \sqrt{2\xi}\,\frac{\partial f}{\partial\xi}\right]\mu U_e - \frac{\sqrt{2\xi}}{\rho}\frac{\partial f}{\partial\eta}\left[\frac{\eta}{U_e}\frac{dU_e}{dx} - \frac{\eta\rho\mu U_e}{2\xi}\right] \qquad (20.6.80)$$

The momentum equation (20.6.71) becomes

$$\frac{\partial^3 f}{\partial\eta^3} + f\frac{\partial^2 f}{\partial\eta^2} + \beta\left[1 - \left(\frac{\partial f}{\partial\eta}\right)^2\right] - 2\xi\left[\frac{\partial f}{\partial\eta}\frac{\partial^2 f}{\partial\eta\,\partial\xi} - \frac{\partial^2 f}{\partial\eta^2}\frac{\partial f}{\partial\xi}\right] = 0 \qquad (20.6.81)$$

where $\beta(\xi)$ is the dimensionless pressure gradient parameter

$$\beta = \frac{2\xi}{U_e}\frac{dU_e}{d\xi} \qquad (20.6.82)$$

The boundary conditions [Eq. (20.6.72)] become

$$\frac{\partial f}{\partial\eta} = 0 \qquad \text{at } \eta = 0$$
$$f = 0 \qquad \text{at } \eta = 0 \qquad (20.6.83)$$
$$\frac{\partial f}{\partial\eta} = 1 \qquad \text{as } \eta \to \infty$$

Keller Box Scheme

Given the pressure gradient parameter $\beta(\xi)$, the momentum equation (20.6.81) can be solved by using Keller's box scheme (Keller, 1975). A grid of points is defined by

$$\xi_i = \xi_{i-1} + \Delta\xi_i \qquad \text{for } i = 2, \dots, m \qquad (20.6.84)$$

$$\eta_j = \eta_{j-1} + \Delta\eta_j \qquad \text{for } j = 2, \dots, n \qquad (20.6.85)$$

where $\xi_1 = \eta_1 = 0$ (Fig. 20.6.11). Note that $\Delta\xi_i$ and $\Delta\eta_j$ may be taken to be arbitrary positive values consistent with adequate resolution of the boundary layer in the streamwise and normal directions. The boundary layer momentum equation (20.6.81) is rewritten as a system of first-order partial differential equations

$$\frac{\partial f}{\partial\eta} = g \qquad (20.6.86)$$

$$\frac{\partial g}{\partial\eta} = h \qquad (20.6.87)$$

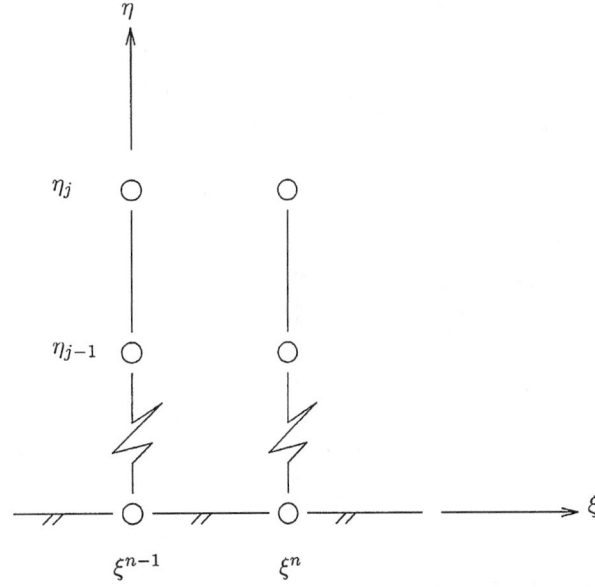

Fig. 20.6.11 Finite-difference grid.

$$\frac{\partial h}{\partial\eta} + fh + \beta(1 - g^2) - 2\xi\left(g\frac{\partial g}{\partial\xi} - h\frac{\partial f}{\partial\xi}\right) = 0 \qquad (20.6.88)$$

where Eq. (20.6.88) is recognized as the boundary layer momentum equation (20.6.81). The function $f(\xi, \eta)$, $g(\xi, \eta)$, and $h(\xi, \eta)$ are defined at the discrete grid points by using the notation $f_{i,j} = f(\xi_i, \eta_j)$, etc. Equations (20.6.86) and (20.6.87) are discretized as

$$f_{i,j} - f_{i,j-1} - \tfrac{1}{2}\Delta\eta_j(g_{i,j} + g_{i,j-1}) = 0 \qquad (20.6.89)$$

$$g_{i,j} - g_{i,j-1} - \tfrac{1}{2}\Delta\eta_j(h_{i,j} + h_{i,j-1}) = 0 \qquad (20.6.90)$$

Equation (20.6.88) is discretized about the centroid of the cell (Fig. 20.6.12) to yield

$$h_{i,j} + h_{i-1,j} - h_{i,j-1} - h_{i-1,j-1} + \tfrac{1}{2}\Delta\eta_j\,\overline{fh}_{i,j} + \Delta\eta_j\,\overline{\beta}_i\left(1 - \tfrac{1}{4}\,\overline{g^2}_{i,j}\right)$$
$$-\frac{1}{4}\frac{\Delta\eta_j}{\Delta\xi_i}\overline{\xi}_i[\,\overline{g}_{i,j}(g_{i,j} + g_{i,j-1} - g_{i-1,j} - g_{i-1,j-1})$$
$$- \overline{h}_{i,j}(f_{i,j} + f_{i,j-1} - f_{i-1,j} - f_{i-1,j-1})] = 0 \qquad (20.6.91)$$

where

$$\overline{fh}_{i,j} = f_{i,j}h_{i,j} + f_{i-1,j}h_{i-1,j} + f_{i,j-1}h_{i,j-1} + f_{i-1,j-1}h_{i-1,j-1}$$
$$\overline{\xi}_i = (\xi_i + \xi_{i-1})$$

with similar expressions for $\overline{g}_{i,j}$, $\overline{g^2}_{i,j}$, $\overline{h}_{i,j}$, and $\overline{\beta}_i$. The boundary conditions are

$$f_{i,1} = 0$$
$$g_{i,1} = 0 \qquad (20.6.92)$$
$$g_{i,n} = 1$$

for $i = 2, \dots, m$. The third boundary condition assumes η_n is outside the boundary layer.

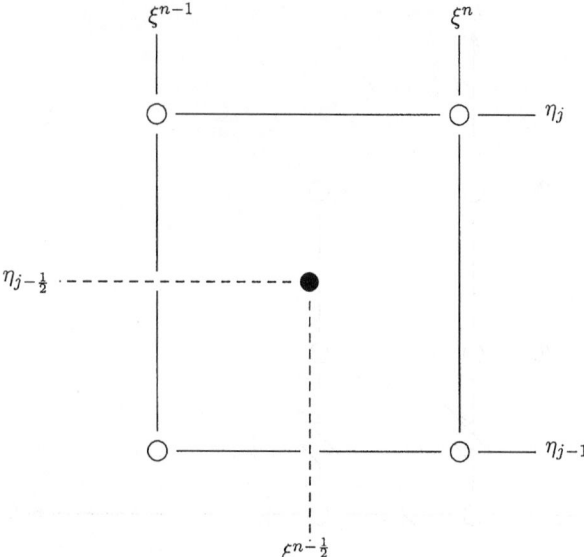

Fig. 20.6.12 Computational cell.

The discrete equations (20.6.89), (20.6.90), and (20.6.91) are second-order accurate in ξ and η, and may be solved by using an implicit technique. Define

$$G_{i,j} = f_{i,j} - f_{i,j-1} - \tfrac{1}{2}\Delta\eta_j\left(g_{i,j} + g_{i,j-1}\right) \qquad (20.6.93)$$

$$H_{i,j} = g_{i,j} - g_{i,j-1} - \tfrac{1}{2}\Delta\eta_j\left(h_{i,j} + h_{i,j-1}\right) \qquad (20.6.94)$$

$$F_{i,j} = h_{i,j} + h_{i-1,j} - h_{i,j-1} - h_{i-1,j-1} + \tfrac{1}{2}\Delta\eta_j \overline{f}\overline{h}_{i,j}$$

$$+ \Delta\eta_j \overline{\beta}_i\left(1 - \tfrac{1}{4}\overline{g^2}_{i,j}\right) - \tfrac{1}{4}\frac{\Delta\eta_j}{\Delta\xi_i}\overline{\xi}_i[\overline{g}_{i,j}(g_{i,j} + g_{i,j-1} - g_{i-1,j} - g_{i-1,j-1})$$

$$- \overline{h}_{i,j}(f_{i,j} + f_{i,j-1} - f_{i-1,j} - f_{i-1,j-1})] \qquad (20.6.95)$$

for $j = 2, \ldots, n$. Then the discrete equations become

$$\begin{aligned}
f_{i,1} &= 0 \\
g_{i,1} &= 0 \\
F_{i,j} &= 0 \qquad \text{for } j = 2, \ldots, n \\
G_{i,j} &= 0 \qquad \text{for } j = 2, \ldots, n \\
H_{i,j} &= 0 \qquad \text{for } j = 2, \ldots, n \\
g_{i,n} - 1 &= 0
\end{aligned} \qquad (20.6.96)$$

for $i = 2, \ldots, m$. These constitute $3n$ equations in $3n$ unknowns and may be solved by Newton's method. Define the vector of equations

$$R_i = (R_{i,1} R_{i,2}, \ldots, R_{i,n})^{\mathrm{T}} \qquad (20.6.97)$$

where ^{T} denotes the vector transpose and $R_{i,j}$ are column vectors of length three defined by

$$\begin{aligned}
R_{i,1} &= (f_{i,1}, g_{i,1}, F_{i,2})^{\mathrm{T}} \\
R_{i,2} &= (G_{i,2}, H_{i,2}, F_{i,3})^{\mathrm{T}} \\
&\cdots \\
R_{i,j} &= (G_{i,j}, H_{i,j}, F_{i,j+1})^{\mathrm{T}} \\
R_{i,n} &= (G_{i,n}, H_{i,n}, g_{i,n} - 1)^{\mathrm{T}}
\end{aligned} \qquad (20.6.98)$$

Define the vector of unknowns

$$U_i = (U_{i,1}, U_{i,2}, \ldots, U_{i,n})^{\mathrm{T}} \qquad (20.6.99)$$

where

$$U_{i,j} = (f_{i,j}, g_{i,j}, h_{i,j})^{\mathrm{T}} \qquad (20.6.100)$$

Equations (20.6.97) are solved by Newton's method. The current iterate for U_i^k is updated:

$$\frac{\partial R_i}{\partial U_i}\delta U_i^k = -R_i \qquad (20.6.101)$$

and

$$U_i^{k+1} = U_i^k + \delta U_i^k \qquad (20.6.102)$$

The Jacobian matrix is block tridiagonal,

$$\frac{\partial R_i}{\partial U_i} = \left\{\begin{array}{cccccc}
A_1 & C_1 & & & & \\
B_2 & A_2 & C_2 & & & \\
& B_3 & A_3 & C_3 & & \\
& & \ddots & \ddots & \ddots & \\
& & & B_{n-1} & A_{n-1} & C_{n-1} \\
& & & & B_n & A_n
\end{array}\right\} \qquad (20.6.103)$$

The diagonal matrices are

$$A_1 = \left\{\begin{array}{ccc}
1 & 0 & 0 \\
0 & 1 & 0 \\
a_{3,1_i} & a_{3,2_i} & a_{3,3_i}
\end{array}\right\} \qquad \text{for } j = 1$$

$$A_j = \left\{\begin{array}{ccc}
1 & -\tfrac{1}{2}\Delta\eta_j & 0 \\
0 & 1 & -\tfrac{1}{2}\Delta\eta_j \\
a_{3,1_i} & a_{3,2_i} & a_{3,3_j}
\end{array}\right\} \qquad \text{for } j = 2, \ldots, n-1$$

$$A_n = \left\{\begin{array}{ccc}
1 & -\tfrac{1}{2}\Delta\eta_n & 0 \\
0 & 1 & -\tfrac{1}{2}\Delta\eta_n \\
0 & 1 & 0
\end{array}\right\} \qquad \text{for } j = n \qquad (20.6.104)$$

where

$$a_{3,1_i} = \tfrac{1}{2}\Delta\eta_{j+1}h_j + \tfrac{1}{4}\frac{\Delta\eta_{j+1}}{\Delta\xi_i}\overline{\xi}_i\overline{h}_{i,j+1}$$

$$a_{3,2_i} = -\tfrac{1}{2}\Delta\eta_j\overline{\beta}_ig_{i,j} - \tfrac{1}{4}\frac{\Delta\eta_{j+1}}{\Delta\xi_i}\overline{\xi}_i(g_{i,j+1} + g_{i,j} - g_{i-1,j+1} - g_{i-1,j} + \overline{g}_{i,j+1})$$

$$a_{3,3_i} = -1 + \tfrac{1}{2}\Delta\eta_{j+1}f_{i,j} + \tfrac{1}{4}\frac{\Delta\eta_{j+1}}{\Delta\xi_i}\overline{\xi}_i(f_{i,j+1} + f_{i,j} - f_{i-1,j+1} - f_{i-1,j})$$

$$(20.6.105)$$

The off-diagonal matrices are

$$B_j = \left\{\begin{array}{ccc}
-1 & -\tfrac{1}{2}\Delta\eta_j & 0 \\
0 & -1 & -\tfrac{1}{2}\Delta\eta_j \\
0 & 0 & 0
\end{array}\right\} \qquad \text{for } j = 2, \ldots, n \qquad (20.6.106)$$

$$C_j = \left\{\begin{array}{ccc}
0 & 0 & 0 \\
0 & 0 & 0 \\
c_{3,1_i} & c_{3,2_i} & c_{3,3_i}
\end{array}\right\} \qquad \text{for } j = 1, \ldots, n-1 \qquad (20.6.107)$$

where

$$c_{3,1_i} = \tfrac{1}{2}\Delta\eta_{j+1}h_{i,j+1} + \tfrac{1}{4}\frac{\Delta\eta_{j+1}}{\Delta\xi_i}\overline{\xi}_i\overline{h}_{i,j+1}$$

$$c_{3,2_i} = -\tfrac{1}{2}\Delta\eta_j\overline{\beta}_ig_{i,j+1} - \tfrac{1}{4}\frac{\Delta\eta_{j+1}}{\Delta\xi_i}\overline{\xi}_i(g_{i,j+1} + g_{i,j} - g_{i-1,j+1}$$
$$- g_{i-1,j} + \overline{g}_{i,j+1})$$

$$c_{3,3_i} = 1 + \tfrac{1}{2}\Delta\eta_{j+1}f_{i,j+1} + \tfrac{1}{4}\frac{\Delta\eta_{j+1}}{\Delta\xi_i}\overline{\xi}_i(f_{i,j+1} + f_{i,j} - f_{i-1,j+1} - f_{i-1,j})$$

$$(20.6.108)$$

The above method is applied to the flow past a flat plate in the absence of a pressure gradient; i.e., $dp/dx = 0$ and $U_e(x) = U_\infty$. For this case, Eqs. (20.6.75) and (20.6.76) become

$$\xi = \rho \mu U_\infty x \qquad (20.6.109)$$

$$\eta = y\sqrt{\frac{U_\infty}{2vx}} \qquad (20.6.110)$$

and a self-similar solution (the *Blasius solution*) is obtained, wherein the dimensionless function f depends only on η, and

$$u = U_\infty \frac{df}{d\eta} \qquad (20.6.111)$$

$$v = U_\infty \left(\frac{U_\infty x}{v}\right)^{-1/2} \frac{1}{\sqrt{2}}(-f + g\eta) \qquad (20.6.112)$$

The computed streamwise and normal velocities are displayed in Figs. 20.6.13 and 20.6.14, respectively, for a grid of 50 points in

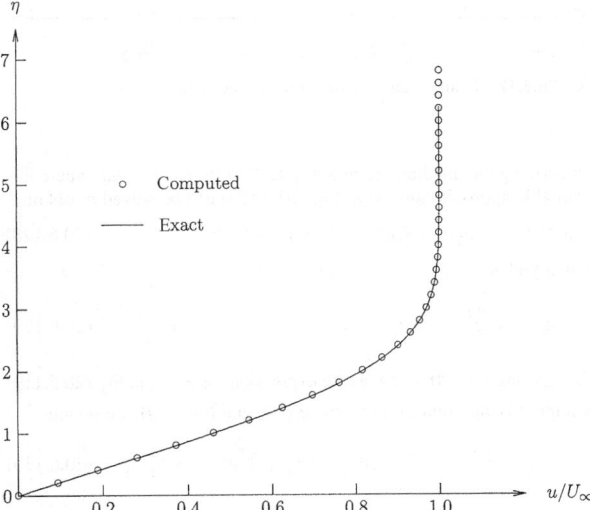

Fig. 20.6.13   Velocity at x direction.

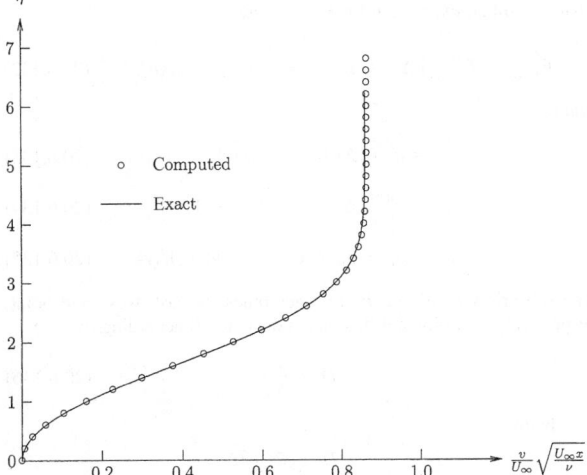

Fig. 20.6.14   Velocity in y direction.

the η direction with $\Delta\eta = 0.2$, together with the exact solution (Schlichting, 1968).

NAVIER-STOKES FLOW

The unsteady Navier-Stokes equations are (with $\tilde{\ }$ omitted for clarity):

$$\frac{\partial u}{\partial x} + \frac{\partial v}{\partial y} = 0 \qquad (20.6.113)$$

$$\frac{\partial u}{\partial t} + u\frac{\partial u}{\partial x} + v\frac{\partial u}{\partial y} = -\frac{1}{\rho}\frac{\partial p}{\partial x} + \hat{v}\left(\frac{\partial^2 u}{\partial x^2} + \frac{\partial^2 u}{\partial y^2}\right) \qquad (20.6.114)$$

$$\frac{\partial v}{\partial t} + u\frac{\partial v}{\partial x} + v\frac{\partial v}{\partial y} = -\frac{1}{\rho}\frac{\partial p}{\partial y} + \hat{v}\left(\frac{\partial^2 v}{\partial x^2} + \frac{\partial^2 v}{\partial y^2}\right) \qquad (20.6.115)$$

where $\hat{v} = \mu/\rho$ is the kinematic viscosity. The first boundary condition is

$$u = u_b \qquad \text{and} \qquad v = v_b \qquad \text{on solid boundary} \qquad (20.6.116)$$

where u_b and v_b are the x and y components of the velocity of the boundary. Equation (20.6.116) is the *no-slip condition*. The second boundary condition is

u and v approach their freestream values away from the solid boundary

$$(20.6.117)$$

SIMPLE Algorithm

The Navier-Stokes equations (20.6.113) to (20.6.115) can be solved by using Patankar's SIMPLE algorithm (Patankar, 1980). A staggered grid (Harlow and Welch, 1965) is used to avoid unphysical oscillations in the pressure. Consider a uniform grid with constant spacing Δx and Δy, as illustrated in Fig. 20.6.15. The x and y components of velocity are defined on adjacent sides of the computational cell, and the pressure is defined at the center of the cell.

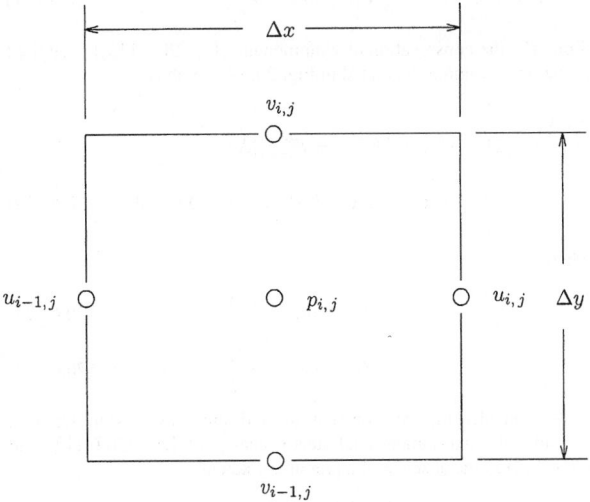

Fig. 20.6.15   Staggered grid.

The conservation of mass is applied to the finite volume in Fig. 20.6.15 to obtain

$$\left(u_{i,j}^{n+1} - u_{i-1,j}^{n+1}\right) + \left(v_{i,j}^{n+1} - v_{i-1,j}^{n+1}\right) = 0 \qquad (20.6.118)$$

where i and j indicate the indices in the x and y directions, respectively, and n indicates the time level. The conservation of x momentum

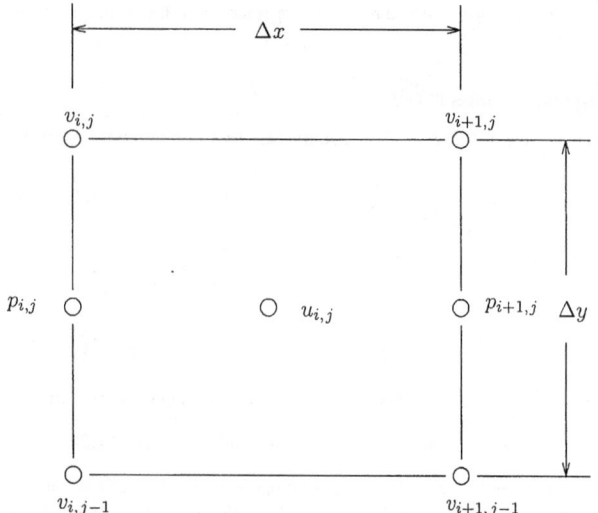

Fig. 20.6.16 Finite volume for x-momentum equation.

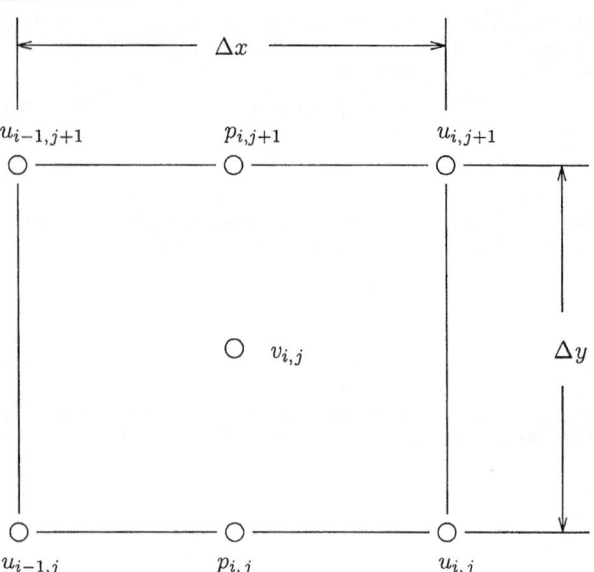

Fig. 20.6.17 Finite volume for y-momentum equation.

[Eq. (20.6.114)] is applied to the finite volume illustrated in Fig. 20.6.16 to obtain

$$\frac{\Delta x\,\Delta y}{\Delta t}\left(u_{i,j}^{n+1} - u_{i,j}^{n}\right) + \left(F_{i+\frac{1}{2},j}^{(1)} - F_{i-\frac{1}{2},j}^{(1)}\right)\Delta y$$
$$+ \left(G_{i,j+\frac{1}{2}}^{(1)} - G_{i,j-\frac{1}{2}}^{(1)}\right)\Delta x + \left(p_{i+1,j}^{n+1} - p_{i,j}^{n}\right)\Delta y = 0 \qquad (20.6.119)$$

where

$$F^{(1)} = u^2 - \hat{\nu}\frac{\partial u}{\partial x} \qquad (20.6.120)$$

$$G^{(1)} = uv - \hat{\nu}\frac{\partial u}{\partial y} \qquad (20.6.121)$$

Similarly, the conservation of y momentum [Eq. (20.6.115)] is applied to the finite volume illustrated in Fig. 20.6.17 to obtain

$$\frac{\Delta x\,\Delta y}{\Delta t}\left(v_{i,j}^{n+1} - v_{i,j}^{n}\right) + \left(F_{i+\frac{1}{2},j}^{(2)} - F_{i-\frac{1}{2},j}^{(2)}\right)\Delta y$$
$$+ \left(G_{i,j+\frac{1}{2}}^{(2)} - G_{i,j-\frac{1}{2}}^{(2)}\right)\Delta x + \left(p_{i,j+1}^{n+1} - p_{i,j}^{n}\right)\Delta x = 0 \qquad (20.6.122)$$

where

$$F^{(2)} = uv - \hat{\nu}\frac{\partial v}{\partial x} \qquad (20.6.123)$$

$$G^{(2)} = v^2 - \hat{\nu}\frac{\partial v}{\partial y} \qquad (20.6.124)$$

We consider the development of a discrete expression for $F^{(1)}$. Consider the one-dimensional steady analog of Eqs. (20.6.113) and (20.6.114) in the absence of a pressure gradient,

$$\frac{\partial u}{\partial x} = 0 \qquad (20.6.125)$$

$$\frac{\partial}{\partial x}\left(u^2 - \hat{\nu}\frac{\partial u}{\partial x}\right) = 0 \qquad (20.6.126)$$

with boundary conditions

$$\begin{aligned} u &= u_1 \quad \text{at } x = x_1 \\ u &= u_2 \quad \text{at } x = x_2 \end{aligned} \qquad (20.6.127)$$

Linearizing the quadratic term in Eq. (20.6.126) as $u^2 \approx \bar{u}u$, where $\bar{u}$ is a suitable approximation to u, Eq. (20.6.126) can be solved to obtain

$$u = u_1 + (u_2 - u_1)\left[e^{\bar{u}(x-x_1)/\hat{\nu}} - 1\right]\left[e^{\bar{u}(x_2-x_1)/\hat{\nu}} - 1\right]^{-1} \qquad (20.6.128)$$

which yields

$$\bar{u}u - \hat{\nu}\frac{\partial u}{\partial x} = \bar{u}\left\{u_1 + (u_1 - u_2)\left[e^{\bar{u}(x_2-x_1)} - 1\right]^{-1}\right\} \qquad (20.6.129)$$

Considering Eq. (20.6.129) as the expression for $F_{i+\frac{1}{2},j}^{(1)}$ in Eq. (20.6.119) and identifying points 1 and 2 to be (i, j) and $(i + 1, j)$, we obtain

$$F_{i+\frac{1}{2}}^{(1)} = u_{i+\frac{1}{2},j}^{n+1}\left[u_{i,j}^{n+1} + \left(u_{i,j}^{n+1} - u_{i+1,j}^{n+1}\right)\left[e^{R_{i+\frac{1}{2},j}} - 1\right]^{-1}\right] \qquad (20.6.130)$$

where

$$R_{i+\frac{1}{2},j} = \frac{u_{i+\frac{1}{2},j}^{n+1}\Delta x}{\hat{\nu}} \qquad (20.6.131)$$

Using a similar expression for $F_{i-\frac{1}{2},j}^{(1)}$ yields

$$\left(F_{i+\frac{1}{2},j}^{(1)} - F_{i-\frac{1}{2},j}^{(1)}\right)\Delta y = a_P u_{i,j}^{n+1} - a_E u_{i+1,j}^{n+1} - a_W u_{i-1,j}^{n+1} \qquad (20.6.132)$$

where

$$a_E = u_{i+\frac{1}{2},j}^{n+1}\Delta y\left[e^{R_{i+\frac{1}{2},j}} - 1\right]^{-1} \qquad (20.6.133)$$

$$a_W = u_{i-\frac{1}{2},j}^{n+1}\Delta y\left[1 + \left[e^{R_{i-\frac{1}{2},j}} - 1\right]^{-1}\right] \qquad (20.6.134)$$

$$a_P = a_E + a_W + \left(u_{i+\frac{1}{2},j}^{n+1} - u_{i-\frac{1}{2},j}^{n+1}\right)\Delta y \qquad (20.6.135)$$

The subscripts E, W, and P are mnemonics for east, west, and point, respectively. Defining a diffusion conductance D according to

$$D = \frac{\hat{\nu}}{\Delta x} \qquad (20.6.136)$$

we obtain

$$a_E = \frac{D_{i+\frac{1}{2},j}R_{i+\frac{1}{2},j}\Delta y}{e^{R_{i+\frac{1}{2},j}} - 1} \qquad (20.6.137)$$

Since the exponential is costly to evaluate computationally, Patankar (1980) developed the *Power Law* approximation

$$
\frac{a_E}{\Delta y D_{i+\frac{1}{2},j}} = \begin{cases} -R_{i+\frac{1}{2},j} & R_{i+\frac{1}{2},j} < 10 \\ \left(1 + 0.1 R_{i+\frac{1}{2},j}\right)^5 - R_{i+\frac{1}{2},j} & -10 \le R_{i+\frac{1}{2},j} < 0 \\ \left(1 - 0.1 R_{i+\frac{1}{2},j}\right)^5 & 0 \le R_{i+\frac{1}{2},j} < 10 \\ 0 & R_{i+\frac{1}{2},j} \ge 10 \end{cases}
$$
$$(20.6.138)$$

Additional approximation (e.g., central difference, hybrid, upwind) are also possible (Patankar, 1980). The coefficients may be written as

$$
a_E = D_{i+\frac{1}{2},j}\, \Delta y \left[\max(0,(1 - 0.1|R_{i+\frac{1}{2},j}|)^5) + \max(-R_{i+\frac{1}{2},j}, 0) \right]
$$
$$(20.6.139)$$

$$
a_W = D_{i-\frac{1}{2},j}\, \Delta y \left[R_{i-\frac{1}{2},j} + \max(0, (1 - 0.1|R_{i-\frac{1}{2},j}|)^5) \right.
$$
$$
\left. + \max(-R_{i-\frac{1}{2},j}, 0) \right]
$$
$$(20.6.140)$$

$$
a_P = a_E + a_W + \left(u_{i+\frac{1}{2},j} - u_{i-\frac{1}{2},j} \right) \Delta y
$$
$$(20.6.141)$$

Similarly,

$$
\left(G^{(1)}_{i,j+\frac{1}{2}} - G^{(1)}_{i,j-\frac{1}{2}} \right) \Delta x = b_P u^{n+1}_{i,j} - b_N u^{n+1}_{i,j+1} - b_S u^{n+1}_{i,j-1}
$$
$$(20.6.142)$$

where

$$
b_N = D_{i,j+\frac{1}{2}}\, \Delta x \left[\max(0,(1 - 0.1|R_{i,j+\frac{1}{2}}|)^5) + \max(-R_{i,j+\frac{1}{2}}, 0) \right]
$$
$$(20.6.143)$$

$$
b_S = D_{i,j-\frac{1}{2}}\, \Delta x \left[R_{i,j-\frac{1}{2}} + \max(0,(1 - 0.1|R_{i,j-\frac{1}{2}}|)^5 \right.
$$
$$
\left. + \max(-R_{i,j-\frac{1}{2}}, 0) \right]
$$
$$(20.6.144)$$

$$
b_P = b_N + b_S + \left(v_{i,j+\frac{1}{2}} - v_{i,j-\frac{1}{2}} \right) \Delta x
$$
$$(20.6.145)$$

with

$$
R_{i,j+\frac{1}{2}} = \frac{v_{i,j+\frac{1}{2}}\, \Delta y}{\hat{\nu}}
$$
$$(20.6.146)$$

The x- and y-momentum equations become

$$
\frac{\Delta x\, \Delta y}{\Delta t} \left(u^{n+1}_{i,j} - u^n_{i,j} \right) + a_P u^{n+1}_{i,j} - a_E u^{n+1}_{i+1,j} - a_W u^{n+1}_{i-1,j}
$$
$$
+ b_P u^{n+1}_{i,j} - b_N u^{n+1}_{i,j+1} - b_S u^{n+1}_{i,j-1} + \Delta y \left(p^{n+1}_{i+1,j} - p^{n+1}_{i,j} \right) = 0 \quad (20.6.147)
$$

$$
\frac{\Delta x\, \Delta y}{\Delta t} \left(v^{n+1}_{i,j} - v^n_{i,j} \right) + c_P v^{n+1}_{i,j} - c_E v^{n+1}_{i+1,j} - c_W v^{n+1}_{i-1,j}
$$
$$
+ d_P v^{n+1}_{i,j} - d_N v^{n+1}_{i,j+1} - d_S v^{n+1}_{i,j-1} + \Delta x \left(p^{n+1}_{i,j+1} - p^{n+1}_{i,j} \right) = 0 \quad (20.6.148)
$$

where the coefficients c_E, c_W, c_P and d_N, d_S, d_P are given by Eqs. (20.6.139) to (20.6.141) and (20.6.143) to (20.6.145) with due consideration to the expressions for $R_{i\pm\frac{1}{2},j}$ and $R_{i,j\pm\frac{1}{2}}$.

Equations (20.6.147) and (20.6.148) are solved by iteration, using the pressure field at the previous time step $p^n_{i,j}$. The resultant expressions for the velocity components, denoted $\hat{u}^{n+1}_{i,j}$ and $\hat{v}^{n+1}_{i,j}$, therefore satisfy

$$
\frac{\Delta x\, \Delta y}{\Delta t} \left(\hat{u}^{n+1}_{i,j} - u^n_{i,j} \right) + a_P \hat{u}^{n+1}_{i,j} - a_E \hat{u}^{n+1}_{i+1,j} - a_W \hat{u}^{n+1}_{i-1,j} + b_P \hat{u}^{n+1}_{i,j}
$$
$$
- b_N \hat{u}^{n+1}_{i,j+1} - b_S \hat{u}^{n+1}_{i,j-1} + \Delta y \left(p^n_{i+1,j} - p^n_{i,j} \right) = 0 \quad (20.6.149)
$$

$$
\frac{\Delta x\, \Delta y}{\Delta t} \left(\hat{v}^{n+1}_{i,j} - v^n_{i,j} \right) + c_P \hat{v}^{n+1}_{i,j} - c_E \hat{v}^{n+1}_{i+1,j} - c_W \hat{v}^{n+1}_{i-1,j} + d_P \hat{v}^{n+1}_{i,j}
$$
$$
- d_N \hat{v}^{n+1}_{i,j+1} - d_S \hat{v}^{n+1}_{i,j-1} + \Delta x \left(p^n_{i,j+1} - p^n_{i,j} \right) = 0 \quad (20.6.150)
$$

However, $\hat{u}_{i,j}$ and $\hat{v}_{i,j}$ do not satisfy conservation of mass [Eq. (20.6.118)]. Consequently, a correction $u'_{i,j}$ and $v'_{i,j}$ must be determined, Writing

$$
u^{n+1}_{i,j} = \hat{u}^{n+1}_{i,j} + u'_{i,j} \qquad (20.6.151)
$$

$$
v^{n+1}_{i,j} = \hat{v}^{n+1}_{i,j} + v'_{i,j} \qquad (20.6.152)
$$

$$
p^{n+1}_{i,j} = p^n_{i,j} + \delta p_{i,j} \qquad (20.6.153)
$$

and subtracting (20.6.149) from (20.6.147) and (20.6.150) from (20.6.148) yields

$$
\left(\frac{\Delta x\, \Delta y}{\Delta t} + a_P + b_P \right) u'_{i,j} - a_W u'_{i-1,j} - a_E u'_{i+1,j} - b_S u'_{i,j-1}
$$
$$
- b_N u'_{i,j+1} + \Delta y \left(\delta p_{i+1,j} - \delta p_{i,j} \right) = 0 \qquad (20.6.154)
$$

$$
\left(\frac{\Delta x\, \Delta y}{\Delta t} + c_P + d_P \right) v'_{i,j} - c_W v'_{i-1,j} - c_E v'_{i+1,j} - d_S v'_{i,j-1}
$$
$$
- d_N v'_{i,j+1} + \Delta x (\delta p_{i,j+1} - \delta p_{i,j}) = 0 \qquad (20.6.155)
$$

These equations are simplified (Patankar, 1980) to

$$
\left(\frac{\Delta x\, \Delta y}{\Delta t} + a_P + b_P \right) u'_{i,j} = -\Delta y (\delta p_{i+1,j} - \delta p_{i,j}) \qquad (20.6.156)
$$

$$
\left(\frac{\Delta x\, \Delta y}{\Delta t} + c_P + d_P \right) v'_{i,j} = -\Delta x (\delta p_{i,j+1} - \delta p_{i,j}) \qquad (20.6.157)
$$

The expressions in Eqs. (20.6.151) and (20.6.152), together with Eqs. (20.6.156) and (20.6.157), are substituted into the conservation of mass [Eq. (20.6.118)] to obtain an equation for the pressure correction:

$$
g_{i,j}\delta p_{i,j} = \left(\hat{u}^{n+1}_{i-1,j} - \hat{u}^{n+1}_{i,j} \right) \Delta y + \left(\hat{v}^{n+1}_{i,j} - \hat{v}^{n+1}_{i,j-1} \right) \Delta x
$$
$$
+ \left(e_{i,j}\delta p_{i+1,j} + e_{i-1,j}\delta p_{i-1,j} \right) \Delta y
$$
$$
+ \left(f_{i,j}\delta p_{i,j+1} + f_{i,j-1}\delta p_{i,j-1} \right) \Delta x \qquad (20.6.158)
$$

where

$$
e_{i,j} = -\Delta y \left[\frac{\Delta x\, \Delta y}{\Delta t} + a_P + b_P \right]^{-1} \qquad (20.6.159)
$$

$$
f_{i,j} = -\Delta x \left[\frac{\Delta x\, \Delta y}{\Delta t} + c_P + d_P \right]^{-1} \qquad (20.6.160)
$$

$$
g_{i,j} = \left(e_{i,j} + e_{i-1,j} \right) \Delta y + \left(f_{i,j} + f_{i,j-1} \right) \Delta x \qquad (20.6.161)
$$

The algorithm is known as SIMPLE (Semi-Implicit Method for Pressure-Linked Equations) (Patankar, 1980). The solution procedure is:

1. The approximate solution $\hat{u}^{n+1}_{i,j}$ and $\hat{v}^{n+1}_{i,j}$ is obtained from Eqs. (20.6.149) and (20.6.150) by iteration. This can be achieved in a two-step procedure. First, each equation is written as scalar tridiagonal system along each y grid line and solved by the Thomas algorithm (Anderson et al., 1984). Second, each equation is written as a tridiagonal system along each x grid line and solved. The two-step procedure is iterated until converged.

2. The correction to the pressure δp_{ij} is obtained by solution of Eq. (20.6.158), using, for example, successive overrelaxation (Anderson et al., 1984).

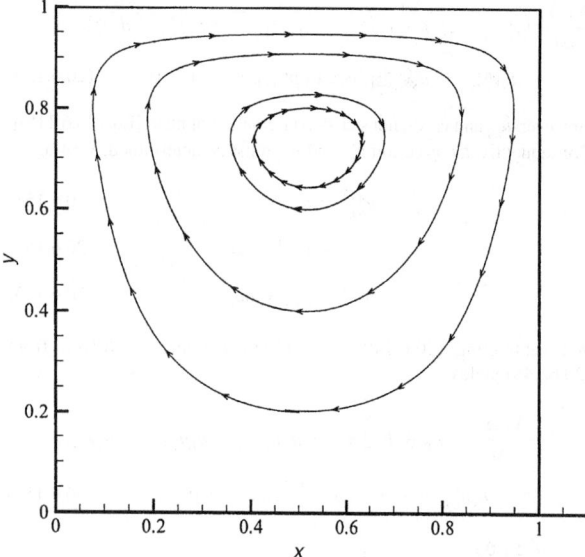

Fig. 20.6.18 Driven flow in cavity.

3. The pressure is updated according to $p_{i,j}^{n+1} = p_{i,j}^{n} + \omega_p\, \delta p_{i,j}$

4. The correction to the velocity, $u_{i,j}'$ and $v_{i,j}'$, is obtained from Eqs. (20.6.156) and (20.6.157) and velocity is updated according to Eqs. (20.6.151) and (20.6.152).

5. The process is repeated until steady state is achieved (if appropriate) or the unsteady solution is obtained for the desired period of time

Several improvements to the SIMPLE algorithm have also been developed as described in Fletcher (1988b).

The above method is applied to the driven flow in a cubic cavity as shown in Fig. 20.6.18. The upper surface moves to the right at a constant speed U. The three remaining walls are stationary. The Reynolds number Re = UL/ν is 100. The streamlines display the convective motion generated by the shearing force exerted by the movement of the upper surface.

REFERENCES

The description in this section is necessarily brief and the interested reader is directed to the numerous texts on aspects of CFD. Fletcher (1988a, b) and Hirsch (1988a, b) provide comprehensive developments of computational fluid dynamics. Laney (1998) and Toro (1997) focus on methods for compressible flows. Cebeci and Cousteix (1999) present a detailed description of boundary layer methods. Wilcox (1988) provides an essential treatment of incompressible and compressible turbulent flows. Quinn (1987) describes methods for efficient computation on parallel computers. Thompson et al. (1999) presents an encyclopedia summary of grid generation methods.

Anderson, D., Tannehill, J., and Pletcher, R. (1984). "Computational Fluid Mechanics and Heat Transfer." McGraw Hill, New York.
Anderson, J. (2001). "Fundamentals of Aerodynamics," 3d ed. McGraw-Hill, New York.
Batchelor, G. (1970). "Introduction to Fluid Dynamics." Cambridge University Press, Cambridge, United Kingdom.

Cebeci, T., and Cousteix, J. (1999). "Modeling and Computation of Boundary-Layer Flows." Horizons Publishing (Springer), Long Beach, CA.
Courant, R., Friedrichs, K., and Lewy, H. (1928). "Uber die partiellen differen-zengleichur-gen der mathematischen physik." Mathematische Annalen, 100:32–74. Also published in IBM Journal, March 1967, pp. 215–234.
Fletcher, C. (1988a). "Computational Techniques for Fluid Dynamics." Vol. I: "Fundamental and General Techniques." Springer-Verlag. Berlin.
Fletcher, C. (1988b). "Computational Techniques for Fluid Dynamics." Vol. II: "Specific Techniques for Different Flow Categories." Springer-Verlag, Berlin.
Harlow, F., and Welch, J. (1965). "Numerical calculation of time-dependent viscous incompressible flow of fluid with a free surface." Physics of Fluids, 8(12):2182–2189.
Hess, J. (1975). "Review of integral equation techniques for solving potential flow problems with emphasis on surface source methods." Computer Methods in Applied Mechanics and Engineering, 5:145–196.
Hess, J., and Smith, A. (1966). "Calculation of potential flow around arbitrary bodies." Progress in Aeronautical Science, 8:1–138.
Hirsch, C. (1988a). "Numerical Computation of Internal and External Flows." Vol. I: "Fundamentals of Numerical Discretization." John Wiley & Sons, New York.
Hirsch, C. (1988b). "Numerical Computation of Internal and External Flows." Vol. II: "Computational Methods for Inviscid and Viscous Flows." John Wiley & Sons, New York.
Houghton, E., and Carpenter, P. (1993). "Aerodynamics for Engineering Students," 4th ed. Arnold, London.
Keller, H. (1975). "Accurate difference methods for nonlinear two-point boundary value problems." SIAM Journal on Numerical Analysis, 11(2):305–320.
Kellogg, O. (1953). "Foundations of Potential Theory." Dover Publications, New York.
Kovenya, V., and Yanenko, N. (1981). "Method of Fractional Steps in Problems in Gas Dynamics." USSR Academy of Sciences, Siberian Division, Novosibirsk, Russia.
Laney, C. (1988). "Computational Gasdynamics." Cambridge University Press, Cambridge.
Lax, P., and Richtmyer, R. (1956). "Survey of the stability of linear finite difference equations." Communications on Pure and Applied Mathematics, 9:267–293.
Milne-Thompson, L. (1958). "Theoretical Aerodynamics." Dover Publications, New York.
Patankar, S. (1980). "Numerical Heat Transfer and Fluid Flow." Hemisphere Publishing Company, New York.
Peyret, R., and Taylor, T. (1983). "Computational Methods for Fluid Flow." Springer-Verlag, Berlin.
Quinn, M. (1987). "Designing Efficient Algorithms for Parallel Computers." McGraw-Hill, New York.
Richardson, L. (1910). "The approximate arithmetical solution by finite differences of physical problems involving differential equations with an application to the stresses in a masonry dam." Transactions of the Royal Society of London, Serial A, 210:307–357.
Richtmyer, R., and Morton, K. (1957). "Difference Methods for Initial-Value Problems." Interscience Publishers, New York.
Roache, P. (1972). "Computational Fluid Dynamics." Hermosa Publishers, Albuquerque, NM.
Roache, P. (1998). "Fundamentals of Computational Fluid Dynamics." Hermosa Publishers, Albuquerque, NM.
Schlichting, H. (1968). "Boundary Layer Theory," 6th ed. McGraw-Hill, New York.
Tani, I. (1977). "History of boundary-layer theory." In "Annual Review of Fluid Mechanics," pp. 87–111. Annual Reviews, Inc., Palo Alto, CA.
Thompson, J., Soni, B., and Weatherill, N. eds. (1999). "Handbook of Grid Generation." CRC Press, Boca Raton, FL.
Toro, E., ed. (1997). "Riemann Solvers and Numerical Methods for Fluid Dynamics." Springer, Berlin.
Wilcox, D. (1997). "Basic Fluid Mechanics." DCW Industries, Inc., La Cañada, CA.
Wilcox, D. (1998). "Turbulence Modeling for CFD." DCW Industries, Inc., La Cañada, CA.

20.7 EXPERIMENTAL FLUID DYNAMICS

by Yiannis Andreopoulos

INTRODUCTION

Experimental fluid dynamics (EFD) is a branch of fluid dynamics that complements the analytical and computational fluid dynamics branches. Although the complete equations describing fluid motions were derived several decades ago, today there are only a few solutions of particular problems of those nonlinear, partial differential equations. For turbulent flows, for instance, these solutions are limited to very low Reynolds number flows, which are not applicable to present data technological flows that are mostly in the high Reynolds number regime. Consequently, physical and numerical experiments will increasingly be employed to provide guidance and solutions to many engineering and technological problems.

The major objective of this section is to introduce the subject of experimental fluid dynamics broadly to readers, to provide an up-to-date view of the fundamentals of some key methods, and to present emerging needs and trends in current technological applications. The field of EFD is rather large and extends over the entire range of fluid dynamics, including not only the customary fields of aeronautics and hydraulics, but also diverse topics as biological and biomedical flows, including insect flight mechanics and pulmonary and blood flows, meteorology, oceanography, granular flows, and metallurgy.

Laboratory experiments require two very important processes: (1) designing or selecting the flow simulation platform and (2) selecting and designing the measurement system. The objectives of the experimental work will determine whether a new experimental setup is needed or an existing facility can be used. The measurement system has to be built around the flow simulation platform by accommodating possible operating restrictions and to fulfill the specific objectives of the experiments. In cases of direct measurements in the field with actual moving objects or in nature, designing the measurement system is the most critical aspect of the work.

Fluid mechanical quantities to be measured can range from forces and flow rates to dissipation rates of kinetic energy and enstrophy. Measurements of lift and drag forces of a body moving in air or water or the power attainable in a turbomachine are typical tasks in engineering applications, while the measurements of the rate-of-strain tensor and vorticity are typical of research-oriented problems that can provide better understanding of the physics of the flows and lead to analytical and computational tools that can predict the efficiency of fluid systems more accurately than existing methods.

EFD encompasses single- and multicomponent fluids, multiphase flows, newtonian and non-newtonian flows, chemically reacting flows, and biological fluid mechanics. We will restrict our description to single-component, chemically nonreactive flows. In that context, issues related to incompressible and compressible flows will be discussed in detail. Particular emphasis will be given to techniques related to applications in emerging technologies and to measurements of quantities that are not in the mainstream, such as spatially and temporally resolved vorticity, rate-of-strain tensor, and dissipation rate of kinetic energy, which are essential in describing the efficiency of fluid systems. The scope of the present contribution excludes the description of facilities used to simulate specific flow phenomena. Extensive coverage of items like wind tunnels and water tunnels, towing tanks and wave tanks, and shock tubes is provided in Refs. 1 to 7 and at NASA Web sites.

BACKGROUND INFORMATION

The relative importance of each of the measurable fluid mechanical quantities and their relationship can be appreciated by considering the application of conservation laws of mass, momentum, and energy, which leads to expressions in the form of integral or differential equations. The continuum approximation will be assumed to be valid in the present description. The fluid flow is inside a control volume V bounded by a control surface ∂V with n, the unit vector, normal in the outside direction. Here $U(x, t) = (U_1, U_2, U_3)$ is the **velocity field** as a function of location x and time t, ρ is the density, E is the **total specific energy** $E_t = (e + \frac{1}{2}U_iU_i)$ and e is the specific internal energy. Einstein's summation convention of repeated indices is applied here.

Conservation of Mass

$$\frac{d}{dt}\int_V \rho\, dV + \int_{\partial V} \rho U_k n_k\, dA = 0 \qquad (20.7.1)$$

$$\frac{\partial\rho}{\partial t} + \frac{\partial}{\partial x_k}(\rho U_k) = 0 \qquad (20.7.2)$$

Conservation of Momentum: Newton's Second Law

$$\frac{d}{dt}\int_V \rho U_i\, dV + \int_{\partial V} \rho(U_k n_k)U_i\, dA = -\int_{\partial V} \sigma_{ki}n_k\, dA + \int_V \rho f_i\, dV$$

$$(20.7.3)$$

where σ_{ki} is the second-order stress tensor and f_i represents the components of the specific external forces acting on this control volume.

$$\frac{\partial}{\partial t}(\rho U_i) + \frac{\partial}{\partial x_k}(\rho U_k U_i) = \frac{\partial \sigma_{ki}}{\partial x_k} + \rho f_i \qquad (20.7.4)$$

where $\sigma_{ki} = -p\delta_{ki} + \tau_{ki} = -p\delta_{ki} + 2\mu S_{ki}$ and S_{ki} is the rate-of-strain tensor.

Conservation of Energy

$$\frac{d}{dt}\int_V \rho E_t\, dV + \int_{\partial V} \rho E_t U_k n_k\, dA = \int_{\partial V} U_i \sigma_{ki}n_k\, dA - \int_{\partial V} q_k n_k\, dA + \int_V \rho U_i f_i\, dV$$

$$(20.7.5)$$

$$\frac{\partial}{\partial t}(\rho E_t) + \frac{\partial}{\partial x_k}(\rho U_k E_t) = \frac{\partial U_i \sigma_{ki}}{\partial x_k} + \rho U_i f_i \qquad (20.7.6)$$

Velocity Gradient Tensor, Strain-Rate, Vorticity, and Dissipation Rate

The velocity gradient tensor A_{ij} can be decomposed, not uniquely, into a symmetric and antisymmetric part. Specifically, these two parts are the rate-of-strain tensor $S_{ij} = \frac{1}{2}(\partial U_i/\partial x_j + \partial U_j/\partial x_i)$ and the-rate-of rotation tensor $R_{ij} = \frac{1}{2}(\partial U_i/\partial x_j - \partial U_j/\partial x_i) = \frac{1}{2}\varepsilon_{ijk}\Omega_k$, where Ω_k is the vorticity vector. Ω_k is defined as the curl of the velocity vector

$$\Omega_k = \epsilon_{ijk}\frac{\partial U_i}{\partial x_j} \qquad (20.7.7)$$

and ϵ_{ijk} is the alternating unit vector. At each point in the flow field we may imagine a spherical fluid particle whose motion can be decomposed into translation, expansion, shearing, and rotation. The vorticity can be interpreted as just twice the instantaneous solid-body-like rotation rate of the fluid particles along the principal axes in the fluid where there exists no shear deformation. We may alternatively interpret the **vorticity** as the circulation per unit area of a surface perpendicular to the vorticity field.

Why is vorticity so important in understanding the dynamics of turbulent fluid flow? Insight into this may be gained by looking at the equation describing the transport of vorticity[8]

$$\frac{d\Omega_i}{dt} = S_{ik}\Omega_k - \Omega_i S_{kk} + \epsilon_{iqn}\frac{1}{\rho^2}\frac{\partial\rho}{\partial x_q}\frac{\partial p}{\partial x_\eta} + \epsilon_{iqn}\frac{\partial}{\partial x_q}\left(\frac{1}{\rho}\frac{\partial\tau_{nj}}{\partial x_j}\right) \quad (20.7.8)$$

This transport equation indicates that, for an observer moving with the flow, the rate of change of vorticity component Ω_i of a particle is provided by four dynamically significant processes: (1) stretching or compression and reorientation or tilting by the strain S_{ik}, (2) vorticity generation through dilatation, (3) baroclinic generation through the interaction of pressure and density gradients, and (4) viscous effects expressed by the viscous stress term. The term $S_{ik}\Omega_k$ consists of one stretching or compression component and two tilting components. This term has been illuminated mostly through numerical work and it appears to play an important role in the various physical processes involved in turbulent flows. If the viscous term can be ignored when its magnitude is small, then the change of vorticity of a fluid element in a lagrangian frame of reference can be entirely attributed to vortex stretching and/or tilting, to dilatational effects, and to baroclinic torque.

If the above equation is multiplied by Ω_i, then the transport equation for the enstrophy $\frac{1}{2}\Omega_i\Omega_i$ can be obtained:

$$\frac{d(\frac{1}{2}\Omega_i\Omega_i)}{dt} = S_{ik}\Omega_k\Omega_i - \Omega_i\Omega_i S_{kk} + \epsilon_{iqn}\frac{1}{\rho^2}\Omega_i\frac{\partial\rho}{\partial x_q}\frac{\partial p}{\partial x_\eta}$$
$$+ \epsilon_{iqn}\frac{\partial}{\partial x_q}\left(\Omega_i\frac{1}{\rho}\frac{\partial\tau_{nj}}{\partial x_j}\right) - \epsilon_{iqn}\frac{1}{\rho}\frac{\partial\Omega_j}{\partial x_q}\frac{\partial\tau_{nj}}{\partial x_j} \quad (20.7.9)$$

The physical meaning of the terms associated with the time-dependent changes of enstrophy of a fluid element in a lagrangian frame of reference are similar with those responsible for the generation or destruction of vorticity. The physical mechanisms involved in these complex processes are not fully understood. Enstrophy is a very significant quantity in fluid dynamics because it is not only related to the solenoidal dissipation, but also to several of the invariants of the rate-of-strain matrix S_{ij}, which may lead to detection of vortex tubes and shear layers in the flow field. In addition, enstrophy is a source term in the **transport equation** of **dilatation** S_{kk}:

$$\frac{dS_{kk}}{dt} = -S_{ik}S_{ki} + \frac{1}{2}\Omega_k\Omega_k + \frac{1}{\rho^2}\frac{\partial\rho}{\partial x_k}\frac{\partial p}{\partial x_k} - \frac{1}{\rho}\frac{\partial^2 p}{\partial x_k\partial x_k}$$
$$+ \frac{\partial}{\partial x_k}\left(\frac{1}{\rho}\frac{\partial\tau_{kq}}{\partial x_q}\right) \quad (20.7.10)$$

This transport equation relates the change of dilatation along a particle path, which can be caused by the straining action of the dissipative motions ($S_{ik}S_{ik}$) as well as by the rotational energy of the spinning motions as it is expressed by the enstrophy $\frac{1}{2}\Omega_i\Omega_i$. Pressure and density gradients as well as viscous diffusion can also affect dilatation. It should be noted that the above transport equation reduces to the well-known Poisson equation for incompressible flows of constant density ($S_{kk} = 0$).

The dissipation rate of kinetic energy, defined as $E = \tau_{ij}S_{ij}$, is a key quantity in determining viscous losses in fluid systems. It can be expanded[9] as:

$$E = \tau_{ij}S_{ij} = 2\mu S_{ij}S_{ij} - \frac{2}{3}\mu S_{kk}S_{kk} \quad (20.7.11)$$

The second term in the right-hand side of the above relation represents the additional contribution of compressibility to the dissipation rate of kinetic energy. This term disappears in the cases of **incompressible**

flow. Since this term is always positive, the negative sign in front may erroneously suggest that compressibility reduces dissipation. This is incorrect because the term $S_{ij}S_{ij}$ also contains contributions from dilatation effects, which can be revealed if one considers that

$$S_{ij}S_{ij} = \frac{1}{2}\Omega_k\Omega_k + \frac{\partial U_i}{\partial x_j}\frac{\partial U_j}{\partial x_i} \quad (20.7.12)$$

The second term in the right-hand side represents the inhomogeneous contribution in the case of incompressible flows. In the case of compressible flows, terms related to dilatation can be extracted, and the instantaneous total dissipation rate then becomes

$$E = \mu\Omega_k\Omega_k + 2\mu\left[\frac{\partial U_i}{\partial x_j}\frac{\partial U_j}{\partial x_i} - S_{kk}S_{kk}\right] + \frac{1}{3}\mu S_{kk}S_{kk} \quad (20.7.13)$$

The third term on the right-hand side describes the direct effects of compressibility, i.e., dilatation on the dissipation rate. The first and the last terms on the right-hand side of the last relation are quadratic with positive coefficients and positive signs and they are, therefore, always positive. The second term on the right-hand side indicates the contributions to the dissipation rate by the purely nonhomogeneous part of the flow. Its time-averaged contribution disappears in homogeneous flows. This term, in principle, can obtain negative values, and thus it can reduce the dissipation rate. This does not violate the second thermodynamic law as long as the total dissipation remains positive at any point in space and time

The above description indicates that velocity, pressure, and temperature are the primary measurable quantities that can determine the local stresses acting on a body inside a flow field and therefore can be used to estimate the resultant forces and possibly torques exerted on various parts of a fluid system. The availability of this type of information will help determine the performance of the fluid system. However, if estimates of the efficiency of the system are required, determination of kinetic energy dissipation rate is essential. This demand is more critical in the case of turbulent flows, where viscous losses are substantially larger than in laminar flows.

CLASSIFICATION OF MEASUREMENTS

A typical block diagram of the measuring system of an experimental technique is shown in Fig. 20.7.1. It includes a probe, a transducer, a signal processing unit with an amplifier, an interface to the data processing unit such as an analog to digital converter, a data processing unit with a CPU, and a storage or a display unit. This type of block diagram is very useful in estimating and reducing uncertainties along the line with the various stages of the measurement system.

The output of instruments measuring fluid mechanical quantities can be purely mechanical, electrical, or optical. Most of the techniques used in EFD include probes and transducers that convert the measured value of a fluid mechanical quantity into an electrical output, which, through an appropriate interface, can feed the CPU of a computer or the local processing unit of the instrument with a display, as shown in Fig. 20.7.1. Substantial effort has been devoted in the past to develop transducers and instrumentation that convert the measured quantity into an electrical signal, which carries the information. Optical

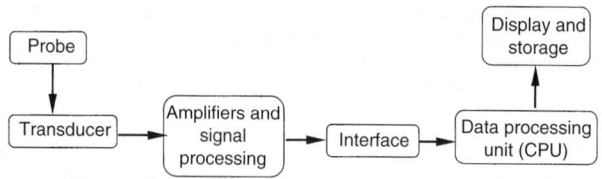

Fig. 20.7.1 Typical block diagram of a measurement system.

carriers and optical techniques, however, have the advantage of faster transmission and processing of information because light travels faster than electricity. In that respect the use of light and the development of optical converters in the future may provide an incentive to replace electrical converters of mechanical quantities.

In most optical techniques the use of flow tracers or flow markers is required. Flow tracers are solid particles that are introduced into the flow and scatter light when they are illuminated. These particles should have a dynamic response comparable to the dynamics of the flow; i.e., they should follow a temporally fluctuating flow very closely, and therefore their density has to match that of the fluid. This problem is more critical in gas flows than in liquid flows. In addition the presence of these particles should not interfere with the flow and change its structure. Flow markers, on the other hand, are molecules of the same fluid of the flow field or foreign molecules that emit light and can be detected when they are excited by photons. In that respect, inertia effects similar to those found in flow tracers are not present here because of the molecular nature of the flow markers.

Flow rate of information from one block to the next (see Fig. 20.7.1) can be hampered by the limited processing speed and storage capability in each of the blocks. This may have an effect on the spatial and/or temporal resolution of the measurement.

Experimental techniques used in EFD can provide qualitative and/or quantitative information about the flow field. Flow visualization techniques, for instance, although qualitative in nature can provide substantially useful information on the flow structure that could not be revealed even through detailed quantitative measurements.

Another way to classify experimental techniques is by considering whether they provide single-point measurement or multipoint measurements. The latter can be on a plane inside the flow field or in a volume that is part of the flow field. These techniques are usually referred as *global measurement techniques.*

GENERAL REQUIREMENTS

The first requirement of every experimental technique is to avoid interference with the flow itself because any disturbance introduced by the probe, the probe-volume or flow tracers needed in the particular technique can alter the flow field substantially. Consider, for instance, an intrusive technique which is using a probe holder to support the sensor. If the sensor is located far upstream of the probe holder and the whole system is designed so that all disturbances introduced into the flow field travel downstream, the flow at the sensor will not be affected. In separated flows, however, where flow reversals take place, disturbances propagate upstream and the use of intrusive measurement techniques is prohibited. In these cases, which are characterized by partial differential equations [see Eq. (20.7.4)] that are elliptical, the use of nonintrusive, mostly optical techniques is recommended.

The second requirement is that the technique should be sensitive to small spatial and temporal changes of the measured quantity in the flow field. This brings the issues of spatial and temporal resolution of the probe. In a steady state laminar flow, for instance, the probe size should be small enough to be able to resolve spatial gradients of the quantity inside the flow field. The case of measurement of a quantity Q_i that varies drastically in the x_j direction is shown in Fig. 20.7.2. If we assume that the size of the sensing area has a linear scale Δx_s, then the change of Q_i along the x_j direction will be $\Delta Q_i = (\partial Q_i/\partial x_j)\Delta x_s$, and since most sensors are not able to measure the variation of Q_i in the x_j direction, the sensor will measure an area-weighed average of $Q_i + \overline{\Delta Q_i}$. The question for the experimentalist is whether this uncertainty in ΔQ_i is acceptable or not. Further reducing the size of the sensing area Δx_s will reduce the error. However, this is not always attainable even if fabrication of small size sensors were possible, because electronic or optical interference noise may increase the uncertainty.

In case of temporal flow variations, the sensor should have adequate frequency response to resolve variations in the measured quantity within the smallest time scale acceptable. Usually separate tests with temporal

Fig. 20.7.2 Spatial resolution of sensing area.

step change in the measured quantity are needed to determine the frequency response of the sensor. For dynamic measurements of pressure, velocity, temperature, or density the shock tube is an ideal platform for frequency response tests.

Spatial and Temporal Resolution in Turbulent Flows

Turbulent flows are characterized by a large range of spatial and temporal scales and the sensors are not only required to have adequate **spatial and temporal response** and resolution but also good response to suddenly changing flow directions (yaws and pitch response).

The spatial and temporal resolution of any probe affects the accuracy of measurements of time-dependent velocity U_i, velocity gradients A_{ij}, or vorticity components Ω_k. This resolution depends on the arrangement of the sensors used in the technique, the hardware used to generate the measuring probe volume and to process the signals, the particles used as flow tracers to scatter light, and their arrival rate in the measuring locations. Thus the technique has a typical characteristic length scale L_p, which is usually associated with the separation distance between the sensing areas, and a time scale T_p, which depends mostly on the speed of data acquisition or burst processor. The technique should be able to resolve the small scales that exist in a given flow, which appear to be the Kolmogorov viscous scales $\eta = \nu^{3/4}\epsilon^{-1/4}$ for length and $T_\varepsilon = \nu^{1/2}\epsilon^{-1/2}$ for time, where ϵ is the dissipation rate of turbulent kinetic energy and ν is the kinematic viscosity. The quantity ϵ is the time-averaged value of the fluctuating part of E in Eqs. (20.7.11) and (20.7.13). L_p and T_p should be less than η and T_ϵ, respectively, in order to capture contributions from the smallest eddies of the flow field. In the case of vorticity or velocity gradients, the small scales are responsible for most of the content of vorticity or velocity gradient fluctuations, since small eddies are uncorrelated. Wyngaard[10] was the first to evaluate the problem of spatial resolution of **hot-wire** based vorticity techniques and found that no attenuation occurred for $L_p/\eta = 1$, while substantial attenuation of vorticity was evident for $L_p/\eta > 3.3$. The review by Wallace and Foss[11] concludes that reasonable estimates of velocity gradients can be obtained without attenuation with $L_p/\eta < 2$ to 4.

Although L_p has to be as small as possible, it cannot be minimized independently from other parameters such as the speckle noise. Errors due to noise in the measurements amplify as the length scale decreases. There is certainly an optimum length intermediate between attenuation and amplification of noise.

Determining η requires a good estimate of turbulent kinetic energy dissipation rate ϵ or the total dissipation rate E, which is very difficult to measure directly in a laboratory. Previous studies obtained the value of E by making assumptions to reduce some of the terms in the expression of the total dissipation rate:[9]

The estimates that usually are provided for E, particularly those based on indirect methods, may be very misleading. An experiment with poor spatial resolution will result in attenuation of all measured turbulent quantities, including the dissipation rate, which will indicate high η and therefore high resolution.

It should be mentioned here, however, that there is increasing evidence provided in Refs. 12 and 13 that the Taylor micro scale λ should be used to determine the spatial resolution in vorticity measurements. It has been shown that in the limit of vanishing separation Δx_j between the measurement locations, the root mean square (rms) of the fluctuations of one velocity gradient $\partial U_i / \partial x_j$ is given by

$$\lim_{\Delta x_j \to 0} \overline{\left[\frac{\Delta U_i}{\Delta x_j}\right]}^2 = \frac{\overline{u_i^2}}{\lambda_{ij}^2}\left[1 + \lambda_{ij}^2\left[\frac{\partial \rho}{\partial(\Delta x_j)}\right]^2\right]$$

where λ_{ij} is the Taylor micro scale. If the gradient $\partial \rho / \partial \Delta x_j$, which indicates the degree of variation of the rms of velocity fluctuations, is much smaller than $1/\lambda_{ij}^2$, then the relation

$$\lim_{\Delta x_j \to 0} \overline{\left[\frac{\Delta U_i}{\Delta x_j}\right]}^2 = \frac{\overline{u_i^2}}{\lambda_{ij}^2}$$

provides the best estimate for ideal resolution when Δx_j tends to zero. In that respect the ratio L_p/λ_{ij} should be less than 1.

An example of the effect of temporal resolution on the statistical results of vorticity flux $\partial \Omega_z / \partial y$, for instance, is obtained from the data of Andreopoulos and Agui.[82] Artificial sampling times Δt were produced by skipping by various numbers of equally sampled data digitized at rates substantially faster than the Kolmogorov frequency $1/T_\epsilon$. The results, plotted in Fig. 20.7.3, show a substantial dependence of the rms of vorticity flux on temporal resolution Δt.

Although statistics of fluctuations of individual velocity components do not require increased temporal resolution, cross-correlation/shear stress and velocity gradients/vorticity measurements by optical techniques using particles as flow tracers require a high degree of coincidence of these particles at different locations inside the probe volume. This can be achieved by providing a high rate of particles, which results in practically continuous signals. The resolution in time depends basically on the rate of arrivals of particles in the probe volume and on the data processing rate of the hardware.

As the Reynolds number of the flow increases, its spatial and temporal scales decrease, and therefore, if the size of the measuring volume and data acquisition rate remain unchanged, then it is expected that the spatial and temporal resolution of the measuring technique will deteriorate.

MEASUREMENTS OF PRESSURE AND TEMPERATURE

There is an extensive description of probes and instrumentation used to measure pressure in textbooks[1,14,15] or in handbooks.[16, 17] Manometers, gages, transducers and microphones are used to measure pressure at single points inside the flow field or at its boundaries.

Pressure-sensitive paint (PSP) can provide useful quantitative and qualitative information on the pressure field on a solid surface, i.e., at more than one point. In that respect, PSP has the advantage of being a field pressure measurement. The technique is based on the phenomena of photoluminescence and oxygen quenching of fluorescent dyes.[18–22]

In its simplest terms, the PSP method is based on the sensitivity of some luminescent dye to molecular oxygen. After absorbing a photon, a luminescent molecule is placed in an excited state. The molecule then returns to its ground state by emitting a new photon at a longer wavelength, as illustrated in Fig. 20.7.4. For some luminescent molecules, however, oxygen quenching allows the molecule to return to a ground state without emitting a photon. Hence, for a given level of excitation, the emitted light intensity from the luminescent molecules varies inversely with the local oxygen partial pressure. Since the mole fraction of oxygen in air is fixed, the oxygen partial pressure is easily converted into the air pressure.

PSP simply consists of luminescent molecules suspended in some type of oxygen permeable binder. Currently, most of these binders are some form of silicone polymer. The vast majority of PSP formulations to date come in a liquid form that is suitable for use with normal spray-painting equipment and methods. Typically, in its simplest application, PSP is the topmost layer of a multilayer coating on a model surface. The PSP is usually applied over a white undercoat, which provides two related benefits. The white undercoating reflects a large portion of the light that is incident upon it, which results in the benefits of amplifying not only the excitation illumination, but the emission illumination as well.

One of the characteristics of PSP is that the response of the luminescent molecules in the coating degrades with time of exposure to the excitation illumination. This degradation occurs because of a photochemical reaction that occurs when the molecules are excited. Eventually, this degradation of the molecules determines the useful life of the PSP coating. This characteristic becomes more important for larger models, as the cost and time of PSP reapplication becomes a significant factor. A second undesirable characteristic of PSP is that the emission intensity is affected by the local temperature. This behavior is due to the effect temperature has on the energy state of the luminescent

Fig. 20.7.3 Effect of temporal resolution on vorticity flux rms data from Andreopoulos and Agui[82,83] are made nondimensional by inner wall scaling.

Fig. 20.7.4 Excitation an illumination in a typical pressure-sensitive paint.

molecules, and the oxygen permeability of the binder. This temperature dependence becomes even more significant in compressible flow tests, where the recovery temperature over the model surface is not uniform.

For the measurement of temperature the classical techniques using thermocouples, resistive and solid-state thermometers, pyrometers, constant-current anemometers, liquid crystals, interferometers, and infrared imaging are described in Refs. 1 and 23, as well as in several textbooks.

Temperature-sensitive paint is used in a global measurement technique that is very similar to PSP except that the coating is temperature sensitive rather than pressure sensitive. In the case of TSP, changes in the free volume in the binder affect the luminescence; as the temperature increases, the luminescence decreases. As in PSP, a ratio of two images provides the measured quantity. In this case, one image is taken at a known, constant temperature, and the other is taken at the unknown conditions. The ratio of the two images is then converted to temperature by using the appropriate calibration. As in PSP testing, during the acquisition of the images, the test surface is exposed to excitation light in only a narrow frequency band. The detector is then notch-filtered to pass only the luminescence band for the particular probe molecule. Of course, these two bands must not overlap.

MEASUREMENTS OF VELOCITY

The measurement of flow velocity at single or multiple points may be accomplished by direct measurement of the speed of the flow or indirectly by a variety of techniques that can be in general classified as pressure-based methods using pitot or pitot static tubes, thermal anemometry, laser Doppler anemometry (LDA), **particle image velocimetry (PIV)**, planar **Doppler velocimetry (PDV)**, **molecular tagging velocimetry (MTV)**, which appear to be the most commonly used techniques. Pressure-based methods are adequately explained in several textbooks and they will be excluded from the present description.

Thermal Anemometry

Thermal anemometry has been used extensively for many years in fluid mechanics to measure instantaneous fluid velocity. The technique depends on the convective heat loss to the surrounding fluid from an electrically heated sensing element or probe. If only the fluid velocity varies, then the heat loss can be interpreted as a measure of that variable. There are several textbook on the subject[24–26] that cover comprehensively the associated techniques.

Cylindrical sensors (hot wires and hot films) are most commonly used to measure the fluid velocity, while flush sensors (hot films) are employed to measure the wall shear stress. Hot-wire sensors are, as the name implies, made from short lengths of resistance wire and are circular in cross section. **Hot-film** sensors consist of a thin layer of conducting material that has been deposited on a nonconducting substrate. Hot-film sensors may also be cylindrical but may also take other forms, such as those that are flush-mounted.

Thermal anemometry is very popular among researchers because the technique involves the use of very small probes that offer very high spatial resolution and excellent frequency response characteristics. The basic principles of the technique are relatively straightforward and the probes are relatively easy to fabricate. The sensors can be operated in the constant-temperature mode (CTA) or constant-current mode (CCA).

Hot-wire anemometry has been used for many years in the study of laminar, transitional, and turbulent boundary layer flows, and much of our current understanding of the physics of boundary layer transition has come solely from hot-wire measurements. Thermal anemometers are also ideally suited to the measurement of unsteady flows such as those that arise behind rotating blade rows when the flow is viewed in the stationary frame of reference. Until the advent of laser anemometry, this was the only available technique for the acquisition of data. In present times, thermal anemometry remains one of the primary tools used in experimental fluid mechanical work.

A hot-wire probe consists of the sensor itself, which is supported on prongs that are embedded in nonconducting (often ceramic) material.

Fig. 20.7.5   Basic hot-wire probe.

Figure 20.7.5 shows the main components of a basic hot-wire probe. The sensing element is heated by a current I under a voltage. A hot-wire sensor has to have several characteristics to make it a useful device:

• Its thermal capacity/inertia should be small so that it provides good frequency response.
• It should have a high temperature coefficient of resistance.
• Its electrical resistance should be such that it can be easily heated with an electrical current at practical voltage and current levels.

The most common wire materials are tungsten, platinum, and a platinum-iridium alloy. Tungsten wires are strong and have a high temperature coefficient of resistance (0.004/°C). However, they cannot be used at high temperatures in many gases because of poor oxidation resistance. Platinum has good oxidation resistance and temperature coefficient (0.003/°C), but is very weak, particularly at high temperatures. The platinum-iridium wire is a compromise between tungsten and platinum with good oxidation resistance, and more strength than platinum, but it has a low temperature coefficient of resistance (0.00085/°C). Tungsten is presently the more popular hot-wire material. The choice of diameter of the sensors, d, depends on the particular application. It can vary from 0.8 μm to 5 μm, while its length is between 0.7 mm and 1 mm. Since these elements are sensitive to heat transfer between the element and its environment, temperature and composition changes can also be sensed. Figure 20.7.6 shows a photograph of a hot-wire probe used for near-wall measurements.

The heat transfer rate from the sensor to the fluid can be, to a certain extent, simulated by considering a heated cylinder at uniform

Fig. 20.7.6   Photograph of boundary-layer-type hot-wire probe.

Fig. 20.7.7   Thermal energy balance in hot wires.

temperature T_w inside a steady flow field. Part of the electrically generated heat along the sensor will be transferred to the fluid by force and free convection, part of it will be transferred to the prongs by conduction, and the remaining will be radiated (see Fig. 20.7.7). For relatively long sensors with $l/d > 200$, heat rate by conduction Q_C can be ignored. For the usual hot-wire anemometry temperature range of a few hundred degrees Celsius, radiation heat transfer Q_R is less than 0.1 percent of the convective heat rate Q_C and may be disregarded. Free convection can be also ignored if $\mathrm{Re} > \mathrm{Gr}^{1/3}$, where Gr is the Grashof number, defined as $\mathrm{Gr} = g\beta(T_w - T)d^3/\nu^2$, where β is the thermal expansion coefficient of the fluid.

The temperature of the fluid in the immediate vicinity of the cylinder will have values between the fluid temperature T and T_w, and the fluid properties will vary accordingly. By convention, gas flow properties such as thermal conductivity, density, kinematic viscosity, and thermal diffusivity will be evaluated at the average film temperature defined as $T_f = \frac{1}{2}(T + T_w)$.

If Q_F is the heat flux convected from the wire to the fluid, the Nusselt number is defined as $\mathrm{Nu} = Q_F/\pi lk(T_w - T)$, with k being the thermal conductivity of the fluid. If one restricts the analysis to air only, the Prandtl number is constant, and then Nu depends on $\mathrm{Re} = \rho Ud/\mu$, $\mathrm{Nu} = \mathrm{Nu} \,(\mathrm{Re})$.

Under the above assumptions Nu appears to be

$$\mathrm{Nu} = (C + D\mathrm{Re}^n)(T_w + T_\infty)^{0.17} \qquad (20.7.14)$$

where C, D, and n are constants. The heat flux rate Q_F is provided by the electrical heating and therefore is $Q_F = IE = I^2/R = E^2/R$, where I and E are the current through and the voltage across the sensor, respectively, and R is the resistance of the wire. The technique then can provide $Q_F = f(E)$ under constant I for the CCA-mode operation or $Q_F = f(R)$ under constant resistance/temperature for is the CTA-mode operation.

In the CTA mode the above relation becomes

$$E^2 \frac{R_w}{(R_w + R_a)^2 (T_w - T)} = X + YU_{\mathrm{eff}}^n \qquad (20.7.15)$$

where U_{eff} is the effective cooling velocity of the fluid. This relation is known as King's law.[27] If we ignore the variation of fluid properties and since T_w is constant, then the partial sensitivities of voltage on velocity and fluid temperature are

$$\frac{\partial E}{\partial U} = \frac{nYU_{\mathrm{eff}}^{n-1}}{2} \left[\frac{(R_w + R_a)^2}{R_w} \frac{T_w - T}{X + YU_{\mathrm{eff}}^n} \right]^{1/2} \qquad (20.7.16)$$

$$\frac{\partial E}{\partial T} = -\frac{1}{2} \left[\frac{(R_w + R_a)^2}{R_w} \frac{X + YU_{\mathrm{eff}}^n}{T_w - T} \right]^{1/2} \qquad (20.7.17)$$

Then $\Delta E = (\partial E/\partial U)\Delta U + (\partial E/\partial T)\Delta T$, and, if the objective is to measure velocity and velocity fluctuations, the partial sensitivity $\partial E/\partial U$ has

to be maximized and the partial derivative $\partial E/\partial T$ has to be minimized in comparison or completely eliminated. In order to achieve this, the operating wire temperature T_w has to be large, since the term $(T_w - T)$ is the nominator of $\partial E/\partial U$ and in the denominator of $\partial E/\partial T$. For accurate measurements an overheat ratio $(R_w - R_a)/R_a = \alpha(T_w - T_a)$ of 0.80 is recommended, where R_a is the cold resistance and α is the coefficient temperature.

Frequency Response   The response of a hot-wire sensor of mass m_w and specific heat c_w is governed by the first-order differential equation given in Ref. 28 as

$$m_w c_w \frac{\partial T_w}{\partial t} = I^2 R_a [1 + \alpha(T_w - T_a)] - \pi k(T_w - T)\mathrm{Nu} \qquad (20.7.18)$$

which is the unsteady heat equation with the first term on the left-hand side representing the accumulated heat rate by the wire mass. The solution of this equation contains the exponential term $e^{-t/M}$, where M is the time constant

$$M = \frac{m_w c_w}{I^2 R_a \alpha - \pi kl\mathrm{Nu}} = \frac{\rho_w c_w d^2}{k\mathrm{Nu}} \frac{R_w}{R_a} \qquad (20.7.19)$$

In order to minimize M, small values of wire diameter d are used; in fact, wire size is the only control a user can have. However, the mechanical strength of the wire under aerodynamic load limits the reduction in d. Reductions in M, at the expense of sensitivity, can be also obtained by reducing the ratio R_w/R_a.

Angular Sensitivity   Unfortunately the output of a hot-wire anemometer depends on the yaw and pitch angles of the incoming velocity vector. This angular response of the probe can be found through appropriate yaw and pitch angle calibration, which is usually carried out at constant free-stream velocity. This angular sensitivity is modeled in the expression of the sensor cooling velocity U_{eff} in various ways by decomposing the local velocity vector into the wire coordinates (see Fig. 20.7.8):

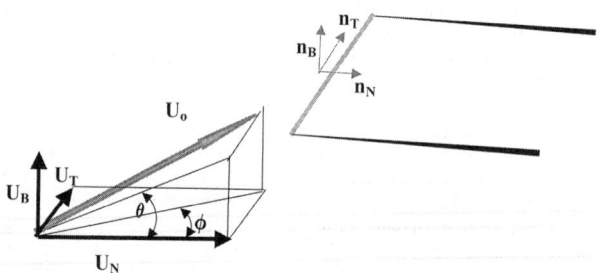

Fig. 20.7.8   Components of velocity vector in wire coordinates.

1. Cosine law with $U_{\mathrm{eff}} = U_o \cos \phi$ (see Ref. 28). This is the simplest of the expressions, and it is valid for small yaw angle ϕ.

2. According to Champagne, Schleicher, and Weherman,[29] U_{eff} includes a fraction of the tangential component of the local velocity vector U_T, and the expression becomes $U_{\mathrm{eff}} = (U_N^2 + k^2 U_T^2)^{1/2}$ where k is a coefficient with values $k = 0.2$ for $l/d = 200$ and $k = 0$ for $ld = 600$.

3. A more general expression for U_{eff}, due to Jörgersen,[30] includes cooling in the binormal direction from the component U_B: $U_{\mathrm{eff}} = (U_N^2 + h^2 U_B^2 + k^2 U_T^2)^{1/2}$ where h is another coefficient with values of the order of 1. Apparently this is the most complete expression for U_{eff} and it can be applied in principle to any three-dimensional flow field with reasonable turbulence intensities.

The coefficients k and h can be determined through individual calibration wire and may include the unavoidable effects of prong and probe holder interference.

Multisensor Probes A single-sensor probe oriented perpendicularly to the flow direction can yield information of one velocity component only, under the assumption that the lateral components are considerably smaller than the longitudinal one. If measurements of two velocity components are required, then a probe with two sensors independently oriented at ±45° each to the mean direction should be used. Triple wires consisting of three sensors are recommended for use in three-dimensional flows.

Laser Doppler Anemometry

The laser Doppler anemometry (LDA) technique, which is also known as laser Doppler velocimetry (LDV), is based on light scattering off flow tracers that exist in the flow or are introduced by the user. The flow field is illuminated by a laser beam and the particle tracer that follows the flow produces a Doppler frequency shift due to its velocity.

This is a nonintrusive optical technique to measure the **Doppler frequency shift** contained in scattered light. The idea of measuring the Doppler shift due to the motion of particles inside a flow was originated by Cummins et al.[31] after the discovery of narrow line width lasers. Cummins, a professor of physics at City College, demonstrated in Yeh and Cummins[32] that the Doppler frequency shift Δv is related to the velocity vector of the light scattering particle $\mathbf{V}$ through

$$\Delta v = v_s - v_i = 1/\lambda (\mathbf{k}_s - \mathbf{k}_i) \cdot \mathbf{V} \qquad (20.7.20)$$

where $\mathbf{k}_s$ and $\mathbf{k}_i$ are the unit vectors in the scattered and incident light directions and λ is the wavelength of the incident light. Because of the inability of existing photodetectors to measure directly the extremely high frequencies of the original laser beam, several arrangements have been used to offset this problem. The most common arrangement is shown in Fig.(20.7.9). The original laser beam is split into two beams that intersect in an interference pattern of standing waves consisting of

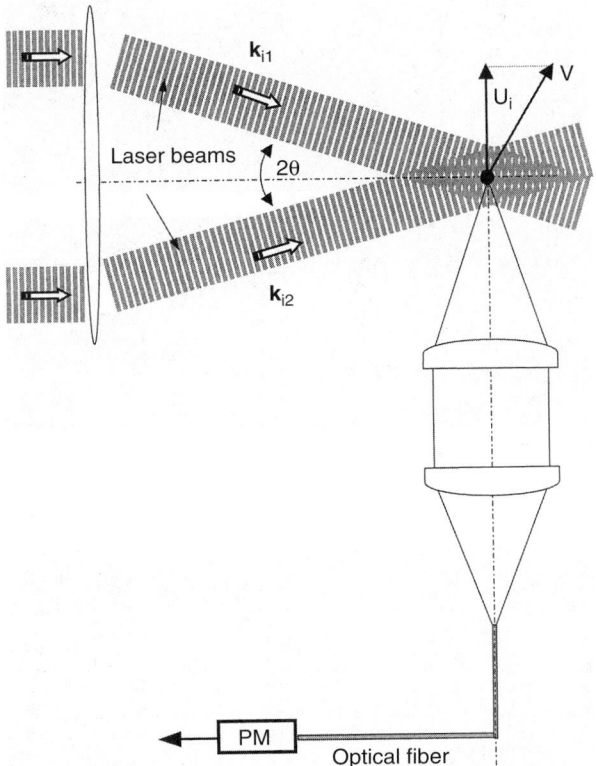

Fig. 20.7.9 Typical laser doppler anemometry arrangement.

high- and low-intensity surfaces called *fringes*. As micrometre-sized liquid or solid particles entrained in a fluid pass through the intersection of the two laser beams, the scattered light received from the particles fluctuates in intensity. The laser frequency of the two beams cancels out and the Doppler frequency is independent of the scattered light direction $\mathbf{k}_s$, which is the observation angle:

$$v_D = v_{s1} - v_{s2} = 1/\lambda (\mathbf{k}_{i1} - \mathbf{k}_{i2}) \cdot \mathbf{V} = 2 \sin\theta \, U_i/\lambda \qquad (20.7.21)$$

This last feature is a great advantage of this arrangement, since it provides some flexibility in the placement of the frequency detector, which is usually a photomultiplier that converts the light intensity into photocurrent. Then Eq. (20.7.21) is transformed into the simple form

$$U_i = v_D \lambda/2 \sin\theta \qquad (20.7.22)$$

Another advantage of linear equation (20.7.22) is that it contains no undetermined constants, thus eliminating the need for calibration and errors due to instrumentation drifting. The proportionality coefficient $\lambda/(2 \sin\theta)$ is the calibration factor, although no calibration is actually necessary (see Refs. 33, 34, and 35 for more details).

The velocity direction can be fixed if one of the laser beams has a frequency slightly different from that of the other. All three components of velocity can be measured by crossing four or six laser beams of different wavelenths or polarization in the same measuring volume, and separating out the scattered light with filters. The frequency is measured by digital counters (if data rates are above 50 per second) or photon correlators (low data rates and signal level) or spectral analyzers (Doppler shift frequency not over a few megahertz).

A laser power source is required, with excellent frequency stability, narrow linewidth, small beam diameter, and a gaussian beam intensity profile (bright at the center). Typically, helium, neon, or argon ion lasers are used, with power levels from 10 mW to 20 W.

LDA techniques have provided spatially precise point measurements, at speeds from millimetres per second to supersonic. In some cases, measurement point distances of up to 7 m have been used in large wind tunnels.

The limitation of this technique is that it provides single-point measurement only. Data rates are dictated by particle arrival in the measuring volume, not when the user wants to sample data. Particles should follow the flow closely. However, particles cannot enter all flows through turbulent diffusion, and simple viscous diffusion is not adequate to provide particles inside the laminar cores of vortices, for instance. Particles should be small enough to follow the flow, but reducing their size may produce a signal below the noise threshold. There is also greater noise at higher speeds. The signal level depends on detector solid angle, posing problems of optical access. Mie scattering intensity is much better in the forward direction, but it is difficult to set up forward receiving optics which remains aligned to the moving measurement volume. Time to alignment of emitting optics can range from 3 hours for a one-component system to a day for a three-component system. The setup may be also sensitive to vibrations. These problems may be alleviated with fiber optic probes, at the cost of signal level because of fiber power loss.

LDA techniques have been used in a variety of flow applications that include low- and high-speed boundary layers, combustion studies, and turbomachinery and rotorcraft flows. They have been also used in atmospheric turbulence measurement, and in air data systems for aircraft. In periodic flows, phase-resolved averaging is used to construct 3-D velocity fields with high accuracy.

Particle Image Velocimetry

Particle image velocimetry (PIV) is a whole-flow-field, nonintrusive optical technique that provides instantaneous velocity vector measurements in a cross-section of a flow. Two velocity components are usually measured on the illumination plane. The use of a stereoscopic arrangement provides all three velocity components to be recorded, resulting in instantaneous 3-D velocity vectors for the whole area. The use of modern CCD cameras and dedicated computing hardware results in real-time velocity maps. In PIV, the velocity vectors are derived from subsections of the target area of the particle-seeded flow by measuring the movement of particles between two light pulses. The technique is nonintrusive and measures the velocities of micrometre-sized particles following the flow. It has been tested in flows with velocity range from zero to supersonic.

The flow is illuminated by a laser sheet in the target area. A camera lens images the target area onto the CCD array of a digital camera. Two sequential images of the flow field are captured by the CCD array, produced by two light pulses in separate image frames. Once a sequence of two light pulses is recorded, the images are divided into small subsections called *interrogation areas*. These interrogation areas from each image frame are cross-correlated with each other, pixel by pixel. The correlation produces a signal peak, identifying the common particle displacement.[36–40] In PIV, the velocity vectors are derived from subsections of the target area of the particle-seeded flow by measuring the movement of particles between two light pulses, $\overline{V} = \overline{\Delta x_i}/\Delta t$.

A typical experimental setup of a PIV system is shown in Fig. 20.7.10. This system has been used to measure the velocity in the near field of a compressible jet at CCNY (see Wang[41]). Talcum flow tracer particles of $1\,\mu\text{m}$ in diameter have been added to the flow. A plane (light sheet), about 1 mm thick, within the flow is illuminated by means of a dual laser system (the time delay between pulses depending on the mean

Fig. 20.7.10   Experimental setup for particle image velocimetry measurements in a compressible jet flow facility.

flow velocity and the magnification at imaging). It is assumed that the tracer particles move with local flow velocity between the two illuminations. The light scattered by the tracer particles is recorded via a high-quality lens either on a single frame or on two separate frames on a special cross-correlation CCD sensor. It is assumed that all particles within one interrogation area have moved homogeneously between the two illuminations. The projection of the vector of the local flow velocity into the plane of the light sheet (two-component velocity vector) is calculated, taking into account the time delay between the two illuminations and the magnification at imaging.

The process of interrogation is repeated for all interrogation areas of the PIV recording. It was possible to capture 100 PIV dual frames per minute with the present 1k × 1k PIVCAM 10-30 cameras. The duration

of processing of one PIV recording with 3600 instantaneous velocity vectors (depending on the size of the recording and of the interrogation area) is of the order of a few seconds with present PCs.

Prior to the detailed velocity measurement in the jet flow facility, the overall structure of the flow was captured by the high-resolution PIVCAM 10-30 camera. This is shown in Fig. 20.7.11 for the case of a nitrogen jet issuing in ambient still air. On the left side of this figure, the 7-mm jet exit nozzle is shown. The flow as it exits the nozzle starts to mix with the ambient air and decelerates. A few diameters downstream of the jet exit, the flow starts to spread out.

The two images shown in Fig. 20.7.12 correspond to the flow described above with the Mach 0.6 nitrogen jet. If these two images were overlapped, the same particle displacement could be revealed. Cross correlation of the two frames provided the velocity vectors shown in Fig. 20.7.13.

Fig. 20.7.11   Flow structure of a nitrogen jet at Mach 0.6.

Fig. 20.7.12   Cross-correlation images for nitrogen jet at Mach 0.6.

Fig. 20.7.13   Resultant velocity for nitrogen jet at Mach 0.6.

Planar Doppler Velocimetry Techniques

Planar Doppler velocimetry (PDV), also known as global Doppler velocimetry (GDV), is a relatively new nonintrusive planar optical technique that utilizes molecular filters to measure the Doppler frequency shift contained in scattered light. As mentioned earlier, the idea of measuring the Doppler shift due to the motion of particles inside a flow was originated by Cummins et al.[31] after the discovery of narrow linewidth lasers. The Doppler frequency shift Δv is related to the velocity vector of the light-scattering particle, $\mathbf{V}$ (see[32]) through

$$\Delta v = v_s - v_i = 1/\lambda(\mathbf{k}_s - \mathbf{k}_i) \cdot \mathbf{V}$$

where $\mathbf{k}_s$ and $\mathbf{k}_i$ are the unit vectors in the scattered and incident light directions and λ is the wavelength of the incident light.

A laser sheet is formed inside the flow field and global measurements can be obtained by recording the light intensity, which is related to the Doppler frequency shift. The measurement of the frequency shift is made spectroscopically by using a high-resolution iodine gas cell, which acts as a frequency-dependent filter with very high sensitivity. The edge of an absorption line in molecular iodine is used to determine the frequency of the Doppler shift of the scattered light. Image data are acquired via digital cameras, which have to be placed appropriately for each axis of interest. The intensity of the light from a given point in the object plane is attenuated in proportion to its Doppler shift. The image plane then contains gray levels proportional to the product of the local intensity of the illuminating light sheet and the Doppler shift. To eliminate the problem of both scattering signal and illumination intensity variations spatially in the measurement window, the Doppler signal intensity map is normalized by a reference intensity map from the same view of the flow. The camera is split into two halves to record the unshifted image field, and the Doppler-shifted image is normalized by the unshifted image, pixel by pixel.

The technique was first described in a U.S. patent by Komine[42] and by Komine et al.[43] and Meyer and Komine,[44] who used an iodine filter. At about the same time Miles and Lempert[45] and Miles et al.[46] used an I_2 molecular filter with Rayleigh scattering to demonstrate the feasibility of measuring mean velocity, density, and temperature in high-speed flows. A good review article on the subject can be found in Samimy and Wernet.[47] Further information can be found in Refs. 48 to 55.

Figure 20.7.14 shows a typical scan of a typical transfer function of the I_2 vapor filter. The laser is tuned at the middle point of the range of the normalized light intensity. Any scattered light of flow tracers inside the flow field will cause a Doppler frequency shift depending on the

Fig. 20.7.15 Experimental setup for one-component micro PDV measurements inside a shock wave in the large-scale shock tube facility at CCNY.

rectangular grids. The major hardware components are a Nd:YAG laser, a CCD camera, and an absorption filter. Presently a Q-switched Quanta Ray GRC-14s Nd:YAG laser with 250-mJ pulse energy is available in the laboratory. This laser is equipped with a LightWave injection seeder. A Princeton Instruments (Roper Scientific) back-illuminated camera, model MicroMax 1024B with 1k × 1k pixels and an ultraviolet ASROMED camera with a chip of 540 × 650 pixels are used for imaging. An I_2 absorption filter provided by Innovative Scientific Solutions Inc. has a typical absorption spectrum shown in Fig. 20.7.16. These hardware pieces are adequate to provide measurements of one-velocity components only. Measurements of three velocity components by using PDV have been already made successfully in supersonic jets by combining the pair of filtered and unfiltered images on the same CCD chip.[47]

Implementation of the PDV techniques to the present flow in the shock-tunnel environment is not straightforward. Particle seeding is one of the major issues in flow measurements based on Doppler frequency shift. In the past, acetone has been used to provide light scatterers in flows

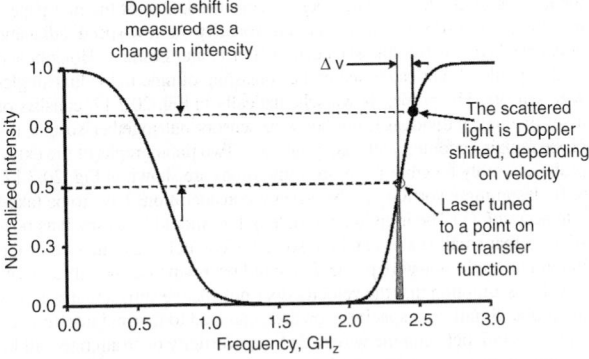

Fig. 20.7.14 Typical transfer function of I_2 vapor filter.

velocity magnitude which corresponds to a normalized light intensity. This spectroscopic transfer function is obtained in advance and it is typical of the filter used and the tuned frequency of the laser.

Figure 20.7.15 shows a typical arrangement currently used at CCNY to measure the velocity field within a shock wave interacting with an isotropic and homogeneous turbulence generated by metallic

Fig. 20.7.16 Absorption spectrum of I_2 vapor filter.

configured in high-speed wind tunnels where temperatures are rather low. In the present flow case, the use of acetone is not certain because the temperature in the shock tunnel is higher than the flow temperature in supersonic wind tunnels. Submicrometre talcum powder particles have been used successfully in our work in the past. The scales of this flow can be adequately resolved with a magnification of about 200, or less. The pixel-to-pixel distance is 9 μm, and with this magnification the flow scales resolution is about 45 nm. For a shock thickness of about 3λ, where λ is the molecular mean free path, and helium as a driven gas at one-fifth atmospheric pressure with $\lambda = 3.75\,\mu$m, the shock thickness can be resolved by 90 pixels. Talcum powder with particle size less than 0.2 μm is used for light scattering.

Unfortunately the pulsed **Nd:YAG lasers** are known for long-term frequency drifts and pulse-to pulse fluctuations in frequency and power. The variation in the power can be remedied by normalization of the filtered signal with the unfiltered one. Frequency variations, however, require a frequency monitoring system. The system described by Mosedale et al.[56] is a good remedy to this problem. This system requires a small cut-out portion of the laser beam to focus on two fast-response photo detectors after it is split in two and after one of the two parts passes through a second I_2 filter. This system can be also used to characterize the frequency-transmission relation of each filter.

Currently, however, PDV is considered to be more suitable for the measurement of high-speed flows (estimated to be above 30 m/s) because of the limit on velocity resolution, which is approximately equal to 1 m/s, which is the uncertainty of a very good DGV system. It is one of the very few Mie-scattering whole-field velocimetry techniques that could be truly regarded as instantaneous with the use of a short-duration (~10 ns) pulsed laser.

Other Techniques

In molecular tagging velocimetry (MTV), the flowing medium is premixed with molecules that can be turned into long-lifetime flow markers upon excitation by photons. Typically a pulsed laser is used to "tag" small regions of interest. The tagged regions are then imaged at two successive instants within the lifetime of the tracer. The measured displacement vectors provide estimates of the average velocity for each small region during the brief interval between exposures. The fluid is water in this case, premixed with a water-soluble phosphorescent triplex compound. The corresponding velocity field is determined by a spatial correlation technique.[57, 58] One advantage of the MTV technique over other commonly used whole-field velocimetry techniques that rely on light scattering from particles is its molecular nature. In that respect, the issues related to flow tracers that are particularly important in gas flows are eliminated. The molecular nature of MTV also makes it suitable for obtaining other flow variables, such as pressure, temperature, and concentration, along with the velocity field.

There are several "unseeded" tagging methods where gas tags are produced from molecules naturally occurring in air (i.e., nitrogen, oxygen, water vapor). Examples of molecular tags produced from air are N_2 + ion,[59] ozone,[60] hydroxyl,[61] nitric oxide,[62,63] and vibrationally excited oxygen.[64, 65] Many of the methods use nonlinear laser excitation[59, 62-65] to produce these tags from air and the tag is produced only in a small region near the laser focus.

MEASUREMENTS OF VORTICITY, RATE OF STRAIN, AND DISSIPATION

The measurement of one or more components of vorticity in turbulent flows of research and technical interest has been a long-held, but elusive, goal of fluid mechanics instrument developers and researchers. The review article by Wallace and Foss[11] provides a detailed account of the available techniques, as of that date, used to measure vorticity and compare existing data. A majority of the techniques are based on thermal anemometry or on optical methods. Optical techniques are nonintrusive, are sensitive to large changes in velocity direction, and most of them do not require any calibration. These characteristics make them ideal for use in separated flows and flows with large turbulence intensities where the use of intrusive methods is prohibited and recirculating

effects are present. Most of the optical techniques are based on detecting light scattered by particles imbedded in the flow. In that respect, seeding the flow with appropriate particles that follow the fluid motion very closely is critical. It should be noted here that, for gas flows in which seed must be added and in which the flow does not recirculate, adding particles can be far more intrusive than a hot-wire probe.

There is a need for a formal definition of the measured quantities, either vorticity components Ω_k or velocity gradient tensor terms A_{ij}. Specifically, there are two common ways to derive vorticity. In the first method, measurements of velocity components at nearby locations are obtained and used in a finite-difference scheme to derive vorticity. In the second method measurements of microcirculation around a small area are obtained and vorticity is computed as circulation density.

The first definition is based on a Taylor's series expansion of velocity in a nearby location, which to a first-order approximation becomes $U_i(r_j + \Delta r_j) = U_i(r_j) + \Delta r_j(\partial U_i/\partial r_j) + \cdots$.

For r_j in the y direction and $\Delta r_j = (0, \Delta y, 0)$, the above equation becomes $U_i(y + \Delta y) \approx U_i(y) + \Delta y(\partial U_i/\partial y)$, which yields the following relation to compute the velocity gradient:

$$\frac{\partial U_i}{\partial y} \approx \frac{U_i(y + \Delta y) - U_i(y)}{\Delta y}$$

This relation will also define the uncertainties in the measurement of the velocity gradient.

The second method of estimating vorticity is due to Foss and his coworkers,[66-68] who have developed a hot-wire technique for the measurement of the transverse vorticity component. The technique involves the measurement of microcirculation Γ around a small finite domain of area ΔS normal to the flow. The estimated vorticity appears to be computed through the relation $< \Omega_z > = \Gamma/\Delta S$.

These two methods are not the only ones that are used. There is no need to have a computational formalism for vorticity when it is estimated explicitly. For instance, direct estimate of vorticity through measurements of fluid rotation does not require vorticity computation through an implicit scheme.[69]

The inherent difficulties of velocity gradient/vorticity measurements, which require high spatial and temporal resolution in turbulent flow fields, have only recently been substantially overcome.

Hot-Wire-Based Vorticity Probes

Several multi-hot-wire probes have been developed that are capable of measuring velocity-gradient-related quantities. The difference among them is the orientation and number of wires used. One of the first papers to provide well-documented data on vorticity in low-speed adiabatic boundary layers is by Vukoslavcevic, Wallace, and Balint.[70] Honkan and Andreopoulos[71] fabricated one probe consisting of nine individual single-wire sensors. The probe, shown schematically in Fig. 20.7.17, consists of a set of three individual triple-hot-wire sensors put together so that the probe remains geometrically axisymmetric. Two photographs of the entire probe assembly together the probe dimensions are shown in Fig. 20.7.17.

In designing a vorticity probe, several considerations have to be taken into account. (1) The individual wire length l_w should be as small as possible so that small scales can be resolved adequately. (2) The size of the individual triple-hot-wire probe, l_p, should be as small as possible to satisfy the assumption that the velocity does not change substantially across the probe. Small wire spacing, however, can lead to thermal interference and cross talk between the wires. (3) Since vorticity or strain rate will be computed from velocity gradients, spacing of the individual probes should be finite so that velocity gradients do not "disappear." If this spacing becomes small, the effect of noise may overwhelm the signal.[11] (4) Each of the wires should be controlled independently of the others. The use of one common prong as in the case of Ref. 70 may create problems such as imbalance under dynamic conditions. (5) The transfer function of the hot-wire anemometer system should be a three-dimensional one. This suggests that the probe should be able to respond to yaw and pitch angle variations of the velocity vector in addition to its magnitude.

Detailed lookup tables were created through velocity, pitch, and yaw calibrations, which are subsequently used to reduce the acquired data.

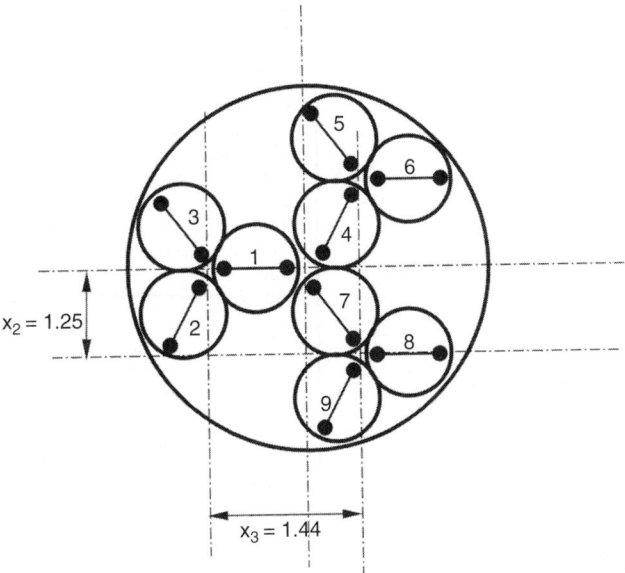

$x_2 = 1.25$

$x_3 = 1.44$

Fig. 20.7.17 Sensor arrangement and photographs of the probe (dimensions in millimetres).

A new vorticity probe designed for use in nonisothermal flows or in compressible flows is described by Agui et al.[72] and Briassulis et al.[73] The probe consists of 12 wires, which is a modification of the original design with nine wires by Honkan and Andreopoulos.[71] The three additional wires were operated in the so-called constant-current mode and used to measure time-dependent total temperature.

Since the probe essentially consists of a set of three modules or arrays (see Fig. 20 7.18), it is necessary to provide several key features of the individual hot-wire modules. Each module contains three hot wires operated in the constant-temperature mode (CTM) and one cold-wire sensor operated in the constant-current mode (CCM). Each wire of

the triple-wire submodule is mutually orthogonal to the others, thus oriented at 54.7° to the probe axis. Each of the 5-μm-diameter tungsten sensors is welded on two individual prongs, which have been tapered at the tips. Each sensor is operated independently, since no common prongs are used. Each of the 2.5-μm-diameter cold wires is located on the outer part of the submodule.

The hot-wire output voltage Ei of the ith sensor is related to the effective cooling velocity, $U_{i,\text{eff}}$ through Eq. (20.7.15), known as King's law[27]:

$$\frac{E_i^2}{T_w - T_0} = A_i \left[\frac{T_0}{T_r} \right]^a + B_i \left[\frac{T_0}{T_r} \right]^b (\rho U_{i,\text{eff}})^n$$

(a)

(b)

Fig. 20.7.18 Vorticity probe: (*a*) probe sensor geometry and arrays; (*b*) close-up view of the probe.

where T_w is the hot-wire temperature, T_0 is the total temperature of the flow, and T_r is a reference temperature, the ambient temperature in the present case. The values of the exponents a and b were taken as suggested by Kovasznay[77]: $a = b = 0.768$. The effective velocity $U_{i,\text{eff}}$ is related to the normal, tangential, and binormal components of the velocity vector in reference to the ith sensor, respectively, by

$$U_{\text{eff}} = (U_N^2 + h^2 U_B^2 + k^2 U_T^2)^{1/2}$$

where k and h are coefficients that, for a given probe, depend on the yaw and pitch angle of the velocity vector. Details of the techniques associated with the use of triple wire arrays can be found in Ref. 74, while estimates of errors related to probe geometry and turbulence intensity are described in Ref. 75.

Mass fluxes ρU_i can be further separated into density and velocity by using the method adopted in Ref. 76. Decoupling density from mass fluxes assumes that static pressure fluctuations are small. This is the so-called weak version of the original "strong Reynolds analogy" hypothesis of Morkovin.[78] The original hypothesis is based on the assumption that pressure and total temperature fluctuations are very small. In the present work, total temperature was measured directly and therefore no corresponding assumptions were needed. The pressure, however, was measured at the wall and not at the location of the hot-wire measurement. The mean value of this pressure signal was used to separate the density and velocity signals since no mean pressure variation has been detected across a given section of the flow. The procedure involves an expression for mass flux m_i in terms of total temperature T_0 and pressure p at the centroid of each array: $m_i = \rho U_i = pU_i/RT = pU_i/[R(T_0 - U_k U_k/2c_p)]$, where U_i is the instantaneous velocity component, $i = 1, 2,$ or 3 and $U_k U_k = U_1^2 + U_2^2 + U_3^2$. The velocity can be decomposed into $U_i = \bar{U}_i + u_i$. An iterative scheme can be used to decouple density and velocity. During the first iteration it was assumed that the quantity $(u_2^2 + u_3^2)/2c_p$, where u_2 and u_3 are the velocity components in the spanwise and normal directions, respectively, is substantially smaller that the quantity $T_0 - U_1^2/2c_p$. Then the above relation can be rearranged to obtain a quadratic equation for U_i, $(Rm_i/2c_p)U_i^2 + pU_i - m_i RT_0 = 0$.

Optical Techniques

Laser speckle strophometry (LSS) or particle image velocimetry (PIV) techniques are nonintrusive but cannot in general provide adequate time-history information. In addition a slight out-of-plane motion of scattering particles due to flow three dimensionality results in different speckle patterns (speckle decorrelation).

Laser speckle strophometric methods have been developed.[79–81] In these techniques, the speckle pattern, which is the result of the diffraction of a plane wave produced by random grating, is basically affected by the flow velocity gradient while the translation of the grating caused by the constant flow velocity will not affect the pattern.

These authors applied their double pulse strophometric technique to obtain direct measurements of the quantities $\partial U/\partial x + \partial U/\partial y$, $\partial U/\partial x - \partial U/\partial y$, and $\partial U/\partial z$ in a channel flow with Reynolds number based on the channel width in the range of 2,250 to 6,700. This information was used to obtain statistics of the $\partial U/\partial y$ gradient.

One of the first direct measurements of vorticity were obtained by measuring the solid rotations of tiny transparent spherical beads of 25 μm in diameter,[69] each containing embedded planar crystal mirrors specially developed for this purpose, which were suspended in a refractive index-matched liquid.

LDA-Based Vorticity Techniques   The techniques used to obtain indirect measurements of vorticity take advantage of the Doppler effect and they have configurations similar to those used in laser Doppler anemometry (LDA). In typical dual-beam configurations for velocity measurements, two laser beams cross each other to form the measuring/probe volume with the fringe pattern. Receiving optics can be aligned in any direction since the Doppler frequency shift Δf_D is independent of the k_s direction. The velocity component normal to the fringe orientation $U_i(x_j)$ can be obtained by measuring the Δf_D as

$$\Delta f_D = \frac{\vec{\text{V}}}{2\pi} \cdot \left[\vec{k}_{01} - \vec{k}_{02} \right]$$

where $\vec{k}_{01}$ and $\vec{k}_{02}$ are the wave number vectors along the two incident beams forming the measurement volume.

A natural extension of the traditional single-point LDA measurement of one velocity component shown in Fig. 20.7.9 to obtain one velocity gradient is possible by forming two independent probe volumes at a distance Δx_j apart to measure two velocity components in the same direction, $U_i(x_j)$ and $U_i(x_j + \Delta x_j)$. The velocity gradient $\partial U_i/\partial x_j$ can be estimated through the finite difference relation

$$\frac{\partial U_i}{\partial x_j} \approx \frac{\Delta U_i}{\Delta x_j} \approx \frac{U_i(x_j + \Delta x_j) - U_i(x_j)}{\Delta x_j}$$

For one component vorticity measurements, simultaneous measurements of the two velocity gradients are required.

The LAVOR method described by Agui and Andreopoulos[82,83] uses two parallel laser beams focused with a long-focal-length lens to generate a moderately narrow probe volume defined by the region of intersection of the two beams. The velocity at two different locations inside the probe volume has been measured. There are several advantages to this approach as compared to the "sheet" beam approach. These include fewer optical components, simple alignment of the optics, smaller windows for optical access, and better use of laser intensity.

A second category of single-point velocity gradient/vorticity measurement techniques exists that are not based on the interference of two laser beams and the scattering light of a particle moving through the fringes that is realized on the surface of the photodetector. This category includes techniques that use laser-Doppler signals from two particles illuminated by two different light beams. The idea of heterodyning of the scattered light from two particles originally proposed by Durst, Melling, and Whitelaw[33] was explored by Hanson[85] and more recently by Ötügen, Su, and Papadopoulos[86] and Yao et al.[87] Hanson's heterodyning arrangement uses an optical filter to define the relative separation of two particles, which requires fabrication of miniature spatial gratings and rings with geometries tailored to the specific flows under investigation. His technique, which has been tested in a cylindrical laminar Couette flow, is insensitive to the sign of velocity gradient. The technique in Ref. 86 resolves the sign of the velocity gradient by using a Bragg cell to shift the frequency of one laser beam, and it has been tested in the laminar flow formed between two concentric cylinders with one of them rotating. The technique of Yao et al.,[87] which is quite similar to that of Ötügen et al., has been tested successfully in a laminar channel flow.

In all these techniques that estimate the velocity gradient through the heterodyning of the scattered light from two different particles, the velocity field cannot be measured simultaneously with the velocity gradient; it requires separate measurements. In that respect, the inability of these methods to provide information of the velocity field at the same time with the velocity gradient measurements may be considered a drawback.

PIV-Based Vorticity Techniques   In particle image velocimetry, crossing of the flowing particle paths in highly fluctuating flows may cause ambiguity in the detection scheme. PIV is also limited by spatial resolution loss due to averaging from the correlation schemes. Although the resolution of a typical CCD camera may be typically 1k × 1k pixels, subregions of 32 × 32 pixels known as *interrogation zones* are usually considered to validate a local velocity measurement. According to Arik,[88] in a typical PIV setup approximately 10 particles are needed to validate a velocity vector measurement and 10 more velocity vectors are needed to evaluate a vorticity vector. Thus approximately 100 particles are needed to evaluate a vorticity vector, a number that results in 30 to 40 particles for each of the three vorticity components.

Lourenco et al.[89] used PIV techniques to obtain one-component vorticity measurements in a jet flow. Meinhart and Adrian[90] used PIV to measure the near wall vortical structure of low-speed boundary layers with reasonable spatial resolution.

Holographic particle image velocimetry (HPIV) was used by Zhang et al.[91] to obtain velocity gradient measurements in a square duct flow with moderate resolution.

A polarization-based approach with dual-plane particle image velocimetry (DSPIV) has been used by Kähler,[92] Ganapathisubramani et al.[93] and Hu et al.[94] to obtain vorticity measurements in boundary layer flows and in lobed jet mixers. A two-frequency DSPIV was applied in jet flows to measure all nine components of the velocity gradient tensor with resolution of about 6 Kolmogorov length scales.[95] This recent work, which is based on PIV techniques, have demonstrated the potential to provide measurements of vorticity with excellent spatial resolution. At the moment, their temporal resolution is rather limited, which, however, does not affect the process of statistical averaging. In the following, a description is provided of optical techniques that have been used to measure one velocity gradient/vorticity at a single point with reasonable success. These techniques can be classified as direct or indirect, depending on whether vorticity is obtained from the measurement of fluid particle rotation or from the measurement of individual velocity gradients inside the flow field.

MICRO- AND NANOSENSORS AND ASSOCIATED TECHNIQUES

Over the past 5 to 12 years, there has been an increasing interest in understanding fluid phenomena at the micrometre- and nanometre-length scales. This has been evident in both basic and applied fluid mechanics research. The application of microscale fluid mechanics to MEMS and BioMEMS represents one of the most significant practical applications. In addition, there is a need to resolve smaller scales in measurements of mainstream macroscopic flows.

These two needs led to the development of microsensors and nanosensors. It is very encouraging to see that, in general, as the physical size of the probes is reduced, their frequency response increases. Reduction in probe dimensions is expected to improve spatial and temporal resolution.

Thermal anemometry–based **micromachined** probes have been explored by Tai[96] and Lofdahl et al.,[97] but they were never ready for practical use. Jiang, Tai, Ho, and Li[98] developed new micrometre-sized polysilicon wires and demonstrated the potential for batch fabrication. Jiang, Tai, Ho, Karan, and Garstenauer[99] have implemented these hot wires into anemometers and performed calibration of them in wind tunnels. The micromachined anemometers (Fig. 20.7.19a) consist of a sensing wire, two parallel support shanks, a silicon beam, and a thick (500-μm) Si handle. The silicon beam is like the probe body and is 1 μm long, 200 μm wide, and 75 μm thick. The two parallel support shanks are 100 μm long, 20 μm wide, and 0.5 μm thick. Various sizes of the sensing wires have been made, and they are 10 to 160 μm long, 1 μm wide, and 0.5 μm thick. This structure is similar to that of the conventional anemometers, so direct comparison can be made. The package of the anemometers is a ceramic tube a few centimetres long and 3 mm in diameter (Fig. 20.7.19b). The anemometer handle is soldered with electrical cables, placed inside the tube, and then epoxy-fixed. During handling, all mechanical vibrations have to be avoided in order to prevent damage of the wires.

Ebefors et al.[100] used doped silicon for both the prongs and the hot-wire sensor, along with a polyamide hinge, to make a three-component sensor, shown schematically in Fig. 20.7.20. An SEM photograph is shown in Fig. 20.7.21.

(a)   (b)

Fig. 20.7.19   (a) SEM of a micromachined hot wire. (b) Micrograph of the anemometer packaging from Jiang et al.[99]

Fig. 20.7.20   Schematitic view of the 3-D hot-wire probe based on a polyamide joint. Silicon chip is 3.5 mm × 3 mm × 0.5 mm and the three wires are 500 mm × 5 mm × 2 mm. From Ebefors et al.[100]

Fig. 20.7.21 SEM photo of the *x*- and *z*-hot-wire resistance elements with electrical connection through the poluamide joint. From Ebefors et al.[100]

Non-silicon-based hot wires with a thermal element size of 2.7 μm $\times$ 50 μm have been fabricated by Chen and Liu[101] (Fig. 20.7.22).

Micrometre-resolution particle image velocimetry (μ**PIV**) refers to the application of PIV to measure velocity fields of fluid motion with length scales of the order 100 μm, and with spatial resolution of individual velocity measurements of order 1 to 10 μm. The PIV technique is limited to maximal spatial resolutions of 0.2 to 1.0 mm because of fundamental physical limitations as well as practical implementation difficulties. In μPIV, emphasis is placed on the ability to accurately and reliably measure fluid velocity with micrometre spatial resolution. In order to achieve microscale velocity measurements, novel developments in PIV image recording hardware, flow-tracing particles, system design, and analysis software are required. The theoretical background of μPIV is based on the interactions between the flow fluid and the seed particles, which occur at nanometre length scales. μPIV can be thought of as a practical, application of nanotechnology to perform science and engineering at micrometre-length scales.

The main component of a typical μPIV setup like that shown in Fig. 20.7.23 is the microscope lens and epifluorescent prism/filter which are typical in microscopy imaging. This system was adopted originally by Meinhart et al.[102] for μPIV measurements. A pulsed monochromatic light source, such as a pulsed Nd:YAG laser system is used. The laser system is similar to those used in PIV applications. It consists of two Nd:YAG laser cavities, beam-combining optics, and a frequency-doubling crystal. The illumination light is delivered to the microscope through beam-forming optics, which can consist of a variety of optical elements that will sufficiently modify the light so that it will fill the back of the objective lens, and thereby broadly illuminate the microfluidic device. The beam-forming optics can consist of a liquid-filled optical fiber to direct the light into the microscope. Recently, light has been delivered into the microfluidic device by using total reflection through a transparent wall. This instrument is called a total internal reflection fluorescence (TIRF) microscope.

Typically, the working fluid is directed through the microfluidic device by pressure, or induced electrokinetically. The working fluid can be of any visibly transparent fluid, such as water. Liquids are most commonly used. It is conceivable to use gaseous fluids, such as air, but μPIV experiments using air have not been demonstrated successfully to date. The working fluid should be seeded with fluorescent particles, which can be manufactured from a variety of materials, such as polystyrene. It is important that the particles have a specific gravity closely matched to the working fluid, and less than, say, 1 μm, preferably 200 to 700 nm.

A sensitive large-format interline-transfer CCD camera is commonly used to record the particle image fields. A large-format CCD array is desirable because it allows for more particle images to be recorded, and increases the spatial dynamic range of the measurements. The interline transfer feature of the CCD camera allows for two particle image frames to be recorded back to back to within a 500-ns time delay.

These techniques can be extended to provide three-dimensional information on the entire flow field by changing the focal plane of the objective lens with piezoelectric positioners to scan the whole depth.

In μPIV, fluorescent particles with diameters typically in the range 200 to 300 nm are used to obtain velocity data with an out-of-plane spatial resolution typically of order 1 μm. Near-wall measurements with this method have so far been obtained no closer than 450 nm from the surface.

Recently, **nano-PIV** (nPIV) has been introduced as a technique specifically designed for near-surface interrogation of flow[103,104,105] where near-wall measurements with μPIV have so far been obtained no closer than 450 nm from the surface. The new nPIV method relies on an evanescent wave illumination generated by the total internal reflection (TIR) of light at the interface between two materials of different refractive indices,[106] e.g., a glass-liquid interface. With this method, the motion of particles within a region of order 100 nm of the surface is measured. In nPIV studies, to date, the size of the seed particles are in the 100- to 300-nm range; i.e., they are of the same order of, or larger than, the illuminated region. It would be desirable to use much smaller seed particles to reduce some of the complications that can be caused by larger particles, such as the modification of the evanescent wave field and particle–fluid interactions that compromise the accuracy of fluid velocity measurement and the ability of particles to move close to the surface.

The use of **quantum dot** (QD) nanoparticles for near-surface measurements of velocity has been introduced by Pouya et al.[107] These particles are an order of magnitude smaller than what has been used to date in nPIV. Nanocrystal QDs are semiconductor nanoparticles that are chemically synthesized with precisely controlled sizes in the range of 1 to 10 nm.[108–112] Among the most developed QDs for fluorescence imaging are the coreshell dots composed of an optically active nanocrystal core of CdSe surrounded by a protective shell of ZnS. The surface of QD is covered with a ligand shell that can be functionalized for broad chemical flexibility. Several properties of QDs make them extremely attractive for fluid flow studies. By modifying the functional groups in the ligand shell, QDs may be dispersed in specific chemical environments such as polar and nonpolar liquids. Since QDs can be solubilized within the fluid, they are expected to behave more like molecules and some of the near-wall issues of "solid" particles should not manifest themselves to QD tracers (e.g., particle-fluid and particle-wall interactions). The emission wavelength of a QD depends on its size, and

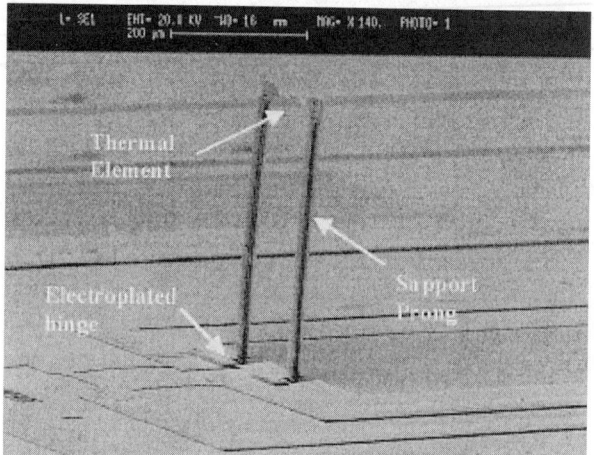

Fig. 20.7.22 SEM micrographs of a hot-wire anemometer after postrelease strengthening by electroplating. From Chen and C. Liu.[101]

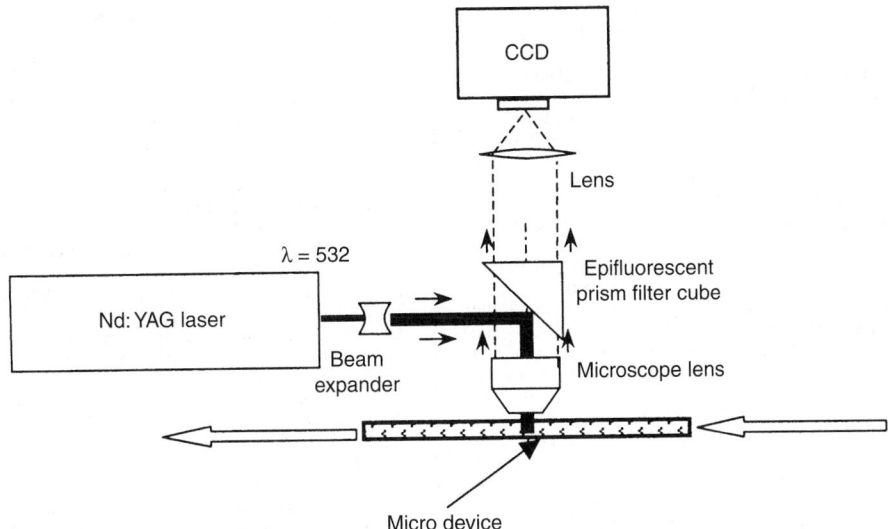

Fig. 20.7.23 Experimental setup for μPIV measurements.

can be tuned across the entire visible spectrum by varying the diameter of the particle. The excitation band is very broad, and the emission is independent of the excitation wavelength. Thus, a size series of QDs with different emissions can be excited with the same light source.

The experimental setup is similar to that of μPIV. Pouya et al.[107] indicated that QDs showed large enough displacement to be discernible with the naked eye.

Nuclear magnetic resonance (NMR), one of the most powerful analytic tools for studying chemical composition, is opening the door to microfluidic applications. NMR is a phenomenon involving a property found in atomic nuclei of almost every molecule called *spin,* which gives rise to a magnetic moment, meaning that the nuclei behave like bar magnets with north and south poles. When a magnetic filed is applied, these magnets align their axes along the lines of the magnetic field. This alignment is not perfect and it results in a wobbly rotation about the magnetic lines that is characteristic for each type of nucleus. Hilty et al.[113] developed a remote NMR technique that provided a strong signal to overcome the sensitivity limitation of NMR by introducing hyperpolarized xenon nuclei, which have a very long spin-relaxation time (several minutes). In this technique, signal encoding and detection are carried out independently while in a conventional setup the two processes take place within a single device. In addition to the magnetic field, the nuclei of the sample are also subjected to a sequence of radio-frequency pulses during the encoding phase, and because of this they absorb and re-emit energy at characteristic frequencies. In the detection phase the frequencies of these encoded signals are used to obtain an NMR spectrum that contains distinct information to characterize the sample.

REFERENCES

1. Tavoularis, S., "Measurement in Fluid Mechanics," Cambridge University Press, 2005.
2. Bradshaw, P., http://vonkarman.stanford.edu/tsd/pbstuff/tunnel/index.html.
3. Bradshaw, P., and Pankhurst, R. C., "Progress in Aeronautical Science," Chap. 1, pp. 1–69. Pergamon Press, Oxford, U.K., 1964.
4. Rae, W.H., and Pope, A., "Low-Speed Wind Tunnel Testing," John Wiley & Sons, New York, 1984.
5. Mehta, R. D., and Bradshaw, P., *Aeron. J.* 73, pp. 443–449, 1979.
6. Gad-el-Hak, M., *Exper. Fluids* 5, pp. 289–297, 1987.
7. Briassulis, G., Agui, J. H., Andreopoulos, J., and Watkins, C.B., *Exper. Thermal Fluid Sci.* 13, pp. 430–446, 1996.
8. Agui, J. H., Briassulis, G., Andreopoulos, Y., *J. Fluid Mech.* 524, pp. 143– 195, 2005.
9. Briassulis G., Agui, J., and Andreopoulos Y., *J. Fluid Mech.* 432, pp. 219– 283, 2001.
10. Wyngaard, J., *J. Phys. E: Sci. Instrum.,* Series 2, pp. 983–987, 1969.
11. Wallace, J. M., and Foss J., *Annu. Rev. Fluid Mech.* 27, pp. 469–514, 1995.
12. Tsinober, A., Kit, E., and Dracos T., *J. Fluid Mech.* 242, pp. 169–192, 1992.
13. Honkan, A., and Andreopoulos, Y., *J. Fluid Mech.* 350, pp. 29–96, 1997.
14. Figliola, R. S., and Beasley, D. E., "Theory and Design of Mechanical Measurements," John Wiley & Sons, New York, 1995.
15. Doebelin, E. O., "Measurement Systems Application and Design," 4th ed., McGraw-Hill, New York, 1990.
16. Tavoularis, S., in "Encyclopedia of Fluid Mechanics," Vol. 1, Chap. 36, Gulf Publishing, Houston, pp. 1207–1255, 1986.
17. Gad-el-Hak, M., Basic Instruments, in "Handbook of Fluid Dynamics," Chap. V, CRC Press, pp. 33-1–33-22, 1998.
18. Peterson, J. I., and Fitzgerald, V. F., *Review of Scientific Instruments,* vol. 51, no. 5, pp. 133–136, 1980.
19. McLachlan, B. G., and Bell, J. H., *Thermal and Fluid Science* 10, no. 5, pp. 470–483, 1995.
20. Morris, M. J., *AIAA Journal* 31, no. 3, pp. 419–421, 1993.
21. McLachlan, B. G., *J. of Aircraft* 32, no. 2, pp. 217–225, 1995.
22. Reda, D. C., Wilder, M. C., Mehta, R., and Zilliac, G., AIAA-97-2489, *32nd Thermo Physics Conference,* Atlanta, GA, June 23–25, 1997.
23. "Handbook of Experimental Fluid Mechanics," Springer Verlag, ed. C. Tropea et al., 2006.
24. Bruun., H. H., "Hot-Wire Anemometry," Oxford University Press, Oxford, U.K., 1995.
25. Lomas, C. G., "Fundamentals of Hot Wire Anemometry," Cambridge University Press, Washington, 1983.
26. Perry, A. E., "Hot-Wire Anemometry," *Clarendon,* Oxford, 1982.
27. King. L. V., *Phil. Trans. Roy. Soc. London* A214, pp. 373–432, 1914.
28. Bradshaw, P., "An Introduction to Turbulence and Its Measurement," Pergamon, Press, Oxford, 1971.
29. Champagne, F. H., Sleicher, C. A., and Wehrmann, O. H., *J. Fluid Mech.* 28, pp. 153–175, 1967.
30. Jorgensen, F. E., *DISA Info.* 11, pp. 31–37, 1971.
31. Cummins, H. Z., Knable, N., and Yeh, Y., *Phys. Review Letters* 12, no 6, pp. 51–53, 1964.
32. Yeh, Y., and Cummins, H. Z., *Applied Physics Letters* 4, no. 10, pp. 176–178, 1964.
33. Durst, F., Melling, A., and Whitelaw, J. H., "Principles and Practice of Laser Doppler Anemometry," 2nd ed., Academic Press, New York, 1981.
34. Durst, F., "Measurements and Predictions of Complex Turbulent Flows: Principles of Laser Doppler Anemometers," Von Karman Institute for Fluid Dynamics, Lecture Series 1980–1983, 1980.
35. Buchave, P., George, W. K., and Lumley, J. L., "Annual Review of Fluid Mechanics," Vol. 11, pp. 443–503, 1979.
36. Adrian, R. J., *Annual Rev. Fluid Mechanics* 23, pp. 261–304, 1991.

37. Adrian, R. J., and Yao, C. S., *Appl. Optics* 24, pp. 42–52, 1985.
38. Grant, I., *Proceedings of Institute of Mechanical Engineers* 211, pp. 55–76, 1997.
39. Hinsch, K.D., "Particle Image Velocimetry," in "Speckle Metrology," ed. R. S. Sirohi, Marcel Dekker, New York, pp. 235–323, 1993.
40. Raffel, M., Willert, C. E., and Kompenhans, J. *Particle Image Velocimetry,*. Springer, 1998.
41. Wang, Z., "Experimental Investigation of Compressible Turbulent Jets," Ph.D. thesis, City University of New York, 2003.
42. Komine, H., U.S. Patent 4,919,536, 1990.
43. Komine, H., Brosnan, S., Litton, A., and Stappaers, E., AIAA Paper 91-0337, 1991.
44. Meyers, J. F., and Komine, H., *Laser Anemometry* 1, pp. 289–296, 1991.
45. Miles, R. B., and Lempert,W. R., AIAA Paper 90–0624, 1990.
46. Miles, R. B., Lempert, W. R., and Forkey, J. N., AIAA Paper 91–0357, 1991.
47. Samimy, M., and Wernet, M. P., *AIAA Journal* 38, no. 4, pp. 553–574, 2000.
48. Miles, R. B., Forkey, J. N., and Lempert, W. R., AIAA Paper 92–3894, 1992.
49. Forkey, J., Finkelstein, N. D., Lempert, W. R., and Miles, R. B., AIAA Paper 95-0298, 1995.
50. Meyers, J. F., *Meas. Sci. Technol.* 6, pp.769–783, 1995.
51. Arnette, S. A., Samimy, M., and Elliott, G. S., *AIAA Journal* 33, pp. 430–438, 1995.
52. Miles, R. B., Lempert, W. R., and Forkey, J., AIAA-91–0367, 1991.
53. McKenzie, R. L., *Appl. Optics* 35, no. 6, pp. 948–964, 1996.
54. Elliott, G. S., Samimy, M., and Arnette, S. A., AIAA Paper 93–0512, 1993.
55. Elliott, G. S., Samimy, M., and Arnette, S. A., *Experiments in Fluids* 18, pp. 107–118, 1994.
56. Mosedale, A. D., Elliot, G. S., Carter, C. D., and Beutner, T. J., *J. AIAA* 38, no. 6, pp. 1010–1024, 2000.
57. Gendrich, C. P., Koochesfahani, M. M., and Nocera, C. G., *Experiments in Fluids* 23, no. 5, pp. 361–372, 1997.
58. Stier, B., and Koochesfahani, M. M., *Exp. Fluids* 26, pp. 297–304, 1999.
59. Ress, J. M., Laufer, G., and Krauss, R. H., *AIAA J.* 33, pp. 296–301, 1995.
60. Pitz, R. W., Wehrmeyer, J. A., Ribarov, L. A., Oguss, D. A., Batliwala, F., DeBarber, P. A., Deusch, S., and Dimotakis, P. E., *Meas. Sci. Tech.* 11, p. 1259, 2000.
61. Ribarov, L. A., Wehrmeyer, J. A., Hu, S., and Pitz, R. W., *Experiments in Fluids* 37, pp. 65–74, 2004.
62. Dam, N. J., Klein-Douwel, J. H., Sijtsema, N. M., and ter Meulen, J. J., *Optics Letters* 26, pp. 36–38, 2001.
63. Sijtsema, N. M., Dam, N. J., Klein-Douwel, J. H., and ter Meulen, J. J., *AIAA J.* 40, pp. 1061–1064, 2002.
64. Miles, R. B., and Lempert, W. R., *Ann. Rev. Flu. Mech.* 29, pp. 285–326, 1997.
65. Noullez, A., Wallace, G., Lempert, W. R., Miles, R. B., and Frisch, U., *J. Fluid Mech.* 339, pp. 287–307, 1997.
66. Foss, J. F., *Bull. Am. Phys. Soc.* 21, p. 1237 (abstract), 1976.
67. Foss, J. F., *Proc. Biennial Symp. on Turbulence,* 7th Missouri–Rolla, pp. 208–218, 1981.
68. Foss, J. F., *Exp. Therm. Fluid Sci.* 8, pp. 260–270, 1994.
69. Frish, M. B., and Webb, W. W., *J. Fluid Mech.* 107, pp. 173–185, 1981.
70. Vukoslavoevic, P., Wallace, J. M., and Balint, J., *J. Fluid Mech.* 228, pp. 25–51, 1991.
71. Honkan, A., and Andreopoulos, Y., *J. Fluid Mech.* 350, pp. 29–96, 1997.
72. Agui, J. H., Briassulis, G., Andreopoulos, Y., *J. Fluid Mech.* 524, pp. 143–195, 2005.
73. Briassulis, G., Agui, J., and Andreopoulos, Y., *J. Fluid Mech.* 432, pp. 219–283, 2001.
74. Andreopoulos, J., *Rev. Sci. Instr* 54, no. 6, pp. 733–740, 1983.
75. Andreopoulos, J., *Physics E. Sci. Instr.* 16, pp. 1264–1271, 1983.
76. Briassulis G., Agui J., Andreopoulos, J., and Watkins C. B., *Exp. Thermal and Fluid Science* 13, pp. 430–446, 1996.
77. Kovasznay, L. S. G., *J. Aeronaut. Sci.* 17, p. 565, 1950.
78. Morkovin, M.V., *AGARDOgraph* 24, 1956.
79. Breyer, H., Kriegs, H., Schmidt, K., and Staude, W., *Experiments in Fluids* 15, pp. 200–208, 1993.
80. Kriegs, H., Schulz, R., and Staude, W., *Experiments in Fluids* 15, pp. 240–246, 1993.
81. Kriegs, H., and Staude, W., *Measurement Science and Technology* 6, p. 653, 1995.
82. Agui, J. H., and Andreopoulos, J., *International Symposium on Laser Anemometry,* FED vol. 191, pp. 11–19, June 20–24, 1994.
83. Agui, J. H., and Andreopoulos, Y., *Experiments in Fluids* 34, pp. 192–205, 2003.
84. Kolmogorov, A. A., *C. R. Akad. Sci. SSSR* 30, pp. 301–306, 1941.
85. Hanson, S. G., *2nd Symposium on Applications of Laser Anemometry to Fluid Mechanics,* July 2–5, Lisbon, 1984.
86. Ötügen, M. V., Su, W-J., and Papadopoulos, G., *Meas. Science Technol.* 9, pp. 267–274, 1998.
87. Yao, S., Tong, P., and Ackerson, B., *Applied Optics* 40, no. 24, pp. 4022–4027, 2001.
88. Arik, E. B., "Flows at Ultra High Reynolds and Rayleigh Numbers: A Status Report," eds. R. J. Donnelly and K. R. Sreenivasan, Springer-Verlag, New York, pp. 138–158, 1998.
89. Lourenco, L., and Krothapalli, A., *Experiments in Fluids* 5, pp. 23–31, 1987.
90. Mcinhart, C. D., and Adrian, R. J., AIAA paper 95-0789, 1995.
91. Zhang, J., Tao, B., and Katz, J., *Experiments in Fluids* 23, pp. 373–381, 1997.
92. Kähler, C. J., *Experiments in Fluids* 36, pp. 114–130, 2004.
93. Ganapathisubramani, B., Longmire, E. K., Marusic, I., and Pothos, S., *Experiments in Fluids* 32, no. 2, pp. 222–231, 2004.
94. Hu, H., Saga, T., Kobayashi, T., Taniguchi, N., and Yasuki, M., *Experiments in Fluids* 31, pp. 277–293, 2001.
95. Mullin, J. A., and Dahm, W. J. A., *Experiments in Fluids* (to appear) 2006.
96. Tai, Y. C., *Dig. Tech. Papers, Transducers* 85, pp. 354–357, 1985.
97. Lofdahl, L., Stemme, G., and Johansson, B., *Experiments in Fluids* 12, pp. 270–276, 1992.
98. Jiang, F., Tai, Y. C., Ho, C. H., and Li, W. J., *Tech. Digest 1994 Solid-State Sensors & Actuator Workshop,* Hilton Head, SC, pp. 264–267, 1994.
99. Jiang, F., Tai, Y. C., Ho, C. M., Karan, R., and Garstenauer, M., *Technical Digest, International Electron Device Meeting* (IEDM-94), San Francisco, pp. 139–142, Dec. 1994.
100. Ebefors, E., Kalvesten, E., and Stemme, G., *Proc. MEMS,* New York, pp. 93–98, 1998.
101. Chen, J., and Liu, C., *J. Microelectromechanical Systems* 12, no. 6, pp. 979–988, December 2003.
102. Meinhart, C. D., Wereley, S. T., and Santiago, J. G., *Exp. in Fluids* 27, pp. 414–419, 1999.
103. Zettner, C. M., and Yoda, M., *Exp. Fluids* 34, p. 115, 2003.
104. Jin, S., Huang, P., Park, J., Yoo, J. Y., and Breuer, K. S., *Exp. Fluids* 37, p. 825, 2004.
105. Sadr, R., Yoda, M., Zheng, Z., and Conlisk, A. T., *J. Fluid Mech.* 506, p. 357, 2004.
106. Axelrod, D., Burghardt, T. P., and Thompson, N. L., *Annu. Rev. Biophys. Bioeng.* 13, p. 247, 1984.
107. Pouya, S., Koochesfahani, M., Snee, P., Bawendi, M., and Nocera, D., *Experiments in Fluids* 39, pp. 784–786, 2005.
108. Murray, C. B., Norris, D. J., and Bawendi, M. G., *J. Am. Chem. Soc.* 115, p. 8706, 1993.
109. Dabbousi, B. O., Rodriguez-Viejo, J., Mikulec, F. V., Heine, J. R., Mattoussi, H., Ober, R., Jensen, K. F., and Bawendi, M. G., *J. Phys. Chem. B* 101, p. 9463, 1997.
110. Bruchez, M., Jr., Moronne, M., Gin, P., Weiss, S., and Alivisatos. A. P., *Science* 281, p. 2013, 1998.
111. Mattoussi, H., Mauro, J. M., Goodman, E., Anderson, G. P., Sundar, V. C., Mikulec, F. V., and Bawendi, M. G., *J. Am. Chem. Soc.* 122, p. 12142, 2000.
112. Murray, C. B., Kagan, C. R., and Bawendi, M. G., *Rev. Mat. Sci.* 30, p. 545, 2000.
113. Hilty, C., McDonnell, E. E., Granwehr, J., Pierce, K. L., Han, S. I., and Pines, A., *PNAS* 142, no. 42, pp. 14960–14963, 2005.

20.8 INTRODUCTION TO BIOMECHANICS
by Ali M. Sadegh

GLOSSARY

In describing the relationship of body parts or the location of an external object with respect to the body, the use of directional terms is necessary. The following are commonly used directional terms; see Fig. 20.8.1.

Superior: Closer to the head
Inferior: Farther away from the head
Anterior: Toward the front of the body
Posterior: Toward the back of the body
Medial: Toward the midline of the body
Lateral: Away from the midline of the body
Proximal: Closer to the trunk
Distal: Away from the trunk
Abduction: Motion away from the body midline
Adduction: Motion toward the body midline
Range of motion: Range of translation and rotation of a joint for each of its six degrees of freedom
In vivo: Within a living organism
In vitro: In an artificial environment outside the living organism

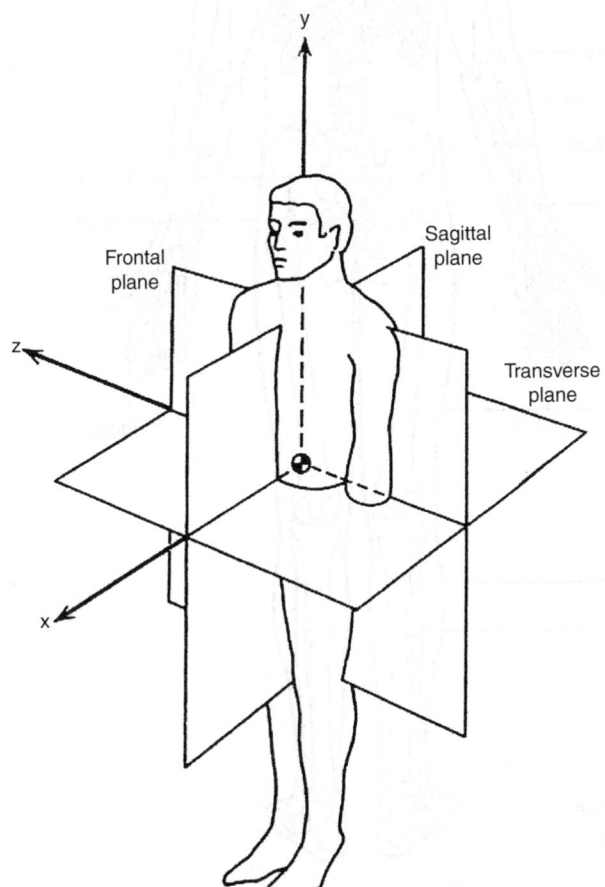

Fig. 20.8.1 The body spatial coordinate system.

INTRODUCTION—WHAT IS BIOMECHANICS?

The concepts of biomechanics were implicitly reflected in medical literatures since the nineteenth century and later in mechanics literatures. For example, von Helmholtz, who was a professor of physiology, pathology, and anatomy at Bonn and later professor of physics in Berlin (1871), had contributed to the mechanism of hearing and theory of color vision and optics and velocity of nerve pulses. Then, in the twentieth century, some researchers employed mechanics principles to explain some phenomena in human physiology. In the early 1970s, the international community adopted the term *biomechanics* to describe biological systems from mechanical perspectives.

By definition, **biomechanics** is the application of mechanics in biology and physiology, i.e., utilization of physics laws to study and to analyze the anatomical and functional aspects of living organisms.

Biomechanics has been involved in virtually every facet of modern medical science and technology. Biomechanics has helped in solving all kinds of clinical problems associated with virtually all human organs from the cardiovascular system to prosthetic devices. Even the molecular biology field, which may seem far from biomechanics, utilizes physics principles to understand the mechanics of the formation, design, function, and production of the molecules. Orthopedics has profoundly relied on biomechanics research to treat patients with musculoskeletal problems. Orthopedic surgeons employ biomechanics as their clinical tools. Rehabilitation and healing are intimately related to the stress and strain in the tissues. Injury biomechanics in automotive accidents and other trauma is becoming more important to modern society since it has an economic impact. Therefore, it is no surprise that biomechanics researchers may have academic backgrounds in all engineering and medical disciplines, including orthopedic, cardiac, sports medicine, physical therapy, and many other fields.

Similar to an engineering approach, biomechanics problems should be solved through the following steps.

1. Acquire knowledge of the geometry and configuration of an organ; anatomy, morphology, and histology of the organ should be studied first.

2. Determine the mechanical properties. Unlike with engineering materials, the mechanical properties of living tissues are very difficult to ascertain due to the size of the tissue, or large deformation and nonlinearity of the constitutive equation.

3. Derive the governing equation. The fundamental laws of physics, e.g., conservation of mass, momentum, and energy, should be employed, and the governing differential or integral equation for the living tissue should be determined.

4. Determine proper boundary conditions of the organ.

5. Solve the boundary value problem analytically or numerically.

6. Similar to engineering, physiological experiments including animals (in vivo or in vitro) are needed to verify or to compare with the analytical/numerical solutions.

The body spatial coordinate system, which is frequently referred to in this section, is shown in Fig. 20.8.1.

BIOMECHANICS OF BONE

Bones are the main load-bearing tissue in the skeletal system, the purpose of which is to protect internal organs, to provide rigid links and muscle attachment sites, and to facilitate muscle action and body movement. There are 206 bones in the human body that are categorized in four groups according to their general shapes and functions: Short bones provide gliding motion and serve as shock absorbers (e.g., carpals); flat bones provide protection for underlying organs (e.g., scapulae or shoulder blades);

irregular bones have different shapes to fulfill specific functions in the body (e.g., vertebrae in the spine); and long bones, provide movements for upper and lower extremities (Fig. 20.8.2).

After the enamel of the teeth, bone is the second-hardest structure of the body. Bone is a highly vascular tissue with excellent remodeling capability. That is, it can repair, alter its properties, and change its geometry in response to changes in mechanical loading. For example, when bone is not subjected to mechanical loading, its density changes; or when it is subjected to new and repeated mechanical loads, then its shape, architecture, and density change.

Bone is composed of inorganic and organic phases and water. Based on the weight, bone is approximately 60 percent inorganic, 30 percent organic, and 10 percent water. On a volume basis, bone is approximately 40 percent inorganic, 35 percent organic, and 25 percent water. The inorganic portion of the bone, which is in the form of small ceramic crystalline-type mineral, consists of calcium and phosphate, resembling synthetic hydroxyapatite $[Ca_{10}(PO_4)_6(OH)_2]$. The organic phase of the bone consists primarily of type I collagen (90 percent by weight), some other collagen of types III and IV, and a variety of noncollagenous proteins.

Fig. 20.8.2 The human skeleton. (*From Hall, 2003.*)

Bone Structure

Macroscopically, there are two types of bone: cortical, or compact, bone and cancellous, or trabecular, bone. This is called the highest level of bone porosity. Cortical bone forms the outer shell or cortex of the bone and has a dense structure, while cancellous bone is within this shell and is composed of thin plates, or rods (known as trabeculae), in a mesh structure (Fig. 20.8.3). The interstices between the trabeculae are filled with red marrow. The porosity of cortical bone ranges from 5 to 30 percent, and that of cancellous (trabecular) bone ranges from 30 to 90 percent. Except for the joint surfaces, which are covered with articular cartilage, a dense fibrous membrane called the periosteum surrounds all bones.

Microscopically, the osteon or haversian system is the main structural unit of bone. This is called the second level of bone porosity. At the center of each osteon is a small channel, called the haversian canal, that contains blood vessels and nerve fibers. The osteon itself consists of a concentric series of layers (lamellae) of mineralized matrix surrounding the central canal, similar to growth rings on a tree (Fig 20.8.4). Along the boundaries of each layer, or lamella, are small cavities known as

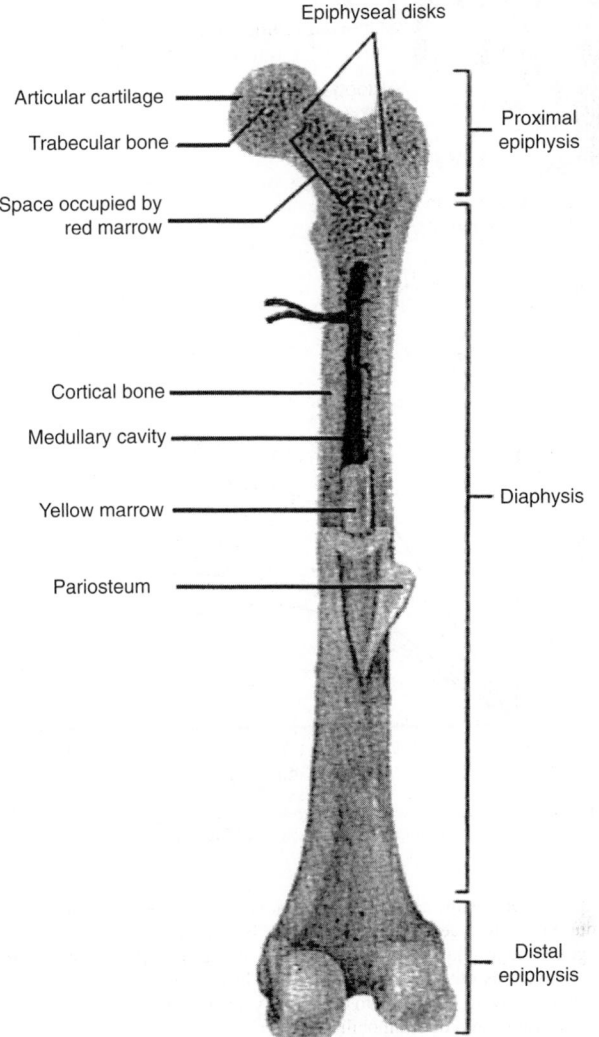

Epiphyseal disks

Articular cartilage

Trabecular bone

Space occupied by red marrow

Cortical bone

Medullary cavity

Yellow marrow

Pariosteum

Proximal epiphysis

Diaphysis

Distal epiphysis

Femur

Fig. 20.8.3 Femur, showing cortical and trabecular bones. (*From "Hole's Human Anatomy and Physiology" and Hall, 2003.*)

lacunae, each containing one bone cell, or osteocyte. Osteon is typically about 200 μm in diameter, which varies with age, and has a length of 1 to 3 mm. The lacunae of adjacent lamellae are connected by numerous small channels, or canaliculi, which ultimately reach the haversian canal.

Mechanical Properties of Bone

CORTICAL BONE

Based on the anisotropy of bone microstructure, the elastic and strength properties of human cortical bone are anisotropic. In diaphysis of long bones, cortical bone is both stronger and stiffer in the longitudinal direction than in the radial or circumferential directions (Table 20.8.1). Cortical bones are stronger in compression than in tension (Table 20.8.1). Mechanical properties of human cortical bone change (decrease) as age increases (Fig. 20.8.6). While the modulus does not reduce much, if at all, the strength is reduced at a rate of about 2 percent per decade.

As with all biological materials, bone is viscoelastic and nonlinear. It has been shown that cortical bone is moderately strain-rate-dependent. That is, over a six order of magnitude increase in strain rate, modulus changes by only a factor of 2 and strength by a factor of 3 (Fig. 20.8.5). The majority of physiological activities occur in the range of 0.01 to 1.0 percent strain per second, where the changes in bone strength are negligible.

In some simple failure analyses of bone, cortical bone has been considered as simple isotropic materials and the von Mises criterion has been employed. This approach is not capable of describing multiaxial failure of cortical bone. However, the Tsai-Wu criterion, commonly used for composite materials, has been applied to cortical bone using transversely isotropic and orthotropic properties of bone.

TRABECULAR BONE

The modulus of elasticity and the strength of trabecular bone vary depending on the anatomic site, loading direction, loading mode, and age and health of the donor. In compression, the anisotropy of trabecular bone, both modulus and strength, decreases with age, falling approximately 10 percent per decade (Fig. 20.8.6). The trabecular bone is stronger in compression and is weaker in shear loading mode. Both the modulus and the strength of the trabecular bone depend heavily on the apparent density, yet they depend on the anatomic site. Note that the density of the bone is the ratio of mass to volume of the actual bone tissue; however, the **apparent density** is defined as the ratio of the mass of the bone tissue to the bulk volume of the specimen, including the volume associated with the vascular channels. The density of the trabecular bone also decreases with age. The apparent density of cortical bone is about 1.85 g/cm^3, which does not vary much with anatomic sites or species. However, the apparent density of trabecular bone is as low as 0.10 g/cm^3 for the spine (vertebra) (Kopperdahl and Keaveny, 1998), about 0.3 g/cm^3 for the human tibia (Linde et al., 1989), and up to 0.6 g/cm^3 for the load-bearing portion of the proximal femur (Morgan and Keaveny, 2001).

Bone Remodeling

Bone, as a living tissue, has the ability to remodel, by altering its size, shape, architecture, density, and structure to accommodate the mechanical loadings placed on it. According to Wolff's law, bone remodeling is a function of the magnitude and direction of the mechanical stresses that act on the bones (Wolff, 1892). In other words, Wolff's law states that bone is laid down (deposition) where needed and is resorbed where not needed. Bone remodeling particularly occurs around implants and prostheses where the mechanical stresses to the bone have been altered (Sadegh et al., 1993; Cowin et al., 1992; Luo et al., 1995). In a bone fracture repair process, an implant is firmly attached to the bone to stabilize the fracture site. After the fracture has healed, if the implant is not extracted, the strength and stiffness of the bone may be diminished. This is due to the fact that the material of implants is about 10 times stiffer than the bone, and thus the majority of the loading will be carried

Fig. 20.8.4 Diagram of a sector of the shaft of a long bone showing structural details of cortical bone. (*From Kutz, 2003.*)

Table 20.8.1 (*a*) **Elastic moduli of cortical bone.**
(*b*) **Ultimate strength of cortical bone**

(*a*)	
Longitudinal modulus, MPa	17,900 (3,900)*
Transverse modulus, MPa	10,100 (2,400)
Shear modulus, MPa	3,300 (400)
Longitudinal Poisson's ratio	0.40 (0.16)
Transverse Poisson's ratio	0.62 (0.26)

*Standard deviations are given in parentheses.

(*b*)	
Longitudinal, MPa	
Tension	135 (15.6)*
Compression	205 (17.3)
Transverse, MPa	
Tension	53 (10.7)
Compression	131 (20.7)
Shear, MPa	65 (4.0)

*Standard deviations are given in parentheses.
SOURCES: Reilly and Burstein (1975) and Kutz (2003).

Fig. 20.8.5 Strain rate sensitivity of cortical bone for longitudinal tensile loading (*From Kutz, 2003; data from McElhaney and Byars, 1965.*)

out by the implant. This is called **stress shielding** where an implant may cause bone resorption.

BIOMECHANICS OF ARTICULAR CARTILAGE

Articular cartilage is a tough, dense, white connective elastic tissue, about 1 to 5 mm thick, that covers the ends of bones in joints and enables the bones to move smoothly over one another. Articular cartilage is an isolated tissue without blood vessels, lymph channels, and

nerves; yet it withstands the rigorous joint environment without failing during the average lifetime. However, when articular cartilage is damaged through injury or a lifetime of use, it does not heal as rapidly or effectively as other tissues in the body. Instead, the damage tends to spread, allowing the bones to rub directly against each other and resulting in pain and reduced mobility.

The primary function of articular cartilage is to distribute joint loads over a wide area and to reduce the stress concentration on the joint. Its secondary function is to provide a contacting surface with minimal friction and wear.

Fig. 20.8.6 Variation of ultimate stress of trabecular bone (from the human vertebra and femur) with age. (*From Kutz, 2003; data from Mosekilde et al., 1987 and McCalden et al., 1997.*)

Table 20.8.2 Representative Properties of the Human Articular Cartilage Taken from the Lateral Condyle of the Femur

Property	Value
Poisson's ratio	0.10
Compressive modulus, MPa	0.70
Permeability coefficient, $m^4/(N \cdot s)$	1.18×10^{-15}

SOURCES: Mankin et al. (1994) and Kutz (2003).

BIOMECHANICS OF TENDONS AND LIGAMENTS

Tendons and ligaments, along with the joint capsule, surround, connect, and stabilize the joints of the skeletal system. Ligaments and joint capsules, which connect bone to other bones, provide stability to joints by guiding joint motion and preventing excessive joint motion. The function of the tendons is to attach muscle to bone and to transmit tensile loads from muscle to bone. Tendons and ligaments are sparsely vascularized and are composed of a combination of collagen and elastin fibers arranged primarily in parallel along the axis of loading. They have relatively few fibroblast cells and abundant extracellular matrix. In general, the cellular material occupies about 20 percent of the total tissue volume, while the extracellular matrix accounts for the remaining 80 percent. About 70 percent of the matrix consists of water, and approximately 30 percent is solids.

Both tendons and ligaments are passive tissues, and they do not have the ability to contract as muscle tissue does; however, they are extensible in tensile loading, and will return to their original length after being stretched, when they are extended within their elastic limits. In the unloaded state, the fibers are slightly crimpled, and thus at the initial stage of tensile loading the structure acts only to straighten out the component fibers and exhibits low stiffness, i.e., in the toe region in Fig. 20.8.8a. The straightened fibers exhibit linear deformation under further tensile loading, resulting in this load-deformation curve of Fig. 20.8.8a. Typical tensile properties of ligament and tendon are shown in Table 20.8.3. Tendons are generally stiffer than ligament due to the higher concentration of collagen. Tendons and ligaments are highly viscoelastic and rate-dependent; i.e., they will fail at lower extensions when loaded at high rates. That is, rapid motion (stretching) of ligaments or tendons may result in tear or rupture. For the whole ligaments and tendons the linear response can extend above 20 percent (Fig. 20.8 8b).

BIOMECHANICS OF MUSCLES

Skeletal muscle is the most abundant tissue in the human body, accounting for 40 to 45 percent of the total body weight. There are 430 skeletal muscles, found in pairs on the right and left sides of the body. There are three types of muscles: (1) cardiac muscles which compose the heart,

Chondrocytes are mature cartilage cells. They are sparsely distributed and account for less than 10 percent of the tissue's volume. Chondrocytes are arranged in three zones through the thickness of cartilage (Fig. 20.8.7). The first zone is close to the articular surface where chondrocytes' long axes are parallel to the articular surface. In the second, middle zone, the chondrocytes are round and randomly distributed. In the third, deep zone, the chondrocytes are arranged in columnar fashion perpendicular to the articular surface.

The organic matrix of cartilage is composed of a dense network of fine collagen fibrils enmeshed in a concentrated solution of proteoglycans (PGs). The collagen content of cartilage tissue ranges from 10 to 30 percent by net weight, and the PG content from 3 to 10 percent by wet weight; the remaining 60 to 87 percent is water, inorganic salts, and small amounts of other matrix proteins, glycoproteins, and lipids.

Cartilage is predominantly loaded in compression and is biphasic (solid-fluid) and viscoelastic. When cartilage is compressively loaded, the water within the proteoglycan matrix is extruded. The stiffness of cartilage is a function of its permeability, and the fluid pressure within the matrix supports approximately 20 times more load than the underlying material during physiological loading (Mankin et al., 1994). Cartilage material properties are given in Table 20.8.2.

Fig. 20.8.7 Collagen fibrils of chondrocyte arrangement through the thickness of cartilage, with three distinct zones: surface tangential, middle, and deep. (*From Nordin and Frankel, 1989, by permission.*)

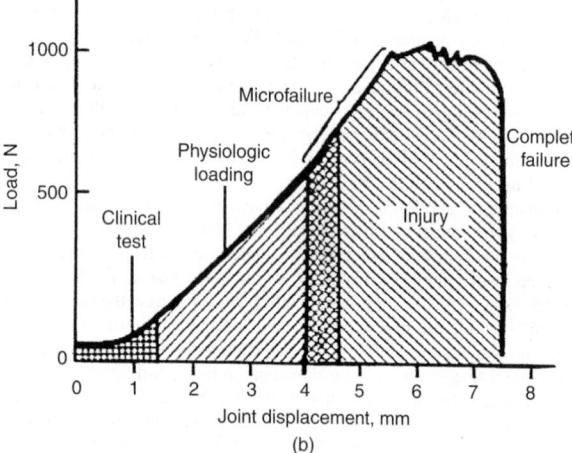

Fig. 20.8.8 (*a*) Typical load deformation of collagenous soft tissue such as ligament and tendon. (*Kutz, 2003.*) (*b*) Tensile testing of a human anterior cruciate ligament in vitro. (*Nordin and Frankel, 1989, by permission.*)

Table 20.8.3   Representative Properties of Human Tendon and Ligament under Tensile Loading

Tissue	Property	Value
Tendon	Elastic modulus, GPa	1.2–1.8
	Ultimate strength, MPa	50–105
	Ultimate strain, %	9–35
	Energy absorbed during elastic deformation	4–10% per cycle
Ligament	Stiffness, N/mm	150
	Ultimate load, N	368
	Energy absorbed to failure, N · mm	1,330

SOURCES: Woo et al. (1994, 1997) and Kutz (2003).

(2) smooth or involuntary muscles which line the hollow internal organs, and (3) skeletal or voluntary muscles which are attached to the skeleton through the tendons, and cause it to move.

Muscles are molecular machines that convert chemical energy to a force (mechanical energy). The structural unit of skeletal muscle is the fiber, a long cylindrical cell with hundreds of nuclei, ranging from about 10 to 100 μm in thickness and about 1 to 30 cm in length. Individual muscle fibers are connected by three levels of collagenous tissue: endomysium, which surrounds individual muscle fibers; perimysium, which collects bundles of fibers into fascicles; and epimysium, which encloses the entire muscle belly (see Fig. 20.8.9). The functional unit of skeletal muscle is the motor unit, which is the smallest part of the muscle that can be made to contract independently. The motor unit consists of a single motor neuron and all the muscle fibers that are to be stimulated by it. Note that when stimulated, all muscle fibers in the motor unit respond as one (Fig. 20.8.10).

The tendons and the connective tissues in and around the muscle belly are viscoelastic structures that help determine the mechanical characteristics of whole muscle during contraction and passive extension. In simulating musculotendon actuation, Hill (1970) showed that the tendons represent a springlike elastic component located in series with the contractile component (element) CC (or CE), which is in parallel with parallel elastic element (PEE) and in series with the series elastic element (SEE) (Fig. 20.8.11).

The mechanism of muscle contraction is as follows. After stimulation by the motor nerve, there is an interval of a few milliseconds, known as the latency period, before the tension in the muscle fibers begins to rise and contraction starts. Physiologically, according to the sliding filament theory, active shortening of the muscle results from the relative movement of the actin and myosin filaments past one another. This mechanical response of muscle fibers is known as twitching. Some muscle fibers contract with a speed of only 10 ms, while others may take 100 ms or longer.

During muscle contraction, a tension load is exerted on the bone to overcome the external load of the body. The muscle tension is generally applied at a distance from the center of rotation of a joint, to create a moment (torque) on the involved joint.

There are three types of muscle contraction. First, when muscles develop sufficient tension to overcome the resistance of the body segment, the muscles shorten and cause joint movement in the same direction as the net torque generated by the muscles. This contraction is called a **concentric.** Second, when muscle tension is developed but no change in muscle length occurs, the contraction is said to be **isometric.** In this case due to the developed contraction the diameter of the muscles is increased. Third, when a muscle cannot develop sufficient tension and the opposing joint torque exceeds that produced by the tension, the muscle progressively lengthens instead of shortens. In this case the muscle is said to contract **eccentrically.** Note that concentric and eccentric contractions involve dynamic work, in which the muscle moves a joint or controls the movement, while isometric contractions involve static work, in which the joint position is maintained (Fig. 20.8.12).

BIOMECHANICS OF JOINTS

Anatomically, there are three types of joints: immovable joints, such as sutures of the skull; slightly movable joints, such as cartilaginous joints (discs) where applied forces are attenuated to the adjacent bones; and freely movable (synovial) joints such as articular joints in the upper and lower extremities. In this section the biomechanics of articular joints in upper and lower extremities is discussed.

Biomechanics of the Knee

The human knee is the largest and one of the most complex joints in the body. The knee joint is situated between the body's two longest bones, the femur and tibia, and bears tremendous loads. The fact that the knee joint provides the mobility required for locomotor activities makes it one of the high weight-bearing joints of the body. Anatomically, the knee joint comprises (1) two condylar articulations of the tibiofemoral, with the third articulation being the patellofemoral joint; (2) medial and lateral menisci cartilaginous disc located between the tibial and femoral condyles; (3) articular capsule which is a double-layered membrane that surrounds the joint; (4) the synovial fluid, which is a clear, slightly yellow liquid that provides lubrication inside the articular capsule at the joint; and (5) the patella, also known as kneecap (Fig. 20.8.13). In addition, there are several ligaments in the knee joint; the two important

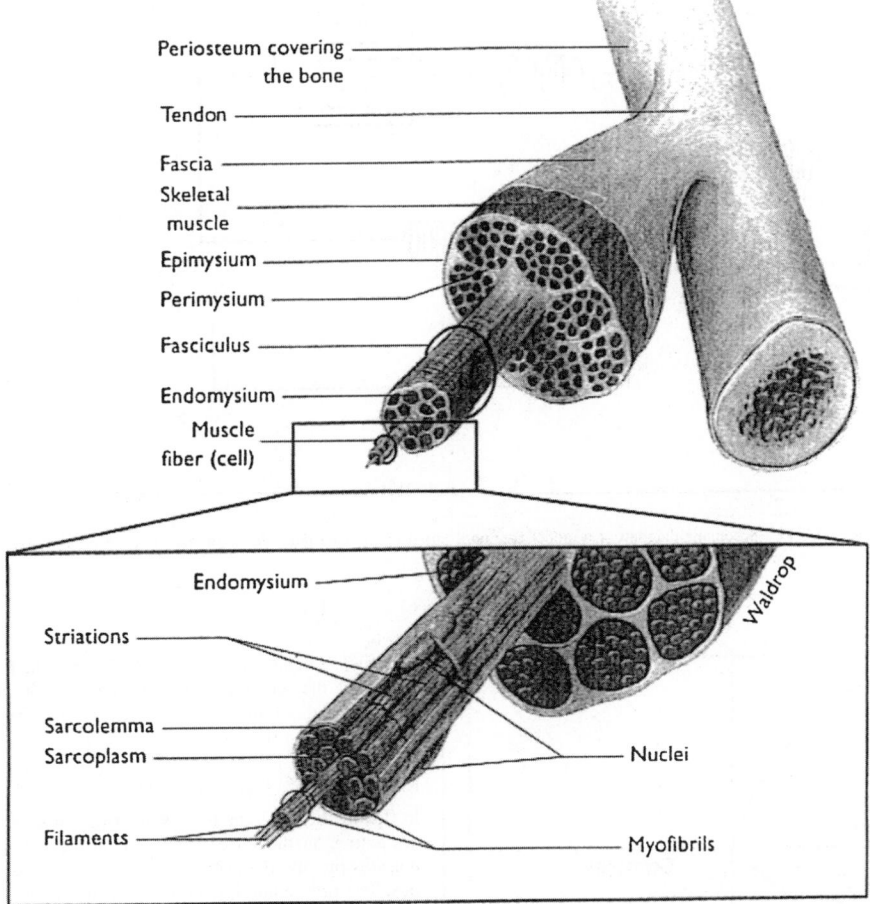

Fig. 20.8.9 Schematic drawings of the structural organization of muscle. (*From "Hole's Human Anatomy and Physiology" and Hall, 2003.*)

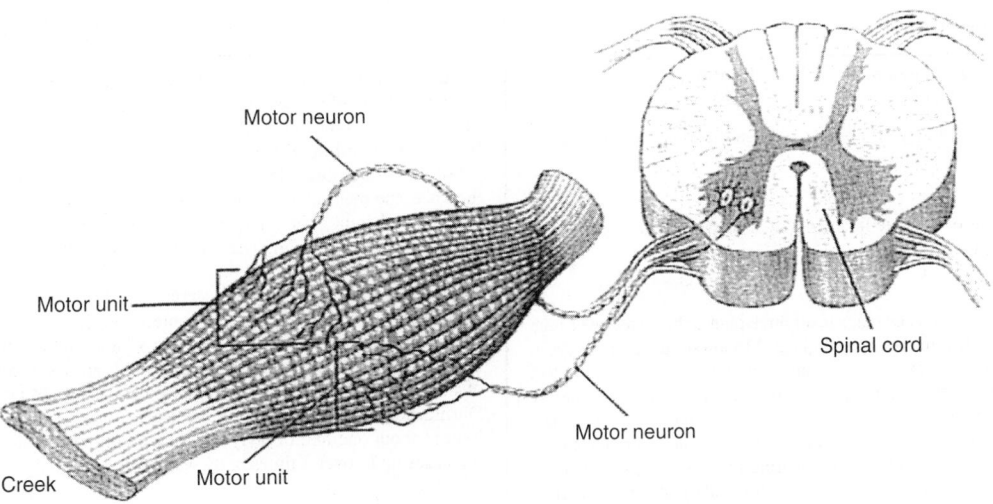

Fig. 20.8.10 A motor unit of muscle with its neuron. (*From "Hole's Human Anatomy and Physiology" and Hall, 2003.*)

Fig. 20.8.11 Schematic diagram of a model commonly used to simulate musculotendon actuation. (*Modified from Zajac and Gordon, 1989 and Pandy et al., 1990.*)

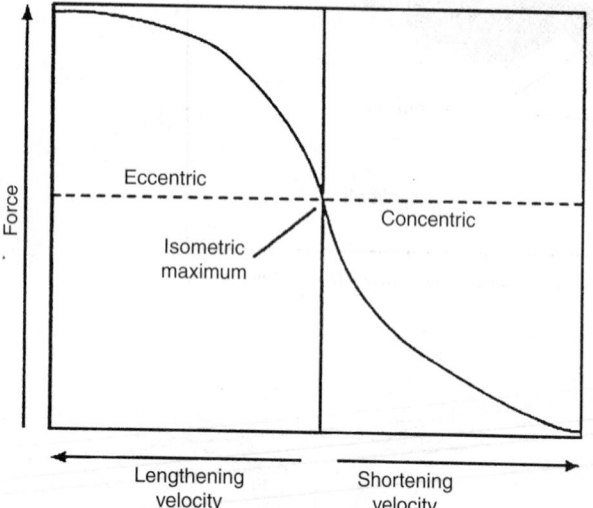

Fig. 20.8.12 The three types of muscle contraction are shown schematically. (*From Hall, 2003.*)

ones are the anterior cruciate ligament (ACL) and the posterior cruciate ligament (PCL) (Fig. 20.8.13).

KNEE KINEMATICS

The knee joint motion takes place in all three planes; however, the range of motion is greatest in the sagittal plane. Motion in this plane is from full extension to full flexion of the knee, which is approximately 140°. When the knee is fully extended, its rotation is almost completely restricted by the interlocking of the femoral and tibial condyles. However, when the knee reaches about 90° flexion, the external rotation ranges from 0° to approximately 45°. During flexion/extension, in addition to the rolling contact between the femoral and tibial condyles, there is tangential gliding between the two condyles as well. Also, in flexion/extension motion, the patella glides inferiorly and superiorly against the distal end of the femur in the vertical direction. The anterior and poste-

rior cruciate ligaments (ACL and PCL) limit the forward and backward sliding of the femur on the tibial plateaus during knee flexion and extension and also limit knee hyperextension. They both play an important role in knee stability.

KNEE LOADS

In flexing the knee, the three hamstring muscles are the primary flexors acting on the knee. However, during extension the quadriceps muscles provide the primary force for the knee. During daily activities, the tibiofemoral joint is subjected to compression and shear forces. Specifically, during the stance phase of gait, the compressive force at the tibiofemoral joint exceeds 3 times the body weight; or during stair climbing the compressive force is approximately 4 times the body weight. Figure 20.8.14 shows the static load diagram of the knee during stair climbing. Figure 20.8.15 shows the joint reaction force, in terms of body weight transmitted through the tibial plateau during walking, in one gait cycle. During the gait cycle the joint reaction force shifts from the medial to the lateral tibial plateau. In stance phase, when the force reached its peak value, the medial plateau mainly carried the load; and in the swing phase, when the force was minimal, the load was carried by the lateral plateau. The contact area of the medial tibial plateau is approximately 50 percent larger than that of the lateral tibial plateau.

Although the tibial plateaus are the main load-bearing structures in the knee, the menisci act as force absorption and distribute loads at the tibiofemoral joint over a larger area, thus reducing the magnitude of contact stress at the joint. The joint cartilages and ligaments also contribute to load bearing.

Biomechanically, the patella serves two important functions. First, it aids knee extension by producing anterior displacement of the quadriceps tendon throughout the entire range of motion, thereby lengthening the lever arm of the muscle force. Second, it allows wider distribution of compressive stress on the femur by increasing the area of contact. During normal walking gait, compressive force at the patellofemoral joint is about one-half of body weight, and during the stair climbing, it increases up to over 3 times body weight.

Biomechanics of the Hip

The hip joint is a ball-and-socket joint and is one of the largest and most stable joints (less likely to be dislocated) in the body with a great deal

Fig. 20.8.13 The knee joint. (*a*) Sagittal cross-sectional view (*"Hole's Human Anatomy and Physiology," 1996; Hall, 2003.*) (*b*) Posterior view. (*Hall, 2003.*) (*c*) Top view. (*Hall, 2003.*)

Fig. 20.8.14 Free-body diagram of the knee during stair climbing. (*From Nordin, and Frankel, 1989, by permission.*)

Fig. 20.8.15 The knee reaction force in terms of body weight transmitted through the tibial plateau during walking in one cycle. (*Adapted from Morrison, 1970; Nordin and Frankel, 1989, by permission.*)

Fig. 20.8.16 The hip joint. (*From Hall, 2003.*)

Fig. 20.8.17 The range of motion of the hip joint in sagittal plane during one gait cycle. (*From Nordin and Frankel, 1989, by permission.*)

Fig. 20.8.18 Body segment length expressed as a fraction of body height *H*. (*From Drillis and Contini, 1966.*)

of mobility. Anatomically, the hip joint is composed of the head of the femur, which forms approximately two-thirds of a sphere, and the acetabulum of the pelvis, which is a socket (see Fig. 20.8.16).

The acetabular surface and the head are covered with articular cartilages. The femur is a major weight-bearing bone and is the longest, largest, and strongest bone in the body. The femoral neck is its weakest component and is primarily composed of trabecular bone. The hip joint cartilage covers both articulating surfaces. Several large and strong ligaments provide the stability of the hip.

HIP KINEMATICS

The hip joint, due to its ball-and-socket nature, moves in all three planes. In the sagittal plane it flexes or extends, in the transverse plane it rotates internally and externally, and in the frontal plane it has abduction and adduction motions. The maximum range of motion is in the sagittal plane where the range of flexion is from 0° to approximately 140° and the range of extension is from 0° to 15°. The range of abduction is from 0° to 30°, whereas that of adduction is from 0° to 25°. Its external rotation ranges from 0° to 90°, and its internal rotation ranges from 0° to 70° when the hip is flexed. The average range of hip joint motion in the sagittal plane during normal walking, in one gait cycle, is shown in Fig. 20.8.17. The joint was maximally flexed during late swing phase of gait, as the limb moved forward for heel strike. Maximum tension was reached at heel-off.

HIP LOADS

During flexion six muscles that cross the joint anteriorly (the iliacus, psoas major, pectineus, rectus femoris, sartorius, and tensor fascia latae) are primarily responsible for the motion. However, during extension, the three hamstring muscles (the biceps femoris, semitendinosus, and semimembranosus) and the gluteus provide the motion. These muscles are the knee flexors as well.

The hip is the major weight-bearing joint. During upright standing on both feet the body weight is evenly distributed to each of the hip joints. To calculate the forces and torques in each joint or muscle of the body, the dimensions and length of each body segment are needed. The length between each joint of the human body varies with body build, gender, and racial origin. However, an average set of segment lengths expressed as a percentage of body height was prepared by Drillis and Contini (1966) and is shown in Fig. 20.8.18. To determine the mechanism of balance during quiet standing, Winter et al. (1998) modeled the body as 14 segments with 21 markers, as shown in Fig. 20.8.19. The location of the markers and the accompanying table give the definition of each of the 14 segments, along with the mass fraction of each segment.

During a single-leg stance, the force of the abductor muscle is calculated as follows. The weight of the body above the hip joint is approximately equal to five-sixths of the total body weight. The moment arising from the weight above the hip joint must be balanced by a moment occurring from the force of the abductor muscles, as shown in

Segment	Mass Fraction	Definition of Segment COM
Head	0.081	(13 + 14)/2
Trunk 4	0.136	(9 + 12 + 21)/3
Trunk 3	0.078	((19 +20)/2 + 21)/2
Trunk 2	0.065	(17 + 18 + 19 + 20)/4
Trunk 1	0.076	(17 + 18 + 15 + 16)/4
Pelvis	0.142	(15+13)/2
Thighs	0.100 (2)	0.433 x 3 + 0.567 x 5 and 0.433 x 4 + 0.567 x 6
Legs & feet	0.060 (2)	0.606 x 1 + 0.394 x 3 and 0.606 x 2 + 0.394 x 4
Upper arms	0.028 (2)	0.436 x 8 + 0.564 x 9 and 0.436 x 11+ 0.564 x 12
Lower arms	0.022 (2)	0.682 x 7 + 0.318 x 8 and 0.682 x 10 + 0.3186 x 11
Total	1.00	

Fig. 20.8.19 Estimation of the mass fraction and the center of mass (COM) of 14 segments of the body with 21 markers. (*From Winter et al., 1998.*)

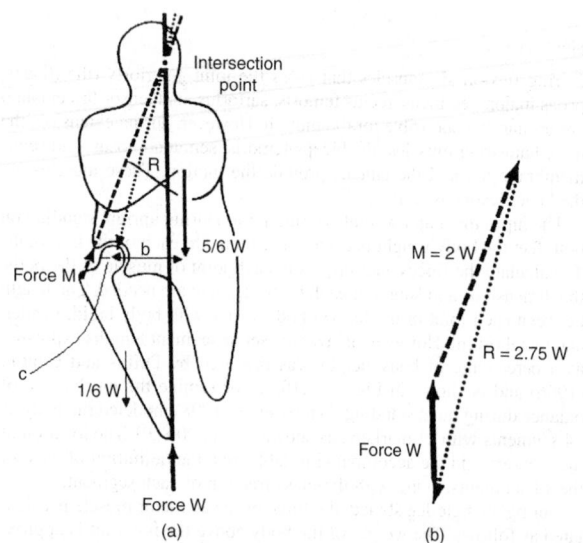

Fig. 20.8.20 Free-body diagram of hip joint during a single-leg stance. The abductor muscle is counterbalance to the moment arising from gravitational force of the body. (*From Nordin and Frankel, 1989, by permission.*)

Fig. 20.8.20. A simple balance of moment with respect to the center of the femoral head reveals that the abductor muscle force M is $M = \frac{5}{6}Wb/c$, where c is the radius of the ball (femoral head) and b is the distance between the center of the ball and the center of gravity (CG) of mass of the body, without the right hip (minus the one leg). Substitution of the values of b and c leads to the abductor muscle force, which is about 2 times the body weight. The reaction force of the hip joint during a single-leg stance is approximately 3 times the body weight. This magnitude varies as the position of the upper body changes.

During dynamic activities several investigators [Paul, 1967; Andriacchi et al., 1980; Bergmann et al., 1997 and 2001] have studied the loads on the hip joint. Figure 20.8.21 shows the hip joint reaction force in units of body weight during walking, one gait cycle. In both men and women, the largest peak of about 7 times the body weight was reached just before toe-off. An increase in gait velocity increases the magnitude of the hip joint reaction force in both swing and stance phases.

BIOMECHANICS OF THE SPINE

The human spine is a complex and functionally significant structure of the human body. The principal function of the spine is to protect the spinal cord and transfer loads from the upper body, head, and trunk to the lower extremities. The spine is capable of motion in all three planes

Fig. 20.8.21 Hip joint reaction forces in units of body weight during walking one gait cycle. Shaded area indicates variations among subjects. (*Adapted from Paul 1967; Nordin and Frankel, 1989, by permission.*)

through its 24 vertebrae which are articulated by intervertebral discs. The discs, surrounding ligaments and muscles, gain the spine stability. The spinal column is structurally divided into four regions. Starting from superior to inferior, there are 7 cervical vertebrae (C1 to C7), 12 thoracic vertebrae (T1 to T12), 5 lumbar vertebrae (L1 to L5), and 5 fused sacral vertebrae (S1 to S5) (Fig. 20.8.22).

The spine contains four normal curves. The thoracic and sacral curves, which are concave anteriorly, are present at birth and are referred to as primary curves. The lumbar and cervical curves, which are concaved posteriorly, are developed from supporting the body in an upright position when young children begin to sit up and stand. These curves are known as secondary spinal curves.

Because of the anatomical changes of vertebrae throughout the spinal column, the amount of movement between adjacent vertebrae is varied. In this subsection, the biomechanics of the lumbar spine and the cervical spine are discussed.

Biomechanics of the Lumbar Spine

The vertebral bodies are designed to bear mainly compressive loads. Anatomically, the vertebral bodies of the lumbar region are thicker and wider than those in the thoracic and cervical regions. A typical vertebra consists of a body, a hollow ring known as the neural arch, and several bony processes (Fig. 20.8.23). The spinous and transverse processes serve as outriggers to improve the mechanical advantage of the attached muscles.

The functional unit of the spine, known as the motion segment, consists of two vertebrae and their intervening soft tissues (Fig. 20.8.24). The intervertebral discs in adults account for approximately one-fourth of the height of the spine. The intervertebral disc bears and distributes loads and restrains excessive motion. The intervertebral disc is composed of two functional structures: a thick outer ring composed of fibrous cartilage, called annulus fibrosus, and a central gelatinous material known as the nucleus pulposus, or nucleus

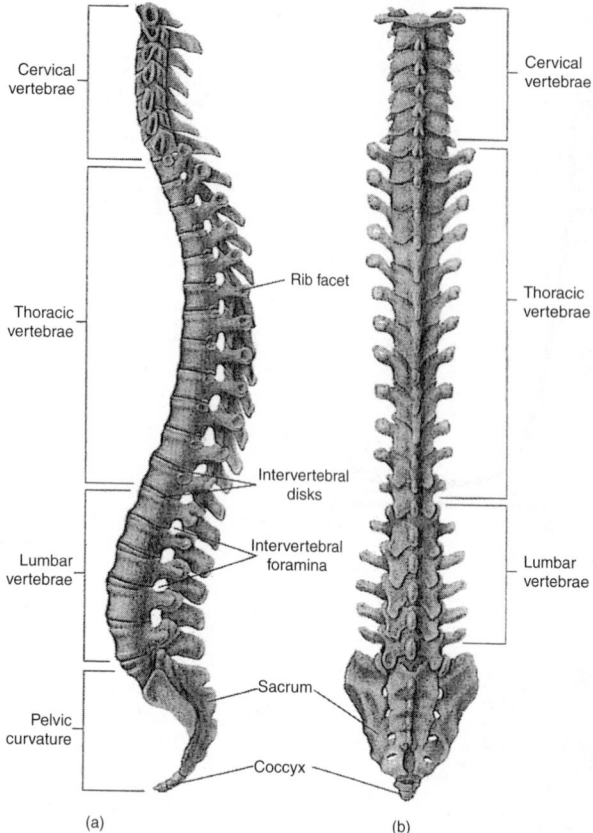

Fig. 20.8.22 The spine. (*a*) Lateral view; and (*b*) the posterior view. (*From "Hole's Human Anatomy and Physiology," 1996; Hall, 2003.*)

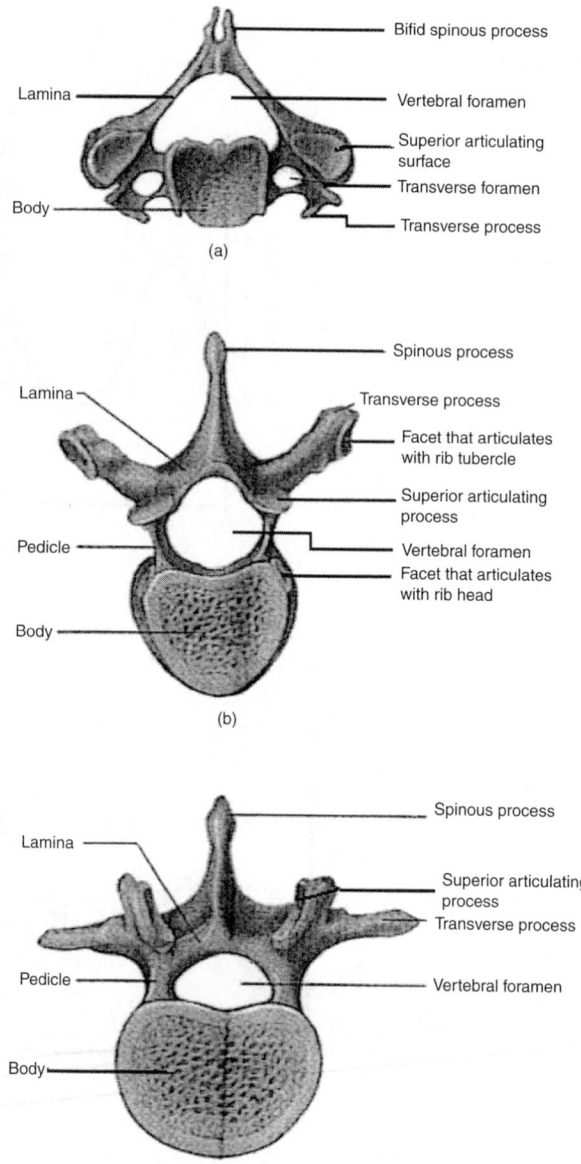

Fig. 20.8.23 Superior views of the typical vertebrae. (*a*) Cervical vertebra, (*b*) thoracic vertebra, and (*c*) lumbar vertebra. (*From "Hole's Human Anatomy and Physiology," 1996; Hall, 2003.*)

(Fig. 20.8.25). The annulus surrounds the nucleus pulposus. The annulus fibrosus, composed of fibrocartilage which is in crisscross arrangement of the coarse collagen fiber bundles within the fibrocartilage, allows the annulus fibrosus to withstand high bending and torsional loads. The nucleus pulposus lies directly in the center of all discs except those in the lumbar segment, where it has a slightly posterior position (Fig. 20.8.25*c*). In young adults, the nucleus is rich in hydrophilic (water-binding) glycosaminoglycans, and it diminishes in glycosaminoglycan content with age and becomes less hydrated. The disc is separated from the vertebral body by the end plate, composed of hyaline cartilage.

The vertebral body is comprised of a thin outer cortical shell and the encompassed cancellous or trabecular bone. The cortical shell and the cancellous bone with the ratio of 45:55 share the load transmitted through the vertebral body, respectively, for an adult of 45 years or younger (Goal and Weinstein, 2000). For subjects over 40 years, the cortical bone shell transmits about 65 percent of the total load exerted on the body, due to an increase in the porosity of the cancellous bone. Table 20.8.4 shows the mechanical properties of the cortical shell and the cancellous bone in the vertebral body. The mechanical properties of ligaments surrounding the vertebrae are given in Table 20.8.5.

SPINE KINEMATICS

The ranges of motion of individual motion segments of the spine have been measured by investigators. The represented values on the relative amount of motion at different levels of the spine are presented by White and Panjabi (1978) in Fig. 20.8.26. The range of flexion and extension is approximately 4° in the upper thoracic motion segments, about 6° in

the middle thoracic region, and about 12° in the two lower thoracic segments. This range progressively increases in the lumbar motion segments, reaching a maximum of 20° at the lumbosacral level. The range of motion is strongly age-dependent, decreasing by about 50 percent from youth to old age. The first 50° to 60° of spine flexion occurs in the lumbar spine, mainly in the lower motion segments. The thoracic spine contributes little to flexion of the total spine because of the oblique orientation of the facets. Significant axial rotation occurs at the thoracic and lumbosacral levels.

SPINAL LOAD

During daily activities the disc is subjected to a combination of compression, bending, and torsional loads. While the facets appear to

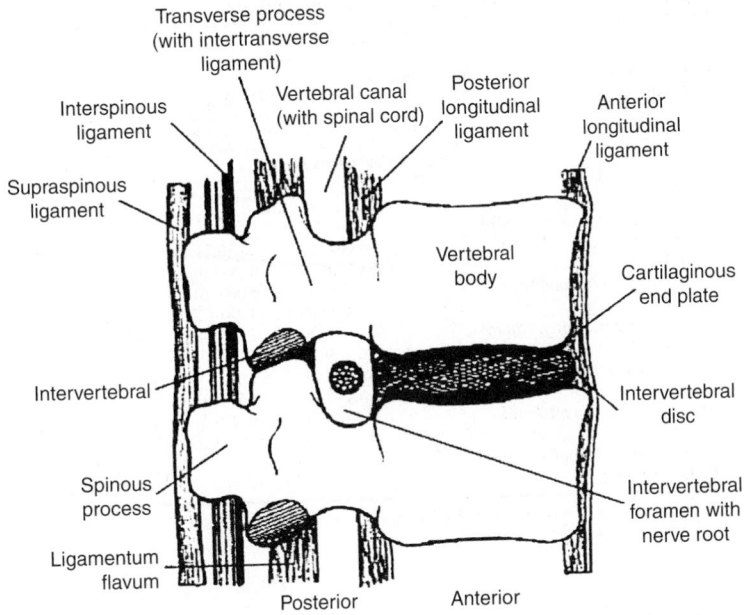

Fig. 20.8.24   The vertebral motion segment and the associated soft tissues. (*From Hall, 2003.*)

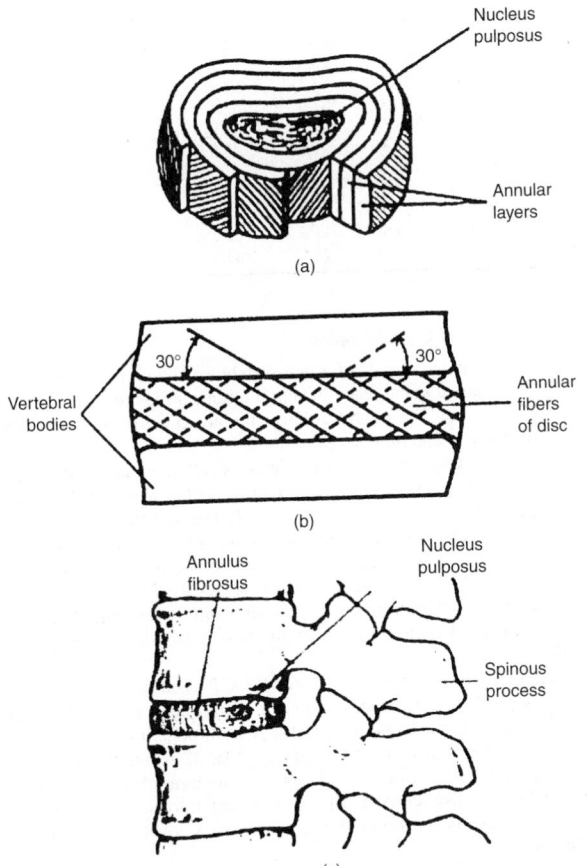

Fig. 20.8.25   Schematic drawings of the intervertebral disc. (*a*) Concentric layers of the annulus fibrosus. (*From Nordin and Frankel, 1989, by permission.*) (*b*) Orientation of the annulus fibrosus. (*From Nordin and Frankel, 1989, by permission.*) (*c*) The lateral view of the lumbar vertebral motion segment. (*From Hall, 2003.*)

function as a guide for the relative movement of adjacent vertebrae, they in fact carry about 30 percent of the total load when the spine is hyperextended (El-Bohy and King, 1986). The vertebral arches and intervertebral joints play an important role in resisting shear forces. The abdominal muscles and the vertebral portion of the psoas muscle initiate flexion. The weight of the upper body produces further flexion.

During the standing position, the center of gravity of the total body is anterior to the spinal column (passes through L4), which makes the postural muscles active. That is, the spine is under a constant forward bending moment. Therefore, the extensor muscles in the back provide a tension, which counteracts the torque (Fig. 20.8.27). The relative loads on the third lumbar disc for various postures are shown in Fig. 20.8.28. As the body continues to flex, the moment arm of the trunk, arm, head, and neck is increased which leads to the increase in tension in the back extensor muscles. Figure 20.8.29 illustrates the back muscle tension, with approximately a 6-cm lever arm counter to torque created by the weight of the body and the external load. When the body is flexed (forward inclination), the spine makes the disc bulge on the concave side of the spine and retracts on the convex side. That is, the disc protrudes anteriorly and is retracted posteriorly.

When a disc is subjected to a compressive load, in a standing position, the compressive stress predominates in the inner portion of the disc, the nucleus pulposus, and tensile stress predominates in the annulus fibrosus. While a person is lifting an object, the increase in abdominal pressure, contributes to the stiffness of the lumbar spine and thus helps in lifting. While a person is in a sitting position, the loads on the lumbar spine are less if the back is supported, since part of the weight of the upper body is supported by the backrest. The trunk muscles involved in motion contribute to the stability of the spine and maintenance of the posture in an upright position.

During slow walking the increase in loads on the lumbar spine is relatively modest. However, the loads increase with an increase in speed of walking. In his study, Cappozzo (1984) showed that the loads were maximal around toe-off and increased approximately linearly with walking speed. Figure 20.8.30 revealed the axial load on the L3-L4 motion segment in terms of body weight during walking at four speeds. The solid horizontal line in the figure shows the upper body weight (UBW), and the maximum load occurred about the left heel strike (LHS) and right heals strike (RHS).

Table 20.8.4 Mechanical Properties of a Vertebral Body

Cortical bone		
Modulus of elasticity, N/mm² or MPa	Shear modulus, N/mm² or MPa	Poisson's ratio (ν)
12,000	4,615.0	0.30[a]

Spongy or Cancellous Bone				
	Males[b]	Females[b]	a	c
Tensile strength, N/mm² or MPa	—	—	—	1.18
Compressive strength, N/mm² or MPa	4.6(0.3)[d] (0.2–10.5)[e]	2.7(0.2) (0.3–7.0)	—	1.37–1.86
Compression at rupture, %	9.5(0.4) (5.3–14.4)	9.00(0.6) (3.2–14.7)	—	2.5
Limit of proportionality, N/mm² or MPa	4.0(0.1) (0.1–9.7)	2.2(0.1) (0.2–6.0)	—	
Compression at limit of proportionality, %	6.7–(0.2) (4.1–8.6)	6.1(0.4) (2.6–10.0)	—	
Modulus of elasticity, N/mm² or MPa	55.6(0.7) (1.1–139.1)	35.1(0.6) (5.2–103.6)	100	68.7–88.3
Shear modulus, N/mm² or MPa			41	
Poisson's ratio ν			0.2	0.14

[a]Shirazi-Adl et al. (1984). [b]Yamada (1970). [c]Lindhal (1976). [d]Mean (SD). [e]Range.
SOURCE: Goal and Weinstein (2000).

Table 20.8.5 Mechanical Properties of Spinal Elements

Property	Ligament parameter	ALL	TL	CL	LF	ISL	SSL
Structural	X-area, mm²	63.7 (11.3)	3.0 (0.4)	20.9 —	56.0 (7.0)	75.7 (41.3)	30.1 (8.8)
	Load at failure, N	177.0 (30.0)	15.0 (1.0)	191.5 (58.2)	87.0 (24.7)	90.5 (44.5)	118.3 (68.9)
	Stiffness, N/mm	164.0 (10.3)	3.7 (0.9)	49.7 (21.9)	66.5 (43.5)	41.5 (1.5)	55.1 (45.9)
Material	Stress at failure, MPa	1.5 (0.1)	3.6 (1.3)	(3.0) —	1.3 (0.1)	1.3 (0.1)	3.6 (2.2)
	Stress at failure, %	8.2 (0.2)	17.1 (1.2)	9.0 (0.8)	13.0 (1.0)	7.8 (1.4)	7.8 (0.8)
	Young's modulus, MPa	20.1 (2.2)	58.7 (26.5)	32.9 —	19.5 (0.5)	11.6 (3.5)	62.7 (37.5)

SOURCE: Goal and Weinstein (2000) with permission.

Biomechanics of Cervical Spine

The cervical spine consists of seven vertebrae. The first two vertebrae, C1 and C2, are atypical, each having a unique structure and role; and the rest of them, C3 through C7, have similar structure and function to those of thoracic and lumbar spine.

The first vertebra in the cervical spine is called the **atlas**, which has no vertebral body and is composed of a ring within which an oval fossa articulates anteriorly with the dens of C2. The superior facets of C1, which form the base of the atlanto-occipital joint, bear the weight of the skull. The second vertebra is called *axis* and is composed of dens, superior facets, pedicle, and spinous process (Fig. 20.8.31). The dens is an articular process that protrudes superiorly from the vertebral body and around which C1 rotates. This configuration allows significant rotation of the head.

The five typical cervical vertebrae C3-C7, each has a body that is elliptical with a transversely concave upper surface, two pedicles, two laminae, and a spinous process. A motion segment consists of two adjacent vertebrae is shown in Fig. 20.8.32 where the ligaments and the intervertebral disc are shown. The intervertebral disc, similar to the lumbar disc, consists of a central gelatinous mass, known as nucleus pulposus, surrounded by a tough outer covering, the annulus fibrosus.

CERVICAL SPINE KINEMATICS

The cervical spine is the most mobile region of the spine. The range of motion of the cervical spine in flexion-extension is about 145°, about

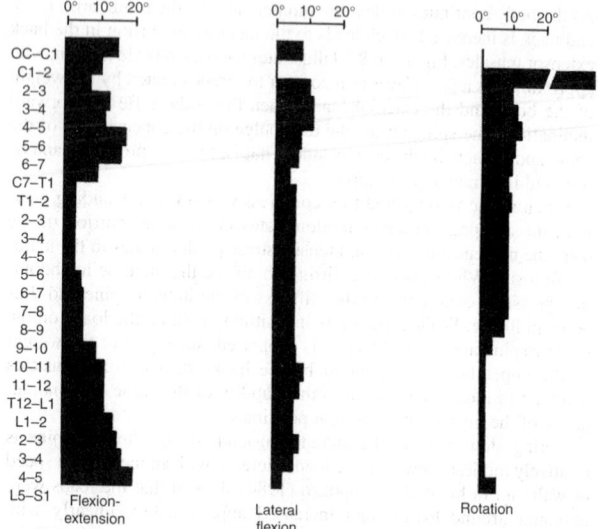

Fig. 20.8.26 Representative values for type and range of motion at different levels of spine. (*Adapted from White and Panjabi, 1978; Nordin and Frankel, 1989, by permission.*)

Fig. 20.8.27 Schematic lateral view of the human body; the line of gravity passes through L4. (*From Hall, 2003.*)

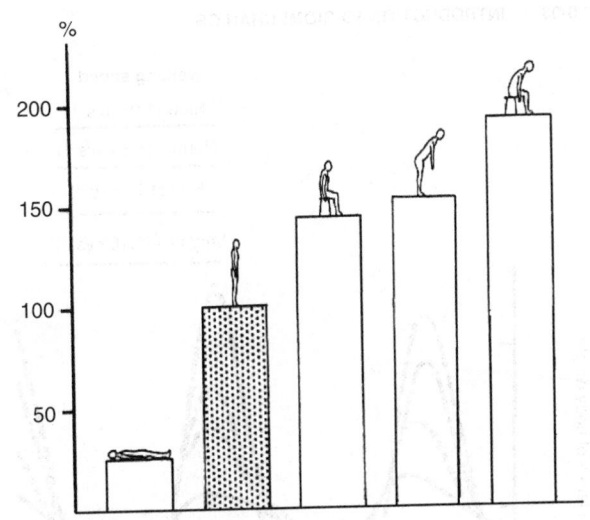

Fig. 20.8.28 The relative loads on the third lumbar disc for various postures. (*From Nordin and Frankel, 1989, by permission.*)

Fig. 20.8.29 Free-body diagram of spine and body loads when subject lifts an external load. The back muscle with a moment of arm of approximately 6 cm must support the moment produced by the body and the load. (*From Hall, 2003.*)

Walking speed

Slow (1.05 m/s⁻¹)

Normal (1.38 m/s⁻¹)

Fast (1.72 m/s⁻¹)

Very fast (2.16 m/s⁻¹)

Fig. 20.8.30 Variation of axial load on L3-L4 motion segment based on the body weight during walking at four speeds. The horizontal line shows the upper body weight (UBW), which represents the gravitational component of the load. (*From Nordin and Frankel, 1989, by permission.*)

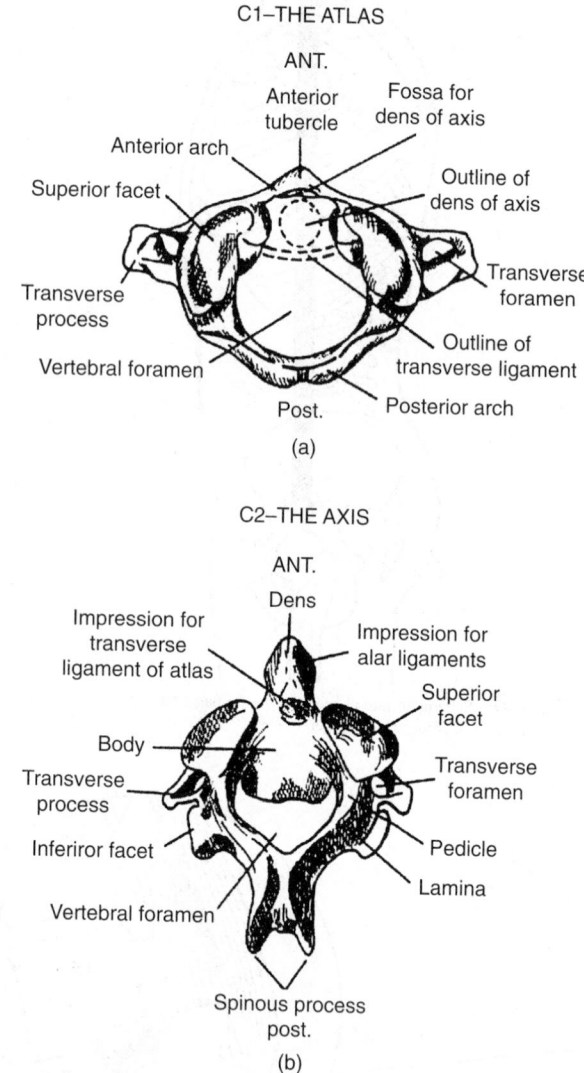

Fig. 20.8.31 Superior views of the atypical cervical vertebrae (*a*) C1 and (*b*) C2. (*From Nordin and Frankel, 1989, by permission.*)

180° in axial rotation, and about 90° in lateral bending. Specifically, motion at the atlanto-occipital articulation, between C1 and the skull, consists of 10° to 15° of flexion-extension and 8° of lateral flexion, with no rotation. The skull rotation is transferred into motion at the C1-C2 articulation, the atlantoaxial joint. The C1-C2 joint is the most mobile segment of the cervical spine, having up to 47° of axial rotation, representing 50 percent of the axial rotation of the whole cervical spine. However, at this joint, about 10° of flexion and extension takes place with minimal lateral flexion (White and Panjabi, 1978). Flexion, extension, lateral flexion, and axial rotation occur at the joints below C2. A total of +/−45° of axial rotation takes place in C3-C7. Note that the range of motions varies widely among individuals.

The cervical spine stability has been discussed extensively in White and Penjabi's publications. Instability is defined as excessive relative movements of adjacent vertebrae where there is a need for clinical procedure to return the vertebrae to their normal positions. They suggested that the adult cervical spine is unstable, or on the brink of instability, when any of the following conditions is present: more than 3.5 mm of transverse (horizontal) relative displacement of two adjacent vertebrae, or more than an 11° difference in rotation from that of either adjacent vertebra. These cervical conditions need clinical attention to be corrected.

CERVICAL SPINE LOAD

Distribution of loads throughout the vertebra is a complex issue, which has been studied by several investigators. The results of these studies indicate that the facets bear a portion of the load (King et al., 1975; Miller et al., 1978). In particular, in extension, the facets bear up to

33 percent of the total load, reducing substantially the load of the discs. Harms-Ringdahl (1986) calculated the bending moment generated around the axes of motion of the atlanto-occipital joint (Occ-C1) and at the C7-T1 motion segments. Figure 20.8.33 shows the extension and flexion moments at these joints in a variety of head positions. During extreme flexion the maximum bending moment in the neck occurs at the C7-T1 joint, while the moment at the Occ-C1 joint is relatively high. As the head moves to its natural position, the moment at the Occ-C1 joint slightly decreases while the moment at C7-T1 significantly decreases so that both joints carry approximately equal amounts of load. As the head continues to rotate and approaches the extreme extension, the moment on the Occ-C1 is further reduced while that of C7-T1 increases in the negative direction.

The head and neck, due to their structures, i.e., a large mobile ball atop a column, are particularly vulnerable to dynamic injuries. Excessive loading and motion of the head can easily create stresses and bending moments that exceed the strength of the structure, which leads into subluxation and destabilization of the neck.

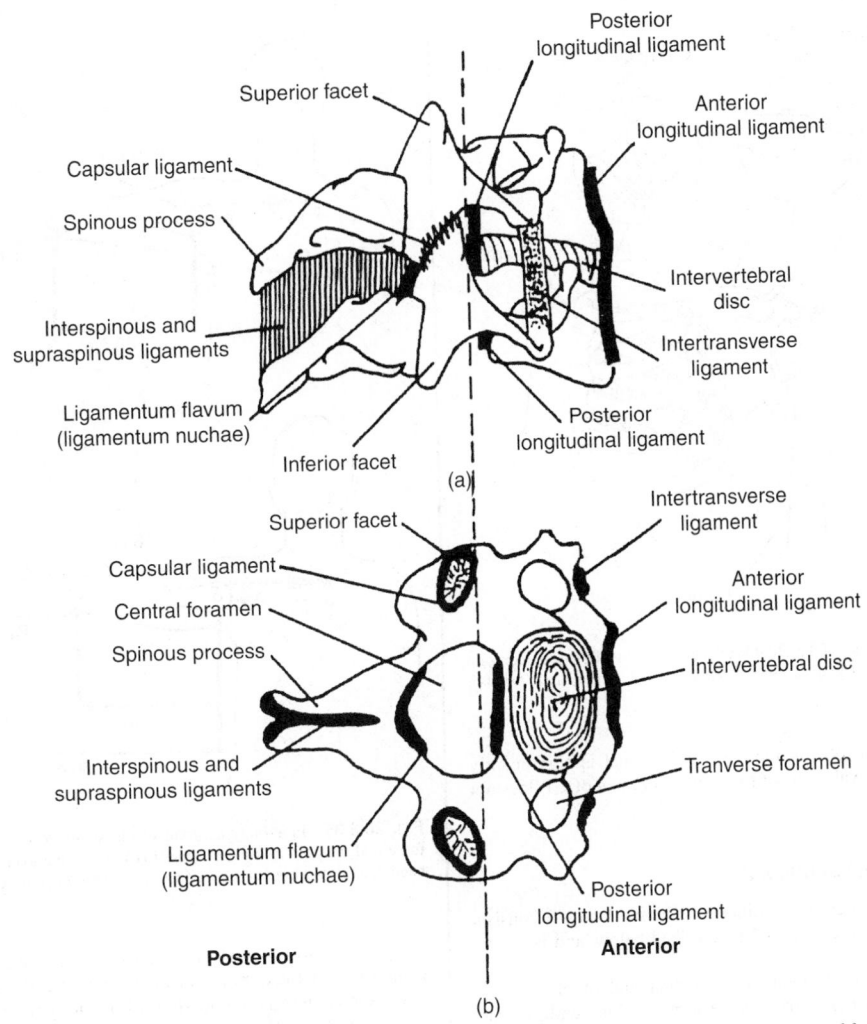

Fig. 20.8.32 Lateral and superior schematic views of a cervical motion segment of a typical vertebra. (*Adapted from White et al., 1975; Nordin and Frankel, 1986, by permission.*)

Extreme flexion	Slight flexion	Neutral	Chin tucked in	Extreme extension
OCC–C1	OCC–C1	OCC–C1	OCC–C1	OCC–C1
1.5 N·m	1.4 N·m	1.3 N·m	1.0 N·m	−0.2 N·m
(0.9–1.8)	(0.8–1.7)	(0.6–1.5)	(0.4–1.5)	(−0.4–0.3)
C7–T1	C7–T1	C7–T1	C7–T1	C7–T1
4.3 N·m	3.7 N·m	1.2 N·m	−0.4 N·m	−1.7 N·m
(3.7–6.5)	(3.0–6.2)	(0.5–2.2)	(−0.8–0.9)	(−2.4– −1.1)

Fig. 20.8.33 Extension-flexion moments at atlanto-occipital (Occ-C1) joint and C7-T1 motion segment (marked with X's) for five positions of the head. The arrow represents the force vector produced by the weight of the head. (*Adapted from Harms-Ringdahl, 1986; Nordin and Frankel, 1989, by permission.*)

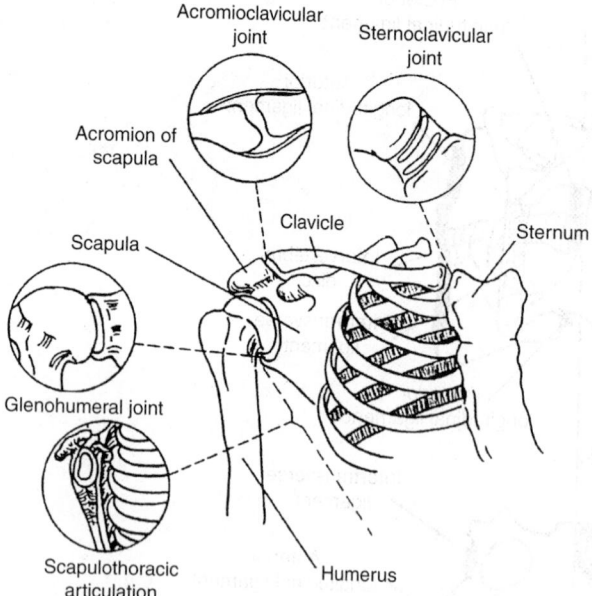

Fig. 20.8.34 Biomechanics of the shoulder. Schematic view of the shoulder joint and the four articulations. (*Adapted from DePalma, 1983; Nordin and Frankel, 1989, by permission.*)

During normal axial rotation the size of the canal is significantly decreased because the oval shape of the central foramen of C1 does not coincide with that of C2.

BIOMECHANICS OF THE SHOULDER

The shoulder is the most complex joint in the human body. It consists of four major separate articulations, as described below, and is shown in Fig. 20.8.34.

1. At the sternoclavicular joint, the proximal end of the clavicle articulates with the clavicular notch of the sternum along with the cartilage of the first rib.

2. At the acromioclavicular joint, the acromion process of the scapula (shoulder blade) articulates with the distal end of the clavicles.

3. At the glenohumeral joint, the hemispherical head of the humerus articulates with the glenoid fossa of the scapula. This is a ball-and-socket configuration joint and is the most freely moving joint in the human body, enabling flexion, extension, hyperextension, abduction, adduction, horizontal abduction and adduction, and medial and lateral rotation of the humerus. There are four major muscles on the posterior, superior, and anterior sides of the joint, providing the joint movements; they are called rotator cuff muscles. The capsule of the joint is surrounded and is stabilized by several ligaments and the tendons of four muscles, as well as the biceps. Negative pressure within the capsule of the glenohumeral joint also helps to stabilize the joint.

4. At the scapulothoracic joint, the anterior scapula is articulated with the thoracic wall. This is due to the movement of the scapula in the sagittal and frontal planes.

Shoulder Loads

The ball-and-socket configuration of glenohumeral joint provides mainly the rotation of the shoulder; however, it includes some rolling and gliding motion as well. The rotator cuff muscles are oriented such that their tendons and muscle masses may push on the head of the humerus, thereby stabilizing the joint.

Fig. 20.8.35 Free-body diagram of the shoulder joint, showing the reaction force and the deltoid muscle force. (*a*) Without the load; (*b*) with a load in the hand. (*Adapted from Nordin and Frankel, 1989, by permission.*)

When the arm is in 90° of abduction (extended horizontally in the frontal plane), the deltoid muscle (superior of the shoulder joint) provides a tension of about one-half of the body weight. This can be calculated by a simple static equilibrium of the joint, as shown in Fig. 20.8.35. When a light weight of about 0.025 times the body weight (BW) is held in the hand, the muscle force is approximately equal to the body weight. In these cases the joint reaction force, which is equal and opposite to the muscle force, is as high as the body weight. Therefore, the glenohumeral joint is a major load-bearing joint.

THE ELBOW

At the elbow joint, three bones are joined: humerus, ulna, and radial. The elbow joint consists of three articulations: the humeroulnar, humeroradial, and proximal radioulnar. All these articulations are enclosed in the same joint capsule, which is reinforced by the anterior and posterior radial collateral and ulnar collateral ligaments. The elbow joint allows 2° of freedom in motion: flexion-extension and pronation-supination (rotation of the radius around the ulna).

Elbow Load

The elbow is not considered a weight-bearing joint; however, it is regularly subjected to large loads during our daily activities. Two simple diagrams of Fig. 20.8.36 show the muscle forces and joint reaction forces, which could be calculated by a simple static equilibrium. The research has shown that the compressive loads during activities such as dressing and eating could reach as high as 300 N (67 lb), and during rising from a chair when the body is supported by the arms, the load could reach as high as 1,700 N (382 lb). During push-up exercises peak forces on the elbow could reach as high as 45 percent of the body weight.

Fig. 20.8.36 Free-body diagram of the elbow joint showing (*a*) the reaction force in flexion and the tendons of the biceps and the brachialis muscle forces and (*b*) the tendon of the extensor (triceps) muscle tendon. (*From Nordin and Frankel, 1989, by permission.*)

BIOMATERIALS

By definition, **biomaterial** is a biocompatible material suitable for inclusion, augmentation, or replacement of the function of bodily tissues or organs. Biomaterials science is a multidisciplinary field that brings together researchers from diverse academic backgrounds, such as bioengineering, chemistry, chemical engineering, electrical engineering, mechanical engineering, materials science, biology, microbiology, and medicine. Biomaterials can be synthetic and natural, solid and sometimes liquid. Solid biomaterials can be metals, ceramics, polymers, glasses, carbons, and composite materials. Such materials are used as molded or machined parts, coatings, fibers, films, foams, and fabrics. Biomaterials are generally integrated into medical devices and implants and are rarely used on their own.

Historically, in the early 1900s bone plates were successfully implemented to stabilize bone fractures and to accelerate their healing, and since 1960s blood vessel replacement and artificial heart valves and joint replacements have been used. Since then, with the advent of new materials and prostheses, the field of biomaterials has rapidly expanded and has saved many lives.

Key factors in the design and use of biomaterials are biocompatibility, toxicology, biofunctionality, and, to a lesser extent, availability of biomaterials. Except for their brittle characteristics, ceramic materials are ideal candidates as biomaterials. Most biomaterials and medical devices perform satisfactorily, improving the quality of life for the recipient or saving lives; however, they have exhibited some failures. This is due to the fact that manufactured materials and devices have a failure rate. Other factors such as genetic differences, gender, body chemistries, living environment, and physical activity as well as physicians' "talent" for implanting devices may contribute to the success or failure of the biomaterials and implants.

Biocompatibility

The understanding and measurement of the biocompatibility of biomaterials are not fully understood. It has been accepted that no foreign material placed within a living body is completely compatible. Only those materials (natural tissues) manufactured by the body itself (autogenous) are compatible, and any other substance that is recognized as foreign initiates some type of reaction (host-tissue response). Natural material that is obtained from an individual and will be implanted into the same person is called an **autograft.** However, if the donor is a different individual, the material is called an **allograft.**

Toxicity

Toxicology in biomaterials has evolved into a sophisticated science, which deals with the substances that migrate out of biomaterials. For example, for polymers, many low-molecular-weight materials have

shown some level of physiologic activity and cell toxicity. It is reasonable to say that a biomaterial should not give off anything from its mass unless it is specifically designed to do so, e.g., cancer drug delivery systems to destroy specific cells.

In terms of tissue response to the biomaterials and implants, there are three types of biomaterials: bioinert, bioresorbable, and bioactive. (Buddy, Ratner et al., 1996).

Bioinert Any material that once placed in the human body has minimal interaction with its surrounding tissue is called **bioinert.** For example, stainless steel, titanium, alumina, partially stabilized zirconia, and ultrahigh-molecular-weight polyethylene (UHMWPE) (to some degree) are bioinert. A fibrous capsule that might form around bioinert implants may compromise the biofunctionality of bioinert materials. In addition, the tissue remodeling around the implant (bone ingrowth or architectural changes such as resorption or depositon of bone) may alter the functionality of the implant.

Bioresorbable Materials that upon placement within the human body start to dissolve (are resorbed) and slowly are replaced by advancing tissue (such as bone) are **bioresorbable.** Common examples of bioresorbable materials are tricalcium phosphate [$Ca_3(PO_4)_2$] and polylactic–polyglycolic acid copolymers. Calcium oxide, calcium carbonate, and gypsum are other common materials that have been utilized during the last three decades.

Bioactive Materials that upon placement in the human body interact with the surrounding bone and in some cases even soft tissue are **bioactive.** This occurs through a time-dependent kinetic modification of the surface, triggered by their implantation within the living bone. An ion-exchange reaction between the bioactive implant and the surrounding body fluids results in the formation of a biologically active carbonate apatite layer on the implant. For example, coating an implant with synthetic hydroxyapatite [$Ca_{10}(PO_4)_6(OH)_2$] triggers bone ingrowth.

OCCUPATIONAL BIOMECHANICS

Human Factors in Design

In interacting with machines and devices, the human musculoskeletal system often has to provide forces to power a product or actuate a controller. The designs of machines and devices are influenced by humans' ability to supply the force and their ability to interact with the machine. The interface between humans and machines requires that humans sense the state of the device and be able to control it. That is, a product must be comfortable for a person to use. Thus products and devices must be designed with human factors in mind [see Bridger (1995)]. The geometric properties of humans (i.e., their height, reach, and seating requirements; the size of the holes they can fit through; and the forces that they can exert in specific directions with their body segments) are called anthropometric data. The armed forces in Military Standard MIL-STD-1472 have collected the majority of these data. A sample of these data is shown in Fig. 20.8.37 and Table 20.8.6. In addition, occupational biomechanics researchers have studied measures for humans' ability to perform various activities. Figure. 20.8.38 shows the average human strength for differing body positions (Dreyfuss, 1967). For example, in the standing position, the push and pull force of the hand in the position slightly lower than the torso region is higher than that of the hand position at the neck level. More detailed information is available in MIL-HDBK (Military Handbook) 759A and books in occupational biomechanics.

PROSTHESES AND IMPLANTS

There are implants and prostheses for virtually every member and organ of the human body. Being among the oldest implants, hip and knee prostheses are discussed in this subsection.

Hip Implants

The hip prosthesis has been the most active area of joint replacement research since **total hip** joint **replacement (THR)** among the most common orthopedic procedures. The hip joint is called a ball-and-socket

Fig. 20.8.37   Anthropometric man. (*From MIL-STD-1472D.*)

Table 20.8.6   Anthropometric Data

| | Percentile values, cm | | | | | |
| | 5th percentile | | | 95th percentile | | |
	Ground troops	Avia-tors	Women	Ground troops	Avia-tors	Women
Weight, kg	65.5	60.4	46.6	91.6	96.0	74.5
Standing body dimensions						
1. Stature	162.8	164.2	152.4	185.6	187.7	174.1
2. Eye height (standing)	151.1	152.1	140.9	173.3	175.2	162.2
3. Shoulder (acromial) height	133.6	133.3	123.0	154.2	154.8	143.7
4. Chest (nipple) height*	117.9	120.8	109.3	136.5	138.5	127.6
5. Elbow (radiate) height	101.0	104.8	94.9	117.8	120.0	110.7
6. Fingertip (dactylion) height		61.5			73.2	
7. Waist height	96.6	97.6	93.1	115.2	115.1	110.3
8. Crotch height	76.3	74.7	68.1	91.8	92.0	83.9
9. Gluteal furrow height	73.3	74.6	66.4	87.7	88.1	81.0
10. Kneecap height	47.5	46.8	43.8	58.6	57.8	52.5
11. Calf height	31.1	30.9	29.0	40.6	39.3	36.6
12. Functional reach	72.6	73.1	64.0	90.9	87.0	80.4
13. Functional reach, extended	84.2	82.3	73.5	101.2	97.3	92.7
	Percentile values, in					
Weight, lb	122.4	133.1	102.3	201.9	211.6	164.3
Standing body dimensions						
1. Stature	64.1	64.6	60.0	73.1	73.9	68.5
2. Eye height (standing)	59.5	59.9	55.5	68.2	69.0	63.9
3. Shoulder (acromial) height	52.6	52.5	48.4	60.7	60.9	56.6
4. Chest (nipple) height*	46.4	47.5	43.0	53.7	54.5	50.3
5. Elbow (radiate) height	39.8	41.3	37.4	46.4	47.2	43.6
6. Fingertip (dactylion) height		24.2			28.8	
7. Waist height	38.0	38.4	36.6	45.3	45.3	43.4
8. Crotch height	30.0	29.4	26.8	36.1	36.2	33.0
9. Gluteal furrow height	28.8	29.4	26.2	34.5	34.7	31.9
10. Kneecap height	18.7	18.4	17.2	23.1	22.8	20.7
11. Calf height	12.2	12.2	11.4	16.0	15.5	14.4
12. Functional reach	38.6	28.8	25.2	35.8	34.3	31.7
13. Functional reach, extended	33.2	32.4	28.9	39.8	38.3	36.5

*Bus point height for women.
SOURCE: Dreyfuss (1967).

Fig. 20.8.38 *Average human strength for different tasks. (From H. Dreyfuss, 1967.)*

joint because the spherical head of the femur moves inside the cup-shaped hollow socket (acetabulum) of the pelvis. To duplicate this action, a total hip replacement implant has three parts: the stem, which fits into the femur and provides stability; the ball, which replaces the spherical head of the femur; and the cup, which replaces the worn-out hip socket (Fig. 20.8.39). There are many designs of hip implant parts based on age, weight, bone quality, activity level, and

health of the recipient. The stem portions of most hip implants are made of titanium- or cobalt/chromium-based alloys; they come in different shapes and degrees of roughness. Cobalt/chromium-based alloys or ceramic materials (aluminum oxide or zirconium oxide) are used in making the ball portions, which are polished smooth to allow easy rotation within the prosthetic socket. The acetabular socket can be made of metal, with a liner made of ultrahigh-molecular-weight

Fig. 20.8.39   Total hip implant.

Fig. 20.8.40   Components of the total knee arthroplasty implant.

polyethylene (UHMWPE) or a combination of polyethylene backed by metal. All together, these components weigh between 14 and 18 oz, depending on the size needed. The chance of a hip replacement lasting 20 years is about 80 percent.

There are many hip implant designs (such as modular stems) and a variety of surface textures (such as sintered beads, meshes, and etched). For more information the reader is referred to manufacturers' Web sites.

IMPLANT FIXATION

To hold the femoral and acetabular components in place, two methods, cemented and cementless (bone ingrowth or osseointegration), are used. The cement, an acrylic polymer called polymethylmethacrylate (PMMA), is used for older patients. However, the cementless method is used for younger patients (under 50 years of age) who are active and have good bone quality. Cemented fixation relies on a stable interface between the prosthesis and the cement and a solid mechanical bond between the cement and the bone.

Cementless implants were introduced in the 1980s, and the implant is directly attached to bone without the use of cement. The surface topography of the implant is textured or has porous regions, and the implant is generally coated with hydroxyapatite so that the new bone actually grows into the surface of the implant. In general, these designs are larger and longer than those used with cement. The surgeon's accuracy in placing the implant is an important factor in the stability of the implant since the channel (bone cavity) must match the shape of the implant precisely. Intimate contact between the component and bone is crucial to permit bone ingrowth. New bone growth cannot bridge gaps larger than 1 to 2 mm.

Hybrid THR, which was introduced in the early 1980s, has one component, usually the acetabular socket, inserted without cement and the other component, usually the femoral stem, inserted with cement. A hybrid hip takes advantage of the excellent track records of cementless hip sockets and cemented stems.

Loosening of hip implants may occur due to several factors.

1. Cracks (fatigue fractures) in the cement that occur over time can cause the prosthetic stem to loosen and become unstable. This is more often the case with patients who are very active or very heavy.

2. The action of the metal ball against the polyethylene cup of the acetabular component creates polyethylene wear debris. The microscopic debris particles are absorbed by cells around the joint and initiate an inflammatory response from the body, which tries to remove them.

3. Micromotion between the implant and the bone prevents bone ingrowth into the implant.

The hip implants could be made of two or three types of materials, namely, metal, ceramic, and polyethylene. The crucial part of the implant is the contact between the ball and the acetabular liner. The ball could be made of metal or ceramic, and the liner could be made of ceramic, metal, and UHMWPE. The metal and plastic implants wear at a rate of about 0.1 mm/yr. Metal-on-metal implants wear at a rate of about 0.01 mm/yr about 10 times less than that of metal and plastic. There are also concerns about the wear debris generated from the metal-on-metal implants. Ceramics are the most resistant to wear of all available hip replacement implants. However, due to low fracture toughness, ceramic implants can break inside the body.

The Knee Implants

The **total knee arthroplasty (TKA)** implants are more complex because of the articulation of three bones—distal femur, proximal tibia, and patella (the kneecap)—where the surfaces actually roll and glide as the knee bends. There are many knee implant designs in the market having the following major components (Fig. 20.8.40).

1. **Femoral component.** The metal femoral component curves around the distal end of the femur and has an interior groove so the kneecap can move up and down smoothly against the bone as the knee bends and straightens. In some designs, a smaller piece may be used (unicompartmental knee replacement) to resurface just the medial or lateral part of the condyle.

2. **Tibial component.** The tibial component is a flat metal platform.

3. **Polyethylene cushion.** The cushion may be part of the platform (fixed) or separate (mobile).

4. **Patellar component.** The patellar component is a dome-shaped piece of polyethylene that duplicates the shape of the kneecap anchored to a flat metal plate.

The metal parts of the implant are made of titanium- or cobalt/chromium-based alloys. The plastic parts are made of ultrahigh-molecular-weight polyethylene. All together, the components weigh between 15 and 20 oz, depending on the size selected.

Due to its continuous sliding and rolling, knee implant wear is a major problem where microscopic particulate debris are generated.

Similar to the hip implants, the knee implants may be cemented, cementless, or hybrid, depending on the type of fixation used to hold the implant in place.

Other Implants

The majority of orthopedic implants are made of metals. There is considerable mismatch between the metal and bone properties, which can lead to stress shielding and bone resorption. New composite implants or graded materials are being developed to closely match the bone properties and thereby reduce the risk of bone resorbtion.

There are many other orthopedic implants and prostheses for virtually every joint in the body as well as plates and rods for bone repair and fracture fixations. In addition, there are many types of screws for different locations and varieties of bones. For these topics the reader is referred to specialized orthopedic books, journals, and manufacturers' Web sites.

Dental implants have been widely used in recent years. There are two categories of dental implants: endosteal implants, which enter the bone tissue, and subperiosteal systems, which contact the exterior bone surfaces (Lemons, 1988).

Mechanical or tissue **prosthetic heart valves** are used to replace the natural heart valves (miteral, tricuspid, aortic, and pulmonic) when these no longer perform their normal functions because of disease (Gidden et al., 1993). Mechanical valves are of two types: caged ball and tilting disk. The materials most widely used are silicone elastomer, cobalt/chrome-based alloys, titanium, and pyrolytic carbon.

A **stent** is a device made of inert materials and designed to serve as a temporary or permanent internal scaffold to maintain or increase the lumen of a vessel. For further information see Schatz (1989).

REFERENCES

Andriacchi, T., et al. 1980. A Study of Lower-Limb Mechanics during Stair-Climbing, *Jour. Bone Joint Surg.* **62A**:749–757.

Aydin, Tozeren, 2000. "Human Body Dynamics," Springer-Verlag.

Berger, S. A., et al., 2000. "Introduction to Bioengineering," Oxford Press.

Bergmann, G., et al., 1997. Hip Joint Forces during Load Carrying, *Clin. Orthop.* **335**:190.

Bergmann, G., et al., 2001. Hip Contact Forces and Gait Patterns from Routine Activities, *Jour. Biomech.* **34**:859

Bogduk, N., and S. R. Mercer, 2000. Biomechanics of the Cervical Spine, I. Normal Kinematics, *Clin. Biomech.* **15**:633.

Brault, J., et al., 2000. Cervical Muscle Response during Whiplash: Evidence of a Lengthening Muscle Contraction, *Clin. Biomech.* **15**:426.

Bridger, R. S., 1995. "Introduction to Ergonomics," McGraw-Hill.

Buddy, Ratner, et al., 1996. "Biomaterials Science: An Introduction to Materials in Medicine," Academic Press.

Cappozzo, A., 1984. Compressive Load in the Lumbar Vertebral Column during Normal Level Walking, *Jour. Orthop. Res.* **1**:292.

Chaffin, Don, et al., 1999. "Occupational Biomechanics," Wiley.

Cholewicki, J., et al., 1999. Intra-abdominal Pressure Mechanism for Stabilizing the Lumbar Spine, *Jour. Biomech.* **32**:13.

Cowin, S., et al, 1992. An Evolutionary Wolff's Law for Trabecular Architecture, *Jour. of Biomech. Eng.* **116**:129–136.

Davis, K. G., and W. S. Marras, 2000. The Effect of Motion on Trunk Biomechanics, *Clin. Biomech.* **15**:703.

DePalma, A. F., 1983. Biomechanics of the Shoulder, in "Surgery of the Shoulder," 3d ed., J. B. Lippincott, Philadelphia, pp. 65–85.

Dreyfuss, H., 1967. "The Measure of Man: Human Factors in Design," Whitney Library of Design, New York. This is a looseleaf book of 30 anthropometric and biomedical charts. A classic.

Drillis, R., and R. Contini, 1966. "Body Segment Parameters," Rep. 1163-03, Office of Vocational Rehabilitation, Dept. of Health, Education, and Welfare, New York.

El-Bohy, A., and A. King, 1986. Intervertebral Disc and Facet Contact Pressure in Axial Torsion, in "Adv. in Bioeng." Lantz and King, (eds.), ASME, pp. 26–27.

Fung, Y. C., 1993. "Biomechanics of Mechanical Properties of Living Tissues," Springer-Verlag.

Gibson, L., and M. Ashby, 1997. "Cellular Solids: Structures and Properties," 2d ed., Pergamon Press, Oxford, U.K.

Gidden, Yagonathan, and F. J. Schoen, 1993. Prosthetic cardiac valves, Special supplement to *Cardiovascular Pathology*, **2**(3), PP167S-177S.

Goal, Vijay, and James, Weinstein, 2000. "Biomechanics of the Spine: Clinical and Surgical Perspective," CRC Press.

Hall, Susan, 2003. "Basic Biomechanics," 4th ed., McGraw-Hill.

Hall, S. J., 1985. Effect of Attempted Lifting Speed on Forces and Torque Exerted on the Lumbar Spine, *Med. Sci. Sports Exerc.* **17**:440.

Harms-Ringdahl, K., 1986. On Assessment of Shoulder Exercise and Load-Elicited Pain in the Cervical Spine. Thesis. Karolinska Institute, University of Stockholm.

Hill, A. V., 1970. "First and Last Experiments in Muscle Mechanics," Cambridge University Press, Cambridge.

Hodges, P. W., et al., 2001. In Vivo Measurement of the Effect of Intra-abdominal Pressure on the Human Spine, *Jour. Biomech.* **34**:347.

"Hole's Human Anatomy and Physiology," 1996. McGraw-Hill.

"Human Factors Engineer Design for Army Material," MIL-HDBK 759A. Almost 700 pages of information on all aspects of human factors.

"Human Factors Engineer Design Criterion for Military Systems, Equipment, and Facilities," MIL-STD 1472 D.

Imwold, Denis, and Janet Parker, 2000. "Anatomica, the Complete Home Medical Reference," Global Book Publishing.

King, A. I., P. Prasad, and C. L. Ewing, 1975. Mechanism of Spinal Injury due to Caudocephalad Acceleration, *Othop. Clin. North Am.* **6**:19.

Kopperdahl, D., and T. Keaveny, 1998. Yield Strain Behavior of Trabecular Bone, *Jour. Biomech.* **31**(7):601–608.

Kutz, Myer (ed.) 2003. "Standard Handbook of Biomedical Engineering and Design," McGraw-Hill.

Lemons, J. E., 1988. "Dental Implant Retrieval Analyses, *Jour. Dent. Ed.* **52**:748–757.

Linde, F., et al., 1989. Energy Absorptive Properties of Human Trabecular Bone Specimens during Axial Compression, *Jour. Orthop. Res.* **7**(3): 432–439.

Lindhal, O., 1976. Mechanical Properties of Dried Defatted Spongy Bone, *Acta Orthop. Scand.*, **47**:11.

Luo, G., et al., 1995. Implementation of Strain Rate as a Bone Remodeling Stimulus, *Jour. Biomech. Eng.* **117**:329–338.

Mankin, H., et al., 1994. Form and function of Articular Cartilage, in S. R. S. Simon (ed.), Orthopedic Basic Science, American Academy of Orthopedic Surgeons, Chicago.

McCalden, R., et al., 1997. Age-Related Changes in the Compressive Strength of Cancellous Bone: The Relative Importance of Changes in Density and Trabecular Architecture, *Jour. Bone Joint Surg.* **79A**(3):421–427.

McElhaney, J., and E. Byars, 1965. Dynamic Response of Biological Materials, *Proc. Amer. Soc. Mech. Eng.* ASME 65-WA/HUF-9:8, Chicago.

Miller, M. D., et. al., 1978. Significant New Observations on Cervical Spine Trauma, *Am. J. Roentgenol.* **130**:659.

Morgan, E., and T. Keaveny, 2001. Dependence of Yield Strain of Human Trabecular Bone on Anatomic Site, *Jour. Biomech.* **34**(5):69–577.

Morrison, J., 1970. The Mechanics of the Knee Joint in Relation to Normal Walking, *Jour. Biomech.* **3**:164–171.

Mosekilde, L., 1986. Normal Vertebral Body Size and Compressive Strength: Relations to Age and to Vertebral and Iliac Trabecular Bone Compressive Strength, *Bone* **7**:207–212.

Mosekilde, L., et al., 1987. Biomechanical Competence of Vertebral Trabecular Bone in Relation to Ash Density and Age in Normal Individuals, *Bone* **8**(2): 79–85.

Mow, V. C., et al., 1974. Some Surface Characteristics of Articular Cartilage, I. A Scanning Electron Microscopy Study and a Theoretical Model for the Dynamic Interaction of Synovial Fluid and Articular Cartilage, *Jour. Biomech.* **7**:449–1974.

Murray, M., 1967. Gait as a Total Pattern of Movement, *Am. J. Phys. Med.* **46**:290.

Nachemson, A., et al., 1986. Valsalva Maneuver Biomechanics "Effects on Lumbar Trunk Loads of Elevated Intra-abdominal Pressure," *Spine* **11**:476.

Nachemson, A., 1975. Towards a Better Understanding of Back Pain: a Review of the Mechanics of the Lumbar Disc, *Rheumatol. Rehabil.* **14**:129.

Nahum, Alan, and John Melvin, 1993. "Accidental Injury, Biomechanics and Prevention," Springer-Verlag.

Nordin, Margareta, and Victor Frankel, 1989. "Basic Biomechanics of the Musculoskeletal System," 2d ed., Williams and Williams.

Noyes, F. R., 1977. Functional Properties of Knee Ligaments and Alteration Induced by Immobilization, *Clin. Orthop.* **123**:210.

Pandy, M., et al., 1990. An Optimal Control Model for Maximum-Height Human Jumping, *Jour. Biomech.* **23**:1185–1198.

Paul, J., 1967. Forces at the Human Hip Joint, Ph.D. thesis, University of Chicago.

Reilly, D., and A. Burstein, 1975. The Elastic and Ultimate Properties of Compact Bone Tissue, *Jour. Biomech.* **8**:393–405.

Sadegh, A., et al., 1993. Bone Ingrowth: An Application of the Boundary Element Method to Bone Remodeling at the Implant Interface, *Jour. Biomech.* **26**(2): 167–182.

Schatz, R. A, 1989. A View of Vascular Stents, *Jour. Am. Coll. Cardiol.* **13**: 445–447.

Shirazi-Adl, A., S. C. Shrivastva, and A. M. Ahmed, 1989. Stress Analysis of the Lumbar Disc-Body Unit in Compression, *Spine*, **9**:120.

Ullman, David, 1997. "The Mechanical Design Process," McGraw-Hill.

White, A., and M. Panjabi, 1978. "Clinical Biomechanics of the Spine," J. B. Lippincott, Philadelphia.

White, A., et al., 1975. Biomechanical Analysis of Clinical Stability in the Cervical Spine, *Clin. Orthop.* **109**:85.

Williams, P., and R. Warwick, "Gray's Anatomy," 36th ed., Churchill Livingston, Edinburgh, pp. 506–515.

Winter, D., et al., 1998, Stiffness Control of Balance in Quiet Standing, *Jour. Neurophys.* **80**:1211–1221.

Winter, David, 2005. "Biomechanics and Motor Control of Human Movement," Wiley.

Wolff, J., 1892. "Das Gesetz der Transformation der Knochen," A. Hirschwald, Berlin.

Woo, S. L., et al., 1997. Structure and Function of Tendons and Ligaments, in V. C. Mow and W. C. Hayes (eds.), "Basic Orthopedic. Biomechanics," Lippincott-Raven, Philadelphia.

Woo, S. L., et al., 1994. Anatomy, Biology and Biomechanics of Tendon, Ligament and Meniscus, in S. R. Simon (ed.), "Orthopedic Basic Science," American Academy of Orthopedic Surgeons, Chicago.

Yamada, H., 1970. "Strength of Biological Materials," F. G. Evans, Ed., Williams & Wilkins, Baltimore.

Zajac, F., and M. Gordon, 1989. Determining Muscle's Force and Action in Multi-articular Movement, *Exercise and Sport Sci. Rev.* **17**:187–230.

20.9   HUMAN INJURY TOLERANCE AND ANTHROPOMETRIC TEST DEVICES
by Srirangam Kumaresan and Anthony Sances, Jr.

HUMAN INJURY TOLERANCE RELATED TO AUTOMOTIVE SAFETY

This section discusses an overview of injury tolerance of various body components relevant to the design of automotive vehicles. For the sake of brevity, the chapter provides a brief definition of injury criteria and tabulates the tolerance values proposed by the **National Highway Safety Traffic Administration** (NHTSA) at different time periods. The injury criteria are proposed both for adult and pediatric populations. The injury values for the pediatric population were based on scaling procedures.[1]

Head Injury Criteria

A blunt force or load applied to the head may result in injury to the skull, the brain, or both. Early work sought to establish a relationship between skull fracture and brain injury through head acceleration. Lissner et al. found that the time required to produce a fracture was smaller for higher levels of acceleration, using four cadavers.[2] Later on, Gadd developed the **severity index** (SI) to fit published literature to assess **head injury**.[3] Versace noted some shortcomings of SI and proposed an alternative index, the **head injury criterion** (HIC).[4] This index was adopted by **NHTSA,** and in 1972 a proposal was advanced to replace the SI with the following expression:

$$\text{HIC} = \left[\left(\frac{1}{t_2 - t_1} \right) \int_{t_1}^{t_2} a(t)\, dt \right]_{\text{maximum}}^{2.5} (t_2 - t_1) \le 1,000$$

where $t_2 - t_1$ is the time for integration and $a(t)$ is the acceleration at the center of the head of the dummy. A value of 1,000 over a maximum period of 36 ms was chosen as the threshold of injury in the **Federal Motor Vehicle Safety Standards** (FMVSS 208). Table 20.9.1 summarizes the head injury values.

Neck Injury Criteria

The force applied to the head may result in an injury to the neck. The neck injuries occur when the range of motion or force exceeds the limit of human tolerance. The neck injury results in the form of fracture or dislocation of vertebra, ligamentous damage, or a combination. The current **FMVSS** No. 208 injury criteria for the neck using the alternative sled test include individual tolerance limits for compression (4,000 N),

tension (3,300 N), shear (3,100 N), flexion (190 N · m), and extension (57 N · m). The concept that neck injury criteria should be a linear combination of axial tension loads and extension-bending moments was developed by Prasad and Daniel and was based on their results from experimental tests on porcine subjects.[8] In 1996, Klinich et al. presented a composite neck injury indicator based on a linear combination of neck forces and moments.[9] The neck injury criterion, called N_{ij}, incorporates critical limits for all possible modes of neck loading: tension or compression combined with either flexion (forward) or extension (rearward) bending moment:

$$N_{ij} = F_z/F_{\text{int}} + M_y/M_{\text{int}}$$

where F_z is the axial load, F_{int} is the critical intercept value of load used for normalization, M_y is the flexion/extension bending moment, and M_{int} is the critical intercept value of moment used for normalization. Tables 20.9.2 to 20.9.4 summarize the neck injury tolerance.

Chest Injury Criteria

The acceleration and deflection applied individually or in combination to the chest in excess would injure the rib cage and underlying organs. The current FMVSS No. 208 stipulates that during the specified crash test, the chest deflection of the 50th percentile **Hybrid III** male dummy cannot exceed 76 mm (3 in) and its chest acceleration cannot exceed 60 g. Using the probability of injury curve reported in literature, lines of equal probability of injury for the linear combination of deflection and spinal acceleration were generated and the **combined thoracic injury criteria** (CTI) was proposed in 1998 with the following equation:

$$\text{CTI} = A_{\text{max}}/A_{\text{int}} + D_{\text{max}}/D_{\text{int}}$$

where A_{max} is the maximum value of 3-ms clip spinal acceleration (A_s), D_{max} is the maximum value of the dummy deflection D, and A_{int} and D_{int} are the respective intercepts. Table 20.9.5 summarizes the chest injury tolerance values.

Lower Extremity Criteria

The femur load tolerance is proposed to represent the lower extremity injuries. Currently the tolerance limit is recommended for the adult population. Table 20.9.6 summarizes the limit values.

Table 20.9.1   Head Injury Criteria

Source	Criterion	1 year	3 year	6 year	5th percentile female	50th percentile male	95th percentile male
Kleinberger, et al.,[5] 1998	HIC 36 ms	660	900	1,000	1,000	1,000	—
Eppinger, et al.,[6] 1999	HIC 15 ms	390	570	700	700	700	700
Eppinger, et al.,[7] 2000	HIC 15 ms	390	570	700	700	700	700

Table 20.9.2   Neck Injury Criteria, N_{ij}

Criterion	1 year	3 year	6 year	5th percentile female	50th percentile male	95th percentile male
Kleinberger et al., 1998						
N_{ij}	1.4	1.4	1.4	1.4	1.4	1.4
Tension/compression, N	2,200	2,500	2,900	3,200	3,600	—
Flexion, N · m	85	100	125	210	410	—
Extension, N · m	25	30	40	60	125	—
Eppinger et al., 1999						
N_{ij}	1.0	1.0	1.0	1.0	1.0	1.0
Tension/compression, N	1,465	2,120	2,800	3,370	4,500	5,440
Flexion, N · m	43	68	93	155	310	415
Extension, N · m	17	27	39	62	125	166

Table 20.9.3   Neck Peak Injury Values

Value	1 year	3 year	6 year	5th percentile female	50th percentile male	95th percentile male
In position						
Tension, N				2,620	4,170	5,030
Compression, N				2,520	4,000	4,830
Out of position						
Tension, N	780	1,130	1,490	2,070		
Compression, N	960	1,380	1,820	2,520		

SOURCE: Eppinger, et al., 2000.

Table 20.9.4   Neck Injury Criteria, N_{ij}

	1 year	3 year	6 year	5th percentile female	50th percentile male	95th percentile male
N_{ij} critical intercept values	1.0	1.0	1.0	1.0	1.0	1.0
In position						
Tension, N				4,287	6,806	8,216
Compression, N				3,880	6,106	7,440
Flexion, N · m				155	310	415
Extension, N · m				67	135	179
Out of position						
Tension, N	1,460	2,120	2,800	3,880		
Compression, N	1,460	2,120	2,800	3,880		
Flexion, N · m	43	68	93	155		
Extension, N · m	17	27	37	61		

SOURCE: Eppinger et al., 2000.

Table 20.9.5   Chest Injury Criteria

Criterion	1 year	3 year	6 year	5th percentile female	50th percentile male	95th percentile male
Kleinberger et al., 1998						
Combined thoracic index (CTI)	1.0	1.0	1.0	1.0	1.0	—
Critical acceleration, g	55	70	85	85	85	—
Critical deflection, mm	49	57	63	83	102	—
Peak acceleration, g	40	50	60	60	60	—
Peak deflection, mm	37	42	47	62	76	—
Eppinger et al., 1999						
Peak acceleration, g	50	55	60	60	60	55
Peak deflection, mm	30	34	40	52	63	70
Eppinger et al., 2000						
Peak acceleration, g	50	55	60	60	60	55
Peak deflection, mm	30	34	40	52	63	70

Table 20.9.6   Lower Extremity Criteria—Femur Load, kN

	1 year	3 year	6 year	5th percentile female	50th percentile male	95th percentile male
Kleinberger et al., 1998	—	—	—	6.8	10	—
Eppinger et al., 1999	—	—	—	6.8	10	12.7
Eppinger et al., 2000	—	—	—	6.8	10	12.7

BIOMECHANICAL BASIS TO DEVELOP ANTHROPOMETRIC TEST DUMMIES FOR AUTOMOBILE CRASH TESTS

This section focuses on the biomechanical basis for anthropometric test dummies to assess head and neck injuries in accident environments with specific attention to the development of **Federal Motor Vehicle Safety Standard** (FMVSS).

Head Injury Assessment

In 1966 Gadd proposed a **severity index** (SI) based upon the Wayne State Tolerance Curve (WSTC).[1] In 1971, the **head injury criterion** (HIC) was proposed by Versace.[2] The HIC attempted to incorporate the observation that higher peak acceleration was tolerable if presented in a shorter period of time. Versace emphasized that the WSTC represented only the boundary between acceptable and unacceptable levels, and that it did not provide a scaling for injury. In 1984, Mertz again emphasized that injury assessment values should be used only to interpret the response of dummies and thereafter only as a guide to design.[3] Nevertheless, a graph relating HIC to the percentage of population expected to experience life-threatening brain injury was advanced. The basis for the curve was a series of 54 cadaver impacts.[4] The result of the impact was scored on the presence or absence of a skull fracture. By using the lowest **HIC** for which a fracture was observed (450) and the highest HIC resulting in a nonfracture (2,351), a normal curve was generated between the two. The appearance of a skull fracture was equated to life-threatening brain injury. Kleinberger reviewed the literature on head injury and the origin of HIC for the **National Highway Transportation Safety Administration** (NHTSA).[5] Since the agents directly responsible for brain injury (e.g., pressure, force, stress, inertial movement) are difficult to measure, a surrogate was sought. It was apparent that peak acceleration and impulse-area measurements, although they are related to head injury, are not easily related to experimental data. Energy absorption calculations were found to be difficult to apply, and, therefore, a simple relationship between energy absorption and brain injury was not possible. Sances has given a review of the various theories of skull fracture and human head injury tolerance.[6] A review of the recent HIC for child and adult dummies has also been given.[7] The value of HIC $\leq$ 1,000 for an integration duration of up to 36 ms is recommended in the **Federal Motor Vehicle Safety Standards** (FMVSS) Standard 208 for the midsized male (50% **Hybrid III**). The value of 1,000 has been adopted for the midsized male, small female, and 6-year-old child. This decision was made after reviewing a large quantity of test data available in the NHTSA databases.[8] It has been noted that the presence or absence of a skull fracture does not have much significance on the course of brain injury.[9] The question of the survivability of a fall may be addressed by observing the results of documented accidental falls onto hard surfaces. The variability of humans to the effects of falls onto unyielding surfaces was investigated by Snyder in 1977.[10] It was found that falls from low heights can be fatal, while falls from significant heights are not necessarily so.

The HIC is useful in setting standards and comparing the relative effectiveness of different methods of head protection to differing levels of translational acceleration and the occurrence of skull fracture. However, the HIC does not measure the presence or absence of other injury mechanisms such as rotational acceleration, internal pressure, and differential movement of tissue within the brain leading to shearing or tearing of tissue.

Neck Injury Assessment

The early studies by Patrick and Mertz are used to determine the injury values for tension and flexion of the Hybrid III 50% male dummy, which is routinely used for rear impact and frontal flexion analysis.[11,12] However, the cadavers that were used by Patrick and Mertz in the 1960 and 1970 studies had been used prior to the studies in windshield and steering wheel studies or for chest experiments.[13] Because CT and MRI equipment was not available, plain X-rays were used to evaluate injury. Consequently, the ligamentous damage could not be evaluated. In another study involving vehicle rollover tests with dummies, the authors have indicated that injury occurs at the neck when the Hybrid III 50% male dummy (Fig. 20.9.1) neck transducer measured 2,000 N.[14] In this study, measured axial neck loads above 7,000 N were 4 for the production vehicle and 0 for the roll cage vehicle. There were five force levels for the production vehicle and two for the roll cage above 6,000 N. At 5,000 N, there were six for the production vehicle and three for the roll cage. But this study is severely limited by the inability to predict injury values. In a laboratory study, a clinical-burst-type fracture was produced when forces of approximately 8,000 N were applied at speeds of 4.0 to 4.5 m/s to the head of a cadaver.[15] This study shows that four potentially injurious neck loads occurred in the production vehicles when the threshold is set at 7,000 N for neck injury and none in the roll cage vehicles.

Human head and neck axial tension studies in the intact human cadaver indicate injury values in the region of 3,400 N.[16] There are marked differences between the Hybrid III neck and the human neck. One important missing characteristic of the Hybrid III is the lack of head lag when the neck goes into flexion and/or extension.[17]

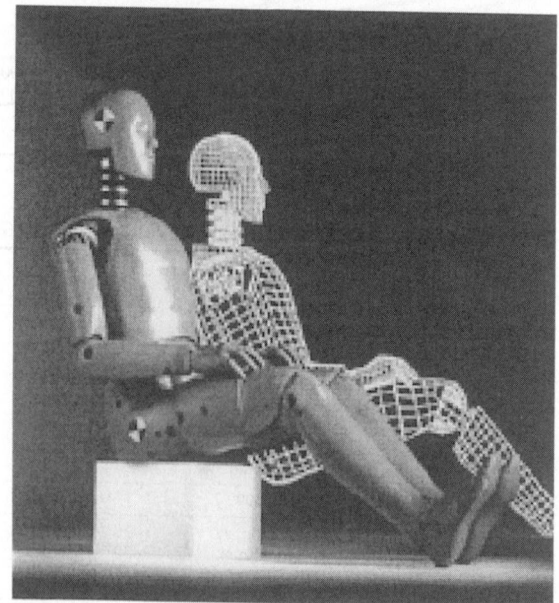

Fig. 20.9.1   Picture of 50th percentile male dummy. (*Source*: http://www.ftss.com.)

Differences in axial compression, dynamic, and static loading of the **Hybrid III** head and neck system has shown that the dynamic and static characteristics of the Hybrid III are 3 to 5 times stiffer than those of the human. A comparison of the Hybrid III 50% male and the cadaveric neck at slow rates showed that dummy necks had a stiffness of about 5,600 N/mm while the human head neck, which averaged about 2,000 N/mm in the axial compression mode.[18] Parallel axial compression studies in the intact human cadaver head-neck system at rates of 3 to 8 m/s were done.[19] Weights represented the muscles; impacts were made on the top of the Hybrid III 50% male head-neck and the cadaveric head-neck system. Forces and moments were measured at the C7/T1 level. This preparation produced clinical-type burst and flexion compression injuries to the cervical spine. The flexion and extension bending moments measured at C7/T1 were 2 to 3 times greater in the human tissue compared with the Hybrid III. Studies on 20 human cervical columns in vertical compression at velocities from 2.5 to 8 m/s showed that the initial force deflection curve was very low until the soft tissue loaded up to about 12 mm of compression.[20] The average level of injury was at 3,326 N with an average stiffness of 555 N/mm. Subsequent dynamic studies of the Hybrid III 50% neck showed markedly greater stiffness compared with the human cadaveric neck with values up to 2,160 N/mm at velocities of 7.6 m/s. The Hybrid III does not exhibit head lag similar to the human.[21]

Scaling approaches have been done to provide injury assessment values for the 6-month, 12-month, 3-year, and 6-year old, the small female, the midsized male, and large male dummies.[22] The major source for the child biomechanical data is a series of tests conducted on pigs by General Motors and the Ford Motor Company with a simulated airbag inflating in front of an out-of-position occupant in the passenger seat. Scaling from these studies, assuming that a 10-week-old pig represents a 3-year-old child, matched tests were then conducted with the 3-year old airbag dummy. These studies were then advanced to the child and adult dummies. Unfortunately, little information is available for the material properties of pediatric tissue except for a study performed over 100 years ago by Duncan.[23] It is well known that there are substantial differences between the 3-year old child and the porcine geometry and development.[24]

REFERENCES

Human Injury Tolerance

1. Kleinberger, M., Yoganandan, N., and Kumaresan, S., Biomechanical considerations for child occupants. *42nd Meeting, Association for the Advancement of Automotive Medicine*, pp. 115–136, 1998.
2. Lissner, H. R., Lebow, M., and Evans, P. G., Experimental studies on the relation between acceleration and intracranial pressure changes in man, *Surg. Gynecol. Obstet.*, September 1960, pp. 329–338.
3. Gadd, C. W., Use of a weighted-impulse criterion for estimating injury hazard, *Proc. 10th Stapp Car Crash Conference*, 1966.
4. Versace, J., A Review of the Severity Index, *Proc. 15th Stapp Car Crash Conf.*, Society of Automotive Engineers, New York, 1971.
5. Kleinberger, M., Sun, E., Eppinger, R., Kuppa, S., and Saul, R., "Development of Improved Injury Criteria for the Assessment of Advanced Automotive Restraint Systems," NHTSA, 1998.
6. Eppinger, R., Sun, E., Bandak, F., Haffner, M., Khaewpong, N., Maltese, M., Kuppa, S., Nguyen, T., Takhounts, E., Tannous, R., Zhang, A., and Saul, R., "Development of Improved Injury Criteria for the Assessment of Advanced Automotive Restraint Systems—II," NHTSA, 1999.
7. Eppinger, R., Sun, E., Kuppa, S., and Saul, R., "Supplement: Development of Improved Injury Criteria for the Assessment of Advanced Automotive Restraint Systems—II," NHTSA, 2000.
8. Prasad P., and Daniel, R. E., A Biomechanical analysis of head, neck, and torso injuries to child surrogates due to sudden torso acceleration, Society of Automotive Engineers, 1984; Paper No 84 1656.
9. Klinich K. D., Saul, R. A., Kleinberger, M., Backaitis, S. H., and August, G., Techniques for developing child dummy protection reference values, Docket No.74-14 N97, Item 069, October 1996.

Anthropometric Test Devices

1. Gadd, C. W., Use of a Weighted-impulse Criterion for Estimating Injury Hazard, *Proc. 10th Stapp Car Crash Conf*, Society of Automotive Engineers, New York, 1966.
2. Versace, J., "A Review of the Severity Index," *Proc. 15th Stapp Car Crash Conf.*, Society of Automotive Engineers, New York, 1971.
3. Mertz, Harold J., "Injury Assessment Values Used to Evaluate Hybrid III Response Measurements," USG 2284 Part III, Attachment I, Enclosure 2, February 1984.
4. Prasad, Priya, and Mertz, Harold J., The Position of the United States Delegation to the ISO Working Group 6 on the Use of HIC in the Automotive Environment, *Government/Industry Meeting & Exposition*, Washington, DC, May 20–23, 1985. Society of Automotive Engineers, Inc. Warrendale, PA. Paper 851246.
5. Kleinberger, M., Sun, E., Eppinger, R., Kuppa, S., and Saul, R., "Development of Improved Injury Criteria for the Assessment of Advanced Automotive Restraint Systems," NHTSA, 1998.
6. Sances, A., Jr., and Yoganandan, N., Human Head Injury Tolerance, in Sances, A., Jr., Thomas, D. J., Ewing, C. L., Larson, S. J., and Unterharnscheidt, F. eds., "Mechanisms of Head and Spine Trauma, Aloray Publisher, Goshen, NY, pp. 189–218, 1986.
7. Sances, A., Jr., Biomechanical Analysis of Injury Criterion for Child and Adult Dummies. *Critical Reviews in Biomedical Engineering*, 1999, pp. 361–365.
8. Kleinberger, M., Sun; E., Eppinger, R., Kuppa, S., and Saul, R., "Development of Improved Injury Criteria for the Assessment of Advanced Automotive Restraint Systems," NHTSA, 1998.
9. Prasad, P., Melvin, J. W., Huelke, D. F., King, A. I., and Nyquist, G. W., "Head" in Melvin, J. W., and Weber, Kathleen eds., "Review of Biomechanical Impact Response and Injury in the Automotive Environment," University of Michigan Transportation Research Institute, UMTRI-85-3.
10. Snyder, Richard G., Foust, David R., and Bowman, Bruce M., "Study of Impact Tolerance Through Free-Fall Investigation." University of Michigan Highway Safety Research Institute, Report VM-HSRI-77-8, 15 Dec, 1977.
11. Mertz H. J., and Patrick, L. M., Investigation of the kinematics and kinetics of whiplash, *Proc. 11th Stapp Car Crash Conf.*, Anaheim, CA, pp. 267–317. 1967.
12. Mertz H. J., and Patrick, L. M., Strength and response of the human neck, *Proc. 15th Stapp Car Crash Conf.*, Coronado, CA, pp. 207–255, 1971.
13. Yoganandan, N. Biomechanical Assessment of Whiplash, in "Frontiers in Head and Neck Trauma," eds. Yoganandan N., Pintar F., Larson S. J., Sances, A., Jr., et al., IOS Press, 1998.
14. Bahling G. S., Bundorf R. T., et al. Rollover and Drop Tests—The Influence of Roof Strength on Injury Mechanics Using Belted Dummies, Society of Automotive Engineers, Paper No. 902314, 1990.
15. Pintar, F., Sances, A., Jr., Yoganandan, N., Reinartz, J., Maiman, D. J., Suh, J. K., et al. Biodynamics of the Total Human Cadaveric Cervical Spine. *Proc. 34th Stapp Car Crash Conf.*, 1990.
16. Yoganandan, N., Maiman, D. J., Cusick, J. F., Pintar F. A., Sances, A., Jr., Walsh, P. R., and Larson, S. J., "Human head-neck biomechanics under axial tension, *Me. Eng. Physics*, 18, no. 4, pp. 289–294.
17. Sances, A., Jr., Experimental System to Investigate Whiplash Injury Biomechanics, in "Frontiers in Head and Neck Trauma," eds. Yoganandan, N., Pintar, F. A., Larson, S. J., and Sances, A., Jr., IOS Press, Burke, VA, pp. 308–312, 1998.
18. Yoganandan, N., Sances, A., Jr., and Pintar, F., Biomechanical evaluation of the axial compressive responses of the human cadaveric and manikin necks, *J. Biomech. Eng.* 1 11, no. 3, 250–255, 1989.
19. Pintar, F., Sances, A., Jr., Yoganandan, N., Reinartz, J., Maiman, D. J., Suh, J. K., et al. Biodynamics of the Total Human Cadaveric Cervical Spine. *Proc. 34th Stapp Car Crash Conf.*, 1990.
20. Pintar, F., Yoganandan, Y., Voo, L., Cusick, J., et al., Dynamic Characteristics of the Human Cervical Spine, Society of Automotive Engineers, *39th Stapp Car Crash Conf.*, 195, 1995.
21. Ewing C. L., and Thomas D. J., Human head and neck response to impact acceleration. Monograph 21, USAARL 73-1, Naval Aerospace and Regional Medical Centre, Pensacola, 1972.
22. Mertz, H. J., Prasad, P., and Irwin, A. J, 1997, "Injury Risk Curves for Children and Adults in Frontal and Rear Collisions," *41st Stapp Conf.* 973318, pp. 13–30.
23. Kleinberger, M., Yoganandan, N., and Kumaresan, S., Biomechanical considerations for child occupant protection, *42nd AAAM Conf.*, Charlottesville, VA, October 1998.
24. Kumaresan S., Yoganandan N., and Pintar, F. A., Age-specific pediatric cervical spine biomechanical responses, *SAE Trans.* (*41st Stapp Car Crash Conf.*) 106, pp. 3581–3611, 1997.

20.10   AIR-INFLATED FABRIC STRUCTURES

by Paul V. Cavallaro and Ali M. Sadegh

INTRODUCTION

Air-inflated fabric structures fall within the category of tensioned structures and provide unique advantages in their use over traditional structures. These advantages include lightweight designs, rapid and self-erecting deployment, enhanced mobility, large deployed-to-packaged volume ratios, fail-safe collapse, and possible rigidification.

Most of the research and development pursued in air-inflated structures can be traced to space, military, commercial, marine engineering, and recreational applications. Examples include airships, weather balloons, inflatable antennas and radomes, temporary shelters, pneumatic muscles and actuators, inflatable boats, temporary bridging, and energy absorbers such as automotive air bags. However, the advent of today's high-performance fibers combined with continuous textile manufacturing processes has produced an emerging interest in air-inflated structures. Air-inflated structures can be designed as viable alternatives to conventional structures.

Because these structures combine both the textile and the structural engineering disciplines, the structural designer should become familiar with the terminology used in textile materials and their manufacturing processes. A glossary is provided in Ref. 1.

DESCRIPTION OF AIR-INFLATED FABRIC STRUCTURES

Air-inflated fabric structures are constructed of lightweight fabric skins that enclose a volume of pressurized air. The fabric is typically formed in a variety of textile architectures including those shown in Fig. 20.10.1. Each architecture has its own design, manufacturing, tooling, and cost implications. Structurally, these architectures will behave differently when subjected to loads.

The plain-weave architecture provides orthogonal yarn placement, resulting in extensional stiffness along the two **yarn** axes; however, it lacks shear stiffness for off-axis loads. While the braided architecture provides the fabric with shear stiffness due to the nonorthogonality of the yarns, it lacks extensional stiffness. The angle between the braid axis and the yarns θ is referred to as the braid angle or bias angle. Both the triaxial braid and axial strap reinforced braid architectures behave similarly in that they impart to the fabric extensional and shear stiffnesses.

Plain Weave   Braid

Triaxial Braid   Strap-Reinforced Braid

Fig. 20.10.1   Various fabric architectures used in air-inflated fabric structures.

The air pressure develops a biaxial pretensioning stress throughout the fabric. This pretensioning stress enables the structure to generate its intended shape, provides stiffness to resist deflections, and affords stability against collapse from external loads. Fabric materials can often be idealized as tension-only materials for design purposes; i.e., their in-plane compressive moduli and bending moduli are considered negligible.

Stiffness of the structure is primarily a function of the inflation pressure. As the inflation pressure increases, the pretensioning stresses throughout the fabric increase and, in turn, stiffen the structure. Once external loads are applied, stresses from these loads superimpose with the pretensioning stresses in the fabric. As a result, a complete redistribution of stress occurs. This stress redistribution balances the loads and maintains the structure in a state of static equilibrium. Depending upon the type of air-inflated fabric structure (i.e., beam, arch, etc.) and applied loads (i.e., tension, compression, shear, bending, torsion, etc.), the redistribution of stresses can either increase or decrease the net tension stresses in the fabric skin. However, stability of the structure is only ensured when no regions of the fabric experience a net loss in tensile stress. Otherwise, if the stresses from applied loads begin to relax the pretensioning stress (i.e., the tension approaches zero), the onset of wrinkling is said to have occurred within the structure. Wrinkling decreases the structure's load-carrying capability, and upon further loading, eventual buckling or collapse will result.

There are two significant and unique characteristics that air-inflated fabric structures provide over conventional structures. First, upon an overload condition, a collapse of an air-inflated structure does not necessarily damage the fabric membrane. When the overload condition is removed, the air-inflated fabric structure simply restores itself to its design load configuration. Second, since wrinkling can be visually detected, it can serve as a warning indicator prior to collapse.

FIBER MATERIALS AND YARN CONSTRUCTIONS

Proper selection of fiber materials and yarn constructions are important factors that must be considered in the design of air-inflated fabric structures. Both should be optimized together to achieve the desired performance characteristics at the fabric and structural levels.

Many of today's air-inflated fabric structures use yarns constructed of high-performance continuous fibers such as **Vectran**® (thermoplastic liquid-crystal polymer), **PEN** (polyethylene naphthalate), DSP (dimensionally stable polyester), and others. These fibers provide improved structural performance (high strength, low elongation, fatigue, flexfold, cyclic loadings, creep, etc.) and enhanced environmental resistance (ultraviolet rays, heat, humidity, moisture, abrasion, chemicals, etc.). Other fibers used in air-inflated fabric structures include Kevlar, Dacron, nylon, Spectra, and polyester. Hearle[2] provides additional information on fiber materials and their mechanical properties.

Yarns are constructed from fibers that may be aligned unidirectionally or arranged in a number of twisted bundles. Twist is used to improve the handling susceptibility of the yarns by grouping the fibers together, especially during textile processing. Twist, which is measured in turns per unit yarn length, affects the yarn tensile properties. For discontinuous or staple fiber yarns, twist can increase the yarn breaking strength because the internal forces at the ends of a fiber can transfer to neighboring fibers via interfiber shear forces. However, twist in continuous fiber yarns can reduce the yarn breaking strength as observed by Hearle et al.[3] Therefore, a minimal twist is recommended for providing adequate handling protection to continuous fiber yarns.

Hearle et al.[3] experimentally investigated the effects of twist on the tensile behavior of several continuous fiber yarns. Results showed that

Fig. 20.10.2   Yarn tensile testing using Instron textile grips.

yarn tenacity (defined as tensile strength measured in grams force per denier or grams force per tex) decreased with increasing twist for three prescribed yarn tensions used during twist formation. In general, the yarn modulus decreased with increasing twist, yarn elongation at break increased with increasing twist, and yarn elongation decreased with increasing yarn tension. A difference in the load extension behavior of twisted and nontwisted yarns is that a twisted yarn subjected to tension will undergo compaction of its cross section through migration of its fibers and will develop greater interfiber frictional forces than a nontwisted yarn. As with fabrics, the structural performance of yarns can be tailored by changes in their architecture and processing.

Once the yarns are processed into the fabric (i.e., by weaving, braiding, etc.), it is recommended that tensile tests be performed on sample yarns removed from the fabric. This will allow the as-processed yarn properties to be compared to the design requirements. For example, the as-woven tensile properties of continuous fiber, nontwisted yarns removed from a **plain-woven fabric** air beam were measured[4] using an Instron machine configured with textile grips, as shown in Fig. 20.10.2. The cross-sectional areas of the yarns were computed based on fiber diameter and quantity. The tests revealed that the average breaking stress of the weft yarns was nearly 20% less than that of the warp yarns. The reduction in breaking stress of the weft yarns was due to fiber damage caused by the use of higher tensions in these yarns during weaving.

EFFECTS OF FABRIC CONSTRUCTION ON STRUCTURAL BEHAVIOR

Fabric materials are constructed from yarns that cross over and under each other in a repetitive, undulating pattern. The undulations shown in Fig. 20.10.3 are referred to as *crimp,* which is based on Pierce's geometric fabric model.[5]

Pierce geometric model relates these parameters as they are coupled among yarn families. The crimp height h is related to the crimp angle α

Bidirectional crimp

Unidirectional crimp

Fig. 20.10.3   Examples of Pierce's geometric model for plain-woven fabrics with bi-directional and unidirectional crimp.

and yarn length L, as measured between yarns, and the sum of yarn diameters at the crossover points by the following equations, described by Hearle et al.[3]

$$
\begin{aligned}
p &= (L - D\alpha)\cos\alpha + D\sin\alpha \\
h &= (L - D\alpha)\sin\alpha + D\,(L - \cos\alpha) \qquad (20.10.1) \\
D &= h_1 + h_2
\end{aligned}
$$

These equations were based on idealized geometry and assumptions such as restricting the yarns to circular cross sections and no consideration of force or stiffness effects.

Plain-woven and braided fabrics behave differently under load because their yarn families are aligned at different angles. Plain-woven fabrics utilize a nearly orthogonal yarn placement of **warp** and **weft** (or **fill**) yarns. By general textile definitions, **warp yarns** are identified as those yarns running parallel to the selvage and are virtually unlimited in their length. **Weft yarns** are at right angles to the selvage and are limited in length by the width of the weaving equipment. On the other hand, braided fabrics have a $+\theta/-\theta$ yarn placement, where θ is commonly referred to as the *bias angle,* with respect to the braid axis.

The biaxial stress-strain behavior of plain-woven fabrics is initially dominated by **crimp interchange** rather than yarn elasticity. Because the factors of safety used in air-inflated fabric structures are typically within the range of 4 to 6, the operating stresses are designed to be low with respect to fabric strength. Therefore, the structural performance of fabrics must address the influences of crimp interchange. The relative yarn motions (slip and rotation) affect the stiffness properties. The **crimp ratio,** which is denoted as C, is defined as the waviness of the yarns and is obtained by measuring the length of a yarn in its fabric state L_{fabric} and the length of that same yarn after it has been extracted from the fabric L_{yarn} and straightened out according to Eq. (20.10.2).

$$
C = \frac{L_{\text{yarn}} - L_{\text{fabric}}}{L_{\text{fabric}}} \qquad (20.10.2)
$$

The following equation described by Hearle et al. relates the crimp height to C.

$$
\frac{h}{p} = \frac{4}{3}\left(\frac{L}{p} - 1\right)^{1/2} \equiv \frac{4}{3}C^{1/2} \qquad (20.10.3)
$$

Consider a plain-woven fabric subject to a tensile load along the direction of one yarn family. The loaded yarns will attempt to straighten, decrease their crimp heights, and elongate their effective lengths. However, the yarns of the crossing family are forced to increase their crimp heights, resulting in contractions of their effective lengths. This coupling effect associated with changes in crimp is referred to as **crimp interchange** and is similar to Poisson's effect exhibited in metals. In tests of plain-woven fabrics along the direction of a given yarn family, crimp interchange can be easily observed by reducing the width of the specimen. Crimp interchange is a source of nonlinear load extension behavior for fabrics.

Hearle et al.[3] describe a limiting phenomenon to crimp interchange. As the biaxial tensile loads continually increase for a given loading ratio, a configuration results in which yarn kinematics (i.e., slip at the crossover points) cease and the spacing between yarns converges to minimum values. This configuration is referred to as the **extensional jamming point,** which can prevent a family of yarns from straightening and thereby not achieve its full strength. Crimp interchange depends upon the ratio of initial crimp between yarn axes and the ratio of stress between yarn axes rather than the levels of stress. Crimp interchange introduces significant nonlinear effects in the mechanical response of the fabric.

Now consider the plain-woven fabric subjected to shearing by applying a uniaxial load at $\pm45°$ to either yarn family. The yarn families will rotate at the crossover points with respect to each other and become increasingly skewed as the angle between the yarns changes. The change in angle is referred to as the **shear angle**. At larger shear angles, the available space between yarn families decreases, and rotational jamming (locking) of the yarn families occurs. This phenomenon is known as shear jamming, and the angle at which the yarn families become jammed is referred to as the **shear-jamming** angle. The shear-jamming angle decreases with increasing yarn density ratio and can be estimated from Pierce's geometric fabric model or obtained experimentally with various trellising or biaxial test fixtures. Continued loading beyond the onset of shear jamming will produce shear wrinkles leading to localized out-of-plane deformations.

It is important to determine the extension and **shear jamming** points for both fabric manufacturing and structural stiffness concerns. In general, jamming is related to the maximum number of weft yarns that can be woven into a fabric for a given warp yarn size and spacing.

OPERATION

Today's air-inflated fabric structures can be operated at various pressure levels depending upon the fabric architecture, service loads, and ambient temperatures. Woven structures are generally designed to operate at pressures up to 20 psi while triaxial braided and strap-reinforced braided structures are generally capable of higher inflation pressures. Blowers, which may be used for inflation pressures up to 3 psi, provide a high volumetric airflow rate. However, air compressor systems will be required for higher inflation pressures. Air compressors are available in single- and dual-stage configurations. The single-stage air compressor delivers a high-pressure capability but at a low volumetric airflow rate. The volumetric flow rate is measured in standard cubic feet per minute (scfm). Dual-stage air compressors are designed to provide an initial high volumetric airflow rate at low pressures in the first stage and can be followed by a low volumetric airflow rate at high pressure. When a dual-stage air compressor is used, the first stage is most beneficial to erecting the braided structure. At this time, no appreciable resistance to the design loads is available. However, the second-stage mode of the compressor is used to fully inflate the braided structure to its proper operating pressure so that the design loads can be supported. Because the dual-stage compressor can achieve the desired inflation pressure much more rapidly than the single-stage compressor, it is more appropriate for those applications in which the need for rapid deployment justifies the additional cost.

INFLATION AND PRESSURE RELIEF VALVES

Valves designed for use in air-inflated fabric structures must be chosen according to several criteria including locations, orifice size, pressure relief controls, and potential strength degradation in the surrounding fabric regions. Ideally, valves should be positioned in structural elements that interface with the fabric and bladder, such as metal clamps, discs, or end plates at the end termination zones of the structure. This avoids the need for cutting additional penetration holes through the fabric or repositioning the yarns to accommodate holes, thus preventing stress concentrations. When fibers must be removed to accommodate insertion of valves, fabric reinforcement layers should be bonded and stitched to the main fabric in the surrounding region. The valves should be readily accessible for user access but should not jeopardize the integrity of the fabric when the structure is subjected to handling events (such as drops and impacts). Valve locations should be further optimized to mitigate the effects of interference with other objects when the structure is deployed or stowed.

A variety of pneumatic valve designs including threaded valves, quick connect disconnect valves, check valves (one-way, two-way), and pressure relief valves are readily available. The use of pressure relief valves is recommended to avoid accidental overpressures and pressure increases due to changes in ambient temperatures. Valves can also be configured to manifold inflation lines in air-inflated structures containing multiple pressure chambers. This also provides capabilities for self-erecting and controlled deployments by sequencing the inflation timing of multiple chambers. Sizing of the valve orifices should be matched to provide the optimal inflation and deflation times for the air volumes and operating pressures required by the structure.

CONTINUOUS MANUFACTURING AND SEAMLESS FABRICS

Prior to continuous circular weaving and braiding processes in use today, air-inflated fabric structures were constructed by using adhesively bonded, piece-cut manufacturing methods. These methods were limited to relatively low pressures because of fabric failures and air leakage through the seams. Continuous weaving and braiding processes can eliminate or minimize the number of seams, resulting in improved reliability, increased pressure capacities, and greater structural load-carrying capability. However, when seams cannot be avoided, the seam construction should be designed in such a way that failure of the surrounding fabric occurs rather than of the seam. For the safe and reliable use of air-inflated fabric structures, sufficient factors of safety against burst and seam failures must be provided. To guard against burst, e.g., a minimum factor of safety of 6 is used on yarn strength for each yarn family.

High fiber densities within the fabric and rip-stop weave constructions afford improved resistance to punctures, impacts, tears, and abrasion. Rip-stop fabrics have periodic inclusions of stronger, larger yarns along both yarn axes that contain the tear or rupture.

Like traditional composite materials, fabrics can be tailored to meet specific structural performance requirements. Fiber placement can be optimized for air-inflated fabric structures by varying the denier of the yarns (defined as the mass in grams of a 9,000-m yarn) and yarn counts along each direction. For instance, consider a pressurized fabric cylinder. The ratio of hoop stress per unit length to axial stress per unit circumference is 2:1. One can ensure equal factors of safety against yarn burst in the weft (hoop) and warp (longitudinal) yarns by weaving twice as many weft yarns per unit length of air beam as the number of warp yarns per unit circumference. Alternatively, the same can be accomplished by doubling the denier of the weft yarns in comparison to the denier of the warp yarns.

IMPROVED DAMAGE TOLERANCE

Assorted methods are used to enhance the reliability of air-inflated fabric structures against various damage mechanisms. Resistance to punctures, impacts, tears, and abrasion can be improved by using high-density weaves, rip-stop fabrics, and coatings. High-density weaves are less susceptible to penetrations and provide greater coverage protection for bladders. Rip-stop fabrics have periodic inclusions of high-strength yarns woven in a cellular arrangement used to prevent fractures of the basic yarns from propagating across cells.

Additionally, coatings protect the fabric against environmental exposure to ultraviolet rays, moisture, fire, chemicals, etc. Coating such as urethane, PVC (polyvinyl chloride), neoprene, EPDM (ethylene propylene diene monomer) are commonly used. Additives such as Hypalon further enhance a coating's resistance to ultraviolet light and abrasion. Coatings can be applied in two stages. First, coating the yarns prior to forming the fabric by a liquid bath immersion provides the best treatment to the fibers. Second, coatings can be applied by spraying, painting or laminating directly to the fabric after forming. This stage coats the yarns and bridges the gaps formed between adjacent yarns. The maximum protection is achieved when both stages are utilized. While protective coatings have been shown to increase the stiffness of the fabric by restricting relative yarn motions, they remain flexible enough to not adversely impact the stowing and packaging operations of the structure. Additional information on coating materials and processes is provided by Fung.[6]

RIGIDIFICATION

Air-inflated fabric structures can be rigidified by coating the fabric with resins such as thermoplastics, thermosets, and shape memory polymers. Prior to inflation, these particular coatings are applied to the fabric and remain initially uncured, thus acting as flexible coatings. After the structure is inflated and properly erected, a phase change is triggered in the coating by a controlled chemical reaction (curing process) activated by exposure to elevated temperature, ultraviolet light, pressure, diffusion, etc. Once the phase change is fully developed, the coatings bind the yarns together; stiffen the fabric in tension, compression, and shear; and behave similar to a matrix material found in traditional fiber-reinforced composites. The air-inflated fabric structure is now rigidified and no longer requires the inflation pressure to maintain its shape and stiffness. Depending upon the coating used, the transition process may be permanent or reversible (except where thermoset plastics are used). Reversible rigidification is especially suited for those applications requiring multiple long-term deployments.

The performance of the rigidified fabric structure can be assessed by using **classical lamination theory (CST)** as described by Jones.[7] Unlike in the inflated version of the structure, the elastic and shear moduli of the rigidified structure are based on the constituent properties of the fibers and cured coating. As such, the elastic and shear moduli can be readily estimated by using CST. However, the rigidified structure may have different failure modes than its pressurized counterpart. The designer must guard against new failure modes including localized shell buckling and compression rather than wrinkling.

Rigidification is of particular interest for space structures because of restrictions on payload weights and stowage volumes. Cadogan et al.[8,9] have pursued the use of shape memory composites for rigidizing deployable space frames.

AIR BEAMS

Fabric air beams are examples of air-inflated structural elements that are capable of supporting a variety of loads, similar to conventional beams. To date, seamless air beams have been constructed using continuous manufacturing methods that have produced diameters ranging up to 42 in. They are constructed of an outer fabric skin that contains an internally sealed film or bladder made of an impermeable elastomerlike material. The bladder contains the air, prevents leakage, and transfers the pressure to the fabric. The air beam has a cylindrical cross section, and its length can be configured to a straight or curved shape such as an arch. The ends are closed using a variety of end termination methods consisting of bonding, stitching, mechanical clamps, etc., depending upon the inflation pressure and loading requirements. Clamping methods are available to permit disassembly, repair, and replacement of the bladder and fabric layers.

Once an air beam is sufficiently pressurized, the fabric becomes pretensioned and provides the air beam with a plurality of stiffnesses including axial, bending, shear, and torsion. Upon inflation, the ratio of hoop (cylindrical) stress per unit length of beam to the longitudinal stress per unit circumference is 2:1, as shown in Fig. 20.10.4.

In the context of plain-woven air beams, the warp yarns are aligned parallel to the longitudinal axis of the air beam and resist axial and bending loads. The weft yarns spiral through the weave and are located at nearly 90° to the warp axis, thus lying along the hoop direction of the air beam. Weft yarns provide stability against collapse by maintaining the circular cross section of the air beam.

For braided air beams, the braid axis is aligned with the longitudinal axis of the air beam. If the ends of a braided air beam are unconstrained

from moving in the longitudinal direction, the fibers will exhibit a scissoring effect, causing the length of the beam to expand or shorten with pressure depending upon the selection of θ. Eventually, the braided yarns will achieve a maximum rotation, and the fabric will become fully jammed. This phenomenon can be easily demonstrated with the well-known Chinese finger trap toy. Unlike in plain-woven fabric structures, the bias angle of braided fabric structures can be controlled to allow expansion or contraction of the structure when inflated. For an unconstrained braided air beam to resist axial tension, longitudinal reinforcements such as distributed axial yarns (i.e., triaxial braid) or axial straps, webbing, belts, etc., must bonded on at multiple locations around the circumference. Various design methods for constructing strap-reinforced and axial yarn-reinforced braided fabric air beams were developed by Brown[10] and Brown and Sharpless.[11,12]

The bending stiffness EI for metal beams is written as the product of elastic modulus E and area moment of inertia I and has units of (force × distance^2). However, in air-inflated fabric structures, the fabric elastic modulus E_f is commonly denoted in units of force per unit length, and therefore, $E_f I$ is denoted in units of (force × distance^3). Likewise, fabric strengths are denoted in units of force per unit length.

Now, consider the air beam subjected to transverse loads. Upon loading, the pretension stresses and the bending stresses will algebraically add. That is, the compressive bending stresses will subtract (relax) from the pretension on the compressive surface of the air beam while the ten-

$$M = \sum_{i=1}^{N} |F_{ib} \cdot y_i|$$

where: $\dfrac{F_{Nb}}{y_N} = \dfrac{F_{ib}}{y_i}$

and: $y_N = \dfrac{d}{2}$

therefore: $F_{ib} = \dfrac{\dfrac{2 \cdot M}{d}}{\displaystyle\sum_{i=1}^{N} (\sin(\alpha(i)))^2}$

Fig. 20.10.5 Idealized distribution of warp yarn forces due to bending of a plain-woven fabric air beam.

sile bending stresses will add to the pretension along the tensile surface, as shown in Fig. 20.10.5. The instant at which any point along the length of the air beam develops a net zero longitudinal tensile stress, the structure is said to have reached the onset of wrinkling. The corresponding bending moment is referred to as the **wrinkling moment** M_w. Prior to wrinkling, the moment-curvature relationship is expectedly linear in the absence of both material nonlinearities and notable changes in air pressure and volume. A stress balance analysis based on strength-of-materials equations can generally be used to compute M_w. Such a stress balance analysis was used to show the variation of warp yarn tensions for a 2-in-diameter plain-woven air beam subjected to four-point bending loads, as depicted in Fig. 20.10.6. Once wrinkling has occurred, the air beam moment-curvature relation behaves nonlinearly because with further loading, the cross section loses bending stiffness, the neutral axis displaces away from the centroidal axis of the cross section, and eventual collapse occurs. The spread of wrinkling around the circumference is similar to the flow of plasticity in metal beams subjected to bending. Loading beyond the wrinkling onset will result in an

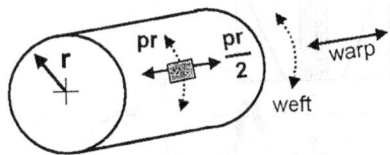

Weft yarn tension per unit length of cylinder = pr

Warp yarn tension per unit circumference = pr/2

Yarn density ratio (YDR) = $\dfrac{\text{\# weft yarns per unit length of cylinder}}{\text{\# warp yarns per unit circumference}}$

Fig. 20.10.4 Yarn tensions in a plain-woven pressurized fabric cylinder, and definition of yarn density ratio.

Fig. 20.10.6 Combined pressure- and bending-induced forces in warp yarns at various distances along a plain-woven air beam based on a simple stress balance analysis.

2" dia, 63 Warp yarns, L_{total} = 96", $L_{support}$ = 44.375", A = 14.625", Pressure = 20 psi, P_{wr} = 4.3 lb, Moment$_{wr}$ = 31.44 in-lb

Fig. 20.10.8 Four-point flexure test on a 6-in-diameter plain-woven air beam constructed of 3,000-denier, 2:1 YDR Vectran fabric.

increase of pressure due to loss of volume, and the work done on the air will affect the postwrinkled bending stiffness. Thus, the postwrinkled response becomes nonlinear. The superposition of the pressure and bending-induced forces is shown for each architecture in Fig. 20.10.7.

Consider now a plain-woven air beam subjected to inflation pressure P. The force in the warp yarns F_{warp}, is shown in Eq. (20.10.4) as simply the pressure times the cross-sectional area divided by the number of warp yarns N_{warp}. The force in the weft yarns F_{weft} was idealized by the assumption of orthogonal yarn placement and is shown in Eq. (20.10.5). The Wrinkling moment M_W for a woven air beam was derived from a simple force balance of the pressure- and bending-induced yarn forces

and is shown in Eq. (20.10.6).

$$F_{warp} = P\pi r^2/N_{warp} \qquad (20.10.4)$$

$$F_{weft} = P\pi rL/N_{weft} \qquad (20.10.5)$$

$$M_W = P\pi r^3/2 \qquad (20.10.6)$$

Note that M_W is independent of the fabric material properties.

The global flexure response of plain-woven air beams was experimentally investigated for use in inflatable military shelters.[13,14] The air beams were constructed of nontwisted yarns and were inflated to various operating pressures. Operating pressures were considered to be safe when the resulting yarn tensions did not exceed 30 percent of their breaking strength. The experimental setup is shown in Fig. 20.10.8. A plot of the experimental load versus deflection is shown in Fig. 20.10.9.

Fig. 20.10.7 Superposition of pressure- and bending-induced yarn forces in plain-woven air beam, a triaxial braided air beam, and a dual-strap reinforced braided air beam.

Fig. 20.10.9 Experimental load versus deflection plot for an uncoated 6-in-diameter air beam constructed of 3,000-denier nontwisted Vectran yarns in a plain-woven 2:1 yarn density ratio fabric using a 37-in span between load points and an 85-in span between support points.

Fig. 20.10.11 Comparison of 2-in-diameter Vectran and PEN plain-woven air beams subjected to four-point bending tests at various load point displacement rates.

Fig. 20.10.10 Plot of total load versus midspan deflection for a 2-in-diameter plain-woven air beam constructed of 1,500-denier, 2:1 YDR Vectran fabric.

Fig. 20.10.12 Section view of an example drop-stitch construction for air-inflated fabrics.

A second series of four-point bending tests[4] was conducted on a 2-in-diameter, uncoated plain-woven Vectran air beam to measure the midspan deflections as a function of inflation pressures ranging from 10 to 100 psi. The load versus midspan deflection curves are shown in Fig. 20.10.10.

In a third series of tests,[4] the flexure behavior of uncoated plain-woven Vectran and PEN air beams was compared for two pressures using a constant load point displacement rate. Both air beams had identical crimp in the warp and weft yarns, yarn density ratios (YDRs), deniers, diameters, and lengths. [The **yarn density ratio** is defined as the number of weft (hoop) yarns per unit beam length divided by the number of warp (longitudinal) yarns per unit circumference.] The graph of Fig. 20.10.11 shows the total applied load versus the midspan deflection responses.

Although the ratio of the as-woven elastic moduli of the Vectran and PEN yarns obtained through as-woven yarn tensile tests was 4.7:1.0, there were minimal differences in deflections of these air beams at the inflation pressures used. It was observed that the global bending behavior of the air beams was (1) invariant with E_{yarn} for the range of inflation pressures considered, (2) dominated by the kinematics of crimp interchange, and (3) subjected to considerable transverse shearing deformations. The effects of crimp interchange and shearing deformations precluded the use of traditional strength-of-materials design methods for plain-woven fabric air beams.

DROP-STITCH FABRICS

Drop-stitch technology was originally pursued by the aerospace community and is now frequently used in inflatable boats. It extends the available shapes and configurations that air-inflated fabric structures can achieve to include flat and curved panels having moderate to large aspect ratios and variable thickness. Drop-stitch construction consists of external skins laminated to a pair of intermediate woven fabric layers separated by a length of perpendicularly aligned fibers as shown in Fig. 20.10.12. During weaving of the intermediate layers, fibers are "dropped" between these layers. Upon inflation, the skins separate and form a panel of thickness controlled by the effective stitch lengths. Air-inflated fabric structures that incorporate drop-stitch technologies include floors for inflatable boats, energy-absorbing walls, temporary barriers, lightweight foundation forms, and a variety of recreational products.

EFFECTS OF AIR COMPRESSIBILITY ON STRUCTURAL STIFFNESS

The load-deflection response of air-inflated fabric structures may depend upon additional stiffening influences other than the initial inflation pressure. These may include contributions from nonlinearities in the fabric stress-strain behavior, the work done on the air by external loads, and other phenomena. If appreciable changes in pressure or volume occur during loading, as in the case of energy absorbers, work is performed on the air through compressibility, which stiffens the structure. Air compressibility can be modeled from thermodynamic principles in accordance with the ideal gas law shown in Eq. (20.10.7).

$$PV = mRT \qquad (20.10.7)$$

where P = absolute pressure
V = volume
m = mass
R = gas constant for air
T = temperature, K

For a quasi-static, isothermal process, the work done on the air by compression is

$$W_{air} = -\int_{V_1}^{V_2} P\, dV = -\int_{V_1}^{V_2} \frac{mRT}{V}\, dV = -mRT \ln \frac{V_2}{V_1} \quad (20.10.8)$$

The total energy of the structure E_{total} is the work done by all external forces W_{ext}, which is related to the total strain energy of the fabric U_f and W_{air} as shown in the energy balance of Eq. (20.10.8).

$$E_{total} = W_{ext} = U_f - W_{air} \qquad (20.10.9)$$

For homogeneous films and membranes, the elastic and shear moduli are readily determined by standardized tests. However, for the case of plain-woven and braided fabrics, the elastic and shear moduli can vary not only with pressure but also with fabric architecture (plain weave, harness weave, braid, yarn densities, etc.), external loads (**crimp interchange**), and coatings, if present. With increasing pressure, slack within the yarns is removed, the yarns begin to straighten (**decrimp**), and contact between intersecting yarns at the crossover points causes compaction.

EXPERIMENTS ON PLAIN-WOVEN FABRICS

Hearle[3] identified three distinct regions of stiffness for tensile loading of a plain-woven fabric, as shown in Fig. 20.10.13. For safe practice, air-inflated fabric structures are designed to operate with yarn forces ranging between regions I and II.

This ensures that the yarns are loaded well below their breaking strengths to provide sufficient factors of safety against burst. Once the fabric is biaxially stressed and subjected to an in-plane shear stress, the yarns will shear (rotate) with respect to their original orientations. The shear stiffness (i.e., resistance to yarn rotations) results from interyarn friction and compaction at the crossover points. Hence, the shear modulus is actually a system property rather than a constitutive (i.e., material) property. As the shear rotations increase, an upper bound is reached when yarns of both directions become kinematically locked, resulting in what is referred to as **shear jamming**. For pure shear loading, the trellising fixture shown in Fig. 20.10.14 can be used. Figure 20.10.15 shows

Fig. 20.10.14   Picture frame test fixture for pure shear loading of fabrics.

the shear force versus shear angle plot of an uncoated Vectran fabric. Further shear loading induces localized shear wrinkling instabilities that lead to increased out-of-plane deformations.

Farboodmanesh et al.[15] conducted shear tests on rubber-coated, plain-woven fabrics and established that the initial shear response was dominated by the coating, and with increased shearing the behavior of a coated fabric transitions to that of an uncoated fabric.

Unlike in fabric structures, the homogeneity of membranes excludes the kinematic motions associated with crimp interchange, and strain energy is developed throughout all stages of biaxial tensile loading. However, membrane structures composed of homogeneous materials are also susceptible to both bending and shear wrinkling instabilities.

Many testing methods have been developed for evaluating the mechanical properties of yarns and fabrics including ASTM standards, Kawabata Evaluation System, military specifications (MIL-Spec), British standards, etc. It is recommended that the structural engineer become familiar with the applicability of these standards to the design and testing of fabric materials as applicable to air-inflated fabric structures. Saville[16] provides a description of many standard tests used in evaluating fiber, yarns, and fabrics.

Recently a multiaxial tension and shear test fixture[13] was developed to permit simultaneous measuring of E_{warp}, E_{weft}, and G_f for structural fabrics. The fixture, as shown in Fig. 20.10.16, was designed for use with conventional tension/torsion machines to characterize the elastic and shear moduli of fabrics as functions of biaxial loads. It utilizes a cruciform-shaped specimen and was designed to evaluate both strength

Fig. 20.10.13   Stages of axial stiffness for woven fabric subjected to tension.

Fig. 20.10.15   Stages of shear stiffness for pure shear loading of a 2:1 yarn density ratio woven fabric.

Fig. 20.10.16 Combined biaxial tension and in-plane shear test fixture. (U.S. Patent No. 6,860,156.)

and stiffness properties of various fabric architectures such as weaves, braids, and knits subjected to biaxial loads, shear loads, or **combined biaxial** and shear loads. For fabrics constructed of two principal fiber directions, the fixture utilizes two rhombus-shaped frames connected with rotary joints, as shown in Fig. 20.10.16 with the biaxial tension and shear modes illustrated for a plain-woven fabric. For triaxial braided fabrics, the fixture uses a third rhombus-shaped frame with additional rotary joints.

A STRAIN ENERGY–BASED DEFLECTION SOLUTION FOR BENDING OF AIR BEAMS WITH SHEAR DEFORMATIONS

The air beam bending experiments exhibited that deflections are functions of internal pressure and transverse shear deformations may be significant. If shearing deformations become appreciable, they will increase the total deflection of air-inflated fabric structures. Therefore, a shear-deformable theory such as that developed by Timoshenko[17] must be employed in place of the Euler-Bernoulli[18] theory to compute air beam deflections.

This section simulates the four-point bending experiments and analytically estimates the fabric shear modulus when changes in pressure and volume are negligible. Fabric shearing deformations, which were evident for each of the four-point bending tests shown in Figs. 20.10.9 through 20.10.11, were observed to decrease with increasing pressure. It will be shown that the shear modulus of the fabric is not a function of the warp yarn's elastic modulus. This was accomplished by deriving a shear-deformable air beam deflection equation for the four-point loading arrangement described by Fig. 20.10.17. From that equation, and using experimentally obtained air beam deflections, the fabric shear modulus G_f can be computed. This equation captured both the bending and transverse shear deformations that comprise the total midspan deflection.

The nearly orthogonal yarn directions were assumed to be truly orthogonal, with the warp axis being parallel to the longitudinal axis of the air beam. This uncoupled the bending stresses from the weft (hoop) yarns. The warp yarns were assumed to exclusively support the bending stresses. In a truly orthogonal fabric air beam, the weft yarns can be considered as parallel rings rather than continuously spiraled yarns with small lead angles. Also an equivalent cylinder having a cross-sectional area equal to the total cross-sectional areas of all the warp yarns A_{total} was used to represent the actual air beam. The inner radius of the equivalent cylinder r, was taken as the nominal air beam radius, and the outer radius of the equivalent cylinder r_o was computed by

$$r_o = \sqrt{\frac{A_{total}}{\pi} + r_i^2} \qquad (20.10.10)$$

Castigliano's second theorem[18] was used and considered the strain energies from both bending and shearing for the equivalent cylinder. The total strain energy U_{total} was expressed as the sum of the bending and shearing strain energies:

$$U_{total} = \int_0^L \frac{M(x)^2}{2E_f I_{cyl}} \, dx + \int_0^L FS \frac{V(x)^2}{2A_{cyl} G_f} \, dx \qquad (20.10.11)$$

where A_{cyl} = cross-sectional area of equivalent cylinder (equal to total area of warp yarns A_{total})
E_f = fabric elastic modulus along warp (longitudinal) axis measured from biaxial tests
FS = shear strain correction factor (for tube = 2.0)
G_f = pressure-dependent fabric shear modulus
I_{cyl} = area moment of inertia of equivalent cylinder about bending axis
L = length of air beam
$M(x)$ = bending moment
$V(x)$ = transverse shear force along longitudinal axis
x = longitudinal position along air beam

For thin-walled cylinders of outer radius r_o and thickness t_{cyl}, I_{cyl} is computed as shown in Eq. (20.10.10) by neglecting terms containing higher-order forms of t_{cyl}.

$$I_{cyl} = \pi r_o^3 t_{cyl} \qquad (20.10.12)$$

The total midspan deflection δ_{total} due to the total applied P_{total} is derived from minimizing the total strain energy U_{total}. This leads to

$$\delta_{total} = \frac{1}{48} \frac{8a^3 P_{total} + 12 ca^2 P_{total} + 3c^2 a P_{total}}{E_f I_{cyl}}$$
$$+ \frac{1}{4} FS \frac{2 a P_{total}}{G_f A_{cyl}} \qquad (20.10.13)$$

where a = distance between support point and adjacent load point and c = distance between load points. The first and second terms of Eq. (20.10.13) represent the bending component δ_{bend} and the shearing component δ_{shear} of the total midspan deflection, respectively. The fabric elastic modulus E_f is measured from biaxial tension tests performed on the fabric using the 2:1 weft-to-warp loading ratio. The magnitudes of the warp yarn tension used in the biaxial tests must match the warp yarn

Fig. 20.10.17   Four-point bending arrangement for shear-deformable beam deflection equation.

forces of the air beam at the inflation pressure of interest. By using the second term d_{shear} the fabric shear modulus G_f may be calculated from

$$G_f = \frac{aP_{total}FS}{2\delta_{shear}A_{cyl}} \qquad (20.10.14)$$

Equation (20.10.14) indicates that G_f is not a function of the fabric elastic modulus E_f whereas P_{total} and δ_{shear} are functions of inflation pressure. Alternatively, by determining the shear jamming angle of the fabric and the corresponding δ_{shear}, a lower bound value of G_f can be determined.

ANALYTICAL AND NUMERICAL MODELS

A variety of analytical and numerical models have been developed to predict the load deflection behavior for inflatable structures[4,13,19–26] and for the load extension behavior of the fabrics themselves.[27,28]

Unit Cell Numerical Models

One could create a finite element model of an air beam with all its discrete yarns, bladder, contact surfaces, friction etc.; however, research has shown that solutions to such a model would exceed the computational limits of today's hardware. The need to obtain a computationally efficient model that would not exceed hardware limits or encounter convergence problems led to the development of localized yarn interaction models for plain-woven fabrics. These models are commonly referred to as unit cell models.

Unit cell models are used to develop the constitutive behavior on the material scale rather than the structural scale. All sources of influence on constitutive behavior are present on the material scale. Once this behavior is known, it can be applied to the structural scale by using homogenization methods. **Homogenization methods** simply employ the as-established constitutive behavior from the unit cell models in global structural models constructed of homogeneous membranes. This two-step approach reduces the necessary computations and runtime for full-structure modeling while preserving the effects of yarn interactions on material behavior.

A unit cell of a plain-woven fabric generally consists of several rows of warp and weft yarns. Since the yarns exhibit relative displacements and rotations with respect to each other, researchers have investigated two options for treating yarn kinematics: (1) rotation-only (pinned centers at the yarn crossover points) and (2) rotation and translation at the yarn crossover points, as shown in Fig. 20.10.18. In the rotation-only condition, the centers of the contact areas are hinged and remain coincident. Friction is due only to the relative rotation of the yarns and no sliding is allowed. This kinematic condition is appropriate for modeling pneumatic muscles[29] where the fabric is braided and the relative angle of rotation at the yarn crossover points is significantly large. In the rotation and translation condition, no kinematic constraints are imposed at the yarn crossover points. This enables interacting yarns to slide and rotate with respect to

Fig. 20.10.19   Example unit cell model and loading procedure.

each other. Therefore, contact-induced friction forces at the yarn crossover points provide resistance to both relative rotations and translations.

The bladder generally does not contribute any structural stiffness to the fabric, nor is it necessary to include the bladder as a contacting medium. In addition, the inclusion of a bladder contact surface was found to create unnecessary convergence difficulties and added computational expenses. Since the bladder was eliminated from these models, the unit cell should be subjected to a uniform pressure representing the air beam pressure. There are several methods of applying the internal pressure to the unit cell model, depending on individual preferences.

Example of a Unit Cell Model

In a specific example of the **unit cell**, each yarn was represented by an isotropic thin shell with material properties of either Vectran or PEN, depending on the fabric being studied. The shell unit cell model consisted of four warp yarns and four weft yarns, as shown in Fig. 20.10.19. The model was subjected to internal pressures ranging from 0.0138 MPa (1 psi) to 0.276 MPa (20 psi). Contact surfaces were permitted to allow yarn translation and rotations as described by option 2. Two compliant membrane elements ($E_{membrane} \ll E_{yarn}$) connecting the center points of the contact regions were created. These membrane elements were used to determine the overall stress and strain of the shell unit cell model and did not contribute to the structural stiffness of the model.

A three-step computational procedure was employed to determine the elastic and shear moduli of the unit cell. First, the vertical yarns were restrained and the horizontal yarns (x direction) were subjected to +1 percent displacement. The model was solved, and the stress in the x direction and the strains in the x and y directions were determined. Then the elastic modulus $E_1 = \sigma_1/\varepsilon_1$ and Poisson's ratio $\nu_{21} = -\varepsilon_2/\varepsilon_1$ were determined. Second, the horizontal yarns were restrained, and the vertical yarns (y direction) were subjected to +1 percent displacement in the y direction. The model was solved, and the stress in the y direction and the strains in the x and y directions were determined. Then the elastic modulus $E_2 = \sigma_2/\varepsilon_2$ and Poisson's ratio $\nu_{21} = -\varepsilon_1/\varepsilon_2$ were determined.

Finally, the third step was to calculate the shear modulus. This is difficult because the elastic and shear moduli of a plain-woven fabric structure are highly dependent on the internal pressure; furthermore, the elastic and shear moduli are independent. Two methods are suggested. The first method is for the case where the fabric is coated and there is no relative motion between the yarns. Then the following approach is recommended.

In this approach the fabric is assumed to be a continuum with orthotropic material properties. The unit cell model in this case was subjected to a +1 percent nodal displacement at a 45° angle (not shown).

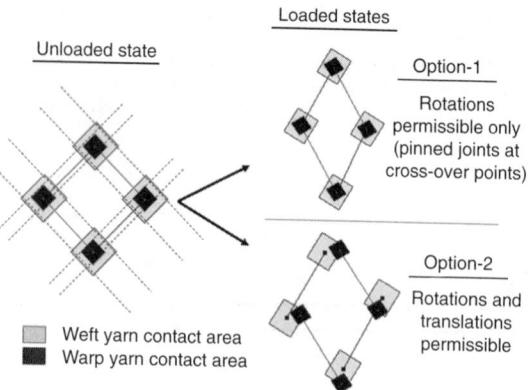

Fig. 20.10.18   Treatment of yarn kinematics in unit cell models.

The model was solved, and the stresses and strains of the unit cell at the 45° angle were determined. The shear modulus was calculated according to Jones[7] by

$$G_f = \left(\frac{4}{E_x} - \frac{1}{E_1} - \frac{1}{E_2} + \frac{2v_{12}}{E_1} \right)^{-1} \qquad (20.10.15)$$

where E_x is the elastic modulus in the direction of the 45° load. This expression may be appropriate for fabrics with stiff coatings.

The second approach to calculating G_f assumes that the woven fabric is not a continuum and that each yarn independently responds to the external load. In this case, the horizontal yarns were subjected to a shear force F_{shear}. The shear modulus was then calculated based on the equilibrium of the horizontal yarn and its shear deformation as

$$G_f = G_{12} = \frac{\tau}{\gamma} \qquad (20.10.16)$$

where $\tau_{shear} = \Sigma F_{reaction}/\Sigma A$, $\gamma = \arctan(\delta_{shear}/L)$, and $\Sigma F_{reaction}$ is the sum of reaction forces at the support, δ_{shear} is the nodal transverse displacement, A is the yarn cross section, and L is the yarn length.

It is recommended that a number of tests on the unit cell model be performed in which the coefficient of friction is varied and the changes in the elastic and shear moduli are determined. Therefore, the results of the shell unit cell model will be parametric in both friction and pressure.

Structural Air Beam Models

Once the elastic and shear moduli of the membrane elements are determined from the unit cell model, a global structural finite element model of the air beam, using membrane elements, can be created such as the one shown in Fig. 20.10.20. Here, the fabric skin would be discretized using membrane elements. However, nine elastic material constants (E_1, E_2, E_3, v_{12}, v_{13}, v_{23}, G_{12}, G_{13}, and G_{23}) are required, of which only E_1, E_2, and G_{12} had been calculated from the beam unit cell model. Because a membrane formulation is used for the fabric skin, the elastic modulus E_3 would not influence the global beam deflection and was therefore chosen to be of the same order of magnitude as E_1 and E_2 (in the order of $E_3 = 64,000$ MPa for the current example). It was also noted that any load applied in the direction of one yarn is decoupled from all orthogonal yarns. That is, the Poisson's ratio is nearly 0; therefore, the condition that $v_{12} = v_{13} = v_{23} = 0.0$ may be applied. No method was developed to determine G_{13} and G_{23}. It is hypothesized that neither value had a significant effect on the beam deflection. One could examine this hypothesis by a parametric study of values for G_{13} and G_{23}.

In the above-mentioned example, using the analytical strain energy solution, the equivalent in-plane fabric shear modulus G_f for the 50.8-mm (2.0-in) diameter Vectran beam pressurized to 0.689 MPa (100 psi) was calculated as 86.06 MPa (12,480 psi). The computed bending and shear components of the midspan deflection were 12.39 mm (0.488 in) and 39.54 mm (1.557 in), respectively. The elastic and shear moduli in the beam unit cell model were calculated for an internal pressure of 0.137 MPa (20 psi) as $E_1 = 64,033$ MPa, $E_2 = 64,098$ MPa, and $G_f = G_{12} = 60.96$ MPa. Similarly, for an internal pressure of 0.207 MPa (30 psi), the elastic moduli were calculated as $E_1 = 64,045$ MPa, $E_2 = 64,049$ MPa, and $G_f = G_{12} = 80.64$ MPa.

Fig. 20.10.20 Example of an air beam global finite-element model subjected to four-point bending.

Concluding Remarks

The unit cell modeling method was used to determine the influence of pressure, weave architecture (including yarn density ratio), crimp interchange, coefficients of friction and biaxial loading ratios on the effective fabric elastic and shear moduli. These moduli were shown to increase with increasing pressure. The coefficient of friction between yarns did not influence the elastic and shear moduli significantly because the relative displacements of the interacting yarns were small compared to their cross sections. The shear modulus of plain-woven fabrics was shown to be a system property; it is a kinematic property of the yarn assemblage in fabric form and is independent of the yarn elastic moduli. The use of yarns with higher elastic modulus may not significantly influence the deflection of air beams because crimp interchange can prevent the as-woven yarns from developing elastic behavior. It has also been shown that coatings can stiffen the fabric material, resulting in smaller air beam deflections.

Homogenization is particularly suitable for developing computationally efficient global models of air-inflated fabric structures in conjunction with unit cell methods. The unit cell method captures the nonlinear constitutive behavior of the plain-woven fabric in response to applied loads. This behavior is transferred to a homogeneous material in a global model of the air-inflated fabric structure.

The studies have shown that air-inflated fabric structures differ fundamentally from conventional metal structures and fiber/matrix composite structures. While the plain-woven fabric may appear to be an orthotropic material, its mechanical behavior indicated otherwise. It does not behave as a continuum, but rather as a discrete yarn assembly.

REFERENCES
1. Celanese Acetate LLC, "Complete Textile Glossary," 2001.
2. J. W. S. Hearle,"High Performance Fibres," Woodhead Publishing, 2001.
3. J. W. S. Hearle, P. Grosberg, and S. Backer, "Structural Mechanics of Fibers, Yarns and Fabrics," Wiley, New York, 1969.
4. P. V. Cavallaro et al., Mechanics of Plain-Woven Fabrics for Inflated Structures, *Composite Structures Jour.* 61:375–393 (2003).
5. F. T. Pierce, *Jour. Textile Institute*, 28:T45 (1937).
6. W. Fung, "Coated and Laminated Textiles," Woodhead Publishing, 2002.
7. R. M. Jones, "Mechanics of Composite Materials," Hemisphere Publishing Co., 1975.
8. D. P. Cadogan et al., "Shape Memory Composite Development for Use in Gossamer Space Inflatable Structures," American Institute of Aeronautics and Astronautics, AIAA 2002-1372, 2002.
9. U.S. Patent No. 6,910,308 by Cadogan et al.
10. U.S. Patent No. 5,677,023 by G. Brown.
11. U.S. Patent No. 5,735,083 by G. Brown and G. Sharpless.
12. U.S. Patent No. 5,421,128 by G. Brown and G. Sharpless.
13. P. V. Cavallaro et al., Effects of Coupled Biaxial Tension and Shear Stresses on Decrimping Behavior in Pressurized Woven Fabrics, 2004 ASME International Mechanical Engineering Congress and Exposition, Anaheim, Calif., Nov. 13, 2004, IMECE2004-59848.
14. Army Center of Excellence for Inflatable Composite Structures, http://nsc.natick.army.mil/media/fact/facilities/ICS.htm.
15. S. Farboodmanesh et al., "Effect of Construction on Mechanical Behavior of Fabric Reinforced Rubber," Rubber Division Meeting, American Chemical Society, Pittsburgh, Pa., Oct. 8–11, 2002.
16. B. P. Saville, "Physical Testing of Textiles," Woodhead Publishing, 1999.
17. S., Timoshenko, "Strength of Materials," D. Van Nostrand Company, 1946.
18. A. C. Ugural and S. K. Fenster, "Advanced Strength and Applied Elasticity," 2d ed., Elsevier North-Holland, New York, 1977.
19. P. S. Bulson, Design Principles of Pneumatic Structures, *The Structural Engineer*, 51 (6), June 1973.
20. E. C. Steeves, Fabrication and Testing of Pressurized Rib Tents, Tech. Rep. 79-008, U.S. Army, Soldier Biological Chemical Command, Natick, Mass., 1979.
21. E. C. Steeves, The Structural Behavior of Pressure Stabilized Arches, Tech. Rep. 78/018 (AD-A063263), U.S. Army, Soldier Biological Chemical Command, Natick, Mass., 1978.
22. E. C. Steeves, Behavior of Pressure Stabilized Beams under Load, Tech. Rep. 75-47-AMEL, U.S. Army, Soldier Biological Chemical Command, Natick, Mass., 1975.
23. E. C. Steeves, A Linear Analysis of the Deformation of Pressure Stabilized Beams, Tech. Rep. AND-006-493, U.S. Army, Soldier Biological Chemical Command, Natick, Mass., 1975.

24. E. C. Steeves, Pressure Stabilized Beam Finite Element, Tech. Rep. 79-002, U.S. Army, Soldier Biological Chemical Command, Natick, Mass., 1978.
25. W. B. Fichter, A Theory for Inflated Thin-Wall Cylindrical Beams, NASA Tech. Note D-3466, National Aeronautics and Space Administration, Washington, 1966.
26. M. Stein, and J. M. Hedgepeth, Analysis of Partly Wrinkled Membranes, Tech. Note D-813, NASA Langley Research Center, July 1961.
27. W. D. Freeston et al., Mechanics of Elastic Performance of Textile Materials, Part XVIII, Stress-Strain Response of Fabrics under Two-Dimensional Loading, *Textile Res. Jour.*, **37**:948–975 (1967).
28. R. M. Sidhu et al., Finite Element Analysis of Textile Composite Preform Stamping, *Jour. Composite Structures*, **52**:483–497 (2001).
29. K. G. Klute and B. Hannaford, Finite Element Modeling of McKibben Artificial Muscle Actuators, *IEEE/ASME Trans. on Mechatronics*, March 1998.

20.11   ROBOTICS, MECHATRONICS, AND INTELLIGENT AUTOMATION
by Andrew A. Goldenberg

OVERVIEW

Description of the Fields

In the 1950s engineering schools used to teach students the subjects related to mechanical and electrical engineering in the same department, usually named electromechanical engineering. This approach was considered suitable to provide the required level of professional knowledge that would allow graduates to work effectively in this branch of engineering.

As the technology, in particular electronics, begun to unravel significant advances, and the industrial requirements became more focused, engineering schools started splitting up the electromechanical departments as independent entities into electrical and mechanical, respectively, with scope, aims, and objectives that were not necessarily similar.

In 1954 a U.S. patent was filed by G. Devol for a programmable universal manipulator arm (PUMA). It constituted the first registered patent in the field of robotics. Soon thereafter, together with John Engelberger and others, they set up Unimation Inc., acquired the rights to this patent, and started a business of manufacturing robots. Unimation was the first company solely dedicated to the business of robots. The first Unimation robot was used at GM, Trenton, New Jersey, in 1961. Eventually, an entire industry as we know it today emerged.

The paradox was that as the departments of electrical and mechanical engineering became independent, the new field of robotics was calling for a merger of these two now established solitudes. Furthermore, the third major disciplinary component of robotics was emerging, that is, the computer science, which provided the capability of software development that is required in robotic systems.

In the 1970s a large number of important industrial projects in the field of robotics took off. The most prominent was the Shuttle Space Remote Manipulator System, developed by Rockwell, NASA, Spar Aerospace of Toronto, Ontario, and National Research Council in Canada. In parallel, the automotive industry embraced the use of robots in earnest. As a result, the industry began requiring skills that combined electronics, mechanical, and software engineering. Many engineering schools set up robotics programs in late 1970s and early 1980s. Usually these programs were hosted by one of the three constituent departments, and the programs they offered reflected the departmental basic discipline. One of the few institutions that started offering a multidisciplinary program in robotics was the Institute of Robotics at Carnegie-Mellon University in the early 1980s.

In parallel an industrial association was formed, the Robotics Institute of America (RIA). RIA had the mandate to standardize the use of robotics. It issued a formal definition that is still used widely to qualify automatic equipment as a **robot**—a mechanical device that can be programmed to perform some task of manipulation or locomotion under automatic control. This definition provided a basis for the interdisciplinary nature of robotics.

At that time the above definition and the use of robotics in general were limited to manipulator arms. All robotic manufacturers in the 1980s were offering manipulator arms. The teaching and research were also confined to robotic arms, and major applications such as in the space sector were also focused entirely on manipulator arms. The robotic arm was generically an open kinematic chain consisting of a base, shoulder, elbow joints, and a three-dimensional wrist, ending with an end effector in the form of a gripper. There were many variations to this basic robot configuration, nonetheless they were all classified as arms.

Robotic applications in late 1980s and early 1990s evolved from automotive welding and painting to assembly in electronics manufacturing. Also, new applications emerged in the medical, security, mining, and other sectors. These applications required more complex robotics structures than the common manipulator arms. In terms of design and development, they were developed using the same basic disciplines and engineering design principles. These applications led to the emergence of a new field: mechatronics.

The term was actually used in the 1970s in Japan. **Mechatronics** is defined as the synergistic combination of precision mechanical engineering, electronic control, and systems thinking in the design of products and manufacturing processes. It extended the scope of automation from robotic arms to intelligent systems. Mechatronics was integrating mechanical, electronics, and software engineering in a more aggressive way than robotics. The major difference was the emphasis on microelectronics, nonexistent in the 1960s, and the use of new materials in mechanical design of small, miniature, portable devices, leading to the emergence of ultraminiature or microsystems, and their fabrication methods, as defined by the new field that emerged in the 1980s: MEMS, or microelectromechanical systems. MEMS is a branch of mechatronics that addresses microsystems. It has further evolved into systems designed based on particle components: nanotechnology.

Today, the teaching and research of mechatronics and robotics often overlap, as more people have embraced both fields of study and professional work. To attempt to precisely differentiate between the two entities is fruitless at this point of technological development. The expertise is overlapping, but most practitioners see robotics as a subentity of mechatronics. However, there are peculiar issues in each of the fields requiring distinct specialization, and both fields belong to the area of intelligent automation.

Robotics

Based on the single-axis servomechanism methods of control system design that were well developed by late 1960s, the development of industrial automation for machine tools began with **numerically controlled (NC) machines.** These tools were designed in the 1960s to perform unstaffed operations of cutting, lathing, drilling, and polishing using primarily two-dimensional positioning devices. The devices were the result of integrating two or three axes of motion, usually perpendicular to each other. Each axis was driven by a standard servomechanism. The NC machines were programmed using logic circuits such that the motions of the servomechanisms were coordinated. The automatic units were machining profiles in two and three dimensions.

The typical **servomechanism** was operated using available hardware (analog) and software (logic circuits). The underlined control system design was based on PID controller and later on optimal and adaptive control. The coordination of the motion of two or three axes was realized by effecting simultaneously very small steps of motion of the axial servomechanisms. The frequency of these steps was selected to correspond to the required speed of motion. The coordination was provided by ensuring that all axes start and stop at the same instant of time. By ensuring that arbitrarily small steps of motion can be performed, limited only by the servomechanism's ability to monitor its own axial motion online by feedback, high-precision motion was generated. In modern machining the repeatability can be guaranteed to a hundredth of a thousand of an inch, or 0.00254 mm.

The control of NC machining along with design, programming, and language used for programming have dominated the industrial automation sector until the mid-1960s. At that time it was recognized that the operations of the NC machine are limited to standard machining operations. Any task that would require more than two or three axes of motion, including translational and rotational motion, could not be executed. At this stage the notion of articulated motion, i.e., the handling of a tool, tracking a path with the tool, and also orienting the tool arbitrarily in three-dimensional space, became intriguing enough, such that the impetus started shifting from the linear NC machines to new devices that could be programmed to deliver the tool motion over an arbitrary path and with arbitrary orientation.

This generated the birth of robotics as a domain of engineering research and design. The robotics were multidegree-of-freedom devices with a capability similar to that of NC machines, but with a higher number of axes active to move based on coordinated motion plans. The purpose was to handle and operate a tool over a workpiece while following specific paths (contours). The robotics provided the domain of development of radically more complex mechanisms. Furthermore, the advent of microelectronics provided a new technology of hardware and software that became the cornerstone in the proliferation of robotics in industry.

The field of mechatronics started to become acknowledged in the 1980s. It was providing a different approach to automation, and it generalized the robotics technology. The mechatronics approach is different from that of robotics in the sense that it does not regard the arm as the central piece of an automated system. Both domains are regarded as an all-encompassing effort to eliminate human involvement in order to provide the industry with the capability of supplying more consistent, more economical, products and compete worldwide on equal footing without concern that the cost of production is significantly dependent on local wages. This reason for introducing automation is more powerful today than it was late in the last century.

Mechatronics

Mechatronics is a multidisciplinary field that is the synergistic combination of mechanical systems, electronics, sensors, actuators, and software. Mechatronics is in essence the integration of these areas. In practice as well as education the division along the classical disciplines is no longer effective, as this emerging branch of engineering generates the most appropriate vehicle for interdisciplinary work. Integration must be done *concurrently,* unlike in the classical approach to design whereby the work in each domain is done in solitude, and the integration is referred to as the last stage before testing and verification of whole system. In mechatronics, integration is elevated to the position of most relevant stage of design. Mechatronics as a theoretical subject deals with the interaction between various disciplines and their expression in the context of system design. In principle, mechatronics is a methodology for product design. Therefore, education of mechatronics must be based on real problems, product design, and teamwork.

There has been already a certain amount of soul-searching to answer the question, What is mechatronics? It is recognized that there are many designs in which electronics and mechanical systems are integrated, but poorly from a technical perspective and thus ineffectively. In mechatronics the design of an intelligent system begins with the conceptual design that involves the participation of all the members by discipline in an integrated effort to formulate solutions that are feasible and efficient based on considerations of all technological branches involved in the project. The next stage of development is to delegate the various members by disciplines to define their subsystems specifications in coordination with the other members, to ensure that the various discipline's input can be related and associated effectively. Last, but not least, the detailed design phase is undertaken, which is concerned with the independent work of each discipline team. Following this stage and the manufacturing, the assembled team again resumes the collaborative work on integration and testing. At all times the development work demands interaction, and it actually provides the basis for intense collaboration between the various disciplines involved.

Mechatronics identity as a collection of disciplines required to formulate a common approach to development of intelligent systems is also a novel but resurrected way of providing education. A mechatronics engineer is expected to be proficient in mechanical design, electronics design, software development, testing, and integration. It provides the best opportunity for a true technology-oriented training of systems engineers. Whereas the classical version of system engineering was limited to concepts and system specifications, the modern interpretation is also embedding the subsystem development in the context of system design. In other words, there is less of a hierarchy, or vertical integration, and more of an equally valued teamwork, or horizontal integration. The mechatronics engineering is inherently interdisciplinary.

The emergence of this paradigm facilitates innovation and, more importantly, transfer of innovative technology to the use of humanity. Innovation is the product of out-of-the-ordinary ideas that usually emerge in contradiction with the state-of-the-art way of thinking. Many innovative ideas do not progress to becoming relevant ideas because they do not cover the necessary ground to identify their usefulness. In reality, ideas are better appreciated when there is a correlation to their use. However, usefulness is not necessarily associated with immediacy. But the modern society is less inclined to serve as a recipient if the usefulness is hidden. This would indicate that ideas that can be shown to have the capability to become products will be better received. Mechatronics is an excellent vehicle to provide innovation in intelligent automation with a much better way of determining the benefits to the society that are emerging from supporting endeavors related to this field.

In terms of formal definition of mechatronics, it may be wiser not to seek to refine the definition while the field is still emerging. Too early revision would jeopardize future attempts to encompass new paradigms. Moreover, one could point out that the robotics domain has been struggling with its own definition since mid-1980s. Robotics was defined by the RIA so rigidly that immediate repercussions followed with regard to the interpretation of "what mechanism qualifies as a robot." Japan, the United States, and the rest of the world had different views of what constituted a robot. As a result, statistics that were compiled to assess the number of robots worldwide became political tools in the hands of those who benefited from claiming that the numbers in the United States are much lower than those in Japan. However, they were using different definitions of robots; hence the comparisons of robot populations were meaningless.

In terms of expertise, know-how, and education, mechatronics covers a variety of topics in the aforementioned disciplines. There is a danger, already prevailing, that research and development would begin in the basic disciplines, allowing the field to evolve as an integrative one, instead of consolidating a basis that would allow the field to be defined independent of its constituent disciplines. Again from robotics, some lessons could be learned. Early on in the 1980s most of, if not all, the research had connotations with the basic disciplines that were contributing to its integrative nature. In particular, control systems theory, mechanical CAD and FEM design, computer architecture, and electronic hardware design have contributed most to robotics research and development. There were many attempts to open forums for discussing the fundamentals of robotics, but they failed to make convincing arguments. Today, there is a large community of researchers who identify themselves as robotics fundamentalists, as well as at least as large a community in each of the constituting disciplines, who are arguing that robotics is an integration of fundamental engineering science disciplines and thus does not qualify to be regarded as a fundamental field. This conundrum does not seem to get abated by the progress of the field, yet one could argue that some stagnation is perceivable, and the opportunity of explaining it is provided by the fact that advances in the constituent technologies have not been created equally. Some domains are further off, such as microelectronics and computer architectures; others are somewhat behind, such as materials and mechanical design.

It is timely to consider these historical facts in the context of mechatronics. Undoubtedly, if no effort is allocated to these issues, one could argue that at least, similar to robotics, mechatronics is an integrative field, whose progress will be dependent on the advances of the constituent technologies. This issue has an impact on education as well as on training of mechatronics engineers. The argument has already been stated. It is necessary to build a basis that would identify unequivocally the field as fundamental. The most assured way of achieving this goal is to educate in the context of a, yet to be formulated, fundamental discipline. Many schools have taken serious steps in this direction. In parallel many journals and conferences bearing the word *mechatronics* in their brand name have been started; some have already achieved notoriety and are regarded as "archival journals." Unfortunately, a quick review of teaching curriculum and journals content would only reemphasize the ambiguity that already existed at the onset. Many schools teach mechatronics as a reproduction of earlier curricula in robotics and controls. As to some journals, they do not publish content that is different from that of robotics, control, and control technology publications. It begs the question, What is the identity?

Clearly the identity could be found in the interdisciplinary interaction between subsystems. The field identity is strongly related to these issues; however, they have not achieved a predominant role in the literature and education. To attain this goal, it is necessary to shift the teaching from the fundamental topics of electronics, sensors, actuators, control, and mechanical structure design to teaching integration and product design. This implies focusing on implementation and means of fulfilling required system specifications, instead of development of subsystems. This would lead to education based on problem solving, case studies, and real industrial product development. But this approach is impeded by the inherent tendency to teach subjects close to the basic disciplines. It is necessary to promote education in new departments of mechatronics where equal emphasis can be allocated to all the constituent disciplines in order to homogenize their contribution toward a systematic approach to this emerging discipline. A possible approach to this goal is to educate by deengineering "black boxes," thus allowing the unknown to be explored and basic science be understood by exploration of existing benchmark systems. They offer the opportunity to correlate rapidly the basic science to the practice, as well as to unveiling the interactions between the various disciplines.

Another issue is the prevalence of opportunity presented by the Internet. It is well known that many terms of reference—books, papers, parts, devices, components—are available on the Internet. This provides a phenomenal source of information that was not available at the onset of robotics. It is possible to carry many courses in engineering by reviewing material published on the Web, perhaps with some qualified guidance from a more experienced person in the field. In terms of understanding basic concepts as well as learning about new devices and components, the Web provides the best source. It contains and provides what is typically taught in the classroom. It would be beneficial to focus the teaching on such aspects that do not have adequate representation on the Web. This category includes mechatronics integration and subsystems interactions, which by and large are dependent on the professional experience and less so on material available in textbooks. It would be ideal if mechatronics were taught by practitioners instead of nonpracticing Ph.D.s, similar to the medical profession where surgery is taught by surgeons instead of biologists.

The mechatronics field presents a unique opportunity to engage the professional community as a whole comprising experts, practitioners, academics, and the socioeconomic sector as well.

Intelligent Automation

Originating in the fields of robotics, mechatronics, and automation, the research and industrial communities have generated a new term of reference: **intelligent automation.** There have been some initial attempts to define this new entity formally, but not all efforts have generated a satisfactory result because of terms of reference that are ambiguous and significant overlap, mainly with the predecessors robotics and mechatronics.

A close and detailed investigation of systems that are classified as robotics or mechatronics would reveal that the boundaries between these fields are not at all easily and clearly defined. Based on common fundamental physics and engineering science principles that provide the underpinning of these fields, it could be argued that a complete abolition of the need to define them separately is in order. However, the research community seems to be more inclined to maintain distinctions between these domains, whereas the practical/professional domain is less concerned with the definitions as long as the systems that are created work and fulfill required specifications.

Intelligent automation could be understood as an extension of mechatronics and robotics. There is a great deal of uncertainty as to where systems that could be defined as robotics or mechatronics begin their journey as "intelligent." Some serious arguments were made to distinguish between the mechanical design of ordinary robotics and mechatronics and that of intelligent automation systems. It was found that the inclusion of new technology such as micro- and nanotechnology could provide the way to render them intelligent. Other attempts focused on software. It is common practice to consider adaptive and learning systems as intelligent. An example is fuzzy-logic-based expert systems, a new domain of research and business activities.

In the next section examples of robotic-based and mechatronic systems are given. No attempt was made to underline their intelligence, although they could be considered "practically intelligent."

TYPICAL APPLICATIONS

In an attempt to illustrate the emerging technologies of robotics and mechatronics, and their combined by-product intelligent automation, a number of examples of niche applications are presented in this section. The purpose is to facilitate the understanding of these fields via examples instead of taking the usual route of presenting the underlining theory and tools that support design and developments in these areas. The examples should elucidate two issues: (1) complexity of the design of some robotics and mechatronics applications and (2) proliferation of intelligent automation in new areas that have not been addressed so far.

The examples are given in two sections: robotics and mechatronics. As indicated in the previous section, it was not considered useful to segregate some of the systems to the intelligent automation category. The design of most of the provided examples is of high complexity, leading to advanced methods of control, microelectronics, sensing, and custom-developed software. These are the main elements of intelligent automation. Furthermore, the distinction between robotics and mechatronics could be also difficult to draw if functionality is the only concern. Although commonly perceived as "arms" or "manipulators," robotics could really encompass any system that has a number of degrees of freedom either activated in automatic closed loop or when the loop is closed by the human operator. The existence of an arm that may or may not be the main element of the system does not warrant a separate classification. However, to follow the accepted norms, in some of the examples although there is a robotic arm involved, the system is much more complex, and therefore it was categorized as mechatronic.

The provision for separate categories seems to hinder rather than support the efforts in technology development in both academic and industrial domains. Some see mechatronics and intelligent automation as new fields, whereas practically they are significantly overlapping.

Robotics

In robotics the applications usually involve a manipulator arm. They encompass industrial operations such as welding and painting in car manufacturing, pick-and-place of parts and subassemblies, parts mating and assembly in the electronics industry, and machine shop operations. They are based on stationary robotics. In the late 1980s, mobile robotics emerged mainly as remote-controlled devices for operations in hazardous environments. The mobility issues rendered robotics as more complex systems, but as integration of such systems began to dominate, they were viewed as mechatronics. We are addressing this field in the section "Mechatronics."

Another domain that has been gaining popularity in the last 10 to 15 years is *modular* robotics. They are ordinary robotics in terms of functionality; are offered in kits including joints, links, and controllers; and are differentiated by providing ways of reconfiguring the robot's arms in different configurations for a wide variety of tasks.

The most notorious application of robotic arms is in the space sector. The most common use is in car manufacturing (welding, painting). Applications in electronics assembly of components and systems are also growing, pushing up the number of industrial robots in use today.

Applications of robotic arms encountered in the car manufacturing sector are well documented in the literature from the late 1970s to the mid-1980s. New applications in electronic industry are also well documented in the later publications.

To illustrate the use of robotics, we present a number of niche applications. The purpose is to highlight the use of robotics in areas that are less publicized. These examples could provide an impetus for consideration of robotic-based applications in new areas. There are no reasons for limiting the proliferation of robotics in all sectors of society, but the technology has still not matured, such that development and use of new applications in some areas are not economical and reliable.

SPACE SHUTTLE REMOTE MANIPULATOR SYSTEM

The **shuttle remote manipulator system (SRMS)** or "Canadarm" was a joint venture between the governments of the United States and Canada to supply the NASA Space Shuttle program with a robotic arm for the deployment/retrieval of space hardware from the payload bay of the orbiter. A schematic view of the SRMS is shown in the Fig. 20.11.1.

The SRMS is a robotic arm consisting of a shoulder, elbow, and wrist separated by an upper and lower arm boom, giving it a total of 6 degrees of freedom (shoulder pitch and yaw, elbow pitch and wrist pitch, yaw and roll). At a total weight of approximately 431 kg, the SRMS is capable of maneuvering payloads of up to 60,000 lb at a rate of 0.06 m/s with a maximum contingency operation payload weight of 250,000 lb. Under unloaded conditions the SRMS can achieve a maximum translational rate of 0.6 m/s. However, the SRMS is incapable

of supporting its own weight on Earth. It must be supported by specialized ground handling equipment during its testing and packaging for shipment. Although the SRMS can handle very heavy payloads, movement of the tip is very accurately controlled, allowing precise handling of delicate payloads. The length of the SRMS is approximately 15 m. A control system is used to deploy payloads in automatic mode to a positional accuracy of ± 2.0 in and $\pm 1.0°$ of a preprogrammed target point and orientation at the aforementioned rates and load conditions. The SRMS may also be operated manually (remote control) by the astronauts to the same accuracy with the use of hand controllers and closed-circuit televisions (CCTVs) mounted on the manipulator arm. The SRMS was designed to have a life of 10 years or 100 missions. The SRMS is comprised of the following component subsystems:

Joints There are two basic joints (Fig. 20.11.2) in the shoulder, which allow the whole arm to pitch (up-and-down motion) and yaw (side-to-side motion); there is one joint in the elbow to allow the lower arm to pitch; and there are three joints in the wrist to allow the tip of the arm to pitch, yaw, and roll (rotating motions). SRMS can accomplish very complex maneuvers. The motors are equipped with their own brakes, encoder, and speed control.

Fig. 20.11.2 SRMS's basic joint (Spar).

Fig. 20.11.1 SRMS mechanical subsystem. (*From Spar Aerospace Ltd., Toronto, Ontario.*)

Fig. 20.11.3 SRMS's end effector. (*From Spar.*)

Booms   Linking the shoulder, elbow, and wrist joints are the upper and lower arm booms. These booms are constructed of graphite epoxy. The upper arm boom is approximately 5 m long by 33 cm in diameter comprising 16 plies of graphite epoxy (each ply is 0.013 cm thick) for a total weight of just under 23 kg. The lower arm boom is approximately 5.8 m long by 33 cm in diameter comprising 11 plies of graphite epoxy for a total weight of just over 22.7 kg. Each boom is protected with a Kevlar bumper (the same material used in bulletproof vests) to preclude the possibility of dents or scratches on the carbon composite.

End Effector   The end effector (Fig. 20.11.3) of the SRMS allows the arm to capture stationary or free-flying payloads (e.g., orbiting satellites) by providing a large capture envelope (a cylinder 20.3 cm in diameter and 10 cm deep) and a mechanism/structure capable of soft docking and rigidizing the payload/end effector interface. This action is accomplished by a two-stage mechanism in the end effector which closes three cables (like a snare) around a grapple probe (knobbed pin) bolted onto the payload and then draws it into the device until close contact is established and a load of approximately 499 kg is imparted to the grapple probe.

Control System   The hand controllers used by the astronauts (one for arm linear motion and one for rotary motion) are generating the commands representing what the astronauts would like the arm to do. Built-in software calculates which joints to move, what direction to move them in, how fast to move them, and what angle to move to. The computer issues the commands every 80 ms. Any changes inputted by the astronauts to the initial trajectory commanded are reexamined and recalculated, and updated commands are then sent out to each of the joints.

Thermal Protection   The SRMS is covered over its entire length with a multilayer insulation thermal blanket system, which provides passive thermal control. This material consists of alternate layers of Kapton, Dacron scrim cloth, and a beta cloth outer covering. In extreme cold conditions, thermostatically controlled electric heaters (resistance elements) attached to critical mechanical and electronic hardware can be powered on to maintain a stable operating temperature.

A REMOTE-CONTROLLED HYDRAULIC ROBOT FOR DIFFICULT AND HAZARDOUS TASKS

Utility companies must provide tree trimming in the vicinity of live aerial electrical distribution lines, to avoid outages due to ruptured cables which are caused by erratic motion of tree branches during storms and other natural causes. The operation of tree trimming is usually done with **hydraulic wood-cutting** tools by a trained operator stationed in a bucket mounted at the end of an aerial boom of approximately 45 ft in length (Fig. 20.11.4). The boom is a two-link hydraulic arm with up to four joints that is mounted on a specially retrofitted truck. Two sets of controllers for the aerial boom are used in parallel: one in the operator bucket and one on a panel aboard the truck, where a second operator (usually the driver) provides guidance and support from the ground. The operator has the ability to lift the bucket to a height of approximately 45 ft.

Fig. 20.11.4 Slave manipulator and aerial boom.[1]

The operator carries in the bucket a set of tools: **circular saw, linear saw, grappling hook,** etc. The operator can connect one tool at a time to the source of hydraulic power available on the bucket. The tools are handheld, and the operator can perform the cutting operations as needed.

The above-mentioned operations have proved to be dangerous for workers. Occasionally the operator's tool touches live cables, and due to grounding of the boom and bucket, serious injuries of the operator result. There are also cases where the bucket, the upper boom, or the operator come into contact with the cables. Note that in the majority of cases the cables carry 22.5 kV. While there is insulation, and grounding is assumed to be avoided, there are instances where the configuration of the system and the work of the operator generate electric shocks to the operator. In response to concerns for worker safety, a remote master-slave hydraulic manipulator was developed with the intent of relocating workers away from the hazardous tasks. The operator was provided with a handheld controller (the master) located on the platform of the track from where the operator can control the slave (a new five-joint hydraulic manipulator) that is attached at the end of the aerial boom, replacing the bucket where the operator used to be located. An assortment of hydraulic tools can be attached to the slave to enable its use in a variety of high-risk tasks.

The slave is a 5-degree-of-freedom **hydraulic manipulator arm,** and the master is a 5-degree-of-freedom electric arm shaped as a reduced-size replica of the slave. The master is instrumented to provide the operator with torque feedback with respect to two axes of rotation. This allows the operator to feel the forces and moments along and about certain directions at the contact between the tool and branches. Such capability could be extended to a complete (6-degree-of-freedom) force and moment reflection onto the master. The communication between the master and slave is via a fiber-optic cable to ensure electric isolation between the tools and the ground equipment.

The operator is comfortably seated on a swivel chair located on the track bed. The operator has two hand controllers. One is the master of the attached slave, operated with the right hand, and the other is a simpler master that allows the operator to control the aerial boom, operated by the left hand. To achieve this configuration, the aerial boom hydraulic control has been converted from the original hydraulic valves lever controller to a replica master to facilitate operation with only one hand.

The control system of the slave is an open-loop resolved rate controller that transforms commands from the task space (relative to a coordinate frame located at the interface between the tool and a hydraulic coupler that constitutes the end effector of the slave robot). The commands in task space are generated by the master in accordance with the operator's right-hand movements.

To perform the type of tasks described above, there is a need of a hydraulic coupler/adapter designed for quick connect and disconnect of

Fig. 20.11.5 Hydraulic tools.[1]

Fig. 20.11.6 Tip of arm reaching the opening of the cavity.[1]

tools from hydraulic lines with one fitting. The coupler features two lines that can handle a flow at 3,000 psi in either direction and can withstand this pressure when disconnected. A hydraulic tool with tool adapter can be connected or disconnected automatically with a single linear motion simply by pulling or pushing on a locking ring. The same ring can be used for the automatic load/unload operation of the tools. Various hydraulic tools have been retrofitted with an adapter to allow the operator to connect the tools with minimum effort (Fig. 20.11.5). The only drawback is that the operator has to bring the slave down to ground every time a change of tooling is required. This is considered unproductive, and means to alleviate it are being researched.

The slave joints are the shoulder turret, shoulder lift, elbow, wrist pitch, and wrist roll. The base of the slave manipulator is located on a small platform attached to the end of the aerial boom. The platform is controlled in closed loop to maintain its plane perpendicular to the gravity vector. This allows the operator and controller to generate and control the arm movements, respectively, in an unambiguous way.

The slave is designed to carry a radial and axial load of up to 67.5 kg (150 lb). The coupler and the adapter of each tool are locked together with balls and locking ring. For safety, the coupler has an extra security pin which prevents the accidental disengagement of the adapter.

LONG-REACH MANIPULATOR ARM

In nuclear reactor installations, there is a need to inspect the structural integrity of supports of critical subsystems, such as the heavy-water boiler. Such a system is located in a closed vault that is not accessible by humans. The walls have a thickness of 7 ft to ensure that no radiation emanates from the vault. Any repair and maintenance operation inside the vault can be done with long-reaching tools that are transferred to the vault via a few cylindrical openings whose diameter is 10 in (25 cm).

To perform operations in the vault through the available openings, specially designed long-reach arms are required and must be fitted with suitable tools. An array of CCTV cameras are also required, mounted on the arm, to provide views of the vault interior, in order to help the operator guide the robot arm by remote control.

A long-reach manipulator arm was successfully designed for this purpose. The robot is modular in the sense that the long-reach arm is formed by a tandem of short linkages that are coupled together to provide different lengths for various tasks and sizes of tools. The maximum length is 21 ft. As the robot has to pass through 7 ft of wall thickness, and it has a base length outside the vault of about 2 ft, it provides 12 ft inside the vault. This was seen as sufficient for most tasks. The tools include various grippers, cameras, grinders, water blasters, and other modules.

In the specific application reported, the manipulator was used to carry ultrasonic equipment that was used to check the structural integrity of some of the support columns. The robot with the sensors attached was inserted into the vault. The robot had to perform two operations. One was to put a rubber cap on an opening at the bottom of the

column. This enables filling with water the cylindrical cavity surrounding the support column; it is necessary for ultrasonic inspection. The task of inserting the cup to close the cavity hermetically was very tedious. The operator required hours of training to perform this task. Six cameras were mounted on the arm to help guide the operator movements. After the cavity was sealed and water was used to fill the cavity, the actual inspection took place.

In terms of arm reach, the bottom of the column was the farthest point to be reached by the robot tool. As seen in the schematic Fig. 20.11.6, the robot had to be curled around the column. The operation of inspection also required the robot to traverse the surface of the column, starting from the bottom and advancing toward the top (Fig. 20.11.7). At least four columns needed inspection for every boiler.

Although the portion of the arm inside the cylindrical tunnel was supported by the wall structure, the 12 ft of arm inside the vault was in effect "nearly cantilevered" to the inside edge of the cylindrical opening. However, this cantilever was not rigid, as the arm was free to move around the pivot point at the edge of the opening. This presented a formidable challenge for the designers as well as operators. The designers have oriented each coupler, attaching two adjacent links in such a way that the moments around the axis of the coupler would be balanced by the structure, thus rigidizing as much as possible the arm against structural vibrations caused by the payload and arm links' mass moments of inertia. The result was visible in the camera's images. The robot arm

Fig. 20.11.7 Arm carrying the ultrasonic equipment.[1]

was stiff, and the operator could control very well the tip of the tool to perform the dexterous operations of inserting the plug and carrying the ultrasonic equipment at a fixed distance from the surface of the column. Note that if a surface scan was insufficient, then the alternative—drilling into the support to remove a sample—had to be carried out.

The long-reach arm has five joints with zero-backlash harmonic drives: joint 1, roll 360°; joint 2, pitch ±120°; joint 3, pitch ±120°; joint 3, pitch ±120°; joint 4, roll ±175°. Every link has a length between 0.25 and 1.8 m. The links are connected to the joints with quick-connect fittings. The cross section of the link is constrained to fit in a 0.2-m (8-in) diameter. The reach is 1.35 m from joint 2 axis and 4.00 m from the edge of the in-vault opening. The total weight of the arm is 55 kg, and the weight of the largest member of the modular arm is 23 kg.

The lift capacity of the gripper is 7 kg and of the arm at full extension is 10 kg, and the maximum lift is 25 kg. The maximum speed of arm joints 1 through 3 is 15°/s, and of the joints 4 through 6 is 90°/s. The arm motion resolution is 0.5 mm, which is very high for such a long arm.

MODULAR AND RECONFIGURABLE ROBOT

Industrial robots are used in many sectors of manufacturing and service industries. Other sectors are laboratory automation, mobile robots for security, and aerospace. In all these applications the robots have standard configurations, i.e., articulated, gantry, SCARA, cylindrical, etc. They possess any number of joints, usually between three and six. The mechanical structure is fixed; therefore the acquisition process must take into account present needs as well as future ones, to ensure that the robot can perform required future tasks. In this process it is inherent that the designer of the automatic robotic-based task must consider the requirements conservatively. Often the result is that the acquired robot may have more joints (degrees of freedom) than necessary for the task at hand, or the supplier cannot provide a suitable configuration without exceeding the minimum number of required joints.

As a solution to this problem, technology developers have introduced a new concept: **modular robots.** These robots can be disassembled into a set of modules: joints and links. The joints are usually each a rotary or linear 1 degree of freedom, and the links can have a multitude of cross sections and lengths. These robots can be had as one, two, three, etc., joint arms depending on the application. Implicitly the cost will be determined by the number of such joints that are acquired. Thus, the user can trim the cost to the minimum necessary. This concept allows also acquiring more joints at a later stage in the event that a new task is being automated that may require higher complexity of motion.

There are several published results on this technology. Some providers are limiting their scope to strictly mechanical modularity. Others are also introducing modularity in the computer architecture and embedded electronics. More recently software has been also modularized to allow independent integration of joint modules consisting of electromechanical structure (motor, drive, gear, encoder, brake, torque sensor, embedded controller) that can be easily coupled with links, thus forming an open- or closed-loop kinematic chain.

One technology of the **modular joint (MRJ)** is shown in Fig. 20.11.8 and is described below. The basic module is a single rotary joint. There are also multiple joint modules: three- and two-dimensional wrists.

Fig. 20.11.8   Modular joint.

Fig. 20.11.9   Modular and reconfigurable robot (MRR).

The robot configurations that can be obtained from the modules are SCARA (Fig. 20.11.9), articulated, gantry, and various custom configurations of one to five joints. The modules can be assembled and reassembled, thus providing reconfigurability. For example, an articulated four-joint arm can be reassembled as a SCARA robot.

The joints are controlled using embedded software based on PID controls. The assembled robot can be taught points and paths for playback, or it can be programmed in closed loop in the task space, using an inverse kinematic routine that can be run online or based on a lookup table. The kinematics runs on a host PC that is interfaced with the embedded controller via RS-232. The basic specifications of some of the MRJ modules are shown in Table 20.11.1.

The modular wrist specification are shown in Table 20.11.2

Table 20.11.1   Modular Joint Specifications

Specs/Module	MRJ-0	MRJ-1	MRJ-2
Gear ratio	80–160	50–160	50–160
Torque range, N·m	250–500	60–200	30–100
Speed range, deg/s	50–100	150–480	150–480
Motor power, W	1,200	830	370
Motion range, deg	340	340	340
Weight, kg	15.2	9.2	7.2
Size, mm—box/turret diam.	260 × 200 × 150/240	152 × 200 × 152/202	126 × 180 × 126/170
Repeatability, deg	0.0015	0.0015	0.0015

Table 20.11.2   Modular Wrist Specifications

Specs/module	MRW	MRW-S
Type	Roll-Pitch-Roll	Linear roll
Nominal payload, kg	6	8
Max. payload, kg	8	12
Max. speed, deg/s	360–360–360	1.1 m/s and 360 deg/s
Motor power, W	168–168–70	168–70
Motion range, deg	340–220–340	200 mm and 340°
Weight, kg	8.8	6
Size, mm	126 × 126 × 680	103 × 600 × 80
Repeatability, deg	0.0015	0.002 mm and 0.0015°

Mechatronics

The technology of mechatronics is currently present in the following sectors: military, law enforcement, service food, manufacturing, etc. A few interesting examples are given in this section.

The systems presented are only a small set of niche applications that may inspire new applications to be considered. The main motivation in developing such systems was the need or desire to improve the execution, economics, and quality of human life. The systems presented are focused on hazardous environments where the human may be exposed to danger. The examples cover meat processing industry, tire industry, utilities, biotechnology, and security and defense.

The example from the meat processing industry is representative of situations where the human working conditions generate a reduction in human capability. The example from the tire industry expresses situations where the job difficulty may lead to poor performance. In both cases the result of the deterioration of the human performance on the job may lead to further problems, even to catastrophes. It was imperative to consider some form of automation in such cases.

The example from the domain of utilities provides a number of opportunities to consider automation. In this example repairs to underground gas pipelines are considered. Such operations involve serious disruptions because of excavations, hazard to the human operator because of gas leaks, and major costs that lead to increases in the price of gas distributed to consumers. All these factors could be alleviated by automation.

The examples from the security and defense are the most eloquent in terms of hazard encountered by human operators. The need of remote control tools to maintain standoffs is very timely in dealing with some of the terrorism-related threats in the world today.

AUTOMATIC ROBOTIC-BASED MEAT GRADER SYSTEM

In commercial abattoirs, **grading of meat** is performed routinely to establish the thickness of fat. The grading is performed at a federally legislated point on the back of the carcass. In the case of hogs, that is at 7 cm from the spine between the third and fourth ribs below the neck. The grading is performed with widely used electronic equipment that measures the thickness. The equipment has a long needle that gets inserted manually by the grader. The equipment measures the impedance of the tissue, and the differentiation between the impedance of the fat and that of the meat generates a reading of the thickness. The grading device, operated by a qualified grader, has to be inserted in every single carcass. In the abattoir the carcass are transported attached to an overhead conveyor. The operator must perform the task while the conveyor is moving, at a rate of maximum 700 to 800 carcasses per hour shift in large abattoirs. This work is performed along the cutting line, as one of several steps of preparing the meat for packaging. The work is tedious; the environment is hardened by odors and generally leads to serious fatigue. The operator works continuously only 20 min at a time and rest the next 20 min. Thus two operators are working in every 8-h shift. Large abattoirs have two or even three shifts. The fatigue of the operator leads to inconsistent grading. The insertion of the needle must be done in one stroke, at a certain speed, and the needle must be inserted perpendicular to the surface of the meat. When it arrives at the grading station, the carcass is already split along the spine from the tail down about halfway. The operator inserts an arm between the two parts of the carcass (the carcass back is facing the operator), finds the third rib, and presses a hand against the third and fourth ribs while he or she inserts the needle. Clearly, experience helps the operator to find the ribs, insert the needle between them, maintain the needle orthogonal to the surface, maintain the speed of insertion, and perform the insertion in one stroke. The equipment is connected to a PC, and it registers the carcass number and the grade automatically.[1]

The operation described above measures only the thickness of fat. The meat processing industry has been asked to perform additional measurements indicating the quality of the meat. There are developments currently underway to develop measurement devices for all sorts of meat properties. It is also desirable to effect grading in a noninvasive way. This would preserve the integrity of the carcass surface.

Grading operator fatigue, inconsistency in grading, the need to perform additional measurements, provision for both invasive or noninvasive grading, and the high cost of grading operations have led to considerations of robotic-based automation. A prototype system was built and labeled *Automatic Pork Grading System* (APGS). The choice of developing this system for hogs was the result of collaboration between the developers and a pork processing plant. The approach and methodology are applicable to the red meat industry as well.

APGS replaces the grader in the repetitive and tedious manual task of grading pork carcasses. Only manual grading of fat/lean thickness is performed. However, automatic grading can be performed for fat/lean thickness, PSE (paleness, softness, and exudativeness) characteristics, and marbling content of a fresh pork carcass.

Automatic grading can be performed (1) invasively, by the insertion of a grading probe, at a designated spot identified using ultrasound technology; (2) noninvasively, using ultrasound and infrared technology; and (3) as a combination of both. Manual grading of fat/lean thickness is representative of invasive grading. Noninvasive grading ensures that no cross contamination occurs.

APGS comprises three subsystems: robotic subsystem (Fig. 20.11.10), sensing and probing subsystem, and clamping subsystem (Fig. 20.11.11). The robotic subsystem carries the sensing and probing subsystem which scans the pork carcass, as the carcass is moving along the conveyor line, to (1) locate the designated spot for probing (invasive or noninvasive) and (2) perform the grading. The clamping subsystem automatically clamps the pork carcass as it enters the grading station and presents the carcass at a constant position and orientation during the grading operation performed by the robotic and probing subsystems. The sensor probing subsystem uses a dual-echo ultrasonic technology to detect the probing site and perform the probing.

The APGS is a modular robot. It provides simultaneous multiple probing at several sites on the carcass and/or complete continuous scanning of the carcass. It is flexible enough to be used with any type of commercially available grading probe. The system can work at a very high rate of 1,200 carcasses per hour or 3 s per carcass. The system does not require large floor space. The system is readily expandable to other species such as beef or lamb.

Fig. 20.11.10   Robotic subsystem of APGS.[1]

Fig. 20.11.11   Clamping subsystem of APGS.[1]

AUTOMATIC ROBOTIC-BASED TIRE
CASING ANALYZER

Truck tire wear and tear generates a need for replacing them as often as needed. The industry of used tire retreading is growing for a number of reasons, in particular environmental concerns. Tire retreading is a growing business that requires effective detection of used tires that are candidates for retreading.

An **automatic** robotic-based **tire casing analyzer** (ATCA) has been developed. It detects defects and wear. It marks the locations of defects, and it provides a report on the state of the tire. The operator then decides if the tire could be retreaded. For many years, the tire industry has been searching for a nondestructive, simple way to inspect tires and tire casings for flaws. Good casings are presently being discarded and disposed of in landfill sites, causing a detrimental effect on the environment. Or, tires are often prepared for retreading only to discover that the casing has major irreparable flaws. The new technology of ATCA eliminates these problems, resulting in significant economic benefits. The technology is based on an intelligent controller that uses two robotic systems and carries 28 ultrasonic sensors. The controller provides autonomous positioning and guidance of the sensors which are delivered into a tested tire by a very compact foldable robot. The sensor's manifold conforms to the interior shape of the tested tire. Based on the ultrasonic signal, the sensors detect and evaluate defects in the tire. An intelligent algorithm has been developed to evaluate defects; classify them according to size, shape, etc.; and graphically zoom in on a detected defect to be displayed on a computer screen.

The robot arms and sensory arrays have been specifically designed to handle a variety of small and large tires. A sensor protection mechanism has been developed to prevent collisions of delicate sensor arrays with nails that may have penetrated the tire walls. Special low-noise material was used to achieve quality ultrasonic inspection.

The robotic arms are performing the ultrasonic inspection from within the cavity of the tire (carrying the emitters) and over the external surface (carrying the receivers) (Fig. 20.11.12). Loading and unloading of tires are done automatically by the operator using a lifting device.

After loading, the unit starts operating by pressing a set of disks onto the sides of the tire to enclose its inner cavity. Then a foldable robot arm that was stowed inside one of the disks gets deployed internally. The robot arm carries a manifold of ultrasonic sensors that have three segments. The segments get deployed in such a way that the center segment faces the inner side of the tire tread, and the two other face the inner sides of the tire. The manifold positions itself at a suitable distance from the internal surface of the tire to enable ultrasonic waves to penetrate and reach the receivers that are connected on an external manifold mounted on the other robot arm external to the tire. Emitters and receivers are automatically aligned for maximum efficiency. The operation of the ultrasonic sensors, including the positioning and alignment, is fully automatic.

After both the emitters and the receivers are positioned accurately, the tire starts rotating at a predetermined speed to perform the inspection (Fig. 20.11.13).

After the inspection of the tire is finished, the tire stops rotating, and the robotic arms retrieve the sensor manifolds to allow the tire to be removed. The inner robot folds itself into the cavity of the disk, and the outer robot arm lifts the sensors from the proximity of the tire. Immediately the tire is lowered to the load/unload plate, and the disks dislodge themselves from the tire. The operator now rolls the tire off the unit.

ATCA inspects all regular truck tire sizes, regardless of tread depth and pattern, design, or wear. It automatically detects various defects: air leaks, nail holes, ply separation. It can identify internal ply separation of 1 in. The robotic unit is fully automated and operator-friendly and has a short cycle (2.5 min per tire on average). The operator can define the flaw size to be detected, graphically displayed on the computer monitor, and printed out.

Defects are automatically marked on the tire using specially designed paint-based marking unit. Safety and protective devices are incorporated to prevent injury. The outer robot is designed to be compact, to enable fitting the system into a transportation container.

Fig. 20.11.12   View of internal and external robot arms carrying ultrasonic sensors.[1]

ROBOT FOR INTERNAL OPERATIONS IN UNDERGROUND GAS PIPES

Underground cast iron gas pipes are made out of 12-ft-long sections connected by bell-and-spigot joints. A joint is generally filled with jute packing and sealed with lead. The joint can develop leaks over time and must be repaired. One current most frequently done repair procedure is to inject an anaerobic sealant into the jute packing. This procedure requires a 4-ft × 6-ft excavation at each joint to allow the utility to reach the pipe (in the northeastern United States and Canada the pipes are 8 ft deep in the ground), drill a hole through the bell into the jute, and inject a measured quantity of anaerobic sealant. Such a process is hazardous, laborious, and costly, and it interferes with transportation and pedestrians in urban areas.

To alleviate the above problems, a novel technique was developed to perform cast iron bell-and-spigot joint sealing from inside the pipe, using a remotely controlled robot. The robot is launched into the pipe through a special opening, travels inside the pipe until a desired joint is reached, and drills a hole into the joint spigot at the highest point. Then anaerobic sealant is injected into the jute packing to reseal the joint. This approach replicates the repair procedure currently done externally. The sealant flows under gravity and saturates the jute. The operation is done while the main is kept in service.

The new internal sealing system is capable of sealing several joints from a single excavation while keeping the main in service. The system has been designed for 6-in- (15-cm-) diameter pipes. It can be inserted into the main up to 150 ft (over 45 m) in each direction from the entry

Fig. 20.11.13   General view of ATCA.

Fig. 20.11.14 View of the gas pipe robot end effector.[1]

point. Up to 24 joints can therefore be sealed from a single excavation. Adaptation of the concept to larger pipe diameters (up to 36 in) has also been performed using a different methodology of moving inside the pipe.

The joint sealing system consists of a small robotic working head (Fig. 20.11.14) equipped with a video camera for search and joint identification, an umbilical cable, cable insertion unit, system storage reel with the tool control components, and an operator's station that includes a control panel and video monitor. In operation, the robot head is inserted into the main through a small tap, and pushed along the pipe by its motorized umbilical cable. The operator visually locates the desired joint and positions the robot at the joint. The robot head is then raised into drilling position, and a small hole is drilled through the spigot into the joint cavity. Next the sealant head is rotated into position, and a measured amount of sealant is injected into the joint. The head is retrieved, and the unit is then moved to the next joint.

The robotic tool head is inserted through a small tap, using a custom-designed saddle. Semirigid umbilical cable is used to push and control the tool head position. The umbilical cable is stored on a custom-designed reel. A miniature video camera mounted on the tool head is used to locate each joint for sealing.

The operator's station is provided with a control panel and video monitor, allowing remote control of the tool head for drilling and sealant injection.

The operation of the robot is as follows. The excavation is performed by a crew designated by the gas company. Then the robot system operator crew takes over the operation. First it drills an opening of about 3-in in diameter into the pipe. This drilling is done nonrobotically with specialized equipment that prevents gas emission. This equipment has been also developed by the robotic design group. The drill unit is held within a cylindrical guide that is attached to a saddle that covers the pipe. The saddle has been designed for this task. The saddle is mounted first, followed by the cylindrical guide. The guide has a tap that closes one end of the guide to prevent the gas from emanating during the drilling. The same guide and valve are used to guide the robot head into the pipe and avoid emission during the operation of the robot.

Two openings are drilled side by side. Each is tilted toward one or the other side of the pipe. The tilt helps insertion of the robot head and umbilical cord. After the drilling operation is finished, the drilling equipment is removed, and the robot is inserted. The robot has skids to slide at the bottom of the pipe. These skids are folded during the insertion into the pipe, and they deploy automatically after the head has entered the pipe fully. The head is pushed into the pipe by the automatic umbilical cord. The cord is stiff longitudinally, and compliant radially,

allowing it to curl on a storage drum for transportation. When the robot reaches a joint that needs to be sealed, the operator stops the robot and lifts a small drill vertically up. The operator positions the drill bit at a location convenient for anaerobic insertion. The drill bit is automatically lifted and pressed against the surface of the pipe. The drilling into the spigot takes approximately 2 to 3 min. The operator maintains a visual observation of the process. The drilling is monitored in closed loop by a force sensor. When the drill has penetrated the wall, the force decreases and the drill stops. Then the system retrieves the drill, rotates the end effector, and brings the anaerobic flow line (activated pneumatically) in line with the hole. The system lifts and presses the anaerobic flow line end onto the opening secured by a specially designed gasket that prevents backflowing, and the anaerobic starts flowing. The flow is controlled in closed loop by a pressure differential sensor. The sensor indicates that the pressure has reached a maximum which implies that the joint is filled with anaerobic material.

DNA MICROARRAYER

A **DNA array** is a generic name covering different molecular biology products and techniques. It can be described as a manifold of DNA fragments (spots) of oligonucleotides at low, medium, or high density. DNA arrays are used as research or diagnostic tools. The array of DNA fragments on a solid surface allows detection of the expression of thousands of genes in a single experiment. The gene array technology is becoming one of the most common techniques used in all molecular biology laboratories.

The advantages of DNA array technology include the simultaneous analysis of a large number of genes in a single experiment; quantitative and reproducible results; speeding up of basic biological research and disease diagnosis; reduction of time, cost, and risks associated with discovery and development.

In terms of technology used to make DNA arrays, there are three methods: (1) photolithography method that is based on site oligonucleotide synthesis; (2) microspotting with quill pins; and (3) ink jetting. The spotting tools are split/quill pins, solid pins, piezoelectric pins, capillary, and ring-and-pin systems. The support used for DNA arrays can be nylon membrane, polypropylene membrane, or glass slides. The methods of DNA binding to the support are based on electrostatic and hydrophobic interactions with covalent links. For detection of gene expression, first cDNA is spotted onto the slides, then the target DNA is hybridized with the cDNA, and the expression is identified through probe labeling: radioactivity (^{33}P) and fluorescence (CY 3 and CY 5), and subsequent detection (reading of color intensity).

The applications of the DNA array are in research and clinical use, such as mapping and genotyping of single-nucleotide polymorphism in the human genome; disease diagnosis: tracking mutation in genes involved in many cancers (p53, BRCA1, BRCA2, etc.); viral, bacterial, and fungal infections: species assignment and detection of drug-resistant strains (HIV, *Mycobacterium* sp, *Candida* sp, etc.).

Robot spotters have been developed for DNA spotting on glass slides. A particular spotter has a large spotting surface (75 or 126 slides), and modular and reconfigurable structure, is able to spot up to 83,000 spots per 25-mm × 75-mm slide, and is capable of performing other biolaboratory tasks such as arraying, gridding, rearraying, and pipetting. The robot is a cell based on a high-quality three-axis gantry robot, with 1.25-μm resolution of motion along each axis, and impedance control to avoid high impact forces at the contact between the slide and the pin. The system has a repeatability of 3 μm and high bandwidth communication with the controller. There is no heating or vibration, and the speed is 1 spot/slide/pin (61,000 genes are spotted in 3.5 h). The slides can be 25 × 75 mm, 20 × 25 mm, or 50 × 75 mm. Also 1 to 8 microtiter source plates and one to six small membrane holders can be used along with up to 126 slides. The cleaning of the pins between loads is done in a water bath that uses active water pump for cleaning.

A vacuum chamber is used to dry out excess material from the quill pins and dry off the water after washing them. A blotting pad is used to eliminate excess material from the pins before spotting. The quill pins have long life (1,000,000 spots). They generate 75- or 90-μm-diameter spots. Each pin can be replaced individually, and up to 250 dots per sample/one dip can be obtained.

Environmental control is provided with a positive-pressure HEPA Fun Filter and PPHC humidity controller (limits evaporation and fast drying of the spots).

The motion controller is based on an embedded PC-104/Pentium. A graphical-user interface is used to program the spotter and monitor the execution. The host computer is based on Pentium/Windows NT. The host controller module can be at a remote location. The graphical-user interface is intuitive and Windows-based for easy control of number of slides, rows, columns, array patterns, dots per row or column, speed, spotting order, type and direction of source plate or membranes, number of pins, etc. Simple robotic language is used for programming the robot.

There are specialized options such as electronic pipetting, colony picking, stacker for multiple source plates, etc.

The robot (Fig. 20.11.15) is designed to collect DNA from 96- and 384-well source plates; 1 to 48 pins (Fig. 20.11.16) can be used simultaneously.

Fig. 20.11.15   DNA microarrayer.[1]

Fig. 20.11.16   Manifold of up to 48 quill pins.

Each pin collects 250 nL of solution and spots 0.6 nL per dot on 75 to 126 microscope glass slides. The center-to-center distance is 120 μm.

Gridding can be performed on up to 15 nylon filters. The filter is standard size equal to 96/384-well plate size. The number of spots per standard filter is maximum 6,144.

Pipetting to and from plates 96/384 can be executed with electronic pipetters in volumes from 0.2 to 200 mL with an accuracy of 10 percent. The pipetters are assembled on manifolds containing four pipetters.

Rearraying can also be performed on 96 and 384 plates also in volumes of 0.2 to 200 mL.

MOBILE ROBOT FOR EXPLOSIVE
AND ORDNANCE DISPOSAL

Police, military, ERT, fire, nuclear, and other hazardous response personnel require remote-controlled equipment for standoff in dealing with explosives, ordnance, and other hazardous materials. The remote-controlled equipment consists of a mobile platform, video equipment, a robot arm mounted on the platform, and a series of auxiliary operational tools such as firearms, disrupters, and X-ray units and associated means of attaching them to the arm or platform. As well, such a system includes an **operator control unit (OCU)** that includes cable and wireless communication with the platform, video monitor, means of controlling the platform and arm, as well as activating the operational tools. The communication covers data, video, and audio links.

The platforms and arms come in various sizes. The platforms are usually anywhere from 60 to 135 cm in length and from 45 to 66 cm in width. The arms can have a reach from 0.8 cm to 2 to 3 m. The payload capacity varies from 5 to over 75 kg depending on the configuration of the arm. The total weight varies from 25 to 350 kg. The speed of the platform varies from 2 km/h to more than 10 km/h.

This class of robots is subdivided in three major groups: small, medium, and large. There are commercial suppliers for each category. The small category is mainly used for reconnaissance and surveillance. It is usually not used with an arm. The large category addressed large payloads, ammunition firing, and other hazardous tasks involving heavy payloads. The medium category is used for a mix of operations, some pertinent to the small robots and others to the large ones.

These **mobile robots** could be used for surveillance, reconnaissance, handling of hazardous items, manipulation of suspected packages, neutralizing and handling such items as improvised explosive devices (IEDs), hazardous chemicals, and radioactive materials. They are all-weather, all-terrain mobile robots that can be used indoors (buildings,

Fig. 20.11.17 Mobile robot with arm deployed.[1]

public institutions, airports, homes, etc.) and outdoor environments (terrain cluttered with obstacles, ditches, gravel, snow, mud, etc.).

The mobile robots have special features: some use only wheels for traction, others use tracks, yet others use quick-remove tracks that can be mounted or dismounted very easily. The tracks, permanent or removable, are necessary for climbing stairs and obstacles.

Other special features are related to precise independent joint control, high dexterity of arm (Fig. 20.11.17), long reach, high payload, open computer architecture for integrating biochemical sensors, and options for autonomous navigation. In terms of control, the operator has the choice to control the arm in joint or task space. The platform is always controlled in task space, by coordinating the motion of the wheels. Typically the platform has two motors, actuating two wheels, one on each side. The other wheels are rotated by chain or other means of transmission.

In terms of mobility the platform is equipped with four or six wheels (Fig. 20.11.18) and a set of fixed or removable tracks. The tracks are used for rough terrain and stair or slope climbing. The robot is able to turn on the spot and maneuver precisely in an obstacle-cluttered area.

Fig. 20.11.18 Mobile robot with arm stowed.[1]

The drives also feature a drive hub, which can disconnect the drive from the wheels to facilitate handling of the robot if necessary.

The robots are remotely controlled by a wireless or cable link; cable link has an onboard automatic winding cable system or an independent reeled cable system. Two-way data and audio and one-way video links are usually available. Up to four cameras and a microphone provide the operator with images and sounds of the environment. A surveillance camera can be mounted on the arm (low to the ground, high in the air, close or far from the gripper) or on a separate articulated boom (usually positioned vertically up).

The wrist/gripper of the arm can be configured to perform operations such as aiming and firing disrupters, pushing heavy payloads, reaching under cars or through car windows, and breaking windows. The robot can insert a car key to open car doors and turn on the car engine, unlock and open building doors, manipulate sensory devices for inspection and surveillance, and handle delicate or hazardous materials. A force-controlled high-quality gripper provides dexterous handling capability.

The robots are water-resistant. Some are even amphibious. All exposed surfaces are covered with anodizing or protective paint. All detector and electronics compartments are protected against water and contaminants. They can operate in all types of weather conditions, day and night, and various types of wet and dry surfaces. Operational temperature range is usually from -20 to $+45°C$. Environmental humidity can be up to 90 percent at 25°C. Storage temperature range is from 0 to 45°C.

This type of mobile robot was further developed as semi- and full autonomous vehicles. As well, certain arm movements can be automated. In terms of platform, features have been developed such as programmed motion (straight line, circular, pivoting), movement in reference to compass, GPS, and land markers. These movements can be controlled in closed-loop or in resolved rate control mode. For the arm, there are automatic movements such as deployment from stowage configuration and vice versa.

MOBILE ROBOT FOR IN-GROUND MINE DETECTION

An articulated remote-controlled mobile manipulator has been developed to scan autonomously off-road and unstructured terrain for buried land mines and explosives. The robot consists of a dual-arm manipulator mounted on a six-wheel mobile platform, a mine or explosive detector, range sensors, terrain-mapping software, and hardware and software for communication, data transfer, and control. The platform can be used also with tracks mounted over the wheels. The tracks help in climbing steep slopes and moving over very rough terrain. The platform can be operated in remote control, semiautonomous, and autonomous modes. The terrain map is generated with respect to a range sensor frame to minimize the computational load in real time. Sensors scan and measure the range with respect to the terrain and relative to the reference coordinate frame. Also the normal direction to the terrain is computed in real time. Based upon the terrain map that is generated in real time, the mine detector's position and orientation are controlled such that a desired (constant) separation and orientation with respect to the terrain is maintained.

One of the dual arms, the *detector arm* (DA), holds the mine or explosive detector (e.g., metal detector, ground penetrating radar, or a nuclear quadripole resonance instrument) that is manipulated autonomously. A plurality of mine detectors can be mounted on the DA. The autonomous motion of the DA is synthesized and controlled online based on the three-dimensional map of the terrain. The map is generated in real time by fusion of sensory data acquired from a scanning laser range finder, an array of four ultrasonic sensors, and joint encoders of the dual arm.

The range finder, a motorized scanning mirror, and ultrasonic sensors are mounted on the other articulated arm, the *sensor Arm* (SA). The SA has a pan joint that allows it to be positioned relative to the DA such that it always scans "ahead" of the DA (Fig. 20.11.19).

The DA holding the detector can maintain it at a desired constant vertical distance from the ground irrespective of undulations of the a priori

Fig. 20.11.19   Mine detector arms: DA and SA.

unknown terrain profile. As well, the DA can maintain the detector orientation constant with respect to the unknown terrain.

The terrain map is generated in real time by using a laser range finder, an array of four ultrasonic sensors, and some joint encoders of the dual arm. The map is generated with respect to the range finder's reference frame (Fig. 20.11.20). The map generation is preceded by image registration that is a fusion of data acquired from the sensors. The map is used to generate the path of DA. Before being used, the map is filtered and enhanced algorithmically (statistically and via fuzzy logic). The fuzzy logic uses a database of terrain signatures. The path of the DA is used to generate the commands of the DA joints that would maintain the detector at the desired separation and orientation relative to the terrain. The DA performs the scanning autonomously.

The performance and reliability of the detection process largely depend on the scanner distance from the terrain, orientation with respect to the terrain, size, scanning speed, and associated filtering and data acquisition software. The scanning involves motion of the sensor manifold (usually a relatively large plate) at constant speed, while maintaining the plate at a constant distance from the ground and in fixed orientation relative to the ground (the detector plate's normal is parallel to the local terrain normal).

The scanning is difficult to perform autonomously because of uncertainties involved in detection of terrain undulations. The uncertainties are estimated in real time (map filtering).

A variety of mine or explosive detectors and terrain scanning sensors can be mounted and interfaced. The entire system (sensors, dual arms, and operational software) is portable to various mobile platforms and vehicles.

The detector arm is an articulated arm with 5 degrees of freedom. Four degrees of freedom are sufficient to move the detector along an arc (i.e., a side-to-side scan), at a constant distance from the ground and parallel to the terrain profile (i.e., the normal of the detector plate is parallel to the local normal of the terrain). The fifth degree of freedom corresponds to a linear motion that supersedes the mobile platform motion when the arms are advanced while the platform is stationary.

The sensor arm carries a laser range finder, two ultrasonic range finders mounted vertically, and two range finders mounted horizontally. The measurements of the latter are used to prevent the collision of the arms with sidewalls when the robot is scanning hallways and narrow passages. The laser range finder and the two vertical ultrasonic range finders measure the distance of the sensors from the ground. These measurements are fused to the joint encoder data to generate a range image of the terrain map. The map is generated locally and online by using a periodic map-building procedure. The ultrasonic range finders move along the arc of the side-to-side scan but do not scan the terrain across the arc. Moreover, the ultrasonic range finders measure the minimum distance of the terrain from the sensors in a region covered by the ultrasonic cones. Thus, the measurements of the ultrasonic range finders cannot be directly used in the range image, but they can be fused to the laser measurements to reduce the uncertainty.

TECHNOLOGICAL UPDATE ON SOME EMERGING APPLICATIONS

Emerging applications of intelligent automation can be found in several sectors such as space, medical, security, and biotechnology. They utilize (1) technology of robotic arms that are embedded in automatic cells, (2) cells that are integrated based on mechatronics technology, and (3) the capability of machine adaptation and learning, thus providing intelligent automation.

In the domain of space, a consortium of major players in the sector is currently developing the Space Mobile Station (SMS). On the station various docking ports and remote-controlled manipulators will be used. One such manipulator is the Special Purpose Dexterous Manipulator (SPDM). This has dual arms with two similar branches, each consisting of seven joints. Seven-joint arms are labeled in the literature as *redundant*. They provide excess degrees of freedom that allow one to effect a task in many ways (configurations). This property can be utilized to perform the tasks in an optimal way by selecting from the many configurations one that optimizes a certain criterion. For example, minimum energy consumption could be used as a criterion in selecting the robot configuration.

The SPDM will be used on the SMS to perform manipulation, assembly, and other tasks that require two arms working in coordination. The domain of coordinated arms is not new in the context of industrial robots, but it presents a new challenge in space robotics.

In the medical field, intelligent automation began being utilized in the 1980s in two domains: surgery and orderly help in hospitals. The latter is based on various mobile autonomous guided vehicles (AGVs), some with an arm mounted for delivery of food, medication, and other necessities to the hospital beds. The former is growing very rapidly by introduction of newly designed robotics in operating rooms. Initially robots were used to perform life-threatening procedures such as brain biopsy in neurological surgery. Later, knee and hip replacement surgery with robotics became standard. Currently laparoscopic surgery is performed in abdomen and thoracic-related surgery. As well, methodologies of surgery with robotic tools are being developed for the brain, heart, prostate, and breast. Some of these procedures will be performed under MRI inside close-bore MRI for best imaging. This requires the development of MRI-compatible equipment, as well as remote control of the surgical tools because of the location of the surgical team in a room adjacent to the MRI, and control of the procedure through computer monitors and master controllers, such as joysticks and pushbuttons.

In the security domain, new technology for standoff and inspection is being developed. There are developments related to inspecting, by

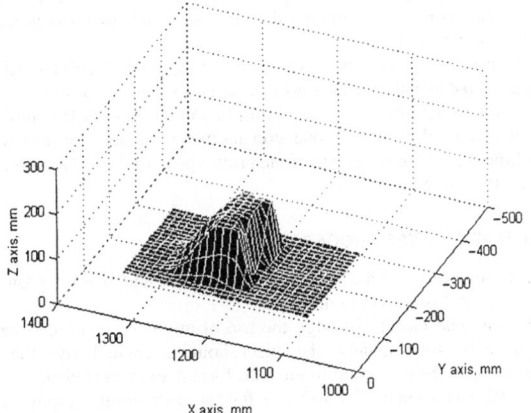

Fig. 20.11.20   Terrain map generated by the robot.

remote control, the interior of passenger cars and trucks. Other developments address the need to negotiate in the event of terrorist-related hostage situations. In the military, there is an acute need for remote devices that can detect and neutralize improvised explosive devices (IEDs). Various unstaffed ground and aerial devices are being developed that could detect IEDs, in-ground mines, and unexploded ammunition. Other technology is being developed for removing the hazardous material after detection.

In the domain of biotechnology, drug discovery and structural proteomics are leading the effort in research and development. There is a need for a wide range of automated equipment to accelerate the drug discovery processes via research in proteomics. Structural proteomics is a crucial domain of molecular biology in the context of drug discovery. Equipment that can perform reliably tens of thousands of experiments each day is being developed using robotic arms, mechatronics cells, and intelligent automation, such as fuzzy-logic-based identification of patterns in the genetic map of humans for detection of predisposition to certain chronic diseases.

In the general domain of intelligent automation, there are developments of mobile robots that climb walls to wash high-rise windows, autonomous vacuuming of floors, robot arms used in the construction industry, modular laboratory equipment that can be carried home by students for studying intelligent automation, automatic guidance of the visually impaired, etc.

TRENDS AND SUMMARY

This section has discussed the related fields of robotics, mechatronics, and intelligent automation, and illustrated them by presenting some selected niche applications. Many other applications would serve the same purpose. The aim was to support the view that these fields have a great impact on humanity, and the opportunities for further developments in these emerging domains are tremendous. Opportunities abound in this domain both scientifically and commercially. There is a great entrepreneurial spirit in conjunction with these areas, as the technological progress made in the last decade has generated extraordinary scientific and commercial activities. Education and research in colleges and universities also have been harnessed onto this wave of technological development, leading to a significant increase in the number of students, publications, patents, and commercial spinoffs.

The domain has achieved significant results, and it continues to evolve as new applications are emerging. The new applications mentioned in the section "Technological Update on Some Emerging Applications" are only representative of the thrust. They are only illustrating this exciting domain of science and engineering. Furthermore, they are demonstrating the interdisciplinary nature of this area of technology.

Nonetheless, there are some main issues that require more thorough consideration as to the best way to accomplish the proliferation of intelligent automation in every sector of life. Already, medicine, security, manufacturing, education, all major sectors of human life and economy, are subscribing to the era of intelligent automation. There is a need to extend this list to services, rehabilitation, entertainment, and other items. This may be possible only if opportunities are provided, human resources are available, and the economics are favorable for developing new technology to serve humanity in the most effective way.

REFERENCE
1. *Engineering Services Inc.,* www.esit.com.

20.12   RAPID PROTOTYPING

by Ali M. Sadegh

In the process of product development in a **computer-aided design** (CAD) environment, when a design concept has been tested and a virtual prototype (solid model) has been created, a physical prototype is needed for testing. This stage of the process is called **alpha prototype** testing. The alpha prototype is a replica of the part showing principal geometric features, such as all the dimensions, slots, cavities, and surface texture. In the case of a device or a machine, the interaction between parts of the device or the machine could be tested with the alpha prototype. The next step in the product development is **beta prototype** testing, where limited full-scale parts are made with planned manufacturing processes and are examined and tested by volunteers or customers in actual operation condition of the device. In the end, the parts or the devices are made with the final materials and processes and are tested in an independent laboratory or limited (selective) market.

Traditional alpha prototype manufacturing utilizes common materials including metal, woods, clay, ceramics, construction papers, and polymers with machining and metal forming. This method requires craftsmanship; it is time-consuming, labor-intensive, and costly.

During the last decade, a class of technologies known as **rapid prototyping** has emerged that can automatically construct physical models, the alpha prototype, from computer-aided design (CAD) data. Rapid prototyping is an alternative technology that uses computer-controlled equipment to automatically and rapidly fabricate prototypes. These technologies have dramatically reduced the time needed for the design and product development.

Rapid prototyping (RP), also referred to as **three-dimensional printing,** or 3-D printing, allows designers to quickly create tangible prototypes of their designs. These prototypes are excellent visual aids for communicating ideas with coworkers or customers and even testing the functionality of devices or machines. In addition to prototypes, RP techniques can also be used to make tooling for a manufacturing process (referred to as **rapid tooling**) and even production-quality parts (**rapid manufacturing**). For small production runs and complicated objects, rapid prototyping is often the best manufacturing process available.

Clearly the word *rapid* is a relative term, since most of the 3-D printing machines take from 1 h to more than 48 h to build, depending on the size and complexity of the part or the device. There is significant time saving compared to the few days or few months that is required for traditional prototyping.

The principal operation of a majority of rapid prototyping machines is the **layered manufacturing process.** That is, a software package "slices" the CAD model into a number of thin (about 0.003-in or 0.1-mm) layers, which are then built up one atop another. Of course, because of the machine dimensions, the size of the prototype is limited. Part volume is generally limited to 5,000 in^3.

RAPID PROTOTYPE PROCESS

The following steps are required to construct and build a prototype through rapid prototyping machines.

1. **The solid model.** Through the use of any computer-aided design software, the solid model of the object should be created. Then the solid object should be converted to the STL format, explained next.

2. **The STL format.** The software format of stereolithography (STL), which was the first RP technology, has been adopted as the standard of the rapid prototyping industry.

This format represents a three-dimensional surface as an assembly of planar triangles. The file contains the coordinates of the vertices and the direction of the outward normal of each triangle. The CAD file should be transferred to the **STL format** with the extension .stl, which is universal for all of the RP machines.

3. **Creating layers from STL file.** Rapid prototyping machines generally have a preprocessing program that slices the STL file. Each RP machine has its own solid modeler, which allows users to alter the scale or some dimensions of the model as well as place the object in a desired orientation to be built. The preprocessing software then slices the STL model into a number of layers from 0.01 to 0.7 mm thick, depending on the RP machine's precision. In case the solid model has overhangs, internal cavities, or very thin-walled sections, the program may also generate an auxiliary structure to support the model during the build. The materials used for the structural support are generally very fragile and could be taken out easily through a cleaning process.

4. **Construction of the object.** Through several techniques (presented below), each sliced layer will be constructed from different materials such as paper, plaster, polymer, or powdered metal. Each layer is glued, fused, or sintered to the previous layer. Finally the object is built when the construction of all the layers are completed. Almost all the RP machines are autonomous and build the object without intervention.

5. **Cleaning.** After the object is built, it is removed from the machine by detaching any support for the object. Some photosensitive materials need to be fully cured before use. Prototypes may also require detaching the structural supports through cleaning and surface treatment. In some cases, sanding, sealing, and/or painting the model will improve its appearance and durability.

RAPID PROTOTYPING TECHNIQUES

Since the early 1980s, several technologies have been patented and developed. However, most commercially available rapid prototyping machines use one the following techniques.

• **Stereolithographic apparatus (SLA).** The stereolithography technique, which was first patented in 1986 by 3D Systems of Valencia, CA, revolutionized the rapid prototyping technology. The commercial machine is called stereolithographic apparatus (SLA). It uses a high power laser to selectively solidify a liquid photopolymer, layer by layer, into the shape of the finished prototype. In essence, the SLA machine is a **material additive process** in nature where the polymer solidifies when exposed to ultraviolet laser. The model is built upon a platform situated just below the surface in a vat of liquid epoxy or acrylate resin.

Through a series of servocontrolled mirrors, as shown in Fig. 20.12.1, the laser is guided and shone on the liquid photosensitive polymers to solidify the top layer, leaving excess areas liquid. Then, an elevator incrementally lowers the platform into the liquid polymer. A sweeper recoats the solidified layer with liquid, and the laser traces the second layer atop the first. This process is repeated until the prototype is complete. Once the very top layer of the object is completed, the elevator raises the platform and the part is removed from the vat and rinsed clean of excess liquid. Supports are broken off and the model is then placed in an ultraviolet oven for complete curing. Figure 20.12.2 shows samples of final products.

• **Fused deposition modeling (FDM).** This is a process that deposits thin beads of melted (fused) thermoplastic materials (filament), and is generally extruded from a tip that moves in a horizontal plane. As the material solidifies, a prototype model is built, layer by layer, typically up to 10 in. by 10 in. in cross section and up to 16 in. high.

The platform is maintained at a lower temperature, so that the thermoplastic quickly hardens. After the platform lowers, the extrusion head deposits a second layer upon the first. Support structures are built along the way. Materials used in this process are high-strength ABS plastic, standard ABS, elastomers (96 durometer), polycarbonate, and polyphenylsulfone. To remove the supporting structure, parts are postprocessed by mechanical agitation in a warm water solution.

• **Laminated object manufacturing (LOM).** In this method each slice of the model is cut (from thin layers of paper, polymer, or sheet metal) using a numerically controlled laser. The slices are bonded together with adhesive materials to form a prototype. After each layer is put on top of the previous layer, a heated roller applies pressure to bond the material (paper) to the base. Then, the platform lowers the object to prepare for the next layer. This process is repeated as needed to build the part. LOM prototypes can be sanded to reduce jagged edges; however, they cannot be rigorously tested. This technique was developed by Helisys of Torrance, CA.

• **Selective laser sintering (SLS).** This technique employs a laser beam to selectively sinter together fusible materials such as powdered metal, nylon, and elastomer. **Sintering** is the heating and fusing of

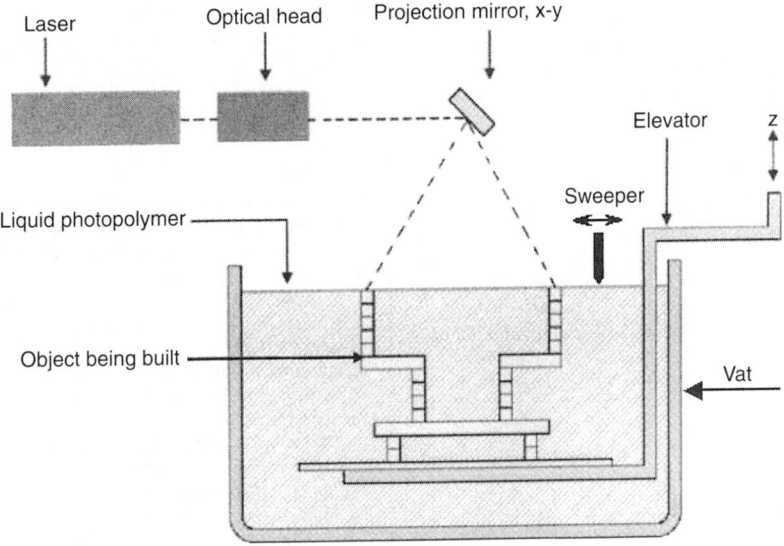

Fig. 20.12.1 Schematic diagram of stereolithography.

Fig. 20.12.2 Samples of products built by stereolithography.

small particles resulting in a hard bonded layer of material. The laser beam traces the pattern of the first layer, sintering it together. Then, the platform is lowered by the height of the next layer and powder is reapplied, and once again the laser beam traces the next layer, which is fused to the previous layer. This process continues layer by layer until the part is complete. The unsintered powder supports the part as the layers are sintered. The University of Texas first patented this method in 1989.

• **3-D Inkjet Printing.** In this process an inkjet printing head deposits or "prints" glue-like binder on a thin layer of powdered material. The binder fluid fuses the powder together in a desired area, to form a layer of the part. As the platform is lowered, more powder is added and leveled, and the inkjet head prints (despites) the binder to form another layer of the part. Layer by layer, the part is built. Unbounded powder remains to support the part. When the part is finished, it is removed from the container and the unbound (excess) powder is blown off. The resolution of each layer is about 0.1 mm. This process is relatively fast, and produces parts that are somewhat fragile, unless the part is coated with wax, Cyanoacrylate (CA) glue, or other sealants to improve durability and surface finish. This process was first developed by MIT. The ZCorp 3D printer, produced by Z Corp., uses two different materials: a starch-based powder (not as strong, but can be burned out, for investment casting applications) and a ceramic powder.

APPLICATIONS OF RAPID PROTOTYPING

Because of the time and cost savings, rapid prototyping machines have revolutionized product development processes in many industries such as automotive, aerospace, medical, and consumer products. Although the rapid prototyping applications are vast, almost all fall into one of the following categories: prototyping, rapid tooling, or rapid manufacturing.

Prototyping Rapid prototyping (RP) plays an important role in quickly making prototypes for communication and alpha prototyping testing purposes. While solid modeling of an object in computers and engineering drawings helps one to understand the function of an object, clearly, touching and feeling the actual prototype of the object dramatically improve communication among people, including engineers.

Rapid Tooling *Rapid tooling* (RT) is a process that employs rapid prototyping models, directly or indirectly, to create a mold quickly. The mold could be used for a limited volume of alpha or beta prototypes. Rapid tooling is less expensive and requires less time than conventional fabrication of tooling. Note that conventional tooling is one of the slowest and most expensive steps in the manufacturing process, since it requires a high level of precision. Typically with RT, the time needed for the fabrication of the first articles is below one-fifth that of conventional tooling and the cost is about 5 percent that of the conventional tooling. The disadvantage of rapid tooling is that the tool life is considerably less than that of a conventional tool.

Indirect Tooling RP parts are also used as patterns for making molds and dies. RP models can be indirectly used in a number of manufacturing processes:

Vacuum casting. This is the simplest and oldest rapid tooling technique. A RP positive pattern is suspended in a vat of liquid silicone or room temperature vulcanizing (RTV) rubber. Specifically, silicone liquid is cast around a master model that is typically made by a rapid prototyping machine (preferably SLAs). A partial vacuum is applied to avoid air bubbles that may be trapped in between the master and silicone. After curing, the mold is cut open and the master (SLA) is removed, leaving a cavity to make copies. The flexibility of the silicone allows limited undercuts, which might avoid slides or additional parting planes. Vacuum casting is a copying technique typically used for the production of small series (10 to 20) of functional plastic prototypes. An alternative approach, known as the Keltool powder metal sintering process, uses rubber molds to produce metal tools. When the binder cures, the "green" metal tool is removed from the rubber mold and then sintered. At this stage the metal is only 70 percent dense, so it is infiltrated with copper to bring it close to its theoretical maximum density. Note that vacuum molding is a different process where, through suction, a sheet of plastic or metal is forced to conform to a preset shaped mold.

Injection molding. To prepare a mold for an injection-molding machine, first a stereolithography machine is used to make a match-plate positive pattern of the desired molding. To form the mold, the SLA pattern is plated with nickel, which is then reinforced with a stiff ceramic material. The two mold halves are separated to remove the pattern, leaving a matched die set that can produce tens of thousands of injection moldings.

Other methods such as casting, including sand casting, may be employed for rapid tooling. In casting, an RP model is used as the positive pattern around which the sand mold is built.

Direct Tooling This technique, developed at Sandia National Laboratories and Stanford University, can create metal tools from CAD data. As in the rapid prototyping process, a laser beam melts the top layer of the part in areas where material is to be added. Powder metal is injected into the molten pool, which then solidifies. This process is repeated layer by layer until the part is built. Unlike traditional powder metal processing, this method produces fully dense parts, since the metal is melted, and not merely sintered. The resulting parts have exceptional mechanical properties. This method is appropriate for parts with simple geometry or a more uniform cross section.

In a different process, an RP machine selectively sinters polymer-coated steel pellets together to produce a metal mold. The mold is then placed in a furnace where the polymer binder is burned off and the part is infiltrated with copper.

Rapid Manufacturing When a limited number of products is needed, rapid manufacturing (RM) is much cheaper, since it does not require tooling. RM is also ideal for producing custom parts tailored to the user's exact specifications. Rapid manufacturing is a natural extension of rapid prototyping. RM will never completely replace other manufacturing techniques, especially in large production runs where mass production is more economical. However, when an object cannot be made by subtractive (machining, grinding) or compressive (forging, etc.) processes, such as internal voids or complex structures, rapid manufacturing is the solution.

Conclusion

Rapid prototyping has changed product development processes and the way companies design and build products. The most attractive feature of RP is the short prototyping time, which saves tremendous cost during product development.

As a result of the advent of computers, RP technology is constantly improving. For example the accuracy of the surface finish is improving and the laser optics and motor control (used in RP) have increased their accuracy in all three directions. In addition, RP companies are developing new polymers that will be less prone to curing and temperature-induced warpage.

RP machines using nonpolymeric materials, such as metals, ceramics, and composites, produce functional parts that could be used for testing. These machines expand the range of products that can be made by rapid manufacturing. Much of the RP research is focused on the development of new materials for the RP machines that make functional parts.

REFERENCES

Ashley, Steven, From CAD Art to Rapid Metal Tools, *Mechanical Engineering*, March 1997, p. 82.
Ashley, Steven, Rapid Prototyping Is Coming of Age, *Mechanical Engineering*, July 1995, p. 63.
Eggert, Rudolph, "Engineering Design," Prentice Hall, 2005.
Griffith, Michelle, and Lamancusa, John S., Rapid Prototyping Technologies, *Rapid Prototyping*, 1998.
Hartwig, Glenn, Rapid 3D Modelers, *DE*, March 1997, pp. 38–39.
Hilton, Peter, Making the Leap to Rapid Tool Making, *Mechanical Engineering*, July 1995, p. 75.
Langdon, Ray, A Decade of Rapid Prototyping, *Automotive Engineer*, May 1997, pp. 44–45.
Otto, Kevin, and Wood, Kristin, "Product Design," Prentice Hall, 2001.
Waterman, Pamela J., Rapid Prototyping, *DE* March 1997, p. 30.

20.13 MISCELLANY

by E. A. Avallone

PREFERRED NUMBERS

REFERENCE: ANSI Standard Z17.1.

Many manufactured articles are made in several sizes which may be designated by some dimension, speed, capacity, or other feature. Each such series of products may be paralleled by a series of numbers.

It is generally agreed that such number series should be **geometric progressions;** i.e., each term should be a fixed percentage larger than the preceding. A geometric series provides small steps for small numbers, large steps for large numbers, and this best meets most requirements. The small steps in the diameter of the numbered twist drills would be absurd in drills of 1 in diameter and larger.

In the case of sized objects that are used principally as raw material, e.g., steel rod, an arithmetic progression may be preferable because it tends to reduce the cost of machining. It is desirable to be able to buy raw material a fixed amount (rather than a fixed percentage) larger than the finished article.

Preferred numbers is the name given to various series proposed for general use. These are either geometric progressions or approximations thereto. A geometric series is defined by one term and the ratio of each term to the preceding one. On the choice of these elements for a preferred number series, there is as yet no general agreement. The same value would hardly be satisfactory for all cases. The idea of preferred numbers is to provide a master series from which terms can be chosen to suit any needs. This would ultimately lead to a comprehensive plan in all fields of manufacture, so that, for example, the sizes of shafting would be in accord with the sizes of bearings, and indeed with all manner of cylindrical machine elements.

An advantage of a geometric series is that if linear dimensions are chosen in the series, areas, volumes, and other functions of powers of dimensions are also members of the same series.

In one of the most carefully considered systems of preferred numbers the base term is 1, and the ratio is $\sqrt[80]{10}$. In this series, the 81st term is 10, and accordingly the series from 10 to 100 or from 0.01 to 0.1, or, in general, from 10^n to 10^{n+1} is identical with the series from 1 to 10 with the decimal point shifted. This series will rarely be used in full; some will choose alternate terms, some every fourth, fifth, tenth, or twentieth term. The index of the root, 80, has as factors, 2^4 and 5, so that the series readily yields subseries having as ratios the roots of 10 with indices 2, 4, 8, 16, 5, 10, 20, 40, thus giving a wide range of choice. See Table 20.13.1.

The strict logic of this series has been somewhat impaired by the adoption of rounded values that are slightly different in the 1-to-10 and 10-to-100 intervals. For the United States, ANSI has adopted a Table of Preferred Numbers (ANSI Z17.1) which differs slightly from the system described in the preceding paragraph.

Another type of series is the **semigeometric series** consisting of a basic geometric series with 1 as the base term and a ratio of 2, giving a series . . . 1/8, 1/4, 1/2, 1, 2, 4, ... Between consecutive terms are inserted arithmetic series of 2, 4, 8, or 16 terms, in general using different numbers of terms in different intervals.

A similar procedure is used to establish the numbers of teeth in a prescribed number of gears that are intended for use in a gear train to provide a stepped gradation of rotational speeds within an upper and lower bound.

Table 20.13.1   Basic Series of Preferred Numbers: R 80 Series

1.00	1.80	3.15	5.60
1.03	1.85	3.25	5.80
1.06	1.90	3.35	6.00
1.09	1.95	3.45	6.15
1.12	2.00	3.55	6.30
1.15	2.06	3.65	6.50
1.18	2.12	3.75	6.70
1.22	2.18	3.87	6.90
1.25	2.24	4.00	7.10
1.28	2.30	4.12	7.30
1.32	2.36	4.25	7.50
1.36	2.43	4.37	7.75
1.40	2.50	4.50	8.00
1.45	2.58	4.62	8.25
1.50	2.65	4.75	8.50
1.55	2.72	4.87	8.75
1.60	2.80	5.00	9.00
1.65	2.90	5.15	9.25
1.70	3.00	5.30	9.50
1.75	3.07	5.45	9.75

SOURCE: American National Standard Preferred Numbers Z17.1, reproduced with permission of ANSI.

Preferred sizes are different from preferred numbers, but when reduced to linear dimensions, they enjoy the convenience and simplicity of "whole" numbers. Metallic bar stock, e.g., may be supplied in diameters (in inches) of, say, 1, 1¼, 1½, 1¾, 2, ... , 8, 8½, 9.

PROFESSIONAL ENGINEERING PRACTICE

The regulatory engineering environment may be broadened to include the professional working environment in which engineers conduct their practice. Ultimately, the engineers' endeavors result in a product of some sort, be it a tangible item or drawings, specifications, inspection documents, research reports, and the like. Keeping that in mind, the subjects which follow address briefly some aspects pertinent to the practice of engineering.

Personnel

The engineer often will employ other persons, and as an employer, the engineer is subject to all applicable laws and regulations related to the employer-employee relationship. The administration of personnel matters will involve the engineer directly or will be delegated to a surrogate within the organization. Among the things which relate to personnel are worker's compensation insurance; accident and health insurance; antidiscrimination laws related to hiring, termination, and promotion; general liability insurance; and so on. Depending upon the staffing and the type of work which is performed in the engineer's office, it may prove prudent to carry employer's liability insurance to cover situations which are excluded from worker's compensation and general liability insurance. In these regards, it is wise to seek legal counsel.

If the employees are affiliated with organized labor, collective bargaining will be involved with the one or more unions representing the employees, and it will encompass matters such as pay, fringe benefits, and working conditions.

The objective of good personnel administration is to have the entity run smoothly, with maximum interpersonal harmony and good will so that the organization will prosper.

Some other pertinent details related to personnel administration are discussed briefly in Sec. 17.1.

Professional Engineer Licensing and Registration

Many engineers, whether self-employed or employed by others, may seek to obtain a license as a registered professional engineer. Generally, the terms *licensed professional engineer* and *registered professional engineer* are synonymous.

The licensing of professional engineers (and of those in other professions and trades) falls under the aegis of the individual state education departments, which control and administer professional licensing within the state's jurisdiction. In years past, some states granted the license as a professional engineer to a few persons who, although lacking formal credentials (education and otherwise), had a demonstrated record of performance of the duties of a professional engineer. The licenses were granted under "grandfather" clauses. This is now very seldom the case.

The usual route to obtain a license as a professional engineer is via a combination of education, experience, and written examination. The successful attainment of an undergraduate degree in engineering will admit the applicant to a first examination, the successful completion of which results in the applicant being granted a certificate as Engineer in Training (EIT). Additional suitable experience for another 4 years admits the EIT into the final written examination for professional engineer. The successful completion of the final examination results in the award of a license as Professional Engineer.

Federal, state, and most local municipalities require that their key engineering personnel be licensed professional engineers. Engineering documents submitted to those bodies, likewise, will require that the signatory thereto be a licensed professional engineer.

There are minor variations from state to state as to acceptable experience, the degree to which academic experience may be acceptable as engineering experience, etc.; but by and large, there is a degree of uniformity among the state education departments in the matter of licensing professional engineers.

Professional engineers active in several state jurisdictions may be required to obtain a license as professional engineer in those different states. To that end, most states have established procedures to grant a license as professional engineer by comity, but usually this will be possible only if the original license was granted on the basis of a written examination. Comity is often initiated through the offices of the National Society of Professional Engineers as a service to its members.

The work bearing the professional engineer's seal is considered creative work and is so addressed in most stated education laws. The work carries an implicit copyright, even though a statement to that effect may not be placed on the documents. Said work is the property of the professional engineer; unauthorized copying of such documents is illegal, and violators are subject to legal action and redress.

The basic intent of formal licensing of professional engineers is to protect the public from incompetent practice. A licensed professional engineer whose competence is called into question is subject to formal investigative procedures, and when the circumstances warrant it, the license may be revoked or suspended.

Detailed information regarding license as a professional engineer can be obtained from the state education department having jurisdiction, and it will set forth the requirements, privileges, and responsibilities inherent in the granting of the license.

In recent years, there has been recognition that a licensed professional engineer provides services in a rapidly changing environment of evolving technology, with the concomitant necessary continual development of knowledge and skills. To that end, most jurisdictions now require that, as a prerequisite to license renewal, professional engineers accrue specified minimal amounts of continuing education, usually in the form of contact hours or continuing education units. Acceptable continuing education includes formal coursework and/or educational activities; teaching a course in an acceptable subject area; authoring an article in a peer-reviewed journal or a published book; making a technical presentation at an approved technical conference; obtaining a patent related to the practice of engineering; and completion of an approved self-study program.

Inquiries along these lines are made to the entity having jurisdiction over granting licenses and renewals.

Product Liability

In the current consumer and/or user climate, the matter of product liability has assumed a large role and often impacts upon the engineer's input to certain products. In some cases, the engineer may be held personally responsible and liable for some outcome even though the engineer may have been employed by others. The case reports of litigation involving product liability are voluminous, and it is not surprising that identical circumstances will result in verdicts different from one jurisdiction to another. The laws related to product liability are based principally on two concepts: negligence and strict liability. The details are not addressed here; suffice it to say that the engineer who faces product liability litigation must retain competent legal counsel to prepare and implement an appropriate defense.

Professional Liability Insurance

This type of insurance is often carried by engineers in private practice to protect the engineer as well as to provide relief to an injured party, when the latter may result from error(s) and/or omission(s) in the documents prepared by the engineer; hence, the term *errors and omissions insurance,* or E&O insurance. The nature of professional liability insurance is quite complex and is usually handled between the engineer and the insurer, often with the active participation of the engineer's legal counsel. Essentially, in obtaining professional liability insurance, the engineer enlists a broader financial entity (the insurance company) to absorb a portion of the costs of claims in exchange for a premium paid to the insurance company. Actions related to professional liability insurance evolve from the management decisions made by the engineer (or the firm), as part of the overall conduct of the engineering practice and consideration of the financial risks incurred thereby.

In many cases, before any work proceeds, the client will require that the engineer provide evidence that professional liability insurance is in effect, and that evidence often is incorporated into the contractual agreement between engineer and client.

Professional liability insurance policies are traditionally of the claims-made type. Claims arising out of work undertaken and performed while the policy is in effect will be covered. Claims arising from that work at a later date will **not** be covered unless the insurance has continued in effect to the time the claim is made. All coverage ceases when the policy is canceled or not renewed.

Insurance premiums depend on several factors: limits of liability, deductibles, type of work performed and the services rendered in connection therewith, experience of engineer and his or her "track record," and the geographic location of the work (there are low-risk states and high-risk states). There are other factors which may be subject to consideration by the insurer.

PUBLISHING AND EDITING

Sizes of Type

The unit of height of a line of printer's type is the "point." One point = $\frac{1}{72}$ in. Six-point type is consequently $\frac{6}{72} = \frac{1}{12}$ in high. The sizes generally employed in books and periodicals are 6-, 7-, 8-, 9-, 10-, 11-, and 12-point. These are shown in Fig. 20.13.1.

6-point   10-point
7-point
8-point   11-point
9-point   12-point

Fig. 20.13.1 Examples of type sizes.

Copy Preparation and Copy Fitting

It is recommended that manuscript material be printed double-spaced on 8½- by 11-in paper with 1½-in margins at the top and left side, 1-in margins on the bottom and left side, and with each page containing 25 lines, double-spaced. Thus a printed line will be 60 characters long. Nearly all printers can print at 10 characters to the inch and double-spaced (three lines to the inch). For convenience in copyediting, printed copy should be printed double-spaced. One manuscript page so printed is equivalent to about 250 words (assuming that each word averages six characters).

One way to determine approximately how many lines of manuscript copy are equivalent to how many lines of finished book type is to print, to the specifications given above, a number of lines from a book which has been printed in the desired type font. This will show the relationship between the number of manuscript-copy lines and lines set in type.

Nearly all word-processing programs will produce satisfactory manuscript. Authors working directly with a publisher will usually make arrangements to submit their manuscripts in computer-readable form. Some publishers require it. Publishers have very comprehensive format-conversion facilities. Authors working in groups or with intermediate editors, however, face the problem that others of their associates or editors may not have much conversion capability. In such cases, the editors or group leaders will make arrangements to accept such computer-readable copy as they can. In most cases, a printed copy should accompany all computer-readable material.

Proofreading

When correcting proof, use ink or pencil of a color different from the marks already on the proof. Put your marks in the right or left margin, whichever is nearer to the word corrected. If there are several corrections in a single line, place them in order from left to right, separated by a slant line (e.g., tr/cap/,). If the same correction is made several times in the same line with no intervening correction, make your correction once in the margin, followed by an appropriate number of slant lines. For instance, if you add an *s* to three words in a line, write in the margin s///. When you wish to insert words, put a caret at the point of insertion and write the additional words in the margin. To delete material without substituting anything, cross it out and put a delete sign in the margin. When you delete words and substitute other words, cross out the unwanted material, use a caret within the line, and write the new material in the margin; the delete sign is then unnecessary. Standard proofreader's marks are shown in Fig. 20.13.2.

Fig. 20.13.2 Proofreader's marks.

Occasionally you may decide that material you have deleted should be restored. Place a row of dots under the crossed-out material, cross off the delete sign in the margin, and write "stet" (let it stand).

Special Alphabets

The primary alphabets used as symbols are Greek, script, and German. Examples of these are shown in Figs. 20.13.3 through 20.13.6.

A B Γ Δ E Z H Θ I K Λ M N Ξ O Π P Σ Τ Υ φ X Ψ Ω
α β γ δ ε ζ η θ ι κ λ μ ν ξ ο π ρ σ τ υ φ χ ψ ω

Fig. 20.13.3   Greek alphabet.

A B Γ Δ E Z H Θ I K Λ M N Ξ O Π P Σ Τ Υ Φ X Ψ Ω
α β γ δ ε ζ η θ ι κ λ μ ν ξ ο π ρ σ τ υ φ χ ψ ω

Fig. 20.13.4   Boldface Greek alphabet.

Note that the capital Greek letters used by the printer are easily mistaken for the Roman letters A, B, E, Z, H, I, K, M, N, O, P, T, and X and may not be suitable as symbols. If they are used in a manuscript, they should be identified with a marginal note so they will not be improperly set. The form of many Greek boldface symbols is markedly different from the style of the lightface characters.

Lightface: 𝒜 𝓑 𝒞 𝒟 �ℰ 𝓕 𝒢 𝒦 𝒥 𝒥 𝒦 𝓛
𝓜 𝓝 𝒪 𝒫 𝒬 𝓡 𝒮 𝒥 𝒰 𝒱 𝒲 𝒳 𝒴 𝒵
Boldface: 𝓑 𝓔 𝓔 𝓕 𝓚 𝓙 𝓜

Fig. 20.13.5   Script alphabet.

𝔄 𝔅 ℭ 𝔇 𝔈 𝔉 𝔊 ℌ ℑ 𝔍 𝔎 𝔏 𝔐 𝔑 𝔒 𝔓 𝔔 ℜ 𝔖 𝔗 𝔘 𝔙 𝔚 𝔛 𝔜
a b c d e f g h i j k l m n o p q r ſ t u v w x y z
ä ü ö 𝔄 𝔘 𝔒 ch ck st ß ſt tz

Fig. 20.13.6   German Fraktur alphabet.

Script and German symbols (Figs. 20.13.5 and 20.13.6) may be indicated by circling in a distinctive color the corresponding English letter.

Index

RXPREP COURSE BOOK
2015 EDITION

KAREN SHAPIRO, PHARMD, BCPS

SHERRY A. BROWN, PHARMD, MBA, BCPS

With Stephanie D. Garrett, PharmD, BCPS

CONTRIBUTORS

AMINE ALE-ALI
PHARMD, BCOP

PAUL BERINGER
PHARMD

HEATHER R. BREAM-ROUWENHORST
PHARMD, BCPS

KATIE E. CARDONE
PHARMD, BCACP, FNKF, FASN

GEORGE DEMAAGD
PHARMD, BCPS

PAULINA DEMING
PHARMD, PHC

CATRINA DERDERIAN
BS, PHARMD

JEFFREY FUDIN
BS, PHARMD, FCCP

MUOI GI
PHARMD, BCPS, BCOP

JEFF GOAD
PHARMD, MPH, FAPHA, FCPHA, FCSHP

SUSAN E. GORMAN
PHARMD, MS, DABAT, FAACT

ERIC GUPTA
PHARMD, FNLA, BCPS, CLS

BRETT HEINTZ
PHARMD, BCPS-ID, AAHIVE

JAN D. HIRSCH
BSPHARM, PHD

DOUG HUMBER
PHARMD

KIM JONES
PHARMD, BCPS

JEFF LEE
PHARMD, FCCP

KELVIN LI
RPH

JOSEPH D. MA
PHARMD

JOEL C. MARRS
PHARMD, FCCP, FNLA, BCPS-AQ CARDIOLOGY, BCACP, CLS

CRAIG MARTIN
PHARMD, BCPS

JESS MARTINEZ
PHARMD

CYNTHIA L. MORRIS-KUKOSKI
PHARMD, DABAT, FAACT

TIEN NG
PHARMD, FCCP, BCPS (AQ-C)

NANCY N. NGUYEN
PHARMD, BCPS, AAHIVP

NATHAN PAINTER
PHARMD, CDE

ALLISON PROVINE
PHARMD, BCPS

RENU F. SINGH
PHARMD, BCACP, CDE

KIMBERLY B. TALLIAN
PHARMD, BCPP, FASHP, FCCP, FCSHP

STACY A. VOILS
PHARMD, MSC, BCPS

ROBIN WACKERNAH
PHARMD, BCPP

D. RAYMOND WEBER
PHARMD, BSPHARM, BCOP, BCPS, RPH